John Klein

THE CIVIL ENGINEERING HANDBOOK

Editor-in-Chief

W. F. CHEN

Purdue University
West Lafayette, Indiana

CRC Press
Boca Raton New York London Tokyo

Library of Congress Cataloging-in-Publication Data

The civil engineering handbook / editor-in-chief, W.F. Chen
p. cm.
Includes bibliographical references and index.
ISBN 0-8493-8953-4 (acid-free paper)
1. Civil engineering—Handbooks, manuals, etc. I. Chen, Wai-Fah, 1936–
TA151.C57 1995
624—dc20 94-45572
CIP

Direct all inquiries to CRC Press LLC, 2000 N.W. Corporate Blvd., Boca Raton, Florida 33431.

International Standard Book Number 0-8493-8953-4
Library of Congress Card Number 94-45572
Printed in the United States of America 4 5 6 7 8 9 0
Printed on acid-free paper

Foreword

This is the beginning of a new era in the engineering of civil works. Recent emphasis on sustainable development, rehabilitation of the civil infrastructure, and the environment is bringing much attention and change to the civil engineering profession. The practicing civil engineer is faced with new challenges, such as environmental sensitivity, new technologies, new materials, and new ways of getting the project completed (design/build, fast track, etc.). Projects are becoming more complicated and involved. The computer is now an essential tool for every project. For effective engineering in this age, a firm grasp of basic fundamentals is more essential than ever before. *The Civil Engineering Handbook* provides very crisp reviews of the fundamentals. It is written to enable the practicing engineer to move from the fundamentals to handling typical modern-day problems, including rehabilitation and retrofitting, in the whole spectrum of civil engineering. It should be an indispensable tool for every practicing civil engineer.

The Editor-in-Chief, Dr. W. F. Chen, is the George E. Goodwin Distinguished Professor of Civil Engineering. He and many of the associate editors and the contributors are or have been affiliated with Purdue University's School of Civil Engineering. Their expertise and experience is widely acknowledged. This handbook is an example of how these persons are fulfilling our mission to work with the profession and to have an impact on the practice of civil engineering. Enrollments in civil engineering programs nationwide are at all-time highs. The undergraduate and graduate programs in civil engineering are among the largest in the United States and have over 9000 living alumni. The School has extensive education, research, and outreach programs, many of which are done in cooperation with engineering practitioners in private practice, industry, and government. This interaction enhances the credibility and value of this handbook to the practicing engineer.

Vincent P. Drnevich
Professor and Head
School of Civil Engineering
Purdue University

Preface

The Civil Engineering Handbook is a comprehensive reference work covering the broad spectrum of civil engineering. It has been written with the practicing civil engineer in mind. The ideal reader will be a B.S.-level engineer with a need for a single reference source to keep abreast of new techniques and practices, as well as review standard practices.

The *Handbook* stresses professional applications; emphasis has been placed on ready-to-use material. It contains many formulas and tables that give immediate answers to questions arising from practical work. It also contains a brief description of the essential elements of each subject, thus enabling the reader to understand the logical background of these results and to think beyond them. Traditional as well as new and innovative practices are covered.

As a result of rapid advances in computer technology, a revolution has occurred in civil engineering research and practice. A third aspect, *computing,* has been added to the *theoretical* and *experimental* aspects of the field to form the basis of civil engineering. Thorough coverage of computers and software as used in connection with many new computational and design methods is essential. Thus, computational aspects of civil engineering form the main focus of several chapters. *The Civil Engineering Handbook* is the first comprehensive handbook published to use a modern CAD/CAE approach in advancing civil engineers into the 21st century.

The *Handbook* is organized into eight sections, covering the eight traditional areas of civil engineering:

- Construction engineering
- Environmental engineering
- Geotechnical engineering
- Hydraulic engineering
- Materials engineering
- Structural engineering
- Surveying engineering
- Transportation engineering

The subdivision of each section into several chapters is made by the associate editors and is somewhat arbitrary, as the many subjects of the individual chapters are cross-linked in many ways and cannot be arranged in a definite sequence. To this end, in addition to the complete table of contents presented at the front of the book, an individual table of contents precedes each of the eight sections and gives a general outline of the scope of the subject area covered. Finally, each chapter begins with its own table of contents. The reader should look over these tables of contents to become familiar with the structure, organization, and content of the book. In this way, the book can also be used as a survey of the field of civil engineering and by the student or civil engineer to find the topics that he or she wants to examine in depth. It can be used as an introduction to or a survey of a particular subject in the field, and the references at the end of each chapter can be consulted for more detailed studies.

The chapters of the *Handbook* have been written by many authors, all experts in their fields, and the eight sections have been carefully edited and integrated by the various associate editors, almost all heads of their areas in the School of Civil Engineering at Purdue University. This handbook is a testimonial to the dedication of the associate editors, the publisher, and the editorial associates. I wish to thank all the authors for their contributions and the reviewers for their constructive comments. I also wish to acknowledge at CRC Press Joel Claypool, Publisher, Kristen Peterson, Development Editor, and Carol Whitehead, Full Service Coordinator.

W. F. Chen
Editor-in-Chief

About the Editor

W. F. Chen is George E. Goodwin Distinguished Professor of Civil Engineering and Head of the Department of Structural Engineering, School of Civil Engineering at Purdue University. He received his B.S. in civil engineering from the National Cheng-Kung University, Taiwan, in 1959, M.S. in structural engineering from Lehigh University, Bethlehem, Pennsylvania in 1963, and Ph.D. in solid mechanics from Brown University, Providence, Rhode Island in 1966.

Dr. Chen's research interests cover several areas, including constitutive modeling of engineering materials, soil and concrete plasticity, structural connections, and structural stability. He is the recipient of numerous engineering awards, including the AISC T. R. Higgins Lectureship Award, the ASCE Raymond C. Reese Research Prize, and the ASCE Shortridge Hardesty Award. Most recently, he was elected to the National Academy of Engineering.

Dr. Chen is a member of the Executive Committee of the Structural Stability Research Council, the Specification Committee of the American Institute of Steel Construction, and the editorial board of six technical journals. He has worked as a consultant for Exxon's Production and Research Division on offshore structures, for Skidmore, Owings and Merrill on tall steel buildings, and for World Bank on the Chinese University Development Projects.

A widely respected author, Dr. Chen's works include *Limit Analysis and Soil Plasticity* (Elsevier, 1975), the two-volume *Theory of Beam-Columns* (McGraw-Hill, 1976–77), *Plasticity in Reinforced Concrete* (McGraw-Hill, 1982), *Stability Design of Steel Frames* (CRC Press, 1991), and *Plasticity for Structural Engineers* (Springer-Verlag, 1988). He is the editor of two book series, one in structural engineering and the other in civil engineering. He has authored or coauthored more than 430 papers in journals and conference proceedings. He is the author or coauthor of 17 books, has edited 11 books, and has contributed chapters to 27 other books. His most recent books are *Plastic Design and Second-Order Analysis of Steel Frames* (Springer-Verlag, 1994), and the two-volume *Constitutive Equations for Engineering Materials* (Elsevier, 1994).

Dr. Chen has taught at Lehigh University as well as Purdue University.

Associate Editors

W. F. Chen
Purdue University
West Lafayette, Indiana

Jacques W. Delleur
Purdue University
West Lafayette, Indiana

W. L. Dolch
Purdue University (Emeritus)
West Lafayette, Indiana

Donn E. Hancher
University of Kentucky
Lexington, Kentucky

Milton E. Harr
Purdue University
West Lafayette, Indiana

Robert B. Jacko
Purdue University
West Lafayette, Indiana

Edward M. Mikhail
Purdue University
West Lafayette, Indiana

Kumares C. Sinha
Purdue University
West Lafayette, Indiana

Contributors

Arch Alexander
Purdue University
West Lafayette, Indiana

David Bernstein
Massachusetts Institute of Technology
Cambridge, Massachusetts

J. S. Bethel
Purdue University
West Lafayette, Indiana

Mark D. Bowman
Purdue University
West Lafayette, Indiana

Jonathan D. Bray
University of California, Berkeley
Berkeley, California

C. B. Burke
Christopher B. Burke Engineering, Ltd.
Rosemont, Illinois

T. T. Burke, Jr.
Purdue University
West Lafayette, Indiana

Susan E. Burns
Georgia Institute of Technology
Atlanta, Georgia

Michael J. Cassidy
University of California, Berkeley
Berkeley, California

W. F. Chen
Purdue University
West Lafayette, Indiana

Menashi D. Cohen
Purdue University
West Lafayette, Indiana

Wesley G. Crawford
Purdue University
West Lafayette, Indiana

Jacques W. Delleur
Purdue University
West Lafayette, Indiana

Richard J. Deschamps
Purdue University
West Lafayette, Indiana

Sidney Diamond
Purdue University
West Lafayette, Indiana

W. L. Dolch
Purdue University (Emeritus)
West Lafayette, Indiana

Said M. Easa
Lakehead University
Ontario, Canada

Bengt H. Fellenius
University of Ottowa
Ontario, Canada

Patrick J. Fox
Purdue University
West Lafayette, Indiana

J. David Frost
Georgia Institute of Technology
Atlanta, Georgia

Peter G. Furth
Northeastern University
Boston, Massachusetts

T. F. Fwa
National University of Singapore
Kent Ridge, Singapore

Aldo Giorgini
Purdue University
West Lafayette, Indiana

Amy Grider
Purdue University
West Lafayette, Indiana

Donn E. Hancher
University of Kentucky
Lexington, Kentucky

Milton E. Harr
Purdue University
West Lafayette, Indiana

Robert W. Holden II
Purdue University
West Lafayette, Indiana

R. D. Holtz
University of Washington
Seattle, Washington

M. H. Houck
George Mason University
Fairfax, Virginia

Dana N. Humphrey
University of Maine
Orono, Maine

Roy E. Hunt
Geotechnical Consultant
Bricktown, New Jersey

Tom Iseley
Louisiana Tech University
Ruston, Louisiana

Robert B. Jacko
Purdue University
West Lafayette, Indiana

Steven D. Johnson
Purdue University
West Lafayette, Indiana

Matthew G. Karlaftis
Purdue University
West Lafayette, Indiana

Vasiliki Keramida
Keramida Environmental, Inc.
Indianapolis, Indiana

Anton J. Kleywegt
Purdue University
West Lafayette, Indiana

Raghavan Kunigahalli
University of Wisconsin
Madison, Wisconsin

Jesus Larralde
California State University, Fresno
Fresno, California

J. Y. Richard Liew
National University of Singapore
Kent Ridge, Singapore

E. M. Lui
Syracuse University
Syracuse, New York

D. A. Lyn
Purdue University
West Lafayette, Indiana

Guy A. Meadows
University of Michigan
Ann Arbor, Michigan

Edward M. Mikhail
Purdue University
West Lafayette, Indiana

Austin D. Pan
Purdue University
West Lafayette, Indiana

Egor P. Popov
University of California, Berkeley
Berkeley, California

Julio A. Ramirez
Purdue University
West Lafayette, Indiana

A. R. Rao
Purdue University
West Lafayette, Indiana

Pedro C. Repetto
Woodward-Clyde Consultants
Plymouth Meeting, Pennsylvania

J. Rhodes
University of Strathclyde
Glasgow, United Kingdom

David V. Rosowsky
Clemson University
Clemson, South Carolina

James E. Rowings, Jr.
Iowa State University
Ames, Iowa

Jeffrey S. Russell
Unviersity of Wisconsin, Madison
Madison, Wisconsin

Rodrigo Salgado
Purdue University
West Lafayette, Indiana

John F. Senft
Purdue University
West Lafayette, Indiana

N. E. Shanmugam
National University of Singapore
Kent Ridge, Singapore

Kumares C. Sinha
Purdue University
West Lafayette, Indiana

Gary R. Smith
Pennsylvania State University
University Park, Pennsylvania

Yorgos J. Stephanedes
University of Minnesota
Minneapolis, Minnesota

Robert M. Sykes
Ohio State University
Columbus, Ohio

Mang Tia
University of Florida
Gainesville, Florida

Jolyon D. Thurgood
Leica, Inc.
Englewood, Colorado

B. H. W. van Gelder
Purdue University
West Lafayette, Indiana

Roger L. Wayson
University of Central Florida
Orlando, Florida

Leo Weitzman
LVW Associates, Inc.
West Lafayette, Indiana

Robert K. Whitford
Purdue University
West Lafayette, Indiana

Thomas F. Wolff
Michigan State University
East Lansing, Michigan

William L. Wood
Purdue University
West Lafayette, Indiana

J. R. Wright
Purdue University
West Lafayette, Indiana

Young Mook Yun
Purdue University
West Lafayette, Indiana

Contents

SECTION I Construction

SECTION II Environmental Engineering

SECTION III Geotechnical Engineering

SECTION IV Hydraulic Engineering

SECTION VI Structural Engineering

SECTION VIII Transportation Engineering

THE CIVIL ENGINEERING HANDBOOK

Above is the base, or jacket, of the world's tallest fixed offshore platform, the Bullwinkle platform. It is located 150 miles southwest of New Orleans in the Gulf of Mexico and stands in 1353 feet of water. With the platform and drilling rigs in place, the Bullwinkle structure stands at about 1615 feet, which is 161 feet taller than the world's tallest building, the Sears Tower.

Twenty-eight piles were used to secure the jacket to the ocean floor. Each pile is 84 inches in diameter and up to 541 feet long. The piles were driven to penetrations of up to 437 feet with underwater hammers. The weight of the jacket is about 49,375 tons and the total structural weight of Bullwinkle is over 77,000 tons. The structure was designed to withstand a wave height of 72 feet, a storm tide of 5 feet, and a wind velocity of 140 miles per hour.

Since October 1993, Bullwinkle has been producing more than 54,000 barrels of oil and 95 million cubic feet of natural gas per day. (Photo courtesy of Shell Oil Company.)

I

Construction

Donn E. Hancher
University of Kentucky

THE CONSTRUCTION INDUSTRY is one of the largest segments of business in the United States, with the percentage of the gross national product spent in construction over the last several years averaging about 10 percent. For 1993 the total amount spent on construction in the U.S. is estimated at $375 billion [*Engineering News Record*, Jan. 25, 1993]. Of this total about $105 billion is estimated for public projects, $105 billion for private nonresidential projects, and the rest for residential projects.

Construction is the realization phase of the civil engineering process, following conception and design. It is the role of the constructor to turn the ideas of the planner and the detailed plans of the designer into physical reality. The owner is the ultimate consumer of the product and is often the general public for civil engineering projects. Not only does the constructor have an obligation to the contractual owner, or client, but an ethical obligation to the general public to perform the work so the final product will serve its function economically and safely.

0-8493-8953-4/95/$0.00 + $.50

The construction industry is typically divided into specialty areas with each area requiring different skills, resources, and knowledge to participate effectively in it. The area classifications typically used are residential (single- and multifamily housing), building (all buildings other than housing), heavy/highway (dams, bridges, ports, sewage-treatment plants, highways), utility (sanitary and storm drainage, water lines, electrical and telephone lines, pumping stations), and industrial (refineries, mills, power plants, chemical plants, heavy manufacturing facilities). Civil engineers can be heavily involved in all of these areas of construction, although fewer are involved in residential. Due to the differences in each of these market areas, most engineers specialize in only one or two of the areas during their careers.

Construction projects are complex and time-consuming undertakings which require the interaction and cooperation of many different persons to accomplish. All projects must be completed in accordance with specific project plans and specifications, along with other contract restrictions that may be imposed on the production operations. Essentially, all civil engineering construction projects are unique. Regardless of the similarity to other projects, there are always distinguishing elements of each project which make it unique, such as the type of soil, the exposure to weather, the human resources assigned to the project, the social and political climate, and so on. In manufacturing raw resources are brought to a factory with a fairly controlled environment; in construction the “factory” is set up on site and production is accomplished in an uncertain environment.

It is this diversity among projects that makes the preparation for a civil engineering project interesting and challenging. Although it is often difficult to control the environment of the project, it is the duty of the contractor to predict the possible situations that may be encountered and to develop contingency strategies accordingly. The dilemma of this situation is that the contractor who allows for contingencies in project cost estimates will have a difficult time competing against other less competent or less cautious contractors. The failure rate in the construction industry is the highest in the U.S.; one of the leading causes for failure is the inability to manage in such a highly competitive market and to realize a fair return on investment.

Participants in the Construction Process

There are several participants in the construction process, all with an important role in developing a successful project. The owner, either private or public, is the party that initiates the demand for the project and ultimately pays for its completion. The owner’s role in the process varies considerably; however, the primary role of the owner is to effectively communicate the scope of work desired to the other parties. The designer is responsible for developing adequate working drawings and specifications, in accordance with current design practices and codes, to communicate the product desired by the owner upon completion of the project. The prime contractor is responsible for managing the resources needed to carry out the construction process in a manner which ensures that the project will be conducted safely, within budget, and on schedule, and that it meets or exceeds the quality requirements of the plans and specifications. Subcontractors are specialty contractors who contract with the prime contractor to conduct a specific portion of the project within the overall project schedule. Suppliers are the vendors who contract to supply required materials for the project within the project specifications and schedule. The success of any project depends on the coordination of the efforts of all the parties involved, hopefully to the financial advantage of all. In recent years these relationships have become more adversarial with much conflict and litigation, often to the detriment of the projects.

Construction Contracts

Construction projects are done under a variety of contract arrangements for each of the parties involved. They range from a single contract for a single element of the project to a single contract for the whole project, including the financing, design, construction, and operation of the facility. Typical contract types include lump sum, unit price, cost plus, and construction management.

These contract systems can be used with either the competitive bidding process or with negotiated processes. A contract system becoming more popular with owners is design-build, in which all the responsibilities can be placed with one party for the owner to deal with. Each type of contract impacts the roles and responsibilities of each of the parties on a project. It also impacts the management functions to be carried out by the contractor on the project, especially the cost engineering function.

A new development in business relationships in the construction industry is *partnering*. Partnering is an approach to conducting business that confronts the economic and technological challenges in industry in the twenty-first century. This new approach focuses on making long-term commitments with mutual goals for all parties involved to achieve mutual success. It requires changing traditional relationships to a shared culture without regard to normal organizational boundaries. Participants seek to avoid the adversarial problems typical for many business ventures. Most of all, a relationship must be based upon trust. Although partnering in its pure form relates to a long-term business relationship for multiple projects, many single-project partnering relationships have been developed, primarily for public owner projects. Partnering is an excellent vehicle to attain improved quality on construction projects.

Partnering is not to be construed as a legal partnership with the associated joint liability. Great care should be taken to make this point clear to all parties involved in a partnering relationship.

Partnering is not a quick fix or panacea to be applied to all relationships. It requires total commitment, proper conditions, and the right chemistry between organizations for it to thrive and prosper. The relationship is based upon trust, dedication to common goals, and an understanding of each other's individual expectations and values. The partnering concept is intended to accentuate the strength of each partner and will be unable to overcome fundamental company weaknesses; in fact, weaknesses may be magnified. Expected benefits include improved efficiency and cost effectiveness, increased opportunity for innovation, and the continuous improvement of quality products and services. It can be used by either large or small businesses, and it can be used for either large or small projects. Relationships can develop among all participants in construction: owner-contractor, owner-supplier, contractor-supplier, contractor-contractor. (Contractor refers to either a design firm or construction company.)

Goals of Project Management

Regardless of the project, most construction teams have the same performance goals:

Cost:	Complete the project within the cost budget, including the budgeted costs of all change orders
Time:	Complete the project by the scheduled completion date or within the allowance for work days
Quality:	Perform all work on the project meeting or exceeding the project plans and specifications.
Safety:	Complete the project with zero lost-time accidents.
Conflict:	Resolve disputes at the lowest practical level and have zero disputes.
Project start-up:	A successful start-up of the completed project by the owner with zero rework.

Basic Functions of Construction Engineering

The activities involved in the construction engineering for projects include the following basic functions:

Cost engineering:	The cost estimating, cost accounting, and cost control activities related to a project, plus the development of cost databases.

Project planning and scheduling:	The development of initial project plans and schedules, project monitoring and updating, and the development of as-built project schedules.
Equipment planning and management:	The selection of needed equipment for projects, productivity planning to accomplish the project with the selected equipment in the required project schedule and estimate, and the management of the equipment fleet.
Design of temporary structures:	The design of temporary structures required for the construction of the project, such as concrete formwork, scaffolding, shoring, and bracing.
Contract management:	Managing the activities of the project to comply with contract provisions, document contract changes, and minimize contract disputes.
Human resource management:	The selection, training, and supervision of the personnel needed to complete the project work within schedule.
Project safety:	The establishment of safe working practices and conditions for the project, the communication of these safety requirements to all project personnel, the maintenance of safety records, and the enforcement of these requirements.

Innovations in Construction

There are several innovative developments in technological tools which have been implemented or are being considered for implementation for construction projects. New tools such as CAD systems, expert systems, bar coding, and automated equipment offer excellent potential for improved productivity and cost effectiveness in industry. Companies who ignore these new technologies will have difficulty competing in the future.

Scope of This Section of the Handbook

The scope of Section I, Construction, in this handbook is to present the reader with the essential information needed to perform the major construction engineering functions on today's construction projects. Examples are offered to illustrate the principles presented, and references are offered for further information on each of the topics covered.

1

Construction Estimating

James E. Rowings, Jr.
Iowa State University

1.1 Introduction

The preparation of estimates represents one of the most important functions performed in any business enterprise. In the construction industry, the quality of performance of this function is paramount to the success of the parties engaged in the overall management of capital expenditures for construction projects. The estimating process, in some form, is used as soon as the idea for a project is conceived. Estimates are prepared and updated continually as the project scope and definition develops and, in many cases, throughout construction of the project or facility.

The parties engaged in delivering the project continually are asking themselves "What will it **cost**?" To answer this question, some type of estimate must be developed. Obviously, the precise answer to this question cannot be determined until the project is completed. Posing this type of question elicits a finite answer from the estimator. This answer, or estimate, represents only an approximation or expected value for the cost. The eventual accuracy of this approximation depends on how closely the actual conditions and specific details of the project match the expectations of the estimator.

Extreme care must be exercised by the estimator in the preparation of the estimate to subjectively weigh the potential variations in future conditions. The estimate should convey an assessment of the accuracy and risks.

1.2 Estimating Defined

Estimating is a complex process involving collection of available and pertinent information relating to the **scope** of a project, expected resource consumption, and future changes in resource costs. The process involves synthesis of this information through a mental process of visualization of the constructing process for the project. This visualization is mentally translated into an approximation of the final cost.

0-8493-8953-4/95/$0.00 + $.50

At the outset of a project the estimate cannot be expected to carry a high degree of accuracy since little information is known. As the design progresses more information is known and the accuracy should improve.

Estimating at any stage of the project cycle involves considerable effort to gather information. The estimator must collect and review all of the detailed plans, specifications, available site data, available resource data (labor, materials, and equipment), contract documents, resource cost information, pertinent government regulations, and applicable owner requirements. Information gathering is a continual process by estimators due to the uniqueness of each project and constant changes in the industry environment.

Unlike the production from a manufacturing facility, each product of a construction firm represents a prototype. Considerable effort in planning is required before a cost estimate can be established. Most of the effort in establishing the estimate revolves around determining the approximation of the cost to produce the one-time product.

The estimator must systematically convert information into a forecast of the component and collective costs that will be incurred in delivering the project or facility. This synthesis of information is accomplished by mentally building the project from the ground up. Each step of the building process should be accounted for along with the necessary support activities and imbedded temporary work items required for completion.

The estimator must have some form of systematic approach to ensure that all cost items have been incorporated and that none have been duplicated. Later in this chapter is a discussion of alternate systematic approaches which are used.

The quality of an estimate depends on the qualifications and abilities of the estimator. In general, an estimator must demonstrate the following capabilities and qualifications:

- Extensive knowledge of construction
- Knowledge of construction materials and methods
- Knowledge of construction practices and contracts
- Ability to read and write construction documents
- Ability to sketch construction details
- Ability to communicate graphically and verbally
- Strong background in business and economics
- Ability to visualize work items
- Broad background in design and code requirements

Obviously, from the qualifications cited, estimators are not born but are developed through years of formal or informal education and experience in the industry. The breadth and depth of the requirements for an estimator lend testimony to the importance and value of the individual in the firm.

1.3 Estimating Terminology

There are a number of terms that are used in the estimating process that should be understood. The American Association of Cost Engineers has developed a glossary of terms and definitions to have a uniform technical vocabulary. Several of the more common terms and the definitions are given below.

1.4 Types of Estimates

There are two broad categories for estimates: conceptual (or approximate) estimates and detailed estimates. Classification of an estimate into one of these types depends on the available information, the extent of effort dedicated to preparation, and the use for the estimate. The classification of an estimate into one of these two categories is an expression of the relative confidence in the accuracy of the estimate.

Conceptual Estimates

At the outset of the project, when the scope and definition are in the early stages of development, little information is available, yet there is often a need for some assessment of the potential cost. The owner needs to have a rough or approximate value for the project's cost for purposes of determining the economic desirability of proceeding with design and construction. Special quick techniques are usually employed utilizing minimal available information at this point to prepare a conceptual estimate. Little effort is expended to prepare this type of estimate, which often utilizes only a single project parameter, such as square feet of floor area, span length of a bridge, or barrels per day of output. Using available, historical cost information and applying like parameters, a quick and simple estimate can be prepared. These types of estimates are valuable in determining the order of magnitude of the cost for very rough comparisons and analysis, but are not appropriate for critical decision making and commitment.

Many situations exist which do not warrant or allow expenditure of the time and effort required to produce a detailed estimate. Feasibility studies involve elimination of many alternatives prior to any detailed design work. Obviously, if detailed design were pursued prior to estimating, the cost of the feasibility study would be enormous. Time constraints may also limit the level of detail which can be employed. If an answer is required in a few minutes or a few hours, then the method must be a conceptual one, even if detailed design information is available.

Conceptual estimates have value, but many limitations as well. Care must be exercised to choose the appropriate method for conceptual estimating based on the available information. The estimator must be aware of the limitations of his estimate and communicate these limitations so that the estimate is not misused. Conceptual estimating relies heavily on past cost data, which is adjusted to reflect current trends and actual project economic conditions.

The accuracy of an estimate is a function of time spent in its preparation, the quantity of design data utilized in the evaluation, and the accuracy of the information used. In general, more effort and more money produce a better estimate, one in which the estimator has more confidence regarding the accuracy of his or her prediction. To achieve a significant improvement in accuracy requires a larger-than-proportional increase in effort. Each of the three conceptual levels of estimating has several methods which are utilized depending on the project type and the availability of time and information.

Order of Magnitude

The order-of-magnitude estimate is by far the most uncertain estimate level which is used. As the name implies, the objective is to establish the order of magnitude of the cost, or more precisely the cost within a range of plus 30 or minus 50%.

Various techniques can be employed to develop an order-of-magnitude estimate for a project or portion of a project. Presented below are some examples and explanations of various methods used.

Rough Weight Check. When the object of the estimate is a single criterion, such as a piece of equipment, the order of magnitude cost can be estimated quickly based on the weight of the object. For the cost determination, equipment can be grouped into three broad categories:

1. Precision/computerized/electronic
2. Mechanical/electrical
3. Functional

Precision equipment includes electronic or optical equipment such as computers and surveying instruments. Mechanical/electrical equipment includes pumps and motors. Functional equipment might include heavy construction equipment, automobiles, and large power tools. Precision equipment tends to cost ten times more per pound than mechanical/electrical equipment, which in turn costs ten times per pound more than functional equipment. Obviously, if you know the average cost per pound for a particular class of equipment (e.g, pumps), this information is more useful than a broad category estimate. In any case, the estimator should have a feel for the approximate cost per pound for the three categories so that quick checks can be made and order-of-magnitude estimates performed with minimal information available. Similar approaches using the capacity of equipment, such as flow rate, can be used for order-of-magnitude estimates.

Cost Capacity Factor. This quick method is tailored to the process industry. It represents a quick shortcut to establish an order-of-magnitude estimate of the cost. Application of the method involves four basic steps:

1. Obtain information concerning the cost (C_1 or C_2) and the input/output/throughput or holding capacity (Q_1 or Q_2) for a project similar in design or characteristics to the one being estimated.
2. Define the relative size of the two projects in the most appropriate common units of input, output, throughput, or holding capacity. As an example, a power plant is usually rated in kilowatts of output, a refinery in barrels per day of output, a sewage treatment plant in tons per day of input, and a storage tank in gallons or barrels of holding capacity.
3. Using the three known quantities (the sizes of the two similar plants in common units and the cost of the previously constructed plant), the following relationship can be developed:

$$C_1/C_2 = (Q_1/Q_2)^x$$

   where x is the appropriate cost capacity factor. With this relationship the estimate of the cost of the new plant can be determined.
4. The cost determined in the third step is adjusted for time and location by applying the appropriate construction cost indices. (The use of indices is discussed later in this chapter.)

The cost capacity factor approach is also called the *six-tenths rule* since, in the original application of the exponential relationship, x was determined to be equal to about 0.6. In reality, the factors for various processes vary from 0.33 to 1.02 with the bulk of the values for x around 0.6.

Example. Assume that we have information on an old process plant which has the capacity to produce 10,000 gallons per day of a particular chemical and the cost today to build the plant would be $1,000,000. The appropriate cost factor for this type of plant is 0.6. An order-of-magnitude estimate of the cost is required for a plant with a capacity of 30,000 gallons per day.

$$C = \$1{,}000{,}000(30{,}000/10{,}000)^{0.6} = \$1{,}930{,}000$$

Comparative Cost of Structure. This method is readily adaptable to virtually every type of structure, including bridges, stadiums, schools, hospitals, and offices. Very little information is required about the planned structure except that the following general characteristics should be known:

1. Use—school, office, hospital, and so on.
2. Kind of construction—wood, steel, concrete, and so on.
3. Quality of construction—cheap, moderate, top grade.
4. Locality—labor and material supply market area.
5. Time of construction—year.

By identifying a similar completed structure with nearly the same characteristics, an order-of-magnitude estimate can be determined by proportioning cost according to the appropriate unit for the structure. These units might be as follows:

Bridges—span in feet (adjustment for number of lands)
Schools—pupils
Stadium—seats
Hospital—beds
Offices—square feet
Warehouses—cubic feet

Example. Assume that the current cost for a 120-pupil school constructed of wood frame for a city is \$1,200,000. We are asked to develop an order-of-magnitude estimate for a 90-pupil school.
Solution. The first step is to separate the per-pupil cost.

$$\$1{,}200{,}000/120 = \$100{,}000/\text{pupil}$$

Apply the unit cost to the new school.

$$\$100{,}000/\text{pupil} \times 90 \text{ pupils} = \$900{,}000$$

Feasibility Estimates

This level of conceptual estimate is more refined than the order-of-magnitude estimate and should provide a narrower range for the estimate. These estimates, if performed carefully, should be within plus or minus 20 to 30%. To achieve this increase in accuracy over the order-of-magnitude estimate requires substantially more effort and more knowledge about the project.

Plant Cost Ratio. This method utilizes the concept that the equipment proportion of the total cost of a process facility is about the same regardless of the size or capacity of the plant for the same basic process. Therefore, if the major fixed equipment cost can be estimated, the total plant cost can be determined by factor multiplication. The plant cost factor or multiplier is sometimes called the Lang factor (after the man who developed the concept for process plants).

Example. Assume that a historical plant with the same process cost 2.5 million dollars with the equipment portion of the plant costing 1 million dollars. Determine the cost of a new plant if the equipment has been determined to cost 2.4 million dollars.

$$C = 2.4/(1.0/2.5)$$
$$C = 6 \text{ million dollars}$$

Floor Area. This method is most appropriate for hospitals, stores, shopping centers, and residences. Floor area must be the dominant attribute of cost (or at least it is assumed to be by the estimator). There are several variations of this method, a few of which are explained below.

Total Horizontal Area. For this variation it is assumed that cost is directly proportional to the development of horizontal surfaces. It is assumed that the cost of developing a square foot of ground-floor space will be the same as a square foot of third-floor space or a square foot of roof space. From historical data a cost per square foot is determined and applied uniformly to the horizontal area which must be developed to arrive at the total cost.

Example. Assume that a historical file contains a warehouse building that cost \$200,000 that was 50 ft × 80 ft with a basement, three floors, and an attic. Determine the cost for a 60 ft × 30 ft warehouse building with no basement, two floors, and an attic.
Solution. Determine the historical cost per square foot.

Basement area	4,000
1st floor	4,000
2nd floor	4,000
3rd floor	4,000
Attic	4,000
Roof	4,000
	24,000

$$\$200{,}000/24{,}000 = \$8.33/\text{ft}^2$$

Next, calculate the total cost for the new project.

1st floor	1800
2nd floor	1800
Attic	1800
Roof	1800
	7200

$$7200\ \text{ft}^2 \times \$8.33/\text{ft}^2 = \$60{,}000$$

Finished Floor Area. This method is by far the most widely used approach for buildings. With this approach, only those floors which are finished are counted when developing the historical base cost and when applying the historical data to the new project area. With this method the estimator must exercise extreme care to have the same relative proportions of area to height to avoid large errors.

Example. Same as the preceding example.
Solution. Determine historical base cost.

1st floor	4,000
2nd floor	4,000
3rd floor	4,000
	12,000 ft^2fa

$$\$200{,}000/12{,}000 = \$16.67/\text{ft}^2\text{fa}$$

where ft^2fa is square feet of finished floor area.

Next, determine the total cost for the new project.

1st floor	1800
2nd floor	1800
	3600 ft^2fa

$$3600\ \text{ft}^2\text{fa} \times \$16.67/\ \text{ft}^2\text{fa} = \$60{,}000$$

As can be seen, little difference exists between the finished floor area and total horizontal area methods; however, if a gross variation in overall dimensions had existed between the historical structure and the new project, a wider discrepancy between the methods would have appeared.

Cubic Foot of Volume Method. This method accounts for an additional parameter which affects cost: floor-to-ceiling height.

Example. The same as the preceding two examples except that the following ceiling heights are given:

	Old Structure	New Structure
1st floor	14	12
2nd floor	10	12
3rd floor	10	—

Solution. Determine the historical base cost.

$$
\begin{aligned}
14 \times 4000 &= 56{,}000\ \text{ft}^3\\
10 \times 4000 &= 40{,}000\ \text{ft}^3\\
10 \times 4000 &= \underline{40{,}000\ \text{ft}^3}\\
& \quad\ 136{,}000\ \text{ft}^3
\end{aligned}
$$

$$\$200{,}000/136{,}000\ \text{ft}^3 = \$1.52/\text{ft}^3$$

Next, determine the total cost for the new warehouse structure.

1st floor	$1800\ \text{ft}^2 \times 12\ \text{ft}$	$= 21{,}600\ \text{ft}^3$
2nd floor	$1800\ \text{ft}^2 \times 12\ \text{ft}$	$= \underline{21{,}600\ \text{ft}^3}$
		$43{,}200\ \text{ft}^3$

$$43{,}200\ \text{ft}^3 \times \$1.52/\text{ft}^3 = \$65{,}664$$

Appropriation Estimates

As a project scope is developed and refined it progresses to a point where it is budgeted into a corporate capital building program budget. Assuming the potential benefits are greater than the estimated costs a sum of money is set aside to cover the project expenses. From this process of appropriation comes the name of the most refined level of conceptual estimate. This level of estimate requires more knowledge and effort than the previously discussed estimates.

These estimating methods reflect a greater degree of accuracy. Appropriation estimates should be between plus or minus 10 to 20%. As with the other forms of conceptual estimates, several methods are available to prepare appropriation estimates.

Parametric Estimating/Panel Method. This method employs a database where key project parameters, project systems, or panels (as in the case of buildings) are priced from past projects using appropriate units. The costs of each parameter or panel are computed separately and multiplied by the number of panels of each kind. Major unique features are priced separately and included as separate line items. Numerous parametric systems exist for different types of projects. For process plants the process systems and piping are the parameters. For buildings, various approaches have been used, but one approach to illustrate the method is as follows:

Parameter	Unit of Measure
Site work	Square feet of site area
Foundations and columns	Building square feet
Floor system	Building square feet
Structural system	Building square feet
Roof system	Roof square feet
Exterior walls	Wall square feet minus exterior windows
Interior walls	Wall square feet (interior)
HVAC	Tons or BTU
Electrical	Building square feet
Conveying systems	Number of floor stops
Plumbing	Number of fixture units
Finishes	Building square feet

Each of these items would be estimated separately by applying the historical cost for the appropriate unit for similar construction and multiplying times the number of units for the current project. This same approach is used on projects such as roads. The units or parameters used are often the same as the bid items and the historical prices are the average of the low-bid unit prices received in the last few contracts.

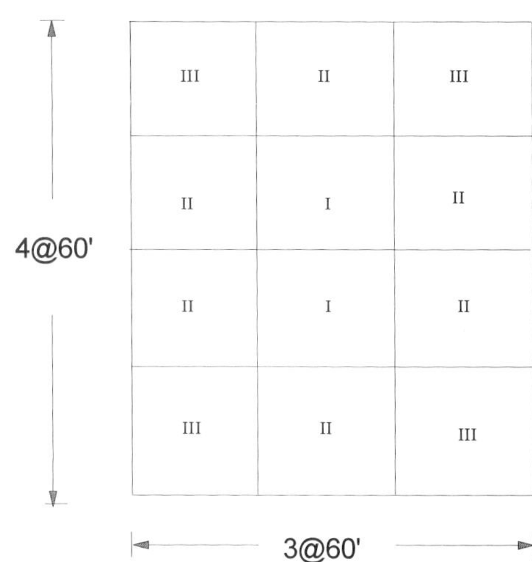

FIGURE 1.1 Plan view—warehouse building.

Bay Method. This method is appropriate for buildings or projects that consist of a number of repetitive or similar units. In the plan view of a warehouse building shown in Fig. 1.1, the building is made up of three types of bays. The only difference between them is the number of outside walls. By performing a definitive estimate of the cost of each of these bay types an appropriation estimate can be made by multiplying this bay cost times the number of similar bays and totaling for the three bay types.

Example. We know from a definitive estimate that the cost of the three bay types is as follows:

Type I = \$40,000
Type II = \$50,000
Type III = \$60,000

Determine the cost for the building structure and skin (outer surface).
Solution.

2 Type I @ 40,000	= \$ 80,000
6 Type II @ 50,000	= \$ 300,000
4 Type III @ 60,000	= \$ 240,000
	\$ 620,000

After applying the bay method for the overall project, the estimate is modified by making special allowances (add-ons) for end walls, entrances, stairs, elevators, and mechanical and electrical equipment.

Plant Component Ratio. This method requires a great deal more information than other methods used in the process industry. Definitive costs of the major pieces of equipment are needed. These can be determined from historical records or published data sources. Historical records also provide the data which identifies the relative percentage of all other items. The total project cost is then estimated as follows:

$$\text{TPC} = \frac{\text{ET}}{1 - \text{PT}}$$

where

TPC = total plant cost
ET = total estimated equipment cost
PT = sum of percentages of other items or phases (major account divisions)

Example. The total equipment cost for a plant is estimated to be \$500,000. The following percentages represent the average expenditures in other cost phases:

Engineering, overhead, and fees	22%
Warehousing	5%
Services	2%
Utilities	6%
Piping	20%
Instrumentation	5%
Electrical	6%
Buildings	4%
	70% = PT

$$\frac{500{,}000}{(1.0 - 0.70)} = \$1{,}670{,}000$$

While the solution here appears simple, in fact, the majority of time and effort is spent collecting the equipment cost and choosing the appropriate percentages for application.

Time and Location Adjustments

It is often desirable when preparing conceptual estimates to utilize cost data from a different period of time or from a different location. Costs vary with both time and location and it is therefore necessary to adjust the conceptual estimate for the differences of time and location from the historical base. A construction cost indexing system is used to identity the relative differences and permit adjustment.

Cost Indexing

A cost index is a dimensionless number associated with a point in time and/or location which illustrates the cost at that time or location relative to a base point in time or base location. The cost index provides a comparison of cost or cost change from year to year and/or location to location for a fixed quantity of services and commodities. The concept is to establish cost indices to avoid having to estimate all of the unique features of every project when it is reasonable to assume that the application of relative quantities of resources is constant or will follow the use of historical data on a proportional basis without knowledge of all of the design details. If the cost index is developed correctly, the following simple relationship will exist:

$$\text{New cost/New index} = \text{Historical cost/Historical index}$$

An example of the way in which a cost index might be computed is given below. The cost elements which were used for developing a cost index for concrete in 1982 are as follows:

C_1 = four hours for a carpenter = \$60
C_2 = one cubic yard concrete = \$40
C_3 = three hours for laborer = \$30
C_4 = 100 fbm lumber (2 × 10) = \$20
C_5 = 100 # rebar = \$20
C_6 = one hour from an ironworker = \$15

$$C_a = 60 + 40 + 30 + 20 + 20 + 15 = 185$$

Calculating C_b similarly for another time or location involves the following steps.

$$
\begin{aligned}
C_1 &= \text{four hours for a carpenter} &&= \$56\\
C_2 &= \text{one cubic yard concrete} &&= \$42\\
C_3 &= \text{three hours for laborer} &&= \$27\\
C_4 &= \text{100 fbm lumber } (2 \times 10) &&= \$19\\
C_5 &= \text{100 \# rebar} &&= \$20\\
C_6 &= \text{one hour from an ironworker} &&= \$12
\end{aligned}
$$

$$C_a = 56 + 42 + 27 + 19 + 20 + 12 = 176$$

Using the CI_a as the base with an index equal to 100, the CI_b index can be calculated as follows:

$$CI_b = (C_b/C_a) \times 100 = (176/185) \times 100 = 95.14$$

The key to creating an accurate and valid cost index is not the computational approach but the correct selection of the cost elements. If the index will be used for highway estimating, the cost elements should include items such as asphalt, fuel oil, paving equipment, and equipment operators. Appropriately, a housing cost index would include timber, concrete, carpenters, shingles, and other materials common to residential construction.

Most of the cost indices are normalized periodically to a base of 100. This is done by setting the base calculation of the cost for a location or time equal to 100 and converting all other indices to this base with the same divisor or multiplier.

While it is possible to develop specialized indices for special purposes, there are numerous indices which are published. These include several popular indices, such as the *Engineering News Record* building cost index and construction cost index and the Mean's construction cost index and historical cost index. These indices are developed using a wide range of cost elements. For example, the Mean's construction cost index is composed of 84 construction materials, 24 building crafts' labor hours, and 9 different equipment rental charges that correspond to the labor and material items. These cost indices are tabulated for the major metropolitan areas 4 times each year and for the 16 major UCI construction divisions. Additionally, indices dating back to 1913 can be found to adjust costs from different periods of time. These are referred to as historical cost indices.

Application of Cost Indices. These cost indices can have several uses:

Comparing costs from city to city (construction cost indices)
Comparing costs from time to time (historical cost indices)
Modifying costs for various cities and times (both)
Estimating replacement costs (both)
Forecasting construction costs (historical cost indices)

The cost index is only a tool and must be applied with sound judgment and common sense.

Comparing Costs from City to City. The construction cost indices can be used to compare costs between cities since the index is developed identically for each city. The index is an indicator of the relative difference. The cost difference between cities for identical buildings or projects in a different city can be found by using the appropriate construction cost indices (CCI). The procedure is as follows:

$$\text{Cost, city A} = \frac{\text{CCI for city A}}{\text{CCI for city B}}(\text{Known cost, city B})$$

$$(\text{Known cost, city B}) - (\text{Cost, city A}) = \text{Cost difference}$$

Comparing Costs from Time to Time. The cost indices can be used to compare costs for the same facility at different points in time. Using the historical cost indices of two points in time, one can calculate the difference in costs between the two points in time. It is necessary to know the cost and the historical index for time B and the historical cost index for time A.

$$\text{Cost, time A} = \frac{\text{HCI for time A}}{\text{HCI for time B}}(\text{Cost, time B})$$

$$(\text{Known cost, time B}) - (\text{Cost, time A}) = \text{Cost difference}$$

Modifying Costs for Various Cities and Times. The two prior uses can be accomplished simultaneously where it is desired to use cost information from another city and time for a second city and time estimate. Care must be exercised to establish the correct relationships. The following example illustrates the principle.

Example. A building cost $2,000,000 in 1980 in South Bend. How much will it cost to build in Boston in 1984?

Given:	HCI, 1984	= 114.3
	HCI, 1980	= 102.2
	CCI, S. Bend	= 123.4
	CCI, Boston	= 134.3

$$\frac{(\text{HCI, 1984})(\text{CCI, Boston})}{(\text{HCI, 1980})(\text{CCI, S. Bend})}(\text{Cost, S. Bend}) = (\text{Cost, Boston})$$

$$\frac{(114.3)(134.3)}{(102.2)(123.4)}(2{,}000{,}000) = \$2{,}430{,}000$$

Estimating Replacement Costs. The historical cost index can be used to determine replacement cost for a facility built a number of years ago or one that was constructed in stages.

Example. A building was constructed in stages over the last 25 years. It is desired to know the 1985 replacement cost for insurance purposes. The building has had two additions since the original 1961, $300,000 portion was built. The first addition was in 1970 at a cost of $200,000, and the second addition came in 1974 at a cost of $300,000. The historical cost indices are as follows:

1965	100.0	=	HCI
1974	49.8	=	HCI
1970	34.6	=	HCI
1961	23.9	=	HCI

Solution. The cost of the original building is

$$\frac{100}{23.9}\$300{,}000 = \$1{,}255{,}000$$

The cost of addition A is

$$\frac{100}{34.6}\$200{,}000 = \$578{,}000$$

The cost of addition B is

$$\frac{100}{49.8}\$300{,}000 = \$602{,}000$$

So,

$$\text{Total replacement cost} = \$2{,}435{,}000$$

Construction Cost Forecasting. If it is assumed that the future changes in cost will be similar to the past changes, the indices can be used to predict future construction costs. By using these past indices, future indices can be forecast and in turn used to predict future costs. Several approaches are available for developing the future index. Only one will be presented here.

The simplest method is to examine the change in the last several historical cost indices and use an average value for the annual change in the future. This averaging process can be accomplished by determining the difference between historical indices each year and finding the average change by dividing by the number of years.

Detailed Estimates

Estimates classified as detailed estimates are prepared after the scope and definition of a project are essentially complete. To prepare a detailed estimate requires considerable effort in gathering information and systematically forecasting costs. These estimates are usually prepared for bid purposes or definitive budgeting. Because of the information available and the effort expended, detailed estimates are usually fairly accurate projections of the costs of construction. A much higher level of confidence in the accuracy of the estimate is gained through this increased effort and knowledge. These types of estimates are used for decision making and commitment.

The Estimating Process

Estimating to produce a detailed construction cost estimate follows a rigorous process made up of several key steps. These key steps are explained below.

Familiarization with Project Characteristics. The estimator must be familiar with the project and evaluate the project from three primary avenues: scope, constructibility, and risk. Having evaluated these three areas in a general way the estimator will decide whether the effort to estimate and bid the work has a potential profit or other corporate goal potential (long-term business objective or client relations). In many cases, investigation of these three areas may lead to the conclusion that the project is not right for the contractor. The contractor must be convinced that the firm's competitive advantage will provide the needed margin to secure the work away from competitors.

Scope: Just because a project is available for bidding does not mean that the contractor should invest the time and expense required for the preparation of an estimate. The contractor must carefully scrutinize several issues of scope for the project in relation to the company's ability to perform. These scope issues include the following:

1. The technological requirements of the project
2. The stated milestone deadlines for the project
3. Required material and equipment availability
4. The staffing requirements

The contractor must honestly assess the technological requirements of the project to be competitive and the internal or subcontractor technological capabilities which can be employed. This is especially true on projects requiring fleets of sophisticated or specialized equipment or on projects whose duration dictates employment of particular techniques such as slipforming. On these types of projects, the contractor must have access to the fleet, as in the case of an interstate highway project, or access to a knowledgeable subcontractor, as in the case of high-rise slipforming.

The contractor must examine closely the completion date for the project as well as any intermediate contractual milestone dates for portions of the project. The contractor must feel comfortable that these dates are achievable and probably that there exists some degree of time allowance for contingencies which might arise. Failure to complete a project on time can seriously damage the reputation of a contractor and has the potential to inhibit future bidding opportunities with the client. If the contract time requirements are not reasonable in the contractor's mind after having estimated the required time by mentally sequencing the controlling work activities, two choices exist. The obvious first choice is to not **bid** the project. Alternatively, the contractor may choose to reexamine the project for other methods or sequences which will allow earlier completion. The contractor should not proceed with the estimate without a plan for timely completion of the project.

A third issue which must be examined in relation to the project's scope is availability to satisfy the requirements for major material commodities and equipment to support the project plan. Problems in obtaining structural steel, timber, quality concrete, or other materials can have pronounced effects on both the cost and schedule of a project. If these problems can be foreseen, then solutions should be sought or the project should not be considered for bidding.

Staffing requirements, including staffing qualifications as well as required numbers, must be evaluated to determine if sufficient levels of qualified manpower will be available when required to support project needs. This staffing evaluation must include supervisory and professional support and the various crafts which will be required. While the internal staffing (supervisory and professional support) is relatively simple to analyze, the craft availability is extremely uncertain and to some degree uncontrollable. With the craft labor in much of the construction industry (union sector) having no direct tie to any one construction company, it is virtually impossible to determine how many workers of a particular craft will be available during a particular month or week. With the current trend away from union construction (and the predominant use of hiring halls) the problem of prediction is compounded. Workers needed for various tasks may or may not be from a particular craft and they may or may not belong to the union. The ability to predict craft labor availability today is a function of construction economy prediction. When there is a booming construction market some shortfalls in craft labor supply can be expected.

Constructibility: A knowledgeable contractor, having made a preliminary review of the project documents, can assess the constructibility of the project. Constructibility evaluations include examination of construction quality requirements, allowable tolerances, and the overall complexity of the project. The construction industry has general norms of quality requirements and tolerances for the various types of projects. Contractors tend to avoid bidding for projects where the quality or tolerances specified are outside those norms. The alternative for the contractor is to overcompensate for the risk associated with achieving the requirements by inflating the bid.

Complexity of a project is viewed in terms of the relative technology requirement for the project execution compared with the technology in common practice in the given area. Where the project documents indicate an unusual method to the contractor, the contractor must choose to either accept the new technology or not bid. The complexity may also come about because of dictated logistical or scheduling requirements which must be met. Where the schedule does not allow flexibility in sequence or pace the contractor may deem the project unsuitable to pursue through bidding.

The flexibility left to the contractor in choosing methods creates interest in bidding the project. The means and methods of work are the primary ways that contractors achieve competitive advantage. This flexibility challenges the contractor to develop a plan and estimate for the work that will be different and cheaper than the competition's.

Risk: The contractor must also evaluate the myriad of potential problems that might be encountered on the project. These risks can include the following:

Material and workmanship requirements not specified
Contradictory clauses interpreted incorrectly
Impossible specifications

Unknown or undiscovered site conditions
Judgment error during the bidding process
Assumption of timely performance of approvals and decisions by the owners
Interpretation and compliance requirements with the contract documents
Changes in cost
Changes in sequence
Subcontractor failure
Suspension of work
Weather variations
Environmental issues
Labor and craft availability
Strikes and labor disputes
Utility availability

This list represents a sample of the risks rather than an inclusive listing. In general, a construction firm faces business risks, project risks, and operational risks, which must be offset in some way.

Examine the Project Design. Another aspect of the information which is important to the individual preparing the estimate is the specific design information which has been prepared. The estimator must be able to read, interpret, and understand the technical specifications, the referenced standards and any project drawings, and documents. The estimator must closely examine material specifications so that an appropriate price for the quality and characteristics specified can be obtained. The estimator must use sound judgment when pricing substitute materials for providing an assumption of "or-equal" quality for a material to be used. A thorough familiarity and technical understanding is required for this judgment. The same is also true for equipment and furnishings which will be purchased. The estimator must have an understanding of referenced documents which are commonly identified in specifications. Standards of testing and performance are made a part of the specifications by a simple reference. These standards may be client or more universal standards such as State Highway Specifications or ASTM (American Society for Testing and Materials) documents. If a specification is referenced that the estimator is not familiar with, he or she must make the effort to locate and examine it prior to bid submittal.

In some cases the specifications will identify prescribed practices which are to be followed. The estimator must assess the degree to which these will be rigidly enforced and where allowances will be made or performance criteria will be substituted. Use of prescriptive specifications can choke off innovation by the contractor but may also protect the contractor from performance risks. Where rigid enforcement can be expected the estimator should follow the prescription precisely.

The drawings contain the physical elements, their location, and their relative orientation. These items and the specifications communicate the designer's concept. The estimator must be able to examine the drawings and mentally visualize the project as it will be constructed to completion. The estimator relies heavily on the information provided in the drawings for determination of the quantity of work required. The drawings provide the dimensions so that lengths, widths, heights, areas, volumes, and numbers of items can be developed for pricing the work. The drawings show the physical features which will be a part of the completed project but do not show the items which may be required to achieve completion (such as formwork). It is also common that certain details are not shown on the drawings for the contractor but are developed by shop fabricators at a later time as shop drawings.

The estimator must keep a watchful eye for errors and omissions in the specifications and drawings. Discrepancies are often identified between drawings, between specifications, or between drawings and specifications. The discrepancies must be resolved either by acceptance of a risk or through communication with the designer. The best choice of solution depends on the specifics of the discrepancy and the process or the method for award of contract.

Structuring the Estimate. The estimator either reviews a plan or develops a plan for completing the project. This plan must be visualized during the estimating process; it provides the logical flow of the project from raw materials to a completed facility. Together with the technical specifications, the plan provides a structure for the preparation of the detailed estimate. Most estimators develop the estimate around the structure of the technical specifications. This increases the likelihood that items of work are covered without duplication in the estimate.

Determine the Elements of Cost. This step involves the development of the quantities of work (a **quantity survey**) to be performed and their translation into expected costs. Translating a design on paper into a functioning, completed project involves the transformation and consumption of a multitude of resources. These basic ingredients or resources which are utilized and incorporated in a project during construction can be classified into one of the following categories:

1. Labor
2. Material
3. Equipment
4. Capital
5. Time

Associated with the use or consumption of each of these resources is a cost. It is the objective of the estimator performing a detailed estimate to identify the specific types of resources which will be used, the quantity of such resources, and the cost of the resource. Every cost item within an estimate is either one or a combination of these five basic resources. The common unit used to measure the different types of resources is dollars. Although overhead costs may not be broken down into the component resource costs, overhead items are a combination of several of these basic resources.

Labor Resources

Labor resources refer to the various human craft or skill resources which actually build a project. Through the years, large numbers of crafts have evolved to perform specialized functions and tasks in the construction industry. The specialties or crafts have been defined through a combination of collective bargaining agreements, negotiation and labor relations, and accepted extensions of trade practices. In most cases, the evolutionary process of definition of work jurisdiction has followed a logical progression; however, there are limited examples of bizarre craftwork assignment. In all, there are over 30 different crafts in the construction industry. Each group or craft is trained to perform a relatively narrow range of construction work either differentiated by material type, construction process, or type of construction project. Where union construction is dominant, the assignment of work to a particular craft can become a significant issue with the potential for stopping or impeding progress. Usually in nonunion construction, jurisdictional disputes are nonexistent, and much more flexibility exists in the assignments of workers to tasks. In union construction, it is vital that the estimator acknowledge the proper craft for a task since labor wage rates can vary substantially between crafts. In nonunion construction, there exists more managerial flexibility, and the critical concern to the estimator must be that a sufficient wage rate is used which will attract the more productive craftsworkers without hindering the chances of competitive award of the construction contract.

The source of construction labor varies between localities. In some cities, the only way of performing construction is through union construction. This, however, has been changing, and will most likely continue to change over the next few years. Open-shop or nonunion construction is becoming the predominant form.

With union construction, the labor source is the hiring hall. The usual practice is for the superintendent to call the craft hiring hall for the type of labor which is needed and request the number of craftsworkers which are needed for the project. The craftsworkers are then assigned to projects in the order in which they became available for work (were released from other projects).

This process, while fair to all craftsworkers, has some drawbacks for the contractor since the personnel cannot be selected based on particular past performance.

These union craftsworkers in construction have their primary affiliation with the union, and only temporarily are affiliated with a particular company, usually for the duration of a particular project. The training and qualifications for these craftsworkers must, therefore, be a responsibility of the union. This training effort provided through the union is financed through a training fund established in the collective bargaining agreement. Apprenticeship programs are conducted by union personnel to develop the skills needed by the particular craft. A second avenue for control is through admission into the union and acceptance after a trial period by the employer. The training for the craftsworker for this approach may have been in another vocational program, on-the-job experience, or a military training experience. The supply of craftsworkers in relation to the demands is thus controlled partially through the admissions into the training or apprenticeship program.

Open-shop or nonunion construction does not yet have well-established training programs or clearly defined and organized sources of labor. For the most part, the open-shop contractor relies on the other training sources (union apprenticeship, vocational schools, and military training) for preparation of the craftsworker, and most exercise considerable effort in screening and hiring qualified labor. Typically, craftsworkers are hired for primary skill areas, but can be utilized on a much broader range of tasks. A trial period for new employees is used to screen craftsworkers for the desired level of skill required for the project. Considerably more effort is required for recruiting and maintaining a productive work force in the open-shop mode.

Cost of Labor. For a detailed estimate, it is imperative that the cost of labor resources be determined with precision. This is accomplished through a three-part process from data in the construction bidding documents which identifies the nature of work and the physical quantity of work. The first step in the process involves identifying the craft which will be assigned the work and determining the hourly cost for that labor resource. This is termed the *labor rate.* The second part of the process involves estimation of the expected rate of work accomplishment by the chosen labor resource. This is termed the labor **productivity.** The third step involves combining this information by dividing the labor rate by the labor productivity to determine the labor resource cost per physical unit of work. The labor cost can be determined by multiplying the quantity of work by the unit labor resource cost. This entire process will be illustrated later in this chapter; however, an understanding of labor rate and labor productivity measurement must first be developed.

Labor Rate: The labor rate is the total hourly expense or cost to the contractor for providing the particular craft or labor resource for the project. This labor rate includes **direct costs** and **indirect costs.** Direct labor costs include all payments made directly to the craftsworkers. The following is a brief listing of direct labor cost components:

1. Wage rate
2. Overtime premium
3. Travel time allowance
4. Subsistence allowance
5. Show-up time allowance
6. Other work or performance premiums

The sum of these direct labor costs is sometimes referred to as the *effective wage rate.* Indirect labor costs include those costs which are incurred as a result of use of labor resources but which are not paid directly to the craftsworker. The components of indirect labor cost include the following:

1. Vacation fund contributions
2. Pension fund contributions
3. Group insurance premiums

4. Health and welfare contributions
5. Apprenticeship and training programs
6. Workers' compensation premiums
7. Unemployment insurance premiums
8. Social security contribution
9. Other voluntary contribution or payroll tax

It is the summation of direct and indirect labor costs that is termed the labor rate—the total hourly cost of providing a particular craft labor resource. Where a collective bargaining agreement is in force, most of these items can be readily determined on an hourly basis. Others are readily available from insurance companies or from local, state, and federal statutes. Several of the direct cost components must be estimated based on past records to determine the appropriate allowance to be included. These more difficult items include overtime, show-up time, and performance premiums. A percentage allowance is usually used to estimate the expected cost impact of such items.

Labor Productivity: Of all the cost elements which contribute to the total project construction cost, labor productivity ranks at the top for variability. Because labor costs represent a significant proportion of the total cost of construction, it is vital that good estimates of productivity be made relative to the productivity which will be experienced on the project. Productivity assessment is a complex process and not yet fully understood for the construction industry.

The following example illustrates the calculation of a unit price from productivity data.

Example. To form 100 square feet of wall requires 6 hours of carpenter time and 5 hours of common laborer time. This assumption is based on standards calculated as averages from historical data. The wage rate with burdens for carpenters is \$21.00/hr. The wage rate with burdens for common laborers is \$17.00/hr.

Solution. The unit cost may be calculated as follows:

Carpenter—6 hours at \$22.00/hr	= \$132.00
Laborer—5 hours at \$18.00/hr	= 90.00
Total labor cost for 100 ft^2	= \$222.00

$$\text{Labor cost per ft}^2 = \$222.00/100\ \text{ft}^2 = \$2.22/\text{ft}^2$$

This labor cost is adjusted for the following conditions:

Weather adjustment	1.05
Job complexity	1.04
Crew experience	0.95
Management	1.00

$$\text{Adjusted unit cost} = 2.22 \times 1.05 \times 1.04 \times 0.95 \times 1.00 = \$2.30/\text{ft}^2$$

Equipment Resources

One of the most important decisions which a contractor makes involves the selection of construction equipment. Beyond very simple construction projects, a significant number of the activities require some utilization of major pieces of equipment. This equipment may either be purchased by the contractor or leased for the particular project at hand. The decision for selection of a particular type of equipment may be the result of an optimization process or may be based solely on the fact that the contractor already owns a particular piece of equipment which should be put to use. This decision must be anticipated or made by the estimator, in most cases, to forecast the expected costs for equipment on a project being estimated.

Equipment Selection Criteria. It is important for the estimator to have a solid background in and understanding of various types of construction equipment. This understanding is most important when making the decisions about equipment. The estimator, having recognized the work to be performed, must identify the most economical choice for equipment. There are four important criteria which must be examined to arrive at the best choice:

1. Functional performance
2. Project flexibility
3. Companywide operations
4. Economics

The functional performance is only one criterion, but an important one, for the selection of construction equipment. For each activity, there is usually a clear choice based on the most appropriate piece of equipment to perform the task. The functional performance is usually examined solely from the perspective of functional performance. The usual measures are capacity and speed. These two parameters also give rise to the calculation of production rates.

A second criterion which must be used is project flexibility. Although each task has a most appropriate piece of equipment based on functional performance, it would not be prudent to mobilize a different piece of equipment for each activity. Equipment selection decisions should consider the multiple uses which the item of equipment possesses for the particular project. The trade-off between mobilization expense and duration versus efficiency of the operation must be explored to select the best fleet of equipment for the project.

Companywide usage of equipment becomes an important factor when determining whether to purchase a particular piece of equipment for a project application. If the investment in the equipment cannot be fully justified for the particular project, then an assessment of future or concurrent usage of the equipment is necessary. This whole process necessarily influences selection decisions by the estimator since the project cost impacts must be evaluated. Equipment which can be utilized on many of the company projects will be favored over highly specialized single project oriented equipment.

The fourth, and probably most important, criterion which the estimator considers is the pure economics of the equipment selection choices. Production or hourly costs of the various equipment alternatives should be compared to determine the most economical choice for the major work tasks involving equipment. A later section in this chapter explains and illustrates the process of determining equipment costs which the estimator should follow.

Production Rates

Equipment production rates can be determined in a relatively simple fashion for the purposes of the estimator. Most manufacturers produce handbooks for their equipment that provide production rates for tasks under stated conditions.

Equipment Costs

Equipment costs represent a large percentage of the total cost for many construction projects. Equipment represents a major investment for contractors and it is necessary that the investment generate a return to the contractor. The contractor must not only pay for the equipment purchased but pay the many costs associated with the operation and maintenance of the equipment. Beyond the initial purchase price, taxes, and set-up costs, the contractor has costs for fuel, lubricants, repairs, and so on, which must be properly estimated when preparing an estimate. A system must be established to measure equipment costs of various types to provide the estimator with a data source to use when establishing equipment costs.

The cost associated with equipment can be broadly classified as direct equipment costs and indirect equipment costs. Direct equipment costs include the ownership costs and operating expenses while indirect equipment costs are the costs which occur in support of the overall fleet of equipment but which cannot be specifically assigned to a particular piece of equipment. Each of the broad cost categories will be discussed in greater detail in the following sections.

Direct Equipment Costs

Direct equipment expenses are costs which can be assigned to a particular piece of equipment, and are usually divided into ownership and operating expenses for accounting and estimating. The concept behind this separation is that the ownership costs occur regardless of whether the equipment is used on a project.

Ownership Costs. Ownership costs include depreciation, interest, insurance, taxes, set-up costs, and equipment enhancements. There are several views which are taken of ownership costs relating to loss in value or depreciation. One view is that income must be generated to build a sufficient reserve to replace the equipment at the new price when it becomes obsolete or worn out. A second view is that ownership of a piece of equipment is an investment, and, as such, must generate a monetary return on that investment equal to or larger than the investment made. A third view is that the equipment ownership charge should represent the loss in value of the equipment from the original value due purely to ownership assuming some arbitrary standard loss in value due to use. These three views can lead to substantially different ownership costs for the same piece of equipment depending on the circumstances. For simplicity, ownership will be viewed as in the third view. The depreciation component of ownership cost will be discussed separately in the following section.

Depreciation Costs. Depreciation is the loss in value of the equipment due to use and/or obsolescence. There are several different approaches for calculating depreciation either based on hours of operation or on real-time years of ownership. In both cases, some arbitrary useful life is assumed for the particular piece of equipment based on experience with similar equipment under similar use conditions. The simplest approach for calculating depreciation is the straight-line method. Using the useful life, either hours of operation or years, the equipment is assumed to lose value uniformly over the useful life from its original value down to its salvage value. The salvage value is the expected market value of the equipment at the end of its useful life.

Operating Expenses. Operating costs are items of cost which are directly attributable to the use of the equipment. Operating costs include such items as fuel, lubricants, filters, repairs, tires, and sometimes operator's wages. Obviously, the specific project conditions will greatly influence the magnitude of the operating costs. It is, therefore, important that on projects where the equipment is a significant cost item, such as large civil works projects such as dams or new highway projects, attention must be given to the job conditions and operating characteristics of the major pieces of equipment.

Equipment Rates

The equipment rates that are used in an estimate represent an attempt to combine the elements of equipment cost that have been explained above. The pricing of equipment in an estimate is also influenced by market conditions. On very competitive projects, the contractor will often discount the actual costs to win the project. In other cases, even though the equipment has been fully depreciated, a contractor may still include an ownership charge in the estimate because the market conditions will allow the cost to be included in the estimate.

Materials Costs

Materials costs can represent the major portion of a construction estimate. The estimator must be able to read and interpret the drawings and specifications and develop a complete list of the materials required for the project. With this quantity takeoff, the estimator then identifies the cost of these materials. The materials costs include several components: the purchase price, shipping and packaging, handling, and taxes.

There are two types of materials: **bulk materials** and engineered materials. Bulk materials are those materials that have been processed or manufactured to industry standards. Engineered materials are those that have been processed or manufactured to project standards. Examples of bulk materials are sand backfill, pipe, and concrete. Examples of engineered materials are compressors, handrailing, and structural steel framing. The estimator must get unit price quotes on bulk materials and must get quotes on the engineered materials that include design costs as well as the processing and other materials costs.

Subcontractor Costs

The construction industry continues to become more specialized. The building sector relies almost entirely on the use of specialty contractors to perform different trade work. The heavy/highway construction industry subcontracts a smaller percentage of work. The estimator must communicate clearly with the various subcontractors to define the scope of intended work. Each subcontractor furnishes the estimator with a quote for the defined scope of work with exceptions noted. The estimator must then adjust the numbers received for items that must be added in and items that will be deleted from their scope. The knowledge of the subcontractor and any associated risk on performance by the subcontractor must also be assessed by the estimator. The estimator often receives the subcontractor's best estimate only a few minutes before the overall bid is due. The estimator must have an organized method of adjusting the overall bid up to the last minute for changes in the subcontractor's prices.

Example. 15,000 cubic yards of material must be hauled onto a job site for use as structural fill. As the material is excavated it is expected to swell. The swell factor is 0.85. The material will be hauled by four 12-yd^3 capacity trucks. The trucks will be loaded by a 1.5-yd^3 excavator. Each cycle of the excavator will take about 30 seconds. The hauling time will be 9 minutes, the dumping time 2 minutes, the return time 7 minutes, and the spotting time 1 minute. The whole operation can be expected to operate 50 minutes out of every hour. The cost of the trucks is \$66/hour and the excavator will cost about \$75/hour. What is the cost per cubic yard for this operation?
Solution.

$$\begin{aligned} \text{Excavator capacity} &= 1.5\ \text{yd}^3 \times 0.85 = 1.28\ \text{yd}^3/\text{cycle} \\ \text{Hauler capacity} &= 12 \times 0.85 = 10.2\ \text{yd}^3/\text{cycle} \\ \text{Number of loading cycles} &= 10.2/1.28 = 8\ \text{cycles} \end{aligned}$$

Truck cycle time:

Loading	8 cycles × 0.5 minute =	4 minutes
Haul		9 minutes
Dump		2 minutes
Return		7 minutes
Spot		1 minute
		23 minutes

Fleet production:

$$\begin{aligned} 4 \times (50/23) \times 10.2 &= 89.75\ \text{yd}^3/\text{hour} \\ 15{,}000/89.75 &= 168\ \text{hours} \end{aligned}$$

Cost:

168 × 66 × 4	= \$44,352
168 × 75	= \$12,600
	\$56,952

$$56{,}952/15{,}000 = \$3.80/\text{yd}^3$$

Example. 90 cubic yards of concrete need to be placed. Site conditions dictate that the safest and best method of placement is to use a crane and a 2-cubic-yard bucket. It is determined that to perform the task efficiently, five laborers are needed—one at the concrete truck, three at the point of placement, and one on the vibrator. It is assumed that supervision is done by the superintendent. The wage rate for laborers is $18.00/hr.

Time needed:	
Set-up	30 minutes
Cycle:	
Load	3 minutes
Swing, dump, and return	6 minutes
	9 minutes

No. of cycles	90/2	= 45 cycles
Total cycle time	45 × 9	= 405 minutes
Disassembly subtotal		= 15 minutes
Inefficiency (labor, delays, etc.) 10% of cycle time		= 41 minutes
Total operation time	405 + 15 + 41	= 461 minutes
Amount of time needed (adjusted to workday)		= 8 hours
Laborers—five for 8 hours at $18.00/hr		= $720.00
Cost per 90 yd^3		= $720.00
Cost per cubic yard	$720/90 yd^3	= $8/yd^3

Example. A small steel-frame structure is to be erected, and you are to prepare an estimate of the cost based on the data given below and the assumptions provided. The unloading, erection, temporary bolting, and plumbing will be done by a crew of 1 foreman, 1 crane operator, and 4 structural steel workers with a 55-ton crawler crane. The bolting will be done by 2 structural-steel workers using power tools. The painting will be done by a crew of 3 painters (structural-steel) with spray equipment. For unloading at site, erection, temporary bolting, and plumbing, allow 7 labor-hours per ton for the roof trusses, and allow 5.6 labor-hours per ton for the remaining steel. Assume 60 crew hours will be required for bolting. Allow 1.11 labor-hours per ton for painting.

Materials:		
A 36 structural	Steel trusses	15 tons
	Columns, etc.	50 tons
Costs:		
Structural steel supply:		22¢/pound
Fabrication:		$450/ton—trusses
		$250/ton—other steel
Freight cost:		$1.65/100 pounds
Field bolts:		250 @ 60¢ each
Paint:		41 gallons @ $16.00/gallon
Labor costs:	Assume payroll taxes and insurance are 80% of labor wage; use the following wages:	

Foreman	$	24.10
Crane operator	$	21.20
Structural steel worker	$	22.10
Painter	$	20.20

Equipment costs:		
	Crane	\$ 915.00/day
	Power tools	\$ 23.40/day
	Paint equipment	\$ 68.00/day
Move in/out:		\$ 300.00
Overhead:		40% of field labor cost
Profit:		12% of all costs

Solution.

Materials:		
Structural steel: 65 × 2000 × .22	=	\$ 28,600
Freight: 65 × 2000/100 × 1.65	=	2,145
Field bolts: 250 × .60	=	150
Paint: 41 × 16	=	656
		\$ 31,551
Fabrication:		
Truss: 15 × 450	=	\$ 6,750
Frame: 50 × 250	=	12,500
		\$ 19,250
Labor crew costs:		
Erection:		
1 foreman:		24.10
1 crane operator:		21.20
4 structural steel workers:		88.40
		\$ 133.40
Paint:		
3 painters:		\$ 60.60
Bolting:		
2 structural steel workers:		\$ 44.20
Erection:		
Frame:		
(50 × 5.6)/6	= 46.7 crew hours—6 days	
	46.7 × \$133.40 =	\$ 6,239
Trusses:		
(15 × 7)/6	= 17.5 crew hours—2 days	
	17.5 × 133.40 =	\$ 2,340
Paint:		
(65 × 1.11)/3	= 24 crew hours—3 days	
	24 × 60.60 =	\$ 1,455
Bolting:		
	60 × 44.20 =	\$ 2,652
	Total labor =	\$12,686
Equipment		
Crane: 8 days × 915/day	=	\$ 7,320
Power tools: 8 days × 23.40/day	=	187
Paint equipment: 3 days × 68/day	=	204
Move in/out	=	300
		\$ 8,011

Summary

Materials:	$ 31,551
Fabrication:	19,250
Labor:	12,686
Equipment:	8,011
Payroll taxes and insurance:	
80% of 12,686	10,149
Overhead:	
40% of (12,686+10,149)	9,134
	$ 90,781
Profit:	
12% of 90,781	10,894
Bid =	$101,675

Project Overhead

Each project requires certain items of cost that cannot be identified with a single item of work. These items are referred to as project overhead. These items are normally described in the general conditions of the contract. The items which are a part of the project overhead include but are not limited to the following:

- Bonds
- Permits
- Mobilization
- Professional services (such as scheduling)
- Safety equipment
- Small tools
- Supervision
- Temporary facilities
- Travel and lodging
- Miscellaneous costs (e.g., cleanup, punch list)
- Demobilization

Each of these types of items should be estimated and included in the cost breakdown for a project.

Markup

Once the direct project costs are known, the estimator adds a sum of money to cover a portion of the general overhead for the firm and an allowance for the risk and investment made in the project—the profit. Each of these elements of **markup** is in large part determined by the competitive environment for bidding the project. The more competition, the less the markup.

General Overhead

Each business has certain expenses that are not variable with the amount of work that they have under contract. These expenses must be spread across the projects. The typical method for spreading the general overhead is to assign it proportionally according to the size of the project in relation to the expected total volume of work for the year. General overhead costs typically include the following:

- Salaries (home office)
- Employee benefits
- Professional fees
- Insurance
- Office lease or rent

Office stationery and supplies
Maintenance
Job procurement and marketing
Home office travel and entertainment
Advertising

The only restriction on the items of general overhead is that they must have a legitimate business purpose. The estimator typically will start with the proportional amount and then add a percentage for profit.

Profit

The profit that is assigned to a project should recognize the nature of risk that the company is facing in the project and an appropriate return on the investment being made in the project. The reality is that the profit is limited by the competition. A larger number of bidders requires that a smaller profit be assigned to have a chance at having the low bid. This process of assigning profit is usually performed at the last minute by the senior management for the company submitting the bid.

1.5 Contracts

The estimator prepares the estimate in accordance with the instructions to bidders. There are numerous approaches for buying construction services which the estimator must respond to. These various approaches can be classified by three characteristics: the method of award, the method of bidding/payment, and incentives/disincentives which may be attached.

Method of Award

There are three ways in which construction contracts are awarded: competitive awards, negotiated awards, and combination competitive-negotiated awards. With a purely competitive award the decision is made solely on the basis of price. The lowest bidder will be awarded the project. Usually, public work is awarded in this manner, and all who meet the minimum qualifications (financial) are allowed to compete. In private work the competitive method of award is used extensively; however, more care is taken to screen potential selective bidders.

The term *selective bid process* describes this method of competitive award. At the opposite extreme from competitive awards are the negotiated awards. In a purely negotiated contract, the contractor is the only party asked to perform the work. Where a price is required prior to initiating work, this price is negotiated between the contractor and the client. Obviously, this lack of competition relieves some of the tension developed in the estimator through the competitive bid process since there is no need to be concerned with the price another contractor might submit. The contractor must still, if asked, provide a firm price which is acceptable to the client and may have to submit evidence of cost or allow an audit. As the purely competitive and purely negotiated method of contract awards represent the extremes, the combination competitive-negotiated award may fall anywhere in between. A common practice for relatively large jobs is to competitively evaluate the qualifications of several potential constructors and then select and negotiate with a single contractor a price for the work.

Method of Bidding/Payment

Several methods of payment are used to reimburse contractors for the construction services which they provide. These methods of payment include lump sum or firm price, unit-price, and cost-plus. Each of these methods of payment requires an appropriate form of bidding which recognizes the unique incentive and risk associated with the method. The requirements for completeness of design and scope definition vary for the various types. The lump-sum or firm price contract is widely used for well-defined projects with completed designs. This method allows purely competitive bidding. The contractor assumes nearly all of the risk, for quantity and quality. The comparison for bidding is based entirely on the total price submitted by the contractors, and payment for the work is limited to the agreed-upon contract price with some allowance for negotiated changes. The lump sum is the predominant form used for most building projects.

The unit-price contract is employed on highway projects, civil works projects, and pipelines. For these projects the quality of the work is defined but the exact quantity is not known at the time of bidding. The price per unit is agreed upon at the time of bidding but the quantity is determined as work progresses and is completed. The contractor therefore assumes a risk for quality performance but the quantity risk is borne by the owner. There is a strong tendency, by contractors, to overprice or front-load those bid items which will be accomplished first and compensate with lower pricing on items of work which will be performed later. This allows contractors to improve their cash flow and match their income closer to their expenses. Each unit-price given must include a portion of the indirect costs and profits which are a part of the job. Usually, quantities are specified for bidding purposes so that the prices can be compared for competitive analysis. If contractors "unbalance" or front-load certain bid items to an extreme, they risk being excluded from consideration. The unit-price approach is appropriate for projects where the quantity of work is not known, yet where competitive bidding is desirable.

A third method, with many variations, which is used is the cost-plus method of bidding/payment. With this method the contractor is assured of being reimbursed for the costs involved with the project plus an additional amount to cover the cost of doing business and an allowance for profit. This additional amount may be calculated as a fixed fee, a percent of specified reimbursable costs, or a sliding-scale amount. The cost to the owner with this method of bidding/payment is open-ended and thus the risks lie predominantly with the owner. This method is used in instances where it is desired to get the construction work underway prior to completion of design or where it is desired to protect a proprietary process or production technology and design. Many of the major power plant projects, process facilities, and other long-term megaprojects have used this method in an attempt to shorten the overall design/construct time frame and realize earlier income from the project.

Of the several variations which are used most relate to the method of compensation for the "plus" portion of the cost and the ceiling which is placed on the expenditures by the owner. One of the variations is the cost plus a fixed fee. With this approach it is in the contractor's best interest to complete the project in the least time with the minimum nonreimbursable costs so that his profits during a given time period will be maximized. Where the scope, although not defined specifically, is generally understood this method works well. The owner must still control and monitor closely actual direct costs. A second variation is the cost plus a percentage. This method offers little protection for the owner on the cost of the project or the length of performance. This method, in fact, may tempt the contractor to prolong project completion to continue a revenue stream at a set return. The sliding scale approach is a third approach. This method of compensation is a combination of the two approaches described above. With this approach a target amount for the project cost is identified. As costs exceed this amount the fee portion decreases as a percentage of the reimbursable portion. If the costs are less than this target figure, there may be a sliding scale which offers the contractor an increased fee for good cost containment and management.

In addition to the method of calculations of the plus portion for a cost-plus method there may be a number of incentives attached to the method. These typically take the form of bonuses and penalties for better time or cost performance. These incentives may be related to the calendar or working day allowed for completion in the form of an amount per day for early completion. Similarly, there may be a penalty for late completion. The owner may also impose or require submittal of a guaranteed maximum figure for a contract to protect the owner from excessive costs.

1.6 Computer-Assisted Estimating

The process of estimating has not changed but the tools of the estimator are constantly evolving. The computer has become an important tool for estimators, allowing them to produce more estimates in the same amount of time and with improved accuracy.

Today the computer is functioning as an aid to the estimator by using software and digitizers to read the architect/engineer's plans; by retrieving and sorting historical cost databases; by analyzing information and developing comparisons; and by performing numerous calculations without error and presenting the information in a variety of graphical and tabular ways.

The microcomputer is only as good as the programmer and data entry person. The estimator must still use imagination to create a competitive plan for accomplishing the work. The computer estimating tools assist and speed the estimator in accomplishing many of the more routine tasks.

Defining Terms*

Bid: To submit a price for services; a proposition either verbal or written, for doing work and for supplying materials and/or equipment.

Bulk materials: Material bought in lots. These items can be purchased from a standard catalog description and are bought in quantity for distribution as required.

Cost: The amount measured in money, cash expended, or liability incurred, in consideration of goods and/or services received.

Direct cost: The cost of installed equipment, material, and labor directly involved in the physical construction of the permanent facility.

Indirect cost: All costs that do not become a part of the final installation but which are required for the orderly completion of the installation.

Markup: Includes the percentage applications such as general overhead, profit, and other indirect costs.

Productivity: Relative measure of labor efficiency, either good or bad, when compared to an established base or norm.

Quantity survey: Using standard methods to measure all labor and material required for a specific building or structure and itemizing these detailed quantities in a book or bill of quantities.

Scope: Defines the materials and equipment to be provided and the work to be done.

**Source: AACE Recommended Practices and Standards.* American Association of Cost Engineers (AACE, Inc.), November 1991.

References

Adrian, J. J. 1982. *Construction Estimating.* Reston Publishing, Reston, VA.

Bauman, H. C. 1964. *Fundamentals of Cost Engineering in the Chemical Industry.* Reinhold Publishing, Florence, KY.

Collier, K. F. 1974. *Fundamentals of Construction Estimating and Cost Accounting.* Prentice-Hall, Englewood Cliffs, NJ.

Cost Engineer's Notebook. American Association of Cost Engineers, Morgantown, WV.

Helyar, F. W. 1978. *Construction Estimating and Costing.* McGraw-Hill Ryerson Ltd., Scarborough, Ontario.

Gooch, K. O. and Caroline, J. 1980. *Construction for Profit.* Reston Publishing, Reston, VA.

Hanscomb, R. *et al.* 1983. *Yardsticks for Costing.* Southam Business Publications, Ltd., Toronto.

Humphreys, ed. 1984. *Project and Cost Engineers' Handbook.* Marcel Dekker, New York.

Hunt, W. D. 1967. *Creative Control of Building Costs.* McGraw-Hill, New York.

Landsdowne, D. K. 1983. *Construction Cost Handbook.* McGraw-Hill Ryerson Ltd., Scarborough, Ontario.

Neil, J. M. 1982. *Construction Cost Estimating for Project Control.* Prentice-Hall, Englewood Cliffs, NJ.

Peurifoy, R. L., 1975. *Estimating Construction Costs.* McGraw-Hill, New York.

Seeley, I. H., 1978. *Building Economics.* The Macmillan Press, Ltd., London.

Vance, M. A., 1979. *Selected List of Books on Building Cost Estimating.* Vance Bibliographies, Monticello, IL.

Walker, F. R., 1980. *The Building Estimator's Reference Book.* Frank R. Walker Publishing, Chicago, IL.

For Further Information

For more information on the subject of cost estimating one should contact the following professional organizations which have additional information and recommended practices.

AACE, International (formerly the American Association of Cost Engineers), 209 Prairie Ave, Suite 100, Morgantown, West Virginia 26507, 800-858-COST.

American Society of Professional Estimators, 11141 Georgia Ave, Suite 412, Wheaton, Maryland 20902, 301-929-8848.

There are numerous textbooks on the subject of cost estimating and construction cost estimating available. Cost engineering texts usually have a large portion devoted to both conceptual estimating and detailed estimating. The following reference materials are recommended:

Process Plant Construction Estimating Standards. Richardson Engineering Services, Mesa, AZ.

Contractor's Equipment Cost Guide. Data quest—The Associated General Contractors of America (AGC).

The Building Estimator's Reference Book. Frank R. Walker, Lisle, IL.

Means Building Construction Cost Data. R.S. Means, Duxbury, MA.

Estimating Earthwork Quantities. Norseman Publishing, Lubbock, TX.

Caterpillar Performance Handbook, 24th ed. Caterpillar, Peoria, IL.

Means Man-Hour Standards. R.S. Means, Duxbury, MA.

Rental Rates and Specifications. Associated Equipment Distributors.

Rental Rate Blue Book. Data quest—The Dun & Bradstreet Corporation, New York.

Historical Local Cost Indexes. AACE—Cost Engineers Notebook, Vol. 1

Engineering News Record. McGraw-Hill, New York.

U.S. Army Engineer's Contract Unit Price Index. U.S. Army Corps of Engineers.

Chemical Engineering Plant Cost Index. McGraw-Hill, New York.

Bureau of Labor Statistics. U.S. Department of Labor.

2

Construction Planning and Scheduling

Donn E. Hancher
University of Kentucky

2.1 Introduction

One of the most important responsibilities of construction **project** management is the planning and scheduling of construction projects. The key to successful profit making in any construction company is to have successful projects. Therefore, for many years, efforts have been made to plan, direct, and control the numerous project activities to obtain optimum project performance. Since every construction project is a unique undertaking, project managers must plan and schedule their work utilizing their experience with similar projects, and applying their judgment to the particular conditions of the current project.

Until just a few years ago, there was no generally accepted formal procedure to aid in the management of construction projects. Each project manager had a different system, which usually included the use of the Gantt chart, or bar chart. The bar chart was, and still is, quite useful for illustrating the various items of work, their estimated time durations, and their position in the work schedule as of the report date represented by the bar chart. However, the relationship which exists between the identified work items is by implication only. On projects of any complexity, it is difficult, if not virtually impossible, to identify the interrelationships between the work items, and there is no indication of the criticality of the various activities in controlling the **project duration**. A sample bar chart for a construction project is shown in Fig. 2.1.

The development of the critical path method (CPM) in the late 1950s has provided the basis for a more formal and systematic approach to project management. Critical path methods involve a graphical display (network diagram) of the activities on a project and their interrelationships,

0-8493-8953-4/95/$0.00 + $.50

WORK DESCRIPTION	SCHEDULED DATES			
	JUNE	JULY	AUGUST	SEPT.
CLEARING & LAYOUT				
EXCAVATE				
FORMWORK & REBAR				
CONCRETE FOUNDATIONS				
STRUCTURAL STEEL				
MASONRY				
PLUMBING				
ELECTRICAL				
HVAC				
ROOFING				
CARPENTRY				
LATH & PLASTER				
DOORS & WINDOWS				
TERRAZZO				
GLAZING				
HARDWARE & MILLWORK				
PAINTING				
EXTERIOR CONCRETE				

FIGURE 2.1 Sample Gantt or bar chart.

and an arithmetic procedure which identifies the relative importance of each activity in the overall project schedule. These methods have been applied with notable success to project management in the construction industry and several other industries, when applied earnestly as dynamic management tools. Also, they have provided a much-needed basis for performing some of the other vital tasks of the construction project manager, such as **resource** scheduling, financial planning, and cost control. Today's construction manager who ignores the use of critical path methods is ignoring a useful and practical management tool.

Planning and Scheduling

Planning for construction projects involves the logical analysis of a project, its requirements, and the plan (or plans) for its execution. This will also include consideration of the existing constraints and available resources which will affect the execution of the project. Considerable planning is required for the support functions for a project, material storage, worker facilities, office space, temporary utilities, and so on. Planning, with respect to the critical path method, involves the identification of the activities for a project, the ordering of these activities with respect to each other, and the development of a network logic diagram which graphically portrays the activity planning. Fig. 2.2 is an I-J CPM logic diagram.

The planning phase of the critical path method is by far the most difficult, but also the most important. It is here that the construction planner must actually build the project on paper. This can only be done by becoming totally familiar with the project plans, specifications, resources, and constraints and then looking at various plans for feasibly performing the project and selecting the best one.

The most difficult planning aspect to consider, especially for beginners, is the level of detail needed for the activities. The best answer is to develop the minimum level of detail required to enable the user to schedule the work efficiently. For instance, general contractors will normally consider two or three activities for mechanical work to be sufficient for their schedule. However,

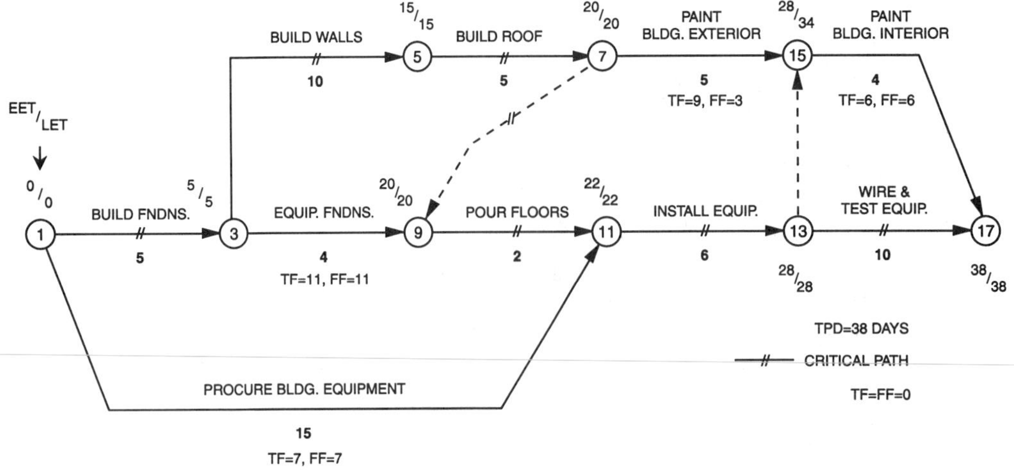

FIGURE 2.2 I-J CPM logic diagram.

to mechanical contractors, this would be totally inadequate since they will need a more detailed breakdown of their activities to schedule their work. Therefore, the level of activity detail required depends on the needs of the user of the plan, and only the user can determine his or her needs after gaining experience in the use of critical path methods.

Once the activities have been determined, they must be arranged into a working plan in the network logic diagram. Starting with an initial activity in the project, one can apply known constraints and reason that all remaining activities must fall into one of three categories:

1. They must precede the activity in question.
2. They must follow the activity in question.
3. They can be performed concurrently with the activity in question.

The remaining planning function is the estimation of the time durations for each activity shown on the logic diagram. The estimated activity time should reflect the proposed method for performing the activity, plus consideration of the levels at which required resources are supplied. The estimation of activity times is always a tough task for the beginner in construction because it requires a working knowledge of the production capabilities of the various crafts in the industry, which can only be acquired through many observations of actual construction work. Therefore, the beginner will have to rely on the advice of superiors for obtaining time estimates for work schedules.

Scheduling of construction projects involves the determination of the timing of each work item, or activity, in a project within the overall time span of the project. Scheduling, with respect to the critical path methods, involves the calculation of the starting and finishing times for each activity and the project duration, the evaluation of the available float for each activity, and the identification of the critical path or paths. In a broader sense it also includes the more complicated areas of construction project management such as financial funds, flow analyses, resource scheduling and leveling, and inclement weather scheduling.

The planning and scheduling of construction projects using critical path methods has been discussed as two separate processes. Although the tasks performed are different, the planning and scheduling processes normally overlap. The ultimate objective of the project manager is to develop a working plan with a schedule which meets the completion date requirements for the project. This requires an interactive process of planning and replanning, and scheduling and rescheduling, until a satisfactory working plan is obtained.

Controlling

The *controlling* of construction projects involves the monitoring of the expenditure of time and money in accordance with the working plan for the project, as well as the resulting product quality or performance. When deviations from the project schedule occur, remedial actions must be determined which will allow the project to be finished on time and within budget if at all possible. This will often require replanning the order of the remaining project activities.

If there is any one factor for the unsuccessful application of the critical path method to actual construction projects, it is the lack of project **monitoring** once the original schedule is developed. Construction is a dynamic process; conditions often change during a project. The main strength of the critical path method is that it provides a basis for evaluating the effects of unexpected occurrences (such as delivery delays) on the total project schedule. The frequency for performing updates of the schedule depends primarily on the job conditions, but are usually needed most as the project nears completion. For most projects, monthly updates of the schedule are adequate. At the point of 50% completion, a major update should be made to plan and schedule the remaining work. The control function is an essential part of successful CPM scheduling.

Critical Path Methods

The **critical path** technique was developed from 1956 to 1958 in two parallel but different problems of planning and control in projects in the U.S.

In one case, the U.S. Navy was concerned with the control of contracts for its Polaris missile program. These contracts compromised research and development work as well as the manufacture of component parts not previously made. Hence, neither cost nor time could be accurately estimated, and completion times therefore had to be based upon probability. Contractors were asked to estimate their operational time requirements on three bases: optimistic, pessimistic, and most likely dates. These estimates were then mathematically assessed to determine the probable completion date for each contract, and this procedure was referred to as the program evaluation and review technique (PERT). It is therefore important to understand that the PERT systems involve a probability approach to the problems of planning and control of projects and are best suited to reporting on works in which major uncertainties exist.

In the other case, the E.I. du Pont de Nemours Company was constructing major chemical plants in America. These projects required that both time and cost be accurately estimated. The method of planning and control that was developed was originally called project planning and scheduling (PPS), and covered the design, construction, and maintenance work required for several large and complex jobs. PPS requires realistic estimates of cost and time and is thus a more definitive approach than PERT. It is this approach that was developed into the critical path method, which is used frequently in the construction industry. Although there are some uncertainties in any construction project, the cost and time required for each operation involved can be reasonably estimated and all operations may then be reviewed by CPM in accordance with the anticipated conditions and hazards that may be encountered on the site.

There are several variations of CPM used in planning and scheduling work, but these can be divided into two major classifications: (1) activity-on-arrows, or I-J CPM; and (2) activity-on-nodes, especially the precedence version. The original CPM system was the I-J system, with all others evolving from it to suit the needs and desires of the users. There is a major difference of opinion as to which of the two systems is the best to use for construction planning and scheduling. There are pros and cons for both systems, with neither system having a significant edge over the other. The only important thing to consider is that both systems be evaluated thoroughly before deciding which one to use. This way, even though both systems will do a fine job, you'll never have to wonder if your method is inadequate.

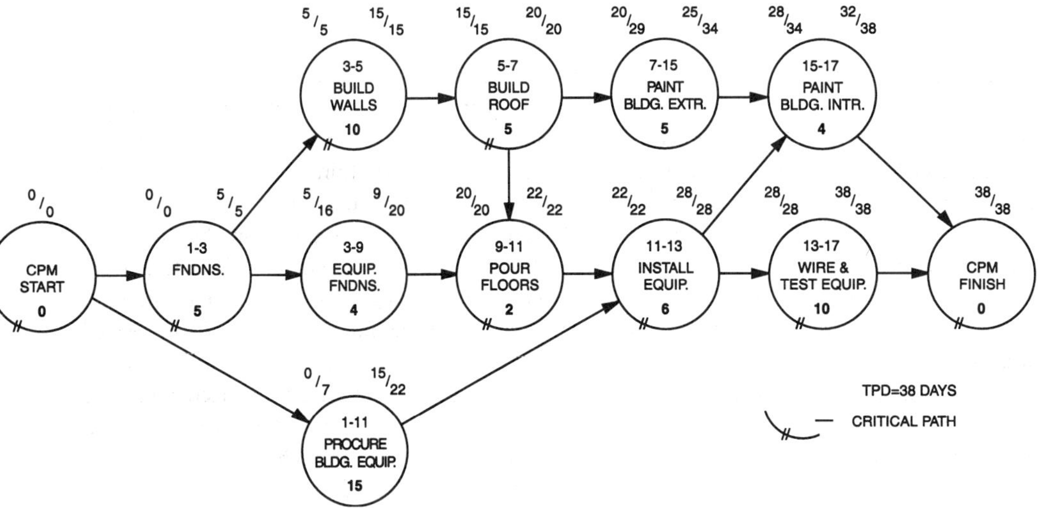

FIGURE 2.3 Activity-on-node CPM diagram.

The two CPM techniques used most often for construction projects are the I-J and precedence techniques. As was mentioned earlier, the I-J CPM technique was the first one developed. It is also, therefore, the technique used most widely in the construction industry. It is often called activity-on-arrows, and sometimes referred to as PERT. This last reference is a misnomer since PERT is a distinctly different technique, as noted previously; however, many people don't know the difference. An example of an I-J CPM diagram is shown in Fig. 2.2, complete with calculated event times.

The other CPM technique often used in construction planning and scheduling is the precedence method. It is actually a more sophisticated version of the activity-on-nodes system, initiated by John W. Fondahl of Stanford University. A diagram of an activity-on-nodes system is shown in Fig. 2.3. Notice that the activities are now the nodes (or circles) on the diagram, and the arrows simply show the **constraints** which exist between the activities. The time calculations represent the activity's early and late start and finish times.

The precedence technique was developed to add flexibility to the activity-on-nodes system. The only constraint used for activity-on-nodes is the finish-to-start relationship, which implies that one activity must finish before its following activity can start. In the precedence system, there are four types of relationships which can be used; also, the activities are represented by rectangles instead of circles on the logic diagram. A complete precedence network plus calculations is shown in Fig. 2.4.

Advantages of CPM

The critical path methods have been used for planning and scheduling construction projects for over twenty years. The estimated worth of their use varies considerably from user to user, with some contractors feeling CPM is a waste of time and money. It is difficult to believe that anyone would feel that detailed planning and scheduling work is a waste! Most likely, the unsuccessful applications of CPM resulted from trying to use a level of detail far too complicated for practical use, or the schedule was developed by an outside firm with no real input by the user, or the CPM diagram was not reviewed and updated during the project.

Regardless of past uses or misuses of CPM, the basic question is still the same: "What are the advantages of using CPM for construction planning and scheduling?" Experience with the application of CPM on several projects has revealed the following observations:

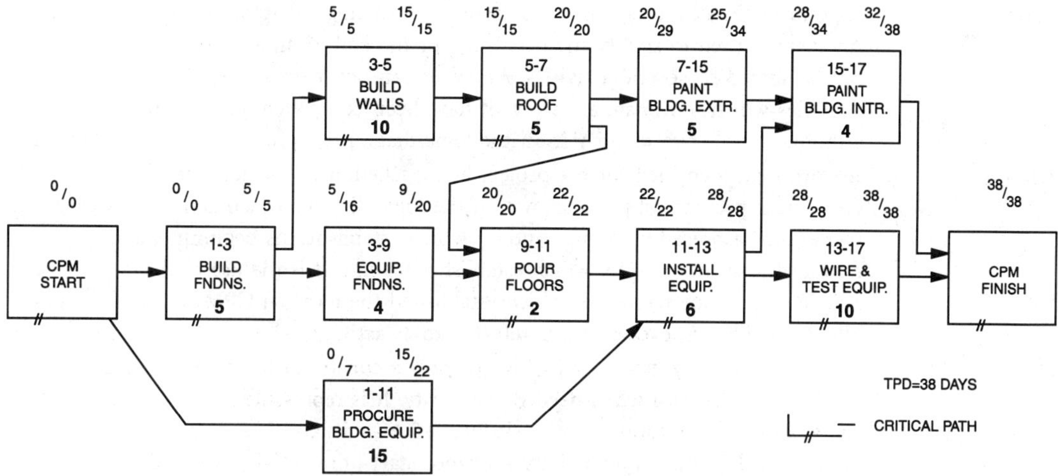

FIGURE 2.4 Precedence CPM logic diagram.

1. CPM encourages a logical discipline in the planning, scheduling, and control of projects.
2. CPM encourages more long-range and detailed planning of projects.
3. All project personnel get a complete overview of the total project.
4. CPM provides a standard method of documenting and communicating project plans, schedules, and time and cost performances.
5. CPM identifies the most critical elements in the plan, focusing management's attention to the 10 to 20% of the project that is most constraining on the scheduling.
6. CPM provides an easy method for evaluating the effects of technical and procedural changes which occur on the overall project schedule.
7. CPM enables the most economical planning of all operations to meet desirable project completion dates.

An important point to remember is that CPM is an open-ended process that permits different degrees of involvement by management to suit their various needs and objectives. In other words, you can use CPM at whatever level of detail you feel is necessary. However, one must always remember that you only get out of it what you put into it. It will be the responsibility of the user to choose the technique which is the best. They are all good and they can all be used effectively in the management of construction projects; just pick the one best liked and use it.

2.2 I-J Critical Path Method

The first CPM technique developed was the I-J CPM system and it is, therefore, widely used in the construction industry. It is often called activity-on-arrows, and sometimes referred to as PERT (which is a misnomer). The objective of this section is to instruct the reader on how to draw I-J CPM diagrams, how to calculate the event times, activity and float times, and how to handle the overlapping work schedule.

Basic Terminology for I-J CPM

There are several basic terms used in I-J CPM which need to be defined before trying to explain how the system works. A sample I-J CPM diagram is shown in Fig. 2.5 and will be referred to while defining the basic terminology.

Event (node): A point in time in a schedule, represented on the logic diagram by a circle. An event is used to signify the beginning or the end of an activity, and can be shared by several activities. An event can occur only after all the activities which terminate at the event have been completed. Each event has a unique number to identify it on the logic diagram.

Activity (A_{ij}): A work item identified for the project being scheduled. The activities for I-J CPM are represented by the arrows on the logic diagram. Each activity has two events: a preceding event (*i*-node) which establishes its beginning, and a following event (*j*-node) which establishes its end. It is the use of the *i*-node and *j*-node references which established the term I-J CPM. In Fig. 2.5 activity A, excavation, is referred to as activity 1-3.

Dummy: A fictitious activity used in I-J CPM to show a constraint between activities on the logic diagram when needed for clarity. It is represented as a dashed arrow and has a duration of zero. In Fig. 2.5, activity 3-5 is a dummy activity used to show that activity E cannot start until activity A is finished.

Activity duration (T_{ij}): Duration of an activity, expressed in working days, usually eight-hour days based on a five-day work week.

EET_i: Earliest possible occurrence time for event *i*, expressed in project workdays, cumulative from the beginning of the project.

LET_i: Latest permissible occurrence time for event *i*, expressed in project workdays, cumulative from the end of the project.

ACTIVITY	DESCRIPTION	DURATION	PREDECESSOR
A	EXCAVATION	2	----
B	BUILD FORMS	3	----
C	PROCURE REINF. STEEL	1	----
D	FINE GRADING	2	A
E	ERECT FORMWORK	2	A, B
F	SET REINF. STEEL	2	D, E, C
G	PLACE/FINISH CONCRETE	1	F

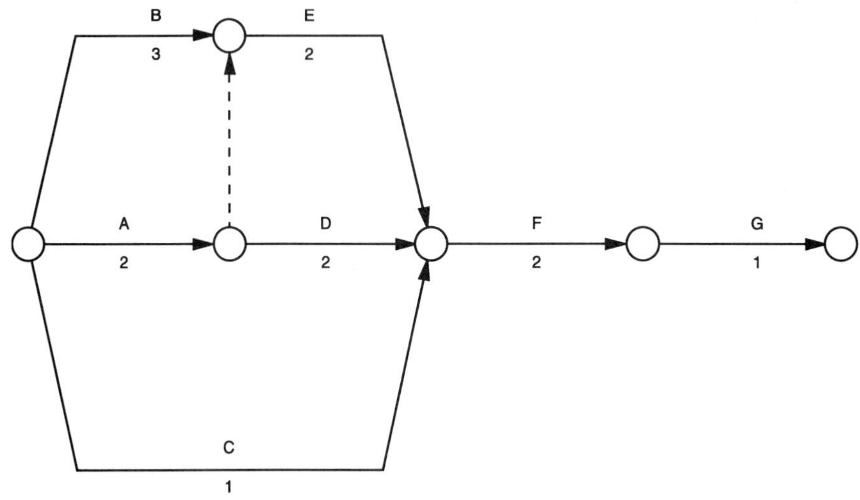

FIGURE 2.5 Sample I-J CPM activities and diagram.

Developing the I-J CPM Logic Diagram

The initial phase in the utilization of CPM for construction planning is the development of the CPM logic diagram, or **network model**. This will require that the preparer first become familiar with the work to be performed on the project and constraints, such as resource limitations, which may govern the work sequence. It may be helpful to develop a list of the activities to be scheduled and their relationship to other activities. Then it is time to draw the logic diagram. This is not an exact science, but an interactive process of drawing and redrawing until a satisfactory diagram is attained.

A CPM diagram must be a closed network in order for the time and float calculations to be completed. Thus, there is a single starting node or event for each diagram and a single final node or event. In Fig. 2.5, the starting node is event 1 and the final node is event 11. Also, notice in Fig. 2.5 that event 11 is the only event which has no activities following it. If any other event in the network is left without an activity following it, then it is referred to as a *dangling node* and will need to be closed back into the network for proper time calculations to be made.

The key to successful development of CPM diagrams is to concentrate on the individual activities to be scheduled. By placing each activity on the diagram in the sequence desired with respect to all other activities in the network, the final logic of the network will be correct. Each activity has a variety of relationships to other activities on the diagram. Some activities must precede it, some must follow it, some may be scheduled concurrently, and others will have no relationship to it. Obviously, the major concern is to place the activity in a proper sequence with those which must precede it and those which must follow it. In I-J CPM these relationships are established via the activity's preceding event (i-node) and following event (j-node).

The key controller of logic in I-J CPM is the event. Simply stated, all activities shown starting from an event are preceded by all activities which terminate at that event, and cannot start until all the preceding activities are completed. Therefore, one of the biggest concerns to watch for is to not carelessly construct the diagram and needlessly constrain activities when not necessary. There are several basic arrangements of activities in I-J CPM; some of the simple relationships are shown in Fig 2.6. Sequential relationships are the name of the game—it is just a matter of taking care to show the proper sequences.

The biggest problem for most beginners in I-J CPM is the use of the dummy activity. As defined earlier, the dummy activity is a special activity used to clarify logic in I-J CPM networks, is shown as a dashed line, and has a duration of zero workdays. The dummy is used primarily for two logic cases, the complex logic situation and the unique activity number problem. The complex logic situation is the most important use of the dummy activity to clarify the intended logic. The proper use of a dummy is depicted in Fig 2.7 where it is desired to show that activity A_{rs} needs to be completed before both activities A_{st} and A_{jk}, and that activity A_{ij} precedes only A_{jk}. The incorrect way to show this logic is depicted in Fig. 2.8. It is true that this logic shows that A_{rs} precedes both A_{st} and A_{jk}, but it also implies that A_{ij} precedes both A_{jk} and A_{st}, which is not true. Essentially, the logic diagram in Fig. 2.7 was derived from the one in Fig. 2.8 by separating event j into two events, j and s, and connecting the two with the dummy activity A_{st}.

The other common use of the dummy activity is to ensure that each activity has a unique i-node and j-node. It is desirable in I-J CPM that any two events may not be connected by more than one activity. This situation is depicted in Fig. 2.9. This logic would result in two activities with the same identification number, i-j. This is not a fatal error in terms of reading the logic, but it is confusing and will cause problems if utilizing a computer to analyze the schedule. This problem can be solved by inserting a dummy activity at the end of one of the activities, as shown in Fig. 2.10. It is also possible to add the dummy at the front of the activity, which is the same logic.

Each logic diagram prepared for a project will be unique if prepared independently. Even if the same group of activities is included, the layout of the diagram, the number of dummy activities, the event numbers used, and several other elements will differ from diagram to diagram. The truth is that they are all correct if the logic is correct. When preparing a diagram for a project, the

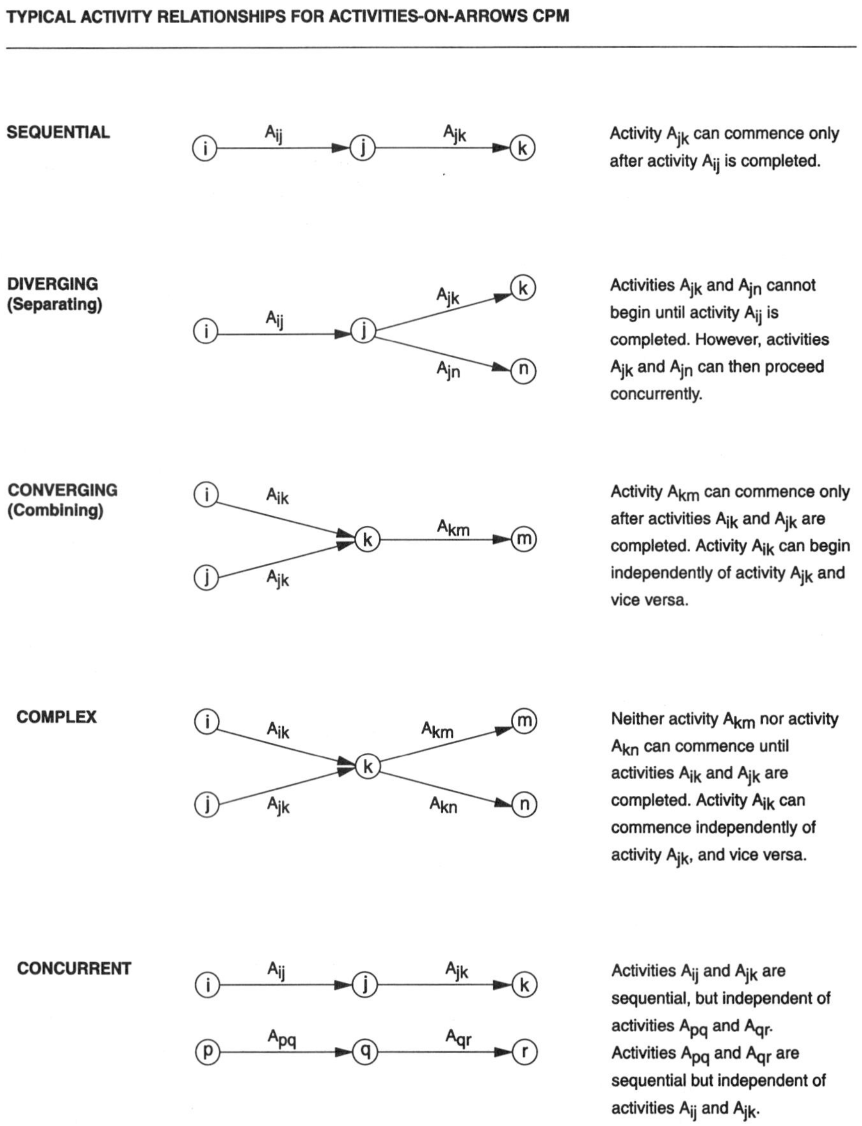

TYPICAL ACTIVITY RELATIONSHIPS FOR ACTIVITIES-ON-ARROWS CPM

SEQUENTIAL	(i) —A_{ij}→ (j) —A_{jk}→ (k)	Activity A_{jk} can commence only after activity A_{ij} is completed.
DIVERGING (Separating)	(i) —A_{ij}→ (j); (j) —A_{jk}→ (k); (j) —A_{jn}→ (n)	Activities A_{jk} and A_{jn} cannot begin until activity A_{ij} is completed. However, activities A_{jk} and A_{jn} can then proceed concurrently.
CONVERGING (Combining)	(i) —A_{ik}→ (k); (j) —A_{jk}→ (k); (k) —A_{km}→ (m)	Activity A_{km} can commence only after activities A_{ik} and A_{jk} are completed. Activity A_{ik} can begin independently of activity A_{jk} and vice versa.
COMPLEX	(i) —A_{ik}→ (k); (j) —A_{jk}→ (k); (k) —A_{km}→ (m); (k) —A_{kn}→ (n)	Neither activity A_{km} nor activity A_{kn} can commence until activities A_{ik} and A_{jk} are completed. Activity A_{ik} can commence independently of activity A_{jk}, and vice versa.
CONCURRENT	(i) —A_{ij}→ (j) —A_{jk}→ (k); (p) —A_{pq}→ (q) —A_{qr}→ (r)	Activities A_{ij} and A_{jk} are sequential, but independent of activities A_{pq} and A_{qr}. Activities A_{pq} and A_{qr} are sequential but independent of activities A_{ij} and A_{jk}.

FIGURE 2.6 Typical I-J CPM activity relationships.

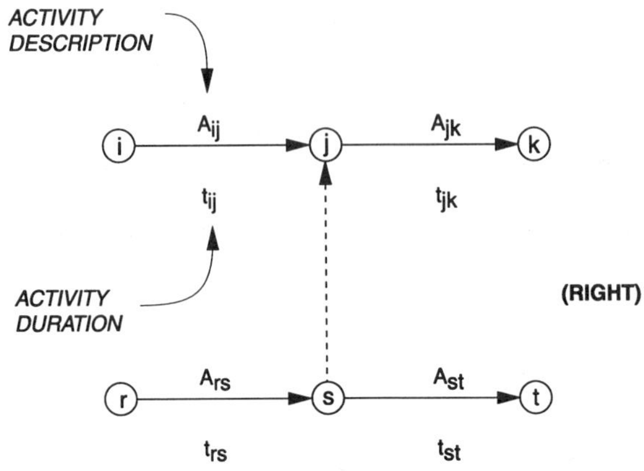

FIGURE 2.7 Correct use of I-J dummy activity.

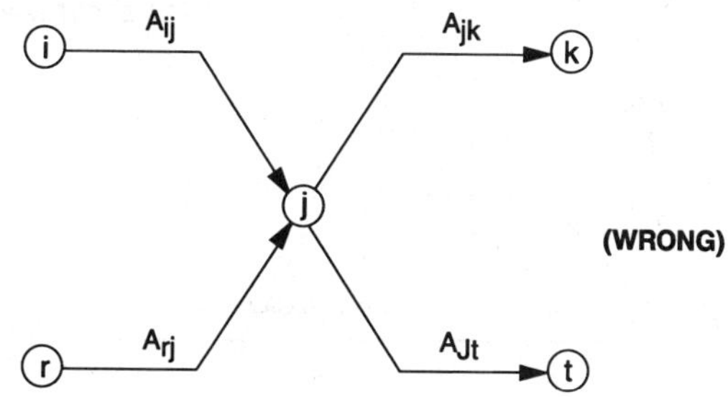

FIGURE 2.8 Incorrect use of I-J dummy activity.

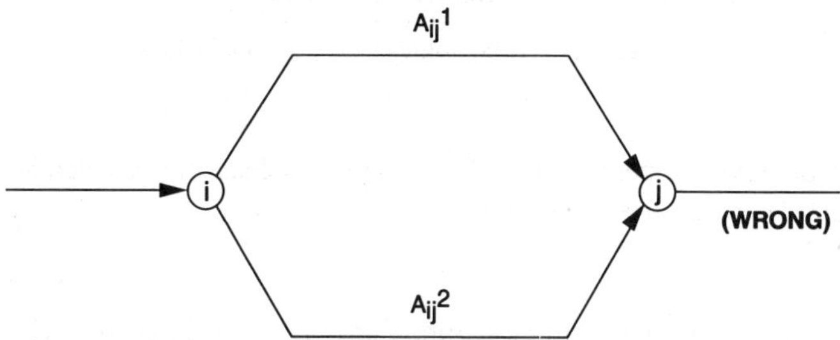

FIGURE 2.9 Incorrect nodes for parallel activities.

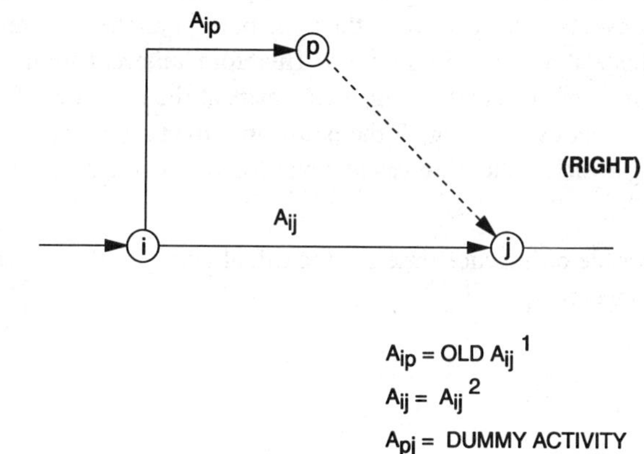

FIGURE 2.10 Correct nodes for parallel activities.

scheduler should not worry about being too neat on the first draft, but should try to include all the activities in the proper order. The diagram can be fine-tuned later after the original schedule is checked.

I-J Network Time Calculations

An important task in the development of a construction schedule is the calculation of the network times. In I-J CPM this involves the calculation of the event times, from which the activity times of interest are then determined. Each event on a diagram has two event times: the early event time (EET) and the late event time (LET); these are depicted in Fig. 2.11. Each activity has two events: the preceding event, or the i-node, and the following event, or the j-node. Therefore, each activity has four associated event times: EET_i, LET_i, EET_j, and LET_j . A convenient methodology

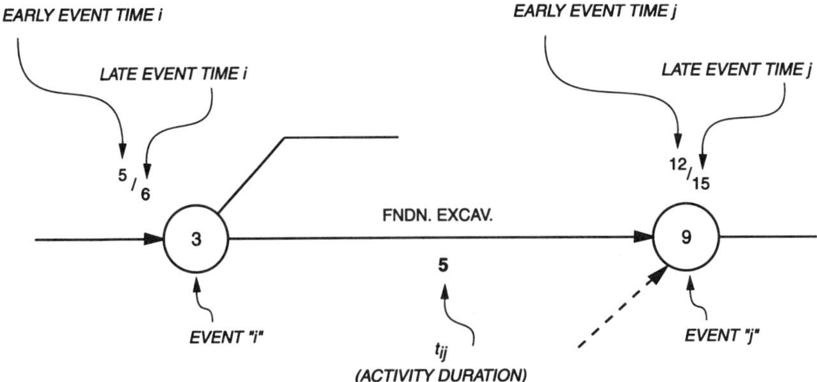

FIGURE 2.11 Terminology for I-J CPM activities.

for determining these event times involves a forward pass to determine the early event times and a reverse pass to determine the late event times.

Forward Pass

The objective of the forward pass is to determine the early event times for each of the events on the I-J diagram. The process is started by setting the early event time of the initial event on the diagram (there is to be only one initial event) equal to zero (0). Once this is done, then all the other early event times can be calculated; this will be explained by the use of Fig. 2.12. The early event time of all other events is determined as the maximum of all the early finish times of all the activities which terminate at an event in question. Therefore, an event should not be considered until the early finish times of all activities which terminate at the event have first been calculated. The forward pass is analogous to trying all the paths on a road network, finding the maximum time that it takes to get each node. The calculations for the forward pass can be summarized as follows:

1. The earliest possible occurrence time for the initial event is taken as zero [$\text{EET}_i = 0 = \text{EST}_{ij}$ (i = initial event)].

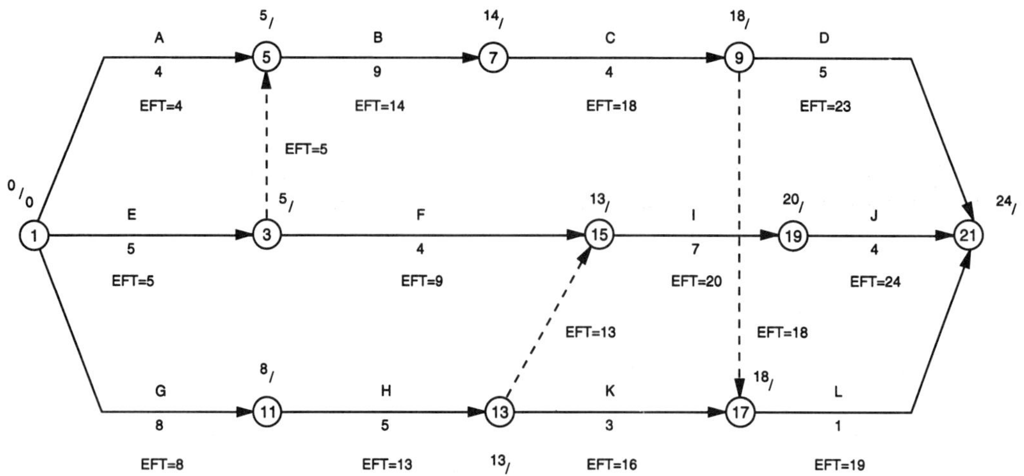

FIGURE 2.12 Forward pass calculations for I-J CPM.

2. Each activity can begin as soon as its preceding event (i-node) occurs ($EST_{ij} = EET_i$; $EFT_{ij} = EET_i + T_{ij}$).
3. The earliest possible occurrence time for an event is the largest of the early finish times for those activities which terminate at the event [$EET_j = \max EFT_{pj}$ (p = all events that precede event j)].
4. The total project duration is the earliest possible occurrence time for the last event on the diagram [$TPD = EET_j$ (j = terminal event on diagram)].

The early event time of event 1 in Fig. 2.12 was set equal to 0; i.e., EET(1) = 0. It is then possible to calculate the early finish times for activities 1-5, 1-3, and 1-11 as noted. Activity early finish times are not normally shown on an I-J CPM diagram, but are shown here to help explain the process. Since both events 3 and 11 have only a single activity preceding them, their early event times can be established as 5 and 8, respectively. Event 5 has two preceding activities; therefore, the early finish times for both activities 1-5 (4) and 3-5 (5) must be found before establishing that its early event time equals 5. Note that dummy activities are treated just as regular activities for calculations. Likewise, the early finish time for event 15 cannot be determined until the early finish times for activities 3-15 (9) and 13-15 (13) have been calculated. EET(15) is then set as 13. The rest of the early event times on the diagram are thus wisely calculated resulting in an estimated total project duration of 24 days.

Reverse Pass

The objective of the reverse pass is to determine the late event time for each event on the I-J diagram. This process is started by setting the late event time of the terminal, or last, event on the diagram (there is to be only one terminal event) equal to the early event time of the event; i.e., $LET_j = EET_j$. Once this is done, then all the other late event times can be calculated; this will be explained by the use of Fig. 2.13. The late event time of all other events is determined as the minimum of all the late start times of all the activities which originate at an event in question. Therefore, an event should not be considered until the late event times of all activities which originate at the event have first been calculated. The calculations for the reverse pass can be summarized as follows:

1. The latest permissible occurrence time for the terminal event is set equal to the early event time of the terminal event. This also equals the estimated project duration [$LET_j = EET_j = TPD$ (j = terminal event on diagram)].

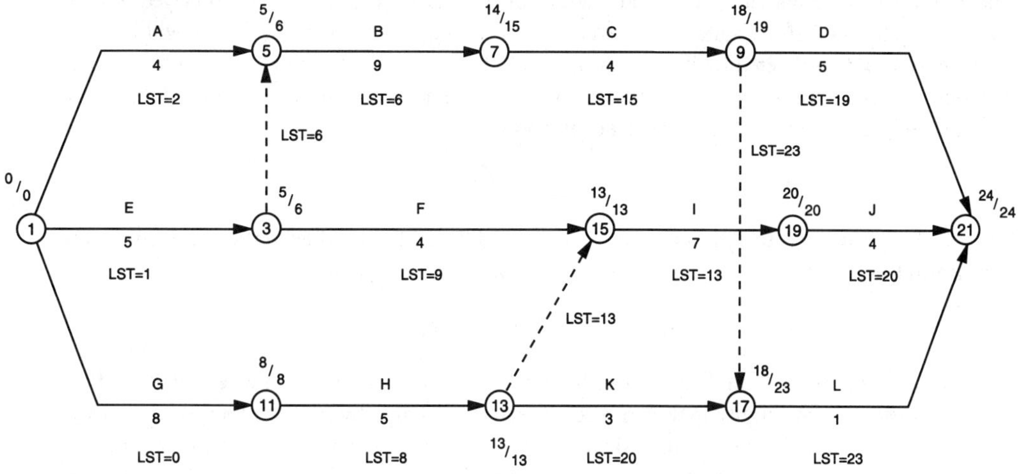

FIGURE 2.13 Reverse pass calculations for I-J CPM.

2. The latest permissible finish time for an activity is the latest permissible occurrence time for its following event (j-node) ($LFT_{ij} = LET_j$; $LST_{ij} = LET_j - T_{ij}$).
3. The latest permissible occurrence time for an event is the minimum (earliest) of the latest start times for those activities which originate at the event [$LET_i = \min LST_{ip}$ (p = all events that follow event i)].
4. The latest permissible occurrence time of the initial event should equal its earliest permissible occurrence time (zero). This provides a numerical check [$LET_i = EET_i = 0$ (i = initial event on diagram)].

The late event time of event 21 in Fig. 2.13 was set equal to 24. It is then possible to calculate the late start times for activities 9-21, 19-21, and 17-21 as noted. Activity late start times are not normally shown on an I-J CPM diagram, but are shown here to help explain the reverse pass process. Since both events 17 and 19 have only a single activity which originates from them, their late event times can be established as 23 and 20, respectively. Event 9 has two originating activities; therefore, the late start times for both activities 9-17 (23) and 9-21 (19) must be found before establishing that its late event time equals 19. Note that dummy activities are treated just as regular activities for calculations. Likewise, the late event time for event 3 cannot be determined until the late start times for activities 3-5 (6) and 3-15 (9) have been calculated. LET(3) is then set as 6. The rest of the late event times on the diagram are thus calculated resulting in the late event time of event 1 checking in as zero.

Activity Float Times

One of the primary benefits of the critical path methods is the ability to evaluate the relative importance of each activity in the network by its calculated **float**, or **slack**, time. In the calculation procedures for CPM, an allowable time span is determined for each activity; the boundaries of this time span are established by the activity's early start time and late finish time. When this bounded time span exceeds the activity's duration, the excess time is referred to as float time. Float time can exist only for noncritical activities; activities with zero float are **critical activities** and make up the critical path(s) on the network. There are three basic float times for each activity: **total float**, **free float**, and interfering float. It should be noted that once an activity is delayed to finish beyond its early finish time, then the network event times must be recalculated for all following activities before evaluating their float times.

Total Float

The total float for an activity is the total amount of time that the activity can be delayed beyond its early finish time before it delays the overall project completion time. This delay will occur if the activity is not completed by its late finish time. Therefore, the total float is equal to the time difference between the activity's late finish time and its early finish time. The total float for an activity can be calculated by the following expression:

$$TF_{ij} = LET_j - EETI - T_{ij}$$

Since the LET_j for any activity is its late finish time, and the EET_i plus the duration, T_{ij}, equals the early finish time, then the above expression for the total float of an activity reduces to

$$TF_{ij} = LFT_{ij} - EFT_{ij}$$

The calculation of the total float can be illustrated by referring to Fig. 2.14 where the total float and free float for each activity is shown below the activity. For activity B the total float = 15 − 5 − 9 = 1, and for activity K the total float = 23 − 13 − 3 = 7. The total float for activities G, H, I, and J is zero, thus these activities make up the critical path for this diagram. Float times are not usually noted on dummy activities; however, dummies do have float since they are activities.

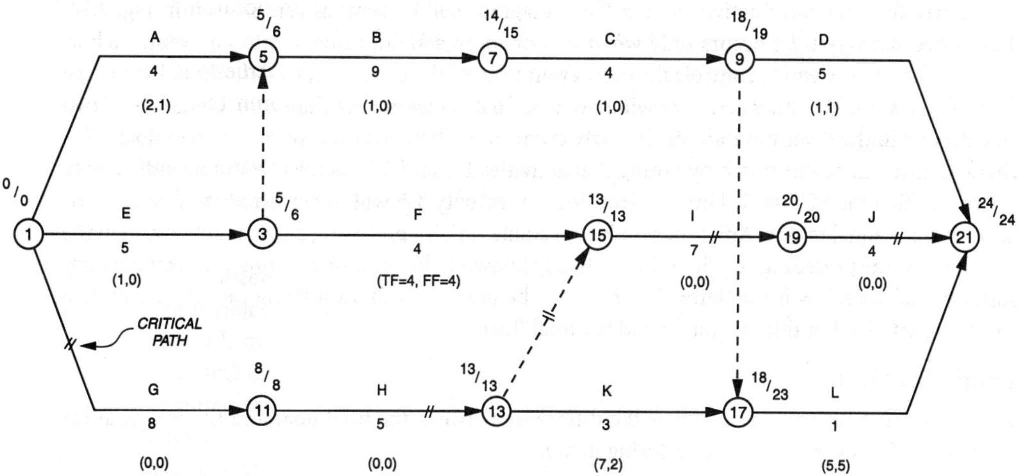

FIGURE 2.14 I-J CPM diagram with float times.

Activity 13-15 connects two critical activities and has zero total float, thus it is on the critical path and should be marked as such.

One of the biggest problems in the use of CPM for scheduling is the misunderstanding of float. Although the total float for each activity is determined independently, it is not an independent property, but is shared with other activities which precede or follow it. The float value calculated is good only for the event times on the diagram. If any of the event times for a diagram change, then the float times must be recalculated for all the activities affected. For example, activity B has a total float $= 1$; however, if either activity A or E is not completed until CPM day 6, then the EET(5) must be changed to 6 and the total float for activity B is now equal to zero. Since activity B had one day of float originally, then the project duration is not affected. However, if the EET(5) becomes greater than 6, then the project duration will be increased.

A chain of activities is a series of activities linked sequentially in a CPM network. Obviously, there are many different possible chains of activities for a given network. Often, a short chain will have the same total float, such as the chain formed by activities B, C, and D in Fig. 2.14. The total float for each of these activities is 1 day. Note that if the total float is used up by an earlier activity in a chain, then it will not be available for following activities. For instance, if activity B is not finished until day 15, then the total float for both activities C and D becomes zero, making them both critical. Thus, one should be very careful when discussing the float time available for an activity with persons not familiar with CPM. This problem can be avoided if it is always a goal to start all activities by their early start time, if feasible, and save the float for activities which may need it when problems arise.

Free Float

Free float is the total amount of time that an activity can be delayed beyond its early finish time before it delays the early start time of a following activity. This means that the activity must be finished by the early event time of its j-node, thus the free float is equal to the time difference between the EET_j and its early finish time. The free float can be calculated by the following expressions:

$$\text{FF}_{ij} = \text{EET}_j - \text{EET}_i - \text{T}_{ij} \quad \text{or} \quad \text{FF}_{ij} = \text{EET}_j - \text{EFT}_{ij}$$

The calculation of free float can be illustrated by referring to Fig. 2.14 where the total float and free float for each activity are shown below the activity. For activity A the free float $= 14 - 5 - 9 = 0$, and for activity D the free float $= 24 - 18 - 5 = 1$.

The free float for most activities on a CPM diagram will be zero, as can be seen in Fig. 2.14. This is because free float occurs only when two or more activities merge into an event, such as event 5. The activity which controls the early event time of the event will, by definition, have a free float of zero while the other activities will have free float values greater than zero. Of course, if two activities tie in the determination of the early event time, then both will have zero free float. This characteristic can be illustrated by noting that activities 1-5 and 3-5 merge at event 5, with activity 3-5 controlling the EET = 5. Thus the free float for activity 3-5 will be zero, and the free float for activity 1-5 is equal to one. Any time you have a single activity preceding another activity, then the free float for the preceding activity is immediately known to be zero, since it must control the early start time of the following activity. Free float can be used up without hurting the scheduling of a following activity, but this cannot be said for total float.

Interfering Float

Interfering float for a CPM activity is the difference between the total float and the free float for the activity. The expression for interfering float is

$$\mathrm{IF}_{ij} = \mathrm{LET}_j - \mathrm{EET}_j \quad \text{or} \quad \mathrm{IF}_{ij} = \mathrm{TF}_{ij} - \mathrm{FF}_{ij}$$

The concept of interfering float comes from the fact that if one uses the free float for an activity, then the following activity can still start on its early start time, so there is no real interference. However, if any additional float is used, then the following activity's early start time will be delayed. In practice, the value of interfering float is seldom used; it is presented here since it helps one to better comprehend the overall system or float for CPM activities. The reader is encouraged to carefully review the sections on activity float, and refer to Figs. 2.14 and 2.15 for graphic illustrations.

Activity Start and Finish Times

One of the major reasons for the utilization of CPM in the planning of construction projects is to estimate the schedule for conducting various phases (activities) of the project. Thus it is essential that one know how to determine the starting and finishing times for each activity on a CPM diagram. Before explaining how to do this, it is important to note that any starting or finish time determined is only as good as the CPM diagram and will not be realistic if the diagram is not kept up to date as the project progresses. If one is interested only in general milestone planning, this is not as critical. However, if one is using the diagram to determine detailed work schedules and delivery dates, then updating is essential. This topic will be discussed in "Updating the CPM Network" later in this chapter.

The determination of the early and late starting and finish times for I-J CPM activities will be illustrated by reference to Fig. 2.14. There are four basic times to determine for each activity: early start time, late start time, early finish time, and late finish time. Activity F on Fig. 2.14, as for all other activities on an I-J CPM diagram, has four event times:

$$\mathrm{EET}_i = 5, \quad \mathrm{LET}_i = 6, \quad \mathrm{EET}_j = 13, \quad \mathrm{LET}_j = 13$$

A common mistake is to refer to these four times as the early start time, late start time, early finish time, and late finish time for activity F, or Activity 3-15. Although this is true for all critical activities and may be true of some other activities, this is not true of many of the activities on an I-J diagram, and this procedure should not be used.

The basic activity times can be determined by the following four relationships:

Early start time, $\mathrm{EST}_{ij} = \mathrm{EET}_i$ (5 for activity F)
Late start time, $\mathrm{LST}_{ij} = \mathrm{LET}_j - \mathrm{T}_{ij}$ $(13 - 4 = 9$ for activity F)
Early finish time, $\mathrm{EFT}_{ij} = \mathrm{EST}_{ij} + \mathrm{T}_{ij}$ $(5 + 4 = 9$ for activity F)
Late finish time, $\mathrm{LFT}_{ij} = \mathrm{LET}_j$ (13 for activity F)

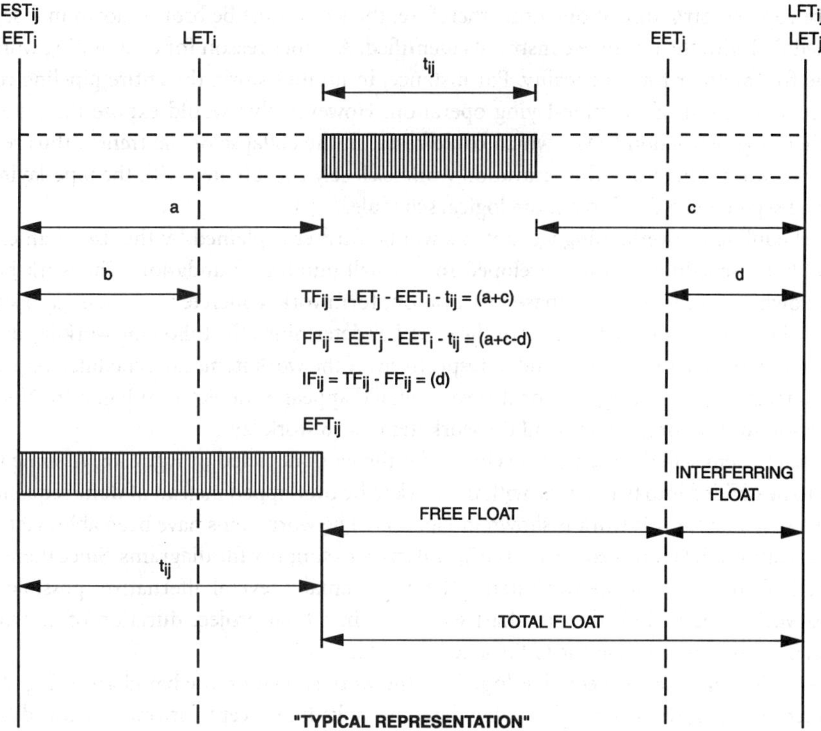

FIGURE 2.15 Graphic representation of I-J CPM float.

As can be seen, the early finish time for activity F is 9, not 13, and the late start time is 9, not 6. The four basic activity times can be found quickly and easily, but cannot be read directly off the diagram. For many construction projects today the starting and finishing times for all activities are shown on a computer printout, not only in CPM days, but in calendar days also. As a further example, the EST, LST, EFT, and LFT of activity A on Fig. 2.14 are 0, 2, 4, and 6, respectively. These times are in terms of CPM days; instructions will be given later for converting activity times in CPM days into calendar dates.

Overlapping Work Items in I-J CPM

One of the most difficult scheduling problems encountered is the overlapping work items problem. This occurs often in construction and requires careful thought by the scheduler whether using I-J CPM or the precedence CPM technique (precedence will be discussed in the following section). The overlapping work situation occurs when two or more work items which must be sequenced will take too long to perform end to end, thus the following items are started before their preceding work items are completed. Obviously, the preceding work items must be started and worked on sufficiently in order for the following work items to begin. This situation occurs often with construction work such as concrete wall (form, pour, cure, strip, finish) and underground utilities (excavate, lay pipe, test pipe, backfill). Special care must be taken to show the correct logic to follow on the I-J diagram, while not restricting the flow of the work as the field forces use the CPM schedule.

The overlapping work item problem is also encountered for several other reasons in construction scheduling. A major reason is the scheduling required to optimize scarce or expensive resources, such as concrete forms. It is usually too expensive or impractical to purchase enough forms to form an entire concrete structure at one time; therefore, the work must be broken down into segments and scheduled with the resource constraints identified. Another reason for overlapping work items could be for safety or for practicality. For instance, in utilities work, the entire pipeline could be excavated well ahead of the pipe-laying operation. However, this would expose the pipe trench to weather or construction traffic which could result in the collapse of the trench, thus requiring expensive rework. Therefore, the excavation work is closely coordinated with the pipe-laying work in selected segments to develop a more logical schedule.

The scheduling of overlapping work items will be further explained by the use of an example. Assume that a schedule is to be developed for a small building foundation. The work has been broken down into four separate phases: excavation, formwork, concrete placement, and stripping and backfilling. A preliminary analysis of the work has determined the following workday durations of the four work activities: 4, 8, 2, and 4, respectively. If the work items are scheduled sequentially, end to start, the I-J CPM diagram for this work would appear as depicted in Fig. 2.16. Notice that the duration for the completion of all the work items is 18 workdays.

A more efficient schedule can be developed for the work depicted in Fig. 2.16. Assume that the work is to be divided into two halves, with the work to be overlapped instead of done sequentially. A bar chart schedule for this work is shown in Fig. 2.17. The work items have been abbreviated as E1 (start excavation), E2 (complete excavation), and so on to simplify the diagrams. Since there is some float available for some of the work items, there are actually several alternatives possible. Notice that the work scheduled on the bar chart will result in a total project duration of 13 workdays, which is five days shorter than the CPM schedule of Fig. 2.16.

An I-J CPM schedule has been developed for the work shown on the bar chart of Fig. 2.17 and is depicted in Fig. 2.18. At first glance the diagram looks fine except for one obvious difference: the project duration is 14 days, instead of 13 days for the bar chart schedule. Closer review reveals that there are serious logic errors in the CPM diagram at events 7 and 11. As drawn, the second half of the excavation (E2) must be completed before the first concrete placement (CP1) can start. Likewise, the first wall pour cannot be stripped and backfilled (S/BF1) until the second half of the formwork is completed (F2). These are common logic errors caused by the poor development of I-J activities interrelationships. The diagram shown in Fig. 2.19 is a revised version of the I-J CPM

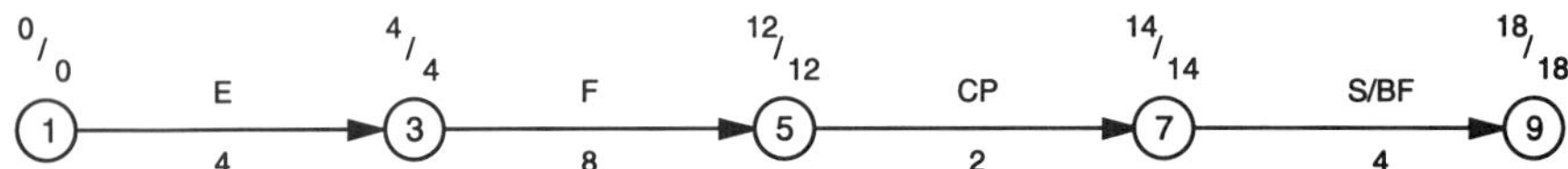

FIGURE 2.16 I-J CPM for sequential activities.

Proj. Work Day →	1	2	3	4	5	6	7	8	9	10	11	12	13	14	15
Proj. CPM Day →	0	1	2	3	4	5	6	7	8	9	10	11	12	13	14
EXCAVATION	E1		E2												
FORMWORK			F1				F2								
POUR CONCRETE							P1				P2				
STRIP/BACKFILL								S/BF1				S/BF2			

FIGURE 2.17 Bar chart for overlapping work items.

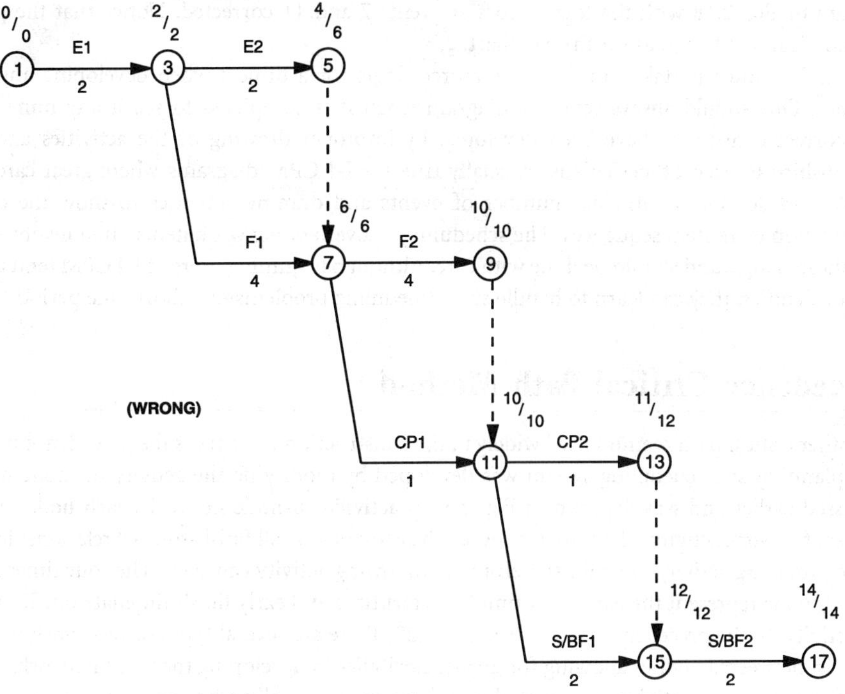

FIGURE 2.18 I-J CPM no. 1 for overlapping work items.

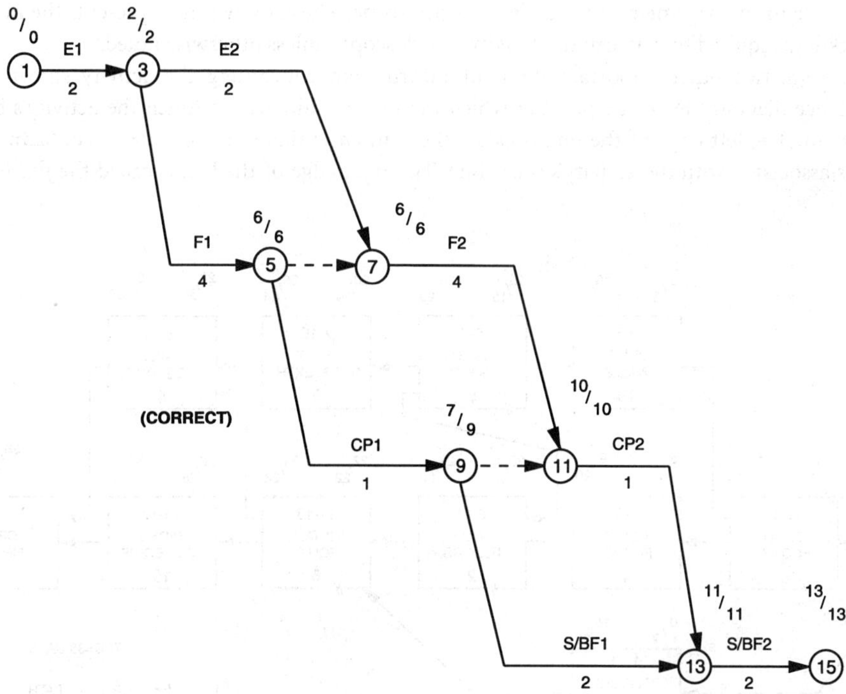

FIGURE 2.19 I-J CPM no. 2 for overlapping work items.

diagram of Fig. 2.18 with the logic errors at events 7 and 11 corrected. Notice that the project duration is now 13 days, as for the bar chart.

Great care must be taken to show the correct logic for a project when developing any CPM diagram. One should always review a diagram when it is completed to see if any unnecessary or incorrect constraints have been developed by improper drawing of the activities and their relationships to each other. This is especially true for I-J CPM diagrams where great care must be taken to develop a sufficient number of events and dummy activities to show the desired construction work item sequences. The scheduling of overlapping work items often involves more complicated logic and should be done with care. Although beginning users of I-J CPM tend to have such difficulties, they can learn to handle such scheduling problems in a short time period.

2.3 Precedence Critical Path Method

The other critical path method used widely in the construction industry is the precedence method. This planning and scheduling system was developed by modifying the activity-on-node method discussed earlier and was depicted in Fig. 2.3. In activity-on-node networks each node or circle represents a work activity. The arrows between the activities are all finish-to-start relationships; that is, the preceding activity must finish before the following activity can start. The four times shown on each node represent the early start time/late start time and early finish time/late finish time for the activity. In the precedence system (see Fig. 2.20) there are several types of relationships which can exist between activities allowing for greater flexibility in developing the CPM network.

The construction activity on a precedence diagram is typically represented as a rectangle (see Fig. 2.21). There are usually three items of information placed within the activity's box: the activity number, the activity description, and the activity time (or duration). The activity number is usually an integer number, although alphabetical characters are often added to denote the group responsible for management of the activity's work scope. The activity time represents the number of workdays required to perform the activity's work scope, unless otherwise noted.

There are two other important items of information concerning the activity shown on a precedence diagram. First, the point at which the relationship arrows touch the activity's box is important. The left edge of the box is called the *start edge;* therefore, any arrow contacting this edge is associated with the activity's start time. The right edge of the box is called the *finish edge;*

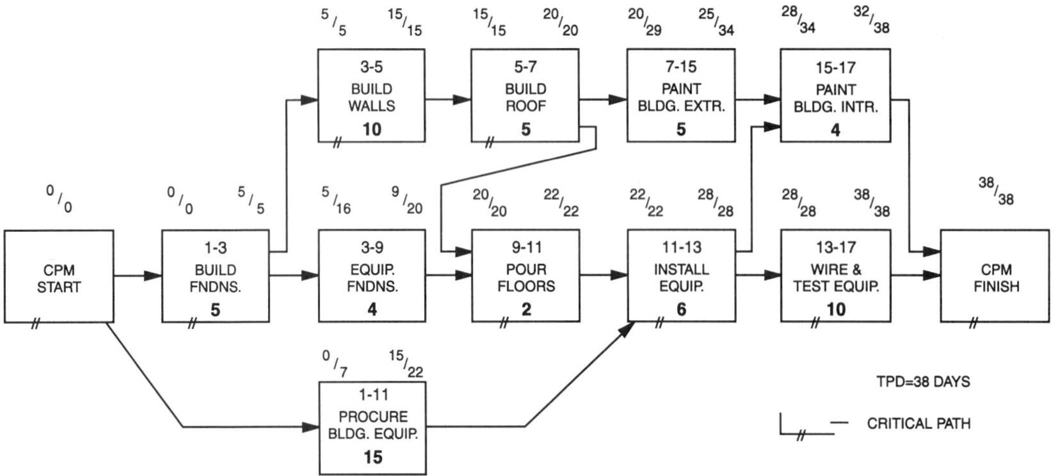

FIGURE 2.20 Precedence CPM diagram.

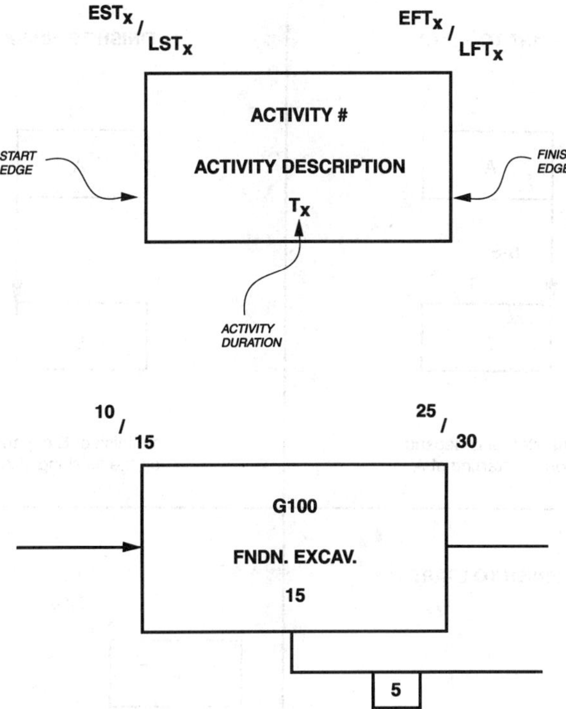

FIGURE 2.21 Precedence CPM activity information.

therefore, any arrow contacting this edge is associated with the activity's finish time. Second, the calculated numbers shown above the box on the left represent the early start time and the late start time of the activity, and the numbers above the box on the right represent the early finish time and the late finish time of the activity. This is different from I-J CPM where the calculated times represent the event times and not activity times.

Precedence Relationships

The arrows on a precedence diagram represent the relationships that exist between different activities. There are four basic relationships which are used, as depicted in Fig. 2.22. The *start-to-start* relationship states that activity B cannot start before activity A starts; that is, EST(B) is greater than or equal to EST(A). The greater-than situation will occur when another activity which precedes activity B has a greater time constraint than EST(A).

The *finish-to-start* relationship states that activity B cannot start before activity A is finished; that is, EST(B) is greater than or equal to EFT(A). This relationship is the one most commonly used on a precedence diagram. The *finish-to-finish* relationship states that activity B cannot finish before activity A finishes; that is, EFT(B) is greater than or equal to EFT(A). This relationship is used mostly in precedence networks to show the finish-to-finish relationships between overlapping work activities.

The fourth basic relationship is the lag relationship. Lag can be shown for any of the three normal precedence relationships and represents a time lag between the two activities. The lag relationship shown in Fig. 2.22 is a start-to-start (S-S) lag. It means that activity B cannot start until X days of work are done on activity A. Often when there is an S-S lag between two activities, there is a corresponding F-F lag because the following activity will require X days of work to complete after the preceding activity is completed. The use of lag time on the relationships allows much

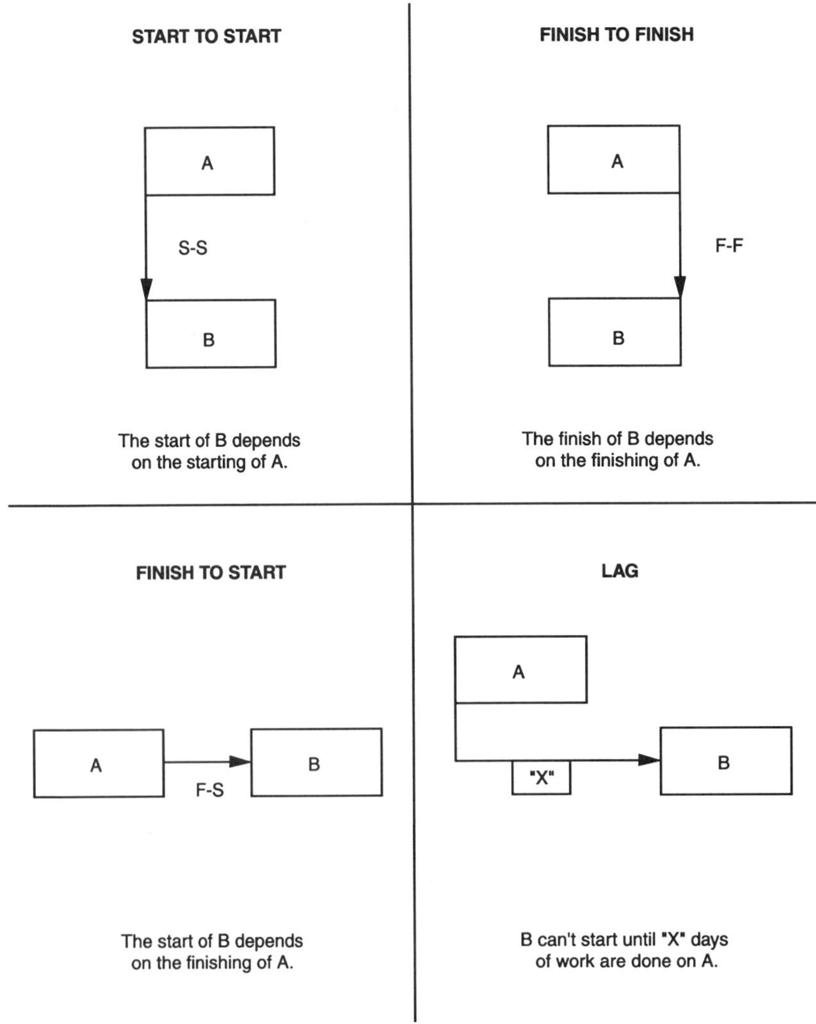

FIGURE 2.22 Precedence CPM activity relationships.

greater flexibility in scheduling delays between activities (such as curing time) and for scheduling overlapping work items (such as excavation, laying pipe, and backfilling). A precedence with all three basic relationships and three lags is shown in Fig. 2.23.

Precedence Time Calculations

The time calculations for precedence are somewhat more complex than for I-J CPM. The time calculation rules for precedence networks are shown in Fig. 2.24. These rules are based on the assumption that all activities are continuous in time; that is, once started, they are worked through to completion. In reality this assumption simplifies interpretation of the network and the activity float times. It ensures that the EFT equals the EST plus the duration for a given activity.

The time calculations involve both a forward pass and a reverse pass, as in I-J CPM. However, for precedence, one is calculating activity start and finish times directly and not event times as in I-J. In the forward pass one evaluates all activities preceding a given activity to determine its EST.

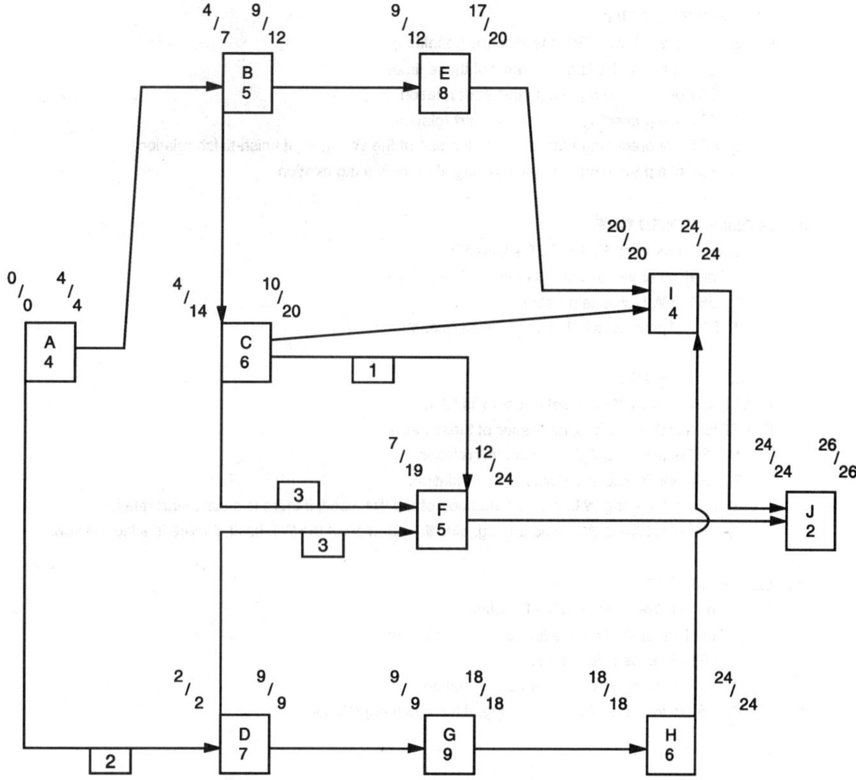

FIGURE 2.23 Precedence diagram with activity times.

For activity F in Fig. 2.23 there are three choices: C-to-F (start-to-start), EST(F) = 4 + 3 = 7; C-to-F (finish-to-finish), EST(F) = 10 + 1 − 5 = 6; D-to-F (finish-to-start), EST(F) = 2 + 3 = 5. Since 7 is the largest EST, the C-to-F (S-S) relationship controls. The EFT is then determined as the EST plus the duration; for activity F, the EFT = 12.

For the reverse pass in precedence one evaluates all activities following a given activity to determine its LFT. For activity F the only following activity is activity J; therefore, the LFT(F) is 24 − 0 = 24. For activity D there are two choices for LFT: D-to-F (start-to-start, with lag) LFT(D) = 19 − 3 + 7 = 23; D-to-G (finish-to-start), LFT(D) = 9 − 0 = 9. Since 9 is the smallest LFT, the D-to-G relationship controls. The LST is then determined as the LFT minus the duration; for activity D, the LST is 2.

A major advantage of the precedence system is that the times shown on a precedence activity truly represent actual start and finish times for the activity. Thus, less-trained personnel can more quickly read these times from the precedence network than from an I-J network. As for I-J CPM, the activity times represent CPM days, which relate to workdays on the project. The conversion from CPM days to calendar dates will be covered in the next section.

Precedence Float Calculations

The float times for precedence activities have the same meaning as for I-J activities; however, the calculations are different. Before float calculations can be made, all activity start and finish times must be calculated as just described. Remember again that all activities are assumed to be continuous in duration. The float calculations for precedence networks are depicted in Fig. 2.25.

I. **EARLIEST START TIME**

A. EST of first Work Item (W.I.) is zero (by definition).

B. EST of all other W.I.'s is the **greater** of these times:

1) EST of a preceding W.I. if start-start relation.
2) EFT of a preceding W.I. if finish-start relation.
3) EFT of a preceding W.I., less the duration of the W.I. itself, if finish-finish relation.
4) EST of a preceding W.I., plus the lag, if there is a lag relation.

II. **EARLIEST FINISH TIME**

A. For first Work Item, EFT = EST + Duration.

B. EFT of all other W.I.'s is the **greater** of these times:

1) EST of W.I. plus its duration.
2) EFT of preceding W.I. if finish-finish relation.

III. **LATEST FINISH TIME**

A. LFT for last Work Item is set equal to its EFT.

B. LFT for all other W.I.'s is the **lesser** of these items:

1) LST of following W.I. if finish-start relation.
2) LFT of following W.I. if finish-finish relation.
3) LST of following W.I., plus the duration of the W.I. itself, if there is a start-start relation.
4) LST of following W.I., less the lag, plus the duration of the W.I. itself, if there is a lag relation.

IV. **LATEST START TIME**

A. LST of first Work Item = LFT - Duration.

B. LST of all other W.I.'s is the **lesser** of these items:

1) LFT of W.I. less its duration.
2) LST of following W.I. if start-start relation.
3) LST of following W.I. less the lag, if there is a lag relation.

EST = Early Start Time LST = Late Start Time
EFT = Early Finish Time LFT = Late Finish time

Lag = Number of days of lag time associated with a relationship.

FIGURE 2.24 Precedence activity times calculation rules.

The total float of an activity is the total amount of time that an activity can be delayed before it affects the total project duration. This means that the activity must be completed by its late finish time; therefore, the total float for any precedence activity is equal to its LFT minus its EFT. For activity F in Fig. 2.23 its total float is $24 - 12 = 12$.

The free float of an activity is the total amount of time that an activity can be delayed beyond its EFT before it delays the EST of a following activity. In I-J CPM this is a simple calculation equal to the EET of its j-node minus the EET of its i-node minus its duration. However, for a precedence activity it is necessary to check the free float existing between it and each of the activities which follows it, as depicted in Fig. 2.25. The actual free float for the activity is then determined as the minimum of all the free float options calculated for the following activities. For most precedence activities, as for I-J, the free float time is normally equal to zero. For activity C in Fig. 2.23 there are three choices: C-to-F (start-to-start), FF $= 7 - 4 - 3 = 0$; C-to-F (finish-to-finish), FF $= 12 - 10 - 1 = 1$; C-to-I (finish-to-start), FF $= 20 - 10 = 10$. Since the smallest is 0, then the C-to-F (S-S) relationship controls and the FF $= 0$.

Overlapping Work Items

A major reason that many persons like to use the precedence CPM system for construction scheduling is its flexibility for overlapping work items. Fig. 2.26 depicts the comparable I-J and precedence diagrams necessary to show the logic and time constraints shown in the bar chart at the

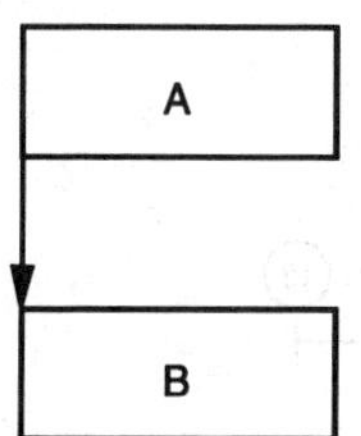

$TF_A = LFT_A - EFT_A$
$FF_A = EST_B - EST_A$

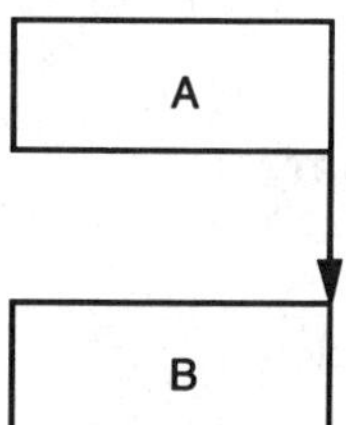

$TF_A = LFT_A - EFT_A$
$FF_A = EFT_B - EST_A$

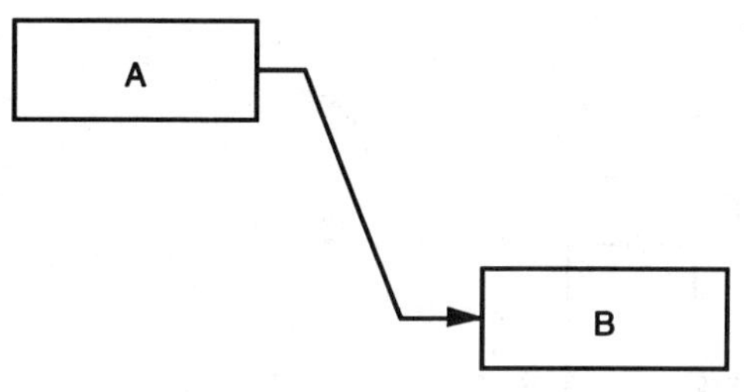

$TF_A = LFT_A - EFT_A$
$FF_A = EST_B - EFT_A$

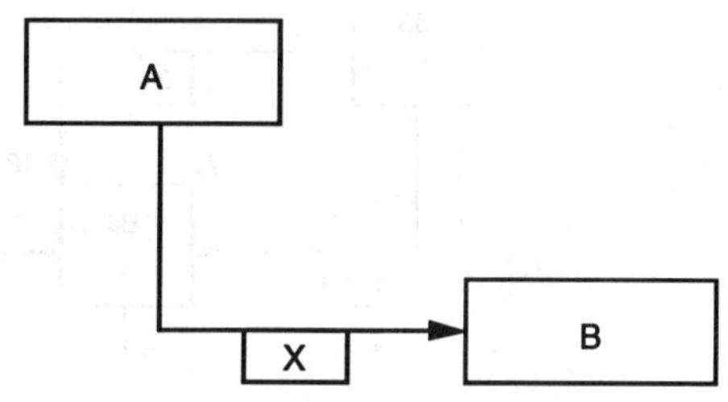

$TF_A = LFT_A - EFT_A$
$FF_A = EST_B - EST_A - "X"$

NOTES:

(1) All activities assumed continuous in duration.

(2) The Total Float for all activities is always equal to their Late Finish Time minus their Early Finish Time.

(3) The Free Float calculations shown are for each type of relationship. If an activity has several following activities, then the Free Float for the activity is the **smallest** of the Free Float calculations made for each following activity.

FIGURE 2.25 Precedence float calculation rules.

top of the figure. Although the precedence version is somewhat easier to draw, one does have to be careful in calculating the activity times. There is also a tendency for all of the precedence activities to be critical due to the continuous time constraint for the activities. If one understands how to use either CPM system, the network development will not be difficult; therefore, it is mostly a matter of preference.

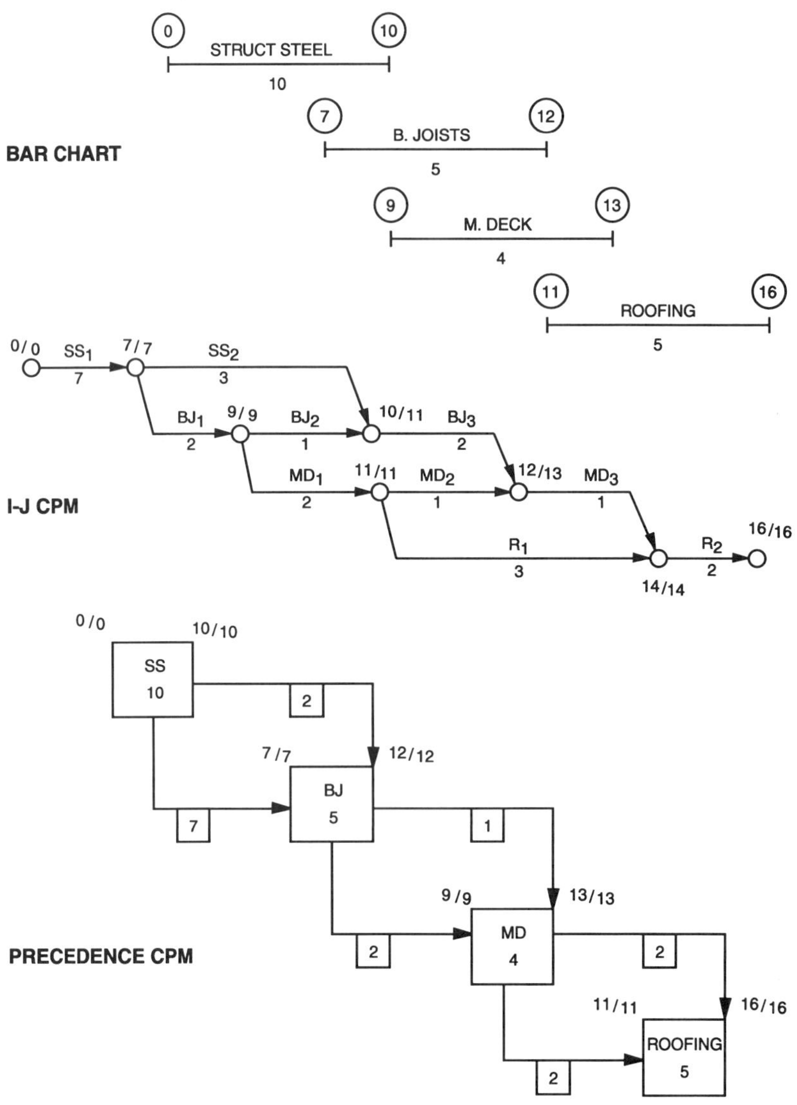

FIGURE 2.26 Precedence overlapping work schedules.

2.4 CPM Day to Calendar Day Conversion

All CPM times on the logic diagrams are noted in CPM days, which are somewhat different from project workdays and decidedly different from calendar days or dates. Since most persons utilizing the activity times from a CPM diagram need to know the starting and finishing time requirements in calendar dates, one needs to know how to convert from CPM days to calendar dates. To illustrate this relationship refer to Fig. 2.27, which depicts a typical activity from an I-J CPM activity, plus a CPM day to CAL day conversion chart.

The CPM/CAL conversion table represents CPM days on the left and regular calendar days on the right. For the sample project shown the project start date is February 3, 1993. Accordingly, the first CPM day is 0 and is shown on the left side. CPM days are then noted consecutively, skipping weekends, holidays, and other nonworkdays for the project. CPM days are referenced to

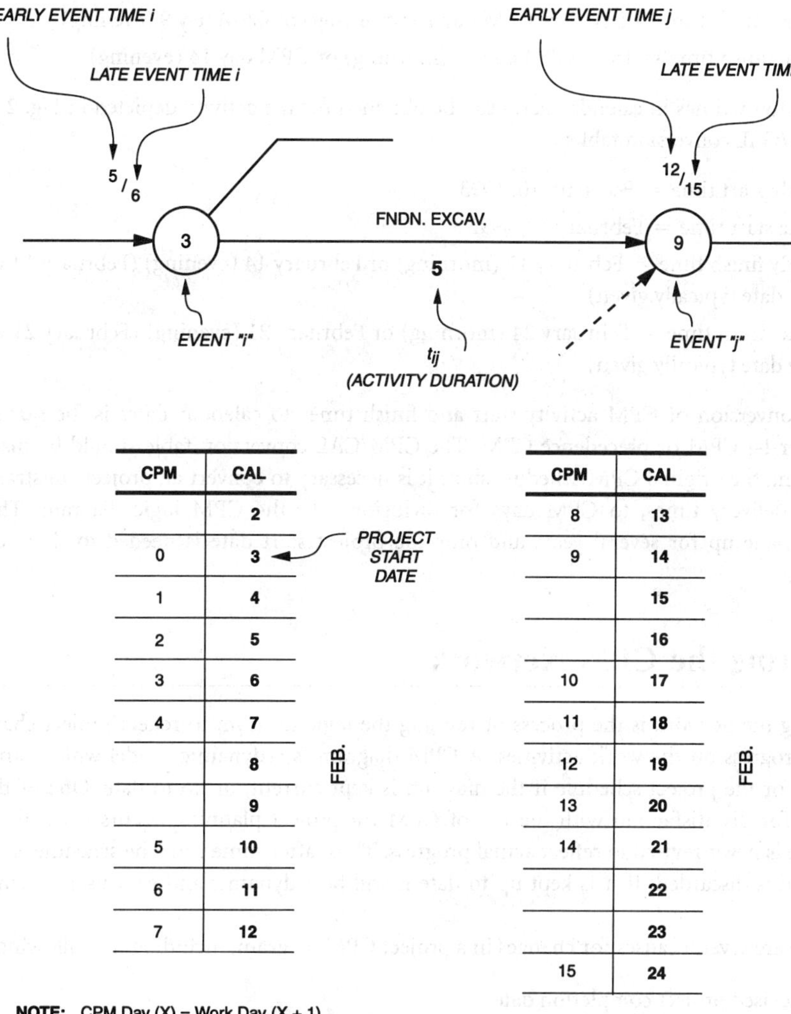

CPM	CAL
	2
0	3
1	4
2	5
3	6
4	7
	8
	9
5	10
6	11
7	12

CPM	CAL
8	13
9	14
	15
	16
10	17
11	18
12	19
13	20
14	21
	22
	23
15	24

NOTE: CPM Day (X) = Work Day (X + 1)
(i.e. CPM Day 0 = Project Work Day 1)

FIGURE 2.27 CPM/CAL conversion chart.

the morning of a workday; therefore, CPM day 0 is equal to the morning of project workday 1 and also equal to the morning of February 3, 1993.

Activity start dates are read directly from the conversion table since they start in the morning. For instance, if an activity has an early start on CPM day 5, then it would start on February 10, 1993. This is also the morning of workday 6. Finish times can also be read directly from the table, but one must remember that the date is referenced to the morning of the day. For instance, if an activity has an early finish of CPM day 8, then it must finish on the morning of February 13. However, most persons are familiar with finish times referenced to the end of the day; therefore, unless the project crew is working 24-hour days, one must back off by one CPM day to give the finish date of the evening of the workday before. This means the finish time for the activity finishing by the morning of February 12 would be given as February 12, 1993. This process is followed for both the early finish time and the late finish time for an activity.

The activity shown in Fig. 2.27 has the following activity times:

- Early start time = CPM day 5
- Late start time = 15 − 5 = CPM day 10

- Early finish time = 5 + 5 = CPM day 10 (morning) or CPM day 9 (evening)
- Late finish time = 15 = CPM day 15 (morning) or CPM day 14 (evening)

The activity times in calendar dates can be obtained for the activity depicted in Fig. 2.27 using the CPM/CAL conversion table:

- Early start time = February 10, 1993
- Late start time = February 17, 1993
- Early finish time = February 17 (morning) or February 14 (evening) (February 14 would be the date typically given)
- Late finish time = February 24 (morning) or February 21 (evening) (February 21 would be the date typically given)

The conversion of CPM activity start and finish times to calendar dates is the same process for either I-J CPM or precedence CPM. The CPM/CAL conversion table should be made when developing the original CPM schedule since it is necessary to convert all project constraint dates, such as delivery times, to CPM days for inclusion into the CPM logic diagram. The charts can be made up for several years and only the project start date is needed to show the CPM days.

2.5 Updating the CPM Network

Updating the network is the process of revising the logic diagram to reflect project changes and actual progress on the work activities. A CPM diagram is a dynamic model which can be used to monitor the project schedule if the diagram is kept current, or up to date. One of the major reasons for dissatisfaction with the use of CPM for project planning occurs when the original schedule is never revised to reflect actual progress. Thus, after some time the schedule is no longer valid and is discarded. If it is kept up to date it will be a dynamic and very useful management tool.

There are several causes for changes in a project CPM diagram, including the following:

1. Revised project completion date
2. Changes in project plans, specifications, or site conditions
3. Activity durations not equal to the estimated durations
4. Construction delays (e.g., weather, delivery problems, subcontractor delays, labor problems, natural disasters, owner indecision)

In order to track such occurrences, the project schedule should be monitored and the following information collected for all activities underway and those just completed or soon to start:

1. Actual start and finish dates, including actual workdays completed
2. If not finished, workdays left to complete and estimated finish date
3. Reasons for any delays or quick completion times
4. Lost project workdays and the reasons for the work loss

Frequency of Updating

A major concern is the frequency of updates required for a project schedule. The obvious answer is that updates should be frequent enough to control the project. The major factor is probably cost, since monitoring and updating are expensive and do cause disturbances, no matter how slight, for the project staff. Other factors are the management level of concern, the average duration of most activities, total project duration, and the amount of critical activities. Some general practices followed for updating frequencies include the following:

1. Updates may be made at uniform intervals (daily, weekly, monthly).
2. Updates may be made only when significant changes occur on the project schedule.
3. Updates may be made more frequently as the project completion draws near.
4. Updates may be made at well-defined milestones in the project schedule.

In addition to keeping up with the actual project progress, there are other reasons that make it beneficial to revise the original project network:

- To provide a record for legal action or for future schedule estimates
- To illustrate the impact of changes in project scope or design on the schedule
- To determine the impact of delays on the project schedule
- To correct errors or make changes as the work becomes better defined

Methods for Revising the Project Network

If it is determined that the current project network is too far off the actual progress on a project, then there are three basic methods to modify the network. These methods will be illustrated for a small network where the original schedule is as shown in Fig. 2.28 and the project's progress is evaluated at the end of CPM day 10.

I. Revise existing network (see Fig. 2.29).
   A. Correct diagram to reflect actual durations and logic changes for the work completed on the project schedule.
   B. Revise logic and duration estimates on current and future activities as needed to reflect known project conditions.

II. Revise existing network (see Fig. 2.30).
   A. Set durations equal to zero for all work completed and set the EET (first event) equal to the CPM day of the schedule update.
   B. Revise logic and duration estimates on current and future activities as needed to reflect known project conditions.

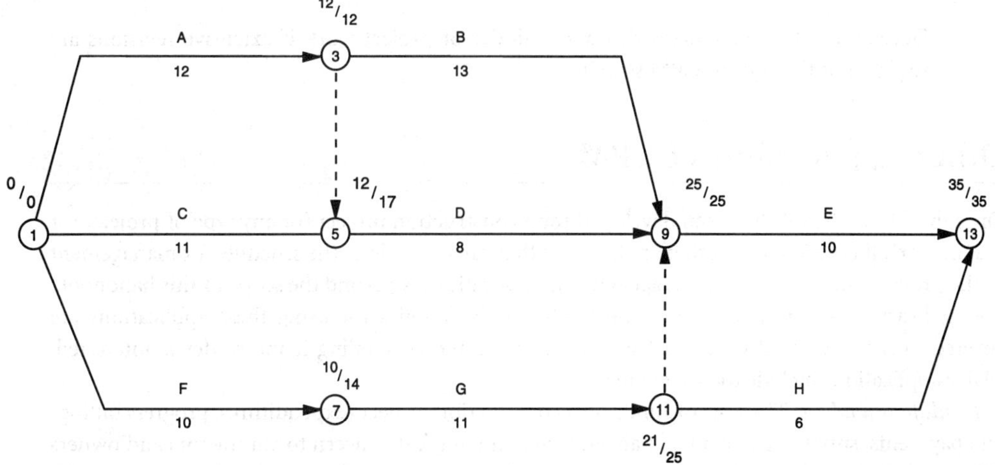

FIGURE 2.28 I-J CPM diagram for updating example.

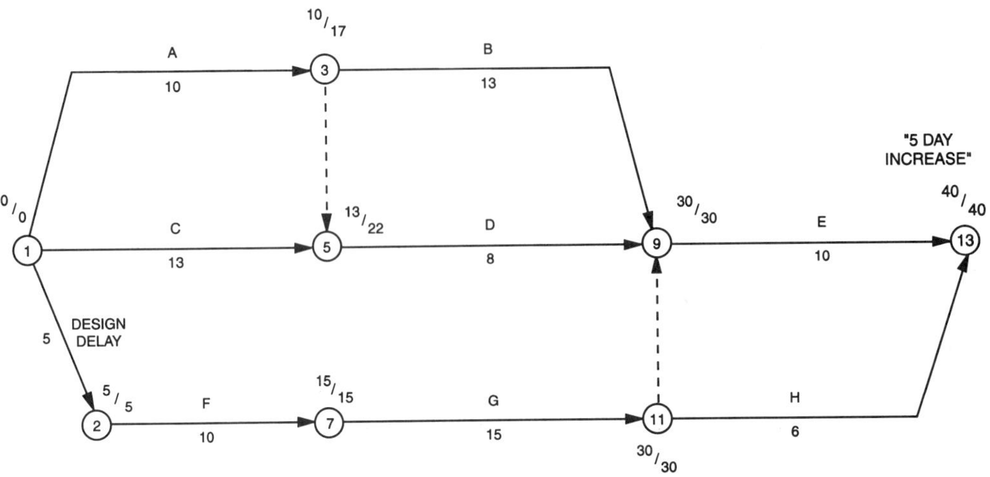

FIGURE 2.29 Updated I-J CPM diagram (version I).

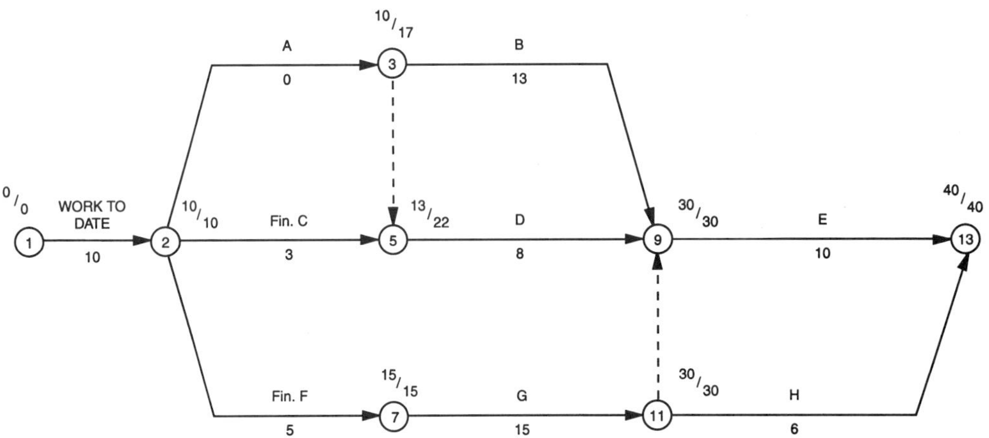

FIGURE 2.30 Updated I-J CPM diagram (version II).

III. Develop a totally new network for remainder of project work if extensive revisions are required of the current CPM schedule.

2.6 Other Applications of CPM

Once the CPM schedule has been developed for a construction project (or any type of project, for that matter), there are several other applications that can be made of the schedule for management of the project. Since complete coverage of these applications is beyond the scope of this handbook, the applications are only mentioned here. The methodologies for using these applications are covered in most textbooks on construction planning and scheduling if the reader is interested. Major applications include the following:

Funds flow analysis: The flow of funds on a construction project (expenditures, progress billings and payments, supplier payments, retainages, etc.) are of great concern to contractors and owners alike for their financing projections for projects. CPM activities can be costed and used, along with other project cost conditions, to predict the flow of funds over the life of the project.

Resource allocation and analysis: The efficient utilization of resources is a very important problem in construction project management, especially for the scheduling of scarce resources. By identifying the resource requirements of all activities on a CPM network, the network provides the basis for evaluating the resource allocation needs of a project. If the demand is unsatisfactory over time, then several methods are available to seek a more feasible project schedule to minimize the resource problem. The controlling factor in most cases is the desire to minimize project cost.

Network compression: This process is sometimes referred to as the *time-cost trade-off problem.* In many projects the need arises to reduce the project duration to comply with a project requirement. This can be planned by revising the CPM network for the project to achieve the desired date. This is accomplished by reducing the durations of critical activities on the network and usually is a random process with minimal evaluation of the cost impact. By developing a cost utility curve for the network's activities, especially the critical or near-critical activities, the network compression algorithm can be used to seek the desired reduced project duration at minimal increased cost.

2.7 Summary

Construction project planning and scheduling is a key element of successful project management. Time spent in planning prior to a project start, or early in the project, will most always pay dividends for all participants on the project. There are several methods available to utilize for such planning, including the bar chart and the critical path methods. Key information on these methods has been presented in this chapter of the handbook. Use of any of the methods presented will make project planning a much easier and more logical process. They all also provide the basis for other management applications on the project. The reader is encouraged to seek more information on the methods presented and investigate other methods available. Finally, there are many computer software packages now available to facilitate the planning process. The use of these systems is encouraged, but the reader should take care to purchase a system that provides the services desired at a reasonable cost. Take time to investigate several systems before selecting one, and be sure to pick a reliable source that is likely to be in business for some time in the future.

Defining Terms

Activity: A distinct and identifiable operation within a project, whose performance will consume one or more resources. The concept of the distinct activity is fundamental in network analysis. The level of detail at which distinct activities are identified in planning depends largely on the objectives of the analysis. An activity may also be referred to as an operation, or work item. A dummy activity, which does not consume a resource, can be used to identify a constraint which is not otherwise apparent.

Activity duration: The estimated time required to perform the activity; the allocation of the time resource to each activity. It is customary to express the activity duration in work-time units; that is, workday, shift, week, and so on. An estimate of activity duration implies some definite allocation of other resources (labor, materials, equipment, capital) necessary to the performance of the activity in question.

Constraints: Limitations placed on the allocation of one or several resources.

Critical activities: Activities which have zero float time. This includes all activities on the critical path.

Critical path: The connected chain, or chains, of critical activities (zero float), extending from the beginning of the project to the end of the project, whose summed activity duration give the minimum project duration. Several may exist in parallel.

Float (slack): In the calculation procedures for any of the critical path methods, an allowable time span is determined for each activity to be performed within. The boundaries of this time span for an activity are established by its early start time and late finish time. When this bounded time span exceeds the duration of the activity, the excess time is referred to as float time. Float can be classified according to the delayed finish time available to an activity before it affects the starting time of its following activities. It should be noted that once an activity is delayed to finish beyond its early finish time, then the network calculations must be redone for all following activities before evaluating their float times.

Free float: The number of days that an activity can be delayed beyond its early finish time without causing any activity which follows it to be delayed beyond its early start time. The free float for many activities will be zero since it only exists when an activity doesn't control the early start time of any of the activities which follow it.

Management constraints: Constraints on the ordering of activities due to the wishes of management. For instance, which do you install first, toilet partitions or toilet fixtures? It doesn't usually matter, but they can't be done easily at the same time. Therefore, one will be scheduled first and the other constrained to follow; this is a management constraint.

Monitoring: The periodic updating of the network schedule as the project progresses. For activities which have already been performed, their estimated durations can be replaced by their actual durations. The network can then be recalculated. It will often be necessary to replan and reschedule the remaining activities as necessary to comply with the requirement that the project duration remain the same.

Network model: The graphical display of interrelated activities on a project, showing resource requirements and constraints; a mathematical model of the project and the proposed methods for its execution. A network model is actually a logic diagram which has been prepared in accordance with established diagramming conventions.

Physical constraints: Constraints on the ordering of activities due to physical requirements. For instance, the foundation footings cannot be completed until the footing excavation work is done.

Planning: The selection of the methods and the order of work for performing the project. (Note that there may be feasible methods and, perhaps, more than one possible ordering for the work. Each feasible solution represents a plan.) The required sequence of activities (preceding, concurrent, or following) is portrayed graphically on the network diagram.

Project: Any undertaking which has a definite point of beginning and a definite point of ending, and which requires one or more resources for its execution. It must also be capable of being divided into interrelated component tasks.

Project duration: The total duration of the project, based on the network assumptions of methods and resource allocations. It is obtained as the linear sum of activity durations along the critical path.

Resource constraints: Constraints on the ordering of activities due to an overlapping demand for resources which exceeds the available supply of the resources. For instance, if two activities can be performed concurrently but each requires a crane, and only one crane is available, then one will have to be done after the other.

Resources: Those things which must be supplied as input to the project. They are broadly categorized as manpower, material, equipment, money, time, and so on. It is frequently necessary to identify them in greater detail (draftsmen, carpenters, cranes, etc.).

Scheduling: The process of determining the time of a work item or activity within the overall time span of the construction project. It also involves the allocation of resources (men, material, machinery, money, time) to each activity, according to its anticipated requirements.

Total Float: The total time available between an activity's early finish time and late finish time as determined by the time calculations for the network diagram. If the activity's finish is delayed more than its number of days of total float, then its late finish time will be exceeded and the total project duration will thus be delayed. Total float also includes any free float which is available for the activity.

References

Callahan, M. T., Quackenbush, D. G., and Rowings, J. E. 1992. *Construction Project Scheduling,* 1st ed. McGraw-Hill, New York.

Harris, R. B. 1978. *Precedence and Arrow Networking Techniques for Construction,* 1st ed. John Wiley & Sons, New York.

For Further Information

There are hundreds of books, papers, and reports available on the subject of construction planning and scheduling in the U.S. alone. The two references cited above are only two such publications which are often used by the author. In addition there are many computer software packages available which give in-depth details of basic and advanced applications of planning and scheduling techniques for construction project management. Some of the commonly used packages are Primavera, Open Plan, Harvard Project Manager, and Microsoft Project. Another excellent source of information on new developments in scheduling is the *ASCE Journal of Construction Engineering and Management,* which is published quarterly.

3

Equipment Productivity

Tom Iseley
Louisiana Tech University

3.1 Introduction

Whether a construction contract is unit price, lump sum, or cost-plus; whether the construction project is to be linear (i.e., concept → design → procurement → construction) or fast-tracked (i.e., design/build), the cost of construction is a major factor in all projects. The major factors that impact construction costs are materials, labor, equipment, overhead, and profit. The cost of equipment for civil engineering construction projects can range from 25% to 40% of the total project cost.

Figure 3.1 illustrates the ability to influence the construction cost of a project; the greatest influence is near the front end of the project. Many decisions made by design engineers near the beginning of the project, during the conceptual and design phase, dictate the equipment that will be acceptable for a particular project. The design may, in fact, restrict the best and most cost-effective solution from being utilized. For example, many sewer projects are designed around traditional specifications and equipment when microtunneling would be safer, more environmentally and socially acceptable, and more cost-effective when total life-cycle costing is applied.

It is important for design engineers and construction engineers to be knowledgeable about construction equipment. Construction equipment is an integral part of the construction process. The cost of construction is a function of the design of the construction operation.

This chapter will provide an overview of the construction equipment selection and utilization process. It will describe typical equipment spreads associated with two major classifications of civil engineering construction projects: heavy/highway and municipal/utility. Methods for determining equipment productivity and cost will be discussed.

3.2 Heavy/Highway Construction Projects

These projects include new road construction, dams, airports, waterways, rehabilitation of existing roadways, marine construction, bridges, and so on. Each project can be segmented into various phases or operations. The equipment spread selected for a specific construction operation is critical to the success of a project.

0-8493-8953-4/95/$0.00 + $.50

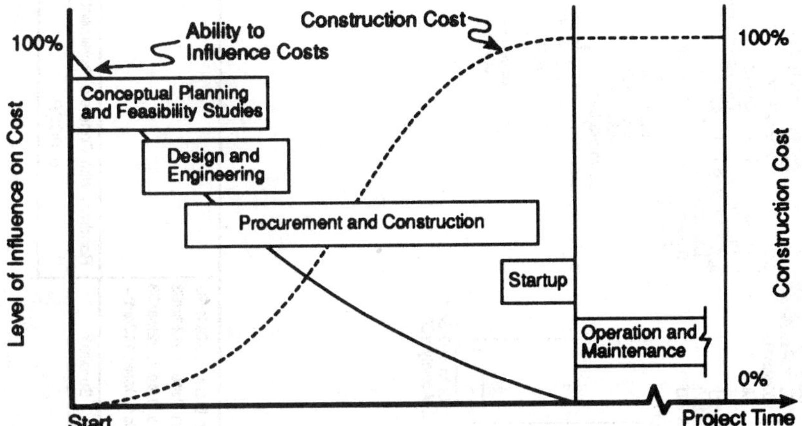

FIGURE 3.1 Ability to influence construction cost over time. (*Source:* Hendrickson, C. and Au, T. *Project Management for Construction.* Prentice Hall, Englewood Cliffs, NJ.)

Figure 3.2 (a) illustrates a typical sequence of activities for the construction of a new highway. This highway project contains culverts and a bridge. While there will be some overlap in the equipment utilization, each activity must be evaluated carefully to identify all operations in the activity and to ensure that the equipment selected for each operation is compatible with the tasks to be completed. Figure 3.2 (b) lists the activities associated with the project, the duration of each, and whether they are critical or noncritical. The intent is not to provide a detailed description of each activity, but to illustrate how equipment selection and utilization are a function of the associated variables. For example, the first activity (clearing and grubbing) is extremely critical as no activity can begin until the area is cleared. Even though clearing land is often considered to be a basic, straightforward activity, it is still more an art than a science. The production rates of clearing land are difficult to forecast because they depend on the following factors:

- The quantity and type of vegetation
- Purpose of the project
- Soil conditions
- Topography
- Climatic conditions
- Local regulations
- Project specifications
- Selection of equipment
- Skill of operators

To properly address these variables requires research and a thorough evaluation of the site to determine the following:

- Density of vegetation
- Percent of hardwood present
- Presence of heavy vines
- Average number of trees by size category
- Total number of trees

A method for quantifying the density of vegetation and the average number of trees per size category is described in Caterpillar [1993, p. 22-2] and Peurifoy and Ledbetter [1985, p. 158].

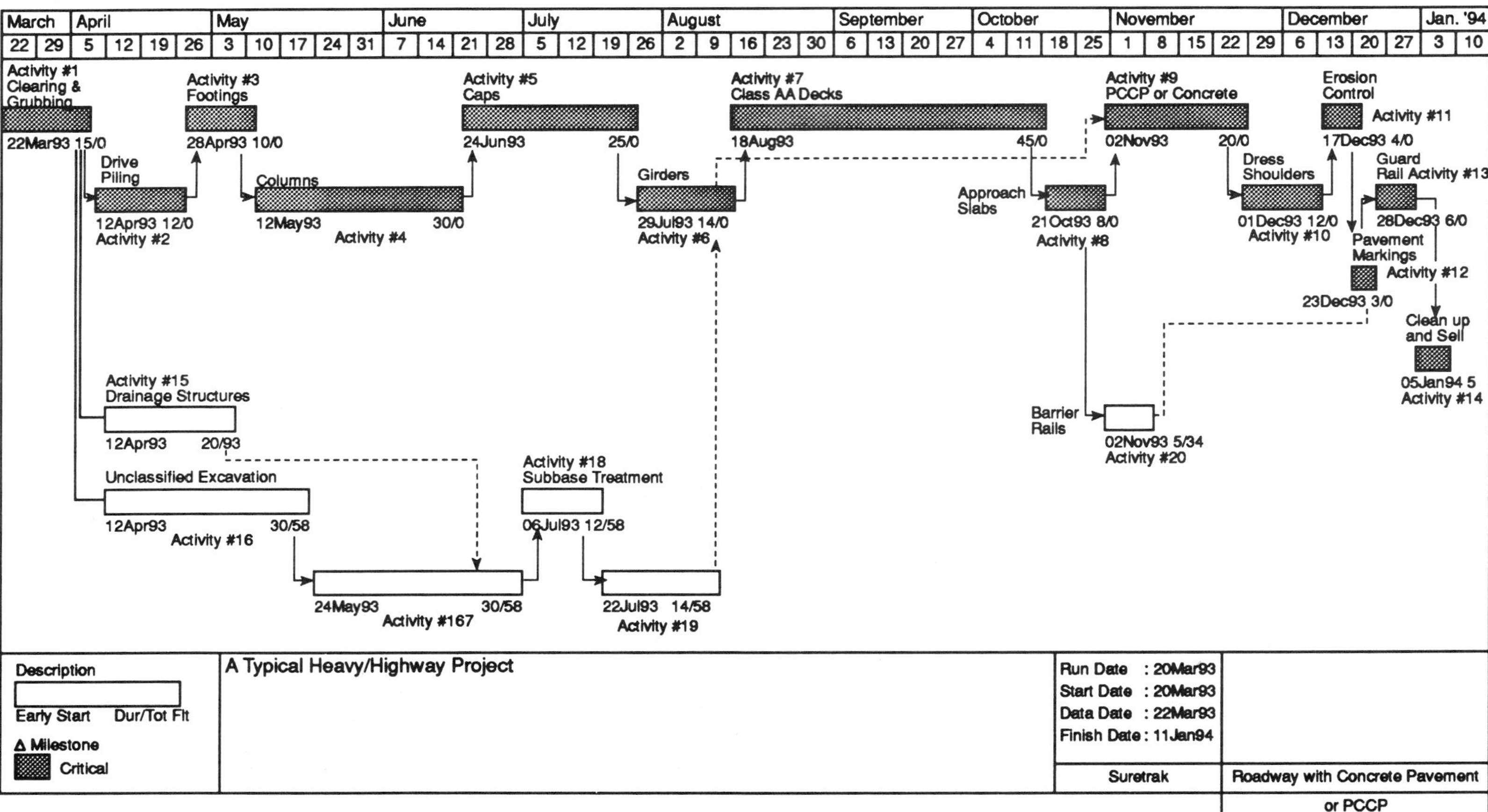

FIGURE 3.2 (a) Typical sequence of activities for a new highway project.

Activity Number	Activities on Critical Path	Duration (Days)
1.	Clearing and Grubbing	15
2.	Drive Piling for Bridge	12
3.	Construct Footings for Bridge Foundation	10
4.	Construct Columns for Bridge Foundation	30
5.	Construction Column Caps	25
6.	Construct Girders for Bridge Deck	14
7.	Construct Class AA Bridge Deck	45
8.	Construct Approach Slabs	8
9.	Place Pavement	20
10.	Dress Road Shoulders	12
11.	Erosion Control	4
12.	Pavement Markings	3
13.	Guard Rail	6
14.	Project Clean Up and Final Inspection	5
Activity Number	**Activities Not on Critical Path**	**Duration (Days)**
15.	Drainage Structures	20
16.	Unclassified Excavation	30
17.	Borrow	30
18.	Subbase Treatment	12
19.	Asphaltic Base	14
20.	Barrier Rails	5

FIGURE 3.2 (continued) (b) Activities associated with a new highway project.

The most common type of construction equipment used for clearing and grubbing activities is a bulldozer. The term *bulldozer* is used to define a tractor mounted with a dozing blade. Tractor size can vary from less than 70 flywheel horsepower (FWHP) to more than 775 FWHP. Dozing blades are available in many types. The appropriate blade depends on the job to be accomplished. For example, types of blades include the following:

U—universal
SU—semiuniversal
S—straight
P—power angle and tilt
A—angling
V—tree cutter

Caterpillar [1993, pp. 1-39–1-41] provides a detailed description and illustration of various blades commonly used. To maximize production, it is important to ensure that the tractor and dozer blade are properly matched. The two major factors to be considered in selecting the proper blade and tractor are material to be moved and tractor limitations. For example, the weight and horsepower of the tractor determine its ability to move material. Particle size, particle shape, voids, and water content are the main factors affecting the level of difficulty of moving material.

Caterpillar [1993] and Peurifoy and Ledbetter [1985] utilize the following relationship to estimate the time required to perform various operations, such as felling trees with bulldozers and piling in windows with bulldozers:

$$T = B + M_1N_1 + M_2N_2 + M_3N_3 + M_4N_4 + DF \tag{3.1}$$

where

T = time required per acre, minutes

B = base time required for a bulldozer to cover an acre with no trees requiring splitting or individual treatment, minutes

M = time required per tree in each diameter range, minutes

N = number of trees per acre in each diameter range, obtained from field survey

D = sum of diameter in feet of all trees per acre, if any, larger than 6 feet in diameter at ground level

F = time required per foot of diameter for trees larger than 6 feet in diameter, minutes

While industry average data exists for the variables in Eq. (3.1) which correlate operation duration with FWHP of bulldozers, a firm's historic database should provide improved accuracy in forecasting production. Such a database can be developed by project owner representatives and constructors. This database could then be used for future projects and comparison with industry averages.

Many large constructors do not maintain such detailed databases. For example, the equipment selected for the clearing and grubbing activity for the project illustrated in Fig. 3.2 was based on estimator experience. When questioned about the selection process, the estimator stated that he made his selection based on the fact that the project consisted of more than 25 acres of timber with trees larger than 24 inches in diameter. If there had been less than 25 acres with less than 24-inch diameter trees he would have selected smaller equipment.

The type of production estimating used by this constructor is extremely effective when experienced estimators are responsible for making the necessary decisions. However, a more detailed database would permit decisions to be made at a lower level by less experienced people without sacrificing accuracy. This would allow the more experienced estimator to utilize his or her time more effectively. In addition, a more detailed database would provide better information to substantiate the impact if a change in conditions should develop.

Table 3.1 illustrates major equipment requirements for each activity shown on Fig. 3.2. The equipment listed can be categorized as follows: bulldozers, excavators, compactors, graders, scrapers, spreaders, cranes, loaders, trucks, and miscellaneous (asphalt spreaders, screeds, water trucks, power brooms, farm tractors, generators).

Although it is beyond the scope of this chapter to provide an in-depth analysis of the estimated productivity of each of the above classes of equipment, this analysis is provided in the references cited at the end of this chapter. Several categories of machines are discussed in general terms. Equipment manufacturers also provide reliable productivity information.

It is important to be able to segment a project into its basic activities, as illustrated in Fig. 3.2. Each activity must then be further segmented into its basic operations. Individual operations within an activity are unique and require specific combinations of equipment to be executed in a cost-effective manner. Each machine selected will have a unique operation within an activity; the operation is a function of the cycle time of the machines needed to execute it. All the machines must be selected so that their productivities balance. The duration of the activities listed in Fig. 3.2 are a function of all the operations necessary to accomplish the activity.

Duration estimating is extremely important, but, because of the many variables involved, it is not an easy task. Excellent simulation methods and computer software programs are available to help project managers evaluate more precisely the impact of variables of operations and processes [Halpin and Riggs, 1992].

It is important to ensure that the project is designed for constructibility. The project designers must be knowledgeable concerning construction processes and the variables impacting total life-cycle costs.

Table 3.1 Equipment Selected for Typical Heavy/Highway Project (See Fig. 3.2)

Activity Number	Equipment Description: No. of Units	Size	Type	Equipment Cost/Unit	Remarks
1	2	285–340 FWHP	Bulldozer	$385,000	Provide: Track-type
					2—U blades
					1—Rake
	1	160–180 FWHP	Bulldozer	$143,000	Provide: Track-type w/P blade
	1	140–160 FWHP	Hydraulic excavator	$400,000	
2	1	65–75 ton	Crane	$312,000	Provide: 1—pile driving hammer
				$75,000	1—set of cable leads
				$40,000	1—drill and power pack
				$120,000	
3,4,5	1	50–75 ton	Crane	$285,000	
	1	small	Generator	$30,000	50 KW
6	2	75–100 ton	Cranes	$600,000	
7	1	50–75 ton	Crane	$285,000	
	1	Medium	Generator	$100,000	150 KW
	1	150 CFM	Air compressor	$12,000	
	1		Bridge Screed	$40,000	
	1		Concrete pump truck	$175,000	
8	1	140–160 FWHP	Hydraulic excavator	$400,000	
	1		Clarey screed	$7,500	
	1	63–70 FWHP	Bulldozer	$65,000	
9	1		Compactor	$75,000	Smooth drum roller
	1		Compactor	$60,000	Pneumatic roller
	1		Power broom	$18,000	Broce model T-20
	1		Asphalt spreader	$175,000	Barber-Green BFS-185
10	1	75–90 FWHP	Bulldozer	$75,000	Track-type
	1	135–145 FWHP	Motor grader	$115,000	12G
	1	175–180 FWHP	Scraper	$215,000	Provide: elevating equipment/model 615
11	3	50–70 FWHP	Farm tractors	$40,000	Provide: miscellaneous attachments
	2	1800–3000 gal	Water trucks	$3,000	
12					Subcontract work
13					Subcontract work
14	1	135–145 FWHP	Motor grader	$115,000	
	1	75–90 FWHP	Bulldozer	$75,000	Track-type
	1	118–133 FWHP	Hydraulic excavator	$160,000	
	1		Dump truck	$62,000	
15	1	118–133 FWHP	Hydraulic excavator	$160,000	
	1	75–90 FWHP	Bulldozer	$75,000	Track-type
	1	145–170 FWHP	Front-end loader	$75,000	Provide: rubber-tired type
	4		Compactors	$75,000	Provide: vibra-plate tamps
	1	Small	Compactor	$20,000	Provide: self-propelled roller
16,17	6	350–450 FWHP	Scrapers	$300,000	Model 631E
	1	320–370 FWHP	Bulldozer	$550,000	Push tractor for scrapers
	2	120–140 FWHP	Bulldozers	$75,000	To level and spread material
	1	140–160 FWHP	Hydraulic excavator	$400,000	
	1	180–200 FWHP	Motor grader	$225,000	16G
	1	175–215 FWHP	Compactor	$180,000	Provide: self-propelled roller
	1	10,000 gal	Water truck	$40,000	
	1	250–300 FWHP	Farm tractor	$105,000	Provide: 24–28 inch plow
18	3		Soil stabilizers	$150,000	Type: Raygo
	4	1800–3000 gal	Water trucks	$25,000	
	1		Compactor	$150,000	Provide: self-propelled
	1		Compactor	$75,000	Sheep-foot roller
	2	135–145 FWHP	Motor grader	$75,000	Provide: pneumatic roller
			Truck	$40,000	
19	1		Compactor	$75,000	Smooth drum roller
	1		Compactor	$60,000	Pneumatic roller
	1		Power broom	$18,000	Broce model T-20
	1		Asphalt spreader	$175,000	Barber Green BFS-185

Note: Equipment costs represent estimated market value. They illustrate the size of investment that must be recovered.

Equipment Productivity

Once the equipment needs for an activity have been identified, the next step is to conduct an equipment productivity analysis to select the optimum size. The objective is to determine the number of units and the size of equipment that would permit the constructor to accomplish the activity with a duration resulting in the lowest cost.

Since most civil engineering construction projects are awarded based on lowest cost, it is of utmost importance to the constructor to select the proper equipment spread providing the lowest construction cost for the project. The project is segmented into various activities; therefore, the lowest cost must be determined for each activity.

Bulldozer Productivity

When a constructor lacks reliable historic data, most bulldozer equipment manufacturers will provide production information. Manufacturer's production information is useful to conduct a comparative analysis when developing a conceptual estimate or schedule. Production data provided by manufacturers (or any other source) must be applicable to a particular situation.

For example, Caterpillar [1993] contains excellent production curves for estimating dozing production in units of loose cubic yards (LCY) of materials per hour (LCY/hr). Figure 3.3 illustrates a typical production curve from Caterpillar [1993, 1-58]. This information is based on numerous field studies made under varying job conditions. These production curves provide the maximum uncorrected production based on the following conditions:

1. 100% efficiency (60 minute hour—level cycle)
2. Power shift machines with 0.05 minute fixed times
3. Machine cuts for 50 feet (15 m), then drifts blade load to dump over a high wall. (Dump time—0 seconds)
4. Soil density of 2300 lb/LCY (1370 kg/m^3)
5. Coefficient of traction: track machines—0.5 or better, wheel machines—0.4 or better
6. Hydraulic controlled blades are used
7. Dig 1F; carry 2F; return 2R (1F—first forward gear, etc.)

As long as the conditions that the production curves are based on are understood, they can be used for other conditions by applying the appropriate correction factors using the following relationship:

$$\text{Production (LCY/hr)} = \text{Maximum production (from production curve)} \times \text{Correction factor} \qquad (3.2)$$

Common correction factors are provided for such variables as operator skill, type of material being handled, method of dozing (i.e., dozing in a slot or side-by-side dozing), visibility, time efficiency (actual minutes per hour of production), transmission type, dozer blade capacity, and grades. Caterpillar [1993, pp. 1-59–1-60] is an excellent source for correction factors and provides an excellent example of how to use production curves and correction factors.

When easy-to-use production curves do not apply to a particular situation, the basic performance curves must be utilized. Manufacturers provide a performance curve for each machine. Track-type tractor performance curves are in the form of drawbar pull (DBP) versus ground speed, and rubber-tired type machine performance curves are in the form of rimpull (RP) versus ground speed.

The DBP vs. speed curves will be discussed in this section and the RP vs. speed curves will be discussed later in connection with rubber-tired scrapers.

Drawbar horsepower is the power available at the tractor drawbar for moving the tractor and its towed load forward. Figure 3.4 illustrates the transfer of power from the flywheel to the drawbar.

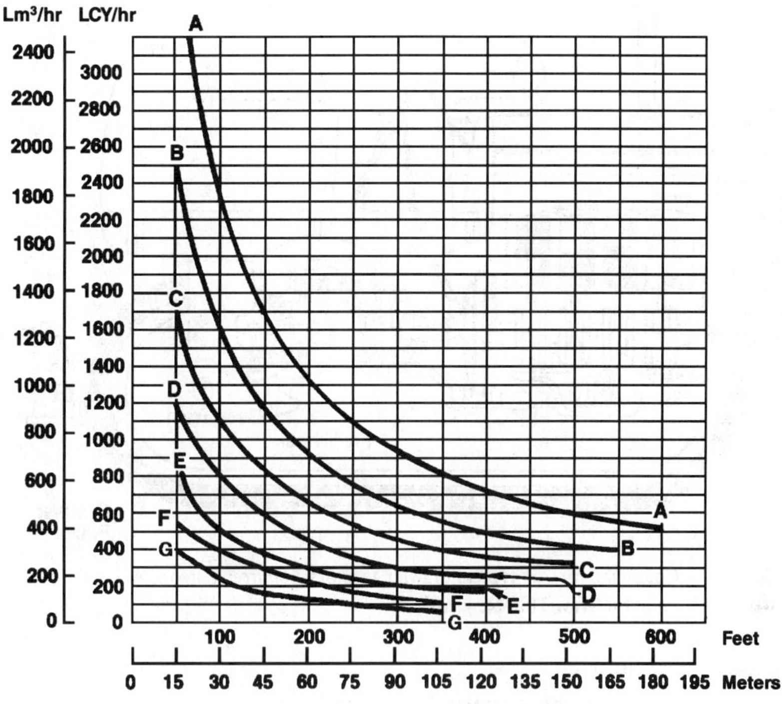

FIGURE 3.3 Typical production curve for estimation of dozing production. A-D11N-11SU, a bulldozer as manufactured by Caterpillar, Inc. with a model number D11N utilizing an SU type of blade designed for use with a D11 machine can be expected to move the volume of earth per hour as indicated by the A curve on this chart for a specific distance. For comparative purposes the FWHP of each machine is D11N—770, D10N—520, D9N—370, D8N—285, D7H—215, and D6H—165. (*Source: Caterpillar Performance Handbook,* 24th ed., 1993. Caterpillar, Peoria, IL, p. 1-58.)

The relationship between DB horsepower, DB pull, and speed can be expressed as

$$\text{DBHP} = \frac{\text{DBP (lb)} \times \text{speed (mph)}}{375} \tag{3.3}$$

where

DBHP = drawbar horsepower
DBP = drawbar pull (pounds)
Speed = ground speed (miles per hour)
375 = conversion factor

For example, to obtain the compaction required in activity 17 in Fig. 3.2, tillage may be necessary to accelerate drying of the soil. This tillage could be accomplished with a 24 to 28 inch agricultural plow pulled behind a track-type tractor. Production requirements could demand that this tillage operation needs to move at a speed of 3 mph and would impose a 22,000 lb DBP. Thus, the tractor

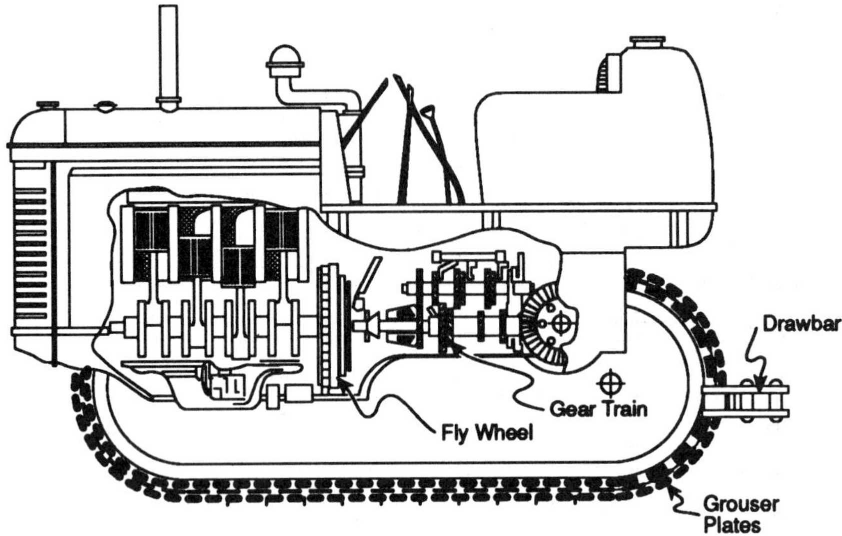

FIGURE 3.4 Characteristics of typical crawler tractors. (*Source:* Carson, A. B. 1961. *General Excavation Methods.* McGraw-Hill, New York.)

must be able to apply at least

$$\frac{22{,}000 \text{ lb DBP} \times 3 \text{ mph}}{375} = 176 \text{ DBHP}$$

However, as can be seen in Table 3.1, for Activity 17 a 250 to 300 FWHP rubber-tired farm tractor was selected by the constructor. Obviously, this decision for a more powerful, rubber-tired machine was made because a higher production rate was needed to keep all operations in balance.

Figure 3.5 is a DBP vs. speed performance curve for a model D8N track-type tractor as manufactured by Caterpillar, Inc. The same type curves are available from all other manufacturers,

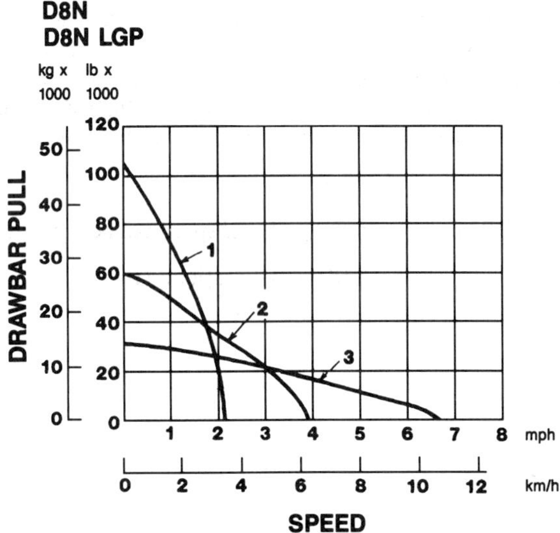

FIGURE 3.5 Drawbar pull vs. ground speed power shift. 1—1st gear, 2—2nd gear, 3—3rd gear. (*Source: Caterpillar Performance Handbook,* 24th ed., 1993. Caterpillar, Peoria, IL, p. 1-58.)

whether domestic or foreign. As illustrated, once the demand has been defined in terms of the necessary DBP required to perform the task, the gear range and ground speed are determined for a specific machine. Obviously this speed is critical information when trying to determine the cycle times for each machine at the work face. The cycle times are essential in determining how long each operation will take, and the length of each operation defines the duration of each activity shown in Fig. 3.2.

Excavator Production

Excavators are a common and versatile type of heavy construction equipment. As in the project illustrated by Fig. 3.2, excavators are used for accomplishing many activities, such as lifting objects, excavating for trenches and mass excavation, loading trucks and scrapers, and digging out stumps and other buried objects.

The production of an excavator is a function of the digging cycle, which can be divided into the following segments:

1. Time required to load the bucket
2. Time required to swing with a loaded bucket
3. Time to dump the bucket
4. Time to swing with an empty bucket

This cycle time depends on the machine size and the job conditions. For example, a small excavator can usually cycle faster than a large one, but it will handle less payload per cycle. As the job conditions become more severe, the excavator will slow down. As the soil gets harder and as the trench gets deeper, it takes longer to fill the bucket.

Other factors that greatly impact excavator production are digging around obstacles such as existing utilities, having to excavate inside a trench shield, or digging in an area occupied by workers.

Many excavator manufacturers provide cycle time estimating data for their equipment. This information is an excellent source when reliable historical data are not available. If an estimator can accurately predict the excavator cycle time and the average bucket payload, the overall production can be calculated as follows:

$$\frac{\text{Production}}{\text{(LCY/hr)}} = \frac{\text{Cycles}}{\text{Hr}} \times \frac{\text{Average bucket payload (LCY)}}{\text{Cycle}} \qquad (3.4)$$

Caterpillar [1993, pp. 4-106–4-107] provides estimated cycle times for common excavators with bucket size variations. These values are based on no obstructions, above average job conditions, above average operator skill, and a 60 to 90° angle of swing. Correction factors must be applied for other operating conditions.

Table 3.2 illustrates the level of detailed information available from manufacturers on specific machines. This table presents information on four Caterpillar excavators commonly used on civil engineering projects.

Once the cycle time is determined, either by measuring or estimating, the production can be determined by the following relationship:

$$\text{LCY/60-min hr} = \text{Cycles/60-min hr} \times \text{Average bucket payload (LCY)} \qquad (3.5)$$

where

$$\text{Average bucket payload} = \text{Heaped bucket capacity} \times \text{Bucket fill factor} \qquad (3.6)$$

This production is still based on production occurring the full 60 minutes of each hour. Since this does not occur over the long term, job efficiency factors are presented at the lower left corner of Table 3.3 and applied as follows:

$$\text{Actual Production (LCY/hr)} = \text{LCY/60-min hr} \times \text{Job efficiency factor} \qquad (3.7)$$

Table 3.2 Cycle Time Estimating Chart for Excavators

MODEL		E70B	311	312	E140	320	E240C	325	330	235D	350	375
Bucket Size L (yd³)		280 0.37	450 0.59	520 0.68	630 0.82	800 1.05	1020 1.31	1100 1.44	1400 1.83	2100 2.75	1900 2.5	2800 3.66
Soil Type		← Packed Earth →				← Hard Clay →						
Digging Depth	(m) (ft)	1.5 5	1.5 5	1.8 6	1.8 6	2.3 8	3.2 10	3.2 10	3.4 11	4.0 13	4.2 14	5.2 17
Load Bucket	(min)	0.08	0.07	0.07	0.09	0.09	0.09	0.09	0.09	0.11	0.10	0.11
Swing Loaded	(min)	0.05	0.06	0.06	0.06	0.06	0.07	0.06	0.07	0.10	0.09	0.10
Dump Bucket	(min)	0.03	0.03	0.03	0.03	0.03	0.05	0.04	0.04	0.04	0.04	0.04
Swing Empty	(min)	0.06	0.05	0.05	0.05	0.05	0.06	0.06	0.07	0.08	0.07	0.09
Total Cycle Time	(min)	0.22	0.21	0.21	0.23	0.23	0.27	0.25	0.27	0.33	0.30	0.34

Source: Caterpillar, 1993. *Caterpillar Performance Handbook,* 24th ed., p. 4-106. Caterpillar, Peoria, IL.

Example. Determine the actual production rate for a Cat 225D hydraulic excavator (150 FWHP) as required for activity #16, unclassified excavation, for the project represented in Fig. 3.2. It is estimated that the realistic productive time for the excavator will be 50 minutes per hour. Thus, the job efficiency factor will be 50/60 = 0.83. The soil type is a hard clay.
Solution.

$$\text{Average bucket payload} = \text{Heaped bucket capacity} \times \text{Bucket fill factor}$$

Enter Table 3.2 and select

1. 1.78 LCY bucket capacity for a Cat 225D
2. 0.25 minute total cycle time

$$1.51\ \text{LCY} = 1.78\ \text{LCY} \times 0.85$$

Enter Table 3.3 and select average production rate based on a work time of 60 minutes per hour. Select the column headed by a 1.5 LCY bucket payload and the row that represents a 0.25 minute cycle time. From this the average production is determined to be 360 LCY per 60-minute hour.

$$\text{Actual production} = \text{LCY/60-min hr} \times \text{Job efficiency factor}$$

$$299\ \text{LCY/hr} = 360\ \text{LCY/60-min hr} \times 0.83$$

When an excavator is used for trenching, often the desired rate of production needs to be expressed in lineal feet excavated per hour. The trenching rate depends on the earth-moving production of the excavator being used and the size of the trench to be excavated.

Scraper Production

Scrapers provide the unique capability to excavate, load, haul, and dump materials. Scrapers are available in various capacities by a number of manufacturers with options such as self-loading with elevators, twin engines, or push-pull capability.

Scrapers are usually cost-effective earthmovers when the haul distance is too long for bulldozers yet too short for trucks. This distance typically ranges from 400 to 4000 feet; however, the economics should be evaluated for each project.

The production rate of a scraper is a function of the cycle time required to load, haul the load, dump the load, and return to the load station. The times required to load and dump are usually uniform once established for a specific project while the travel times can vary a significant amount

Table 3.3 Cubic Yards per 60-Minute Hour[a]

Estimated Cycle Times		Estimated Bucket Payload[b]—Loose Cubic Yards																		Estimated Cycle Times	
Seconds	Minutes	0.25	0.50	0.75	1.00	1.25	1.50	1.75	2.00	2.25	2.50	2.75	3.00	3.25	3.50	3.75	4.00	4.50	5.00	Cycles per Min.	Cycles per Hr.
10.0	.17																			6.0	360
11.0	.18																			5.5	330
12.0	.20	75	150	225	300	375														5.0	300
13.3	.22	**67**	**135**	**202**	**270**	337	404	472	540	607	675	742	810	877	945	1012	1080	1215	1350	4.5	270
15.0	.25	**60**	**120**	**180**	**240**	**300**	**360**	**420**	480	540	600	660	720	780	840	900	960	1080	1200	4.0	240
17.1	.29	**52**	**105**	**157**	**210**	**262**	**315**	**367**	**420**	**472**	**525**	**577**	630	682	735	787	840	945	1050	3.5	210
20.0	.33	**45**	**90**	**135**	**180**	**225**	**270**	**315**	**360**	**405**	**450**	**495**	**540**	**585**	**630**	**675**	**720**	810	900	3.0	180
24.0	.40	**37**	**75**	**112**	**150**	**187**	**225**	**262**	**300**	**337**	**375**	**412**	**450**	**487**	**525**	**562**	**600**	675	750	2.5	150
30.0	.50	30	60	**90**	**120**	**150**	**180**	**210**	**240**	**270**	**300**	**330**	**360**	**390**	**420**	**450**	**480**	510	600	2.0	120
35.0	.58	26	51	77	102	128	**154**	**180**	**205**	**231**	**256**	**282**	**308**	**333**	**360**	**385**	**410**	462	613	1.7	102
40.0	.67					112	135	157	180	202	**225**	**247**	**270**	**292**	**315**	**337**	360	405	450	1.5	90
45.0	.75									180	200	220	240	260	280	300	320	360	400	1.3	78
50.0	.83																			1.2	72

Job Efficiency Estimator

Work Time/Hour	Efficiency
60-min	100%
55	91%
50	83%
45	75%
40	67%

Average Bucket Payload = (Heaped Bucket Capacity) × (Bucket Fill Factor)

Material	Fill Factor Range (Percent of Heaped Bucket Capacity)
Moist loam or sandy clay	A—100–110%
Sand and gravel	B—95–110%
Hard, tough clay	C—80–90%
Rock—well blasted	60–75%
Rock—poorly blasted	40–50%

[a] Actual hourly production = (60 min. hr. production) × (Job efficiency factor)
[b] Estimated bucket payload = (Amount of material in the bucket) = (Heaped bucket capacity) × (Bucket fill factor)
Numbers in boldface indicate average production.
Source: Caterpillar. 1993. *Caterpillar Performance Handbook,* 24th ed. Caterpillar, Peoria, IL.

during the project due to variation of the travel distance. The load time can be decreased by preripping the tight soils, prewetting the soil, and designing the operation to load down grade.

It is common practice for a push tractor during the loading operation to add the necessary extra power. The pattern selected for the tractor-assisted loading operation is important in the design of the operation to maximize production. The standard patterns are back tracking, chain, and shuttle. A thorough description of these patterns is provided in Peurifoy and Ledbetter [1985, p. 187].

The performance of a scraper is the function of the power required for the machine to negotiate the job site conditions and the power that is available by the machine. The power required is a function of rolling resistance (RR) and the effect of grade (EOG). RR is the force that must be exerted to roll or pull a wheel over the ground. It is a function of the internal friction of bearings, tire flexing, tire penetration into the surface, and the weight on the wheels.

Each ground surface type has a rolling resistance factor (RR_F) associated with it. However, as a general rule, the RR_F consists of two parts. First, it takes at least a 40 lb force per each ton of weight just to move a machine. Second, it takes at least a 30 lb force per each ton of weight for each inch of tire penetration. Therefore the RR_F can be determined as follows:

$$RR_F = 40 \text{ lb/ton} + 30 \text{ lb/ton/inch of penetration} \tag{3.8}$$

Rolling resistance is then calculated by using the RR_F and the gross vehicle weight (GVW) in tons:

$$RR = RR_F \times GVW \tag{3.9}$$

RR can be expressed in terms of pounds or percent. For example, a resistance of 40 pounds per ton of equipment weight is equal to a 2% RR.

The EOG is a measure of the force due to gravity which must be overcome as the machine moves up an incline, but is recognized as grade assistance when moving downhill. Grades are generally measured in percent slope. It has been found that for each 1% increment of adverse grade an additional 20 lb of resistance must be overcome for each ton of machine weight. Therefore, the effect of grade factor (EOG_F) can be determined by

$$EOG_F = (20 \text{ lb/ton/\% grade}) \times (\text{\% of grade}) \tag{3.10}$$

The EOG is then calculated by

$$EOG = EOG_F \times GVW \tag{3.11}$$

The total resistance (TR) associated with a job site can be calculated by

$$\text{Machine moving uphill: } TR = RR + EOG \tag{3.12}$$

$$\text{Machine moving on level ground: } TR = RR \tag{3.13}$$

$$\text{Machine moving downhill: } TR = RR - EOG \tag{3.14}$$

Once the power requirements are determined for a specific job site, a machine must be selected that has adequate power available. Available power is a function of horsepower and operating speed. Most equipment manufacturers provide user-friendly performance charts to assist with evaluating the influence of GVW, TR, speed, and rimpull. Rimpull is the force available between the tire and the ground to propel the machine.

The relationship of the power train to rimpull for a rubber-tired tractor can be expressed by the following relationship:

$$\text{Rimpull} = \frac{375 \times HP \times \text{Efficiency}}{\text{Speed (mph)}} \tag{3.15}$$

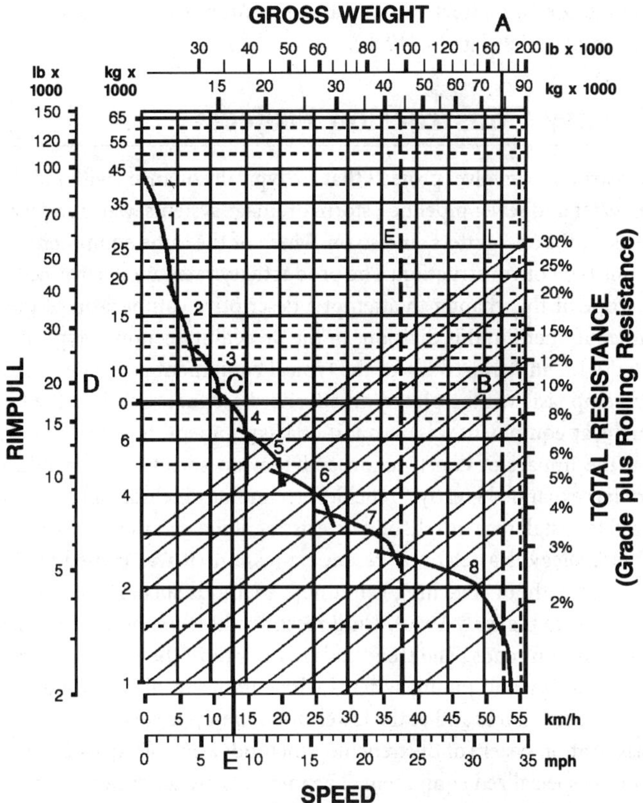

KEY
1 — 1st Gear Torque Converter Drive
2 — 2nd Gear Torque Converter Drive
3 — 3rd Gear Direct Drive
4 — 4th Gear Direct Drive
5 — 5th Gear Direct Drive
6 — 6th Gear Direct Drive
7 — 7th Gear Direct Drive
8 — 8th Gear Direct Drive

KEY
A — Loaded 77 965 kg (171,880 lb)
B — Intersection with 10% total resistance line
C — Intersection with rimpull curve (4th gear)
D — Required rimpull 7756 kg (17,100 lb)
E — Speed 12.9 km/h (8 mph)

FIGURE 3.6 Rimpull-speed-gradeability curves. (*Source: Caterpillar Performance Handbook*, 24th ed., 1993. Caterpillar, Peoria, IL.)

Figure 3.6 illustrates information available from a typical performance chart. The following example illustrates how this information can be utilized.

Example. A scraper with an estimated payload of 34,020 kg (75,000 lb) is operating on a total effective grade of 10%. Find the available rimpull and maximum attainable speed.

$$\text{Empty weight} + \text{payload} = \text{Gross weight}$$

$$43{,}945 \text{ kg} + 34{,}020 \text{ kg} = 77{,}965 \text{ kg}$$

$$(96{,}880 \text{ lb} + 75{,}000 \text{ lb} = 171{,}880 \text{ lb})$$

Solution. Using Fig. 3.6, read from 77,965 kg (171,880 lb) on top of gross weight scale down (line B) to the intersection of the 10% total resistance line (point *C*).

Go across horizontally from *C* to the Rimpull Scale on the left (point *D*). This gives the required rimpull: 7593 kg (16,740 lb).

Where the line *CD* cuts the speed curve, read down vertically (point *E*) to obtain the maximum speed attainable for the 10% effective grade: 13.3 km/h (8.3 mph).

The vehicle will climb the 10% effective grade at a maximum speed of 13.3 km/h (8.3 mph) in 4th gear. Available rimpull is 7593 kg (16,740 lb).

3.3 Municipal/Utility Construction Projects

Municipal/utility construction involves projects that are typically financed with public funds and include such things as water and sewer pipelines, storm drainage systems, water and wastewater treatment facilities, streets, curbs and gutters, and so on. Much of the same equipment listed in Fig. 3.2 will be utilized for this type of construction. The productivity rates are determined the same way.

It is beyond the scope of this chapter to attempt a descriptive comparison of the various types of construction and equipment in this division. In the preceding section, a typical heavy/highway project was presented with an itemized list of the typical equipment associated with each activity. In this segment, the emphasis will be placed on advanced technology while the emphasis in the section on heavy/highway equipment was on traditional equipment. In recent years, more concern has been placed on the impact of construction activities on society. As a result, the trenchless technology industry has expanded greatly. Trenchless technology includes all methods, equipment, and materials utilized to install new or rehabilitate existing underground infrastructure systems.

While *trenchless technology* is a relatively recent expression (it was coined in the mid-1980s), the ability to install pipe without trenching is not new. Methods such as auger boring and slurry boring have been used since the early 1940s. Until recently, these methods were used primarily to cross under roadways and railroads. The trend today is to utilize the trenchless concept to install complete underground utility and piping systems with minimum disruption and destruction to society and the environment, safely and at the lowest total life-cycle cost.

Figure 3.7 is a classification system of the trenchless methods available to install new systems. Each method involves unique specialized equipment. The methods are described in detail in Trenchless Technology Center [1993]. No one method is compatible with all installations. Each project should be evaluated separately; the method selected should be compatible, safe, and cost-effective, and should provide a high probability of success.

Only the microtunneling technique will be described in this chapter. This technique is well suited for installing sanitary and storm sewer pipelines, which require a high degree of accuracy for alignment and grade.

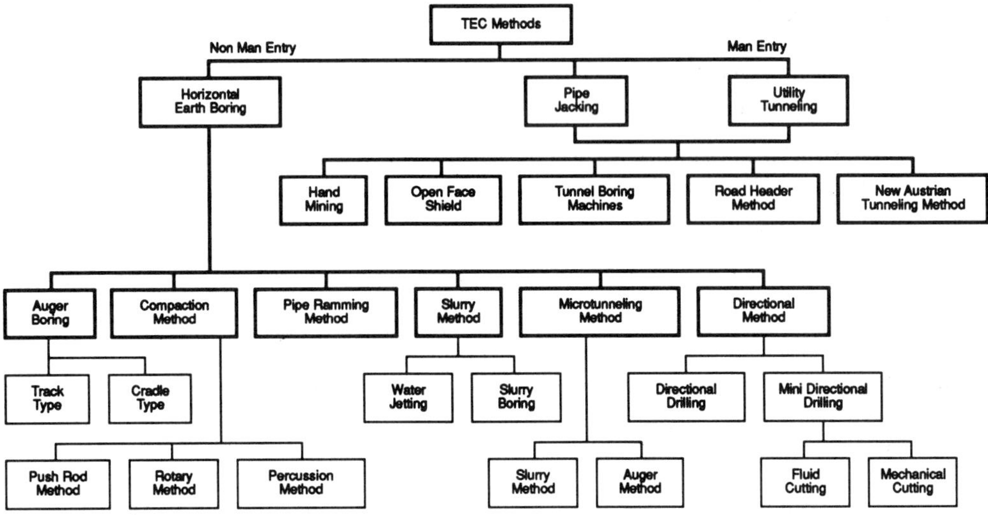

FIGURE 3.7 Trenchless excavation construction (TEC) classification system.

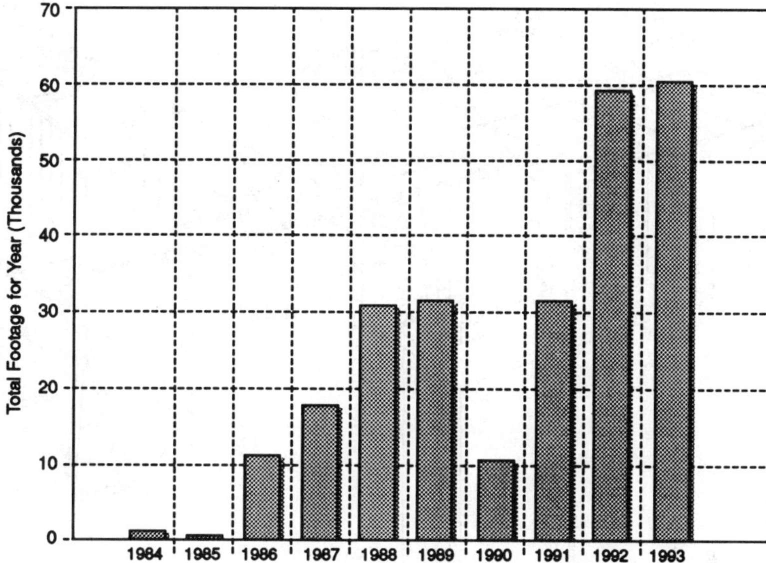

FIGURE 3.8 Total U.S. microtunneling footage by year.

Microtunneling systems are laser-guided, remote control, pipejacking systems. In most instances, because of their high accuracy, the product pipe is installed in one pass. Most machines have the capability to counterbalance the earth pressure at the work face continuously so that dewatering is not required.

These systems were developed in Japan in the mid-1970s, introduced in Germany in the early 1980s, and first used in the U.S. in 1984. Figure 3.8 shows the growth of the microtunneling industry in the U.S. By mid-1993, more than 50 miles of pipe had been installed by this method, and it continues to grow in demand at an increasing rate because of its extraordinary capability. For example, in a residential Houston area, microtunneling was used in 1987 to install almost four miles of gravity sewer lines (10 inch to 24 inch diameter) because the residents did not want their neighborhood torn apart by traditional methods. In 1989, in Staten Island, New York, this method was used to install a five-foot diameter gravity sewer at a depth of 80 feet, under 60 feet of groundwater, with the longest single drive 1600 linear feet. It was installed at an accuracy of ±1 inch horizontal and vertical. In 1992–93, two raw water intake lines were installed, one above the other, in Jordan Lake, near Carey, North Carolina. These examples and many others are helping engineers realize the unique capability of microtunneling to solve complex problems safely, cost-effectively, and with minimum environmental impact.

Figure 3.9 illustrates the two basic types of systems. They provide similar capabilities but are differentiated by their spoil removal system. One provides a slurry spoil transportation system and the other provides an auger spoil removal system.

Figure 3.10 is a schematic drawing which illustrates the basic components and systems of the microtunneling methods.

The microtunneling process consists of five independent systems: the mechanized excavation system, the propulsion system, the spoil removal system, the guidance control system, and the pipe lubrication system.

The Mechanized Excavation System

The cutter head is mounted on the face of the microtunnel boring machine and is powered by electric or hydraulic motors located inside the machine. Cutting heads are available for a variety

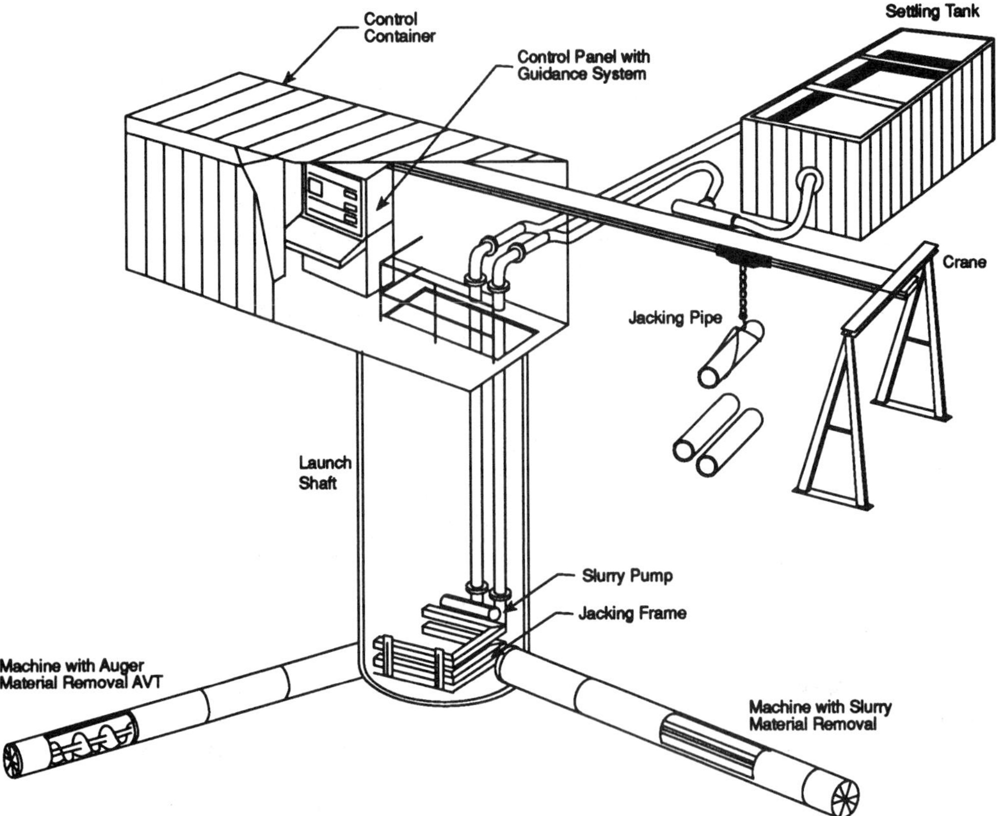

FIGURE 3.9 Auger and slurry microtunneling systems.

of soil conditions, ranging from soft soils to rock, including mixed-face conditions and boulders. The microtunnel machines may operate above or below the groundwater. Each manufacturer produces cutting heads which are unique. These machines have been used successfully on projects where rock is encountered with unconfined compressive strength up to 30,000 psi. Also, they can handle boulders and other obstructions that are up to 30% of the diameter of the machine by incorporating crushing capability in the head. This crushing mechanism reduces the boulder to 3/4 inch to 1 inch particles that can be removed by an auger or by the slurry spoil removal system.

The boring machine also houses the articulating steering unit with steering jacks and the laser control target. Additional components which may be located in the microtunnel boring machine, depending on the type of machine, include the rock crusher, mixing chamber, pressure gauges, flow meters, and control valves.

Most machines have the capability of counterbalancing the actual earth pressure and the hydrostatic pressure independently of each other. The actual earth pressure is counterbalanced by careful control over the propulsion system and spoil removal system. This force is carefully regulated to stay higher than the actual earth pressure but lower than the passive earth pressure so that subsidence and heave are avoided. The groundwater can be maintained at its original level by counterbalancing with slurry pressure or compressed air.

The Guidance Control System

The heart of the guidance control system is the laser. The laser provides the alignment and grade information for the machine to follow. The laser beam must have an unobstructed pathway from the drive shaft to the target located in the machine. The laser must be supported in the drive shaft

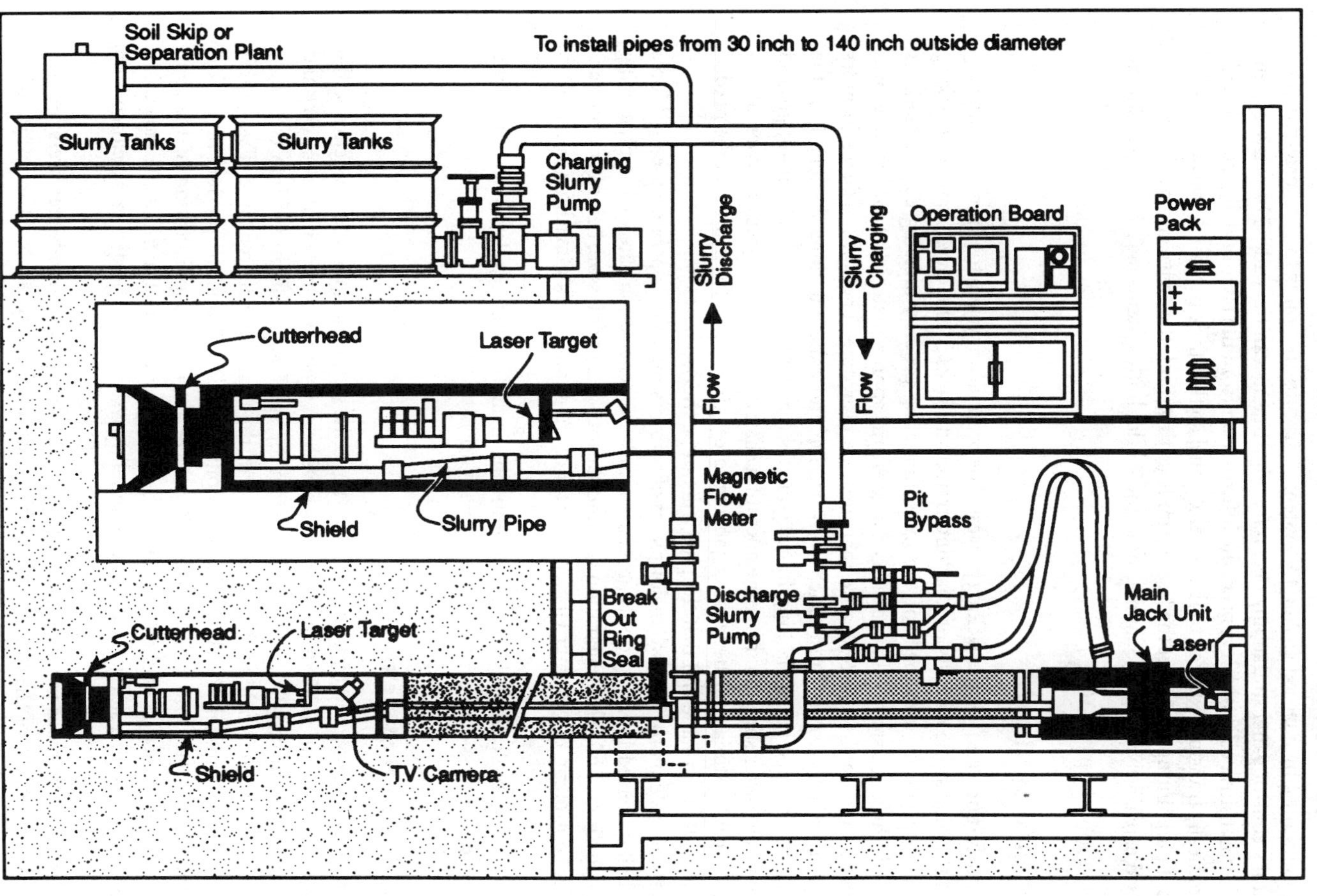

FIGURE 3.10 Slurry microtunneling system.

so that it is independent of any movement that may take place as a result of forces that are being created by the propulsion system. The target receiving the laser information can be an active or passive system. The passive system consists of a target that receives the light beam from the laser; the target is monitored by a closed circuit TV system. This information is then transferred back to the operator's control panel so that any necessary adjustments can be made. The active system consists of photosensitive cells which convert the information from the laser into digital data. This data is electronically transmitted back to the control panel so that the operator is provided with a digital readout pinpointing the target the laser beam is hitting. Both the active and passive systems have been used extensively in the U.S. and around the world; both systems have been found to be reliable.

The Propulsion System

The microtunneling process is a pipe-jacking process. The propulsion system for the microtunneling machine and the pipe string consists of a jacking frame and jacks in the drive shaft. The jacking units have been specifically designed for the microtunneling process, offering compactness of design and high thrust capacity. That capacity ranges from approximately 100 tons to well over 1000 tons, depending on the soil resistance that must be overcome. The soil resistance includes resistance from face pressure and resistance from friction and adhesion along the length of the steering head and the pipe string. Jacking force estimates may be based on drive length, ground conditions, pipe characteristics, and machine operating characteristics, specifically on overcut and lubrication. A reliable estimate of the required jacking force is important to ensure that the needed thrust capacity will be available and that the pipe will not be overloaded.

The propulsion system provides two major pieces of information to the operator: the total force or pressure being exerted by the propulsion system and the penetration rate of the pipe being pushed through the ground. The penetration rate and the total jacking pressure being generated are important for controlling the counterbalancing forces of the tunnel boring machine to maintain safe limits.

The types of pipe typically used for microtunneling include concrete, clay, steel, PVC, and centrifugally cast fiberglass-reinforced polyester pipe (GRP).

The Spoil Removal System

Microtunneling spoil removal systems can be divided into the slurry transportation system and the auger transportation system. Both of these systems have been used extensively in this country and abroad and have been successful. In the slurry system the spoil is actually mixed into the slurry in a chamber that is located behind the cutting head of the tunnel boring machine. The spoil is hydraulically removed through the slurry discharge pipes installed inside the product pipe. This material is then discharged into a separation system. The degree of sophistication of the spoil separation system is a function of the type of spoil being removed. The effluent of the separation system becomes the charging slurry for the microtunneling system; thus the system is a closed loop system.

Because the slurry chamber pressure is used to counterbalance the groundwater pressure, it is important that the velocity of the flow as well as the pressure be closely regulated and monitored. Regulation is accomplished by variable speed charging and discharging pumps, bypass piping, and control valves. As a result of this capability to counterbalance the hydrostatic head accurately, these machines have worked successfully in situations with extremely high hydrostatic pressure. The machine can be completely sealed off from external water pressure, allowing underwater retrieval, as was successfully accomplished on two recent projects at Corps of Engineer lakes. The auger spoil removal system utilizes an independent auger system in an enclosed casing inside the product pipe for spoil removal. The spoil is augered to the drive shaft, collected in a skip, and hoisted to a surface storage facility near the shaft. Water may be added to the spoil in the machine to facilitate spoil removal. However, one of the advantages of the auger system is that the spoil does not have to reach pumping consistency for removal.

The Control System

All microtunneling systems rely on remote control capability, allowing operators to be located in a safe and comfortable control cabin typically at the surface immediately adjacent to the drive shaft so the operator can visually monitor activities in the shaft. If the control cabin cannot be set up adjacent to the drive shaft for visual monitoring, a CCTV system can be set up in the shaft to allow the operator to monitor activities using a TV monitor. A key ingredient to a successful project is the operator's skill. The operator must monitor numerous bits of information which are continuously fed to the control panel, evaluate this information, and make decisions regarding future actions. Information relayed back to the control panel is audible as well as visual, as sounds generated in the microtunneling machine are sent to the operator. Other information that must be monitored includes the line and grade of the machine, cutter-head torque, jacking thrust, steering pressures, slurry flow rates and pressures for slurry systems, and the rate of advancement.

The sophistication of the control system varies from totally manual to completely automatic. With the manual system, the operator evaluates all the information and makes necessary decisions regarding correcting actions. It is the responsibility of the operator to record all information at appropriate intervals during the pipe-jacking procedure. All monitoring and recording of data is automated; the computer provides a printout on the condition of the various systems at selected time intervals. Systems using fuzzy logic are available for making necessary corrections in the operational process. This allows the machine to automatically acquire, evaluate, and compare the data to corrections typically utilized for the existing condition. The machine will then make those corrections. With this system, the operator monitors the actions to ensure that the automatic corrections are those that the operator thinks are appropriate. Manual override of the automatic corrections is also possible.

The Pipe Lubrication System

The pipe lubrication system consists of a mixing tank and the necessary pumping equipment, which transmits the lubricant from a reservoir near the shaft either to the application points inside the machine or along the inside barrel of the pipe. Pipe lubrication is optional but recommended for most installations, particularly for lines of substantial length. The lubricant can be either a bentonite or polymer-based material. For pipe systems less than 36 inches in diameter, the application point is at the machine steering head at the face of the tunnel. For sizes greater than 36 inches, application points can be installed at intervals throughout the pipe. Lubrication can substantially reduce the total thrust required to jack the pipe.

Equipment Cost

The cost of the project must include the cost of equipment needed to build the project. The constructor must be able to determine as accurately as possible the duration of each piece of equipment required for each activity of the project, and he or she must be able to apply cost factors to this time commitment. The cost factor should represent the actual equipment cost experienced by the constructor. If the cost is too low the equipment will not pay for itself. If the rate is too high it may result in not being competitive. To know the true equipment cost requires accurate record keeping.

The constructor can (1) lease the equipment or (2) purchase the equipment. If the equipment is leased, determining equipment cost is straightforward because the rental rate will be established. If the equipment is to be purchased, the anticipated owning and operating (O&O) cost will need to be determined.

Associated Equipment Distributors publishes an annual compilation of nationally averaged rental rates for construction equipment. The following need to be taken into consideration when considering the leasing option:

1. *Time basis of the rates quoted.* It is common practice in the industry to base rates on one shift of 8 hours per day, 40 hours per week, or 176 hours per month. If these hours are exceeded, an extra fee can be charged.
2. *Cost of repairs.* The lessor usually bears the cost of repairs due to normal wear and tear, and the lessee bears all other costs. Normal wear and tear is that which would be expected to result from the use of the equipment under normal circumstances. This can lead to disputes because in many cases normal wear and tear is difficult to distinguish.
3. *Operator.* Unless specifically stated otherwise, the operator is not included in the rental rates.
4. *Fuel and lubricants.* Unless specifically stated otherwise, the lessee is responsible for the cost of fuel, lubricants, and all preventive maintenance work while the equipment is being rented.
5. *Condition of equipment.* It is standard practice for the equipment to be delivered to the lessee in good operational condition, and to be returned to the lessor in the same condition less normal wear and tear.
6. *Freight charges.* Unless specifically stated otherwise, the rental rates are f.o.b. the lessor's shipping point.
7. *Payment and taxes.* Normally, rental rates are payable in advance, and no license, sale, or use taxes are included in the rates.
8. *Insurance.* It is standard practice for the lessee to furnish the lessor a certificate of insurance prior to equipment delivery.

The factors influencing the calculation of owning and operating costs are investment and depreciation (ownership costs) and maintenance, repairs, lubrication, and fuel (operating costs). If a firm has similar equipment they should have reliable historical data to help forecast the cost that should be applied to a specific piece of equipment. Many times, however, this is not the case. Therefore, the constructor must use an approximation based on assumed cost factors. Most equipment manufacturers can provide valuable assistance in selecting cost factors that should apply to the type of work being considered.

Whether rental rates or O&O costs are being utilized, they should eventually be expressed as total hourly equipment cost without the operator cost. This facilitates the determination of the machine performance in terms of cost per units of material. For example,

$$\text{Top machine performance} = \frac{\text{Lowest possible equipment hourly cost}}{\text{Highest possible hourly productivity}}$$

$$\text{Top machine performance} = \frac{\text{Cost/hr}}{\text{Units of material/hr}} = \frac{\text{Cost(\$)}}{\text{Units of material}}$$

Caterpillar [1985], Peurifoy and Ledbetter [1985], and *Production and Cost Estimating* [1981] contain detailed descriptive information on how to develop O&O costs. These references contain numerous examples that show how to apply specific factors.

The following is a summary of the principles presented in Peurifoy and Ledbetter [1985]:

I. Ownership costs (incurred regardless of the operational status)
   A. Investment costs
      1. Interest (money spent on equipment which could have been invested at some minimum rate of return)
      2. Taxes (property, etc.)

3. Insurance
4. Storage

Investment costs can be expressed as a percentage of an average annual value of the equipment ($\overline{p}$). For equipment with no salvage value:

$$\overline{p} = \frac{p(N+1)}{2N}$$

where p is the total initial cost and N is the useful life in years. For equipment with salvage value:

$$\overline{p} = \frac{p(N+1) + S(N-1)}{2N}$$

Example.

Interest on borrowed money	= 12%
Tax, insurance, storage	= 8%
Total	= 20%
Investment cost	= $0.20\overline{p}$

B. Depreciation (the loss in value of a piece of equipment over time due to wear, tear, deterioration, obsolescence, etc.)

II. Operation costs
   A. Maintenance and repair
      1. Depends on type of equipment, service, care
      2. Usually taken into consideration as a ratio or percentage of the depreciation cost
   B. Fuel consumed
      1. Gas engine = 0.06 gal/FWHP-hr
      2. Diesel engine = 0.04 gal/FWHP-hr
   C. Lubricating oil

FWHP-hr is the measure of work performed by an engine based on average power generated and duration. Two major factors that impact the FWHP-hr are the extent to which the engine will operate at full power and the actual time the unit will operate in an hour.

$$\text{TF} = \text{Time factor} = \frac{50 \text{ min}}{60} \times 100 = 83.3\%$$

$$\text{EF} = \text{Engine factor} = \frac{\%\text{ of time at full load}}{\%\text{ of time at less than full load}}$$

$$\text{OF} = \text{Operating factor} = \text{TF} \times \text{EF}$$

$$\text{Fuel consumed} = \text{OF} \times \text{Rate of consumption}$$

The amount of lubricating oil consumed includes the amount used during oil changes plus oil required between changes.

$$q = \frac{\text{FWHP} \times \text{OF} \times 0.006\ \#/\text{FWHP-hr}}{7.4\ \#/\text{gal}} = \frac{c}{t} = \frac{\text{Gal}}{\text{Hr}}$$

where OF is the operating factor, c is the crankcase capacity in gallons, t is the number of hours between changes, and # is pound.

Example. Hydraulic excavator.

160 FWHP—diesel engine
Cycle time = 20 s
Filling the dipper = 5 s at full power
Remainder of time = 15 s at half power
Assume shovel operates 50 min/hr.

$$\text{TF} = \frac{50}{60} \times 100 = 83.3\%$$

Engine factor:

Filling	5/20 × 1	= 0.25
Rest of Cycle	15/20 × .50	= 0.375
	Total	0.625

$$\text{OF} = \text{TF} \times \text{EF} \times 0.625 \times 0.833 = 0.520$$

$$\frac{\text{Fuel consumed}}{\text{Hr}} = 0.52 \times 160 \times 0.04 = 3.33 \text{ gal/hr}$$

3.4 Preventive Maintenance

Preventive maintenance (PM) is necessary for sound equipment management and protection of a company's assets. Minimum corporate PM standards should be established. Specific maintenance procedures should be available from the equipment department on most major pieces of equipment. If specific standards are not available, the manufacturer's minimum maintenance recommendations need to be used. A functioning PM program will comprise at least the following:

1. The PM program will be written and have specific responsibilities assigned. Your company division and/or area managers will have responsibility to see that the program actually works as designed.
2. Periodic service and inspections on all equipment in operation will be performed, documented, and reported (in writing). Each division/area will implement the service and inspection using the equipment manufacturers' recommendations as guidelines. For most major pieces of equipment, this will be defined by the equipment department.
3. A systematic method of scheduling and performing equipment repairs will be implemented.
4. A fluid analysis program with regular sampling (including, but not limited to, testing for aluminum, chromium, copper, iron, sodium, silicon, plus water and fuel dilution) will be implemented.
5. All necessary permits will be acquired.
6. Federal, state, and local laws that affect the trucking industry will be followed.

3.5 Mobilization of Equipment

The following are factors that should be taken into consideration to facilitate and expedite mobilization of equipment:

1. Type and size of equipment
2. Number of trucks and trailers needed to make the move
3. Rates (company charges or rental charges)
4. Equipment measurements (weight, height, width, length)
5. Permits (vary with state)
6. Federal, state, and local laws affecting the trucking industry

The purpose of mobilization is to maximize efficiency and minimize cost using either rental or company trucks. This requires research on the above items by using equipment dealer support, appropriate law enforcement agencies and so on.

Acknowledgments

The author would like to express his sincere appreciation to Mr. Danny A. Lott for his input. Mr. Lott has over 18 years of professional management experience with two leading corporations in construction and communications. For the past eight years he has been the equipment maintenance and truck operations manager for T. L. James and Company, Inc., Ruston, LA. Mr. Lott provided a wealth of insight into the approach and substance of this chapter.

References

Associated Equipment Distributors. Undated. *Rental Rates and Specifications.* Associated Equipment Distributor, Chicago, IL.

Carson, A. B. 1961. *General Excavation Methods.* McGraw-Hill, New York.

Caterpillar. 1985. *Caterpillar Performance Handbook,* 22nd ed. Caterpillar, Peoria, IL.

Halpin, D. W. and Riggs, L. S. 1992. *Planning and Analysis of Construction Operation.* John Wiley & Sons, New York.

Hendrickson, C. and Au, T. *Project Management for Construction.* Prentice Hall, Englewood Cliffs, NJ.

The Herrenknecht Microtunneling System. 1993. Herrenknecht Corporation, Greenville, SC 29615.

The Iseki Microtunneling System. 1993. Iseki, Inc., San Diego, CA.

Peurifoy, R. L. and Ledbetter, W. 1985. *Construction Planning, Equipment and Methods,* 4th ed. McGraw-Hill, New York.

Production and Cost Estimating of Material Movement with Earthmoving Equipment. 1981. Terex Corporation, Hudson, Ohio.

Trenchless Technology Center. 1993. *Trenchless Excavation Construction Equipment and Methods Manual,* 2nd ed., ed. D. T. Iseley and R. Tanwani, p. 1-3. National Utility Contractors Association, Arlington, VA.

For Further Information

A good introduction to practical excavation methods and equipment is in *General Excavation Methods* by Carson.

Construction Planning, Equipment and Methods by Peurifoy and Ledbetter is particularly helpful for practical techniques of predicting equipment performance and production rates.

An excellent introduction to trenchless techniques that are used to install new underground utility and piping systems is *Trenchless Excavation Construction Equipment and Methods Manual* developed by the Trenchless Technology Center at Louisiana Tech University.

4

Design and Construction of Concrete Formwork

Arch Alexander
Purdue University

4.1 Introduction

A supporting structure or element that is erected and used during construction, and is not incorporated into the finished structure, is considered to be a temporary structure. The actual time the temporary structure may be used can vary from a few hours to as long as several months or more. Common examples of temporary structures, but by no means a complete listing, include concrete formwork, trench excavation sheeting and bracing, braces and guys for steel erection, and access scaffolding.

Design of the members of temporary structures generally must follow the same codes and regulations as the members of permanent structures. Some codes may allow increased allowable loads and stresses since temporary structures are used for a shorter time. The Occupational Safety and Health Act (OSHA) of the U.S. government contains criteria that the designer of temporary structures must follow. State and local safety codes (which must be followed by the designer) may also exist to regulate job site safety.

For the materials ordinarily used in the construction of temporary structures, the governing codes commonly follow and incorporate by reference the basic technical codes published by national organizations. These national organizations include the American Concrete Institute (ACI), the American Institute of Steel Construction (AISC), the American Society for Testing of Materials (ASTM), and the National Forest Product Association (NFPA). These organizations have developed specifications and standards for wood, concrete, and steel. They are listed here.

0-8493-8953-4/95/$0.00 + $.50

- ACI Standard 318 *Building Code Requirements for Reinforced Concrete*
- AISC *Specification for Design, Fabrication and Erection of Structural Steel for Buildings*
- AISC *Code of Standard Practice*
- NFPA *National Design Specification (NDS) for Wood Construction*
- *Design Values for Wood Construction,* supplement to the *National Design Specifications for Wood Construction*

In addition to the above technical specifications, the following publications may provide useful information to the designer of temporary structures.

- *Plywood Design Specification (PDS),* American Plywood Association (APA)
- *Specifications for Aluminum Structures,* The Aluminum Association, Inc.

There exist many technical manuals and publications used for both temporary and permanent structures. A list of those most commonly encountered includes *Formwork for Concrete* published by ACI, *Concrete Forming* published by APA, *Manual of Steel Construction* published by AISC, *Manual of Concrete Practice* published by ACI, *CRSI Handbook* published by the Concrete Reinforcing Steel Institute, *Timber Construction Manual* published by the American Institute of Timber Construction (AITC), *Concrete Manual* published by the U.S. Department of the Interior Bureau of Reclamation, *Recommended Practice for Concrete Formwork* by ACI, *Wood Handbook: Wood as an Engineering Material* published by the U.S. Department of Agriculture, *Standard Specifications and Load Tables for Open Web Steel Joists* published by the Steel Joist Institute, *Light Gage Cold Formed Steel Design Manual* published by the American Iron and Steel Institute, *Minimum Design Loads for Buildings and Other Structures* by the American National Standards Institute (ANSI), and *Formwork, Report of the Joint Committee* published by the Concrete Society as Technical Report No. 13 (Great Britain).

4.2 Concrete Formwork

The designer of concrete forms must choose appropriate materials and utilize them so that the goals of safety, economy, and quality are met. The formwork when constructed and used properly should produce a concrete product that has the desired size and finish. The formwork should be easily built and stripped so that it saves time for the contractor. It should have sufficient strength and stability to safely carry all live and dead loads encountered before, during, and after the placing of the concrete, and it should be sufficiently resistant to deformations such as sagging or bulging in order to produce concrete which satisfies requirements for straightness and flatness.

Since formwork costs can exceed 50% of the total cost of the concrete structure, economy is an important consideration. Formwork cost savings should ideally begin with the architect and engineer. They should choose the sizes and shapes of the elements of the structure after considering the forming requirements and framework cost, as well as the usual design requirements of appearance and strength. Keeping constant dimensions where possible, using dimensions that match standard material sizes, and avoiding complex shapes for elements in order to save concrete are some examples of how the designer can reduce forming costs.

Concrete forms that do not perform to produce satisfactory concrete elements are not economical. Forms not carefully designed, constructed, and used probably will not provide the surface finish or the dimensional tolerance required by the specifications for the finished concrete work. To correct defects which occur in the concrete from improperly constructed forms may require considerable patching, rubbing, grinding, or in extreme cases, demolition and rebuilding.

4.3 Materials

Most concrete forms are either constructed using basic materials such as lumber, plywood, and steel, or are prefabricated panels sold or leased to contractors by the panel manufacturers. Use of the prefabricated panels may save labor costs on jobs where forms are reused many times. Panel manufacturers will provide layout drawings, and sometimes they provide supervision of the construction where prefabricated forms are used. Even where prefabricated forms are chosen, there are usually parts of the concrete structure that must be formed using lumber and plywood job-built forms.

Lumber

Lumber suitable for constructing concrete forms is available in a variety of sizes, grades, and species groups. The contractor should determine what is economically available before specifying a particular grade or species of lumber for constructing the forms.

Some of the most widely available species of lumber include Douglas fir, southern pine, ponderosa pine, and spruce. Douglas fir and southern pine are among the strongest woods available and are often chosen for use in formwork. The strength and stiffness of lumber varies widely with different species and grades. Choice of species and grade will greatly affect size and spacing of formwork components.

Most lumber has been planed on all four sides to produce a uniform surface and consistent dimensions, and is referred to as S4S (surfaced on four sides) lumber. The sizes produced have minimum dimensions specified in the *American Softwood Lumber Standard,* PS 20-70. Lumber is usually referred to by its nominal dimensions (e.g., 2 × 4, 2 × 6, etc.) and the actual dimensions are somewhat smaller for both finished and rough-sawn lumber. Rough-sawn lumber will have dimensions about 1/8-inch larger than finished S4S lumber. Lumber sizes commonly used, along with their section properties, are given in Table 4.1.

Lumber used in forming concrete must have a predictable strength. Predictable strength is influenced by many factors. Lumber that has been inspected and sorted during manufacturing

Table 4.1 Properties of Dressed Lumber

Standard Size Width × Depth	S4S Dressed Size Width × Depth	Cross-Sectional Area *A* (in.^{2})	Moment of Inertia *I* (in.^{4})	Section Modulus *S* (in.^{3})	Weight in Pounds per Lineal Foot*
1 × 4	3/4 × 3 1/2	2.63	2.68	1.53	.64
1 × 6	3/4 × 5 1/2	4.13	10.40	3.78	1.00
1 × 8	3/4 × 7 1/4	5.44	23.82	6.57	1.32
1 × 12	3/4 × 11 1/4	8.44	88.99	15.82	2.01
2 × 4	1 1/2 × 3 1/2	5.25	5.36	3.06	1.28
2 × 6	1 1/2 × 5 1/2	8.25	20.80	7.56	2.01
2 × 8	1 1/2 × 7 1/4	10.88	47.64	13.14	2.64
2 × 10	1 1/2 × 9 1/4	13.88	98.93	21.39	3.37
2 × 12	1 1/2 × 11 1/4	16.88	177.98	31.64	4.10
4 × 2	3 1/2 × 1 1/2	5.25	.98	1.31	1.28
4 × 4	3 1/2 × 3 1/2	12.25	12.51	7.15	2.98
4 × 6	3 1/2 × 5 1/2	19.25	48.53	17.65	4.68
4 × 8	3 1/2 × 7 1/4	25.38	111.15	30.66	6.17
6 × 2	5 1/2 × 1 1/2	8.25	1.55	2.06	2.01
6 × 4	5 1/2 × 3 1/2	19.25	19.65	11.23	4.68
6 × 6	5 1/2 × 5 1/2	30.25	76.26	27.73	7.35
6 × 8	5 1/2 × 7 1/4	41.25	193.36	51.53	10.03
8 × 2	7 1/4 × 1 1/2	10.88	2.04	2.72	2.64
8 × 4	7 1/4 × 3 1/2	25.38	25.90	14.80	6.17
8 × 6	7 1/4 × 5 1/2	41.25	103.98	37.81	10.03
8 × 8	7 1/4 × 7 1/4	56.25	263.67	70.31	13.67

* Weights are for wood with a density of 35 pounds per cubic foot.

will carry a grade stamp indicating the species, grade, moisture condition when surfaced, and perhaps other information. Grading is accomplished by following rules which were established by recognized grading agencies and are published in the *American Softwood Lumber Standard.* Lumber can be graded visually by a trained technician or by a machine. Visually graded lumber has its design values based on provisions of ASTM-D245, *Methods for Establishing Structural Grades and Related Allowable Properties for Visually Graded Lumber.* Machine stress rated (MSR) lumber has design values based on nondestructive stiffness testing of individual pieces. Some visual grade requirements also apply to MSR lumber. Lumber which has a grade established by a recognized agency should always be used for formwork where strength is important.

Allowable Stresses for Lumber

The National Design Specification for Wood Construction contains allowable stresses and makes comprehensive recommendations for engineered wood construction for stress-graded lumber. These stress values tabulated in the NDS correspond to the species and grades of lumber produced in the U.S.

The allowable stresses depend not only on species and grade but on other factors as well. These include moisture content, size of cross section, and duration of loading. As an example of how the NDS presents allowable stresses, Table 4.2 contains values for allowable stresses for southern pine and Douglas fir. These stress values represent lumber used in a seasoned condition (less than 19% moisture) and for normal load duration.

Table 4.2 Allowable Stresses for Selected Grades and Species of Lumber

		Stress Values in Pounds per Square Inch (psi)					
	Size	Bending		Horizontal	Compression		Modulus of
Species and Grade	Classification	Single	Repetitive	Shear	Perpendicular	Parallel	Elasticity
Douglas Fir—Larch (Surfaced Dry or Green, Used at 19% Maximum Moisture)							
No. 1	2″ to 4″	1750	2050	95	625	1250	1,800,000
No. 2	thick	1450	1650	95	625	1000	1,700,000
No. 3	2″ to 4″	800	925	95	625	600	1,500,000
Stud	wide	800	925	95	625	600	1,500,000
Construction		1050	1200	95	625	1150	1,500,000
Standard		600	675	95	625	925	1,500,000
Utility		275	325	95	625	600	1,500,000
No. 1	2″ to 4″	1500	1750	95	625	1250	1,800,000
No. 2	thick	1250	1450	95	625	1050	1,700,000
No. 3	5″ and	725	850	95	625	675	1,500,000
Stud	wider	725	850	95	625	675	1,500,000
Southern Pine (Surfaced Green, Used Any Condition)							
No. 1	2″ to 4″	1350	1550	95	375	825	1,500,000
No. 2	thick	1150	1300	85	375	650	1,400,000
No. 3	2″ to 4″	625	725	85	375	400	1,200,000
Stud	wide	625	725	85	375	400	1,200,000
Construction		825	925	95	375	725	1,200,000
Standard		475	525	85	375	600	1,200,000
Utility		200	250	85	375	400	1,200,000
No. 1	2″ to 4″	1200	1350	85	375	825	1,500,000
No. 2	thick	975	1100	85	375	675	1,400,000
No. 3	5″ and	550	650	85	375	425	1,200,000
Stud	wider	575	675	85	375	425	1,200,000

Source: Design Values for Wood Construction, A Supplement to the 1986 Edition National Design Specification, National Forest Products Association, Washington, D.C.

Table 4.3 Load Duration Factors for Lumber

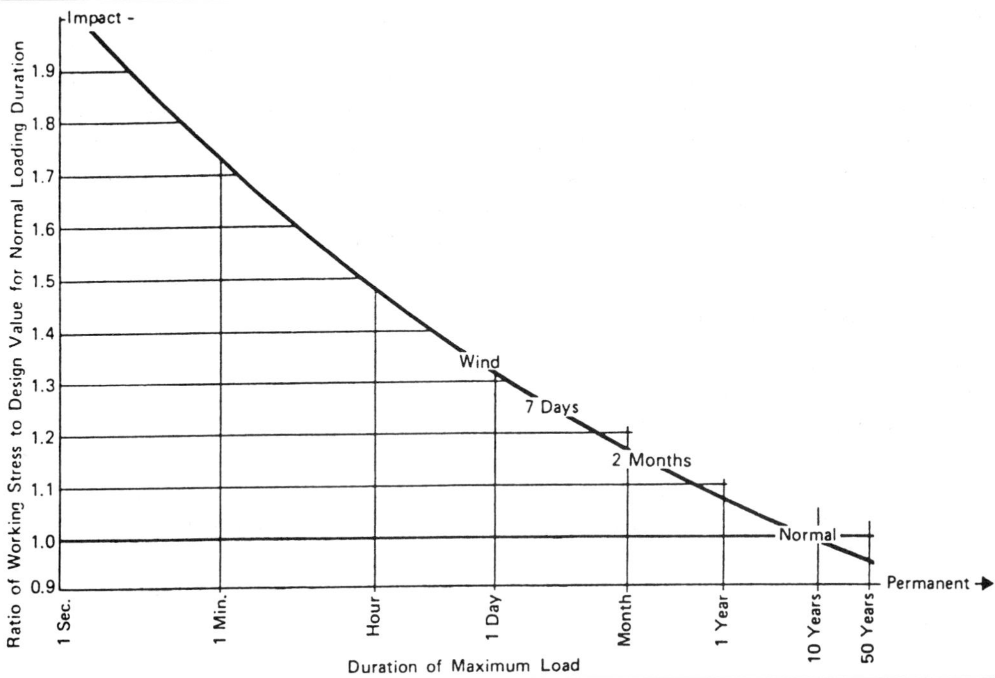

Source: Wood Construction, 1986 Edition, National Design Specification, p. 62. National Forest Products Association, Washington, D.C.

A normal load duration is defined as a load acting for ten years. For shorter load durations, the stresses can be increased for permanent structures using a load duration factor (LDF). Table 4.3 contains values for load duration factors taken from the NDS.

ACI Committee 347 in the *Guide to Formwork for Concrete* states:

> For formwork materials with limited reuse, allowable stresses specified in the appropriate design codes or specifications for temporary structures may be used. Where there will be a considerable number of formwork reuses or where formwork is fabricated from materials such as steel, aluminum, or magnesium, it is recommended that the formwork be designed as a permanent structure carrying permanent loads.

ACI Committee 347 does not explain what is considered limited reuse so the designer is left to use his or her own judgment in deciding whether to use increased stresses for short durations of load. It would seem reasonable that for forms used a few times, and which have not been damaged or weakened by repeated stripping and handling, the increased stresses could be used in their design. The appropriate increase used is 25% above the normal allowable stresses. This load duration factor is recommended for a load applied for seven days.

The NDS also allows bending stresses to be increased for beams which share their loads with other beams. The increased allowable stress is referred to as a repetitive member stress. For a beam to qualify as a repetitive member, it must be one of at least three members which are spaced no further apart than two feet and are joined by a load-distributing element such as plywood sheathing. These repetitive stresses may be applied to some formwork. Since the intent of allowing increased stress for repetitive flexural members is to take advantage of the load sharing provided by continuity, gang panels assembled securely by bolting or nailing and intended for multiple reuse would seem to qualify for this increase. They should not be used where the stresses have already been increased by 25% for short duration loads.

Plywood

Plywood is used extensively for concrete forms and provides the following advantages:

- It is economical in large panels.
- It is available in various thicknesses.
- It creates smooth, finished surfaces on concrete.
- It has predictable strength.
- It is manufactured in more than 40 surface textures that can provide various architectural finishes.

Plywood is available in two types: exterior and interior. The exterior type is made with waterproof glue and has all plies made with veneers of at least C grade. While many exterior plywood panels could be used, the plywood industry produces a special product intended for concrete forming called Plyform. This panel has two smooth sides (usually B grade veneer on both front and back) and is available in three classes—Class I, Class II, and Structural I. Class I is stronger than Class II because of the higher grade of veneers used in the panel. Structural I is the strongest of the three classes and is specially intended for applications where high strength and stiffness or maximum reuse is desired. Plyform is also available with a surface treatment of thermosetting, resin-impregnated material that is bonded to the panel surfaces. This abrasion-resistant surface, which gives an extremely smooth finish to concrete and allows more reuses of gang forms, is called a high-density overlay (HDO). Table 4.4 summarizes plywood grades and uses for concrete forms.

Engineering Properties of Plywood

Plywood is manufactured by peeling veneers from a log in thin layers, then gluing these veneers together to form plywood panels. Depending on the panel thickness, different numbers of veneer layers are used. To produce a panel that has desirable properties in both directions, the grain direction in different layers of veneer is oriented perpendicular to adjacent layers. Laying panels

Table 4.4 Plywood Grade-Use Guide for Concrete Forms

Terms for Specifying	Description	Typical Trademarks	Veneer Grade Faces	Veneer Grade Inner Plies	Veneer Grade Backs
APA B-B Plyform Class I & II	Specifically manufactured for concrete forms. Many reuses. Smooth, solid surfaces	APA PLYFORM B-B CLASS I EXTERIOR 000 PS 1-83	B	C	B
APA High Density Overlaid Plyform Class I & II	Hard, semi-opaque resin-fiver overlay, heat fused to panel faces. Smooth surface resists abraision. Up to 200 reuses	HDO PLYFORM I EXT-APA (Edge grade stamp)	B	C-Plugged	B
APA Structural I Plyform	Especially designed for engineered applications. All group ! species. Stronger and stiffer than Plyform Class I & II. Recommended for high pressures. Also available in HDO faces.	APA STRUCTURAL I PLYFORM B-B CLASS I EXTERIOR 000 PS 1-83	B	C or C plugged	B
APA B-C EXT	Sanded panel often used for concrete forming where only one smooth side is required	APA B-C GROUP 1 EXTERIOR 000 PS 1-83	B	C	C

Source: Concrete Forming, APA Design/Construction Guide, p. 6. American Plywood Association, Tacoma, Washington.

with veneer grain in perpendicular directions in alternate layers is called cross-banding. Because of cross-banding, the mechanical properties of adjacent veneers are not the same.

The section properties such as moment of inertia and section modulus cannot be calculated using the same formulas used for other common materials. The calculations of these section properties involve using a transformed area approach with the veneers. The form designer does not have to make these calculations. For plywood which conforms to the *U.S. Product Standard*, the engineering data are published in the *Plywood Design Specification (PDS)* from the American Plywood Association. Section properties for Plyform Class I, Class II, and Structural I are shown in Table 4.5. These section properties have been adjusted to account for cross-banded veneers. These values have been calculated by transforming all plies to the properties of the face ply. In using these values, the designer needs only the allowable stresses for the face ply and does not have to be concerned with the actual construction of the panel. The tabulated section properties are used for calculating flexural stresses, shear stresses, and deflections. For types of plywood other than Plyform panels, section properties can be found in the *PDS*.

Table 4.5 Section Properties for Plywood

		Properties for Stress Applied Parallel with Face Grain			Properties for Stress Applied Perpendicular to Face Grain		
Thickness (Inches)	Approximate Weight (psf)	Moment of Inertia I (in.^{4})	Effective Section Modulus KS (in.^{3})	Rolling Shear Constant Ib/Q (in.^{2})	Moment of Inertia I (in.^{4})	Effective Section Modulus KS (in.^{3})	Rolling Shear Constant Ib/Q (in.^{2})
			Class I				
15/32	1.4	0.066	0.244	4.743	0.018	0.107	2.419
1/2	1.5	0.077	0.268	5.153	0.024	0.130	2.739
19/32	1.7	0.115	0.335	5.438	0.029	0.146	2.834
5/8	1.8	0.130	0.358	5.717	0.038	0.175	3.094
23/32	2.1	0.180	0.430	7.009	0.072	0.247	3.798
3/4	2.2	0.199	0.455	7.187	0.092	0.306	4.063
7/8	2.6	0.296	0.584	8.555	0.151	0.422	6.028
1	3.0	0.427	0.737	9.374	0.270	0.634	7.014
1 1/8	3.3	0.554	0.849	10.430	0.398	0.799	8.419
			Class II				
15/32	1.4	0.063	0.243	4.499	0.015	0.138	2.434
1/2	1.5	0.075	0.267	4.891	0.020	0.167	2.727
19/32	1.7	0.115	0.334	5.326	0.025	0.188	2.812
5/8	1.8	0.130	0.357	5.593	0.032	0.225	3.074
23/32	2.1	0.180	0.430	6.504	0.060	0.317	3.781
3/4	2.2	0.198	0.454	6.631	0.075	0.392	4.049
7/8	2.6	0.300	0.591	7.990	0.123	0.542	5.997
1	3.0	0.421	0.754	8.614	0.220	0.812	6.987
1 1/8	3.3	0.566	0.869	9.571	0.323	1.023	8.388
			Structural I				
15/32	1.4	0.067	0.246	4.503	0.021	0.147	2.405
1/2	1.5	0.078	0.271	4.908	0.029	0.178	2.725
19/32	1.7	0.116	0.338	5.018	0.034	0.199	2.811
5/8	1.8	0.131	0.361	5.258	0.045	0.238	3.073
23/32	2.1	0.183	0.439	6.109	0.085	0.338	3.780
3/4	2.2	0.202	0.464	6.189	0.108	0.418	4.047
7/8	2.6	0.317	0.626	7.539	0.179	0.579	5.991
1	3.0	0.479	0.827	7.978	0.321	0.870	6.981
1 1/8	3.3	0.623	0.955	8.841	0.474	1.098	8.377

Source: Concrete Forming, APA Design/Construction Guide, p. 14. American Plywood Association, Tacoma, Washington.

The section properties in Table 4.5 are for a 12-inch-wide strip of plywood. The values are given for both possible orientations of the face grain of the panel with the direction of stress. These two orientations are sometimes called the "strong direction" and the "weak direction." When the panel is supported so that the stresses are in a direction parallel to the face grain, it is said to be oriented in the strong direction. This is sometimes described as having the supports perpendicular to the grain or having the span parallel to the grain. All of these describe the strong direction orientation.

The three different section properties found in Table 4.5 are moment of inertia (I), effective section modulus (KS), and rolling shear constant (Ib/Q).

The moment of inertia is used in calculating deflections in plywood. Deflections due to both flexure and shear are calculated using I. Standard formulas can be used to find bending and shear deflections.

The effective section modulus (KS) is used to calculate bending stress (f_b) in the plywood:

$$f_b = \frac{M}{KS} \tag{4.1}$$

It should be noted that because of the cross-banded veneers and the fact that veneers of different strengths may be used in different plies, the bending stress cannot be correctly calculated using the moment of inertia. That is, KS is not equal to I divided by half the panel thickness:

$$KS \neq \frac{I}{c} \tag{4.2}$$

Therefore,

$$f_b \neq \frac{Mc}{I} \tag{4.3}$$

The rolling shear constant is used to calculate the rolling shear stress in plywood having loads applied perpendicular to the panel. The name *rolling shear stress* comes from the tendency of the wood fibers in the transverse veneer plies to roll over one another when subjected to a shear stress in the veneer plane. Horizontal shear stress in a beam is

$$f_v = \frac{VQ}{Ib} = \frac{V}{Ib/Q} \tag{4.4}$$

where V is the shear force in the beam at the section and Ib/Q is the section property depending on size and shape of the cross section.

Rolling shear stress, f_s, in a plywood beam is

$$f_s = \frac{V}{Ib/Q} \tag{4.5}$$

where Ib/Q is the rolling shear constant. The rolling shear constant for Plyform is shown in Table 4.5.

Allowable Stresses for Plywood

Table 4.6 shows the values of allowable stresses and moduli of elasticity for the three classes of Plyform. It also has load tables showing allowable pressures on Class I and Structural I Plyform. Plywood stresses can be modified in the same way as lumber stresses to account for short duration loads. The load duration factor may be applied to increase the allowable stresses if the plywood loads have a duration of not more than one week and if the forms are not for multiple reuse. Stresses in Table 4.6 have been reduced for "wet use" since fresh concrete will be in contact with

Table 4.6 Allowable Stresses and Pressures for Plyform Plywood

Allowable Stresses	Plyform Class I	Plyform Class II	Structural I Plyform
Modulus of elasticity, E^* (psi)	1,500,000	1,300,000	1,500,000
Bending stress, F_b (psi)	1930	1330	1930
Rolling shear stress F_s (psi)	72	72	102

*Use when shear deflection is not computed separately. Note: All stresses have been increased by 25% for short-term loading.

Recommended Maximum Pressures on Plyform Class I (Pounds per Square Foot, psf)
Face Grain Parallel to Supports, Plywood Continuous across Two or More Spans

	Plywood Thickness (Inches)													
	15/32		1/2		19/32		5/8		23/32		3/4		1 1/8	
Deflection limit	L/360	l/270	L/360	l/270	L/360	l/270	L/360	l/270	L/360	l/270	L/360	l/270	L/360	l/270
Support spacing														
4	2715	2715	2945	2945	3110	3110	3270	3270	4010	4010	4110	4110	5965	5965
8	885	885	970	970	1195	1195	1260	1260	1540	1540	1580	1580	2295	2295
12	335	395	405	430	540	540	575	575	695	695	730	730	1370	1370
16	150	200	175	230	245	305	265	325	345	390	370	410	740	770
20	—	115	100	135	145	190	160	210	210	270	225	285	485	535
24	—	—	—	—	—	100	—	110	110	145	120	160	275	340
32	—	—	—	—	—	—	—	—	—	—	—	—	130	170

Recommended Maximum Pressures on Plyform Class I (Pounds per Square Foot, psf)
Face Grain across Supports, Plywood Continuous across Two or More Spans

	Plywood Thickness (Inches)													
	15/32		1/2		19/32		5/8		23/32		3/4		1 1/8	
Deflection limit	L/360	l/270	L/360	l/270	L/360	l/270	L/360	l/270	L/360	l/270	L/360	l/270	L/360	l/270
Support spacing														
4	1385	1385	1565	1565	1620	1620	1770	1770	2170	2170	2325	2325	4815	4815
8	390	390	470	470	530	530	635	635	835	835	895	895	1850	1850
12	110	150	145	195	165	225	210	280	375	400	460	490	1145	1145
16	—	—	—	—	—	—	—	120	160	215	200	270	710	725
20	—	—	—	—	—	—	—	—	115	125	145	155	400	400
24	—	—	—	—	—	—	—	—	—	—	—	100	255	255

Source: Concrete Forming, APA Design/Construction Guide, p. 14. American Plywood Association, Tacoma, Washington.

the plywood form sheathing. The stresses in Table 4.6 also have been increased by 25% for load duration.

These allowable stresses can be used without regard to the direction of the grain. Grain orientation is accounted for in the calculations of the section properties in Table 4.5 that are used to calculate actual stresses.

Ties

Ties are devices used to hold the sides of concrete forms together against the fluid pressure of fresh concrete. Ties are loaded in tension and have an end connector that attaches them to the sides of the form. In order to maintain the correct form width, some ties are designed to spread the forms and hold them at a set spacing before the concrete is placed. Some ties are designed to be removed from the concrete after it sets and after the forms have been removed. These ties take the form of tapered steel rods which are oiled or greased so that they can be extracted from one side of the wall. They usually have a high strength and are used in heavier panel systems where it is desirable to minimize the total number of ties.

The removal of ties allows the concrete to be patched. Patching allows a smoother surface on the concrete and helps to eliminate the potential for staining from rusting of steel tie ends. One type of

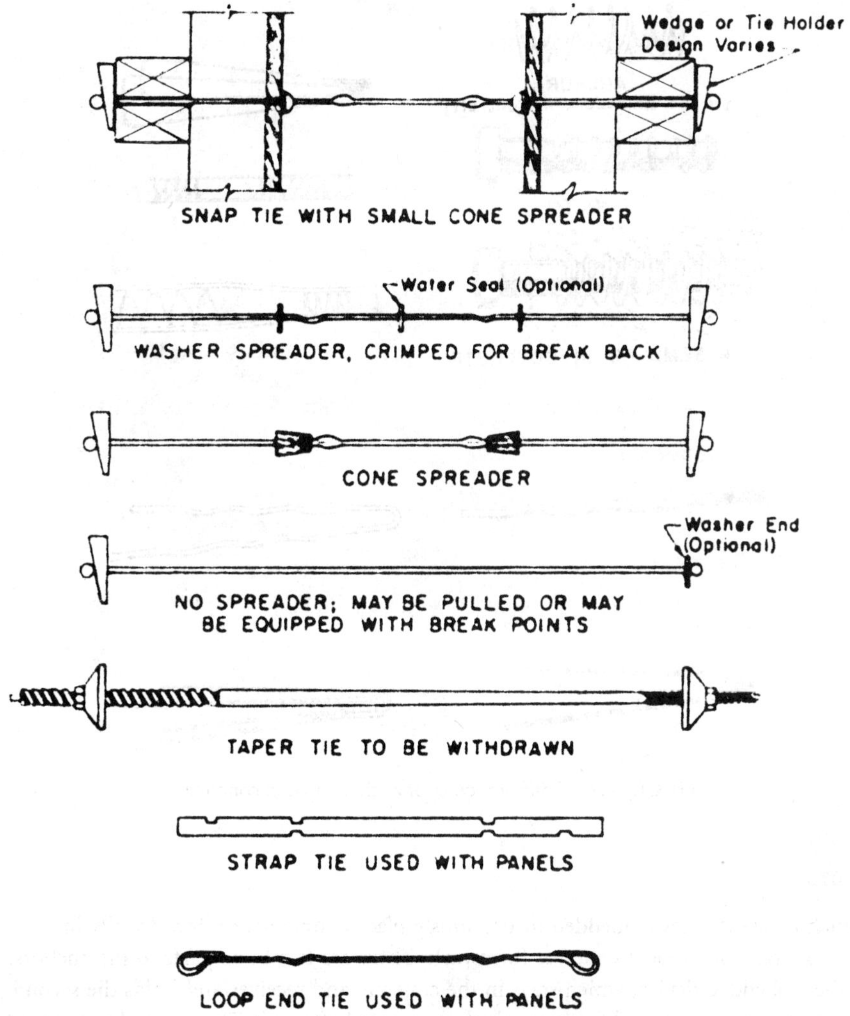

FIGURE 4.1 A drawing showing common types of form ties.

tie is designed to be partially removed by either unscrewing the tie ends from a threaded connector that stays in the concrete, or by breaking the tie ends back to a point weakened by crimping. There is another type of tie that has water seals attached which may be used if ordinary grout patching will not provide a tight seal. Figure 4.1 shows some of these types of ties.

Nonmetallic ties have recently been introduced. They are produced of materials such as a resin-fiber composite and are intended to reduce tie-removal and concrete-patching labor costs. Since they are of nonmetallic materials, they do not rust or stain concrete that is exposed to the elements.

The rated strength of ties should include a factor of safety of 2.0. ACI Committee 347 has revised previous recommendations that called for a factor of safety of 1.5 for ties. The new rating matches the factors of safety applied to the design of other form components. When specifying ties, the manufacturer's data should be carefully checked to ensure that the required factor of safety is incorporated within the rated capacity. If older ratings are found that use a 1.5 factor of safety, then use the ties at only 75% of that rated capacity.

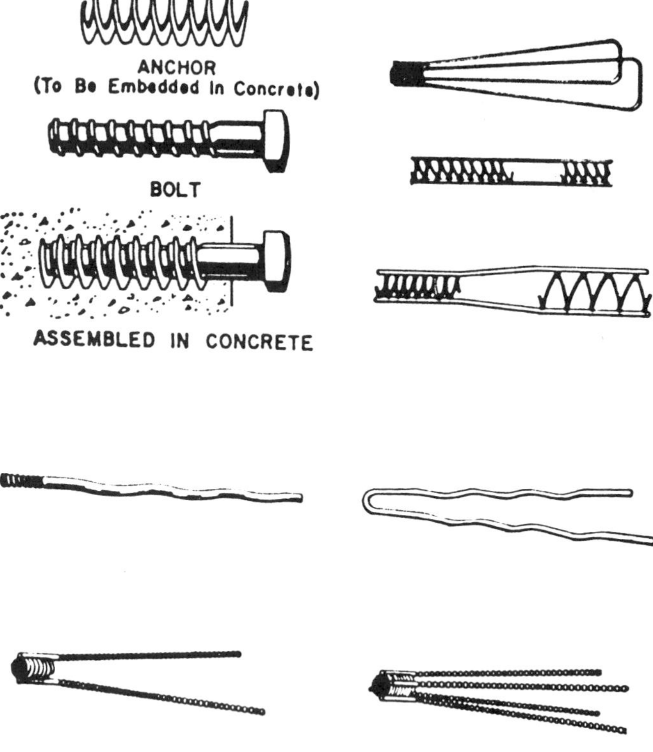

FIGURE 4.2 Common types of anchors used in concrete.

Anchors

Form anchors are devices embedded in previously placed concrete, or occasionally in rock, that are used to attach or support concrete formwork. There are two basic parts to the anchors. One part is the embedded device, which stays in the concrete and receives and holds the second part. The second part is the external fastener, which is removed after use. The external fastener may be a bolt or other type of threaded device, or it may have an expanding section that wedges into the embedded part. Figure 4.2 shows some typical anchors.

ACI Committee 347 recommends two factors of safety for form anchors. For anchors supporting only concrete and dead loads of the forms, a factor of safety of 2 is used. When the anchor also supports construction live loads and impact loads, a factor of safety of 3 should be used.

The rated capacity of various anchors is often given by the manufacturers. Their holding power depends not only on the anchor strength but on the strength of the concrete in which they are embedded. The depth of embedment and the area of contact between the anchor and concrete are also important in determining capacity. It is necessary to use the data provided by the manufacturers, which are based on actual load tests for various concrete strengths, to determine the safe anchor working load for job conditions. This will require an accurate prediction of the concrete strength at the time the anchor is loaded. Estimated concrete strengths at the age when anchor loads will be applied should be used to select the type and size of anchor required.

Hangers

Hangers are used to support concrete formwork by attaching the formwork to the structural steel or precast concrete structural members. Various designs are available and each manufacturer's safe

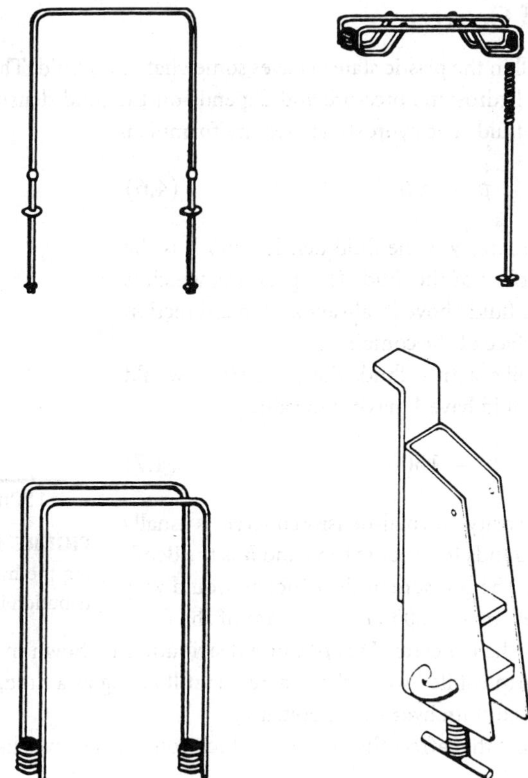

FIGURE 4.3 Examples of types of form hangers used to support formwork from existing structural members.

load rating should be used when designing hanging assemblies. For hangers having more than one leg, the form designer should carefully check to see if the rated safe load is for the entire hanger or each leg. ACI Committee 347 recommends a safety factor of 2 for hangers. Figure 4.3 shows some types of hangers.

Column Clamps

Devices that surround a column form and withstand the lateral pressure of the fresh concrete are called column clamps. They may be loaded in tension or flexure or in a combination of both. There are available several commercial types of column clamps designed to fit a range of sizes of column forms. Care should be taken to read the manufacturer's instructions for using their clamps. Rate of placement of the concrete and maximum height of the form may be restricted. Deflection limits for the forms may be exceeded if only the strength of the clamps is considered. Where deflection tolerances are important, either previous satisfactory experience with the form system or additional analysis of the clamp and form sheathing is suggested.

4.4 Loads on Concrete Formwork

Concrete forms must be designed and built so that they will safely carry all the live and dead loads applied to them. These loads include the weight and pressure of concrete, the weight of reinforcing, the weight of the form materials and any stored construction materials, the construction live loads imposed by workers and machinery applied to the forms, and loads from wind or other natural forces.

Lateral Pressure of Concrete

Fresh concrete that is still in the plastic state behaves somewhat like a fluid. The pressure produced by a true fluid is called hydrostatic pressure and depends on the fluid density and on the depth below the surface of the fluid. The hydrostatic pressure formula is

$$p = \gamma h \tag{4.6}$$

where p is the fluid pressure, γ is the fluid density, and h is the depth below the free surface of the fluid. This pressure, which is due to the weight of the fluid above it, always acts in a direction perpendicular to the surface of the container.

If concrete behaved like a true fluid, the pressure it would produce on the forms would have a maximum value of

$$p = 150h \tag{4.7}$$

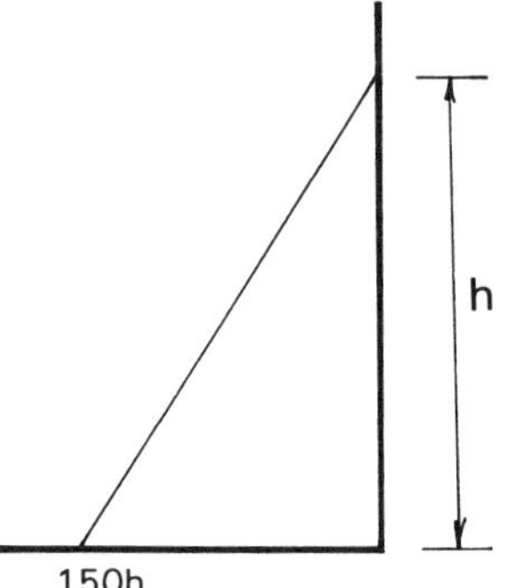

FIGURE 4.4 A diagram showing the hydrostatic pressure distribution in fresh concrete.

for concrete of normal density. Normal-density concrete is usually assumed to weigh 150 pounds for a cubic foot, and h is the depth of concrete in the form. The pressure of the concrete would vary from the maximum pressure of $150h$ at the bottom of the form to zero at the surface of the concrete. This pressure distribution is shown in Fig. 4.4. For placing conditions where the form is filled while the concrete is still acting as a fluid, the form should be designed to resist this maximum hydrostatic pressure.

There are several factors that affect the degree to which concrete behaves like a true fluid.

- Fresh concrete is a mixture of aggregate, water, cement, and air, and can only approximate fluid behavior.
- The internal friction of the particles of solids in the mixture against each other and against the forms and reinforcing tends to reduce the actual pressure to below hydrostatic levels. This internal friction tends to be reduced in a concrete mixture that is wetter, that is, has a high slump; therefore, the pressure will more closely approach the hydrostatic level in high slump mixes.
- The bridging of the aggregate from one side of the form to the opposite in narrow forms also tends to limit pressure. The weight of the concrete tends to be partly supported by the bridging, which prevents the pressure from increasing to true fluid levels.
- In addition to the interaction of the solids in the mixture, the stiffening of the concrete due to the setting of the cement has a marked influence on the pressure. As the concrete stiffens, it tends to become self-supporting, and the increase in pressure on the forms is reduced as the filling continues. Since the setting of the cement can begin in as little as 30 minutes or in as long as several hours after mixing, the lateral pressure may or may not be changed from the hydrostatic condition for any given filling of a form. The things that most directly affect the stiffening of the concrete include the concrete temperature, the use of admixtures such as retarders, the amount and type of cement, and the amount and types of cement substitutes such as fly ash or other pozzolans.

Considering the effective lateral pressure in the concrete to be the modified hydrostatic pressure from Eq. (4.6), the variables that most directly influence this pressure are the density of the mixture, the temperature of the mixture, the rate of placement of the concrete in the forms, the use of admixtures or cement replacements, and the effect of vibration or other consolidation methods.

Other factors which have been considered as influences on pressure include aggregate size and shape, size of form cross section, consistency of the concrete, amount and location of the reinforcement, and the smoothness of the surface of the forms. Studies have shown that the effect

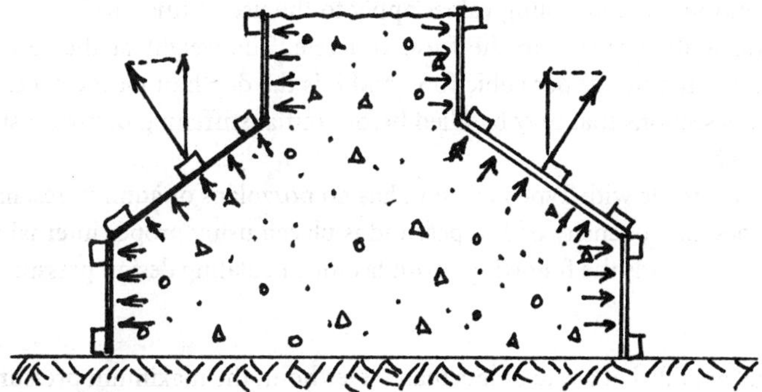

FIGURE 4.5 A diagram showing how concrete pressure creates uplift in forms with sloping sides.

of these other factors is usually small when conventional placement practices are used, and that the influence of these other factors is generally ignored.

The temperature of the concrete has an important effect on pressure because it affects the setting time of the cement. The sooner the setting occurs, the sooner the concrete will become self-supporting, and the sooner the pressure in the form will cease to increase with increasing depth of concrete.

The rate of placement (usually measured in feet of rise of concrete in the form per hour) is important because the slower the form is filled, the slower the hydrostatic pressure will rise. When the concrete stiffens and becomes self-supporting, the hydrostatic pressure ceases to increase. At low rates of placement, the hydrostatic pressure will have reached a much lower value before setting starts and the maximum pressure reached at any time during the filling of the forms will be reduced.

Internal vibration is the most common method of consolidating concrete in formwork. When the probe of the vibrator is lowered into the concrete, it liquefies the surrounding mixture and produces a full fluid pressure to the depth of vibration. This is why proper vibration techniques are so important in avoiding excessive pressure on concrete forms. Vibrating the concrete below the level necessary to eliminate voids between lifts could reliquefy concrete that has started to stiffen and increase pressure beyond expected design levels. The pressures in concrete placed using proper internal vibration usually exceed by 10 to 20% those pressures from placement where consolidation is by simple spading. When vibration is to be used, the forms should be constructed with additional care to avoid leaking.

In some cases, it is acceptable to consolidate concrete using vibrators attached to the exterior of the forms, or to revibrate the concrete to the full depth of the form. These techniques produce greater pressures than those from normal internal vibration and usually require specially designed forms.

Because the pressure of the concrete acts perpendicular to the form surfaces, it is possible that sloping forms such as shown in Fig. 4.5 could have significant uplift forces applied to them. Uplift forces should be considered in the design of the form and should be resisted by proper tie-downs or deadweights.

Recommended Design Values for Form Pressure

ACI Committee 347, after reviewing data from field and laboratory investigations of formwork pressure, has published recommendations for calculating design pressure values. The basic lateral pressure value for freshly placed concrete is

$$p = wh \tag{4.8}$$

No maximum or minimum controlling values apply to the use of this formula, which represents the equivalent hydrostatic pressure in the fresh concrete. The weight of the concrete, w, is the weight of the concrete in pounds per cubic foot, and h is the depth of plastic concrete in feet. For forms of small cross sections that may be filled before initial stiffening occurs, h should be taken as the full form height.

For concrete that is made with Type I cement, has no pozzolans or admixtures, has a maximum slump of four inches, has a density of 150 pcf, and is placed using proper internal vibration, ACI Committee 347 recommends the following formulas for calculating design pressure values.

Wall Forms

For wall forms that are filled at a rate of less than 7 feet/hour, the maximum pressure is

$$p = 150 + 9000\frac{R}{T} \tag{4.9}$$

where p is the maximum lateral pressure of concrete in the form, R is the rate of placement of the concrete in feet/hour, and T is the temperature of the concrete in the form in degrees F. For wall forms filled at a rate of between 7 feet/hour and 10 feet/hour, the maximum pressure is

$$p = 150 + \frac{43{,}400}{T} + 2800\frac{R}{T} \tag{4.10}$$

Design values from both Eq. (4.9) and Eq. (4.10) should not exceed $150h$ or 2000 psf. Both of these formulas predict a maximum value of pressure in the wall form during placing. From these maximum values, no prediction should be made about what the pressure distribution is at any given time. An envelope of maximum pressure can be found by considering the concrete to be fully fluid to the depth in the form where the maximum pressure is reached. This depth where the maximum pressure is reached is the depth where the concrete has become self-supporting and the pressure has ceased to increase.

Example 4.1. A wall form 12 feet high is filled with normal-weight concrete having a temperature of 70°F. The concrete rises during placement at a rate of 5 feet per hour.

Using Eq. (4.9) the maximum pressure is

$$p = 150 + 9000\frac{R}{T}$$

$$p = 150 + 9000\frac{5}{70} = 793 \text{ psf}$$

The hydrostatic pressure for the 12-foot depth is 150(12) = 1800 psf.

Comparing the pressure from the formula to the fully hydrostatic pressure, or the limiting value of 2000, shows that the maximum pressure expected in the wall form at any time during placement is 793 psf.

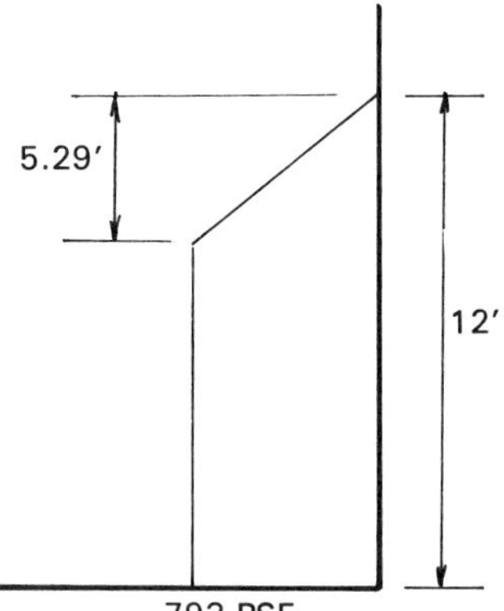

FIGURE 4.6 A diagram showing the envelope of maximum pressure in Example 4.1.

The envelope of maximum pressure will show a hydrostatic pressure to a depth below the top of the form of 793/150 = 5.29 feet. Figure 4.6 shows the envelope of maximum pressure.

Column Forms

Column forms usually have a smaller total volume than wall forms and will fill faster for a given delivery rate of concrete. When the column form fills in a relatively short period of time, it is likely that the concrete will act as a fluid and the pressure will be fully hydrostatic. Equation (4.9) gives the pressure for this condition.

According to ACI Committee 347 recommendations, when the concrete is made with Type I cement weighing no more than 150 pcf, having a slump of not more than four inches, and having no pozzolans or admixtures in it, and is consolidated using internal vibration to a depth of not more than four feet below the surface, Eq. (4.9) may be used to calculate the maximum lateral pressure in the form.

$$p = 150 + 9000\frac{R}{T} \tag{4.11}$$

where p is the pressure in psf, R is the rate of placement in feet per hour, and T is the temperature of the concrete in degrees F. This formula is limited to columns where lifts do not exceed 18 feet. The pressure from the formula has a minimum recommended value of 600 psf and a maximum value of 3000 psf.

The pressure distribution in a column form is hydrostatic until the time the concrete begins to stiffen. Maximum pressure is assumed to increase by 150 psf per foot of depth until the maximum value given by Eq. (4.9) is reached. The pressure then remains constant to the bottom of the form.

Example 4.2. A 16-foot-high concrete column form is filled at a rate of 10 feet per hour with 80 degree concrete. The maximum pressure, using Eq. (4.11), is

$$p = 150 + 9000\frac{10}{80} = 1275 \text{ psf}$$

and will occur at a depth of

$$1275/150 = 8.5 \text{ feet}$$

below the top of the form.

The envelope of maximum pressure to be used to design the column form is shown in Fig. 4.7.

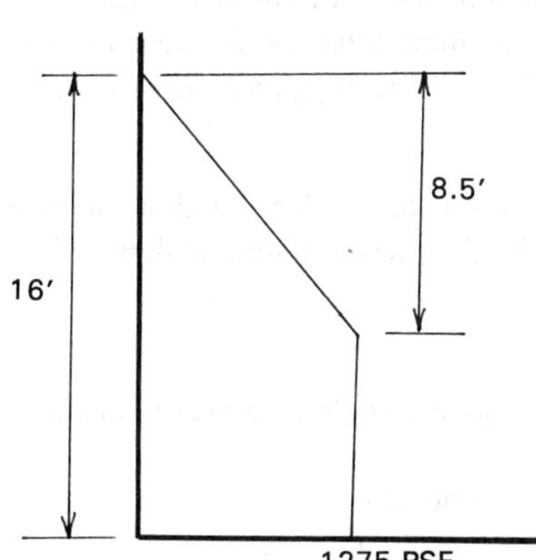

FIGURE 4.7 A diagram showing the envelope of maximum pressure in Example 4.2.

The formulas recommended by the ACI Committee 347 report for pressure in wall and column forms apply when the conditions stated above exist. In some cases, conditions may vary from these standard conditions. Adjustments to the design pressures can be made to account for some conditions other than those specified. These adjustments are explained below.

Consolidation by Spading. Lower pressures result when concrete is spaded rather than consolidated using internal vibration. The pressures from the formulas may be reduced 10% in this situation.

Mixtures Containing Retarders, Fly Ash, Superplasticizers. Concrete that uses retarders, fly ash or other pozzolans, or superplasticizers will have its initial stiffening delayed. Since this will increase the pressure of the concrete on the forms, the forms should be designed using the assumption of a full liquid head [Eq. (4.8)].

Concrete Density. For concrete having densities ranging between 100 pcf and 200 pcf, the pressure can be found by multiplying the pressure from normal-weight concrete (150 pcf) by the ratio of the densities. If the concrete has a density of 200 pcf, its pressure would be 200/150 or 1.33 times the pressure for normal-weight concrete.

Gravity Loads on Formwork

Gravity loads are from all the live loads and deadweights that are applied to and supported by the formwork. These include the weight of the concrete and reinforcing steel, the weight of the forms themselves, and any construction loads from workers, equipment, or stored materials. Loads from upper floors may also be transferred to lower-level forms in multistory construction. The largest loads generally are due to the weight of the concrete element being formed and to the construction live load from workers and equipment. Since the majority of concrete used has a weight of around 140 to 145 pounds per cubic foot, it is common to use a value for design of 150 pounds per cubic foot, which includes an allowance for the reinforcing steel. Where lightweight or heavy-weight concrete is used, the density of that particular mix should be used in calculating formwork loads.

Trying to predict what value should be used for construction loads to account for the weight of workers and equipment is difficult. The weights of the workers and equipment would have to be estimated for each situation and their locations taken into account when trying to determine worst-case loadings. As a guide to the designer in ordinary and usual conditions, ACI Committee 347 recommends using a minimum construction live load of 50 psf of horizontal projection when no motorized buggies are used for placing concrete. When motorized buggies are used, a minimum construction live load of 75 psf should be used. The dead load of the concrete and forms should be added to the value of the live load.

The dead load of the concrete depends on the thickness of the concrete element. For every inch of thickness of the concrete, 150/12 or 12.5 psf should be used. The dead load of the forms can vary from a value of 4 or 5 psf to as much as 15 to 18 psf. In many cases, the weight of the forms is small when compared to the other loads and can be safely neglected. When the design of the form is complete and member sizes are known, the assumption should be checked if the form weight is neglected. ACI Committee 347 recommends that a minimum total load for design of 100 psf be used (regardless of concrete thickness) without use of motorized buggies, or 125 psf when motorized buggies are used.

Example 4.3. A reinforced concrete slab with a thickness of 9 inches is placed with a concrete pump. For normal weight concrete, find the gravity loads the slab forms should be designed to support.

Dead load of concrete slab = 150(9/12) = 112.5 psf
Dead load of forms (estimate) = 10 psf (check this after design is completed, and after member sizes and weights are known)
Construction live load = 50 psf (Committee 347 recommendations)
Total design load for slab form = 172.5 psf

Lateral Loads

In addition to fluid pressures, formwork must also resist lateral loads caused by wind, guy cable tensions, starting and stopping of buggies, bumping by equipment, and uneven dumping of concrete. Since many formwork collapses can be attributed to bracing inadequate for handling lateral loads, it is important that these loads be properly resisted by an adequate bracing system.

The first step in choosing what bracing is required is in determining the magnitude of the lateral loads created by the effects listed above. When lateral loads cannot be easily or precisely determined, minimum lateral loads recommended by Committee 347 of the American Concrete Institute may be used.

Slab Forms

Bracing of the slab forms should be provided to resist the greater of 100 pounds per lineal foot of slab edge, or 2% of the total dead load on the form distributed as a uniform load on the slab edge. When considering the dead load, use only the area of the slab formed in a single pour. If slab forms are enclosed, as might be the case in cold weather operations, the lateral load produced by the wind acting on the forms, enclosure, and any other windbreaks attached to the forms, should be considered.

Wall Forms

Bracing for wall forms should resist a minimum load of 100 pounds per foot of wall applied at the top of the form, or should resist the wind load prescribed by the local code. The wind load used should be at least 15 pounds per square foot. If the wall forms are located below grade and are not subjected to wind loads, bracing should be designed to be adequate to hold the panels in alignment during concrete placement.

Considerations for Multistory Construction

Formwork loads in multistory construction are transferred by shores and reshores to the floors below. The shores directly support the slab forms and carry the loads to the level below. When the shores are removed from the slab and the slab forms are stripped, new shores are placed under that slab. These new shores are called reshores. The reshores transfer any additional loads applied to the floor slab directly to the level below. The loads in the shores and reshores from the slab may be carried all the way to the ground when the building is only a few stories high, or when work for a taller building is still only a few levels above the ground. Otherwise, the loads have to be supported by the lower floors of the building. Depending on the speed of the construction, the loads may be imposed on floors that have not yet reached their full design strength. This understrength, due to lack of curing time combined with the possibility that loads from several floors above may be applied to the floor, can create dangerous overloads or even failures. Most of the failures of concrete buildings occur during the construction phase [Chen and Mosallam, 1991]. To avoid a disastrous accident during the construction of a multistory building, careful analysis should be made of the loads in the forms, the shores, the reshores, and the concrete floor systems for each step in the building process. According to ACI Committee 347, the structure's capacity to carry these construction loads should be reviewed or approved by the engineer-architect. The plan for the forms and shoring remain the responsibility of the contractor.

The basic scheme for ensuring that loads applied to any floor level in the building do not exceed that floor's capacity is to use as many levels of shores and reshores as necessary to distribute the loads. With this system, the loads can either be carried through all levels to the ground below, or when the building has too many levels to make that arrangement practical, the latest-applied loads can be distributed through the shores and reshores to several floors simultaneously so that no one level has a load that exceeds its capacity. The ability of any floor to support construction loads depends on the service capacity that the engineer designed the floor to carry, as well as the age of the concrete when the construction loads are applied. Since the service capacity is based on the specified minimum 28-day strength, the floor capacity prior to the concrete reaching that strength will be less than this service capacity.

When preparing a forming system for multistory buildings, not only must the sizes and specifications for the forming elements be selected, the schedule for when the forms will be erected and stripped must be carefully prepared. In planning this kind of a system and its schedule of use, consider:

- The dead load of the concrete and reinforcing, and the dead load of the forms when significant
- The construction live loads (workers, equipment, storage of materials)
- Design strength of concrete specified

- Cycle time for placing of floors of building
- Structural design load for the floor element supporting construction loads (slab, beam, girder, etc.)—include all loads the engineer designed the slab to carry
- The rate at which the concrete will gain strength under job conditions—use to find the strength of the concrete when loads are applied to it
- The way the loads applied at the different levels are distributed to the floors at the different levels by the elements of the form system

Because of the complex ways the elements of forms for multistory buildings interact with the building elements, and because these interactions may change from one building to another (due to cycle times, concrete properties, weather, different ratios of dead-to-live loads, and many other variables), no single method or procedure for forming and shoring will be satisfactory for all projects.

4.5 Analysis and Design for Formwork

The objective for the formwork designer is to choose a system that will allow concrete elements that meet the requirements of the job to be cast in a safe and economical way. The designer has to choose not only the materials for constructing the forms but also has to determine the size of the form components, the spacing of all supporting members, and the best way to properly assemble the forms to produce a stable and usable structure. After the materials for constructing the form have been selected and the loads that the form must withstand are determined, the designer must determine how to make the form strong enough to carry the stresses from these loads. The forms must also be proportioned so that all deformations are less than those allowable under the specifications and conditions of the job. Many times, the form designer can rely on past experience with other jobs with similar requirements and simply use the same forming system as used before. However, the current job requirements, the materials to be used, or the loading conditions are sometimes different enough to make it necessary that the forms be designed for this particular situation. Form design for the particular situation can be handled in two ways: by relying on tables that give allowable loads for various materials such as lumber and plywood, or by following a rational design procedure similar to those used in designing permanent structures. In the rational design procedure, known elastic properties of the materials are used to determine the required member sizes, spacings, and other details of the form system, and established engineering principles are followed.

Simplifying Assumptions for Design

Because of the many assumptions that can be made about loads, job conditions, workmanship, and quality of materials for formwork, too much refinement in design calculations may be a waste of time. The designer's effort to attain a high degree of precision in calculations is usually negated by the accuracy of the field construction and by uncertainties about loads and materials. A simplified approach for the rational design of formwork members is usually justified. The need to make the forms convenient to construct makes choosing modular spacing desirable even when larger dimensions are indicated by calculations. For example, the strength of the plywood sheathing for a wall form may be sufficient to allow a stud spacing of 14 inches. For ease of construction and to make sure the edges of panels are supported, a spacing of 12 inches for the studs would probably be chosen even though it is conservative.

For concrete formwork that supports unusually heavy loads, requires a high degree of control of critical dimensions, or presents unusual danger to life or property, a complete and precise structural design may be essential. In this situation, a qualified structural engineer experienced in this kind of design should be engaged.

Table 4.7 Beam Formulas

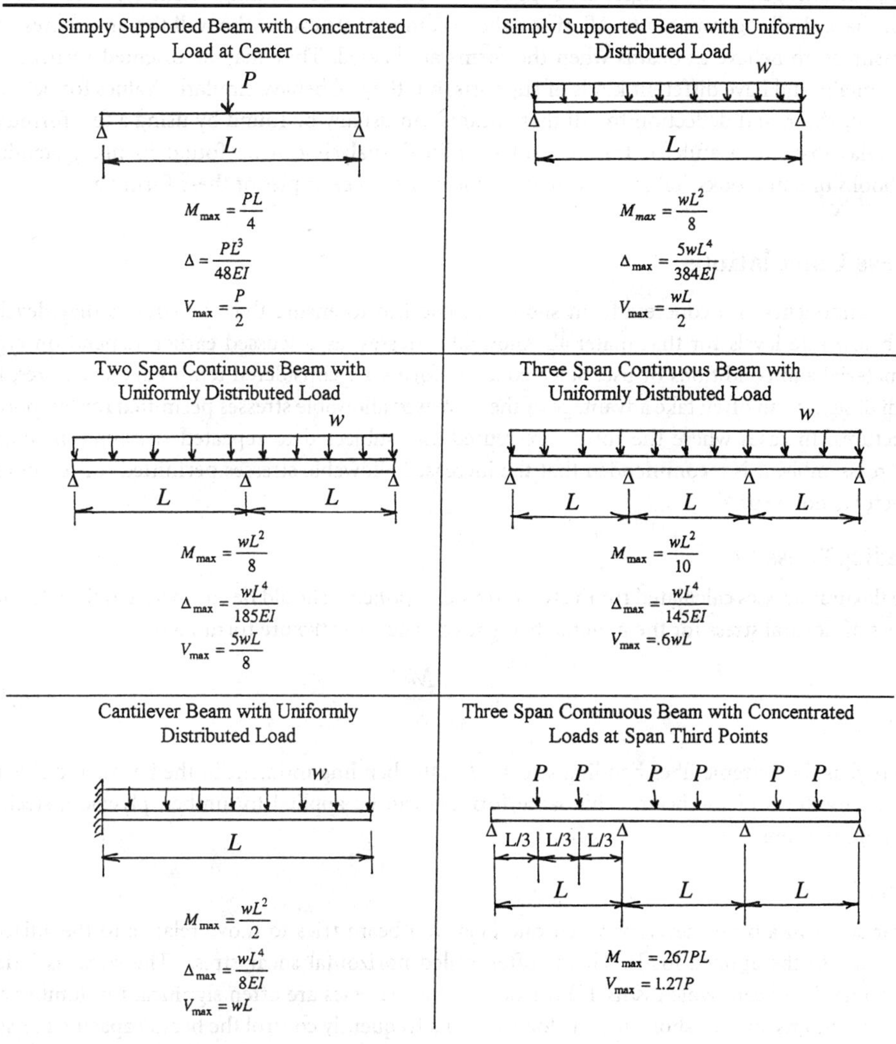

Simply Supported Beam with Concentrated Load at Center	Simply Supported Beam with Uniformly Distributed Load
$M_{max} = \frac{PL}{4}$	$M_{max} = \frac{wL^2}{8}$
$\Delta = \frac{PL^3}{48EI}$	$\Delta_{max} = \frac{5wL^4}{384EI}$
$V_{max} = \frac{P}{2}$	$V_{max} = \frac{wL}{2}$
Two Span Continuous Beam with Uniformly Distributed Load	**Three Span Continuous Beam with Uniformly Distributed Load**
$M_{max} = \frac{wL^2}{8}$	$M_{max} = \frac{wL^2}{10}$
$\Delta_{max} = \frac{wL^4}{185EI}$	$\Delta_{max} = \frac{wL^4}{145EI}$
$V_{max} = \frac{5wL}{8}$	$V_{max} = .6wL$
Cantilever Beam with Uniformly Distributed Load	**Three Span Continuous Beam with Concentrated Loads at Span Third Points**
$M_{max} = \frac{wL^2}{2}$	$M_{max} = .267PL$
$\Delta_{max} = \frac{wL^4}{8EI}$	$V_{max} = 1.27P$
$V_{max} = wL$	

The following simplifying assumptions are commonly used in many situations to allow more straightforward design calculations.

- Assume all loads are uniformly distributed. Loads on sheathing, joists, and studs are in fact distributed, though not necessarily uniformly. Loads on other form members may be point loads but can usually be approximated by equivalent distributed loads. In cases where the spacing of the point loads is large compared to the member span, bending stresses and deflections should be checked for actual load conditions.
- Beams continuous over three or more spans have values for moment, shear, and deflection approximated by the formulas shown in Table 4.7 for continuous beams.

Beam Formulas for Analysis

The components that make up a typical wall form include the sheathing to contain the concrete, studs that support the sheathing, and wales that support the studs. The wales are usually supported by wall ties. For a slab form, the components are similar and are commonly called sheathing, joists, and stringers. The stringers are supported by shores or scaffolding. Other types of concrete forms have similar components. All of the above components except the wall ties and shores may be assumed to behave as beams when the forms are loaded. They may be oriented vertically or horizontally and have different kinds of supports, but they all behave similarly. Values for bending moment, shear, and deflection for all these beams can usually be found by using a few formulas. Formulas that are useful for this type of simplified analysis can be found in many standard textbooks or handbooks. Table 4.7 contains some common examples of these formulas.

Stress Calculations

The components of a concrete form should be checked to ensure that the stresses they develop are below safe levels for that material. Allowable stresses, as discussed earlier, depend on types of material and conditions of use. Since concrete forms are considered temporary structures, the form designer can often take advantage of the increased allowable stresses permitted for temporary structures. In cases where the forms are reused and subjected to repeated stripping, handling, and reassembly, it is recommended that the increased allowable stresses permitted for temporary structures not be used.

Bending Stress

The flexural stresses calculated for the formwork components should be compared to the allowable values of flexural stress for the material being used. The basic flexure formula is

$$f_b = \frac{M}{S} \tag{4.12}$$

where f_b is the extreme fiber bending stress, M is the bending moment in the beam, and S is the section modulus of the beam. This same formula can be applied to lumber, plywood, steel, or aluminum beams.

Shear Stress

Shear stress in a beam develops when one fiber in a beam tries to move relative to the adjacent fiber due to the applied loads. This is often called horizontal shear stress. The stress is indeed horizontal in a beam which is itself horizontal. These stresses are often significant in lumber and plywood beams and for short, heavily loaded spans frequently control the beam capacity. For steel and aluminum beams they are a controlling stress less frequently.

The formula for calculating shear stress is

$$f_v = \frac{VQ}{Ib} \tag{4.13}$$

where V is the shear force in the beam at that cross section, I is the centroidal moment of inertia, b is the beam width at the level in the cross section where f_v is desired, and Q is the moment of the area of the part of the beam cross section above or below the plane where stress is being calculated taken about the beam centroid. Since the maximum value of shear stress is usually what is of interest in formwork design, this formula can be simplified for specific materials. For lumber beams which have a rectangular cross section with width b and height d, the maximum shear stress will occur at the point in the span where the shear force, V, is a maximum at the center of the cross section, and will have the value

$$f_v = \frac{3V}{2bd} \tag{4.14}$$

For plywood, the shear stress is called rolling shear stress and is calculated using Eq. (4.5).

For steel W beams, the shear stress is

$$f_v = \frac{V}{d\,t_w} \tag{4.15}$$

where V is the shear force, d is the depth of the steel beam, and t_w is the thickness of the web.

Bearing Stress

Wood is relatively weak when subjected to compression stresses perpendicular to the grain. Since this is the direction of bearing stresses for lumber beams, these stresses should be compared to allowable values. The bearing stress in a lumber beam is

$$f_{brg} = \frac{R}{A_{brg}} \tag{4.16}$$

where R is the beam reaction and A_{brg} is the bearing area.

Deflections

For the required dimensions and finish to be achieved using the concrete forms, they must resist excessive deformations. The allowable amount of deformation will depend on the job specifications and the type of concrete elements being formed. Concrete that will be exposed usually will have to be cast with a form system that is rigid enough to prevent unsightly bulges or waves. When choosing the sizes and spacing of elements of the concrete forms, deflections must be controlled and kept below the specified limits. A common way to specify limits of deflection in a formwork member is to require it to be less than some fraction of its span. A frequently used value is 1/360 of the span. A deflection limit may also be given as a fraction of an inch such as 1/16 inch for sheathing and 1/8 for other form members. If limits for deflections are given by job specifications, the individual members of the form system should be sized to meet these limits as well as the strength requirements. If no deflection limits are specified, form deformations should not be so large that the usefulness of the cast concrete member is compromised.

Deflections for the members of a concrete form can usually be calculated using formulas such as those in Table 4.7.

Example 4.4—Wall Form Design. A reinforced concrete wall 10 feet 9 inches tall and 16 inches thick will be formed using job-built panels that are 12 feet high and 16 feet long. The wall will be poured in sections 80 feet long, and the concrete will be placed with a pump having a capacity of 18 cubic yards per hour. The expected temperature of the concrete is 75°F. The form panels will have sheathing that is ¾-inch-thick B-B Plyform, Class I. The sheathing will be supported by 2 × 4 lumber studs that are in turn supported by horizontal double 2 × 6 lumber wales. The wales are held against spreading by wall ties that pass through the form. The lumber has allowable stresses as follows: bending = 1100 psi, shear = 190 psi. The modulus of elasticity for the lumber is 1,500,000 psi. Calculate lateral pressure from concrete.

The rate of placement is R = (volume placed in form (yd^3/hr))/volume of form 1 ft high. The volume of form (1 ft.) = (1)(80)(16/12) = 106.7 cubic feet = 3.95 yd^3. R = 18/3.95 = 4.56 ft per hour. Use Eq. (4.9) to calculate pressure, p:

$$p = 150 + \frac{9000R}{T} = 150 + \frac{9000(4.56)}{75}$$

$$p = 697 \text{ psf}$$

To check full hydrostatic condition: $150h$ = 150(10.75)1613 psf; therefore, use 697 psf from formula depth from top of wall to maximum pressure = 697/150 = 4.65 ft.

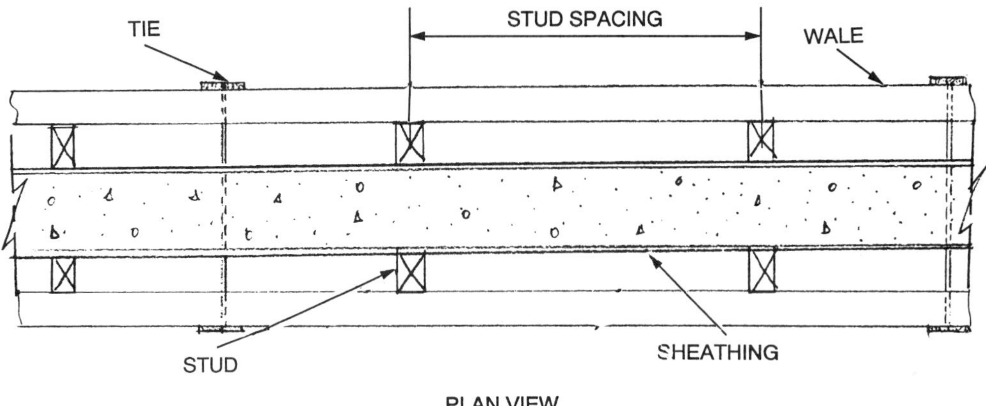

To find the support spacing for plywood sheathing, assume the sheets of plywood are oriented in the form panel as shown in the following sketch. The face grain is across the supports (strong direction). Table 4.6 shows that the plywood will carry a pressure of 730 psf with a supports spacing of 12 inches. This is also a convenient modular spacing for support of panel edges.

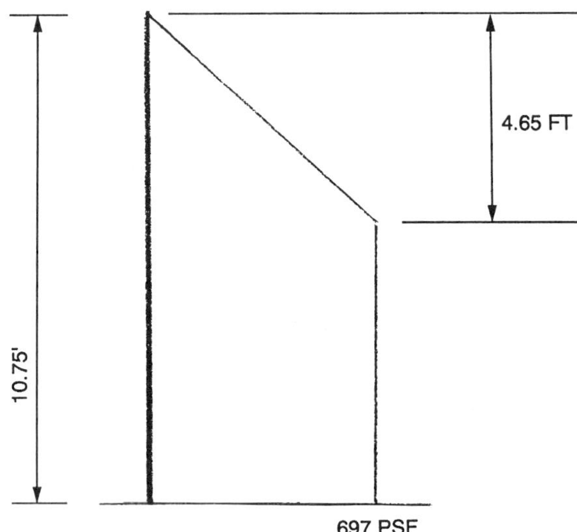

To determine the wale spacing: with a 12 inch stud spacing, the load on each stud will be 697 plf. The studs must be supported by the wales so that the bending stress, the shear stress, and the deflection in the studs do not exceed allowable levels.

If the studs have three or more spans the maximum bending moment can be approximated from Table 4.7 as

$$M = \frac{WL^2}{10}$$

Solving for L gives

$$L = \sqrt{\frac{10M}{w}}$$

The allowable bending stress, F_b, gives the allowable bending moment when multiplied by the section modulus, S.

$$S(2 \times 4) = bd^2/6 = 1.5(3.5)^2/6 = 3.06 \text{ in.}^3$$

$$M_{\text{all}} = F_b S$$

Substituting the allowable moment for L,

$$L = \sqrt{\frac{10F_b S}{w}} = \sqrt{\frac{10(1100)(3.06)(12 \text{ in./ft})}{697}} = 24.1 \text{ inches}$$

For convenience in layout, use a 2-foot wale spacing.

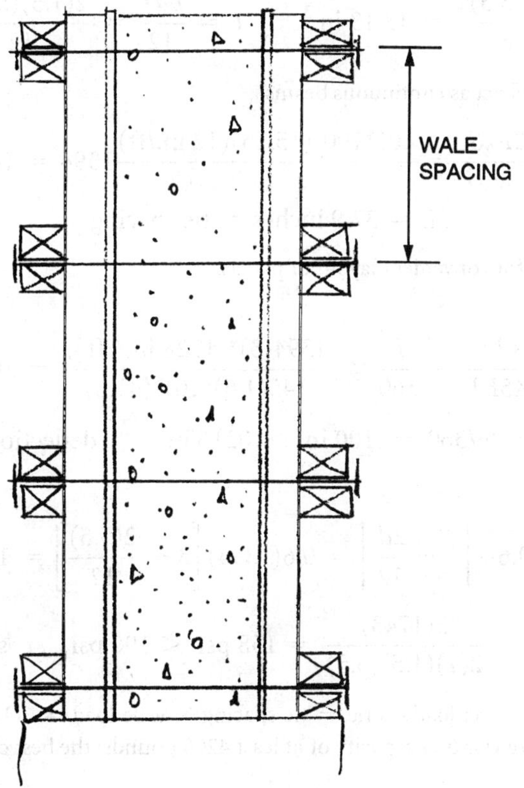

To determine deflection, assume the specifications limit deflections to l/360:

$$\Delta = \frac{wL^4}{145EI} \leq \frac{L}{360} \Rightarrow \frac{697(2)^4(1728 \text{ in.}^3/\text{ft}^3)}{145(1.5)(10)^6(5.36)} = \frac{2(12)}{360}$$

$$0.0165 < 0.0667 \qquad \text{deflection is OK}$$

To determine shear: from Table 4.7 for a continuous beam the maximum shear is

$$V = 0.6w\left[L - \frac{2d}{12}\right]$$

$$V = 0.6(697)\left[2 - \frac{2(3.5)}{12}\right] = 592 \text{ pounds}$$

$$f_v = \frac{3V}{2bd} = \frac{2(592)}{2(1.5)(3.5)} = 169 \text{ psi} < 190 \text{ psi} \qquad \text{shear is OK}$$

A 2-foot wale spacing is therefore OK.

For wale, determine tie spacing and size of tie required. Loads on wale are actually point loads from studs but are often treated as distributed loads. For a 2-foot wale spacing the equivalent distributed wale load, w, is

$$2(697) = 1394 \text{ plf}$$

For 2 × 6 double wales the section properties are

$$S = \frac{bd^2}{6} = \frac{1.5(5.5)^2}{6} = 15.13 \text{ in.}^3 \qquad I = \frac{bd^3}{12} = \frac{2(1.5)(5.5)^3}{12} = 41.6 \text{ in.}^3$$

Wales are 16 ft long and act as continuous beams:

$$L^2 = \frac{10F_bS}{w} = \frac{10(1100)(15.13)(12 \text{ in./ft})}{}1394 = 1432 \text{ in.}^2$$

$$L = 37.9 \text{ inches} = \text{tie spacing}$$

Use 3-foot tie spacing for convenient layout of panels.

Check deflection:

$$\Delta = \frac{wL^4}{145EI} \leq \frac{L}{360} \Rightarrow \frac{1394(3)^4(1728 \text{ in.}^3/\text{ft}^3)}{145(1.5)(10)^6(41.6)} = .0216 \text{ in.}$$

$$L/360 = 36/360 = .100 \text{ in.} > .0216 \text{ in.} \qquad \text{deflection is OK}$$

Check shear:

$$V = 0.6w\left[L - \frac{2d}{12}\right] = 0.6(1394)\left[3 - \frac{2(5.5)}{12}\right] = 1734 \text{ lb}$$

$$f_v = \frac{3V}{2bd} = \frac{3(1743)}{2(2)(1.5)(5.5)} = 158 \text{ psi} < 190 \text{ psi} \qquad \text{shear is OK}$$

Find the required tie size: load on tie = tie spacing × wale load = 3(1394) = 4182 pounds. The tie must have a safe working capacity of at least 4200 pounds; the best choice is probably a 5K (5000 lb) tie.

References

American Plywood Association. 1988. *Concrete Forming.* Tacoma, WA.

American Plywood Association. *Plywood Design Specification.* Tacoma, WA.

Brand, R. E. 1975. *Falsework and Access Scaffolds in Tubular Steel.* McGraw-Hill, New York.

Chen, W. F. and Mosallam, K. H. 1991. *Concrete Buildings, Analysis for Safe Construction.* CRC Press, Boca Raton, FL.

Hurd, M. K. 1989. *Formwork for Concrete,* 5th ed. American Concrete Institute, Detroit, MI.

Moore, C. E. 1977. *Concrete Form Construction.* Van Nostrand Reinhold, New York.

National Forest Products Association. 1986. *National Design Specification.* Washington, D.C.

Peurifoy, P. E. 1976. *Formwork for Concrete Structures.* McGraw-Hill, New York.

Ratay, R. T. 1984. *Handbook of Temporary Structures in Construction.* McGraw-Hill, New York.

For Further Information

For information about designing and using concrete forms in a safe and economical way, see *Guide to Formwork for Concrete,* reported by Committee 347 of the American Concrete Institute. This is included in the fifth edition of the comprehensive work by M. K. Hurd, *Formwork for Concrete,* published by the American Concrete Institute.

For recommendations and guidance in using plywood for concrete forming, see *Concrete Forming,* a design/construction guide published by the American Plywood Association.

For information about safety requirements in concrete operations, including design and construction of formwork, see *Safety and Health Regulations for Construction* as it appears in the United States Occupational Safety and Health Act 1988 revision.

For a comprehensive discussion of the latest techniques for analysis and design of concrete formwork, see *Concrete Buildings, Analysis for Safe Construction* by W. F. Chen and K. H. Mosallam. This book is especially helpful with its discussion of the treatment of formwork for multistory concrete buildings. It includes both a matrix structural analysis procedure and a simple hand calculation procedure to estimate the distribution of loads between floor slabs during construction.

5

Contracts and Claims

Gary R. Smith
Pennsylvania State University

5.1 Introduction

Engineers and architects excel in their mastery of the technical aspects of planning and design, while contractors are highly proficient in identifying cost-effective processes to build complex modern structures. However, when evaluated on the basis of their knowledge of **contracts**, many of these professionals do not understand the importance of the contract language that forms the basis for their relationship with the owner or with each other. Even small contracts have complex contract relationships, due to increased regulation of the environment and safety. Few would argue that the proliferation of contract claims consultants and attorneys reflects positively on the ability of designers and contractors to deliver quality products without litigation. While it is commonly heard that contractors actively seek claims for profit, few reputable contractors would pursue a claim that is frivolous or subjective. Owners and design professionals reflect their heightened awareness of the potential for claims by using restrictive contract language.

This section will focus on the basics: elements of contracts, contract administration, interpretation of some key clauses, the common causes of claims, and resolution alternatives. The type of contract is an important indication of how the contracting parties wish to distribute the financial risks in the project. The discussion on interpretation of contracts presents common interpretation practices and is not intended to replace competent legal advice. Good contract administration and

0-8493-8953-4/95/$0.00 + $.50

interpretation practices are needed to ensure proper execution of the project contract requirements. In the event that circumstances do not evolve as anticipated, a claim may be filed to settle disputed accounts. Claims are often viewed by owners and engineers as the contractor's strategy to cover bidding errors or omissions. Those who have successfully litigated a claim are not likely to agree that claims are "profitable" undertakings. A claim is a formalized complaint by the contractor, and the contractor's right to file for the claim is an important element of contract law. In many situations claims help to define new areas of contract interpretation. These disputes often relate to some particularly troublesome clause interpretation and serve to provide contract administrators additional guidance on contract interpretation.

5.2 Contracts

Sweet [1989, p.4] describes contract formation as follows:

> Generally, American law gives autonomy to contracting parties to choose the substantive content of their contracts. Since most contracts are economic exchanges, giving parties autonomy allows each to value the other's performance. To a large degree autonomy assumes and supports a marketplace where participants are free to pick the parties with whom they deal and the terms upon which they will deal.

The terms of a contract will be enforced, no matter how harshly some language treats one of the parties. **Equity** or fairness is occasionally used as the basis for a claim, but it is seldom used by the courts to settle a dispute ensuing from a contract relationship. The most common contract relationships created by modern construction projects are

- The owner and contractor(s)
- The owner and design professional
- The contractor and subcontractor(s)
- The contractor and the **surety**

If the owner hires a construction manager, this creates an additional contract layer between the owner and the designer or contractor. These contracts form the primary basis of the relationship among the parties. It is important that project-level personnel as well as corporate managers understand the importance of the contract and how properly to interpret the contract as a whole.

A contract is a binding agreement between the parties to exchange something of value. Contracts are generally written, but unless there is a statutory requirement that prohibits their use, oral contracts are valid agreements. The basic elements of a valid contract are

- Competent parties
- Offer and acceptance
- Reasonable certainty of terms
- Proper subject matter
- **Consideration**

Competent parties must be of a proper age to enter into a contract and must have sufficient mental capacity to understand the nature of the agreement. *Offer and acceptance* indicates that there has been a meeting of the minds or mutual assent. A contract cannot be formed if there is economic duress, fraud, or mutual mistakes. The *terms* of the contract should be clear enough that an independent third party can determine whether the two parties performed as promised. While this is rarely a problem in public construction contracts, the private industry sector has a greater potential for problems, due to more informal exchanges in determining boundaries of a contract. Contract *subject matter* must not be something that is illegal.

The last element of a valid contract is *consideration*. Contracts are generally economic exchanges; therefore, something of value must be exchanged. Consideration need not be an equal exchange. Courts will uphold seemingly unbalanced consideration if all the elements of a contract are met and there is no evidence of fraud or similar problems.

Form of Agreement

The actual form of agreement, which describes the contracting parties' authority, the work in general, the consideration to be paid, penalties or bonuses, and time for performance, is often a very brief document containing under a dozen pages. This document is seldom the issue of concern in a dispute. More commonly the documents that detail the relationships and project requirements are the source of disagreement. Primarily, these documents for a construction project are the general conditions, special conditions, technical specifications, and plans.

Contract types require a division or separation among the wide variety of contracts used in the industry. In keeping with the economic exchange concept, contracts can be identified as either *fixed price* or *cost reimbursable*. Fixed price contracts establish a fixed sum of money for the execution of a defined quantity of work. These contracts are often termed *hard dollar contracts*. Fixed price contracts fall into two major categories: lump sum and unit price. *Lump sum contracts* require the contractor to assume all risks assigned by the contract for their stated price. Adjustments to costs and extensions of time require a modification to the original agreement. *Unit price contracts* permit more flexibility by establishing costs relative to a measurable work unit; cubic yards and square feet are examples of work units.

Reimbursable contracts allow for contract adjustments relative to project scope as determined by the cost and do not, generally, address a final fixed price. Fixed price contracts allocate more risk to the contractor and thus require more effort, money, and time on design documentation before construction is initiated. Cost reimbursable contracts require greater risk sharing between the owner and contractor and often require more owner personnel for contract administration during the construction phase to enforce cost and schedule. Cost reimbursable contracts are more easily used for fast-tracking of design and construction. Reimbursable contracts are also very flexible for changing design or scope of work and establish the basis for a less adversarial relationship between the owner and contractor. [Contracts Task Force, 1986, p. 8.] Figure 5.1, from the *Construction Industry Cost Effectiveness* (CICE) *Report*, portrays the time advantages associated with cost reimbursable contracts when the owner has a demand for a facility that is highly schedule-driven [CICE, 1982, p. 9]. Often both forms of contracts exist on a project simultaneously. Prime contractors will often have cost reimbursable contracts with the owner and fixed price contracts with their subcontractors.

5.3 Contract Administration

The contractor must concentrate on constructing the project and concurrently attend to the terms of the contract documents. Contract administration involves numerous daily decisions that are based on interpretation of the contract documents. A record of these deliberations is important to both parties. The primary tools for controlling a project contract are the cost and schedule report updates. In addition, quality and safety reports are indicative of project administration success. Administration of the contract requires that accurate records be maintained as a permanent record of the contract process. In the event that the project manager would need to negotiate a change order, prepare a claim, or reconstruct specific events, the project data from records and correspondence are often needed. Richter and Mitchell [1982], Fig. 5.2 emphasize the importance of accurate records and documents. The relative priority of documents would be determined by the nature of the dispute.

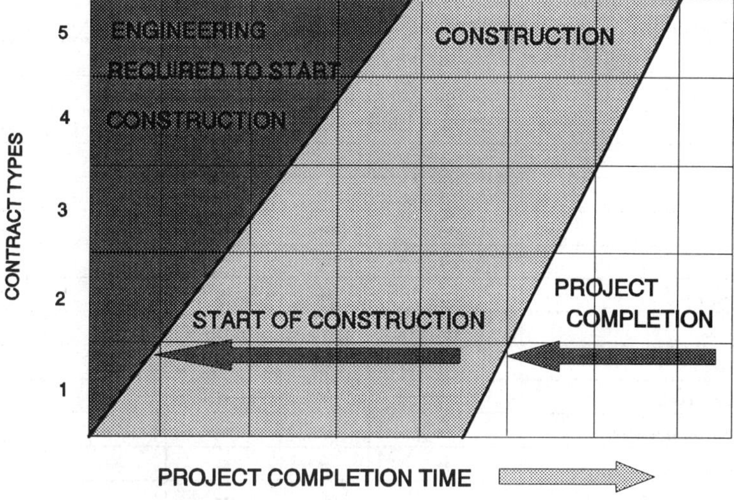

FIGURE 5.1 Contract time and type comparison. *Source:* CICE (Construction Industry Cost Effectiveness Project Report). 1982. *Contractual Arrangement, Report A-7.* The Business Round Table, New York, p. 9.

Trauner [1993] places emphasis on professional information management as a necessary and cost-effective measure for reducing risk on the project. The following list highlights the importance of information management:

1. Appropriate documentation permits future users to verify how the project was built.
2. Lessons learned on the project are recorded for the benefit of future projects.
3. Continuous, contemporaneous documentation reduces the chance of misunderstanding of day-to-day concerns.
4. Records prevent the loss of information otherwise left to memory.
5. Project personnel turnover problems can be reduced with a complete project history.
6. Written reports are the best means of keeping multiple parties informed of project progress.
7. Written reports reduce oral communications and number of meetings.
8. Documentation and monitoring of the project are supported by information management.
9. Establishing defined documentation requirements assists the manager in focusing on the most important aspects of the project.

Progress Reports

Performance documentation covers a wide variety of reports and charts. The project schedule is essential for determining the status of the project at any given point in time, and it can also be used to estimate the time impact of disruptions at the project site. It is important, therefore, that the schedule be updated at frequent intervals to ensure that the actual start dates, finish dates, and percent complete are recorded.

TYPE OF DOCUMENT	AUTHORIZATION	ISSUES	SCHEDULES	PAYMENT	AUDIT
Agreement	**		**	**	**
General Conditions	**	**	**	**	**
Special Conditions	**	**	**	**	**
Technical Specifications	**	**	**		
Bid Invitation		**		**	
Addenda	**	**	**	**	**
Drawings	**	**		**	
Bid Proposal		**		**	**
Subject Files	**	**	**	**	
Chronological Files		**	**	**	
A/E Correspondence	**	**	**	**	
Contractor Correspondence	**	**	**	**	**
Owner Correspondence	**	**	**	**	**
Conference Notes		**	**		
Shop Drawing Logs		**	**		
Survey Books		**	**		
Inspection Reports	**	**	**	**	
Pay Requisitions			**	**	**
Delivery Schedules		**	**		
Test Reports		**			
Daily Reports	**	**	**	**	**
Subcontracts	**	**	**	**	**
Purchase Orders	**	**	**	**	**
Schedules		**	**		**
Photographs		**			
Technical Reports		**			
Cost Records				**	**
Estimates		**		**	**
Change Order Files	**	**	**	**	**
Extra Work Orders	**	**	**	**	**
Payrolls				**	**
Building Codes	**	**			

FIGURE 5.2 Contract document use in claims. *Source:* Richter, I. and Mitchell, R. 1982. *Handbook of Construction Law and Claims.* Reston Publishing Company, Inc., Reston, VA.

Progress should be recorded in daily and weekly reports. Daily reports should be prepared by personnel who can report on field and office activities. Weather information, subcontractor performance, workforce data, equipment use, visitor data, meeting notations, and special or unusual occurrences are entered into a standard diary form, which is filed on-site and in the home office.

Progress reporting should include a photographic progress journal. A log of photograph dates and locations is needed to preserve the specific nature of the photograph. Photographs provide strong visual evidence of the site conditions reported in the progress reports.

The personal project diaries of superintendents also record daily activity. These records summarize key events of the day including meetings, oral agreements or disagreements, telephone discussions, and similar events. Diaries also record drawing errors, provide notations on differing site conditions observed on the site, and other discrepancies. Personal project diaries should be collected at the end of the project and stored with project records.

Quality Records

Complete records of all quality tests performed on materials and reports from inspections should be retained. In addition to test results, plots or statistical analyses performed on the data should also be stored for later use. Inspection reports should be retained as an integral part of the quality recordation and documentation. Rework should be noted, and the retest results should be noted. Problems with quality and notes on corrective procedures applied should be evident in the records.

Change Order Records

Changes should be tracked by a change order record system separate from other project records. Careful attention is needed to ensure compliance with notice requirements, proper documentation of costs, and estimation of the anticipated time impact. An understanding beforehand of the change order process and the required documentation will reduce the risk of a change order request not being approved. Change orders can have a significant impact on the progress of remaining work as well as on the changed work. Typical information included in a change request includes the specification and drawings affected, the contract clauses that are appropriate for filing the change, and related correspondence. Once approved, the change order–tracking system resembles traditional cost and schedule control.

Correspondence Files

Correspondence files should be maintained in a chronological order. The files may cover the contract, material suppliers, subcontracts, minutes of meetings, and agreements made subsequent to meetings. It is important that all correspondence, letters, and memorandums be used to clarify issues, not for the self-serving purpose of preparing a claim position. If the wrong approach in communications is employed, the communications may work against the author in the eventual testimony on their content. Oral communications should be followed by a memorandum to file or to the other party to ensure that the oral communication was correctly understood. Telephone logs, fax transmissions, or other information exchanges also need to be recorded and filed.

Drawings

Copies of the drawings released for bidding and those ultimately released for construction should be archived for the permanent project records. A change log should be maintained to record the issuance or receipt of revised drawings. Obsolete drawings should be properly stamped and all copies recovered. Without a master distribution list, it is not always possible to maintain control of drawing distribution. Shop drawings should also be filed and tracked in a similar manner. Approval dates, release dates, and other timing elements are important to establishing the status of the project design and fabrication process.

5.4 Reasoning with Contracts

The contract determines the basic rules that will apply to the contract. However, unlike many other contracts, construction contracts usually anticipate that there will be changes. *Changes* or *field variations* are created from many different circumstances. Most of these variations are successfully negotiated in the field, and once a determination is made on the cost and time impact, the contracting parties modify the original agreement to accommodate the change. When the change order negotiation process fails, the change effectively becomes a dispute. The contractor commonly will perform a more formal analysis of the items under dispute and present a formal claim document to the owner to move the negotiations forward. When the formal claim analysis fails to yield results,

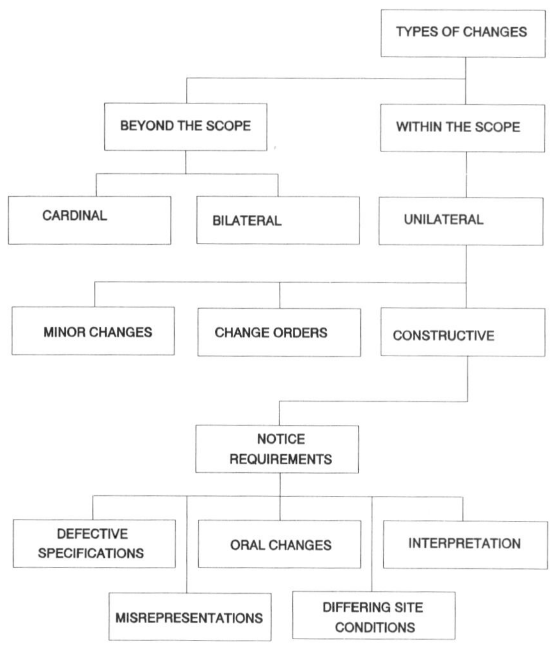

FIGURE 5.3 Types of changes.

the last resort is to file the claim for litigation. Even during this stage, negotiations often continue in an effort to avoid the time and cost of litigation. Unfortunately, during the maturation from a dispute to a claim, the parties in the dispute often become entrenched in positions and feelings and lose their ability to negotiate on the facts alone. Contract wording is critical, and fortunately, most standard contracts have similar language. It is important to understand the type of dispute that has developed. Figure 5.3 was developed to aid in understanding the basic relationships among the major types of changes.

5.5 Changes

Cardinal and **bilateral** changes are beyond the scope of the contract. *Cardinal* changes describe either a single change or an accumulation of changes that are beyond the general scope of the contract. Exactly what is beyond the scope of a particular contract is a case-specific determination based on circumstances and the contract; there is no quick solution or formula to determine what constitutes a cardinal change. Cardinal changes require very thorough claim development.

A bilateral change is generated by the need for a change that is recognized as being outside the contract scope and, therefore, beyond the owner's capability to issue a unilateral change. A bilateral change permits the contractor either to consent to performing the work required by the change or to reject the change and not perform the additional work. Bilateral changes are also called *contract modifications*. Obviously, the gray area between what qualifies as a unilateral change and a bilateral change requires competent legal advice before a contractor refuses to perform the work.

Several distinctions can be made among *unilateral* changes. *Minor changes* that do not involve increased cost or time can be ordered by the owner or the owner's representative. Disputes occasionally arise when the owner believes that the request is a minor change, but the contractor believes that additional time and/or money is needed. Minor changes are also determined by the specific circumstances. *Change orders* are those changes conducted in accordance with the change order clause of the contract, and unless the change can be categorized as a cardinal change, the contractor is obligated to perform the requested work. *Constructive changes* are unilateral changes that are not considered in the changes clause; they can be classified as oral changes, defective specifications, misrepresentation, contract interpretation, and differing site conditions. However, before constructive changes can be considered in more detail, contract notice requirements must be satisfied.

5.6 Notice Requirements

All contracts require the contractor to notify the owner as a precondition to claiming additional work. The reason for a written notice requirement is that the owner has the right to know the extent of the liabilities accompanying the bargained-for project. Various courts that have reviewed notice cases agree that the notice should allow the owner to investigate the situation to determine the character and scope of the problem, develop appropriate strategies to resolve the problem, monitor the effort, document the contractor resources used to perform the work, and remove interferences that may limit the contractor in performing the work.

Contracts often have several procedural requirements for filing the notice. Strict interpretation of the notice requirements would suggest that where the contract requires a written notice, only a formal writing will satisfy the requirement. The basic elements in most contracts' change order clauses are the following:

- Only persons with proper authority can direct changes.
- The directive must be in writing.
- The directive must be signed by a person with proper authority.
- Procedures for communicating the change are stated.
- Procedures for the contractor response are defined.

Figure 5.4 is a decision analysis diagram for disputes involving notice requirements.

The applicability of the clause should be at issue only if the contract has been written such that the notice clause is only effective for certain specific situations. Written notice implies that a formal letter has been delivered that clearly defines the problem, refers to the applicable contract provisions, and states that the contractor expects to be compensated for additional work and possibly given additional time to complete the work. However, notice can also be delivered in other ways. Verbal statements have been found to constitute notice to satisfy this requirement. The principal issues are owner knowledge of events and circumstances; owner knowledge that the contractor expects compensation or a time extension under some provision of the contract; and timing of the communication.

Owner knowledge is further divided into *actual knowledge* and *constructive knowledge*. Actual knowledge is knowledge that is clear, definite, and unmistakable. Constructive knowledge can be divided into *implied* knowledge and *imputed* knowledge. Implied knowledge is communicated by deduction from the circumstances, job site correspondence, or conduct of the parties. While this may not be complete, it is generally sufficient to alert the owner that additional investigation is warranted. Evidence of owner knowledge is more compelling if it involves a problem that is caused by the owner or that is within the owner's control. Imputed knowledge refers to situations in which proper notice is given to an individual who has the duty to report it to the person affected.

Knowledge that the contractor is incurring additional expense is not sufficient to make the owner liable for the costs. If the owner is unaware that the contractor expects payment for the additional cost, the owner may not be held liable for payment.

Notice Timing

Timing of the notice is important. If the notice is given too late for the owner to control the extent of its liability for additional costs, the court may not find that the notice requirement was satisfied.

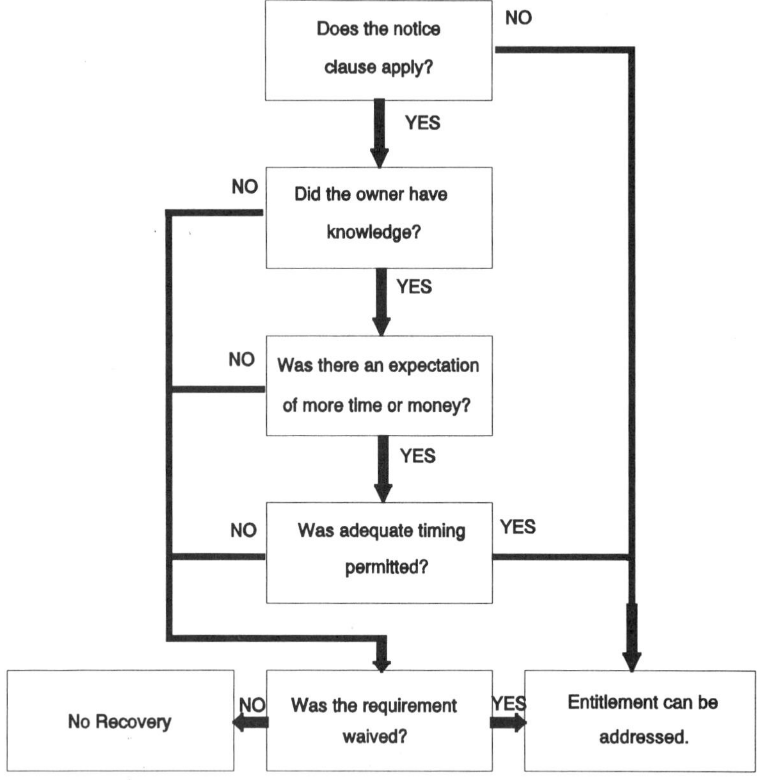

FIGURE 5.4 Notice disputes flowchart.

Generally contracts will specify a time limit for submission of the notice. Slippage of time may not be meaningful if the character of the problem cannot be ascertained without passage of time. However, in some cases the passage of time obscures some of the information, which will prevent the owner from verifying information or controlling costs.

Form of Notice

If notice was not given and evidence of constructive notice is not clear, the remaining recourse is for the contractor to show that the requirement was waived. The owner cannot insist on compliance with the contract where the owner's actions have conflicted with the same requirements. If a statute requires written notice, the requirement cannot be waived. Waiver can only occur by the owner or the owner's representative.

The form of the communication is usually a formal letter. Notice can occur in job site correspondence, letters, memos, and other site documents. Project meeting minutes that summarize discussions about project situations may be sufficient, provided they are accurately drafted. In some instances CPM (critical path method) updates that show delay responsibilities have been found to constitute notice of delay, since they kept the owner fully informed of progress.

5.7 Oral Changes

Oral communication is very common on construction projects. In most cases the oral instructions are clearly understood, and no problems result from the exchange. Oral modifications may be valid even though there may be specific contract language prohibiting oral change orders. Through their consent or mutual conduct, the parties to a contract may waive the written change requirement. Therefore, the owner must be consistent in requiring that all changes be written. The contractor must also be consistent if submitting written changes; failure to provide the written change may indicate that it was a minor change and therefore no additional time or payment was expected. Any inconsistent conduct in the handling of changes will often eliminate the written requirement.

While the actions of the parties may waive a contract clause, the requirement will be upheld when there are statutory requirements for written directives. The owner must be aware of incurring additional liability. The owner may understand that the contractor is accruing additional cost but may not know the contractor is expecting the owner to pay for the additional cost. This may happen when the contractor, in some fashion, indicates that the work is being completed on a voluntary basis. However, when the owner has made an express or implied promise to pay the contractor for the work, recovery is likely. The contractor must make the owner aware at the time of the change that the owner will be expected to pay for additional costs. Acceptance of completed work is not sufficient to show that the owner agreed to pay for the work.

The person approving the change must also have the authority to act for the owner and incur the liability for the owner on the extra work. Generally, the authority is clearly written, but there are cases in which the conduct of an individual implies that he or she has authority. Contractors need to know who has the authority to direct changes at the site. Owners, on the other hand, may appear to extend authority to someone they know does not have explicit authority, but fail to correct the action directed by the unauthorized person. Waiver of the requirements is caused by works, actions, or inactions of the owner that result in abandonment of a contract requirement. The owner must consistently require that the changes be in writing; any deviation from this requirement will result in abandonment of the clause that specifies that all changes be in writing.

5.8 Contract Interpretation

The rules for contract interpretation are well established in common law. The rules are split into two major divisions: procedural and operational. *Procedural rules* are the rules within which the court must operate. *Operational rules* are applied to assist in the interpretation of the facts in the case.

Procedural rules establish the objective of interpretation; measures for the admissibility of evidence; controls on what interpretation can be adopted; and standards for evaluating interpretations. The objective of interpretation focuses on determining the intent of the parties in the contract. Courts will not uphold hidden agendas or secret intentions. The admissibility of evidence provides the court the opportunity to look at separate contracts, referenced documents, oral agreements, and **parol** evidence (oral evidence provided to establish the meaning of a word or term). Courts have no right to modify the contract of the parties, and they cannot enforce contracts or provisions that are illegal or against public policy or where there is evidence of fraud [Thomas and Smith, 1993]. The last function of interpretation controls is to incorporate existing law. Generally, the laws where the contract was made will govern the contract. However, in the construction business the performance of the contract is governed by the law where the contracted work is performed.

Operational interpretation rules are primarily those applied to ascertain the meaning of the contract. The "plain meaning rule" establishes the meaning of words or phrases that appear to have an ambiguous or unclear meaning. Generally, the words will be assigned their common meaning unless the contracting parties had intended to use them differently. A *patent ambiguity* is an obvious conflict within the provisions of the contract. When a patent ambiguity exists, the court will look to the parties for good faith and fair dealing. Where one of the parties recognizes an ambiguity, a duty to inquire about the ambiguity is imposed on the discovering party. Practical construction of a contract's terms is based on the concept that the intentions of the contracting parties are best demonstrated by their actions during the course of the contract.

Another common rule is to interpret the contract as a whole. A frequent mistake made by contract administrators in contract interpretation is to look too closely at a specific clause to support their position. The court is not likely to approach the contract with the same narrow viewpoint. All provisions of the contract should be read in a manner that promotes harmony among the provisions. Isolation of specific clauses may work in a fashion to render a part of the clause or another clause inoperable. When a provision may lead to more than one reasonable interpretation, the court must have a tie breaker rule. A common tie breaker is for the court to rule against the party that wrote the contract, since they failed to clearly state their intent.

When the primary rules of interpretation are not sufficient to interpret a contract, additional rules can be applied. When language is ambiguous, the additional interpretation guides suggest that technical words be given their technical meaning with the viewpoint of a person in the profession and that all words be given consistent meaning throughout the agreement. The meaning of the word may also be determined from the words associated with it.

In the case of ambiguities occurring because of a physical defect in the structure of the contract document, the court can reconcile the differences looking at the entire contract; interpret the contract so that no provision will be treated as useless; and where a necessary term was omitted inadvertently, supply it to aid in determining the meaning of the contract. Some additional guidance can be gained by providing that specific terms govern over general terms; written words prevail over printed words; and written words are chosen over figures. Generally, where words conflict with the drawings, the words will normally govern. It is possible, in some cases, that the drawings will be interpreted as more specific if they provide more specific information to the solution of the ambiguity.

The standards of interpretation for choosing between meanings are the following:

- A reasonable interpretation is favored over an unreasonable one.
- An equitable interpretation is favored over an inequitable one.
- A liberal interpretation is favored over a strict one.
- An interpretation that promotes the legality of a contract is favored.
- An interpretation that upholds the validity of a contract is favored.
- An interpretation that promotes good faith and fair dealing is favored.
- An interpretation that promotes performance is favored over one that would hinder performance.

5.9 Defective Specifications

Defective specifications are not a subject area of the contract like a differing site condition or notice requirement. However, there is an important area of the law that considers the impact of defective specifications under *implied warranties*. The theory of implied warranty can be used to resolve disputes originating in either the specifications or the plans; the term *defective specification* will refer to both. The contract does contemplate defects in the plans and specifications and requires the contractor to notify the designer when errors, inconsistencies, or omissions are discovered.

Defective specifications occur most frequently when the contractor is provided a *method specification*. A method specification implies that the information or method is sufficient to achieve the desired result. Since many clauses are mixtures, it is imperative to identify what caused the failure. For example, was the failure caused by a poor concrete specification or poor workmanship? Another consideration in isolating the cause of the failure is to identify who had control over the aspect of performance that failed. When the contractor has a performance specification, the contractor controls all aspects of the work. If a method specification was used, it must be determined that the contractor satisfactorily followed the specifications and did not deviate from the work. If the specification is shown to be commercially impractical, the contractor may not be able to recover if it can be shown that the contractor assumed the risk of impossibility. Defective specifications are a complex area of the law, and competent legal advice is needed to evaluate all of the possibilities.

5.10 Misrepresentation

Misrepresentation is often used in subsurface or differing site condition claims where the contract does not have a differing site conditions clause. In the absence of a differing site conditions clause, the owner assigns the risk for unknown subsurface conditions to the contractor [Jervis and Levin, 1988]. To prove misrepresentation the contractor must demonstrate that he or she was justified in relying on the information, the conditions were materially different from conditions indicated in the contract documents, the owner erroneously concealed information that was material to the contractor's performance, and the contractor had an increase in cost due to the conditions encountered. More commonly, a differing site condition clause is included in the contract.

5.11 Differing Site Conditions

One of the more common areas of dispute involves differing site conditions. However, it is also an area in which many disputes escalate due to misunderstandings of the roles of the soil report, disclaimers, and site visit requirements. The differing site condition clause theoretically reduces the cost of construction, because the contractors do not have to include contingency funds to cover the cost of hidden or latent subsurface conditions [Stokes and Finuf, 1986]. The federal differing site conditions (DSC) clause, or a slightly modified version, is used in most construction contracts. The clause is divided into two parts, commonly called Type I and Type II conditions. A *Type I* condition allows additional cost recovery if the conditions differ materially from those indicated in the contract documents. A *Type II* condition allows the contractor additional cost recovery if the actual conditions differ from what could have been reasonably expected for the work contemplated in the contract. Courts have ruled that when the wording is similar to the federal clause, federal precedent will be used to decide the dispute. More detailed discussions of the clause can be found elsewhere [Parvin and Araps, 1982; Currie *et al.*, 1971].

Type I Conditions

A Type I condition occurs when site conditions differ materially from those indicated in the contract documents. With a DSC clause, the standard of proof is an indication or suggestion that may be established through association and inference. Contract indications are normally found in the plans and specifications and may be found in borings, profiles, design details, contract clauses, and sometimes in the soil report. Information about borings, included in the contract documents, is a particularly valuable source, since they are commonly held to be the most reliable reflection of the subsurface conditions. While the role of the soil report is not consistent, the courts are often willing to go beyond the contract document boundaries to examine the soil report when a DSC clause is present. This situation arises when the soil report is referred to in the contract documents but not made a part of the contract documents. Groundwater is a common problem condition in DSC disputes, particularly where the water table is not indicated in the drawings. Failure to indicate the groundwater level has been interpreted as an indication that the water table exists below the level of the borings or that it is low enough not to affect the anticipated site activities.

The contractor must demonstrate, in a DSC dispute, that he or she was misled by the information. To show that he or she was misled, the contractor must show where his or her bid incorporated the incorrect information and how the bid would have been different if the information had been correct. These proofs are not difficult for the contractor to demonstrate. However, the contractor must also reasonably interpret the contract indications. The contractor's reliance on the information may be reduced by other contract language, site visit data, other data known to the contractor, and previous experience of the contractor in the area. If these reduce the contractor's reliance on the indications, the contractor will experience more difficulty in proving the interpretation.

Owners seek to reduce their exposure to unforeseen conditions by disclaiming responsibility for the accuracy of the soil report and related information. Generally, this type of disclaimer will not be effective. The disclaimers are often too general and nonspecific to be effective in overriding the DSC clause—particularly when the DSC clause serves to reduce the contractor's bid.

Type II Conditions

A Type II DSC occurs when the physical conditions at the site are of an unusual nature, differing materially from those ordinarily encountered and generally recognized as inherent in work. The conditions need not be bizarre, but simply unknown and unusual for the work contemplated. A Type II condition would be beyond the conditions anticipated or contemplated by either the owner or the contractor. As in the Type I DSC, the contractor must show that he or she was reasonably misled by the information provided. The timing of the DSC may also be evaluated in Type II conditions. The contractor must establish that the DSC was discovered after contract award.

5.12 Claim Preparation

Claim preparation involves the sequential arrangement of project information and data to the extent that the issues and costs of the dispute are defined. There are many methods to approach development and cost of a claim, but all require a methodical organization of the project documents and analysis. Assuming that it has been determined that there is entitlement to a recovery, as determined by consideration of interpretation guidelines, the feasibility of recovery should be determined. Once these determinations are complete, claims are generally prepared by either a total-cost approach or an actual-cost approach.

An actual-cost approach, also called a *discrete approach*, will allocate costs to specific instances of modifications, delays, revisions, and additions where the contractor can demonstrate a cost increase. Actual costs are considered to be the most reliable method for evaluating a claim. Permissible costs are direct labor, payroll burden costs, materials, equipment, bond and insurance premiums, and subcontractor costs. Indirect costs that are recoverable include labor inefficiency, interest and financing costs, and profit. Impact costs include time impact costs, field overhead costs, home office overheads, and wage and material escalation costs. Pricing the claim requires identification and pricing of recoverable costs. The recoverable costs depend primarily on the type of claim and the specific causes of unanticipated expenses. Increased labor costs and losses of productivity can occur under a wide variety of circumstances. Increased costs for bonding and insurance may be included when the project has been delayed in completion or the scope has changed. Material price escalation may occur in some circumstances. In addition, increased storage costs or delivery costs can be associated with many of the common disputes. Equipment pricing can be complicated if a common schedule of values cannot be determined.

Total cost is often used when the cost overrun is large, but no specific item or areas can be identified as independently responsible for the increase. Stacked changes and delays often leave a contractor in a position of being unable to relate specific costs fully to a particular cause. The total-cost approach is not a preferred approach for demonstrating costs. A contractor must demonstrate that the bid and actual costs incurred were reasonable; costs increased because of actions by the defendant; and the nature of the losses make it impossible or highly impractical to determine the costs accurately. Good project information management will improve the likelihood that the contractor can submit an actual-cost claim rather than a total-cost claim. However, due to the complexity of some projects, the total-cost approach may be the most appropriate method.

5.13 Dispute Resolution

Alternate dispute resolution (ADR) techniques have slowly gained in popularity. High cost, lost time, marred relationships, and work disruptions characterize the traditional litigation process.

However, many disputes follow the litigation route as the main recourse if a significant portion of the claim involves legal issues. The alternatives—dispute review boards, **arbitration**, **mediation**, and mini trials—are usually established in the contract development phase of the project.

The traditional litigation process is the primary solution mechanism for many construction claims. This is particularly important if the dispute involves precedent-setting issues and is not strictly a factual dispute. The large expense of trial solutions often is associated with the cost of recreating the events on the project that created the original dispute. Proof is sought from a myriad of documents and records kept by contractors, engineers, subcontractors, and suppliers in some cases. Once filing requirements have been met, a pretrial hearing is set to clarify the issues of the case and to establish facts agreeable to the parties.

The discovery phase of litigation is the time-consuming data-gathering phase. Requests for and exchange of documents, depositions, and interrogatories are completed during this time period. Evidence is typically presented in a chronological fashion with varying levels of detail, depending on the item's importance to the case. The witnesses are examined and cross-examined by the lawyers conducting the trial portion of the claim. Once all testimony has been presented, each side is permitted to make a summary statement. The trier of the case, a judge or jury, deliberates on the evidence and testimony and prepares the decision. Appeals may result if either party feels there is an error in the decision. Construction projects present difficult cases, since they involve both technological issues and terminology issues for the lay jury or judge. The actual trial time may last less than a week after several years of preparation. Due to the high cost of this procedure, the alternative dispute resolution methods have continued to gain in popularity.

Dispute review boards have gained an excellent reputation for resolving complex disputes without litigation. Review boards are a real-time, project-devoted dispute resolution system. The board, usually consisting of three members, is expected to stay up to date with the project progress. This alone relieves the time and expense of the traditional document requests and timeline reconstruction process of traditional discovery and analysis. The owner and contractor each appoint one member of the dispute review board. The two appointees select the third member, who typically acts as the chairman. The cost of the board is shared equally. Typically the board members are highly recognized experts in the type of work covered by the contract or design. The experience of the board members is valuable because they quickly grasp the scope of a dispute and can provide their opinion on liability. Damage estimates are usually left to the parties to work out together. However, the board may make recommendations on settlement figures as well. Board recommendations are not binding but are admissible as evidence in further litigation.

Arbitration hearings are held before a single arbitrator or, more commonly, before an arbitration panel. A panel of three arbitrators is commonly used for more complex cases. Arbitration hearings are usually held in a private setting over a period of one or two days. Lengthy arbitrations meet at convenient intervals when the arbitrators' schedules permit the parties to meet; this often delays the overall schedule of an arbitration. Information is usually presented to the arbitration panel by lawyers, although this is not always the case. Evidence is usually submitted under the same administrative rules the courts use. Unless established in the contract or by a separate agreement, most arbitration decisions are binding. An arbitrator, however, has no power to enforce the award. The advantages of arbitration are that the hearings are private; small claims can be cost-effectively heard; knowledge of the arbitrator assists in resolution; the proceedings are flexible; and results are quickly obtained.

Mediation is essentially a third party–assisted negotiation. The neutral third party meets separately with the disputing parties to hear their arguments and meets jointly with the parties to point out areas of agreement where no dispute exists. A mediator may point out weaknesses and unfounded issues that the parties have not clarified or that may be dropped from the discussion. The mediator does not participate in settlements but acts to keep the negotiations progressing to settlement.

Mediators, like all good negotiators, recognize resistance points of the parties. A primary role of a mediator is to determine whether there is an area of commonality where agreement may be reached. The mediator does not design the agreement. Confidentiality of the mediator's discussions with the parties is an important part of the process. If the parties do agree on a settlement, they sign an agreement contract. The mediator does not maintain records of the process or provide a report to the parties on the process.

A major concern that can be expressed about the ADR system is that it promotes a private legal system specifically for business, where few if any records of decisions are maintained, yet decisions may affect people beyond those involved in the dispute. ADR may also be viewed as a cure-all. Each form is appropriate for certain forms of disputes. However, when the basic issues are legal interpretations, perhaps the traditional litigation process will best match the needs of both sides.

5.14 Summary

Contract documents are the framework of the working relationship of all parties to a project. The contracts detail technical as well as business relationships. Claims evolve when either the relationship or the technical portion of the contract fails. While it is desirable to negotiate settlement, often disputes cannot be settled and a formal resolution is necessary. If the contracting managers had a better understanding of the issues considered by the law in contract interpretation, perhaps there would be less of a need to litigate.

Defining Terms

Arbitration: The settlement of a dispute by a person or persons chosen to hear both sides and come to a decision.

Bilateral: Involving two sides, halves, factions; affecting both sides equally.

Consideration: Something of value given or done in exchange for something of value given or done by another, in order to make a binding contract; inducement for a contract.

Contract: An agreement between two or more people to do something, esp. one formally set forth in writing and enforceable by law.

Equity: Resort to general principles of fairness and justice whenever existing law is inadequate; a system of rules and doctrines, as in the U.S., supplementing common and statute law and superseding such law when it proves inadequate for just settlement.

Mediation: The process of intervention, usually by consent or invitation, for settling differences between persons, companies, etc.

Parol: Spoken evidence given in court by a witness.

Surety: A person who takes responsibility for another; one who accepts liability for another's debts, defaults, or obligations.

References

Contracts Task Force. 1986. *Impact of Various Construction Contract Types and Clauses on Project Performance(5-1),* The Construction Industry Institute, Austin, TX.

CICE (Construction Industry Cost Effectiveness Project Report). 1982. *Contractual Arrangements, Report A-7.* The Business Round Table, New York.

Currie, O. A., Ansley, R. B., Smith, K. P., and Abernathy, T. E. 1971. Differing Site (Changed) Conditions. *Briefing Papers* No. 71-5. Federal Publications, Washington, D.C.

Jervis, B.M. and Levin, P. 1988. *Construction Law Principles and Practice.* McGraw-Hill, New York.

Parvin, C. M. and Araps, F. T. Highway construction claims—A comparison of rights, remedies, and procedures in New Jersey, New York, Pennsylvania, and the southeastern states. *Public Contract Law.* 12(2).

Richter, I. and Mitchell, R. 1982. *Handbook of Construction Law and Claims.* Reston Publishing, Reston, VA.

Stokes, M. and Finuf, J. L. 1986. *Construction Law for Owners and Builders.* McGraw-Hill, New York.

Sweet, J. 1989. *Legal Aspects of the Architecture, Engineering, and Construction Process,* 4th ed. West Publishing, St. Paul, MN.

Thomas, R. and Smith, G. 1993. *Construction Contract Interpretation.* The Pennsylvania State University, Department of Civil Engineering.

Trauner, T. J. 1993. *Managing the Construction Project.* John Wiley & Sons, New York.

For Further Information

A good practical guide to construction management is *Managing the Construction Project* by Theodore J. Trauner, Jr. The author provides good practical advice on management techniques that can avoid the many pitfalls that are found in major projects.

A comprehensive treatment of the law can be found in *Legal Aspects of Architecture, Engineering, and the Construction Process* by Justin Sweet. This book is one of the most comprehensive treatments of construction law that has been written.

The *Handbook of Modern Construction Law* by Jeremiah D. Lambert and Lawrence White is another comprehensive view of the process, but more focused on the contractor's contract problems.

6

Construction Automation

Jeffrey S. Russell
University of Wisconsin–Madison

Raghavan Kunigahalli
University of Wisconsin–Madison

6.1 Introduction

The construction industry plays a vital role in a nation's economy. In the U.S., for example, this industry annually contributes over $400 billion and accounts for 55 to 65% of the country's total investment in gross fixed capital formation. In addition, the U.S. construction industry employed over 6.7 million workers in 1991, making it the economic sector employing the largest number of manual workers [Perini, 1991; Wright, 1991].

Decline in construction productivity has been reported by many studies conducted throughout the world. In the U.S., construction productivity, defined as gross product originating per person-hour in the construction industry, has shown an average annual net decrease of nearly 1.7% since 1969. The average of all industries for the same period has been a net annual increase of 0.9%, while the manufacturing industry has posted an increase of 1.7% [Groover *et al.*, 1989]. This decline in construction productivity is a matter of global concern because of its impact on the economy's health. Recent trends have made availability of capital and innovation through the application of automated technologies the defining parameters of competitiveness in today's global economy. These trends enable projects to be constructed with improved quality, shorter construction schedules, increased site safety, and lower construction costs.

Much of the increase in productivity in the manufacturing industry can be attributed to the development and application of automated manufacturing technologies. This, combined with concern about declining construction productivity, has motivated many industry professionals and researchers to investigate the application of automation technology to construction. As a

0-8493-8953-4/95/$0.00 + $.50

practical matter, these efforts have recognized that complete automation of construction works is not presently technically and economically feasible. Because of frequently reconfigured operations often under severe environmental conditions, the construction industry has been slower than the manufacturing industry to adopt automation technology [Paulson, 1985].

Use of automation in construction faces larger challenges compared with other industries, such as manufacturing. The physical nature of any construction project is a primary obstacle to meaningful work automation. In batch manufacturing, the work object is mobile though the production facility and work tools can be stationary. In construction, the "work object" is stationary, of large dimensions, and constantly changing as work progresses, while tools are mobile, whether hand-held or mechanized. In addition, construction processes are usually performed in dusty and noisy environments, preventing the use of fragile, high-precision, and sensitive electronic devices.

Construction has traditionally been very resistant to technical innovation. Past efforts to industrialize construction in the U.S. were undertaken at the time when industrial automation technology was at its infancy. Additionally, engineering and economic analyses of prefabrication processes and systems were lacking.

On the other hand, numerous construction tasks have or will become more attracted to automated technologies based upon the following characteristics: (1) repetitiveness, (2) tedious and boring, (3) hazardous to health, (4) physically dangerous, (5) unpleasant and dirty, (6) labor intensiveness, (7) vanishing skill area, (8) high skill requirement, (9) precision dexterity requirement, and (10) critical to productivity [Kangari and Halpin, 1989]. For example, some construction tasks have been historically noted for their arduous, repetitive nature, with relatively little dynamic decision making required on the part of a human laborer. Such tasks may include placing of concrete (see Fig. 6.1), placing of drywall screws, finishing of concrete (see Fig. 6.2), and placing of masonry block, among others. The work involved in these tasks is rather unattractive for humans. Robots, however, are applicable to these types of work tasks provided that the technology and economics are feasible.

In some cases, the work site itself poses a significant health hazard to humans involved. Hazards are associated with work in undersea areas, underground, at high elevations, on chemically or radioactively contaminated sites (see Fig. 6.3), and in regions with prevailing harsh temperatures. In the construction industry, work-related illness and injuries, including fatalities, occur at a rate 54% higher than the aggregate rate for all other industries. Accidents involving persons

FIGURE 6.1 Concrete placement. (Photo courtesy of Oscar J. Boldt Construction Co.)

FIGURE 6.2 Concrete finishing. (Photo courtesy of Oscar J. Boldt Construction Co.)

on the job site are estimated to account for 6.5% (both direct and indirect cost to a contractor) of the total contract cost [Business Roundtable, 1982]. Consequently, liability insurance for most types of construction work is very costly. Replacing humans with robots for dangerous construction tasks can contribute to the reduction of these costs.

Many industrial nations, including Japan, France, Germany, and to some extent the U.S., suffer from a shortage of skilled construction labor. Representatives of the National Concrete Masonry

FIGURE 6.3 Hazardous waste site. (Photo courtesy of RMT, Inc.)

Association have indicated that there will be a need for an additional 130,000 masons in the U.S. over the next 10 years [130,000 More, 1988]. This trend of worker shortages in many traditional construction trades will most likely continue into the future. This will result, as it has over the past two decades, in an increase in the real cost of construction labor. These facts, together with the rapid advancement in automation and robotics technology, indicate a promising potential for gradual automation and robotization of construction work.

Construction automation refers to the use of a mechanical/electrical/computer-based system to operate and control construction equipment/devices. There are two types of construction automation:

1. Fixed construction automation
2. Programmable construction automation

Fixed construction automation involves a sequence of operations performed by equipment fixed in their locations. In other words, an automated facility, whether it is permanently indoors or temporarily on the construction site, is set up specifically to perform only one function or produce one product. In programmable construction automation, equipment has the ability to change its sequence of operations easily to accommodate a wide variety of products.

6.2 Fixed Construction Automation

Fixed construction automation is useful in mass production or prefabrication of building components such as:

1. Reinforcing steel
2. Structural steel
3. Exterior building components (e.g., masonry, granite stone, precast concrete)

Examples of Fixed Construction Automation

In this section, selected examples of fixed construction automation are highlighted.

Automated Rebar Prefabrication System

The automated rebar prefabrication system places reinforcing bars for concrete slab construction. The system consists of a NEC PC98000XL high-resolution-mode personal computer that uses AutoCAD™, DBASE III Plus™, and BASIC™ software. The information regarding number, spacing, grade and dimension, and bending shapes of rebars is found from the database generated from an AutoCAD file. This information is used by an automatic assembly system to fabricate the rebar units.

The assembly system consists of two vehicles and a steel rebar arrangement support base. Of the two vehicles, one moves in the longitudinal direction and the other in the transverse direction. The longitudinally moving vehicle carries the rebars forward until it reaches the preset position. Then it moves backward and places the rebars one by one at preset intervals on the support base. Upon completion of placement of the rebars by the longitudinally moving vehicle, the transversely moving vehicle places the rebars in a similar manner. The mesh unit formed by such a placement of rebars is tied together automatically [Miyatake and Kangari, 1993].

Automated Brick Masonry

The automated brick masonry system, shown in Fig. 6.4, is designed to spread mortar and place bricks for masonry wall construction. The system consists of:

1. Mortar-spreading module
2. Brick-laying station

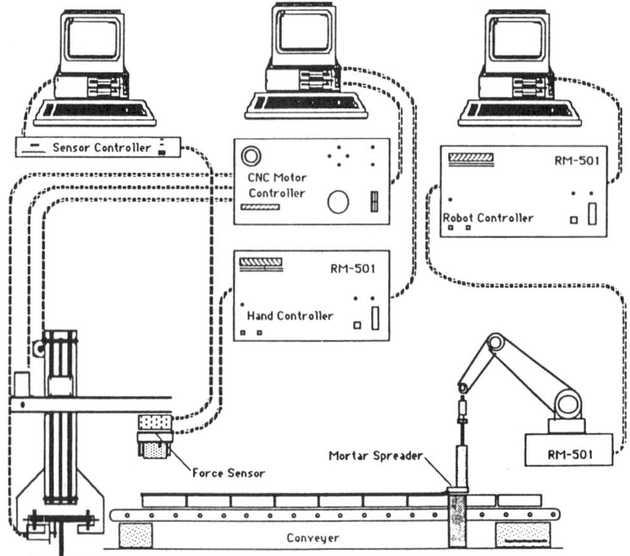

FIGURE 6.4 Automated brick masonry. (*Source:* Bernold *et al.* 1992. Computer-controlled brick masonry. *ASCE Journal of Computing in Civil Engineering.* 6(2):147–161. Reproduced by permission of ASCE.)

The controls of the system are centered around three personal computers responsible for:

1. Collecting and storing data in real time
2. Interfacing a stepping motor controller and a robot controller
3. Controlling the mortar-spreading robot

A Lord 15/50 force-torque sensor is used to determine the placing force of each brick. The system is provided with an integrated control structure that includes a conveyor for handling the masonry bricks [Bernold *et al.*, 1992].

Automated Stone Cutting

The purpose of the automated stone-cutting facility is to precut stone elements for exterior wall facings. The facility consists of the following subsystems:

1. Raw material storage
2. Loading
3. Primary workstation
4. Detail workstations
5. Inspection station
6. End-product inventory

A special lifting device has been provided for automated material handling. The boom's rigidity enables the computation of exact location and orientation of the hook. Designs for the pallets, the primary saw table, the vacuum lift assembly, and the detail workstation have also been proposed [Bernold *et al.*, 1992].

6.3 Programmable Construction Automation

Programmable construction automation includes the application of the construction robots and numerical control machines described below.

Construction Robots

The International Standards Organization (ISO) defines a robot as "an automatically controlled, re-programmable, multi-purpose, manipulative machine with several re-programmable axes, which may be either fixed in place or mobile for use in industrial automation applications" [Rehg, 1992]. For construction applications, robots have been categorized into three types [Hendrickson and Au, 1988]:

1. Tele-operated robots in hazardous or inaccessible environments
2. Programmed robots as commonly seen in industrial applications
3. Cognitive or intelligent robots that can sense, model the world, plan, and act to achieve working goals

The important attributes of robots from a construction point of view are their (1) **manipulators,** (2) **end effectors,** (3) **electronic controls,** (4) **sensors,** and (5) **motion systems** [Warszawski, 1990]. For further explanation of these attributes, refer to the definitions section at the end of this chapter.

Applications of Construction Robots

Table 6.1 presents a partial list of construction robot prototypes developed in the U.S. and in other countries. Brief summaries of several of these prototypes are provided below. Several of these descriptions have been adapted from Skibniewski and Russell [1989].

John Deere 690C Excavator. The John Deere 690C excavator is a tele-operated machine; that is, it is fully controlled by a human operating from a remote site. It is equipped with a model 60466T, six-cylinder, four-stroke turbocharged diesel engine, producing a maximum net torque of 450 ft-lb (62.2 kgf-m) at 1300 revolutions per minute (rpm) [*Technical Specifications*, 1985]. The engine propels the excavator at traveling speeds ranging from 0 to 9.8 mph (15.8 km/h).

The arm on the 690C excavator has a lifting capacity of 11,560 lb (5243 kgf) over side and 10,700 lb (4853 kgf) over end. The rated arm force is 15,900 lb (7211 kgf), and the bucket digging force is 25,230 lb (11,442 kgf) [*Technical Specifications*, 1985].

The John Deere 690C excavator has been implemented in a cooperative development program with the U.S. Air Force within the Rapid Runway Repair (RRR) project. The major task of the RRR is the repair of runways damaged during bombing raids. The Air Force is currently investigating other areas in which the 690C could be implemented, including heavy construction work, combat earthmoving in forward areas, mine field clearing, and hazardous material handling.

Robot Excavator (REX). The primary task of the robot excavator (REX) is to remove pipelines in areas where explosive gases may be present. This robot, shown in Fig. 6.5, is an autonomous machine able to sense and adjust to its environment. REX achieves its autonomous functions by incorporating three elements into its programming [Whittaker, 1985a]:

1. Subsurface premapping of pipes, structures, and other objects using available utility records and ground penetrating sensors. Magnetic sensing is the leading candidate for premapping metallic pipes.
2. Primary excavation for gross access near target pipes. Trenching and augering are the leading candidates for this operation.
3. Secondary excavation, the fine and benign digging that progresses from the primary excavation to clear piping. Secondary excavation is accomplished with the use of a supersonic air jet.

Table 6.1 Example Construction Robotic Prototypes

System Description	Application	Research Center
	Excavation	
John Deere 690C	Tele-operated excavation machine	John Deere, Inc., Moline, IL
Robot excavator (REX)	Autonomous excavation, sandblasting, spray washing, and wall finishing	The Robotics Institute, Carnegie-Mellon Univ., Pittsburgh, PA
Super hydrofraise excavation control system	Excavate earth	Obayashi Co., Japan
Haz-Track	Remotely controlled excavation	Kraft Telerobotics
Hitachi RX2000	Pile driving	Hitachi Construction Machinery Co., Japan
Remote core sampler (RCS)	Concrete core sampling for radiated settings	The Robotics Institute, Carnegie-Mellon Univ., Pittsburgh, PA
Laser-aided grading system	Automatic grading control for earthwork	Gradeway Const. Co. and Agtex Dev. Co., San Francisco, CA; Spectra-Physics, Dayton, OH
	Tunneling	
Shield machine control system	Collect and analyze data for controlling tunneling machine	Obayashi Co. and Kajima Co., Japan
Microtunneling machine	Tele-operated microtunneling	American Augers, Wooster, OH
Tunnel wall lining robot	Assemble wall liner segments in tunnels for sewer systems and power cables	Ishikawajima-Harima Heavy Industries, Tokyo Electric Power Co., Kajima Co., Japan
	Concrete	
Automatic concrete distribution system	Carry concrete from batching plant to the cable crane	Obayashi Co., Japan
Automatic slipform machines	Placement of concrete sidewalks, curbs, and gutters	Miller Formless Systems Co., McHenry, IL; Gomaco, Ida Grove, IA
Concrete placing robot for slurry walls	Place and withdraw tremie pipes and sense upper level of concrete as it is poured	Obayashi Co., Japan
Horizontal concrete distributor (HCD)	Place concrete for horizontal slabs	Takenaka Komuten Co., Japan
Shotcrete robot	Spray concrete tunnel liner	Kajima Co, and Obayashi Co., Japan
Automatic concrete vibrator tamper	Vibrate cast-in-place concrete	Obayashi Co., Japan
Automatic laser beam-guided floor robot	Finish surface of cast-in-place concrete	Obayashi Co., Japan
Slab-finishing robot	Finish surface of cast-in-place concrete	Kajima Co., Japan
	Structural Members	
Auto-claw, auto-clamp	Erect structural steel beams and columns	Obayashi Co., Japan
Structural element placement	Place reinforcing steel	Kajima Co., Japan
Structural element welding	Weld large structural blocks for cranes and bridges	Mitsubishi Heavy Industries Co., Japan

Table 6.1 (continued) Example Construction Robotic Prototypes

System Description	Application	Research Center
	Structural Members	
Shear stud welder	Weld shear connectors in composite steel/concrete construction	Massachusetts Institute of Technology, Cambridge
Automatic carbon fiber wrapper	Wrap existing structures with carbon steel	Obayashi Co., Japan
Automated pipe construction	Pipe bending, pipe manipulation, and pipe welding	University of Texas, Austin
Blockbots	Construction of concrete masonry walls	Massachusetts Institute of Technology, Cambridge
Wallbots	Construction of interior partitions, metal track studs	Massachusetts Institute of Technology, Cambridge
Interior finishing	Building walls and partitions, plastering, painting, and tiling walls and ceilings	Israel Institute of Technology, National Building Research Institute
	Non-Concrete Spraying	
Fireproof spraying robot	Fireproof structural steel	Shimizu Co., Japan
Paint spraying robot	Paint balcony rails in high-rise buildings	Shimizu Co., Japan
	Inspection	
Wall inspection robot (KABEDOHDA)	Inspect reinforced concrete walls	Obayashi Co., Japan
Wall inspection robot	Inspect facade	Kajima Co., Shimizu Construction Co., and Taisei Co., Japan
Bridge inspection robot	Inspect structural surface of a bridge	University of Wales
GEO robot	Finish facade/surface	Eureka, France
	Other	
Clean room inspection and monitoring robot (CRIMRO)	Inspect and monitor the amount of particles in the air	Obayashi Co., Japan
Integrated surface patcher (ISP) material handling	Hot resurfacing on highways, pick and distribute construction materials (e.g., prefabricated concrete materials and pipe)	Secmar Co., France; Tokyo Construction Co. and Hitachi Construction Co., Japan
Autonomous pipe mapping	Mapping subsurface pipes	The Robotics Institute, Carnegie-Mellon Univ., Pittsburgh, PA
Terregator	Autonomous navigation	The Robotics Insitute, Carnegie-Mellon Univ., Pittsburgh, PA
Remote work vehicle (RWV)	Nuclear accident recovery work, wash contaminated surfaces, remove sediments, demolish radiation sources, apply surface treatment, package and transport materials	The Robotics Institute, Carnegie-Mellon Univ., Pittsburgh, PA
ODEX III	Inspection, surveillance, material transport	Odetics, Inc., French Commissariat a l'Energie Atomique

FIGURE 6.5 Robotic excavator. (Photo courtesy of The Robotics Institute, Carnegie-Mellon University.)

The hardware that REX uses for primary excavation is a conventional backhoe retrofitted with servo valves and joint resolvers that allow the computer to calculate arm positions within a three-dimensional space. The manipulator arm can lift a 300 lb (136 kg) payload at full extension and over 1000 lb (454 kg) in its optimal lifting position.

REX uses two primary sensor modes: tactile and acoustic. The tactile sensor is an instrumented compliant nozzle. The instrumentation on the nozzle is an embedded tape switch that is activated when the nozzle is bent. The second sensor employed in excavation is an acoustical sensor, allowing for three-dimensional imaging.

Haz-Trak. Haz-Trak, developed by Kraft Telerobotics, is a remotely controlled excavator that can be fitted with a bulldozer blade for grading, backfilling, and leveling operations [Jaselskis and Anderson, 1994]. Haz-Trak uses force feedback technology allowing the operator to actually feel objects held by the robot's manipulator. The operator controls the robot's arm, wrist, and grip movements through devices attached to his or her own arm. Thus, the robot arm instantly follows the operator's movements.

Pile-Driving Robot. The Hitachi RX2000, shown in Fig. 6.6, is a pile-driving machine directed by a computer-assisted guiding system. It consists of a piling attachment (such as an earth auger or a vibratory hammer) directly connected to the tip of a multijointed pile driver arm. The pile driver arm uses a computer-assisted guiding system called an "arm tip locus control." Coordinates of arm positions are calculated using feedback from angle sensors positioned at joints along the arm. A control lever operation system is provided to increase efficiency. The compactness of the RX2000 and its leaderless front attachment enable efficient piling work even in congested locations with little ground stabilization. Further, the vibratory hammer has a center hole chuck that firmly chucks the middle part of a sheet pile or an H-steel pile. Hence, pile length is not limited by the base machine's dump height [Uchino *et al.*, 1993].

Laser-Aided Grading System. Spectra-Physics of Dayton, Ohio, has developed a microcomputer-controlled, laser-guided soil-grading machine (see Fig. 6.7). A laser transmitter creates a plane of light over the job site. Laser light receptors mounted on the equipment measure the height of the blade relative to the laser plane. Data from the receiver are then sent to the microcomputer

FIGURE 6.6 Pile-driving robot. (*Source:* Uchino *et al.* 1993. Multi-jointed pile driving machine with a computer-assisted guiding system. *Proceedings, Tenth International Symposium on Automation and Robotics for Construction*, Houston TX, pp. 363–370.)

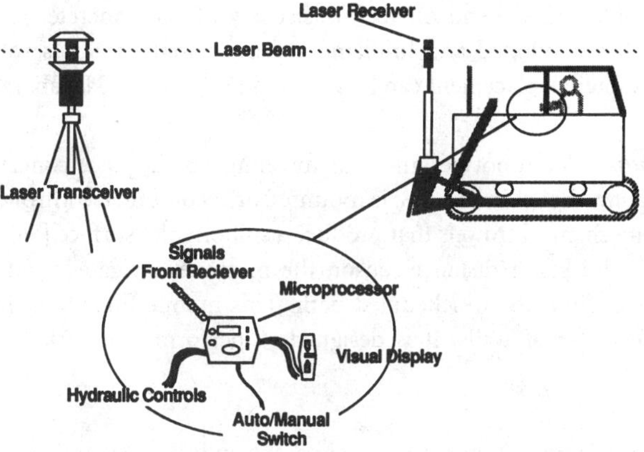

FIGURE 6.7 Laser-aided grading system. (*Source:* Tatum, C. B., and Funke, A. T. 1988. Partially automated grading: Construction process innovation. *ASCE Journal of Construction Engineering and Management.* 114(1):19–35. Reproduced by permission of ASCE.)

that controls the height of the blade through electronically activated valves installed in the machine's hydraulic system. A similar device has been developed by Agtek Company in cooperation with a construction contractor in California [Paulson, 1985]. An automated soil-grading process implemented by these machines relieves the operator from having to manually position and control the grading blades, thus increasing the speed and quality of grading, as well as work productivity [Tatum and Funke, 1988].

Automatic Slipform Machines. Miller Formless Systems Company has developed four automatic slipform machines—M1000, M7500, M8100, and M9000—for sidewalk and curb and gutter construction [*Technical Specifications*, 1988]. All machines are able to pour concrete closer to obstacles than is possible with alternative forming techniques. They can be custom-assembled for the construction of bridge parapet walls, monolithic sidewalk, curb and gutter, barrier walls, and other continuously formed elements commonly used in road construction.

The M1000 machine is suitable for midrange jobs, such as the forming of standard curb and gutter, sidewalks to four feet, and cul-de-sacs. The M7500 is a sidemount-design machine for pouring barrier walls, paved ditches, bridge parapets, bifurcated walls, and other types of light forming jobs. The M8100 is a midsize system with a sidemount design combined with straddle-paving capabilities. The machine can be extended to 16-ft (4.88-m) slab widths with added bolt-on expansion sections. The M9000 multidirectional paver is designed for larger-volume construction projects. It can perform an 18-ft (5.49-m) wide paving in a straddle position. Options are available for wider pours, plus a variety of jobs from curbs to irrigation ditches in its sidemount mode.

Horizontal Concrete Distributor. The HCD, developed by Takenaka Company, is a hydraulically driven, three-boom telescopic arm that cantilevers from a steel column. The boom can extend 66 ft (20 m) in all directions over an 11,000-ft (1000-m) surface area. A cockpit located at the end of the distributor houses the controls for an operator to manipulate the boom direction and flow of concrete. The weight of the robot is 4.97 tons (4508 kgf), and it can be raised along the column by jacks for the next concrete pour. On average, the relocation procedure takes only 1.5 hours [Sherman, 1988].

Shotcrete Robot. Traditionally, in tunneling work, a skilled operator has been needed to regulate the amount of concrete to be sprayed on a tunnel surface and the quality of the hardening agent to be added, both of which depend on the consistency of the concrete. Kajima Construction Company of Japan has developed and implemented a semi-autonomous robotic applicator by which high-quality shotcrete placement can be achieved [Sagawa and Nakahara, 1985].

Slab-Finishing Robot. The robot designed for finishing cast-in-place concrete slabs by Kajima Construction Company, shown in Fig. 6.8, is mounted on a computer-controlled mobile platform and equipped with mechanical trowels that produce a smooth, flat surface [Saito, 1985]. By means of a gyrocompass and a linear distance sensor, the machine navigates itself and automatically corrects any deviation from its prescheduled path. This mobile floor-finishing robot is able to work to within one meter of walls. It is designed to perform the work of at least six skilled workers.

Auto-Claw and Auto-Clamp. Two robotic devices, shown in Fig. 6.9 and used for steel beam and column erection on construction sites, have been developed by Obayashi Construction Company of Japan. Both construction robots have been developed to speed up erection time and to minimize the risks incurred by steel workers. Both have been implemented on real job sites.

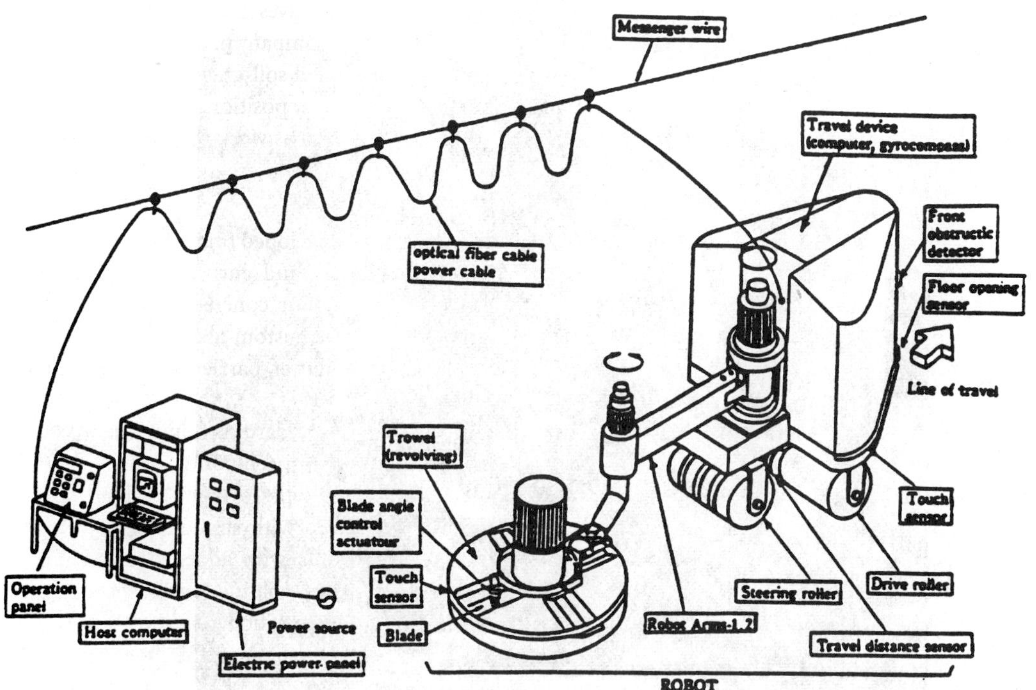

FIGURE 6.8 Slab-finishing robot. (*Source:* Skibniewski, M. J. 1988. *Robotics in Civil Engineering.* Van Nostrand Reinhold, New York.)

The auto-claw consists of two steel clamps extended from a steel-encased unit containing a DC battery pack, electrical panel, and microprocessor unit, which is in turn suspended from a standard crane. The two clamps have a rated capacity of two tons (1.824 kgf) each and can be adjusted to fit beam flanges from 8 to 12 inches (203.2 to 304.8 mm). The clamps are automatically released by remote radio control once the beam is securely in place. Fail-safe electronic circuitry prevents the accidental release of the clamps during erection by keeping the circuit broken at such times. The steel beams require no special preparation for using this robot.

The auto-clamp's essential purpose and mechanics are the same as for the auto-claw except that the auto-clamp uses a special electro-steel cylinder tube to secure and erect columns. A steel appendage plate with a hole in the center must be welded to one end of the column. The steel cylinder is electrically inserted and locked into the hole by remote control, whereupon the column can be erected. The auto-clamp has a rated lifting capacity of 15 tons (13,605 kg). The appendage plates must be removed after the columns are erected. Like the auto-claw, the auto-clamp is equipped with a fail-safe system preventing the cylinder from retracting from the hole during erection [Sherman, 1988].

Automated Pipe Construction. Research into automated pipe construction is under way at the University of Texas at Austin [O'Connor *et al.*, 1987]. Research efforts are focused on developing and integrating three pipe production technologies: bending, manipulation, and welding. The pipe manipulator, shown in Fig. 6.10, was adapted from a 20-ton rough-terrain hydraulic crane with an attachment to the main boom [Hughes *et al.*, 1989]. The attachment includes an elevating, telescoping, auxiliary boom with a wrist and pipe-gripping jaws. Associated research has concentrated on improving productivity through automated lifting and manipulating of horizontal piping [Fisher and O'Connor, 1991].

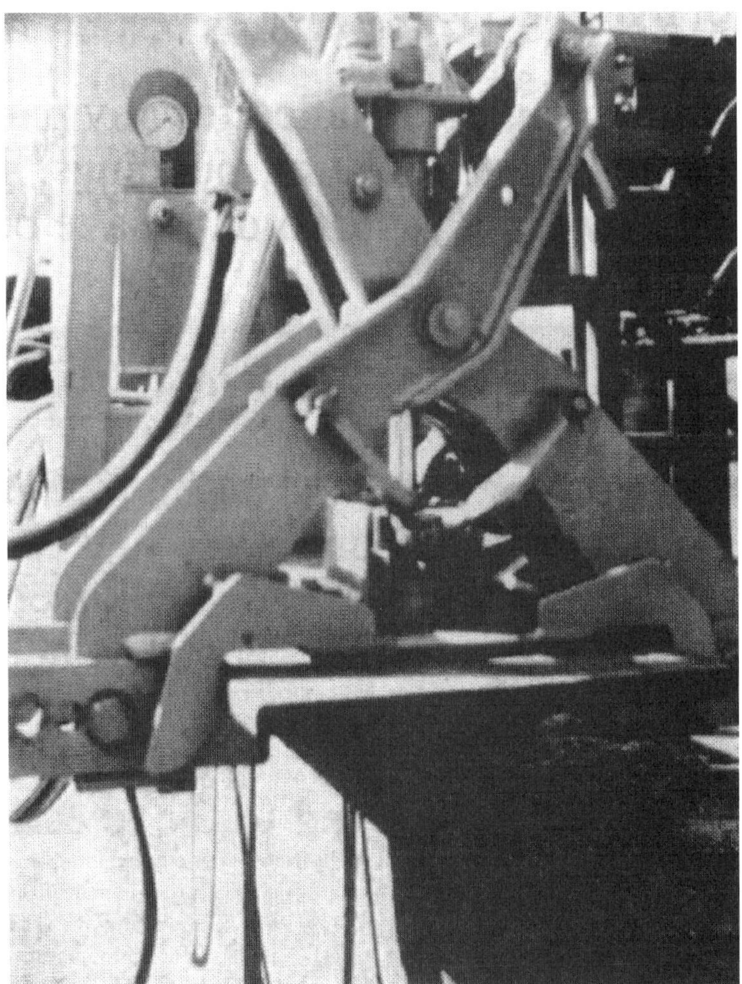

FIGURE 6.9 Auto-claw and auto-clamp. (*Source:* Sherman, P. J. 1988. Japanese construction R&D: Entrée into U.S. market, *ASCE Journal of Construction Engineering and Management.* 114(1):138–142. Reproduced by permission of ASCE.)

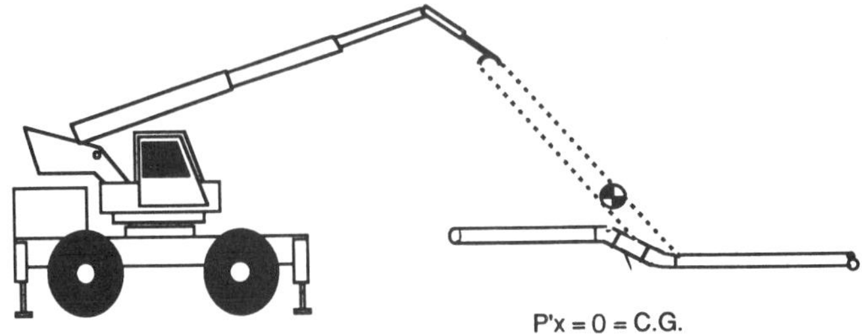

FIGURE 6.10 Pipe manipulator. (*Source:* Fisher, D. J., and O'Connor, J. T. 1991. Constructability for piping automation: Field operations. *ASCE Journal of Construction Engineering and Management.* 117(3):468–485. Reproduced by permission of ASCE.)

Blockbots. Another application involves the design, development, and testing of the "blockbot" robot intended to automate the placement of masonry blocks to form walls. The complete wall assembly consists of four major components [Slocum *et al.*, 1987]:

1. A six-axis "head" that will actually place the blocks on the wall
2. A 20- to 30-ft (6- to 9.1-m) hydraulic scissors lift used to roughly position the placement head both vertically and longitudinally
3. A large-scale metrology system, sensors, and other related computer control equipment
4. A block-feeding system/conveyor to continually supply the placement head

To facilitate construction, the blocks are stacked upon each other with no mortar between the levels. The wall is then surface-bonded using Surewall^{TM}, a commercial fiberglass-reinforced bonding cement. This process produces a wall with strength comparable to that of a traditional mortar wall.

Wallbots. Researchers at the Massachusetts Institute of Technology (MIT) are engaged in the Integrated Construction Automation Design Methodology (ICADM) project [Slocum *et al.*, 1987]. This work attempts to integrate the efforts of material suppliers, architects, contractors, and automated construction equipment designers.

The process of building interior wall partitions is divided between two separate robots: a trackbot and a studbot. Circumventing the need for complex navigational systems, the trackbot is guided by a laser beacon that is aligned manually by a construction worker. The trackbot is separated into two parallel workstations: an upper station for the ceiling track and a lower station for the floor track. Detectors are mounted on the ends of the effector arms to ensure that the laser guidance system achieves the necessary precision. The placement of the track consists of four steps: (1) the effector arm grabs a piece of track, (2) the effector arm positions the track, (3) two pneumatic nail guns fasten the track, and (4) the trackbot moves forward, stopping twice to add additional fasteners.

Once the trackbot has completed a run of track, the studbot can begin placing studs. Location assessment is made by following the track and employing an encoding wheel or an electronic distance measuring (EDM) instrument. The studbot then references a previously sorted floor plan to ascertain locations of studs to be placed. The stud is removed from its bin and placed into position. The positioning arm then spot-welds the stud into place.

Interior Finishing Robot. An interior finishing robot, shown in Fig. 6.11, can execute the following tasks: (1) building walls and partitions, (2) plastering walls and ceilings, (3) painting walls and ceilings, and (4) tiling walls. The arm of the robot has six degrees of freedom with a nominal reach of 5.3 ft (1.6 m) and a lifting capacity of 66 lb (145 kgf). The robot is designed to perform interior finishing work in residential and commercial buildings with single or multiple floor levels and interior heights of 8.5 to 8.8 ft (2.60 to 2.70 m). A three-wheel mobile carriage measuring 2.8 × 2.8 ft (0.85 × 0.85 m) enables motion of the robot between static workstations [Warszawski and Navon, 1991; Warszawski and Rosenfeld, 1993].

Fireproofing Spray Robot. Shimizu Company has developed two robot systems for spraying fireproofing material on structural steel [Yoshida and Ueno, 1985]. The first version, the SSR-1, was built to (1) use the same materials as in conventional fireproofing, (2) work sequentially and continuously with human help, (3) travel and position itself, and (4) have sufficient safety functions for the protection of human workers and of building components. The second robot version, the SSR-2, was developed to improve some of the job site functions of SSR-1. The SSR-2 can spray faster than a human worker but requires time for transportation and setup. The SSR-2 takes about 22 minutes for one work unit whereas a human worker takes about 51 minutes. The SSR-2 requires

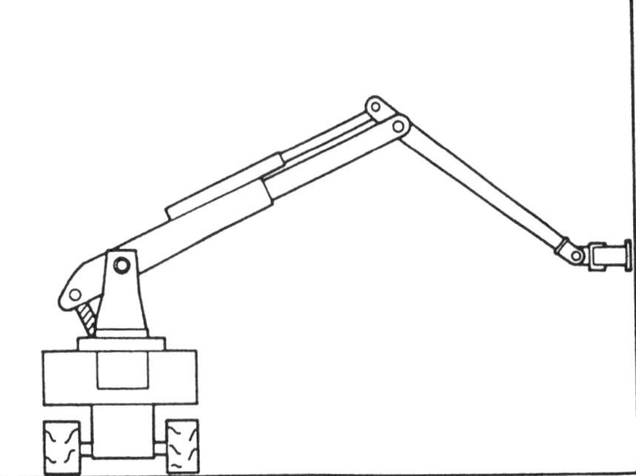

FIGURE 6.11 Interior finishing robot. (*Source:* Warszawski, A. and Navon, R. 1991. Robot for interior-finishing works. *ASCE Journal of Construction Engineering and Management,* 117(3)402–422.)

relatively little manpower for the spraying preparation—only some 2.1 person-days compared with 11.5 for the SSR-1. As the positional precision of the robot and supply of the rock wool feeder were improved, the SSR-2 could achieve the same quality of dispersion of spray thickness as for that applied by a human worker.

Exterior Wall Painting Robot. The exterior wall painting robot, shown in Fig. 6.12, paints walls of high-rise buildings, including walls with indentations and protrusions. The robot is mounted on mobile equipment that permits translational motion along the exterior wall of a building. The robot consists of the following:

1. Main body that sprays paint
2. Moving equipment to carry the robot main body to the proper work position
3. Paint supply equipment
4. A controller

The robot main body consists of the following:

1. Main frame
2. Painting gun
3. Gun driver
4. Control unit

The painting gun is driven in three principal translational directions (x, y, and z). The painting gun is also provided with two rotational degrees of freedom. The robot moving equipment consists of the following:

1. A transporter that propels the moving equipment along the outside of the building being painted
2. A work stage on which the robot main body is mounted
3. A mast that serves as a guide for raising and lowering the work stage

The top of the mast is attached to a travel fitting, and the fitting moves along a guide rail mounted on the top of the building [Terauchi *et al.*, 1993].

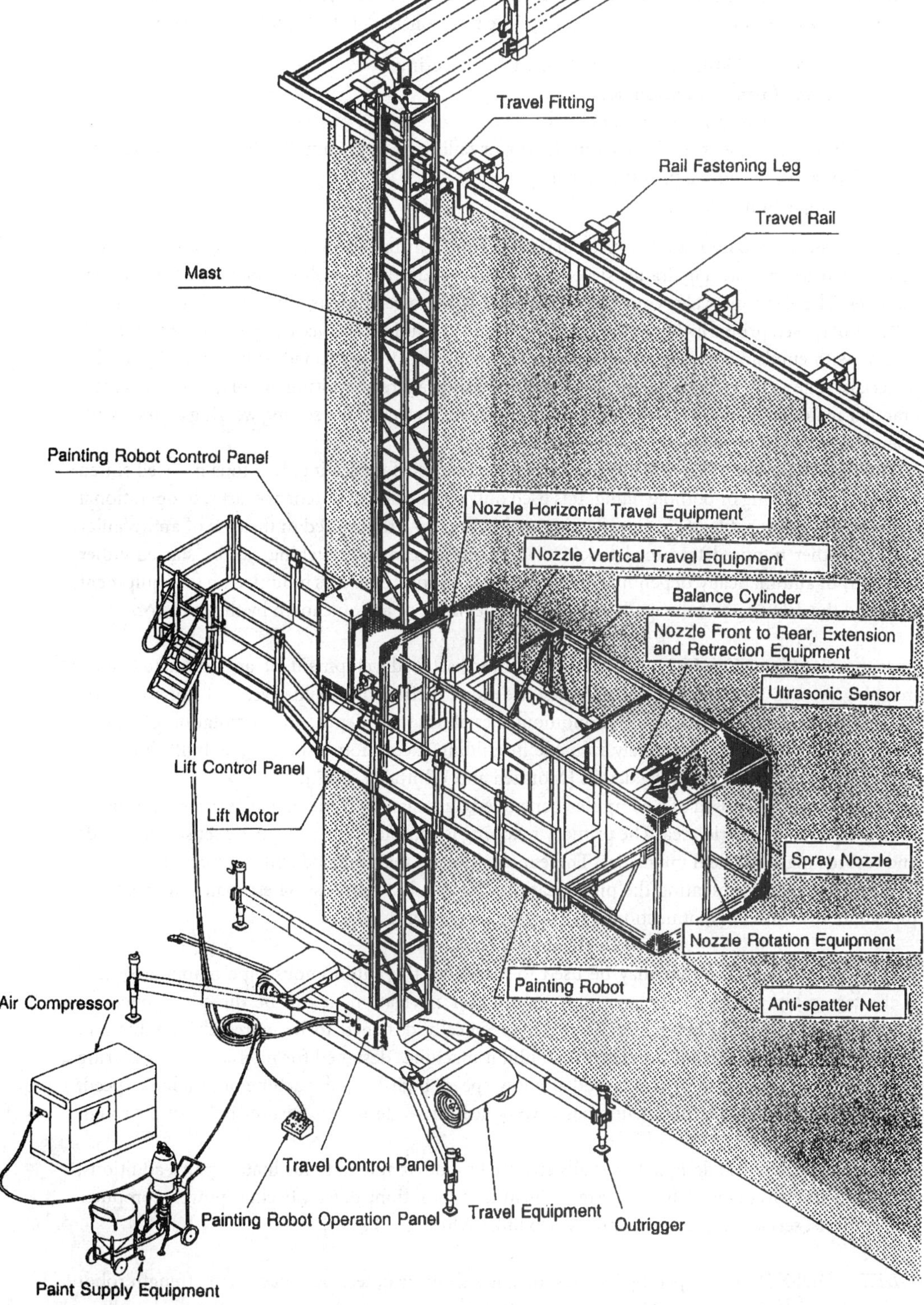

FIGURE 6.12 Exterior wall painting robot. (*Source:* Terauchi *et al.* 1993. Development of an exterior wall painting robot—Capable of painting walls with indentations and protrusions. *Proceedings, Tenth International Symposium on Automation and Robotics for Construction.* Houston, TX, pp. 363–370.)

Integrated Surface Patcher (ISP). Secmar Company of France developed a prototype of the integrated surface patcher (ISP) [Point, 1988]. The unit consists of the following components:

1. A 19-ton (17, 234-kgf) carrier with rear wheel steering
2. A 3.9-yd^3 (3-m^3) emulsion tank
3. A 5.2-yd^3 (4-m^3) aggregate container
4. A built-in spreader working from the tipper tailboard (a pneumatic chip spreader with 10 flaps and a 10-nozzle pressurized bar)
5. A compaction unit

The ISP unit has a compressor to pressurize the emulsion tank and operate the chip-spreading flaps. The machine uses a hydraulic system driven by an additional motor to operate its functional modules. The electronic valve controls are operated with power supplied by the vehicle battery.

The ISP is used primarily for hot resurfacing repairs, including surface cutting, blowing and tack coating with emulsion, as well as for repairs requiring continuous treated or nontreated granular materials. The unit is suitable for deep repairs using aggregate-bitumen mix, cement-bound granular materials, and untreated well-graded aggregate, as well as for sealing wearing courses with granulates.

The current design of the ISP allows only carriageway surface sealing. It is thus not well suited for surface reshaping or pothole filling. It is used only for routine maintenance tasks. In operational terms, ISP is not capable of on-line decision making on how to proceed in the case of an irregular crack or other nonpredetermined task. However, the automated patching can be started either manually or automatically, depending on the presence of optical readers mounted on the equipment that read the delimiters of the work area, and on the mode of action chosen by the operator.

Autonomous Pipe Mapping. Another application is the development of an automated pipe-mapping system. Current manual methods are slow, inefficient, qualitative, and nonrepetitive. The intention of the system is to autonomously establish size, depth, and orientation of buried pipes. This knowledge is extremely valuable in guiding excavation, validating as-built drawings, and building databases of piping details [Motazed and Whittaker, 1987].

The system is composed of a computer-controlled Cartesian *x-y* table that allows various sensors to be swept across an arid area. The primary mapping is completed by a magnetic sensor that reads and records magnetic field intensities. These intensities are manipulated and interpreted, resulting in a line drawing representing the pipe locations. Higher-level processing estimates the depth of pipes and identifies interconnections such as elbows, tees, and crosses.

Terregator. A machine that may be used to transport the autonomous pipe mapping system is the terregator. Designed for autonomous outdoor navigation, it can be directly applied on a construction site. The terregator has been specifically designed to be extremely durable and powerful in order to prevent problems that inhibit machines designed for interior use. Its gearing is adjustable to allow it to be configured as a low-speed, high-torque machine or as a high-speed, low-torque machine. The terregator has a six-wheel-drive design to ensure mobility on rough terrain.

The terregator is also designed as a fully enclosed modular system to facilitate repairs, additions, or system improvements. The subsystems include locomotion, power, backup power, computer and controls, serial links, sensors, and a video link [Whittaker, 1985b].

ODEX. ODEX III, developed by Odetics, Inc., is a six-legged, tele-operated, high-strength robot designed for inspection, surveillance, and material handling in nuclear power plants and outdoor hazardous environments [Jaselskis and Anderson, 1994]. ODEX III has telescoping legs that can extend it to a full height of 7.9 ft (2.4 m), a manipulator arm, and sensors on each foot to determine proper foot placement.

Numerical Control

Numerical control refers to control of construction equipment using numbers [Luggen, 1984]. Questions such as "What numbers are used to control a piece of equipment?" and "In what format are they presented to the equipment?" are basic to understanding numerical control. Numerically controlled equipment consists of a machine control unit (MCU) and a machine tool (such as an end effector). The MCU cannot think, judge, or reason in relation to the environment in which it works. The machine accepts and responds to commands from the control unit [Luggen, 1984]. For example, a numerically controlled pumped-concrete placement system may use numbers corresponding to (1) position (x, y, z) of the discharging end of the placement pipe, (2) pumping pressure, and (3) the speed at which the discharging end of the placement pipe travels.

Numerical Control (NC) Programs

The numerically controlled tool concept is based on textual programming methods to describe the structural components with the help of control surfaces. The description of the structural component is taken from the architectural drawing, converted to a code, and entered on a code carrier such as a computer disk. The format of the control data and the equipment commands need to be defined in detail. The control program consists of a sequence of commands in a standardized symbolic format. The control program is transferred to the MCU, which translates the program to equipment-level instructions. The equipment-level instruction may be coded on perforated paper tape (NC tape), computer cards, magnetic tape, or floppy disks [Rembold *et al.*, 1985].

Computers are used to derive equipment-level instructions using information from the control program. For a computer to accept and process the NC program data, the input programs must conform to the exacting requirements of the programming language of the computer. Hence, the general-purpose computer must be primed to handle the specific input program. The general-purpose computer is converted to a special-purpose computer through insertion of the NC program [Maynard, 1971].

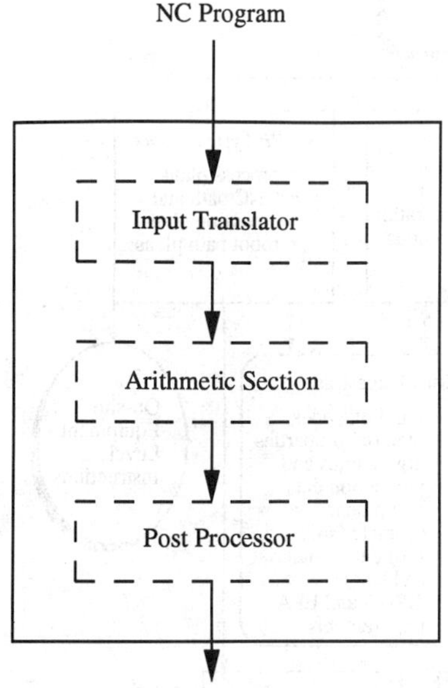

FIGURE 6.13 Software sections for NC programming processing.

The NC program, when processed by a computer, passes through three modules, as shown in Fig. 6.13. The input translator converts the NC program into a binary-coded system called *machine language*. Next the machine language instructions are passed on to the arithmetic section, which performs the required mathematical and geometric computations to calculate the path of the numerically controlled equipment. The postprocessor checks the limitations of a particular piece of equipment (such as maximum pumping pressure or maximum velocity of the placing boom). The final output corresponds to the equipment-level instructions [Maynard, 1971].

Computer Numerical Control (CNC)

A CNC system performs control functions similar to those of the NC system. However, CNC systems can have a microcomputer or multiprocessor architecture that is highly flexible. Logic control, geometric data processing, and NC program executions are supervised by a central processing unit (CPU). Hence, CNC is a software control system that performs the following tasks using a microcomputer: (1) system management, (2) data input/output, (3) data correction, (4) control of the NC program, (5) processing of operator commands, and (6) output of the NC process variables to the display [Rembold *et al.*, 1985].

6.4 Computer-Integrated Construction (CIC)

Computer-integrated construction (CIC) is defined as "a strategy for linking existing and emerging technologies and people in order to optimize marketing, sales, accounting, planning, management, engineering, design, procurement and contracting, construction, operation and maintenance, and support functions" [Miyatake and Kangari, 1993]. Computer-aided design/computer-aided construction (CAD/CAC) systems are a major subset of CIC that focus on design and construction issues [Kunigahalli and Russell, 1994]. Fig. 6.14 presents the architecture of CAD/CAC systems. The development of CAD/CAC systems requires multidisciplinary research efforts in a variety of areas such as:

1. Computer-aided design (CAD) and geometric modeling
2. Algorithms and data structures
3. Artificial intelligence
4. Computer numerical control (CNC) and robotics
5. Group technology (GT)
6. Computer-aided process planning (CAPP)

The implementation of CIC requires technologies related to (1) computer-aided engineering, (2) automatic material handling and data identification systems, (3) network communications, (4) object-oriented programming, (5) knowledge-based systems (KBS), and (6) database manage-

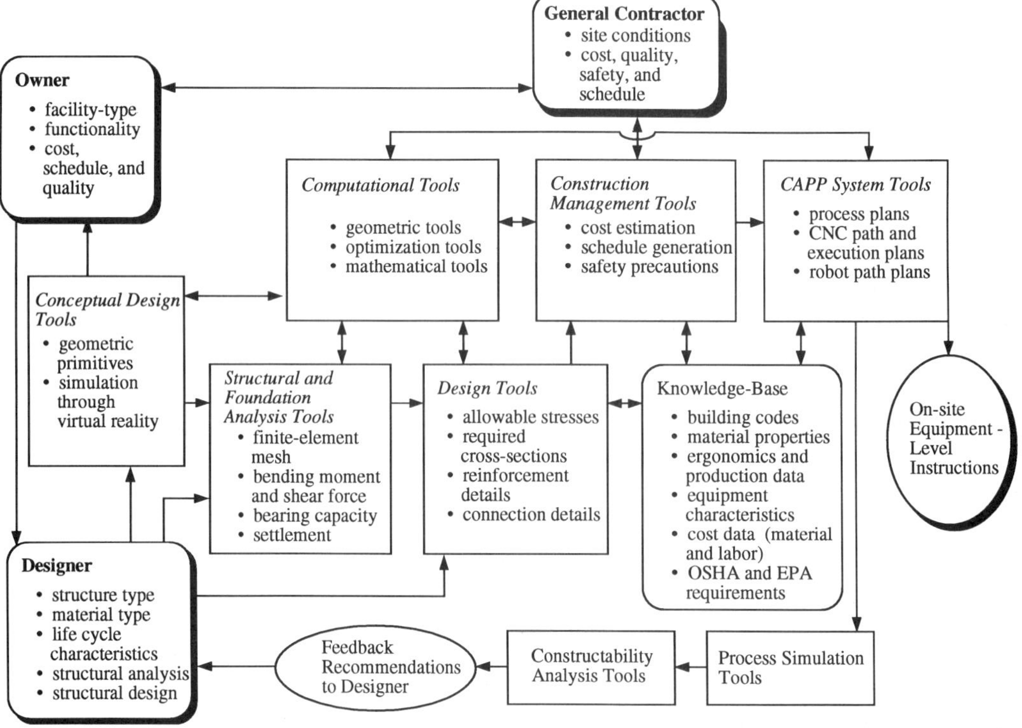

FIGURE 6.14 Framework of CAD/CAC systems.

ment systems [Miyatake and Kangari, 1993]. Three of these areas are discussed below, followed by an example application of CIC.

Computer-Aided Design (CAD) and Geometric Modeling

Computer-aided design (CAD) can be described as "using a computer in the design process." A CAD model requires graphical data processing that comprises many techniques to process and generate data in the form of lines and figures. Thus, the input representation of textual or pictorial data is performed with techniques of character and pattern recognition [Rembold *et al.*, 1985].

Models are used to represent physical abstract entities and phenomena, not just for the purpose of making pictures (creating sectional views) but to represent their structure and behavior [Foley and Van Dam, 1982]. CAD software modeling can be classified into the following three categories: (1) basic 2-D and 3-D wire-frame modeling, (2) surface modeling, and (3) solid modeling.

Basic 2-D and 3-D Wire-Frame Modeling

In 2-D and 3-D wire-frame models, lines are stored as edges in an edge table, with each line pointing to its two end vertices stored in a vertex table. Wire-frame CAD models are not capable of recognizing the faces delineated by lines and vertices of the object being represented. Wire-frame CAD models are generally used as a substitute for manual drafting.

Surface Modeling

Surface modeling allows users to add faces to geometric models. Hence, hidden surface removal is possible in surface models. However, surface models do not contain information on the interior and exterior of the object.

Solid Modeling

A solid geometric model is an unambiguous and informationally complete mathematical representation of the physical shape of an object in a form that a computer can easily process [Mortenson, 1985]. Topology and algebraic geometry provide the mathematical foundation for solid modeling. Solid modeling's computational aspects include (1) data structures and algorithms from computer science and (2) application considerations from design and construction of engineering projects.

The following techniques are available for solid modeling of civil engineering facilities [Requicha, 1980]:

1. Primitive instancing
2. Cell decompositions
3. Spatial occupancy enumeration (SOE)
4. Constructive solid geometry (CSG)
5. Sweep representations
6. Boundary representation (B-Rep)

Primitive Instancing. The primitive instancing modeling technique consists of an independent approach to solid-object representation in the context of the group technology (GT) paradigm. The modeling approach is based on the notion of families of objects, with each member of the family being distinguishable by a few parameters. For example, columns, beams, and slabs can be grouped as separate families in the case of general buildings. Each object family is called a *generic primitive*, and individual objects within a family are referred to as *primitive instances* [Requicha, 1980].

Cell Decompositions. Cell decompositions are generalizations of triangulations. Using the cell decomposition modeling technique, a solid may be represented by decomposing it into cells and representing each cell in the decomposition. This modeling technique can be used for analysis of trusses and frames in industrial and general buildings, bridges, and other civil engineering facilities. In fact, the cell decomposition technique is the basis for finite-element modeling [Mortenson, 1985].

Spatial Occupancy Enumeration (SOE). The spatial occupancy enumeration (SOE) technique is a special case of the cell decomposition technique. A solid in the SOE scheme is represented using a list of spatial cells occupied by the solid. The spatial cells, called *voxels*, are cubes of a fixed size lying in a fixed spatial grid. Each cell may be represented by the coordinates of its centroid. Cell size determines the maximum resolution. This modeling technique requires large memory space, leading to inefficient space complexity. However, this technique may be used for motion planning of automated construction equipment under complete-information models [Requicha, 1980].

Constructive Solid Geometry (CSG). Constructive solid geometry (CSG), often referred to as *building-block geometry,* is a modeling technique that defines a complex solid as a composition of simpler primitives. Boolean operators are used to execute the composition. CSG concepts include (1) regularized Boolean operators, (2) primitives, (3) boundary evaluation procedures, and (4) point membership classification. CSG representations are ordered binary trees. Operators specify either rigid motion, regularized union, intersection, or difference and are represented by nonterminal nodes. Terminal nodes are either primitive leaves that represent subsets of 3-D Euclidean space or transformation leaves that contain the defining arguments of rigid motions. Each subtree that is not a transformation leaf represents a set resulting from the application of the motional and combinational operators to the sets represented by the primitive leaves.

The CSG modeling technique can be adopted to develop computer-aided design and drafting (CADD) systems for civil engineering structures. It can be combined with primitive instancing that incorporates the group technology paradigm to assist the designer. Although CSG technique is most suitable for design engineering applications, it is not suitable for construction engineering applications, as it does not store topological relationships required for construction process planning [Requicha, 1980].

Sweep Representation. The sweep representation technique is based on the idea of moving a point, curve, or surface along a given path; the locus of points generated by this process results in 1-D, 2-D, and 3-D objects, respectively. Two basic ingredients are required for sweep representation: an object to be moved and a trajectory to move it along. The object can be a curve, surface, or solid. The trajectory is always an analytically definable path. There are two major types of trajectories: translational and rotational [Mortenson, 1985].

Boundary Representation (B-Rep). The boundary representation modeling technique involves representing a solid's boundary by decomposing it into a set of faces. Each face is then represented by its bounding edges and the surface in which it lies. Edges are often defined in the 2-D parametric space of the surface as segments of piecewise polynomial curves. A simple enumeration of a solid's faces is sufficient to unambiguously separate the solid from its complement. However, most boundary representation schemes store additional information to (1) aid feature extraction and (2) determine topological relationships. The additional information enables intelligent evaluation of CAD models for construction process planning and automated equipment path planning required in CAD/CAC systems [Requicha and Rossignac, 1992; Kunigahalli *et al.*, 1994; Kunigahalli and Russell, 1994].

The boundary representation technique, storing topological relationships among geometric entities, is most suitable for computer-aided generation of construction process plans. However, primitive instancing, sweep representation, and CSG techniques are useful in developing user-friendly CAD software systems for the design of civil engineering structures. Hence, CAD systems that incorporate CSG or primitive instancing techniques during interactive design processes and that employ boundary representation techniques for internal storage of design information are efficient for use in CAD/CAC systems [Kunigahalli and Russell, 1994].

CAD Applications in Civil Engineering

AutoCAD. AutoCAD is the most widely used CAD software in civil engineering applications. In an effort toward computer-integrated construction (CIC), researchers have developed a link between AutoCAD and a knowledge-based planning program [Cherneff *et al.*, 1991].

CATIA. CATIA is a 3-D solid modeling software marketed by IBM Corporation. Stone & Webster Engineering Corporation, in cooperation with IBM, has developed an integrated database for engineering, design, construction, and facilities management. The system uses the DB2 relational database management system and the CATIA computer-aided-design software system [Reinschmidt *et al.*, 1991].

Walkthrough™. Bechtel Corporation has developed a 3-D simulation system called Walkthrough to aid in marketing, planning, and scheduling of construction projects. Walkthrough was developed to replace the use of plastic models as a design tool [Cleveland and Francisco, 1988]. It was designed to allow users to interact with a 3-D computer model as they would with a plastic model. The system uses 3-D, real-time animation that lets the user visually move through the computer model and observe visual objects. Graphics of the system are presented such that objects are recognizable to users not accustomed to typical CAD images. This includes the use of multiple colors and shading. Walkthrough uses a Silicon Graphics IRIS workstation with specialized processors facilitating the high-speed graphics required for real-time animation. This visualization and simulation system supports files from IGDS (Intergraph CAD system) and 3DM [Morad *et al.*, 1992].

Object-Oriented CAD Model. An object-oriented CAD model for the design of concrete structures that uses EUROCODE2, a European standard for concrete structures, has been developed by German researchers. The primitive instancing solid modeling technique was employed in the development of this object-oriented model [Reymendt and Worner, 1993]. A committee, entitled "NEW TECCMAR," formed under the Japanese construction ministry, has developed a three-dimensional finite-element method (FEM) program with an extended graphical interface to analyze general buildings [Horning and Kinura, 1993].

Automated Material Management

Automated material management systems are another important function of CIC. They comprise automated material identification systems and automated material handling systems.

Automated Material Identification Systems

When construction materials arrive at CIC job sites, they are identified at the unloading area, and the job site inventory database in the central computer is updated. CIC requires very tight control on inventory and integrated operation of automated equipment. Further, all construction materials must be tracked from the time of their arrival at the job site to their final position in the finished facility. Such tracking of construction materials may be done by employing automated identification systems.

There are two means of tracking construction materials: direct and indirect. Direct tracking involves identifying a construction material by a unique code on its surface. This method of tracking can be employed with the use of large prefabricated components. Indirect tracking involves identifying construction material by a unique code on the material handling equipment. This method of tracking can be employed for tracking bulk materials such as paints [Rembold *et al.*, 1985]. Select automatic identification systems for construction materials are described below.

Bar Coding. The U.S. Department of Defense (DOD) was the first organization to implement bar coding technology. The Joint Steering Group for Logistics Applications of Automated Marking and Reading Symbols (LOGMARS) spearheaded the DOD's effort in the implementation of bar coding technology. The symbology of bar codes conveys information through the placement of wide or narrow dark bars that create narrow or wide white bars. With the rise of the LOGMARS project, code 39 (also called "3 of 9" coding) has become a standard for bar coding. To date, most construction bar code applications have used the code 39 symbology [Teicholz and Orr, 1987; Bell and McCullough, 1988].

Laser beams and magnetic foil code readers are two basic technologies available for reading bar codes. Lasers offer the ability to read bar codes that move rapidly. Magnetic code readers are among the most reliable identification systems. It is possible to transmit the code without direct contact between the code reader and the write head on the code carrier. When the workpiece passes the read head, the code is identified by the code reader [Teicholz and Orr, 1987; Rembold *et al.*, 1985].

Voice Recognition. Voice recognition provides computers the capability of recognizing spoken words, translating them into character strings, and sending these strings to the central processing unit (CPU) of a computer. The objective of voice recognition is to obtain an input pattern of voice waveforms and classify it as one of a set of words, phrases, or sentences. This requires two steps: (1) analyze the voice signal to extract certain features and characteristics sequentially in time and (2) compare the sequence of features with the machine knowledge of a voice, and apply a decision rule to arrive at a transcription of the spoken command [Stukhart and Berry, 1992].

Vision Systems. A vision system takes a two-dimensional picture by either the vector or the matrix method. The picture is divided into individual grid elements called pixels. From the varying grey levels of these pixels, the binary information needed for determining the picture parameters is extracted. This information allows the system, in essence, to see and recognize objects. The process is shown in Fig. 6.15.

The vector method is the only method that yields a high picture resolution with currently available cameras. The vector method involves taking picture vectors of the scanned object and storing them at constant time intervals. After the entire cycle is completed, a preprocessor evaluates the recomposed picture information and extracts the parameters of interest [Rembold *et al.*, 1985].

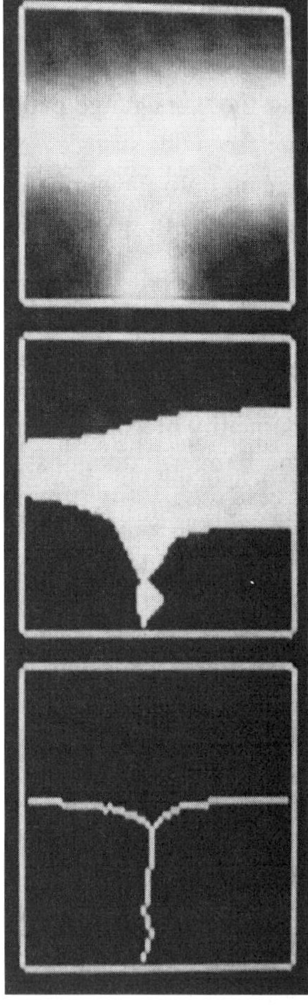

FIGURE 6.15 Vision system. (Photo courtesy of The Robotics Institute, Carnegie-Mellon University.)

Automated Material Handling Systems

Automated material handling systems play an important role in CIC. Efficient handling of construction materials such as prefabricated and precast components is possible through an effective automated material handling system operating in conjunction with an automated material identification system.

Towlines. A towline consists of a simple track with a powered chain that moves carts of other carriers from pickup points to assigned destinations. Towlines can be controlled by sophisticated computer electronic techniques. Towlines interface efficiently with other automated material handling systems. Automated material identification systems can be easily integrated with towline material handling systems. Optical scanners or photoelectric readers can be used at important locations and intersections along the track to read the bar-coded information attached to the cart and relay signals to the control system. The control system then routes the cart to its destination [Considine and Considine, 1986].

Underhang Cranes. There are two types of motor-driven underhang cranes: (1) single-bridge overhead cranes that can operate on multiple runways and (2) double-bridge overhead cranes that can achieve higher hook lifts with greater load-carrying capacity. A motor-driven crane consists of: (1) a track used for crane runways and bridge girders, (2) the end trucks, (3) the control package, (4) a drive assembly, (5) drive wheels, (6) a drive line shaft, (7) a traveling pushbutton control, and (8) runway and cross-bridge electrification.

Power and Free Conveyor Systems. Conveyor systems allow precast or prefabricated components to be carried on a trolley or on multiple trolleys propelled (1) by conveyors through some part of the system and (2) by gravity or manually through another part of the system. Conveyor systems provide the high weight capacities that are normally required in construction.

Inverted Power and Free Conveyor Systems. An inverted power and free conveyor system is an upside-down configuration of the power and free conveyor system. The load is supported on a pedestal-type carrier for complete access.

Track and Drive Tube Conveyors. Track and drive tube conveyors can be employed to transport components for prefabrication. This conveyor system consists of a spinning tube (drive tube) mounted between two rails. Carriers of prefabricated units need to be equipped with a drive wheel capable of moving between 0° and 45°. This drive wheel is positioned against the spinning drive tube. Speed of the moving component can be controlled by varying the angle of the drive wheel. When the drive wheel is in the 0° position, the carrier remains stationary. As the angle between the drive wheel and the drive tube is increased, the carrier accelerates forward.

Automatic Vertical Transport System (AVTS). Fujita Corp.'s AGVS is a system under development that is designed to deliver material throughout the job site. The system uses an automated elevator system that automatically loads material onto a lift, hoists the material to the designated floor, and automatically unloads the lift [Webster, 1993].

Interlocks. Interlocks allow transfer of hoist carriers between adjacent crane runways, thereby maximizing the area covered by the overhead material handling systems. Interlocks also eliminate duplicate handling. Cross- connected, double-locking pins help to ensure that the safety stops will not operate until the crane and connecting track are in proper alignment.

Automatically Guided Vehicles. Automatically guided vehicles (AGVs) are the most flexible of all material handling equipment. AGVs can be controlled either by (1) programmable controllers, (2) on-board microprocessors, or (3) a central computer. Because of their lack of dependence on manual guidance and intervention, AGVs can also be categorized as construction robots. AGVs have their own motive power aboard. The steering system is controlled by signals emanating from a buried wire [Considine and Considine, 1986].

Autonomous Dump Truck System. The autonomous dump truck system, shown in Fig. 6.16, enables driverless hauling operations such as hauling of earth and gravel by dump trucks on heavy construction sites. The two major functions of this system are autonomous driving function and advanced measurement function.

The driving distance and velocity of the vehicle are detected by encoder sensors attached to the truck tires. Direction of the vehicle is detected by a fiber-optic gyroscope. Positions of the vehicle are determined using data from the encoder sensors and fiber-optic gyroscope.

A laser transmitter/receiver is equipped at the left side of the test vehicle, and laser reflectors are installed along the driving route at a spacing of approximately 50 m. Positional errors accumulated in long-distance driving are corrected using the feedback information from the laser transmitter/receiver. The autonomous vehicle system recognizes the workers wearing helmets by utilizing a color image processor [Sugiura *et al.*, 1993].

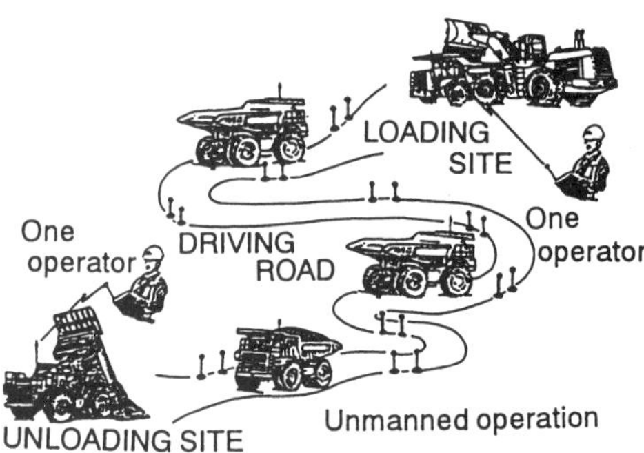

FIGURE 6.16 Autonomous dump truck system. (*Source:* Sugiyura *et al.* 1993. Autonomous dump truck systems for transporting and positioning heavy-duty materials in heavy construction sites. *Proceedings, Tenth International Symposium on Automation and Robotics for Construction.* Houston, TX, pp. 253–260.)

Network Communication

Communication technology, transferring information from one person or computer system to another, plays a vital role in the implementation of CIC. Establishment of an effective communications network such that (1) originating messages receive the correct priority and (2) accurate data arrive at the final destination is a very difficult task [Miyatake and Kangari, 1993]. To ensure smooth operations in CIC, many automated devices and computers must be linked. Computer networking techniques enable a large number of computers to be connected. Computer networks can be classified as (1) wide-area networks (WANs), which serve geometric areas larger than 10 km; (2) local-area networks (LANs), which are confined to a 10-km distance, and (3) high-speed local networks (HSLNs), which are confined to a distance less than 1 km.

Wide-area networks such as ARPANET can be employed to connect the construction company's corporate office to various automated project sites. LANs combined with HSLNs can be employed to facilitate efficient data exchange among automated construction equipment (such as a CNC concrete placement machine, floor leveling robots, and wall painting robots) operating on job sites. Various gateways (computers that transfer a message from one network to another) can be used to link networks.

Three types of commonly used network arrangements—ring, star, and bus—are shown in Fig. 6.17. In a *ring* network arrangement, the connecting coaxial cable must be routed back to where it begins. This results in network breakdown whenever the ring breaks. The *star* network arrangement is easily expanded, but the network relies on a server at the center of the star. Further, all communications between nodes must pass through the center. The *bus* network arrangement is open-ended, and hence a node can be added easily to the network [Chang *et al.*, 1991].

The efficiency of a network system depends on the following parameters [Rembold *et al.*, 1985]:

1. Transmission speed and maximum transmission distance
2. Time delay necessary to respond to interrupts and data requests
3. Additional hardware and software needed for expansion
4. Reliability, fault tolerance, and availability
5. Unique logic structure
6. Standard plug-in principle

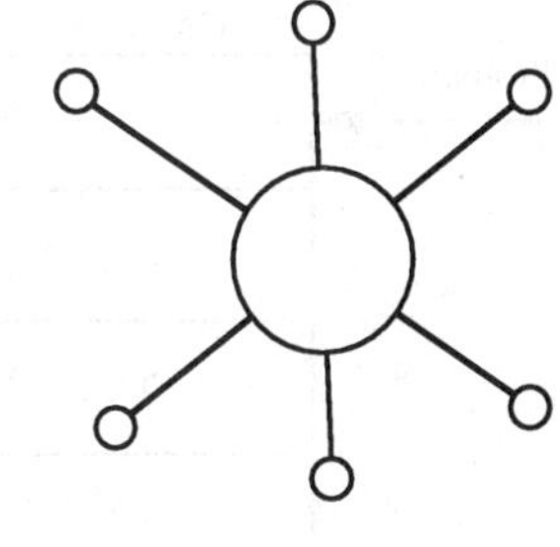

(a) Ring

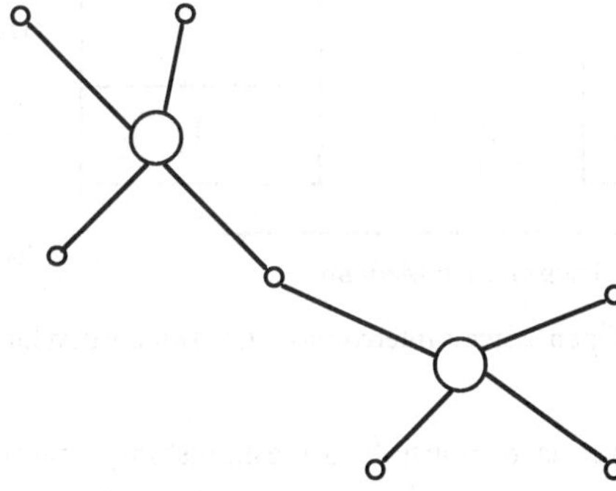

(b) Star

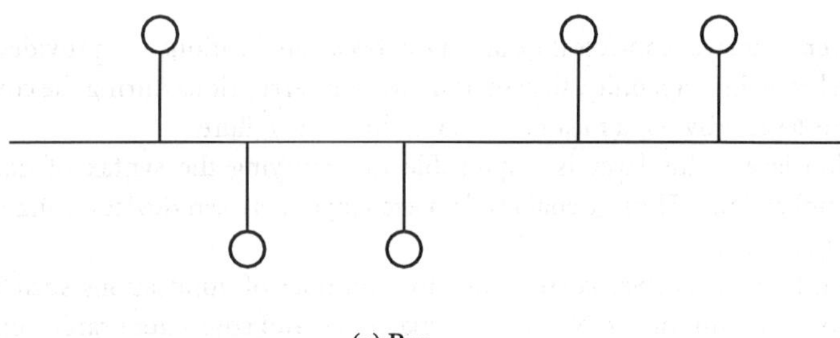

(c) Bus

FIGURE 6.17 Three common network arrangements.

7. Possible geographic distribution of communication processes
8. Cost of the system components

In an attempt to enable network communications using computers and devices from different vendors, the International Standards Organization (ISO) has developed a model for LANs called the Open Systems Interconnect (OSI) model, which is shown in Fig. 6.18. The OSI splits the communication process into seven layers as described below.

1. Physical layer. The physical layer corresponds to electrical and mechanical means of data transmission. It includes coaxial cable, connectors, fiber optics, and satellite links.
2. Data link layer. Functions of this layer include resolution of contention for use of the shared transmission medium, delineation and selection of data addressed to this node, detection of noise, and error correction.

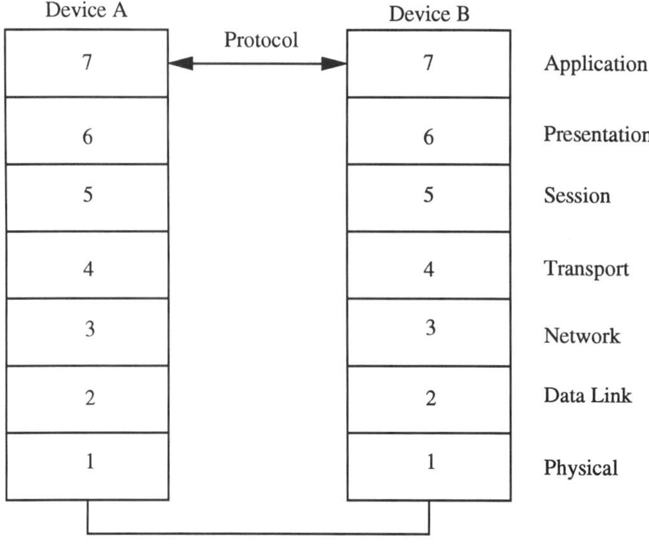

FIGURE 6.18 Open Systems Interconnect (OSI) model developed by ISO.

3. Network layer. This layer is responsible for establishing, maintaining, and terminating connections. Further, this layer enables internetwork routing using a global standard for assigning addresses to nodes.
4. Transport layer. This layer provides a network-independent service to the session, presentation, and application layers. Loss or duplication of information is also checked by this layer.
5. Session layer. This layer controls the dialogue between applications and provides a checkpoint and resynchronizing capability. In case of network interruptions during the communication session, this layer provides a means to recover from the failure.
6. Presentation layer. This layer is responsible for verifying the syntax of data exchanged between applications. Thus, it enables data exchanges between devices using different data encoding systems.
7. Application layer. This layer corresponds to a number of applications such as CAD/CAC systems, construction robots, NC or CNC machines, and computer graphic interfaces. This is the most complex layer and ensures that data transferred between any two applications are clearly understood [Chang *et al.*, 1991].

Example Application of Computer-Integrated Construction

The Shimizu Manufacturing by Advanced Robotics Technology (SMART) system, shown in Fig. 6.19, is part of an overall CIC strategy. The SMART system integrates high-rise construction processes such as (1) erection and welding of steel frames, (2) placement of precast concrete floor slabs, and (3) exterior and interior wall panel installation.

In the SMART system, steel-frame columns and beams are automatically conveyed to designated locations. Assembly of these structural components is greatly simplified by using specially designed joints.

The SMART system consists of (1) operating platform, (2) jacking towers, (3) vertical lifting crane, and (4) weather protection cover. The operating platform is an automated assembly system and consists of a computer-control room, monorail hoists for automated material handling,

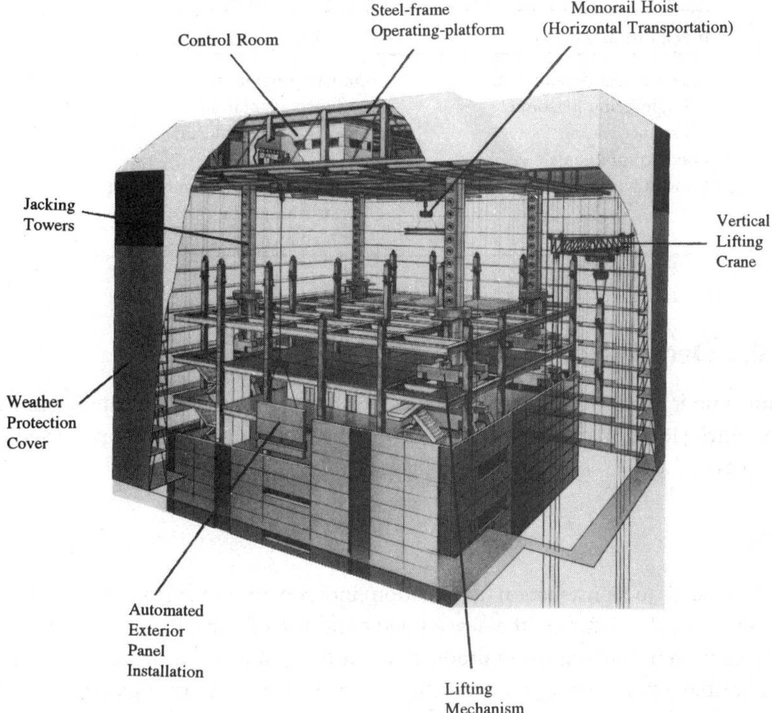

FIGURE 6.19 Shimizu's Manufacturing by Advanced Robotic Technology (SMART) system. (*Source:* Miyatake, Y., and Kangari, R. 1993. Experiencing computer integrated construction. *ASCE Journal of Construction Engineering and Management.* 119(2): 307–323. Reproduced by permission of ASCE.)

and a structural steel frame that eventually becomes the top roof of the building. Upon completion of the foundation activity, the operating platform is assembled and mounted on top of the four jacking towers of the lifting mechanism. After completion of construction work in each floor of the building, the entire automated system is lifted up by vertical jacks. Thus, the automated construction work is performed, floor by floor, until the entire building is completed [Miyatake and Kangari, 1993].

6.5 Economics

Three major factors contributing to economic benefits of construction automation are: productivity, quality, and savings in skilled labor [Kangari and Halpin, 1989]. These benefits must be weighed against the costs of automation, including initial investment and operating costs; these are further described in Table 6.2. Economic data resulting from analyses of several robot applications are described below.

Automated Stone Cutting

Benefits of a partially automated stone-cutting mill were assessed through computer simulation [Hijazi *et al.*, 1988]. In comparison to traditional stone-cutting methods, simulation of the automated system resulted in a 74% increase in productivity and 42% less time to process identical orders.

Table 6.2 Costs Associated with Construction Robotics

Initial Investment Costs	Robot Operating Costs
Research and development	On-site programming
Engineering personnel	Software adaptation
Product testing	Labor cost (technicians)
Robot components	Electricity
Control hardware	System dismantling and re-setup
	Robot transporting costs
	Maintenance and repair

Steel Bridge Deck Welding

The economic implications of using robot welders in steel bridge deck fabrication was studied by Touran and Ladick [1988]. Using the robot welders in the fabrication shop was predicted to reduce fabrication costs by 5.6%.

Excavation

Most reports related to construction automation indicate improvement in productivity and quality. In one particular instance, the development and use of partially automated laser-guided grading equipment not only improved productivity of fine grading operations, but increased the contractor's flexibility in managing its fleet. The contractor gained a major competitive advantage that contributed to the contractor's growth from an annual volume of $500,000 in 1976 to over $50,000,000 in 1984 [Tatum and Funke, 1988].

Large-Scale Manipulators

Use of large-scale manipulators (LSMs) can reduce the amount of non-value-added tasks and increase productivity. According to a case study by the Construction Industry Institute, LSMs have high potential for industrial construction. They can be used for elevated concrete placement, painting, and sandblasting, as well as pipe, cable tray, and structural steel erection. These tasks on average constitute 33% of total project work-hours [Hsieh and Haas, 1993].

Results from a productivity analysis performed by videotaping identical pipe-handling operations by a pipe manipulator and a telescopic rough-terrain crane indicate a shorter cycle time for the pipe manipulator [Hsieh *et al.*, 1993].

Interior Finishing Robot

A performance study of an interior-finishing robot indicated that the net productivity of the robot can reach 10–19 m^2/h in a one-layer coating and 8–8.5 m^2/h in a dry (mortarless) building. These figures are four to five times higher than for an average construction worker. Wages of $25 per hour, 1500 to 2000 hours of robot usage per year, suitable site conditions, and proper organization of material packaging can result in savings of 20 to 50% in the cost of interior finishing work [Warszawski and Navon, 1991; Warszawski and Rosenfeld, 1993].

Exterior Building Finishers

Results from an outdoor experiment using a tile-setting robot indicate a setting efficiency of 14 m^2/day, with an average adhesive strength of 17.2 kg/cm, representing improved productivity and quality [Kikawada *et al.*, 1993].

Automated Slab Placing and Finishing

According to a study on automated concrete placement and finishing [Moselhi *et al.*, 1992], automation of both placing and finishing concrete slabs would require a minimum annual work volume of 144,321 m^2 (1,600,000 ft^2) of pavement in order to be more economical than the conventional manual process. Thus, at present, the sizable capital cost of the initial investment precludes smaller paving contractors from considering automation.

Shimizu's SMART System

Shimizu Corporation's experience with the SMART system includes (1) improved productivity, (2) attractive working environment, (3) all-weather protection, (4) higher quality and durability, (5) reduced construction schedule duration, and (6) reduced amount of waste and damage to materials. Upon further advancement of the SMART system, a 50% reduction in construction duration is expected [Miyatake and Kangari, 1993].

6.6 Summary

A brief description of construction industry characteristics followed a discussion on the importance of construction automation. Fixed construction automation was defined, and selected examples of fixed construction automation were provided. Following this, programmable automation including robotic and numerical control applications were described. Computer-integrated construction (CIC), which provides an intelligent approach to planning, design, construction, and management of facilities, requires emerging technology that encompasses research efforts from a variety of engineering and computer science disciplines. A detailed description of CIC and supporting areas that play important roles in implementing CIC was provided.

Defining Terms

Electronic controls: Computer-based hardware units designated to control and coordinate the positions and motions of manipulator arms and effectors. A controller is always equipped with manipulator control software, enabling an operator to record a sequence of manipulator motions and subsequently play back these motions a desired number of times. More sophisticated controllers may plan entire sequences of motions and tool activators given a desired work task.

End Effectors: Tools and devices on automated construction equipment, including discharge nozzles, sprayers, scrapers, grippers, and sensors. The robot tools are usually modified compared with tools used by human workers, or even specially designed to accommodate unique characteristics of the working machine.

Manipulators: Stationary, articulated arms that are essential components of industrial robotics. The role of a manipulator arm is to move an effector tool to the proper location and orientation relative to a work object. To achieve sufficient dexterity, arms typically require six axes of motions (i.e., six degrees of freedom): three translational motions (right/left, forward/back, up/down) and three rotational motions (pitch, roll, and yaw).

Motion Systems: Systems that enable the essential features of mobility and locomotion for construction equipment. A variety of mobile platforms can support stationary manipulator arms for performance of required tasks. An example selection of automatically guided vehicle (AGV) platforms is presented in Skibniewski [1988]. However, most automated tasks supported by AGVs in construction will require modified control systems and larger payloads than those in automated factories.

Sensors: A device for converting environmental conditions into electrical signals. An environmental condition might be a mechanical, optical, electrical, acoustic, magnetic, or other physical effect. These effects may occur with various levels of intensity and can be assessed quantitatively by more sophisticated sensors. These measurements are used to control robot movements and, in advanced robots, to plan operations. Sensors are very important to robotics in construction since they instantaneously convey elements of the building environment to the control unit.

Source: Adapted from Hunt, V. D. 1983. *Industrial Robotics Handbook.* Industrial Press, New York.

References

Bell, L. C. and McCullough, B. G. 1988. *Bar Code Applications in Construction.* A report to Construction Industry Institute, SD-33, University of Texas, Austin. February.

Bernold, L. E., Altobelli, F. R., and Taylor, H. 1992. Computer-controlled brick masonry. *J. Comput. Civ. Eng., ASCE.* 6(2):147–161.

Business Roundtable. 1982. *Improving Construction Safety Performance*, p. 3. Construction industry cost effectiveness report. Business Roundtable, New York.

Chang, T., Wysk, R. A., and Wang, S. 1991. *Computer-Aided Manufacturing.* Prentice Hall, Englewood Cliffs, NJ.

Cherneff, J., Logcher, R., and Sriram, D. 1991. Integrating CAD with construction schedule generation. *J. Comput. Civ. Eng., ASCE.* 5(1):64–86.

Cleveland, A. B., and Francisco, V. 1988. Use of real-time animation in the construction process. *Transp. Res. Rec.* 1186:5–39.

Considine, D. M. and Considine, G. D. 1986. *Standard Handbook of Industrial Automation.* Chapman & Hall, New York.

Fisher, D. J. and O'Connor, J. T. 1991. Constructability for piping automation: Field operations. *J. Constr. Eng. Manage., ASCE.* 117(3):468–485.

Foley, J. D. and Van Dam, A. V. 1982. *Fundamentals of Interactive Computer Graphics.* Addison-Wesley, Reading, MA.

Groover, M. P., Parriera, N. D., Doydum, C., and Smith, R. 1989. A survey of robotics technology in construction. *SME Tech. Pap.* May 7–11, pp. 1–20.

Hendrickson, C. and Au, T. 1988. *Project Management for Construction: Fundamental Concepts for Owners, Engineers, Architects and Builders,* p. 486. Prentice Hall, Englewood Cliffs, NJ.

Hijazi, A., AbouRizk, S., and Halpin, D. W. 1988. *Impact of Automation on Construction Fixed Plant Operations.* Technical Report. Purdue University, West Lafayette, IN.

Horning, C. and Kinura, H. 1993. Three-dimensional FEM analysis with graphical interface. *Proc. 5th Int. Conf. Comput. Civ. Build. Eng.* Anaheim, CA, pp. 1283–1290.

Hsieh, T. and Haas, C. 1993. Applications of large-scale manipulators in the construction environment. *Proc. 10th Int. Symp. Automat. Robot. Constr.* Houston, TX, pp. 55–62.

Hsieh, T. Y., Fulton, C., Gibson, G. E., and Haas, C. T. 1993. An evaluation of the pipe manipulator performance in a material handling yard. *Proc. 10th Int. Symp. Automat. Robot. Constr.*, Houston, TX, pp. 55–62.

Hughes, P. J., O'Connor, J. T., and Traver, A. E. 1989. Pipe manipulator enhancements for increased automation. *J. Constr. Eng. Manage., ASCE.* 115(3):412–423.

Hunt, V. D. 1983. *Industrial Robotics Handbook.* Industrial Press, New York.

Jaselskis, E. J. and Anderson, M. R. 1994. On hazardous waste and robotic applications. *J. Environ. Eng., ASCE.* 120(2):359–378.

Kangari, R. and Halpin, D. W. 1989. Potential robotics utilization in construction. *J. Constr. Eng. Manage., ASCE.* 115(1):126–143.

Kikawada, K., Ashikaga, S., Ishikawa, S., and Nishigaki, S. 1993. Practical tile-setting robot for exterior walls. *Proc. 10th Int. Symp. Automat. Robot. Constr.* Houston, TX, pp. 71–75.

Kunigahalli, R. and Russell, J. S. 1994. Framework for development of CAD/CAC systems. Under review.

Kunigahalli, R., Russell, J. S., and Veeramani, D. 1994. An algorithm to extract topological relationships from a wire-frame CAD model of concrete floors. *J. Comput. Civ. Eng., ASCE.* In press.

Luggen, W. W. 1984. *Fundamentals of Numerical Control.* Delmar, Albany, NY.

Maynard, H. B. 1971. *Industrial Engineering Handbook.* McGraw-Hill, New York.

Miyatake, Y. and Kangari, R. 1993. Experiencing computer integrated construction. *J. Constr. Eng. Manage., ASCE.* 119(2):307–323.

Morad, A. A., Cleveland, A. B., Beliveau, Y. J., Francisco, V. D., and Dixit, S. S. 1992. Path finder: AI-based path planning system. *J. Comput. Civ. Eng., ASCE.* 6(2):114–128.

Mortenson, M. E., 1985. *Geometric Modeling.* John Wiley & Sons, New York.

Moselhi, O., Fazio, P., and Hanson, S. 1992. Automation of concrete slab-on-grade construction. *J. Constr. Eng. Manage., ASCE.* 118(4):731–748.

Motazed, B. and Whittaker, W. L. 1987. *Automated Pipe Mapping.* The Robotics Institute, Carnegie-Mellon University, Pittsburgh, PA. March.

O'Connor, J. T., Traver, A. E., and Tucker, R. L. 1987. Research into automated piping construction. *Proc. 4th Int. Symp. Robot. Artif. Intel. Build. Constr.* Building Research Station, Haifa, Israel. June, pp. 268–273.

130,000 more concrete masons needed in next 10 years. 1988. *Constr. Labor Relations.* Vol. 33, Feb. 24, p. 1391.

Paulson, B. C. 1985. Automation and robotics for construction. *J. Constr. Eng. Manage., ASCE.* 111(3):190–207.

Perini, P. B. 1991. *CCRE and the Construction Industry: An Overview.* Center for Construction Research and Education, Massachusetts Institute of Technology, Cambridge, p. 1.

Point, G. 1988. Two major innovations in current maintenance: The multi-purpose vehicle and the integrated surface patcher. *67th TRB Annu. Meet.,* Transportation Research Board, Washington, D.C., p. 29.

Rehg, J. A. 1992. *Introduction to Robotics in CIM Systems.* Prentice Hall, Englewood Cliffs, NJ.

Reinschmidt, K. F., Griffs, F. H., Browner, P. L. 1991. Integrated engineering, design, and construction. *J. Constr. Eng. Manage., ASCE.* 117(4):756–772.

Rembold, V., Bleme, C., and Dillmann, R. 1985. *Computer-Integrated Manufacturing Technology and Systems.* Marcel Dekker, New York.

Requicha, A. A. G. 1980. Representations of rigid solids: Theory, methods, and systems. *ACM Comp. Surv.* 12(4):439–465.

Requicha, A. A. G. and Rossignac, J. R. 1992. Solid modeling and beyond. *IEEE Comp. Graph. Appl.* September, pp. 31–44.

Reymendt, J. and Worner, J. D. 1993. Object-oriented modeling of concrete structures. *Proc. 5th Int. Conf. Comput. Civ. Build. Eng.* Anaheim, CA, pp. 86–93.

Sagawa, Y. and Nakahara, Y. 1985. Robots for the Japanese construction industry. *IABSE Proceedings.* May, p. 86.

Saito, M. 1985. *The Development of a Mobile Robot for Concrete Slab Finishing.* Technical Report. Mechanical Engineering Development Dept., Kajima Co., Tokyo, Japan.

Sherman, P. J. 1988. Japanese construction R&D: Entrée into U.S. market. *J. Constr. Eng. Manage., ASCE.* 114(1):138–142.

Skibniewski, M. J. 1988. *Robotics in Civil Engineering,* p. 65. Van Nostrand Reinhold, New York.

Skibniewski, M. J. and Russell, J. S. 1989. Robotic applications to construction. *Cost Eng.* 31(6): 10–18.

Slocum, A. H., Laura, D., Levy, D., Schena, B., and Ziegler, A. 1987. Construction automation research at the Massachusetts Institute of Technology. *Proc. 4th Int. Symp. Robot. Artif. Intel. Build. Constr.* Building Research Station, Haifa, Israel. June, pp. 222–233.

Stukhart, G. and Berry, W. D. 1992. *Evaluation of Voice Recognition Technology.* Report to Construction Industry Institute, SD-76. University of Texas, Austin. June.

Sugiyura, H., Yute, S., Nishide, K., Hatekeyama, O., and Nishigaki, S. 1993. Autonomous dump trucks systems for transporting and positioning heavy-duty materials in heavy construction sites. *Proc. 10th Int. Symp. Automat. Robot. Constr.* Houston, TX, pp. 253–260.

Tatum, C. B. and Funke, A. T. 1988. Partially automated grading: Construction process innovation. *J. Constr. Eng. Manage., ASCE.* 114 (1):19–35.

Technical Specifications for Miller Formless Systems, Automated Slip Forming Systems, Bulk Cement and Unloading Systems. 1988. Miller Formless Systems Co., McHenry, IL.

Technical Specifications of the John Deere 690C Excavator. 1985. John Deere, Inc., Moline, IL. August.

Teicholz, E. and Orr, J. N. 1987. *Computer-Integrated Manufacturing Handbook.* McGraw-Hill, New York.

Terauchi, S., Miyajima, T., Mirjamoto, T., Arai, K., and Takiwaza, S. 1993. Development of an exterior wall painting robot—Capable of painting walls with indentations and protrusions. *Proc. 10th Int. Symp. Automat. Robot. Contr.* Houston, TX, pp. 285–292.

Touran, A. and Ladick, D. 1988. Applications of robotics on bridge deck fabrication. *J. Constr. Eng. Manage., ASCE.* 115(1):35–52.

Uchino, T., Narisawa, J. Sato, Y., and Kumazawa, K. 1993. Multi-jointed pile driving machine with a computer-assisted guiding system. *Proc. 10th Int. Symp. Automat. Robot. Constr.* Houston, TX, pp. 363–370.

Warszawski, A. 1990. *Industrialization and Robotics in Building: A Managerial Approach.* Harper & Row, New York.

Warszawski, A. and Navon, R., 1991. Robot for interior-finishing works. *J. Constr. Eng. Manage., ASCE.* 117(3):402–422.

Warszawski, A. and Rosenfeld, Y. 1993. Feasibility analysis of robotized vs. manual performance of interior finishing tasks. *Proc. 10th Int. Symp. Automat. Robot. Constr.* Houston, TX, pp. 383–390.

Webster, A. C. 1993. Japanese building design and construction technology. *J. Prof. Issues Eng. Educ. Prac., ASCE.* 119(4):358–377.

Whittaker, W. L. 1985a. *REX: A Robot Excavator.* The Robotics Institute, Carnegie-Mellon University, Pittsburgh, PA. August.

Whittaker, W. L. 1985b. *Terregator.* Technical Report. The Robotics Institute, Carnegie-Mellon University, Pittsburgh, PA. August.

Wright, R. N. 1991. Competing in for construction. *Constr. Bus. Rev.* 1(3):36–39.

Yoshida, T. and Ueno, T. 1985. *Development of a Spray Robot for Fireproof Treatment,* pp. 48–63. Shimizu Technical Research Bulletin 4. Tokyo, Japan.

For Further Information

Proceedings of the International Symposium on Automation and Robotics in Construction (ISAARC) provide information on the latest developments in the field of construction automation. The *International Journal of Automation in Construction,* the *ASCE Journal of Computing in Civil Engineering,* and the *ASCE Journal of Construction Engineering and Management* document the advances in computer-integrated construction.

The following books provide a basic introduction to construction robotics: *Robotics in Civil Engineering* by M. J. Skibniewski and *Industrialization and Robotics in Building: A Managerial Approach* by A. Warszawski.

In 1994, Citizens Gas & Coke Utility in Indianapolis, Indiana completed construction of a $35 million ammonia destruct/desulfurization facility. Citizens Gas & Coke Utility operates one of the largest merchant coke plants in the U.S., producing high-quality foundry and blast furnace coke. The recently completed two-phase project has eliminated approximately 5000 pounds a day of ammonia that previously had been discharged to the City of Indianapolis advanced wastewater treatment plants. The ammonia destruct system (shown above) is over 90% efficient, and there are no releases of pollutants to the air of the generation of any hazardous waste. The ammonia molecule is destroyed at high temperature using nickel catalyst, and even the resulting hydrogen gas is used as an energy source for the Citizens Gas & Coke Utility coke oven batteries.

1994 also saw the completion of the Citizens Gas desulfurization plant (right) that has eliminated in one year over 4000 tons of hazardous waste. 99.99%-pure elemental sulfur is recovered for sale.

The ammonia destruct facility was the first of its kind in the Western Hemisphere and, coupled with its sister desulfurization facility, represents the only combination ammonia destruct/desulfurization facility like it in the world. Krupp Wilputt designed and built the facility for Citizens Gas & Coke Utility. (Photo courtesy of Citizens Gas & Coke Utility.)

Environmental Engineering

Robert B. Jacko
Purdue University

DURING THE EVOLUTION OF THE U.S., the water, air, and land resources available to our forefathers were immeasurably vast. So vast, in fact, that they appeared to be of infinite proportions, and their use and consumption were taken for granted. However, as the population grew it became clear that these resources, particularly a clean and abundant water supply, were not infinite and in some cases not even available. A case in point is the water supply problem that confronted New York almost from its inception. A visitor to New York in 1748 declared, "There is no good water to be met within the town itself" [Koeppel, 1994]. In 1774 the city authorized a water system, but it was not until 1841, when the Croton Aqueduct was completed, that New Yorkers could experience cool, clean water for drinking, bathing, and fire fighting. They could even dream about the luxury of indoor plumbing. Four years prior to 1841, a son was born to a humble British family in the Yorkshire town of Thorne who was to make a major contribution regarding the handling of human waste

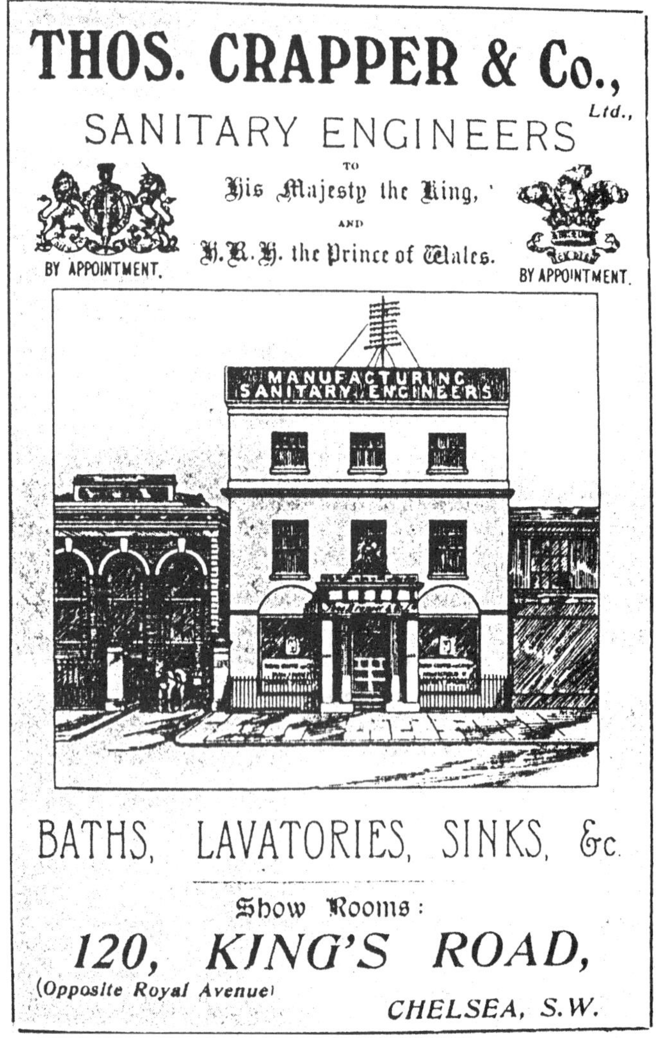

FIGURE II.1 Thomas Crapper invented many improvements to indoor flush toilets.

products. The child's name was Thomas Crapper. Figure II.1 shows an advertisement for Thomas Crapper & Company. Crapper was an entrepreneurial sanitary engineer and the inventor of many improvements to indoor flush toilets [Rayburn, 1989].

By 1850 there were only 83 public water supplies in the U.S., but the demand was growing, and by 1870 there were 243 [Fuhrman, 1984]. With these burgeoning public water supplies came the need to consider the disposal of the "used" water. In Europe during the Middle Ages, people simply threw their excreta out the window, as the woodcut in Fig. II.2 demonstrates [Rayburn, 1989]. Word has it that some sport was involved in this process involving the passersby in the street below [Alleman, 1994].

Recognition at about this time that water supplies, disease, and disposal of human waste were interconnected led to the requirement that used water and excrement be discharged to sewers. In 1850, a member of the Sanitary Commission of Massachusetts, Lemuel Shattuck, reported the relationship between water supply, sewers, and health. He recommended the formation of a State Board of Health, which would include a civil engineer, a chemist or physicist, two physicians, and two others. During this time, a French chemist by the name of Louis Pasteur was initiating research that was to found the field of bacteriology and connect bacteria with disease. In addition, Pasteur

FIGURE II.2 "Sanitation" in the Middle Ages. (From an old woodcut.)

was to demonstrate the benefits of utilizing bacteria in industrial processes. The use of bacteria to stabilize municipal waste was coming to the fore.

In 1887, the Massachusetts State Board of Health established an experiment station at Lawrence for investigating water treatment and water pollution control. This station was similar to others which had been established in England and Germany and was the forerunner of eight others established throughout the U.S. Topics investigated were primary wastewater treatment, secondary treatment via trickling filters, and activated sludge.

As the population of the U.S. continues to grow, greater demand is being placed on our natural resources. What were once adequate treatment and disposal methods now require far greater levels of cleanup before waste is discharged to water courses, the atmosphere, or onto the land. In essence, the use of water, air, and land is no longer a free economic good, as has been assumed for so many years. The cost of using water, air, and land resources is the cleanup cost prior to their return to the environment. This section will deal with those broader topics in water treatment, wastewater treatment, air pollution, landfills, and incineration.

References

Alleman, J. E. 1994. Personal communication.

Fuhrman, R. E. 1984. History of water pollution control. *J. Water Pollut. Control Fed.* 56(4):306–313.

Koeppel, G. 1994. A struggle for water. *Invent. Technol.* 9(3).

Rayburn, W. 1989. *Flushed with Pride.* Pavilion, London.

7

Water and Wastewater Planning

Robert M. Sykes
Ohio State University

7.1 Treatment Standards

Water works, water distribution systems, sewerage, and sewage treatment works constitute an integrated system. The primary purpose of this system is to protect the public health and to prevent nuisances. This is achieved as follows:

- *Water works* produce potable waters that are free of pathogens and poisons.
- *Water distribution* systems prevent the posttreatment contamination of potable water while storing it and delivering it to users on demand.
- *Sewerage systems* efficiently and safely collect contaminated used water, thereby preventing disease transmission and nuisance, and transmit it to sewage treatment works without loss or contamination of the surrounding environment.
- *Sewage treatment works* remove contaminants from the used water prior to its return to its source, thereby preventing contamination of the source and nuisance.

Overall, this system has been successful in controlling waterborne disease, and such disease is now rare in modern industrial economies.

The secondary purpose of sewage treatment is to preserve wildlife and to maintain an ambient water quality sufficient to permit recreational, industrial, and agricultural uses.

Water Treatment

Potable water quality in the U.S. is regulated by the U.S. EPA under authority of the "Safe Drinking Water Act" of 1974 (PL 93-523) and its amendments. The Act applies to any piped water supply that either has at least 15 connections or regularly serves at least 25 people. Day-to-day administration of the Act is delegated to the states by the U.S. EPA. The fundamental obligations of the U.S. EPA are to establish primary regulations for the protection of the public health; to establish secondary

0-8493-8953-4/95/$0.00 + $.50

regulations relating to taste, odor, color, and appearance of drinking water; to protect underground drinking water supplies; and to assist the states via technical assistance, personnel training, and money grants. Regulations include criteria for water composition, treatment technologies, system management, and statistical and chemical analytical techniques.

Maximum Contaminant Limits

The U.S. EPA has established both "Maximum Contaminant Limits (MCL)," which are legally enforceable standards of quality, and "Maximum Contaminant Limit Goals (MCLG)," which are nonenforceable health-based targets. The MCL are summarized in Table 7.1. For comparative purposes, the earlier standards of the U.S. Public Health Service and the standards of the World Health Organization are included. The standards apply at the consumer's tap, not at the treatment plant nor at any point in the distribution system.

It should be noted that the bacterial limits are no longer given as most probable numbers (MPN) or as membrane filter counts (MFC) but rather as the fraction of 100 mL samples that test positive in any month. The limits on lead and copper are thresholds that require implementation of specific treatment processes to inhibit corrosion and scale dissolution. Some substances are not yet subject to regulation, but in the interim they must be monitored. Some substances must be monitored by all facilities; others must be monitored only if, in the judgment of the state authority, monitoring is warranted.

Violations of Drinking Water Regulations

Water supply systems must notify the people they serve whenever:

- A violation of a National Primary Drinking Water Regulation or monitoring requirement occurs.
- Variances or exemptions are in effect.
- Noncompliance with any schedule associated with a variance or exemption occurs.

If the violation involves an MCL, a prescribed treatment technique, or a variance/exemption schedule, a notice must be published in the local newspapers within 14 days. If there are no local newspapers, the notice must be given by hand delivery or posting. In any case, notification by mail or hand delivery must occur within 45 days, and notification must be repeated quarterly as long as the problem persists.

If (1) the violation incurs a severe risk to human health as specified by a state agency, (2) the system violates the MCL for nitrate, (3) the system violates the MCL for total coliform when fecal coliform or *Escherichia coli* is known to be present, or (4) there is an outbreak of waterborne disease in an unfiltered supply, notification must be made by television and radio within 72 hours.

Wastewater Treatment

Wastewater discharges are regulated under the Federal Water Pollution Control Act of 1972 (PL 92-500), as amended.

Stream Standards

Terms like *pollution* and *contamination* require quantitative definition before abatement programs can be undertaken. Quantification permits engineering and economic analysis of projects. In the U.S., the water bodies are first classified as to suitability for "beneficial uses," which include the following:

- Wildlife preservation—warm water habitats, exceptional warm water habitats, cold water habitats
- Historic and/or scenic preservation

Table 7.1 Maximum Contaminant Concentrations Allowable in Drinking Water (Action Levels) *(continues)*

Parameter	Authority		
	U.S. PHS[a]	U.S. EPA[b,c]	WHO[d]
Pathogens and parasites:			
Total coliform bacteria (number/100mL)	1	<5% positive samples in a set of ≥40 per month, or <1 sample positive in a set of <40 per month	0
Inorganic Positions (mg/L):			
Antimony	—	0.006	—
Arsenic	0.05	0.05 (interim)	0.05
Asbestos (million fibers > 10μm per liter)	—	7.0	—
Barium	1	2	—
Beryllium	—	0.004	—
Cadmium	0.01	0.005	0.005
Chromium	0.05	0.1 (total)	0.05
Copper	—	1.3 90th percentile action level, requires corrosion control	—
Cyanide	0.2	0.2	0.1
Fluoride	See Nuisances	4.0	—
Lead	0.05	0.015 90th percentile action level, requires corrosion control	0.05
Mercury (inorganic)	—	0.002	0.001
Nickel	—	0.1	—
Nitrate (as N)	10	10	10
Nitrite (as N)	—	1	—
Nitrate plus nitrite (as N)	—	10	—
Selenium	0.01	0.05	0.01
Sulfate	—	Deferred (400 to 500?)	—
Thallium	—	0.002	—
Organic poisons (μg/L, except as noted):			
Acrylamide	—	Use in treatment, storage, and distribution restricted	—
Alachor	—	2	—
Aldicarb	—	Deferred (3?)	—
Aldicarb sulfoxide	—	Deferred (4?)	—
Aldicarb sulfone	—	Deferred (2?)	—
Aldrin and dieldrin	—	—	0.03
Atrazine	—	3	—
Benzene	—	5	10
Benzo[a]pyrene	—	0.2	0.01
Bromobenzene	—	Monitor	—
Bromochloromethane	—	Monitor if ordered	—
Bromodichloromethane	—	Monitor	—
Bromoform	—	Monitor	—
Bromomethane	—	Monitor	—
n-Butylbenzene	—	Monitor if ordered	—
sec-Butylbenzene	—	Monitor if ordered	—
tert-Butylbenzene	—	Monitor if ordered	—
Carbofuran	—	40	—

Table 7.1 (continued) Maximum Contaminant Concentrations Allowable in Drinking Water (Action Levels)

Parameter	Authority		
	U.S. PHS[a]	U.S. EPA[b,c]	WHO[d]
Carbon chloroform extract	200	—	—
Carbon tetrachloride	—	5	—
Chlordane	—	2	0.3
Chlorobenzene	—	100	—
Chlorodibromomethane	—	Monitor	—
Chloromethane	—	Monitor	—
Chloroform	—	Monitor	30
Chloromethane	—	Monitor	—
m-Chlorotoluene	—	Monitor	—
p-Chlorotoluene	—	Monitor	—
2,4-D	—	70	100
Dalapon	—	200	—
DDT	—	—	1
1,2-Dibromo-3-chloropropane (DBCP)	—	0.2	—
Dibromomethane	—	Monitor	—
m-Dichlorobenzene	—	Monitor	—
o-Dichlorobenzene	—	600	—
p-Dichlorobenzene	—	75	—
Dichlorodifluoromethane	—	Monitor if ordered	—
1,1-Dichloroethane	—	Monitor	—
1,2-Dichloroethane	—	5	10
1,1-Dichloroethylene	—	7	0.3
cis-1,2-Dichloroethylene	—	70	—
trans-1,2-Dichloroethylene	—	100	—
Dichloromethane	—	5	—
1,2-Dichloropropane	—	5	—
1,3-Dichloropropane	—	Monitor	—
2,2-Dichloropropane	—	Monitor	—
1,1-Dichloropropene	—	Monitor	—
1,3-Dichloropropene	—	Monitor	—
Di(2-ethylhexyl)adipate	—	400	—
Di(2-ethylhexyl)phthalate	—	6	—
Dinoseb	—	7	—
Dioxin (2,3,7,8-TCDD)	—	30×10^{-6}	—
Diquat	—	10	—
Endothall	—	100	—
Endrin	—	2	—
Epichlorhydrin	—	Use in treatment, storage, and distribution	
Ethylbenzene	—	700	—
Ethylene dibromide (EDB)	—	0.05	—
Fluorotrichloromethane	—	Monitor if ordered	—
Glyphosate (aka Rodeo™ and Roundup™)	—	700	—
Heptachlor	—	0.4	0.1
Heptachlor epoxide	—	0.2	—
Hexachlorobenzene	—	1	0.01
Hexachlorobutadiene	—	Monitor if ordered	—
Hexachlorocyclopentadiene (Hex)	—	50	—
Isopropylbenzene	—	Monitor if ordered	—
p-Isopropyltoluene	—	Monitor if ordered	—
Lindane	—	0.2	3
Methoxychlor	—	40	30
Naphthalene	—	Monitor if ordered	—
Oxamyl (Vydate)	—	200	—

Table 7.1 (continued) Maximum Contaminant Concentrations Allowable in Drinking Water (Action Levels) *(continues)*

Parameter	Authority		
	U.S. PHS[a]	U.S. EPA[b,c]	WHO[d]
Pentachlorophenol	—	1	10
PCB (polychlorinated biphenyl)	—	0.5	—
Picloram	—	500	—
n-Propylbenzene	—	Monitor if ordered	—
Silvex (2,4,5-TP)	—	10	—
Simazine	—	4	—
Styrene	—	100	—
2,3,7,8-TCDD (dioxin)	—	30×10^{-6}	—
1,1,1,2-Tetrachloroethane	—	Monitor	—
1,1,2,2-Tetrachloroethane	—	Monitor	—
Tetrachloroethylene	—	5	—
Toluene	—	1,000	—
Toxaphene	—	3	—
1,2,3-Trichlorobenzene	—	Monitor if ordered	—
1,2,4-Trichlorobenzene	—	70	—
1,1,1-Trichloroethane	—	200	—
1,1,2-Trichloroethane	—	5	—
Trichloroethylene	—	5	—
1,2,3-Trichloropane	—	Monitor	—
2,4,6-Trichlorophenol	—	—	10
Trihalomethanes (total)	—	100	—
1,2,4-Trimethylbenzene	—	Monitor if ordered	—
1,3,5-Trimethylbenzene	—	Monitor if ordered	—
Vinyl chloride	—	2	—
Xylene (total)	—	10,000	—
m-Xylene	—	Monitor	—
o-Xylene	—	Monitor	—
p-Xylene	—	Monitor	—
Radioactivity (pCi/L, except as noted):			
Gross alpha	—	Deferred (15?)	2.7
Gross beta	1000	—	27
Gross beta/proton (mrem/yr)	—	Deferred (4?)	—
Radium-226	10	Deferred (5?)	—
Radium-228	—	Deferred (5?)	—
Radon-222	—	Deferred (200?)	—
Strontium-90	3	—	—
Uranium (μg/L)	—	Deferred (20?)	—
Nuisances (mg/L, except as noted):			
Alkyl benzene sulfonate	0.5	—	—
Aluminum	—	—	0.2
Chloride	250	250	250
Color (Pt-Co units)	15	15	15
Copper	1	See above	1
Corrosivity (Langelier Index)	—	—[aa]	—
Fluoride	depending on air temperature 0.8–1.7	See above	—
Hardness (as $CaCO_3$)	—	—	500
Hydrogen sulfide	—	—	—[bb]
Iron	0.3	0.3	0.3
Manganese	0.05	0.05	0.1
Methylene blue active substances	—	0.5	—
Odor (threshold odor number)	3	3	—[cc]
pH	—	6.5/8.5	6.5/8.5
Phenol (μg/L)	1	—	—

Table 7.1 (continued) Maximum Contaminant Concentrations Allowable in Drinking Water (Action Levels)

Parameter	Authority		
	U.S. PHS[a]	U.S. EPA[b,c]	WHO[d]
Silver	0.05	0.05	—
Sodium	—	—[aa]	200
Sulfate	250	Deferred	400
Taste	—	—	—[cc]
Total dissolved solids	500	500	1000
Turbidity	5	All samples ≤ 5; 95% of samples ≤ 0.5	5
Zinc	5	5	5

[a] Hopkins, O. C. 1962. *Public Health Service Drinking Water Standards 1962.* U.S. Department of Health, Education, and Welfare, Public Health Service, Washington, D.C.

[b] Pontius, F. W. 1990. Complying with the new drinking water quality regulations. *J. Am. Water Works Assn.*, 82(2): 32.

[c] Auerbach, J. 1994. Costs and benefits of current SDWA regulations. *J. Am. Water Works Assn.*, 86(2): 69.

[d] Anonymous. 1984. *Guidelines for Drinking Water Quality: Volume 1. Recommendations.* World Health Organization, Geneva.

[aa] To be monitored and reported to appropriate agency and/or public.

[bb] Not detectable by consumer.

[cc] Not offensive for most consumers.

- Recreation—primary or contact recreation (e.g., swimming) and secondary or noncontact recreation (e.g., boating)
- Fisheries—commercial and sport
- Agricultural usage—crop irrigation and stock watering
- Industrial usage—process water, steam generation, cooling water
- Navigation
- Hydropower
- Public water supply source (which assumes treatment prior to use)

Once a water body has been classified, the water volume and composition needed to sustain that usage can be specified. Such a specification is called a *stream standard.* A water body is contaminated if any one of the various volume and composition specifications is violated. Stream standards are revised every three years. Standards are issued by competent state agencies subject to review and approval by the U.S. EPA. Commonly recommended standards for various beneficial uses are given in Tables 7.2 through 7.6.

Rules of Thumb. The determination of whether a particular discharge will cause a stream standard violation is difficult, particularly when the contaminants undergo physical, chemical, or biological transformations and when competing processes (e.g., BOD decay and reaeration) occur. There are, however, a few rules of thumb that are useful guides to permit specification. The rules of thumb restrict BOD_5 and settleable solids, which adversely affect stream dissolved oxygen levels, and ammonia, which is toxic to many fish.

Fuller [1912] reviewed several field studies conducted in France, Massachusetts, and Ohio on the effect of sewage discharges on rivers and reached the following conclusions:

- The sewage should be settled and skimmed to remove settleable and floatable material.
- The flow in the receiving stream should be at least 4 to 7 cfs per 1000 persons.

Table 7.2 Maximum Contaminant Concentrations Allowable in Sources of Public Water Supplies *(continues)*

Parameter	Authority		
	FWPCA[a]	U.S. EPA[b]	CEC[c]
Pathogens and parasites (number/100 mL):			
Total coliform bacteria	10,000	—	5000
Fecal coliform bacteria	2,000	—	2000
Fecal streptococci	—	—	1000
Salmonella	—	—	—[d]
Inorganic poisons (mg/L):			
Arsenic	0.05	0.05	0.05
Barium	1	1	1
Boron	1	—	1
Cadmium	0.01	0.01	0.005
Chromium	0.05	0.05	0.05
Cyanide	0.2	—	0.05
Fluoride	0.8–1.7	—	0.7/1.7
Lead	0.05	0.05	0.05
Mercury	—	0.002	0.001
Nitrate (as N)	10	10	11.3
Selenium	0.01	0.01	0.01
Silver	0.05	0.05	—
Organic poisons (μg/L)			
Aldrin and dieldrin	34	—[e]	—
Chlordane	3	—[e]	—
Chloroform extract	—	—	200
2,4-D (see herbicides)	—	100	—
DDT	42	—[e]	—
Endrin	1	0.2	—
Ether-soluble hydrocarbons	—	—	200
Heptachlor	18	—[e]	—
Heptachlor epoxide	18	—	—
Herbicides (total)	100	—	—
Lindane	56	4	—
Methoxychlor	35	100	—
Organophosphates and carbamates	100	—	—
Pesticides	—	—	2.5
Polycyclic aromatic hydrocarbons	—	—	0.2
Silvex (see herbicides)	—	10	—
Toxaphene	—	5	—
Radioactivity (pCi/L):			
Gross beta	1,000	—	—
Radium-226	3	—	—
Strontium-90	10	—	—
Nuisances (mg/L, except as noted):			
Aesthetic qualities	—	—[f]	—
Ammonia (as NH^{4+})	0.64	—	1.5
Biochemical oxygen demand	—	—	5
Chloride	250	—	200
Color (Pt-Co units)	75	75	100
Conductivity (μs/L)	—	—	1000
Copper	1	1	0.05
Dissolved oxygen	>4	—	>50.% sat
Iron	0.3	0.3	2
Manganese	0.05	0.05	0.1
Methylene blue–active substances	0.5	—	0.2
Odor (threshold odor number)	—[g]	—	10

Table 7.2 (continued) Maximum Contaminant Concentrations Allowable in Sources of Public Water Supplies

Parameter	Authority		
	FWPCA[a]	U.S. EPA[b]	CEC[c]
Oil and grease	—[h]	—[h]	—
pH (pH units)	6–8.5	5–9	5.5–9
Phenol	0.001	0.001	0.005
Phosphate (as P_2O_5)	—	—	0.7
Sulfate	250	—	250
Tainting substances	—	—[i]	—
Temperature (°C)	30	—	25
Total dissolved solids	500	250	—
Total Kjeldahl nitrogen	—	—	2
Uranyl ion (as UO_2^{2-})	5	—	—
Zinc	5	5	5

[a] Ray, H. C., *et al.* 1968. *Water Quality Criteria, Report of the National Technical Advisory Committee to the Secretary of the Interior.* U.S. Department of the Interior, Federal Water Pollution Control Administration, Washington, D.C.

[b] Anonymous. 1976. *Quality Criteria for Water.* U.S. Environmental Protection Agency, Office of Water Planning and Standards, Criteria and Standards Division, Criteria Branch, Washington, D.C.

[c] Council of the European Communities. 1975. Council directive of 16 June 1975. *Official Journal of the European Communities: Legislation,* 19(L31), 1–37.

[d] Absent in 1000 mL.

[e] Human exposure to be minimized.

[f] To be free from substances attributable to wastewaters or other discharges that (1) settle to form objectionable deposits; (2) float as debris, scum, oil, or other matter to form nuisances; (3) produce objectionable color, odor, taste, or turbidity; (4) injure or are toxic or produce adverse physiological response in humans, animals or plants; and (5) produce undesirable or nuisance aquatic life.

[g] Not objectionable.

[h] Virtually free.

[i] Substances should not be present in concentrations that produce undesirable flavors in the edible portions of aquatic organisms.

Fuller's recommendations result in an increase of 3 to 5 mg/L in the stream BOD_5. The requirement for sedimentation is especially important, because settleable solids have a disproportionate impact on stream oxygen values. Floatables should be removed to avoid nuisance.

In 1913 the Royal Commission on Sewage Disposal published its studies on the development of the biochemical oxygen demand test procedure and its applications. The Commission concluded that the BOD_5 of rivers should be held to less than 4 mg/L. This recommendation is supported by the work of Sladacek and Tucek [1975], who concluded that a BOD_5 of 4 mg/L produced a stream condition called *beta-mesosaprobic* [Kolkwitz and Marrson, 1909], in which the stream benthos (bottom dwelling) contains predominantly clean-water fauna.

The unprotonated ammonia molecule, NH_3, is toxic to fish at concentrations about 0.02 mg NH_3/L [Anonymous, 1976]. Ammonia reacts with water to form the ammonium ion, NH_4^+, which is the predominant form in most natural waters:

$$NH_3 + H_2O = NH_4^+ + OH^- \tag{7.1}$$

The equilibrium constant for this reaction is called the *base ionization constant* and is defined as

$$K_b = \frac{[NH_4^+][OH^-]}{[NH_3]} \tag{7.2}$$

Table 7.3 Maximum Contaminant Concentrations Allowable in Recreational Waters

Parameter	Authority	
	FWPCA[a]	CED[b]
Aesthetics	—[c]	—[d]
General recreational use		
Fecal coliform bacteria (number/100mL)	2000	—
Miscellaneous	—[e]	—
Primary contact recreation		
Total coliform bacteria (number/100mL)	—	10,000
Fecal coliform bacteria (number/100mL)	200	2,000
Fecal streptococci (number/100mL)	—	100
Salmonella (number/1L)	—	0
Enteroviruses (PFU/10L)	—	0
pH (pH units)	6.5–8.3	6/9
Temperature (°C)	30	—
Clarity (Secchi disc, ft)	4	3.28
Color (Pt-Co units)	—	—[f]
Dissolved oxygen (% sat.)	—	80–120
Mineral oils (mg/L)	—	0.3
Methylene blue–active substances (mg/L)	—	0.3
Phenols (mg/L)	—	0.005
Miscellaneous	—[e]	—[g]

[a] Ray, H. C., *et al.* 1968. *Water Quality Criteria, Report of the National Technical Advisory Committee to the Secretary of the Interior.* U.S. Department of the Interior, Federal Water Pollution Control Administration, Washington, D.C.

[b] Council of the European Communities. 1975. Council directive of 16 June 1975. *Official Journal of the European Communities: Legislation,* 19(L31), 1–7.

[c] All surface waters should be capable of supporting life forms of aesthetic value. Surface waters should be free of substances attributable to discharges or wastes as follows: (1) materials that will settle to form objectionable deposits; (2) floating debris, oil, scum, and other matter; (3) substances producing objectionable color, odor, taste, or turbidity; (4) materials, including radionuclides, in concentrations or in combinations that are toxic or that produce undesirable physiological responses in human, fish, and other animal life and plants; (5) substances and conditions or combinations thereof that produce undesirable aquatic life.

[d] Tarry residues and floating materials such as wood; plastic articles; bottles; containers of glass, rubber, or any other substances; and waste or splinters shall be absent.

[e] Surface waters, with specific and limited exceptions, should be of such quality as to provide for the enjoyment of recreational activities based on the utilization of fishes, waterfowl, and other forms of life without reference to official designation of use. Species suitable for harvest by recreation users should be fit for human consumption.

[f] No abnormal change.

[g] If the quality of the water has deteriorated or if their presence is suspected, competent authorities shall determine the concentrations of pesticides, heavy metals, cyanides, nitrates, and phosphates. If the water shows a tendency toward eutrophication, competent authorities shall check for ammonia and total Kjeldahl nitrogen.

It is also possible to write what is called an *acid ionization constant:*

$$K_a = \frac{[H^+][NH_3]}{[NH_4^+]} \tag{7.3}$$

Table 7.4 Maximum Contaminant Concentrations Allowable in Aquatic Habitats[a]

Parameter	Habitat		Parameter	Habitat	
	Freshwater	Marine		Freshwater	Marine
Inorganic poisons (mg/L, except as noted)			Organic poisons (μg/L, except as noted)		
Alkalinity (as $CaCO_3$)	≥20	—	Aldrin and dieldrin	0.003	0.003
Ammonia (un-ionized)	0.02	—	Chlordane	0.01	0.004
Beryllium	—	—	DDT	0.001	0.001
Hard water	1.1	—	Demeton	0.1	0.1
Soft water	0.011	—	Endosulfan	0.003	0.001
Cadmium	—	5	Endrin	0.004	0.004
Hard water	0.0012–0.012	—	Guthion	0.01	0.01
Soft water	0.0004–0.004	—	Heptachlor	0.001	0.001
Chlorine (total residual)	0.01	0.01	Lindane	0.01	0.004
Salmonid fish	0.002	0.002	Malathion	0.1	0.1
Chromium	0.1	—	Methoxychlor	0.03	0.03
Copper (X96-hr LC_{50})	0.1	0.1	Mirex	0.001	0.001
Cyanide	0.005	0.005	Oil and grease (times the 96-hr LC_{50})[b]	0.01	0.01
Dissolved oxygen	≥5	—	Parathion	0.04	0.04
Hydrogen sulfide (undissociated)	0.002	0.002	Phthalate esters	3	—
Iron	1	—	Polychlorinated biphenyls	0.001	0.001
Lead (times the 96-hr LC_{50})	0.01	—	Toxaphene	0.005	0.005
Mercury (μg/L)	0.05	0.1			
Nickel (times the 96-hr LC_{50})	0.01	0.01	Miscellaneous (mg/L, except as noted)		
pH (pH units)	6.5/9.0	6.5/8.5	Temperature increase (°C)	—[c]	1
Phosphorus (elemental, μg/L as P)	—	0.1	Total phosphate (as P)	0.025	—
Selenium (times the 96-hr LC_{50})	0.01	0.01	Turbidity	—[d]	—
Silver (times the 96-hr LC_{50})	0.01	0.01			
Total dissolved gases (% sat.)	110	110			

[a] Anonymous. 1976. *Quality Criteria for Water.* U.S. Environmental Protection Agency, Office of Water Planning and Standards, Criteria and Standards Division, Criteria Branch, Washington, D.C.

[b] Oils or petrochemical levels that cause deleterious effects to the biota should not be allowed in the sediments, and surface waters should be virtually free from floating nonpetroleum oils of vegetable or animal origin, as well as petroleum-derived oils.

[c] In cooler months, maximum plume temperatures must be such that important species will not die if the plume temperature falls to the ambient water temperature. In warmer months, the maximum plume temperature may not exceed the optimum temperature of the most sensitive species by more than one-third of the difference between that species' optimum and ultimate upper incipient lethal temperatures. During reproductive seasons, the plume temperature must permit migration, spawning, egg incubation, fry rearing, and other reproductive functions of important species.

[d] The compensation point for photosynthesis may not be reduced by more than 10% of the seasonally adjusted norm.

This corresponds to the reaction

$$NH_4^+ = NH_3 + H^+ \tag{7.4}$$

The two constants are related by the ionization product of water:

$$K_a K_b = [H^+][OH^-] = K_w \tag{7.5}$$

The molar fraction of the total ammonia concentration that is unprotonated ammonia is

$$f = \frac{[NH_3]}{[NH_3] + [NH_4^+]} = \frac{1}{1 + \dfrac{[H^+]}{K_a}} = \frac{1}{1 + \dfrac{K_b}{OH^-}} \tag{7.6}$$

Table 7.5 Maximum Contaminant Concentrations Allowable in Irrigation Water[a]

Parameter	Concentration	Parameter	Concentration
Pathogens and parasites (number/100mL)		Selenium	0.05
Total coliform bacteria	5000	Vanadium	10
Fecal coliform bacteria	1000	Zinc	5.0
Inorganic poisons (mg/L, except as noted)		Herbicides (mg/L vs. corn):	
Aluminum	1.0	Acrolein	60
Arsenic	1.0	Amitrol-T	>3.5
Beryllium	0.5	Dalapon	<0.35
Boron	0.75	Dichlobenil	>10
Cadmium	0.005	Dimethylamines	>25
Chloride (meq/L)	20	Diquat	125
Chromium	5.0	Endothall	25
Cobalt	0.2	Fenac	10
Copper	0.2	Pichloram	>10
Electrical conductivity (mmhos/cm)	1.5–18	Radionuclides (pCi/L):	
Lead	5.0	Gross beta	1000
Lithium	5.0	Radium-226	3
Manganese	2.0	Strontium-90	10
Molybdenum	0.005	Soil deflocculation (units as noted):	
Nickel	0.5	Exchangeable sodium ratio (%)	10–15
pH (pH units)	4.5–9.0	Sodium absorption ratio [$(meq/L)^{0.5}$]	4–18

[a] Ray, H. C., *et al.* 1968. *Water Quality Criteria, Report of the National Technical Advisory Committee to the Secretary of the Interior.* U.S. Department of the Interior, Federal Water Pollution Control Administration, Washington, D.C.

Table 7.6 Maximum Contaminant Concentrations in Surface Waters That Have Been Used for Cooling Water[a]

	Source			
	Freshwater		Brackish	
Parameter	Once-through	Recycle	Once-through	Recycle
Inorganic substances (mg/L, except as noted):				
Acidity (as $CaCO_3$)	0	200	0	0
Alkalinity (as $CaCo_3$)	500	500	150	150
Aluminum	3	3	—	—
Bicarbonate	600	600	180	180
Calcium	500	500	1,200	1,200
Chloride	600	500	22,000	22,000
Hardness (as $CaCO_3$)	850	850	7,000	7,000
Hydrogen Sulfide	—	—	4	4
Iron	14	80	1	1
Manganese	2.5	10	0.02	0.02
Nitrate (as NO_3)	30	30	—	—
pH (pH units)	5–8.9	3–9.1	5–8.4	5–8.4
Phosphate (as PO_4)	4	4	5	5
Sulfate	680	680	2,700	2,700
Suspended solids	5,000	15,000	250	250
Total dissolved solids	1,000	1,000	35,000	35,000
Organic substances (mg/L):				
Carbon-chloroform extract	—[b]	100	—[b]	100
Chemical oxygen demand	—	100	—	200
Methylene blue–active substances	1.3	1.3	—	1.3
Miscellaneous (units as noted):				
Color (Pt-Co units)	—	1,200	—	—
Temperature (°F)	100	120	100	120

[a] Ray, H. C., *et al.* 1968. *Water Quality Criteria, Report of the National Technical Advisory Committee to the Secretary of the Interior.* U.S. Department of the Interior, Federal Water Pollution Control Administration, Washington, D.C.

[b] No floating oil.

Table 7.7 Total Ammonia Concentrations Corresponding to 0.20 mg/L of Unprotonated (Free) Ammonia

Temp	Receiving Water pH								
(°C)	6.0	6.5	7.0	7.5	8.0	8.5	9.0	9.5	10.0
0	241	76.2	24.1	7.64	2.43	0.782	0.261	0.0962	0.0441
5	160	50.7	16.0	5.09	1.62	0.527	0.180	0.0707	0.0360
10	108	34.1	10.8	3.42	1.10	0.360	0.128	0.0540	0.0308
15	73.3	23.2	7.35	2.34	0.753	0.252	0.0933	0.0432	0.0273
20	50.3	15.9	5.04	1.61	0.522	0.179	0.0702	0.0359	0.0250
25	32.8	10.4	3.30	1.06	0.348	0.124	0.0528	0.0304	0.0233
30	24.8	7.85	2.50	0.803	0.268	0.0983	0.0448	0.0278	0.0225
35	17.7	5.62	1.79	0.580	0.197	0.0760	0.0377	0.0256	0.0218
40	12.8	4.06	1.30	0.424	0.148	0.0604	0.0328	0.0240	0.0213
45	9.37	2.98	0.955	0.316	0.114	0.0496	0.0294	0.0230	0.0209
50	6.94	2.21	0.712	0.239	0.0892	0.0419	0.0269	0.0222	0.0207

Consequently, a rule-of-thumb estimate of the allowable total ammonia concentration is

$$\frac{0.02 \text{ mg NH}_3\text{/L}}{f} = \left(1 + \frac{[\text{H}^+]}{K_a}\right) \times 0.02 \text{ mg NH}_3\text{/L} \tag{7.7}$$

Values of the total ammonia concentration that correspond to an unprotonated ammonia concentration of 0.02 mg/L are given in Table 7.7. These differ slightly from the values given by Thurston et al. [1974], because of rounding and differences in values of the acid ionization constants employed.

Nondegradation. Finally, in some instances it may be desirable to preserve existing water quality. In fact, the Water Quality Act of 1972 includes a "nondegradation" provision. This has been interpreted by the Ohio Environmental Protection Agency as meaning that there should be no measurable increase in contaminant levels, which as a practical matter means that no contaminant concentration may be increased by more than 10%.

Effluent Standards

Because of the difficulty in assigning legal responsibility for stream standard violations when more than one discharge occurs, it is administratively easier to impose an *effluent standard.* Each discharger is required to obtain a National Pollutant Discharge Elimination System (NPDES) permit from the competent state authority. The permit specifies the location, times, volume, and composition of the permitted discharge. The permit specifications are set by the state agency so as to prevent any violation of stream standards. However, any violation of a permit condition is prosecutable regardless of the impact on stream conditions.

Permit conditions for conservative contaminants, for poisons, for the traditional rules of thumb, and for antidegradation conditions are easily established by calculating a steady state mass balance on the receiving stream at the point where the outfall meets it:

$$\overline{Q}_R \cdot \overline{C}_{QR} = \overline{Q}_W \overline{C}_{QW} + (\overline{Q}_R + \overline{Q}_W)\overline{C}_o \tag{7.8}$$

where

$\overline{C}_o$ = the stream standard for the most stringent beneficial use, generally a one-day, four-day, or seven-day average (kg/m^3 or lbm/ft^3)

$\overline{C}_{QR}$ = the flow-weighted (flow-composited) contaminant concentration in the river upstream of the outfall (kg/m^3 or lbm/ft^3)

$\overline{C}_{QW}$ = the flow-weighted (flow-composited) contaminant concentration in the wastewater (kg/m^3 or lbm/ft^3)

$\overline{Q}_R$ = the time-averaged (steady state) river flow for the critical period, generally the seven-day-average low flow with a ten-year return period (m^3/s or ft^3/s)

$\overline{Q}_W$ = the time-averaged (steady state) wastewater flow for the critical period, generally the maximum 7-day or 30-day average wastewater flow rate (m^3/s or ft^3/s)

The unknown in this case is the contaminant concentration in the treated effluent, $\overline{C}_{QW}$, and it becomes the NPDES permit condition. Sometimes the competent regulatory agency will "reserve" some stream assimilation capacity by using a $\overline{C}_o$ value that is lower than the relevant stream standard.

It should be noted that for the purpose of judging compliance with an NPDES permit, the Water Quality Act defines the effluent load to be the product of the arithmetically averaged contaminant concentration and the arithmetically averaged flow. This product will be larger or smaller than the true load depending on whether concentrations and flows are positively or negatively correlated.

For nonconservative contaminants, the permit conditions must be determined via water quality simulations. The simulations generally must be verified by in situ field data, and the process is time-consuming and expensive.

The minimum requirements of a NPDES discharge permit are shown in Table 7.8. These requirements define secondary biological treatment. While such requirements might be imposed on a small discharge to a large body of water, most permits are much more stringent, in order to meet relevant water quality standards. More comprehensive permits include:

- Separate restrictions for summer and winter conditions
- Limits on ammonia
- Limits on specific substances known to be in the influent wastewater
- More stringent sampling and monitoring requirements, perhaps at various points throughout the treatment facility or tributary sewers instead of only the outfall
- Specification of particular treatment processes

The more important parameters, such as $CBOD_5$, are subject to the more stringent sampling programs. Some parameters, such as flow, oxygen, and chlorine, are easily monitored continuously. The discharger may be requested to monitor parameters that have no apparent environmental impact in order to develop databases for future permits.

Table 7.8 Secondary Treatment Information[a]

	Averaging period	
Parameter	7 days	30 days
Five-day biochemical oxygen demand (BOD_5)		
Maximum effluent concentration (mg/L)	45	30
Removal efficiency (%)	—	85
Suspended solids		
Maximum effluent concentration (mg/L)	45	30
Removal efficiency (%)	—	85
pH (pH units):		
Minimum	—	6.0
Maximum	—	9.0

[a] Environmental Protection Agency. 1976. Secondary treatment information: biochemical oxygen demand, suspended solids and pH. *Federal Register,* 41(144):30786–30789.

7.2 Planning

The problem of projecting future demands may be subdivided into two parts, namely selection of (1) the planning period and (2) the projection technique.

Selection of Planning Period

In order to assess all the impacts of a project, the planning period should be at least as long as the economic life of the facilities. Estimates of the economic life of equipment, buildings, etc., are published by the U.S. Internal Revenue Service. This is especially important for long-lived facilities, because they tend to attract additional demand beyond that originally planned for.

Buildings have economic lives on the order of 20 years. For large pipelines the economic life might be 50 years, and dams might have economic lives as long as 100 years. The U.S. Government usually requires a planning period of 20 years for federally subsidized projects. Usefully accurate projections, however, cannot be made for periods much longer than 10 years. The demands projected for the economic life of a project cannot, therefore, be regarded as really likely to occur. Rather, the projected demands set the boundaries of the problem. That is, they provide guides to the maximum plant capacity, storage volume, land area, and other resources that might be needed. Preliminary facilities designs are made using these guides, but the actual facilities construction is staged to meet shorter-term projected demands. The longer-term projections and plans serve to guarantee that the various construction stages will produce an integrated, efficient facility, and the staging itself permits reasonably accurate tracking of the actual demand evolution.

Optimum Construction Staging

The basic question is, "How much capacity should be constructed at each stage?" This is a problem in cost minimization. Consider Fig. 7.1. The smooth curves represent the projected demand over the economic life of some sort of facility, say a treatment plant, and the stepped lines represent the installed capacity. Note that installed capacity always exceeds projected demand. Public utilities usually set their prices to cover their costs. This means that the consumers generally will be paying for capacity they cannot use, and it is desirable to minimize these excess payments on the grounds of both equity and efficiency.

The appropriate procedure is to minimize the present worth of the costs for the entire series of construction stages. This is done as follows. The present worth of any future cost is calculated using the prevailing interest rate and the time elapsed between now and the actual expenditure:

$$\text{PW} = \sum_{j=0}^{n} C_j \exp\{-i\,t_j\} \tag{7.9}$$

where

PW = the present worth of the jth cost
C_j = the jth cost at the time it is incurred
i = the prevailing interest rate (per year)
t_j = the elapsed time between now and the date of the jth cost

Manne [1967] and Srinivasan [1967] have shown that for either linearly or exponentially increasing demands, the time between capacity additions, Δt, should be constant. The value of Δt is computed by minimizing PW in Eq. (7.9). To do this it is first necessary to express the cost C_j as a function of the added capacity ΔQ_j. Economies of scale reduce the average cost per unit of capacity as the capacity increases, so the cost-capacity relationship may be approximated functionally as

$$C = kQ^a \tag{7.10}$$

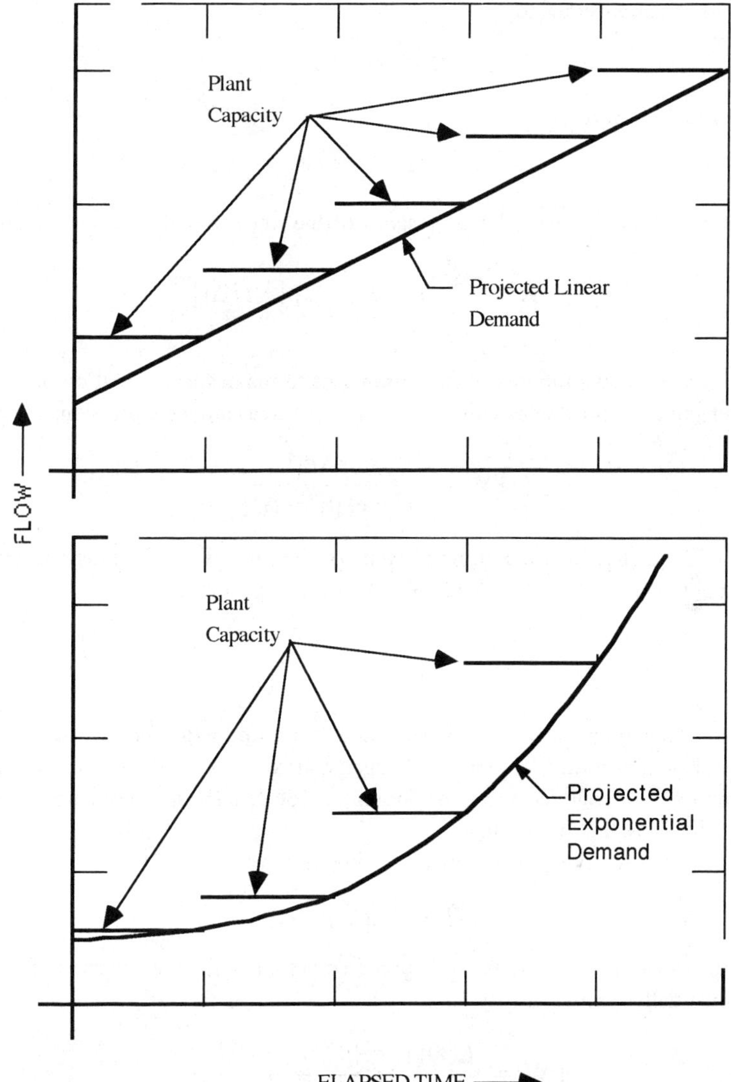

FIGURE 7.1 Projected demand-capacity relationship.

where

a, k = empirical coefficients determined by regression of costs on capacity
C = the cost to construct a facility of capacity Q (\$)
Q = the capacity (m^3/s or ft^3/sec)

For water treatment plants, a is usually between 0.5 and 0.8 [Clark and Dorsey, 1982].

Suppose that the demand is projected to increase linearly over time:

$$Q = Q_o + rt \tag{7.11}$$

where

Q_o = the demand at the beginning of the planning period (m^3/s or ft^3/sec)
r = the linear rate of increase of demand (m^3/s per year or ft^3/sec per year)

The capacity additions are equal to

$$\Delta Q = r\Delta t \tag{7.12}$$

and the cost of each addition is

$$C_j = k(r\Delta t)^a \tag{7.13}$$

Substituting Eq. (7.13) into Eq. (7.9), the present worth of the series of additions becomes

$$\text{PW} = \sum_{j=0}^{n} k(r\Delta t)^a \exp\{-ij\Delta t\} \tag{7.14}$$

The problem is somewhat simplified if the number of terms in the summation, n, is allowed to become infinite, because for that case the series converges to a simpler expression:

$$\text{PW} = \frac{k(r\Delta t)^a}{1 - \exp\{-i\Delta t\}} \tag{7.15}$$

The value of Δt that yields the minimum value of the present worth is determined by taking logarithms of both sides of Eq. (7.15) and differentiating with respect to Δt:

$$a = \frac{i\Delta t}{\exp\{i\Delta t\} - 1} \tag{7.16}$$

It should be noted that for linearly increasing demand the optimum time between capacity additions (i.e., the time that minimizes the discounted expansion costs) depends only on the interest rate i and the economies of scale factor a. The annual rate of demand increase r does not affect the result, and neither does the coefficient k.

It is more usual to project exponentially increasing demands:

$$Q = Q_0 \exp\{rt\} \tag{7.17}$$

In this case the capacity additions are still spaced equally in time, but the sizes of the additions increase exponentially [Srinivasan, 1967]:

$$\text{PW} = \frac{k(Q_o[\exp\{r\Delta t\} - 1])^a}{1 - \exp\{-(1 - ar)\Delta t\}} \tag{7.18}$$

$$a = \frac{(1 - ar)[1 - \exp\{-r\Delta t\}]}{r[\exp\{(1 - ar)\Delta t\} - 1]} \tag{7.19}$$

In this case the exponential demand growth rate does enter the calculation of Δt.

An exponential growth rate of 2% per year is about twice the national average for the U.S., so Eq. (7.16), which assumes linear growth and is easier to solve, may provide adequate accuracy for most American cities. However, urban areas in some developing countries are growing much faster than 2% per year, and Eq. (7.19) must be used in those cases.

Different components of water supply systems have different cost functions [Rachford, Scarato, and Tchobanoglous, 1969; Clark and Dorsey, 1982], and their capacities should be increased at different time intervals. The analysis just described can also be applied to the system components. In this case, the various component expansions need to be carefully scheduled so that bottlenecks are not created.

Finally, it should be remembered that the future interest rates and growth rates used to estimate the optimum expansion time interval are uncertain. Allowing for these uncertainties by adopting high estimates of these rates in order to calculate "conservative" values for the expansion capacity and costs is not the economically optimal strategy [Berthouex and Polkowski, 1970]. Considering first the interest rate, if the range of possible interest rates can be estimated, then the optimum strategy is to adopt the midpoint of the range. The same strategy should be used when selecting growth rates: if the projected growth is exponential, adopt the midrange value of the estimated exponential growth rates. However, for linear growth a different strategy is indicated: the minimal cost of expansion is achieved by using a linear growth rate somewhat less than the midrange value of the estimated rates. The "under design" capacity increment should be about 5 to 10% less than the capacity increment calculated using the midrange rate.

Population Projections

Because water demand is proportional to population, projections of water demand are reduced to projections of population. Engineers in the U.S. no longer make population projections. Population projection is the responsibility of designated agencies, and any person, company, or agency planning future public works is required to base those plans on the projections provided by the designated agency.

Projection errors of about 10% can be expected for planning periods less than 10 years, but errors in excess of 50% can be expected if the planning period is 20 years or more [McJunkin, 1964]. Because nearly all projections are for periods of 20 years or more, all projections are nonsense if one regards them as predictions of future conditions. Their real function is regulatory: They force the engineer to design and build facilities that may be easily modified, either by expansion or by decommissioning.

The four most commonly used methods of population projection are

- Extrapolation of historical census data for the community's total population
- Analysis of components of change (also known as cohort analysis, projection matrix, and Leslie matrix)
- Correlation with the total population of larger, surrounding regions
- Estimation of ultimate development

The first method was the principal one employed prior to World War II. Nowadays, cohort analysis is nearly the only method used.

Population Extrapolation Methods

Extrapolation procedures consist of fitting some assumed function to historical population data for the community being studied. The procedures differ in the functions fitted, the method of fitting, and the length of the population record employed. The functions usually fitted are the straight line, the exponential, and the logistic:

$$P = P_o + rt \tag{7.20}$$

$$P = P_o \exp\{rt\} \tag{7.21}$$

$$P = \frac{P_{max}}{1 + \left(\frac{P_{max}}{P_o} - 1\right)\exp\{-rt\}} \tag{7.22}$$

where

P = the population at time t (capita)
P_{max} = the maximum possible number of people (capita)
P_o = the population at the beginning of the fitted record (capita)

r = the linear or exponential growth rate (number of people per year or per year, respectively)

t = the elapsed time from the beginning of fitted record (years)

The preferred method of fitting Eqs. (7.20) through (7.22) to historical population data is least squares regression. Because the exponential and logistic equations are nonlinear, they should be fitted using a nonlinear least squares procedure. Nonlinear least squares procedures are iterative and require a computer. Suitable programs that are available for IBM™-compatible and Macintosh™ personal computers are Minitab™, SAS™, SPSS™, and SYSTAT™.

Components of Change

At the present time, the preferred method of population projection is the method of components of change. The method is also known as cohort analysis, the projection matrix, and the Leslie [1945] matrix. It is a discretized version of a continuous model originally developed by Lotka [1956].

In this method the total population is divided into males and females, and the two sexes are subdivided into age groups. For consistency's sake, the time step used in the model must be equal to the age increment. The usual subdivision is a five-year increment; therefore, the age classes considered are 0/4, 5/9, 10/14, 15/19, etc. This duration is chosen because it is exactly half the interval between censuses, so that every other time step corresponds to a census.

Consider first the female age classes beginning with the second class (i.e., 5/9); the 0/4 class will be considered later. For these higher classes, the only processes affecting the number of females in each age class are survival and migration. The relevant equation is

$$\left\{\begin{array}{l}\text{Number of females}\\ \text{in age class } i+1\\ \text{at time } t+\Delta t\end{array}\right\} = \left\{\begin{array}{l}\text{Number of females}\\ \text{in age class } i\\ \text{at time } t\\ \text{that survive } \Delta t\end{array}\right\} + \left\{\begin{array}{l}\text{Number of female}\\ \text{migrants during } \Delta t\\ \text{of age class } i+1\end{array}\right\}$$

$$P_f(i+1, t+\Delta t) = l_f(i,t)P_f(i,t) + M_f(i+1, t+\Delta t) \qquad (7.23)$$

where

$l_f(i,t)$ = the fraction of females in age class i (i.e., aged $5i$ through $5i + 4$ years) at time t that survive Δt years (here 5)

$M_f(i+1, t+\Delta t$ = the net number of female migrants (immigrants positive, emigrants negative) that join the age class $i + 1$ between times t and $t + \Delta t$

$P_f(i,t)$ = the number of females in the age class i at time t

$P_f(i+1, t+\Delta t)$ = the number of females in age class $i + 1$ (i.e., aged $5(i + 1)$ through $5(i + 1) + 4$ years) at time $t + \Delta t$

The survival fractions $l_f(i,t)$ can be calculated independently using the community's death records. The migration rates, however, are calculated using Eq. (7.23). This is done by using census data for $P_f(i+1, t+\Delta t)$ and $P_f(i,t)$ and the independently calculated $l_f(i,t)$. Consequently, any errors in the census data or death records are absorbed into the migration rate.

It is a matter of convenience how the migration is expressed, and it is most convenient to express it as a correction to the calculated number of survivors:

$$M_f(i+1, t+\Delta t) = m_f(i,t)l_f(i,t)P_f(i,t) \qquad (7.24)$$

where

$m_f(i,t)$ = the female migration rate for age class $i + 1$, in units of number of females joining age class $i + 1$ during the period t to $t + \Delta t$ per survivor of age class i

With this definition, Eq. (7.23) can be written as

$$P_f(i+1, t+\Delta t) = l_f(i,t)[1 + m_f(i,t)]P_f(i,t) = l_{\text{ef}}(i,t)P_f(i,t) \qquad (7.25)$$

where

$$l_{\mathrm{ef}}(i,t) = \text{the effective fraction of females in age class } i \text{ that survive from time } t \text{ to } t+\Delta t$$
$$= l_f(i,t)[1+m_f(i,t)]$$

Now consider the first female age class, which is comprised of individuals aged 0 to 4 years. The processes affecting this class are birth and migration. What is needed is the net result of births and migrations for the five-year period Δt. However, birth rates are usually given in terms of the number of births per female per year. Therefore, a one-year rate must be converted into a five-year rate. Furthermore, over a five-year period the number of females in any given age class will change due to death and migration. The birth rate itself changes, also. These changes are handled by averaging the beginning and end of period rates:

$$\text{Total births for age class } i \text{ from } t \text{ to } t+\Delta t$$
$$= \frac{5}{2}[b_f(i,t)P_f(i,t) + b_f(i,t+\Delta t)P_f(i,t+\Delta t)] \tag{7.26}$$

where

$b_f(i,t)$ = the female birth rate for females in age class i at time t, in units of female babies per female per year

Note that the factor 5 is needed to convert from one-year to five-year rates, and the factor $\frac{1}{2}$ is needed to average the two rates.

The total number of births is obtained by summing Eq. (7.26) over all the age classes. If this is done, and if terms containing like age classes are collected, the total number of births is

$$P_f(0,t+\Delta t) = \frac{5}{2}\sum_{i=0}^{k}[b_f(i,t) + b_f(i+1,t+\Delta t)l_{\mathrm{ef}}(i,t)]P_f(i,t)$$
$$= \sum_{i=0}^{k}\bar{b}_f(i,t)P_f(i,t) \tag{7.27}$$

where

$\bar{b}_f(i,t)$ = the average female birth rate for females in the age class i for the period t to $t+\Delta t$
$= \frac{5}{2}[b_f(i,t) + b_f(i+1,t+\Delta t)l_{\mathrm{ef}}(i,t)]$

Equations (7.26) and (7.27) are most conveniently written in matrix form as follows:

$$\begin{bmatrix} P_f(0,t+\Delta t) \\ P_f(1,t+\Delta t) \\ \vdots \\ P_f(k,t+\Delta t) \end{bmatrix} = \begin{bmatrix} \bar{b}_f(0,t) & \bar{b}_f(1,t) & \cdots & \bar{b}_f(k,t) \\ l_{\mathrm{ef}}(0,t) & 0 & \cdots & 0 \\ 0 & l_{\mathrm{ef}}(1,t) & \cdots & 0 \\ 0 & 0 & l_{\mathrm{ef}}(k-1,t) & 0 \end{bmatrix} \begin{bmatrix} P_f(0,t) \\ P_f(1,t) \\ \vdots \\ P_f(k,t) \end{bmatrix} \tag{7.28}$$

The first row in the coefficient matrix has numerous zeros (because very young and very old women are not fertile), and the main diagonal is entirely zeros.

The model for males is slightly more complicated. First, for all age classes other than the first (i.e., for $i = 1, 2, 3, \ldots, k$), the number of males depends solely on the fraction of the previous male age class that survives plus any migration that occurs:

$$\begin{aligned} P_m(i+1, t+\Delta t) &= l_m(i,t)[1+m_m(i,t)]P_m(i,t) \\ &= l_{em}(i,t)P_m(i,t) \end{aligned} \tag{7.29}$$

where

$l_{em}(i,t)$ = the effective fraction of males in age class i that survives from time t to $t+\Delta t$
$= l_m(i,t)[1+m_m(i,t)]$

$l_m(i,t)$ = the fraction of males in the age class i (i.e., aged $5i$ through $5i+4$ years) at time t that survives Δt years (here 5)

$m_m(i,t)$ = the male migration rate for age class $i+1$, in units of number of males joining age class $i+1$ during the period t to $t+\Delta t$ per survivor of age class i

$P_m(i,t)$ = the number of males in the age class i at time t

$P_m(i+1, t+\Delta t)$ = the number of males in age class $i+1$ (i.e., aged $5(i+1)$ through $5(i+1)+4$ years) at time $t+\Delta t$

The male birth rate depends on the number of females, not males, and the number of births must be written in terms of the population vector for women. The number of births for men (aged 0 to 4) is

$$P_m(0, t+\Delta t) = \sum_{i=0}^{k} \bar{b}_m(i,t)P_f(i,t) \tag{7.30}$$

where

$b_m(i,t)$ = the *male* birth rate for *females* in age class i at time t, in units of *male* babies per *female* per year

$\bar{b}_m(i,t)$ = the average *male* birth rate for *females* in age class i at time t, in units of the total number of *male* babies born during the interval Δt (here 5 years) per *female* in age class i
$= \frac{5}{2}[b_m(i,t) + b_m(i+1, t+\Delta t)l_{em}(i,t)]$

$P_m(0, t+\Delta t)$ = the number of males in age class 0 (i.e., aged 0 to 4 years) at time $t+\Delta t$

Consequently, the projection for males involves the population vectors for both males and females and coefficient matrices for each and is written as

$$\begin{bmatrix} P_m(0,t+\Delta t) \\ P_m(1,t+\Delta t) \\ P_m(k,t+\Delta t) \end{bmatrix} = \begin{bmatrix} \bar{b}_m(0,t) & \bar{b}_m(1,t) & \cdots & \bar{b}_m(k,t) \\ 0 & 0 & \cdots & 0 \\ \vdots & \vdots & & 0 \\ 0 & 0 & \cdots & 0 \end{bmatrix} \begin{bmatrix} P_f(0,t) \\ P_f(1,t) \\ \vdots \\ P_f(k,t) \end{bmatrix} + \cdots$$

$$+ \begin{bmatrix} 0 & 0 & \cdots & 0 \\ l_{em}(0,t) & 0 & \cdots & 0 \\ 0 & l_{em}(1,t) & & 0 \\ 0 & 0 & l_{em}(k-1,t) & 0 \end{bmatrix} \begin{bmatrix} P_m(0,t) \\ P_m(1,t) \\ \vdots \\ P_m(k,t) \end{bmatrix} \tag{7.31}$$

Correlation Methods

In the correlation method, one attempts to use the projected population of a larger, surrounding area—say, a state or county—to estimate the future population of the community being studied. This method is sometimes preferable to a direct projection of the community's population, because there is usually more accurate and more extensive data available for the larger area. This is especially true for smaller cities. The projected population of the larger, surrounding area is usually obtained by the method of components just described.

The correlation method entails plotting the available population data from the surrounding area and from the community against one another. It is assumed that the community is following the pattern of growth exhibited by the surrounding area, or something close to it. The correlation takes the form

$$P_c(t) = b_1 \cdot P_a(t) + b_0 \tag{7.32}$$

where

b_0, b_1 = regression coefficients
$P_a(t)$ = the population of the larger, surrounding area at time t
$P_c(t)$ = the population of the community at time t

Ultimate Development

The concept of *ultimate* or *full development* is actually somewhat nebulous but nonetheless useful, at least for short-term projections. Almost all communities have zoning regulations that control the use of both developed and undeveloped areas within the communities' jurisdictions. Consequently, mere inspection of the zoning regulations suffices to determine the ultimate population of the undeveloped areas.

There are several problems with this method. First, it cannot assign a date for full development. In fact, because population trends are not considered, the method cannot determine whether full development will ever occur. Second, the method cannot account for the chance that zoning regulations may change; for example, an area zoned for single-family, detached housing may be rezoned for apartments. Zoning changes are quite common in undeveloped areas, because there is no local population to oppose them. Third, the method cannot account for annexations of additional land into the jurisdiction. However, it should be noted that the second and third problems are under the control of the community, so it can force the actual future population to conform to the full development projection.

Siting and Site Plans

Flood Plains

Water and wastewater treatment plants are frequently built in the flood plain. Of course, the intakes of surface water treatment plants and the outfalls of wastewater treatment plants are necessarily in the flood plain. However, the remainder of the facilities are often built in risky areas, either to minimize the travel time of maintenance crews or because, in the case of wastewater plants, the site is the low point of a drainage district.

The Wastewater Committee requires that wastewater treatment plants be fully operational during the 25-year flood and that they be protected from the 100-year flood [Wastewater Committee, 1990]. The intake pumping stations of water treatment plants should be elevated at least 3 feet above the highest level of either the largest flood of record or the 100-year flood, or it should be protected to those levels [Water Supply Committee, 1987].

Permits

Water and wastewater treatment plants require a number of permits and must conform to a variety of design, construction, and operating codes.

Federal Permits. The National Environmental Policy Act of 1969 (PL 91-190) requires an *Environmental Impact Statement (EIS)* for "major federal actions significantly affecting the quality of the human environment." The EIS should consider alternatives to the proposed action, both long-term and short-term effects, irreversible and irretrievable commitments of resources, and unavoidable adverse impacts. An EIS may be required even if the only federal action is the issuance of a permit. An EIS is not required if the relevant federal agency (here the U.S. Army Corps of Engineers or U.S. EPA) certifies that there is no significant environmental impact.

The fundamental federal permit is the *National Pollution Discharge Elimination System (NPDES)* permit, which the Federal Water Pollution Control Act of 1972 requires for each discharge to a navigable waterway. Authority to grant NPDES permits is vested in the U.S. EPA; however, it has delegated that authority to the various states, and as a practical matter, dischargers apply to the relevant state agency. Treatment plants may also be required to obtain a *404 permit* from the U.S. Army Corps of Engineers if the proposed facility creates an obstruction in or over a navigable waterway. A *broadcasting license* may be required by the Federal Communications Commission if radio signals are used for telemetry and remote control.

Besides these permits and licenses, treatment plants must be designed in conformance with a number of federal regulations, including:

- Clean Air Act of 1990 requirements governing emissions from storage tanks for chemicals and fuels (Title 40, Parts 50 to 99 and 280 to 281)
- Americans with Disabilities Act of 1990 (PL 101-336) requirements regarding access for handicapped persons
- Occupational Health and Safety Act requirements regarding workplace safety

State and Local Permits. The design of water and wastewater treatment plants is subject to review and regulation by state agencies. Such reviews are normally conducted upon receipt of an application for a NPDES permit. The states generally are required to make use of U.S. EPA guidance documents regarding the suitability of various treatment processes and recommended design criteria. Some states also explicitly require adherence to the so-called "Ten States Standards," which are more correctly referred to as the Great Lakes–Upper Mississippi River Board of State Public Health and Environmental Managers' *Recommended Standards for Water Works* and *Recommended Standards for Wastewater Facilities*. States also require permits for construction in floodplains, because large facilities will alter flood heights and durations. Highway departments require permits for pipelines that cross state roads. Generally, such crossings must be either bored or jacked so that traffic is not disrupted.

Facilities must also conform to local codes, including:

- State building codes, usually based on the Building Official and Code Administrators (BOCA) Building Code
- State electrical codes, usually based on the National Electrical Code as specified by the National Fire Protection Association (NFPA-70)
- State plumbing code

These are usually administered locally by county, municipal, or township agencies. Most commonly, the agencies require a permit to build (which must be obtained before construction and which requires submission of plans and specifications to the appropriate agencies) and a permit to occupy (which requires a postconstruction inspection).

In many cases, either design engineers voluntarily adhere to, or local regulations require adherence to, other specifications, such as the following:

- General materials specifications, sampling procedures, and analytical methods: American Society for Testing and Materials (1916 Race Street, Philadelphia, PA 19103-1187)
- Specifications for water treatment chemicals and equipment: American Water Works Association (6666 West Quincy Avenue, Denver, CO 80235)
- Specifications for fire control systems and chemical storage facilities: National Fire Protection Association (1 Batterymarch Park, PO Box 9101, Quincy, MA 02269-9191)

- Specifications for chemical analytical methods: Water Environment Federation (601 Wythe Street, Alexandria, VA 22314-1994) and American Water Works Association (6666 West Quincy Avenue, Denver, CO 80235)

Other Permits. Railroad crossings require the permission of the railroad. As with highways, pipelines must be either bored or jacked under the road bed so that traffic is not disrupted.

Excavation almost always occurs near other utilities and should be coordinated with them.

7.3 Design Flows and Loads

Flow and Load Averaging

Designers must account for the inherent variability in flows and loads. This is done via the definition of various duration averages, maxima, and minima. The average flow for some specified time period T is, for continuous and discrete data, respectively,

$$\overline{Q} = \frac{1}{T}\int_0^T Q(t)dt \tag{7.33}$$

$$\overline{Q} = \frac{\sum_i Q_i}{n} \tag{7.34}$$

where

n = the number of discrete flow measurements during the specified period (dimensionless)
$Q(t)$ = the flow at time instant t (m^3/s or ft^3/s)
$\overline{Q}$ = the arithmetic mean volumetric flow for the specified period (m^3/s or ft^3/s)
Q_i = the ith flow measurement during the specified period (m^3/s or ft^3/s)
T = the duration of the specified period (s)
t = the time elapsed since the beginning of the period (s)

The usual averaging periods (T) are

- The annual average, otherwise known as average day
- The maximum (or minimum) 1-hour average
- The maximum (or minimum) 24-hour average
- The maximum (or minimum) 3-day average
- The maximum (or minimum) 7-day average
- The maximum (or minimum) 30-day average

In each case, the database consists of 12 consecutive months of flow and load records. A maximum average flow or load for some specified period is computed by identifying the continuous time interval in the annual record that is equal in duration to the specified period and that produces the maximum total volume of flow or mass of contaminant load. The minimum averages are determined similarly. If the NPDES permit is written with seasonal limits, the maxima and minima should be determined using seasonal rather than annual data.

The *instantaneous* contaminant load is the product of the instantaneous contaminant concentration and the instantaneous wastewater volumetric flow rate:

$$W(t) = Q(t)C(t) \tag{7.35}$$

$$W_i = Q_i C_i \tag{7.36}$$

where

$C(t)$ = the contaminant concentration at time instant t (kg/m^3 or lbm/ft^3)
C_i = the ith measurement of the contaminant concentration (kg/m^3 or lbm/ft^3)
$W(t)$ = the load at time instant t (kg/s or lb/s)
W_i = the ith load measurement (kg/s or lb/s)

The *average* contaminant load during some interval T is calculated as

$$\overline{W} = \frac{1}{T}\int_0^T W(t)dt \tag{7.37}$$

$$\overline{W} = \frac{\sum_i W_i}{n} \tag{7.38}$$

where

$\overline{W}$ = the average contaminant load for the specified period (kg/s or lb/s)

The Average Concentration

The most natural and useful definition of an average concentration is the *flow-weighted (flow-composited) concentration*:

$$\overline{C}_Q = \frac{\int_0^T Q(t)C(t)dt}{\int_0^T Q(t)dt} = \frac{\overline{W}}{\overline{Q}} \tag{7.39}$$

$$\overline{C}_Q = \frac{\sum_i Q_i C_i}{\sum_i Q_i} = \frac{\overline{W}}{\overline{Q}} \tag{7.40}$$

where $\overline{C}_Q$ is the flow-weighted (flow-composited) average concentration (kg/m^3 or lb/ft^3). Note that the (arithmetic) average load is the product of the (arithmetic) average flow and the flow-weighted average concentration.

The simple arithmetic average concentration is frequently encountered:

$$\overline{C} = \frac{1}{T}\int_0^T C(t)dt \tag{7.41}$$

$$\overline{C} = \frac{\sum_i C_i}{n} \tag{7.42}$$

where $\overline{C}$ is the arithmetic average contaminant concentration (kg/m^3 or lbm/ft^3).

The general relationship between the average load and the arithmetic average concentration is

$$\overline{W} = \overline{Q}\overline{C} + \mathrm{covar}(Q, C) = \overline{Q}\overline{C} + R_{QC}S_Q S_C \tag{7.43}$$

where

$$\mathrm{covar}(Q, C) = \text{the covariance of } Q \text{ and } C \text{ (kg/s or lb/s)}$$
$$= \frac{\sum_i (Q_i - \overline{Q})(C_i - \overline{C})}{n - 1}$$

$$R_{QC} = \text{the Pearson product-moment correlation coefficient (dimensionless)}$$
$$= \frac{\sum_i (Q_i - \overline{Q})(C_i - \overline{C})}{\sqrt{\sum_i (Q_i - \overline{Q})^2}\sqrt{\sum_i (C_i - \overline{C})^2}}$$

S_C = the standard deviation of the contaminant concentration measurements (kg/m^3 or lbm/ft^3)

$$= \sqrt{\frac{\Sigma_i (C_i - \overline{C})^2}{n-1}}$$

S_Q = the standard deviation of the flow measurements (m^3/s or ft^3/s)

$$= \sqrt{\frac{\Sigma_i (Q_i - \overline{Q})^2}{n-1}}$$

Note that the average load is equal to the product of the average flow and arithmetic average concentration only if the flows and concentrations are uncorrelated ($R_{QC} = 0$). This is generally false for raw wastewaters, but it may be true for biologically treated effluents. In most raw wastewaters the flows and concentrations tend to be positively correlated; therefore the product $\overline{QC}$ underestimates the mean load. The underestimate can be substantial.

It should be noted that for regulatory purposes the U.S. EPA *defines* the average load to be the product of the arithmetic average flow and the arithmetic average concentration, $\overline{QC}$. Generally, the EPA formula, $\overline{W} = \overline{QC}$, underestimates the true loads.

Annual Average per Caput Flows and Loads

The annual average per caput (per capita) water demand is the total volume of water supplied during a complete year divided by the midyear population. The usual units are either gallons per caput per day (gpcd) or liters per caput per day (Lpcd). Sometimes the volumes are expressed as cubic feet or cubic meters. The calculation proceeds as follows:

$$\overline{Q}_{pc} = \frac{1}{T}\int_0^T \frac{Q(t)dt}{P(t)} = \frac{1}{n}\sum_{i=1}^{n} \frac{Q_i}{P_i} \tag{7.44}$$

$$\overline{W}_{pc} = \frac{1}{T}\int_0^T \frac{Q(t)C(t)dt}{P(t)} = \frac{1}{n}\sum_{i=1}^{n} \frac{Q_i C_i}{P_i} \tag{7.45}$$

where

$\overline{Q}_{pc}$ = the annual average per caput water demand (m^3/cap·s or ft^3/cap·s)
$\overline{W}_{pc}$ = the annual average per caput load (kg/cap·s) or lb/cap·s

Because the rate of population increase is usually only a few percent per year or less, the integral and sum are evaluated by assuming P is constant and equal to the midyear population. P is then factored out, and the integral and sum become the total volume of water supplied in either gallons or liters during the year.

Automatic Samplers

Some automatic samplers capture a constant sample volume at uniformly spaced time increments: the data from these samplers are arithmetically averaged concentrations. Other samplers capture sample volumes at uniformly spaced time increments but the sample volumes are proportional to the simultaneous flow rate: the data from these samplers are flow-weighted average concentrations. Flow-weighted concentrations are also produced by samplers that collect constant sample volumes but adjust the sampling time increments so that each sample represents the same total flow volume.

Water Treatment

The flow through water treatment plants is controlled by the operator, and this simplifies the design process. The designer must still allow for variations in consumption, but these can be satisfied by system storage of finished water.

Annual Average per Caput Flows and Loads

In public water supplies, the water demand is the sum of the volumes supplied to households and to commercial and industrial enterprises and public agencies. An approximate breakdown by use of the water withdrawn for public supplies in 1980 is given in Table 7.9. The estimates were compiled from several independent sources, and they may each be in error by several gpcd. The data are national averages, and particular cities will exhibit substantial divergences from them. In 1980, public water supplies in the U.S. served 186 million people (81% of the total population), and their total withdrawal from both surface waters and groundwater amounted to 183 gpcd [Solley *et al.*, 1983]. Of the total withdrawal, 66% was obtained from surface waters, and 34% came from groundwater; 21% of the withdrawal went to consumptive uses, most of which was lawn sprinkling.

These national averages mask a wide range of local variations [Solley *et al.*, 1983]. For example, the amount of water withdrawn in 1980 varied from 63 gpcd in the Virgin Islands to 575 gpcd in Utah, and the consumptive use varied from none in Delaware to 71% in Wyoming. In general, per caput water withdrawals are greatest in the hot, arid Colorado River valley and the Great Basin (341 gpcd, 45% consumptive) and least in the wetter, cooler New England states (148 gpcd, 10% consumptive).

About one-eighth of the withdrawal is lost either as treatment byproducts (i.e., sludges and brines) or as unaccounted-for water. The latter category includes such things as unmetered (i.e., free) water, leakage, fire control, and metering errors.

One-third of the withdrawal is supplied to commercial and industrial enterprises, although many industrial plants have independent supplies.

Somewhat less than half the withdrawal is supplied to households. The principal uses there are lawn sprinkling, toilet flushing, and bathing. The particular estimates of aggregate household use cited in Table 7.9 are the weighted average of the demands reported by Linaweaver *et al.* [1966] for detached single-family houses and apartments that had metered water and sewerage. It was assumed that detached single-family homes comprised 67% of the total number of housing units and that the remainder were apartments. For detached single-family homes with metered water and septic tanks, a better estimate of usage exclusive of lawn sprinkling is 44 gpcd [Schmidt

Table 7.9 Estimated Breakdown, by Use, of the Annual Average Water Withdrawal by Public Supplies in 1980

Use	Rate (gpcd)	Percent of Total Withdrawal
Total withdrawal	183	100[a]
Commercial/industrial	63	34[a]
Public	35	19[b]
Unaccounted for	20	11[c,d]
Treatment losses	8	4[e]
Metered agencies	7	4[b]
Households	85	46[f,g]
Lawn sprinkling	25	14[f,g]
Toilet flushing	25	14[h]
Bathing	20	11[h]
Laundry	8	4[h]
Kitchen	6	3[h]
Drinking	0.5	0.2[h]

[a] Solley, W. B., Chase, E. B., and Mann, W. B., IV. 1983. *Estimated Water Use in the United States in 1980,* Circular No. 1001. U.S. Geological Survey, Text Products Division, Distribution Branch, Alexandria, VA.

[b] By difference, but see: Schneider, M. L. 1982. *Projections of Water Usage and Water Demand in Columbus, Ohio: Implications for Demand Management.* M. S. Thesis. The Ohio State University, Columbus.

[c] Seidel, H. F. 1985. Water utility operating data: An analysis. *J. Am. Water Works Assn.* 77(5):34.

[d] Keller, C. W. 1976. Analysis of unaccounted-for water. *J. Am. Water Works Assn.* 68(3):159.

[e] Lin, S., Evans, R. L., Schnepper, D., and Hill, T. 1984. *Evaluation of Wastes from the East St. Louis Water Treatment Plant and Their Impact on the Mississippi River.* Circular No. 160. State of Illinois, Department of Energy and Natural Resources, State Water Survey, Champaign.

[f] Linaweaver, F. P., Jr., Geyer, J. C., and Wolff, J. B. 1966. *Final and Summary Report on the Residential Water Use Research Project,* Report V, Phase 2. The Johns Hopkins University, Department of Environmental Engineering Science, Baltimore, MD.

[g] King, G. W., *et al.*, 1984. *Statistical Abstract of the United States 1985.* 105th ed. U.S. Government Printing Office, Washington, D.C.

[h] Bailey, J. R., Benoit, R. J., Dodson, J. L., Robb, J. M., and Wallman, H. 1969. *A Study of Flow Reduction and Treatment of Waste Water from Households.* Water Pollution Control Research Series 11050FKE 12/69. U.S. Department of the Interior, Federal Water Quality Administration, Washington, D.C.

et al., 1980] and lawn sprinkling might be reduced by two-thirds [Linaweaver *et al.*, 1966]. For high-income areas, the lawn sprinkling usage might be doubled, but in-house usage would increase by only about 10%.

The per caput annual average withdrawal rate for public supplies rose steadily at about 1% per year for over 100 years; some statistics are given in Table 7.10. The rate of increase from 1870 is 1.09% per year, and the rate since 1950 is 0.78% per year. The annual increase in withdrawals may have declined or even ceased since the mid-1980s because of conservation efforts.

Because it is probable that no city in the U.S. has a usage breakdown like that in Table 7.9, local usage projections should always be based on local data. This is not a very severe requirement, because all American cities already have water supply systems, and the records maintained by these systems can be used as the basis of design. National average data are more useful for making adjustments to local usage projections to account for factors such as income levels, water prices, and climate.

Table 7.10 Per Caput Rates of Water Withdrawal by Public Supplies in the U.S. from 1870 to 1980

Year	Withdrawal (gpcd)
1870	55[a]
1907	121[b]
1950	145[c]
1960	151[c]
1970	166[c]
1980	183[c]

[a]Amount supplied, not withdrawn. Fanning, J. T. 1882. *A Practical Treatise on Hydraulic and Water-Supply Engineering: Relating to the Hydrology, Hydrodynamics, and Practical Construction of Water-Works in North America,* 3rd ed. Van Nostrand, New York.

[b]Amount supplied, not withdrawn. Metcalf, L., Gifford, F. J., and Sullivan, W. F. 1913. Report of Committee on Water Consumption Statistics and Records. *J. New England Water Works Assn.*, 27(1):29.

[c]King, G. W., *et al.*, 1984. *Statistical Abstract of the United States 1985,* 105th ed. U.S. Government Printing Office, Washington, D.C.

Peak Demand

Some reported peaking factors for entire cities are given in Table 7.11. The 1882 data given by Fanning are for large cities in the Northeast, and the data given by Metcalf *et al.* [1913] are for cities and towns in Massachusetts. The data for 1932 and 1936 were obtained from numerous cities of all sizes throughout the country. The peak data for 1936 were obtained during a severe

Table 7.11 Reported Ratios of Peak to Average per Caput Demands for American Cities

	Averaging Period					
Year	30 days	7 days	3 days	1 day	4 hr	1 hr
1882[a]	1.017	1.027	—	—	—	1.731
1913[b]	1.28	1.47	—	1.98	—	—
1932[c]						
< 10,000 cap.	—	—	—	—	—	3.30
> 10,000 cap.	—	—	—	—	—	2.98
5,000/20,000 cap.	—	—	—	1.66	—	—
> 122,000 cap.	—	—	—	1.53	—	2.34
1936[d]	—	—	1.75	1.82	2.40	2.79
1957[e]	—	—	—	≤2.09	—	3.30

[a]Fanning, J. T. 1882. *A Practical Treatise on Hydraulic and Water-Supply Engineering: Relating to the Hydrology, Hydrodynamics, and Practical Construction of Water-Works in North America,* 3rd ed. Van Nostrand, New York.

[b]Metcalf, L., Gifford, F. J., and Sullivan, W. F. 1913. Report of Committee on Water Consumption Statistics and Records, *J. New England Water Works Assn.* 27(1):29.

[c]Folwell, A. P. 1932. Maximum daily and hourly water consumption in American cities. *Public Works,* 63(10):13.

[d]Anonymous. 1936. Record-breaking consumptions: established by severe drought and high summer temperatures. *Water Works Eng.* 89(19):1236.

[e]Wolff, J. B. 1957. Forecasting residential requirements. *J. Am. Water Works Assn.* 49(3):225.

drought, and the peak-to-average ratio was calculated using the average demand for June of 1935. Wolff's data were obtained from cities in the Northeast.

The detailed data reported in the cited papers indicate that the peak demands for the 4-hour averaging period are only slightly smaller than those for the 1-hour averaging period. The 3-day averages are also nearly equal to the 1-day averages. However, the ratio of peak to average demands generally declines as the peak averaging period increases, as it must.

The detailed data also show a weak tendency for the peak-to-average ratio to decline as the population rises. This tendency is not apparent in the data for 1932 and 1936, which represent about 400 cities nationwide. However, for 70 cities in the Northeast, Wolff [1957] reported the ratio of peak 1-day to average day demands as, for purely residential areas,

$$\frac{\text{Maximum 1-day average demand}}{\text{Annual average demand}} = 2.51P^{-0.0855} \tag{7.46}$$

and for combined residential and industrial areas,

$$\frac{\text{Maximum 1-day average demand}}{\text{Annual average demand}} = 2.09P^{-0.0574} \tag{7.47}$$

where P is the service population in thousands.

In view of the insensitivity of peaking factors to population, the peaking factors in Table 7.12 are recommended for American communities. These ratios are somewhat higher than those recommended a generation ago (e.g., Fair *et al.*, 1954], but the amount of lawn sprinkling has increased substantially since then, and lawn sprinkling is a major component of the peak demands. The same ratios may be used for residential neighborhoods, with two exceptions [Linaweaver *et al.*, 1966]. The exceptions are newly developed lots, where new lawns are being established, and very large lots in high-income areas. In these areas, the ratio of peak hour to average day demands reaches 15 to 20.

In the late 19th century, peak demands in northern cities usually occurred in the winter as a result of the need to discharge water through hydrants and household taps to prevent freezing, because the water mains often were shallow and the houses poorly heated. By the early 20th century, lawn sprinkling had shifted the peak demand to summer. By 1930, 55% of American cities reported that peak demand occurred in July, and 25% reported that it occurred in August [Folwell, 1932]. Only 15% of the cities experienced peak demands in winter.

The peaking factors just cited are weighted toward Eastern cities, which receive significant rainfall throughout the summer, somewhat mitigating lawn sprinkling. Many Western cities, however, receive little or no rain in the summer, and residents often water their lawns several times a week. The practical result of this practice is that the so-called maximum daily average demand occurs nearly every day during the Western summer [Linaweaver *et al.*, 1966].

Table 7.12 Recommended Peaking Factors for Water Demand

	Averaging Period					
Type of Community	30 days	7 days	3 days	1 day	4 hr	1 hr
Population greater than 5000, with commerce and industry	1.3	1.5	2.0	2.0	3.3	3.3
Purely residential and resort [b,c]	1.9	2.2	—	2.9	5.0	6.2

[a] Henderson, A. D. 1956. The sprinkling load—Long Island, N.Y., and Levittown, Pa. *J. Am. Water Works Assn.* 48(4):361.

[b] Critchlow, H. T. 1951. Discussion: S. K. Keller (1951) , "Seasonal water demands in vacation areas." *J. Am. Water Works Assn.* 43(9):701.

Finally, there is the issue of fire demand. The amount of water used for fire control over an entire year is very small, and it might not affect the peak demand statistics for a whole city. However, in the neighborhood of an ongoing fire, the fire demand dominates the local water requirements and controls the design of the local storage and distribution systems. In the U.S., water supply systems are usually designed in accordance with the recommendations of the Insurance Services Office (ISO). These recommendations are used to set fire insurance premiums for new or small cities for which there is little data on fire insurance claims. For old, large cities that have ample statistics on claims, premiums are usually set on an actuarial basis.

The ISO calculates a *needed fire flow* (NFF_i) at each of several representative locations in the city. It is assumed that only one building is on fire at any time. The formula for the NFF_i is [ISO, 1980]:

$$NFF_i = C_i O_i (X + P)_i \tag{7.48}$$

where

NFF_i = the needed fire flow (gpm)
C_i = the fire flow based on the size of the building and its construction (ranging from 500 gpm to 8000 gpm)
O_i = a correction factor for the kinds of materials stored in the building (ranging from 0.75 to 1.25)
$(X + P)_i$ = a correction factor for the proximity of other buildings (ranging from 0.0 to 1.75)

Criteria for assigning values to the various factors in Eq. (7.48) are given in the ISO recommendations. In any case, the NFF_i must be not be greater than 12,000 gpm nor less than 500 gpm. For one- and two-family dwellings, NFF_i ranges from 500 gpm to 1500 gpm; for larger dwellings, NFF_i is increased up to a maximum of 3500 gpm. The water supply system should be able to deliver to the fire location the total of the NFF_i and the maximum 24-hour average demand while maintaining a system pressure of at least 20 psig everywhere. For an NFF_i of 2500 gpm or less, the total flow must be sustainable for 2 hours; for an NFF_i between 3000 and 3500 gpm, for 3 hours; and for an NFF_i greater than 3500 gpm, for 4 hours.

These recommendations refer to the water distribution system itself. The treatment plant design flow need not be the total of the NFF_i and the maximum 24-hour average demand, so long as there is sufficient storage capacity to make up the difference between plant capacity and the total flow.

Factors Affecting Household Demand

The principal factors affecting household water demand are household income, water price, climate, and conservation regulations. These factors primarily affect lawn sprinkling, and a very strong climatic effect distinguishes the humid East from the arid West.

In general, lawn sprinkling demand amounts to about 60 percent of the net evapotranspiration rate [Linaweaver *et al.*, 1966]:

$$Q_{ls} = 0.60 A_l (E_{pot} - P_{eff}) \tag{7.49}$$

where

Q_{ls} = the lawn sprinkling demand (annual, monthly, or daily average, as appropriate) in m^3/s
A_l = the lawn area in hectares
E_{pot} = the potential evapotranspiration rate (annual, monthly, or daily average) to conform to Q_{ls} in m^3/ha-s
P_{eff} = the effective precipitation rate (annual, monthly, or daily average) to conform to Q_{ls} in m^3/ha-s

The lawn sprinkling demand is modified by both household income and water price. The lawn sprinkling demand elasticities for income and price are calculated as

$$\eta_I = \frac{dQ_{ls}/Q_{ls}}{dI/I} \tag{7.50}$$

$$\eta_C = \frac{dQ_{ls}/Q_{ls}}{dC/C} \tag{7.51}$$

where

C = water cost (\$/m^3 or \$/ft^3)
I = household income (\$/yr)
Q_{ls} = lawn sprinkling demand (m^3/s or ft^3/s)
η_C = the elasticity of lawn sprinkling demand with respect to water cost (dimensionless)
η_I = the elasticity of lawn sprinkling demand with respect to household income (dimensionless)

In metered, sewered Eastern communities, the income and price elasticities for the summer average lawn sprinkling demand are about 1.5 and -1.6, respectively [Howe and Linaweaver, 1967]. This means that if household income were increased by 10%, lawn sprinkling demand would be increased by 15%; if water prices were increased by 10%, lawn sprinkling demand would be reduced by 16%. In metered, sewered Western communities, however, the income and price elasticities are only 0.4 and -0.7, respectively.

These differences in response result from climate. In the humid East, there are frequent summer rains, and reductions in lawn sprinkling will not result in loss of lawns. In the arid West, however, summer rains are infrequent or nonexistent, and lawns will die without regular irrigation.

Prices and incomes have less effect on peak daily lawn sprinkling demands than on the summer average. This is an artifact of utility billing practices [Howe and Linaweaver, 1967]. Because consumers are billed only monthly or quarterly for the total amount of water used, the cost of a single afternoon's watering is not a major portion of the bill and does not deter water use. If necessary, peak daily demands can be curtailed by use prohibitions [Heggie, 1957].

Wastewater Treatment

It has been the custom in the U.S. to use the annual average flows and loads as the basis of process design, exceptions being made for process components that are especially sensitive to maximum or minimum flows and loads. However, nowadays NPDES permits generally are written in terms of 30-day and 7-day average limits, and this suggests that the design basis ought to be either a maximum 30-day average or maximum 7-day average load or flow [Joint Task Force, 1992]. If other averages are written into the permit, they should become the design basis. If the permit conditions are seasonal, the averages should be calculated on a seasonal basis (a maximum 30-day average load for the winter, a maximum 30-day average load for the summer). Consequently, the design of each component in a treatment plant must consider several different loading and flow conditions, and the most severe condition will control its design.

Table 7.13 is a compilation of recommended design flows and loads for various treatment plant components. The basic design period is assumed to be either the maximum 7-day or 30-day average flow or load, depending on which is most severe given the associated temperature and composition. Most regulatory authorities still use the annual average load or flow in their regulations.

Typical peaking factors are included in Table 7.13. There is substantial variation in peaking factors between communities, especially between large and small communities, and designs should be based on local data.

Table 7.13 Recommended Design Flows and Loads for Wastewater Treatment Plant Components[a] *(continues)*

Process or Material	Design Period	Peaking Factor (Peak Value/Annual Average)
Operating costs:		
Chemicals	Annual average	1.0
Electricity	Annual average	1.0
Ultimate hydraulic capacity of tanks, conduits, pumps, weirs, screens, etc.	Transmit the instantaneous annual maximum flow with one of each parallel unit out of service; check solids deposition at minimum flows	Maximum and minimum rated capacities of influent sewer or pumping station
Preliminary treatment:		
Coarse screen size	Max. 7- or 30-d avg flow*	2.0 or 1.4
Screenings storage volume	Max. 3-d avg load	4.0
Grit chamber size	Max. 1-h avg flow	3.0
Grit storage volume	Max. 3-d avg load	4.0
Comminutor size	Max. 1-h avg flow	3.0
Influent pumping station	Max. 1-h avg flow and min. 1-h avg flow	3.0 and 0.33
Influent sewer	Max. 1-h avg flow and min. 1-h avg flow	3.0 and 0.33
Flow meters	Max. 1-h avg flow and min. 1-h avg flow	3.0 and 0.33
Primary treatment:		
Primary clarifier overflow rate	Max. 1-h avg flow and max. 7-day or 30-day avg flow*	3.0 and 2.0 or 1.4
Outlet weir length	Max. 1-h avg flow	3.0
Outlet trough capacity	Max. 1-h avg flow and $\frac{1}{2}$ max. 7-d or 30-d max. avg flow	3.0 and 2.0 or 1.4
Sludge volume	Max. 30-d avg load	1.3
Scum volume	Max. 7-d or 30-d avg load	1.6 or 1.3
Activated sludge:		
Food-to-microorganism ratio *(F/M)* or SRT	50 to 150% of value for max. 7-d or max. 30-d avg load*	2.4 or 1.6
Aeration volume	Max SRT or min. *F/M* at max. 7-d or 30-d avg. BOD load*	2.4 or 1.6
Aeration tank weirs and troughs	Max. instantaneous flow plus max. recycle	3.0
Air supply	Max. 1-h avg BOD load; include NOD for nitrifying systems	2.8
Temperature (nitrification)	30-d avg preceding critical period	—
Alkalinity (nitrification)	Min. 7-d or 30-d avg load	—
Waste sludge volume and dry solids	Min. SRT or max *F/M* at max. 7-d or 30-d avg BOD load*	1.6 or 1.3
Return sludge pumps	Max. 7-d or 30 d avg flow*	2.0 or 1.4 and
Final clarifier size	Max. 1-h avg and max. 7-d or 30-d avg flow* and max. 24-h avg solids load	$2.8(Q + Q_r)$ MLSS
Trickling filters:		
Media volume	Max. 24-h BOD load	2.4
Distributor and recirculation system	Max. 7-d or 30-d avg flow* and min. 1-h avg flow	2.0 or 1.4 and 0.33
Underdrains	Max. 7-d or 30-d avg flow* plus recirculation	$(2.0 \text{ or } 1.4)Q + Q_r$
Intermediate clarifiers	Max. 1-h avg flow	3.0
Final clarifier size	Max. 1-h avg flow	3.0

Table 7.13 (continued) Recommended Design Flows and Loads for Wastewater Treatment Plant Components[a]

Process or Material	Design Period	Peaking Factor (Peak Value/Annual Average)
Wastewater ponds:		
Facultative pond area	Max. 7-d or 30-d avg BOD load	1.6 or 1.3
Facultative pond volume	Ann. avg flow	1.0
Aerated pond volume	Ann. avg flow	1.0
Aerated pond air supply	Max. 7-d or 30-d avg BOD load	1.6 or 1.3
Effluent sand filters:		
Filter area	Max. 1-h avg flow	3.0
Disinfection:		
Chlorine contactor volume	Max. 1-h avg flow	3.0

[a] Compiled from:

Fair, G. M., Geyer, J. C., and Morris, J. C. 1954. *Water Supply and Waste-Water Disposal.* John Wiley & Sons, Inc., New York.

Joint Task Force. 1992. *Design of Municipal Wastewater Treatment Plants: Volume 1—Chapters 1–12.* Water Environment Federation, Alexandria, VA; American Society of Civil Engineers, New York.

Metcalf & Eddy, Inc. 1991. *Wastewater Engineering: Treatment, Disposal, and Reuse,* 3rd ed., revised by G. Tchobanoglous and F. L. Burton. McGraw-Hill, Inc., New York.

Mohlman, F. W., Chairman, Thomas, H. A. Jr,. Secretary, Fair, G. M., Fuhrman, R. E., Gilbert, J. J., Heacox, R. E., Norgaard, J. C., and Ruchhoft, C. C. 1946. Sewage treatment at military installations. Report of the Subcommittee on Sewage Treatment of the Committee on Sanitary Engineering, National Research Council Division of Medical Science, Washington, D.C. *Sewage Works J.* 18(5):791.

Wastewater Committee of the Great Lakes–Upper Mississippi River Board of State Public Health and Environment Managers. 1990. *Recommended Standards for Wastewater Facilities, 1990 Edition.* Health Education Services, Albany, NY.

Note: The Wastewater Committee (1990) uses the annual average flow wherever the Table recommends the maximum 7-day or 30-day average flow.

Flow and Load Equalization

The design flows and load recommendations in Table 7.13 generally do not account for hourly flow variations within a 24-hour period. Normal daily peak flows and loads do not affect average activated sludge process performance unless the daily 3-hour average maximum flow exceeds 160% of the 24-hour average flow (140% of the average for the 6-hour average maximum flow). If higher peak flows do occur, they can be accounted for by increasing the size of the aeration tank. An approximate rule for the volume increase is [Joint Committee, 1977]:

$$\Delta V = 60(\mathrm{PF}_{3\mathrm{h}} - 1.6) \tag{7.52}$$

where

ΔV = the additional aeration tank volume needed to account for excessive peak flows (%)

$\mathrm{PF}_{3\mathrm{h}}$ = the peaking factor for the daily 3-hour average maximum flow (decimal fraction)

For the daily 6-hour average maximum flow, the rule is

$$\Delta V = 60(\mathrm{PF}_{6\mathrm{h}} - 1.4) \tag{7.53}$$

where $\mathrm{PF}_{6\mathrm{h}}$ is the peaking factor for the daily 6-hour average maximum flow (decimal fraction).

Flow equalization is generally not recommended at municipal wastewater treatment plants unless the daily 3-hour average maximum flow exceeds 210% of the average (180% of the average for the daily 6-hour average maximum flow). The principal cause of performance degradation during peak hydraulic loads is secondary clarifier failure. In activated sludge plants, this seems to become a problem at overflow rates of about 1000 gpd/ft^2 [Ongerth, 1979].

Table 7.14 Population Equivalents for Municipal Sewage

Source	Volume (L/cap d)	TSS [VSS] (g/cap d)	$CBOD_5$ (g/cap d)	Total *N* (g/cap d)	Total *P* (g/cap d)
Households[c,d,e,f]	220	38 [34]	45	15	4
Sanitary sewage, no industry[b]	—	—	54	—	—
Combined sewage, little industry[b]	—	—	73	—	—
Average of 31 cities[a]	400	95 [64]	100	14	—
Combined sewage, much industry[b]	—	—	150	—	—

[a] Committee of the Sanitary Engineering Division on Sludge Digestion. 1937. Standard practice in separate sludge digestion. *Proc. ASCE,* 63(1):39–106.

[b] Theriault, E. J. 1927. *The Oxygen Demand of Polluted Waters,* Bulletin No. 173. U.S. Public Health Service, Washington.

[c] Vollenweider, R. A. 1968. *Scientific Fundamentals of the Eutrophication of Lakes and Flowing Waters, with Particular Reference to Nitrogen and Phosphorus as Factors in Eutrophication,* OECD Report No. DAS/CSI/68.27. Organization for Economic Cooperation and Development, Directorate for Scientific Affairs, Paris.

[d] Watson, K. S., Farrel, R. P., and Anderson, J. S., 1967. The contribution from the individual home to the sewer system. *J. Water pollution Cont. Fed.* 39(12):2039.

[e] Webb, P., ed. 1964. *Bioastronautic Data Book,* NASA SP- 3006. National Aeronautics and Space Administration, Scientific and technical Information Division, Washington.

[f] Zanoni, A. E. and Rutkowski, R. J. 1972. Per capita loadings of domestic wastewater. *J. Water Pollution Cont. Fed.* 44(9):1756.

Population Equivalents

The concentrations of various substances in raw sewage varies widely, because (1) water consumption rates vary widely, yielding differing degrees of dilution, and (2) potable waters start out with different amounts of minerals depending on local geology. However, the rates of contaminant mass generation per caput are fairly uniform, even across cultures. Table 7.14 presents a summary of American and some European data.

The organic fraction is reported variously as 5-day biochemical oxygen demand (BOD_5), (dichromate) chemical oxygen demand (COD), volatile solids (VS), and total organic carbon (TOC). Approximate ratios among some of these parameters are given in Table 7.15. The tabulated ratios are weighted towards the results of the U.S. EPA study [Burns and Roe, Inc., 1977], because of the large number of plants included.

7.4 Intakes and Wells

Natural waters are almost never homogeneous, either in space or in time. Even aquifers often show some vertical and seasonal variation in water composition. There are, however, patterns to the variations, and a judicious selection of the exact point and timing of water withdrawals can simplify and improve treatment plant design, operation, and performance.

The factors requiring consideration by the designer of an intake or well field are [Babbitt *et al.,* 1967; Burdick, 1930]:

- Source of supply (e.g., river, lake, impounding reservoir, spring, or well)
- Nature of the withdrawal site, including depth of water, character of bottom (silt, gravel, cobbles, etc.), wind fetch and wave protection, flow stratification and seasonal turnover,

Table 7.15 Relationships among Organic Composition Indicators for Municipal Sewage[a]

Parameter Ratio	Raw or Settled Sewage	Conventional, Nonnitrified Effluent	Nitrified Effluent
BOD/COD	0.50	0.30	0.15
TOC/COD	0.40	0.40	——
COD/VSS	2.0?	1.4	1.4
$\frac{\text{Suspended BOD}}{\text{VSS}}$	——	0.50	——
$\frac{\text{Suspended BOD}}{\text{TSS}}$	——	0.40	——

[a] Sources:

Austin, S., Yunt, F., and Wuerdeman, D. 1981. *Parallel Evaluation of Air- and Oxygen-Activated Sludge,* Order No. PB 81-246 712, National Technical Information Service, Springfield, VA.

Bishop, D. F., Heidman, J. A., and Stamberg, J. B. 1976. Single-stage nitrification-denitrification. *J. Water Pollution Cont. Fed.* 48(3):520.

Boon, A. G. and Burgess, D. R. 1972. Effects of diurnal variations in flow of settled sewage on the performance of high-rate activated-sludge plants. *Water Pollution Cont.* 71:493.

Burns and Roe, Inc. 1977. *Federal Guidelines—State Pretreatment Programs—Vol. II—Appendixes 1–7,* EPA-430/9-76-017b, U.S. Environmental Protection Agency, Office of Water Program Operations, Municipal Construction Division, Washington.

Dixon, H., Bell, B., and Axtell, R. J. 1972. Design and operation of the works of the Basingstoke Department of Water Pollution Control. *Water Pollution Cont.* 71:167.

Ford, D. L. 1969. Application of the total carbon analyzer for industrial wastewater evaluation. In *Proc. 23rd Industrial Waste Conf., May 7, 8, and 9, 1968.* Engineering Bulletin 53(2), Engineering Extension Series No. 132, ed. D. Bloodgood, p. 989. Purdue University, Lafayette, IN.

Hoover, S. R. and Porges, N. 1952. Assimilation of dairy wastes by activated sludge: II. The equation of synthesis and rate of oxygen consumption. *Sewage and Industrial Wastes.* 24(3):306.

Hunter, J. V. and Heukelekian, H. 1965. The composition of domestic sewage fractions. *J. Water Pollution Cont. Fed.* 37(8):1142.

LaGrega, M. D. and Keenan, J. D. 1974. Effects of equalizing wastewater flows. *J. Water Pollution Cont. Fed.* 46(1):123.

Nash, N., Krasnoff, P. J., Pressman, W. B., and Brenner, R. C. 1977. Oxygen aeration at Newton Creek. *J. Water Pollution Cont. Fed.* 49(3):388.

Nicholls, H. A. 1975. Full scale experimentation on the new Johannesburg extended aeration plants. *Water SA.* 1(3):121.

Reinhart, D. R. 1979. Nitrification treatability study for Carrollton, Ga. *J. Water Pollution Cont. Fed.* 51(5):1032.

Rickert, D. A. and Hunter, J. V. 1971. Effects of aeration time on soluble organics during activated sludge treatment. *J. Water Pollution Cont. Fed.* 43(1):134.

Schaffer, R. B., Van Hall, C. E., McDermott, G. N., Barth, D., Stenger, V. A., Sebesta, S. J., and Griggs, S. H. 1965. Application of a carbon analyzer in waste treatment. *J. Water Pollution Cont. Fed.* 37(11):1545.

Torpey, W. N. and Lang, M. 1958. Effects of aeration period on modified aeration. *J. Sanitary Eng. Div., ASCE.* 84(SA 3): Paper 1681.

salt fronts and intrusion, consolidated or unconsolidated aquifer materials, artesian or non-artesian aquifer

- Relationship of site to possible or actual sources of contamination
- Permanence of existing conditions; that is, to what extent such events as floods, droughts, or other development might change the quantity and quality of the water, the site topography, or access to the water source

- Special requirements of the site; that is, the need to provide special facilities to cope with moving sand bars, fish and wildlife, seasonal debris, ice, navigation, violent storms, high flood stages and low drought stages, accessibility of site, availability and reliability of electrical power, and distance to pumping station
- Site properties that affect the stability of the proposed structures (e.g., the character of the foundation soils)

The source of supply, nature of the withdrawal site, special requirements of the site, and relationship of the site to sources of contamination all have direct impacts on the quality of the raw water and the treatment it requires. Each of these is discussed in this section.

River Intakes

Hazards

River intakes are exposed to a variety of hazards, including floating and suspended debris, fish and wildlife, ice, variable water stages due to floods and droughts, and shipping and pleasure boating. Although these hazards primarily affect the longevity and reliability of the structures involved, there are some indirect effects on water quality.

Debris and Wildlife. The principal concern raised by debris and wildlife is that they might enter the plant piping and damage mechanical equipment, such as pumps and valves. Small and fragile debris such as fish, leaves, floating oils, cloth, and paper may pass through pumps and valves without damaging them. Nevertheless, such materials are objectionable, because they are macerated by the mechanical equipment and increase the amount of organic matter dissolved in the raw water. High concentrations of dissolved organic matter produce tastes, odors, color, and turbidity; reduce the efficiency of disinfection; and promote the regrowth of microbes in the distribution system. This regrowth increases both the threat of disease transmission and system corrosion.

The defense against debris and wildlife is screens. For intakes these come in two general classes: bar racks and traveling screens. Screen design is discussed in section 7.5, "Screens."

The screenings, collected by manual or mechanical cleaning, undergo some compaction and decay during collection and storage, and they should not be returned to the raw water source. The debris has been concentrated from a large volume of water, and its physical, chemical, and biological properties may have been changed. The mere fact of concentration means that if the debris is returned to the raw water source, a high local debris concentration will be created, which may be a nuisance. The changes in properties may cause the debris to settle in the water source, forming sludge deposits, or the dissolved organic matter concentration in the source may be increased. The preferred route of disposal consists of incineration, landfill, or both.

Fish require special considerations, which are listed in section 7.5, "Screens."

Ice. A major hazard to all surface water intakes in the temperate zone is ice. The principal ice forms are surface, anchor, frazil, and slush [Baylis and Gerstein, 1948; Gisiger, 1947].

Frazil ice consists of fine needlelike ice crystals, which form suspended in the water and later deposit on various surfaces. It sometimes accumulates in the water as slushy masses. Because the crystals are small and the slush is fragile, they can penetrate deeply into the intake piping system, and large amounts of ice may coat intake screens, the inside of conduits, and even pump impellers. This generally results in very serious clogging of the intake system. Frazil ice occurs under turbulent conditions in a very narrow temperature range, maybe only $-0.05°C$ to $+0.10°C$. The condition is usually temporary, starting in the late evening to early morning and lasting until noon. Its formation is prevented by restricting the velocity through intake screens and conduits to less than 30 fpm and by very mild heating of the flow to about 2°F above the ambient water temperature [Pankratz, 1988].

Anchor ice forms on submerged objects. This is especially troublesome in the guideways for slidegates. The usual preventive measures are some sort of heating device, which may be as simple as an air bubble curtain or as involved as hot steam lines.

Surface ice damages structures by impact, static horizontal pressure, and static lifting. The last occurs when ice forms on the structure during low water stages and the water levels subsequently rise. Drifting surface ice and ice floes may also enter the intakes, clogging them. In large lakes and rivers, ice floes may pile up against structures or be stranded in shallow water. At the Chicago intakes in Lake Michigan, stranded ice has been observed reaching from the lake bottom, some 37 ft below the water surface, to 25 ft above the water surface [Burdick, 1946]. Consequently, even bottom intakes are threatened. Because of these possibilities, intakes are usually designed to incorporate barriers and deflectors to keep ice away from the exposed structures.

Floods and Droughts. Floods threaten intakes both by simple submergence, which may damage electric equipment and furnishings, and by structural damage due to hydraulic forces and the impacts of entrained debris such as logs and large rocks. Floods also move sand bars and may either bury intakes or isolate them from the water flow. Droughts can lower water stages to the point that the intakes no longer contact the water. Very low water stages, if they happen fast enough, may destabilize river banks, causing the banks to slump and threaten intake foundations and screens.

Flood protection consists of barriers and deflectors such as pilings, river training to stabilize the banks, and levees to protect buildings and equipment. Regulatory authorities may require that pumps, controls, and electrical switching equipment be installed above the expected maximum flood stage for some rare design flood. Drought protection generally involves the capability to withdraw water from several different elevations. This may be done by constructing multiple intakes or by constructing floating or moving intakes. The former are built on moored barges; the latter are built on rails that permit the pump suction bell to be moved up and down the bank as the river stage changes.

Both floods and droughts exacerbate the problem of suspended debris, especially silts and sands. Floods themselves resuspend stream bottoms and carry increased land erosion, and droughts often force water intakes to be placed so close to the bottom that the intake suction resuspends some sediments.

All of these problems require some estimate of flood and drought stages. First, each time series (not the ranked values) is analyzed for trends and serial correlation. If none are found, the series may be fitted to some extreme value distribution, usually either Gumbel's distribution or the log Pearson Type III distribution.

Next, the maximum annual flood and minimum annual drought water surface elevations are ranked according to severity. For floods, this means highest ($m = 1$) to lowest ($m = n$), but for droughts it means lowest ($m = 1$) to highest ($m = n$). Each elevation is then assigned an "exceedance" probability and a return period based on its rank:

$$P = \frac{m}{n+1} = \frac{1}{T} \tag{7.54}$$

where

m = the rank of a flood or a drought, beginning with most severe (highest stage for flood, lowest for drought) and ending with least severe (lowest stage for flood, highest for drought) (dimensionless)

n = the number of records of maximum annual flood or minimum annual drought (dimensionless)

P = the "exceedance" probability of either (1) the probability of a flood stage *greater* than or equal to the given elevation or (2) the probability of a drought stage *less* than or equal to the given elevation (per yr)

T = the return period of either (1) a flood stage equal to or larger than the given elevation or (2) a drought stage less than or equal to the given elevation (yr)

For small data sets, the probabilities are sometimes assigned according to Eq. (7.55) [Taylor, 1990]:

$$P = \frac{m - 0.5}{n} = \frac{1}{T} \tag{7.55}$$

The exceedance probability is the probability that a flood or drought equal to or worse than the given event will occur in any randomly chosen year. The return period is the average length of time between such occurrences.

Each elevation can also be assigned a nonexceedance probability and an associated return period:

$$P + P_o = 1 \tag{7.56}$$

$$\frac{1}{T} + \frac{1}{T_o} = 1 \tag{7.57}$$

where

P_o = the "nonexceedance" probability, which is either (1) the probability of a flood stage elevation *less* than the given elevation or (2) the probability of a drought stage *greater* than the given elevation (per yr)

T_o = the return period of either (1) a flood stage *less* than the given elevation or (2) a drought stage *greater* than the given elevation (yr)

If the Gumbel equation gives an adequate fit (i.e., a plot of the flows vs. the equivalent flood return period on Gumbel probability paper is linear), the design flood and drought stages for some design return period T may be calculated from Eqs. (7.58) and (7.59), respectively:

$$z_F(T) = \bar{z}_F - s_F \frac{\sqrt{6}}{\pi}[\gamma + \ln(\ln T_o)] \tag{7.58}$$

$$z_D(T) = \bar{z}_D - s_D \frac{\sqrt{6}}{\pi}[\gamma + \ln(\ln T)] \tag{7.59}$$

where

s_D = the standard deviation of the observed minimum annual drought stages (m or ft)
s_F = the standard deviation of the observed maximum annual flood stages (m or ft)
$\bar{z}_D$ = the arithmetic average of the observed minimum annual drought stages (m or ft)
$z_D(T)$ = the estimated drought stage elevation with a return period of T years (m or ft)
$\bar{z}_F$ = the arithmetic average of the observed maximum annual flood stages (m or ft)
$z_F(T)$ = the estimated flood stage elevation with a return period of T years (m or ft)
γ = 0.57722..., Euler's constant

Note that in each case, the return period is that for a stage lower than the given stage.

Although the Gumbel distribution often provides a satisfactory fit, other distributions may be needed in other cases. A collection of these may be found in Chow (1964). The U.S. Water Resources Council has adopted the log Pearson Type III distribution for flood analyses on federally funded projects [Hydrology Committee, 1977].

Navigation. The principal hazards due to navigation are collisions between boats and the intake structure; anchors snagging parts of the intake; and contaminants discharged either accidentally or deliberately. The recommended defense against all these hazards is avoidance of navigation channels. However, this conflicts with the need for access to deep water during droughts and cannot always be achieved. Once again, various kinds of barriers and deflectors are constructed, and intake depths are selected to avoid water contaminated by waste discharges from boats. This normally means selection of deeper waters.

In the U.S., all navigable waters are under the jurisdiction of the U.S. Army Corps of Engineers, and construction within navigable waters requires a Corps-issued permit.

Contaminant Distribution

Rivers are frequently stratified both horizontally and vertically. This is important for two reasons:

1. Water quality in rivers can be expected to vary significantly with the distance from the bank. Most importantly, contaminants will tend to cling to the bank from which they are discharged, so that water intakes are best placed on the opposite bank.
2. Contaminant dilution does not always or even usually occur. What is needed is a method to predict when and where contaminants will occur. It is especially desirable to locate zones of low contamination and to predict when known spills will arrive at the intake.

To predict where contaminants are at any time, it is necessary to consider how they are transported by the water source. Four mechanisms need to be accounted for: advection (convection), shear flow dispersion, turbulent diffusion, and molecular (Fickian) diffusion. *Advection* is the transport achieved by the average motion of the whole body of water. In the case of rivers, this is the transport due to the mean velocity. *Shear flow dispersion* is caused by the vertical and horizontal variations in velocity over the cross section. If velocity measurements are taken at small time intervals at a fixed point in the cross section, it will be discovered that the velocity there fluctuates. These short-period fluctuations about the predicted velocity distribution give rise to another transport mode: *turbulent diffusion*. Finally, *molecular diffusion* arises because of the random thermal motions of contaminant molecules.

These four transport mechanisms are additive. However, one or more of them may sometimes be ignored. For example, by definition there is no advection or shear flow dispersion vertically or laterally in a river; transport in these directions is due entirely to turbulent diffusion and molecular diffusion. In river mouths and in estuaries and lakes, the mean velocity is often small and advection is negligible, even in the "downstream" direction. In these cases, only shear flow dispersion, turbulent diffusion, and molecular diffusion are active. But in the mainstem of most rivers, only advection need be considered.

Longitudinal Transport. Longitudinal transport in rivers is dominated by advection. Shear flow dispersion is a minor component, but it becomes important during unsteady phenomena like spills. Turbulent diffusion and molecular diffusion are negligible contributors to longitudinal transport.

Consider a spill of a conservation contaminant some distance upstream of an intake. If the flow is uniform and steady, the predicted areally weighted cross-sectional average contaminant concentration at the intake is [Crank, 1975; Fischer *et al.*, 1979]

$$\overline{C}_A(x,t) = \frac{M/A}{\sqrt{4\pi K t}} \exp\left[-\frac{(x-Ut)^2}{4Kt}\right] \tag{7.60}$$

where

A = the cross-sectional area of the stream (m^2 or ft^2)

$\overline{C}_A(x,t)$ = the cross-sectional average contaminant concentration at a distance x downstream from the apparent origin at time t after the spill (kg/m^3 or lbm/ft^3)

K = the longitudinal shear flow diffusivity (m^2/day or ft^2/day)

M = the mass of the contaminant spilled (kg or lbm)
t = the time elapsed since the spill (s)
U = the (uniform flow) areally weighted, cross-sectional average stream velocity between the spill and the intake (m/day or ft/day)
x = the effective distance the spill has traveled from its apparent origin (m or ft)

A commonly used semiempirical formula for K is [Fischer *et al.*, 1979]

$$K = \frac{0.011U^2W^2}{Hv_*} \tag{7.61}$$

where

g = the acceleration due to gravity (9.806650 m/s^2 or 32.1740 ft/s^2)
H = the mean depth (m or ft)
R = the hydraulic radius (m or ft)
S = the energy gradient (m/m or ft/ft)
v_* = the shear velocity (m/s or ft/s)
= $\sqrt{gRS}$
W = the stream (top) width (m or ft)

The minimum distance downstream of a spill or an outfall that the water must travel before the contaminant has spread sufficiently for Eq. (7.60) to apply is called the *initial period* and is given approximately by [Fischer *et al.*, 1979]

$$L_{\text{ip}} = \frac{0.67UW^2}{Hv_*} \tag{7.62}$$

where L_{ip} is the initial period (m or ft). The error in Eq. (7.62) is on the order of 50%. At L_{ip} from the spill, the contaminant cloud looks as if it had traveled a shorter distance.

The apparent origin of the spill is L_{ao} downstream from its actual location [Fischer *et al.*, 1979]

$$L_{\text{ao}} = \frac{0.12UW^2}{Hv_*} \tag{7.63}$$

where L_{ao} is the distance from the actual spill to its apparent origin (m or ft). This equation also has an error of about 50%.

Lateral Distribution. If a spill occurs within L_{ip} of an intake, the effect of lateral spreading of the contaminant will be important. The transport problem is now two-dimensional, and the important processes are advection and turbulent diffusion. Vertical mixing can be ignored, because streams are very much wider than deep, and the vertical velocity gradients are steeper than the horizontal velocity gradients, so they become vertically mixed long before they are laterally mixed. For steady, uniform flow with a uniform velocity profile and isotropic turbulence, the longitudinal and lateral contaminant concentrations due to a spill can be estimated from

$$\overline{C}_z(x, y, t) = \frac{M/H}{4\pi t\sqrt{KE}} \exp\left[-\frac{(x - Ut)^2}{4Kt}\right] \times \cdots$$
$$\times \sum_{j=-\infty}^{\infty}\left\{\exp\left[-\frac{(y - 2jW - y_o)^2}{4Et}\right] + \exp\left[-\frac{(y - 2jW + y_o)^2}{4Et}\right]\right\} \tag{7.64}$$

where

$\overline{C}_z(x, y, t)$ = the depth-averaged contaminant concentration at distances of x downstream from the apparent origin and y from the right bank, at a time t after the spill (kg/m^3 or lb/ft^3)

E = the turbulent diffusivity (m^2/s or ft^2/s)

H = the mean depth (m or ft)

M = the mass of the contaminant spilled (kg or lbm)

t = the time elapsed since the spill (s)

U = the (uniform flow) areally weighted, cross-sectional average stream velocity between the spill and the intake (m/day or ft/day)

x = the effective distance the spill has traveled from its apparent origin (m or ft)

y = the distance across a stream measured from the right bank (m or ft)

y_o = the location of a spill measured from the right bank (m or ft)

The error in this equation is approximately 50%. As a practical matter, we are interested in peak concentrations, which occur when $x = Ut$; the first exponential term is unity. As for the summation, only a few terms centered on $j = 0$ need to be computed.

A commonly used estimate of E is [Fischer *et al.*, 1979]

$$E = 0.6H v_* \tag{7.65}$$

If the source of the contaminant is a steady discharge rather than a spill, the steady-state depth-averaged prediction is [Fischer *et al.*, 1979]

$$\overline{C}_{z,t}(x, y) = \frac{W\overline{Q}_W\overline{C}_{QW}/(\overline{Q}_R + \overline{Q}_W)}{\sqrt{4\pi E\tau}} \times \cdots$$
$$\times \sum_{j=-\infty}^{\infty}\left\{\exp\left[-\frac{(y - 2jW - y_o)^2}{4E\tau}\right] + \exp\left[-\frac{(y - 2jW + y_o)^2}{4E\tau}\right]\right\} \tag{7.66}$$

where

$\overline{C}_{QR}$ = the flow-weighted (flow-composited) contaminant concentration in the river upstream of the outfall (kg/m^3 or lbm/ft^3)

$\overline{C}_{QW}$ = the flow-weighted (flow-composited) contaminant concentration in the wastewater (kg/m^3 or lb/ft^3)

$\overline{C}_{z,t}(x, y)$ = the depth- and time-averaged concentration at distance x from the apparent origin and y from the right bank (kg/m^3 or lb/ft^3)

$\overline{Q}_R$ = the time-averaged (steady state) river flow for the critical period, generally the 7-day average low flow with a 10-year return period (m^3/s or ft^3/s)

$\overline{Q}_W$ = the time-averaged (steady state) wastewater flow for the critical period, generally the maximum 7-day or 30-day average wastewater flow rate (m^3/s or ft^3/s)

τ = the time of travel from the apparent origin (s)

Note that the time of travel from the source, τ, replaces the elapsed time, t.

Adsorbed Contaminants. Because streams mix vertically much more quickly than they mix laterally, it is usually assumed that contaminant concentrations are uniform with respect to depth.

There are, however, important exceptions to this rule: settleable solids (especially sand, silt, and clay) and floatable solids such as grease.

Sands, silts, and clays are important for two reasons. First, they themselves are contaminants, causing turbidity. However, this is not a major issue. Sands, silts, and clays are quite reactive, and they adsorb a large number of other contaminants, particularly pesticides, volatile organic substances, and heavy metals. Consequently, the presence of sand, silt, and clay in a finished water raises the possibility that other, more serious contaminants are present. There is one advantage in all this: they are easily removed, and so, consequently, are the poisons adsorbed by them.

The amount of contaminant adsorbed onto silt and clay particles can be expressed in terms of a partition coefficient [Thomann and Mueller, 1987]:

$$\Pi = \frac{C_p}{C_d} \tag{7.67}$$

where

C_d = the dissolved contaminant concentration based on total volume of liquid and solids (kg/m^3 or lb/ft^3)

C_p = the concentration of contaminant adsorbed to suspended solids (kg contaminant/kg clay or lb contaminant/lb clay)

Π = the partition coefficient (m^3/kg or ft^3/lb)

The partition coefficient may be used to calculate the fraction of the contaminant that is dissolved or adsorbed [Thomann and Mueller, 1987]:

$$\frac{C_d}{C_t} = \frac{1}{1 + \Pi C_c} \tag{7.68}$$

$$\frac{C_p}{C_t} = \frac{\Pi C_c}{1 + \Pi C_c} \tag{7.69}$$

where

C_t = the total concentration of contaminant, based on the total volume of liquid and solids (kg/m^3 or lb/ft^3)

$= C_d + C_p C_c$

Table 7.16 Typical Partition Coefficients for Silt/Water Systems[a]

Contaminant	Partition Coefficient (L/kg)
Heavy metals	10^4 to 10^5
Benzo(a)pyrene	10^4 to 10^5
Methoxychlor	10^4
Napthalene	10^3

[a] Thomann, R. V. and Mueller, J. A. 1987. *Principles of Water Quality Modeling and Control.* Harper & Row, Publishers, New York.

Typical values of the partition coefficient are given in Table 7.16.

If there is no resuspension of sediment, then a simple steady, uniform flow model for contaminant loss due to sedimentation of silts and clays is

$$C_t(x) = C_{to}\frac{1 + \Pi C_{co}\exp\left(-\frac{v_s x}{HU}\right)}{1 + \Pi C_{co}} \tag{7.70}$$

where

C_{co} = the initial silt and clay concentration at a point source of sorbable contaminants (kg/m^3 or lb/ft^3)

C_{to} = the initial total concentration of contaminant at a point source (kg/m^3 or lb/ft^3)

$C_t(x)$ = the total concentration of contaminant at distance x from a point source (kg/m^3 or lb/ft^3)

v_s = the settling velocity of the sediment (m/s or ft/sec)

x = the distance from the point source (m or ft)

Lake and Reservoir Intakes

Hazards

All of the hazards that threaten river intakes also threaten lake intakes: namely, debris and wildlife, ice, floods and droughts, and shipping. There are, of course, differences. Ice and debris in rivers are often carried by strong currents that may produce strong impacts. In lakes, the currents and impacts are less severe. Rivers in flood can move sizable boulders, but the debris in lakes is limited to floatable objects. Floods and droughts in rivers are generally short-term phenomena, driven by the recent weather and lasting a few days to a season or two, whereas floods and droughts in lakes often last for years and are driven by long-cycle weather patterns. There are, however, two hazards that are peculiar to lakes: surface waves and seiches. Both are wind-driven phenomena.

Surface Waves. Waves are formed initially by fluctuations in the wind stress on the water surface, but once formed, they interact with each other and the wind to form a large variety of wave sizes. Wave power is transmitted to structures when the waves are (1) reflected by the structure, (2) break on the structure, or if it is submerged, (3) pass over the structure.

For reflected, nonbreaking waves, the simplest analysis is Minikin's [1950]. If a wave is reflected by a vertical wall, it is supposed to rise up the wall above the still water level a distance equal to 1.66 times the wave height. This produces a momentary pressure gradient like that shown in Fig. 7.2, with a wave pressure at the still water elevation of $\rho g H_w$. The total horizontal force due to the wave alone is

$$F = \underbrace{\rho g W H H_w}_{\substack{\text{Wave force below} \\ \text{still water depth}}} + \underbrace{\rho g W \frac{H_w^2}{2}}_{\substack{\text{Wave force above} \\ \text{still water depth}}} \tag{7.71}$$

where

F = the total wave force (N or lbf)
g = the acceleration due to gravity (9.80665 m/s^2 or 32.174 ft/s^2)
H = the still water depth (m or ft)
H_w = the wave height measured crest to trough (m or ft)
W = the width of the wall (m or ft)
ρ = the density of water (kg/m^3 or slug/ft^3)

The wave moment is

$$M = \rho g W \left[\frac{1.66 H_w^2}{2} \left(H + \frac{1.66 H_w}{3} \right) + \frac{H^2 H_w}{2} \right] \tag{7.72}$$

where M is the turning moment (N-m or lbf-ft). These estimates should be doubled for ratios of water depth to wave length less than about 0.20 [King and Brater, 1963]. The forces and moments due to the still water level should be added to the wave forces and moments.

Breaking waves would develop horizontal forces on the order of [King and Brater, 1963]

$$F = 87.8 W H_w^2 \tag{7.73}$$

where

F = the horizontal wave force (lbf)
H_w = the wave height (ft)
W = the width of the structure (ft)

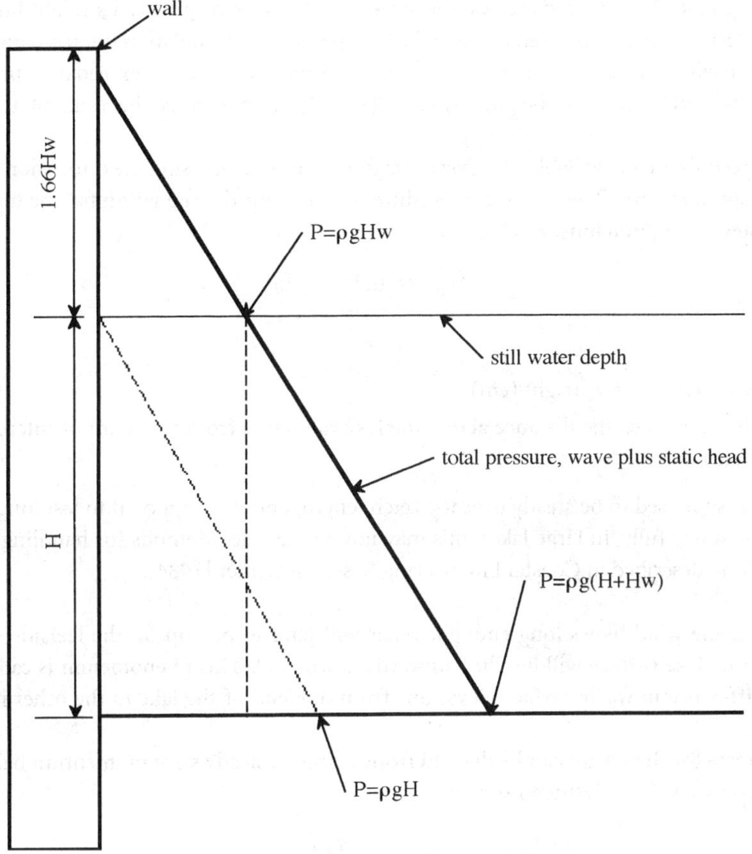

FIGURE 7.2 Reflected wave pressure on vertical wall according to Minikin [1950].

Note that this is an empirical equation written in customary American units. The force would be centered at the still water level and be distributed approximately $\frac{1}{2}H_w$ above and below the still water level.

Forces on submerged structures depend upon the depth of submergence and the dimensions of the structure relative to the wave length of the waves. Both horizontal and vertical forces can be developed.

In many locales, wave heights are routinely recorded, and the needed information can be obtained from the records. Designs will generally be based on some rare, large wave, and it will be necessary to interpolate within or extrapolate from the recorded waves to get the design wave height, H_{wd}. Interpolation/extrapolation is normally done using the Rayleigh distribution [Coastal Engineering Research Center, 1984]:

$$P = \text{Prob}(H_w \geq H_{\text{wd}}) = \exp\left(-\frac{H_{\text{wd}}^2}{H_{\text{rms}}^2}\right) \tag{7.74}$$

$$H_{\text{rms}} = \sqrt{\frac{\sum_1^n H_w^2}{n}} \tag{7.75}$$

where

H_{wd} = the design wave height (m or ft)

H_{rms} = the root-mean-square wave height (m or ft)

A semilog plot of the exceedance vs. the square of the wave heights is a straight line, so linear regression techniques may be used to make the interpolation/extrapolation unbiased and objective. The wave heights are rank-ordered from largest to smallest, and the exceedance probability is calculated for each recorded height using $m/(n+1)$, where m is the rank of the recorded height.

If wave records are unavailable, then wave heights are estimated using the dimensions of the lake and wind speed records. The simplest procedure uses an empirical equation for the highest waves given by Stevenson [Hutchinson, 1975]:

$$H_w = 0.105\sqrt{L_f} \tag{7.76}$$

where

H_w = the highest wave height (cm)
L_f = the *fetch* (i.e., the distance across the lake measured from the point of interest upwind) (cm)

The wind is supposed to be steady over the fetch length, and it is supposed to last long enough to develop the waves fully. In large lakes, this may not be the case. Methods for handling these cases and others are described in Coastal Engineering Research Center [1984].

Seiches. If the wind blows long enough, water will tend to pile up on the lee side of the lake, and the entire lake surface will be tilted upwards downwind. This phenomenon is called a *seiche*, and the difference in water surface elevations from one end of the lake to the other is called the *set-up*.

An estimate for the set-up can be derived from a simple, steady state momentum balance on an arbitrary control volume [Hutchinson, 1975]:

$$H_{su} = \frac{\tau_w L}{\rho g H} \tag{7.77}$$

As a first-order approximation, the distance L may be identified as the length of the lake, H as its mean depth, and H_{su} its set-up. This implies that the sloping lake surface is planar; it is usually curved. The effects of varying lake depth and thermal stratification have also been ignored, but their incorporation requires a complete two- or three-dimensional hydrodynamic model, and what is wanted here is an order-of-magnitude estimate of the set-up.

The wind shear stress may be estimated by a variety of semiempirical formulas, but a common estimate is [Hutchinson, 1975]:

$$\tau_w = 3.5 \times 10^{-6} v_{10}^2 \tag{7.78}$$

where

v_{10} = the wind velocity 10 m above the surface (cm/s)
τ_w = the wind shear stress (dynes/cm^2)

Note that Eq. (7.78) gives the stress in cgs units, so if it is used in Eq. (7.77), cgs units must be used there, too.

Another empirical equation for the set-up is [King and Brater, 1963]:

$$\frac{H_{su}}{H} = 4.88 \times 10^{-5}\left(\frac{L_f}{H}\right)^{1.66}\left(\frac{v_{10}^2}{gL_f}\right)^{2.20(L_f/H)^{-0.0768}} \tag{7.79}$$

Note that this equation is written in dimensionless ratios.

Lake Stratification. During the summer, temperate zone lakes tend to be stratified vertically. The stratified lake may be divided into three zones: an *epilimnion*, a *mesolimnion* (alias *thermocline* and *metalimnion*), and a *hypolimnion*. The epilimnion is the top portion of the lake. It is generally well-mixed and somewhat turbulent because of the wind; it is also warm. The bottom part of the lake is the hypolimnion. It, too, is homogeneous, but it is poorly mixed and nonturbulent or even laminar; it is cool. The mesolimnion is the transition zone between the other layers, normally defined as the region in which the temperature changes.

The stratification is due to the density differences between the warm, light surface water and the cold, heavy bottom water. It develops as follows: In the early spring, the lake is still cold from the previous winter, and it is well-mixed from top to bottom because of the seasonally strong winds. As the spring and early summer progress, solar heating increases and the winds die down. The heating tends to produce a warm surface layer that is lighter than the cold, deep waters. If the winds are strong enough, they will mix the warm and cold waters, and the whole lake will warm up uniformly. However, at some point, the increasing heating rate will overcome the decreasing wind shear, and the warm surface waters will not be mixed down into the whole lake. The lake is then stratified. The epilimnion will continue to warm up during the summer, but the hypolimnion will remain cool.

The stratification breaks down in the fall as solar heating declines and the winds pick up. This generally results in complete mixing of the lake. If the lake surface does not freeze during the winter, the mixing will continue until the next spring, when the lake again stratifies. Such lakes are called *monomictic*, meaning one mixing per year. If the lake surface does freeze, wind mixing is impossible, and the lake becomes stagnant. It may exhibit an inverse stratification, with near 0°C water just under the ice and denser 4°C water below. Such lakes are called *dimictic*, meaning two mixings per year.

Many variations are possible. First, lakes might not stratify. High-latitude and high-altitude lakes may cycle between frozen, stagnant periods and well-mixed periods, whereas equatorial lakes may be mixed continuously throughout the year. Second, in temperate lakes, local conditions may result in weak stratification that comes and goes during the summer, or the lake may stratify into more than three zones. Third, deep bottom waters may remain permanently segregated from the rest of the lake, not participating in the general mixing. Such lakes are called *meromictic*, meaning mixed only in part. The bottom waters of these lakes are always significantly denser than the remainder of the lake, especially the hypolimnion, because they contain elevated concentrations of dissolved solids.

Intake Location. Because lakes and reservoirs stratify, the depth of intakes requires special consideration. The temperature difference between the epilimnion and hypolimnion is usually large enough to affect the kinetics and hydraulics of the various treatment processes, and tank volumes, chemical dosages, and power consumption will depend on whether upper or lower waters are withdrawn. Particularly in the cases of tastes, odors, and turbidity, there may be major differences between the epilimnion and hypolimnion that may dictate the kinds of treatment processes used, not merely their sizing. Many of the characteristics of the epilimnion and hypolimnion change during the year, and a single depth of withdrawal will not always be preferred. Consequently, the intakes should be designed with multiple inlet ports at various depths.

The effects of seiches also deserve consideration. When strong winds set-up a lake, it is the epilimnion waters that pile up on the lee end. This means that at the lee end the epilimnion becomes deeper, and the hypolimnion is compressed and becomes shallower. In fact, the epilimnion may become deep enough that the hypolimnion is completely displaced. The opposite occurs at the windward end of the lake: the epilimnion may be completely blown away, and the hypolimnion may surface. The result of these effects is that the plant operator may not be able to control the kinds of water withdrawn from the lake, and the plant must be designed to accommodate any raw water.

Stratification also affects the fate of contaminant discharges and spills. These inputs will tend to seek water of their own density rather than simply disperse, and they may be found at any elevation in the lake. The elevation may vary seasonally as the input's density varies and the lake temperatures change.

Algal Growth in the Epilimnion. The principal water quality problems associated with stratification are (1) the presence of algae and their metabolic products in the epilimnion and (2) reduced oxygen concentrations and algal decay products in the hypolimnion. These problems are connected.

Algae grow by absorbing nutrients and water and using light energy to convert them to more algae. Except for diatoms, the synthesis stoichiometry for algae growing on nitrate and phosphorus is approximately as follows [Jewell, 1968]:

$$106\,CO_2 + 16\,HNO_3 + H_3PO_4 + \frac{162}{2}H_2O \rightarrow C_{106}H_{180}O_{45}N_{16}P + \frac{599}{4}O_2 \quad (7.80)$$

The algae have a formula weight of 2428, and they are, on a dry weight basis, 52% carbon, 30% oxygen, 9.2% nitrogen, 7.4% hydrogen, and 1.3% phosphorus. Diatoms have an exterior shell of silica, which comprises about one-fourth of the dry weight. However, the material inside the shell has the composition just given. If the synthesis begins from ammonia rather than nitrate, the stoichiometry is

$$106\,CO_2 + 16\,NH_3 + H_3PO_4 + \frac{129}{2}H_2O \rightarrow C_{106}H_{180}O_{45}N_{16}P + \frac{471}{4}O_2 \quad (7.81)$$

Note that the formation of 1 g of algae results in the release of either 1.97 g of oxygen or 1.55 g, depending on whether the synthesis starts with nitrate or ammonia. Conversely, when the algae decay, the same amount of oxygen is consumed. Algae grow in the warm, illuminated epilimnion and settle into the cool, dark hypolimnion. This leads to the fundamental problem of lakes: oxygen release occurs in the epilimnion, whereas oxygen consumption occurs in the hypolimnion, and because the two do not mix, the hypolimnion can become anoxic.

The effects of algae may be estimated by a somewhat simplified analysis of the spring bloom. The algae growing in the epilimnion may be either nutrient limited or light limited. They are most likely nutrient limited if the epilimnion is clear and shallow. In this case, the algae grow until some nutrient is exhausted. Supposing that the nutrient is phosphorus, the maximum possible algal concentration is

$$X_{max} = Y_P P_i \quad (7.82)$$

where

P_i = the initial available orthophosphate concentration (kg/m^3 or lb/ft^3)

X_{max} = the maximum possible algal concentration in the epilimnion due to nutrient limitation (kg/m^3 or lb/ft^3)

Y_P = the algal true growth yield coefficient on phosphorus (kg algae/kg P or lb algae/lb P)

Other nutrients can be limiting. Diatoms are often limited by silica.

Limnologists prefer to work in terms of the areal concentration rather than the volumetric concentration. This can be calculated by multiplying X_{max} by the volume of the epilimnion and dividing it by the surface area of the lake:

$$X_{max,A} = \frac{X_{max} V_{epi}}{A} = Y_P P_i H_{epi} \quad (7.83)$$

where

A = the epilimnion plan area (m^2 or ft^2)
H_{epi} = the epilimnion depth (m or ft)
V_{epi} = the epilimnion volume (m^3 or ft^3)
$X_{max,A}$ = the maximum areal algal concentration in the epilimnion due to nutrient (here P) limitation (kg/m^2 or lb/ft^2)

Note that the areal concentration is simply proportional to the epilimnion depth and the initial phosphorus concentration. Of course, if nitrogen were the limiting nutrient, the slope would depend on nitrogen rather than phosphorus.

The light-limited case is more complicated, because the maximum algal concentration is the result of a balancing of rates rather than simple nutrient exhaustion. The simplest mass balance for the algae in the epilimnion is

$$\underbrace{V_{epi}\frac{dX}{dt}}_{\text{Rate of algal accumulation}} = \underbrace{\mu X V_{epi}}_{\text{Rate of algal growth}} - \underbrace{k_D X V_{epi}}_{\text{Rate of algal loss}} \tag{7.84}$$

where

k_D = the algal decay rate (per second)
t = elapsed time (s)
X = the algal concentration (kg/m^3 or lb/ft^3)
μ = the algal specific growth rate (per second)

In this simplified model, the specific growth rate is supposed to depend only on light intensity and temperature, and the decay rate is supposed to depend only on temperature.

The dependency of the specific growth rate on light can be represented by Steele's [1965] equation:

$$\mu(z) = \mu_{max}\frac{I(z)}{I_{sat}}\exp\left(1 - \frac{I(z)}{I_{sat}}\right) \tag{7.85}$$

where

I_{sat} = the saturation light intensity for the algae (W/m^2 or Btu/ft^2-s)
$I(z)$ = the light intensity at depth z from the surface (W/m^2 or Btu/ft^2-s)
μ_{max} = the maximum algal specific growth rate (per second)
$\mu(z)$ = the algal specific growth rate at depth z (per second)
z = the depth from the surface (m or ft)

The light intensity declines exponentially with depth according to the Beer-Lambert Law:

$$I(z) = I_o \exp[-(a + bX)z] \tag{7.86}$$

where

a = the water extinction coefficient (per m or per ft)
b = the algal extinction coefficient (m^2/kg or ft^2/lb)
I_o = the surface light intensity (W/m^2 or Btu/ft^2-s)

Substituting Eq. (7.86) into Eq. (7.85) and averaging over depth and time yields [Thomann and Mueller, 1987]:

$$\overline{\mu}_{zt} = \frac{ef\mu_{\max}}{(a + bX)H_{\text{epi}}}\left\{\exp\left[-\frac{\overline{I}_o}{I_{\text{sat}}}\exp(-[a + bX]H_{\text{epi}})\right] - \exp\left(-\frac{\overline{I}_o}{I_{\text{sat}}}\right)\right\} \tag{7.87}$$

where

e = 2.71828..., the base of the natural logarithms
f = the fraction of the 24-hour day that is sunlit (dimensionless)
$\overline{I}_o$ = the average sunlight intensity during daylight (W/m^2 or Btu/ft^2-s)
$\overline{\mu}_{z,t}$ = the depth- and time-averaged algal specific growth rate (per second)

During the summer, the average surface light intensity is usually much larger than the saturation light intensity, and the exponential of their ratio is nearly zero. Also, the light intensity at the bottom of the epilimnion will be close to zero, because the algal population will increase until nearly all the light has been absorbed. The exponential of 0 is 1, so Eq. (7.88) reduces to [Lorenzen and Mitchell, 1973 and 1975]:

$$\overline{\mu}_{zt} \approx \frac{ef\mu_{\max}}{(a + bX)H_{\text{epi}}} \tag{7.88}$$

The algal growth rate declines as its population increases (which is called "self-shading"), and at some point it just equals the decay rate. No further population increase is possible, so this becomes the maximum algal concentration. Substituting Eq. (7.88) into (7.84) and solving for the steady-state algal concentration yields

$$X_{\max,A}H_{\text{epi}} = \frac{ef\mu_{\max}}{bk_D} - \frac{a}{b}H_{\text{epi}} \tag{7.89}$$

Equation (7.89) predicts that the maximum areal algal concentration under light-limited conditions will be a linear function of epilimnion depth with a negative slope. Under nutrient-limited conditions, the maximum areal concentration is also linear, but it has a positive slope. These two lines form the boundaries to a triangular region, shown in Fig. 7.3, that contains all possible areal algal concentrations. Any given lake will have some specific epilimnetic depth. The maximum algal concentration that can occur in that lake can be determined by projecting a vertical line from the depth axis to the first line intersected. That line will give the concentration. If the epilimnetic depth is so deep that it lies outside the triangular region, no algae will be found in the lake. In fact, it is the reduction in mixed depth in the spring, due to stratification, that permits algal blooms to occur.

Algal Death in the Hypolimnion. As a first-order approximation, it may be assumed that the quantity of algae that settle into the hypolimnion is equal to the amount of algae that can be formed from the limiting nutrient in the epilimnion. This will be true even in light-limited lakes, because the algal concentration given by Eq. (7.89) is a dynamic equilibrium, and algae will be continuously formed and removed at that concentration until the limiting nutrient is exhausted. Algal decay results in oxygen consumption, so the oxygen balance in the hypolimnion is

$$\beta_X X_{\max} V_{\text{epi}} = (C_o - C)V_{\text{hypo}} \tag{7.90}$$

where

C = the final oxygen concentration in the hypolimnion (kg/m^3 or lb/ft^3)
C_o = the initial oxygen concentration in the hypolimnion kg/m^3 or lb/ft^3)

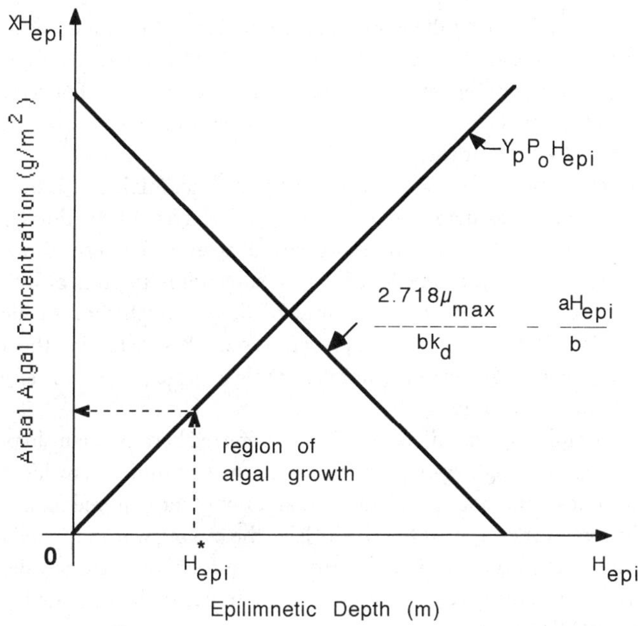

FIGURE 7.3 Permissible algal growth region in epilimnion.

V_{hypo} = the volume of the hypolimnion (m^3 or ft^3)
β_X = the algal oxygen demand (kg O_2/kg algae or lb O_2/lb algae)

When algae decay, their nitrogen content is first released as ammonia, and the ammonia is subsequently converted to nitrate. This means that the algal "carbonaceous" oxygen demand of 1.55 g O_2/g algae is satisfied first, and it proceeds as long as C is greater than zero. The ammonia released during this process is oxidized only if oxygen is left over from the first step.

Algal Control. Figure 7.3 and Eq. (7.90) provide the means of estimating the impacts of algae on lakes and reservoirs and clues to the control of excessive algal blooms:

- The initial nutrient concentration may be reduced by changes in land use or by restrictions on used water discharges.
- The algal decay rate could be increased by adding some sort of poison.
- The lake may be artificially mixed to increase the epilimnetic depth, thereby shifting the algal concentration rightwards and downwards along the light-limited line.

Restriction of nutrient input is the method of choice, but the underlying assumption is either that the algae are nutrient-limited to begin with or that inputs can be reduced to the point that they become nutrient-limited. If the algae are or remain light-limited, control of nutrient inputs will not affect the concentration of algae in the epilimnion. The effect of nutrient restrictions is to reduce the slope of the nutrient line in Fig. 7.3, so if the algae are nutrient-limited, the reduction in algae is proportional to the reduction in nutrient input. Nutrient reductions will always reduce the amount of oxygen consumption in the hypolimnion and the amount of algal excretion product in the epilimnion. Nutrient input reductions are relatively easily attained when the inputs come from point sources such as sewage treatment plants, but reductions are difficult to come by when the sources are diffuse like agriculture, forests, or lawns.

Poisons such as copper sulfate [$CuSO_4 \cdot 5H_2O$], calcium hypochlorite [$Ca(OCl)_2$], and various organic algicides act by increasing the algal death rate. This has the effect of reducing the vertical intercept of Eq. (7.90) and lowering the whole light-limited line. Normal dosages are a few tenths of a mg/L of either copper sulfate or calcium hypochlorite, but the dosages are species-specific and vary substantially depending on the kinds of algae present [Hale, 1957]. For reasons of economy, the epilimnion should be small. The poison dosages needed to kill algae are often fatal to other wildlife, and this may prevent their use.

The reduction of epilimnetic algal concentrations by artificial mixing of lakes and reservoirs to control algae has been achieved many times [Cooley and Harris, 1954; Riddick, 1958; Symons, 1969; Lorenzen and Mitchell, 1975]. Furthermore, cyanobacteria (formerly known as "blue-green algae"), which are especially objectionable, are more susceptible to control by mixing than are other kinds of algae. To be effective, the lake must be deep enough that the algae either are or will become light limited. Referring to Fig. 7.3, it can be seen that increasing the depth of shallow, nutrient-rich systems will only increase the number of algae present, and this effect also has been observed [Hooper *et al.*, 1952; Symons, 1969].

The usual method of mixing is air diffusion. This is preferred because many hypolimnetic waters are oxygen-depleted, and mixing low-oxygen water into the epilimnion may kill or injure wildlife and may contribute to taste and odor problems. The amount of air required for satisfactory mixing is dependent on the total work required to mix the lake, the mixing work supplied by the wind, and the offsetting heat input from the sun. The resulting air requirement is strongly dependent on local conditions, but a rough and ready rule of thumb is 20 scfm per million square feet of lake surface [Lorenzen and Fast, 1977].

Littoral vs. Pelagic Zones. Lakes also stratify horizontally and may be divided into a *littoral*, or near-shore, zone and a *pelagic*, or central, zone. Contaminant levels are generally very much higher in the littoral than in the pelagic zone. The kinds of contaminants present depend largely on lake size and the intensity of the currents and waves that develop in the lake as a consequence of its size.

Small lakes have weak currents and little or no waves, except during storms. The littoral zone in these lakes is usually defined to be the zone of rooted aquatic plants, and it is subdivided into zones of emergent plants, floating plants, and submerged plants [Wetzel, 1975]. The emergent plants are generally limited to water depths less than about 1.5 m; the floating plants can exist in depths between 0.5 and 3.0 m; and the submerged plants can maintain themselves at virtually any depth that has sufficient sunlight.

Conditions in the small lake littoral are good for plant growth. Land drainage, with its generally high nutrient load, must first pass through the littoral, and the weak currents there ensure that much of the nutrient load is retained. The littoral is shallow and well-illuminated, and the rooted plants provide a large surface area for the attachment and growth of microscopic algae. The net result is that plant productivity in the littoral is very high and may, in some cases, comprise the bulk of the net photosynthesis in the whole lake [Wetzel, 1975]. The net result of all this activity often is a water that is high in turbidity, color, taste, and odor.

The littoral of large lakes is defined to be the region between the shore line and the breakers [Coastal Engineering Research Center, 1984]. Within this zone, there are few aquatic plants, the bottom is coarse-grained, and the water has a high turbidity because of suspended silt and sand. There is almost always a *littoral drift:* a current parallel to the shore and tending in the same direction as the mean wind. Somewhat farther offshore there may be a *coastal jet:* a fairly strong current parallel to the coast [Csanady, 1975].

Water quality in littoral zones is generally not suitable for water supplies, and better-quality water may be had farther offshore. The likelihood of contamination from spills on land is also greatly reduced, because they are intercepted by the drift and coastal jet. Placement of the intake in the open water will also provide the opportunity to choose between epilimnetic and hypolimnetic waters, although there are more hazards from boating and shipping.

Wells

Aquifer Choice

Groundwaters occur in a variety of geological strata, including unconsolidated materials, porous rocks, fractured rocks, and limestones with solution channels. Any of these formations can be developed into an acceptable source as long as:

- The water can be brought up to acceptable quality standards by treatment.
- The permeability of the formation permits adequate flows into the wells.
- The volume and recharge rate of the aquifer permit useful long-term withdrawal rates.

Except where igneous or other impermeable rocks lie close to the surface, usable groundwater supplies are found everywhere that has a net runoff of precipitation. They are even found in deserts, but these supplies are usually either fossil deposits from an earlier climatological epoch (and nonrenewable) or fed from distant precipitation and subject to low renewal rates.

It is commonly believed that well waters are pure because they have been filtered through great thicknesses of fine-pored material. This, of course, is false. First, some common contaminants are derived from the weathering of the formations that produce the well waters. Second, there are a variety of ways by which wells become contaminated. And third, it is not true that all wells draw on naturally filtered water: in heavily weathered limestone regions, the wells may tap actual underground pools or streams.

The composition of groundwater supplies frequently exhibits strong vertical stratification. In part, this is because the waters in different strata may be separated by impermeable materials, which prevent vertical mixing of the supplies. The separated aquifers may be in formations with different mineralogies. In this case the weathering products would be different, both in amount and in kind.

Shallow aquifers near the surface are replenished by local precipitation and are vulnerable to contamination by spills on the surface and drainage from local agriculture, lawns, transportation modes, and even land treatment of used water. Most contaminants enter the aquifer with water percolating from the surface, and they tend to form a layer on top of the supply. Vertical mixing deeper into the supply will proceed very slowly unless the groundwater table undergoes seasonal variations in elevation. These variations move the surface contaminant layer up and down through the soil and so distribute the contaminant in depth.

Deep aquifers may be fed from afar and are subject to similar contamination at their origin.

The common contaminants due to surface spills and land use practices are gasoline and other fuels (from leaking storage tanks), tetraethyl lead (in the gasoline), road salts ($NaCl$ and $CaCl_2$), ammonia, and nitrate. The occurrence of these contaminants is predictable, and well fields can be sited to avoid them. Where this is not possible, suitable regulations can sometimes mitigate the contamination.

The common contaminants from formation weathering are total dissolved solids, calcium, magnesium, sulfate, sodium, chloride, ferrous iron, manganous manganese, and sulfide. Occasionally, one finds arsenic, copper, lead, or mercury, which are derived from local ore deposits. Reliable treatment processes exist for the removal of all of these contaminants. However, they can frequently be avoided altogether by switching to a different aquifer, and this is often a matter of simply changing the well depth.

Groundwaters can also be contaminated *by* wells. The most common problem is the well casing itself. The placement of the casing frequently leaves small channels between the casing wall and the surrounding formation, and these channels serve as routes by which surface drainage may enter the aquifer. These channels may be closed by grouting operations and by paving the ground near the well to deflect surface drainage. The inside of the well casing is an even better route of contamination, and the tops of the casings must be sealed to prevent the entry of surface drainage, soil, and dust.

The same problems exist with other kinds of wells that may penetrate the aquifer, even if they do not draw from it. These other wells include gas and oil wells and deep well injection facilities. In these cases, contamination from below the aquifer is also possible, because these systems are usually pressurized. The kinds of contaminants associated with gas and oil wells include various hydrocarbons, hydrogen sulfide, and brines associated with the deposits. Virtually anything can be found in deep well injection systems, and because such systems are often used to dispose of especially hazardous materials, the contamination from them may pose severe public health problems.

Well Construction and Operation

The construction and operation of wells can also contribute to water supply contamination. This occurs three ways:

- failure to develop the well properly
- corrosion of the well materials
- precipitate formation during pumping

When wells are constructed, the formation next to the casing normally becomes plugged with fines [Johnson Division, 1975]. This occurs in two general ways. First, if the formation consists of unconsolidated materials, the well-drilling operation will compact the materials near the casing, reducing both their porosity and permeability. Second, if the well is drilled in rock, the drilling muds used to lubricate the bits and remove the cuttings will deposit clays in the pores of the rock. The compactions and deposits are eliminated by *developing* the well. This generally consists of either rapidly changing the direction of flow through the well screen or flushing the screens with high-velocity water jets. Well development will also remove fine material from the formation itself, at least near the casing, and this will increase well yields. Failure to develop wells generally results in reduced water flow rates and continuing problems with sands, silts, and clays in the water supply.

Well casings and screens are subject to corrosion at varying rates and ultimately must be replaced. However, corrosion rates can be greatly accelerated if unsuitable materials are used in certain waters. Groundwater conditions leading to excessive rates of corrosion include [Johnson Division, 1975]:

- A pH less than 7
- A dissolved oxygen concentration greater than 2.0 mg/L
- Almost any hydrogen sulfide concentration, even below 1 mg/L
- Total dissolved solids in excess of 1000 mg/L
- A carbon dioxide concentration above 50 mg/L
- A chloride concentration greater than 500 mg/L

The well screen is more vulnerable to corrosion than the casing, because its function is to screen out the sands and gravels surrounding the well and prevent their entry into the pump, and any small failure of the screen will allow large amounts of material into the casing.

The corrosion products are also of interest. Most of the product will be ferrous or ferric iron. However, most older wells also have a lead packer that closes the opening between the casing itself and the well screen. This packer is also subject to corrosion, which will introduce lead into the supply. Many submersible pumps have brass or bronze fittings, which leach lead.

Finally, there is precipitate formation. This occurs in the screen openings and the pores of the formation near the well. It is a problem associated with hard, alkaline waters. The flow through the formation pores and screen can be rapid, and this results in a low-pressure zone. Any dissolved gases in the formation are at equilibrium with the static formation pressure, so they tend to come out of solution in the low-pressure zone. In the case of carbon dioxide, its loss from solution causes bicarbonate to decompose, forming both new carbon dioxide to replace that lost and carbonate.

$$2\,HCO_3^- \rightarrow CO_2 + H_2O + CO_3^{2-} \tag{7.91}$$

The increased carbonate will precipitate both calcium and ferrous iron as carbonates:

$$Ca^{2+} + CO_3^{2-} \rightarrow CaCO_3 \tag{7.92}$$

$$Fe^{2+} + CO_3^{2-} \rightarrow FeCO_3 \tag{7.93}$$

Both these materials can enter the water distribution system, causing nuisances, and in large quantities, both can clog the screen and the surrounding formation. They are generally removed with acids, but acids can corrode the screen and casing and should be used cautiously.

References

Section 7.1

Anonymous. 1976. *Quality Criteria for Water*, U.S. Environmental Protection Agency, Office of Water Quality Planning and Standards, Division of Criteria and Standards, Washington, D.C.

Fuller, G. W. 1912. *Sewage Disposal.* McGraw-Hill Book Co., Inc., New York.

Kolkwitz, R. and Marrson, M. 1909. Ecology of animal saprobia. *International Revue der Gesamten Hydrobiologie und Hydrographie,* 2:126. (trans.: United States Joint Publications Research Service, U.S. Department of Commerce, 1967, In *Biology of Water Pollution: A Collection of Selected Papers on Stream Pollution, Waste Water and Water Treatment,* Pub. No. CWA-3, Keup, L. E., Ingram, W. M., and Mackenthum, K. M., Eds., U.S. Dept. Interior, Federal Water Pollution Control Adm., Cincinnati, Ohio)

Royal Commission on Sewage Disposal. 1913. Eighth report. *Parliamentary Session Papers, Reports from Commissioners, Inspectors et al.,* vol. 25, 10 March to 15 August, 1913.

Sladacek, V. and Tucek, F. 1975. Relation of the saprobic index to BOD_5. *Water Research,* 9(9):791.

Thurston, R.V., *et al.* 1974. *Aqueous Ammonia Equilibrium Calculations,* Tech. Rept. No. 74-1. Montana State University, Fisheries Bioassay Laboratory, Bozeman, MT.

Section 7.2

Berthouex, P. M. and Polkowski, L. B. 1970. Design capacities to accommodate forecast uncertainties. *J. Sanit. Eng. Div. ASCE.* 96(SA5):1183.

Clark, R. M. and Dorsey, P. 1982. A model of costs for treating drinking water. *J. Am. Water Works Assn.* 74(12):618.

Leslie, P. H. 1945. On the use of matrices in certain population mathematics. *Biometrika.* 33(Part III):183.

Lotka, A. J. 1956. *Elements of Mathematical Biology.* Dover Publications, Inc., New York. (Note: This is a corrected reprint of the 1st edition, published in 1924 under the title *Elements of Physical Biology* by The Williams and Wilkins Co., Inc., of Baltimore.)

Manne, A. S. 1967. Calculations for a single producing area. In *Investments for Capacity Expansion: Size, Location, and Time-Phasing,* ed. A. S. Manne, p. 28. The MIT Press, Cambridge, MA.

McJunkin, F. E. 1964. Population forecasting by sanitary engineers. *J. Sanit. Eng. Div. ASCE.* 90(SA4):31.

Rachford, T. M., Scarato, R. F., and Tchobanoglous, G. 1969. Time-capacity expansion of waste treatment systems. *J. Sanit. Eng. Div. ASCE.* 95(SA6):1063.

Srinivasan, T. N. 1967. Geometric rate of growth of demand. In *Investments for Capacity Expansion: Size, Location, and Time-Phasing,* ed. A. S. Manne, p. 151. The MIT Press, Cambridge, MA.

Wastewater Committee, Great Lakes–Upper Mississippi River Board of State Public Health and Environmental Managers. 1990. *Recommended Standards for Wastewater Facilities, 1990 Edition.* Health Education Services, Albany, NY.

Water Supply Committee, Great Lakes–Upper Mississippi River Board of State Public Health and Environmental Managers. 1987. *Recommended Standards for Water Works.* Health Education Services, Albany, NY.

Section 7.3

Burns and Roe, Inc. 1977. *Federal Guidelines—State Pretreatment Programs—Vol. II—Appendixes 1–7,* EPA-430/9-76-017b. U.S. Environmental Protection Agency, Office of Water Program Operations, Municipal Construction Division, Washington, D.C.

Fair, G. M., Geyer, J. C., and Morris, J. C. 1954. *Water Supply and Waste-Water Disposal.* John Wiley & Sons, Inc., New York.

Folwell, A. P. 1932. Maximum daily and hourly water consumption in American cities. *Public Works.* 63(10):13.

Heggie, G. D. 1957. Effects of sprinkling restrictions. *J. Am. Water Works Assn.* 49(3):267.

Howe, C. W. and Linaweaver, F. P., Jr. 1967. The impact of price on residential water demand and its relation to system design and price structure. *Water Resources Res.* 3(1):13.

ISO (Insurance Services Office). 1980. *Fire Suppression Rating Schedule,* ed. 6-80. Insurance Services Office, New York.

Joint Committee of the Water Pollution Control Federation and the American Society of Civil Engineers. 1977. *Wastewater Treatment Plant Design,* Manual of Practice No. 8, Water Pollution Control Federation, Alexandria, VA; American Society of Civil Engineers, New York.

Joint Task Force. 1992. *Design of Municipal Wastewater Treatment Plants: Vol. I—Chapters 1–12,* WEF Manual of Practice No. 8, ASCE Manual and Report on Engineering Practice No. 76, Water Environment Federation, Alexandria, VA; American Society of Civil Engineers, New York.

Linaweaver, F. P., Jr., Geyer, J. C., and Wolff, J. B. 1966. *Final and Summary Report on the Residential Water Use Research Project, Report V, Phase Two,* The Johns Hopkins University, Department of Environmental Engineering Science, Baltimore, MD.

Ongerth, J. E. 1979. *Evaluation of Flow Equalization in Municipal Wastewater Treatment,* EPA-600/2-79-096. U.S. Environmental Protection Agency, Municipal Environmental Research Laboratory, Cincinnati, OH.

Schmidt, C. J., *et al.* 1980. *Design Manual: Onsite Wastewater Treatment and Disposal Systems,* U.S. Environmental Protection Agency, Office of Research and Development, Municipal Environmental Research Laboratory, Cincinnati, OH.

Solley, W. B., Chase, E. B., and Mann, W. B., IV. 1983. *Estimated Water Use in the United States in 1980,* Circular No. 1001. U.S. Geological Survey, Distribution Branch, Text Products Division, Alexandria, VA.

Wolff, J. B. 1957. Forecasting residential requirements. *J. Am. Water Works Assn.* 49(3):225.

Section 7.4

Babbitt, H. E., Doland, J. J., and Cleasby, J. L. 1967. *Water Supply Engineering,* 6th ed., McGraw-Hill, New York.

Baylis, J. A. and Gerstein, H. H. 1948. Fighting frazil ice at a waterworks. *Engineering News-Record.* 104(April 15):80 (vol. p. 562).

Burdick, C. B. 1930. Water-works intakes of the Middle West. *Engineering News-Record.* 104(May 22):834.

Burdick, C. B. 1946. Water works intakes. *J. Am. Water Works Assn.* 38(3):315.

Chow, V. T. 1964. *Handbook of Hydrology.* McGraw-Hill, New York.

Coastal Engineering Research Center. 1984. *Shore Protection Manual,* Vol. 1 and 2, Department of the Army, U.S. Army Corps of Engineers, Waterways Experiment Station, Vicksburg, MS.

Cooley, P. and Harris, S. L. 1954. The prevention of stratification in reservoirs. *J. Inst. Water Eng.* 8(7):517.

Crank, J. 1975. *The Mathematics of Diffusion.* Oxford University Press, London.

Csanady, G. T. 1975. The coastal jet conceptual model in the dynamics of shallow seas. In *The Sea: Ideas and Observation on Progress in the Study of the Seas, Vol. 6, Marine Modeling,* ed. E. D. Goldberg, I. N. McCave, J. J. O'Brien, and J. H. Steele, p. 117. John Wiley & Sons. New York.

Fair, G. M. and Geyer, J. C. 1954. *Water Supply and Waste-Water Disposal,* John Wiley & Sons, New York.

Fischer, H. B., List, E. J., Koh, R. C. Y., Imberger, J., and Brooks, N. H. 1979. *Mixing in Inland and Coastal Waters.* Academic Press, New York.

Gisiger, P. E. 1947. Safeguarding hydro plants against the ice menace. *Civil Eng.* 17(1):24.

Hale, F. E. 1957. *The Use of Copper Sulfate in Control of Microscopic Organisms.* Phelps Dodge Refining Corp., New York.

Hooper, F. F., Ball, R. C., and Tanner, H. A. 1952. An experiment in the artificial circulation of a small Michigan lake. *Trans. Am. Fisheries Soc.: 82nd Annual Meeting, September 8, 9, 10, Dallas,* 52:222.

Hutchinson, G. E. 1975. *A Treatise on Limnology: Vol. 1, Geography, Physics, and Chemistry—Part 1, Geography and Physics of Lakes.* John Wiley & Sons, New York.

Hydrology Committee. 1977. *Guidelines for Determining Flow Frequency, Bull. No. 17A.* U.S. Water Resources Council, Washington.

Jewell, W. J. 1968. *Aerobic Decomposition of Algae and Nutrient Regeneration,* Ph.D. dissertation. Stanford University, Stanford, CA.

Johnson Division. 1975. *Ground Water and Wells: A Reference Book for the Water-Well Industry.* UOP Inc., St. Paul, MN.

King, H. W. and Brater, E. F. 1963. *Handbook of Hydraulics,* 5th ed. McGraw-Hill Book Co., Inc., New York.

Lorenzen, M. and Fast, A. 1977. *A Guide to Aeration/Circulation Techniques for Lake Management,* Cat. No. PB-264 126. U.S. Department of Commerce, National Technical Information Service, Springfield, VA.

Lorenzen, M. and Mitchell, R. 1973. Theoretical effects of artificial destratification on algal production in impoundments. *Env. Sci. Techn.* 7(10):939.

Lorenzen, M. W. and Mitchell, R. 1975. An evaluation of artificial destratification for control of algal blooms. *J. Am. Water Works Assn.* 67(7):373.

Minikin, R. R. 1950. *Winds, Waves and Maritime Structures.* Charles Griffin & Co., Ltd., London.

Riddick, T. M. 1958. Forced circulation of large bodies of water. *J. San. Eng. Div. ASCE.* 84(SA4): paper 1703.

Steele, J. H. 1965. Notes on some theoretical problems in production ecology. In *Primary Productivity in Aquatic Environments, Memoria di Istituto Italiano di Idrobiologia,* 18 Suppl., ed. C. R. Goldman, p. 383. University of California Press, Berkeley.

Symons, J. M. 1969. *Water Quality Behavior in Reservoirs: A Compilation of Published Research Papers.* U.S. Department of Health, Education, and Welfare, Public Health Service, Consumer Protection and Environmental Health Service, Environmental Control Administration, Bureau of Water Hygiene, Cincinnati, OH.

Taylor, J. K. 1990. *Statistical Techniques for Data Analysis.* Lewis Publishers, Boca Raton, FL.

Thomann, R. V. and Mueller, J. A. 1987. *Principles of Surface Water Quality Modeling and Control.* Harper & Row, New York.

Wetzel, R. G. 1975. *Limnology,* W. B. Saunders Co., Philadelphia, PA.

8

Physical Water Treatment Processes

Robert M. Sykes
Ohio State University

8.1 Screens

The important kinds of screening devices are bar screens, coarse screens, comminutors and in-line grinders, fine screens, and microscreens [Pankratz, 1988].

Bar Screens

Bar screens may be subdivided into (1) trash racks, (2) mechanically cleaned bar screens, and (3) manually cleaned bar screens.

Trash Racks

Trash racks are frequently installed in surface water treatment plant intakes to protect traveling screens from impacts by large debris and to prevent large debris from entering combined/storm

0-8493-8953-4/95/$0.00 + $.50

water sewerage systems. Typical openings are 1 to 4 inches. The bars themselves are made of steel, and their shape and size depend on the expected structural loads, which are both static (due to the headloss through the rack) and dynamic (due to the impacts of moving debris). The racks are cleaned intermittently by mechanically driven rakes that are drawn across the outside of the bars. The raking mechanism should be able to lift and move the largest expected object.

Mechanically Cleaned Bar Screens

Mechanically cleaned bar screens are usually installed in the headworks of sewage treatment plants to intercept large debris. They may be followed by coarse screens, comminutors, and in-line grinders. The clear openings between the bars are usually ½ to 1¾ inches wide [Hardenbergh and Rodie, 1960; Pankratz, 1988; Wastewater Committee, 1990]. In sewage treatment plants, the approach channel should be perpendicular to the plane of the bar screen and straight. Approach velocities should lie between 1.25 ft/s (to avoid grit deposition) and 3.0 ft/s (to avoid forcing material through the openings) [Wastewater Committee, 1990]. Velocities through the openings should be limited to 2 to 4 ft/s [Joint Task Force, 1992].

Several different designs are offered [Pankratz, 1988].

Inclined and Vertical Multirake Bar Screens. Multirake bar screens are used wherever intermittent or continuous heavy debris loads are expected. The spaces between the bars are kept clear by several rows of rakes mounted on continuous belts. The rake speed and spacing is adjusted so that any particular place on the screen is cleaned at intervals of less than one minute. The bars themselves may be either vertical or inclined, although the latter facilitates debris lifting.

The raking mechanism may be placed in front of the bars, behind them, or it may loop around them. In the most common arrangement, the continuous belt and rakes are installed in front of the screen, and the ascending side of the belt is the cleaning side. At the top of the motion, a high-pressure water spray dislodges debris from the rake and deposits it into a collection device.

If the belt and rakes are installed so that the descending side is behind the screen (and the ascending side is either in front of or behind the screen) there will be some debris carryover.

Catenary, Multirake Bar Screens. Standard multirake bar screens support and drive the rake belt with chain guides, shafts, and sprockets at both the top and bottom of the screen. Catenary bar screens dispense with the bottom chain guide, shaft, and sprocket. This avoids the problem of interference by deposited debris in front of the screen. The rakes are weighted so that they drag over debris deposits, and the bar screen is inclined so that the weighted rakes lie on it.

Reciprocating-Rake Bar Screens. Reciprocating-rake bar screens have a single rake that is intermittently drawn up the face of the screen. Because of their lower solids-handling capacity, they are used only in low-debris loading situations. The reciprocating mechanism also requires more headroom than multirake designs. However, they are intrinsically simpler in construction, have fewer submerged moving parts, and are less likely to jam.

Arc, Single-Rake Bar Screens. In these devices, the bars are bent into circular arcs, and the cleaning rake describes a circular arc. The rake is normally cleaned at the top of the arc by a wiper. The flow is into the concave face of the screen, and the rakes are upstream of it.

Manually Cleaned Bar Screens

Manually cleaned bar screens are sometimes installed in temporary bypass channels for use when the mechanically cleaned bar screen is down for servicing. The bars should slope at 30 to 45 degrees from the horizontal, and the total length of bars from the invert to the top must be reachable by the rake. The opening between the bars should not be less than 1 in., and the velocity through it should be between 1 and 2 ft/s. The screenings will usually be dragged up over the top of the bars and deposited into some sort of container. The floor supporting this container should be drained or grated.

Bar Screen Headlosses

The maximum headloss allowed for dirty bar racks is normally about 2.5 ft [Fair *et al.*, 1954]. For clean bar racks the minimum headloss can be calculated from Kirschmer's formula [Fair and Geyer, 1954]:

$$h_L = \beta \left(\frac{w}{b}\right)^{4/3} h_v \sin\theta \tag{8.1}$$

where

b = the minimum opening between the bars (m or ft)
h_L = the headloss through the bar rack (m or ft)
h_v = the velocity head of the approaching flow (m or ft)
w = the maximum width of the bars facing the flow (m or ft)
β = a dimensionless shape factor for the bars
θ = the angle between the facial plane of the bar rack and the horizontal

Some typical values of the shape factor β are given in Table 8.1.

The angle θ is an important design consideration, because sloping the bar rack increases the open area exposed to the flow and helps keep the velocity through the openings to less than the desired maximum. Slopes as flat as 30° from the horizontal make manual cleaning of the racks easier, although nowadays racks are always mechanically cleaned.

Coarse Screens

Coarse screens may be constructed as traveling screens, rotating drum screens, or rotating disk screens [Pankratz, 1988].

Traveling Screens

Traveling screens are the most common type of coarse screen. They are used in water intakes to protect treatment plant equipment from debris and at wastewater treatment plants to remove debris from the raw sewage. They consist of flat panels of woven wire mesh supported on steel frames. The panels are hinged together to form continuous belt loops that are mounted on motor-driven shafts and sprockets. Traveling screens are used to remove debris smaller than 2 in., and the mesh openings are generally about 1/8 in. to 3/4 in., typically 1/4 to 3/8 in. Traveling screens are often preceded by bar racks to prevent damage by large objects. When used in water intakes in cold or temperate climates, the approach velocity to traveling screens is normally kept below 0.5 ft/s in

Table 8.1 Kirschmer's Shape Factors for Bars[1]

Bar Cross Section	Shape Factor (Dimensionless)
Sharp-edged, rectangular	2.42
Rectangular with semicircular upstream face	1.83
Circular	1.79
Rectangular with semisircular faces upstream and downstream	1.67
Tear-drop with wide face upstream	0.76

[1]Fair, G. M., Geyer, J. C., and Morris, J. C. 1954. *Water Supply and Waste-Water Disposal,* John Wiley & Sons, New York.

order to prevent the formation of frazil ice, to prevent resuspension of sediment near the intake, and to permit fish to swim away.

Traveling screens are cleaned intermittently by advancing the continuous belt so that dirty panels are lifted out of the water. The panels are cleaned by high-pressure water sprays, and the removed debris is deposited into a drainage channel for removal.

Traveling screens may be installed so that their face is either perpendicular to the flow or parallel to it, and either one or both sides of the continuous belt loop may be used for screening. The alternatives are as follows [Pankratz, 1988]:

- Direct, through, or single flow—the screen is installed perpendicular to the flow, and only the outer face of the upstream side of the belt screens the flow. The chief advantage to this design is the simplicity of the inlet channel.
- Dual flow—the screen is installed so that both the ascending and descending sides of the belt screen the flow. This is accomplished either by arranging the inlet and outlet channels so that the flow enters through the outside face of the loop and discharges along the center axis of the loop (dual entrance, single exit) or so that the flow enters along the central axis of the loop and discharges through the inner face of the loop (single entrance, dual exit). The chief advantage to this design is that both sides of the belt loop screen the water, and the required screen area is half that of a direct flow design.

Rotating Drum Screens

In drum screens, the wire mesh is wrapped around a cylindrical framework, and the cylinder is partially submerged in the flow, typically to about two-thirds to three-fourths of the drum diameter. As the mesh becomes clogged, the drum is rotated, and high-pressure water sprays mounted above the drum remove the debris.

The flow may either enter the drum along its central axis and exit through the inner face of the drum, or enter the drum through the outer face of the mesh and exit along its central axis. The flow along the central axis may either be one way, in which case one end of the drum is blocked, or two way.

Rotating Disk Screens

In disk screens, the wire mesh is supported on a circular disk framework, and the disk is partially submerged in the flow, typically to about two-thirds to three-fourths of the disk diameter. As the mesh becomes clogged, the disk is rotated, and the mesh is cleaned by a high-pressure water spray above it. The disks may be mounted so that the disk plane is either vertical or inclined.

For reasons of economy, disk screens are normally limited to flows less than 20,000 gpm [Pankratz, 1988].

Fish Screens

Surface water intakes must be designed to minimize injury to fish by the intake screens. This generally entails several design and operating features and may require consultation with fisheries biologists [Pankratz, 1988]:

- Small mesh sizes—mesh sizes should be small enough to prevent fish from becoming lodged in the openings.
- Low intake velocities—the clean screen should have an approach velocity of less than 0.5 fps to permit fish to swim away.
- Continuous operation—this minimizes the amount of debris on the screen and the local water velocities near the screen surface, which enables fish to swim away.
- Escape routes—the intake structure should be designed so that the screen is not the downstream end of a channel. This generally means that the intake channel should direct flow parallel or at an angle to the screens, with an outlet passage downstream of the screens.

- Barriers—the inlet end of the intake channel should have some kind of fish barrier, such as a curtain of air bubbles.
- Fish pans and two-stage cleaning—the screen panels should have a tray on their bottom edge that will hold fish in a few inches of water as the panels are lifted out of the flow. As the screen rotates over the top sprocket, the fish tray should dump its contents into a special discharge channel, and the screen and tray should be subjected to a low-pressure water spray to move the fish through the channel back to the water source. A second, high-pressure water spray is used to clean the screen once the fish are out of the way.

Wire Mesh Headlosses

Traveling screens are usually cleaned intermittently when the headloss reaches 3 to 6 in. The maximum design headloss for structural design is about 5 ft [Pankratz, 1988].

The headloss through a screen made of vertical, round, parallel wires or rods is [Blevins, 1984]

$$h_L = 0.52 \cdot \frac{1-\varepsilon^2}{\varepsilon^2} \cdot \frac{U^2}{2g} \tag{8.2}$$

if

$$\mathrm{Re} = \frac{\rho U d}{\varepsilon \mu} > 500 \tag{8.3}$$

and

$$0.10 < \varepsilon < 0.85 \tag{8.4}$$

where

d = the diameter of the wires or rods in a screen (m or ft)
g = the acceleration due to gravity (9.80665 m/s^2 or 32.174 ft/s^2)
s = the distance between wire or rod centers in a screen (m or ft)
U = the approach velocity (m/s or ft/s)
ε = the screen porosity, that is, the ratio of the open area measured at the closest approach of the wires to the total area occupied by the screen (dimensionless)
= $(s - d)/s$
μ = the dynamic viscosity of water ($N \cdot s/m^2$ or $lbf \cdot s/ft^2$)
ρ = the water density (kg/m^3 or lb/ft^3)

Blevins [1984] gives headloss data for a wide variety of other screen designs.

Comminutors and In-line Grinders

Comminutors and in-line grinders are used to reduce the size of objects in raw wastewater. They are supposed to eliminate the need for coarse screens and screenings handling and disposal. They require upstream bar screens for protection from impacts from large debris.

Comminutors consist of a rotating, slotted drum that acts as a screen and peripheral cutting teeth and shear bars that cut down objects too large to pass through the slots and that are trapped on the drum surface. The flow is from outside the drum to its inside. Typical slot openings are ¼ in. to ⅜ in. Typical headlosses are 2 to 12 in.

In-line grinders consist of pairs of counterrotating, intermeshing cutters that shear objects in the wastewater. The product sizes also are typically ¼ in. to ⅜ in., and the headlosses are typically 12 to 18 in.

Both in-line grinders and comminutors tend to produce "ropes" and "balls" from cloth, which can jam downstream equipment. If the wastewater contains large amounts of rags and solids, in-line grinders may require protection by upstream coarse screens, which defeats their function. Comminutors and in-line grinders also chop up plastics and other nonbiodegradable material, which end up in wastewater sludges and which may prevent disposal of the sludges on land because of esthetics. Comminutors and in-line grinders are also subject to wear from grit and require relatively frequent replacement. Comminutors and, to a lesser extent, in-line grinders are nowadays not recommended [Joint Task Force, 1992].

Fine Screens

Fine screens are sometimes used in place of clarifiers in scum dewatering and centrate and sludge screening. In wastewater treatment, they are preceded by bar screens for protection from impact by large debris, but not by comminutors, because screen performance depends on the development of a "precoat" of solids. Fine screens generally remove fewer solids from raw sewage than do primary settling tanks, say 15 to 30% of suspended solids for openings of 6 to 1 mm [Joint Task Force, 1992]. There are three kinds [Hazen and Sawyer, Inc., 1975; Metcalf & Eddy, Inc., 1991; Pankratz, 1988].

Continuous-belt Fine Screens

Continuous-belt fine screens consist of stainless steel wedgewire elements mounted on horizontal supporting rods and forming a continuous belt loop. As the loop moves, the clogged region of the screen is lifted out of the water. Either the supporting rods themselves or the upper head sprocket mount blades that fit between the wires and dislodge accumulated debris as the wires are carried over the head sprocket. A supplementary brush or doctor blade may be used to removed sticky material.

The openings between the wires are generally between $\frac{3}{16}$ in. and $\frac{1}{2}$ in. The openings in continuous-belt fine screens are usually too coarse for use as primary sewage treatment devices, although they may be satisfactory for some industrial wastewaters containing fibrous or coarse solids.

Rotary Drum Fine Screens

In rotary drum fine screens, stainless steel wedgewire is wrapped around a horizontal cylindrical framework that is partially submerged. Generally, about 75% of the drum diameter and 66% of its mesh surface area are submerged. As the drum rotates, dirty wire is brought to the top, where it is cleaned by high-pressure water sprays and doctor blades.

The flow direction may be either in along the drum axis and outward through the inner surface of the wedgewire or in through the outer surface of the wedgewire and out along the axis.

Common openings are 0.01 to 0.06 in. (0.06 in. is preferred for raw wastewater), and the usual wire diameter is 0.06 in. Typical hydraulic loadings are 16 to 112 gpm/ft^2, and typical suspended solids removals for raw municipal sewage are 5 to 25%.

Inclined, Self-cleaning Static Screens

Inclined, self-cleaning static screens consist of inclined panes of stainless steel wedgewire. The wire runs horizontally. The flow is introduced at the top of the screen, and it travels downward along the screen surface. Solids are retained on the surface, and screened water passes through it and is collected underneath the screen. As solids accumulate on the screen surface, they impede the water flow, which causes the water to move the solids downward to the screen bottom.

Common openings are 0.01 to 0.06 in. (determined by in situ tests), and the usual wire diameter is 0.06 in. Typical hydraulic loadings are 4 to 16 gpm per inch of screen width, and typical suspended solids removals for raw municipal sewage are 5 to 25%.

Disk Fine Screens

Disk fine screens consist of flat disks of woven stainless steel wire supported on steel frameworks and partially submerged in the flow. As the disk rotates, the dirty area is lifted out of the flow and cleaned by high-pressure water sprays.

The mesh openings are generally about 1/32 in. Disk fine screens are limited to small flows, generally less than 20,000 gpm for reasons of economy [Pankratz, 1988].

Microscreens

Microscreens are used as tertiary suspended solids removal devices following biological wastewater treatment and secondary clarification [Hazen and Sawyer, Engineers, 1975]. Typical mesh openings are 20 to 25 μm and range from about 15 to 60 μm. The hydraulic loading is typically 5 to 10 gpm/ft^2 of submerged area. The suspended solids removal from secondary clarifier effluents is about 40 to 60%. Effluent-suspended solids concentrations are typically 5 to 10 mg/L.

Most microscreens are rotary drums, but there are some disk microscreens. These are similar to rotary drum and disk fine screens, except for the mesh size and material, which is usually a woven polyester fiber.

Microscreen fabrics gradually become clogged despite the high-pressure water sprays, and the fabric must be removed from the drum or disk for special cleaning every few weeks.

Orifice Walls

Orifice walls are sometimes installed in the inlet zones of sedimentation tanks to improve the lateral and vertical distribution of the flow. Orifice walls will not disperse longitudinal jets, and if jet formation cannot be prevented, it may be desirable to install adjustable vertical vanes to redirect the flow over the inlet cross section.

The relationship between head across an orifice and the flow through it is [King and Brater, 1963]:

$$Q = C_D A \sqrt{2gh_L} \tag{8.5}$$

Empirical discharge coefficients for sharp-edged orifices of any shape lie between about 0.59 and 0.66, as long as the orifice Reynolds number ($\text{Re} = d\sqrt{gh_L}/\nu$) is larger than about 10^5, which is usually the case [Lea, 1938; Smith and Walker, 1923]. Most of the results are close to 0.60.

In Hudson's [1981] design examples, the individual orifices are typically 15 to 30 cm in diameter, and they are spaced 0.5 to 1.0 m apart. The jets from these orifices will merge about 6 orifice diameters downstream from the orifice wall—which would be about 1 to 2 m in Hudson's examples—and that imaginary plane should be taken as the boundary between the inlet zone and settling zone.

8.2 Biological and Chemical Reactors

Hydraulic Retention Time

Regardless of tank configuration, mixing condition, or the number or volume of recycle flows, the average residence for water molecules in a tank is [Wen and Fan, 1975]

$$\theta_h = \frac{V}{Q} \tag{8.6}$$

where

θ_h = the hydraulic retention time (s)
V = the active liquid volume in the tank (m^3 or ft^3)
Q = the volumetric flow rate through the tank not counting any recycle flows (m^3/s or ft^3/s)

Note that Q does *not* include any recycle flows. The hydraulic *retention* time is also called the hydraulic *detention* time and, by chemical engineers, the *space time*. It is often abbreviated HRT.

Reaction Order

The reaction order is the apparent number of reactant molecules participating in the reaction. Mathematically, it is the exponent on the reactant concentration in the rate expression:

$$r_{1,2} = kC_1^p C_2^q \tag{8.7}$$

where

C_1 = the concentration of substance 1 (mol/L)
C_2 = the concentration of substance 2 (mol/L)
k = the reaction rate coefficient (units vary)
p = the reaction order of substance 1 (dimensionless)
q = the reaction order of substance 2 (dimensionless)
$r_{1,2}$ = the rate of reaction of components 1 and 2 (units vary)

In this case, the reaction rate is pth order with respect to substance 1 and qth order with respect to substance 2.

Many biological reactions are represented by the Monod equation [Monod, 1942]:

$$r = \frac{r_{\max} \cdot C}{K_C + C} \tag{8.8}$$

where

$r_{\max}$ = the maximum reaction rate (units vary, same as r)
K_C = the affinity constant (same units as C)

The Monod function is an example of a "mixed"-order rate expression, because the rate varies from first order to zero order as the concentration increases.

The units of the reaction rate vary depending on the mass balance involved. In general, one identifies a control volume (usually one compartment or differential volume element in a tank) and constructs a mass balance on some substance of interest. The units of the reaction rate will then be the mass of the substance per unit volume per time. The units of the rate constant will be determined by the need for dimensional consistency.

It is important to note that masses in chemical reaction rates have identities and they do not cancel. Consequently, the ratio kg COD/kg VSS·s does *not* reduce to 1/s. This becomes obvious if it is remembered that the mass of a particular organic substance can be reported in a variety of ways: kg, mol, BOD, COD, TOC, and so on. The ratios of these various units are not unity, and the actual numerical value of a rate will depend on the method of expression of the mass.

Many reactions in environmental engineering are represented satisfactorily as first order. This is a consequence of the fact that the substances are contaminants, and the goal of the treatment process is to reduce their concentrations to very low levels. In this case, the Maclaurin series representation of the rate expression may be truncated to the first-order terms:

$$r(C_1) = \underbrace{r(0)}_{=0} + \underbrace{(C_1 - 0)\frac{\partial r}{\partial C_1}\bigg|_{C_1=0}}_{\text{First order in } C_1} + \underbrace{\text{Higher-order terms}}_{\text{Truncated}} \tag{8.9}$$

Many precipitation and oxidation reactions are first order in the reactants. Disinfection is frequently first order in the microbial concentration, but the order of the disinfectant may vary. Flocculation is second order. Substrate removal reactions in biological processes generally are first order at low concentration and zero order at high concentration.

Effect of Tank Configuration on Removal Efficiency

The general steady state removal efficiency is

$$E = \frac{C_o - C}{C_o} \tag{8.10}$$

This is affected by both the reaction kinetics and the hydraulic regime in the reactor.

Completely Mixed Reactors

The completely mixed reactor (CMR) is also known as the continuous-flow, stirred-tank reactor (CFSTR or, more commonly, CSTR). Because of mixing, the contents of the tank are homogeneous, and a mass balance yields

$$\frac{dVC}{dt} = QC_o - QC - kC^nV \tag{8.11}$$

where

C = the concentration in the homogeneous tank and its effluent flow (kg/m^3 or slug/ft^3)
C_o = the concentration in the influent liquid (kg/m^3 or slug/ft^3)
k = the reaction rate constant (here, m^{3n}/kg$^n \cdot$ s or ft^{3n}/slug$^n \cdot$ s)
n = the reaction order (dimensionless)
Q = the volumetric flow rate (m^3/s or ft^3/s)
V = the tank volume (m^3 or ft^3)
t = elapsed time (s)

The steady state solution is

$$\frac{C}{C_o} = \frac{1}{1 + kC^{n-1}\theta_h} \tag{8.12}$$

For first-order reactions, this becomes

$$\frac{C}{C_o} = \frac{1}{1 + k\theta_h} \tag{8.13}$$

In the case of zero-order reactions, the steady state solution is

$$C_o - C = k\theta_h \tag{8.14}$$

If the mixing intensity is low, CSTRs may develop "dead zones" that do not exchange water with the inflow. If the inlets and outlets are poorly arranged, some of the inflow may pass directly to the outlet without mixing with the tank contents. This latter phenomenon is called *short-circuiting*. In the older literature, short-circuiting and complete mixing were often confused. They are opposites. Short-circuiting cannot occur in a tank that is truly completely mixed. (Short-circuiting in clarifiers is discussed later in this chapter.)

The analysis of short-circuiting and dead zones in mixed tanks is due to Cholette and Cloutier [1959] and Cholette *et al.* [1960]. For a "completely mixed" reactor with both short-circuiting and dead volume, the mass balance of an inert tracer *on the mixed volume* is

$$\underbrace{f_m V \frac{dC_m}{dt}}_{\substack{\text{Accumulation in} \\ \text{mixed volume}}} = \underbrace{(1 - f_s)QC_o}_{\substack{\text{Fraction of influent} \\ \text{entering mixed volume}}} - \underbrace{(1 - f_s)QC_m}_{\substack{\text{Flow leaving} \\ \text{mixed volume}}} \tag{8.15}$$

where

C_m = the concentration of tracer in the mixed zone (kg/m^3)

f_m = the fraction of the reactor volume that is mixed (dimensionless)

f_s = the fraction of the influent that is short-circuited directly to the outlet (dimensionless)

The observed effluent is a mixture of the short-circuited flow and the flow leaving the mixed zone:

$$QC = f_s QC_o + (1 - f_s)QC_m \tag{8.16}$$

Consequently, for a slug application of tracer (in which the influent momentarily contains some tracer and is thereafter free of it), the observed washout curve is

$$\frac{C}{C_i} = (1 - f_s)\exp\left\{-\frac{(1 - f_s)Qt}{f_m V}\right\} \tag{8.17}$$

where C_i is the apparent initial concentration (kg/m^3).

Equation (8.17) provides a convenient way to determine the mixing and flow conditions in completely mixed tanks. All that is required is a slug tracer study. The natural logarithms of the measured effluent concentrations are then plotted against time, and the slope and intercept yield the values of the fraction short-circuited and the fraction mixed.

Mixed-cells-in-series

Consider the rapid mixing tank shown in Fig. 8.1. This particular configuration is called mixed-cells-in-series or tanks-in-series, because the liquid flows sequentially from one cell to the next.

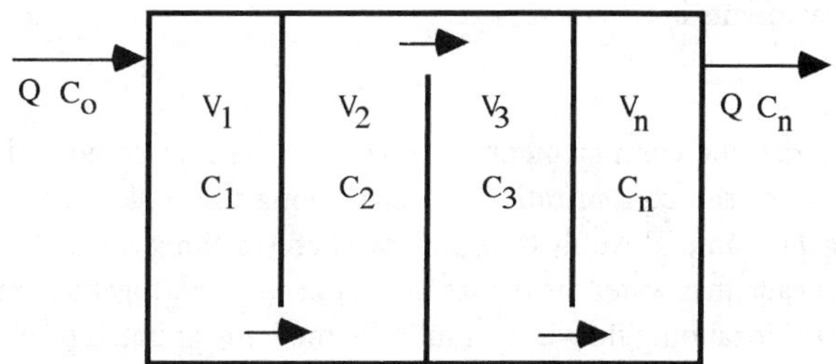

FIGURE 8.1 Mixed-cells-in-series configuration.

Each cell has a mixer, and each cell is completely mixed. Mixed-cells-in-series is the usual configuration for flocculation tanks and activated sludge aeration tanks.

The flow is assumed to be continuous and steady, and each compartment has the same volume, that is, $V_1 = V_2 = V_3 = \ldots = V_n$. The substance concentrations in the first through last compartments are $C_1, C_2, C_3, \cdots, C_n$, respectively. The substance mass balance for each compartment has the same mathematical form as Eq. (8.11) and the steady state concentration in each compartment is given by Eq. (8.12) [Hazen, 1904; MacMullin and Weber, 1935; Kehr, 1936; Ham and Coe, 1918; Langelier, 1921].

Because of the mixing, the concentration in the last compartment is also the effluent concentration. Consequently, the ratio of effluent to influent concentrations is

$$\frac{C_n}{C_o} = \frac{C_1}{C_o} \cdot \frac{C_2}{C_1} \cdot \frac{C_3}{C_2} \cdot \quad \cdots \quad \cdot \frac{C_n}{C_{n-1}} \tag{8.18}$$

Because all compartments have the same volume and process the same flow, all their HRTs are equal. Therefore, for a first-order reaction, Eq. (8.18) becomes

$$\frac{C_n}{C_o} = \left(\frac{1}{1 + k\theta_{h1}}\right)^n \tag{8.19}$$

where

θ_{h1} = the HRT of a single compartment (s)
$= V/nQ$
n = the number of mixed-cells-in-series

Equation (8.19) has significant implications for the design of all processing tanks used in natural and used water treatment. Suppose that all the internal partitions in Fig. 8.1 are removed, so that the whole tank is one completely mixed, homogeneous compartment. Because there is only one compartment, its HRT is n times the HRT of a single compartment in the partitioned tank. Now, divide Eq. (8.12) by Eq. (8.19) and expand the bracketed term in Eq. (8.19) by the binomial theorem:

$$\frac{C(1\text{ cell})}{C_n(n\text{ cells})} = \left(1 + nk\theta_{h1} + \frac{n(n-1)}{2!}(k\theta_{h1})^2 + \cdots + (k\theta_{h1})^n\right) \Big/ (1+nk\theta_{h1}) > 1 \tag{8.20}$$

Therefore, the effect of partitioning the tank is to reduce the concentration of reactants in its effluent, that is, to increase the removal efficiency.

The effect of partitioning on higher-order reactions is even more pronounced. Partitioning has no effect in the case of zero-order reactions [Levenspiel, 1972].

The efficiency increases with the number of cells. It also increases with the hydraulic retention time. This means that total tank volume can be traded against the number of cells. A tank can be made smaller—more economical—and still achieve the same degree of particle destabilization, if the number of compartments in it is increased.

Ideal Plug Flow

As n becomes very large, the compartments approach differential volume elements, and, if the partitions are eliminated, the concentration gradient along the tank becomes continuous. The result is an *ideal plug flow* tank. The only transport mechanism along the tank is advection: there is no dispersion. This means that water molecules that enter the tank together stay together and exit together. Consequently, ideal plug flow is hydraulically the same as batch processing. The distance traveled along the plug flow tank is simply proportional to the processing time in a batch reactor, and the coefficient of proportionality is the average longitudinal velocity.

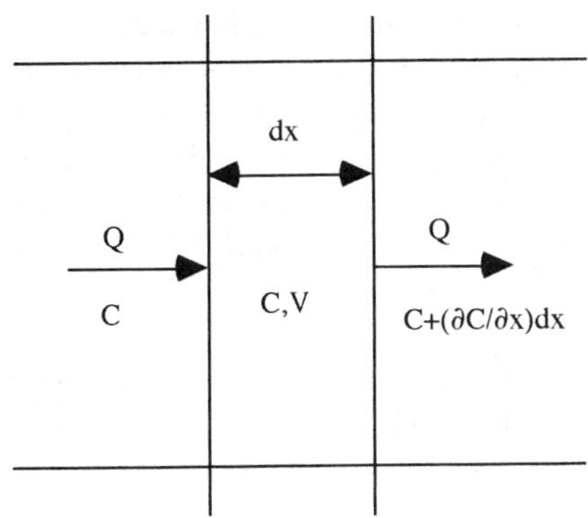

FIGURE 8.2 Differential control volume for an ideal plug flow reactor.

Referring to Fig. 8.2, the mass balance on a differential volume element is

$$\frac{\partial C}{\partial t} = -U\frac{\partial C}{\partial x} - kC^p \tag{8.21}$$

For steady state conditions, this becomes, for reaction orders greater than 1,

$$C^{1-p} = C_o^{1-p} - (1-p)k\theta_h \tag{8.22}$$

For first-order reactions, one gets

$$C = C_o \exp(-k\theta_h) \tag{8.23}$$

The relative efficiencies of ideal plug flow tanks and tanks that are mixed-cells-in-series are easily established. In the case of first-order reactions, the power series expansion of the exponential is

$$\exp(-k\theta_h) = 1 - k\theta_h + \frac{(k\theta_h)^2}{2!} - \frac{(k\theta_h)^3}{3!} + \cdots < \lim_{n\to\infty}\left(\frac{1}{1+k\theta_{h1}}\right)^n \tag{8.24}$$

Consequently, ideal plug flow tanks are the most efficient; completely mixed tanks are the least efficient; and tanks consisting of mixed-cells-in-series are intermediate.

Plug Flow with Axial Dispersion

Of course, no real plug flow tank is ideal. If it contains mixers, they will break down the concentration gradients and tend to produce a completely mixed tank. Furthermore, the cross-sectional velocity variations induced by the wall boundary layer will produce a longitudinal mixing called *shear-flow dispersion*. Shear-flow dispersion shows up in the mass balance equation as a "diffusion" term, so Eq. (8.21) should be revised as follows:

$$\frac{\partial C}{\partial t} = K\frac{\partial^2 C}{\partial x^2} - U\frac{\partial C}{\partial x} - kC^p \tag{8.25}$$

where K is the (axial) shear flow dispersion coefficient (m^2/s or ft^2/s). Langmuir's [1908] boundary conditions for a tank are as follows. Inlet:

$$x = 0 \qquad QC_o = QC - KA\frac{dC}{dx} \tag{8.26}$$

Outlet:

$$x = L \qquad \frac{dC}{dx} = 0 \tag{8.27}$$

where

A = the cross-sectional area of the reactor (m^2 or ft^2)

L = the length of the reactor (m or ft)

The first condition is a mass balance around the tank inlet. The rate of mass flow (kg/s) approaching the tank inlet in the influent pipe must equal the mass flow leaving the inlet in the tank itself. Transport within the tank is due to both advection and dispersion, but dispersion in the pipe is assumed to be negligible, because the pipe velocity is high. The outlet condition assumes that the reaction is nearly complete—which is, of course, the goal of tank design—so the concentration gradient is nearly zero and there is no dispersive flux. The Langmuir boundary conditions produce a formula that reduces in the limit to ideal plug flow as K approaches zero and to ideal complete mixing as K approaches infinity. A correct formula must do this. No other set of boundary conditions produces this result [Wehner and Wilhem, 1956; Pearson, 1959; Bishoff, 1961; Fan and Ahn, 1963]. The general solution is [Danckwerts, 1953]:

$$\frac{C}{C_o} = \frac{4a \cdot \exp\left\{\frac{1}{2}(1+a)\mathrm{Pe}\right\}}{(1+a)^2 \cdot \exp\{a\mathrm{Pe}\} - (1-a)^2} \tag{8.28}$$

where

a = part of the solution to the characteristic equation of the differential equation (1/m or 1/ft)
$= \sqrt{1 + (4kK/U^2)}$
Pe = the turbulent Peclet number (dimensionless)
$= UL/K$

A tank with axial dispersion has an efficiency somewhere between ideal plug flow and ideal complete mixing. Consequently, a real plug flow tank with axial dispersion behaves as if it were compartmentalized. The equivalence of the number of compartments and the shear flow diffusivity can be represented by [Levenspiel and Bischoff, 1963]

$$\frac{1}{n} = 2 \cdot \mathrm{Pe}^{-1} - 2 \cdot \mathrm{Pe}^{-2} \cdot \left[1 - \exp\left\{-\mathrm{Pe}^{-1}\right\}\right] \tag{8.29}$$

The axial dispersion—and, consequently, the Peclet number—in pipes and ducts has been extensively studied, and the experimental results can be summarized as follows [Wen and Fan, 1975]. For Reynolds numbers less than 2000:

$$\frac{1}{\mathrm{Pe}} = \frac{1}{\mathrm{Pe} \cdot \mathrm{Sc}} + \frac{\mathrm{Re} \cdot \mathrm{Sc}}{192} \tag{8.30}$$

For Reynolds numbers greater than 2000:

$$\frac{1}{\mathrm{Pe}} = \frac{30 \times 10^6}{\mathrm{Re}^{2.1}} + \frac{1.35}{\mathrm{Re}^{1/8}} \tag{8.31}$$

where

D = the molecular diffusivity of the substance (m^2/s or ft^2/s)
d = the pipe or duct diameter (m or ft)
Pe = the duct Peclet number (dimensionless)
$= Ud/K$
Re = the duct Reynolds number (dimensionless)
$= Ud/\nu$
ν = the kinematic viscosity (m^2/s or ft^2/s)
Sc = the Schmidt number (dimensionless)
$= \nu/D$

These formulas assume that the dispersion is generated entirely by the shear flow of the fluid in the pipe or duct. When mixers are installed in tanks, Eqs. (8.30) and (8.31) no longer apply, and the axial dispersion coefficient must be determined experimentally.

More important, the use of mixers generally results in very small Peclet numbers, and the reactors tend to approach completely mixed behavior, which is undesirable because the efficiency is reduced. Thus, there is an inherent contradiction between high turbulence and ideal plug flow, both of which are wanted in order to maximize tank efficiency. The usual solution to this problem is to construct the reactor as a series of completely mixed cells. This allows the use of any desired mixing power and preserves the reactor efficiency.

Sequencing Batch Reactors

Another way to combine high turbulence with ideal plug flow is to construct a *sequencing batch reactor* (SBR), which operates as follows:

- Starting out empty, the tank is first filled; any needed chemicals are added during the filling.
- The full tank is then stirred and aerated, if necessary, and the reactions proceed.
- After mixing and reacting, the tank is allowed to stand quiescently, if necessary, to settle out any precipitate that has formed.
- The tank supernatant is drained off.
- The tank may sit idle between the draining and filling operation while valves and pumps are switched.

This is essentially the mode of operation of what used to be called "contact beds," a form of sewage treatment employed around the turn of the century and a predecessor of the trickling filter [Dunbar, 1908; Metcalf and Eddy, 1916]. It is also the fill-and-draw mode used by Ardern and Lockett and many others studying the activated sludge process. The design problem here is scheduling the filling, stirring, draining, and idle phases of the cycle so as to meet the plant design flow rate.

If water production is to be continuous, at least one tank must be filling and one draining at each moment. Consider the schedule shown in Fig. 8.3. The total cycle time for a single tank is the sum of the times for filling, processing, draining, and idling. The plant flow first fills tank no. 1. Then the flow is diverted to tank no. 2, and so on. Tank no. 1 is again available after its cycle is completed. Raw water will be available at that moment, if another tank has just completed filling. This means that the cycle time for a single tank must be equal to or less than the product of the filling time and the number of available tanks:

$$m\,\theta_f \geq \theta_f + \theta_r + \theta_s + \theta_d + \theta_i \tag{8.32}$$

where

m = the number of tanks (dimensionless)
θ_d = the draining time (s)

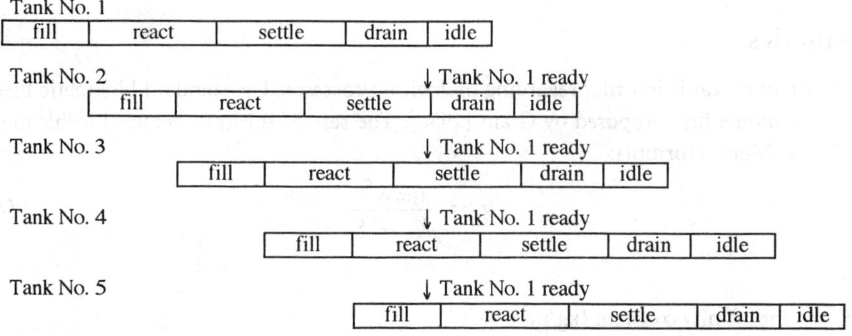

FIGURE 8.3 Operating schedule for sequencing batch reactors.

θ_f = the filling time (s)

θ_i = the idle time (s)

θ_r = the reaction time (s)

θ_s = the settling time (s)

The idle time is freely adjustable, but the other times are set by the hydraulic capacity of the inlet and outlet devices and the time required for reaction and settling. A similar analysis of the draining operation leads to the conclusion that the product of the number of tanks and the draining time must be equal to the cycle time:

$$m\theta_d \geq \theta_f + \theta_r + \theta_s + \theta_d + \theta_i \tag{8.33}$$

The number of tanks in both equations must be the same, so the filling time must equal the draining time. They are otherwise freely adjustable within the limits of the hydraulic system.

An SBR is not a viable option for rapid mixing. The principal problem is the short processing time employed in rapid mixing, generally 30 seconds or less. With such short running times, electric motors do not achieve steady state operation. Instead, they are always in the start-up mode, during which they draw electric currents much higher than their steady state demand. Under these conditions, the motors are prone to overheating and burnout. Also, the fill-process-draw-idle sequence of an SBR rapid mixing tank would require valve and pump switching to occur within a few seconds. This is intrinsically difficult because of the weights of the equipment parts, and it imposes high mechanical stresses on them. Finally, even if quick switching could be achieved, it would not be desirable because of water hammer in the conduits.

In any application of SBRs, the headloss is a problem. In continuous-flow tanks, the headloss amounts to several centimeters at most. However, in SBRs the headloss is the difference between the high water level and low water level, and this may amount to a few meters, even in small tanks. In flat country, these headlosses become a significant problem, because they must be met by pumping.

The SBR is suitable where the processing time is long, say, several hours, so that only a few cycles are needed each day. This is the case in many activated sludge plants.

Ideal plug flow tanks are more efficient than SBR systems. The total tankage in an SBR system is the product of the number of tanks, the filling time, and the flow rate. The tankage required for an ideal plug flow reactor is the product of the flow and processing time. If the same conversion efficiency is required of both systems, the processing times will be equal, and the ratio of the system volumes is

$$\frac{V_{\text{SBR}}}{V_{\text{PF}}} = \frac{\theta_f + \theta_r + \theta_s + \theta_d + \theta_i}{\theta_r + \theta_s} > 1 \tag{8.34}$$

Plug flow reactors do not incorporate settling—a separate clarifier is required for that—but SBRs do. Consequently, the settling time of the clarifier associated with the plug flow reactor must be included in the denominator.

Biokinetics

An important caveat is in order regarding biological processes. The standard biokinetic theory is based on a model first proposed by Gram [1956]. The rate of substrate removal in his model is given by the Monod formula:

$$q = \frac{q_{\max} S}{K_s + S} \tag{8.35}$$

where

K_s = the affinity constant (kg/m^3)

q = the specific uptake rate (kg substrate/kg biomass·s)

q_{max} = the maximum specific uptake rate (kg substrate/kg biomass·s)

S = the substrate concentration (kg/m^3)

Equation (8.35) is well verified for pure cultures of microbes growing on pure substances. From the preceding discussion, it might be assumed that organic matter removal (i.e., total influent substrate minus soluble effluent substrate and product) in plug flow reactors would be greater than in mixed-cells-in-series reactors, which, in turn, would be more efficient than completely mixed reactors. Unfortunately, this is not true. In the case of mixed microbial populations consuming synthetic or natural wastewater, the effluent soluble organic matter concentration is not affected by reactor configuration [Badger *et al.*, 1975; Chudoba *et al.*, 1991; Haseltine, 1961; Kroiss and Ruider, 1977; Toerber *et al.*, 1974]. This is true regardless of how the organic matter is measured: biochemical oxygen demand, chemical oxygen demand, or total organic carbon.

Biological processes do not violate the laws of reaction kinetics. Instead, it is the simplistic application of these laws that is at fault. The soluble organic matter concentration in the effluents of biological reactors is not residual substrate (i.e., a reactant); rather, it is a microbial product [Baskir and Spearing, 1980; Erickson, 1980; Grady *et al.*, 1972; Hao and Lau, 1988; Rickert and Hunter, 1971].

In the case of suspended growth processes, the appropriate design procedure is to assume that all reactors are completely mixed, regardless of any internal baffling. Equation (8.35) can then be used to estimate the effluent soluble BOD, COD, or TOC, but it must be regarded as a purely statistical correlation and not as a mechanistic hypothesis.

Fixed film processes are designed according to other empirical procedures.

It should not be assumed that reactor configuration has no effect. Sequencing batch reactors and mixed-cells-in-series reactors with very short compartmental detention times (less than 10 minutes) suppress activated sludge filamentous bulking and are the preferred configuration for that reason.

8.3 Mixers and Mixing

The principal objects of mixing are (1) blending different liquid streams, (2) suspension of particles, and (3) mass transfer. The main mass transfer operations are treated in separate sections below. This section focuses on blending and particle suspension.

Mixing Devices

Mixing devices are specialized to either laminar or turbulent flow conditions.

Laminar/High Viscosity

Low-speed mixing in high-viscosity liquids is done in the laminar region with impeller Reynolds numbers below about 10. The mechanical mixers most commonly used are:

- Gated anchors and horseshoes—large U-shaped mixers that fit up against or near the tank wall and bottom, usually with cross members, called *gates*, running between the upright limbs of the U
- Helical ribbons
- Helical screws
- Paddles
- Perforated plates—usually in stacks of several separated plates having an oscillatory motion with the flat portion of the plate normal to the direction of movement

Turbulent/Low to Medium Viscosity

High-speed mixing in low- to moderate-viscosity fluids is done in the turbulent region with impeller Reynolds numbers above 10,000. The commonly used agitators are:

- Anchors and horseshoes
- Disk—either with or without serrated or sawtooth edge for high shear
- Jets
- Propellers
- Static in-line mixers—tubing with internal vanes fixed to the inner tubing wall, set at an angle to the flow to induce crosscurrents
- Radial flow turbines—blades are mounted either to a hub on the drive shaft or to a flat disk (Rushton turbines) attached to the drive shaft; blades are oriented radially and may be flat, with the flat side oriented perpendicular to the direction of rotation, or curved, with the convex side oriented perpendicular to the direction of rotation
- Axial flow turbines—blades are mounted either to a hub on the drive shaft or to a flat disk (Rushton turbines) attached to the drive shaft; blades flat and pitched with the flat side oriented at an angle (usually 45°) to the direction of rotation
- Smith turbines—turbines specialized for gas transfer with straight blades having a C-shaped cross section and oriented with the open part of the C facing the direction of rotation
- fluidfoil—having hub-mounted blades with winglike cross sections to induce axial flow

Power Dissipation

Fluid Deformation Power

Energy is dissipated in turbulent flow by the internal work due to volume element compression, stretching, and twisting [Lamb, 1932]:

$$\frac{\Phi}{\mu} = 2\left(\frac{\partial u}{\partial x}\right)^2 + 2\left(\frac{\partial v}{\partial y}\right)^2 + \left(\frac{\partial w}{\partial z}\right)^2 + \left(\frac{\partial u}{\partial y} + \frac{\partial v}{\partial x}\right)^2 + \left(\frac{\partial u}{\partial z} + \frac{\partial w}{\partial x}\right)^2 + \left(\frac{\partial v}{\partial z} + \frac{\partial w}{\partial y}\right)^2 \tag{8.36}$$

where

u, v, w = the local velocities in the x, y, z directions, respectively (m/s or ft/s)
Φ = Stokes' [1845] energy dissipation function (W/m^3 or ft lbf/ft^3·s)
μ = The absolute or dynamic viscosity (N s/m^2 or lbf s/ft^2)

The first three terms on the right-hand side of Eq. (8.36) are compression and stretching terms, and the last three are twisting terms. The left-hand side of Eq. (8.36) can also be written as

$$\frac{\Phi}{\mu} = \frac{\varepsilon}{\nu} = \Gamma^2 \tag{8.37}$$

where

ε = the local power dissipation per unit mass (watts/kg or ft·lbf/s·slug)
ν = the kinematic viscosity (m^2/s or ft^2/s)
Γ = the characteristic strain rate (per second)

Equation (8.37) applies only at a point. If the total energy dissipated in a mixed tank is needed, it must be averaged over the tank volume. If the temperature and composition are uniform everywhere in the tank, the kinematic viscosity is a constant, and one gets

$$\overline{\Gamma}^2 = \frac{1}{\nu V}\int\int\int \varepsilon \cdot dx\,dy\,dz \tag{8.38}$$

where

V = the tank volume (m^3 or ft^3)
$\overline{\Gamma}$ = the spatially averaged (root-mean-square) characteristic strain rate (per second)

With this definition, the total power dissipated by mixing is

$$P = \Phi V = \mu\overline{\Gamma}^2 V \tag{8.39}$$

where P is the mixing power (W or ft·lbf/s).

Camp–Stein Theory

Camp and Stein [1943] assumed that the axial compression/stretching terms can always be eliminated by a suitable rotation of axes so that a differential volume element is in pure shear. This may not be true for all three-dimensional flow fields, but there is numerical evidence that it is true for some [Clark, 1985]. Consequently, the power expenditure per unit volume is

$$\frac{dP}{dV} = \Phi = \mu\left[\left(\frac{\partial u}{\partial y} + \frac{\partial v}{\partial x}\right)^2 + \left(\frac{\partial u}{\partial z} + \frac{\partial w}{\partial x}\right)^2 + \left(\frac{\partial v}{\partial z} + \frac{\partial w}{\partial y}\right)^2\right] = \mu G^2 \tag{8.40}$$

where

dP = the total power dissipated deforming the differential volume element (N·m/s, ft·lbf/s)
dV = the volume of the element (m^3 or ft^3)
G = the absolute velocity gradient (per second)
Φ = Stokes' [1845] dissipation function (watts/m^3 or lbf/ft^2·s)

If the velocity gradient is volume averaged over the whole tank, one gets

$$P = \mu\overline{G}^2 V \tag{8.41}$$

where $\overline{G}$ is the root-mean-square velocity gradient (per second).

The Camp–Stein rms velocity gradient is numerically identical to the rms characteristic strain rate, but (because of the unproved assumption of pure shear embedded in $\overline{G}$) $\overline{\Gamma}$ is the preferred concept. Camp and Stein used the assumption of pure shear to derive a formula for the flow around a particle and the resulting particle collision rate, thereby connecting the flocculation rate to $\overline{G}$ and P. However, it is also possible to derive the collision rate based on $\overline{\Gamma}$ [Saffman and Turner, 1956]; this yields a better physical representation of the flocculation process and is the preferred method.

Energy Spectrum and Eddy Size

Mixing devices are pumps, and they create macroscopic, directed currents. As the currents flow away from the mixer, they rub against and collide with the rest of the water in the tank. These collisions and rubbings break off large eddies from the current. The large eddies repeat the process of collision/shear, producing smaller eddies, and these do the same until there is a spectrum of eddy sizes. The largest eddies in the spectrum are of the same order of size as the mixer. The large eddies move quickly and contain most of the kinetic energy of the turbulence, and their motion is controlled by inertia rather than viscosity. The small eddies move slowly and are affected by viscosity.

The size of the smallest eddies is called the *Kolmogorov length scale* [Landahl and Mollo-Christensen, 1986]:

$$\eta = \left(\frac{\nu^3}{\varepsilon}\right)^{1/4} \tag{8.42}$$

where

ε = the power input per unit mass (watts/kg or ft·lbf/s·slug)
η = the Kolmogorov length (m or ft)

The Kolmogorov wave number is defined by custom as [Landahl and Mollo-Christensen, 1986]

$$k_K = \frac{1}{\eta} \tag{8.43}$$

where k_K is the Kolmogorov wave number (per m or per ft). The general formula for the wave number is

$$k = \frac{2\pi}{L} \tag{8.44}$$

where L is the wavelength (m or ft).

The kinetic energy contained by an eddy is one-half the square of its velocity times its mass. It is easier to calculate the energy per unit mass, because this is merely one-half the square of the velocity. The mean water velocity in mixed tanks is small, so nearly all the kinetic energy of the turbulence is in the velocity fluctuations, and the kinetic energy density function can be defined in terms of these fluctuations:

$$E(k)\,dk = \tfrac{1}{2}(u_k^2 + v_k^2 + w_k^2)n(k)\,dk \tag{8.45}$$

where

$E(k)\,dk$ = the total kinetic energy contained in the eddies between wave numbers k and $k + dk$ (m^2/s^2 or ft^2/s^2)
$n(k)\,dk$ = the number of eddies between the wave numbers k and $k + dk$ (dimensionless)
u_k, v_k, w_k = the components of the velocity fluctuation in the x, y, and z directions for eddies in the wavelength interval k to $k + dk$ (m/s or ft/s).

A plot of $E(k)\,dk$ versus k is called the *energy spectrum*. An energy spectrum plot can have a wide variety of shapes, depending on the power input and the system geometry [Brodkey, 1967]. However, if the power input is large enough, all spectra contain a range of small-sized eddies that are a few orders of magnitude larger than the Kolmogorov length scale η. The turbulence in this range of eddy sizes is isotropic and independent of the geometry of the mixing device, although it does depend on the power input. Consequently, it is called the *universal equilibrium range*. At very-high-power inputs, the universal equilibrium range subdivides into a class of larger eddies that are influenced only by inertial forces and a class of smaller eddies that are influenced by molecular viscous forces. These subranges are called the *inertial convective subrange* and the *viscous dissipation subrange*, respectively.

When the energy density, $E(k)$, is measured, the inertial convection subrange is found to occur at wave numbers less than about one-tenth the Kolmogorov wavelength, and the viscous dissipation subrange lies entirely between about $0.1k_K$ and k_K [Grant *et al.*, 1962; Stewart and Grant, 1962]. Similar results have been obtained theoretically [Matsuo and Unno, 1981]. Therefore, η is the diameter of the smallest eddy in the viscous dissipative subrange, and the largest eddy in the viscous dissipation subrange has a diameter of about $20\pi\eta$.

The relative sizes of floc particles and eddies is important in understanding how they interact. If the eddies are larger than the floc particles, they entrain the flocs and transport them. If the eddies are smaller than the flocs, the only interaction is shearing of the floc by the eddies. It is also important whether the flocs interact with the inertial convective subrange eddies or the viscous dissipative subrange eddies, because the formulas connecting eddy diameter and velocity with mixing power are different for the two subranges. In particular, a collision rate formula based on $\overline{\Gamma}$ would be correct only in the viscous dissipative subrange [Cleasby, 1984].

Typical recommended $\overline{\Gamma}$ values are on the order of 900 per second for rapid mixing tanks and 75 per second for flocculation tanks [Joint Committee, 1969]. At 20°C, the implied power inputs per unit volume are about 0.81 m^2/s^3 for rapid mixing and 0.0056 m^2/s^3 for flocculation. The diameter of the smallest eddy in the viscous dissipative subrange in rapid mixing tanks is 0.030 mm, and the diameter of the largest eddy is 1.9 mm. The sizes of flocculated particles generally range from a few hundredths of a millimeter to a few millimeters, and the sizes tend to decline as the mixing power input rises [Boadway, 1978; Lagvankar and Gemmel, 1968; Parker *et al.*, 1972; Tambo and Watanabe, 1979; Tambo and Hozumi, 1979]. Therefore, they are usually contained either within the viscous dissipative subrange or they are smaller than any possible eddy and lie outside the universal equilibrium range. In water treatment, only the viscous dissipative subrange processes need to be considered.

Turbines

An example of a typical rapid mixing tank is shown in Fig. 8.4. Such tanks approximate cubes or right cylinders; the liquid depth approximates the tank diameter. The impeller is usually a flat disk with several short blades mounted near the disk's circumference. The blades may be flat and perpendicular to the disk, as shown, or they may be curved or pitched at an angle to the disk. The number of blades varies, but a common choice is six. Almost always, there are several baffles mounted along the tank wall to prevent vortexing of the liquid. The number of baffles and their width are design choices, but most commonly there are four baffles.

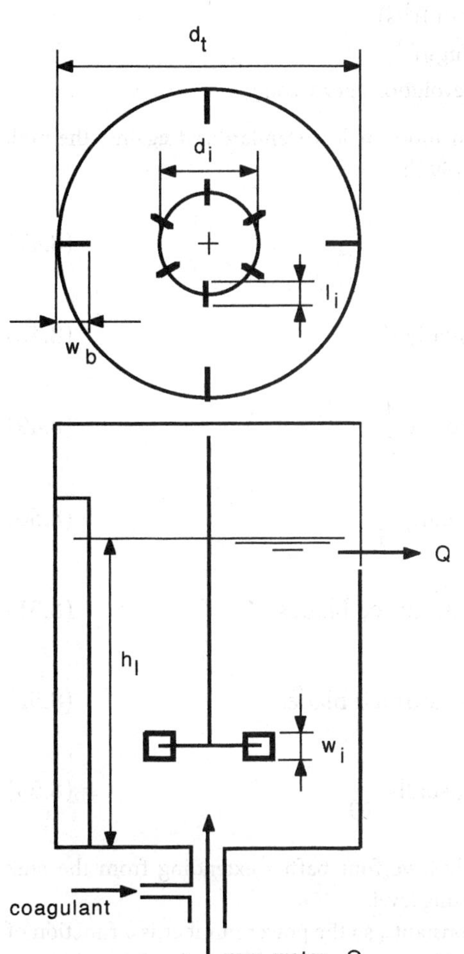

FIGURE 8.4 Turbine definition sketch.

Turbine Power. The power dissipated by the turbulence in a tank is related to the geometry of the tank and mixer and the rotational speed of the mixer. Dimensional analysis suggests an equation of the following form [Rushton *et al.*, 1950]:

$$\mathrm{Ru} = f(\mathrm{Re}, \mathrm{Fr}, d_i, d_t, h_i, h_l, l_i, N_b, N_i, n_b, n_i, p_i, w_b, w_i) \quad (8.46)$$

where

d_i = the impeller diameter (m or ft)

d_t = the tank diameter (m or ft)

Fr = the Froude number of the impeller (dimensionless) $= \omega^2 d_i / g$

g = the acceleration due to gravity (m/s^2 or ft/s^2)

h_l = the depth of liquid in the tank (m or ft)
h_i = the height of the impeller above the tank bottom (m or ft)
l_i = the impeller blade length (m or ft)
N_b = the baffle reference number, that is, the number of baffles in some arbitrarily chosen "standard" tank (dimensionless)
n_b = the number of baffles in the tank (dimensionless)
N_i = the impeller reference number, that is, the number of impeller blades on some arbitrarily chosen "standard" impeller (dimensionless)
n_i = the number of impeller blades (dimensionless)
P = the power dissipated by the turbulence (watts or ft·lbf/s)
p_i = the impeller blade pitch (m or ft)
Re = the impeller Reynolds number (dimensionless)
= $\omega d_i^2/\nu$
Ru = the Rushton power number (dimensionless)
= $P/\rho\omega^3 d_i^5$
w_b = the width of the baffles (m or ft)
w_i = the impeller blade width (m or ft)
ν = the kinematic viscosity of the liquid (m^2/s or ft^2/s)
ρ = the mass density of the liquid (kg/m^3 or $slugs/ft^3$)
ω = the rotational speed of the impeller (Hz, revolutions per second)

The geometry of turbine/tank systems has been more or less standardized against the tank diameter [Holland and Chapman, 1966; Tatterson, 1994]:

$$\frac{h_l}{d_t} = 1 \tag{8.47}$$

$$\frac{1}{6} \leq \frac{h_i}{d_t} \leq \frac{1}{2}; \quad \text{usually } d_i \tag{8.48}$$

$$\frac{1}{4} \leq \frac{d_i}{d_t} \leq \frac{1}{2}; \quad \text{usually } \frac{1}{3} \tag{8.49}$$

$$\frac{1}{6} \leq \frac{w_i}{d_i} \leq \frac{1}{4}; \quad \text{usually } \frac{1}{5} \tag{8.50}$$

$$\frac{l_i}{d_i} = \frac{1}{4}; \quad \text{for hub-mounted blades} \tag{8.51}$$

$$\frac{l_i}{d_i} = \frac{1}{8}; \quad \text{for disk-mounted blades} \tag{8.52}$$

$$\frac{1}{12} \leq \frac{w_b}{d_t} \leq \frac{1}{10}; \quad \text{usually } \frac{1}{10} \tag{8.53}$$

Turbines usually have six blades, and tanks usually have four baffles extending from the tank bottom to somewhat above the highest liquid operating level.

For any given tank, all the geometric ratios are constants, so the power number is a function of only the Reynolds number and the Froude number. Numerous examples of such relationships are given by Holland and Chapman [1966].

For impeller Reynolds numbers below 10, the hydraulic regime is laminar, and Eq. (8.46) is found experimentally to be

$$\text{Re} \cdot \text{Ru} = \text{Constant} \tag{8.54}$$

The value of the constant is typically about 300, but it varies between 20 and 4000 [Tatterson, 1994].

Equation (8.54) indicates that the power dissipation is proportional to the viscosity, the square of the impeller rotational speed, and the cube of the impeller diameter:

$$P \propto \mu \omega^2 d_i^3 \tag{8.55}$$

For impeller Reynolds numbers above 10,000, the hydraulic regime is turbulent, and the experimental relationship for baffled tanks is

$$\text{Ru} = \text{Constant} \tag{8.56}$$

Typical values of the constant are [Tatterson, 1994]:

- Hub-mounted flat blades, 4.0
- Disk-mounted flat blades, 5.0
- Pitched blades, 1.27
- Propellers, 0.6

The power number for any class of impeller varies significantly with the details of the design. Impeller design and performance are discussed by Oldshue and Trussell [1991].

Equation (8.56) indicates that the power dissipation is proportional to the liquid density, the cube of the impeller rotational speed, and the fifth power of the impeller diameter:

$$P \propto \rho \omega^3 d_i^5 \tag{8.57}$$

The typical turbine installation operates in the turbulent region. Operation in the transition region between the laminar and turbulent zones is not recommended, because mass transfer rates in the transition region tend to be lower and less predictable than in the other regions [Tatterson, 1994].

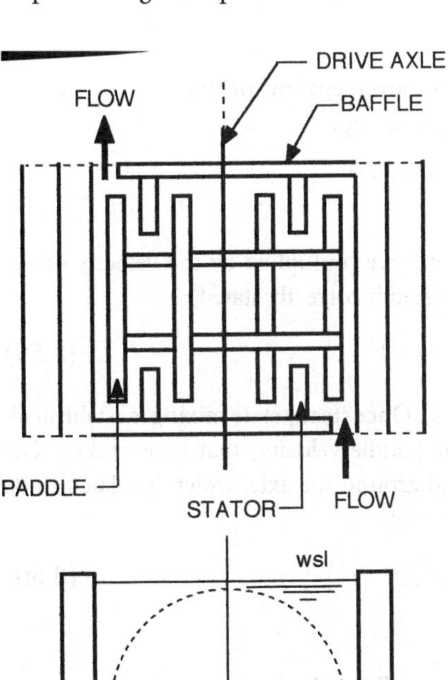

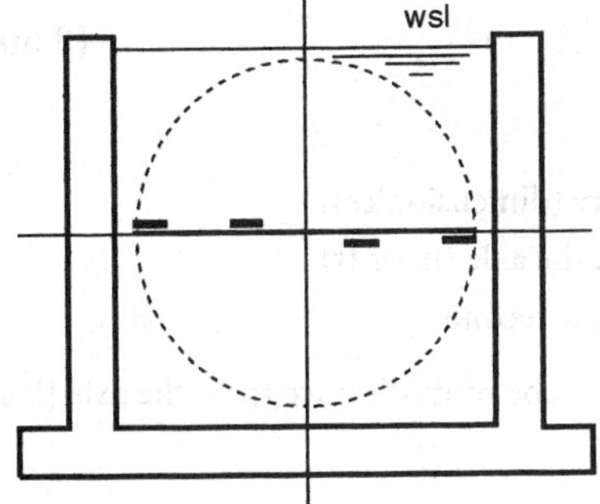

FIGURE 8.5 Flocculation paddles and compartment.

Paddle Wheel Flocculators

A typical flocculation tank compartment is depicted in Fig. 8.5. Paddle flocculators similar to this design, but without the stators and baffles and with the axles transverse to the flow, were first introduced by Smith [1932]. A set of flocculation paddles is mounted on a drive axle, which runs along the length of the compartment parallel to the flow. The axle may be continuous throughout the whole tank or it may serve only one or two compartments. Alternatively, the axle may be mounted vertically in the compartment or horizontally but transverse to the flow. In these cases, each compartment has its own axle. The paddles are mounted parallel to the drive axle. The number of paddles may be the same in each compartment or it may vary. The compartments in the flocculator are separated from each other by cross walls called *baffles*. The baffles are not continuous across the tank; there are openings between the baffles and tank walls so that water can flow from one compartment to the next. In Fig. 8.5, an opening is shown at one end of each baffle, and the openings alternate from one side of the tank to the other so they do not line up. This arrangement minimizes short-circuiting. The spaces are sometimes put at the top or bottom of the baffles to force an over-and-under flow pattern. The compartments also contain *stators*, boards fixed to the baffle walls that are intended to prevent the start of a vortex in the compartment.

Although flocculator performance is usually correlated with tank-average parameters such as $\overline{\Gamma}$ and HRT, it should be borne in mind that the actual flocculation process occurs in the immediate vicinity of the paddles and their structural supports. The flow around the paddles and supports is sensitive to their exact geometry and rotational speed. This means that precise prediction of flocculator performance requires the testing of full-scale units. Facilities may require redesign and reconstruction of the paddle system in the light of operating experience. Some engineers will prefer to specify commercially available paddle systems, which have demonstrated satisfactory performance on similar waters. Contracts with vendors should include performance specifications and guarantees.

Paddle geometry can be connected to power input by Camp's [1955] method. The drag force on the paddle is given by

$$F_D = \tfrac{1}{2} C_D \rho A (v_p - v)^2 \tag{8.58}$$

where

F_D = the drag force on the paddle (N or lbf)
C_D = the drag coefficient (dimensionless)
A = the area of the paddle normal to the direction of movement (m^2 or ft^2)
v_p = the velocity of the paddle relative to the tank (m/s or ft/s)
v = the velocity of the water relative to the tank (m/s or ft/s)
ρ = the density of the water (kg/m^3 or $slug/ft^3$)

The power dissipated by the paddle is simply the drag force multiplied by the velocity of the paddle relative to the *tank* (*not* relative to the water, as is often incorrectly stated):

$$P_p = \tfrac{1}{2} C_D \rho A v_p (v_p - v)^2 \tag{8.59}$$

where P_p is the power dissipated by the paddle (W or ft·lb/s). Once steady state mixing is established, the water velocity will be some constant fraction of the paddle velocity; that is, $v = k v_p$. The paddle speed itself is taken to be the speed of its centroid around the axle, which is related to its radial distance from the axle. Making these substitutions yields

$$P_p = 4\pi^3 C_D \rho (1 - k)^2 A r^3 \omega^3 \tag{8.60}$$

where

k = the ratio of the water velocity to the paddle velocity (dimensionless)
r = the radial distance of the centroid of the paddle to the axle (m or ft)
ω = the rotational velocity of the paddle (revolutions per second)

If the paddle is wide, the area should be weighted by the cube of its distance from the axle [Fair *et al.*, 1968].

$$r^3 A = \int_{r_0}^{r_1} r^3 L \, dB = \tfrac{1}{4}(r_1 - r_0)^4 L \tag{8.61}$$

where

B = the width of the paddle (m or ft)
L = the length of the paddle (m or ft)
r_1 = the distance of the outer edge of the paddle from the axle (m or ft)
r_0 = the distance of the inner edge of the paddle from the axle (m or ft)

This refinement changes Eq. (8.60) to

$$P_p = \pi^3 C_D \rho (1-k)^2 (r_1 - r_0)^4 L\omega^3 \tag{8.62}$$

Equation (8.60) or (8.62) should be applied to each paddle and support in the tank, and the results should be summed to obtain the total power dissipation:

$$P = 4\pi^3 C_D \rho (1-k)^2 \omega^3 \sum_{i=1}^{n} A_i r_i^3 \tag{8.63}$$

$$P = \pi^3 C_D \rho (1-k)^2 \omega^3 \sum_{i=1}^{n} (r_{1,i} - r_{0,i})^4 L_i \tag{8.64}$$

Equations (8.63) and (8.64) provide the needed connection to the volume-averaged characteristic strain rate:

$$\overline{\Gamma}^2 = \frac{\varepsilon}{\nu} = \frac{P}{\mu V} \tag{8.65}$$

It is generally recommended that the strain rate taper downward from the inlet chamber to the outlet chamber, say, from 100 per second to 50 per second. The tapering of $\overline{\Gamma}$ that is required can be achieved by reducing, from inlet to outlet, either the rotational speed of the paddle assemblies or the paddle area.

Another criterion sometimes encountered is the *Bean number* [Bean, 1953]. This is defined as the volume swept out by the elements of the paddle assembly per unit time—called the *displacement*—divided by the flow through the flocculation tank:

$$\text{Be} = \frac{2\pi\omega \sum_{i=1}^{n} r_i A_i}{Q} \tag{8.66}$$

where Be is the Bean number (dimensionless). Bean's recommendation, based on a survey of actual plants, is that Be should be kept between 30 and 40, if it is calculated using all the paddle assemblies in the flocculation tank [Bean, 1953]. For a given facility, the Bean number is proportional to the spatially averaged characteristic strain rate and the total water power. However, the ratio varies as the square of the rotational speed.

The mixing conditions inside a tank compartment are usually turbulent, so the drag coefficient is a constant.

Camp reports that the value of k for paddle flocculators *with stators* varies between 0.32 and 0.24 as the rotational speed increases [Camp, 1955].

The peripheral speed of the paddle assembly is usually kept below 2 ft/s, and speeds of less than 1 ft/s are recommended for the final compartment [Bean, 1953; Hopkins and Bean, 1966]. In older plants, peripheral speeds were generally below 1.8 ft/s [Bean, 1953]. This practice appears to have been based on laboratory data developed using 1-gallon jars without stators. The laboratory data indicated impaired flocculation at peripheral speeds above 1.8 ft/s, but the results may have been caused by vortexing, which would actually reduce the velocity gradients in the liquid [Leipold, 1934].

Paddles are usually between 4 and 8 in. wide, and the spacing between paddles should be greater than this [Bean, 1953]. The total paddle area should be less than 25% of the plan area of the compartment.

Jets

The power expended by a jet is simply the kinetic energy of the mass of liquid injected into the tank:

$$P = \tfrac{1}{2}\dot{m}v^2 \tag{8.67}$$

where $\dot{m}$ is the mass flow rate in the jet at its inlet (kg/s or slug/s). The mass flow rate is determined by the mixing requirements.

Static In-line Mixers

The energy dissipated by static mixers is determined by the pressure drop through the unit:

$$P = Q \cdot \Delta p \tag{8.68}$$

where

Δp = the pressure drop (N/m^2 or lbf/ft^2)
Q = the volumetric flow rate (m^3/s or ft^3/s)

The pressure drop depends on the details of the mixer design.

Gas Sparging

The power dissipated by gas bubbles is

$$P = \rho g Q H \tag{8.69}$$

where

g = the acceleration due to gravity (9.80665 m/s^2 or 32.174 ft/s^2)
H = the depth of bubble injection (m or ft)
Q = the gas flow rate (m^3/s or ft^3/s)
ρ = the liquid density (kg/m^3 or slug/ft^3)

Blending

The principal purpose of all mixing is blending two or more different liquid streams.

Batch Mixing Times

The time required to blend two or more liquids to some acceptable level of macroscopic homogeneity is determined by batch blending tests. The test results are usually reported in terms of a homogenization number that is defined to be the number of impeller revolutions required to achieve homogenization [Tatterson, 1994]:

$$\text{Ho} = \omega\theta_m \tag{8.70}$$

where

Ho = the homogenization number (dimensionless)
θ_m = the blending time (s)
ω = the rotational speed of the impeller (Hz, revolutions per second)

The degree of mixing is often determined from tracer data by calculating the *fractional unmixedness*, which is defined in terms of measured concentration fluctuations [Godfrey and Amirtharajah, 1991; Tatterson, 1994]:

$$X_t = \left| \frac{C_t - C_\infty}{C_\infty - C_o} \right| \tag{8.71}$$

where

C_o = the initial tracer concentration in the tank, if any, prior to tracer addition and mixing (kg/m^3 or lb/ft^3)

C_t = the maximum tracer concentration at any point in the tank at time t (kg/m^3 of lb/ft^3)

C_∞ = the calculated tracer concentration for perfect mixing (kg/m^3 or lb/ft^3)

X_t = the fractional unmixedness (dimensionless)

Blending is considered to be complete when X_t falls to 0.05, meaning that the observed concentration fluctuations are within 5% of the perfectly mixed condition.

Turbines. The Prochazka–Landau correlation [Tatterson, 1994] for a turbine with six flat, disk-mounted blades in a baffled tank is

$$\omega \theta_m = 0.905 \left(\frac{d_t}{d_i} \right)^{2.57} \log \left(\frac{X_o}{X_f} \right) \tag{8.72}$$

For a turbine with four pitched blades, their correlation is

$$\omega \theta_m = 2.02 \left(\frac{d_t}{d_i} \right)^{2.20} \log \left(\frac{X_o}{X_f} \right) \tag{8.73}$$

And for a three-bladed marine propeller, it is

$$\omega \theta_m = 3.48 \left(\frac{d_t}{d_i} \right)^{2.05} \log \left(\frac{X_o}{X_f} \right) \tag{8.74}$$

where

d_i = the impeller diameter (m or ft)

d_t = the tank diameter (m or ft)

X_o = the initial fractional unmixedness, typically between 2 and 3 in the cited report (dimensionless)

X_f = the final fractional unmixedness, typically 0.05

θ_m = the required batch mixing time (s)

ω = the impeller rotational speed (Hz, revolutions per second)

Note that impeller diameter is more important than impeller speed, because the mixing time is inversely proportional to the impeller diameter raised to a power greater than 2, whereas it is inversely proportional to the speed to the first power.

The gist of Eqs. (8.72) through (8.74) is that for any particular design the homogenization number is a constant. These equations are best used to convert data from one impeller size to another.

The homogenization number is also a constant in the laminar region, but the constant is independent of impeller size. Turbines are seldom used in laminar conditions, but propellers mounted in draft tubes are. Reported homogenization numbers for propeller/tube mixers range from 16 to 130 [Tatterson, 1994].

Static In-line Mixers. A wide variety of proprietary static in-line mixers is available. Some designs are specialized to either laminar or turbulent flow conditions and other designs are general-purpose mixers. Static mixers are liable to clogging from large suspended solids and require filtered or screened feed streams.

The usual specification is that the coefficient of variation of concentration measurements in the mixer outlet be equal to or less than 0.05 [Godfrey and Amirtharajah, 1991; Tatterson, 1994]. Mixer unit lengths are typically 5 to 50 times the pipe diameter, depending on the design.

The standard deviation of concentration measurements over any mixer cross section declines exponentially with mixer length and may be correlated by [Godfrey and Amirtharajah, 1991]

$$\frac{s}{s_o} = 2\exp\left(-\frac{1.54 f^{0.5} L}{D}\right) \tag{8.75}$$

where

D = the mixer's diameter

f = the mixer's Darcy–Weisbach friction factor (dimensionless)

L = the mixer's length (m or ft)

s = the standard deviation of the concentration measurements over the outlet cross section (kg/m^3 or lb/ft^3)

s_o = the standard deviation of the concentration measurements over the inlet cross section (kg/m^3 or lb/ft^3)

Jets. Jet mixers have relatively high power requirements, but they are low-maintenance devices. They are restricted to turbulent, medium- to low-viscosity liquid mixing, and the jet Reynolds number at the inlet should be above 2100 [Tatterson, 1994]. The required pump may be either inside or outside the tank, depending on equipment design.

Jets may enter the tank either axially on the tank bottom or radially along the tank side. Axial entry can be used for deep tanks in which the liquid depth to tank diameter ratio is between 0.75 and 3.0, and radial entry can be used for shallow tanks in which the depth/diameter ratio is between 0.25 and 1.25 [Godfrey and Amirtharajah, 1991]. If the depth/diameter ratio exceeds 3.0, multiple jets are required at different levels. Radial inlets are frequently angled upward.

Mixing of the jet and surrounding fluid does not begin until the jet has traveled at least 10 inlet diameters, and effective mixing occurs out to about 100 inlet diameters. Oldshue and Trussell [1991] recommend that the tank diameter to jet inlet diameter ratio be between 50 and 500.

If the initial jet Reynolds number is 5000 or more, the batch mixing time is given by [Godfrey and Amirtharajah, 1991]

$$\frac{\theta_m v_j}{d_j} = 6\left(\frac{d_t}{d_j}\right)^{3/2}\left(\frac{d_l}{d_j}\right)^{1/2} \tag{8.76}$$

For initial jet Reynold's numbers below 5000, the mixing time is given by

$$\frac{\theta_m v_j}{d_j} = \frac{30{,}000}{\text{Re}}\left(\frac{d_t}{d_j}\right)^{3/2}\left(\frac{d_l}{d_j}\right)^{1/2} \tag{8.77}$$

where

d_j = the jet's inlet diameter (m or ft)

d_l = the liquid depth (m or ft)

d_t = the tank diameter (m or ft)

Re = the jet's inlet Reynold's number (dimensionless)
$= v_j d_j / \nu$
v_j = the jet's inlet velocity (m/s of ft/s)
θ_m = the required mixing time (s)

Other correlations are given by Tatterson [1994].

Continuous Flow

In order to account for their exponential residence time distributions, the required hydraulic retention time in continuous flow tanks ranges from 50 to 200 times the batch mixing time, and is typically about 100 times θ_m [Tatterson, 1994]:

$$\theta_h \approx 100\theta_m \tag{8.78}$$

Particle Suspension

Settleable Solids

Settleable solids are usually put into suspension using downward-directed axial flow turbines, sometimes with draft tubes. Tank bottoms should be dished, and the turbine should be placed relatively close to the bottom, say, between one-sixth and one-fourth of the tank diameter [Godfrey and Amirtharajah, 1991]. Antivortex baffles are required. Sloping side walls, bottom baffles, flat bottoms, radial flow turbines, and large tank diameter to impeller diameter ratios should be avoided as they permit solids accumulation on the tank floor [Godfrey and Amirtharajah, 1991; Tatterson, 1994].

The impeller speed required for the suspension of settleable solids is given by the Zweitering [1958] correlation:

$$\omega_{js} = \frac{S\nu^{0.1} d_p^{0.2} \left[g(\rho_p - \rho)/\rho\right]^{0.45} X_p^{0.13}}{d_i^{0.85}} \tag{8.79}$$

where

d_p = the particle diameter (m or ft)
d_i = the turbine diameter (m or ft)
g = the acceleration due to gravity (9.80665 m/s^2 or 32.174 ft/s^2)
S = the impeller/tank geometry factor (dimensionless)
X_p = the weight fraction of solids in the suspension (dimensionless)
ω_{js} = the impeller rotational speed required to just suspend the particles (Hz, revolutions per second)
ρ = the liquid density (kg/m^3 or slug/ft^3)
ρ_p = the particle density (kg/m^3 or slug/ft^3)

Equation (8.79) is the "just suspended" criterion. Lower impeller speeds will allow solids to deposit on the tank floor. The impeller/tank geometry factor varies significantly. Typical values for many different configurations are given by Zweitering [1958].

The Zweitering correlation leads to a prediction for power scale-up that may not be correct. In the standard geometry, the liquid depth, tank diameter, and impeller diameter are proportional to one another. Combining this geometry with the power correlation for the turbulent regime [Eq. (8.57)], one gets

$$\frac{P}{V} \propto \frac{\omega^3 d_i^5}{d_l d_t^2} \propto \frac{d_i^{-2.55} d_i^5}{d_l d_t^2} \propto d_t^{-0.55} \tag{8.80}$$

Some manufacturers prefer to scale power per unit volume as $d_t^{-0.28}$, and others prefer to keep the power per unit volume constant.

If solids are to be distributed throughout the whole depth of the liquid, then the modified Froude number must be greater than 20 [Tatterson, 1994]:

$$\mathrm{Fr} = \frac{\rho\omega^2 d_i^2}{g(\rho_p - \rho)d_p}\left(\frac{d_p}{d_i}\right)^{0.45} > 20 \tag{8.81}$$

The concentration profile can be estimated from [Tatterson, 1994]:

$$\frac{X_p(z)}{\overline{X}_p} = \frac{\mathrm{Pe}\cdot\exp(-\mathrm{Pe}\cdot z/d_l)}{1-\exp(-\mathrm{Pe})} \tag{8.82}$$

$$\mathrm{Pe} = 330\left(\frac{\omega d_i}{v_p}\right)^{-1.17}\left(\frac{\varepsilon d_p^4}{\nu^3}\right)^{-0.095} \tag{8.83}$$

where

d_i = the impeller's diameter (m or ft)
d_l = the total liquid depth (m or ft)
d_p = the particle's diameter (m or ft)
Pe = the solid's Peclet number (dimensionless)
= $v_{ps}d_l/d_p$
v_p = the particle's free settling velocity in still liquid (m/s or ft/s)
v_{ps} = the particle's free settling velocity in stirred liquid (m/s or ft/s)
$\overline{X}_p$ = the mean particle mass fraction in the tank (dimensionless)
$X_p(z)$ = the particle mass fraction at elevation z above the tank bottom (dimensionless)
z = the elevation above the tank bottom (m or ft)
ε = the power per unit mass (W/kg or ft·lbf/slug·s)
ρ = the density of the liquid (kg/m^3 or slug/ft^3)
ρ_p = the density of the particle (kg/m^3 or slug/ft^3)
ν = the kinematic viscosity (m^2/s or ft^2/s)
ω = the rotational speed of the impeller (Hz, revolutions per second)

Floatable Solids

The submergence of low-density, floating solids requires the development of a vortex, so only one antivortex baffle or very narrow baffles ($w_b = d_t/50$) should be installed [Godfrey and Amirtharajah, 1991]. The axial flow turbine should be installed close to the tank bottom and perhaps off-center. The tank bottom should be dished. The minimum Froude number for uniform mixing is given by

$$\mathrm{Fr}_{\min} = 0.036\left(\frac{d_i}{d_t}\right)^{-3.65}\left(\frac{\rho_p-\rho}{\rho}\right)^{0.42} \tag{8.84}$$

where

$\mathrm{Fr}_{\min}$ = the required minimum value of the Froude number (dimensionless)
Fr = the impeller Froude number (dimensionless)
= $\omega^2 d_i/g$

8.4 Rapid Mixing and Flocculation

Rapid Mixing

Rapid or flash mixers are required to blend treatment chemicals with the water being processed. Chemical reactions also occur in the rapid mixer, and the process of colloid destabilization and flocculation begins there.

Particle Collision Rate

Within the viscous dissipation subrange, the relative velocity between two points along the line connecting them is [Saffman and Turner, 1956; Spielman, 1978]

$$|\overline{u}| = \sqrt{\frac{2\varepsilon}{15\pi\nu}} \cdot r \tag{8.85}$$

where

r = the radial distance between the two points (m or ft)

$|\overline{u}|$ = the average of the absolute value of the relative velocity of two points in the liquid along the line connecting them (m/s or ft/s)

ε = the power input per unit mass (W/kg or ft·lbf/slug·s)

ν = the kinematic viscosity (m^2/s or ft^2/s)

One selects a target particle of radius r_1 and number concentration C_1 and a moving particle of radius r_2 and number concentration C_2. The moving particle is carried by the local eddies, which may move either toward or away from the target particle at velocities given by Eq. (8.85). A collision will occur whenever the center of a moving particle crosses a sphere of radius $r_1 + r_2$ centered on the target particle. The volume of liquid crossing this sphere is

$$Q = \frac{1}{2}|\overline{u}|4\pi(r_1 + r_2)^2 = \sqrt{\frac{8\pi}{15}}\sqrt{\frac{\varepsilon}{\nu}}(r_1 + r_2)^3 \tag{8.86}$$

where Q is the volumetric rate of flow of liquid into the collision sphere (m^3/s or ft^3/s).

The Saffman–Turner collision rate is, therefore,

$$R_{1,2} = \sqrt{\frac{8\pi}{15}}\sqrt{\frac{\varepsilon}{\nu}}C_1C_2(r_1 + r_2)^3 \tag{8.87}$$

A similar formula has been derived by Delichatsios and Probstein [1975].

Particle Destabilization

The destabilization process may be visualized as the collision of colloidal particles with eddies containing the coagulants. If the diameter of the eddies is estimated to be η, then Eq. (8.87) can be written as

$$\sqrt{\frac{\pi}{120}}\sqrt{\frac{\varepsilon}{\nu}}C_1C_\eta(d_1 + \eta)^3 \tag{8.88}$$

where

C_η = the "concentration" of eddies containing the coagulant (dimensionless)

d_1 = the diameter of the colloidal particles (m or ft)

$R_{1,\eta}$ = the rate of particle destabilization (no./m^3·s or no./ft^3·s)

η = the diameter of the eddies containing the coagulant (m or ft)

Equation (8.88) is the Amirtharajah–Trusler [1986] destabilization rate formula. The "concentration" C_η may be regarded as an unknown constant that is proportional to the coagulant dosage.

The effect of a collision between a stable particle and an eddy containing coagulant is a reduction in the net surface charge of the particle. The particle's surface charge is proportional to its zeta potential, which in turn is proportional to its electrophoretic mobility. Consequently, $R_{1,\eta}$ is the rate of reduction of the average electrophoretic mobility of the suspension.

Equation (8.88) has unexpected implications. First, the specific mixing power may be eliminated from Eq. (8.88) by means of the definition of the Kolmogorov length scale, yielding [Amirtharajah and Trusler, 1986]

$$R_{1,\eta} = \sqrt{\frac{\pi}{120}} C_\eta \frac{(d_1 + \eta)^3}{\eta^2} C_1 \tag{8.89}$$

The predicted rate has a minimum value when the Kolmogorov length scale is twice the particle diameter (i.e., $\eta = 2d_1$), and the mixing power per unit mass required to produce this ratio is

$$\varepsilon = \frac{\nu^3}{16d_1} \tag{8.90}$$

In strongly mixed tanks, the power dissipation rate varies greatly from one part of the tank to another. It is highest near the mixer, and ε should be interpreted as the power dissipation at the mixer.

The predicted minimum is supported by experiment, and it does occur at about $\eta = 2d_1$, if η is calculated for the conditions next to the mixer [Amirtharajah and Trusler, 1986]. For a given tank and mixer geometry, the energy dissipation rate near the mixer is related to the average dissipation rate for the whole tank. In the case of completely mixed tanks stirred by turbines, the minimum destabilization rate will occur at values of $\overline{\Gamma}$ between about 1500 and 3500 per second, which is substantially above the usually practice.

Second, the destabilization rate is directly proportional to the stable particle concentration:

$$R_{1,\eta} = kC_1 \tag{8.91}$$

where k is the first-order destabilization rate coefficient (per second).

The rate coefficient, k, applies to the conditions near the mixer. However, what is needed for design is the rate $R_{1,\eta}$ volume averaged over the whole tank. If the tank is completely mixed, C_1 is the same everywhere and can be factored out of the volume average. Thus, the volume average of the rate, $R_{1,\eta}$, becomes a volume average of the rate coefficient, k. Throughout most of the volume, the power dissipation rate is much smaller than the rate near the mixer, and in general $\eta \gg d_1$. Therefore, the particle diameter can be neglected, and to a first-order approximation one has

$$\overline{k} = \sqrt{\frac{\pi}{120}} \sqrt{\frac{\nu^3}{\overline{\Gamma}}} C_\eta \tag{8.92}$$

Equation (8.92) is limited to power dissipation rates less than 1500 per second.

Inspection of Eq. (8.92) shows that the rate of destabilization should *decrease* if either the temperature or the mixing power *increases*. This is counterintuitive. Experience with simple chemical reactions and flocculation indicates that reaction rates always rise whenever the temperature or mixing power is increased. However, in chemical reactions and particle flocculation, the driving forces are the velocity of the particles and their concentrations. In destabilization, the driving forces are these plus the size of the eddies, and eddy size dominates.

The prediction that the destabilization rate decreases as the mixing power increases has indirect experimental support, although it may be true only for completely mixed tanks [Amirtharajah and Trusler, 1986; Camp, 1968; Vrale and Jorden, 1971]. The prediction regarding temperature has not been tested.

Optimum Rapid Mixing Time

It is known that there is a well-defined optimum rapid mixing time [Letterman *et al.,* 1973; Camp, 1968]. This optimum is not predicted by the Amirtharajah–Trusler formula. If rapid mixing is continued beyond this optimum, the flocculation of the destabilized particles and their settling will be impaired, or at least there will be no further improvement. For the flocculation of activated carbon with filter alum, the relationship between the rms strain rate, batch processing time, and alum dose is [Letterman *et al.*, 1973]

$$\overline{\Gamma} t C^{1.46} = 5.9 \times 10^6 (\text{mg/L})^{1.46} \tag{8.93}$$

where

C = the dosage of filter alum, $Al_2(SO_4)_3(H_2O)_{18}$ (mg/L)
t = the duration of the batch mixing period (in seconds)

Equation (8.93) does not apply to alum dosages above 50 mg/L, because optimum rapid mixing times were not always observed at higher alum dosages. The experiments considered powdered activated carbon concentrations between 50 and 1000 mg/L, alum dosages between 10 and 50 mg/L, and values of $\overline{\Gamma}$ between 100 and 1000 per second. At the highest mixing intensities, the optimum value of t ranged from 14 seconds to 2.5 minutes. These results have not been tested for other coagulants or particles. Consequently, it is not certain that they are generally applicable. Also, the underlying cause of the optimum is not known.

It is also clear that very high mixing powers in the rapid mixing tank impair subsequent flocculation in the flocculation tank. For example, Camp [1968] reports that rapid mixing for 2 minutes at a $\overline{\Gamma}$ of 12,500 per second prevents flocculation for at least 45 minutes. If $\overline{\Gamma}$ is reduced to 10,800 per second, flocculation is prevented for 30 minutes. The period of inhibition was reduced to 10 minutes when $\overline{\Gamma}$ was reduced to 4400 per second. Strangely enough, if the particles were first destabilized at a $\overline{\Gamma}$ of 1000 per second, subsequent exposure to 12,500 per second had no effect. Again, the cause of the phenomenon is not known.

Design Criteria for Rapid Mixers

Typical engineering practice calls for an rms characteristic strain rate in the rapid mixing tank of about 600 to 1000 per second and a hydraulic retention time of 1 to 3 seconds [Joint Task Force, 1990]. Impeller tip speeds should be limited to less than 5 m/s to avoid polymer shear.

Flocs begin to form within 2 seconds, and conduits downstream of the rapid mixing chamber should be designed to minimize turbulence. Typical conduit velocities are 1.5 to 3.0 ft/s.

Flocculation

Quiescent and Laminar Flow Conditions

In quiescent water, the Brownian motion of the destabilized colloids will cause them to collide and agglomerate. Eventually, particles large enough to settle will form, and the water will be clarified. The rate of agglomeration is increased substantially by mixing. This is due to the fact that mixing creates velocity differences between neighboring colloidal particles, which increases their collision frequency.

Perikinetic Flocculation. Coagulation due to the Brownian motion alone is called "perikinetic" flocculation. The rate of perikinetic coagulation was first derived by Smoluchowski in 1916–1917 as follows [Levich, 1962]. Consider a reference particle having a radius r_1. A collision will occur with another particle having a radius r_2 whenever the distance between the centers of the two particles is reduced to $r_1 + r_2$. (Actually, because of the van der Waals attraction, the collision will occur even if the particles are somewhat farther apart.) The collision rate will be the rate at which particles with radius r_2 diffuse across a sphere of radius $r_1 + r_2$ centered on the reference particle. For a spherically symmetrical case like this, the mass conservation equation becomes [Crank, 1975]:

$$\frac{\partial C_2(r,t)}{\partial t} = \frac{D_{1,2}}{r^2}\frac{\partial}{\partial r}\left[r^2\frac{\partial C_2(r,t)}{\partial t}\right] \tag{8.94}$$

where

$C_2(r,t)$ = the number concentration of the particle with radius r_2 at a point a distance r from the reference particle at time t (no./m^3 or no./ft^3)
$D_{1,2}$ = the mutual diffusion coefficient of the two particles (m^2/s or ft^2/s)
= $D_1 + D_2$;
D_1 = the diffusion coefficient of the reference particle (m^2/s or ft^2/s)
D_2 = the diffusion coefficient of the particle with radius r_2 (m^2/s or ft^2/s)
r = the radial distance from the center of the reference particle (m or ft)
t = elapsed time (seconds)

Only the steady state solution is needed, so the left-hand side of Eq. (8.94) is zero. The boundary conditions are

$$r \to \infty \qquad C_2(r) = C_2 \tag{8.95}$$

$$r = r_1 + r_2 \qquad C_2(r) = 0 \tag{8.96}$$

Therefore, the particular solution is

$$C_2(r) = C_2\left(1 - \frac{r_1 + r_2}{r}\right) \tag{8.97}$$

The rate at which particles with radius r_2 diffuse across the spherical surface is

$$4\pi(r_1 + r_2)^2 D_{1,2}\left.\frac{dC_2(r)}{dr}\right|_{r=r_1+r_2} = 4\pi D_{1,2}C_2(r_1 + r_2) \tag{8.98}$$

If the concentration of reference particles is C_1, the collision rate will be

$$R_{1,2} = 4\pi D_{1,2}C_1C_2(r_1 + r_2) \tag{8.99}$$

where $R_{1,2}$ is the rate of collisions between particles of radius r_1 and radius r_2 due to Brownian motion (collisions/m^3·s or collisions/ft^3·s).

Orthokinetic Flocculation. If the suspension is gently stirred, so as to produce a laminar flow field, the collision rate is greatly increased. The mixing is supposed to produce a velocity gradient near the reference particle, $G = du/dy$. The velocity, u, is perpendicular to the axis of the derivative, y. The collisions now are due to the velocity gradient, and the process is called *gradient* or *orthokinetic* flocculation. As before, a collision is possible if the distance between the centers of two particles is less than the sum of their radii, $r_1 + r_2$. Smoluchowski selects a reference particle with a radius of r_1, and, using the local velocity gradient, one calculates the total fluid flow through a circle centered on the reference particle with radius $r_1 + r_2$ [Freundlich, 1922]:

$$Q = 4\int_0^{r_1+r_2} Gy(r^2 - y^2)^{1/2}dy = \tfrac{4}{3}G(r_1 + r_2)^3 \tag{8.100}$$

where

G = the velocity gradient (per second)
Q = the total flow through the collision circle (m^3/s or ft^3/s)
y = the distance from the center of the reference particle normal to the local velocity field (m or ft)

The total collision rate between the two particle classes will be

$$R_{1,2} = \tfrac{4}{3} G C_1 C_2 (r_1 + r_3)^3 \tag{8.101}$$

The ratio of the orthokinetic to perikinetic collision rates for equal size particles is

$$\frac{R_{1,2}(\text{ortho})}{R_{1,2}(\text{peri})} = \frac{4\mu G N_A r^3}{RT} \tag{8.102}$$

where

N_A = Avogadro's constant (6.022137×10^{23} particles per mol)
R = the gas constant (8.3243 J/mol·K or 1.987 Btu/lb·°R)
T = the absolute temperature (K or °R)

Einstein's [1956] formula for the diffusivity of colloidal particles has been used to eliminate the joint diffusion constant. Note that the effect of mixing is very sensitive to the particle size, varying as the cube of the radius. In fact, until the particles have grown by diffusion to some minimal size, mixing appears to have little or no effect. Once the minimal size is reached, however, flocculation is very rapid [Freundlich, 1922].

Corrections to Smoluchowski's analysis to account for van der Waals forces and hydrodynamic effects are reviewed by Spielman [1978].

Turbulent Flocculation and Deflocculation

The purposes of the flocculation tank are to complete the particle destabilization begun in the rapid mixing tank and to agglomerate the destabilized particles. Flocculation tanks are operated at relatively low power dissipation rates, and destabilization proceeds quite rapidly; it is probably completed less than one minute after the water enters the tank. Floc agglomeration is a much slower process, and its kinetics are different. Furthermore, the kinds of mixing devices used in flocculation are different from those used in rapid mixing. The effects of tank partitioning on the degree of agglomeration achieved are quite striking, and flocculation tanks are always partitioned into at least four mixed-cells-in-series.

Flocculation Rate. First, consider the kinetics of flocculation itself. Floc particles form when smaller particles collide and stick together. The basic collision rate is given by the Saffman–Turner equation [Harris *et al.*, 1966; Argaman and Kaufman, 1970; Parker *et al.*, 1972]:

$$R_{1,2} = \sqrt{\frac{8\pi}{15}} \sqrt{\frac{\varepsilon}{\nu}} C_1 C_2 (r_1 + r_2)^3 \tag{8.103}$$

This needs to be adapted to the situation in which there is a wide variety of particle sizes, not just two. This may be done as follows [Harris *et al.*, 1966]. It is first supposed that the destabilized colloids, also called the *primary particles*, may be represented as spheres, each having the same radius r. The volume of such a sphere is

$$V_1 = \tfrac{4}{3}\pi r^3 \tag{8.104}$$

Floc particles are aggregations of these primary particles, and the volume of the aggregate is equal to the sum of the volumes of the primary particle it contains. An aggregate consisting of i primary particles is called an *i-fold* particle and has a volume equal to

$$V_i = \tfrac{4}{3}\pi i r^3 \tag{8.105}$$

From this it is seen that the i-fold particle has an effective radius given by

$$r_i = \sqrt[3]{i r^3} \tag{8.106}$$

The total floc volume concentration is

$$\Phi = \tfrac{4}{3}\pi r^3 \textstyle\sum_{i=1}^{p} i C_i \tag{8.107}$$

where

C_i = the number concentration of aggregates containing primary particles (no./m^3 or no./ft^3)
p = the number of primary particles in the largest floc in the suspension (dimensionless)
Φ = the floc volume concentration (m^3 floc/m^3 water or ft^3 floc/ft^3 water)

A k-fold particle can arise in several ways. All that is required is that the colliding flocs contain a total of k particles, so the colliding pairs may be a floc having $k - 1$ primary particles and a floc having 1 primary particle, or a floc having $k - 2$ primary particles and a floc having 2 primary particles, or a floc having $k - 3$ primary particles and a floc having 3 primary particles, and so on. The k-fold particle will disappear, and form a larger particle, if it collides with anything except a p-fold particle. Collisions with p-fold particles cannot result in adhesion, because the p-fold particle is supposed to be the largest possible. In both kinds of collisions, it is assumed that the effective radius of a particle is somewhat larger than its actual radius, because of the attraction of van der Waals forces. Each kind of collision can be represented by a summation of terms like Eq. (8.103), and the net rate of formation of k-fold particles is the difference between the summations [Harris *et al.*, 1966]:

$$R_k = \bar{\Gamma}\sqrt{\frac{8\pi}{15}} \cdot \left[\frac{1}{2}\sum_{i=1,\, j=k-1}^{k} (ar_i + ar_j)^3 C_i C_j - \sum_{i=1}^{p-1} (ar_k + ar_i)^3 C_k C_i\right] \tag{8.108}$$

where

R_k = the net rate of formation of k-fold particles (no./s·m^3 or no./s·ft^3)
a = the ratio of the effective particle radius to its actual radius (dimensionless)

Note that the factor $\frac{1}{2}$ in the first summation is required to avoid double counting of collisions.

If there is no formation process for primary particles, Eq. (8.108) reduces to

$$R_1 = -\bar{\Gamma}\sqrt{\frac{8\pi}{15}} \cdot \left[\sum_{i=1}^{p-1} (ar + ar_i)^3 C_i\right] \cdot C_1 \tag{8.109}$$

where R_1 is the rate of loss of primary particles due to flocculation (no./s·m^3 or no./s·ft^3). The summation may be rearranged by factoring out r, which requires use of Eq. (8.106) to eliminate r_i, and by using Eq. (8.107) to replace the resulting r^3:

$$R_1 = -\sqrt{\frac{3}{10\pi}}\bar{\Gamma} a^3 \Phi \sigma C_1 \tag{8.110}$$

where

σ = the particle size distribution factor (dimensionless)

$$= \frac{\sum_{i=1}^{p-1} C_i(1+i^{1/3})^3}{\sum_{i=1}^{p-1} iC_i}$$

Except for a factor reflecting particle stability (which is not included above) and substitution of the Saffman–Turner collision rate for the Camp–Stein collision rate, Eq. (8.110) is the Harris–Kaufman–Krone flocculation rate formula for primary particles.

In any given situation, the resulting rate expression involves only two variables: the number concentration of primary particles, C_i, and the particle size distribution factor, σ. The power dissipation rate, kinematic viscosity, and floc volume concentration will be constants. Initially, all the particles are primary, and the size distribution factor has a value of 8. As flocculation progresses, primary particles are incorporated into ever larger aggregates, and σ declines in value, approaching a lower limit of 1.

If the flocculation tank consists of mixed-cells-in-series, σ will approach a constant but different value in each compartment. If the mixing power in each compartment is the same, the average particle size distribution factor will be [Harris *et al.*, 1966]:

$$\overline{\sigma} = \frac{\left(\dfrac{C_{1,o}}{C_{1,e}}\right)^{1/n} - 1}{\sqrt{\dfrac{3}{10\pi}}\overline{\Gamma} a^3 \Phi} \tag{8.111}$$

where

$\overline{\sigma}$ = the average particle size distribution factor for a flocculation tank consisting of n mixed-cells-in-series (dimensionless)

$C_{1,o}$ = the number concentration of primary particles in the raw water (no./m^3 or no./ft^3)

$C_{1,e}$ = the number concentration of primary particles in the flocculated water (no./m^3 or no./ft^3)

An equation of the same form as Eq. (8.109) has been derived by Argaman and Kaufman [1970] by substituting a formula for a turbulent eddy diffusivity into Smoluchowski's quiescent rate formula, Eq. (8.99). They make two simplifications. First, they observe that the radius of a typical floc is much larger than that of a primary particle, so r can be eliminated from Eq. (8.99). Second, ignoring the primary particles and the largest flocs, the total floc volume concentration, which is a constant everywhere in the flocculation tank, would be

$$\Phi \approx \frac{4}{3}\sum_{i=2}^{p-1} C_i r_i^3 \tag{8.112}$$

Therefore,

$$R_1 \approx -\sqrt{\frac{3}{10\pi}}\overline{\Gamma}\Phi a^3 C_1 \tag{8.113}$$

These two substitutions eliminate any explicit use of the particle size distribution function. However, early in the flocculation process, the primary particles comprise most of the particle volume, so Eq. (8.113) is at some points of the process grossly in error.

Equations (8.110) and (8.113) are devices for understanding the flocculation process. The essential prediction of the models is that flocculation of primary particles can be represented as a first-order reaction with an apparent rate coefficient that depends only on the total floc volume concentration, the mixing power, and the water viscosity.

Experiments in which a kaolin/alum mixture was flocculated in a tank configured as one to four mixed-cells-in-series were well described by the model, as long as the value of $\overline{\Gamma}$ was kept below about 60 per second [Harris *et al.*, 1966]. The average particle size distribution factor, $\overline{\sigma}$, was observed to vary between about 1 and 4, and it appeared to decrease as the mixing power increased.

Deflocculation Rate. If the mixing power is high enough, the collisions between the flocs and the surrounding liquid eddies will scour off some of the floc's primary particles, and this scouring will limit the maximum floc size that can be attained.

The maximum size can be estimated from the Basset–Tchen equation for the sedimentation of a sphere in a turbulent liquid [Basset, 1888; Hinze, 1959]:

$$\rho_p V \frac{d u_p}{dt} = \rho_p V g - \rho V g - 3\pi \rho \nu d_p (u - u_p) + \rho V \frac{du}{dt} - \frac{1}{2}\rho \nu \frac{d(u - u_p)}{dt}$$
$$- \frac{9\rho V \sqrt{\nu/\pi}}{d_p} \int_0^t \frac{d[u(t-\tau) - u_p(t-\tau)]}{\sqrt{t-\tau}} \tag{8.114}$$

where

d_p = the floc diameter (m or ft)
g = the acceleration due to gravity (9.80665 m/s^2 or 32.174 ft/s^2)
t = the elapsed time from the beginning of the floc's motion (s)
u = the velocity of the liquid near the floc (m/s or ft/s)
u_p = the velocity of the floc (m/s or ft/s)
V = the floc volume (m^3 or ft^3)
ν = the kinematic viscosity (m^2/s or ft^2/s)
ρ = the density of the liquid (kg/m^3 or slug/ft^3)
ρ_p = the density of the floc (kg/m^3 or slug/ft^3)
τ = the variable of integration (s)

The terms in this equation may be explained as follows. The term on the left-hand side is merely the rate of change of the floc's downward momentum. According to Newton's second law, this is equal to the resultant force on the floc, which is given by the terms on the right-hand side. The first term is the floc's weight. The second term is the buoyant force due to the displaced water. The density of the floc is assumed to be nearly equal to the density of the liquid, so the gravitational and buoyant forces nearly cancel. The third term is Stoke's drag force, which is written in terms of the difference in velocity between the floc and the liquid. Stoke's drag force would be the only drag force experienced by the particle, if the particle were moving at a constant velocity with a Reynold's number much less than 1, and the liquid were stationary. However, the moving floc displaces liquid, and when the floc is accelerated some liquid must be accelerated also, and this gives rise to the remaining terms. The fourth term is an acceleration due to the local pressure gradients set up by the acceleration of the liquid. The fifth and sixth terms represent the additional drag forces resulting from the acceleration of the displaced liquid. The fifth term is the "virtual inertia" of the floc. It represents the additional drag due to the acceleration of displaced liquid in the absence of viscosity. The sixth term is the so-called Basset term. It is a further correction to the virtual inertia for the viscosity of the liquid.

Parker *et al.* [1972] used the Basset–Tchen equation to estimate the largest possible floc diameter in a turbulent flow. First they absorb the last term into the Stoke's drag force. Then they subtract from both sides the quantity $\rho_p V(du/dt) + \frac{1}{2}\rho V(du/dt)$, getting

$$\left(\rho_p + \frac{1}{2}\rho\right)V\frac{d(u - u_p)}{dt} = (\rho_p - \rho)V\frac{du}{dt} - 3b\pi\rho\nu d_p(u - u_p) \tag{8.115}$$

where b is a coefficient greater than 1 that reflects the contribution of the Basset term to the drag force (dimensionless). They argue that the relative acceleration of the liquid and floc, which is the term on the left-hand side, is small relative to the acceleration of the liquid itself:

$$3b\pi\rho\nu d_p(u - u_p) \approx (\rho_p - \rho)V\frac{du}{dt} \tag{8.116}$$

The time required by an eddy to move a distance equal to its own diameter is approximately

$$t \approx \frac{d}{u} \tag{8.117}$$

This is also approximately the time required to accelerate the eddy from zero to u, so the derivative in the right-hand side of Eq. (8.116) is

$$\frac{du}{dt} \approx \frac{u^2}{d} \tag{8.118}$$

Therefore,

$$3b\pi\rho\nu d_p(u - u_p) \approx (\rho_p - \rho)V\frac{u^2}{d} \tag{8.119}$$

Finally, it is assumed that the effective eddy is the same size as the floc; this also means the distance between them is the floc diameter. The eddy velocity is estimated as its fluctuation, which is given by the Saffman–Turner formula, yielding

$$3b\pi\nu d_p(u - u_p) \approx (\rho_p - \rho)V\frac{2\varepsilon d_p^2}{15\pi\nu d_p} \tag{8.120}$$

Now, the left-hand side is the total drag force acting on the floc. Ignoring the contribution of the liquid pressure, an upper bound on the shearing stress on the floc surface can be calculated as

$$\tau_s 4\pi d_p^2 \approx 3b\pi\rho\nu d_p(u - u_p) = (\rho_p - \rho)\frac{\pi d_p^3}{6}\frac{2\varepsilon d_p^2}{15\pi\nu d_p} \tag{8.121}$$

Note that the shearing stress τ_s increases as the square of the floc diameter. If τ_s is set equal to the maximum shearing strength that the floc surface can sustain, without loss of primary particles, then the largest possible floc diameter is [Parker *et al.*, 1972]

$$d_{p\max} \approx \left[\frac{180\pi\tau_{s\max}}{(\rho_p - \rho)(\varepsilon/\nu)}\right]^{1/2} \tag{8.122}$$

The rate at which primary particles are eroded from flocs by shear may now be estimated; it is assumed that only the largest flocs are eroded. The largest flocs are supposed to comprise a constant fraction of the total floc volume:

$$\frac{4}{3}\pi r_p^3 C_p = f\frac{4}{3}\pi r^3 \sum_{i=1}^{p} iC_i \tag{8.123}$$

$$C_p = \frac{f}{p}\sum_{i=1}^{p} iC_i \tag{8.124}$$

where

f = the fraction of the total floc volume contained in the largest flocs (dimensionless)
r_p = the effective radius of the largest floc (m or ft)
$= p^{1/3}r$

Each erosion event is supposed to remove a volume ΔV_p from the largest flocs:

$$\Delta V_p = f_e 4\pi r_p^2 \cdot \Delta r_p \tag{8.125}$$

where

ΔV_p = the volume eroded from the largest floc surface in one erosion event (m^3 or ft^3)
f_e = the fraction of the surface that is eroded per erosion event (dimensionless)

If the surface layer is only one primary particle thick, Δr_p is equal to the diameter of a primary particle, that is, $2r$. The number of primary particles contained in the eroded layer is equal to the fraction of its volume that is occupied by primary particles divided by the volume of one primary particle [Parker *et al.*, 1972]:

$$n = \frac{f_p f_e 4\pi r_p^2 2r}{\frac{4}{3}\pi r^3} \tag{8.126}$$

$$n = 6 f_p f_e p^{2/3} \tag{8.127}$$

where

n = the number of primary particles eroded per erosion event (no. primary particles/floc·erosion)
f_p = the fraction of the eroded layer occupied by primary particles (dimensionless)
p = the number of primary particles in the largest floc (dimensionless)

The frequency of erosion events is approximated by dividing the velocity of the effective eddy by its diameter. This represents the reciprocal of the time required for an eddy to travel its own diameter and the reciprocal of the time interval between successive arrivals. If it is assumed that the effective eddy is about the same size as the largest floc, then the Saffman–Turner formula gives

$$f = \left(\frac{2\varepsilon}{15\pi\nu}\right)^{1/2} \tag{8.128}$$

Combining these results, one obtains the primary particle erosion rate [Parker *et al.*, 1972]:

$$R_{1e} = \overline{\Gamma}\sqrt{\frac{24}{5\pi}} \cdot \left(\frac{f \cdot f_p \cdot f_e}{p^{1/3}}\right) \cdot \sum_{i=1}^{p} iC_i \tag{8.129}$$

The radius of the largest floc may be eliminated from Eq. (8.129) by use of Eq. (8.122), producing

$$R_{1e} = \overline{\Gamma}^2\sqrt{\frac{3}{200}} \cdot \frac{f \cdot f_p \cdot f_e \Phi \sqrt{\rho_p - \rho}}{\pi\sqrt{\tau_s r}} \tag{8.130}$$

$$R_{1e} = k_e \Phi \overline{\Gamma}^2 \tag{8.131}$$

where k_e is the erosion rate coefficient (no. primary particles·s/m^3 floc).

The rate of primary particle removal by flocculation, given by Eq. (8.110), is

$$R_{1f} = k_f \Phi \overline{\Gamma} C_1 \tag{8.132}$$

where k_f is the flocculation rate coefficient (m^3 water/m^3 floc). The derivation makes it obvious that k_f depends on the floc size distribution. It should be kept in mind, however, that the coefficient k_e contains f, which is the fraction of the floc volume contained in the largest flocs. Consequently, k_e is also a function of the floc size distribution. Also, note that the rate of primary particle loss due to flocculation varies as the square root of the mixing power, while the rate of primary particle production due to erosion varies directly as the mixing power. This implies that there is a maximum permissible mixing power.

If the products $k_e\Phi$ and $k_f\Phi$ are constants, then a steady state particle balance on a flocculator consisting of equal-volume mixed-cells-in-series yields [Argaman and Kaufman, 1970]

$$\frac{C_{1,e}}{C_{1,o}} = \frac{1 + \dfrac{k_e \Phi \overline{\Gamma}^2 V_1}{C_{1,o} Q} \displaystyle\sum_{i=1}^{n=1} \left(1 + \dfrac{k_f \Phi \overline{\Gamma} V_1}{Q}\right)^i}{\left(1 + \dfrac{k_f \Phi \overline{\Gamma} V_1}{Q}\right)^n} \tag{8.133}$$

where

n = the number of mixed-cells-in-series (dimensionless)

V_1 = the volume of one cell (m^3 or ft^3)

Equation (8.133) was tested in laboratory flocculators consisting of four mixed-cells-in-series with either turbines or paddle mixers [Argaman and Kaufman, 1970]. The raw water fed to the flocculators contained 25 mg/L kaolin that had been destabilized with 25 mg/L of filter alum. The total hydraulic retention times varied from 8 to 24 minutes, and the rms characteristic strain rate varied from 15 per second to 240 per second. The concentration of primary particles was estimated by allowing the flocculated water to settle quiescently for 30 minutes and measuring the residual turbidity. The experimental data for single-compartment flocculators was represented accurately by the model. With both kinds of mixers, the optimum value of $\overline{\Gamma}$ varied from about 100 per second down to about 60 per second as the hydraulic retention time was increased from 8 to 24 minutes. The observed minimum in the primary particle concentration is about 10 to 15% lower than the prediction, regardless of the number of compartments in the flocculator.

Flocculation Design Criteria. The degree of flocculation is determined by the dimensionless number $\overline{\sigma}\Phi\overline{\Gamma}\theta_h$. The floc volume concentration is fixed by the amount and character of the suspended solids in the raw water, and the particle size distribution factor is determined by the mixing power, flocculator configuration, and raw water suspended solids. For a given plant, then, the dimensionless number can be reduced to $\overline{\Gamma}\theta_h$, which is sometimes called the *Camp number* in honor of Thomas R. Camp, who promoted its use in flocculator design.

The Camp number is proportional to the total number of collisions that occur in the suspension as it passes through a compartment. Because flocculation is a result of particle collisions, the Camp number is a performance indicator and a basic design consideration. In fact, specification of the Camp number and either the spatially averaged characteristic strain rate or hydraulic detention time suffices to determine the total tankage and mixing power required. It is commonly recommended that flocculator design be based on the product $\overline{\Gamma}\theta_h$ and some upper limit on $\overline{\Gamma}$ to avoid floc breakup [Camp, 1955; James M. Montgomery, Inc., 1985; Joint Task Force, 1990]. Many regulatory authorities require a minimum HRT in the flocculation tank of at least 30 minutes [Water Supply Committee, 1987]. In this case, the design problem is reduced to selection of $\overline{\Gamma}$.

Another recommendation is that flocculator design requires only the specification of the product $\overline{\Gamma}\Phi\theta_h$; sometimes Φ is replaced by something related to it, such as raw water turbidity of coagulant dosage [O'Melia, 1972; Ives, 1968; Culp/Wesner/Culp, Inc., 1986]. This recommendation applies mainly to upflow contact clarifiers in which the floc volume concentration can be manipulated.

The average absolute velocity gradient employed in the flocculation tanks studied by Camp ranged from 20 to 74 per second, and the median value was about 40 per second; hydraulic retention times ranged from 10 to 100 minutes, and the median value was 25 minutes [Camp, 1955]. Both the $\overline{\Gamma}$ values and HRTs are somewhat smaller than current practice. Following the practice of Langelier [1921], who introduced mechanical flocculators, most existing flocculators are designed with tapered power inputs. This practice is supposed to increase the settling velocity of the flocs produced. In a study of the coagulation of colloidal silica with alum, TeKippe and Ham [1971] showed that tapered flocculation indeed produced the fastest settling floc. Their best results were obtained with a flocculator divided into four equal compartments, each having a hydraulic retention time of 5 minutes, and $\overline{\Gamma}$ values of 140, 90, 70, and 50 per second, respectively. The $\overline{\Gamma}\theta_h$ product was 105,000. A commonly recommended design for flocculators that precede settlers calls for a Camp number between 30,000 and 60,000, and HRT of at least 1000 to 1500 seconds (at 20°C and maximum plant flow) and a tapered rms characteristic strain rate ranging from about 60 per second in the first compartment to 10 per second in the last compartment [Joint Task Force, 1990]. For direct filtration, smaller flocs are desired, and the Camp number is increased to 40,000 to 75,000, the detention time is between 900 and 1500 seconds, and the rms characteristic strain rate is tapered from 75 per second to 20 per second.

None of these recommendations is fully in accord with the kinetic model or the empirical data. They ignore the effect of the size distribution factor, which causes the flocculation rate for primary particles to vary by a factor of at least four, and which is itself affected by mixing power and configuration. The consequence of this omission is that different flocculators designed for the same $\overline{\Gamma}\theta_h$ or $\overline{\Gamma}\Phi\theta_h$ will produce different results if either the mixing power distributions or tank configurations are different. Also, pilot data obtained at one set of mixing powers or tank configurations cannot be extrapolated to others. The recommendations quoted above merely indicate in a general way the things that require attention. In every case, flocculator design requires a special pilot plant study to determine the best combination of coagulant dosage, tank configuration, and power distribution.

Finally, the data of Argaman and Kaufman [1970] suggest that at any given average characteristic strain rate there is an optimum flocculator hydraulic retention time, and conversely. The existence of an optimum HRT has also been reported by Hudson [1973] and by Griffith and Williams [1972]. This optimum HRT is not predicted by the Argaman–Kaufman model; Eq. (8.133) predicts that the degree of flocculation will increase uniformly as θ_h increases. Using an alum/kaolin suspension and a completely mixed flocculator, Andreu-Villegas and Letterman [1976] showed that the conditions for optimum flocculation were approximately

$$\overline{\Gamma}^{2.8} C\,\theta_h = 44 \times 10^5 (\text{mg min/L s}^{2.8}) \tag{8.134}$$

The Andreu-Villegas–Letterman equation gives optimum $\overline{\Gamma}$ and HRT values that are low compared to most other reports. In one study, when the $\overline{\Gamma}$ values were tapered from 182 to 16 per second in flocculators with both paddles and stators, the optimum mixing times were 30 to 40 minutes [Wagner, 1974].

8.5 Sedimentation

Kinds of Sedimentation

Four distinct kinds of sedimentation processes are recognized:

- **Free settling.** When discrete particles settle independently of each other and the tank walls, the process is called *free, unhindered, discrete, type I*, or *class I* settling. This is a limiting case for very dilute suspensions of noninteracting particles. It is unlikely that free settling ever occurs in purification plants, but its theory is simple and serves as a starting point for more realistic analyses.

- **Flocculent settling.** In *flocculent, class II*, or *type II* settling, the particle agglomeration process continues in the clarifier. Because the velocity gradients in clarifiers are small, the particle collisions are due primarily to the differences in the particle settling velocities. Aside from the collisions, and the resulting flocculation, there are no interactions between particles or between particles and the tank wall. This is probably the most common settling process in treatment plants designed for turbidity removal.
- **Hindered settling.** *Hindered, type III, class III*, or *zone* settling occurs whenever the particle concentration is high enough that particles are influenced either by the hydrodynamic wakes of their neighbors or by the counterflow of the displaced water. When observed, the process looks much like a slowly shrinking lattice, with the particles representing the lattice points. The rate of sedimentation is dependent on the local particle concentration. Hindered settling is the usual phenomenon in lime/soda softening plants, upflow contact clarifiers, secondary clarifiers in sewage treatment plants, and sludge thickeners.
- **Compressive settling.** *Compressive* settling is the final stage of sludge thickening. It occurs in sludge storage lagoons. It also occurs in batch thickeners, if the sludge is left in them long enough. In this process, the bottom particles are in contact with the tank or lagoon floor, and the others are supported by mutual contact. A slow compaction process takes place as water is exuded from between and within the particles, and the particle lattice collapses.

Kinds of Settling Tanks

There are several kinds of settlement tanks in use:

- **Conventional.** The most common are the *conventional rectangular* and *conventional circular* tanks. *Rectangular* and *circular* refer to the tank's shape in plan sections. In each of the designs, the water flow is horizontal, and the particles settle vertically relative to the water flow (but at an angle relative to the horizontal). The settling process is either free settling or flocculent settling.
- **Tube, tray**, or **high-rate.** Sometimes, sedimentation tanks are built with an internal system of baffles, which are intended to regulate the hydraulic regime and impose ideal flow conditions on the tank. Such tanks are called *tube, tray*, or *high-rate* settling tanks. The water flow is parallel to the plane of the baffles, and the particle paths form some angle with the flow. The settling process is either free settling or flocculent settling.
- **Upflow.** In *upflow contact clarifiers*, the water flow is upward, and the particles settle downward. The rise velocity of the water is adjusted so that it is equal to the settling velocity of the particles, and a *sludge blanket* is trapped within the clarifier. The settling process in an upflow clarifier is hindered settling, and the design methodology is more akin to that of thickeners.
- **Thickeners.** *Thickeners* are tanks designed to further concentrate the sludges collected from settling tanks. They are employed when sludges require some moderate degree of dewatering prior to final disposal, transport, or further treatment. They look like conventional rectangular or circular settling tanks, except they contain mixing devices. The settling process is hindered settling.
- **Flotation.** Finally, there are *flotation tanks*. In these tanks the particles rise upward through the water, and they may be thought of as upside-down settling tanks. Obviously, the particle density must be less than the density of water, so the particles can float. Oils and greases are good candidates for flotation. However, it is possible to attach air bubbles to almost any

particle, so almost any particle can be removed in a flotation tank. The process of attaching air bubbles to particles is called *dissolved air flotation*. Flotation tanks can be designed for mere particle removal or for sludge thickening.

Floc Properties

The most important property of the floc is its settling velocity. Actually, coagulation/flocculation produces flocs with a wide range of settling velocities, and plant performance is best judged by the velocity distribution curve. The slowest flocs control settling tank design. In good plants, one can expect the slowest 5% by weight of the flocs to have settling velocities less than about 2 to 10 cm/min. The higher velocity is found in plants with high raw water turbities, because the degree of flocculation increases with floc volume concentration. The slowest 2% by weight will have velocities less than roughly 0.5 to 3 cm/min. Poor plants will produce flocs that are very much slower.

Alum/clay floc sizes range from a few hundredths of a mm to a few mm [Boadway, 1978; Dick, 1970; Lagvankar and Gemmell, 1968; Parker *et al.*, 1972; Tambo and Hozumi, 1979; Tambo and Watanabe, 1979]. A typical median floc diameter for alum coagulation might be a few tenths of a millimeter; the largest diameter might be 10 times as large. Ferric iron/clay flocs are generally larger than alum flocs [Ham and Christman, 1969; Parker *et al.*, 1972]. Floc size is correlated with the mixing power and the suspended solids concentration. Relationships of the following form have been reported [Boadway, 1978; François, 1987]:

$$d_p \propto \frac{X^b}{\overline{\Gamma}^a} \tag{8.135}$$

where

a = a constant ranging from 0.2 to 1.5 (dimensionless)
b = a constant (dimensionless)
d_p = the minimum, median, mean, or maximum floc size (m or ft)
X = the suspended solids concentration (kg/m^3 or lb/ft^3)
$\overline{\Gamma}$ = the rms characteristic strain rate (per second)

Equation (8.135) applies to all parts of the floc size distribution curve, including the largest observed floc diameter, the median floc diameter, and so on. The floc size distribution is controlled by the forces in the immediate vicinity of the mixer, and these forces are dependent on the geometry of the mixing device [François, 1987]. This makes the reported values of the coefficients highly variable, and, although good correlations may be developed for a particular facility or laboratory apparatus, the correlations cannot be transferred to other plants or devices unless the conditions are identical.

When flocs grown at one rms velocity gradient are transferred to a higher one, they become smaller. The breakdown process takes less than a minute [Boadway, 1978]. If the gradient is subsequently reduced to its former values, the flocs will regrow, but the regrown flocs are weaker and smaller than the originals [François, 1987].

Flocs consist of a combination of silt/clay particles, the crystalline products of the coagulant and entrained water. The specific gravities of aluminum hydroxide and ferric hydroxide crystals are about 2.4 and 3.4, respectively, and the specific gravities of most silts and clays are about 2.65 [Hudson, 1972]. However, the lattice of solid particles is very loose, and nearly all of the floc mass is due to entrained water. Consequently, the mass density of alum/clay flocs ranges from 1.002 to 1.010 g/cm^3, and the density of iron/clay flocs ranges from 1.004 to 1.040 g/cm^3 [Lagvankar and Gemmell, 1968; Tambo and Watanabe, 1979]. With both coagulants, density decreases with floc diameter and mixing power. Typically,

$$\rho_p - \rho \propto \frac{1}{d_p^a} \tag{8.136}$$

Free Settling

Free settling includes both nonflocculent and flocculent settling.

Calculation of the Free, Nonflocculent Settling Velocity

Under the quiescent conditions in settling tanks, particles quickly reach a constant, so-called terminal settling velocity, and Bassett's [1888] equation for the force balance on a particle becomes:

$$\underbrace{0}_{\text{Change in momentum}} = \underbrace{\rho_p V_p g}_{\text{Particle weight}} - \underbrace{\rho V_p g}_{\text{Buoyant force}} - \underbrace{C_D \rho A_p \frac{v_s^2}{2g}}_{\text{Drag force}} \tag{8.137}$$

$$v_s = \sqrt{\frac{2g V_p(\rho_p - \rho)}{C_D \rho A_p}} \tag{8.138}$$

where

A_p = the cross-sectional area of the particle normal to the direction of fall (m^2 or ft^2)
C_D = the drag coefficient (dimensionless)
g = the acceleration due to gravity (9.80665 m/s^2 or 32.174 ft/s^2)
V_p = the volume of the particle (m^3 or ft^3)
v_s = the terminal settling velocity of the particle (m/s or ft/s)
ρ = the density of the liquid (kg/m^3 or $slug/ft^3$)
ρ_p = the density of the particle (kg/m^3 or $slug/ft^3$)

For a sphere, Eq. (8.138) becomes

$$v_s = \sqrt{\frac{4g d_p(\rho_p - \rho)}{3C_D \rho}} \tag{8.139}$$

where d_p is the particle's diameter (m or ft).

Newton assumed that the drag coefficient was a constant, and indeed if the particle is moving very quickly it is a constant, with a value of about 0.44 for spheres. However, in general the drag coefficient depends on the size, shape, and velocity of the particle. It is usually expressed as a function of the particle Reynolds number. For spheres, the definition is

$$\text{Re} = \frac{\rho v_s d_p}{\mu} \tag{8.140}$$

The empirical correlations for C_D and Re for spheres are shown in Fig 8.6 [Rouse, 1937]. Different portions of the empirical curve may be represented by the following theoretical and empirical formulas.

Theoretical Formula. Stokes [1856] (for Re < 0.1),

$$C_D = \frac{24}{\text{Re}} \tag{8.141}$$

For spheres, the Stokes terminal settling velocity is

$$v_s = \frac{g(\rho_p - \rho)d_p^2}{18\mu} \tag{8.142}$$

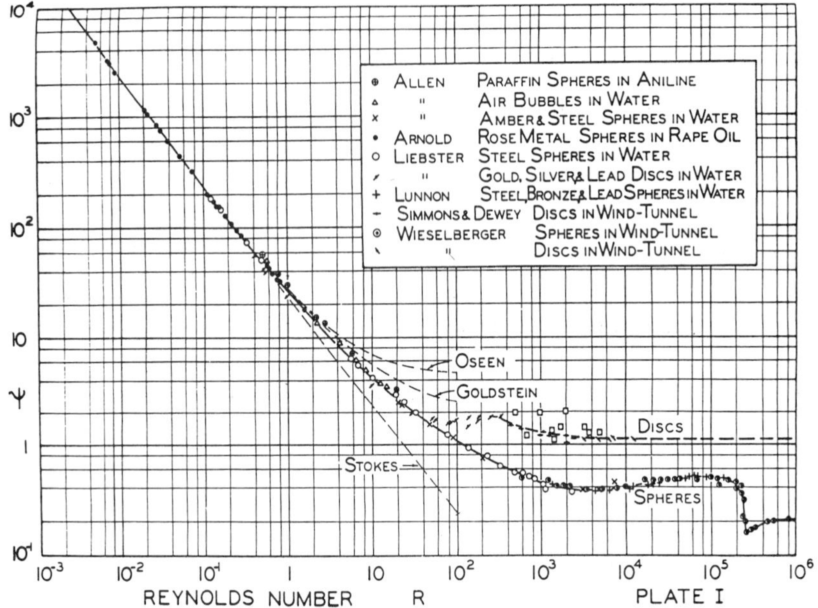

FIGURE 8.6 Drag coefficients for sedimentation. (*Source:* Rouse, H. 1937. Nomogram for the settling velocity of spheres. In *Report of the Committee on Sedimentation*, p. 57, P. D. Trask, chm., National Research Council, Division of Geology and Geography, Washington, D.C.)

Oseen [1913]–Burgess [1916] (for Re < 1),

$$C_D = \frac{24}{\text{Re}} \cdot \left(1 + \frac{3\text{Re}}{16}\right) \tag{8.143}$$

Goldstein [1929] (for Re < 2),

$$C_D = \frac{24}{\text{Re}} \cdot \left[1 + \frac{3\text{Re}}{16} - \frac{19\text{Re}^2}{1280} + \frac{71\text{Re}^3}{20{,}480} - \frac{30{,}179\text{Re}^4}{34{,}406{,}400} + \frac{122{,}519\text{Re}^5}{560{,}742{,}400} - \cdots\right] \tag{8.144}$$

Empirical Formulas. Allen [1900] (for $10 < \text{Re}_{\text{eff}} < 200$),

$$C_D = \frac{10.7}{\sqrt{\text{Re}_{\text{eff}}}} \tag{8.145}$$

The Reynolds number in Eq. (8.145) is based on an effective particle diameter:

$$\text{Re}_{\text{eff}} = \frac{\rho v_s d_{\text{eff}}}{\mu} \tag{8.146}$$

where

d_p = the actual particle diameter (m or ft)
d_2 = the diameter of a sphere that settles at a Reynold's number of 2 (m or ft)
d_{eff} = the effective particle diameter (m or ft)
$= d_p - 0.40d_2$.

The effective diameter was introduced by Allen to improve the curve fit. The definition of d_2 was arbitrary: Stokes' law does not apply at a Reynolds number of 2. For spheres, the Allen terminal settling velocity is

$$v_s = 0.25 d_{\text{eff}} \left[\frac{g(\rho_p - \rho)}{\sqrt{\rho\mu}} \right]^{2/3} \tag{8.147}$$

Shepherd [Anderson, 1941] (for 1.9 < Re < 500),

$$C_D = \frac{18.5}{\text{Re}^{0.60}} \tag{8.148}$$

For spheres, the Shepherd terminal settling velocity is

$$v_s = 0.153 d_p^{1.143} \left[\frac{g(\rho_p - \rho)}{\rho^{0.40} \mu^{0.60}} \right]^{0.714} \tag{8.149}$$

McGaughey [1956]: Examination of Fig. 8.6 shows that the slope of the curve varies from −1 to 0 as the Reynolds number increases from about 0.5 to about 4000. This is the transition region between the laminar Stokes' law and the fully turbulent Newton's law. For this region, the drag coefficient may be generalized as follows:

$$C_D = k\text{Re}^{n-2} \tag{8.150}$$

$$v_s^n = \frac{4g}{3k} d_p^{3-n} \left(\frac{\rho_p - \rho}{\rho^{n-1} \cdot \mu^{2-n}} \right) \tag{8.151}$$

where

k = a dimensionless curve-fitting constant ranging in value from 24 to 0.44

n = a dimensionless curve-fitting constant ranging in value from 1 to 2

Equation (8.150) represents the transition region as a series of straight line segments. Each segment will be accurate for only a limited range of Reynolds numbers.

Fair–Geyer [1954] (for Re < 10^4),

$$C_D = \frac{24}{\text{Re}} + \frac{3}{\sqrt{\text{Re}}} + 0.34 \tag{8.152}$$

Newton [Anderson, 1941] (for 500 < Re < 200,000),

$$C_D = 0.44 \tag{8.153}$$

For spheres, the Newton terminal settling velocity is

$$v_s = 1.74 \sqrt{\frac{g d_p (\rho_p - \rho)}{\rho}} \tag{8.154}$$

Referring to Fig. 8.6, it can be seen that there is a sharp discontinuity in the drag coefficient for spheres at a Reynolds number of about 200,000. The discontinuity is caused by the surface roughness of the particles and turbulence in the surrounding liquid. It is not important, because Reynolds numbers this large are never encountered in water treatment.

A sphere with the properties of a median alum floc (d_p = 0.50 mm and S_p = 1.005) would have a settling velocity of about 0.5 mm/s and a Reynolds number of about 0.2 (at 10°C). Most floc particles are smaller than 0.5 mm, so they will be slower and have smaller Reynolds numbers. This means that Stokes' law is an acceptable approximation in most cases of alum/clay floc sedimentation.

For sand grains, the Reynolds number is well into the transition region, and the Fair–Geyer formula is preferred. If reduced accuracy is acceptable, one of the Allen or Shepherd formulas may be used.

Except for the Stokes, Newton, Allen, and Shepherd laws, the calculation of the terminal settling velocity is iterative, because the drag coefficient is a polynomial function of the velocity. Graphical solutions are presented by Camp [1936a], Fair and Geyer [1954], and Anderson [1941].

Nonspherical particles can be characterized by the ratios of their diameters measured along their principal axes of rotation. Spherical and nonspherical particles are said to be equivalent if they have the same volume and weight. If Re is less than 100 and the ratios are 1:1:1 (as in a cube), the nonspherical particles settle at 90 to 100% of the velocity of the equivalent sphere [Task Committee, 1975]. For ratios of 4:1:1, 4:2:1, or 4:4:1, the velocity of the nonspherical particle is about two-thirds the velocity of the equivalent sphere. If the ratios are increased to 8:1:1, 8:2:1, or 8:4:1, the settling velocity of the nonspherical particles falls to a little more than half that of the equivalent sphere. The shape problem is lessened by the fact that floc particles are formed by the drag force into roughly spherical or teardrop shapes.

In practice, Eqs. (8.142) through (8.154) are almost never used to calculate settling velocities. The reason for this is the onerous experimental and computational workload their use requires. Floc particles come in a wide range of sizes, and the determination of the size distribution would require an extensive experimental program. Moreover, the specific gravity of each size class would be needed. In the face of this projected effort, it is much easier to measure settling velocities directly using a method like Seddon's, which is described below.

The settling velocity equations are useful when experimental data obtained under one set of conditions must be extrapolated to another. For example, terminal settling velocities depend on water temperature, because temperature strongly affects viscosity. The ratio of water viscosities for 0°C and 30°C, which is the typical range of raw water temperatures in the temperature zone, is about 2.24. This means that a settling tank designed for winter conditions will be between 1.50 and 2.24 times as big as a tank designed for summer conditions, depending on the Reynolds number.

The terminal velocity also varies with particle diameter and specific gravity. Because particle size and density are inversely correlated, increases in diameter tend to be offset by decreases in specific gravity, and some intermediate particle size will have the fastest settling velocity. This is the reason for the traditional advice that pinhead flocs are best.

Settling Velocity Measurement

The distribution of particle settling velocities can be determined by the method first described by Seddon [1889] and further developed by Camp [1945]. Tests for the measurement of settling velocities must be continued for at least as long as the intended settling zone detention time. Furthermore, samples must be collected at several time intervals in order to determine whether the concentration trajectories are linear or concave downward.

A vertical tube is filled from the bottom with a representative sample of the water leaving the flocculation tank, or any other suspension of interest. The depth of water in the tube should be at least equal to the expected depth of the settling zone, and there should be several sampling ports at different depths. The tube and the sample in it should be kept at a constant, uniform temperature to avoid the development of convection currents. A tube diameter at least 100 times the diameter of the largest particle is needed to avoid measurable "wall effects" [Dryden *et al.*, 1956]. The effect of smaller tube/particle diameter ratios can be estimated using McNown's [Task Committee, 1975] formula:

$$\frac{v_s}{v_t} = 1 + \frac{9d_p}{4d_t} + \left(\frac{9d_p}{4d_t}\right)^2 \tag{8.155}$$

where

d_p = the particle's diameter (m or ft)
d_t = the tube's diameter (m or ft)
v_s = the particle's free terminal settling velocity (m/s or ft/s)
v_t = the particle's settling velocity in the tube (m/s or ft/s)

The largest expected flocs are on the order of a few mm in diameter, so the minimum tube diameter will be tens of cm.

Initially, all the various particle velocity classes are distributed uniformly throughout the depth of the tube. Therefore, at any particular sampling time after the settling begins, say, t_i, the particles sampled at a distance h below the water surface must be settling at a velocity less than

$$v_i = \frac{h_i}{t_i} \tag{8.156}$$

where

h_i = the depth below the water's surface of the sampling port for the ith sample (m or ft)
t_i = the ith sampling time (s)
v_i = the limiting velocity for the particles sampled at t_i (m/s or ft/s)

The weight fraction of particles that are slower than this is simply the concentration of particles in the sample divided by the initial particle concentration:

$$P_i = \frac{X_i}{X_o} \tag{8.157}$$

where

P_i = the weight fraction of the suspended solids that settle more slowly than v_i (dimensionless)
X_i = the suspended solids concentration in the sample collected at t_i (kg/m^3 or lb/ft^3)
X_o = the initial, homogeneous suspended solids concentration in the tube (kg/m^3 or lb/ft^3)

The results of a settling column test would look something like the data in Fig. 8.7, which are taken from Seddon's paper. The data represent the velocities of river muds, which are slower than alum or iron flocs.

Flocculent versus Nonflocculent Settling

In nonflocculent settling, the sizes and velocities of the particles do not change. Consequently, if the trajectory of a particular concentration is plotted on depth–time axes, a straight line is obtained. In flocculent settling, the particles grow and accelerate as the settling test progresses, and the concentration trajectories are concave downward. These two possibilities are shown in Fig. 8.8. The same effect is seen in Fig. 8.7, where the samples from 8 ft, which represent longer sampling times, yield higher settling velocities than the 2-ft samples.

Design of Rectangular Clarifiers

What follows is the Hazen–Camp [Hazen, 1904; Camp, 1936a, 1936b] theory of free settling in rectangular tanks. Refer to Fig. 8.9. The liquid volume of the tank is divided into four zones: (1) an inlet or dispersion zone, (2) a settling zone, (3) a sludge zone, and (4) an outlet or collection zone [Camp, 1936b].

The inlet zone is supposed to be constructed so that each velocity class of the incoming suspended particles is uniformly distributed over the tank's transverse vertical cross section. A homogeneous

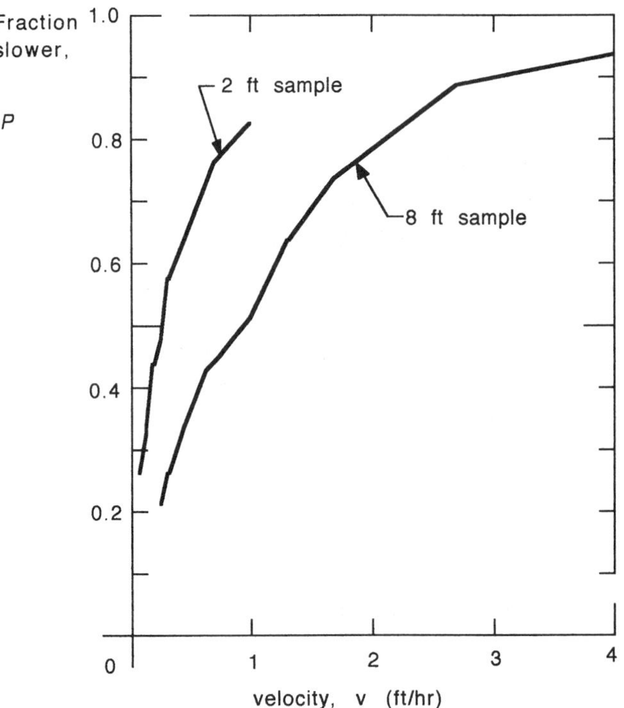

FIGURE 8.7 Settling velocity distributions for Mississippi River sediments. (Data from Seddon, J. A. 1889. Clearing water by settlement, *Journal of the Association of Engineering Societies*, 8(10):477.)

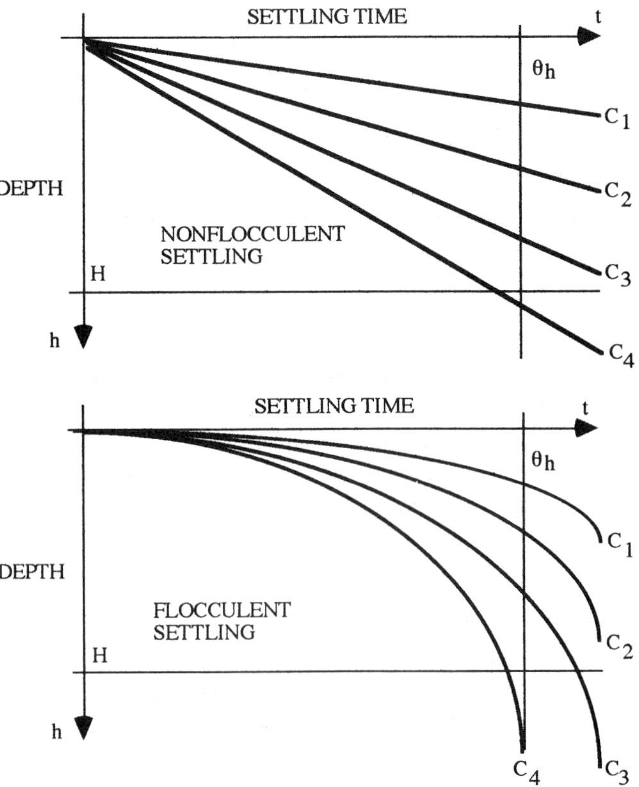

FIGURE 8.8 Particle depth–time trajectories for flocculent and nonflocculent free settling.

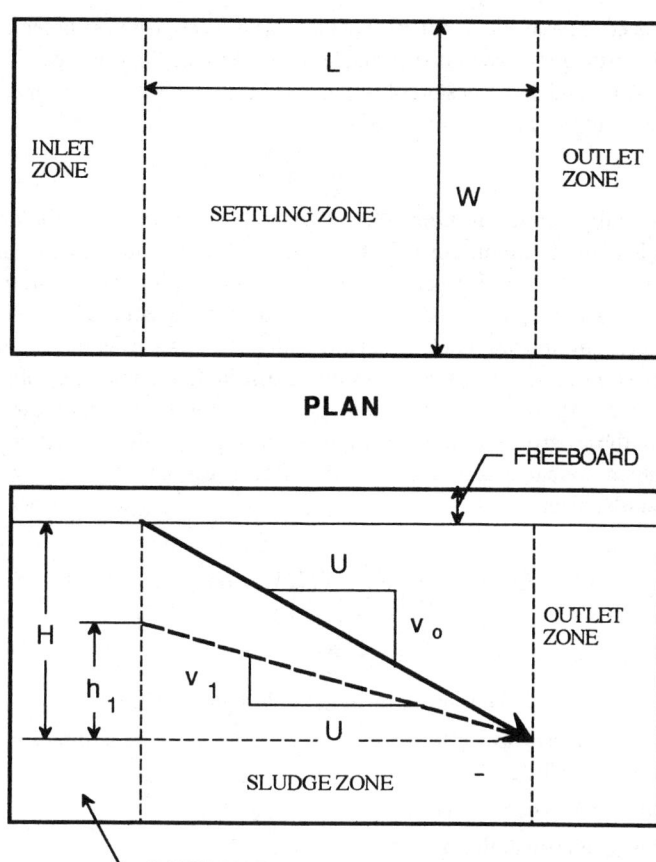

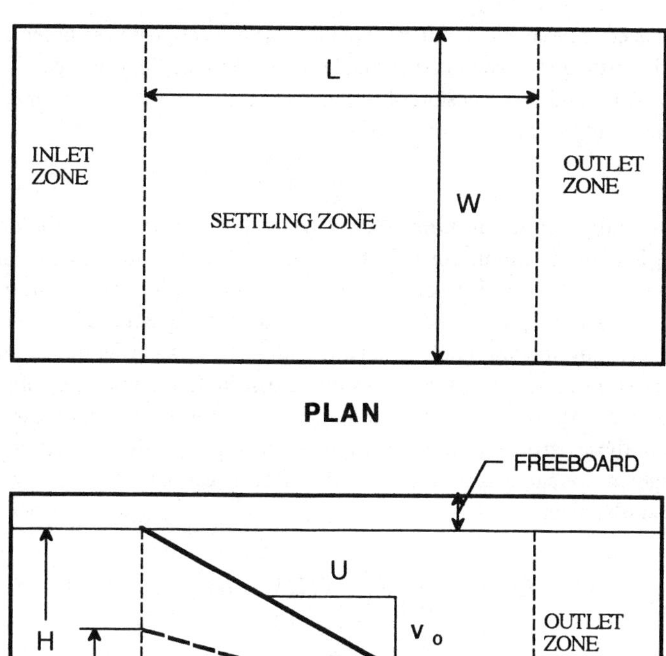

FIGURE 8.9 Settling tank definition sketch. (After Camp, T. R. 1936a. A study of the rational design of settling tanks, *Sewage Works Journal*, 8(5):742.)

distribution is achieved in many water treatment plants as a consequence of the way flocculation tanks and rectangular settling tanks must be connected. It is not achieved in circular tanks or in rectangular tanks in sewage treatment plants, unless special designs are adopted.

The sludge zone contains accumulated sludge and sludge removal equipment. Particles that enter the sludge zone are assumed to remain there until removed by the sludge removal equipment. Scour of the sludge must be prevented. Sludge thickening and compression also occur in the sludge zone, but this is not normally considered in clarifier design.

The water flow through the settling zone is supposed to be laminar and horizontal, and the water velocity is supposed to be the same at each depth. Laminar flow is readily achievable by baffling, but the resulting water velocities are not uniform, and often they are not horizontal. The corrections needed for these conditions are discussed below.

The outlet zone contains the mechanisms for removing the clarified water from the tank. Almost always, water is removed from near the water surface, so the water velocity in the outlet zone has a net upward component. This upward component may exceed the settling velocities of the slowest particles, so it is assumed that any particle that enters the outlet zone escapes in the tank effluent.

Five problems need to be considered in rectangular clarifier design: (1) sedimentation of the floc particles out of the settling zone and into the sludge zone, (2) the settling velocity distribution of the floc particles, (3) scour and resuspension of settled particles from the sludge zone, (4) turbulence in the settling zone, and (5) short-circuiting of the flow.

Sedimentation

Suppose that the settling occurs in Camp's ideal rectangular clarifier, and the horizontal water velocity is everywhere steady and uniform. In this case, column test sampling time may be equated with time of travel through the settling zone, and the depth–time plot of concentration trajectories in a settling column is also a plot of the trajectories in the settling zone. The settling zone can be represented on Fig. 8.8 by drawing a horizontal line at the settling zone depth, H, and a vertical line at its hydraulic retention time, θ_h. Trajectories that cross the horizontal line before θ_h represent those particles that are captured and enter the sludge zone. Trajectories that cross the vertical line above H represent those particles that escape capture and appear in the tank effluent.

The flow-weighted average concentration in the effluent can be estimated by a simple depth average over the outlet plane:

$$\overline{C}_H = \frac{1}{Q}\int_0^H C \cdot dQ = \frac{1}{UWH}\int_0^H CUW \cdot dh = \frac{1}{H}\int_0^H C \cdot dh \tag{8.158}$$

where

C = the suspended solids' concentration at depth h (kg/m^3 or lb/ft^3)
$\overline{C}_H$ = the depth-averaged suspended solids' concentration over the settling zone outlet's outlet plane (kg/m^3 or lb/ft^3)
H = the settling zone's depth (m or ft)
L = the settling zone's length (m or ft)
Q = the flow through the settling zone (m^3/s or ft^3/s)
U = the horizontal water velocity through the settling zone (m/s or ft/s)
W = the settling zone's width (m or ft)

The settling zone particle removal efficiency is [San, 1989]

$$E = \frac{C_i - \overline{C}_H}{C_i} = \frac{1}{H}\int_0^H f \cdot dh \tag{8.159}$$

where

C_i = the suspended solids' concentration in the influent flow to the settling zone (kg/m^3 or lb/ft^3)
E = the removal efficiency (dimensionless)
f = the fraction removed at each depth at the outlet plane of the settling zone (dimensionless)

The removal efficiency can also be calculated from the settling velocity distribution, such as Fig. 8.7 [Sykes, 1993]:

$$E = 1 - P_o + \frac{1}{v_o}\int_0^{P_o} v \cdot dP \tag{8.160}$$

where

P = the weight fraction of the suspended solids that settles more slowly than v (dimensionless)
P_o = the weight fraction of the suspended solids that settles more slowly than v_0 (dimensionless)

v = any settling velocity between zero and v_0 (m/s of ft/s)
v_o = the tank overflow rate (m/s or ft/s)

Note that the integration is performed along the vertical axis, and the integral represents the area between the velocity distribution curve and the vertical axis. The limiting velocities are calculated using Eq. (8.156) by holding the sampling time t_i constant at θ_h and using concentration data from different depths. Equation (8.160) reduces to the formulas used by Zanoni and Blomquist [1975] to calculate removal efficiencies for flocculating suspensions.

Nonflocculent Settling. Certain simplifications occur if the settling process is nonflocculent. However, nonflocculent settling is probably limited to grit chambers [Camp, 1953]. Referring to Fig. 8.9, the critical particle trajectory goes from the top of the settling zone on the inlet end to the bottom of the settling zone on the outlet end. The settling velocity that produces this trajectory is v_o. Any particle that settles at v_o or faster is removed from the flow. A particle that settles more slowly than v_o, say, v_1, is removed only if it begins its trajectory within h_1 of the settling zone's bottom. Similar triangles show that this critical settling velocity is also the tank overflow rate and the ratio of the settling zone depth to hydraulic retention time:

$$v_o = \frac{Q}{WL} = \frac{H}{\theta_h} \tag{8.161}$$

where v_o is the critical particles' settling velocity (m/s or ft/s).

In nonflocculent free settling, the settling velocity distribution does not change with settling time, and the efficiency predicted by Eq. (8.160) depends only on overflow rate, regardless of the combination of depth and detention time that produces it. Increasing the overflow rate always reduces the efficiency, and reducing the overflow rate always increases it.

Flocculent Settling. This is not the case in flocculent settling. Consider a point on one of the curvilinear contours shown in Fig. 8.8. Equation (8.156) now represents the average velocity of the selected concentration up to the selected instant. In the case of the contour that exits the settling zone at its bottom, the slope of the chord is v_o. The average velocity of any given contour increases as the sampling time increases. This has several consequences:

- Fixing v_0 and increasing both H and θ_h increases the efficiency.
- Fixing v_0 and reducing both H and θ_h reduces the efficiency.
- Increasing v_0 by increasing both H and θ_h may increase, decrease, or not change the efficiency, depending on the trajectories of the contours; if the curvature of the contours is large, efficiency increases will occur for new overflow rates that are close to the original v_0.
- Reducing v_0 by reducing both H and θ_h may increase, decrease, or not change the efficiency, depending on the trajectories of the contours; if the curvature of the contours is large, efficiency reductions will occur for new overflow rates that are close to the original v_0.

Any other method of either increasing or decreasing v_0 yields the same results as those obtained in nonflocculent settling.

Scour

Settling zone depth is important. The meaning of Eq. (8.161) is merely that if a particle's settling velocity is equal to the overflow rate it will reach the top of the sludge zone. However, once there, it still may be scoured from the sludge zone and resuspended in the flow. Repeated depositions and scourings would gradually transport the particle through the tank and into the effluent flow, resulting in clarifier failure. Whether or not this happens depends on the depth-to-length ratio of the settling zone.

Camp [1942] assumes that the important variable is the average shearing stress in the settling zone/sludge zone interface. In the case of steady, uniform flow, the accelerating force due to the weight of the water is just balanced by retarding forces due to the shearing stresses on the channel walls and floor [Yalin, 1977]:

$$\tau_o = \gamma H S \tag{8.162}$$

The shearing stress depends on both the water velocity and the settling zone depth. The critical shearing stress is that which just initiates mass movement in the settled particles. Dimensional analysis suggests a correlation of the following form [Task Committee, 1975]:

$$\frac{\tau_c}{(\gamma_p - \gamma)d_p} = f(\mathrm{Re}_*) \tag{8.163}$$

where

d_p = the particle's diameter (m or ft)
Re_* = the boundary Reynolds number (dimensionless)
= $v_* d_p / \nu$
v_* = the shear velocity (m/s or ft/s)
= $\sqrt{\tau_c/\rho}$
γ = the specific weight of the liquid (N/m^3 or lbf/ft^3)
γ_p = the specific weight of the particle (N/m^3 or lbf/ft^3)
ν = the kinematic viscosity of the liquid (m^2/s or ft^2/s)
ρ = the density of the liquid (kg/m^3 or slug/ft^3)
τ_c = the critical shearing stress that just initiates particle movement (N/m^2 or lbf/ft^2)

Equation (8.163) applies to granular, noncohesive materials such as sand. Alum and iron flocs are cohesive. The cohesion is especially well developed on aging, and individual floc particles tend to merge together. However, a conservative assumption is that the top layer of the deposited floc, which is fresh, coheres weakly.

For the conditions in clarifiers, Equation (8.163) becomes [Mantz, 1977]

$$\frac{\tau_c}{(\gamma_p - \gamma)d_p} = \frac{0.1}{\mathrm{Re}_*^{0.30}} \tag{8.164}$$

This leads to relationships between the settling zone's depth and length. For example, the horizontal velocity is related to the critical shearing stress by [Chow, 1959]

$$U_c = \sqrt{8/f}\,\sqrt{\tau_c/\rho} \tag{8.165}$$

where

f = the Darcy–Weisbach friction factor (dimensionless)
U_c = the critical horizontal velocity that just initiates particle movement (m/s or ft/s)

Consequently,

$$\frac{H}{L} \geq \frac{v_o}{\sqrt{8/f}\,\sqrt{\tau_c/\rho}} = \text{Constant} \tag{8.166}$$

Equation (8.166) is a lower limit on the depth-to-length ratio.

Ingersoll *et al.* [1956] assume that the turbulent fluctuations in the water velocity at the interface cause scour. They hypothesize that the deposited flocs will be resuspended if the vertical fluctuations in the water velocity at the sludge interface are larger than the particle settling velocity. Using the data of Laufer [1950], they concluded that the vertical component of the root-mean-square velocity fluctuation of these eddies is approximately equal to the shear velocity, and they suggested that scour and resuspension will be prevented if

$$\frac{v_o}{\sqrt{\tau_o/\rho}} > 1.2 \text{ to } 2.0 \tag{8.167}$$

This leads to

$$\frac{H}{L} > (1.2 \text{ to } 2.0)\sqrt{f/8} \tag{8.168}$$

The Darcy–Weisbach friction factor for a clarifier is about 0.02, so the right-hand side of Eq. (8.168) is between 0.06 and 0.1, which means that the length must be *less than* 10 to 16 times the depth. Camp's criterion would permit a horizontal velocity that is 2 to 4 times as large as the velocity permitted by the Ingersoll–McKee–Brooks analysis.

Short-Circuiting

A tank is said to "short-circuit" if a large portion of the influent flow traverses a small portion of the tank's volume. In extreme cases, some of the tank's volume is a "dead zone" that neither receives nor discharges liquid. Two kinds of short-circuiting occur: *density currents* and *streaming.*

Density Currents. Density currents develop when the density of the influent liquid is significantly different from the density of the tank's contents. The result is that the influent flow either floats over the surface of the tank or sinks to its bottom. The two common causes of density differences are differences in (1) temperature and (2) suspended solids concentrations.

Temperature differences arise because the histories of the water bodies differ. For example, the influent flow may have been drawn from the lower portions of a reservoir, while the water in the tank may have been exposed to surface weather conditions for several hours. Small temperature differences can be significant.

Suspended solids have a similar effect. The specific gravity of a suspension can be calculated as follows [Fair *et al.*, 1954]:

$$S_s = \frac{S_p S_w}{f_w S_p + (1 - f_w) S_w} \tag{8.169}$$

where

f_w = the weight fraction of water in the suspension (dimensionless)
S_p = the specific gravity of the particles (dimensionless)
S_s = the specific gravity of the suspension (dimensionless)
S_w = the specific gravity of the water (dimensionless)

By convention all specific gravities are referenced to the density of pure water at 3.98°C, where it attains its maximum value. Equation (8.169) can be written in terms of the usual concentration units of mass/volume as follows:

$$S_s = S_w + \frac{X}{\rho} \cdot \left(1 - \frac{S_w}{S_p}\right) \tag{8.170}$$

where

X = the concentration of suspended solids (kg/m^3 or slug/ft^3)
ρ = the density of water (kg/m^3 or slug/ft^3)

Whether or not the density current has important effects depends on its location and its speed [Eliassen, 1946; Fitch, 1957]. If the influent liquid is lighter than the tank contents and spreads over the tank surface, any particle that settles out of the influent flow enters the stagnant water lying beneath the flow and settles vertically all the way to the tank bottom. During their transit of the stagnant zone, the particles are protected from scour, so once they leave the flow they are permanently removed from it. If the influent liquid spreads across the entire width of the settling zone, then the density current itself may be regarded as an ideal clarifier. In this case, the depth of the settling zone is irrelevant. Clearly, this kind of short-circuiting is desirable.

If the influent liquid is heavier than the tank contents, it will settle to the bottom of the tank. As long as the flow uniformly covers the entire bottom of the tank, the density current itself may be treated as an ideal clarifier. Now, however, the short-circuiting flow may scour sludge from the tank bottom. The likelihood of scour depends of the velocity of the density current. According to von Karman [1940], the density current velocity is

$$U_{dc} = \left[\frac{2gQ(\rho_{dc} - \rho)}{\rho W}\right]^{1/3} \tag{8.171}$$

where

g = the acceleration due to gravity (9.80665 m/s^2 or 32.174 ft/s^2)
Q = the flow rate of the density current (m^3/s or ft^3/s)
U_{dc} = the velocity of the density current (m/s or ft/s)
W = the width of the clarifier (m or ft)
ρ = the density of the liquid in the clarifier (kg/m^3 or slug/ft^3)
ρ_{dc} = the density of the density current (kg/m^3 or slug/ft^3)

Streaming. Density currents divide the tank into horizontal layers stacked one above the other. An alternative arrangement would be for the tank to be divided into vertical sections placed side by side. The sections would be oriented to run the length of the tank from inlet to outlet. Fitch [1956] calls this flow arrangement *streaming*.

Suppose that the flow consists of two parallel streams, each occupying one-half of the tank width. A fraction f of the flow traverses the left section, and a fraction $1 - f$ traverses the right section. Within each section, the flow is distributed uniformly over the depth. The effective overflow rate for the left-hand section is v_1 and a weight fraction P_1 settles slower than this. For the right-hand section the overflow rate is v_2, and the fraction slower is P_2. The flow-weighted average removal efficiency would be

$$\overline{E} = f \cdot \left(1 - P_1 + \frac{1}{v_1}\int_0^{P_1} v \cdot dP\right) + (1-f)\cdot\left(1 - P_2 + \frac{1}{v_2}\int_0^{P_2} v \cdot dP\right) \tag{8.172}$$

This can also be written as

$$\begin{aligned}\overline{E} &= 1 - P_o + \frac{1}{v_o}\int_0^{P_o} v \cdot dP + f(P_0 - P_1) \\ &\quad - \frac{1}{2v_o}\int_{P_1}^{P_o} v \cdot dP - (1-f)(P_2 - P_o) + \frac{1}{2v_o}\int_{P_o}^{P_2} v \cdot dP\end{aligned} \tag{8.173}$$

The first three terms on the right-hand side is the removal efficiency without streaming. The remaining terms represent corrections to the ideal removal efficiency due to streaming. Fitch shows that both corrections are always negative, and the effect of streaming is to reduce tank efficiency.

The degree of reduction depends on the shape of the velocity distribution curve and the design surface loading rate.

The mere fact that tanks have walls means that some streaming is inevitable: the drag exerted by the walls causes a lateral velocity distribution. However, the more important point is that the inlet and outlet conditions must be designed to achieve and maintain uniform lateral distribution of the flow. One design criterion used to achieve uniform lateral distribution is a length-to-width ratio. The traditional recommendation was that the length-to-width ratio should lie between 3:1 and 5:1 [Joint Committee, 1990]. However, length-to-width ratio restrictions are no longer recommended [Joint Task Force, 1969]. It is more important to prevent the formation of longitudinal jets. This can be done by proper design of inlet details.

The specification of a surface loading rate, a length-to-depth ratio, and a length-to-width ratio uniquely determines the dimensions of a rectangular clarifier and accounts for the chief hydraulic problems encountered in clarification.

Traditional Rules of Thumb

Some regulatory authorities specify a minimum hydraulic retention time, a maximum horizontal velocity, or both: for example, for water treatment [Water Supply Committee, 1987],

$$\theta_h = \frac{WHL}{Q} \geq 4 \text{ hr} \tag{8.174}$$

$$U = \frac{Q}{WH} \leq 0.5 \text{ fpm} \tag{8.175}$$

Hydraulic detention times for primary clarifiers in wastewater treatment tend to be 2 hours at peak flow [Joint Task Force, 1992].

Some current recommendations for the overflow rate for both rectangular and circular clarifiers for water treatment are as follows [Joint Task Force, 1990]. For lime/soda softening:

Low magnesium, v_o = 1700 gpd/ft^2
High magnesium, v_o = 1400 gpd/ft^2

For alum or iron coagulation:

Turbidity removal, v_o = 1000 gpd/ft^2
Color removal, v_o = 700 gpd/ft^2
Algae removal, v_o = 500 gpd/ft^2

The recommendations for wastewater treatment are as follows [Wastewater Committee, 1990]. For primary clarifiers:

Average flow, v_o = 1000 gpd/ft^2
Peak hourly flow, v_o = 1500 to 3000 gpd/f^2

For activated sludge clarifiers, at peak hourly flow, not counting recycles:

Conventional, v_o = 1200 gpd/ft^2
Extended aeration, v_o = 1000 gpd/ft^2
Second stage nitrification, v_o = 800 gpd/ft^2

For trickling filter humus tanks, at peak hourly flow:

v_o = 1200 gpd/ft^2

The traditional rule of thumb overflow rates for conventional rectangular clarifiers in alum coagulation plants are 0.25 gal/min·ft^2 in regions with cold winters and 0.38 gal/min·ft^2 in regions with mild winters. The Joint Task Force [1992] summarizes an extensive survey of the criteria used by numerous engineering companies for the design of wastewater clarifiers. The typical practice appears to be about 800 gpd/ft^2 at average flow for primary clarifiers and 600 gpd/ft^2 for secondary clarifiers. The latter rate is doubled for peak flow conditions.

Typical side water depths for all clarifiers is 10 to 16 ft. Secondary wastewater clarifiers should be designed at the high end of the range.

Hudson [1981] has reported that in manually cleaned, conventional clarifiers the sludge deposits often reach to within 30 cm of the water surface near the tank inlets, and scour velocities range from 3.5 to 40 cm/s. The sludge deposits in manually cleaned tanks are often quite old, at least beneath the surface layer, and the particles in the deposits are highly flocculated and "sticky." Also, the deposits are well compacted, because there is no mechanical collection device to stir them up. Consequently, Hudson's data represent the upper limits of scour resistance.

The limit on horizontal velocity, Camp's shearing stress criterion, and the Ingersoll–McKee–Brook velocity fluctuation criterion are different ways of representing the same phenomenon. In principle, all three criteria should be consistent and produce the same length-to-depth ratio limit. However, different workers have access to different data sets and draw somewhat different conclusions. Because scour is a major problem, a conservative criterion should be adopted. This means a relatively short length-to-depth ratio, which means a relatively deep tank.

The limits on tank overflow rate, horizontal velocity, length-to-depth ratio, and hydraulic retention time overdetermine the design; only three of them are needed to specify the dimensions of the settling zone. They may also be incompatible.

Design of Circular Tanks

In most designs, the suspension enters the tank at the center and flows radially to the circumference. Other designs reverse the flow direction, and in one proprietary design the flow enters along the circumference and follows a spiral path to the center.

The collection mechanisms in circular tanks have no bearings under water and are less subject to corrosion, reducing maintenance. However, center-feed tanks tend to exhibit streaming, especially in tank diameters larger than about 125 ft. Streaming is reduced in peripheral feed tanks and in center feed tanks with baffles [Joint Task Force, 1990].

The design principles used for rectangular tanks also apply, with a few exceptions, to circular tanks.

Free, Nonflocculent Settling

The analysis of particle trajectories for circular tanks is given by Fair *et al.* [1954] for center-feed, circular clarifiers. The trajectory of a freely settling, nonflocculent particle curves downward, because as the distance from the center-feed increases the horizontal water velocity decreases. At any point along the trajectory, the slope of the trajectory is given by the ratio of the settling velocity to the water velocity:

$$\frac{dz}{dr} = \frac{v_s}{U_r} = -\frac{2\pi H v_s}{Q} \tag{8.176}$$

where

H = the water depth in the settling zone (m or ft)
Q = the flow through the settling zone (m^3/s or ft^3/s)
r = the distance from the center of the tank (m or ft)
U_r = the horizontal water velocity at r (m/s or ft/s)
v_s = the particle's settling velocity (m/s or ft/s)
z = the elevation of the particle about the tank's bottom (m or ft)

The minus sign on the right-hand side is needed because the depth variable is positive upward and the particle is moving down.

Equation (8.176) can be solved for the critical case of a particle that enters the settling zone at the water surface and just reaches the bottom of the settling zone at its outlet cylinder:

$$v_o = \frac{Q}{\pi(r_o^2 - r_i^2)} \tag{8.177}$$

where

r_i = the radius of the inlet baffle (m or ft)

r_o = the radius of the outlet weir (m or ft)

The denominator in Eq. (8.177) is the plan area of the settling zone. Consequently, the critical settling velocity is equal to the settling zone overflow rate.

The analysis leading to Eq. (8.177) also applies to tanks with peripheral feed and central takeoff. The only change required is the deletion of the minus sign in the right-hand side of Eq. (8.176), because the direction of the flow is reversed, and the slope of the trajectory is positive in the given coordinate system. Furthermore, the analysis applies to spiral flow tanks, if the integration is performed along the spiral stream lines. Consequently, all horizontal flow tanks are governed by the same principle.

Free, Flocculent Settling

The trajectories of *nonflocculating* particles in a circular clarifier are curved in space. However, if the horizontal coordinate were the time of travel along the settling zone, the trajectories would be linear. This can be shown by a simple change of variable:

$$\frac{dz}{dt} = \frac{dz}{dr} \cdot \frac{dr}{dt} = \frac{v_s}{U_r} \cdot U_r = -v_s \tag{8.178}$$

This means that Fig. 8.8 also applies to circular tanks, if the horizontal coordinate is the time of travel. Furthermore, it applies to both flocculent and nonflocculent settling in circular tanks.

Equation (8.160) applies to circular tanks with uniform feed over the inlet depth, whether they be center-feed, peripheral-feed, or spiral-flow tanks.

Unfortunately, the inlet designs of most circular clarifiers do not produce a vertically uniform feeding pattern. Usually, all of the flow is injected over a small portion of the settling zone depth. As long as the flow is injected at the top of the settling zone, this does not change matters, but other arrangements may.

The one case where Camp's formula for clarifier efficiency does not apply is the upflow clarifier. However, this is also a case of hindered settling, and it will be discussed later.

The plan area of the circular settling zone is determined by the overflow rate, and this rate will be equal to the one used for a rectangular tank having the same efficiency. The design is completed by choosing either a settling zone depth or a detention time. No general analysis for the selection of these parameters has been published, and designers are usually guided by traditional rules of thumb. The traditional rule of thumb in the U.S. is a detention time of 2 to 4 hr [Joint Committee, 1990]. Many regulatory agencies insist on a 4-hr detention time [Water Supply Committee, 1987].

Design of High-Rate, Tube or Tray Clarifiers

High-rate settlers, also known as *tray* or *tube* settlers, are laminar flow devices. They eliminate turbulence, density currents, and streaming—and the problems associated with them. Their behavior is nearly ideal and predictable. Consequently, allowances for nonideal and uncertain behavior can be eliminated, and high-rate settlers can be made much smaller than conventional rectangular and circular clarifiers (hence their name).

Hayden [1914] published the first experimental study of the efficiency of high-rate clarifiers. His unit consisted of a more-or-less conventional rectangular settler containing a system of corrugated steel sheets. This is a form of Camp's tray clarifier. The sheets were installed 45° from the horizontal, so that particles that deposited on them would slide down the sheets into the collection hoppers below. The sheets were corrugated for structural stiffness. The high-rate clarifier had removal efficiencies that were 40 to 100% higher than simple rectangular clarifiers having the same geometry and dimensions and treating the same flow.

Nowadays, Hayden's corrugated sheets and the Hazen–Camp trays are replaced by modules made out of arrays of plastic tubes. The usual tube cross section is an area-filling polygon, such as the isosceles triangle, the hexagon, the square, and the chevron. Triangles, squares, and chevrons are preferred because alternate rows of tubes can be sloped in different directions, which stiffens the module and makes it self-supporting. When area-filling hexagons are used, the alternate rows interdigitate and must be strictly parallel. Alternate rows of hexagons can be sloped at different angles if the space-filling property is sacrificed.

Sedimentation

Consider a particle being transported along a tube that is inclined at an angle θ from the horizontal. Yao [1970, 1973] analyzes the situation as follows. Refer to Fig. 8.10. The coordinate axes are parallel and perpendicular to the tube axis. The trajectory of the particle along the tube will be the resultant of the particle's own settling velocity and the velocity of the water in the tube:

$$\frac{dx}{dt} = v_x = u - v_s \sin\theta \tag{8.179}$$

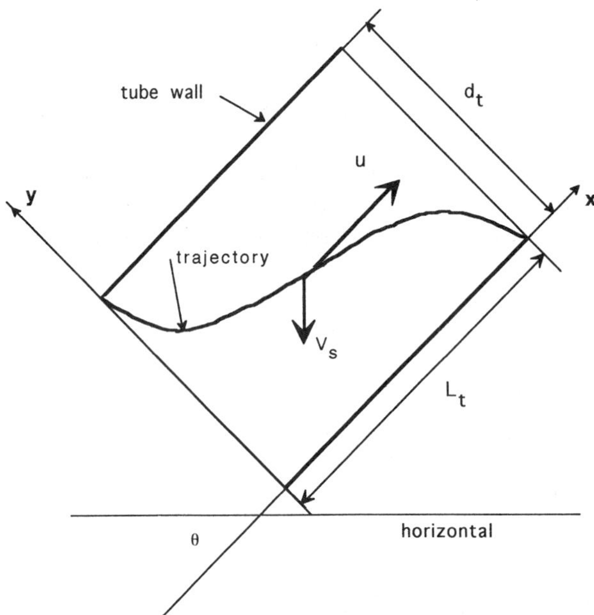

FIGURE 8.10 Trajectory of the critical particle in a tube. [After Yao, K. M. 1970. Theoretical study of high-rate sedimentation, *Journal of the Water Pollution Control Federation,* 42(2, part 1):218.]

$$\frac{dy}{dt} = v_y = -v_s \cos\theta \tag{8.180}$$

where

t = time (s)
u = the water velocity at a point in the tube (m/s or ft/s)
v_s = the particle's settling velocity (m/s or ft/s)
v_x = the particle's resultant velocity in the x direction (m/s or ft/s)
v_y = the particle's resultant velocity in the y direction (m/s or ft/s)
θ = the angle of the tube with the horizontal (rad)

Equations (8.179) and (8.180) may be combined to yield the differential equation of the trajectory of a particle transmitting the tube:

$$\frac{dy}{dx} = -\frac{v_s \cos\theta}{u - v_s \sin\theta} \tag{8.181}$$

The water velocity u varies from point to point across the tube cross section. For fully developed laminar flow in a circular tube, the distribution is a parabola [Rouse, 1978]:

$$\frac{u}{U_t} = 8\left(\frac{y}{d_t} - \frac{y^2}{d_t^2}\right) \tag{8.182}$$

where

d_t = the tube diameter (m or ft)
U_t = the mean water velocity in the tube (m/s or ft/s)
y = the distance from the tube invert (m or ft)

The critical trajectory begins at the crown of the tube and ends at its invert. Substituting for u and integrating between these two points yields the following relationship between the settling velocity of a particle that is just captured and the mean water velocity in a circular tube [Yao, 1970]:

$$\frac{v_s}{U_t} = \frac{4}{3}\Big/\left(\sin\theta + \frac{L_t}{d_t}\cdot\cos\theta\right) \tag{8.183}$$

where L_t is the effective length of the tube (m or ft). Yao [1970] has performed the same analysis for other cross section shapes, and the equations differ only in the numerical value of the numerator on the right-hand side. For laminar flow in square tubes:

$$\frac{v_s}{U_t} = \frac{11}{8}\Big/\left(\sin\theta + \frac{L_t}{a_t}\cdot\cos\theta\right) \tag{8.184}$$

where a is the length of one side of the square cross section (m or ft).

For laminar flow between parallel plates, for laminar flow over a plane, or for an idealized flow that has a uniform velocity everywhere:

$$\frac{v_s}{U_t} = 1\Big/\left(\sin\theta + \frac{L_t}{h_p}\cdot\cos\theta\right) \tag{8.185}$$

where h_p is the thickness of the flow (m or ft).

This critical velocity may be related to the tube overflow rate. The definition of the overflow rate for a square tube is

$$\text{Tube overflow rate} = \frac{\text{Flow through tube}}{\text{Horizontally projected area}}$$
$$v_{ot} = \frac{U_t a_t^2}{L_t a_t \cos\theta} = \frac{U_t}{(L_t/a_t)\cos\theta} \tag{8.186}$$

where v_{ot} is the tube overflow rate (m/s or ft/s). Therefore,

$$v_s = \frac{11}{8} v_{ot} \bigg/ \left(1 + \frac{L_t}{a_t} \cdot \tan\theta\right) \tag{8.187}$$

In the ideal Hazen–Camp clarifier, the critical settling velocity is equal to the tank overflow rate. This is not true for tubes. Tubes are always installed as arrays, and the critical velocity for the array is much less than the tank overflow rate. The number of tubes in such an array can be calculated as follows. First, the fraction of the settling zone surface occupied by the tube ends is less than the plan area of the zone, because the inclination of the tubes prevents full coverage. The area occupied by open tube ends that can accept flow is

$$A_o = W(L - L_t \cdot \cos\theta) \tag{8.188}$$

where

A_o = the area of the settling zone occupied by tube ends (m^2 or ft^2)
L = the length of the settling zone (m or ft)
L_t = the length of the tubes (m or ft)
W = the width of the settling zone (m or ft)
θ = the angle of the tubes with the horizontal plane (rad)

The number of square tubes in this area is

$$n_t = \frac{W(L - L_t \cos\theta)}{a_t^2/\sin\theta} \tag{8.189}$$

The effective plan area of the array of square tubes is simply the number of tubes times the plan area of a single tube:

$$A_{\text{eff}} = n_t L_t a_t \cos\theta = W(L - L_t \cos\theta)\frac{L_t}{a_t}\sin\theta\cos\theta \tag{8.190}$$

The overflow rate of the array of square tubes is

$$v_o = Q \bigg/ \left[W(L - L_t \cdot \cos\theta) \cdot \frac{L_t}{a_t} \cdot \sin\theta \cdot \cos\theta\right] \tag{8.191}$$

This is also the overflow rate of each tube in the array, if the flow is divided equally among them.

Assuming equal distribution of flow, the critical settling velocity may be related to the overflow rate of the array of square tubes by substituting Eq. (8.191) into (8.187):

$$v_s = \frac{11}{8} Q \bigg/ \left[\left(1 + \frac{L_t}{a_t}\tan\theta\right) W(L - L_t\cos\theta)\frac{L_t}{a_t}\sin\theta\cos\theta\right] \tag{8.192}$$

For parallel plates, the equivalent formula is

$$v_s = Q \bigg/ \left[\left(1 + \frac{L_t}{h_p}\tan\theta\right) W(L - L_t\cos\theta)\frac{L_t}{h_p}\sin\theta\cos\theta\right] \quad (8.193)$$

Tube Inlets

Near the tube inlet the velocity distribution is uniform, not parabolic:

$$u = U_t \quad (8.194)$$

The relationships just derived do not apply in the inlet region, and the effective length of the tubes should be diminished by the inlet length.

The distance required for the transition from a uniform to a parabolic distribution in a circular tube is given by the Schiller–Goldstein equation [Goldstein, 1965]:

$$L_{en} = 0.0575 r_t \mathrm{Re}_t \quad (8.195)$$

where

L_{en} = the length of the inlet region (m or ft)
Re_t = the tube's Reynold's number (dimensionless)
= $U_t d_t / \nu$
r_t = the tube's radius (m or ft)

Tube diameters are typically 2 in., and tube lengths are typically 2 ft (Culp/Wesner/Culp, Inc., 1986). The transition zone length near the inlet of a circular tube will be about 0.15 to 0.75 ft. Tubes are normally 2 ft long, depending on water temperature and velocity.

If the velocity remained uniform everywhere for the whole length of the tube, the formula for the critical velocity in a circular tube would be

$$\frac{v_s}{U_t} = 1 \bigg/ \left(\sin\theta + \frac{L_t}{d_t}\cdot\cos\theta\right) \quad (8.196)$$

Comparison with Eq. (8.183) shows that this is actually one-fourth less than the critical velocity for a fully developed parabolic velocity profile, so the effect of a nonuniform velocity field is to reduce the efficiency of the tube.

Performance Data

Long, narrow tubes with low water velocities perform best [Hansen and Culp, 1967]. However, water velocity is more important than tube diameter, and tube diameter is more important than tube length. For tubes that were 2 in. in diameter and 2 ft long, turbidity removal was seriously degraded when the water velocity exceeded 0.0045 ft/s.

Recommended tank overflow rates for tube settlers range from 2.5 to 4.0 gpm/ft^2 (Culp/Wesner/Culp, Inc., 1986). It is assumed that the tubes are 2 in. in diameter, 2 ft long, and inclined at 60°. For conventional rapid sand filters, a settled water turbidity of less than 3 TU is preferred, and the overflow rate should not exceed 2.5 gpm/ft^2 at temperatures above 50°F. A settled water turbidity of up to 5 TU is acceptable to dual media filters, and the overflow rate should not exceed 2.5 gpm/ft^2 at temperatures below 40°F and 3.0 gpm/ft^2 at temperatures above 50°F.

Clarifier Inlets

Conventional Clarifiers

The inlet zone should distribute the influent flow uniformly over both the depth and width of the settling zone. If both goals cannot be met, then uniform distribution across the width has priority, because streaming degrades tank performance more than density currents do. Several design principles should be followed:

- In order to prevent floc breakage, the rms characteristic strain rates in the influent channel, piping, and appurtenances must not exceed the velocity gradient in the final compartment of the flocculation tank. Alum/clay flocs and sludges should not be pumped or allowed to free-fall over weirs at any stage of their handling. This rule does not apply to the transfer of settled water to the filters [Hudson and Wolfner, 1967; Hudson, 1981].
- The influent flow should approach the inlet zone *parallel* to the longitudinal axis of the clarifier. Avoid side-overflow weirs. The flow in channels feeding such weirs is *normal* to the axis of the clarifier, and its momentum across the tank will cause a nonuniform lateral distribution. At low flows, most of the water will enter the clarifier from the upstream end of the influent channel, and at high flows most of it will enter the clarifier from the downstream end of the channel. These maldistributions occur even if baffles and orifice walls are installed in the inlet zone [Yee and Babb, 1985].
- A simple inlet pipe is unsatisfactory, even if it is centered on the tank axis and even if there is an orifice wall downstream of it, because the orifice wall will not dissipate the inlet jet. The inlet pipe must end in a tee that discharges horizontally, and an orifice wall should be installed downstream of the tee [Kleinschmidt, 1961]. A wall consisting of adjustable vertical vanes may be preferable to an orifice wall.

The orifice wall should have about 30 to 40% open area [Hudson, 1981]. The orifice diameters should be between 1/64th and 1/32nd of the smallest dimension of the wall, and the distance between the wall and the inlet tee should be approximately equal to the water depth. The headloss through the orifice wall should be about four times the approaching velocity head.

High-Rate Clarifiers

In a high-rate clarifier, the settling zone is the tube modules. The inlet zone consists of the following regions and appurtenances:

- A region near the tank inlet that contains the inlet pipe and tee, a solid baffle wall extending from above the water surface to below the tube modules, and, perhaps, an orifice wall below the solid baffle
- The water layer under the settling modules and above the sludge zone

The inlet pipe and tee are required for uniform lateral distribution of the flow. The solid baffle wall must deflect the flow under the tube modules. Horizontally, it extends completely across the tank. Vertically, it extends from a few inches above the maximum water level to the bottom of the tube modules. If the raw water contains significant amounts of floatables, the baffle wall should extend sufficiently above the maximum water surface to accommodate some sort of skimming device. The orifice wall extends across the tank and from the bottom of the baffle wall to the tank floor. The extension of the orifice wall to the tank bottom requires that special consideration be given to the sludge removal mechanism.

Most tube modules are installed in existing conventional clarifiers to increase their hydraulic capacity. The usual depth of submergence of tube modules in retrofitted tanks is 2 to 4 ft, in order to provide room for the outlet launders, and the modules are generally about 2 to 3 ft thick.

The water layer under the tube modules needs to be thick enough to prevent scour of the sludge deposited on the tank floor. Conley and Hansen [1978] recommend a minimum depth under the modules of 4 ft.

Outlets

Outlet structures regulate the depth of flow in the settling tank and help maintain a uniform lateral flow distribution. In high-rate clarifiers, they also control the flow distribution among the tubes, which must be uniform.

Launders

The device that collects the clarified water usually is called a *launder*. There are three arrangements:

- Troughs with side-overflow weirs (the most common design)
- Troughs with submerged perforations along the sides
- Submerged pipes with side perforations

Launders may be constructed of any convenient, stable material; concrete and steel are the common choices. A supporting structure is needed to hold them up when the tank is empty and down when the tank is full. Launders are usually provided with invert drains and crown vents to minimize the loads on the supports induced by tank filling and draining. The complete outlet structure consists of the launder plus any ancillary baffles and the supporting beams and columns.

Clarified water flows into the launders either by passing over weirs set along the upper edges of troughs or by passing through perforations in the sides of troughs or pipes. The launders merge downstream until only a single channel pierces the tank's end wall.

The outlet structures in some very old plants consist of simple overflow weirs set in the top of the downstream end wall. This design is unacceptable.

If troughs-with-weirs are built, the weir settings will control the water surface elevation in the tank. *V-notch* weirs are preferred over horizontal sharp-crested weirs, because they are more easily adjusted. V-notch weirs tend to break up large, fragile alum flocs. This may not be important.

Perforated launders are built with the perforations set 1 to 2 ft below the operating water surface [Hudson, 1981; Culp/Wesner/Culp, Inc., 1986]. This design is preferred when the settled water contains significant amounts of scum or floating debris, when surface freezing is likely, and when floc breakage must be minimized [Hudson, 1981; Culp/Wesner/Culp, Inc., 1986]. Perforated launders permit significant variations in the water surface elevation, which may help to break up surface ice.

Finger launders [James M. Montgomery, Consulting Engineers, Inc., 1985] consist of long troughs or pipes that run the length of the settling zone and discharge into a common channel or manifold at the downstream end of the tank. Finger launders are preferred for all rectangular clarifiers, with or without settling tube modules, because they maximize floc removal efficiency. There are several reasons for the superiority of finger launders [James M. Montgomery, Consulting Engineers, Inc., 1985]:

- By drawing off water continuously along the tank, they reduce tank turbulence, especially near the outlet end.
- They dampen wind-induced waves. This is especially true of troughs with weirs, because the weirs protrude above the water surface.
- If a diving density current raises sludge from the tank bottom, the sludge plume is concentrated at the downstream end of the tank. Therefore, most of the launder continues to draw off clear water near the center and upstream end of the tank.
- Finger launders impose a nearly uniform vertical velocity component everywhere in the settling zone, which produces a predictable, uniform velocity field. This uniform velocity field eliminates many of the causes of settler inefficiency, including bottom scour, streaming, and gradients in the horizontal velocity field.
- Finger launders eliminate the need for a separate outlet zone. Settled water is collected from the top of the settling zone, so the outlet and settling zones are effectively merged.

The last two advantages are consequences of Fisherström's [1955] analysis of the velocity field under finger launders. The presence or absence of settling tube modules does not affect the analysis, nor does it change the conclusions. The modules merely permit the capture of particles that would otherwise escape.

The outlet design should include so-called hanging or cross baffles between the launders. Hanging baffles run across the width of the clarifier, and they extend from a few inches above the maximum water surface elevation to a few feet below it. If settling tube modules are installed in the clarifier, the hanging baffle should extend all the way to the top of the modules. In this case, it should be better called a cross baffle. The baffles are pierced by the launders. The purpose of the baffles is to promote a uniform vertical velocity component everywhere in the settling/outlet zone. They do this by suppressing the longitudinal surface currents in the settling/outlet zone that are induced by diving density currents and wind.

The number of finger launders is determined by the need to achieve a uniform vertical velocity field everywhere in the tank. There is no firm rule for this. Hudson [1981] recommends that the center-to-center distance between launders be 1 to 2 tank depths. The number of hanging baffles is likewise indeterminate. Slechta and Conley [1971] successfully suppressed surface currents by placing the baffles at the quarter points of the settling/outlet zone. However, this spacing may be too long. The clarifier in question also had tube modules, which helped to regulate the velocity field below the launders.

Other launder layouts have serious defects and should be avoided. The worst choice for the outlet of a conventional rectangular clarifier consists of a weir running across the top of the downstream end wall, which was a common design 50 years ago. The flow over the weir will induce an upward component in the water velocity near the end wall. This causes strong vertical currents that carry all but the fastest particles over the outlet weir.

Regulatory authorities often attempt to control the upward velocity components in the outlet zone by limiting the so-called weir loading or weir overflow rate. This number is defined to be the ratio of the volumetric flow rate of settled water to the total length of weir crest or perforated wall. If water enters both sides of the launder, the lengths of both sides may be counted in calculating the rate.

A commonly used upper limit on the weir rate is the "Ten States" specification of 20,000 gal/ft·day for peak hourly flows less than or equal to 1 mgd and 30,000 gal/ft·day for peak hourly flows greater than 1 mgd [Wastewater Committee, 1990]. Babbitt *et al.* [1967] recommend an upper limit of 5000 gal/ft·day. Walker Process Equipment, Inc. [Joint Task Force, 1969], recommends the following limits, which are based on coagulant type:

- Low raw water turbidity/alum—8 to 10 gpm/ft
- High raw water turbidity/alum—10 to 15 gpm/ft
- Lime/soda softening—15 to 18 gpm/ft

The Ten States weir loading yields relatively short launders and high upward velocity components. Launders designed according to the Ten States regulation require a separate outlet zone, which would be defined to be all the water under the horizontal projection of the launders. A commonly followed recommendation [Joint Committee, 1969] is to make the outlet zone one-third of the total tank length, and to cover the entire outlet zone with a network of launders. More recently, it is recommended that the weirs cover enough of the settling zone surface so that the average rise rate under them not exceed 1 to 1.5 gpm/ft^2 [Joint Task Force, 1992].

Weir/Trough Design

The usual effluent launder weir consists of a series of V-notch or triangular weirs. The angle of the notch is normally 90° because this is the easiest angle to fabricate, it is less likely to collect trash than narrower angles, and the flow through it is more predictable than wider angles. Individual weir plates are typically 10 cm wide, and the depth of the notch is generally around 5 cm. The spacing between notches is about 15 cm, measured bottom point to bottom point. The flat surface between notches is for worker safety. If the sides of the notches merged in a point, the point would be hazardous to people working around the launders. The sides of the V are beveled at 45° to produce a sharp edge, the edge being located on the weir inlet side. The stock from which the weirs are cut is usually either hot-dipped galvanized steel or aluminum sheet 5 to 13 mm thick. Fiberglass also has been used. Bolt slots permit vertical adjustment of the weir plates.

The usual head-flow correlation for 90° V-notch weirs is King's equation [King and Brater, 1963]:

$$Q = 2.52H^{2.47} \tag{8.197}$$

where

Q = the flow over the weir in ft^3/s
H = the head over the weir notch in ft

Adjacent weirs behave nearly independently of one another as long as the distance between the notches is at least 3.5 times the head [Barr, 1910a; Barr, 1910b; Rowell, 1913]. Weir discharge is independent of temperature between 39 and 165°F [Switzer, 1915].

Because finger launders are supposed to produce a uniform vertical velocity field in the settling zone, each weir must have the same discharge. This means that the depth of flow over each notch must be the same. The hydraulic gradient along the tank also is small, and the water surface may be regarded as flat, at least for design purposes. Wind setup may influence the water surface more than clarifier wall friction.

The water profile in the effluent trough may be derived by writing a momentum balance for a differential cross-sectional volume element. The result is the so-called Hinds–Favre equation [Camp, 1940; Chow, 1959; Favre, 1933; Hinds, 1926]:

$$\frac{dy}{dx} = \frac{S_o - S_f(2nq_wQ_x/gA_x^2)}{1 - (Q_x^2/gA_x^2H_x)} \tag{8.198}$$

where

A_x = the cross-sectional area of the flow at x (m^2 or ft^2)
B_x = the top width of the flow at x (m or ft)
g = the acceleration due to gravity (9.80665 m/s^2 or 32.174 ft/s^2)
H_x = the mean depth of flow at x (m or ft)
= A_x/B_x
n = the number of weir plates attached to the trough (dimensionless)
= 1, if flow enters over one edge only
= 2, if flow enters over both edges
Q_x = the flow at x (m^3/s or ft^3/s)
q_w = the weir loading rate ($m^3/m{\cdot}s$ or $ft^3/ft{\cdot}s$)
S_f = the energy gradient (dimensionless)
S_o = the invert slope (dimensionless)
x = the distance along the channel (m or ft)
y = the depth above the channel invert at x (m or ft)

For a rectangular cross section, which is the usual trough shape, the mean depth is equal to the depth.

If it is assumed that the energy gradient is caused only by wall friction, then it may be replaced by the Darcy–Weisbach formula (or any other wall friction formula):

$$S_f = \frac{fU^2}{8gR} \tag{8.199}$$

where

f = the Darcy–Weisbach friction factor (dimensionless)
R = the hydraulic radius (m or ft)
U = the mean velocity (m/s or ft/s)

An approximate solution to Eq. (8.198) may be had by substituting the average values of the depth and the hydraulic radius into the integral. The information desired is the depth of water at the upstream end of the trough, because this will be the point of highest water surface elevation (even if not greatest depth in the trough). Camp's [1940] solution for the upstream depth is

$$H_o = \sqrt{H_x^2 + \frac{2Q_x^2}{gb^2 H_x} - 2x\overline{H}(S_o - \overline{S}_f)} \tag{8.200}$$

where

$\overline{H}$ = the mean depth along the channel (m or ft)
H_o = the depth of flow at the upstream end of the channel (m or ft)
$\overline{S}_f$ = the average energy gradient along the channel (dimensionless)

H_o can be calculated if the depth of flow is known at any point along the trough. The most obvious and convenient choice is the depth at the free overflow end of the channel, where the flow is critical. The critical depth for a rectangular channel is given by [King and Brater, 1963]:

$$H_c = \sqrt[3]{\frac{Q_c^2}{gb^2}} \tag{8.201}$$

In smooth channels, the critical depth section is located at a distance of about $4H_c$ from the end of the channel [Rouse, 1936; O'Brien, 1932]. In a very long channel, the total discharge can be used with little error.

The actual location of the critical depth section is not important, because the overflow depth is simply proportional to the critical depth [Rouse, 1936, 1943; Moore, 1943]:

$$H_e = 0.715 H_c \tag{8.202}$$

Estimation of the mean values of the depth, hydraulic radius, and energy gradient for use in Eq. (8.200) requires knowledge of H_o, so an iterative calculation is required. An initial estimate for H_o can be obtained by assuming that the trough is flat and frictionless and that the critical depth occurs at the overflow. This yields

$$H_o \cong \sqrt{3} \cdot \sqrt[3]{\frac{Q^2}{gb^2}} \tag{8.203}$$

A first estimate for the average value of the depth of flow may now be calculated. Because the water surface in the trough is nearly parabolic, the best estimators for the averages are [Camp, 1940]:

$$\overline{H} = \frac{2H_o + H_e}{3} \tag{8.204}$$

$$\overline{R} = \frac{b\overline{H}}{b + 2\overline{H}} \tag{8.205}$$

The average energy gradient can be approximated using any of the standard friction formulas, such as the Manning equation. The side inflow has the effect of slowing the velocity in the trough, because the inflow must be accelerated, and a somewhat higher than normal friction factor is needed.

Combining-Flow Manifold Design

A perforated trough may be treated as a trough with weirs, if the orifices discharge freely to the air. In this case, all the orifices have the same diameter and the orifice equation may be used to calculate the required diameters. If the orifices are submerged, the launder is a combining-flow manifold, and a different design procedure is required.

Consider a conduit with several perpendicular laterals. The hydraulic analysis of each junction involves eight variables [McNown, 1945, 1954]:

- The velocity in the conduit upstream of the junction
- The velocity in the lateral
- The velocity of the combined flow in the conduit downstream of the junction
- The pressure difference in the conduit upstream and downstream of the junction
- The pressure difference between the lateral exit and the conduit downstream of the junction
- The conduit diameter (assumed to be the same upstream and downstream of the junction)
- The lateral diameter
- The density of the fluid

There are three equations connecting these variables:

- Continuity
- Headloss for the lateral flow
- Headloss for the conduit flow

Besides these equations, there is the requirement that all laterals deliver the same flow. The conduit diameter is usually kept constant, also.

The continuity equation for the junction is

$$Q_d = Q_u + Q_l \tag{8.206}$$

where

Q_d = the flow downstream from the lateral (m^3/s or ft^3/s)
Q_l = the flow entering from the lateral (m^3/s or ft^3/s)
Q_u = the flow upstream from the lateral (m^3/s or ft^3/s)

If the conduit has the same diameter above and below the junction with the lateral, the headloss for the conduit flow may be represented as a sudden contraction [McNown, 1945; Naiz, 1954]:

$$h_{Lc} = K_c \left(\frac{U_d^2}{2g} - \frac{U_u^2}{2g} \right) \tag{8.207}$$

where

h_{Lc} = the headloss in the conduit at the lateral (m or ft)
K_c = the headloss coefficient (dimensionless)
U_d = the velocity downstream of the lateral (m/s or ft/s)
U_u = the velocity upstream of the lateral (m/s or ft/s)

The headloss coefficient, K_c, depends on the ratio of the lateral and conduit diameters [Soucek and Zelnick, 1945; McNown, 1954; Naiz, 1954; Powell, 1954]. Niaz's [1954] analysis of McNown's data yields the following approximate relationships:

$$\begin{aligned} \frac{d_l}{d_c} &= \frac{1}{4} \rightarrow K_c = 1.4 \\ \frac{d_l}{d_c} &= \frac{1}{2} \rightarrow K_c = 1.0 \\ \frac{d_l}{d_c} &= 1 \rightarrow K_c = 0.5 \end{aligned} \tag{8.208}$$

where

d_c = the diameter of the conduit (m or ft)
d_l = the diameter of the lateral (m or ft)

The situation with respect to the lateral headloss is more complicated. The headloss may be expressed either in terms of the lateral velocity or in terms of the downstream conduit velocity [McNown, 1954]:

$$h_{Ll} = K_{ll} \cdot \frac{U_l^2}{2g} \tag{8.209}$$

$$h_{Ll} = K_{lc} \cdot \frac{U_d^2}{2g} \tag{8.210}$$

where

h_{Ll} = the headloss in the lateral (m or ft)
K_{lc} = the lateral's headloss coefficient based on the conduit's velocity (dimensionless)
k_{ll} = the lateral's headloss coefficient based on the lateral's velocity (dimensionless)
U_d = the conduit's velocity downstream of the lateral (m/s or ft/s)
U_l = the lateral's velocity (m/s or ft/s)

The headloss coefficients depend on both the ratio of the lateral and conduit diameters and the ratio of the lateral and conduit flows (or velocities). If the lateral velocity is much larger than the conduit velocity, all of the lateral velocity head is lost, and K_{ll} is equal to 1. The situation here is similar to that of a jet entering a reservoir. If the lateral velocity is much smaller than the conduit velocity, the lateral flow loses no energy. In fact, the headloss calculated from the Bernoulli equation will be *negative*, and its magnitude will approach the velocity head in the conduit downstream of the junction. For this case, K_{lc} will approach -1. The negative headloss is an artifact caused by the use of cross-sectional average velocities in the Bernoulli equation. If the lateral discharge is very small relative to the conduit flow, it enters the conduit boundary layer, which has a very small velocity. Some empirical data on the variation of the headloss coefficients are given by McNown [1954] and Powell [1954].

The energy equation is written for an arbitrary element of water along its path from the clarifier to the outlet of the launder. To simplify matters, it is assumed that the launder discharges freely to the atmosphere. The water element enters the launder through the *j*th lateral, counting from the downstream end of the launder:

$$\frac{U_o^2}{2g} + \frac{p_o}{\gamma} + z_o = \frac{U_e^2}{2g} + \frac{p_e}{\gamma} + z_e + h_{Li}(j) + h_{Ll}(j) + \sum_{i=1}^{j-1} h_{Lc}(i) \tag{8.211}$$

where

p_e = the pressure at the conduit's exit (N/m^2 or lbf/ft^2)
p_o = the pressure in the clarifier (N/m^2 or lbf/ft^2)
U_e = the velocity at the conduit's exit (m/s or ft/s)
U_o = the velocity in the clarifier (m/s or ft/s)
z_e = the elevation at the conduit's exit (m or ft)
z_o = the elevation in the clarifier (m or ft)

Wall friction losses in the launder and its laterals are ignored, because they are usually small compared to the other terms. The velocity of the water element at the beginning of its path in the clarifier will be small and can be deleted. The sum of the pressure and elevation terms at the beginning is simply the water surface elevation. The pressure of a free discharge is zero (gauge pressure). Consequently, Eq. (8.211) becomes

$$h_{Li}(j) + h_{Ll}(j) + \sum_{i=1}^{j-1} h_{Lc}(i) = z_{\text{cws}} - z_{\text{ews}} - \frac{U_e^2}{2g} \tag{8.212}$$

where

z_{cws} = the clarifier's water surface (m or ft)
z_{ews} = the conduit exit's water surface (m or ft)

Equation (8.212) assumes that the laterals do not interact. This will be true as long as the lateral spacing is at least six lateral diameters [Soucek and Zelnick, 1945].

The right-hand side of Eq. (8.212) is a constant and the same for each lateral. All the water elements begin with the same total energy, and they all end up with the same total energy, so the total energy loss for each element must be the same.

The flows into laterals far from the outlet of the launder experience more junction losses than those close to the outlet. This means that the head available for lateral entrance and exit is reduced for the distant laterals. Consequently, if all the lateral diameters are equal, the lateral discharge will decrease from the launder outlet to its beginning [Soucek and Zelnick, 1945]. The result will be a nonuniform vertical velocity distribution in the settling zone of the clarifier. This can be overcome by increasing the lateral diameters from the launder outlet to its beginning.

Most perforated launders are built without lateral tubes. The headloss data quoted above do not apply to this situation, because the velocity vectors for simple orifice inlets are different from the velocity vectors for lateral tube inlets. Despite this difference, some engineers use the lateral headloss data for orifice design [Hudson, 1981].

Data are also lacking for launders with laterals or orifices on each side. These data deficiencies make perforated launder design very uncertain. It is usually recommended that the design be confirmed by full-scale tests.

The usual reason given for perforated launders is their relative immunity to clogging by surface ice. However, the need for a uniform velocity everywhere in the settling zone controls the design. If freezing is likely, it would be better to cover the clarifiers. The launders could then be designed for V-notch weirs or freely discharging orifices, which are well understood.

Sludge Zone

Sludge Collection

The sludge collection zone lies under the settling zone. It provides space for the sludge removal equipment and, if necessary, for temporary sludge storage.

The most common design consists of a bottom scraper and a single hopper. Periodically, the solids deposited on the clarifier floor are scraped to the hopper set into the tank floor. The solids are collected as a sludge that is so dilute it behaves hydraulically like pure water. Periodically, the solids are removed from the hopper via a discharge line connected to the hopper bottom.

An alternative scheme is sometimes found in the chemical and mining industries and in dust collection facilities. In this case, the entire tank floor is covered with hoppers. No scraper mechanisms are required. However, the piping system needed to drain the hoppers is more complex.

A third system consists of perforated pipes suspended near the tank floor and lying parallel to it. A slight suction head is put on the pipes, and they are drawn over the entire tank floor sucking up the deposited solids. This system eliminates the need for hoppers, but it produces a very dilute sludge.

Sludge Composition

The composition of the sludges produced by the coagulation and sedimentation of natural waters is summarized in Table 8.2, and by wastewater treatment in Table 8.3. Alum coagulation is applied to surface waters containing significant amounts of clays and organic particles, so the sludges produced also contain significant amounts of these materials. Lime softening is often applied to groundwaters, which are generally clear. Consequently, lime sludges consist mostly of calcium and magnesium precipitates.

Sludge Collectors/Conveyors

There are a variety of patented sludge collection systems offered for sale by several manufacturers. The two general kinds of sludge removal devices are *flights*, or *squeegees*, and *suction manifolds*.

The first consists of a series of boards called flights, or squeegees, that extend across the width of the tank. The boards may be constructed of water-resistant woods, corrosion-resistant metals, or engineering plastics. In traditional designs, the flights are attached to continuous chain loops, which are mounted on sprockets and moved by a drive mechanism. The flights, chains, and sprockets are submerged. The drive mechanism is placed at ground level and connected to the sprocket/chain system by some sort of transmission. In addition to the primary flight system, which moves sludge

Table 8.2 Table Range of Composition of Water Treatment Sludges[1]

Sludge Component	Alum Sludges	Iron Sludges	Lime Sludges
Total suspended solids (% by wt)	0.2 to 4.0	0.25 to 3.5	2.0 to 15.0
Aluminum			
(% by wt of TSS, as Al)	4.0 to 11.0	—	—
(mg/L, as Al)	295 to 3750	—	—
Iron			
(% by wt of TSS, as Fe)	—	4.6 to 20.6	—
(mg/L, as Fe, for 2% TSS)	—	93.0 to 4120	—
Calcium (% by wt of TSS, as Ca)	—	—	30 to 40
Silica/ash (% by wt of TSS)	35 to 70	—	3 to 12
Volatile suspended solids (% by wt of TSS)	15 to 25	5.1 to 14.1	7 (as carbon)
BOD_5 (mg/L)	30 to 300	—	Little or none
COD (mg/L)	30 to 5000	—	Little or none
pH (standard units)	6 to 8	7.4 to 8.5	9 to 11
Color (sensory)	Gray-brown	Red-brown	White
Odor (sensory)	None	—	None to musty
Absolute viscosity (g/cm·s)	0.03 (nonnewtonian)	—	—
Dewaterability	Concentrates to 10% solids in 2 days on sand beds, producing a spongy semisolid	—	Compacts to 50% solids in lagoons, producing a sticky semisolid; dewatering impaired if Ca/Mg ratio is 2 or less
Settleability	50% in 8 hr	—	50% in 1 week
Specific resistance (s^2/g)	1.0×10^9 to 5.4×10^{11}	4.1×10^8 to 2.0×10^{12}	0.20×10^7 to 26×10^7
Filterability	Poor	—	Poor

[1]Compiled from Culp/Wesner/Culp, Inc. [1986]; James M. Montgomery, Inc. [1985]; and J. T. O'Connor [1971].

Table 8.3 Range of Composition of Wastewater Sludges[1]

Sludge Component	Primary Sludge	Waste-Activated Sludge	Trickling Filter Humus
pH	5 to 8	6.5 to 8	—
Higher heating value (Btu/lb)	6800 to 10,000	6500	—
Specific gravity of particles	1.4	1.08	1.3 to 1.5
Specific gravity of sludge	1.02 to 1.07	$1 + 7 \times 10^{-8} X$	1.02
Color	Black	Brown	Grayish brown to black
COD/VSS	1.2 to 1.6	1.4	—
C/N	—	3.5 to 14.6	—
C (% by wt of TSS)	—	17 to 44	—
N (% by wt of TSS)	1.5 to 4.0	2.4 to 6.7	1.5 to 5.0
P as P_2O_5 (% by wt of TSS)	0.8 to 2.8	2.8 to 11	1.2 to 2.8
K as K_2O (% by wt of TSS)	0.4	0.5 to 0.7	—
VSS (% by wt of TSS)	60 to 93	61 to 88	64 to 86
Grease and fat (% by wt of TSS)	7 to 35	5 to 12	—
Cellulose (% by wt of TSS)	4 to 15	7	—
Protein (% by wt of TSS)	20 to 30	32 to 41	—

[1]Anonymous 1979. *Process Design Manual for Sludge Treatment and Disposal,* EPA 625/1-79-011. U.S. Environmental Protection Agency, Municipal Environmental Research Laboratory, Technology Transfer, Cincinnati, OH.

to one end of the tank, there may be a secondary flight system, which moves the sludge collected at the tank end to one or more hoppers.

In some newer designs, the flights are suspended from a traveling bridge, which moves along tracks set at ground level along either side of the clarifier. Bridge-driven flights cannot be used with high-rate settlers, because the flight suspension system interferes with the tube modules. Bridge systems also require careful design to ensure compatibility with effluent launders. If finger launders extending the whole length of the settling zone are used, bridge systems may not be feasible.

In either system, the bottom of the tank is normally finished with a smooth layer of grout, and two or more longitudinal rails are placed along the length of the tank to provide a relatively smooth bearing surface for the flights. The tops of the rails are set slightly above the smooth grout layer. The grout surface is normally pitched toward the sludge hopper to permit tank drainage for maintenance. The recommended minimum pitch is $1/16$ in. per ft (0.5%) [Joint Committee, 1969].

The flights are periodically dragged along the bottom of the tank to scrape the deposited sludge into the sludge hoppers. The scraping may be either continuous or intermittent, depending on the rate of sludge deposition. Flight speeds are generally limited to less than 1 ft/min to avoid solids resuspension [Joint Task Force, 1969].

In some traveling bridge designs, the flights are replaced by perforated pipes, which are subjected to a slight suction head. The pipes suck deposited solids off the tank floor and transfer them directly to the sludge processing and disposal systems. Suction manifolds dispense with the need for sludge hoppers, which simplifies and economizes tank construction, but they tend to produce dilute sludges, and they may not be able to collect large, dense flocs.

The sludge zone should be deep enough to contain whatever collection device is used and to provide storage for sludge solids accumulated between removal operations. Generally 12 to 18 in. of additional tank depth is provided.

Sludge Hoppers

Sludge hoppers serve several purposes. First, they store the sludge until it is removed for processing and storage. For this reason, the hoppers should have sufficient volume to contain all the solids deposited on the tank floor between sludge removals. Second, they channel the flow of the sludge to the inlet of the drain pipe. To facilitate this, the bottom of the hoppers should be square and only somewhat larger than the inlet pipe bell. Third, the hoppers provide sufficient depth of sludge over the pipe inlet to prevent short-circuiting of clear water. There is no general rule regarding the minimum hopper depth. Fourth, hoppers prevent resuspension of solids by diving density currents near the inlet and by the upflow at the outlet end of the tank. Fourth, hoppers provide some sludge thickening.

The number of hoppers and their dimensions are determined by the width of the tank and the need, if any, for cross collectors. For example, the length of a hopper side at the top of the hopper will be the minimum width of the cross collector. At the bottom of the hopper, the minimum side length will be somewhat larger than the diameter of the pipe inlet bell. The minimum side wall slopes are generally set at 45° to prevent sludge adhesion.

Freeboard

The tops of the sedimentation tank walls must be higher that the maximum water level that can occur in the tank. Under steady flow, the water level in the tank is set by the backwater from the effluent launder. In the case of effluent troughs with V-notch weirs, the maximum steady flow water level is the depth over the notch for the maximum expected flow, which is the design flow. From time to time, waves caused by hydraulic surges and wind will raise the water above the expected backwater. In order to provide for these transients, some *freeboard* is provided. The freeboard is defined to be the distance between the top of the tank walls and the predicted water surface level for the (steady) design flow. The actual amount of freeboard is somewhat arbitrary, but a common choice in the U.S. is 18 in. [Joint Task Force, 1969].

The freeboard also may be set by safety considerations. In order to minimize pumping, the operating water level in most tanks is usually near the local ground surface elevation. Consequently, the tops of the tank walls also will be near ground level. In this situation, the tanks require some sort of guardrail and curb in order to prevent pedestrians, vehicles, and debris from falling in. The top of the curb becomes the effective top of the tank wall. Curbs are usually at least 6 in. above the local ground level.

Hindered Settling

In hindered settling, particles are close enough to be affected by the hydrodynamic wakes of their neighbors, and the settling velocity becomes a property of the suspension. The particles move as a group, maintaining their relative positions like a slowly collapsing lattice: large particles do not pass small particles. The process is similar to low-Reynold's-number bed expansion during filter backwashing, and the Richardson–Zaki [Richardson and Zaki, 1954] correlation would apply, if the particles were of uniform size and density and their settling velocity were known. Instead, the velocity must be measured.

Hindered settling is characteristic of activated sludge clarifiers, many lime-soda sludge clarifiers, and gravity sludge thickeners.

Settling Column Tests

If the particle concentration is large enough, a well-defined interface forms between the clear supernatant and the slowly settling particles. Formation of the interface is characteristic of hindered settling; if a sharp interface does not form, the settling is free.

Hindered settling velocities are frequently determined in laboratory-scale settling columns. Vesilind [1974] identifies several deficiencies in laboratory-scale units that do not occur in field units:

- The initial settling velocity depends on the liquid depth [Dick and Ewing, 1967].
- "Channeling"—in narrow-diameter cylinders, the water tends to flow along the cylinder wall, which is the path of least resistance, and the measured settling velocity is increased.
- "Volcanoing"—in the latter part of the settling process, during compression, small columns of clear liquid erupt at various places across the sludge/water interface, which increases the measured interface velocity; this is a form of channeling that occurs in wide, unmixed columns.
- At high solids concentrations, narrow cylinders also permit sludge solids bridging across the cylinder, which inhibits settling.
- Narrow cylinders dampen out liquid turbulence, which prevents flocculation and reduces measured settling velocities.

Vesilind [1974] recommends the following procedure for laboratory settling column tests:

- The minimum column diameter should be 8 in., but larger diameters are preferred.
- The depth should be that of the proposed thickener, but at least 3 ft.
- The column should be filled from the bottom from an aerated, mixed tank.
- The columns less than about 12 in. in diameter should be gently stirred at about 0.5 rpm.

In any settling test, the object is to produce a plot of the batch flux (the rate of solids transport settling across a unit area) versus the solids concentration. There are two procedures in general use.

Kynch's Method. Kynch's [1952] batch settling analysis is frequently employed. The movement of the interface is monitored and plotted as in Fig. 8.11. The settling velocity of the interface is obviously the velocity of the particles in it, and it can be calculated as the slope of the interface height–time plot:

$$v_X = \frac{z' - z}{t} \tag{8.213}$$

where

t = the sampling time (s)

v_X = the settling velocity of the particles (which are at concentration X) in the interface (m/s or ft/s)

z = the height of the interface at time t (m or ft)

z' = the height of the vertical intercept of the tangent line to the interface height–time plot (m or ft)

Initially the slope is linear, and the calculated velocity is the velocity of the suspension's initial concentration. As settling proceeds, the interface particle concentration increases and its settling velocity decreases, which is indicated by the gradual flattening of the interface/time plot. The interface concentration at any time can be calculated from Kynch's [1952] formula:

$$X = \frac{X_o z_o}{z'} \tag{8.214}$$

where

X = the interface suspended solids concentration (kg/m^3 or lb/ft^3)

X_o = the initial, homogeneous suspended solids' concentration (kg/m^3 or lb/ft^3)

z_o = the initial interface height and the liquid depth (m or ft)

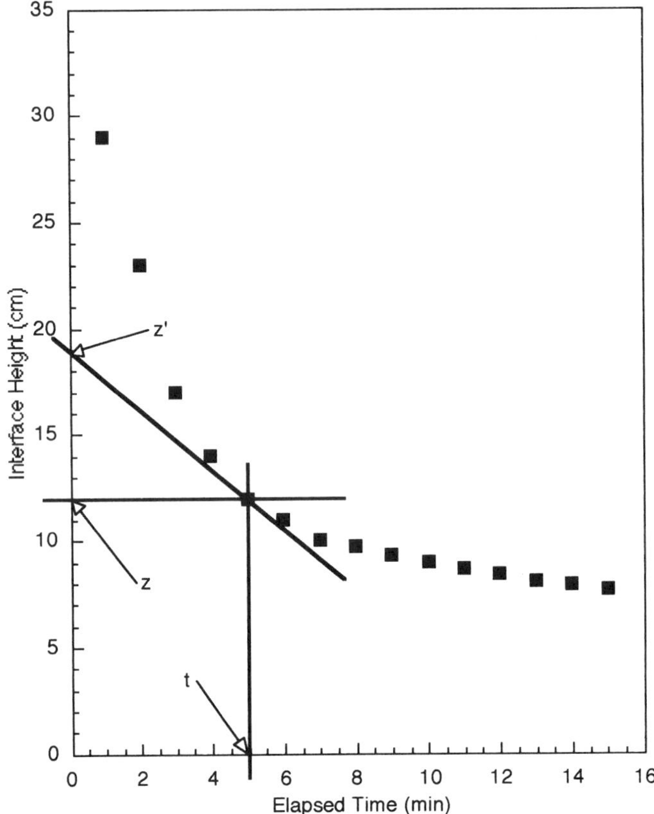

FIGURE 8.11 Typical interface height–time plot for hindered settling.

Note that the elevation in the denominator is the intercept of the extrapolated tangent to the interface height/time curve; it is not the elevation of the interface.

The batch flux for any specified solids concentration is

$$F = v_X X \tag{8.215}$$

where F is the flux of solids settling through a horizontal plane in a batch container (kg/m^2·s or lb/ft^2·s).

In the derivation of Eq. (8.214) Kynch shows that if the water is stationary a concentration layer that appears on the bottom of the settling column travels at a constant velocity upward until it intersects the interface. The concentration exists momentarily at the interface and then is replaced by another higher concentration. Dick and Ewing [1967] reviewed earlier studies and concluded that there were several deficiencies with the Kynch analysis:

- Concentration layers do not travel at constant velocities, at least in clay suspensions.
- Stirring the bottom of a suspension increases the rate of subsidence of the interface.
- The Talmadge–Fitch [Talmadge and Fitch, 1955] procedure, which is an application of Kynch's method for estimating interface concentrations, underestimates the settling velocities at high sludge concentrations [Fitch, 1962; Alderton, 1963].

Vesilind [1974] recommends that this method not be used for designing thickeners for wastewater sludges or for other highly compressible sludges.

Initial Settling Velocity Method. Most engineers prefer to prepare a series of dilutions of the sludge to be tested and to determine only the initial settling velocity that occurs during the linear portion of the interface height/time curve. The resulting correlation between the initial settling velocities and the initial solids concentration can usually be represented by the simple decaying exponential proposed by Duncan and Kawata [1968]:

$$v_{X_o} = aX_o^{-b} \tag{8.216}$$

where

a = a positive constant (units vary)
b = a positive constant (dimensionless)
X_o = the initial suspended solids concentration in the column (kg/m^3 or lb/ft^3)
v_{X_o} = the settling velocity for a suspended solids concentration of X_o (m/s or ft/s)

The batch flux for each initial concentration is calculated using Eq. (8.215) and it is plotted versus the suspended solids concentration. An example is shown in Fig. 8.12. This method assumes that the concentration in the interface is not changing as long as the height–time plot is linear, and by doing so it necessarily entails Kynch's theory.

Thickener Design

Thickeners can be conceptually divided into three layers. On the top is a clear water zone in which free, flocculent settling occurs. Below this is a sludge blanket. The upper portion of the sludge blanket is a zone of hindered settling. The lower portion is a zone of compression. Many engineers believe that the particles in the compression zone form a self-supporting lattice, which must be broken down by gentle mixing. The compression zone may not exist in continuous flow thickeners.

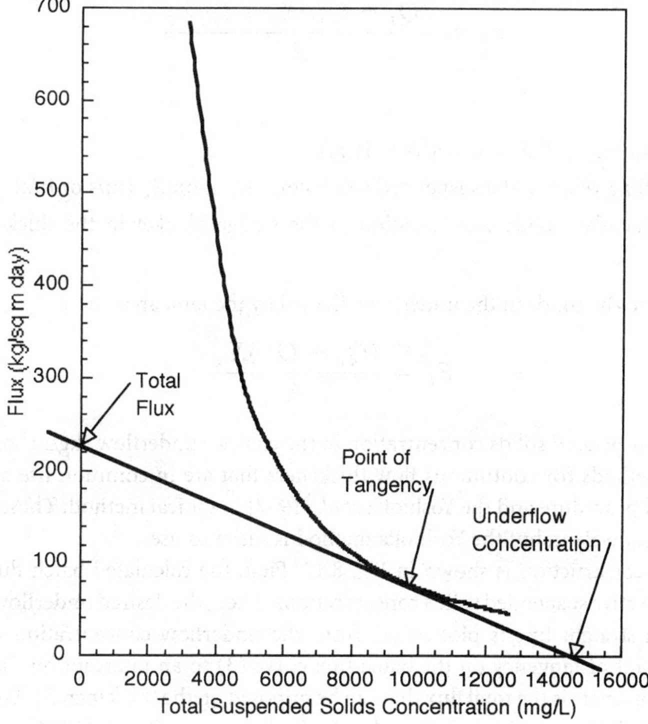

FIGURE 8.12 The Yoshioka construction for the total flux.

Hindered Settling Zone

In free settling, the critical loading parameter is the hydraulic flow per unit plan area (e.g., $m^3/m^2{\cdot}s$ or $ft^3/ft^2{\cdot}s$). In hindered settling the critical loading parameter is the total solids flux, which is the solids mass loading rate per unit area (e.g., $kg/m^2{\cdot}s$ or $lbm/ft^2{\cdot}s$). The result of a hindered settling analysis is the required thickener cross-sectional area. The depth must be determined from a consideration of the clarification and compression functions of the thickener.

The total solids flux can be expressed three ways for a perfectly efficient thickener, each of which yields the same numerical value. First, it is the total solids loading in the influent flow divided by the plan area of the clarifier. For an activated sludge plant or a lime-soda plant with solids recycling, this would be calculated as

$$F_t = \frac{(Q + Q_r)X_i}{A} \tag{8.217}$$

where

A = the thickener plan area (m^2 or ft^2)
F_t = the total solids flux ($kg/m^2{\cdot}s$ or $lb/ft^2{\cdot}s$)
Q = the design water or wastewater flow rate (m^3/s or ft^3/s)
Q_r = the recycle flow rate (m^3/s or ft^3/s)
X_i = the suspended solids concentration in the flow entering the thickener (kg/m^3 or lb/ft^3)

Second, it can be calculated as the flux through the sludge blanket inside the thickener divided by the plan area. In a continuous flow thickener, this consists of the flux due to the water movement through the tank plus the flux due to the settling of particles through the moving water. If sludge is wasted from the thickener underflow, the total water flow through the sludge blanket and in the underflow is $Q_r + Q_w$:

$$F_t = \frac{(Q_r + Q_w)X_c + v_{X_c}X_c}{A} \tag{8.218}$$

where

Q_w = the waste sludge flow rate (m^3/s or ft^3/s)
v_{X_c} = the settling velocity at a suspended solids concentration X_c (m/s or ft/s)
X_c = the suspended solids concentration in the sludge blanket in the thickener (kg/m^3 or lb/ft^3)

Third, it is equal to the solids in the underflow divided by the tank area:

$$F_t = \frac{(Q_r + Q_w)X_u}{A} \tag{8.219}$$

where X_u is the suspended solids concentration in the clarifier underflow (kg/m^3 or lb/ft^3).

Two design methods for continuous flow thickeners that are in common use are the Coe and Clevenger [1916] procedure and the Yoshioka *et al.* [1957] graphical method. These procedures are mathematically equivalent, but the Yoshioka method is easier to use.

The Yoshioka construction is shown in Fig. 8.12. First, the calculated batch fluxes are plotted against their respective suspended solids concentrations. Then, the desired underflow concentration is chosen, and a straight line is plotted (1) from the underflow concentration on the abscissa (2) through a point of tangency on the batch flux curve (3) to an intercept on the ordinate. The intercept on the ordinate is the total flux that can be imposed on the thickener, F_t. Equation (8.217), (8.218), or (8.219) is then used to calculate the required plan area. Parker [1983] recommends that the peak hydraulic load rather than the average hydraulic load be used in the calculation.

If the right-hand side of Eq. (8.218) is plotted for all possible values of X_c, it will be found that F_t is the minimum of the function [Dick, 1970]. Consequently, Eq. (8.216) may be used to eliminate v_{X_c} from Eq. (8.218) and the minimum of the total flux formula may be found by differentiating with respect to X_c [Dick and Young, no date]:

$$F_t = [a(b-1)]^{1/b}\left(\frac{b}{b-1}\right)\left(\frac{Q_r + Q_w}{A}\right)^{(b-1)/b} \tag{8.220}$$

Clarification Zone

Clarification is impaired if the sludge blanket comes too close to the water surface and if the overflow rate is too high. If effluent suspended solids concentrations must be consistently below about 20 mg/L, the clarifier side water depth must be at least 16 ft, and the overflow rate must be less than 600 gpd/ft^2 [Parker, 1983].

Compression Zone

If very high sludge solids concentrations are required, a compression zone may form. Its depth can be estimated from the Roberts–Behn formula either in terms of the suspension dilution, D, which is defined to be the mass of water in the sludge divided by the mass of particles [Roberts, 1949; Behn, 1957],

$$D_t - D_\infty = (D_0 - D_\infty)e^{-Kt} \tag{8.221}$$

where

D_0 = the initial dilution (kg water/kg solids or lb water/lb solids)
D_t = the dilution at time t (kg water/kg solids or lb water/lb solids)
D_∞ = the ultimate dilution (kg water/kg solids or lb water/lb solids)
K = the rate constant (per second)
t = the elapsed compression time (s)

or, alternatively, in terms of the interface height, H [Behn and Liebman, 1963],

$$H_t - H_\infty = (H_o - H_\infty)e^{-Kt} \tag{8.222}$$

where

H_o = the initial interface height (m or ft)
H_t = the interface height at time t (m or ft)
H_∞ = the ultimate interface height (m or ft)

These are connected by

$$D_t = \frac{\rho H_t}{X_o H_o} - \frac{\rho}{\rho_p} \tag{8.223}$$

where

ρ = the water density (kg/m^3 or lb/ft^3)
ρ_p = the density of the particles (kg/m^3 or lb/ft^3)

The unknown here is the time required to achieve the desired compressive thickening. The compression parameters are D_o (or H_o), D_∞ (or H_∞), and K, and these are determined from a batch settling test. The computational procedure is iterative. One selects a trial value for D_∞ (or H_∞) and plots the differences on semilog paper. If the data come from the compression region, a value of the ultimate dilution factor or depth can be found that will produce a straight line.

The volume of the compression zone that produces the required underflow dilution factor, D_u, is given by [Behn and Liebman, 1963]

$$V_{cz} = (Q + Q_r)X_i\left(\frac{t_u}{\rho_p} + \frac{D_0 - D_u}{\rho K} + \frac{D_\infty t_u}{\rho}\right) \tag{8.224}$$

where

t_u = the compression time required to produce a dilution ratio of D_u (s)
V_{cz} = the volume of the compression zone (m^3 or ft^3)
X_i = the suspended solids concentration in the influent flow (kg/m^3 or lb/ft^3)

The compression time to reach the underflow dilution factor, t_u, is calculated from Eq. (8.224) once D_u is selected. The depth of the compression zone is calculated by dividing the compression zone volume by the plan area determined from the hindered settling analysis.

In Behn's [1957] soil consolidation theory of compression, the parameter K depends on the depth at which compression begins, H_o:

$$K = \frac{k(\rho_p - \rho)g}{\gamma H_o} \tag{8.225}$$

where k is Darcy's permeability coefficient (m/s or ft/s). This relationship arises because the force that expels water from the sludge is the net weight of the solids. Consequently, one can expect the parameter K to vary with the depth of the compression zone in the thickener. An iterative solution is required. The value of K determined from the batch settling test is used to get a first estimate of the compression zone volume and depth. The calculated depth is then compared with the value of H_o that occurred in the test, and K is adjusted accordingly until the calculated depth matches H_o.

Rules of Thumb

Commonly used limits on total solids fluxes on activated sludge secondary clarifiers for the maximum daily flow and maximum return rate are [Wastewater Committee, 1990]:

- Conventional: $F_t = 50$ lb/ft^2·day
- Extended aeration: $F_t = 35$ lb/ft^2·day
- Second stage nitrification: $F_t = 35$ lb/ft^2·day

Current practice for average flow conditions is around 20 to 30 lb/ft^2·day [Joint Task Force, 1992].

8.6 Filtration

Granular Media Filters

Particle Removal Mechanisms

The possible removal mechanisms are [Ives, 1975; Tien, 1989] as follows:

- *Mechanical straining.* Straining occurs when the particles are larger than the local pore. It is important only if the ratio of particle diameter to pore size is larger than about 0.2 [Herzig *et al.*, 1970]. Straining is undesirable, because it concentrates removed particles at the filter surface and reduces filter runs.
- *Sedimentation.* The effective horizontal surface in sand filters is roughly 3% of the surface area of the sand grains, and this amounts to nearly 400 times the plan area of the bed for *each* meter of sand depth [Fair and Geyer, 1954]. Consequently, filters act in part like very large sedimentation basins [Hazen, 1904].

- *Inertial impact.* Particles tend to persist moving in a straight line, and if they are large enough their momentum may overcome the liquid drag forces and lead to a collision with the media.
- *Hydrodynamic diffusion.* Because of the liquid velocity variation across the particle diameter, there is a net hydrodynamic force on the particle normal to the direction of flow. If the particles are spherical and the velocity gradient is linear (laminar flow), this force moves the particle toward high velocities and away from any media surface. However, if particle shape is irregular and the flow field nonuniform and unsteady (turbulent flow), the particle drift appears to be random. In the first case, removals are reduced. In the second case, random movements produce a turbulent diffusion, with transport from high-concentration to low-concentration areas. Particle adsorption produces a low-concentration region near the media surface, so the hydrodynamic diffusion transports particles to the media.
- *Interception.* If the liquid stream lines bring a particle center to within one particle radius of the media surface, the particle will strike the surface and may adhere.
- *Brownian diffusion.* Brownian diffusion will cause particles to move from high-concentration zones to low-concentration zones. Adsorption to the media surface creates a low-concentration zone near the media, so Brownian diffusion will transport particles from the bulk flow toward the media.
- *Electrostatic and London–Van der Waals attraction.* Electrical fields, due either to electrostatic charges or to induced London–Van der Waals fields, may attract or repel particles and media.
- *Adhesion.* Once the particles collide with the media surface, they must adhere. Adhesion occurs both with the clean media surface and with particles already collected on the surface. Consequently, particles that are to be removed must be colloidally unstable [Gregory, 1975].

Performance

In general, granular media filters remove particles that are much smaller than the pore opening. Also, transport to the media surface depends on the local particle concentration in the flow. The result is that particle removal is more or less exponential with depth. This is usually expressed as Iwasaki's [1937] law:

$$\frac{X}{X_o} = e^{-\lambda H} \tag{8.226}$$

where

H = the depth of filter media (m or ft)
X = the suspended solids concentration leaving the filter (kg/m^3 or lb/ft^3)
X_o = the suspended solids concentration entering the filter (kg/m^3 or lb/ft^3)
λ = the filter coefficient (per m or per ft)

Ives and Sholji [1965] and Tien [1989] have summarized empirical data for the dependence of the Iwasaki filter coefficient on the diameter of the particle to be removed, d_p, the diameter of the filter media, d_m, the filtration rate, U_f, and the water viscosity, μ. The dependencies would have the following form:

$$\lambda \propto \frac{d_p^{\alpha}}{U_f^{\beta} d_m^{\gamma} \mu^{\delta}} \tag{8.227}$$

The exponents on the variables are highly uncertain and system specific, but α appears to be on the order of 1 to 1.5; β appears to be on the order of 0.5 to 2; γ appears to be on the order of 0.5 to 2; and δ appears to be on the order of 1. A worst-case scenario may be estimated by adopting the most unfavorable exponent for a suggested process change.

For practical design purposes, γ is sometimes taken to be 1.0, and filter depth and media size are traded off according to [James M. Montgomery, Consulting Engineers, Inc., 1985]

$$\frac{H_1}{d_{m1}} = \frac{H_2}{d_{m2}} \tag{8.228}$$

The Committee of the Sanitary Engineering Division on Filtering Materials [1936] reported that the depth of penetration of suspended solids (silts and flocs) into filter media was more or less proportional to the square of the media particle diameter. In the Committee's studies, the coefficient of proportionality ranged from about 10 to 25 in./mm^2, and it appears to be a characteristic of the specific sand source. They recommended Allan's procedure for formulating sand specifications, which reduces to

$$H_{\min} = \sum_{i=1}^{n} H_i \propto \frac{1}{\sum_{i=1}^{n} \frac{f_i}{d_{mi}^2}} \tag{8.229}$$

where

d_{mi} = the grain diameter of media size class i (m or ft)
f_i = the fraction by weight of media size d_{mi} (dimensionless)
H_i = the depth of filter media size d_{mi} (m or ft)
$H_{\min}$ = the minimum depth of media (m or ft)

Equation (8.229) was derived by assuming that the penetration of suspended particles into the sand mixture should be equal to the penetration into any single grade.

Because the removal rate is exponential, there is *no* combination of filtration rate or media depth that will remove all particles. A minimum filtered water turbidity of 0.1 to 0.2 seems to be the best that can be expected; this minimum will be proportional to the influent turbidity.

For many years, the standard filtration rate in the USA was 2 gpm/ft^2, which is usually attributed to Fuller's (1898) studies at Louisville. Cleasby and Baumann [1962] have shown that increasing the filtration rate to 6 gpm/ft^2 triples the effluent turbidity, but that turbidities less than 1 TU were still achievable at the higher rate. Their filters held 30 in of sand, either with an effective size of 0.5 mm and a uniformity coefficient of 1.89 or sieved to lie between 0.59 and 0.84 mm. The suspended particles were ferric hydroxide flocs. Reviewing a number of filtration studies, Cleasby (1990) concluded that filtration rates up to 4 gpm/ft^2 were acceptable as long as coagulation of the raw water is nearly complete and no sudden increases in filtration rate occur. Higher filtration rates required filter aids, but effluent turbidities of less than 0.5 NTU are achievable.

Fair and Geyer (1954) summarized whole-plant coliform removal data reported by the U.S. Public Health Service as follows:

$$C_f = aC_o^b \tag{8.230}$$

where

a, b = empirical coefficients
C_o = the concentration of total coliform bacteria in the raw, untreated water in no./100 mL
C_f = the finished water concentration of total coliform bacteria in no./100 mL

Values of the coefficients for different plant designs are shown in Table 8.4. Also shown is the concentration of bacteria in the filter influent, C_{pi}, that will permit achievement of a finished water concentration of 1/100 mL. The filtration rates, bed depths, and grain sizes are not specified.

Table 8.4 Coliform Removal Efficiencies for Different Water Treatment Plant Configurations[1]

	Turbid River Water			Clear Lake Water		
Treatment Process	a	b	C_{pi}	a	b	C_{pi}
Chlorination only	0.015	0.96	80	0.050	0.76	50
Flocculation, settling, filtration	0.070	0.60	80	0.087	0.60	60
Flocculation, settling, filtration, postchlorination	0.011	0.52	6000	0.040	0.38	4500

[1]Fair, G. M. and Geyer, J. C. 1954. *Water Supply and Waste-Water Disposal,* John Wiley & Sons, New York.

Hazen's [1896] empirical rule for the proportion of bacteria passing through filters is

$$P_{\text{pass}} = \frac{1}{2}\frac{U_f^2 d_{\text{es}}}{\sqrt{H}} \tag{8.231}$$

where

d_{es} = the effective size on the filter media in mm
H = the media depth in in.
P_{pass} = the percentage of the bacteria passing through the filter
U_f = the filtration rate in mgad

Hazen's formula appears to approximate the median of a log-normal distribution. Approximately 40% of the cases will lie between one-half and twice the predicted value. In somewhat less than one-fourth of the cases, the percentage passing will be greater than twice the predicted value. One-sixth of the cases will exceed 4 times the predicted value. The proportion of cases exceeding 10 times the predicted value is less than 1%.

Immediately after backwashing, the water in the filter pores is very turbid, and the initial product following the startup of filtration must be discarded. Approximately 10 bed volumes must be discarded to obtain stable effluent suspended solids concentrations [Cleasby and Baumann, 1962].

Water Balance and Number of Filter Boxes

The water balance (in units of volume) for a rapid sand filter for a design period T is

$$\underbrace{U_f A t_f}_{\text{Gross product}} = \underbrace{U_b A t_b}_{\text{Wash water}} + \underbrace{F_p P Q_{\text{pc}} T}_{\text{Community demand}} \tag{8.232}$$

where

A = the plan area of the filter media (m^2 or ft^2)
F_p = the peaking factor for the design period (dimensionless)
P = the projected service population (capita)
Q_{pc} = the per caput water demand (m^3/s·cap or ft^3/s·cap)
T = the design period (s)
t_b = the time spent backwashing during the design period (s)
t_f = the time spent filtering during the design period (s)
U_b = the backwash rate (m/s or ft/s)
U_f = the filtration rate (m/s or ft/s)

The design period T is determined by a balancing of the costs of storage vs. filtration capacity. It is typically on the order of one to two weeks. The peaking factor F_p corresponds to the design period. Typical values are given in Chapter 7, Tables 7.11 and 7.12. The service population P is the projected population, usually 20 years hence.

A commonly used filtration rate is 4 gpm/ft^2. The duration of the filter run is determined by the required effluent quality. A rough rule of thumb is that the filter must be cleaned after accumulating 0.1 lbm suspended solids per sq ft of filter area per ft of head loss [Cleasby, 1990]. Actual filters may capture between one-third and three times this amount before requiring cleaning.

Backwash rates depend on the media and degree of fluidization required. Full fluidization results in a bed expansion of 15 to 30%. Backwashing is continued until the wash water is visibly clear, generally about 10 min.

The economical number of filters is often estimated using the Morrill–Wallace [1934] formula:

$$n = 2.7\sqrt{Q_{\text{des}}} \tag{8.233}$$

where

n = the number of filter boxes (dimensionless)
Q_{des} = the plant design flow in mgd

Filter Box Design

The usual requirement is that there be at least two filters, each of which must be capable of meeting the plant design flow [Water Supply Committee, 1987]. If more than two filters are built, then they must be capable of meeting the plant design flow with one filter out of service. The dimensions of rapid sand filters are more or less standardized and may be summarized as follows [Babbitt, Doland and Cleasby, 1967; Joint Task Force, 1990; Culp/Wesner/Culp, Inc., 1986; Kawamura, 1991]:

Vertical Cross Section. The vertical cross section of a typical rapid sand filter box is generally at least 8 to 10 ft deep (preferably deeper to avoid air binding) above the floor and consists of the following layers:

- At least 12 and more likely 16 to 18 in. of gravel and torpedo sand
- 24 to 30 in. of filter media, either fine or coarse sand or sand and crushed anthracite coal
- At least 6 ft of water, sufficient to avoid negative or zero gauge pressures anywhere in the filter and consequent air binding
- At least 6 in. of freeboard, and preferably more to accommodate surges during backwashing

The backwash troughs are placed sufficiently high that the top of the expanded sand layer is at least 18 in. below the trough inlets.

Plan Dimensions. The plan area of the sand beds of gravity filters generally does not exceed 1000 ft^2, and larger sand beds are constructed as two subunits with separate backwashing.

Filters are normally arranged side by side in two parallel rows with a pipe gallery running between the rows. The pipe gallery should be open to daylight to facilitate maintenance.

Wash Water Effluent Troughs. The design of wash water effluent troughs is discussed earlier in this chapter under "Sedimentation." The troughs are submerged during filtration.

Air Scour. A major problem with air scour systems is disruption of the gravel layer and gravel/sand mixing. The gravel layer is eliminated in the nozzle/strainer and porous plate false bottom systems, and special gravel designs are used in the perforated block with gravel systems.

In some European designs [Degrémont, 1965], wash water overflow troughs are dispensed with, and the wash water is discharged over an end wall. The sand bed is not fluidized, and air scour and backwash water are applied simultaneously. An additional, simultaneous cross-filter surface flow of clarified water is used to promote movement of the dirty wash water to the wash water effluent channel.

Floor Design. Filter floors serve three essential purposes: (1) they support the filter media; (2) they collect the filtrate; and (3) they distribute the backwash water. In some designs, they also distribute the air scour. The usual designs are as follows [Cleasby, 1990; Joint Task Force, 1990; Kawamura, 1991]:

- *Pipe laterals with gravel* consist of a manifold with perforated laterals placed on the filter box floor in at least 18 in. of gravel. The coarsest gravel must be deeper than the perforated laterals (at least 6 in. total depth), and there must be at least 10 to 16 in. of finer gravel and torpedo sand above the coarsest gravel. Precast concrete inverted-V laterals are also available. The lateral orifices are drilled into the pipe bottoms or V lateral sides and are normally about ¼ to ¾ in. in diameter and 3 to 12 in. apart.

  The usual rules of thumb regarding sizing are (1) a ratio of orifice area to filter plan area of 0.0015 to 0.005, (2) a ratio of lateral cross-sectional area to total orifice area served of 2 to 4, and (3) a ratio of manifold cross-sectional area to lateral cross-sectional area served of 1.5 to 3.

  These systems generally exhibit relatively high headlosses and inferior backwash distribution and are not recommended. Poor backwash distribution can lead to gravel/sand mixing. This is aggravated by air scour, and air scour should not be applied through the laterals.
- *Blocks with gravel* consist of ceramic or polyethylene blocks overlain by at least 12 in. of gravel. The blocks are grouted onto the filter box flow and to each other. The blocks are usually about 10 in. high by 11 in. wide by 2 ft long. The tops are perforated with 5⁄32 to 5⁄16 in. orifices typically numbering about 45 per sq ft. The filtrate and backwash water flow along channels inside the block that run parallel to the lengths. Special end blocks discharge vertically into collection channels built into the filter box floor.

  Special gravel designs at least 12 to 15 in. deep are required if air scour is applied through the blocks. Porous plates and nozzle/strainer systems are preferred for air scour.
- *False bottoms with gravel* consist of precast or cast-in-place plates that have inverted pyramidal ¾ in. orifices at 1-ft intervals horizontally that are filled with porcelain spheres and gravel. The plates are installed on 2-ft-high walls that sit on the filter box floor, forming a crawl space. The gravel layer above the false bottom is at least 12 in. deep. The flow is through the crawl space between the filter box floor and the false bottom.
- *False bottoms without gravel* consist of precast plates or cast-in-place monolithic slabs set on short walls resting on the filter box floor. The walls are at least 2 ft high to provide a crawl space for maintenance and inspection. Cast-in-place slabs are preferred, because they eliminate air venting through joints. The plates are perforated, and the perforations contain patented nozzle/strainers that distribute the flow and exclude the filter sand. These systems are commonly used with air scour, which is applied through the nozzle/strainers.

  The nozzle/strainer openings are generally small, typically on the order of 0.25 mm, and are subject to clogging from construction debris, rust, and fines in the filter media and/or gravel. The largest available opening size should be used, and the effective size of the filter media should be twice the opening size. Some manufacturers recommend that a 6-in. layer of pea stone be placed over the nozzles to avoid sand and debris clogging of the nozzle openings.

  Plastic nozzle/strainers are easily broken during installation and placement of media. Nozzle-strainer materials must be carefully matched to avoid differential thermal expansion and contraction.

- *Porous plates* consist of sintered aluminum oxide plates mounted on low walls or rectangular ceramic saddles set on the filter box floor. These systems are sometimes used when air scour is employed.

  The plates are fragile and easily broken during installation and placement of media. They are subject to the same clogging problems as nozzle/strainers. They are not recommended for hard, alkaline water, lime-soda softening installations, iron/manganese waters, or iron/manganese removal installations.

Good backwash distribution requires that headloss of the orifices or pores exceed all other minor losses in the backwash system.

Hydraulics

Filter hydraulics are concerned with the clean filter headloss, which is needed to select rate-of-flow controllers, and the backwashing headloss.

Clean Filter Headloss. For a uniform sand, the initial headloss is given by the Ergun [1952] equation:

$$\frac{\Delta p}{\gamma} = h_L = f_m \left(\frac{H}{d_{\text{eq}}}\right)\left(\frac{1-\varepsilon}{\varepsilon^3}\right)\left(\frac{U_f^2}{g}\right) \tag{8.234}$$

where

A_p = the cross-sectional area of a sand grain (m^2 or ft^2)
d_{eq} = the equivalent diameter of a nonspherical particle (m or ft)
= $6V_p/A_p$
g = the acceleration due to gravity (9.80665 m/s^2 or 32.174 ft/s^2)
f_m = the MacDonald–El-Sayed–Mow–Dullien [1979] friction factor (dimensionless)
H = the thickness of the media layer (m or ft)
h_L = the headloss (m or ft)
Δp = the pressure drop due to friction (Pa or lbf/ft^2)
U_f = the filtration rate (m/s or ft/s)
V_p = the volume of a sand grain (m^3 or ft^3)
ε = the bed porosity (dimensionless)
γ = the specific weight of the water (N/m^3 or lbf/ft^3)

The friction factor f_m is given by the MacDonald–El-Sayed–Mow–Dullien [1979] equation:

$$f_m \geq 180\frac{1-\varepsilon}{\text{Re}} + 1.8, \text{ for smooth media} \tag{8.235}$$

$$f_m \leq 180\frac{1-\varepsilon}{\text{Re}} + 4.0, \text{ for rough media} \tag{8.236}$$

$$\text{Re} = \frac{\rho U_f d_{\text{eq}}}{\mu} \tag{8.237}$$

where

μ = the dynamic viscosity of water ($N{\cdot}s/m^2$ or $lbf{\cdot}s/ft^2$)
ρ = the water density (kg/m^3 or lbm/ft^3)

For filters containing several media sizes, the Fair and Geyer [1954] procedure is employed. It is assumed that the different sizes separate after backwashing and that the bed is stratified. The headloss is calculated for each media size by assuming that the depth for that size is

$$H_i = f_i H \tag{8.238}$$

where

f_i = the fraction by weight of media size class i (dimensionless)
H = the depth of the settled filter media (m or ft)
H_i = the depth of filter media size class i (m or ft)

The Fair–Geyer method yields a lower bound to the headloss. Filters do not fully stratify unless there is a substantial difference in terminal settling velocities of the media sizes. In a partially stratified filter, the bed porosity will be reduced because of intermixing of large and fine grains, and the headloss will be larger than that predicted by the Fair–Geyer method. An upper bound can be found by assuming that the entire bed is filled with grains equal in size to the effective size.

Backwashing. The headloss required to fluidize a bed is simply the net weight of the submerged media:

$$\frac{\Delta p}{\gamma} = h_L = H(\rho_s - \rho)(1 - \varepsilon) \tag{8.239}$$

where ρ_p is the particle density (kg/m^3 or lbm/ft^3). During backwashing, the headloss increases according to Eq. (8.234) until the headloss specified by Eq. (8.239) is reached; thereafter, the backwashing headloss is a constant regardless of flow rate [Cleasby and Fan, 1981].

The only effect of changing the flow rate in a fluidized bed is to change the bed porosity. The expanded bed porosity can be estimated by the Richardson–Zaki [1954] equation:

$$\varepsilon_e = \left(\frac{U_b}{v_s}\right)^{\mathrm{Re}^{0.03}/4.35} \quad \text{for } 0.2 < \mathrm{Re} < 1 \tag{8.240}$$

$$\varepsilon_e = \left(\frac{U_b}{v_s}\right)^{\mathrm{Re}^{0.1}/4.45} \quad \text{for } 1 < \mathrm{Re} < 500 \tag{8.241}$$

where

U_b = the backwashing rate (m/s or ft/s)
v_s = the free (unhindered) settling velocity of a media particle (m/s or ft/s)
ε_e = the expanded bed porosity (dimensionless)

The Reynold's number is that of the media grains in free, unhindered settling:

$$\mathrm{Re} = \frac{\rho v_s d_{eq}}{\mu} \tag{8.242}$$

The free settling velocity can be estimated either by Stokes' [1856] equation,

$$C_D = \frac{24}{\mathrm{Re}} \quad \text{for Re} < 0.1 \tag{8.141}$$

$$v_s = \frac{g(\rho_p - \rho)d_{eq}^2}{18\mu} \tag{8.142}$$

or by Shepherd's equation [Anderson, 1941],

$$C_D = \frac{18.5}{\mathrm{Re}^{0.60}} \quad \text{for } 1.9 < \mathrm{Re} < 500 \tag{8.148}$$

$$v_s = 0.153 d_{\mathrm{eq}}^{1.143} \left[\frac{g(\rho_p - \rho)}{\rho^{0.40} \mu^{0.60}} \right]^{0.714} \tag{8.149}$$

Generally, the particle size greater than 60% by weight of the grains, d_{60}, is used. The Reynold's number for common filter material is generally close to 1 and because of the weak dependence of bed expansion on the Reynold's number (raised to the one-tenth power), it is often set arbitrarily to 1.

The depth of the expanded bed may be calculated from

$$H_e = H \frac{1 - \varepsilon}{1 - \varepsilon_e} \tag{8.243}$$

where H_e is the expanded bed depth (m or ft).

The maximum backwash velocity that does *not* fluidize the bed, U_{mf}, can be estimated by setting the expanded bed porosity in Eq. (8.240) or (8.241) equal to the settled bed porosity. Alternatively, one can use the Wen–Yu [1966] correlation:

$$\mathrm{Ga} = \frac{\rho(\rho_p - \rho) g d_{\mathrm{eq}}^3}{\mu} \tag{8.244}$$

$$U_{mf} = \frac{\mu (33.7^2 + 0.0408\mathrm{Ga})^{0.5}}{\rho d_{\mathrm{eq}}} - \frac{33.7\mu}{\rho d_{\mathrm{eq}}} \tag{8.245}$$

where

Ga = the Galileo number (dimensionless)

U_{mf} = the minimum bed fluidization backwash rate (m/s or ft/s)

In dual-media filters, the expansion of each layer must be calculated separately and totaled. Each layer will experience the same backwash flow rate, but the minimum flow rates required for fluidization will differ.

The optimum expansion for cleaning without air scour is about 70%, which corresponds to maximum hydraulic shear [Cleasby *et al.*, 1975].

Fluidization alone is not a very effective cleaning mechanism; fluidization plus air scour is preferred. Air scour is either applied alone followed by bed fluidization or applied simultaneously with a backwash flow rate that does not fluidize the bed. The latter method is the most effective cleaning procedure.

For typical fine sands with an effective size of about 0.5 mm, air scour and backwash are applied separately. Generally, air is applied at a rate of about 1 to 2 scfm/ft^2 for 2 to 5 min followed by a water rate of 5 to 8 gpm/ft^2 (nonfluidized) to 15 to 20 gpm/ft^2 (fully fluidized) [Cleasby, 1990]. Backwash rates greater than about 15 gpm/ft^2 may dislodge the gravel layer.

Simultaneous air scour and backwash may be applied to dual-media filters and to coarse-grain filters with effective sizes of about 1.0 mm or larger. Air flow rates are about 2 to 4 scfm/ft^2, and water flow rates are about 6 gpm/ft^2 [Cleasby, 1990].

Amirtharajah [1984] has developed a theoretical equation for optimizing air flow and backwash rates:

$$0.45 Q_a^2 + 100 \left(\frac{U_b}{U_{mf}} \right) = 41.9 \tag{8.246}$$

where

Q_a = the air flow rate ($scfm/ft^2$)

U_b = the backwash rate (m/s or ft/s)

U_{mf} = the minimum fluidization velocity based on the d_{60} grain diameter (m/s or ft/s)

Equation (8.246) may overestimate the water flow rates [Cleasby, 1990].

Water Treatment

U.S. EPA Surface Water Rule

Enteric viruses and the cysts of important pathogenic protozoans like *Giardia lamblia* are highly resistant to the usual disinfection processes. The removal of these organisms from drinking water depends almost entirely on coagulation, sedimentation, and filtration. Where possible, source protection is helpful in virus control.

The U.S. Environmental Protection Agency [U.S. EPA, 1989; Malcolm Pirnie, Inc. and HDR Engineering, Inc., 1990] has issued treatment regulations and guidelines for public water supplies that use surface water sources and groundwater sources that are directly influenced by surface waters. Such systems are required to employ filtration and must achieve 99.9% ("three log") removal or inactivation of *G. lamblia* and 99.99% ("four log") removal or inactivation of enteric viruses. The performance requirements for filtration are that the filtered water turbidity never exceed 5 TU and that individual filter systems meet the following turbidity limits at least 95% of the time:

- Conventional rapid sand filters preceded by coagulation, flocculation, and sedimentation—0.5 TU
- Direct filtration—0.5 TU
- Cartridge filters and approved package plants—0.5 TU
- Slow sand filters—1.0 TU
- Diatomaceous earth filters—1.0 TU

Public supplies can avoid the installation of filters if they meet the following basic conditions [Pontius, 1990]:

- The system must have an effective watershed control program.
- Before disinfection, the fecal and total coliform levels must be less than 20/100 mL and 100/100 mL, respectively, in at least 90% of the samples.
- Before disinfection, the turbidity level must not exceed 5 TU in samples taken every 4 hours.
- The system must practice disinfection and achieve 99.9 and 99.99% removal or inactivation of *G. lamblia* cysts and enteric viruses, respectively, daily.
- The system must submit to third-party inspection of its disinfection and watershed control practices.
- The system cannot have been the source of a water-borne disease outbreak, unless it has been modified to prevent another such occurrence.
- The system must be in compliance with the total coliform rule requirements.
- The system must be in compliance with the total trihalomethane regulations.

The *Guidance Manual for Compliance with the Surface Water Treatment Requirements for Public Water Systems* [Malcolm Pirnie, Inc. and HDR Engineering, Inc., 1990] lists other requirements and options.

Media

The generally acceptable filtering materials for water filtration are silica sand, crushed anthracite coal, and granular activated carbon [Water Supply Committee, 1987; Standards Committee on Filtering Material, 1989]:

- Silica sand particles shall have a specific gravity of at least 2.50, shall contain less than 5% by weight acid-soluble material and shall be free (less than 2% by weight of material smaller than 0.074 mm) of dust, clay, micaceous material, and organic matter.
- Crushed anthracite coal particles shall have a specific gravity of at least 1.4, a Mohr hardness of at least 2.7, and shall contain less than 5% by weight of acid-soluble material and be free of shale, clay, and debris.

In addition, filters normally contain a supporting layer of gravel and torpedo sand that retains the filtering media. In general, the supporting gravel and torpedo sand shall consist of hard, durable, well-rounded particles with less than 25% by weight having a fractured face, less than 2% by weight being elongated or flat pieces, less than 1% by weight smaller than 0.074 mm, and less than 0.5% by weight consisting of coal, lignin, and organic impurities. The acid solubility shall be less than 5% by weight for particles smaller than 2.36 mm, 17.5% by weight for particles between 2.36 and 25.4 mm, and less than 25% for particles equal to or larger than 25.4 mm. The specific gravity of the particles shall be at least 2.5.

The grain size distribution is summarized in terms of an *effective size* and a *uniformity coefficient*:

- The effective size(s) is the sieve opening that passes 10% by weight of the sample.
- The uniformity coefficient (UC) is the ratio of the sieve opening that passes 60% by weight of the sample to the opening that passes 10% by weight of the same sample.

Typical media selections for various purposes are given in Table 8.5.

Single Media Filters. The typical filter cross section for a single-media filter (meaning either sand, coal, or GAC used alone) is given in Table 8.6. The various media are placed coarsest to finest from bottom to top, and the lowest layer rests directly on the perforated filter floor. Various manufacturers recommend different size distributions for their underdrain systems; two typical recommendations are given in Table 8.7. Kawamura [1991] recommends a minimum gravel depth of 16 in. in order to avoid disruption during backwashing.

Dual Media Filters. The principal problem with single-media filters is that the collected solids are concentrated in the upper few inches of the bed. This leads to relatively short filter runs. If the solids can be spread over a larger portion of the bed, the filter runs can be prolonged. One way of doing this is dual-media filters. Dual-media filters consist of an anthracite coal layer on top of a sand layer. The coal grains are larger than the sand grains, and suspended solids penetrate more deeply into the bed before being captured.

The general sizing principle is to minimize media intermixing following backwashing. The critical size ratio for media with different densities is that which produces equal settling velocities [Conley and Hsiung, 1969]:

$$\frac{d_1}{d_2} = \left(\frac{\rho_2 - \rho_m}{\rho_1 - \rho_m}\right)^{0.625} \tag{8.247}$$

where

d_1, d_2 = the equivalent grain diameters of the two media (m or ft)
ρ_1, ρ_2 = the densities of the two media grains (kg/m^3 or lb/ft^3)
ρ_m = the density of a fluidized bed (kg/m^3 or lbm/ft^3)

Table 8.5 Media Specification for Various Applications[1–3]

Process	Media	Effective Size (mm)	Depth (ft)
Coagulation, settling, filtration; filtration rate less than or equal to 4 gpm/ft^2	Medium sand alone	0.45 to 0.55	2.0 to 2.5
Coagulation, settling, filtration; filtration rate greater than 4 gpm/ft^2 with polymer addition	Coarse sand alone	0.8 to 2.0	2.6 to 7.0
Coagulation, settling, filtration; filtration rate greater than 4 gpm/ft^2	Coarse sand	0.9 to 1.1	1.0 to 1.5
	Anthracite	0.9 to 1.4	1.5 to 2.0
Direct filtration of surface water	Coarse sand	1.5	1.0 to 1.5
	Anthracite	1.5	1.5 to 2.0
Iron and manganese removal	Coarse sand alone	<0.8	2.5 to 3.0
Iron and manganese removal	Coarse sand	0.9 to 1.1	1.0 to 1.5
	Anthracite	0.9 to 1.4	1.5 to 2.0
Coarse single media with simultaneous air scour and backwash for coagulation, settling, and filtration	Coarse sand alone	1.4 to 1.6	3.5 to 7.0
Coarse single media with simultaneous air scour and backwash for iron and manganese removal	Coarse sand alone	1.0 to 2.0	5.0 to 10.0

[1]Cleasby, J. L. 1990. Filtration. In *Water Quality and Treatment: A Handbook of Community Water Supplies,* 4th ed., p. 455, F. W. Pontius, tech. ed., McGraw-Hill, New York.

[2]Kawamura, S. 1991. *Integrated Design of Water Treatment Facilities,* John Wiley & Sons, New York.

[3]*Note:* Uniformity coefficient 1.6 to 1.7 for medium sand and 1.5 for coarse sand or coal.

Table 8.6 Typical Single-Media, Rapid Sand Filter Cross Section and Media Specifications[1]

Material	Effective Size (mm)	Uniformity Coefficient (Dimensionless)	Depth (in.)
Silica Sand	0.45 to 0.55	≤1.65	24 to 30
Anthracite coal			
Surface water turbidity removal	0.45 to 0.55	≤1.65	24 to 30
Groundwater Fe/Mn removal	≤0.8		24 to 30
Granular activated carbon	0.45 to 0.55	≤1.65	24 to 30
Torpedo sand	0.8 to 2.0	≤1.7	3
Gravel	3/16 to 3/32 in.	—	2 to 3
	1/2 to 3/16 in.	—	2 to 3
	3/4 to 1/2 in.	—	3 to 5
	1 3/4 to 3/4 in.	—	3 to 5
	2 1/2 to 1 3/4 in.		5 to 8

[1]Water Supply Committee. 1987. *Recommended Standards for Water Works,* Health Research, Albany, NY.

Table 8.7 Gravel Specifications for Specific Filter Floors

Gravel Size Range (U.S. Standard Sieve Sizes, mm)	Layer Thickness (in.) General[1]	Leopold Dual-Parallel Lateral Block[1,2]	Wheeler[2]
1.70 to 3.35	3	3	—
3.35 to 6.3	3	3	—
4.75 to 9.5	—	—	3
6.3 to 12.5	3	3	—
9.5 to 16.0	—	—	3
12.5 to 19.0	3	3	—
16.0 to 25.0	—	—	3
25.0 to 31.5	—	—	To cover drains
19.0 to 37.5	4 to 6	—	—

[1]Kawamure, S. 1991. *Integrated Design of Water Treatment Facilities,* John Wiley & Sons, New York.
[2]Williams, R. B. and Culp, G. L., eds. 1986. *Handbook of Public Water Systems,* Van Nostrand Reinhold, New York.

The fluid density is that of the fluidized bed, which is a mixture of water and solids:

$$\rho_m = \varepsilon\rho + \tfrac{1}{2}(\rho_1 + \rho_2)(1 - \varepsilon) \tag{8.248}$$

Kamamura [1991] uses the density of water, ρ, instead of ρ_m and an exponent of 0.665 instead of 0.625.

The media are generally proportioned so that the ratio of the weight fractions is equal to the ratio of the grain sizes:

$$\frac{d_1}{d_2} = \frac{f_1}{f_2} \tag{8.249}$$

where f_1, f_2 are the weight fractions of the two media (dimensionless).

Some intermixing of the media at the interface is desirable. Cleasby and Sejkora [1975] recommend the following size distributions of coal and sand:

- Sand—an effective size 0.46 mm and a uniformity coefficient of 1.29
- Coal—an effective size of 0.92 mm and a uniformity coefficient of 1.60
- Ratio of the diameter larger than 90% of the coal to the diameter larger than 10% of the sand of 4.05

An interfacial size ratio of coal to sand of at least 3 is recommended [Joint Task Force, 1990].

The coal layer is usually about 18 in. deep, and the sand layer is about 8 in. deep. A typical specification for the media is given in Table 8.8 [Culp and Culp, 1974].

Table 8.8 Size Specifications for Dual-Media Coal Sand Filters[1]

U.S. Sieve Size (mm)	Percentage by Weight Passing Coal	Sand
4.75	99 to 100	—
3.35	95 to 100	—
1.40	60 to 100	—
1.18	30 to 100	—
1.00	0 to 50	—
0.850	0 to 5	96 to 100
0.600	—	70 to 90
0.425	—	0 to 10
0.297	—	0 to 5

[1]Culp, G. L. and Culp, R. L. 1974. *New Concepts in Water Purification,* Van Nostrand Reinhold, New York.

Operating Modes

Several different filter operating modes are recognized [AWWA Filtration Committee, 1984]:

- *Variable-control, constant-rate.* In this scheme, plant flow is divided equally among the filters in service, and each filter operates at the same filtration rate and at a constant water level in the filter box. Each filter has an effluent rate-of-flow controller that compensates for the changing headloss in the bed.

This is a highly automated scheme that requires only minimal operator surveillance. It requires the most instrumentation and flow control equipment, because each filter must be individually monitored and controlled.

- *Flow control from filter water level.* In this scheme, the plant flow is divided equally among the filters in service by a flow-splitter. Each filter operates at a constant water level that is maintained by a butterfly valve on the filtered water line. The water levels differ among boxes in accordance with the differences in media headloss due to solids captured.

  This is an automated scheme that requires little operator attention. Individual filters do not require flow measurement devices, so less hardware is needed than in the variable-control, constant-rate scheme.

- *Inlet flow splitting (constant rate, rising head).* In this scheme, the plant flow is divided equally among the filters in service by a flow-splitter that discharges above the highest water level in each filter. Each filter operates at a constant filtration rate, but the rates differ from one filter to the next. Instead of a rate-of-flow controller, the constant filtration rate is maintained by allowing the water level in each filter box to rise as solids accumulate in the media. When a filter reaches its predetermined maximum water level, it is removed from service and backwashed. Automatic overflows between boxes are needed to prevent spillage. The filter discharges at an elevation above the media level, so negative pressures in the media are impossible.

  The monitoring instrumentation and flow control devices are minimal, but the operator must attend to filter cleanings.

- *Variable declining rate.* In this scheme, all filters operate at the same water level (maintained by a common inlet channel) and discharge at the same level (which is maintained above the media to avoid negative pressure). The available head is the same for each filter, and the filtration rate varies from one filter to the next in accordance with the relative headloss in the media. When the inlet water level reaches some predetermined elevation, the dirtiest filter is taken out of service and cleaned. The flow rate on each filter declines stepwise as clean filters are brought on line that take up a larger share of the flow.

  This scheme produces a better filtrate, because the filtration rate automatically falls as the bed becomes clogged, which reduces shearing stresses on the deposit. It also makes better use of available head, because most of the water is always going through relatively clean sand. The control system is simplified and consists mainly of a flow rate limiter on each filter to prevent excessive filtration rates in the clean beds. The operator must monitor the water level in the filters and plant flow rate.

- *Direct filtration.* Direct filtration eliminates flocculation and settling but not chemical addition and rapid mixing. The preferred raw water for direct filtration has the following composition [Direction Filtration Subcommittee, 1980; Cleasby, 1990]:

  - Color less than 40 Hazen (platinum-cobalt) units
  - Turbidity less than 5 formazin turbidity units (FTU)
  - Algae (diatoms) less than 2000 asu/mL (1 asu = 400 μm^2 projected cell area)
  - Iron less than 0.3 mg/L
  - Manganese less than 0.05 mg/L

  The coagulant dosage should be adjusted to form small, barely visible pinpoint flocs; large flocs shorten filter runs by increasing headlosses and promoting early breakthrough [Cleasby, 1990]. The optimum dosage is determined by filter behavior and is the smallest that achieves the required effluent turbidity.

  Dual-media filters are required to provide adequate solids storage capacity. Design filtration rates range from 4 to 5 gpm/ft^2, but provision for operation at 1 to 8 gpm/ft^2 should be made [Joint Task Force, 1990].

Wastewater Treatment

Granular-media filters are used in wastewater treatment plants as tertiary processes following secondary clarification of biological treatment effluents. A wide variety of designs are available, many of them proprietary. The major problems in wastewater filters are the relatively high solids concentrations in the influent flow and the so-called stickiness of the suspended solids. These problems require that special consideration be given to designs that produce long filter runs and to effective filter cleaning systems.

Wastewater filters are almost always operated with the addition of coagulants. Provision should be made to add coagulants to the secondary clarifier influent and the filter influent. Rapid mixing is required; it may be achieved via tanks and turbines or static inline mixers.

Filters may be classified as follows [Metcalf & Eddy, Inc., 1991]:

- *Stratified or unstratified media.* Backwashing alone tends to stratify monomedia beds with the fines on top. Simultaneous air scour and backwash produces a mixed, unstratified bed. Deep bed filters almost always require air scour and are usually unstratified.
- *Mono, dual, or multimedia.* Filter media may be either a single layer of one material such as sand or crushed anthracite, two separate layers of different material such as sand and coal, or multilayer filters (usually five) of sand, coal and garnet, or ilmenite.
- *Continuous or discontinuous operation.* Several proprietary cleaning systems are available that permit continuous operation of the filter: downflow moving bed, upflow moving bed, traveling bridge, and pulsed bed.

  In downflow moving beds, both the water and the filter media move downward. The sand is removed below the discharge point of the water, subjected to air scouring, and returned to the top of the filter.

  In upflow beds, the water flows upward through the media and the media moves downward. The media is withdrawn continuously from the bottom of the filter, washed, and returned to the top of the bed.

  In traveling bridge filters, the media is placed in separate cells in a tank, and the backwashing system is mounted on a bridge that moves from one cell to the next for cleaning. At any moment, only one of the cells is being cleaned and the others are in service.

  Pulsed bed filters are really semicontinuous filters. Air is diffused just above the media surface to keep solids in suspension, and periodically air is pulsed through the bed, which resuspends the surface layer of the media, releasing collected solids. Solids do migrate into the bed, and eventually the filtering process must be shut down for backwashing.

  Conventional filters, which are taken out of service for periodic cleaning, are classified as discontinuous.

The most common designs for new facilities specify discontinuous service, dual-media filters with coagulation and flow equalization to smooth the hydraulic, and solids loading. Older plants with space restrictions are often retrofitted with continuous service filters that can operate under heavy and variable loads.

Performance and Media

The liquid applied to the filter should contain less than 10 mg/L TSS [Metcalf & Eddy, Inc., 1991; Joint Task Force, 1991]. Under this condition, dual-media filters with chemical coagulation can produce effluents containing turbidities of 0.1 to 0.4 JTU. If the coagulant is aluminum or ferric iron salts, orthophosphate will also be removed, and effluent orthophosphate concentrations of about 0.1 mg/L as PO_4 can be expected. As the influent suspended solids concentration increases, filter efficiency falls off. At an influent load of 50 mg/L TSS, filtration with coagulation can be expected to produce effluents of about 5 mg/L TSS.

Commonly used media are described in Table 8.9.

Table 8.9 Media for Wastewater Effluent Filtration[1–3]

Filter Type and Typical Filtration Rate	Media	Effective Size (mm)	Media Depth (in.)
Monomedia, shallow bed, 3 gpm/ft^2	Sand	0.35 to 0.60	10 to 12
Monomedia, shallow bed, 3 gpm/ft^2	Anthracite	0.8 to 1.5	12 to 20
Monomedia, conventional, 3 gpm/ft^2	Sand	0.4 to 0.8	20 to 30
Monomedia, conventional, 4 gpm/ft^2	Anthracite	0.8 to 2.0	24 to 36
Monomedia, deep bed, 5 gpm/ft^2	Sand	2.0 to 3.0	36 to 72
Monomedia, deep bed, 5 gpm/ft^2	Anthracite	2.0 to 4.0	36 to 84
Dualmedia, 5gpm/ft^2	Sand	0.4 to 0.8	6 to 12
	Anthracite	0.8 to 2.0	12 to 30
Trimedia, trilayer 5 gpm/ft^2	Anthracite (top)	1.0 to 2.0	8 to 20
	Sand (middle)	0.4 to 0.8	8 to 16
	Garnet/ilmenite (bottom)	0.2 to 0.6	2 to 6

[1]Metcalf & Eddy, Inc. 1991. *Waster Engineering: Treatment, Disposal, and Reuse,* 3rd ed., rev. by G. Tchobanoglous and F. L. Burton, McGraw-Hill, New York.

[2]Joint Task Force of the Water Environment Federation and the American Society of Civil Engineers, 1991. *Design of Municipal Wastewater Treatment Plants: Volume II—Chapters 13–20,* WEF Manual of Practice No. 8, ASCE Manual and Report on Engineering Practice No. 76, Water Environment Federation, Alexandria, VA; American Society of Civil Engineers, New York.

[3]*Note:* Uniformity coefficient: 1.5 for sand and 1.6 for other materials.

Backwashing

The general recommendation is that the backwashing rate expand the bed by 10% so that the media grains have ample opportunity to rub against one another [Joint Task Force, 1991]. Either preliminary or concurrent air scour should be provided. The usual air scour rate is 3 to 5 scfm/ft^2. If two or more media are employed, a Baylis-type rotary water wash should be installed at the expected media interface heights of the expanded bed. The rotary water washes are operated at pressures of 40 to 50 psig and water rates of 0.5 to 1.0 gpm/ft^2 for single-arm distributors and 1.5 to 2.0 gpm/ft^2 for double-arm distributors. Distributor nozzles should be equipped with strainers to prevent clogging.

8.7 Activated Carbon

Preparation and Regeneration

Pure carbon occurs as crystals of either diamond or graphite or as fullerene spheres. The atoms in diamond are arranged as tetrahedra, and the atoms in graphite are arranged as sheets of hexagons. Activated carbon grains consist of random arrays of microcrystalline graphite. Such grains can be made from many different organic substances, but various grades of coal are most commonly used, because the product is hard, dense, and easy to handle.

The raw material is first carbonized at about 500°C in a nonoxidizing atmosphere. This produces a char that contains some residual organic matter and many small graphite crystals. The char is then subjected to a slow oxidation either in air at about 500°C or steam or carbon dioxide at 800 to 950°C. The preferred activation atmospheres are steam and carbon dioxide, because the oxidation reactions are endothermic and more easily controlled. The slow oxidation removes residual organic matter and small graphite crystals and produces a network of microscopic pores. The higher temperature promotes the formation of larger graphite crystals, which reduces the randomness of their arrangement.

It is almost always economical to recover and regenerate spent carbon, unless the quantities are very small. This is true even of powdered carbon, if it can be easily separated from the process stream. The economic break point for onsite regeneration depends on carbon usage and varies from 500 to 2000 lb/day [Snoeyink, 1990]. Larger amounts should be shipped to commercial regeneration plants.

Spent carbon can be regenerated as it was made by heating in a rotary kiln, fluidized bed, multiple hearth, or infrared furnaces [von Dreusche, 1981; McGinnis, 1981]. In multiple hearth furnaces, the required heat is supplied by the hot combustion gases of a fuel. The fuel is oxidized with only small excess air to minimize the amount of oxygen fed to the furnace. In infrared furnaces, heating is supplied by electrical heating elements. This method of heating provides somewhat better control of oxygen concentration in the furnace, but some oxygen enters with the carbon itself through the feedlock.

The usual regeneration stages in any furnace are as follows [McGinnis, 1981; Snoeyink, 1990]:

- A drying stage—usually operated at about 200 to 700°C, depending on furnace type
- A pyrolysis stage—usually operated at about 500 to 800°C
- A coke oxidation stage—usually operated at about 700°C for water treatment carbon and at about 900°C for wastewater treatment carbons [Culp and Clark, 1983].

The pyrolysis stage vaporizes and cokes the adsorbed organics. The vaporized material diffuses into the hot gases in the furnace and either reacts with them or is discharged as part of the furnace stack gas. The coked material is oxidized either by water vapor or by carbon dioxide, and the oxidation products exit in the furnace stack gases:

$$C + CO_2 \rightarrow 2\,CO \tag{8.250}$$

$$C + H_2O \rightarrow CO + H_2 \tag{8.251}$$

For water treatment carbons, the pyrolysis and activation stages are operated at temperatures below that required to graphitize the adsorbate char. Consequently, the newly formed char is more reactive than the original, largely graphitized activated carbon, and it is selectively removed by the oxidation process. Wastewater treatment carbons are oxidized at temperatures that graphitize as well as oxidize the adsorbate char. Significant amounts of the original carbon are also oxidized at this temperature.

Regeneration losses are generally on the order of 5 to 10% by weight. This is due mostly to spillage and other handling and transport losses, not to furnace oxidation. There is usually a reduction in grain size, which affects the hydraulics of fixed bed adsorbers and the adsorption rates. However, the adsorptive capacity of the original carbon is little changed [McGinnis, 1981].

Characteristics

The properties of commercially available activated carbons are given in Table 8.10. For water treatment, the American Water Works Association [AWWA, 1990a,b] requires the following:

- The carbon should contain no soluble organic or inorganic substances in quantities capable of producing health-damaging effects on those consuming the water or that would otherwise render the water unfit for public use.
- The carbon should not impart to the water any contaminant that exceeds the limits as defined by the U.S. Environmental Protection Agency.
- The moisture content should not exceed 8% by weight when packed or shipped.
- The apparent density should not be less than 0.36 g/cm^3.
- The average particle size should be at least 70% of its original size after subjection to either the stirred abrasion test or the Ro-Tap abrasion test (described in the standard).
- The adsorptive capacity as determined by the iodine number should not be less than 500.

Table 8.10 Properties of Commercially Available Activated Carbons[1–4]

Property	Granular Activated Carbon	Powdered Activated Carbon
Effective size (mm)	0.6 to 0.9	<0.01 (65 to 90% <0.044)
Uniformity coefficient (dimensionless)	<1.9	Not applicable
Real density of carbon, excluding pores (g/cm^3)	2.0 to 2.1	Same
Dry density of grains (g/cm^3)	0.9 to 1.4	Same
Wet density of Grains (g/cm^3)	1.5 to 1.6	Same
Dry, bulk density of packed bed (g/cm^3) (apparent density)	0.40 to 0.50	Same
Pore volume of grains (cm^3/g)	0.60 to 0.95	Same
Specific surface area of grains (m^2/g)	600 to 1100	Same
Iodine number (mg I_2/g C at 0.02 NI_2 at 20 to 25°C)	600 to 1100	Same

[1]Joint Task Force. 1990. *Water Treatment Plant Design,* 2nd ed., McGraw-Hill, New York.

[2]Culp/Wesner/Culp, Inc. 1986. *Handbook of Public Water Supply Systems,* R. B. Williams and G. L. Culp, eds., Van Nostrand Reinhold, New York.

[3]Joint Task Force of the Water Environment Federation and the American Society of Civil Engineers. 1991. *Design of Municipal Wastewater Treatment Plants: Volume II—Chapters 13–20,* WEF Manual of Practice No. 8, ASCE Manual and Report on Engineering Practice No.76, Water Environment Federation, Alexandria, VA; American Society of Civil Engineers, New York.

[4]Snoeyink, V. L. 1990. Adsorption of organic compounds. In *Water Quality and Treatment: A Handbook of Community Water Supplies,* 4th ed., p. 781, American Water Works Association, F. W. Pontius, ed., McGraw-Hill, New York.

Uses

Taste and Odor

The traditional use of activated carbon is for taste and odor control. The usage is generally seasonal, and PAC slurry reactors (actually flocculation tanks) have often been the preferred mode of application.

The principal sources of tastes and odors are [Thimann, 1963]

- Manufacturing—generally due to oils, phenols and cresols, ammonia and amines (especially diamines), sulfides and mercaptans, and volatile fatty acids (especially propionic, butyric and isobutyric, valeric and isovaleric, and caproic acids)
- Putrefaction of proteinaceous and fatty material—generally diamines (skatol, indole, γ-amino butyric acid, cadaverine, putrescine, tryptamine, and tyramine), sulfides and mercaptans, and volatile fatty acids
- Microbial and metazoan blooms—especially cyanobacterial blooms

The odors associated with algal and other blooms have been classified as shown in Table 8.11.

Odors are normally measured by the threshold odor number (TON). This is defined as the sample with the smallest diluted volume that still retains an odor. The samples are diluted with distilled water, and the TON is calculated as [Joint Editorial Board, 1992],

$$\mathrm{TON} = \frac{V_{\mathrm{of}} + V_s}{V_s} \tag{8.252}$$

where

TON = the threshold odor number (dimensionless)

V_{of} = the volume of odor-free water used in the threshold odor test (mL)

V_s = the volume of original sample used in the threshold odor test (mL)

Table 8.11 Taste and Odor-Producing Organisms[1–3]

Organism	Odor When Numbers Are Moderate	Odor When Numbers Are Abundant
	Cyanobacteria	
Anabaena circinalis	Grassy	Pigpen
Anabaena flosaquae	Moldy, grassy, nasturtium	Pigpen
Anabaenopsis	—	Grassy
Anacystis	Grassy	Septic
Aphanizomenon	Moldy, grassy	Septic
Clathrocystis	Sweet, grassy	—
Cylindrospermum	Grassy	Septic
Coeloshaerium	Sweet, grassy	—
Gloeotrichia	—	Grassy
Gomphosphaeria	Grassy	Grassy
Nostoc	Musty	Septic
Oscillatoria	Grassy	Musty, spicy
Rivularia	Grassy	Musty
	Green Algae	
Actinastrum	—	Grassy, musty
Chara	Skunk, garlic	Garlic, spoiled
Chlorella	—	Musty
Cladophra	—	Septic
Closterium	—	Grassy
Cosmarium	—	Grassy
Dictyosphaerium	Grassy, nasturtium	Fishy
Gleocystis	—	Septic
Hydrodictyon	—	Septic
Nitella	Grassy	Grassy, septic
Pandorina	—	Grassy
Pediastrum	—	Grassy
Scendesmus	—	Grassy
Spirogyra	—	Grassy
	Diatoms	
Asterionella	Geranium, spicy	Fishy
Cyclotella	Geranium	Fishy
Diatoma	—	Aromatic
Fragilaria	Geranium, aromatic	Musty
Melosira	Geranium	Musty
Meridion	—	Spicy
Pleurosigma	—	Fishy
Synedra	Earthy, grassy	Musty
Stephanodiscus	Aromatic geranium	Fishy
Tabellaria	Aromatic geranium	Fishy
	Flagellated Algae and Protozoa	
Bursaria	Irish moss, salt marsh	Fishy
Ceratium	Fishy	Septic, vile stench
Chlamydomonas	Musty, grassy	Fishy, septic
Chrysosphaerella	—	Fishy
Cryptomonas	Violet	Violet
Codonella	Fishy	—
Dinobryon	Violet	Fishy
Eudorina	—	Fishy
Euglena	—	Fishy
Glendinium	—	Fishy
Gonium	—	Fishy
Mallomonas	Violet	Fishy

Table 8.11 (continued) Taste and Odor-Producing Organisms[1–3]

Organism	Odor When Numbers Are Moderate	Odor When Numbers Are Abundant
Pandorina	—	Fishy
Peridnium	Cucumber	Fishy
Synura	Cucumber, muskmelon	Fishy
Uroglenopsis	—	Fishy, cod liver oil, oily
	Rotifera	
Anuraea	Fishy	—
	Crustacea	
Cyclops	Fishy	—
Daphnia	Fishy	—

[1]Palmer, C. M. 1967. *Algae in Water Supplies: An Illustrated Manual on the Identification, Significance, and Control of Algae in Water Supplies,* Public Health Service Publication No. 657, U.S. Department of Health, Education, and Welfare, Public Health Service, Division of Water Supply and Pollution Control, Washington, D.C.

[2]Turre, G. J. 1942. Pollution from natural sources, particularly algae. In *Taste and Odor Control in Water Purification,* p. 17, West Virginia Pulp and Paper Company, Industrial Chemical Sales Division, New York.

[3]Rohlich, G. A. and Sarles, W. B. 1949. Chemical composition of algae and its relationship to taste and odor, *Taste and Odor Journal,* 18(10):1.

Conventional water treatment processes consisting of coagulation, settling, and rapid sand filtration do not substantially reduce the TON. In fact, coagulation with lime or soda ash frequently increases it [Anonymous, 1942]. Coagulant chemicals sometimes change the character of the odor. Rapid sand filtration is effective in odor removal.

THM Precursors

The naturally occurring humic substances in surface water and groundwater react with chlorine to form chloroform ($CHCl_3$), which is a known carcinogen [Joint Task Force, 1990]. This problem presents itself continuously, so GAC-packed columns are the preferred mode of application. GAC will remove both the trihalomethanes and their precursor humic substances. The removal efficiency of the precursors is somewhat greater than that of the THMs. THM removal is best on fresh carbon, and removals are impaired by competitive adsorption on older carbons. The useful life of GAC replacement media in rapid sand filter boxes is about 2 months [Ohio River Valley Water Sanitation Commission, 1980].

Organic Poisons and Priority Pollutants

Water supply intakes on large rivers are subject to transport and processing spills of most industrial organic chemicals. Large spills are often detected early on and tracked, so that utilities can shut down operations while they pass. However, many smaller spills go undetected, and medically significant concentrations of organic poisons can enter public water supplies. This problem is similar to the THM problem, and it requires continuous water treatment by GAC in packed columns. GAC filter-adsorbers do not remove priority pollutants reliably, and desorption of small chlorinated organic molecules sometimes occurs [Ohio River Valley Water Sanitation Commission, 1980; Graese *et al.*, 1987]. GAC filter adsorbers also exhibit lower carbon capacities and earlier breakthrough due to pore clogging by water treatment chemicals and flocs.

Wastewater SOC

Stringent NPDES effluent permits may require GAC-packed columns to remove soluble organic carbon (SOC) such as residual soluble BOD_5, priority pollutants, or THM precursors [Joint Task

Force, 1991]. The suspended solids content of biologically treated effluents usually must be removed by chemical coagulation, sedimentation, and filtration prior to carbon treatment. The Lake Tahoe results indicate that empty bed contacting times of 15 to 30 min can produce an effluent BOD_5, COD, and TOC of less than 1, 12, and 3 mg/L, respectively, if clear, biologically treated water is fed to the columns [Culp and Culp, 1971].

Equilibria

A nonpolar adsorbent, such as activated carbon, will adsorb nonpolar solutes from a polar solvent, such as water [Adamson, 1982]. Furthermore, Traube's rule [Freundlich, 1922] states that the adsorption of a homologous series of organic substances from an aqueous solution increases as the chain length increases.

Adsorption is an equilibrium process in which organic molecules dissolved in water displace water molecules from the surface of the activated carbon. The equilibrium depends on the water phase activities of the organic molecules (a_o), the water phase activity of the water itself (a_w), the mole fraction of the organic molecule in the adsorbed layer (x_o), and the mole fraction of water in the adsorbed layer (x_w), and it is characterized by a free energy of adsorption [Adamson, 1982]:

$$K = \frac{a_w x_o}{a_o x_w} = e^{-\Delta G^o/RT} \tag{8.253}$$

where

a_o = water phase activity of organic molecules (mol/L)
a_w = water phase activity of water itself (mol/L)
ΔG^o = the standard free energy of a reaction (J/mol or ft-lbf/lb-mol)
K = the equilibrium constant (units vary)
R = the gas constant (8.314 510 J/K·mol or 1545 ft·lbf/°R·lb-mol)
T = the absolute temperature (K or °R)
x_o = the mole fraction of organic matter in the adsorbed layer (dimensionless)
x_w = the mole fraction of water in the adsorbed layer (dimensionless)

Langmuir

The Langmuir isotherm follows directly from the equilibrium constant if all adsorption sites on the carbon have equal free energies of adsorption; the solution is dilute so that concentrations are nearly equal to activities; the mole fraction of water is eliminated, using the fact that the sum of the mole fractions ($x_o + x_w$) must be unity; the number of adsorption sites per unit mass of carbon can be estimated so that the mole fraction of adsorbed organic matter can be expressed as mass of organic matter per unit mass of carbon:

$$q_e = \frac{q_{\max} C_e}{K_C + C_e} \tag{8.254}$$

where

C_e = the equilibrium concentration of organic matter in water (kg/m^3 or lb/ft^3)
K_C = the affinity constant for the Langmuir isotherm (kg/m^3 or lb/ft^3)
q_e = the mass of adsorbed organic matter per unit mass of activated carbon at equilibrium (kg/kg or lbm/lbm)
$q_{\max}$ = the maximum capacity of activated carbon to adsorb organic matter (kg/kg or lbm/lbm)

Freundlich

It is generally not true that the free energy of adsorption is equal for all sites, so the Langmuir isotherm fails to fit many data sets. However, good fits are usually obtained with the Freundlich isotherm. This can be derived from the Langmuir isotherm if it is assumed that the distribution of adsorption energies is a decaying exponential [Halsey and Taylor, 1947]:

$$q_e = kC_e^{1/n} \tag{8.255}$$

where

k = a positive empirical coefficient (units vary)
n = a positive empirical exponent (dimensionless)

The representation of the exponent on the concentration as a fraction is traditional and has no special meaning.

Three-Parameter Isotherm

Occasionally it is convenient to use a purely empirical three-parameter adsorption isotherm formula [Crittenden and Weber, 1978]:

$$q_e = \frac{q_{\max} C_e}{K_C + C_e^{\beta}} \tag{8.256}$$

where β is an empirical exponent (dimensionless).

Freundlich Parameter Values

Freundlich parameter values for a number of important organic contaminants are given in Table 8.12.

Kinetics

During operation a packed column first becomes saturated at its inlet end, and the carbon grains are in equilibrium with the influent solute concentration, C_o. Downstream of the saturated zone is the mass transfer zone, in which the carbon is adsorbing solute from the pore water. In this zone, the pore water solute concentration varies from its influent value to some very low nonzero value. The mass transfer zone moves through the bed at a constant velocity toward the outlet end, and as it does the effluent concentration begins to rise. The so-called breakthrough curve is a plot of the effluent concentration vs. service time, and it is a mirror image of the concentration profile in the mass transfer zone. The breakthrough concentration, C_b, is set by the required product water quality, and when it is reached the carbon must be taken out of service and replaced or regenerated.

Empirical Column Tests

In pilot studies, several small columns are connected in series, and sampling taps are installed in each column's outlet line [Wagner and Julia, 1981; Hutchins, 1981]. The test solution is pumped through the columns at some controlled rate, and a plot of effluent concentration versus service time is prepared for each column. The columns will become saturated in sequence from first to last. The sampling taps represent distances along the intended field column's depth, and the use of several columns in series permits determination of the rate at which the mass transfer zone moves through the bed. The service time, t_b, required to reach the specified breakthrough concentration, C_b, is determined for each column, and it is plotted against the cumulative hydraulic retention time [Hutchins, 1981]:

$$t_b = a_1\theta_h + a_o \tag{8.257}$$

Table 8.12 Approximate Values of Freundlich Adsorption Isotherm Parameters for Various Organic Substances Sorbed onto Filtrasorb 300™ Activated Carbon from Distilled Water at 22°C[1,2]

Contaminant	k (mg/g)(L/mg)$^{1/n}$	$1/n$
Acenaphthene, pH 5.3	190	0.36
Acenaphthylene, pH 5.3	115	0.37
Acetophenone, pH 3.0 to 9.0	74	0.44
2-Acetylaminofluorene, pH 7.1	318	0.12
Acridine orange, pH 3.0 to 7.0	180	0.29
Acridine orange, pH 9.0	210	0.38
Acridine yellow, pH 3.0	210	0.14
Acridine yellow, pH 7 to 9	230	0.12
Acrolein, pH 5.2	1.2	0.65
Acrylonitrile, pH 5.8	1.4	0.51
Adenine, pH 3.0	38	0.38
Adenine, pH 7 to 9	71	0.38
Adipic acid, pH 3.0	20	0.47
Aldrin, pH 5.3	651	0.92
4-Aminobiphenyl, pH 7.2	200	0.26
Anethole, pH 3 to 9	300	0.42
o-Anisidine, pH 3.0	20	0.41
o-Anisidine, pH 7 to 9	50	0.34
Anthracene, pH 5.3	376	0.70
Benzene, pH 5.3	1.0	1.6
Benzidine dihydrochloride, pH 3.0	110	0.35
Benzidine dihydrochloride, pH 7 to 9	220	0.37
Benzoic acid, pH 3.0	51	0.42
Benzoic acid, pH 7.0	0.76	1.8
Benzoic acid, pH 9.0	0.0008	4.3
3,4-Benzofluoranthene [benzo(b)fluoranthene], pH 7.0	57	0.37
Benzo(k)fluoranthene, pH 7.1	181	0.57
Benzo(ghi)perylene, pH 7.0	10.7	0.37
Benzo(a)pyrene, pH 7.1	33.6	0.44
Benzothiazole, pH 3 to 9	120	0.27
α-BHC [hexachloro-cyclohexane], pH 5.4	303	0.43
β-BHC [hexachloro-cyclohexane,], pH 5.4	220	0.49
γ-BHC [hexachloro-cylcohexane, Lindane], pH 5.3	256	0.49
Bromoform, pH 5.3	19.6	0.98
4-Bromophenyl phenyl ether, pH 5.3	144	0.68
5-Bromouracil, pH 3 to 7	44	0.47
5-Bromouracil, pH 9.0	21	0.56
Butylbenzyl phthalate, pH 5.3	1,520	1.26
n-Butyl phthalate, pH 5.3	220	0.45
Carbon tetrachloride, pH 5.3	11.1	0.83
Chlorobenzene, pH 7.4	91	0.99
Chlordane, pH 5.3	245	0.38
Chloroethane, pH 5.3	0.59	0.95
bis(2-Chloroethoxy) methane, pH 5.8	11	0.65
bis(2-Chloroethyl) ether, pH 5.3	3.9	0.80
Chloroform, pH 5.3	2.6	0.73
bis(2-Chloroisopropyl) ether, pH 5.4	24	0.57
p-Chloro-*m*-cresol, pH 3.0	122	0.29
p-Chloro-*m*-cresol, pH 5.5	124	0.16
p-Chloro-*m*-cresol, pH 9.0	99	0.42
2-Chloronaphthalene, pH 5.5	280	0.46
1-Chloro-2-nitrobenzene, pH 3 to 9	130	0.46
2-Chlorophenol, pH 3 to 9	51.0	0.41
4-Chlorophenyl phenyl ether, pH 5.3	111	0.26
5-Chlorouracil, pH3 to 7	25	0.58
5-Chlorouracil, pH 9.0	7.3	0.90
Cyclohexanone, pH 7.3	6.2	0.75

Table 8.12 (continued) Approximate Values of Freundlich Adsorption Isotherm Parameters for Various Organic Substances Sorbed onto Filtrasorb 300™ Activated Carbon from Distilled Water at 22°C[1,2]

Contaminant	k (mg/g)(L/mg)$^{1/n}$	$1/n$
Cytosine, pH 3.0	0.63×10^{-12}	13.67
Cytosine, pH 7 to 9	1.1	1.6
DDE, pH 5.3	232	0.37
DDT, pH 5.3	322	0.50
Dibenzo (a, h) anthracene, pH 7.1	69.3	0.75
Dibromochloromethane, pH 5.3	4.8	0.34
1,2-Dibromo-3-chloropropane, pH 5.3	53	0.47
1,2-Dichlorobenzene, pH 5.5	129	0.43
1,3-Dichlorobenzene, pH 5.1	118	0.45
1,4-Dichlorobenzene, pH 5.1	121	0.46
3,3 Dichlorobenzidine, pH 7.2 to 9.1	300	0.20
Dichlorobromomethane, pH 5.3	7.9	0.61
1,1-Dichloroethane, pH 5.3	1.79	0.53
1,2-Dichloroethane, pH 5.3	3.57	0.83
1,2-*trans*-Dichloroethane, pH 6.7	3.05	0.51
1,1-Dichloroethene, pH 5.3	4.91	0.54
2,4-Dichlorophenol, pH 3.0	147	0.35
2,4-Dichlorophenol, pH 5.3	8.21	0.46
2,4-Dichlorophenol, pH 9.0	141	0.29
1,2-Dichloropropane, pH 5.3	5.86	0.60
1,2-Dichloropropene, pH 5.3	8.21	0.46
Dieldrin, pH 5.3	606	0.51
Diethyl phthalate, pH 5.4	110	0.27
4-Dimethylaminobenzene, pH 7.0	249	0.24
n-Dimethylnitrosamine, pH 7.5	68×10^{-6}	6.6
2,4-Dimethlyphenol, pH 3.0	78	0.44
2,4-Dimethylphenol, pH 5.8	70	0.44
2,4-Dimethylphenol, pH 9.0	108	0.33
Dimethylphenylcarbinol, pH 3.0	110	0.60
Dimethylphenylcarbinol, pH 7 to 9	210	0.34
Dimethyl phthalate, pH 3 to 9	97	0.41
4,6-Dinitro-*o*-cresol, pH 3.0	237	0.32
4,6-Dinitro-*o*-cresol, pH 5.2	169	0.35
4,6-Dinitro-*o*-cresol, pH 9.0	42.74	0.90
2,4-Dinitrophenol, pH 3.0	160	0.37
2,4-Dinitrophenol, pH 7.0	33	0.61
2,4-Dinitrophenol, pH 9.0	41	0.25
2,4-Dinitrotoluene, pH 5.4	146	0.31
2,6-Dinitrotoluene, pH 5.4	145	0.32
Diphenylamine, pH 3 to 9	120	0.31
1,1-Diphenylhydrazine, pH 7.5	135	0.16
1,2-Diphenylhydrazine, pH 5.3	16,000	2.0
α-Endsulfan, pH 5.3	194	0.50
β-Endosulfan, pH 5.3	615	0.83
Endosulfan sulfate, pH 5.3	686	0.81
Endrin, pH 5.3	666	0.80
Ethylbenzene, pH 7.4	53	0.79
EDTA [ethylenediaminetetraacetic acid], pH 3 to 9	0.86	1.5
bis (2-Ethylhexyl) phthalate, pH 5.3	11,300	1.5
Fluranthene, pH 5.3	664	0.61
Fluorene, pH 5.3	330	0.28
2,4-Dinitrophenol, pH 3 to 7	5.5	1.0
2,4-Dinitrophenol, pH 9.0	1.3	1.4
Guanine, pH 3.0	75	0.48
Guanine, pH 7 to 9	120	0.40
Heptachlor, pH 5.3	1,220	0.95
Heptachlor epoxide, pH 5.3	1,038	0.70

Table 8.12 (continued) Approximate Values of Freundlich Adsorption Isotherm Parameters for Various Organic Substances Sorbed onto Filtrasorb 300™ Activated Carbon from Distilled Water at 22°C[1,2]

Contaminant	k (mg/g)(L/mg) $^{1/n}$	$1/n$
Hexachlorbenzene, pH 5.3	450	0.60
Hexachlorobutadiene, pH 5.3	258	0.45
Hexachlorocyclopentadiene, pH 5.3	96.5	0.38
Hydroquinone, pH 3.0	90	0.25
Isophorone, pH 5.5	32	0.39
Methylene chloride, pH 5.8	1.30	1.16
4,4′-Methylene-*bis* (2-chloroaniline), pH 7.5	190	0.64
Naphthalene, pH 5.6	132	0.42
α-Naphthol, pH 3 to 9	180	0.32
β-Naphthol, pH 3 to 9	200	0.26
α-Naphthylamine, pH 3.0	140	0.25
α-Naphthylamine, pH 7 to 9	160	0.34
β-Naphthylamine, pH 7.5	150	0.30
p-Nitroaniline, pH 3 to 9	140	0.27
Nitrobenzene, pH 7.5	68	0.43
4-Nitrobiphenyl, pH 7.0	370	0.27
2-Nitrophenol, pH 3.0	101	0.26
2-Nitrophenol, pH 5.5	99	0.34
2-Nitrophenol, pH 9.0	85	0.39
4-Nitrophenol, pH 3.0	80.2	0.17
4-Nitrophenol, pH 5.4	76.2	0.25
4-Nitrophenol, pH 9.0	71.2	0.28
N-Nitrosodiphenylamine, pH 3 to 9	220	0.37
N-Nitrosodi-*n*-propylamine, pH 3 to 9	24.4	0.26
p-Nonylphenol, pH 3.0	53	1.04
p-Nonylphenol, pH 7.0	250	0.37
p-Nonylphenol, pH 9.0	150	0.27
PCB 1221, pH 5.3	242	0.70
PCB 1232, pH 5.3	630	0.73
Pentachlorophenol, pH 3.0	260	0.39
Pentachlorophenol, pH 7.0	150	0.42
Pentachlorophenol, pH 9.0	100	0.41
Phenanthrene, pH 5.3	215	0.44
Phenol, pH 3 to 9	21	0.54
Phenylmercuric acetate, pH 3 to 7	270	0.44
Phenylmercuric acetate, pH 9.0	100	0.41
Styrene, pH 3 to 9	120	0.56
1,1,2,2,-Tetrachloroethane, pH 5.3	10.6	0.37
Tetrachloroetheane, pH 5.3	50.8	0.56
1,2,3,4-Tetrahydronaphthalene, pH 7.4	74	0.81
Thymine, pH 3 to 9	27	0.51
Toluene, pH 5.6	26.1	0.44
1,2,4-Trichlorobenzene, pH 5.3	157	0.31
1,1,1-Trichloroethane, pH 5.3	2.48	0.34
1,1,2-Trichloroethane, pH 5.3	5.81	0.60
Trichloroethane, pH 5.3	28.0	0.62
Trichlorofluoromethane, pH 5.3	5.6	0.24
2,4,6-Trichlorophenol, pH 3.0	219.0	0.29
2,4,6-Trichlorophenol, pH 6.0	155.1	0.40
2,4,6-Trichlorophenol, pH 9.0	130.1	0.39
Uracil, pH 3 to 9	11	0.63
p-Xylene, pH 7.3	85	0.19

[1]Dobbs, R. A. and Cohen, J. M. 1980. *Carbon Adsorption Isotherms for Toxic Organics*, EPA-600/8-80-023, U.S. Environmental Protection Agency, Office of Research and Development, Municipal Environmental Research Laboratory, Cincinnati, OH.

[2]Filtrasorb 300™ is a trademark of Calgon Corporation, Inc.

where

a_o = an empirical constant (s)
a_1 = an empirical coefficient (dimensionless)
C_b = the specified breakthrough concentration (kg/m^3 or lb/ft^3)
t_b = the service time to breakthrough (s)
θ_h = the hydraulic retention of the carbon columns (s)

Column hydraulic retention times are usually reported as empty bed contacting times (EBCT), which is simply the hydraulic retention time calculated assuming the bed is empty. The ratio of the true HRT to the EBCT is equal to the bed porosity:

$$\varepsilon = \frac{\text{HRT}}{\text{EBCT}} = \frac{\varepsilon V/Q}{V/Q} \tag{8.258}$$

The abscissal intercept, $-a_o/a_1$, is the smallest possible EBCT that produces the required effluent quality. The ordinal intercept, a_o, represents that portion of the bed that is not exhausted when breakthrough occurs. It represents the length of the mass transfer zone. If the service time to column saturation, t_{sat}, is determined, the length of the MTZ is

$$H_{\text{mtz}} = \frac{(t_{\text{sat}} - t_b)H}{t_{\text{sat}}} = \frac{a_o H}{t_b} \tag{8.259}$$

where

H = the bed length (m or ft)
H_{mtz} = the length of the mass transfer zone (m or ft)
t_{sat} = the service time to bed saturation (s)

It is desirable to minimize the MTZ so as to maximize the efficiency of carbon usage. The length of the MTZ can be reduced by reducing the flow rate, but this increases the required column size.

The carbon usage rate for conventional GAC columns can be estimated as

$$R_C = \frac{\rho_C(1-\varepsilon)V}{t_b} \tag{8.260}$$

where

R_C = the carbon usage rate (kg/s or lb/s)
V = the empty bed volume (m^3 or ft^3)
ε = the bed porosity when filled with carbon (dimensionless)
ρ_C = the bulk density of the dry carbon (kg/m^3 or lb/ft^3)

In this case, a portion of the carbon removed from the adsorber (equal to the volume of the MTZ) is not saturated with adsorbate and does not require regeneration. Pulsed column operation overcomes this deficiency, and the base carbon usage rate for a pulsed column is calculated as

$$R_{\text{bcu}} = \frac{\rho_C(1-\varepsilon)Q}{a_1} \tag{8.261}$$

where

Q = the water flow rate (m^3/s or ft^3/s)
R_{bcu} = the base carbon usage rate (kg/s or lb/s)

The Logistic Model

A purely empirical procedure has been developed by Oulman [1980]. His model is based on the observation that the breakthrough curve is sigmoidal and can be approximated by the logistic equation:

$$\frac{C}{C_o} = \frac{1}{1 + e^{-(a+bt)}} \tag{8.262}$$

where

a = an empirical constant (dimensionless)
b = an empirical coefficient (per second)
C = the effluent contaminant concentration at time t (kg/m^3 or lb/ft^3)
C_o = the influent contaminant concentration (kg/m^3 or lb/ft^3)
t = the service time (s)

This is actually a rearrangement of the Bohart–Adams equation, and the parameters in the exponential can be related to column properties and hydraulic loading as follows:

$$a = -\frac{k_a a_b H}{v_i \varepsilon} \tag{8.263}$$

$$b = k_a C_o \tag{8.264}$$

where

H = the bed depth (m or ft)
k_a = the Bohart–Adams adsorption rate coefficient (per second)
q_b = the adsorption capacity of a carbon bed at bed saturation (kg substance/kg C or lb substance/lb C)
v_i = the interstitial water velocity in a packed bed (m/s or ft/s)

The logistic equation can be linearized by rearrangement, and the coefficients a and b are the intercept and slope of the plot:

$$\ln\left(\frac{C/C_o}{1 - C/C_o}\right) = a + bt \tag{8.265}$$

The only unknowns are the adsorption rate coefficient, k_a, and the bed capacity, q_b, and these may be calculated from the slope and intercept of the plot.

Clark *et al.* [1986] derived a modified logistic formula by assuming that mass transfer to the carbon was controlled by transport through the water film and that the solute concentration at the grain surface was in equilibrium with the quantity of adsorbed material. Freundlich's isotherm was used to describe the equilibrium. The effluent concentration vs. service time plot is,

$$C = \left(\frac{C_o^{n-1}}{1 + Ae^{-rt}}\right)^{1/(n-1)} \tag{8.266}$$

$$A = \left(\frac{C_o^{n-1}}{C_b^{n-1}} - 1\right)e^{rt_b} \tag{8.267}$$

$$r = \frac{(n - 1)k_l v_{az}}{v_i \varepsilon} \tag{8.268}$$

where

k_l = the liquid film mass transfer coefficient (m/s or ft/s)
n = an empirical exponent in the Freundlich adsorption isotherm (dimensionless)
r = an empirical coefficient (per second)
v_{az} = the velocity of the adsorption zone through an activated carbon bed (m/s or ft/s)

Equation (8.266) can be linearized as follows for curve fitting:

$$\ln\left[\left(\frac{C_o}{C}\right)^{n-1} - 1\right] = \ln A - rt \tag{8.269}$$

The logistic equation assumes that the breakthrough curve is symmetrical. When this is not the case, the modified logistic equation may give a better fit. The derivation of the modified logistic model shows that the coefficients A and r will vary with the liquid velocity in the bed, and they do [Clark, 1987]. However, because the model ignores the effect of surface diffusion with the carbon grain pores, which is probably the rate-limiting process, Eqs. (8.267) and (8.268) cannot be used to calculate A and r and they should be determined empirically.

Packed Column Theory

A commonly used model is the homogeneous surface diffusion model (HSDM) developed by Crittenden and Weber [1978a,b,c]. This model was developed for the adsorption of single components.

It is assumed that the rate of adsorption is controlled by the rate of diffusion of adsorbed molecules along the walls of the pores inside the carbon grains. The water phase is assumed to be dilute everywhere, so that bulk flow in the pores and concentration effects on diffusivities can be ignored. The pore surface area is supposed to be uniformly distributed throughout the carbon particles and the mass balance for adsorbate adsorbed to the carbon surface is written in spherical coordinates as

$$\frac{\partial q}{\partial t} = D_s\left(\frac{\partial^2 q}{\partial r^2} + \frac{2}{r}\cdot\frac{\partial q}{\partial r}\right) \tag{8.270}$$

where

D_s = the surface diffusivity (m^2/s or ft^2/s)
q = the mass of adsorbate per unit mass of carbon (kg adsorbate/kg C or lb adsorbate/lb C)
r = the radial distance from the center of the carbon grain (m or ft)
t = elapsed time (s)

The water phase at any level in the bed is assumed to be of uniform composition (except for the boundary layer) and without channeling. Mass transfer from the interstitial water to the carbon grains is represented as transport across a surface film. The mass balance for the adsorbate dissolved in the water flowing through the bed is [James M. Montgomery, Consulting Engineers, Inc., 1985]

$$\underbrace{\frac{\partial C}{\partial t}}_{\text{Accumulation in water}} = \underbrace{D_p\frac{\partial^2 C}{\partial z^2}}_{\text{Dispersive flux}} - \underbrace{v_i\frac{\partial C}{\partial z}}_{\text{Advective flux}} - \underbrace{\frac{3k_l(1-\varepsilon)}{a_p\phi\varepsilon}(C - C_s)}_{\text{Transfer to carbon}} \tag{8.271}$$

where

a_p = the radius of an activated carbon grain (m or ft)
C = the bulk water phase adsorbate concentration (kg/m^3 or lb/ft^3)
C_s = the adsorbate concentration at the water-carbon interface (kg/m^3 or lb/ft^3)
D_p = the water phase dispersion coefficient (m^2/s or ft^2/s)
k_l = the liquid film mass transfer coefficient (m/s or ft/s)
v_i = the interstitial water velocity in a packed bed (m/s or ft/s)
z = the distance from the inlet of a granular activated carbon bed (m or ft)
ε = the filter bed porosity (dimensionless)
ϕ = the sphericity, the ratio of the surface area of an equivalent volume sphere to the surface area (not counting internal pores) of the carbon grain (dimensionless)

At the interface between the carbon grain and the interstitial water, there is an equilibrium between the adsorbed material and the dissolved material, which can be represented by some suitable isotherm such as the Freundlich:

$$q_s = kC_s^{1/n}$$

The interfacial area per unit bed volume is $3(1-\varepsilon)/a_p\phi$.

Equations (8.270) and (8.271) are connected by the requirement that solute disappearing from the water appear as adsorbate in the carbon. The transfer of solute to the carbon can be represented on either an areal-flux basis,

$$D_s\rho_p \left.\frac{\partial q}{\partial r}\right|_{r=a_p} = k_l(C - C_s) \tag{8.272}$$

where ρ_p is the particle density (kg/m^3 or lbm/ft^3), or a bed-volume basis,

$$\frac{3k_l(1-\varepsilon)(C-C_s)}{a_p\phi} = \rho_p(1-\varepsilon)\frac{\partial}{\partial t}\left(4\pi\int_0^{a_p} qr^2 dr\right) \tag{8.273}$$

The value of $q(r,z,t)$ is given by Eq. (8.273) with the condition that $q(a_p,z,t)$ be q_s. The relevant boundary conditions are

$$C(z,t) - C_o + \frac{D_p}{v_i}\cdot\frac{\partial C}{\partial z} \quad \text{at } z = 0 \tag{8.274}$$

$$\frac{\partial C}{\partial z} = 0 \quad \text{at } z = H \tag{8.275}$$

$$\frac{\partial q}{\partial r} = 0 \quad \text{at } r = 0 \tag{8.276}$$

$$q(r,t) = 0 \quad \text{at } t = 0 \tag{8.277}$$

$$C(z,t) = 0 \quad \text{at } t = 0 \tag{8.278}$$

Numerical procedures for solving these equations are presented by Weber and Crittenden [1975] and by Roy *et al.* [1993].

A number of parameters must be evaluated before the model can be used:

- The Freundlich isotherm constants
- The liquid film mass transfer coefficient

- The liquid film diffusivity
- The surface diffusivity

These parameters can be estimated using various published correlations, which are listed below, or by the analysis of batch or continuous flow adsorption tests.

The Freundlich isotherm parameters must be determined experimentally for the substance of interest and the intended carbon. Typical values for a variety of substances are given in Table 8.12.

The mass transfer coefficient can be estimated using any one of several correlations. Williamson *et al.* [1963]:

$$\frac{k_l}{\varepsilon v_i} \cdot \mathrm{Sc}^{0.58} = 2.40 \cdot \mathrm{Re}^{-0.66} \qquad 0.08 \leq \mathrm{Re} \leq 125 \tag{8.279}$$

where

D_l = the molecular diffusivity of the adsorbate in the bulk water (m^2/s or ft^2s)
d_p = the diameter of a carbon grain (m or ft)
Re = the particle Reynolds number (dimensionless)
= $\rho v_i d_p/\mu$
Sc = the Schmidt number (dimensionless)
= $\mu/\rho D_l$
μ = the dynamic viscosity of water ($N{\cdot}s/m^2$ or $lbf{\cdot}s/ft^2$)
ρ = the water density (kg/m^3 or lbm/ft^3)

Wakao and Funazkri [1978]:

$$\frac{k_l d_p}{D_l} = 2 + 1.1 \cdot \mathrm{Re}^{0.6}\mathrm{Sc}^{1/3} \qquad 3 \leq \mathrm{Re} \leq 10^4 \tag{8.280}$$

where

Re = the modified particle Reynolds number (dimensionless)
= $\rho\varepsilon v_i d_p/\mu$

The liquid phase solute diffusivities can be estimated in several ways. Stokes–Einstein [Einstein, 1956]:

$$D_l = \frac{kT}{6\pi a_p \mu} \tag{8.281}$$

where

k = Boltzmann's constant (1.380658×10^{-23} J/K or approximately 0.5658×10^{-23} ft·lbf/°R)
T = the absolute temperature (K or °R)

Polson [1950]:

$$D_l = 2.74 \times 10^{-5} \cdot M_r^{-1/3} \tag{8.282}$$

where

D_l = the diffusivity in cm^2/s
M_r = the relative molecular weight (dimensionless)

Wilke–Chang [Wilke and Chang, 1955; Hayduk and Laudie, 1974]:

$$\frac{D_l \mu}{T} = 7.4 \times 10^{-8} \times \frac{\sqrt{2.6 M_{rl}}}{v_{\mathrm{mbp}}^{0.6}} \tag{8.283}$$

where

D_{li} = the diffusivity of species i in water in cm^2/s
M_{rl} = the molecular weight of water in g
= 18 g
v_{mbp} = the molar volume of the liquid phase of species i at its boiling point in cm^3/g-mole
μ = the viscosity of water in centipoise (cp)

Note: 1cp = 0.001 N·s/m^2 (exactly). Perry and Chilton [1973] recommend the Le Bas volumes in place of v_{mbp}, if values for the latter cannot be found. Table 8.13 gives the Le Bas volumes for common structural components of organic substances.

Surface diffusivities of volatile organic substances and polycyclic aromatic substances can be estimated by the Suzuki–Kawazoe [1975] correlation:

$$D_s = 1.1 \times 10^{-4} \cdot e^{-.532T_b/T} \tag{8.284}$$

where

D_s = the surface diffusivity in cm^2/s
T = the adsorption temperature in K
T_b = the boiling point of the organic substance in K

Table 8.13 Components of the Molar Volume of Liquids

Structural Component	Molar Volumes[1] (cm^3/g-mol)	Molar Volumes[2] (cm^3/g-mol)
Bromine, add	27.0	—
Carbon, add	14.8	16.5
Chlorine, end (R-Cl), add	21.6	19.5
Chlorine, inner (R-Cl-R), add	24.6	19.5
Fluorine, add	8.7	—
Hydrogen, add	3.7	1.98
Iodine, add	37.0	—
Nitrogen, double bonded, add	15.6	5.69
Nitrogen, triple bonded, add	16.2	5.69
Nitrogen, primary amines (RNH_2), add	10.5	5.69
Nitrogen, secondary amines (R_2NH), add	12.0	5.69
Nitrogen, tertiary amines (R_3N), add	10.8	5.69
Oxygen, if not combined as follows, add	7.4	5.48
Oxygen, in methyl esters, add	9.1	5.48
Oxygen, in methyl ethers, add	9.9	5.48
Oxygen, in other esters or ethers, add	11.0	5.48
Oxygen, in acids, add	12.0	5.48
Oxygen, combined with N, P, or S, add	8.3	5.48
Phosphorus, add	27.0	—
Sulfur, add	25.6	17.0
3-member rings, deduct	6.0	—
4-member rings, deduct	8.5	—
5-member rings, deduct	11.5	—
6-member rings (benzene, cyclohexane), deduct	15.0	20.2
Anthracene ring, deduct	47.5	20.2
Naphthalene ring, deduct	30.0	20.2

[1]Le Bas. 1915. *The Molecular Volumes of Liquid Chemical Compounds*, Longmans, London. (Quoted in Perry, R. H. and Chilton, C. H. 1973. *Chemical Engineer's Handbook*, 5th ed., McGraw-Hill, New York.)

[2]Bennet, C. O. and Myers, J. E. 1974. *Momentum, Heat, and Mass Transfer*, 2nd ed., McGraw-Hill, New York.

It is expected that the coefficient in front of the exponential will depend only on the carbon used and not on the adsorbate. The correlation was developed using Takeda HGR 513™, which is derived from coconut shell and has a mean grain size of 3.41 mm and a specific surface area of 1225 m^2/g.

A simplified approach has been developed by Hand *et al.* [1984] that eliminates the need for computer simulations. They have, rather, fitted an empirical formula to the results of many computer solutions. The equations and boundary conditions are first nondimensionalized by introducing the following dimensionless variables and groups:

$$\overline{C} = \frac{C}{C_o} \tag{8.285}$$

$$\overline{C}_s = \frac{C_s}{C_o} \tag{8.286}$$

$$\overline{q} = \frac{q}{q_o} \tag{8.287}$$

$$\overline{r} = \frac{r}{a_p} \tag{8.288}$$

$$\overline{z} = \frac{z}{H} \tag{8.289}$$

$$\text{Bi} = \frac{k_l a_p (1-\varepsilon)}{D_s D_g \varepsilon \phi} \tag{8.290}$$

$$D_g = \frac{q_o \rho_p (1-\varepsilon)}{C_o \varepsilon} \tag{8.291}$$

$$E_d = \frac{D_s D_g \theta_h}{a_p^2} \tag{8.292}$$

$$\text{St} = \frac{k_l \theta_h (1-\varepsilon)}{a_p \varepsilon \phi} \tag{8.293}$$

$$T = \frac{t}{\theta_h (1 + D_g)} \tag{8.294}$$

where

Bi = the Biot number (dimensionless)
C_o = the adsorbate's influent concentration (kg/m^3 or lb/ft^3)
D_g = the solute distribution parameter (dimensionless)
E_d = the diffusivity modulus, the ratio of the Stanton to Biot numbers (dimensionless)
H = the depth of the carbon bed (m or ft)
q_o = the mass of adsorbate per unit mass of activated carbon at equilibrium with the influent concentration C_o (kg adsorbate/kg C or lb adsorbate/lb C)
St = the modified Stanton number (dimensionless)
T = the throughput, the ratio of the adsorbate fed to the amount of adsorbate that may be adsorbed at equilibrium (dimensionless)

The resulting transport equations and boundary conditions are [Hand *et al.*, 1984]

$$\frac{1}{1+D_g} \cdot \frac{\partial \overline{C}}{\partial T} = -\frac{\partial \overline{C}}{\partial \overline{z}} - 3 \cdot \mathrm{St} \cdot (\overline{C} - \overline{C}_s) \tag{8.295}$$

$$\frac{\partial \overline{q}}{\partial T} = \left(1 + \frac{1}{1+D_g}\right) \cdot \frac{E_d}{\overline{r}^2} \cdot \frac{\partial}{\partial \overline{r}}\left(\overline{r}^2 \frac{\partial \overline{q}}{\partial \overline{r}}\right) \tag{8.296}$$

$$\left.\frac{\partial \overline{q}}{\partial \overline{r}}\right|_{\overline{r}=1} = \mathrm{Bi} \cdot \left(\overline{C} - \overline{C}_s\right) \tag{8.297}$$

$$\left.\frac{\partial \overline{q}}{\partial \overline{r}}\right|_{\overline{r}=0} = 0 \tag{8.298}$$

$$\overline{C}(\overline{z} = 0, T \geq 0) = 1 \tag{8.299}$$

$$\overline{C}\left[T(1+D_g) < \overline{z} \leq 1, T < \frac{1}{1+D_g}\right] = 0 \tag{8.300}$$

$$\overline{q}(0 \leq \overline{r} \leq 1, 0 \leq \overline{z} \leq 1, T = 0) = 0 \tag{8.301}$$

The nondimensional Freundlich isotherm is

$$\overline{q}_s = \overline{C}_s^{1/n} \tag{8.302}$$

If the solute distribution parameter, D_g, is greater than about 50, the Biot number, Bi, is greater than about 0.5, and the Freundlich exponent, $1/n$, is less than 1, the numerical solution is controlled by Bi, $1/n$, and the Stanton number, St. Hand *et al.* [1984] have summarized the results of numerous computer simulations as parameters in the following empirical formula:

$$T = A_o + A_1\left(\frac{C}{C_o}\right)^{A_2} + \frac{A_3}{1.10 - \left(\frac{C}{C_o}\right)^{A_4}} \tag{8.303}$$

Equation (8.303) gives the processing time in terms of the throughput parameter, T, required to achieve a specified effluent concentration, C. The breakthrough ratio, C/C_o, must lie between about 0.02 and 0.99. Values for the parameters are given in Table 8.14. There are several apparent discontinuities in the parameter values, which makes interpolation hazardous. However, a three-dimensional plot of the coefficients reveals a smoother pattern, and it should be used as the basis of interpolation.

Equation (8.303) is valid only if the bed contacting time is deep enough to develop a mass transfer zone. Minimum bed contacting times depend on the minimum Stanton number, which can be estimated from empirical correlation:

$$\mathrm{St}_{\min} = B_o + B_1 \cdot \mathrm{Bi} \tag{8.304}$$

Values of the correlation parameters are given in Table 8.15. The minimum bed contacting time is given by

$$\theta_{h,\min} = \frac{\mathrm{St}_{\min} a_p \varepsilon \phi}{k_l(1-\varepsilon)} \tag{8.305}$$

Table 8.14 Parameter Values for the Hand–Crittenden–Thacker Breakthrough Formula[1] *(continues)*

$1/n$	Bi	Equation (8.303) Parameters				
		A_o	A_1	A_2	A_3	A_4
0.05	0.5	−5.447	6.599	0.02657	0.01938	20.45
0.05	2.0	−5.468	6.592	0.004989	0.004988	0.5023
0.05	4.0	−5.531	6.585	0.02358	0.009019	0.2731
0.05	6.0	−5.607	6.582	0.02209	0.01313	0.2145
0.05	8.0	−5.606	6.505	0.02087	0.01708	0.1895
0.05	10.0	−5.664	6.457	0.01816	0.01994	0.1493
0.05	14.0	−0.6628	1.411	0.06071	0.02023	0.1439
0.05	25.0	−0.6628	1.351	0.03107	0.02035	0.1300
0.05	≥100.0	0.6659	0.7113	2.987	0.01678	0.3610
0.10	0.5	−1.920	3.055	0.005549	0.02428	15.31
0.10	2.0	−2.279	3.394	0.04684	0.004751	0.3847
0.10	4.0	−2.337	3.378	0.04399	0.00865	0.2434
0.10	6.0	−2.407	3.374	0.04132	0.01255	0.1966
0.10	8.0	−2.478	3.371	0.03899	0.01628	0.1764
0.10	10.0	−2.566	3.371	0.03500	0.01939	0.1508
0.10	16.0	−2.567	3.306	0.02097	0.01948	0.1368
0.10	30.0	−2.569	3.242	0.009595	0.01961	0.1218
0.10	≥100.0	−2.568	3.191	0.001555	0.01968	0.1101
0.20	0.5	−1.441	2.569	0.06092	0.002333	0.3711
0.20	2.0	−1.474	2.558	0.05848	0.005026	0.2413
0.20	4.0	−1.507	2.519	0.05552	0.008797	0.1875
0.20	6.0	−1.035	1.983	0.06928	0.01230	0.1679
0.20	8.0	−0.1692	1.078	0.01449	0.01550	0.1681
0.20	10.0	−1.403	2.188	0.05219	0.01842	0.1336
0.20	13.0	−1.369	2.119	0.03949	0.01845	0.1276
0.20	25.0	−1.514	2.209	0.01794	0.01851	0.1152
0.20	≥100.0	0.6803	0.649	2.570	0.01495	0.3698
0.30	0.5	−1.7587	2.847	0.04953	0.003022	0.1568
0.30	2.0	−1.6579	2.689	0.04841	0.005612	0.1409
0.30	4.0	−0.5657	1.538	0.08445	0.008808	0.1391
0.30	6.0	−0.1971	1.119	0.1179	0.01153	0.1359
0.30	8.0	−0.1971	1.069	0.1198	1.1392	0.1327
0.30	10.0	−0.1734	1.000	0.1203	0.01594	0.1340
0.30	15.0	−0.1734	0.9194	0.07177	0.01416	0.08627
0.30	35.0	0.6665	0.4846	1.719	0.01344	0.2595
0.30	≥100.0	0.6962	0.5170	2.055	0.01296	0.3032
0.40	0.5	−0.5343	1.604	0.09406	0.004141	0.1378
0.40	2.0	−0.1663	1.191	0.1223	0.006261	0.1348
0.40	5.0	−0.1663	1.132	0.1155	0.008634	0.1268
0.40	6.0	−0.1663	1.090	0.1123	0.01046	0.1243
0.40	9.0	0.4919	0.4918	0.4874	0.01137	0.1477
0.40	12.0	0.5641	0.4192	0.6398	0.01154	0.1490
0.40	15.0	0.6407	0.4325	1.048	0.01162	0.2127
0.40	25.0	0.6724	0.3970	1.153	0.01128	0.2169
0.40	≥100.0	0.7414	0.4481	1.930	0.01020	0.3064
0.50	0.5	−0.04080	1.100	0.1590	0.005467	0.1391
0.50	4.0	−0.04080	0.9828	0.1116	0.008072	0.1114
0.50	10.0	0.09460	0.7549	0.09207	0.009877	0.09076
0.50	14.0	0.02300	0.8021	0.05754	0.009662	0.08453
0.50	25.0	0.02300	0.7937	0.03932	0.009326	0.08275
0.50	≥100.0	0.5292	0.2918	0.08243	0.008317	0.07546

Table 8.14 (continued) Parameter Values for the Hand–Crittenden–Thacker Breakthrough Formula[1]

		Equation (8.303) Parameters				
$1/n$	Bi	A_0	A_1	A_2	A_3	A_4
0.60	0.5	0.3525	0.6921	0.2631	0.005482	0.1218
0.60	2.0	0.5220	0.5042	0.3273	0.005612	0.1287
0.60	6.0	0.6763	0.3346	0.4823	0.005898	0.1389
0.60	14.0	0.7695	0.2595	0.7741	0.005600	0.1655
0.60	50.0	0.8491	0.2158	1.343	0.004725	0.2238
0.60	≥100.0	0.8312	0.2273	1.175	0.004961	0.2121
0.70	0.5	0.575	0.4491	0.2785	0.004122	0.1217
0.70	4.0	0.7153	0.3072	0.4421	0.004371	0.1384
0.70	12.0	0.7879	0.2435	0.6616	0.004403	0.1626
0.70	25.0	0.8295	0.2041	0.7845	0.004050	0.1790
0.70	≥100.0	0.8470	0.1907	0.9317	0.003849	0.1832
0.80	0.5	0.7089	0.3141	0.3575	0.003276	0.1193
0.80	4.0	0.7846	0.2397	0.4844	0.003206	0.1350
0.80	14.0	0.8394	0.1890	0.6481	0.003006	0.1577
0.80	100.0	0.8827	0.1462	0.808	0.002537	0.1745
0.90	0.5	0.8654	0.1576	0.4450	0.001650	0.1408
0.90	4.0	0.8548	0.1714	0.4950	0.001910	0.1423
0.90	16.0	0.8662	0.1640	0.5739	0.001987	0.1576
0.90	100.0	0.8932	0.1330	0.6241	0.001740	0.1642

[1]Hand, D. W., Crittendon, J. C., and Thacker, W. E. 1984. Simplified models for design of fixed-bed adsorption systems, *Journal of Environmental Engineering*, 110(2): 440.

The minimum processing time before regeneration or replacement is required is calculated from the throughput:

$$t_{\min} = \theta_{h,\min}\left(1 + D_g\right)T \tag{8.306}$$

If longer processing times are desired, the hydraulic retention time must be increased proportionately.

Table 8.15 Parameter Values for Estimating Minimum Stanton Number for Establishment of Mass Transfer Zone[1]

	$0.5 \le \text{Bi} \le 10$		$\text{Bi} \ge 10$	
Freundlich Exponent, $1/n$	B_o	B_1	B_o	B_1
0.05	1.990	0.02105	0.0	0.22
0.10	2.189	0.02105	0.0	0.24
0.20	2.379	0.04211	0.0	0.28
0.30	2.547	0.1053	0.0	0.36
0.40	2.684	0.2316	0.0	0.50
0.50	2.737	0.5263	0.0	0.80
0.60	3.421	1.158	0.0	1.50
0.70	7.105	1.789	0.0	2.50
0.80	13.16	3.684	0.0	5.00
0.90	56.84	6.316	0.0	12.0

[1]Hand, D. W., Crittendon, J. C., and Thacker, W. E. 1984. Simplified models for design of fixed-bed adsorption systems, *Journal of Environmental Engineering*, 110(2):440.

Batch and Ideal Plug Flow Reactors

The Biot number is the ratio of the mass flux through the liquid film surrounding a carbon grain to the mass flux due to surface diffusion within the grain. For Biot numbers greater than about 100, mass transfer is controlled by surface diffusion, whereas liquid film diffusion controls at small Biot numbers [Traegner and Suidan, 1989].

For large Biot numbers, the HSDM has an exact power series solution in closed, batch systems [Crank, 1956]:

$$\frac{q(t)}{q_e} = 1 - \frac{6}{\pi^2} \cdot \sum_{i=1}^{\infty} \frac{1}{i^2} \cdot \exp\left(-\frac{i^2 \pi^2 D_s t}{a_p^2}\right) \tag{8.307}$$

where q_e is the mass of adsorbate per unit mass of activated carbon at equilibrium (kg adsorbate/kg C or lb adsorbate/lb C). Note that the solution depends only on the grain size and the surface diffusivity. Particle size does not affect either the equilibrium position or the surface diffusivity, but small particles adsorb solute more quickly [Najm *et al.*, 1990]. Consequently, batch tests are more quickly performed if powdered carbon is used.

Numerical solutions to the HSDM model for large Biot numbers have been summarized in the form of power series by Hand *et al.* [1983].

If both liquid film transport and surface diffusion are important, only a numerical solution is possible. Traegner and Suidan [1989] have shown how the Levenberg–Marquardt nonlinear optimization procedure can be used to adjust both D_s and k_l so that model predictions match experimental data.

PAC Slurry Reactors

The batch adsorption solution may also be applied to turbulent, completely mixed reactors. Taking into account the residence time distribution for a continous flow stirred tank, one obtains [Nakhla *et al.*, 1989; Najm *et al.*, 1990; Adham *et al.*, 1993]

$$C = C_o - X_{\mathrm{PAC}} \cdot kC^{1/n} \cdot \left[1 - \frac{6}{\pi^2} \cdot \sum_{i=1}^{\infty} \frac{1}{i^2\left(1 + \dfrac{i^2 \pi^2 D_s \theta_h}{a_p^2}\right)}\right] \tag{8.308}$$

Application

Powdered Activated Carbon

Powdered activated carbon is usually applied by adding it to the water or wastewater to be treated and mixing the carbon and water in a slurry reactor for the requisite contacting time. The carbon is subsequently removed from the liquid by settling or filtration. Unless the treated water is nearly free of other suspended solids, the spent carbon cannot be regenerated and is wasted. Carbon wasting is practicable only for intermittent, short-duration treatment.

If the treated water is clear, settling and/or filtration will recover a product suitable for regeneration. Furthermore, it is possible to operate a countercurrent flow of carbon and water so as to minimize carbon requirement. In such a system, several slurry reactors are constructed and operated in series. Each reactor is equipped with a filter or settler, and the solids captured are pumped to the next upstream reactor [Hutchins, 1981].

The required contacting times must be empirically determined; the homogeneous surface diffusion model described above is widely used as a design aid. The adsorption rate is inversely proportional to the square of the carbon grain size, so the required contacting time for equal removal efficiencies is proportional to the square of the grain diameter [Najm *et al.*, 1990]:

$$\theta_h \propto a_p^2 \tag{8.309}$$

The adsorptive capacity and surface diffusivity are independent of grain size [Randtke and Snoeyink, 1983; Najm *et al.*, 1990].

Most reported carbon adsorption studies have employed pure solutions of single substances. Studies in complex natural waters indicate that the background organic matter may reduce the sorptive capacity of the test substance by about 50% [Najm *et al.*, 1990].

Granular Activated Carbon

Granular activated carbon systems are usually preferred when continuous carbon treatment is required, because carbon recovery for regeneration is easier. Plant surveys have indicated several design deficiencies that require special attention [Culp and Clark, 1983; Graese *et al.*, 1987; Akell, 1981]:

- Wet carbon forms electrolytic cells and is corrosive.
- Wet carbon is abrasive to steel.
- Dry, fresh carbon in large piles adsorbs oxygen and can spontaneously ignite; the ignition temperature depends strongly on how the carbon was manufactured and is about 300°C for steam-activated carbons.
- Idle carbon beds loaded with ketones, aldehydes, and some organic acids are subject to "temperature excursions" and occasional bed fires.
- Carbon dust is explosive if it contains more than 8% by weight organic matter; most steam-activated carbons are well below this limit.
- Adsorbers are generally devoid of oxygen and cannot be entered without respirators and/or ventilation.
- Wastewater adsorbates may undergo anaerobic decomposition, producing explosive gases such as methane and poisonous gases such as hydrogen sulfide; control of column dissolved oxygen levels is necessary.
- GAC and slurry transfer and feed equipment are often undersized.
- Backwash lines must be vented.
- Backwash nozzles subject to breakage or clogging must be avoided.
- Regeneration furnace feeds must be uniform to avoid temperature fluctuations and inconsistent reactivation.
- Control systems and drive motors must be protected from the weather and from furnace emissions and radiant heat.

Granular activated carbon systems come in three general types: packed columns, pulsed columns, and rapid sand filters.

Packed Columns. Packed columns are large steel or reinforced concrete tanks, generally pressurized to maximize throughput rates [Culp and Clark, 1983; Culp/Wesner/Culp, Inc., 1986; Snoeyink, 1990; Joint Task Force, 1990]. Prefabricated tanks must be less than 12 ft in diameter for transport. Larger tanks are constructed on-site.

In water treatment plants, the columns are installed after the rapid sand filters and operated downflow. Filtration rates between 3 and 7 gpm/ft^2 and depths of 2.5 to 15 ft are employed. The governing variable is contact time. This should be determined by pilot studies and model studies as described above. The most important restriction is that the contacting time be long enough to establish a mass transfer zone. Contacting times longer than this are needed to provide adequate service life, and the actual value chosen depends on the economics of the particular application. An empty bed contacting time (EBCT) of 10 to 20 min is common; 5 min suffices for taste and odor control. However, some applications may require an EBCT of a few hours.

Pulsed Columns. Pulsed columns are designed to permit intermittent replacement of carbon in operating columns [Hutchins, 1981; Snoeyink, 1990]. Liquid flow is upward, and carbon is withdrawn from the column bottom at specified intervals. Fresh carbon is added at the top of the column. The columns have no freeboard, so the carbon cannot move during operation or cleaning, unless it is being withdrawn. This prevents vertical mixing of the bed.

The carbon discharged from pulsed beds is always saturated with adsorbate at the raw water solute concentration. This is the most efficient way to use carbon. Carbon usage rates are calculated as "base rate," discussed above [Hutchins, 1981].

Rapid Sand Filters. It has become common to remove the sand from rapid sand filters and replace it with activated carbon [Culp and Clark, 1983; Graese *et al.*, 1987; Joint Task Force, 1990]. Unless the filter box is extended, the contacting time will be short (usually less than 9 min) and may not be sufficient to establish a mass transfer zone. Such installations are normally restricted to taste and odor control. Taste and odor control systems may be in service for 1 to 5 years before the carbon must be replaced.

Because the carbon also acts as a particle filter, several operational problems must be considered:

- Backwashing rates sufficient to remove accumulated solids and prevent mudball formation may lead to carbon loss; careful carbon sizing and backwash control are needed.
- Carbon grain pores may be fouled by deposition of calcium carbonate, magnesium hydroxide, and ferric and manganic oxides, all of which reduce adsorptive capacity.

8.8 Aeration and Gas Exchange

Equilibria and Kinetics of Unreactive Gases

Equilibria

According to Henry's law, at equilibrium the concentration of a dissolved gas in a liquid is proportional to the partial pressure of the gas in the gas phase [Brezonik, 1994]:

$$C = K_H p \tag{8.310}$$

where

C = the concentration of dissolved gas in the liquid (kg/m^3 or lb/ft^3)
K_H = the Henry's law constant ($kg/m^3 \cdot Pa$ or $lb \cdot ft/lbf$)
p = the partial pressure of the dissolved gas in the gas phase (Pa or lbf/ft^2)

Equation (8.310) can be written in a variety of ways. The pressure on the right-hand side can be replaced by a concentration, in which case the Henry's law constant takes on the units of a volume ratio (e.g., m^3 gas/m^3 liquid). This is frequently (and incorrectly) called the dimensionless Henry's law constant. Also, it is often convenient to work in moles or pound-moles instead of kilograms or pounds. Other variants are discussed below under "Air Stripping of Volatile Organic Substances."

Equation (8.310) is often written with the constant on the left-hand side:

$$p = HC \tag{8.311}$$

where H is the (reciprocal) Henry's law constant (Pa·m^3/kg or lbf/lb·ft). Equation (8.310) will be used in order that the symbol H may be used without confusion for depth in various mass transfer formulas.

Mass Transfer Kinetics

The simplest model of mass transfer between two phases is the Lewis–Whitman two-film theory [Brezonik, 1994]. It is assumed that a "stagnant" film exists on each side of the phase interface. Each film exists only statistically. Its fluid properties may differ from the bulk fluid only on average and not at each instant. Its thickness depends on the details of the flow regime in the bulk fluid. Mass transport through each film is supposed to occur by molecular diffusion, and Fick's law applies. If the films are in steady state (which does not mean that the bulk fluids are, although they may be), the total mass flux through each film will be the same, and the fluxes will be proportional to the pressure or concentration differences across the films:

$$\underbrace{F}_{\text{Total flux}} = \underbrace{k_g\left(\frac{p - p_i}{RT}\right)}_{\substack{\text{Flux through}\\\text{gas film}}} = \underbrace{k_l(C_i - C)}_{\substack{\text{Flux through}\\\text{liquid film}}} \tag{8.312}$$

where

C = the gas concentration in the bulk liquid (mol/m^3 or lb-mol/ft^3)
C_i = the gas concentration in the liquid at the gas/liquid interface (mol/m^3 or lb-mol/ft^3)
F = the total flux through the two films (mol/m^2·s or lb-mol/ft^2·s)
k_g = the gas film mass transfer coefficient (m/s or ft/s)
k_l = the liquid film mass transfer coefficient (m/s or ft/s)
p = the partial pressure of the dissolving gas in the bulk gas phase (Pa or lbf/ft^2)
p_i = the partial pressure of the dissolving gas at the gas/liquid interface (Pa or lbf/ft^2)
R = the gas constant (8.314510 J/mol·K or 1545.356 ft·lbf/lb-mol·°R)
T = the absolute temperature (K or °R)

Henry's law connects the interfacial partial pressure and the interfacial dissolved gas concentration, and it also allows the bulk gas phase partial pressure to be expressed as the equivalent gas solubility:

$$C_i = K_H p_i \tag{8.313}$$

$$C_s = K_H p \tag{8.314}$$

where C_s is the dissolved gas concentration equivalent to the gas's partial pressure in the gas phase, its equilibrium concentration (kg/m^3 or lb/ft^3).

The overall concentration difference, $C_s - C$, may be partitioned as follows:

$$C_s - C = (C_s - C_i) + (C_i - C) \tag{8.315}$$

$$C_s - C = K_H(p - p_i) + (C_i - C) \tag{8.316}$$

The pressures and concentrations may be eliminated via the flux relationships in Eq. (8.312), and an overall liquid film mass transfer coefficient may be defined:

$$\frac{1}{K_l} = \frac{RTK_H}{k_g} + \frac{1}{k_l} \tag{8.317}$$

where

K_l = the overall liquid film mass transfer coefficient (m/s or ft /s)
$= F/(C_s - C)$

A similar analysis, which starts by writing the overall pressure difference analogously to Eq. (8.315), yields

$$\frac{RT}{K_g} = \frac{RT}{k_g} + \frac{1}{k_l K_H} \tag{8.318}$$

where

K_g = the overall gas film mass transfer coefficient (m/s or ft/s)
$= F/(p - p_{eq})$
p_{eq} = the gas partial pressure equivalent to the bulk liquid phase concentration (Pa or lbf/ft^2)
$= K_H C$

If the Henry's law constant and the flux are defined in terms of molar concentration in both the gas and liquid phases, the product RT does not appear in Eqs. (8.338), (8.344), or (8.345).

Note that the two overall transfer coefficients are connected by the Henry's law constant:

$$K_H = \frac{K_g}{K_l} = \frac{RT(C_s - C)}{p - p_{eq}} \tag{8.319}$$

It is largely a matter of convenience which overall mass transfer coefficient is employed. If K_H is very small, then K_l is approximately equal to k_l. Similarly, if K_H is very large, then K_g is nearly equal to k_g. Furthermore, a large Henry's law constant means that K_g is larger than K_l, and the mass transfer rate is limited by diffusion through the water film, and the concentration gradient across the water film is large. Conversely, a small Henry's law constant means that mass transfer is controlled by diffusion through the gas film, and the pressure gradient across the gas film is large.

The underlying film mass transfer coefficients are supposed to depend on the molecular diffusivities of the substances. Depending on whether the fluid regime is laminar, transitional, or turbulent (which affects the film structure), the dependency is either to the first power of the diffusivity or to its square root. The diffusivity itself depends on the reciprocal square root of the relative molecular weight. Consequently, if the mass transfer coefficient for one substance is known, the coefficient for another can be estimated [Brezonik, 1994]:

$$\frac{k_{l1}}{k_{l2}} = \left(\frac{D_{l1}}{D_{l2}}\right)^n = \left(\frac{M_{r2}}{M_{r1}}\right)^{n/2} \tag{8.320}$$

$$\frac{k_{g1}}{k_{g2}} = \left(\frac{D_{g1}}{D_{g2}}\right)^n = \left(\frac{M_{r2}}{M_{r1}}\right)^{n/2} \tag{8.321}$$

where

D_{ij} = the diffusivity of the *j*th substance in the *i*th phase (m^2/s or ft^2/s)
M_{rj} = the relative molecular weight of the *j*th substance (g/mol or lb/lb-mol)

The exponent n in Eq. (8.321) varies from 0.5 under turbulent conditions to 1.0 under laminar, and must be determined experimentally [Brezonik, 1994].

The diffusivities can be estimated by using Eqs. (8.281), (8.282), or (8.283).

K_g and K_l are areal mass transfer coefficients, because the flux must be multiplied by the interfacial area to get the total rate of mass transfer. In many systems, the interfacial area cannot be determined, but the gas or liquid volumes can. It is then easier to use an overall volumetric mass transfer coefficient. The areal and volumetric coefficients are related by

$$K_l a = \frac{K_l A}{V} \tag{8.322}$$

where

A = the (possibly unknown) interfacial area (m^2 or ft^2)
$K_l a$ = the overall volumetric mass transfer coefficient (per second)

Oxygen Transfer

Oxygen is an unreactive, high Henry's law constant gas, whose gas transfer kinetics are liquid film limited.

Oxygen Solubility

The solubility of oxygen depends on the temperature of the water, the humidity of the air, and the mean submergence of the gas bubbles as they rise from the diffusers to the tank surface. Henry's law for oxygen solubility is

$$C_s = K_H p_{O_2} \tag{8.323}$$

where

C_s = the solubility of oxygen in water (kg/m^3 or lb/ft^3)
K_H = the Henry's law equilibrium constant (kg/m^3·Pa or lb/ft·lbf)
p_{O_2} = the partial pressure of oxygen in the gas phase (Pa or lbf/ft^2)

Because *dry* air is 20.946% by volume oxygen [Weast *et al.*, 1983], Henry's law can also be written as

$$C_s = 0.20946 K_H p_{da} \tag{8.324}$$

where p_{da} is the pressure of dry air (Pa or lbf/ft^2). Air usually contains some water vapor, and Henry's law for humid air is

$$C_s = 0.20946 K_H (p_{ha} - p_v) \tag{8.325}$$

where p_v is the pressure of the water vapor (Pa or lbf/ft^2).

By convention, tabulated values of oxygen solubility are given for standardized conditions of (1) 1 atmosphere (101.325 kPa or 2116.22 lbf/ft^2) total humid air pressure and (2) 100% relative humidity in the gas phase:

$$C_s' = 0.20946 K_H (p_{sa} - p_{vsat}) \tag{8.326}$$

where

C_s' = oxygen's solubility in pure water under standard conditions or 20°C and 1 atm total pressure of water-saturated air (kg/m^3 or lb/ft^3)
p_{vsat} = the vapor pressure of water at 20°C (Pa or lbf/ft^2)

Because the vapor pressure of water depends only on temperature, one can also write

$$p_{sa} - p_{vsat} = c \cdot p_{sa} \tag{8.327}$$

$$C_s' = 0.20946 \cdot c \cdot K_H p_{sa} \tag{8.328}$$

The value of c varies from about 0.993 to 0.905 as the temperature increases from 0 to 45°C.

Table 8.16 Solubilities of Oxygen in Pure Water in Contact with Water-saturated Air at 1 Atm Total Pressure[1]

Temperature (°C)	Oxygen Solubility (mg/L) for Salinity (g/kg)					
	0.0	0.5	1.0	1.5	2.0	2.5
0.	14.62	14.57	14.52	14.47	14.42	14.37
5.	12.77	12.73	12.69	12.64	12.60	12.56
10.	11.29	11.25	11.22	11.18	11.14	11.11
15.	10.08	10.05	10.02	9.99	9.96	9.93
20.	9.09	9.07	9.04	9.01	8.99	8.96
25.	8.26	8.24	8.22	8.19	8.17	8.15
30.	7.56	7.54	7.52	7.50	7.48	7.46
35.	6.95	6.93	6.91	6.89	6.88	6.86
40.	6.41	6.40	6.38	6.36	6.35	6.33
45.	5.93	5.92	5.90	5.89	5.87	5.86
50.	5.49	5.48	5.47	5.45	5.44	5.43

[1]Computed from formula of Joint Editorial Board. 1992. *Standard Methods for the Examination of Water and Wastewater,* 18th ed. American Public Health Association, Washington, D.C.

Standard values of the oxygen solubility are given in Table 8.16. These concentrations may be calculated from [Joint Editorial Board, 1992]:

$$\begin{aligned} \ln C_s' &= -139.44411 + \frac{157.5701 \times 10^3}{T} - \frac{66.42308 \times 10^6}{T^2} + \frac{12.43800 \times 10^9}{T^3} \\ &\quad - \frac{862.1949 \times 10^9}{T^4} - S\,(‰)\left[0.017674 - \frac{10.754}{T} + \frac{2140.7}{T^2}\right] \end{aligned} \tag{8.329}$$

where

S (‰) = the salinity (g "dissolved solids"/kg water)

T = the absolute temperature in K

The salinity is discussed below.

Salinity and Chlorinity. The solubilities of dissolved gases are reduced by the presence of dissolved salts, which is called the "salting out" effect [Glasstone, 1947].

The original definition of "salinity" was "mass of salt per unit mass of seawater." This proved to be a very difficult chemical determination, because simple drying leads to loss of hydrogen chloride and carbon dioxide and the residue readily absorbs moisture from the atmosphere [Riley and Chester, 1971]. The definition due to Knudsen is that the salinity is the grams of dissolved inorganic matter in 1 kg of water after the carbonate has been replaced by oxide and after bromide and iodide have been replaced by chloride. This would be approximately equivalent to the ash left in a volatile solids test of fresh water.

It is simpler to measure the total halides by a silver titration. The "chlorinity" is defined to be the grams of silver needed to precipitate all the halides in 328.5233 g of water. With this definition, the empirical relationship between salinity and chlorinity is [Riley and Chester, 1971]

$$S\,(‰) = 0.03 + 1.805Cl\,(‰) \tag{8.330}$$

Nowadays, salinity is determined via conductance measurements. The currently accepted formula is [Lewis, 1980; Perkin and Lewis, 1980]

$$S\,(‰) = \sum_{j=0}^{5} \left[a_j + b_j f(T)\right] R_T^{j/2} \tag{8.331}$$

$$R_T = \frac{R}{R_p r_T} \tag{8.332}$$

$$R_p = 1 + \frac{A_1 p + A_2 p^2 + A_3 p^3}{1 + B_1 T + B_2 T^2 + B_3 R + B_4 RT} \tag{8.333}$$

$$r_T = c_0 + c_1 T + c_2 T^2 + c_3 T^3 + c_4 T^4 \tag{8.334}$$

where

$A_1, A_2, A_3 = 2.070 \times 10^{-5}, -6.370 \times 10^{-10}, 3.989 \times 10^{-15}$

$a_0, a_1, a_2, a_3, a_4, a_5 = 0.0080, -0.1692, 25.3851, 14.0941, -7.0261, 2.7081$

$B_1, B_2, B_3, B_4 = 3.426 \times 10^{-2}, 4.464 \times 10^{-4}, 4.215 \times 10^{-1}, -3.107 \times 10^{-3}$

$b_0, b_1, b_2, b_3, b_4, b_5 = 0.0005, -0.0056, -0.0066, -0.0375, 0.0636, -0.0144$

$c_0, c_1, c_2, c_3, c_4 = 0.6766097, 0.0200564, 1.104259 \times 10^{-4}, -6.6698 \times 10^{-7}, 1.0031 \times 10^{-9}$

$f(T) = \dfrac{T - 15}{1 + 0.0162(T - 15)}$

p = the gauge pressure of the in situ sample in decibars

R = the ratio of the measured in situ conductivity of the sample to the conductivity of a solution containing 32.4356 g KCl in 1 kg solution at 15°C and 1 atm total pressure

R_P = a pressure correction factor (dimensionless)

r_T = a temperature correction factor (dimensionless)

T = the in situ sample temperature in °C

Equation (8.331) is valid for salinities ranging from 2 to 39 g/kg, temperatures from −2 to 35°C, and pressures from 0 to about 2000 decibars. The pressure correction factor can be ignored in most treatment plant designs.

For the lower salinities normally encountered in water and wastewater treatment (0 to 1 g/kg), the following formula may be used [Hill and Dauphinee, 1986]:

$$S\,(‰) = \sum_{j=0}^{5} \left[a_j + b_j f(T)\right] R_T^{j/2} - \frac{0.0080}{1 + 600 R_T + 1.6 \times 10^5 R_T^2} - \frac{0.0005 f(T)}{1 + 10 R_T + 1000 R_T^{3/2}} \tag{8.335}$$

Effect of Pressure. An air bubble just released from a diffuser is subjected to a total pressure equal to the local barometric pressure plus the hydrostatic head due to the submergence, $p_b + \gamma h$. Air bubbles in contact with water quickly reach the water temperature and become saturated with water vapor, so the dry air pressure in the bubble is $p_b + \gamma h - p_{vsat}$. The partial pressure of oxygen near the diffuser would be $0.20946(p_b + \gamma h - p_{vsat})$.

The hydrostatic head falls as the bubble rises. Moreover, the volume percentage of oxygen in the bubble declines as oxygen is absorbed by the surrounding water. Consequently, one needs to average both the hydrostatic pressure and the bubble oxygen pressure to get an average aeration tank solubility. The result is

$$C_s = 0.20946 \bar{f} K_H (p_b + \tfrac{1}{2}\gamma h - p_{vsat}) \tag{8.336}$$

where

$\bar{f}$ = the depth-averaged ratio of the oxygen partial pressure at any depth to the initial oxygen partial pressure (dimensionless)

h = the depth of submergence of the diffuser (m or ft)
γ = the weight density of water (N/m^3 or lbf/ft^3)

The oxygen transfer efficiency is typically around 10%, so $\overline{f}$ is typically around 0.95.

The mean oxygen solubility in an aeration tank, corrected for all these effects, is related to the standard oxygen solubility by

$$\overline{C}_s = C_s' \cdot \frac{\overline{f}}{c} \cdot \frac{p_b + \frac{1}{2}\gamma h - p_{vsat}}{p_{sa}} \tag{8.337}$$

where $\overline{C}_s$ is the depth-averaged oxygen solubility (kg/m^3 or lb/ft^3). The ratio $\overline{f}/c$ ranges from about 0.96 to 1.05 as the temperature increases from 0 to 45°C, but under summer conditions in the temperate zone—which usually control the design—the ratio is about 0.99.

Density and vapor pressure data for water are given in Tables 8.17 and 8.18, respectively. Air properties versus elevation above or below sea level are given in Table 8.19.

Gas Diffusers

There are two broad classes of air diffusion devices, based on bubble size—coarse bubble and fine bubble. Coarse bubble diffusers produce air bubbles that are 6 to 10 mm in diameter, and *new* fine bubble diffusers produce bubbles that are 2 to 5 mm in diameter.

The standard oxygen transfer efficiencies of several kinds of diffusers are given in Table 8.20. It should be noted that the fine bubble systems are generally two to three times as efficient as coarse bubble systems. The cost of the increased efficiency is additional maintenance; the benefit is reduced energy usage. Nowadays, fine bubble systems are cost-effective.

Coarse Bubble Diffusion. The most common coarse bubble diffusers consist of perforated pipes and valved orifices. Such devices have low transfer efficiencies, but they resist clogging. Coarse bubble diffusers are generally installed along the bottom of one or both walls to develop spiral flow.

Table 8.17 Densities of Pure Water at Selected Temperatures[1]

Temperature (°C)	Mass Density (kg/m^3)	Weight Density (N/m^3)	Temperature (°C)	Mass Density (kg/m^3)	Weight Density (N/m^3)
0	999.84	9805.1	25	997.04	9777.6
5	999.96	9806.2	30	995.65	9764.0
10	999.70	9803.7	35	994.03	9748.1
15	999.09	9797.7	40	992.02	9728.4
20	998.20	9789.0	45	990.21	9710.6

[1]Recalculated from Dean, J. A. 1992. *Lange's Handbook of Chemistry*, 14th ed. McGraw-Hill, New York.

Table 8.18 Vapor Pressures of Pure Water in Contact with Air at Selected Temperatures[1]

Temperature (°C)	Vapor Pressure (kPa)	Vapor Pressure (lbf/in.^{2})	Temperature (°C)	Vapor Pressure (kPa)	Vapor Pressure (lb/in.^{2})
0	0.615	0.0892	25	3.192	0.4629
5	0.879	0.1275	30	4.275	0.6201
10	1.237	0.1794	35	5.666	0.8218
15	1.717	0.2490	40	7.432	1.0780
20	2.356	0.3417	45	9.623	1.3956

[1]Recalculated from Dean, J. A. 1992. *Lange's Handbook of Chemistry*, 14th ed. McGraw-Hill, New York.

Table 8.19 Properties of the Standard Atmosphere versus Altitude[1]

Altitude above Mean Sea Level (m)	Pressure (kPa)	Density (kg/m^3)	Absolute Viscosity ($N \cdot s/m^2$)	Temperature (K)
−1000	113.93	1.3470	1.8206×10^{-5}	294.65
−500	107.47	1.2849	1.8050×10^{-5}	291.40
0	101.325	1.2250	1.7894×10^{-5}	288.15
500	95.461	1.1673	1.7737×10^{-5}	284.90
1000	89.876	1.1117	1.7579×10^{-5}	281.65
1500	84.559	1.0581	1.7420×10^{-5}	278.40
2000	79.501	1.0066	1.7260×10^{-5}	275.15
2500	74.691	0.9570	1.7099×10^{-5}	271.91
3000	70.121	0.9092	1.6938×10^{-5}	268.66

[1]Weast, R. C., Astle, M. J., and Beyer, W. H. 1983. *CRC Handbook of Chemistry and Physics,* 64th ed. CRC Press, Boca Raton, FL.

The oxygen transfer rate (OTR) for a completely mixed reactor is

$$R = \alpha \cdot K_l a \cdot (\beta \cdot C_s - C) \cdot V \tag{8.338}$$

where

C = the oxygen concentration in the aeration tank (kg/m^3 or lb/ft^3)

C_s = the oxygen solubility in the aeration tank under the given temperature and diffusers submergence (kg/m^3 or lb/ft^3)

$K_l a$ = the volumetric mass transfer coefficient (per second)

R = the oxygen transfer rate (kg/s or lb/s)

V = the aeration tank volume (m^3 or ft^3)

Table 8.20 Comparative Efficiencies of Diffusers in Clean Water at 15 ft Submergence[1]

Diffuser System	Air Flow Rate (scfm/diffuser)	Standard Oxygen Transfer Efficiency (%)
Ceramic plate grids	2.0 to 5.0 ($scfm/ft^2$)	26 to 33
Ceramic disk grids	0.4 to 3.4	25 to 40
Ceramic dome grids	0.5 to 2.5	27 to 39
Porous plastic disk grids	0.6 to 3.5	24 to 35
Perforated membrane grids	0.5 to 20.5	16 to 38
Rigid porous tube grids	2.4 to 4.0	28 to 32
Rigid porous tube in dual-spiral roll	3 to 11	17 to 28
Rigid porous tube in single-spiral roll	2 to 12	13 to 25
Nonrigid porous tube grid	1 to 7	26 to 36
Nonrigid porous tube in single-spiral roll	2 to 7	19 to 37
Perforated membrane grid	1 to 4	22 to 29
Perforated membrane midwidth	2 to 6	16 to 19
Perforated membrane midwidth	2 to 12	21 to 31
Perforated membrane in single-spiral roll	2 to 6	15 to 19
Coarse bubble in dual-spiral roll	3.3 to 9.9	12 to 13
Coarse bubble midwidth	4.2 to 45	10 to 13
Coarse bubble in single-spiral roll	10 to 35	9 to 12

[1]ASCE Committee on Oxygen Transfer. 1989. *Design Manual: Fine Pore Aeration Systems,* EPA/625/1-89/023, U.S. Environmental Protection Agency, Office of Research and Development, Center for Environmental Research Information, Risk Reduction Engineering Laboratory, Cincinnati, OH.

α = the ratio of the mass transfer coefficient under conditions in the aeration tank to the mass transfer coefficient under standard conditions (dimensionless)

β = the ratio of the oxygen solubility for the salinity in the aeration tank to the oxygen solubility in pure water (dimensionless)

Both the alpha and beta values depend on water composition, and alpha also depends on the aeration equipment. They should be determined by field testing.

Beta is generally near 1.0, unless the salinity of the water is high. The effect is accounted for by Eq. (8.315) and Table 8.16, and beta should be set to unity, if oxygen solubilities are calculated from Eq. (8.315).

Some aeration field data are given in Table 8.21 [Joint Task Force, 1988]. For diffused air and oxygen systems, the alpha value for raw and settled municipal wastewater is about 0.2 to 0.3; it rises to about 0.5 to 0.6 for conventional, unnitrified effluents and to about 0.8 to 0.9 for highly treated, nitrified effluents. Alpha values for fine bubble diffusers are smaller than those for coarse bubble systems. For surface aeration systems, the alpha value for raw and settled municipal wastewater is about 0.6; it may rise to 1.2 for clean water.

The volumetric mass transfer coefficient is temperature dependent [Stenstrom and Gilbert, 1981]:

$$\frac{K_l a(T_1)}{K_l a(T_o)} = 1.024^{T_1 - T_o} \tag{8.339}$$

where

T_o = the reference temperature in °C

T_1 = the aeration tank temperature in °C

The usual reference temperature is 20°C.

Table 8.21 Observed and Standard Oxygen Transfer Efficiencies at Selected Wastewater Treatment Plants[1]

Location	System	Observed Transfer Efficiency (%)	Average Alpha Values	Range of Alpha Values	Expected Efficiency under Standard Conditions (%)
Madison, WI	Ceramic grid, step feed	17.8	0.64	0.42 to 0.98	25 to 37
Whittier Narrows, CA	Ceramic grid, plug flow	11.2	0.45	0.35 to 0.60	25 to 37
Seymour, WI	Ceramic grid, plug flow	16.5	0.66	0.49 to 0.75	25 to 37
Lakewood, OH	Ceramic grid, plug flow	14.5	0.51	0.44 to 0.57	25 to 37
Lakewood, OH	Ceramic grid, plug flow	8.9	0.31	0.26 to 0.37	25 to 37
Madison, WI	Ceramic and plastic tubes, step feed	11.0	0.62	0.46 to 0.85	13 to 32
Madison, WI	Wide-band, fixed orifice, nonporous diffusers, step feed	10	1.07	0.83 to 1.19	9 to 13
Orlando, FL	Wide-band, fixed orifice, nonporous diffusers, complete mix	7.6	0.75	0.67 to 0.83	9 to 13
Nassau, Co., NY	Flexible membrane tubes, plug flow	7.6	0.36	0.27 to 0.42	15 to 29
Whittier Narrows, CA	Jet aerators, plug flow	9.4	0.58	0.48 to 0.72	15 to 24
Brandon, WI	Jet aerators, complete mix	10.9	0.45	0.40 to 0.50	15 to 24
Brandon, WI	Jet aerators, complete mix	7.5	0.47	0.46 to 0.48	15 to 24

[1]Joint Task Force of the Water Pollution Control Federation and the American Society of Civil Engineers. 1988. *Aeration: A Wastewater Treatment Process,* WPCF Manual of Practice No. FD-13, ASCE Manuals and Reports on Engineering Practice No. 68. Water Pollution Control Federation, Alexandria, VA; American Society of Civil Engineers, New York.

The so-called standard oxygen transfer rate (SOTR) of equipment is usually reported under standard conditions of (1) clean water, (2) zero dissolved oxygen, (3) 20°C, (4) 1 standard atmosphere (101.325 kPa or 2116.22 lbf/ft^2) of ambient air pressure, and (4) a specified depth of submergence. The SOTR is calculated as

$$R_{\text{std}} = K_l a \cdot C_s \cdot V \tag{8.340}$$

where R_{std} is the standard oxygen transfer rate (kg/s or lb/s). The conversion of SOTRs to OTRs is

$$\frac{R}{R_{\text{std}}} = \cdot \frac{\alpha \cdot 1.024^{T-20} \cdot [\beta \cdot C_s(T) - C]}{C_s(20°\text{C})} \tag{8.341}$$

The oxygen transfer efficiency (OTE) is the ratio of the oxygen absorbed to the oxygen supplied in the air flow through the diffuser:

$$E_{\text{O}_2} = \frac{R}{0.20946 Q_a \rho_a M_{r\text{O}_2}} \tag{8.342}$$

where

E_{O_2} = the oxygen transfer efficiency (dimensionless)
$M_{r\text{O}_2}$ = the relative molecular weight of oxygen (31.998, either g/mol or lb/lb-mol)
Q_a = the air flow rate under standard conditions of 1 atm pressure and 0°C (m^3/s or ft^3/s)
ρ_a = the molar oxygen density (mol/m^3 or lb-mol/ft^3)

The standard oxygen transfer efficiency (SOTE) is the efficiency for clean water at 20°C and 1 atm pressure at a specified submergence.

The molar density of oxygen can be calculated from the ideal gas law:

$$\rho_a = \frac{n}{V} = \frac{P}{RT} \tag{8.343}$$

Fine Bubble Diffusers. There are three types of fine bubble diffusers [ASCE Committee on Oxygen Transfer, 1989]:

- Sintered ceramic plates made from alumina, aluminum silicate, and silica
- Rigid porous plastic plates and tubes, usually made from high density polyethylene (HDPE) and styrene-acrylonitrile
- Nonrigid porous plates and tubes made from rubber and HDPE
- Perforated plastic membranes, both disks and tube sheaths, usually made from polyvinyl chloride (PVC) with added plasticizer

Presently, only the first and last types are in widespread use. The diffusers are installed as (1) plate holders and air manifolds set *on* the aeration tank floor, or (2) disk, dome, or tube diffusers attached to air manifolds that cover the tank floor and are set somewhat above it.

Fine bubble diffusers gradually lose oxygen transfer efficiency. This is usually modeled as [ASCE Committee on Oxygen Transfer, 1989]

$$R = \alpha \cdot F \cdot K_l a \cdot (\beta \cdot C_s - C) \cdot V \tag{8.344}$$

where F is the fouling factor (dimensionless). The fouling factor decreases with time of service. This occurs because either the diffuser pores accumulate airborne particulates and precipitates from the water (type I) or because of biofilm growth on the diffuser surface (type II). For design purposes, the rate of fouling may be taken as constant so that the fouling factor may be modeled as

$$F = 1.0 - k_f t \tag{8.345}$$

Table 8.22 Effect of Air Flow Rate on Diffuser Oxygen Transfer Efficiency[1]

Diffuser System	m
Ceramic dome grid	0.150
Ceramic disk grid	0.133
Ceramic disk grid	0.126
Rigid porous disk grid	0.097
Rigid porous tube in double-spiral roll	0.240
Nonrigid tube in spiral roll	0.276
Perforated membrane disk grid	0.195
9-in. perforated membrane disk grid	0.11
EPDM perforated membrane tube grid	0.150

[1]ASCE Committee on Oxygen Transfer. 1989. *Design Manual: Fine Pore Aeration Systems,* EPA/625/1-89/023. U.S. Environmental Protection Agency, Office of Research and Development, Center for Environmental Research Information, Risk Reduction Engineering Laboratory, Cincinnati, OH.

where

$$k_f = \text{the fouling rate (per second)}$$
$$t = \text{the service time (s)}$$

Typical values for k_f range from 0.03 to 0.07 per month.

The standard oxygen transfer efficiencies of fine bubble diffusers depend somewhat on air flow rate [ASCE Committee on Oxygen Transfer, 1989]:

$$\frac{E_{\text{stdO}_2,2}}{E_{\text{stdO}_2,1}} = \left(\frac{Q_{ad1}}{Q_{ad2}}\right)^m \tag{8.346}$$

Values of m for various diffusers are given in Table 8.22.

Air Piping. The volumetric air flow rate for the piping temperature and pressure can be estimated from the ideal gas law:

$$\frac{p_2 Q_{a_2}}{T_2} = \frac{p_{\text{sa}} Q_{\text{Ra}}}{T_{\text{sc}}} \tag{8.347}$$

where

p_2 = the blower outlet pressure (Pa or lbf/ft^2)
p_{sa} = the pressure of the standard atmosphere (101.325 kPa or 2116.22 lbf/ft^2)
Q_a = the air flow rate (m^3/s or ft^3/s)
Q_{Ra} = the required process air flow rate at standard conditions of 1 atm pressure and 0°C (m^3/s or ft^3/s)
T_2 = the blower outlet temperature (K or °R)
T_{sc} = the temperature for standard conditions for gases (273.15 K or 491.67°R)

The piping inlet temperature may be estimated from the adiabatic polytropic process [Perry and Chilton, 1973]:

$$T_2 = T_1 \left(\frac{p_2}{p_1}\right)^{(n-1)/n} \tag{8.348}$$

where

n = the exponent for the polytropic process (dimensionless)
= 1.40, for air
p_1 = the blower intake pressure (N/m^2 or lbf/ft^2)
p_2 = the blower outlet pressure (N/m^2 or lbf/ft^2)
T_1 = the blower intake temperature (K or °R)
T_2 = the blower outlet temperature (K or °R)

The blower power requirement is [Perry and Chilton, 1973]

$$P = \left(\frac{n}{n-1}\right)\rho_a Q_a R T_1 \left[\left(\frac{p_2}{p_1}\right)^{(n-1)/n} - 1\right] \tag{8.349}$$

where

P = the blower power (W or ft-lbf/s)
Q_a = the air flow rate (m^3/s or ft^3/s)
R = the gas constant
= 8.314510 J/mol·K or 1545.356 ft·lbf/lb-mol·°R)
ρ_a = the mass density of the air (mol/m^3 or lb-mol/ft^3)

The blower power is the power in the compressed air, and the power required by the blower motor will be larger in inverse proportion to the blower efficiency.

For small temperature and pressure changes (less than 10%), air may be treated as an incompressible fluid [Metcalf & Eddy, Inc., 1991]. The principal difference from water distribution is that the blower exit gas is hot (about 140°F to 180°F) and allowances must be made for piping expansion and contraction.

The headloss may be calculated using the Darcy–Weisbach equation and friction factors from the Moody diagram. The air density may be calculated from the ideal gas law. The viscosity of air is given by the Chapman–Enskog formula [Blevins, 1984]:

$$\mu_a = 26.69 \times 10^{-7} \frac{M_{\text{ra}}^{1/2} T^{1/2}}{\sigma^2 \Omega_v} \tag{8.350}$$

$$\Omega_v = 1.147 \left(\frac{T}{T_e}\right)^{-0.145} + \left(\frac{T}{T_e} + 0.5\right)^{-2.0} \tag{8.351}$$

where

M_{ra} = the relative molecular weight for air (28.966 g/mol, for air)
T = the air temperature in K
T_e = the effective temperature of the force potential in K (78.6 K, for air)
μ_a = the dynamic viscosity of air in N·s/m^2
σ = the collision diameter in Å(3.711 Å, for air)

Surface Aerators

The common surface aerators are the Kessener brush and its derivatives (brushes or disks mounted on a horizontal shaft and normally found in oxidation ditches), low-speed radial flow turbines, and high-speed axial flow turbines (usually equipped with draft tubes and found in aerated lagoons).

Surface aerators are rated in terms of mass of oxygen transferred per unit time per unit power, for example, kg/kW·s or lb/hp·hr. The design equation is [Joint Task Force, 1988]

$$\frac{R/P}{R_{\text{std}}/P} = \frac{\alpha \cdot 1.024^{T-20} \cdot [\beta \cdot C_s(T) - C]}{C_s(20°\text{C})} \tag{8.352}$$

Both alpha and beta values are often reported to be about 1 in municipal wastewater. The usual theta value, 1.024, applies. Standard aeration rates for low-speed radial flow systems range from about 2.0 to 5.0 lb O_2/hp·hr; for high-speed axial flow aerators from 2.0 to 3.6 lb O_2/hp·hr; and for brushes from 1.5 to 3.6 lb O_2/hp·hr [Metcalf & Eddy, 1991].

Absorption of Reactive Gases

Some gases react either with water or with other solutes in it. If the reactions are fast (as are most acid/base reactions), the reaction increases both the amount of gas that is absorbed and its absorption rate.

Equilibria

Henry's law applies only to the dissolved gas molecule and not to any of its reaction products.

Ammonia. Ammonia chemistry is outlined in Chapter 7. Ammonia reacts with water to form the ammonium ion, NH_4^+:

$$NH_3 + H_2O \leftrightarrow NH_4^+ + OH^- \tag{8.353}$$

The molar fraction of the total ammonia concentration that is unprotonated ammonia is

$$f = \frac{[NH_3]}{[NH_3] + [NH_4^+]} = \frac{1}{1 + \dfrac{[H^+]}{K_a}} = \frac{1}{1 + \dfrac{K_b}{[OH^-]}} \tag{8.354}$$

where

K_a = the acid ionization constant (mol/L)
K_b = the base ionization constant (mol/L)

Values of the acid ionization constant for various temperatures are given in Table 8.23.

Carbon Dioxide. Atmospheric carbon dioxide goes into solution, forming aqueous carbon dioxide, and this reacts with water to form carbonic acid [Stumm and Morgan, 1970]:

$$CO_2(g) \leftrightarrow CO_2(aq) \tag{8.355}$$

$$CO_2(aq) + H_2O \leftrightarrow H_2CO_3 \tag{8.356}$$

The reaction in Eq. (8.355) goes virtually to completion, and dissolved carbon dioxide and carbonic acid are not distinguished by the usual analytical methods. Therefore, it is customary to define a *composite carbonic acid* concentration:

$$H_2CO_3^* \leftrightarrow CO_2(aq) + H_2CO_3 \tag{8.357}$$

Table 8.23 Acid Ionization Constants for Selected Gases—pK_a Values[1]

	Temperature (°C)									
Acid	0	5	10	15	20	25	30	35	40	50
NH_4^+	10.081	9.904	9.731	9.564	9.400	9.425	9.093	8.947	8.805	8.539
$H_2CO_3^*$	6.577	6.517	6.465	6.429	6.382	6.352	6.327	6.309	6.296	6.285
HCO_3^-	10.627	10.558	10.499	10.431	10.377	10.329	10.290	10.250	10.220	10.172
HCl	—	—	—	—	—	−6.2	—	—	—	—
HCN	—	—	9.63	9.49	9.36	9.21	9.11	8.99	8.88	—
HOCl	7.82	7.75	7.69	7.63	7.58	7.54	7.50	7.46	—	7.05
H_2S	—	7.33	7.24	7.13	7.05	6.97	6.90	6.82	6.79	6.69
HS^-	—	13.5	—	13.2	—	12.90	12.75	12.6	—	—
H_2SO_3	1.63	—	1.74	—	—	1.89	—	1.98	—	2.12
HSO_3^-	—	—	—	—	6.91 (at 18°C)	7.20	—	—	—	—
pK_w	14.938	14.727	14.528	14.340	14.163	13.995	13.836	13.685	13.542	13.275

[1]Mostly from Dean, J. A., ed. 1992. *Lange's Handbook of Chemistry,* 14th ed. McGraw-Hill, New York. Supplemented by Weast, R. C., Astle, M. J., and Beyer, W. H. 1983. *CRC Handbook of Chemistry and Physics,* 64th ed. CRC Press, Boca Raton, FL; and Blaedel, W. J. and Meloche, V. W. 1963. *Elementary Quantitative Analysis,* 2nd ed. Harper and Row, New York.

At equilibrium, the ratio of $CO_2(aq)$ to H_2CO_3 is constant, depending only on temperature, so this convention merely introduces a constant factor into the equilibrium constants. The composite carbonic acid ionizes to produce bicarbonate, and bicarbonate ionizes to make carbonate:

$$H_2CO_3^* \leftrightarrow H^+ + HCO_3^- \tag{8.358}$$

$$HCO_3^- \leftrightarrow H^+ + CO_3^{2-} \tag{8.359}$$

The system is completed by the ionization of water:

$$H_2O \leftrightarrow H^+ + OH^- \tag{8.360}$$

The equilibrium constants for this system may be written as:

$$K_{HCO_2}^* = \frac{[H_2CO_3^*]}{p_{CO_2}} \tag{8.361}$$

$$K_1 = \frac{[H^+][HCO_3^-]}{[H_2CO_3^*]} \tag{8.362}$$

$$K_2 = \frac{[H^+][CO_3^{2-}]}{[HCO_3^-]} \tag{8.363}$$

$$K_w = [H^+][OH^-] \tag{8.364}$$

where

$[X]$ = the activity of species X (moles/L), usually approximated as the molar concentration in dilute solutions

p_{CO_2} = the partial pressure of carbon dioxide in the atmosphere (atm)

Numerical values for these constants at several temperatures are given in Table 8.23.

The molar concentrations of the various species can be written in terms of the proton concentration and the equilibrium constants by using Eqs. (8.361) through (8.364) to eliminate variables:

$$f_o = \frac{[H_2CO_3^*]}{[H_2CO_3^*] + [HCO_3^-] + [CO_3^{2-}]} = \frac{1}{1 + \frac{K_1}{[H^+]} + \frac{K_1K_2}{[H^+]^2}} \tag{8.365}$$

$$f_1 = \frac{[HCO_3^-]}{[H_2CO_3^*] + [HCO_3^-] + [CO_3^{2-}]} = \frac{1}{1 + \frac{[H^+]}{K_1} + \frac{K_2}{[H^+]}} \tag{8.366}$$

$$f_2 = \frac{[CO_3^{2-}]}{[H_2CO_3^*] + [HCO_3^-] + [CO_3^{2-}]} = \frac{1}{1 + \frac{[H^+]}{K_2} + \frac{[H^+]^2}{K_1K_2}} \tag{8.367}$$

where

f_o = the molar fraction of the carbonate species in the form of the composite acidity (dimensionless)

f_1 = the molar fraction of the carbonate species in the form of bicarbonate (dimensionless)

f_2 = the molar fraction of the carbonate species in the form of carbonate (dimensionless)

Table 8.24 Distribution of Alkaline Species for the Carbonate System[1]

Ratio of Phenolphthalein to Total Alkalinity, P/T	Bicarbonate (meq/L)	Carbonate (meq/L)	Hydroxide (meq/L)
0	T	0	0
$<½$	$T - 2P$	$2P$	0
½	0	$2P$ or T	0
$>½$	0	$2(T - P)$	$2P - T$
1	0	0	T

[1]Hardenbergh, W. A. and Rodie, E. B. 1963. *Water Supply and Waste Disposal,* International Textbook, Scranton, PA.

The rather wide separation in the values of K_1 and K_2, and the fact that significant quantities of bicarbonate and hydroxide do not occur together make possible a simple procedure for determining the distribution of hydroxide, carbonate, and bicarbonate. The alkalinity titrated to pH 8.3 is called the *phenolphthalein alkalinity* (because phenolphthalein is used to detect the end point), and the total quantity of acid needed to reduce the pH from the original sample value past pH 8.3 down to pH 4.5 is called traditionally the *methyl orange alkalinity*, or total alkalinity. (The latter term is better, because bromcresol green is the preferred indicator.) The ratio of these two values determines the distribution of carbonate forms, as indicated in Table 8.24.

Chlorine. Chlorine gas readily dissolves in water according to Henry's law [Stover *et al.*, 1986]:

$$[\mathrm{Cl_2(aq)}] = K_H p_{\mathrm{Cl_2}} \tag{8.368}$$

$$K_H = 4.805 \times 10^{-6} e^{2818.48/T} \tag{8.369}$$

where

$[\mathrm{Cl_2(aq)}]$ = the concentration of *molecular* chlorine in mol/L
K_H = Henry's law constant in mol/L·atm
$p_{\mathrm{Cl_2}}$ = the partial pressure of molecular chlorine in the gas phase in atm
T = the absolute temperature in K

Dissolved chlorine reacts strongly with water to form hypochlorous acid [White, 1986]:

$$\mathrm{CL_2(aq)} + \mathrm{H_2O} \rightarrow \mathrm{HOCl} + \mathrm{H^+} + \mathrm{Cl^-} \tag{8.370}$$

$$K_o \frac{[\mathrm{H^+}][\mathrm{Cl^-}][\mathrm{HOCl}]}{[\mathrm{Cl_2(aq)}]} \tag{8.371}$$

$$pK_o = -0.579 + \frac{1190.7}{T} \tag{8.372}$$

where T is the absolute temperature in K. In pure water, the total concentration can be written [Stover *et al.*, 1986]:

$$[\mathrm{Cl_2(aq)}] + [\mathrm{HOCl}] = K_H p_{\mathrm{Cl_2}} + (K_O K_H p_{\mathrm{Cl_2}})^{1/3} \tag{8.373}$$

In buffered water with significant background chloride concentrations, the total solubility is

$$[\mathrm{Cl_2(aq)}] + [\mathrm{HOCl}] = K_H p_{\mathrm{Cl_2}} \left[1 + \frac{K_o}{[\mathrm{H^+}][\mathrm{Cl^-}]} + \frac{K_o K_a}{[\mathrm{H^+}]^2[\mathrm{Cl^-}]}\right] \tag{8.374}$$

The hypochlorous acid ionizes to form the hypochlorite ion:

$$HOCl \leftrightarrow H^+ + OCl^- \tag{8.375}$$

$$K_a = \frac{[H^+][OCl^-]}{[HOCl]} \tag{8.376}$$

$$pK_a = -10.069 + 0.025T - \frac{3000}{T} \tag{8.377}$$

where T is kelvin. The molar fraction of hypochlorous acid depends strongly on pH and can be estimated from

$$f = \frac{[HOCl]}{[HOCl] + [OCl^-]} = \frac{1}{1 + \dfrac{K_a}{[H^+]}} \tag{8.378}$$

Below pH 7, a mixture of hypochlorous acid and hypochlorite is nearly all un-ionized acid, but about pH 9 it is nearly all the ion. Values of the acid ionization constants are given in Table 8.23.

Chlorine Dioxide. The solubility of chlorine dioxide follows Henry's law [Haas, 1990]:

$$C_{ClO_2} = K_H p_{ClO_2} \tag{8.379}$$

$$\ln K_H = 58.84621 + \frac{47.9133}{T} - 11.0593 \ln T \tag{8.380}$$

where

C_{ClO_2} = the mole fraction of chlorine dioxide in water (dimensionless)
K_H = the Henry's law constant in atm
p_{ClO_2} = the partial pressure of chlorine dioxide in the gas phase in atm
T = the absolute temperature in K

Above pH 9, chlorine dioxide disproportions according to

$$2ClO_2 + 2OH^- \leftrightarrow ClO_3^- + ClO_2^- + H_2O \tag{8.381}$$

An equilibrium constant is not yet available.

Hydrogen Chloride. The solubility of hydrogen chloride in pure water ranges from 823 g/L at 0°C to 633 g/L at 40°C [Dean, 1992].

Hydrogen chloride is a strong acid with an acid ionization constant of about 1.3×10^6 [Dean, 1992] and is nearly completely ionized in water:

$$HCl \leftrightarrow H^+ + Cl^- \tag{8.382}$$

Hydrogen Cyanide. The solubility of hydrogen cyanide is virtually unlimited. The cyanide ion forms complexes with many metals, which further enhances the solubility.

Hydrogen cyanide is a weak acid:

$$HCN \leftrightarrow H^+ + CN^- \tag{8.383}$$

$$K_a = \frac{[H^+][CN^-]}{[HCN]} \tag{8.384}$$

Values of the acid ionization constant for several temperatures are given in Table 8.23.

Hydrogen Sulfide. The solubility of hydrogen sulfide in pure water ranges from about 7.1 g/L at 0°C to about 1.9 g/L at 50°C [Dean, 1992].

Hydrogen sulfide is a weak acid, and it dissociates twice as the pH is raised:

$$\mathrm{H_2S} \leftrightarrow \mathrm{HS^-} + \mathrm{H^+} \tag{8.385}$$

$$K_1 = \frac{[\mathrm{H^+}][\mathrm{HS^-}]}{[\mathrm{H_2S}]} \tag{8.386}$$

$$\mathrm{HS^-} \leftrightarrow \mathrm{S^{2-}} + \mathrm{H^+} \tag{8.387}$$

$$K_2 = \frac{[\mathrm{H^+}][\mathrm{S^{2-}}]}{[\mathrm{HS^-}]} \tag{8.388}$$

The sulfide ion forms highly insoluble precipitates with many metals, which greatly enhances its solubility. Values of the acid ionization constant are given in Table 8.23.

Ozone. The solubility of ozone follows Henry's law [Joint Task Force, 1990]:

$$C_{\mathrm{O_3}} = K_H p_{\mathrm{O_3}} \tag{8.389}$$

$$K_H = \frac{1.29 \times 10^6}{T} - 3720.5 \tag{8.390}$$

where

$C_{\mathrm{O_3}}$ = the ozone concentration in mg/L
K_H = the Henry's law constant in mg/L·%
$p_{\mathrm{O_3}}$ = the partial pressure of ozone in the gas phase in % of 1 atm
T = the temperature in K

The units of pressure are somewhat peculiar. For example, if the exit gas from the generator is at 1 atm total pressure and 20°C, and if it contains 0.05% ozone, the solubility of ozone is 0.34 mg/L. Ozone decomposes spontaneously, even in the absence of reductants.

Sulfur Dioxide. Sulfur dioxide is used as a reductant to remove excess chlorine following disinfection.

The solubility of sulfur dioxide in pure water ranges from 798 g/L at 0°C to about 188 g/L at 45°C [Dean, 1992].

$$\mathrm{SO_2(g)} \leftrightarrow \mathrm{SO_2(aq)} \tag{8.391}$$

$$C_{\mathrm{SO_2}} = K_H p_{\mathrm{SO_2}} \tag{8.392}$$

The Henry's law constant is approximately 247, 75.0, and 22.9 g/L·atm at 4, 16, and 27°C, respectively [Stover *et al.*, 1986].

Sulfur dioxide reacts with water to form sulfurous acid, which ionizes stepwise to form bisulfite and sulfite ions [Stover *et al.*, 1986]:

$$\mathrm{SO_2} + \mathrm{H_2O} \leftrightarrow \mathrm{H_2SO_3} \tag{8.393}$$

$$K_0 = \frac{[\mathrm{H_2SO_3}]}{[\mathrm{SO_2(aq)}]} \tag{8.394}$$

$$\mathrm{H_2SO_3} \leftrightarrow \mathrm{HSO_3^-} + \mathrm{H^+} \tag{8.395}$$

$$K_1 = \frac{[H^+][HSO_3^-]}{[H_2SO_3]} \tag{8.396}$$

$$HSO_3^- \leftrightarrow SO_3^{2-} + H^+ \tag{8.397}$$

$$K_2 = \frac{[H^+][SO_3^{2-}]}{[HSO_3^-]} \tag{8.398}$$

Acid ionization constants are given in Table 8.23.

Kinetics

The effect of fast reactions is to increase the concentration gradient in the interfacial layers and, consequently, the rate of gas absorption. This is expressed mathematically and experimentally by an increase in the liquid film mass transfer coefficient, k_l [Sherwood *et al.*, 1975]. The increase depends on the details of the reaction, including the other reactants, and must be determined empirically.

Air Stripping of Volatile Organic Substances

Nowadays, an important treatment process is the removal of volatile organic substances—many of which are toxic and/or carcinogenic—from water by air-stripping either in packed towers or by air diffusion or surface aeration in tanks.

Packed Towers

The common packing materials are 1-in. to 2-in. Berl saddles, Pall rings, Raschig rings, plastic rods, spheres, and plastic Tellerettes. The air and water flow rates are countercurrent, with the water generally tricking down over the packing and the air forced upward through the bed by blowers.

If the contaminant concentration in water is dilute so that Henry's law governs its solubility, the required height of packing is given by [Sherwood *et al.*, 1975]

$$H = \left(\frac{Q_l/A}{K_l a}\right) \cdot \left(\frac{\dfrac{K_{HC} Q_a}{Q_l}}{\dfrac{K_{HC} Q_a}{Q_l} - 1}\right) \cdot \ln\left[\frac{\left(C_{l,i} - \dfrac{C_{a,i}}{K_{HC}}\right)\left(\dfrac{K_{HC} Q_a}{Q_l} - 1\right) + 1}{\left(C_{l,e} - \dfrac{C_{a,i}}{K_{HC}}\right)\left(\dfrac{K_{HC} Q_a}{Q_l}\right)}\right] \tag{8.399}$$

where

A = the cross-sectional area of the packed tower (m^2 or ft^2)
$C_{a,i}$ = the concentration of contaminant in the influent air (kg/m^3 or lb/ft^3)
$C_{l,e}$ = the concentration of contaminant in the effluent water (kg/m^3 or lb/ft^3)
$C_{l,i}$ = the concentration of contaminant in the influent water (kg/m^3 or lb/ft^3)
H = the height of the packing (m or ft)
K_{HC} = the so-called dimensionless Henry's law constant (m^3 water/m^3 air or ft^3 water/ft^3 air) $= C_a/C_l$
K_l = the overall areal liquid phase mass transfer coefficient for clean water (m/s or ft/s)
Q_a = the air flow rate (m^3/s or ft^3/s)
Q_l = the water flow rate (m^3/s or ft^3/s)

In environmental engineering practice, the influent air has no contaminant in it, so $C_{a,i}$ in Eq. (8.399) is zero. Henry's law constants are given in a variety of units, and one must be careful to use the appropriate numerical value.

The principal problem in using Eq. (8.399) is evaluation of the overall mass transfer coefficient. This has been the subject of numerous studies, and the Onda correlations are currently preferred [Roberts *et al.*, 1985; Lamarche and Droste, 1989; Staudinger *et al.*, 1990]:

$$\frac{A_w}{A_v} = 1 - \exp\left\{-1.45 \cdot \left(\frac{\sigma_c}{\sigma_l}\right)^{0.45} \left(\frac{\rho_l Q_l / A}{A_v \mu_l}\right)^{0.1} \left[\frac{\rho_l^2 g}{A_v (\rho_l Q_l / A)^2}\right]^{0.05} \left[\frac{(\rho_l Q_l / A)^2}{\rho_l \sigma_l A_v}\right]^{0.2}\right\} \tag{8.400}$$

$$\frac{k_a}{A_v D_a} = 5.23 \cdot \left(\frac{\rho_a Q_a / A}{A_v \mu_a}\right)^{0.7} \left(\frac{\mu_a}{\rho_a D_a}\right)^{1/3} \tag{8.401}$$

$$k_l \left(\frac{\rho_l}{\mu_l g}\right)^{1/3} = 0.0051 \left(\frac{\rho_l Q_l}{A_w \mu_l}\right)^{2/3} \left(\frac{\mu_l}{\rho_l D_l}\right)^{0.5} \left(A_v d_p\right)^{0.4} \tag{8.402}$$

where

A = the plan area of the packed tower (m^2 or ft^2)
A_v = the specific surface area of the packing (m^2/m^3 or ft^2/ft^3)
A_w = the wetted specific surface area of the packing (m^2/m^3 or ft^2/ft^3)
D_a = the diffusivity of the contaminant vapor in air (m^2/s or ft^2/s)
D_l = the diffusivity of the contaminant in water (m^2/s or ft^{2}s)
d_p = the nominal packing size (m or ft)
g = the acceleration due to gravity (9.80665 m/s^2 or 32.174 ft/s^2)
k_a = the air film mass transfer coefficient (m/s or ft/s)
k_l = the water film mass transfer coefficient (m/s or ft/s)
Q_a = the air flow rate (m^3/s or ft^3/s)
Q_l = the water flow rate (m^3/s or ft^3/s)
μ_a = the dynamic viscosity of air (N/m^2·s or lbf/ft^2·s)
μ_l = the dynamic viscosity of water (N/m^2·s or lbf/ft^2·s)
ρ_a = the mass density of air (kg/m^3 or lb/ft^3)
ρ_l = the mass density of water (kg/m^3 or lb/ft^3)
σ_c = the critical surface tension of the packing (N/m or lbf/ft)
σ_l = the surface tension of water (N/m or lbf/ft)

Staudinger *et al.* [1990] have estimated that the Onda correlations yield a ±30% error in the mass transfer rate, $K_l a$. LaMarche and Droste [1989] also found that the Onda predictions were uniformly too high, but their error was smaller. For design purposes, use 70% of the predicted $K_l a$.

Properties of Packings. Physical properties of common packing materials are given in Table 8.25.

The critical surface tension is that which produces a contact angle of zero between the solid surface and the liquid film surface that is in contact with air [Adamson, 1982]. Table 8.26 lists some critical surface tensions.

Henry's Law Constants. Henry's law can be written in several different ways, and each way assigns different units to the constant [Munz and Roberts, 1987]:

$$p_i = K_{HXp} \cdot X_{li} \tag{8.403}$$

$$p_i = K_{HCp} \cdot C_{li} \tag{8.404}$$

$$C_{ai} = K_{HC} \cdot C_{li} \tag{8.405}$$

Table 8.25 Physical Properties of Packings[1]

Packing	Nominal Size (in.)	Bulk Density (lb/ft³)	Specific Surface Area (ft²/ft³)	Porosity (%)
Berl saddles, ceramic	2	39	32	72
	1 ½	40	46	71
	1	45	76	68
	¾	49	87	66
	½	54	142	62
	¼	56	274	60
Intalox saddles, ceramic	3	37	28	80
	2	42	36	79
	1 ½	42	59	80
	1	44	78	77
	¾	44	102	77
	½	45	190	78
	¼	54	300	75
Pall rings, ceramic	3	40	20	74
	2	38	29	74
Pall rings, polypropylene	3 ½	4.25	26	92
	2	4.5	31	92
	1 ½	4.75	39	91
	1	5.5	63	90
	⅝	7.25	104	87
Pall rings, steel	2	24	31	96
	1 ½	26	39	95
	1	30	63	94
	⅝	37	104	93
Raschig rings, carbon	3	23	19	78
	2	27	28	74
	1 ½	34	38	67
	1	27	57	74
	¾	34	75	67
	½	27	114	74
	¼	46	212	55
Raschig rings, ceramic	4	36	14	80
	3	37	19	75
	2	41	28	74
	1 ½ (¼-in. wall)	46	36	68
	1 ½ (3/16-in. wall)	43	37	73
	1	42	58	74
	¾	50	74	72
	½	55	112	64
	¼	60	217	62
Raschig rings, steel	3	25	20	95
	2	37	29	92
	1 ½	49	39	90
	1	71	56	86
	¾ (1/16-in. wall)	94	75	80
	¾ (1/32-in. wall)	52	81	89
	⅝	62	103	87
	½ (1/16-in. wall)	132	111	73
	½ (1/32-in. wall)	75	122	85
Tellerites, LDPE	1	10	76	83

[1] Perry, R. H. and Chilton, C. H. 1973. *Chemical Engineer's Handbook,* 5th ed. McGraw-Hill, New York.

The different constants are connected by

$$K_{HC} = K_{HXp} \cdot \frac{v_l M_{rl}}{RT} \tag{8.406}$$

$$K_{HCp} = K_{HXp} \cdot v_l \tag{8.407}$$

where

$C_{a,i}$ = the concentration of contaminant in the air (kg/m^3 or lbm/ft^3)
$C_{l,i}$ = the concentration of contaminant in the water (kg/m^3 or lbm/ft^3)
K_{HC} = the so-called dimensionless Henry's law constant (m^3 water/m^3 air or ft^3 water/ft^3 air)
= C_a/C_l
K_{HCp} = the Henry's law constant (Pa·m^3/kg or lbf·ft/lb)
K_{HXp} = the Henry's law constant (Pa in lbf/ft^2)
M_{rl} = the molecular weight of water (18 kg/mol or 18 lb/lb-mol)
p_i = the partial pressure of species i in air (Pa or lbf/ft^2)
R = the gas constant (8.314 J/mol·k or 1545 ft-lbf/lb-mol·°R)
T = the absolute temperature (K or °R)
v_l = the specific volume of water (m^3/kg or ft^3/lb)
X_{li} = the mole fraction of species i in water (dimensionless)

Table 8.26 Critical Surface Tensions for Typical Materials[1]

Material	Critical Surface Tension (dyne/cm)
Carbon	56
Ceramic	61
Glass	73
Paraffin	20
Polyethylene	33
Polyvinyl chloride	40
Steel	75

[1]Onda, T. and Koyama, 1967. *Kagaku Kogaku,* 31:126.

Henry's law constants for several important volatile organic compounds are listed in Table 8.27. The temperature dependency is represented by an empirical fit to the van't Hoff relationship. This is done separately for each form of the constant; for example,

$$\log K_{HXp} = a - \frac{b}{T} \tag{8.408}$$

$$\log K_{HC} = a' - \frac{b'}{T} \tag{8.409}$$

$$\log K_{HCp} = a'' - \frac{b''}{T} \tag{8.410}$$

where

a, a', a'' = empirical constants (dimensionless)
b, b', b'' = empirical constants (K or °R)

Diffused Air and Surface Aeration

Volatile organic contaminants can be removed by diffused and surface aeration also. The laws governing oxygen transfer also apply. The gas and liquid film mass transfer coefficients depend on molecular diffusivities, which in turn depend on molecular weights. Therefore, if one knows k_l and k_g for one substance in a particular mass transfer system, one can in principle calculate for any other substance. One needs to distinguish between surface aerators and diffused air, because the air bubbles accumulate contaminant as they rise through the liquid.

Surface Aerators. In the case of surface aerators, there is no contaminant concentration in the ambient air, so the solubility of the contaminant is zero. For a completely mixed reactor, the steady state mass balance is [Roberts *et al.*, 1984]

$$Q_l(C_{l,i} - C_{l,e}) = K_l a_j \cdot C_{l,e} V \tag{8.411}$$

where

$C_{l,e}$ = the concentration of contaminant in the effluent water (kg/m^3 or lb/ft^3)
$C_{l,i}$ = the concentration of contaminant in the influent water (kg/m^3 or lb/ft^3)
$K_l a_j$ = the overall volumetric liquid phase mass transfer coefficient for species j (per second)
Q_l = the water flow rate (m^3/s or ft^3/s)
V = the tank volume (m^3 or ft^3)

In the more general case of mixed-cells-in-series and in ideal plug flow one gets

$$\frac{C_{l,e}}{C_{l,i}} = \left[\frac{1}{1 + K_l a_j (V_1/Q_l)}\right]^n \tag{8.412}$$

Table 8.27 Henry's Law Constants for Volatile Organic Substances *(continues)*

Substance	K_{HC} at 20°C $\left(\frac{\text{m}^3\text{ water}}{\text{m}^3\text{ air}}\right)$	K_{HXp} at 20°C (atm)	a, a'	b, b' (K)	Reference
Benzene	—	—	8.68	1852	3
	0.306	—	—	—	7
Bromoform	—	35	—	—	3
	0.017	—	4.729	1905	6
Carbon tetrachloride	—	1290	10.06	2038	3
	0.98	—	5.853	1718	6
	0.936	—	—	—	7
Chlorobenzene	0.131	—	—	—	7
Chloroform	0.13	170	9.10	2013	3
	0.15	—	1.936	809.1	4
	0.12	—	4.990	1729	6
Chloromethane	—	480	6.93	1248	3
p-Dichlorobenzene	0.078	—	—	—	7
1, 1-Dichloroethane	—	—	8.87	1902	3
	0.22	—	2.080	803.8	4
1, 2-Dichloroethane	—	61	—	—	3
	0.046	—	5.156	1904	4
cis-1, 2-Dichloroethylene	0.181	—	—	—	7
trans-1, 2-Dichloroethylene	0.375	—	—	—	7
Dichlorodifluoromethane	—	—	818	1470	3
	11	—	5.811	1399	6
1, 2-Dichloromethane	—	—	7.92	1822	3
Dieldrin	—	0.0094	—	—	3
Diethyl ether	0.039	—	5.953	2158	4
Hexachloroethane	0.12	—	6.982	2320	6
Methylene chloride	0.077	—	—	—	7
Naphthalene	0.015	—	—	—	7
Pentachlorophenol	—	0.12	—	—	3
Phenol	11×10^{-6}	—	—	—	1
Tetrachloroethylene	—	1100	10.38	2159	3
	0.12	—	5.920	1802	6
	0.535	—	—	—	7

Table 8.27 (continued) Henry's Law Constants for Volatile Organic Substances

Substance	K_{HC} at 20°C $\left(\frac{m^3 \text{ water}}{m^3 \text{ air}}\right)$	K_{HXp} at 20°C (atm)	a, a'	b, b' (K)	Reference
Toluene	—	340 (25°C)	—	—	3
	0.15	—	11.18	3518	4
	0.244	—	—	—	7
Toxaphene	—	3500	—	—	3
1, 1, 1-Trichloroethane	—	430	9.39	1993	3
	0.55	—	5.327	1636	6
	0.645	—	—	—	7
Trichloroethylene	—	550	8.59	1716	3
	0.37	—	2.189	767.8	4
	0.32	—	6.026	1909	6
	0.430	—	—	—	7
1, 2, 4-Trimethylbenzene	—	353	—	—	3
	0.195	—	—	—	7
o-xylene	0.175	—	—	—	7
Vinyl chloride	—	3.55×10^5	—	—	3

[1]Berger, B. B. 1983. *Control of Organic Substances in Water and Wastewater,* EPA-600/8-83-011. U.S. Environmental Protection Agency, Office of Research and Development, Washington, D. C.

[2]Gosset, J. M. 1987. Measurement of Henry's law constants for C_1 and C_2 chlorinated hydrocarbons, *Environmental Science and Technology,* 21:202.

[3]Kavanaugh, M. C. and Trussell, R. R. 1980. Design of aeration towers to strip volatile organic contaminants for drinking water, *Journal of the American Water Works Association,* 71(12):684.

[4]LaMarche, P. and Droste, R. L. 1989. Air-stripping mass transfer correlation for volatile organics, *Journal of the American Water Works Association,* 81(1):78.

[5]McKinnon, R. J. and Dyksen, J. E. 1984. Removing organics from groundwater through aeration plus GAC, *Journal of the American Water Works Association,* 76(5):42.

[6]Munz, C. and Roberts, P. V. 1987. Air-water phase equilibria, *Journal of the American Water Works Association,* 79(5):62.

[7]Yuteri, C., Ryan, D. F., Callow, J. J., and Gurol, M. D. 1987. The effect of chemical composition of water on Henry's law constant, *Journal of the Water Pollution Control Federation,* 51(10):950.

Note: Reference 3 used Eq. (8.408) and calculated K_{HXp}. References 4 and 6 used Eq. (8.409) and calculated K_{HC}.

$$\frac{C_{l,e}}{C_{l,i}} = \exp\left(-\frac{K_l a_j V}{Q_l}\right) \tag{8.413}$$

where n is the number of mixed-cells-in-series (dimensionless).

For relatively insoluble substances such as oxygen and many volatile organics, the gas film resistance is small and may be neglected [Kavanaugh and Trussel, 1980]. In that case, K_l is nearly equal to k_l. The organic vapor mass transfer coefficient for a particular system may be estimated from the coefficient for oxygen, if it is known, via Eq. (8.321).

Air Diffusion. Assuming a completely mixed reactor, the ratio of the liquid effluent to the liquid influent concentrations of the volatile organic substance removal is given by [Roberts *et al.,* 1984]

$$\frac{C_{l,e}}{C_{l,i}} = \frac{1}{1 + \frac{Q_a K_{HCj}}{Q_l}\left[1 - \exp\left(-\frac{K_l a_j V}{K_{HCj} Q_a}\right)\right]} \tag{8.414}$$

The concentration of the contaminant in the off-gas is

$$C_{a,e} = K_{HCj} \cdot C_{l,e}\left[1 - \exp\left(-\frac{K_l a_j V}{K_{HCj} Q_a}\right)\right] \tag{8.415}$$

where

$C_{a,e}$ = the concentration of contaminant in the effluent air (kg/m^3 or lb/ft^3)
K_{HCj} = the dimensionless Henry's law constant for species j (m^3 water/m^3 air or ft^3 water/ft^3 air)
$K_l a_j$ = the overall volumetric liquid phase mass transfer coefficient for species j (per second)
Q_a = the air flow rate (m^3/s or ft^3/s)

References

Section 8.1

Babbitt, H. E., Doland, J. J., and Cleasby, J. L. 1967. *Water Supply Engineering,* 6th ed. McGraw-Hill, New York.

Blevins, R. D. 1984. *Applied Fluid Dynamics Handbook.* Van Nostrand Reinhold, New York.

Fair, G. M., Geyer, J. C., and Morris, J. C. 1954. *Water Supply and Waste-Water Disposal.* John Wiley & Sons, New York.

Hardenbergh, W. A. and Rodie, E. B. 1960. *Water Supply and Waste Disposal.* International Textbook, Scranton, PA.

Hazen and Sawyer, Engineers. 1975. *Process Design Manual for Suspended Solids Removal,* EPA 625/1-75-003a. U.S. Environmental Protection Agency, Technology Transfer, Washington.

Hudson, H. E., Jr. 1981. *Water Clarification Processes: Practical Design and Evaluation.* Van Nostrand Reinhold, New York.

Joint Task Force of the Water Environment Federation and the American Society of Civil Engineers. 1992. *Design of Municipal Wastewater Treatment Plants: Volume I—Chapters 1–12, WEF Manual of Practice No. 8, ASCE Manual and Report on Engineering Practice No. 76.* Water Environment Federation, Alexandria, VA; American Society of Civil Engineers, New York.

King, H. W. and Brater, E. F. 1963. *Handbook of Hydraulics for the Solution of Hydrostatic and Fluid-flow Problems,* 5th ed. McGraw-Hill, New York.

Lea, F. C. 1938. *Hydraulics for Engineers and Engineering Students,* 6th ed. Edward Arnold, London.

Metcalf & Eddy, Inc. 1991. *Wastewater Engineering: Treatment, Disposal and Reuse,* 3rd ed., revised by G. Tchobanoglous and F. L. Burton. McGraw-Hill, New York.

Pankratz, T. M. 1988. *Screening Equipment Handbook for Industrial and Municipal Water and Wastewater Treatment.* Technomic Publishing, Lancaster, PA.

Smith, D. and Walker, W. J. 1923. Orifice flow, *Proceedings of the Institution of Mechanical Engineers,* 1(1):23.

Wastewater Committee, Great Lakes–Upper Mississippi River Board of State Public Health and Environmental Managers. 1990. *Recommended Standards for Wastewater Facilities, 1990 Edition.* Health Education Services, Albany, NY.

Section 8.2

Badger, R. B., Robinson, D. D., and Kiff, R. J. 1975. Aeration plant design: derivation of basic data and comparative performance studies, *Water Pollution Control,* 74(4):415.

Baskir, C. I. and Spearing, G. 1980. Product formation in the continuous culture of microbial populations grown on carbohydrates, *Biotechnology and Bioengineering,* 22(9):1857.

Bishoff, K. B. 1961. A note on boundary conditions for flow reactors, *Chemical Engineering Science,* 16(1/2):131.

Cholette, A. and Cloutier, L. 1959. Mixing efficiency determinations for continuous flow systems, *Canadian Journal of Chemical Engineering,* 37(6): 105.

Cholette, A., Blanchet, J., and Cloutier, L. 1960. Performance of flow reactors at various levels of mixing, *Canadian Journal of Chemical Engineering,* 38(2):1.

Chudoba, J., Strakova, P., and Kondo, M. 1991. Compartmentalized versus completely-mixed biological wastewater treatment systems, *Water Research,* 25(8):973.

Danckwerts, P. V. 1953. Continuous flow systems—Distribution of residence times, *Chemical Engineering Science,* 2(1):1.

Dunbar, W. P. 1908. *Principles of Sewage Treatment,* trans. H. T. Calvert, Charles Griffin, London.

Erickson, L. E. 1980. Analysis of microbial growth and product formation with nitrate as nitrogen source, *Biotechnology and Bioengineering,* 22(9):1929.

Fan, L.-T. and Ahn, Y.-K. 1963. Frequency response of tubular flow systems, *Process Systems Engineering, Chemical Engineering Progress Symposium No. 46,* 59(46):91.

Grady, C. P. L., Jr., Harlow, L. J., and Riesing, R. R. 1972. Effects of growth rate and influent substrate concentration on effluent quality from chemostats containing bacteria in pure and mixed culture, *Biotechnology and Bioengineering,* 14(3):391.

Gram, A. L., III. 1956. *Reaction Kinetics of Aerobic Biological Processes,* Report No. 2, I.E.R. Ser. 90. University of California at Berkeley, College of Engineering, Sanitary Engineering Research Laboratory, Berkeley, CA.

Ham, A. and Coe, H. S. 1918. Calculation of extraction in continuous agitation, *Chemical and Metallurgical Engineering,* 19(9):663.

Hao, O. J. and Lau, A. O. 1988. Kinetics of Microbial By-Product Formation in Chemostat Pure Cultures, *Journal of the Environmental Engineering Division, ASCE,* 114(5):1097.

Haseltine, T. R. 1961. Sludge reaeration in the activated sludge process—A survey, *Journal of the Water Pollution Control Federation,* 33(9):946.

Hazen, A. 1904. On Sedimentation, *Transactions of the American Society of Civil Engineers,* 53:45.

Kehr, R. W. 1936. Detention of Liquids being Mixed in Continuous Flow Tanks, *Sewage Works Journal,* 8(6):915.

Kroiss, H. and Ruider, E. 1977. Comparison of the plug-flow and complete mix activated sludge process, *Progress in Water Technology,* 8(6):169.

Langelier, W. F. 1921. Coagulation of water with alum by prolonged agitation, *Engineering News-Record,* 86(22):924.

Langmuir, I. 1908. The velocity of reactions in gases moving through heated vessels and the effect of convection and diffusion, *Journal of the American Chemical Society,* 30(11):1742.

Levenspiel, O. 1972. *Chemical Reaction Engineering,* 2nd ed., John Wiley & Sons, New York.

Levenspiel, O. and Bischoff, K. B. 1963. Patterns of flow in chemical process vessels, *Advances in Chemical Engineering,* 4:95.

MacMullin, R. B. and Weber, M. 1935. The theory of short-circuiting in continuous-flow mixing vessels in series and the kinetics of chemical reactions in such systems, *Transactions of the American Institute of Chemical Engineers,* 31(2):409.

Metcalf, L. and Eddy, H. P. 1916. *American Sewerage Practice: Vol. III, Disposal of Sewage, 2nd Impression, with Appendix on Activated Sludge and Minor Revisions,* McGraw-Hill, New York.

Monod, J. 1942. Recherches sur la croissance des cultures bacteriennes, *Actualitiés scientifiques et industrielles,* No. 911, Hermann & Cie., Paris.

Pearson, J. R. A. 1959. A note on the "Danckwerts" boundary conditions for continuous flow reactors, *Chemical Engineering Science,* 10(4):281.

Rickert, D. A. and Hunter, J. V. 1971. Effects of aeration time on soluble organics during activated sludge treatment, *Journal of the Water Pollution Control Federation,* 43(1):134.

Toerber, E. D., Paulson, W. L., and Smith, H. S. 1974. Comparison of completely mixed and plug flow biological systems, *Journal of the Water Pollution Control Federation,* 46(8):1995.

Wehner, J. F. and Wilhem, R. H. 1956. Boundary conditions of flow reactor, *Chemical Engineering Science,* 6(2):89.

Wen, C. Y. and Fan, L. T. 1975. *Models for Flow Systems and Chemical Reactors,* Marcel Dekker, New York.

Section 8.3

Bean, E. L. 1953. Study of physical factors affecting flocculation, *Water Works Engineering,* 106(1):33 and 65.

Boadway, J. D. 1978. Dynamics of growth and breakage of alum floc in presence of fluid shear, *Journal of the Environmental Engineering Division, Proceedings of the American Society of Civil Engineers,* 104(EE5):901.

Brodkey, R. S. 1967. *The Phenomena of Fluid Motions,* 2nd printing with revisions. Private Printing, Columbus, OH. (First printing: Addison-Wesley Publishing Co., Inc., Reading, MA.)

Camp, T. R. 1955. Flocculation and flocculation basins, *Transactions of the American Society of Civil Engineers,* 120:1.

Camp, T. R. and Stein, P. C. 1943. Velocity gradients and internal work in fluid motion, *Journal of the Boston Society of Civil Engineers,* 30(4):219.

Clark, M. M. 1985. Critique of Camp and Stein's velocity gradient, *Journal of Environmental Engineering,* 111(6):741.

Cleasby, J. L. 1984. Is velocity gradient a valid turbulent flocculation parameter? *Journal of Environmental Engineering,* 110(5):875.

Fair, G. M., Geyer, J. C., and Okun, D. A. 1968. *Water and Wastewater Engineering: Vol. 2, Water Purification and Wastewater Treatment and Disposal.* John Wiley & Sons, New York.

Godfrey, J. C. and Amirtharajah, A. 1991. Mixing in liquids. In *Mixing in Coagulation and Flocculation,* A. Amirtharajah, M. M. Clark, and R. R. Trussell, eds., p. 35. American Water Works Association, Denver.

Grant, H. L., Stewart, R. W., and Moilliet, A. 1962. Turbulence spectra from a tidal channel, *Journal of Fluid Mechanics,* 12(part 2):241.

Holland, F. A. and Chapman, F. S. 1966. *Liquid Mixing and Processing in Stirred Tanks.* Reinhold, New York.

Hopkins, E. S. and Bean, E. L. 1966. *Water Purification Control,* 4th ed. Williams & Wilkins, Baltimore.

Joint Committee of the American Society of Civil Engineers, the American Water Works Association and the Conference of State Sanitary Engineers. 1969. *Water Treatment Plant Design.* American Water Works Association, New York.

King, H. W. and Brater, E. F. 1963. *Handbook of Hydraulics for the Solution of Hydrostatic and Fluid-flow Problems,* 5th ed. McGraw-Hill, New York.

Lagvankar, A. L. and Gemmell, R. S. 1968. A size-density relationship for flocs, *Journal of the American Water Works Association,* 60(9):1040. See errata: *Journal of the American Water Works Association,* 60(12):1335.

Lamb, H. 1932. *Hydrodynamics,* 6th ed. Cambridge University Press, Cambridge.

Landahl, M. T. and Mollo-Christensen, E. 1986. *Turbulence and Random Processes in Fluid Mechanics.* Cambridge University Press, Cambridge.

Leipold, C. 1934. Mechanical agitation and alum floc formation, *Journal of the American Water Works Association,* 26(8):1070.

Matsuo, T. and Unno, H. 1981. Forces acting on floc and strength of floc, *Journal of the Environmental Engineering Division, Proceedings of the American Society of Civil Engineers,* 107(EE3):527.

Oldshue, J. Y. and Trussell, R. R. 1991. Design of impellers for mixing. In *Mixing in Coagulation and Flocculation,* A. Amirtharajah, M. M. Clark, and R. R. Trussell, eds., p. 309. American Water Works Association, Denver.

Parker, D. S., Kaufman, W. J., and Jenkins, D. 1972. Floc breakup in turbulent flocculation processes, *Journal of the Sanitary Engineering Division, Proceedings of the American Society of Civil Engineers,* 98(SA1):79.

Rushton, J. H., Costich, E. W., and Everett, H. J. 1950. Power characteristics of mixing impellers, part I, *Chemical Engineering Progress,* 46(8):395.

Saffman, P. G. and Turner, J. S. 1956. On the collision of drops in turbulent clouds, *Journal of Fluid Mechanics,* 1(part 1):16.

Smith, M. C. 1932. Improved mechanical treatment of water for filtration, *Water Works and Sewerage,* 79(4):103.

Stewart, R. W. and Grant, H. L. 1962. Determination of the rate of dissipation of turbulent energy near the sea surface in the presence of waves, *Journal of Geophysical Research,* 67(8):3177.

Stokes, G. G. 1845. On the theories of internal friction of fluids in motion, etc., *Transactions of the Cambridge Philosophical Society,* 8:287.

Streeter, V. L. 1948. *Fluid Dynamics.* McGraw-Hill, New York.

Tambo, N. and Hozumi, H. 1979. Physical characteristics of flocs—II. Strength of floc, *Water Research,* 13(5):421.

Tambo, N. and Watanabe, Y. 1979. Physical characteristics of flocs—I. The floc density function and aluminum floc, *Water Research,* 13(5):409.

Tatterson, G. B. 1994. *Scaleup and Design of Industrial Mixing Processes.* McGraw-Hill, New York.

Zweitering, T. N. 1958. Suspending solid particles in liquids by agitation, *Chemical Engineering Science,* 8(3/4):244.

Section 8.4

Amirtharajah, A. and Trusler, S. L. 1986. Destabilization of particles by turbulent rapid mixing, *Journal of Environmental Engineering,* 112(6):1085.

Andreu-Villegas, R. and Letterman, R. D. 1976. Optimizing flocculator power input, *Journal of the Environmental Engineering Division, Proceedings of the American Society of Civil Engineers,* 102(EE2):251.

Argaman, Y. and Kaufman, W. J. 1970. Turbulence and flocculation, *Journal of the Sanitary Engineering Division, Proceedings of the American Society of Civil Engineers,* 96(SA2):223.

Basset, A. B. 1888. *A Treatise on Hydrodynamics: With Numerous Examples.* Deighton, Bell, Cambridge, UK.

Camp, T. R. 1955. Flocculation and flocculation basins, *Transactions of the American Society of Civil Engineers,* 120:1.

Camp, T. R. 1968. Floc volume concentration, *Journal of the American Water Works Association,* 60(6):656.

Crank, J. 1975. *The Mathematics of Diffusion,* 2nd ed. Oxford University Press, Clarendon Press, Oxford.

Culp/Wesner/Culp, Inc. 1986. *Handbook of Public Water Systems.* R. B. Williams and G. L. Culp, eds. Van Nostrand Reinhold, New York.

Delichatsios, M. A. and Probstein, R. F. 1975. Scaling laws for coagulation and sedimentation, *Journal of the Water Pollution Control Federation,* 47(5):941.

Einstein, A. 1956. *Investigations on the Theory of the Brownian Movement,* R. Furth, ed., trans. A. D. Cowper. Dover, New York.

Freundlich, H. 1922. *Colloid and Capillary Chemistry,* trans. H. S. Hatfield. E. P. Dutton, New York.

Griffith, J. D. and Williams, R. G. 1972. Application of jar-test analysis at Phoenix, Ariz., *Journal of the American Water Works Association,* 64(12):825.

Harris, H. S., Kaufman, W. J., and Krone, R. B. 1966. Orthokinetic flocculation in water purification, *Journal of the Sanitary Engineering Division, Proceedings of the American Society of Civil Engineers,* 92(SA6):95.

Hinze, J. O. 1959. *Turbulence: An Introduction to Its Mechanism and Theory.* McGraw-Hill, New York.

Hudson, H. E., Jr. 1973. Evaluation of plant operating and jar-test data, *Journal of the American Water Works Association,* 65(5):368.

Ives, K. J. 1968. Theory of operation of sludge blanket clarifiers, *Proceedings of the Institution of Civil Engineers,* 39(2):243.

James M. Montgomery, Inc. 1985. *Water Treatment Principles and Design.* John Wiley & Sons, New York.

Joint Committee of the American Society of Civil Engineers, the American Water Works Association and the Conference of State Sanitary Engineers. 1969. *Water Treatment Plant Design.* American Water Works Association, New York.

Joint Task Force. 1990. *Water Treatment Plant Design,* 2nd ed. McGraw-Hill, New York.

Langelier, W. F. 1921. Coagulation of water with alum by prolonged agitation, *Engineering News-Record,* 86(22):924.

Letterman, R. D., Quon, J. E., and Gemmell, R. S. 1973. Influence of rapid-mix parameters on flocculation, *Journal of the American Water Works Association,* 65(11):716.

Levich, V. G. 1962. *Physicochemical Hydrodynamics,* trans. Scripta Technica, Inc. Prentice-Hall, Englewood Cliffs, NJ.

O'Melia, C. R. 1972. Coagulation and flocculation. In *Physicochemical Processes: for Water Quality Control,* W. J. Weber, Jr., ed., p. 61, John Wiley & Sons, Wiley-Interscience, New York.

Parker, D. S., Kaufman, W. J., and Jenkins, D. 1972. Floc breakup in turbulent flocculation processes, *Journal of the Sanitary Engineering Division, Proceedings of the American Society of Civil Engineers,* 98(SA1):79.

Saffman, P. G. and Turner, J. S. 1956. On the collision of drops in turbulent clouds, *Journal of Fluid Mechanics,* 1(part 1):16.

Spielman, L. A. 1978. Hydrodynamic aspects of flocculation. In *The Scientific Basis of Flocculation,* K. J. Ives, ed., p. 63, Sijthoff & Noordhoff International Publishers B. V., Alphen aan den Rijn, the Netherlands.

TeKippe, J. and Ham, R. K. 1971. Velocity-gradient paths in coagulation, *Journal of the American Water Works Association,* 63(7):439.

Vrale, L. and Jorden, R. M. 1971. Rapid mixing in water treatment, *Journal of the American Water Works Association,* 63(1):52.

Wagner, E. G. 1974. Upgrading existing water treatment plants: Rapid mixing and flocculation, p. IV-56 in *Proceedings AWWA Seminar on Upgrading Existing Water-Treatment Plants.* American Water Works Association, Denver, CO.

Water Supply Committee, Great Lakes–Upper Mississippi River Board of State Public Health and Environmental Managers. 1987. *Recommended Standards for Water Works,* 1987 edition. Health Research Inc., Albany, NY.

Section 8.5

Alderton, J. L. 1963. Discussion of "Analysis of thickener operation," by V. C. Behn and J. C. Liebman, *Journal of the Sanitary Engineering Division, Proceedings of the American Society of Civil Engineers,* 89(SA6):57.

Allen, H. S. 1900. The motion of a sphere in a viscous fluid, *The London, Edinburgh, and Dublin Philosophical Magazine and Journal of Science,* 50, 5th Series, no. 304, p. 323, and no. 306, p. 519.

Anderson, E. 1941. Separation of Dusts and Mists, p. 1850 in *Chemical Engineers Handbook,* 2nd ed., 9th imp., J. H. Perry, ed., McGraw-Hill, New York. The source of Shepherd's formula is cited as a personal communication.

Babbitt, H. E., Doland, J. J., and Cleasby, J. L. 1967. *Water Supply Engineering,* 6th ed., McGraw-Hill, New York.

Barr, J. 1910a. Experiments on the flow of water over triangular notches, *Engineering,* 89(8 April): 435.

Barr, J. 1910b. Experiments on the flow of water over triangular notches, *Engineering,* 89(15 April):470.

Basset, A. B. 1888. *A Treatise on Hydrodynamics: With numerous Examples,* Deighton, Bell, Cambridge, UK.

Behn, V. C. 1957. Settling behavior of waste suspensions, *Journal of the Sanitary Engineering Division, Proceedings of the American Society of Civil Engineers,* 83(SA5): Paper No. 1423.

Behn, V. C. and Liebman, J. C. 1963. Analysis of thickener operation, *Journal of the Sanitary Engineering Division, Proceedings of the American Society of Civil Engineers,* 89(SA3):1.

Boadway, J. D. 1978. Dynamics of growth and breakage of alum floc in presence of fluid shear, *Journal of the Environmental Engineering Division, Proceedings of the American Society of Civil Engineers,* 104(EE5):901.

Burgess, R. W. 1916. The uniform motion of a sphere through a viscous liquid, *American Journal of Mathematics,* 38:81.

Camp, T. R. 1936a. A study of the rational design of settling tanks, *Sewage Works Journal,* 8(5):742.

Camp, T. R. 1936b. Discussion: "Sedimentation in quiescent and turbulent basins," by J. J. Slade, Jr., Esq., *Proceedings of the American Society of Civil Engineers,* 62(2):281.

Camp, T. R. 1940. Lateral Spillway Design, *Transactions of the American Society of Civil Engineers,* 105:606.

Camp, T. R. 1942. Grit chamber design, *Sewage Works Journal,* 14(2):368.

Camp, T. R. 1945. Sedimentation and the design of settling tanks, *Proceedings of the American Society of Civil Engineers,* 71(4, part 1):445.

Camp, T. R. 1953. Studies of sedimentation basin design, *Sewage and Industrial Wastes,* 25(1):1.

Chow, V. T. 1959. *Open-Channel Hydraulics,* McGraw-Hill, New York.

Coe, H. S. and Clevenger, G. H. 1916. Methods for determining the capacities of slime settling tanks, *Transactions of the American Institute of Mining Engineers,* 55:356.

Conley, W. P. and Hansen, S. P. 1978. Advanced Techniques for Suspended Solids Removal, p. 299 in *Water Treatment Plant Design for the Practicing Engineer,* R. L. Sanks, ed., Ann Arbor Science Publishers, Ann Arbor, MI.

Culp/Wesner/Culp, Inc. 1986. *Handbook of Public Water Systems,* R. B. Williams and G. L. Culp., eds., Van Nostrand Reinhold, New York.

Dick, R. I. 1970. Role of activated sludge final settling tanks, *Journal of the Sanitary Engineering Division, Proceedings of the American Society of Civil Engineers,* 96(SA2):423.

Dick, R. I. 1970. Discussion: "Agglomerate size changes in coagulation," *Journal of the Sanitary Engineering Division, Proceedings of the American Society of Civil Engineers,* 96(SA2):624.

Dick, R. I. and Young, K. W. no date. Analysis of thickening performance of final settling tanks, p. 33 in *Proceedings of the 27th Industrial Waste Conference, May 2, 3 and 4, 1972,* Engineering Extension Series No. 141, J. M. Bell, ed., Purdue University, Lafayette, IN.

Dick, R. I. and Ewing, B. B. 1967. Evaluation of activated sludge thickening theories, *Journal of the Sanitary Engineering Division, Proceedings of the American Society of Civil Engineers,* 93(SA4):9.

Dryden, H. L., Murnaghan, F. D., and Bateman, H. 1956. *Hydrodynamics,* Dover Press, New York.

Duncan, J. W. K. and Kawata, K. 1968. Discussion of "Evaluation of activated sludge thickening theories," by R. I. Dick and B. B. Ewing, *Journal of the Sanitary Engineering Division, Proceedings of the American Society of Civil Engineers,* 94(SA2):431.

Eliassen, R. 1946. Discussion: "Sedimentation and the design of settling tanks by T.R. Camp," *Proceedings of the American Society of Civil Engineers,* 42(3):413.

Fair, G. M., Geyer, J. C., and Morris, J. C. 1954. *Water Supply and Waste-Water Disposal,* John Wiley & Sons, New York.

Favre, H. 1933. *Contribution a l' Étude des Courants Liquides,* Dunod, Paris.

Fisherström, C. N. H. 1955. Sedimentation in rectangular basins, *Proceedings of the American Society of Civil Engineers,* 81(Separate No. 687):1.

Fitch, E. B. 1956. Flow path effect on sedimentation, *Sewage and Industrial Wastes,* 28(1):1.

Fitch, E. B. 1957. The significance of detention in sedimentation, *Sewage and Industrial Wastes,* 29(10):1123.

Fitch, E. B. 1962. Sedimentation process fundamentals, *Transactions of the American Institute of Mining Engineers,* 223:129.

François, R. J. 1987. Strength of aluminum hydroxide flocs, *Water Research,* 21(9):1023.

Goldstein, S. 1929. The steady flow of viscous fluid past a fixed spherical obstacle at small Reynolds numbers, *Proceedings of the Royal Society of London Ser. A,* 123:225.

Goldstein, S., ed. 1965. *Modern Developments in Fluid Dynamics: An Account of Theory and Experiment Relating to Turbulent Boundary Layers, Turbulent Motion and Wakes, Composed by the Fluid Motion Panel of the Aeronautical Research Committee and Others, in Two Volumes,* Vol. I. Dover Publications, New York.

Ham, R. K. and Christman, R. F. 1969. Agglomerate size changes in coagulation, *Journal of the Sanitary Engineering Division, Proceedings of the American Society of Civil Engineers,* 95(SA3):481.

Hansen, S. P. and Culp, G. L. 1967. Applying shallow depth sedimentation theory, *Journal of the American Water Works Association,* 59(9):1134.

Hayden, R. 1914. Concentration of slimes at Anaconda, Mont., *Transactions of the American Institute of Mining Engineers,* 46:239.

Hazen, A. 1904. On sedimentation, *Transactions of the American Society of Civil Engineers,* 53:45.

Hinds, J. 1926. Side channel spillways: Hydraulic theory, economic factors, and experimental determination of losses, *Transactions of the American Society of Civil Engineers,* 89:881.

Hudson, H. E., Jr. 1972. Density considerations in sedimentation, *Journal of the American Water Works Association,* 64(6):382.

Hudson, H. E., Jr. 1981. *Water Clarification Processes: Practical Design and Evaluation,* Van Nostrand Reinhold, New York.

Hudson, H. E., Jr., and Wolfner, J. P. 1967. Design of mixing and flocculating basins, *Journal of the American Water Works Association,* 59(10):1257.

Ingersoll, A. C., McKee, J. E., and Brooks, N. H. 1956. Fundamental concepts of rectangular settling tanks, *Transactions of the American Society of Civil Engineers,* 121:1179.

James M. Montgomery, Consulting Engineers, Inc. 1985. *Water Treatment Principles and Design,* John Wiley & Sons, New York.

Joint Committee of the American Society of Civil Engineers, the American Water Works Association and the Conference of State Sanitary Engineers. 1990. *Water Treatment Plant Design,* 2nd ed. McGraw-Hill, New York.

Joint Task Force of the American Society of Civil Engineers, the American Water Works Association and the Conference of State Sanitary Engineers. 1969. *Water Treatment Plant Design,* American Water Works Association, Inc., New York.

Joint Task Force of the Water Environment Federation and the American Society of Civil Engineers. 1992. *Design of Municipal Wastewater Treatment Plants: Volume I. Chapters 1–12,* WEF Manual of Practice No. 8, ASCE Manual and Report on Engineering Practice No. 76. Water Environment Federation, Alexandria, VA; American Society of Civil Engineers, New York.

King, H. W. and Brater, E. F. 1963. *Handbook of Hydraulics for the Solution of Hydrostatic and Fluid-flow Problems,* 5th ed., McGraw-Hill, New York.

Kleinschmidt, R. S. 1961. Hydraulic Design of Detention Tanks, *Journal of the Boston Society of Civil Engineers,* 48(4, sect. 1):247.

Kynch, G. J. 1952. A theory of sedimentation, *Transactions of the Faraday Society,* 48:166.

Lagvankar, A. L. and Gemmell, R. S. 1968. A size-density relationship for flocs, *Journal of the American Water Works Association,* 60(9):1040. See errata: *Journal of the American Water Works Association,* 60(12):1335 (1968).

Laufer, J. 1950. Some recent measurements in a two-dimensional turbulent channel, *Journal of the Aeronautical Sciences,* 17(5):277.

Mantz, P. A. 1977. Incipient transport of fine grains and flakes by fluids—extended Shields diagram, *Journal of the Hydraulics Division, Proceedings of the American Society of Civil Engineers,* 103(HY6):601.

McGauhey, P. M. 1956. Theory of sedimentation, *Journal of the American Water Works Association,* 48(4):437.

McNown, J. S. 1945. Discussion of "Lock manifold experiments," by E. Soucek and E. W. Zelnick, *Transactions of the American Society of Civil Engineers,* 110:1378.

McNown, J. S. 1954. Mechanics of manifold flow, *Transactions of the American Society of Civil Engineers,* 119:1103.

Moore, W. L. 1943. Energy loss at the base of a free overflow, *Transactions of the American Society of Civil Engineers,* 108:1343.

Naiz, S. M. 1954. Discussion of "Mechanics of manifold flow," by J. S. McNown, *Transactions of the American Society of Civil Engineers,* 119:1132.

O'Brien, M. P. 1932. Analyzing hydraulic models for the effects of distortion, *Engineering News-Record,* 109(11):313.

O'Connor, J. T. 1971. Management of water-treatment plant residues, p. 625 in *Water Quality and Treatment: A Handbook of Public Water Supplies,* 3rd ed., P. D. Haney *et al.*, ed. McGraw-Hill, New York.

Oseen, C. W. 1913. Über den gültigkeitsbereich der stokesschen widerstandformel, *Arkiv för Matematik, Astonomi och Fysik,* 9(16):1.

Parker, D. S. 1983. Assessment of secondary clarification design concepts, *Journal of the Water Pollution Control Federation,* 55(4):349.

Parker, D. S., Kaufman, W. J., and Jenkins, D. 1972. Floc breakup in turbulent flocculation processes, *Journal of the Sanitary Engineering Division, Proceedings of the American Society of Civil Engineers,* 98(SA1):79.

Powell, R. W. 1954. Discussion of "Mechanics of manifold flow," by J. S. McNown, *Transactions of the American Society of Civil Engineers,* 119:1136.

Richardson, J. F. and Zaki, W. N. 1954. Sedimentation and fluidization, *Transactions of the Institute of Chemical Engineers: Part I,* 32:35.

Roberts, E. J. 1949. Thickening—Art or Science? *Mining Engineering,* 101:763.

Rouse, H. 1936. Discharge characteristics of the free overfall, *Civil Engineering,* 6(4):257.

Rouse, H. 1937. Nomogram for the settling velocity of spheres. In *Report of the Committee on Sedimentation,* p. 57, P. D. Trask, chm., National Research Council, Division of Geology and Geography, Washington, D.C.

Rouse, H. 1943. Discussion of "Energy loss at the base of a free overflow," by W. L. Moore, *Transactions of the American Society of Civil Engineers,* 108:1343.

Rouse, H. 1978. *Elementary Mechanics of Fluids,* Dover Publications, New York.

Rowell, H. S. 1913. Note on James Thomson's V-notches, *Engineering,* 95(2 May):589.

San, H. A. 1989. Analytical approach for evaluation of settling column data, *Journal of Environmental Engineering,* 115(2):455.

Seddon, J. A. 1889. Clearing water by settlement, *Journal of the Association of Engineering Societies,* 8(10):477.

Slechta, A. F. and Conley, W. R. 1971. Recent experiences in plant-scale application of the settling tube concept, *Journal of the Water Pollution Control Federation,* 43(8):1724.

Soucek, E. and Zelnick, E. W. 1945. Lock manifold experiments, *Transactions of the American Society of Civil Engineers,* 110:1357.

Stokes, G. G. 1856. On the effect of the internal friction of fluids on the motion of pendulums, *Transactions of the Cambridge Philosophical Society,* 9(II) 8.

Switzer, F. G. 1915. Tests on the effect of temperature on weir coefficients, *Engineering News,* 73(13):636.

Sykes, R. M. 1993. Flocculent and nonflocculent settling, *Journal of Environmental Science and Health, Part A, Environmental Science and Engineering,* A28(1):143.

Talmadge, W. P. and Fitch, E. B. 1955. Determining thickener unit areas, *Industrial and Engineering Chemistry,* 47:38.

Tambo, N. and Hozumi, H. 1979. Physical characteristics of flocs. II. Strength of floc, *Water Research,* 13(5):421.

Tambo, N. and Watanabe, Y. 1979. Physical characteristics of flocs. I. The floc density function and aluminum floc, *Water Research,* 13(5):409.

Task Committee for the Preparation of the Manual on Sedimentation. 1975. *Sedimentation Engineering,* ASCE Manuals and Reports on Engineering Practice No. 54, V. A. Vanoni, ed. American Society of Civil Engineers, New York.

U.S. EPA. 1979. *Process Design Manual for Sludge Treatment and Disposal,* EPA 625/1-79-011. U.S. Environmental Protection Agency, Municipal Environmental Research Laboratory, Technology Transfer, Cincinnati.

Vesilind, P. A. 1974. *Treatment and Disposal of Wastewater Sludges,* Ann Arbor Science Publishers, Ann Arbor, MI.

von Karman, T. 1940. The engineer grapples with nonlinear problems, *Bulletin of the American Mathematical Society,* 46(8):615.

Water Supply Committee of the Great Lakes–Upper Mississippi River Board of State Sanitary Engineers. 1987. *Recommended Standards for Water Works,* 1987 ed., Health Research, Albany, NY.

Wastewater Committee, Great Lakes–Upper Mississippi River Board of State Public Health and Environmental Managers. 1990. *Recommended Standards for Wastewater Facilities, 1990 Edition.* Health Education Services, Albany, NY.

Yalin, M. S. 1977. *Mechanics of Sediment Transport,* 2nd ed., Pergamon Press, Oxford.

Yao, K. M. 1970. Theoretical study of high-rate sedimentation, *Journal of the Water Pollution Control Federation,* 42(2, part 1):218.

Yao, K. M. 1973. Design of high-rate settlers, *Journal of the Environmental Engineering Division, Proceedings of the American Society of Civil Engineers,* 99(EE5):621.

Yee, L. Y. and Babb, A. F. 1985. Inlet design for rectangular settling tanks by physical modeling, *Journal of the American Water Works Association,* 57(12):1168.

Yoshioka, N., Hotta, Y., Tanaka, S., Naito, S., and Tsugami, S. 1957. Continuous thickening of homogeneous flocculated slurries, *Chemical Engineering,* 21(2):66.

Zanoni, A. E. and Blomquist, M. W. 1975. Column settling tests for flocculant suspensions, *Journal of the Environmental Engineering Division, Proceedings of the American Society of Civil Engineers,* 101(EE3):309.

Section 8.6

Amirtharajah, A. 1984. Fundamentals and theory of air scour, *Journal of the Environmental Engineering Division, Proceedings of the American Society of Civil Engineers,* 110(3):573.

Anderson, E. 1941. Separation of dusts and mists. In *Chemical Engineers Handbook,* 2nd ed., 9th imp., p. 1850, J. H. Perry, ed., McGraw-Hill, New York.

AWWA Filtration Committee. 1984. Committee report: Comparison of alternative systems for controlling flow through filters, *Journal of the American Water Works Association,* 76 (1):91.

Babbitt, H. E., Doland, J. J., and Cleasby, J. L. 1967. *Water Supply Engineering,* 6th ed., McGraw-Hill, New York.

Cleasby, J. L. 1990. Filtration. In *Water Quality and Treatment: A Handbook of Community Water Supplies,* 4th ed., p. 455, F. W. Pontius, tech. ed., McGraw-Hill, New York.

Cleasby, J. L., Amirtharajah, A., and Baumann, E. R. 1975. Backwash of granular filters. In *The Scientific Basis of Filtration,* p. 255, K. J. Ives, ed., Noordhoff, Leyden.

Cleasby, J. L. and Baumann, E. R. 1962. Selection of sand filtration rates, *Journal of the American Water Works Association,* 54(5):579.

Cleasby, J. L. and Fan, K. S. 1981. Predicting fluidization and expansion of filter media, *Journal of the Environmental Engineering Division, Proceedings of the American Society of Civil Engineers,* 107(3):455.

Cleasby, J. L. and Sejkora, G. D. 1975. Effect of media intermixing on dual media filtration, *Journal of the Environmental Engineering Division, Proceedings of the American Society of Civil Engineers,* 101(EE4):503.

Committee of the Sanitary Engineering Division on Filtering Materials for Water and Sewage Works. 1936. Filter sand for water purification plants, *Proceedings of the American Society of Civil Engineers,* 62(10):1543.

Conley, W. R. and Hsiung, K.-Y. 1969. Design and application of multimedia filters, *Journal of the American Water Works Association,* 61(2):97.

Culp, G. L. and Culp, R. L. 1974. *New Concepts in Water Purification,* Van Nostrand Reinhold, New York.

Culp/Wesner/Culp, Inc. 1986. *Handbook of Public Water Supplies,* R. B. Williams and G. L. Culp, eds., Van Nostrand Reinhold, New York.

Degrémont, S. A. 1965. *Water Treatment Handbook,* 3rd English ed., trans. D. F. Long, Stephen Austin & Sons, Hertford, UK.

Direct Filtration Subcommittee of the AWWA Filtration Committee. 1980. The status of direct filtration, *Journal of the American Water Works Association,* 72(7):405.

Ergun, S. 1952. Fluid flow through packed columns, *Chemical Engineering Progress,* 48(2):89.

Fair, G. M. and Geyer, J. C. 1954. *Water Supply and Waste-Water Disposal,* John Wiley & Sons, New York.

Fuller, G. W., 1898. *The Purification of the Ohio River Water at Louisville, Kentucky,* D. Van Nostrand, New York.

Gregory, J. 1975. Interfacial phenomena. In *The Scientific Basis of Filtration,* K. J. Ives, ed., p. 53, Noordhoff International Publishing, Leyden.

Hazen, A. 1896. *The Filtration of Public Water-Supplies,* John Wiley & Sons, New York.

Hazen, A. 1904. On sedimentation, *Transactions of the American Society of Civil Engineers,* 53:63.

Herzig, J. R., Leclerc, D. M., and Le Goff, P. 1970. Flow of suspension through porous media, application to deep bed filtration, *Industrial and Engineering Chemistry,* 62(5):8.

Ives, K. J. 1975. Capture mechanisms in filtration. In *The Scientific Basis of Filtration,* K. J. Ives, ed., p. 183, Noordhoff International Publishing, Leyden.

Ives, K. J. and Sholji I. 1965. Research on variables affecting filtration, *Journal of the Sanitary Engineering Division, Proceedings of the American Society of Civil Engineers,* 91(SA4):1.

Iwasaki, T. 1937. Some notes on filtration, *Journal of the American Water Works Association,* 29(10):1591.

James M. Montgomery, Consulting Engineers, Inc. 1985. *Water Treatment Principles and Design,* John Wiley & Sons, New York.

Joint Task Force of the American Society of Civil Engineers and the American Water Works Association. 1990. *Water Treatment Plant Design,* 2nd ed., McGraw-Hill, New York.

Joint Task Force of the Water Environment Federation and the American Society of Civil Engineers. 1992. *Design of Municipal Wastewater Treatment Plants: Volume II—Chapters 13–20,* WEF Manual of Practice No. 8, ASCE Manual and Report on Engineering Practice No. 76, Water Environment Federation, Alexandria, VA; American Society of Civil Engineers, New York.

Kawamura, S. 1991. *Integrated Design of Water Treatment Facilities,* John Wiley & Sons, New York.

MacDonald, I. F., El-Sayed, M. S., Mow, K., and Dullien, F. A. L. 1979. Flow through porous media—the Ergun equation revisited, *Industrial & Engineering Chemistry Fundamentals,* 18(3):199.

Malcolm Pirnie, Inc. and HDR Engineering, Inc. 1990. *Guidance Manual for Compliance with the Surface Water Treatment Requirements for Public Water Systems.* American Water Works Association, Denver.

Metcalf & Eddy, Inc. 1991. *Wastewater Engineering: Treatment, Disposal, and Reuse,* 3rd ed., rev. by G. Tchobanoglous and F. L. Burton, McGraw-Hill, New York.

Morrill, A. B. and Wallace, W. M. 1934. The design and care of rapid sand filters, *Journal of the American Water Works Association,* 26(4):446.

Pontius, F. W. 1990. Complying with the new drinking water quality regulations, *Journal of the American Water Works Association,* 82(2):32.

Richardson, J. F. and Zaki, W. N. 1954. Sedimentation and fluidization, *Transactions of the Institute of Chemical Engineers: Part I,* 32:35.

Standards Committee on Filtering Material. 1989. *AWWA Standard for Filtering Material,* AWWA B100-89, American Water Works Association, Denver, CO.

Stokes, G. G. 1856. On the Effect of the Internal Friction of Fluids on the Motion of Pendulums, *Transactions of the Cambridge Philosophical Society,* 9(II):8.

Tien, C. 1989. *Granular Filtration of Aerosols and Hydrosols,* Butterworth, Boston.

U.S. EPA. 1989. Filtration and disinfection: Turbidity, *Giardia lamblia*, viruses *Legionella* and heterotrophic bacteria: final rule, *Federal Register*, 54(124):27486.

Water Supply Committee of the Great Lakes–Upper Mississippi River Board of State Public Health and Environmental Managers. 1987. *Recommended Standards for Water Works, 1987 Edition,* Health Research, Albany, NY.

Wen, C. Y. and Yu, Y. H. 1966. Mechanics of fluidization, *Chemical Engineering Progress,* Symposium Series 62, American Institute of Chemical Engineers, New York.

Section 8.7

Adamson, A. W. 1982. *Physical Chemistry of Surfaces,* 4th ed. John Wiley & Sons, New York.

Adham, S. S., Snoeyink, V. L., Clark, M. M., and Anselme, C. 1993. Predicting and verifying TOC removal by PAC in pilot-scale UF systems, *Journal of the American Water Works Association,* 85(12):59.

Akell, R. B. 1981. Safety aspects of activated carbon technology. In *Activated Carbon for Wastewater Treatment,* p. 223, J. R. Perrich, ed., CRC Press, Boca Raton, FL.

Anonymous. 1942. Effect of water treatment processes on tastes and odors. In *Taste and Odor Control in Water Purification,* p. 35, West Virginia Pulp and Paper Company, Industrial Chemical Sales Division, New York.

AWWA Standards Committee on Activated Carbon. 1990a. *AWWA Standard for Granular Activated Carbon,* B604-90. American Water Works Association, Denver, CO.

AWWA Standards Committee on Activated Carbon. 1990b. *AWWA Standard for Powdered Activated Carbon,* B600-90. American Water Works Association, Denver, CO.

Bennet, C. O. and Myers, J. E. 1974. *Momentum, Heat, and Mass Transfer,* 2nd ed. McGraw-Hill, New York.

Clark, R. M. 1987. Modeling TOC removal by GAC: The general logistic equation, *Journal of the American Water Works Association,* 79(1):33.

Clark, R. M., Symons, J. M., and Ireland, J. C. 1986. Evaluating field scale GAC systems for drinking water, *Journal of Environmental Engineering,* 112(4):744.

Crank, J. 1956. *The Mathematics of Diffusion,* Oxford University Press, London.

Crittenden, J. C. and Weber, W. J., Jr. 1978a. Predictive model for design of fixed-bed adsorbers: Parameter estimation and model development, *Journal of the Environmental Engineering Division, Proceedings of the American Society of Civil Engineers,* 104(EE2):185.

Crittenden, J. C. and Weber, W. J., Jr. 1978b. Predictive model for design of fixed-bed adsorbers: Single component model verification, *Journal of the Environmental Engineering Division, Proceedings of the American Society of Civil Engineers,* 104(EE3):433.

Crittenden, J. C. and Weber, W. J., Jr. 1978c. A model for design of multicomponent systems, *Journal of the Environmental Engineering Division, Proceedings of the American Society of Civil Engineers,* 104(EE6):1175.

Culp, R. L. and Clark, R. M. 1983. Granular activated carbon installations, *Journal of the American Water Works Association,* 75(8):398.

Culp, R. L. and Culp, G. L. 1971. *Advanced Wastewater Treatment.* Van Nostrand Reinhold, New York.

Culp/Wesner/Culp, Inc., 1986. *Handbook of Public Water Supply Systems,* R. B. Williams and G. L. Culp, eds., Van Nostrand Reinhold, New York.

Dobbs, R. A. and Cohen, J. M. 1980. *Carbon Adsorption Isotherms for Toxic Organics,* EPA-600/8-80-023, U.S. Environmental Protection Agency, Office of Research and Development, Municipal Environmental Research Laboratory, Cincinnati.

Einstein, A. 1956. *Investigations on the Theory of the Brownian Movement,* R. Fürth, ed., A. D. Cowper, trans. Dover Publications, New York.

Freundlich, H. 1922. *Colloid and Capillary Chemistry,* H. S. Hatfield, trans. E. P. Dutton, New York.

Graese, S. L., Snoeyink, V. L., and Lee, R. G. 1987. *GAC Filter-Adsorbers.* American Water Works Association, Research Foundation, Denver, CO.

Halsey, G. and Taylor, H. S. 1947. The adsorption of hydrogen on tungsten, *The Journal of Chemical Physics,* 15(9):624.

Hand, D. W., Crittenden, J. C., and Thacker, W. E. 1983. User-oriented batch reactor solutions to the homogeneous surface diffusion model, *Journal of Environmental Engineering,* 109 (1):82.

Hand, D. W., Crittenden, J. C., and Thacker, W. E. 1984. Simplified models for design of fixed-bed adsorption systems, *Journal of Environmental Engineering,* 110(2):440.

Hayduk, W. and Laudie, H. 1974. Prediction of diffusion coefficients for nonelectrolytes in dilute aqueous solutions, *Journal of the American Institute of Chemical Engineers,* 20(3):611.

Hutchins, R. A. 1981. Development of design parameters. In *Activated Carbon for Wastewater Treatment,* p. 61, J. R. Perrich, ed., CRC Press, Boca Raton, FL.

James M. Montgomery, Consulting Engineers, Inc. 1985. *Water Treatment: Principles and Design,* John Wiley & Sons, New York.

Joint Editorial Board. 1992. *Standard Methods for the Examination of Water and Wastewater,* 18th ed. American Public Health Association, Washington, D.C.

Joint Task Force of the American Society of Civil Engineers and the American Water Works Association. 1990. *Water Treatment Plant Design,* 2nd ed. McGraw-Hill, New York.

Joint Task Force of the Water Environment Federation and the American Society of Civil Engineers. 1991. *Design of Municipal Wastewater Treatment Plants: Volume II—Chapters 13–20,* WEF Manual of Practice No. 8, ASCE Manual and Report on Engineering Practice No. 76. Water Environment Federation, Alexandria, VA; American Society of Civil Engineers, New York.

Le Bas. 1915. *The Molecular Volumes of Liquid Chemical Compounds,* Longmans, London.

Liu, K.-T. and Weber, W. J., Jr. 1981. Characterization of mass transfer parameters for adsorber modeling and design, *Journal of the Water Pollution Control Federation,* 53(10):1541.

McGinnis, F. K. 1981. Infrared furnaces for reactivation. In *Activated Carbon for Wastewater Treatment,* p. 155, J. R. Perrich, ed. CRC Press, Boca Raton, FL.

Najm, I. N., Snoeyink, V. L., Suidan, M. T., Lee, C. H., and Richard, Y. 1990. Effect of particle size on background natural organics on the adsorption efficiency of PAC, *Journal of the American Water Works Association,* 82(1):65.

Nakhla, G. F., Suidan, M. T., and Traegner-Duhr, U. K. 1989. Steady state model for the expanded bed anaerobic GAC reactor operating with GAC replacement and treating inhibitory wastewaters, *Proceedings of the Industrial Waste Symposium, 62nd Annual Conference of the Water Pollution Control Federation, San Francisco.* Water Pollution Control Federation, Alexandria, VA.

Ohio River Valley Water Sanitation Commission. 1980. *Water Treatment Process Modifications for Trihalomethane Control and Organic Substances in the Ohio River,* EPA-600/2-80-028. U.S. Environmental Protection Agency, Office of Research and Development, Municipal Environmental Research Laboratory, Cincinnati, OH.

Oulman, C. S. 1980. The logistic curve as a model for carbon bed design, *Journal of the American Water Works Association,* 72(1):50.

Palmer, C. M. 1967. *Algae in Water Supplies: An Illustrated Manual on the Identification, Significance, and Control of Algae in Water Supplies,* Public Health Service Publication No. 657 (reprint, 1967). U.S. Department of Health, Education, and Welfare, Public Health Service, Division of Water Supply and Pollution Control, Washington, D.C.

Perry, R. H. and Chilton, C. H. 1973. *Chemical Engineer's Handbook,* 5th ed. McGraw-Hill, New York.

Polson, A. 1950. Some aspects of diffusion in solution and a definition of a colloidal particle, *Journal of Physical Colloid Chemistry,* 54:649.

Randtke, S. J. and Snoeyink, V. L. 1983. Evaluating GAC adsorptive capacity, *Journal of the American Water Works Association,* 75(8):406.

Rohlich, G. A. and Sarles, W. B. 1949. Chemical composition of algae and its relationship to taste and odor, *Taste and Odor Journal,* 18(10):1.

Roy, D., Wang, G.-T., and Adrian, D. D. 1993. Simplified calculation procedure for carbon adsorption model, *Water Environment Research,* 65(6):781.

Snoeyink, V. L. 1990. Adsorption of organic compounds. In *Water Quality and Treatment: A Handbook of Community Water Supplies,* 4th ed., p. 781, American Water Works Association, F. W. Pontius, ed., McGraw-Hill, New York.

Suzuki, M. and Kawazoe, K. 1975. Effective surface diffusion coefficients of volatile organics on activated carbon during adsorption from aqueous solution, *Journal of Chemical Engineering of Japan,* 8(5):379.

Thimann, K. V. 1963. *The Life of Bacteria: Their Growth, Metabolism, and Relationships,* 2nd ed. Macmillan, New York.

Traegner, U. K. and Suidan, M. T. 1989. Parameter evaluation for carbon adsorption, *Journal of Environmental Engineering,* 115(1):109.

Turre, G. J. 1942. Pollution from natural sources, particularly algae. In *Taste and Odor Control in Water Purification,* p. 17, West Virginia Pulp and Paper Company, Industrial Chemical Sales Division, New York.

von Dreusche, C. 1981. Regeneration systems. In *Activated Carbon for Wastewater Treatment,* p. 137, J. R. Perrich, ed., CRC Press, Boca Raton, FL.

Wagner, N. J. and Julia, R. J. 1981. Activated carbon adsorption. In *Activated Carbon for Wastewater Treatment,* p. 41, J. R. Perrich, ed., CRC Press, Boca Raton, FL.

Wakao, N. and Funazkri, T. 1978. Effect of fluid dispersion coefficients in dilute solutions, *Chemical Engineering Science,* 33:1375.

Weber, W. J., Jr. and Crittenden, J. C. 1975. MADAM I: A numeric method for design of adsorption systems, *Journal of the Water Pollution Control Federation,* 47(5):924.

Wilke, C. R. and Chang, P. 1955. Correlation of diffusion coefficients in dilute solutions, *American Institute of Chemical Engineering Journal,* 1:264.

Williamson, J. E., Bazaire, K. E., and Geankoplis, C. J. 1963. Liquid-phase mass transfer at low Reynolds numbers, *Industrial and Engineering Chemistry Fundamentals,* 2(2):126.

Section 8.8

Adamson, A. W. 1982. *Physical Chemistry of Surfaces,* 4th ed. John Wiley & Sons, New York.

ASCE Committee on Oxygen Transfer. 1989. *Design Manual: Fine Pore Aeration Systems,* EPA/625/1-89/023, U.S. Environmental Protection Agency, Office of Research and Development, Center for Environmental Research Information, Risk Reduction Engineering Laboratory, Cincinnati, OH.

Bennet, C. O. and Myers, J. E. 1974. *Momentum, Heat, and Mass Transfer,* 2nd ed. McGraw-Hill, New York.

Berger, B. B. 1983. *Control of Organic Substances in Water and Wastewater,* EPA-600/8-83-011. U.S. Environmental Protection Agency, Office of Research and Development, Washington, D.C.

Blevins, R. D. 1984. *Applied Fluid Dynamics Handbook,* Van Nostrand Reinhold, New York.

Brezonik, P. L. 1994. *Chemical Kinetics and Process Dynamics in Aquatic Systems,* Lewis Publishers, CRC Press, Boca Raton, FL.

Dean, J. A., ed. 1992. *Lange's Handbook of Chemistry,* 14th ed. McGraw-Hill, New York.

Glasstone, S. 1947. *Thermodynamics for Chemists,* D. Van Nostrand, Princeton, NJ.

Gosset, J. M. 1987. Measurement of Henry's law constants for C_1 and C_2 chlorinated hydrocarbons, *Environmental Science and Technology,* 21:202.

Haas, C. N. 1990. Disinfection. In *Water Quality and Treatment: A Handbook of Community Water Supplies,* 4th ed., p. 877, F. W. Pontius, ed. McGraw-Hill, New York.

Hardenbergh, W. A. and Rodie, E. B. 1963. *Water Supply and Waste Disposal,* International Textbook, Scranton, PA.

Hill, K. D. and Dauphinee, T. M. 1986. The extension of the "Practical Salinity Scale 1978" to low salinities, *IEEE Journal of Oceanic Engineering,* OE-11(1):109.

Joint Editorial Board. 1992. *Standard Methods for the Examination of Water and Wastewater,* 18th ed. American Public Health Association, Washington, D.C.

Joint Task Force of the American Society of Civil Engineers and American Water Works Association. 1990. *Water Treatment Plant Design,* 2nd ed. McGraw-Hill, New York.

Joint Task Force of the Water Pollution Control Federation and the American Society of Civil Engineers. 1988. *Aeration: A Wastewater Treatment Process,* WPCF Manual of Practice No. FD-13, ASCE Manuals and Reports on Engineering Practice No. 68. Water Pollution Control Federation, Alexandria, VA; American Society of Civil Engineers, New York.

Kavanaugh, M. C. and Trussel, R. R. 1980. Design of aeration towers to strip volatile contaminants from drinking water, *Journal of the American Water Works Association,* 72(12):684.

Lamarche, P. and Droste, R. L. 1989. Air-stripping mass transfer correlation for volatile organics, *Journal of the American Water Works Association,* 81(1):78.

Lewis, E. L. 1980. The "Practical Salinity Scale 1978" and its antecedents, *IEEE Journal of Oceanic Engineering,* OE-5(1):3.

Metcalf & Eddy, Inc. 1991. *Wastewater Engineering: Treatment, Disposal, and Reuse,* 3rd ed., revised by G. Tchobanoglous and F. L. Burton. McGraw-Hill, New York.

McKinnon, R. J. and Dyksen, J. E. 1984. Removing organics from groundwater through aeration plus GAC, *Journal of the American Water Works Association,* 76(5):42.

Morris, J. C. 1966. The acid ionization constant of HOCl from 5 to 35 C, *Journal of Physical Chemistry,* 70:3798.

Munz, C. and Roberts, P. V. 1987. Air-water phase equilibria of volatile organic solutes, *Journal of the American Water Works Association,* 79(5):62.

Perkin, R. G. and Lewis, E. L. 1980. The "Practical Salinity Scale 1978": Fitting the data, *IEEE Journal of Oceanic Engineering,* OE-5(1):9.

Perry, R. H. and Chilton, C. H., eds. 1973. *Chemical Engineer's Handbook,* 5th ed. McGraw-Hill, New York.

Riley, J. P. and Chester, R. 1971. *Introduction to Marine Chemistry.* Academic Press, New York.

Roberts, P. V., Hopkins, G. D., Munz, C., and Riojas, A. H. 1985. Evaluating two-resistance models for air stripping of volatile organic contaminants in a countercurrent, packed column, *Environmental Science and Technology,* 19(2):164.

Roberts, R. V., Munz, C., and Dändliker, P. 1984. Modeling volatile organic solute removal by surface and bubble aeration, *Journal of the Water Pollution Control Federation,* 56(2):157.

Sherwood, T. K., Pigford, R. L., and Wilke, C. R. 1975. *Mass Transfer,* McGraw-Hill, New York.

Staudinger, J., Knocke, W. R., and Randall, C. W. 1990. Evaluating the Onda mass transfer correlation for the design of packed-column air stripping, *Journal of the American Water Works Association,* 82(1):73.

Stenstrom, M. K. and Gilbert, R. G. 1981. Effects of alpha, beta and theta factor upon the design, specification and operation of aeration systems, *Water Research,* 15(6):643.

Stover, E. L., Haas, C. N., Rakness, K. L., and Scheible, O. K. 1986. *Design Manual: Municipal Wastewater Disinfection,* EPA/625/1-86-021. U.S. Environmental Protection Agency, Office of Research and Development, Water Engineering Research Laboratory, Center for Environmental Research Information, Cincinnati, OH.

Stumm, W. and Morgan, J. J. 1970. *Aquatic Chemistry,* Wiley-Interscience, John Wiley & Sons, New York.

Weast, R. C., Astle, M. J., and Beyer, W. H. 1983. *CRC Handbook of Chemistry and Physics,* 64th ed. CRC Press, Boca Raton, FL.

White, G. 1986. *Handbook of Chlorination,* 2nd ed. Van Nostrand, Princeton, NJ.

Yuteri, C., Ryan, D. F., Callow, J. J., and Gurol, M. D. 1987. The effect of chemical composition of water on Henry's law constant, *Journal of the Water Pollution Control Federation,* 51(10):950.

9

Chemical Water Treatment Processes

Robert M. Sykes
Ohio State University

9.1 Coagulation

Surface waters contain a variety of suspended colloidal solids that have aesthetic, economic, or health impacts. Simple sedimentation and direct, unaided filtration are impractical in the case of clays and organic detritus because the overflow and filtration rates required for their removal lead to facilities that are 100 to 200 times larger than those built today [Fanning, 1887; Fuller, 1898]. Consequently, all surface water treatment plants incorporate processes that destabilize and agglomerate colloids into larger, fast-settling particles.

Colloids

Properties

Colloidal systems (dispersoids, colloidal dispersions, colloidal suspensions, colloidal solutions, and sols) consist of particles suspended in some sort of medium. The chemical composition of the particles is usually different from that of the medium, but examples where they are the same are known. Colloidal systems are distinguished from true solutions and mechanical suspensions by the following criteria [Voyutsky, 1978]:

Opalescence. Colloidal systems scatter visible light. If a light beam is passed through a suspension of colloidal particles, some of the light beam will be scattered at right angles and a cloudy streak will be seen running along its path. This is called the *Tyndall cone* after its discoverer.

A consequence of scattering is that colloidal systems do not transmit images of objects; when the transmitted light is viewed along its path, only a uniform glow is seen. This property is called *turbidity,* and colloidal systems are said to be *turbid.* By contrast, true solutions, even if they are colored, transmit clear images of objects.

0-8493-8953-4/95/$0.00 + $.50

Opalescence is the basis of colloid measurement. If light intensity measurements are made collinearly with the beam, the procedure is called *turbidimetry.* If the measurements are made at right angles to the path of the beam, the procedure is called *nephelometry.*

Turbidimetry requires subtraction of the light intensity leaving the sample from the light intensity entering the sample. For low turbidities this difference is small and its measurement is inherently inaccurate. Nephelometry is preferred at low turbidities because only the intensity of the scattered light need be known, and very low light intensities can be measured accurately.

In Rayleigh's theory the scattered light intensity is given by [Jirgensons and Straumanis, 1956]:

$$H_s = 24\pi^3 \cdot \left(\frac{n_p^2 - n_m^2}{n_p^2 + 2n_m^2}\right) \cdot \frac{n V_p^2}{\lambda^4} \cdot I_o \tag{9.1}$$

where

H_s = the total scattered light intensity summed over all angles from nonconducting, spherical particles (W or ft·lbf/s)
I_o = the irradiance of the incident beam (W/m^2 or $ft \cdot lbf/ft^2 \cdot s$)
n = the concentration of particles ($number/m^3$ or $number/ft^3$)
n_m = the refractive index of the suspending medium (dimensionless)
n_p = the refractive index of the particle (dimensionless)
λ = the wavelength of the incident beam (m or ft)

The observed intensity of scattered light varies with (a) the angle at which the light is measured, (b) the size and properties of the particles, (c) the properties of the suspending medium, (d) the wavelengths in the incident light, and (e) whether or not the light is polarized. Consequently, the units of turbidity are somewhat arbitrary, and the weight concentrations of particles in different waters may be different, even if the turbidities are the same.

The turbidity units used in environmental engineering are based on several different but related standards. The earliest standard was based on the silica frustules of diatoms [Committee on Standard Methods of Water Analysis, 1901]. The shells were cleaned of organic matter and ground and sieved through a 200 mesh screen, so the particles were smaller than 74 μm. This means the original standard suspension includes some particles that were larger than colloids. A suspension containing 1 mg/L of these prepared particles was defined to have a turbidity of 1.

Nowadays, the clay kaolin, the organic colloid formazin, and styrene divinylbenzene beads are used instead of diatomaceous earth, but the concentrations of these materials are adjusted so that one turbidity unit of any of them produces approximately the same degree of scattering as 1 mg/L of diatomaceous earth [Joint Editorial Board, 1992]. All these modern standards also include particles that are larger than colloids. However, many of the suspended particles in surface waters are themselves supracolloidal, so the use of standards containing supracolloidal particles is not an error.

If the turbidity is between 25 and 1000 units, it is often measured using a *Jackson tube.* This is an example of turbidimetry. In this device a standardized candle is viewed through a layer of sample contained in a glass tube with opaque sides and a clear bottom. Sample can be added to or withdrawn from the tube until the image of the candle disappears and a uniformly illuminated field remains. The depth of sample is correlated with the turbidity. For example, if the turbidity is 100 units, the image of the candle disappears at a sample depth of 21.5 cm; it disappears at 39.8 cm if the turbidity is 50 units. Turbidity measurements performed this way are reported in Jackson turbidity units (JTU).

For turbidities less than 25 units the scattered light intensity at 90° from the incident path is measured. Various commercial instruments are used, and they are calibrated against standard suspensions. Measurement at 90° is called *nephelometry*, and the instruments are called *nephelometers*. The measured turbidity is reported in nephelometric turbidity units (NTU).

The treatment goal for potable waters is to produce a final turbidity less than 0.5 units.

Dialysis. Colloidal particles can be dialyzed. This means that they cannot pass through a semipermeable membrane. True solutes of low molecular weight will. Consequently, if a system containing water, true solutes, and colloidal particles is placed on one side of a semipermeable membrane and pure water is placed on the other side of the membrane, the true solutes pass through the membrane, equilibrating their concentrations on either side, but the colloids do not. This is one way of purifying colloidal systems from dissolved salts. The process is evidently dependent on the sizes of the membrane's pores, and so this is another size classification scheme: Colloids are larger than true solutes. The traditional membranes were animal tissues such as bulls' bladders and parchment, but nowadays various synthetic membranes are used, and the pore sizes can be specified [Voyutski, 1978].

Osmotic Pressure. If a colloidal system is dialyzed, it will exhibit an osmotic pressure on the dialysis membrane, just like a true solution. Osmotic pressure is proportional to the number of particles suspended in the dispersing medium. It does not depend on the size of the particles, so it does not matter whether the particles are single atoms, large molecules, or sols.

The osmotic pressure of the colloidal system is calculated using Einstein's formula, which is the same as the osmotic pressure equation for true solutes [Einstein, 1956]:

$$p = \frac{RTn}{VN_A} \tag{9.2}$$

where

N_A = Avogadro's number (6.0221367×10^{23} particles/mole)
n = the concentration of particles (number/m^3 or number/ft^3)
p = the osmotic pressure of the colloidal system (N/m^2 or lbf/ft^2)
R = the gas constant (8.314510 J/mol·K or 1545.356 ft·lbf/lb-mol·°R)
T = the absolute temperature (K or °R)

It is estimated that a gold sol 0.5% by weight (about the highest concentration that can be achieved) consisting of particles about 1 nm in diameter would develop an osmotic pressure of only 1 to 2 mm water head [Svedberg, 1924].

Brownian Movement. Colloidal particles exhibit so-called Brownian movement, which is visible in the case of the larger particles under a microscope. The Brownian movement is due to the momentum transmitted to the colloidal particles by the thermal motion of the suspending medium. The resulting paths of the colloidal particles consist of connected, broken straight lines oriented at random and having random lengths. The result is that the particles diffuse according to Fick's law. Einstein's formula for the diffusivity of colloidal particles is as follows [Einstein, 1956]:

$$D = \frac{RT}{6\pi N_A \mu r} \tag{9.3}$$

where

D = the diffusivity of the suspended particles (m^2/s or ft^2/s)
r = the radius of the suspended particle (m or ft)
μ = the absolute viscosity of the suspending medium (N·s/m^2 or lbf·s/ft^2)

Colloidal particles have a small settling velocity, which may be estimated from Stoke's law. Consequently, in a perfectly quiescent container, the particles will tend to settle out. This will establish a concentration gradient, with higher concentrations toward the bottom, and the resulting upward diffusion will at some point balance the sedimentation. At equilibrium the particles are distributed vertically in the container according to the *hypsometric law*, which was first derived for the distribution of gases in a gravitational field [Svedberg, 1924]:

$$\frac{n_2}{n_1} = \exp\left\{\frac{gN_A V_p(\rho_p - \rho)(z_2 - z_1)}{RT}\right\} \tag{9.4}$$

where

g = the acceleration due to gravity (9.80665 m/s^2 or 32.174 ft/s^2)
n_1 = the number (or concentration) of particles at height z_1
n_2 = the number (or concentration) of particles at height z_2
V_p = the volume of a single particle (m^3 or ft^3)
z_1 = the elevation of particle concentration n_1 (m or ft)
z_2 = the elevation of particle concentration n_2 (m or ft)
ρ = the mass density of the suspending medium (kg/m^3 or $slugs/ft^3$)
ρ_p = the mass density of the particles (kg/m^3 or $slugs/ft^3$)

Electrophoresis. If an electric field is applied to a colloidal system, *all* the particles will migrate slowly to *one* electrode. This means that the particles are charged and that they all have the same kind of charge—positive or negative—although the absolute values of the charges may differ. This should be contrasted with solutions of electrolytes, which contain equal numbers of positive and negative charges: When true solutions are electrolyzed, particles are attracted to both electrodes.

Stability. Many colloidal systems are unstable, and the particles can be coagulated in a variety of ways. In fact, one of the main problems of colloid chemistry involves devising means to make the particles stay in suspension.

Composition. The particles themselves usually have a different composition from the suspending medium. Therefore, the systems consist of more than one chemical phase—usually two but sometimes more—and they are heterogeneous. True solutions consist of a single phase.

Particle Size. The traditional range of sizes of colloidal particles was set by Zsigmondy [1914] at 1 to 100 nm. The upper size limit was chosen because it is somewhat smaller than the smallest particle that can be seen under a light microscope. Also, particles smaller than 100 nm do not settle out of suspension, even under quiescent conditions, whereas particles around 1 μm—the size of bacteria—do. The lower limit is somewhat smaller than can be detected by an ultramicroscope. Consequently, these are operational limits determined by the available instrumentation; they are not fundamental properties of colloidal systems.

These sizes may be compared to those of other particles:

- Atoms: 0.1 to 0.6 nm
- Small molecules: 0.2 to 5.0 nm
- Small polymers: 0.5 to 10 nm
- Colloids: 1.0 to 100 nm
- Clay: $<$ 2000 nm (smaller clays are colloidal)
- Bacteria: 250 to 10,000 nm (these and larger particles are settleable)

- Silt: 2000 to 50,000 nm
- Visible particles: > 50,000 nm
- Very fine sand: 50,000 to 100,000 nm

Dispersions of particles with diameters between 100 and 1000 nm are sometimes called *fine dispersions*; if the diameters are larger than 1000 nm, the dispersion is called *coarse*. Fine and coarse dispersions are maintained by turbulence in the suspending medium, not by the random thermal motion of its molecules.

Kinds of Colloidal Dispersions. Colloids can also be classified according to chemistry. The simplest scheme, from Ostwald [1915], is as follows:

- Gas in gas (impossible)
- Liquid in gas (fogs, mists, clouds)
- Solid in gas [smokes, fumes (ammonium chloride)]
- Gas in liquid (foams)
- Liquid in liquid (emulsions, cream)
- Solid in liquid (colloidal gold)
- Gas in solid (meerschaum, pumice)
- Liquid in solid [metallic mercury in ointments, opal (water in amorphous silica)]
- Solid in solid [ruby glass (gold in glass), cast iron (carbon in iron)]

The important colloids in water and sewage treatment are foams, emulsions, and solids in liquids. Smokes, fumes, fogs, and mists are important in air pollution.

Classification by Stability. Colloidal systems are traditionally divided into two broad groups:

- Reversible, lyo(hydro)philic, or emulsoid
- Irreversible, lyo(hydro)phobic, or suspensoid

The various terms used to describe each class are not exact synonyms because they emphasize different aspects of colloidal stability. Furthermore, they are probably best thought of as endpoints on a continuous spectrum rather than separate groups. Lyophilic colloids are typically organic materials, especially naturally occurring ones, and lyophobic colloids are primarily inorganic materials.

The dichotomy reversible/irreversible, proposed by Zsigmondy [1914], is based on the idea of thermodynamic spontaneity. A colloidal system is called reversible if after drying it can be reformed simply by adding the dispersion medium. It is called irreversible it does not reform spontaneously.

The distinctions lyophilic/lyophobic, introduced by Neumann [Ostwald, 1915], and hydrophilic/hydrophobic, introduced by Perrin [Ostwald, 1915], refer to the sensitivity of the system to the addition of electrolytes. A lyo(hydro)philic colloidal particle remains in suspension and uncoagulated over relatively wide ranges of electrolyte concentration, but lyo(hydro)phobic colloidal particles are stable only over narrow ranges of electrolyte concentration.

Suspensoid comprehends both the ideas of irreversibility and electrolyte sensitivity.

Emulsoid comprehends both reversibility and insensitivity to electrolytes.

Because these definitions are not fully equivalent, they sometimes lead to contradictory classifications. For example, clays spontaneously form stable suspensions when mixed with natural waters, so they can be classified as reversible and, by extension, hydrophilic [Fridrikhsberg, 1986]. On the other hand, the stability of clay suspensions is sensitive to electrolyte concentration, so they can also be classified as hydrophobes [James M. Montgomery, Consulting Engineers, Inc., 1985]. Aluminum hydroxide and ferric hydroxide, which are discussed below, also exhibit these contrary tendencies [Voyutski, 1978]. Furthermore, some colloid scientists maintain that organic substances such as cellulose and protein are not properly classified as any kind of colloid; they are actually high-molecular-weight molecules in true solution [Voyutski, 1978]. By implication, the only true colloids are suspensoids. Most workers, however, continue to include organic materials among the colloids.

Stability

Colloidal dispersions are said to be stable if the particles remain separated from one another for long times. If the particles coalesce, the dispersion is unstable. There are two phenomena that affect stability: solvation and surface charge [Kruyt, 1930].

Hydrophilic colloids are naturally stabilized by both solvation and surface charge. Hydrophobic colloids are not solvated and depend entirely on surface charge for stability. Clays and metallic hydroxides are partially solvated but are stabilized in part by surface charges.

A particle is solvated if its surface bonds to water. The particular kind of bonding involved is called *hydrogen bonding*. This is a sort of weak electrostatic bonding. It occurs whenever the system contains surfaces that have strongly electronegative atoms like O, N, or F. Even when these atoms are covalently bonded into molecules, their attraction for electrons is so strong that the electron cloud is distorted and concentrated in their vicinity. This produces a region of excess negative charge. Hydrogen atoms tend to be attracted to these zones of excess negativity, and this attraction leads to a weak bonding between molecules. For example, water molecules hydrogen bond both to each other and to ammonia:

$$\begin{matrix} H-\underset{\substack{| \\ H}}{O}\cdots \underset{\substack{| \\ H}}{H}-O \end{matrix} \quad \text{and} \quad \begin{matrix} H-\underset{\substack{| \\ H}}{O}\cdots H-\underset{\substack{| \\ H}}{N}-H \end{matrix}$$

Here the solid lines indicate normal covalent bonds, and the three dots indicate hydrogen bonds. Typical hydrogen bond energies are about 5 kcal/mole, compared to about 50 to 100 kcal/mole for covalent bonds.

Hydrogen bonding leads to a competition between water molecules and other colloids for the particle surface, and in the case of hydrophilic colloids the water wins. Consequently, the particles are prevented from coalescing because they are coated with a film of water that cannot be displaced.

The surface charge on colloids is developed in three ways. Some colloids contain surface groups that readily ionize in water or take up protons [Stumm and Morgan, 1970]. For example,

$$—OH \rightarrow —O^-$$

$$—NH_2 \rightarrow —NH_3^+$$

$$—COOH \rightarrow —COO^-$$

These ionizations and protonations are strongly pH dependent, and the resulting surface potential varies with the pH.

The second method involves selective bonding of ions from the surrounding solution. This occurs because the plane of the crystal lattice that forms the particle surface has unsatisfied electrostatic and covalent bonds that ions in the suspending medium can complete. The bonding is very specific and depends on both the detailed chemistry of the particle surface and the kinds of ions dissolved in the water.

The third method is ion exchange. Some ions, usually cations, diffuse out of the crystal lattice of the colloidal particle into the surrounding medium, and they are replaced by other ions that diffuse from the medium into the particle. If the two ions have different charges, the lattice will acquire or lose charge. Ion exchange is not very specific, except that small, highly charged ions tend to replace large, weakly charged ions.

All three mechanisms may occur on a single particle. The net result is the surface potential, ψ_o.

The surface potential influences the remaining ions in the suspending medium by electrostatic repulsion and attraction; the result is the so-called electrical double layer [Voyutski, 1978]. If the particle has a net negative charge (which is typical of clays), positive ions in the suspending medium adsorb electrostatically to the exterior of the particle surface in a layer one or more ions thick. This adsorbed layer is called the *Stern layer,* and it reduces the net potential on the particle from ψ_o to ψ_δ. The reduced electrostatic field repels and attracts ions in the suspending medium depending on whether they are of like or unlike sign, respectively. Consequently, the solution is not electrically neutral near the particle surface, and any given thin layer of solution will contain an excess of charge opposite in sign to ψ_δ. The ions in the solution layer bearing the charges of opposite sign are called the *counterions,* and the layer itself is called the *diffuse* or *Gouy* layer. Away from the particle surface, the net observed charge is the sum of the charges due to the particle surface charge, the Stern layer, and the intervening Gouy layer. It falls off with distance, until at large distances the system appears to be electrically neutral.

The charge on colloids is usually determined by measuring their velocity in an electrical field. Because moving particles have an attached boundary layer of water, what is actually determined is the net of the voltage on the particle itself and the counterions in the boundary layer. The result is called the *electrokinetic potential* or the *zeta potential,* and it is calculated using the Helmholtz–Smoluchowski equation [Voyutski, 1978]:

$$\zeta = \frac{k\pi\mu v}{\varepsilon \cdot \mathbf{E}} \tag{9.5}$$

where

$\mathbf{E}$ = the imposed potential gradient (V/m)

k = a constant in the Helmholtz-Smoluchowski equation that depends on the particle shape and imposed electric field, generally between 4 and 8 (dimensionless)

v = the particle velocity (m/s or ft/s)

ε = the dielectric constant of the suspending medium (dimensionless)

μ = the absolute viscosity of the suspending medium ($N \cdot s/m^2$ or $lbf \cdot s/ft^2$)

ζ = the zeta potential (V)

The ratio $v/\mathbf{E}$ is called the *electrophoretic mobility.*

The thickness of the boundary layer will depend on the velocity of the particle. Consequently, the volume of water and the number of counterions associated with a moving colloid vary with the imposed electric field and other factors. This means that ζ also varies with these conditions, and it cannot be identified with ψ_δ. Nevertheless, as long as the experimental conditions are standardized, the zeta potential remains a useful index of the surface potential on the particles. Furthermore, many colloids coagulate spontaneously if the Stern layer potential, ψ_δ, is near zero; the zeta potential is also zero in this case.

Coagulation Chemistry

Coagulation Mechanisms

Colloidal particles can be coagulated in four ways:

- Surface potential reduction
- Compression of the Gouy layer
- Interparticle bridging
- Enmeshment

Reduction of surface potential is effective only against hydrophobic colloids. If the surface potential arises because of ionization or protonation of surface groups, a change in pH via the addition of acid or base will eliminate it. Addition of counterions that adsorb to the surface of the particles also can reduce the surface potential.

Compression of the Gouy layer permits colloidal particles to approach each other closely before experiencing electrostatic repulsion, and their momentum may overcome the residual repulsion and cause collision and adhesion. The Gouy layer can be compressed by the addition of counterions that *do not* adsorb to the particles. The compression is greatest for highly charged ions because the electrostatic attraction per ion increases with its charge. According to the Schulze-Hardy rule [Voyutski, 1978] the molar concentration of an ion required to coagulate a colloid is proportional to the reciprocal of its charge raised to the sixth power. Consequently, the relative molar concentrations of mono-, di-, tri-, and tetravalent ions required to coagulate a colloid are in the ratios $1{:}(1/2)^6{:}(1/3)^6{:}(1/4)^6$ or 1:0.016:0.0013:0.00024.

Interparticle bridging is accomplished by adding microscopic filaments to the suspension. These filaments are long enough to bond to more than one particle surface, and they entangle the particles, forming larger masses. The filaments may be either uncharged in water (nonionic), positively charged (cationic), or negatively charged (anionic).

Enmeshment occurs when a precipitate is formed in the water by the addition of suitable chemicals. If the precipitate is voluminous it will surround and trap the colloids, and they will settle out with it.

Coagulant Dosage

A typical example of coagulation by aluminum and iron salts is shown in Fig. 9.1, in which residual turbidity after settling is plotted against coagulant dose. At low coagulant dosages, nothing happens. However, as the dosage is increased a point is reached at which rapid coagulation and settling occurs. This is called the *critical coagulation concentration* (CCC). Coagulation and settling also occur at somewhat higher concentrations of coagulant. Eventually, increasing the dosage fails to coagulate the suspension, and the concentration marking this failure is called the *critical restabilization concentration* (CSC). At still higher coagulant dosages, turbidity removal again occurs. This second turbidity removal zone is called the *sweep zone*.

Figure 9.1 can be explained as follows. Between the CCC and the CSC, aluminum and iron salts coagulate silts and clays by surface charge reduction [Dentel and Gossett, 1988; Mackrle, 1962; Stumm and O'Melia, 1968]. Aluminum and iron form precipitates of aluminum hydroxide [$Al(OH)_3$] and ferric hydroxide [$Fe(OH)_3$], respectively. These precipitates are highly insoluble and hydrophobic, and they adsorb to the silt and clay surfaces. The net charge on the aluminum hydroxide precipitate is positive at pH values less than about 8.0; the ferric hydroxide precipitate is positive at pH values less than about 6.0 [Stumm and Morgan, 1970]. The result of the hydroxide adsorption is that the normally negative surface charge of the silts and clays is reduced, and so is the

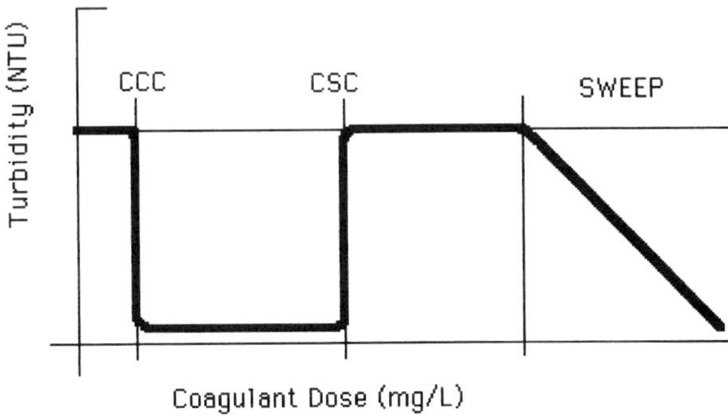

FIGURE 9.1 Supernatant turbidity vs. coagulant dose.

zeta potential. Coagulation and precipitation of the silts and clays occurs when enough aluminum or iron has been added to the suspension to reduce the zeta potential to near zero; this is the condition between the CCC and the CSC.

At dosages below the CCC the silts and clays retain enough negative charge to repel each other electrostatically.

As the aluminum or iron dosage approaches the CSC, aluminum and ferric hydroxide continue to adsorb to the silts and clays, the silts and clays become positive, and they are stabilized again by electrostatic repulsion, although the charge is positive.

If large amounts of aluminum or iron salts are used, the quantity of hydroxide precipitate formed will exceed the adsorption capacity of the silt and clay surfaces, and free hydroxide precipitate will accumulate in the suspension. This free precipitate will enmesh the silts and clays and remove them when it settles out. This is the sweep zone.

In water treatment the coagulant dosages employed lie between the CCC and the CSC. This is done to minimize coagulant costs. In this region the coagulation mechanism is surface charge reduction via adsorption of aluminum or ferric hydroxides to the particle surfaces. Consequently, there should be a relationship between the raw water turbidity and the dosage required to destabilize it. Examples of empirical correlations are given in Stein [1915], Hopkins and Bean [1966], Langelier *et al.* [1953], and Hudson [1965]. For particles of uniform size, regardless of shape, the surface area is proportional to the two-thirds power of the concentration. This rule is also true for different suspensions having the same size distribution. Hazen's [1890] rule of thumb, Eq. (9.6), follows this rule very closely:

$$C_{\text{Alum}} = 0.349 + 0.0377 \cdot C_{\text{TU}}^{2/3} \qquad R^2 = 0.998 \tag{9.6}$$

where

C_{Alum} = the filter alum dosage in grains/gallon
C_{TU} = the raw water turbidity in JTU

However, when waters from several different sources are compared it is found that the required coagulant dosages do not follow Hazen's rule of thumb. The divergences from the rule are probably due to differences in particle sizes in the different waters. For constant turbidity the required coagulant dosage varies inversely with particle size; the required dosage nearly triples if the particle size is reduced by a factor of about 10 [Langelier *et al.*, 1953].

The Jar Test. Although Eq. (9.6) is useful as a guideline, coagulant dosages must in practice be determined experimentally. The determination must be repeated on a frequent basis, at least daily but often once or more per work shift, because the quantities and qualities of the suspended solids in surface waters vary. The usual method is the *jar test.*

The jar test attempts to simulate the intensity and duration of the turbulence in key operations as they are actually performed in the treatment plant: chemical dosing (rapid mixing), colloid destabilization and agglomeration (coagulation/flocculation), and particle settling. Because each plant is different, the details of the jar test procedure will vary from facility to facility, but the general outline, developed by Camp and Conklin [1970], is as follows:

- Two-liter aliquots of a representative sample are placed into each of several standard two-liter laboratory beakers. Typically, six beakers are used because the common laboratory mixing apparatus has space for six beakers. Beakers with stators are preferred because they provide better control of the turbulence. The intensity of the turbulence is measured by the root-mean-square velocity gradient G. (The rms characteristic strain rate, $\overline{\Gamma}$, is nowadays preferred.)
- The mixer is turned on and the rotational speed is adjusted to produce the same rms velocity gradient as that produced by the plant's rapid mixing tank.

- A known amount of the coagulant is added to each beaker, usually in the form of a concentrated solution, and the rapid mixing is allowed to continue for a time equal to the hydraulic detention time of the plant's rapid mixing tank.
- The mixing rate is slowed to produce an rms velocity gradient equal to that in the plant's flocculation tank, and the mixing is continued for a time equal to the flocculation tank's hydraulic detention time.
- The mixer is turned off and the flocculated suspension is allowed to settle quiescently for a period equal to the hydraulic detention time of the plant's settling tanks.
- The supernatant liquid is sampled and analyzed for residual turbidity.

Hudson and Singley [1974] recommend sampling the contents of each beaker for residual suspended solids as soon the turbulence dies out in order to develop a settling velocity distribution curve for the flocculated particles.

The supernatant liquid should be clear, and the floc particles should be compact and dense: "pinhead" floc, called as such because of its size. Large, feathery floc particles are undesirable, because they are fragile and tend to settle slowly, and they may indicate dosage in the sweep zone, which is uneconomic.

If the suspension does not coagulate or if the result is "smokey" or "pin-point" floc, one of the following applies:

- More coagulant is needed.
- The raw water has insufficient alkalinity, and the addition of lime or soda ash is required. (This necessitates a more elaborate testing program to determine the proper ratios of coagulant and base.)
- The water is so cold that the reactions are delayed. (The test should be conducted at the temperature of the treatment plant.)

The jar test is also used to evaluate various coagulant aids.

Finally, it should be noted that the jar test simulates an ideal plug flow reactor. This means that it will not accurately simulate the performance of the flocculation and settling tanks unless they exhibit ideal plug flow, too. In practice, flocculation tanks must be built as mixed-cells-in-series, and settling tanks should incorporate tube modules.

Aluminum and Iron Chemistry

The chemistries of aluminum and ferric iron are very similar. Both cations react strongly with water molecules to form hydroxide precipitates and release protons:

$$AL^{3+} + 3H_2O \rightarrow Al(OH)_3(s) + 3H^+ \tag{9.7}$$

$$Fe^{3+} + 3H_2O \rightarrow Fe(OH)_3(s) + 3H^+ \tag{9.8}$$

Furthermore, at high pH both precipitates react with the hydroxide ion and redissolve, forming aluminate and ferrate ions:

$$Al(OH)_3(s) + OH^- \rightarrow Al(OH)_4^- \tag{9.9}$$

$$Fe(OH)_3(s) + OH^- \rightarrow Fe(OH)_4^- \tag{9.10}$$

Both cations also form a large number of other dissolved ionic species, some of which are polymers, and many of yet unknown structure.

The dissolution of aluminum and ferric hydroxide at high pH is not a significant problem in water treatment because the high pH levels required do not normally occur. However, the hydrolysis reactions of Eqs. (9.7) and (9.8) are. Both reactions liberate protons, and unless these protons are removed from solution only trace amounts—if any—of the precipitates are formed. In fact, if aluminum salts are added to pure water, no visible precipitate is formed. There are also many natural waters in which precipitate formation is minimal. These generally occur in granitic or basaltic regions.

Filter Alum. The most commonly used coagulant is *filter alum,* also called *aluminum sulfate.* Filter alum is made by dissolving bauxite ore in sulfuric acid. The solution is treated to remove impurities, neutralized, and evaporated to produce slabs of aluminum sulfate. The product is gray- to yellow-white in color, depending on the impurities present, and the crystals include variable amounts of water of hydration: $Al_2(SO_4)_3(H_2O)_n$, with n taking the values 0, 6, 10, 16, 18, and 27. It is usually specified that the water-soluble alumina $[Al_2O_3]$ content exceed 17% by weight. This implies an atomic composition of $Al_2(SO_4)_3(H_2O)_{14.3}$. The commercial product also should contain less than 0.5% by weight insoluble matter and less than 0.75% by weight iron, reported as ferric oxide $[Fe_2O_3]$ [Hedgepeth, 1934; Sidgwick, 1950].

Filter alum can be purchased as lumps ranging in size from ¾ to 3 in., as granules smaller than the NBS no. 4 sieve, as a powder, or as a solution. The solution is required to contain at least 8.5% by weight alumina. The granules and powder must be dissolved in water prior to application, and the lumps must be ground prior to dissolution. Consequently, the purchase of liquid aluminum sulfate, sometimes called *syrup alum,* eliminates the need for grinders and dissolving apparatus, and these savings may offset the generally higher unit cost and increased storage volumes and costs.

The dissolution of filter alum and its reaction with alkalinity to form aluminum hydroxide may be described by

$$Al_2(SO_4)_3(H_2O)_{14.3}(s) + 6HCO_3^- \rightarrow 2Al(OH)_3(s) + 6CO_2 + 3SO_4^{2-} + 14.3H_2O \tag{9.11}$$

The aluminum hydroxide precipitate is white, and the carbon dioxide gas produced appears as small bubbles in the water and on the sides of the jar test beaker. The sulfate released passes through the treatment plant and into the distribution system. One mole of filter alum releases six moles of protons, so its equivalent weight is 1/6 of 600 g, or 100 g. The alkalinity consumed is six equivalents or 300 g (as $CaCO_3$). This is the source of the traditional rule of thumb that **1 g of filter alum consumes 0.5 g of alkalinity.**

The acid-side and base-side equilibria for the dissolution of aluminum hydroxide are [Hayden and Rubin, 1974]:

$$Al^{3+} + H_2O \leftrightarrow Al(OH)_3(s) + 3H^+ \tag{9.12}$$

$$K_{s1} = \frac{[H^+]^3}{[Al^{3+}]} = 10^{-10.40} \quad (25°C) \tag{9.13}$$

$$Al(OH)_3(s) + OH^- \leftrightarrow Al(OH)_4^- \tag{9.14}$$

$$K_{s2} = \frac{[Al(OH)_4^-]}{[OH^-]} = 10^{1.64} \quad (25°C) \tag{9.15}$$

Substituting the ionization constant for water into Eq. (9.15) produces:

$$K_{s2} \cdot K_w = [Al(OH)_4^-] \cdot [H^+] = 10^{-12.35} \quad (25°C) \tag{9.16}$$

Equations (9.13) and (9.16) plot as straight lines on log/log coordinates. Together they define a triangular region of hydroxide precipitation, which is shown in Fig. 9.2. Figure 9.2 shows the

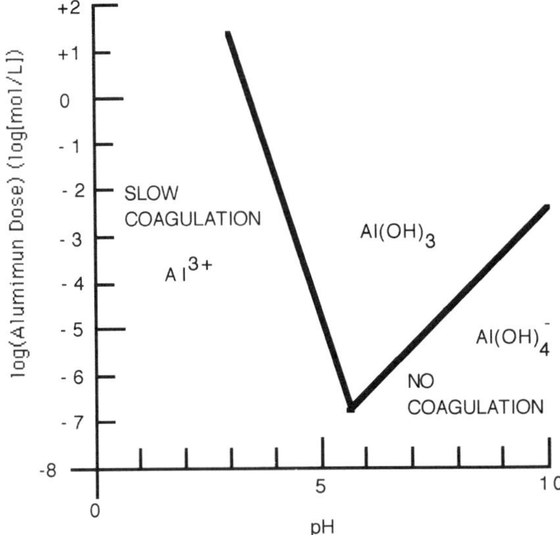

FIGURE 9.2 Aluminum hydroxide precipitation zone.

conditions for the precipitation of aluminum hydroxide itself. However, it is known that this triangular region also corresponds to the region in which silts and clays are coagulated [Dentel and Gossett, 1988; Hayden and Rubin, 1974]. The rectangular region in the figure is the usual range of pH levels and alum dosages seen in water treatment. It corresponds to Hazen's recommendations.

Other aluminum species not indicated in Fig. 9.2 are $AlOH^{2+}$ and $Al_8(OH)_{20}^{4+}$. These species are significant under acid conditions. The general relationship among the aluminum species may be represented as [Rubin and Kovac, 1974]:

$$Al^{3+} \leftrightarrow \left\{ \begin{array}{c} AlOH^{2+} \\ Al_8(OH)_{20}^{4+} \end{array} \right\} \leftrightarrow Al(OH)_3(s) \leftrightarrow Al(OH)_4^-$$

$$\updownarrow$$

$$AlOOH(s)$$

$$\updownarrow$$

$$Al_2O_3(s)$$

The aluminum ion, Al^{3+}, is the dominant species below pH 4.5. Between pH 4.5 and 5.0 the cations $AlOH^{2+}$ and $Al_8(OH)_{20}^{4+}$ are the principal species. Aluminate, $Al(OH)_4^-$, is the major species above pH 9.5 to 10. Between pH 5 and 10 various solids are formed. The fresh precipitate is aluminum hydroxide, but as it ages it gradually loses water, eventually becoming a mixture of bayerite and gibbsite. As the solid phase ages the equilibrium constants given in Eqs. (9.13) and (9.15) change. The values given are for freshly precipitated hydroxide.

The usual aluminum coagulation operating range intersects the restabilization zone, so coagulation difficulties are sometimes experienced. These can be overcome by the following methods [Rubin and Kovac, 1974]:

- Increasing the alum dosage to get out of the charge reversal zone; this is really a matter of increasing the sulfate concentration, which compresses the Gouy layer.
- Decreasing the alum dosage to get below the CSC; the floc is generally less settleable, and the primary removal mechanism is filtration, which is satisfactory as long as the total suspended solids concentration is low.

- Adding lime to raise the pH and move to the right of the restabilization zone.
- Adding polyelectrolytes to flocculate the positively charged colloids or coagulant aids such as bentonite or activated silica, which are negatively charged and reduce the net positive charge on the silt-clay-hydroxide particles by combining with them. Bentonite and activated silica also increase the density of the floc particles, which improves settling, and activated silica improves flow toughness.

The same problems arise with iron coagulants, and the same solutions may be employed.

Ferrous and Ferric Iron. Iron salts are seldom used in water treatment, but they are efficient coagulants nonetheless. The three forms usually encountered are ferric chloride [$FeCl_3 \cdot 6H_2O$], ferric sulfate [$Fe_2(SO_4)_3(H_2O)_9$], and ferrous sulfate [$FeSO_4(H_2O)_7$]. Anhydrous forms of the ferric salts are available.

Ferrous sulfate also is known in the trade as *copperas, green vitriol, sugar sulfate,* and *sugar of iron.* Ferrous sulfate occurs naturally as the ore copperas, but it is more commonly manufactured. Copperas can be made by oxidizing iron pyrites [FeS_2]. The oxidation yields a solution of copperas and sulfuric acid, and the acid is neutralized and itself converted to copperas by the addition of scrap iron or iron wire. However, the major source of copperas is waste pickle liquor. This is a solution of ferrous sulfate that is produced by soaking iron and steel in sulfuric acid to remove mill scale. Again, residual sulfuric acid in the waste liquor is neutralized by adding scrap iron or iron wire. The solutions are purified and evaporated, yielding pale green crystals. Although the heptahydrate [$FeSO_4(H_2O)_7$] is the usual product, salts with 0, 1, or 4 waters of hydration may also be obtained. Copperas is sold as lumps and granules. Because of its derivation from scrap steel, it should be checked for heavy metals.

Ferrous iron forms precipitates with both hydroxide and carbonate [Stumm and Morgan, 1970]:

$$Fe(OH)_2(s) \leftrightarrow Fe^{2+} + 2OH^- \tag{9.17}$$

$$K_{OH^-} = [Fe^{2+}] \cdot [OH^-] = 2 \times 10^{-15} \quad (25°C) \tag{9.18}$$

$$FeCO_3(s) \leftrightarrow Fe^{2+} + CO_3^{2-} \tag{9.19}$$

$$K_{CO_3^{2-}} = [Fe^{2+}] \cdot [CO_3^{2-}] = 2.1 \times 10^{-11} \quad (25°C) \tag{9.20}$$

In most waters the carbonate concentration is high enough to make ferrous carbonate the only solid species, if any forms.

The alkalinity of natural and used waters is usually composed entirely of bicarbonate, and copperas will not form a precipitate in them. This difficulty may be overcome by using a mixture of lime and copperas, the so-called lime-and-iron process. The purpose of the lime is to convert bicarbonate to carbonate so a precipitate may be formed:

$$Fe^{2+} + HCO_3^- + OH^- \rightarrow FeCO_3 + H_2O \tag{9.21}$$

If the raw water has a significant carbonate concentration, less lime will be needed because the original carbonate will react with some of the copperas. If the raw water has a significant carbonic acid concentration, additional lime will be required to neutralize it. In either case enough lime should be used to form an excess of carbonate and, perhaps, hydroxide, which forces the reaction to completion. Hazen's [1890] rule of thumb for raw water with carbonate (pH > 8.3) is as follows:

$$\{CaO\} = 1.16 \cdot \{FeSO_4(H_2O)_7\} - \{\text{phenolphthalein alkalinity}\} + 0.6 \tag{9.22}$$

For raw water without carbonate (pH < 8.3),

$$\{CaO\} = 1.16 \cdot \{FeSO_4(H_2O)_7\} + 0.50 \cdot \{H_2CO_3^*\} + 0.6 \tag{9.23}$$

The curly braces indicate that the concentration units are equivalents per liter. One mole of ferrous sulfate is two equivalents.

The lime-and-iron process produces waters with a pH around 9.5, which may be excessive and may require neutralization prior to distribution or discharge. Consequently, copperas is normally applied in combination with chlorine. The intention is to oxidize the ferrous iron to ferric iron, and the mixture of copperas and chlorine is referred to as *chlorinated copperas:*

$$FeSO_4(H_2O)_7 + \tfrac{1}{2}Cl_2 \rightarrow Fe^{3+} + Cl^- + SO_4^{2-} + 7H_2O \tag{9.24}$$

The indicated reaction ratio is about 7.84 g ferrous sulfate per g of chlorine, but dissolved oxygen in the raw water will also convert ferrous to ferric iron, and the practical reaction ratio is more like 7.3 g ferrous sulfate per g chlorine [Hardenbergh, 1940].

Ferric sulfate $[Fe_2(SO_4)_3 \cdot 9H_2O]$ is prepared by oxidizing copperas with nitric acid or hydrogen peroxide. Evaporation produces a yellow crystal. Besides the usual nonahydrate, ferric sulfates containing 0, 3, 6, 7, 10, and 12 waters of hydration may be obtained [Sidgwick, 1950].

Ferric chloride is made by mixing hydrochloric acid with iron wire, ferric carbonate, or ferric oxide. The solid is red-yellow in color. The usual product is $FeCl_3 \cdot 6H_2O$, but the anhydrous salt and salts with 2, 2.5, and 3.5 waters of hydration are also known [Hedgepeth, 1934; Sidgwick, 1950].

Both ferric salts produce ferric cations upon dissolution:

$$Fe_2(SO_4)_3(H_2O)_9 \rightarrow 2Fe^{3+} + 3SO_4^{2-} + 9H_2O \tag{9.25}$$

$$FeCl_3(H_2O)_6 \rightarrow Fe^{3+} + 3Cl^- + 6H_2O \tag{9.26}$$

The chemistry of the ferric cation is similar to that of the aluminum cation, except for the species occurring under acidic conditions [Rubin and Kovac, 1974]:

$$Fe^{3+} \leftrightarrow \left\{ \begin{array}{c} FeOH^{2+} \\ Fe(OH)_2^+ \\ Fe_2(OH)_2^{4+} \end{array} \right\} \leftrightarrow \begin{array}{c} Fe(OH)_3(s) \leftrightarrow Fe(OH)_4^- \\ \updownarrow \\ FeOOH(s) \\ \updownarrow \\ Fe_2O_3(s) \end{array}$$

In the case of iron the equilibrium involving $Fe(OH)_2^+$ must be considered, as well as Eqs. (9.27) and (9.29) [Stumm and Morgan, 1970]:

$$Fe^{3+} + 3H_2O \leftrightarrow Fe(OH)_3(s) + 3H^+ \tag{9.27}$$

$$K_{s0} = \frac{[H^+]^3}{[Fe^{3+}]} = 10^{-3.28} \qquad (25°C) \tag{9.28}$$

$$Fe(OH)_3(s) + OH^- \leftrightarrow Fe(OH)_4^- \tag{9.29}$$

$$K_{s1} = \frac{[Fe(OH)_4^-]}{[OH^-]} = 10^{-4.5} \qquad (25°C) \tag{9.30}$$

$$Fe(OH)_3(s) \leftrightarrow Fe(OH)_2^+ + OH^- \tag{9.31}$$

$$K_{s2} = \left[Fe(OH_2^+)\right] \cdot [OH^-] = 10^{-16.6} \qquad (25°C) \tag{9.32}$$

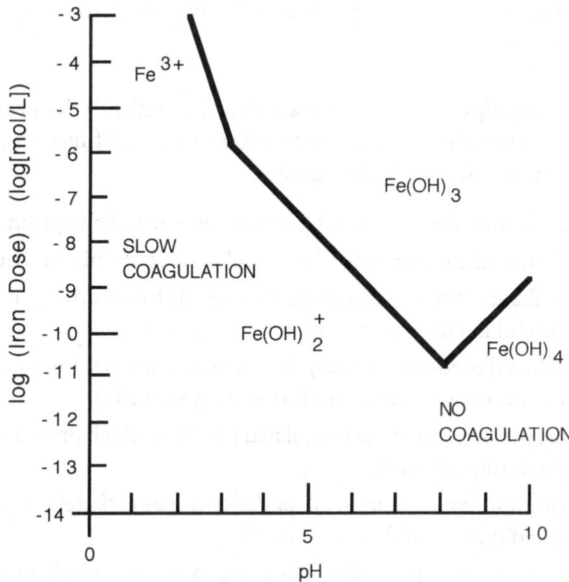

FIGURE 9.3 Ferric hydroxide precipitation zone.

The lines defined by Eqs. (9.28), (9.30), and (9.32) are plotted on log/log coordinates in Fig. 9.3. They define a polygon that indicates the region where ferric hydroxide precipitate may be expected.

Lime. Lime is usually purchased as *quicklime* [CaO] or *slaked lime* [$Ca(OH)_2$]. The former is available as lumps or granules; the latter is a white powder. Synonyms for quicklime are *burnt lime, chemical lime, unslaked lime,* and *calcium oxide.* If it is made by calcining limestone or lime/soda softening sludges, quicklime may contain substantial amounts of clay. The commercial purity is 75 to 99% by weight CaO. The synonyms for slaked lime are *hydrated lime* and *calcium hydroxide.* The commercial product generally contains 63 to 73% CaO.

Lime is used principally either to raise the pH to change the surface charge on the colloids and precipitates, or to provide the alkalinity needed by aluminum and iron.

The lime dosage required for the formation of aluminum and ferric hydroxide, in equivalents per liter, is simply the aluminum or ferric iron dosage minus the original total alkalinity:

$$\{CaO\} + \{Ca(OH)_2\} = \{Al^{3+}\} + \{Fe^{3+}\} - \{\text{total alkalinity}\} \tag{9.33}$$

All concentrations are in meq/L.

As a practical matter, both the coagulant and the lime dosages are determined experimentally by the jar test; the calculation suggested by Eq. (9.33) cannot be used to determine lime dosages, although it may serve as a check on the reasonableness of the jar test results.

The jar test is also used to determine the lime dosage needed to reduce the surface charge. The charge reduction is usually checked by an electrophoresis experiment to measure the zeta potential associated with the particles.

It is possible to coagulate many waters simply by adding lime. In a few instances, chiefly in anoxic groundwaters, the coagulation occurs because the raw water contains substantial amounts of ferrous iron, and one is actually employing the lime-and-iron process. Usually, the precipitate formed with lime is calcium carbonate, perhaps with some magnesium hydroxide. This is the lime/soda softening process.

Coagulant Aids. The principal coagulant aids are lime, bentonite, fuller's earth, activated silica, and various organic polymers. These aids are used to perform several functions, although any given aid will perform only one or a few of the functions:

- They may change the pH of the water, which alters the surface charge on many colloids.
- They may provide the alkalinity needed for coagulation by aluminum and ferric iron.
- They may reduce the net surface charge on the colloids by adsorbing to the colloid surface; this is especially useful in escaping the restabilization zone.
- They may link colloidal particles together, forming larger masses; in some cases the coagulant aid may totally replace the coagulant, but this is often expensive.
- They may increase the strength of the flocculated particles, which prevents floc fragmentation and breakthrough during filtration.
- They may increase the concentration of particles present, thereby increasing the rate of particle collision and the rate of flocculation.
- They may increase the density of the flocculated particles, which improves settling tank efficiencies.

Bentonite and fuller's earth are clays, and both are members of the montmorrillonite-smectite group. The clays are used during periods of low turbidity to increase the suspended solids concentration, the rate of particle collision, and the density of the floc particles. Because they carry negative charges when suspended in water, bentonite and fuller's earth may also be used to reduce surface charges in the restabilization zone. Clay dosages as high as 7.0 g/gal (120 mg/L) have been used [Babbitt *et al.*, 1967]. The required coagulant dosages are also increased by the added clay, and voluminous, fluffy flocs are produced, which, however, settle more rapidly than the flocs formed from aluminum hydroxide alone.

Activated silica is an amorphous precipitate of sodium silicate [Na_2SiO_3]. Sodium silicate itself is sold as a solution containing about 30% by weight SiO_2. This solution is very alkaline, having a pH of about 12. The precipitate is formed by diluting the commercial solution to about 1.5% by weight SiO_2 and reducing the alkalinity of the solution to about 1100 to 1200 mg/L (as $CaCO_3$) with sulfuric acid. Chlorine and sodium bicarbonate have also been used as acids. The precipitate is aged for 15 min to 2 hr and diluted again to about 0.6% by weight SiO_2. This second dilution stops the polymerization reactions within the precipitate. The usual application rate is 1:12 to 1:8 parts of silica to parts of aluminum hydroxide. The activated silica precipitate bonds strongly to the coagulated silts/clays/hydroxides and strengthens the flocs. This reduces floc fragmentation due to hydraulic shear in sand filters and limits floc "breakthrough" [Vaughn *et al.*, 1971; Kemmer, 1988].

The organic polymers used as coagulant aids may be classified as nonionic, anionic, or cationic [James M. Montgomery, Consulting Engineers, Inc., 1985; O'Melia, 1972; Kemmer, 1988]:

- *Nonionic.*
  - Polyacrylamide, $[\text{—}CH_2\text{—}CH(CONH_2)\text{—}]_n$, mol. wt. over 10^6.
  - Polyethylene oxide, $[\text{—}CH_2\text{—}CH_2\text{—}]_n$, mol. wt. over 10^6.
- *Anionic.*
  - Hydrolyzed polyacrylamide, $[\text{—}CH_2\text{—}CH(CONH_2)CH_2CH(CONa)\text{—}]$, mol. wt. over 10^6.
  - Polyacrylic acid, $[\text{—}CH_2\text{—}CH(COO^-)\text{—}]_n$, mol. wt. over 10^6.
  - Polystyrene sulfonate, $[\text{—}CH_2\text{—}CH(\varnothing SO_3^-)\text{—}]_n$, mol. wt. over 10^6.
- *Cationic.*

- Polydiallyldimethylammonium, mol. wt. below 10^5.

```
 [    H    H         H    H     ]
 [    |    |         |    |     ]
 [  —C—  C———— C—  C—           ]
 [    |    |         |    |     ]
 [    H    C         C    H     ]
 [       /   \     /   \        ]
 [     H       N        H       ]
 [           / + \              ]
 [         C       C            ]
 [       /           \          ]
 [     H3              H3       ]n
```

- Polyamines, [—CH_2—CH_2—NH_2—]$_n$, mol. wt. below 10^5.
- Quarternized polyamines, [—CH_2—CH(OH)—CH_2—$N(CH_3)_2$—]$_n$, mol. wt. below 10^5.

These materials are available from several manufacturers under a variety of trade names. The products are subject to regulation by the U.S. EPA. They are sold as powders, emulsions, and solutions.

All polymers function by adsorbing to the surface of colloids and metal hydroxides. The bonding may be purely electrostatic, but hydrogen bonding and van der Waals bonding occur, too, and may overcome electrostatic repulsion.

Anionic polymers will bind to silts and clays despite the electrostatic repulsion if their molecular weight is high enough. The bonding is often very specific, and some polymers will not bind to some colloids.

The charges on cationic and anionic polymers are due to the ionization and protonation of amino [—NH_2], carboxyl [—COOH], and amide [—$CONH_2$] groups and are pH dependent. Consequently, cationic polymers are somewhat more effective at low pHs, and anionic polymers are somewhat more effective at high pHs.

Cationic polymers both reduce the surface charge on silts and clays and form interparticle bridges, which literally tie the particles together. Cationic polymers are sometimes used as the sole coagulant. Anionic and nonionic polymers generally function by forming interparticle bridges. Anionic and nonionic polymers are almost always used in combination with a primary coagulant. Dosages are generally on the order of one to several mg/L and are determined by jar testing.

Coagulant Choice

The choice of the coagulants to be employed and their dosages is determined by their relative costs and the jar test results. The rule is to choose the least-cost combination that produces satisfactory coagulation, flocculation, settling, and filtration. This rule should be understood to include the minimization of treatment chemical leakage through the plant and into the distribution system. The important point here is that the choice is an empirical matter, and as such it is subject to change as the raw water composition changes and as relative costs change.

Nevertheless, there are some differences between filter alum and iron salts that appear to have general applicability:

- Iron salts react more quickly to produce hydroxide precipitate than does alum, and the precipitate is tougher and settles more quickly [Babbitt *et al.*, 1967].
- Ferric iron precipitates over a wider range of pHs than does alum, 5.0 to 11.0 versus 5.5 to 8.0. Ferrous iron precipitates between pH 8.5 and 11.0 [Committee on Water Works Practice, 1940].
- Alum sludges dewater with difficulty, especially on vacuum filters: The floc is weak and breaks down, so solids are not captured; the sludge is slimy, requiring frequent shutdowns for fabric

cleaning; and the volume of sludge requiring processing is larger than with iron salts [Rudolfs, 1940; Joint Committee, 1959].

- Iron salts precipitate more completely, and the iron carryover into the distribution system is less than the aluminum carryover. The median iron concentration in surface waters coagulated with iron salts is about 80 μg/L, and the highest reported iron concentration is 0.41 mg/L [Miller *et al.*, 1984]. The median aluminum concentration in finished waters treated with alum is about 90 to 110 μg/L, and nearly 10% of all treatment plants report aluminum concentrations in their product in excess of 1 mg/L [Letterman and Driscoll, 1988; Miller *et al.*, 1984]. Aluminum carryover is a problem because of indications that aluminum may be involved in some brain and bone disorders in humans, including Alzheimer's disease [Alfrey *et al.*, 1976; Crapper *et al.*, 1973; Davison *et al.*, 1982; Kopeloff *et al.*, 1942; Klatzo *et al.*, 1965; Platts *et al.*, 1977]. The aluminum may be present as Al^{3+}, and its concentration in finished waters appears to increase with increases in fluoride (caused by fluoridation) and dissolved organic matter, both of which form soluble complexes with Al^{3+} [Driscoll and Letterman, 1988].
- Alum is easy to handle and store, but iron salts are difficult to handle. Iron salts are corrosive, and they absorb atmospheric moisture, caking. This precludes feeding the dry compound [Rudolfs, 1940; Babbitt *et al.*, 1967].
- There is no significant cost advantage accruing to either alum or iron salts.

Iron salts are often recommended for the coagulation of cold, low-turbidity waters. However, a recent study suggests there is no substantial advantage for iron coagulation under these conditions [Haarhoff and Cleasby, 1988]. There is an optimum aluminum dosage of about 0.06 mmol/L (1.6 mg Al^{3+}/L), and dosages above this degrade settling and removal. Iron does not exhibit such an optimum.

It would appear from these comparisons that iron would be the coagulant of choice. However, in the U.S., filter alum is nearly always the primary or sole coagulant in water treatment. On the other hand, iron salts predominate in sewage treatment: Of 107 sewage treatment plants using chemical precipitation in operation in the U.S. in 1938, only eight used alum, and two of those used it in conjunction with iron salts [Donaldson, 1938]. Furthermore, all of the nine chemical precipitation plants then under construction were designed to use either ferric chloride or copperas.

The reasons for this contrast include historical accident, economics, and performance. The preference for filter alum in water treatment appears to be due to historical accident and the way water treatment sludges are handled—or, rather, not handled. Alum is the easiest coagulant to come by; its use goes back to antiquity; and well-documented studies of alum coagulation are found throughout the eighteenth and nineteenth centuries [Baker, 1948]. When mechanical filters (also known as *rapid sand filters*) were first being developed in the nineteenth century, their inventors adopted alum as a coagulant almost without reflection, and the choice stuck. The choice was reinforced by the fact that the principal available competitor for alum was copperas, which required high lime dosages to be effective and was relatively difficult to handle [Committee on Water Works Practice, 1940]. Chlorinated copperas was not then known. On the other hand, Hazen [1890] concluded in an influential report that ferric salts were best for coagulating sewage.

Sludge handling practices are also important. Water treatment sludges usually do not require processing prior to disposal, so the defects of alum sludges are not always apparent. This situation arises because water treatment sludges consist of relatively inert inorganic materials, and in the past (but no longer) plants were often permitted to dispose of their sludges by discharge to the nearest stream. Nowadays the sludges are simply lagooned or placed in landfills. In contrast, sewage treatment sludges are putrescible and require processing prior to disposal. This usually involves a dewatering step, and the comparative ease and economy of dewatering iron sludges is decisive. Ferric iron also precipitates the sulfides found in sewage sludges, reducing their nuisance potential.

There are important disadvantages to ferric iron. First, ferric iron solutions are acidic and corrosive and require special materials of construction and operational practices. Second, any carryover of ferric hydroxide into a water distribution system is immediately obvious and undesirable. Alum hydroxide carryover would not be noticed.

9.2 Softening, Stabilization, and Demineralization

Hardness

The natural weathering of limestone, dolomite, and gypsum produces waters that contain elevated levels of calcium and magnesium (and bicarbonate):

$$CO_2 + H_2O + CaCO_3 \rightarrow Ca^{2+} + 2HCO_3^- \tag{9.34}$$

$$2CO_2 + 2H_2O + CaMg(CO_3)_2 \rightarrow Ca^{2+} + Mg^{2+} + 4HCO_3^- \tag{9.35}$$

$$CaSO_4(H_2O)_2 \rightarrow Ca^{2+} + SO_4^{2-} + 2H_2O \tag{9.36}$$

In the case of limestone and dolomite, weathering is an acid/base reaction with the carbon dioxide dissolved in the percolating waters. In the case of gypsum it is a simple dissolution that occurs whenever the percolating water is unsaturated with respect to calcium sulfate.

Waters that contain substantial amounts of calcium and magnesium are called "hard." Waters that contain substantial amounts of bicarbonate are called "alkaline." Hard waters are usually also alkaline. For reasons connected to Clark's lime/soda softening process, the "carbonate hardness" is defined as that portion of the calcium and magnesium that is equal to (or less than) the sum of the concentrations of bicarbonate and carbonate, expressed in meq/L. The "noncarbonate hardness" is defined as the excess of calcium and magnesium over the sum of the concentrations of bicarbonate and carbonate, expressed in meq/L. Except for desert evaporite ponds, the concentrations of carbonate and hydroxide are negligible in natural waters.

Hardness is undesirable for two reasons:

- Hard waters lay down calcium and magnesium carbonate on hot surfaces, which reduces the heat transfer capacity of boilers and heaters and the hydraulic capacity of water and steam lines.
- Hard waters precipitate natural soaps; the precipitation consumes soaps uselessly, which increases cleaning costs, and the precipitate itself accumulates on surfaces and in fabrics, which requires additional cleaning and which reduces the useful life of fabrics.

It is generally believed that these costs become high enough to warrant municipal water softening when the water hardness exceeds about 100 mg/L (as $CaCO_3$).

Modern steam boilers require feedwaters that have mineral contents much lower than what can be achieved via lime/soda softening. Feedwater demineralization is usually accomplished via ion exchange or reverse osmosis.

Some scale deposition is desirable in water distribution systems in order to minimize lead and copper solubility.

Lime/Soda Chemistry

The excess lime process for the removal of carbonate hardness and the practice of reporting hardness in units of calcium carbonate were introduced by Thomas Clark in 1841 and 1856, respectively [Baker, 1981]. The removal of noncarbonate hardness via the addition of soda ash or potash was introduced by A. Ashby in 1876.

The underlying principle is that calcium carbonate and magnesium hydroxide are relatively insoluble. Magnesium hydroxide is more insoluble than magnesium carbonate. The solubility products for dilute solutions of calcium carbonate are [Shock, 1984]:

$$K_{sp} = [Ca^{2+}] \cdot [CO_3^{2-}] \tag{9.37}$$

Calcite:

$$\log K_{sp} = -171.9065 - 0.077993T + \frac{2839.319}{T} + 71.595 \log T \tag{9.38}$$

Aragonite:

$$\log K_{sp} = -171.9773 - 0.077993T + \frac{2903.293}{T} + 71.595 \log T \tag{9.39}$$

Vaterite:

$$\log K_{sp} = -172.1295 - 0.077993T + \frac{3074.688}{T} + 71.595 \log T \tag{9.40}$$

where T is the absolute temperature in K. For the magnesium solutions one has [Stumm and Morgan, 1970]

$$\begin{aligned} K_{sp} &= [Mg^{2+}] \cdot [OH^-]^2 \\ &= 10^{-9.2} \text{ (active, 25°C)} \\ &= 10^{-11.6} \text{ (brucite, 25°C)} \end{aligned} \tag{9.41}$$

$$\begin{aligned} K_{sp} &= [Mg^{2+}] \cdot [CO_3^{2-}] \\ &= 10^{-4.9} \text{ (magnesite, 25°C)} \\ &= 10^{-5.4} \text{ (nesquehonite, 25°C)} \end{aligned} \tag{9.42}$$

The solubility product for calcite varies nearly linearly with temperature from $10^{-8.09}$ at 5°C to $10^{-8.51}$ at 40°C. This is the basis of the "hot lime" process [Powell, 1954].

The reactions involved can be summarized as follows. First a slurry of calcium hydroxide is prepared, either by slaking quicklime or by adding slaked lime to water. When this slurry is mixed with hard water, the following reactions occur in sequence:

1. Reaction with carbon dioxide and carbonic acid:

$$Ca(OH)_2 + H_2CO_3 \rightarrow CaCO_3 + H_2O \tag{9.43}$$

2. Reaction with bicarbonate:

$$Ca(OH)_2 + 2HCO_3^- \rightarrow CaCO_3 + CO_3^{2-} + H_2O \tag{9.44}$$

3. Reaction with raw water calcium:

$$Ca^{2+} + CO_3^{2-} \rightarrow CaCO_2 \tag{9.45}$$

4. Reaction with magnesium:

$$Ca(OH)_2 + Mg^{2+} \rightarrow Mg(OH)_2 + Ca^{2+} \tag{9.46}$$

5. Reaction with soda ash:

$$Na_2CO_3 + Ca^{2+} \rightarrow CaCO_3 + 2Na^+ \tag{9.47}$$

The reaction with carbon dioxide is a nuisance because it consumes lime and produces sludge but does not result in any hardness removal. Carbon dioxide concentrations are generally negligible in surface waters but may be substantial in groundwaters. In that case it may be economical to remove the carbon dioxide by aeration prior to softening.

Equations (9.44) and (9.45) are the heart of the Clark process. First, hydrated lime reacts with bicarbonate to form an equivalent amount of free carbonate. The calcium in the lime precipitates out as calcium carbonate, and the free carbonate formed reacts with an equivalent amount of the raw water's original calcium, removing it as calcium carbonate, too. The net result is a reduction in the calcium concentration.

Equation (9.46) is called the "excess lime" reaction because a substantial concentration of free, unreacted hydroxide is required to drive the precipitation of magnesium hydroxide. The net result is the replacement of magnesium ions by calcium ions. If there is any free carbonate left over from Eq. (9.45), it will react with an equivalent amount of the calcium removing it.

Equation (9.47) is Ashby's process for the removal of noncarbonate hardness. Any calcium left over from Eq. (9.45) is precipitated with either soda ash or potash.

Calcium Removal Only

Unless the magnesium concentration is a substantial portion of the total hardness, say, more than one-third, or if the total hardness is very high, say, more than 300 mg/L (as $CaCO_3$), only calcium is removed. The traditional rule of thumb is that cold water softening can reduce the calcium concentration to about 0.8 meq/L (40 mg/L as $CaCO_3$) [Tebbutt, 1992]. The magnesium concentration is unchanged.

The reactions are most easily summarized as a bar chart. First, all the ionic concentrations and the concentration of carbon dioxide/carbonic acid are converted to meq/L. Carbon dioxide/carbonic acid acts like a diprotic acid in Eq. (9.43), so its equivalent weight is one-half its molecular weight. The bar chart is drawn as two rows with the cations on top and the anions on the bottom. Carbon dioxide/carbonic acid is placed in a separate box to the left. The sequence of cations from left to right is calcium, magnesium, and all others. The sequence of anions from left to right is carbonate, bicarbonate, and all others:

<table>
<tr><td rowspan="2">CO_2
+
H_2CO_3</td><td colspan="3">Ca^{2+}</td><td>Mg^{2+}</td><td>other cations</td></tr>
<tr><td>CO_3^{2-}</td><td>HCO_3^-</td><td colspan="3">other anions</td></tr>
</table>

The lime requirement for calcium removal is the sum of the carbon dioxide/carbonic acid demand and the lime required to convert bicarbonate to carbonate. If the calcium concentration exceeds the carbonate and bicarbonate concentrations combined (as shown), then all the bicarbonate is converted. However, if the sum of carbonate and bicarbonate is greater than the calcium concentration, only enough bicarbonate is converted to remove the calcium. The calculation is

$$\{CaO\} = \{CO_2 + H_2CO_3\} + \min\left[\{\text{original } Ca^{2+} - CO_3^{2-}\} \text{ or } \{HCO_3^-\}\right] \quad (9.48)$$

where $\{x\}$ is the concentration of species x in meq/L.

The soda ash requirement is calculated as the calcium that cannot be removed by the original carbonate plus the bicarbonate converted to carbonate:

$$\{Na_2CO_3\} = \{\text{original } Ca^{2+}\} - \{CO_3^{2-}\} - \{HCO_3^-\} \quad (9.49)$$

For the bar diagram shown the soda ash requirement is zero.

The sludge solids produced consist of calcium carbonate. The concentration in suspension just prior to settling is

$$X_{ss} = 50.04\left[\{CaO\} + \{\text{original } Ca^{2+}\}\right] \quad (9.50)$$

where X_{ss} is the concentration of suspended solids in mg/L. If the raw water contains any suspended solids, these will be trapped in the precipitate, and they must be included.

Calcium and Magnesium Removal

If magnesium must be removed, excess lime treatment is required. The traditional rule of thumb is that cold water excess lime softening will reduce the calcium concentration to about 0.8 meq/L and the magnesium concentration to about 0.2 meq/L.

The bar chart relevant to excess lime treatment would look as follows:

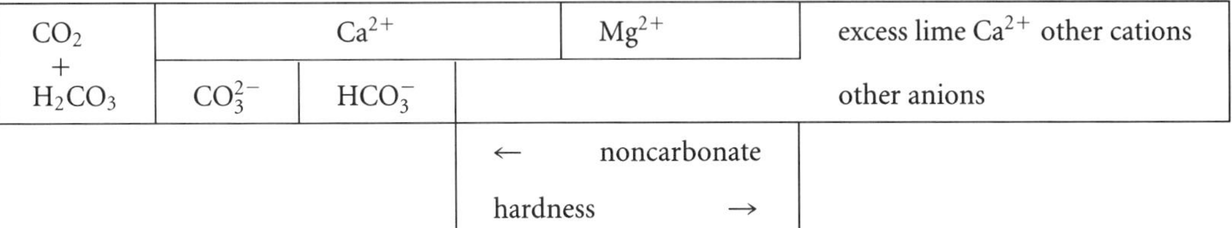

The required lime dosage is

$$\{CaO\} = \{CO_2 + H_2CO_3\} + \{HCO_3^-\} + \{Mg^{2+}\} + \{\text{excess lime}\} \tag{9.51}$$

where $\{x\}$ is the concentration of species x in meq/L. Note that all the carbon dioxide/carbonic acid, all the bicarbonate, and all the magnesium must be reacted. The quantity of excess lime influences the concentration of magnesium that can be achieved and the rate of reaction. In cold water a free hydroxide concentration of about 1 meq/L is required to reduce the magnesium concentration to 50 mg/L (as $CaCO_3$), and about 1.4 meq/L is required to produce a magnesium concentration of 8 mg/L (as $CaCO_3$) [Powell, 1954]. The equivalent pHs are 11 and 11.2, respectively.

The soda ash requirement is

$$\begin{aligned}\{Na_2CO_3\} &= \{Ca^{2+}\} + \{Mg^{2+}\} + \{\text{excess lime Ca}\} - \{CO_3^{2-}\} - \{HCO_3^-\}\\ &= \{\text{noncarbonate hardness}\} + \{\text{excess lime Ca}\}\end{aligned} \tag{9.52}$$

The solids formed consist of calcium carbonate and magnesium hydroxide and any silts and clays in the raw water. The total suspended solids concentration in mg/L prior to settling is

$$X_{ss} = 50.04\left[\{CaO\} + \left\{\text{original } Ca^{2+}\right\}\right] + 29.16\{\text{original } Mg^{2+}\} \tag{9.53}$$

Any silts and clays in the raw water will be trapped in the precipitates and should be added to get the total suspended solids.

Recarbonation

When only calcium is removed, the settled water has a pH of between 10 and 10.6, is supersaturated with respect to calcium carbonate, and contains suspended calcium carbonate crystals that did not settle out. The traditional rule of thumb is that the residual calcium hardness is about 0.8 meq/L and the magnesium hardness is unchanged. In the case of excess lime treatment, the settled water is supersaturated with respect to both calcium carbonate and magnesium hydroxide and contains suspended particles of both crystals. The final pH is between 11 and 11.5. Such waters will deposit hard scales in sand filters and distribution systems and corrode lead and copper. The usual practice is to convert carbonate and hydroxide to bicarbonate by reaction with carbon dioxide gas.

Consider the following bar chart, which represents excess lime-softened water after sedimentation:

Ca^{2+}	Mg^{2+}	other cations	
CO_3^{2-}	OH^-	excess lime OH^-	other anions

It is convenient to use concentration units of mmol/L because the carbonate and carbon dioxide are both converted to bicarbonate and so behave as a monoprotic base and acid. The carbon dioxide requirement is in mmol/L:

$$[CO_2] = [CO_3^{2-}] + [OH^-] + [\text{excess lime } OH^-] \tag{9.54}$$

In the case of excess lime softening, the carbonate and hydroxide concentrations associated with calcium and magnesium would be about 0.4 and 0.2 mmol/L, respectively, and the excess lime hydroxide (which is associated with other cations) would be about 1 to 1.5 mmol/L. Therefore, the expected carbon dioxide dosage is about 1.6 to 2.1 mmol/L. In the case of calcium removal only there is no hydroxide residual, so the expected carbon dioxide dosage would be about 0.4 mmol/L.

In either case the final water pH will be 8.3 or a little higher. Whether or not such waters will deposit scale can be estimated by using the Larson–Buswell [1942] correction to Langelier's [Langelier, 1936, 1946; Hoover, 1938] equilibrium pH_{eq}:

$$pH_{eq} = -\log\left\{\frac{K_2(Ca^{2+})(\text{Alkalinity})}{K_{sp}}\right\} + 9.30 + \frac{2.5\sqrt{\mu}}{1 + 5.3\sqrt{\mu} + 5.5\mu} \tag{9.55}$$

or the formula derived by Singlely *et al.* [1985]:

$$pH_{eq} = -\log\left\{\frac{K_2(Ca^{2+})(\text{Alkalinity})}{K_{sp}}\right\} + \frac{2.5\sqrt{\mu} + 3.63\mu}{1 + 3.30\sqrt{\mu} + 2.61\mu} \tag{9.56}$$

Langelier's saturation index, I_{sat}, is calculated as

$$I_{sat} = pH_{act} - pH_{eq} \tag{9.57}$$

where

I_{sat} = Langelier's saturation index
pH_{act} = the measured pH of the water
pH_{eq} = the equilibrium pH at which a water neither deposits nor dissolves calcium carbonate
K_2 = the second acid dissociation constant for carbonic acid
K_{sp} = the solubility product of calcium carbonate
(Ca^{2+}) = the calcium ion concentration in mg/L as $CaCO_3$
$[Ca^{2+}]$ = the calcium ion concentration in mol/L
(Alkalinity) = the total alkalinity in mg/L as $CaCO_3$
[Alkalinity] = the total alkalinity in eq/L
μ = the ionic strength in mol/L
$\approx$ Total dissolved solids (mg/L)/40,000

The values of the calcium carbonate solubility product were given above. The second acid ionization constant of carbonic acid can be estimated by [Shock, 1984]:

$$\log K_2 = -107.8871 - 0.03252849T + \frac{5151.79}{T} + 38.92561\log T - \frac{563713.9}{T^2} \tag{9.58}$$

where T is the absolute temperature in K.

A water is stable if its measured pH is equal to pH_{eq}. It will deposit calcium carbonate scale if its measured pH is greater than pH_{eq}, and it will dissolve calcium carbonate scale if its measured pH is less than pH_{eq}.

More information is provided by Ryznar's [1944] stability index, I_{stab}:

$$I_{stab} = 2pH_{eq} - pH_{act} \tag{9.59}$$

A stability index between about 6.3 and 6.8 will result in virtually no calcium carbonate scale or iron corrosion. Indices below 6.0 result in scale deposition from hot water, and the deposition is heavy below an index of 5.0. Indices above 7.0 produce iron corrosion, and indices above 8.0 indicate very corrosive water.

Lead and Copper Control

The action level for lead is 0.015 mg/L, and the maximum contaminant level goal (MCLG) for lead is zero. The action level for copper is 1.3 mg/L, and the MCLG for copper is also 1.3 mg/L. If 10% of the taps sampled exceed either of the action levels, the water system must do one or more of the following [EPA, 1991]:

- Install corrosion control for the distribution system
- Treat the raw water to remove lead and copper
- Replace lead service lines
- Conduct public education to help the public reduce their exposure to lead

The general corrosion reaction is an oxidation in which the metallic element becomes an ion which may react with other solution ions:

$$Me \rightarrow Me^{n+} + ne^- \tag{9.60}$$

Corrosion can be discussed under three heads [Schock, 1990]:

- *Immunity.* Immunity to corrosion means that the water chemistry is such that the metal is thermodynamically stable and will not corrode and go into solution. However, lead metal in contact with water is unstable and will corrode. Immunity generally requires some kind of cathodic protection, and the protected metal survives only as long as the sacrificial metal exists. In potable waters, zinc, aluminum, steel, and iron provide cathodic protection to lead, lead-tin solders, and copper.
- *Passivation.* Metal surfaces may be passivated by the deposition of a stable, nonporous film that protects the thermodynamically unstable metal from corrosion. In waters containing carbonate ion, the lead surface may be coated with a layer of plumbous carbonate [cerrusite, $PbCO_3$), hydroxyplumbous carbonate [hydrocerrusite, $Pb(CO_3)_2(OH)_2$] or plumboacrite [$Pb_{10}(CO_3)_6(OH)_6O$] [Schock, 1990]. The minimum equilibrium lead concentration is 0.069 mg/L, and it occurs at pH 9.8 and a dissolved carbonate concentration of 4.8 mg/L as C [Schock, 1990]. The equilibrium lead concentrations for pHs between 8.0 and 9.0 and carbonate concentrations of less than 10 mg/L are less than 0.20 mg Pb/L. At pH 9.8, lead concentrations increase as carbonate concentrations increase or decrease. At a carbonate concentration of 4.8 mg C/L, lead concentrations increase as the pH increases or decreases.

  All of these lead concentrations are above the action level of 0.015 mg/L, but water in distribution systems is seldom in equilibrium with the piping and appurtenances, and measured lead concentrations may be less than the equilibrium values.

  Lead surfaces can also be passivated by zinc orthophosphate. Typical dosages are 0.4 to 0.6 mg/L as PO_4^{3-} [Lee *et al.*, 1989]. Theoretically, an equilibrium lead concentration of 0.01 mg/L can be achieved if the pH is adjusted to 7.6, the phosphate dosage is 4.5 mg/L as PO_4^{3-}, and the bicarbonate concentration is 5 to 10 mg/L as C [Schock, 1989]. Zinc concentrations should not exceed 5 mg/L (for aesthetics), and this may limit the usefulness of zinc orthophosphate.

Polyphosphate solubilizes lead, but it is slowly hydrolyzed to orthophosphate.

In low alkalinity waters, sodium silicate at concentrations above 20 mg/L as SiO_2 and pHs around 8.2 slowly forms protective coatings on lead surfaces.

- *Protection.* Protection occurs when a layer of material is deposited that reduces the diffusion of material to and from the metal surface. The most common protective layer is calcium carbonate scale. This would require recarbonation to stop at a positive Langelier saturation index or a Ryznar stability index less than 6.

It should be noted that corrosion is not inhibited at Langelier indexes as high as 2. The corrosion rate is influenced more strongly by dissolved oxygen and total dissolved solids than by water stability. For mild steel the empirical formula for the rate of penetration is [Singley *et al.*, 1985]

$$R_{pen} = \frac{C_{TDS}^{0.253} \cdot C_{DO}^{0.820}}{10^{0.0876 I_{sat}} \cdot t^{0.373}} \tag{9.61}$$

where

C_{DO} = the dissolved oxygen concentration in mg/L
C_{TDS} = the total dissolved solids concentration in mg/L
I_{sat} = Langelier's saturation index (dimensionless)
R_{pen} = the mild steel corrosion penetration rate in mils per year
t = the exposure time in days

Ion Exchange

Materials

An ion exchanger is a solid that absorbs certain dissolved ions from solution and replaces them with other ions. Some materials exchange cations, and others exchange anions. The exchange maintains the electroneutrality of both the exchanger and the solution, so the number of equivalents of cations or anions released is equal to the number of equivalents of cations or anions absorbed.

The property of ion exchange is widespread among many kinds of naturally occurring and synthetic solid materials. However, the most useful ion exchange materials are synthetic organic polymers consisting of polystyrene chains crosslinked with divinylbenzene and various attached functional groups. The functional groups may be divided into four classes. Letting R represent the resin lattice, the groups are [Abrams and Benezra, 1967]:

- *Strong acid cation* (SAC) exchangers that contain the sulfonate ($R{-}SO_3^-$) group. These exchangers operate over a wide pH range and can remove all cations from solution (replacing them with protons) or all doubly and triply charged cations (replacing them with sodium).
- *Weak acid cation* (WAC) exchangers that contain the carboxylate ($R{-}COO^-$) group. These exchangers operate in neutral to alkaline pHs where the carboxylate group is ionized and can remove doubly charged and triply charged ions if carbonate or bicarbonate is present in sufficient quantities to neutralize the protons released.
- *Strong base anion* (SBA) exchangers that contain the quarternary amine [$R{-}N(CH_3)_3^+$] group. These exchangers can operate over a wide range of pHs and can remove all anions, replacing them with hydroxide or chloride ions.
- *Weak base anion* (WBA) exchangers that contain a primary, secondary, or tertiary amine ($R{-}NH_3^+$, $R{=}NH_2^+$, $R{\equiv}NH^+$) group. These exchangers operate in the acidic range where the amino group is protonated.

Most commercial resins are homogeneous gels [Abrams and Benezra, 1967]. The solid consists of an open crystalline matrix that adsorbs water and permits the free diffusion of ions. The charge of the functional groups is balanced by the diffusing ions. The degree of cross-linking influences the ionic diffusion rate, and large ions may be excluded from the gel.

Macroporous, highly cross-linked resins are available [Abrams and Benezra, 1967]. In these materials, ion exchange is limited to the interior surfaces of the macropores. However, ionic diffusion in the macropores is rapid. Macroporous resins are largely limited to nonaqueous processes.

Equilibria

The exchange process can be described as a two-phase equilibrium involving ion X with charge n and ion Y with charge m:

$$\text{R—}X^n_m + nY^m \leftrightarrow \text{R—}Y^m_n + mX^n \tag{9.62}$$

$$K = \frac{[\text{R—}Y^m_n]^n [X^n]^m}{[\text{R—}X^n_m]^m [Y^m]^n} \tag{9.63}$$

Note that the terms $[\text{R—}Y^n_m]$ and $[\text{R—}Y^m_n]$ refer to the activities of ions in the solid phase, while the terms $[X^n]$ and $[Y^m]$ refer to the ionic activities in water. In practice, concentrations rather than activities are used and the equilibrium constant is called a *selectivity coefficient.* The selectivity coefficient varies with the ionic strength of the solution. Concentrations of ions in the exchanger are generally reported as a mass-to-mass ratio, whereas concentrations in water are reported as a mass-to-volume ratio.

Sometimes another constant called the *separation factor* is used:

$$\alpha = \frac{[\text{R—}Y^m_n][X^n]}{[\text{R—}X^n_m][Y^m]} = K \cdot \frac{[\text{R—}X^n_m]^{m-1}[Y^m]^{n-1}}{[\text{R—}Y^m_n]^{n-1}[X^n]^{m-1}} \tag{9.64}$$

Note that the separation factor is not an equilibrium constant and will vary significantly with water composition. Some separation factors relative to sodium are given in Table 9.1.

Table 9.1 Separation Factors Relative to Sodium and Chloride for Various Ions

Strong Acid Cation Exchangers		Strong Base Anion Exchangers	
Cation	α	Anion	α
Ra^{2+}	13.0	CrO_4^{2-}	100.
Ba^{2+}	5.8	SeO_4^{2-}	17.0
Pb^2	5.0	SO_4^{2-}	9.1
Sr^2	4.8	HSO_4^-	4.1
Cu^{2+}	2.6	NO_3^-	3.2
Ca^{2+}	1.9	Br^-	2.3
Zn^{2+}	1.8	$HAsO_4^{2-}$	1.5
Fe^{2+}	1.7	SeO_3^{2-}	1.3
Mg^{2+}	1.67	HSO_3^{3-}	1.2
K^+	1.67	NO_2^-	1.1
Mn^2	1.6	Cl^-	1.0
NH_4^+	1.3	HCO_3^-	.27
Na^+	1.0	CH_3COO^-	.14
H^+	0.67	F^-	.07

Source: Clifford, D. A. 1990. Ion exchange and inorganic adsorption. In *Water Quality and Treatment: A Handbook of Community Water Supplies,* 4th ed., p. 561. F. W. Pontius, ed. McGraw-Hill, New York.

Note: N = 0.01; TDS = 500 mg/L as $CaCO_3$.

In general, ion exchangers preferentially adsorb more highly charged ions over less highly charged ions, and smaller ions over larger ions. The general preference sequence for cation exchangers is [Kemmer, 1988]

$$Fe^{3+} > Al^{3+} > Pb^{2+} > Ba^{2+} > Sr^{2+} > Cd^{2+}$$

$$> Zn^{2+} > Cu^{2+} > Fe^{2+} > Mn^{2+} > Ca^{2+} > Mg^{2+}$$

$$> K^{+} > NH_4^{+} > Na^{+} > H^{+} > Li^{+}$$

In most natural waters, ferric iron and aluminum form precipitates, which should be removed prior to the ion exchange bed to prevent clogging. Ferrous iron may be oxidized by dissolved oxygen after exchange and precipitate in or on the resin beads. This can be prevented by applying a reductant such as sodium sulfite to the raw water or mixed with the regenerant.

For anion exchangers the preference sequence is

$$CrO_4^{2-} > SO_4^{2-} > SO_3^{2-} > HPO_4^{2-} > CNS^{-} > CNO^{-} > NO_3^{-}$$

$$> NO_2^{-} > Br^{-} > Cl^{-} > CN^{-}HCO_3^{-} > HSiO_3^{-} > OH^{-} > F^{-}$$

These sequences are affected by both the ionic strength of the solution and the chemical composition of the ion exchanger.

Operating parameters and important resin properties are summarized in Table 9.2. Note that the operating exchange capacity varies with the concentration of the regenerant solution. This is a consequence of the equilibrium nature of the process.

Sodium Cycle Softening

Health and Ecology Notes

The sodium concentration in the finished water is equal to the original hard water sodium concentration plus the sodium required to replace the calcium and magnesium hardness removed:

$$C_{Naf} = C_{Nao} + \frac{23}{50} C_{HCao} + \frac{23}{50} C_{HMgo} \tag{9.65}$$

where

C_{HCao} = the original calcium hardness [eq/m^3 or gr (as $CaCO_3$)/ft^3]
C_{HMgo} = the original magnesium hardness [eq/m^3 or gr (as $CaCO_3$)/ft^3]
C_{Naf} = the final sodium concentration [eq/m^3 or gr (as $CaCO_3$)/ft^3]
C_{Nao} = the original sodium concentration [eq/m^3 or gr (as $CaCO_3$)/ft^3]

If a very hard water is softened the resulting sodium concentration will be high, and drinking water may compose a significant fraction of the dietary sodium intake. This may be of concern for people on restricted sodium diets.

A zero-hardness water is corrosive because of the lack of scale-forming calcium ions. Such waters are not suitable for household use without further treatment because lead and copper will be dissolved from plumbing fixtures. Zinc orthophosphate additions and pH adjustment may be required. Partially softened waters may be acceptable if the final pH and carbonate concentration are carefully adjusted.

Sodium cycle softening produces a waste brine that may adversely affect fresh water biota. Many states have restrictions on the increase in total dissolved solids concentration that they will permit in receiving waters. Common restrictions are an increase in TDS of 100 mg/L and an absolute upper limit of 750 mg/L.

Table 9.2 Ion Exchange Resin Properties[1–4]

Parameter	Strong Acid Cation Exchanger	Strong Base Anion Exchanger
Effective size (mm)	0.45 to 0.55	0.45 to 0.55
Uniformity coefficient	1.7	1.7
Specific gravity of wet grains (dimensionless)	$\leq$1.3	$\geq$1.07
Moisture content of wet grains (%)	43 to 45	43 to 49
Iron tolerance (mg/L)	5	0.1
Chlorine tolerance (mg/L)	1	0.1
Silica tolerance (mg/L)	—	10 (<30% total anions)
Servive flow rate (gpm/ft^3)	$\leq$5	2 to 3
Minimum depth (in.)	30	30
Backwash flow rate (gpm/ft^3)	5 to 8	2 to 3
Backwash expansion (%)	50	50 to 75
Backwash duration (min)	5 to 15	5 to 20
Flushing flow rate (gpm/ft^3)	1.0 to 1.5	0.5
Flushing volume (empty bed volumes)	2 to 5	2 to 10
Flushing duration (min)	30 to 70	30 to 150
Operating ion exchange capacity (kg $CaCO_3$/ft^3)	9 to 25	9 to 17
Regenerant concentration (% by wt)		
NaCl	3 to 12	1.5 to 12
H_2SO_4	2 to 4	—
NaOH	—	2 to 4
Regenerant dose (lb/ft^3)		
NaCl	5 to 20	5 to 20
H_2SO_4	2.5 to 10	—
NaOH	—	3.5 to 8
Regenerant efficiency (%)		
NaCl	30 to 50	—
H_2SO_4	20 to 40	—
NaOH	—	—
Regenerant application rate (gpm/ft^3)	0.5	0.5
Regenerant contact time (min)	50 to 80	60 to 90

[1]Clifford, D. A. 1990. Ion exchange and inorganic adsorption. In *Water Quality and Treatment: A Handbook of Community Water Supplies*, 4th ed., p. 561. F. W. Pontius, ed. McGraw-Hill, New York.
[2]Kemmer, F. N. 1988. *The Nalco Water Handbook,* 2nd ed. McGraw-Hill, New York.
[3]Powell, S. T. 1954. *Water Conditioning for Industry.* McGraw-Hill, New York.
[4]Çulp/Wesner/Culp, Inc. 1986. *Handbook of Public Water Systems.* R. B. Williams and G. L. Culp, ed. Van Nostrand Reinhold, New York.

Waste brines do not adversely affect biological wastewater treatment processes (including septic tanks) at chloride concentrations up to several thousand mg/L [Ludzack and Noran, 1965].

Operating Cycle

The sodium cycle ion exchange softening process consists of the following cycle:

- Hard water is passed through a bed of fresh ion exchange resin that is preloaded with sodium ions; calcium and magnesium ions are absorbed from solution and replaced by a charge-equivalent amount of sodium ions.
- At the end of the ion exchange service run (which is indicated either by preset timer or by effluent monitoring), the ion exchange bed is backwashed to remove sediment, and the washwater is run to waste.
- The ion exchange material is then regenerated by slowly pumping through it a sodium chloride brine; the required brine volume normally exceeds the pore spaces in the bed; the

spent regenerant brine is run to waste; the concentration of the brine determines the exchange capacity of the resin.

- The bed is then flushed with several empty bed volumes of hard water to remove the spent regenerant brine, and the flushing water is run to waste; the bed is put back in service.

The design problem is to determine:

- The required bed volume and dimensions
- The duration of a cycle
- The mass of salt required for regeneration each cycle
- The volume of regenerant brine required each cycle
- The volume of waste brine produced each cycle
- The composition of the waste brine

Bed Volume and Salt Requirement

For all intents and purposes, the removal of calcium and magnesium is nearly complete. Thus the required ion exchange bed volume, V, can be calculated as

$$V = \frac{Q t_s C_{Ho}}{q_{iec}} \tag{9.66}$$

where

C_{Ho} = the hardness of the raw water [eq/m^3 or gr (as $CaCO_3$)/ft^3]
Q = the raw water flow rate (m^3/s or ft^3/s)
q_{iec} = the capacity of the ion exchange resin [eq/m^3 or gr (as $CaCO_3$)/ft^3]
t_s = the time in service of the bed (s)
V = the volume of the bed (m^3 or ft^3)

The time in service, t_s, is a design choice. The hard water flow rate, Q, the raw water hardness, C_{Ho}, and the bed exchange capacity, q_{iec}, are determined by the design problem.

The service flow rate in Table 9.2 yields a minimum bed volume, service time, and hydraulic detention time. The minimum depth requirement yields a maximum cross-sectional area, which is intended to control short circuiting. Sometimes the service flow rate is given as an areal rate (approach velocity). In that case the service flow rate and minimum depth combine to yield a minimum bed volume, service time, and hydraulic detention time.

Regeneration Scheduling

The cycle time is the sum of the required service time, backwash time, regeneration time, flushing time, and down time for valve opening and closing and pump startup and shutdown:

$$t_c = t_s + t_b + t_r + t_f + t_d \tag{9.67}$$

where

t_b = the backwashing time (s)
t_c = the cycle time (s)
t_d = the down time for valve and pump adjustments (s)
t_f = the flushing time (s)
t_r = the regeneration time (s)
t_s = the time in service of the bed (s)

The last four components of the cycle time are more or less fixed, whereas the service time is freely adjustable (by adjusting the bed volume) and can be chosen so that the cycle time fits comfortably into a convenient work schedule.

Partial Softening

It is usually desirable to produce a finished hardness C_{Hf} greater than zero in order to facilitate corrosion control. This is accomplished by bypassing some of the raw water around the exchanger and mixing it with softened water, as shown in Fig. 9.4. The finished water hardness is

$$C_{Hf} = \frac{Q_b C_{Ho}}{Q_s + Q_b} = \frac{Q_b C_{Ho}}{Q} \tag{9.68}$$

where

C_{Ho} = the hardness of the raw water [eq/m^3 or gr (as $CaCO_3$)/ft^3]
C_{Hf} = the hardness of the finished water [eq/m^3 or gr (as $CaCO_3$)/ft^3]
Q = the raw water flow rate (m^3/s or ft^3/s)
Q_b = the raw water flow bypassed around the softener (m^3/s or ft^3/s)
Q_s = the raw water flow processed through the softener (m^3/s or ft^3/s)

The fractions of the raw water flow that are softened and bypassed are

$$f_s = \frac{Q_s}{Q} = 1 - \frac{C_{Hf}}{C_{Ho}} \tag{9.69}$$

$$f_b = \frac{Q_b}{Q} = \frac{C_{Hf}}{C_{Ho}} \tag{9.70}$$

where

f_b = the fraction of the raw water that is bypassed (dimensionless)
f_s = the fraction of the raw water that is softened (dimensionless)

The bed volume is sized based on the volume of water softened, Q_s.

The salt requirement per cycle, N_{NaCl}, is the manufacturer's recommended dosage rate per unit bed volume, m_{NaCl}, times the bed volume:

$$M_{NaCl} = m_{NaCl} V \tag{9.71}$$

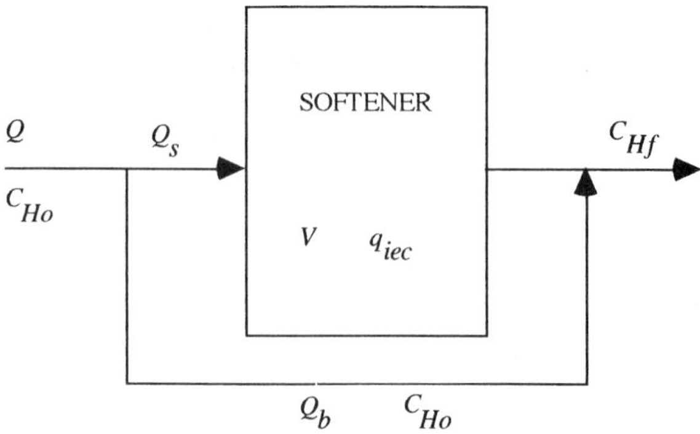

FIGURE 9.4 Flow scheme for partial softening.

where

M_{NaCl} = the mass of salt required to regenerate an ion exchange bed per cycle (kg or lb)
m_{NaCl} = the salt dosage per unit bed volume (kg/m^3 or lb/ft^3)

This will vary with the desired bed ion exchange capacity. The salt efficiency is defined as the sodium exchanged divided by the sodium supplied. If equivalents are used to express masses, the efficiency is

$$E_{Na} = \frac{Vq_{iec}}{M_{NaCl}/58.5} \quad (9.72)$$

where E_{Na} is the salt efficiency (dimensionless). Large ion exchange capacities require disproportionate salt dosages, which significantly reduces the salt efficiency.

Waste Brine

The waste brine is composed of the wash water, the spent regenerant brine, and the flushing water.

The wash water volume is merely the backwash rate, U_b, times the backwash duration and bed cross-sectional area, A:

$$V_{ww} = U_b A t_b \quad (9.73)$$

where

A = the cross-sectional area of the bed (m^2 or ft^2)
t_b = the backwash duration (s)
U_b = the backwash rate (m/s or ft/s)
V_{ww} = the wash water volume (m^3 or ft^3)

The regenerant brine volume is equal to the mass of salt that is required for regeneration, divided by the weight fraction of salt in the brine and the density of the brine:

$$V_{rb} = \frac{M_{NaCl}}{\rho_{rb} f_{NaCl}} \quad (9.74)$$

where

f_{NaCl} = the mass fraction of salt in regenerant brine (dimensionless)
M_{NaCl} = the mass of salt required to regenerate an ion exchange bed per cycle (kg or lb)
ρ_{rb} = the density of regenerant brine (kg/m^3 or lbm/ft^3)

Densities for various brine concentrations and compositions are given in standard handbooks such as the *CRC Handbook of Chemistry and Physics*.

The amount of flushing water is equal to the empty bed volume times the number of bed volumes used for flushing:

$$V_f = nV \quad (9.75)$$

where

n = the number of empty bed volumes of flushing water needed to remove spent brine from an ion exchange bed (dimensionless)
V = the empty bed volume (m^3 or ft^3)
V_f = the flushing water volume (m^{3}or ft^3)

The total waste brine volume per cycle is

$$V_{wb} = V_{ww} + V_{rb} + V_f \tag{9.76}$$

The waste brine contains the following:

- All the calcium and magnesium removed
- All the chloride in the regenerant brine
- All the sodium not exchanged during regeneration

The composition is most easily calculated if the units are equivalents. If the equivalents fraction of calcium in the hard water is f_{Ca}, the masses of the various ions in the waste brine are

$$M_{\text{Ca}} = f_{\text{Ca}} V q_{iec} \left(\frac{20 \text{ g Ca}}{\text{eq}}\right)\left(\frac{1 \text{ kg}}{10^3 \text{ g}}\right) \quad \text{(as Ca)} \tag{9.77}$$

$$M_{\text{Mg}} = (1 - f_{\text{Ca}}) V q_{iec} \left(\frac{12.16 \text{ g Mg}}{\text{eq}}\right)\left(\frac{1 \text{ kg}}{10^3 \text{ g}}\right) \quad \text{(as Mg)} \tag{9.78}$$

$$M_{\text{Na}} = M_{\text{NaCl}} \left(\frac{23 \text{ g Na}}{58.5 \text{ g NaCl}}\right)\left(\frac{1 \text{ kg}}{10^3 \text{ g}}\right) - V q_{iec} \left(\frac{23 \text{ g Na}}{\text{eq}}\right)\left(\frac{1 \text{ kg}}{10^3 \text{ g}}\right) \quad \text{(as Na)} \tag{9.79}$$

$$M_{\text{Cl}} = M_{\text{NaCl}} \left(\frac{35.5 \text{ g Cl}}{58.5 \text{ g NaCl}}\right)\left(\frac{1 \text{ kg}}{10^3 \text{ g}}\right) \quad \text{(as Cl)} \tag{9.80}$$

where

f_{Ca} = the equivalents fraction of calcium in the hard water (dimensionless)
M_{Ca} = the mass of calcium in waste brine from ion exchange in kg
M_{Cl} = the mass of chloride in waste brine from ion exchange in kg
M_{Mg} = the mass of magnesium in waste brine from ion exchange in kg
M_{Na} = the mass of sodium in waste brine from ion exchange in kg

Chloride Cycle Dealkalization and Desulfurization

Strong base anion exchange resins will release chloride ions and absorb bicarbonate, carbonate, and sulfate. The regenerant is a sodium chloride brine, but the chloride is the exchangeable ion and the sodium ion is passive. The calculations parallel those for sodium cycle softening.

Demineralization

Waters may be nearly completely demineralized by a combination of strong acid hydrogen cycle and strong base hydroxide cycle resins.

Raw waters are first processed through the hydrogen cycle resin, which removes all cations and replaces them with protons. The result is a dilution solution of mineral acids and carbonic acid. The pH will depend on the total amount of cations removed: Each equivalent of cations is replaced by one equivalent of protons.

The carbonic acid is derived from the carbonate and bicarbonate originally present in the raw water. Carbonic acid decomposes to form carbon dioxide gas until it is in equilibrium with the gas phase partial pressure of CO_2. If the initial alkalinity is high enough a vacuum degasifier will be required after the hydrogen cycle exchanger to remove the carbon dioxide gas.

The initial bicarbonate and carbonate ions are removed by conversion to carbon dioxide gas and degasification. The remaining anions (Cl^-, SO_4^{2-}, NO_3^-, etc.) can be removed by hydroxide cycle ion exchange. Note that the hydrogen cycle exchanger must precede the hydroxide cycle exchanger in order to avoid the formation of calcium carbonate and magnesium hydroxide deposits.

The design calculations for hydrogen cycle and hydroxide cycle exchangers parallel those for sodium cycle softening, differences being that H^+ and OH^- are being released to the water rather than Na^+ and that the regenerants are strong acids (H_2SO_4 or HCl) and strong bases (NaOH or KOH). The waste products are fairly strong acid and base solutions. The wastes do not exactly neutralize each other because the acid and base efficiencies are usually different.

Strong acid hydrogen cycle exchangers are usually regenerated with sulfuric acid. However, the acidity required to drive the regeneration is generally on the order of 2 to 8% acid by weight, and in this range sulfuric acid is only partially ionized. The result is a low acid efficiency, generally on the order of 30%. Furthermore, in hard waters the calcium sulfate solubility limit in the waste acid may be exceeded, and a gypsum sludge may be produced. Both problems can be avoided using hydrochloric acid, but it is expensive and HCl vapors pose a venting problem.

If the removal of weak acid anions such as HCO_3^-, $HSiO_4^-$, or CH_3COO^- is not required or if such anions are absent, weak base anion exchanger resins can be used instead of strong base anion resins. WBA resins can be regenerated with a wide variety of bases.

Mixed bed exchangers contain both SAC and SBA resins. Both cation exchange and anion exchange occur at every level of the bed. The proportions of the two resin types depend on their relative ion exchange capacities. SBA resins have much lower densities than SAC resins, so they can be separated by backwashing. Regeneration is accomplished by down-flowing base through the upper SBA resin and up-flowing acid through the lower SAC resin. The wastes are drawn off at the interface between the separated resins. A mechanism for remixing the beads after regeneration must be included in the bed design.

9.3 Disinfection

Waterborne Diseases

The principal waterborne diseases in the U.S. are listed in Table 9.3. Almost all of these diseases are transmitted via fecal contamination of water and food, but a few are more commonly transmitted by direct person-to-person contact or by inhalation of microbially contaminated air. Most of these diseases are actually quite rare, there being only one or a few cases a year, and some—such as cholera—have not been observed for decades. However, all these diseases have permanent reservoirs among either humans or wild animals both in the U.S. and abroad, and all of them have the potential to cause outbreaks unless careful sanitation is maintained.

For the last 30 years, the U.S. has averaged about 33 outbreaks of waterborne disease per year, and each outbreak has involved about 220 cases [Craun, 1988]. The majority of the outbreaks occur in noncommunity and individual systems, but the majority of the cases occur in community systems. About one-fifth of the outbreaks occur in surface water systems that either do not treat the raw water or that provide only disinfection [Craun, 1988]. About two-fifths of the outbreaks occur in groundwater systems that do not provide any treatment [Craun and McCabe, 1973]. Only half the outbreaks can be attributed to a specific organism; in recent years *Giardia lamblia* is the organism most often identified, displacing *Salmonella*. However, *Cryptosporidium parvum*–induced diarrhea may be as common as giardiasis [Rose, 1988]. In recent years, the U.S. has experienced nearly one proven *Cryptosporidium* outbreak per year [Pontius, 1993]. From 1989 to 1990 there were 26 outbreaks of waterborne disease reported in the U.S. Of these the causative organism was not identified in 14 outbreaks. *Giardia* was responsible in 7 outbreaks, hepatitis A in 2, whereas Norwalk-like viruses, *E. coli* 0157:H7, and cyanobacteria were responsible for 1 each [Herwaldt *et al.*, 1992].

Table 9.3 Principal Waterborne Diseases in the U.S.[1,2]

Group/Organism	Disease	Annual Number of Outbreaks	Annual Number of Cases
Unknown Etiology	acute gastroenteritis	16.	3353.
	Bacteria		
Salmonella typhi	typhoid fever	1.4	10.
Salmonella paratyphi	paratyphoid fever	—	—
Other Salmonella	salmonellosis	1.	171.
Shigella	bacillary dysentery	2.	647.
Vibrio cholerae	cholera	—	—
Enteropathogenic *Escherichia coli*	gastroenteritis	0.1	125.
Yersinia enterocolitica	gastroenteritis	—	—
Campylobacter jejuni	gastroenteritis	0.1	375.
Legionella pneumophilia et al.	acute respiratory illness	—	—
Mycobacterium tuberculosis et al.	tuberculosis	—	—
Atypical mycobacteria	pulmonary illness	—	—
Misc. opportunistic bacteria	varies	—	—
	Viruses		
Polioviruses	poliomyelitis	—	—
Coxsackieviruses A	aseptic meningitis	—	—
Coxsackieviruses B	aseptic meningitis	—	—
Echoviruses	aseptic meningitis	—	—
Other enteroviruses	AHC; encephalitis	—	—
Reoviruses	mild upper respiratory and gastrointestinal illness	—	—
Rotaviruses	gastroenteritis	—	—
Adenoviruses	mild upper respiratory and gastrointestinal illness	—	—
Hepatitis A virus	infectious hepatitis	1.5	50.
Norwalk and related G. I. viruses	gastroenteritis	0.5	154.
	Protozoans		
Acanthamoeba castellani	amebic meningoencephalitis	—	—
Balantidium coli	balantidiasis (dysentery)	—	—
Cryptosporidium parvum	—	—	—
Entamoeba histolytica	amebic dysentery	0.3	4.
Giardia lamblia	giardiasis (gastroenteritis)	3.3	2228.
Naegleria flowleri	primary amebic meningoencephalitis	—	—
	Helminths		
Ascaris lumbricoides	ascariasis	—	—
Trichuris trichiura	trichuriasis	—	—
Ancylostoma duodenale	hookworm disease	—	—
Necator americanus	hookworm disease	—	—
Strongyloides stercoralis	threadworm disease	—	—
	Cyanobacteria		
Anabaena flos-aquae	gastroenteritis	—	—
Microcystis aeruginosa	gastroenteritis	—	—
Aphanizomenon flos-acquae	gastroenteritis and neurotoxins	—	—
Schizothrix calcicola	gastroenteritis	—	—

[1]Sobsey, M. and Olson, B. 1983. Microbial agents of waterborne disease. *Assessment of Microbiology and Turbidity Standards for Drinking Water: Proceedings of a Workshop, December 2–4, 1981.* EPA 570-9-83-001. P. S. Berger and Y. Argaman, ed. U.S. Environmental Protection Agency, Office of Drinking Water, Washington, DC.

[2]Craun, G. F. and McCabe, L. J. 1973. Review of the causes of water-borne disease outbreaks. *Journal of the American Water Works Association.* 65(1):74.

The Total Coliform Rule

In the past the sanitary quality of drinking was judged by either a most probable number (MPN) or membrane filter count (MFC) of coliform bacteria, and the maximum contaminant level was set at 1 total coliform per 100 mL. The U.S. Environmental Protection Agency [1989b] has abandoned this method and now uses the following rule [Pontius, 1990]:

- Total coliform must be measured in 100 mL samples using the multiple tube fermentation technique, the membrane filter technique, or the presence-absence coliform test [all of which are described in *Standard Methods for the Examination of Water and Wastewater,* 18th ed., American Public Health Association, Washington, DC. (1992)] or the Colilert™ System [MMO-MUG test, approved *Federal Register,* 57:24744 (1992)].
- The number of monthly samples depends on system size as prescribed in Table 9.4.

Table 9.4 Sampling Requirements of the Total Coliform Rule[1]

Population Served	Minimum Number of Routine Samples per Month[a]	Population Served	Minimum Number of Routine Samples per Month[a]
25–1000[b]	1[c]	59,001–70,000	70
1001–2500	2	70,001–83,000	80
2501–3300	3	83,001–96,000	90
3301–4100	4	96,001–130,000	100
4101–4900	5	130,001–220,000	120
4901–5800	6	220,001–320,000	150
5801–6700	7	320,001–450,000	180
6701–7600	8	450,001–600,000	210
7601–8500	9	600,001–780,000	240
8501–12,900	10	780,001–970,000	270
12,901–17,200	15	970,001–1,230,000	300
17,201–21,500	20	1,230,001–1,520,000	330
21,501–25,000	25	1,520,001–1,850,000	360
25,001–33,000	30	1,850,001–2,270,000	390
33,001–41,000	40	2,270,001–3,020,000	420
41,001–50,000	50	3,020,001–3,960,000	450
50,001–59,000	60	3,960,001 or more	480

[1] Environmental Protection Agency. 1989. Total coliforms. Final rule. *Federal Register.* 54(124): 27544.

[a] In lieu of the frequency specified in this table, a noncommunity water system using groundwater and serving 1000 persons or fewer may monitor at a lesser frequency specified by the state until a sanitary survey is conducted and the state reviews the results. Thereafter, noncommunity water systems using groundwater and serving 1000 persons or fewer must monitor each calendar quarter during which the system provides water to the public unless the state determines that some other frequency is more appropriate and notifies the system (in writing). Five years after promulgation, noncommunity water systems using groundwater and serving 1000 persons or fewer must monitor at least once per year. A noncommunity water system using surface water or groundwater under the direct influence of surface water, regardless of the number of persons served, must monitor at the same frequency as a like-sized public water system. A noncommunity water system using groundwater and serving more than 1000 persons during any month must monitor at the same frequency as a like-sized community water system, except that the state may reduce the monitoring frequency for any month the system serves 1000 persons or fewer.

[b] Includes public water systems that have at least 15 service connections but serve fewer than 25 persons.

[c] For a community water system serving 25 to 10,000 persons, the state may reduce this sampling frequency if a sanitary survey conducted in the last five years indicates that the water system is supplied solely by a protected groundwater source and is free of sanitary defects. However, in no case may the state reduce the sampling frequency to less than once per quarter.

Table 9.5 Monitoring and Repeat-Sample Frequency after a Total Coliform-Positive Routine Sample[1]

Number of Routine Samples per Month	Number of Repeat Samples[a]	Number of Routine Samples Next Month[b]
1 or fewer	4	5
2	3	5
3	3	5
4	3	5
5 or greater	3	Table 9.4

[1]Environmental Protection Agency. 1989. Total coliforms. Final rule. *Federal Register.* 54(124):27544.

[a]Number of repeat samples for the same month for each total-coliform-positive routine sample.

[b]Except where the state has invalidated the original routine sample, substitutes an on-site evaluation of the problem, or waives the requirement on a case-by-case basis.

- No more than 5.0% of the monthly samples can test positive if 40 or more samples are analyzed each month.
- No more than 1 sample can test positive if less than 40 samples are analyzed each month.
- Unfiltered surface water systems must sample the first service connection each day that the final turbidity exceeds 1 NTU, and service connection samples must be included in the positive and negative counts.
- Coliform-positive samples must be analyzed to determine whether fecal coliform are present.
- Repeat samples must be collected within 24 hours of the laboratory report of a coliform-positive result; the number of repeat samples depends on system size as prescribed in Table 9.5.

When the total coliform rule is violated, immediate corrective action is required. The public must be notified of the violation. The method of notification includes both the public media and mail. The notification itself must conform to federal regulations regarding language and must include a description of the total coliform rule, the public health significance of the violation, precautions consumers should take, a description of the corrective actions being taken by the utility, and telephone numbers for additional information. Corrective measures include flushing mains, increasing disinfectant doses, and improvements to filtration performance.

Community water systems collecting fewer than five samples per month are required to conduct a sanitary survey within five years of the promulgation of the rule and every five years thereafter. Noncommunity water systems are required to conduct a sanitary survey within ten years after the rule is promulgated and every five years thereafter. If the noncommunity water system uses protected and disinfected groundwater, the sanitary survey may be repeated every ten years instead of every five years.

Disinfectants

The solubilities of the gaseous disinfectants and pH effects on their forms are discussed under "Aeration and Gas Exchange, Absorption of Reactive Gases," Chapter 8.

Chlorine

Chlorine can be purchased either as pressurized liquid chlorine or as solid hypochlorite salts of calcium or sodium. Liquid chlorine is preferred for reasons of economy, but solid hypochlorite salts are preferred for reasons of safety.

Water Chemistry of Chlorine. Both hypochlorous acid and hypochlorite are two-electron acceptors like chlorine, so they preserve the full oxidizing potential of chlorine:

$$Cl_2 + 2e^- \rightarrow 2Cl^- \tag{9.81}$$

$$HOCl + H^+ + 2e^- \rightarrow H_2O + Cl^- \tag{9.82}$$

$$OCl^- + 2H^+ + 2e^- \rightarrow H_2O + Cl^- \tag{9.83}$$

The Breakpoint Curve. The breakpoint curve is a plot of the measured chlorine residual versus the chlorine dose. It consists of three regions (see Fig. 9.5):

- An initial region at low chlorine dosages where there is no measurable residual. This represents the immediate chlorine demand (ICD) due to the oxidation of reactive substances such as ferrous iron and sulfide.
- A second region at higher doses in which the measured residual first increases with chlorine dose and then decreases. This region occurs when the raw water contains ammonia or organic nitrogen. The measured residual consists mostly of monochloramine, and it is called the *combined residual.* Combined residual rises as ammonia is converted to monochloramine, and it declines as monochloramine is converted to nitrogen gas and nitrate. The minimum point of the falling limb of the curve is the "breakpoint," which marks the elimination of the combined residual.
- A final region in which the measured residual increases linearly 1:1 with chlorine dose. The measured residual is a mixture of molecular chlorine, hypochlorous acid, and hypochlorite and is called the *free available chlorine* (FAC) residual. (See "Aeration and Gas Exchange, Absorption of Reactive Gases," Chapter 8.)

The existence of an ICD and a combined residual depends on raw water composition, and either or both may be absent. Ferrous iron and sulfide are common in groundwater but rare in surface water. Unless it has been nitrified, wastewater contains significant amounts of ammonia, but surface waters do not.

Monochloramine, hypochlorite, and hypochlorous acid differ greatly in their disinfecting power, and determination of the breakpoint curve is needed to know which form is present.

Combined Residual Chlorine. Hypochlorous acid reacts with ammonia to form monochloramine [Palin, 1977; Wei and Morris, 1974; Saunier and Selleck, 1979]:

$$HOCl + NH_3 \rightarrow NH_2Cl + H_2O \tag{9.84}$$

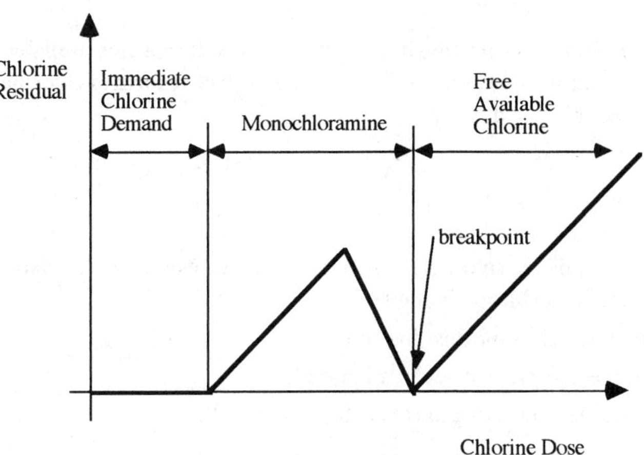

FIGURE 9.5 The breakpoint curve.

This is the only product formed until the ammonia is nearly exhausted. Then, hypochlorous acid reacts with monochloramine to form dichloramine. Small amounts of nitrogen trichloride may also be formed:

$$HOCl + NH_2Cl \rightarrow NHCl_2 \tag{9.85}$$

$$HOCl + NHCl_2 \leftrightarrow NCl_3 \tag{9.86}$$

Saunier and Selleck [1979] propose that mono- and dichloramine react with water and/or hydroxide to form hydroxylamine:

$$NH_2Cl + OH^- \rightarrow NH_2OH + Cl^- \tag{9.87}$$

$$NHCl_2 + H_2O + OH^- \rightarrow NH_2OH + HOCl + Cl^- \tag{9.88}$$

They further propose that the hydroxylamine is oxidized by hypochlorous acid to nitrosyl hydride. This intermediate was also suggested by Wei and Morris. However, nitrosyl hydride has never been observed, and its existence may be impossible [Sidgwick, 1950].

$$HOCl + NH_2OH \rightarrow NOH + H_2O + H^+ + Cl^- \tag{9.89}$$

If nitrosyl hydride does exist, it would either react with mono- and dichloramine to form nitrogen gas or with hypochlorous acid to form nitrate:

$$NOH + NHCl_2 \rightarrow N_2 + HOCl + H^+ + Cl^- \tag{9.90}$$

$$NOH + NH_2Cl \rightarrow N_2 + H_2O + H^+ + Cl^- \tag{9.91}$$

$$NOH + HOCl \rightarrow HNO_2 + H^+ + Cl^- \tag{9.92}$$

$$HNO_2 + HOCl \rightarrow HNO_3 + H^+ + Cl^- \tag{9.93}$$

The net stoichiometric would be a mixture of Eqs. (9.94) and (9.95):

$$3HOCl + 2NH_3 \rightarrow N_2 + 3H_2O + 3H^+ + 3Cl^- \tag{9.94}$$

$$4HOCl + NH_3 \rightarrow HNO_3 + H_2O + 4H^+ + 4Cl^- \tag{9.95}$$

It appears that the formation of nitrogen gas predominates, and the molar ratio of hypochlorous acid to ammonia at the breakpoint is about 1.6 to 1.7 [Wei and Morris, 1974]. This amounts to about 8.1 to 8.6 kg chlorine per kg of total kjeldahl nitrogen (TKN).

Required Chlorine Dose. Generally, it is desirable to produce a free available chlorine residual consisting of hypochlorous acid because this is the most effective form of chlorine. The free residual can be related to the chlorine dose by

$$D_{Cl_2} = \mathrm{ICD} + 8.6\mathrm{TKN} + C_{FAC} \tag{9.96}$$

where

C_{FAC} = the free, available chlorine consisting of a mixture of dissolved chlorine, hypochlorous acid, and hypochlorite in mg/L

D_{Cl_2} = the required chlorine dose in mg/L

ICD = the immediate chlorine demand in mg/L

TKN = the total kjeldahl nitrogen concentration in mg/L

The pH dependency of the distribution of hypochlorous acid and hypochlorite is discussed under "Aeration and Gas Exchange, Absorption of Reactive Gases," Chapter 8.

Dechlorination. Chlorine is highly toxic to fish: The 96-hour LC_{50} is about 10 μg/L for most freshwater fish and about 2 μg/L for salmonids [Criteria Branch, 1976]. High concentrations are also undesirable in drinking water for aesthetic reasons.

The usual method of dechlorination is reaction with sulfur dioxide. Sulfur dioxide reacts with both hypochlorous acid and monochloramine:

$$SO_2 + HOCl + H_2O \rightarrow 3H^+ + Cl^- + SO_4^{2-} \tag{9.97}$$

$$SO_2 + NH_2Cl + H_2O \rightarrow NH_4^+ + 2H^+ + Cl^- + SO_4^{2-} \tag{9.98}$$

Chlorine can also be removed by reaction with activated carbon. The initial reaction with fresh carbon results in the formation of carbon oxides (·C=O) on the carbon surfaces [Stover *et al.*, 1986]:

$$\cdot C + HOCl \rightarrow \cdot C{=}O + H^+ + Cl^- \tag{9.99}$$

The surface oxides may eventually evolve carbon monoxide or carbon dioxide. Monochloramine undergoes a similar reaction:

$$\cdot C + NH_2Cl + H_2O \rightarrow \cdot C{=}O + NH_4^+ + Cl^- \tag{9.100}$$

Chlorine Dioxide

Chlorine dioxide is an unstable greenish-yellow gas. Because of its instability, it cannot be stored or transported and is prepared on site immediately prior to use. The usual methods are the reaction between hydrochloric acid and sodium chlorite and the reaction of sodium chlorite with chlorine, both in aqueous solution [Katz, 1980; Haas, 1990]:

$$4HCl + 5NaClO_2 \rightarrow 4ClO_2 + 5NaCl + 2H_2O \tag{9.101}$$

$$Cl_2 + 2NaClO_2 \rightarrow 2ClO_2 + 2NaCl \tag{9.102}$$

Excess chlorine is used to drive the chlorine/chlorite reaction to completion. The normal recommendation is 1 kg pure chlorine per kg pure sodium chlorite (1.3 mol/mol), but even this excess produces conversion of only about 60%. Some facilities use more chlorine, and others acidify the reaction. Chlorine dioxide can also be formed from sodium hypochlorite and sodium chlorite in aqueous solution:

$$NaOCl + 2NaClO_2 + H_2O \rightarrow 2ClO_2 + 3Na^+ + Cl^- + 2OH^- \tag{9.103}$$

In all these reactions, chlorine, chlorite, and chlorate are formed in side reactions.

Chlorine dioxide is an oxidant, and it acts as a one-electron acceptor forming chlorite:

$$ClO_2 + e^- \rightarrow ClO_2^- \tag{9.104}$$

Large doses of chlorite cause anemia, and the total residual of chlorine dioxide, chlorite, and chlorate should be kept below 1 mg/L [Haas, 1990].

The relative advantages and disadvantages of chlorine dioxide as a disinfectant are [Katz, 1980]:

- It destroys phenols rather than chlorinating them and producing taste and odor products.
- It destroys some algal taste and odor compounds.
- It does not hydrolyze, so its disinfecting power is not affected by pH.
- Because it does not hydrolyze, it can be removed from water by air stripping.
- It does not react with ammonia.

Chlorine dioxide reacts rapidly with both manganous manganese and ferrous iron, forming hydroxide precipitates. Chlorine dioxide itself does not form trihalomethanes [Rav-Acha, 1984].

However, the excess chlorine used to make it will form THMs. Also, some chlorine is released during chlorine dioxide reactions, and this chlorine will form trace amounts of THMs.

Ozone

Ozone is prepared by passing a dry, oil-free, particulate-free, oxygen-containing gas between two high-voltage electrodes (generally 7500 to 20,000 volts) [Katz, 1980].

$$3O_2 \leftrightarrow 2O_3 \tag{9.105}$$

The yields are generally 1 to 3% ozone by volume in air and 2 to 6% in oxygen. The reaction liberates heat, and decomposition of ozone to oxygen is favored at high temperatures, so that the ozone generators must be cooled.

Ozone decomposes in water to yield the free radicals $HO^{\cdot}$ and $HO_2^{\cdot}$. These are very strong oxidants and are one to two orders of magnitude more effective than chlorine. The free radicals will react with each other to form oxygen, so ozone residuals are unstable and spontaneously disappear:

$$O_3 + H_2O \rightarrow 2HO_2^{\cdot} \tag{9.106}$$

$$O_3 + HO_2^{\cdot} \rightarrow HO^{\cdot} + O_2 \tag{9.107}$$

$$HO^{\cdot} + HO_2^{\cdot} \rightarrow H_2O + O_2 \tag{9.108}$$

The rate of decomposition can be estimated by [Joint Task Force, 1990]:

$$C_{O_3}(t) = C_{O_3}(0) \times 10^{-kt} \tag{9.109}$$

$$k = 10^{0.63\text{pH}-6.97} \tag{9.110}$$

where

$C_{O_3}(0)$ = the initial ozone concentration in mg/L
$C_{O_3}(t)$ = the ozone concentration at time t in mg/L
k = a pH-dependent rate constant in minutes
t = the elapsed time in minutes

Disinfection Kinetics

The Chick–Watson Law

In batch disinfection systems the die-away of microbes is often exponential, which implies a first-order decay law called the Chick–Watson law [Chick, 1908; Watson, 1908]:

$$\frac{dN}{dt} = KC^nN \tag{9.111}$$

$$N(t) = N_o e^{-KC^nt} \tag{9.112}$$

where

C = the disinfectant concentration (kg/m^3 or lb/ft^3)
K = the coefficient of specific lethality (m^{3n}/kg^n·s or ft^{3n}/lb^n·s)
N = the number concentration of microbes (number/m^3 or number/ft^3)
n = the coefficient of dilution (dimensionless)
t = elapsed time (s)

Table 9.6 Specific Lethality Coefficients for 5°C

	K(L/mg·min)			
Disinfectant	Enteric Bacteria	Amoebic Cysts	Viruses	Spores
Ozone	500.	0.5	5.	2.
Hypochlorous Acid	20.	0.05	≥1.0	0.05
Hypochlorite	0.2	0.0005	<0.02	<0.0005
Monochloramine	0.1	0.02	0.005	0.001

Source: Morris, J. C. 1975. Aspects of the quantitative assessment of germicidal efficiency. In *Disinfection: Water and Wastewater*, p. 1. J. D. Johnson, ed. Ann Arbor Science Publishers, Ann Arbor, MI.

The coefficient of dilution is frequently near 1, and Morris [1975] argues that quality of the data typically obtained does not warrant any other value. Some specific lethality coefficients reported by Morris are given in Table 9.6.

Effect of Microbial Clusters

Microorganisms frequently exist clustered together or with other suspended solids. This offers protection for those organisms embedded within the cluster and changes the pattern of inactivation. Die-away curves may exhibit an initial lag (called a "shoulder") and/or a sharp reduction in slope as the contacting time increases. This situation can be modeled as a multipoisson process in which microorganisms in individual clusters are inactivated sequentially [Wei and Chang, 1975]:

$$S_t = \sum_{i=1}^{n} \left[X_i^o e^{-kt} \sum_{r=0}^{i-1} \frac{(kt)^r}{r!} \right] \tag{9.113}$$

where

i = the number of viable microorganisms in a cluster (dimensionless)
k = the inactivation rate for a single microorganism (per second)
n = the maximum number of microorganisms in a cluster (dimensionless)
S_t = the concentration of clusters containing viable microorganisms at time t (number/m^3 or number/ft^3)
X_i^o = the initial concentration of clusters containing i viable microorganisms (dimensionless)
t = the contacting time (s)

Equation (9.113) fits all of the observed die-away patterns.

Ct Products

The results of disinfection studies are frequently reported as Ct products. These products can be derived directly from the Chick–Watson law. Their use implies that the Chick–Watson law accurately describes the die-away curve, and that the dilution coefficient is 1:

$$Ct = \frac{\ln N_o/N(t)}{K} \tag{9.114}$$

The value of Ct depends on the ratio of the initial to final microorganism concentrations. Generally, the products are reported for ratios that are multiples of 10, and the reduction in microbial counts is referred to as 1-log (10:1 ratio), 2-log (100:1 ratio), 3-log (1000:1), and so on. Alternatively, one speaks of viable count reductions of 90%, 99%, 99.9%, and so on, respectively.

Strictly speaking, the concept of the Ct product does not apply if the dilution coefficient is not 1. When this situation occurs, reported Ct products vary with the disinfectant concentration, and the

Table 9.7 Pathogen Inactivation Requirements of the Surface Water Treatment Rule[1]

Process	Removal Credited to Process (logs)		Additional Removal by Disinfection (logs)	
	G. lamblia	Virus	*G. lamblia*	Virus
Convention treatment	2.5	2.0	0.5	2.0
Direct filtration	2.0	1.0	1.0	3.0
Slow sand filters	2.0	2.0	1.0	2.0
Diatomaceous earth filters	2.0	1.0	1.0	3.0

[1]Environmental Protection Agency. 1989. Drinking water; national primary drinking regulations; filtration, disinfection, turbidity, *Giardia lamblia*, viruses, *Legionella*, and heterotrophic bacteria. Final rule. *Federal Register.* 54(124):27485.

applicable concentration must be specified. The result is a specification of the required contacting time.

Ct products also do not apply when there is microbial clustering. However, this situation can be handled by estimating the initial lag period. The die-away curve is plotted semilogarithmically, and the linear portion is extrapolated back to the initial microorganism concentration. The initial lag period can then be read off the time axis. During the lag period there is no inactivation, so any disinfection system must provide a contacting time longer than this.

Finally, it must be remembered that *Ct* products are obtained from batch disinfection experiments. If they are to be used in the design of continuous flow systems, the systems must achieve nearly ideal plug flow. In this regard the U.S. Environmental Protection Agency requires that the contacting time for any system be calculated as the time that at least 90% of the water is held [EPA, 1990]. This can be determined by tracer studies.

Ct products can be determined for any microorganism, but as a practical matter only those microbes most resistant to disinfection are of interest. For design purposes the critical organisms are the cysts of the protozoan *Giardia lamblia* and viruses. The U.S. Environmental Protection Agency [1989a] requires that water treatment plants achieve an overall reduction of 99.9% (3-log) of *G. lamblia* cysts and 99.99% (4-log) of viruses. The agency credits coagulation-flocculation-settling filtration with most of the removal of these indicator organisms and requires chemical disinfection for the remainder of the 3-log and 4-log removals. Their requirements are shown in Table 9.7. The *Ct* products for chemical disinfection of cysts and viruses are given in Tables 9.8 and 9.9.

Chlorine

Because chlorine forms a weak acid in water and forms weak oxidants with ammonia, chlorination kinetics exhibit special features.

Table 9.8 *Ct* Products for One-Log *Giardia lamblia* Inactivation[1]

Disinfectant	pH	*Ct* Product (L/mg·min)			
		Temperature (°C)			
		0.5	5.0	10.	15.
Free available chlorine (at 2 mg/L)	6.	49.	35.	26.	19.
	7.	70.	50.	37.	28.
	8.	101.	72.	54.	36.
	9.	146.	146.	78.	59.
Ozone	—	0.97	0.63	0.48	0.32
Chlorine dioxide	—	21.	8.4	7.4	6.3
Chloramine (preformed)	—	1270.	730.	620.	500.

[1]Environmental Protection Agency. 1989. Drinking water; national primary drinking water regulations; filtration, disinfection, turbidity, *Giardia lamblia*, viruses, *Legionella*, and heterotrophic bacteria. Final rule. *Federal Register.* 54(124):27485.

Table 9.9 *Ct* Products for Virus Inactivation between pH 6 and 9[1]

		Ct Product (L/mg·min)			
		Temperature (°C)			
Disinfectant	Log Inactivation	0.5	5.0	10.	15.
Free available chlorine	2.	6.	4.	3.	2.
	3.	9.	6.	4.	3.
Ozone	2.	0.9	0.6	0.5	0.3
	3.	1.4	0.9	0.8	0.5
Chlorine dioxide	2.	8.4	5.6	4.2	2.8
	3.	25.6	17.1	12.8	8.6
Chloramine	2.	1243.	857.	843.	428.
	3.	2063.	1423.	1067.	712.

[1]Environmental Protection Agency. 1989. Drinking water; national primary drinking water regulations; filtration, disinfection, turbidity, *Giardia lamblia,* viruses, *Legionella,* and heterotrophic bacteria. Final rule. *Federal Register.* 54(124):27485.

pH Effects. As Table 9.6 shows, the hypochlorite ion is a very poor disinfectant compared with undissociated hypochlorous acid. Consequently, in a mixture of hypochlorite and hypochlorous acid nearly all the inactivation is achieved by the undissociated acid. This means that the specific lethality coefficient that is reported for FAC will depend on pH.

A pH-independent specific lethality coefficient based on HOCl can be calculated from the following:

$$K_{\mathrm{FAC}} C_{\mathrm{FAC}}^n = K_{\mathrm{HOCl}} C_{\mathrm{HOCl}}^n \tag{9.115}$$

where

C_{FAC} = the free available chlorine concentration ($\mathrm{kg/m^3}$ or $\mathrm{lb/ft^3}$)
C_{HOCl} = the hypochlorous acid concentration ($\mathrm{kg/m^3}$ or $\mathrm{lb/ft^3}$)
K_{FAC} = the coefficient of specific lethality based on the free available chlorine concentration ($\mathrm{m}^{3n}/\mathrm{kg}^n\cdot\mathrm{s}$ or $\mathrm{ft}^{3n}/\mathrm{lb}^n\cdot\mathrm{s}$)
K_{HOCl} = the coefficient of specific lethality based on the hypochlorous acid concentration ($\mathrm{m}^{3n}/\mathrm{kg}^n\cdot\mathrm{s}$ or $\mathrm{ft}^{3n}/\mathrm{lb}^n\cdot\mathrm{s}$)
n = the dilution coefficient (dimensionless)

The molar fraction of hypochlorous acid depends strongly on pH and can be estimated from the following:

$$pK_a = -10.069 + 0.025T - \frac{3000}{T} \tag{9.116}$$

$$f = \frac{[\mathrm{HOCl}]}{[\mathrm{HOCl}] + [\mathrm{OCl^-}]} = \frac{1}{1 + [K_a/\mathrm{H^+}]} \tag{9.117}$$

where T is in kelvin. Consequently, the relationship of the HOCl-based coefficient to the pH-dependent FAC-based coefficient is

$$K_{\mathrm{HOCl}} = K_{\mathrm{FAC}} \left\{ \frac{K_a + [\mathrm{H^+}]}{[\mathrm{H^+}]} \right\}^n \tag{9.118}$$

Temperature Effects. The analysis of Fair and Geyer [1954] indicates that the Streeter–Phelps theta value for coliform inactivation by free available chlorine is approximately 1.04 at pHs below 8.5 and 1.08 at pHs above 9:

$$K_{\mathrm{HOCl}}(T_1) = K_{\mathrm{HOCl}}(T_0) \cdot \theta^{T_1 - T_0} \tag{9.119}$$

where

$K_{\mathrm{HOCl}}(T_0)$ = the specific lethality coefficient at the reference temperature T_0 ($\mathrm{m}^{3n}/\mathrm{kg}^n \cdot \mathrm{s}$ or $\mathrm{ft}^{3n}/\mathrm{lb}^n \cdot \mathrm{s}$)
$K_{\mathrm{HOCl}}(T_1)$ = the specific lethality coefficient at the chlorine contactor temperature T_1 ($\mathrm{m}^{3n}/\mathrm{kg}^n \cdot \mathrm{s}$ or $\mathrm{ft}^{3n}/\mathrm{lb}^n \cdot \mathrm{s}$)
T_0 = the reference temperature, frequently 5°C (K or °R)
T_1 = the chlorine contactor temperature (K or °R)
θ = the Streeter–Phelps temperature coefficient (dimensionless)
= 1.04 for pH less than 8.5
= 1.08 for pH greater than 9

Chloramines. Monochloramine occurs naturally in the disinfection of many wastewaters because the wastewaters contain substantial amounts of ammonia. In water treatment, monochloramines are formed when anhydrous ammonia is added along with chlorine. Monochloramine, dichloramine, and nitrogen trichloride are oxidants, and they are usually reported as equivalent molecular chlorine and called the *combined available chlorine* or *combined residual chlorine.*

Monochloramine is also a poor disinfectant. However, disinfection with monochloramine is sometimes practiced in order to reduce trihalomethane and phenolic odor formation. Best results are obtained with low pHs around 6 and rather low ammonia-to-chlorine weight ratios (around 2:1), which enhance the formation of dichloramine [Haas, 1990].

The Streeter–Phelps theta value for monochloramine inactivation is about 1.08.

Contactor Design

The *surface water treatment rule* requires that the specified inactivation of cysts and viruses be completed by the time the water reaches the first customary service at the peak hourly flow. The Ct product may be calculated using the detention times of all the tanks and piping between the point of application of the disinfectant and the first service. For the purpose of the calculation the disinfectant residuals are measured at tank outlets and the downstream end of pipeline sections. The hydraulic detention time of pipe sections can be estimated by assuming ideal plug flow. The hydraulic detention of tanks is estimated as the time required for 10% of a tracer applied to the tank inlet to reach the outlet, T_{10}. In the case of cysts the calculation is

$$\frac{\sum_{i=1}^{n} (C_{\mathrm{eff}} T_{10})_i}{(Ct)_{99.9}} \geq 1 \tag{9.120}$$

where

C_{eff} = the disinfectant concentration at the effluent end of the ith sequential unit ($\mathrm{kg/m^3}$ or $\mathrm{lb/ft^3}$)
$(Ct)_{99.9}$ = the Ct product for 99.9% inactivation ($\mathrm{m^3 \cdot s/kg}$ or $\mathrm{ft^3 \cdot s/lb}$)
T_{10} = the detention time for 90% of the flow entering the ith sequential unit (s)

The detention should be determined by tracer studies. The best method of tracer application is step input because pulse-input tests require more work [Teefy and Singer, 1990]. Suitable tracers are chloride (at drinking water concentrations less than 250 mg/L), fluoride (at drinking water concentrations less than 2 mg/L), and Rhodamine WT (at drinking water concentrations less than 0.1 μg/L). Fluoride adsorbs to aluminum hydroxide and ferric hydroxide flocs, which reduces its usefulness. Rhodamine B is a carcinogen and should not be used.

Mixed-Cells-in-Series

In the past, little attention has been paid to the hydraulic design of disinfectant contactors, and the residence time distributions of existing contactors are generally unknown. In this situation tracer studies are mandatory in order to satisfy the requirements of the surface water treatment rule.

It is possible, however, to produce contactor designs that have predictable residence distributions and that achieve predictable degrees of inactivation. The easiest way to do this is to build contactors as mixed-cells-in-series. Each cell should have a mixer and a disinfectant dosing device so that both the hydraulic regime and the disinfectant concentration are controllable.

For a contactor consisting of n mixed-cells-in-series—each cell having the same volume V_1—the ratio of the outlet to inlet microorganism concentrations is

$$\frac{N}{N_o} = \left[\frac{1}{1 + KC(V_1/Q)}\right]^n \tag{9.121}$$

where

C = the disinfectant concentration (kg/m^3 or lb/ft^3)
K = the specific lethality coefficient (m^3/kg·s or ft^3/lb·s)
N = the required effluent microbial concentration (number/m^3 or number/ft^3)
N_o = the influent microbial concentration (number/m^3 or number/ft^3)
n = the number of mixed-cells-in-series (dimensionless)
Q = the flow rate through the contactor (m^3/s or ft^3/s)
V_1 = the volume of one cell (m^3 or ft^3)

It is assumed that the coefficient of dilution is 1, and that the cells are equivolume.

The ratio N/N_o is determined by the required log inactivation: The ratio is 0.10 for a 1-log inactivation, 0.01 for a 2-log inactivation, and so on. The specific lethality coefficient can be calculated from Ct product data by Eq. (9.114). In the case of chlorine, if the disinfectant concentration is reported in terms of free available chlorine $\{[Cl_2] + [HOCl] + [OCl^-]\}$, the specific lethality coefficient for the contactor pH must be used. If the concentration of the undissociated hypochlorous acid is reported, no pH correction is needed, but the specific lethality coefficient for HOCl itself is needed. The fraction of hypochlorous acid in the free available chlorine is given by Eq. (9.116). In any case the specific lethality coefficient must be corrected for temperature using Eq. (9.117).

For a reactor consisting of mixed-cells-in-series, the value of T_{10} can be calculated theoretically. The exit age distribution for n mixed-cells-in-series is given by Eq. (9.122) [Wen and Fan, 1975]:

$$E = \frac{n^n}{(n-1)!}\left(\frac{t}{nV_1/Q}\right)^{n-1}\exp\left(-\frac{t}{V_1/Q}\right) = \frac{C(t)nV_1}{M} \tag{9.122}$$

In a pulse-input tracer test E is also the ratio of the tracer concentration in the effluent times the tank volume to the mass of tracer injected. The cumulative exit age distribution is given by Eq. (9.123) [Wen and Fan, 1975]:

$$F = 1 - \exp\left(-\frac{t}{V_1/Q}\right)\sum_{i=1}^{n}\frac{1}{(i-1)!}\left(\frac{t}{V_1/Q}\right)^{i-1} = \frac{C(t)}{C_o} \tag{9.123}$$

This is also the dimensionless tracer response obtained in a step-input tracer test. In either equation the value of T_{10} is the time for the first 10% of the tracer to appear in the effluent.

The computation recommended by the surface water treatment rule underestimates the inactivation actually achieved by mixed-cells-in-series if the dimensionless contacting number $KC(nV_1/Q)$ is less than 14 and the number of compartments is 4 or more [Lawler and Singer, 1993]. For larger values of the parameter or fewer compartments, the SWTR overestimates the inactivation achieved.

Ultraviolet Irradiator Design

Ultraviolet radiation destroys nucleic acids, which inactivates microorganisms [Stover *et al.*, 1986]. Optimum disinfection occurs at UV wavelengths about 254 nm, and most commercial lamps are designed to emit near the optimum.

Ultraviolet disinfection units consist of parallel arrays of long emission lamps. The lamps are generally 0.9 to 1.6 m long and 1.5 to 1.9 cm in diameter. They are usually spaced several cm apart (center line to center line) and are operated submerged. The liquid flow may be either parallel to the lamp axes or normal to them. The tanks containing the arrays are designed with inlet and outlet baffles to control flow distribution.

The inactivation rate is directly proportional to the UV light intensity, generally expressed as $\mu W/cm^2$. However, both water and suspended solids absorb UV light, so in any real reactor the light intensity declines with distance from the lamp. Inactivation rates are expressed in terms of the average intensity, $\bar{I}$. Because the details of the intensity distribution vary from one design to the next, different designs achieve different inactivation rates even if they produce the same average intensity.

For batch systems the inactivation rate would be represented by [Stover *et al.*, 1986]:

$$\frac{dN}{dt} = -a\bar{I}^b N \tag{9.124}$$

where

a = an empirical rate coefficient ($cm^{2b}/\mu W^b \cdot s$)
b = an empirical exponent (dimensionless)
$\bar{I}$ = the volume-averaged light intensity ($\mu W/cm^2$)
N = the number concentration of microbes (number/m^3)
t = exposure time (s)

Values of a are strongly dependent on the suspended solids content of the water being irradiated. Reported wastewater treatment plant values vary from 40×10^{-9} $(cm^2/\mu W)^{-b} \cdot s^{-1}$ to 0.10×10^{-3} $(cm^2/\mu W)^{-b} \cdot s^{-1}$. Reported values of b are less variable and range from 1.09 to 2.2.

Typical average UV light intensities are on the order of several thousand $\mu W/cm^2$, which corresponds to a UV power density of several W/L. Tubular arrays of lamps generally produce lower average intensities for the same power density than do uniform arrays.

The disinfection units may be analyzed as plug flow reactors with dispersion [Stover *et al.*, 1986]:

$$N = N_o \exp\left[\frac{UL}{2E}\left(1 - \sqrt{1 + \frac{4a\bar{I}^b E}{U^2}}\right)\right] + 0.26X_{SS}^{1.96} \tag{9.125}$$

where

a = an empirical rate coefficient ($cm^{2b}/\mu W^b \cdot s$)
b = an empirical exponent (dimensionless)

E = longitudinal shear dispersion coefficient (m^2/s)
$\bar{I}$ = the volume-averaged light intensity ($\mu W/cm^2$)
L = the length of the irradiation chamber (m)
N = the effluent number concentration of microbes (coli/100 mL)
N_o = the influent number concentration of microbes (coli/mL)
U = the mean velocity in the irradiator (m/s)
X_{SS} = the suspended solids concentration in mg/L

The first term on the right-hand side represents the residual microbial concentration that would be in the irradiator effluent if all the microorganisms were dispersed as single cells or cysts. The second term represents the microorganisms that are incorporated into suspended solids and that escape irradiation.

The coefficient and exponent on the suspended solids concentration X_{SS} represent the experience in disinfection of coliform bacteria at the Port Richmond Water Pollution Control Plant on Staten Island, New York. If the suspended solids concentration is given as mg/L, the resulting fecal coliform count is given as cells per 100 mL. Fecal coliform data from other wastewater treatment facilities are scattered widely around the Port Richmond correlation. Correlations for water treatment plants are not available.

Sludge Disinfection

The general requirements for pathogens in sewage sludges applied to land are given in Table 9.10 [EPA, 1993]. The Class A requirements apply to sewage sludges used on home gardens and lawns. Class A specifications also may be used for sludges applied to agricultural land, forest, public contract sites, or reclamation sites. Class B specification may be used for sludges applied to agricultural land, forests, public contract sites, and reclamation sites if certain site usage, cropping, and pasturing restrictions are met.

Sewage sludges can be graded Class A if they are subjected to certain pathogen reduction treatments. There are six alternative treatments.

Class A

Alternative 1. This alternative requires heat inactivation of pathogens. Specific testing for enteric viruses and helminth ova is not required.

The minimum temperature for heat inactivation regardless of contacting time is 50°C [EPA, 1993]. The contacting time required depends on the solids content of the sludge, the temperature, and the mode of heating. Required contacting times for disinfection temperatures of 50°C

Table 9.10 General Pathogen Restrictions for the Disposal of Sewage Sludges[1]

Pathogen	Class A Requirements (lawn and garden)	Class B Requirements (agricultural land)
Fecal coliform	All alternatives: <1000 MPN per g total solids	$<2 \times 10^6$ per g total solid.
Salmonella	All alternatives: <3 MPN per 4 g total solids	—
Enteric viruses	Alternatives 3 through 6: <1 PFU per 4 g total solids	—
Helminth ova	Alternatives 3 through 6: <1 ovum per 4 g total solids	—

[1]Environmental Protection Agency. 1993. Standards for the use or disposal of sewage sludge. *Federal Register.* 58(32):9248.

and above are given by the following equations. Note that each method of heating and sludge concentration has an absolute minimum contacting time regardless of temperature.

- If the sludge solids concentration is 7% by weight or higher, heating of the bulk sludge is by heat exchanger; use Eq. (9.126) below. The minimum contacting time regardless of disinfection temperature is 20 min.

$$t_c = \frac{131{,}700{,}000}{10^{0.1440T}} \tag{9.126}$$

where

t_c = the contacting time in days
$\geq$ 20 min regardless of temperature
T = the inactivation temperature in °C
$\geq$ 50°C

- If the sludge solids concentration is 7% by weight or higher, sludge heating is by diffused hot gas or hot immiscible liquid; use Eq. (9.126), with the proviso that the minimum contacting time is 15 s, regardless of disinfection temperature.
- If the sludge concentration is less than 7% solids by weight, use Eq. (9.127) to determine contacting times, with the proviso that the minimum contacting time regardless of inactivation temperature is 30 min:

$$t_c = \frac{50{,}070{,}000}{10^{0.1440T}} \tag{9.127}$$

where

t_c = the contacting time in days
$\geq$ 30 min regardless of temperature
T = the inactivation temperature in °C
$\geq$ 50°C

Alternative 2. This method combines pH and heat inactivation:

- The pH of the sludge shall be kept above 12 for 72 hours.
- During the period of elevated pH, the sludge temperature shall be kept at 52°C or higher for at least 12 hours.
- After heat/pH treatment, the sludge shall be air-dried to a solids content of at least 50% by weight.

Alternative 3. No treatment method is specified; rather, the sludge quality is further constrained. The sewage sludge shall be tested periodically for enteric viruses and viable helminth ova as well as fecal coliform and *Salmonella,* and it will be graded Class A as long as the conditions of Table 9.10 are met. Failure to meet the tabulated requirements necessitates pathogen treatment and retesting of the treated sludge.

Alternative 4. No pathogen reduction treatment is required. However, the sludge must meet the conditions in Table 9.10 at the time of sale or usage.

Alternative 5. The sludge must meet the fecal coliform and *Salmonella* standards in Table 9.10 at the time of sale or usage and must be subjected to pathogen reduction treatments specified in Appendix B of the rule.

Alternative 6. The sludge must meet the fecal coliform and *Salmonella* standards in Table 9.10 at the time of sale and usage and must be subjected to pathogen reduction treatments approved by the permitting authority.

Class B

Alternative 1. The sludge must meet the fecal coliform standard in Table 9.10 at the time of usage.

Alternative 2. The sludge must be subjected to a pathogen reduction treatment specified in Appendix B of the rule.

Alternative 3. Sludges graded Class B must be subjected to pathogen reduction processes approved by the permitting authority. Also, there are site usage restrictions:

- Food crops that touch the sludge/soil mixture and are not entirely above ground may not be harvested for 14 months after sludge application.
- Food crops with harvested parts below the ground may not be harvested for 20 months after sludge application if the sludge remains on the land surface for 4 months or longer prior to incorporation into the soil.
- Food crops with harvested parts below the ground may not be harvested for 38 months after sludge application when the sludge lies on the land surface for less than 4 months before incorporation into the soil.
- Food crops, feed crops, and fiber crops shall not be harvested for 30 days after sludge application.
- Animals may not be pastured on the land for 30 days after sludge application.
- Turf may not be harvested for 1 year after sludge application if the turf will be used as a lawn or if there is a high potential for public contact with the turf.
- Public access to land with a high potential to public exposure shall be restricted for 1 year after sludge application.
- Public access to land with a low potential for public exposure shall be restricted for 30 days after sludge application.

Lime Stabilization

Lime stabilization requires that the sludge pH be raised above 12 and preferably above 12.5. At these pHs there is substantial hydrolysis of organic molecules, and the lime dosage is controlled by the amount of solids to be treated. Some guidelines for hydrated lime dosages are as follows [Joint Task Force, 1992]:

Primary sludge: 0.10 to 0.15 kg $Ca(OH)_2$ per kg suspended solids
Activated sludge: 0.30 to 0.50 kg $Ca(OH)_2$ per kg suspended solids
Septage: 0.10 to 0.30 kg $Ca(OH)_2$ per kg suspended solids

At pH 12 a minimum contacting time of at least 72 hours is required for stabilization (Class A, alternative 2). The hydrolysis reactions consume lime, so additional dosages of lime may be required during the contacting period.

Fecal streptococci are not inactivated unless the contacting time is at least 1 h, and nematodes, protozoa, and mites are not affected by high pHs.

Lime stabilization releases organic nitrogen as ammonia, and some means of controlling the emitted ammonia gas is required.

Trihalomethane Formation

The chlorination of natural waters containing humic and fulvic acids results in the formation of chloroform and other halogenated methane derivatives. In test animals, chloroform causes central nervous system depression, hepatotoxicity, nephrotoxicity, teratogenicity, and carcinogenicity [Symons *et al.*, 1981]. Chloroform in drinking water is suspected to cause cancer in humans.

Trihalomethane formation can be avoided either by removing the organic precursors with activated carbon prior to chlorination or by switching to chlorine dioxide or ozone [Symons *et al.*, 1981].

The U.S. Environmental Protection Agency is preparing a disinfection byproducts (DBP) rule to regulate the concentrations of trihalomethanes (THM) and haloacetic acids (HAA) that are formed during chlorination. The substances likely to be regulated are chloroform ($CHCl_3$), bromoform ($CHBr_3$), bromodichloromethane ($CHCl_2Br$), dibromochloromethane ($CHClBr_2$), dichloroacetic acid ($Cl_2CHCOOH$), trichloroacetic acid (Cl_3CCOOH), monochloroacetic acid ($ClCH_2COOH$), monobromoacetic acid ($BrCH_2COOH$), bromochloroacetic acid ($BrClCHCOOH$), and dibromoacetic acid ($Br_2CHCOOH$).

Predictive, empirical models for the resulting THM or HAA have been developed by Watson [1993]. In these models the concentration of the DBP is given in μg/L; all other concentrations are in mg/L; the extinction coefficient for UV light at 254 nm is given in reciprocal cm; and the contacting time is in hours. Chloroform:

$$\begin{aligned} CHCl_3 &= 0.064(\text{TOC})^{0.329}(\text{pH})^{1.161}(^\circ\text{C})^{1.018}(\text{Cl}_2\text{ dose})^{0.561} \times \cdots \\ &\times (\text{Br}^- + 0.01)^{-0.404}(k_{\text{ext},254})^{0.874}(\text{h})^{0.269} \end{aligned} \tag{9.128}$$

Bromodichloromethane, for $Cl_2/Br^- < 75$:

$$CHCl_2Br = 0.0098(\text{pH})^{2.550}(^\circ\text{C})^{0.519}(\text{Cl}_2\text{ dose})^{0.497}(\text{Br}^-)^{0.181}(\text{h})^{0.256} \tag{9.129}$$

Bromodichloromethane, for $Cl_2/Br^- > 75$:

$$CHCl_2Br = 1.325(\text{TOC})^{-0.725}(^\circ\text{C})^{1.441}(\text{Cl}_2\text{ dose})^{0.632}(\text{Br}^-)^{0.794}(\text{h})^{0.204} \tag{9.130}$$

Dibromochloromethane, for $Cl_2/Br^- < 50$:

$$CHClBr_2 = 14.998(\text{TOC})^{-1.665}(^\circ\text{C})^{0.989}(\text{Cl}_2\text{ dose})^{0.729}(\text{Br}^-)^{1.241}(\text{h})^{0.261} \tag{9.131}$$

Dibromochloromethane, for $Cl_2/Br^- > 50$:

$$\begin{aligned} CHClBr_2 &= 0.028(\text{TOC})^{-1.078}(\text{pH})^{1.956}(^\circ\text{C})^{0.596}(\text{Cl}_2\text{ dose})^{1.072} \times \cdots \\ &\times (\text{Br}^-)^{1.573}(k_{\text{ext},254})^{-1.175}(\text{h})^{0.200} \end{aligned} \tag{9.132}$$

Bromoform:

$$CHBr_3 = 6.533(\text{TOC})^{-2.031}(\text{pH})^{1.603}(\text{Cl}_2\text{ dose})^{1.057}(\text{Br}^-)^{1.388}(\text{h})^{0.136} \tag{9.133}$$

Monochloroacetic acid, for $h > 12$:

$$\begin{aligned} ClCH_2COOH &= 1.634(\text{TOC})^{0.753}(\text{pH})^{-1.124}(\text{Cl}_2\text{ dose})^{0.509} \times \cdots \\ &\times (\text{Br}^- + 0.01)^{-0.085}(\text{h})^{0.300} \end{aligned} \tag{9.134}$$

Dichloroacetic acid:

$$\begin{aligned} Cl_2CHCOOH &= 0.605(\text{TOC})^{0.291}(^\circ\text{C})^{0.665}(\text{Cl}_2\text{ dose})^{0.480} \times \cdots \\ &\times (\text{Br}^- + 0.01)^{-0.568}(k_{\text{ext},254})^{0.726}(\text{h})^{0.239} \end{aligned} \tag{9.135}$$

Trichloroacetic acid:

$$\begin{aligned} Cl_3CCOOH &= 87.182(\text{TOC})^{0.355}(\text{pH})^{-1.732}(Cl_2\ \text{dose})^{0.881} \times \cdots \\ &\times (\text{Br}^- + 0.01)^{-0.679}(k_{\text{ext},254})^{0.901}(\text{h})^{0.264} \end{aligned} \tag{9.136}$$

Monobromoacetic acid:

$$\begin{aligned} BrCH_2COOH &= 0.176(\text{TOC})^{1.664}(\text{pH})^{-0.927}(^\circ\text{C})^{0.450}(\text{Br}^-)^{0.795} \times \cdots \\ &\times (k_{\text{ext},254})^{-0.624}(\text{h})^{0.145} \end{aligned} \tag{9.137}$$

Dibromoacetic acid:

$$\begin{aligned} Br_2CHCOOH &= 84.945(\text{TOC})^{-0.620}(^\circ\text{C})^{0.657}(Cl_2\ \text{dose})^{-0.200} \times \cdots \\ &\times (\text{Br}^-)^{1.073}(k_{\text{ext},254})^{0.651}(\text{h})^{0.120} \end{aligned} \tag{9.138}$$

where

Br^- = the bromide concentration in mg/L
°C = the reaction temperature in °C
Cl_2 dose = the chlorine dose in mg/L
DBP = the disinfection byproduct in μg/L
h = the contacting time in hours
$k_{\text{ext},254}$ = the extinction coefficient for ultraviolet light at 254 nm in cm^{-1}
pH = the contacting pH
TOC = the total organic carbon concentration in mg/L

References

Section 9.1

Adamson, A. W. 1982. *Physical Chemistry of Surfaces,* 4th ed. John Wiley & Sons, New York.

Alfrey, A. C., LeGendre, G. R., and Kaehny, W. D. 1976. The dialysis encephalopathy syndrome. A possible aluminum intoxification. *New England Journal of Medicine.* 294(1):184.

Babbitt, H. E., Doland, J. J., and Cleasby, J. L. 1967. *Water Supply Engineering,* 6th ed. McGraw-Hill, New York.

Baker, M. N. 1948. *The Quest for Pure Water.* American Water Works Association, New York.

Camp, T. R. and Conklin, G. F. 1970. Towards a rational jar test for coagulation. *Journal of the New England Water Works Association.* 84(3):325.

Committee on Standard Methods of Water Analysis. 1901. Second report of progress. *Public Health Papers and Reports.* 27:377.

Committee on Water Works Practice. 1940. *Manual of Water Quality and Treatment,* 1st ed. American Water Works Association, New York.

Crapper, D. R., Krishnan, S. S., and Dalton, A. J. 1973. Brain aluminum in Alzheimer's disease and experimental neurofibrillary degeneration. *Science.* 180(4085):511.

Davison, A. M., Walker, G. S., Oli, H., and Lewins, A. M. 1982. Water supply aluminum concentration, dialysis dementia, and effect of reverse osmosis water treatment. *The Lancet.* II(8302):785.

Dentel, S. K. and Gossett, J. M. 1988. Mechanisms of coagulation with aluminum salts. *Journal of the American Water Works Association.* 80(4):187.

Donaldson, W. 1938. Chemical treatment of sewage. In *Modern Sewage Disposal: Anniversary Book of the Federation of Sewage Works Associations,* p. 85. L. Pearse, ed. Federation of Sewage Works Associations, New York.

Driscoll, C. T. and Letterman, R. D. 1988. Chemistry and fate of Al(III) in treated drinking water. *Journal of Environmental Engineering.* 114(1):21.

Einstein, A. 1956. *Investigations on the Theory of the Brownian Movement,* trans. A. D. Cowper. R. Fürth, ed. Dover Publications, New York.

Fanning, J. T. 1887. *A Practical Treatise on Hydraulic and Water-Supply Engineering.* Van Nostrand, New York.

Fridrikhsberg, D. A. 1986. *A Course in Colloid Chemistry,* trans. G. Leib. Mir, Moscow.

Fuller, G. W. 1898. *Report on the Investigations into the Purification of the Ohio River Water at Louisville, Kentucky, made to the President and Directors of the Louisville Water Company.* Van Nostrand, New York.

Haarhoff, J. and Cleasby, J. L. 1988. Comparing aluminum and iron coagulants for in-line filtration of cold water. *Journal of the American Water Works Association.* 80(4):168.

Hardenbergh, W. A. 1940. *Operation of Water-Treatment Plants.* International Textbook, Scranton, PA.

Hayden, P. L. and Rubin, A. J. 1974. Systematic investigation of the hydrolysis and precipitation of aluminum(III). In *Aqueous-Environmental Chemistry of Metals,* p. 317. A. J. Rubin, ed. Ann Arbor Science, Ann Arbor, MI.

Hazen, A. 1890. Report of experiments upon the chemical precipitation of sewage made at the Lawrence Experiment Station during 1889. In *Experimental Investigations by the State Board of Health of Massachusetts, upon the Purification of Sewage by Filtration and by Chemical Precipitation, and upon the Intermittent Filtration of Water. Made at Lawrence, Mass., 1888–1890: Part II of Report on Water Supply and Sewerage,* p. 735. Wright & Potter, State Printers, Boston, MA.

Hedgepeth, L. L. 1934. Coagulants used in water purification and why. *Journal of the American Water Works Association.* 26(9):1222.

Hopkins, E. S. and Bean, E. L. 1966. *Water Purification Control,* 4th ed. Williams & Wilkins, Baltimore, MD.

Hudson, H. E., Jr. 1965. Physical aspects of flocculation. *Journal of the American Water Works Association.* 57(7):885.

Hudson, H. E., Jr., and Singley, J. E. 1974. Jar testing and utilization of jar-test data. *Proceedings AWWA Seminar on Upgrading Existing Water-Treatment Plants.* p. VI-79. American Water Works Association, Denver, CO.

James M. Montgomery, Consulting Engineers, Inc. 1985. *Water Treatment Principles and Design.* John Wiley & Sons, New York.

Jirgensons, B. and Straumanis, M. E. 1956. *A Short Textbook of Colloid Chemistry.* John Wiley & Sons, New York.

Joint Committee of the American Society of Civil Engineers and the Water Pollution Control Federation. 1959. *Sewage Treatment Plant Design.* ASCE Manual of Engineering Practice No. 36, WPCF Manual of Practice No. 8. American Society of Civil Engineers and Water Pollution Control Federation, New York.

Joint Editorial Board. 1992. *Standard Methods for the Examination of Water and Wastewater,* 18th ed. American Public Health Association, Washington, DC.

Kemmer, F. N., ed. 1988. *The NALCO Water Handbook,* 2nd ed. McGraw-Hill, New York.

Klatzo, I., Wismiewski, H., and Streicher, E. 1965. Experimental production of neurofibrillary degeneration. *Journal of Neuropathology and Experimental Neurology.* 24(1):187.

Kopeloff, L. M., Barrera, S. E., and Kopeloff, N. 1942. Recurrent conclusive seizures in animals produced by immunologic and chemical means. *American Journal of Psychiatry.* 98(4):881.

Kruyt, H. R. 1930. *Colloids: A Textbook,* 2nd ed., trans. H. S. van Klooster. John Wiley & Sons, New York.

Langelier, W. F., Ludwig, H. F., and Ludwig, R. G. 1953. Flocculation phenomena in turbid water clarification. *Transactions of the American Society of Civil Engineers.* 118:147.

Letterman, R. D. and Driscoll, C. T. 1988. Survey of residual aluminum in filtered water. *Journal of the American Water Works Association.* 80(4):154.

Mackrle, S. 1962. Mechanism of coagulation in water treatment. *Journal of the Sanitary Engineering Division, Proceedings of the American Society of Civil Engineers.* 88(SA3):1.

Miller, R. G., Kopfler, F. C., Kelty, K. C., Stober, J. A., and Ulmer, N. S. 1984. The occurrence of aluminum in drinking water. *Journal of the American Water Works Association.* 76(1):84.

O'Melia, C. R. 1972. Coagulation and flocculation. In *Physicochemical Processes for Water Quality Control,* p. 61. W. J. Weber, Jr., ed. John Wiley & Sons, New York.

Ostwald, W. 1915. *A Handbook of Colloid-Chemistry,* trans. M. H. Fischer. P. Blaikston's Son, Philadelphia, PA.

Platts, M. M., Goode, G. C., and Hislop, J. S. 1977. Composition of domestic water supply and the incidence of fractures and encephalopathy in patients on home dialysis. *British Medical Journal.* 2(6088):657.

Rubin, A. J. and Kovac, T. W. 1974. Effect of aluminum(III) hydrolysis on alum coagulation. In *Chemistry of Water Supply, Treatment, and Distribution,* p. 159. A. J. Rubin, ed. Ann Arbor Science, Ann Arbor, MI.

Rudolfs, W. 1940. Chemical treatment of sewage. *Sewage Works Journal.* 12(6):1051.

Sidgwick, N. V. 1950. *The Chemical Elements and Their Compounds, Vols. I and II.* Oxford University Press, London.

Stein, M. F. 1915. *Water Purification Plants and Their Operation.* John Wiley & Sons, New York.

Stumm, W. and Morgan, J. J. 1970. *Aquatic Chemistry: An Introduction Emphasizing Equilibria in Natural Waters.* John Wiley & Sons, New York.

Stumm, W. and O'Melia, C. R. 1968. Stoichiometry of coagulation. *Journal of the American Water Works Association.* 60(5):514.

Svedberg, T. 1924. *Colloid Chemistry: Wisconsin Lectures.* Chemical Catalog, New York.

Vaughn, J. C., Turre, G. J., and Grimes, B. L. 1971. Chemicals and chemical handling. In *Water Quality and Treatment: A Handbook of Public Water Supplies,* 3rd ed., p. 526. P. D. Haney *et al.,* ed. McGraw-Hill, New York.

Voyutsky, S. 1978. *Colloid Chemistry,* trans. N. Bobrov. Mir, Moscow.

Zsigmondy, R. 1914. *Colloids and the Ultramicroscope: A Manual of Chemistry and Ultramicroscopy,* trans. J. Alexander. John Wiley & Sons, New York.

Section 9.2

Abrams, I. M. and Benezra, L. 1967. Ion exchange polymers. In *Encyclopedia of Polymer Science and Technology,* p. 692. Vol. 7. John Wiley & Sons, New York.

Baker, M. N. 1981. *The Quest for Pure Water: The History of Water Purification from the Earliest Records to the Twentieth Century, Volume I,* 2nd ed. American Water Works Association, Denver, CO.

Clifford, D. A. 1990. Ion exchange and inorganic adsorption. In *Water Quality and Treatment: A Handbook of Community Water Supplies,* 4th ed., p. 561. F. W. Pontius, ed. McGraw-Hill, New York.

Culp/Wesner/Culp, Inc. 1986. *Handbook of Public Water Systems.* R. B.Williams and G. L. Culp, ed. Van Nostrand Reinhold, New York.

Environmental Protection Agency. 1991. Drinking water regulations: Maximum contaminant level goals and national primary drinking water regulations for lead and copper. *Federal Register.* 56(110):26460.

Hoover, C. P. 1938. Practical application of the Langelier method. *Journal of the American Water Works Association.* 30(11):1802.

Kemmer, F. N., ed. 1988. *The Nalco Water Handbook,* 2nd ed. McGraw-Hill, New York.

Langelier, W. F. 1936. The analytical control of anti-corrosion water treatment. *Journal of the American Water Works Association.* 28(10):1500.

Langelier, W. F. 1946. Chemical equilibria in water treatment. *Journal of the American Water Works Association.* 38(2):169.

Larson, T. E. and Buswell, A. M. 1942. Calcium carbonate saturation index and alkalinity interpretations. *Journal of the American Water Works Association.* 34(11):1667.

Ludzack, F. J. and Noran, D. K. 1965. Tolerance of high salinities by conventional wastewater treatment processes. *Journal of the Water Pollution Control Federation.* 37(10):1404.

Lee, R. G., Becker, W. C., and Collins, D. W. 1989. Lead at the tap: Sources and control. *Journal of the American Water Works Association.* 81(7):52.

Powell, S. T. 1954. *Water Conditioning for Industry.* McGraw-Hill, New York.

Ryznar, J. W. 1944. A new index for determining amount of calcium carbonate scale formed by a water. *Journal of the American Water Works Association.* 36(4):472.

Schock, M. R. 1984. Temperature and ionic strength corrections to the Langelier index—revisited. *Journal of the American Water Works Association.* 76(8):72.

Schock, M. R. 1989. Understanding corrosion control strategies for lead. *Journal of the American Water Works Association.* 81(7):88.

Schock, M. R. 1990. Internal corrosion and deposition control. In *Water Quality and Treatment: A Handbook of Community Water Supplies,* 4th ed., p. 997. F. W. Pontius, ed. McGraw-Hill, New York.

Singley, J. E., Pisigan, R. A., Jr., Ahmadi, A., Pisigan, P. O., and Lee, T.-Y. 1985. *Corrosion and Carbonate Saturation Index in Water Distribution Systems.* EPA/600/S2-85/079. U.S. Environmental Protection Agency, Water Engineering Research Laboratory, Center for Environmental Research Information, Cincinnati, OH.

Stumm, W. and Morgan, J. J. 1970. *Aquatic Chemistry: An Introduction Emphasizing Chemical Equilibria in Natural Waters.* John Wiley & Sons, New York.

Tebbutt, T. H. Y. 1992. *Principles of Water Quality Control,* 4th ed. Pergamon Press, Oxford, UK.

Section 9.3

Chick, H. 1908. An investigation of the laws of disinfection. *Journal of Hygiene.* 8:92.

Craun, G. F. 1988. Surfacewater supplies and health. *Journal of the American Water Works Association.* 80(2):40.

Craun, G. F. and McCabe, L. J. 1973. Review of the causes of water-borne disease outbreaks. *Journal of the American Water Works Association.* 65(1):74.

Criteria Branch, ed. 1976. *Quality Criteria for Water.* Environmental Protection Agency, Office of Water Planning and Standards, Division of Criteria and Standards, Washington, DC.

Environmental Protection Agency. 1989a. Drinking water; national primary drinking water regulations; filtration, disinfection, turbidity, *Giardia lamblia,* viruses, *Legionella,* and heterotrophic bacteria. Final rule. *Federal Register.* 54(124):27485.

Environmental Protection Agency. 1989b. Total coliforms. Final rule. *Federal Register.* 54(124):27544.

Environmental Protection Agency. 1990. *Guidance Manual for Compliance with the Filtration and Disinfection Requirements for Public Water Systems Using Surface Water Sources,* Washington, D.C.

Environmental Protection Agency. 1993. Standards for the use or disposal of sewage sludge. *Federal Register.* 58(32):9248.

Fair, G. M. and Geyer, J. C. 1954. *Water Supply and Waste-water Disposal.* John Wiley & Sons, New York.

Haas, C. N. 1990. Disinfection. In *Water Quality and Treatment: A Handbook of Community Water Supplies,* 4th ed., p. 877. F. W. Pontius, ed. McGraw-Hill, New York.

Herwaldt, B. L., Craun, G. F., Stokes, S. L., and Juranek, D. D. 1992. Outbreaks of waterborne disease in the United States: 1989–1990. *Journal of the American Waterworks Association.* 84(4):129.

Joint Task Force of the American Society of Civil Engineers and American Water Works Association. 1990. *Water Treatment Plant Design,* 2nd ed. McGraw-Hill, New York.

Joint Task Force of the Water Environment Federation and the American Society of Civil Engineers. 1992. *Design of Municipal Wastewater Treatment Plants: Volume II.* Chapters 13–20. Water Environment Federation, Alexandria, VA; American Society of Civil Engineers, New York.

Katz, J., ed. 1980. *Ozone and Chlorine Dioxide Technology for Disinfection of Drinking Water.* Noyes Data Corporation, Park Ridge, NJ.

Lawler, D. F. and Singer, P. C. 1993. Analyzing disinfection kinetics and reactor design: A conceptual approach versus the SWTR. *Journal of the American Water Works Association.* 85(11):67.

Morris, J. C. 1975. Aspects of the quantitative assessment of germicidal efficiency. In *Disinfection: Water and Wastewater,* p. 1. J. D. Johnson, ed. Ann Arbor Science, Ann Arbor, MI.

Palin, A. T. 1977. Water disinfection: Chemical aspects and analytical control. In *Disinfection: Water and Wastewater,* p. 67. J. D. Johnson, ed. Ann Arbor Science, Ann Arbor, MI.

Pontius, F. W. 1990. New regulations for total coliforms. *Journal of the American Water Works Association.* 82(8):16.

Pontius, F. W. 1993. Protecting the public against Cryptosporidium. *Journal of the American Water Works Association.* 85(8):18.

Rav-Acha, C. 1984. The reactions of chlorine dioxide with aquatic organic materials and their health effects. *Water Research.* 18(11):1329.

Rose, J. B. 1988. Occurrence and significance of *Cryptosporidium* in water. *Journal of the American Water Works Association.* 80(2):53.

Saunier, B. M. and Selleck, R. E. 1979. The kinetics of breakpoint chlorination in continuous flow systems. *Journal of the American Water Works Association.* 71(3):164.

Sidgwick, N. V. 1950. *The Chemical Elements and Their Compounds, Volume I.* Oxford University Press, London.

Sobsey, M. and Olson, B. 1983. Microbial agents of waterborne disease. In *Assessment of Microbiology and Turbidity Standards for Drinking Water: Proceedings of a Workshop, December 2–4, 1981.* EPA 570-9-83-001. P. S. Berger and Y. Argaman, ed. Environmental Protection Agency, Office of Drinking Water, Washington, DC.

Stover, E. L., Haas, C. N., Rakness, K. L., and Scheible, O. K. 1986. *Design Manual: Municipal Wastewater Disinfection.* EPA/625/1-86-021. Environmental Protection Agency, Office of Research and Development, Water Engineering Research Laboratory, Center for Environmental Research Information, Cincinnati, OH.

Symons, J. M., Stevens, A. A., Clark, R. M., Geldreich, E. E., Love, O. T., Jr., and DeMarco, J. 1981. *Treatment Techniques for Controlling Trihalomethanes in Drinking Water.* EPA-600/2-81-156. Environmental Protection Agency, Office of Research and Development, Municipal Environmental Research Laboratory, Drinking Water Research Division, Cincinnati, OH.

Teefy, S. M. and Singer, P. C. 1990. Performance and analysis of tracer tests to determine compliance of a disinfection scheme with the SWTR. *Journal of the American Water Works Association.* 82(12):88.

Watson, H. E. 1908. A note on the variation of the rate of disinfection with change in the concentration of the disinfectant. *Journal of Hygiene.* 8:536.

Watson, M. 1993. *Final Report: Mathematical Modeling of the Formation of THMs and HAAs in Chlorinated Natural Waters.* American Water Works Association, Denver, CO.

Wei, J. and Chang, S. L. 1975. A multi-poisson distribution model for treating disinfection data. In *Disinfection: Water and Wastewater,* p. 11. J. D. Johnson, ed. Ann Arbor Science, Ann Arbor, MI.

Wei, I. W. and Morris, J. C. 1974. Dynamics of breakpoint chlorination. In *Chemistry of Water Supply, Treatment, and Distribution,* p. 297. A. J. Rubin, ed. Ann Arbor Science, Ann Arbor, MI.

Wen, C. Y. and Fan, L. T. 1975. *Models for Flow Systems and Chemical Reactors.* Marcel Dekker, New York.

10

Biological Water Treatment Processes

Robert M. Sykes
Ohio State University

10.1 Activated Sludge

The principal wastewater treatment scheme is the activated sludge process. Its various modifications are capable of removing organic matter, converting ammonia to nitrate, converting nitrate to nitrogen gas, and achieving high removals of phosphorus.

Process Parameters and Variables

Gram's Model Parameters

There are two sets of variables used in activated sludge design and operation. These variables are chosen because there are empirical correlations between them and the effluent quality, and because they relate directly to process sizing and operation.

The first set is derived from Gram's [1956] theory. Refer to Fig. 10.1. *Solids retention time* (SRT) can be calculated as follows:

$$\theta_c = \frac{VX}{Q_w X_w + (Q - Q_w) X_e} \tag{10.1}$$

where

Q = the settled sewage flow rate (m^3/s or ft^3/s)
Q_w = the waste activated sludge flow rate (m^3/s or $ft^3 s$)
V = the aeration tank volume (m^3 or ft^3)
X = the concentration of volatile suspended solids in the aeration tank mixed liquor, MLVSS (kg VSS/m^3 or lb VSS/ft^3)
X_e = the concentration of volatile suspended solids in the final effluent (kg VSS/m^3 or lb VSS/ft^3)

0-8493-8953-4/95/$0.00 + $.50

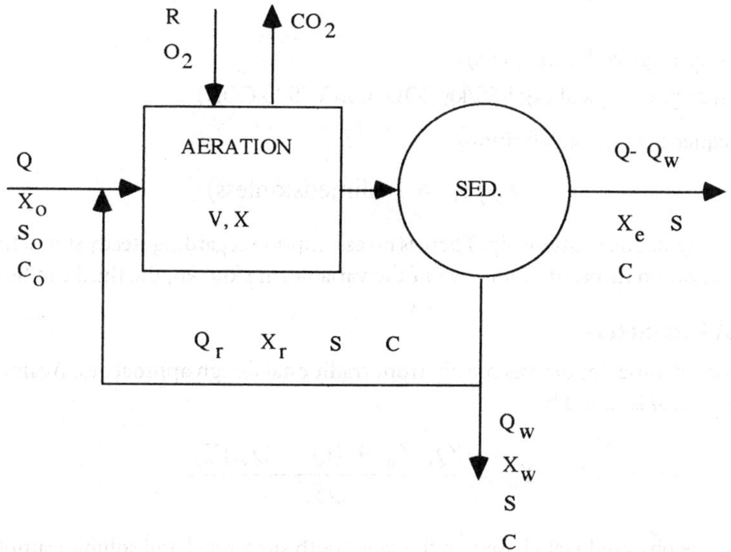

FIGURE 10.1 Definition sketch for activated sludge variables—conventional, non-nitrifying process.

X_w = the concentration of volatile suspended solids in the waste activated sludge (kg VSS/m^3 or lb VSS/ft^3)

θ_c = the solids retention time (s)

The symbol θ_c was introduced by Rolf Eliasson.

Synonyms for SRT are *cell residence time* (CRT), *mean cell residence time* (MCRT), and *sludge age. Sludge age* has been used in the past to mean several different things: the ratio of aeration tank biomass to influent suspended solids loading [Torpey, 1948], the reciprocal food-to-microorganism ratio [Heukelekian, *et al.* 1951], the reciprocal specific uptake rate [Fair and Thomas, 1950], the aeration period [Keefer and Meisel, 1950], and some undefined relationship between system biomass and the influent BOD and suspended solids loading [Eckenfelder, 1956].

Observed yield can be found as follows:

$$Y_o = \frac{Q_w X_w + (Q - Q_w) X_e}{Q(S_o - S)} \tag{10.2}$$

where

Y_o = the observed yield based on the net reduction in organic matter (kg VSS/kg COD of lb VSS/lb COD)

S_o = the *total* (both suspended and soluble) organic matter concentration in the settled sewage (kg COD/m^3 or lb COD/ft^3)

S = the *soluble only* organic matter concentration in the final effluent (kg COD/m^3 or lb COD/ft^3)

Specific uptake (utilization) rate is given as

$$q = U = \frac{Q(S_o - S)}{VX} \tag{10.3}$$

where $q = U =$ the specific uptake (or utilization) rate (kg COD/kg VSS·s or lb COD/lb VSS·s). A related concept is the *specific growth rate*,

$$\mu = Yq \tag{10.4}$$

where

μ = the specific growth rate (per s)
Y = the true growth yield (kg VSS/kg COD or lb VSS/lb COD)

As a consequence of these definitions,

$$Y_o q \theta_c \equiv 1 \text{ (dimensionless)} \tag{10.5}$$

This is purely a semantic relationship. There is no assumption regarding steady states, time averages, or mass conservation involved. If any two of the variables are known, the third can be calculated.

Traditional Parameters

The second set of variables derives mostly from traditional design approaches. Again, refer to Fig. 10.1. *Observed yield* is found by

$$Y'_o = \frac{Q_w X_w + (Q - Q_w) X_e}{Q S_o} \tag{10.6}$$

where Y'_o = the observed yield based on the *total* (both suspended and soluble) supply of organic matter (kg VSS/kg COD or lb VSS/lb COD).

Food-to-microorganism ratio (F/M or F:M) is given by

$$F = \frac{Q S_o}{V X} \tag{10.7}$$

where F = the food-to-microorganism ratio, F/M or F:M (kg COD/kg VSS·s or lb COD/lb VSS·s). Synonyms for F/M are *loading, BOD loading, BOD loading factor, biological loading, organic loading, plant load,* and *sludge loading.* McKinney's [1962] original definition of F/M as the ratio of the BOD and VSS concentrations is still sometimes encountered. Synonyms for McKinney's original meaning are *loading factor* and *floc loading.*

The SRT is defined above in Eq. (10.1), so

$$Y'_o F \theta_c \equiv 1 \text{ (dimensionless)} \tag{10.8}$$

These two sets of variables are related through the removal efficiency:

$$E = \frac{S_o - S}{S_o} = \frac{q}{F} = \frac{Y'_o}{Y_o} \tag{10.9}$$

where E = the removal efficiency (dimensionless).

Return Sludge Flow Rate

For both sets of variables, a steady state suspended solids balance on the secondary clarifier gives the return activated sludge flow rate, Q_r:

$$\frac{Q_r}{Q} = \frac{1 - (V/Q)/\theta_c}{(X_r/X) - 1} \cong \frac{1}{(X_r/X) - 1} \tag{10.10}$$

where

Q = the settled sewage flow rate (m^3/s or ft^3/s)
Q_r = the return activated sludge (RAS) flow rate (m^3/s or ft^3/s)
V = the aeration tank volume (m^3 or ft^3)
X = the concentration of volatile suspended solids in the aeration tank mixed liqour, MLVSS (kg VSS/m^3 or lb VSS/ft^3)

X_r = the concentration of volatile suspended solids in the return (or recycle) activated sludge (kg VSS/m^3 or lb VSS/ft^3)
θ_c = the solids retention time (s)

Steady-State Oxygen Consumption

The steady-state total oxygen demand balance on the whole system applicable to any activated sludge process is

$$R = Q(S_o - S) + 4.57Q(N_o - N) - 2.86G_{N_2} - 1.98\frac{VX}{\theta_c} \tag{10.11}$$

where

R = the oxygen utilization rate (kg O_2/s or lb O_2/s)
N_o = the *total* (both suspended and soluble) TKN of the settled wastewater (kg N/m^3 or lb N/ft^3)
N = the *soluble only* TKN of the final effluent (kg N/m^3 or lb N/ft^3)
G_{N_2} = the rate of production of nitrogen gas by denitrifying bacteria (kg N/s or lb N/s)
4.57 = the oxygen demand of the TKN for the conversion of TKN to HNO_3 (kg O_2/kg TKN or lb O_2/lb TKN)
2.86 = the oxygen demand of nitrogen gas for the conversion of nitrogen gas to HNO_3 (kg O_2/kg N_2 or lb O_2/lb N_2)
1.98 = the oxygen demand of the VSS, assuming complete oxidation to CO_2, H_2O and HNO_3 (kg O_2/kg VSS) or lb O_2/lb VSS

If there is no denitrification, the term for nitrogen gas production, G_{N_2}, is zero. In municipal wastewaters there is no denitrification unless there is first nitrification because all the influent nitrogen is in the reduced forms of ammonia or amines. However, some industrial wastewaters (principally explosives and some agricultural chemicals) do contain significant amounts of nitrates.

If there is no nitrification, the influent TKN is merely redistributed between the soluble TKN output, QN, and the nitrogen incorporated into the waste solids, VX/θ_c. The oxygen demand of the nitrogen in the waste solids exactly cancels the oxygen demand of the TKN removed, and the oxygen utilization rate becomes

$$R = Q(S_o - S) - 1.42\frac{VX}{\theta_c} \tag{10.12}$$

$$R = (1 - 1.42Y_o)Q(S_o - S) \tag{10.13}$$

$$R = [(1 - 1.42Y)q + 1.42k_d] \cdot VX \tag{10.14}$$

where 1.42 = the oxygen demand of the VSS assuming oxidation to CO_2, H_2O and NH_3 (kg O_2/kg VSS). Equation (10.14) is written in terms of Gram's model. Many engineers prefer this form, although all three versions of Eqs. (10.12), (10.13), and (10.14) reduce to one another identically.

In mixed cells in series, the oxygen uptake rate is highest in the inlet compartment and lowest in the outlet compartment. Consequently, the rate of oxygen supply must be "tapered." A commonly recommended air distribution is given later in this chapter (see Table 10.7). This should be checked against the mixing requirements.

Carbonaceous BOD Removal

Designs can be based on either pilot plant data or on the traditional rules of thumb. The traditional rules of thumb are acceptable only for municipal wastewaters that consist primarily of domestic

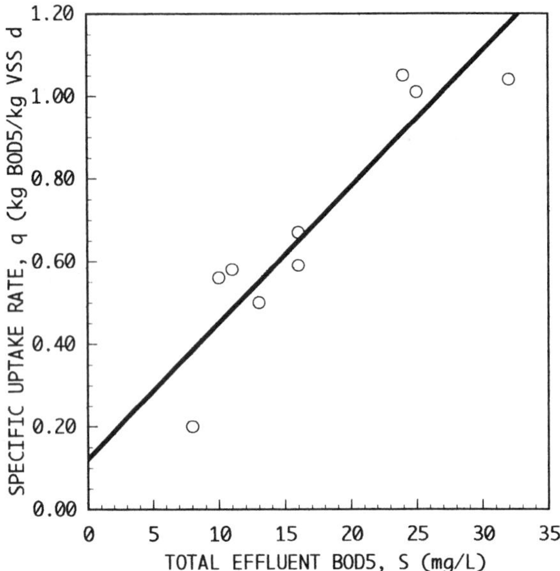

FIGURE 10.2 Typical plot of specific uptake rate vs. effluent BOD_5. (Data from Wuhrmann, 1954.)

wastes. The design of industrial treatment facilities requires pilot testing and careful wastewater characterization.

Pilot Plant Data for Non-nitrifying Activated Sludge

Both the specific uptake rate, $q = U$, and the food-to-microorganism ratio, F, are correlated with the composition of the final effluent. Examples of such correlations are shown in Figs. 10.2 and 10.3. Furthermore, both q and F are correlated with the solids retention time, θ_c. Examples are shown in Figs. 10.4 and 10.5.

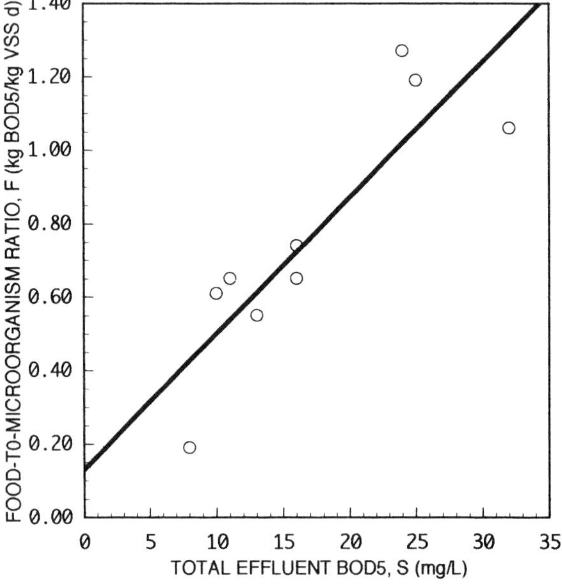

FIGURE 10.3 Typical plot of foot-to-microorganism ratio vs. effluent BOD_5. (Data from Wuhrmann, 1954.)

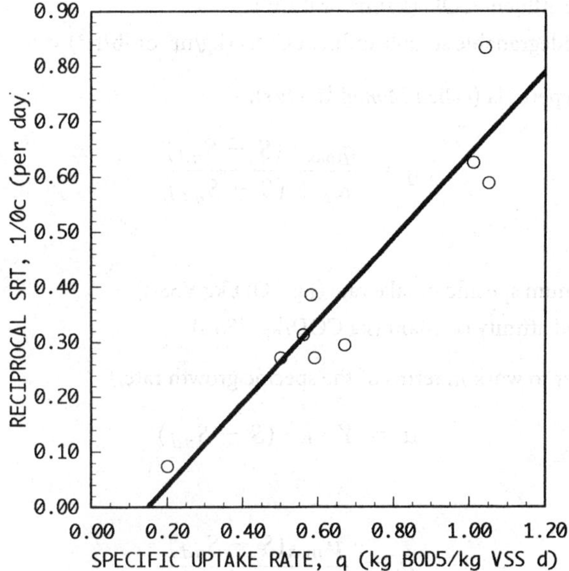

FIGURE 10.4 Typical plot of reciprocal solids retention time vs. specific uptake rate. (Data from Wuhrmann, 1954.)

Pilot plant data are often summarized as formulas. As long as the COD of the settled wastewater is constant, the correlation between effluent COD and q can be approximated as either a straight line (called *first order kinetics*),

$$q = k \cdot (S - S_{nd}) \tag{10.15}$$

where

k = the first order rate constant (m^3/kg VSS·s)
q = the specific uptake rate (kg COD/kg VSS·s or lb COD/lb VSS·s)

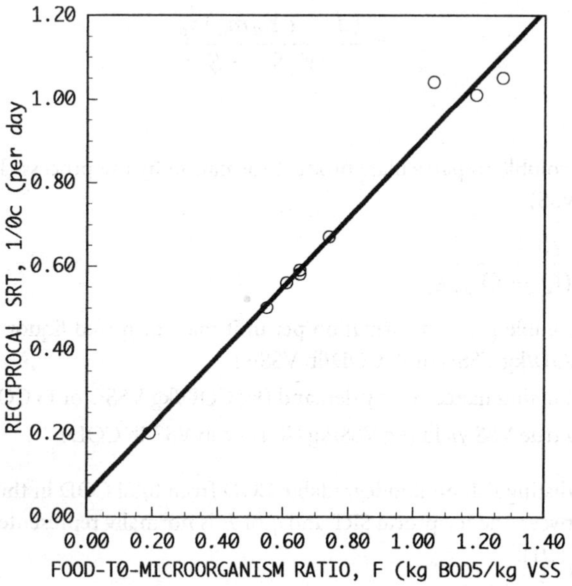

FIGURE 10.5 Typical plot of reciprocal solids retention time vs. food-to-microorganism ratio. (Data from Wuhrmann, 1954.)

S = the soluble effluent COD (kg/m^3 or lb/ft^3)
S_{nd} = the nonbiodegradable soluble effluent COD (kg/m^3 or lb/ft^3)

or as a rectangular hyperbola (called *Monod kinetics*),

$$q = \frac{q_{\max} \cdot (S - S_{nd})}{K_s + (S - S_{nd})} \tag{10.16}$$

where

$q_{\max}$ = the maximum specific uptake rate (kg COD/kg VSS s)
K_s = the Monod affinity constant (kg COD/kg VSS s)

Some people prefer to work in terms of the specific growth rate,

$$\mu = Y \cdot k \cdot (S - S_{nd}) \tag{10.17}$$

or

$$\mu = \frac{\mu_{\max}(S - S_{nd})}{K_s + (S - S_{nd})} \tag{10.18}$$

where

μ = the specific growth rate (per s)
$\mu_{\max}$ = the maximum specific growth rate (per s)
= $Yq_{\max}$

The nonbiodegradable COD, S_{nd}, is defined to be the intercept of the q versus S plot on the COD axis. It is generally on the order of 10 to 30 mg/L and comprises most of the effluent COD in efficient plants.

If the settled sewage exhibits a large variation in BOD_5 and COD, better correlations may be obtained with [Sykes, 1991],

$$F = \frac{(d - cY_a m_a)S_o}{cY_a S_o - S} \tag{10.19}$$

where

c = the ratio of soluble to particulate product formation by the process (kg COD/kg VSS or lb COD/lb VSS)

$$= \frac{QS}{Q_w X_w + (Q - Q_w)X_e} \tag{10.20}$$

d = the rate of soluble product formation per unit mass of mixed liquor volatile suspended solids (kg COD/kg VSS·s or lb COD/lb VSS·s)
m_a = the adjusted maintenance energy demand (kg COD/kg VSS·s or lb COD/lb VSS·s)
Y_a = the adjusted true VSS yield (kg VSS/kg COD or lb VSS/lb COD)

It is not necessary to distinguish nonbiodegradable COD from total COD in this formula.

The correlation between the reciprocal SRT and q or F is normally represented by a straight line [Gram, 1956; Sykes, 1991]:

$$\frac{1}{\theta_c} = Yq - k_d \tag{10.21}$$

or

$$\frac{1}{\theta_c} = Y_a F - Y_a m_a \tag{10.22}$$

where

Y = the true growth yield (kg VSS/kg COD or lb VSS/lb COD)
k_d = the "decay" rate (per second)

These relationships can be rearranged into formulas for the observed yields:

$$Y_o = \frac{Y}{1 + k_d \theta_c} \tag{10.23}$$

or

$$Y'_o = \frac{Y_a}{1 + Y_a m_a \theta_c} \tag{10.24}$$

It should be noted that the "decay" rate is *not* the endogenous respiration rate of the sludge. The endogenous respiration rate of the sludge is determined by measuring the oxygen consumption rate of sludge solids suspended in a solution of mineral salts. The measured rate includes the respiration of bacteria and fungi oxidizing cellular food reserves (true endogenous respiration) and the respiration of predators feeding on their prey (technically exogenous respiration). The "decay rate" is determined by regression of the reciprocal solids retention time on the BOD or COD specific uptake rate. It represents a variety of VSS loss processes including at least:

- Viral lysis of microbial and metazoan cells
- Hydrolysis of solids by exocellular bacterial and fungal enzymes
- Hydrolysis of solids by "intestinal" protozoan, rotiteran, and nematodal enzymes
- Simple dissolution
- Abiotic hydrolysis
- Respiration

The sludge endogenous respiration rate and the decay are uncorrelated. The former decreases as the sludge age increases, but the latter is independent of sludge age.

Finally, q and F generally exhibit a strong linear correlation, which is often convenient. Figure 10.6 is an example. The theoretical formula for the correlation [Sykes, 1991] is:

$$q = (1 - cY_a)F + (cY_a m_a - d) \tag{10.25}$$

Typical values for the parameters are given in Table 10.1. For any given data set, the parameter values are connected because they are determined via regression on the data. This means that the absolute values of the errors of the estimates are about the same. However, the values of the true yields, Y and Y_a, are about an order of magnitude larger than the values of the parameters k_d, m_a, and d. Consequently, while the values of the true yields are quite accurate, the values of the other parameters are fairly uncertain. Moreover, efficient activated sludge plants produce effluents with very little soluble organic matter, and the experimental procedures for COD and BOD yield data with large variances. These large errors result in poor correlations between q and S or F and S. This means that the design procedure must make allowance for substantial uncertainty.

Temperature. Field and laboratory data indicate that the BOD removal efficiency increases from about 80–85 to 90–95% as the temperature increases from about 5 to 30°C [Benedict and Carlson, 1973; Hunter *et al.*, 1966; Keefer, 1962; Ludzack *et al.*, 1961; Sawyer, 1942; Sayigh and Malina,

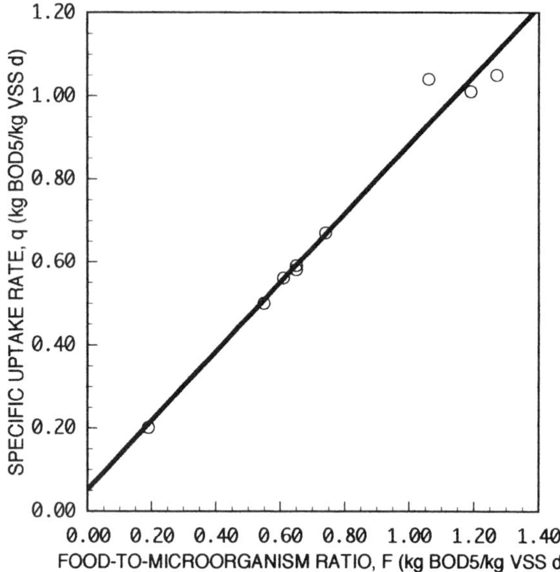

FIGURE 10.6 Typical plot of specific uptake rate vs. food-to-microorganism ratio. (Data from Wuhrmann, 1954.)

1978]. Increases above 30°C do not improve BOD removal efficiency, and increases above 45°C degrade BOD removal efficiency [Hunter *et al.*, 1966].

The true yield coefficients, Y and Y_a, do not vary with temperature.

The values of d, k_d and m_a are so uncertain that temperature adjustments may not be warranted; however, most engineers do make temperature corrections to k_d. This is justified by the reduction in predation and consequent increase in solids production that occurs at low temperatures [Ludzack *et al.*, 1961]. Low-temperature operation also results in poorer flocculation and greater amount of dispersed fine solids. This is also due to reduced predation. The temperature correction to the decay rate would be

$$\frac{k_d(T_1)}{k_d(T_o)} = 1.100^{T_1 - T_o} \tag{10.26}$$

The parameters of the Monod function vary with temperature approximately as follows [Novak, 1974; Giona *et al.*, 1979]:

$$\frac{q_{\max}(T_1)}{q_{\max}(T_o)} \cong 1.100^{T_1 - T_o} \tag{10.27}$$

$$\frac{K_s(T_1)}{K_s(T_o)} \cong 1.075^{T_1 - T_o} \tag{10.28}$$

Lin and Heinke [1977] analyzed 26 years of data from each of 13 municipal plants and concluded that the temperature dependence of the first-order rate coefficient was

$$\frac{k(T_1)}{k(T_o)} \cong 1.125^{T_1 - T_o} \tag{10.29}$$

pH. The optimum pH for the activated sludge process lies between 7.0 and 7.5, but BOD removal is not affected substantially by pHs between about 6.0 and 9.0 [Keefer and Meisel, 1951]. BOD removal falls off sharply outside that range and is reduced by about 50% at pHs of 5 or 10.

Table 10.1 Typical Parameter Values for the Conventional, Non-nitrifying Activated Sludge Process for Municipal Wastewater at Approximately 15 to 25°C[1–7]

Parameter	Symbol	Units	Typical Value	Range (%)
True growth yield	Y	$\frac{\text{kg VSS}}{\text{kg COD}}$	0.4	±10
		$\frac{\text{kg VSS}}{\text{kg CBOD}_5}$	0.7	±10
Adjusted true VSS yield	Y_a	$\frac{\text{kg VSS}}{\text{kg COD}}$	0.4	±10
		$\frac{\text{kg VSS}}{\text{kg CBOD}_5}$	0.6	±10
Decay rate	k_d	per day	0.05	±100
Adjusted maintenance energy demand	m_a	$\frac{\text{kg COD}}{\text{kg VSS d}}$	0.2	±50
		$\frac{\text{kg CBOD}_5}{\text{kg VSS d}}$	0.07	±50
Maximum uptake rate	$q_{\max}$	$\frac{\text{kg COD}}{\text{kg VSS d}}$	10	±50
		$\frac{\text{kg CBOD}_5}{\text{kg VSS d}}$	6.0	±50
Maximum specific growth rate	$\mu_{\max}$	per day	4	±50
Affinity constant	K_s	mg COD/L	500 (total COD)	±50
		mg COD/L	50 (biodegradable COD)	±50
		mg $CBOD_5$/L	100	±50
First-order rate constant	k	$\frac{\text{L}}{\text{mgVSS} \cdot \text{d}}$	0.02 (total COD)	±100
			0.2 (biodegradable COD)	±100
			0.06 (BOD)	±100
Ration soluble to particulate product information	c	$\frac{\text{kg COD}}{\text{kg VSS}}$	0.7	±50
		$\frac{\text{kg CBOD}_5}{\text{kg VSS}}$	0.2	±50
Rate of soluble product formation per unit biomass	d	$\frac{\text{kg COD}}{\text{kg VSS d}}$	0.1	±100
		$\frac{\text{kg CBOD}_5}{\text{kg VSS d}}$	0.04	±100

[1]Joint Task Force of the Water Environment Federation and the American Society of Civil Engineers. 1992. *Design of Municipal Wastewater Treatment Plants: Volume I.* Chapters 1–12. WEF Manual of Practice No. 8; ASCE Manual and Report on Engineering Practice No. 76. Water Environment Federation, Alexandria, VA; American Society of Civil Engineers, New York.

[2]Goodman, B. L. and Englande, A. J., Jr. 1974. A unified model of the activated sludge process. *Journal of the Water Pollution Control Federation.* 46(2):312.

[3]Lawrence, A. W. and McCarty, P. L. 1970. Unified basis for biological treatment design and operation. *Journal of the Sanitary Engineering Division, Proceedings of the American Society of Civil Engineers.* 96(3):757.

[4]Jordan, W. J., Pohland, F. G., and Kornegay, B. H. 1971. Evaluating treatability of selected industrial wastes. *Proceedings of the 26th Industrial Waste Conference,* p. 514. May 4, 5, and 6, 1971. Engineering Extension Series No. 140. ed. J. M. Bell, Purdue University, West Lafayette, IN.

[5]Peil, K. M. and Gaudy, A. J., Jr. 1971. Kinetic constants for aerobic growth of microbial populations selected with various single compounds and with municipal wastes as substrates. *Applied Microbiology.* 21:253.

[6]Sykes, R. M. 1991. The product-maintenance theory of the activated sludge process. *Journal of Environmental Science and Health, Part A, Environmental Science and Engineering.* A26(6):855.

[7]Wuhrmann, K. 1954. High-rate activated sludge treatment and its relation to stream sanitation: I. Pilot plant studies. *Sewage and Industrial Wastes.* 26(1):1.

Table 10.2 Approximate Composition of Growing Bacteria and Nutrient Requirements for Biological Treatment

Component	Weight Percentage[1,2,3]		Mole Ratio	Eckenfelder's Guidelines[4] $\left(\frac{\text{mg substance}}{\text{kg BOD}}\right)$
	Total Weight	Dry Weight		
Water	80	—	—	—
Solids	20	100	—	—
Ash	—	7	—	—
Volatile solids	—	93	—	—
C	—	54	6.5	—
O	—	23	2.1	—
N	—	9.6	1.0	[a]
H	—	7.4	10.7	—
P	—	3	0.14	[b]
K	—	1	0.04	4500
Mg	—	0.7	0.04	3000
Na	—	0.5	0.03	50
S	—	0.5	0.02	—
Ca	—	0.5	0.02	6200
Fe	—	0.025	—	12,000
Cu	—	0.004	—	146
Mn	—	0.004	—	100
Co	—	—	—	130
CO_3	—	—	—	2700
Mo	—	—	—	430
Se	—	—	—	0.0014
Zn	—	—	—	160
Proteins	—	50	—	—
RNA	—	20	—	—
Carbohydrates	—	10	—	—
Lipids	—	7	—	—
DNA	—	3	—	—
Inorganic ions	—	3	—	—
Small molecules	—	3	—	—

[1]Bowen, H. J. M. 1966. *Trace Elements in Biochemistry.* Academic Press, New York.
[2]Porter, J. R. 1946. *Bacterial Chemistry and Physiology.* John Wiley & Sons, New York.
[3]Watson, J. D. 1965. *Molecular Biology of the Gene.* W. A. Benjamin, New York.
[4]Eckenfelder, W. W., Jr. 1980. Principles of biological treatment. In *Theory and Practice of Biological Wastewater Treatment,* ed. K. Curi and W. W. Eckenfelder, Jr., p. 49. Sijthoff & Noordhoff BV., Germantown, MD.

[a] $N(kg/d) = [0.123 \cdot f + 0.07 \cdot (0.77 - f)] \cdot \Delta MLVSS(kg/d)/0.77$

[b] $P(kg/d) = [0.026 \cdot f + 0.01 \cdot (0.77 - f)] \cdot \Delta MLVSS(kg/d)/0.77$

where f = the biodegradable fraction of the MLVSS ≈ 0.6 to 0.4 at SRTs of 1 to 4 days, respectively.

Nutrients. Most municipal and many industrial wastewaters have a proper balance of nutrients for biological waste treatment. However, some industrial wastes may be deficient in one or more required elements. The traditional rule of thumb is that the BOD:N and BOD:P ratios should be less than 20:1 and 100:1, respectively [Helmers *et al.*, 1951]. Sludge yields and treatment efficiencies fall when the BOD:P ratio exceeds about 220 [Greenberg *et al.*, 1955; Verstraete and Vissers, 1980].

Some industrial wastes are deficient in metals, especially potassium. Table 10.2 is the approximate composition of bacterial cells and may be used as a guide to nutrient requirements.

Waste activated sludge is typically about 75% volatile solids, and the volatile solids contain about 7% N and 3% P. See Table 10.3.

Oxygen. The rate of carbonaceous BOD removal is reduced at oxygen concentrations below about 0.5 mg/L [Orford *et al.*, 1963]. However, most authorities require higher aeration tank DOs.

Table 10.3 Typical Activated Sludge Compositions

Parameter	Typical Value	Range	Sources
Volatile solids (% of TSS), municipal	70	65 to 75	2, 6, 4
Volatile solids (% of TSS), industrial	—	up to 92	3
Nitrogen (% of VSS)	7.0	—	1, 6, 7
Phosphorus (% of VSS), conventional plants	2.6	1.1 to 3.8	1, 5, 6, 7, 8, 9
Phosphorus (% of VSS), enhanced P-removal plants	5.5	4.5 to 6.8	10, 11

[1]Ardern, E. and Lockett, W. T. 1914. Experiments on the oxidation of sewage without the aid of filters. *Journal of the Society of Chemical Industry.* 33(10):523.

[2]Babbitt, H. E. and Baumann, E. R. 1958. *Sewage and Sewage Treatment,* 8th ed. John Wiley & Sons, New York.

[3]Eckenfelder, W. W., Jr., and O'Connor, D. J. 1961. *Biological Waste Treatment.* Pergamon Press, New York.

[4]Joint Committee of the Water Pollution Control Federation and the American Society of Civil Engineers. 1977. *Wastewater Treatment Plant Design.* Manual of Practice No. 8. Water Pollution Control Federation, Washington, D.C.; American Society of Civil Engineers, New York.

[5]Levin, G. V., Topol, G. J., Tarnay, A. G., and Samworth, R. B. 1972. Pilot-plant tests of a phosphate removal process. *Journal of the Water Pollution Control Federation.* 44(10):1940.

[6]Levin, G. V., Topol, G. J., and Tarnay, A. G. 1975. Operation of full-scale biological phosphorus removal plant. *Journal of the Water Pollution Control Federation.* 47(3):577.

[7]Martin, A. J. 1927. *The Activated Sludge Process.* Macdonald and Evans, London.

[8]Metcalf, L. and Eddy, H. P. 1916. *American Sewage Practice: Vol. III Disposal of Sewage,* 2nd ed. McGraw-Hill, New York.

[9]Mulbarger, M. C. 1971. Nitrification and Denitrification in Activated Sludge Systems. *Journal of the Water Pollution Control Federation.* 43(10):2059.

[10]Scalf, M. R., Pfeffer, F. M., Lively, L. D., Witherow, J. L., and Priesing, C. P. 1969. Phosphate Removal at Baltimore, Maryland. *Journal of the Sanitary Engineering Division, Proceedings of the American Society of Civil Engineers.* 95(SA5):817.

[11]Vacker, D., Connell, C. H., and Wells, W. N. 1967. Phosphate Removal through Municipal Wastewater Treatment at San Antonio, Texas. *Journal of the Water Pollution Control Federation.* 39(5):750.

For non-nitrifying systems, the Joint Task Force [1988] recommends a DO of 2.0 mg/L under average conditions and 0.5 mg/L during peak loads. The Wastewater Committee [1990] and the Technical Advisory Board [1980] require a minimum DO of 2.0 mg/L at all times.

Poisons. Carbonaceous BOD removal is not affected by salinity up to that of seawater [Stewart *et al.*, 1962]. Approximate concentrations at which some poisons become inhibitory are indicated in Tables 10.4 and 10.5.

Traditional Rules of Thumb

Nowadays, most regulatory authorities have approved certain rules of thumb. An example is shown in Table 10.6 [Wastewater Committee, 1990]. The rules of thumb should be used in the absence of pilot plant data or when the data are suspect.

The Conventional Process

The aeration tank volume should be adjusted so that it can carry between 150 and 250% of the solids required to treat the design flow and load. The capacity of the aeration system should be based on the maximum one-hour average oxygen demand load.

Reactor Configuration. The principal operational problem of efficient activated sludge processes is filamentous bulking. The filamentous bacteria responsible for bulking are strict aerobes (some are microaerophilic) and are limited to the catabolism of small organic molecules (simple sugars, volatile fatty acids, and short-chain alcohols). They grow faster than the zoogloeal bacteria at low concentrations of oxygen, nutrients, and substrates and come to dominate the activated sludge community under those conditions. Completely mixed aeration tanks are especially prone to bulking because the substrate concentration is low everywhere in such a tank. High-rate processes, which produce relatively high effluent BODs, may not bulk.

Table 10.4 Approximate Threshhold Concentrations for Inhibition of Activated Sludges by Inorganic Substances[1,2]

Substance		Threshhold Concentration for Inhibition (mg/L)		
		Non-nitrifying	Nitrifying	Denitrifying
Ammonia	(NH_3)	480	—	—
Arsenic	(As)	0.1	—	—
Arsenate	(AsO_2)	—	—	1.0
Barium	(Ba)	—	—	0.1
Borate	(BO_4)	10	—	—
Cadmium	(Cd)	1	5.0	1.0
Calcium	(Ca)	2500	—	—
Chromium	(Cr VI)	1	0.25	0.05
Chromium	(Cr III)	50	—	0.01
Copper	(Cu)	0.1	0.05	20
Cyanide	(CN)	0.5	0.3	0.1
Iron	(Fe)	1000	—	—
Lead	(Pb)	0.1	—	0.05
Magnesium	(Mg)	—	50	—
Manganese	(Mn)	10	—	—
Mercury	(Hg)	0.1	—	0.006
Nickel	(Ni)	1	0.5	5.0
Silver	(Ag)	5	—	0.01
Sulfate	(SO_4)	—	500	—
Sulfide	($S^=$)	25	—	—
Zinc	(Zn)	0.1	0.1	0.1

[1]EPA. 1977. *Federal Guidelines: State and Local Pretreatment Programs. Volume I.* (EPA-430/9-76-017a) *and Volume II. Appendices 1–7* (EPA-430/9-76-017b). Environmental Protection Agency, Office of Water Programs Operations, Municipal Construction Division, Washington, D.C.

[2]Knoetze, C., Davies, T. R., and Wiechers, S. G. 1980. Chemical inhibition of biological nutrient removal processes. *Water SA.* 6(4):171.

The bulking problem is often controlled by designing the aeration tank to be an ideal plug flow reactor. Such a design is called a *selector*. This can actually be done if the facility is a sequencing batch reactor (fill-and-draw reactor). It can be closely approximated if the aeration tank is baffled so as to be *mixed cells in series*. In either case the design objective is to create a zone of relatively high substrate concentration near the aeration tank inlet, which favors the growth of zoogloeal species.

Substrate removal is quick and is probably complete in less than 30 min. Consequently, if a method of mixed cells in series is designed, the hydraulic detention time in each cell should be very short, say 5 to 10 min. Even then, a dilute sewage may favor the filamentous bacteria.

The Joint Task Force [1992] summarizes a number of design recommendations and notes that there is no consensus on the design details for selectors. It should be noted that the semiaerobic process described later is also an effective inhibitor of filamentous bacteria because it denies them access to oxygen.

Air Distribution. Modern aeration systems are usually tapered aeration systems; even in sequencing batch reactors the air supply may be tapered in time. Recommendations for the tapering are given in Table 10.7.

The Contact-Stabilization Process

An important modification of the conventional, non-nitrifying activated sludge process is *contact stabilization* [Zablatsky *et al.*, 1959]. Synonyms are *sludge reaeration* [Ardern and Lockett, 1914a, 1914b], *bio flocculation* [Martin, 1927], *bio sorption* [Ullrich and Smith, 1951] and *step aeration* [Torpey, 1948]. Variants are the *Hatfield process* and the *Kraus process* [Haseltine, 1961].

Table 10.5 Approximate Threshhold Concentrations for Inhibition of Activated Sludges by Organic Substances[1,2]

Substance	Threshhold Concentration for Inhibition (mg/L)		
	Non-nitrifying	Nitrifying	Denitrifying
Acetone	—	840	—
Allyl alcohol	—	19.5	—
Allyl chloride	—	180	—
Allyl isothiocyanate	—	1.9	—
Analine	—	0.65	—
Benzidine	500	—	—
Benzyl thiuronium chloride	—	49	—
Carbamate	0.5	0.5	—
Carbaryl	—	—	10
Carbon disulfide	—	35	—
CEEPRYN™	100	—	—
CHLORDANE™	—	0.1	10
2-chloro-6-trichloro-methyl-pyridine	—	100	—
Creosol	—	4	—
Diallyl ether	—	100	—
Dichlorophen	—	50	—
Dichlorophenol	0.5	5.0	—
Dimethyl ammonium dimethyl dithiocarbamate	—	19.3	—
Demethyl paranitroso aniline	—	7.7	—
Dithane	0.1	0.1	10
Dithiooxamide	—	1.1	—
EDTA	25	—	—
Ethyl urethane	—	250	—
Guanadine carbonate	—	19	—
Hydrazine	—	58	—
8-Hydroxyquinoline	—	73	—
Mercaptothion	10	10	10
Methylene blue	—	100	—
Methylisothiocyanate	—	0.8	—
Methyl thiuronium sulfate	—	6.5	—
NACCONOL™	200	—	—
Phenol	—	4	0.1
Piperidinium cyclopentamethylene dithiocarbamate	—	57	—
Potassium thiocyanate	—	300	—
Pyridine	—	100	—
Skatole	—	16.5	—
Sodium cyclopentamethylene dithiocarbamate	—	23	—
Sodium dimethyl dithiocarbamate	—	13.6	—
Streptomycin	—	400	—
Strychnine hydrochloride	—	175	—
Tetramethyl thiuram disulfide	—	30	—
Tetramethyl thiuram monosulfide	—	50	—
Thioacetamid	—	0.14	—
Thiosemicarbazide	—	0.18	—
Thiourea	—	0.075	—
Trinitrotoluene	20	—	—

[1]EPA. 1977. *Federal Guidelines: State and Local Pretreatment Programs. Volume I.* (EPA-430/9-76-017a) *and Volume II. Appendices 1–7* (EPA-430/9-76-017b). Environmental Protection Agency, Office of Water Programs Operations, Municipal Construction Division, Washington.

[2]Knoetze, C., Davies, T. R., and Wiechers, S. G. 1980. Chemical inhibition of biological nutrient removal processes. *Water SA.* 6(4):171.

Table 10.6 Summary of Recommended Standards for Wastewater Facilities[1]

Treatment Scheme	Food-to-Micro-organism Ratio F $\left(\frac{\text{kg BOD}_5}{\text{kg VSSday}}\right)$	Mixed Liquor Total Suspended Solids X $\left(\frac{\text{mg TSS}}{\text{L}}\right)$	Aeration Tank Load $\left(\frac{\text{kg BOD}_5}{\text{m}^3 \cdot \text{day}}\right)$	Air Supply $\left(\frac{\text{m}^3}{\text{kg BOD}_5}\right)$	Return Sludge Flow % Design Ave. Flow		Secondary Clarifier Overflow Rate $\left(\frac{\text{L}}{\text{m}^2 \cdot \text{s}}\right)$	Secondary Clarifier Solids' Flux $\left(\frac{\text{kg TSS}}{\text{m}^2 \cdot \text{d}}\right)$
					min	max		
Conventional, nonnitrifying, plug flow	0.2 to 0.5	1000 to 3000	0.64	94	15	100	0.56	245
Conventional, nonnitrifying, complete mix	0.2 to 0.5	1000 to 3000	0.64	94	15	100	0.56	245
Step aeration	0.2 to 0.5	1000 to 3000	0.64	94	15	100	0.56	245
Contact-stabilization	0.2 to 0.6	1000 to 3000	0.8	94	50	150	0.56	245
Extended aeration	0.05 to 0.1	3000 to 5000	0.24	128	50	150	0.47	171
Single-stage nitrification	0.05 to 0.1	3000 to 5000	0.24	—	—	—	0.47	171
Two-stage nitrification, carbonaceous stage	—	—	—	—	15	100	0.56	245
Two-stage nitrification, nitrification stage	—	—	—	—	50	200	0.38	171

[1]Wastewater Committee of the Great Lakes–Upper Mississippi River Board of State Public Health and Environmental Managers. 1990. *Recommended Standards for Wastewater Facilities.* Health Education Services, Albany, NY.

Note: The design waste sludge flow shall range from 0.5 to 25% of the design average flow, but not less than 10 gpm.

Table 10.7 Distribution of Oxygen Consumption along Plug Flow and Mixed-Cells-in-Series Aeration Tanks[1]

Aeration Tank Volume	Carbonaceous Demand (%)	Carbonaceous Plus Nitrogenous Demand (%)
First fifth	60	46
Second fifth	15	17
Third fifth	10	14
Fourth fifth	10	13
Last fifth	5	10

[1]Boon, A. G. and Chambers, B. 1985. Design protocol for aeration systems—U.K. perspective. *Proceedings—Seminar Workshop on Aeration System Design, Testing, Operation, and Control.* EPA 600/9-85-005. W. C. Boyle, ed. Environmental Protection Agency, Risk Reduction Engineering Laboratory, Cincinnati, OH.

The distinguishing feature of contact stabilization is the inclusion of a tank for the separate aeration of the return activated sludge. (See Fig. 10.7.) The practical effect is that a much smaller total aeration tank volume is needed to hold the required system biomass because the biomass is held at a higher average concentration.

The basic design principle proposed by Haseltine [1962] is that the distribution of activated sludge solids between the contacting and stabilization tanks does not affect the required system biomass. Haseltine's data, collected from 36 operating facilities, is shown in Fig. 10.8. Haseltine's rule may be expressed mathematically as

$$V_c X_c + V_s X_s = \frac{Q(S_o - S)}{q} = \frac{QS_o}{F} \tag{10.30}$$

where

F = the required food-to-microorganism ratio (kg COD/kg VSS·s or lb COD/lb VSS·s)
Q = the settled sewage flow rate (m^3/s or ft^3/s)
q = the required specific uptake rate (kg COD/kg VSS·s or lb COD/lb VSS·s)
S = the required soluble effluent organic matter concentration (kg COD/m^3 or lb COD/ft^3)
S_o = the settled sewage total organic matter concentration (kg COD/m^3 or lb COD/ft^3)
V_c = the volume of the contacting tank (m^3)
V_s = the volume of the stabilization tank (m^3)

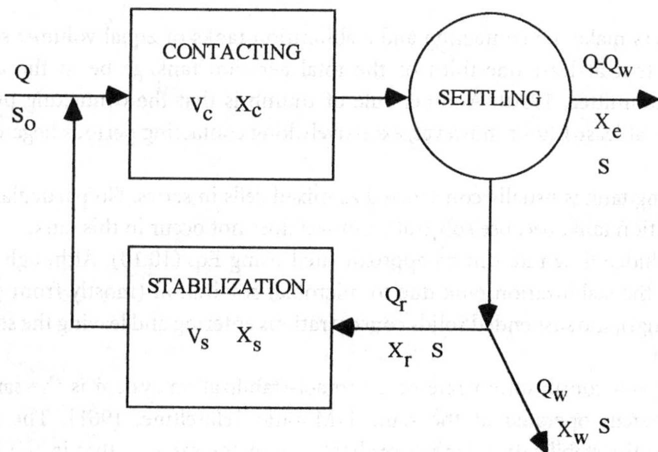

FIGURE 10.7 Flow schematic for the contact-stabilization process.

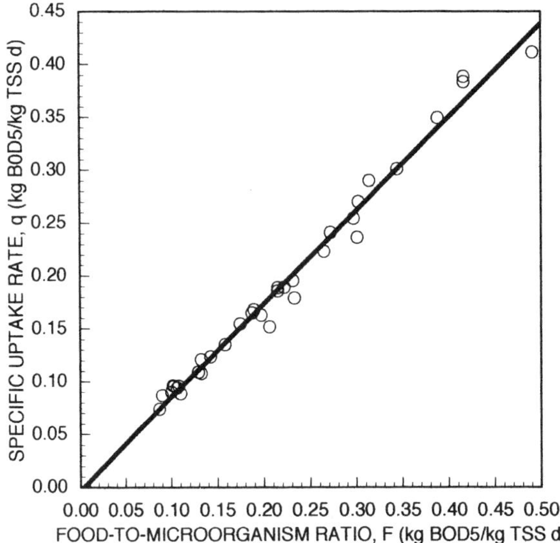

FIGURE 10.8 Correlation of F and q for contact-stabilization plants. (Data from Haseltine, 1961.)

X_c = the concentration of VSS in the contacting tank (mg/L)
X_s = the concentration of VSS in the stabilization tank (mg/L)

The usual relationship between SRT, specific uptake rate, and decay rate applies, except the decay rate should be corrected for the distribution of solids between the contacting and stabilization tanks. Jenkins and Orhon [1973] recommend the following relationship:

$$k_d = \beta \cdot \left[\frac{0.74q}{(1-\beta)3.2 + q} \right] \tag{10.31}$$

where

k_d = the specific decay rate (per second)
q = the specific uptake rate (kg COD/kg VSS·day or lb COD/lb VSS·day)
β = the weight fraction of the total activated sludge inventory that is in the stabilization basin (dimensionless)

Most engineers make the contacting and stabilization tanks of equal volume; some regulatory bodies require that at least one-third of the total aeration tankage be in the contacting tank [Wastewater Committee, 1990]. Another rule of thumb is that the contacting period, counting recycle flows, be at least 1 hour; however, excessively long contacting periods negate the benefits of the process.

The contacting tank is usually constructed as mixed cells in series. No particular design is used for the stabilization tank, because substrate removal does not occur in this tank.

The return sludge flow rate can be approximated using Eq. (10.10). Although there are some solids losses in the stabilization tank due to microbial respiration (mostly from predation upon bacteria and fungi), the suspended solids concentrations entering and leaving the stabilization tank are nearly equal.

The total oxygen consumption rate of a contact-stabilization system is the same as that of a conventional system operated at the same F/M ratio [Haseltine, 1961]. The rate of oxygen consumption in the stabilization tank is probably about the same as that in the final 60 to 80% of the conventional, mixed-cells-in-series aeration tank. The recommendations of Boon and

Chambers [1985], given in Table 10.7, would suggest that the stabilization tank could account for 25 to 40% of the total oxygen consumption. However, the uncertainties in this process require considerable flexibility in the capacity of the air distribution system. It is probably desirable to size the air piping and diffuser systems so that either the contacting tank or the stabilization tank could receive 100% of the estimated total air requirement.

The Extended Aeration Process

The extended aeration process was discovered about 1947 by F. S. Barckhoff, Jr., the plant operator in East Palestine, Ohio [Knox, 1958]. Originally it was called aerobic *digestion* because its purpose was to stabilize waste activated sludge in the aeration tank rather than in separate sludge digestion facilities. In older designs there is no intentional wasting of activated sludge. The only solids to leave the system are those in the final effluent. The result is SRTs on the order of months to a year or more.

The chief advantage of the extended aeration process is the elimination of waste sludge–processing facilities. The chief disadvantages are the increase in oxygen consumption (needed to decompose the activated sludge solids) and the lack of kinetic control.

Nowadays, extended aeration plants frequently incorporate solids wasting facilities, and the term really designates plants with long aeration periods (usually 24 hours) and long SRTs (about 30 days).

Full-scale plants fed milk wastes have in fact operated for periods up to a year without sludge wasting [Forney and Kountz, 1959; Kountz and Forney, 1959]. Laboratory units fed glucose and soluble nutrients and managed so that no solids left the systems (other than those removed in sampling) have been operated for periods of three years (SRT of 500 days) without net solids accumulation [Gaudy *et al.*, 1970; Gaudy *et al.*, 1971]. However, the MLSS in the laboratory units fluctuated widely during this time.

Most wastewaters contain significant amounts of clay and other inert inorganic solids, and these gradually accumulate in the system. Extended aeration plants accumulate suspended solids until the secondary clarifier fails, producing periods of very high effluent suspended solids concentrations [Morris *et al.*, 1963]. These slug discharges usually are associated with high flows or bulking. Bulking is always a hazard because most extended aeration plants incorporate completely mixed aeration tanks.

In terms of Gram's theory the ideal extended aeration plant is operated so that sludge growth just equals sludge decay. Consequently, there is no net sludge production, and the SRT is infinite, as follows:

$$\mu = k_d \tag{10.32}$$

$$S = \frac{k_d K_s}{\mu_{\max} - k_d} \tag{10.33}$$

$$Y_o = 0 \tag{10.34}$$

According to the product-maintenance model, the influent organic matter is expended, satisfying the maintenance energy demand of the biomass and insoluble organic product synthesis; there is no VSS production. The product-maintenance model for extended aeration yields [Sykes, 1991]

$$S = \frac{S_o d}{m_a} \tag{10.35}$$

$$Y_o' = 0 \tag{10.36}$$

No actual extended aeration facility can operate at infinite SRT because activated sludge flocs are always lost in the settled effluent. And even at SRTs of 500 days or more the soluble effluent COD amounts to 5 to 10% of the influent COD (Gaudy *et al.*, 1970; Gaudy *et al.*, 1971].

Table 10.8 Typical Parameter Values for the Extended Aeration Activated Sludge Process at Approximately 20°C[1–4]

Parameter	Symbol	Units	Value
True yield	Y	$\frac{\text{kg VSS}}{\text{kg COD}}$	0.4
		$\frac{\text{kg VSS}}{\text{kg BOD}_5}$	0.7
Decay rate	k_d	per day	0.012
Maximum specific growth rate	μ_{max}	per day	0.04
Maximum uptake rate	q_{max}	$\frac{\text{kg COD}}{\text{kg VSS d}}$	0.05
Affinity constant	K_s	mg BOD_5/L	10

[1]Goodman, B. L. and Englande, A. J., Jr. 1974. A unified model of the activated sludge process. *Journal of the Water Pollution Control Federation.* 46(2):312.

[2]Middlebrooks, E. J. and Garland, C. F. 1968. Kinetics of model and field extended-aeration wastewater treatment units. *Journal of the Water Pollution Control Federation.* 40(4):586.

[3]Middlebrooks, E. J., Jenkins, D., Neal, R. C., and Phillips, J. L. 1969. Kinetics and effluent quality in extended aeration. *Water Research.* 3(1):39.

[4]Morris, G. L., Van Den Berg, L., Culp, G. L., Geckler, J. R., and Porges, R. 1963. *Extended Aeration Plants and Intermittent Watercourses.* Pub. Health Service Pub. No. 999-WP-8. Department of Health, Education and Welfare, Public Health Service, Division of Water Supply and Pollution Control, Cincinnati, OH.

The SRTs employed in extended aeration are sufficient to permit year-round nitrification in most locales; however, traditional designs did not provide sufficient aeration capacity to support nitrification.

Extended aeration plants are frequently built either as oxidation ditches or as aerated lagoons. Both of these designs may incorporate anoxic zones. In the case of oxidation ditches that employ surface aerators for both aeration and mixing, the bottom may accumulate a deposit of activated sludge. In the case of lagoons, the placement of surface aerators may result in unaerated zones [Nicholls, 1975]. These anoxic zones can be the sites of denitrification, and extended aeration plants may accomplish substantial nitrogen removal even in the absence of sludge wasting.

Some values of the Gram kinetic model parameters are given in Table 10.8. It should be noted that the decay rate appears to decline at long SRTs. For SRTs up to about 40 days Goodman and Englande [1974] recommend

$$k_d = 0.48 \times 0.75^{\ln \theta_c / \ln 2} \times 1.075^{T-20} \tag{10.37}$$

where

k_d = the decay rate in per day
T = the temperature in °C
θ_c = the solids retention time in days

For SRTs on the order of hundreds of days, the decay rate predicted by Eq. (10.37) may be high by a factor of about two.

Extended aeration plants have usually been recommended for small or isolated installations where regular, trained operating staffs are difficult to obtain. This practice results in serious operation and maintenance problems, most typically (Guo *et al.*, 1981):

- Clogging of comminutors
- Clogging of sludge return lines
- Clogging of air diffusers
- Failure of skimmers

- Icing of clarifier outlet weirs
- Insufficient biomass
- Insufficient aeration
- Offensive odors

Extended aeration plants are also subject to bulking and scum formation, but this is a result of the combination of completely mixed aeration tanks and long SRTs.

Nitrification

Microbiology

Nitrification is the biological conversion of ammonia to nitrate. It is carried out by aerobic, chemoautotrophic bacteria in two steps [Stanier *et al.*, 1976]. *Nitrosomonas, nitrosospira, nitrosococcus*, and *nitrosolobus* activate the first step:

$$NH_3 + \tfrac{3}{2}O_2 \rightarrow HNO_2 + H_2O \tag{10.38}$$

Nitrobacter, nitrospina, and *nitrococcus* bacteria activate the second step:

$$HNO_2 + \tfrac{1}{2}O_2 \rightarrow HNO_3 \tag{10.39}$$

The overall reaction is given as

$$NH_3 + 2O_2 \rightarrow HNO_3 + H_2O \tag{10.40}$$

The first step is slow and controls the overall rate of conversion. Consequently, in most systems only small amounts of nitrite are observed, and the process can be represented as a one-step conversion of ammonia to nitrate.

Kinetics and Yield. At low ammonia concentrations, the rate of growth of the *Nitroso-* genera can be correlated with the ammonia concentration using Monod kinetics:

$$\mu_N = \frac{\mu_{\max N} N}{K_N + N} \tag{10.41}$$

where

K_N = the ammonia affinity constant (kg NH_3-N/m^3 or lb NH_3-N/ft^3)
N = the total ammonia concentration (kg NH_3-N/m^3 or lb NH_3-N/ft^3)
μ_N = the specific growth rate of the *Nitroso-* genera (per second)
$\mu_{\max N}$ = the maximum specific growth rate of the *Nitroso-* genera (per second)

The values of the kinetic parameters are usually estimated as follows [Parker *et al.*, 1975; Scheible and Heidman, 1993]: For pH less than 7.2,

$$\mu_{\max N} = 0.47 \cdot \exp\{0.098(T - 15)\} \cdot [1 - 0.833(7.2 - \text{pH})] \cdot \left(\frac{C}{C + 1.3}\right) \tag{10.42}$$

For pH between 7.2 and 9.0,

$$\mu_{\max N} = 0.47 \cdot \exp\{0.098(T - 15)\} \cdot \left(\frac{C}{C + 1.3}\right) \tag{10.43}$$

For any pH,

$$K_N = 10^{0.051T - 1.158} \tag{10.44}$$

where

$$
\begin{aligned}
C &= \text{the aeration tank dissolved oxygen concentration in mg/L} \\
K_N &= \text{the affinity constant in mg NH}_3\text{-N/L} \\
\text{pH} &= \text{the aeration tank pH (standard units)} \\
T &= \text{the aeration tank temperature in }^\circ\text{C} \\
\mu_{\max N} &= \text{the maximum growth rate of the } \textit{Nitroso-} \text{ genera (per day)}
\end{aligned}
$$

Scheible and Heidman [1993] recommend a constant value of the affinity constant of 1.0 mg NH_3-N/L because of the high degree of variability in the reported values.

Note that the temperature, pH, and oxygen effects are included in the estimators for the biokinetic constants. The effective Streeter-Phelps theta values for this process are

$$
\begin{aligned}
&\text{Maximum growth rate, } \mu_{\max N} \\
&\qquad \theta = e^{0.098} = 1.10; \\
&\text{Affinity coefficient, } K_N \\
&\qquad \theta = 10^{0.051} = 1.125;
\end{aligned}
$$

which are nearly the same as that of the heterotrophic population. However, the maximum growth rate of the nitrifiers is much smaller than that of the heterotrophs, so nitrification is very sensitive to SRT at cold temperatures.

At 20°C the affinity coefficient is predicted to be about 0.73 mg N/L. This means that the rate of nitrification is independent of ammonia concentration down to concentrations on the order of 1 mg NH_3-N/L. The practical consequence of this is that reactor configuration has little effect on overall ammonia removal. Aeration tanks are sized assuming complete mixing, which is very conservative, because the effective ammonia specific uptake rate in mixed cells in series is $\mu_{\max N}/Y_N$.

The maximum specific growth rates of the *Nitroso-* genera fall off sharply above pH 9.0, and the estimators do not apply above that pH.

The nitrification process produces nitric acid, and in poorly buffered waters this may cause a significant decline in pH. The effect can be estimated as follows, noting that ammonia exists as the ammonium cation at normal pH values:

$$NH_4^+ + 2O_2 \rightarrow 2H^+ + N_3^- + H_2O \tag{10.45}$$

Thus, 1 g NH_3-N reduces the alkalinity by 7.14 g (as $CaCO_3$).

The affinity constant for the DO correction, given here as 1.3 mg/L, may be as high as 2.0 mg/L in some systems. Scheible and Heidman [1993] recommend a value of 1.0 mg/L.

The yield of nitrifiers normally includes both the *Nitroso-* and *Nitro-* genera and may be approximated [Parker *et al.*, 1975] by

$$NH_4^+ + 1.83O_2 + 1.98HCO_3^- \rightarrow 0.021C_5H_7O_2N + 1.88H_2CO_3 + 0.98NO_3^- + 1.041H_2O \tag{10.46}$$

From this, the estimated true yield coefficient for the nitrifiers as a whole is 0.17 g VSS/g NH_3-N or 0.037 g VSS/g NBOD.

The nitrifier "decay" rate is usually assumed to be the same as that of the heterotrophs. This is reasonable because the so-called decay rate is really a predation effect.

Ammonia Inhibition. Un-ionized ammonia—that is, NH_3 itself—is inhibitory to both the *Nitroso-* and the *Nitro-* genera. The inhibition threshold concentrations of un-ionized ammonia for the *Nitroso-* group is about 10 to 150 mg NH_3/L; for the *Nitro-* group it is about 0.1 to 1.0 mg NH_3/L. Inhibition of the *Nitro-* group results in the accumulation of nitrite.

The usual way to handle these effects is to adopt the *Haldane* kinetic model for the specific growth rate [Haldane, 1930]:

$$\mu = \frac{\mu_{\max N} \cdot N}{K_N + N + (N^2/K_i)} \tag{10.47}$$

where K_i = the Haldane inhibition constant (kg NH_3-N/m^3 or lb NH_3-N/ft^3).

In heterotroph-free cultures, the inhibition constant has been reported to be about 20 mg N/L at 19°C and pH 7.0 [Rozich and Castens, 1986]. The basis here is the total ammonia concentration, both NH_3-N and NH_4^+-N. Under the given conditions the ratio of total ammonia concentration to un-ionized ammonia would be about 250:1. At a full-scale nitrification facility treating landfill leachate, the observed inhibition constant was 36 mg/L of total ammonia nitrogen [Keenan *et al.*, 1979]. The wastewater temperature varied from 0 to 29°C, and the pH varied from 7.3 to 8.6. Another report gave the inhibition constant a value of 9,000 mg N/L at pH 8.0 and 23°C [Gee *et al.*, 1990]. Again the basis was the total ammonia concentration. The ratio of total to un-ionized ammonia would have been about 20:1.

Other Inhibitors. A list of other inhibitors and the approximate threshold concentration for nitrification inhibition is given in Tables 10.4 and 10.5. The reduction in nitrification rate in some industrial wastewaters due to inhibitors can be severe. Adams and Eckenfelder [1977] give some laboratory data for nitrification rates for pulp and paper, refinery and phenolic wastes that are only about 0.1% of the rates in municipal wastewater. The reported rates are low by an order of magnitude even if ammonia inhibition is accounted for.

Nitrifier Biomass

The nitrifier population is normally only a small portion of the MLSS. If it is assumed that the heterotrophs are carbon limited and the nitrifiers are NH_3-N limited, then the nitrifier biomass can be estimated from a simple nitrogen balance:

$$\frac{VX_N}{Y_{oN} \cdot \theta_c} = \underbrace{Q(N_o - N)}_{\text{Total N removed}} - \underbrace{f_{nH} \cdot \frac{VX_H}{\theta_c}}_{\text{N removed by heterotrophs}} \tag{10.48}$$

where

f_{nH} = the fraction of nitrogen in the heterotrophs (kg N/kg VSS)
$\approx$ 0.070 kg N/kg VSS (see Table 10.3)
N = the soluble TKN (not ammonia) of the final effluent (kg TKN/m^3 or lb TKN/ft^3)
N_0 = the TKN (soluble plus particulate) of the settled wastewater (kg TKN/m^3 or lb TKN/ft^3)
Q = the settled sewage flow rate (m^3/s ft^3/s)
V = the volume of the aeration tank (m^3 or ft^3)
X_H = the concentration of heterotrophs in the aeration tank (kg VSS/m^3 or lb VSS/ft^3)
X_N = the concentration of nitrifiers in the aeration tank (kg VSS/m^3 or lb VSS/ft^3)
Y_{0N} = the observed yield for nitrifier growth on ammonia (kg VSS/kg NH_3-N)
θ_c = the solids retention time (s)

Organic Nitrogen Production

NPDES permits are written in terms of ammonia-nitrogen because of its toxicity. However, a substantial portion of the effluent TKN in nitrifying facilities is soluble organic nitrogen. As a rough guide, the ratio of TKN to NH_3-N in settled effluents is about 2:1 to 3:1 [Barth *et al.*, 1968;

Beckman *et al.*, 1972; Clarkson *et al.*, 1980; Lawrence and Brown, 1976; Mulbarger, 1971; Prakasam *et al.*, 1979; Stankewich, 1972].

Two-Stage Nitrification

In the two-stage nitrification process, the nitrification step is preceded by a roughing step—either activated sludge or trickling filter—which is designed to remove about one-half to three-quarters of the settled sewage CBOD. The benefits of this scheme are that the total biomass that must be carried in the plant and the oxygen consumption are reduced. The costs are an additional clarifier and increased waste sludge production. The aeration tankage requirements of two-stage nitrification are comparable to those of non-nitrifying facilities, which suggests that most conventional plants can be readily upgraded to nitrification without major expense. Mulbarger [1971] estimates the increase in capital cost above that of a non-nitrifying facility to be about 10%.

The first stage of the process is also known as *modified aeration* or *high-rate activated sludge.*

The BOD and TKN loads to the second stage are the expected effluent quality of the first stage. The first stage effluent BOD is a semi-free design choice; it determines the specific uptake rate for the first stage. The first stage effluent TKN is the TKN *not* incorporated into the first stage's waste activated sludge. It may be estimated by

$$\underbrace{QN}_{\text{1st stage soluble effluent TKN}} = \underbrace{QN_o}_{\text{Settled sewage total TKN}} - \underbrace{f_n \cdot \frac{VX}{\theta_c}}_{\text{TKN in 1st stage waste-activated sludge}} \tag{10.49}$$

where f_n = the weight fraction of nitrogen in the waste activated sludge solids (kg N/kg VSS or lb N/lb VSS).

The nitrification kinetics of the second stage can be represented by the EPA model described above. The kinetics of CBOD removal in the second stage are not well known. The second stage influent CBOD is a microbial product formed in the first stage—not residual settled sewage CBOD—and its removal kinetics may not be well represented by the data in Tables 10.1 and 10.8.

Denitrification

A variety of proprietary denitrification processes are being marketed, including A/O™ (U.S. Patent No. 4,056,465), BARDENPHO™ (U.S. Patent No. 3,964,998), and BIODENITRO™ (U.S. Patent No. 3,977,965). Other patents are listed by the Joint Task Force [1992].

Microbiology

Many aerobic heterotrophic bacteria, especially pseudomonads, can utilize nitrate and nitrite as a terminal electron acceptor. The half cell reactions are

$$NO_2^- + 3e^- + 4H^+ \rightarrow \tfrac{1}{2}N_2 + 2H_2O \tag{10.50}$$

$$NO_3^- + 5e^- + 6H^+ \rightarrow \tfrac{1}{2}N_2 + 3H_2O \tag{10.51}$$

The oxygen reduction half cell is

$$O_2 + 4e^- + 4H^+ \rightarrow 2H_2O \tag{10.52}$$

Consequently, the electronic equivalent weights of nitrite and nitrate are 1.713 and 2.857 g O_2/g NO_3-N, respectively. The nitrate equivalent weight is confirmed by Wuhrmann's [1968] data.

Growth Kinetics and Yield. The energy available to denitrification is sharply reduced, from about 1 ATP per electron pair for oxygen based oxidations to about 0.4 ATP per electron pair [Sykes, 1975]. This substantially reduces waste activated sludge production.

The theoretical growth stoichiometry for the anoxic growth of heterotrophic bacteria growing on methanol and nitrate plus nutrient salts is given by Smarkel [1977] as

$$13.7CH_3OH + 11.8HNO_3 \rightarrow C_5H_7O_2N + 8.67CO_2 + 5.40N_2 + 29.8H_2O \quad (10.53)$$

This closely approximates the empirical formula given by McCarty *et al.* [1969]:

$$1.08CH_3OH + NO_3^- + H^+ \rightarrow 0.065C_5H_7O_2N + 0.76CO_2 + 0.47N_2 + 2.44H_2O \quad (10.54)$$

The calculated true growth yields for organic matter and nitrate nitrogen are 0.172 g VSS/g COD and 0.684 g VSS/g NO_3-N.

The denitrifier growth rate can be limited by either the nitrate concentration or the COD concentration, and values of the Gram model kinetic parameters have been reported for both cases. Laboratory results for some of these parameters are summarized in Table 10.9. Additional data are given by Scheible and Heidman [1993]. While the values for the true growth yields, the microbial decay, and perhaps the affinity coefficient for nitrate are reliable, the maximum specific growth rate and uptake rate are questionable. Large-scale pilot studies have produced maximum specific uptake rates that range from about 0.1 to 0.4 g NO_3-N/g MLVSS d at 20°C [Ekama and Marais, 1984; Parker *et al.*, 1975].

Table 10.9 Typical Parameter Values for Laboratory-Scale Denitrifying Activated Sludge Processes Using Methanol at Approximately 20°C[1–4]

Parameter	Symbol	Units	Typical Value
True yield	Y_{COD}	$\frac{\text{kg VSS}}{\text{kg COD}}$	0.17
	Y_{NO_3}	$\frac{\text{kg VSS}}{\text{kg NO}_3\text{-N}}$	0.68
Decay rate	k_d	per day	0.04
Maximum specific growth rate	μ_{max}	per day	0.3
Maximum uptake rate	q_{max}	$\frac{\text{kg COD}}{\text{kg VSS d}}$	2
		$\frac{\text{kg NO}_3\text{-N}}{\text{kg VSS d}}$	0.5
Affinity constant	K_s	$\frac{\text{mg BOD}_5}{\text{L}}$	150
		$\frac{\text{mg COD}_{total}}{\text{L}}$	75
		$\frac{\text{mg COD}_{degrad}}{\text{L}}$	10
		$\frac{\text{mg NO}_3\text{-N}}{\text{L}}$	0.08

[1]Johnson, W. K. 1972. Process kinetics for denitrification. *Journal of the Sanitary Engineering Division, Proc. ASCE.* 98(SA4):623.

[2]McClintock, S. A., Sherrard, J. H., Novak, J. T., and Randall, C. W. 1988. Nitrate versus oxygen respiration in the activated sludge process. *Journal of the Water Pollution Control Federation.* 60(3):342.

[3]Moore, S. F. and Schroeder, E. D. 1970. An investigation of the effects of residence time on anaerobic bacterial denitrification. *Water Research.* 4(10):685.

[4]Moore, S. F. and Schroeder, E. D. 1971. The effect of nitrate feed rate on denitrification. *Water Research.* 5(7):445.

[5]Stensel, H. D., Loehr, R. C., and Lawrence, A. W. 1973. Biological kinetics of suspended-growth denitrification. *Journal of the Water Pollution Control Federation.* 45(2):249.

Oxygen. The threshold for oxygen inhibition of denitrification is about 0.1 mg/L, and the denitrification rate is reduced by 50% at oxygen concentrations around 0.2 to 0.3 mg/L [Focht and Chang, 1975]. At 2.0 mg/L of oxygen the denitrification rate is reduced by 90%.

Inhibitors. The reduction of nitrite is inhibited by nitrite itself. In unacclimated cultures the nitrite reduction rate falls by about 80% when the nitrite nitrogen concentration exceeds about 8 mg/L [Beccari *et al.*, 1983]. This may be due to the accumulation of free nitrous acid because there is an unusually sharp fall in the rate of nitrite reduction as the pH falls below 7.5.

Other reported inhibitors are [Painter, 1970]:

- Metal chelating agents (e.g., sodium diethyldithiocarbamate, orthophenanthroline, potassium cyanide, and 4-methyl-1:2-dimercaptobenzene)
- Cytochrome inhibitors (e.g., 2-*n*-heptyl-4-hydroxyquinoline-N-oxide)
- *p*-chloromercuribenzoate
- Hydrazine
- Chlorate
- Copper

Tables 10.4 and 10.5 summarize some quantitative data on inhibition.

Temperature. Focht and Chang [1975] have reviewed Q_{10} data for a variety of nitrification processes (including mixed and pure cultures and suspended and film systems), and Lewandowski [1982] has reviewed theta values for waste water. The average theta for all these processes is about 1.095, and the type of reductant does not appear to affect the value.

pH. The optimum pH for the reduction of nitrate is about 7.0 to 7.5 [Beccari et al., 1983; Focht and Chang, 1975; Parker *et al.*, 1975]. The rate of nitrate reduction falls off sharply outside that range and falls to about half its maximum value at pH's of 6.0 and 8.0.

The optimum range of pH for nitrite reduction appears to be narrowly centered at 7.5. At pH 7.0, it is only 20% of its maximum value, and at pH 8.0 it is about 70% of its maximum [Beccari *et al.*, 1983].

Semiaerobic Denitrification

The simplest denitrification facility is the *semiaerobic process* developed by Ludzack and Ettinger [1962]. This process is suitable for moderate removals of nitrate.

A schematic of the process train is shown in Fig. 10.9. In this process, mixed liquor from the effluent end of a nitrifying aeration tank is recirculated to an anoxic tank ahead of the aeration tank where it is mixed with settled sewage. The anoxic tank is mixed but not aerated. The heterotrophs of the activated sludge utilize the nitrates in the mixed liquor to oxidize the CBOD of the settled sewage, and the nitrates are reduced to nitrogen gas.

The nitrate removal efficiency may be approximated by

$$E_{NO_3} = \frac{Q_m + Q_r}{Q + Q_m + Q_r} \tag{10.55}$$

where

E_{NO_3} = the nitrate removal efficiency (dimensionless)
Q = the settled sewage flow (m^3/s or ft^3/s)
Q_m = the mixed liquor return flow (m^3/s or ft^3/s)
Q_r = the return sludge flow (m^3/s or ft^3/s)

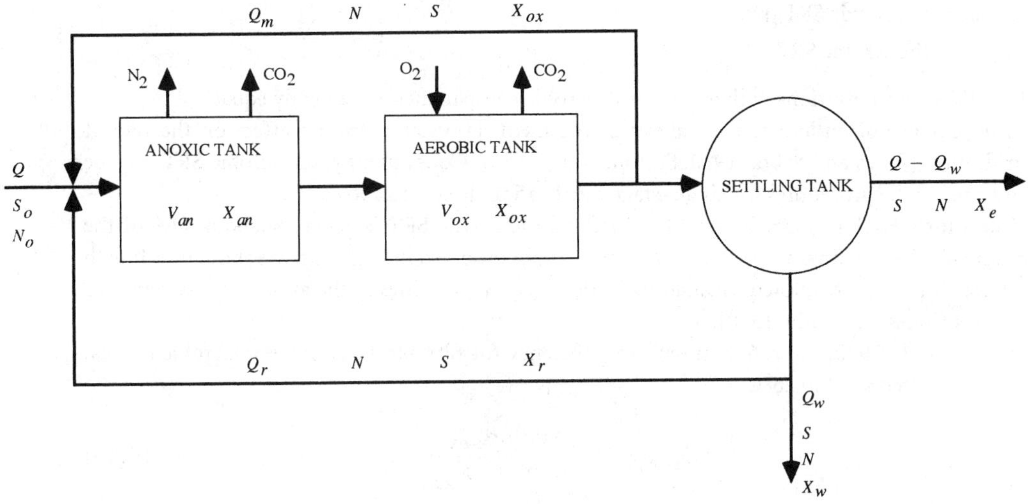

FIGURE 10.9 The Ludzack–Ettinger semiaerobic process for nitrogen removal.

Because of economic limitations on the amount of return flows, the practical removal efficiency is limited to maybe 80%. If high removal efficiencies are needed, a separate stage denitrification facility must be built.

It is assumed that the BOD of the settled sewage is sufficient to reduce the nitrate produced in the aerobic tank, and this is usually the case. However, it may be necessary to add additional reductant in some cases.

It is important to distinguish the anoxic SRT from the aerobic SRT because nitrification occurs only under aerobic conditions. These may be estimated as follows:

$$\theta_c = \theta_{c,an} + \theta_{c,ox} = \frac{V_{an}X_{an} + V_{ox}X_{ox}}{Q_wX_w + (Q - Q_w)X_e} \tag{10.56}$$

$$\theta_{c,an} = \frac{V_{an}X_{an}}{Q_wX_w + (Q - Q_w)X_e} \tag{10.57}$$

$$\theta_{c,ox} = \frac{V_{ox}X_{ox}}{Q_wX_w + (Q - Q_w)X_e} \tag{10.58}$$

where

Q = the settled sewage flow (m^3/s or ft^3/s)
Q_m = the mixed liquor return flow (m^3/s or ft^3/s)
Q_r = the return sludge flow (m^3/s or ft^3/s)
X_{an} = the volatile suspended solids concentration in the anoxic tank (kg/m^3 or lb/ft^3)
X_e = the volatile suspended solids concentration in the settled effluent (kg/m^3 or lb/ft^3)
X_{ox} = the volatile suspended solids concentration in the aerobic (oxic) tank (kg/m^3 or lb/ft^3)
X_w = the volatile suspended solids concentration in the waste activated sludge (kg/m^3 or lb/ft^3)
V_{an} = the volume of the anoxic compartment (m^3 or ft^3)
V_{ox} = the volume of the aerobic (oxic) compartment (m^3 or ft^3)
θ_c = the systems' solids retention time, SSRT (s)

$\theta_{c,an}$ = the anoxic SRT (s)

$\theta_{c,ox}$ = the aerobic SRT (s)

The MLVSS concentrations in the anoxic and aerobic compartments are nearly equal.

The presence of an anoxic zone ahead of the nitrification zone has no effect on the rate of nitrification [Jones and Sabra, 1980; Sutton, *et al.*, 1980]. Consequently, the aerobic SRT may be chosen using the procedures for single-stage nitrification discussed above.

The anoxic HRT is normally 1 to 4 hours, and the anoxic SRT is generally about 30% of the system SRT. The maximum specific nitrate uptake rates observed in field units are about one-fourth to one-half of the rate quoted in Table 10.9. Also, $q_{\max,NO_3}$ declines as the anoxic and system SRT increases [Jones and Sabra, 1980].

Sutton *et al.* [1980] indicate that removal efficiency for filtrable TKN in a semiaerobic process can be estimated at 24 to 26°C:

$$E_{\text{TKN,filt}} = \frac{0.98\theta_{c,ox}}{0.26 + \theta_{c,ox}} \tag{10.59}$$

at 14 to 16°C:

$$E_{\text{TKN,filt}} = \frac{1.05\theta_{c,ox}}{1.00 + \theta_{c,ox}} \tag{10.60}$$

and at 7 to 8°C

$$E_{\text{TKN,filt}} = \frac{1.11\theta_{c,ox}}{5.04 + \theta_{c,ox}} \tag{10.61}$$

where $E_{\text{TKN,filt}}$ = the removal efficiency for filtrable total kjeldahl nitrogen (dimensionless). These results are supported by the data of Sutton, *et al.* [1980] and by the data of Jones and Sabra [1980].

Kinetic data for the removal of nitrate from municipal wastewater in the anoxic zone at 20°C are provided by Burdick *et al.* [1982]:

$$q_{\text{NO}_3\text{-N}} = 0.03F_{an} + 0.029 \tag{10.62}$$

$$q_{\text{NO}_3\text{-N}} = 0.12\left(\frac{1}{\theta_c}\right)^{0.706} \tag{10.63}$$

where

$q_{\text{NO}_3\text{-N}}$ = the anoxic zone specific nitrate consumption rate in kg NO_3-N/kg MLTSS·d

F_{an} = the anoxic zone food-to-microorganism ratio in kg BOD/kg MLTSS·d

θ_c = the total system solids retention time in days

Typical design values for the A/O™ and BARDENPHO™ processes are given in Table 10.10. Recent BARDENPHO™ designs are tending towards the lower limits tabulated.

A nitrogen/COD balance on the system and anoxic tank produces the following:

$$\begin{aligned} Q(N_{\text{TKN},o} - N_{\text{TKN}}) &= (Q + Q_m + Q_r)\left(N_{\text{NO}_3} - N_{\text{NO}_3an}\right) + \dots \\ &\quad + f_N \cdot \left\{ Y_{o,\text{COD},ox} \cdot (S_o - S) + \dots \right. \end{aligned} \tag{10.64}$$

$$\left. Y_{o,\text{NO}_3,an} \cdot \left(1 - \frac{Y_{o,\text{COD},ox}}{Y_{o,\text{COD},an}}\right) \cdot \left[(Q_m + Q_r)N_{\text{NO}_3} - (Q + Q_m + Q_r)N_{\text{NO}_3,an}\right] \right\}$$

Table 10.10 Typical Design Values for Commercial Semiaerobic Processes[1–4]

Design Parameter	A/O™ Value	BARDENPHO™ Value
Anaerobic HRT (h)	0.5 to 1.0	0.6 to 2.0
First anoxic HRT (h)	0.5 to 1.0	2.2 to 5.2
Oxic HRT (h)	3.5 to 6.0	6.6 to 19.0
Second anoxic HRT (h)	—	2.2 to 5.7
Reaeration HRT (h)	—	0.5 to 2.0
Oxic SRT (d)	—	>10
F/M (kg BOD/kg MLVSS · d)	0.15 to 0.25	—
MLVSS (mg/L)	3000 to 5000	3000
Return sludge flow (% settled sewage flow)	20 to 50	—
Mixed liquor recycle flow (% settled sewage flow)	100 to 300	400

[1]Barnard, J. L. 1974a. Cut P and N without chemicals. *Water & Wastes Engineering.* 11(7):33.

[2]Barnard, J. L. 1974b. Cut P and N without chemicals. *Water & Wastes Engineering.* 11(8):41.

[3]Roy, F. Weston, Inc. 1983. *Emerging Technology Assessment of Biological Phosphorus Removal: 1. PHOSTRIP PROCESS; 2. A/O PROCESS; 3. BARDENPHO PROCESS.* Environmental Protection Agency, Wastewater Research Division, Municipal Environmental Research Laboratory, Cincinnati, OH.

[4]Weichers, H. N. S. *et al.*, ed. 1984. *Theory, Design and Operation of Nutrient Removal Activated Sludge Process.* Water Research Commission, Pretoria, South Africa.

where

f_N = the weight fraction of nitrogen in the biomass (kg N/kg VSS)

N_{NO_3} = the final effluent nitrate-nitrogen concentration (kg N/m^3)

$N_{NO_3,an}$ = the nitrate-nitrogen concentration in the effluent of the anoxic tank (kg N/m^3)

$N_{TKN,o}$ = the total TKN in the settled wastewater (kg N/m^3)

N_{TKN} = the soluble TKN in the final effluent (kg N/m^3)

Q = the settled sewage flow rate (m^3/s)

Q_m = the mixed liquor recirculation rate (m^3/s)

Q_r = the return sludge flow rate (m^3/s)

S = the soluble effluent COD (kg COD/m^3)

S_o = the total COD of the settled sewage (kg COD/m^3)

$Y_{o,COD,an}$ = the observed heterotrophic yield from COD consumption under anoxic (denitrifying) conditions (kg VSS/kg COD)

$Y_{o,COD,ox}$ = the observed heterotrophic yield from COD consumption under aerobic conditions (kg VSS/kg COD)

$Y_{o,NO_3,an}$ = the observed heterotrophic yield from nitrate nitrogen consumption under anoxic (denitrifying) conditions (kg VSS/kg NO_3-N)

This ignores the nitrifying biomass on the grounds that it is small.

In Eq. (10.64) the settled sewage flow rate, the influent TKN, and the target effluent nitrate-nitrogen concentrations are known. The effluent soluble TKN can be estimated as in the designs for single-stage nitrification; it is two or three times the effluent ammonia-nitrogen concentration. Both the effluent ammonia-nitrogen concentration and the effluent soluble BOD and COD are fixed by the aerobic SRT. The nitrate-nitrogen concentration in the effluent of the anoxic stage is probably small. The return sludge flow rate may be estimated using Equation (10.10). The observed anoxic and aerobic heterotrophic yields must be calculated using the anoxic and aerobic SRT, respectively.

The unknown is the required mixed liquor recirculation rate. Generally, recirculation rates of 100 to 400% of the settled sewage flow are employed. Higher rates produce more complete nitrate removal.

Once the flow rates are known, simple mass balances around the system can be used to calculate the various mass conversions. For anoxic consumption of nitrate and production of nitrogen gas,

$$R_{NO_3,an} = R_{N_2,an} = (Q_m + Q_r)N_{NO_3} - (Q + Q_m + Q_r)N_{NO_3,an} \tag{10.65}$$

For anoxic consumption of COD,

$$R_{COD,an} = QS_o + (Q_m + Q_r)S - (Q + Q_m + Q_r)S_{an} \tag{10.66}$$

$$R_{COD,an} = \frac{Y_{o,NO_3,an}}{Y_{o,COD,an}} \cdot R_{NO_3,an} = 3.98 \cdot R_{NO_3,an} \tag{10.67}$$

For aerobic consumption of COD,

$$R_{COD,ox} = Q(S_o - S) - R_{COD,an} \tag{10.68}$$

where

N_{NO_3} = the concentration of nitrate-nitrogen in the settled effluent(kg/m^3 or lb/ft^3)
$N_{NO_3,an}$ = the concentration of nitrate-nitrogen in the effluent of the anoxic tank (kg/m^3 or lb/ft^3)
$R_{COD,an}$ = the rate of COD consumption in the anoxic tank (kg/s or lb/s)
$R_{COD,ox}$ = the rate of COD consumption in the aerobic tank (kg/s or lb/s)
$R_{N_2,an}$ = the rate of nitrogen gas production in the anoxic tank (kg/s or lb/s)
$R_{NO_3,an}$ = the rate of nitrate-nitrogen consumption in the anoxic tank (kg/s or lb/s)
S_{an} = the COD in the effluent of the anoxic tank (kg/m^3 or lb/ft^3)

Note that the effluent COD is fixed by the aerobic SRT. The COD in the effluent of the anoxic tank can be calculated using Eq. (10.66). If it is negative a supplemental reductant, usually methanol, is required.

The waste sludge production rate is given by

$$Q_w X_w \leq \frac{V_{an}X_{an} + V_{ox}X_{ox}}{\theta_c} = Y_{o,COD,ox} \cdot R_{COD,ox} + Y_{o,COD,an} \cdot R_{COD,an} \tag{10.69}$$

The SRT is fixed by the specified effluent ammonia-nitrogen concentration, and the MLSS is a semi-free choice, so the only unknown is the total tankage.

Two- and Three-Stage Denitrification

If low effluent nitrate concentrations are needed, a separate stage denitrification process fed supplementary reductant is required. A schematic process train for separate stage denitrification is shown in Fig. 10.10, which is Mulbarger's [1971] three-sludge process. This scheme was modified and built by the Central Contra Costa Sanitary District and Brown and Caldwell, Engineers [Horstkotte *et al.*, 1974]. The principal changes are (1) the use of lime in the primary clarifier to remove metals and (2) the substitution of a single-stage nitrification process for the two stage process used by Mulbarger.

The nitrification stage can be designed using the procedures described above.

The denitrification stage can be designed using the methanol and nitrate kinetic data in Table 10.9.

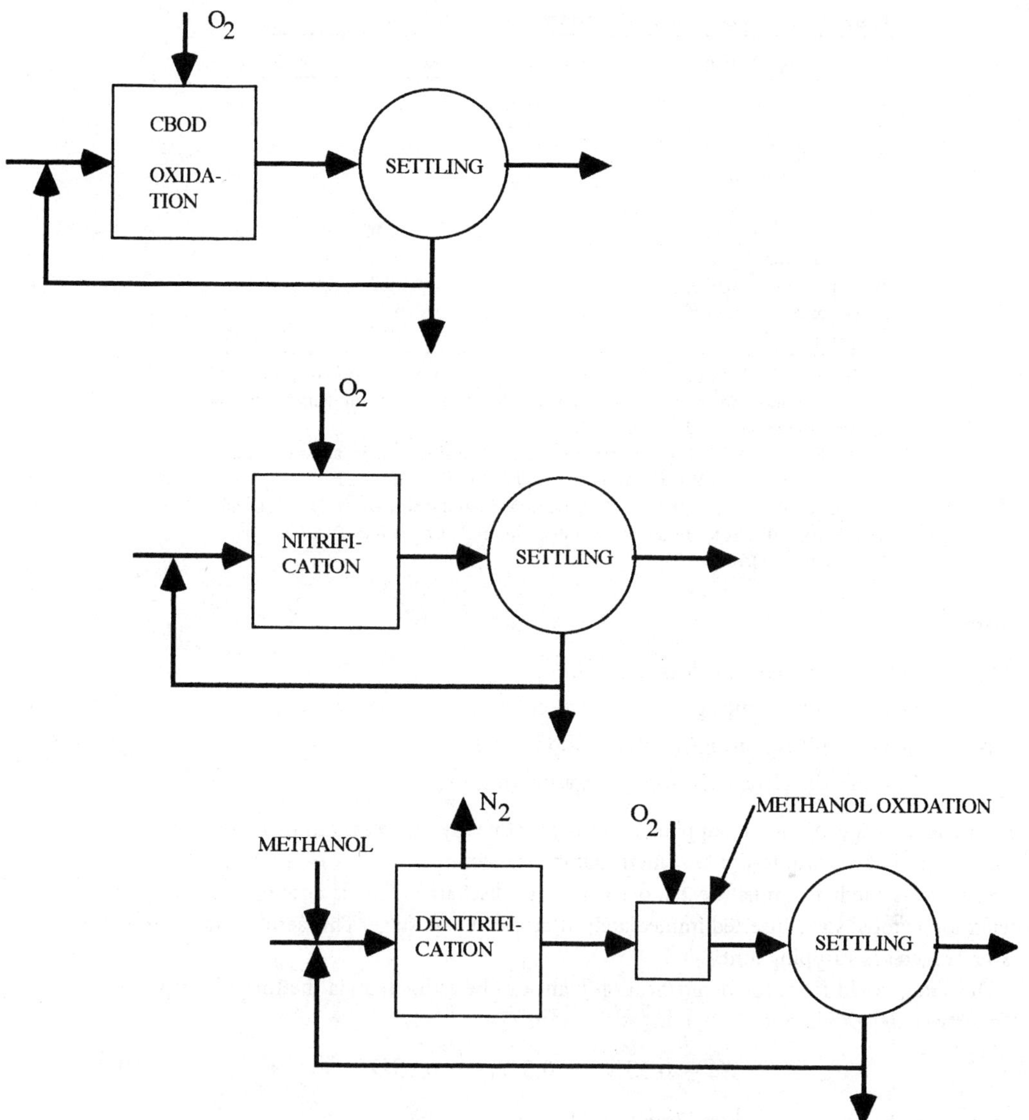

FIGURE 10.10 The three-sludge nitrogen removal scheme.

Any nonfermentable organic substance can be used as the reductant for nitrate. The usual choice is methanol because it is usually the cheapest. Fermentable substances like sugar or molasses are not used because a substantial portion of the reductant can be dissipated as gaseous end products like hydrogen.

Disregarding microbial growth, the stoichiometry of methanol oxidation is

$$CH_3OH + \tfrac{6}{5}HNO_3 \rightarrow CO_2 + \tfrac{3}{5}N_2 + \tfrac{13}{5}H_2O \tag{10.70}$$

This suggests a methanol dosage of at least 1.90 g methanol per g nitrate-nitrogen. In order to account for growth and the presence of oxygen and nitrite in the feed, McCarty *et al.*, [1969] recommend a dosage of

$$C_m = 2.47N_o + 1.53N_1 + 0.87D_o \tag{10.71}$$

Table 10.11 Separate Stage Denitrification Design Values[1–3]

Design Variable	Reference 1	Reference 2	Reference 3
Anoxic HRT (h)	3.0	1.6	0.82
Oxic HRT (h)	0.4	0.3	0.79
MLVSS (mg/L)	1390	2460	2600
F/M (kg COD/kg VSS · d)	0.73	—	0.79
Anoxic SRT (d)	38	7.6	29
Methanol nitrate ratio (kg Methanol/kg NO_3-N)	5.83	3.07	3.54
Effluent NO_3-N (mg/L)	0.9	0.7	0.8
System nitrogen removal (%)	91	>69	—
Temperature (°C)	12 to 22	10	16 to 19

[1] Barth, E. F., Brenner, R. C., and Lewis, R. F. 1968. Chemical-biological control of nitrogen and phosphorus in wastewater effluent. *Journal of the Water Pollution Control Federation.* 40(12):2040.

[2] Mulbarger, M. C. 1971. Nitrification and denitrification in activated sludge systems. *Journal of the Water Pollution Control Federation.* 43(10):2059.

[3] Horstkotte, G. A., Niles, D. G., Parker, D. S., and Caldwell, D. H. 1974. Full-scale testing of a water reclamation system. *Journal of the Water Pollution Control Federation.* 46(1):181.

where

C_m = the required methanol dosage (mg/L)
N_o = the initial nitrate-nitrogen concentration (mg/L)
N_1 = the initial nitrite-nitrogen concentration (mg/L)
D_o = the initial dissolved oxygen concentration (mg/L)

In general, excess methanol is supplied to ensure that nearly complete nitrate removal is achieved. This means that the nitrate-limited kinetic parameters control.

The excess methanol must be removed prior to discharge. This is accomplished via a short detention aerobic stage inserted immediately after the anoxic stage. This aerobic stage also serves as a nitrogen gas stripping unit.

The sludge yield from the denitrification plant can be estimated via the formula suggested by McCarty *et al.* [1969]:

$$X = 0.53N_o + 0.32N_1 + 0.19D_o \tag{10.72}$$

where X = the biomass produced (mg VSS/L).

The design parameter values used by Mulbarger [1971] and Horstkotte *et al.* [1974] are given in Table 10.11.

Phosphorus Removal

Except in the PHOSTRIP™ process, phosphorus is removed from wastewater by incorporation into the biomass and subsequent biomass wastage from the system. The steady state system phosphorus balance is

$$P = P_o - f_P \frac{VX}{Q\theta_c} \tag{10.73}$$

where

f_P = the concentration of phosphorus in the volatile suspended solids (kg P/kg VSS or lb P/lb VSS)
P = the soluble effluent phosphorus concentration (kg/m^3 or lb/ft^3)

P_o = the total influent phosphorus concentration (kg/m^3 or lb/ft^3)
Q = the settled sewage flow rate (m^3/s or ft^3/s)
V = the aeration tank volume (m^3 or ft^3)
X = the mixed liquor volatile suspended solids concentration (kg/m^3 or lb/ft^3)
θ_c = the solids retention time (s)

Growing bacteria typically consist of about 3% by weight phosphorus—of which 65% is incorporated into deoxyribonucleic acid (DNA) and ribonucleic acid (RNA), 15% into phospholipids, and the remainder as acid-soluble phosphate esters [Roberts *et al.*, 1955]. Conventional activated sludge plants typically remove about 40% of the influent phosphorus [EPA, 1977].

Nearly all microbes can accumulate more than 3% phosphorus [Harold, 1966], and the bacterium *Acinetobacter* has been reported to contain as much as 16% phosphorus [Eagle *et al.*, 1989]. The increased phosphorus content consists of volutin granules. Volutin is a long chain polymer of orthophosphate ions. The polymer is highly charged and is associated with K^+, Ca^{2+}, and Mg^{2+} ions; RNA; poly-betahydroxybutyric acid (PHB); proteins; and phospholipids [Eagle *et al.*, 1989].

The phosphorus content of activated sludges with volutin-containing bacteria generally consists of 4 to 7% by weight of the total suspended solids, and the plants generally achieve better than 90% phosphorus removal [Scalf *et al.*, 1969; Vacker *et al.*, 1967].

A number of proprietary phosphorus removal processes are being marketed that induce volutin accumulation in bacteria, including A^2/O™(U.S. Patent No. 4,056,465), BARDENPHO™(U.S. Patent No. 3,964,998), PHOSTRIP™and PHOSTRIP II™(U.S. Patent No. 4,042,493; 4,141,822; 4,183,808; and 4,956,094), and VIP™(U.S. Patent No. 4,867,883). The Joint Task Force [1992] lists several other patents and processes.

Microbiology

The underlying mechanism of volutin accumulation is the derepression of the synthesis of three enzymes [Harold, 1966]: alkaline phosphatase (which hydrolyzes extracellular phosphate esters), polyphosphate kinase (which synthesizes polyphosphate chains by transferring orthophosphate from adenosine triphosphate), and polyphosphatase (which hydrolyzes polyphosphate chains). Bacteria that are subjected to repeated anaerobic/aerobic cycles may have three to five times the normal amount of these enzymes. The biochemical process is called *polyphosphat uberKompensation* or "polyphosphate overplus" [Harold, 1966]. This is not to be confused with *luxury uptake*, which occurs when growth is inhibited by lack of an essential nutrient.

The jargon of the profession now distinguishes among aerobic, anoxic, and anaerobic conditions. *Aerobic* means that dissolved oxygen is present. Both anoxic and anaerobic mean that dissolved oxygen is absent. However, *anaerobic* also means that there is no alternative electron acceptor present—especially nitrite or nitrate—whereas *anoxic* means that alternative electron acceptors are present, usually nitrate.

Volutin formation occurs in activated sludge when it is exposed to an anaerobic/aerobic cycle [Wells, 1969]. The currently accepted biochemical model for polyphosphate overplus states that during the anaerobic phase volutin is hydrolyzed to orthophosphate, and the hydrolysis energy is used to absorb short-chain, volatile fatty acids (VFA, principally acetic acid) that are polymerized into PHB and related substances like poly-betahydroxyvaleric acid (PHV) [Bowker and Stensel, 1987]. Orthophosphate is released to the surrounding medium, and cellular phosphorus levels fall to less than one-half normal [Smith *et al.*, 1954]. Under subsequent aerobic conditions, the PHB and PHV are oxidized for energy and used for cell synthesis; the volutin is resynthesized by using some of the oxidation energy to reabsorb and polymerize the released orthophosphate. Consequently, there is an alternation between volutin-rich aerobic and PHB-rich anaerobic states. Peak volutin levels are reached in about one hour after reexposure to oxygen or nitrate. At peak volutin content, the cells contain several times the normal amount of total phosphorus. Under prolonged aeration, the volutin is broken down and incorporated into DNA, RNA, and phospholipids.

The anaerobic phase is supposed to serve two purposes. First, and most importantly, it shuts down the metabolism of most aerobic bacteria. The exception is *Acinetobacter* species. This permits *Acinetobacter* species (which are supposed to be able to capture the energy of volutin hydrolysis) to absorb volatile fatty acids without competition from other aerobic bacteria. The result is an activated sludge enriched in volutin-accumulating *Acinetobacter* bacteria. Second, if the wastewater does not contain appreciable amounts of VFA, then a prolonged anaerobic phase will encourage the growth of anaerobic and facultatively anaerobic bacteria that can ferment sewage organic matter to VFA and other small molecules.

It should be noted that while this scheme is supported by circumstantial evidence, such as concurrent orthophosphate release and acetate uptake, the enzymes required have not yet been demonstrated to exist [Harold, 1966].

Kinetics. The kinetics of phosphorus uptake and release are very poorly established in the public literature. In the early laboratory studies it was reported that the release and subsequent uptake of orthophosphate each require about three hours for completion [Fuhs and Chen, 1975; Shapiro, 1967]. The anaerobic phase in A/O™ systems generally has a hydraulic retention time of 30 to 90 minutes, BARDENPHO™ systems generally incorporate an anaerobic HRT or 1 to 2 hours, and PHOSTRIP™ plants operate with an anaerobic HRT of 5 to 20 hours [Bowker and Stensel, 1987]. In PHOSTRIP™ the anaerobic phase is applied to settled sludge solids in a sidestream operation.

Soluble Organic Matter Requirement. If effluent phosphorus concentrations less than 1 mg/L are desired, the settled wastewater should have a soluble BOD to soluble P ratio of at least 15 or a total BOD to total P ratio of at least 35 [Bowker and Stensel, 1987].

Effect of Oxygen and Nitrate. Oxygen and nitrate (in nitrifying plants) are carried into the anaerobic zone by sludge and mixed liquor recycles. Air-lift pumps, which are sometimes used for sludge recycle, saturate water with oxygen. This permits some growth of aerobic bacteria in the anaerobic zone and reduces the competitive advantage of *Acinetobacter* spp. The net effect of oxygen and nitrate is to reduce the effective BOD:P ratio below that required for optimum phosphorus uptake. The reduction can be estimated from the initial oxygen or nitrate concentrations. The fraction of the removed COD that is actually oxidized is $1 - 1.42Y_0$. For rapidly growing bacteria, the observed yield is nearly equal to the true growth yield of 0.4 g VSS/g COD, so the fraction oxidized is about 43%. In the case of denitrification the theoretical ratio is 3.53 kg COD removed for every kg of nitrate-nitrogen that is reduced [McCarty *et al.*, 1969]; reported values of the ratio range from 2.2 to 10.2 [Bowker and Stensel, 1987].

Empirical Design

Weichers *et al.* [1984] recommend the following semi-empirical formula for estimating phosphorus uptake:

$$\Delta P_s = S_o\left\{\left[\frac{(1-f_{nbs}-f_{nbp})Y_h}{1+k_{dh}\theta_c}\right](\gamma + f_P\cdot f_{nbX_a}\cdot k_{dh}\cdot\theta_c) + f_P\left(\frac{f_{nbp}}{\beta_X}\right)\right\} \tag{10.74}$$

where

f_{anX} = the anaerobic mass fraction of the MLSS (dimensionless)

f_{nbp} = the nonbiodegradable fraction of the particulate influent COD (dimensionless) ≈ 0.13 [according to Weichers *et al.*, 1984]

f_{nbs} = the nonbiodegradable fraction of the soluble influent COD (dimensionless) ≈ 0.05 [according to Weichers *et al.*, 1984]

f_{nbX_a} = the nonbiodegradable fraction of the active biomass (dimensionless)
$\approx$ 0.20 [according to Weichers *et al.*, 1984]

f_P = the phosphorus concentration in the nonbiodegradable volatile solids, in mg P/mg VSS
$\approx$ 0.015 mg P/mg VSS [according to Weichers *et al.*, 1984]

k_{dh} = the decay coefficient of the heterotrophic bacteria, in per day
$\approx$ 0.24/day [according to Weichers *et al.*, 1984]

P_f = the excess phosphorus removal propensity factor (dimensionless)
$= (S_{anrb} - 25) f_{anX}$

ΔP_s = the phosphorus removal by incorporation into the sludge, in mg/L

S_{anrb} = the readily biodegradable COD in the anaerobic reactor, in mg/L

S_o = the total influent COD, in mg/L

Y_h = the true growth yield of the heterotrophic bacteria, in mg VSS/mg COD
$\approx$ 0.45 mg VSS/mg COD [according to Weichers *et al.*, 1984]

β_X = the COD of the VSS, in mg COD/mg VSS
$\approx$ 1.48 mg COD/mg VSS [according to Weichers *et al.*, 1984]

γ = the coefficient of excess phosphorus removal by the active biomass, in mg P/mg VSS
$= 0.35 - 0.29 \exp(-0.242 P_f)$

θ_c = the solids retention time, in days

The formula distinguishes between the active biomass—which absorbs and releases phosphorus—and inert ("endogenous") biomass. The phosphorus content of the active biomass is supposed to vary between about 3 and 35% by weight of the VSS. The phosphorus content of the inert biomass is estimated to be about 1.5% of VSS.

The anaerobic mass fraction is the fraction of the system biomass held in the anaerobic zone. Because MLSS concentrations do not vary substantially from one reactor to the next in a series of tanks, this is also approximately the fraction of the total system reactor volume in the anaerobic zone.

The readily available COD is estimated from oxygen uptake rates (OUR). First the OUR of the mixed liquor is measured under steady loading and operating conditions. This should be done at several different times to establish that the load is steady. Then the influent load is shut off, and the OUR is measured at several different times again. There should be an immediate drop in OUR within a few minutes followed by a slow decline. The immediate drop represents the readily biodegradable COD. The calculation [Weichers *et al.*, 1984] is as follows:

$$S_{rb} = \frac{(R_o - R_{nl})V}{f_{ox} Q} \tag{10.75}$$

where

f_{ox} = the fraction of the consumed COD that is oxidized by rapidly growing bacteria (dimensionless)
$= 1 - \beta_X Y_h$
$\approx$ 0.334 [according to Weichers *et al.*, 1984]

R_{nl} = the oxygen uptake rate immediately after the load is removed (kg/s or lb/s)

R_o = the oxygen uptake rate during the steady load (kg/s or lb/s)

Q = the wastewater flow rate during loading (m^3/s or ft^3/s)

S_{rb} = the readily degradable COD (kg/m^3 or lb/ft^3)

V = the reactor volume (m^3 or ft^3)

β_X = the COD of the VSS (kg COD/kg VSS or lb COD/lb VSS)
$\approx$ 1.48 kg COD/kg VSS [according to Weichers *et al.*, 1984]

A/O™

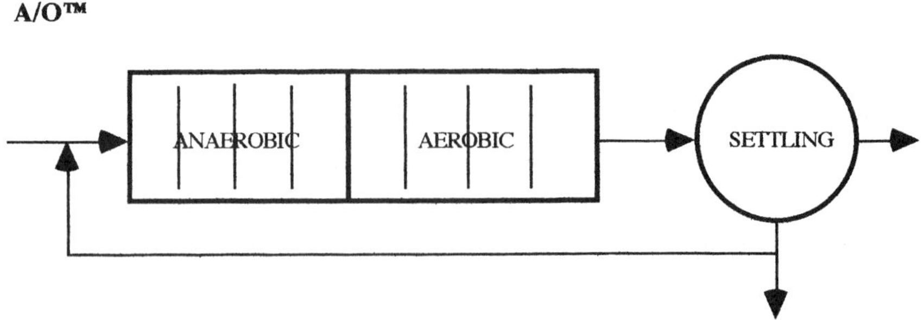

MODIFIED BARDENPHO™

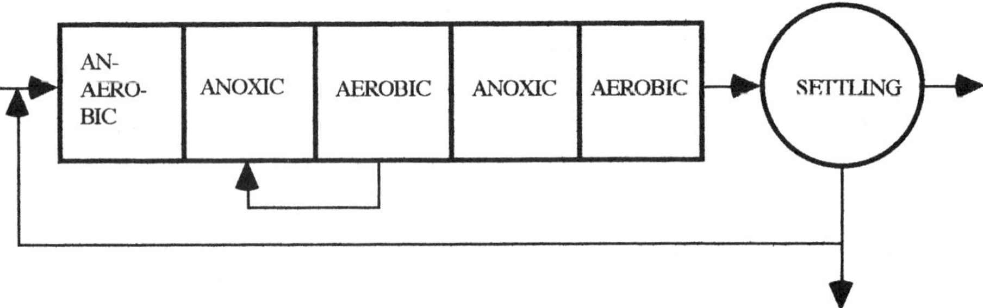

MODIFIED UCT™

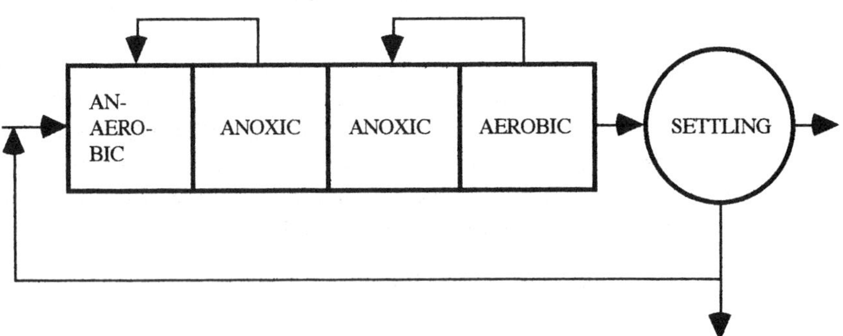

FIGURE 10.11 Schemes for biological phosphorus removal. *(continues)*

Flow Schematics and Performance

Flow schematics of the principal phosphorus removal processes are shown in Fig. 10.11. The PHOSTRIP™ process removes phosphorus by subjecting a portion of the recycled activated sludge to an anaerobic holding period in a thickener. The phosphorus-rich supernatant is then treated with lime, and the phosphorus leaves the system as a hydroxylapatite [$Ca_5(PO_4)_3OH$] sludge. The phosphorus-depleted activated sludge solids in the thickener are returned to aeration. The PHOSTRIP™ process consistently produces soluble effluent phosphorus concentrations under 1 mg/L and usually produces total effluent phosphorus concentrations of about 1 mg/L, although excursions of effluent suspended solids sometimes degrade performance [Roy F. Weston, Inc., 1983]. The process cannot be used to remove phosphorus from a nitrifying sludge, but it can be applied to the first stage of a two-stage nitrification plant. The PHOSTRIP™ process is best suited to

PHOSTRIP™

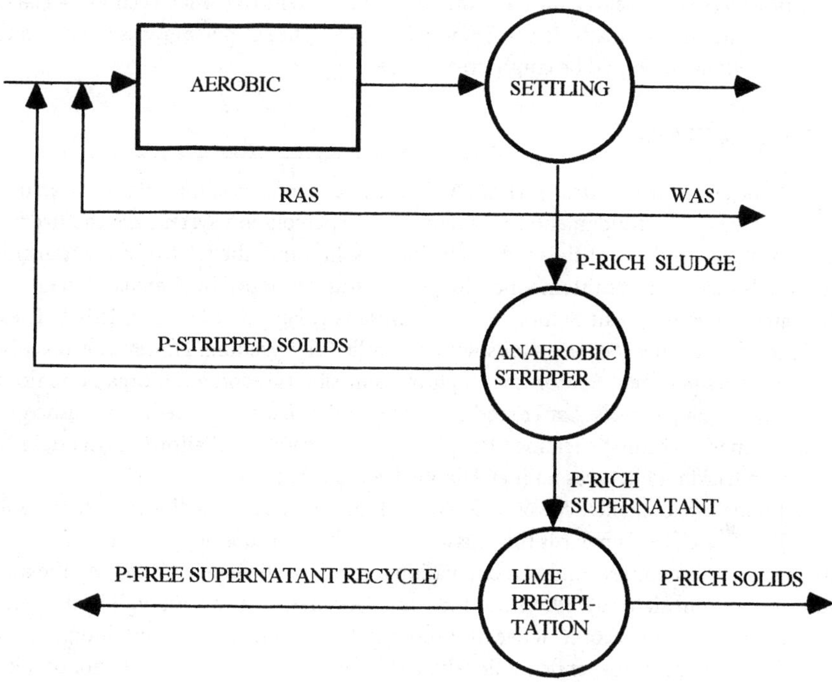

GENERIC PROCESS

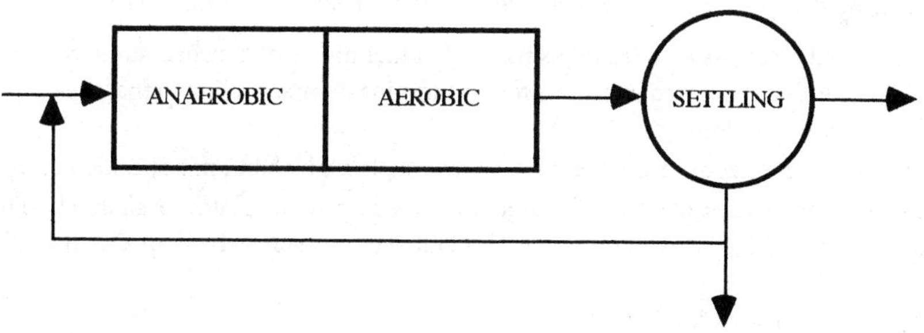

FIGURE 10.11 (continued) Schemes for biological phosphorus removal.

low hardness waters so that the lime treatment step does not produce unwanted calcium carbonate sludges.

The A/O™ process nitrifies, partially denitrifies, and partially removes phosphorus. Total effluent phosphorus concentrations are generally 1.5 to 3.0 mg/L [Roy F. Weston, Inc., 1983]. Phosphorus is removed from the system in the waste activated sludge.

The BARDENPHO™ process also nitrifies, partially denitrifies, and partially removes phosphorus. Supplemental methanol for denitrification and alum for phosphate precipitation may be required if consistently high removals of nitrogen and phosphorus are needed [Roy F. Weston, Inc., 1983]. Phosphorus is removed from the system in the waste activated sludge.

Except for the PHOSTRIP™ process, all the schemes for biological phosphorus removal produce phosphorus-rich waste activated sludges. If these sludges are subjected to digestion, phosphorus-rich supernatants are produced, and the supernatants require a phosphorus removal treatment prior to recycle.

All the processes show degradation of total phosphorus removal when secondary clarifiers are subjected to hydraulic overloads. If consistently low total effluent phosphorus concentrations are required, effluent filters should be considered.

10.2 Trickling Filters

Trickling filters (also known as *bacteria beds, biological filters,* and *percolating filters*) are an outgrowth of sewage farming. The fundamental problems of any percolation system are the hydraulic and pneumatic transmissibilities of the media. The initial solution to the relatively low transmissibility of natural soils was the intermittent sand filter, which was developed by Frankland in Great Britain and the Lawrence Experiment Station in Massachusetts [Bruce and Hawkes, 1983]. These filters are constructed of relatively coarse sands with controlled size grading. However, it was discovered at the Lawrence Experiment Station that high transmissibilities and good organic removals were obtainable by utilizing gravels. Latter studies increased the size of the media to crushed rock of a few inches diameter. The first trickling filter plant was constructed at Salford, England, in 1892; the first American trickling filter was built at Atlanta, Georgia, in 1903.

The traditional trickling filter classification by hydraulic and BOD loading is given in Table 10.12. The Joint Task Force [1992] regards this classification scheme as obsolete.

Low-rate filters sometimes employ recirculation during low-flow periods to keep the media wet. Low-rate and intermediate-rate filters produce nuisance flies. They also slough in the early spring when the filter fly larvae become active and disrupt the biofilm. During the sloughing, the filter effluent will contain more suspended solids than its influent. High-rate filters do not usually harbor filter flies, and they discharge solids at low rates more or less continuously. The low rate of solids discharge is not sloughing, because it is not larva induced, seasonal, or flashy.

Media

Trickling filter media (packings) can be classified as random, stacked, or modular:

- Random media consist of relatively small individual pieces of crushed stone or ceramic or plastic shapes that are placed into the filter by dumping or by shoveling so that the orientations of the pieces are random.
- Stacked media consist of individual pieces that are placed by hand in a specified orientation.
- Modular media consist of relatively large integrated arrays of plates or channels. These are placed by hand in fixed orientations, and individual modules may be shaped in the field to fit.

Table 10.12 Trickling Filter Classification[1,2]

Class	Media	Hydraulic Loading (mgd/ac)	BOD_5 Loading (lb/d/1000 ft^3)	Depth (ft)	BOD_5 Removal[a] (%)	Recirculation/ Nitrification
Low	stone	1 to 4	5 to 20	5 to 10	75 to 85	no/fully
Intermediate	stone	4 to 10	15 to 30	6 to 8	50 to 70	yes/partial
High	stone	10 to 40	30 to 60	3 to 8	40 to 80	yes/none
Super	plastic	15 to 90	30 to 150	<40	65 to 85	yes/some
Roughing	stone/plastic	60 to 180	<300	3 to 20	40 to 85	no/none

[1]Joint Task Force of the Water Environment Federation and the American Society of Civil Engineers. 1992. *Design of Municipal Wastewater Treatment Plants: Volume I.* Chapters 1–12. WEF Manual of Practice No. 8; ASCE Manual and Report on Engineering Practice No. 76. Water Environment Federation, Alexandria, VA; American Society of Civil Engineers, New York.

[2]Schwinn, D. E. and Gassett, R. B., ed. 1974. *Process Design Manual for Upgrading Wastewater Treatment Plants.* Environmental Protection Agency, Technology Transfer, Washington, D.C.

[a]After settling.

Older filters built before 1960 consist almost entirely of crushed stone placed by hand shovel and randomly oriented. Hand placement is required to avoid damage by dumping to the underdrain system and to the media itself. The stone is usually sieved and washed on site to remove fines and long or flat shapes. The usual size range is 3 to 4½ in. (76 to 114 mm).

Stone must be durable against freeze-thaw cycles and chemically and biologically inert to the wastewater and microbes. Some furnace slags and burnt clays leach iron, which discolors the effluent. The usual test of durability is a loss of less than 10% by weight after 20 cycles of soaking for 16 to 18 hours in a saturated sodium sulfate solution at 70°F followed by drying at 230°F (ASCE, 1936; ASTM Committee C-9, 1993).

Other random packings include ceramic, carbon, metal or plastic rings (Raschig, Lessing, cross-partition, and spiral), and saddles (Berl and Intalox). Rings are short, hollow cylinders, different kinds having different designs for partition walls within the cylinder. For example, Rashig rings have no inner partitions, Lessing rings have a single cross-wall running down the axis of the cylinder, and cross-partitioned rings have two intersecting walls running down the axis. Saddles are strips of material that have been bent into a semicircular curve; the surface of the strip may be grooved or folded.

Stacked media are almost never used because they are expensive to install (because of the careful hand labor required) and because they are prone to channeling.

Modular media consist of arrays of parallel sheets of plastic joined together. Early designs used corrugated sheets with the corrugation channels aligned vertically. Two kinds are in use: vertical-flow, fully corrugated bundles (VFC) and vertical-flow, semicorrugated bundles (VFS). The latter alternate flat sheets with corrugated sheets. In recent years the corrugations have been arranged so that the flow is diagonal through a bundle, with alternating layers directing the flow to opposite sides. This arrangement is called *cross flow* (XF). VFC is preferred for strong wastewaters and filters with high organic loadings. XF is preferred for weak wastewaters and low organic loadings [Joint Task Force, 1992].

Plastic media must be resistant to ultraviolet radiation, disintegration, erosion, aging, all common acids and alkalis, organic compounds, and fungus and biological attack [Wastewater Committee, 1990]. The media must not be toxic to the biofilm or leach toxic materials. The media should be able to support foot traffic for distributor maintenance. If it cannot, walkways must be provided.

An important property of any media is the specific surface area, A_v, which is defined as the total area available for biofilm attachment divided by the bulk volume of the media. The surface area actually covered by biofilm and wastewater depends on the hydraulic loading, and media-specific minimum flows are required to achieve the maximum treatment capacity. Some typical values of A_v are given in Table 10.13.

In order to promote free air flow, the Wastewater Committee [1990] specifies maximum specific surface areas of 30 ft^2/ft^3 (100 m^2/m^3) for carbonaceous BOD removal and 45 ft^2/ft^3 (150m^2/m^3) for nitrification.

Table 10.13 Physical Properties of Trickling Filter Media[1]

Media	Unit Size (in.)	Specific Weight (lb/ft^3)	Specific Surface Area (ft^2/ft^3)	Void Space (%)
Modular plastic	24 × 24 × 48	2 to 6	25 to 35	94 to 97
Random plastic	varies	2 to 5	25 to 50	>94
Redwood	47½ × 47½ × 35¾	10	14	76
Blast furnace slag	2 to 3	68	29	49
Rock (granite)	1 to 3	90	19	46
Rock (granite)	4	70	13	60

[1]Schwinn, D. E. and Gassett, R. B., ed. 1974. *Process Design Manual for Upgrading Existing Wastewater Treatment Plants.* Environmental Protection Agency, Technology Transfer, Washington, D.C.

Rock media filters that depend on natural draft for aeration are usually limited to a maximum media depth of 10 ft [Wastewater Committee, 1990]; the usual practice is to build them no deeper than 6 ft. Greater depths can be used if forced aeration is provided. Natural draft modular plastic media filters are typically 16 to 26 ft deep [Wastewater Committee, 1990].

Hydraulics and Pneumatics

Hydraulic Retention Time and Film Thickness

The current model for the hydraulic regime in trickling filters is as follows [Suschka, 1987]:

- There is a biofilm layer attached to the media surface.
- On the liquid side of the biofilm there is a heterogeneous layer of porous biofilm and stagnant water.
- A wastewater film flows freely over the heterogeneous layer.
- The remainder of the media voids is filled with air.

It appears that most of the liquid on the media surface is in the heterogeneous layer rather than in the freely flowing liquid layer.

For clean modular media without a biofilm the thickness of the freely flowing water film can be estimated by

$$h_f = \sqrt[3]{\frac{3q}{\gamma \sin \alpha}}; \tag{10.76}$$

where

h_f = the thickness of the free liquid (m or ft)
q = the wastewater flow rate per unit width of media surface ($m^3/m \cdot s$ or $ft^3/s \cdot ft$)
α = the inclination of the media surface measured from the horizontal (rad)
γ = the unit weight of the liquid (N/m^3 or lbf/ft^3)

However, once the biofilm and its associated heterogeneous layer develop, the effective water layer thickness is determined by experiment to be approximately

$$h_w = 3.4 \times \sqrt[3]{\frac{3q}{\gamma \sin \alpha}} \tag{10.77}$$

where h_w = the effective water film thickness (m or ft).

An order of magnitude estimate of the HRT of trickling filters with modular media is given by Suschka [1987] as

$$\theta_h = \frac{2H}{(Q/V)^{0.75}} \tag{10.78}$$

where

H = the depth of media in m
Q = the total wastewater flow rate including recirculated flow in m^3/h
V = the media volume in m^3
θ_h = the HRT in min

Equation (10.78) appears to be most representative of media with specific surface areas of about 100 to 120 m^2/m^3. The HRT of media with specific surface areas of about 40 m^2/m^3 is only about one-third of the HRT predicted by Eq. (10.78), and the HRT of media with specific surface areas around 200 m^2/m^3 is about three times the equation value.

Distributor Systems

The preferred distributor systems consist of rotary sprinkler arms driven by variable speed electric motors [Joint Task Force, 1992]. The wastewater should be distributed as uniformly as possible over the filter cross section; the maximum areal application rate should not vary by more than 10% from one point to another on the cross section [Wastewater Committee, 1990].

Impulse-driven arms are not desirable because they fail to achieve adequate wetting. Fixed nozzles require high maintenance, and uniform distribution of the wastewater is nearly impossible [Joint Task Force, 1992].

Traditional designs attempted to provide a steady flow to each area of the filter cross section, but recent studies suggest that such designs results in poor wetting of the media surface, with perhaps as little as one-third of the media being used [Joint Task Force, 1992]. Recent recommendations [Joint Task Force, 1992] focus on the instantaneous dosing intensity (*Spülkraft*, in German) on a subarea due to the passage of a distributor arm. This is normally expressed as the total depth deposited on the area per passage of an arm and is calculated as

$$I_f = \frac{Q + Q_r}{n\omega A} \tag{10.79}$$

where

A = the plan area of the filter (m^2 or ft^2)
I_f = the instantaneous dosing intensity (m/pass or ft/pass)
Q = the settled sewage flow rate (m^3/s or ft^3/s)
Q_r = the flow rate of the recirculated filter effluent (m^3/s or ft^3/s)
n = the number of distributor arms (dimensionless)
ω = the rotational speed of the distributor (Hz, rev/s)

Current practice results in dosing intensities on the order of 2 to 10 mm/pass, but field experience suggests it should be somewhere between a few hundred and several hundred mm/pass [Joint Task Force, 1992]. The flushing intensity may be site-, media-, and/or application-specific, and the selection of the distributor motor and controller should allow maximum flexibility in the selection of rotation speeds.

It may be noted in passing that the recommendation for high flushing intensities is supported by chemical engineering experience with packed beds for liquid/gas mass transfer operations [Leva, 1953]. Because of improved wetting, the efficiency of such units appears to increase with hydraulic loading nearly up to the point of flooding. Packed towers are often operated in what is called *loading region*, which occurs when the liquid rate is high enough that the gas pressure drop across the packing varies as a power of the gas flow rate that is greater than 2. In this region Eqs. (10.80) and (10.81) do not apply.

Drainage

The usual requirements are [Wastewater Committee, 1990]:

- The underdrains should collect wastewater from the entire floor of the filter.
- The inlet openings to the drains must cover at least 15% of the filter cross section area.
- The drains themselves must be only half full at instantaneous peak flows (to permit air flow), and they should be designed to provide a minimum velocity of 2 ft/s (0.61 m/s) at an invert slope of 1%.

Aeration

The trickling filter is an aerobic process, and the oxygen needed is supplied by air flowing through the packed bed. This air flow is developed either by natural convection due to density differences between the air inside and outside the bed or by blowers.

As long as the pressure drop is small, say 10% or less, air may be treated as an incompressible fluid, and the usual headloss equations apply. Piping losses may be estimated using the Darcy–Weisbach equation, and minor losses may be estimated as multiples of the velocity head. Conversion of headlosses to pressure drops requires the specific weight of *air*:

$$\Delta p = \gamma h_L \tag{10.80}$$

where

g = the acceleration due to gravity (m/s^2 or ft/s^2)
h_L = the headloss experienced by the flowing air (m or ft)
M_r = the molecular weight of air (28.96 g/mol or 28.96 lb/lb-mol)
Δp = the pressure drop (Pa or lbf/ft^2)
p_{atm} = local atmospheric pressure (Pa or lbf/ft^2)
R = the gas constant (8.314 510 J/mol·K or 1,545.356 ft·lbf/lb-mol·°R)
T = the absolute temperature (K or °R)
γ = the specific weight of air (N/m^3 or lbf/ft^3)
= $p_{atm} M_r g/RT$

The pressure drop in the filter bed itself may be estimated via the Ergun [1952] equation, which is discussed under "Filtration, Granular Media Filters, Hydraulics":

$$\frac{\Delta p}{\gamma} = h_L = f_m \left(\frac{H}{\phi d}\right)\left(\frac{1-\varepsilon}{\varepsilon^3}\right)\left(\frac{v_s^2}{g}\right) \tag{10.81}$$

For random and modular plastic media, this can be rewritten as

$$\frac{\Delta p}{\gamma} = h_L = \frac{1}{6} f_m (H A_v)\left(\frac{1}{\varepsilon^3}\right)\frac{v_s^2}{g} \tag{10.82}$$

where

A_p = the surface area of a single media "particle"
A_v = the specific surface area of the filter media—that is, the total media surface area divided by the total volume of the filter bed *voids* (m^2/m^3 or ft^2/ft^3)
= $6(1-\varepsilon)/d_{eq}$, for granular media
d_{eq} = the equivalent particle diameter (m or ft)
= $6(V_p/A_p)$
f_m = MacDonald-El-Sayed-Mow-Dullien [1979] friction factor (dimensionless)
$\geq 180 \cdot (1-\varepsilon/\text{Re}) + 1.8$, for smooth media
$\leq 180 \cdot (1-\varepsilon/\text{Re}) + 4.0$, for rough media
H = the height of the filter (m or ft)
Re = the bed Reynold's number (dimensionless)
= $\rho v_s d_{eq}/\mu$ or $\rho v_s/\mu \cdot 6(1-\varepsilon)/A_v$
V_p = the volume of a single media particle (m^3 or ft^3)
v_s = the superficial (approach) velocity of the air (m/s or ft/s)

ε = the bed porosity (dimensionless)
μ = the dynamic viscosity (N·s/m^2 or lbf·s/ft^2)

In these relations

$$RA_v = \varepsilon \tag{10.83}$$

$$AA_v = P \tag{10.84}$$

where

A = the total cross-sectional plan area of the bed, including both voids and media (m^2 or ft^2)
P = the wetted perimeter of the media in any planar cross section (m or ft)
R = the bed hydraulic radius—that is, the void volume divided by the surface area of the media (m or ft)

The void volume in the bed must be reduced by the liquid and biomass volumes in the filter, and, in the case of crushed stone, the grain diameter should be increased by twice the film thicknesses.

If no blower is installed, the air flow is developed by the natural draft. The natural draft is the gas phase pressure difference developed between the air in the bottom of the trickling and the surrounding air. It may be estimated from the following equation [Schroeder and Tchobanoglous, 1976]:

$$\Delta p = \frac{p_{atm} M_r g H}{R}\left(\frac{1}{T_o} - \frac{1}{T_1}\right) \tag{10.85}$$

where

Δp = the natural draft developed by the temperature difference (Pa or lbf/ft^2)
T_o = the cooler absolute temperature (K or °R)
T_1 = the warmer absolute temperature (K or °R)

If there is a significant temperature gradient within the filter, the logarithmic mean filter air temperature may be estimated by

$$T_m = \frac{T_2 - T_1}{\ln(T_2/T_1)} \tag{10.86}$$

where

T_2 = the warmer absolute temperature (K or °R)
T_1 = the cooler absolute temperature (K or °R)
T_m = the logarithmic mean temperature (K or °R)

The air flow through natural draft filters may stagnate for temperature differences of less than 3°C [Joint Task Force, 1992]. If temperature differences this small are expected to be common, forced air circulation is needed.

In natural draft systems the air passages consist of the pore spaces in the media, the underdrains, and "chimneys" placed around the periphery of the bed and connecting the underdrains with the atmosphere. The underdrains and effluent piping must have at least 50% of their cross-sectional area open to air movement at the peak instantaneous hydraulic flow rate (including the recirculated flow) [Wastewater Committee, 1990].

Plastic media filters for municipal wastewater should have at least 1 ft^2 of ventilation area for each 10 to 15 ft of filter circumference and 1 to 2 m^2 of ventilation area per 1000 m^3 of media [Joint Task Force, 1992].

The minimum air flows recommended by the Joint Task Force [1992] for roughing filters located at 120 to 320 kg BOD_5 $m^3 \cdot d$ (75 to 200 lb BOD_5/d/1000 ft^3) are

$$Q_{\text{air}} = 0.9 \cdot PF \cdot K \cdot Q(S_i - S) \tag{10.87}$$

Minimum air flows recommended for standard rate filters loaded at 40 to 80 kg $BOD_5/m^3 \cdot d$ (25 to 50 lb BOD_5/d/1000ft^3) are:

$$Q_{\text{air}} = PF \cdot K \cdot Q(S_i - S) \tag{10.88}$$

where

K = the volumetric air requirement per unit mass of BOD_5 removed [75 m^3 air (SC)/kg BOD_5 or 1200 ft^3 air (SC)/lb BOD_5]
PF = the load peaking factor (dimensionless)
Q = total hydraulic flow rate applied to the filter, including recirculated flow (m^3/s or ft^3/s)
Q_{air} = the air flow rate [m^3 air (SC)/s or ft^3 air (SC)/s]
S_i = the total BOD_5 in the flow applied to the filter (kg/m^3 or lb/ft^3)
S = the soluble BOD_5 in the filter effluent (kg/m^3 or lb/ft^3)

The suggested volumetric air requirement amounts to an oxygen to BOD_5 mass ratio of about 21 kg/kg and an implied oxygen transfer efficiency of 10%.

Carbonaceous BOD Removal

There are many empirical and semitheoretical kinetic formulas for the design of trickling filters [Roberts, 1973]. The more commonly encountered ones are listed below in chronological order. Designers should bear in mind that the validity of these formulas is under active review, and it appears that all of them fail to account for (1) the fraction of the media that is wetted, and (2) the liquid dosing pattern. These points are discussed below. It is recommended that trickling filter designs be based on large-scale pilot studies. In this regard the older literature for stone and plastic media contains mainly data from filters that may have been hydraulically *underloaded*, and these reports may seriously underestimate filter capacities.

It should also be borne in mind that all the models assume the effluent BOD is merely residual influent BOD that the microbes did not remove and metabolize. However, it has been shown that in laboratory scale filters fed known, soluble substrates about 85% of the soluble organic matter in the effluent is a high molecular weight ($M_r > 1000$) microbial product [Namkung and Rittman, 1986]. This means that all trickling filter formulas, including those derived theoretically, have no mechanistic meaning, and the model parameter values can be expected to vary with wastewater composition and media configuration. All models should be treated as regression models.

National Research Council Formula

The NRC (Mohlman *et al.*, 1946; Fair *et al.*, 1948) formula was derived from correlations using data from stone media trickling filters at U.S. Army bases during World War II:

$$\frac{S_{se}}{S_o} = \frac{0.0085\sqrt{\dfrac{W}{VF}}}{1 + 0.0085\sqrt{\dfrac{W}{VF}}} \tag{10.89}$$

where

F = the Mountfort [1924] recirculation factor (dimensionless)
$= (1 + R)/[1 + (1 - f)R]^2$

f = the "available" fraction of the BOD_5, usually assumed to be about 0.90 (dimensionless)
R = the recirculation ratio—that is, the ratio of the settled sewage flow to the recycled treated flow (dimensionless)
S_o = the BOD_5 of the effluent of the primary settling tank in mg/L
S_{se} = the BOD_5 of the settled, final effluent in mg/L
V = the volume of the filter bed in ac·ft
W = the BOD_5 load on the filter in lb/day

The scatter about this formula is very large, and it is only a rough guide to trickling filter performance.

Velz–Howland–Schulze Contacting Time Model

Velz's [1948] law for the removal of organic matter by trickling filters is that the rate of removal per unit depth is proportional to the remaining organic matter concentration, which yields an exponential decay versus depth:

$$\frac{S}{S_o} = e^{-K_V H} \tag{10.90}$$

where

H = the depth of media (m or ft)
K_V = the Velz decay rate (1/m or 1/ft)
S = the remaining organic matter concentration (kg/m^3 or lb/ft^3)
S_o = the concentration of biodegradable organic matter in the filter influent (kg/m^3 or lb/ft^3)

Trickling filter depth is proportional to the hydraulic retention time of the bed as long as the media are uniformly coated with biofilm, so Velz's formula can also be expressed in terms of the HRT. The HRT of a trickling filter packed with spheres was derived theoretically by Howland [1958]:

$$\theta_h = 2^{2/3} \cdot 1.3 \cdot \left(\frac{3\nu}{g}\right)^{1/3} \frac{f^{2/3} H}{(Q/A)^{2/3}} \tag{10.91}$$

where

A = the plan area of the filter (m^2 or ft^2)
g = the acceleration due to gravity (m/s^2 or ft/s^2)
r = the radius of a piece of filter media (m or ft)
ν = the kinematic viscosity (m^2/s or ft^2/s)
θ_h = the hydraulic retention of the filter (s)

Schulze [1960] applied the Howland's formula to Velz's equation and fitted it to data from an experimental low-rate filter consisting of vertical wire grids. The curve fits were good, even though no correction for biodegradability was made. The final form of the equation was

$$\frac{S}{S_o} = \exp\left\{-\frac{K_1 H}{(Q/A)^{2/3}}\right\} \tag{10.92}$$

where K_1 = an empirical constant (m$^{-1/3}$·s$^{-2/3}$ or ft$^{-1/3}$·s$^{-2/3}$).

The Velz–Howland–Schulze trickling filter formula is sometimes further modified to take into account the specific surface area of the packing material and the variation of biofilm thickness and activity with depth [Eckenfelder, 1961; Eckenfelder and Barnhart, 1963; Kehrberger and Busch, 1969]:

$$\frac{S}{S_o} = \exp\left\{-\frac{KA_v^l H^m}{(Q/A)^n}\right\} \tag{10.93}$$

where

K = an empirical constant ($\mathrm{m}^{1-l-m-n}\cdot s^{-n}$ or $\mathrm{ft}^{1-l-m-n}\cdot s^{-n}$)

l, m, n = empirical exponents that account for the nonuniform distribution of biofilm in the bed (dimensionless)

This equation frequently gives very good fits to experimental data, and it is the most commonly cited formula, but see below. The most common form of the equation, especially in plastic media studies, is the Germain [1966] formula, which absorbs the specific surface area into the rate constant and assigns values of 1 to the exponent m and 0.5 to the exponent n:

$$\frac{S_{se}}{S_o} = \exp\left(-\frac{K_G H}{(Q/A)^{0.5}}\right) \tag{10.94}$$

where

A = the plan area of the trickling filter (m^2 or ft^2)

K_G = the Germain treatability factor [$\mathrm{s}^{-0.5}\cdot\mathrm{m}^{-0.5}$ or $(\mathrm{gpm})^{0.5}/\mathrm{ft}^2$]

$\approx 0.088\ (\mathrm{gpm})^{0.5}/\mathrm{ft}^2$ for settled domestic sewage

Q = the flow rate of the settled sewage not including any recirculated flow (m^3/s or gpm)

S_o = the BOD_5 in the settled sewage not including the recirculated flow ($\mathrm{kg/m}^3$ or $\mathrm{lb/ft}^3$)

S_{se} = the BOD_5 in the settled effluent ($\mathrm{kg/m}^3$ or $\mathrm{lb/ft}^3$)

Equations (10.93) and (10.94) are often erroneously called *Eckenfelder's equation*.

Eckenfelder's Retardant Model

The Eckenfelder [1961] retardant formula was derived from the assumption that the rate of BOD removal decreases as the contacting time increases:

$$\frac{dS}{dt} = \frac{kS}{1 + kt} \tag{10.95}$$

where

k = a rate constant (1/s)

S = the residual organic matter concentration ($\mathrm{kg/m}^3$)

t = the duration of the contacting time (s)

Integrating over the depth of the bed, substituting the Howland formula for the bed HRT, and accounting for biofilm distribution and media-specific surface area, one gets

$$\frac{S}{S_o} = \frac{1}{1 + \dfrac{KA_v H^m}{(Q/A)^n}} \tag{10.96}$$

Eckenfelder applied his model to data from stone filters treating domestic sewage and obtained the following:

$$\frac{S_{se}}{S_o} = \frac{1}{1 + \dfrac{2.5H^{0.67}}{(Q/A)^{0.50}}} \tag{10.97}$$

where

A = the plan area of the filter in acres
H = the depth of media in ft
Q = the settled sewage flow rate in mgd
S_{se} = the settled trickling filter effluent BOD_5 in mg/L
S_o = the settled sewage BOD_5 in mg/L

This formula appears to be forgotten, although it gives good curve fits.

Galler–Gotaas Correlation

Galler and Gotaas [1964] performed multiple linear (logarithmically transformed) regression on a large sample of published operating data and obtained the following formula, which is one of several versions they derived:

$$S_{se} = \frac{0.464 S_i^{1.19} \left(\frac{Q + Q_r}{Q} \right)^{0.28} (Q + Q_r)^{0.13}}{(1 + H)^{0.67} T^{0.15}} \tag{10.98}$$

where

H = the depth of media in ft
Q = the settled sewage flow rate in mgad
Q_r = the recirculation rate in mgad
S_i = the BOD_5 in the total flow applied to the filter in mg/L
S_{se} = the BOD_5 in the settled, final effluent in mg/L
T = the sewage temperature in °C

The multiple correlation coefficient for the logarithmic form of Eq. (10.98) is 0.974.

Equation (10.98) was modified by Blain and McDonnell [1965] in order to account for strong correlations among the data set:

$$S_{se} = \frac{0.860 S_o^{1.31} Q^{0.11}}{\left(\frac{Q + Q_r}{Q} \right)^{0.35} H^{0.68} T^{0.57}} \tag{10.99}$$

where S_o = the BOD_5 of the settled sewage applied to the filter not including the recirculated flow or BOD_5 in mg/L. The logarithmic form of Eq. (10.99) has a multiple linear correlation coefficient of 0.869.

Oxygen-limitation Model

The steady state oxygen balance on a non-nitrifying trickling filter is [Schroeder and Tchobanoglous, 1976]

$$Q_i (S_i - S) = \frac{Q_{air} (p_{O_2 i} - p_{O_2 e}) M_{rO_2}}{RT} \tag{10.100}$$

where

M_{rO_2} = the relative molecular weight of oxygen (kg/mol or lb/lb-mol)
$p_{O_2 e}$ = the partial pressure of oxygen in the exit air stream (Pa or lbf/ft^2)
$p_{O_2 i}$ = the partial pressure of oxygen in the inlet air stream (Pa or lbf/ft^2)
Q_i = the wastewater flow applied to filter, including any recycle (m^3/s or ft^3/s)
Q_{air} = the volumetric flow rate of the air through the filter (m^3/s or ft^3/s)

R = the gas constant (8.314 510 J/mol·K or 1,545.356 ft·lbf/lb-mol·°R)
S_i = the total COD of the wastewater applied to the filter including any recycle (kg/m^3 or lb/ft^3)
S = the total COD of the wastewater leaving the filter including any solids (kg/m^3 or lb/ft^3)
T = the absolute temperature (K or °R)

Oxygen transfer efficiencies are generally small, so the air flow rates must be fairly high.

Oxygen will be the limiting factor when the concentrations of oxygen and soluble COD satisfy the relationship derived by Williamson and McCarty [1976a, 1976b]:

$$\frac{C_{O_2}}{S} = \frac{C_{O_2 i}}{S_i} < \frac{D_{fCOD_s} \cdot \nu_{O_2} \cdot M_{rO_2}}{D_{fO_2} \cdot \nu_{COD_s} \cdot M_{rCOD_s}} \tag{10.101}$$

where

C_{O_2} = the oxygen concentration in the bulk liquid film (kg/m^3 or lb/ft^3)
$C_{O_2 i}$ = the oxygen concentration at the interface of the liquid and biofilm (kg/m^3 or lbm/ft^3)
D_{fCOD_s} = the diffusivity of soluble COD in the biofilm (m^2/s or ft^3/s)
D_{fO_2} = the diffusivity of oxygen in the biofilm (m^2/s or ft^2/s)
M_{rCOD_s} = the relative molecular weight of soluble COD (kg/mol or lb/mol)
M_{rO_2} = the relative molecular weight of oxygen (kg/mol or lb/mol)
S = the concentration of soluble COD in the bulk liquid film (kg/m^3 or lb/ft^3)
S_i = the soluble COD concentration at the interface of the liquid and biofilm (kg/m^3 or lb/ft^3)
ν_{COD_s} = the stoichiometric coefficient for soluble COD for heterotroph growth (dimensionless)
ν_{O_2} = the stoichiometric coefficient for oxygen for heterotroph growth (dimensionless)

Williamson and McCarty estimate that a biofilm will be oxygen limited whenever the soluble BOD in the liquid film exceeds 40 mg/L (about 80 mg/L soluble COD).

If there is colloidal COD in the wastewater, there will be a mass transfer to the film via adsorption and flocculation. Consequently, the soluble COD flux and liquid concentration that causes oxygen limitation in the film will be reduced. An approximate value for the total COD that results in oxygen limitation is probably the Williamson-McCarty estimate for soluble COD, that is, 40 mg/L.

A model of oxygen-limited COD removal that is based on Higbie's [1935] penetration theory has been developed by Mehta *et al.*, [1972] and discussed by Logan [1993]. Their analysis leads to the following COD removal for a *single* module of plastic media:

$$S_i - S_e = \sqrt{\frac{6D_l g^{1/3}}{3^{1/2}\pi\nu^{1/3}}} \cdot \frac{A_v^{2/3}\sqrt{h}}{(Q/A)^{2/3}} \cdot (C_s - C_i) \tag{10.102}$$

where

A = the plan area of the filter (m^2 or ft^2)
A_v = the specific surface area of the media (m^2/m^3 or ft^2/ft^3)
C_i = the (uniform) concentration of oxygen in the water film entering the module (kg/m^3 or lb/ft^3)
$\approx$ 0 mg/L for oxygen limited conditions
C_s = the solubility of oxygen in the water film (kg/m^3 or lb/ft^3)
D_l = the diffusivity of oxygen in the water film (m^2/s or ft^2/s)
h = the height of one module of the plastic media (m or ft)

Q = the flow rate (m^3/s or ft^3/s)
S_e = the COD in the liquid film at the bottom of the module (kg/m^3 or lb/ft^3)
S_i = the COD in the liquid film at the top of the module (kg/m^3 or lb/ft^3)
ν = the kinematic viscosity of water in (m^2/s or ft^2/s)

The total removal for a bed would be the number of module layers times the single module removal. If the biomass is oxygen limited, the oxygen concentration in the liquid film should be small relative to the solubility.

There is little or no public information regarding the ability of Eq. (10.102) to describe field and laboratory data. Furthermore, the Mehta oxygen transfer model does not take into account the two liquid films that probably exist—the freely flowing surface film and the stagnant film in the heterogeneous layer. Logan [1993] points out that the Mehta model produces oxygen profiles across the liquid film that violate the underlying assumption of Higbie's penetration theory. Higbie assumed that the quantity of oxygen absorbed into the liquid film is negligible, so that the oxygen concentration in the main body of the film is nearly unchanged. This assumption yields a simple boundary condition ($C = C_o$ as $z \to \infty$) and permits a closed form solution for oxygen mass transfer coefficient.

Logan *et al.* [1987] dispensed with the Higbie assumption and developed a computer simulation model that takes into account the two-dimensional concentration and velocity distributions in a laminar film flowing over a plate. This is a general BOD/O_2 uptake model that can be applied to oxygen-limited and BOD-limited filters as well as nitrifying filters. The numerical routine automatically makes the adjustment via comparison of biofilm boundary conditions.

The Logan–Hermanowicz model assumes that mass transfer from the liquid film controls BOD removal; there is no biokinetic component of the model. The mass transport equation in the liquid film for both oxygen and soluble COD can be written as follows (two separate equations result):

$$\frac{\partial C}{\partial t} = D_l \frac{\partial^2 C}{\partial x^2} - v(x)\frac{\partial C}{\partial z} \tag{10.103}$$

where

C = the O_2 or soluble COD concentration at x, z, and t (kg/m^3 or lb/ft^3)
D_l = the O_2 or soluble COD diffusivity (m^2/s or ft^2/s)
g = the acceleration due to gravity (9.80665 m/s^2 or 32.174 ft/s^2)
t = elapsed time (s)
$v(x)$ = the (laminar) liquid velocity parallel to the plate at x = (m/s or ft/s)
$= v_{max}(1 - x/\delta_f)$
v_{max} = the maximum liquid velocity at the surface (m/s or ft/s)
w = the plate width normal to the flow (m or ft)
x = the distance from the air-water interface towards the biofilm (m or ft)
z = the distance along the plate in the direction of flow measured from the point of establishment of the boundary layer (m or ft)
δ_f = the liquid film thickness (m or ft)
$= \sqrt{3Q\nu/gw\cos\theta}$
ν = the kinematic viscosity (m^2/s or $ft^2 s$)
θ = the angle of inclination measured from the horizontal (rad)

The appropriate boundary conditions are

$z = 0$; both oxygen and soluble COD, with different values, $C = C_o$
$x = 0$; for oxygen, $C = C_s$; for soluble COD, $\partial C/\partial x = 0$

$x = \delta_f$; for both oxygen and soluble COD, with equal values,

$$D_l \frac{\partial C}{\partial x} = D_l C \sqrt{\frac{3E_b(1-\varepsilon)}{a_c^2}} \tag{10.104}$$

where

a_c = the radius of a bacterial cell (m or ft)
$\approx 1\ \mu\text{m}$
E_b = the collector efficiency of a single bacteria cell (dimensionless)
≈ 0.0035
ε = the biofilm porosity (dimensionless)
≈ 0.8

Under steady state conditions and in the long-run average, the oxygen and COD fluxes into the biofilm must be equal. The two possible exceptions to this rule would be fermentable substrates, like glucose, and alternative electron acceptors, like sulfate or nitrate. In the absence of these exceptions the possible oxygen and COD fluxes are calculated from the biofilm boundary condition, and the smaller flux is used for both components.

The numerical simulations gave good fits to data from a variety of plastic media [Logan *et al.*, 1987]. Various modular media are distinguished by their HRTs, liquid film velocities, and liquid film thicknesses. These parameters can be calculated from standard laminar flow theory using Eqs. (10.76) and (10.78) given (1) the uninterrupted flow lengths, (2) the specific surface area, and (3) the plate angle from the horizontal.

Bruce–Merkens Correlation

An empirical formula developed by Bruce and Merkens [1973] fits a variety of plastic and stone media:

$$\frac{S}{S_i} = \exp\left(-\frac{K_{15}\theta^{T-15}A_v}{Q/V}\right) \tag{10.105}$$

where

A_v = the specific surface area of the media in m^2/m^3
K_{15} = the reaction rate coefficient at 15°C
≈ 0.037 m/d
Q = the settled sewage flow rate in m^2/d
V = the media volume in m^3
θ = the Streeter–Phelps temperature correction coefficient (dimensionless)
≈ 1.08

Institution of Water and Environmental Managers

The formula recommended by the Institution of Water and Environmental Management [Joint Task Force, 1992] is the following:

$$\frac{S}{S_i} = \frac{1}{1 + \dfrac{K\theta^{T-15}A_v^m}{(Q/V)^n}} \tag{10.106}$$

where

A_v = the specific surface area of the filter media in m^2/m^3
m = 1.407 for stone and random media and 0.7324 for modular plastic media

K = 0.0204 m^{m+n}/d^n for stone and random media and 0.400m^{m+n}/d^n for modular plastic media

n = 1.249 for stone and random media and 1.396 for modular plastic media

Q = the settled sewage flow rate in m^3/d

S = the settled effluent BOD_5 in mg/L

S_i = the influent BOD_5 in mg/L

V = the filter volume in m^3

θ = the Streeter–Phelps temperature coefficient—1.111 for stone and random media and 1.089 for modular plastic media

Note that media depth is not a factor in this model.

Relative Reliability of Formulas

Benjes [1978] has compared the predictions of the NRC formula, Eckenfelder's equation, and the Galler–Gotaas correlation to data from 20 treatment plants. Data from these treatment plants were not used in the original studies, so Benjes's work is a test of the predictive capabilities of the models. In general the coefficient of determination (the correlation coefficient squared) was 0.50 for both the NRC formula and the Galler–Gotaas correlation; it was 0.56 for the Eckenfelder equation. All of the formulas tend to predict better effluents than are actually achieved, especially when the effluent BODs are large; the Eckenfelder formula yields the smallest predicted effluent BODs. The divergences from observed data are greatest for effluent BODs above 30 mg/L. Below 30 mg/L the three formulas are about equally accurate, but errors of plus or minus 50% should be expected.

Effect of Bed Geometry, Recirculation, and Hydraulic Load

All trickling filter models that incorporate Howland's HRT formula predict that if the media volume is held constant and the media depth is increased, the BOD_5 removal efficiency will improve. This occurs because the liquid film thickness increases when Q/A does, and the HRT increases because the total volume of liquid in the bed is larger. These models also predict that increasing recirculation will improve removal efficiency for the same reason.

The NRC formula also predicts that recirculation will improve removal efficiency. Again, this is because of increased contacting between the wastewater and the media.

In the case of empirical equations like the Galler–Gotaas correlation, the prediction falls out of the regression and no mechanism is implied.

It is now believed that increased hydraulic loading does in fact improve removal efficiency, but the improvement is due to more complete wetting of the media surface and a resulting increase in the effective media surface area. The high-intensity, low-frequency dosing recommended by the Joint Task Force [1992] is intended to provide maximum wetting of the surface area.

The effects of wetting can be reported as changes in the rate coefficient of the Germain version of the Velz–Howland–Schulze formula. In general the Germain treatability constant, K_G, increases with hydraulic loading up to some maximum value, at which it is supposed that the media is completely wet [Joint Task Force, 1992]. The same result was obtained by Dow Chemical for plastic media filters [Joint Task Force, 1992]. In the Dow study the reaction rate constant increased with hydraulic load up to about 0.75 gpm/ft^2, above which the rate coefficient was constant [Joint Task Force, 1992]. Albertson [Joint Task Force, 1992] has correlated the data from 18 plastic media filters, with the following results:

$$K_G = 0.074\left(\frac{Q+Q_r}{A}\right)^3 - 0.333\left(\frac{Q+Q_r}{A}\right)^2 + 0.442\left(\frac{Q+Q_r}{A}\right) + 0.0633, \qquad R = 0.76 \tag{10.107}$$

where

A = the plan area of the filter in m^2

K_G = the Germain treatability constant for a 6.1 m filter at 20°C with an applied BOD_5 of 150 mg/L, in $L^{0.5}/s^{0.5}\cdot m^2$

= 0.210 $L^{0.5}/s^{0.5}\cdot m^2$ at 0.5 $L/m^2\cdot s$ (0.75 gpm/ft^2)

Q = the settled sewage flow rates in L/s

Q_r = the recirculated effluent flow rate in L/s

There was no difference between vertical flow and cross-flow media. The Albertson correlation also indicates that a minimum total hydraulic loading rate of about 0.75 gpm/ft^2 is needed to completely wet the media. Albertson uses a Germain treatability constant of 0.203 $L^{0.5}/s^{0.5}\cdot m^2$ for fully wetted media at the reference conditions of 20°C, 6.1 m depth and 150 mg/L influent BOD. This is slightly smaller than the value predicted by Eq. (10.107).

The Dow study also indicated that BOD removal per unit media volume was independent of media depth for any given hydraulic load that achieved complete wetting of the media. This result was supported by Bruce and Merkens [1973] and by others [Joint Task Force, 1992]; the relationship between the Germain treatability coefficient and the media depth can be represented as

$$\frac{K_{G1}}{K_{G2}} = \sqrt{\frac{H_2}{H_1}} \tag{10.108}$$

so that $K_G\sqrt{H}$ is a constant. The consequence of this is that the removal efficiency should depend on the volumetric rather than the areal hydraulic loading rate:

$$\frac{S}{S_i} \exp\left(-\frac{K}{\sqrt{Q/V}}\right) \tag{10.109}$$

where K is equal to the constant $K_G\sqrt{H}$. Note the similarity to the Bruce–Merkens model.

Albertson [Joint Task Force, 1992] recommends that the Germain constant also be corrected for the strength of the applied sewage:

$$K_G(20°C, H, S_o) = K_G\left(\frac{6.1}{H}\right)^{0.5}\left(\frac{150}{S_o}\right)^{0.5} \tag{10.110}$$

where

K_G = the Germain treatability constant at 20°C calculated from Eq. (10.107), in $L^{0.5}/(s^{0.5}\cdot m^2)$

$K_G(20°C, H, S_o)$ = the Germain treatability constant at 20°C for the filter depth and influent BOD, in $L^{0.5}/(s^{0.5}\cdot m^2)$

H = the media depth in m

S_o = the BOD_5 of the settled sewage in mg/L, not including the recirculated flow

Equation (10.110) only applies to vertical-flow, modular plastic media. The Germain treatability constant for shallow cross-flow media is significantly smaller.

Tertiary Nitrification

A trickling filter may be classified as *tertiary* or *nitrifying only* as long as the soluble BOD_5 in the applied flow is less than about 12 mg/L and the BOD_5:TKN ratio is less than about 1.0 [Joint Task Force, 1992].

Nitrification in trickling filters is regarded as mass transport limited as long as the ammonia concentration is greater than a few mg/L. The limiting flux may be either that of oxygen or that of ammonia. According to the Williamson–McCarty [1976a, 1976b] analysis, the oxygen flux is rate limiting whenever

$$\frac{C_{O_2}}{S_N} = \frac{C_{O_2 i}}{S_{Ni}} < \frac{D_{fN} \cdot \nu_{O_2} \cdot M_{rO_2}}{D_{fO_2} \cdot \nu_n \cdot M_{rN}} \tag{10.111}$$

where

D_{fN} = the diffusivity of ammonia nitrogen in the biofilm (m^2/s or ft^3/s)

M_{rN} = the relative molecular weight of ammonia nitrogen (kg/mol or lbm/mol)

S_N = the ammonia nitrogen concentration in the bulk liquid film (kg NH_3-N/m^3 or lbm NH_3-N/ft^3)

S_{Ni} = the ammonia nitrogen concentration at the interface of the liquid and biofilm (kg/m^3 or lbm/ft^3)

ν_N = the stoichiometric coefficient of ammonia nitrogen for nitrifier growth (dimensionless)

Williamson and McCarty estimate that nitrification will be oxygen limited whenever the oxygen-to-ammonia ratio in the liquid film is less than 2.7 kg O_2/kg NH_3-N. As a practical matter this means that tertiary nitrifying filters are oxygen limited whenever the ammonia nitrogen concentration is larger than about 2 to 4 mg/L. In this range the nitrification rate is also zero order with respect to ammonia.

Parker–Lutz–Dahl–Bernkopf Oxygen Limitation Model

The oxygen-limited case has been modeled by Parker *et al.* [1989] by adapting a theoretical model of Gujer and Boller [1986]. The Logan–Hermanowicz model is used to calculate oxygen fluxes into the biofilm, and the nitrification rate is set equal to the oxygen flux. An empirical correction is introduced to account for the observed decline in nitrification rates as the filter depth increases:

$$S_{No} - S_{Ne} + K_N \ln\left(\frac{S_{No}}{S_{Ne}}\right) = \frac{E_b A_v J_{O_2}}{4.57k(Q/A)}\left(1 - e^{-kH}\right) \tag{10.112}$$

where

E_b = the Logan–Hermanowicz media effectiveness factor (dimensionless)

J_{O_2} = the maximum oxygen flux for the particular media as determined by the Logan–Hermanowicz model (kg $O_2/m^2 \cdot s$ or lbm $O_2/ft^2 \cdot s$)

k = the Gujer–Boller biofilm patchiness factor (1/m or 1/ft)

K_N = the Monod affinity constant for nitrification (kg NH_3-N/m^3 or lbm NH_3-N/ft^3)

S_{Ne} = the ammonia nitrogen concentration in the filter effluent (kg NH_3-N/m^3 or lbm NH_3-N/ft^3)

S_{No} = the ammonia nitrogen concentration in the filter influent (kg NH_3-N/m^3 or lbm NH_3-N/ft^3)

In a strict application of kinetic concepts, the term containing the Monod affinity constant on the left-hand side would not appear in either oxygen-limited or zero-order nitrification; it causes an underestimate of the removal efficiency.

Both Gujer and Boller and Parker *et al.* adopted the Wezernak–Gannon [1967] estimate of the oxygen demand exerted by ammonia during nitrification (4.33 kg O_2/kg NH_3-N), but the growth stoichiometry cited in the *Process Design Manual for Nitrogen Control* by Parker *et al.* [1975] indicates that nearly 98% of the ammonia nitrogen is oxidized for energy and reducing power. Therefore, the theoretical oxygen equivalent of ammonia has been used in Eq. (10.112).

The modified Gujer–Boller model was able to describe operating data from a variety of modular plastic media filters. The biofilm patchiness factor varied from about 0 to about 0.16/m. The lower value occurred with higher influent ammonia concentrations, which caused a more uniform biofilm growth over the media. The media effectiveness factor varied from about 0.39 to 1.30, with typical values around 0.9. Cross-flow media yielded higher removal efficiencies because of their higher oxygen transfer rates.

Gujer–Boller Biofilm Model

The semitheoretical Gujer–Boller [1986] model for the ammonia-limited case is

$$S_{No} - S_{Ne} + K_N \ln\left(\frac{S_{No}}{S_{Ne}}\right) = \frac{A_v J_{N\max}}{K(Q/A)}\left(1 - e^{-kH}\right) \tag{10.113}$$

where

D_{fN} = the ammonia diffusivity in the biofilm, approximately 0.8 to 0.9 the diffusivity in water (m^2/s or ft^2/s)

$J_{N\max}$ = the maximum ammonia flux from the liquid film to the biomass at the top of the filter (kgNH_3-N/m^2·s or lbm NH_3-N/ft^2· s)

= $\sqrt{2D_{fN} \cdot \mu_{\max} \cdot \gamma \cdot S_{N0}/Y_N}$

S_{Ne} = the ammonia nitrogen concentration in the filter effluent (kg NH_3-N/m^3 or lbm NH_3-N/ft^3)

S_{No} = the ammonia nitrogen concentration in the filter influent (kg NH_3-N/m^3 or lbm NH_3-N/ft^3)

Y_N = the true growth yield of nitrifiers on ammonia (kg VSS/kg NH_3-N or lbm VSS/lbm NH_3-N)

γ = the density of nitrifiers in the biofilm (kg VSS/m^3 or lbm VSS/ft^3)

$\mu_{\max}$ = the maximum specific growth rate of the nitrifiers (1/s)

Equation (10.113) is only valid for ammonia nitrogen concentrations less than about 2 to 4 mg/L, with the lower limit applying around 5°C and the higher limit around 25°C. Okey and Albertson [1989] reviewed the operating results from more than 150 tertiary filters and concluded that the transition from oxygen limitation to ammonia limitation occurred at a total nitrogen loading rate per unit plan area of filter (including recirculated flow) of about 1.2 g TKN/m^2·d. Below this loading rate the nitrogen removal was approximately first order with a Velz reaction rate coefficient of about 0.3 to 0.4 per meter (20°C). The Velz rate coefficient was independent of media type. Above a loading of 1.2 g TKN/m^2·d, nitrogen removal was oxygen limited, highly variable, and strongly dependent on media type.

Okey–Albertson Procedure

Okey and Albertson [Joint Task Force, 1992] have developed an empirical procedure for tertiary nitrifying filters. They first define a transition ammonia concentration for the boundary between oxygen limitation and ammonia limitation of nitrification. This varies with oxygen saturation of the liquid film approximately as shown in Table 10.14 [Okey and Albertson, 1989]. They then assume the filter depth can be divided into an upper oxygen-limited region and a lower ammonia-limited region. The volumes of the two regions are calculated separately and then added to get the total media volume. The filter area is calculated by assuming a minimum hydraulic loading rate, which includes recirculated flow. This procedure is restricted to applied wastewaters containing less than 12 mg/L of soluble BOD_5, a BOD_5:TKN ratio of less than 1.0 and a combined BOD_5 plus TSS concentration of 30 mg/L. It is assumed that modular media having a specific surface area of 138 m^2/m^3 is employed.

Table 10.14 Ammonia Concentrations for the Transition from Oxygen to Ammonia Limitation of Nitrification in Trickling Filters[1]

Oxygen Saturation (%)	Transition Ammonia Concentration (mg NH_3-N/L)		
	10°C	20°C	30°C
25	1.0	0.9	0.8
50	2.2	1.8	1.5
75	3.3	2.6	2.1
100	4.3	3.5	2.9

[1]Okey, R. W. and Albertson, O. E. 1989. Diffusion's role in regulating rate and masking temperature effects in mixed film nitrification. *Journal of the Water Pollution Control Federation.* 61(4):500.

Oxygen-Limited Media Volume. Calculate the oxygen-limited media volume by assuming a nitrification rate of 1.2 g NH_3-N/m^2·d and a media specific surface area of 138 m^2/m^3. The rate is assumed to be constant between 10 and 30°C, and it is reduced below 10°C by using a Streeter–Phelps theta value of 1.035:

$$K_o(10°\text{C} \leq T \leq 30°\text{C}) = 1.2\frac{\text{g NH}_3\text{-N}}{\text{m}^2\cdot\text{d}}$$

$$K_o(T < 10°\text{C}) = 1.2\frac{\text{g NH}_3\text{-N}}{\text{m}^2\cdot\text{d}}\times 1.035^{T-10} \tag{10.114}$$

The oxygen-limited media volume will be

$$V_o = \frac{(Q + Q_r)(C_{\text{TKN}_o} - S_{NT})}{K_o A_v} \tag{10.115}$$

where

A_v = 138 m^2/m^3
C_{TKN_o} = the total kjeldahl nitrogen concentration in the applied flow, including the TKN of the recirculated flow in g TKN/m^3
$Q + Q_r$ = the applied flow in m^3/d
S_{NT} = the transition ammonia nitrogen concentration in g NH_3-N/m^3 obtained from Table 10.14
V_o = the oxygen-limited media volume in m^3

Ammonia-Limited Media Volume. Use the empirical ammonia oxidation rate,

$$K_n(7°\text{C} \leq T \leq 30°\text{C}) = 1.2\left(\frac{S_{Ne}}{S_{NT}}\right)^{0.75}\frac{\text{NH}_3\text{-N}}{\text{m}^2\cdot\text{d}} \tag{10.116}$$

and calculate the required ammonia-limited volume by

$$V_n = \frac{(Q + Q_r)(S_{NT} - S_{Ne})}{K_n A_v} \tag{10.117}$$

where

S_{Ne} = the required effluent ammonia nitrogen concentration in g NH_3-N/m^3
V_n = the ammonia-limited media volume in m^3

Filter Depth. The filter depth is the total media volume divided by the required minimum hydraulic load, here taken to be 0.54 L/m^2·s :

$$H = \frac{V_n + V_o}{(Q + Q_r)/L_Q} \tag{10.118}$$

where

L_Q = the required hydraulic loading rate in m^3/m^2·d
$\geq 0.54 \text{L/m}^2\cdot\text{s} = 47 \text{m}^3/\text{m}^2\cdot\text{d}$

Aeration and Dosing. Okey and Albertson recommend that forced ventilation be employed with a minimum oxygen supply of 50 kg O_2 per kg O_2 consumed—that is, an oxygen transfer efficiency of 2%.

They also recommend instantaneous dosing intensity of 25 to 250 mm/pass and occasional flushing at greater than 300 mm/pass.

Combined BOD and TKN Removal

Heterotrophic bacteria grow more rapidly than nitrifiers and are less sensitive to low oxygen concentrations, so when the influent BOD is high, the heterotrophs tend to displace the nitrifiers, which become limited to the deep, low-BOD region of the filter. TKN removal rates decrease both as the BOD:TKN ratio increases and as the temperature increases. At 15°C the TKN oxidation rate is approximately

$$K_{\text{TKN}}(15^{\circ}\text{C}) \cong 1.086\left(\frac{C_{\text{BOD}_5 i}}{C_{\text{TKN}i}}\right)^{-0.44} \tag{10.119}$$

where

$K_{\text{TKN}}(15^{\circ}\text{C})$ = the total kjeldahl nitrogen removal rate in g TKN/m^2·d
$C_{\text{BOD}_5 i}$ = the total BOD_5 concentration in the applied flow in g TKN/m^3
$C_{\text{TKN}i}$ = the total TKN concentration in the applied flow in g TKN/m^3

At 25°C the TKN oxidation rate is likely to be less than half that indicated by Eq. (10.119); at 10°C the rate may be 50% higher. The scatter about Eq. (10.119) is very large; the standard deviation of the estimate is 0.17 g TKN/m^2·d.

The Albertson–Okey [Joint Task Force, 1992] procedure for combined BOD and ammonia removal is as follows.

Media Volume

Assuming nitrification controls, determine the influent BOD_5 and TKN and estimate the TKN oxidation rate for the indicated BOD_5:TKN ratio. Reduce this estimate by 1 standard deviation—that is, 0.17 g TKN/m^2·d. Use the corrected oxidation rate to calculate the required media volume:

$$K_{\text{TKNdes}} \cong 1.086\left(\frac{C_{\text{BOD}_5 i}}{C_{\text{TKN}i}}\right)^{-0.44} - 0.17 \tag{10.120}$$

$$V_{\text{TKN}} = \frac{(Q + Q_r)(C_{\text{TKN}i} - S_{\text{TKN}e})}{K_{\text{TKNdes}} A_v} \tag{10.121}$$

where

A_v = the specific surface area of the media in m^2/m^3
= 98 m^2/m^3, for design purposes
K_{TKNdes} = the design TKN oxidation rate for 15°C in g TKN/m^2·d
$S_{\text{TKN}e}$ = the required settled effluent TKN in g TKN/m^3
V_{TKN} = the total media volume required for combined BOD and TKN removal in m^3

Aeration and Distribution

Powered ventilation of the towers is required to prevent oxygen limitation. The oxygen requirement (at 2% transfer efficiency) is 50 kg O_2 per kg oxygen demand removed:

$$R_{req} = 50(Q + Q_r)\left[1.2\left(C_{BOD_5 i} - C_{BOD_5 e}\right) + 4.57\left(C_{TKNi} - C_{TKNe}\right)\right] \quad (10.122)$$

where

$C_{BOD_5 e}$ = the total BOD_5 in the filter underflow (kg/m^3 or lb/ft^3)
$C_{BOD_5 i}$ = the total BOD_5 in the applied flow (kg/m^3 or lb/ft^3)
C_{TKNe} = the total TKN in the filter underflow (kg/m^3 or lb/ft^3)
C_{TKNi} = the total TKN in the applied flow (kg/m^3 or lb/ft^3)
$Q + Q_r$ = the applied flow (m^3/s or ft^3/s)
R_{req} = the oxygen requirement in (kg/s or lb/s)

The minimum hydraulic application rate to ensure adequate wetting is 0.54 $L/m^2 \cdot s$. The wastewater application should be intermittent with an instantaneous dosing rate of 15 to 300 mm/pass.

Predator Control

Trickling filters mimic (by design) the physical ecology of brooks. Consequently, they become populated with significant numbers of insect larvae, worms, and snails that are adapted to eating microbial films. No steady state is possible in such an ecosystem, and the biofilm and its predators exhibit oscillations that are out of phase. Lotka's [1925] theory of predation indicates that the period of small oscillations can be approximated by

$$T = \frac{2\pi}{\sqrt{rd}} \quad (10.123)$$

where

T = the period of the (small) oscillation (s)
r = the net (birth minus death) growth rate of the prey (per s)
d = the death rate of the predator (per s)

Lotka's theory also predicts that long-period oscillations have large amplitudes. All trickling filters exhibit such oscillations, but the oscillations in nitrifying filters are more pronounced because the nitrifying bacteria grow so much more slowly than the heterotrophic bacteria. In fact predator infestations can denude significant portions of the media surface, thereby preventing nitrification. Such denudation is followed by predator starvation and population decline, which permit reestablishment of the nitrifying film.

Parker *et al.* [1989] recommend that tertiary nitrifying filters be flooded and then backwashed about once per week to kill and/or remove insect larvae, worms, and snails that consume the nitrifying biomass.

Temperature, pH, and Inhibition

Effect of Temperature on Carbonaceous BOD Removal

The BOD removal efficiency of trickling filters depends on the temperature of the slime layer. In the case of continuously dosed, high-rate filters, this is probably the temperature of the applied wastewater. However, the slime layer in intermittently dosed, low-rate filters spends most of its time in contact with the air circulating through the filter, and the slime temperature will be somewhere between the air temperature and the water temperature. If recirculation is practiced, the applied wastewater temperature tends to approach the ambient air temperature.

Schroepfer *et al.* [1952] derived the following formulas for the effects of temperature on the BOD removal efficiency of rock filters. For low-rate, intermittently dosed filters:

$$E_2 - E_1 = 0.62(T_2 - T_1); \qquad R \cong 0.7 \tag{10.124}$$

For high-rate, continuously dosed filters:

$$E_2 - E_1 = 0.34(T_2 - T_1); \qquad R \cong 0.6 \tag{10.125}$$

where

E = the percentage BOD removal efficiency of the combined filter and secondary clarifier based on the total BOD load (%)

T = the wastewater temperature (°F)

Note that the low-rate, intermittently dosed filters are more sensitive to temperature changes. This is due to the fact that ambient air and wastewater temperatures are correlated and vary together seasonally. However, the variation in air temperature is much larger than the variation in wastewater temperature, so when the filter is standing idle the slime is exposed to air that is either much cooler or much warmer than the applied wastewater.

Recirculation greatly increases the seasonal variation in BOD removal efficiency. In a study of 17 stone filters in Michigan, Benzie *et al.* [1963] found that the summer efficiency was 21 percentage points higher than the winter efficiency when recirculation was practiced but only 6 points higher without recirculation.

The Joint Task Force [1992] recommends that the Germain treatability constant be corrected for temperature by the following formula:

$$K_G(T_1) = K_G(T_o) \times 1.035^{T_1 - T_o} \tag{10.126}$$

where $K_G(T)$ = the Germain treatability constant at temperature T in °C. The indicated temperatures are those of the wastewater in the filter, not the ambient air.

Effect of Temperature on Nitrification

The Joint Task Force concluded that temperature had little effect on the performance of tertiary nitrifying filters. The effects that are commonly reported were attributed instead to deficiencies in wetting, biofilm consumption by predators, overgrowth by heterotrophs, ammonia concentrations, and oxygen transfer limitations. In fact, if filters are liquid film mass transfer limited as assumed in the Logan–Hermanowicz model, the controlling rate process is diffusion, and the Streeter–Phelps [1925] temperature correction factor for the process is about 1.024, which leads to relatively small temperature effects.

The data for combined BOD removal and nitrification is very limited. It appears that the uncorrected TKN oxidation given by Eq. (10.119) should be *decreased* by about 50% if the temperature is 25°C or more, and that it may be *increased* by about 50% if the temperature is 10°C.

Effect of pH

The pH requirements of activated sludge apply BOD removal in trickling filters as well. Tertiary nitrifying filters may develop low pH's as a result of nitric acid formation. Limited data presented by Huang and Hopson [Parker *et al.*, 1975] suggest that the nitrification rate actually increases up to about pH 8.4; at pH 7.1 the rate is only 50% of the maximum and at pH 7.8 it is 90% of the maximum. This is somewhat contrary to the pH effect reported by Downing and Knowles for activated sludge, in which they indicated no increase in nitrification rate above pH 7.2.

Inhibitors

The list of inhibitors given for activated sludge in Tables 10.4 and 10.5 probably applies to trickling filter nitrification as well.

10.3 Sludge Digestion

The stabilization of wastewater sludges can occur either aerobically (with molecular oxygen) or anaerobically (without molecular oxygen). Stabilization means that the bulk of the organic solids is converted to metabolic end products that resist further degradation and do not produce nuisance odors. In aerobic digestion these products are carbon dioxide, water, various inorganic salts, and humic/fulvic materials. In anaerobic digestion the products are methane, carbon dioxide, various inorganic salts, and humic/fulvic materials.

Aerobic digestion consumes energy because of the aeration requirement. Anaerobic digestion is a net energy producer, because of the methane formed, but often the only economic use of the methane is in digester and space heating. Aerobic digestion seldom produces offensive odors (the usual smell being mustiness), but failed anaerobic digesters can be a nuisance because of foul odors.

Sludge digestion substantially reduces the numbers of pathogens and parasites, but it does not qualify as a sludge disinfection process. If digested sludges are to be applied to land, disinfection by heat treatment and/or lime stabilization is required.

Anaerobic Digestion

Facilities Arrangement

According to the ASCE survey of treatment plants [Leininger *et al.*, 1983], about three-fourths of the plants employing anaerobic digestion have single-stage, heated (95°F), mixed tanks. About one-fourth of the plants have two-stage systems with heated, mixed primary tanks followed by unheated, unmixed secondary tanks. The secondary tanks are intended to capture and thicken digested solids. However, the suspended solids produced by the primary tanks settle poorly because of (a) gas flotation, (b) particle size reduction due to digestion, and (c) particle size reduction due to mixing [Brown and Caldwell, Consulting Engineers, 1979]. In full-scale field units, only about one-third of the suspended solids entering the secondary digester will settle out [Fan, 1983]. The construction of secondary digesters does not appear to be warranted.

Plants that attempt to concentrate digested sludges by gravity thickening (in secondary digesters or otherwise) frequently recycle the "supernatant" liquor to the primary settlers. This practice results in digester feed sludges that contain substantial portions of previously digested, inert organic solids, perhaps 25 to 50% of the VS in the feed. Consequently, the observed volatile solids destructions are significantly reduced, frequently to about 30 to 40%. Such low destructions merely indicate the presence of recirculated inert volatile solids and do not imply that incomplete digestion is occurring. However, if substantial fractions of inert organics are being recirculated either the digesters are oversized or their hydraulic retention time and treatment capacity is reduced.

Typical digestion tanks are circular in plan with conical floors for drainage. Typical tank diameters are 24 m and typical sidewall depths are 8 m [Leininger *et al.*, 1983]. Almost half of the digesters are mixed by recirculated digester gas, nearly one-fourth by injection from the roof via gas lances. About two-fifths of the digesters are mixed by external pumps, which generally incorporate external heat exchangers.

Digesters are almost always heated by some kind of external heat exchanger fueled with digester gas. In cold climates, gas storage is most often practiced, usually in floating gasholder covers or flexible membrane covers. Hydrogen sulfide is usually removed by iron sponge scrubbers.

Microbiology and Pattern of Digestion

Many of the bacteria responsible for anaerobic digestion are common intestinal microbes [Kirsch and Sykes, 1971]. They are fastidious anaerobes; molecular oxygen kills them. The preferred temperature is 37°C (human body temperature), and the preferred pH range is about 6.5 to 7.5. Facultatively anaerobic bacteria such as the coliforms comprise only a few percent of the population or less.

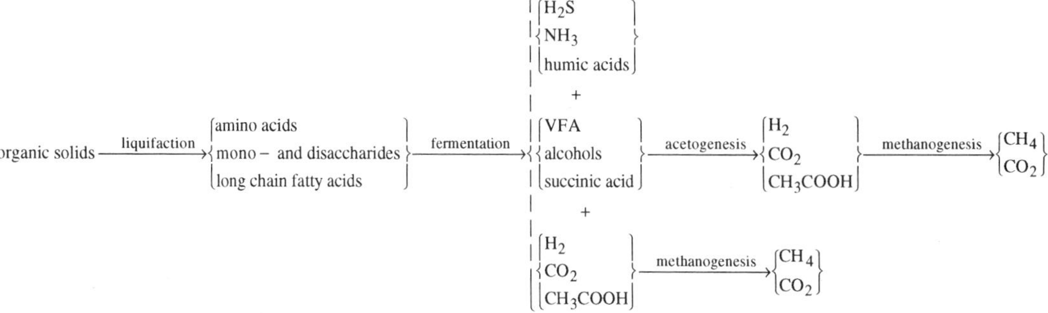

FIGURE 10.12 Pattern of organic solids decomposition in anaerobic digestion.

The pattern of anaerobic decomposition of wastewater sludges is indicated in Fig. 10.12. The primary ecological division for bacteria in digesters is between acidogens and methanogens. The acidogen population (as a whole) hydrolyzes the cellulose and other complex carbohydrates, proteins, nucleotides, and lipids to simple organic molecules and ferments them to hydrogen gas, carbon dioxide, acetic acid and other volatile fatty acids (VFA), other organic acids, alcohols, ammonia, hydrogen sulfide, and humic and fulvic acids. Three-carbon and larger VFA, other organic acids, and alcohols are converted to acetic acid, possibly hydrogen gas, and carbon dioxide by a subgroup of the acidogenic population called the *acetogens.*

The methanogenic population (as a whole) converts acetic acid, carbon dioxide and hydrogen, formic acid, methanol, and tri-, di-, and mono-methylamine to methane [Whitman *et al.*, 1992].

$$CH_3COOH \rightarrow CH_4 + CO_2 \tag{10.127}$$

$$CO_2 + 4H_2 \rightarrow CH_4 + 2H_2O \tag{10.128}$$

$$4HCOOH \rightarrow CH_4 + 3CO_2 + 2H_2O \tag{10.129}$$

$$4CH_3OH \rightarrow 3CH_4 + CO_2 + H_2O \tag{10.130}$$

$$4CH_3NH_2 + 3H_2O \rightarrow 3CH_4 + CO_2 + 4NH_3 \tag{10.131}$$

No other substrates are known to support growth of the methanogens. Nearly all the known methanogens (except for *Methanothrix soehngenii,* which grows only on acetic acid) grow by reducing carbon dioxide with hydrogen, but the largest source of methane in digesters—about 70%—is derived from the lysis of acetic acid [McCarty, 1964]. The methanogens are autotrophs and derive their cell carbon from carbon dioxide.

The composition of typical digested municipal sludge is given in Table 10.15, and the fate of various wastewater sludge components during digestion is indicated in Table 10.16 [Woods and Malina, 1965]. Overall, about 60% of the sludge organics is converted to gas, about 30% ends up as soluble organics, and 10% ends up in the residual humus solids. Over 80% of the influent carbohydrates and lipids is gasified, and most of the remainder ends up as soluble products. Only about half the proteins and other nitrogenous organics is gasified, and nearly one-third is converted to soluble products.

Thermal Classification of Bacteria. Bacteria are classified with respect to their optimum growth temperature [Brock, 1970; Schlegel, 1992] as:

- *Psychrophiles or cryophiles.* Can grow at 0°C, with fastest growth at 15°C or less and maximum growth temperature of 20°C.

Table 10.15 Fate of Wastewater Sludge Constituents During Anaerobic Digestion

Item	Distribution of Feed Material in Products (% by wt)		
	Gas	Liquid	Solid
Carbon	54[a]	26	10
Nitrogen (as N_2)	11	70	9
Volatile solids	60	30	10
Carbohydrate	86	9	5
Fats/lipids	80	14	6
Protein	55	32	13

Source: Woods, C. E. and Malina, J. F., Jr. 1965. Stage digestion of wastewater sludge. *Journal of the Water Pollution Control Federation.* 37(11): 1495.

[a] Experimental error.

- *Mesophiles.* Fastest growth at temperatures between 25 and 40°C.
- *Thermophiles.* Fastest growth at temperatures between 45 and 70°C.
- *Extreme thermophiles.* Fastest growth at temperatures between 65 and 80°C.
- *Hyperthermophiles.* Fastest growth at temperatures above 80°C.

Some sanitary engineers set the upper limit of the psycrophilic range and the lower limit of the mesophilic range at 10°C, but microbiologists use the division points given above. Fungi do not grow at temperatures above 60°C, and protozoa do not grow at temperatures above 50°C.

Of the 29 known species of methanogens, 3 are psychrophiles, 17 are mesophiles, 7 are thermophiles, and 2 are unclassified [Whitman, 1985]. There are only 2 thermophiles that might be present in sludge digesters: *Methanobacterium thermoautotrophicum* and an unnamed species of *Methanomicrobium*. Both *M. thermoautotrophicum* and the *Methanomicrobium* species grow by reducing carbon dioxide with hydrogen. *M. thermoautotrophicum* cannot grow on acetic acid alone but will consume it.

Effect of Temperature on Digestion. Nearly all municipal digesters are heated to about 35°C (or 95°F), which is in the mesothermal range to which the intestinal microbes that dominate the process are adapted. Anaerobic lagoons used to treat packing house wastes may not be heated directly, but much of the process wastewater is hot.

Psycrophilic digesters operate below 30°C. Psychrophilic digestion is quite common in anaerobic lagoons, septic tanks, and wetlands. Psycrophilic digestion is slower than mesophilic or thermophilic digestion. Also, below about 30°C the fraction of the sludge solids converted to gas is reduced [Malý and Fadrus, 1971]. The reduction is roughly linear, and at 10°C the fraction of solids converted to gas is only about 60% of the conversion at 30°C. Long-chain fatty acids accumulate below about 18°C, which causes foaming [Fan, 1983].

Thermophilic digesters operate above 40°C, usually at 50°C or somewhat higher. The chief advantages of thermophilic digestion [Buhr and Andrews, 1977] are as follows:

- More rapid completion of the digestion process, as indicated by the cumulative gas production
- Better dewatering characteristics
- Disinfection of pathogens and parasites, if operated above 50°C

There are, however, several reports of disadvantage [Pohland, 1962; Kirsch and Sykes, 1971]:

- Accumulation of volatile fatty acids and reduction of pH
- Accumulation of long-chain fatty acids
- Reduced gas production per unit volatile solids fed
- Reduced methane content of the gas

Table 10.16 Approximate Composition of Digested Sludge

Item	Primary Digester Sludge		Secondary Digester Settled Sludge Solids with Associated Interstitial Water		Secondary Digester Decanted Supernatant Liquor	
	Median	Range	Median	Range	Median	Range
Total solids, TS (% by wt)	3.1	3.0 to 4.0	4.0	2.5 to 5.5	1.5	1.0 to 5.0
Volatile solids (% of TS)	58.0	49.0 to 65.0	51.0	44.0 to 60.0	50.0	1.0 to 71.0
Total suspended solids (mg/L)	—	—	—	—	(2205.)	(143 to 7772)
Volatile suspended solids (mg/L)	—	—	—	—	(1660.)	(118 to 3176)
pH	7.0	6.9 to 7.1	—	—	(7.2)	(7.0 to 8.0)
Alkalinity (mg/L as $CaCO_3$)	2751	1975 to 3800	—	—	—	(1349 to 3780)
Volatile fatty acids (mg/L as HAc)	220	116 to 350	—	—	—	(250 to 322)
NH_3-N (mg/L)	998	300 to 1100	—	—	—	(480 to 853)
BOD (mg/L)	—	—	—	—	2282	600 to 2650
COD	—	—	—	—	—	11 to 7000
Total P (mg/L)	—	—	—	—	—	(63 to 143)

Sources: Number in parentheses taken from Brown and Caldwell, Consulting Engineers, Inc. 1979. *Process Design for Sludge Treatment and Disposal,* Environmental Protection Agency, Municipal Environmental Research Laboratory, Office of Research and Development, Center for Environmental Research Information, Technology Transfer, Cincinnati, OH. The other numbers are taken from Leininger, K. V., Sailor, M. K., and Apple, D. K. 1983. *A Survey of Anaerobic Digester Operations,* Final Draft Report, ASCE Task Committee on Design and Operation of Anaerobic Digesters, American Society of Civil Engineers, New York. The range is the first and third quartile points of the distribution and represents the limits of the middle 50% of the reported data.

- Offensive odors
- Reduced volatile solids destruction
- Impaired dewatering characteristics
- Increased heating loss rates (although the tanks are smaller and the smaller surface area somewhat offsets the higher heat flux)

The systems reporting impaired digestion were not operating stably, and there is a general opinion that thermophilic digestion is difficult to establish and maintain. Part of the problem may lie in the reduced suite of microbes that are able to grow thermophilically. Naturally occurring thermophilic environments are rare, and there are few thermophiles among the acidogens and methanogens in primary sewage sludge, which is derived entirely from psycrophilic and mesophilic environments. In any particular plant the conversion from mesophily to thermophily may require some time for proper seeding by the very few thermophiles in the influent sludge. If the digester goes sour in the interim, the seeding may fail to take. It should be noted that many of the failed thermophilic digestion experiments were conducted using laboratory-scale units. The seeding problem here is compounded by the Poisson nature of the sludge-sampling process; it is likely that none of the small samples needed for laboratory work will contain any thermophiles. It should be noted that the full-scale thermophilic plant at Los Angeles was a stable process [Garber, 1954].

Gas Stoichiometry

The principal benefit of anaerobic digestion is the methane gas produced, which can be used as a fuel. Typical municipal digesters produce a gas that is approximately 65% by volume methane, 30% carbon dioxide, 2.6% nitrogen, 0.7% hydrogen, 0.4% carbon monoxide, 0.3% hydrogen sulfide, and about 0.2% other illuminants [Pohland, 1962].

Digester gas is saturated with water vapor at the digestion temperature (42 mmHg at 35°C) and contains an aerosol of small grease and sludge particles that is formed as gas bubbles burst at the liquid surface. The aerosols and hydrogen sulfide must be removed prior to transmission and burning.

The fuel value of the gas resides in the methane content. At 25°C (77°F) methane has a lower heating value (product water remains a vapor) of 21,500 Btu/lbm (959 Btu/ft^3) and a higher heating value (product water condenses) of 23,900 Btu/lbm (1,064 Btu/ft^3) [Van Wylen, 1963]. Methane has an autoignition temperature of 650°C and lower and upper explosion limits in air of 5.3 and 15% by volume, respectively [Dean, 1992].

The quality and quantity of this gas is determined by the chemical composition of the volatile solids that are destroyed. This can be estimated using the following modification of Buswell's [1965] stoichiometry:

$$C_vH_wO_xN_yS_z + \left(v - \frac{w}{4} - \frac{x}{2} + \frac{3y}{4} + \frac{z}{2}\right)H_2O \rightarrow$$

$$\left(\frac{v}{2} + \frac{w}{8} - \frac{x}{4} - \frac{3y}{8} - \frac{z}{4}\right)CH_4 + \left(\frac{v}{2} - \frac{w}{8} + \frac{x}{4} + \frac{3y}{8} + \frac{z}{4}\right)CO_2 + yNH_3 + zH_2S \tag{10.132}$$

Organic nitrogen is released as ammonia, which reacts with water to form ammonium hydroxide and trap some of the carbon dioxide produced:

$$NH_3 + CO_2 + H_2O \rightarrow NH_4^+ + HCHO_3^- \tag{10.133}$$

The net result is that each mole of ammonia traps one mole of carbon dioxide. Accounting for this effect, the expected mole fractions of methane, carbon dioxide, and hydrogen sulfide are:

$$f_{CH_4} = \frac{4v + w - 2x - 3y - 2z}{8(v - y + z)} \tag{10.134}$$

$$f_{CO_2} = \frac{4v - w + 2x - 5y + 2z}{8(v - y + z)} \tag{10.135}$$

$$f_{H_2S} = \frac{z}{v - y + z} \tag{10.136}$$

The estimate for hydrogen sulfide given in Eq. (10.136) is a maximum. There are three general processes that reduce its gas phase concentration. First, hydrogen sulfide is a fairly soluble gas—about 100 times as soluble as oxygen—and much of it remains in solution. The Henry's law constant is [Dean, 1992]:

$$\begin{aligned} K_H &= \frac{p_{H_2S}}{[H_2S]} \\ &= 9.822 \frac{\text{atm·L}}{\text{mol}} \text{ (at 25°C)} \\ &= 12.24 \frac{\text{atm·L}}{\text{mol}} \text{ (at 35°C)} \end{aligned} \tag{10.137}$$

Second, hydrogen sulfide is also a weak acid, and the two-step ionization, which liberates bisulfide and sulfide, is pH-dependent:

$$H_2S \leftrightarrow HS^- + H^+ \tag{10.138}$$

$$\begin{aligned} K_1 &= \frac{[H^+][HS^-]}{[H_2S]} \\ &= 1.07 \times 10^{-7} \text{ (at 25°C)} \\ &= 1.51 \times 10^{-7} \text{ (at 35°C)} \end{aligned} \tag{10.139}$$

$$HS^- \leftrightarrow S^{2-} + H^+ \tag{10.140}$$

$$\begin{aligned} K_2 &= \frac{[H^+][S^{2-}]}{[HS^-]} \\ &= 1.26 \times 10^{-13} \text{ (at 25°C)} \\ &= 2.51 \times 10^{-13} \text{ (at 35°C)} \end{aligned} \tag{10.141}$$

Consequently, at 35°C and pH 7 most of the hydrogen sulfide exists as bisulfide, which further increases the amount of sulfide that remains in the sludge. Third, sulfide forms highly insoluble precipitates with many metals and can be trapped in the digested sludge as a metallic sulfide. The most common form is ferrous sulfide:

$$FeS \leftrightarrow Fe^{2+} + S^{2-} \tag{10.142}$$

$$\begin{aligned} K_{sp} &= [Fe^{2+}][S^{2-}] \\ &= 6.3 \times 10^{-18} \text{(about 20 to 25°C)} \end{aligned} \tag{10.143}$$

This is a black solid and gives off a characteristic "rotten egg" odor when dissolved in hydrochloric acid.

The net effect of all three processes yields a hydrogen sulfide partial pressure that can be estimated by

$$p_{H_2S} = \frac{K_H K_{sp} [H^+]^2}{K_1 K_2 [Fe^{2+}]} \quad (10.144)$$

High pH's and high ferrous iron concentrations both reduce the hydrogen sulfide partial pressure. The usual concentration in digester gas is about 1% by volume, but it is variable [Joint Task Force, 1992].

If the hydrogen sulfide partial pressure is low enough to be ignored, which is usually the case, then the mole fractions of methane and carbon dioxide in the product gas are

$$f_{CH_4} = \frac{4v + w - 2x - 3y - 2z}{8(v - y)} \quad (10.145)$$

$$f_{CO_2} = \frac{4v - w + 2x - 5y + 2z}{8(v - y)} \quad (10.146)$$

Carbohydrates and acetic acid produce gases that are 50/50 methane and carbon dioxide. Proteins and long-chain fatty acids produce gases that are closer to 75/25 methane and carbon dioxide.

Equation (10.132) indicates that anaerobic digestion is a pseudohydrolysis reaction and that the weight of the gases produced may exceed the weight of the solids destroyed because of the incorporation of water. However, in the case of carbohydrates and acetic acid, there is no water incorporation, and the weight of the gases equals the weight of the carbohydrate/acetic acid destroyed. There is significant water incorporation in the destruction of long-chain fatty acids, and the weight of the gases formed may be 50% greater than the weight of the fatty acid destroyed. In the case of protein fermentation there is significant water incorporation, but the trapping of carbon dioxide by ammonia yields a gas weight that is lower than the protein weight.

Primary sludges tend to contain more fats and protein and less carbohydrate than secondary sludges. Consequently, secondary sludges produce less gas but with a somewhat higher methane content. On the basis of volatile solids destroyed, the gas yields are as follows. For primary sewage solids [Buswell and Boruff, 1932]:

1.25 g total gas per g VS destroyed
1.16 L total gas (SC) per g VS destroyed
CH_4:CO_2::67:33, by volume

For waste-activated sludge and trickling filter humus [Fair and Moore, 1932c]:

0.71 g total gas per g VS destroyed
0.66 L total gas (SC) per g VS destroyed
CH_4:CO_2::71:29, by volume

It is easier to estimate the methane production from a COD balance on a digester. Because it is an anaerobic process, all the COD removed from the sludge ends up in the methane produced. The COD:CH_4 ratio can be estimated from

$$CH_4 + 2O_2 \rightarrow CO_2 + 2H_2O \quad (10.147)$$

Consequently, the COD of 1 mole of methane is 64 g or 4 g COD/g CH_4. Because 1 mole of any gas occupies 22.414 L at standard conditions (0°C, 1 atm), the ratio of gas volume to COD is 0.350 L CH_4 (SC)/g COD removed.

Digestion Kinetics

Batch Digestion. Most of the early research on anaerobic digesters was conducted in batch reactors that did not receive any seed or sludge solids after the initial loading. Much work is still done in fill-and-draw reactors fed once per day. This latter practice is sometimes justified by the observation that many full-scale units are operated in the fill-and-draw mode.

In batch tests, when the ratio of well-digested seed to fresh solids is large (greater than 10:1), the conversion of organic matter to gas is a pseudo first-order process limited by the amount of undigested solids, and the cumulative gas production can be represented by the monomolecular equation [Fair and Moore, 1932b]:

$$\frac{dV}{dt} = K(V_{\max} - V) \tag{10.148}$$

$$V(t) = V_{\max}[1 - \exp(-Kt)] \tag{10.149}$$

where

K = the rate constant (per s)
t = reaction time (s)
$V(t)$ = the cumulative volume of gas produced up to time t (m^3 or ft^3)
$V_{\max}$ = the maximum volume of gas that can be produced from the sample (m^3 or ft^3)

At low ratios of well-digested seed to fresh solids (less than 1:10), the process is limited initially by the number of microbes and is autocatalytic. The pattern of gas production is sigmoidal and can be represented as the sum of a monomolecular rate term and an autocatalytic rate term [Fair and Moore, 1932b]:

$$\frac{dV}{dt} = \underbrace{K_i(V_{\max} - V)}_{\text{Monomolecular}} + \underbrace{K_2 V(V_{\max} - V)}_{\text{Autocatalytic}} \tag{10.150}$$

$$V(t) = \frac{K_1 V_{\max}\left\{1 - \exp\left[-(K_1 + K_2 V_{\max})t\right]\right\}}{K_1 + K_2 V_{\max} \exp\left[-K_1 + K_2 V_{\max})t\right]} \tag{10.151}$$

where

K_1 = the monomolecular rate coefficient (per s)
K_2 = the autocatalytic rate coefficient ($1/m^3{\cdot}s$ or $1/ft^3{\cdot}s$)

Equation (10.151) may be used to represent the course of digestion between loadings in fill-and-draw digesters fed once per day [Sykes, 1970]. In this mode, peak gas production rates occur about three to six hours after feeding, and half the gas production occurs within about nine hours of feeding.

Continuous-Flow Digesters. Sludge digesters are usually designed to be completely mixed, single-pass reactors without recycle. This is due to the fact that digested sludges are only partially settleable, and solids capture by sedimentation is impractical. Consequently, the hydraulic retention time of the system is also the solids retention time:

$$\theta_c = \frac{VX}{QX} = \frac{V}{Q} = \theta_h \tag{10.152}$$

where

Q = the raw sludge flow rate (m^3/s or ft^3/s)
V = the digester's volume (m^3 or ft^3)
X = the volatile solids concentration in the digester (kg/m^3 or lb/ft^3)
θ_c = the solids retention time (s)
θ_h = the hydraulic retention time (s)

The applicability of Gram's [1956] model for the activated sludge process to anaerobic digestion was first demonstrated by Stewart [1958] and latter confirmed by Agardy *et al.* [1963]. Gram's model is discussed in this chapter under "Activated Sludge, Process Parameters and Variables." The important relationships are:

$$\mu = Yq \tag{10.153}$$

$$q = \frac{q_{max} S}{K_s + S} \tag{10.154}$$

$$\mu = \frac{\mu_{max} S}{K_s + S} \tag{10.155}$$

$$\frac{1}{\theta_c} = \mu - k_d \tag{10.156}$$

where

K_s = Monod's affinity constant (kg COD/m^3 or lb COD/ft^3)
k_d = the "decay" rate (per s)
q = the specific uptake (or utilization) rate (kg COD/kg VSS·s or lb COD/lb VSS·s)
q_{max} = the maximum specific uptake (or utilization) rate (kg COD/kg VSS·s or lb COD/lb VSS·s)
S = the kinetically limiting substrate's concentration (kg COD/m^3 or lb COD/ft^3)
μ = the specific growth rate (per s)
μ_{max} = the maximum specific growth rate (per s)
θ_c = the solids retention time (s)

The rate-limiting step in the conversion of organic solids to methane is the fermentation of saturated long-chain fatty acids to acetic acid [Fan, 1983; Novak and Carlson, 1970; O'Rourke, 1968]. Kinetic constants for growth on selected volatile fatty acids, long-chain fatty acids, and hydrogen are given in Table 10.17. The minimum solids retention time for satisfactory digestion is determined by using the kinetic parameters for long-chain fatty acid fermentation.

The kinetic parameters apply to the "biodegradable" fraction of the lipids in municipal wastewater sludges. "Biodegradable" fraction really means the fraction converted to gas; the remaining lipid is conserved in other soluble and particulate microbial products. O'Rourke [1968] estimates that 72% of the lipids in municipal sludges can be gasified at 35°C. The gasifiable fraction falls to 66% at 25°C and to 59% at 20°C. Lipids are not gasified at 15°C.

The Wastewater Committee [1990] limits the solids loading to primary anaerobic digesters to a maximum of 80 lb VS/1000 ft^3·day (1.3 kg/m^3·d), providing the units are completely mixed and heated to 85 to 95°F. The volume of secondary digesters used for solids capture and storage may not be included in the loading calculation. If the feed sludges have a VS content of 2%, the resulting hydraulic retention time of the primary digester is 16 days.

The median hydraulic retention time (HRT) employed for mixed, heated primary digesters treating primary sludge only is 20 to 25 days [Joint Task Force, 1992]. The implied safety factor

Table 10.17 Gram Model Kinetic Parameters for Anaerobic Digestion at 35°C[1–4]

Gram Model Parameter	Substrate				
	H_2	Acetic Acid	Propionic Acid	Butyric Acid	Long-Chain Fatty Acids[1]
μ_{max} (per day)	1.06	0.324	0.318	0.389	0.267 (0.267)
q_{max} (kg COD/kg VSS · d)	24.7	6.1	9.6	15.6	6.67 (6.67)
K_s (mg COD/L)	569 (mm Hg)	164	71	16	2,000 (1800)
k_d (per day)	−0.009	0.019	0.010	0.027	0.038 (0.030)
Y (kg VS/kg COD)	0.043	0.041	0.042	0.047	0.054 (0.040)

[1]Lawrence, A. W. and McCarty, P. L. 1967. *Kinetics of Methane Fermentation in Anaerobic Waste Treatment.* Tech. Rept. No. 75 Stanford University, Department of Civil Engineering, Stanford, CA.

[2]O'Rourke, J. T. 1968. *Kinetics of Anaerobic Waste Treatment at Reduced Temperatures.* Ph.D. Dissertation. Stanford University, Stanford, CA.

[3]Shea, T. G., Pretorius, W. A., Cole, R. D., and Pearson, E. A. 1968. *Kinetics of Hydrogen Assimilation in Methane Fermentation.* SERL Rept. No. 68-7. University of California, Sanitary Engineering Research Laboratory, Berkeley.

[4]Speece, R. E. and McCarty, P. L. 1964. Nutrient requirements and biological solids accumulation in anaerobic digestion: advances in water pollution research. *Proceedings of the International Conference, London, September, 1962, Vol. II.* W. W. Eckenfelder, Jr., ed. Pergamon Press, New York.

Note: Preferred design values are shown in parentheses.

for 90% conversion of volatile solids to gas is about 2 to 2.5. Excluding HRTs less than 10 days, the median HRT for mixed, heated digesters fed a blend of primary and waste activated sludges is about 30 to 35 days. For either sludge the modal HRT is 20 to 25 days.

Temperature Effects. The varying amounts of seed used by different investigators makes it difficult to interpret the early work on digestion kinetics. Fair and Moore [1934, 1937] attempted to deal with this problem by using as a rate measure the incubation time required for a given batch test to produce 90% of its ultimate gas yield. The rate constant is proportional to the reciprocal of the incubation time. For the monomolecular formula the relationship is

$$K = \frac{\ln(10/9)}{t_{90}} \tag{10.157}$$

The data available to Fair and Moore exhibit a great deal of scatter above 25°C, but there is a clear trend of increasing rates of digestion up to about 55°C. Fair and Moore believed that the data showed a local minimum in digestion rate at about 42°C, and that this minimum marked the boundary between the mesophilic range and the thermophilic range. However, if the smooth curves they drew through the data are removed, the scatter in the data becomes more apparent, and the existence of a local minimum at 42°C becomes problematic, although it may be real.

The data collected by Fair and Moore indicate that between about 10°C and 30°C, the Streeter–Phelps theta value for the gasification rate is about 1.070; between 25°C and 55°C, it is about 1.032. Fan's [1983] data for a full-scale plant yield a theta value of 1.084 for the range 18 to 30°C.

The effect of temperature on digestion can also be represented in terms of Gram's model. For digestion temperatures in the range 20 to 35°C, Parkin and Owen [1986] recommend:

$$q_{max} = 6.67 \times 1.035^{T-35} \text{ (kg COD/kg VSS·d)} \tag{10.158}$$

$$K_s = 1.8 \times 0.8993^{T-35} \text{ (g COD/L)} \tag{10.159}$$

$$k_d = 0.030 \times 1.035^{T-35} \text{ (per day)} \tag{10.160}$$

$$Y = 0.040 \text{ (g VSS/g COD)} \tag{10.161}$$

The base values are referenced to 35°C and are derived from O'Rourke's [1968] work. True growth yields do not vary significantly with temperature. The microbial decay rate is so poorly known that temperature adjustments may not be warranted. Note that the effective theta value for the combined temperature effect on the maximum uptake rate and the affinity constant is about 1.15, which is much larger than the values reported for gasification rates.

Effect of pH. Methane production only occurs between pH 5 and 9; the optimum pH is very near 7 and falls off rapidly as the pH increases or decreases. Price's [1963] data may be summarized as:

- Peak rate of methane formation at pH 7.0
- 90% of peak rate at pH 6.5 and 7.5
- 75% of peak rate at pH 6.0 and 8.0
- 50% of peak rate at pH 5.8 and 8.4
- 25% of peak rate at pH 5.4 and 8.8

Inhibitors. Digestion inhibitors are listed in Table 10.18. Some of the metals listed are also nutrients. The optimum concentration for Na^+, K+, or NH_4^+ is 0.01 mol/L; the optimum concentration for Ca^{2+} or Mg^{2+} is 0.005 mol/L [Kugelman and McCarty, 1965].

Poisons must be in soluble form to be effective. Cobalt, copper, iron, lead, nickel, zinc, and other heavy metals form highly insoluble sulfides ranging in solubility from 10^{-5} to 10^{-11} mg/L, which eliminates the metal toxicity [Lawrence and McCarty, 1965]. Heavy metals may compose as much as 10% of the volatile solids without impairing digestion if they are precipitated as sulfides. The requisite sulfide may be fed as sodium sulfide or as various sulfate salts, which are reduced to sulfide.

It should be noted that methanogens will reduce mercury to mono- and dimethylmercury, which is volatile and insoluble and which may be present in the digester gas. Alkyl mercurials are very toxic.

Halogenated methane analogs like chloroform, carbon tetrachloride, and freon are very toxic to methanogens at concentrations on the order of several mg/L [Kirsch and Sykes, 1971].

Moisture Limitation. The stoichiometry of anaerobic digestion indicates that water is consumed and may be limiting in low-moisture environments. Anaerobic digestion proceeds normally at total solids concentrations up to 20 to 25% by weight [Wujcik, 1980]. Above about 30% TS the rate of methane production is progressively reduced and ceases at 55% TS. In this range the methanogens appear to be water limited rather than salt, ammonia, or VFA limited. Above 55% TS acid production is inhibited.

Mixing

For high-rate digesters fed unthickened sludges, the required mixing power is about 0.2 to 0.3 hp/ft^3 [Joint Task Force, 1992]. Alternatively, the required rms velocity gradient is 50 to 80 per second, and the turnover time is 30 to 45 min.

In conventional digesters mixing becomes impaired at VS loading rates above 0.3 $lb/ft^3 \cdot day$ (4.0 $kg/m^3 \cdot d$) [Metcalf & Eddy, Inc., 1991]. Ammonia toxicity limits VS loadings to about 0.2 $lb/ft^3 \cdot day$ (3.2 $kg/m^3 \cdot d$) [Joint Task Force, 1992].

Sludge pumping becomes a problem at about 8 to 12% TS [Brisbin, 1957]. In this range the Hazen–Williams *C* coefficient should be reduced by 60 to 75%.

Heat Balance

The higher heating value of either raw or digested sludge solids is given as follows [Fair and Moore, 1932a]. For primary sludge:

$$\Delta H = 29P^{4/3}(\text{Btu/lb TS}) \tag{10.162}$$

Table 10.18 Anaerobic Digestion Inhibitors[1]

Substance	Effect	Concentration Units	Concentration
	Inorganic		
Ammonia-nitrogen	moderate	mg/L	1500 to 3000
	strong	mg/L	3000
Calcium	moderate	mg/L	2500 to 4500
	strong	mg/L	8000
Chromium (III)	strong	mg/L	180 to 420 (total)
Chromium (VI)	strong	mg/L	3.0 (soluble)
	strong	mg/L	200 to 260 (total)
Copper	strong	mg/L	0.5 (soluble)
	strong	mg/L	50 to 70 (total)
Magnesium	moderate	mg/L	1000 to 1500
	strong	mg/L	3000
Nickel	strong	mg/L	1.0 (soluble)
	strong	mg/L	30. (total)
Potassium	moderate	mg/L	2500 to 4500
	strong	mg/L	12,000
Sodium	moderate	mg/L	3500 to 5500
	strong	mg/L	8,000
Zinc	strong	mg/L	1.0 (soluble)
	Organic		
Acetaldehyde	50% activity	mmol/L	10.
Acrylic acid	50% activity	mmol/L	12.
Acrylonitrile	50% activity	mmol/L	4.
Acrolein	50% activity	mmol/L	0.2
Aniline	50% activity	mmol/L	26.
Catechol	50% activity	mmol/L	24.
Chloroform	"inhibitory"	mg/L	0.5
3-Chloro-1,3-propandiol	50% activity	mmol/L	6.
1-Chloropropane	50% activity	mmol/L	1.9
1-Chloropropene	50% activity	mmol/L	0.1
2-Chloropropionic acid	50% activity	mmol/L	8.0
Crotonaldehyde	50% activity	mmol/L	6.5
Ethyl acetate	50% activity	mmol/L	11.0
Ethyl benzene	50% activity	mmol/L	3.2
Ethylene dichloride	"inhibitory"	mg/L	5.
Formaldehyde	50% activity	mmol/L	2.4
Kerosene	"inhibitory"	mg/L	500.
Lauric acid	50% activity	mmol/L	2.6
Linear alkylbenzene sulfonate	"inhibitory"	mg/L	1% of dry solids
Nitrobenzene	50% activity	mmol/l	0.1
Phenol	50% activity	mmol/L	26.
Propanal	50% activity	mmol/L	90.
Resorcinol	50% activity	mmol/L	29.
Vinyl acetate	50% activity	mmol/L	8.

[1]Parkin, G. F. and Owen, W. F. 1986. Fundamentals of anaerobic digestion of wastewater sludge. *Journal of Environmental Engineering*. 112(5):867.

For waste-activated sludge:

$$\Delta H = 25P^{4/3}(\text{Btu/lb TS}) \tag{10.163}$$

where P = the percentage of volatile solids in the total solids. Note that the heating value does not vary linearly with the volatile solids content of the sludge solids. For primary sludges the heating value extrapolated to 100% VS is 13,500 Btu/lb; for waste activated sludges the heating value extrapolated to 100% VS is 11,600 Btu/lb.

Table 10.19 Heat Transfer Coefficients for Various Materials[1]

Material	Transfer Coefficient, k (Btu · in/ft²·hr·°F)
Air	0.17
Brick	3.0 to 6.0
Concrete	2.0 to 3.0
Earth, dry	10
Earth, wet	30
Mineral wool insulation	0.26 to 0.29
Steel	5.2 to 6.0

[1]Joint Task Force of the Water Environment Federation and the American Society of Civil Engineers. 1992. *Design of Municipal Wastewater Treatment Plants: Volume II.* Chapters 13–20. WEF Manual of Practice No. 8; ASCE Manual and Report on Engineering Practice No. 76. Water Environment Federation, Alexandria, VA; American Society of Civil Engineers, New York.

The principal use of digester gas is digester heating. The steady state heat transfer due to conduction through material is directly proportional to (a) the temperature difference and (b) the area normal to the heat flow, and it is inversely proportional to (c) the thickness of the material. A heat transfer coefficient, k, may be defined for any pure substance by the following equation:

$$\Delta H = \frac{kA\Delta T}{L} \tag{10.164}$$

where

A = the area normal to the heat flux (m^2 or ft^2)
L = the thickness of the medium conducting the heat (m or ft)
ΔH = the heat flow (J/s or Btu/s)
k = the heat transfer coefficient (J/m·s·K or Btu·in./ft^2·s·°F)
ΔT = the temperature difference across the conducting medium (K or °R)

The units of the transfer coefficient would be either J/m·s·K or Btu·in./ft^2·s·°F. Heat transfer coefficients for some materials are given in Table 10.19.

If a wall or roof is made up of several layers of different substances, then an overall heat transfer coefficient, K, can be calculated by summing the temperature drops across each component and noting that each component transmits the same heat flux and has the same area:

$$\Delta T = (T_1 - T_2) + (T_2 - T_3) + \ldots + (T_{n-2} - T_{n-1}) + (T_{n-1} - T_n)$$

$$\frac{\Delta H}{KA} = \frac{L_1 \Delta H}{k_1 A} + \frac{L_2 \Delta H}{k_2 A} + \ldots + \frac{L_{n-2}\Delta H}{k_{n-2} A} + \frac{L_{n-1}\Delta H}{k_{n-1} A} \tag{10.165}$$

$$\frac{1}{K} = \frac{L_1}{k_1} + \frac{L_2}{k_2} + \ldots + \frac{L_{n-2}}{k_{n-2}} + \frac{L_{n-1}}{k_{n-1}}$$

where K = the overall heat transfer coefficient (J/m^2·s·K or Btu/ft^2·s·°F). As a matter of convenience, the thickness of the composite is adsorbed into the definition of the overall transfer coefficient.

A complete heat balance on an anaerobic digester is

$$\underbrace{\Delta H_{\text{req}}}_{\text{Heat required}} = \underbrace{C_p \rho Q \left(T_{\text{dig}} - T_{\text{slu}}\right)}_{\text{Raw sludge heating}} + \underbrace{K_r A_r \left(T_{\text{dig}} - T_{\text{air}}\right)}_{\text{Heat through roof}} + \underbrace{K_w A_w \left(T_{\text{dig}} - T_{\text{grd}}\right)}_{\text{Heat through wall}}$$

$$+ \underbrace{K_f A_f \left(T_{\text{dig}} - T_{\text{grd}}\right)}_{\text{Heat through floor}} - \underbrace{H_{\text{met}} Q \left(X_{vo} - X_{ve}\right)}_{\text{Metabolic heat}} \tag{10.166}$$

where

A = area normal to heat flux (m^2 or ft^2)
C_p = constant pressure specific heat of water (J/kg or Btu/lb)

H_{met} = metabolic heat release (J/kg·VS or Btu/lb·VS)
ΔH_{req} = heat requirement (J/s or Btu/s)
K = overall heat transfer coefficient ($J/m^2 \cdot s \cdot K$ or $Btu/ft^2 \cdot s \cdot {}^\circ F$)
Q = sludge flow rate (m^3/s or ft^3/s)
T_{air} = air temperature (K or °R)
T_{dig} = digester temperature (K or °R)
T_{grd} = ground temperature (K or °R)
T_{slu} = sludge temperature (K or °R)
X_{ve} = effluent VSS (kg/m^3 or lb/ft^3)
X_{vo} = influent VSS (kg/m^3 or lb/ft^3)
ρ = mass density of water (kg/m^3 or lb/ft^3)

When VS are destroyed approximately 80% of their fuel value is retained in the methane formed, and 20% is liberated to the digesting sludge as metabolic heat [Fan, 1983]. A somewhat conservative estimate of the metabolic heat release is 2000 Btu/lb VS destroyed (1100 cal/g VS destroyed). This raises the possibility of autothermal anaerobic digestion. Ignoring the heat losses by conduction and setting the heat requirement to zero, one gets

$$X_{vo} - X_{ve} = \frac{C_p \rho (T_{dig} - T_{slu})}{H_{met}} \tag{10.167}$$

Assuming 60% VS destruction and a sludge temperature increase of 40°F, the influent VS concentration for autothermal mesophilic digestion is about 5% by weight. This is equivalent to about 8% by weight TS, which is near the pumping limit.

Aerobic Digestion

Treatment Goals

The treatment goal is the production of a stabilized sludge that will not produce offensive odors or attract disease vectors. A sludge is considered to be stabilized when its specific oxygen uptake rate is reduced to 1.5 mg O_2/g TS·h at 20°C [Environmental Protection Agency, 1993]. Federal regulations require minimum solids retention times of 40 days at 20°C and 60 days at 15°C and a minimum volatile solids destruction of 38% [Environmental Protection Agency, 1993].

Typical compositions of aerobic digester supernatants are summarized in Table 10.20.

Table 10.20 Properties of Aerobic Digester Supernatant[1]

Parameter (units)	Mean Value	Range
BOD (mg/L)	500	9 to 1700
Soluble BOD (mg/L)	51	4 to 183
COD (mg/L)	2600	228 to 8140
Suspended solids (mg/L)	3400	46 to 11,500
Alkalinity (mg/L, as $CaCo_3$)	—	473 to 514
pH	7.0	5.9 to 7.7
TKN (mg/L)	170	10 to 400
Total P (mg/L)	98	19 to 241
Soluble P (mg/L)	26	2.5 to 64

[1]Schwinn, D. E. and Gassett, R. B., ed. 1974. *Process Design Manual for Upgrading Existing Wastewater Treatment Plants.* Environmental Protection Agency, Technology Transfer, Washington, D.C.

Facilities

Aerobic digestion is usually restricted to smaller facilities where the cost of aeration is offset by operational and facilities simplicity [Joint Task Force, 1992].

The digesters are constructed as open, unheated tanks. A variety of plan geometries have been built, including rectangular, circular, and annular tanks [Joint Task Force, 1992]. Side water depths range from 10 to 25 ft. Aerobic digesters are susceptible to foaming, and freeboard heights of 1.5 to 4 ft are required to retain the foam. Aeration and mixing is usually provided by diffused air systems, either coarse or fine bubble. Air flow rates of 20 to 40 scfm/1000 cf are needed for mixing. Diffused air permits better control of dissolved oxygen and reduces heat losses, which is important in cold climates. Mechanical surface aerators have lower maintenance costs, but they produce greater heat losses, increase foam production, and are more susceptible to reduction in oxygen transfer efficiency due to foam.

Typical process loadings are 24 to 140 lb VS/1000 ft^3·day, and reactor volume allowances are 3 to 4 ft^3/cap [Schwinn and Gassett, 1974]. Many small facilities store digested sludge in the digester for substantial time periods prior to disposal (e.g., because of seasonal land application), and allowances must be made for this additional storage.

Microbiology

The process of aerobic sludge digestion is a continuation of phenomena occurring in the activated sludge process. In some installations soluble substrate levels are low, and heterotrophic growth of bacteria is small. Initially, there may also be some endogenous respiration by bacteria starved for substrate. The principal digestion process is the predation and scavenging by "worms"—rotifers and protozoa of the bacteria and other sludge solids. Some bacteria may hydrolyze particulate matter, and the digestion processes of the predators and scavengers may release soluble substrates that support some heterotrophic growth and the growth of their phages. Additional soluble substrate may be released when the phages lyze the cells of their bacterial hosts.

Kinetics

The usual assumptions are that the volatile solids may be divided into an inert fraction and a biodegradable fraction whose destruction obeys first-order kinetics [Adams *et al.*, 1974]. For a completely mixed digester the volatile solids destruction may be modeled as

$$\frac{X_{vd} - X_{vi}}{X_{vo} - X_{vi}} = \frac{1}{1 + k_d\,\theta_h} \tag{10.168}$$

where

k_d = the decay rate (per s)
X_{vd} = the VSS in the digester (kg/m^3 or lb/ft^3)
X_{vi} = the inert or nonbiodegradable VSS (kg/m^3 or lb/ft^3)
X_{vo} = influent VSS (kg/m^3 or lb/ft^3)
θ_c = the solids retention time (s)

A typical decay rate at 20°C is 0.08 to 0.12 per day [Brown and Caldwell, 1979]. The decay may decline with increasing suspended solids concentrations. For bench-scale units Reynolds [1973] reported a decline from 0.72/d at a TSS of 8400 mg/L to 0.34/d at a TSS of 22,700 mg/L. The sludge was digested at room temperature. Reynolds decay rates are substantially higher than other reported rates; his sludges were obtained from a contact-stabilization plant.

At very long HRTs, the digester VSS concentration, X_{vd}, approaches the inert or nonbiodegradable VSS concentration, X_{vi}. Typically, about 50 to 60% of the volatile solids in waste activated sludge are biodegradable [Reynolds, 1973].

It should be noted that the suspended ash (TSS minus VSS) is solubilized during digestion, and its concentration declines in parallel with the decline in VSS [Eckenfelder, 1956; Reynolds, 1973]. However, the solubilized solids remain in the liquid as part of the sludge and are not removed unless a dewatering process is applied to the sludge.

For temperatures above 15°C, SRTs range from 10 to 15 days for waste activated sludge alone and 15 to 20 days for primary sludge alone and for mixtures of waste activated and primary sludges [Schwinn and Gassett, 1974].

Temperature Effects. The variation of the decay rate with temperature is given approximately by

$$k_d = 0.332\{1 - \exp[-0.403(T - 8)]\}; \qquad R^2 = 0.53 \tag{10.169}$$

where

k_d = the decay rate per day
T = the digestion temperature in °C

Equation (10.169) was derived from the data summarized by Brown and Caldwell [1979] using the Thomas graphical method for fitting the BOD curve. All the data were used, and the data span the temperature range 10 to 64°C. The derived curve lies somewhat above the hand-drawn curve presented in the report. The scatter about either line is very large, and digestion rates should be based upon pilot studies.

An examination of the plotted data suggests that the digestion rate reaches a maximum of 0.23/d at a digestion temperature of 40°C. There is no clear thermophilic digestion range, which may reflect the limited number of thermophilic eucaryotes. There are no thermophilic rotifers or worms, which are the dominant predators effecting aerobic digestion.

pH. The comments on the activated sludge process apply here as well. Organic solids destruction is not appreciably affected between pH 6 and 9. However, as long as the dissolved oxygen concentration is adequate (above 2 mg/L), aerobic digesters will nitrify and the pH will fall in poorly buffered waters.

Inhibitors. See Tables 10.4 and 10.5. If nitrification is desired, the special requirements of the nitrifying bacteria will control.

Oxygen Requirements

The general oxygen balance for activated sludge also applies to aerobic digestion. For nonnitrifying digesters,

$$R = 1.42Q(X_{vo} - X_{ve}) \tag{10.170}$$

and for nitrifying digesters,

$$R = 1.98Q(X_{vo} - X_{ve}) \tag{10.171}$$

See the comments about oxygen requirements in the activated sludge process, especially the requirements of the nitrifying bacteria.

Mixing Requirements

The power required to mix thickened sludges may be estimated from the following [Zwietering, 1958; Reynolds, 1973]:

$$P = 0.00475\mu^{0.3} X_{\mathrm{TSS}}^{0.298} V \tag{10.172}$$

where

$P_{\min}$ = minimum required mixing power, in hp
V = digester volume, in 1000 gal
X_{TSS} = the TSS concentration, in mg/L
μ = the liquid viscosity, in centipoises

Autothermal Thermophilic Digestion

The heat balance given earlier for anaerobic digesters also applies to aerobic digesters. However, in aerobic digestion all the higher heating value of the destroyed volatile solids is released as metabolic heat, so the break-even point for autothermal digestion is a feed sludge containing between 1 and 2% VS. This is equivalent to about 2 to 3% TS, which is well within the limits for good mixing.

In European practice, waste activated sludges are first thickened to at least 2.5% by weight VSS [Joint Task Force, 1992]. The digesters are cylindrical with a height-to-diameter ratio of 0.5 to 1.0. They are operated in the fill-and-draw mode with two temperature phases per cycle. The first temperature phase is 35 to 50°C and is intended to stabilize the sludge. The second phase is 50 to 65°C and is intended to reduce pathogens. The HRT for the digester is 5 to 6 days, with a minimum HRT of 20 hours in either temperature phase. Aeration is generally by diffused air. Substantial foaming occurs; foam cutters are needed to control foam accumulation. Nitrification does not occur at thermophilic temperatures, which reduces the oxygen requirement.

References

Section 10.1

Adams, C. E. and Eckenfelder, W. W., Jr. 1977. Nitrification design approach for high strength ammonia wastewaters. *Journal of the Water Pollution Control Federation.* 49(3):413.

Ardern, E. and Lockett, W. T. 1914a. Experiments on the oxidation of sewage without the aid of filters. *Journal of the Society of Chemical Industry.* 33(10):523.

Ardern, E. and Lockett, W. T. 1914b. The oxidation of sewage without the aid of filters, part II. *Journal of the Society of Chemical Industry.* 33(23):1122.

Babbitt, H. E. and Baumann, E. R. 1958. *Sewerage and Sewage Treatment,* 8th ed. John Wiley & Sons, New York.

Barnard, J. L. 1974a. Cut P and N without chemicals. *Water & Wastes Engineering.* 11(7):33.

Barnard, J. L. 1974b. Cut P and N without chemicals. *Water & Wastes Engineering.* 11(8):41.

Barth, E. F., Brenner, R. C., and Lewis, R. F. 1968. Chemical-biological control of nitrogen and phosphorus in wastewater effluent. *Journal of the Water Pollution Control Federation.* 40(12):2040.

Beccari, M., Passino, R., Ramadori, R., and Tandoi, V. 1983. Kinetics of dissimilatory nitrate and nitrite reduction in suspended growth culture. *Journal of the Water Pollution Control Federation.* 55(1):58.

Beckman, W. J., Avendt, R. J., Mulligan, T. J., and Kehrberger, G. J. 1972. Combined carbon oxidation-nitrification. *Journal of the Water Pollution Control Federation.* 44(9):1917.

Benedict, A. H. and Carlson, D. A. 1973. Temperature acclimation in aerobic bio-oxidation systems. *Journal of the Water Pollution Control Federation.* 45(1):10.

Boon, A. G. and Chambers, B. 1985. Design protocol for aeration systems—U.K. perspective. In *Proceedings—Seminar Workshop on Aeration System Design, Testing, Operation, and Control,* ed. W. C. Boyle. EPA 600/9-85-005. Environmental Protection Agency, Risk Reduction Engineering Laboratory, Cincinnati, OH.

Bowen, H.J.M. 1966. *Trace Elements in Biochemistry.* Academic Press, New York.

Bowker, R.P.G., and Stensel, H. D. 1987. *Design Manual: Phosphorus Removal.* EPA/625/1-87-001. Environmental Protection Agency, Center for Environmental Research Information, Water Engineering Research Laboratory, Cincinnati, OH.

Burdick, C. R., Refling, D. R., and Stensel, H. D. 1982. Advanced biological treatment to achieve nutrient control. *Journal of the Water Pollution Control Federation.* 54(7):1078.

Clarkson, R. A., Lau, P. J., and Krichten, D. J. 1980. Single-sludge pure-oxygen nitrification and phosphorus removal. *Journal of the Water Pollution Control Federation.* 52(4):770.

Eagle, L. M., Heymann, J. B., Greben, H. A., and Potgeiger, D.J.J. 1989. The isolation and characterization of volutin granules as subcellular components involved in biological phosphorus removal. *Water Science and Technology.* 21:397.

Eckenfelder, W. W., Jr. 1956. Studies on the oxidation kinetics of biological sludges. *Sewage and Industrial Wastes.* 28(8):983.

Eckenfelder, W. W., Jr. 1980. Principles of biological treatment. In *Theory and Practice of Biological Wastewater Treatment,* ed. K. Curi and W. W. Eckenfelder, Jr., p. 49. Sijthoff & Noordhoff BV., Germantown, MD.

Eckenfelder, W. W., Jr. and O'Connor, D. J. 1961. *Biological Waste Treatment.* Pergamon Press, New York.

Ekama, G. A. and Marais, G.v.R. 1984. Biological nitrogen removal. In *Theory, Design and Operation of Nutrient Removal Activated Sludge Processes,* ed. H.N.S. Weichers *et al.*, Water Research Commission, Pretoria, South Africa.

EPA. 1977. *Federal Guidelines: State and Local Pretreatment Programs. Volume I.* (EPA-430/9-76-017a) *and Volume II. Appendices 1–7.* (EPA-430/9-76-017b). Environmental Protection Agency, Office of Water Programs Operations, Municipal Construction Division, Washington, D.C.

Fair, G. M. and Thomas, H. A., Jr. 1950. The concept of interface and loading in submerged aerobic biological sewage treatment systems. *Journal and Proceedings of the Institution of Sewage Purification.* 3:235.

Focht, D. D. and Chang, A. C. 1975. Nitrification and denitrification processes related to waste water treatment. *Advances in Applied Microbiology.* 19:153.

Forney, C. and Kountz, R. R. 1959. Activated sludge total oxidation metabolism. *Proceedings of the 13th Industrial Waste Conference.* 43(3):313. May 5, 6, and 7, 1958. Extension Series No. 96, Engineering Bulletin, ed. D. E. Bloodgood. Purdue University, Engineering Extension Department, West Lafayette, IN.

Fuhs, G. W. and Chen, M. 1975. Microbiological basis of phosphate removal in the activated sludge process for the treatment of wastewater. *Microbial Ecology.* 2:119.

Gaudy, A. F., Ramanathan, M., Yang, P. V., and DeGeare, T. V. 1970. Studies of the operational stability of the extended aeration process. *Journal of the Water Pollution Control Federation.* 42(2):165.

Gaudy, A. F., Yang, P. V., and Obayashi, A. W. 1971. Studies of the total oxidation of activated sludge with and without hydrolytic pretreatment. *Journal of the Water Pollution Control Federation.* 43(1):40.

Gee, C. S., Suidan, M. T., Pfeffer, J. T. 1990. Modeling of nitrification under substrate-inhibiting conditions. *Journal of Environmental Engineering.* 116(1):18.

Giona, A. R., Annesini, M. C., Toro, L., and Gerardi, W. 1979. Kinetic parameters for municipal wastewater. *Journal of the Water Pollution Control Federation.* 51(5):999.

Goodman, B. L. and Englande, A. J., Jr. 1974. A unified model of the activated sludge process. *Journal of the Water Pollution Control Federation.* 46(2):312.

Gram, A. L., III. 1956. *Reaction Kinetics of Aerobic Biological Process.* Rept. No. 2, I.E.R. Ser. 90. University of California at Berkeley, Department of Engineering, Sanitary Engineering Research Laboratory, Berkeley.

Greenberg, A. E., Klein, G., and Kaufman, W. J. 1955. Effect of phosphorus on the activated sludge process. *Sewage and Industrial Wastes.* 27(3):277.

Guo, P.H.M., Thirumurthi, D., and Jank, B. E. 1981. Evaluation of extended aeration activated sludge package plants. *Journal of the Water Pollution Control Federation.* 53(1):33.

Haldane, J.B.S. 1930. *Enzymes.* Longmans, Green, London.

Harold, F. M. 1966. Inorganic phosphates in biology: Structure, metabolism and function. *Bacteriological Reviews.* 30:772.

Haseltine, T. R. 1961. Sludge reaeration in the activated sludge process—A survey. *Journal of the Water Pollution Control Federation.* 33(9):946.

Helmers, E. N., Frame, J. D., Greenberg, A. E., and Sawyer, C. N. 1951. Nutritional requirements in the biological stabilization of industrial wastes—II. Treatment with domestic sewage. *Sewage and Industrial Wastes.* 23(7):884.

Heukelekian, H., Orford, H. E., and Manganelli, R. 1951. Factors affecting the quantity of sludge production in the activated sludge process. *Sewage and Industrial Wastes.* 23(8):945.

Horstkotte, G. A., Niles, D. G., Parker, D. S., and Caldwell, D. H. 1974. Full-scale testing of a water reclamation system. *Journal of the Water Pollution Control Federation.* 46(1):181.

Hunter, J. V., Genetelli, E. J., and Gilwood, M. E. 1966. Temperature and retention time relationships in the activated sludge process. *Proceedings of the 21st Industrial Waste Conference,* 50(2):953. May 3, 4, and 5, 1966. Engineering Extension Series No. 121. Engineering Bulletin, ed. D. E. Bloodgood. Purdue University, West Lafayette, IN.

Jenkins, D. and Orhon, D. 1973. The mechanism and design of the contact stabilization activated sludge process. *Advances in Water Pollution Research: Proceedings of the Sixth International Conference,* p. 353. Jerusalem, June 18–23, 1972, ed. S. H. Jenkins, Pergamon Press, New York.

Johnson, W. K. 1972. Process kinetics for denitrification. *Journal of the Sanitary Engineering Division, Proceedings ASCE.* 98(SA4):623.

Joint Committee of the Water Pollution Control Federation and the American Society of Civil Engineers. 1977. *Wastewater Treatment Plant Design.* Manual of Practice No. 8. Water Pollution Control Federation, Washington, D.C.; American Society of Civil Engineers, New York.

Joint Task Force of the Water Pollution Control Federation and the American Society of Civil Engineers. 1988. *Aeration: A Wastewater Treatment Process.* WPCF Manual of Practice No. FD-13; ASCE Manuals and Reports on Engineering Practice No. 68. Water Pollution Control Federation, Alexandria, VA; American Society of Civil Engineers, New York.

Joint Task Force of the Water Environment Federation and the American Society of Civil Engineers. 1992. *Design of Municipal Wastewater Treatment Plants: Volume I,* chapters 1–12, WEF Manual of Practice No. 8; ASCE Manual and Report on Engineering Practice No. 76. Water Environment Federation, Alexandria, VA; American Society of Civil Engineers, New York.

Jones, P. H. and Sabra, N. M. 1980. Effect of systems solids retention time (SSRT or sludge age) on nitrogen removal from activated-sludge systems. *Water Pollution Control.* 79:106.

Jordan, W. J., Pohland, F. G., and Kornegay, B. H. (1971). Evaluating treatability of selected industrial wastes. *Proceedings of the 26th Industrial Waste Conference,* p. 514. May 4, 5, and 6, 1971. Engineering Extension Series No. 140, ed. J.M. Bell, Purdue University, West Lafayette, IN.

Keefer, C. E. 1962. Temperature and efficiency of the activated sludge process. *Journal of the Water Pollution Control Federation.* 34(11):1186.

Keefer, C. E. and Meisel, J. 1950. Activated sludge studies—I. Effect of sludge age on oxidizing capacity. *Sewage and Industrial Wastes.* 22(9):1117.

Keefer, C. E. and Meisel, J. 1951. Activated sludge studies—III. Effect of pH of sewage on the activated sludge process. *Sewage and Industrial Wastes.* 23(8):982.

Keenan, J. D., Steiner, R. L., and Fungaroli, A. A. 1979. Substrate inhibition of nitrification. *Journal of Environmental Science and Health, Part A, Environmental Science and Engineering.* A14(5):377.

Knoetze, C., Davies, T. R., and Wiechers, S. G. 1980. Chemical inhibition of biological nutrient removal processes. *Water SA.* 6(4):171.

Knox, H. 1958. *Progress Report on Aerobic Digestion Plants in Ohio.* Presented at the 5th Annual Wastes Engineering Conference. University of Minnesota, Minneapolis-St. Paul, MN. [Read in manuscript. Files kept by Ohio Environmental Protection Agency, Columbus, Ohio.]

Kountz, R. R. and Forney, C. 1959. Metabolic energy balances in a total oxidation activated sludge system. *Sewage and Industrial Wastes.* 31(7):819.

Lawrence, A. W. and Brown, C. G. 1976. Design and control of nitrifying activated sludge systems. *Journal of the Water Pollution Control Federation.* 48(7):1779.

Lawrence, A. W. and McCarty, P. L. 1970. Unified basis for biological treatment design and operation. *Journal of the Sanitary Engineering Division, Proceedings of the American Society of Civil Engineers.* 96(3):757.

Levin, G. V., Topol, G. J., Tarnay, A. G., and Samworth, R. B. 1972. Pilot-plant tests of a phosphate removal process. *Journal of the Water Pollution Control Federation.* 44(10):1940.

Levin, G. V., Topol, G. J., and Tarnay, A. G. 1975. Operation of full-scale biological phosphorus removal plant. *Journal of the Water Pollution Control Federation.* 47(3):577.

Lewandowski, Z. 1982. Temperature dependency of biological denitrification with organic materials addition. *Water Research.* 16(1):19.

Lin, K.-C. and Heinke, G. W. 1977. Plant data analysis of temperature significance in the activated sludge process. *Journal of the Water Pollution Control Federation.* 49(2):286.

Ludzack, F. J., and Ettinger, M. B. 1962. Controlling operation to minimize activated sludge effluent nitrogen. *Journal of the Water Pollution Control Federation.* 34(9):920.

Ludzack, F. J., Schaffer, R. B., and Ettinger, M. B. 1961. Temperature and feed as variables in activated sludge performance. *Journal of the Water Pollution Control Federation.* 33(2):141.

Martin, A. J. 1927. *The Activated Sludge Process.* Macdonald and Evans, London.

McCarty, P. L., Beck, L., and St. Amant, P. 1969. Biological denitrification of wastewaters by addition of organic materials. *Proceedings of the 24th Industrial Waste Conference,* p. 1271. May 6, 7, and 8, 1969. Engineering Extension Series No. 135. ed. D. E. Bloodgood, Purdue University, West Lafayette, IN.

McClintock, S. A., Sherrard, J. H., Novak, J. T., and Randall, C. W. 1988. Nitrate versus oxygen respiration in the activated sludge process. *Journal of the Water Pollution Control Federation.* 60(3):342.

McKinney, R. E. 1962. Mathematics of complete-mixing activated sludge. *Journal of the Sanitary Engineering Division, Proceedings of the American Society of Civil Engineers.* 88(SA3):87.

Metcalf, L. and Eddy, H. P. 1916. *American Sewerage Practice: Vol. III Disposal of Sewage,* 2nd ed. McGraw-Hill, New York.

Middlebrooks, E. J. and Garland, C. F. 1968. Kinetics of model and field extended-aeration wastewater treatment units. *Journal of the Water Pollution Control Federation.* 40(4):586.

Middlebrooks, E. J., Jenkins, D., Neal, R. C., and Phillips, J. L. 1969. Kinetics and effluent quality in extended aeration. *Water Research.* 3(1):39.

Moore, S. F. and Schroeder, E. D. 1970. An investigation of the effects of residence time on anaerobic bacterial denitrification. Water Research. 4(10):685.

Moore, S. F., and Schroeder, E. D. 1971. The effect of nitrate feed rate on denitrification. *Water Research.* 5(7):445.

Morris, G. L., Van Den Berg, L., Culp, G. L., Geckler, J. R., and Porges, R. 1963. *Extended Aeration Plants and Intermittent Watercourses.* Public Health Service Pub. No. 999-WP-8. Department of Health, Education and Welfare, Public Health Service, Division of Water Supply and

Pollution Control, Technical Services Branch, Robert A. Taft Sanitary Engineering Center, Cincinnati, OH.

Mulbarger, M. C. 1971. Nitrification and denitrification in activated sludge systems. *Journal of the Water Pollution Control Federation.* 43(10):2059.

Nicholls, H. A. 1975. Full scale experimentation on the new Johannesburg extended aeration plants. *Water SA.* 1(3):121.

Novak, J. T. 1974. Temperature-substrate interactions in biological treatment. *Journal of the Water Pollution Control Federation.* 46(8):1984.

Orford, H. E., Heukelekian, H., and Isenberg, E. 1963. Effect of sludge loading and dissolved oxygen on the performance of the activated sludge process. *Air and Water Pollution.* 5(2/4):251.

Painter, H. A. 1970. A review of literature on inorganic nitrogen metabolism in microorganisms. *Water Research.* 4(6):393.

Parker, D. S., Stone, R. W., Stenquist, R. J., and Culp, G. 1975. *Process Design Manual for Nitrogen Control.* Environmental Protection Agency, Technology Transfer, Washington, D.C.

Peil, K. M. and Gaudy, A. J., Jr. 1971. Kinetic constants for aerobic growth of microbial populations selected with various single compounds and with municipal wastes as substrates. *Applied Microbiology.* 21:253.

Porter, J. R. 1946. *Bacterial Chemistry and Physiology.* John Wiley & Sons, New York.

Prakasam, T.B.S., Lue-Hing, C., Bogusch, E., and Zenz, D. R. 1979. Pilot-scale studies of single-stage nitrification. *Journal of the Water Pollution Control Federation.* 51(7):1904.

Roberts, R. B., Abelson, P. H., Cowie, D. B., Bolton, E. T., and Britten, R. J. 1955. *Studies of Biosynthesis in* Escherichia coli. Bull. No. 607. Carnegie Institution, Washington, D.C.

Roy F. Weston, Inc. 1983. *Emerging Technology Assessment of Biological Phosphorus Removal:* 1. PHOSTRIP Process; 2. A/O Process; 3. BARDENPHO Process. Environmental Protection Agency, Wastewater Research Division, Municipal Environmental Research Laboratory, Cincinnati, OH.

Rozich, A. F. and Castens, D. J. 1986. Inhibition kinetics of nitrification in continuous-flow reactors. *Journal of the Water Pollution Control Federation.* 58(3):220.

Sawyer, C. N. 1942. Activated sludge oxidations: VI. Results of feeding experiments to determine the effect of the variables temperature and sludge concentration. *Sewage Works Journal.* 12(2):244.

Sayigh, B. A. and Malina, J. F., Jr. 1978. Temperature effects on the activated sludge process. *Journal of the Water Pollution Control Federation.* 50(4):678.

Scalf, M. R., Pfeffer, F. M., Lively, L. D., Witherow, J. L., and Priesing, C. P. 1969. Phosphate removal at Baltimore, Maryland. *Journal of the Sanitary Engineering Division, Proceedings of the American Society of Civil Engineers.* 95(SA5):817.

Scheible, O. K. and Heidman, J., (ed.) 1993. *Manual: Nitrogen Control.* EPA/625/R-93/010. Environmental Protection Agency, Office of Research and Development, Center for Environmental Research Information, Risk Reduction Engineering Laboratory, Cincinnati, OH.

Shapiro, J. 1967. Induced release and uptake of phosphate by microorganisms. *Science.* 155:1269.

Smarkel, K. S. 1977. Personal communication.

Smith, I. W., Wilkinson, J. F., and Duguid, J. P. 1954. Volutin production in *Aerobacter aerogenes* due to nutrient imbalance. *Journal of Bacteriology.* 68:450.

Stanier, R. Y., Adelberg, E. A., and Ingraham, J. L. 1976. *The Microbial World,* 4th ed. Prentice Hall, Engelwood Cliffs, NJ.

Stankewich, M. J. 1972. Biological nitrification with the high purity oxygen process. *Proceedings of the 27th Industrial Waste Conference,* p. 1. May 2, 3, and 4, 1972. Engineering Extension Series No. 141, ed. J. M. Bell, Purdue University, West Lafayette, IN.

Stensel, H. D., Loehr, R. C., and Lawrence, A. W. 1973. Biological kinetics of suspended-growth denitrification. *Journal of the Water Pollution Control Federation.* 45(2):249.

Stewart, M. J., Ludwig, H. F., and Kearns, W. H. 1962. Effects of varying salinity on the extended aeration process. *Journal of the Water Pollution Control Federation.* 34(11):1161.

Sutton, P. M., Jank, B. E., and Vachon, D. 1980. Nutrient removal in suspended growth systems without chemical addition. *Journal of the Water Pollution Control Federation.* 52(1):98.

Sykes, R. M. 1975. Theoretical heterotrophic yields. *Journal of the Water Pollution Control Federation.* 47(3):591.

Sykes, R. M. 1991. The product-maintenance theory of the activated sludge process. *Journal of Environmental Science and Health, Part A, Environmental Science and Engineering.* A26(6):855.

Technical Advisory Board of the New England Interstate Water Pollution Control Commission. 1980. *Guides for the Design of Wastewater Treatment Works, WT-3* (formerly TR-16). New England Interstate Environmental Training Center, South Portland, ME.

Torpey, W. N. 1948. Practical results of step aeration. *Sewage Works Journal.* 20(5):781.

Ullrich, A. H. and Smith, M. W. 1951. The biosorption process of sewage and waste treatment. *Sewage and Industrial Wastes.* 23(10):1248.

Vacker, D., Connell, C. H., and Wells, W. N. 1967. Phosphate removal through municipal wastewater treatment at San Antonio, Texas. *Journal of the Water Pollution Control Federation.* 39(5):750.

Verstraete, W. and Vissers, W. 1980. Relationship between phosphate stress, effluent quality, and observed cell yield in a pure-oxygen activated-sludge plant. *Biotechnology and Bioengineering.* 22:2591.

Wastewater Committee of the Great Lakes–Upper Mississippi River Board of State Public Health and Environmental Managers. 1990. *Recommended Standards for Wastewater Facilities.* Health Education Services, Albany, NY.

Watson, J. D. 1965. *Molecular Biology of the Gene.* W.A. Benjamin, New York.

Weichers, H.N.S. *et al.*, (ed.) 1984. *Theory, Design and Operation of Nutrient Removal Activated Sludge Processes.* Water Research Commission, Pretoria, South Africa.

Wells, W. W. 1969. Differences in phosphate uptake rates exhibited by activated sludges. *Journal of the Water Pollution Control Federation.* 41(5):765.

Wuhrmann, K. 1954. High-rate activated sludge treatment and its relation to stream sanitation: I. Pilot plant studies. *Sewage and Industrial Wastes.* 26(1):1.

Wuhrmann, K. 1968. Research developments in regard to concept and base values of the activated sludge system. In *Advances in Water Quality Improvement,* ed. E. F. Gloyna and W. W. Eckenfelder, Jr., p. 143. University of Texas at Austin.

Zablatsky, H. R., Cornish, M. S., and Adams, J. K. 1959. An application of the principles of biological engineering to activated sludge treatment. *Sewage and Industrial Wastes.* 31(11):1281.

Section 10.2

ASCE. 1936. *Filtering Materials for Sewage Treatments.* Manual of Engineering Practice No. 13. American Society of Civil Engineers, New York.

ASTM Committee C-9 on Concrete and Concrete Aggregates. 1993. Standard test method for soundness of aggregates by use of sodium sulfate or magnesium sulfate. *Annual Book of ASTM Standards, Sect. 4—Construction,* Vol. *04.02/Concrete and Aggregates. ASTM Designation C88-90.* American Society for Testing and Materials, Philadelphia, PA.

Benjes, H. H., Jr. 1978. Small community wastewater treatment facilities—biological treatment systems. *Design Seminar Handout: Small Wastewater Treatment Facilities.* Environmental Protection Agency, Environmental Research Information Center, Technology Transfer, Cincinnati, OH.

Benzie, W. J., Larkin, H. O., and Moore, A. F. 1963. Effects of climatic and loading factors on trickling filter performance. *Journal of the Water Pollution Control Federation.* 35(4):445.

Blain, W. A. and McDonnell, A. J. 1965. Discussion: "Analysis of biological filter variables" by W. S. Galler and H. B. Gotaas. *Journal of the Sanitary Engineering Division, Proc. ASCE.* 91 (SA4):57.

Bruce, A. M. and Hawkes, H. A. 1983. Biological Filters. In *Ecological Aspects of Used-Water Treatment, Volume 3, The Processes and Their Ecology,* ed. C. R. Curds and H. A. Hawkes, p. 1. Academic Press, New York.

Bruce, A. M. and Merkens, J. C. 1973. Further studies of partial treatment of sewage by high-rate biological filtration. *Water Pollution Control.* 72(5):499.

Ergun, S. 1952. Fluid flow through packed columns. *Chemical Engineering Progress.* 48(2):89.

Eckenfelder, W. W., Jr. 1961. Trickling filtration design and performance. *Journal of the Sanitary Engineering Division, Proc. ASCE.* 87(SA4):33.

Eckenfelder, W. W., Jr., and Barnhart, E. L. 1963. Performance of a high rate trickling filter using selected media. *Journal of the Water Pollution Control Federation.* 35(12):1535.

Fair, G. M., Fuhrman, R. E., Ruchhoft, C. C., Thomas, H. A., Jr., and Mohlman, F. W. 1948. Sewage treatment at military installations—Summary and conclusions, by the NRC Subcommittee on Sewage Treatment. *Sewage Works Journal.* 20(1):52.

Galler, W. S. and Gotaas, H. B. 1964. Analysis of biological filter variables. *Journal of the Sanitary Engineering Division, Proc. ASCE.* 90(SA4):59.

Germain, J. E. 1966. Economical treatment of domestic waste by plastic-medium trickling filters. *Journal of the Water Pollution Control Federation.* 38(2):192.

Gujer, W. and Boller, M. 1986. Design of a nitrifying tertiary trickling filter based on theoretical concepts. *Water Research.* 20(11):1353.

Higbie, R. 1935. The rate of absorption of a pure gas into a still liquid during short periods of exposure. *Transactions of American Institute of Chemical Engineers.* 31(2):365.

Howland, W. E. 1958. Flow over porous media as in a trickling filter. *Proceedings of the Twelfth Industrial Waste Conference,* p. 435. May 13, 14 and 15, 1957. Extension Series No. 94, Engineering Bulletin 42(3). ed. D. E. Bloodgood, Purdue University, West Lafayette, IN.

Joint Task Force of the Water Environment Federation and the American Society of Civil Engineers. 1992. *Design of Municipal Wastewater Treatment Plants: Volume I.* Chapters 1-12. WEF Manual of Practice No. 8; ASCE Manual and Report on Engineering Practice No. 76. Water Environment Federation, Alexandria, VA; American Society of Civil Engineers, New York.

Kehrberger, G. J. and Busch, A. W. 1969. The effects of recirculation on the performance of trickling filter beds. *Proceedings of the 24th Industrial Waste Conference,* p. 37. May 6, 7, and 8, 1969. Engineering Extension Ser. No. 135. ed. D. E. Bloodgood, Purdue University, West Lafayette, IN.

Leva, M. 1953. *Tower Packings and Packed Tower Design,* 2nd ed. United States Stoneware, Akron, OH.

Logan, B. E. 1993. Oxygen transfer in trickling filters. *Journal of Environmental Engineering.* 119(6):1059.

Logan, B. E., Hermanowicz, S. W., and Parker, D. S. 1987. A fundamental model for trickling filter process design. *Journal of the Water Pollution Control Federation.* 59(12):1029.

Lotka, A. J. 1925. *Elements of Physical Biology.* Williams & Wilkins, Baltimore, MD. [Reprinted as *Elements of Mathematical Biology.* 1956. Dover Publications, New York.]

MacDonald, I. F., El-Sayed, M. S., Mow, K., and Dullien, F. A. L. 1979. Flow through porous media—the Ergun equation revisited. *Industrial & Engineering Chemistry Fundamentals.* 18(3):199.

Mehta, D. S., Davis, H. H., and Kingsbury, R. P. 1972. Oxygen theory in biological treatment process design. *Journal of the Sanitary Engineering Division, Proc. ASCE.* 98(SA3):471.

Mohlman, F. W., Thomas, H. A., Jr., Fair, G. M., Fuhrman, R. E., Gilbert, J. J., Heacox, R. E., Norgaard, J. C., and Ruchhoft, C. C. 1946. Sewage treatment at military installations, report of the Subcommittee on Sewage Treatment of the Committee on Sanitary Engineering, National Research Council Division of Medical Science, Washington, D.C. *Sewage Works Journal.* 18(5)791.

Mountfort, L. F. 1924. Correspondence on: A. J. Martin (1924), *The Bio-Aeration of Sewage*, (Paper No. 4490, Dec. 11, 1923). *Minutes of Proceedings of the Institution of Civil Engineers.* vol. CCXVII, session 1923–1924, part I, pp. 190–196.

Namkung, E. and Rittman, B. E. 1986. Soluble microbial products (SMP) formation kinetics by biofilms. *Water Research.* 20(6):795.

Okey, R. W. and Albertson, O. E. 1989. Diffusion's role in regulating rate and masking temperature effects in fixed film nitrification. *Journal of the Water Pollution Control Federation.* 61(4):500.

Parker, D., Lutz, M., Dahl, R., and Bernkopf, S. 1989. Enhancing reaction rates in nitrifying trickling filters through biofilm control. *Journal of the Water Pollution Control Federation.* 61(5):618.

Parker, D. S., Stone, R. W., Stenquist, R. J., and Culp, G. 1975. *Process Design Manual for Nitrogen Control.* Environmental Protection Agency, Technology Transfer, Washington, D.C.

Roberts, J. 1973. Towards a better understanding of high rate biological film flow reactor theory. *Water Research.* 7(11):1561.

Schroeder, E. D. and Tchobanoglous, G. 1976. Mass transfer limitations on trickling filter design. *Journal of the Water Pollution Control Federation.* 48(4):771.

Schroepfer, G. J., Al-Hakim, M. B., Seidel, H. F., and Ziemke, N. R. 1952. Temperature effects on trickling filters. *Sewage and Industrial Wastes.* 22(6):705.

Schulze, K. L. 1960. Load and efficiency of trickling filters. *Journal of the Water Pollution Control Federation.* 32(2):245.

Streeter, H. W. and Phelps, E. B. 1925. *A Study of the Pollution and Natural Purification of the Ohio River.* Public Health Bulletin No. 146. U.S. Public Health Service, Washington, D.C. [Reprinted 1958 by Department of Health, Education, & Welfare, Washington, D.C.]

Suschka, J. 1987. Hydraulic performance of percolating biological filters and consideration of oxygen transfer. *Water Research.* 21(8):865.

Velz, C. J. 1948. A basic law for the performance of biological filters. *Sewage Works Journal.* 20(4):607.

Wastewater Committee of the Great Lakes–Upper Mississippi River Board of State Public Health and Environmental Managers. 1990. *Recommended Standards for Wastewater Facilities.* Health Education Services, Albany, NY.

Wezernak, C. T. and Gannon, J. J. 1967. Oxygen-nitrogen relationships in autotrophic nitrification. *Applied Microbiology.* 15:1211.

Williamson, K. and McCarty, P. L. 1976a. A model of substrate utilization by bacterial films. *Journal of the Water Pollution Control Federation.* 48(1):9.

Williamson, K. and McCarty, P. L. 1976b. Verification studies of the biofilm model for bacterial substrate utilization. *Journal of the Water Pollution Control Federation.* 48(2):281.

Section 10.3

Adams, C. E., Eckenfelder, W. W., Jr., and Stein, R. M. 1974. Modifications to aerobic digester design. *Water Research.* 8:213.

Agardy, F. J., Cole, R. D., and Pearson, E. A. 1963. *Kinetic and Activity Parameters of Anaerobic Fermentation Systems: First Annual Report.* SERL Rept. 63-2. University of California, Sanitary Engineering Research Laboratory, Berkeley, CA.

Brisbin, S. G. 1957. Flow of concentrated raw sewage sludges in pipes. *Journal of the Sanitary Engineering Division, Proceedings of the American Society of Civil Engineers.* 83(SA3):1274.

Brock, T. D. 1970. *Biology of Microorganisms,* 2nd ed. Prentice Hall, Englewood Cliffs, NJ.

Brown and Caldwell, Consulting Engineers and Environmental Technology Consultants. 1979. *Process Design Manual for Sludge Treatment and Disposal.* EPA 625/1-79-011. Environmental Protection Agency, Center for Environmental Research Information, Technology Transfer, Cincinnati, OH.

Buhr, H. O. and Andrews, J. F. 1977. Thermophilic anaerobic digestion process. *Water Research.* 11(2):129.

Buswell, A. W. 1965. Methane fermentation. *Proceedings of the 19th Industrial Waste Conference,* p. 508. May 5, 6, and 7, 1964, Engineering Bulletin, 49(1a,b), ed. D. E. Bloodgood, Purdue University, West Lafayette, IN.

Buswell, A. W. and Boruff, C. B. 1932. The relation between the chemical composition of organic matter and the quality and quantity of gas production during digestion. *Sewage Works Journal.* 4(3):454.

Dean, J. A. (ed.) 1992. *Lange's Handbook of Chemistry,* 14th ed. McGraw-Hill, New York.

Eckenfelder, W. W., Jr. 1956. Studies on the oxidation kinetics of biological sludge. *Sewage and Industrial Wastes.* 28(8):983.

Environmental Protection Agency. 1993. Final rule: Standards for the use or disposal of sewage sludge. *Federal Register.* 58(32):9248.

Environmental Protection Agency. 1994. Final rule: Standards for the use or disposal of sewage sludge [amendments]. *Federal Register.* 59(38):9095.

Fair, G. M. and Moore, E. W. 1932a. Heat and energy relations in the digestion of sewage solids: I. The fuel value of sewage solids. *Sewage Works Journal* 4(2): 242.

Fair, G. M. and Moore, E. W. 1932b. Heat and energy relations in the digestion of sewage solids: II. Mathematical formulation of the course of digestion. *Sewage Works Journal* 4(3):428.

Fair, G. M. and Moore, E. W. 1932c. Heat and energy relations in the digestion of sewage solids: IV. Measurement of heat and energy interchange. *Sewage Works Journal.* 4(5):755.

Fair, G. M. and Moore, E. W. 1934. Time and rate of sludge digestion, and their variation with temperature. *Sewage Works Journal.* 6(1):3.

Fair, G. M. and Moore, E. W. 1937. Observation on the digestion of a sewage sludge over a wide range of temperatures. *Sewage Works Journal.* 9(1):3.

Fan, K.-S. 1983. *Full-Scale Field Demonstration of Unheated Anaerobic Contact Stabilization.* Ph.D. Dissertation. Ohio State University, Columbus, OH.

Garber, W. F. 1954. Plant-scale studies of thermophilic digestion at Los Angeles. *Sewage and Industrial Wastes.* 26(10):1202.

Gram, A. L., III. 1956. *Reaction Kinetics of Aerobic Biological Processes.* Rept. No. 2, I. E. R. Series 90. University of California, Sanitary Engineering Research Laboratory, Berkeley.

Joint Task Force of the Water Environment Federation and the American Society of Civil Engineers. 1992. *Design of Municipal Wastewater Treatment Plants: Volume II.* Chapters 13–20. WEF Manual of Practice No. 8; ASCE Manual and Report on Engineering Practice No. 76. Water Environment Federation, Alexandria, VA; American Society of Civil Engineers, New York.

Kirsch, E. J. and Sykes, R. M. 1971. Anaerobic digestion in biological waste treatment. *Progress in Industrial Microbiology.* 9:155.

Kugelman, I. J. and McCarty, P. L. 1965. Cation toxicity and stimulation in anaerobic digestion. *Journal of the Water Pollution Control Federation.* 37(1):97.

Lawrence, A. W. and McCarty, P. L. 1965. The role of sulfide in preventing heavy metal toxicity in anaerobic treatment. *Journal of the Water Pollution Control Federation.* 37(3):392.

Lawrence, A. W. and McCarty, P. L. 1967. *Kinetics of Methane Fermentation in Anaerobic Waste Treatment.* Tech. Rept. No. 75. Stanford University, Department of Civil Engineering, Stanford, CA.

Leininger, K. V., Sailor, M. K., and Apple, D. K. 1983. *A Survey of Anaerobic Digester Operations.* Final Draft Report. ASCE Task Committee on Design and Operation of Anaerobic Digesters. American Society of Civil Engineers, New York.

Malý, J. and Fadrus, H. 1971. Influence of temperature on anaerobic digestion. *Journal of the Water Pollution Control Federation.* 43(4):641.

McCarty, P. L. 1964. Anaerobic treatment fundamentals: I. Chemistry and microbiology. *Public Works.* 95(9):107.

Metcalf & Eddy, Inc. 1991. *Wastewater Engineering: Treatment, Disposal, and Reuse,* 3rd ed. G. Tchobanglous and F. L. Burton, ed. McGraw-Hill, New York.

Novak, J. T. and Carlson, D. A. 1970. The kinetics of anaerobic long fatty acid degradation. *Journal of the Water Pollution Control Federation.* 42(11):1932.

O'Rourke, J. T. 1968. *Kinetics of Anaerobic Waste Treatment at Reduced Temperatures.* Ph.D. Dissertation. Stanford University, Stanford, CA.

Parkin, G. F. and Owen, W. F. 1986. Fundamentals of anaerobic digestion of wastewater sludge. *Journal of Environmental Engineering.* 112(5):867.

Pohland, F. G. 1962. *General Review of the Literature on Anaerobic Sewage Sludge Digestion.* Engineering Bulletin, 46(5). Purdue University, West Lafayette, IN.

Price, R. H. 1963. *Rate of Methane Production in Acetate-Acclimated Culture Derived from an Anaerobic Digester,* M.S.C.E. Thesis. Purdue University, West Lafayette, IN.

Reynolds, T. D. (1973). Aerobic digestion of thickened waste activated sludge. *Proceedings of the 28th Industrial Waste Conference,* p. 12. May 1, 2, and 3, 1973. Engineering Extension Series No. 142. J. M. Bell, ed. Purdue University, West Lafayette, IN.

Schlegel, H. G. 1992. *General Microbiology,* 7th ed. M. Kogut, trans. Cambridge University Press, Cambridge, UK.

Schwinn, D. E. and Gassett, R. B., ed. 1974. *Process Design Manual for Upgrading Existing Wastewater Treatment Plants.* Environmental Protection Agency, Technology Transfer, Washington, D.C.

Shea, T. G., Pretorius, W. A., Cole, R. D., and Pearson, E. A. 1968. *Kinetics of Hydrogen Assimilation in Methane Fermentation.* SERL Rept. No. 68-7. University of California, Sanitary Engineering Research Laboratory, Berkeley.

Speece, R. E. and McCarty, P. L. 1964. Nutrient requirements and biological solids accumulation in anaerobic digestion: Advances in water pollution research. *Proceedings of the International Conference, London, September, 1962, Vol. II.* W. W. Eckenfelder, Jr., ed. Pergamon Press, New York.

Stewart, M. J. 1958. *Reaction Kinetics and Operational Parameters of Continuous-Flow Anaerobic-Fermentation Processes.* Rept. No. 4, I. E. R. Series 90. University of California, Sanitary Engineering Research Laboratory, Berkeley.

Sykes, R. M. 1970. *Hydrogen Production in the Anaerobic Digestion of Sewage Sludge.* Ph.D. Dissertation. Purdue University, West Lafayette, IN.

Van Wylen, G. J. 1963. *Thermodynamics.* John Wiley & Sons, New York.

Wastewater Committee of the Great Lakes–Upper Mississippi River Board of State Public Health and Environmental Managers. 1990. *Recommended Standards for Wastewater Facilities.* Health Education Services, Albany, NY.

Whitman, W. B. 1985. Methanogenic bacteria. In *The Bacteria: A Treatise on Structure and Function—Vol. VIII. The Archaebacteria,* ed. C. R. Woese and R. S. Wolfe, p. 3. Academic Press, New York.

Whitman, W. B., Bowen, T. L., and Boone, D. R. 1992. The Methanogenic Bacteria. In *The Prokaryotes: A Handbook on the Biology of Bacteria—Ecophysiology, Isolation, Identification, Applications—Volume I,* ed. A. Balows, H. G. Truper, M. Dworkin, W. Harder, and K.-H. Schliefer, p. 719. Springer-Verlag, New York.

Woods, C. E. and Malina, J. F., Jr. 1965. Stage digestion of wastewater sludge. *Journal of the Water Pollution Control Federation.* 37(11):1495.

Wujcik, W. J. 1980. *Dry Anaerobic Fermentation to Methane of Organic Residues.* Ph. D. Dissertation. Cornell University, Ithaca, NY.

Zwietering, T. N. 1958. Suspending of solid particles in liquid by agitators. *Chemical Engineering Science.* 8: 244.

11

Air Pollution

Robert B. Jacko
Purdue University

Robert W. Holden II
Purdue University

11.1 Introduction

The quality of the ambient air is an issue that is a common denominator among all people throughout the world. This statement is based on the simple fact that to live everyone must breathe. Despite this fact, air quality is an issue that has been historically ignored until it deteriorates to a point where breathing is uncomfortable or even to where life itself is threatened. However, this approach to air quality is changing rapidly as no aspect of the environment has recently received greater attention than that of air pollution and its effects on our health and well-being.

In the U.S. this attention is illustrated by the Clean Air Act Amendments of 1990. This legislation is being heralded as the most comprehensive piece of environmental legislation ever enacted. The scope of this legislation's effects can be illustrated by the cost of compliance with its provisions. The cost of compliance, by industry, is being estimated at 25 to 30 billion dollars annually. The level of attention the issue of air quality management is receiving is indeed significant, but the field is often misunderstood.

The focus of this chapter is to provide a synopsis of the various aspects involved in air quality engineering and management. The chapter will begin by presenting an overview of the major air quality regulations and pollutants of concern. This discussion will be followed by descriptions of methods used in estimating and quantifying emissions, methods of controlling typical emission sources, a discussion of the meteorology affecting dispersion of emitted pollutants, and conclude with a discussion of the models used to estimate the effects of emission of pollutants on the ambient atmosphere.

0-8493-8953-4/95/$0.00 + $.50

11.2 Regulations

Historical Perspective

The regulation of air pollution has evolved from a level of local ordinances in the late 1800s to the federally driven regulatory efforts of today. In 1881, Cincinnati and Chicago became the first American cities to pass smoke control ordinances. This type of local ordinance was the primary means of air quality regulation until the federal government began addressing the issue with the passage of the Air Pollution Control–Research and Technical Assistance Act in 1955. However, this act was not a means of federal regulation, but only a means of providing funds for federal research and technical assistance for an issue that, at the time, was felt to be a state and local problem.

In 1963, the president of the U.S. pushed for the passage of the first Clean Air Act. At that time Congress recognized that air pollution "resulted in mounting dangers to the public health and welfare, including injury to agricultural crops and livestock, damage to and deterioration of property, and hazards to air and ground transportation" [Cooper and Alley, 1990]. This act was the first to address interstate air pollution problems. Further regulations on air pollution were introduced in 1965 with the first set of amendments to the Clean Air Act. These amendments were divided into two provisions addressing air pollution prevention and air pollution resulting from motor vehicles. This act set a national standard for emissions from automobiles to prevent automobile manufacturers from having to comply with 50 different sets of emission standards.

Regulation of ambient air quality was first addressed with the Air Quality Act of 1967. This act was also significant in that for the first time the federal government was granted enforcement authority and was required to develop and promulgate air quality criteria based on scientific studies.

The foundations of the air quality regulations that are in effect today were laid in 1970 with the second set of amendments to the Clean Air Act. These amendments grouped areas of the country into two classes based on the quality of their ambient air in relation to established standards. Separate regulations were developed to apply to the areas based on the air quality in that particular area. This set of amendments also set a time frame in which the areas of the country not in compliance with established ambient air standards would come into compliance with these standards. The authority of the federal government over air quality issues took a giant step forward with this act, and a giant leap forward when this act was coupled with the National Environmental Policy Act that established the Environmental Protection Agency (U.S. EPA) in 1970. This provided for air quality regulation that could be developed and managed at the federal level but implemented by the individual states.

Despite the new level of federal enforcement over the Clean Air Act, the deadlines for compliance with the ambient air standards were not met and in 1977, the Clean Air Act was amended for the third time. The Amendments of 1977 took a proactive stance toward ambient air quality with provisions to prevent areas currently meeting ambient air standards from deteriorating, while at the same time requiring those areas not in compliance with ambient air standards to come into compliance. The amendments of 1977 further required review of air quality criteria and regulations every five years by the U.S. EPA.

Again, despite the new regulations, air quality did not improve. However, federal regulatory efforts plateaued until 15 November 1990 when the Clean Air Act was revised for the fourth time and created the air quality regulations in effect today.

Regulatory Overview

Air Pollution Sources

Air pollution is defined as the intentional or unintentional release of various compounds into the atmosphere. These compounds consist of both gases (vapors and fumes) and solids (particulates and aerosols) which can be emitted from natural and/or human sources. Typically, pollution arising from human sources, such as manufacturing and automobiles, far outweighs the contribution of compounds arising from natural sources, such as volcanoes, forest fires, and decay of natural compounds [Environmental Resources Management, 1992].

When evaluating the regulatory effects of the emission of various pollutants, the source of the pollutant is always considered. However, the term *source* takes on several meanings when used in different situations. In this chapter, it will be used to relate to human sources that are stationary in nature.

There are two types of stationary sources that must be considered when addressing emissions: point and nonpoint sources. Point sources include such things as stacks, vents, and other specific points where gas streams are designed to be emitted. Nonpoint sources, or fugitive or secondary sources, include releases of compounds from leaking valves, flanges, and pumps, or release of compounds from wastewater treatment plants [Environmental Resources Management, 1992].

Regulation of Ambient Air Quality

National Ambient Air Quality Standards. National Ambient Air Quality Standards (NAAQS) have been established for criteria pollutants. These consist of six primary pollutants and one secondary pollutant. The six primary criteria pollutants, or pollutants that are emitted directly to the atmosphere, are carbon monoxide (CO), nitrogen oxides (NO_x), particulates, sulfur oxides (SO_x), volatile organic compounds (VOCs), and lead. The secondary criteria pollutant is ground-level ozone, and is called a secondary pollutant because it is formed through photochemical reactions between VOCs, NO_x, and sunlight. Therefore, ground-level ozone is not emitted directly to the atmosphere, but formed only after its precursors have been emitted and photochemically react.

NAAQS were set by the U.S. EPA based on two criteria: primary standards for the protection of human health and secondary standards for the protection of the public well-being (such as vegetation, livestock, and other items that can be related to nonhealth effects). These standards differ in that the primary standards are designed to directly protect human health, while the secondary standards are designed to protect the quality of life. Table 11.1 lists the NAAQS for each of the criteria pollutants and the time frame the standard is applied over. For further definitions of a regulated air pollutant the reader is advised to contact a state environmental regulatory office for the most current definitions from the U.S. EPA.

Attainment and Nonattainment. The U.S. EPA monitors concentrations of the criteria pollutants through a national monitoring network. If the monitoring data show that the NAAQS levels have

Table 11.1 National Ambient Air Quality Standards

Criteria Pollutant	Averaging Period	Primary NAAQS ($\mu g/m^3$)	Secondary NAAQS ($\mu g/m^3$)
PM-10	Annual	50	50
	24 hours	150	150
Sulfur dioxide (SO_2)	Annual	80	
	24 hours	365	
	3 hours		1,300
Nitrogen dioxide (NO_2)	Annual	100	100
Ozone	1 hour	235	235
Carbon monoxide (CO)	8 hours	10,000	10,000
	1 hour	40,000	40,000
Lead	Quarterly	1.5	1.5

Table 11.2 Ozone Nonattainment Area Classifications

Ozone Concentration (ppm)	Nonattainment Classification
0.120–0.138	Marginal
0.139–0.160	Moderate
0.161–0.180	Serious
0.181–0.280	Severe
Above 0.280	Extreme

been exceeded then that area of the country is in nonattainment. If the monitoring shows that NAAQS levels have not been exceeded then the area is in attainment.

The attainment/nonattainment designation applies to each criteria pollutant. As a result an area may have exceeded the NAAQS and be in nonattainment for ozone, but may not have exceeded the NAAQS for SO_2 and is therefore still an attainment area for SO_2.

With the implementation of the Clean Air Act Amendments of 1990, the nonattainment provisions were amended to expand nonattainment designations based on the air quality in the area. While the previous regulations only considered areas to be attainment or nonattainment, the new regulations have established classes of nonattainment that range from marginal to extreme. Table 11.2 lists the new definitions of nonattainment for ground-level ozone. The new regulations also include differing requirements for areas in various classes of nonattainment in an effort to bring these areas into compliance with the NAAQS.

Perhaps the most common requirement that will affect the greatest number of sources is the requirement for the installation of Reasonably Available Control Technology (RACT). RACT will be defined at the state level of enforcement authority and will be a specified level of control for major sources in nonattainment areas.

Regulation of Emission Rates

The NAAQS set acceptable concentrations for pollutants in the ambient atmosphere but do not enforce emission rates for sources such that these levels are met. Regulation of emission rates from stationary sources to control ambient concentrations arise from four programs: Prevention of Significant Deterioration (PSD), New Source Review (NSR), New Source Performance Standards (NSPS), and Hazardous Air Pollutants (HAPs). Each is described below.

Prevention of Significant Deterioration. When the Clean Air Act was amended in 1977, provisions were included to prevent areas in attainment with the NAAQS from being polluted up to the level of the NAAQS. These provisions are the major regulatory program for attainment areas and are known as the PSD provisions. PSD regulates new major sources and major modifications to existing sources.

Under PSD, a major source is defined as a source that has the potential to emit more than 100 tons per year (tpy) if the source is one of the 28 listed sources in the program, or has the potential to emit 250 tpy if the source is not among the listed sources. A major modification is the expansion of an existing source that increases emissions beyond a specific de-minimis amount.

Any new source that is regulated by PSD must apply for a PSD permit prior to beginning construction. A PSD permit application requires the preparation of an extensive amount of information on not only the process but also the impacts of the project [Environmental Resources Management, 1992]. To comply with the PSD provisions, an applicant must demonstrate the use of Best Available Control Technology (BACT) and demonstrate that the project will have no adverse effects on ambient air quality through ambient monitoring and/or dispersion modeling.

BACT specifies a level of emissions control a process must have. BACT can be a piece of add-on control equipment such as a catalytic incinerator or baghouse, or can involve process modifications or work-practice standards such as the use of water-based paints as opposed to solvent-based paints, or ensuring solvent storage tanks are covered when not in use. A control technology review is done in the preparation of a PSD permit to determine what other, similar sources have used as a BACT level of control. This ensures that suggested BACT is at least as effective as what has been previously used.

New Source Review. The NSR provisions were established at the same time as the PSD provisions and regulate new major sources and major modifications in nonattainment areas. The NSR provisions are more stringent than the PSD provisions. The goal of the NSR program is to improve the ambient air quality in areas that do not meet the NAAQS.

NSR requires that each new major source or major modification install a Lowest Achievable Emission Rate (LAER) level of emissions control, obtain emissions offsets equal to the source's emission rate plus a penalty for cleaner air, and investigate alternate sites for the proposed expansion. Unlike the PSD provisions, a major source under NSR depends on the classification of the nonattainment area the source is to be constructed in. For example, a major source in an extreme ozone nonattainment area is any source emitting more than 10 tpy of VOCs. However, in a moderate nonattainment area a major source is any source emitting more than 100 tpy of VOCs.

Under NSR, a LAER level of control is required to be installed. This level of control is similar to BACT in that it is at least as stringent, but often is more stringent and is related to process modification. The LAER level of control is determined on a case-by-case basis as is BACT emissions control.

Emissions offsets are also required under the NSR provisions. Emissions offsets are a method of lowering total emissions in a nonattainment area by requiring new sources to first reduce emissions from an existing operating source. This is done by a ratio such that for the total new emissions a greater amount of existing emissions will be offset or eliminated. This reduction in existing emissions can result from adding new controls on existing sources, shutting existing sources down, or purchasing "banked" offsets from another company that has previously shut a source down and has documented these emissions.

New Source Performance Standards. NSPS are based on the premise that new sources should be able to operate with lesser amounts of emissions than older sources. As a result, the NSPS establish emission rates for specific pollutants for specific sources that have been constructed since 1971. NSPS standards have been established for 64 different types of sources.

Hazardous Air Pollutants. The emission of HAPs was originally regulated in 1970 when Congress authorized the U.S. EPA to establish standards for HAPs not regulated under the NAAQS. The National Emission Standards for Hazardous Air Pollutants (NESHAP) program was developed for this purpose. However, this program was ineffective and managed to regulate only seven hazardous compounds by 1990: asbestos, benzene, mercury, beryllium, vinyl chloride, arsenic, and radionuclides.

In 1990, a new HAPs program was established to regulate a new list of 189 hazardous compounds. Table 11.3 lists the 189 regulated HAPs. This program will regulate major sources of HAPs by requiring the initial installation of Maximum Achievable Control Technology (MACT) and following this with the requirement for facilities to address residual risk after the application of MACT.

A major source for the emission of HAPs is any source with the potential to emit 10 tpy of any single HAP or 25 tpy of any combination of three or more HAPs. MACT for these sources will be promulgated by the U.S. EPA with emissions limits for categories for industrial sources.

State and Local Air Quality Programs. In addition to the federal air quality programs described above, many state and local governments have their own air quality regulations. These regulations are required to be at least as stringent as the federal programs; many are far more stringent.

Table 11.3 Hazardous Air Pollutants

CAS Number	Chemical Name	CAS Number	Chemical Name
75070	Acetaldehyde	123911	1,4-Dioxane (1,4-Diethyleneoxide)
60355	Acetamide	122667	1,2-Diphenylhydrazine
75058	Acetonitrile	106898	Epichlorohydrin (1-Chloro-2,3-epoxypropane)
98862	Acetophenone	106887	1,2-Epoxybutane
53963	2-Acetylaminofluorene	140885	Ethyl acrylate
107028	Acrolein	100414	Ethylbenzene
79061	Acrylamide	51796	Ethyl carbamate (Urethane)
79107	Acrylic acid	75003	Ethyl chloride (Chloroethane)
107131	Acrylonitrile	106934	Ethylene dibromide (Dibromoethane)
107051	Allyl chloride	107062	Ethylene dichloride (1,2-Dichloroethane)
92671	4-Aminobiphenyl	107211	Ethylene glycol
62533	Aniline	151564	Ethyleneimine (Aziridine)
90040	*o*-Anisidine	75218	Ethylene oxide
1332214	Asbestos	96457	Ethylene thiourea
71432	Benzene (including benzene from gasoline)	75343	Ethylidene dichloride (1,1-Dichloroethane)
92875	Benzidine	50000	Formaldehyde
98077	Benzotrichloride	76448	Heptachlor
100447	Benzyl chloride	118741	Hexachlorobenzene
92524	Biphenyl	87683	Hexachlorobutadiene
117817	Bis(2-ethylhexyl)phthalate (DEHP)	77474	Hexachlorocyclopentadiene
542881	Bis(chloromethyl)ether	67721	Hexachloroethane
75252	Bromoform	822060	Hexamethylene-1,6-diisocyanate
106990	1,3-Butadiene	680319	Hexamethylphosphoramide
156627	Calcium cyanamide	110543	Hexane
105602	Caprolactam	302012	Hydrazine
133062	Captan	7647010	Hydrochloric acid (hydrogen chloride) (gas only)
63252	Carbaryl	7664393	Hydrogen fluoride (Hydrofluoric acid)
75150	Carbon disulfide	123319	Hydroquinone
56235	Carbon tetrachloride	78591	Isophorone
463581	Carbonyl sulfide		1,2,3,4,5,6-Hexachlorocyclohexane (all stereoisomers, including lindane)
120809	Catechol		
133904	Chloramben	108316	Maleic anhydride
57749	Chlordane	67561	Methanol
7782505	Chlorine	72435	Methoxychlor
79118	Chloroacetic acid	74839	Methyl bromide (Bromomethane)
532274	2-Chloroacetophenone	74873	Methyl chloride (Chloromethane)
108907	Chlorobenzene	71556	Methyl chloroform (1,1,1-Trichloroethane)
510156	Chlorobenzilate	78933	Methyl ethyl ketone (2-Butanone)
67663	Chloroform	60344	Methylhydrazine
107302	Chloromethyl methyl ether	74884	Methyl iodide (Iodomethane)
126998	Chloroprene	108101	Methyl isobutyl ketone (Hexone)
1319773	Cresols/Cresylic acid (isomers and mixture)	624839	Methyl isocyanate
95487	*o*-Cresol	80626	Methyl methacrylate
108394	*m*-Cresol	1634044	Methyl tert-butyl ether
106445	*p*-Cresol	101144	4,4′-Methylenebis(2-chloroaniline)
98828	Cumene	75092	Methylene chloride (Dichloromethane)
94757	2,4-D (2,4-Dichlorophenoxyacetic acid, including salts and esters)	101688	4-4′ Methylenediphenyl diisocyanate (MDI)
		101779	4,4,-Methylenedianiline
2559	DDE (1,1-dichloro-2,2-bis(*p*-chlorophenyl) ethylene)	91203	Naphthalene
		98953	Nitrobenzene
334883	Diazomethane	92933	4-Nitrobiphenyl
132649	Dibenzofuran	100027	4-Nitrophenol
96128	1,2-Dibromo-3-chloropropane	79469	2-Nitropropane
84742	Dibutylphthalate	684935	N-Nitroso-N-methylurea
106467	1,4-Dichlorobenzene	62759	N-Nitrosodimethylamine
91941	3,3-Dichlorobenzidene	59892	N-Nitrosomorpholine
111444	Dichloroethyl ether (Bis(2-chloroethyl)ether)	56382	Parathion
542756	1,3-Dichloropropene	82688	Pentachloronitrobenzene (Quintobenzene)
62737	Dichlorvos	87865	Pentachlorophenol
111422	Diethanolamine	108952	Phenol
121697	N,N-Dimethylaniline	106503	*p*-Phenylenediamine
64675	Diethyl sulfate	75445	Phosgene
119904	3,3′-Dimethoxybenzidine	7803512	Phosphine
60117	Dimethyl aminoazobenzene	7723140	Phosphorus
119937	3,3′,-Dimethylbenzidine	85449	Phthalic anhydride
79447	Dimethylcarbamoyl chloride	1336363	Polychlorinated biphenyls (Aroclors)
68122	N,N-Dimethylformamide	1120714	1,3-Propane sulfone
57147	1,1-Dimethylhydrazine	57578	β-Propiolactone
131113	Dimethyl phthalate	123386	Propionaldehyde
77781	Dimethyl sulfate	114261	Propoxur (Baygon)
	4,6-Dinitro-*o*-cresol, and salts	78875	Propylene dichloride (1,2-Dichloropropane)
51285	2,4-Dinitrophenol	75569	Propylene oxide
121142	2,4-Dinitrotoluene	75558	1,2-Propylenimine (2-Methyl aziridine)

Table 11.3 (continued) Hazardous Air Pollutants

CAS Number	Chemical Name	CAS Number	Chemical Name
91225	Quinoline	75014	Vinyl chloride
106514	Quinone	75354	Vinylidene chloride (1,1-Dichloroethylene)
100425	Styrene	1330207	Xylenes (isomers and mixture)
96093	Styrene oxide	95476	*o*-Xylene
1746016	2,3,7,8-Tetrachlorodibenzo-*p*-dioxin	108383	*m*-Xylene
79345	1,1,2,2-Tetrachloroethane	106423	*p*-Xylene
127184	Tetrachloroethylene (Perchloroethylene)	0	Antimony compounds
7550450	Titanium tetrachloride	0	Arsenic compounds (inorganic including arsine)
108883	Toluene	0	Beryllium compounds
	2,4-Toluenediamine	0	Cadmium compounds
584849	2,4-Toluene diisocyanate	0	Chromium compounds
95534	*o*-Toluidine	0	Cobalt compounds
8001352	Toxaphene (chlorinated camphene)	0	Coke oven emissions
120821	1,2,4-Trichlorobenzene	0	Cyanide compounds[1]
79005	1,1,2-Trichloroethane	0	Glycol ethers[2]
79016	Trichloroethylene	0	Lead compounds
95954	2,4,5-Trichlorophenol	0	Manganese compounds
88062	2,4,6-Trichlorophenol	0	Mercury compounds
121448	Triethylamine	0	Fine mineral fibers[3]
1582098	Trifluralin	0	Nickel compounds
540841	2,2,4-Trimethylpentane	0	Polycyclic organic matter[4]
108054	Vinyl acetate	0	Radionuclides (including radon)[5]
593602	Vinyl bromide	0	Selenium compounds

Note: For all listings above which contain the word "compounds" and for glycol ethers, the following applies: Unless otherwise specified, these listings are defined as including any unique chemical substance that contains the named chemical (antimony, arsenic, etc.) as part of that chemical's infrastructure.

[1]X′CN where X = H′ or any other group where a formal dissociation may occur. For example, KCN or $Ca(CN)_2$.

[2]Includes mono- and di-ethers of ethylene glycol, diethylene glycol, and triethylene glycol R—(OCH2CH2)$_n$—OR′ where $n = 1, 2, \text{ or } 3$, R = akyl or aryl groups, R′ = R, H, or groups which, when removed, yield glycol ethers with the structure: R—(OCH2CH2)$_n$—OH. Polymers are excluded from the glycol category.

[3]Includes mineral fiber emissions from facilities manufacturing or processing glass, rock, or slag fibers (or other mineral derived fibers) of average diameter 1 micrometer or less.

[4]Includes organic compounds with more than one benzene ring, and which have a boiling point greater than or equal to 100°C. Limited to, or refers to, products from incomplete combustion of organic compounds (or material) and pyrolysis processes having more than one benzene ring, and which have a boiling point greater than or equal to 100°C.

[5]A type of atom which spontaneously undergoes radioactive decay.

One of the most common requirements at the state and local levels of regulation is the requirement for all new sources, regardless of size, to obtain an air pollution construction permit prior to beginning construction of the source. In many cases, small sources are determined to be exempt from the permitting requirements, or are merely given registration status as opposed to a full construction and operating permit. Nonetheless, even small sources are required to give notification prior to beginning construction or face serious penalties for not doing so.

11.3 Emissions Estimation

Estimation of emissions from a source is a process which involves the qualification and quantification of pollutants that are generated by the source. This process is of paramount importance as the emission estimates will be used to describe the applicability of various regulations, and thus influence how the source is constructed and operated.

To begin the process of estimation, the source must be reviewed to determine its size and nature. This includes the quantification of all raw material inputs (existing or planned), production steps, and release points to qualify what types of emissions might possibly exist. In this step, review of similar sources is imperative as this information can provide a vast array of information that is easily overlooked. The reader is referred to the Air and Waste Management Association's *Air Pollution Engineering Manual* or the U.S. EPA's AP-40 as references for the review of similar sources. These texts provide an overview of a variety of industrial processes and the types and quantities of emissions they generate and emissions controls they employ.

After the source has been reviewed and the potential emissions qualified, the process of assessing the quantities of pollutants that are or can be emitted can begin.

Typically, emissions estimates are generated in two ways: by mass balance or by the use of emission factors. A mass balance is a process based on the fact that because mass is neither created nor destroyed, the mass of raw material into an operation must come out of the operation somewhere. From this, the amount of raw materials into the operation can be quantified and proportioned as to their amounts in either the finished product or in a waste stream. As a result, the portion of the raw materials released into the air can be quantified. However, an appropriate mass balance is a difficult task as there are typically many different raw material inputs into a facility, making the process very complex. Further, many types of raw material inputs result in emissions that are not readily apparent. Because of these factors, the mass balance approach to estimating emissions is not recommended.

The second method of quantifying emissions is the use of an emission factor. An emission factor is a relation between a common operation and the average emissions it generates. For example, the combustion of 1,000,000 standard cubic feet of natural gas in a small industrial boiler results in the emission of 35 pounds of CO. Therefore, the emission factor would be 35 lb CO/10^6 scf of natural gas combusted. Emission factors are generated simply by relating emissions to a representative operating variable, and are often a single number that is the weight of pollutant divided by a unit weight of the activity that generates the pollutant [U.S. EPA, 1993]. Further, when operating variables become complex or contain a number of variables, the emission factor might consist of a series of equations encompassing the variables to determine emissions.

The use of emission factors is common, and the U.S. EPA generates emission factors for almost every conceivable process. These factors are based on statistical averages of all acceptable data from a variety of sources. These emission factors are published in a manual entitled the *Compilation of Air Pollutant Emission Factors*, or AP-42, available from the National Technical Information Service (NTIS). Examples of typical emission factors are given in Tables 11.4, 11.5, and 11.6.

The use of emissions factors is relatively simple and typically consists of the process of unit cancellation once the quantity of operational variable has been determined. For example, the

Table 11.4 Emission Factors for Particulate Matter (PM) from Natural Gas Combustion[a,b]

Combustor Type (Size, 10^6 Btu/hr heat input) [SCC][c]	Filterable PM[d]			Condensible PM[e,f]		
	kg/10^6 m^3	lb/10^6 ft^3	Rating	kg/10^6 m^3	lb/10^6 ft^3	Rating
Utility/large industrial boilers (>100) [10106001, 10100604]	16–80	1–5	B	ND[g]	ND	
Small industrial boilers (10–100) [10200602]	99	6.2	B	120	7.5	D
Commercial boilers (0.3–<10) [10300603]	72	4.5	C	120	7.5	C
Residential furnaces (<0.3) [no SCC]	2.8	0.18	C	180	11	D

[a] Expressed as weight pollutant/volume natural gas fired. All factors represent uncontrolled emissions.

[b] Based on an average natural gas higher heating value of 8270 kcal/m^3 (1000 Btu/scf). The emission factors in this table may be converted to other natural gas heating values by multiplying the given emission factor by the ratio of the specified heating value to this average heating value.

[c] SCC = Source Classification Code.

[d] Filterable PM is that particulate matter collected on or prior to the filter of an EPA method 5 (or equivalent) sampling train.

[e] Condensible PM is that particulate matter collected in the impinger portion of an EPA method 5 (or equivalent) sampling train.

[f] Total PM is the sum of the filterable PM and condensible PM. All PM emissions can be assumed to be less than 10 microns in aerodynamic equivalent diameter.

[g] ND = no data.

Source: U.S. EPA. 1993. *Compilation of Air Pollutant Emission Factors.* Research Triangle Park.

Table 11.5 Emission Factors for Sulfur Dioxide (SO_2), Nitrogen Oxides (NO_x), and Carbon Monoxide (CO) from Natural Gas Combustion[a,b]

Combustor Type (Size, 10^6 Btu/hr heat input) [SCC][c]	SO_2[d]			NO_x[e]			CO		
	kg/10^6 m^3	lb/10^6 ft^3	Rating	kg/10^6 m^3	lb/10^6 ft^3	Rating	kg/10^6 m^3	lb/10^6 ft^3	Rating
Utility/large industrial boilers (>100) [10100601, 10100604]									
Uncontrolled	9.6	0.6	A	8800	550[f]	A	640	40	A
Controlled—low NO_x burners	9.6	0.6	A	1300	81[f]	D	ND[g]	ND	
Controlled—flue gas recirculation	9.6	0.6	A	850	53[f]	D	ND	ND	
Small industrial boilers (10–100) [10200602]									
Uncontrolled	9.6	0.6	A	2240	140	A	560	35	A
Controlled—low NO_x burners	9.6	0.6	A	1300	81[f]	D	980	61	D
Controlled—flue gas recirculation	9.6	0.6	A	480	30	C	590	37	C
Commercial boilers (0.3–<10) [10300603]									
Uncontrolled	9.6	0.6	A	1600	100	B	330	21	C
Controlled—low NO_x burners	9.6	0.6	A	270	17	C	425	27	C
Controlled—flue gas recirculation	9.6	0.6	A	580	36	D	ND	ND	
Residential furnaces (<0.3) [no SCC]									
Uncontrolled	9.6	0.6	A	1500	94	B	640	40	B

[a] Expressed as weight pollutant/volume natural gas fired.

[b] Based on an average natural gas higher heating value of 8270 kcal/m^3 (1000 Btu/scf). The emission factors in this table may be converted to other natural gas heating values by multiplying the given emission factor by the ratio of the specified heating value to this average heating value.

[c] SCC = Source Classification Code.

[d] Based on average sulfur content of natural gas, 4600 g/10^6 Nm^3 (2000 gr/10^6 scf).

[e] Expressed as NO_2. For tangentially fired units, use 4400 kg/10^6 m^3 (275 lb/10^6 ft^3). At reduced loads, multiply factor by load reduction coefficient. Note that NO_x emissions from controlled boilers will be reduced at low load conditions.

[f] Emission factors apply to packaged boilers only.

[g] ND = no data.

Source: U.S. EPA. 1993. *Compilation of Air Pollutant Emission Factors.* Research Triangle Park.

Table 11.6 Emission Factors for Carbon Dioxide (CO_2) and Total Organic Compounds (TOC) from Natural Gas Combustion[a,b]

Combustor Type (Size, 10^6 Btu/hr heat input) [SCC][c]	CO_2			TOC		
	kg/10^6 m^3	lb/10^6 ft^3	Rating	kg/10^6 m^3	lb/10^6 ft^3	Rating
Utility/large industrial boilers (>100) [10100601, 10100604]	ND[g]	ND		28[f]	1.7[d]	C
Small industrial boilers (10–100) [10200602]	1.9E06	1.2E05	D	92[g]	5.8[e]	C
Commercial boilers (0.3–<10) [10300603]	1.9E06	1.2E05	C	92[h]	5.8[f]	C
Residential furnaces (<0.3) [no SCC]	2.0E06	1.3E05	D	180[h]	11[h]	D

[a] Expressed as weight pollutant/volume natural gas fired. All factors represent uncontrolled emissions.
[b] Based on an average natural gas higher heating value of 8270 kcal/m^3 (1000 Btu/scf). The emission factors in this table may be converted to other natural gas heating values by multiplying the given emission factor by the ratio of the specified heating value to this average heating value.
[c] SCC = Source Classification Code.
[d] Methane comprises 17% of organic compounds.
[e] Methane comprises 52% of organic compounds.
[f] Methane comprises 34% of organic compounds.
[g] ND = no data.
Source: U.S. EPA. 1993. *Compilation of Air Pollutant Emission Factors.* Research Triangle Park.

determination of annual emissions from a 20 MMBtu/hr (MMBtu denotes a million British thermal units) natural-gas-fired boiler consists of the following process.

Example 11.1 Natural gas higher heating value (HHV) = 1,000 Btu/scf. Therefore, the boiler uses

$$\left(\frac{20{,}000{,}000\ \text{Btu/hr}}{1000\ \text{Btu/ft}^3}\right) = 20{,}000\ \text{ft}^3\text{/hr} \tag{11.1}$$

or for operation over 8760 hours/year, annual use is

$$20{,}000\frac{\text{ft}^3}{\text{hr}} \times 8760\frac{\text{hr}}{\text{year}} = 1.752 \times 10^8\frac{\text{ft}^3}{\text{year}} \tag{11.2}$$

Thus, the operational variable for the emission factor has been quantified for annual use. Now, turning to Tables 11.4 through 11.6, the emissions factors are used for a small, uncontrolled industrial boiler. Emissions are determined in the following manner. For filterable particulate matter, the emission factor = 6.2 lb/10^6 ft^3 (Table 11.4):

$$\frac{6.2\ \text{lb}}{10^6\ \text{ft}^3} \times \left(\frac{1.752 \times 10^8\ \text{ft}^3\text{/year}}{1{,}000{,}000\ \text{ft}^3/10^6\text{ft}^3}\right) = 1086.2\ \text{lb part./year} \tag{11.3}$$

For condensible particulate matter, the emission factor = 7.5 lb/10^6 ft^3 (Table 11.4):

$$\frac{7.5\ \text{lb}}{10^6\ \text{ft}^3} \times \left(\frac{1.752 \times 10^8\ \text{ft}^3\text{/year}}{1{,}000{,}000\ \text{ft}^3/10^6\ \text{ft}^3}\right) = 1314\ \text{lb part./year} \tag{11.4}$$

For sulfur dioxide, the emission factor = 0.6 lb/10^6 ft^3 (Table 11.5):

$$\frac{0.6\ \text{lb}}{10^6\ \text{ft}^3} \times \left(\frac{1.752 \times 10^8\ \text{ft}^3\text{/year}}{1{,}000{,}000\ \text{ft}^3/10^6\text{ft}^3}\right) = 105.1\ \text{lb SO}_2\text{/year} \tag{11.5}$$

For nitrogen oxides, the emission factor $= 140\ \text{lb}/10^6\ \text{ft}^3$ (Table 11.5):

$$\frac{140\ \text{lb}}{10^6\ \text{ft}^3} \times \left(\frac{1.752 \times 10^8\ \text{ft}^3/\text{year}}{1{,}000{,}000\ \text{ft}^3/10^6\ \text{ft}^3}\right) = 24{,}528\ \text{lb NO}_x/\text{year} \tag{11.6}$$

For carbon monoxide, the emission factor $= 35\ \text{lb}/10^6\ \text{ft}^3$ (Table 11.5):

$$\frac{35\ \text{lb}}{10^6\ \text{ft}^3} \times \left(\frac{1.752 \times 10^8\ \text{ft}^3/\text{year}}{1{,}000{,}000\ \text{ft}^3/10^6\ \text{ft}^3}\right) = 6132\ \text{lb CO/year} \tag{11.7}$$

For total organic carbon, the emission factor $= 5.8\ \text{lb}/10^6\ \text{ft}^3$:

$$\frac{5.8\ \text{lb}}{10^6\ \text{ft}^3} \times \left(\frac{1.752 \times 10^8\ \text{ft}^3/\text{year}}{1{,}000{,}000\ \text{ft}^3/10^6\ \text{ft}^3}\right) = 1016\ \text{lb TOC/year} \tag{11.8}$$

or, because the TOC emission factor encompasses methane, which is not a VOC, the emission quantity can be corrected to obtain VOC emissions. This is done by the fact that the emission factor documentation (Table 11.6) reports that methane comprises 52% of the emissions. Therefore, the VOC portion of the TOC is 48% of the total. This may be determined by

$$1016\frac{\text{lb}}{\text{year}}\text{TOC} \times 0.48 = 487.8\frac{\text{lb}}{\text{year}}\text{VOC} \tag{11.9}$$

11.4 Stack Sampling

In the field of air pollution, the process of quantifying emissions is often referred to as air or stack sampling. This process consists of examining a sample of gas from the emission stream to determine both the physical characteristics of the stream and the concentrations of pollutants contained therein. While this seems relatively easy, the process is somewhat more complicated because of the nature of the medium being sampled.

As opposed to a liquid sample that can be contained, transported, and examined in a remote location with relative ease, a gas sample obtained on-site must either be quantified directly or be altered such that the constituents contained within the sample are immobilized. Immobilization is necessary because it is impractical to transport an actual quantity of the gas sample for later analysis. However, even though the sample has been transformed for evaluation at a separate location, the sample must still provide an accurate depiction of the pollutants in the gas stream being emitted. As the pollutants of concern consist of both the solid and gaseous states, sampling methods consist of a wide variety of procedures that are specific to the pollutant of concern. These methods vary from sampling for entrained particulate to the detection of multitudes of different organics and inorganics.

As a wide variety of procedures exist, the U.S. EPA has standardized these procedures and codified them such that the data resulting from their application are precise and accurate if appropriate methods are used in specific sampling scenarios. These procedures or methods refer directly to the analysis of one or more pollutants. Table 11.7 is a listing of the currently approved U.S. EPA methods with their title and appropriate *Code of Federal Regulations* reference. It should be noted that the references for the technical corrections should be reviewed in addition to the original citation for a complete description of the relevant sampling methodology.

All of the methods listed in Table 11.7 employ similar initial methods to measure the basic characteristics of the gas stream. For instance, sampling methods 1 through 3 consist of the measurement of the physical dimensions of the duct or stack, velocity, and CO_2 and O_2 concentrations, respectively. These methods are used to reflect on the appropriate locations and sample volumes

Table 11.7 Summary of U.S. EPA Emission Test Methods

Method	Reference	Description
		40 CFR Part 60 Appendix A
1–8	42 FR 41754	Velocity, Orsat, PM, SO_2, NO_x, etc.
	43 FR 11984	Corrections and amendments to methods 1 through 8
1/24	52 FR 34639	Technical corrections
	52 FR 42061	Corrections
2–25	55 FR 47471	Technical amendments
1	48 FR 45034	Reduction of number of traverse points
1	51 FR 20286	Alternative procedure for site selection
1A	54 FR 12621	Traverse points in small ducts
2A	48 FR 37592	Flow rate in small ducts—volumetric meters
2B	48 FR 37594	Flow rate–stoichiometry
2C	54 FR 12621	Flow rate in small ducts—standard pitot
2D	54 FR 12621	Flow rate in small ducts—rate meters
2E P	56 FR 24468	Flow rate from landfill wells
2F	Tentative	3D pitot for velocity
3	55 FR 05211	Molecular weight
3/3B	55 FR 18876	Method 3B applicability
3A	51 FR 21164	Instrumental method for O_2 and CO_2
3B	55 FR 05211	Orsat for correction factors and excess air
3C P	56 FR 24468	Gas composition from landfill gases
3	48 FR 49458	Addition of quality assurance/quality control
4	48 FR 55670	Addition of quality assurance/quality control
5	48 FR 55670	Addition of quality assurance/quality control
5	45 FR 66752	Filter specification change
5	48 FR 39010	Dry gas meter revision
5	50 FR 01164	Incorporation of dry gas meter and probe calibration procedures
5	52 FR 09657	Use of critical orifices as calibration standards
5	52 FR 22888	Corrections
5A	47 FR 34137	Particulate matter from asphalt roofing (prop. as M-26)
5A	51 FR 32454	Addition of quality assurance/quality control
5B	51 FR 42839	Nonsulfuric acid PM
5C	Tentative	PM from small ducts
5D	49 FR 43847	PM from baghouses
5D	51 FR 32454	Addition of quality assurance/quality control
5E	50 FR 07701	PM from fiberglass plants
5F	51 FR 42839	PM from FCCU
5F	53 FR 29681	Barium titration procedure
5G	53 FR 05860	PM from woodstove—dilution tunnel
5H	53 FR 05860	PM from woodstove—stack
6	49 FR 26522	Addition of quality assurance/quality control
6	48 FR 39010	Dry gas meter revision
6	52 FR 41423	Use of critical orifices for FR/vol meas.
6A	47 FR 54073	SO_2/CO_2
6B	47 FR 54073	Auto SO_2/CO_2
6A/B	49 FR 09684	Incorporation of collection test changes
6A/B	51 FR 32454	Addition of quality assurance/quality control
6C	51 FR 21164	Instrumental method for SO_2
6C	52 FR 18797	Corrections
7	49 FR 26522	Addition of quality assurance/quality control
7A	48 FR 55072	Ion chromatograph NO_x analysis
7A	53 FR 20139	ANPRM
7A	55 FR 21752	Revisions
7B	50 FR 15893	UV NO_x analysis for nitric acid plants
7A/B	Tentative	High SO_2 interference
7C	49 FR 38232	Alkaline permanganate/colorimetric for NO_x
7D	49 FR 38232	Alkaline permanganate/IC for NO_x
7E	51 FR 21164	Instrumental method for NO_x
8	36 FR 24876	Sulfuric acid mist and SO_2
8	42 FR 41754	Addition of particulate and moisture
8	43 FR 11984	Miscellaneous corrections
9	39 FR 39872	Opacity
9A	46 FR 53144	LIDAR opacity; called alternative 1
10	39 FR 09319	CO
10	53 FR 41333	Alternative trap
10A	52 FR 30674	Colorimetric method for PS-4
10A	52 FR 33316	Correction notice
10B	53 FR 41333	GC method for PS-4
11	43 FR 01494	H_2S
12	47 FR 16564	Pb
12	49 FR 33842	Incorporation method of additions
13A	45 FR 41852	F—colorimetric method
13B	45 FR 41852	F—SIE method

Table 11.7 (continued) Summary of U.S. EPA Emission Test Methods *(continues)*

Method	Reference	Description
13A/B	45 FR 85016	Correction to methods 13A and 13B
14	45 FR 44202	*F* from roof monitors
15	43 FR 10866	TRS from petroleum refineries
15	54 FR 46236	Revisions
15	54 FR 51550	Correction notice
15A	52 FR 20391	TRS alternative/oxidation
16	43 FR 07568	TRS from kraft pulp mills
16	43 FR 34784	Amendment to method 16, H_2S loss after filters
16	44 FR 02578	Amendment to method 16, SO_2 scrubber added
16	54 FR 46236	Revisions
16	55 FR 21752	Correction of figure ($\pm10\%$)
16A	50 FR 09578	TRS alternative
16A	52 FR 36408	Cylinder gas analysis alternative method
16B	52 FR 36408	TRS alternative/GC analysis of SO_2
16A/B	53 FR 02914	Correction 16A/B
17	43 FR 07568	PM, in-stack
18	48 FR 48344	VOC, general gas chromatograph method
18	49 FR 22608	Corrections to method 18
18	52 FR 51105	Revisions to improve method
18	52 FR 10852	Corrections
18	59 FR 19308	Revisions to improve quality assurance/quality control
19	44 FR 33580	*F*-factor, coal sampling
19	52 FR 47826	Method 19A incorporation into M-19
19	48 FR 49460	Corr. to *F*-factor equations and F_c value
20	44 FR 52792	NO_x from gas turbines
20	47 FR 30480	Corr. and amend.
20	51 FR 32454	Clarifications
21	48 FR 37598	VOC leaks
21	49 FR 56580	Corrections to method 21
21	55 FR 25602	Clarifying revisions
22	47 FR 34137	Fugitive VE
22	48 FR 48360	Add smoke emission from flares
23	56 FR 5758	Dioxin/dibenzofuran
24	45 FR 65967	Solvent in surface coatings
24A	47 FR 50644	Solvent in ink (proposed as method 29)
24	Tentative	Solvent in water-borne coatings
24	57 FR 30654	Multicomponent coatings
24	Tentative	Radiation-cured coatings
25	45 FR 65956	Total gaseous nonmethane organics
25	53 FR 04140	Revisions to improve method
25	53 FR 11590	Correction notice
25A	48 FR 37595	Total organic carbon/flame ionization detector
25B	48 FR 37597	TOC/NDIR
25C P	56 FR 24468	VOC from landfills
25D	59 FR 19311	VO from TSDF—purge procedure
25E P	56 FR 33555	VO from TSDF—vapor pressure procedure
26	56 FR 5758	HCl
26	57 FR 24550	Corrections to method 26
26	59 FR 19309	Expand method 26 to HCl, halogens, and other hydrogen halides
26A	59 FR 19309	Isokinetic HCl, halogens, and other hydrogen halides method
27	48 FR 37597	Tank truck leaks
28	53 FR 05860	Wood stove certification
28A	53 FR 05860	Air-to-fuel ratio
29	Tentative	Multiple metals
		Part 60 Appendix B
PS-1	48 FR 13322	Opacity
PS-1	Tentative	Revisions
PS-2	48 FR 23608	SO_2 and NO_x
PS 1-5	55 FR 47471	Technical amendments
PS-3	48 FR 23608	CO_2 and O_2
PS-4	50 FR 31700	CO
PS-4A	56 FR 5526	CO for MWC
PS-5	48 FR 32984	TRS
PS-6	53 FR 07514	Velocity and mass emission rate
PS-7	55 FR 40171	H_2S
		Part 60 Appendix F
Prc 1	52 FR 21003	Quality assurance for CEMS
Prc 1 P	54 FR 52207	Revision

Table 11.7 (continued) Summary of U.S. EPA Emission Test Methods

Method	Reference	Description
		Part 60 Appendix J
App-J	55 FR 33925	Wood stove thermal efficiency
		Alternative Procedures and Miscellaneous
	48 FR 44700	S-factor method for sulfuric acid plants
	48 FR 48669	Corrections to S-factor publication
	49 FR 30672	Add fuel analysis procedures for gas turbines
	51 FR 21762	Alternative PST for low-level concentrations
	54 FR 46234	Misc. revisions to Appendix A, 40 CFR Part 60.
	55 FR 40171	Monitoring revisions to Subpart J (Petr. Ref.)
		Part 60
	54 FR 06660	Test methods and procedures rev. (40 CFR 60)
	54 FR 21344	Correction notice
	54 FR 27015	Correction notice
		Part 61 Appendix B
101	47 FR 24703	Hg in air streams
101A	47 FR 24703	Hg in sewage sludge incinerators
101	49 FR 35768	Corrections to methods 101 and 101A
102	47 FR 24703	Hg in H_2 streams
103	48 FR 55266	Revised Be screening method
104	48 FR 55268	Revised beryllium method
105	40 FR 48299	Hg in sewage sludge
105	49 FR 35768	Revised Hg in sewage sludge
106	47 FR 39168	Vinyl chloride
107	47 FR 39168	VC in process streams
107	52 FR 20397	Alternative calibration procedure
107A	47 FR 39485	VC in process streams
108	51 FR 28035	Inorganic arsenic
108A	51 FR 28035	Arsenic in ore samples
108B	55 FR 22026	Arsenic in ore alternative
108C	55 FR 22026	Arsenic in ore alternative
108B/C	55 FR 32913	Correction notice
111	50 FR 05197	Polonium-210
114 P	54 FR 09612	Monitoring of radionuclides
115 P	54 FR 09612	Radon-222
		Part 61
	53 FR 36972	Corrections
		Part 51 Appendix M
201	55 FR 14246	PM-10 (EGR procedure)
201A	55 FR 14246	PM-10 (CSR procedure)
201/A	55 FR 24687	Correction of equations
201	55 FR 37606	Correction of equations
202	56 FR 65433	Condensible PM
203 P	57 FR 46114	Transmissometer for compliance
203A P	58 FR 61640	Visible emissions—2–6 min avg.
203B P	58 FR 61640	Visible emissions—time exception
203C P	58 FR 61640	Visible emissions—instantaneous
204	Tentative	VOC capture efficiency
204A	Tentative	VOC capture efficiency
204B	Tentative	VOC capture efficiency
204C	Tentative	VOC capture efficiency
204D	Tentative	VOC capture efficiency
204E	Tentative	VOC capture efficiency
204F	Tentative	VOC capture efficiency
205	Tentative	Dilution calibration verification
206	Tentative	Ammonia (NH_3)
		Part 63 Appendix A
301	57 FR 61970	Field data validation protocol
302	Tentative	Generic GC/MS procedure
303	58 FR 57898	Coke oven door emissions
304A	59 FR 19590	Biodegradation rate (vented)
304B	59 FR 19590	Biodegradation rate (enclosed)
305	59 FR 19590	Compound specific liquid waste

Table 11.7 (continued) Summary of U.S. EPA Emission Test Methods

Method	Reference	Description
306 P	58 FR 65768	Hexavalent chromium
306A P	58 FR 65768	Simplified chromium sampling
306B P	58 FR 65768	Surface tension of chromium suppressors
307 P	58 FR 62593	Solvent degreaser VOC
308 P	58 FR 66079	Methanol
309	Tentative	Aerospace solvent recovery material balance
		Part 64 Appendix A
101 P	58 FR 54648	VOC CEMS performance specification
102 P	58 FR 54648	GC CEMS performance specification

P = proposal.
Tentative = under evaluation.
Source: U.S. EPA, Office of Air Quality Planning and Standards.

that must be withdrawn to provide a representative sample of the gas stream. As a result, many other sampling methods employ these basic methods during their trials.

Of all the sampling methods, the first five are typically employed in most sampling scenarios. As a result of the great number and variation between all of the individual methods, this discussion will focus on the basic procedures and hardware of method 5, as this is the most common sampling method. The reader is referred to 40 CFR Part 60 for the specific sampling procedures for method 5 and other methods. Additionally, the reader is referred to *Methods of Air Sampling and Analysis* for further information on the analysis of specific compounds.

The U.S. EPA method 5 sampling train is used in the determination of particulate in gas streams. The method 5 sampling train is composed of a heated sampling probe, a sample case, and a control case. A schematic of the assembly is illustrated in Fig. 11.1.

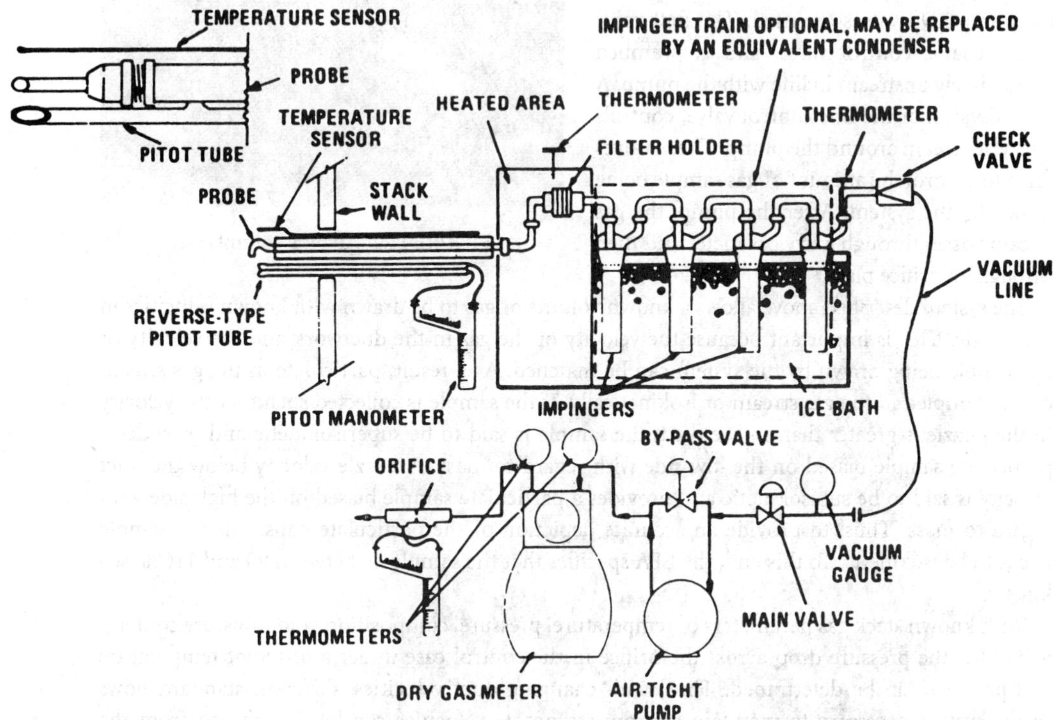

FIGURE 11.1 Method 5 sampling hardware schematic.

In Fig. 11.1, the heated sampling probe is attached to the sample case. The probe consists of a nozzle of known inner diameter, a thermistor to determine stack temperature, another thermistor to determine probe temperature, and a pitot tube. Ending in a stainless steel or glass ball joint depending on the probe liner material, the probe is joined to the filter housing in the sample case by a ground glass joint. The pitot tube and thermistors are connected to the control case through the "umbilical cord" running from the sample case to the control case. The umbilical cord houses both a section of tubing the gas stream is drawn through and a wire harness connecting the control case and the sample case. This configuration results in the sample being drawn from the gas stream through the probe, the sample case, and finally the control case.

Inside the sample case, the gas stream is passed through a heated filter housing to remove particulate. The housing is heated to prevent the gas stream from falling below the dew point, and fouling the filter with moisture and most importantly control the particulate formation temperature at 250°F. After the filter housing, the gas stream is passed through a set of four impingers immersed in an ice bath. The first two impingers are filled with a liquid-absorbing reagent (dependent on the pollutant being sampled) to remove a pollutant from the gas stream. These are followed by an empty third impinger serving as a moisture trap, and a fourth impinger filled with silica gel to adsorb any remaining moisture. As a result of passing through the sample case, the gas stream being sampled has had the particulate filtered from it, and the moisture and pollutant removed. To sample other pollutants, the contents of the first two impingers are altered to remove the specific pollutant of concern.

After being drawn through the sample case, the gas stream passes through the umbilical cord to the control case. In Fig. 11.1, the control case flow path is shown, while Fig. 11.2 is a photograph of a typical control case. A vacuum gauge indicates negative pressure in the line downstream of the filter. To control flow to the pump, two valves are used. The first valve is the coarse control valve, and is plumbed immediately upstream in line with the pump. A second valve, or the fine control valve, controls a recycle stream around the pump. These valves serve to control the amount of gas sample being drawn by the system. After the pump, the gas stream passes through a dry gas meter and then through an orifice plate.

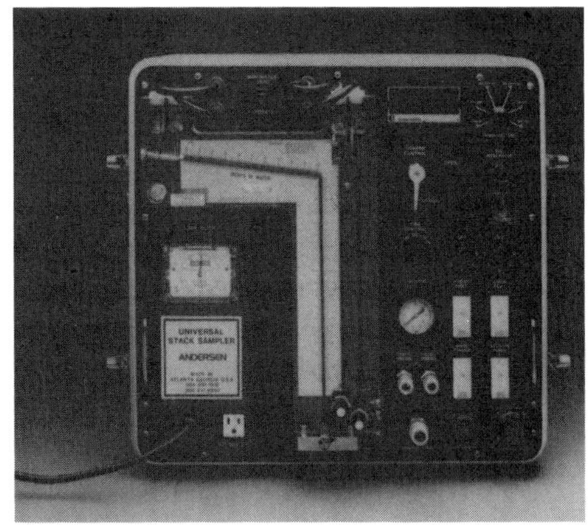

FIGURE 11.2 Typical control case.

The system described above allows a known volume of gas to be drawn with known velocities in the nozzle. This is important because the velocity of the gas in the ductwork and the velocity of the sample being drawn by the system can be matched. As a result, particulate in the gas stream can be sampled in the gas stream or isokinetically. If the sample is collected such that the velocity in the nozzle is greater than in the duct, the sample is said to be superisokinetic and provides a particulate sample biased on the low side with regard to mass. A nozzle velocity below the duct velocity is said to be subisokinetic and provides a particulate sample biased on the high side with regard to mass. Thus, to provide an accurate depiction of the particulate emissions the sample needs to be isokinetic. To this end, the EPA specifies that the sample be between 90 and 110% isokinetic.

With known stack gas parameters of temperature, pressure, composition, and moisture content, values for the pressure drop across the orifice in the control case under a different temperature and pressure can be determined. Thus, with changing duct velocities, different standard flows in the system necessary to maintain appropriate nozzle velocities can be determined from the calibration graphs for the control case. As mentioned previously, standard flows are then used to determine the appropriate ΔH (pressure drop across the orifice meter) value for a specific velocity pressure. The method 5 sampling train is unique in that both a rate meter (orifice plate) and totalizing meter (dry gas meter) allow for a post-test check of the isokinetic percentage.

11.5 Emissions Control

Particulates

Aerodynamic Diameter

Engineers who are concerned with the removal of solid particles from gas streams are less concerned with the physical shape of the particle and more interested in the particles' aerodynamic behavior in the gas stream. As such, the term *aerodynamic diameter* is widely used in the design and selection of air pollution control hardware for particulate control. Aerodynamic diameter can be defined as an equivalent diameter of a nonspherical particle whose actual shape can be spherical but is usually nonspherical and whose aerodynamic behavior is identical to a unit density sphere in stokes flow.

The measurement of aerodynamic diameter is best performed using isokinetic sampling procedures directly in the gas stream with a multistage impactor which has been calibrated with unit density spheres. One such impactor is the Anderson® Impactor seen in Fig. 11.3. As the particles proceed through the multistage impactor, their velocities are stepwise increased at each impaction stage. Immediately following this acceleration, the conveying gas stream is routed through 90 degree turns at each stage of the impactor. Since the particles have much greater inertia than the gas molecules, the particles cannot negotiate the 90 degree turns; that is, they deviate from the streamlines of gas flow and impact a collection surface or stage.

Each collection stage is gravimetrically analyzed and the data is presented graphically on a log-probability plot seen in Fig. 11.4. Most particulates generated by abrasion, fracturing, or condensation phenomena have aerodynamic diameters that are log-normally distributed as is reflected by the straight line shown in Fig. 11.4. This plot is very useful to the engineer either designing or selecting the appropriate control hardware. For example, particles with an extremely wide aerodynamic diameter distribution would be represented by an almost vertical line or an infinite slope in Fig. 11.4. On the other hand, particles which did not vary widely in their aerodynamic diameters would be represented by an almost horizontal line or a line with near-zero slope. Therefore, a quick glance at the slope immediately tells the engineer whether he or she is dealing with an almost infinite variability in particle sizes or a near mono-dispersion of particles.

Additionally, the intersection of the line in Fig. 11.4 with the 50% probability value on the abscissa is the mass median diameter of the particle distribution. This mass median diameter immediately tells the engineer what type of particulate control hardware probably will be needed. For example, particle diameter distributions having submicron mass median diameters require relatively high-energy devices for removal while super-micron-diameter particles require lesser amounts of energy.

Typically, for particles larger than 40 microns in aerodynamic diameter, gravity force is utilized for removal. Obviously this force is very cost-effective and if properly combined with low transport velocities and subsequently high residence times respectable removal efficiencies can result.

As the particles become smaller, greater forces must be brought into play for their removal. For particles between 10 to 40 microns in aerodynamic diameter, centrifugal forces are brought into play through the use of cyclones. For the removal of particles smaller than 10 microns in aerodynamic diameter, fabric filtration, electrostatic precipitators, and high-energy wet scrubbing are employed.

Settling Chambers

Particles larger than 40 microns in aerodynamic diameter settle readily under the influence of gravity. If the particulate matter is being carried in an exhaust gas stream as opposed to fugitive

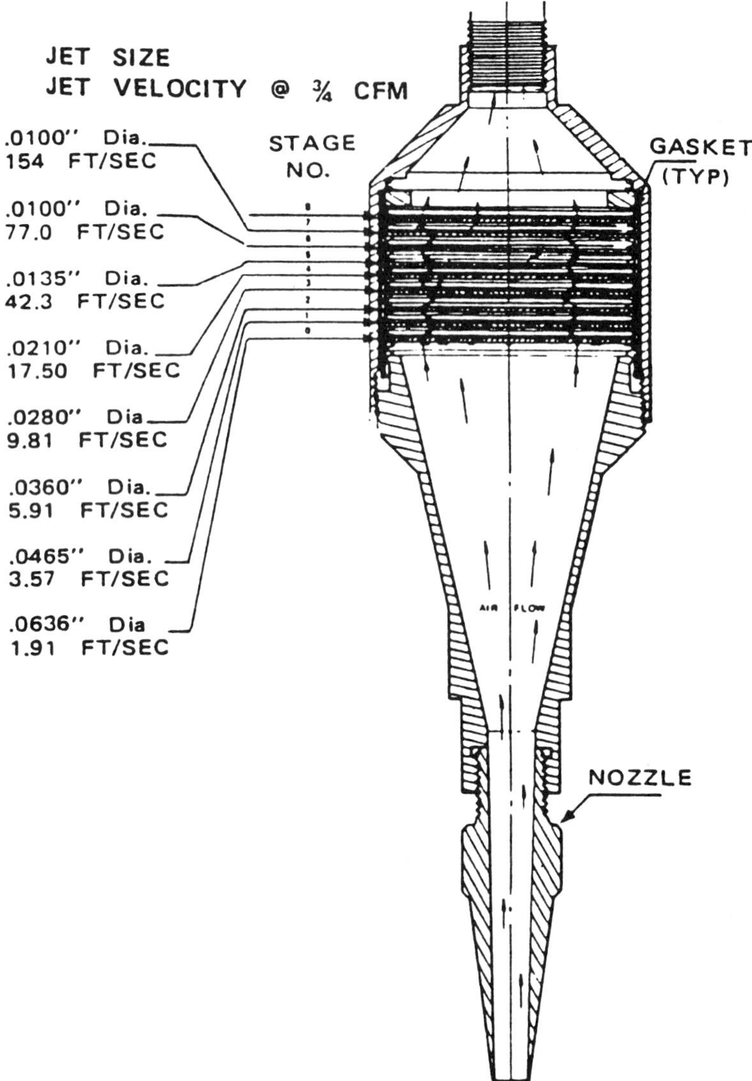

FIGURE 11.3 Anderson® impactor.

dust in the atmosphere, a settling chamber is a very cost-effective device for their removal. A settling chamber is essentially a wide spot in a duct which significantly reduces the gas velocity and, therefore, the particulate transport velocity, allowing enough residence time for gravity to act on the particle and separate it from the gas stream.

A side view of a settling chamber is seen in Fig. 11.5. As the entering particle decelerates due to the increased cross-sectional area for flow, gravity force accelerates the particle to its terminal settling velocity. If the residence time in the chamber is sufficient such that the particle falls at least a distance h, the particle will be captured in the chamber. In other words, for 100% capture, a particle's fallout time must at least be equal to or less than its transport time (residence time):

$$T_f = T_t \tag{11.10}$$

If particle fallout time is expressed as

$$T_f = \frac{h}{V_t} \tag{11.11}$$

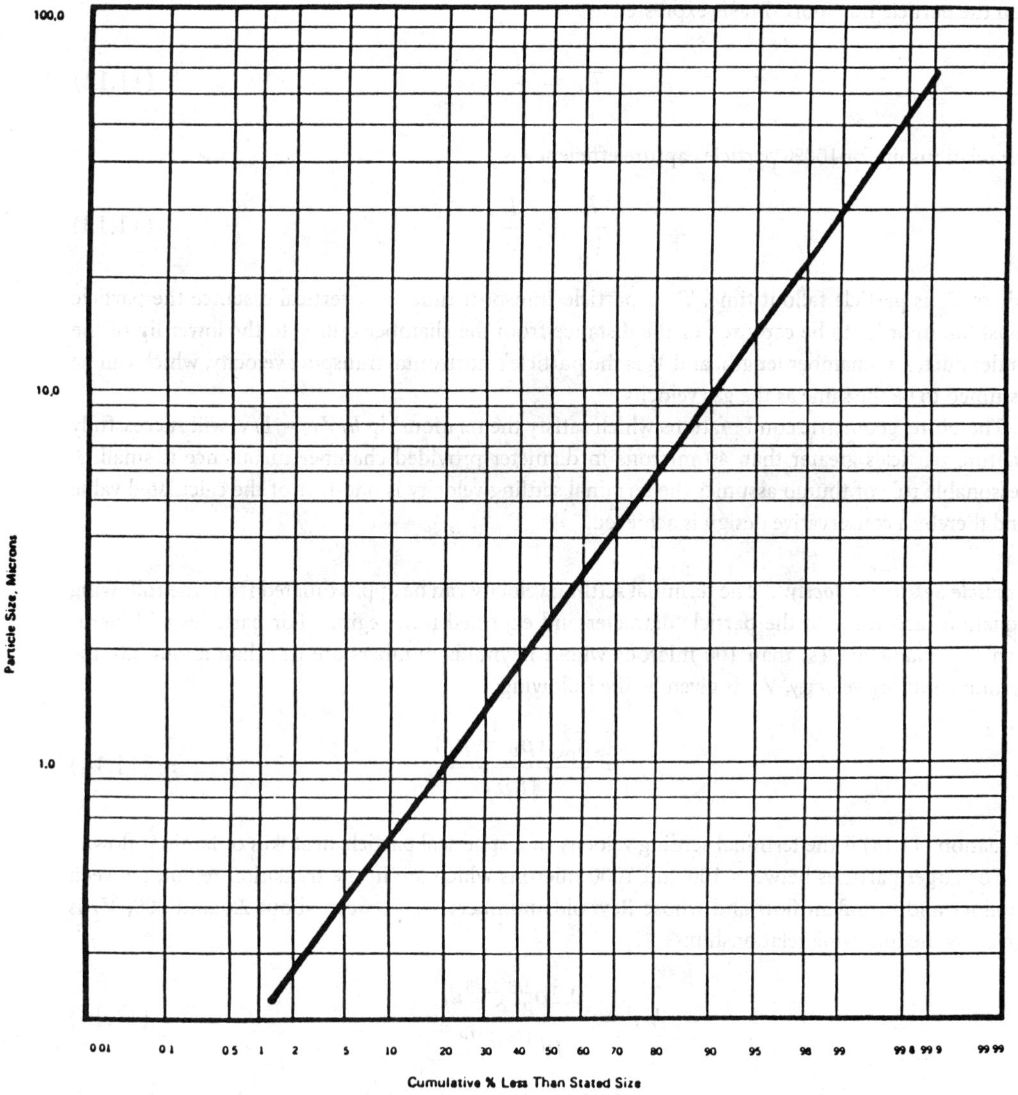

FIGURE 11.4 Log probability plot.

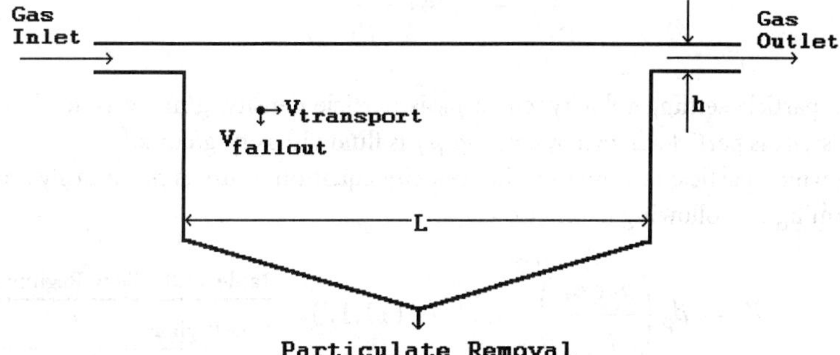

FIGURE 11.5 Settling chamber.

and the particle transport time is expressed as

$$T_t = \frac{L}{V} \tag{11.12}$$

the relationship for 100% particle capture efficiency is

$$\frac{h}{V_t} = \frac{L}{V} \tag{11.13}$$

where T_f is particle fallout time, T_t is particle transport time, h is vertical distance the particle must fall in order to be captured or the distance from the chamber ceiling to the lower lip of the outlet duct, L is chamber length, and V is the particle's horizontal transport velocity, which can be assumed to be the same as the gas velocity.

Therefore, geometric combinations which satisfy the relationship $h/V_t = L/V$ will successfully capture particles greater than 40 microns in diameter provided chamber turbulence is small. A reasonable rule of thumb assumes the terminal settling velocity is one-half of the calculated value and thereby a conservative design is achieved.

Particle Settling Velocity. The terminal settling velocity can be approximated from the following equations according to the particle diameter and expected flow regime. For particles with aerodynamic diameters less than 100 microns whose Reynolds numbers are less than about 2.0, the terminal settling velocity, V_t, is given by the following:

$$V_t = \frac{d_p^2 g(\rho_p - \rho_f)}{18\mu_f} \tag{11.14}$$

Equation (11.14) is the terminal settling velocity of a spherical particle in stokes or laminar flow.

For larger particles between 100 and 1000 microns which are in the transition region between laminar and turbulent flow and whose Reynolds numbers are between about 2.0 and 500, V_t is given by the following relationship:

$$V_t = \frac{0.2\rho_p^{2/3} g^{2/3} d_p}{\rho f^{1/3} \mu_f^{1/3}} \tag{11.15}$$

For particles larger than 1000 microns which are in turbulent flow regime and whose Reynolds numbers are between 500 and 10^5, V_t is given by the following:

$$V_t = 1.74\left(\frac{\rho_p g d_p}{\rho_f}\right)^{1/2} \tag{11.16}$$

where V_t is particle settling velocity, cm/s; ρ_p is particle density, g/m^3; g is acceleration due to gravity, cm/s^2; d_p is particle diameter, cm; and μ_f is fluid viscosity, g/cm-s.

Selecting which particle terminal settling velocity equation to use is more easily and accurately done by forming the following K criterion:

$$K = d_p\left(\frac{\rho_p g \rho_f}{\mu_f^2}\right)^{1/3} \tag{11.17}$$

and selecting the flow regime according to the value of K in Table 11.8.

Table 11.8 Flow Regime K Values

Flow Regime	K Range
Stokes	$K \leq 3.3$
Intermediate	$3.3 \leq K \leq 43.6$
Turbulent	$43.6 \leq K \leq 2360$

Cyclones

Particle aerodynamic diameters from 40 down to 10 microns are usually removed from a gas stream by cyclonic separation. In this smaller particle diameter range, an additional force, namely centrifugal force, must be applied to effect their removal. In the case of the settling chamber for larger particles, one g force was utilized. With the cyclone, many equivalent g forces are utilized via centrifugal force. A ratio of the centrifugal to gravitational force is called the separation factor of the cyclone:

$$\text{Separation factor} = \frac{\text{Centrifugal force}}{\text{Gravity force}} = \frac{mV^2/r}{mg} = \frac{v^2}{rg} \qquad (11.18)$$

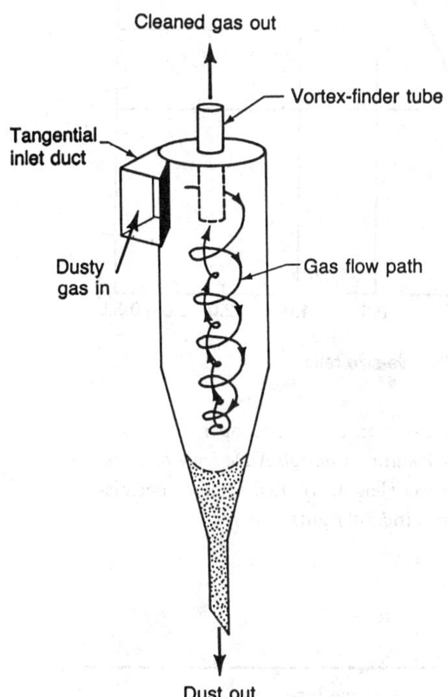

FIGURE 11.6 Cyclone. (*Source:* Cooper, C. D. and Alley, F. C. 1990. *Air Pollution Control: A Design Approach.* Waveland Press, Prospect Heights, IL. Reprinted by permission of Waveland Press, Inc. All rights reserved.)

For cyclones this separation factor can be as high as 2500, which means 2500 gs of force are applied to the particles as compared to 1 g in a settling chamber. A cyclonic separator can be thought of as a settling chamber of revolution. In other words, the cyclone is also a wide spot in the exhaust gas duct which provides sufficient residence time for the centrifugal forces to act and remove the particles from the gas stream.

A conventional cyclone consists of a tangential entry, a main cylinder section, a conical lower section with provision for particle removal, and a gas outlet tube, as shown in Fig. 11.6. The gas enters the main cylinder tangentially and spirals downward (forming the primary vortex) into the conical section, thus imposing a centrifugal force on the entrained particles which move radially outward, impacting the side walls and falling to the bottom of the cyclone. Since there is no gas exit at the bottom of the cyclone, the primary vortex downward movement stops, the spiral tightens (now called the secondary vortex) and moves vertically up and out of the top of the cyclone through the outlet tube. The size of these cyclones can vary from 2 feet in diameter up to 12 to 15 feet in diameter. Figure 11.7 shows the dimensional labeling for a typical tangential entry cyclone. The conventional cyclone design geometry dimensional relationships as given by Lapple are shown in Table 11.9. Note that each of the lengths are proportioned to the main body diameter of the cyclone. Once the main body diameter is chosen, the remaining lengths are set.

Since Lapple's original work on the conventional cyclone, other researchers have suggested geometry ratios slightly different from that of Lapple. These geometries are shown in Table 11.9, and have been identified as "high efficiency" and "high throughput."

Conventional Cyclone Design Approach. In the 1950s, Lapple correlated the collection efficiency data from many sizes of conventional cyclones using a "cut-diameter" variable. This correlation, seen in Fig. 11.8, is the classic approach to designing and sizing a conventional cyclone. The

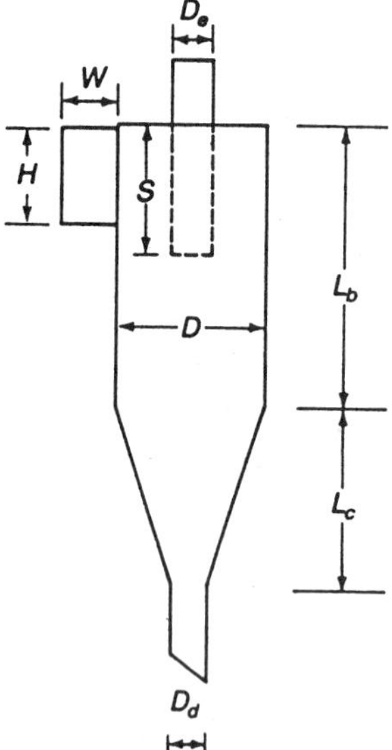

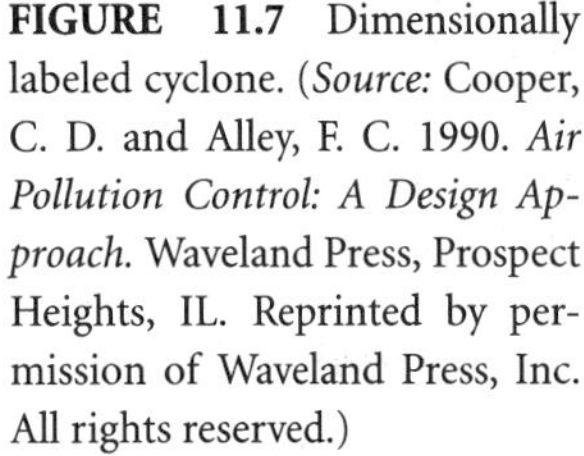

FIGURE 11.7 Dimensionally labeled cyclone. (*Source:* Cooper, C. D. and Alley, F. C. 1990. *Air Pollution Control: A Design Approach.* Waveland Press, Prospect Heights, IL. Reprinted by permission of Waveland Press, Inc. All rights reserved.)

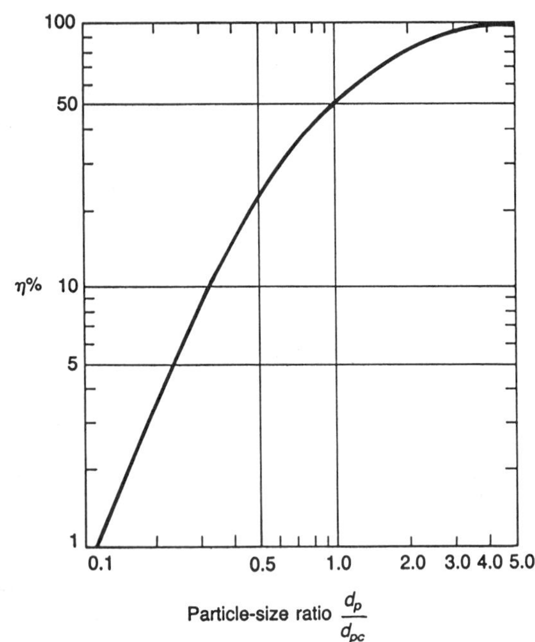

FIGURE 11.8 Cut diameter. (*Source:* Cooper, C. D. and Alley, F. C. 1990. *Air Pollution Control: A Design Approach.* Waveland Press, Prospect Heights, IL. Reprinted by permission of Waveland Press, Inc. All rights reserved.)

Table 11.9 Lapple Design Values for Cyclones

	Cyclone Type		
	High Efficiency	Conventional	High Throughput
Body diameter, D/D	1.0	1.0	1.0
Height of inlet, H/D	0.5	0.5	0.75
Width of inlet, W/D	0.2	0.25	0.375
Diameter of gas exit, D_e/D	0.5	0.5	0.75
Length of vortex finder, S/D	0.5	0.625	0.875
Length of body, L_b/D	1.5	2.0	1.5
Length of cone, L_c/D	2.5	2.0	2.5
Diameter dust outlet, D_d/D	0.375	0.25	0.375

Source: Cooper, C. D. and Alley, F. C. 1990. *Air Pollution Control: A Design Approach.* Waveland Press, Prospect Heights, IL. Reprinted by permission of Waveland Press, Inc. All rights reserved.

cut-diameter is that diameter of particle captured at a 50% efficiency in the cyclone. It is calculated from the following equation:

$$d_p = \left(\frac{9\mu W_i}{\pi N v_i (\rho_p - \rho_g)} \right)^{1/2} \tag{11.19}$$

where d_{pc} is the cyclone cut-diameter, μ is the gas velocity, W_i is the inlet width of the cyclone, v_i is the inlet gas velocity, ρ_p and ρ_g are the particle and gas densities, respectively, and N is the number of primary vortex revolutions. N can be estimated by the ratio of L_b/H (refer to Fig. 11.6).

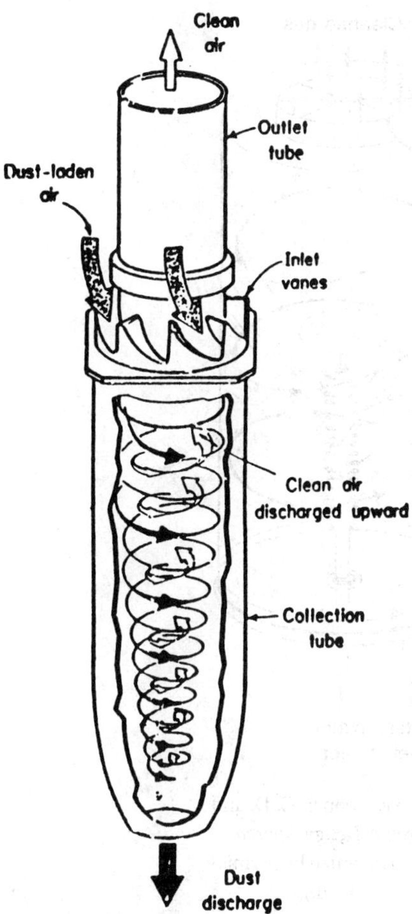

FIGURE 11.9 Axial flow cyclone.

Once the cut-diameter d_{pc} is calculated, the Lapple nondimensional ratio, D_p/D_{pc} can be formed and Fig. 11.8 used to determine the specific collection efficiency of the particles of interest. The overall collection efficiency of the cyclone is the summation of the specific efficiency from each size range chosen weighted according to the mass of particles in each size range.

Another version of a cyclonic separator is the axial flow cyclone shown in Fig. 11.9. These units are usually much smaller than the conventional unit and are usually ganged together in parallel to increase the amount of gas they process. These units vary in size from a few inches to 18 inches in diameter.

Wet Scrubbers

Where gravity or centrifugal forces fail to remove smaller particles, wet scrubbers can be employed. However, keep in mind that the wet scrubber merely converts an air pollution problem into a water pollution problem. The wet scrubber presents a liquid, usually water, into a particulate-laden gas stream where the liquid droplets and particles are brought together. Upon contact, the particle is surrounded by liquid and the apparent mass of the original particle is increased. The particles, now with a greater apparent mass, can be removed from the gas stream using straightforward impaction or cyclonic mechanisms.

Generally, the smaller the particle to be removed, the greater the energy input into the scrubber for effective particulate removal. The water- and particulate-laden gas stream can be brought together in a number of different ways. One common way is to inject the water through spray nozzles into the relatively low-velocity gas stream. These low-energy wet scrubbers are referred to as *spray chamber scrubbers* and are usually effective only for relatively large particles greater than 5 microns. The spray nozzles relative to the gas stream are usually arranged cocurrent, countercurrent, or crosscurrent using a cone-type spray pattern with liquid requirements from 15 to 25 gallons/1000 ft^3 at a liquid delivery pressure of about 40 psi. There are literally hundreds of spray chamber scrubber configurations on the market. Figure 11.10 shows a typical configuration.

Other methods of bringing the water and particulate matter together involve impaction of a liquid droplet–gas mixture on a solid surface. In orifice scrubbers, the gas impinges on a liquid surface and then the droplet–gas mixture must negotiate a series of tortuous turns through a baffle arrangement. In impingement scrubbers the liquid droplet–gas mixture flows upward through perforated trays containing liquid and froth and then impacts on plates above the trays.

High-efficiency scrubbers for submicron particulate matter accelerate the gas stream in a venturi and present the liquid to the gas near the throat of the venturi. These venturi scrubbers are very effective for submicron particulate removal and operate with relatively high gas–side pressure drops from approximately 20 to 120 inches–water column. As a result they are energy intensive and have high operational costs. However, they can achieve collection efficiencies of 93% or greater on particles larger than 0.3 microns [Cooper and Alley, 1990].

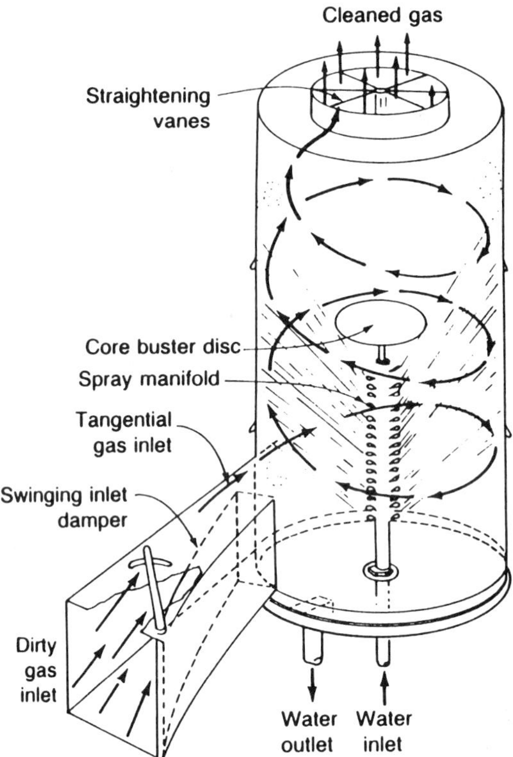

FIGURE 11.10 Wet scrubber. (*Source:* Cooper, C. D. and Alley, F. C. 1990. *Air Pollution Control: A Design Approach.* Waveland Press, Prospect Heights, IL. Reprinted by permission of Waveland Press, Inc. All rights reserved.)

The general equation for overall wet scrubber particulate removal efficiency is given by

$$N = 1 - e^{-f_{(\text{system})}} \tag{11.20}$$

where $f_{(\text{system})}$ is a term representing some function of the particular scrubber system variables. The form of this equation predicts an exponentially increasing collection efficiency with particle diameter and asymptotically approaches 100%.

Wet Scrubber Design Approach. Calvert suggests a design approach based on a single droplet target efficiency, impaction parameter and the concept of particle penetration [Calvert, 1972]. Penetration is defined as the converse of collection efficiency:

$$P = 1 - N \tag{11.21}$$

where N is the particle collection efficiency fraction and P is the fraction of particles that escape (or penetrate) the collection device.

Combining the general efficiency and penetration equations yields

$$P = 1 - (1 - e^{-f_{(\text{system})}}) \tag{11.22}$$

or $P = e^{-f_{(\text{system})}}$. Calvert defines $f_{(\text{system})}$ as an empirical function dependent on the particle aerodynamic diameter, d_p. Now:

$$f_{(\text{system})} = A_{\text{cut}} d_p^{B_{\text{cut}}} \tag{11.23}$$

where A_{cut} is a parameter which characterizes the particulate aerodynamic size distribution and B_{cut} is an empirical constant of 2.0 for plate towers and venturi scrubbers and 0.7 for centrifugal scrubbers. The penetration is now:

$$P = e^{-A_{cut} d_p^{B_{cut}}} \tag{11.24}$$

which yields the penetration for one particle size distribution. The overall penetration is the summation of the penetration from each chosen size range weighted according to the mass of particles in each size range.

By integrating the penetration, P, over a lognormal size distribution of particles and various geometric standard deviations and mass median diameters, Calvert developed the curves in Fig. 11.11. Knowing the required overall collection efficiency of the scrubber and the mass median diameter of the particle size distribution challenging the scrubber, the cut size of the scrubber can be determined.

Consider this example: an in situ aerodynamic particle size sample and subsequent gravimetric analysis indicated that the mass median diameter of the distribution was 10 microns and the geometric standard deviation of the distribution was 2.0. If a collection efficiency of 98% is required to meet the emission standards what must the cut diameter of the scrubber be?

$$\begin{aligned} P &= 1 - N \\ &= 1 - 0.98 \\ &= 0.02 \end{aligned} \tag{11.25}$$

Now from Fig. 11.11, for $P = 0.02$ and SD $= 2.0$ microns, A_{cut}/MMD $= 0.19$ and, therefore, $A_{cut} = \text{MMD} \times 0.19 = 10 \times 0.19 = 1.9$ microns. In this example, the scrubber must collect the 1.9-micron particles with an efficiency of at least 50% to meet the overall scrubber efficiency of 98%.

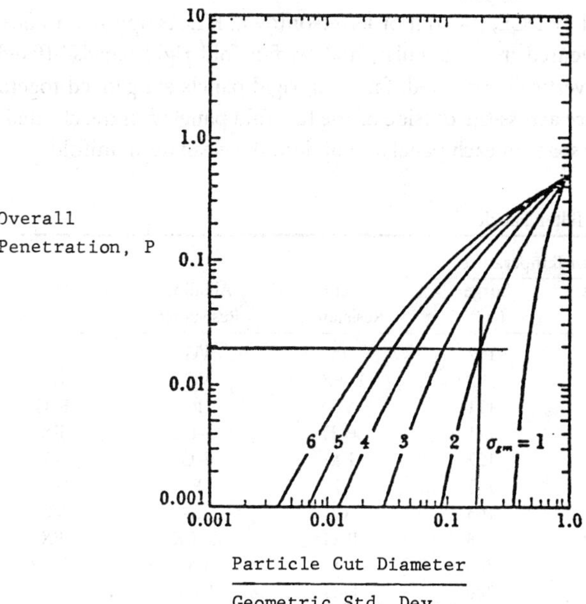

FIGURE 11.11 Calvert design curves.

Fabric Filters

Where it is not desired to create a water pollution problem from an air pollution problem, dry collection of fine particulate matter in gas streams can be very effectively accomplished using baghouse filtration. Baghouses have been around for quite some time and their usefulness is continually being extended due to advances in the media or substrate material onto which the particles are being collected. The "heart" of a baghouse filtration system is the media which acts as a substrate for collection of a filter "cake" which ultimately does the "filtering."

Baghouses are typically grouped into three classes based on the method employed to clean the filtration media: reverse air, shaker, and pulse jet. The reverse air baghouse is cleaned by taking the baghouse compartment to be cleaned out of operation and reversing the air flow through the filter, thus removing the cake from the filtration bags. A shaker baghouse also has a compartment taken out of service and mechanically shakes the bags to dislodge the filter cake physically. The most recent innovation in baghouse cleaning consists of the pulse-jet baghouse. This type of baghouse is cleaned by pulsing high pressure blasts of air through the interior of the bag while the bag is in service.

Filtration Media. Many different types of fabrics are used in baghouse filters. Fiberglass and even ceramic fibers are used in high-temperature applications. More commonly used are cotton, polyesters, and teflon coatings. Matching the proper bag material with the gas stream characteristics is extremely important for long bag life. Table 11.10 is an example of some of the more popular fabric materials with their temperature limitations and acid, alkali, and flexure/abrasion resistance.

As can be seen from Table 11.10, most conventional fibers are limited to operating temperatures less than 260°C (500°F). Nextel® is a relatively new high-temperature fiber developed by 3M which has an upper temperature limit of 760°C (1400°F). The fiber is made of continuous individual ceramic filaments. The technology is based on a "sol-gel" process whereby a chemical sol is extruded through a spinneret and then fired. The resulting metal oxide fibers are polycrystalline and nonporous. The filtration performance of Nextel® is very similar to that of conventional woven fiberglass [Hansen, 1994].

Teflon(PTFE)-coated polyethylene beads sintered into a rigid matrix is a newer technology applied as a baghouse filter media. The media has the trade name Sintamatic, is imported from Europe, and is sold by DCE, Inc. out of Jeffersontown, KY. As opposed to flexible fabric material, this media is configured in rectangular, hollow, fan-fold rigid panels affixed to an upper clean air manifold. The vertically oriented, fan-fold, rigid panels are ganged together in parallel units. Filtration takes place across the outside of the fan-fold panel with the cleaned gas proceeding into the hollow interior space in each panel and up into the clean air manifold.

Table 11.10 Properties of Filter Fabrics

	Maximum Temperature					
Material	Continuous (°C)	Surge (°C)	Acid Resistance	Alkaline Resistance	Flex Abrasion	Relative Cost
Cotton	82	107	P	VG	VG	2
Polypropylene	88	93	P–EX	VG	EX	1.5
Wool	93–102	121	VG	P	F–G	3
Nylon	93–107	121	P–F	G–EX	EX	2.5
Orlon	116	127	EX	F–G	G	2.75
Acrylic	127	137	G	F	G	3
Dacron	135	163	G	G	VG	2.8
Nomex	204	218	P–G	G–EX	EX	8
Teflon	204–232	260	EX	EX	F	25
Fiberglass	260	288	F–G	F–G	F	6
Nextel®	760	1200	G–EX	EX–G	EX	High

P—poor, F—fair, G—good, VG—very good, EX—excellent.

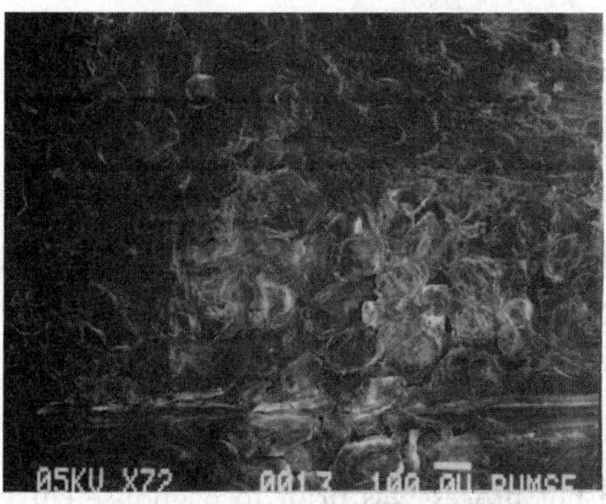

FIGURE 11.12 Sintamatic filter media.

Since the media is composed of polyethylene and PTFE it is very resistant to acids and bases. However, the major hindrance is temperature; it has an upper limit of about 140°F. A scanning electron microscope was used to examine a cross section of the media [Pederson, 1993]. Figure 11.12 shows the media magnified 72 times with the horizontal white bar in the lower right of the photo representing 100 microns. Most of the polyethylene beads are between approximately 50 and 200 microns, with the average being 100 microns. Most of the visible pores appear to be smaller than 100 microns. An analysis of the pore structure indicated the media porosity to be about 0.054 [Pederson, 1993]. This author has sampled a Sintamatic baghouse handling toxic metal dusts in an ambient air stream and found the unit to perform extremely well.

Air-to-Cloth Ratio. The primary design variable for a baghouse filter is the air-to-cloth ratio (A/C) which is the actual air flow rate through the baghouse divided by the filtration or cloth area. The A/C has units of meters/min and ranges from 1.0 to 7.7 cm/s (2.0 to 15.0 ft/min). A very conservative design has A/C ratios around 1.0 cm/s (2.0 ft/min) while most industrial applications are in the 2.5 to 3.0 cm/s (5 to 6 ft/min) A/C range. Applications in aggregate crushing, sand reclaiming operations, and other applications where the particulate matter is relatively large in diameter and easily cleaned off the surface of the filter media can use A/C ratios between 3.0 to 7.7 cm/s (6 to 15 ft/min).

The selection of an optimum A/C ratio depends on the cleaning method, type of filter media, temperature, and the characteristics of the particles. Also involved are the trade-offs between initial cost and future operation and maintenance costs. A lower A/C means lower power cost, less maintenance, and higher dust collection efficiency. However, a low A/C also means more filter area and higher initial cost. Assuming no short-circuiting of inlet gas is occurring around the bags within the baghouse, poor collection efficiencies usually result from too high an air-to-cloth ratio for the application. Returning to conservative A/C ratios in the range of 2.0 or even 1.0 solves most former baghouse failure problems.

Baghouses typically have overall collection efficiencies greater than 98% for particles less than 10 microns in aerodynamic diameter. Therefore, they are widely used when high collection efficiencies for relatively small particulate matter are required. Their high collection efficiency is due to three mechanisms of impaction, interception, and diffusion acting simultaneously to remove particles down into the submicron size range.

For the environmental engineer, the challenge is to properly match one of the various types of baghouses available with the gas stream and the particulate matter to be collected.

Gas Conditioning. Since most baghouses are not very tolerant of high temperatures and moisture condensation, conditioning of the challenge gas stream is an extremely important consideration. High temperatures can cause bag media failure and even fires or baghouse dust explosions; gas-entrained sparks or static electricity buildup on the bags can also cause fires and explosions. If water vapor within the gas stream is allowed to condense on the filter cake, the cake turns to mud, the pressure drop quickly approaches very high values, and the baghouse blinds. High gas temperatures and moisture condensation must be avoided in baghouse systems. Three methods of reducing baghouse inlet gas temperatures are ambient air dilution, radiation and free convection, and water injection. A starting point common to all three cooling methods is the calculation of the heat energy that must be removed from the gas stream prior to baghouse entry.

Gas Stream Cooling Requirements. The following heat transfer rate equation is used to calculate the heat energy that must be removed from the gas stream:

$$q = mc_p(T_p - T_{bi}) \tag{11.26}$$

where q is the heat removed in Btu/hr; m is the mass flow rate of gas in lb/hr; c_p is the specific heat of the gas at constant pressure in Btu/lb, °F; T_p is the process gas temperature; and T_{bi} is the baghouse gas inlet temperature in degrees Fahrenheit. Regardless of the method of gas cooling that is employed prior to baghouse entry, the amount of heat to be removed from the gas stream is the same for the three methods.

Ambient Air Dilution. Mixing ambient air with the hot process gas prior to baghouse entry is an effective method of cooling. Even in some systems employing other methods of cooling, emergency cooling is usually effected by dilution with ambient air. Usually, a simple tee employing a butterfly valve in the ambient air leg is inserted in the duct just upstream of the baghouse. A temperature-controlled modulating valve driver can be used to position the butterfly valve and maintain the desired inlet baghouse gas temperature.

The following heat transfer rate equation can be solved for the desired mass flow rate of dilution air to maintain the inlet baghouse gas temperature below desired limits:

$$m_{\text{air}} = \frac{q}{c_{p_{\text{air}}}(T_{bi} - T_{\text{amb}})} \tag{11.27}$$

where q is the heat energy to be removed from the process gas stream, Btu/hr; c_p is the specific heat of the ambient air, 0.24 Btu/lb, °F; T_{bi} is the baghouse inlet gas temperature; and T_{amb} is the ambient dilution air temperature, °F.

Now the dilution air volume flow rate, Q, is calculated from

$$Q = \frac{m_{\text{air}}}{\rho_{bi_{\text{air}}}} \tag{11.28}$$

where $\rho_{bi_{\text{air}}}$ is the density of the gas entering the baghouse, lb/ft^3.

While this method is effective for cooling, it must be pointed out that the dilution air volume flow rate is a significant portion of the total gas the baghouse must now handle. In some situations, depending on the original process gas temperature, the dilution air flow rate will exceed the process gas flow rate. Therefore, the main disadvantage with this method of cooling is the added gas that the baghouse must now process.

Radiation and Free Convection. This cooling method employs a long duct for radiation and free convection gas cooling between the process and the baghouse. With this method, the heat energy needed to be removed from the gas stream can be calculated from the following equation:

$$q = q_{\text{free convection}} + q_{\text{radiation}} \tag{11.29}$$

or

$$q = U_{\text{combined}}A(\text{LMTD}) \tag{11.30}$$

where U_{combined} is the overall heat transfer coefficient of the cooling duct, Btu/(hr ft^2 °F); A is the exterior duct area for heat transfer; and LMTD is the log mean temperature difference between the cooling duct and the ambient air. Since we are interested in the diameter and length required for the cooling duct, the above equation is solved for the heat transfer area, A, as

$$A = \pi DL = \frac{q}{U_{\text{combined}}\text{LMTD}} \tag{11.31}$$

and for a chosen diameter,

$$L = \frac{q}{U_{\text{combined}} \text{LMTD} \pi D} \tag{11.32}$$

Radiation and free convection cooling has the distinct advantage over the dilution cooling scheme in that the gas handled by the baghouse contains no added dilution air and, therefore, is significantly less in volume flow. There is the added expense of the ductwork, but this is usually more than offset by the reduction in the amount of gas passing through the baghouse. Since the controlling thermal resistance in this cooling method is the internal duct wall convective heat transfer coefficient, schemes to spray water on the external surface of the pipe to significantly reduce the required pipe length are usually not successful.

This cooling method can require up to a few hundred feet of cooling duct, so there must be ample room available at the site for successful implementation. Where space is available, long horizontal duct runs have worked successfully; where space is at a premium, serpentine configurations can be used effectively.

Evaporative Cooling. Utilization of the latent heat of vaporization of water of about 1000 Btu/(lb water) is an extremely effective method of cooling process gas prior to baghouse entry. This method requires a contact chamber, spray nozzles, and a dew point sensing and feedback control system to modulate the water flow to the spray nozzles. This method has the advantages of small space requirement and lower capital cost, but it does require a reasonably sophisticated water flow control system to prevent dew point problems (liquid drops forming in the baghouse).

The total heat energy to be removed by the injected water is the sum of the sensible and latent heat absorbed by the water. This includes the sensible heat raising the liquid water from its injection temperature up to its vaporization point, the latent heat during vaporization, and the sensible heating of the water vapor up to the desired baghouse inlet gas temperature. In equation form:

$$q_{\text{total}} = q_{\text{sensible}} + q_{\text{latent}} \tag{11.33}$$

where

$$\begin{aligned} q_{\text{sensible}} &= m c_p \Delta T_{\text{liquid}} + m c_p \Delta T_{\text{vapor}} \\ q_{\text{latent}} &= m h_{fg} \end{aligned} \tag{11.34}$$

c_p is the specific heat at constant pressure of liquid water or vapor and h_{fg} is the latent heat of vaporization of water changing phase from a liquid to a vapor, 1000 Btu/lb.

Example 11.2. It is desired to cool 50,000 acfm of process gas at 1500°F down to 400°F prior to baghouse entry using water injection. Compute the amount of heat needed to be removed from the gas stream and the needed water injection rate as well as the rise in dew point temperature after injection. Heat to be removed from gas stream:

$$\begin{aligned} q &= m c_p (T_{\text{process gas}} - T_{\text{baghouse inlet}}) \\ q &= (50{,}000)(0.02)(0.25)(1500 - 400) \\ q &= 275{,}000 \, \frac{\text{Btu}}{\text{min}} \end{aligned} \tag{11.35}$$

Mass flow rate of water assuming initial water temperature is 70°F and the vaporization point of water occurs at 212°F:

$$q = m c_p \Delta T_{\text{liquid}} + m c_p \Delta T_{\text{vapor}} + m h_{fg} \tag{11.36}$$

substituting:

$$275{,}000\frac{\text{Btu}}{\text{min}} = m\left[(1.0)(212-70)+(0.46)(400-212)+\left(1000\,\frac{\text{Btu}}{\text{lb}}\right)\right] \quad (11.37)$$

solving for m:

$$m = 224\,\frac{\text{lb water}}{\text{min}} \quad \text{or} \quad 27\,\frac{\text{gal}}{\text{min}} \quad (11.38)$$

Therefore, 27 gpm of water must be injected directly into the gas stream to cool it from 1500°F down to 400°F. This injection of liquid adds 7800 acfm of vapor to the process gas stream at 400°F which must be considered as part of the baghouse air to cloth ratio.

The vapor equivalent of the 27 gpm liquid water injected is calculated as follows:

$$\text{Vapor volume (ft}^3) = 10.1\left(\frac{T}{P}\right)L \quad (11.39)$$

where L is in gallons, T is temperature in Rankine (i.e., 400 + 460), and P is pressure in Hg. Substituting:

$$\text{Vapor volume (ft}^3) = 10.1\left(\frac{860}{29.92}\right)27 \quad (11.40)$$

$$= 7838 \text{ acfm at } 400°\text{F}$$

This is a 36% increase in gas flow that the baghouse will have to handle (over that of the cooled process gas of 21,938 acfm at 400°F).

Effect on Dew Point

The added water vapor to the process gas stream will increase the specific humidity of the mixture. The specific humidity (SH) of the mixture is

$$\text{SH} = 224\,\frac{\text{lb water/min}}{(21{,}938 \text{ acfm})}\left(0.046\,\frac{\text{lb}}{\text{ft}^3}\right) \quad (11.41)$$

$$= 0.22\,\frac{\text{lb water}}{\text{lb dry air}}$$

This specific humidity corresponds to a dew point of 151°F and compares to 42°F prior to water injection. Therefore, water injection of 27 gpm elevates the dew point of the baghouse inlet gas from 42°F to 151°F. Caution must be exercised so that the 151°F dew point temperature of the inlet gas is never approached or liquid droplets will blind the baghouse.

Sulfur Dioxide

The emission of sulfur dioxide (SO_2) is a common occurrence from the combustion of fossil fuels for power generation and from other industrial processes that involve the use of sulfur. Combustion processes, however, generate up to 75% of all SO_2 emissions with the fuel itself being the source of sulfur. As a result, this discussion will focus on the control of SO_2 from combustion sources.

SO_2 is formed when fuel-bound sulfur is combusted, and in the process converted to SO_2 by combining with the combustion air at a molecular weight ratio of 2:1. As a result, one gram of fuel-bound sulfur results in 2 grams of SO_2 and necessitates substantial control measures for flue gas treatment.

The nature of formation of SO_2 leaves two options in its emissions control. One, control emissions by eliminating the formation of SO_2, or two, remove SO_2 from the combustion flue gas after its formation. The first process is referred to as fuel conversion and the second as flue gas desulfurization (FGD).

Fuel Conversion

The process of fuel conversion seems relatively simple as it consists of the removal of sulfur from the fuel to be combusted or the use of a different, lower-sulfur fuel. This can involve the use of natural gas, distillate oils with low sulfur contents, or low-sulfur coals in place of higher-sulfur counterparts.

These alternatives pose simple answers to a more complex issue, in that many combustion units are designed to burn a specific fuel type, making switching to a different fuel cost-prohibitive. Additionally, the use of low-sulfur coal is often cost-prohibitive as the transportation costs are often extreme. Further, fuel conversion involving the removal of sulfur from crude oils that can be distilled is far simpler than the removal of sulfur from coals.

As a result of the cost issues and the complexity of the removal of sulfur from fuel sources, controlling emissions by fuel conversion is typically a good method of reducing SO_2 emissions from sources burning distillate oils or sources that have exposure to a supply of low-sulfur coal. Often fuel conversion is used to lower total control costs or to reduce SO_2 emissions below regulated levels.

Flue Gas Desulfurization

Flue gas desulfurization (FGD), in this discussion, is a process that focuses on the removal of SO_2 from a combustion gas stream. On a broader scale, flue gas desulfurization can be applied to control sulfurous compounds in many different types of industrial emissions. FGD systems consist of two different types—throwaway and regenerative—that operate on either a wet or dry basis. Throwaway and regenerative systems differ in that throwaway systems convert the sulfur compounds into a form that is disposed of as a solid waste, while regenerative systems convert the sulfur into either an elemental sulfur or sulfuric acid form that can be reused or sold.

Throwaway Systems. Throwaway systems operate by the scrubbing of the gas stream where the scrubber water contains a reactant, typically lime or limestone, such that the SO_2 present reacts and is removed in a solid form in the scrubber blowdown. Often a simple scrubber configuration is used and is followed by some sort of gravity thickener to remove the solids from the effluent scrubber water. The overall stoichiometry for the reaction can be represented by the following [Cooper and Alley, 1990]:

$$\begin{aligned} CaO + H_2O &\rightarrow Ca(OH)_2 \\ SO_2 + H_2O &\rightarrow H_2SO_3 \\ H_2SO_3 + Ca(OH)_2 &\rightarrow CaSO_3 \cdot 2H_2O \\ CaSO_3 \cdot 2H_2O + \tfrac{1}{2}O_2 &\rightarrow CaSO_4 \cdot 2H_2O \end{aligned} \tag{11.42}$$

Limestone and lime systems operated in this manner are capable of 90 and 95% removal efficiencies, respectively.

A variation on the scrubbing method is the spray injection of a lime slurry into the hot gas stream. As the slurry is injected into the hot gas stream the SO_2 reacts with the lime in the aqueous phase, dries, and is removed in the solid phase by filtration in a baghouse. This type of operation results in lower maintenance, energy use, and operating costs [Cooper and Alley, 1990].

Dry lime or limestone injection systems operate by injecting the lime or limestone into the flue gas, causing the SO_2 present to adsorb and react. Final reaction products are then removed by filtration. These systems are limited by the fact that the site of reaction is only on the surface of the reactant, and is thus hindered in a manner that spray injection is not.

Regenerative Systems. Regenerative FGD systems operate in a manner that is similar to throwaway systems with the exception of an additional step that either transforms or reclaims the reaction products. An example of this is magnesium oxide scrubbing. This process employs the use of MgO as the scrubbing agent, wherein SO_2 is absorbed and forms magnesium sulfite (or sulfate). This reaction by-product is then recalcined, forming MgO and SO_2, with the MgO being returned to the scrubber and the SO_2 being captured in a concentrated form that can be utilized as a feedstock in sulfuric acid production. Disadvantages to the system lie in the heat required to recalcine the magnesium sulfite [Wark and Warner, 1981]. Other examples of regenerative FGD systems are single alkali scrubbing, double alkali scrubbing, citric acid scrubbing, the Sulf-X process, and the Wellman-Lord process.

Nitrogen Oxides

The formation of nitrogen oxides (NO_x) is similar to SO_2 in that NO_x is formed as the result of combustion processes. NO_x is composed of several different forms of nitrogenous compounds. Included among these are nitrous oxide (N_2O), nitric oxide (NO), nitrogen dioxide (NO_2), nitrogen trioxide (N_2O_3), nitrogen tetroxide (N_2O_4), and nitrogen pentoxide (N_2O_5). While all of these components can exist in a combustion gas stream at some point, NO_x typically refers to NO and NO_2.

The formation of NO_x begins with the formation of NO in the flame zone of a combustion process where fuel-bound nitrogen combines with the oxygen present in combustion air to form NO and is termed fuel NO_x. Additionally, atmospheric nitrogen in the combustion air also combines with oxygen to form NO and is known as thermally generated NO_x. After formation, the NO generated rapidly cools with the gas stream where a majority (approximately 95%) converts to NO_2. As a result, NO_2 is the principal component of NO_x that is emitted to the atmosphere. For a more complete discussion of combustion by-product formation, the reader is referred to the chapter on incineration.

NO_x can be controlled by a variety of methods. These control methods can be grouped in three forms: fuel conversion, combustion modifications, and flue gas treatment. Fuel conversion is the process of changing fuels to take advantage of lower-nitrogen fuels. This consists primarily of the use of natural gas over fuel oils or coals, and suffers from the same types of limitations as fuel conversion to control SO_2, described above. Therefore, fuel conversion will be omitted from this discussion. The second form is combustion modifications that lower the potential for formation of NO_x during combustion processes. The third form is the use of downstream controls to remove NO_x from the flue gas.

Combustion Modifications

Combustion modifications take advantage of the characteristics of NO_x formation in an effort to minimize it. These efforts focus on the combination of nitrogen and oxygen in the region of combustion (flame zone) at a high enough temperature to form NO. Modifications employed for this include combustion air variations, low NO_x burners (LNB), and fuel reburning [Makansi, 1988].

Combustion Air Variations. One of the most simple combustion modifications is to alter the manner in which the combustion air is supplied to the flame. This is done in an effort to lower both the peak flame temperature and oxygen concentrations in the regions of highest temperature. Typical methods consist of the following [Makansi, 1988]:

1. Placing burners out of service (BOOS) and fuel biasing that provide combustion regions that are fuel rich followed by regions that are fuel lean to stretch the combustion zone, lowering peak temperatures and oxygen concentrations. NO_x reductions of up to 20% are possible with these methods.
2. Low excess air firing (LEA) to reduce the excess combustion air from typical levels of 10–20% to 2–5%. This reduces oxygen concentrations in the flame zone and results in decreased NO_x formation, up to 20%, and a more efficient flame.
3. Overfire air is a means of air staging, or elongating the combustion zone by forcing a portion, 10–20%, of the combustion air to a set of ports above the burners. This in essence creates fuel-rich and fuel-lean zones and results in NO_x emission reductions of 15–30%.
4. Flue gas recirculation (FGR) is a process that recycles a portion of the combustion gases back into the virgin combustion air to reduce combustion temperatures and thereby reduces thermally generated NO_x up to 20–30%.

Low* NO_x *Burners. Low NO_x burners are a technology that has developed in order to retrofit existing, NO_x intensive, combustion units with a burner that will allow the combustion unit to operate at its design level, but with substantially lower NO_x emissions. This is done within the burner itself by combining the combustion air and fuel in different manners (this varies from vendor to vendor) such that oxygen levels are reduced in the critical NO_x formation zones. Low NO_x burners themselves are capable of NO_x reductions of up to 20–30%, and when coupled with overfire air systems NO_x reductions of up to 50% are possible [Smith, 1993]. Installation of these types of burners is limited by the design of the combustion unit, and in some cases will require modification of existing fuel-and air-handling equipment.

Fuel Reburning. Fuel reburning is a method of fuel staging wherein a portion of the fuel for the combustion unit is fed into the unit downstream of the initial combustion zone. This action creates a second combustion zone which is operated substoichiometrically. In the second zone, the NO created in the first zone is kept at temperature for longer periods of time and thereby allowed to convert back to elemental N_2. The fuel added for reburning must be of high enough volatility to allow continued combustion and therefore this method favors oil- and gas-fired units. NO_x reductions of 75–90% are possible with this type of configuration [Makansi, 1988].

Downstream Processes

Downstream controls for the removal of NO_x from combustion gas streams consist of two types. The first type is the addition of urea or ammonia to the hot combustion stream, causing the NO_x in the gas stream to be converted into water and nitrogen. This first type includes both selective catalytic and noncatalytic reduction. The second type of downstream control is the removal of NO_x from a gas stream by scrubbing.

Selective Catalytic Reduction. Selective catalytic reduction (SCR) of NO_x utilizes a catalytic transition metal grid in combination with ammonia at temperatures of 600–700°F to convert the NO_x present in the gas stream back to elemental N_2 and water. This process is governed by the NO_x concentration in the flue gas with the injection of ammonia being a function of this concentration. Removal efficiencies for this type of operation range from 80–90% [Makansi, 1988].

However, drawbacks to this type of removal system include a poor catalyst life of only one to five years and emission of unreacted ammonia. The portion of ammonia passing through the system is referred to as ammonia slip.

Selective Noncatalytic Reduction. Selective noncatalytic reduction (SNCR) is similar to SCR in that the injection of ammonia or urea is utilized to convert NO_x emissions into elemental N_2 and water. However, this system does not utilize a catalyst for this reaction; instead, the ammonia is injected in a higher-temperature region of the gas stream taking advantage of the heat as a catalyst. The temperature required, typically 1600–2000°F, necessitates that the injection location be either physically in the combustion unit or immediately downstream in the ductwork. Removal efficiencies of up to 80% are possible with SNCR.

Flue Gas Denitrification. Flue gas denitrification (FGDN) is the process of scrubbing NO_x from a gas stream. However, while the principles of operation for the scrubbing system are the same for its particulate removal and FGD counterparts, the scrubbing liquid is substantially different. This is due to the fact that many of the NO_x constituents vary in their degree of water solubility. As a result, scrubbing systems for the removal of NO_x typically employ a series of individual scrubbers whose makeup liquid varies with intended removal. Operation of the scrubbers is relatively maintenance free with the requirement of continued chemical addition for the makeup water. Removal efficiencies can approach 90% with FGDN.

Volatile Organic Compounds

The control of VOCs is a complex issue as there are a great number of organic compounds being emitted, either directly or indirectly, with all compounds having various structures and properties. The issue is further complicated by the fact that a majority of the air toxics list, provided earlier in the chapter, is composed of VOCs and requires a MACT level of control as opposed to the nontoxic VOCs RACT level of control requirement for most cases of VOC emission. As a result, there are a great number of VOC emissions controls that are in place or are being recommended for various situations.

However, the description of all methods of VOC control is beyond the scope of this chapter. The reader is referred to the Air and Waste Management Association's *Air Pollution Engineering Manual* and the chapter in this handbook on incineration for further information on the subject of VOC emissions control for those points not covered in this discussion. Emissions control focused on in this discussion will consist of the standard methods of adsorption and incineration that are capable of high removal efficiencies and are most likely to be considered in control technology reviews.

Carbon Absorption

Carbon absorption is the use of the physicochemical process of adsorption to remove dilute organics from a gas stream. Adsorption itself is the interphase accumulation or concentration of substances at a surface or interface of the adsorbent, in this case activated carbon [Weber, 1972]. Accumulation or concentration of molecules at the surface of the adsorbent is the result of the molecule being attached to the surface by van der Waals forces, physical adsorption, chemical interaction with the adsorbent, or chemical adsorption. Both types of adsorption are reversible, although chemical adsorption requires a greater driving force to desorb the molecules.

The adsorption process itself consists of the use of an adsorbent medium, typically activated carbon, that is placed in the gas stream such that the gas has to pass through the medium. As the gas passes through the carbon, the constituents of the gas stream make contact with it and adsorb. After a period of time, the carbon can no longer adsorb organics and is removed from service. This carbon can then be reactivated by steam stripping or incineration to remove the organics, or disposed of as a solid or hazardous waste. Steam stripping of the carbon provides the option of reclaiming the organics previously lost to the atmosphere. Activated carbon is composed of nutshell or coal that has been charred in the absence of oxygen and activated by steam stripping or various other methods.

In operation, a carbon adsorber consists of a bed of carbon whose dimensions are a function of the VOCs being removed. The bed usually operates in a passive mode, as described earlier, with the carbon that is first exposed being the first expended. As the carbon is expended or spent, a front develops that demarks spent carbon from carbon that is still active. Through the life of the carbon bed, the front progresses from one end of the bed to the other. The end of the operational life of the bed is realized when the front "breaks through" the opposite end of the bed and the bed has lost its ability to remove organics. In operation, the bed is removed from service and replaced prior to breakthrough such that continued emissions control is ensured.

Carbon adsorption is employed for a wide variety of uses due to its versatility and the potential to reclaim organics. Operation of carbon beds is a relatively simple process that is maintenance nonintensive. Typically, maintenance of the beds involves only the servicing of the air-handling equipment and exchange or recharge of the bed once the carbon is spent. In many cases, exchange of the carbon is handled by the vendor. Lifetimes of the carbon bed and the removal efficiency it will provide are functions of the design of the bed. However, in most cases carbon adsorbers are designed for removal efficiencies of 99% or greater.

The most serious drawback to adsorption is the fact that it is not a means of ultimate disposal. Adsorption provides only the ability to concentrate and transfer a pollutant from a gas stream to a medium where ultimate disposal is possible.

Incineration

The second type of control commonly employed for the emission of VOCs is incineration. This control oxidizes the organics to CO_2 and water by combustion or through the use of a catalyst. Incineration for the control of VOCs is a means of ultimate disposal and is typically employed for more concentrated gas streams where the composition of organics in the gas stream is known and reasonably constant. Destruction efficiencies in excess of 99% are possible with incineration. However, large variations in the organic feed stream result in inefficiencies during combustion and can produce products of incomplete combustion (PICs) and result in poor destruction efficiencies. The degree to which this is possible is dependent on the type of incinerator employed. Again, the reader is referred to the chapter on incineration for a more complete discussion of combustion and operation of incinerators in general.

There are three types of incinerators: direct, thermal, and catalytic. These vary by the manner in which the oxidation of the organics takes place, or by the manner in which the organics are combusted.

Direct Incineration. Direct incineration is the combustion of the gas stream of organics itself. This is done by igniting the gas stream and allowing the organics to combust instantaneously on their own and requires concentrations of organics in the gas stream that will support this type of combustion. As a result, this type of incineration is used on gas streams with concentrated organics. Flaring is a form of direct incineration.

Thermal Oxidation. Thermal oxidation is a process that oxidizes organics by introducing the organics around a flame such that oxidation of the organics is the result of the elevated temperature in the chamber. Unlike direct incineration, the organics themselves are never the primary fuel for combustion; they are a secondary combustion process. In fact the concentrations of organics must be below that required for combustion on their own. Oxidation in this case is a function of the temperature, 1200–2000°F, and residence time, 0.2–2.0 seconds, of the organics in the combustion chamber.

Catalytic Oxidation. Catalytic oxidation is the third type of oxidation process commonly used in the control of VOCs. This process utilizes the ability of a transition metal catalyst to oxidize organic compounds to CO_2 and water at relatively low temperatures. For this type of oxidation, organic concentrations below those that require thermal or direct incineration are used. Operation of this type of incinerator consists of the introduction of the organic-laden gas stream into a bed of catalyst at temperatures of 650–800°F. As the gas stream passes through the bed, the organics react in the presence of the catalyst and are oxidized.

11.6 Odor

Odor is defined as the characteristic of a substance that makes it perceptible to the sense of smell. This definition includes all odors regardless of their hedonic tone. In the field of air pollution, concern over odor is limited to those compounds or mixtures thereof that result in the annoyance of an individual.

In human terms, compounds producing displeasure due to their odor are typically not threatening to human health but do produce a great deal of physiological stress. In fact, in Metcalf and Eddy's text on wastewater engineering, one of the principal characteristics of wastewater considered is the odor. This is due to the fact that offensive odors can cause decreased appetites, lowered water consumption, impaired respiration, nausea, and mental perturbation. The Metcalf and Eddy text further reports that in extreme situations offensive odors from wastewater treatment facilities can have substantial impacts on a communitywide basis [Tchobanoglous, 1979].

Sense of Smell

The sense of smell is a sensation that is produced when a stimulant comes into contact with the olfactory membranes located high in the nasal passages. As the stimulants contact the olfactory membranes they are absorbed and excite the membranes. There are thought to be seven primary classes of stimulants that affect the olfactory membranes. These seven stimulants are as follows.

1. Camphoraceous
2. Musky
3. Floral
4. Peppermint
5. Ethereal
6. Pungent
7. Putrid

The seven stimulants are felt to be the primary sensations; however, this is a topic of debate as others feel that there are perhaps 50 or more classes of primary stimulants. The primary stimulants illustrate the complexity of the sense of smell in relation to the other senses in that there are only three primary color sensations of the eye and four primary sensations associated with taste.

While there are seven stimulants of the olfactory membranes, reaction to a stimulant varies from person to person: a rose may not smell as sweet to one person as it does to another. This results in great difficulty in assessing the magnitude of an odor.

Characteristics of Odor

Human response to odor depends on the characteristics of the property being assessed. The odor intensity, detectability, character, and hedonic tone all influence the response to a particular compound [Prokop, 1992].

The odor intensity is the strength of the perceived sensation. Intensity of the odor is a function of the concentration of the odiferous compound coupled with a human response. Intensities of odors can be the same for different compounds at different concentrations.

Detectability is the minimum concentration of an odiferous compound that produces an olfactory response. The detectable limit for odors is referred to as the odor threshold. Measurement of this characteristic of odor is difficult as the threshold varies from person to person and is further complicated as the detectable concentration varies with previous exposure. Odor threshold limits are often reported as the concentration that produces an olfactory response in 50% of the test population.

The character of an odor refers to the associations of the person sensing the odor. This is the characteristic that separates the odors of different compounds that are presented in similar

Table 11.11 Odor Descriptions of Various Compounds

Compound	Formula	Molecular Weight	50 Percent Detection Thresholds (μg/m³)	Odor Descriptions
Acetaldehyde	CH_3CHO	44	90	Pungent, fruity
Ammonia	NH_3	17	3700	Pungent, irritating
Dimethyl sulfide	$(CH_3)_2S$	62	51	Decayed cabbage
Hydrogen sulfide	H_2S	34	5.5	Rotten eggs
Methyl mercaptan	CH_3SH	48	2.4	Rotten cabbage
Pyridine	C_5H_5N	79	1500	Pungent, irritating
Trimethylamine	$(CH_3)_3N$	59	5.9	Pungent, fishy

Sources: Adapted from Prokop, W. H. 1992. *Air Pollution Engineering Manual—Odors.* Air and Waste Management Association. Van Nostrand Reinhold, New York; and Nagy, G. Z. 1991. The odor impact model. *J. Air Waste Manage. Assoc.* 41(10):1360–1362.

intensities. Table 11.11 presents a list of several different compounds with their odor thresholds and associated characteristics.

Similar to the character of an odor is the hedonic tone of the odor. The hedonic tone is a reflection on the degree of pleasantness or unpleasantness associated with an odor. Hedonic tone is assessed by the response of different individuals.

Odorous Compounds

Odorous compounds are emitted from a variety of sources and consist of both organic and inorganic compounds that exist primarily in the gas phase, but can also exist as solids. Table 11.11 lists examples of both organic (such as acetaldehyde) and inorganic (such as ammonia) compounds. Typically odorous compounds are emitted from processes which involve anaerobic decomposition of organic matter [Prokop, 1992].

Most odorous compounds are significantly volatile compounds with molecular weights from 30 to 150. Typically the lower molecular weight compounds have higher vapor pressures and thus are more volatile [Prokop, 1992]. A positive example of the use of an odiferous compound is the addition of mercaptans to nonodiferous natural gas supplies. This process is done in order to provide a method of detection for leaks of natural gas, and thus, hopefully, prevent catastrophes.

Measurement

The measurement of odor is subjective as the process necessitates the assessment of odor characteristics by noninstrumental means. Intensity, detectability, character, and hedonic tone are typically assessed by sensory methods involving the use of an odor panel. An odor panel is a group of people that are presented a series of gas streams and asked to qualify and quantify what is presented.

The particulars of the measurement of odor are beyond the scope of this chapter. For further information, the reader is referred to the chapter on odors in the Air and Waste Management Association's *Air Pollution Engineering Manual* for a synopsis of odor measurement involving odor panels.

Odor Control Techniques

Odor control techniques involve essentially the same elements as the methods employed in the control of VOCs. Controls that are typically used consist of the following:

1. Process modification
2. Masking agents or odor modification

Table 11.12 Process Modification

Advantages	Disadvantages
Process changes alone may reduce odors enough to minimize odor complaints	The effect of modifications is difficult to predict without extensive testing
Changes can be made relatively quickly	Process modifications may not be able to provide a suitable reduction
Costs should be lower than the installation of downstream controls	Modifications could result in decreased production capacity

3. Carbon adsorption
4. Absorption/chemical oxidation
5. Incineration

The first control to be considered is prevention of odor formation. This is an attempt to reduce the generation of odors through changes in process equipment design and/or operating procedures. This process would typically involve the identification and elimination of the area in the process that creates the compound of concern. In some instances, this can simply involve ensuring that a waste stream receives proper aeration. However, another situation would be one in which the compound in question is an integral part of the process and as a result necessitates the use of a downstream control. Advantages and disadvantages of process modification are listed in Table 11.12.

A second consideration in the reduction of an emissions odor is odor masking or modification. This type of control method attempts to modify the hedonic tone of the odor or the emissions stream such that the offending odor is less unpleasant. Odor masking does not result in a reduction of the offending compound; it merely attempts to alter the intensity or character of the offending compound. Usually these types of controls are not very effective and are expensive in relation to other methods of control.

Carbon adsorption involves the removal of the odiferous compounds by passing the emission stream through a carbon bed. The carbon is stationary in the bed, and as the gas stream passes through, the organics adsorb onto the carbon through a physicochemical process. As the carbon reaches a maximum capacity it is either exchanged with new carbon or regenerated. Regeneration of the carbon is typically done by passing a steam through the bed to desorb the organics. A few of the advantages and disadvantages are listed in Table 11.13.

Absorption is the removal of organics by wet scrubbing with various oxidizing agents to remove odors from emission streams. A typical scrubbing system would consist of a venturi scrubber followed by a packed-tower scrubber. The venturi scrubber serves to remove particulate and presaturate the gas stream. After the venturi, the gas stream is passed through a packed-tower scrubber where the odiferous organics are removed. Sodium hypochlorite and potassium

Table 11.13 Carbon Adsorption

Advantages	Disadvantages
Control efficiencies can range from 95 to 98%	Extensive preconditioning of the gas stream may be required
A relatively small amount of auxiliary fuel is required since carbon beds are typically regenerated every 8 to 24 hours	Adsorption beds typically require large amounts of space
	Specialized maintenance may be required
	Even with regenerating systems, periodic replacement of carbon is needed every three to five years
	Regeneration off gas may be another source of odor

Table 11.14 Absorption

Advantages	Disadvantages
Respectable odor control efficiencies are possible	Control efficiencies may not meet required levels for odor control
No auxiliary fuel is needed	Particulate control is needed upstream of the scrubber
	Corrosion of the scrubber is a problem
	Creates a wastewater disposal problem

permanganate are two of the more commonly used oxidants in packed-tower scrubbers. Table 11.14 lists a few of the advantages and disadvantages of absorption for odor control.

The final control option generally used for odors is incineration. Thermal and catalytic incinerators are employed in controlling odors; however, thermal incineration is more common. Temperatures and residence times for thermal incineration vary depending on the nature of the gas stream. Catalytic incineration is a method to lower operating temperatures and thus lower operating costs; however, catalyst poisoning is a significant problem. Both types of incineration offer the most effective means of odor control, but do so at the highest capital and operating costs.

11.7 Air Pollution Meteorology

After an emission source has been characterized as to type and size, the question arises as to how the source and the resulting emissions will affect the quality of the downwind ambient air. This determination requires that the source further be described by modeling, or the estimation of the atmospheric dispersion of the various components of the gas stream. The modeling step is often what determines an appropriate emission rate.

Once the gas stream is emitted into the atmosphere, a complex series of events begins to unfold. The components (or pollutants) in the gas stream join the environment and are transported, dispersed, and eventually removed from the atmosphere according to the physical nature of the component. However, before examining the dispersion of the gas stream, a basic understanding of the environment which governs the dispersion is required.

Wind (Advection)

The atmosphere is an ocean of air that is in continual motion primarily as a result of solar heating and the rotation of the earth. As the sun heats different parcels of air in different areas, the heated air responds by expanding and increasing in pressure. Subsequently, differential pressure areas arise and result in large-scale air motions, or winds. This is illustrated on a global basis by the large-scale winds created as a result of differential heating between the equator and the poles. Figure 11.13 illustrates the various winds that would be created by differential heating on a global scale if the earth were smooth and of homogenous composition [Cooper and Alley, 1990].

Wind Variations

As described previously, movement of air (or wind) is a result of the pressure gradient created by high- and low-pressure regions interacting. This pressure gradient results in a force, F_p, that creates a velocity vector from the high-pressure region to the low-pressure region and, ideally, is perpendicular to both pressure regions or isobars. This is shown in Fig. 11.14(a), where V is the velocity vector or wind created by the pressure gradient force.

However, as the earth is rotating the effects of the Coriolis forces, F_{Cor}, are realized on the movement of air as well. These forces result in the deflection of the wind from the ideal perpendicular gradient. In the northern hemisphere, these forces deflect the wind to the right, relative

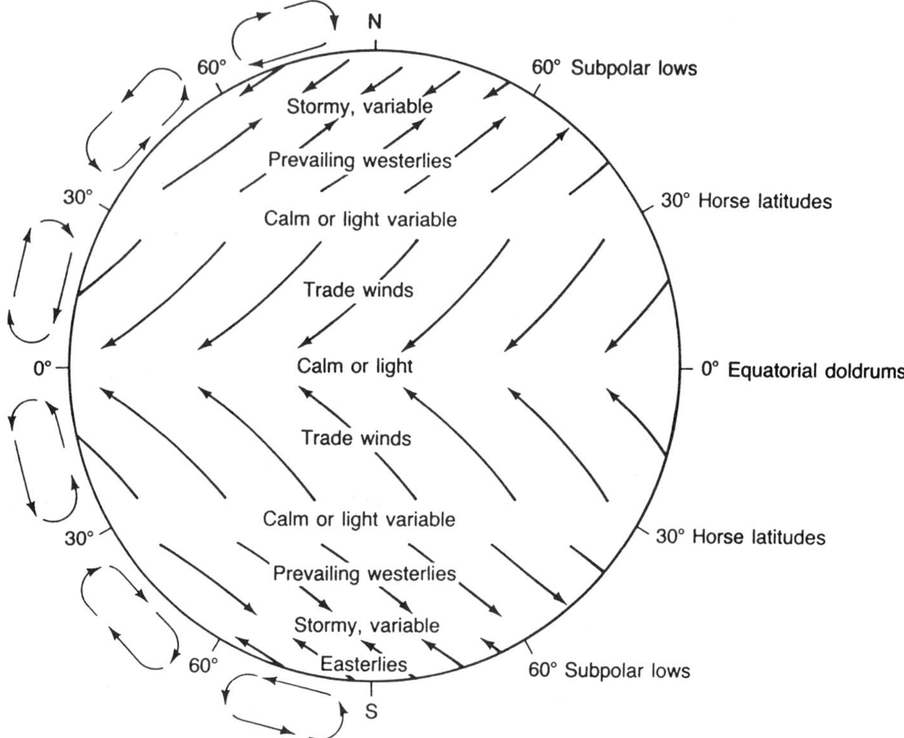

FIGURE 11.13 General global air circulation. (*Source:* Cooper, C. D. and Alley, F. C. 1990. *Air Pollution Control: A Design Approach.* Waveland Press, Prospect Heights, IL. Reprinted by permission of Waveland Press, Inc. All rights reserved.)

to the surface, while in the southern hemisphere the force deflects the wind to the left. The magnitude of the Coriolis force is a function of the velocity of the air, latitude, and the earth's rotational speed. These forces are at a maximum at the poles and zero at the equator [Wark and Warner, 1981]. When the pressure gradient and Coriolis forces are combined, the result, depicted in Fig. 11.14(b), is a velocity vector whose angle is a function of the relative magnitude of the two forces to each other. In the ideal case, the pressure gradient and Coriolis forces are balanced such that the resultant velocity vector is parallel to the isobars. This is depicted in Fig. 11.14(c). The resulting wind always moves such that the pressure gradient force is in the direction of the low-pressure region and is balanced by the Coriolis force in the direction of the high-pressure region. The corresponding velocity vector is at a right angle to both. Movement of the wind is to the right of the Coriolis force and is therefore moving such that the low-pressure region is to the left of the vector. This idealized wind is referred to as the geostrophic wind and approximates conditions a few hundred meters above the earth's surface [Wark and Warner, 1981].

While the geostrophic wind is associated with parallel isobars, the gradient wind is associated with curved isobars. The gradient wind differs from the geostrophic wind by the centripetal acceleration, a_c, associated with the movement of a parcel of air in a curvilinear motion. The gradient wind is evident around centers of high and low pressure [Wark and Warner, 1981].

Winds at the earth's surface are further complicated by the fact that the earth is not smooth and homogenous. As a result several other factors need to be considered when discussing the magnitude and direction of the wind. Among these are [Cooper and Alley, 1990]:

1. Topography
2. Diurnal and seasonal variation in surface heating
3. Variation in surface heating from the presence of ground cover and large bodies of water

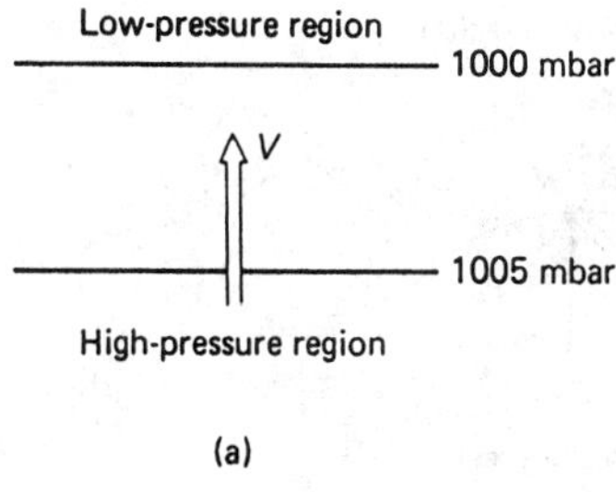

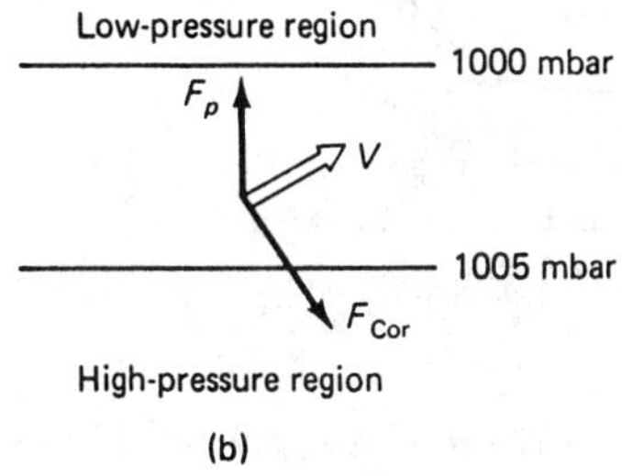

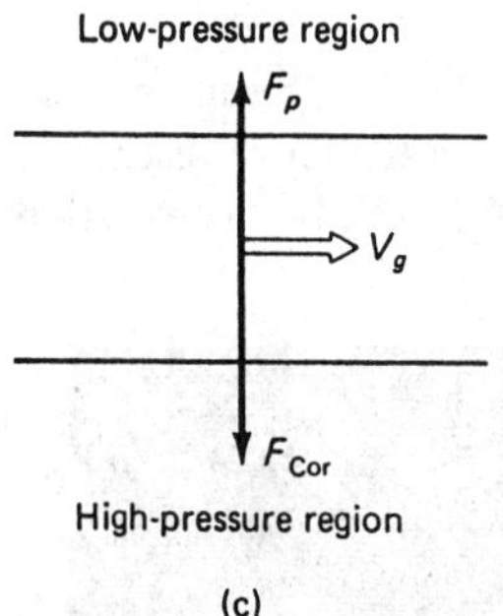

FIGURE 11.14 Effects of various forces on wind direction relative to pressure isobars. (a) Pressure gradient force only. (b) Pressure gradient force with Coriolis force. (c) Balanced pressure gradient and Coriolis forces. (*Source:* Wark, K. and Warner, C. F. 1981. *Air Pollution, Its Origin and Control.* Harper & Row, New York. Used by permission.)

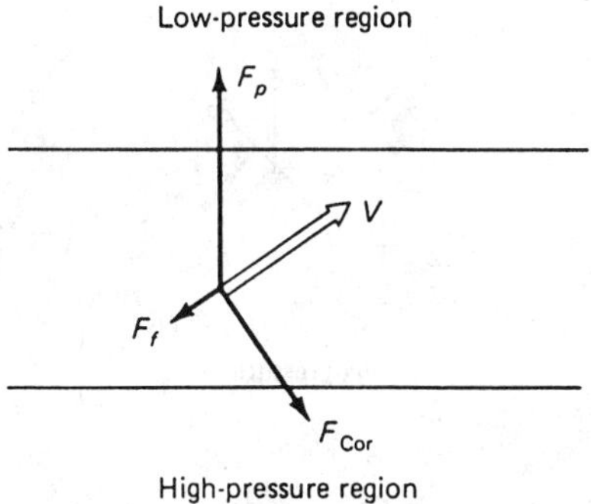

FIGURE 11.15 Frictional force effect on the magnitude and direction of the wind. (*Source:* Wark, K. and Warner, C. F. 1981. *Air Pollution, Its Origin and Control.* Harper & Row, New York. Used by permission.)

Principal among these factors is the frictional force, F_f, arising from surface roughness or the effect of the earth's topography. Additionally, variation in heating arises from daily and seasonal changes that affect the movement of air on a local basis. Both of the factors are combined as surface heating is a function of incoming solar radiation and the local surface characteristics.

The region of the atmosphere between the earth's surface and the upper reaches of the atmosphere is referred to as the planetary boundary layer [Wark and Warner, 1981]. In this layer, all of the factors mentioned above result in this frictional force that combines with the pressure and Coriolis forces to alter the direction and magnitude of the wind at a slight angle toward the low-pressure region. This is illustrated in Fig. 11.15. The frictional force acts opposite to the direction of the velocity vector, which in turn acts at a right angle to the Coriolis force. The resulting magnitude of the wind velocity is the sum of the components of the individual force vectors, with the direction being at a slight angle toward the low-pressure region. This surface wind is of a magnitude less than the geostrophic wind.

The combination of the pressure gradient, Coriolis, centripetal acceleration, and frictional forces is demonstrated in the clockwise and counterclockwise flow around high- and low-pressure systems, respectively. This pattern of flow is shown in Fig. 11.16.

Wind Rose

The wind in a specific location varies with the movement of pressure systems and heating patterns and produces characteristic patterns that can be represented by statistical diagram called a *wind rose* [Turner, 1979]. A wind rose is a polar diagram that plots the frequency of the observed direction of a wind as a spoke. Additionally, the magnitude of the wind from a particular direction is included in the diagram as the length of the individual segments of the spoke. The observed direction of the

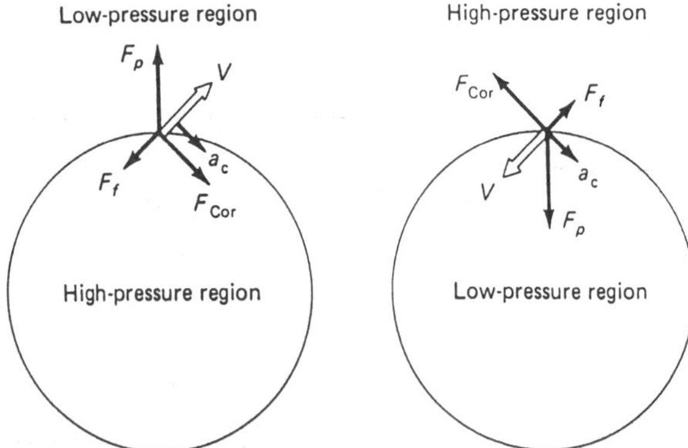

FIGURE 11.16 Force balances around (a) a high-pressure region, and (b) a low-pressure region. (*Source:* Wark, K. and Warner, C. F. 1981. *Air Pollution, Its Origin and Control.* Harper & Row, New York. Used by permission.)

wind is the direction from which the wind is blowing. Figure 11.17 is a wind rose generated from AIRS data for 1988 in Indianapolis, Indiana.

Vertical Velocity Profile

When all of the factors influencing the magnitude and direction combine, the resulting wind is found to vary in magnitude with elevation above the earth's surface. In the lower reaches of the atmosphere the magnitude of the wind is retarded by the frictional forces, but as elevation

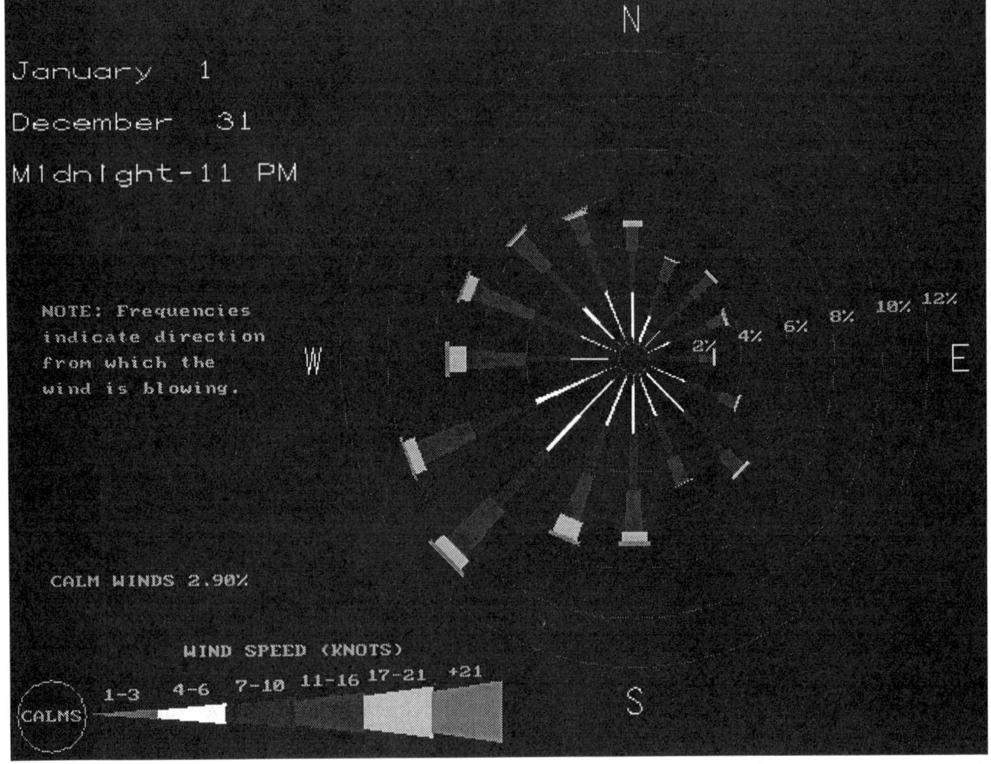

FIGURE 11.17 1988 wind rose, Indianapolis, Indiana. (Courtesy Outarra Fatagoma.)

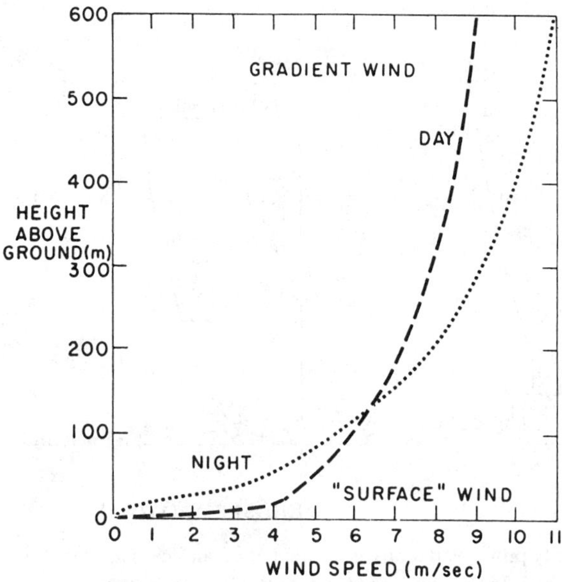

FIGURE 11.18 Diurnal variations in the vertical velocity profile. (*Source:* Turner, D. B. 1979. Meteorological fundamentals. In *Recommended Guide for the Prediction of the Dispersion of Airborne Effluents,* ed. J. R. Martin. The American Society of Mechanical Engineers, New York. Used by permission.)

increases the frictional forces diminish and the magnitude of the wind increases. This relationship is described as the vertical velocity profile of the wind, and can be approximated numerically by

$$\frac{u}{u_1} = \left(\frac{z}{z_1}\right)^p \tag{11.43}$$

where u is the wind velocity at an elevation of z, u_1 is the wind velocity at an altitude of z_1, and p is a constant based on the stability of the atmosphere. The constant p ranges from 0 to 1. Figures 11.18 and 11.19 demonstrate the effects of the local surface characteristics and atmospheric stability on the vertical velocity profile of the wind.

Stability

As the wind varies with differential heating, elevation, and surface characteristics, the stability of the atmosphere, or its ability to enhance or impede vertical motion, also varies [Turner, 1979]. The turbulence of the lower atmosphere is a function of the vertical temperature gradient, wind speed and direction, and surface characteristics. Stability can be classified into six categories based on general atmospheric conditions. These categories, included in Table 11.15, decrease in dispersional ability as the letter designation increases. Further, actual atmospheric conditions are assigned a stability class by several different methods, with each method varying in degree of complexity. A general description of the atmospheric stability, letter designation, and associated vertical velocity profile constant are given in Table 11.16.

Perhaps the method most often used is based on reference to a given vertical temperature gradient a parcel of air should encounter. The reference gradient is the dry adiabatic lapse rate.

Lapse Rates

The dry adiabatic lapse rate is the theoretical rate of cooling that should predict the temperature of a parcel of air as it rises in the atmosphere. This rate is derived by the situation that when a

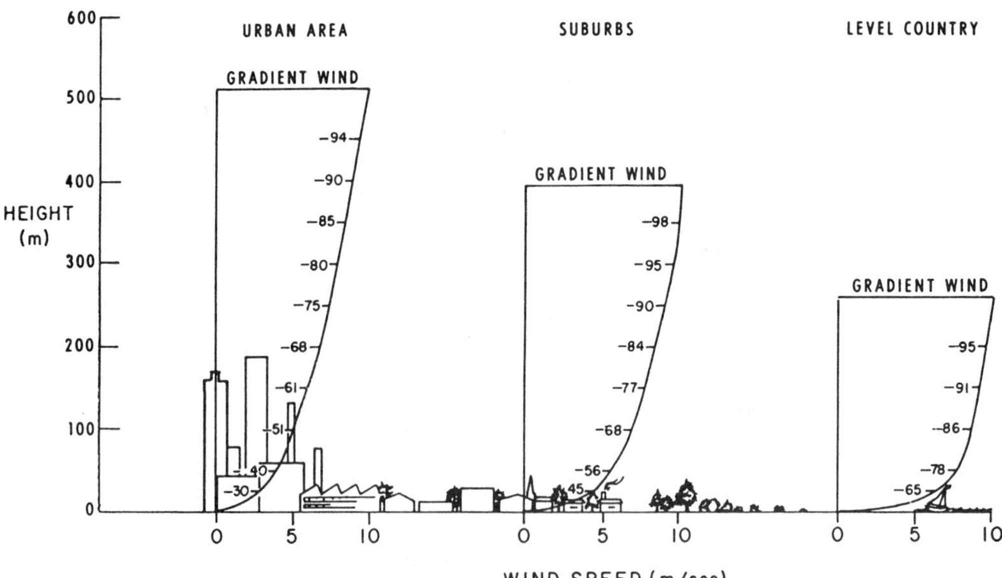

FIGURE 11.19 Vertical velocity profile variations as a function of surface characteristics. (*Source:* Turner, D. B. 1979. Meteorological fundamentals. In *Recommended Guide for the Prediction of the Dispersion of Airborne Effluents,* ed. J. R. Martin. The American Society of Mechanical Engineers, New York. Used by permission.)

Table 11.15 Key to Stability Categories

Surface Wind at 10 meters (m/s)	Day: Incoming Solar Radiation[c] Strong[a]	Day: Incoming Solar Radiation[c] Moderate	Day: Incoming Solar Radiation[c] Slight[b]	Night: Thinly Overcast or Low Cloud Cover (≥ 1/2)	Night: Cloud (≤ 3/8)
<2	A	A-B	B		
2-3	A-B	B	C	E	F
3-5	B	B-C	C	D	E
5-6	C	C-D[d]	D	D	D
>6	C	D	D	D	D

[a] Strong incoming solar radiation corresponds to a solar altitude greater than 60° with clear skies.
[b] Slight insolation corresponds to a solar altitude from 15° to 35° with clear skies.
[c] Incoming radiation that would be strong with clear skies can be expected to be reduced to moderate with broken (5/8 to 7/8 cloud cover) middle clouds and to slight with broken low clouds.
[d] The neutral class, D, should be assumed for overcast conditions during both the night and day.
Source: Turner, D. B. 1969. *Workbook Atmospheric Dispersion Estimates.* U.S. Public Health Service, Cincinnati, OH.

Table 11.16 Stability Category Descriptions

Stability Descriptions[a]	Letter Designations[a]	Velocity Profile Constants (*P* Values)
Very unstable	A	0.12
Moderately unstable	B	0.16
Slightly unstable	C	0.20
Neutral	D	0.25
Moderately stable	E	0.30
Very stable	F	0.40

[a] *Source:* Turner, D. B. 1979. Meteorological fundamentals. In *Recommended Guide for the Prediction of the Dispersion of Airborne Effluents,* ed. J. R. Martin, pp. 1–15. The American Society of Mechanical Engineers, New York. Used by permission.

parcel of air rises in the atmosphere it expands and cools as the surrounding pressure decreases. If the parcel is assumed to expand and cool under adiabatic conditions (that is, assuming there is no heat exchange with its surroundings) the parcel of air should cool at a rate of −5.4°F per 1000 feet (−1°C/100 meters) increase in elevation.

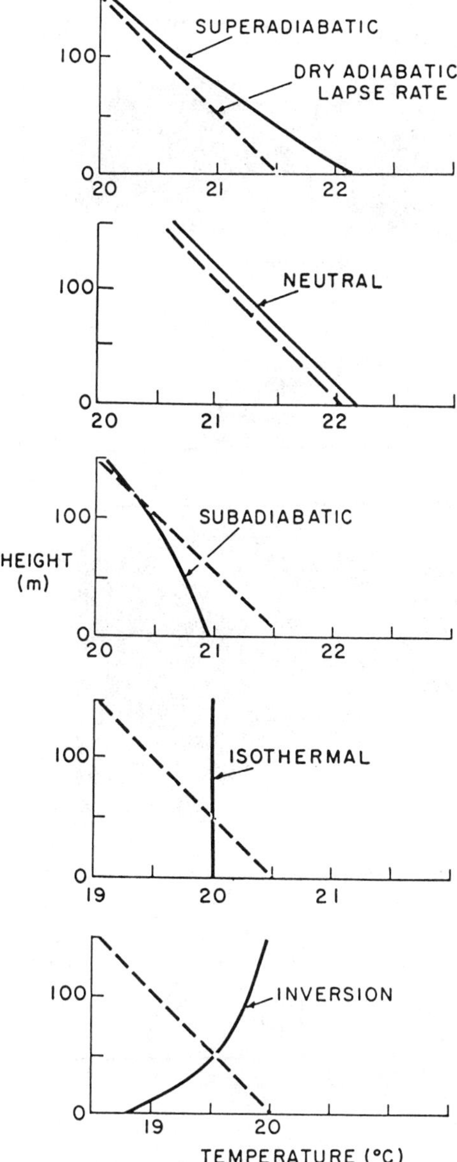

FIGURE 11.20 Atmospheric lapse rates in comparison to the dry, adiabatic lapse rate for (a) superadiabatic, (b) neutral, (c) subadiabatic, (d) isothermal, and (e) inversion conditions. (*Source:* Turner, D. B. 1979. Meteorological fundamentals. In *Recommended Guide for the Prediction of the Dispersion of Airborne Effluents,* ed. J. R. Martin. The American Society of Mechanical Engineers, New York. Used by permission.)

Stability Classes

Actual lapse rates in the atmosphere can be determined readily and correspond to atmospheric conditions at that particular time. There are five general lapse rates found in the atmosphere.

1. *Superadiabatic.* Shown in Fig. 11.20(a), this condition occurs on days when strong solar heating is present, and results in a lapse rate greater than −1°C/100 meters. Superadiabatic conditions are typically found only in the lower 200 meters of the atmosphere and are associated with a stability class of A, or very unstable conditions.
2. *Neutral.* This condition is associated with overcast days and strong to moderate wind speeds. The lapse rate approximates the dry adiabatic lapse rate very closely, as shown in Fig. 11.20(b). Stability classes associated with neutral conditions are B, C, or D indicating moderately unstable to neutral conditions.
3. *Subadiabatic.* This condition is associated with relatively calm days without strong solar heating. The lapse rate is below the −1°C/100 meters rate, as shown in Fig. 11.20(c). Subadiabatic conditions correspond to stability classes of D or E for neutral to moderately stable.
4. *Isothermal.* Shown in Fig. 11.20(d), the temperature in the atmosphere is constant with height, and as a result, there is no lapse rate. Isothermal conditions coincide with a stability class of D.
5. *Inversion.* When the temperature gradient increases with height, often found in the evenings on days with strong solar heating, the condition is referred to as a thermal inversion. This condition demonstrates extremely stable conditions, with a stability class of F, and results in a positive lapse rate. This is shown in Fig. 11.20(e).

Plume Characteristics

For each of the atmospheric stability classes listed above, there is a plume type that is characteristic in both plume shape and downwind concentration. That is, as the gas stream is emitted from the source, the gas stream (or plume) initially rises and then begins to dissipate in the atmosphere.

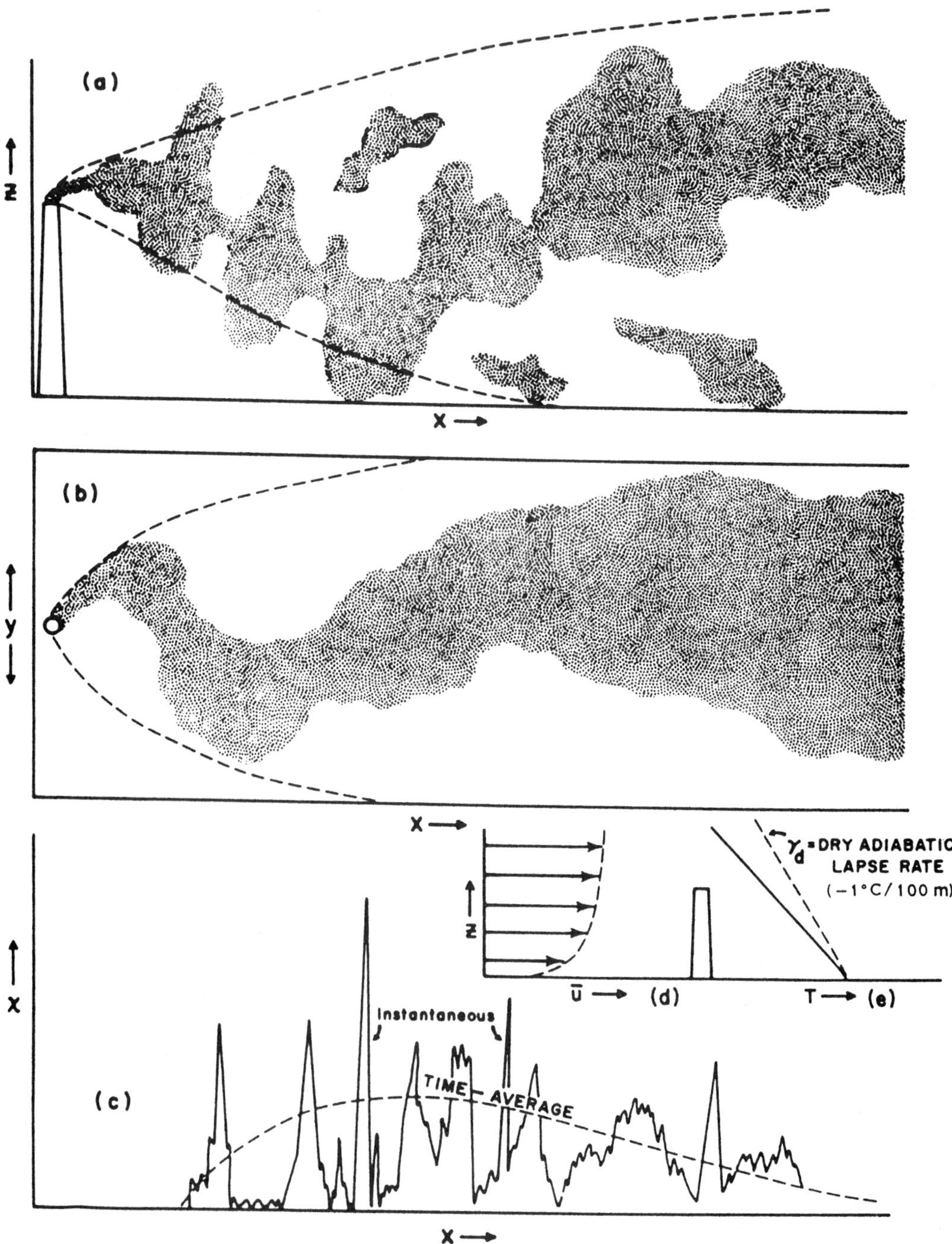

FIGURE 11.21 Looping plume. (*Source:* Pack, D. H. and Halitsky, J. 1979. Behaviour of airborne effluents. In *Recommended Guide for the Prediction of the Dispersion of Airborne Effluents*, ed. J. R. Martin. The American Society of Mechanical Engineers, New York. Used by permission.)

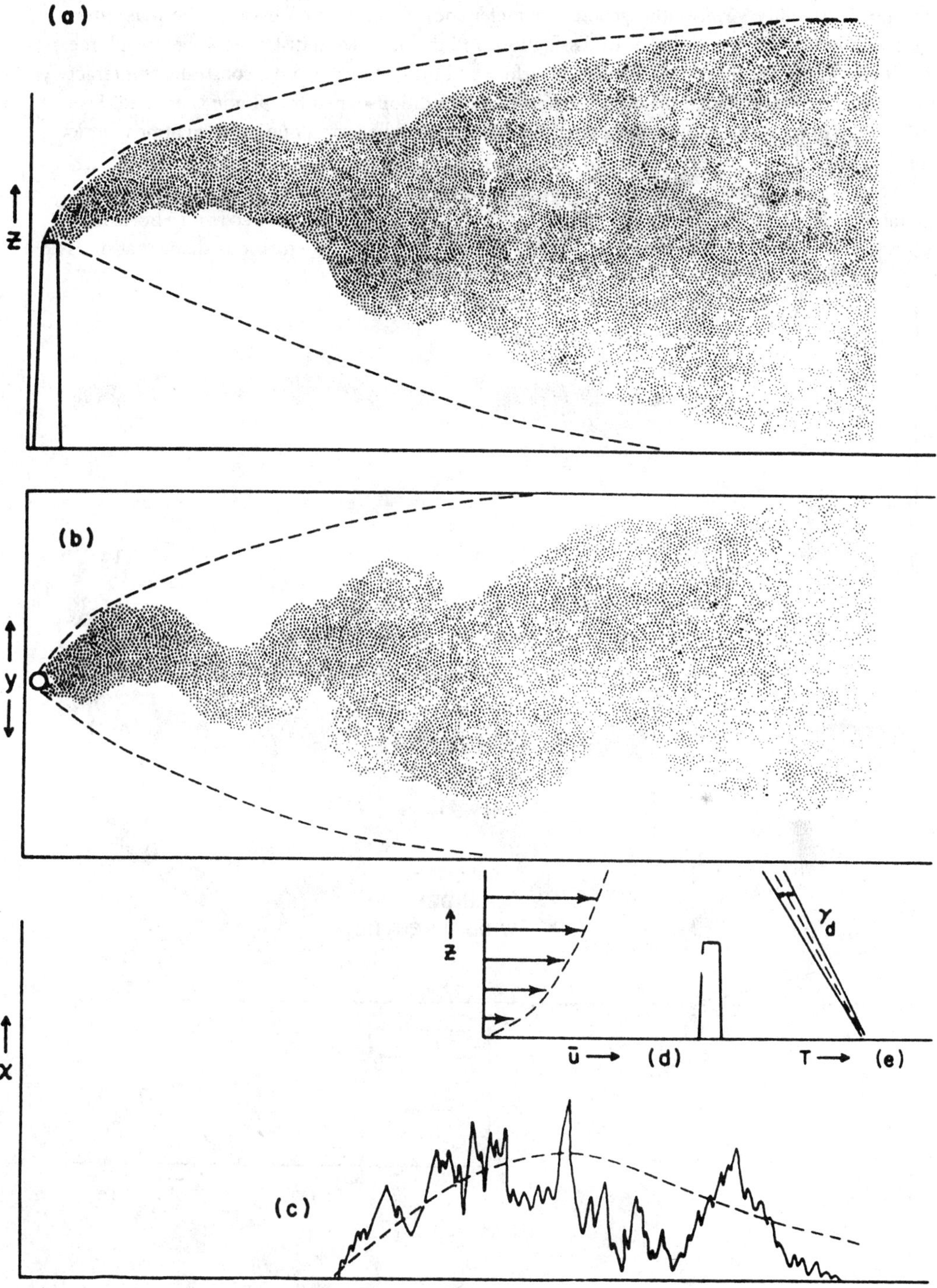

FIGURE 11.22 Coning plume. (*Source:* Pack, D. H. and Halitsky, J. 1979. Behaviour of airborne effluents. In *Recommended Guide for the Prediction of the Dispersion of Airborne Effluents*, ed. J. R. Martin. The American Society of Mechanical Engineers, New York. Used by permission.)

Atmospheric stability governs the general characteristics of any plume because the maximum height the plume rises to, the degree of dissipation, and the downwind distances where the plume constituents first come into contact with the surface and the distance that the constituents interact with the surface are all functions of the atmospheric conditions when the plume is emitted. Five general types of plumes can be visually identified, each generally corresponding to an atmospheric stability class.

The first type of plume is the *looping plume* associated with great instability in the atmosphere, or a stability class of A. The name of the plume arises from its shape when viewed from a horizontal perspective. A looping plume, shown in Fig. 11.21, is characterized by a tortuous shape that finds

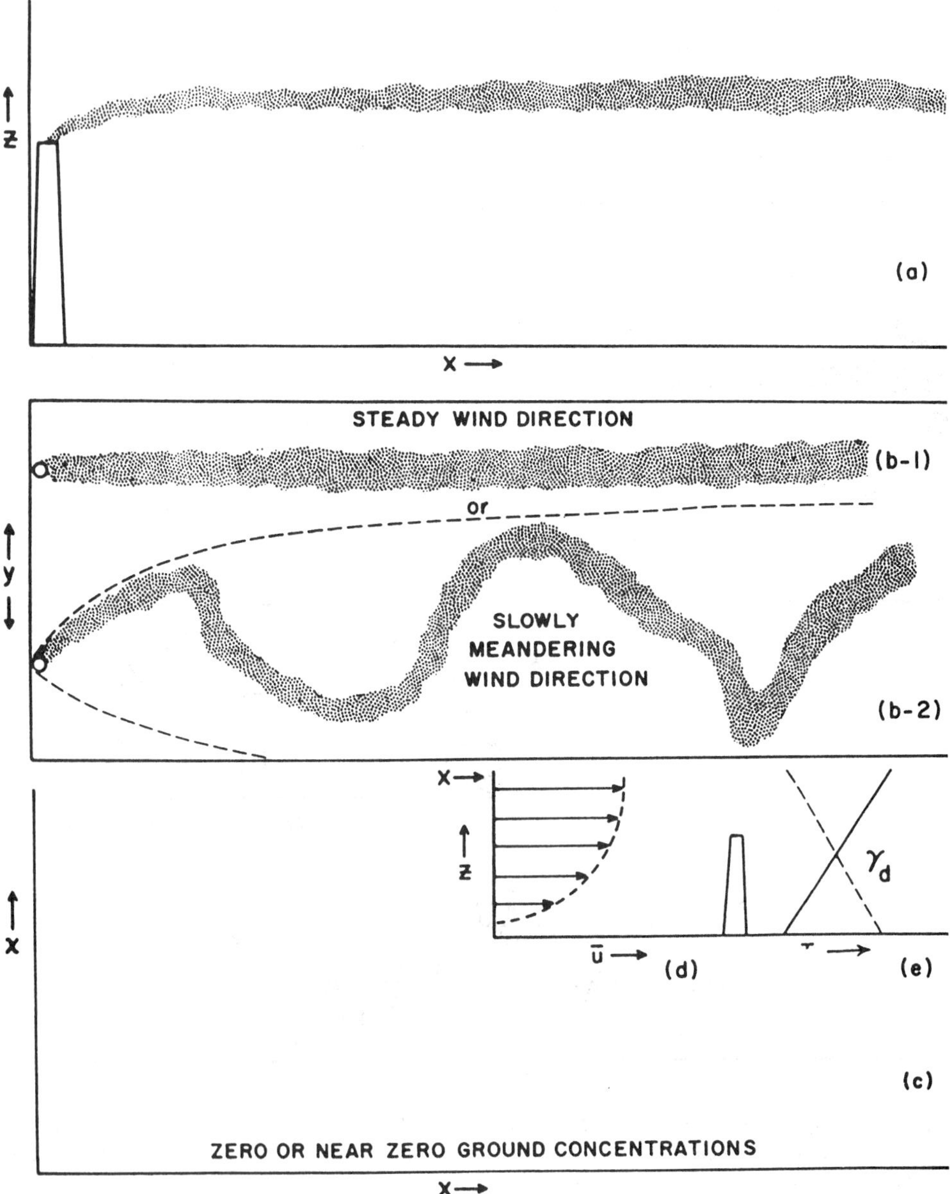

FIGURE 11.23 Fanning plume. (*Source:* Pack, D. H. and Halitsky, J. 1979. Behaviour of airborne effluents. In *Recommended Guide for the Prediction of the Dispersion of Airborne Effluents*, ed. J. R. Martin. The American Society of Mechanical Engineers, New York. Used by permission.)

the plume rising and falling. This type of plume has a very short downwind initial contact distance with plume constituents resulting in a gradual concentration profile over a long downwind distance. Strong instability in the atmosphere creates this situation as the plume is influenced only by great mixing and no forces preventing surface contact. Additionally, when viewed vertically, the plume meanders a great distance from its original centerline, and thus has the potential for the plume constituents to affect a large surface area.

The second type of plume is the *coning plume*, shown in Fig. 11.22, and corresponds to the near-neutral conditions or atmospheric conditions in classes B, C, or D. Initially, the plume follows the same pattern as the looping plume when viewed vertically. However, when viewed horizontally the

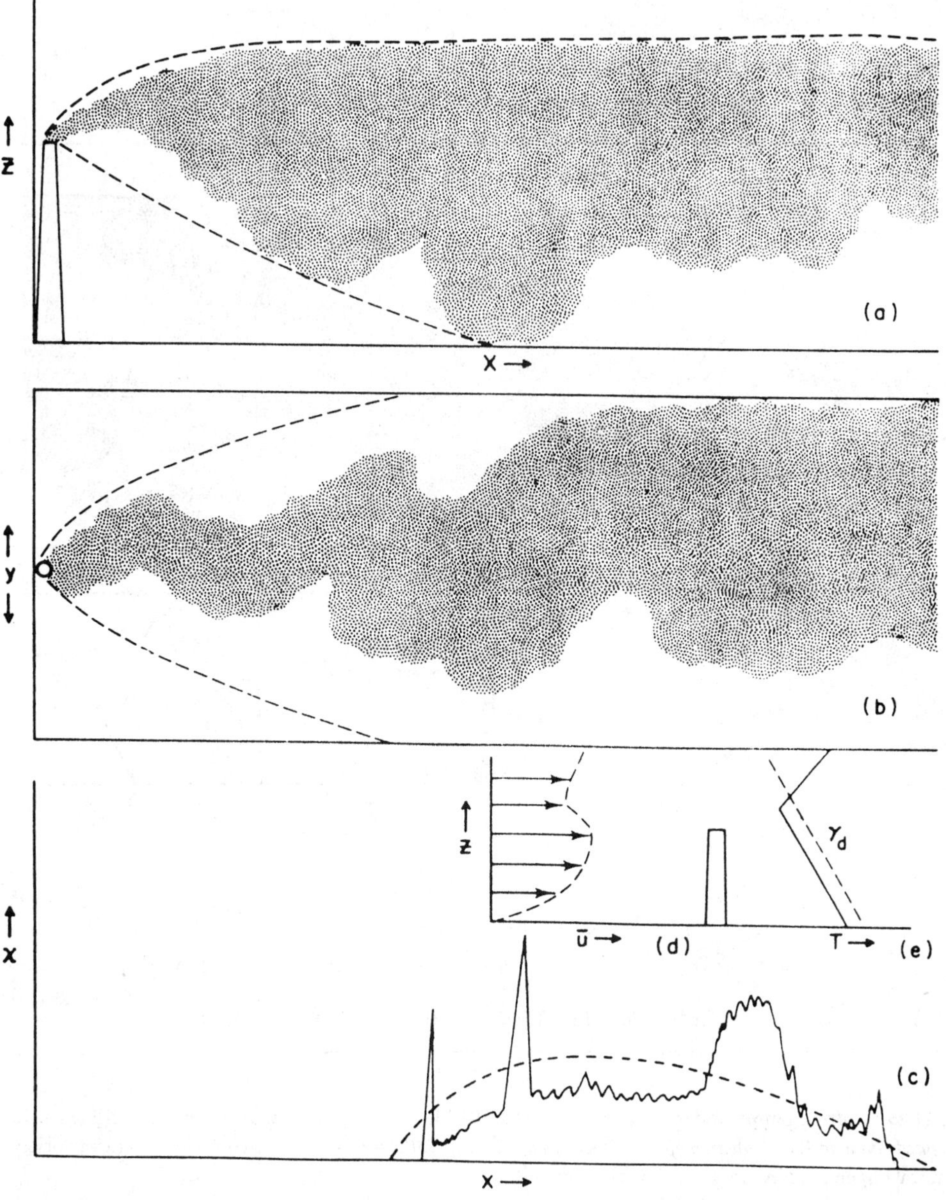

FIGURE 11.24 Fumigation plume. (*Source:* Pack, D. H. and Halitsky, J. 1979. Behaviour of airborne effluents. In *Recommended Guide for the Prediction of the Dispersion of Airborne Effluents*, ed. J. R. Martin. The American Society of Mechanical Engineers, New York. Used by permission.)

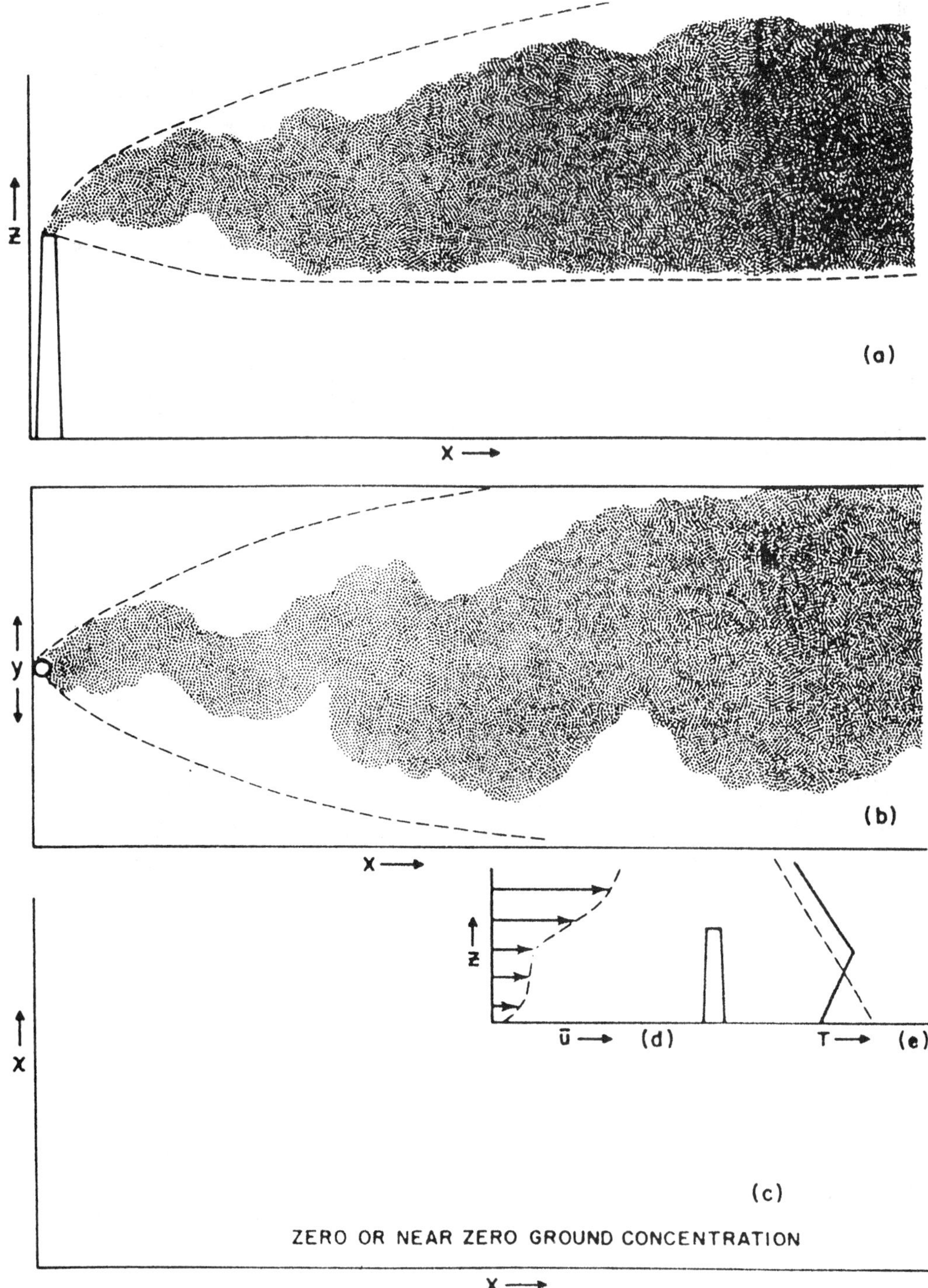

FIGURE 11.25 Lofting plume. (*Source:* Pack, D. H. and Halitsky, J. 1979. Behaviour of airborne effluents. In *Recommended Guide for the Prediction of the Dispersion of Airborne Effluents*, ed. J. R. Martin. The American Society of Mechanical Engineers, New York. Used by permission.)

plume has a much more gradual pattern of dissipation. This is due to the decreased mixing ability of the atmosphere. As a result of the decreased mixing, the plume is carried over a much greater distance before its initial surface contact. Downwind distances these constituents are deposited over are less than the looping plume, and result in higher surface concentrations of the constituents.

The third type of plume is the *fanning plume*, the plume type associated with inversion conditions. These conditions, with a stability class of F, result in the plume being held in a vertical plane by the cooler air below the stack and the warmer air above it, as is illustrated in Fig. 11.23. As a result, the downwind concentrations are virtually nonexistent as the plume constituents have a much greater time to dissipate before contacting the earth's surface.

The *fumigation plume* is the fourth type of plume and is the result of an elevated inversion. The warmer air above the stack holds the plume rise to a maximum elevation at the lower limit of the inversion. This type of plume, shown in Fig. 11.24, is similar to a coning with the exception that surface concentrations tend to be higher as a result of the limited upper mixing boundary.

The final type of plume is the *lofting plume*, shown in Fig. 11.25. This plume is the result of a surface inversion that restricts the plume's ability to reach the surface. Concentration profiles for this plume are identical to the fanning plume; none of the plume's constituents are detectable until the plume has traveled through the surface inversion.

11.8 Dispersion Modeling

The estimation of downwind concentrations of pollutants emitted from a source involves approximating the cumulative effect of many atmospheric and emission variables. This modeling effort is done mainly to determine the effects the source presently has, or will have in the future, on its surrounding environment. Types of air pollution models vary from very basic models describing a single source to regional airshed models that describe multitudes of sources over large metropolitan areas.

All of the models employed for air quality analysis are derived from four basic types of models. These are the Gaussian, numerical, statistical or empirical, and physical. Gaussian models are the most widely used models for the estimation of nonreactive pollutants. Numerical models are favored in the analysis of urban areas with reactive pollutants, but are a great deal more complex. Statistical or empirical models are typically used in situations where data is lacking, and physical models are actual simulations in wind tunnels or other fluid-modeling facilities [U.S. EPA, 1993].

Because of the wide range of models, the description of the fundamentals of each individual model is beyond the scope of this chapter. As a result, this discussion will focus on the basics of the most common, the double Gaussian or normal dispersion model. This model will be described in an effort to cover the basic approaches to modeling a single source in noncomplex terrain for both gaseous and particulate pollutants.

In describing the dispersion of these pollutants the source, often a stack, is referred to as having an effective height. This is the description of the height where the plume loses its influence from the emission source and begins to be dispersed in the atmosphere.

Plume Rise

Coordinate System

The first step in describing the plume rise, or effective stack height, is the development of a system of coordinates to describe the dispersion of the plume in three dimensions. The system employed assumes the origin to be at the base of the source. From the origin the x axis extends in the direction of the mean wind, the y axis extends perpendicular to the x axis, and the z axis extends vertically (along the stack) from the origin, perpendicular to both the x and y axes.

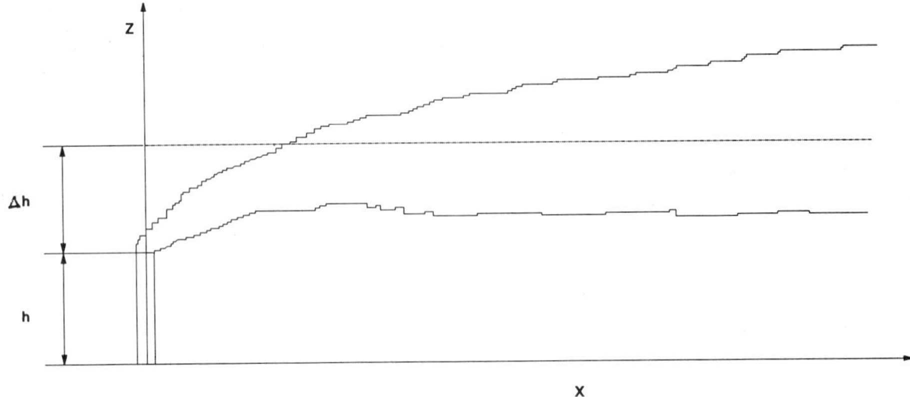

FIGURE 11.26 Plume rise.

Effective Stack Height

Plume height (H), or effective stack height, is the elevation in the z axis the plume rises to. This is defined by the following equation:

$$H = h + \Delta h \tag{11.44}$$

where the plume height H is the sum of the height of the actual stack h and the additional height Δh the plume rises under the effects of the momentum and buoyancy forces of the gas stream being emitted. Figure 11.26 illustrates the plume rise.

The determination of Δh is done by attributing this variable to the major force affecting it. From this plumes are grouped into two categories: momentum sources and buoyant sources. Momentum sources are small-volume sources having appreciable exit speeds with little temperature excess. These are conditions described by exit speeds of greater than or equal to 10 m/s with temperatures less than 50°C above the ambient temperature. Buoyant sources are larger-volume sources with higher temperatures. Typically these sources will have volumetric flow rates greater than 50 m^3/s and temperatures greater than 50°C above ambient temperature [Frizzola *et al.*, 1979]. In describing these situations, only one force is considered in determining the plume rise.

Momentum Sources. Plume rise from a momentum source is described mathematically as a function of the exit conditions and the wind speed at the stack height by the following relationship.

$$\Delta h = D\left(\frac{V_s}{u_s}\right)^{1.4} \tag{11.45}$$

where

Δh is the height of the plume above the stack (m)

D is the diameter of the stack (m)

V_s is the emission velocity (m/s)

u_s is the mean wind speed at the stack height (m/s).

Buoyant Plume Sources. Buoyant plume sources, as described previously, can be described mathematically only by inclusion of atmospheric conditions at the time of emission. As plume rise varies according to the atmosphere, three different equations are used to describe this situation. The three equations apply to stable conditions, unstable conditions, and both stable and unstable conditions in the absence of appreciable wind.

Under stable conditions, or stability classes of E or F, the following equations describe the plume rise if there is appreciable wind present.

$$\Delta h = 2.6\left(\frac{F}{u_s S}\right)^{1/3} \quad \text{and} \quad F = gV_s\left(\frac{D}{2}\right)^2\left(\frac{\rho_a - \rho_s}{\rho_a}\right) \tag{11.46}$$

where

g is the acceleration due to gravity (9.807 m/s^2)
V_s is the emission velocity (m/s)
D is the diameter of the stack (m)
ρ_a is the density of the ambient air (g/m^3)
ρ_s is the stack gas density (g/m^3)
S is the stability parameter (1/s^2)

The stability parameter is further defined as

$$S = \left(\frac{g}{T}\right)\left(\frac{\partial\Theta}{\partial z}\right) \tag{11.47}$$

where T is the ambient temperature (K) and $\partial\Theta/\partial z$ is the vertical potential temperature gradient. For stability categories E and F, $\partial\Theta/\partial z$ is 0.02 and 0.04 K/m, respectively.

In unstable atmospheres, the plume rise is described by

$$\Delta h = \frac{1.6F^{1/3}(3.5x)^{2/3}}{\overline{u}_s} \tag{11.48}$$

where the downwind distance the plume rises to an elevation of Δh is $3.5x$. x is determined by the following equations.

$$\begin{aligned} x &= 14F^{5/8} \quad \text{when} \quad F \le 55\left(\frac{\text{m}^4}{\text{s}^3}\right) \\ x &= 34F^{2/5} \quad \text{when} \quad f > 55\left(\frac{\text{m}^4}{\text{s}^3}\right) \end{aligned} \tag{11.49}$$

During both stable and unstable conditions, if there is an absence of wind there is an almost vertical rise in the plume. This rise can be described by the following equation.

$$\Delta h = 5.0\frac{F^{1/4}}{S^{3/8}} \tag{11.50}$$

After the plume type and rise have been determined, the effective stack height should be determined by adding the plume rise to the physical stack height. It should be noted that synonymous units are required in the calculation of effective stack height and that the plume rise equations listed above are all in S.I. units.

Gaseous Dispersion

Currently, worldwide use is being made of the double normal Gaussian mathematical dispersion model for estimating downwind concentrations of pollutants from point, area, and fugitive sources. The Gaussian model is used throughout the U.S. by the Environmental Protection Agency and the respective state and local air pollution control agencies. Its development over the past 25 years has

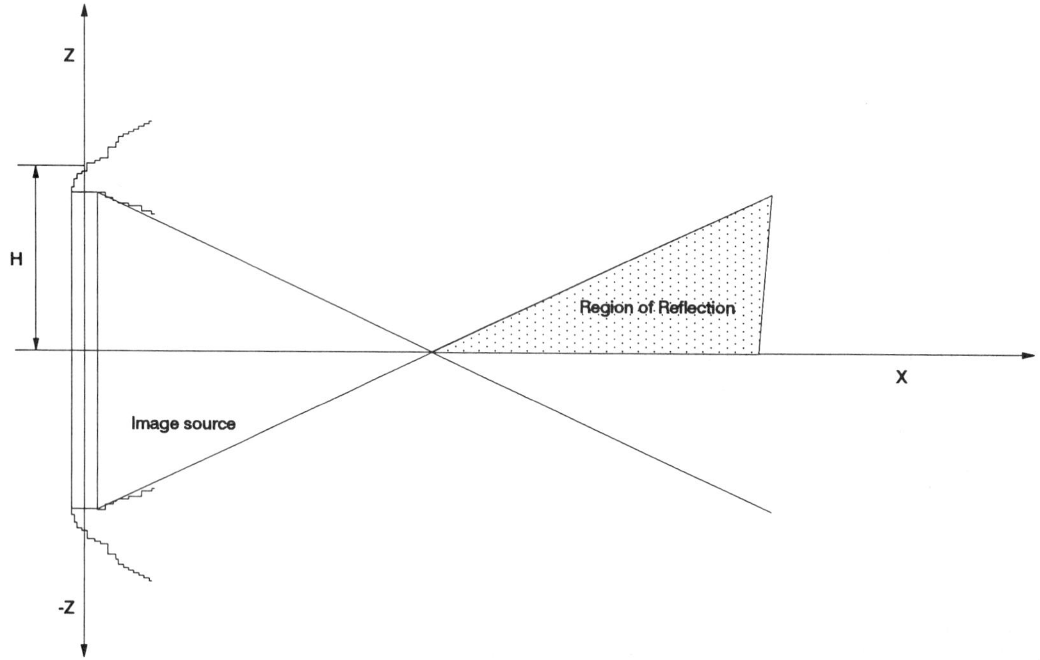

FIGURE 11.27 Region of reflection.

resulted in a rather powerful and cost-effective tool for regulatory planning and implementation purposes.

The model is a combination of empirical and theoretical considerations. It is certainly not a perfect tool but through validations and sometimes calibrations, it is widely accepted. The model uses an exponential decay term (e^{-t}) to account for downwind dilution effects and borrows the Gaussian distribution from statistics to simulate the vertical and crosswind distribution of a pollutant at any point downwind. Furthermore, the model recognizes plume reflection at the earth's surface and provides for a superposition solution or mirror-image stack below the ground plane to account for pollutant reflection at the surface. Figure 11.27 illustrates the plume reflection effect. Elevated sources such as tall stacks are handled by shifting the plume centerline above the surface.

For a surface which absorbs and does not reflect pollutants Eq. (11.51) can be used.

$$C(x, y, z) = \frac{Q}{2\pi u_s \sigma_y \sigma_z} \exp\left[-\frac{1}{2}\left(\frac{y^2}{\sigma_y^2} + \frac{(z-H)^2}{\sigma_z^2}\right)\right] \tag{11.51}$$

where

Q is the mass emission rate of pollutant (g/s)
u_s is the wind speed at stack height at the time of emission (m/s)
$\sigma_y$is a stability constant described in Table 11.1 (m)
H is the effective stack height (m)
$\sigma_z$is a stability constant described in Table 11.1 (m)

For situations which require the consideration of surface reflection, which occurs for most cases, the double normal Gaussian model of Eq. (11.52) must be used.

$$C(x, y, z) = \frac{Q}{2\pi u_s \sigma_y \sigma_z}\left[\exp\left(-\frac{y^2}{2\sigma_y^2}\right)\right]\left[\exp\left(\frac{-(z-H)^2}{2\sigma_z^2}\right) + \exp\left(\frac{-(z+H)^2}{2\sigma_z^2}\right)\right] \tag{11.52}$$

Table 11.17 σ_y and σ_z Determinations for Open-country Conditions with Downwind Distances between 100 and 10,000 Meters

Stability Class	σ_y (m)	σ_z (m)
A	$0.22x(1 + 0.0001x)^{-1/2}$	$0.20x$
B	$0.16x(1 + 0.0001x)^{-1/2}$	$0.12x$
C	$0.11x(1 + 0.0001x)^{-1/2}$	$0.08x(1 + 0.0001x)^{-1/2}$
D	$0.08x(1 + 0.0001x)^{-1/2}$	$0.06x(1 + 0.0001x)^{-1/2}$
E	$0.06x(1 + 0.0001x)^{-1/2}$	$0.03x(1 + 0.0001x)^{-1}$
F	$0.04x(1 + 0.0001x)^{-1/2}$	$0.016x(1 + 0.0001x)^{-1}$

Source: Smith, M. E., *et al.* 1979. Calculations of dispersion and deposition. In *Recommended Guide for the Prediction of the Dispersion of Airborne Effluents*, ed. John R. Martin. The American Society of Mechanical Engineers, New York. Used by permission.

The σ_y and σ_z values are borrowed from the Gaussian distribution statistics and are analogized here as pollutant dispersion variables instead of standard deviations as we usually think of them in a statistical sense. σ_y is the pollutant dispersion coefficient in the crosswind direction and σ_z is the dispersion coefficient in the vertical. These dispersion coefficients were originally derived from diffusion experiments on the O'Neill, Nebraska, grass flats following World War II. Since then they have been modified significantly with experiments in St. Louis, Missouri, and elsewhere giving rise to both an urban and a rural version of the model. The dispersion coefficients can be determined from Table 11.17 for various downwind distances and atmospheric stability classes [Smith *et al.*, 1979].

It is interesting to note that Eq. (11.52) does not contain a σ_x or downwind dispersion coefficient. The reason for this is that this model simulates a continuous point source release as opposed to a "puff" or instantaneous release. The resulting downwind pollutant concentration gradient along a limited segment of the plume centerline is insignificant compared to the crosswind and vertical gradients. Hence, no σ_x or downwind dispersion coefficient is used in the continuous release model.

Note that the exponential term in Eq. (11.52) can never be greater than unity and that all parameters with units must cancel out. The engineering units of concentration are formed from the first term in the equation. The numerator has the units of mass per time and the denominator volume per time, yielding mass per volume or pollutant concentration units. A worst-case scenario for the model would be the situation in which the stack height is zero, as in the case of a burning pile of leaves, and the concentration is desired directly downwind on the ground along the plume centerline. In this situation, note that $y = 0$, $z = 0$, and $H = 0$, resulting in the exponential term going to a unity value (1.0) leaving the final term to be $Q/(\pi u_s \sigma_y \sigma_z)$.

Particulate Dispersion

The double normal Gaussian model just discussed is valid for gases and suspended particulate matter. In addition, the model can be modified to handle settleable particulate matter. This is usually defined as those particles greater than 40 microns in diameter. (A typeset period on this page is about 1000 microns in diameter.) We are interested in having a model consider particulate settling because in some instances surface deposition of a toxic particle from an elevated plume is very important to know. Instead of the model calculating pollutant concentration, we may be more interested in mass of deposition per unit area per unit time.

Examining the dynamics of particles settling out of an elevated plume, we can differentiate the centerline of the gaseous portion of the plume from the centerline of the settleable particulate portion of the plume. Considering only the settleable particulate plume centerline, we note that for each size class of particle there is a distance downwind beyond which no more particles of that size will be found because they have all settled to the ground. In other words, the plume centerline for that particle size has disappeared. We can represent the disappearance of that particle size class with a plume centerline that tilts downward from near the top of the stack and eventually disappears downwind into the surface. This type of Gaussian model is referred to as a *tilted plume* model.

The amount of plume tilt is determined from the terminal settling velocities of the various sizes of particles in the plume, the distance downwind where they have settled out, and the mean wind speed. If the terminal settling velocity of the particle is defined as V_t, the downwind distance as x, and the wind speed as u, the ratio $V_t x/u$ can be used to shift the plume centerline downward. Equation (11.53) is the familiar Gaussian model with the plume-shift term incorporated.

$$C(x, y, z) = \frac{Q_P}{2\pi u_s \sigma_y \sigma_z} \exp\left[-\frac{1}{2}\left\{\left(\frac{y}{\sigma_y}\right)^2 + \left(\frac{z - [H - (V_t x/u_s)]}{\sigma_z}\right)^2\right\}\right] \tag{11.53}$$

Additionally, the mass deposition rate of particulate in a given area can be determined by multiplying the concentration deposited by the terminal settling velocity. This is given by the following equation.

$$\omega(x, y, z) = \frac{Q_P V_t}{2\pi u_s \sigma_y \sigma_z} \exp\left[-\frac{1}{2}\left\{\left(\frac{y}{\sigma_y}\right)^2 + \left(\frac{z - [H - (V_t x/u_s)]}{\sigma_z}\right)^2\right\}\right] \tag{11.54}$$

Table 11.18 Preferred Air Quality Models

Preferred Models	
Buoyant line and point source dispersion model (BLP)	BLP is a Gaussian plume dispersion model for plume rise and downwash effects from stationary line sources.
CALINE3	CALINE3 is used to estimate nonreactive pollutants from highway traffic.
Climatological dispersion model (CDM 2.0)	CDM is a climatological steady-state Gaussian plume model for seasonal or annual arithmetic average pollutant concentrations in an urban area.
Gaussian-plume multiple source air quality algorithm (RAM)	RAM is a Gaussian plume model for estimating relatively stable pollutants from point and area sources in level terrain.
Industrial source complex model (ISC2)	ISC2 is a Gaussian plume model to describe an industrial source complex in either short- or long-term modes with limited terrain adjustment.
Multiple point Gaussian dispersion algorithm with terrain adjustment (MPTER)	MPTER is a multiple point source algorithm to the Gaussian model.
Single source (CSTER) model	CSTER is a steady-state, Gaussian dispersion model for single source, single location modeling in rural or urban locations.
Urban airshed model (UAM)	UAM is an urban simulation model for estimating ozone concentrations from short-term conditions.
Offshore and coastal dispersion model (OCD)	OCD is a Gaussian model developed to estimate the effect of offshore sources on coastal regions.
Emissions and dispersion model system (EDMS)	EDMS is a combined emissions/dispersion model for assessing pollution at civilian airports and military air bases.
Complex terrain dispersion model plus algorithms for unstable conditions (CTDMPLUS)	CTDMPLUS is a refined point source Gaussian air quality model for use in all stability conditions for complex terrain situations.

Source: U.S. Environmental Protection Agency. 1993. *Guideline on Air Quality Models.* Research Triangle Park.

Table 11.19 Alternative Air Quality Models *(continued)*

	Alternative Models
Air quality display model (AQDM)	AQDM is a climatological steady-state Gaussian plume model to determine annual average sulfur dioxide and particulate concentrations in urban areas.
Air resources regional pollution assessment model (ARRPA)	ARRPA is a segmented plume model to estimate surface dry mass deposition of sulfur dioxide and sulfate.
APRAC-3	APRAC-3 is a model which computes hourly average carbon monoxide concentrations for any urban location.
Compter	Compter is a Gaussian steady-state model applicable to both urban and rural areas for open and elevated terrain.
ERT visibility model	The ERT visibility model is designed to address visibility impairment for arbitrary lines of sight.
HIWAY-2	HIWAY-2 can estimate concentrations of nonreactive pollutants from highway traffic.
Integrated model for plumes and atmospheric chemistry in complex terrain (IMPACT)	IMPACT is a finite-difference grid model used to estimate impacts of pollutants in simple or complex terrain from point or area sources.
LONGZ	LONGZ is a steady-state Gaussian plume model for urban and rural areas to estimate annual average concentrations of pollutants from up to 14,000 sources.
Maryland power plant siting program model (PPSP)	PPSP is a Gaussian plume model applicable to tall stacks in either rural or urban area with simple terrain.
Mesoscale puff model (MESOPUFF II)	MESOPUFF II is a short-term, regional scale puff model used to estimate concentrations of up to five pollutant species.
Mesoscale transport diffusion and deposition model for industrial sources (MTDDIS)	MTDDIS is a variable trajectory Gaussian puff model for long-range transport of point source emissions of level or rolling terrain.
Models 3141 and 4141	Models 3141 and 4141 are modifications of CRSTER for complex terrain.
MULTIMAX	MULTIMAX is a Gaussian plume model applicable to both urban and rural areas.
Multi-source model (SCSTER)	SCSTER is a modified version of CRSTER to account for multiple sources.
Pacific gas and electric plumes model (PLUME5)	PLUME5 is a Gaussian steady-state plume model applicable to both urban and rural areas in uneven terrain.
PLMSTAR air quality simulation model	PLMSTAR is a mesoscale Lagrangian photochemical model designed to predict atmospheric concentrations of photochemical products of reactive hydrocarbons.
Plume visibility model (PLUVUE II)	PLUVUE II is used for estimating visual range reduction and atmospheric discoloration from a single emission source.
Point, area, line source algorithm (PAL-DS)	PAL-DS is an algorithm for estimating short-term dispersion.
Random-walk advection and dispersion model (RADM)	RADM uses the random-walk method to simulate atmospheric dispersion.
Reactive plume model (RPM-II)	RPM-II is used for estimating short-term concentrations of primary and secondary pollutants from point or area sources.

Table11.19 (continued) Alternative Air Quality Models

Alternative Models	
Regional transport model (RTM-II)	RTM-II is used to estimate the dispersion of air pollutants from multiple sources over large distances.
SHORTZ	SHORTZ uses the steady-state Gaussian plume model to estimate ground-level concentrations in urban and rural terrain for both simple and complex terrain.
Simple line-source model (GMLINE)	GMLINE is used to determine exhaust concentrations 100 meters from a roadway in flat terrain.
Texas climatological model (TCM-2)	TCM is used for determining seasonal or annual arithmetic average pollutant concentrations of nonreactive pollutants.
Texas episodic model (TEM-8)	TEM is a short-term Gaussian plume model for determining short-term concentrations of nonreactive pollutants.
AVACTA II	AVACTA II is a Gaussian model in which atmospheric dispersion phenomena are described by the evolution of plume elements.
Shoreline dispersion model (SDM)	SDM is multipoint Gaussian model that calculates source impacts from fumigation events.
WYNDvalley model	WYNDvalley is a multilayer grid dispersion model that permits flexibility in modeling.
Dense gas dispersion model (DEGADIS)	DEGADIS is used to model the transport of toxic chemical releases into the atmosphere.

Source: U.S. Environmental Protection Agency. 1993. *Guideline on Air Quality Models.* Research Triangle Park.

Regulatory Air Models

From the first basic Gaussian air models, an entire library of models has developed with each new model being employed for new uses or being a refinement of an existing model. As a result of continuous updating and advancement of air models, the use of specific models in regulatory applications changes as new models become available.

In regulatory efforts the use of a specific model is based on the desired result of the modeling. For example, the use of a simple Gaussian model to describe the impact of a major source in complex terrain is beyond the capabilities of the model and is therefore not recommended. However, if only rough estimates are required for a single source in open terrain, the simple Gaussian model would probably suffice. Because of the existence of these types of situations, individual models are recommended for specific applications. Tables 11.18 and 11.19 are a listing of the regulatory air models and their applications currently preferred or recommended as alternatives by the U.S. EPA.

References

Calvert, S., Goldschmid, J., Leith, D., and Mehta, D. 1972. Wet scrubber system study. *Scrubber Handbook,* Volume I. U.S. Department of Commerce, NTIS. PB-213016.

Cooper, C. D. and Alley, F. C. 1990. *Air Pollution Control: A Design Approach.* Waveland Press, Prospect Heights, IL.

Environmental Resources Management. 1992. *Clean Air Act Primer.*

Frizzola, J. A., *et al.* 1979. Calculation of effective stack height. In *Recommended Guide for the Prediction of the Dispersion of Airborne Effluents*, ed. J. R. Martin, pp. 38–44. The American Society of Mechanical Engineers, New York.

Hansen, L. D. 1994. A critical evaluation of the Nextel advantage. Class Report, C.E. 457, Purdue University.

Lodge, J. P. Jr., ed. 1989. *Methods of Air Sampling and Analysis.* Lewis Publishers, Chelsea, MI.

Makansi, J. 1988. Reducing NO_x emissions: A special report. *Power.* September.

Nagy, G. Z. 1991. The odor impact model. *J. Air Waste Manage. Assoc.* 41 (10):1360–1362.

Pack, D. H. and Halitsky, J. 1979. Behaviour of airborne effluents. In *Recommended Guide for the Prediction of the Dispersion of Airborne Effluents*, ed. J. R. Martin, pp. 16–37. The American Society of Mechanical Engineers, New York.

Pedersen, E. 1993. An analysis of sintamatic baghouse filter media. Class Report, C.E. 457, Purdue University.

Prokop, W. H. 1992. *Air Pollution Engineering Manual—Odors.* Air and Waste Management Association. Van Nostrand Reinhold, New York.

Smith, D. J. 1993. Low-NO_x burners lead technologies to meet CAA's Title IV. *Power Engineering.* June.

Smith, M. E., *et al.* 1979. Calculations of dispersion and deposition. In *Recommended Guide for the Prediction of the Dispersion of Airborne Effluents,* ed. J. R. Martin, pp. 43–55. The American Society of Mechanical Engineers, New York.

Tchobanoglous, G. 1979. *Wastewater Engineering: Treatment, Disposal, and Reuse.* Metcalf and Eddy, Inc., McGraw-Hill, New York.

Turner, D. B. 1969. *Workbook Atmospheric Dispersion Estimates.* U.S. Public Health Service, Cincinnati, OH.

Turner, D. B. 1979. Meteorological fundamentals. In *Recommended Guide for the Prediction of the Dispersion of Airborne Effluents,* ed. J. R. Martin, pp. 1–15. The American Society of Mechanical Engineers, New York.

U.S. EPA. 1993. *Guideline on Air Quality Models.* Research Triangle Park.

U.S. EPA. 1993. *Compilation of Air Pollutant Emission Factors.* Research Triangle Park.

Wark, K. and Warner, C. F. 1981. *Air Pollution, Its Origin and Control.* Harper & Row, New York.

Weber, J. 1972. Adsorption. In *Physicochemical Processes for Water Quality Control*, pp. 199–259. John Wiley & Sons, New York.

The great London Smog Episode of December 1952 as seen looking west along the Thames River toward Oakley Street and the Prince Albert Bridge. Nearly 4000 deaths were attributed to the five-day episode of smog. It was one of the most disastrous and lethal occurrences of air pollution to date. This picture was taken at 4 P.M. by the now-retired Purdue University professor, William Patterson, while studying at University College in London. (Photo courtesy of Robert B. Jacko, Purdue University.)

12

Incinerators

Leo Weitzman
LVW Associates, Inc.

Many types of devices are used for incineration. The most obvious are incinerators, which are furnaces especially designed and built to burn wastes. However, wastes, especially hazardous wastes, are also burned in boilers and industrial furnaces, mainly cement and aggregate kilns. Approximately 50% of the incinerable hazardous wastes produced in the U.S. in 1993 were burned in cement kilns. The important point to consider is that irrespective of the device, as soon as it is used to burn wastes, it becomes subject to all appropriate laws and regulations which govern the handling, storage, and combustion of wastes.

When properly performed, incineration is highly efficient, destroying virtually all organic contaminants, reducing the volume of material to be landfilled, and producing extremely low levels of air emissions. Incineration facilities frequently encounter opposition from neighbors and from political groups, and it can be argued that such opposition represents the greatest barrier to their use. Incineration is also heavily regulated by federal, state, and local statutes and regulations. The regulations govern every facet of the design, construction, testing, and operation

0-8493-8953-4/95/$0.00 + $.50

of all waste combustion facilities, and a thorough understanding of the legal aspects is essential to successful operation of an incineration facility. The regulatory requirements are briefly discussed below, but because of their complexity and the fact that they are subject to frequent changes, the reader is strongly urged to contact all appropriate regulatory agencies to obtain the latest regulatory standards and requirements before proceeding with any facet of waste management.

The following discussion restricts itself to the laws currently in effect in the United States of America. Virtually every environmental law in effect can apply to an incineration facility. Many facilities are subject to state and local statutes, but these are, in the U.S., similar to the federal laws; in many cases, the laws are identical. Obviously, the same cannot be said for other countries. Nevertheless, all environmental laws have an inherent similarity which makes an understanding of one set of comprehensive laws applicable to an understanding of any other set. The discussion is to be considered descriptive of the general concepts of the laws.

12.1 Regulations and Regulatory Background

The following four U.S. laws are the most important to waste combustion:

1. Clean Air Act (CAA) of 1972 and its subsequent amendments and reauthorizations (most recent being 1990), which specify ambient concentrations of a variety of air pollutants and which limit the emissions of hazardous or toxic air pollutants from all sources, including some nonhazardous waste incinerators.
2. Toxic Substances Control Act (TSCA), which bans the use of polychlorinated biphenyls (PCBs) and sets strict regulations for their incineration and disposal by other means. This law applies only to incinerators burning PCBs.
3. Resource Conservation and Recovery Act (RCRA) of 1976 and its successors, the Hazardous and Solid Waste Amendments (HSWA), are the basis for regulation of all wastes and specifically of incineration.
4. Clean Water Act (for all effluents to any waterway or wastewater treatment plant).

It is the author's experience that even though specific laws in other jurisdictions vary, they are similarly organized and this discussion is still applicable from a technical point of view. An incineration facility includes units for receiving, storing, pretreating (if necessary), transfer, and burning of the wastes. It usually also includes laboratory facilities for testing wastes received and samples of discharge streams, and other facilities related to record keeping. The laws require that extensive and detailed records be kept. Most of these activities, including incineration of hazardous and medical wastes are regulated under solid waste (in the U.S., RCRA) laws. The incineration of nonhazardous waste is regulated under the clean air laws as well as under RCRA and, in the U.S., the incineration of PCB wastes is regulated under TSCA. The application of the clean water legislation to incinerators is equivalent to that for any wastewater discharge and is not covered further herein.

Incineration and Clean Air Laws

The first significant environmental regulations for incinerators in the U.S. were promulgated under the Clean Air Act of 1972. The CAA required that states set up regulatory programs to reduce the ambient concentrations of the following five general categories of air pollution, called criteria pollutants:

- Particulate
- Sulfur oxides (SO_2 and SO_3)
- Nonmethane hydrocarbons (HC)
- Nitrogen oxides (NO_x)
- Carbon monoxide (CO)

All sources of air pollution, including incinerators, are required to meet emission standards for particulate. While in theory all sources were also required to meet hydrocarbon and CO standards, the CO limits in many cases were set at levels greater than that typically found in most incinerators. Note that more recently waste combustion laws have set more stringent limits on CO emissions.

The limits on SO_x and NO_x and total hydrocarbons are generally imposed only when the facility is in an area that does not meet the ambient air quality standards specified by the CAA ("nonattainment" areas) and when the potential emissions of these contaminants exceeds a specified value. Many large incinerators, boilers, cement kilns, and other industrial furnaces used for hazardous waste destruction can be major sources of contaminants and, hence, CAA regulations influence their normal operation. Some of the regulatory restrictions under the CAA and subsequent amendments established that may impact incinerators and BIFs include the following:

- Prevention of significant deterioration (PSD) requires that any major emission source in an area where the ambient air quality is currently being met must have sufficiently low emissions so that it will not significantly worsen it. The intent was to ensure that the air quality in areas that had better than the minimum specified by the CAA not be degraded to the minimum. All new major sources built in nonattainment areas are subject to a new source review (NSR) to ensure that they will not contribute to a significant further decrease in ambient air quality.
- National emission standard hazardous air pollutant (NESHAP) regulations regulate the emissions of specific, toxic pollutants.
- State and local restrictions on the emission of specific compounds and substances classified as toxic.
- Restrictions on the emissions of metals, HCl and toxic organics based on risk to health and the environment (as determined by a risk assessment) in addition to the limits specified in regulations.

The Clean Air Act was extensively amended and expanded in its 1990 reauthorization (PL0101-549, 101st Congress, Nov. 15, 1990). These amendments resulted in major changes to the types of contaminants regulated and to the procedures to be followed for their regulation. The most significant change from the perspective of the combustion of hazardous waste appears to be the CAA's establishment of ambient air concentration limits for a large number of toxic materials. It appears likely that these limits will affect hazardous waste combustors, which are classified as significant sources of contaminants. It appears clear that the impact of these hazardous waste combustors of all types on the ambient air quality will have to be routinely determined in virtually all jurisdictions. Such an evaluation may result in more stringent emission limits than those required on a technology basis.

Emission standards for municipal waste combustors under the Clean Air Act are published in 40 CFR part 60, subpart Ca, 40 CFR § 60.30a through § 60.39a. These standards place limits on the emissions of particulate and the other criteria pollutants, acid gas (especially hydrogen chloride), chlorodibenzodioxins (CDDs), and chlorodibenzofurans (CDFs). They also specify the procedures for compliance testing, operator training, and reporting and record keeping.

Virtually all boilers and industrial furnaces, whether they burn hazardous waste or not, are subject to significant regulation under the Clean Air Act. These regulations still apply even when the units burn wastes or engage in other activities subject to RCRA standards. When more than one law applies, the most stringent of the applicable regulations must be obeyed.

Incinerator Regulation under the Toxic Substances Control Act

The Toxic Substances Control Act (TSCA) governs the incineration of polychlorinated biphenyls. PCBs are a class of compounds which had been used extensively for many industrial applications, especially as dielectric fluids, and which were banned under TSCA. Such separation of PCB handling from the handling of other wastes is unique to the U.S. The PCB incinerator regulations are codified in 40 CFR § 761.70, Annex I, and require the following:

- For a solids incinerator, PCB emissions must not exceed 1 mg per kg of PCB fed to the incinerator—this corresponds to a **destruction and removal efficiency (DRE)** of 99.9999%.
- Particulate emissions must be controlled to a level specified by the **Environmental Protection Agency (EPA)** regional administrator or, for systems to be operated in more than one region, the director of the exposure evaluation division of the Office of Pesticides and Toxic Substances (OPTS) of the EPA.
- The HCl must be controlled and the level of control for each facility must be specified by the EPA regional administrator or, for systems to be operated in more than one region, the director of the exposure evaluation division of the OPTS.
- The incinerator must satisfy the combustion conditions in 40 CFR § 761.70: 1200°C (±100°C) with 3% oxygen and a 2-second gas residence time in the combustion chamber or 1600°C (±100°C) with 2% oxygen and a 1 ½-second gas residence time in the combustion chamber; and a combustion efficiency (CE) greater than 99% as calculated by the following equation:

$$\%CE = 100\% \left[CO_2/(CO_2 + CO)\right] \tag{12.1}$$

  where CO_2 is the molar or volume fraction of carbon dioxide in the exit gas from the combustion chamber and CO is the molar or volume fraction of carbon monoxide in the exit gas from the combustion chamber.
- The incinerator must be tested (trial burn) prior to use, and it must contain sufficient monitors and safety interlocks to automatically shut off the PCB feed if minimum operating conditions are not met.

For the purpose of incineration, a waste is restricted under TSCA if it contains more than 50 ppm PCB. If it contains over 500 ppm PCB it may only be destroyed in an approved incinerator or equivalent pursuant to 40 CFR § 761.60(e). The PCB Regulations also allow the disposal of "mineral oil dielectric fluid" and other liquids contaminated with less than 500 ppm PCB in a "high efficiency boiler" defined as one meeting the requirements in 40 CFR § 761.60:

1. The boiler is rated at a minimum of 50 MMBtu/hr.
2. For natural gas and oil fired boilers the concentration of CO $\leq$ 50 ppm and $O_2 \geq 3\%$.
3. For coal fired boilers the concentration of CO$\leq$ 100 ppm and $O_2 \geq 3\%$.
4. The PCB-contaminated fluid does not make up more than 10% on a volume basis of the fuel feed rate.
5. The boiler must be at operating temperature when the PCB-contaminated material is fed. No waste feed during start-up or shutdown.
6. The owner and operator must comply with monitoring and record-keeping requirements described in 40 CFR § 760.60(6).

Incinerator Regulation under the Resource Conservation and Recovery Act

The Resource Conservation and Recovery Act of 1976 (RCRA) as amended in 1986 under the name Hazardous and Solid Waste Amendments (HSWA) requires that the EPA promulgate regulations governing the handling of hazardous wastes. Note that RCRA is commonly used to refer to RCRA, HSWA, and subsequent amendments. The term will be similarly used herein. In the U.S. the RCRA regulations are given in the Code of Federal Regulations, 40 CFR § 260 through § 280, which set design and performance standards for hazardous waste generation, storage, transport, disposal, and treatment. The RCRA regulations set standards for all aspects of waste management, including standards for generators and transporters of hazardous wastes, and owners and operators of treatment, storage, and disposal facilities (waste combustion devices fall under this heading).

The general requirements for all treatment, storage, and disposal (TSD or TSDF) facilities are described in "Standards for Owners of Hazardous Waste Treatment, Storage, and Disposal Facilities," 40 CFR § 264, which specifies that an owner or operator satisfy requirements such as the following:

- Construct all facilities such as storage tank farms, drum storage areas, waste-receiving areas, and landfills in a manner which minimizes the environmental impact of both routine operations and upset conditions.
- Develop a contingency plan and emergency procedures to cope with spills, fires, and other accidents.
- Maintain records of waste produced, treated, and disposed, and identify their fate or disposition.
- Develop a closure and postclosure plan which includes costs, and show that money will be available to implement the plan.
- Meet financial requirements that verify ability to clean up in case of an accident and ability to close the facility after its useful life is over.
- Manage containers, tanks, surface impoundments, waste piles, and landfills properly.

All hazardous waste incinerators are subject to the general standards for storage and disposal as well as the specific standards dealing with the incineration process. The reader is referred to *Permit Applicants' Guidance Manual for the General Facility Standards of 40 CFR § 264*, SW-968, October 1983 for details on the general standards.

The RCRA regulations require owners and operators of all TSD facilities to obtain an operating permit from the appropriate regulatory agency, the EPA regional office, or, if authority has been so transferred, a state agency. To obtain a permit, the applicant submits the following general information as well as process (e.g., container storage, tank treatment, land disposal, incineration) information:

- Description of the facility
- Description of the waste
- Security procedures and inspection schedule
- Contingency plan
- Description of preventive maintenance procedures
- Personnel training program
- A closure plan including cost estimates
- Assurance that the operator of the facility is financially able to assume this responsibility

Definition of Solid, Hazardous, and Medical Waste

RCRA classifies a waste as any material which has no value and which is commonly disposed. It further specifically excludes from this definition any waste material discharged to the air (and regulated under the Clean Air Act) or to a wastewater treatment plant or waterway (and regulated under the Clean Water Act). This is a simplistic definition, but it is functionally reasonably adequate. See 40 CFR § 260.10 for the legal definitions of the terms related to hazardous waste.

A solid waste can be further classified as a nonhazardous waste, a medical waste, or a hazardous waste. The classification governs the regulations for the waste's incineration and, hence, defines the types and operating conditions for the combustors that may be used. A nonhazardous waste is any solid waste which does not meet the requirements for a medical waste or a hazardous waste. Certain wastes, such as household and commercial refuse, are classified as nonhazardous by law. Similarly, certain high-volume industrial wastes, such as mine tailings and ash from combustion of coal, are classified by law as nonhazardous. Medical wastes are those which might contain some form of pathogen such as waste from hospitals. As can be seen, the classifications are generally made on the basis of exclusions, that is, (1) a waste material is a solid waste if it is not an air pollutant or a wastewater, and (2) a solid waste is a nonhazardous waste if it does not meet the definition of a hazardous or medical waste. The definition of a hazardous waste thus becomes critical to that of the other types of wastes. A waste is classified as hazardous under RCRA if:

- It exhibits any of the following characteristics:
  - *Ignitability.* It is ignitable, detonates readily, or it is an oxidizing agent which causes other materials to ignite or burn as defined in 40 CFR § 261.21 (waste category D001).
  - *Corrosivity.* It has a pH of less than 2 or more than 12.5, or is corrosive to steel as defined in 40 CFR § 261.22 (waste category D002).
  - *Reactivity.* It is unstable and readily undergoes change without detonating; it will react with air or water or will spontaneously react under shock or friction as described in 40 CFR § 261.23 (waste category D003).
  - *Toxicity.* It is toxic, it leaches specific contaminants at an excessive rate as shown (formerly by the EP toxicity 40 CFR § 261.24 and Appendix II and now) by the TCLP (40 CFR § 261.24 and Appendix II), which has superseded the EP toxicity test.
- It is specifically listed in 40 CFR § 261.11 subpart C as a hazardous waste because of any of the following:
  - It exhibits any of the hazardous waste characteristics described above.
  - It has been found to be fatal to humans in low doses or, in the absence of data on human toxicity, it has been shown to be toxic to animals at specified low doses [40 CFR § 261.11 (a)2].
  - It contains a toxic constituent listed in Appendix VIII to part 261 (Appendix VIII is a list of compounds and classes of compounds, e.g., lead and compounds not otherwise specified, which have been determined by the EPA administrator to be toxic) unless the administrator determines that the waste does not pose a present or potential hazard (40 CFR § 261.111).

Wastes classified as hazardous solely because they are ignitable, reactive, or corrosive (but not toxic) and contain no Appendix VIII constituents are exempt from those hazardous waste incinerator regulations which relate to performance; however, they are still subject to waste analysis and closure requirements of the regulations and must have a permit [40 CFR § 264.340(b)].

Regulation of Nonhazardous Waste Incinerators

At present, incinerators burning refuse and other nonhazardous wastes are regulated only under RCRA and the Clean Air Act. The regulations under RCRA for solid wastes are codified in 40 CFR § 240 and for medical wastes in § 259. The regulation of solid wastes and especially medical wastes includes a substantial amount of record keeping to track the wastes' ultimate disposal or destruction. Emission limits on nonhazardous waste incinerators are set on particulate, hydrogen chloride gas, sulfur dioxide, chlorodibenzodioxins, and chlorodibenzofurans.

Particulate emissions are also called flyash. The allowable particulate emissions vary between 0.015 and 0.1 gr/scf (34 and 229 mg/scm) corrected to 7% oxygen, depending on the age of the furnace and the particular regulations that are applicable in the jurisdiction. Particulate control devices are required in almost all cases. The HCl gas forms from the combustion of chlorinated materials, most commonly polyvinyl chloride (PVC). Sulfur dioxide (SO_2) forms from the combustion of sulfur in fuel or waste and from the decomposition of sulfur-bearing minerals in the waste. Because of the relatively low sulfur content of the fuel and waste, their combustion is usually only a minor source of sulfur dioxide emissions from incinerators. Waste gypsum (containing calcium sulfate) from building materials is the main source of sulfur emissions. Operation of refuse incinerators requires that the waste sorting and acceptance program keep gypsum products from entering the combustor.

Physically, medical wastes are similar to general refuse, but these wastes may be contaminated with pathogens. They often also contain large quantities of PVC, which results in elevated hydrogen chloride emissions. A major concern when operating an incinerator for medical wastes is maintenance of a sufficiently high temperature for a sufficiently high solids and gas residence time to ensure that the waste is sterilized. Chlorodibenzodioxin and chlorodibenzofuran emissions are regulated to a level of 30 ng/m^3, although this can be less if a site-specific risk assessment indicates that such a change is necessary.

Regulation of Hazardous Waste Incinerators

The regulations governing hazardous waste incinerators are covered in 40 CFR § 264, subpart 0, which requires that the owners or operators also perform the following:

- Analyze wastes as specified in the permit.
- Meet the following minimum performance standards: 99.99% destruction and removal efficiency of selected principal organic hazardous constituents (POHCs) and 99.9999% DRE for dioxin-listed wastes; 99% removal of hydrochloric acid or 1.8 kg/h HCl emissions, whichever is greater; chlorodibenzodioxin and chlorodibenzofuran emissions are regulated to a level of 30 ng/m^3, although this can be less if a site-specific risk assessment indicates that such a change is necessary; and particulate emissions of 180 mg/dscm corrected to 7% O_2.
- Operate the incinerator within limits specified in the permit for (as a minimum) the following control parameters: carbon monoxide level in exhaust, waste feed rate and composition, combustion temperature, indicator of combustion gas velocity, and other requirements necessary to meet performance standards.
- Control fugitive emissions.
- Install automatic waste feed cutoffs to stop the flow of hazardous waste if the limits are exceeded.
- Perform inspections and monitoring (I&M).
- Remove wastes upon closure.

The hazardous waste combustion regulations were expanded in 1991 to include regulation of boilers and industrial furnaces. (Federal Register, Vol. 56, No. 35, Feb. 21, 1991, p. 7134; Vol. 56, No. 166, August 27, 1991, p. 42504 and subsequent amendments). These boiler and industrial furnace (BIF) regulations made the following changes:

- Expanded the regulation of the combustion of hazardous wastes to include destruction in boilers and industrial furnaces.
- Placed limits on carbon monoxide and, in some cases, total hydrocarbons as a mechanism to minimize the emissions of hazardous products of combustion.
- Placed limits on toxic metal emissions on the basis of site-specific risk determinations.
- Limited HCl and Cl_2 emissions to the lesser of the following: (1) 4 lb/hr, (2) 99% removal efficiency, (3) risk-based limit.

The major change in performance requirements placed on hazardous waste combustion devices by the BIF regulations was risk-based limits on ambient concentrations of noncarcinogenic metals, HCl and Cl_2, specified as reference air concentrations (RACs) and for carcinogenic metals based on the risk-specific dose (RSD) for each of the metals. The RAC for a constituent is defined as the maximum annual average concentration of that constituent at any point in the vicinity of the combustor. The value is specified in the regulation and is based on acute toxicity data. The RSD is defined as the maximum ambient concentration that results in a 10^{-5} increase in cancer risk. The total emissions of carcinogenic metals must, therefore, be such that the combined risk from their emission will result in a risk below this value. Methods for calculating the emission limits are given in the EPA's implementation document for the BIF regulations [EPA, 1992].

In addition, a guidance document on metals and hydrogen chloride controls for hazardous waste incinerators [EPA, 1989c] has been prepared to assist in this effort. It gives calculation techniques and worksheets which can be used to set risk-based emission limits. It allows the limits to be set on the basis of feed rate levels of the contaminants or on the emissions achieved by the total system, combustor and air pollution control equipment. Since metal emissions tend to increase with increased combustor temperature, assurance that the emissions during operation do not exceed those during the trial burn may require additional permit conditions, such as maximum temperature.

EPA has published numerous guidance documents describing specific procedures to be followed in various aspects of incinerator permitting and operation. The reader is referred to the *Engineering Handbook for Hazardous Waste Incineration* [Bonner, 1981] for discussion of incineration equipment and ancillary systems. This book has been updated and is scheduled to be published in 1994 under the title *Engineering Handbook for Combustion of Hazardous Waste.* The *Guidance Manual for Hazardous Waste Incinerator Permits* [EPA, 1983] and *Handbook, Guidance on Setting Permit Conditions and Reporting Trial Burn Results* [EPA, 1989b] along with the implementation document for the BIF Regulations [EPA, 1992] provide necessary information for permitting a hazardous waste combustor and operating it in compliance with applicable regulations.

Oxygen Correction Factors

RCRA (and some other) regulations require reporting of CO, chlorodibenzodioxin, chlorodibenzofuran, and particulate concentrations in the flue gas corrected to 7% oxygen. The revised correction factor (Federal Register, Vol. 55, No. 82, April 27, 1990, p. 17918) is

$$CO_c = CO_m \times 14/(E - Y)$$

where CO_c is the corrected CO concentration, CO_m is the measured CO concentration, E is the enrichment factor, percentage of oxygen used in the combustion air (21% for no enrichment), and Y is the measured oxygen concentration in the stack (by Orsat).

Regulatory Requirements for Risk Assessments

In 1993, through a series of memos and policy statements, the U.S. Environmental Protection Agency expanded the review of incinerators to include estimation of risk through a site-specific risk assessment. The details of this estimating procedure are beyond the scope of this manual. Briefly, however, this policy requires that the EPA perform a risk assessment on all waste combustion facilities to determine whether the statutory emission limits are adequate for protecting health and the environment. If the risk assessment indicates that they are not, it requires that the emission limits for the facility be set at a lower value. The risk assessment methodology to be followed by EPA is described in *Technical Implementation Document for EPA's Boiler and Industrial Furnace Regulations* [EPA, 1992] and amendments. The methodology is being constantly updated. The latest version of the risk assessment procedure can be obtained from the Environmental Protection Agency.

12.2 Principles of Combustion and Incineration Thermodynamics

The physical and chemical processes of combustion are the same whether the fuels are burned in an open fire, an engine, or a refractory-lined chamber such as a boiler or incinerator. Combustion requires the presence of organic matter, oxygen (usually air), and an ignition source. The word "fuel" in the context of combustion is used to designate any organic material which releases heat in the combustion chamber regardless of whether it is a virgin fuel, such as natural gas or fuel oil, or a waste material. When organic matter containing the combustible elements carbon, sulfur and hydrogen is raised to a high enough temperature (order of 600–800°F, 300–400°C) the chemical bonds are excited and the compounds break down. If there is insufficient oxygen present for the complete oxidation of the compounds, the process is termed pyrolysis. If sufficient oxygen is present, the process is termed combustion.

Pyrolysis is a necessary first step in the combustion of most solids and many liquids. The rate of pyrolysis is controlled by three mechanisms. The first mechanism is the rate of heat transfer into the fuel particle. Clearly, therefore, the smaller the particle or the higher the temperature, the greater the rate of heating and the faster the pyrolytic process. The second mechanism is the rate of the pyrolytic process itself. The third mechanism is the diffusion of the combustion gases away from the pyrolyzing particles. Clearly, the last mechanism is likely to be a problem only in combustion systems which pack the waste material into a tight bed and provide very little gas flow.

At temperatures below approximately 500°C (900°F) the pyrolysis reactions appear to be rate controlling for solid particles less than 1 cm in diameter. Above this temperature, heat and mass transfer appears to limit the rate of the pyrolysis reaction. For larger pieces of solid under most incinerator conditions, heat and mass transfer are probably the rate-limiting step in the pyrolysis process [Niessen, 1978]. Since pyrolysis is the first step in the combustion of most solids and many liquids, heat and mass transfer is also the rate-limiting step for many combustion processes. Pyrolysis produces a wide variety of complex organic molecules which form by two mechanisms: cracking and recombination. In cracking, the constituent molecules of the fuel break down into smaller portions. In recombination, the original molecules or cracked portions of the molecule recombine to form larger, often new, organic compounds such as benzene. Pyrolysis is also termed destructive distillation. The products of pyrolysis are commonly referred to as **products of incomplete combustion (PICs).** PICs include a large number of different organic molecules.

Pyrolysis is only the first step in combustion. To complete combustion, a properly designed incinerator, boiler, or industrial furnace mixes the pyrolysis off-gases with oxygen and exposes the mixture to high temperatures. The resulting chemical reactions, those of combustion, destroy the organics present to very high levels. Combustion is a spontaneous chemical reaction which takes place between any type of organic compound (and many inorganic materials) and oxygen. Combustion liberates energy in the form of heat and light. The combustion process is so violent and releases so much energy that the exact organic compounds involved become relatively unimportant. The vast majority of the carbon, hydrogen, oxygen, sulfur, and nitrogen will behave, in many ways, like a mixture of the elements. This is a very important concept since for all but the most detailed calculations, the combustion process can be evaluated on an elemental basis.

The process of combustion can be viewed as taking place in three primary zones: (1) the volatilization or pyrolysis zone, referred to here as the preflame zone; (2) the flame zone; and (3) the postflame or burnout zone. In the first zone, the organic material in the gaseous, liquid, or solid fuel is vaporized and mixes with air or another source of oxygen. Those organic compounds which do not vaporize typically pyrolize, forming a combustible mixture of organic gases. Volatilization is endothermic (heat absorbing) while pyrolysis is, at best, only slightly exothermic. As a result, this step in the combustion process requires a heat source to get the process started.

The source of the initial heat, the ignition source, the match or pilot flame for example, provides the energy to start the combustion reaction. Once started, the reaction will be self-sustaining as long as fuel and oxygen are replenished at a sufficiently high rate to maintain the temperature above that needed to ignite the next quantity of fuel. If this condition is met, the ignition source can be removed. The energy released from the initial reaction will activate new reactions and the combustion process will continue. A material that can sustain combustion without the use of an external source of ignition is termed *autogenous.*

In order to speed up the phase change to the vapor a liquid is usually atomized by a nozzle which turns it into fine droplets. The high surface-to-volume ratio of the droplets increases the rate at which the liquid absorbs heat, increases its rate of vaporization or decomposition, and produces a flammable gas which then mixes with oxygen in the combustion chamber and burns. Atomization, while usually desirable, is not always necessary. In certain combustion situations the gas temperatures may be high enough or the gas velocities in the combustion chamber large enough to allow the fuel or waste to become a gas without being atomized.

The phase change is speeded for solids by agitating them to expose fresh surface to the heat source and improve volatilization and pyrolysis of the organic matter. Agitation of solids also increases the rate at which heat and oxygen transfer into the bed and of combustion gases out of the bed. In practice this is done in many different ways, such as tumbling the solids in a kiln, raking the solids over a hearth, agitating with a hot solid material that has a high heat capacity (as in a fluidized bed), burning the solids in suspension, burning the solids in a fluidized bed, or, if the combustion is rapid enough, drawing some of the air that feeds the flame through a grate holding the solids.

It is important that the amount of fuel charged to the burning bed not exceed the heating capacity of the heat source. If this occurs, the burning mass will require more heat to vaporize or pyrolize than the heat source (the flame zone) can supply and the combustion reactions will not continue properly. In most cases such overloading will also result in poor distribution of air to the burning bed and improper flow of combustion gases away from the flame. The combined result will be that the flame will be smothered.

Consider one of the simplest forms of combustion, that of a droplet of liquid or a particle of solid (the fuel) suspended in a hot oxidizing gas, as shown in Fig. 12.1. The fuel contains a core of solid or liquid whose temperature will be below its boiling or pyrolysis point. That temperature is shown as T_B. The liquid or solid core is surrounded by a vapor shell which consists of the

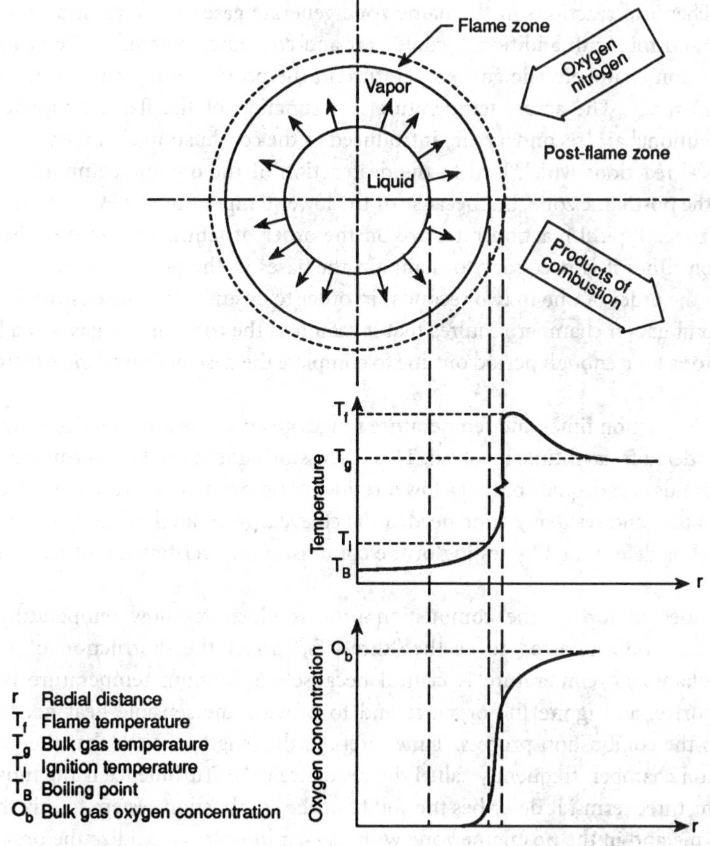

FIGURE 12.1 Combustion around a droplet of fuel.

vaporized liquid and the products of pyrolysis of the fuel. The fuel and its surrounding vapor are the first zone of combustion. The vapor cloud surrounding the liquid core is continually expanding or moving away from the core. As it expands it heats up and mixes with oxygen diffusing inward from the bulk gases. At some distance from the core the mixture reaches the proper temperature (T_i) and oxygen/fuel ratio to ignite. The actual distance from the core where the expanding vapor cloud ignites is a complex function of the following factors: the bulk gas temperature, the vapor pressure of the liquid and latent heat of vaporization, the temperature at which the material begins to pyrolyze, the turbulence of the gases around the droplet (which affects the rate of mixing of the outflowing vapor and the incoming oxygen), the amount of oxygen needed to produce a stable flame for the liquid, and the heat released by the combustion reaction.

Ignition creates the second zone of the combustion process, the flame zone. The flame zone has a small volume compared to that of the preflame and postflame zones in most combustors, and a molecule of material will only be in it for a very short period of time, on the order of milliseconds. Here, the organic vapors rapidly react with air (the chemical reactions are discussed below) to form the products of combustion. The temperature in the flame zone, T_f, is very high, usually well over 3000°F. At these elevated temperatures, the atoms in the molecule are very reactive and the chemical reactions are rapid. Reaction rates are on the order of milliseconds.

The very high temperature in the flame zone is the main reason one can consider the major chemical reactions that occur in combustion to be functions of the elements involved and not of the specific compounds. The vast majority (on the order of 99% or more) of the organic constituents released from the waste and fuel are destroyed in the flame zone.

The flame around a droplet can be viewed as a balance between the rate of outward flow of the combustible vapors against the inward flow of heat and oxygen. In a stable flame, these two flows are balanced and the flame appears to the eye to be stationary.

The rapid chemical reactions in the flame zone generate gaseous combustion products which flow outward and mix with additional, cooler, air and combustion gases in the postflame region of the combustion chamber. The gas temperature in the postflame region is in the 600–1200°C (1200–2200°F) range. The actual temperature is a function of the flame temperature and the amount of additional air (secondary air) introduced to the combustion chamber.

The chemical reactions which lead to the destruction of the organic compounds continue to take place in the postflame zone, but because of the lower temperatures they are much slower than in the flame zone. Typical reaction rates are on the order of tenths of a second. Because of the longer reaction times it is necessary to maintain the gases in the postflame zone for a relatively long time (on the order of one to two seconds) in order to ensure adequate destruction. Successful design of a combustion chamber requires that it maintain the combustion gases at a high enough temperature for a long enough period of time to complete the destruction of the hazardous organic constituents.

Note that the reaction times and temperature ranges given above are intended only to provide a sense of the orders of magnitude involved. This discussion should not be interpreted to mean that one or two seconds are adequate or that a lower residence time or temperature is not acceptable. The actual temperature and residence time needed to achieve a given level of destruction is a complex function which is determined by testing of the combustor and verification of its performance by the trial burn.

The above description of the combustion process illustrates how temperature, time, and turbulence, commonly referred to as the "three Ts," affect the destruction of organics in a combustion chamber. Temperature is critical because a minimum temperature is required to pyrolyze, vaporize, and ignite the organics and to provide the sensible heat needed to initiate and maintain the combustion process. Time refers to the length of time that the gases spend in the combustion chamber, frequently called the residence time. Turbulence is the most difficult to measure of the three terms. It describes the ability of the combustion system to mix the gases both within the flame and in the postflame zone with oxygen in order to oxidize the organics released from the fuel.

The following three points illustrate the importance of turbulence:

1. The process of combustion consumes oxygen in the immediate vicinity of pockets of fuel-rich vapor.
2. The destruction of organic compounds occurs far more rapidly and cleanly under oxidizing conditions.
3. In order to achieve good destruction of the organics, it is necessary to mix the combustion gases moving away from the oxygen-poor pockets of gas with the oxygen-rich gases in the bulk of the combustion chamber.

Turbulence can, therefore, be considered to be the ability of the combustor to keep the products of combustion mixed with oxygen at an elevated temperature. The better this ability (up to a point), the higher the degree of destruction of the organics.

Complex flames behave in an analogous manner to that described above. The major difference is that the flame is often shaped by the combustion device to optimize the three Ts. To illustrate, consider a Bunsen burner flame. The fuel is introduced through the bottom of the burner's tube and accelerated by a nozzle in the tube to increase turbulence. Openings on the side of the tube permit air to enter and mix with the fuel. This air is called primary air since it mixes with the fuel prior to ignition. The flow rate of the fuel and air are adjusted so that the mixture is slightly too rich (too much fuel or not enough oxygen) to maintain combustion. When the mixture hits the ambient air at the mouth of the burner, it mixes with additional oxygen and ignites. The flame of a properly adjusted Bunsen burner will be hollow. The core will contain a mixture of fuel and air which is too rich to burn, the preflame zone. The flame zone is well defined. In it the rapid flow of primary air and fuel increases turbulence. The postflame zone is virtually nonexistent for an open burner since there is no combustion chamber to maintain the elevated temperature.

The Bunsen burner is designed for gaseous fuels. The fuel is premixed with air to minimize the amount of oxygen that must diffuse into the flame to maintain combustion. Premixing the fuel with air also increases the velocity of the gases exiting the mouth of the burner, increasing turbulence in the flame and producing a flame with a higher temperature than that of a simple gas flame in air. Liquid combustion adds a level of complexity. Liquid burners consist of a nozzle, whose function is to atomize the fuel, mounted into a burner, burner tile, or burner block which shapes the flame so that it radiates heat properly backward and provides good mixing of the fuel and air. The whole assembly is typically called the burner. The assembly may be combined into a single unit, or the burner and nozzle may be independent devices. The fundamental principles of operation of liquid burners are the same as those of a Bunsen burner, with the added complexity of atomizing the fuel so that it will vaporize readily. In all liquid fuel burners, the fuel is first atomized by a nozzle to form a finely dispersed mist in air. Heat radiating back from the flame vaporizes the mist. The nozzle mixes the vapor with some air, but not enough to allow ignition. The mixture is now equivalent to the gas mixture in the tube of the Bunsen burner; it is a mixture of combustible gases and air at a concentration too rich (too much fuel or not enough oxygen) to ignite.

As the fuel-air mixture moves outward, it mixes with additional air, either by its impact with the oxygen-rich gases in the combustion or by the introduction of air through ports in the burner. As the gases mix with air, they form a flame front. The flame radiates heat backwards to the nozzle where it vaporizes the fuel. Since most nozzles cannot tolerate flame temperatures, the nozzle and burner must be matched so that the cooling effect of the vaporizing fuel prevents radiation from overheating the nozzle. Similarly, if the liquid does not evaporate in the appropriate zone (if, for example, it is too viscous to be atomized properly), it will not vaporize and mix adequately with air. Proper balance of the various factors results in a stable flame. Clearly, all liquid burned in a nozzle must have properties within the nozzle's design limits. A flame which flutters a lot and has numerous streamers is "soft." One which has a sharp, clear, spearlike (as in a Bunsen burner flame) or spherical appearance is "hard." Hard flames tend to be hotter than soft flames.

Nozzles operate in many ways. Some nozzles operate like garden hoses—the pressure of the liquid fed to them is used to atomize the liquid fuel. Others use compressed air, steam, or nitrogen to atomize the liquid. Nitrogen is used in those cases when the liquid fuel is reactive with steam or air. A third form of nozzle atomizes the liquid by firing the liquid against a rotating plate or cup. The type of nozzle used for any given application is a function of the properties of the liquid.

A great deal of information about the fuel and about the combustion process can be gained by looking at the flame in a furnace. **CAUTION! Protective lenses must always be worn when examining the flame.** The flame's color is a good indicator of its temperature. But this indicator must be used with caution as the presence of metals can change the flame color. In the absence of metals, yellow flames are the coolest. As the color moves up the spectrum (red, orange, yellow, green, blue, indigo, and violet) the flame temperature increases. One will often see different colors in different areas of the flame. A sharp flame formed by fuel with high heating values will typically have a blue to violet core surrounded by a yellow to orange zone. Such a flame would be common in a boiler or industrial furnace where coal, fuel oil, or some other "hot" fuel was being burned.

Another useful piece of operating information is the shape of the flame. A very soft (usually yellow or light orange) flame with many streamers may indicate that the fuel is inhomogeneous and probably has a heating value approaching the lower range for maintaining combustion. This is acceptable unless a large amount of soot is observed. Soot (black smoke) released from a flame is indicative of localized lack of oxygen. While a small amount (a few fine streamers) of soot are common in a soft flame, large amounts of soot or a steady stream of soot from one point indicates some form of burner maladjustment.

Large amounts of soot emanating from the flame can occur because of two reasons. First, the burner may not be supplying enough air to the flame. The system should be shut down and the burner inspected for blockage in the air supply. Second, the fuel could contain too much water or other material with a low heating value such as a heavily halogenated organic. In this case, improved fuel (waste) blending may resolve the issue. Production of large amounts of soot are usually associated with a rapid rise in the concentration of CO and hydrocarbon in the flue gas. A CO monitor is often a useful tool for ensuring that burners are properly adjusted.

12.3 Combustion Chemistry

The combustion of a compound is normally represented as a single chemical reaction. In fact, numerous chemical reactions can occur. Consider, for example, one of the simplest combustion processes, the burning of methane in the presence of air. The equation is

$$CH_4 + 2O_2 \rightarrow CO_2 + 2H_2O \tag{12.2}$$

In fact, many more reactions are possible. If the source of the oxygen is air, nitrogen will be carried along with the oxygen at a ratio of approximately 79 moles (or volumes) of nitrogen for each 21 of oxygen. The nitrogen will mainly be a diluent for the combustion process but a small (but important) fraction will also participate to form trace quantities of NO_x. In addition, if the combustion is less than complete, some of the carbon will form CO rather than CO_2. Because of the presence of free radicals in a flame, molecular fragments can coalesce and form larger organic molecules. When the material being burned contains elements such as chlorine, numerous other chemical reactions are possible. For example, the combustion of carbon tetrachloride with methane can result in the following products:

$$\begin{aligned} CH_4 + CCl_4 + O_2 + N_2 \rightarrow\ & CO_2 + H_2O + HCl + N_2 + CO + Cl_2 + CH_3Cl \\ & + CH_2Cl_2 + CHCl_3 + C_2H_5Cl + C_2H_4Cl_2 + 2H_3Cl_3 + ? \end{aligned} \tag{12.3}$$

The goal of a well-designed combustor is to minimize the release of the undesirable products and convert as much of the organics to CO_2, water, and other materials which may safely be released after treatment with an APCD. The combustion products of a typical properly operating combustor will contain on the order of 5–12 percent CO_2, 20–100 ppm CO, 10–25 percent H_2O, ppb quantities of POHCs and PICs, ppm quantities of NO_x, and ppm quantities of SO_x.

The lower concentrations of CO appear to be a normal product of combustion and are, most likely, limited by chemical equilibrium. If the combustion is poor (poor mixing of the oxygen and fuel or improper atomization of the fuel) localized pockets of gas will form where there is insufficient oxygen to complete the combustion. CO will form in these localized pockets, and since the reaction of CO to CO_2 is slow outside the flame zone (on the order of seconds at the postflame zone conditions) it will not be completely destroyed. This mode of failure is commonly termed *kinetics limiting* because the rate at which the chemical reactions occur was less than the time that the combustor kept the constituents at the proper conditions of oxygen and temperature to destroy the intermediate compound.

Similar explanations can be offered for the formation of other PICs. Many are normal equilibrium products of combustion (usually in minutely small amounts) at the conditions of some point in the combustion process. Because of the similarity, PIC formation is commonly associated with CO emissions. Test data [EPA, 1991, 1992], have shown that PICs rarely if ever occur when the CO level is less than 100 ppm (dry and adjusted to 7% O_2). They sometimes occur at CO levels over 100 ppm. It must be noted that PICs occur during the combustion of all fuels, including wood, petroleum products, and coal. Their formation is not characteristic just of the combustion of hazardous wastes.

Hydrogen forms two major products of combustion depending on whether chlorine is present or not. If chlorinated organic wastes are burned, the hydrogen will preferentially combine with the chlorine and form HCl. The thermodynamics of HCl formation are such that all but a small fraction (order of 0.1%) of the chlorine will form HCl. The balance is chlorine gas. The reaction between free chlorine and virtually any form of hydrogen found in the combustion chamber is so rapid that the Cl_2:HCl ratio will be equilibrium limiting in virtually all cases. Organically bound oxygen will behave like a source of oxygen for the combustion process.

Behavior of the halogens is similar to that of chlorine discussed above. Fluorine, which is a more electronegative compound than chlorine, will be converted to HF during the combustion process. Like chlorine, it will form an equilibrium between the element and the acid, but the thermodynamics dictates that this equilibrium will result in a lower F_2:HF ratio than the Cl_2:HCl. Bromine and iodine tend to form more of the gas than the acid. Combustion of a brominated or iodinated material will result in significant releases of bromine or iodine gas. This fact is important to incinerator design since Br_2 and I_2 will not be removed by simple aqueous scrubbers. Furthermore, because the production of the elemental gases is equilibrium limiting, modifications to the combustion system will not reduce their concentration in the flue gas significantly. It is sometimes possible to increase the amount of acid formed by adding salts, but the level of conversion of the salt to acids is uncertain.

Organic sulfur forms the di- and trioxides during combustion. The vast majority of the sulfur will form SO_2 with trace amounts of SO_3 also forming. The ratio of the two is equilibrium limiting. SO_3 forms a strong acid (H_2SO_4, sulfuric acid) when dissolved in water. It is thus readily removed by a scrubber designed to remove HCl. SO_2 forms sulfurous acid (H_2SO_3), a weak acid which is not controlled well by a typical acid gas scrubber designed just for HCl removal.

Nitrogen enters the combustion process both as the element, with the combustion air, and as chemically bound oxygen in the waste or fuel. During combustion, the nitrogen forms a variety of oxides. The ratio between the oxides is governed by a complex interaction between kinetic and equilibrium relationships and is highly temperature dependent. The reaction kinetics are such that the reactions to create, destroy, and convert the various oxides from one to the other occur at a reasonable rate only at the high temperatures of the flame zone. Nitrogen oxides are, therefore, controlled by modifying the shape or temperature distribution of the flame and by the addition of ammonia to lower the equilibrium NO_x concentration, and to hence decrease NO_x emissions. NO_x formation and control as well as the concept of equilibrium is discussed below.

Particulate and Metal Fume Formation

The term *particulate matter* refers to any solid and condensible matter emitted to the atmosphere. Particulate emissions from combustion are composed of varying amounts of soot, unburned droplets of waste or fuel, and ash. Soot consists of unburned carbonaceous residue, consisting of the high molecular weight portion of the POM. The soot can and does condense both on its own and on the other particulate, other inorganic salts such as sodium chloride, and metals. The formation of particulate in a combustor is intimately related to the physical and chemical characteristics of the wastes, fuels, combustion aerodynamics, the mechanisms of waste/fuel/air mixing, and the effects of these factors on combustion gas temperature-time history. The reader is referred to the EPA guide on metals and HCl controls for hazardous waste incinerators [EPA, 1989c] for further information on this subject.

Particulate can form by the following three fundamental mechanisms: abrasion, ash formation, and volatilization. Abrasion is the simplest mechanism; it forms particulate by the straightforward action of the solid mass rubbing against itself and against the walls of the combustion chamber. Since abrasion tends to form coarse particulate which can be readily removed by a reasonably well-designed APCD, it is not usually of primary concern when evaluating combustors and will not be discussed further.

The formation of particulate by ash formation occurs when ash-containing liquid wastes and fuels are burned, the organics are destroyed, and the inorganic ash remains behind. The size of the resulting particulate is a function of the concentration of the inorganics in the fuel, the size of the droplets formed by the nozzle, and the ash's physical characteristics. When a droplet of atomized liquid leaves the nozzle and hits the combustion chamber the organic portions vaporize and burn, leaving the ash behind in suspension. Depending on the composition of the ash, a fraction may volatilize as well [Barton *et al.,* 1988]. Regardless of whether this occurs or not, nearly all of the inorganic fraction is typically entrained by the gas flow.

The particulate formed by volatilization/condensation is especially germane to the emission of metals [EPA, 1989c, 1991, 1992]. Many metals and their salts will form vapors at the temperatures of the flame and the postflame zones of a combustion chamber. When the vapors cool, they condense to form very fine (<1 micron in diameter) particulate that tends to be relatively difficult to capture in an air pollution control device. Volatilization of metals and other inorganics can occur whether the waste or fuel is a solid or a liquid. As long as it contains an inorganic fraction, it may volatilize. This phenomenon is of particular concern when the inorganic fraction includes toxic metals (e.g., Cr, Cd) or their compounds.

The diameter of the entrained particles can range from 1 micron to over 50 microns, but typically does not exceed 20 microns [Petersen, 1984; Goldstein and Siegmund, 1976]. The particle size distribution can significantly affect the collection efficiency of air pollution control systems and the ability to meet RCRA particulate standards. Generally, particles less that 1 micron in diameter are more difficult to collect, requiring higher-energy wet or dry control devices.

Another source of flue gas entrained particulate is the dissolved salts in quench water and sometimes the scrubber. Particulate forms in the quench when the water is evaporated by the hot flue gases. The solids dissolved and suspended in the water could form particulate. Particulate formation in the scrubber is less common but can occur if the quench does not cool the flue gas completely or if gas velocities in the scrubber are high enough to entrain liquid droplets. The salts, such as NaCl, can escape to the flue gas along with entrained mist. Although mist elimination equipment provides an added measure of control, some of the salts can escape to the stack and will eventually be measured as particulate by conventional sampling methods. The contribution of these salts to the overall particulate loading, while not large, could result in the failure to meet the particulate emission standard.

A model to predict the partitioning of toxic metals is under development [Barton *et al.,* 1988]. Toxic metal emissions from hazardous waste incinerators can result from the following mechanisms:

- Relatively volatile metals, such as lead (Pb) and cadmium (Cd), in the waste feed vaporize and enter the gas stream. The higher the incinerator temperature, the higher the vaporization rate. As the gas cools, the vapors condense homogeneously to form new, less than 1 micron, particles, and heterogeneously on the surface of entrained ash particles, preferentially to fine particulate, because of their large surface area.
- Under combustion conditions, metals react to form compounds such as chlorides and sulfides. Because these compounds (especially many metal chlorides) are often more volatile than the original metal species, they more readily vaporize and enter the gas stream. Upon secondary oxidative reactions these metals can return to their original form and will proceed to condense into new or existing particles.
- Metals and their reactive species can also remain trapped in the ash bed of solid waste incinerators, contributing to the metal loading of the residual ash. Entrainment of particulate from the ash bed into the gas stream can, however, contribute to the flue gas loading of toxic metals.

Material and Energy Balances

Stoichiometry and thermochemistry form the basis for most design calculations for incinerators and, in fact, all combustors. One of the simplest forms of combustion is that of charcoal or coke, which may, for the purpose of this calculation, be assumed to be pure carbon. The example could just as well be based on the combustion of diamond, but budgets are tight. The major reaction is

$$C + O_2 \rightarrow CO_2 \tag{12.4}$$

One mole of carbon (12 lb) combines with one molecule or mole of oxygen gas (32 lb) to form one molecule or mole of CO_2 (44 lb) gas. If the source of oxygen for the combustion is air (approximately 21 mole or volume percent oxygen and 79 mole or volume percent nitrogen), the 1 mole of oxygen brings with it 79/21 = 3.76 moles or 105 lb of nitrogen. Note that for combustion calculations the other components of air (except for possibly the water vapor) are usually ignored.

By the ideal gas law, 1 lb-mole of any gas at STP occupies 387 ft^3, the combustion of 12 lb of carbon will produce 387 scf of CO_2, and will consume 387 scf of oxygen gas. If air is used as the source of oxygen, 1456 scf of nitrogen gas will be moved through the system.

The amount of oxygen or air required to exactly burn all of the fuel available is called the stoichiometric oxygen, stoichiometric air, theoretical oxygen, or theoretical air. Virtually all combustion devices operate with an excess of combustion air, that is, at over 100% theoretical air with the amount fed specified by the equivalence ratio, ER, defined by

$$\text{ER} = \frac{\text{Volume of combustion air}}{\text{Volume of theoretical air required}} \tag{12.5}$$

An alternative method of specifying the amount of air actually fed to the combustor is the percent excess air, which is defined by

$$\text{Percent excess air} = \%\text{EA} = (\text{ER} - 1) \times 100\% \tag{12.6}$$

If the combustion of carbon were performed using 120% excess air, ER = 2.2, then the same amount of carbon and CO_2 would be involved, but the combustor would be fed 2.2 times the theoretical amount of air than under stoichiometric conditions, hence 2.2 lb-moles or 851 scf of oxygen and 8.27 lb-moles (3203 scf) of nitrogen would pass through the system. The mass and volumetric flow rates of the combustion air and flue gas streams can thus be computed by using stoichiometry and the concept of excess air. The same calculations can be applied to complex systems.

The percent excess air (also called excess oxygen) can be readily calculated for an actual combustor from the analysis of oxygen and carbon dioxide in the combustion gases. Such a determination is typically performed using an Orsat apparatus and the following equation:

$$\%EA = \frac{O_2}{0.266N_2 - O_2} \tag{12.7}$$

Equation (12.7) assumes that the amount of nitrogen in the wastes and fuel and the CO in the flue gas are negligible. Both assumptions are usually acceptable for incinerator applications. The value 0.266 in Eq. (12.7) is simply the quotient 21/79, the ratio of oxygen to nitrogen in air. This equation can be used along with a mass balance and energy balance to cross-check the flue gas flow rate, composition, and temperature against the feed rate, composition, and heating value of the wastes and fuels fed to the incinerator to determine whether the measurements performed are mutually consistent.

Mass balances on more complex systems are performed in the same way. No matter how complex the system, the calculations can be based on the elemental feed rates of the streams. Consider the combustion of a mixture of methane and carbon tetrachloride in air. The balanced chemical reaction for the combustion is

$$CH_4 + CCl_4 + 2O_2 + 7.52N_2 \rightarrow 2CO_2 + 4HCl + 7.52N_2 + \Delta H_c$$

One mole of methane reacts with one mole of carbon tetrachloride and two moles of oxygen to form two moles of carbon dioxide and four moles of HCl. The nitrogen does not participate in the main combustion reaction; it is carried along with the oxygen. The coefficient, 7.52, for nitrogen is obtained by multiplying the number of moles of oxygen participating in the chemical reaction (2) by the volumetric ratio of nitrogen and oxygen in air (79/21). The combustion produces $(2 + 4 + 7.52) = 13.52$ moles of flue gas. If the calculations are performed in lb-moles, the combustor will produce $13.52 \times 387 = 5232$ scf of flue gas. If the calculation were performed in metric units the 13.52 g-moles of flue gas would occupy (at 0.02404 scm/g-mole) 0.325 m^3 at STP.

Stoichiometric calculations allow one to determine the amount of gas formed by the combustion process. Thermodynamic calculations allow one to determine the temperature of the gases. The combined material and energy balance is used to determine the temperature, gross composition, and flow rates in an incinerator or combustor. The heat balance is completely analogous to the mass balance that was performed above. The mass balance takes advantage of the fact that matter is neither created nor destroyed in an incinerator; all of the elements that enter the combustor must come out, although the materials react chemically to form different compounds. The heat balance is based on the fact that all of the energy that goes into the incinerator will be released at some point. Because energy can flow into and out of a system in many ways, a rigorous energy balance can be very time-consuming and complex. Fortunately, many simplifying assumptions can be made.

The first step in the heat balance is to calculate the heat input to the combustor. The heat input is simply the sum of the sensible heat released or absorbed by each stream. The heating values of pure compounds and for fuels such as natural gas or fuel oil can be determined by calorimetric methods or, more frequently, tabulated from the heat of formation of the major constituents of the wastes and auxiliary fuels. Direct measurement of the wastes' heating value can be performed by a calorimetric techniques on samples of the waste. Unfortunately, this is often not practical. The typical laboratory calorimeter is a small device, and it is usually difficult to obtain a representative sample of the waste small enough to fit into it. While multiple samples can be tested, the cost becomes high and evaluation of the data becomes complex. It is recommended that for critical applications the heat of formation be used to calculate the heat of combustion whenever the chemical composition of the waste is known.

The heat of combustion or heating value of any compound can be calculated from the heat of formation by the following formula:

$$\Delta H_{c,\,298\,\text{K}} = \sum n_p(\Delta H_f)_p - \sum n_r(\Delta H_f)_r \tag{12.8}$$

where $\Delta H_{c,\,298\,\text{K}}$ is the heat of combustion at 298 K, n is the stoichiometric coefficient, ΔH_f is the heat of formation of each compound, and the subscripts p and r refer to the products and reactants. The following example shows how the heat of combustion of 1,3-dichloropropane (liquid) is calculated from the heat of formation. Table 12.1 gives its heat of formation as 388 Btu/lb and its molecular weight at 179. The combustion equation is

$$C_3H_6Cl_2(L) + 4O_2 \rightarrow 3CO_2 + 2H_2O(L) + 2HCl$$

The heat of formation for each of these compounds is given per unit weight and per mole, and each compound's molecular weight is given in the following:

	$C_3H_6Cl_2(L)$	+	$4O_2$	→	$3CO_2$	+	$2H_2O(L)$	+	2HCl
H_f (Btu/lb)	−3,880		0		−3,848		−6,832		−1,088
Mol. wt.	179		32		44		18		36.5
H_f (Btu/lb-mole)	−69,500		0		−169,000		−123,000		−39,700

The heat of combustion of 1,3-dichloropropane liquid is

$$\begin{aligned} &3(-169{,}300) + 2(-123{,}000) + 2(-39{,}700) - (-69{,}500) \\ &\quad = -594{,}500\ \text{Btu/lb-mole or} - 1{,}530\ \text{Btu/lb} \end{aligned}$$

That is, 1,530 Btu are released when liquid 1,3-dichloropropane is burned and the water in the combustion products is condensed. What happens if the water is not condensed and its latent heat released, as when the LHV is computed? The combustion of one mole (179 pounds) of 1,3-dichloropropane forms 2 moles (36 pounds) of water. The latent heat of vaporization of water at ambient temperature is approximately 9500 cal/g-mole, 528 cal/g, 17,100 Btu/lb-mole, or 950 Btu/lb.

The LHV of 1,3-dichloropropane, when the water formed in the reaction does not condense and release its latent heat, is 594,500 − 2(17,100) = 560,300 Btu/lb-mole of 1,3-dichloropropane burned. This corresponds to 3,130 Btu/lb.

This calculation illustrates the higher heating value (HHV) and lower heating value (LHV) for a material. The HHV corresponds to the heat released by the combustion of a compound when the water formed condenses and gives up its latent heat. This is heating value measured in a calorimeter. The LHV corresponds to the heat released on combustion when the water leaves the combustion chamber in the vapor form, as occurs in an incinerator. As can be seen, the difference is not insignificant.

A published heating value for highly halogenated compounds must also be used with caution. The reason for this is that calorimetric determinations are made by placing the compound in question with oxygen into a calorimeter. Highly chlorinated compounds, such as CCl_4, $CHCl_3$, or highly chlorinated benzenes, do not contain sufficient hydrogen to form HCl. When a system contains more moles of chlorine than hydrogen, a significant fraction of the chlorine is converted to Cl_2 gas rather than to HCl. This condition is counter to the one that occurs in an incinerator, where sources of hydrogen are plentiful. For example, the combustion of chloroform in a calorimeter will occur in the following way (the heats of formation for each compound are given directly beneath the formula):

$$\begin{aligned} &CHCl_3 \;+ O_2 \rightarrow \quad CO_2 \quad + \quad HCl \quad + Cl_2 \\ \Delta H_f &= (-24{,}200) + 0 + (-94{,}052) + (-22{,}063) + 0 \\ \Delta H_c &= -91{,}915\ \text{cal/g-mole} \end{aligned}$$

Table 12.1 Heats of Formation and Combustion of Pure Compounds

			Reference condition, 1 atmosphere, 25°C, 77°F											
			Heat of Formation				Higher Heating Value				Lower Heating Value			
Compound	Formula	Mol. Weight	Cal/ g-mol	Cal/g	Btu lb-mole	Btu/lb	Cal/ g-mol	Cal/g	Btu/ lb-mole	Btu/lb	Cal/ g-mol	Cal/g	Btu/ lb-mole	Btu/lb
Saturated, Parafins, Alkanes														
Methane	CH_4	16	17,889	1,118	23,200	2,013	212,797	13,300	383,035	23,940	191,759	11,985	345,166	21,573
Ethane	C_2H_6	30	20,234	674	36,421	1,214	372,821	12,427	671,078	22,369	341,264	11,375	614,275	20,476
Propane	C_3H_8	44	24,820	564	44,676	1,015	530,604	12,059	955,087	21,707	488,528	11,103	879,350	19,985
n-Butane	C_4H_{10}	58	30,150	520	54,270	936	687,643	11,856	1,237,757	21,341	635,048	10,949	1,143,086	19,708
n-Pentane	C_5H_{12}	72	35,000	486	63,000	875	845,162	11,738	1,521,292	21,129	782,048	10,862	1,407,686	19,551
n-Hexane	C_6H_{14}	86	39,960	465	71,928	836	1,002,571	11,658	1,804,628	20,984	928,938	10,802	1,672,088	19,443
Each C past C_6	CH_2	14	4,925	352	8,865	633	157,444	11,246	283,399	20,243	146,925	10,495	264,465	18,890
Unsaturated Compounds, Alkenes														
Ethylene	C_2H_4	28	(12,496)	(446)	(22,493)	(803)	337,234	12,044	607,021	21,679	316,196	11,293	569,153	20,327
Propylne	C_3H_6	28	(4,879)	(116)	(8,782)	(209)	491,986	11,714	885,575	21,085	460,429	10,963	828,772	19,733
1-Butene	C_4H_8	56	30	1	54	1	649,446	11,597	1,169,003	20,875	607,370	10,846	1,093,266	19,523
1-Pentene	C_5H_{10}	70	5,000	71	9,000	129	806,845	11,526	1,452,321	20,747	754,250	10,775	1,357,650	19,395
1-Hexene	C_6H12	84	9,650	115	17,370	207	964,564	11,483	1,736,215	20,669	901,450	10,732	1,622,610	19,317
Each C past C_6	CH_2	14	4,925	352	8,865	633	157,444	11,246	283,399	20,243	146,925	10,495	264,465	18,890
Other Organic Compounds														
Acetaldehyde	C_2H_2OH	43	36,760	925	71,568	1,664	25,820	5,833	451,475	10,499	235,041	5,466	423,074	9.839
Acetic Acid	C_2H_4COOH	73	116,400	1,595	209,520	2,870	242,497	3,322	436,494	5,979	216,199	2,962	389,158	5,331
Acetylene	C_2H_2	26	(54,194)	(2,084)	(97,549)	(3,752)	310,615	11,947	559,107	21,504	300,096	11,542	540,173	20,776
Benzene	C_6H_6	78	(19,820)	(254)	(35,676)	(457)	789,083	10,116	1,420,349	18,210	757,526	9,712	1,363,547	17,481
Benzene (L)	C_6H_6	78	(11,720)	(150)	(21,096)	(270)	780,983	10,013	1,405,769	18,023	749,426	9,608	1,348,967	17,294
1,3-Butadiene	C_4H_6	54	(26,330)	(488)	(47,394)	(878)	607,489	11,250	1,093,480	20,250	575,932	10,665	1,036,678	19,198
Cyclohexane	C_6H_{12}	84	37,340	445	67,212	800	936,874	11,153	1,686,373	20,076	873,760	10,402	1,572,768	18,723
Ethanol	C_2H_5OH	98	46,240	574	101,232	1,033	336,815	3,437	606,267	6,186	305,258	3,115	549,464	5,607
Ethanol (L)	C_2H_5OH	46	66,356	1,443	119,441	2,597	326,669	7,102	588,058	12,784	285,142	6,416	531,256	11,549
Ethylbenzene	$C_6H_5C_2H_5$	106	(7,120)	(67)	(12,816)	(121)	1,101,121	10,388	1,982,018	18,698	1,048,526	9,892	1,887,347	17,805
Ethylene glycol (L)	$C_2H_4(OH)_2$	62	(108,580)	(1,751)	(195,444)	(3,152)	501,635	8,091	902,943	14,564	470,078	7,582	846,140	13,647
Methanol	CH_3OH	32	12,190	381	21,942	686	218,496	6,828	393,293	12,290	197,458	6,171	355,424	11,107
Methanol (L)	CH_3OH	32	48,100	1,503	86,580	2,706	182,586	5,706	328,655	10,270	161,548	5,048	290,786	9,087
Methylcyclohexane	$C_6H_{11}CH_3$	98	57,036	582	102,665	1,048	1,079,547	11,016	1,943,185	19,828	1,005,914	10,264	1,810,645	18,476
Methylcyclohexane (L)	$C_6H_{11}CH_3$	98	36,990	377	66,582	679	1,099,593	11,220	1,979,267	20,197	1,025,960	10,469	1,846,728	18,844
Styrene	$C_6H_5C_2H_3$	104	45,450	437	81,810	787	980,234	9,425	1,764,421	16,966	938,158	9,021	1,688,684	16,237
Toluene (methylbenzene)	$C_6H_5CH_3$	92	(11,950)	(130)	(21,510)	(234)	943,582	10,256	1,698,448	18,461	901,506	9,799	1,622,711	17,628
Toluene (methylbenzene) (L)	$C_6H_5CH_3$	92	(2,820)	(31)	(5,076)	−55.2	934,452	10,157	1,682,014	18,283	892,376	9.700	1,606.277	17,460

Inorganic Compounds														
Ammonia	NH_3	17	11,040	649	19,872	1,169	91,436	5,379	164,584	9,681	75,657	4,450	136,183	8,011
Carbon dioxide	CO_2	44	94,052	2,138	169,294	3,848	0	0	0	0	0	0	0	0
Carbon monoxide	CO	28	26,416	943	47,549	1,698	67,636	2,416	121,745	4,348	67,636	2,416	121,745	4,348
Hydrogen chloride	HCl	36.5	22,063	604	39,713	1,088	0	0	0	0	0	0	0	0
Hydrogen sulfide	H_2S	34	4,815	142	8,667	255	134,462	3,955	424,032	7,119	123,943	3,645	223,097	6,562
Nitrogen oxides	NO	30	(21,600)	(720)	(38,880)	(1,296)	21,600	720	28,880	1,296	21,600	720	38,880	1,296
	N_2O	44	(8,041)	(183)	(14,474)	(329)	8,041	183	14,474	329	8,041	183	14,474	329
	NO_2	46	(19,490)	(424)	(35,082)	(763)	19,490	424	35,082	763	19,490	424	35,082	763
	N_2O_4	92	(2,309)	(25)	(4,156)	(45.2)	2,309	25	4,156	45	2,309	25	4,156	45
Sulfur dioxide	SO_2	46	70,960	1,543	127,728	2,777	0	0	0	0	0	0	0	0
Sulfur trioxide	SO_3	80	94,450	1,181	170,010	2,125	(23,490)	(294)	(42,282)	(529)	(23,490)	(294)	(42,282)	(529)
Sulfur trixoide (L)	SO_3	80	104,800	1,310	188,640	2,358	(33,840)	(423)	(60,912)	(761)	(33,840)	(423)	(60,912)	(761)
Water	H_2O	18	57,798	3,211	104,036	5,780	10,519	584	18,934	1,052	0	0	0	0
Water (L)	H_2O	18	68,317	3,795	122,971	6,832	0	0	0	0	(10,519)	(584)	(18,934)	(1,052)
Chlorinated Organics														
Methylene chloride (L)	CH_3CL	50.5	22,630	448	40,734	807	161,802	3,204	291,244	5,767	151,283	2,996	272,309	5,392
Dichloromethane (L)	CH_2Cl_2	85	22,800	268	41,040	483	115,378	1,357	207,680	2,443	115,378	1,357	207,680	2,443
Chloroform (L)	$CHCl_3$	119.5	24,200	203	43,560	365	67,724	567	121,903	1,020	67,724	567	121,903	1,020
Carbontetrachloride (L)	CCl_4	154	24,000	156	43,200	281	21,670	141	39,006	253	21,670	141	39,006	253
Ethyl chloride	C_2H_5Cl	64.5	26,700	414	48,060	745	320,101	4,963	576,182	8,933	299,063	4,637	538,313	8,346
1,1-Dichlorethane (L)	$C_2H_4Cl_2$	99	31,050	314	55,890	565	269,497	2,722	485,095	4,900	258,978	2,616	466,160	4,709
1,1,2,2-Tetrachloroethane (L)	$C_2H_2Cl_4$	168	36,500	217	65,700	391	171,539	1,021	308,770	1,838	171,539	1,021	308,770	1,838
Monochloro n-propane (L)	C_3H_7Cl	78.5	31,100	396	55,980	713	478,070	6,091	860,526	10,962	446,513	5,688	803,723	10,239
1,3-Dichloropropane (L)	$C_3H_6Cl_2$	179	38,600	216	69,480	388	424,316	2,370	763,769	4,267	403,278	2,253	725,900	4,055
1-Chlorobutane (L)	C_4H_9Cl	92.5	35,200	381	63,360	685	636,339	6,879	1,145,410	12,383	594,263	6,424	1,069,673	11,564
1-Chloropentane (L)	$C_5H_{11}Cl$	106.5	41,800	392	75,240	706	792,108	7,438	1,425,794	13,388	739,513	6,944	1,331,123	12,499
1-Chloroethylene (L)	C_2H_3Cl	62.5	8,400	134	15,120	242	270,084	4,321	486,151	7,778	259,565	4,153	467,217	7,475
trans-1,2-Dichloro ethylene (L)	$C_2H_6Cl_2$	101	1,00	10	1,800	17.8	367,864	3,642	662,155	6,556	346,826	3,434	624,287	6,181
Trichloroethylene (L)	C_2HCl_3	131.5	1,400	11	2,520	19.2	184,576	1,404	332,237	2,527	184,576	1,404	332,237	2,527
Tetrachloroethylene (L)	C_2Cl_4	166	3,400	20	6,120	36.9	136,322	821	245,380	1,478	136,322	821	245,380	1,478
3-Chloro-1-propene (L)	C_3H_5Cl	76.5	150	2	270	3.53	440,703	5,761	793,265	10,369	419,665	5,486	755,397	9,874
monochlorobenzene (L)	C_6H_5Cl	112.5	12,390	110	22,302	198	710,619	6,317	1,279,114	11,370	689,581	6,130	1,241,246	11,033
p-dichlorobenzene (L)	$C_6H_4Cl_2$	147	5,500	37	9,900	67.3	671,255	4,566	1,208,259	8,219	660,736	4,495	1,189,325	8,091
Hexachlorobenzene (L)	C_6Cl_6	285	8,100	28	14,580	51.2	483,639	1,697	870,550	3,055	483,639	1,697	870,550	3,055
Benzylchloride	$C_6H_5CH_2Cl$	126.5		0	0	0.0	885,378	6,999	1,593,680	12,598	853,821	6,750	1,536,878	12,149

Heat of combustion of chlorinated compounds assume that all chlorine goes to HCl with addition of liquid water if needed.
Values in parentheses are negative.
Positive values indicate heat released from combustor.

Table 12.2 Selected Properties of Waste Constituents and Fuels

	Waste Constituents												
	Proximate Analysis (as-received) Weight%				Ultimate Analysis (Dry) Weight%						Higher Heating Value (kcal/kg)		
Waste Component	Moisture	Volatile Matter	Fixed Carbon	Non-Comb.	C	H	O	N	S	Non-Comb.	As Received	Dry	Moisture and Ash Free
Paper and Paper Products													
Paper, mixed	10.25	75.94	7.44	5.38	43.31	5.82	44.32	0.25	0.20	6.00	3778	4207	4475
Newsprint	5.97	81.12	11.48	1.43	49.14	6.10	43.03	0.05	0.16	1.52	4430	4711	4778
Brown paper	5.83	83.92	9.24	1.01	44.90	6.08	47.34	0.00	0.11	1.07	4031	4281	4333
Trade magazine	4.11	66.39	7.03	22.47	32.91	4.95	38.55	0.07	0.09	23.43	2919	3044	3972
Corrugated boxes	5.20	77.47	12.27	5.06	43.73	5.70	44.93	0.09	0.21	5.34	3913	4127	4361
Plastic-coated paper	4.71	84.20	8.45	2.64	45.30	6.17	45.50	0.18	0.08	2.77	4078	4279	4411
Waxed milk cartons	3.45	90.92	4.46	1.17	59.18	9.25	30.13	0.12	0.10	1.22	6293	6518	6606
Paper food cartons	6.11	75.59	11.80	6.50	44.74	6.10	41.92	0.15	0.16	6.93	4032	4294	4583
Junk mail	4.56	73.32	9.03	13.09	37.87	5.41	42.74	0.17	0.09	13.72	3382	3543	4111
Food and Food Waste													
Vegetable food waste	78.29	17.10	3.55	1.06	49.06	6.62	37.55	1.68	0.20	4.89	997	4594	4833
Citrus rinds and seeds	78.70	16.55	4.01	0.74	47.96	5.68	41.67	1.11	0.12	3.46	948	4453	4611
Meat scraps (cooked)	38.74	56.34	1.81	3.11	59.59	9.47	24.65	1.02	0.19	5.08	4235	6913	7283
Fried fats	0.00	79.64	2.36	0.00	73.14	11.54	14.82	0.43	0.07	0.00	9148	9148	9148
Mixed garbage I	72.00	20.26	3.26	4.48	44.99	6.43	28.76	3.30	0.52	16.00	1317	4719	5611
Mixed garbage II	—	—	—	—	41.72	5375	27.62	2.97	0.25	21.81	—	4026	5144
Trees, Wood, Brush, Plants													
Green logs	50.00	42.25	7.25	0.50	50.12	6.40	42.26	0.14	0.08	1.00	1168	2336	2361
Rotten timbers	26.80	55.01	16.13	2.06	52.30	5.5	39.0	0.2	1.2	2.8	2617	2528	2644
Demolition softwood	7.70	77.62	13.93	0.75	51.0	6.2	41.8	0.1	<.1	0.8	4056	4398	4442
Waste hardwood	12.00	75.05	12.41	0.54	49.4	6.1	43.7	0.1	<.1	0.6	3572	4056	4078
Furniture wood	6.00	80.92	11.74	1.34	49.7	6.1	42.6	0.1	<.1	1.4	4083	4341	4411
Evergreen shrubs	69.00	25.18	5.01	0.81	48.51	6.54	40.44	1.71	0.19	2.61	1504	4853	4978
Balsam spruce	74.35	20.70	4.13	0.82	53.30	6.66	35.17	1.49	0.20	3.18	1359	5301	5472
Flowering plants	53.94	35.64	8.08	2.34	46.65	6.61	40.18	1.21	0.26	5.09	2054	4459	4700
Lawn grass I	75.24	18.64	4.50	1.62	46.18	5.96	36.43	4.46	0.42	6.55	1143	4618	4944
Lawn grass II	65.00	—	—	2.37	43.33	6.04	41.68	2.15	0.05	6.75	1494	4274	4583
Ripe leaves I	9.97	66.92	19.29	3.82	52.15	6.11	30.34	6.99	0.16	4.25	4436	4927	5150
Ripe leaves II	50.00	—	—	2.37	43.33	6.04	41.68	2.15	0.05	8.20	1964	3927	4278
Wood and bark	20.00	67.89	11.31	0.80	50.46	5.97	42.37	0.15	0.05	1.00	3833	4785	4833
Brush	40.00	—	—	5.00	42.52	5.90	41.20	2.00	0.05	8.33	2636	4389	4778
Mixed greens	62.00	26.74	6.32	4.94	40.31	5.64	39.00	2.00	0.05	13.00	1494	3932	4519
Grass, dirt, leaves	21–62	—	—	—	36.20	4.75	26.61	2.10	0.26	30.08	—	3491	4994

Domestic Wastes													
Upholstery	6.9	75.96	14.52	2.62	47.1	6.1	43.6	0.3	.1	2.8	3867	4155	4272
Tires	1.02	64.92	27.51	6.55	79.1	6.8	5.9	0.1	1.5	6.6	7667	7726	8278
Leather	10.00	68.46	12.49	9.10	60.00	8.00	11.50	10.00	0.40	10.10	4422	4917	5472
Leather, shoe	7.46	57.12	14.26	21.16	42.01	5.32	22.83	5.98	1.00	22.86	4024	4348	5639
Shoe, heel & sole	1.15	67.03	2.08	29.74	53.22	7.09	7.76	0.50	1.34	30.09	6055	6126	8772
Rubber	1.20	83.98	4.94	9.88	77.62	10.35	—	—	2.00	10.00	6222	6294	7000
Mixed plastics	2.0	—	—	1.00	60.00	7.20	22.60	—	—	10.20	7833	7982	8889
Plastic film	3–20	—	—	—	67.21	9.72	15.82	0.46	0.07	6.72	—	7692	8261
Polyethylene	0.20	98.54	0.07	1.19	84.54	14.18	0.00	0.06	0.03	1.19	10,932	10,961	11,111
Polystyrene	0.20	98.67	0.68	0.45	87.10	8.45	3.96	0.21	0.02	0.45	9122	9139	9172
Polyurethane	0.20	87.12	8.30	4.38	63.27	6.26	17.65	5.99	0.02	4.38(a)	6554	6236	6517
Polyvinyl chloride	0.20	86.89	10.85	2.06	45.14	5.61	1.56	0.08	0.14	2.06(b)	5419	5431	5556
Linoleum	2.10	64.50	6.60	26.80	48.06	5.34	18.70	0.10	0.40	27.40	4528	4617	6361
Rags	10.00	84.34	3.46	2.20	55.00	6.60	31.20	4.12	0.13	2.45	3833	4251	4358
Textiles	15–31	—	—	—	46.19	6.41	41.85	2.18	0.20	3.17	—	4464	4611
Oils, paints	0		—	16.30	66.85	9.62	5.20	2.00	—	16.30	7444	7444	8889
Vacuum cleaner dirt	5.47	55.68	8.51	30.34	35.69	4.73	20.08	6.25	1.15	32.09	3548	3753	5533
Household dirt	3.20	20.54	6.26	70.00	20.62	2.57	4.00	0.50	0.01	72.30	2039	2106	7583
Municipal Wastes													
Street sweepings	20.00	54.00	6.00	20.00	34.70	4.76	35.20	0.14	0.20	25.00	2667	3333	4444
Mineral (c)	2–6	—	—	—	0.52	0.07	0.36	0.03	0.00	99.02	—	47	—
Metalic (c)	3–11	—	—	—	4.54	0.63	4.28	0.05	0.01	90.49	—	412	4333
Ashes	10.00	2.68	24.12	63.2	28.0	0.5	0.8	—	0.5	70.2	2089	2318	7778

Fuels

	Percent by Weight							Higher Heating Value		sp. gr.
	C	H	O	S	N	Water	Ash	kcal/kg	kcal/liter	kg/l
Fuel Oil										
#1, #2	84.7	15.3	nil	0.02	nil	nil	<.05	11,061	9,070	0.82
#3	85.8	12.1	nil	1.2	nil	nil	<0.1	10,528	9,474	0.90
#5	87.9	10.2	nil	1.1	nil	0.05	<1.0	10,139	9,733	0.96
#6	88.3	9.5	nil	1.2	nil	0.05	<2.0	10,000	10,000	1.00
Natural Gas	69.3	22.7	nil	nil	0.08	nil	nil	13.21	9.435	0.000714

Sources: Properties of waste constituents from Niessen, W. R. 1978. *Combustion and Incineration Processes.* Marcel Dekker, New York. Properties of fuel oil from Danielson, 1973. Properties of gas from Theodore, 1987.

When sufficient hydrogen is present, as in an incinerator, the combustion follows the following reaction, with the corresponding free energies of formation given beneath the compounds:

$$\begin{array}{lcccccccc} & CHCl_3 & + \frac{1}{2}O_2 + & H_2O & \rightarrow & CO_2 & + & 3HCl \\ \Delta H_c = & (-24{,}200) + & 0 & + (-57{,}598) & + & (-94{,}052) & + & (-22{,}063) \end{array}$$

$$\Delta H_c = -140{,}837 \text{ cal/g-mole}$$

As can be seen, the heats of combustion are significantly different. It is, therefore, recommended that the heats of combustion be calculated from the higher and lower heating values of the pure components whenever possible. See Theodore [1987, pp. 143, 146] for a discussion of this subject.

Waste streams can have a negative heat release in a combustion chamber. For example, if a waste stream is pure water, it will absorb approximately 1050 Btu/lb (its latent heat of vaporization) when injected into an incinerator. The latent heat of vaporization of organic constituents in the waste is much smaller than that of water (typically one-quarter to one-half that of water) and is usually included in the tabulated heating values.

The heat of combustion of complex materials such as paper, leather, fuel oil, and so on cannot be determined from heats of formation. Table 12.2 lists typical heating values for these materials.

Table 12.3 Enthalpies of Gases

°C	N_2	O_2	Air	H_2	CO	CO_2	H_2O
			Standard Condition: 20° C, 293.16 K (cal/scm)				
16	−1.55	−1.55	−1.60	−1.55	−1.60	−1.95	−1.80
21	0.00	0.00	0.00	0.00	0.00	0.00	0.00
25	0.66	0.66	0.69	0.66	0.69	0.83	0.77
38	3.73	3.73	3.68	3.66	3.68	4.76	4.27
93	16.86	17.08	16.81	16.65	16.81	22.24	19.47
149	29.99	30.56	29.94	29.85	30.01	40.72	34.88
204	43.18	44.33	43.28	42.98	43.28	60.06	50.51
260	56.54	58.39	56.70	56.18	56.77	80.25	66.13
316	69.95	72.66	70.26	69.02	70.33	101.09	82.47
371	83.58	87.15	84.03	82.44	84.10	122.57	98.89
427	97.28	101.85	97.94	95.57	98.02	144.62	115.58
482	111.20	116.83	112.00	108.84	112.14	167.17	132.64
538	125.25	131.96	126.27	122.33	126.42	190.14	149.98
593	139.60	147.23	140.69	135.46	140.90	213.62	167.60
649	153.94	162.71	155.24	148.87	155.53	237.45	185.58
704	168.58	178.27	170.01	162.22	170.37	261.57	203.85
760	183.27	193.97	184.86	175.92	185.28	286.05	222.40
816	198.11	209.81	199.91	189.69	200.41	310.95	241.39
871	213.24	225.72	215.18	196.82	215.61	335.93	260.58
927	228.29	241.78	230.31	216.95	230.88	361.19	279.99
982	243.49	257.76	245.72	230.50	246.29	386.80	299.76
1,038	258.76	273.96	261.14	244.28	261.92	412.42	319.73
1,093	274.25	290.30	276.62	258.40	277.55	438.39	340.07
1,149	289.87	306.50	292.25	273.10	293.18	464.37	360.55
1,204	305.29	322.91	308.02	287.95	309.02	490.56	381.24
1,260	321.27	339.46	323.86	302.86	324.79	516.89	402.37
1.316	337.11	356.02	339.70	317.49	340.84	543.43	423.42
1,371	352.88	372.57	355.68	332.54	356.83	569.98	444.89
1,649	433.08	456.42	436.46	406.75	437.60	704.20	554.64
1,927	514.43	541.69	518.38	484.68	519.31	841.13	668.38
2,204	596.35	628.46	600.79	564.88	601.79	980.06	785.12
2,482	678.98	716.58	684.21	645.51	684.82	1119.78	904.15
2,760	762.25	806.21	768.41	729.29	768.55	1260.57	1025.17
3,038	846.03	896.76	853.11	813.42	852.54	1402.14	1147.51
3,316	944.21	995.59	938.38	898.69	936.88	1544.85	1271.28
3,593	1014.36	1081.07	1024.22	985.67	1021.44	1688.35	1395.51

Table 12.3 (continued) Enthalpies of Gases

°F	N_2	O_2	Air	H_2	CO	CO_2	H_2O
Standard Condition: 70° F, 530° R (Btu/scf)							
60	−0.22	−0.22	−0.22	−0.22	−0.22	−0.27	−0.25
70	0	0	0	0	0	0	0
77	0.09	0.09	0.10	0.09	0.10	0.12	0.11
100	0.52	0.52	0.52	0.51	0.52	0.67	0.60
200	2.36	2.39	2.36	2.33	2.36	3.12	2.17
300	4.20	4.28	4.20	4.18	4.21	5.71	4.89
400	6.05	6.21	6.07	6.02	6.07	8.42	7.08
500	7.92	8.18	7.95	7.87	7.96	11.25	9.27
600	9.80	10.18	9.85	9.67	9.86	14.17	11.56
700	11.71	12.21	11.78	11.55	11.79	17.18	13.86
800	13.63	14.27	13.73	13.39	13.74	20.27	16.20
900	15.58	16.37	15.70	15.25	15.72	23.43	18.59
1,000	17.55	18.49	17.70	17.14	17.72	26.65	21.02
1,100	19.56	20.63	19.72	18.98	19.75	29.94	23.49
1,200	21.57	22.80	21.76	20.86	21.80	33.28	26.01
1,300	23.62	24.98	23.83	22.73	23.88	36.66	28.57
1,400	25.68	27.18	25.91	24.65	25.97	40.09	31.17
1,500	27.76	29.40	28.02	26.58	28.09	43.58	33.83
1,600	29.88	31.63	30.16	27.58	30.22	47.08	36.52
1,700	31.99	33.88	32.28	30.40	32.36	50.62	39.24
1,800	34.12	36.12	34.44	32.30	34.52	54.21	42.01
1,900	36.26	38.39	36.60	34.23	36.71	57.80	44.81
2,000	38.43	40.68	38.77	36.21	38.90	61.44	47.66
2,100	40.62	42.95	40.96	38.27	41.09	65.08	50.53
2,200	42.78	45.25	43.17	40.35	43.31	68.75	53.43
2,300	45.02	47.57	45.39	42.44	45.52	72.44	56.39
2,400	47.24	49.89	47.61	44.49	47.77	76.16	59.34
2,500	49.45	52.21	49.85	46.60	50.01	79.88	65.35
3,000	60.69	63.96	61.17	57.00	61.33	98.69	77.73
3,500	72.09	75.91	72.65	67.92	72.78	117.88	93.67
4,000	83.57	88.07	84.20	79.16	84.34	137.35	110.03
4,500	95.15	100.42	95.89	90.46	95.99	156.93	126.71
5,000	106.82	112.98	107.69	102.20	107.17	176.66	143.67
5,500	118.56	125.67	119.56	113.99	119.48	196.50	160.82
6,000	132.32	139.52	131.51	125.94	131.30	216.50	178.16
6,500	142.15	151.50	143.54	138.13	143.15	236.61	195.57

Enthalpies are for a gaseous system and do not include latent heat of vaporization of water.

L_v = 1059.9 BTU/lb or 50.34 BTU/scf of H_2O vapor at 60° F and 14.696 psia.

Source: Danielson, 1973.

Having established the heat input rate to the incinerator, it is necessary to then determine the resultant gas temperature. When performing material and energy calculations by hand, this is done by using the gas composition calculated from the material balance and determining through iteration the resultant temperature based on the composition and the enthalpy (heat content) of gases as given in Table 12.3. When doing the calculation on a computer, one normally uses correlations of heat capacity and the resulting enthalpy of each gas constituent.

Two correlations are typically used for calculating the enthalpy versus temperature of the components of a flue gas. One is based on the heat capacity of the gases:

$$C_p = \mathrm{A} + \mathrm{B}T + \mathrm{C}T^{-2} \tag{12.9}$$

$$\mathrm{H} = \mathrm{A}(T - T_0 + \tfrac{1}{2}\mathrm{B}(T^2 - T_0^2) - \mathrm{C}[(1/T) - (1/T_0)] \tag{12.10}$$

Table 12.4 Coefficients for Mean Heat Capacity Equations

Coef.	N2	O2	H2	CO	CO2	H2Oa	NO	NO2	CH4	C2H4	C2H6	C3H8	C4H10
A	9.3355	8.9465	13.5050	16.526	−0.89286	34.190	14.169	11.005	−160.820	−22.800	1.648	−0.966	0.945
B	−122.56	4.8044E−03	−167.96	−0.6841	7.2967	−43.868	−0.40861	51.650	105.100	29.433	4.124	7.279	8.873
C	256.38	−42.679	278344	−47.985	−0.980174	19.778	−16.877	−86.916	−5.9452	−8.5185	−0.153	−0.3755	−0.438
D	−196.08	56.615	−134.01	42.246	5.7835E−03	−0.88407	−17.899	55.580	77.408	43.683	1.74E−03	7.58E−03	8.36E−03
b	−1.5	1.5	−0.75	0.75	0.5	0.25	0.5	−0.5	0.25	0.5	1	1	1
c	−2	−1.5	−1	−0.5	1	0.5	−0.5	−0.75	0.75	0.75	2	2	2
d	−3	−2	−1.5	−0.75	2	1	−1.5	−2	−0.5	−3	3	3	3
Max T	6,300	6,300	6,300	6,300	6,300	6,300	6,300	6,300	3,600	3,600	2,700	2,700	2,700
Max T	3,500	3,500	3,500	3,500	3,500	3,500	3,500	3,500	2,000	2,000	1,500	1,500	1,500
Max.	0.43%	0.30%	0.60%	0.42%	0.19%	0.43%	0.34%	0.26%	0.15%	0.07%	0.83%	0.40%	0.54%

and the second is based on the mean heat capacities over a specified temperature range:

$$C_{p,\,\mathrm{mean}} = \mathrm{A} + \mathrm{B}t^b + \mathrm{C}t^c + \mathrm{D}t^d \tag{12.11}$$

with the enthalpy calculated by

$$H = \int C_{p,\,\mathrm{mean}}(T)(T - T_0) \tag{12.12}$$

The coefficients for Eqs. (12.11) and (12.12) are given in Table 12.4. The coefficients for Eqs. (12.9) and (12.10) are given in Table 12.5.

The full power of a material and energy balance is illustrated in Tables 12.6 and Table 12.7, which show the complete material and energy balance for an incinerator burning three waste streams (solid waste, high Btu liquid, and low Btu liquid) and a supplemental fuel stream of #2 fuel oil. The liquid waste streams are synthetic mixtures of components which may be used as part of a trial burn. The waste stream compositions are shown in the upper half of Table 12.6. As can be seen, the solid waste stream consists of refuse contaminated with chloroform, 1,1-dichloroethane, ethylene glycol, and ethanol. The low BTU liquid stream is water-contaminated ethylene glycol and ethanol, and the liquid waste consists of fuel oil doped with the specific organic compounds.

The compositions of each stream are given as ultimate analyses and as mole fractions. For the pure compounds (such as chloroform) one does not normally have an ultimate analysis. In that case, one can use number of atoms of the elements (carbon, hydrogen, etc.) in the compound.

Table 12.5 Coefficients for Heat Capacity and Gas Enthalpy vs. Temperature Equation

$$H = \mathrm{A}(T - T_0) + 0.5\mathrm{B}(T^2 - T_0^2) - \mathrm{C}[(1/T - (1/T_0)]$$
$$C_p = \mathrm{A} + \mathrm{B}T + \mathrm{C}T^2$$

	Temperature input as K, *H* calculated as (cal/g-mole)			Temperature input as K, *H* calculated as (cal/g)		
	A	B	C	A	B	C
O_2	7.168	1.002E–03	–4.000E+04	0.2240	3.131E–05	–1.250E+03
N_2	6.832	8.988E–04	1.201E+04	0.2400	3.210E–05	–1.289E+02
CO_2	10.570	2.100E–03	–2.059E+05	0.2402	4.773E–05	–4680E+03
HCl	6.278	1.241E–03	3.000E+04	0.1720	3.400E–05	8.219E+02
H_2O	7.308	2.466E–03	0	0.4060	1.370E–04	0
	Temperature input as °R, *H* calculated as (BTU/lb-mole)			Temperature input as °R, *H* calculated as (BTU/lb)		
	A	B	C	A	B	C
O_2	7.168	5.57E–04	–2.33E+05	0.2240	1.704E–05	–7.290E+03
N_2	6.832	4.99E–04	–7.00E+04	0.2440	1.783E–05	–2.502E+03
CO_2	10.570	1.17E–03	–1.20E+06	0.2402	2.652E–05	–2.729E+04
HCl	6.278	6.89E–04	1.75E+05	0.1720	1.889E–05	4.793E+03
H_2O	7.308	1.37E–03	0	0.4060	7.611E–05	0
	Temperature input as K, *H* calculated as (calscm)			Temperature input as °R, *H* calculated as (BTUscf)		
	A	B	C	A	B	C
O_2	2.982E+02	4.168E–02	–1.664E+06	1.852E–02	1.438E–06	–6.028E+02
N_2	2.842E+02	3.739E–02	–4.996E+05	1.765E–02	1.290E–06	–1.810E+02
CO_2	4.397E+02	8.735E–02	–8.565E+06	2.731E–02	3.015E–06	–3.103E+03
HCl	2.611E+02	5.162E–02	1.248E+06	1.622E–02	1.782E–06	4.521E+02
H_2O	3.040E+02	1.026E–01	0	1.888E–02	3.540E–06	0

Temperature range: 20 to 2000°C.

Temperature must be input in K to obtain the appropriate values for C_p in (cal/g-mole-k) or (BTU/lb-mole-°R); °R = 1.8 x K; (BTU/lb-mole) = (cal/g-mole) x 1.8.

Table 12.6 Waste Feed Rates and Composition for System with Heat Exchanger

				Chloroform $CH(Cl)_3$		1,1-Dichloroethane $C_2H_4(Cl)_2$		Ethyleneglycol $C_2H_4(OH)_2$		Ethanol C_2H_5OH		Water H_2O		Fuel Gas or Oil		Refuse	
Heat of formation (cal/g-mole)					24,200		31,050		(103,580)		66,356						
Stream			lb/hr	%	lb/hr	%	lb/hr	%	lb/hr	%	lb/hr	%	lb/hr	%	lb/hr	%	lb/hr
Solid waste			1,500	8.0%	120.0	8.0%	120.0	8.0%	120.0	16.0%	240.0		0.0		0.0	60.0%	900.0
High BTU liquid			2,000	3.0%	60.0	4.0%	80.0	10.0%	200.0	50.0%	1,000.0		0.0	33.0%	660.0		0.0
Low BTU liquid			2,000		0.0	0.1%	2.0	1.0%	20.0	2.0%	40.0	96.9%	1,938.0		0.0		0.0
Supplemental fuel			100		0.0		0.0		0.0		0.0		0.0	100.0%	100.0		0.0
Nat'l gas		0			0.0		0.0		0.0		0.0		0.0		0.0		0.0
Fuel oil		1			0.0		0.0		0.0		0.0		0.0		0.0		
Total (lb/hr)			5,600		180		202		340		1,280		1,938		760		900
		MW	lb/hr														
Ultimate	C	12	1,759	10.04%	18	24.24%	49	38.71%	132	52.17%	668	0.00%	0	84.41%	642	27.8%	251
Analysis	H	1	574	0.84%	2	4.04%	8	9.68%	33	13.04%	167	11.11%	215	15.19%	115	3.7%	34
	O	16	2,550	0.00%	0	0.00%	0	51.61%	175	34.78%	445	88.89%	1,723	0.00%	0	23.0%	207
Wet basin	N	14	7	0.00%	0	0.00%	0	0.00%	0	0.00%	0	0.00%	0	0.10%	1	0.7%	6
	Cl	35.5	305	89.12%	160	71.72%	145	0.00%	0	0.00%	0	0.00%	0	0.00%	0	0.0%	0
	S	32	1	0.00%	0	0.00%	0	0.00%	0	0.00%	0	0.00%	0	0.10%	1	0.1%	1
	H_2O	18	199	0	0	0	0	0	0	0	0	0	0	0.10%	1	22.0%	198
	Ash	–	204	0	0	0	0	0	0	0	0	0	0	0.10%	1	22.6%	204
Molecular weight					119.5		99		62		46		18				
			Total Moles per Hour	Number Moles	Moles per Hour	Number Moles	Moles per Hour	Number Moles	Moles per Hour	Number Moles	Moles per Hour	Number Moles	Moles per Hour	Mole Fraction	Moles per Hour	Mole Fraction	Moles per Hour
Formula	C		146.55	1	1.51	2	4.08	2	10.97	2	55.65	0	0.00	31.64%	53.46	20.68%	20.89
or mole	H		596.11	1	1.51	4	8.16	6	32.90	6	166.96	2	215.33	68.32%	115.53	5.15%	55.69
fraction	O		170.77	0	0.00	0	0.00	2	10.97	1	27.83	1	107.67	0.00%	0.04	23.71%	24.25
(water as	N		0.51	0	0.00	0	0.00	0	0.00	0	0.00	0	0.00	0.03%	0.05	0.45%	0.45
elements)	Cl		8.60	3	4.52	2	4.08	0	0.00	0	0.00	0	0.00	0.00%	0.00	0.00%	0.00
	S		0.05	0	0.00	0	0.00	0	0.00	0	0.00	0	0.00	0.01%	0.02	0.02%	0.02
			922.58		1.51		2.04		5.48		27.83		107.67	100.00%	169.32	100.00%	130.52
Lower heating value (BTU/lb)					1,179		4,709		13,647		11,549		(1,060)		19,000		4,263
(BTU/lb-mole)					140,837		466,160		846,140		531,256		104,036	Oil =	19,000 BTU/lb		
(cal/g-mole)					78,243		258,978		470,078		295,142		57,798	Gas =	950 BTU/scf		
Higher heating value (BTU/lb)					1,179		4,719		13,696		11,614		5,835				
(BTU/lb-mole)					140,837		467,162		849,144		534,259		105,038		2,04E+04		4,600
(cal/g-mole)					78,243		259,534		471,747		296,811		58,354				
Heat input (BTU/hr)			3.68E+07		2.12E+05		9.51E+05		4.64E+06		1.48E+07		–2.05E+06		1.44E+07		3.84E+06

This corresponds to the molecular structure of the compound. For example, chloroform has one carbon, one hydrogen, and three chlorine atoms, and this is shown in the "Number moles" column for chloroform. The other compounds are similarly filled in.

For complex mixtures, such as refuse, natural gas, or fuel oil, one would normally enter the ultimate analysis. Table 12.6 shows both the ultimate and formula analyses for all streams to illustrate how one calculates one from the other. It also shows the LLV and HHV for the components of each of the waste streams. The LHV for the pure compounds were calculated from their heats of formation. The LHV for the refuse was obtained from Niessen [1978, p. 48].

Note that, while heats of combustion are usually given as negative numbers in the literature, they are shown to be positive in Table 12.6. This change in sign is consistent with the standard

Table 12.7 Flue Gas Properties for System with Heat Exchanger

	@ 0% Excess Air		Exit from Combustion Chamber and Heat Exchanger @ % Excess Air → 120%					Exit of Quench		Exit of Scrubber		
	Moles/Hr	Mole %	SCFM	Moles/Hr	% Wet	% Dry	lb/hr	SCFM	% Wet	SCFM	% Wet	Lb/Hr
CO_2	146.55	11.9%	945	146.55	6.1%	6.9%	6,448	945	4.6%	945	4.6%	6,448
H_2O	293.75	23.9%	1,895	293.75	12.1%	0.0%	5,288	6,711	32.9%	6,711	32.9%	18,728
HCl	8.60	0.7%	55	8.60	0.4%	0.4%	314	55	0.3%	0.28*	0.0%	2
N_2	782.64	63.5%	11,106	1,721.82	71.1%	81.0%	48,211	11,106	54.4%	11,106	54.4%	48,211
O_2	0	0.0%	1,160	249.65	10.3%	11.7%	7,989	1,610	7.9%	1,610	7.9%	7,989
SO_2	0.046649365	0.0%	0	0.00	0.0%	0.0%	0	0	0.0%	0	0.00%	0
O_2 req'd	208.04	—		457.70								
	1,231.55		15,611	2,420.38		1.00		20,427		20,372		
			Molecular weight = 28.2 Wet					MW =	25.8	MW =	25.7	

Combustor operating conditions input/output table Enter operating conditions in this table	INPUT		CALCULATED		LHV (Btu/lb)
Scrubber efficiency*	99.5%				
Solid waste feed (lb/hr)	1,500		8.95E+06	(Btu/hr)	5,968
High Btu liquid waste feed (lb/hr)	2,000		2.73E+07	(Btu/hr)	13,633
Low Btu liquid waste feed (lb/hr)	2,000		−1,31E+06	(Btu/hr)	(655)
Suppl. fuel (oil lb/hr or gas scfm)	100		1.90E+06	(Btu/hr)	19,000
("0" if oil, "1" if gas)	0				
% excess air	120%	120%	by temperature		
% oxygen in stack (dry)	11.7%	11.7%			
Comb. chamber temp. °F	1,872	1,872	1,295	K	
Heat exch. thermal duty (Btu/hr)			2.15E+07		
% Heat loss or boiler duty	5%	or	1.84E+06	(Btu/hr)	
Total input to incin. (excl losses)	5,600 lb/hr		3.68E+07	(Btu/hr)	
		In	Out		AV—by feed (including losses), Spreadsheet 4A
Enth. both ways (10^3 Btu/hr)		34,968	34,968		AX by gas flow, from spreadsheet 3
Comb. gas flow, SCFM, dry, no HCl)			13,661		
Temp. @ heat exch. outlet (°F)	800		700	K	
Temp @ quench outlet (°F)	161	161	345	K	
Water evaporated in quench	13,441 lb/hr		1,612	gal/hr	
Gas flow rate leaving quench (ACFM, wet)			23,916		1,001 Btu/lb latent of heat water @ quench *T*
Gas flow rate leaving quench (SCFM, wet)			20,427		
% moisture			32.9%		
% mosture @ saturation			32.9%		

*Based on scrubber efficiency input.

nomenclature of positive when heat flows into a system and negative when it flows out of it. The compounds are the system when defined in the literature; as a result the heat of combustion is shown as negative. The heat balance calculations are performed using the incinerator as the point of reference; hence, heat released from the waste (a negative quantity) is heat absorbed by the combustion chamber (a positive quantity).

The negative value for the heat of combustion of water needs to be mentioned. Since the water is fed to the incinerator as a liquid and exits the combustion chamber as a gas, its LHV is equal to its latent heat of vaporization. Its HHV is zero since it is based on the water emerging from the system as a liquid rather than a gas. The -1060 Btu per pound is the heat of vaporization of water at 70°F.

Table 12.7 shows how the composition, size, and temperature of the output stream are determined. The elements fed to the incinerator form the compounds shown in the upper left of Table 12.7. The amounts of the compounds formed are determined by applying the rules for a mass balance described above. This computes the amount of the combustion products formed. The moles of oxygen required to form these combustion products is the sum of the oxygen required to form all of the oxides (CO_2, H_2O from hydrogen, and SO_2 minus the moles of oxygen fed to the combustor). In this example, the stoichiometric (0% excess air) quantity of oxygen required is 208.04 moles/hr. Since each mole of oxygen carries with it 79/21 or 3.76 moles of nitrogen, the moles of nitrogen in the flue gas can readily be calculated by multiplying 208.04 by 3.76. This results in the analysis given in the upper first columns of Table 12.7, identified by "@ 0% excess air."

It is now necessary to compute the composition of the gases at the excess air ratio of the incinerator. To do this, one multiplies the excess air ratio (1.20 in this case, 120% excess air) by the O_2 required value to obtain the moles of oxygen that are fed. This oxygen similarly carries nitrogen with it at the ratio of 3.76 moles of nitrogen for each mole of oxygen, and this nitrogen is added further to the nitrogen calculated at 0% excess air to obtain the actual number of moles of nitrogen in the exit of the combustor. Once the molar flow rate of each of the major component gases of the flue gas is known, its SCFM can be computed by multiplying the moles per hour by 387 scf/lb-mole and dividing by 60 minutes per hour. The percent of each component on a wet and dry basis can also be calculated by dividing that component's flow rate by the total gas flow rate including and excluding the water, respectively.

The heat input to the incinerator is obtained by summing the heat input from each of the streams. Either the LHV or the HHV can be used to perform the energy calculations. The usage is a matter of personal preference. This computation uses the LHV. If the HHV is used, the appropriate corrections for the latent heat of vaporization of the water entering and leaving the combustor must be made. The bottom row of Table 12.6 shows the amount of sensible heat released by combustion of each of the waste streams. To illustrate, the fuel oil's LHV is 19,000 Btu/lb and it is fed at a rate of 760 lb/hr in both the high Btu liquid and supplemental fuel streams. The total heat contribution is, therefore, 14.4 MMBtu/h. The other streams' heat contributions are similarly calculated. The total heat input from the fuel is 36.8 MMBtu/h.

The calculation further assumes that the heat loss to the surroundings is 5%. This is conservative but useful for the purpose. As a result, the heat available to raise the temperature of the 2420 lb-moles/h of gas is 34.968 MMBtu/h for a sensible heat content of 14,450 Btu/lb. It is now possible to compute the temperature of the flue gas by solving either Eq. 12.10 or 12.12 for temperature. The form of these equations requires that their solution for temperature involve an iteration, readily accomplished with a computer. If a computer is not available, the calculation can be time-consuming; however, one can use the enthalpy of gases given in Table 12.3 to estimate the temperature for the calculated enthalpy. Table 12.3 indicates that the gas temperature is, therefore, approximately 1850°F, well within the level of error of the calculation to 1872°F calculated by iteration. Please note that this is an unusually good agreement between the two methods of calculation. The typical difference is usually in the range ± 150°F.

The heat loss through the walls of the combustion chamber is rarely known. It can, however, be estimated during shakedown or during the trial burn by comparing the measured temperature of the flue gas to that calculated by a mass and energy balance. The heat loss from the incinerator as a whole will not vary from this value by more than a few percent. If necessary, even this small variation can be taken into account by assuming that the rate of heat loss is proportional to the difference between combustion chamber or duct temperatures and the ambient temperature. The temperature of the gases in the combustion chamber or duct is usually constant, and the ambient temperature will not normally vary by more than about 100°F between the seasons.

Assume that these results were obtained during the summer, when the temperature was 100°F. The heat loss was 1.84 MMBtu/hr. The combustion chamber temperature was 1872°F, so the temperature difference was 1772°. On a cold winter day when the ambient temperature is 0°F, the temperature difference would be 1872°. The heat loss in the winter could then be estimated to be

$$1.84\,\text{MMBtu/h} \times (1872/1772) = 1.94\,\text{MMBtu/h}$$

Since, in this case, the heat loss was assumed to be only 5% of the incinerator's thermal duty, the difference is only 6% of the heat loss, and the impact of temperature variation on the incinerator's thermal duty is a negligible 0.3%.

Obviously, factors such as heavy winds, rain, or snow hitting the incinerator surface can increase the heat loss, but usually a small correction for the measured heat loss will adequately take such considerations into account. Localized cooling can have an impact on the incinerator's operation. For example, unusually cold weather or a large amount of precipitation can cool a portion of ductwork or refractory to the point where ash can solidify at that point, causing a blockage. Similarly, cold weather can result in the condensation of acid gases onto the walls of an air pollution control device corroding the materials of construction. However, these localized pockets of cooling will rarely impact the mean temperature of the combustion chamber to a degree where the destruction efficiency of the system is jeopardized.

A note about significant figures. The calculations are carried out here to a large number of significant figures. This is done for illustrative purposes only. It makes it easier for the reader to follow the calculations using an electronic calculator. The number of significant figures shown should in no way be construed as a reflection on the accuracy of the calculations, which are good estimates, but not substitutes for actual test data. In addition, these calculations do not consider the trace constituents, such as CO and the POHCs, which must be considered as part of the overall incinerator evaluation. Since their concentrations in the exit of most incinerators and other combustors is low compared to the gases shown here, they can be ignored for the purpose of the overall mass and energy balance.

12.4 Incineration and Combustion Systems

The following three categories of devices are used to incinerate hazardous wastes: incinerators, boilers, and industrial furnaces. The three categories are differentiated only by their primary function, not by the fundamental concepts associated with waste combustion. Incinerators are specifically designed to burn waste materials, including hazardous wastes. Boilers and industrial process furnaces are not specifically designed and built to burn wastes. Boilers are intended to generate steam and industrial furnaces are intended to produce a product such as cement or lime. A BIF which is used to burn wastes must provide the high-temperature combustion environment needed to destroy organic materials. Burning wastes in BIFs destroys the waste and utilizes its heating value to replace fossil fuel. Incinerators are broken down into two categories by the type of waste they burn: nonhazardous waste (termed here "refuse incinerators") and hazardous waste. Each type is subject to different regulations. Both types are increasingly required to meet the same very low emission limits. The main design difference between them relates to the fact that hazardous waste incinerators must achieve very high levels of destruction for toxic contaminants while refuse incinerators must handle very large quantities of highly inhomogeneous solids.

Nonhazardous Waste Incinerators

While nonhazardous wastes can be incinerated in a wide variety of different types of furnaces, the vast majority of municipal refuse and nonhazardous commercial and industrial wastes are, currently, incinerated in mass-burn waste-to-energy plants similar to the one shown in Fig. 12.2. The term mass burn refers to the fact that the waste is not pretreated prior to being fed to the incinerator, although large objects such as large appliances (white goods) or construction debris (especially gypsum board) will usually be removed.

Incineration consists of the following processes: (1) waste receiving, (2) waste storage and segregation, (3) waste burning, (4) ash discharge, (5) heat recovery, (6) acid gas control, (7) particulate control, and (8) fan and stack.

Wastes are received on the tipping floor where trucks arrive and dump their loads of refuse into the holding pit. During this process, an operator will typically identify objects and materials which are unsuitable for incineration such as large objects, major appliances (white goods), or noncombustible construction debris, especially gypsum board and other gypsum products. The waste feed-crane operator typically segregates these wastes from the waste that will be incinerated. Large objects can result in jamming of the waste feed and transport mechanism. If they are combustible, they will, if possible, be broken up prior to incineration. White goods have very few combustible components and are typically segregated. Refrigerators and air conditioners are a particular concern since they may contain chlorofluorocarbons (CFCs), which can produce highly corrosive hydrogen fluoride during combustion. CFC incineration is not legal in most nonhazardous waste incinerators. Gypsum is kept out of the combustion chamber because it will release sulfur dioxide (SO_2) in excess of the environmentally acceptable limits for the combustor when heated to combustion temperatures.

The waste from the refuse holding pit is fed by a crane into a chute leading to a grate. The grate may be a moving screen or a set of reciprocating or fixed grates. The important factor is to

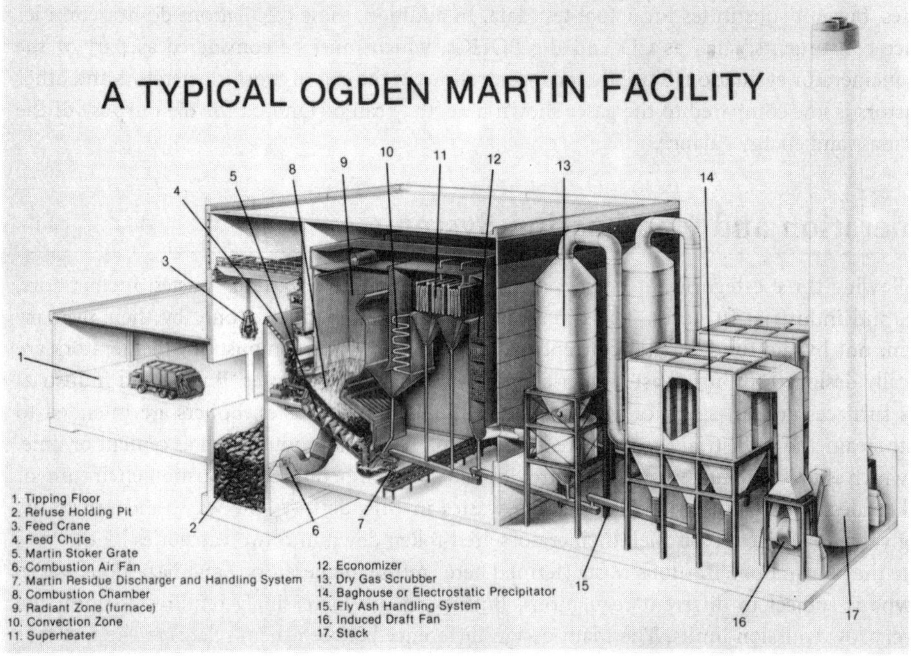

FIGURE 12.2 Waste-to-energy facility. (Reproduced courtesy of Ogden Martin Systems, Inc.)

move the burning mass through the combustion chamber and to discharge the ash into a suitable receiver. The grate must allow a sufficient solids residence time for the combustibles in the waste to burn down completely. Typical solids residence times range from thirty minutes to several hours for municipal refuse. Shorter solids residence times occur for highly flammable light materials, longer residence times for heavy materials such as flammable furniture. Waste consisting of paper and other light materials could have an even shorter solids residence time. The ash remaining after incineration falls into a collection and discharge system. The ash may drop into a water-filled tank or trough, where it is cooled and the slurry discharged, or onto a dry conveyor, where it is air cooled prior to discharge. The steam and hot gases from either type of ash-handling system are typically drawn back into the furnace.

The combustion chamber itself can be refractory lined or it may (as shown in Fig. 12.2) be lined with steam tubes similar to those in a boiler. Steam-tube-lined combustion chambers are often called water-wall furnaces. The gases then pass through a series of heat exchangers similar in concept to the heat recovery parts of any steam boiler. The gases are thus cooled and energy converted to useful steam. Some incinerator designs include a quench (not shown) where the gases are further cooled with water sprays. The cooled gases enter the air pollution control system.

The air pollution control system on a refuse incinerator is designed to control acid gas and particulate. Typical acid gas control devices are dry or wet scrubbers. Typical particulate removal devices are fabric filter baghouses or electrostatic precipitators. Venturi scrubbers have been used for both particulate and acid gas removal; however, the high pressure drop inherent in the design, coupled with the high gas flows of large refuse incinerators, tends to result in higher energy usage than with the fabric filter or electrostatic precipitator.

The principal emissions of concern from a refuse incinerator are particulates, mercury and other toxic metals, chlorodibenzodioxins and chlorodibenzofurans (dioxins and furans), and acid gases. The particulate forms by the entrainment of ash and other inert material and by the vaporization of salts or metal compounds in the hot flame zones followed by their condensation in the cooler zones of the system. Mercury has a significant vapor pressure at even the low temperatures encountered in the incineration system. It can pass through the system as a vapor. Fortunately, it is usually present in only minute quantities in refuse and does not require special removal techniques in most cases. Chlorodibenzodioxins and chlorodibenzofurans can form in the combustion chamber from the combustion of chlorinated materials. The most common chlorinated material in refuse is polyvinyl chloride (vinyl plastic). There is increasing evidence [Bruce *et al.*, 1990] that chlorodibenzodioxins and chlorodibenzofurans can form by the recombination of other compounds in the air pollution control system. Temperatures in the range of 230 to 400°C (450 to 750°F) [EPA, 1992] appear to favor the formation of these compounds. At this point, the information is preliminary, but if verified, it would tend to encourage the use of fabric filters (which operate at temperatures below 230°C).

Hazardous Waste Incinerators

There are many types of hazardous waste incinerators in use today. The following are examples of six types: liquid injection, rotary kiln, fluidized bed, fixed and moving hearth, infrared furnace, and plasma arc furnace. The liquid injection, rotary kiln, and hearth systems are traditional designs that have been in use for many years. Fluidized and circulating bed and infrared combustors represent newer-generation designs for treatment of a variety of solid wastes, often with claims of higher efficiency at lower operating costs for specific applications. Infrared and plasma arc furnaces are highly experimental at this time and are not discussed further herein. Numerous tests performed in support of regulatory development and for RCRA permitting [EPA, 1986a] have shown that, when properly designed and operated, hazardous waste incinerators can meet virtually any emission and destruction standard required.

A relatively recent development in incineration is the availability of many designs in modular and transportable systems designed to treat contaminated materials and wastes directly at the site of contamination. Portable incinerators refer to those units whose major components (kiln, SCC, APCE, etc.) are mounted on trailers, and on-site assembly consists largely of bolting the modules together and connecting utilities. The capacity of portable incinerators is limited by restrictions on the size of over-the-road loads. Transportable incinerators are built in prefabricated pieces which are assembled on-site to form the major units. Assembly of a transportable incinerator typically requires a foundation and significant on-site construction, which can take several months. The operating units can be as large as desired and, hence, have a waste treatment capacity approaching that of fixed-site systems.

Liquid Injection Incinerators

Liquid injection incinerators are currently the most common types used. As the name implies, these units are designed to incinerate liquid or pumpable slurry and sludge wastes, usually in a single, refractory-lined cylindrical combustion chamber positioned in a horizontal or vertical arrangement. Often, process vent gases are incinerated in liquid injection incinerators as well. The secondary combustion chamber of a multiple chamber incinerator is similar in design to a liquid injection incinerator. Most liquid injection incinerators consist of only a primary combustion chamber; however, for some cases, it may prove necessary to stage the combustion in multiple chambers.

Typical liquid injection incinerator combustion chamber mean gas residence time and temperature ranges are 0.5 to 2.0+ seconds and 700 to 1200°C (1300 to 2200°F). The combustion chambers vary in dimensions with length-to-diameter ratios in the range of 2 or 3 to 1 and a diameter less than 12 ft. Liquid injection feed rates are as high as 1500 gal/h (6000 l/h) of organic liquid and 4000 gal/h (16,000 l/h) of aqueous liquid.

The primary advantages of liquid incinerators are their ability to incinerate a wide range of gases and pumpable liquid wastes, and to operate with a minimum of moving parts. Because the wastes are injected through atomizing nozzles, the physical properties of the waste are very important to the safe and efficient operation of these units. The primary waste properties that must be considered when evaluating a liquid injection incinerator (or any combustor burning liquid wastes) are the viscosity and solids content. Viscosity is important because the liquids must be atomized into fine droplets for adequate vaporization or pyrolysis. The viscosity should be, typically, less than 10,000 Saybolt seconds (SSU). High solids content can lead to nozzle erosion, plugging, and caking, which can result in poor atomization of the wastes and less efficient combustion. Wastes are often blended and pretreated to meet burner and nozzle specifications.

Liquid injection incinerators can be positioned either horizontally or vertically. Horizontally fired incinerators are the simplest design in that the waste feed systems, the air inlets, the combustion gas exhaust, and the air pollution control systems are at ground level and readily accessible. Piping and ductwork runs tend to be shorter than in other designs as well. Such incinerators, however, have relatively large footprints. In addition, any ash in the waste will collect in the combustion chamber so that, unless the waste has a very low ash content, the furnaces have to be shut down for manual ash removal on a regular basis.

Vertical liquid injection incinerators may be downfired or upfired depending primarily on the type of wastes being incinerated. Upfired incinerators require relatively little space but they require that the hot flue gas be brought down to the APCD in refractory-lined ducts. Like the horizontally fired units, they also tend to accumulate ash in the combustion chamber. Their principal use is for burning clean solvents with a low ash content under conditions which may not require the use of air pollution control devices. Upfired incinerators are relatively less common for hazardous waste applications and when used they are limited to small installations—less than about 1000 lb/h waste feed.

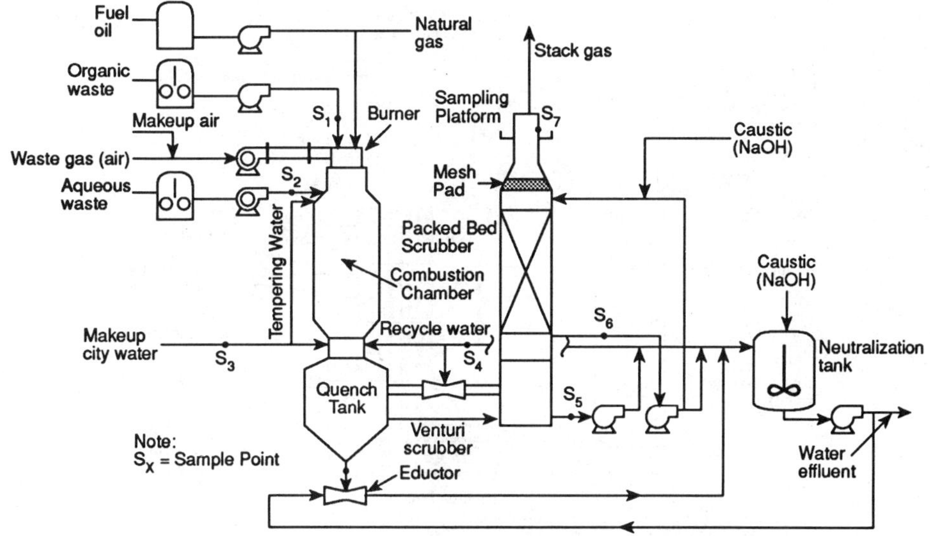

FIGURE 12.3 Typical downfired liquid incinerator arrangement.

Downfired incinerators have the burner(s) at the top of the combustion chamber. Figure 12.3 illustrates a forced draft, downfired liquid injection incinerator. In this particular system the combustion gases move downward and impinge on a wet quench located at the bottom of the chamber called a wetbottom. In other variations on the design, the wastes, fuel, and air are introduced into the combustion chamber tangentially, causing a swirling "vortex" gas-flow pattern, shown in Fig. 12.4. The vortex pattern increases the velocity of the gas and hence tends to increase the mixing of the gases with air. The swirling, or vortex, flow is imparted by the shape of the air inlets at the bottom of the combustor and by the arrangement of the air tuyeres in a cyclonic combustor, which are aimed to inject the combustion air with a tangential component to its flow.

The combination of vortex flow and wetbottom is useful when wastes contain significant inorganics with a melting point in the combustor's temperature range. The melting inorganic materials form a slag that can stick to the refractory, building up on it and eventually reducing the combustor's cross-sectional area. Some slags can corrode or otherwise damage the refractory. Cyclonic flow combined with a wetbottom tends to sweep the slag and ash from the walls into the wetbottom, thereby reducing buildup.

The organic compound destruction achieved by a liquid injection incinerator is largely determined by how well the particular design atomizes the waste materials, converts the organic constituents to a vapor by vaporization or pyrolysis, maintains a stable flame, and provides adequate mixing of the combustion gases with oxygen at temperatures sufficiently high to destroy the organics.

Some of the factors that need to be considered are particle size distribution of the droplets produced by the nozzle, droplet impaction on walls or other surfaces, presence of cool zones in the organics destruction zone, and presence of stagnant zones where poor mixing of organics and oxygen (localized pyrolyzing conditions) may occur. The first two factors relate to the precombustion and combustion zones of the incinerator. If the waste droplets do not completely evaporate prior to exiting the flame zone, there is a risk that a fraction of the waste may evaporate and not be completely destroyed. Improper evaporation can occur because the droplet size is too large, the heating value of the waste drops, or because the nozzle is spraying a fraction of the liquid onto a cool surface, preventing its ignition. The atomizing nozzle must be selected to match the properties of the fluids involved and they must be properly mounted so that no spray hits any surfaces. In addition, it is necessary to continuously test the viscosity, heating value, and (for some

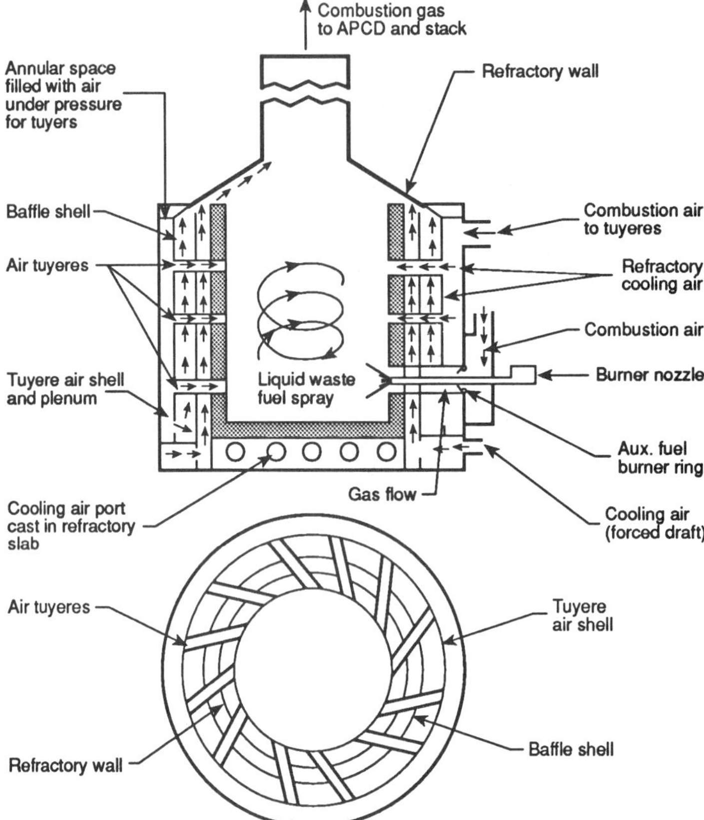

FIGURE 12.4 Tangentially fired vortex combustor. [*Source:* Bonner, T. A. 1981. *Engineering Handbook for Hazardous Waste Incineration.* EPA-SW-889 (NTISPB81-248163)].

types of nozzles) vapor pressure of the waste stream to ensure that it remains within the nozzle's design specifications.

Rotary Kiln Incinerators

The rotary kiln incinerator is the single most common design for the large-scale combustion of solid hazardous waste. The typical system consists of a solid waste feed system, a pumpable waste feed system, an auxiliary fuel combustor, a kiln, a transition section, an ash drop and dump system, a secondary combustion chamber with auxiliary fuel and pumpable waste feed systems, and air pollution control equipment. Most rotary kilns incinerators are the first stage of a two (or more) chamber incinerator. The secondary combustion chamber (sometimes called an afterburner) destroys the remainder of the volatilized combustible matter released from the heating of the solid wastes. Exceptions do exist; most industrial furnace kilns that have been tested with hazardous waste appear to perform well without a secondary combustion chamber, as do a few hazardous waste incinerator kilns. For the most part, rotary kiln hazardous waste incinerators require the secondary combustion chamber to complete the destruction of the organic constituents released from the solids. The secondary combustion chamber is essentially identical to a liquid injection incinerator.

A rotary kiln is a cylindrical shell mounted on a slight incline from a horizontal plane. It is usually refractory lined, although it may, if the temperature is low enough, be simply a metal shell. Its design is an adaptation of industrial process kilns used in the manufacture of lime, cement, and aggregate

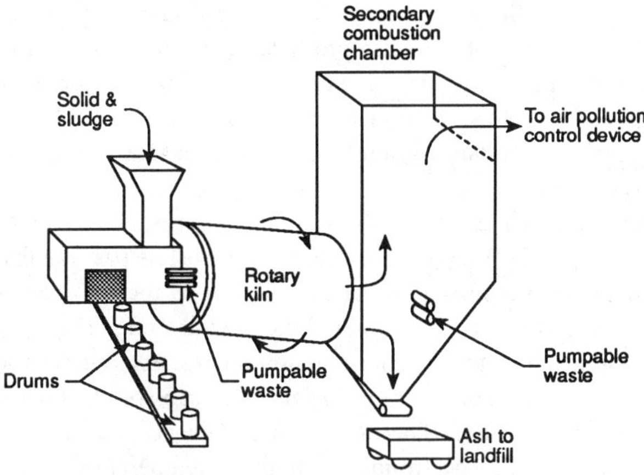

FIGURE 12.5 Rotary kiln incinerator schematic. (*Source:* ASME, 1988. *Hazardous Waste Incineration, A Resource Document.* American Society of Mechanical Engineers, New York.)

materials. The rotary kiln is used extensively for the incineration of bulk and containerized solids, sludges, liquids, and gases. Most rotary kiln incinerators consist of two combustion chambers. The first chamber is the kiln. The second chamber is similar to a liquid injection incinerator. It often burns pumpable hazardous wastes along with the off-gases from the kiln.

Figure 12.5 is a schematic of a rotary kiln incinerator system which illustrates how the two combustion chambers treat wastes. The kiln consists of a firing end (shown on the left) containing a solids feed system and a series of nozzles and lances to feed the wastes and supplementary fuels if needed.

Solid materials, either in containerized or bulk form, are fed to the kiln by many methods. Four common methods of feeding are gravity feed, conveyors, augur, and shredder. With gravity feed, a chute is built into the front of the kiln (the front is defined as the point at which the wastes are fed into the kiln) and the wastes are simply dumped through the chute and into the kiln. In larger kilns, the chute may be left open during operation. Combustion air is pulled through the chute by the induced draft system. Alternatively, the chute may be fitted with a door, a ram feeder, or some other device which reduces air infiltration but allows solids to be fed into the kiln.

A somewhat more automated method of feeding the solids to the kiln is a conveyor, which transports the waste to either a chute (gravity feeder) or directly into the kiln. In the latter case, the conveyor must be made of a material capable of withstanding the kiln temperatures. Screw augers are also used to transport bulk wastes into the kiln; however, their use is limited to wastes that will not jam the auger and that can be moved by such a device. They can successfully be used for feeding powdered materials in many cases.

A recent development in feeders is a blender-feeder, which combines a hammer mill with a solids feed system. These devices can be large and powerful enough to shred steel drums of waste material. Their advantage is that the blended and shredded wastes burn more smoothly and with a lower probability of flaring up and causing puffing or explosions in the combustion chamber.

Pumpable wastes can be fed to a kiln in two ways. Those wastes which can be atomized (usually with viscosity of under about 740 SSU) are fed through atomizing nozzles. It is highly desirable that an autogenous waste or some auxiliary fuel is pumped into the kiln in this way at all times to maintain temperature and to properly ignite the other wastes. The considerations for the nozzle are similar to those for a nozzle in a liquid injection incinerator. Those pumpable wastes which cannot be atomized are typically fed to the kiln through a lance, a pipe discharging directly into the kiln with a minimum of bends or constrictions. Steam, air, or nitrogen is sometimes injected with the waste into the lance to facilitate the waste's flow and to provide limited atomization.

As the kiln rotates, it mixes the solid wastes with combustion air and moves the wastes toward the discharge end. The constant rotation also promotes exposure of waste surface to the radiant heat from the flames and hot refractory to enhance heat transfer efficiency and release combustible organics into the gas. As the solids and liquids burn, they produce combustion gases which are swept down the kiln into a secondary combustion chamber or (for a single chamber system) the air pollution control equipment.

Operation of a rotary kiln incinerator involves several concerns. First is the seals at the front and rear ends. As can be seen in Fig. 12.5, the rotating kiln must slide past the fixed wall at the front where the waste feed and burner nozzles are and at the rear by the ash drop and combustion gas exhaust. There is no practical way to seal a rotary kiln to withstand positive pressure at the points where the rotating equipment meets the stationary components. As a result, a rotary kiln is operated under negative pressure. The system is designed to draw air at a specified maximum rate through the seals and other openings. If the seals become worn or damaged, the air infiltration can become excessive and the incinerator will have trouble maintaining temperature at an acceptable gas flow rate. A properly operating incinerator must include routine inspections and a regular maintenance program of the seals.

A second potential problem that is of particular concern with rotary kiln incinerators is that of "puffing." Normally, the gases leaving the kiln are "pulled" into the secondary combustion chamber by the pressure differential between the two. If there is a sudden increase in the gas production rate in the kiln (due to sudden explosion, combustion, or volatilization of a chunk of waste, for example) or a draft decrease in the secondary combustion chamber (due, for example, to a problem with the fan) the gas flow rate may exceed the capacity of the downstream equipment and an overpressure could result. Flue gas from the kiln, potentially containing unburned POHCs and PICs, could thus be released as a "puff" from the rotating juncture between the kiln and the secondary combustion chamber. Normally, the seals can contain the gas from a specified level of overpressure. When the level is exceeded or the seals are damaged or worn, however, puffing can occur.

This problem is of special concern in incinerators burning munitions or other explosive wastes. In these cases, puffing can occur when a shell or piece of explosive detonates suddenly. These incinerators are designed to withstand explosions, but puffing can frequently occur. The burning of drummed wastes can also lead to puffing. In this case, if the contents of a drum burn rapidly, the effect could be similar to (although usually not as severe as) an explosion discussed above. The resultant overpressure can produce puffing. When puffing may occur, incinerators are equipped with an emergency vent stack, water column, or other emergency relief vents.

The length-to-diameter ratios of rotary kilns can range from 2 to 10. Outside shell diameters are generally limited to less than 15 ft to allow shipping of the cylinder sections. Rotational speeds of the kiln are usually measured as a linear velocity at the shell. Typical values are on the order of 0.2 to 1 inch per second. Temperatures for burning vary between 800 to 1600°C (1500 to 3000°F). Bulk gas residence times in the kiln are generally maintained at 2 seconds or higher.

The solids retention time in a rotary kiln is a function of the length-to-diameter ratio of the kiln, the slope of the kiln, and its rotational velocity. The functional relationship between these variables is given by the following rough approximation [Bonner, 1981]:

$$t_{\text{solids}} = 0.19(L/D)/SN \tag{12.13}$$

where t_{solids} is the retention time in minutes, L is the length of the kiln in feet, D is the diameter of the kiln in feet, S is the slope of the kiln, and N is the revolutions per minute. Typical ranges of these parameters are $L/D = 2$–10, $S = 0.03$–0.09 ft/ft, and the rotational speed is 1–5 ft/min (which can be converted to rpm by dividing by the kiln circumference measured in feet). The retention time requirements for burnout of any particular solid waste should be determined experimentally or extrapolated from operating experience with similar wastes. In a movable grate furnace, the retention time is given by the ratio of the length of the grate, L, and its speed, S.

Air/solids mixing in the kiln is primarily a function of the kiln's rotational velocity, assuming a relatively constant gas flow rate. As rotational velocity is increased, the solids are carried up higher along the kiln wall and showered down through the air/combustion gas mixture. Since solids retention time is also affected by rotational velocity, there is a trade-off between retention time and air/solids mixing. Mixing is improved to a point by increased rotational velocity, but the solids retention time is reduced. Mixing is also improved by increasing the excess air rate, but this reduces the operating temperature. Thus, there is a trade-off between gas and solids retention time and mixing.

The longer the solid waste is kept in the kiln, the cleaner the bottom ash becomes. The additional cost of prolonging the solid waste retention time in the kiln is small compared to the total cost of incineration. To extend the solid waste retention time, one or more of the following steps can be used.

1. Rotate the kiln slower.
2. Reduce the kiln slope (when designing) unless slope is variable.
3. Reduce the waste (ash) feed rate into the kiln.
4. Install a longer kiln.

Rotary kiln incinerators can be operated in two modes: slagging and nonslagging. If the kiln is operated in the nonslagging mode, the ash temperature is kept below its fusion point. In the slagging mode, the ash temperature is allowed to rise above its fusion temperature and the ash forms a liquid (or more accurately semiliquid) mass in the kiln. The type of kiln refractory (or other wall material) and the type of ash removal equipment used will be influenced by whether the kiln is a slagging or nonslagging type. Many types of molten ash, called slags, are corrosive, and will dissolve an improperly chosen refractory.

Fluidized Bed Incinerators

The fluidized bed incinerator consists of a bed of sand, limestone, or other mineral and the combustion residue in a refractory chamber. The bed is fluidized by blowing air and, if additional gas is needed, recirculated combustion gases into the bottom through a set of tuyeres. The flowing gas agitates the bed material, turning it into an expanded turbulent mass which has properties similar to a fluid (hence the name). Bed depth typically varies between 1.5 and 10 ft and gas velocities are typically in the range of 2.5 to 8.0 ft/s. It is generally desirable to maintain the depth of the bed as small as possible, consistent with complete combustion and excess air levels, to minimize the pressure drop and power consumption. Figure 12.6 is a schematic of a fluidized bed incinerator. Trenholm *et al.* [1984] give performance data on fluidized bed incinerators burning industrial wastes.

The combustion chamber of a fluidized bed is lined with brick or castable refractory. It has two distinguishable zones: a fluidized bed zone composed of inert granular bed material that is fluidized by directing air upward through the bed, and a freeboard area extending from the top of the bed to the exit of the combustor. Wastes and fuel, if needed, are fed either directly into the fluidized bed or into the windbox beneath the bed, where they ignite. Fuel nozzles, for start-up or supplemental heat, if needed, are located in either the windbox (hot windbox) or in the freeboard (cold windbox). The combustion of the wastes and fuel heats the bed material to temperatures high enough that it, rather than the flame, acts as the combustion source.

The freeboard serves two major functions. First, its larger cross-sectional area slows the fluidizing gas velocity and keeps the larger bed particles from escaping. Second, it acts as a secondary combustion chamber for the off-gases from the bed. Higher heating value liquid wastes or auxiliary fuels can be burned in the freeboard area in a manner analogous to the secondary combustion chamber of a two-chamber incinerator.

The fluidized bed combustor is especially appropriate for burning tars and other sticky materials. In fact, it has been used for many years in the petroleum industry to burn still bottoms and

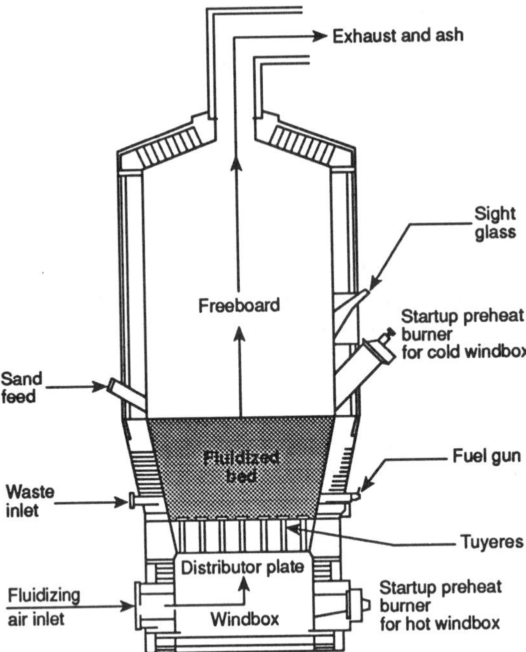

FIGURE 12.6 Fluidized bed incinerator schematic. (*Source:* Battelle, 1972. *Fluidized-Bed Incineration of Selected Carbonaceous Industrial Wastes.* Battelle Columbus Laboratories, final report prepared under grant #12120 FYF for the state of Ohio Department of Natural Resources.)

other high-molecular-weight residues. The tars coat the bed particles and increase particle size and weight. The enlarged particles tend to remain in the bed until the waste is burned off. The residence time for wastes in the fluidized bed can be as much as 12 to 14 seconds. The rapid mixing of the bed also provides good agitation, exposing new surfaces to the hot combustion gases.

The fluidized bed has a very high thermal mass, which helps even out fluctuations in the combustion of highly heterogeneous wastes. As discussed for rotary kilns, heterogeneous wastes can burn unevenly. When a highly flammable clump of waste ignites, it can release a puff of gas which would overload the combustion gas-handling system, resulting in a potentially dangerous condition. The thermal mass of the fluidized bed reduces such uneven burning, making this type of incinerator a likely candidate for such wastes.

While sand is the commonly used bed material for fluidized bed incinerators, other materials that participate in the chemical reaction can be used as well. For example, lime or limestone is sometimes used in fluidized bed boilers that burn high-sulfur coal. The bed material absorbs the sulfur oxides formed in the combustion. A similar method is used to absorb the acid gases formed during incineration.

Fluidized bed combustors operate at relatively low bed temperatures, 425 to 800°C (800 to 1500°F) and freeboard temperatures up to about 1000°C (1800°F) [EPA, 1971]. At start-up, the bed is preheated by a burner located above and impinging down onto it. Because of its high thermal mass and excellent ability to transfer heat from the bed to the incoming waste, a fluidized bed incinerator is capable of burning materials whose heating value is low. Normal incinerators require that the mean heating value of the combined wastes and fuels be a minimum of about 3300 to 4400 kcal/kg (6000 to 8000 Btu/lb). Any one waste stream could have a lower heating value, but the total of all of them should be above this minimum. A fluidized bed incinerator requires a minimum gross heating value of 2500 kcal/kg (4500 Btu/lb) and as little as 550 kcal/kg (1000 Btu/lb) if no water is present.

The rapid motion of the fluidized bed can cause attrition of the bed particles and of the refractory. This creates particulate that will be carried over into the flue gas. As a result, fluidized bed incinerators often place a greater load on the APCD than similarly sized conventional incinerators. Attrition requires that the operators keep tight control of the gas flow. The flow must be great enough to fluidize the bed, but it should not be much greater than is required for this purpose. Note that particulate typically formed by attrition, while fine enough to be carried out in the flue gas, is rarely sufficiently fine to cause difficulty with most types of air pollution control devices. The concern relates to the quantity of particulate captured overloading the flyash-handling capability of the system, rather than with the collection of this relatively coarse particulate matter by the APCD.

The fluidized bed combustor can provide several advantages, including the ability to incinerate a wide variety of wastes. It operates at relatively low and uniform temperatures, thereby tending to have lower NO_x emissions than standard combustors. It also achieves a higher combustion efficiency because of the high mixing and large surface area for reaction. The large mass of the bed makes it tolerant of wide variations in waste heating values. Finally, proper use of additives such as limestone or lime gives this type of incinerator the potential to neutralize acids in the bed.

The fluidized bed combustor does have limitations in its applicability. First of all, it is mechanically complex. Second, it cannot typically burn wastes whose ash forms particulate much larger than the bed material. Large ash particles will fall to the bottom of the bed and will eventually cause defluidization. Defluidization is a phenomenon where the bed settles down and is not blown about by the combustion gas. Third, the ash formed must have a fusion temperature (melting point) greater than the bed temperature. If the ash should melt, it would agglomerate and cause defluidization.

The composition of the ash that a waste material will produce must thus be carefully controlled to keep the particle size small and prevent the fusion temperature of the bed material from dropping below the bed temperature. The fusion temperature of the bed material places an upper limit on the combustion chamber temperature. The fusion temperature of sand, for example, is 900°C (1650°F). Another disadvantage is poor combustion efficiency under low loads.

Multiple Hearth Incinerators

The multiple hearth incinerator was originally developed for ore roasting and drying of wet materials. It is typically used today to incinerate sewage sludges and liquid combustible wastes, but is rarely used for solid wastes. Its design is most appropriate to wastes containing large amounts of water and wastes requiring long solids residence times. It is illustrated in Fig. 12.7. The furnace is a refractory-lined vertical steel shell containing a series of flat hearths. Each hearth has a hole in the middle. A rotating shaft with rabble arms attached at each hearth runs the length of the cylindrical shell. The incinerator is also equipped with an air blower, burners, an ash removal system, and a waste feeding system. Liquid wastes and auxiliary fuel can be injected at points in the furnace to assist the combustion of the solids or simply to destroy the liquid. Multiple hearth furnaces that burn hazardous waste are usually equipped with a secondary combustion chamber (afterburner).

Solid waste is fed to the incinerator in a continuous stream, usually from an auger, onto the top hearth, where it is plowed by the rotating rabble arms, which slowly move the waste across the hearth and into a hole leading to the lower hearth, where another rabble arm plows the waste. The waste thus falls from hearth to hearth until it is discharged from the bottom. The bottommost hearth is usually the only one supplied with overfire air; the other hearths are fed underfire air. The gases in the multiple hearth incinerator flow upward, countercurrent to the waste. The hot gases from the lower hearths dry the waste fed to the upper hearths and eventually ignite the dried solids.

Because a multiple hearth incinerator operates in a countercurrent mode, it has the same difficulties in dealing with volatile hazardous organic waste as a countercurrent kiln. The initial drying zone typically operates at moderate temperatures; any volatile materials in the solid will evaporate and leave the combustor without being exposed to a flame. It is, therefore, necessary to

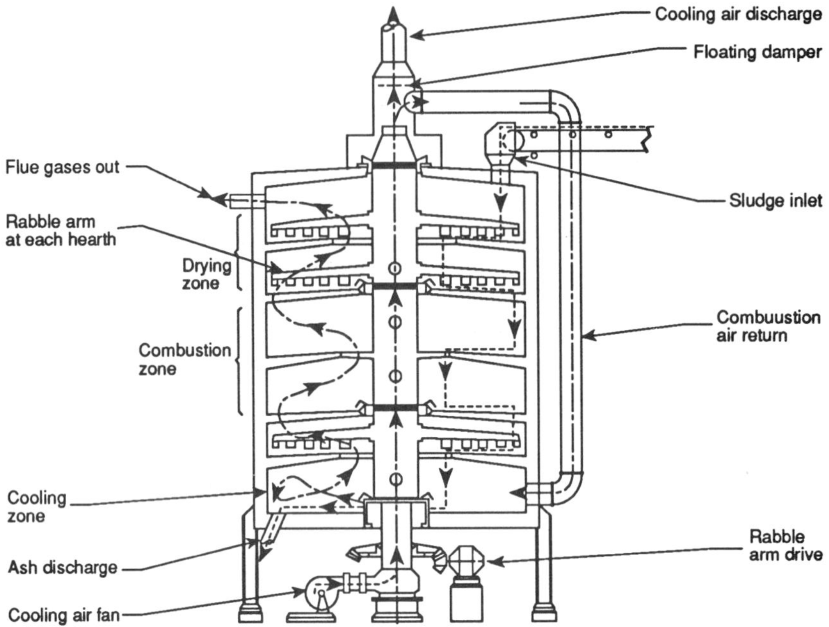

FIGURE 12.7 Typical multiple hearth incinerator. (*Source:* EPA, 1980. Environmental Protection Agency, Air Pollution Training Institute Course #427. *Combustion Evaluation.* Student Manual EPA 450/2-80-063, Student Workbook EPA 450/2-80-063.)

duct the off-gases to a secondary combustion chamber in order to ensure efficient incineration of the volatilized organics.

The incineration process in a multiple hearth furnace occurs in three stages. In the uppermost sets of hearths, the incoming wastes are dried at moderate temperatures. In this zone, volatile organics can also be released into the gas stream. Incineration of combustible matter takes place in the middle sets of hearths. The final set of hearths serves to cool the waste prior to discharge. Discharge of the solids is usually by means of a second auger.

Multiple hearth incinerators are rarely used for hazardous waste destruction. Their principle application is for the combustion of sewage sludge. The wet, homogeneous nature of this waste is well suited to the multiple hearth design. Little test data are available on their efficiency of destruction of organics. One set of published data on tests of four multiple hearth incinerators conducted by the EPA [Bostian *et al.,* 1988; Bostian and Crumple, 1989] indicated the presence of trace amounts of common organic POHCs in the emissions even when they were not detected in the sewage sludge. The reason for this is not given, but the measured levels were very low. The incinerators were equipped with afterburners or secondary combustion chambers and inlet/outlet tests were conducted at one of the four sites. Destruction efficiencies of 90 to 99+% were observed.

The metal emissions tests conducted as part of the same program showed that small amounts of a variety of metals (from the sewage sludge) were released from these incinerators. The size of the metal particulate was not given, but based on the poor removal efficiency of the scrubber at one of the sites (less than 90% and as low as 50% for beryllium), one can probably assume that the particulate was very fine.

Commercial multiple hearth incinerators come in several sizes ranging from 6 to 25 ft in diameter and from 12 to 75 ft in height. Upper-zone temperatures, depending on the heat content of the wastes and supplementary fuel firing, range from 350 to 550°C (650 to 1000°F); midzone temperatures range from 800 to 1000°C (1500 to 1800°F); and lower-zone temperatures range from 300 to 550°C (600 to 1000°F). Generally, multiple hearth units are not common for hazardous wastes incineration because capital and maintenance costs are high. Also, the applications of these units are limited to incineration of wastes with high water content and relatively uniform size (e.g., sludges). The incineration of bulk solids can jam moving parts, resulting in significant maintenance and operational problems.

Controlled Air Incinerators

In the 1970s an incinerator design termed *controlled air* or *starved air* was developed. Figure 12.8 is a schematic of such a unit. The design offers the advantages of lower fuel requirements and lower particulate formation than similarly sized fixed hearth incinerators. However, they are technically more complex and require that the equipment and operation be designed and maintained to prevent random air infiltration. Their use, at present, is generally limited to the burning of hospital and pathological wastes.

Controlled air systems are batch units. Waste is continually fed into them through an air lock or ram feeder, and the ash accumulates. Periodically, the system is shut down and cooled and the ash is removed. Combustion air to the primary chamber is tightly controlled to maintain an oxygen level which is close to, or even slightly below, stoichiometric conditions. This results in pyrolysis of the waste. The combustion gases from the primary chamber go into the secondary chamber where they mix with additional air to complete the combustion process. Such staged combustion results in very low gas flows in the primary chamber and minimization of particulate releases from the solids.

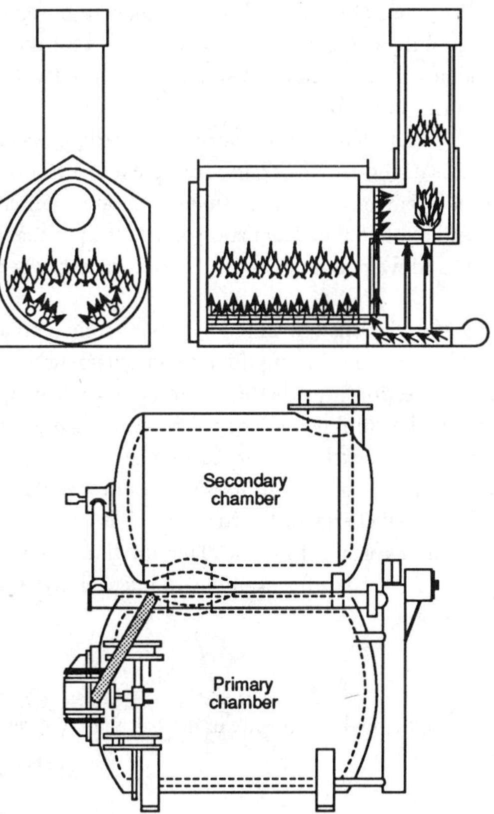

FIGURE 12.8 Two controlled air incinerator configurations. (*Source:* EPA, 1980. Environmental Protection Agency, Air Pollution Training Institute Course #427. *Combustion Evaluation.* Student Manual EPA 450/2-80-063, Student Workbook EPA 450/2-80-063.)

Boilers and Industrial Furnaces

Many boilers and high-temperature industrial process furnaces (BIFs) operate at conditions which are suitable for the destruction of selected types of hazardous wastes, and their use for this purpose is common. The practice achieves two things. First, it destroys the waste. Second, the heating value of the waste replaces fossil fuel with an economic benefit. The practice must, however, be approached with caution.

Wastes, including hazardous wastes, by their very nature, have a highly variable composition. Boilers and industrial furnaces, on the other hand, are primarily operated to produce energy or a product and are relatively intolerant of impurities or "off-specification" raw materials. Furthermore, the systems may not be capable of withstanding some of the products of combustion of certain types of wastes. For example, HCl from the combustion of chlorinated wastes may damage refractory. Many boilers or high-temperature industrial processes are not equipped with scrubbers or other air pollution control devices. Even when the processes do include an APCD, the ash or combustion products from the waste may damage the components and reduce their operating efficiency. Clearly, any program of cofiring wastes into such a combustion device must include a high level of waste selectivity and quality control. This fact must be considered when evaluating the feasibility of using any of these processes to burn a waste.

A second factor which must be considered is the temperature/time/oxygen regime to which a given BIF exposes the waste. Many furnaces have high flame temperatures, but the mean gas temperature in the combustion chamber may be low. For example, most fire-tube boilers have the flame virtually surrounded by the tube walls. The flame may be hot enough to achieve organic destruction, but the walls cool the combustion gases quickly so that the postflame zone may not provide the extended residence time at elevated temperature needed to destroy those organics which escape destruction in the flame zone.

A third factor that must be considered before burning hazardous waste in a process furnace is the point at which the waste is introduced. Take, for example, the introduction of wastes at a point other than the flame zone of a cement kiln. Since cement kilns are countercurrent kilns (where the solids flow in the opposite direction to the combustion gases), the problem is of particular concern. The waste may be vaporized and swept out of the combustion chamber before it has sufficient time to be destroyed.

The U.S. hazardous waste regulations (40 CFR § 260.10) define a boiler as an enclosed device using controlled flame combustion and having three major characteristics. First, the combustion chamber and energy recovery section must be of integral design. This means that a boiler must have heat transfer surfaces in the combustion chamber rather than as an add-on in the duct. A combustion chamber that has a heat exchanger or economizer mounted on the flue gas outlet and no tubes or other steam-generating surfaces in its walls is not a boiler. Second, the boiler must recover at least 60% of the thermal energy of the fuel. Third, at least 75% of the recovered energy must be utilized to perform some form of function. The function can be internal (for example, to preheat air or pump liquids) or external (for example, to make electricity). The following is a list of different types of incinerators.

1. Cement kilns
2. Lime kilns
3. Aggregate kilns (e.g., lightweight aggregate or asphalt plant aggregate drying kilns)
4. Phosphate kilns
5. Coke ovens
6. Blast furnaces
7. Smelting, melting, and refining furnaces
8. Titanium dioxide chloride process oxidation reactors
9. Ethane reforming furnaces
10. Pulping liquor recovery furnaces
11. Waste sulfuric acid furnaces and halogen acid furnaces

The *Technical Implementation Document for EPA's Boiler and Industrial Furnace Regulations* [EPA, 1992] gives the legal description of the specific types of furnaces which are regulated. There are other furnace types which are not listed in the regulations whose operating temperatures and other conditions may be adequate for the destruction of hazardous waste and which appear to meet the criteria of the regulations. They are not regulated at present but may be subject to hazardous waste combustion regulation if used for the purpose. Examples are glass melt furnaces, carbon black furnaces, and activated carbon retort kilns.

The most common industrial furnaces used for waste disposal are cement kilns, lime kilns, lightweight aggregate plants, blast furnaces, and spent acid recovery furnaces. The main factor that differentiates a boiler or industrial furnace from an incinerator is the purpose of its use. An incinerator is built and operated mainly for the purpose of destroying waste materials. Heat may be recovered from this operation, but that is ancillary to the system's main function, which is waste destruction. A boiler or industrial furnace, on the other hand, is operated for the purpose of making a physical product or producing energy. Waste materials may be utilized in doing this, but that is an ancillary purpose to the operation.

Both the boiler and industrial furnace equipment categories can achieve the high organic compound destructions of incinerators. Emissions test data on three boilers have shown a largely unimpaired performance under a wide operating window that extends well beyond normal operating practices for the units [Wool *et al.*, 1989]. However, other data have shown that cofiring, if not properly done, can result in increased emissions [EPA, 1987]. In 1991, the EPA expanded the hazardous waste incinerator regulations (40 CFR § 260) to include boilers and industrial furnaces [FR Vol. 56, No. 35, p. 7134, and subsequent amendments]. These BIF regulations require incinerators and BIFs to meet the same performance standards with operating requirements adjusted for the unique characteristics of the individual types of furnaces.

In order to fire hazardous wastes in existing boilers and industrial furnaces it may be necessary to add equipment and modify the system. Examples of such modifications include the following:

- Install waste storage, blending, pretreatment, and handling facilities.
- Set up sampling and analytical facilities to characterize the wastes and ensure that the waste composition is acceptable.
- Add burners, guns, nozzles, or other types of equipment to feed the wastes to the furnace.
- Upgrade combustion controls to handle the wastes.
- Add monitors for waste feed rate, oxygen, CO, and so on which are required for a facility burning hazardous waste.
- Upgrade or modify the air pollution control equipment to meet the BIF emission requirements.

Waste storage and handling equipment will generally have to be added to a boiler or industrial furnace which is being modified to burn hazardous waste. The facilities may consist of little more than a tank truck which is hooked directly to the waste feed system; however, in most cases, permanent tankage will be preferable. Provisions for blending the wastes to maintain a consistent composition must usually be made. Waste blending is usually needed for incinerators and is often more important for boilers and furnaces since boilers will typically have tighter waste specifications.

A testing program to verify that the wastes received satisfy the specifications for combustion in the boiler or furnace must be included in the waste management program. The legal requirements for such a waste analysis plan are beyond the scope of this type of technical handbook. See the EPA's technical implementation document for the BIF rules [EPA, 1992] and related guidance for legal requirements. Operationally, it is important for a boiler or industrial furnace to keep a tight control on the wastes fed to the furnace. Undesirable waste properties or components could result in damage to the system or to the production of off-specification products. As a minimum the testing program should include the following properties:

- Heating value—large amounts of low heating value materials could have a deleterious impact on flame stability and furnace performance and could cause a drop in temperature.
- Viscosity, density, vapor pressure—this information is needed to ensure that the wastes can be atomized satisfactorily by the nozzles and burners. Poor waste atomization was the probable cause of unsatisfactory performance in a number of tests conducted on boilers and industrial furnaces burning hazardous wastes discussed below.
- Halogen, sulfur, and nitrogen content—these values should be within bounds to keep acid gas emissions within acceptable limits, minimize corrosion, and maintain product quality.
- Ash and metals—these values need to be controlled to prevent the formation of fumes and keep the release of particulate and, specifically, toxic metals to the environment to within acceptable limits. Ash constituents with a low fusion temperature can cause an unacceptable rate of deposition on system components, especially boiler tubes.

Clearly, the decision to burn hazardous waste in a boiler or industrial furnace results in modifications to the equipment and operating procedures. Process changes that will usually be required in order to legally burn hazardous waste include the following:

- Addition of waste feed nozzles and guns—it is generally not recommended to feed the wastes through the same firing guns as the primary fuel because the wastes will have chemical and physical properties that are significantly different from those of the fuel.
- Addition of automatic waste feed cutoffs—these are legally required for any combustor burning hazardous waste.
- Addition or upgrading of combustion control equipment—combustion of hazardous waste requires tight control of operating parameters. For example, a wide temperature excursion may be acceptable for normal operation of a boiler, but it is unacceptable when hazardous waste is burned. Existing controllers may not be able to maintain the tighter bounds on the control parameters legally required when burning hazardous wastes.
- Addition of continuous monitors are required for combustion gas, CO, CO_2, possibly O_2, and possibly HCl. Other required monitors are temperatures of the flue gases, indicator of combustion gas flow, and indicators of proper operation of the waste nozzles and of the air pollution control equipment. The waste flame must be equipped with a cutoff that stops the hazardous waste flow in case any operating parameter goes outside its permit limits.

The existing indicators for the primary fuel feeds must be wired into the waste feed system so that both the primary fuel and waste fuel are shut off in case of primary flame failure. The control system should incorporate a means of maintaining the primary flame if the problem lies with the waste combustion system only. In this way, residual hazardous waste in the combustion chamber will be destroyed and operation of the system will continue.

The final general area of consideration that will be discussed here is the impact of the waste feed on the air pollution control device. Very few boilers and industrial furnaces are presently equipped with acid gas control devices, so the impact will be restricted to particulate. Tests on boilers and cement kilns have shown that burning wastes containing halogens in cement kilns or (for any type of combustor) containing metals or salts can increase the amount of fine particulate that is formed. In addition to having very small diameters, the particulate can have a lower resistivity than that produced when only primary fuel is used. These changes can have an impact on the performance of the air pollution control device. The impact is especially of concern for an electrostatic precipitator or ionizing wet scrubber since their performance is especially susceptible to both size and resistivity, although the finer particulate could reduce the collection efficiency of any type of device. The impact on all air pollution control equipment should, therefore, be considered when evaluating the feasibility of burning hazardous waste in a boiler or process furnace.

12.5 Air Pollution Control and Gas Conditioning Equipment for Incinerators

Incineration is one of the most difficult applications of air pollution control systems. The gas must be cooled, and multiple pollutants must be controlled with very high efficiencies. Hydrogen chloride, hydrogen fluoride, and sulfuric acid vapors can be highly corrosive if handled improperly. Particulate matter generated in the incinerators is primarily in the submicron range. Pollutants having special toxicity problems such as metal compounds and dioxin/furan compounds generally enter the systems at very low concentrations and must be reduced to concentrations which challenge the limits of analytical techniques. The air pollution control system typically can consist of a heat recovery boiler, a quench, a particulate removal device, and an acid gas removal device. The devices are often mounted in series. For example, the off-gas treatment system for a typical high-performance rotary kiln incinerator in hazardous waste duty can include: (1) a boiler to partially cool the combustion gas and recover steam; (2) a dry scrubber, also called a spray dryer, to remove some acid gas and reduce downstream caustic requirements; (3) a quench to further cool the gases; (4) a reheater to prevent condensation in the fabric filter; (4) a fabric filter; (5) a packed column for further acid gas removal; (6) a HydroSonics™ scrubber, which includes a steam ejector to provide the motive force for the flue gases and to act as a backup air pollution control device in case of failure of an upstream component; and (7) the stack. As can be seen, the air pollution control system represents a major fraction of the total system cost.

A variety of particulate and acid gas control devices are used for incinerators. They typically are broken down into two categories: wet and dry. The wet devices include Venturi and Venturilike scrubbers, "wet" electrostatic precipitators, and proprietary wet scrubbers. Dry particulate removal devices include fabric filters, electrostatic precipitators, and high-efficiency particulate absolute (HEPA) filters. HEPA filters are used in particulate control where a very high level of particulate removal is required, such as radioactive waste incinerators. They are similar in operation to fabric filters, although the nature of the fabric is such that a high level of particulate control is achieved. Wet devices remove acid gases as well as particulate, and the two functions are often combined in one device.

Acid gas removal in incinerators is performed through the use of a packed bed scrubber, Venturi scrubber, or ionizing wet scrubber. The ionizing wet scrubber includes a packed bed in its design, and the acid gas removal largely occurs in the packed bed. Dry scrubbing is a technique whereby lime or another lime-based sorbent is injected into the hot zone of the incinerator. The sorbent removes a fraction of the acid gas in suspension and is collected on the particulate removal device, usually a fabric filter. Dry scrubbers are usually used in conjunction with another form of acid gas removal device.

Quench

The first step in the air pollution control system is the gas cooling, which is usually accomplished by means of a water quench sometimes preceded by a boiler. The boiler is a noncontact heat exchanger which utilizes the heat from the combustion gases to make steam. The quench is a portion of the duct, or a separate chamber, in which water is sprayed into the duct to cool it by evaporative cooling. The quench increases the water vapor content of the gas stream substantially, and drops the gas temperature by adiabatic evaporation. Typically, the gas stream exiting the quench reaches the adiabatic saturation temperature, and further cooling by evaporation is impossible. Any additional cooling must occur due to sensible heat transfer to the exiting liquid stream.

The solids content of the quench liquor is very important. Submicron particles can be formed by droplet evaporation if the aqueous stream contains dissolved solids when injected into the hot portions of the evaporative cooler. Suspended solids in the aqueous stream will also be released to the gas stream on evaporation of the water; however, this particulate will tend to be larger and, hence, more easily removed by the particulate control system. Particulate noncompliance conditions can be created if the potential particulate formation in the quench is not considered.

Because of the critical nature of the quench system, a backup, emergency quench system is desirable. The emergency quench consists of a separate set of nozzles and water feed lines from an independent source. The water supply for the emergency quench should activate even in case of a complete power failure. A fire supply or elevated tank supplying water by gravity feed are possible sources. The system's response time must be very short since catastrophic damage can occur within seconds, especially if downstream ductwork or control devices are constructed of fiber-reinforced plastic (FRP) materials. Due to the short response times necessary, the emergency quench system should not depend on manual activation by the operator.

Heat Recovery Systems

Many incinerator designs incorporate some form of heat recovery systems. Systems designed to recover heat from the burning of municipal waste (waste-to-energy systems) recover heat by the inclusion of steam tubes in the combustion chamber walls; these are termed water-wall incinerators. Such water-wall units are not commonly used for the combustion of hazardous wastes since the relatively cold walls can reduce the incinerator's combustion zone temperature and, hence, the level of organic compound destruction which is achieved. A notable exception to this is the use of large industrial or utility boilers for waste destruction. In this case, the amount of waste burned is small compared to the amount of fuel used. In addition, the large size of the furnace minimizes the effect of the cold walls.

In general, however, incinerators burning hazardous wastes do not utilize water-wall designs. They do, however, often include a boiler which is mounted in the exhaust gas immediately after the final combustion chamber. The boiler serves two functions. First, it recovers heat from the flue gas in the form of usable steam. Second, it cools the flue gas without increasing its volume and mass by injecting steam as in a quench.

Heat recovery is an important aspect of the boiler. The high temperature and volume of the combustion gas lends itself to the production of usable steam to drive steam ejectors, provide steam for atomization in the nozzles, drive turbine pumps, or for numerous other applications. The steam coming directly from a hazardous waste incinerator is generally not at a high enough pressure to drive turbines for electrical generation; however, if the plant includes a cogeneration plant, then the steam could be fed to an external, secondary furnace to boost its pressure for such a use. In most cases, however, an incinerator does not produce enough steam to warrant the expense of an electricity cogeneration system.

The second purpose for the heat recovery system, gas cooling, makes it a significant component in the incinerator design. The temperature of the combustion gas leaving the secondary combustion chamber is typically on the order of 1000°C (2000°F). It must be cooled to below about 70°C (160°F) prior to entering the air pollution control system. A quench cools the gas by evaporating water, which then increases the mass of the combustion gas stream; a boiler cools the combustion gas by indirect heat transfer without increasing its mass. If the boiler can cool the combustion gas from 1100 to 250°C (2000 to 500°F), it will reduce the amount of water that the downstream quench will produce by approximately 75%. The quench, the air pollution control system, the fan or ejector, and stack can thus be proportionately smaller.

There are a number of factors which must be considered prior to including a boiler in an incineration system. First, a boiler, by its very nature, is designed to cool the gas stream and, hence, it must be placed downstream from all combustion chambers which destroy hazardous constituents. If it is placed in the combustion chamber, it will cool the gases and change the temperature-residence time regime to which the combustion gases are exposed, potentially causing a lesser destruction of organics. Second, in some cases the ash particles present in the secondary combustion chamber exhaust gas may be at least partially melted. Sodium chloride, a very common component of wastes, melts at flame temperatures. The molten particles can solidify on the cool walls of the boiler tubes, fouling them and even restricting gas flow. The boiler may thus be impractical for those incinerators burning wastes with high salt content or whose ash has a fusion temperature (melting point) below that found in the combustion chamber.

Another factor is the presence of corrosive gases in the gas stream. The boiler must be operated to ensure that it has no spots that are below the combustion gas stream's dew point. Acid gases (such as SO_2, HCl, or HF) in the combustion gas will dissolve in the condensed aqueous liquid, corroding the tubes. HCl is especially corrosive to metals, including most forms of stainless steel. HF is also corrosive to metals, and it also attacks silica-based refractories. SO_2 is relatively insoluble in water and is relatively noncorrosive; however, it lowers the dew point of water substantially, and its presence requires the boiler to be operated at a higher temperature to prevent condensation.

To minimize the risk of condensation, boilers are typically sized so that the temperature of the exit gas is above 450°F. Furthermore, because of the particularly corrosive nature of HF on refractory materials, the presence of more than trace amounts of fluorides may prevent the use of boilers.

Electrostatic Precipitators

Electrostatic precipitators (ESPs) can be used for two entirely different types of service. "Wet" precipitators can be used as the principal particulate control device within a wet scrubbing system including a gas cooler and an acid gas absorber. While basic operating principles of the wet and conventional ESP are similar, the two different styles are subject to quite different operating problems. They both use the principle of electrostatic attraction. The incoming particulate is ionized and then collected on charged plates. In a dry precipitator, the plates are periodically rapped or shaken and the accumulated dust collected in hoppers at the bottom. In a wet precipitator, a continuous stream of recirculated liquor drains over the plates and removes the accumulated particulate matter. A wet ESP can be used to control acid gas as well as particulate. Wet ESPs operate at relatively low gas temperatures, and the precipitators are limited to one or two electrical fields in series.

Figure 12.9 is a drawing of a conventional dry electrostatic precipitator. In electrostatic precipitators, particles are electrically charged during passage through a strong, nonuniform electrical field. The field is generated by a transformer-rectifier set which supplies pulsed DC power to a set of small diameter electrodes suspended between grounded collection plates. Corona discharges on these electrodes generate electron flow which in turn leads to the formation of negative ions as the electrons travel on the electrical field lines toward the grounded plates. The negative ions also continue on the field lines toward the plates. Within the corona itself, positive ions form, and these travel back to the high-voltage electrodes.

Performance of an ESP is a function of the ability of the particulate matter to receive and maintain a charge, the velocity of the particulate migration to the collection plates, the ability of the particulate to adhere to the plates after they are captured, and the ability of the system to minimize reentrainment of the particulate during the cleaning or rapping cycle.

The ability of the particulate to maintain a charge and to adhere to the collection plates is a function of the resistivity of the particulate and of the flue gas. The resistivity is a measure of the ability of electrical charges on the particles to pass through the dust layer to the grounded collection plates. Dust layer resistivity is expressed in units of ohm-centimeters and the values can range from 1×10^8 to more than 1×10^{13}. The effective migration velocity increases nonlinearly as the resistivity decreases.

Dust layer resistivity at any given site in a precipitator is the net effect caused by two different paths of charge conduction. When there are conductive compounds adsorbed on the surfaces of the particles, the electrons can pass along the particle surfaces within the dust layer to reach the grounded collection plates. The most common conductive materials on the surfaces of particles

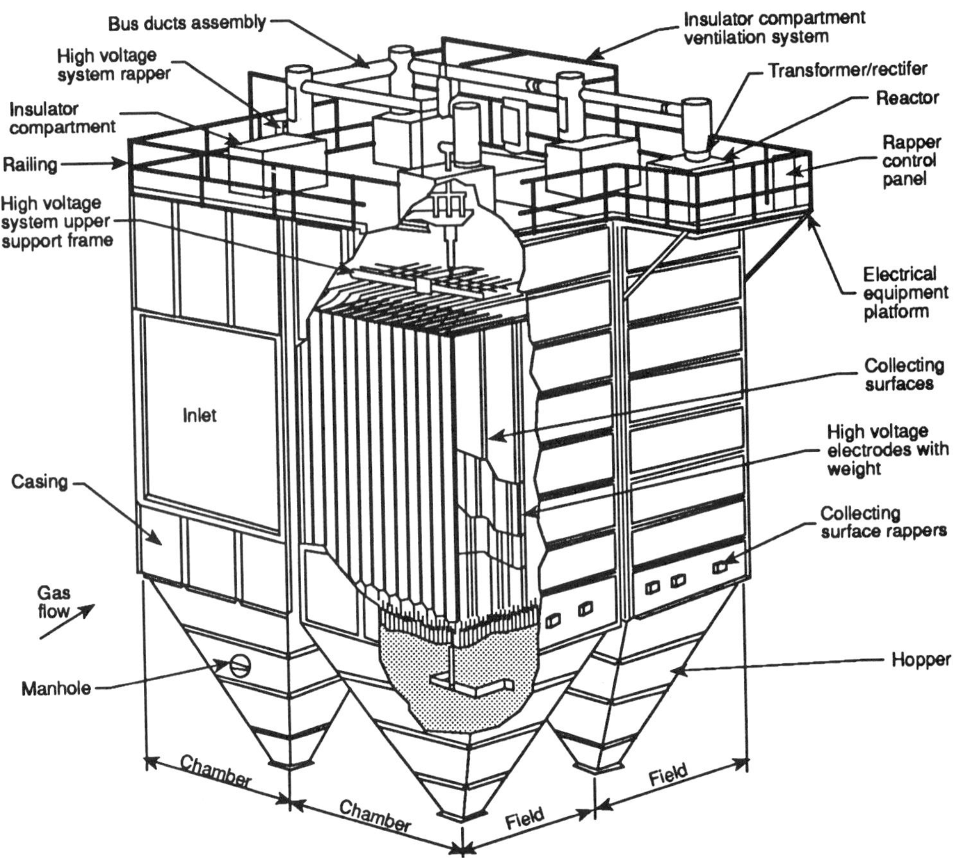

FIGURE 12.9 Conventional dry electrostatic precipitator.

include sulfuric acid and water vapor. When there are conductive materials in the particles themselves, there can be an electrical path directly through the bulk material. The most common conductive compounds within the precipitated particles include carbonaceous materials, sodium compounds, and potassium compounds. The overall particle size distribution affects the resistivity because, in both types of conductivity, the electrical current must pass through a number of separate particles in the dust layer prior to reaching the grounded surface. Smaller particles form a dust layer with less voids and greater particle-to-particle contact.

Dust layer resistivity is not a constant value throughout a precipitator at any given time, and the average resistivity can vary substantially over time at a given incinerator. This temporal and spatial variability is due to the extreme sensitivity of resistivity values to the gas temperature, the presence of vapors such as sulfuric acid mist, and the dust chemical composition. The rate of electrical charge conduction through the dust layer on the collection plates determines the electrostatic voltage drop across the layer, and this can affect both the ability of the dust to adhere to the plates and the ability of the plates to attract additional dust. When conductivity by either path is high, the dust layer voltage drop is low. Some of the weakly held particulate matter can be reentrained due to rapping with excessive force. Also, localized high gas velocity zones through the precipitator can scour off some of the dust layer. For applications where the dust shows a high resistivity, the precipitator will have to be larger. It is sometimes possible to modify resistivity by injecting small amounts of sulfuric acid into the gas stream entering the ESP as a conditioning agent.

One of the first steps in precipitator design is to determine the necessary collection plate area for the efficiency desired. The efficiency should increase as the specific collection area is increased. However, the cost and mechanical complexity of the precipitator also increase with the specific collection area. Also, the vulnerability to malfunctions can increase as the size increases. The optimum size which provides for high-efficiency performance without excessive costs and reliability problems must be determined. Equipment sizing must take into account the numerous nonideal factors which are difficult to express mathematically but nevertheless have important effects on performance. This has generally been done by the determination of "effective" migration velocities which include theoretical parameters plus the effects due to nonuniform particle size distribution, nonuniform gas flow distribution, nonuniform gas temperature distribution, and rapping reentrainment. During the design stage, the anticipated dust layer resistivity range should be carefully evaluated. For existing units, the resistivity variability can be measured by a variety of instruments such as the cyclonic probe and the point-to-plane probes. For new units, the resistivity range in similar incinerators handling similar wastes should be reviewed.

High-efficiency precipitators have more than one electrical field. Two or more fields are normally provided in the direction of the gas flow. For large incinerators, the gas flow can be split into two or more chambers, each of which has several fields in series. The sectionalization of the precipitators improves both precipitator performance and reliability. For this reason, sectionalization should be considered in the preparation of precipitator specifications. As in the case of precipitator sizing, there is an optimum balance between the number of independent fields and the cost. One of the underlying reasons for sectionalization is the significant particle concentration gradient and dust layer thickness gradients between the inlet and outlet of the precipitator. At the inlet of the precipitator, the dust layer accumulates rapidly since 60 to 80% of the mass is collected rapidly. This makes this field more prone to electrical sparking due to the nonuniformities in the dust layer electrical fields. Also, the fine particles which are initially charged in the inlet field but not collected create a space charge in the interelectrode zone. This space charge inhibits current flow from the discharge electrodes and collection plates. By sectionalizing the precipitator, the inherent electrical disturbances in the inlet field do not affect the downstream fields. Another reason for sectionalization is the differences in electrical sparking which are normally moderate to frequent in the inlet fields and low to negligible in the outlet fields. The number of fields in series and the number of chambers in parallel are generally determined empirically based on the performance of previous commercial units. In the case of hazardous waste incinerators, there are some practical limits to the number of fields used simply because the gas flow rates are relatively small.

Because of the high voltage, sparks will occur in a precipitator. The sparking rate needs to be maintained at a specified level, and the ESP includes automatic voltage controllers to quench the electrical sparks. These are electronic devices which reduce the applied voltage to zero for milliseconds and then increase it in several steps to a level close to the one at which sparking occurred. When sparking rates increase in a field, the net effect is to lower the peak voltages and the applied time of the electrical power.

Rapping Techniques. For dry electrostatic precipitators, there are two major approaches to rapping: external roof-mounted rappers and internal rotating hammer rappers. The external rappers are connected to groups of collection plates or an individual high voltage frame by means of rapper shafts, insulators, and shaft seals. The advantage of these roof-mounted rappers is that there is access to the rapper during operation and the intensities can be adjusted for variations in dust layer resistivity. The disadvantage is that the large number of rapper shaft components can attenuate rapping energy and become bound to the hot or cold decks.

The internal rotating hammer rappers have individual rappers for each collection plate. Due to the greater rapping forces possible, these can be used for moderately high resistivity dusts. The disadvantages of these rappers are the inability to adjust the frequency and intensity in various portions of the precipitator and the inaccessibility for maintenance. Also, the internal rotating hammer rappers can be vulnerable to maintenance problems such as shearing of the hammer bolts, distortion of the hammer anvils, bowing of the support shafts, and failure of the linkages.

The type of rapping system chosen should be based on the anticipated resistivity range, the frequency of routine incinerator outages, and the cost of the equipment. For both styles of rappers, the high-voltage frame rapper shafts must include high-voltage insulators to prevent transmission of high DC voltages to external, accessible equipment. These insulators must be kept clean and dry. Purge air blowers with electrical resistance heaters are used for this purpose.

Hoppers and Solids Discharge Equipment. For dry electrostatic precipitators used in dry scrubbing systems, proper design of the hopper and solids discharge equipment is especially important. These units handle high mass loadings, and some of the reaction products are hygroscopic and prone to bridging. Hopper heaters and thermal insulation are important to avoid the hopper overflow conditions which could cause an undervoltage trip of a field and which could possibly cause serious collection plate to discharge electrode alignment problems.

Gas Distribution. One of the most important steps in ensuring adequate gas distribution is to allow sufficient space for gradual inlet and outlet transition sections. Units with very sharp duct turns before and after the transition are also prone to gas distribution problems.

Proper gas distribution is achieved by the use of one or more perforated gas distribution screens at the inlet and outlet of the precipitator. These are generally hung from the top and cleaned by means of externally mounted rappers. Location of the gas distribution screens (and ductwork turning vanes) is usually based on either 1/16th scale flow models or gas distribution computer models.

Fabric Filters

The fabric filter consists of a series of filter bags made of fabric and hung from a frame in a "bag house." The gas to be treated enters the bag house, passing through the fabric of the bags, which filter out particulate. The cleaned gas exits the bag house to subsequent further cleaning or discharge. The parameters which influence bag house construction are type of filter bag material, gas to cloth ratio (also called facial velocity), direction of gas flow through the bags, and type of bag cleaning employed.

Particulate collection on a fabric filter occurs primarily by the combined action of impaction and interception within the dust cake supported on the fabric. Without a well-developed dust cake, the filtration efficiencies would be very low. Particulate capture efficiency in a typical bag house is normally very high. Penetration of particles through a fabric filter is due not to inefficient impaction/interception, but rather to dust seepage through the dust cake and fabric, gas flow through gaps or tears in the fabric, gas flow through gaps between the bags and the tube sheet, and gas flow through gaps in tube sheet welds. The emissions are minimized by operating the unit at proper air-to-cloth ratios and by ensuring that the bags have not deteriorated due to chemical attack, high-temperature damage, flex/abrasion, or other mechanical damage. The efficiency of fabric filters is not highly sensitive to the inlet particle size distribution. The particle size distribution of the material penetrating the bag house is similar to the inlet gas stream particulate due to the emission mechanisms.

The typical bag house is arranged in compartments with dampers that permit isolation of each compartment from the rest with poppet or butterfly dampers. In case of failure of one or more bags in one compartment, the compartment can be isolated and the system can keep operating until the bags are replaced or repaired. For some systems the individual compartments are closed prior to rapping or shaking in order to minimize the release of particulate during the cleaning cycle.

The two basic styles of fabric filters used on incinerators are pulse jet units and reverse air units. Reverse air units are identified by bags that are suspended under tension from tube sheets directly above the hopper. The particulate-laden gas stream enters the interior of the bags through the tube sheet and collects as a dust on the inside of the bags. The bags are cleaned by blowing filtered gas from the exhaust through the bags in a reverse direction to normal gas flow.

The reverse flow dislodges the dust cake into the hopper. In order to accommodate the reverse flow of air necessary for cleaning, reverse air bag houses have multiple compartments. For cleaning, each compartment is isolated from the flow gas stream with dampers.

Pulse jet bag houses are the most common since these are well suited for the relatively small gas-flow rates in hazardous waste incinerators. In pulse jet units, the fabric is supported on a cage which is suspended from a tube sheet near the top of the collector. The particulate-laden gas stream enters through a duct on the side of the bag house or in the upper portions of the hoppers. The dust cake accumulates on the exterior surfaces of the bags. The bags are cleaned by the combined action of a compressed air pulse and the reverse air flow induced by this pulse. A set of solenoid-valve-controlled diaphragm valves are used to supply compressed air cleaning flow to each row of bags. Some units have multiple compartments to allow for bag cleaning off-line or to allow for maintenance of a portion of the unit during incinerator operation.

The following section on fabric filters emphasizes pulse jet units. Two sketches of pulse jet bag houses are shown in Figs. 12.10 and 12.11. The first of these is an isometric sketch which shows the locations of the clean side access hatches on the top and the filtered gas exit duct. The second sketch is a side elevation showing the bag support and the bag-cleaning apparatus.

The air-to-cloth ratio is the total fabric area divided by the total gas flow rate in actual cubic feet per minute (ACFM) (wet basis). It is defined in Eq. (12.14). The units should be specified along with the air-to-cloth ratio dimensions of length/time since a value of x in English units is quite different from a value of x in metric units.

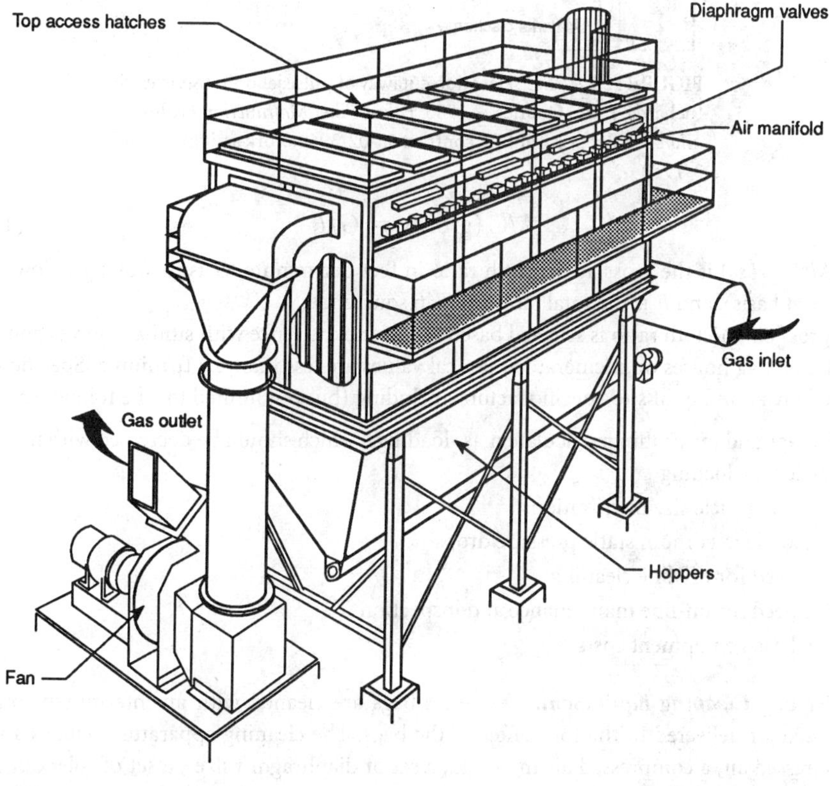

FIGURE 12.10 Isometric sketch of a top-access-type pulse jet unit. (*Source:* Richards, J., and Quarles, P. 1986. *Fabric Filter Malfunction Evaluation.* Final report for EPA contract 68-02-3960, work assignment 3-131.)

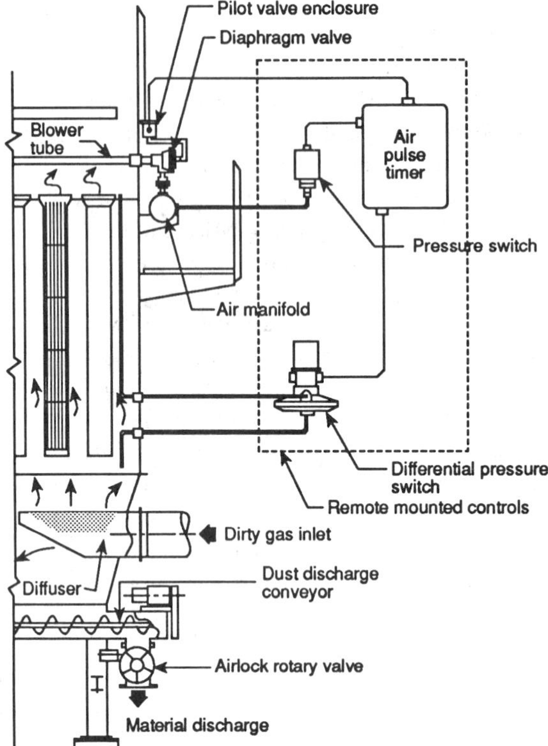

FIGURE 12.11 Side elevation cutaway of pulse jet unit. (*Source:* Richards, J. and Quarles, P. 1986. *Fabric Filter Malfunction Evaluation.* Final report for EPA contract 68-02-3960, work assignment 3-131.)

$$\text{A/C (gross)} = G/F \tag{12.14}$$

where A/C (gross) is the gross air-to-cloth ratio in feet per minute, G is the total gas flow rate in ACFM (wet basis), and F is the total fabric area in square feet.

The gross air-to-cloth ratio is selected based on prior experience with similar sources and dusts. For pulse jet bag houses on incinerators, typical values are less than 4.0 ft/minute. Specific values are based on a number of site-specific factors, including (but not limited to) the following:

- Average and maximum particulate mass loadings, which should be decreased with increasing particulate loading
- Typical particle size distributions
- Allowable maximum static pressure drops
- The need for off-line cleaning
- The need for on-line maintenance and inspection
- Purchased equipment costs

Pulse Jet Bag Cleaning Equipment. Pulse jet bags are cleaned with an intermittent pulse of compressed air delivered to the top center of the bags. The cleaning apparatus includes a source of compressed air, a compressed air manifold, a set of diaphragm valves, a set of solenoid valves, and compressed air delivery tubes. Cleaning is done on a row-by-row basis. On-line cleaning means that the gas stream is continuing to flow through the compartments as the rows of bags are cleaned one by one. During off-line cleaning, a compartment is isolated while the rows are cleaned one by one. Off-line cleaning minimizes dust cake discharge problems since the falling solids from the outsides of the bags are not being opposed by an upward-flowing unfiltered gas stream.

Table 12.8 Fabric Capabilities

Material	Temperature Limits		Chemical Resistance		Flex/Abrasion Resistance
	Long-Term	Short-Term	HCl and H_2SO_4	HF	
Acrylic copolymer	225	250	Good	—	Fair
Modacrylic	275	300	Good	—	Fair
Polyester	250	275	Fair	Fair	Good
Polyphenylene sulfide	375	400	Good	—	Good
Nylon arimid	400	425	Poor	Poor	Good
Flourocarbon	450	500	Good	Good	Good
Polyimide	450	500	Good	Good	Good
Fiberglass	500	550	Fair	Poor	Fair
Stainless steel	1200	1300	Fair	—	Good

Sources: Richards, J. and Segall, R. 1985. *Inspection Techniques for Evaluation of Air Pollution Control Equipment,* Vol. II. EPA Report 340/1-85-022b; PEI. 1986. PEI Associates, Inc. *Operation and Maintenance Manual for Fabric Filters.* EPA Report EPA-625/1-86-020.

As a general rule, the minimum bag house gas temperature should be 50°F above the acid dewpoint to take into account the gas temperature spatial variability in the bag house and the short-term fluctuations in the average gas temperature. Sufficient thermal insulation should be provided so that the gas temperature drop across the bag house does not exceed 25 to 40°F (depending on the inlet gas temperature). Air infiltration should be minimized by selection of proper hatch gaskets and latches, proper solids discharge valves, and proper shell welding practices.

Fabrics and Support Cages. The selection of fabric materials must be based on the expected gas stream temperatures and acid gas concentrations. A summary of the general capabilities of commonly used fabrics is provided in Table 12.8. The long-term temperature limits presented in Table 12.8 are slightly below the general temperature limits often stated for the various types of materials. The reduced long-term temperature values increase the service life of the bags. The short-term maximum temperature limits specified in Table 12.8 are slightly higher than general temperature values. However, these values should not be exceeded for more than 15 minutes. Severe gas temperature spikes will lead to premature bag failure even if the long-term temperatures are maintained in the proper range.

It should be noted that for many units there is only a narrow optimum gas temperature range. There can be only 100 to 150°F between the long-term upper gas temperature limit and the acid-dewpoint-related lower gas temperature limits. Proper process control and conscientious maintenance are necessary to maintain the narrow gas temperature range throughout the bag house.

In addition to the temperature and acid sensitivities, the abrasion and flex resistance of the material should be considered. Materials which are vulnerable to flex and abrasion problems should be used only on cages which provide the maximum support. Also, the cages should not have exposed sharp edges which could cut the fabric.

Hoppers and Solids Discharge Equipment. The proper design of the hopper and solids discharge equipment is important in ensuring long-term reliable operation. Hopper heaters and thermal insulation are important to prevent the hygroscopic, acidic ash from cooling. This can result in bridging of solids and hopper overflows. Due to the gas entry ducts in the upper portions of the hoppers, dust reentrainment blasting of the pulse jet bags occurs as the hopper solids levels increase.

Fabric Filter System Instruments. Table 12.9 summarizes the categories of instruments often used on pulse jet fabric filters. The gas temperature monitors are especially important since they provide indications of incinerator upset and bag house air infiltration. The pulse jet fabric filter static pressure drop gauges should be mounted in accessible locations since they are prone to pluggage

Table 12.9 Pulse Jet Fabric Filter System Instruments

Vessel	Parameter Measured	Process/Equipment Controlled	Portable Instrument Port or Sampling Tap
Stack	Opacity	—	—
Fan	Fan motor current	—	—
	Fan vibration	—	—
Pulse jet	Fabric filter		
	Inlet gas temperature	Emergency bypass	Yes
	Outlet gas temperature	—	Yes
	Compressed air pressure (each separate header)	—	—
	Static pressure drop, overall bag house	Cleaning system	Yes
	Static pressure drop, each compartment	—	Yes

of one or both of the lines. Also, compressed air gauges are necessary on each separate header to identify units with leak problems.

High-Efficiency Particulate Absolute Filters

These very high efficiency collectors are conceptually similar to fabric filters. They are used for the control of relatively small gas volume incinerators firing low-level radioactive wastes. The filter elements are composed of thick fiberglass mats with radiation-resistant binders. The filters are constructed in 2-foot-square panels approximately 1 foot deep. The filter is pleated within each of the panels in order to increase the filtering area. A prefilter is often used to reduce the frequency of replacement of the expensive HEPA filters. This usually consists of a set of low-efficiency panel filters. The average approach velocities range from 300 to 500 meters (1000 to 1500 feet) per minute. Particles are collected by the combined action of impaction and Brownian diffusion on the surfaces of the filter mat. Unlike fabric filters, the accumulated material (or dust cake) is not the main filtering element in HEPA filters. The particulate removal efficiencies are rated at least 99.97% efficient for DOP droplets with a mean size of 0.3 microns.

The initial static pressure drop across a set of new HEPA filter panels is between 1 and 1.5 inches water-column (W.C.). The units are replaced whenever the pressure drop reaches a preset maximum limit which is normally approximately 2.5 inches. High-pressure drops are not desirable since this increases the risks of leakage through the seals around the panels and the risks of particle seepage through the filter elements.

Gas-Atomized (Venturi) Scrubbers

There are a large number of quite different devices which are included in the general category of gas-atomized scrubbers. These include but are not limited to adjustable-throat venturis, rod decks, and collision scrubbers. The common element of all of these devices is the utilization of a high-velocity gas stream to atomize a relatively slow-moving injected liquid stream. One commercial type of adjustable-throat venturi is shown in Fig. 12.12. This includes a wedge which moves up and down within the diverging section of the throat in order to vary the cross-sectional area. The wedge is moved by means of a hydraulic actuator below the elbow of the venturi diverging section. The particular scrubber shown in Fig. 12.12 has a cyclonic demister for collection of the liquor droplets formed in the venturi throat.

A second type of venturi scrubber is the rod deck. The deck consists of one or more horizontal rows of rods across the throat of the scrubber. Liquor is introduced by means of downward-oriented

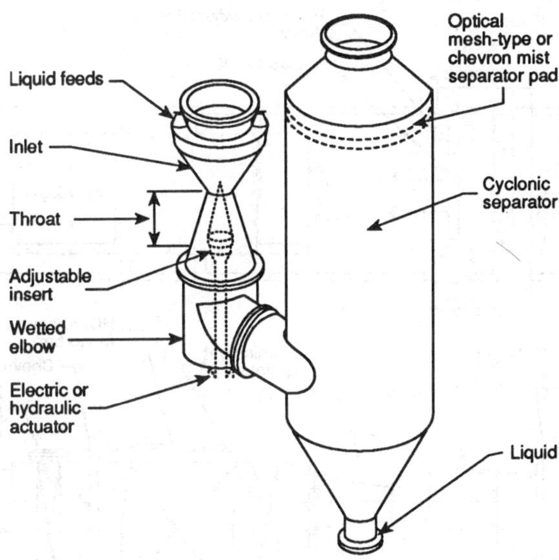

FIGURE 12.12 Adjustable-throat venturi scrubber.

nozzles above the rod decks. Some models include the provision for movement of one of the decks in order to vary the cross-sectional area and the operating static pressure drop.

A schematic of a collision scrubber is shown in Fig. 12.13. The gas stream is split into two equal streams which are then directed against each other. Impaction occurs due to the significant differences in relative velocities of the water droplets and the particles in the colliding streams. This particular unit has a chevronlike demister for collection of the water droplets.

All of the fundamental mechanisms employed in gas-atomized scrubbers are particle-size dependent. The two most commonly used mechanisms are impaction and Brownian diffusion. Impaction is a very effective means of capture for particles larger than 0.5 microns, and Brownian diffusion is the primary capture mechanism for particles in the less than 0.1 micron range. In the 0.1 to 0.5 micron range, both collection mechanisms can be active, but neither is especially effective.

The rate of Brownian diffusion is inversely proportional to particle diameter. As particle size decreases, Brownian diffusion increases. It also increases as the gas temperature increases due to the increased kinetic energy of the gas molecules striking the small particles. Due to the combined action of impaction and Brownian diffusion, penetration of particulate matter in gas-atomized scrubbers is low in the greater than 1.0 micron range and in the less than 0.10 micron range. However, there is a peak in the penetration curve (penetration $= 1 -$ collection efficiency/100) at approximately 0.2 to 0.5 microns. Gas-atomized scrubbers and other air pollution control devices using impaction and Brownian diffusion are least effective in the submicron particle size range.

Scrubbing systems that can achieve high particulate removal efficiencies in the submicron particle size range utilize a flux force/condensation mechanism to aid capture. Flux force/condensation conditions are initiated by removing the sensible heat from the gas stream downstream of the quench so that a portion of the gas stream water vapor condenses on the particles to be removed. The two primary physical mechanisms active in flux force/condensation are diffusiophoresis and heterogeneous condensation [Calvert *et al.,* 1973; Calvert and Jhaveri, 1974]. Diffusiophoresis is the net force due to nonequal molecular collisions around the surface of a particle. The conditions which favor diffusiophoresis occur when the particle is near another particle or surface undergoing condensation. The mass flux of water vapor toward the condensation surface creates the nonequal molecular forces on the second particle. Diffusiophoresis is important only for very small submicron particles, which are affected by molecular collisions.

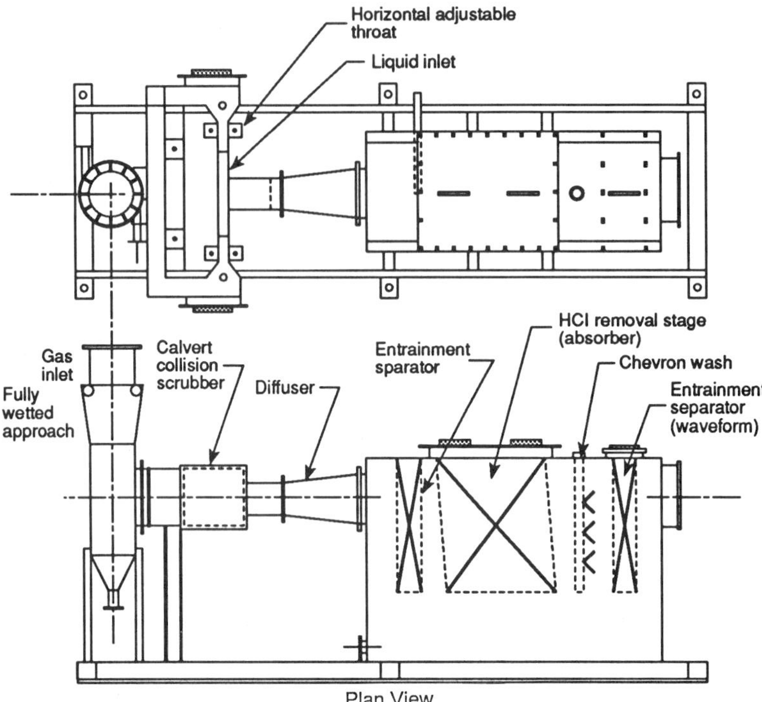

FIGURE 12.13 Collision scrubber.

The static pressure drop through a venturi scrubber can be estimated either by means of the pilot scale tests or by one of the published theoretical equations. The pressure drop calculation [Yung, 1977] is presented in the following equation:

$$\gamma p = 0.005(L/G)V^2 \tag{12.15}$$

where γp is the static pressure drop in W.C., V is the gas velocity in ft/sec, and (L/G) is the liquid-to-gas ratio (liters/min)/(liters/min). This equation indicates that the pressure drop is a strong function of the gas velocity through the throat used to accelerate and atomize the liquid. Also, the pressure drop is directly proportional to the liquid-to-gas ratio. This equation does not take into account the relatively small dry frictional energy losses of the gas stream passing through the restricted throat.

The static pressure drop for most gas-atomized scrubbers on hazardous waste incinerators is in the range of 25 to 60 inches W.C. However, some units operate at pressure drops as high as 100 inches W.C. [Anderson, 1984]. The liquid-to-gas ratios for most commercial gas-atomized scrubbers is in the range of 4 to 15 gallons per thousand ACFM. At liquor rates less than 4 gallons per thousand, efficiency drops off rapidly due to an insufficient number of liquor droplet "targets" in the throat. This can be a problem in relatively arid areas where make-up water is limited and there is a need to have high purge rates to control solids levels in the recirculated liquor. The scrubber system efficiency decreases at high liquor flow rates due to a change in the droplet size distribution formed in the scrubber. A liquid-to-gas ratio of 10 gallons per thousand ACFM is generally considered optimal.

Due to the complexity of most wet scrubber systems, numerous instruments are necessary to monitor performance. Table 12.10 summarizes the categories of instruments usually necessary and the types of units used.

Table 12.10 Gas-Atomized Scrubber System Instruments

Vessel	Parameter Measured	Process/Equipment Controlled	Portable Instrument Port or Sampling Tap
Quench	Inlet gas temperature	Emergency quench	—
	Outlet gas temperature	Emergency quench	Yes
	Makeup water flow rate	—	—
	Makeup water pressure	Incinerator trip	—
	Emergency water pressure	Incinerator trip	—
	Recirculation liquor flow	—	—
	Recirculation liquor pH	Alkali feed rate	Yes
	Inlet static pressure	Induced draft fan, gas recirc. damper	—
Venturi	Pressure drop	Adjustable throat	Yes
	Recirculation liquor flow	—	—
	Liquor inlet header pressures	—	—
	Recirculation liquor pH	Alkali feed rate	Yes
Demister	Pressure drop	Flush water sprays	Yes
Recirculation pumps	Discharge pressure	Emergency quench	—
Fan	Inlet gas temperature	—	Yes
	Fan motor current	—	—
	Fan vibration	Fan trip	—

Hydrosonics Scrubbers

This is a group of ejector-type scrubbers. Most of the units used for high-efficiency particulate collection use a fan as a source of motive power. However, one of the designs using a supersonic steam (or compressed air nozzle) can be used without a fan. The scrubbers consist of a cyclonic pretreatment chamber, one or more converging section nozzles for flue gas, a ring of liquor spray nozzles around the flue gas converging sections, a gas-liquor mixing section, a long contact throat, and a mist eliminator. The units rely on a combination of particle condensation growth and particle impaction. Accordingly, the relationships presented earlier between pressure drop and scrubber performance also apply to this category of scrubbers.

Two of the most common types of hydrosonics systems used are shown in Figs. 12.14 and 12.15. The unit in Fig. 12.14 is a Tandem Nozzle design and the unit in Fig. 12.15 is the SuperSub™unit having the steam or compressed air ejector nozzle. For both types of units, the gas stream from the incinerator initially enters a cyclonic chamber where the temperature is reduced to approximately the adiabatic saturation temperature. This chamber also serves as a cyclonic precleaner for the removal of large particles emitted from the incinerator.

For the unit shown in Fig. 12.15, a compressed air nozzle operating at supersonic velocities is used for the initial atomization of scrubber liquor and for the generation of suction. The flue gas and atomized liquor then pass through a subsonic nozzle. A ring of spray nozzles around the subsonic nozzle injects an additional cocurrent stream of liquor. The flue gas is then accelerated in a long throat where particle growth by condensation and particle capture by impaction occur. Water droplets are collected in a low-pressure drop cyclonic collector or in a horizontally oriented chevron demister vessel.

The compressed air requirements for the supersonic ejector unit are 0.04 to 0.05 pounds per pound of flue gas [Zink, 1988a]. For a flue gas stream of 50,000 ACFM saturated at 180°F, this is equivalent to approximately 125 to 160 pounds per minute or 1600 to 2100 SCFM. Compressor

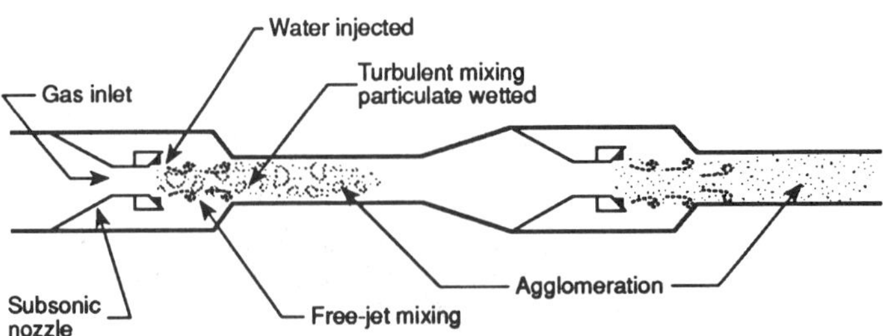

FIGURE 12.14 Tandem nozzle Hydro-Sonics scrubber. (*Source:* Holland, O. and Means, J. 1988. *Utilization of Hydro-Sonic Scrubbers for the Abatement of Emissions from Hazardous, Industrial, Municipal, and Bio-Medical Wastes.* John Zink Co., Technical Paper 7802.)

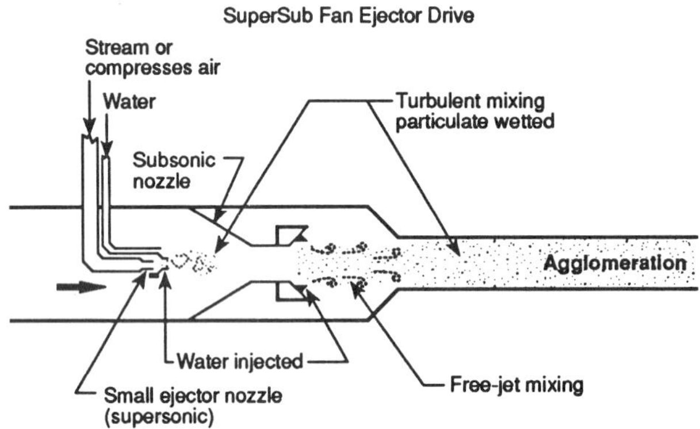

FIGURE 12.15 SuperSub Hydro-Sonics scrubber. (From Zink, 1988a. *Hydro-Sonic Systems, Gas Cleaning Equipment Tandem Nozzle—Series TN.* Bulletin HSS 0003A. John Zink Co.)

horsepower requirements based on the manufacturer's data are 0.06 to 0.07 HP per pound of flue gas per minute [Zink, 1988b]. For the 50,000 ACFM example, the horsepower required is approximately 185 to 220. The compressor must be equipped with a means of removing condensed oil so that this material is not volatilized and recondensed as nonwettable particles. If low-pressure steam is available from a waste heat boiler or other source, it can be used instead of compressed air in the nozzle. The steam requirements are typically on the order of 0.03 pounds per pound of flue gas [Zink, 1988b].

The fan horsepower requirements are a function of the static pressure drop across the entire system. The common range is 10 to 35 inches of water. The fan horsepower requirements range between 2.1 HP/1000 ACFM for a 10 inch pressure drop to 7.3 HP/1000 ACFM for a 35 inch pressure drop [Zink, 1988b].

The instrumentation requirements for a hydrosonics scrubber are similar to those for a venturi scrubber.

Ionizing Wet Scrubbers

The ionizing wet scrubber (IWS) utilizes electrostatic charges for capture of particulate matter. The initial part of the control device is an ionizer section which functions like an electrostatic precipitator. Instead of grounded precipitator collection plates within the nonuniform electric field, the ionizer scrubber uses a packed bed scrubber downstream of the electric field for particle capture. The operating principles of an ionizing wet scrubber are similar to those of a conventional electrostatic precipitator with two major exceptions. Due to the removal of collected particulate matter on wetted packing, resistivity is not a major factor. Also, the migration velocities of the charged particles are lower since the applied electric field does not extend into the packed bed. However, this is offset by the much shorter migration velocities to the collection surfaces.

The power source for an IWS consists of a standard transformer-rectifier (TR) set identical to the type used on electrostatic precipitators. A separate TR set is used for each ionizer/packed bed module in series. For incinerators, the number of modules in series can range from one to four depending on the particle size distribution, the mass emission regulatory limits, and the presence of an upstream adjustable-throat venturi section. The overall power requirements of the units range between 0.2 and 0.4 kVA per 1000 ACFM [Ceilcote, undated].

The ionizer is a set of small-diameter negatively charged wires centered between grounded metallic plates. Alignment is maintained at 3 inches plus or minus 0.25 inches to ensure maximum operating voltages. The actual operating voltage normally varies between secondary voltages of 20 and 25 kilovolts and secondary amperages of 25 to 100 milliamps [Ceilcote, undated]. The operating voltages are controlled by an automatic voltage controller which utilizes spark rate as the monitored variable. The spark rate is a function of the particle size distribution, the particulate matter loading, and the ionizer electrode alignment. The spark rate is usually maintained between 50 and 100 sparks per minute [Ceilcote, 1975].

Until the secondary voltage reaches the onset point, there is insufficient voltage to initiate a sustained corona on the negatively charged discharge electrodes. Once this voltage is exceeded, the secondary current rises rapidly as the voltage is increased. Generally, the maximum currents and voltages are a function of the spark rate set by the operator, with 50 to 100 sparks per minute being the manufacturer's recommendation for most facilities. However, the controller also has maximum primary current, secondary current, and primary voltage limits to protect the TR set. There are also undervoltage limits with a short time delay to protect against short circuits due to broken wires or failure of an insulator.

The size of the ionizing wet scrubber is based primarily on the actual gas flow rate. The cross-sectional area of the ionizer is based on the desired superficial velocity necessary to achieve adequate particle charging. The packed bed size is based primarily on the acid gas removal requirements, not the charged particle removal requirements. The turndown capability of the IWS system is very good. As long as particle size distribution and loadings remain relatively constant, the performance for both gas and particulate removal should improve as the gas flow rate is decreased.

The ionizer is cleaned using a programmable controller. During washing, the ionizer is shut down for approximately 3 minutes to prevent electrical-spark-related damage to the small-diameter discharge wires and to the TR set. Generally, cleaning is done on a 4- to 8-hour schedule. However, this depends strongly on the particulate loading.

Acid gas removal is accomplished within a packed bed immediately downstream of the ionizer. This is usually a 4-foot irrigated bed of 2-inch-diameter Tellerette packing. A set of sprays is used to maintain recirculation liquor flow across the packing. An internal sump is used as part of the recirculation loop. The pH of the liquor is maintained between 6 and 8 by means of alkaline addition. The liquor recirculation rates within an IWS stage are approximately 10 gallons per 1000 ACFM. This includes the deluge water used on a routine basis to clean the electrodes. Make-up water requirements are generally in the range of 2 gallons per 1000 ACFM per stage.

Packed Bed and Tray Tower Scrubbers

Hydrogen chloride, hydrogen fluoride, and sulfur dioxide are the main pollutants collected in packed bed and tray tower scrubbers. These are relatively soluble gases which can be collected with high efficiencies in a variety of units. The most common type of packed bed scrubber is the vertical tower with randomly stacked packing. The counterflow arrangement inherent in the vertical tower design has a performance advantage in that the driving forces for absorption are maximized by this flow arrangement. The gas stream encounters progressively cleaner liquor as it approaches the scrubber outlet at the top of the vessel. Another advantage of this approach is that it requires little plant area. The main disadvantage is the length of the ductwork from the outlet at the top to the inlet of ground-mounted fans. The tray tower scrubbers share the advantages and limitations of the vertical packed bed scrubbers.

Another common scrubber style is the horizontal crossflow packed bed. The configuration of this unit is very compatible with rod deck and ionizing particulate wet scrubbers. Demisters installed in the exhaust ends of the horizontal vessels are also slightly more efficient than demisters in vertical towers since the collected liquor drains off without being opposed by the gas stream. The main disadvantage is the possible absorption performance problems caused by the driving force gradient across the packed bed. Removal efficiency can be very high for gas passing across the top of the bed and somewhat lower for gas passing through the bottom. This is due to the reduction in pH levels as the liquor flows downward through the bed.

Gas and vapor collection in air pollution control devices is achieved by either absorption or adsorption. Absorption is the dissolving of a soluble component into droplets or sheets of liquid. Adsorption is the physical bonding of molecules to the surface of dry particles entrained in the gas stream or contained within a bed. Some of the newer air pollution control systems use both absorption and adsorption in separate control devices arranged in series.

Absorption is the transfer of a gas or vapor phase compound into a liquid phase. In the gas of hazardous waste incinerators, the gas or vapor phase compounds include primarily hydrogen chloride, hydrogen fluoride, and sulfur dioxide. The liquid streams which receive these contaminants generally consist of recirculated liquids containing sufficient alkali to maintain the design pH for the system. Numerous chemical engineering texts and handbooks give the design procedures for absorbers. See, for example, Perry's *Chemical Engineering Handbook*, McGraw-Hill, a book which is updated at regular intervals. The key operating conditions for an absorber are the types of packing material, the liquid-to-gas ratio for the system, and the height and diameter of the absorber. Table 12.11 lists some common types of packing material.

Table 12.11 Packing Material Data

Packing Material	Size	Packing Factor ft^3/ft^2
Raschig rings, ceramic and porcelain	1.0	155
	2.0	65
	3.0	37
Raschig rings, metal	1.0	137
	2.0	57
Tellerettes	1.0	40
	2.0	20
	3.0	15
Intalox saddles, ceramic	1.0	98
	2.0	40
Intalox saddles, plastic	1.0	30
	2.0	20
	3.0	15
Glitsch ballast saddles, plastic	1.0	33
	2.0	21
	3.0	16

The pressure drop through the packed tower can be estimated using manufacturer-supplied data relating the pressure drop per foot of packing as a function of the type of packing, how it is placed in the column (random packed or stacked in a pattern), and the liquid loading rate in terms of gallons per square foot per minute. Typical static pressure drops per foot of packing are generally in the range of 0.1 to 1.50 inches of water [Schifftner and Hesketh, 1983].

Dry Scrubbing Systems

There are two basic styles of dry scrubbing systems in use for hazardous incinerators: spray atomizer systems and dry injection systems. The spray atomizer systems generally include two fluid nozzles and rotary atomizers. Spray atomizer systems utilize an evaporating alkali slurry for absorption and adsorption of acid gases. The systems generally consist of a large atomizer vessel followed by either a fabric filter or an electrostatic precipitator. Dry injection systems use a finely divided alkali solid for adsorption of the acid gases. The dry injection systems require more alkali reagent due to the lower collection efficiency inherently involved. However, the dry injection systems are less expensive and easier to operate. Both spray atomizer and dry injection dry scrubbing systems use recycle loops to increase the utilization of the alkali materials.

A number of physical processes combine in the removal of acid gases in dry scrubbing systems. Those systems using atomized liquid droplets initially have gas-phase-controlled absorption into the drops as they are beginning to evaporate. The factors which influence mass transfer rates are similar to those for absorption. These include the diffusivity of the acid gas molecule in air, the gas temperature, the liquor spray rate, the droplet size distribution, and the flue gas distribution around each of the nozzles or rotary atomizers used in the reaction vessel.

As droplet evaporation continues, the accumulating reaction products lower the droplet water vapor pressure and reduce the evaporation rate. Mass transfer of pollutants into the droplets begins to be controlled by the diffusion of water molecules through the precipitating matrix of reaction products and the undissolved reagent. Eventually, mass transfer of acid gases and volatile metals to the dried alkali particles is limited by the vapor pressures and diffusion rates of the pollutants within the drying particles. The important operating variables during this phase include the gas temperature, the size distribution of the adsorbent particles, the quantity of adsorbent available, and the residence time. These factors also limit the mass transfer rates of systems using only dry alkali particles.

Dry scrubber system vessels are designed by the equipment suppliers. The information necessary to select the most appropriate and economical unit for a specific incinerator should be based on visits to operating dry scrubbing systems, on available performance data for existing systems, and on information supplied by the suppliers. In a typical spray dryer dry scrubber the incinerator flue gas initially enters a cyclonic chamber for removal of the large particles. The cyclone outlet gas is then treated in an upflow quench reactor for removal of HCl and other acid gases.

The reagent is usually a calcium hydroxide slurry at 5 to 15% by weight atomized with compressed air [Dhargalkar and Goldbach, 1988b]. The atomizer vessel outlet gas temperature is carefully controlled to ensure that it does not approach the saturation temperature so closely that the solids are difficult to handle. The acid gas neutralization reactions in the quench reactor are shown below. The efficiency of acid gas removal is primarily a function of the stoichiometric ratio of alkali (such as calcium hydroxide) to the combined quantities of acid gas. Typical operating stoichiometric ratios are in the range of 2.5 to 3.5 moles of alkali per mole of acid gas.

The size of the atomizing vessel is based on the evaporating rates of the slurry droplets at the prevailing gas stream temperatures. Generally, the residence time is between 6 and 8 seconds. Nozzle operating pressures and spray angles are selected by the manufacturer to achieve the necessary initial droplet size populations for proper evaporation. An atomizer vessel can have one or more spray nozzles.

Calcium hydroxide is the most common alkali used since it is relatively inexpensive and easy to handle. Calcium oxide (quicklime) is less expensive. However, a lime slaker is necessary in order to prepare the atomizer feed slurry. Improper operation of the lime slaker can result in reduced effectiveness of the absorption step. Other possible alkali materials include soda ash and sodium bicarbonate.

To increase acid gas removal efficiency, an alkali dry injection system can be installed downstream of the atomizer vessel. The manufacturer of this type of system uses a mixture of waste alkali materials termed "TESISORB" for dry adsorption. It is also claimed that this material improves the dust cake properties in the downstream fabric filter used for particulate and adsorbent collection [Dhargalkar, 1988a].

12.6 Trial Burn and Compliance Test for Hazardous Waste Incinerators

This section summarizes how to design a trial burn for an incinerator or BIF. The discussion focuses on setting the operating procedures for the unit. Guidance documents on the various aspects of trial burn design are available from the EPA [EPA, 1983, 1986a, 1986b, 1989a, 1989b, 1992]. An especially important reference is *Test Methods for Evaluating Solid Waste* [EPA, 1986c], a multivolume description of the sampling and analytical methods. This reference is continuously expanded and updated to reflect the latest EPA procedures and is incorporated into the RCRA regulations by reference.

Limits on operating conditions for incinerators can be set as either an absolute or a rolling average limit. The absolute limit is based on the mean measured value of the control parameter during the trial burn. It is easy to determine and to monitor but is conservative to the point where it may not be usable for many combustors. If the absolute limit is unacceptable, one can use an hourly rolling average (HRA) as an alternative. The HRA is the average of the instantaneous measurements taken over the past hour of operation. Every minute, the average is computed by dropping the minus 60 minute measurement (counting the present at time zero) and adding the most recent measurement to the calculation.

The BIF regulations require that a maximum temperature also be set to limit the formation of metal fumes. The regulations specify that this maximum be set as the mean of the maximum HRA from each of the three runs of the trial burn or compliance test. To illustrate, consider a compliance test consisting of three runs. The maximum value of the HRA (note that this is not the same as the maximum temperature observed) during each run was 1900, 2100, and 2030°F. The limit on the maximum HRA temperature would thus be the mean of these three values, or 2010°F. The trial burn or the compliance test serves the following two main purposes:

1. It demonstrates that the combustor can meet all applicable regulations.
2. It establishes the conditions under which the combustor can meet the applicable regulations.

Assuming the combustor is capable of meeting the applicable regulations, the trial burn should use measurement methods which are adequate to demonstrate the requirements. Methods have been developed for the vast majority of measurements required by the trial burn. If these methods, and the associated quality assurance/quality control (QA/QC) procedures are adhered to, the trial burn will be capable of demonstrating compliance or noncompliance. Achievement of the second purpose requires the combustor to operate during the test at the worst-case conditions that will be encountered during normal operation. If the combustor satisfies the regulatory, health, and safety demands under these worst-case conditions, it will satisfy them under less severe operation.

Trial Burn Operating Conditions

Worst-case conditions for the combustor are defined by a series of limits (absolute or rolling, maxima or minima) on the parameters summarized in Table 12.12. Most of the parameters listed in Table 12.12 are reasonably independent of one another. Changes in one will not affect other

Table 12.12 Control Parameters for Incinerators

Group	Parameter
Group A Continuously monitored parameters are interlocked with the automatic waste feed cutoff. Interruption of waste feed is automatic when specified limits are exceeded. The parametera are applicable to all facilities.	1. Minimum temperature measured at each combustion chamber exit 2. Maximum CO emmissions measured at the stack or other appropriate location 3. Maximum flue gas flow rate or velocity measured at the stack or other appropriate location 4. Maximum pressure in PCC and SCC 5. Maximum feed rate of *each* waste type to *each* combustion chamber 6. The following as applicable to the facility: • Minimum differential pressure across particulate venturi scrubber • Minimum liquid-to-gas ratio and pH to wet scrubber • Minimum caustic feed to dry scrubber • Minimum kVA settings to ESP (wetdry) and kV for ionized wet scrubber (IWS) • Minimum pressure differential accross bag house • Minimum liquid flowrate to IWS
Group B Parameters do *not* require continuous monitoring and are thus *not* interlocked with the waste feed cutoff systems. Operating records are nevertheless required to ensure that trial burn worst-case conditions are not exceeded.	7. POHC incinerability limits 8. Maximum total halides and ash feed rate to the incinerator system 9. Maximum size of batches or containerized waste 10. Minimum particulate scrubber blowdown or total solids content of the scrubber liquid
Group C Limits on these parameters are set independently of trial burn test conditions. Instead, limits are based on equipment manufacturers' design and operating specifications and are thus considered good operating practices. Selected parameters do *not* require continuous monitoring and are *not* interlocked with the waste feed cutoff.	11. Minimum/maximum nozzle pressure to scrubber 12. Maximum total heat input capacity for each chamber 13. Liquid injections chamber burner settings: • Maximum viscosity of pumped waste • Maximum burner turndown • Minimum atomization fluid pressure • Minimum waste heating value (only applicable when a given waste provides 100% heat input to a given combustion chamber) 14. APCE inlet gas temperature

Items 5 and 9 are closely related; therefore these are discussed under group A parameters.

Item 14 can be a group B or C parameter. See text.

Source: EPA, 1989. Environmental Protection Agency, Center for Environmental Research Information. *Handbook, Guidance on Setting Permit Conditions and Reporting Trial Burn Results,* Volume II of the Hazardous Waste Incineration Guidance Series. EPA/625/6-89/019.

parameters to a significant extent, so they can be set as extremes fairly easily. But, as combustion calculations show, the following parameters are highly interdependent:

1. Primary and secondary chamber (PCC, SCC) temperatures
2. Flue gas velocity
3. Waste feed rates
4. Waste composition
5. Oxygen or excess air—not always a permit condition per se, but set by some regulatory agencies

Their values must be set by a series of iterative combustion calculations to find the waste compositions, waste feed rates, and air flows that result in the desired temperature and flue gas flow rates. A great deal of combustion calculation can result if they are not set in an orderly process.

The first step in establishing the conditions for the trial burn is to specify the temperatures of each of the combustion chambers. In a multichamber system, the temperature of the secondary combustion chamber is usually more critical and should be set first. Maximum temperatures are set from equipment limits and metals emission considerations. Minimum temperatures are set on the basis of operating experience or the experience of the vendor as to the minimum temperature at which a given design successfully destroyed organic constituents.

The next factor that is set is the maximum gas flow rates. The goal here is to come as close as practical to the capacity of the fans, ducts, and air pollution control equipment. The maximum gas flow rate that is desired during actual operation, which may be lower than the theoretical maximum based on equipment capacity, should be the goal. The permit will place an upper limit on this value; maximizing it will thus give the operators as much flexibility as possible.

Waste feed rates for the trial burn should also be considered as a relatively inflexible condition. The permit will set an upper limit on the feed rate of each waste category to each combustion chamber. A common mistake in many permit applications is to overly categorize the waste. Normally, the following waste categorization should prove adequate:

1. High heating value (or Btu) liquid waste (greater than 5000 Btu/lb)
2. Low heating value liquid wastes (less than 5000 Btu/lb)
3. Solid wastes

The values are maximized by varying the composition of each waste stream. Typically one does this by setting the type and amount of POHCs required, then adding ash (as a soil or as a flyash, for example). Heating values are increased by then adding fuel oil. They are decreased by replacing a portion of the fuel oil with an oxygenated fuel such as methanol or by adding water. Combustion calculations are useful for finding the waste compositions and quantities which give the required temperatures, halogen inputs, and gas flow rates.

It is necessary that the operating conditions during the trial burn include the desired maximum or minimum values for the Group A and B control parameters. EPA [1989b] specifies that the permit condition on PCC and SCC exit gas temperatures be set at the mean temperature measured during the trial burn. Assume that the mean during a trial burn was 1800±100°F. If the permit condition were determined by the first, simpler, condition, the absolute minimum operating temperature specified by the permit condition would be 1800°F. If one desired that the incinerator be allowed to burn hazardous waste at a mean temperature of 1800°F with a variation of ±100°F, it would be necessary to conduct the trial burn at a lower mean temperature, 1700°F. Then during operation, the normal 100°F temperature fluctuation would not result in frequent waste feed cutoff. Similarly, the trial burn would have to be conducted at somewhat lower excess air (oxygen concentration), higher gas flow rate, higher waste feed rates, and so on than the minimum or maximum (as the case may be) during operation.

POHC Selection—Incinerability Ranking

Because of the wide range of organic compounds present in most wastes it is impossible to test for each one. The approach taken is to select the most difficult to destroy compounds (POHCs) that will occur in the waste and demonstrate during the trial burn that they can be properly destroyed under the worst-case operating conditions of the system. The POHCs are chosen on the basis of the types of organic hazardous compounds that will be in the waste and incinerability ranking or rankings that are most suited for this application.

The POHC selection process begins with examination of the waste stream that will be burned during operation and identification of those hazardous organic compounds which occur in significant quantities. There is no specific guidance available at present to specify what constitutes a significant quantity. The decision must be made on a case-by-case basis, evaluating anticipated concentrations of organic compounds, their potential impact on health or the environment, or the public concern they generate. Those organic compounds so identified are then grouped into categories such as aliphatics, aromatics, and chlorinated aliphatics and the POHCs selected from the most refractory (as established by an appropriate incinerability ranking) compound from each category. This procedure establishes the POHCs as representing the worst-case conditions of organic hazardous compounds in the waste feed.

A point in POHC selection is summarized by the glib statement "Don't choose a POHC that's a PIC." Certain compounds such as chloroform, carbon tetrachloride, and methane can be normal products of combustion. Their formation appears to be dominated by a quasi-equilibrium so that their presence cannot be reduced without radically changing the incinerator. The reader is referred to Dellinger *et al.* [1988] for information that could be used to identify which compounds could be PICs and to EPA [1986b] for further guidance on the limits of measurement methods to assist in determining the amount of POHC that should be fed to an incinerator. If a given compound forms in the incinerator, its presence in the stack would decrease its apparent DRE. As a result, such compounds have been avoided in the past when possible. If, due to sampling and analysis, compound availability, or other constraints, a POHC must be selected which is also a PIC, the compound should be spiked up to levels high enough to override the "PIC effect" on DRE.

The amount of each POHC that must be fed to the incinerator during a trial burn is determined by the sensitivity of the measurement method that will be used. The quantity should be sufficient so the measurement method used can show 99.99% (or 99.9999% for PCB or dioxin-listed wastes) DRE. The amount fed must be greater than 10^4 (10^6 for PCB and dioxin-listed wastes) times the method's detection limit to ensure that the test does indeed show greater than 99.99% (99.9999% for PCB or dioxin-listed wastes) DRE.

For example, if the sampling and analytical method used to measure the emissions of a given POHC has a detection limit of 10 grams emitted from the stack, then one must feed a minimum of 10×10^4 g or 100 kg of that POHC for each run of the trial burn to establish a DRE of 99.99% or 10×10^6 mg (1000 kg) to establish 99.9999%. Note that, since the sampling train removes only a small fraction of the total gas emitted, the determination must be based on the minimum sensitivity of the combined sampling and analytical methods for measuring the incinerator's total emissions, not just on the sensitivity of the analytical methods.

Incinerability ranking is a concept that was developed to identify those organic compounds which are the most difficult to destroy. The incinerability ranking allows the incinerator operator the flexibility to burn other wastes which are less difficult to destroy than those tested. Without this ability, incinerator operation might be limited to only those specific compounds that were burned during the trial burn. The RCRA regulations [40 CFR §170.62(b)(4)] require that the "Director will specify as trial Principal Organic Hazardous Constituents (POHCs).... based on his estimate of the difficulty of incineration of the constituents identified in the waste analysis, their concentration or mass in the waste feed, and for wastes listed in Part 261, Subpart D (listed wastes) the hazardous waste organic constituent or constituents identified in Appendix VII of that part as the basis for listing."

A number of incinerability rankings have been proposed [Dellinger *et al.*, 1986] and any of them may be appropriate for a given application. Each ranking strives to correlate a measurable property of the compound to its "incinerability"—how readily it is destroyed in an incinerator. Each ranking is based on a different property, such as the compound's heat of combustion or how readily it is destroyed under substoichiometric oxygen conditions. The difficulty in using such a ranking lies in the fact that while each ranking method can be tied to a specific destruction mechanism, any properly operating incinerator subjects the waste to a combination of destruction mechanisms. As a result, each ranking system lists specific compounds in somewhat different order. There is no single recommended ranking system at present, but virtually all incinerator and BIF tests are conducted on POHCs selected on the basis of both the heat of combustion ranking [EPA, 1983] and the thermal stability ranking [EPA, 1992]. The recommended procedure in the selection process is as follows:

1. Examine the wastes that will be incinerated during actual operation and identify those compounds that are likely to be present in significant amounts.
2. Classify the compounds that will be burned into broad categories such as aliphatics, aromatics, chlorinated aliphatics, and chlorinated aromatics.
3. Select the POHCs by choosing at least one representative compound from each category. The compound should be the most difficult to destroy of those present within the category by the above two incinerability ranking schemes. Until guidance on the subject has been published, a conservative approach would be to use compounds that are difficult to destroy according to both ranking schemes.

General Waste Stream Composition

Having identified the POHCs, the next step in the trial burn design is to establish the overall composition of the waste streams that will be burned during the trial burn. Here again, the fact that the trial burn must represent the worst-case operating conditions should be kept in mind. Even "omnivores" have ranges of acceptable waste which are achieved by an extensive blending program, mixing wastes of different property to achieve the desired ranges of heating value, viscosity, size of waste particles, ash content, metals content, halogen content, water content, and other parameters that may be important for a specific application.

The waste streams for the test are prepared by blending the major components into a base material. If the POHCs or metal spikes are soluble and otherwise compatible in the base material, they can be blended into the waste stream so produced. In many cases, blending is not possible. The spikes may be immiscible with the base material, they may be chemically incompatible, or the act of blending may cause them to evaporate prior to being fed to the combustor. In those cases, the spike can be added as a side stream to the base material just prior to injection into the combustor. For example, solid waste streams can be spiked by adding jars or sealed plastic bags of the POHC or metal spikes to the solid just before it is fed to the combustor. Organic streams (of POHCs, for example) which are immiscible in a liquid stream may be metered into the liquid waste feed line for the combustor.

The waste streams burned during the test should have the worst-case properties of each of the above parameters, and they should be fired at maximum rates and at maximum sizes of charges or containers. Liquid wastes for the test can be created by mixing individual components to achieve the desired blend of properties. Solid waste streams are more difficult to mimic. In general, all waste streams that contain combustible material should include combustibles for the trial burn.

Defining Terms

Destruction and removal efficiency (DRE): The percentage of a material that is released from the stack versus that which is fed to the combustor. For example, 99.9999% DRE means 1 mg of a material emitted from the stack per kg of the same material fed to the combustor.

Environmental Protection Agency (EPA): While the term refers to the U.S. Environmental Protection Agency, it normally encompasses all appropriate regulatory agencies that have jurisdiction.

Products of incomplete combustion (PICs): Organic compounds which form in an incinerator or combustor and which are released to the atmosphere.

References

Anderson, 1984. *Anderson 200 Inc. Scrubbing and Filtration Systems to Control Gaseous and Particulate Emissions from Hazardous Waste Incinerators.* Bulletin TR82-9000145. Anderson 2000, Inc.

ASME, 1988. *Hazardous Waste Incineration, A Resource Document.* American Society of Mechanical Engineers, New York.

Barton, R. G., Maly, P. M., Clark, W. D., Seeker, W. R., and Lanier, W. S. 1988. Prediction of the fate of toxic metals in hazardous waste incinerators. Energy and Environmental Research Corporation, Final Report, pp. 12–191.

Battelle, 1972. *Fluidized-Bed Incineration of Selected Carbonaceous Industrial Wastes.* Battelle Columbus Laboratories, final report prepared under grant #12120 FYF for the state of Ohio Department of Natural Resources.

Bonner, T. A. 1981. *Engineering Handbook for Hazardous Waste Incineration.* EPA-SW-889 (NTISPB81-248163).

Bostian, H. E., Crumpler, E. P., Palazzolo, M. A., Barnett, K. W., and Dykes, R. M. 1988. Emissions of metals and organics from four municipal wastewater sludge incinerators—Preliminary data. Paper presented at the Conference on Municipal Sewage Treatment Plant Sludge Management, Palm Beach, FL, June 28–30.

Bostian, H. E. and Crumpler, E. P. 1989. Metals and organic emissions at four municipal wastewater sludge incinerators. Portion of paper presented on waste incineration presented at meeting of Pacific Basin Consortium for Hazardous Waste Management, Singapore, April 3–6.

Brady, J. 1982. Understanding venturi scrubbers for air pollution control. *Plant Engineering,* September 30.

Bruce, K. R., Beach, L. O., and Gullett, B. K. 1990. The role of gas-phase Cl_2 in the formation of PCDD/PCDF in municipal and hazardous waste combustion. *Proc. Univ. Calif. Irvine Incineration Conf.*, San Diego, CA, May 14–18.

Calvert, S., Richards, J., and Jhaveri, N. 1973. *Feasibility of Flux Force/Condensation Scrubbing for Fine Particulate Collection.* Environmental Protection Technology Series. U.S. EPA Report EPA-650/2-73-036.

Calvert, S. and Jhaveri, N. 1974. Flux force/condensation scrubbing. *J. Air Pollut. Control Assoc.* 24 (10): 946–951.

Ceilcote. Undated. Bulletin 12-19. Ceilcote Co.

Ceilcote. 1975. Particle charging aids wet scrubber's submicron efficiency. *Chemical Engineering,* July 21. Ceilcote Co.

Danielson, J. A. (ed.). 1973. *Air Pollution Engineering Manual,* AP-40, second edition. Environmental Protection Agency, Office of Air Quality Planning and Standards, Research Triangle Park, NC.

Dhargalkar, P. 1988. Control of emissions from municipal solid waste incinerators. Paper presented at the Conference on Incineration of Wastes sponsored by the New England Section of the Air Pollution Control Assocation, April 12–13.

Dhargalkar, P. and Goldbach, K. 1988. Control of heavy metal emissions from waste incinerators. Paper presented at the NATO Advanced Research Workshop, Control and Fate of Atmospheric Heavy Metals, Oslo, Norway, September 12–16.

Dellinger, B., Rubey, W., Hall, D., and Graham, J. 1986. Incinerability of hazardous waste and hazardous materials. *Hazardous Waste and Hazardous Materials* 3(2):139–150.

Dellinger, B., Tirey, D. A., Taylor, P. H., Pan, J., and Lee, C. C. 1988. Products of incomplete combustion from the high temperature pyrolysis of chlorinated methanes. *Proc. 3rd Chemical Congr. North America 195th Nat. Meet. Am. Chem. Soc.*

EPA, 1971. Environmental Protection Agency. *Fluid Bed Incineration of Petroleum Refinery Wastes.* Prepared by American Oil Company, Mandane Refinery, Mandane, ND, Superintendent of Documents Number 5501-0052.

EPA, 1980. Environmental Protection Agency, Air Pollution Training Institute Course #427. *Combustion Evaluation.* Student Manual EPA 450/2-80-063, Student Workbook EPA 450/2-80-063.

EPA, 1983. Environmental Protection Agency. *Guidance Manual for Hazardous Waste Incinerator Permits.* Mitre Corp., NTIS PB84-100577.

EPA, 1986a. Environmental Protection Agency. *Handbook, Permit Writer's Guide to Test Burn Data.* Hazardous Waste Incineration Series, EPA/625/6-86/012.

EPA, 1986b. Environmental Protection Agency. *Practical Guide—Trial Burns for Hazardous Waste Incinerators.* Midwest Research Institute. EPA Publication No. 600/2-86-050.

EPA, 1986c (updated). Environmental Protection Agency. *Test Methods for Evaluating Solid Waste. Physical/Chemical Method,* SW-846, Third Edition. Office of Solid Waste and Emergency, Washington, D.C.

EPA, 1987. Environmental Protection Agency. *Background Information Document for the Development of Regulations to Control the Burning of Hazardous Wastes in Boilers and Industrial Furnaces,* Boilers, Volume 1, Industrial Furnaces, Volume II. FA035/50A-E, Final Report Submitted by Engineering-Science.

EPA, 1989a. Environmental Protection Agency. *Hazardous Waste Incineration Measurement Guidance Manual,* Volume III of the Hazardous Waste Incineration Guidance Series. EPA/625/6-89/021.

EPA, 1989b. Environmental Protection Agency, Center for Environmental Research Information. *Handbook, Guidance on Setting Permit Conditions and Reporting Trial Burn Results,* Volume II of the Hazardous Waste Incineration Guidance Series. EPA/625/6-89/019.

EPA, 1989c. Environmental Protection Agency. *Guidance on Metal and HCl Controls for Hazardous Waste Incinerators,* Volume IV of the Hazardous Waste Incineration Guidance Series. EPA, Office of Solid Waste.

EPA, 1990. Environmental Protection Agency. *Methodology for Assessing Health Risks Associated with Indirect Exposure to Combustor Emissions.* Office of Health and Environmental Assessment, EPA/600/6-90/003.

EPA, 1991. Environmental Protection Agency. *Federal Register.* Vol. 56, No. 35, Feb. 21, 1991, p. 7134; Vol. 56, No. 166, August 27, 1991, p. 425042.

EPA, 1992. Environmental Protection Agency. *Technical Implementation Document for EPA's Boiler and Industrial Furnace Regulations.* EPA530-R-92-011, NTIS No. PVB92-154 947.

Goldstein, H. L. and Siegmund, C. W. 1976. Influence of heavy fuel oil composition and boiler combustion conditions on particulate emissions. *Environmental Science and* (12):1109, 12–181.

Hinshaw, G. D., Klamm, S. W., Huffman, G. L., and Lin, P. C. L. 1990. Sorption and desorption of POHCs and PICs in a full-scale boiler under sooting conditions. In *Proc. 16th Ann. Res. Symp. Remedial Action, Treatment and Disposal of Hazardous Waste,* Cincinnati, OH, April 3–5; EPA/600/9-90 037, August.

Holland, O. and Means, J. 1988. *Utilization of Hydro-Sonic Scrubbers for the Abatement of Emissions from Hazardous, Industrial, Municipal, and Bio-Medical Wastes.* John Zink Co., Technical Paper 7802.

Niessen, W. R. 1978. *Combustion and Incineration Processes.* Marcel Dekker, New York.

PEI. 1986. PEI Associates, Inc. *Operation and Maintenance Manual for Fabric Filters.* EPA Report EPA-625/1-86-020.

Petersen, H. H. 1984. Electrostatic precipitators for resource recovery plants. *Conf. Am. Soc. Mechanical Engineers,* Orlando, Florida, pp. 12–171.

Richards, J. and Segall, R. 1985. *Inspection Techniques for Evaluation of Air Pollution Control Equipment,* Vol. II. EPA Report 340/1-85-022b.

Richards, J. and Quarles, P. 1986. *Fabric Filter Malfunction Evaluation.* Final report for EPA contract 68-02-3960, work assignment 3-131.

Schifftner, K. and Hesketh, H. 1983. *Wet Scrubbers.* The Environment and Energy Handbook Series. Ann Arbor Science Publishers, Ann Arbor, MI.

Schifftner, K. 1989. Wet scrubbers for air pollution control. Paper presented at the Regional Conference Series on Incineration of Medical Wastes. Orlando, Florida, October 10–11.

Theodore, L. and Reynolds, J. 1987. *Introduction to Hazardous Waste Incineration.* John Wiley & Sons, New York.

Trenholm, A., Gorman, P., and Jungclaus, G. 1984. *Performance Evaluation of Full-Scale Hazardous Waste Incinerators.* EPA-600/2-84-181, PB85 129518.

Western Precipitation. Undated. *Electrostatic Precipitator Operating Manual.*

Wool, M., Castaldini, C., and Lips, H. 1989. *Engineering Assessment Report: Hazardous Waste Cofiring in Industrial Boilers under Nonsteady Operating Conditions.* Acurex Draft Report TR-86-103/ESD, EPA Risk Reduction Engineering Laboratory, Cincinnati, Ohio.

Yung, S., *et al.* 1977. *Venturi Scrubber Performance Model.* EPA publication EPA-600/2-77-172.

Zink, 1988a. *Hydro-Sonic Systems, Gas Cleaning Equipment Tandem Nozzle—Series TN.* Bulletin HSS 0003A. John Zink Co.

Zink, 1988b. *Hydro-Sonic Systems, Gas Cleaning Equipment SuperSub Series.* Bulletin HSS 0004A. John Zink Co.

13

Solid Waste/Landfills

Vasiliki Keramida

Keramida Environmental, Inc.

13.1 Introduction

The proper management of solid waste is now, more than ever, a matter of national and international concern. As a nation, we are generating more solid waste than ever before. At the same time, we are finding that there are limitations to traditional solid waste management practices. As the generation of solid waste continues to increase, the capacity to handle it is decreasing. Many landfills and incinerators have closed, and new disposal facilities are often difficult to cite.

Even though municipal solid waste (MSW) constitutes only a portion of the solid waste stream, the rate of its generation is staggering. The U.S. Environmental Protection Agency's (EPA) most recent data show that in 1988, 180 million tons, or 4.0 pounds per person per day of MSW, were generated in the U.S. [EPA, 1990a]. By the year 2000, generation of MSW is projected by the EPA to reach 216 million tons, or 4.4 pounds per person per day. Based on current trends and information, EPA anticipates that 20 to 28% of MSW will be recovered annually by 1995. Exceeding this projected range will require fundamental changes in government programs, technology, and corporate and consumer behavior. According to EPA data [EPA, 1990a], recovery of MSW materials for recycling and composting was 13% in 1988, combustion was 14% of total generation, and the remaining 73% of the MSW stream was taken to landfills.

In response to the growing national concern about the solid waste disposal crisis, EPA developed an "agenda for action" and a national strategy for addressing the MSW management problems [EPA, 1989a]. The cornerstone of the strategy is "integrated waste management," where source reduction (i.e., reduction of the quantity and toxicity of materials and products entering the solid waste stream) followed by recycling are the first steps of an effective solid waste management system, and are complemented by environmentally sound combustion and landfilling.

13.2 Solid Waste

Regulatory Framework

Solid waste is regulated under Subtitle D of the Resource Conservation and Recovery Act (RCRA) and the corresponding federal regulations found in 40 Code of Federal Regulations (CFR) Parts

0-8493-8953-4/95/$0.00 + $.50

257 and 258. Subtitle D solid waste is not subject to the hazardous waste regulations under Subtitle C of RCRA. Solid waste is defined in 40 CFR 257 as "any garbage, refuse, sludge from waste treatment plant, water supply treatment plant, or air pollution control facility and other discarded material, including solid, liquid, semisolid, or contained gaseous material resulting from industrial, commercial, mining, and agricultural operations, and from community activities."

Household hazardous wastes and hazardous small quantity generator (SQG) wastes from businesses and industry generating less than 100 kilograms of hazardous waste per month are exempt from RCRA's Subtitle C regulations for hazardous waste and thus are regulated as solid waste under RCRA's Subtitle D. In accordance with the RCRA definition of solid waste, the following categories of Subtitle D solid waste have been identified by EPA:

- Municipal solid waste
- Household hazardous waste
- Municipal sludge
- Municipal waste combustion ash
- Industrial nonhazardous process waste
- Small quantity generator hazardous waste
- Agricultural waste
- Oil and gas waste
- Mining waste

Subtitle D of RCRA establishes a framework of federal, state, and local government cooperation in controlling the management of nonhazardous solid waste. The federal role in this arrangement is to establish the overall regulatory direction, by providing minimum nationwide standards for protecting human health and the environment from the disposal practices of solid waste. The actual planning, direct implementation, and enforcement of solid waste programs under Subtitle D, however, remain largely state and local functions.

Solid Waste Characteristics

To analyze the characteristics of solid waste, EPA has conducted numerous studies to determine the weight, volume, characteristics, and management methods of wastes regulated under Subtitle D of RCRA [EPA, 1990a; EPA, 1990b; EPA, 1989a; EPA, 1988a; EPA, 1986]. These studies revealed that based on data collected up to 1988, more than 11 billion tons of solid waste are generated each year, including 7.6 billion tons of industrial nonhazardous waste, which includes about 56 million tons of electric utility waste and 240 million tons of solid wastes generated by three industrial categories: iron and steel, inorganic chemicals, and plastics and resins. The studies also recorded 2 to 3 billion tons of oil and gas waste (including both drilling and produced wastes), more than 1.4 billion tons of mining waste, and nearly 180 million tons of municipal solid waste.

The Subtitle D solid waste regulations in 40 CFR Parts 257 and 258 focus on municipal solid waste, including household hazardous waste and small quantity generator hazardous waste, as well as industrial nonhazardous process wastes and municipal sludge. Several Subtitle D wastes, in particular oil and gas wastes, utility wastes, and mining wastes, are being considered separately for rule making by EPA. In addition, EPA has been closely evaluating, in a separate effort, the characteristics and management practices for municipal waste combustion ash.

Municipal Solid Waste

EPA's definition of municipal solid waste states that MSW comes from residential, commercial, institutional, and industrial sources and includes durable goods, nondurable goods, containers and packaging, food waste, yard wastes, and miscellaneous inorganic wastes [EPA, 1990a; EPA, 1989a]. Examples of wastes from these categories include appliances, newspapers, clothing, food

Table 13.1 Materials Generated in Municipal Solid Waste (MSW) by Weight, 1988

Material	Weight Generated (in Million Tons)	Percent of Total MSW
Paper and paperboard	71.8	40.0
Glass	12.5	7.0
Metals		
Ferrous	11.6	6.5
Aluminum	2.5	1.4
Other nonferrous	1.1	0.6
Plastics	14.4	8.0
Rubber and leather	4.6	2.5
Textiles	3.9	2.2
Wood	6.5	3.6
Food wastes	13.2	7.4
Yard wastes	31.6	17.6
Other	5.8	3.1
Total MSW	179.5	100

scraps, boxes, disposable diapers, disposable tableware, office and classroom paper, wood pallets, and cafeteria wastes.

Generation of MSW in 1988, the latest year for which information is available, totaled approximately 180 million tons [EPA, 1990a]. The EPA 1990 MSW characterization report provides detailed information on the generation of MSW and other projections for its future production [EPA, 1990a]. Table 13.1 provides a breakdown by weight of the materials generated in MSW in 1988. Paper and paperboard products were the largest component of MSW by weight (40%) and yard wastes were the second largest component (about 18%).

The various materials in MSW make up the many individual products that enter the MSW stream. The products generated in MSW by weight, grouped into major product categories, are

Table 13.2 Products Generated in Municipal Solid Waste (MSW) by Weight, 1988

Product	Weight Generated (in Million Tons)	Percent of Total MSW
Containers/packaging (boxes, bottles, cans, bags, etc., made of glass, steel, aluminum, paper, and plastic)	56.8	31.6
Nondurable goods (newspapers, office paper, disposable tableware, diapers, books/printed material, clothing)	50.4	28.1
Yard wastes (grass, leaves, etc.)	31.6	17.6
Durable goods (appliances, furniture, tires, batteries, electronics)	24.9	13.9
Food wastes	13.2	7.4
Other (stones, concrete, dirt, demolition, etc.)	2.7	1.4
Total MSW	179.5	100

shown in Table 13.2 for 1988. Containers and packaging, including all types of packaging materials, were the largest product category generated in MSW by weight (about 32%). Nondurable goods, such as newspapers, disposable diapers, and disposable tableware, were the second largest category by weight (28%).

Although solid waste is usually characterized by weight, information about volume is important for such issues as determining landfill capacity uptake. Volume estimates of MSW, however, are far more difficult to make than weight estimates. Wide ranges for the volume occupied by solid waste are reported in the literature [Salvato, 1972; Bond *et al.*, 1973]. Loose refuse can weigh from 100 to 240 lb/yd^3, while refuse compacted in a landfill can weigh 700 to 1250 lb/yd^3, depending on the compaction applied. EPA has attempted to estimate the volume of materials as they would typically be found in a landfill, after a significant amount of compaction [EPA, 1990a]. These estimates were largely based on empirical data that were used to estimate density factors (pounds per cubic yard) with corroboration from actual landfill studies. Table 13.3 compares 1988 volume and weight figures for discarded materials in MSW as estimated by EPA. Discarded materials constitute the MSW remaining after recovery for recycling and composting has taken place. The paper and paperboard category ranked first in both volume and weight, at about 34% of the discarded MSW for the 1988 data. Plastics ranked second in volume at 20%, while yard waste ranked second in weight at 20%. Four materials constitute significantly larger proportions by volume of the discarded MSW than by weight: plastics, rubber and leather, textiles, and aluminum. By contrast, three materials constitute significantly smaller proportions by volume than by weight: yard wastes, food, and glass. The remaining four materials, namely paper and paperboard, ferrous metals, wood, and miscellaneous wastes, make up almost the same portion of the discarded MSW, either by weight or volume.

Trends in Municipal Solid Waste Generation

Generation of municipal solid waste grew steadily between 1960 and 1988, from 88 million to almost 180 million tons per year. The per capita generation of MSW for the same period saw an increase from 2.7 to 4.0 pounds per person per day. By 2000, projected per capita MSW generation is 4.4 pounds per person per day, for a total of 216 million tons per year. The projection for MSW generation in the year 2010 is over 250 million tons, or approximately 4.9 pounds per person per day [EPA, 1990a], marking an increase of more than 21% over the 1988 MSW generation.

Table 13.3 Weight and Volume of Materials Discarded in Municipal Solid Waste (MSW), 1988

Material	Solid Waste Weight Generated (in Million Tons)	Solid Waste Weight Recovered by Recycling/ Composting (in Million Tons)	Solid Waste Weight Discarded (in Million Tons)	Solid Waste Weight (% of Total Discarded MSW)	Solid Waste Volume-Compacted in Landfill (% of Total Discarded MSW)
Paper and paperboard	71.8	18.4	53.4	34.2	34.1
Glass	12.5	1.5	11.0	7.1	2.0
Ferrous metals	11.6	0.7	10.9	7.0	9.8
Aluminum	2.5	0.8	1.7	1.1	2.3
Plastics	14.4	0.2	14.2	9.2	19.9
Rubber and leather	4.6	0.1	4.5	2.9	6.4
Textiles	3.9	0.0	3.9	2.5	5.3
Wood	6.5	0.0	6.5	4.2	4.1
Food wastes	13.2	0.0	13.2	8.5	3.3
Yard wastes	31.6	0.5	31.1	19.9	10.3
Miscellaneous	6.9	1.4	5.5	3.6	2.5
Total	179.5	23.6	155.9	100.0	100.0

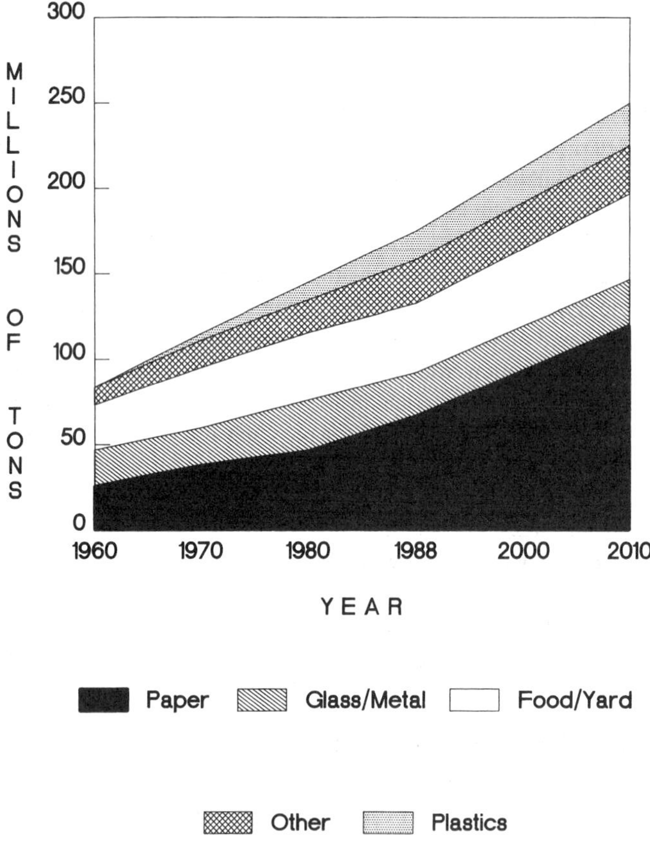

FIGURE 13.1 U.S. municipal solid waste generation, 1960–2010.

Table 13.4 Projected Per Capita Generation of Municipal Solid Waste (MSW), by Material, 1988 to 2010 (in Pounds per Person per Day)

Material	1988	1995	2000	2010
Paper and paperboard	1.60	1.80	1.96	2.35
Glass	0.28	0.23	0.21	0.18
Metals	0.34	0.34	0.35	0.34
Plastics	0.32	0.39	0.43	0.50
Rubber and leather	0.10	0.10	0.11	0.11
Textiles	0.09	0.09	0.09	0.09
Wood	0.14	0.16	0.17	0.20
Other	0.07	0.06	0.06	0.06
Food wastes	0.29	0.28	0.27	0.27
Yard wastes	0.70	0.70	0.70	0.70
Miscellaneous	0.06	0.06	0.06	0.06
Inorganic wastes				
Total MSW generated	4.00	4.21	4.41	4.86

Source: U.S. EPA, OSWER. 1990a. *Characterization of Municipal Solid Waste in the United States: 1990 Update,* p. 61. EPA/530-SW-90-042.

Figure 13.1 shows the generation, in millions of tons, of materials in MSW between 1960 and 1988, with projections to 2010. Paper and plastics have shown the most remarkable increase over the years and a continued increase in their generation is projected to the year 2010. Food and yard wastes as well as glass and metal, on the other hand, have had only a nominal increase over the period 1960 to 1988, and no marked increased is projected for the generation of these wastes.

Some further insight into projected generation of materials in MSW can be gained from Table 13.4, which presents detailed projected per capita generation of MSW by material category. Paper and plastics are projected to grow substantially in per capita generation. Other materials are projected to decline in per capita generation or to increase only slightly.

13.3 Landfills

A landfill is an area of land or an excavation in which wastes are placed for permanent disposal. While alternative waste disposal methods, such as incineration, along with the advent of recycling, composting, and pollution prevention, are scaling back the numbers of active landfills, the engineering, construction, and operation of landfills are now more complex than ever. Driven by public pressure and subsequent regulatory requirements, landfill design and operation now have to conform to strict federal standards.

The EPA had mandated double liners, environmental controls for groundwater protection, and other design standards for hazardous waste landfills since 1984, under the Subtitle C requirements of RCRA. Since 1993, the nation's 6000 municipal solid waste landfills have been required to comply with similar strict standards under new federal regulations issued in accordance with Subtitle D requirements of RCRA.

Under the authority of Subtitle D of RCRA, EPA first promulgated the Criteria for Classification of Solid Waste Disposal Facilities and Practices (40 CFR Part 257) in 1979 [EPA, 1979]. Those Subtitle D criteria established minimum national performance standards necessary to ensure that "no reasonable probability of adverse effects on health or the environment" will result from solid waste disposal facilities. A facility that meets the criteria is classified in 40 CFR 257 as a "sanitary landfill." A facility that fails to satisfy any of the criteria is considered an "open dump." The criteria under 40 CFR 257 include general environmental performance standards addressing eight major topics: flood plains, endangered species, surface water, groundwater, land application, disease, air, and safety.

In 1984, Congress made significant modifications to Subtitle D of RCRA through the Hazardous and Solid Waste Amendments (HSWA). To fulfill its responsibilities under HSWA, EPA conducted a comprehensive study of solid waste characteristics, waste disposal practices, and environmental and public health impacts resulting from solid waste disposal. The results of this study were submitted to Congress in 1988 [EPA, 1988a]. The 1984 HSWA provisions, furthermore, required EPA to revise the criteria, then in existence, for solid waste disposal facilities that may receive household hazardous waste or hazardous waste from small quantity generators as part of the municipal solid waste stream disposed of at those facilities. As a result of these requirements, EPA promulgated the Criteria for Municipal Solid Waste Landfills, a broad new solid waste regulation [EPA, 1991a] in 1991. The new rule, which took effect in 1993, imposes minimum national criteria for **municipal solid waste landfills (MSWLFs)** in the areas of location, operation, design, groundwater monitoring and corrective action, closure and postclosure, and financial assurance. Under the new rule, states may incorporate the new federal requirements into state solid waste permitting programs and, with EPA approval, assume primary responsibility for implementing and enforcing them. "Approved" states have flexibility in applying EPA criteria, so that state-specific environmental conditions can be accommodated. In states that fail to gain approval, MSWLFs are subject to the federal criteria as specified in 40 CFR 258, with no allowance to alter the criteria for special environmental considerations.

Minimum Federal Regulatory Criteria for Municipal Solid Waste Landfills

The Subtitle D federal criteria for MSWLFs [EPA, 1991a] apply to all existing and new MSWLFs, as well as to lateral expansions, with few exceptions for some small landfills, under certain conditions. A summary of the Subtitle D minimum criteria imposed on MSWLFs [EPA, 1991a] is presented below.

Location Restrictions

There are six restricted areas. Sites located near airports, floodplains, unstable areas, wetlands, seismic impact zones, and fault areas are unsuitable for operating MSWLFs, unless the MSWLF can meet the specific criteria stated under the rule for each location.

Operating Criteria

Ten new operating criteria for MSWLFs have been mandated, to address procedures for excluding the receipt of regulated quantities of hazardous wastes; daily cover material requirements; disease vector control; explosive gases control; air criteria; public access requirements; run-on/run-off control systems; control of discharges to surface waters; liquid disposal restrictions; and recordkeeping requirements.

Design Criteria

The design criteria of the solid waste disposal rule [EPA, 1991a] have as their goal the protection of groundwater and apply to new MSWLFs and lateral expansions. The design requirements do not apply to MSWLF units in existence as of October 9, 1993. The rule provides owners of MSWLFs with two design options. These two design options are depicted in Fig. 13.2, as presented in the federal regulations (40 CFR Part 258).

The first option is a **composite liner** system, whose minimum requirements are specified in the rule (40 CFR Part 258), and is believed to ensure a protective uniform design standard for MSWLFs in all locations [EPA, 1988b]. It consists of a flexible membrane liner and a two-foot, compacted

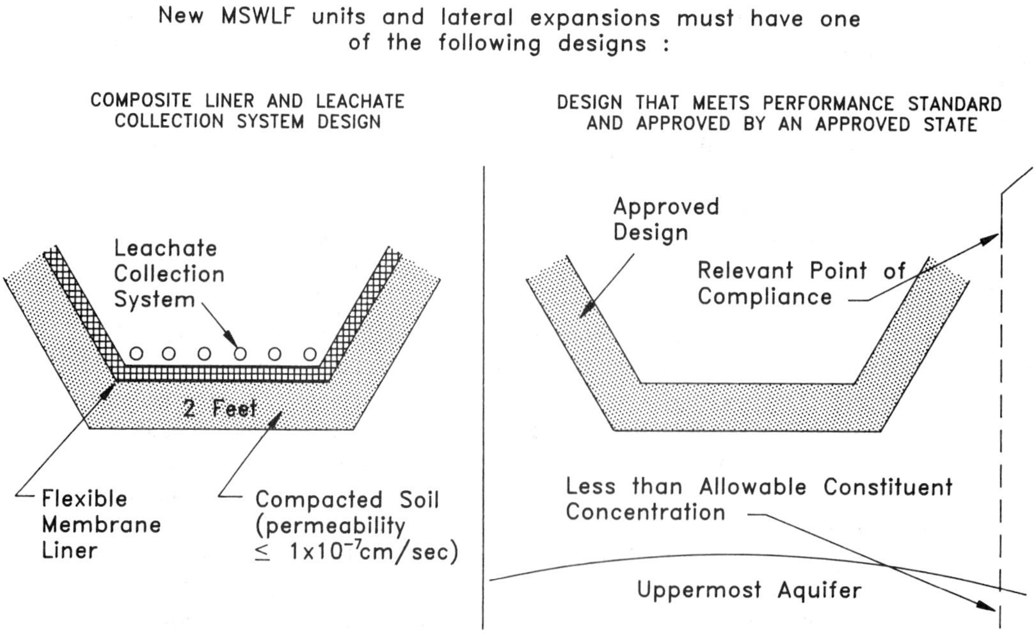

FIGURE 13.2 Municipal solid waste landfill design criteria mandated by federal regulations.

soil component, and a **leachate** collection and removal system that is designed to maintain less than a 30-cm depth of leachate over the liner.

The second option, which is available in states with an EPA-approved solid waste program, allows owners of MSWLFs to consider site-specific conditions in developing a design. This design must be approved by the state environmental agency and must meet the performance standard set in the federal regulations (40 CFR Part 258). Under the regulatory performance standard, the design of an MSWLF must ensure that the **maximum contaminant level (MCL)** values will not be exceeded in the groundwater of the **uppermost aquifer** at the **relevant point of compliance** at the landfill. Factors a state will consider when evaluating a site-specific landfill design include, at a minimum, the following: hydrogeologic characteristics of the site and surrounding land; climatic factors of the area; volume and physical and chemical characteristics of the leachate; and public health effects.

Groundwater Monitoring and Corrective Action

Extensive groundwater monitoring systems and a detection groundwater monitoring program are standard requirements for MSWLFs under the federal solid waste regulations (40 CFR Part 258). The groundwater monitoring system must consist of wells able to provide information on the quality of unaffected, background groundwater as well as of groundwater at the MSWLF's relevant point of compliance. If any of the detection monitoring parameters are found at a statistically significant level over background concentrations, the landfill owner is required to proceed to more intensive groundwater monitoring requirements and, subsequently, to investigative studies to define the extent of the groundwater contamination. If contamination has migrated off-site, the owner of the landfill has to develop and implement a corrective action plan. The remedy selected by the MSWLF owner has to be approved by the state prior to its implementation. The regulations provide explicit procedures for sampling wells and methods for the statistical analysis, and provide detailed analytical requirements for the groundwater under the various possible monitoring phases.

Closure and Postclosure Requirements

A final cover system designed to minimize infiltration and erosion is required under federal regulations (40 CFR Part 258). The infiltration layer must be a minimum of 18 inches of soil with permeability less than or equal to the permeability of the bottom liner system or natural subsoils, or no greater than 1×10^{-5} cm/s, whichever is less. The erosion layer must be a minimum of six inches of soil, able to sustain native plant growth. Written closure and postclosure plans, with provisions for maintenance of final cover, groundwater and methane gas monitoring, and leachate management are required by the regulations. Closure must commence within 30 days of last receipt of waste at the MSWLF, and postclosure activities must continue for a period of 30 years.

Financial Assurance

Under the federal solid waste regulations (40 CFR Part 258) the owner of a MSWLF is required to demonstrate financial responsibility for the cost of the landfills closure, postclosure care, and corrective action for known releases in an amount equal to the cost of a third part conducting these activities. The financial assurance requirements must be met by the owner before a construction permit is approved for the landfill by the regulatory agency. The cost estimates must be updated annually for inflation and operational changes.

Environmental Effects of Municipal Landfills

Old municipal landfills typically did not conform to today's location restrictions. In addition, several had poor operational records and minimal environmental controls. As a result of ever tightening regulations, many old landfills were literally abandoned by their owners. Today, abandoned municipal landfills compose approximately 20% of the sites on the Superfund Program's National Priorities List (NPL). Landfill sites on the NPL contain a combination of principally municipal and to a lesser extent co-disposal and hazardous waste landfills and range in size from one acre to over 600 acres. Superfund municipal landfills are primarily composed of municipal solid waste, therefore, they typically pose a low-level environmental threat rather than a principal threat [EPA, 1991b].

Potential concerns stemming from old, unprotected municipal landfills include leachate generation and possible groundwater contamination; soil contamination; landfill contents; landfill gas; and contamination of surface waters, sediments, and adjacent wetlands. Because these sites share similar characteristics, they lend themselves to remediation by similar technologies. The EPA, after considerable study of these landfills due to their high presence in the Superfund NPL, has developed methods and tools to streamline their investigation and selection of remedy process through a simplified Remedial Investigation/Feasibility Study (RI/FS) approach [EPA, 1991b].

Evaluation and selection of appropriate remedial action alternatives for Superfund municipal landfill sites is a function of a number of factors, including:

- Sources and pathways of potential risks to human health and the environment
- Potential Applicable or Relevant and Appropriate Requirements (ARARs) for the landfill as the landfill's cleanup standards under the Comprehensive Environmental Response, Compensation and Liability Act (CERCLA), commonly known as Superfund; significant ARARs might include RCRA and/or state closure requirements, and federal or state requirements pertaining to landfill gas emissions
- Waste characteristics
- Site characteristics (including surrounding area)
- Regional surface water (including wetlands) and groundwater characteristics and potential uses

Remedial Alternatives for Superfund Municipal Landfills

In general, the remedial actions implemented at most Superfund municipal landfill sites include:

- Containment of landfill contents (i.e., landfill cap)
- Remediation of hot spots
- Control and treatment of leachate
- Control and treatment of landfill gas

Other areas that may require remediation include groundwater, surface waters, sediments, and adjacent wetlands. A summary of the issues associated with the main remedial actions commonly applied to Superfund municipal landfills is presented below.

Landfill Contents

Characterization of a landfill's contents is generally not necessary because containment of the landfill contents by a cap, which is often the most practical remedial alternative, does not require such information. Certain data, however, are necessary to evaluate capping alternatives and should be collected in the field. For instance, certain landfill properties such as the fill thickness, lateral extent, and age will influence landfill settlement and gas generation rates, which will thereby have an influence on the cover type at a site.

The main purpose of a cap is to prevent vertical infiltration of surface water. Lateral migration of water or gases into and out of the landfill can be prevented by a perimeter trench-type barrier. The type of cap would likely be either a native soil cover, single-barrier cap, or composite-barrier cap. The appropriate type of cap to be considered will be based on remedial objectives for the site. For example, a soil cover may be sufficient if the primary objective is to prevent direct contact and minimize erosion. A single barrier or composite cap may be necessary where infiltration is also a significant concern. Similarly, the type of trench will be dependent on the nature of the contaminant to be contained. Impermeable trenches, such as slurry walls, may be constructed to contain liquids while permeable trenches may be used to collect gases.

Hot Spots

More extensive characterization activities and development of remedial alternatives (such as thermal treatment or stabilization) may be appropriate for known or suspected hot spots within a landfill. Hot spots consist of highly toxic and/or highly mobile material and present a potential principal threat to human health or the environment. Hot spots should be characterized if documentation or physical evidence exists to indicate the presence and approximate location of the hot spots. Hot spots may be delineated using geophysical techniques or soil gas surveys and typically are confirmed by excavating test pits or drilling exploratory borings. Excavation or treatment of hot spots is generally practicable where the waste type or mixture of wastes is in a discrete, accessible location of a landfill. A hot spot should be large enough that its remediation would significantly reduce the risk posed by the overall site, but small enough that it is reasonable to consider removal or treatment. Consolidation of hot spot materials under a landfill cap is a potential alternative in cases when treatment is not practical or necessary.

Leachate

Characterization of a site's geology and hydrogeology will affect decisions on capping options as well as on extraction and treatment systems for leachate and possibly groundwater. Although leachate quality is different in each municipal landfill, generally the variables affecting it are the age of the landfill, climate variables such as annual rainfall and ambient temperature, final cover, and factors such as permeability, depth, composition, and compaction of the waste in the landfill. New landfills typically have leachates high in biodegradable organics. As a landfill ages, its contents degrade and produce more complex organics, not so readily amenable to biodegradation, and inorganics.

Characteristics of leachate produced, as well as differences in the quality of leachate generated, by municipal, codisposal, and hazardous waste landfills have been documented [EPA, 1988c]. In general, the collected data show that although the same chemicals are routinely detected at both municipal and hazardous waste landfills, considerably higher concentrations of many chemicals are found at the leachate of hazardous waste facilities. In particular, chemicals such as 1,1,1-trichloroethane, trichloroethene, vinyl chloride, chloroform, pesticides, and PCBs occur with greater frequency and at higher concentrations in leachates at hazardous waste landfills than at municipal facilities. Typical chemical constituents in leachate from municipal landfills are shown in Tables 13.5 and 13.6 [EPA, 1988c].

Leachate generation is of special concern when investigating municipal landfill sites. The principal factors contributing to the leachate quantity are precipitation and recharge from groundwater and surface water. In many landfills, leachate is perched within the landfill contents, above the water table. Placing a limited number of leachate wells in the landfill could be an efficient way to gather information regarding the depth, thickness, and types of wastes present; the moisture content and degree of decomposition of the waste; leachate composition and head levels; and the elevation of the underlying natural soil layer. Leachate wells, in addition, provide good access for landfill gas sampling. It is important to note, however, that without extreme precautions, placing wells into the landfill contents may create health and safety risks. Such installation, furthermore, may create conduits through which leachate can migrate to lower geologic strata, thus contaminating previously nonimpacted groundwater.

Extraction and treatment of leachate may be required to control off-site migration of wastes. Collection and treatment may be necessary indefinitely because of continued contaminant loadings from the landfill. Biological processes are one possible step in the treatment of leachate, given its usually high organic matter manifested as biochemical oxygen demand (BOD). Chemical and physical processes are also applicable, as well as combinations of the three. A cost-effective alternative

Table 13.5 Municipal Landfill Leachate Data—Indicator Parameters and Inorganic Compounds

	Municipal Landfills Leachate Concentration Reported (ppm)	
Indicator Parameters	Minimum	Maximum
Alkalinity	470	57,850
Ammonia	0.39	1,200
Biological oxygen demand	7	29,200
Calcium	95.5	2,100
Chemical oxygen demand	42	50,450
Chloride	31	5,475
Fluoride	0.11	302
Iron	0.22	2,280
Phosphorus	0.29	117.18
Potassium	17.8	1,175
Sulfate	8	1,400
Sodium	12	2,574
Total dissolved solids	390	31,800
Total suspended solids	23	17,800
Total organic carbon	20	14,500
Inorganic Compounds (ppm)		
Aluminum	0.01	5.8
Antimony	0.0015	47
Arsenic	0.0002	0.982
Barium	0.08	5
Beryllium	0.001	0.01
Cadmium	0.0007	0.15
Chromium (total)	0.0005	1.9
Cobalt	0.04	0.13
Copper	0.003	2.8
Cyanide	0.004	0.3
Lead	0.005	1.6
Manganese	0.03	79
Magnesium	74	927
Mercury	0.0001	0.0098
Nickel	0.02	2.227
Vanadium	0.009	0.029
Zinc	0.03	350

Source: U.S. EPA. OSWER. 1988c. *Summary of Data on Municipal Solid Waste Landfill Leachate Characteristics.* EPA/530-SW-88-038.

to on-site leachate treatment, if available, is the discharge of the leachate into the municipal sewer system and the eventual treatment of the leachate by the city's wastewater treatment plant.

Landfill Gas

Several gases typically are generated by decomposition of organic materials in a landfill. The composition, quantity, and generation rates of the gases depend on such factors as refuse quantity and composition, placement characteristics, landfill depth, refuse moisture content, and amount of oxygen present. The principal gases generated (by volume) are carbon dioxide, methane, trace thiols, and, occasionally, hydrogen sulfide. Volatile organic compounds may also be present in landfill gases, particularly at codisposal facilities. Data generated during the landfill gas characterization should include, in addition to the landfill gas characteristics, the role of on-site and off-site surface emissions, and the geologic and hydrogeologic conditions of the site. Constructing an active landfill gas collection and treatment system should be considered where (1) existing or planned homes or buildings may be adversely affected through either explosion or inhalation hazards, (2) final

Table 13.6 Municipal Landfill Leachate Data—Organic Compounds (*continues*)

Indicator Parameters	Municipal Landfills Leachate Concentration Reported (ppb)	
	Minimum	Maximum
Acetone	8	11,000
Acrolein	270	270
Aldrin	NA	NA
α-Chlordane	NA	NA
Aroclor-1242	NA	NA
Aroclor-1254	NA	NA
Benzene	4	1,080
Bromomethane	170	170
Butanol	10,000	10,000
1-Butanol	320	360
2-Butanone (methyl ethyl ketone)	110	27,000
Butyl benzyl phenol	21	150
Carbazole	NA	NA
Carbon tetrachloride	6	397.5
4-Chloro-3-methylphenol	NA	NA
Chlorobenzene	1	685
Chloroethane	11.1	860
Bis(2-chloroethyoxy)methane	18	25
2-Chloroethyl vinyl ether	2	1,100
Chloroform	7.27	1,300
Chloromethane	170	400
Bis(chloromethyl)ether	250	250
2-Chloronaphthalene	46	46
p-Cresol	45.2	5,100
2,4-D	7.4	220
4,4′-DDE	NA	NA
4,4-DDT	0.042	0.22
Dibromomethane	5	5
Di-N-butyl phthalate	12	150
1,2-Dichlorobenzene	3	21.9
1,4-Dichlorobenzene	1	52.1
3,3-Dichlorobenzidine	NA	NA
Dichlorodifluoromethane	10.3	450
1,1-Dichloroethane	4	44,000
1,2-Dichloroethane	1	11,000
1,2-Dichloroethylene (Total)	NA	NA
cis-1,2-Dichloroethylene	190	470
trans-1,2-Dichloroethylene	2	4,800
1,2-Dichloropropane	0.03	500
1,3-Dichloropropene	18	30
Diethyl phthalate	3	330
2,4-Dimethyl phenol	10	28
Dimethyl phthalate	30	55
Endrin	0.04	50
Endrin ketone	NA	NA
Ethanol	23,000	23,000
Ethyl acetate	42	130
Ethyl benzene	6	4,900
Ethylmethacrylate	NA	NA
Bis(2-ethylhexyl)phthalate	16	750
2-Hexanone (methyl butyl ketone)	6	690
Isophorone	4	16,000
Lindane	0.017	0.023
4-Methyl-2-pentanone (methyl isobutyl ketone)	10	710
Methylene chloride (dichloromethane)	2	220,000
2-Methylnaphthalene	NA	NA

Table 13.6 (continued) Municipal Landfill Leachate Data—Organic Compounds

Indicator Parameters	Municipal Landfills Leachate Concentration Reported (ppb)	
	Minimum	Maximum
2-Methylphenol	NA	NA
4-Methylphenol	NA	NA
Methoxychlor	NA	NA
Naphthalene	2	202
Nitrobenzene	4	120
4-Nitrophenol	17	17
Pentachlorophenol	3	470
Phenanthrene	NA	NA
Phenol	7.3	28,800
1-Propanol	11,000	11,000
2-Propanol	94	26,000
Styrene	NA	NA
1,1,2,2-Tetrachloroethane	210	210
Tetrachloroethylene	2	620
Tetrahydrofuran	18	1,300
Toluene	5.55	18,000
Toxaphene	1	1
2,4,6-Tribromophenol	NA	NA
1,1,1-Trichloroethane	1	13,000
1,1,2-Trichloroethane	30	630
Trichloroethylene	1	1,300
Trichlorofluormethane	4	150
1,2,3-Trichloropropane	230	230
Vinyl chloride	8	61
Xylenes	32	310

Source: U.S. EPA. OSWER. 1988c. *Summary of Data on Municipal Solid Waste Landfill Leachate Characteristics.* EPA/530-SW-88-038.

use of the site includes allowing public access, (3) the landfill produces excessive odors, or (4) it is necessary to comply with ARARs. Most landfills will require at least a passive gas collection system (that is, venting) to prevent buildup of pressure below the cap and to prevent damage to the vegetative cover.

FIGURE 13.3 Geosynthetic fabric liner installation conforms to composite liner federal requirements for MSWLFs at the expansion site of Caldwell Sanitary Landfill. (*Source:* Caldwell Sanitary Landfill. With permission.)

Landfills—Present Status

Several landfills across the country have already incorporated the federal criteria for MSWLF into their expansion process. A typical landfill expansion under the new federal criteria is shown in Fig. 13.3 where the newly installed geosynthetic fabric liner, part of a composite liner system for the expansion of the existing MSWLF, is adjoining the active site of the landfill.

Even though the open dumps of the past have given way to high-tech landfill facilities where advanced engineering principles are applied to ensure environmentally safe and aesthetically pleasing conditions, public opposition to citing new landfills has not diminished. As a result, lateral and vertical expansions of existing MSWLF units are today the predominant means of generating new landfill space.

Defining Terms

Composite liner: A liner system for municipal solid waste landfills which, according to 40 CFR 258, consists of two components with the following specifications: The upper component is a minimum 30 mil flexible membrane liner (FML), and the lower component is at least a two-foot layer of compacted soil with a hydraulic conductivity of no more than 1×10^{-7} cm/sec. FML components consisting of high density polyethylene (HDPE) shall be at least 60 mil thick.

Leachate: A liquid that has passed through or emerged from solid waste in a landfill and contains soluble, suspended, or miscible materials removed from such waste.

Maximum contaminant levels (MCLs): Enforceable, allowable concentrations of contaminants in public drinking water supplies, protective of human health, under the federal Safe Drinking Water Act.

Municipal solid waste landfill (MSWLF): A discrete area of land that receives household waste, and that is not a land application unit, surface impoundment, injection well, or waste pile, as those terms are defined in 40 CFR 257. An MSWLF may also receive other types of RCRA Subtitle D wastes, such as commercial solid waste, nonhazardous sludge, and industrial solid waste. Such a landfill may be publicly or privately owned and may be a new MSWLF, an existing MSWLF, or a lateral expansion.

Relevant point of compliance: Under the federal solid waste regulations, this is the point where an MSWLF's impact on groundwater is evaluated. This point, which is specified by a state with an EPA approved solid waste program, cannot be more than 150 meters from the waste management unit boundary and has to be located on land owned by the MSWLF owner.

Uppermost aquifer: The geologic formation nearest the natural ground surface that is an aquifer, including lower aquifers interconnected with this aquifer within an MSWLF's property boundary.

References

Bond, R. G., Straub, C. P., and Prober, R., eds. 1973. *Handbook of Environmental Control,* Vol. 2. CRC Press, Boca Raton, FL.

Salvato, J. A. 1972. *Environmental Engineering and Sanitation,* 2nd ed. Wiley-Interscience, New York.

U.S. EPA. 1991a. 40 Code of Federal Regulations (CFR) Part 258, Criteria for Municipal Solid Waste Landfills.

U.S. EPA, OSWER. 1991b. *Conducting Remedial Investigation/Feasibility Studies for CERCLA Municipal Landfill Sites.* EPA/540-P-91-001.

U.S. EPA, OSWER. 1990a. *Characterization of Municipal Solid Waste in the United States: 1990 Update.* EPA/530-SW-90-042.

U.S. EPA, OSWER. 1990b. *Report to Congress, Methods to Manage and Control Plastic Wastes.* EPA/530-SW-89-051.

U.S. EPA, OSWER. 1989a. *The Solid Waste Dilemma: An Agenda for Action.* EPA/530-SW-89-019.

U.S. EPA, OSWER. 1989b. *Characterization of Products Containing Lead and Cadmium in Municipal Solid Waste in the United States, 1970 to 2000.* EPA/530-SW-89-015.

U.S. EPA, OSWER. 1988a. *Report to Congress, Solid Waste Disposal in the United States.* EPA/530-SW-88-011B.

U.S. EPA, RQEL. 1988b. *Lining of Waste Containment and Other Impoundment Facilities.* EPA/600/2-88/052.

U.S. EPA, OSWER. 1988c. *Summary of Data on Municipal Solid Waste Landfill Leachate Characteristics—Criteria for Municipal Solid Waste Landfills (40 CFR Part 258).* EPA/530-SW-88-038.

U.S. EPA, OSWER. 1986. *Subtitle D Study Phase I Report.* EPA/530-SW-054.

U.S. EPA. 1979. 40 Code of Federal Regulations (CFR) Part 257, Criteria for Classification of Solid Waste Disposal Facilities and Practices, as amended in 1981 and 1991.

The photo shows the foundation construction under way for the new Ottawa National Hockey League stadium. The stadium will become the home of the Ottawa Senators and will be called The Palladium. The stadium is founded on more than one thousand 12-in. steel pipe piles driven through soft sensitive clay into bearing in a dense glacial till at a depth of 25 m below the original ground surface. Activities shown in the photo include pile driving with dynamic monitoring, static loading tests, and excavations. The $200 million construction project was begun in early 1994 and is planned to be completed by the end of 1995.

The Palladium will seat 18,500 people and is designed to be a multipurpose stadium. Hockey will be the primary event, but the Palladium will also host concerts, circus shows, tennis tournaments, and a variety of other events. It will be a tourist attraction that will rival the Canadian Parliament buildings and the Rideau Canal. (Photo courtesy of Bengt H. Fellenius.)

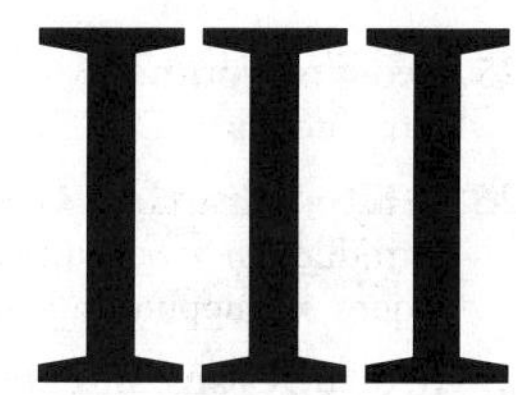

Geotechnical Engineering

Milton E. Harr
Purdue University

CIVIL ENGINEERS ARE IN THE MIDST of a construction revolution. Heavy structures are being located in areas formerly considered unsuitable from the standpoint of the supporting power of the underlying soils. Earth structures are contemplated that are of unprecedented height and size; soil systems must be offered to contain contaminants for time scales for which past experience is either inadequate or absent. Designs must be offered to defy the ravages of floods and earthquakes that so frequently visit major population centers.

All structures eventually transmit their loads into the ground. In some cases this may be accomplished only after circuitous transfers involving many component parts of a building; in other cases, such as highway pavements, contact is generally direct. Load transfer may be between soil and soil or, as in retaining walls, from soil through masonry to soil. Of fundamental importance is the response that can be expected due to the imposed loadings. It is within this framework that *geotechnical engineering* is defined as *that phase of civil engineering that deals with the state of rest or motion of soil bodies under the action of force systems.*

Soil bodies, in their general form, are composed of complex conglomerations of discrete particles, in compact arrays of varying shapes and orientations. These may range in magnitude from the microscopic elements of clay to the macroscopic boulders of a rock fill. At first glance, the task of establishing a predictive capability for a material so complicated appears to be overwhelming.

Although man's use of soil as a construction material extends back to the beginning of time, only within very recent years has the subject met with semiempirical treatment. In large measure, this change began in 1925 when Dr. Karl Terzaghi published his book *Erdbaumechanik*. Terzaghi demonstrated that soils, unlike other engineering materials, possess a mechanical behavior highly dependent on their prior history of loading and degree of saturation and that only a portion of the boundary energy is effective in producing changes within the soil body. Terzaghi's concepts transferred foundation design from a collection of rules of thumb to an engineering discipline. The contents of the present section offer, in a concise manner, many of the products of this and subsequent developments.

Had the section on geotechnical engineering in this handbook been written a mere decade or two ago, the table of contents would have been vastly different. Although some of the newer subjects might have been cited, it is unlikely that their relative importance would have precipitated individual chapters such as contained in the present section, namely: Chapter 15, "Accounting for Variability (Reliability)"; Chapter 23, "Geosynthetics"; Chapter 24, "Geotechnical Earthquake Engineering"; Chapter 25, "Geo-Environment"; Chapter 26, "In Situ Subsurface Characterization"; and Chapter 27, "In Situ Testing and Field Instrumentation." These make up approximately half the chapters in the present section on geotechnical engineering in the *Handbook*. Necessity *does* give birth to invention.

14

Soil Relationships and Classification

Thomas F. Wolff
Michigan State University

14.1 Soil Classification

There are two soil classification systems in common use for engineering purposes. The **Unified Soil Classification System** [ASTM D 2487-93] is used for virtually all geotechnical engineering work except highway and road construction, where the **AASHTO classification system** [AASHTO M 145-87] is used. Both systems use the results of **grain-size analysis** and determinations of **Atterberg limits** to determine a soil's classification. Soil components may be described as **gravel, sand, silt,** or **clay.** A soil comprising one or more of these components is given a descriptive name and a designation consisting of letters or letters and numbers which depend on the relative proportions of the components and the **plasticity** characteristics of the soil.

Grain-Size Characteristics of Soils

Large-grained materials such as **cobbles** and **boulders** are sometimes considered to be soil. The differentiation of cobbles and boulders depends somewhat on local practice, but boulders are generally taken to be particles larger than 200 to 300 mm or 9 to 12 inches. The Unified Soil Classification System suggests that boulders be defined as particles that will not pass a 12-in. (300 mm) opening. Cobbles are smaller than boulders and range down to particles that are retained on a 3-inch (75 mm) sieve. Gravels and sands are classified as **coarse-grained** soils; silts and clays are **fine-grained** soils. For engineering purposes, gravel is defined as soil that passes a 3-inch (75 mm) sieve and is retained by a No. 4 sieve (4.75 mm or 0.187 in.) or No. 10 sieve (2.00 mm or 0.078 in.), depending on the classification system. Sand is defined as soil particles smaller than gravel but retained on a No. 200 sieve (0.075 mm or about 0.003 in.). Soils passing the No. 200 sieve may be silt or clay. Although grain-size criteria were used in some older classification systems to differentiate silt from clay, the two systems described herein make this differentiation based on plasticity rather than grain size.

0-8493-8953-4/95/$0.00 + $.50

Table 14.1 Opening Sizes of Commonly Used Sieves

Inches	Millimeters
1.5	37.5
1	25
0.75	19
0.5	12.5
Sieve No.	**Millimeters**
4	4.75
10	2.00
20	0.850
40	0.425
70	0.212
100	0.150
200	0.075

The grain-size characteristics of soils that are predominantly coarse grained are evaluated by a **sieve analysis.** A **nest of sieves** is prepared by stacking sieves one above the other with the largest opening at the top followed by sieves of successively smaller openings and a catch pan at the bottom. Opening sizes of commonly used sieves are shown in Table 14.1. A sample of dry soil is poured onto the top sieve, the nest is covered, and it is then shaken by hand or mechanical shaker until each particle has dropped to a sieve with openings too small to pass, and the particle is *retained.* The cumulative weight of all material larger than each sieve size is determined and divided by the total sample weight to obtain the *percent retained* for that sieve size, and this value is subtracted from 100% to obtain the *percent passing* that sieve size. Results are displayed by plotting the percent passing (on a linear scale) against the sieve opening size (on a log scale) and connecting the plotted points with a smooth curve referred to as a **grain-size distribution curve.** A sample of some grain-size distribution curves is presented in Fig. 14.1.

The notation D_{xx} refers to the size D, in mm, for which xx percent of the sample by weight passes a sieve with an opening equal to D. The $\mathbf{D_{10}}$ **size,** sometimes called the **effective grain size,** is the grain diameter for which 10% of the sample (by weight) is finer. It is determined from the grain-size distribution curve at the point where the curve crosses a horizontal line through the 10% passing value on the y axis. Other D sizes are found in a similar manner. The D_{50} size, called the **median grain size,** is the grain diameter for which half the sample (by weight) is smaller and half is larger.

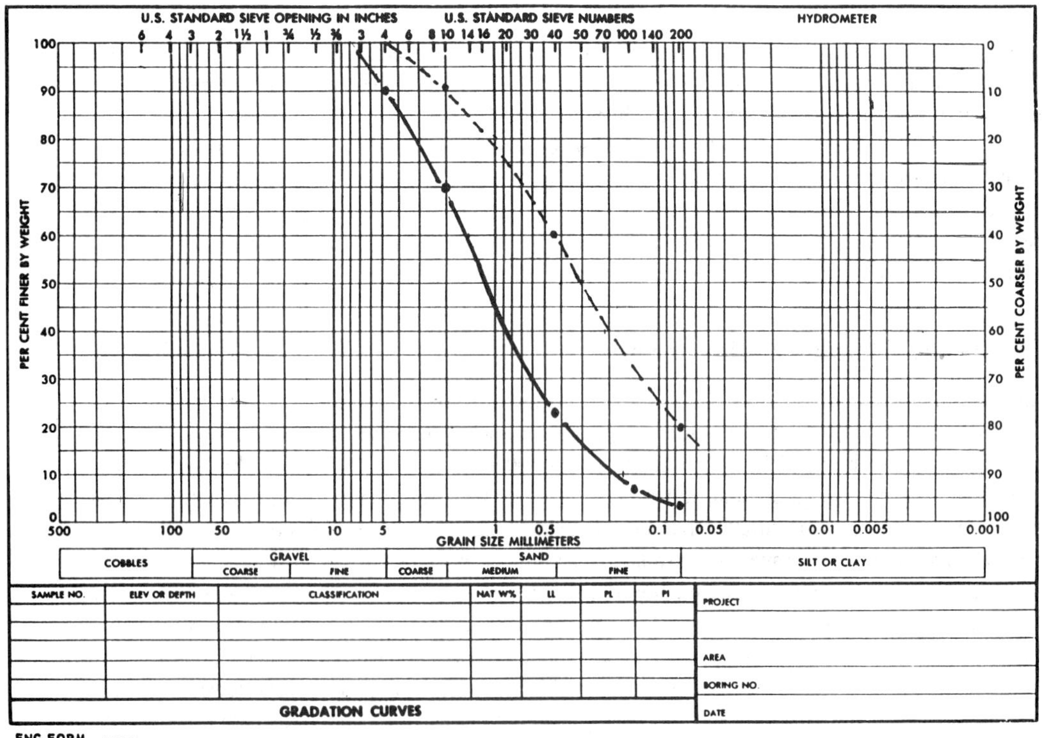

FIGURE 14.1 Typical grain size distribution curves. (From U.S. Army, 1970.)

Two parameters are used to describe the general shape of the grain-size distribution curve. The **coefficient of uniformity,** C_u, is:

$$C_u = \frac{D_{60}}{D_{10}} \tag{14.1}$$

The **coefficient of curvature,** C_c, is:

$$C_c = \frac{D_{30}^2}{D_{60} \times D_{10}} \tag{14.2}$$

Atterberg Limits and Plasticity

Atterberg limits, named after the Swedish soil scientist A. Atterberg, are water content values at which notable changes in soil behavior occur. The **liquid limit,** denoted LL or w_L, marks the transition between liquid and plastic behavior. At water contents above the liquid limit the soil behaves as a viscous liquid; below the liquid limit the soil behaves as a plastic solid. The liquid limit is determined in the laboratory by partly filling a standard brass cup with wet soil and cutting a groove of a standard dimension in the soil. The liquid limit is taken as the water content at which the groove closes a specified amount when the cup is lifted and dropped 1 cm exactly 25 times. The details of the test are given in AASHTO T 89 and ASTM D 4318-93. The **plastic limit,** denoted PL or w_P, is the transition between plastic and brittle behavior. It is determined in the laboratory as the water content at which a 1/8-inch diameter thread of soil begins to crumble when rolled under the palm of the hand. Details of the liquid limit and plastic limit tests are provided by AASHTO T 90 and ASTM D 4318-93. The **shrinkage limit,** denoted SL or w_S, is the water content below which the soil no longer reduces in volume when the water content is reduced. Although Atterberg limits are water contents and are properly decimals or percentages, they are usually expressed as an integer percentage without a percent sign. Thus, a liquid limit of 40% is usually reported as LL = 40.

The **plasticity index,** denoted PI or I_P, is the difference of the liquid limit and the plastic limit:

$$\text{PI} = \text{LL} - \text{PL} \tag{14.3}$$

The **liquidity index,** denoted LI or I_L, is a measure of the natural water content (w) relative to the plastic limit and the liquid limit:

$$\text{LI} = I_L = \frac{w - \text{PL}}{\text{LL} - \text{PL}} \tag{14.4}$$

The Unified Soil Classification System (USCS)

The Unified Soil Classification System is based on the airfield classification system developed by A. Casagrande during World War II. With some modification it was jointly adopted by several U.S. government agencies in 1952. Additional refinements were made and it is currently standardized as ASTM D 2487-93. It is used in the U.S. and much of the world for geotechnical work other than roads and highways.

In the unified system soils are designated by a two-letter symbol: the first identifies the primary component of the soil, and the second describes its grain size or plasticity characteristics. For example, a poorly graded sand is designated SP and a low plasticity clay is CL. Five first-letter symbols are used:

G for gravel
S for sand
M for silt
C for clay
O for organic soil

Clean sands and gravels (having less than 5% passing the No. 200 sieve) are given a second letter P if poorly graded or W if well graded. Sands and gravels with more than 12% by weight passing the No. 200 sieve are given a second letter M if the fines are silty or C if fines are clayey. Sands and gravels having between 5 and 12% are given dual classifications such as SP-SM. Silts, clays, and organic soils are given the second letter H or L to designate high or low plasticity. The specific rules for classification are summarized as follows and described in detail in ASTM D 2487.

Organic soils are distinguished by a dark-brown to black color, an organic odor, and visible fibrous matter.

For soils that are not notably organic the first step in classification is to consider the percentage passing the No. 200 sieve. If less than 50% of the soil passes the No. 200 sieve, the soil is *coarse grained,* and the first letter will be G or S; if more than 50% passes the No. 200 sieve, the soil is *fine grained* and the first letter will be M or C.

For coarse-grained soils, the proportions of sand and gravel in the **coarse fraction** (not the total sample) determine the first letter of the classification symbol. The coarse fraction is that portion of the total sample retained on a No. 200 sieve. If more than half of the coarse fraction is gravel (retained on the No. 4 sieve), the soil is *gravel* and the first letter symbol is G. If more than half of the coarse fraction is sand, the soil is *sand* and the first letter symbol is S.

For sands and gravels the second letter of the classification is based on gradation for clean sands and gravels and plasticity of the fines for sands and gravels with fines. For clean sands (less than 5% passing the No. 200 sieve), the classification is well-graded sand (SW) if $C_u \geq 6$ *and* $1 \leq C_c \leq 3$. *Both* of these criteria must be met for the soil to be SW, otherwise the classification is poorly graded sand (SP). Clean gravels (less than 5% passing the No. 200 sieve) are classified as well-graded gravel (G_W) if $C_u \geq 4$ *and* $1 \leq C_c \leq 3$. If both criteria are not met, the soil is poorly graded gravel (GP).

For sands and gravels where more than 12% of the total sample passes the No. 200 sieve, the soil is a clayey sand (SC), clayey gravel (GC), silty sand (SM), or silty gravel (GM). The second letter is assigned based on whether the fines classify as clay (C) or silt (M) as described for fine-grained soils below.

For sands and gravels having between 5 and 12% of the total sample passing the No. 200 sieve, both the gradation and plasticity characteristics must be evaluated and the soil is given a dual classification such as SP-SM, SP-SC, GW-GC, etc. The first symbol is always based on gradation, whereas the second is always based on plasticity.

For fine-grained soils and organic soils, classification in the unified system is based on Atterberg limits determined by the fraction passing the No. 40 sieve. The liquid limit and plasticity index are determined and plotted on the plasticity chart (Fig. 14.2). The vertical line at LL = 50 separates high-plasticity soils from low-plasticity soils. The *A-line* separates clay from silt. The equation of the A-line is PI = 0.73(LL − 20). The U-line is not used in classification but is an upper boundary of expected results for natural soils. Values plotting above the U-line should be checked for errors.

Inorganic soils with liquid limits below 50 that plot above the A-line and have PI values greater than 7 are **lean clays** and are designated CL; those with liquid limits above 50 that plot above the A-line are **fat clays** and are designated CH. Inorganic soils with liquid limits below 50 that plot below the A-line are silt and are designated ML; those with liquid limits above 50 that plot below the A-line are elastic silts and are designated MH. The plasticity chart has a shaded area; soils that plot in this area (above the A-line with PI values between 4 and 7) are silty clay and are given the dual symbol CL-ML. If the soil under consideration is the fines component of a dually classified sand or gravel, the soil is classified as SM-SC or GM-GC.

Soils with sufficient organic contents to influence properties that have liquid limits below 50 are classified as OL; those with liquid limits above 50 are classified as OH. Soils that are *predominantly* organic, with visible vegetable tissue, are termed **peat** and given the designation Pt.

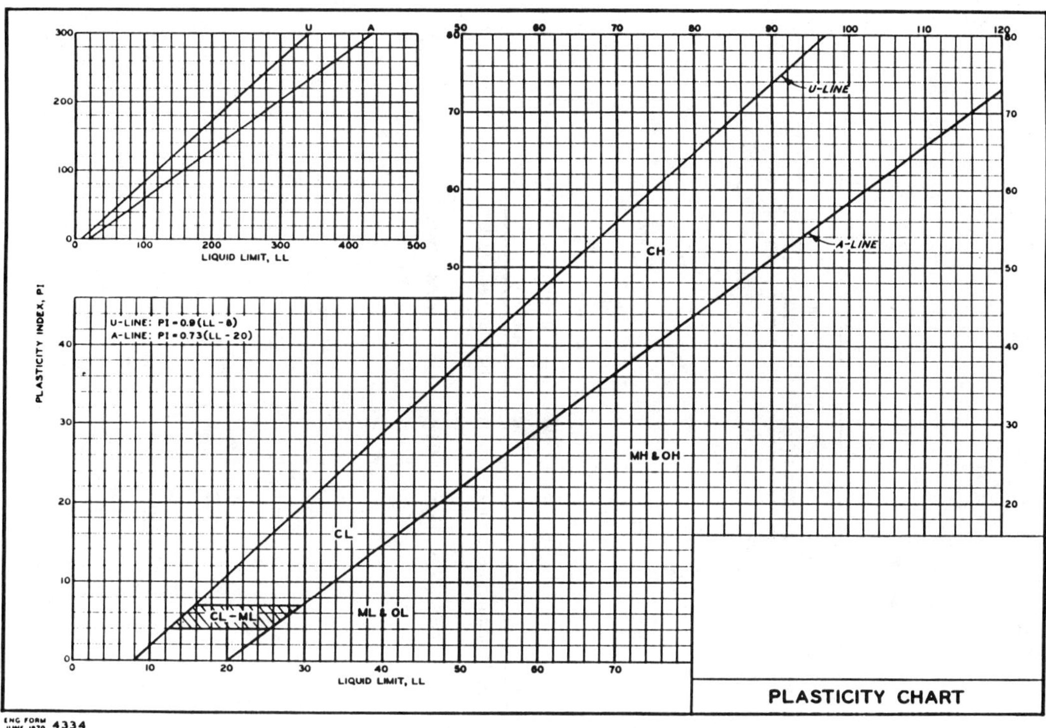

FIGURE 14.2 Plasticity chart for Unified Soil Classification System. (From U.S. Army, 1970.)

The AASHTO Classification System

The AASHTO system classifies soils into seven primary groups, named A-1 through A-7, based on their relative expected quality for road embankments, subgrades, subbases, and bases. Some of the groups are in turn divided into subgroups, such as A-1-a and A-1-b. Furthermore, a *group index* may be calculated to quantify a soil's expected performance within a group.

To determine a soil's classification in the AASHTO system, one first determines the relative proportions of gravel, coarse sand, fine sand, and silt-clay. In the AASHTO system gravel is material smaller than 75 mm (3 in.) but retained on a No. 10 sieve; coarse sand is material passing a No. 10 sieve but retained on a No. 40 sieve; and fine sand is material passing a No. 40 sieve but retained on a No. 200 sieve. Material passing the No. 200 sieve is *silt-clay* and is classified based on Atterberg limits. It should be noted that the division between gravel and sand is made at a smaller size (No. 10 sieve) in the AASHTO system than in the unified system (No. 4 sieve). Secondly, if any fines are present, Atterberg limits are determined and the plasticity index is calculated.

A soil is a **granular material** if less than 35% of the soil by weight passes the No. 200 sieve. Granular materials are classified into groups A-1 through A-3. Soils having more than 35% passing the No. 200 sieve are silt-clay and fall in groups A-4 through A-7.

Having the proportions of the components and the plasticity data, one enters one of the two alternative AASHTO classification tables (Tables 14.2 and 14.3) and checks from left to right until a classification is found for which the soil meets the criteria. It should be noted that, in this scheme, group A-3 is checked before A-2. The AASHTO plasticity criteria are also illustrated in Fig. 14.3.

Soils classified as A-1 are typically well-graded mixtures of gravel, coarse sand, and fine sand. Soils in subgroup A-1-a contain more gravel whereas those in A-1-b contain more sand. Soils in group A-3 are typically fine sands that may contain small amounts of nonplastic silt. Group A-2

Table 14.2 Classification of Soils and Soil-Aggregate Mixtures by the AASHTO System

General Classification	Granular Materials (35% or Less Passing 0.075 mm)			Silt-Clay Materials (More than 35% Passing 0.075 mm)			
Group Classification	A-1	A-3[a]	A-2	A-4	A-5	A-6	A-7
Sieve analysis, percent passing:							
2.00 mm (No. 10)	—	—	—	—	—	—	—
0.425 mm (No. 40)	50 max.	51 min.	—	—	—	—	—
0.075 mm (No. 200)	25 max.	10 max.	35 max.	36 min.	36 min.	36 min.	36 min.
Characteristics of fraction passing 0.425 mm (No. 40)							
Liquid limit	—	—		40 max.	41 min.	40 max.	41 min.
Plasticity index	6 max.	N.P.	[b]	10 max.	10 max.	11 min.	11 min.
General rating as subgrade	Excellent to good			Fair to poor			

[a] The placing of A-3 before A-2 is necessary in the "left to right elimination process" and does not indicate superiority of A-3 over A-2.

[b] See Table 14.3 for values.

From *Standard Specification for Transportation Materials and Methods of Sampling and Testing*. Copyright 1990 by the American Association of State Highway and Transportation Officials, Washington, D.C. Used by permission.

Table 14.3 Classification of Soils and Soil-Aggregate Mixtures

General Classification	Granular Materials (35% or Less Passing 0.075 mm)							Silt-Clay Materials (More than 35% Passing 0.075 mm)			
	A-1			A-2							A-7
Group Classification	A-1-a	A-1-b	A-3	A-2-4	A-2-5	A-2-6	A-2-7	A-4	A-5	A-6	A-7-5, A-7-6
Sieve analysis, percent passing:											
2.00 mm (No. 10)	50 max.	—	—	—	—	—	—	—	—	—	—
0.425 mm (No. 40)	30 max.	50 max.	51 min.	—	—	—	—	—	—	—	—
0.075 mm (No. 200)	15 max.	25 max.	10 max.	35 max.	35 max.	35 max.	35 max.	36 min.	36 min.	36 min.	36 min.
Characteristics of fraction passing 0.425 mm (No. 40)											
Liquid limit	—		—	40 max.	41 min.	40 max.	41 min.	40 max.	41 min.	40 max.	41 min.
Plasticity index	6 max.		N.P.	10 max.	10 max.	11 min.	11 min.	10 max.	10 max.	11 min.	11 min.[a]
Usual types of significant constituent materials	Stone fragments, gravel, and sand		Fine sand	Silty or clayey gravel and sand				Silty soils		Clayey soils	
General rating as subgrade	Excellent to good							Fair to poor			

[a] Plasticity index of A-7-5 subgroup is equal to or less than LL minus 30. Plasticity index of A-7-6 subgroup is greater than LL minus 30.

From *Standard Specification for Transportation Materials and Methods of Sampling and Testing.* Copyright 1990 by the American Association of State Highway and Transportation Officials, Washington, D.C. Used by permission.

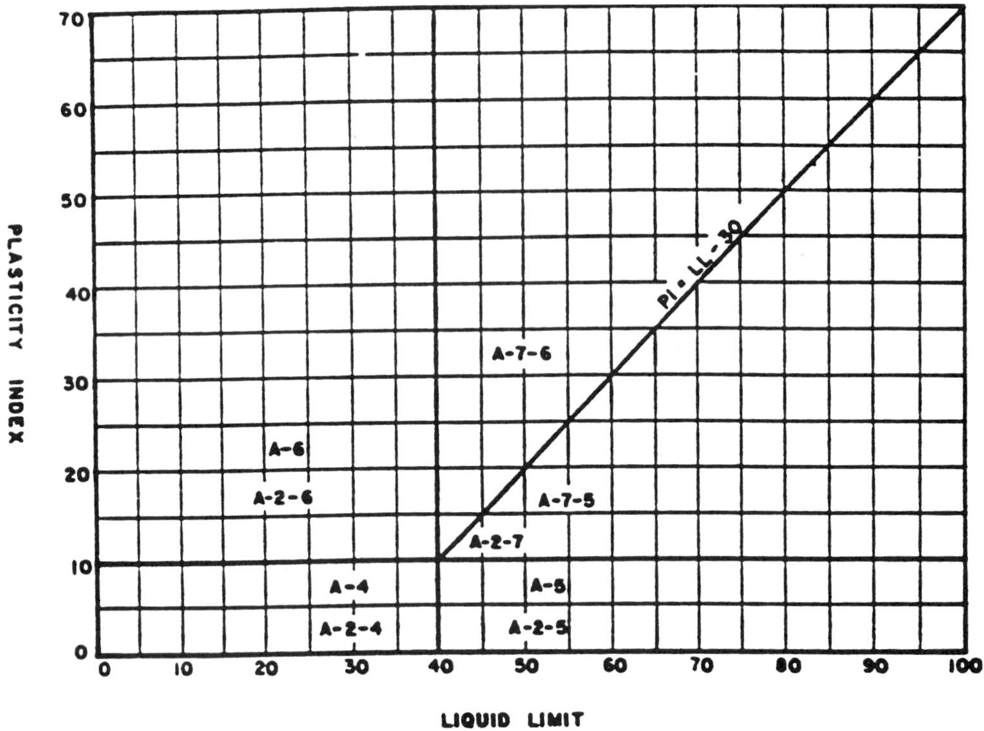

NOTE—A-2 soils contain less than 35 % finer than 200 sieve.

FIGURE 14.3 Plasticity chart for AASHTO Classification System. (From *Standard Specification for Transportation Materials and Methods of Sampling and Testing.* Copyright 1990 by the American Association of State Highway and Transportation Officials, Washington, D. C. Used by permission.)

contains a wide variety of "borderline" granular materials that do not meet the criteria for groups A-1 or A-3.

Soils in group A-4 are silty soils, whereas those in group A-5 are high-plasticity elastic silt. Soils in group A-6 are typically lean clays, and those in group A-7 are typically highly plastic clays.

Within groups containing fines, one may calculate a *group index* to further evaluate relative quality and supporting value of a material as subgrade. The group index is calculated according to the following empirical formula:

$$\text{Group index} = (F - 35)[0.2 + 0.005(\text{LL} - 40)] + 0.01(F - 15)(\text{PI} - 10) \quad (14.5)$$

In this equation F is the percentage of fines (passing the No. 200 sieve) expressed as a whole number. When calculating the group index for A-2-6 and A-2-7, only the PI term is used. The group index is rounded to the nearest whole number and, if negative, it is taken as zero. The expected performance is inversely related to group index. A value of zero indicates a good subgrade material and a value above 20 indicates a very poor material.

Example 14.1. Classify the soil shown by the solid curve in Fig. 14.1. Assume the soil is nonplastic. The following data are obtained:

Percent passing No. 4 sieve:	90%
Percent passing No. 10 sieve:	70%
Percent passing No. 40 sieve:	23%
Percent passing No. 200 sieve:	4%

D_{60} size:	1.50 mm
D_{30} size:	0.61 mm
D_{10} size:	0.18 mm

To classify the soil in the Unified Soil Classification System, the percent passing the No. 200 sieve is first checked. As only 4% passes the No. 200 sieve, the soil is coarse grained. The coarse fraction makes up 96% of the soil, with 10% being gravel (larger than the No. 4 sieve) and 86% being sand (the difference between the Nos. 4 and 200 sieves). As there is more sand than gravel, the soil is a sand and the first letter is S.

As less than 5% passes the No. 200 sieve, the sand is clean (SP or SW). The coefficient of uniformity is $(1.5)/(0.18) = 8.33$. The coefficient of curvature is $0.61^2/(1.5)(0.18) = 1.38$. As the coefficient of uniformity is greater than 6.0 and the coefficient of curvature is between 1 and 3, the soil meets both of the criteria for SW and is so classified.

In the AASHTO system the soil is 30% gravel, 47% coarse sand, 19% fine sand, and 4% silt-clay. Proceeding from left to right on the classification chart, the soil cannot be classified as A-1-a because the soil has 70% passing the No. 10 and that classification permits a maximum of 50%. The soil meets the classification criteria for A-1-b and is so classified.

Example 14.2. Classify the soil represented by the dashed curve in Fig. 14.1. The liquid limit and plastic limit are found to be 30 and 20, respectively. The following data are obtained:

Percent passing No. 4 sieve:	100%
Percent passing No. 10 sieve:	91%
Percent passing No. 40 sieve:	60%
Percent passing No. 200 sieve:	20%
D_{60} size:	0.41 mm
D_{30} size:	0.12 mm
D_{10} size:	<0.074 mm
Liquid limit:	30
Plastic limit:	20

First the soil is classified by the Unified Soil Classification System. As only 20% of the soil is smaller than a No. 200 sieve, the soil is coarse grained. All of the coarse fraction is smaller than the No. 4 sieve, so the soil is sand (first letter S).

As the percentage passing the No. 200 sieve is greater than 12%, the gradation characteristics are not considered, and the Atterberg limits are examined to determine whether the sand is a clayey sand or silty sand. The plasticity index is calculated as $30 - 20 = 10$, and the coordinates $\text{LL} = 30$, $\text{PI} = 10$ are entered on the plasticity chart. As this plots in the CL region, the fines are clay and the soil is a clayey sand SC.

Next the soil is classified according to the AASHTO system. Following the classification table from left to right, group A-1 is eliminated due to too much material passing the No. 40 sieve, and group A-3 is eliminated due to too much material passing the No. 200 sieve. The soil passes the criteria for A-2-4 and is so classified. The group index is calculated as:

$$\text{Group index} = (F - 35)[0.2 + 0.005(\text{LL} - 40)] + 0.01(F - 15)(\text{PI} - 10)$$

where $F = 20$, the percent passing the No. 200 sieve. Thus:

$$\begin{aligned}\text{Group index} &= (20 - 35)[0.2 + 0.005(30 - 40)] + 0.01(20 - 15)(10 - 10)\\ &= (15)[0.2 + (-0.05)] + 0\\ &= 15(0.15)\\ &= 2.25\end{aligned}$$

This is rounded to the nearest whole number, 2, and the soil classification is reported as A-2-4(2).

Example 14.3. A fine-grained soil has the following properties:

Percent passing No. 200 sieve:	65%
Liquid limit:	60
Plastic limit:	28

First the soil is classified according to the unified system. As more than 50% of the sample passes the No. 200 sieve, the soil is fine grained and the plasticity chart is used. The plasticity index is $60 - 28 = 32$. The coordinates (60,32) plot in the CH region, so the soil is a high-plasticity clay (CH).

To classify the soil in the AASHTO system, one notes that more than 35% passes the No. 200 sieve, so the soil is silt-clay. Entering the coordinates (60, 32) on the AASHTO plasticity chart, the classification is A-7-6. The group index is calculated as follows:

$$\begin{aligned}\text{Group index} &= (F - 35)[0.2 + 0.005(\text{LL} - 40)] + 0.01(F - 15)(\text{PI} - 10)\\ &= (65 - 35)[0.2 + 0.005(60 - 40)] + 0.01(65 - 15)(32 - 10)\\ &= 30[0.2 + 0.1] + 0.01(50)(22)\\ &= 9.0 + 11.0\\ &= 20\end{aligned}$$

The complete classification is A-7-6(20), which indicates a poor quality soil for highway construction.

14.2 Weight, Mass, and Volume Relationships

In an engineering context, soil comprises three components: solid particles, water, and air. Many problems in soil mechanics and construction quality control involve making calculations and communicating information regarding the relative proportions of these components and the volumes they occupy, individually or in combination. Given some known values of mass, weight, volume, or density (or relationships among them), it is often necessary to calculate other values. This section defines the terms commonly used in geotechnical engineering to describe such relationships and provides worked examples of typical calculations.

The Phase Diagram

In a typical volume of soil the three components are arranged in a complex mixture. To visualize the relationships among the components when performing weight-volume or mass-volume calculations, drawing a phase diagram is recommended. Much like checking equilibrium with a free body diagram in statics, completing and balancing a phase diagram with numerical values of all quantities permits cross-checking the solution. Figure 14.4 shows an element of soil and the

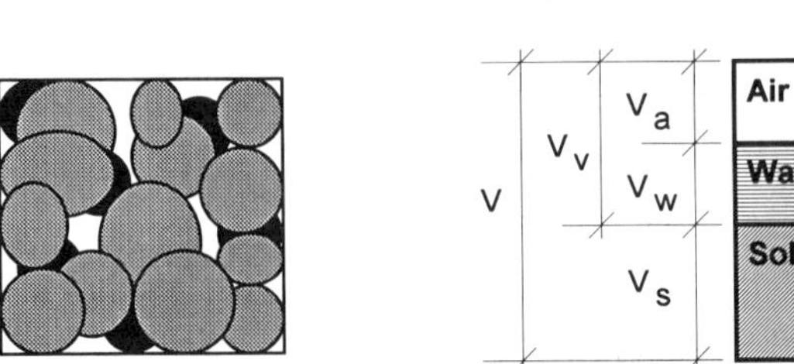

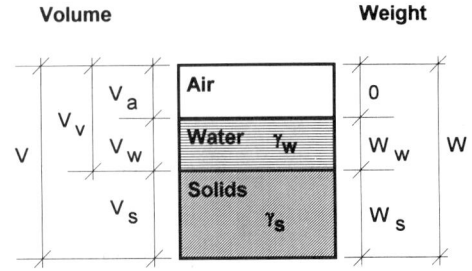

FIGURE 14.4 Soil element and phase diagram.

corresponding phase diagram. All solid particles in the element are taken as an equivalent single mass at the bottom of the diagram. Above the solids is the water—also represented as a single equivalent mass—and above that, the air. On the sides of the diagram are variable names; numerical values are entered here during calculations. On the left side are the total volume V, the volume of solids V_s, the volume of water V_w, and the volume of air V_a. The combined volume of water and air is the volume of voids V_v. On the right side are shown the total weight W, the weight of solids W_s, and the weight of water W_w. The weight of air is negligible. For problems involving masses, mass values are shown on the right instead of weights and given the equivalent designations M, M_s, and M_w.

Volume Relationships

Volume relationships include the void ratio, the porosity, and the degree of saturation. The **void ratio,** denoted e, is the ratio of the volume of voids to the volume of solids:

$$e = \frac{V_v}{V_s} \tag{14.6}$$

The **porosity,** denoted n, is the ratio of the volume of voids to the total volume:

$$n = \frac{V_v}{V} \tag{14.7}$$

The void ratio and porosity are related as follows:

$$n = \frac{e}{1+e} \tag{14.8}$$

$$e = \frac{n}{1-n} \tag{14.9}$$

The **degree of saturation,** denoted S, is the ratio of the volume of water to the volume of voids. It is commonly expressed as a percentage:

$$S = \frac{V_w}{V_v} \times 100\% \tag{14.10}$$

If all the soil voids are filled with water and no undissolved air is present, the soil is said to be **saturated.**

Weight and Mass Relationships

The **water content** (or moisture content), denoted w, is the only relationship involving weights or masses. It is the ratio of the weight of water to the weight of solids or, equivalently, the ratio of the mass of water to the mass of solids:

$$w = \frac{W_w}{W_s} = \frac{M_w}{M_s} \tag{14.11}$$

Unit Weight

The ratio of the weight of a material to its volume is its **unit weight,** sometimes termed *specific weight* or *weight density.* The unit weight of water, γ_w, is 9.81 kN/m^3 in the SI system and 62.4 lb/ft^3 in the English system. The unit weight of solids, γ_s, varies with the mineralogy of the soil particles but is commonly in the range of 26.0 to 27.0 kN/m^3 or 165 to 172 lb/ft^3. The **total unit weight** of a soil (the solids-water-air system), denoted γ, is the ratio of the total weight to the total volume occupied:

$$\gamma = \frac{W}{V} = \frac{W_s + W_w}{V_s + V_w + V_a} \tag{14.12}$$

The **saturated unit weight,** denoted γ_{sat}, is the total unit weight that would be obtained if the air voids were filled with an equal volume of water ($S = 100\%$ and $V_w = V_v$). The **dry unit weight,** denoted γ_d, is often termed the **dry density** and has particular importance in field control of soil compaction. It is the ratio of the weight of solids to the total volume:

$$\gamma_d = \frac{W_s}{V} = \frac{W_s}{V_s + V_w + V_a} \tag{14.13}$$

Note that the dry unit weight matches the weight of a single component—the solids—with the *entire* volume of solids, water, and air. It does not represent the unit weight of any component or consistent set of components, but rather provides a measure of how much solid material by weight is in the total volume of a container, such as an earthmover or a compaction mold. The **buoyant unit weight** or **effective unit weight,** γ', is equal to the saturated unit weight minus the unit weight of water, γ_w:

$$\gamma' = \gamma_{\text{sat}} - \gamma_w \tag{14.14}$$

The buoyant unit weight is sometimes used to directly calculate vertical effective stresses below the water table instead of calculating total stresses and subtracting pore pressures.

Density

The term **density** is used herein to denote the mass-to-volume ratio of a material. However, some references, particularly older ones, use the term to describe unit weight. Density is denoted by ρ. Because $m = W/g$, the unit weight terms defined above can be converted to mass densities as follows:

$$\rho = \frac{\gamma}{g} = \frac{M}{V} \tag{14.15}$$

$$\rho_{\text{sat}} = \frac{\gamma_{\text{sat}}}{g} = \frac{M_{\text{sat}}}{V} \tag{14.16}$$

$$\rho_d = \frac{\gamma_d}{g} = \frac{M_s}{V} \tag{14.17}$$

$$\rho' = \rho_{\text{sat}} - \rho_w \tag{14.18}$$

In the SI system mass densities are commonly expressed in Mg/m^3, kg/m^3, or g/ml. The mass density of water can therefore be expressed $\rho_w = 1000$ kg/m^3 $= 1$ Mg/m^3 $= 1$ g/ml. The mass density of soil solids typically ranges from 2640 to 2750 kg/m^3. Where mass or mass density values (g, kg, or kg/m^3) are given or measured, they must be multiplied by g(9.81m/s^2) to obtain weights or unit weights before performing stress calculations. In the English system mass density values are virtually never used in geotechnical engineering and all work is performed in terms of unit weights (lb/ft^3).

Specific Gravity

To facilitate working problems across different sets of units, it is convenient to express the unit weight and density of solids as a ratio to the unit weight and density of water. This ratio is termed the **specific gravity** and denoted G_s. For most soil minerals G_s is commonly in the range 2.64 to 2.75. Note that:

$$\gamma_s = G_s \gamma_w \tag{14.19}$$

and

$$\rho_s = G_s \rho_w \tag{14.20}$$

Conversion of Unit Weight and Density

A soil sample has a total unit weight of 125 lb/ft^3. It is desired to find its total unit weight in kN/m^3 and its density in kg/m^3. Although the problem can be worked using a chain of conversion factors, a simpler approach is to consider that the unit weight and density of the soil sample have a constant ratio to the unit weight and density of water. Placing the unit weight or density of water in any system of units in both the numerator and denominator of a fraction forms an equality. (It is assumed that the problem is on the planet Earth and a *mass* of 1000 kg *weighs* 9.81 kN!) Thus,

$$\gamma = (125 \text{ lb/ft}^3)\left(\frac{9.81 \text{ kN/m}^3}{62.4 \text{ lb/ft}^3}\right) = 19.7 \text{ kN/m}^3$$

$$\rho = (125 \text{ lb/ft}^3)\left(\frac{1000 \text{ kg/m}^3}{62.4 \text{ lb/ft}^3}\right) = 2003 \text{ kg/m}^3$$

Weight-Volume Problems Involving Defined Quantities

Weight-volume problems may be divided into two categories: those where there is a defined quantity of soil, and those where the quantity of soil is not defined and it is only desired to make conversions among relationships. The solution to problems of the first category is discussed first; discussion of the second category follows. Problems of the first category can be solved in four steps:

1. A blank phase diagram is sketched and known weights, volumes, and unit weights are entered on the diagram.
2. Known volumes are multiplied by their respective unit weights to obtain weights. Known weights are divided by unit weights to obtain volumes. Where values of some relationships are given, additional weight and volume values are calculated using the definitions of Eqs. (14.6), (14.7), (14.10), and (14.11). To numerically balance the phase diagram it is recommended that all calculations be carried to at least four significant digits.
3. Multiplication and division horizontally across the diagram and addition and subtraction vertically along the sides is continued until all weights, volumes, and unit weights are determined and found to numerically balance.
4. All desired values and relationships can now be calculated from the completed and checked diagram.

Example 14.4 (English Units). Assume that a **compaction mold** having a volume of 1/30 ft^3 was filled with moist soil. The total weight of the soil in the mold was found to be 4.10 lb. The soil was oven dried and its weight after drying was 3.53 lb. The specific gravity of solids was known to be 2.70. Water content, void ratio, porosity, degree of saturation, total unit weight, and dry unit

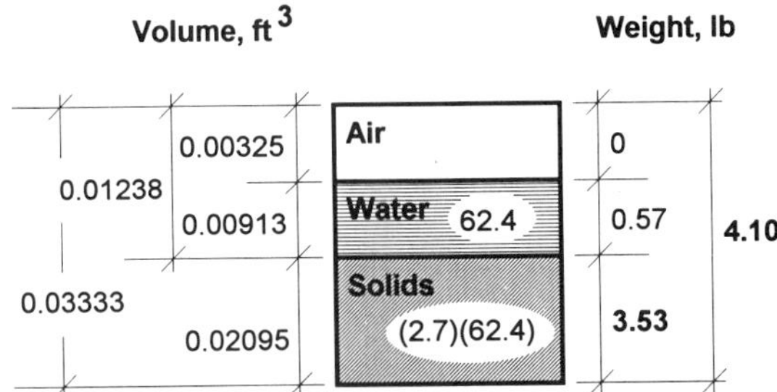

FIGURE 14.5 Phase diagram for Example 14.4.

weight must be determined. A phase diagram is shown in Fig. 14.5, with the known quantities in bold. The weight and volume of water are calculated as

$$W_w = W - W_s = 4.10 - 3.53 = 0.57 \text{ lb}$$

$$V_w = W_w/\gamma_w = 0.57/62.4 = 0.00913 \text{ ft}^3$$

The volume of solids is

$$V_s = W_s/G_s\gamma_w = 3.53/[(2.70)(62.4)] = 0.02095 \text{ ft}^3$$

The volume of air is

$$V_a = V - V_w - V_s = 0.03333 - 0.00913 - 0.02095 = 0.00325 \text{ ft}^3$$

The volume of voids is

$$V_v = V_w + V_a = 0.00913 + 0.00325 = 0.01238 \text{ ft}^3$$

With all quantities now known, all relationships can be determined.

$$w = W_w/W_s = 0.57/3.53 = 0.161 = 16.1\%$$

$$e = V_v/V_s = 0.01238/0.02095 = 0.590$$

$$n = V_v/V = 0.01238/(1/30) = 0.371$$

$$S = (V_w/V_s)100\% = (0.00913/0.01238)100\% = 73.7\%$$

$$\gamma = W/V = 4.10/(1/30) = 123.0 \text{ lb/ft}^3$$

$$\gamma_d = W_s/V = 3.53/(1/30) = 105.9 \text{ lb/ft}^3$$

If the same soil now becomes saturated by the addition of water at constant total volume, the saturated water content and saturated unit weight can be calculated as follows. The new volume of water is the entire void volume, 0.01238 ft^3. Multiplying this value by 62.4 lb/ft^3, the new weight of water is 0.77 lb. The water content at saturation is then

$$w = W_w/W_s = 0.77/3.53 = 0.218 = 21.8\%$$

The total weight is then

$$W = W_s + W_w = 3.53 + 0.77 = 4.30 \text{ lb}$$

The total and dry unit weights are then

$$\gamma = W/V = 4.30/(1/30) = 129.0 \text{ lb/ft}^3$$

$$\gamma_d = W_s/V = 3.53/(1/30) = 105.9 \text{ lb/ft}^3$$

Note that the dry unit weight does not change if water is added without changing total volume.

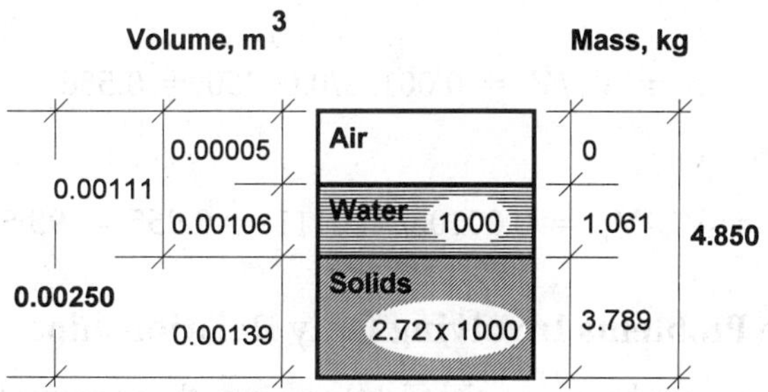

FIGURE 14.6 Phase diagram for Example 14.5.

Example 14.5 (SI units). A soil sample has a volume of 2.5 liters (2.5×10^{-3} m^3) and a total mass of 4.85 kg. A water content test indicates the water content is 28%. Assuming that the specific gravity of solids is 2.72, it is desired to determine the total density, total unit weight, dry density, dry unit weight, void ratio, porosity, and degree of saturation.

A phase diagram is shown in Fig. 14.6, with known values shown in bold.

$$M = M_s + M_w = 4.85 \text{ kg}$$

From the definition of the water content,

$$\begin{aligned} M_s + 0.28M_s &= 4.85 \text{ kg} \\ 1.28M_s &= 4.85 \text{ kg} \\ M_s &= 3.789 \text{ kg} \\ M_w = M - M_s &= 1.061 \text{ kg} \end{aligned}$$

With the mass side of the diagram complete, the masses are divided by density values to obtain volumes:

$$\begin{aligned} V_s &= M_s/\rho_s = 3.789 \text{ kg}/(2720 \text{ kg/m}^3) = 0.00139 \text{ m}^3 \\ V_w &= M_w/\rho_w = 1.061 \text{ kg}/(1000 \text{ kg/m}^3) = 0.00106 \text{ m}^3 \end{aligned}$$

Then

$$V_a = V - V_s - V_w = 0.00250 - 0.00139 - 0.00106 = 0.00005 \text{ m}^3$$

The total density is

$$\rho = M/V = 4.850 \text{ kg}/0.00250 \text{ m}^3 = 1940 \text{ kg/m}^3$$

The total unit weight is

$$\gamma = \rho g = (1940 \text{ kg/m}^3)(9.81 \text{ m/s}^2) = 19{,}031 \text{ N/m}^3 = 19.03 \text{ kN/m}^3$$

The dry density is

$$\rho_d = M/V = 3.789 \text{ kg}/0.00250 \text{ m}^3 = 1515 \text{ kg/m}^3$$

The dry unit weight is

$$\gamma_d = \rho_d g = (1515 \text{ kg/m}^3)(9.81 \text{ m/s}^2) = 14{,}862 \text{ N/m}^3 = 14.86 \text{ kN/m}^3$$

The void ratio is

$$e = V_v/V_s = 0.00111/0.00139 = 0.799$$

The porosity is

$$n = V_v/V = 0.00139/0.00250 = 0.556$$

The degree of saturation is

$$S = V_w/V_v = 0.00106/0.00111 = 0.955 = 95.5\%$$

Weight-Volume Problems Involving Only Relationships

If only relationships (e.g., void ratio or unit weight) are given, the quantity of soil is indefinite and only other relationships can be calculated. Nevertheless, it is convenient to solve such problems using a phase diagram and assuming one fixed weight or volume value. That quantity of solids, water, or soil is "brought to the paper" and used to calculate corresponding quantities of components. Although any one quantity can be assumed to have any value, the following assumptions simplify calculations:

- If a water content is known, assume $W_s = 1.0$ or 100.0 (lb or kN); then $W_w = w$ or $100w$.
- If a void ratio is known, assume the volume of solids $V_s = 1.0\ \text{ft}^3$ or $1.0\ \text{m}^3$; then $V_v = e$.
- If a dry unit weight is known, assume the total volume $V = 1.0\ \text{ft}^3$ or $1.0\ \text{m}^3$; then $W_s = \gamma_d$.
- If a total unit weight is known, assume the total volume $V = 1.0\ \text{ft}^3$ or $1.0\ \text{m}^3$; then $W = \gamma$.

Example 14.6. Assume that a soil has a water content of 30%, a void ratio of 0.850, and a specific gravity of 2.75. It is desired to find the degree of saturation, porosity, total unit weight, and dry unit weight. Any single fixed quantity of soil or water may be assumed in order to start the calculations; it is assumed that the volume of solids $V_s = 1.000\ \text{m}^3$. A phase diagram for the problem is shown in Fig. 14.7. The calculations then proceed as follows:

$$\begin{aligned}
V_v &= eV_s = (0.850)(1.000) = 0.850\ \text{m}^3 \\
V &= V_s + V_v = 1.000 + 0.850 = 1.850\ \text{m}^3 \\
W_s &= V_s G_s \gamma_w = (1.000)(2.75)(9.81) = 26.98\ \text{kN} \\
W_w &= wW_s = (0.30)(26.98) = 8.09\ \text{kN} \\
W &= W_s + W_w = 26.98 + 8.09 = 35.02\ \text{kN} \\
V_w &= W_w/\gamma_w = 8.09/9.81 = 0.825\ \text{m}^3 \\
V_a &= V_v - V_w = 0.850 - 0.825 = 0.025
\end{aligned}$$

At this point the weights and volumes of all components are known for the assumed $1.000\ \text{m}^3$ of solids and all desired relationships can be calculated as follows:

$$\begin{aligned}
S &= (V_w/V_v)100\% = (0.825/0.850)100\% = 97.1\% \\
n &= V_v/V = 0.850/1.850 = 0.459
\end{aligned}$$

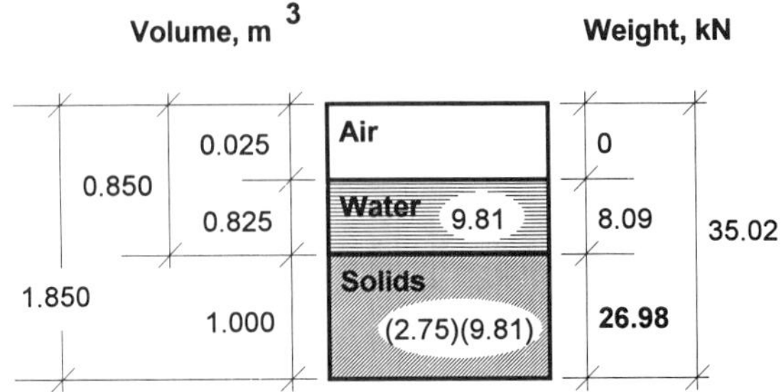

FIGURE 14.7 Phase diagram for Example 14.6.

$$\gamma = 35.02/1.850 = 18.93\ \text{kN/m}^3$$

$$\gamma_d = 26.98/1.850 = 14.58\ \text{kN/m}^3$$

These relationships derive from the given relationships regardless of the quantity of soil considered.

Equations among Relationships

Solving weight-volume problems using a phase diagram provides a visual display of whether sufficient information is available to complete the problem, whether additional assumptions must be introduced, or whether the problem is overconstrained by an unwarranted assumption. For example, without completing a phase diagram, it may not be immediately apparent from the information given whether a soil is saturated. Nevertheless, a few additional equations are sometimes very useful for converting from one relationship to another.

Four distinct relationships combine to form the equation

$$Se = wG_s \tag{14.21}$$

For saturated soils S = 100%, and this can be written

$$e = wG_s \tag{14.22}$$

The total unit weight can be obtained using the following:

$$\gamma = \frac{(G_s + Se)\gamma_w}{1 + e} = \frac{(1 + w)\gamma_w}{w/S + 1/G_s} \tag{14.23}$$

The dry unit weight can be obtained using the following:

$$\gamma_d = \frac{G_s\gamma_w}{1 + e} = \frac{G_s\gamma_w}{1 + (wG_s/S)} \tag{14.24}$$

Example 14.6 can be reworked using the equations as follows:

$$Se = wG_s$$

$$S = wG_s/e = (0.30)(2.75)/0.850 = 0.971 = 97.1\%$$

$$n = \frac{e}{1 + e} = \frac{0.850}{1 + 0.850} = 0.459$$

$$\gamma = \frac{(1 + w)\gamma_w}{w/S + 1/G_s} = \frac{(1.30)(9.81)}{(0.30/0.971) + (1/2.75)} = 18.96\ \text{kN/m}^3$$

$$\gamma_d = \frac{G_s\gamma_w}{1 + e} = \frac{(2.75)(9.81)}{1.85} = 14.58\ \text{kN/m}^3$$

The discrepancy in the fourth decimal place for the unit weight γ is due to rounding and the use of only three significant figures in the input values.

Defining Terms

Section 14.1

AASHTO classification system: A classification system developed by the American Association of State Highway and Transportation Officials that rates soils relative to their suitability for road embankments, subgrades, subbases, and basis.

Atterberg limits: Water contents at which soil changes engineering behavior; the most important ones in classification are the liquid limit and plastic limit.

Boulders: Rock particles larger than 9 to 12 inches or 200 to 300 mm.

Clay: Fine-grained soil that exhibits plasticity.

Coarse grained: Soils that are retained on a No. 200 sieve.

Coarse fraction: In the Unified Soil Classification System, that portion of a soil sample retained on a No. 200 sieve.

Cobbles: Rock particles smaller than a boulder but larger than 3 inches (75 mm).

Coefficient of curvature: A mathematical parameter, $D_{30}^2/(D_{60}D_{10})$, used as a measure of the smoothness of a gradation curve.

Coefficient of uniformity: A mathematical parameter, D_{60}/D_{10}, used as a measure of the slope of a gradation curve.

D_{10} size: The grain size, in mm, for which 10% by weight of a soil sample is finer.

Effective grain size: Another name for the D_{10} size.

Fat clay: Highly plastic clay; clay with a liquid limit greater than 50.

Fine fraction: In the unified soil classification system, that portion of a soil sample passing a No. 200 sieve.

Fine grained: Soil passing a No. 200 sieve.

Grain-size analysis: The determination of the relative proportions of soil particles of each size in a soil sample, performed by passing the sample over a nest of sieves.

Grain-size distribution curve: A plot of percent finer or coarser versus soil-grain size. Grain size is plotted on a logarithmic scale.

Granular material: In the AASHTO classification system, soil with less than 35% passing the No. 200 sieve.

Gravel: Soil or rock particles smaller than 3 inches but retained on a No. 4 sieve (unified soil classification system) or on a No. 10 sieve (AASHTO system).

Lean clay: Clay with low plasticity; clay with a liquid limit less than 50.

Liquid limit: The water content above which soil behavior changes from a plastic solid to a viscous liquid.

Median grain size: The grain size for which one-half of a soil sample, by weight, is larger and half is smaller.

Nest of sieves: A stack of sieves of different sizes, having the largest opening on the top and progressing downward to successively smaller openings.

Plastic limit: The water content above which the soil behavior changes from a brittle solid to a plastic solid.

Plasticity: The ability of a soil, when mixed with water, to deform at constant volume.

Plasticity index: The difference between the liquid and plastic limit.

Peat: A highly organic soil, dark brown to black in color, with noticeable organic odor and visible vegetable matter.

Sand: Soil particles retained on the No. 200 sieve that pass the No. 4 sieve (Unified Soil Classification System) or the No. 10 sieve (AASHTO system).

Shrinkage limit: The water content at which further reduction in water content does not cause a further reduction in volume.

Sieve analysis: A grain-size analysis using a nest of sieves.

Silt: Fine-grained soil having a low plasticity index or not exhibiting plasticity.

Unified Soil Classification System: A descriptive classification system based on Casagrande's airfield system and now standardized by ASTM D 2487-93.

Section 14.2

Buoyant unit weight: The apparent unit weight of a submerged soil, obtained as the total unit weight minus the weight of water.

Compaction mold: A metal mold, typically 1/30 ft^3, used to determine the density of compacted soil.

Degree of saturation: The ratio of the volume of water to the volume of void space in a sample of soil.

Density: The mass per unit volume of a soil or one of its components.

Dry density: The ratio of the mass of solids to the total volume of a soil sample.

Dry unit weight: The ratio of the weight of solids to the total volume of a soil sample.

Effective unit weight: Another term for buoyant unit weight.

Porosity: The ratio of the volume of void spaces to the total volume of a soil sample.

Saturated: The condition in which all of the void spaces in a soil are filled with water and the volume of air is zero.

Saturated unit weight: The unit weight obtained if a soil sample is saturated by adding water at constant total volume.

Specific gravity: The ratio of the density of a material to the density of water; usually refers to the specific gravity of soil solids.

Total unit weight: The total combined weight of solids and water in a unit volume of soil.

Unit weight: The ratio of the weight of a material to its volume.

Void ratio: The ratio of the volume of void space in a soil sample to the volume of solid particles.

Water content: The ratio of the weight of water to the weight of solids of a soil sample.

References

AASHTO Standard M 145-87. The classification of soils and soil-aggregate mixtures for highway construction purposes. *AASHTO Materials, Part I, Specifications.* American Association of State Highway and Transportation Officials, Washington, D.C.

AASHTO Standard T 11-90. Amount of material finer than 75μm sieve in aggregate. *AASHTO Materials, Part I, Specifications.* American Association of State Highway and Transportation Officials, Washington, D.C.

AASHTO Standard T 27-88. Sieve analysis of fine and coarse aggregates. *AASHTO Materials, Part I, Specifications.* American Association of State Highway and Transportation Officials, Washington, D.C.

AASHTO Standard T 88-90. Particle size analysis of soils. *AASHTO Materials, Part I, Specifications.* American Association of State Highway and Transportation Officials, Washington, D.C.

AASHTO Standard T 89-89. Determining the liquid limit of soils. *AASHTO Materials, Part I, Specifications.* American Association of State Highway and Transportation Officials, Washington, D.C.

AASHTO Standard T 90-87. Determining the plastic limit and plasticity index of soils. *AASHTO Materials, Part I, Specifications.* American Association of State Highway and Transportation Officials, Washington, D.C.

ASTM Designation: C 117-90. Test method for materials finer than 75-μm (No. 200) sieve in mineral aggregates by washing. *ASTM Book of Standards.* Sec. 4, Vol. 04.02. American Society for Testing and Materials, Philadelphia, PA.

ASTM Designation: C 136-93. Method for sieve analysis of fine and coarse aggregates. *ASTM Book of Standards.* Sec. 4, Vol. 04.02. American Society for Testing and Materials, Philadelphia, PA.

ASTM Designation: D 422-63(1990). Test method for particle-size analysis of soils. *ASTM Book of Standards.* Sec. 4, Vol. 04.08. American Society for Testing and Materials, Philadelphia, PA.

ASTM Designation: D 653-90. Terminology relating to soil, rock, and contained fluids. *ASTM Book of Standards.* Sec. 4, Vol. 04.08. American Society for Testing and Materials, Philadelphia, PA.

ASTM Designation: D 1140-92. Test method for amount of material in soils finer than the No. 200 (75-μm) sieve. *ASTM Book of Standards.* Sec. 4, Vol. 04.08. American Society for Testing and Materials, Philadelphia, PA.

ASTM Designation: D 2487-93. Standard classification of soils for engineering purposes (unified soil classification system). *ASTM Book of Standards.* Sec. 4, Vol. 04.08. American Society for Testing and Materials, Philadelphia, PA.

ASTM Designation: D 2488-93. Practice for description and identification of soils (visual-manual procedure). *ASTM Book of Standards.* Sec. 4, Vol. 04.08. American Society for Testing and Materials, Philadelphia, PA.

ASTM Designation: D 4318-93. Test method for liquid limit, plastic limit, and plasticity index of soils. *ASTM Book of Standards.* Sec. 4, Vol. 04.08. American Society for Testing and Materials, Philadelphia, PA.

ASTM Designation: E 11-87. Specification for wire-cloth sieves for testing purposes. *ASTM Book of Standards.* Sec. 4, Vol. 04.02. American Society for Testing and Materials, Philadelphia, PA.

Casagrande, A. 1948. Classification and identification of soils. *Transactions ASCE.* 113:901–930.

Das, B. M. 1990. *Principles of Geotechnical Engineering,* 2nd ed. PWS-Kent, Boston, MA.

Holtz, R. D. and Kovacs, W. D. 1981. *An Introduction to Geotechnical Engineering.* Prentice Hall, Englewood Cliffs, NJ.

U.S. Army. 1970. *Laboratory Soils Testing.* Engineer Manual EM 1110-2-1906. Headquarters, Department of the Army, Office of the Chief of Engineers, Washington, D.C.

For Further Information

The development and underlying philosophy of the Unified Soil Classification System was summarized by Casagrande [1948]. The complete and official rules for classifying soil according to the Unified Soil Classification System are given in ASTM Designation D 2487-92. The complete and official rules for classifying soil according to the AASHTO classification system is given in AASHTO Standard M 145-87. The procedures for performing index tests related to soil classification are specified in the other AASHTO and ASTM standards listed in the references.

A more detailed presentation of soil relationships and classification is given in most introductory geotechnical engineering textbooks. Two notable examples are *An Introduction to Geotechnical Engineering* by Holtz and Kovacs and *Principles of Geotechnical Engineering* by B. M. Das.

15

Accounting for Variability (Reliability)

Milton E. Harr
Purdue University

15.1 Introduction

The trend in civil engineering today, more than ever before, is toward providing economical designs at specified levels of safety. Often these objectives necessitate a prediction of the performance of a system for which there exists little or no previous experience. Current design procedures, which are generally learned only after many trial-and-error iterations, lacking precedence, often fall short of expectations in new or alien situations. In addition, there is an increasing awareness that the raw data, on which problem solutions are based, themselves exhibit significant variability. It is the aim of this presentation to demonstrate how concepts of probability analysis may be used to supplement the geotechnical engineer's judgment in such matters.

15.2 Probabilistic Preliminaries

Fundamentals

Within the context of engineering usage there are two primary definitions of the concept of probability: *relative frequency* and *subjective* interpretation. Historically, the measure first offered for the probability of an outcome was its relative frequency. If an outcome A can occur T times in N equally likely trials, the probability of the outcome A is

$$P[A] = \frac{T}{N} \tag{15.1a}$$

Implied in Eq. (15.1a) is that the probability of an outcome A equals the number of outcomes favorable to A (within the meaning of the experiment) divided by the total number of possible

0-8493-8953-4/95/$0.00 + $.50

outcomes, or

$$P[A] = \frac{\text{Favorable outcomes}}{\text{Total possible outcomes}} \tag{15.1b}$$

Example 15.1. Find the probability of drawing a red card from an ordinary well-shuffled deck of 52 cards.
Solution. Of the 52 equally likely outcomes, there are 26 favorable (red card) outcomes. Hence,

$$P[\text{drawing red card}] = \frac{26}{52} = \frac{1}{2}$$

Understood in the example is that if one were to repeat the process a large number of times, a red card would appear in one-half of the trials. This is an example of the relative frequency interpretation. Now, what meaning could be associated with the statement, "The probability of the failure of a proposed structure is 1% ($P[\text{failure}] = 0.01$)"? The concept of repeated trials is meaningless: the structure will be built only once, and it will either fail or be successful during its design lifetime. It cannot do both. Here we have an example of the subjective interpretation of probability. It is a measure of information as to the likelihood of the occurrence of an outcome.

Subjective probability is generally more useful than the relative frequency concept in engineering applications. However, the basic rules governing both are identical. As an example, we note that both concepts specify the probability of an outcome to range from zero to one, inclusive. The lower limit indicates there is no likelihood of occurrence; the upper limit corresponds to a certain outcome.

$$\langle \textbf{Axiom I} \rangle \qquad 0 \leq P[A] \leq 1 \tag{15.2a}$$

The certainty of an outcome C is a probability of unity:

$$\langle \textbf{Axiom II} \rangle \qquad P[C] = 1 \tag{15.2b}$$

Equations (15.2a) and (15.2b) provide two of the three axioms of the theory of probability. The third axiom requires the concept of *mutually exclusive* outcomes. Outcomes are mutually exclusive if they cannot occur simultaneously. The third axiom states the probability of the occurrence of the sum of a number of mutually exclusive outcomes $A(1)$, $A(2)$, $\ldots$, $A(N)$ is the sum of their individual probabilities (addition rules), or

$$\langle \textbf{Axiom III} \rangle \qquad P[A(1)+A(2)+\cdots+A(N)] = P[A(1)]+P[A(2)]+\cdots+P[A(N)] \tag{15.2c}$$

As a very important application of these axioms consider the proposed design of a structure. After construction, only one of two outcomes can obtain in the absolute structural sense: either it is successful or it fails. These are mutually exclusive outcomes. They are also exhaustive in that, within the sense of the example, no other outcomes are possible. Hence, the second axiom, Eq. (15.2b), requires

$$P[\text{success} + \text{failure}] = 1$$

Since they are mutually exclusive, the third axiom specifies that

$$P[\text{success}] + P[\text{failure}] = 1$$

Discrete

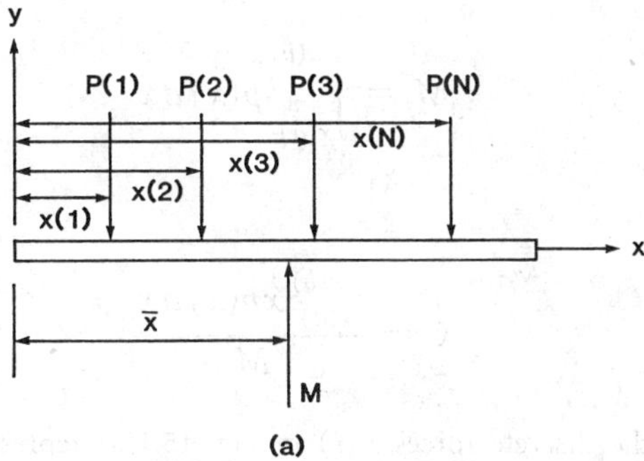

Continuous

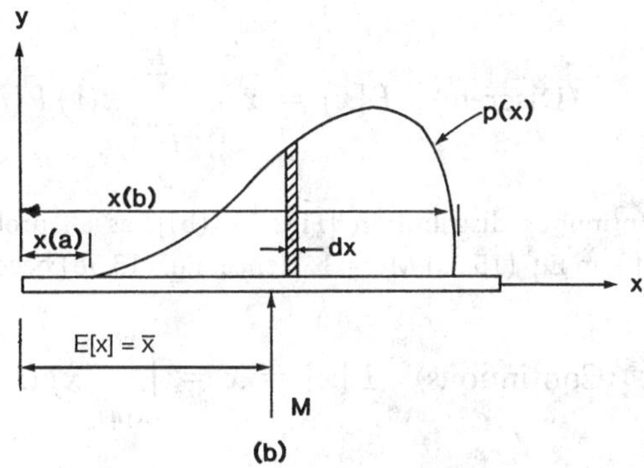

FIGURE 15.1 Equilibrant for discrete and continuous distributions.

The **probability** of the success of a structure is called its **reliability**, R. Symbolizing the probability of failure as $p(f)$, we have the important expression

$$R + p(f) = 1 \tag{15.3}$$

Moments

Consider a system of *discrete* parallel (vertical) forces, $P(1), P(2), \ldots, P(N)$, acting on a rigid beam at the respective distances $x(1), x(2), \ldots, x(N)$, as in Fig. 15.1(a). From statics we have that the magnitude of the *equilibrant*, M, is

$$M = \sum_{i=1}^{N} P(i) \tag{15.4a}$$

and its point of application, $\bar{x}$, is

$$\bar{x} = \frac{\sum_{i=1}^{N} x(i)P(i)}{\sum_{i=1}^{N} P(i)} \tag{15.4b}$$

For a continuously distributed parallel force system [Fig. 15.1(b)] over a finite distance, say from $x(a)$ to $x(b)$, the corresponding expressions are

$$M = \int_{x(a)}^{x(b)} p(x)\,dx \tag{15.5a}$$

and

$$\bar{x} = \frac{\int_{x(a)}^{x(b)} xp(x)\,dx}{M} \tag{15.5b}$$

Suppose now that the discrete forces $P(i)$ in Fig. 15.1(a) represent the frequencies of the occurrence of the N outcomes $x(1)$, $x(2)$, ..., $x(N)$. As the distribution is exhaustive, from axiom II, Eq. (15.2b), the magnitude of the equilibrant must be unity, $M = 1$. Hence, Eq. (15.4b) becomes

$$\langle\text{Discrete}\rangle \quad E[x] = \bar{x} = \sum_{i=1}^{N} x(i)\,P(i) \tag{15.6a}$$

Similarly, for the continuous distribution [Fig. 15.1(b)], as all probabilities $p(x)\,dx$ must lie between $x(a)$ and $x(b)$, in Eq. (15.5a) $M = 1$. Hence, Eq. (15.5b) becomes

$$\langle\text{Continuous}\rangle \quad E[x] = \bar{x} = \int_{x(a)}^{x(b)} xp(x)\,dx \tag{15.6b}$$

The symbol $E[x]$ in Eqs. (15.6) is called the **expected value** or the **expectation** or simply the *mean* of the variable x. As is true of the equilibrant, it is a measure of the central tendency, the *center of gravity* in statics.

Example 15.2. What is the expected value of the number of dots that will appear if a fair die is tossed?

Solution. Here each of the possible outcomes 1, 2, 3, 4, 5, and 6 has the equal probability of $P(i) = 1/6$ of appearing. Hence, from Eq. (15.6a),

$$E[\text{toss of a fair die}] = \tfrac{1}{6}[1 + 2 + 3 + 4 + 5 + 6] = 3.5$$

We note in the above example that the expected value of 3.5 is an impossible outcome. There is no face on the die that will show 3.5 dots; however, it is still the best measure of the central tendency.

Example 15.3. Find the expected value of a continuous probability distribution wherein all values are equally likely to occur (called a *uniform distribution*, Fig. 15.2) between $y(a) = 0$ and $y(b) = 1/2$.

Solution. From Eq. (15.5a), as $M = 1$ and $p(y) = C$ is a constant, we have

$$1 = C\int_0^{1/2} dy, \qquad C = 2$$

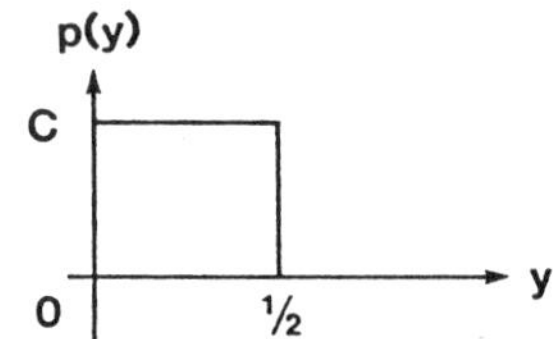

FIGURE 15.2 Uniform distribution.

From Eq. (15.6b), $E[y] = 2\int_0^{1/2} y\,dy = 1/4$, as expected.

As the variables x and y in the above examples are determined by the outcomes of random experiments, they are said to be **random variables.** In classical probability theory, random variables are generally represented by capital letters, such as X and Y. The individual values are customarily denoted by their corresponding lowercase letters, x and y; however, no such distinction will be made here.

The expected value (mean) provides the locus of the central tendency of the distribution of a random variable. To characterize other attributes of the distribution, recourse is had to higher moments. Again, returning to statics, a measure of the dispersion of the distribution of the force system about the centroidal axis, at $x = E[x]$ in Fig. 15.1(b), is given by the *moment of inertia* (the *second central moment*),

$$I(y) = \int_{x(a)}^{x(b)} (x - \bar{x})^2 p(x)\,dx \tag{15.7}$$

The equivalent measure of the scatter (variability) of the distribution of a random variable is called its **variance**, denoted in this text as $v[x]$ and defined as

$$\langle\text{Discrete}\rangle \quad v[x] = \sum_{\text{all}\,x(i)} [x(i) - \bar{x}]^2 p(i) \tag{15.8a}$$

$$\langle\text{Continuous}\rangle \quad v[x] = \int_{x(a)}^{x(b)} (x - \bar{x})^2 p(x)\,dx \tag{15.8b}$$

In terms of the expectation these can be written as

$$v[x] = E[(x - \bar{x})^2] \tag{15.9}$$

which, after expansion, leads to a form more amenable to computations:

$$v[x] = E[x^2] - (E[x])^2 \tag{15.10}$$

This expression is the equivalent of the parallel-axis theorem for the moment of inertia.

Example 15.4. Find the expected value and the variance of the *exponential distribution*, $p(x) = a\exp(-ax)$; $x > 0$, a is a constant.

Solution. We first show that $p(x)$ is a valid probability distribution:

$$\int_0^\infty p(x)\,dx = a\int_0^\infty e^{-ax}\,dx = 1, \quad \text{Q.E.D.}$$

The expected value is

$$E[x] = a\int_0^\infty xe^{-ax}\,dx = \frac{1}{a}$$

Continuing,

$$E[x^2] = a\int_0^\infty x^2 e^{-ax}\,dx = \frac{2}{a^2}$$

whence, using Eq. (15.10),

$$v[x] = \frac{2}{a^2} - \left(\frac{1}{a}\right)^2 = \frac{1}{a^2}$$

It is seen that the variance has the units of the square of those of the random variable. A more meaningful measure of dispersion of a random variable (x) is the positive square root of its variance (compare with radius of gyration of mechanics) called the **standard deviation,** $\sigma[x]$,

$$\sigma[x] = \sqrt{v[x]} \tag{15.11}$$

From the results of the previous example, it is seen that the standard deviation of the exponential distribution is $\sigma[x] = 1/a$.

An extremely useful relative measure of the scatter of a random variable (x) is its *coefficient of variation* $V(x)$, usually expressed as a percentage:

$$V(x) = \frac{\sigma[x]}{E[x]} \times 100(\%) \tag{15.12}$$

For the exponential distribution we found, $\sigma[x] = 1/a$ and $E[x] = 1/a$, hence V (exponential distribution) $= 100\%$. In Table 15.1 are given representative values of the coefficients of variation of some parameters common to civil engineering design. Original sources should be consulted for details.

The coefficient of variation expresses a measure of the reliability of the central tendency. For example, a mean value of a parameter of 10 with a coefficient of variation of 10% would indicate a standard deviation of 1, whereas a similar mean with a coefficient of variation of 20% would demonstrate a standard deviation of 2. The coefficient of variation has been found to be a fairly stable measure of variability for homogeneous conditions. Additional insight into the standard deviation and the coefficient of variation as measures of uncertainty is provided by Chebyshev's inequality [for the derivation see Lipschutz, 1965].

The spread of a random variable is often spoken of as its *range,* the difference between the largest and smallest outcomes of interest. Another useful measure is the range between the mean plus-and-minus h standard deviations, $\bar{x} \pm h\sigma[x]$, called the *h-sigma bounds* (see Fig. 15.3). If x is a random variable with mean value $\bar{x}$ and standard deviation σ, then Chebyshev's inequality states

$$P[(\bar{x} - h\sigma) \leq x \leq (\bar{x} + h\sigma) \geq \frac{1}{h^2} \tag{15.13}$$

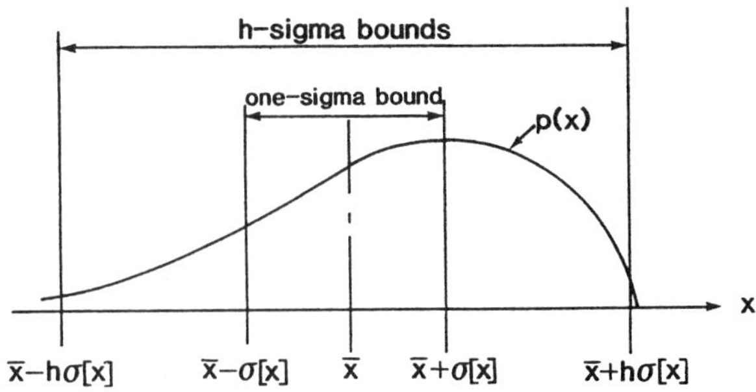

FIGURE 15.3 Range in h-sigma bounds.

Table 15.1 Representative Coefficients of Variation

Parameter	Coefficient of Variation, %	Source
	Soil	
Porosity	10	Schultze [1972]
Specific gravity	2	Padilla and Vanmarcke [1974]
Water content		
Silty clay	20	Padilla and Vanmarcke [1974]
Clay	13	Fredlund and Dahlman [1972]
Degree of saturation	10	Fredlund and Dahlman [1972]
Unit weight	3	Hammitt [1966]
Coefficient of permeability	(240 at 80% saturation to 90 at 100% saturation)	Nielsen *et al.* [1973]
Compressibility factor	16	Padilla and Vanmarcke [1974]
Preconsolidation pressure	19	Padilla and Vanmarcke [1974]
Compression index		
Sandy clay	26	Lumb [1966]
Clay	30	Fredlund and Dahlman [1972]
Standard penetration test	26	Schultze [1975]
Standard cone test	37	Schultze [1975]
Friction angle ϕ		
Gravel	7	Schultze [1972]
Sand	12	Schultze [1972]
c, strength parameter (cohesion)	40	Fredlund and Dahlman [1972]
	Structural Loads, 50-Year Maximum	
Dead load	10	Ellingwood *et al.* [1980]
Live load	25	Ellingwood *et al.* [1980]
Snow load	26	Ellingwood *et al.* [1980]
Wind load	37	Ellingwood *et al.* [1980]
Earthquake load	>100	Ellingwood *et al.* [1980]
	Structural Resistance	
Structural steel		
Tension members, limit state, yielding	11	Ellingwood *et al.* [1980]
Tension members, limit state, tensile strength	11	Ellingwood *et al.* [1980]
Compact beam, uniform moment	13	Ellingwood *et al.* [1980]
Beam, column	15	Ellingwood *et al.* [1980]
Plate, girders, flexure	12	Ellingwood *et al.* [1980]
Concrete members		
Flexure, reinforced concrete, grade 60	11	Ellingwood *et al.* [1980]
Flexure, reinforced concrete, grade 40	14	Ellingwood *et al.* [1980]
Flexure, cast-in-place beams	8–9.5	Ellingwood *et al.* [1980]
Short columns	12–16	Ellingwood *et al.* [1980]
	Ice	
Thickness	17	Bercha [1978]
Flexural strength	20	Bercha [1978]
Crushing strength	13	Bercha [1978]
Flow velocity	33	Bercha [1978]
	Wood	
Moisture	3	Borri *et al.* [1983]
Density	4	Borri *et al.* [1983]
Compressive strength	19	Borri *et al.* [1983]
Flexural strength	19	Borri *et al.* [1983]
Glue-laminated beams		
Live load	18	Galambos *et al.* [1982]
Snow load	18	Galambos *et al.* [1982]

Table 15.2 Probabilities for Range of Expected Values ± h-Sigma Units

h	Chebyshev's Inequality	Gauss's Inequality	Exact Exponential Distribution	Exact Normal Distribution	Exact Uniform Distribution
½	0	0	0.78	0.38	0.29
1	0	0.56	0.86	0.68	0.58
2	0.75	0.89	0.95	0.96	1.00
3	0.89	0.95	0.982	0.9973	1.00
4	0.94	0.97	0.993	0.999934	1.00

In words, it asserts that for any probability distribution (with finite mean and standard deviation) the probability that random values of the variate will lie within h-sigma bounds is at least $\left[1-(1/h^2)\right]$. Some numerical values are given in Table 15.2. It is seen that quantitative probabilistic statements can be made without complete knowledge of the probability distribution function; only its expected value and coefficient of variation (or standard deviation) are required. In this regard, the values for the coefficients of variation given in Table 15.1 may be used in the absence of more definitive information.

Example 15.5. The expected value for the ϕ-strength parameter of a sand is 30°. What is the probability that a random sample of this sand will have a ϕ-value between 20° and 40°?
Solution. From Table 15.1, $V(\phi) = 12\%$; hence, $\sigma[\phi]$ is estimated to be $(0.12)(30) = 3.6°$ and $h = (\phi - \bar{\phi})/\sigma = 10°/3.6° = 2.8$. Hence, $P[20° \leq \phi \leq 40°] \geq 1 - (1/2.8)^2 \geq 0.87$. That is, the probability is at least 0.87 that the ϕ-strength parameter will be between 20° and 40°.

If the unknown probability distribution function is symmetrical with respect to its expected value and the expected value is also its maximum value (said to be *unimodal*), it can be shown [Freeman, 1963] that

$$P[(\bar{x} - h\sigma) \leq x \leq (\bar{x} + h\sigma)] \leq 1 - \frac{4}{9h^2} \tag{15.14}$$

This is sometimes called *Gauss's inequality.* Some numerical values are given in Table 15.2.

Example 15.6. Repeat the previous example if it is assumed that the distribution of the ϕ-value is symmetrical with its maximum at the mean value ($\bar{\phi} = 30°$).
Solution. For this case, Gauss's inequality asserts

$$P[20° \leq \phi \leq 40°] \geq 1 - \frac{4}{9(2.8)^2} \geq 0.94$$

Recognizing symmetry we can also claim $P[\phi \leq 20°] = P[\phi \geq 40°] = 0.03$.

Example 15.7. Find the general expression for the probabilities associated with h-sigma bounds for the exponential distribution, $h \geq 1$.
Solution. From Example 4, we have (with $E[x] = 1/a, \sigma[x] = 1/a$)

$$P[(\bar{x} - h\sigma) \leq x \leq (x + h\sigma)] = \int_0^{(h+1)/a} ae^{-ax}\,dx = 1 - e^{-(h+1)}$$

Some numerical values are given in Table 15.2. The normal distribution noted in this table will be developed subsequently.

The results in Table 15.2 indicate that lacking information concerning a probability distribution beyond its first two moments, from a practical engineering point of view, it may be taken to range within 3-sigma bounds. That is, in Fig. 15.1(b), $x(a) \approx \bar{x} - 3\sigma[x]$ and $x(b) \approx \bar{x} + 3\sigma[x]$.

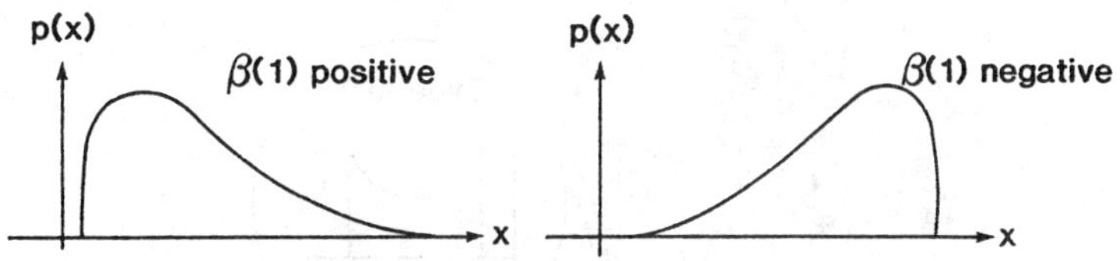

FIGURE 15.4 Coefficient of skewness.

For a symmetrical distribution all moments of odd order about the mean (central moments) must be zero. Consequently, any odd-ordered moment may be used as a measure of the degree of *skewness,* or *asymmetry,* of a probability distribution. The third central moment $E[(x - \bar{x})^3]$ provides a measure of the peakedness (called *kurtosis*) of a distribution.

As the units of the third central moment are the cube of the units of the variable, to provide an absolute measure of skewness, Pearson [1894, 1895] proposed that its value be divided by the standard deviation cubed to yield the dimensionless *coefficient of skewness,*

$$\beta(1) = \frac{E[(x - \bar{x})^3]}{(\sigma[x])^3} \tag{15.15}$$

If $\beta(1)$ is positive, the long tail of the distribution is on the right side of the mean; if it is negative, the long tail is on the left side (see Fig. 15.4). Pearson also proposed the dimensionless *coefficient of kurtosis* as a measure of peakedness:

$$\beta(2) = \frac{E[(x - \bar{x})^4]}{(\sigma[x])^4} \tag{15.16}$$

In Fig. 15.5 are shown the regions occupied by a number of probability distribution types, as delineated by their coefficients of skewness and kurtosis. Examples of the various types are shown schematically.

15.3 Probability Distributions

We note in Fig. 15.5 that the type IV distribution and the symmetrical type VII are unbounded (infinite) below and above. From the point of view of civil engineering applications this represents an extremely unlikely distribution. For example, all parameters or properties (see Table 15.1) are positive numbers (including zero).

The type V (the lognormal distribution), type III (the gamma), and type VI distributions are unbounded above. Hence, their use would be confined to those variables with an extremely large range of possible values. Some examples are the coefficient of permeability, the state of stress at various points in a body, the distribution of annual rainfall, and traffic variations.

The normal (Gaussian) distribution [$\beta(1) = 0, \beta(2) = 3$], even though it occupies only a single point in the universe of possible distributions, is the most frequently used of probability models. Some associated properties were given in Table 15.2. The normal distribution is the well-known symmetrical bell-shaped curve (see Fig. 15.6). Some tabular values are given in Table 15.3. The table is entered by forming the standardized variable z for the normal variate x as

$$z = \left| \frac{x - \bar{x}}{\sigma[x]} \right| \tag{15.17}$$

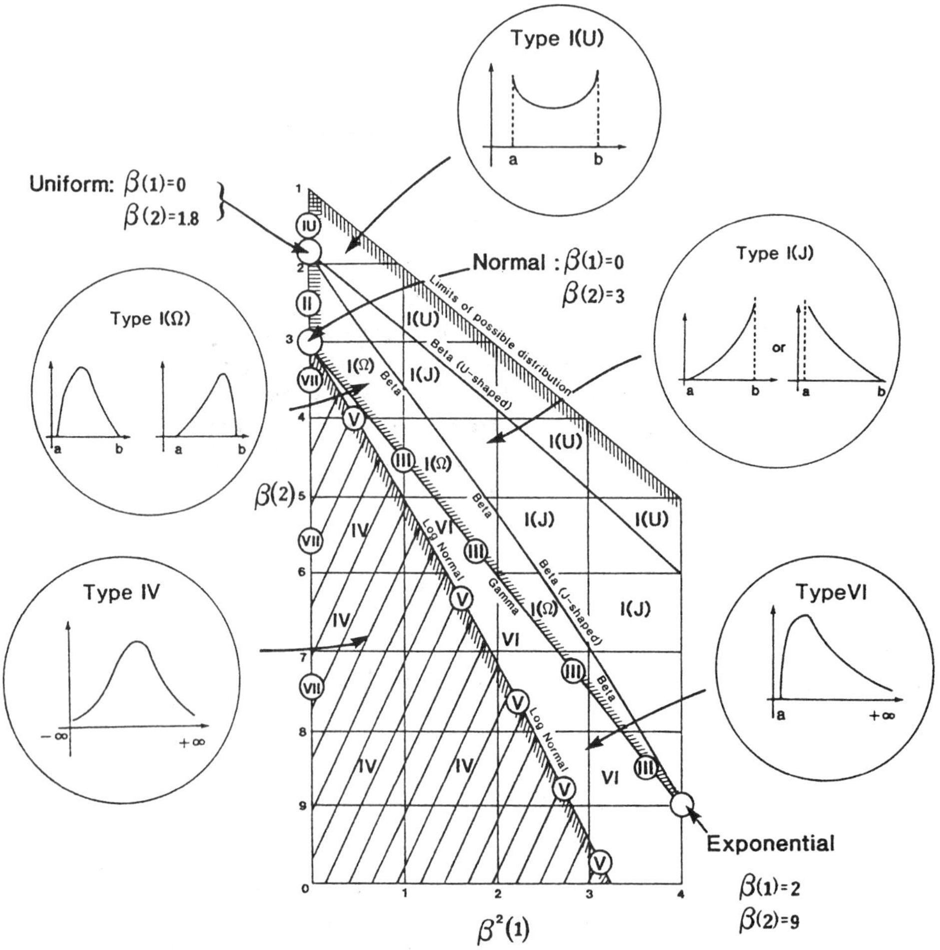

FIGURE 15.5 Space of probability distributions. (After Pearson, E. S. and Hartley, H. O. 1972. *Biometrika Tables for Statisticians,* vol. II. Cambridge University Press, London.)

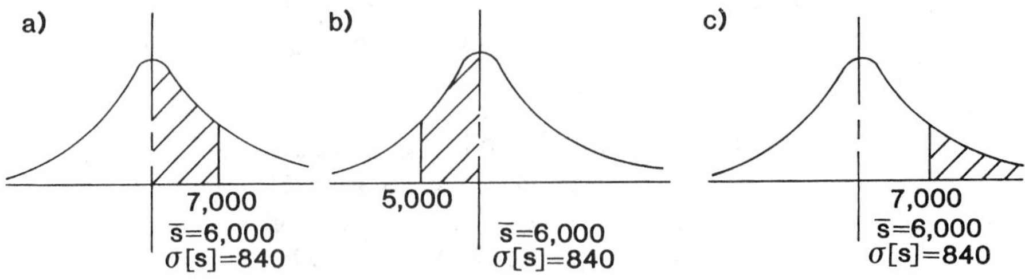

Tabular values yield the probabilities associated with the shaded areas shown in the figure: area $= \psi(z)$.

Example 15.8. Assuming the strength s of concrete to be a normal variate with an expected value of $\bar{s} = 6000$ psi and a coefficient of variation of 14%, find (a) $P[6000 \leq s \leq 7000]$, (b) $P[5000 \leq s \leq 6000]$, and (c) $P[s \geq 7000]$.

Table 15.3 Standard Normal Probability

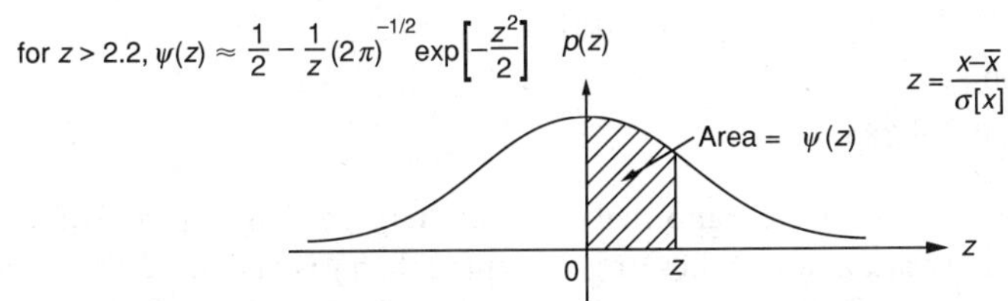

z	0	1	2	3	4	5	6	7	8	9
0	0	.003969	.007978	.011966	.015953	.019939	.023922	.027903	.031881	.035856
.1	.039828	.043795	.047758	.051717	.055670	.059618	.063559	.067495	.071424	.075345
.2	.079260	.083166	.087064	.090954	.094835	.098706	.102568	.106420	.110251	.114092
.3	.117911	.121720	.125516	.129300	.133072	.136831	.140576	.144309	.148027	.151732
.4	.155422	.159097	.162757	.166402	.170031	.173645	.177242	.180822	.184386	.187933
.5	.191462	.194974	.198466	.201944	.205401	.208840	.212260	.215661	.219043	.222405
.6	.225747	.229069	.232371	.235653	.238914	.242154	.245373	.248571	.251748	.254903
.7	.258036	.261148	.264238	.267305	.270350	.273373	.276373	.279350	.282305	.285236
.8	.288145	.291030	.293892	.296731	.299546	.302337	.305105	.307850	.310570	.313267
.9	.315940	.318589	.321214	.323814	.326391	.328944	.331472	.333977	.336457	.338913
1.0	.341345	.343752	.346136	.348495	.350830	.353141	.355428	.357690	.359929	.362143
1.1	.364334	.366500	.368643	.370762	.372857	.374928	.376976	.379000	.381000	.382977
1.2	.384930	.386861	.388768	.390651	.392512	.394350	.396165	.397958	.399727	.401475
1.3	.403200	.404902	.406582	.408241	.409877	.411492	.413085	.414657	.416207	.417736
1.4	.419243	.420730	.422196	.423641	.425066	.426471	.427855	.429219	.430563	.431888
1.5	.433193	.434476	.435745	.436992	.438220	.439429	.440620	.441792	.442947	.444083
1.6	.445201	.446301	.447384	.448449	.449497	.450529	.451543	.452540	.453521	.454486
1.7	.455435	.456367	.457284	.458185	.459070	.459941	.460796	.461636	.462462	.463273
1.8	.464070	.464852	.465620	.466375	.467116	.467843	.468557	.469258	.469946	.470621
1.9	.471283	.471933	.472571	.473197	.473610	.474412	.475002	.475581	.476148	.476705
2.0	.477250	.477784	.478308	.478822	.479325	.479818	.480301	.480774	.481237	.481691
2.1	.482136	.482571	.482997	.483414	.483823	.484222	.484614	.484997	.485371	.485738
2.2	.486097	.486447	.486791	.487126	.487455	.487776	.488089	.488396	.488696	.488989
2.3	.489276	.489556	.489830	.490097	.490358	.490613	.490863	.491106	.491344	.491576
2.4	.491802	.492024	.492240	.492451	.492656	.492857	.493053	.493244	.493431	.493613
2.5	.493790	.493963	.494132	.494297	.494457	.494614	.494766	.494915	.495060	.495201
2.6	.495339	.495473	.495604	.495731	.495855	.495975	.496093	.496207	.496319	.496427
2.7	.496533	.496636	.496736	.496833	.496928	.497020	.497110	.497197	.497282	.497365
2.8	.497445	.497523	.497599	.497673	.497744	.497814	.497882	.497948	.498012	.498074
2.9	.498134	.498193	.498250	.498305	.498359	.498411	.498462	.498511	.498559	.498605
3.0	.498650	.498694	.498736	.498777	.498817	.498856	.498893	.498930	.498965	.498999
3.1	.499032	.499065	.499096	.499126	.499155	.499184	.499211	.499238	.499264	.499289
3.2	.499313	.499336	.499359	.499381	499402	.499423	.499443	.499462	.499481	.499499
3.3	.499517	.499534	.499550	.499566	.499581	.499596	.499610	.499624	.499638	.499651
3.4	.499663	.499675	.499687	.499698	.499709	.499720	.499730	.499740	.499749	.499758
3.5	.499767	.499776	.499784	.499792	.499800	.499807	.499815	.499822	.499828	.499835
3.6	.499841	.499847	.499853	.499858	.499864	.499869	.499874	.499879	.499883	.499888
3.7	.499892	.499896	.499900	.499904	.499908	.499912	.499915	.499918	.499922	.499925
3.8	.499928	.499931	.499933	.499936	.499938	.499941	.499943	.499946	.499948	.499950
3.9	.499952	.499954	.499956	.499958	.499959	.499961	.499963	.499964	.499966	.499967

Solution. The standard deviation is $\sigma[s] = (0.14)(6000) = 840$ psi. Hence (see Fig. 15.6),

- $z = |(7000 - 6000)/840| = 1.19$, $\psi(1.19) = 0.383$.
- By symmetry, $P[5000 \le s \le 6000] = 0.383$.
- $P[s \ge 7000] = 0.500 - 0.383 = 0.117$.

As might be expected from its name, the lognormal distribution (type V) is related to the normal distribution. If x is a normal variate and $x = \ln y$ or $y = \exp(x)$ then y is said to have a lognormal distribution. It is seen that the distribution has a minimum value of zero and is unbounded above. The probabilities associated with lognormal variates can be obtained very easily from those of mathematically corresponding normal variates (see Table 15.3). If $E(y)$ and $V(y)$ are the expected value and coefficient of variation of a lognormal variate, the corresponding normal variate x will have the expected value and standard deviation [Benjamin and Cornell, 1970]:

$$(\sigma[x])^2 = \ln\left\{1 + [V(y)]^2\right\} \tag{15.18a}$$

$$E[x] = \ln E(y) - (\sigma[x])^2/2 \tag{15.18b}$$

Example 15.9. A live load of 20 kips is assumed to act on a footing. If the loading is assumed to be lognormally distributed, estimate the probability that a loading of 40 kips will be exceeded.
Solution. From Table 15.1 we have that the coefficient of variation for a live load, L, can be estimated as 25%; hence, from Eqs. (15.18) we have for the corresponding normal variate, x,

$$\sigma[x] = \sqrt{\ln[1 + (0.25)^2]} = 0.25 \tag{15.19a}$$

and

$$E[x] = \ln 20 - (0.25)^2/2 = 2.96 \tag{15.19b}$$

As $x = \ln L$, the value of the normal variate x equivalent to 40 K is $\ln 40 = 3.69$. We seek the equivalent normal probability $P[3.69 \le x]$. The standardized normal variate is $z = (3.69 - 2.96)/0.25 = 2.92$. Hence, using Table 15.3,

$$P[40 \le L] = 0.50 - \psi(2.92) = 0.500 - 0.498 = 0.002 \tag{15.20}$$

As was noted with respect to Fig. 15.5 many and diverse distributions (as well as the normal, lognormal, uniform, and exponential) can be obtained from the very versatile beta distribution. The beta distribution is treated in great detail by Harr [1977, 1987]. The latter reference also contains FORTRAN programs for beta probability distributions. Additional discussion is given below following Example 15.11.

15.4 Point Estimate Method—One Random Variable

Various probabilistic methods have been developed that yield measures of the distribution of functions of random variables [Harr, 1987]. The simple and very versatile procedure called the *point estimate method* (PEM) is advocated by this writer and will be developed in some detail. This method, first presented by Rosenblueth [1975] and later extended by him in 1981, has since seen considerable use and expansion by this writer and his coauthors (see references). The methodology is presented in considerable detail in Harr [1987].

Consider $y = p(x)$ to be the probability distribution of the random variable x. With analogy to Fig. 15.1, we replace the load on the beam by two reactions, $p(-)$ and $p(+)$, acting at $x(-)$ and $x(+)$, as shown in Fig. 15.7.

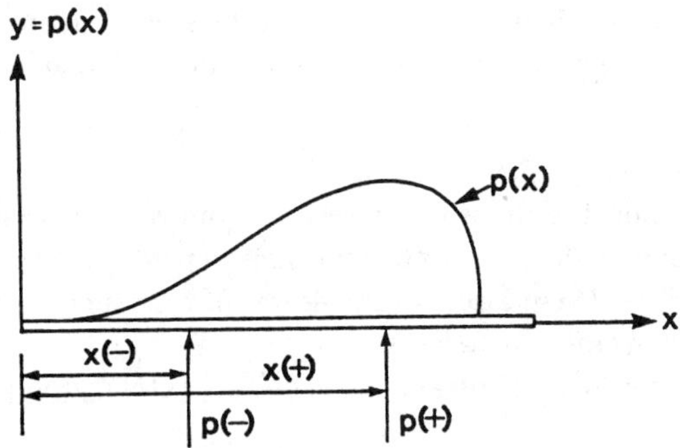

FIGURE 15.7 Point estimate approximations.

Pleading symmetry, probabilistic arguments produce for a random variable x:

$$p(+) = p(-) = \tfrac{1}{2} \tag{15.21a}$$

$$x(+) = \bar{x} + \sigma[x] \tag{15.21b}$$

$$x(-) = \bar{x} - \sigma[x] \tag{15.21c}$$

With the distribution $p(x)$ approximated by the point estimates $p(-)$ and $p(+)$, the moments of $y = p(x)$ are

$$E[y] = \bar{y} = p(-)y(-) + p(+)y(+) \tag{15.22a}$$

$$E[y^2] = p(-)y^2(-) + p(+)y^2(+) \tag{15.22b}$$

where $y(-)$ and $y(+)$ are the values of the function $p(x)$ at $x(-)$ and $x(+)$, respectively. These reduce to the simpler expressions

$$\bar{y} = \frac{y(+) + y(-)}{2} \tag{15.23a}$$

$$\sigma[y] = \left| \frac{y(+) - y(-)}{2} \right| \tag{15.23b}$$

Example 15.10. Estimate the expected value and the coefficient of variation for the well-known coefficient of active earth pressure $K_A = \tan^2(45 - \phi/2)$, if $\bar{\phi} = 30$.
Solution. With the standard deviation of the ϕ-parameter not given, we again return to Table 15.1, $V(\phi) = 12\%$, and $\sigma[\phi] = 3.6°$. Hence, $\phi(+) = 33.6°$, $\phi(-) = 26.4°$. Hence, $K_A(+) = 0.29$, $K_A(-) = 0.38$, and Eqs. (15.23) produce $\bar{K}_A = 0.34$, $\sigma[K_A] = 0.05$; hence, $V(K_A) = 13\%$.

15.5 Regression and Correlation

Thus far, only one-dimensional (*univariate*) random variables have been considered. More generally, concern is directed toward *multivariate formulations,* wherein there are two or more random variables. As an example, consider the *flexure formula*

$$s = Mc/I$$

where s is the stress at the extreme fiber at a distance c from the neutral axis acted on by a bending moment M for a beam in which I is the moment of inertia of the section. If the parameters M, c, and I are random variables (possess uncertainty), what can be said about the unit stress s? Needless to say, granted the probability distribution function of the stress, statements could be made with respect to the reliability of the beam relative to, say, a maximum allowable stress $\hat{s}$; for example,

$$\text{Reliability} = P[s \leq \hat{s}]$$

We first study the functional relationship between random variables called *regression analysis.* It is regression analysis that provides the grist of being able to predict the value of one variable from that of another or of others. The measure of the degree of correspondence within the developed relationship belongs to *correlation analysis.*

Let us suppose we have N pairs of data $[x(1), y(1)], \ldots, [x(N), y(N)]$ for which we postulate the linear relationship

$$y = Mx + B \tag{15.24}$$

where M and B are constants. Of the procedures available to estimate these constants (including best fit by eye), the most often used is the *method of least squares.* This method is predicated on minimizing the sum of the squares of the distances between the data points and the corresponding points on a straight line. That is, M and B are chosen so that

$$\Sigma(y - Mx - B)^2 = \text{Minimum}$$

This requirement is met by the expressions

$$M = \frac{N\,\Sigma xy - \Sigma x\,\Sigma y}{N\,\Sigma x^2 - (\Sigma x)^2} \tag{15.25a}$$

$$B = \frac{\Sigma x^2\,\Sigma y - \Sigma x\,\Sigma xy}{N\,\Sigma x^2 - (\Sigma x)^2} \tag{15.25b}$$

It should be emphasized that a straight line fit was assumed. The reasonableness of this assumption is provided by the **correlation coefficient** ρ, defined as

$$\rho = \frac{\text{cov}[x, y]}{\sigma[x]\,\sigma[y]} \tag{15.26}$$

where $\sigma[x]$ and $\sigma[y]$ are the respective standard deviations and $\text{cov}[x, y]$ is their *covariance.* The covariance is defined as

$$\text{cov}[x, y] = \frac{1}{N}\sum_{i=1}^{N}[x(i) - \bar{x}][y(i) - \bar{y}] \tag{15.27}$$

With analogy to statics the covariance corresponds to the product of inertia.

In concept, the correlation coefficient is a measure of the tendency for two variables to vary together. This measure may be zero, negative, or positive; wherein the variables are said to be *uncorrelated, negatively correlated,* or *positively correlated.* The variance is a special case of the covariance as

$$\text{cov}[x, x] = v[x] \tag{15.28}$$

Application of their definitions produces [Ditlevsen, 1981] the following identities (a, b, and c are constants):

$$E[a + bx + cy] = a + bE[x] + cE[y] \tag{15.29a}$$

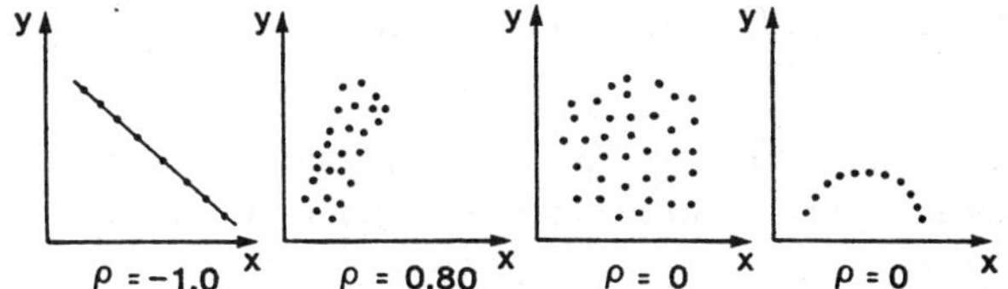

FIGURE 15.8 Example of scatter and correlation coefficients.

$$v[a + bx + cy] = b^2 v[x] + c^2 v[y] + 2bc\ \mathrm{cov}[x, y] \tag{15.29b}$$

$$\mathrm{cov}[x, y] \leq \sigma[x]\sigma[y] \tag{15.29c}$$

$$v[a + bx + cy] = b^2 v[x] + c^2 v[y] + 2bc\sigma[x]\sigma[y]\rho \tag{15.29d}$$

Equation (15.29c) demonstrates that the correlation coefficient, Eq. (15.26), must satisfy the condition

$$-1 \leq \rho \leq +1 \tag{15.30}$$

If there is perfect correlation between variables in the same direction, $\rho = +1$. If there is perfect correlation in opposite directions (one variable increases as the other decreases), $\rho = -1$. If some scatter exists, $-1 < \rho < +1$, with $\rho = 0$ if there is no correlation. Some examples are shown in Fig. 15.8.

15.6 Point Estimate Method—Several Random Variables

Rosenblueth [1975] generalized the methodology for any number of correlated variables. For example, for a function of three random variables—say, $y = y[x(1), x(2), x(3)]$—where $\rho(i, j)$ is the correlation coefficient between variables $x(i)$ and $x(j)$,

$$E[y^N] = p(+++)y^N(+++) + p(++-)y^N(++-) + \cdots + p(---)y^N(---) \tag{15.31a}$$

where

$$y(\pm\pm\pm) = y[\bar{x}(1) \pm \sigma[x1], \bar{x}(2) \pm \sigma[x2], \bar{x}(3) \pm \sigma[x3]] \tag{15.31b}$$

$$p(+++) = p(---) = \frac{1}{2^3}[1 + \rho(1, 2) + \rho(2, 3) + \rho(3, 1)]$$

$$p(++-) = p(--+) = \frac{1}{2^3}[1 + \rho(1, 2) - \rho(2, 3) - \rho(3, 1)]$$

$$p(+-+) = p(-+-) = \frac{1}{2^3}[1 - \rho(1, 2) - \rho(2, 3) + \rho(3, 1)]$$

$$p(+--) = p(-++) = \frac{1}{2^3}[1 - \rho(1, 2) + \rho(2, 3) - \rho(3, 1)] \tag{15.31c}$$

where $\sigma[xi]$ is the standard deviation of $x(i)$. The sign of $\rho(i, j)$ is determined by the multiplication rule of i and j; that is, if the sign of $i = (-)$, and of $j = (+)$, then $(i)(j) = (-)(+) = (-)$.

Equation (15.31a) has $2^3 = 8$ terms, all permutations of the three +s and −s. In general for M variables there are 2^M terms and $M(M-1)/2$ correlation coefficients, the number of combinations of M objects taken two at a time. The coefficient on the right-hand side of Eqs. (15.31c), in general, is $(1/2)^M$.

Example 15.11. The recommendation of the American Concrete Institute [Galambos *et al.*, 1982] for the design of reinforced concrete structures is (in simplified form)

$$R \geq 1.6D + 1.9L$$

where R is the strength of the element, D is the dead load, and L is the lifetime live load. (a) If $\bar{D} = 10$, $\bar{L} = 8$, $V(D) = 10\%$, $V(L) = 25\%$, and $\rho(D, L) = 0.75$, find the expected value and standard deviation of R for the case $R = 1.6D + 1.9L$. (b) If the results in part (a) generate a normal variate and the maximum strength of the element R is estimated to be 40, estimate the implied probability of failure.

Solution. The solution is developed in Fig. 15.9.

Generalizations of the PEM to more than three random variables are given by Harr [1987]. The PEM procedure yields the first two moments of the dependent random functions under consideration. Functional distributions must then be obtained and statements must be made concerning the

a) R=1.6D+1.9L

Variable, x	$\bar{x}$	$\sigma[x]$	x(+)	x(−)
D	10	1	11	9
L	8	2	10	6

$\rho(D,L) = +0.75$

	R(ij)	R(ij)^2
R(++):	36.6	1340
R(+−):	29.0	841
R(−+):	33.4	1116
R(−−):	25.8	666

$p(++) = \frac{1}{4}(1+\rho) = 0.44$

$p(+-) = \frac{1}{4}(1-\rho) = 0.06$

$p(-+) = \frac{1}{4}(1-\rho) = 0.06$

$p(--) = \frac{1}{4}(1+\rho) = 0.44$

$E[R] = \bar{R} = \Sigma R(ij)p(ij)$
$= 0.44(36.6+25.8) + 0.06(29.0+33.4)$
$= \underline{31.20}$

$E[R^2] = \Sigma R(ij)^2 p(ij)$
$= 0.44(1340+666) + 0.06(841+1116)$
$= \underline{1000.06}$

$\upsilon[R] = E[R^2] - (E[R])^2 = 1000.06 - (31.20)^2 = \underline{26.62}$
$\sigma[R] = 5.16$; $V(R) = 16.5\%$

From Eq. (15.29a), the exact solution for $E[R] = 1.6\bar{D} + 1.9\bar{L} = \underline{31.20}$
Eq. (15.29d), the exact solution for $\upsilon[R] = (1.6)^2\upsilon[D] + (1.9)^2\upsilon[L]$
$+2(1.6)(1.9)(1)(2)(0.75) = \underline{26.12}$

Of course, for this example the exact solution is easier to obtain. This is not generally the case.

b) $P_f = P[R \geq 40] = \frac{1}{2} - \psi\left[\frac{40-31.20}{5.16}\right] = \frac{1}{2} - \psi[1.71] = \underline{0.044}$ (Table 3)

The exact solution is $\underline{0.043}$

FIGURE 15.9 Solution to Example 15.11.

probabilities of events. Inherent in the assumption of the form of a particular distribution is the imposition of the limits or range of its applicability. For example, for the normal it is required that the variable range from $-\infty$ to $+\infty$; the range of the lognormal and the exponential is 0 to $+\infty$. Such assignments may not be critical if knowledge of distributions is desired in the vicinity of their expected values and their coefficients of variation are not excessive (say, less than 25%). On the other hand, estimates of reliability (and of the probability of failure) are vested in the tails of distributions. It is in such characterizations that the beta distribution is of great value. If the limits are known, zero is often an option, and probabilistic statements can readily be obtained. In the event that limits are not defined, the specification of a range of the mean plus or minus three standard deviations would generally place the generated beta distribution well within the accuracy required for most geotechnical engineering applications (see Table 15.2).

15.7 Reliability Analysis

Capacity-Demand

The adequacy of a proposed design in geotechnical engineering is generally determined by comparing the estimated resistance of the system to that of the imposed loading. The resistance is the **capacity** C (or strength) and the loading is the induced **demand** D imposed on the structure. In the present writing, because of its greater generality, we shall use a *capacity-demand* concept. Some common examples are the bearing capacity of a soil and the column loads, allowable and computed maximum stresses, traffic capacity and anticipated traffic flow on a highway, culvert sizes and the quantity of water to be accommodated, and structural capacity and earthquake loads.

Conventionally, the designer forms the well-known factor of safety as the ratio of the single-valued nominal values of capacity $\tilde{C}$ and demand $\tilde{D}$ (see Ellingwood *et al.*, 1980), depicted in Fig. 15.10(a),

$$FS = \frac{\tilde{C}}{\tilde{D}} \tag{15.32}$$

For example, if the allowable load is 400 tons per square foot and the maximum calculated load is 250 tons per square foot, the conventional factor of safety would be 1.6. The design is considered satisfactory if the calculated factor of safety is greater than a prescribed minimum value learned from experience with such designs. Thus, in concept, in the above example, if a factor of safety of 1.6 was considered intolerable, the system would be redesigned to decrease the maximum induced load.

In general, the demand function will be the resultant of the many uncertain components of the system under consideration (vehicle loadings, wind loadings, earthquake accelerations, location of the water table, temperatures, quantities of flow, runoff, and stress history, to name only a few). Similarly, the capacity function will depend on the variability of material parameters, testing errors, construction procedures and inspection supervision, ambient conditions, and so on.

A schematic representation of the capacity and demand functions as probability distributions is shown in Fig. 15.10(b). If the maximum demand ($D_{\max}$) exceeds the minimum capacity ($C_{\min}$), the distributions overlap (shown shaded), and there is a nonzero probability of failure.

The difference between the capacity and demand functions is called the safety margin (S); that is,

$$S = C - D \tag{15.33}$$

Obviously, the safety margin is itself a random variable, as shown in Fig. 15.10(c). Failure is associated with that portion of its probability distribution wherein it becomes negative (shown

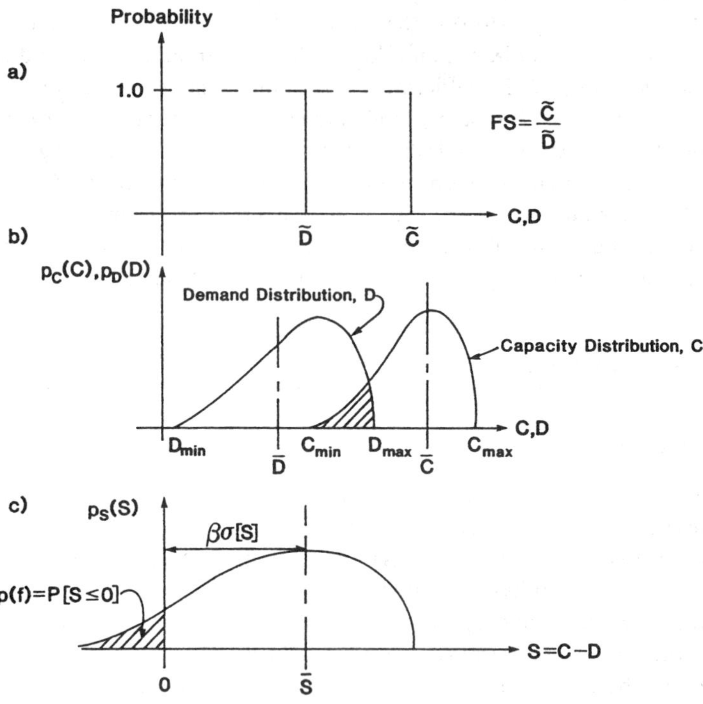

FIGURE 15.10 (a) Conventional factor of safety, (b) capacity–demand model, (c) safety margin.

shaded); that is, that portion wherein $S = C - D \leq 0$. As the shaded area is the probability of failure $p(f)$, we have

$$p(f) = P[(C - D) \leq 0] = P[S \leq 0] \tag{15.34}$$

Reliability Index

The number of standard deviations that the mean value of the safety margin is beyond $S = 0$, Fig. 15.10(c), is called the *reliability index*, β; that is,

$$\beta = \frac{\bar{S}}{\sigma[S]} \tag{15.35a}$$

The reliability index is seen (compare also with h-sigma bounds in Table 15.2) to be the reciprocal of the coefficient of variation of the safety margin, or

$$\beta = \frac{1}{V(S)} \tag{15.35b}$$

Example 15.12. Obtain a general expression for the reliability index in terms of the first two moments of the capacity and the demand functions.

Solution. From Eq. (15.29a) we have $E[S] = E[C] - E[D] = \bar{C} - \bar{D}$. Equation (15.29c) produces $\sigma^2[S] = \sigma^2[C] + \sigma^2[D] - 2\rho\sigma[C]\sigma[D]$. Hence,

$$\beta = \frac{\bar{C} - \bar{D}}{\sqrt{\sigma^2[C] + \sigma^2[D] - 2\rho\sigma[C]\sigma[D]}} \tag{15.36}$$

It is seen that β is a maximum for a perfect positive correlation and a minimum for a perfect negative correlation.

It can be shown that the sum or difference of two normal variates is also a normal variate [Haugen, 1968]. Hence, if it is assumed that the capacity and demand functions are normal variates, it follows directly from Example 15.12 that

$$p(f) = \tfrac{1}{2} - \psi[\beta] \tag{15.37}$$

where $\psi[\beta]$ is standard normal probability as given in Table 15.3.

15.8 Recommended Procedure

We list at this point some desirable attributes of a reliability-based design procedure:

1. It should account for the pertinent capacity and demand factors, their components, and their interactions.
2. It should produce outputs that can be related to the expected performance during the design life of the system under consideration.
3. It should employ as input into formulations quantities, parameters, or material characterizations that can be ascertained within the present state of the art.
4. It should not disregard indices currently considered to be pertinent, such as factor of safety or reliability index. It should serve to supplement this knowledge and reduce uncertainty.
5. Ideally, mathematical computations should be reduced to a minimum.

All of the above can be accommodated by an extension of the point estimate method. The recommended procedure is as follows, where applicable:

1. Using PEM, or an equally valid probabilistic formulation, obtain the expected values and standard deviations of the capacity and demand functions: $E[C]$, $E[D]$, $\sigma[C]$, $\sigma[D]$.
2. Calculate the expected value and standard deviation of the safety margin, $E[S]$, $\sigma[S]$.
3. Fit a beta distribution (and normal distribution, as a check) to the safety margin, using appropriate upper and lower bounds. If unknown, take them as $E[S] \pm 3\sigma[S]$; see Table 15.2.
4. Obtain the probability of failure, $p(f) = P[S \leq 0]$, the reliability, central factor of safety, and reliability index, as appropriate.

Example 15.13. The ultimate bearing capacity Q per unit length of a long footing of width B founded at a depth D below the ground surface, Fig. 15.11(a), is

$$Q = \frac{\gamma B^2}{2} N_\gamma + \gamma DBN_q + cBN_c$$

where γ is the unit weight of the soil, c is the c-parameter of strength (sometimes called the *cohesion*). The dimensionless factors N_γ, N_q, and N_c, called *bearing capacity factors*, are functions of the friction angle ϕ as given by the tabulated values in Fig. 15.11(b).

(a) A very long footing 8 ft wide is founded at a depth of 4 ft in a soil with the parameters

Parameter, x	Expected Value	Standard Deviation	$x(+)$	$x(-)$
γ	110 lb/ft^3	0	110	110
ϕ	35°	5°	40	30
c	200 lb/ft^2	80 lb/ft^2	280	120

a)

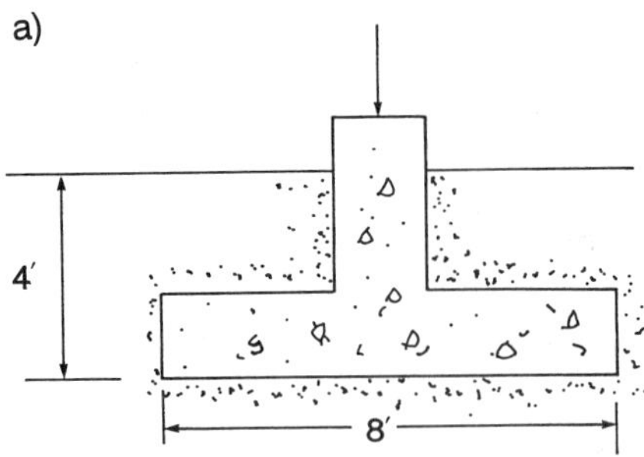

b)

Bearing capacity factors

ϕ	$N\gamma$	N_q	N_c
0	0.00	1.00	5.14
5	0.45	1.57	6.49
10	1.22	2.47	8.34
15	2.65	3.94	10.98
20	5.39	6.40	14.83
25	10.88	10.66	20.72
30	22.40	18.40	30.14
35	48.83	33.30	46.12
40	109.41	64.20	75.31
42.5	170.25	91.90	99.20
45	271.75	133.87	134.87
50	762.86	319.06	266.88

FIGURE 15.11 Example 15.13.

The correlation coefficient $\rho(\phi, c) = -0.50$. Estimate the expected value and standard deviation of the bearing capacity.

(b) If a central factor of safety (CFS) of 4 is required and it is assumed that $V(D) = 50\%$, estimate the probability of failure.

Solution. (a) Forming the required values, as the bearing capacity factors are functions of ϕ only. From Fig. 15.11(b),

$$N_\gamma(+) = 109.41 \qquad N_\gamma(-) = 22.40$$
$$N_q(+) = 64.20 \qquad N_q(-) = 18.40$$
$$N_c(+) = 75.31 \qquad N_c(-) = 30.14$$

Forming the respective values, $Q(\phi, c)$ in tons,

$Q(i, j)$	$Q^2(i, j)$
$Q(++)$: 389.9	152,020
$Q(+-)$: 341.7	116,758
$Q(-+)$: 105.6	11,144
$Q(--)$: 86.3	7,444

and

$$p(++) = p(--) = \tfrac{1}{4}(1 + \rho) = \tfrac{1}{8}$$

$$p(+-) = p(-+) = \tfrac{1}{4}(1 - \rho) = \tfrac{3}{8}$$

and

$$E[Q] = \bar{Q} = \Sigma Q(ij)p(ij) = 227.3 \text{ tons/ft}$$
$$E[Q]^2 = \Sigma Q^2(ij)p(ij) = 67{,}896$$
$$v[Q] = E[Q^2] - (E[Q])^2 = 16{,}231$$

and

$$\sigma[Q] = 127.40 \text{ tons}, \quad V(Q) = 56\%$$

(b) For a CFS = 4, $\bar{D} = 56.8$. As $V(D) = 50\%$, $\sigma[D] = 28.4$. Forming the characteristics of the safety margin, with $\rho(Q, D) = +3/4$, we have $E[S] = \bar{C} - \bar{D} = 170.50$, $\sigma[S] = 143.72$, $S_{\min} = -152.75$, $S_{\max} = 493.75$, $\beta = 170.50/107.75 = 1.58$.

If S is taken as a beta variate $\quad p(f) = 0.059$
If S is taken to be normal $\quad p(f) = 0.057$

Defining Terms

Capacity: The ability to resist an induced demand; resistance or strength of entity.
Correlation coefficient: Measure of the compliance between two variables.
Demand: Applied loading or energy.
Expected value, expectation: Weighted measure of central tendency of a distribution.
Probability: Quantitative measure of a state of knowledge.
Random variable: An entity whose measure cannot be predicted with certainty.
Regression: Means of obtaining a functional relationship among variables.
Reliability: Probability of an entity (or system) performing its required function adequately for a specified period of time under stated conditions.
Standard deviation: Square root of variance.
Variance: Measure of scatter of variable.

References

Benjamin, J. R. and Cornell, C. A. 1970. *Probability, Statistics, and Decision for Civil Engineers.* McGraw-Hill, New York.

Bercha, F. G. 1978. Application of probabilistic methods in ice mechanics. Preprint 3418, ASCE.

Borri, A., Ceccotti, A., and Spinelli, P. 1983. Statistical analysis of the influence of timber defects on the static behavior of glue laminated beams, vol. 1. In *4th Int. Conf. Appl. Stat. Prob. Soil Struct. Eng.* Florence, Italy.

Ditlevsen, O. 1981. *Uncertainty Modeling.* McGraw-Hill, New York.

Ellingwood, B., Galambos, T. V., MacGregor, J. G., and Cornell, C. A. 1980. Development of a probability based load criterion for American National Standard A58. *Nat. Bur. Stand. Spec. Publ. 577,* Washington, D.C.

Fredlund, D. G. and Dahlman, A. E. 1972. Statistical Geotechnical Properties of Glacial Lake Edmonton Sediments. In *Statistics and Probability in Civil Engineering.* Hong Kong University Press (Hong Kong International Conference), ed. P. Lumb, distributed by Oxford University Press, London.

Freeman, H. 1963. *Introduction to Statistical Inference.* John Wiley & Sons, New York.

Galambos, T. V., Ellingwood, B., MacGregor, J. G., and Cornell, C. A. 1982. Probability based load criteria: Assessment of current design practice. *J. Struct. Div., ASCE.* 108(ST5).

Grivas, D. A. and Harr, M. E. 1977. Reliability with respect to bearing capacity failures of structures on ground. *9th Int. Conf. Soil Mech. Found. Eng.* Tokyo, Japan.

Grivas, D. and Harr, M. E. 1979. A reliability approach to the design of soil slopes. *7th Eur. Conf. on S.M.A.F.E.,* Brighton, England.

Hammitt, G. M. 1966. Statistical analysis of data from a comparative laboratory test program sponsored by ACIL, United States Army Engineering Waterways Experiment Station, Corps of Engineers, Miscellaneous Paper No. 4-785.

Harr, M. E. 1976. Fundamentals of probability theory. *Transp. Res. Rec. 575.*

Harr, M. E. 1977. *Mechanics of Particulate Media: A Probabilistic Approach.* McGraw-Hill, New York.

Harr, M. E. 1987. *Reliability-Based Design in Civil Engineering.* McGraw-Hill, New York.

Haugen, E. B. 1968. *Probabilistic Approaches to Design.* John Wiley & Sons, New York.

Lipshutz, S. 1965. *Schaum's Outline of Theory and Problems of Probability.* McGraw-Hill, New York.

Lumb, P. 1972. Precision and Accuracy of Soils Tests. In *Statistics and Probability in Civil Engineering.* Hong Kong University Press (Hong Kong International Conference), ed. P. Lumb, distributed by Oxford University Press, London.

Lumb, P. 1974. Application of Statistics in Soil Mechanics. In *Soil Mechanics—New Horizons,* ed. I. K. Lee. American Elsevier, New York.

Padilla, J. D. and Vanmarcke, E. H. 1974. Settlement of structures on shallow foundations: A Probabilistic analysis. Research Report R74-9, M.I.T.

Pearson, E. S. and Hartley, H. O. 1972. *Biometrika Tables for Statisticians,* vol. II. Cambridge University Press, London.

Pearson, K. 1894, 1895. Skew Variations in Homogeneous Material, Contributions to the Mathematical Theory of Evolution. *Philos. Trans. R. Soc.* Vol. 185 and Vol. 186.

Rosenblueth, E. 1975. Point estimates for probability moments. *Proc. Natl. Acad. Sci., USA.* 72(10).

Rosenblueth, E. 1981. Two-point estimates in probabilities. *Appl. Math. Modeling.* Vol. 5.

Schultze, E. 1972. Frequency Distributions and Correlations of Soil Properties. In *Statistics and Probability in Civil Engineering.* Hong Kong University Press (Hong Kong International Conference), ed. P. Lumb, distributed by Oxford University Press, London.

For Further Information

Ang, A.H.-S. and Tang, W. H. 1975. *Probability Concepts in Engineering Planning and Design, Vol. I—Basic Principles.* John Wiley & Sons, New York.

Guymon, G. L., Harr, M. E., Berg, R. L., and Hromadka, T. V. 1981. A probabilistic-deterministic analysis of one-dimensional ice segregation in a freezing soil column. *Cold Reg. Sci. Tech.* 5: 127–140.

Hahn, G. J. and Shapiro, S. S. 1967. *Statistical Models in Engineering.* John Wiley & Sons, New York.

Jayne, E. T. 1978. Where Do We Stand on Maximum Entropy? In *The Maximum Entropy Formalism,* ed. R. D. Levine and M. Tribus. MIT Press, Cambridge, MA.

Tribus, M. 1969. *Rational Descriptions, Decisions and Designs.* Pergamon Press, New York.

Whitman, R. Y. 1984. Evaluating calculated risk in geotechnical engineering. *J. Geotech. Eng., ASCE.* 110(2).

16

Strength and Deformation

Dana N. Humphrey
University of Maine

16.1 Introduction

The shear strength of soil is generally characterized by the Mohr-Coulomb failure criterion. This criterion states that there is a linear relationship between the shear strength on the failure plane at failure (τ_{ff}) and the normal stress on the failure plane at failure (σ_{ff}) as given in the following equation:

$$\tau_{ff} = \sigma_{ff} \tan \phi + c \tag{16.1}$$

where ϕ is the friction angle and c is the intrinsic cohesion. The strength parameters (ϕ, c) are used directly in many stability calculations, including bearing capacity of shallow footings, slope stability, and stability of retaining walls. The line defined by Eq. (16.1) is called the **failure envelope.** A **Mohr's circle** tangent to a point on the failure envelope (σ_{ff}, τ_{ff}) intersects the x-axis at the **major** and **minor principal stresses** at failure (σ_{1f}, σ_{3f}) as shown in Fig. 16.1. For many soils, the failure envelope is actually slightly concave down rather than a straight line. However, for most situations Eq. (16.1) can be used with a reasonable degree of accuracy provided the strength parameters are determined over the range of stresses that will be encountered in the field problem. For a comprehensive review of Mohr's circles and the Mohr-Coulomb failure criterion, see Lambe and Whitman [1969] and Holtz and Kovacs [1981].

16.2 Strength Parameters Based on Effective Stresses and Total Stresses

The shear strength of soils is governed by **effective stress** ($\boldsymbol{\sigma'}$), which is given by

$$\sigma' = \sigma - u \tag{16.2}$$

0-8493-8953-4/95/$0.00 + $.50

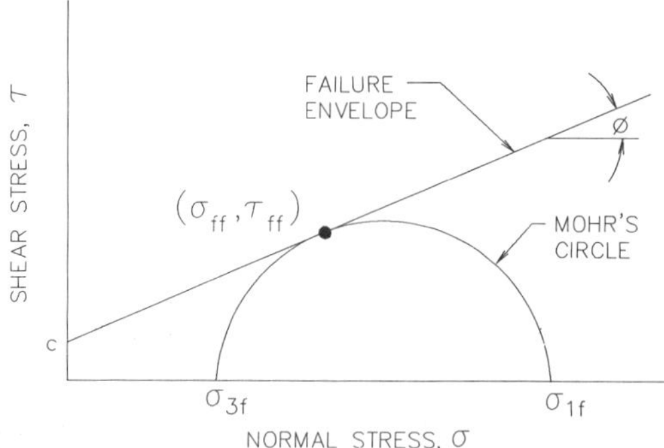

FIGURE 16.1 Mohr-Coulomb failure criteria.

where σ is the **total stress** and u is the pore water pressure. Equation (16.1) written in terms of effective stresses is

$$\tau_{\text{ff}} = (\sigma_{\text{ff}} - u)\tan\phi' + c' = \sigma'_{\text{ff}}\tan\phi' + c' \tag{16.3}$$

where σ'_{ff} is the effective normal stress on the failure plane at failure, and ϕ' and c' are the friction angle and cohesion based on effective stresses. Use of Eq. (16.3) requires knowledge of the pore pressure, but this can be difficult to predict for fine-grained soils. In this case, it is often convenient to assess stability of a structure based on total applied stresses and strength parameters based on total stresses. Determination of strength parameters based on effective and total stresses is discussed further in the following sections.

16.3 Laboratory Tests for Shear Strength

The choice of appropriate shear strength tests for a particular project depends on the soil type, whether the parameters will be used in a total or effective stress analysis, and the relative importance of the structure. Laboratory tests are discussed in this chapter and field tests were discussed in Chapter 14. Common laboratory tests include direct shear, triaxial, direct simple shear, unconfined compression, and laboratory vane. The applicability, advantages, disadvantages, and sources of additional information for each test are summarized in Table 16.1.

Of the available tests, the triaxial test is often used for important projects because of the advantages listed in Table 16.1. The types of triaxial tests are classified according to their drainage conditions during the consolidation and shearing phases of the tests. In a consolidated-drained (CD) test the sample is fully drained during both the consolidation and shear phases of the test. This test can be used to determine the strength parameters based on effective stresses for both coarse- and fine-grained soils. However, the requirement that the sample be sheared slowly enough to allow for complete drainage makes this test impractical for fine-grained soils. In a consolidated-undrained (CU) test the sample is drained during consolidation but is sheared with no drainage. This test can be used for fine-grained soils to determine strength parameters based on total stresses or, if pore pressures are measured during shear, strength parameters based on effective stresses. For the latter use, a CU test is preferred over a CD test because a CU test can be sheared much more quickly than a CD test. In an unconsolidated-undrained (UU) test the sample is undrained during both the consolidation and shear phases. The test can be used to determine the undrained shear strength of fine-grained soils. Further discussion of triaxial tests is given in Holtz and Kovacs [1981] and Head [1982, 1986].

Table 16.1 Summary of Common Shear Strength Tests

Test Type	Applicability	Advantages	Disadvantages	Additional Information
Direct shear	a. Effective strength parameters for coarse-grained and fine-grained soils	a. Simple and inexpensive b. Thin sample allows for rapid drainage of fine-grained soils	a. Only for drained conditions b. Failure plane forced to occur at joint in box c. Nonuniform distribution of stress and strain d. No stress-strain data	a. ASTM D3080* b. U.S. Army, 1970 c. Saada and Townsend, 1981 d. Head, 1982
Triaxial	a. Effective and total strength parameters for coarse-grained and fine-grained soils b. Compared to direct shear tests, triaxial tests are preferred for fine-grained soils	a. Easy to control drainage b. Useful stress-strain data c. Can consolidate sample hydrostatically or to insitu K_0 state of stress d. Can simulate various loading conditions	a. Apparatus more complicated than other types of tests b. Drained tests on fine-grained soils must be sheared very slowly	a. ASTM D2850* b. U.S. Army, 1970 c. Donaghe *et al.*, 1988 d. Head, 1982 e. Head, 1986
Direct simple shear	a. Most common application is undrained shear strength of fine-grained soils	a. K_0 consolidation b. Gives reasonable values of undrained shear strength for design use	a. Nonuniform distribution of stress and strain	a. Bjerrum and Landva, 1966 b. Saada and Townsend, 1981
Unconfined	a. Undrained shear strength of 100% saturated samples of homogenous, unfissured clay b. Not suitable as the only basis for design on critical projects	a. Very rapid and inexpensive	a. Not applicable to soils with fissures, silt seams, varves, other defects, or less than 100% saturation b. Sample disturbance not systematically accounted for	a. ASTM D2166* b. U.S. Army, 1970 c. Head, 1982
Lab vane	Same as for unconfined test	Same as for unconfined test	Same as for unconfined test	Head, 1982

*Designation for American Society of Testing and Materials test procedure.

16.4 Shear Strength of Granular Soils

Granular soil is a frictional material. The friction angle (ϕ') is affected by the grain size distribution and dry density. In general, ϕ' increases as the dry density increases and as the soil become more well graded, as illustrated in Figs. 16.2 and 16.3. Other typical values of ϕ' for granular soils are given in Holtz and Kovacs [1981] and Carter and Bentley [1991]. The friction angle also increases as the angularity of the soil grains increases and as the surface roughness of the particles increases. Wet sands tend to have a ϕ' that is 1° or 2° lower than for dry sands [Holtz and Kovacs, 1981]. The **intermediate principal stress (σ_2)** also affects ϕ'. In triaxial tests σ_2 is equal to either the major principal stress or minor principal stress (σ_1 or σ_3, respectively); however, most field problems occur under **plane strain conditions** where $\sigma_3 \leq \sigma_2 \leq \sigma_1$. It has been found that ϕ' for plane strain conditions (ϕ'_{ps}) is higher than for triaxial conditions (ϕ'_{tx}) [Ladd *et al.*, 1977]. Lade and Lee [1976] recommend the following equation for estimation of ϕ'_{ps}:

$$\phi'_{\text{ps}} = 1.5\phi'_{\text{tx}} - 17^\circ \qquad (\phi'_{\text{tx}} > 34^\circ) \tag{16.4a}$$

$$\phi'_{\text{ps}} = \phi'_{\text{tx}} \qquad (\phi'_{\text{tx}} \leq 34^\circ) \tag{16.4b}$$

In practice ϕ' for granular soils is determined using correlations with results from SPT, CPT, and other in situ tests, as discussed in Chapter 14, or laboratory tests on samples compacted to the same density as the in situ soil. Appropriate laboratory tests are drained direct shear and CD

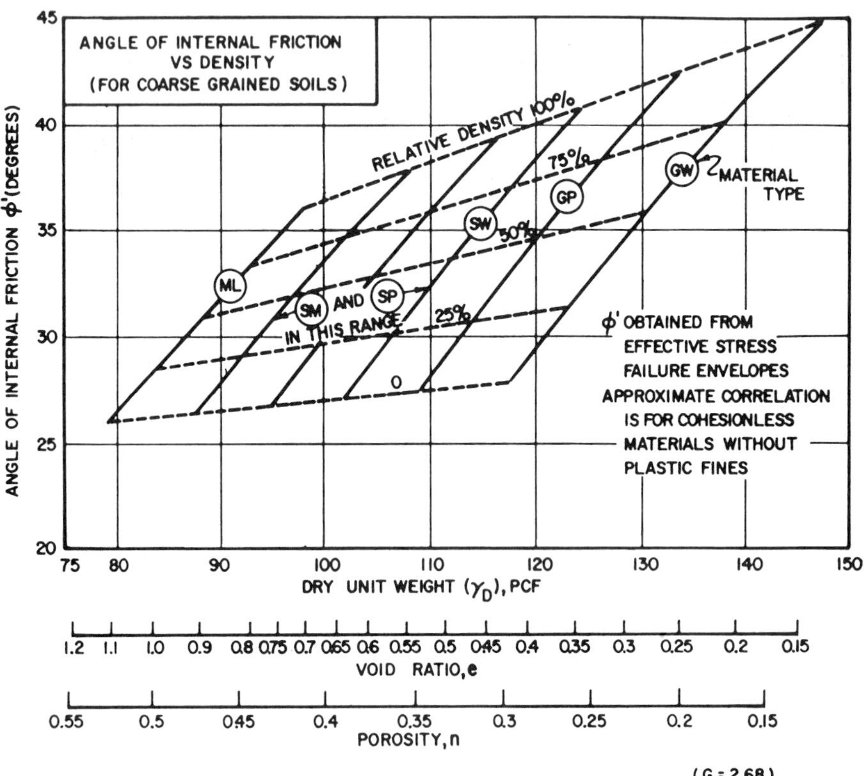

FIGURE 16.2 Correlation of friction angle of granular soils with soil classification and relative density. (*Source:* U. S. Navy. 1986. *Soil Mechanics,* Design Manual 7.1, p. 7.1–149. Naval Facilities Engineering Command, Alexandria, VA.)

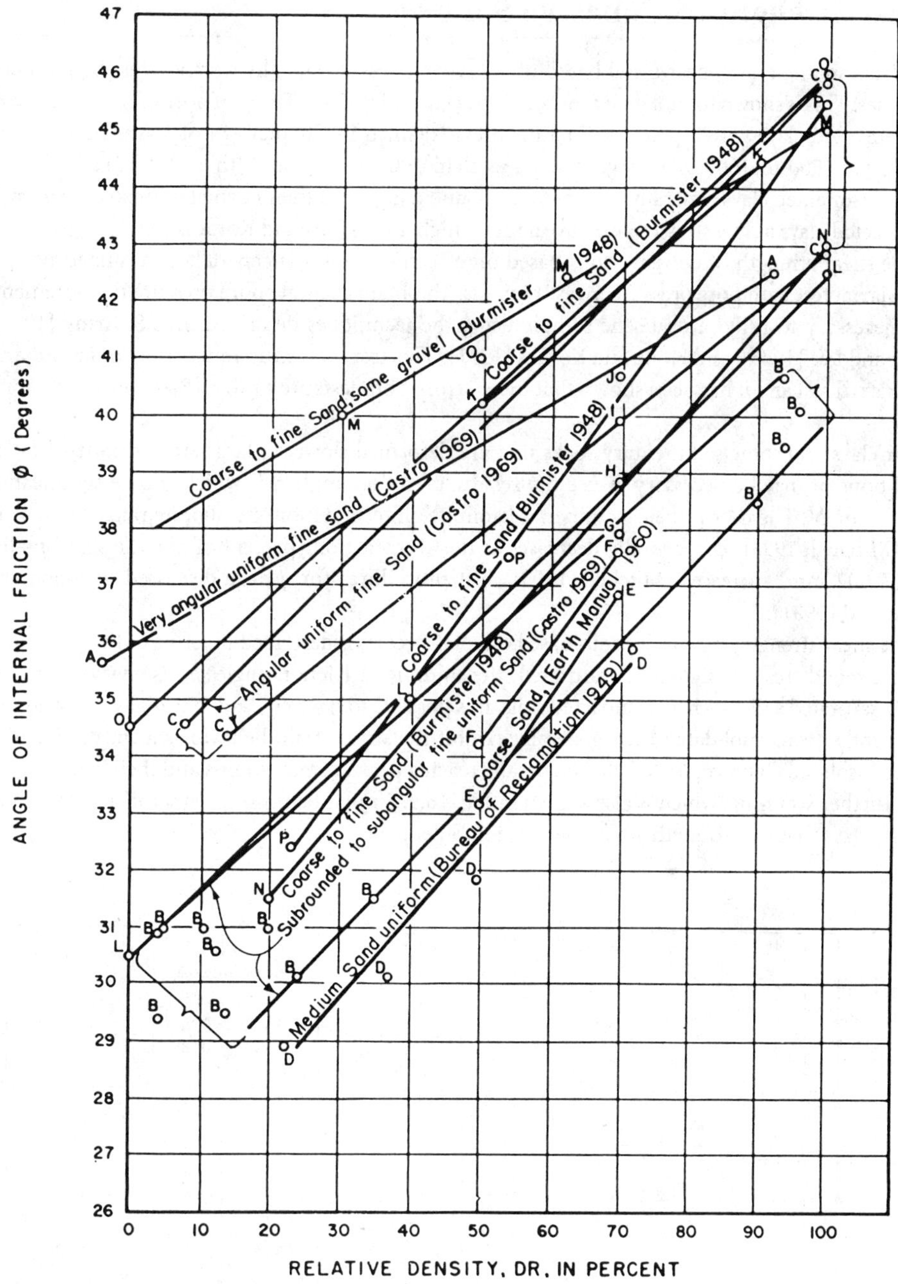

FIGURE 16.3 Friction angle versus relative density for cohesionless soils. (*Source:* Hilf, J. W. 1991. Compacted fill. In *Foundation Engineering Handbook,* ed. H.-Y. Fang, p. 268. Van Nostrand Reinhold, New York. With permission.)

triaxial tests. CU triaxial tests with pore pressure measurements are sometimes used for granular soils with appreciable fines. The method used to prepare the remolded sample and the direction of shearing relative to the direction of deposition has been found to affect ϕ' by up to 2.5° [Oda, 1977; Mahmood and Mitchell, 1974; Ladd *et al.,* 1977]. The c' of granular soils is zero except for lightly cemented soils which can have an appreciable c' [Clough *et al.,* 1981; Head, 1982].

16.5 Shear Strength of Cohesive Soils

The friction angle of cohesive soil based on effective stresses generally decreases as the plasticity increases. This is shown for normally consolidated clays in Fig. 16.4. The c' of normally consolidated, noncemented clays with a preconsolidation stress (defined in Chapter 18) of less than 10,000 to 20,000 psf (500 to 1000 kPa) is generally less than 100 to 200 psf (5 to 10 kPa) [Ladd, 1971]. Overconsolidated clays generally have a lower ϕ' and a higher c' than normally consolidated clays. Compacted clays at low stresses also have a much higher c' [Holtz and Kovacs, 1981].

The shear strength of cohesive soils based on effective stresses is generally determined using a CU triaxial test with pore pressure measurements. To obtain accurate pore pressure measurements it is necessary to fully saturate the sample using the techniques described in U.S. Army [1970], Black and Lee [1973], and Holtz and Kovacs [1981]. This test can be run much more quickly than a CD triaxial test and it has been shown that the ϕ' from both tests are similar [Bjerrum and Simons, 1960].

For clays and some sedimentary rocks that are deformed slowly to large strains under drained conditions, it may be necessary to use the **residual friction angle ϕ'_r**, which can be significantly lower than ϕ'. The ϕ'_r for the clay minerals kaolinite, illite, and montmorillonite range from 4° to 12° [Mitchell, 1993]. ϕ'_r generally decreases as the clay fraction (percent of particle sizes smaller than 0.002 mm) increases [Mitchell, 1993]. Test procedures for ϕ'_r are discussed in Saada and Townsend [1981].

The shear strength of cohesive soils based on total stresses is described in terms of the undrained shear strength (c_u). If the soil is saturated, the undrained friction angle ϕ_u is always zero. For partly saturated soils, such as compacted soils, it is possible to have $\phi_u > 0$. Normally consolidated and lightly overconsolidated clays generally exhibit the stress strain behavior shown in Fig. 16.5. Overconsolidated and cemented clays generally reach a peak at small strains and then lose strength with further straining, which is also shown in Fig. 16.5. Similar behavior occurs for sensitive clays, that is, clays that lose strength when they are remolded.

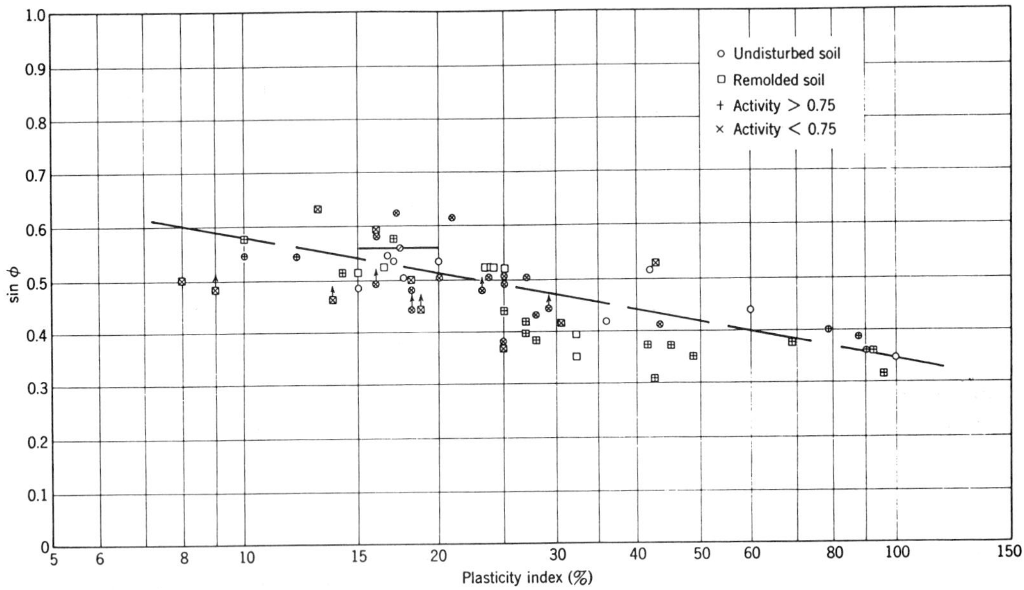

FIGURE 16.4 Friction angle of fine grained soil based on effective stresses versus plasticity index [Kenney, 1959]. (*Source:* Lambe, T. C. and Whitman, R. V. 1969. *Soil Mechanics,* p. 307. John Wiley & Sons, Inc. Copyright © 1969.)

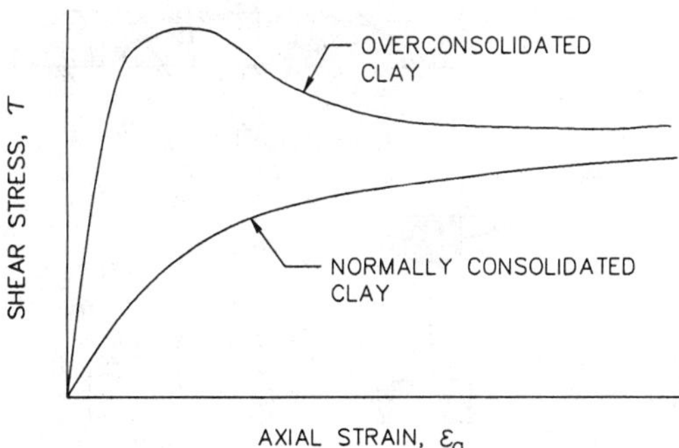

FIGURE 16.5 Stress-strain behavior of normally consolidated and heavily overconsolidated clay.

The undrained shear strength of clay is a function of its stress history. This is often expressed in dimensionless form as the ratio of c_u/σ'_{vc} where σ'_{vc} is the effective vertical consolidation stress. This ratio is empirically related to the overconsolidation ratio (OCR) by [Ladd, 1991]:

$$c_u/\sigma'_{\text{vc}} = S(\text{OCR})^m \tag{16.5}$$

where

S = 0.22 ± 0.03 for sedimentary clay plotting above A-line on plasticity chart
S = 0.25 ± 0.05 for silts and organic clays plotting below A-line
OCR = overconsolidation ratio = $\sigma'_p/\sigma'_{\text{vc}}$ (see Chapter 18)
σ'_p = preconsolidation pressure (see Chapter 18)
m = $0.88(1 - C_s/C_c)$
C_s = swelling index from consolidation test (see Chapter 18)
C_c = compression index from consolidation test (see Chapter 18)

Alternately, the undrained shear strength can be expressed as the ratio of c_u/σ'_p. This is shown for the results from K_0 consolidated triaxial compression (TC), triaxial extension (TE), and direct simple shear (DSS) tests in Fig. 16.6. In a TC test the vertical stress is increased to failure while

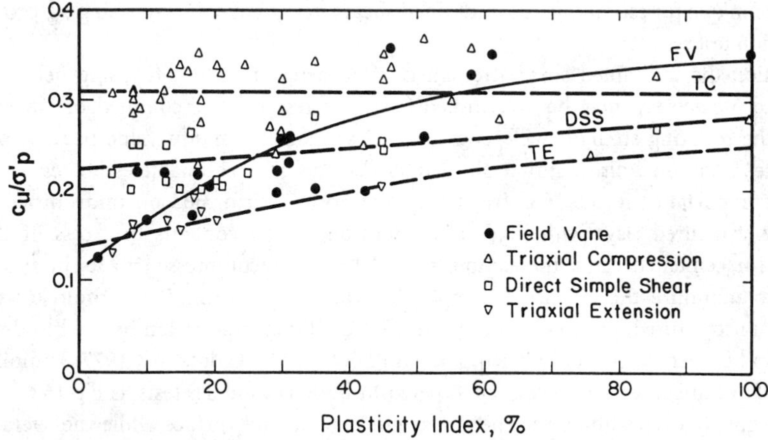

FIGURE 16.6 Undrained shear strength from K_0 consolidated CU triaxial compression, triaxial extension, and direct simple shear tests as well as field vane tests. (*Source:* Mesri, G. 1989. A reevaluation of $s_{u(\text{mob})} = 0.22\sigma'_p$. *Canadian Geotechnical J.* Vol. 26, No. 1, p. 163. With permission.)

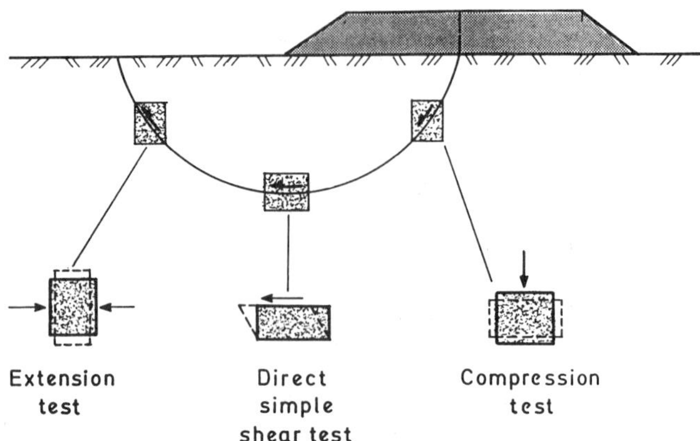

FIGURE 16.7 Relevance of laboratory shear tests to shear strength in the field. (*Source:* Bjerrum, L. 1972. Embankments on soft ground. In *Performance of Earth and Earth-Supported Structures,* Vol. II, p. 16. ASCE, New York. With permission of ASCE.)

in a TE test the vertical stress is decreased to failure. It is seen that TC tests give higher strengths than TE tests while DSS tests give results that are intermediate between the two. Results from field vane (FV) tests are also shown in Fig. 16.6. The applicability of undrained shear strengths from TC, TE, and DSS tests to a typical stability problem is shown in Fig. 16.7. Thus, Mesri [1989] concluded that an average of the results from these three tests would be reasonable for use in design. When this is applied to the data in Fig. 16.6, the following relationship results:

$$c_u = 0.22\sigma_p' \tag{16.6}$$

Mesri [1975] found an identical relationship using results from the FV test and a similar relationship was obtained by Larsson [1980] from a back analysis of 15 embankment failures. It is significant that Eq. (16.6) is independent of the plasticity index of the soil and that the same relationship was obtained using results from laboratory and field tests. This tends to confirm Bjerrum's [1973] conclusion that the "field vane test is the best possible approach for determining the strength for undrained strength stability analysis" [Mesri, 1989, p. 164]. Furthermore, Eq. (16.6) provides a valuable technique for estimating the undrained shear strength of soft clays using σ_p' profiles from consolidation test results.

In practice, the undrained shear strength is often determined in situ using field vane tests. For routine projects, c_u may be determined from the results of unconfined or lab vane tests; however, the resulting strength will generally be less than the in situ value because of sample disturbance. Undrained direct simple shear tests also can give reasonable estimates of c_u [Ladd, 1981]. For important projects, CU triaxial tests are often performed on undisturbed samples. For highly structured clays with high sensitivities and water contents in excess of the liquid limit and for cemented clays, the sample should first be recompressed to its in situ K_0 state of stress to minimize the effects of sample disturbance [Bjerrum, 1973; Jamiolkowski *et al.*, 1985]. For unstructured, uncemented clays, the SHANSEP technique can be used to develop the relationship between c_u/σ_{vc}' and OCR [Ladd and Foote, 1974; Ladd *et al.*, 1977; Jamiolkowski *et al.*, 1985]. For both cases it is necessary to perform both TC and TE tests as Fig 16.6 shows that TC results greatly overestimate the shear strength on a failure surface while the average of TC and TE results yields a more realistic shear strength for use in design. UU triaxial tests do not give meaningful stress-strain data and often give scattered c_u results because of the inability of this test to account for varying degrees of sample disturbance [Jamiolkowski *et al.*, 1985] making the use of this test undesirable for important projects.

Table 16.2 Typical Values of Elastic Modulus and Poisson's Ratio for Granular Soils

	Elastic Modulus, E_s		
Type of Soil	MPa	lb/in.^{2}	Poisson's ratio, μ
Loose sand	10–24	1,500–3,500	0.20–0.40
Medium dense sand	17–28	2,500–4,000	0.25–0.40
Dense sand	35–55	5,000–8,000	0.30–0.45
Silty sand	10–17	1,500–2,500	0.20–0.40
Sand and gravel	69–170	10,000–25,000	0.15–0.35

Source: Das, B. M. 1990. *Principles of Foundation Engineering,* 2nd ed., p. 161. PWS-Kent Publishing Co., Boston. With permission.

16.6 Elastic Modulus of Granular Soils

The **elastic modulus (E_s)** of granular soils based on effective stresses is a function of grain size, gradation, mineral composition of the soil grains, grain shape, soil type, relative density, soil particle arrangement, stress level, and prestress [Lambe and Whitman, 1969; Ladd *et al.,* 1977; Lambrechts and Leonards, 1978]. A granular soil is prestressed if, at some point in its history, it has experienced a stress level that is greater than is currently acting on the soil. This is analogous to overconsolidation of a fine-grained soil, which is discussed in Chapter 18. Of the several factors controlling E_s, the ones having the largest influence are prestress, which can increase E_s by more than a factor of six, and extreme differences in relative density, which can make a fivefold difference in E_s [Lambrechts and Leonards, 1978]. The effect of stress level on modulus is often represented by [Janbu, 1963]

$$E_i = Kp_a\left(\frac{\sigma_3}{p_a}\right)^2 \tag{16.7}$$

where E_i is the initial slope of a stress-strain curve, σ_3 is the minor principal stress, K is a dimensionless modulus number that varies from 300 to 2000, n is an exponent number typically between 0.3 and 0.6, and p_a is atmospheric pressure in the same units as σ_3 and E_i [Mitchell, 1993]. Typical values of K and n are given in Wong and Duncan [1974].

Measuring E_s is very difficult since it is nearly impossible to measure the prestress of an in situ deposit of granular soil or to obtain undisturbed samples for laboratory testing. While CD triaxial tests can be used to measure E_s [Head, 1986], they are restricted to reconstituted samples that cannot duplicate the in situ prestress. For these reasons, E_s is often estimated using in situ tests (Chapter 14). Typical values of E_s and Poisson's ratio (μ) for normally consolidated granular soils are given in Table 16.2.

16.7 Undrained Elastic Modulus of Cohesive Soils

The undrained elastic modulus (E_u) of cohesive soils is a function primarily of soil plasticity and overconsolidation (defined in Chapter 18). It can be determined from the slope of a stress-strain curve obtained from an undrained triaxial test [Holtz and Kovacs, 1981]. However, E_u is very sensitive to sample disturbance, which results in values measured in laboratory tests that are too low [Lambe and Whitman, 1969; Jamiolkowski *et al.,* 1985]. Alternatively, E_u can be measured using in situ tests (Chapter 14) or a crude estimate of E_u can be made from the undrained shear strength using the empirical relations shown in Table 16.3. However, there is significant variability in the ratio E_s/c_u, which has been reported to vary from 40 to more than 3000 [Holtz and Kovacs, 1981].

Table 16.3 Approximate Relationship between Undrained Young's Modulus and Undrained Shear Strength

	E_s/c_u		
OCR*	PI** < 30	30<PI<50	PI>50
<3	600	300	125
3 to 5	400	200	75
>5	150	75	50

OCR = overconsolidation ratio (defined in Chapter 18).
PI = plasticity index (defined in Chapter 14).
Source: U.S. Navy, 1986. *Soil Mechanics,* Design Manual 7.1, p. 7.1-215. Naval Facilities Engineering Command, Alexandria, VA.

Defining Terms

Effective stress (σ'): Intergranular stress that exists between soil particles.
Elastic modulus (E_s): Ratio of the change in stress divided by the corresponding change in strain for an axially loaded sample. Also called Young's modulus.
Failure envelope: A line tangent to a series of Mohr's circles at failure.
Intermediate principal stress (σ_2): In a set of three principal stresses acting at a point in a soil mass, the intermediate principal stress is the one that is less than or equal to the major principal stress but greater than or equal to the minor principal stress.
Major principal stress (σ_1): The largest of a set of three principal stresses acting at a point in a soil mass.
Minor principal stress (σ_3): The smallest of a set of three principal stresses acting at a point in a soil mass.
Mohr's circle: A graphical representation of the state of stress at a point in a soil mass.
Plane strain conditions: A loading condition where the normal strain on one plane is zero as would occur for a long retaining wall or embankment.
Principal planes: A set of three orthogonal (mutually perpendicular) planes that exist at any point in a soil mass on which the shear stresses are zero.
Principal stresses: The normal stresses acting on a set of three principal planes.
Residual friction angle (ϕ_r'): For clays and some sedimentary rocks it is the friction angle that is reached after very large strains.
Total stress (σ): The sum of the effective stress and the pore water pressure.

References

Bjerrum, L. 1973. Problems of soil mechanics and construction on soft clays and structurally unstable soils. In *Proc. 8th Int. Conf. Soil Mech. Found. Eng.* 3:111–159.
Bjerrum, L. and Landva, A. 1966. Direct simple shear tests on Norwegian quick clay. *Geotechnique.* 16(1):1–20.
Bjerrum, L. and Simons, N. E. 1960. Comparison of shear strength characteristics of normally consolidated clays. In *Proc. Res. Conf. Shear Strength Cohesive Soils.* ASCE, New York, pp. 711–726.
Black, D. K. and Lee, K. L. 1973. Saturating laboratory samples by back pressure. *J. Soil Mech. Found. Div., ASCE.* 99(SM1):75–93.
Bowles, J. E. 1992. *Engineering Properties of Soils and Their Measurement.* McGraw-Hill, New York.
Carter, M. and Bentley, S. P. 1991. *Correlations of Soil Properties.* Pentech Press, London.

Clough, G. W., Sitar, N., Bachus, R. C., and Rad, N. S. 1981. Cemented sands under static loading. *J. Geotech. Eng., ASCE.* 107(GT6):799–817.

Donaghe, R. T., Chaney, R. C., and Silver, M. L. 1988. Advanced Triaxial Testing of Soil and Rock, *Am. Soc. Test. Mater., Spec. Tech. Publ. 977.*

Head, K. H. 1982. *Manual of Soil Laboratory Testing, Vol. 2: Permeability, Shear Strength, and Compressibility Tests.* Pentech Press, London.

Head, K. H. 1986. *Manual of Soil Laboratory Testing, Vol. 3: Effective Stress Tests.* John Wiley & Sons, New York.

Holtz, R. D. and Kovacs, W. D. 1981. *An Introduction to Geotechnical Engineering.* Prentice-Hall, Englewood Cliffs, NJ.

Jamiolkowski, M., Ladd, C. C., Germaine, J. T., and Lancellotta, R. 1985. New developments in field and laboratory testing of soils. In *Proc. 11th Int. Conf. Soil Mech. Found. Eng.* A. A. Balkema, Rotterdam. 1:57–153.

Janbu, N. 1963. Soil compressibility as determined by oedometer and triaxial tests. In *Eur. Conf. Soil Mech. Found. Eng.* Weisbaden, Germany. 1:19–25.

Kenney, T. C. 1959. Discussion. *Proc. Am. Soc. Civ. Eng.,* 85(SM3):67–79.

Ladd, C. C. 1971. Strength parameters and stress-strain behavior of saturated clays. *Research Report R71-23.* Soils Publication 278, Department of Civil Engineering Massachusetts Institute of Technology, Cambridge.

Ladd, C. C. 1981. Discussion on laboratory shear devices, *in* Laboratory Shear Strength of Soil, *Am. Soc. Test. Mater., Spec. Tech. Publ. 740:* 643–652.

Ladd, C. C. 1991. Stability evaluations during staged construction. *J. Geotech. Eng., ASCE.* 117(4):540–615.

Ladd, C. C. and Foote, R. 1974. A new design procedure for stability of soft clays. *J. Geotech. Eng., ASCE.* 100(GT7):763–786.

Ladd, C. C., Foote, R., Ishihara, K., Schlosser, F., and Poulos, H. G. 1977. Stress-deformation and strength characteristics. In *Proc. 9th Int. Conf. Soil Mech. Found. Eng.*, Tokyo, 2:421–494.

Lade, P. V. and Lee, K. L., 1976. Engineering properties of soils. *Report UCLA-ENG-7652.* University of California, Los Angeles.

Lambe, T. C. 1951. *Soil Testing for Engineers.* John Wiley & Sons, New York.

Lambe, T. C. and Whitman, R. V. 1969. *Soil Mechanics.* John Wiley & Sons, New York.

Lambrechts, J. R. and Leonards, G. A. 1978. Effects of stress history on deformation of sand. *J. Geotech. Eng., ASCE.* 104(GT11):1371–1387.

Larsson, R. 1980. Undrained shear strength in stability calculations of embankments and foundations on soft clays. *Can. Geotech. J.* 17(4):591–602.

Mahmood, A. and Mitchell, J. K. 1974. Fabric-property relationships in fine granular materials. *Clays Clay Miner.* 22:397–408.

Mesri, G. 1975. Discussion: New design procedure for stability of soft clays. *J. Geotech. Eng., ASCE.* 101(GT4):409–412.

Mesri, G. 1989. A reevaluation of $s_{u(\mathrm{mob})} = 0.22\sigma_p'$. *Can. Geotech. J.* 26(1):162–164.

Mitchell, J. K. 1993. *Fundamentals of Soil Behavior,* 2nd ed. John Wiley & Sons, New York.

Oda, M. 1977. The mechanism of fabric changes during compressional deformation of sand. *Soils Found.* 12(2):1–18.

Saada, A. S. and Townsend, F. C. 1981. State of the art: In Laboratory strength testing of soils. In Laboratory Shear Strength of Soil, *Am. Soc. Test. Mater., Spec. Tech. Publ. 740*: 7–77.

U. S. Army. 1970. Laboratory soils testing. *Engineer Manual EM 1110-2-1906.* Department of the Army, Office of the Chief of Engineers, Washington, D.C.

Wong, K. S. and Duncan, J. M. 1974. Hyperbolic stress-strain parameters for non-linear finite element analyses of stresses and movements in soil masses. *Report TE 73-4.* Department of Civil Engineering, University of California, Berkeley.

For Further Information

Holtz and Kovacs [1981], Lambe and Whitman [1969], and Mitchell [1993] are recommended for a review of the fundamentals of the shear strength of soils.

Laboratory testing procedures are discussed in Head [1982, 1986], U.S. Army [1970], Lambe [1951], and Bowles [1992] as well as the ASTM procedures referenced in Table 16.1.

Major conferences on the shear strength and deformation properties of soil include *Research Conference on Shear Strength of Cohesive Soils,* ASCE, 1960; *Laboratory Shear Testing of Soils,* ASTM STP 361, 1964; *Laboratory Shear Strength of Soil,* STP 740, 1980; and *Advanced Triaxial Testing of Soil and Rock,* ASTM STP 977, 1986.

Relevant state-of-the-art papers include Bjerrum [1973]; Ladd *et al.* [1977]; and Jamiolkowski *et al.* [1985].

17

Groundwater and Seepage

Milton E. Harr
Purdue University

17.1 Introduction

Figure 17.1 shows the pore space available for flow in two highly idealized soil models: *regular cubic* and *rhombohedral.* It is seen that even for these special cases, the pore space is not regular, but consists of cavernous cells interconnected by narrower channels. Pore spaces in real soils can range in size from molecular interstices to cathedral-like caverns. They can be spherical (as in concrete) or flat (as in clays), or display irregular patterns which defy description. Add to this the fact that pores may be *isolated* (inaccessible) or *interconnected* (accessible from both ends) or may be *dead-ended* (accessible through one end only).

In spite of the apparent irregularities and complexities of the available pores, there is hardly an industrial or scientific endeavor that does not concern itself with the passage of matter, solid, liquid, or gaseous, into, out of, or through porous media. Contributions to the literature can be found among such diverse fields (to name only a few) as soil mechanics, groundwater hydrology, petroleum, chemical, and metallurgical engineering, water purification, materials of construction (ceramics, concrete, timber, paper), chemical industry (absorbents, varieties of contact catalysts, and filters), pharmaceutical industry, traffic flow, and agriculture.

The flow of groundwater is taken to be governed by *Darcy's law,* which states that the velocity of the flow is proportional to the **hydraulic gradient.** A similar statement in an electrical system is *Ohm's law* and in a thermal system, *Fourier's law.* The grandfather of all such relations is *Newton's laws of motion.* Table 17.1 presents some other points of similarity.

17.2 Some Fundamentals

The literature is replete with derivations and analytical excursions of the basic equations of steady state groundwater flow [e.g., Polubarinova-Kochina, 1962; Harr, 1962; Cedergren, 1967; Bear, 1972; Domenico and Schwartz, 1990]. A summary and brief discussion of these will be presented below for the sake of completeness.

0-8493-8953-4/95/$0.00 + $.50

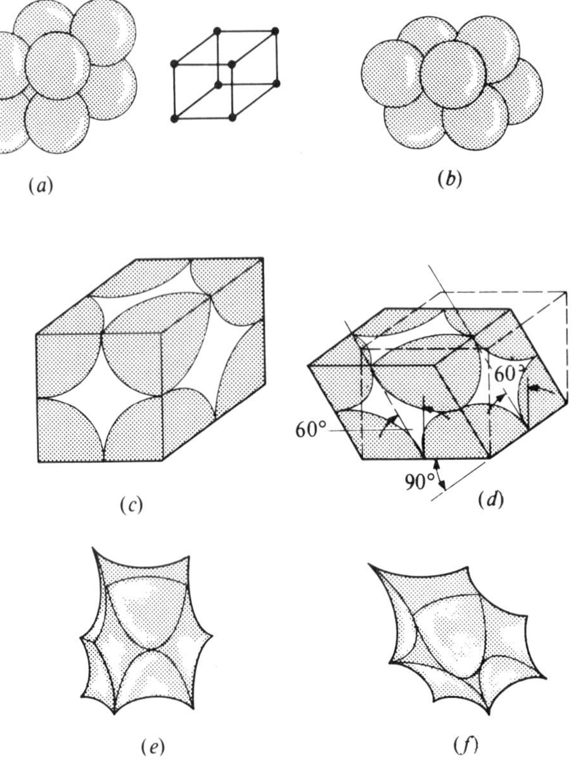

FIGURE 17.1 Idealized void space.

Table 17.1 Some Similarities of Flow Models

Form of Energy	Name of Law	Quantity	Storage	Resistance
Electrical	Ohm's law	Current (voltage)	Capacitor	Resistor
Mechanical	Newton's law	Force (velocity)	Mass	Damper
Thermal	Fourier's law	Heat flow (temperature)	Heat capacity	Heat resistance
Fluid	Darcy's law	Flow rate (pressure)	Liquid storage	Permeability

Bernoulli's Equation

Underlying the analytical approach to groundwater flow is the representation of the actual physical system by a tractable mathematical model. In spite of their inherent shortcomings, many such analytical models have demonstrated considerable success in simulating the action of their prototypes.

As is well known from fluid mechanics, for steady flow of nonviscous incompressible fluids, Bernoulli's equation [Lamb, 1945]

$$\frac{p}{\gamma_w} + z + \frac{\overline{v}^2}{2g} = \text{constant} = h \tag{17.1}$$

where

p = pressure, lb/ft^2

γ_w = unit weight of fluid, lb/ft^3

$\bar{v}$ = seepage velocity, ft/sec

g = gravitational constant, 32.2 ft/s^2

h = total head, ft

demonstrates that the sum of the *pressure head, p/γ_w, elevation head, z,* and *velocity head, $\bar{v}^2/2g$,* at any point within the region of flow is a constant. To account for the loss of energy due to the viscous resistance within the individual pores, Bernoulli's equation is taken as

$$\frac{p_A}{\gamma_w} + z_A + \frac{\bar{v}_A^2}{2g} = \frac{p_B}{\gamma_w} + z_B + \frac{\bar{v}_B^2}{2g} + \Delta h \tag{17.2}$$

where Δh represents the total head loss (energy loss per unit weight of fluid) of the fluid over the distance Δs. The ratio

$$i = -\lim_{\Delta s \to 0} \frac{\Delta h}{\Delta s} = -\frac{dh}{ds} \tag{17.3}$$

is called the *hydraulic gradient* and represents the space rate of energy dissipation per unit weight of fluid (a pure number).

In most problems of interest the velocity heads (the kinetic energy) are so small they can be neglected. For example, a velocity of 1 ft/s, which is large compared to typical seepage velocities through soils, produces a velocity head of only 0.015 ft. Hence, Eq. (17.2) can be simplified to

$$\frac{p_A}{\gamma_w} + z_A = \frac{p_B}{\gamma_w} + z_B + \Delta h$$

and the total head at any point in the flow domain is simply

$$h = \frac{p}{\gamma_w} + z \tag{17.4}$$

Darcy's Law

Prior to 1856, the formidable nature of the flow through porous media defied rational analysis. In that year, Henry Darcy published a simple relation based on his experiments on the flow of water in vertical sand filters in "les fontaines publiques de la ville de Dijon," namely,

$$v = ki = -k\frac{dh}{ds} \tag{17.5}$$

Equation (17.5), commonly called *Darcy's law,* demonstrates a linear dependency between the hydraulic gradient and the *discharge velocity* v. The discharge velocity, $v = n\bar{v}$, is the product of the porosity n and the seepage velocity, $\bar{v}$. The coefficient of proportionality k is called by many names depending on its use; among these are the **coefficient of permeability,** *hydraulic conductivity,* and *permeability constant.* As shown in Eq. (17.5), k has the dimensions of a velocity. It should be carefully noted that Eq. (17.5) states that flow is a consequence of differences in total head and not of pressure gradients. This is demonstrated in Fig. 17.2, where the flow is directed from A to B, even though the pressure at point B is *greater* than that at point A.

Defining Q as the total volume of flow per unit time through a cross-sectional area A, Darcy's law takes the form

$$Q = Av = Aki = -Ak\frac{dh}{ds} \tag{17.6}$$

Darcy's law offers the single parameter k to account for both the characteristics of the medium and the fluid. It has been found that k is a function of γ_f, the unit weight of the fluid, μ, the

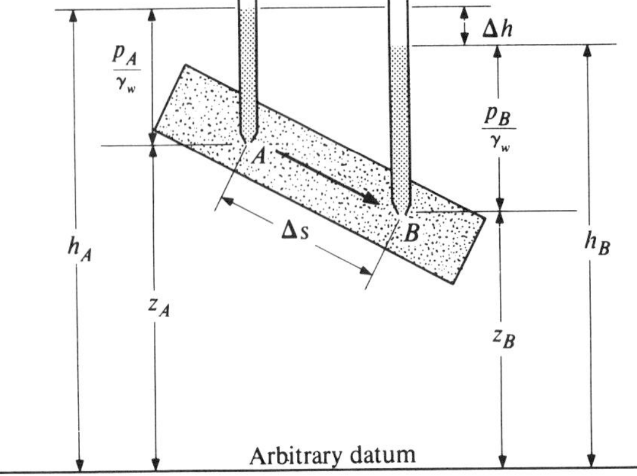

FIGURE 17.2 Heads in Bernoulli's equation.

Table 17.2 Some Typical Values of Coefficient of Permeability

Soil Type	Coefficient of Permeability **k**, cm/s
Clean gravel	1.0 and greater
Clean sand (coarse)	1.0–0.01
Sand (mixtures)	0.01–0.005
Fine sand	0.05–0.001
Silty sand	0.002–0.0001
Silt	0.0005–0.00001
Clay	0.000001 and smaller

coefficient of viscosity, and n, the porosity, as given by

$$k = C\frac{\gamma_f n}{\mu} \tag{17.7}$$

where C (dimensionally an area) typifies the structural characteristics of the medium independent of the fluid properties. The principal advantage of Eq. (17.7) lies in its use when dealing with more than one fluid or with temperature variations. When employing a single relatively incompressible fluid subjected to small changes in temperature, such as in groundwater- and seepage-related problems, it is more convenient to use k as a single parameter. Some typical values for k are given in Table 17.2.

Although Darcy's law was obtained initially from considerations of one-dimensional macroscopic flow, its practical utility lies in its generalization into two or three spatial dimensions. Accounting for the directional dependence of the coefficient of permeability, Darcy's law can be generalized to

$$v_s = -k_s\frac{\partial h}{\partial s} \tag{17.8}$$

where k_s is the coefficient of permeability in the s direction, and v_s and $\partial h/\partial s$ are the components of the velocity and the hydraulic gradient, respectively, in that direction.

Reynolds Number

There remains now the question of the determination of the extent to which Darcy's law is valid in actual flow systems through soils. Such a criterion is furnished by the Reynolds number R (a pure number relating inertial to viscous force), defined as

$$R = \frac{v d \rho}{\mu} \tag{17.9}$$

where

v = discharge velocity, cm/s
d = average of diameter of particles, cm
ρ = density of fluid, g(mass)/cm^3
μ = coefficient of viscosity, g-s/cm^2

The critical value of the Reynolds number at which the flow in aggregations of particles changes from laminar to turbulent flow has been found by various investigators [see Muskat, 1937] to range between 1 and 12. However, it will generally suffice to accept the validity of Darcy's law when the Reynolds number is taken as equal to or less than unity, or

$$\frac{v d \rho}{\mu} \leq 1 \tag{17.10}$$

Substituting the known values of ρ and μ for water into Eq. (17.10) and assuming a conservative velocity of $\frac{1}{4}$ cm/s, we have d equal to 0.4 mm, which is representative of the average particle size of coarse sand.

Homogeneity and Isotropy

If the coefficient of permeability is independent of the direction of the velocity, the medium is said to be *isotropic.* Moreover, if the same value of the coefficient of permeability holds at all points within the region of flow, the medium is said to be *homogeneous* and *isotropic.* If the coefficient of permeability depends on the direction of the velocity and if this directional dependence is the same at all points of the flow region, the medium is said to be *homogeneous* and *anisotropic* (or *aleotropic*).

Streamlines and Equipotential Lines

Physically, all flow systems extend in three dimensions. However, in many problems the features of the motion are essentially planar, with the flow pattern being substantially the same in parallel planes. For these problems, for steady state, incompressible, isotropic flow in the xy plane, it can be shown [Harr, 1962] that the governing differential equation is

$$k_x \frac{\partial^2 h}{\partial x^2} = k_y \frac{\partial^2 h}{\partial y^2} = 0 \tag{17.11}$$

Here the function $h(x, y)$ is the distribution of the total head (of energy available to do work), within and on the boundaries of a flow region, and k_x and k_y are the coefficients of permeability in the x and y directions, respectively. If the flow system is isotropic, $k_x = k_y$, and Eq. (17.11) reduces to

$$\frac{\partial^2 h}{\partial x^2} + \frac{\partial^2 h}{\partial y^2} = 0 \tag{17.12}$$

Equation (17.12), called *Laplace's equation,* is the governing relationship for steady state, laminar-flow conditions (Darcy's law is valid). The general body of knowledge relating to Laplace's equation is called *potential theory.* Correspondingly, incompressible steady state fluid flow is often called *potential flow.* The correspondence is more evident upon the introduction of the *velocity potential* ϕ, defined as

$$\phi(x, y) = -kh + C = -k\left(\frac{p}{\gamma_w} + z\right) + C \tag{17.13}$$

where h is the total head, p/γ_w is the pressure head, z is the elevation head, and C is an arbitrary constant. It should be apparent that, for isotropic conditions,

$$v_x = \frac{\partial \phi}{\partial x} \qquad v_y = \frac{\partial \phi}{\partial y} \tag{17.14}$$

and Eq. (17.12) will produce

$$\nabla^2 \phi = \frac{\partial^2 \phi}{\partial x^2} + \frac{\partial^2 \phi}{\partial y^2} = 0 \tag{17.15}$$

The particular solutions of Eqs. (17.12) or (17.15) that yield the locus of points within a porous medium of equal potential, curves along which $h(x, y)$ or $\phi(x, y)$ are equal to constants, are called *equipotential lines.*

In analyses of groundwater flow, the family of flow paths is given by the function $\psi(x, y)$, called the *stream function,* defined in two dimensions as [Harr, 1962]

$$v_x = \frac{\partial \psi}{\partial y} \qquad v_y = -\frac{\partial \psi}{\partial x} \tag{17.16}$$

where v_x and v_y are the components of the velocity in the x and y directions, respectively.

Equating the respective potential and stream functions of v_x and v_y produces

$$\frac{\partial \phi}{\partial x} = \frac{\partial \psi}{\partial y} \qquad \frac{\partial \phi}{\partial y} = -\frac{\partial \psi}{\partial x} \tag{17.17}$$

Differentiating the first of these equations with respect to y and the second with respect to x and adding, we obtain Laplace's equation:

$$\frac{\partial^2 \psi}{\partial x^2} + \frac{\partial^2 \psi}{\partial y^2} = 0 \tag{17.18}$$

We shall examine the significance of this relationship following a little more discussion of the physical meaning of the stream function. Consider AB of Fig. 17.3 as the path of a particle of water passing through point P with a tangential velocity **v**. We see from the figure that

$$\frac{v_y}{v_x} = \tan\theta = \frac{dy}{dx}$$

and hence

$$v_y dx - v_x dy = 0 \tag{17.19}$$

Substituting Eq. (17.16), it follows that

$$\frac{\partial \psi}{\partial x} dx + \frac{\partial \psi}{\partial y} dy = 0$$

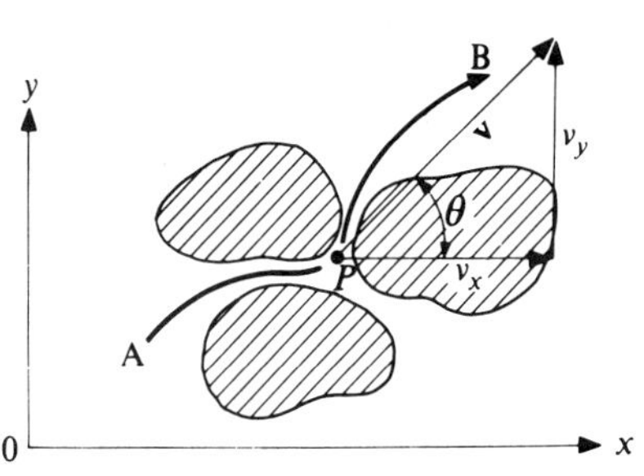

FIGURE 17.3 Path of flow.

which states that the total differential $d\psi = 0$ and

$$\psi(x, y) = \text{constant}$$

Thus we see that the family of curves generated by the function $\psi(x, y)$ equal to a series of constants are tangent to the resultant velocity at all points in the flow region and hence define the path of flow.

The potential $[\phi = -kh + C]$ is a measure of the energy available at a point in the flow region to move the particle of water from that point to the tailwater surface. Recall that the locus of points of equal energy, say, $\phi(x, y) =$ constants, are called *equipotential lines.* The total differential along any curve $\phi(x, y) =$ constant produces

$$d\phi = \frac{\partial \phi}{\partial x}dx + \frac{\partial \phi}{\partial y}dy = 0$$

Substituting for $\partial\phi/\partial x$ and $\partial\phi/\partial y$ from Eqs. (17.16), we have

$$v_x\, dx + v_y\, dy = 0$$

and

$$\frac{dy}{dx} = -\frac{v_x}{v_y} \tag{17.20}$$

Noting the negative reciprocal relationship between their slopes, Eqs. (17.19) and (17.20), we see that, within the flow domain, the families of streamlines $\psi(x, y) =$ constants and equipotential lines $\phi(x, y) =$ constants intersect each other at right angles. It is customary to signify the sequence of constants by employing a subscript notation, such as $\phi(x, y) = \phi_i$, $\psi(x, y) = \psi_j$ (Fig. 17.4).

As only one streamline may exist at a given point within the flow medium, streamlines cannot intersect one another. Consequently, if the medium is saturated, any pair of streamlines act to form a flow channel between them. Consider the flow between the two streamlines ψ and $\psi + d\psi$ in Fig. 17.5; **v** represents the resultant velocity of flow. The quantity of flow through the flow channel per unit length normal to the plane of flow (say, cubic feet per second per foot) is

$$dQ = v_x\, ds \cos\theta - v_y\, ds \sin\theta = v_x\, dy - v_y\, dx = \frac{\partial \psi}{\partial y}dy + \frac{\partial \psi}{\partial x}dx$$

and

$$dQ = d\psi \tag{17.21}$$

Hence the quantity of flow (also called the *discharge quantity*) between any pair of streamlines is a constant whose value is numerically equal to the difference in their respective ψ values. Thus, once a sequence of streamlines of flow has been obtained, with neighboring ψ values differing by

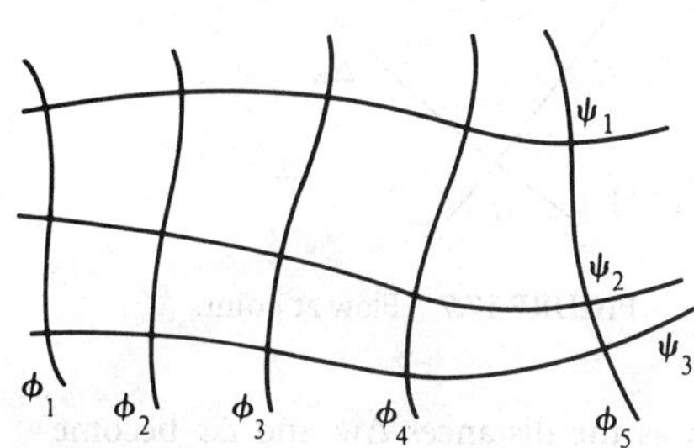

FIGURE 17.4 Streamlines and equipotential lines.

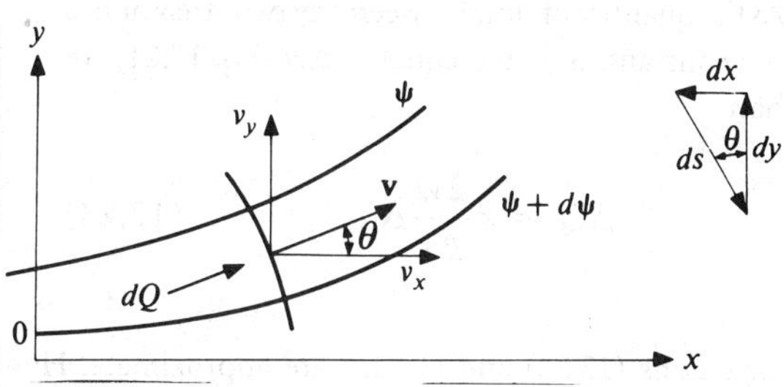

FIGURE 17.5 Flow between streamlines.

a constant amount, their plot will not only show the expected direction of flow but the relative magnitudes of the velocity along the flow channels; that is, the velocity at any point in the flow channel varies inversely with the streamline spacing in the vicinity of that point.

An equipotential line was defined previously as the locus of points where there is an expected level of available energy sufficient to move a particle of water from a point on that line to the tailwater surface. Thus, it is convenient to reduce all energy levels relative to a tailwater datum. For example, a piezometer located anywhere along an equipotential line, say at $0.75h$ in Fig. 17.6, would display a column of water extending to a height of $0.75h$ above the tailwater surface. Of course, the pressure in the water along the equipotential line would vary with its elevation.

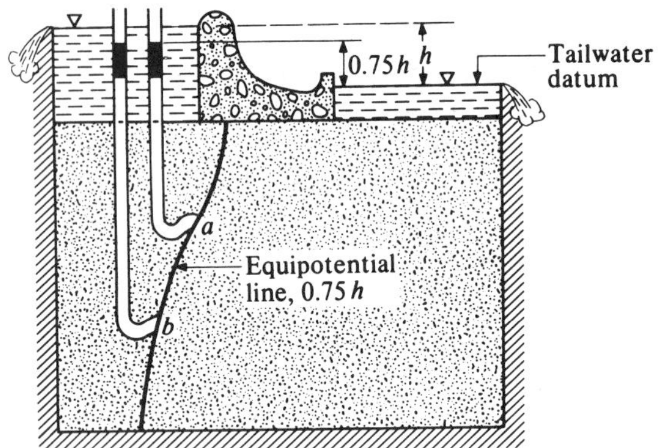

FIGURE 17.6 Pressure head along equipotential line.

17.3 The Flow Net

The graphical representation of special members of the families of streamlines and corresponding equipotential lines within a flow region form a **flow net.** The orthogonal network shown in Fig. 17.4 represents such a system. Although the construction of a flow net often requires tedious trial-and-error adjustments, it is one of the more valuable methods employed to obtain solutions for two-dimensional flow problems. Of additional importance, even a hastily drawn flow net will often provide a check on the reasonableness of solutions obtained by other means. Noting that, for steady state conditions, Laplace's equation also models the action (see Table 17.1) of thermal, electrical, acoustical, odoriferous, torsional, and other systems, the flow net is seen to be a significant tool for analysis.

If, in Fig. 17.7, Δw denotes the distance between a pair of adjacent streamlines and Δs is the distance between a pair of adjacent equipotential lines in the near vicinity of a point within the region of flow, the approximate velocity (in the mathematical sense) at the point, according to Darcy's law, will be

$$v \approx \frac{k\,\Delta h}{\Delta s} \approx \frac{\Delta\psi}{\Delta w} \tag{17.22}$$

As the quantity of flow between any two streamlines is a constant, ΔQ, and equal to $\Delta\psi$ (Eq. 17.21) we have

$$\Delta Q \approx k\frac{\Delta w}{\Delta s}\Delta h \tag{17.23}$$

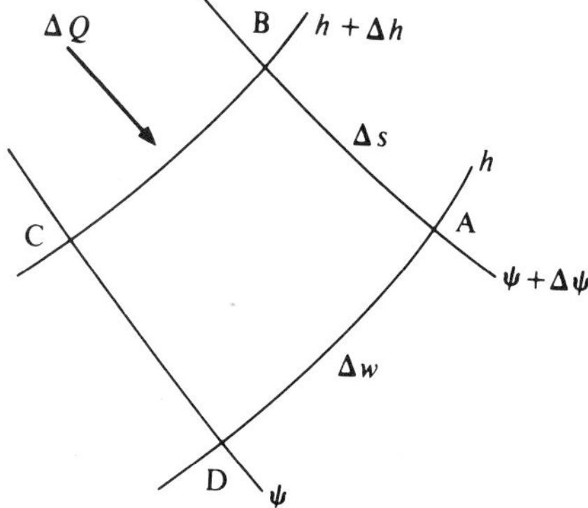

FIGURE 17.7 Flow at point.

Equations (17.22) and (17.23) are approximate. However, as the distances Δw and Δs become very small, Eq. (17.22) approaches the velocity at the point and Eq. (17.23) yields the quantity of discharge through the flow channel.

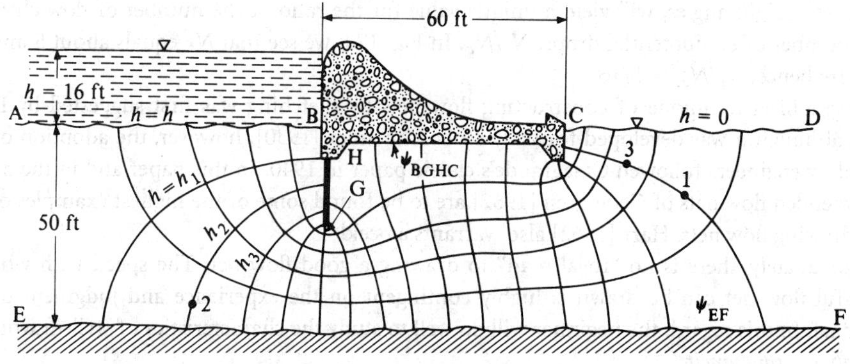

FIGURE 17.8 Example of flow net.

In Fig. 17.8 is shown the completed flow net for a common type of structure. We first note that there are four boundaries: the bottom impervious contour of the structure $BGHC$, the surface of the impervious layer EF, the headwater boundary AB, and the tailwater boundary CD. The latter two boundaries designate the equipotential lines $h = h$ and $h = 0$, respectively. For steady state conditions, the quantity of discharge through the section Q and the head loss ($h = 16$ ft) must be constant. If the flow region is saturated, it follows that the two impervious boundaries are streamlines and their difference must be identically equal to the discharge quantity

$$Q = \psi_{BGHC} - \psi_{EF}$$

From among the infinite number of possible streamlines between the impervious boundaries, we sketch only a few, specifying the same quantity of flow between neighboring streamlines. Designating N_f as the number of flow channels, we have, from above,[1]

$$Q = N_f \, \Delta Q = kN_f \frac{\Delta w}{\Delta s} \Delta h$$

Similarly, from among the infinite number of possible equipotential lines between headwater and tailwater boundaries, we sketch only a few and specify the same drop in head, say, Δh, between adjacent equipotential lines. If there are N_e equipotential drops along each of the channels,

$$h = N_e \, \Delta h \quad \text{and} \quad Q = k \frac{N_f}{N_e} \frac{\Delta w}{\Delta s} h \tag{17.24}$$

If, now, we also require that the ratio $\Delta w / \Delta s$ be the same at all points in the flow region, for convenience, and because a square is most sensitive to visual inspection, we take this ratio to be unity,

$$\frac{\Delta w}{\Delta s} = 1$$

and obtain

$$Q = k \frac{N_f}{N_e} h \tag{17.25}$$

Recalling that Q, k, and h are all constants, Eq. (17.25) demonstrates that the resulting construction, with the obvious requirement that everywhere in the flow domain streamlines and equipotential

[1]There is little to be gained by retaining the approximately equal sign $\approx$.

lines meet at right angles, will yield a unique value for the ratio of the number of flow channels to the number of equipotential drops, N_f/N_e. In Fig. 17.8 we see that N_f equals about 5 and N_e equals 16; hence, $N_f/N_e = 5/16$.

The graphical technique of constructing flow nets by sketching was first suggested by Prasil [1913] although it was developed formally by Forchheimer [1930]; however, the adoption of the method by engineers followed Casagrande's classic paper in 1940. In this paper and in the highly recommended flow nets of Cedergren [1967] are to be found some of the highest examples of the art of drawing flow nets. Harr [1962] also warrants a peek!

Unfortunately, there is no "royal road" to drawing a good flow net. The speed with which a successful flow net can be drawn is highly contingent on the experience and judgment of the individual. In this regard, the beginner will do well to study the characteristics of well-drawn flow nets: *labor omnia vincit.*

In summary, a flow net is a sketch of distinct and special streamlines and equipotential lines that preserve right-angle intersections, satisfy the boundary conditions, and form curvilinear squares.[2] The following procedure is recommended:

1. Draw the boundaries of the flow region to a scale so that all sketched equipotential lines and streamlines terminate on the figure.
2. Sketch lightly three or four streamlines, keeping in mind that they are only a few of the infinite number of possible curves that must provide a smooth transition between the boundary streamlines.
3. Sketch the equipotential lines, bearing in mind that they must intersect all streamlines, including boundary streamlines, at right angles and that the enclosed figures must form curvilinear rectangles (except at singular points) with the same ratio of $\Delta s/\Delta s$ along a flow channel. Except for partial flow channels and partial head drops, these will form curvilinear squares with $\Delta w = \Delta s$.
4. Adjust the initial streamlines and equipotential lines to meet the required conditions. Remember that the drawing of a good flow net is a trial-and-error process, with the amount of correction being dependent upon the position selected for the initial streamlines.

Example 17.1. Obtain the quantity of discharge, Q/kh, for the section shown in Fig. 17.9.

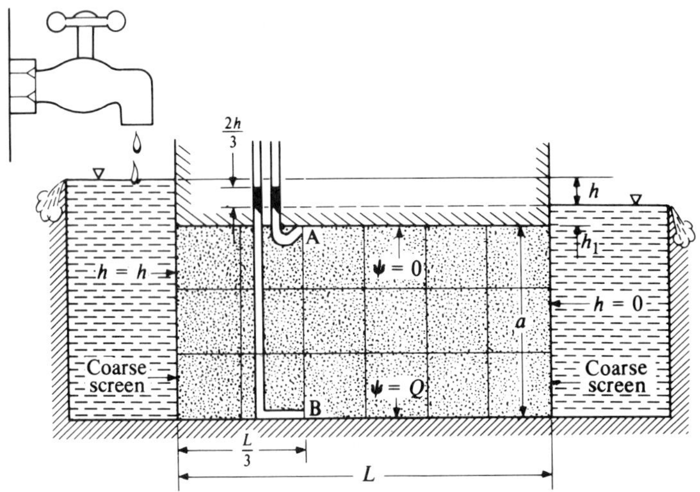

FIGURE 17.9 Example of flow regime.

[2]We except singular squares such as the five-sided square at point H in Fig. 17.8 and the three-sided square at point G. (It can be shown—Harr [1962], p. 84—that a five-sided square designates a point of turbulence). With continued subdividing into smaller squares, the deviations, in the limit, act only at singular points.

Solution. This represents a region of horizontal flow with parallel horizontal streamlines between the impervious boundaries and vertical equipotential lines between reservoir boundaries. Hence the flow net will consist of perfect squares, and the ratio of the number of flow channels to the number of drops will be $N_f/N_e = a/L$ and $Q/kh = a/L$.

Example 17.2. Find the pressure in the water at points A and B in Fig. 17.9.

Solution. For the scheme shown in Fig. 17.9, the total head loss is linear with distance in the direction of flow. Equipotential lines are seen to be vertical. The total heads at points A and B are both equal to $2h/3$ (datum at the tailwater surface). This means that a piezometer placed at these points would show a column of water rising to an elevation of $2h/3$ above the tailwater elevation. Hence, the pressure at each point is simply the weight of water in the columns above the points in question: $p_A = (2h/3 + h_1)\gamma_w, p_B = (2h/3 + h_1 + a)\gamma_w$.

Example 17.3. Using flow nets obtain a plot of Q/kh as a function of the ratio s/T for the single impervious sheetpile shown in Fig. 17.10(a).

Solution. We first note that the section is symmetrical about the y axis, hence only one-half of a flow net is required. Values of the ratio s/T range from 0 to 1, with 0 indicating no penetration and 1 complete cutoff. For $s/T = 1$, $Q/kh = 0$ [see point a in Fig. 17.10(b)]. As the ratio of s/T decreases, more flow channels must be added to maintain curvilinear squares and, in the limit as s/T approaches zero, Q/kh becomes unbounded [see arrow in Fig. 17.10(b)]. If $s/T = 1/2$ [Fig. 17.10(c)], each streamline will evidence a corresponding equipotential line in the

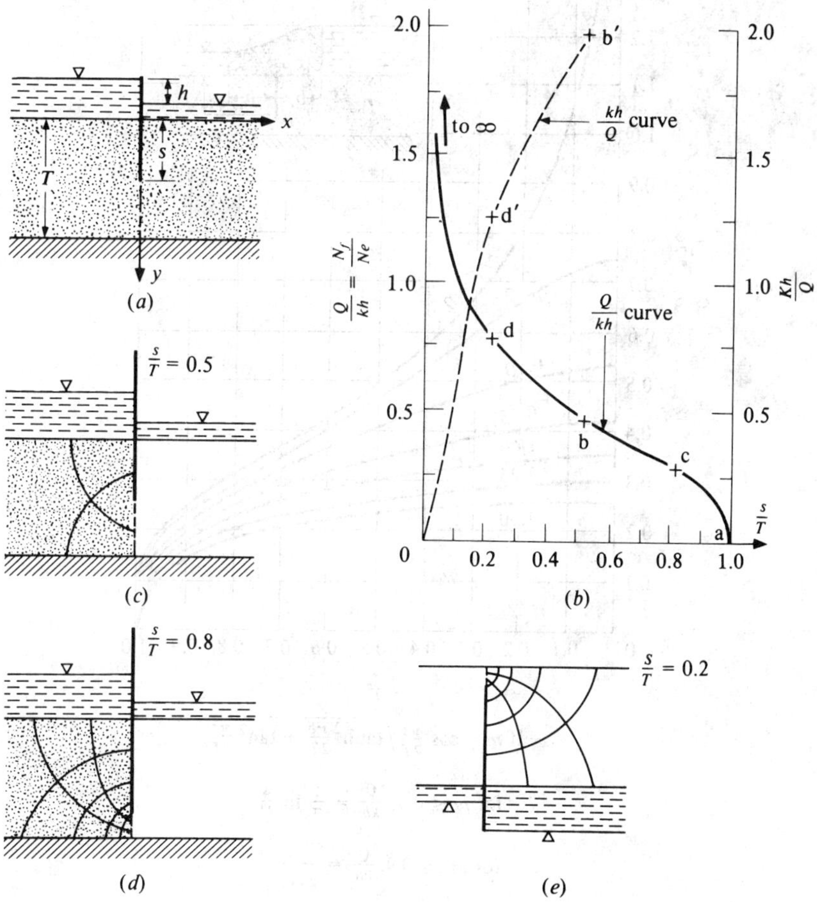

FIGURE 17.10 Example 17.3.

half-strip; consequently, for the whole flow region, $N_f/N_e = 1/2$ for $s/T = 1/2$ [point b in Fig. 17.10(b)]. Thus, without actually drawing a single flow net, we have learned quite a bit about the functional relationship between Q/kh and s/T. If Q/kh was known for $s/T = 0.8$ we would have another point and could sketch, with some reliability, the portion of the plot in Fig. 17.10(b) for $s/T \geq 0.5$. In Fig. 17.10(d) is shown one-half of the flow net for $s/T = 0.8$, which yields the ratio of $N_f/N_e = 0.3$ [point c in Fig. 17.10(b)]. As shown in Fig. 17.10(e), the flow net for $s/T = 0.8$ can also serve, geometrically, for the case of $s/T = 0.2$, which yields approximately $N_f/N_e = 0.8$ [plotted as point d in Fig. 17.10(b)]. The portion of the curve for values of s/T close to 0 is still in doubt. Noting that for $s/T = 0$, $kh/Q = 0$, we introduce an ordinate scale of kh/Q to the right of Fig. 17.10(b) and locate on this scale the corresponding values for $s/T = 0, 0.2$, and 0.5 (shown primed). Connecting these points (shown dotted) and obtaining the inverse, Q/kh, at desired points, the required curve can be had.

A plot giving the quantity of discharge (Q/kh) for symmetrically placed pilings as a function of depth of embedment (s/T), as well as for an impervious structure of width ($2b/T$), is shown in Fig. 17.11. This plot was obtained by Polubarinova-Kochina [1962] using a mathematical approach. The curve labeled $b/T = 0$ applies for the conditions in Example 17.3. It is interesting to note that

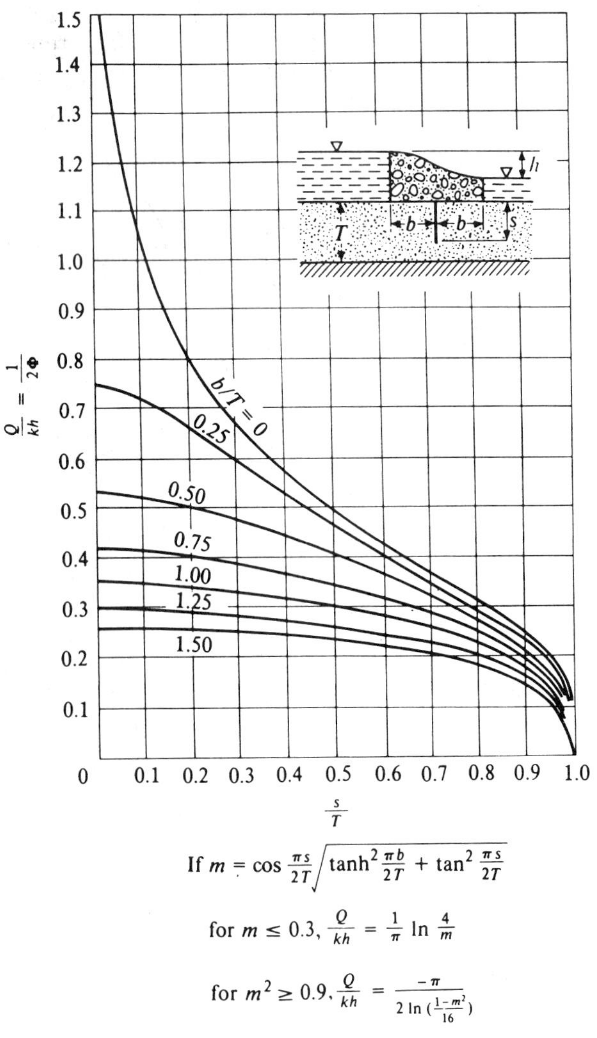

$$\text{If } m = \cos\frac{\pi s}{2T}\sqrt{\tanh^2\frac{\pi b}{2T} + \tan^2\frac{\pi s}{2T}}$$

$$\text{for } m \leq 0.3,\ \frac{Q}{kh} = \frac{1}{\pi}\ln\frac{4}{m}$$

$$\text{for } m^2 \geq 0.9,\ \frac{Q}{kh} = \frac{-\pi}{2\ln\left(\frac{1-m^2}{16}\right)}$$

FIGURE 17.11 Quantity of flow for given geometry.

this whole family of curves can be obtained, with reasonable accuracy (at least commensurate with the determination of the coefficient of permeability, k), by sketching only two additional half flow nets (for special values of b/T).

It was tacitly assumed in the foregoing that the medium was homogeneous and isotropic ($k_x = k_y$). Had isotropy not been realized, a transformation of scale (with x and y taken as the directions of maximum and minimum permeability, respectively) in the y direction of $Y = y(k_x/k_y)^{1/2}$ or in the x direction of $X = x(k_y/k_x)^{1/2}$ would render the system as an equivalent isotropic system [for details, see Harr, 1962, p. 29]. After the flow net has been established, by applying the inverse of the scaling factor, the solution can be had for the anisotropic system. The quantity of discharge for a homogeneous and anisotropic section is

$$Q = \sqrt{k_x k_y}\, h \frac{N_f}{N_e} \tag{17.26}$$

17.4 Method of Fragments

In spite of its many uses, a graphical flow net provides the solution for a particular problem only. Should one wish to investigate the influence of a range of characteristic dimensions (such as is often the case in a design problem), many flow nets would be required. Consider, for example, the section shown in Fig. 17.12, and suppose we wish to investigate the influence of the dimensions A, B, and C on the characteristics of flow, all other dimensions being fixed. Taking only three values for each of these dimensions would require 27 individual flow nets. As noted previously, a rough flow net should always be drawn as a check. In this respect it may be thought of as being analogous to a free-body diagram in mechanics, wherein the physics of a solution can be examined with respect to satisfying conditions of necessity.

An approximate analytical method of solution, directly applicable to design, was developed by Pavlovsky in 1935 [Pavlovsky, 1956] and was expanded and advanced by Harr [1962, 1977]. The underlying assumption of this method, called the **method of fragments,** is that equipotential lines at various critical parts of the flow region can be approximated by straight vertical lines (as, for example, the dotted lines in Fig. 17.13) that divide the region into sections or fragments. The groundwater flow region in Fig. 17.13 is shown divided into four fragments.

Suppose, now, that one computes the discharge in the ith fragment of a structure with m such fragments as

$$Q = \frac{kh_i}{\Phi_i} \tag{17.27}$$

where h_i is the head loss through the ith fragment and Φ_i is the dimensional form factor in the ith fragment, $\Phi_i = N_e/N_f$ in Eq. (17.25). Then, as the discharge through all fragments must

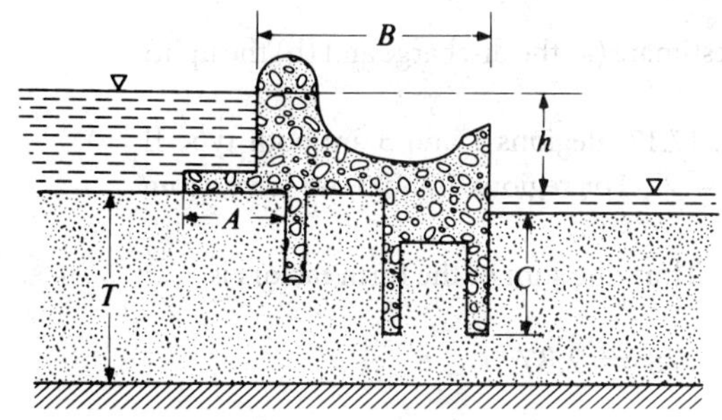

FIGURE 17.12 Example of complicated structure.

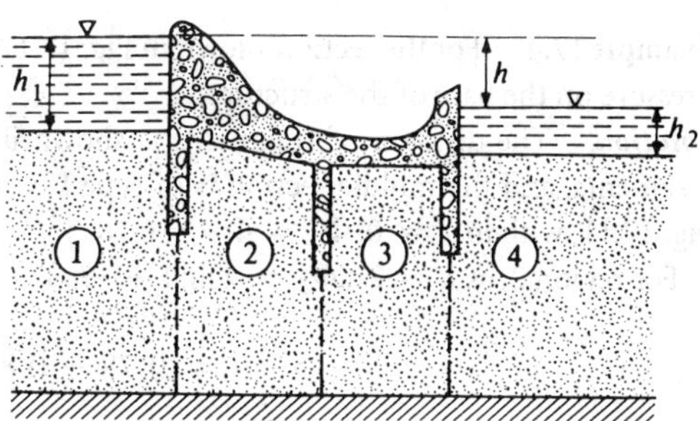

FIGURE 17.13 Four fragments.

be the same,

$$Q = \frac{kh_1}{\Phi_1} = \frac{kh_2}{\Phi_2} = \frac{kh_i}{\Phi_i} = \cdots = \frac{kh_m}{\Phi_m}$$

whence

$$\sum h_i = \frac{Q}{k} \sum \Phi_i$$

and

$$Q = k \frac{h}{\sum_{i=1}^{m} \Phi_i} \tag{17.28}$$

where h (without subscript) is the total head loss through the section. By similar reasoning, the head loss in the ith fragment can be calculated from

$$h_i = \frac{h\Phi_i}{\sum \Phi} \tag{17.29}$$

Thus, the primary task is to implement this method by establishing a catalog of form factors. Following Pavlovsky, the various form factors will be divided into types. The results are summarized in tabular form, Fig. 17.14, for easy reference. The derivation of the form factors is well documented in the literature [Harr 1962, 1977].

Various entrance and emergence conditions for type VIII and IX fragments are shown in Fig. 17.15. Briefly, for the entrance condition, when possible the free surface will intersect the slope at right angles. However, as the elevation of the free surface represents the level of available energy along the uppermost streamline, at no point along the curve can it rise above the level of its source of energy, the headwater elevation. At the point of emergence the free surface will, if possible, exit tangent to the slope [Dachler, 1934]. As the equipotential lines are assumed to be vertical, there can be only a single value of the total head along a vertical line, and, hence, the free surface cannot curve back on itself. Thus, where unable to exit tangent to a slope, it will emerge vertical.

To determine the pressure distribution on the base of a structure (such as that along $C'CC''$) in Fig. 17.16, Pavlovsky assumed that the head loss within the fragment is linearly distributed along the impervious boundary. Thus, in Fig. 17.16, if h_m is the head loss within the fragment, the rate of loss along $E'C'CC''E''$ will be

$$R = \frac{h_m}{L + s' + s''} \tag{17.30}$$

Once the total head is known at any point, the pressure can easily be determined by subtracting the elevation head, relative to the established (tailwater) datum.

Example 17.4. For the section shown in Fig. 17.17(a), estimate (a) the discharge and (b) the uplift pressure on the base of the structure.

Solution. The division of fragments is shown in Fig. 17.17. Regions 1 and 3 are both type II fragments, and the middle section is of type V with $L = 2s$. For regions 1 and 3, we have, from Fig. 17.11, with $b/T = 0$, $\Phi_1 = \Phi_3 = 0.78$.

For region 2, as $L = 2s$, $\Phi_2 = 2\ln(1 + 18/36) = 0.81$. Thus, the sum of the form factors is

$$\sum \Phi = 0.78 + 0.81 + 0.78 = 2.37$$

and the quantity of flow (Eq. 17.28) is $Q/k = 18/2.37 = 7.6$ ft.

Fragment type	Illustration	Form factor, Φ (h is head loss through fragment)
I	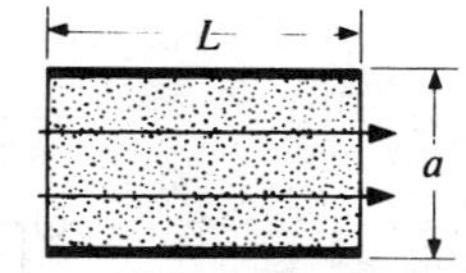	$\Phi = \frac{L}{a}$
II	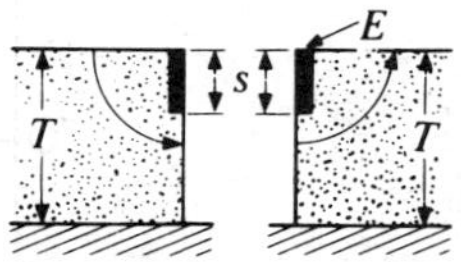	$\Phi = \frac{1}{2}\left(\frac{kh}{Q}\right)$, Fig. 17.11
III	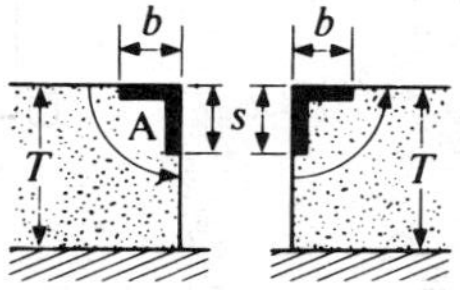	$\Phi = \frac{1}{2}\left(\frac{kh}{Q}\right)$, Fig. 17.11
IV	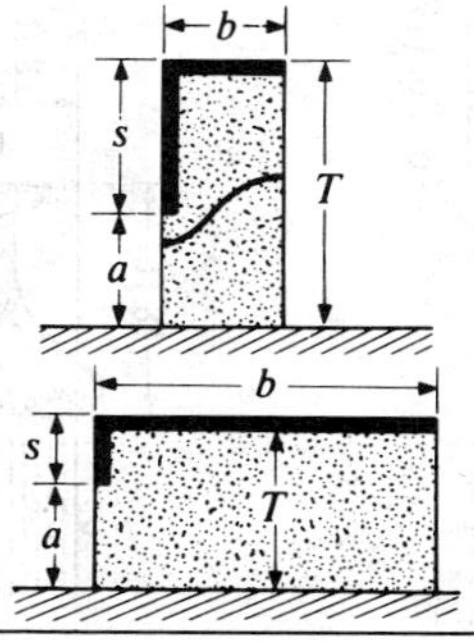	$b \le s$: $\Phi = \ln\left(1 + \frac{b}{a}\right)$ $b \ge s$: $\Phi = \ln\left(1 + \frac{s}{a}\right) + \frac{b-s}{T}$
V	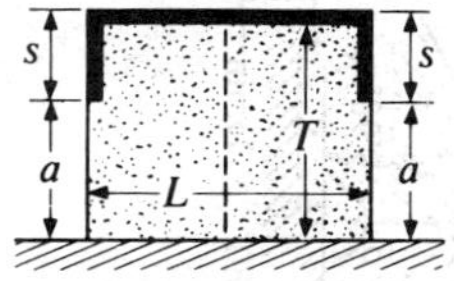	$L \le 2s$: $\Phi = 2\ln\left(1 + \frac{L}{2a}\right)$ $L \ge 2s$: $\Phi = 2\ln\left(1 + \frac{s}{a}\right) + \frac{L-2s}{T}$
VI	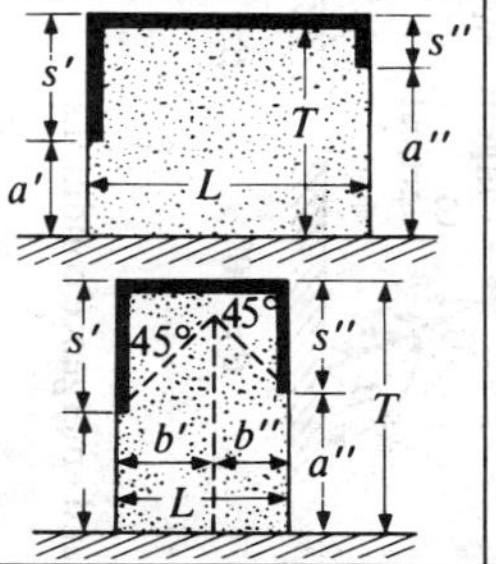	$L \ge s' + s''$: $\Phi = \ln\left[\left(1 + \frac{s'}{a'}\right)\left(1 + \frac{s''}{a''}\right)\right] + \frac{L - (s' + s'')}{T}$ $L \le s' + s''$: $\Phi = \ln\left[\left(1 + \frac{b'}{a'}\right)\left(1 + \frac{b''}{a''}\right)\right]$ where $b' = \frac{L + (s' - s'')}{2}$ $b'' = \frac{L - (s' - s'')}{2}$
VII	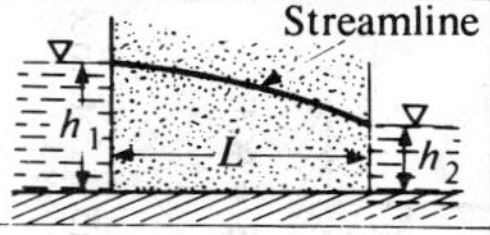	$\Phi = \frac{2L}{h_1 + h_2}$ $Q = k\frac{h_1^2 - h_2^2}{2L}$
VIII	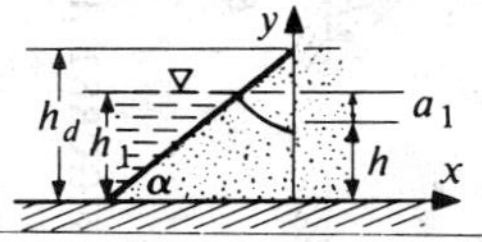	$Q = k\frac{h_1 - h}{\cot\alpha}\ln\frac{h_d}{h_d - h}$
IX	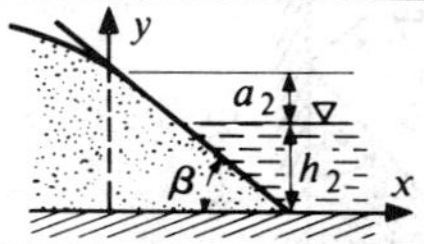	$Q = k\frac{a_2}{\cot\beta}\left(1 + \ln\frac{a_2 + h_2}{a_2}\right)$

FIGURE 17.14 Summary of fragment types and form factors.

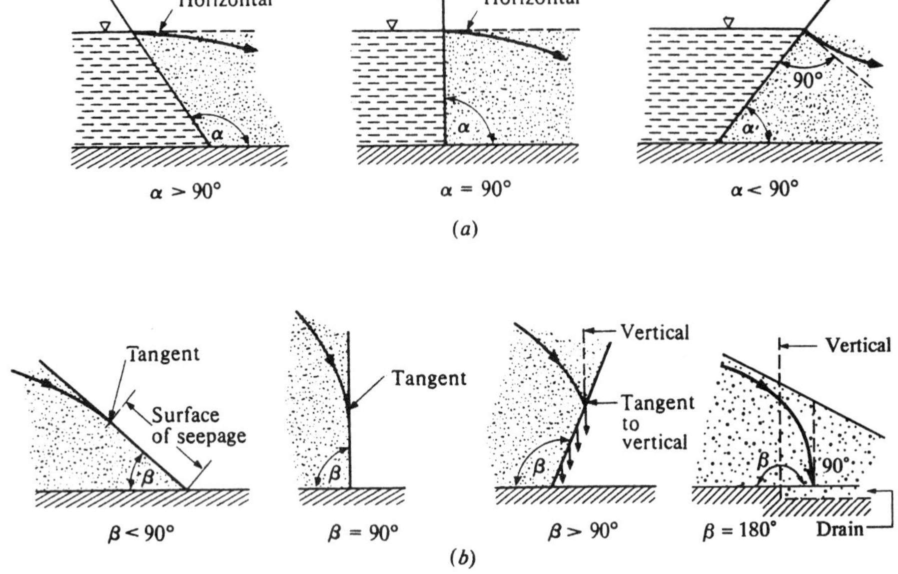

FIGURE 17.15 Entrance and emergence conditions.

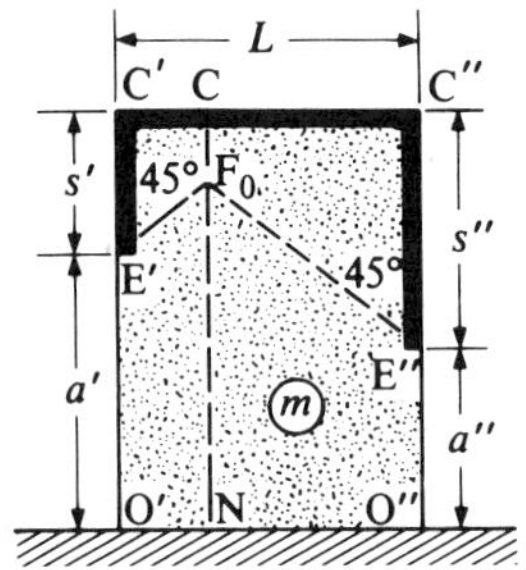

FIGURE 17.16 Illustration of Eq. (17.30).

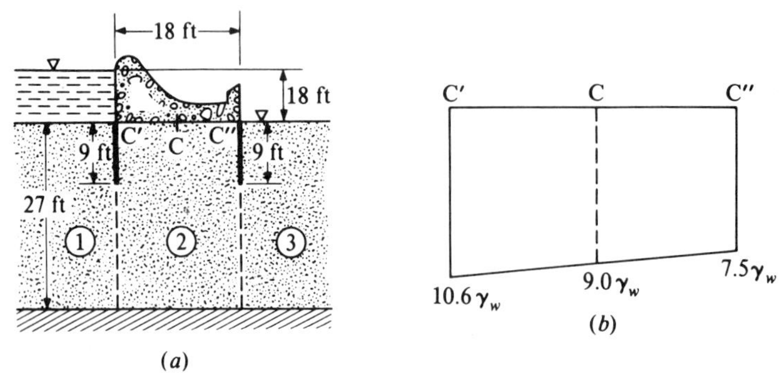

FIGURE 17.17 Example 17.4.

For the head loss in each of the sections, from Eq. (17.29) we find

$$h_1 = h_3 = \frac{0.78}{2.37}(18) = 5.9\,\text{ft}$$

$$h_2 = 6.1\,\text{ft}$$

Hence the head loss rate in region 2 is (Eq. 17.30)

$$R = \frac{6.1}{36} = 17\%$$

and the pressure distribution along $C'CC''$ is shown in Fig. 17.17(b).

Example 17.5.[3] Estimate the quantity of discharge per foot of structure and the point where the free surface begins under the structure (point A) for the section shown in Fig. 17.18(a).
Solution. The line AC in Fig. 17.18(a) is taken as the vertical equipotential line that separates the flow domain into two fragments. Region 1 is a fragment of type III, with the distance B as an unknown quantity. Region 2 is a fragment of type VII, with $L = 25 - B$, $h_1 = 10$ ft, and $h_2 = 0$. Thus, we are led to a trial-and-error procedure to find B. In Fig. 17.18(b) are shown plots of Q/k versus B/T for both regions. The common point is seen to be $B = 14$ ft, which yields a quantity of flow of approximately $Q = 100k/22 = 4.5k$.

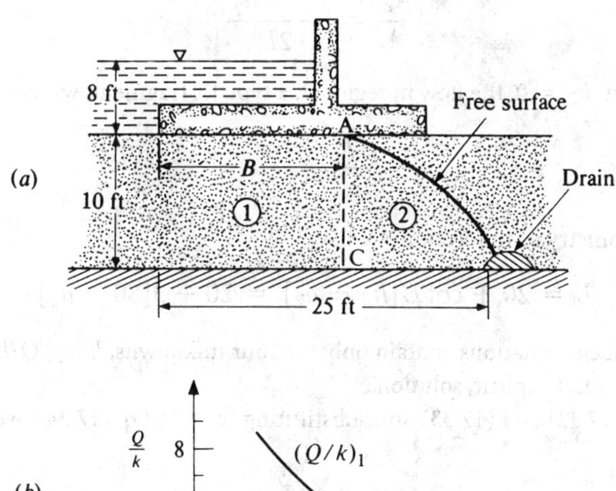

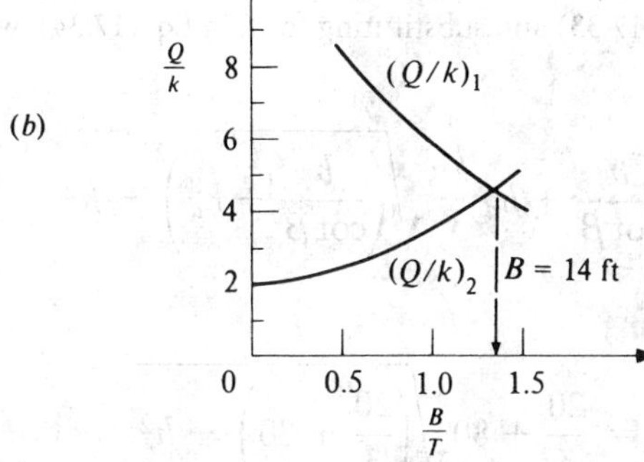

FIGURE 17.18 Example 17.5.

[3]For comparisons between analytical and experimental results for mixed fragments (confined and unconfined flow) see Harr and Lewis [1965].

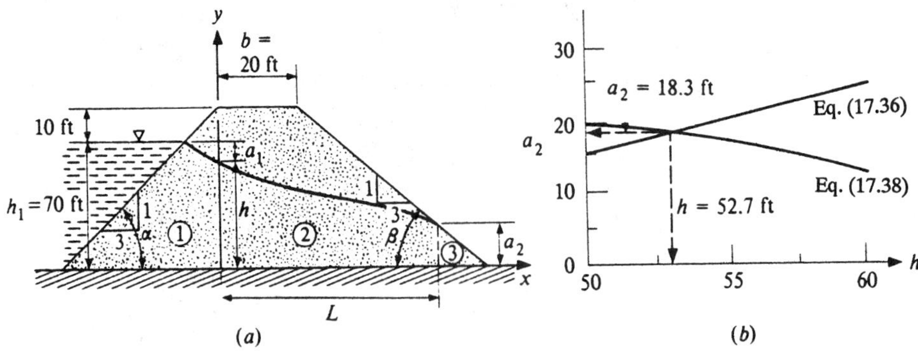

FIGURE 17.19 Example 17.6.

Example 17.6. Determine the quantity of flow for 100 ft of the earth dam section shown in Fig. 17.19(a), where $k = 0.002$ ft/min.

Solution. For this case, there are three regions. For region 1, a type VIII fragment $h_1 = 70$ ft, $\cot\alpha = 3$, $h_d = 80$ ft, produces

$$\frac{Q}{k} = \frac{70 - h}{3} \ln \frac{80}{80 - h} \tag{17.31}$$

For region 2, a type VII fragment produces

$$\frac{Q}{k} = \frac{h^2 - a_2^2}{2L} \tag{17.32}$$

With tailwater absent, $h_2 = 0$, the flow in region 3, a type IX fragment, with $\cot\beta = 3$ produces

$$\frac{Q}{k} = \frac{a_2}{3} \tag{17.33}$$

Finally, from the geometry of the section, we have

$$L = 20 + \cot\beta[h_d - a_2] = 20 + 3[80 - a_2] \tag{17.34}$$

The four independent equations contain only the four unknowns, h, a_2, Q/k, and L, and hence provide a complete, if not explicit, solution.

Combining Eqs. (17.32) and (17.33) and substituting for L in Eq. (17.34), we obtain, in general (b = crest width),

$$a_2 = \frac{b}{\cot\beta} + h_d - \sqrt{\left(\frac{b}{\cot\beta} + h_d\right)^2 - h^2} \tag{17.35}$$

and, in particular,

$$a_2 = \frac{20}{3} + 80\sqrt{\left(\frac{20}{3} + 80\right)^2 - h^2} \tag{17.36}$$

Likewise, from Eqs. (17.31) and (17.33), in general,

$$\frac{a_2 \cot\alpha}{\cot\beta} = (h_1 - h) \ln \frac{h_d}{h_d - h} \tag{17.37}$$

and, in particular,

$$a_2 = (70 - h) \ln \frac{80}{80 - h} \qquad (17.38)$$

Now, Eqs. (17.36) and (17.38), and (17.35) and (17.37) in general, contain only two unknowns (a_2 and h), and hence can be solved without difficulty. For selected values of h, resulting values of a_2 are plotted for Eqs. (17.36) and (17.38) in Fig. 17.19(b). Thus, $a_2 = 18.3$ ft, $h = 52.7$ ft, and $L = 205.1$ ft. From Eq. (17.33), the quantity of flow per 100 ft is

$$Q = 100 \times 2 \times 10^{-3} \times \frac{18.3}{3} = 1.22 \text{ ft}^3/\text{min} = 9.1 \text{ gal/min}$$

17.5 Flow in Layered Systems

Closed-form solutions for the flow characteristics of even simple structures founded in layered media offer considerable mathematical difficulty. Polubarinova-Kochina [1962] obtained closed-form solutions for the two layered sections shown in Fig. 17.20 (with $d_1 = d_2$). In her solution she found a cluster of parameters that suggested to Harr [1962] an approximate procedure whereby the flow characteristics of structures founded in layered systems can be obtained simply and with a great degree of reliability.

The flow medium in Fig. 17.20(a) consists of two horizontal layers of thickness d_1 and d_2, underlain by an impervious base. The coefficient of permeability of the upper layer is k_1 and of the lower layer k_2. The coefficients of permeability are related to a dimensionless parameter ϵ by the expression

$$\tan \pi\epsilon = \sqrt{\frac{k_2}{k_1}} \qquad (17.39)$$

(π is in radian measure). Thus, as the ratio of the permeabilities varies from 0 to ∞, ϵ ranges between 0 and $\frac{1}{2}$. We first investigate the structures shown in Fig. 17.20(a) for some special values of ϵ.

1. $\epsilon = 0$. If $k_2 = 0$, from Eq. (17.39) we have $\epsilon = 0$. This is equivalent to having the impervious base at depth d_1. Hence, for this case the flow region is reduced to that of a single homogeneous layer for which the discharge can be obtained directly from Fig. 17.11.

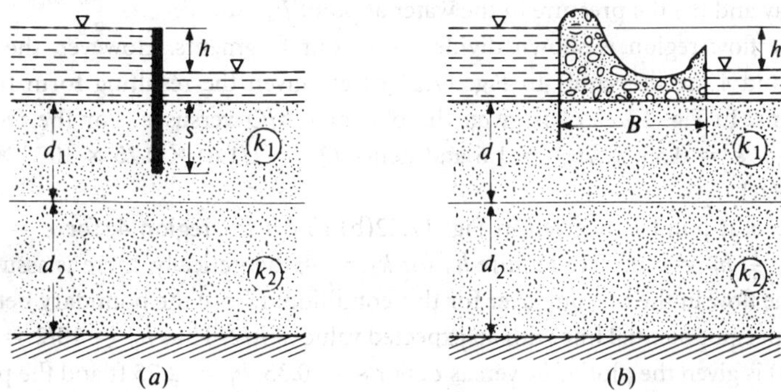

FIGURE 17.20 Examples of two-layered systems.

2. $\epsilon = \frac{1}{4}$. If $k_1 = k_2$, the sections are reduced to a single homogeneous layer, of thickness $d_1 + d_2$, for which Fig. 17.11 is again applicable.
3. $\epsilon = \frac{1}{2}$. If $k_2 = \infty$, $\epsilon = \frac{1}{2}$. This represents a condition where there is no resistance to flow in the bottom layer. Hence, the discharge through the total section under steady state conditions is infinite, or $Q/k_1h = \infty$. However, of greater significance is the fact that the inverse of this ratio equals zero: $k_1h/Q = 0$. It can be shown [Polubarinova-Kochina, 1962] that for $k_2/k_1 \rightarrow \infty$,

$$\frac{Q}{k_1 h} = \sqrt{\frac{k_2}{k_1}} \tag{17.40}$$

Thus, we see that for the special values of $\epsilon = 0$, $\epsilon = \frac{1}{4}$, and $\epsilon = \frac{1}{2}$, measures of the flow quantities can be easily obtained. The essence of the method then is to plot these values, on a plot of k_1h/Q versus ϵ, and connect the points with a smooth curve, from which intermediate values can be had.

Example 17.7. In Fig. 17.20(a), $s = 10$ ft, $d_1 = 15$ ft, $d_2 = 20$ ft, $k_1 = 4k_2 = 1 \times 10^{-3}$ ft/min, $h = 6$ ft. Estimate Q/k_1h.

Solution. For $\epsilon = 0$, from Fig. 17.11 with $s/T = s/d_1 = \frac{2}{3}$, $b/T = 0$, $Q/k_1h = 0.39$, $k_1h/Q = 2.56$.

For $\epsilon = \frac{1}{4}$, from Fig. 17.11 with $s/T = s/(d_1 + d_2) = \frac{2}{7}$, $Q/k_1h = 0.67$, $k_1h/Q = 1.49$.

For $\epsilon = \frac{1}{2}$, $k_1h/Q = 0$.

The three points are plotted in Fig. 17.21, and the required discharge, for $\epsilon = 1/\pi \tan^{-1} \cdot (1/4)^{1/2} = 0.15$, is $k_1h/Q = 1.92$ and $Q/k_1h = 0.52$; whence

$$\begin{aligned} Q &= 0.52 \times 1 \times 10^{-3} \times 6 \\ &= 3.1 \times 10^{-3} \ \text{ft}^3/(\text{min})(\text{ft}) \end{aligned}$$

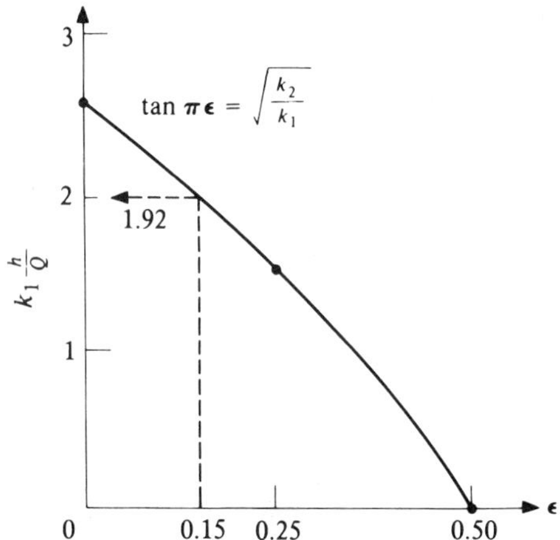

FIGURE 17.21 Example 17.7.

In combination with the method of fragments, approximate solutions can be had for very complicated structures.

Example 17.8. Estimate (a) the discharge through the section shown in Fig. 17.22(a), $k_2 = 4k_1 = 1 \times 10^{-3}$ ft/day and (b) the pressure in the water at point P.

Solution. The flow region is shown divided into four fragments. However, the form factors for regions 1 and 4 are the same. In Fig. 17.22(b) are given the resulting form factors for the listed conditions. In Fig. 17.22(c) is given the plot of k_1h/Q versus ϵ. For the required condition ($k_2 = 4k_1$), $\epsilon = 0.35$, $k_1h/Q = 1.6$ and hence $Q = (1/1.6) \times 0.25 \times 10^{-3} \times 10 = 1.6 \times 10^{-3}$ ft^3/(day)(ft).

The total head at point P is given in Fig. 17.22(b) as Δh for region 4; for $\epsilon = 0$, $h_p = 2.57$ ft and for $\epsilon = \frac{1}{4}$, $h_p = 3.00$. We require h_p for $k_2 = \infty$; theoretically, there is assumed to be no resistance to the flow in the bottom layer for this condition. Hence the boundary between the two layers (AB) is an equipotential line with an expected value of $h/2$. Thus $h_p = 10/2 = 5$ for $\epsilon = \frac{1}{2}$. In Fig. 17.22(d) is given the plot of h_p versus ϵ. For $\epsilon = 0.35$, $h_p = 2.75$ ft and the pressure in the water at point P is $(3.75 + 5)\gamma_w = 8.75\gamma_w$.

The above procedure may be extended to systems with more than two layers.

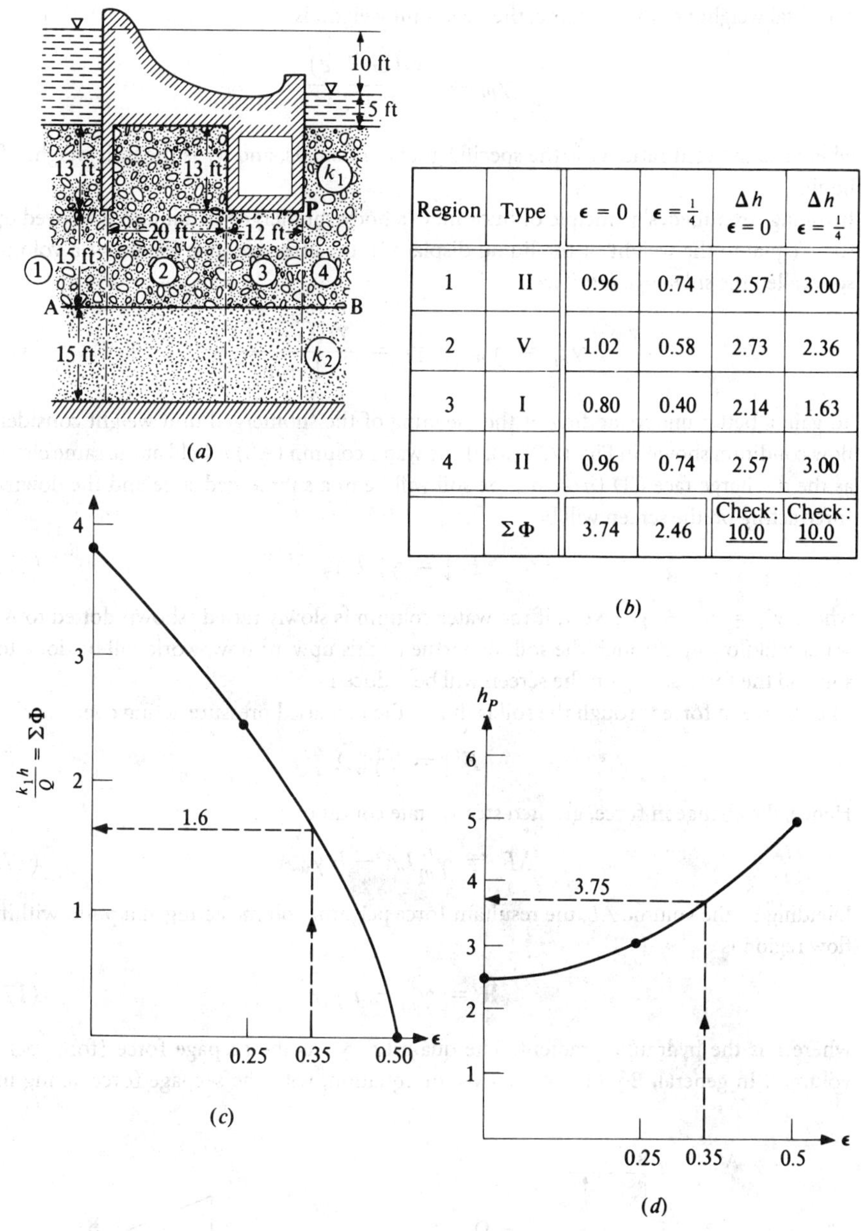

Region	Type	$\epsilon = 0$	$\epsilon = \frac{1}{4}$	Δh $\epsilon = 0$	Δh $\epsilon = \frac{1}{4}$
1	II	0.96	0.74	2.57	3.00
2	V	1.02	0.58	2.73	2.36
3	I	0.80	0.40	2.14	1.63
4	II	0.96	0.74	2.57	3.00
	$\Sigma\Phi$	3.74	2.46	Check: 10.0	Check: 10.0

FIGURE 17.22 Example 17.8.

17.6 Piping

By virtue of the viscous friction exerted on a liquid flowing through a porous medium, an energy transfer is effected between the liquid and the solid particles. The measure of this energy is the head loss h between the points under consideration. The force corresponding to this energy transfer is called the *seepage force.* It is this seepage force that is responsible for the phenomenon known as *quicksand,* and its assessment is of vital importance in stability analyses of earth structures subject to the action of flowing water (seepage).

The first rational approach to the determination of the effects of seepage forces was presented by Terzaghi [1922] and forms the basis for all subsequent studies. Consider all the forces acting on a volume of particulate matter through which a liquid flows.

1. The total weight per unit volume, the mass unit weight, is

$$\gamma_m = \frac{\gamma_l(G+e)}{1+e}$$

where e is the void ratio, G is the specific gravity of solids, and γ_l is the unit weight of the liquid.

2. Invoking Archimedes' principle of buoyancy (a body submerged in a liquid is buoyed up by force equal to the weight of the liquid displaced), the effective unit weight of a volume of soil, called the *submerged unit weight,* is

$$\gamma'_m = \gamma_m - \gamma_l = \frac{\gamma_l(G-1)}{1+e} \tag{17.41}$$

To gain a better understanding of the meaning of the submerged unit weight consider the flow condition shown in Fig. 17.23(a). If the water column (AB) is held at the same elevation as the discharge face CD ($h = 0$), the soil will be in a submerged state and the downward force acting on the screen will be

$$F \downarrow = \gamma'_m LA \tag{17.42}$$

where $\gamma'_m = \gamma_m - \gamma_w$. Now, if the water column is slowly raised (shown dotted to A′B′), water will flow up through the soil. By virtue of this upward flow, work will be done to the soil and the force acting on the screen will be reduced.

3. The change in force through the soil is due to the increased pressure acting over the area, or

$$F \uparrow = h\gamma_w A$$

Hence, the change in force, granted steady state conditions, is

$$\Delta F = \gamma'_m LA - h\gamma_w A \tag{17.43}$$

Dividing by the volume AL, the resultant force per unit volume acting at a point within the flow region is

$$\mathbf{R} = \gamma'_m - i\gamma_w \tag{17.44}$$

where i is the hydraulic gradient. The quantity $i\gamma_w$ is the seepage force (force per unit volume). In general, Eq. (17.44) is a vector equation, with the seepage force acting in the

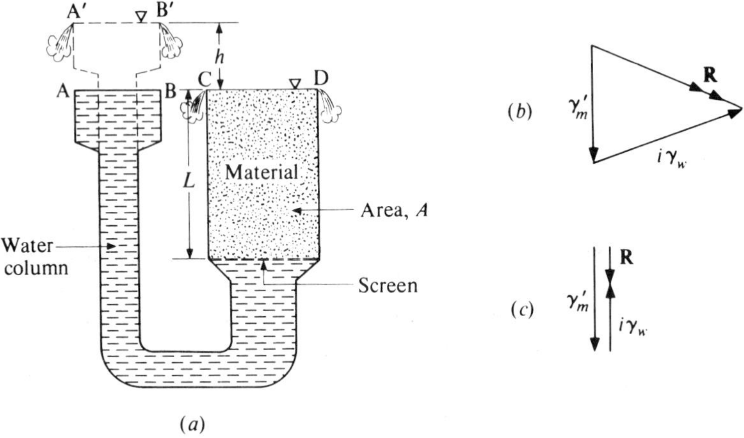

FIGURE 17.23 Piping.

direction of flow [Fig. 17.23(b)]. Of course, for the flow condition shown in Fig. 17.23(a), the seepage force will be directed vertically upward [Fig. 17.23(c)].

If the head h is increased, the resultant force **R** in Eq. (17.44) is seen to decrease. Evidently, should h be increased to the point at which $\mathbf{R} = 0$, the material would be at the point of being washed upward. Such a state is said to produce a **quick** (meaning alive) **condition.** From Eq. (17.44) it is evident that a quick condition is incipient if

$$i_{cr} = \frac{\gamma'_m}{\gamma_w} = \frac{G-1}{1+e} \tag{17.45}$$

Substituting typical values of $G = 2.65$ (quartz sand) and $e = 0.65$ (for sand, $0.57 \leq e \leq 0.95$), we see that as an average value the critical gradient can be taken as

$$i_{cr} \approx 1 \tag{17.46}$$

When information is lacking as to the specific gravity and void ratio of the material, the critical gradient is generally taken as unity.

At the critical gradient, there is no interparticle contact ($\mathbf{R} = 0$); the medium possesses no intrinsic strength, and will exhibit the properties of liquid of unit weight

$$\gamma_q = \left(\frac{G+e}{1+e}\right)\gamma_l \tag{17.47}$$

Substituting the above values for G, e, and $\gamma_l = \gamma_w$, $\gamma_q = 124.8$ lb/ft^3. Hence, contrary to popular belief, a person caught in quicksand would not be sucked down but would find it almost impossible to avoid floating.

Many hydraulic structures, founded on soils, have failed as a result of the initiation of a local quick condition which, in a chainlike manner, led to severe internal erosion called **piping.** This condition occurs when erosion starts at the exit point of a flow line and progresses backward into the flow region forming pipe-shaped watercourses which may extend to appreciable depths under a structure. It is characteristic of piping that it needs to occur only locally and that once begun it proceeds rapidly, and is often not apparent until structural failure is imminent.

Equations (17.45) and (17.46) provide the basis for assessing the safety of hydraulic structures with respect to failure by piping. In essence, the procedure requires the determination of the maximum hydraulic gradient along a discharge boundary, called the *exit gradient,* which will yield the minimum resultant force ($\mathbf{R}_{\text{min}}$) at this boundary. This can be done analytically, as will be demonstrated below, or from flow nets, after a method proposed by Harza [1935].

In the graphical method, the gradients along the discharge boundary are taken as the macrogradient across the contiguous squares of the flow net. As the gradients vary inversely with the distance between adjacent equipotential lines, it is evident that the maximum (exit) gradient is located where the vertical projection of this distance is a minimum, such as at the toe of the structure (point C) in Fig. 17.11. For example, the head lost in the final square of Fig. 17.11 is one-sixteenth of the total head loss of 16 ft, or 1 ft, and as this loss occurs in a vertical distance of approximately 4 ft, the exit gradient at point C is approximately 0.25. Once the magnitude of the exit gradient has been found, Harza recommended that the factor of safety be ascertained by comparing this gradient with the critical gradient of Eqs. (17.45) and (17.46). For example, the factor of safety with respect to piping for the flow conditions of Fig. 17.11 is 1.0/0.25, or 4.0. Factors of safety of 4 to 5 are generally considered reasonable for the graphical method of analysis.

The analytical method for determining the exit gradient is based on determining the exit gradient for the type II fragment, at point E in Fig. 17.14. The required value can be obtained directly from Fig. 17.24 with h_m being the head loss in the fragment.

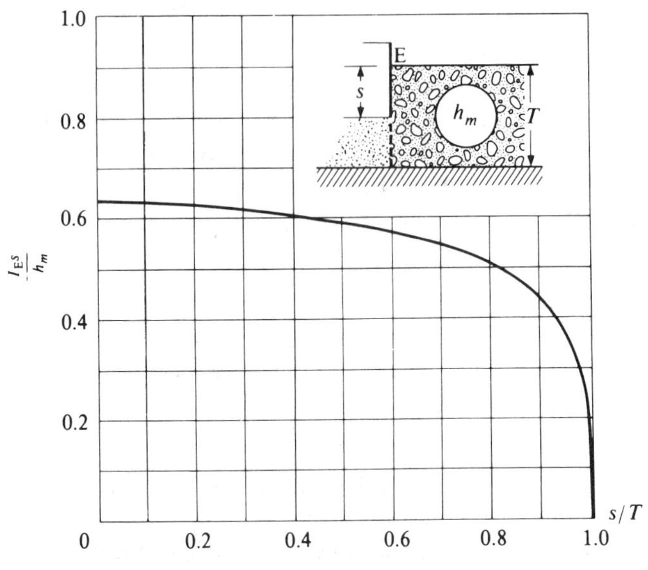

FIGURE 17.24 Exit gradient.

Example 17.9. Find the exit gradient for the section shown in Fig. 17.17(a).

Solution. From the results of Example 17.4, the head loss in fragment 3 is $h_m = 5.9$ ft. With $s/T = \frac{1}{3}$, from Fig. 17.24 we have $I_E s/h_m = 0.62$; hence,

$$I_E = \frac{0.62 \times 5.9}{9} = 0.41 \quad \text{or} \quad \text{FS} = \frac{1}{0.41} = 2.44$$

To account for the deviations and uncertainties in nature, Khosla *et al.* [1954] recommend that the following factors of safety be applied as critical values of exit gradients: gravel, 4 to 5; coarse sand, 5 to 6; and fine sand, 6 to 7.

The use of reverse filters on the downstream surface or where required serves to prevent erosion and decrease the probability of piping failures. In this regard, the Earth Manual of the U.S. Bureau of Reclamation, Washington [1974] is particularly recommended.

Defining Terms

Coefficient of permeability: Coefficient of proportionality between Darcy's velocity and the hydraulic gradient.

Flow net: Trial-and-error graphical procedure for solving seepage problems.

Hydraulic gradient: Space rate of energy dissipation.

Piping: Development of a "pipe" within soil by virtue of internal erosion.

Method of fragments: Approximate analytical method for solving seepage problems.

Quick condition: Condition when soil "liquifies."

References

Bear, J. 1972. *Dynamics of Fluids in Porous Media.* American Elsevier, New York.

Bear, J., Zaslavsky, D., and Irmay, S. 1966. *Physical Principles of Water Percolation and Seepage.* Technion, Israel.

Brahma, S. P. and Harr, M. E. 1962. Transient development of the free surface in a homogeneous earth dam. *Geotechnique.* 12(4).

Casagrande, A. 1940. Seepage Through Dams. In *Contributions to Soil Mechanics 1925–1940.* Boston Society of Civil Engineering, Boston.

Cedergren, H. R. 1967. *Seepage, Drainage and Flow Nets.* John Wiley & Sons, New York.

Dachler, R. 1934. Ueber den Strömungsvorgang bei Hanquellen. *Die Wasserwirtschaft,* no. 5.

Darcy, H. 1856. *Les Fontaines publique de la ville de Dijon.* Paris.

Domenico, P. A. and Schwartz, F. W. 1990. *Physical and Chemical Hydrogeology.* John Wiley & Sons, New York.

Dvinoff, A. H. and Harr, M. E. 1971. Phreatic surface location after drawdown. *J. Soil Mech. Found. ASCE.* January.

Earth Manual 1974. *Water Resources Technology Publication,* 2nd ed. U.S. Department of Interior, Bureau of Reclamation, Washington, D.C.

Forchheimer, P. 1930. *Hydraulik.* Teubner Verlagsgesellschaft, Stuttgart.

Freeze, R. A. and J. A. Cherry. 1979. *Groundwater.* Prentice-Hall, Englewood Cliffs, NJ.

Harr, M. E. 1962. *Groundwater and Seepage.* McGraw-Hill, New York.

Harr, M. E. and Lewis, K. H. 1965. Seepage around cutoff walls. *RILEM,* Bulletin 29, December.

Harr, M. E. 1977. *Mechanics of Particulate Media: A Probabilistic Approach.* McGraw-Hill, New York.

Harza, L. F. 1935. Uplift and seepage under dams on sand. *Trans. ASCE,* vol. 100.

Khosla, R. B. A. N., Bose, N. K. and Taylor, E. M. 1954. *Design of Weirs on Permeable Foundations.* Central Board of Irrigation, New Delhi, India.

Lambe, H. 1945. *Hydrodynamics.* Dover, New York.

Muskat, M. 1937. *The Flow of Homogeneous Fluids through Porous Media.* McGraw-Hill, New York. Reprinted by J. W. Edwards, Ann Arbor, 1946.

Pavlovsky, N. N. 1956. *Collected Works.* Akad. Nauk USSR, Leningrad.

Polubarinova-Kochina, P. Ya. 1941. Concerning seepage in heterogeneous (two-layered) media. *Inzhenernii Sbornik.* 1(2).

Polubarinova-Kochina, P. Ya. 1962. *Theory of the Motion of Ground Water.* Translated by J. M. R. De Wiest. Princeton University Press, Princeton, New Jersey. (Original work published 1952.)

Prasil, F. 1913. *Technische Hydrodynamik.* Springer-Verlag, Berlin.

Scheidegger, A. E. 1957. *The Physics of Flow through Porous Media.* Macmillan, New York.

Terzaghi, K. 1922. Der Grundbruch and Stauwerken und seine Verhütung. *Die Wasserkraft,* p. 445.

For Further Information

This chapter dealt with the conventional analysis of groundwater and seepage problems. Beginning in the mid-1970s another facet was added, motivated by federal environmental laws. These were concerned with transport processes, where chemical masses, generally toxic, are moved within the groundwater regime. This aspect of groundwater analysis gained increased emphasis with the passage of the Comprehensive Environmental Response Act of 1980, the *superfund.* Geotechnical engineers' involvement in these problems is likely to overshadow conventional problems. Several of the above references provide specific sources of information in this regard (cf. Domenico and Schwartz, Freeze and Cherry). The following are of additional interest:

Bear, J. and Verruijt, A. 1987. *Modeling Groundwater Flow and Pollution.* Dordrecht; Boston: D. Reidel Pub. Co.; Norwell, MA.

Mitchell, J. K. 1993. *Fundamentals of Soil Behavior.* John Wiley & Sons, New York.

Nyer, E. K. 1992. *Groundwater Treatment Technology.* Van Nostrand Reinhold, New York.

Sara, M. N. 1994. *Standard Handbook for Solid and Hazardous Waste Facility Assessments.* Lewis Publishers, Boca Raton, FL.

18

Consolidation and Settlement Analysis

Patrick J. Fox
Purdue University

In order to evaluate the suitability of a foundation or earth structure, it is necessary to design against both bearing capacity failure and excessive settlement. For foundations on cohesive soils, the principal design criterion is typically the latter—the control of expected settlements within the limits considered tolerable for the structure. As a result, once allowable foundation displacements have been established, the estimate of total settlement over the service life of the structure is a major factor in the choice of foundation design.

The purpose of this chapter is to present the fundamental concepts regarding settlement analysis for saturated, inorganic, cohesive soils. In addition, the recommended procedure for estimation of foundation settlements is described. Much of this chapter is based on Leonards [1968], Perloff [1975], and Holtz [1991]. Readers may refer to these works for additional information on consolidation and settlement analysis.

18.1 Components of Total Settlement

During construction, surface loads from foundations or earth structures are transmitted to the underlying soil profile. As a result, stresses increase within the soil mass and the structure undergoes a time-dependent vertical settlement. In general, this time-settlement curve can be represented conceptually as shown in Fig. 18.1. The **total settlement,** S, is calculated as the sum of the following three components:

$$S = S_i + S_c + S_s \tag{18.1}$$

where S_i is the **immediate settlement,** S_c is the **consolidation settlement,** and S_s is the **secondary compression settlement.**

Immediate settlement is time-independent and results from shear strains that occur at constant volume as the load is applied to the soil. Although this settlement component is not elastic, it is generally calculated using elastic theory for cohesive soils such as clays. Both consolidation and secondary compression settlement components are time-dependent and result from a reduction of void ratio and concurrent expulsion of water from the voids of the soil skeleton. For consolidation settlement, the rate of void ratio reduction is controlled by the rate at which water can escape from the soil. Therefore, during consolidation, pore water pressure exceeds the steady state condition

0-8493-8953-4/95/$0.00 + $.50

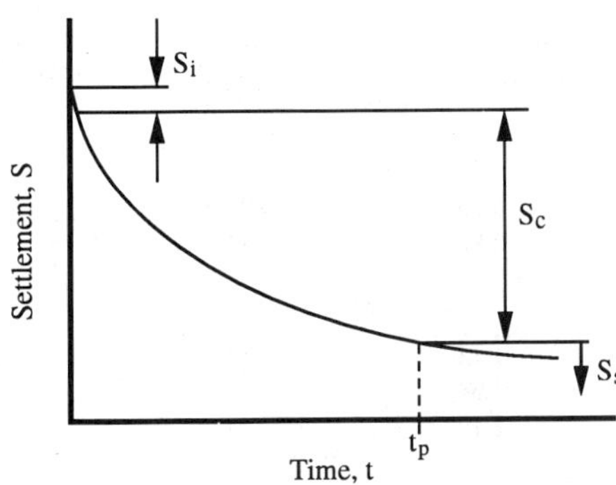

FIGURE 18.1 Time-settlement curve showing total settlement components.

throughout the depth of the layer. Over time, the rate of consolidation settlement continuously decreases as effective stresses increase to approach their equilibrium values. Once the consolidation process is completed at time t_p, settlement continues in the form of secondary compression. During secondary compression, the rate of void ratio reduction is controlled by the rate of compression of the soil skeleton itself. As such, it is essentially a creep phenomenon that occurs at constant vertical effective stress and without sensible excess pressure in the pore water.

The time-settlement relationship shown in Fig. 18.1 is conceptually valid for all soil types. However, large differences exist in the magnitude of the components and the rate at which they occur for different soils. For granular soils, such as sands, the hydraulic conductivity is sufficiently large that consolidation occurs nearly instantaneously with the applied load. In addition, although granular soils do exhibit creep effects, secondary compression is generally insignificant. For cohesive soils, such as clays, hydraulic conductivity is very small and the consolidation of a thick deposit may require years or even decades to complete. Secondary compression can be substantial for cohesive soils. Different from both sands and clays, peats and organic soils generally undergo rapid consolidation and extensive, long-term secondary compression.

The first step in a settlement analysis is a careful study of the changes in applied loads and the selection of appropriate fractions of live load pertinent to each of the three total settlement components. Often, insufficient attention is given to this aspect of the problem. In general, immediate settlements should be computed using 100% of both live and dead loads of the structure. Consolidation and secondary compression settlements should be calculated using 100% of the dead load and permanent live load, but only a reasonable fraction of the transient live load. The proper estimate of this fraction should be made in consultation with the structural engineer on the project [Leonards, 1992].

18.2 Immediate Settlement

For saturated or nearly saturated cohesive soils, a linear elastic model is generally used for the calculation of immediate settlement. Although clays do not behave as linear elastic materials, the rationale for the use of elastic theory has been the availability of solutions for a wide variety of boundary conditions representative of foundation engineering problems. In general, the elastic approximation performs reasonably well in the case of saturated clays under monotonic loading conditions not approaching failure. In addition, for these same conditions, the elastic parameters can generally be assumed as approximately constant throughout an otherwise homogeneous soil mass [Perloff, 1975].

For cohesionless soils, in which the equivalent elastic modulus depends markedly on confinement, the use of linear elastic theory coupled with the assumption of material homogeneity is inappropriate. Immediate settlement on granular soils is most often estimated using the procedure of Schmertmann [1970]. Holtz [1991] reviews this and other available methods in detail.

For those cases in which a linear elastic model is acceptable, solutions for stress distribution and surface deflection under a variety of flexible and rigid surface loading configurations can be found in Harr [1966], Perloff [1975], Poulos and Davis [1974], and Holtz [1991]. One particularly useful relationship is provided herein for the immediate settlement of a circular or rectangular footing at the surface of a deep isotropic stratum. In this case, the immediate vertical displacement is given by

Table 18.1 Values of C_s

Shape and Rigidity	Center	Corner	Edge/Middle of Long Side	Average
Circle (flexible)	1.00		0.64	0.85
Circle (rigid)	0.79		0.79	0.79
Square (flexible)	1.12	0.56	0.76	0.95
Square (rigid)	0.82	0.82	0.82	0.82
Rectangle (flexible): length/width				
2	1.53	0.76	1.12	1.30
5	2.10	1.05	1.68	1.82
10	2.56	1.28	2.10	2.24
Rectangle (rigid): length/width				
2	1.12	1.12	1.12	1.12
5	1.6	1.6	1.6	1.6
10	2.0	2.0	2.0	2.0

Source: Holtz, R. D. 1991. Stress distribution and settlement of shallow foundations. In *Foundation Engineering Handbook,* 2nd ed., ed. H.-Y. Fang. Van Nostrand Reinhold, New York.

$$S_i = C_s q B \left(\frac{1 - \nu^2}{E_u} \right) \tag{18.2}$$

where

S_i = immediate settlement of a point on the surface
C_s = shape and rigidity factor
q = equivalent uniform stress on the footing (total load/footing area)
B = characteristic dimension of the footing
ν = Poisson's ratio
E_u = undrained elastic modulus (Young's modulus)

The coefficient C_s is a function of the shape and rigidity of the loaded area and the point on the footing for which the immediate settlement estimate is desired. Thus, Eq. (18.2) can be used for both rigid and flexible footings, with the appropriate values of C_s given in Table 18.1. The characteristic footing dimension, *B*, is taken by convention as the diameter of a circular footing or the short side length of a rectangular footing. For saturated cohesive soils, constant volume strain is usually assumed and Poisson's ratio, ν, is set equal to 0.5. For soils that are nearly saturated, ν will be less than 0.5. However, using $\nu = 0.5$ is generally acceptable since the magnitude of computed immediate settlement is not especially sensitive to small changes in ν.

Reliable evaluation of the remaining soil parameter, the undrained elastic modulus, E_u, is critical for a good estimate of immediate settlement. In general, E_u is the slope of the undrained stress-strain curve for a stress path representative of the field condition. Figure 18.2 illustrates the measurement of E_u from a plot of principal stress difference, $\Delta\sigma$, versus axial strain, ε_a, as typically obtained from an undrained triaxial test. Principal stress difference is defined as $\sigma_1 - \sigma_3$, where σ_1 and σ_3 are the major and minor principal stresses, respectively. The initial tangent modulus, E_{ui}, is determined from the

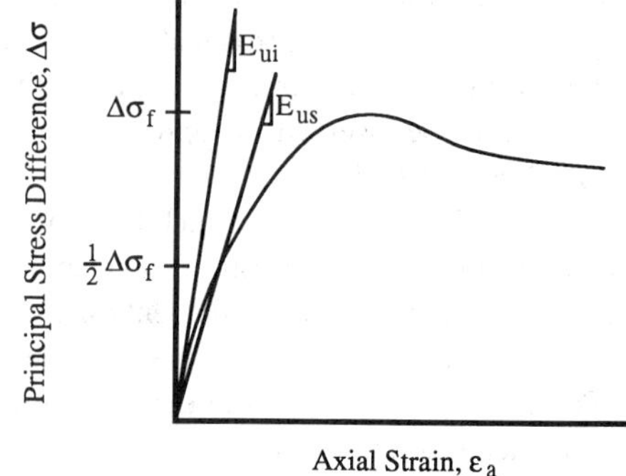

FIGURE 18.2 Definitions of the initial tangent modulus and secant modulus.

initial slope of the curve. The secant modulus, E_{us}, is sometimes used instead of E_{ui} when there is severe nonlinearity in the stress-strain relationship over the stress range of interest. Generally, the secant modulus would be taken at some predetermined stress level, such as 50% of the principal stress difference at failure, $\Delta\sigma_f$, in Fig. 18.2.

As a first approximation, the undrained elastic modulus can be estimated from the undrained shear strength using [Bjerrum, 1972]

$$E_u = 500c_u \quad \text{to} \quad E_u = 1500c_u \tag{18.3}$$

where c_u is the undrained shear strength determined from a field vane shear test. In general, E_u depends strongly on the level of shear stress. The lower value in Eq. (18.3) corresponds to highly plastic clays where the applied stress is relatively large as compared to the soil strength. The higher value is for low plasticity clays under small shear stress. In addition, the E_u/c_u ratio decreases with increasing **overconsolidation ratio** for a given stress level [Holtz and Kovacs, 1981]. Thus, Eq. (18.3) can provide a rough estimate of E_u suitable only for preliminary design computations.

In situations where a field loading test is not warranted, the undrained modulus should be estimated from a consolidated undrained (CU) triaxial test in the laboratory. The following procedure is recommended [Leonards, 1968]:

1. Obtain the highest quality soil samples. If possible, use a large-diameter piston sampler, or excavate blocks by hand from a test pit. Optimally, the laboratory test should be performed the same day as the field sampling operations.
2. Reconsolidate the specimen in the triaxial cell to the estimated initial in situ state of stress. If possible, anisotropic K_o consolidation is preferred to isotropic consolidation. Undrained modulus values determined from unconfined compression tests will significantly underestimate the actual value of E_u, and thereby overestimate the immediate settlement.
3. Load and unload the specimen in undrained axial compression to the expected in situ stress level for a minimum of 5 cycles. For field loading conditions other than a structural foundation, a different laboratory stress path may be needed to better match the actual in situ stress path.
4. Obtain E_u from the fifth (or greater) cycle in similar fashion to that shown in Fig. 18.2.

For sensitive clays of low plasticity, CU triaxial tests will likely yield somewhat low values of E_u, even if the specimens are allowed to undergo appreciable aging and E_u is determined at a low stress level. For highly plastic clays and organic clays, CU tests may yield stress-strain curves that are indicative of in situ behavior. However, it may be difficult to represent the nonlinear behavior with a single modulus value [D'Appolonia *et al.*, 1971].

The undrained elastic modulus is best measured directly from field tests. For near surface clay deposits having a consistency that does not vary greatly with depth, E_u may be obtained from a plate load test placed at footing elevation and passed through several loading-unloading cycles (ASTM D1194). In this case, all the parameters in Eq. (18.2) are known except the factor $(1 - \nu^2)/E$, which can then be calculated. Because of the relatively shallow influence of the test, it may be advisable to use a selection of different size plates and then scale $(1 - \nu^2)/E$ to the size of the prototype foundation. In situations where the loaded stratum is deep or displays substantial heterogeneity, plate load tests may not provide a representative value for E_u. Large-scale loading tests utilizing, for example, an embankment or a large tank of water may be warranted. In this case, the immediate settlement of the proposed foundation is measured directly without requiring Eq. (18.2). Measurement of stress-strain behavior using field tests is preferred to laboratory tests because of the many difficulties in determining the appropriate modulus in the laboratory. The most important of these is the invariable disturbance of soil structure that occurs during sampling and testing. Of the many soil properties defined in geotechnical engineering, E_u is one of the most sensitive to sample disturbance effects [Ladd, 1964].

For many foundations on cohesive soils, the immediate settlement is a relatively small part of the total vertical movement. Thus, a detailed study is seldom justified unless the structure is very sensitive to distortion, footing sizes and loads vary considerably, or the shear stresses imposed by the foundation are approaching a failure condition.

18.3 Consolidation Settlement

Different from immediate settlement, consolidation settlement occurs as the result of volumetric compression within the soil. For granular soils, the consolidation process is sufficiently rapid that consolidation settlement is generally included with immediate settlement. Cohesive soils have a much lower hydraulic conductivity, and, as a result, consolidation requires a far longer time to complete. In this case, consolidation settlement is calculated separately from immediate settlement, as suggested by Eq. (18.1).

When a load is applied to the ground surface, there is a tendency for volumetric compression of the underlying soils. For saturated materials, an increase in pore water pressure occurs immediately upon load application. Consolidation is then the process by which there is a reduction in volume due to the expulsion of water from the pores of the soil. The dissipation of excess pore water pressure is accompanied by an increase in effective stress and volumetric strain. Analysis of the resulting settlement is greatly simplified if it is assumed that such strain is one-dimensional, occurring only in the vertical direction. This assumption of one-dimensional compression is considered to be reasonable when (1) the width of the loaded area exceeds four times the thickness of the clay stratum, (2) the depth to the top of the clay stratum exceeds twice the width of the loaded area, or (3) the compressible material lies between two stiffer soil strata whose presence tends to reduce the magnitude of horizontal strains [Leonards, 1976].

Employing the assumption of one-dimensional compression, the consolidation settlement of a cohesive soil stratum is generally calculated in two steps:

1. Calculate the total (or "ultimate") consolidation settlement, S_c, corresponding to the completion of the consolidation process.
2. Using the theory of one-dimensional consolidation, calculate the fraction of S_c that will have occurred by the end of the service life of the structure. This fraction is the component of consolidation settlement to be used in Eq. (18.1).

In actuality, the total amount of consolidation settlement and the rate at which this settlement occurs is a coupled problem in which neither quantity can be calculated independently from the other. However, in geotechnical engineering practice, total consolidation settlement and rate of consolidation are almost always computed independently for lack of widely accepted procedures to solve the coupled problem. This will also be the approach taken here. The calculation of total consolidation settlement will be presented first, followed by procedures to calculate the rate at which this settlement occurs.

Total Consolidation Settlement

Total one-dimensional consolidation settlement, S_c, results from a change in void ratio, Δe, over the depth of the consolidating layer. The basic equation for calculating the total consolidation settlement of a single compressible layer is

$$S_c = \frac{\Delta e H_o}{1 + e_o} \tag{18.4}$$

where e_o is the initial void ratio and H_o is the initial height of the compressible layer.

Consolidation settlement is sometimes calculated using H_o for the entire consolidating stratum and stress conditions acting at the midheight. This procedure will underestimate the actual settlement, and the error will increase with the thickness of the clay. As Δe generally varies with depth, settlement calculations can be improved by dividing the consolidating stratum into n sublayers for purposes of analysis. Equation (18.4) is then applied to each sublayer and the cumulative settlement is computed using the following equation:

$$S_c = \sum_{i=1}^{n} \frac{\Delta e_i H_{oi}}{1 + e_{oi}} \tag{18.5}$$

where Δe_i is the change in void ratio, H_{oi} is the initial thickness, and e_{oi} is the initial void ratio of the ith sublayer.

The appropriate Δe_i for each sublayer within the compressible soil must now be determined. To begin, both the initial vertical effective stress, σ'_{vo}, and the final vertical effective stress (after excess pore pressures have fully dissipated), σ'_{vf}, are needed. The distribution of σ'_{vo} with depth is usually obtained by subtracting the in situ pore pressure from the vertical total stress, σ_v. Vertical total stress at a given depth is calculated using the following equation:

$$\sigma_v = \sum_{j=1}^{m} \gamma_j z_j \tag{18.6}$$

where

γ_j = unit weight of the jth stratigraphic layer
z_j = thickness of the jth stratigraphic layer
m = number of layers above the depth of interest

It should not be assumed a priori that in situ pore pressures are hydrostatic. Rather, significant upward or downward groundwater flow may be present. For important structures, the installation of piezometers to measure the in situ distribution of pore pressure is warranted. In addition, these piezometers will also provide a valuable check on the estimated initial excess pore pressures and indicate when the consolidation process is complete.

The final vertical effective stress is equal to the initial vertical effective stress plus the change of vertical effective stress, $\Delta\sigma'_v$, due to loading:

$$\sigma'_{vf} = \sigma'_{vo} + \Delta\sigma'_v \tag{18.7}$$

For truly one-dimensional loading conditions, such as a wide fill, $\Delta\sigma'_v$ is constant with depth and equal to the change in total stress applied at the surface of the soil stratum. For situations in which the load is applied over a limited surface area, such as a spread footing, $\Delta\sigma'_v$ will decrease with depth as the surface load is transmitted to increasingly larger portions of the soil mass. In this case, the theory of elasticity can be used to estimate $\Delta\sigma'_v$ as a function of depth under the center of the loaded area.

Once initial and final stress conditions have been established, it is necessary to determine the relationship between void ratio and vertical effective stress for the in situ soils. This information is generally obtained from a laboratory consolidation test (ASTM D2435). The general consolidation testing procedure is to place successive loads on an undisturbed soil specimen (typically 25.4 mm high with a diameter-to-height ratio of at least 2.5 to 1) and measure the void ratio corresponding to the end of consolidation for each load increment. The load increment ratio (LIR) is defined as the added load divided by the previous total load on the specimen. The load increment duration (LID) is the elapsed time permitted for each load increment. For the standard consolidation test, the load is doubled every day, giving LIR = 1 and LID = 24 hours. Detailed procedures for specimen preparation and performance of the laboratory consolidation test are found in Bowles [1992].

A typical laboratory compressibility curve is shown in Fig. 18.3. Void ratio, e, is plotted as a function of vertical effective stress, σ'_v, on a semilogarithmic scale. The open points are void ratios measured at the end of 24 hours for each load increment. They include the contribution from all previous immediate and secondary compression settlement. The solid points represent the sum of changes in void ratio during consolidation alone, and are calculated by subtracting out the immediate and secondary compression from all previous load increments. As indicated in Fig. 18.3, the laboratory compressibility curve is best drawn through the solid points. The reason for this procedure is that S_i, S_c, and S_s are computed separately and then summed to calculate total settlement S using Eq. (18.1). Therefore, immediate and secondary compression settlements should likewise be removed from the laboratory compressibility curve to compute S_c.

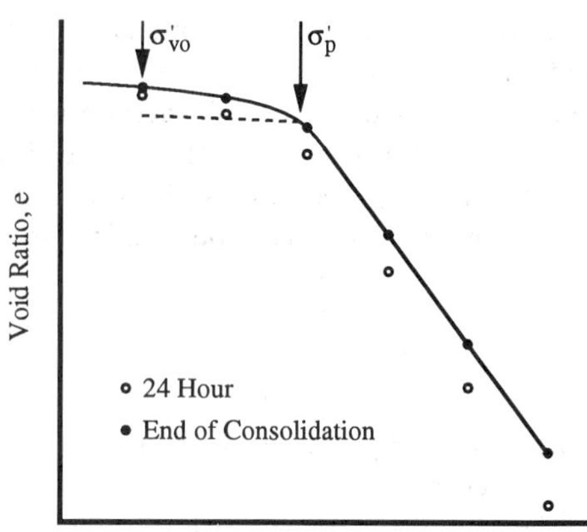

FIGURE 18.3 Typical laboratory compressibility curve.

For the compressibility curve in Fig. 18.3, both the **preconsolidation pressure,** σ'_p, and the in situ initial vertical effective stress are indicated. The stress history of a soil layer is generally expressed by its **overconsolidation ratio** (OCR), which is the ratio of these two values:

$$\text{OCR} = \frac{\sigma'_p}{\sigma'_{vo}} \tag{18.8}$$

Normally consolidated soils have OCR $= 1$, while soils with an OCR > 1 are *preconsolidated* or *overconsolidated*. For the example shown in Fig. 18.3, the soil is overconsolidated. In addition, a soil can be underconsolidated if excess pore pressures exist within the deposit (i.e., the soil is still undergoing consolidation). With the exception of recently deposited materials, soils in the field are very often overconsolidated as a result of unloading, desiccation, secondary compression, or aging effects [Brumund *et al.*, 1976].

The preconsolidation pressure is the stress at which the soil begins to yield in volumetric compression, and it therefore separates the region of small strains ($\sigma'_v < \sigma'_p$) from the region of large strains ($\sigma'_v > \sigma'_p$) on the $e - \log \sigma'_v$ diagram. As a result, for a given initial and final stress condition, the total consolidation settlement of a compressible layer is highly dependent on the value of the preconsolidation pressure. If a foundation applies a stress increment such that the final stress is less than σ'_p, the consolidation settlement will be relatively small. However, if the final stress is larger than σ'_p, much larger settlements will occur. Therefore, accurate determination of the preconsolidation pressure, and its variation with depth, is the most important step in a settlement analysis. The determination of σ'_p is generally performed using the Casagrande graphical construction method [Holtz and Kovacs, 1981].

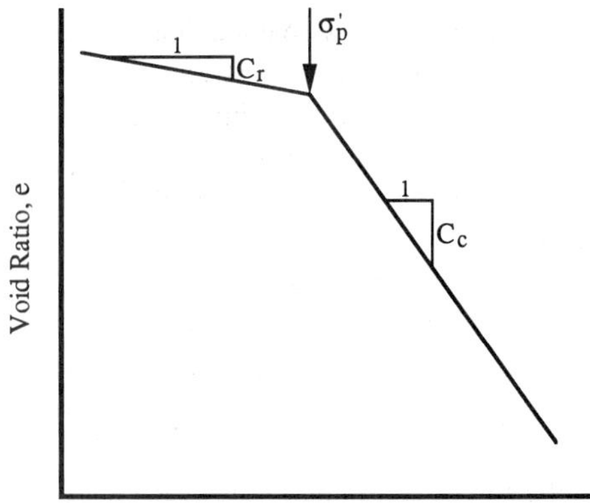

FIGURE 18.4 Simplified approximation of a laboratory compressibility curve.

For analysis purposes, the laboratory compressibility curve is usually approximated as linear (in log scale) for both the overconsolidated and normally consolidated ranges. A typical example is shown in Fig. 18.4. The slope of the overconsolidated range is the *recompression index*, C_r. Although this portion of a compressibility curve is generally not linear on the semilog plot, a line of constant C_r is usually fitted to the data for simplicity. The slope of the normally consolidated portion of the compressibility curve is the *compression index*, C_c. C_c is often constant over typical stress ranges of engineering interest. Both C_c and C_r are calculated using the same formula:

$$C_c = -\frac{\Delta e}{\Delta \log \sigma'_v} \quad \text{normally consolidated stress range} \tag{18.9}$$

$$C_r = -\frac{\Delta e}{\Delta \log \sigma'_v} \quad \text{overconsolidated stress range} \tag{18.10}$$

In many clay deposits, σ'_p, C_c, and C_r vary considerably with depth and the practice of performing one or two consolidation tests to evaluate the entire profile is seldom satisfactory. The following procedure is recommended for performing and interpreting the consolidation test to best obtain these parameters:

1. Obtain the highest quality undisturbed specimens, each representative of one sublayer in the profile, and begin testing the same day if possible. The value of σ'_p determined in the laboratory is very sensitive to sample disturbance and will generally be underestimated from data obtained using poor-quality soil samples.
2. Perform a consolidation test in which each load increment is placed on the specimen at the end of consolidation for the previous increment. The final void ratio obtained from each load will then directly provide the solid points in Fig. 18.3. The end of consolidation can be determined from pore pressure measurements or, less accurately, from graphical procedures such as the Casagrande log time or Taylor square root of time methods. LIR $= 1$ is satisfactory; however, to more accurately measure the value of σ'_p, it may be advantageous to run a second test with smaller load increments in the vicinity of the preconsolidation pressure. If a standard consolidation test is performed using LID $= 24$ hours, plot the cumulative void ratio reduction during consolidation for each increment as shown by the solid dots in Fig. 18.3.
3. Obtain C_r by unloading from σ'_p to σ'_{vo} and then reloading. This path is indicated by the dashed line in Fig. 18.3. Using the initial reloading curve in Fig. 18.3 will yield too large a value of C_r. When the value of C_r is critical to a particular design, a backpressure oedometer should be used for testing. A C_r value obtained by unloading and reloading in a conventional oedometer (without backpressure) is about twice the value obtained in a backpressure oedometer due to expansion of gas within the pore water.
4. Reconstruct the in situ compressibility curve using the method of Schmertmann [1955] as described in Holtz and Kovacs [1981], among others.

Once the in situ compressibility curve has been established for a given sublayer, the change of void ratio can be calculated knowing $\Delta\sigma'_v$. For normally consolidated conditions, the change of void ratio for the ith sublayer is

$$\Delta e_i = C_{ci} \log\left(\frac{\sigma'_{vfi}}{\sigma'_{voi}}\right) \tag{18.11}$$

Substituting Eq. (18.11) into Eq. (18.5), the ultimate consolidation settlement for a normally consolidated soil is

$$S_c = \sum_{i=1}^{n} \frac{C_{ci} H_{oi}}{1 + e_{oi}} \log\left(\frac{\sigma'_{vfi}}{\sigma'_{voi}}\right) \tag{18.12}$$

where the summation is performed over n sublayers.

In the case of overconsolidated clays, the change of void ratio for a given $\Delta\sigma'_v$ is

$$\Delta e_i = C_{ri} \log\left(\frac{\sigma'_{vfi}}{\sigma'_{voi}}\right) \tag{18.13}$$

if $\sigma'_{vf} < \sigma'_p$ and

$$\Delta e_i = C_{ri} \log\left(\frac{\sigma'_{pi}}{\sigma'_{voi}}\right) + C_{ci} \log\left(\frac{\sigma'_{vfi}}{\sigma'_{pi}}\right) \tag{18.14}$$

if $\sigma'_{vf} > \sigma'_p$. Substituting Eqs. (18.13) and (18.14) into Eq. (18.5), the total consolidation settlement of an overconsolidated soil is

$$S_c = \sum_{i=1}^{n} \frac{C_{ri} H_{oi}}{1 + e_{oi}} \log\left(\frac{\sigma'_{vfi}}{\sigma'_{voi}}\right) \tag{18.15}$$

if $\sigma'_{vf} < \sigma'_p$ and,

$$S_c = \sum_{i=1}^{n} \frac{C_{ri} H_{oi}}{1 + e_{oi}} \log\left(\frac{\sigma'_{pi}}{\sigma'_{voi}}\right) + \frac{C_{ci} H_{oi}}{1 + e_{oi}} \log\left(\frac{\sigma'_{vfi}}{\sigma'_{pi}}\right) \tag{18.16}$$

if $\sigma'_{vf} > \sigma'_p$.

As noted earlier, this discussion of consolidation settlement has been limited to conditions of one-dimensional compression. In those cases where the thickness of the compressible strata is large relative to the dimensions of the loaded area, the three-dimensional nature of the problem may influence the magnitude and rate of consolidation settlement. The best approach for problems of this nature is a three-dimensional numerical analysis, but these have not yet become generally accepted in practice. As an alternative, the semiempirical approach of Skempton and Bjerrum [1957] and the stress path method [Lambe, 1967] are more commonly used to take these effects into account.

Rate of Consolidation Settlement

The preceding discussion has described the calculation of ultimate consolidation settlement corresponding to the complete dissipation of excess pore pressure and the return of the soil to an equilibrium stress condition. At any time during the process of consolidation, the amount of settlement is directly related to the proportion of excess pore pressure that has been dissipated. The theory of consolidation is used to predict the progress of excess pore pressure dissipation as a function of time. Therefore, the same theory is also used to predict the rate of consolidation settlement. The one-dimensional theory of Terzaghi is most commonly used for prediction of consolidation settlement rate. The assumptions of the classical Terzaghi theory are as follows:

1. Drainage and compression are one-dimensional.
2. The compressible soil layer is homogenous and completely saturated.
3. The mineral grains and pore water are incompressible.
4. Darcy's law governs the outflow of water from the soil.

5. The applied load increment produces only small strains. Therefore, the thickness of the layer remains unchanged during the consolidation process.
6. The hydraulic conductivity and compressibility of the soil are constant.
7. The relationship between void ratio and vertical effective stress is linear and unique. This assumption also implies that there is no secondary compression settlement.
8. Total stress remains constant throughout the consolidation process.

Accepting these assumptions, the fundamental governing equation for one-dimensional consolidation is

$$\frac{\partial u}{\partial t} = c_v \frac{\partial^2 u}{\partial z^2} \tag{18.17}$$

where

$$c_v = \frac{k(1 + e_o)}{\gamma_w a_v} \tag{18.18}$$

and

k = hydraulic conductivity
γ_w = unit weight of water
e_o = initial void ratio
$a_v = -de/d\sigma'_v$ = coefficient of compressibility

The parameter c_v is called the coefficient of consolidation and is mathematically analogous to the diffusion coefficient in Fick's second law. It contains material properties that govern the process of consolidation and has dimensions of area per time. In general, c_v is not constant because its component parameters vary during the consolidation process. However, in order to reduce Eq. (18.17) to a linear form that is more easily solved, c_v is assumed constant for an individual load increment.

The consolidation equation [Eq. (18.17)] can be solved analytically using the Fourier series method. In the course of the solution, the following dimensionless quantities are defined:

$$Z = \frac{z}{H_{dr}} \tag{18.19}$$

$$T = \frac{c_v t}{H_{dr}^2} \tag{18.20}$$

$$U_z = 1 - \frac{u}{u_i} \tag{18.21}$$

where

z = depth below top of the compressible stratum
H_{dr} = length of the longest pore water drainage path
t = elapsed time of consolidation
u = excess pore pressure at time t and position z
u_i = initial excess pore pressure at position z

Z is a measure of the dimensionless depth within the consolidating stratum, T is the time factor and serves as a measure of dimensionless time, and U_z is the consolidation ratio. U_z is a function of both Z and T, and thus it varies throughout the consolidation process with both time and vertical position within the layer. U_z expresses the progress of consolidation at a

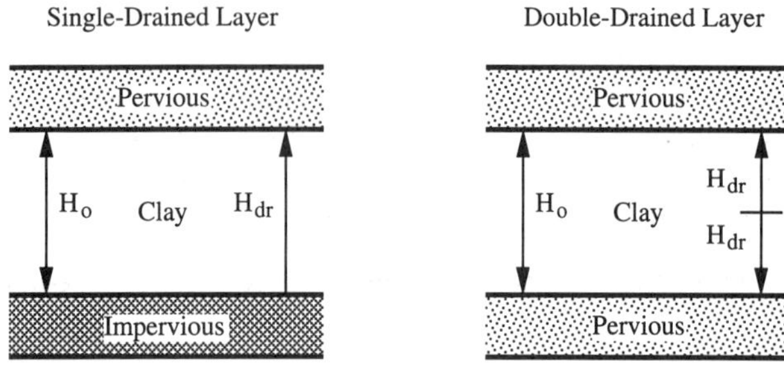

FIGURE 18.5 Boundary drainage conditions for the consolidation problem.

specific point within the consolidating layer. The value of H_{dr} depends on the boundary drainage conditions for the layer. Figure 18.5 shows the two typical drainage conditions for the consolidation problem. A single-drained layer has an impervious and pervious boundary. Pore water can escape only through the pervious boundary, giving $H_{dr} = H_o$. A double-drained layer is bounded by two pervious strata. Pore water can escape to either boundary, and therefore $H_{dr} = H_o/2$.

Figure 18.6 shows the solution to Eq. (18.17) in terms of the above dimensionless parameters. For a double-drained layer, pore pressure dissipation is modeled using the entire figure. However, for a single-drained layer, only the upper or lower half is used. As expected, U_z is zero for all Z at the beginning of the consolidation process ($T = 0$). As time elapses and pore pressures dissipate, U_z gradually increases to 1.0 for all points in the layer and σ_v' increases accordingly. From Fig. 18.6, it is possible to find the consolidation ratio (and therefore u and σ_v') at any time t and any position z within the consolidating layer after the start of loading. The time factor T can be calculated from Eq. (18.20) given the c_v for a particular deposit, the total thickness of the layer, and the boundary drainage conditions.

Figure 18.6 also provides some insight as to the progress of consolidation with time. The isochrones (curves of constant T) represent the percent consolidation for a given time throughout the compressible layer. For example, the percent consolidation at the midheight of a doubly

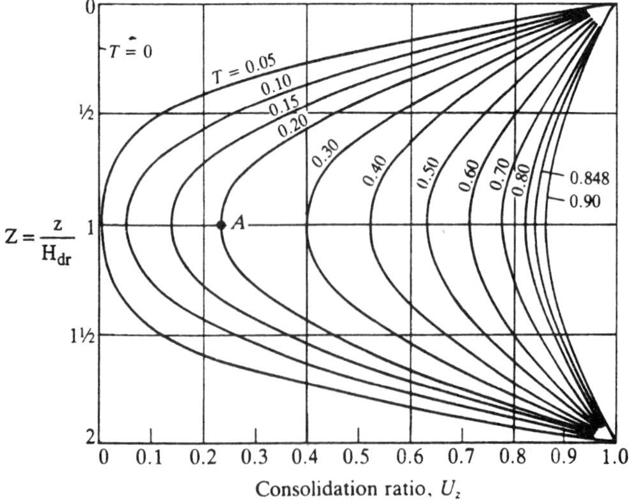

FIGURE 18.6 Consolidation ratio as a function of Z and T. (*Source:* Taylor, D.W. 1948. *Fundamentals of Soil Mechanics.* John Wiley & Sons, New York.)

drained layer for $T = 0.2$ is approximately 23%(see point A in Fig. 18.6). However, at $Z = 0.5$, $U_z = 44\%$ for the same time factor. Similarly, near the drainage surfaces at $Z = 0.1$, the clay is already 86% consolidated. This also means, at that same depth and time, 86% of the original excess pore pressure has dissipated and the effective stress has increased by a corresponding amount.

For settlement analysis, the consolidation ratio U_z is not the quantity of immediate interest. Rather, the geotechnical engineer needs to know the average degree of consolidation for the entire layer, U, defined as

$$U = \frac{s(t)}{S_c} \tag{18.22}$$

where $s(t)$ is the consolidation settlement at time t and S_c is the total (ultimate) consolidation settlement.

The following approximations can be used to calculate U. For $U < 0.60$,

$$T = \frac{\pi}{4}U^2 \tag{18.23}$$

and for $U > 0.60$,

$$T = 1.781 - 0.933\log(100(1 - U)) \tag{18.24}$$

Provided S_c is known, Eqs. (18.22), (18.23), and (18.24) can be used to predict consolidation settlement as a function of time.

18.4 Secondary Compression Settlement

Secondary compression settlement results from the time-dependent rearrangement of soil particles under constant effective stress conditions. For highly compressible soils, such as soft clays and peats, secondary compression is important whenever there is a net increase in σ'_v due to surface loading. Although structures of any consequence would seldom be founded on these soils, highways, for example, must commonly cross areas of compressible soils that are either too deep or too extensive to excavate.

Secondary compression settlement can be predicted using the secondary compression index, C_α, defined as the change of void ratio per log cycle of time:

$$C_\alpha = -\frac{\Delta e}{\Delta \log t} \tag{18.25}$$

Laboratory values for the secondary compression index should be measured at a stress level and temperature corresponding to that expected in the subsurface.

As a first approximation, C_α can be calculated from the compression index using the ratio C_α/C_c. This method has the advantage of not requiring prolonged periods of secondary compression in the laboratory consolidation test. Recommended values for C_α/C_c are [Mesri and Castro, 1987]:

$$\frac{C_\alpha}{C_c} = 0.04 \pm 0.01 \quad \text{for soft inorganic clays} \tag{18.26}$$

$$\frac{C_\alpha}{C_c} = 0.05 \pm 0.01 \quad \text{for highly organic plastic clays} \tag{18.27}$$

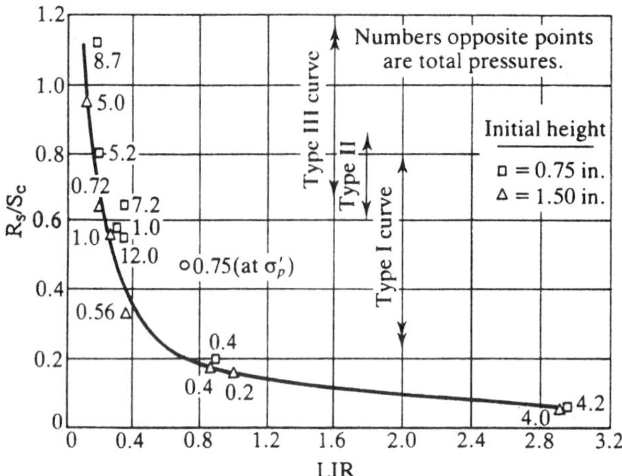

FIGURE 18.7 Effect of LIR on rate of secondary compression for undisturbed Mexico City clay. (*Source:* Leonards, G. A. and Girault, P. 1961. A study of the one-dimensional consolidation test. In *Proc. 5th Int. Conf. Soil Mech. Found. Eng.*, Paris, 1:213–218.)

Once a C_α value has been selected, secondary compression settlement S_s is calculated using the following equation:

$$S_s = \frac{C_\alpha H_o}{1 + e_o} \log \frac{t_f}{t_p} \tag{18.28}$$

where t_p is the time at the end of consolidation, and t_f is the final time for which secondary compression settlement is desired (typically the design life of the structure).

For all loading conditions, including one-dimensional compression, LIR decreases with depth. This is especially true for foundations where the load is spread over a limited surface area. For important structures, it is recommended to account for the effect of LIR on S_s. To begin, the secondary compression settlement per log cycle of time, R_s, is defined as follows,

$$R_s = \frac{\Delta S_s}{\Delta \log t} \tag{18.29}$$

Leonards and Girault [1961] demonstrated that a consistent relationship exists between R_s/S_c and LIR for Mexico City clay, as shown in Fig. 18.7. If a corresponding relationship can be obtained from laboratory consolidation tests, it can be used to estimate secondary compression settlement in the field in the following manner:

1. Divide the compressible strata into sublayers and compute the ultimate consolidation settlement S_{ci} for each ith sublayer using the procedure previously described.
2. Calculate LIR for each sublayer.
3. Obtain R_{si}/S_{ci} using the LIR for each sublayer.
4. Multiply the values of S_{ci} from step 1 by the values of R_{si}/S_{ci} obtained in step 3 to calculate R_{si} for each sublayer.
5. Sum the value of R_{si} for all sublayers to obtain R_s, the total secondary compression settlement per log cycle of time.
6. Calculate the secondary compression settlement using the following equation:

$$S_s = R_s \log \frac{t_f}{t_p} \tag{18.30}$$

Discrepancies of up to 75% may be expected using this procedure. This is indicative of the present state-of-the-art in predicting secondary compression settlement.

In many cases of practical interest, secondary compression is a minor effect relative to the magnitude of consolidation settlement. However, in some instances where very soft soils are involved or where deep compressible strata are subjected to small LIR, secondary compression may account for the majority of total settlement.

Defining Terms

Consolidation settlement: The time-dependent component of total settlement that results from the dissipation of excess pore pressure from within the soil mass.

Immediate settlement: The time-independent component of total settlement that occurs at constant volume as the load is applied to the soil.

Normally consolidated: A condition in which the initial vertical effective stress is equal to the preconsolidation pressure.

Overconsolidated: A condition in which the initial vertical effective stress is less than the preconsolidation pressure.

Overconsolidation ratio: The value of the preconsolidation pressure divided by the initial vertical effective stress.

Preconsolidation pressure: The vertical effective stress at which the soil begins to yield in volumetric compression.

Secondary compression: The time-dependent component of total settlement which occurs after consolidation and results from creep under constant effective stress.

Total settlement: The total vertical displacement of a foundation or earth structure that takes place after construction.

References

ASTM. 1993. *Annual Book of ASTM Standards, Volume 04.08: Soil and Rock; Dimension Stone; Geosynthetics.* American Society for Testing and Materials, Philadelphia, PA.

Bjerrum, L. 1972. Embankments on soft ground. In *Proc.* ASCE Spec. Conf. *Performance Earth Earth-Supported Structures,* Purdue University, 2:1–54.

Bowles, J. E. 1992. *Engineering Properties of Soils and Their Measurement,* 4th ed. McGraw-Hill, New York.

Brumund, W. F., Jonas, J., and Ladd, C. C. 1976. Estimating in situ maximum past (preconsolidation) pressure of saturated clays from results of laboratory consolidometer tests. *Estimation of Consolidation Settlement,* TRB Special Report 163, National Research Council, pp. 4–12.

D'Appolonia, D. J., Poulos, H. G., and Ladd, C. C. 1971. Initial settlement of structures on clay. *J. Soil. Mech. Found., ASCE.* 97(SM10):1359–1377.

Harr, M. E. 1966. *Theoretical Soil Mechanics.* McGraw-Hill, New York.

Holtz, R. D. 1991. Stress distribution and settlement of shallow foundations. In *Foundation Engineering Handbook,* 2nd ed., ed. H.-Y. Fang, pp. 168–222, Van Nostrand Reinhold, New York.

Holtz, R. D. and Kovacs, W. D. 1981. *An Introduction to Geotechnical Engineering.* Prentice-Hall, Englewood Cliffs, NJ.

Ladd, C. C. 1964. Stress-strain modulus of clay in undrained shear. *J. Soil. Mech. Found., ASCE.* 90(SM5):103–134.

Lambe, T. W. 1967. Stress path method. *J. Soil. Mech. Found., ASCE.* 93(SM6):309–331.

Leonards, G. A. 1992. Personal communication.

Leonards, G. A. 1976. Estimating consolidation settlements of shallow foundations on overconsolidated clay. *Estimation of Consolidation Settlement,* TRB Special Report 163, National Research Council, pp. 13–16.

Leonards, G. A. 1968. Predicting settlements of buildings on clay soils. H.F. In *Proc. Soil Mech. Lect. Ser.*, Illinois Section ASCE, Northwestern University.

Leonards, G. A. and Girault, P. 1961. A study of the one-dimensional consolidation test. In *Proc. 5th Int. Conf. Soil Mech. Found. Eng.*, Paris, 1:213–218.

Mesri, G. and Castro, A. 1987. C_α/C_c concept and K_0 during secondary compression. *J. Geotech. Eng., ASCE.* 113:(3):230–247.

Perloff, W. H. 1975. Pressure distribution and settlement. In *Foundation Engineering Handbook*, ed. H. F. Winterkorn and H.-Y. Fang, pp. 148–196, Van Nostrand Reinhold, New York.

Poulos, H. G. and Davis, E. H. 1974. *Elastic Solutions for Soil and Rock Mechanics.* John Wiley & Sons, New York.

Schmertmann, J. H. 1955. The undisturbed consolidation behavior of clay, *Transactions, ASCE.* 120:1201–1233.

Schmertmann, J. H. 1970. Static cone to compute static settlement over sand. *J. Soil. Mech. Found., ASCE.* 96(SM3):1011–1043.

Skempton, A. W., and Bjerrum, L. 1957. A contribution to the settlement analysis of foundations on clay. *Geotechnique,* 7(4):168–178.

Taylor, D. W. 1948. *Fundamentals of Soil Mechanics.* John Wiley & Sons, New York.

19

Stress Distribution

Milton E. Harr
Purdue University

From a microscopic point of view, all soil bodies are composed of discrete particles that are connected to each other by forces of mutual attraction and repulsion. Given an initial state of equilibrium, if an additional force system is applied, deformations may occur; particle arrangements may be altered; and changes in the distribution of the resultant forces may take place. Although the intensity of the generated forces may be high at points of contact, their range of influence is very short. Generally, effects extend only over a distance of molecular size or *in the very near vicinity* of the particles. The internal forces generated at these points by the induced loadings are called **stresses.**

Because stresses reflect distributional changes induced by boundary loadings, they can be thought of as providing a measure of the transmission of induced energy throughout the body. The transformation from real soil bodies, composed of discrete particles, into a form such that useful deductions can be made through the exact processes of mathematics is accomplished by introducing the abstraction of a *continuum*, or *continuous medium*, and pleading *statistical macroscopic equivalents* through the introduction of *material properties*. In concept, the soil is assumed to be continuously distributed in the region of space under consideration. This supposition then brings the continuous space required by mathematical formulations and the material points of real bodies into conformity.

This chapter will present some basic solutions for increases in vertical stresses due to some commonly encountered boundary loadings. Solutions will be presented both from the linear theory of elasticity and from particulate mechanics. Special efforts have been made to present the results in easy-to-use form. Many solutions can also be found in computer software packages.

19.1 Elastic Theory (Continuum)

Three-Dimensional Systems

Soil in the neighborhood of a point is called **isotropic** if its defining parameters are the same in all directions emanating from that point. Isotropy reduces the number of elastic constants at a point to two: E, the *modulus of elasticity*; and μ, *Poisson's ratio* [Harr, 1966]. If the elastic constants are the same at all points within a region of a soil body, that region is said to be **homogeneous**. Invoking

0-8493-8953-4/95/$0.00 + $.50

linear constitutive equations produces

$$
\begin{aligned}
\epsilon_x &= \frac{1}{E}[\sigma_x - \mu(\sigma_y + \sigma_z)] \\
\epsilon_y &= \frac{1}{E}[\sigma_y - \mu(\sigma_x + \sigma_z)] \\
\epsilon_z &= \frac{1}{E}[\sigma_z - \mu(\sigma_x + \sigma_y)] \\
\gamma_{yz} &= \frac{2(1+\mu)}{E}\tau_{yz} \\
\gamma_{xz} &= \frac{2(1+\mu)}{E}\tau_{xz} \\
\gamma_{xy} &= \frac{2(1+\mu)}{E}\tau_{xy}
\end{aligned}
\tag{19.1}
$$

where ϵ_i and σ_i are the normal **strains** and stresses, respectively, and γ_i and τ_i are the shearing strains and stresses in the $i = x, y, z$ directions, respectively.

Force Normal to Surface (Boussinesq Problem)

Assuming that the z direction coincides with the direction of gravity, the vertical stress under a concentrated load P as shown in Fig. 19.1(a), where $R^2 = x^2 + y^2 + z^2$, is [Boussinesq, 1885]

$$\sigma_z = \frac{3Pz^3}{2\pi R^5} \tag{19.2}$$

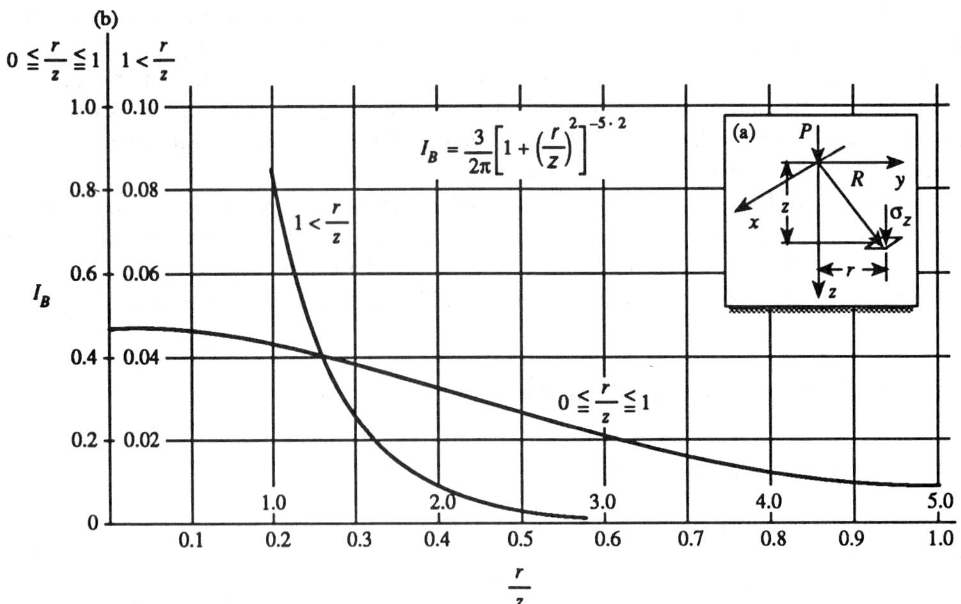

FIGURE 19.1 Concentrated force acting at and normal to surface; $\sigma_z = I_B P/z^2$.

It should be noted that the vertical normal stress (σ_z) is independent of the so-called elastic parameters. Equation (19.2) can also be written as

$$\sigma_z = I_B \frac{P}{z^2} \tag{19.3}$$

where

$$I_B = \frac{3}{2\pi}\left[1 + \left(\frac{r}{z}\right)^2\right]^{-5/2}$$

A plot of the **influence factor** I_B is given in Fig. 19.1(b).

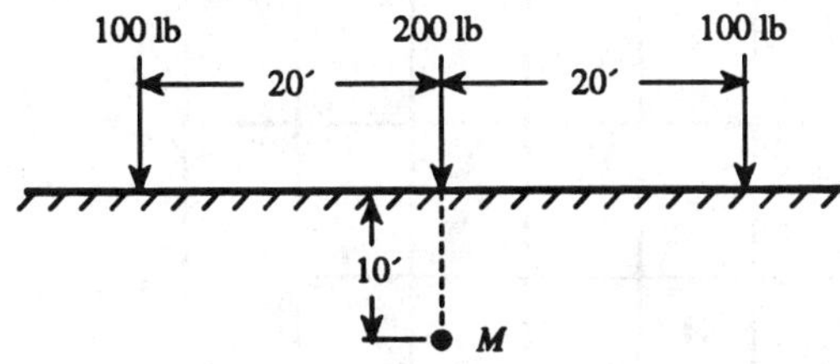

FIGURE 19.2 Example 19.1.

Because superposition is valid, the effects of a number of normal forces acting on the surface of a body can be accounted for by adding their relative influence values.

Example 19.1. Find the vertical stress at point M in Fig. 19.2 due to the three loads shown acting in a line at the surface.

Solution. For the 200 lb force $r/z = 0$ and, from Fig. 19.1(b), $I_B = 0.478$. For the 100 lb forces, $r/z = 20/10 = 2$, $I_B = 0.009$. Hence, the corresponding vertical stress at point M is

$$\sigma_z = \frac{200(0.478)}{100} + \frac{2(100)(0.009)}{100} = 0.974\,\text{lb/ft}^2$$

By applying the principle of superposition, the increased stress due to distributed loadings over flexible areas at the surface can be obtained by dividing the loading into increments (see Fig. 19.3) and treating each increment as a concentrated force.

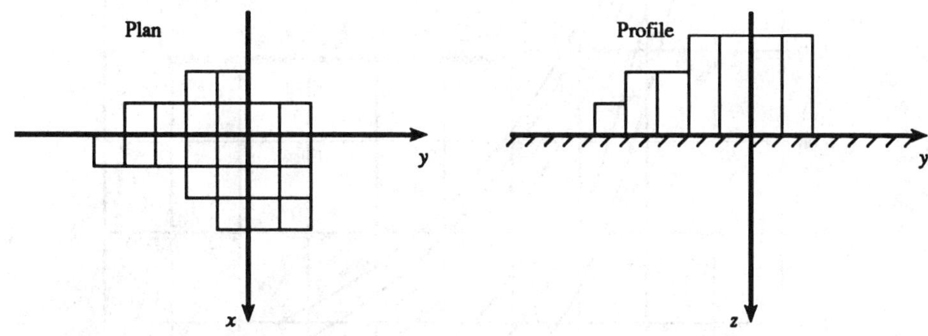

FIGURE 19.3 Distributed loads.

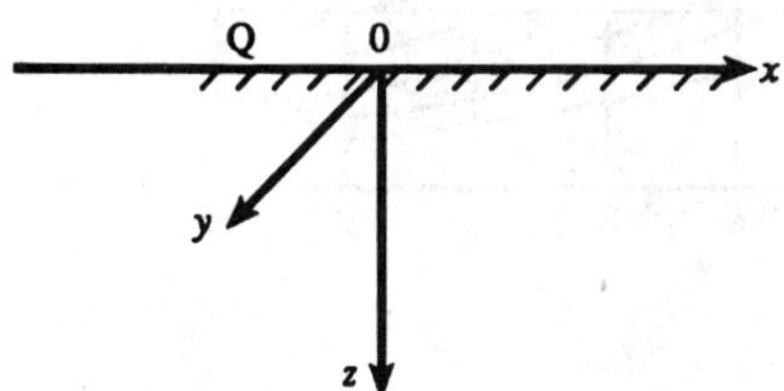

FIGURE 19.4 Concentrated force parallel to surface.

For a concentrated force parallel to the boundary surface (Fig. 19.4), the vertical stress [Cerruti, 1882] is

$$\sigma_z = \frac{3Qxz^2}{2\pi R^5} \tag{19.4}$$

By combining Eqs. (19.2) and (19.4), the increase in vertical stress, consistent with the assumptions of linear elasticity, can be determined for a concentrated force at the surface with any inclination.

Uniform Flexible Load over Rectangular Area

The vertical stress under the corner of a flexible, uniformly loaded, rectangular area with sides a by b, as in Fig. 19.5(a), is

$$\sigma_z = \frac{q}{2\pi}\left[\frac{mn}{\sqrt{1+m^2+n^2}}\frac{1+m^2+2n^2}{(1+n^2)(m^2+n^2)} + \sin^{-1}\frac{m}{\sqrt{m^2+mn^2}\sqrt{1+n^2}}\right] \tag{19.5}$$

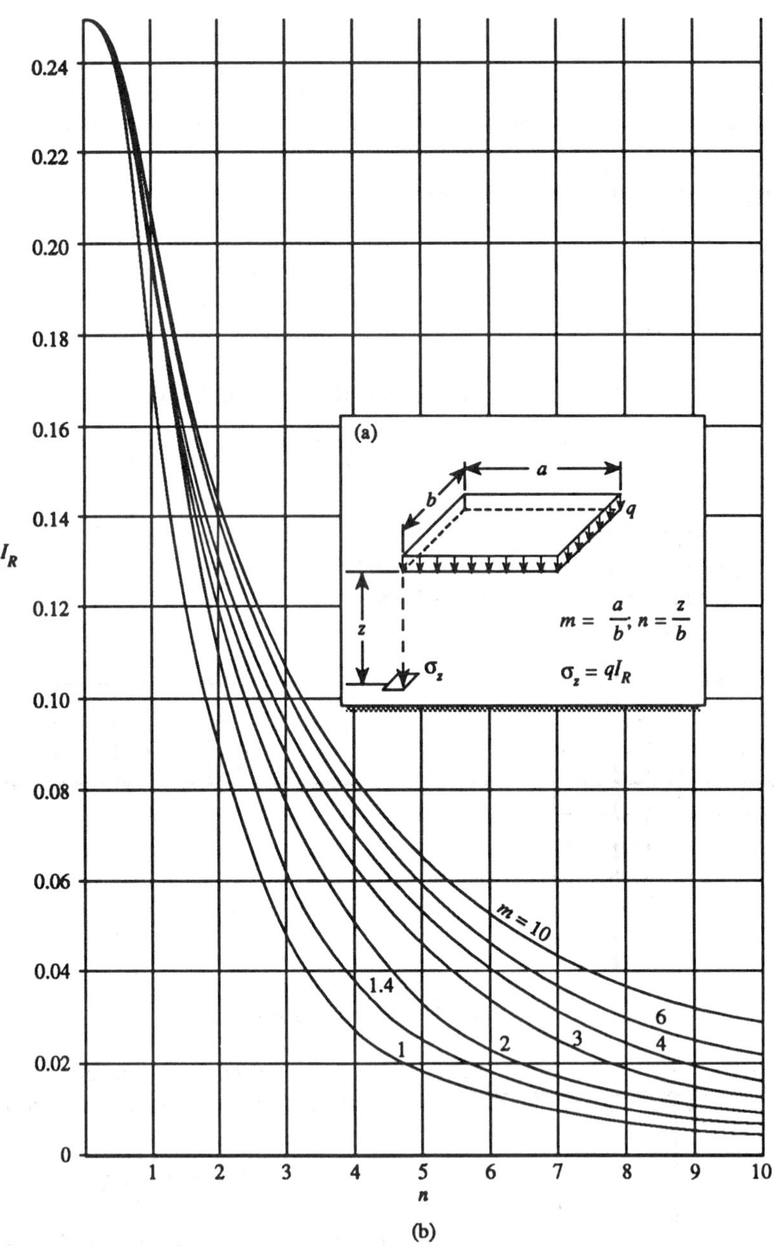

FIGURE 19.5 Normal uniform load over rectangular area. *Source:* After Steinbrenner, W. 1936 A rational method for the determination of the vertical normal stresses under foundations. *Proc. 1st Int. Conf. Soil Mech. Found. Eng.*, vol. 2.

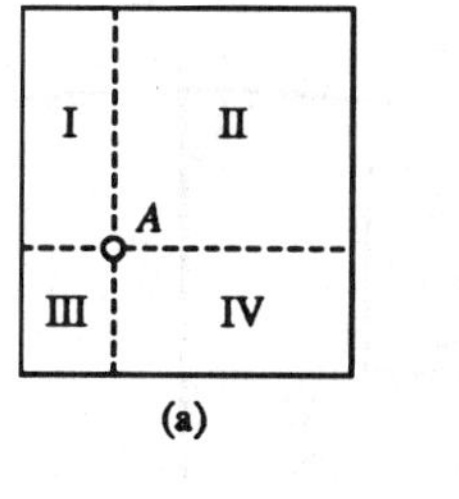

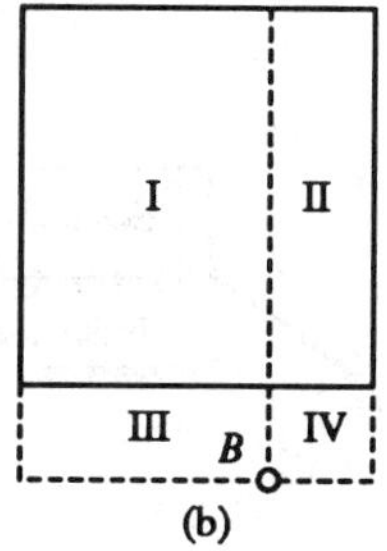

FIGURE 19.6 Vertical stress. (a) Interior. (b) Exterior.

where $m = a/b$ and $n = z/b$. This can also be written as

$$\sigma_z = qI_R$$

where I_R is a dimensionless influence factor. Figure 19.5(b) gives a plot of I_R as a function of the dimensionless ratios m and n. This form of the solution was given by Steinbrenner [1936].

By superposition, the distribution of vertical stress can be obtained anywhere under uniformly loaded, flexible, rectangular loadings. Two cases will be examined.

Vertical Stress for a Point Interior to a Loaded Area. For this case, Fig. 19.6(a), the influence factor I_R is determined from Fig. 19.5(b) for each of the rectangular areas (Roman numerals) with their corresponding m and n values and add them to obtain

$$\sigma_{zA} = q(I_{R\mathrm{I}} + I_{R\mathrm{II}} + I_{R\mathrm{III}} + I_{R\mathrm{IV}})$$

Vertical Stress for a Point Exterior to a Loaded Area. For a point such as B in Fig. 19.6(b), the stress is computed as

$$\sigma_{zB} = q(I_{R\mathrm{I+III}} + I_{R\mathrm{II+IV}} - I_{R\mathrm{III}} - I_{R\mathrm{IV}})$$

Influence Chart Normal Load

Although the above procedure can also be used to approximate irregularly shaped loadings, an influence chart, developed by Newmark [1942] greatly reduces the work required. Such a chart is shown in Fig. 19.7. To use the chart, the shape of the loading is drawn (generally on tracing paper) to scale so that the length AB on the chart represents the depth z at which the vertical stress is desired. The scaled drawing is then oriented so that the point under which the stress is sought is directly over the center of the circles on the chart. The number of blocks covered by the area of loading multiplied by the influence value (each square is $0.001q$ for this chart) yields, for the vertical stress, that part of the uniformly distributed load. By repeating this procedure and varying the size of the drawings, the complete distribution of vertical stresses can be found with depth. A separate drawing of the area is required for each depth. Although the chart was developed for uniform loadings over the whole area, the effects of varying loads can be treated as a series of uniform loadings.

Two-Dimensional Systems

In soil mechanics and foundation engineering, problems such as the analysis of retaining walls or of continuous footings and slopes generally offer one dimension very large in comparison with the other two. Hence, if boundary forces are perpendicular to and independent of this dimension, all cross sections will be the same. In Fig. 19.8 the y dimension is taken to be large, and it is assumed

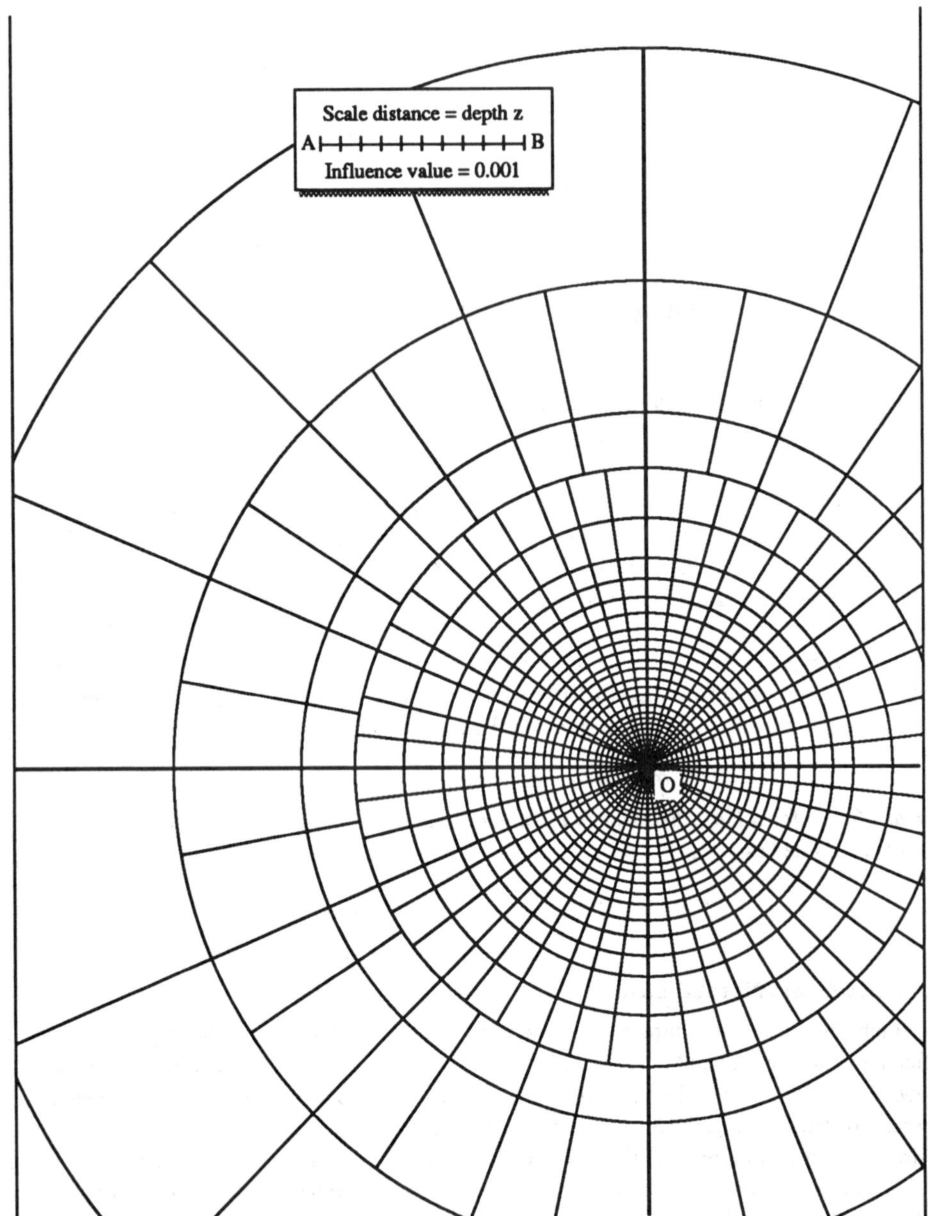

FIGURE 19.7 Influence chart for uniform vertical normal stress. *Source:* After Newmark, N. M. 1942. Influence charts for computation of stresses in elastic foundations. *Univ. of Illinois Bull. 338.*

that the state of affairs existing in the xz plane holds for all planes parallel to it. These conditions are said to define the state of **plane strain.**

Infinite Line Load Normal to Surface (Flamant Problem)

Figure 19.9(a) shows a semi-infinite plane with a concentrated load (line load) of intensity P (per unit run) normal to the surface. The solution, given by Flamant [1892], is

$$\sigma_z = \frac{2Pz^3}{\pi(x^2 + z^2)^2} \tag{19.6}$$

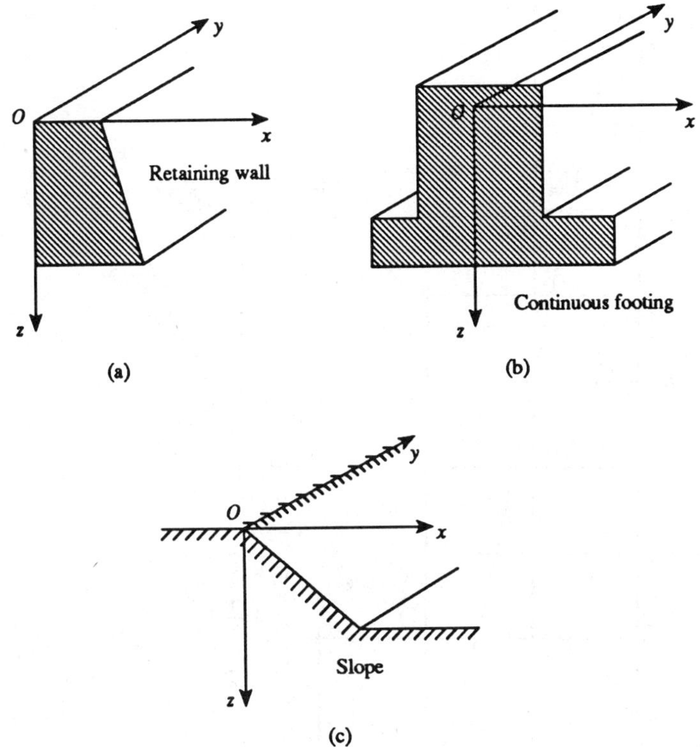

FIGURE 19.8 Examples of plane strain models.

In Fig. 19.9(b) is given a plot of this equation taken as $\sigma_z = I_{\sigma z}(P/z)$. Also plotted are the influence factors for $I_{\sigma x}$ and the tangential stress, I_τ.

Example 19.2. Find the vertical normal stress at a depth corresponding to point A in Fig. 19.10, due to the three normal line forces N_1, N_2, and N_3.
Solution. The influence curve is oriented so that the origin (point O) is directly above point A (as shown). The magnitude of the force multiplied by the ordinate of the σ_z curve immediately under it for $P = 1$ (such as $\overline{B_1C_1}$ under N_1) gives that part of the stress at point A due to the particular force. By superposition the total vertical stress at point A is obtained as the algebraic sum of the contributions of each of the forces. Thus, for the three forces N_1, N_2, and N_3, the increase in vertical stress at point A is $\sigma_{zA} = N_1(\overline{B_1C_1}) + N_2(\overline{B_2C_2}) + N_3(\overline{B_3C_3})$.

Superposition also permits the determination of the stress under any distribution of flexible surface loading. For example, to obtain the vertical stress at point A in Fig. 19.10 under the distributed line load $q(x)$, the load is first divided into a number of increments, and each increment is then treated, as just described, as a concentrated force.

Infinite Strip of Width *b*

An influence chart [Giroud, 1973] provides the stress σ_z at point $P(x, z)$ in Fig. 19.11, due to the distributed vertical load q over a flexible strip of width b in the form

$$\sigma_z = I_{\sigma z} \cdot q \tag{19.7}$$

Plots of this equation as well as the influence factor $I_{\sigma x}$ are also shown.

(a)

P, line load

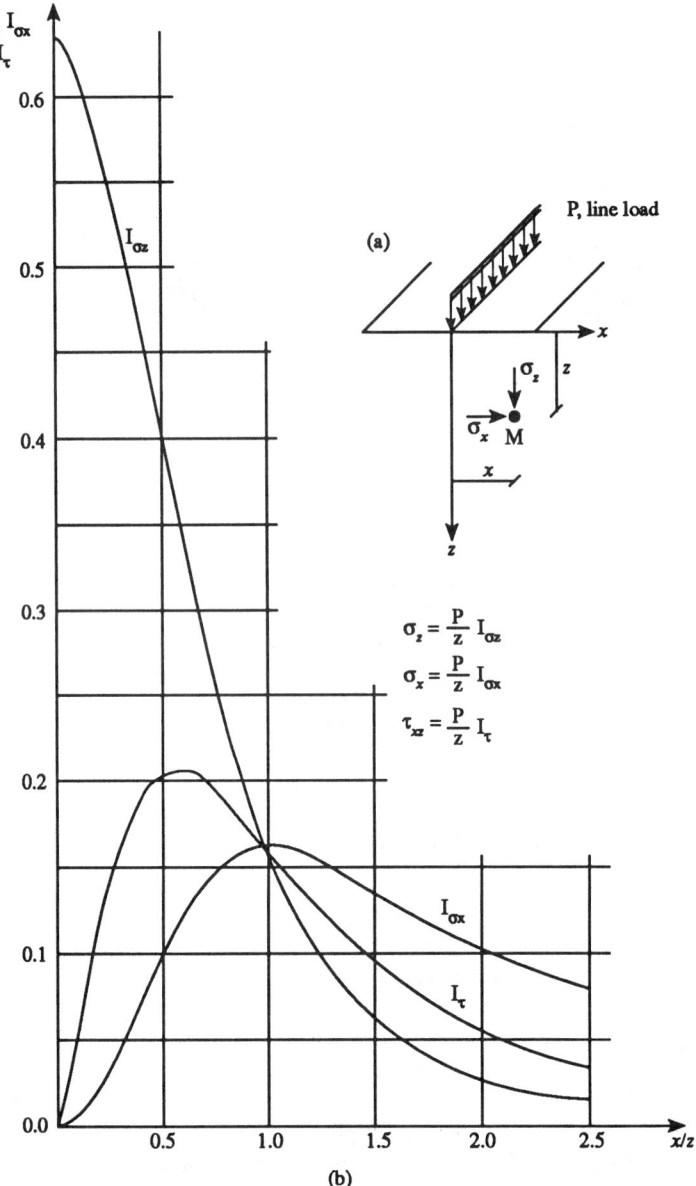

$$\sigma_z = \frac{P}{z} I_{\sigma z}$$

$$\sigma_x = \frac{P}{z} I_{\sigma x}$$

$$\tau_{xz} = \frac{P}{z} I_{\tau}$$

(b)

FIGURE 19.9 Infinite line load, normal to surface.

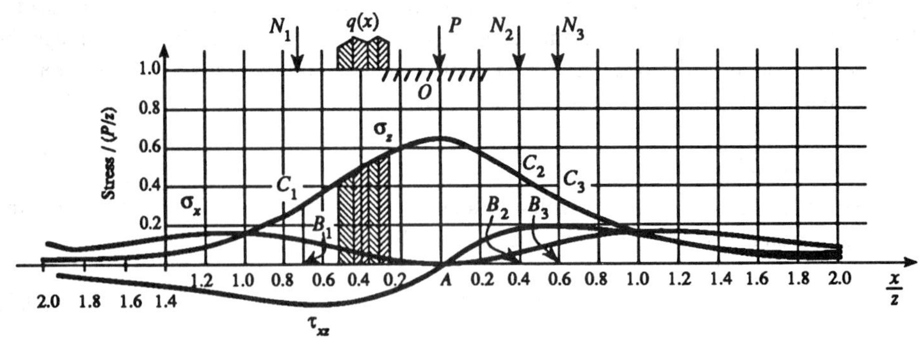

FIGURE 19.10 Example 19.2.

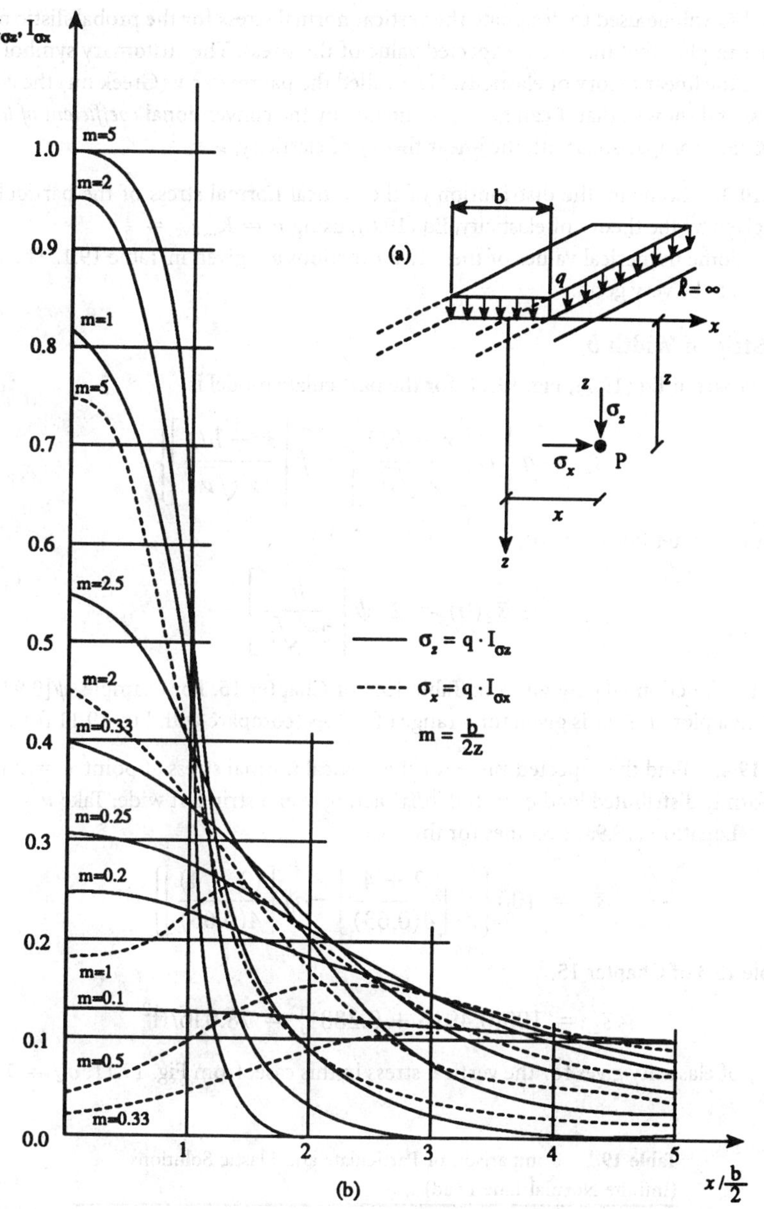

FIGURE 19.11 Stresses under infinite strip loading. *Source:* Giroud, J. P. 1973. *Tables pour le Calcul des Fondations.* 2 vols. Dunod, Paris.

19.2 Particulate Medium

Two-Dimensional Systems

Infinite Line Load Normal to Surface

On the basis of a probabilistic model, Harr [1977] gave the expression for the vertical stress due to a concentrated load (line load) of intensity P (per unit run) normal to the surface (see Fig. 19.9) as

$$\bar{s}_z = \frac{P}{z\sqrt{2\nu\pi}} \exp\left[-\frac{x^2}{2\nu z^2}\right] \tag{19.8}$$

The symbol $\bar{s}_z$ will be used to designate the vertical normal stress for the probabilistic model. The overlaid bar implies that this is the expected value of the stress. The customary symbol σ_z will be reserved for the linear theory of elasticity. Harr called the parameter ν (Greek nu) the *coefficient of lateral stress* and showed that it can be approximated by the conventional *coefficient of lateral earth pressure*, K. For comparisons with the linear theory of elasticity, $\nu \approx \frac{1}{3}$.

Example 19.3. Compare the distribution of the vertical normal stress of the particulate theory with that given by the theory of elasticity, Eq.(19.6), using $\nu = K_{\text{active}} = \frac{1}{3}$.
Solution. Some numerical values of these two functions are given in Table 19.1. The correspondence is seen to be very good.

Infinite Strip of Width b

The counterpart of Eq.(19.7), Fig. 19.11, for the particulate model is

$$\bar{s}_z = q\left\{\psi\left[\frac{x + b/2}{z\sqrt{\nu}}\right] - \psi\left[\frac{x - b/2}{z\sqrt{\nu}}\right]\right\} \tag{19.9a}$$

and under the center line ($x = 0$),

$$\bar{s}_z(0) = 2q\psi\left[\frac{b}{2z\sqrt{\nu}}\right] \tag{19.9b}$$

Values of the function $\psi[\,]$ are given in Table 15.3 of Chapter 15. For example, $\psi[0.92] = 0.321$. In Fig. 19.12 a plot of $\bar{s}_z/q$ is given for a range of values (compare with Fig. 19.11 for $\nu = \frac{1}{3}$).

Example 19.4. Find the expected value for the vertical normal stress at point $x = 2$ ft, $z = 4$ ft for a uniformly distributed load $q = 100$ lb/ft^2 acting over a strip 8 ft wide. Take $\nu = \pi/8$.
Solution. Equation (19.9a) becomes for this case

$$\bar{s}_z = 100\left\{\psi\left[\frac{2+4}{4(0.63)}\right] - \psi\left[\frac{(2-4)}{4(0.63)}\right]\right\}$$

From Table 15.3 of Chapter 15,

$$\bar{s}_z = 100\{0.4916 + 0.2881\} = 78.0\,\text{lb/ft}^2$$

The theory of elasticity gives for the vertical stress in this case, from Fig. 19.11, $\sigma_z = 73\,\text{lb/ft}^2$.

Table 19.1 Comparison of Particulate and Elastic Solutions (Infinite Normal Line Load)

$\frac{x}{z}$	$z\bar{s}_z/P = \left(\frac{1}{2\nu\pi}\right)^{1/2}\exp\left[-\frac{x^2}{2\nu z^2}\right]$	$z\sigma_z/P = \left(\frac{2}{\pi}\right)\left[1 + \frac{x^2}{z^2}\right]^{-2}$
0.0	0.69	0.64
0.1	0.68	0.62
0.2	0.65	0.59
0.3	0.60	0.54
0.4	0.54	0.47
0.5	0.47	0.41
0.6	0.40	0.34
0.8	0.26	0.24
1.0	0.15	0.16
1.2	0.08	0.11
1.5	0.02	0.06
1.8	0.01	0.04
2.0	0.004	0.03

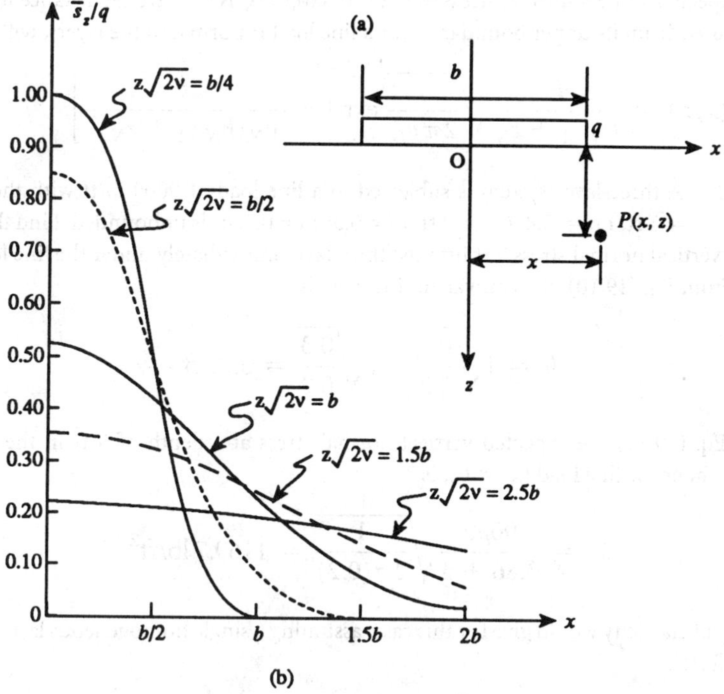

FIGURE 19.12 Uniform normal load over strip.

Multilayer System

Given a system with N layers in which the ith layer has thickness h_i and coefficient of lateral stress ν_i, Fig. 19.13, Kandaurov [1966] showed that the equivalent thickness for the upper $N-1$ layers can be found as

$$\tilde{h}_{N-1} = h_1\sqrt{\frac{\nu_1}{\nu_N}} + h_2\sqrt{\frac{\nu_2}{\nu_N}} + \cdots + h_{N-1}\sqrt{\frac{\nu_{N-1}}{\nu_N}} \tag{19.10}$$

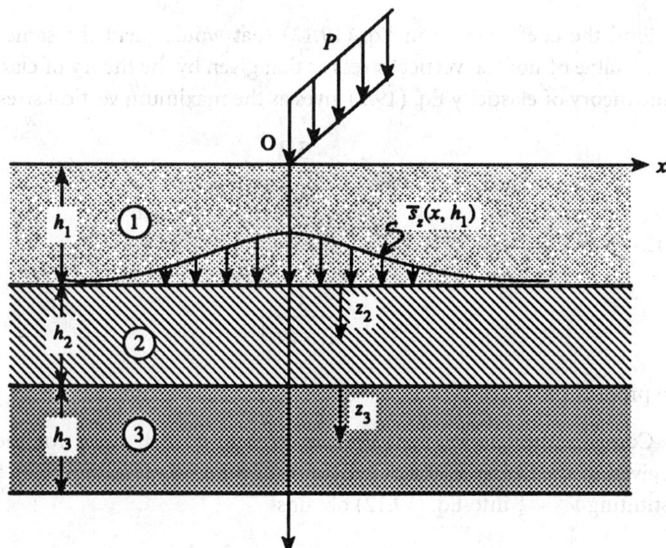

FIGURE 19.13 Multilayer system.

Hence, the expected vertical normal stress in the Nth layer (z_N is the vertical distance into the Nth layer as measured from its upper boundary), for a line load P normal to the layers, will be

$$\bar{s}_z(x,z) = \frac{P}{h_{N-1} + z_N}\sqrt{\frac{1}{2\pi\nu_N}}\exp\left[-\frac{x^2}{2\nu_N(h_{N-1}+z_N)^2}\right] \tag{19.11}$$

Example 19.5. A three-layer system is subjected to a line load of 9000 lb/ft with the following information: $h_1 = 1$ ft, $\nu_1 = 0.4$; $h_2 = 2$ ft, $\nu_2 = 0.3$; $\nu_3 = 0.2$, h_3 is unbounded. Find the expected value for the vertical normal stress 3 ft into the third layer immediately under the line load.
Solution. From Eq. (19.10), the equivalent thickness is

$$\tilde{h} = 1\sqrt{\frac{0.4}{0.2}} + 2\sqrt{\frac{0.3}{0.2}} = 3.86\ \text{ft}$$

Thus, from Eq. (19.11) the expected vertical normal stress at a depth of 3 ft in the third layer, immediately under the line load ($x = 0$), is

$$\bar{s}_z = \frac{9000}{3.86 + 3}\sqrt{\frac{1}{2\pi(0.2)}} = 1169.7\ \text{lb/ft}^2$$

The theory of elasticity would give for this case, assuming a single homogeneous layer, Eq. (19.6), $\sigma_z = 954.9\ \text{lb/ft}^2$.

Three-Dimensional System

Force Normal to Surface

The probabilistic counterpart of the Boussinesq solution, Eq.(19.2), for the expected vertical normal stress is

$$\bar{s}_z = \frac{P}{2\pi\nu z^2}\exp\left[-\frac{r^2}{2\nu z^2}\right] \tag{19.12}$$

where $r^2 = x^2 + y^2$.

Example 19.6. Find the coefficient ν in Eq. (19.12) that would yield the same value for the maximum expected value of normal vertical stress as that given by the theory of elasticity.
Solution. For the theory of elasticity Eq. (19.2) gives as the maximum vertical stress

$$\sigma_{z_{\max}} = \frac{3P}{2\pi z^2}$$

Equation (19.12) gives

$$\bar{s}_{z_{\max}} = \frac{P}{2\pi\nu z^2}$$

Equating the two produces $\nu = \frac{1}{3}$.

Example 19.7. Compare the lateral attenuation of vertical normal stress for the probabilistic theory with that given by the theory of elasticity (for three dimensions) using $\nu = \frac{1}{3}$.
Solution. Substituting $\nu = \frac{1}{3}$ into Eq. (19.12) obtains

$$\bar{s}_z = 3P/2\pi z^2 \exp\left[-3r^2/z^2\right]$$

Table 19.2 Comparison of Particulate and Elastic Solutions (Normal Point Load)

$\frac{r}{z}$	$\frac{z^2\bar{s}_z}{P}$	$\frac{z^2\sigma_z}{P}$
0.0	0.48	0.48
0.1	0.47	0.47
0.2	0.45	0.43
0.4	0.38	0.33
0.6	0.28	0.22
0.8	0.18	0.14
1.0	0.11	0.08
1.2	0.06	0.05
1.5	0.02	0.03
2.0	0.001	0.01

The theory of elasticity, Eq. (19.2), yields $\sigma_z = 3P/2\pi z^2(1 + r^2/z^2)^{-5/2}$. Some values of the two functions are given in Table 19.2. In Fig. 19.14 a nomograph of the expected vertical normal stress is given for a range of ν values.

Example 19.8. Find the expected value of the vertical normal stress at the point $r = 6$ ft, $z = 10$ ft if $\nu = \frac{1}{5}$ under a concentrated vertical force of 100 lb.

Solution. The arrow in Fig. 19.14 indicates that for the given conditions $\nu z^2\bar{s}_z/P = 0.065$. Hence, the expected value of the vertical normal stress is $\bar{s}_z = 0.065(100)(5)(1/100) = 0.33$ lb/ft². The theory of elasticity gives for this case, Fig. 19.1, $\sigma_z = 0.22$ lb/ft².

Distributed Vertical Loads at Surface

Normal Uniform Load over a Rectangular Area

The probabilistic counterpart of Eq. (19.5) for the expected vertical normal stress under the corner of a rectangular area with sides a by b, as in Fig. 19.5(a), is

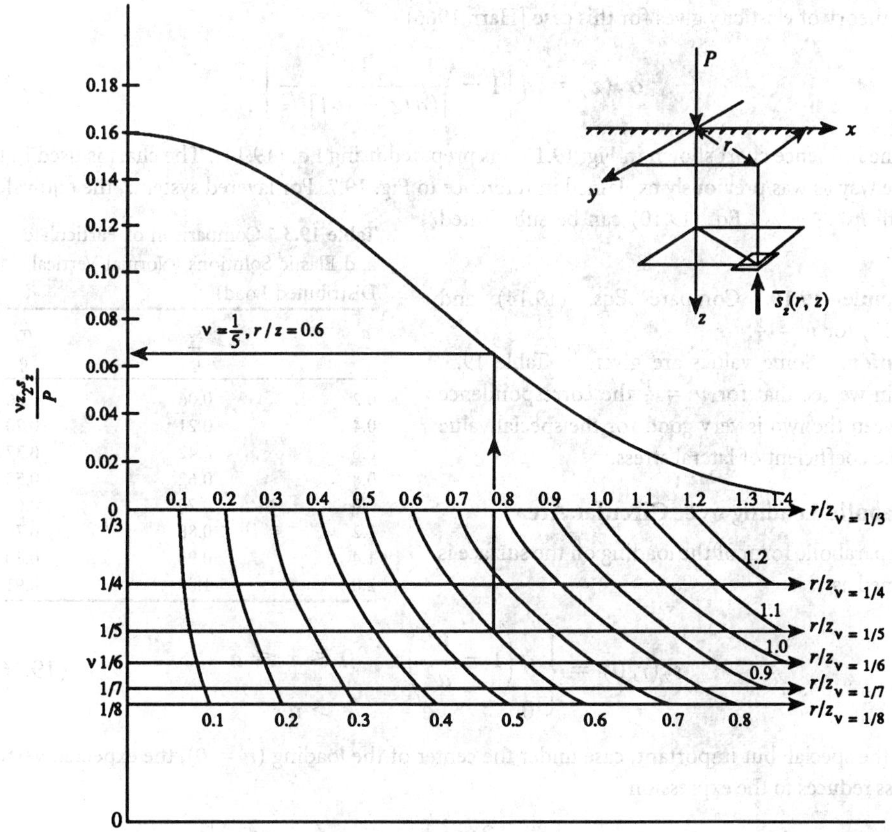

FIGURE 19.14 Expected value of vertical stress, particulate model.

$$\bar{s}_z/q = \psi\left[\frac{a}{z\sqrt{\nu}}\right]\psi\left[\frac{b}{z\sqrt{\nu}}\right] \tag{19.13}$$

Values of the function $\psi[\,]$ are given in Table 15.3 of Chapter 15.

Example 19.9 A uniformly distributed vertical load of magnitude 25 lb/ft^2 acts over an area 4 ft by 8 ft at the surface of a particulate medium. Find the expected vertical normal stress 6 ft below the center of the area. Take $\nu = \frac{1}{3}$.

Solution. From Eq. (19.13), taking the 4 × 8 ft area as four areas, each with sides 2 × 4 ft, we have

$$\bar{s}_z = 4(25)\psi\left[\frac{2}{6\sqrt{1/3}}\right]\psi\left[\frac{4}{6\sqrt{1/3}}\right] = 100\psi[0.58]\psi[1.15] = 100(0.219)(0.375)$$

and $\bar{s}_z = 8.2$ lb/ft^2. The theory of elasticity, as shown in Fig. 5(b), yields for this case $\sigma_z = 7.3$ lb/ft^2.

Uniform Normal Load over Circular Area

The expected vertical normal stress under the center of a circular area, of radius a, subject to a uniform normal load q was obtained by Kandaurov [1959] as

$$\bar{s}_z(z) = q\left[1 - \exp\left(-\frac{a^2}{2\nu z^2}\right)\right] \tag{19.14}$$

The theory of elasticity gives for this case [Harr, 1966]

$$\sigma_z(z) = q\left[1 - \frac{1}{[(a/z)^2 + 1]^{3/2}}\right] \tag{19.15}$$

The influence chart shown in Fig. 19.15 was prepared using Eq. (19.14). The chart is used in the same way as was previously explained in reference to Fig. 19.7. For layered systems, the equivalent depth $\tilde{h}_{N-1} + z_N$, Eq. (19.10) can be substituted for z.

Example 19.10. Compare Eqs. (19.14) and (19.15) for $\nu = \frac{1}{3}$.

Solution. Some values are given in Table 19.3. Again we see that for $\nu = \frac{1}{3}$ the correspondence between the two is very good for the special value of the coefficient of lateral stress.

Table 19.3 Comparison of Particulate and Elastic Solutions (Normal Vertical Distributed Load)

$\frac{a}{z}$	$\frac{\bar{s}_z}{q}$	$\frac{\sigma_z}{q}$
0.2	0.06	0.06
0.4	0.21	0.20
0.6	0.42	0.37
0.8	0.62	0.52
1.0	0.78	0.65
1.2	0.88	0.74
1.4	0.95	0.80
2.0	1.00	0.91

Parabolic Loading over Circular Area

The parabolic form of the loading on the surface is defined as

$$\bar{s}_z(r,0) = \begin{cases} q\left(1 - \dfrac{r^2}{a^2}\right) & 0 \le r \le a \\ 0 & r > a \end{cases} \tag{19.16a}$$

For the special, but important, case under the center of the loading ($r = 0$), the expected vertical stress reduces to the expression

$$\bar{s}_z(0,z) = q\left\{1 - \frac{2\nu z^2}{a^2}\left[1 - \exp\left(-\frac{a^2}{2\nu z^2}\right)\right]\right\} \tag{19.16b}$$

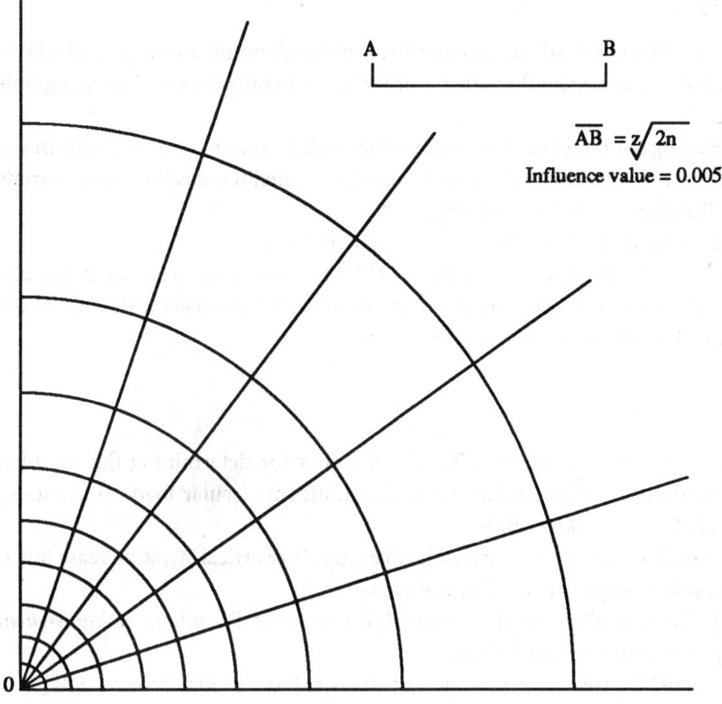

FIGURE 19.15 Influence chart.

Tangential (Horizontal Force at Surface)

For a concentrated force of intensity Q parallel to the boundary surface (the equivalent of Cerruti's solution, Fig. 19.4), Muller [1962] obtained the expression

$$\bar{s}_x = \frac{Q}{2\pi\nu x^2}\exp\left[-\frac{y^2+z^2}{2\nu x^2}\right] \tag{19.17}$$

Example 19.11. Compare the expression for the vertical stress in Eq.(19.17) with that given by Cerruti (Eq. 19.4) for the line $y = 0, z = 1$.

Solution. Cerruti obtained $\sigma_z = 3Qxz^2/2\pi R^5$ where $R^2 = x^2 + y^2 + z^2$. From Eq. (19.17) for $y = 0, z = 1$,

$$\frac{2\pi\bar{s}_z}{Q} = \frac{1}{\nu x^2}\exp\left[-\frac{1}{2\nu x^2}\right]$$

It can be shown that $\lim_{x\to 0}\bar{s}_z = 0$. Some numerical values for two values of ν are given in Table 19.4. It is seen that the vertical stresses for the developed theory are smaller than those given by the theory of elasticity in the near vicinity of the applied load. However, the general patterns of the distributions are quite similar away from the point of loading.

Table 19.4 Comparison of Particulate and Elastic Solutions (Concentrated Force Parallel to Boundary Surface)

		$\frac{2\pi\bar{s}_z}{Q}$	
x	$\frac{2\pi\sigma_z}{Q}$	$\nu = 1/3$	$\nu = 1/2$
0	0	0.00	0.00
0.25	0.64	0+	0+
0.5	0.86	0.03	0.15
0.75	0.74	0.37	0.60
1.0	0.53	0.67	0.74
1.5	0.24	0.68	0.57
2.0	0.11	0.52	0.39
2.5	0.05	0.38	0.27

Defining Terms

Homogeneous: Having elastic parameters the same at all points in a region of a body.

Influence factor: A dimensionless cluster of factors generally obtained from a graphical presentation.

Isotropic: Having defining parameters the same in all directions emanating from a point.

Plane strain: A two-dimensional simplification used when there is little or no variation in strain in the direction normal to the plane.

Strain: Changes in displacements due to changes in stress.

Stress: Intensity of internal forces within a soil body induced by loadings. In the classical theory of elasticity stress is single-valued. The particulate theory deals with a distribution, the best measure of which is the *expected* (mean) value.

References

Ahlvin, R. G. and Ulery, H. H. 1962. Tabulated values for determining the complete pattern of stresses, strains, and deflections beneath a uniform circular load on a homogeneous half space, *Highway Res. Board Bull. 342.*

Barksdale, R. and Harr, M. E. 1966. An influence chart for vertical stress increase due to horizontal shear loadings. *Highway Res. Board Rec. 108.*

Boussinesq, J. 1885. *Applications des Potentiels à l'Étude de l'Équilibre et Mouvement des Solides Élastiques.* Gauthier-Villard, Paris.

Carothers, S. D. 1920. Direct determination of stresses. *Proc. R. Soc., Series A, 97.*

Cerruti, V. 1882. Ricerche intorno al'equilibrio de' corpi elastici isotropi, *Reale Accademia dei Lincei,* serie 3ª Memoria della Classe di scienze fisiche . . . , vol. XIII, Rome.

Egorov, K. E. 1940. Distribution of stresses in base under rigid strip footing. *Mauchn. Issled, Stantsiya Fundamentstroya,* no. 9.

Fadum, R. E. 1948. Influence values for estimating stresses in elastic foundations. *Proc. 2nd Int. Conf. Soil Mech. Found. Eng.,* vol. 3. The Netherlands.

Flamant, A. 1892. *Comptes Rendus,* vol. 114, p. 1465.

Florin, V. A. 1959. *Fundamentals of Soil Mechanics.* 2 vols. Gosstroiizdat, Moscow.

Giroud, J. P. 1973. *Tables pour le Calcul des Fondations.* 2 vols. Dunod, Paris.

Harr, M. E. 1966. *Foundations of Theoretical Soil Mechanics.* McGraw-Hill, New York.

Harr, M. E. 1977. *Mechanics of Particulate Media: A Probabilistic Approach.* McGraw-Hill, New York.

Harr, M. E. 1987. *Reliability-Based Design in Civil Engineering.* McGraw-Hill, New York.

Harr, M. E. and Lovell, C. W., Jr. 1963. Vertical stresses under certain axisymmetrical loading. *Highway Res. Board Rec. 39.*

Highway Research Board. 1962. Stress distribution in earth masses. *Highway Res. Board Bull. 342.*

Kandaurov, I. I. 1959. *Theory of Discrete Distribution of Stress and Deformation . . .* Izd. VATT (in Russian).

Kandaurov, I. I. 1966. *Mechanics of Discrete Media and Its Application to Construction.* Izd. Liter. po Stroitel'stvu (in Russian). Translated into English by Zeidler, R. B. A. A. Balkema, Rotterdam, 1991.

Kandaurov, I. I. 1988. *Mekhanika Zernistykh Sred i Yeye Primeneniye v Stroitel'stve.* Stroyizdat, Leningrad.

Love, A. E. H. 1944. *A Treatise on the Mathematical Theory of Elasticity.* Dover Publications, Inc., New York.

Michell, J. H. 1900. The stress distribution in an aeotropic solid with infinite boundary plane. *Proc. London Math. Soc., 32.*

Mindlin, R. D. 1936. Force at a point in the interior of a semi-infinite solid. *Physics 7.*

Muller, R. A. 1962. Statistical theory of transformation of stress in granular soil bases. *Bases, Foundations, and Soil Mechanics,* no. 4.

Muller, R. A. 1963. Deformation of granular soil. *Proc. All-Union Sci.-Res. Mining Inst.*, Leningrad.

Newmark, N. M. 1940. Stress distributions in soils. *Proc. Purdue Conf. on Soil Mech. and Its Appl.*, Lafayette, IN, July.

Newmark, N. M. 1942. Influence charts for computation of stresses in elastic foundations. *Univ. of Illinois Bull. 338.*

Newmark, N. M. 1947. Influence charts for computation of vertical displacements in elastic foundations. *Univ. of Illinois, Eng. Exp. Stn. Bull. 45.*

Perloff, W. H., Galadi, G. Y., and Harr, M. E. 1967. Stresses and displacements within and under long elastic embankments. *Highway Res. Rec. 181.*

Poulos, H. G. and Davis, E. H. 1974. *Elastic Solutions for Soil and Rock*, John Wiley & Sons, New York. Reprinted in 1991 by Centre for Geotechnical Research, University of Sydney, Sydney, NSW 2006, Australia.

Reimbert, M. L. and Reimbert, A. M. 1974. *Retaining Walls.* Transactions Technological Publishing, Bay Village, OH.

Steinbrenner, W. 1936. A rational method for the determination of the vertical normal stresses under foundations. *Proc. Int. Conf. Soil Mech. Found. Eng.* vol. 2.

Timoshenko, S. P. 1953. *History of Strength of Materials.* McGraw-Hill Book Company, New York.

Todhunter, I. and Pearson, K. 1960. *A History of the Theory of Elasticity and of the Strength of Materials.* Dover Publications, Inc., New York (first published by Cambridge University Press, London, 1886–1893).

Turnbull, W. J., Maxwell, A. A., and Ahlvin R. G. 1961. Stresses and deflections in homogeneous soil masses. *Proc. 5th Intl. Conf. Soil Mech. Found. Eng.*, Paris.

Westergaard, H. M. 1938. A problem of elasticity suggested by a problem in soil mechanics: soft material reinforced by numerous strong horizontal sheets. In *Contributions to the Mechanics of Solids, Dedicated to S. Timoshenko by His Friends on the Occasion of His Sixtieth Birthday Anniversary.* The MacMillan Company, New York.

For Further Information

Many compilations of elastic solutions are available in the literature; primary among these is that of Poulos and Davis [1974], which has recently been reprinted (see references). Historical background can be found in Timoshenko [1953], Todhunter and Pearson [1960], and Love [1944]. Many particulate solutions can be found in Kandaurov [1988] and Harr [1977].

20

Stability of Slopes

Roy E. Hunt
Geotechnical Consultant
Bricktown, New Jersey

Richard J. Deschamps
Purdue University

20.1 Introduction

Slope stability analysis is performed to assess the potential for failure of the slope by rupture. Slopes that are typically assessed fall into a number of categories, as illustrated in Fig. 20.1. The primary objective of a stability analysis is to determine the factor of safety (FS) of a particular slope, to predict when failure is imminent, and to assess remedial treatments when necessary. In many practical situations, an analytical assessment of stability can be made. Other situations do not lend themselves to convenient analytical solutions. Analytical techniques can be applied to slope failures where peak strength occurs essentially simultaneously at every point along single or multiple failure surfaces and the mass moves as a unit or group of units. The sliding surface may be circular, planar, or irregular.

Slope failure forms that cannot be analyzed in the present state of the art include avalanches, flows, failure by lateral spreading, and progressive failure. All of these forms can be initiated by a slide failure at the toe of the mass. The only defense against these failure forms is recognition that their potential exists [Hunt, 1984].

Deformations can be a concern in slopes and embankments due to the effects on surface or buried structures, and because they often precede failure. The finite element method (FEM) has been used to approximate deformations in earth-dam embankments and rock slopes, but only infrequently in natural and cut slopes in soils [Vulliet and Hutter, 1988]. In any case, it is necessary to closely define material properties, slope geometry, and the initial state of stress, which often are difficult to assess accurately. The most common analytical approaches currently used by practitioners to assess slope stability are based on the limiting equilibrium method. It is an approximate solution that considers a state of equilibrium between the forces acting to cause failure (driving forces) and the forces resisting failure (mobilized shear stresses). Two general cases are illustrated in Fig. 20.2.

Limit equilibrium analyses assume the following:

1. The failure surface is of simple geometric shape (planar, circular, log-spiral).
2. The distribution of stresses acting along the failure surface causing failure are determinate.

0-8493-8953-4/95/$0.00 + $.50

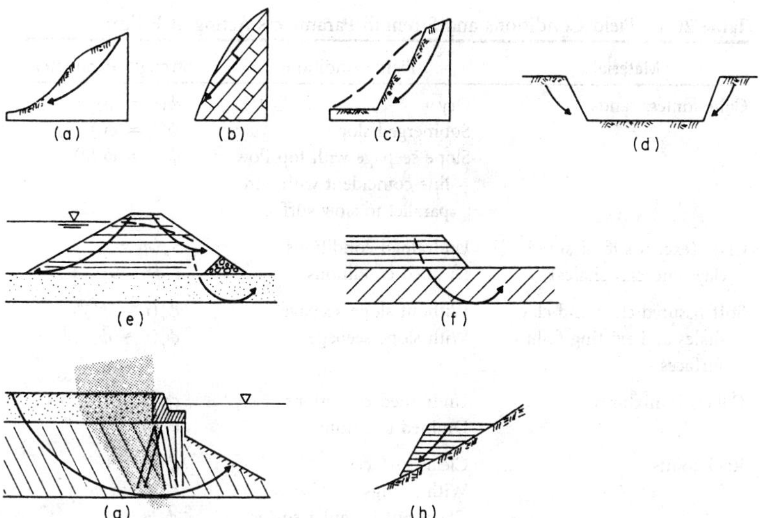

FIGURE 20.1 Categories of slope stability problems: (a) natural soil slope; (b) natural rock slope; (c) cut slope; (d) open excavation; (e) earth dam embankment; (f) embankment over soft soils; (g) waterfront structure; (h) sidehill fill. (*Source:* Hunt, R. E. 1986. Geotechnical Engineering Techniques and Practices. McGraw-Hill, New York. Reprinted with permission of McGraw-Hill Book Co.)

3. The same percentage of mobilized shear strength acts simultaneously along the entire failure surface.

Limit equilibrium methods of analysis typically use the Mohr-Coulomb failure criterion, in which (see Table 20.1)

$$s = c + \sigma_n \tan \phi$$

where s is the shearing resistance, c is the cohesion, σ_n is the normal stress, and ϕ is the angle of internal friction. Appropriate total stress or effective stress strength parameters for specific loading conditions are provided in Table 20.2.

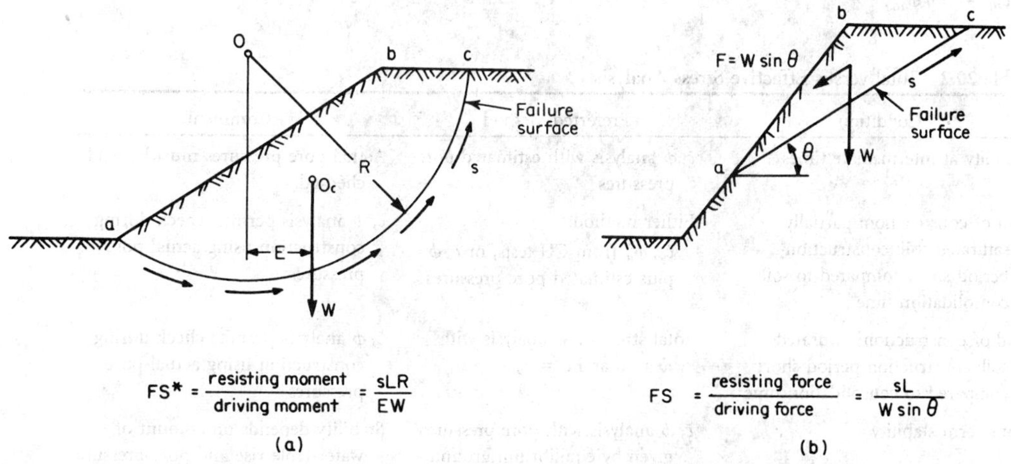

FIGURE 20.2 Forces acting on cylindrical failure surfaces. (a) Rotational cylindrical failure surface with length *L*. Safety factor against sliding FS. (b) Simple wedge failure on planar surface with length *L*. *Note:* The expression for FS is generally considered unsatisfactory (see text).

Table 20.1 Field Conditions and Strength Parameters Acting at Failure

Material	Field Conditions	Strength Parameter[a]
Cohesionless sands	Dry	$\phi(i = \phi)$
	Submerged slope	$\phi(i = \phi')$
	Slope seepage with top flow line coincident with and parallel to slow surface	$\phi(i = \phi'/2)$
Clays (except stiff fissured clays and clay shales)	Undrained conditions	$S_u(\phi = 0)$
	Drained conditions	$c'\phi^b$
Stiff fissured clays and clay shales and existing failure surfaces	Without slope seepage	$\phi_r'(i \approx \phi_r')^b$
	With slope seepage	$\phi_r'(i \approx \phi_r'/2)^b$
Cohesive mixtures	Undrained conditions	c_u, ϕ_u
	Drained coditions	c', ϕ^b
Rock joints	Clean surfaces	ϕ or $\phi + j^c$
	With fillings	$c', \phi,$ or ϕ'
	Clean but irregular surfaces after failure	$\phi_t + j^c$

Source: Hunt, R. E. 1986. *Geotechnical Engineering Techniques and Practices.* McGraw-Hill, New York. Reprinted with permission oif McGraw-Hill Book Co.

[a] i = stable slope angle.

[b] Pore-water presssures: reduce frictional resistance in accordance with $(N - U)\tan\phi$.

[c] j = angle of asperities.

The factor of safety has been expressed in a number of forms, examples of which are given in Fig. 20.2 (see Table 20.3). In slope stability analysis, FS can be considered as the ratio of the total shearing strength available along the sliding surface to the total shearing stresses required to maintain limiting equilibrium, given as

$$\text{FS} = \frac{cL + N\tan\phi}{c_m L + N\tan\phi_m}$$

where L is the length of the failure surface, N is the normal force on the assumed failure surface, c_m is the mobilized cohesion at equilibrium, ϕ_m is the mobilized friction angle at equilibrium, and $c/c_m = \phi/\phi_m$.

Table 20.2 Total versus Effective Stress Analysis

Condition	Preferred Method	Comment
Stability at intermediate times	$\bar{c}, \bar{\phi}$ analysis with estimated pore pressures	Actual pore pressures must be field-checked.
End of construction; partially saturated soil; construction period short compared to soil consolidation time	Either method: $\bar{c}_u, \bar{\phi}_u$ from CU tests, or $\bar{c}, \bar{\phi}$ plus estimated pore pressures	$\bar{c}, \bar{\phi}$ analysis permits check during construction using actual pore pressures
End of construction; saturated soil; construction period short compared to consolidation time	Total stress or s_u analysis with $\phi = 0$ and $c = s_u$	$\bar{c}, \bar{\phi}$ analysis permits check during construction using actual pore pressures
Long-term stability	$\bar{c}, \bar{\phi}$ analysis with pore pressures given by equilibrium ground-water conditions	Stability depends on amount of water-table rise and pore-pressure increase

Source: After Lambe, T. W. and Whitman, R. V. 1969 *Soil Mechanics.* John Wiley & Sons, New York. Adapted by permission of John Wiley & Sons, Inc.

Table 20.3 Minimum Values for FS for Earth and Rock-fill Dams[a,b]

Case	Design Conditions	FS_{min}	Shear Strength[c]	Remarks
I	End of construction	1.3[d]	Q or S[e]	Upstream and downstream slopes
II	Sudden drawdown from maximum pool	1.0[f]	R, S	Upstream slope only, use composite envelope
III	Sudden drawdown from spillway crest	1.2[f]	R, S	Upstream slope only, use composite envelope
IV	Partial pool with steady seepage	1.5	$(R+S)/2$ for $R < S$	Upstream slope only, use intermediate envelope
V	Steady seepage with maximum pool	1.5	$(R+S)/2$ for $R < S$	Downstream slope only, use intermediate envelope
VI	Earthquake (Cases I, IV, and V with seismic loading)	1.0	—[g]	Upstream and downstream slopes

[a] *Source:* From Wilson, S. D. and Marsal, R. J. 1979. *Current Trends in Design and Construction of Embankment Dams.* ASCE, New York.

[b] Not applicable to embankments on clay shale foundations; higher FS values should be used for these conditions.

[c] Q = quick (unconsolidated-undrained test), S = slow (consolidated-drained test), and R = intermediate (consolidated-undrained test).

[d] For embankments more than 50 ft high over relatively weak foundation use $FS_{min} = 1.4$.

[e] In zones where no excess pore-water pressures are anticipated use S strength.

[f] FS should not be less than 1.5 when drawdown rate and pore-water pressures developed from flow nets are used in stability analysis

[g] Use shear strength for case analyzed without earthquake. (Values for FS are based on pseudostatic approach.)

20.2 Factors to Consider

Selection of the proper method to be applied to the analysis of a slope problem requires consideration of a number of factors. Specific details can be found in the referenced works.

- Type of slope to be analyzed, such as natural or cut slope in soil [Bjerrum, 1973; Patton and Hendron, 1974; Brand, 1982; Leonards, 1979, 1982] or rock [Deere, 1976], earth-dam embankments [Lowe, 1967], embankments over soft ground [Chirapuntu and Duncan, 1976; Ladd, 1991], or sidehill fills.
- Location, orientation, and shape of a potential or existing failure surface which is controlled by material type and structural features. The shape of the rupture zone can have single or multiple surfaces, and can be composed of single or multiple wedges. Failure surfaces can be planar, cylindrical or log-spiral, or irregular.
- Material distribution (stratigraphy) within and beneath the slope, divided generally into homogeneous zones wherein the properties are more or less similar in all directions, and nonhomogeneous zones wherein soils are stratified and rock masses contain major discontinuities.
- Material types and representative shear strength parameters; angle of internal friction ϕ, cohesion c, residual strength ϕ_r, and undrained strength s_u (see Table 20.1).
- Drainage conditions; appropriateness for either drained or undrained analysis, which depends on the relative rates of construction and pore pressure dissipation. Often considered as short-term (during construction) or long-term (postconstruction or natural slope) conditions; that is, total versus effective stress analysis (see Table 20.2).
- Distribution of piezometric levels (pore- or cleft-water pressures) along the potential failure surface and an estimate of the maximum value that may prevail.
- Potential earthquake loadings, which are a transient factor.

Hunt [1984, 1986] provides a detailed discussion of these necessary considerations for the various slope types.

20.3 Analytical Approaches

Most slope stability analyses are computationally intensive. Many computer programs have been developed to perform stability analyses on personal computers. Commonly used packages include PCSTABL (Purdue University) and UTEXAS3 (University of Texas). Data is input for the problem geometry, material stratigraphy, and the phreatic surface based on a coordinate system, and material properties are then entered. The programs search for the potential failure surface that produces the lowest value for FS. These programs are used routinely in virtually all geotechnical consulting offices. The following discussion is intended to present the salient features of the assumptions and analytical approach used to perform the stability analysis.

General

Failure, sudden or gradual, results when the mobilized stresses in a slope or its foundation equal the available strength. Limit equilibrium analysis is the basis for most methods available for slope stability evaluations. Consideration is given to a free body of the soil mass bounded by the slope and an assumed "slip" or failure surface. The known or assumed forces acting on the body and the shearing resistance required for stability are estimated. Most practical problems are statically indeterminate and require simplifying assumptions regarding the position and direction of forces to render the problem determinate.

The primary assumption of the limit equilibrium method is that the assumed strength can be mobilized throughout the length of the failure surface simultaneously. Strain compatibility is not considered. This assumption is applicable for stress-strain conditions that can be modeled as perfectly plastic at the failure strength. When soils have some post–peak strength reduction (as with most natural soil), engineering judgment is required to select appropriate strength parameters and safety factors.

Two common free-body assumptions are illustrated in Fig. 20.2. In the cylindrical form shown in Fig. 20.2(a), the mass weight W acting through lever arm E produces a driving moment. This moment is resisted by strength S mobilized along the failure surface of length L that acts through a lever arm R. In the simple wedge [Fig. 20.2(b)], strength S acting along the planar surface of length L resists the driving forces resulting from gravity acting on weight W. A plane strain condition is assumed in most analytical methods currently in use so the potential failure mass is analyzed per unit width. Resistance that would be generated at the lateral extremities of the failure zone are considered insignificant compared to the area of the potential failure surface. When three-dimensional analyses are performed, FS(3-D) $>$ FS(2-D) [Duncan, 1992] for most cases; therefore, two-dimensional analysis normally is conservative.

Common methods of analysis consider a system of forces. The soil mass is divided into a system of "slices," "blocks," or "wedges," and force and/or moment equilibrium conditions are evaluated for the individual components of the soil mass. The soil strength is uniformly adjusted by a scaling factor until the system is in a state of equilibrium. The actual soil strength divided by the strength required to satisfy equilibrium is defined as the factor of safety.

Slices as Free Bodies

In modern force systems the sliding mass is divided into "slices" as shown in Fig. 20.3(a), which illustrates a "finite" slope with a circular failure surface. The forces that act on a slice [Fig. 20.3(b)] include the material weight W, normal force N, and shear force T distributed along an assumed failure surface; water pressure force U; water pressure V acting in a tension crack; and forces $E_r, X_r, E_L,$ and X_L acting along the sides of the slice. A discussion of the treatment of tension cracks

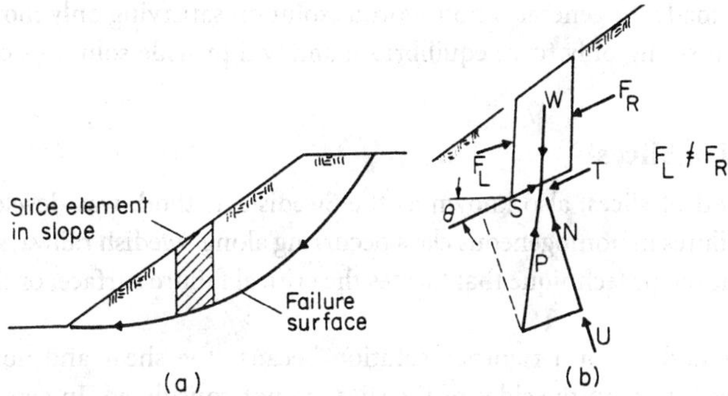

FIGURE 20.3 Finite slope with cylindrical failure surface (a) and forces on an element (b).

can be found in Tschebotarioff [1973] and Spender [1967]. Earthquakes impose dynamic forces in the form of the acceleration of gravity acting on the mass component.

Planar Surfaces, Blocks, and Wedges

Simple planar failures involve a single surface which can define a wedge in soil or rock [Fig. 20.2(b)], a sliding block, or a wedge in rock with a tension crack. The force system is similar to that given in Fig. 20.3, except that the side forces are neglected. Complex planar failures involve a number of planes dividing the mass into two or more blocks, and in addition to the force system for the single block, these solutions provide for interblock forces.

Infinite Slope

The infinite slope pertains where the depth to a planar failure surface is small compared to its length, which is considered as unlimited. Such conditions are found in slopes composed of the following:

- Cohesionless materials, such as clean sands
- Cohesive soils, such as residuum or colluvium, over a sloping rock surface at shallow depth
- OC fissured clays or clay shales with a uniformly deep, weathered zone
- Large slabs of sloping rock layers underlain by a weakness plane

Circular Failure Surfaces

General

In rotational slide failures, methods are available to analyze a circular or log-spiral failure surface, or a surface of any general shape. The location of the critical failure surface is found by determining the lowest value of safety factor obtained from a large number of assumed failure surface positions.

Slice Methods

Common to all slice methods is the assumption that the assumed soil mass and failure surface can be divided into a finite number of slices. Equilibrium conditions are considered for all slices. The problem is strongly indeterminate, requiring several basic assumptions regarding the location of application or resultant directions of applied forces.

The slice methods can be divided into two groups: nonrigorous and rigorous. Nonrigorous methods satisfy either force or moment equilibrium, whereas rigorous methods satisfy both force and moment equilibrium. The factor of safety estimated from rigorous methods is relatively insensitive to the assumptions made to obtain determinacy [Duncan, 1992; Espinoza *et al.*, 1992, 1994]. However, nonrigorous solutions can produce significantly different estimates of safety depending on the assumptions made. In general, a nonrigorous solution satisfying only moment equilibrium is superior to one satisfying only force equilibrium and will provide solutions close to a rigorous method.

Ordinary Method of Slices

The ordinary method of slices, also known as the Swedish method, was developed by Fellenius [1936] to analyze failures in homogeneous clays occurring along Swedish railways in the 1920s. The solution is a trial-and-error technique that locates the critical failure surface, or that circle with the lowest value for FS.

The ordinary method is not a rigorous solution because the shear and normal stresses and pore-water pressures acting on the sides of the slice are not considered. In general the results are conservative. In slopes with low ϕ angles and moderate inclinations, FS may be 10 to 15% below the range of the more exact solutions; with high ϕ and slope inclination, FS can be underestimated by as much as 60%.

For the $\phi = 0$ case, normal stresses do not influence strength, and the ordinary method provides results similar to rigorous methods [Johnson, 1974]. An example analysis using the ordinary method for the $\phi = 0$ case as applied to an embankment over soft ground is given in Fig.20.4.

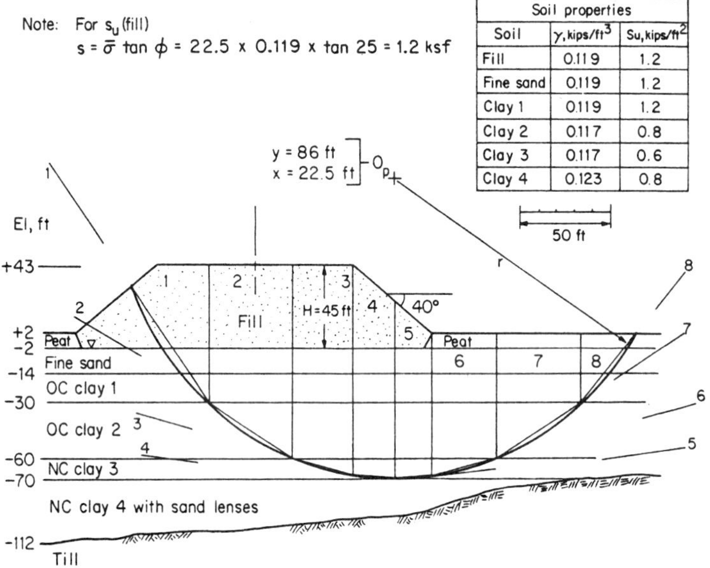

Slice	s_u, ksf	ΔL, ft	$S_u\,\Delta L$	θ, degrees	W, kips*	$W \sin\theta$
1	1.2	76	91.2	56	203.39	168.62
2	0.8	58	46.4	32	359.73	190.63
3	0.6	34	20.4	17	399.74	116.87
4	0.6	24	14.4	5	187.15	16.31
5	0.6	21	12.6	−5	117.60	−10.25
6	0.6	39	23.4	−14	128.46	−31.08
7	0.8	58	46.4	−32	113.40	−60.09
8	1.2	42	50.4	−51	30.24	−23.50
L = 352			Σ = 305.2			Σ = 367.5

$$FS = \frac{\Sigma S_u \Delta L}{\Sigma W \sin\theta}$$

$$FS = \frac{305.2}{367.5} = 0.83$$

FIGURE 20.4 Embankment over soft clay: stability analysis for $\phi = 0$ case using ordinary method of slices. (*Sources:* Hunt, R. E. 1986. *Geotechnical Engineering Techniques and Practices.* McGraw-Hill, New York. Reprinted with permission of McGraw-Hill Book Co.)

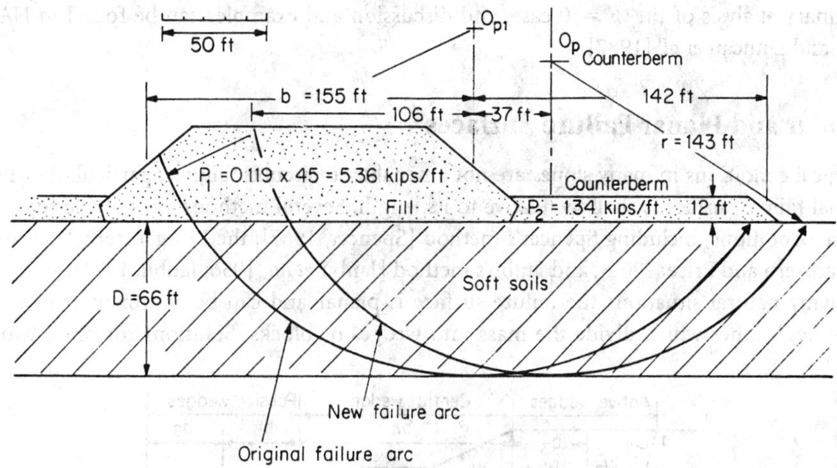

FIGURE 20.5 Counterberm to provide stability for embankment shown in Fig. 20.3. FS = 1.2. (*Source:* Hunt, R. E. 1986. *Geotechnical Engineering Techniques and Practices.* McGraw-Hill, New York. Reprinted with permission of McGraw-Hill Book Co.)

A counterberm added to increase stability (Fig. 20.5) must be of adequate width to cause the critical circle to pass beyond the toe with an acceptable value for FS.

Bishop Slice Methods

The rigorous Bishop method [Bishop, 1955] considers the complete system of forces acting on a slice, as shown in Fig. 20.6. In addition to N_i', U_i, and T_i, included along each side of the slice are shear stresses (x_i with width b_i), effective normal stresses (E_i), and pore-water pressures (U_L and U_r). In analysis, distributions of ($x_i - x_{i+1}$) are found by successive approximation until a number of equilibrium conditions are satisfied. Computations are considerable even with a computer.

The modified (simplified) Bishop method [Bishop, 1955; Janbu *et al.*, 1956] is a simplification of the rigorous method. In the modified method it is assumed that the total influence of the tangential forces on the slice sides is small enough to be neglected.

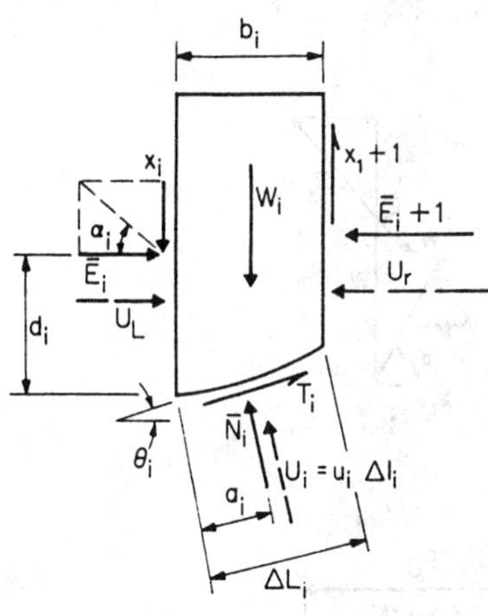

FIGURE 20.6 Complete force system acting on a slice. (*Source:* Lambe, T. W. and Whitman, R. V. 1969. *Soil Mechanics.* John Wiley & Sons, New York. Used with permission of John Wiley & Sons, Inc.)

Other Slice Methods

A number of other methods have been developed that differ in the statics employed to determine FS and the assumptions used to render the problem determinate. Included are Spencer's method [Spencer, 1967], Janbu's rigorous and simplified methods [Janbu *et al.*, 1956], and the Morganstern-Price method [Morganstern and Price, 1965]. Espinoza *et al.* [1992, 1994] present a general framework to evaluate all limit equilibrium methods of stability analysis and illustrate the variability among methods for circular and irregular failure surfaces.

Chart Solutions for Homogeneous Slopes

Various chart solutions have been developed for simple homogeneous slopes, including Taylor [1937, 1948], Janbu [1968], Hunter and Schuster [1968], and Cousins [1978]. They are useful for preliminary analysis of the $\phi = 0$ case, and discussion and examples can be found in NAVFAC [1982] and Duncan *et al.* [1987].

Irregular and Planar Failure Surfaces

Geological conditions in many slopes are not amenable to circular failures, particularly when the potential failure surface is shallow relative to its length. Several methods are suitable for analyses of these conditions, including Spencer's method [Spencer, 1967], the Morganstern-Price method [Morganstern and Price, 1965], and Janbu's method [Janbu *et al.*, 1956; Janbu, 1973].

In many natural situations the failure surface is planar and can be approximated by one or more straight lines which divide the mass into wedges or blocks. Solutions for one-, two-, and

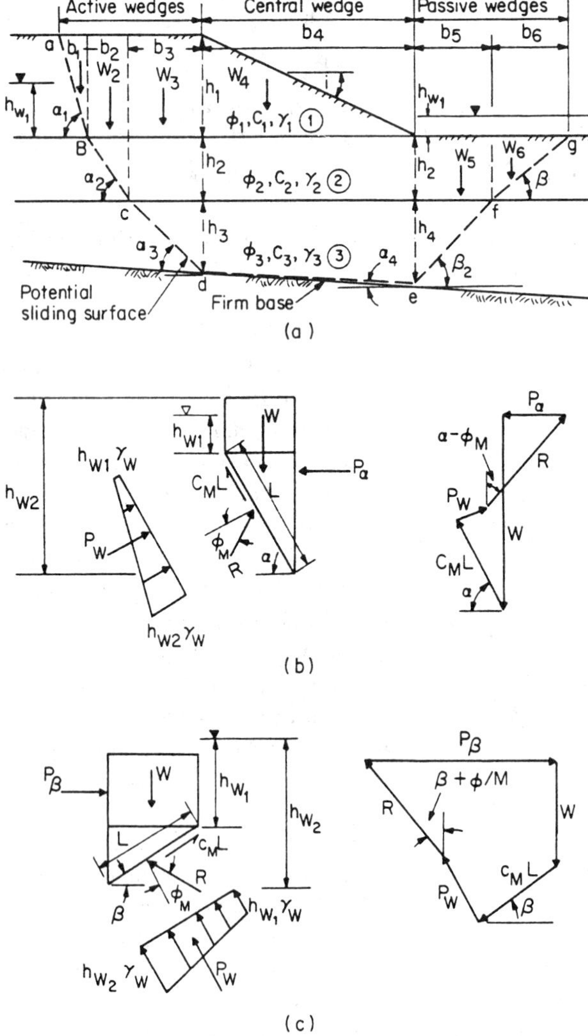

FIGURE 20.7 Stability analysis of translational failure (a). The resultant horizontal force for a wedge sliding along surface *abcde* is P_α (b), and for sliding along surface *efg* is P_β (c). (*Source:* NAVFAC. 1982. *Design Manual, Soil Mechanics, Foundations and Earth Structures, DM-7.* Naval Facilities Engineering Command, Alexandria, VA.)

three-block problems are available in many sources, including Hunt [1986], Huang [1983], and NAVFAC [1982]. A detailed discussion of the analysis of blocks as applied to rock slopes is given in Hoek and Bray [1977].

The translation failure method [NAVFAC, 1982] is based on earth pressures and is suitable for multiple blocks, although interwedge forces are ignored. It is useful where soil conditions consist of several masses with different parameters such as in the case illustrated in Fig. 20.7.

Three-dimensional or tetrahedral wedge failures are common in open-pit mines on heights of one or two benches (60 to 100 ft) but become progressively less prevalent as slope height increases. Failure seems to be associated with weakness planes that are of the same order of size as the slope height involved, with the planes representing three or more intersecting joint sets. The "free blocks" approach tetrahedrons in shape, as shown in Fig. 20.8. A discussion of the analysis of the tetrahedral wedge can be found in Hoek and Bray [1977].

Earthquake Forces

Earthquake forces include cyclic loads which decrease the stability of a slope by increasing shear stresses, pore air and water pressures, and decreasing soil strength. In the extreme case, increases in pore pressure can lead to liquefaction. Sensitive clays and loose fine-grained granular soils above or below groundwater level (GWL), and metastable soils such as loess even when dry, are very susceptible to failure during cyclic loading. The presence of even a thin layer of saturated fine-grained soil, such as silt or clayey silt, can lead quickly to instability in any slope. Embankments over fine-grained saturated soils are particularly susceptible to failure, especially in areas where lateral restraint is limited.

Natural slopes composed of low-sensitivity clay, dense granular soils above or below GWL, or loose coarse-grained soils below GWL generally are stable even during strong ground shaking. Earth-dam embankments can withstand moderate to strong shaking when well-built to modern standards. The greatest risk of damage or failure lies with dams constructed of saturated fine-grained cohesionless materials. The general effect of ground shaking on embankments is slope bulging and crest settlement.

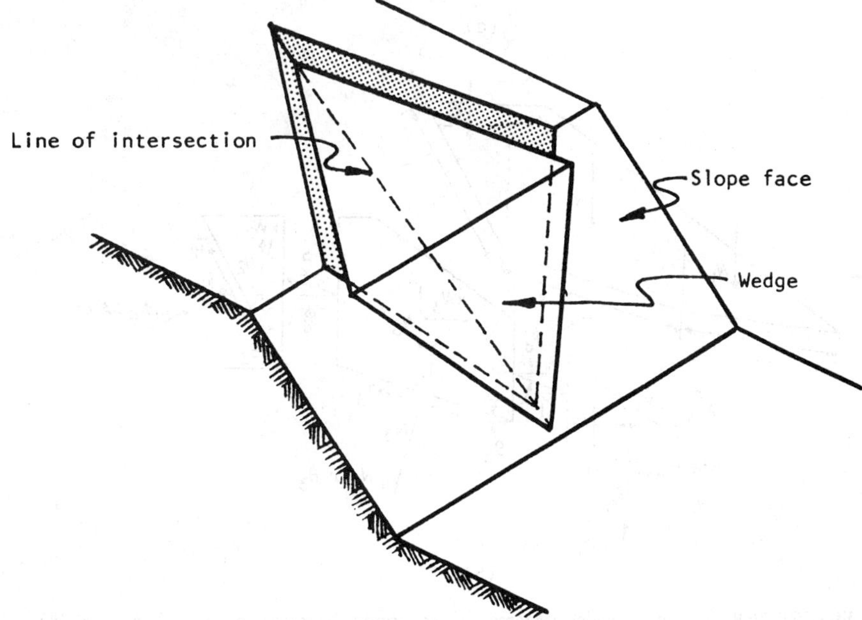

FIGURE 20.8 Geometry of a triangular wedge failure. (*Source:* Hoek, E. and Bray, J. W. 1977. *Rock Slope Engineering*, 2nd ed. The Institute of Mining and Metallurgy, London.)

Pseudostatic Analysis

In the conventional approach, stability is determined as for static loading conditions and the effects of an earthquake are accounted for by including an equivalent horizontal force acting on the mass. The horizontal force, as shown in Fig. 20.9, is expressed as the product of the weight and a seismic coefficient k, which is related to induced accelerations. The effects on pore pressure are not considered, and a decrease in soil strength is accounted for only indirectly. Hall and Newmark [1977] developed design accelerations for horizontal ground motion for slope stability studies as related to magnitude, as given in Table 20.3. Augello *et al.* [1994] provide additional guidance for selecting appropriate design accelerations.

Time-History Dependence

Pseudostatic analysis is based on peak accelerations producing dynamic inertial forces. They may be sufficiently large to drop FS below unity for brief intervals during which displacements occur, but movement ceases when accelerations drop. The time element of the dynamic loadings

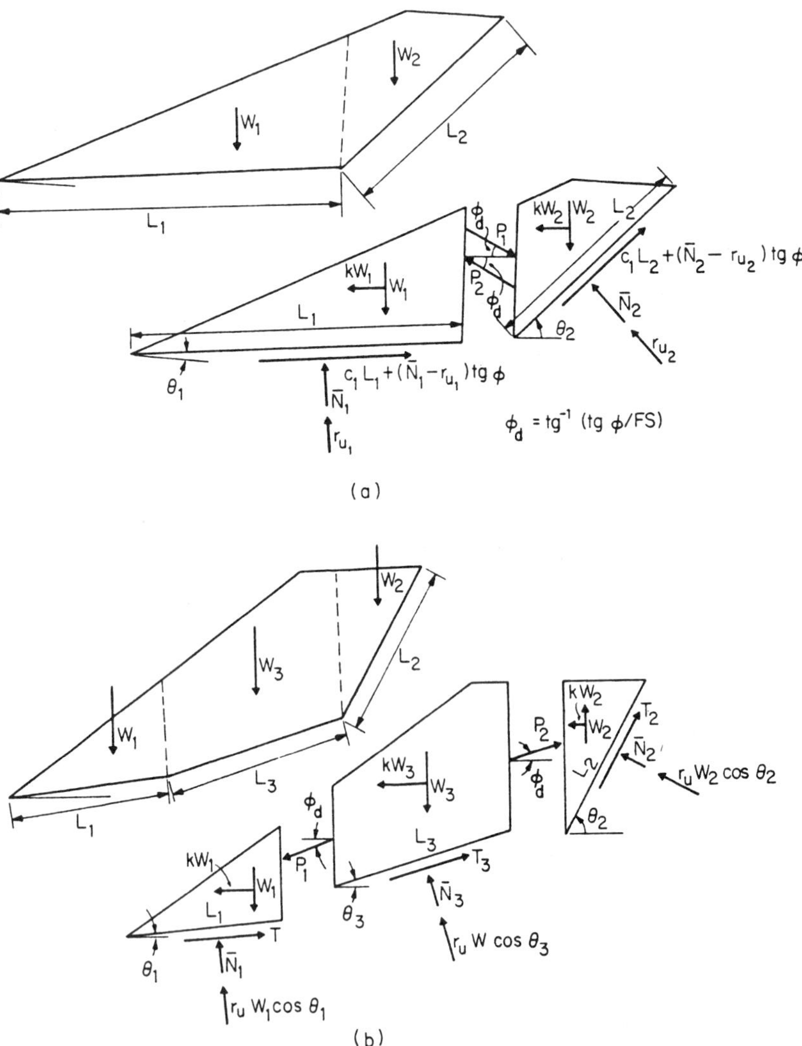

FIGURE 20.9 Complex wedge systems: (a) free-body diagrams for two blocks and (b) free-body diagrams for three blocks. (*Source:* After Huang, Y. H. 1983. *Stability Analysis of Earth Slopes.* Van Nostrand-Reinhold, New York.)

is extremely significant and is not typically considered in analysis. Augello *et al.* [1995] and Bray *et al.* [1995] evaluate the time-history–dependent movements associated with stability of landfill slopes.

If dynamic loads are applied in cycles with small periods, but relatively long duration, soil will not have time to drain between loadings, pore pressures will continue to increase, and failure by liquefaction may occur. The Alaskan event of 1964 lasted almost three minutes, causing disastrous slides at Turnagain Heights where previous earthquakes of equal magnitude but shorter duration had caused little damage.

Dynamic Analysis

Dynamic analysis of embankments is performed employing finite element methods [Newmark, 1965; Seed, 1966; Seed *et al.*, 1975; Byrne, 1991; Finn, 1993]. Embankment response to base excitation is evaluated and the dynamic stresses induced in representative elements of the embankment are computed incorporating nonlinear dynamic material properties by using strain-dependent shear modulus and damping values. Recent work includes the generation of excess pore-water pressure during dynamic loading and the onset of liquefaction. Both the overall deformations and the stability of the embankment section are evaluated.

20.4 Treatments to Improve Stability

General Concepts

Selection Basis

Slope treatments can be placed in one of two broad categories:

- Preventive treatments applied to stable, but potentially unstable natural slopes, slopes to be cut, side-hill fills to be placed, or embankments to be constructed
- Remedial or corrective treatments applied to existing unstable, moving slopes, or to failed slopes

The slope treatment selected is a function of the degree of the hazard and the risk to the public. In natural slopes these factors are very much related to the form of slope failure (fall, slump, avalanche, or flow), the identification of which requires evaluation and prediction by an experienced engineering geologist.

Rating the Hazard and the Risk

Hazard degree relates to the potential failure itself in terms of its possible magnitude and probability of occurrence. Magnitudes can range from a small displacement and material volume, as is common in slump slides, to a large displacement and material volume, such as in a massive debris avalanche. Probability can range from certain to remote. Risk degree relates to the consequences of failure, such as a small volume of material covering portions of a roadway but not endangering lives, to the high risk from the failure of an earth dam resulting in the loss of many lives and much property damage. Safe but economical construction is always the desired result, but the acceptable degree of safety varies with the degree of hazard and risk. An example of a rating system for hazard and risk is given in Hunt [1984].

Treatment Options

Avoid the High-Risk Hazard. There are natural conditions where slope failure is essentially unpredictable and not preventable by reasonable means and the consequences are potentially disastrous. It is best to avoid construction in mountainous terrain subject to massive planar slides or avalanches, slopes in tropical climates subject to debris avalanches, or slopes subject to liquefaction and flows.

Accept the Failure Hazard. In some cases, low to moderate hazards may be acceptable because postfailure cleanup is less costly than some stabilization treatment. Examples are partial temporary closure of a roadway, which is often the approach in underdeveloped countries, or a slide in an open-pit mine where failure is predictable but prevention is considered uneconomical.

In many cases, failures are self-correcting, and eventually a slope may reach a stable condition or work back to where failures do not affect construction. An innovation being used in southern California to protect homes against debris slides, avalanches, and flows is the "A" wall [Hollingsworth and Kovacs, 1981] illustrated in Fig. 20.10. The purpose of the wall is to deflect moving earth masses away from the building.

Eliminate or Reduce the Hazard. Where failure is essentially predictable and preventable, or is occurring or has occurred and is suitable for treatment, slope stabilization methods are applied. For low- to moderate-risk conditions, the approach can be either to eliminate or to reduce the hazard, depending on comparative economics. For high-risk conditions the hazard should be eliminated. The generally acceptable safety factor determined by stability analysis can be taken as follows: low risk, FS = 1.3; significant risk, FS = 1.4; high risk, FS = 1.5.

Slope Stabilization

Slope stabilization methods may be placed in five general categories, as follows (Fig. 20.11):

1. Change slope geometry to decrease the driving forces or increase the resisting forces.
2. Control surface water to prevent erosion, and infiltration to reduce seepage forces.
3. Control internal seepage to reduce the driving forces.
4. Increase material strengths to increase resisting forces.
5. Provide retention to increase the resisting forces.

Where slopes are in the process of failing, the time factor must be considered. Time may not be available for carrying out measures that will eliminate the hazard; therefore, the hazard should be reduced and perhaps eliminated at a later date. The objective is to arrest the immediate movement. Where possible, treatments should be performed during the dry season, when movements will not affect remediation such as breaking horizontal drains.

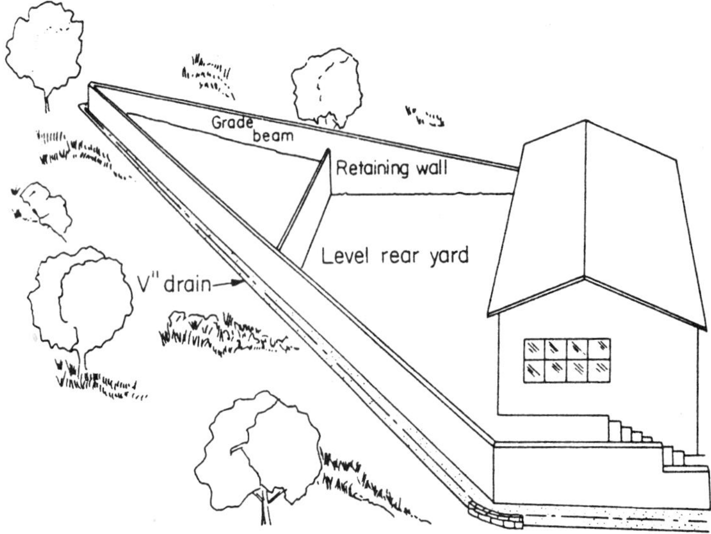

FIGURE 20.10 Typical A-wall layout to deflect debris slides, avalanches, and flows. (*Source:* Hollingsworth, R. and Kovacs, G. S. 1981. Soil slumps and debris flows: Prediction and protection. *Bul. Assoc. Eng. Geol.* 18(1):17–28.)

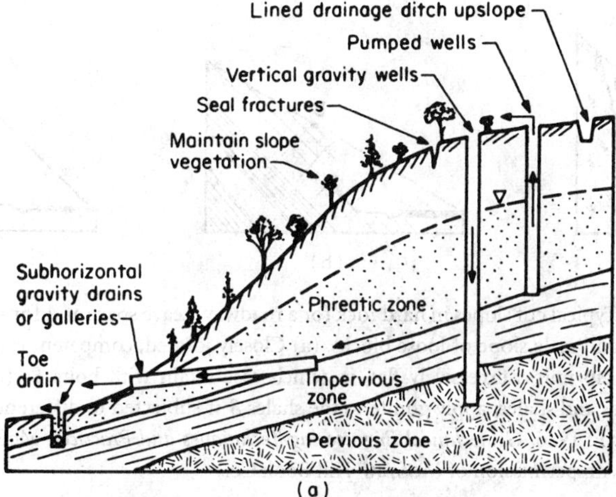

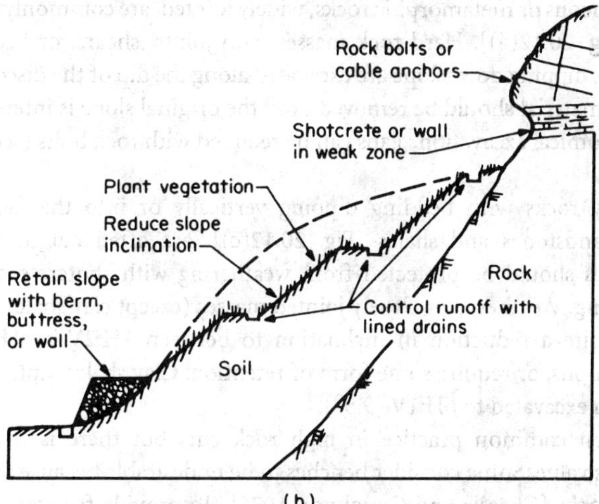

FIGURE 20.11 The general methods of slope stabilization: (a) control of seepage forces, (b) reducing the driving forces and increasing the resisting forces. (*Source:* Hunt, R. E. 1984. *Geotechnical Engineering Investigation Manual.* McGraw-Hill, New York.)

Changing Slope Geometry

Natural Slope Inclinations

In many cases the natural slope represents the maximum long-term inclination, but there are cases where the slope is not stable and is moving. The inclination of existing slopes should be noted during field reconnaissance, since an increase in inclination by cutting may result in failure.

Cut Slopes in Rock

The objective of any cut slope is to form a stable inclination without retention. Controlled blasting procedures are required in rock masses to avoid excessive rock breakage resulting in extensive fracturing. Line drilling and presplitting during blasting operations minimize disturbance of the rock face.

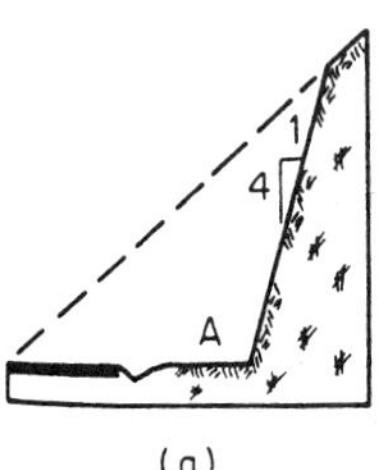

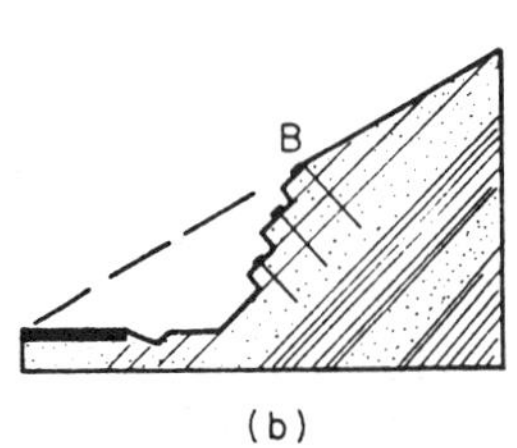

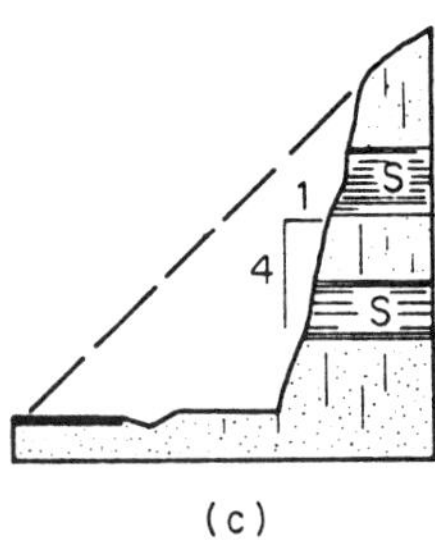

FIGURE 20.12 Typical cut slopes in hard rock for a roadway. Leave space at *A* for storage of block falls and topples and scale slope of loose blocks. (a) Closely jointed, competent rock. (b) Dipping beds; follow the dip unless excessively flat, in which case retain with bolts *B*. (c) Horizontally bedded sandstones and shales; apply gunite to the shales *S* if subjected to differential weathering. (*Source:* Hunt, R. E. 1986. *Geotechnical Engineering Techniques and Practices.* McGraw-Hill, New York. Reprinted with permission of McGraw-Hill Book Co.)

Hard masses of igneous or metamorphic rocks, widely jointed, are commonly cut to 1H:4V (76°) [Terzaghi, 1962] [Fig. 20.12(a)]. Hard rock masses with joints, shears, or bedding representing major discontinuities dipping downslope are excavated along the dip of the discontinuity, as shown in Fig. 20.12(b). All material should be removed until the original slope is intercepted. If the dip is too shallow for economical excavation, slabs can be retained with rock bolts (see "Retention," later in this chapter).

Hard sedimentary rocks with bedding dipping vertically or into the face, or horizontally interbedded hard sandstones and shales [Fig. 20.12(c)], are often cut to 1H:4V, but in the latter case, the shales should be protected from weathering with shotcrete or gunite to retard differential weathering. Weathered or closely jointed masses (except clay shales and dipping major discontinuities) require a reduction in inclination to between 1H:2V to 1H:1V (63° to 45°) depending on conditions, or require some form of retention. Clay shales, unless interbedded with sandstones, are often excavated to 6H:1V (9.5°).

Benching has been common practice in high rock cuts but there is disagreement among practitioners as to its value. Some consider benches to be undesirable because they provide takeoff points for falling blocks [Chassie and Goughnor, 1976]. To provide for storage they must be of adequate width. Block storage space should always be provided at the slope toe with adequate shoulder width to protect the roadway from falls and topples.

Cut Slopes in Soils

Most soil formations are commonly cut to an average inclination of 2H:1V (26°) but consideration must be given to seepage forces and other physical and environmental factors to determine if retention is required. Soil cuts are normally designed with benches, especially for cuts over 25 to 30 ft high. Because the slope angle between benches may be increased, benches reduce the amount of excavation necessary to achieve overall lower inclinations. Drains are installed as standard practice along the slopes and the benches to control runoff, as illustrated in Fig. 20.11(b) and Fig. 20.13.

In soil-rock transition (strong residual soils to weathered rock) such as in Fig. 20.13, cuts are often excavated to between 1H:2V to 1H:1V (63° to 45°) although potential failure along relict discontinuities must be considered. Where there is thin soil cover over rock the soil should be removed or retained as the condition normally will be unstable in cut. Figure 20.13 illustrates an ideal case, often misinterpreted from test boring data, and not present in the slope. In mountainous terrain all of the formations may be dipping, as shown on Fig. 20.14, a potentially very unstable condition. In such conditions, downslope seismic refraction surveys are valuable to define stratigraphy.

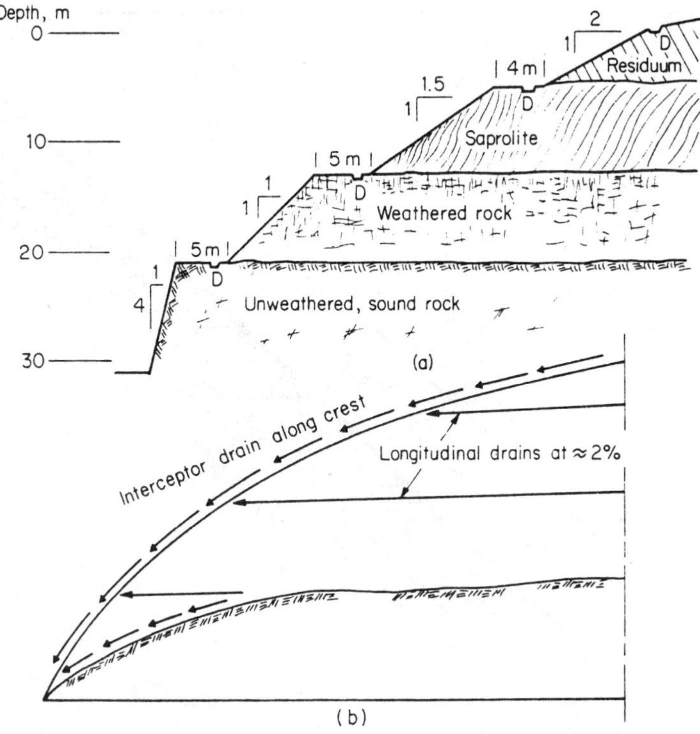

FIGURE 20.13 Benching a cut slope. (a) Typical section with drains located at *D*. Slope in weathered rock varies with rock quality; in saprolite, varies with orientation of foliations. Attempt is made to place benches on contact between material types. (b) General scheme to control runoff and drainage shown in face view. (*Source:* Hunt, R. E. 1986. *Geotechnical Engineering Techniques and Practices.* McGraw-Hill, New York. Reprinted with permission of McGraw-Hill Book Co.)

Failing Slopes

If a slope is failing and undergoing substantial movement, the removal of material from the head to reduce the driving forces can be the quickest method of arresting movement, and benching may be effective in the early stages. Placing material at the toe to form a counterberm increases the resisting forces. An alternative is to remove debris from the toe and permit failure to occur. Eventually the mass may naturally attain a stable inclination.

Changing slope geometry to achieve stability once failure has begun usually requires either the removal of very large volumes or the implementation of other methods. Space is seldom available in critical situations to permit placement of material at the toe, since very large volumes normally are required. As will be discussed, subhorizontal drains are often a very effective intermediate solution.

Surface Water Control

Purpose

Surface water is controlled to eliminate or reduce infiltration and to provide erosion protection. External measures are generally effective, however, only if the slope is stable and there is no internal source of water to cause excessive seepage forces.

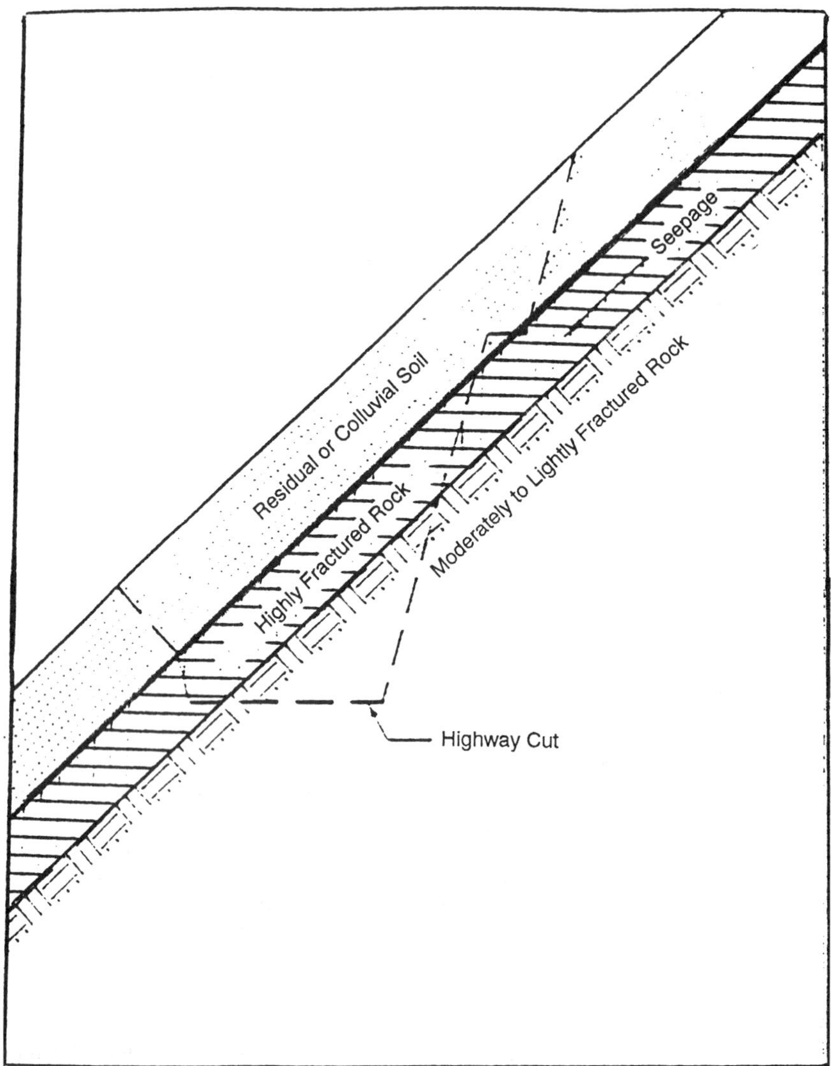

FIGURE 20.14 Slope conditions common along steep slopes in mountainous terrain.

Infiltration and Erosion Protection

Planting the slope with thick, fast-growing native vegetation strengthens the shallow soils with root systems and discourages desiccation, which causes fissuring. Not all vegetation works equally well, and selection requires experience. In the Los Angeles area of California, for example, Algerian ivy has been found to be quite effective in stabilizing steep slopes [Sunset, 1978]. Newly cut slopes should be immediately planted and seeded. Burlap bags or sprayed mulch helps increase growth rate and provide protection against erosion during early growth stages. In addition to plantings, erosion protection along the slope can be achieved with wattling bundles, as illustrated in Fig. 20.15.

Sealing cracks and fissures with asphalt or soil cement will reduce infiltration but will not stabilize a moving slope since the cracks will continue to open. Grading a moving area results in filling cracks with soil, which helps reduce infiltration.

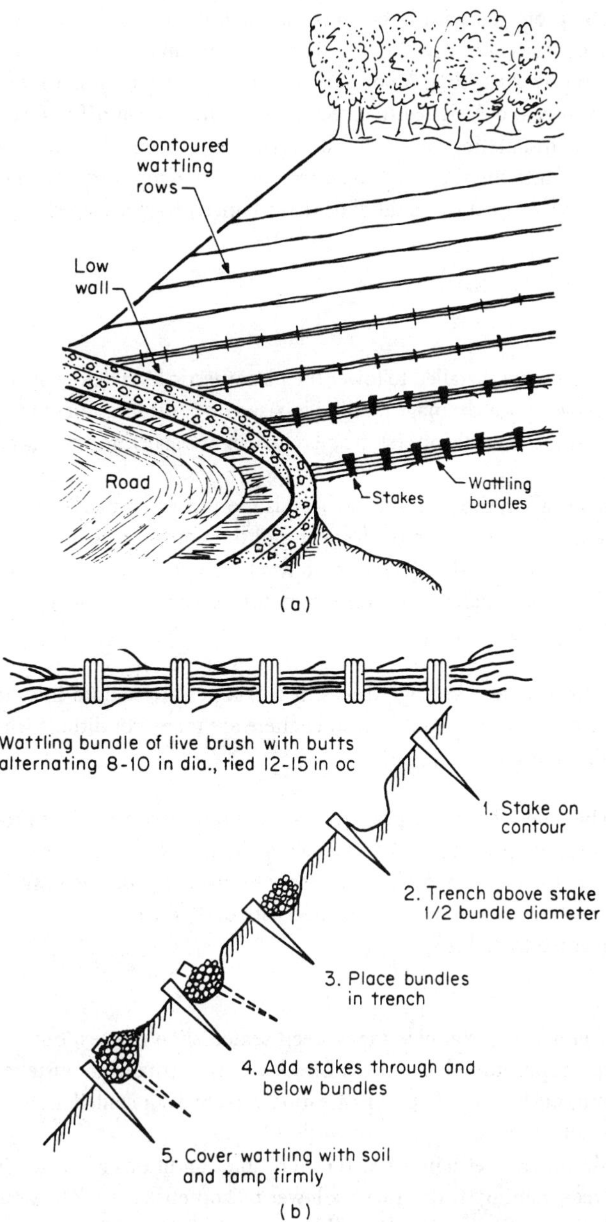

FIGURE 20.15 Erosion protection by installation of wattling bundles along contours of slope face. (a) Contoured wattling on slope face. (b) Sequence of operations for installing wattling on slope face. Work starts at bottom of cut or fill with each contour line proceeding from step 1 through 5. Cigar-shaped bundles of live brush of species which root are buried and staked along slope. They eventually root and become part of the permanent slope cover. (After Gray *et al.* 1980. Adapted with the permission of the American Society of Civil Engineers.)

Surface Drainage Systems

Cut slopes should be protected with interceptor drains installed along the crest of the cut, along benches, and along the toe (Figs. 20.11 and 20.13). On long cuts the interceptors are connected to downslope collectors [Fig. 20.13(b)]. All drains should be lined with nonerodible materials, free of cracks or other openings, and designed to direct all concentrated runoff to discharge offslope.

With failing slopes, installation of an interceptor along the crest beyond the head of the slide area will reduce runoff into the slide. But the interceptor is a temporary expedient, since in time it may break up and cease to function as the slide disturbance progresses upslope.

Internal Seepage Control

General

Internal drainage systems are installed to lower the piezometric level below the potential or existing sliding surface. Selection of the drainage method is based on consideration of the geologic materials, structure, and groundwater conditions (static, perched, or artesian), and the location of the phreatic surface.

As the drains are installed, the piezometric head is monitored by piezometers and the efficiency of the drains is evaluated. The season of the year and the potential for increased flow during wet seasons must be considered, and if piezometric levels are observed to rise to dangerous values (as determined by stability analysis, or from monitoring slope movements), the installation of additional drains is required.

Cut Slopes. Systems to relieve seepage forces in cut slopes are seldom installed in practice, but they should be considered more frequently, since there are many conditions where they would aid significantly in maintaining stability.

Failing Slopes. The relief of seepage pressures is often the most expedient means of stabilizing a moving mass. The primary problem is that, as mass movement continues, the drains will be cut off and cease to function; therefore, it is often necessary to install the drains in stages over a period of time. Installation must be planned and performed with care, since the use of water during drilling could possibly trigger a total failure.

Methods

Deep wells have been used to stabilize many deep-seated slide masses, but they are costly since continuous or frequent pumping is required. Check valves normally are installed so that when the water level rises, pumping begins. Deep wells are most effective if installed in relatively free-draining material below the failing mass.

Vertical gravity drains are useful in perched water-table conditions where an impervious stratum overlies an open, free-draining stratum with a lower piezometric level. The drains permit seepage by gravity through the confining stratum and thus relieve hydrostatic pressures. Clay strata over granular soils, or clays or shales over open-jointed rock, offer favorable conditions for gravity drains where a perched water table exists.

Subhorizontal drains are one of the most effective methods to improve stability of a cut slope, or to stabilize a failing slope. Installed at a slight angle upslope to penetrate the phreatic zone and permit gravity flow, they usually consist of perforated pipe, 2 in. diameter or larger, forced into a predrilled hole of slightly larger diameter than the pipe. Horizontal drains have been installed to lengths of more than 300 ft. Spacing depends on the type of material being drained; fine-grained soils may require spacing as close as 10 to 30 ft, whereas for more permeable materials, 30 to 50 ft may suffice.

Drainage galleries are very effective for draining large moving masses but their installation is difficult and costly. They are used mostly in rock masses where roof support is less of a problem than in soils. Installed below the failure zone to be effective, they are often backfilled with stone. Vertical holes drilled into the galleries from above provide for drainage from the failure zone into the galleries.

Interceptor trench drains can be installed upslope to intercept groundwater flowing into a cut or sliding mass, but they must be sufficiently deep. Perforated pipe is laid in the trench bottom, embedded in sand, and covered with free-draining material, then sealed at the surface. Interceptor trench drains are generally not practical on steep, heavily vegetated slopes because installation of the drains and access roads requires stripping the vegetation, which will tend to decrease stability.

Relief trenches relieve pore pressures at the slope toe. They are relatively simple to install. Excavation should be made in sections and quickly backfilled with stone so as not to reduce the slope stability and possibly cause a total failure. Generally, relief trenches are most effective for small slump slides where high seepage forces in the toe area are the major cause of instability.

Electroosmosis has been used occasionally to stabilize silts and clayey silts, but the method is relatively costly, and not a permanent solution unless operation is maintained.

Increased Strength

Chemicals have sometimes been injected to increase soil strength. In a number of instances the injection of a quicklime slurry into predrilled holes has arrested slope movements as a result of the strength increase from chemical reaction with clays [Handy and Williams, 1967; Broms and Boman, 1979]. Strength increase in saltwater clays, however, was found to be low.

Resistance along an existing or potential failure surface can be increased with drilled piers [Oakland and Chameau, 1989; Lippomann, 1989], shear pins (reinforced concrete dowels), or stone columns. In the latter case the increased resistance is obtained from a significantly higher friction angle obtained in the stone along its width intercepting the failure surface.

Side-Hill Fills

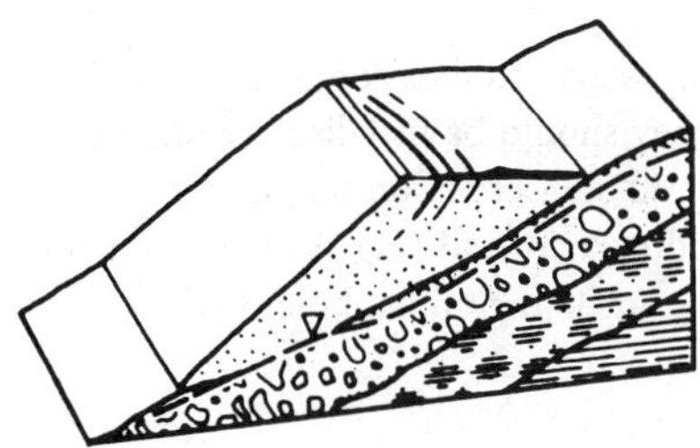

FIGURE 20.16 Early failure stage in side-hill fill as concentric cracks form in the pavement.

Failures

Construction of a side-hill embankment using slow-draining materials can be expected to block natural drainage and evaporation. As seepage pressures increase, particularly at the toe (as shown in Fig. 20.16), the embankment strains and concentric tension cracks form. The movements develop finally into a rotational failure. Fills placed on moderately steep to steep slopes of residual or colluvial soils, in particular, are prone to be unstable unless seepage is properly controlled, or the embankment is supported by a retaining structure.

Preventive Treatments

Interceptor trench drains should be installed along the upslope side of all side-hill fills as standard practice to intercept flow, as shown in Fig. 20.17. Perforated pipe is laid in the trench bottom, embedded in sand, covered by free-draining materials, and then sealed at the surface. Surface flow is collected in open drains and all discharge, including that from the trench drains, is directed away from the fill area.

Where anticipated flows are low to moderate, transverse drains extending downslope and connecting with the interceptor ditches upslope, parallel to the roadway, may provide adequate subfill drainage. Wherever either the fill or the natural soils are slow-draining, however, a free-draining blanket should be installed over the entire area between the fill and the natural slope materials to relieve seepage pressures from shallow groundwater conditions (Fig. 20.17). It is

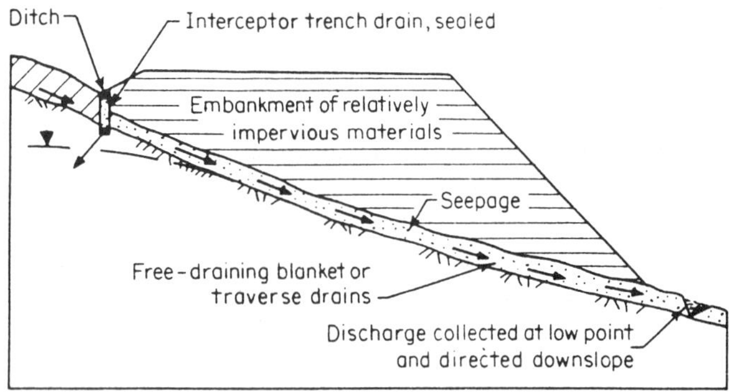

FIGURE 20.17 Proper drainage provisions for a sidehill fill *A* constructed of relatively impervious materials over relatively impervious natural soils *S* subject to surface creep. Upper soils are stripped and replaced with free-draining blanket *B*. Interceptor trench drain *C* installed and sealed with lined ditch *D*. Groundwater discharge collected at low point *E* and directed downslope. (*Source:* Hunt, R. E. 1986. *Geotechnical Engineering Techniques and Practices.* McGraw-Hill, New York. Reprinted with permission of McGraw-Hill Book Co.)

prudent to strip potentially unstable upper soils, which are often creeping on moderately steep to steep slopes, to a depth where stronger soils are encountered. Stepped excavations improve stability. Discharge should be collected at the low point of the fill and drained downslope in a manner that will provide erosion protection.

Retaining structures may be economical on steep slopes that continue for some distance beyond the fill, if stability is uncertain.

Corrective Treatments

If movement downslope has begun in a slow initial failure stage, subhorizontal drains may be adequate to stabilize the embankment if closely spaced. They should be installed during the dry season, if practical, since the use of water to drill holes during the wet season may accelerate total failure. An alternative is to retain the fill with an anchored curtain wall (Fig. 20.18). After total failure, the most practical solutions are either reconstruction of the embankment with proper drainage, or retention with a wall.

Retention

Rock Slopes

Various methods of retaining hard rock slopes are illustrated in Fig. 20.19. They can be described briefly as follows.

- Concrete pedestals are used to support overhangs, where their removal is not practical because of danger to existing construction downslope [Fig. 20.19(a)].
- Rock bolts are used to reinforce jointed rock masses or slabs on a sloping surface [Fig. 20.19(b)]. Ordinary or temporary rock bolts, and fully grouted or permanent rock bolts are described by Lang [1972].
- Concrete straps and rock bolts are used to support loose or soft rock zones or to reduce the number of bolts [Fig. 20.19(c)].
- Cable anchors are used to reinforce thick rock masses [Fig. 20.19(d)]. The reinforcement of a single block by bolts or cables is shown in Fig. 20.20.

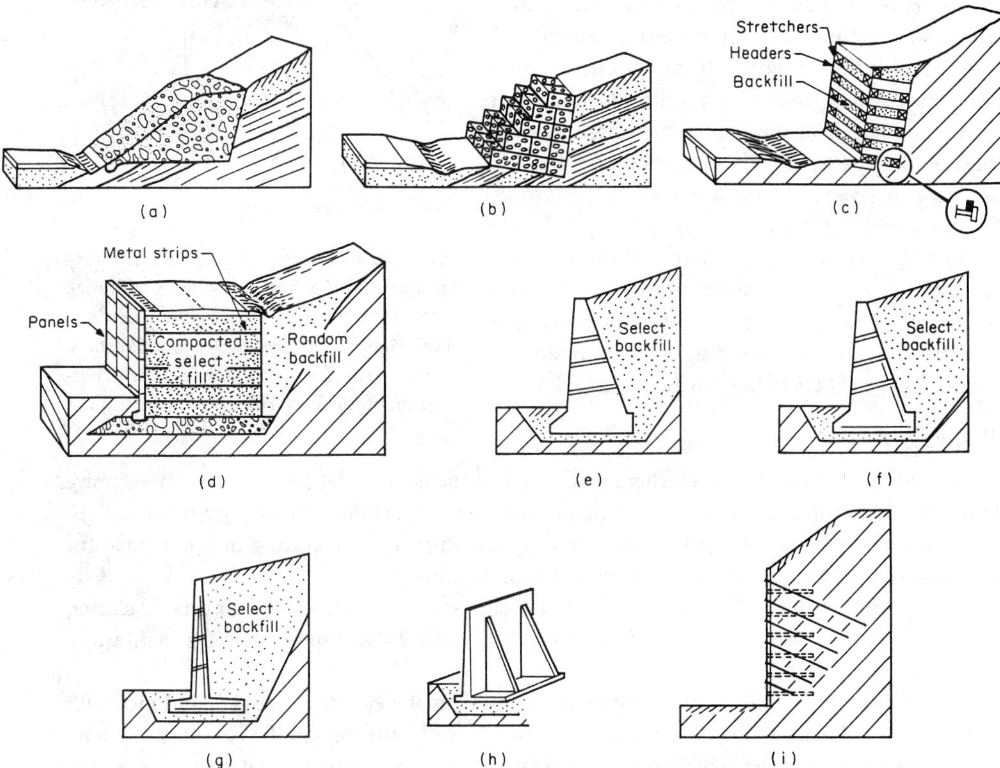

FIGURE 20.18 Various types of retaining walls: (a) rock-filled buttress; (b) gabion wall; (c) crib wall; (d) reinforced earth wall; (e) concrete gravity wall; (f) concrete-reinforced semigravity wall; (g) cantilever wall; (h) counterfort wall; (i) anchored curtain wall. (*Source:* Hunt, R. E. 1986. *Geotechnical Engineering Techniques and Practices.* McGraw-Hill. New York. Reprinted with permission of McGraw-Hill Book Co.)

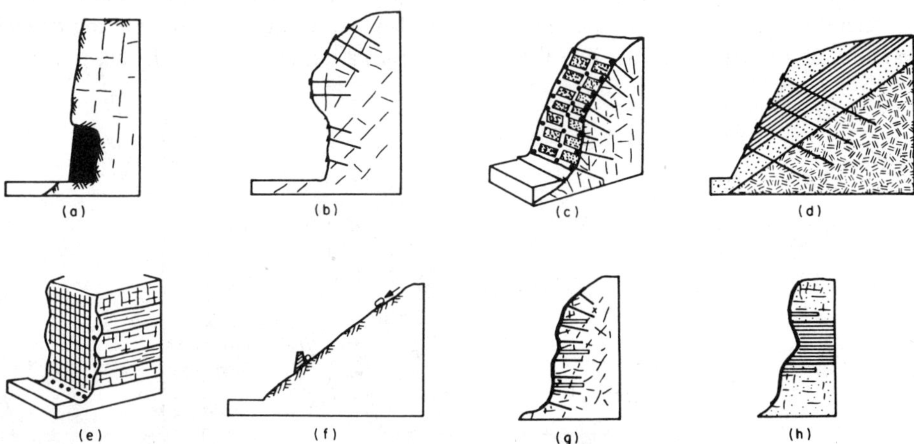

FIGURE 20.19 Various methods of retaining hard rock slopes: (a) concrete pedestals for overhangs; (b) rock bolts for jointed masses; (c) bolts and concrete straps for intensely jointed masses; (d) cable anchors to increase support depth; (e) wire mesh to constrain falls; (f) impact walls to deflect or contain rolling blocks; (g) shotcrete to reinforce loose rock, with bolts and drains; (h) shotcrete to retard weathering and slaking of shales. (*Source:* Hunt, R. E. 1984. *Geotechnical Engineering Investigation Manual.* McGraw-Hill, New York.)

Shotcrete, when applied to rock slopes, usually consists of a wet-mix mortar with aggregate as large as 2 cm (3/4 in.) which is projected by air jet directly onto the slope face. The force of the jet compacts the mortar in place, bonding it to the rock, which first must be cleaned of loose particles and loose blocks. Application is in 3- to 4-in. layers, each of which is permitted to set before application of subsequent layers. Weep holes are installed to relieve seepage pressures behind the face. Since shotcrete acts as reinforcing and not as support, it is used often in conjunction with rock bolts. The tensile strength can be increased significantly by adding 25-mm-long wire fibers to the concrete mix.

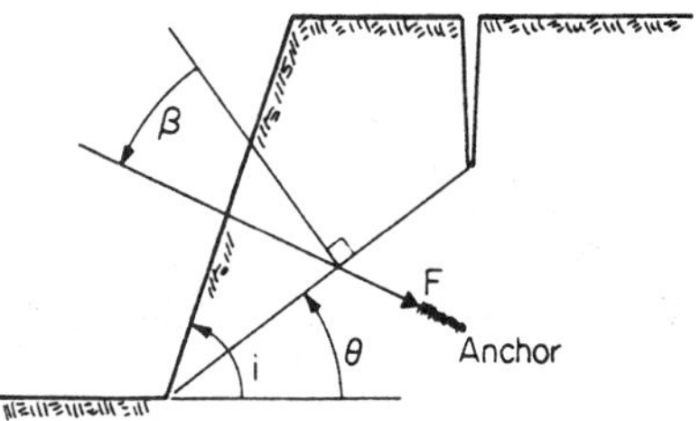

FIGURE 20.20 Definition of angle β for reinforcement of a single rock block with rock bolts. (*Source:* Hunt, R. E. 1986. *Geotechnical Engineering Techniques and Practices.* McGraw-Hill, New York. Reprinted with permission of McGraw-Hill Book Co.)

Soil Slopes

Walls are used to retain earth slopes where space is not available for a flat enough slope or excessive volumes of excavation are required, or to obtain more positive stability under certain conditions. Except for anchored concrete curtain walls, wall types which require cutting into the slope for construction are seldom suitable for retention of a failing slope.

Various types of walls are illustrated in Fig. 20.18. They may be divided into four general classes, with some wall types included in more than one class: gravity walls, nongravity walls, rigid walls, and flexible walls.

Gravity walls provide slope retention by either their weight alone, or their weight combined with the weight of a soil mass acting on a portion of their base or by the weight of a composite system. They are free to move at the top, thereby mobilizing active earth pressure. Included are concrete gravity walls, cantilever walls, counterfort walls, rock-filled buttresses, gabion walls, crib walls, and reinforced earth walls. With the advent of geosynthetics there are many variations of "reinforced earth" [Koerner, 1993], and other innovations such as "soil nailing" [FHWA, 1993].

Nongravity walls are restrained at the top and not free to move. They include basement walls, some bridge abutments, and anchored concrete curtain walls. Anchored concrete curtain walls, such as the one illustrated in Fig. 20.21, can be constructed to substantial heights and have a very high retention capacity. As illustrated in Fig. 20.22, they are constructed from the top down by excavation of a series of benches into the slope and formation of a section of wall, retained by

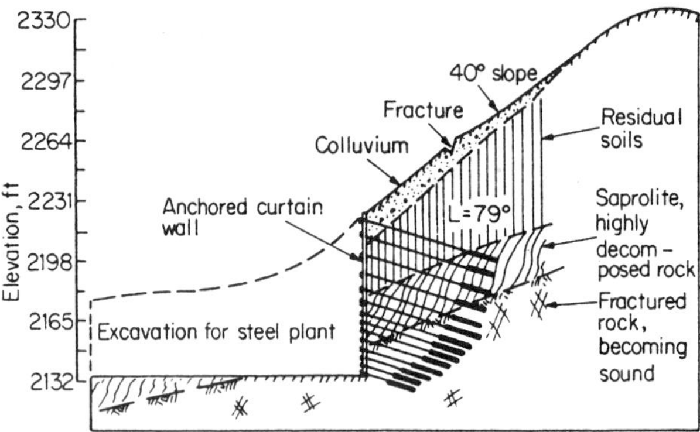

FIGURE 20.21 Section through a 25.5 m (85 ft) high anchored curtain wall constructed to retain a steep slope with a history of landslides. Joao Monlavade, M. G. Brazil. (*Source:* Hunt, R. E. and da Costa Nunes, A. J. 1978. Retaining walls: Taking it from the top. *Civ. Eng., ASCE,* May, 73–75.)

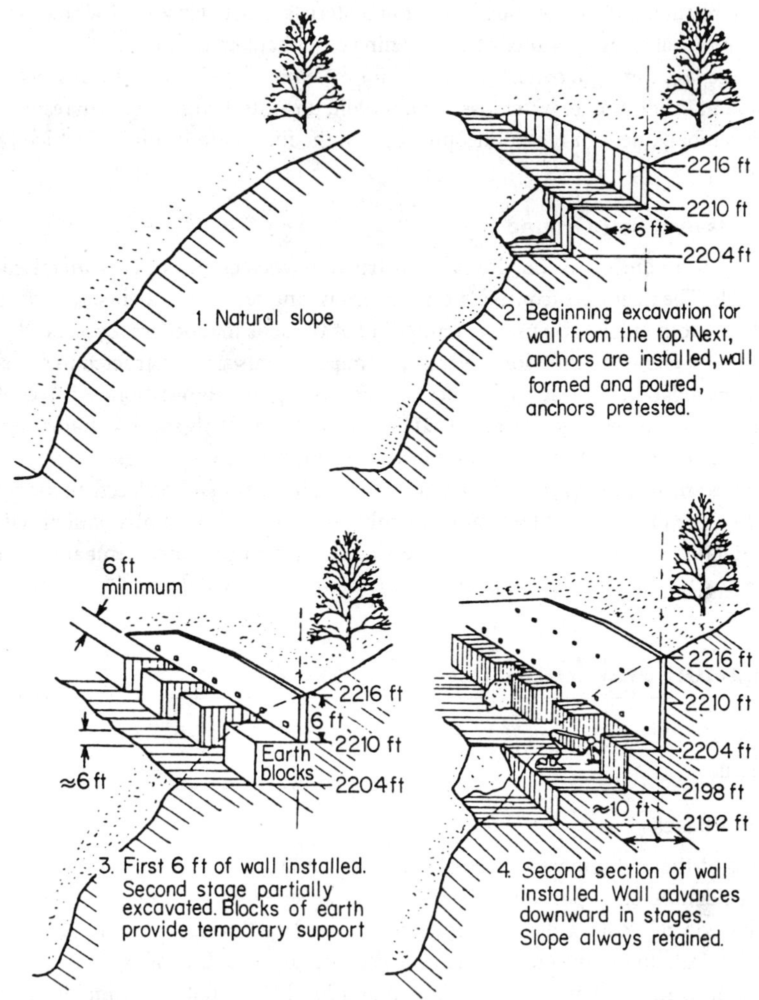
1. Natural slope
2216 ft
2210 ft
≈6 ft
2204ft
2. Beginning excavation for wall from the top. Next, anchors are installed, wall formed and poured, anchors pretested.
6 ft minimum
2216 ft
6 ft
2210 ft
Earth blocks
2204ft
≈6 ft
3. First 6 ft of wall installed. Second stage partially excavated. Blocks of earth provide temporary support
2216 ft
2210 ft
2204 ft
2198 ft
≈10 ft
2192 ft
4. Second section of wall installed. Wall advances downward in stages. Slope always retained.

FIGURE 20.22 The "Brazilian method" of construction of an anchored curtain wall from the top down. (*Source:* Hunt, R. E. and da Costa Nunes, A. J. 1978. Retaining walls: Taking it from the top. *Civ. Eng., ASCE,* May, 73–75.)

anchors, in each bench along the slope. Since the slope is thus retained completely during the wall construction, the system is particularly suited to potentially unstable or unstable slopes. A variation of the anchored curtain wall consists of anchored premolded concrete panels. The advantage of the system is that the wall easily conforms to the slope configuration.

Rigid walls include concrete walls, gravity and semigravity walls, cantilever walls, and counterfort walls. Anchored concrete curtain walls are considered as semirigid. Flexible walls include rock-filled buttresses, gabion walls, crib walls, reinforced earth walls, and anchored sheet-pile walls.

Embankments

Earth Dams

During design and construction of earth dams, stability is provided by controlled compaction of the embankment materials, adequate support by founding materials, and control of seepage through and beneath the embankment. Stable slope inclinations are related to the materials used to construct the embankment, and to the foundation materials. Relatively weak foundation materials either require removal by excavation or the flattening of embankment slopes.

Control of seepage through, beneath, and around the embankment is a critical aspect of design and construction. In addition to water loss in a reservoir, uncontrolled seepage can result in internal erosion of the embankment or high uplift pressures in the foundation, either of which can lead to complete failure.

Embankments over Soft Ground

Failures usually occur during or shortly after construction when embankments are raised over soft foundation soils. The major postconstruction concern is long-term settlements. In most situations it is desirable to avoid failure because the remolding of the soils that occurs significantly decreases their strength, worsening the situation. Therefore, embankments are constructed in stages. During each stage consolidation and strength increase occurs in the foundation soils, enabling the construction of subsequent stages [Ladd, 1991]. Vertical drains, such as wick drains, significantly increase the rate of consolidation and shorten the time interval between stages.

Geosynthetics are used to improve embankment strength. They may be placed on the soft ground prior to placing the first lift, and then subsequently placed within the embankment. Hird [1986] and Low *et al.* [1990] provide methods for assessing stability of reinforced embankments on soft ground. Construction of counterberms (Fig. 20.5) also improves stability.

20.5 Investigation and Monitoring

Exploration

Preliminary Phases

The objectives of the preliminary phases of investigation are to identify potential forms, magnitudes, and incidences of slope failures, and to plan an exploration program. The study scope includes collection of existing data, generation of new data through terrain analysis, and field reconnaissance. Data to be collected include site geology, topographic maps, and remotely sensed imagery, and historical information on regional and local slope failures, climatic conditions of precipitation and temperature, and seismicity.

Using the topographic maps and remotely sensed imagery (air-photo interpretation techniques) terrain analysis is performed to identify unstable and potentially unstable areas, and to establish preliminary conclusions regarding geologic conditions. For a large study area, a preliminary map is prepared on which is shown topography, drainage, active and ancient failures, and geology. The preliminary map is developed into a hazard map after field reconnaissance. At the project location, more detailed maps are prepared illustrating the foregoing items. The methodology identifies the significant features to be examined during field reconnaissance.

The site location is visited and notations are made regarding seepage points, vegetation, creep indications (tilted and bent tree trunks), tension cracks, failure scars, hummocky ground, natural slope inclinations, exposed geology, and prevailing and recent weather conditions.

From the data collected, preliminary evaluations are made regarding slope conditions and an exploration program is planned. The entire slope should be explored, not only the specific area of failure or area to be cut.

Explorations

Seismic refraction profiling is useful to determine the depth to sound rock and the probable groundwater table, and is most useful in differentiating between colluvial or residual soils and the fractured-rock zone. Surveys are made both longitudinal and transverse to the slope. They are particularly valuable on steep slopes with a deep weathering profile where test borings are time-consuming and costly.

Resistivity profiling may be performed to determine the depth to groundwater and to rock. Profiling is generally only applicable to depths of about 15 to 30 ft, but very useful in areas of difficult access. In the soft, sensitive clays of Sweden, the failure surface or potential failure surface is often located by resistivity measurements since the salt content, and therefore the resistivity, often changes suddenly at the slip surface [Broms, 1975].

Test borings are made to confirm the stratigraphy determined by the geophysical explorations, to recover samples of the various materials, and to provide holes for the installation of instrumentation. Borings should extend to adequate penetration below the depth of a cut, and below the depth of any potential failure surface. Sampling should be continuous through a potential or existing rupture zone, and in residual soils and rock masses care should be taken to identify

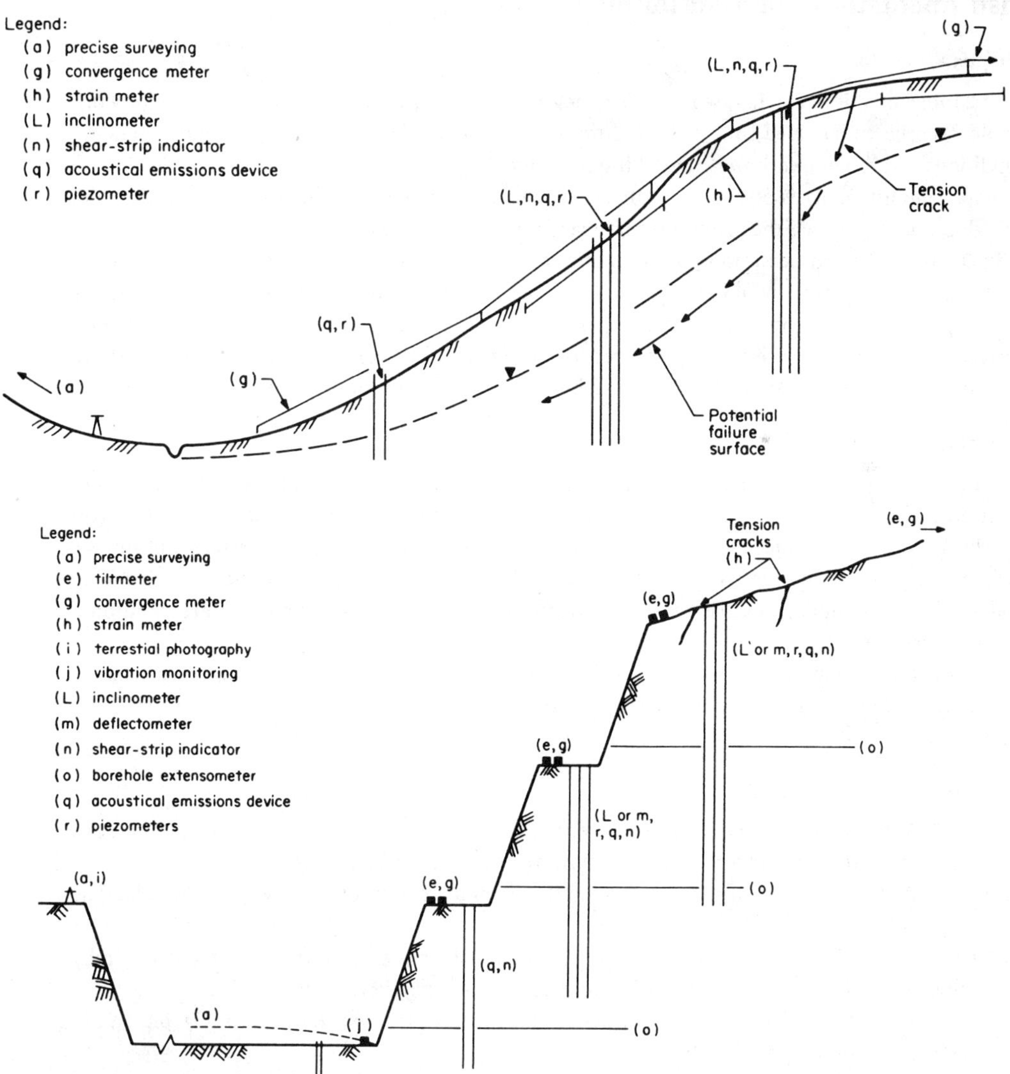

FIGURE 20.23 Slope instrumentation and monitoring: (a) potentially unstable soil slope and (b) rock cut. (*Source:* Hunt, R. E. 1984. *Geotechnical Engineering Investigation Manual.* McGraw-Hill, New York.)

slickensided surfaces. Groundwater conditions must be defined carefully and the static water table, perched, and artesian conditions noted. It is important to remember that the conditions existing at the time of investigation are not likely to be those during failure.

Evaluations

Evaluations are made of the safety factor against total failure on the basis of existing topographic conditions, then under conditions of the imposed cut or fill. For preliminary studies, shear strengths may be estimated from published data, or measured by laboratory or in situ testing. In the selection of the strength parameters, consideration is given to field conditions as well as to changes that may occur with time (reduction from weathering, leaching, solution). Other transient conditions such as weather and earthquakes also require consideration, especially if the safety factor for the entire slope is low and could go below unity with some environmental change.

Instrumentation and Monitoring

Purpose

Where movement is occurring, where safety factors against sliding are low, or where a major work would become endangered by a slope failure, instrumentation is required to monitor changing conditions and provide early warning of impending failure.

Slope-stability analysis is far from an exact science, regardless of the adequacy of the data available, and sometimes the provision for an absolutely safe slope is prohibitively costly.

In cut slopes, instrumentation monitors movements and changing stress conditions to provide early warning and permit invoking contingency plans for remedial measures when low safety factors are accepted in design. In unstable or moving slopes, instrumentation is installed to locate the failure surface and determine pore-water pressures for analysis, and to measure surface and subsurface movements, velocities, and accelerations which provide indications of impending total failure.

Methods Summarized

Instrumentation is discussed in detail in Hunt [1984] and Dunnicliff [1988], and for slopes is illustrated in Fig. 20.23. Surface movements are monitored by survey nets, tiltmeters (on benches), convergence meters, surface extensometers, and terrestrial photography. Subsurface deformations are monitored with inclinometers, deflectometers, shear-strip indicators, steel wire and weights in boreholes, and the acoustical emissions device. Pore-water pressures are monitored with piezometers.

All instruments should be monitored periodically and the data plotted as it is obtained to show changing conditions. Movement accelerations are most significant.

References

Augello, A. J., Bray, J. D., Leonards, G. A., Repetto, P. C., and Byrne, R. J. 1995. Response of landfills to seismic loading. Proceedings, Geoenvironment 2000, *ASCE* (in press).

Bishop, A. W. 1955. The use of the slip circle in the stability analysis of earth slopes. *Geotechnique.* 5(1):7–17.

Bjerrum, L. 1973. Problems of soil mechanics and construction on soft clays and structurally unstable soils. In *Proc. 8th Int. Conf. Soil Mech. Found. Eng.* 3:111–159.

Brand, E. W. 1982. Analysis and design in residual soils. In *Proc. Conf. Eng. Constr. Tropical Soils. ASCE*, Honolulu, pp. 89–143.

Bray, J. D., Augello, A. J., Repetto, P. C., Leonards, G. A., and Byrne, R. J. 1995. Seismic analytical procedures for solid waste landfills. *J. Geotech. Eng., ASCE* (in press).

Broms, B. B. 1975. *Foundation Engineering Handbook*, H. F. Winterkorn and H. Y. Fang, editors, Van Nostrand Reinhold, New York, pp. 373–401.

Broms, B. B. and Boman, P. 1979. Lime columns—A new foundation method. *J. Geotech. Eng., ASCE.* 105(GT4):539–556.

Byrne, P. M. 1991. A cyclic shear-volume coupling and pore pressure model for sand. In *Proc. 2nd Int. Conf. Recent Adv. Geotech. Earth. Eng. Soil Dynamics.* Rolla, I:47–56.

Chassie, R. G. and Goughnor, R. D. 1976. States intensifying efforts to reduce highway landslides. *Civ. Eng.*, Apr., p. 65.

Chirapuntu, S. and Duncan, J. M. 1976. The role of fill strength in the stability of embankments on soft clay. University of California, Berkeley, NITS AD-A027-087.

Cousins, B. F. 1978. Stability charts for simple earth slopes. *J. Geotech. Eng. ASCE.* 104(GT5):267–279.

Deere, D. U. 1976. Dams on rock foundations—Some design questions. In *Proc. Rock Eng. Found. Slopes. ASCE*, New York, pp. 55–85.

Duncan, J. M. 1992. State-of-the-art: Static stability and deformation analysis. *Proc. ASCE, Stability and Performance of Slopes and Embankments—II.* Geotech. Spec. Pub. No. 1(31):222–266.

Duncan, J. M., Buchignani, A. L., and De Wet, M. 1987. *An Engineering Manual for Slope Stability Studies.* Virginia Tech Report.

Dunnicliff, J. 1988. *Geotechnical Instrumentation for Monitoring Field Performance.* John Wiley & Sons, New York.

Espinoza, R. D., Repetto, P. C., and Muhunthan, B. 1992. General framework for stability analysis of slopes. *Geotechnique.* 42(4):603–616.

Espinoza, R. D., Bourdeau, P. L., and Muhunthan, B. 1994. Unified formulation for analysis of slopes with general slip surfaces. *J. Geotech. Eng. ASCE.* 120(7):1185–1204.

Fellenius, W. 1936. Calculation of the stability of earth dams. *Trans. 2nd Cong. Large Dams*, Vol. 4, Washington, D. C.

FHWA. 1993. *FHWA Tour for Geotechnology—Soil Nailing.* U.S. Department of Transportation, Federal Highway Administration, Washington, D.C.

Finn, W. D. L. 1993. Practical studies of the seismic response of a rockfill dam and a tailings impoundment. In *Proc. 3rd Int. Conf. Case Histories Geotech. Eng.* St. Louis.

Gray, D. H., Leiser, A. T., and White, C. A. 1980. Combined vegetative-structural slope stabilization. *Civ. Eng., ASCE.* October, pp. 73–77.

Hall, W. J. and Newmark, N. M. 1977. Seismic design criteria for pipelines and facilities. In *The Current State of Knowledge of Lifeline Earthquake Engineering, Proc. ASCE*, New York, pp. 18–34.

Handy, R. L. and Williams, W. W. 1967. Chemical stabilization of an active landslide. *Civ. Eng.*, August, pp. 62–65.

Hird, C. C. 1986. Stability charts for reinforced embankments on soft ground. *Geotext. Geomembr.* 4(2):107–127.

Hoek, E. and Bray, J. W. 1977. *Rock Slope Engineering*, 2nd ed. The Institute of Mining and Metallurgy, London.

Hollingsworth, R. and Kovacs, G. S. 1981. Soil slumps and debris flows: Prediction and protection. *Bul. Assoc. Eng. Geol.* 18(1):17–28.

Huang, Y. H. 1983. *Stability Analysis of Earth Slopes.* Van Nostrand-Reinhold, New York.

Hunt, R. E. 1984. *Geotechnical Engineering Investigation Manual.* McGraw-Hill, New York.

Hunt, R. E. 1986. *Geotechnical Engineering Techniques and Practices.* McGraw-Hill, New York.

Hunt, R. E. and da Costa Nunes, A. J. 1978. Retaining walls: Taking it from the top. *Civ. Eng.*, May, pp. 73–75.

Hunter, J. H. and Shuster, R. L. 1968. Stability of simple cuttings in normally consolidated clays. *Geotechnique.* 13(3):372–378.

Janbu, N. 1968. *Slope Stability Computations.* Soil Mechanics and Foundation Engineering Report, The Technical University of Norway.

Janbu, N. 1973. *Slope stability computations.* In *Embankment Dam Engineering,* eds. R. C. Hirschfield and S. J. Poulos, pp. 47–86. John Wiley & Sons, New York.

Janbu, N., Bjerrum, L., and Kjaernsli, B. 1956. *Soil Mechanics Applied to Some Engineering Problems,* pp. 5–26. Norwegian Geotechnical Institute, Oslo, Publ. 16.

Johnson, S. J. 1974. Analysis and design relating to embankments. In *Proc. ASCE Conf. Anal. Design Geotech. Eng.* Austin, Texas, Vol. II, pp. 1–48.

Koerner, R. M. 1993. *Designing with Geosynthetics,* 3rd ed. Prentice Hall, Englewood Cliffs, NJ.

Ladd, C. C. 1991. Stability evaluation during staged construction. *J. Geotech. Eng., ASCE.* 117(4): 540–615.

Lang, T. A. 1972. Rock reinforcement. *Bull. Assoc. Eng. Geol.* 9(3):215–239.

Lambe, T. W. and Whitman, R. V. 1969. *Soil Mechanics.* John Wiley & Sons, New York.

Leonards, G. A. 1979. Stability of slopes in soft clays. In *6th Panamerican Conf. Soil Mech. Found. Eng.* 1:225–274.

Leonards, G. A. 1982. Investigation of failures. In *J. Geotech. Eng., ASCE.* 108(GT2):187–246.

Lippomann, R. 1989. Dowelled clay slopes: Recent example. In *Proc. 12th Int. Conf. Soil Mech. Found. Eng.* Rio de Janeiro, 2:1269–1271.

Low, B. K., Wong, K. S., Lim, C., and Broms, B. B. 1990. Slip circle analysis of reinforced embankments on soft ground. *Geotext. Geomembr.* 9(2):165–181.

Lowe, J., III. 1967. Stability analysis of embankments. *J. Soil Mech. Found., ASCE.* 93(SM4):1–34.

Morgenstern, N. R. and Price, V. E. 1965. The analysis of the stability of general slip surfaces. *Geotechnique.* 15(1):79–93.

NAVFAC. 1982. *Design Manual, Soil Mechanics, Foundations and Earth Structures, DM-7.* Naval Facilities Engineering Command, Alexandria, VA.

Newmark, N. M. 1965. Effects of earthquakes on dams and embankments. *Geotechnique.* 15(2): 139–160.

Oakland, M. W. and Chameau, J. L. 1989. Analysis of drilled piers used for slope stabilization. *Trans. Res. Rec.* 1219:21–32.

Patton, F. D., and Hendron, A. J., Jr. 1974. General report on mass movements. In *Proc. 2nd Int. Cong. Int. Assoc. Eng. Geol.* São Paulo, p. V-GR 1.

Seed, H. B. 1966. "A method for earthquake resistant design of earth dams. *J. Soil Mech. Found., ASCE.* 92(SM1):13–41.

Seed, H. B., Idriss, I. M., Lee, K. L., and Makdisi, F. I. 1975. Dynamic analysis of the slide of the lower San Fernando Dam during the earthquake of February 9, 1971. *J. Geotech. Eng., ASCE.* 101(GT9):889–911.

Spencer, E. 1967. A method of analysis of the stability of embankments assuming parallel inter-slice forces. *Geotechnique.* 17(1):11–26.

Sunset. 1978. If hillside slides threaten in southern California, November is planting action month. *Sunset Magazine,* November, pp. 122–126.

Taylor, D. W. 1937. Stability analysis of earth slopes. *Journal of the Boston Society of Civil Engineers,* 24(3). Reprinted in *Contributions to Soil Mechanics 1925–1940,* BSCE, 1940, pp. 337–386.

Taylor, D. W. 1948. *Fundamentals of Soil Mechanics.* John Wiley & Sons, New York.

Terzaghi, K. 1962. Stability of steep slopes on hard, unweathered rock. *Geotechnique.* 12(4):251–270.

Tschebotarioff, G. P. 1973. *Foundations, Retaining and Earth Structures.* McGraw-Hill, New York.

Vulliet, L. and Hutter, K. 1988. Viscous-type sliding laws for landslides. *Can. Geotech. J.* 25(3):467–477.

Wilson, S. D. and Marsal, R. J. 1979. *Current Trends in Design and Construction of Embankment Dams. ASCE,* New York.

21

Retaining Structures

Jonathan D. Bray
University of California at Berkeley

21.1 Introduction

Civil engineering projects often require the construction of systems that retain earth materials. An excavation support system for a cut-and-cover trench for utilities installation is an example of a temporary retaining structure. A reinforced concrete retaining wall utilized in a highway project to accommodate a change in elevation over a limited distance is an example of a permanent retaining structure. Numerous earth retention systems have been developed over the years and a few systems are shown in Fig. 21.1. The design of retaining structures requires an evaluation of the loads likely to act on the system during its design life and the strength, load-deformation, and volume-change response of the materials to the imposed loads. Lateral pressures develop on retaining structures as a result of the adjacent earth mass, surcharge, water, and equipment. The development of lateral earth pressures and the transfer of these pressures to the retaining system are inherently governed by soil-structure interaction considerations. Hence, the analytical procedure should consider the relative rigidity/flexibility of the earth retention system. In this chapter, retaining structures will be broadly classified as either *rigid* or *flexible*. Before applicable design procedures are discussed, **lateral earth pressure** concepts will be reviewed.

21.2 Lateral Earth Pressures

The at-rest lateral earth pressure within a level earth mass of large areal extent can be estimated using the following relationship:

$$\sigma'_h = K_0 \sigma'_v \tag{21.1}$$

where σ'_h is the horizontal effective pressure, σ'_v is the vertical effective pressure, and K_0 is the **lateral earth pressure coefficient at rest**. Note that this relationship is valid in terms of effective stress only. The parameter K_0 is difficult to evaluate. Based on Jaky [1944], K_0 for normally consolidated soils, $K_{0,nc}$, can be approximated as

$$K_{0,nc} \approx 1 - \sin\phi' \tag{21.2}$$

0-8493-8953-4/95/$0.00 + $.50

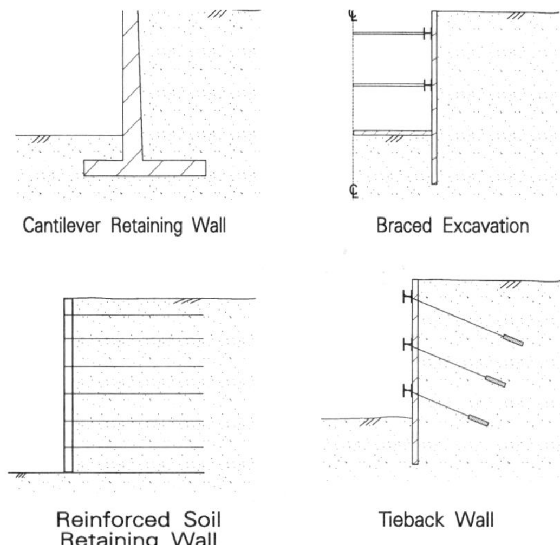

FIGURE 21.1 Examples of earth retention systems.

where ϕ' is the effective friction angle. This correlation appears to be more reasonable for clays and less reasonable for sands, but it is widely used in practice [see data presented in Mayne and Kulhawy, 1982]. A strong case could be made for just using $K_0 \approx 0.4$ for most normally consolidated sands. Schmidt's [1966] relationship as modified by Mayne and Kulhawy [1982] can be used to estimate the coefficient of lateral earth pressure at rest during unloading, $K_{0,u}$, as

$$K_{0,u} \approx K_{0,nc}(\text{OCR})^{\sin\phi'} \tag{21.3}$$

where OCR is the **overconsolidation ratio**. The at-rest lateral earth pressure coefficient during reloading is not known precisely, but it can be approximated as halfway between $K_{0,u}$ and $K_{0,nc}$ [Clough and Duncan, 1991]. K_0 may also be estimated on a site-specific basis using in situ test devices, such as the pressure meter or dilatometer [see Kulhawy and Mayne, 1990].

If a perfectly rigid vertical wall was wished into place (i.e., no lateral deformation occurred), then the at-rest in situ pressure distribution would be preserved. In homogeneous, dry soil with a constant K_0 and unit weight, γ, the horizontal effective stress would increase linearly with depth, z, proportional to the linearly increasing vertical effective stress (i.e., $\sigma'_h = K_0\gamma z$). This would result in a triangular pressure distribution. With the presence of a level water table, the total lateral pressure would consist of two components: horizontal effective earth pressure and water pressure ($u = \gamma_w \cdot z_w$, where u = pore water pressure, γ_w = unit weight of water, and z_w = depth below water table). Layered soil profiles can be easily handled by calculating the vertical effective stress at the depth of interest and multiplying this value by the parameter K_0 that best represents the soil layer at this depth. The compaction of earth fill behind a rigid wall that does not move will lock in higher lateral earth pressures than the at-rest condition [see Duncan and Seed, 1986].

If the rigid vertical retaining wall discussed above translates outward the lateral earth pressure on the back of the wall reduces, since the adjacent earth mass mobilizes strength as it deforms to follow the outward wall movement. The minimum active lateral earth pressure is reached when the soil has mobilized its maximum shear strength. Conversely, if the rigid wall translates inward, the lateral earth pressure on the the back of the wall increases as the adjacent earth mass mobilizes strength and resists the inward wall translation. At the limiting state, the maximum passive lateral earth pressure is attained. Hence, a range of possible magnitudes for the lateral earth pressure on the back of the wall exist, depending on the direction and magnitude of wall movement (see Fig. 21.2). The **minimum active earth pressure** defined by $p'_a = K_a\sigma'_v$ and the **maximum passive earth pressure** defined by $p'_p = K_p\sigma'_v$ represent only the limiting states of the possible lateral

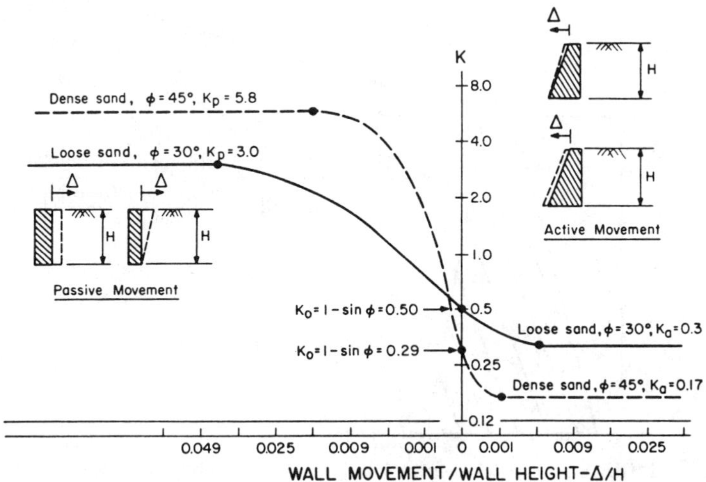

FIGURE 21.2 Relationship between wall movement and earth pressure. (After Clough, G. W. and Duncan, J. M. 1991. Earth Pressures. In *Foundation Engineering Handbook,* ed. H-Y. Fang, pp. 224–235. Van Nostrand Reinhold, New York.)

pressure range. Note that p_a' is attained at relatively low wall displacements, and that an order of magnitude more wall displacement is required to reach p_p'. Consequently, engineers often design retaining structures for the full active state when the earth retaining wall can move outward, but only for a fraction of the full passive state when the retaining wall moves inward.

21.3 Earth Pressure Theories

Rankine Theory

If the vertical wall is frictionless and the retained earth materials are level, homogeneous, isotropic, and characterized by the Mohr-Coulomb strength criterion (i.e., $s = c' + \sigma_n' \tan\phi'$), the limiting states of stress can be estimated using Rankine [1857] earth pressure theory. The minimum active earth pressure is

$$p_a' = K_a \sigma_v' - 2c' \sqrt{K_a} \tag{21.4}$$

where $K_a = \tan^2(45 - \phi'/2)$. Note that the vertical effective stress can include the effects of any applied surcharge as well as the gravity load of the earth materials. Typically, only the destabilizing effects of the active earth pressure (i.e., $p_a' > 0$) are included in developing design pressures, and a tension crack filled with water is conservatively assumed to exist down to the depth at which $p_a' = 0$. Layered soil profiles can be easily handled by calculating p_a' at the top and bottom of each soil layer and realizing that the lateral earth pressure varies linearly between these points. With freely draining cohesionless materials, effective strength parameters should be used and the earth and water pressure distributions should be computed separately. In a short-term undrained analysis of cohesive soils, total strength parameters may be used, but now the earth and water pressures will be calculated together since the total strength parameters include pore water effects.

The Rankine maximum passive earth pressure is

$$p_p' = K_p \sigma_v' + 2c' \sqrt{K_p} \tag{21.5}$$

where $K_p = \tan^2(45 + \phi'/2)$. Rankine theory underestimates the actual maximum passive earth pressure, so engineers often use the full Rankine passive earth pressure in design.

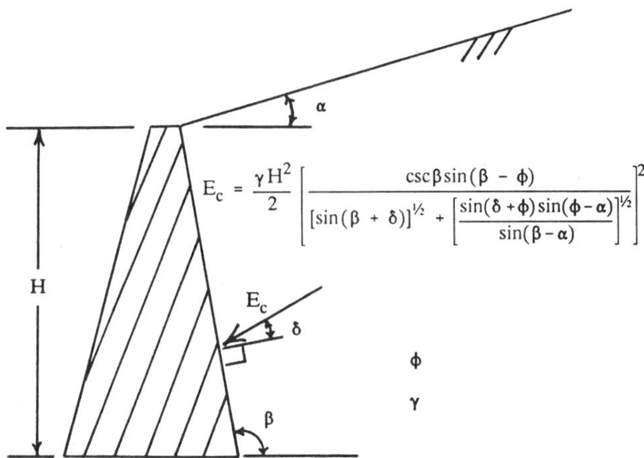

FIGURE 21.3 Closed-form solution for Coulomb minimum active earth force.

Coulomb Theory

Coulomb [1776] solved the lateral earth pressure problem assuming a homogeneous, isotropic material, rough wall, planar failure surface, and Mohr-Coulomb strength criterion. A wide range of earth pressure problems can be solved using the force polygon technique implied in Coulomb's method. The closed-form solution for the minimum active earth force for a general case including dry, cohesionless material, inclined rough wall, and sloping backfill is presented in Fig. 21.3. Coulomb theory can overestimate the actual maximum passive earth pressure acting on a rough wall with the wall friction angle, δ, greater than half of ϕ', so its use should be avoided. The use of log-spiral failure surfaces, as opposed to planar failure surfaces, provides good estimates of minimum active and maximum passive earth pressures, and graphical solution charts are available (see Fig. 21.4).

Example 21.1—Lateral Earth Pressure. Calculate the resultant active earth pressure by Rankine theory for the case shown in Fig. 21.5(a).
Solution.

Sand ($\phi' = 30°, c' = 0$): $K_a = \tan^2(45° - \phi'/2) = \tan^2(45° - 30°/2) = 0.333$;
$p'_a = K_a\gamma' z = K_a\rho' gz$

Clay ($\phi = 0°, c = 24$ kPa): $K_a = \tan^2(45 - 0°/2) = 1.0$; $p_a = \rho_t gz - 2c$

At $z = 3$ m, $p'_a = K_a\rho' gz = (0.333)\,(2.0\text{ Mg/m}^3)\,(9.81\text{ m/s}^2)\,(3\text{ m}) = 19.6$ kPa

At $z = 6^-$ m, $p'_a = 19.6\text{ kPa} + (0.333)\,(2.0 - 1.0\text{ Mg/m}^3)\,(9.81\text{ m/s}^2)\,(3\text{ m}) = 29.4$ kPa

At $z = 6^-$ m, $u = \rho_w g z_w = (1\text{ Mg/m}^3)\,(9.81\text{ m/s}^2)\,(3\text{ m}) = 29.4$ kPa

At $z = 6^+$ m, $p_a = \rho_t gz - 2c = (2\text{ Mg/m}^3)\,(9.81\text{ m/s}^2)\,(6\text{ m}) - 2(24\text{ kPa}) = 69.7$ kPa

At $z = 12^-$ m, $p_a = 69.7\text{ kPa} + (1.8\text{ Mg/m}^3)\,(9.81\text{ m/s}^2)\,(6\text{ m}) = 175.6$ kPa

$$P_{ae} = \tfrac{1}{2}(19.6\text{ kN/m}^2)\,(3\text{ m}) + \tfrac{1}{2}(19.6 + 29.4\text{ kN/m}^2)\,(3\text{ m}) + \tfrac{1}{2}(69.7 + 175.6\text{ kN/m}^2)\,(6\text{ m}) = 839\text{ kN/m}$$

$$P_w = \tfrac{1}{2}(29.4\text{ kN/m}^2)\,(3\text{ m}) = 44\text{ kN/m}$$

See Fig. 21.5(b) for pressure diagram.

21.4 Rigid Retaining Walls

Design lateral active earth pressures for low retaining walls (i.e., height <6 m) are often estimated using conservative design charts [see Department of the Navy, 1982] or using equivalent fluid

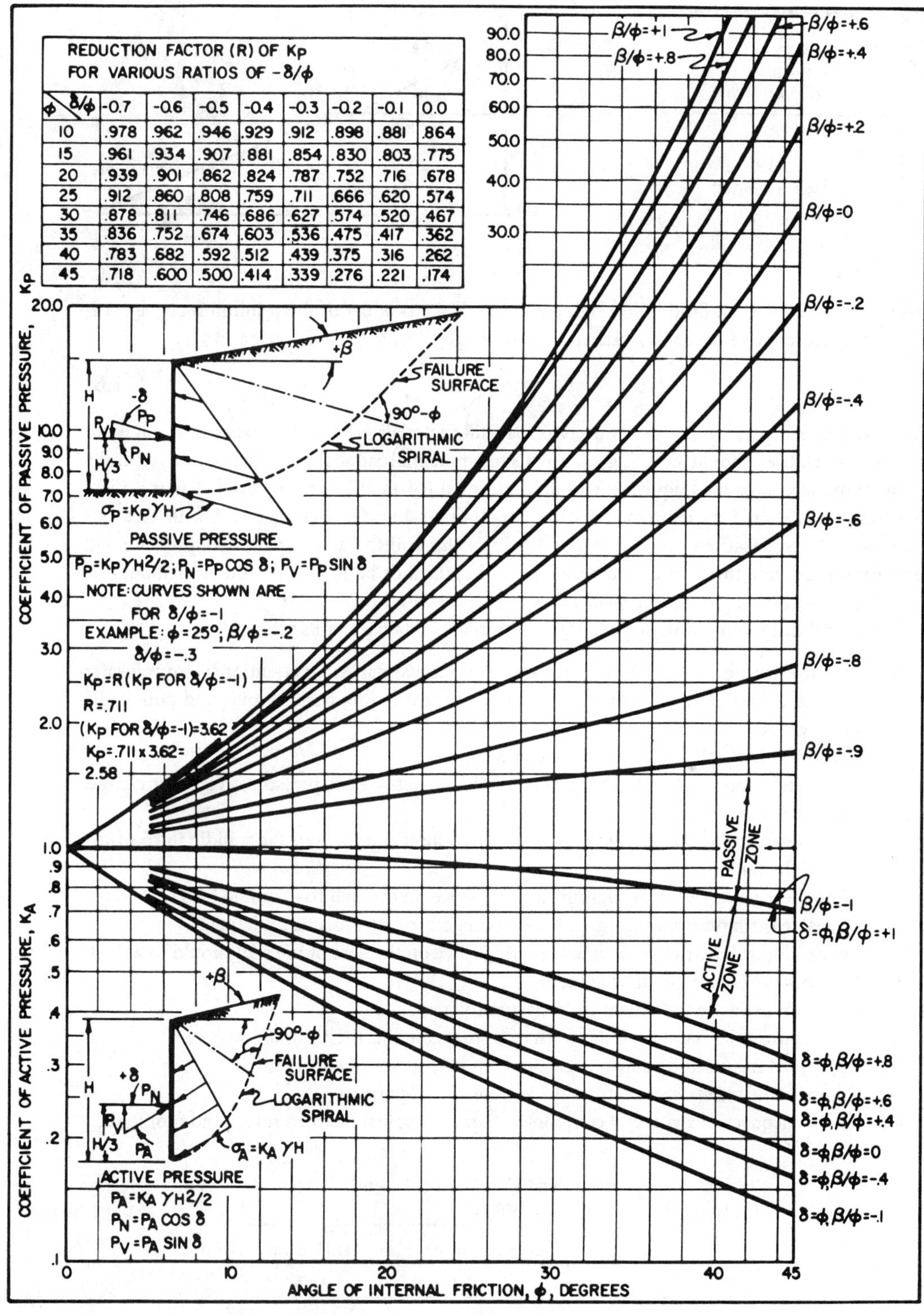

REDUCTION FACTOR (R) OF Kp FOR VARIOUS RATIOS OF −δ/φ

φ \ δ/φ	-0.7	-0.6	-0.5	-0.4	-0.3	-0.2	-0.1	0.0
10	.978	.962	.946	.929	.912	.898	.881	.864
15	.961	.934	.907	.881	.854	.830	.803	.775
20	.939	.901	.862	.824	.787	.752	.716	.678
25	.912	.860	.808	.759	.711	.666	.620	.574
30	.878	.811	.746	.686	.627	.574	.520	.467
35	.836	.752	.674	.603	.536	.475	.417	.362
40	.783	.682	.592	.512	.439	.375	.316	.262
45	.718	.600	.500	.414	.339	.276	.221	.174

FIGURE 21.4 Minimum active and maximum passive lateral earth pressure coefficients developed from log-spiral solution techniques. (After Department of the Navy. 1982. *Foundations and Earth Structures: Design Manual 7.2.* NAVFAC DM-7.2, May.)

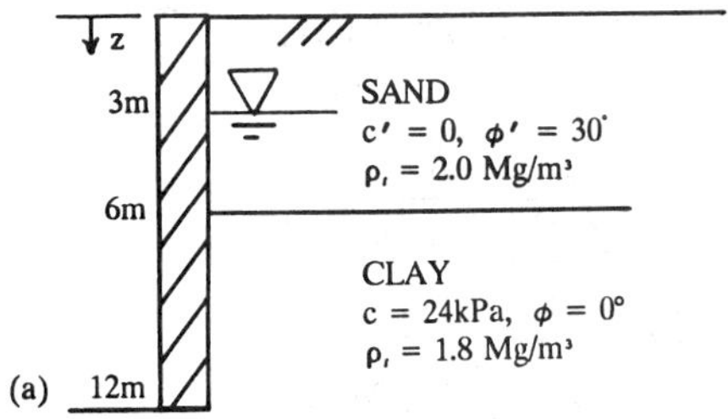

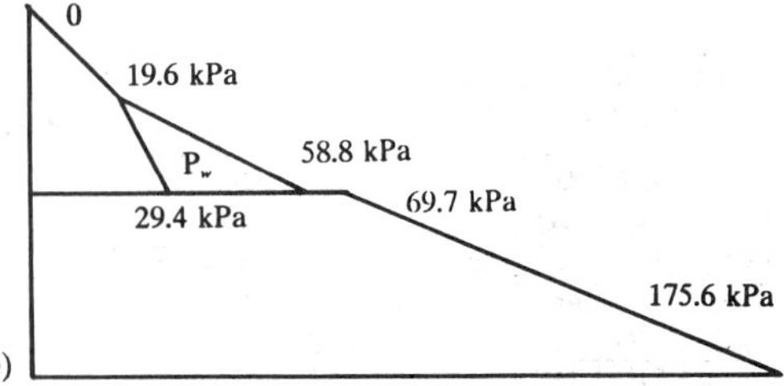

FIGURE 21.5 Example 21.1.

pressures. The equivalent fluid unit weight, γ_{eq}, equals the product of the minimum active earth pressure coefficient and the backfill material's unit weight (i.e., $\gamma_{eq} = K_a \cdot \gamma$), and

$$p_a = \gamma_{eq} \cdot z \tag{21.6}$$

Conservative estimates of γ_{eq} for a variety of backfill materials are listed in Table 21.1. All earth retention structures should be designed to sustain potential surcharge loadings, and typically a minimum surcharge load equivalent to an additional 0.6-m thickness of backfill is specified. Retaining walls should be constructed with free-draining backfill materials and with effective drainage systems, because if water can pond behind the wall the additional water pressure will dramatically increase the load on the wall. If ponding cannot be precluded, the wall should be designed to resist the higher total pressures.

The general design procedure for gravity and cantilever retaining walls follows:

1. Characterize project site and subsurface conditions. Pay particular attention to groundwater and surface water, site geology, availability of free-draining backfill soils, and potentially weak seams.
2. Select tentative wall dimensions (see Fig. 21.6).
3. Estimate the forces acting on the retaining wall (i.e., active earth pressure, weight, surcharge, and resultant).
4. Check overturning stability; the resultant force should act within the middle third of the base of the wall.
5. Check bearing capacity; maximum earth pressure on the wall base should be less than the allowable earth pressure regarding bearing capacity or permissible settlement.
6. Check sliding; horizontal frictional resisting force on the base of the wall should be at least 1.5 times the horizontal driving force.
7. Check for excessive settlement from deeper soil deposits.
8. Check the overall stability of the earth mass that contains the retaining structure.
9. Apply load factors and compute reactions, shears, and moments in the wall.
10. Compute the ultimate strength of the structural components.
11. Check adequacy of structural components against applied factored forces and moments.

Table 21.1 Equivalent Fluid Unit Weights (kN/m³) for Design of Low Retaining Walls

	Level Backfill		2H:1V Backfill
Soil	At-Rest	Active	Active
Clean sand or gravel	7.5	5	6
Silty sand	8.5	6	8
Clayey sand	9.5	7	9
Sandy clay	11	10	11
Fat clay	13	12	14

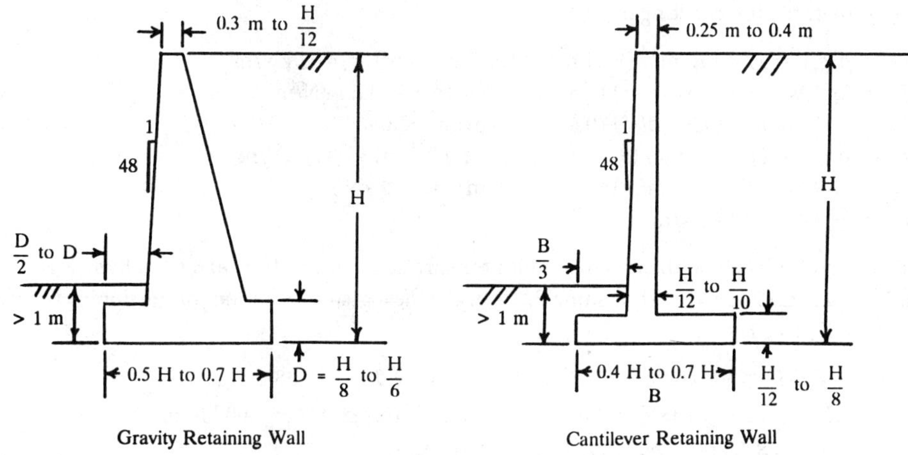

FIGURE 21.6 Tentative gravity and cantilever wall dimensions.

Design procedures differ for retaining systems built of reinforced soil [see Mitchell and Villet, 1987].

Example 21.2—Cantilever Retaining Wall. Check the adequacy of the cantilever retaining wall shown in Fig. 21.7 regarding overturning, bearing capacity, and sliding. The allowable bearing pressure is 360 kPa.

Solution. (a) The overall dimensions of the wall appear to be appropriate (see Fig. 21.6).

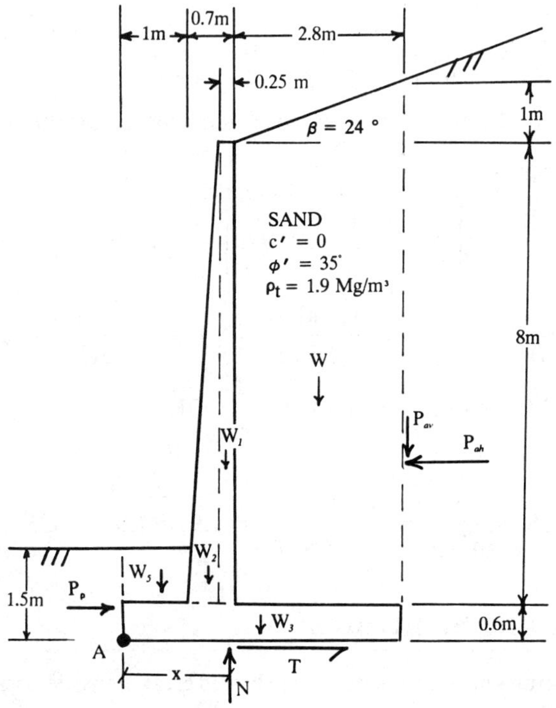

FIGURE 21.7 Example 21.2.

(b) Estimate the forces acting on wall:

$$
\begin{aligned}
W_1 &= \rho_c g A = (2.4\ \text{Mg/m}^3)\ (9.81\ \text{m/s}^2)\ (0.25\ \text{m} \cdot 8\ \text{m}) = 47.1\ \text{kN/m} \\
W_2 &= (2.4\ \text{Mg/m}^3)\ (9.81\ \text{m/s}^2)\ (0.5 \cdot 0.45\ \text{m} \cdot 8\ \text{m}) = 42.4\ \text{kN/m} \\
W_3 &= (2.4\ \text{Mg/m}^3)\ (9.81\ \text{m/s}^2)\ (0.6\ \text{m} \cdot 4.5\ \text{m}) = 63.6\ \text{kN/m} \\
W_4 &= \rho_t g A \approx (1.9\ \text{Mg/m}^3)\ (9.81\ \text{m/s}^2)\ (8.5\ \text{m} \cdot 2.8\ \text{m}) = 444\ \text{kN/m} \\
W_5 &= (1.9\ \text{Mg/m}^3)\ (9.81\ \text{m/s}^2)\ (0.9\ \text{m} \cdot 1.0\ \text{m}) = 16.8\ \text{kN/m} \\
W_T &= \sum W_i = 614\ \text{kN/m}
\end{aligned}
$$

Add 0.6 m of soil behind the wall to account for surcharge; hence, $H = 0.6\ \text{m} + 8\ \text{m} + 1\ \text{m} + 0.6\ \text{m} = 10.2\ \text{m}$. Conservatively assume $P_p \approx 0$. Use the log-spiral solution for the sloping backfill to estimate K_a (see Fig. 21.4).

$$
\begin{aligned}
\beta/\phi &= 24°/35° \cong 0.7 \text{ and } \phi' = 35° \text{ with } \delta = \phi' \rightarrow K_a = 0.38 \\
P_a &= K_a \gamma H^2/2 = (0.38)(1.9\ \text{Mg/m}^3)(9.81\ \text{m/s}^2)(10.2\ \text{m})^2/2 = 368\ \text{kN/m} \\
P_{ah} &= P_a \cos\delta = (368\ \text{kN/m}) \cos 35° = 301\ \text{kN/m} \\
P_{av} &= P_a \sin\delta = (368\ \text{kN/m}) \sin 35° = 211\ \text{kN/m} \\
N &= W_T + P_{av} = 614\ \text{kN/m} + 211\ \text{kN/m} = 825\ \text{kN/m} \\
T &= N \tan\delta_b = (825\ \text{kN/m}) \tan 35° = 578\ \text{kN/m}
\end{aligned}
$$

Location of resultant, N:

$$
\begin{aligned}
\sum M_A = 0 &= (47.1\ \text{kN/m})(1.575\ \text{m}) + (42.4\ \text{kN/m})(1.3\ \text{m}) \\
&\quad + (63.6\ \text{kN/m})(2.25\ \text{m}) + (444\ \text{kN/m})(3.1\ \text{m}) \\
&\quad + (16.8\ \text{kN/m})(0.5\ \text{m}) + (211\ \text{kN/m})(4.5\ \text{m}) \\
&\quad - (301\ \text{kN/m})(3.2\ \text{m}) - (825\ \text{kN/m})(x) \\
x &= 2.0\ \text{m}
\end{aligned}
$$

(c) Check overturning:

$$
\begin{aligned}
&B/3 = 4.5\ \text{m}/3 = 1.5\ \text{m} \qquad 2B/3 = 2 \cdot 1.5\ \text{m} = 3\ \text{m} \\
&1.5\ \text{m} < 2.0\ \text{m} < 3\ \text{m} \qquad \text{OK, since } N \text{ acts within middle third of base}
\end{aligned}
$$

(d) Check bearing capacity:

$$
e = \left|\frac{B}{2} - x\right| = \left|\frac{4.5\ \text{m}}{2} - 2.0\ \text{m}\right| = 0.25\ \text{m}
$$

$$
P_{\max} = \frac{N}{B}\left(1 + \frac{6e}{B}\right) = \frac{825\ \text{kN/m}}{4.5\ \text{m}}\left(1 + \frac{6 \cdot 0.25\ \text{m}}{4.5\ \text{m}}\right) = 245\ \text{kPa}
$$

$$
P_{\max} = 245\ \text{kPa} < q_a = 360\ \text{kPa} \qquad \text{OK}
$$

(e) Check sliding:

$$
\text{FS} = \frac{T}{P_{ah}} = \frac{578\ \text{kN/m}}{301\ \text{kN/m}} = 1.9 > 1.5 \qquad \text{OK}
$$

21.5 Flexible Retaining Structures

Flexible retaining structures include systems used in braced excavations, tie-back cuts, and anchored bulkheads. In this section, braced excavation systems will be discussed. Braced excavation support systems include walls, which may be steel sheetpiles, soldier piles with wood lagging, slurry placed tremie concrete, or secant/tangent piles; and supports, which may be cross-lot struts, rakers, diagonal bracing, tiebacks, or the earth itself in cantilever walls. Active earth pressure theories cannot be used directly to develop estimates of the lateral earth pressure acting on flexible retention structures. The pattern of wall movements during the excavation process does not satisfy Rankine-type assumptions of rigid wall translation or rigid wall rotation about its toe. With respect to the active Rankine state, the movement at the top of the wall is less and the movement at the base of the wall is more. Terzaghi [1943] showed that the resultant force on the flexible retaining structure is about 10% greater than the active Rankine resultant force and that the resultant force is located nearer to midheight of the wall rather than at its lower-third point. Theory is inadequate, since much depends on construction procedures, soil-structure interaction, and stress transfer. Moreover, more conservatism is desirable to guard against a progressive failure of the support system. Consequently, **apparent lateral earth pressure** diagrams, which envelope the maximum strut loads measured for excavation systems (in specific subsurface conditions), are used.

Terzaghi and Peck [1967] have developed the apparent pressure diagrams shown in Fig. 21.8 for sand and clay sites [see Tschebotarioff, 1951 for other diagrams]. Note that for sand, the ratio of the resultant force due to the apparent pressure distribution shown in Fig. 21.8 to that due to active Rankine earth pressures is 1.3. The corresponding ratio for clay is about 1.7. Individual strut loads are computed based on the associated tributary area of the apparent pressure diagram. This is merely a reversal of the procedure used to develop the apparent pressure diagrams. The apparent pressure diagrams are based on field measurements of maximum strut loads, so normal surcharge loads are already included. Some engineers increase the strength of the upper struts by 15% to guard against surcharge overload.

The design wall and wale moments are typically calculated using the assumption that the wall and wale are simply supported between adjacent wales and struts, respectively. If the wall or wale is continuous over at least three supports, then the moment formula for a continuous beam can be used to calculate moments. Since the design of the wall and wale do not require the level of conservatism needed to guard against progressive failure of the struts, only two-thirds of the magnitude of the apparent pressure diagram is used in the computation. Hence, the maximum wall or wale moment, $M_{\max}$, can be calculated by the following formula:

$$M_{\max} = \frac{\frac{2}{3} \cdot \mathrm{AP} \cdot l_{\max}^2}{(8 \text{ or } 10)} \tag{21.7}$$

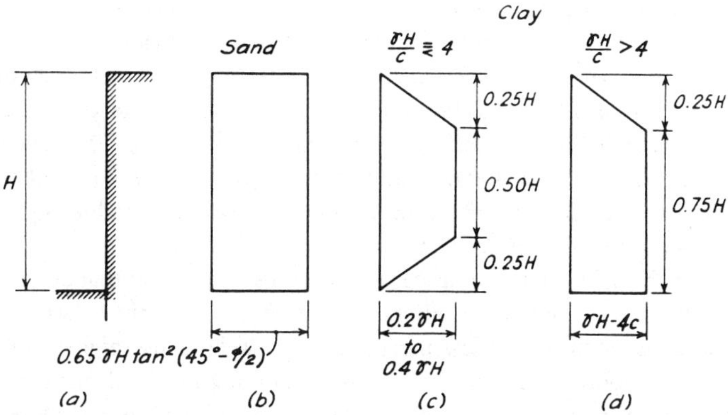

FIGURE 21.8 Lateral pressure distribution for computation of strut loads in braced excavation systems. (After Terzaghi, K. and Peck, R. 1967. *Soil Mechanics in Engineering Practice.* John Wiley & Sons, New York. Copyright © 1967 John Wiley & Sons, Inc. Reprinted by permission of John Wiley & Sons, Inc.)

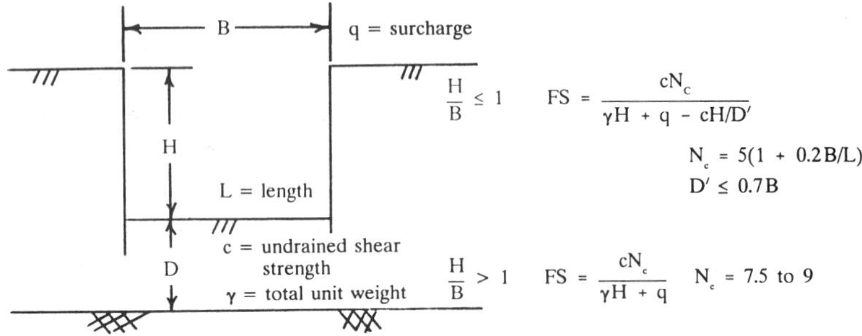

FIGURE 21.9 Factor of safety against base heave.

where AP = apparent distributed load, $l_{\max}$ = maximum span length, and the denominator is 8 for the simply supported condition or 10 for the continuous beam condition. Excavations in deep soft to medium stiff clays may present situations in which the maximum wall moment occurs prior to installing the bottom strut, so wall moments should not just be computed for the final bracing configuration. Lastly, structural details are critically important in braced excavation systems. Typically, stiffeners are added to the wales at strut locations, and long internal struts are braced at points along their length.

The overall stability of the excavation must also be evaluated. Some potential failure mechanisms that should be investigated include **base heave**, bottom blowout, and piping. Calculating the factor of safety against base heave in deep clay deposits is especially important because a low safety factor indicates marginal stability and the potential for excessive movements (see Fig. 21.9). The engineer's primary concern in urban areas or where sensitive structures are near the excavation is often limiting ground movements.

Clough and O'Rourke [1990] present empirical data and analytical results that assist engineers in estimating excavation-induced ground movements. The maximum lateral wall movement, $\Delta_{h,m}$, in well-constructed excavations in stiff clays, residual soils, and sands is typically 0.1% to 0.5% of the height of the excavation, H, with most of the data suggesting $\Delta_{h,m} \approx 0.2\%H$. The stiffness of the support system is not critically important in those soil deposits. Conversely, the support system stiffness is important in controlling movements with excavations in soft to medium stiff clays. In soft to medium stiff clays, the maximum lateral wall movement can range from $0.3\%H$ to over $3\%H$ depending on the factor of safety against base heave and the support system stiffness (see Fig. 21.10). When the factor of safety against base heave is less than 1.5, special care should be exercised in controlling the excavation procedures to minimize movements. Preloading struts, not allowing overexcavation, and employing good construction details (e.g., steel shims) have proven useful in minimizing ground movements. Vertical movements of the ground surface, Δ_v, surrounding the excavation are largely a function of the lateral wall movements, and $\Delta_{v,m} \approx \Delta_{h,m}$. However, vertical ground movements can be much higher if excavation dewatering induces consolidation settlement in underlying clay deposits. Driving sheetpiles in loose sands can also induce significant ground settlement [see Clough and O'Rourke, 1990].

The design of tieback walls used in temporary excavations is similar to that described previously for braced excavations, but there are a number of significant differences [see Juran and Elias, 1991]. Tieback systems often permit less movement because they use higher preloads, positive connections, smaller support spacing, and less overexcavation. The tieback anchor itself, however, is more flexible than an internal strut, and its capacity depends greatly on the bond developed between the soil and grouted anchorage. Tieback systems consequently require good soil conditions and the absence of obstructions in the surrounding ground. Typically, the tieback system is checked against anchor pullout by proof testing each anchor to 100 to 150% of the design load, and the required lengths of the anchors are determined by ensuring their anchorage zones are located

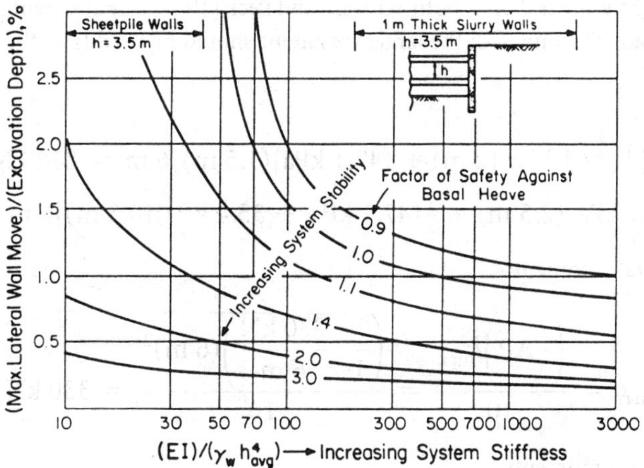

FIGURE 21.10 Design curves to obtain maximum lateral wall movement for soft to medium stiff clays. (After Clough, G. W., Smith, E. M., and Sweeny, B. P. 1989. Movement Control of Excavation Support Systems by Iterative Design. *Proc. ASCE Foundation Engineering: Current Principles and Practices,* Vol. 2, pp. 869–884. Reproduced by permission of ASCE.)

behind potential failure surfaces. Because preloading anchors to roughly 80% of their design load maintains a nearly at-rest stress state in the ground, lateral earth pressures are often assumed to be near at-rest pressures. Numerous other support systems have been developed and, depending on the availability of specialized contractors, these systems may be advantageous. For example, in good soils, soil nailing with a reinforced shotcrete wall has proved to be effective and cost-efficient.

Example 21.3—Braced Excavation. For the braced excavation shown in Fig. 21.11(a), develop estimates of the strut loads, maximum wale moment, maximum wall moment, and maximum excavation-induced ground movements.

Solution. (a) Apparent pressures:

$$\text{Stability number } N = \frac{\gamma H}{c} = \frac{(2\ \text{Mg/m}^3)(9.81\ \text{m/s}^2)(8\ \text{m})}{35\ \text{kPa}} = 4.5$$

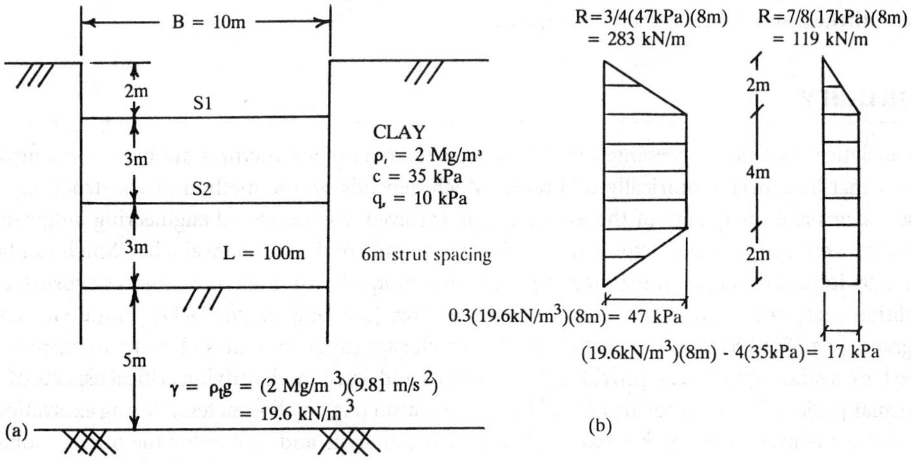

FIGURE 21.11 Example 21.3.

Since N is only slightly larger than 4, both Terzaghi and Peck [1967] clay apparent pressure diagrams should be calculated. The one with the largest resultant should be used (i.e., the left one in which $R = 283$ kN/m).

(b) Strut loads:

$$S1 = 1.15[\tfrac{1}{2}(47.1 \text{ kPa})(2 \text{ m}) + (47.1 \text{ kPa})(1.5 \text{ m})]6 \text{ m} = 810 \text{ kN}$$

$$S2 = [(47.1 \text{ kPa})(2.5 \text{ m}) + \tfrac{1}{2}(47.1 \text{ kPa} + 35.4 \text{ kPa})(0.5 \text{ m})]6 \text{ m} = 830 \text{ kN}$$

(c) Maximum wale moment:

$$M_{max} = \frac{\left(\frac{2}{3}\text{AP}\right)l_{max}^2}{10} = \frac{\left(\frac{2}{3} \cdot \frac{830 \text{ kN}}{6 \text{ m}}\right)(6 \text{ m})^2}{10} = 330 \text{ kN-m}$$

(d) Maximum wall moment:

$$M_{max} = \frac{\left(\frac{2}{3}\text{AP}\right)l_{max}^2}{8} = \frac{\left(\frac{2}{3} \cdot 47.1 \text{ kPa}\right)(3 \text{ m})^2}{8} = 35 \text{ kN-m/m}$$

(e) Estimate ground movements:

$$N_c = 5(1 + 0.2B/L) = 5(1 + (0.2)(10 \text{ m})/100 \text{ m}) = 5.1$$

$$D' = 5 \text{ m} \leq 0.7(10 \text{ m}) = 7 \text{ m}$$

$$\text{FS}_{BH} = \frac{cN_C}{\gamma H + q - cH/D'}$$

$$= \frac{(35 \text{ kPa})(5.1)}{(2 \text{ Mg/m}^3)(9.81 \text{ m/s}^2)(8 \text{ m}) + 10 \text{ kPa} - (35 \text{ kPa})(8 \text{ m})/5 \text{ m}} = 1.6$$

Using Fig. 21.10, for a typical sheetpile wall with $\text{FS}_{BH} \approx 1.6$, $\Delta_{h,m} \approx 1.0\%H$, or

$$\Delta_{h,m} \approx 0.01(8 \text{ m}) = 0.08 \text{ m} = 8 \text{ cm}$$

In urban areas, maximum wall movements should normally be kept less than 5 cm. A heavy sheetpile wall (e.g., PZ 40) or a relatively stiff concrete slurry wall could be used to increase the system stiffness and reduce ground movements.

21.6 Summary

The analytical techniques presented in this section for retaining structures are based on simple models that have been empirically calibrated. Much depends on the method of construction of these systems and the quality of the workmanship involved. Hence, sound engineering judgment should be exercised and local experience in similar ground conditions is invaluable. Much can be gained by implementing an integrated approach that uniquely considers the project's subsurface conditions, site constraints, and excavation procedures [see Bray *et al.*, 1993]. Finite element programs [e.g., SOILSTRUCT, Filz *et al.*, 1990], which capture the unique soil-structure response of each excavation system, can provide salient insights and assist in identifying critical aspects of a particular project. The monitoring of field instrumentation (e.g., inclinometers) during excavation allows the engineer to verify the reasonableness of the analysis and to employ the observational method to optimize the design during construction.

Defining Terms

Apparent lateral earth pressure: Lateral earth pressure acting on tributary area of flexible retaining wall that is necessary to develop measured strut loads.

Base heave: Upward movement of base of excavation and associated inward movement of retaining wall due to bearing capacity-type instability of base soil.

Lateral earth pressure coefficient: Horizontal effective stress divided by vertical effective stress at a point.

Lateral earth pressure coefficient at rest: Lateral earth pressure coefficient when the lateral strain in the soil is zero. Realized for case of 1-D vertical compression (e.g., level ground).

Maximum passive earth pressure coefficient: Maximum value of the lateral earth pressure coefficient. Realized when soil compresses laterally and its full strength is mobilized.

Minimum active earth pressure coefficient: Minimum value of the lateral earth pressure coefficient. Realized when soil expands laterally and its full strength is mobilized.

Overconsolidation ratio: Maximum vertical effective stress in the past divided by the current vertical effective stress.

References

Barker, R. M., Duncan, J. M., Rojiani, V. B., Ooi, P. S. K., Tan, C. K., and Kim, S. G. 1991. *Manuals for the Design of Bridge Foundations.* National Cooperative Highway Research Program Report 343, TRB, Washington, D.C., December.

Bray, J. D., Deschamps, R. J., Parkison, R. S., and Augello, A. J. 1993. Braced excavation at the NIPSCO Bailly Station power plant. In *Proc. 3rd Int. Conf. Case Histories Geotech. Eng.,* pp. 765–774. St. Louis, MO.

Clough, G. W. and Duncan, J. M. 1991. Earth pressures. In *Foundation Engineering Handbook,* ed. H-Y. Fang, pp. 224–235. Van Nostrand Reinhold, New York.

Clough, G. W. and O'Rourke, T. D. 1990. Construction induced movements of in situ walls. In *Proc. ASCE Design Performance Earth Retaining Struct.,* ed. P. C. Lambe and L. A. Hansen, pp. 439–470. Ithaca, NY, June 18–21.

Clough, G. W., Smith, E. M., and Sweeny, B. P. 1989. Movement control of excavation support systems by iterative design. *Proc. ASCE Found. Eng.: Curr. Principles Pract.,* 2:869–884.

Coulomb, C. A. 1776. Essai sur une application des règles des maximis et minimis à quelques problèmes de statique relatifs à l'architecture. *Mèm. Acad. Roy. des Sciences,* Paris. 3:38.

Department of the Navy. 1982. *Foundations and Earth Structures: Design Manual 7.2.* NAVFAC DM-7.2, May.

Duncan, J. M. and Seed, R. B. 1986. Compaction-induced earth pressure under K_0-conditions. *J. Geotech. Eng., ASCE.* 112(1):1–22.

Filz, G., Clough, G. W., and Duncan, J. M. 1990. Draft user's manual for program SOILSTRUCT (isotropic) plane strain with beam element. *Geotech. Eng. Rep.* Virginia Tech, Blacksburg, March.

Jaky, J. 1944. The coefficient of earth pressure at-rest. *J. Soc. Hungarian Architects Engineers.* Budapest, Hungary, pp. 355–358.

Juran, I. and Elias, V. 1991. Ground anchors and soil nails in retaining structures. In *Foundation Engineering Handbook,* ed. H-Y. Fang, pp. 868–905. Van Nostrand Reinhold, New York.

Kulhawy, F. H. and Mayne, P. W. 1990. Manual on Estimating Soil Properties for Foundation Design. *EL-6800, Research Project 1493-6 Final Report.* Electric Power Research Institute, Palo Alto, CA.

Lambe, P. C. and Hansen, L. A., ed. 1990. *Design and Performance of Earth Retaining Structures.* ASCE Geotechnical Special Publication No. 25, New York.

Mayne, P. W. and Kulhawy, F. H. 1982. K_0-OCR relationships in soil. *J. Geotech. Eng., ASCE.* 108(GT6):851–872.

Mitchell, J. K. and Villet, W. C. B. 1987. Reinforcement of Earth Slopes and Embankments. *National Cooperative Highway Research Program Report 290,* TRB, Washington, DC, June.

Peck, R. B., Hanson, W. E., and Thornburn, T. H. 1974. *Foundation Engineering.* John Wiley & Sons, New York.

Rankine, W. J. M. 1857. On the Stability of Loose Earth. *Philos. Trans. R. Soc.,* London, 147 (1):9–27.

Rowe, P. W. 1957. Limit design of flexible walls. *Proc. Midland Soil Mech. Found. Eng. Soc.* 1:29–40.

Schmidt, B. 1966. Discussion of "Earth pressures at-rest related to stress history," *Can. Geotech. J.* 3(4):239–242.

Terzaghi, K. 1943. *Theoretical Soil Mechanics.* John Wiley & Sons, New York.

Terzaghi, K. and Peck, R. 1967. *Soil Mechanics in Engineering Practice.* John Wiley & Sons, New York.

Tschebotarioff, G. P. 1951. *Soil Mechanics, Foundations and Earth Structures.* McGraw-Hill, New York.

For Further Information

Foundation Engineering by Peck *et al.* [1974] provides a good overview of earth pressure theories and retaining structures, with illustrated design examples.

Retaining walls and abutments are discussed in Barker *et al.* [1991], and reinforcement of earth slopes and embankments are discussed by Mitchell and Villet [1987].

The design of anchored bulkheads is presented in Department of the Navy [1982].

State-of-the-art papers on the design and performance of earth retaining structures are presented in Lambe and Hansen [1990].

22 Foundations

Bengt H. Fellenius
University of Ottawa

Before a foundation design can be undertaken, the associated soil profile must be well established. The soil profile is compiled from three cornerstones of information: borehole records with laboratory classification and testing, piezometer observations, and assessment of the overall geology at the site. Projects where construction difficulties, disputes, and litigation arise often have one thing in common: Copies of borehole field records were thought to determine the soil profile.

An essential part of the foundation design is to come up with a foundation type and size which will have acceptable values of deformation (settlement) and an adequate margin of safety to failure (the degree of utilization of the soil strength). Deformation is a function of *change* of effective stress, and soil strength is a *proportional* to effective stress. Therefore, all applications of foundation design start with determining the effective stress distribution of the soil around and below the foundation unit. The distribution then serves as basis for the design analysis.

22.1 Effective Stress

Effective stress is the total stress minus the pore pressure (the water pressure in the voids). The common method of calculating the effective stress, $\Delta\sigma'$, contributed by a soil layer is to multiply the buoyant unit weight, γ', of the soil with the layer thickness, Δh. Usually, the buoyant unit weight is determined as the bulk unit weight of the soil, γ_t, minus the unit weight of water, γ_w, which presupposes that there is no vertical gradient of water flow in the soil.

$$\Delta\sigma' = \gamma'\Delta h \tag{22.1a}$$

The effective stress at a depth, σ'_z, is the sum of the contributions from the soil layers above.

$$\sigma'_z = \Sigma(\gamma'\Delta h) \tag{22.1b}$$

0-8493-8953-4/95/$0.00 + $.50

However, most sites display vertical water gradients: either an upward flow, maybe even artesian (the head is greater than the depth), or a downward flow, and the buoyant unit weight is a function of the gradient, i, in the soil, as follows.

$$\gamma' = \gamma_t - \gamma_w(1 - i) \tag{22.1c}$$

The gradient is defined as the difference in head between two points divided by the distance the water has to flow between these two points. Upward flow gradient is negative and downward flow is positive. For example, if for a particular case of artesian condition the gradient is nearly equal to -1, then, the buoyant weight is nearly zero. Therefore, the effective stress is close to zero, too, and the soil has little or no strength. This is the case of so-called quicksand, which is not a particular type of sand, but a soil, usually a silty fine sand, subjected to a particular pore pressure condition.

The gradient in a nonhydrostatic condition is often awkward to determine. However, the difficulty can be avoided, because the effective stress is most easily determined by calculating the total stress and the pore water pressure separately. The effective stress is then obtained by simple subtraction of latter from the former.

Note the difference in terminology—effective *stress* and pore *pressure*—which reflects the fundamental difference between forces in soil as opposed to in water. Soil stress is directional; that is, the stress changes depending on the orientation of the plane of action in the soil. In contrast, water pressure is omnidirectional, that is, independent of the orientation of the plane. The soil stress and water pressures are determined, as follows.

The *total vertical stress* (symbol σ_z) at a point in the soil profile (also called total overburden stress) is calculated as the stress exerted by a soil column with a certain weight, or bulk unit weight, and height (or the sum of separate values where the soil profile is made up of a series of separate soil layers having different bulk unit weights). The symbol for bulk unit weight is γ_t (the subscript t stands for "total," because the bulk unit weight is also called *total unit weight*).

$$\sigma_z = \gamma_t z \tag{22.2}$$

or

$$\sigma_z = \Sigma\Delta\sigma_z = \Sigma(\gamma_t \Delta h)$$

Similarly, the **pore pressure** (symbol u), if measured in a stand-pipe, is equal to the unit weight of water, γ_w, times the height of the water column, h, in the stand-pipe. (If the pore pressure is measured directly, the head of water is equal to the pressure divided by the unit weight of water, γ_w.)

$$u = \gamma_w h \tag{22.3}$$

The height of the column of water (the head) representing the water pressure is rarely the distance to the ground surface or, even, to the groundwater table. For this reason, the height is usually referred to as the *phreatic height* or the *piezometric height* to separate it from the depth below the groundwater table or depth below the ground surface.

The groundwater table is defined as the uppermost level of zero pore pressure. Notice that the soil can be saturated with water also above the groundwater table without pore pressure being greater than zero. Actually, because of capillary action, pore pressures in the partially saturated zone above the groundwater table may be negative.

The pore pressure distribution is determined by applying that (in stationary situations) the pore pressure distribution is linear in each individual soil layer, and, in pervious soil layers that are "sandwiched" between less pervious layers, the pore pressure is hydrostatic (that is, the gradient is zero).

The *effective overburden stress* (symbol σ'_z) is then obtained as the difference between total stress and pore pressure.

$$\sigma'_z = \sigma_z - u_z = \gamma_t z - \gamma_w h \tag{22.4}$$

Usually, the geotechnical engineer provides the density (symbol ρ) instead of unit weight, γ. The unit weight is then the density times the gravitational constant, g. (For most foundation engineering purposes, the gravitational constant can be taken to be 10 m/s^2 rather than the overly exact value of 9.81 m/s^2.)

$$\gamma = \rho g \tag{22.5}$$

Many soil reports do not indicate the total soil density, ρ_t, only water content, w, and dry density, ρ'_d. For saturated soils, the total density can then be calculated as

$$\rho_t = \rho_d(1 + w). \tag{22.6}$$

The principles of effective stress calculation are illustrated by the calculations involved in the following soil profile: an upper 4 m thick layer of normally consolidated sandy silt, 17 m of soft, compressible, slightly overconsolidated clay, followed by 6 m of medium dense silty sand and a thick deposit of medium dense to very dense sandy ablation glacial till. The groundwater table lies at a depth of 1.0 m. For *original conditions,* the pore pressure is hydrostatically distributed throughout the soil profile. For *final conditions,* the pore pressure in the sand is hydrostatically distributed but artesian with a phreatic height above ground of 5 m, which means that the pore pressure in the clay is non-hydrostatic (but linear, assuming that the final condition is long term). The pore pressure in the glacial till is also hydrostatically distributed. A 1.5 m thick earth fill is to be placed over a square area with a 36 m side. The densities of the four soil layers and the earth fill are: 2000 kg/m^3, 1700 kg/m^3, 2100 kg/m^3, 2200 kg/m^3, and 2000 kg/m^3, respectively.

Calculate the distribution of total and effective stresses as well as pore pressure underneath the center of the earth fill before and after placing the earth fill. Distribute the earth fill by means of the 2:1 method; that is, distribute the load from the fill area evenly over an area that increases in width and length by an amount equal to the depth below the base of fill area.

Table 22.1 presents the results of the stress calculation for the example conditions. The calculations have been made with the Unisettle program [Goudreault and Fellenius, 1993] and the results are presented in the format of a "hand calculation" to ease verifying the computer calculations. Notice that performing the calculations at every meter depth is normally not necessary. The table includes a comparison between the non-hydrostatic pore pressure values and the hydrostatic, as well as the effect of the earth fill, which can be seen from the difference in the values of total stress for original and final conditions.

The stress distribution below the center of the loaded area was calculated by means of the 2:1 method. However, the 2:1 method is rather approximate and limited in use. Compare, for example, the vertical stress below a loaded footing that is either a square or a circle with a side or diameter of B. For the same contact stress, q_0, the 2:1 method, strictly applied to the side and diameter values, indicates that the vertical distributions of stress, $[q_z = q_0/(B + z)^2]$, are equal for the square and the circular footings. Yet, the total applied load on the square footing is $4/\pi = 1.27$ times larger than the total load on the circular footing. Therefore, if applying the 2:1 method to circles and other non-rectangular areas, they should be modeled as a rectangle of an equal size ("equivalent") area. Thus, a circle is modeled as an equivalent square with a side equal to the circle radius times $\sqrt{\pi}$.

More important, the 2:1 method is inappropriate to use for determining the stress distribution along a vertical line below a point at any other location than the center of the loaded area. For this reason, it can not be used to combine stress from two or more loaded areas unless the areas have the same center. To calculate the stresses induced from more than one loaded area and/or below an off-center location, more elaborate methods, such as the Boussinesq distribution (Chapter 19) are required.

Table 22.1 Stress Distribution (2:1 Method) at Center of Earth Fill

	Original Condition (no earth fill)			Final Condition (with earth fill)		
Depth (m)	σ_t (kPa)	u (kPa)	σ' (kPa)	σ_t (kPa)	u (kPa)	σ' (kPa)
		Layer 1 Sandy Silt $\rho = 2000$ kg/m^3				
0.00	0.0	0.0	0.0	30.0	0.0	30.0
1.00 (GWT)	20.0	0.0	20.0	48.4	0.0	48.4
2.00	40.0	10.0	30.0	66.9	10.0	56.9
3.00	60.0	20.0	40.0	85.6	20.0	65.6
4.00	80.0	30.0	50.0	104.3	30.0	74.3
		Layer 2 Soft Clay $\rho = 1700$ kg/m^3				
4.00	80.0	30.0	50.0	104.3	30.0	74.3
5.00	97.0	40.0	57.0	120.1	43.5	76.6
6.00	114.0	50.0	64.0	136.0	57.1	79.0
7.00	131.0	60.0	71.0	152.0	70.6	81.4
8.00	148.0	70.0	78.0	168.1	84.1	84.0
9.00	165.0	80.0	85.0	184.2	97.6	86.6
10.00	182.0	90.0	92.0	200.4	111.2	89.2
11.00	199.0	100.0	99.0	216.6	124.7	91.9
12.00	216.0	110.0	106.0	232.9	138.2	94.6
13.00	233.0	120.0	113.0	249.2	151.8	97.4
14.00	250.0	130.0	120.0	265.6	165.3	100.3
15.00	267.0	140.0	127.0	281.9	178.8	103.1
16.00	284.0	150.0	134.0	298.4	192.4	106.0
17.00	301.0	160.0	141.0	314.8	205.9	109.0
18.00	318.0	170.0	148.0	331.3	219.4	111.9
19.00	335.0	180.0	155.0	347.9	232.9	114.9
20.00	352.0	190.0	162.0	364.4	246.5	117.9
21.00	369.0	200.0	169.0	381.0	260.0	121.0
		Layer 3 Silty Sand $\rho = 2100$ kg/m^3				
21.00	369.0	200.0	169.0	381.0	260.0	121.0
22.00	390.0	210.0	180.0	401.6	270.0	131.6
23.00	411.0	220.0	191.0	422.2	280.0	142.2
24.00	432.0	230.0	202.0	442.8	290.0	152.8
25.00	453.0	240.0	213.0	463.4	300.0	163.4
26.00	474.0	250.0	224.0	484.1	310.0	174.1
27.00	495.0	260.0	235.0	504.8	320.0	184.8
		Layer 4 Ablation Till $\rho = 2200$ kg/m^3				
27.00	495.0	260.0	235.0	504.8	320.0	184.8
28.00	517.0	270.0	247.0	526.5	330.0	196.5
29.00	539.0	280.0	259.0	548.2	340.0	208.2
30.00	561.0	290.0	271.0	569.9	350.0	219.9
31.00	583.0	300.0	283.0	591.7	360.0	231.7
32.00	605.0	310.0	295.0	613.4	370.0	243.4
33.00	627.0	320.0	307.0	635.2	380.0	255.2

Note: Calculations by means of UNISETTLE.

A footing is usually placed in an excavation and fill is often placed next to the footing. When calculating the stress increase from one or more footing loads, the changes in effective stress from the excavations and fills must be included which therefore precludes the use of the 2:1 method (unless all such excavations and fills surround and are concentric with the footing).

A small diameter footing, of about 1 meter width, can normally be assumed to distribute the stress evenly over the footing contact area. However, this cannot be assumed to be the case for wider footings. The Boussinesq distribution assumes ideally flexible footings (and ideally elastic soil). Kany [1959] showed that below a so-called characteristic point, the vertical stress distribution is equal for flexible and stiff footings. The characteristic point is located at a distance of $0.37B$ and $0.37L$ from the center of a rectangular footing of sides B and L and at a radius of $0.37R$ from the center of a circular footing of radius R. When applying Boussinesq method of stress distribution to regularly shaped footings, the stress below this point is normally used rather than the stress below the center of the footing.

Calculation of the stress distribution below a point within or outside the footprint of a footing by means of the Boussinesq method is time-consuming, in particular if the stress from several loaded areas are to be combined. The geotechnical profession has for many years simplified the calculation effort by using nomograms over "influence values for vertical stress" at certain locations within the footprint of a footing. The Newmark influence chart [Newmark, 1935; 1942] is a classic. The calculations are still rather time consuming. However, since the advent of the computer, several computer programs are available which greatly simplify and speed up the calculation effort—for example, Unisettle by Goudreault and Fellenius [1993].

22.2 Settlement of Foundations

A **foundation** is a constructed unit that transfers the load from a superstructure to the ground. With regard to vertical loads, most foundations receive a more or less concentrated load from the structure and transfer this load to the soil underneath the foundation, distributing the load as a stress over a certain area. Part of the soil structure interaction is then the condition that the stress must not give rise to a deformation of the soil in excess of what the superstructure can tolerate.

The amount of deformation for a given contact stress depends on the distribution of the stress over the affected soil mass in relation to the existing stress (the imposed change of effective stress) and the compressibility of the soil layer. The change of effective stress is the difference between the initial (original) effective stress and the final effective stress, as illustrated in Table 22.1. The stress distribution has been discussed in the foregoing. The compressibility of the soil mass can be expressed in either simple or complex terms. For simple cases, the soil can be assumed to have a linear stress-strain behavior and the compressibility can be expressed by an elastic modulus.

Linear stress-strain behavior follows Hooke's law:

$$\varepsilon = \frac{\Delta\sigma'}{E} \tag{22.7}$$

where

ε = induced strain in a soil layer
$\Delta\sigma'$ = imposed change of effective stress in the soil layer
E = elastic modulus of the soil layer

Often the elastic modulus is called *Young's modulus.* Strictly speaking, however, Young's modulus is the modulus for when lateral expansion is allowed, which may be the case for soil loaded by a small footing, but not when loading a larger area. In the latter case, the lateral expansion is constrained. The constrained modulus, D, is larger than the E-modulus. The constrained modulus is also called the *oedometer modulus.* For ideally elastic soils, the ratio between D and E is:

$$\frac{D}{E} = \frac{(1-\nu)}{(1+\nu)(1-2\nu)} \tag{22.8}$$

where ν = Poisson's ratio. For example, for a soil material with a Poisson's ratio of 0.3, a common value, the constrained modulus is 35% larger than the Young's modulus. (Notice also that the concrete inside a concrete-filled pipe pile behaves as a constrained material as opposed to the concrete in a concrete pile. Therefore, when analyzing the deformation under load, use the constrained modulus for the former and the Young's modulus for the latter.)

The deformation of a soil layer, s, is the strain, ε, times the thickness, h, of the layer. The settlement, S, of the foundation is the sum of the deformations of the soil layers below the foundation.

$$S = \sum s = \sum (\varepsilon h) \tag{22.9}$$

However, stress-strain behavior is nonlinear for most soils. The nonlinearity cannot be disregarded when analyzing compressible soils, such as silts and clays; that is, the elastic modulus approach is not appropriate for these soils. Nonlinear stress-strain behavior of compressible soils is conventionally modeled by Eq.(22.10):

$$\varepsilon = \frac{C_c}{1 + e_0} \lg \frac{\sigma'_1}{\sigma'_0} \tag{22.10}$$

where

σ'_0 = original (or initial) effective stress
σ'_1 = final effective stress
C_c = compression index
e = void ratio

The compression index and the void ratio parameters C_c and e_0 are determined by means of oedometer tests in the laboratory.

If the soil is overconsolidated, that is, consolidated to a stress (called "preconsolidation stress") larger than the existing effective stress, Eq.(22.10) changes to

$$\varepsilon = \frac{1}{1 + e_0} \left[C_{cr} \lg \frac{\sigma'_p}{\sigma'_0} + C_c \lg \frac{\sigma'_1}{\sigma'_p} \right] \tag{22.11}$$

where σ'_p = preconsolidation stress and C_{cr} = recompression index.

Thus, in conventional engineering practice of settlement design, two compression parameters need to be established. This is an inconvenience that can be avoided by characterizing the soil with the ratios $C_c/(1 + e_0)$ and $C_{cr}/(1 + e_0)$ as single parameters (usually called compression ratio, CR, and recompression ratio, RR, respectively), but few do. Actually, on surprisingly many occasions, geotechnical engineers only report the C_c parameter—neglecting to include the e_0 value—or worse, report the C_c from the oedometer test and the e_0 from a different soil specimen than used for determining the compression index! This is not acceptable, of course. The undesirable challenge of ascertaining what C_c value goes with what e_0 value is removed by using the Janbu tangent modulus approach instead of the C_c and e_0 approach, applying a modulus number, m, instead.

The Janbu tangent modulus approach, proposed by Janbu [1963; 1965; 1967] and referenced by the Canadian Foundation Engineering Manual, (CFEM) [Canadian Geotechnical Society, 1985], applies the same basic principle of nonlinear stress-strain behavior to all soils, clays as well as sand. By this method, the relation between stress and strain is a function of two nondimensional parameters which are unique for a soil: a stress exponent, j, and a modulus number, m.

In cohesionless soils, $j > 0$, the following simple formula governs:

$$\varepsilon = \frac{1}{mj} \left[\left(\frac{\sigma'_1}{\sigma'_r} \right)^j - \left(\frac{\sigma'_0}{\sigma'_r} \right)^j \right] \tag{22.12}$$

where

ε = strain induced by increase of effective stress
σ'_0 = the original effective stress
σ'_1 = the final effective stress
j = the stress exponent
m = the modulus number, which is determined from laboratory and/or field testing
σ'_r = a reference stress, a constant, which is equal to 100 kPa (= 1 tsf = 1 kg/cm^2)

In an essentially cohesionless, sandy or silty soil, the stress exponent is close to a value of 0.5. By inserting this value and considering that the reference stress is equal to 100 kPa, Eq. (22.12) is simplified to

$$\varepsilon = \frac{1}{5m}\left(\sqrt{\sigma'_1} - \sqrt{\sigma'_0}\right). \tag{22.13a}$$

Notice that Eq. (22.13a) is *not* independent of the choice of units; the stress values must be inserted in kPa. That is, a value of 5 MPa is to be inserted as the number 5000 and a value of 300 Pa as the number 0.3. In English units, Eq. (22.13a) becomes

$$\varepsilon = \frac{2}{m}\left(\sqrt{\sigma'_1} - \sqrt{\sigma'_0}\right). \tag{22.13b}$$

Again, the equation is not independent of units: Because the reference stress converts to 1.0 tsf, Eq. (22.13b) requires that the stress values be inserted in units of tsf.

If the soil is overconsolidated and the final stress exceeds the preconsolidation stress, Eqs. (22.13a) and (22.13b) change to

$$\varepsilon = \frac{1}{5m_r}\left(\sqrt{\sigma'_p} - \sqrt{\sigma'_0}\right) + \frac{1}{5m}\left(\sqrt{\sigma'_1} - \sqrt{\sigma'_p}\right) \tag{22.14a}$$

$$\varepsilon = \frac{2}{m_r}\left(\sqrt{\sigma'_p} - \sqrt{\sigma'_0}\right) + \frac{2}{m}\left(\sqrt{\sigma'_1} - \sqrt{\sigma'_p}\right) \tag{22.14b}$$

where

σ'_0 = original effective stress (kPa or tsf)
σ'_p = preconsolidation stress (kPa or tsf)
σ'_1 = final effective stress (kPa or tsf)
m = modulus number (dimensionless)
m_r = recompression modulus number (dimensionless)

Equation (22.14a) requires stress units in kPa and Eq. (22.14b) stress units in tsf.

If the imposed stress does not result in a new (final) stress that exceeds the preconsolidation stress, Eqs. (22.13a) and (22.13b) become

$$\varepsilon = \frac{1}{5m_r}\left(\sqrt{\sigma'_1} - \sqrt{\sigma'_0}\right) \tag{22.15a}$$

$$\varepsilon = \frac{2}{m_r}\left(\sqrt{\sigma'_1} - \sqrt{\sigma'_0}\right) \tag{22.15b}$$

Equation (22.15a) requires stress units in kPa and Eq. (22.15b) units in tsf.

In cohesive soils, the stress exponent is zero, $j = 0$. Then, in a normally consolidated cohesive soil:

$$\varepsilon = \frac{1}{m} \ln\left(\frac{\sigma'_1}{\sigma'_0}\right) \tag{22.16}$$

and in an overconsolidated cohesive soil

$$\varepsilon = \frac{1}{m_r} \ln\left(\frac{\sigma'_p}{\sigma'_0}\right) + \frac{1}{m} \ln\left(\frac{\sigma'_1}{\sigma'_p}\right). \tag{22.17}$$

Notice that the ratio (σ'_p/σ'_0) is equal to the overconsolidation ratio, OCR. Of course, the extent of overconsolidation may also be expressed as a fixed stress-unit value, $\sigma'_p - \sigma'_0$.

If the imposed stress does not result in a new stress that exceeds the preconsolidation stress, Eq.(22.17) becomes

$$\varepsilon = \frac{1}{m_r} \ln\left(\frac{\sigma'_1}{\sigma'_0}\right) \tag{22.18}$$

By means of Eqs. (22.10) through (22.18), settlement calculations can be performed without resorting to simplifications such as that of a constant elastic modulus. Apart from having knowledge of the original effective stress, the increase of stress, and the type of soil involved, without which knowledge no reliable settlement analysis can ever be made, the only soil parameter required is the modulus number. The modulus numbers to use in a particular case can be determined from conventional laboratory testing, as well as in situ tests. As a reference, Table 22.2 shows a range of normally conservative values, in particular for the coarse-grained soil types, which are typical of various soil types (quoted from the CFEM [Canadian Geotechnical Society, 1985]).

In a cohesionless soil, where previous experience exists from settlement analysis using the elastic modulus approach—Eqs. (22.7) and (22.9)—a direct conversion can be made between E and m, which results in Eq. (22.19a).

$$m = \frac{E}{5(\sigma'_1 + \sigma'_0)} = \frac{E}{10\bar{\sigma}'} \tag{22.19a}$$

Table 22.2 Typical and Normally Conservative Modulus Numbers

Soil Type		Modulus Number	Stress Exp.
Till, very dense to dense		1000–300	$(j = 1)$
Gravel		400–40	$(j = 0.5)$
Sand	dense	400–250	$(j = 0.5)$
	compact	250–150	"
	loose	150–100	"
Silt	dense	200–80	$(j = 0.5)$
	compact	80–60	"
	loose	60–40	"
		Clays	
Silty clay	hard, stiff	60–20	$(j = 0)$
and	stiff, firm	20–10	"
clayey silt	soft	10–5	"
Soft marine clays and organic clays		20–5	$(j = 0)$
Peat		5–1	$(j = 0)$

Equation (22.19a) is valid when the calculations are made using SI units. Notice, stress and E-modulus must be expressed in the same units, usually kPa. When using English units (stress and E-modulus in tsf), Eq. (22.19b) applies.

$$m = \frac{2E}{(\sigma_1' + \sigma_0')} = \frac{E}{\sigma'} \qquad (22.19b)$$

Notice that most natural soils have aged and become overconsolidated with an overconsolidation ratio, OCR, that often exceeds a value of 2. For clays and silts, the recompression modulus, m_r, is often five to ten times greater than the virgin modulus, m, listed in Table 22.2.

The conventional C_c and e_0 method and the Janbu modulus approach are identical in a cohesive soil. A direct conversion factor as given in Eq. (22.20) transfers values of one method to the other.

$$m = \ln 10 \left(\frac{1 + e_0}{C_c} \right) = 2.30 \left(\frac{1 + e_0}{C_c} \right) \qquad (22.20)$$

Designing for settlement of a foundation is a prediction exercise. The quality of the prediction—that is, the agreement between the calculated and the actual settlement value—depends on how accurately the soil profile and stress distributions applied to the analysis represent the site conditions, and how closely the loads, fills, and excavations considered resemble those actually occurring. The quality depends also, of course, on the quality of the soil parameters used as input to the analysis. Soil parameters for cohesive soils depend on the quality of the sampling and laboratory testing. Clay samples tested in the laboratory should be from carefully obtained "undisturbed" piston samples. Paradoxically, the more disturbed the sample is, the less compressible the clay appears to be. The error which this could cause is to a degree "compensated for" by the simultaneous apparent reduction in the overconsolidation ratio. Furthermore, high-quality sampling and oedometer tests are costly, which limits the amounts of information procured for a routine project. The designer usually runs the tests on the "worst" samples and arrives at a conservative prediction. This is acceptable, but only too often is the word "conservative" nothing but a disguise for the more appropriate terms of "erroneous" and "unrepresentative" and the end results may not even be on the "safe side."

Non-cohesive soils cannot easily be sampled and tested. Therefore, settlement analysis of foundations in such soils must rely on empirical relations derived from in situ tests and experience values. Usually, these soils are less compressible than cohesive soils and have a more pronounced overconsolidation. However, considering the current tendency toward larger loads and contact stresses, foundation design must prudently address the settlement expected in these soils as well. Regardless of the methods that are used for prediction of the settlement, it is necessary to refer the results of the analyses back to basics. That is, the settlement values arrived at in the design analysis should be evaluated to indicate what corresponding compressibility parameters (Janbu modulus numbers) they represent for the actual soil profile and conditions of effective stress and load. This effort will provide a check on the reasonableness of the results as well as assist in building up a reference database for future analyses.

Time-Dependent Settlement

Because soil solids compress very little, settlement is mostly the result of a change of pore volume. Compression of the solids is called *initial compression.* It occurs quickly and it is usually considered elastic in behavior. In contrast, the change of pore volume will not occur before the water occupying the pores is squeezed out by an increase of stress. The process is rapid in coarse-grained soils and slow in fine-grained soils. In fine clays, the process can take a longer time than the life expectancy of the building, or of the designing engineer, at least. The process is called *consolidation* and it usually occurs with an increase of both undrained and drained soil shear strength. By analogy with heat dissipation in solid materials, the Terzaghi consolidation theory indicates simple relations for the time required for the consolidation. The most commonly applied theory builds on the assumption that water is able to drain out of the soil at one surface boundary and not at all at the opposite boundary. The consolidation is fast in the beginning, when the driving pore pressures are greater, and slows down with time. The analysis makes use of the relative amount of consolidation obtained at a certain time, called *average degree of consolidation*, which is defined as follows:

$$\bar{U} = \frac{S_t}{S_f} = 1 - \frac{\bar{u}_t}{\bar{u}_0} \tag{22.21}$$

where

$\bar{U}$ = *average* degree of consolidation
S_t = settlement at time t
S_f = final settlement at full consolidation
$\bar{u}_t$ = *average* pore pressure at time t
$\bar{u}_0$ = initial *average* pore pressure (on application of the load; time $t = 0$)

Notice that the pore pressure varies throughout the soil layer and Eq. (22.21) assumes average values. In contrast, the settlement values are not the average, but the accumulated values.

The time for a achieving certain degree consolidation is then, as follows:

$$t = T_v \frac{H^2}{c_v} \tag{22.22}$$

where

t = time to obtain a certain degree of consolidation
T_v = a dimensionless time coefficient
c_v = coefficient of consolidation
H = length of the longest drainage path

The time coefficient, T_v, is a function of the type of pore pressure distribution. Of course, the shape of the distribution affects the average pore pressure values and a parabolic shape is usually assumed. The coefficient of consolidation is determined in the laboratory oedometer test (some in situ tests can also provide c_v values) and it can rarely be obtained more accurately than within a ratio range of 2 or 3. The length of the longest drainage path, H, for a soil layer that drains at both surface boundaries is half the layer thickness. If drainage only occurs at one boundary, H is equal to the full layer thickness. Naturally, in layered soils, the value of H is difficult to ascertain.

Approximate values of T_v for different degrees of consolidation are given below. For more exact values and values to use when the pore pressure distribution is different, see, for example, Holtz and Kovacs [1981].

U(%)	25	50	70	80	90	100
T_v	0.05	0.20	0.30	0.40	0.60	1.00

In partially saturated soils, consolidation determined from observed settlement is initially seemingly rapid, because gas (air) will readily compress when subjected to an increase of pressure. This settlement is often mistaken for the initial compression of the grain solids. However, because the pore pressure will not diminish to a similar degree, initial consolidation determined from observed pore pressures will not appear to be as large. In these soils and in seemingly saturated soils that have a high organic content, gas is present as bubbles in the pore water, and the bubbles will compress readily. Moreover, some of the gas may go into solution in the water as a consequence of the pressure increase. Inorganic soils below the groundwater surface are usually saturated and contain no gas. In contrast, organic soils will invariably contain gas in the form of small bubbles (as well as gas dissolved in the water, which becomes free gas on release of confining pressure when sampling the soil) and these soils will appear to have a fast initial consolidation. Toward the end of the consolidation process, when the pore pressure has diminished, the bubbles will return to the original size and the consolidation process will appear to have slowed.

Generally, the determination—prediction—of the time for a settlement to develop is filled with uncertainty and it is very difficult to reliably estimate the amount of settlement occurring within a specific time after the load application. The prediction is not any easier when one has to consider the development during the build-up of the load. For details on the subject, see Ladd [1991].

The rather long consolidation time in clay soils can be shortened considerably by means of vertical drains. Vertical drains installed at spacings ranging from about 1.2 m through 2.0 m have been very successful in accelerating consolidation from years to months. In the past, vertical drains consisted of sand drains and installation disturbance in some soils often made the drains cause more problems than they solved. However, the sand drain is now replaced by premanufactured band-shaped drains, which do not share the difficulties and adverse behavior of sand drains.

Theoretically, when vertical drains have been installed, the drainage is in the horizontal direction and design formulas have been developed as based on radial drainage. However, vertical drains connect horizontal layers of greater permeability, which frequently are interspersed in natural soils, which make the theoretical calculations quite uncertain. Some practical aspects of the use of vertical drains are described in the CFEM [1985].

The settlement will continue after the end of the consolidation. This type of settlement is called *creep* or *secondary compression* (the consolidation is then called *primary compression*). Creep is a function of a coefficient of secondary compression, C_α, and the ratio of the time considered after full consolidation and the time for full consolidation to develop:

$$\varepsilon_{\text{creep}} = \frac{C_\alpha}{(1 + e_0)} \ln \frac{t_\alpha}{t_{100}} \tag{22.23}$$

where

C_α = coefficient of secondary compression
t_α = time after end of consolidation considered
t_{100} = time for achieving primary compression

In most soils, creep is small in relation to the consolidation settlement and is therefore neglected. However, in organic soils, creep may be substantial.

Magnitude of Acceptable Settlement

For many years, settlement analysis was limited to ascertaining that the expected settlement should not exceed one inch. (Realizing that 25.4 mm is too precise a value—as is 25 mm when transferring this limit to the SI system—some have argued whether the "metric inch" should be 20 mm or 30 mm). Furthermore, both total settlement and differential settlement must be evaluated. The *Canadian Foundation Engineering Manual* [1985] lists allowable displacement criteria in terms of maximum deflection between point supports, maximum slope of continuous structures, and rotation limits for structures. The multitude of limits demonstrate clearly that the acceptable settlement varies with the type and size of structure considered. Moreover, modern structures often have small tolerance for settlement and, therefore, require a more thorough settlement analysis. The advent of the computer and development of sophisticated yet simple to use design software have enabled the structural engineers to be very precise in the analysis of deformations and the effect of deformations on the stress and strain in various parts of a structure. As a not so surprising consequence, requests for "settlement-free" foundations have increased. When the geotechnical engineer is vague on the predicted settlement, the structural designer "plays it safe" and increases the size of footings or changes the foundation type, which may increase the costs of the structure. These days, in fact, the geotechnical engineer cannot just estimate a "less than one inch" value, but must provide a more accurate value by performing a thorough analysis considering soil compressibility, soil layering, and load variations. Moreover, the analysis must be put into the context of the structure, which necessitates a continuous communication between the geotechnical and structural engineers during the design effort. Building codes have started to recognize the complexity of the problem and mandate that the designers collaborate continuously. See, for example, the 1993 Canadian Highway Bridge Design Code.

22.3 Bearing Capacity of Shallow Foundations

When society started building and imposing large concentrated loads onto the soil, occasionally the structure would fail catastrophically. Initially, the understanding of foundation behavior progressed from one failure to the next. Later, tests were run of model footings in different soils and the test results were extrapolated to the behavior of full-scale foundations. For example, loading tests on model size footings in normally consolidated clay showed load-movement curves where the load increased to a distinct peak value—bearing capacity failure—indicating that the capacity (not the settlement) of a footing in clay is independent of the footing size.

The behavior of footing in clay differs from the behavior of footings in sand, however. Figure 22.1 presents results from loading tests on a 150 mm diameter footing in dry sand of densities vary-

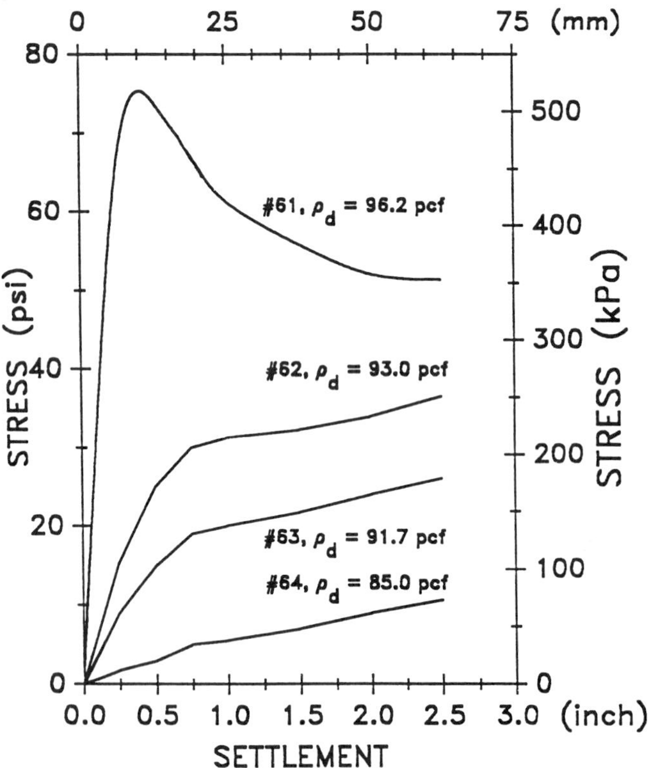

FIGURE 22.1 Contact stress vs. settlement of 150-mm footings. (*Source:* Vesic, 1967.)

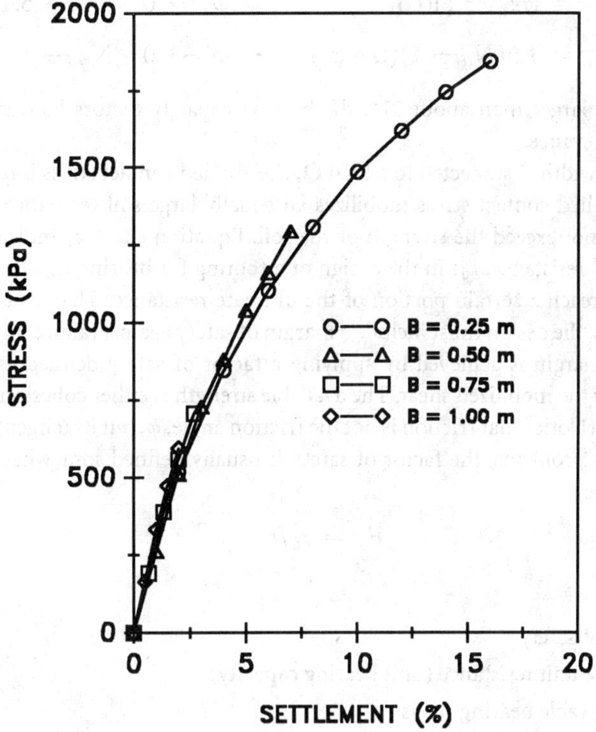

FIGURE 22.2 Stress vs. normalized settlement. (Data from Ismael, 1985.)

ing from very dense to loose. In the dense sand, a peak value is evident. In less dense sands, no such peak is found.

The capacity and the load movement of a footing in sand are almost directly proportional to the footing size. This is illustrated in Fig. 22.2, which shows some recent test results on footings of different size in a fine sand. Generally, eccentric loading, inclined loading, footing shape, and foundation depth influence the behavior of footings. Early on, Terzaghi developed the theoretical explanations to observed behaviors into a "full bearing capacity formula," as given in Eq. (22.24a) and applicable to a continuous footing:

$$r_u = c'N_c + q'(N_q - 1) + 0.5B\gamma'N_\gamma \tag{22.24a}$$

where

r_u = ultimate unit resistance of the footing
c' = effective cohesion intercept
B = footing width
q' = overburden effective stress at the foundation level
γ' = average effective unit weight of the soil below the foundation
N_c, N_q, N_γ = nondimensional bearing capacity factors

The **bearing capacity** factors are a function of the effective friction angle of the soil. Such factors were first originated by Terzaghi, later modified by Meyerhof, Berezantsev, and others. As presented in the *Canadian Foundation Engineering Manual* [1985], the bearing capacity factors are somewhat interrelated, as follows.

$$N_q = \left(e^{\pi \tan\varphi'}\right)\left(\frac{1 + \sin\varphi'}{1 - \sin\varphi'}\right) \qquad \varphi' \to 0 \quad N_q \to 1 \tag{22.24b}$$

$$N_c = (N_q - 1)(\cot \varphi') \qquad \varphi' \to 0 \quad N_c \to 5.14 \tag{22.24c}$$

$$N_\gamma = 1.5(N_q - 1)(\tan \varphi') \qquad \varphi' \to 0 \quad N_\gamma \to 0 \tag{22.24d}$$

For friction angles larger than about 37°, the bearing capacity factors increase rapidly and the formula loses in relevance.

For a footing of width B subjected to a load Q, the applied contact stress is $q(= Q/B)$ per unit length and the applied contact stress mobilizes an equally large soil resistance, r. Of course, the soil resistance can not exceed the strength of the soil. Equation (22.24a) indicates the maximum available (ultimate) resistance, r_u. In the design of a footing for bearing capacity, the applied load is only allowed to reach a certain portion of the ultimate resistance. That is, as is the case for all foundation designs, the design must include a margin of safety against failure. In most geotechnical applications, this margin is achieved by applying a factor of safety defined as the available soil strength divided by the mobilized shear. The available strength is either cohesion, c, friction, $\tan \varphi$, or both combined. (Notice that friction is not the friction angle, φ, but its tangent, $\tan \varphi$). However, in bearing capacity problems, the factor of safety is usually defined somewhat differently and as given by Eq. (22.24e).

$$F_s = r_u / q_{\text{allow}} \tag{22.24e}$$

where

F_s = factor of safety
r_u = ultimate unit resistance (unit bearing capacity)
q_{allow} = the allowable bearing stress

The factor of safety applied to the bearing capacity formula is usually recommended to be no smaller than 3.0, usually equal to 4.0. There is some confusion whether, in the bearing capacity calculated according to Eq. (22.24a), the relation $(N_q - 1)$ should be used in lieu of N_q and, then, whether or not the allowable bearing stress should be the "net" stress, that is, the value exceeding the existing stress at the footing base. More importantly, however, is that the definition of factor of safety given by Eq. (22.24e) is not the same as the factor of safety applied to the shear strength, because the ultimate resistance determined by the bearing capacity formula includes several aspects other than soil shear strength, particularly so for foundations in soil having a substantial friction component. Depending on the details of each case, a value of 3 to 4 for the factor defined by Eq. (22.24e) corresponds, very approximately, to a factor of safety on shear strength in the range of 1.5 through 2.0.

In fact, the bearing capacity formula is wrought with much uncertainty and the factor of safety, be it 3 or 4, applied to a bearing capacity formula is really a "factor of ignorance" and does not always guarantee an adequate safety against failure. Therefore, in the design of footings, be it in clays or sands, the settlement analysis should be given more weight than the bearing capacity formula calculation.

Footings are rarely loaded only vertically and concentrically. Figure 22.3(b) illustrates the general case of a footing subjected to both inclined and eccentric load. Eq. (22.24a) changes to

$$r_u = s_c i_c c' N_c + s_q i_q q' N_q + s_\gamma i_\gamma 0.5 B' \gamma' N_\gamma \tag{22.24f}$$

where factors not defined earlier are

s_c, s_q, s_γ = nondimensional shape factors
i_c, i_q, i_γ = nondimensional inclination factors
B' = equivalent or effective footing width

The shape factors are

$$s_c = s_q = 1 + (B'/L')(N_q/N_c) \tag{22.24g}$$

where L' = equivalent or effective footing length.

$$s_\gamma = 1 - 0.4(B'/L') \tag{22.24h}$$

The inclination factors are

$$i_c = i_q = (1 - \delta/90^\circ)^2 \tag{22.24i}$$

$$i_\gamma = (1 - \delta/\varphi')^2 \tag{22.24j}$$

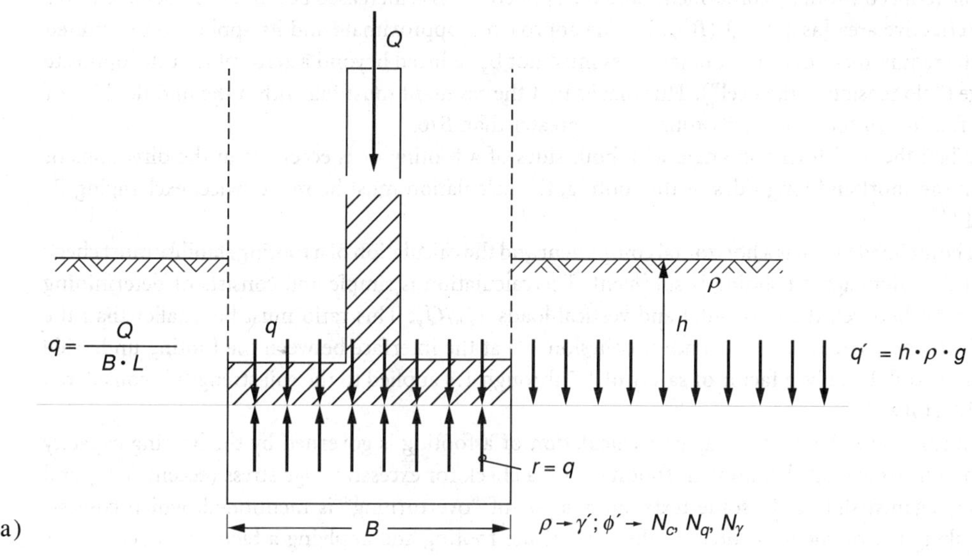

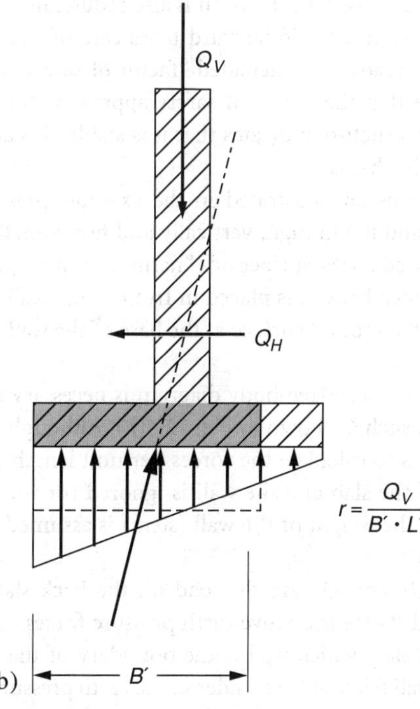

FIGURE 22.3 Input to the bearing capacity formula applied to a strip footing. (*a*) Concentric and vertical loading. (*b*) Vertical and horizontal loading.

An inclined load can have a significant reducing effect on the bearing capacity of a footing. Directly, first, as reflected by the inclination factor and then also because the resultant to the load on most occasions acts off center. An off-center load will cause increased stress, edge stress, on one side and a decreased stress on the opposing side. A large edge stress can be the starting point of a failure. In fact, most footings, when they fail, fail by tilting, which is an indication of excessive edge stress. To reduce the risk for failure, the bearing capacity formula (which assumes a uniform load) applies the term B' in Eq. (22.24f), the effective footing width, which is the width of a smaller footing having the resultant load in its center. That is, the calculated ultimate resistance is decreased because of the reduced width (γ component) and the applied stress is increased because it is calculated over the effective area [as $q = Q/(B'/L')$]. The approach is approximate and its application is limited to the requirement that the contact stress must not be reduced beyond a zero value at the opposite edge ("No tension at the heel"). This means that the resultant must fall within the middle third of the footing, or the eccentricity must not be greater than $B/6$.

When the load forms an angle with both sides of a footing or is eccentric in the directions of both the short and long sides of the footing, the calculation must be made twice, exchanging B' and L'.

The inclined load has a horizontal component and the calculation of a footing stability must check that the safety against sliding is sufficient. The calculation is simple and consists of determining the ratio between the horizontal and vertical loads, Q_h/Q_v. This ratio must be smaller than the soil strength (friction, tan φ', and/or cohesion, c') at the interface between the footing underside and the soil. Usually, a factor of safety of 1.5 through 1.8 applied to the soil strength is considered satisfactory.

In summary, the bearing capacity calculation of a footing is governed by the bearing capacity of a uniformly loaded equivalent footing, with a check for excessive edge stress (eccentricity) and safety against sliding. In some texts, an analysis of "overturning" is mentioned, which consists of taking the moment of forces at the edge of the footing and applying a factor of safety to the equilibrium. This is an incorrect approach, because long before the moment equilibrium has been reached, the footing fails due to excessive edge stress. (It is also redundant, because the requirement for the resultant to be located within the middle third takes care of the "overturning.") In fact, "overturning" failure will occur already at a calculated "factor of safety" as large as about 1.3 on the moment equilibrium. Notice that the factor of safety approach absolutely requires that the calculation of the stability of the structure indicates that it is stable also at a factor of safety very close to unity—theoretically stable, that is.

The bearing capacity calculations are illustrated in the example presented in Fig. 22.4. The example involves a 10.0 m long and 8.0 m high, vertically and horizontally loaded retaining wall (bridge abutment). The wall is placed on the surface of a "natural" coarse-grained soil and backfilled with a coarse material. A 1.0 m thick backfill is placed in front of the wall and over the front slab. The groundwater table lies close the ground surface at the base of the wall. Figure 22.4(a) presents the data to include in an analysis.

In any analysis of a foundation case, a free-body diagram is necessary to ensure that all forces are accounted for in the analysis, such as shown in Fig. 22.4(b). Although the length of the wall is finite, it is normally advantageous to calculate the forces per unit length of the wall. To simplify the computations, the weight of the slab and the wall is ignored (or the slab weight is assumed included in the soil weights, and the weight of the wall (stem) is assumed included in the vertical load applied to the top of the wall).

The vertical forces denoted Q_1 and Q_2 are the load on the back slab of the wall. The two horizontal forces denoted P_1 and P_2 are the active earth pressure forces acting on a fictitious wall rising from the heel of the back slab, which wall is the boundary of the free body. Because this fictitious wall is soil, there is no wall friction to consider in the earth pressure calculation. Naturally, earth pressure also acts on the footing stem (the wall itself). Here, however, wall friction does exist, rotating the earth pressure resultant from the horizontal direction. Because of compaction of the

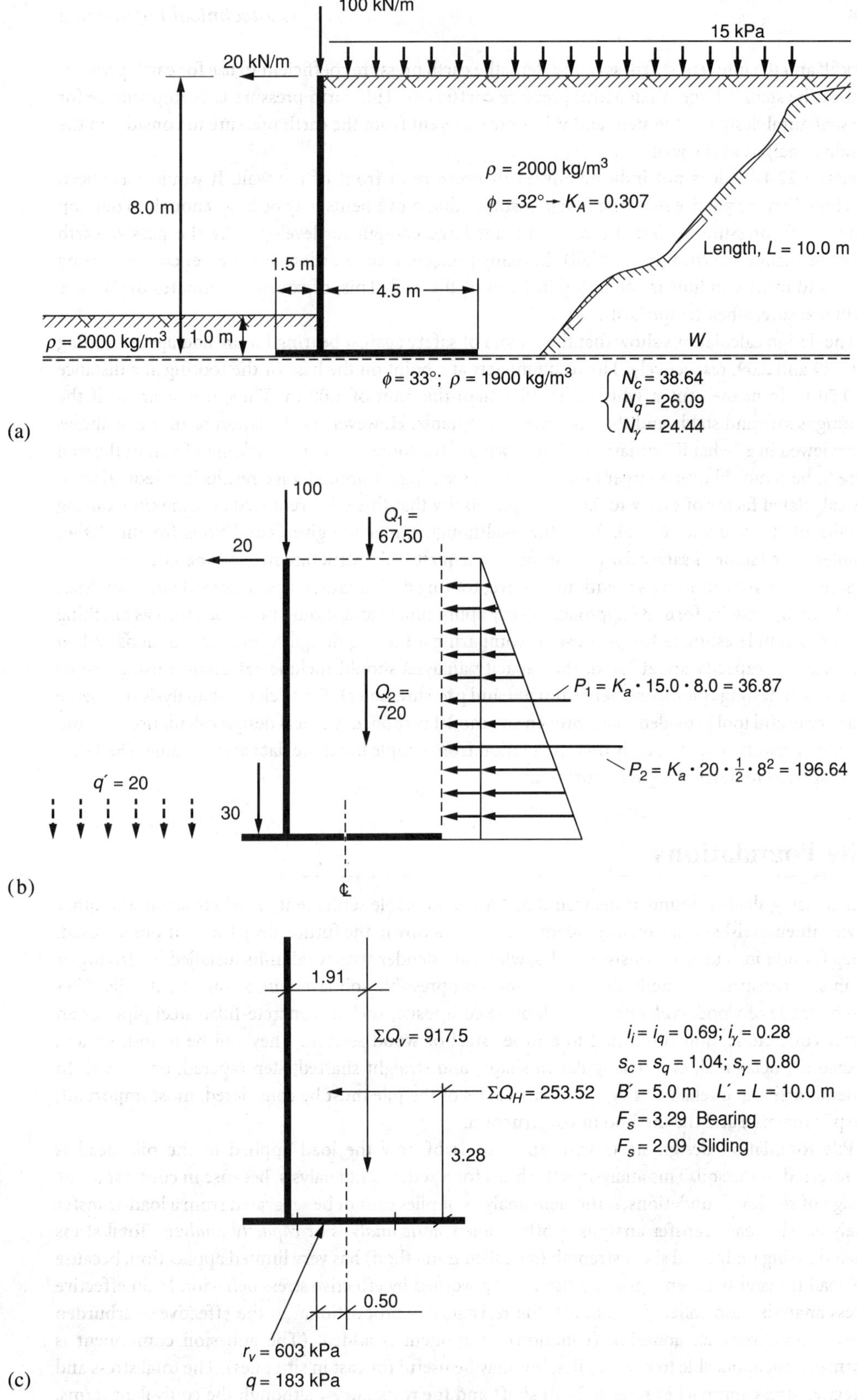

FIGURE 22.4 Bearing capacity example. (*a*) Problem background. (*b*) Free-body diagram. (*c*) Solution data.

backfill and the inherent stiffness of the stem, the earth pressure coefficient to use for earth pressure against the stem is larger than active pressure coefficient. This earth pressure is of importance for the structural design of the stem and it is quite different from the earth pressure to consider in the stability analysis of the wall.

Figure 22.4(b) does not indicate any earth pressure in front of the wall. It would have been developed on the passive side (the design assumes that movements may be large enough to develop active earth pressure behind the wall, but not large enough to develop fully the passive earth pressure against the front of the wall). In many projects a more or less narrow trench for burying pipes and other conduits is often dug in front of the wall. This, of course, eliminates the passive earth pressure, albeit temporarily.

The design calculations show that the factors of safety against bearing failure and against sliding are 3.29 and 2.09, respectively. The resultant acts at a point on the base of the footing at a distance of 0.50 m from the center, which is smaller than the limit of 1.00 m. Thus, it appears as if the footing is safe and stable and the edge stress acceptable. However, a calculation result must always be reviewed in a "what if" situation. That is, what if for some reason the backfill in front of the wall were to be removed over a larger area? Well, this seemingly minor change results in a reduction of the calculated factor of safety to 0.69. The possibility that this fill is removed at some time during the life of the structure is real. Therefore—although under the given conditions for the design problem, the factor of safety for the footing is adequate—the structure may not be safe.

Some words of caution: As mentioned above, footing design must emphasize settlement analysis. The bearing capacity formula approach is very approximate and should never be taken as anything beyond a simple estimate for purpose of comparing a footing design to previous designs. When concerns for capacity are at hand, the capacity analysis should include calculation using results from in situ testing (piezocone penetrometer and pressuremeter). Finite element analysis may serve as a very useful tool provided that a proven soil model is applied. Critical design calculations should never be permitted to rely solely on information from simple borehole data and N values (SPT-test data) applied to bearing capacity formulas.

22.4 Pile Foundations

Where using shallow foundations would mean unacceptable settlement, or where scour and other environmental risks exist which could impair the structure in the future, deep foundations are used. Deep foundations usually consist of piles, which are slender structural units installed by driving or by in situ construction methods through soft compressible soil layers into competent soils. Piles can be made of wood, concrete, or steel, or be composite, such as concrete-filled steel pipes or an upper concrete section connected to a lower steel or wood section. They can be round, square, hexagonal, octagonal, even triangular in shape, and straight shafted, step tapered, or conical. In order to arrive at a reliable design, the particulars of the pile must be considered, most important, the pile material and the method of construction.

Pile foundation design starts with an analysis of how the load applied to the pile head is transferred to the soil. This analysis is the basis for a settlement analysis, because in contrast to the design of shallow foundations, settlement analysis of piles cannot be separated from a load-transfer analysis. The load-transfer analysis is often called *static analysis* or *capacity analysis.* Total stress analysis using undrained shear strength (so-called α-method) has very limited application, because the load transfer between a pile and the soil is governed by effective stress behavior. In an effective stress analysis (also called β-method), the resistance is proportional to the effective overburden stress. Sometimes, an adhesion (cohesion) component is added. (The adhesion component is normally not applicable to driven piles, but may be useful for cast in situ piles). The total stress and effective stress approaches refer to both **shaft** and **toe resistances,** although the equivalent terms, "α-method" and "β-method" usually refer to shaft resistance, specifically.

Shaft Resistance

The general numerical relation for the unit shaft resistance, r_s, is

$$r_s = c' + \beta\sigma'_z \tag{22.25a}$$

The adhesion component, c', is normally set to zero for driven piles and Eq. (22.25a) then expresses that unit shaft resistance is directly proportional to the effective overburden stress.

The accumulated (total) shaft resistance, R_s, is

$$R_s = \int A_s r_s dz = \int A_s(c' + \beta\sigma'_z)dz \tag{22.25b}$$

Table 22.3 Approximate Range of Beta Coefficients

Soil Type	Phi	Beta
Clay	25–30	0.25–0.35
Silt	28–34	0.27–0.50
Sand	32–40	0.30–0.60
Gravel	35–45	0.35–0.80

The beta coefficient varies with soil gradation, mineralogical composition, density, and soil strength within a fairly narrow range. Table 22.3 shows the approximate range of values to expect from basic soil types.

Toe Resistance

Also the unit toe resistance, r_t, is proportional to the effective stress, that is, the effective stress at the **pile toe** ($z = D$). The proportionality coefficient has the symbol N_t. Its value is sometimes stated to be of some relation to the conventional bearing capacity coefficient, N_q, but such relation is far from strict. The toe resistance, r_t, is

$$r_t = N_t\sigma'_{z=D}. \tag{22.26a}$$

The total toe resistance, R_t, acting on a pile with a toe area equal to A_t is

$$R_t = A_t r_t = A_t N_t \sigma'_{z=D}. \tag{22.26b}$$

Table 22.4 Approximate Range of N_t Coefficients

Soil Type	Phi	N_t
Clay	25–30	3–30
Silt	28–34	20–40
Sand	32–40	30–150
Gravel	35–45	60–300

In contrast to the β-coefficient, the toe coefficient, N_t, varies widely. Table 22.4 shows an approximate range of values for the four basic soil types.

Ultimate Resistance—Capacity

The capacity of the pile, Q_{ult} (alternatively, R_{ult}), is the sum of the shaft and toe resistances.

$$Q_{\text{ult}} = R_s + R_t \tag{22.27}$$

When the shaft and toe resistances are fully mobilized, the load in pile, Q_z, (as in the case of a static loading test brought to "failure") varies, as follows:

$$Q_z = Q_u - \int A_s\beta\sigma'_z dz = Q_u - R_s \tag{22.28}$$

Equation (22.28) is also called the *resistance distribution curve.* At the depth $z = D$, Eq. (22.28), of course, states that $Q_z = R_t$.

Notice that the commonly used term "ultimate capacity" is a misnomer and a tautology: a mix of the words "ultimate resistance" and "capacity". Although one cannot be mistaken about the meaning of ultimate capacity, the adjective should not be used, because it makes other adjectives seem proper, such as "load capacity," "allowable capacity," "design capacity," which are at best awkward and at worst misleading, because what is meant is not clear. Sometimes not even the person using these adjectives with "capacity" knows the meaning.

During service conditions, loads from the structure will be applied to the pile head via a pile cap. The loads are normally permanent (or "dead") loads, Q_d, and transient (or "live") loads Q_l. Not generally recognized is that even if soil settlement is small—too small to be noticeable—the soil will in the majority of cases move down in relation to the pile and in the process transfer load to the pile by negative skin friction. (The exception refers to piles in swelling soils and it is then limited to the length of pile in the swelling zone.) Already the extremely small relative movements always occurring between a pile shaft and the soil are sufficient to develop either shaft resistance or negative skin friction. Therefore, every pile develops an equilibrium of forces between, on the one side, the sum of dead load applied to the pile head, Q_d, and dragload, Q_n, induced by negative skin friction in the upper part of the pile, and, on the other side, the sum of positive shaft resistance and toe resistance in the lower part of the pile. The point of equilibrium, called the **neutral plane,** is the depth where the shear stress along the pile changes over from negative skin friction into positive shaft resistance. This is also where there is no relative displacement between the pile and the soil.

The key aspect of the foregoing is that the development of a neutral plane and negative skin friction is an always occurring phenomenon in piles and not only of importance in the context of large settlement of the soil around the piles.

Normally, the neutral plane lies below the midpoint of a pile. The extreme case is for a pile on rock, where the location of the neutral plane is at the bedrock elevation. For a dominantly shaft-bearing pile "floating" in a homogeneous soil with linearly increasing shear resistance, the neutral plane lies at a depth which is about equal to the lower third point of the pile embedment length.

The larger the toe resistance, the deeper the elevation of the neutral plane. And, the larger the dead load, the shallower the elevation of the neutral plane.

The load distribution in the pile during long-term conditions down to the neutral plane is given by the following load-transfer relation. [Below the neutral plane, Q_z follows Eq. (22.28).]

$$Q_z = Q_d + \int A_s q_n dz = Q_d + Q_n \tag{22.29}$$

The transition between the load-resistance curve [Eq. (22.27)] and the load-transfer curve [Eq. (22.28)] is in reality not the sudden kink the equations suggest, but a smooth transition over some length of pile, about 4 to 8 pile diameters above and below the neutral plane. (The length of this transition zone varies with the type of soil at the neutral plane.) Thus, the theoretically calculated value of the maximum load in the pile is higher than the real value. That is, it is easy to overestimate the magnitude of the dragload.

Critical Depth

Many texts suggest the existence of a so-called "critical depth" below which the shaft and toe resistances would be constant and independent of the increasing effective stress. This concept is a fallacy based on past incorrect interpretation of test data and should not be applied.

Effect of Installation

Whether a pile is installed by driving or by other means, the installation affects, disturbs, the soil. It is difficult to determine the magnitude of the shaft and toe resistances existing before the disturbance from the pile installation has subsided. For instance, presence of dissipating excess pore pressures causes uncertainty in the magnitude of the effective stress in the soil, ongoing strength gain due to reconsolidation is hard to estimate, etc. Such installation effects can take a long time to disappear, especially in clays. They can be estimated in an effective stress analysis using suitable assumptions as to the distribution of pore pressure along the pile at any particular time. Usually, to calculate the installation effect, a good estimate can be obtained by imposing excess pore pressures in the fine-grained soil layers, taking care that the pore pressure must not exceed the total overburden stress. By restoring the pore pressure values to the original conditions, which will prevail when the induced excess pore pressures have dissipated, the long-term capacity is established.

Residual Load

The dissipation of induced excess pore pressures is called *reconsolidation.* Reconsolidation after installation of a pile imposes load (residual load) in the pile by negative skin friction in the upper part of the pile, which is resisted by positive shaft resistance in the lower part of the pile and some toe resistance. The quantitative effect of including, as opposed to not to, the residual load in the analysis is that the shaft resistance becomes smaller and the toe resistance becomes larger. If the residual load is not recognized in the evaluation of results from a static loading test, totally erroneous conclusions will be drawn from the test: The shaft resistance appears larger than the true value, while the toe resistance appears correspondingly smaller, and if the resistance distribution is determined from a force gauge in the pile, zeroed at the start of the test, a "critical depth" will seem to exist. For more details on this effect and how to analyze the force gauge data to account for the residual load, see Altaee *et al.* [1992; 1993].

Analysis of Capacity for Tapered Piles

Many piles are not cylindrical or otherwise uniform in shape throughout the length. The most common example is the wood pile, which is conical in shape. Step-tapered piles are also common, consisting of two or more concrete-filled steel pipes of different diameters connected to each other, the larger above the smaller. Sometimes a pile can consist of a steel pipe with a conical section immediately above the pile toe, for example, the Monotube pile, which typically has a 25 feet (7.6 m) long conical end section, tapering the diameter down from 14 inches (355 mm) to 8 inches (203 mm). Piles can have an upper solid concrete section and a bottom H-pile extension.

For the step-tapered piles, obviously each "step" provides an extra resistance point, which needs to be considered in an analysis. (The GRLWEAP wave equation program, for example, can model a pile with one diameter change as having a second pile toe at the location of the step). Similarly, in a static analysis, each such step can be considered as a donut-shaped extra pile toe and assigned a corresponding toe resistance per Eq. (22.26). Each such extra toe resistance value is then added to the shaft resistance calculated using the actual pile diameter.

Piles with a continuous taper (conical piles) are less easy to analyze. Nordlund [1963] suggested a taper correction factor to use to increase the unit shaft resistance in sand for conical piles. The correction factor is a function of the taper angle and the soil friction angle. A taper angle of 1°(0.25 inch/foot) in a sand with a 35° friction angle would give a correction factor of about 4. At an angle of 0.5°, the factor would be about 2.

A more direct calculation method is to "step" the calculation in sub-layers of some thickness and at the bottom of each such sub-layer project the donut-shaped diameter change, which then is treated as an extra toe similar to the analysis of the step-taper pile. The shaft resistance is calculated using the mean diameter of the pile over the same "stepped" length. The shaft resistance over each such particular length consists of the sum of the shaft resistance and the extra-toe resistance. This method requires that a toe coefficient, N_t, be assigned to each soil layer.

The taper does not come into play for negative skin friction. This means that, when determining the dragload, the effect of the taper should be excluded. Below the neutral plane, however, the effect should be included. Therefore, the taper will influence the location of the neutral plane (because the taper increases the positive shaft resistance below the neutral plane).

Factor of Safety

In a pile design, one must distinguish between the design for bearing capacity and design for structural strength. The former is considered by applying a factor of safety to the pile capacity according to Eq. (22.27). The capacity is determined considering positive shaft resistance developed along the full length of the pile plus full toe resistance. Notice that no allowance is given for the dragload. If design is based on only theoretical analysis, the usual factor of safety is about 3.0. If based on the results of a loading test, static or dynamic, the factor of safety is reduced, depending on reliance on and confidence in the capacity value, and importance and sensitivity of the structure to foundation deformations. Factors of safety as low as 1.8 have been used.

Design for structural strength applies to a factor of safety applied to the loads acting at the pile head and at the neutral plane. At the pile head, the loads consist of dead and live load combined with bending at the pile head, but no dragload. At the neutral plane, the loads consist of dead load and dragload, but no live load. Live load and dragload cannot occur at the same time and must, therefore, not be combined in the analysis. It is recommended that for *straight and undamaged piles* the allowable maximum load at the neutral plane be limited to 70% of the pile strength. For composite piles, such as concrete-filled pipe piles, the load should be limited to a value that induces a maximum compression strain of 1.0 millistrain into the pile with no material becoming stressed beyond 70% of its strength. The calculations are interactive inasmuch a change of the load applied to a pile will change the location of the neutral plane and the magnitude of the maximum load in the pile.

The third aspect in the design, calculation of settlement, pertains more to pile groups than to single piles. In extending the approach to a pile group, it must be recognized that a pile group is made up of a number of individual piles which have different embedment lengths and which have mobilized the toe resistance to a different degree. The piles in the group have two things in common, however: They are connected to the same stiff pile cap and, therefore, all pile heads move equally, and the piles must all have developed a neutral plane at the same depth somewhere down in the soil (long-term condition, of course).

Therefore, it is impossible to ensure that the neutral plane is common for the piles in the group, with the mentioned variation of length, etc., unless the dead load applied to the pile head from the cap differs between the piles. This approach can be used to discuss the variation of load within a group of stiffly connected piles. A pile with a longer embedment below the neutral plane or one having mobilized a larger toe resistance as opposed to other piles will carry a greater portion of the dead load on the group. On the other hand, a shorter pile, or one with a smaller toe resistance as opposed to other piles in the group, will carry a smaller portion of the dead load. If a pile is damaged at the toe, it is possible that the pile exerts a negative—pulling—force at the cap and thus increases the total load on the pile cap.

An obvious result of the development of the neutral plane is that no portion of the dead load is transferred to the soil via the pile cap, unless, of course, the neutral plane lies right at the pile cap and the entire pile group is failing.

Above the neutral plane, the soil moves down relative to the pile; below the neutral plane, the pile moves down into the soil. Therefore, at the neutral plane, the relative movement between the pile and the soil is zero, or, in other words, whatever the settlement of the soil that occurs at the neutral plane is equal to the settlement of the pile (the pile group) at the neutral plane. Between the pile head and the neutral plane, only deformation of the pile due to load occurs and this is usually minor. Therefore, settlement of the pile and the pile group is governed by the settlement of the soil at and below the neutral plane. The latter is influenced by the stress increase from the permanent load on the pile group and other causes of load, such as the fill. A simple method of calculation is to exchange the pile group for an equivalent footing of area equal to the area of the pile cap placed at the depth of the neutral plane. The load on the pile group load is then distributed as a stress on this footing calculating the settlement of this footing stress in combination with all other stress changes at the site, such as the earth fill, potential groundwater table changes, adjacent excavations, etc. Notice that the portion of the soil between the neutral plane and the pile toe depth is "reinforced" with the piles and, therefore, not very compressible. In most cases, the equivalent footing is best placed at the pile toe depth (or at the level of the average of the pile toe depths).

Empirical Methods for Determining Axial Pile Capacity

For many years, the N-index of standard penetration test has been used to calculate capacity of piles. Meyerhof [1976] compiled and rationalized some of the wealth of experience available and recommended that the capacity be a function of the N-index, as follows:

$$R = R_t + R_s = mNA_t + nNA_sD \tag{22.30}$$

where

m = a toe coefficient
n = a shaft coefficient
N = N-index at the pile toe
N = average N-index along the pile shaft
A_t = pile toe area
A_s = unit shaft area; circumferential area
D = embedment depth

For values inserted into Eq. (22.30) using base SI units—that is, R in newton, D in meter, and A in m^2—the toe and shaft coefficients, m and n, become:

$m = 400 \cdot 10^3$ for driven piles and $120 \cdot 10^3$ for bored piles(N/m^2)
$n = 2 \cdot 10^3$ for driven piles and $1 \cdot 10^3$ for bored piles (N/m^3)

For values inserted into Eq. (22.30) using English units with R in ton, D in feet, and A in ft^2, the toe and shaft coefficients, m and n, become:

m = 4 for driven piles and 1.2 for bored piles (N/m^2)
n = 0.02 for driven piles and 0.01 for bored piles (N/m^3)

The standard penetration test (SPT) is a subjective and highly variable test. The test and the N-index have substantial qualitative value, but should be used only very cautiously for quantitative analysis. The Canadian Foundation Engineering Manual [1985] includes a listing of the numerous irrational factors influencing the N-index. However, when the use of the N-index is considered with the sample of the soil obtained and related to a site- and area-specific experience, the crude and decried SPT test does not come out worse than other methods of analyses.

The static cone penetrometer resembles a pile. There is shaft resistance in the form of so-called local friction measured immediately above the cone point, and there is toe resistance in the form of the directly applied and measured cone-point pressure.

When applying cone penetrometer data to a pile analysis, both the local friction and the point pressure may be used as direct measures of shaft and toe resistances, respectively. However, both values can show a considerable scatter. Furthermore, the cone-point resistance, (the cone-point being small compared to a pile toe) may be misleadingly high in gravel and layered soils. Schmertmann [1978] has indicated an averaging procedure to be used for offsetting scatter, whether caused by natural (real) variation in the soil or inherent in the test.

The piezocone, which is a cone penetrometer equipped with pore pressure measurement devices at the point, is a considerable advancement on the static cone. By means of the piezocone, the cone information can be related more dependably to soil parameters and a more detailed analysis be performed. Soil is variable, however, and the increased and more representative information obtained also means that a certain digestive judgment can and must be exercised to filter the data for computation of pile capacity. In other words, the designer is back to square one: more thoroughly informed and less liable to jump to false conclusions, but certainly not independent of site-specific experience.

The Lambda Method

Vijayvergiya and Focht [1972] compiled a large number of results from static loading tests on essentially shaft-bearing piles in reasonably uniform soil and found that, for these test results, the mean unit shaft resistance is a function of depth and can be correlated to the sum of the mean overburden effective stress plus twice the mean undrained shear strength within the embedment depth, as follows.

$$r_s = \lambda(\sigma'_m + 2c_m) \tag{22.31}$$

where

r_m = mean shaft resistance along the pile
λ = the lambda correlation coefficient
σ_m = mean overburden effective stress
c_m = mean undrained shear strength

The correlation factor is called "lambda" and it is a function of pile embedment depth, reducing with increasing depth, as shown in Table 22.5.

The lambda method is almost exclusively applied to determining the shaft resistance for heavily loaded pipe piles for offshore structures in relatively uniform soils.

Table 22.5 Approximate Values of λ

Embedment		λ
(ft)	(m)	(−)
0	0	0.50
10	3	0.36
25	7	0.27
50	15	0.22
75	23	0.17
100	30	0.15
200	60	0.12

Field Testing for Determining Axial Pile Capacity

The capacity of a pile is most reliable value when determined in a full-scale field test. However, despite the numerous static loading tests that have been carried out and the many papers that have reported on such tests and their analyses, the understanding of static pile testing in current engineering practice leaves much to be desired. The reason is that engineers have concerned themselves with mainly one question—"Does the pile have a certain least capacity?"—finding little of practical value in analyzing the pile-soil interaction, the load transfer.

A static loading test is performed by loading a pile with a gradually or stepwise increasing force while monitoring the movement of the pile head. The force is obtained by means of a hydraulic jack reacting against a loaded platform or anchors.

The American Society for Testing and Materials, ASTM, publishes three standards, D-1143, D-3689, and D-3966, for static testing of a single pile in axial compression, axial uplift, and lateral loading, respectively. The ASTM standards detail how to arrange and perform the pile test. Wisely, they do not include how to interpret the tests, because this is the responsibility of the engineer in charge, who is the only one with all the site- and project-specific information necessary for the interpretation.

The most common test procedure is the slow maintained load method referred to as the "standard loading procedure" in the ASTM Designation D-1143 and D-3689, in which the pile is loaded in eight equal increments up to a maximum load, usually twice a predetermined allowable load. Each load level is maintained until zero movement is reached, defined as 0.25 mm/hr (0.01 in./hr). The final load, the 200 percent load, is maintained for a duration of 24 hours. The "standard method" is very time-consuming, requiring from 30 to 70 hours to complete. It should be realized that the words "zero movement" are very misleading: The "zero" movement rate mentioned is equal to a movement of more than 2 m (7 ft) per year!

Each of the eight load increments is placed onto the pile very rapidly; as fast as the pump can raise the load, which usually takes about 20 seconds to 2 minutes. The size of the load increment in the "standard procedure"—12.5 percent of the maximum load—means that each such increase of load is a shock to the pile and the soil. Smaller increments that are placed more frequently disturb the pile less, and the average increase of load on the pile during the test is about the same. Such loading methods provide more consistent, reliable, and representative data for analysis.

Tests that consist of load increments applied at constant time intervals of 5, 10, or 15 minutes are called quick maintained-load tests or just "quick tests." In a quick test, the maximum load is not normally kept on the pile longer than any other load before the pile is unloaded. Unloading is done in about ten steps of no longer duration than a few minutes per load level. The quick test allows for applying one or more load increments beyond the minimum number that the particular test is designed for, that is, making use of the margin built into the test. In short, the quick test is, from the technical, practical, and economic points of view, superior to the "standard loading procedure."

A quick test should aim for 25 to 40 increments with the maximum load determined by the amount of reaction load available or the capacity of the pile. For routine cases, it may be preferable to stay at a maximum load of 200 percent of the intended allowable load. For ordinary test arrangements, where only the load and the pile head movement are monitored, time intervals of 10 minutes are suitable and allow for the taking of 2 to 4 readings for each increment. When testing instrumented piles, where the instruments take a while to read (scan), the time interval may have to be increased. To go beyond 20 minutes, however, should not be necessary. Nor is it advisable, because of the potential risk for influence of time-dependent movements, which may impair the test results. Usually, a quick test is completed within three to six hours.

In routine tests, cyclic loading or even single unloading and loading phases must be avoided, as they do little more than destroy the possibility of a meaningful analysis of the test results. There is absolutely no logic in believing that anything of value on load distribution and toe resistance can be obtained from an occasional unloading or from one or a few "resting periods" at certain load levels, when considering that we are testing a unit that is subjected to the influence of several soil types, is subjected to residual stress of unknown magnitude, exhibits progressive failure, etc., and when all we know is what is applied and measured at the pile head.

Interpretation of Failure Load

For a pile that is stronger than the soil, the failure load is reached when rapid movement occurs under sustained or slightly increased load (the pile plunges). However, this definition is inadequate, because plunging requires large movements. To be useful, a definition of failure load must be based on some mathematical rule and generate a repeatable value that is independent of scale relations and the opinions of the individual interpreter. Furthermore, it has to consider the shape of the load-movement curve or, if not, it must consider the length of the pile (which the shape of the curve indirectly does).

Fellenius [1975, 1980] compiled several methods used for interpreting failure or limit loads from a load-movement curve of a static loading test. The most well-known method is the offset limit method proposed by Davisson [1972]. This limit load is defined as the load corresponding to the movement that exceeds the elastic compression of the pile by an offset of 4 mm (0.15 inch) plus a value equal to the diameter of the pile divided by 120. It must be realized, however, that the offset limit load is a deformation limit that is determined taking into account the stiffness and length of the pile. It is not necessarily equal to the failure load of the pile.

The offset limit has the merit of allowing the engineer, when proof testing a pile for a certain allowable load, to determine *in advance* the maximum allowable movement for this load with consideration of the length and size of the pile. Thus, contract specifications can be drawn up including an acceptance criterion for piles proof tested according to quick-testing methods. The specifications can simply call for a test to at least twice the design load, as usual, and declare that at a test load equal to a factor F times the design load the movement shall be smaller than the Davisson offset from the elastic column compression of the pile. Normally, F would be chosen within a range of 1.8 to 2.0. The acceptance criterion could be supplemented with the requirement that the safety factor should also be smaller than a certain minimum value calculated on pile bearing failure defined according to the 80% criterion or other preferred criterion.

Influence of Errors

A static loading test is usually considered a reliable method for determining the capacity of a pile. However, even when using new manometers and jacks and calibrating them together, the applied loads is usually substantially overestimated. The error is usually about 10% to 15% of the applied load. Errors as large as 30% to 40% are not uncommon.

The reason for the error is that the jacking system is required to both provide the load and to measure it, and because load cells with moving parts are considerably less accurate than those without moving parts. For example, when calibrating testing equipment in the laboratory, one ensures that no eccentric loading, bending moments, or temperature variations influence the calibration. In contrast, all of these adverse factors are at hand in the field and influence the test results to an unknown extent, unless a load cell is used.

The above deals with the error of the applied load. The error in movement measurement can also be critical. Such errors do not originate in the precision of the reading—the usual precision is more than adequate—but in undesirable influences, such as heave or settlement of the reference beam during unloading the ground when loading the pile. For instance, one of the greatest spoilers of a loading test is the sun: The reference beam must be shielded from sunshine at all times.

Dynamic Analysis and Testing

The penetration resistance of driven piles provides a direct means of determining bearing capacity of a pile. In impacting a pile, a short-duration force wave is induced in the pile, giving the pile a downward velocity and resulting in a small penetration of the pile. Obviously, the larger the number of blows necessary to achieve a certain penetration, the stronger the soil. Using this basic principle, a large number of so-called pile-driving formulas have been developed for determining pile-bearing capacity. All these formulas are based on equalizing potential energy available for driving in terms of weight of hammer times its height of fall (stroke) with the capacity times penetration (set) for the blow. The set often includes a loss term.

The principle of the dynamic formulas is fundamentally wrong as wave action is neglected along with a number of other aspects influencing the penetration resistance of the pile. Nevertheless, pile-driving formulas have been used for many years and with some degree of success. However, success has been due less to the theoretical correctness of the particular formulas used and more to the fact that the users possessed adequate practical experience to go by. When applied to single-acting hammers, use of a dynamic formula may have some justification. However, dynamic formulas are the epitome of an outmoded level of technology and they have been or must be replaced by modern methods, such as the wave equation analysis and dynamic measurements, which are described below.

Pile-driving formulas or any other formula applied to vibratory hammers are based on a misconception. Vibratory driving works by eliminating resistance to penetration, not by overcoming it. Therefore, records of penetration combined with frequency, energy, amplitudes, and so on can relate only to the resistance not eliminated, not to the static pile capacity after the end of driving.

Pile-driving hammers are rated by the maximum potential energy determined as the ram weight times the maximum ram travel. However, diesel hammers and double-acting air/steam hammers, but also single-acting air/steam hammers, develop their maximum potential energy only during favorable combinations with the pile and the soil. Then, again, the energy actually transferred to the pile may vary due to variation in cushion properties, pile length, toe conditions, etc. Therefore, a relation between the hammer rated energy and measured transferred energy provides only very little information on the hammer.

For reliable analysis, all aspects influencing the pile driving and penetration resistance must be considered: hammer mass and travel, combustion in a diesel hammer, helmet mass, cushion stiffness, hammer efficiency, soil strength, viscous behavior of the soil, and elastic properties of the pile, to mention some. This analysis is made by means of commercially available wave equation programs, such as the GRLWEAP [GRL, 1993].

However, the parameters used as input into a wave equation program are really variables with certain ranges of values and the number of parameters included in the analysis is large. Therefore, the result of an analysis is only qualitatively correct, and not necessarily quantitatively correct, unless it is correlated to observations. The full power of the wave equation analysis is only realized when combined with dynamic measurements during pile driving by means of transducers attached to the pile head. The impact by the pile-driving hammer produces strain and acceleration in the pile which are picked up by the transducers and transmitted via a cable to a data acquisition unit (the Pile Driving Analyzer), which is placed in a nearby monitoring station. The complete generic field-testing procedure is described in the American Society for Testing and Materials, Standard for Dynamic Measurements, ASTM D-4945.

Dynamic testing, also called dynamic monitoring, is performed with the Pile Driving Analyzer (PDA). The PDA measurements provide much more information than just the value of the capacity of the pile, such as the energy transferred into the pile, the stresses in the pile, and the hammer performance. The dynamic data can be subjected to special analyses and provide invaluable information for determining that the piles are installed correctly, that the soil response is what was assumed in the design, and much more. For details, see Rausche *et al.* [1985] and Hannigan [1990]. Should difficulties develop with the pile driving at the site, the dynamic measurements can normally determine the reason for the difficulties and how to eliminate them. In the process, the frequent occurrence is avoided of having difficulties grow into a dispute between the contractor, the engineer, and the owner.

The dynamic measurements provide quantitative information of how the pile hammer functions, the compression and tension stresses that are imposed on the pile during the driving, and how the soil responds to the driving of the pile, including information pile static capacity. The dynamic measurements can also be used to investigate damage and defects in the pile, such as voids, cracks, spalling, local buckling, etc.

Dynamic records are routinely subjected to a detailed analysis called the CAPWAP signal matching analysis [Rausche *et al.*, 1972]. The CAPWAP analysis provides, first of all, a calculated static capacity and the distribution of resistance along the pile. However, it also provides several additional data, for example, the movement necessary to mobilize the full shear resistance in the soil (the quake) and damping values for input in a wave equation analysis.

Pile Group Example: Axial Design

The design approach is illustrated in the following example: A group of 25 piles consisting of 355-mm diameter, closed-end steel pipes are to be driven at equal spacing and in a square configuration at a site with the soil profile equal to that described earlier as background to Table 22.1. The 1.5-m earth fill over 36 m^2 area will be placed symmetrically around the pile group. The pile cap is 9.0 m^2 and placed level with the ground surface.

Each pile will be subjected to dead and live loads of 800 kN and 200 kN, respectively. The soils investigation has established a range of values of the soil parameters necessary for the calculations, such as density, compressibility, consolidation coefficient, as well as the parameters (β and N_t) used in the effective stress calculations of load transfer. A load-transfer analysis is best performed using a range (boundary values) of β and N_t parameters, which differentiate upper and lower limits of reasonable values. The analysis must include several steps in approximately the following order.

Determine first the range of installation length (using the range of effective stress parameters) as based on the required at-least capacity, which is stated, say, to be at least equal to the sum of the loads times a factor of safety of 3.0: 3.0(800 + 200) = 3000 kN.

To obtain a capacity of 3000 kN, applying the lower boundary of β and N_t, the piles have to be installed to a penetration into the sandy till layer of 5 m, that is, to a depth of 32 m. Table 22.6 presents the results of the load-transfer calculations for this embedment depth. The calculations

Table 22.6 Calculation of Pile Capacity

Area A_s = 1.115 m²/m	Live Load, Q_l = 200 kN	Shaft Resistance, R_s = 1817 kN
Area A_t = 0.099 m²	Dead Load, Q_d = 800 kN	Toe Resistance, R_t = 1205 kN
	Total Load = 1000 kN	Total Resistance, R_u = 3021 kN
Factor of Safety = 3.2	Depth to Neutral Plane = 26.51 m	Load at Neutral Plane = 1911 kN

Depth (m)	Total Stress (kPa)	Pore Pres. (kPa)	Eff. Stress (kPa)	Incr. R_s (kN)	$Q_d + Q_n$ (kN)	$Q_U - R_S$ (kN)
	Layer 1	Sandy Silt	ρ = 2000 kg/m³	β = 0.40		
0.00	30.00	0.00	30.00	0.0	800	3232
1.00 (GWT)	48.40	0.00	48.40	17.5	817	3215
4.00	104.30	30.00	74.30	82.1	900	3133
	Layer 2	Soft Clay	ρ = 1700 kg/m³	β = 0.30		
4.00	104.30	30.00	74.30		900	3133
5.00	120.13	43.53	76.60	25.2	925	3108
6.00	136.04	57.06	78.98	26.0	951	3082
7.00	152.03	70.59	81.44	26.8	978	3055
8.00	168.08	84.12	83.96	27.7	1005	3027
9.00	184.20	97.65	86.55	28.5	1034	2999
10.00	200.37	111.18	89.20	29.4	1063	2969
11.00	216.60	124.71	91.89	30.3	1094	2939
12.00	232.88	138.24	94.64	31.2	1125	2908
13.00	249.19	151.76	97.43	32.1	1157	2876
14.00	265.55	165.29	100.26	33.1	1190	2842
15.00	281.95	178.82	103.12	34.0	1224	2808
16.00	298.38	192.35	106.03	35.0	1259	2773
17.00	314.84	205.88	108.96	36.0	1295	2737
18.00	331.33	219.41	111.92	37.0	1332	2701
19.00	347.85	232.94	114.91	37.9	1370	2663
20.00	364.40	246.47	117.93	39.0	1409	2624
21.00	380.97	260.00	120.97	40.0	1449	2584
	Layer 3	Silty Sand	ρ = 2100 kg/m³	β = 0.50		
21.00	380.97	260.00	120.97		1449	2584
22.00	401.56	270.00	131.56	70.4	1519	2513
23.00	422.17	280.00	142.17	76.3	1596	2437
24.00	442.80	290.00	152.80	82.2	1678	2355
25.00	463.45	300.00	163.45	88.2	1766	2267
26.00	484.11	310.00	174.11	94.1	1860	2172
27.00	504.80	320.00	184.80	100.1	1960	2072
	Layer 4	Ablation Till	ρ = 2200 kg/m³	β = 0.55		
27.00	504.80	320.00	184.80		1960	2072
30.00	569.93	350.00	219.93	372.4	2332	1700
32.00	613.41	370.00	243.41	285.1	2617	1205
						N_t = 50

Note: Calculations by means of UNIPILE.

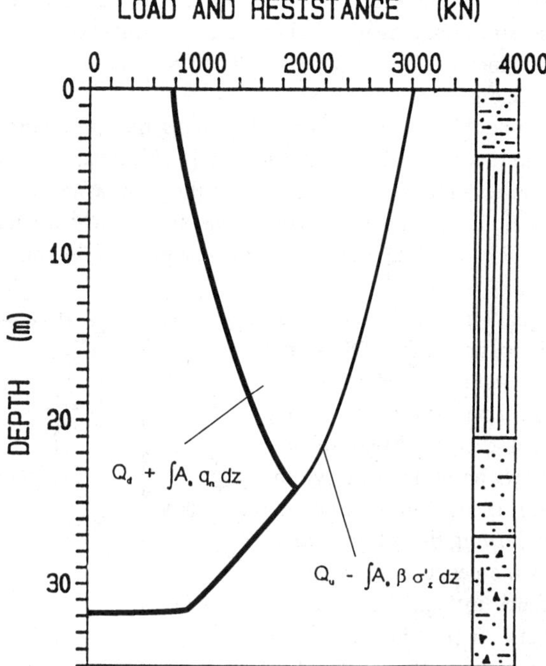

FIGURE 22.5 Load-transfer and resistance curves.

have been made with the Unipile program [Goudreault and Fellenius, 1990] and the results are presented in the format of a hand calculation to ease verifying the computer calculations. The precision indicated by stress values given with two decimals is to assist in the verification of the calculations and does not suggest a corresponding level of accuracy. Moreover, the effect of the 9 m^2 "hole" in the fill for the pile cap was ignored in the calculation. Were its effect to be included in the calculations, the calculated capacity would reduce by 93 kN or the required embedment length increase by 0.35 m.

The calculated values have been plotted in Fig. 22.5 in the form of two curves: a resistance curve giving the load transfer as in a static loading test to failure (3000 kN); and a load curve for long-term conditions, starting at the dead load of 800 kN and increasing due to negative skin friction to a maximum at the neutral plane. The load and resistance distributions for the example pile follow Eqs. (22.27) and (22.28).

Pile Group Settlement

The piles have reached well into the sand layers, which will not compress much for the increase of effective stress. Therefore, settlement of the pile group will be minimal and will not govern the design.

Installation Phase

The calculations shown above pertain to the service condition of the pile and are not quite representative for the installation (construction) phase. First, when installing the piles (these piles will be driven), the earth fill is not yet placed, which means that the effective stress in the soil is smaller than during the service conditions. More importantly, during the pile driving, large excess pore pressures are induced in the soft clay layer and, probably, also in the silty sand, which further reduces the effective stress. An analysis imposing increased pore pressures in these layers suggests that the EOID capacity is about 2200 kN at the end of the initial driving (EOID) using the depth and effective stress parameters indicated in Table 22.6 for the 32 m installation depth. The subsequent dissipation of the pore pressures will result in an about 600-kN soil increase of capacity due to set-up. (The stress increase due to the earth fill will provide the additional about 200 kN to reach the 3000-kN service capacity.)

The design must include the selection of the pile-driving hammer, which requires the use of software for wave equation analysis, called WEAP analysis [Goble *et al.*, 1980; GRL, 1993; Hannigan, 1990]. This analysis requires input of soil resistance in the form as result of static load-transfer analysis. For the installation (initial driving) conditions, the input is calculated considering the induced pore pressures. For restriking conditions, the analysis should consider the effect of soil set-up.

By means of wave equation analysis, pile capacity at initial driving—in particular the EOID—and restriking (RSTR) can be estimated. However, the analysis also provides information on what driving stresses to expect, indeed, even the length of time and the number of blows necessary to drive the pile. The most commonly used result is the bearing graph, that is, a curve showing the ultimate resistance (capacity) versus the penetration resistance (blow count) as illustrated in Fig. 22.6. As in the case of the static analysis, the parameters to input to a wave equation analysis can vary within upper and lower limits, which results in not one curve but a band of curves within envelopes as shown in Fig. 22.6. The input parameters consist of the particular hammer to use with its expected efficiency, the static resistance variation, dynamic parameters for the soil, such as damping and quake values, and many other parameters. It should be obvious that no one should expect a single answer to the analysis. Figure 22.6 shows that at EOID for the subject example, when the predicted capacity is about 2200 kN, the penetration resistance (PRES) will be about 10 blows/25 mm through about 20 blows/25 mm.

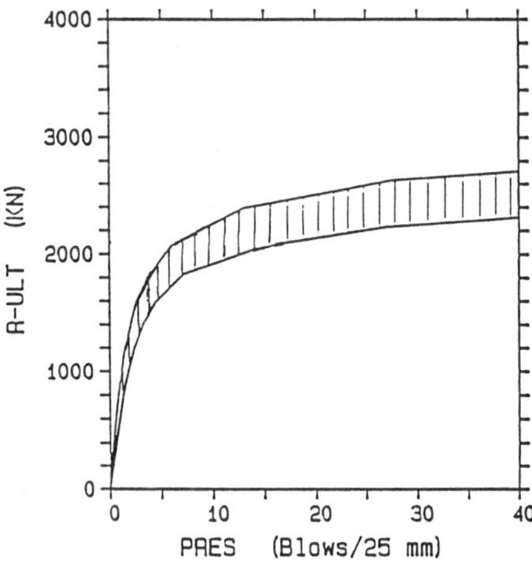

FIGURE 22.6 Bearing graph from WEAP analysis.

Notice that the wave equation analysis postulates observation of the actual penetration resistance when driving the piles, as well as a preceding static analysis. Then, common practice is to combine the analyses with a factor of safety ranging from 2.5 through 3.0.

Figure 22.6 demonstrates that the hammer selected for the driving cannot drive the pile against the 3000-kN capacity expected after full set-up. That is, restriking cannot prove out the capacity. This is a common occurrence. Bringing in a larger hammer may be a costly proposition. It may also be quite unnecessary. If the soil profile is well known, the static analysis correlated to the soil profile and to careful observation during the entire installation driving for a few piles, sufficient information is usually obtained to support a satisfactory analysis of the pile capacity and load transfer. That is, the capacity after set-up is inferred and sufficient for the required factor of safety.

When conditions are less consistent, when savings may result, and when safety otherwise suggests it to be good practice, the pile capacity is tested directly. Conventionally, this is accomplished by means of a static loading test. Since about 1975, dynamic tests have likewise often been performed. Static tests are costly and time-consuming, and are therefore usually limited to one or a few piles. In contrast, dynamic tests can be obtained quickly and economically and can be performed on several piles, thus providing assurance in numbers. For larger projects, static and dynamic tests are often combined. More recently, a new testing method called Statnamic has been proposed [Bermingham and Janes, 1989]. The Statnamic method is particularly intended for high-capacity bored piles (drilled piers).

When the capacity is determined by direct testing, the factor of safety of the design is reduced. The usual range is from 2.0 to 2.2. When the design and the installation are tested by means of static or dynamic proof testing on the site, the factor of safety is often reduced to the range of 1.8 through 2.0. Notice that the results of the tests must not just be given in terms of a capacity value (which can vary depending on how the ultimate resistance is defined); the load transfer should also be included in the analysis.

All analyses for a project must apply the same factor of safety. Therefore, for the subject example, when the designer knows that the capacity will be verified by means of a direct test, the minimum factor of safety to apply reduces to about 2.2, say. That is, the piles need only be driven to a final (after set-up) capacity of 2000 kN or 2200 kN. Considering the initial driving conditions, the capacity at EOID could be limited to about 1600 kN and this should be obtainable at a penetration into the till of slightly less than one meter and a penetration resistance of 4 to 5 blow/25 mm. Then, the subsequent increase due to set-up to 2200 kN is verified in a dynamic of static test *combined* with restriking to verify that the PRES values have increased to beyond about 10 blows/25 mm.

Summary of Axial Design of Piles

In summary, pile design consists of the following steps.

1. Compile soil data and perform a static analysis of the load transfer.
2. Verify that the ultimate pile resistance (capacity) is at least equal to the factor of safety times the sum of the dead and the live load (do not include the dragload in this calculation).
3. Verify that the maximum load in the pile, which is the sum of the dead load and the dragload is smaller than the appropriate factor of safety (usually 1.5) times the structural strength of the pile. (Do not include the live load in this calculation.)
4. Verify that the pile group settlement does not exceed the maximum deformation permitted by the structural design.
5. Perform wave equation analysis to select the pile-driving hammer and to decide on the driving and termination criteria (for driven piles).
6. Observe carefully the pile driving (construction) and verify that the work proceeds as anticipated. Document the observations (that is, keep a complete and carefully prepared log!).
7. When the factor of safety needs to be 2.5 or smaller, verify pile capacity by means of static or dynamic testing.

Design of Piles for Horizontal Loading

Because foundation loads act in many different directions depending on the load combination, piles are rarely loaded in true axial direction only. Therefore, a more or less significant lateral component of the total pile load always acts in combination with an axial load. The imposed lateral component is resisted by the bending stiffness of the pile and the shear resistance mobilized in the soil surrounding the pile.

An imposed horizontal load can also be carried by means of inclined piles, if the horizontal component of the axial pile load is at least equal to and acting in the opposite direction to the imposed horizontal load. Obviously, this approach has its limits, as the inclination cannot be impractically large. It should, preferably, not be greater than 4 (vertical) to 1 (horizontal). Also, only one load combination can provide the optimal lateral resistance.

In general, it is not correct to resist lateral loads by means of combining the soil resistance for the piles (inclined as well as vertical) with the lateral component of the vertical load for the inclined piles. The reason is that resisting an imposed lateral load by means of soil shear requires the pile to move against the soil. The pile will rotate due to such movement and an inclined pile will then either push up against or pull down from the pile cap, which will substantially change the axial load in the pile.

In design of vertical piles installed in a homogeneous soil and subjected to horizontal loads, an approximate and usually conservative approach is to assume that each pile can sustain a horizontal load equal to the passive earth pressure acting on an equivalent wall with depth of $6b$ and width $3b$, where b is the pile diameter or face-to-face distance [Canadian Foundation Engineering Manual, 1985].

Similarly, the lateral resistance of a pile group may be approximated by the soil resistance on the group calculated as the passive earth pressure over an equivalent wall with depth equal to $6b$ and width equal to:

$$L_e = L + 2b \tag{22.32}$$

where

L_e = equivalent width

L = the length, center-to-center, of the pile group in plan perpendicular to the direction of the imposed loads

b = the width of the equivalent area of the group in plan parallel to the direction of the imposed loads

The lateral resistance calculated according to Eq. (22.32) must not exceed the sum of the lateral resistance of the individual piles in the group. That is, for a group of n piles, the equivalent width of the group, L_e, must be smaller than n times the equivalent width of the individual pile, $6b$. For an imposed load not parallel to a side of the group, calculate for two cases, applying the components of the imposed load that are parallel to the sides.

The very simplified approach expressed above does not give any indication of movement. Nor does it differentiate between piles with fixed heads and those with heads free to rotate; that is, no consideration is given to the influence of pile bending stiffness. Because the governing design aspect with regard to lateral behavior of piles is lateral displacement, and the lateral capacity or ultimate resistance is of secondary importance, the usefulness of the simplified approach is very limited in engineering practice.

The analysis of lateral behavior of piles must involve two aspects:

1. *Pile response.* The bending stiffness of the pile, how the head is connected (free head, or fully or partially fixed head).
2. *Soil response.* The input in the analysis must include the soil resistance as a function of the magnitude of lateral movement.

The first aspect is modeled by treating the pile as a beam on an "elastic" foundation, which is accomplished by solving a fourth-degree differential equation with input of axial load on the pile, material properties of the pile, and the soil resistance as a nonlinear function of the pile displacement.

The derivation of lateral stress may make use of a simple concept called *coefficient of subgrade reaction* having the dimension of force per volume [Terzaghi, 1955]. The coefficient is a function of the soil density or strength, the depth below the ground surface, and the diameter (side) of the pile. In cohesionless soils, the following relation is used:

$$k_s = n_h \frac{z}{b} \tag{22.33}$$

where

k_s = coefficient of horizontal subgrade reaction

n_h = coefficient related to soil density

z = depth

b = pile diameter

The intensity of the lateral stress, p_z, mobilized on the pile at Depth z is then as follows:

$$p_z = k_s y_z b \tag{22.34}$$

where y_z = the horizontal displacement of the pile at depth z. Combining Eqs. (22.33) and (22.34), we get

$$p_z = n_h y_z z. \tag{22.35}$$

The relation governing the behavior of a laterally loaded pile is then as follows:

$$Q_h = EI\frac{d^4y}{dx^4} + Q_v\frac{d^2y}{dx^2} - p \tag{22.36}$$

where

Q_h = lateral load on the pile

EI = bending stiffness (flexural rigidity)

Q_v = axial load on the pile

Design charts have been developed that, for an input of imposed load, basic pile data, and soil coefficients, provide values of displacement and bending moment. See, for instance, the Canadian Foundation Engineering Manual [1985].

The design charts cannot consider all the many variations possible in an actual case. For instance, the $p - y$ curve can be a smooth rising curve, can have an ideal elastic-plastic shape, or can be decaying after a peak value. As an analysis without simplifying shortcuts is very tedious and time-consuming, resort to charts has been necessary in the past. However, with the advent of the personal computer, special software has been developed, which makes the calculations easy and fast. In fact, as in the case of pile-driving analysis and wave equation programs, engineering design today has no need for computational simplifications. Exact solutions can be obtained as easily as approximate ones. Several proprietary and public domain programs are available for analysis of laterally loaded piles.

One must not be led to believe that, because an analysis is theoretically correct, the results also predict the true behavior of the pile or pile group. The results must be correlated to pertinent experience and, lacking this, to a full-scale test at the site. If the experience is limited and funds are lacking for a full-scale correlation test, then a prudent choice is necessary of input data, as well as of margins and factors of safety.

Designing and analyzing a lateral test is much more complex than for the case of axial behavior of piles. In service, a laterally loaded pile almost always has a fixed-head condition. However, a fixed-head test is more difficult and costly to perform as opposed to a free-head test. A lateral test without inclusion of measurement of lateral deflection down the pile (bending) is of limited value. While an axial test should not include unloading cycles, a lateral test should be a cyclic test and include a large number of cycles at different load levels. The laterally tested pile is much more sensitive to the influence of neighboring piles than is the axially tested pile. Finally, the analysis of the test results is very complex and requires the use of a computer and appropriate software.

Seismic Design of Lateral Pile Behavior

Seismic design of lateral pile behavior is often taken as being the same as the conventional lateral design. A common approach is to assume that the induced lateral force to be resisted by piles is static and equal to a proportion, usually 10% of the vertical force acting on the foundation. If all the horizontal force is designed to be resisted by inclined piles, and all piles—including the vertical ones—are designed to resist significant bending at the pile cap, this approach is normally safe, albeit costly.

The seismic wave appears to the pile foundation as a soil movement forcing the piles to move with the soil. The movement is resisted by the pile cap; bending and shear are induced in the piles; and a horizontal force develops in the foundation, starting it to move in the direction of the wave. A half period later, the soil swings back, but the foundation is still moving in the first direction, and therefore the forces increase. This situation is not the same as the one originated by a static force.

Seismic lateral pile design consists of determining the probable amplitude and frequency of the seismic wave as well as the natural frequency of the foundation and structure supported by the piles. The first requirement is, as in all seismic design, that the natural frequency must not be the same as that of the seismic wave. Then the probable maximum displacement, bending, and shear induced at the pile cap are estimated. Finally the pile connection and the pile cap are designed to resist the induced forces.

There is at present a rapid development of computer software for use in detailed seismic design.

Defining Terms

Capacity: The maximum or ultimate soil resistance mobilized by a foundation unit.

Capacity, bearing: The maximum or ultimate soil resistance mobilized by a foundation unit subjected to downward loading.

Dragload: The load transferred to a deep foundation unit from negative skin friction.

Factor of safety: The ratio of maximum available resistance or of the capacity to the allowable stress or load.

Foundations: A system or arrangement of structural members through which the loads are transferred to supporting soil or rock.

Groundwater table: The upper surface of the zone of saturation in the ground.

Load, allowable: The maximum load that may be safely applied to a foundation unit under expected loading and soil conditions and determined as the capacity divided by the factor of safety.

Neutral plane: The location where equilibrium exists between the sum of downward acting permanent load applied to the pile and dragload due to negative skin friction and the sum of upward acting positive shaft resistance and mobilized toe resistance. The neutral plane is also where the relative movement between the pile and the soil is zero.

Pile: A slender deep foundation unit, made of wood, steel, concrete, or combinations thereof, which is either premanufactured and placed by driving, jacking, jetting, or screwing, or cast in situ in a hole formed by driving, excavating, or boring. A pile can be a non-displacement, low-displacement, or displacement type.

Pile head: The uppermost end of a pile.

Pile point: A special type of pile shoe.

Pile shaft: The portion of the pile between the pile head and the pile toe.

Pile shoe: A separate reinforcement attached to the pile toe of a pile to facilitate driving, to protect the lower end of the pile, and/or to improve the toe resistance of the pile.

Pile toe: The lowermost end of a pile. (Use of terms such as pile tip, pile point, or pile end in the same sense as pile toe is discouraged).

Pore pressure: Pressure in the water and gas present in the voids between the soil grains minus the atmospheric pressure.

Pore pressure, artesian: Pore pressure in a confined body of water having a level of hydrostatic pressure higher than the level of the ground surface.

Pore pressure, hydrostatic: Pore pressure varying directly with a free-standing column of water.

Pore pressure elevation, phreatic: The elevation of a groundwater table corresponding to a hydrostatic pore pressure equal to the actual pore pressure.

Pressure: Omnidirectional force per unit area. (Compare **stress.**)

Settlement: The downward movement of a foundation unit or soil layer due to rapidly or slowly occurring compression of the soils located below the foundation unit or soil layer, when the compression is caused by an increase of effective stress.

Shaft resistance, negative: Soil resistance acting downward along the pile shaft because of an applied uplift load.

Shaft resistance, positive: Soil resistance acting upward along the pile shaft because of an applied compressive load.

Skin friction, negative: Soil resistance acting downward along the pile shaft as a result of downdrag and inducing compression in the pile.

Skin friction, positive: Soil resistance acting upward along the pile shaft caused by swelling of the soil and inducing tension in the pile.

Stress: Unidirectional force per unit area. (Compare **pressure.**)

Stress, effective: The total stress in a particular direction minus the pore pressure.

Toe resistance: Soil resistance acting against the pile toe.

References

Altaee, A., Evgin, E., and Fellenius, B. H. 1992. Axial load transfer for piles in sand. I: Tests on an instrumented precast pile. *Can. Geotech. J.* 29, (1) 11–20.

Altaee, A., Evgin, E., and Fellenius, B. H. 1993. Load transfer for piles in sand and the critical depth. *Can. Geotech. J.* 30(2): 465–463.

American Society of Civil Engineers. 1989. *Proc. Cong. Found. Eng., Geotech. Eng. Div., ASCE.* Evanston, IL.

American Society of Civil Engineers. 1991. *Proc. Geotech. Eng. Cong. ASCE.* Boulder, CO.

American Society of Civil Engineers. 1994. Geotechnical Engineering Division. *Prof. Spec. Conf. Vert. Hor. Deform. Found. Embank., ASCE.* Houston, TX.

Bermingham, P. and Janes, M. 1989. An innovative approach to loading tests on high capacity piles. *Proc. Int. Conf. Piling Deep Found.,* 1: 409–427. A. A. Balkema, Rotterdam.

Canadian Geotechnical Society, 1985. *Canadian Foundation Engineering Manual* (2nd Ed.). BiTech, Vancouver.

Davisson, M. T., 1972. High capacity piles. *Proc. Innov. in Found. Const., ASCE.* Illinois Section. Chicago, pp. 81–112.

Fang, H.-Y. 1991. *Foundation Engineering* (2nd ed.). Van Nostrand Reinhold, New York.

Fellenius, B. H. 1975. Test loading of piles. Methods, interpretation and new proof testing procedure. *J. Geotech. Eng., ASCE.* 101 (GT9): 855–869.

Fellenius, B. H. 1980. The analysis of results from routine pile loading tests. *Ground Engineering,* London. 13(6):19–31.

Fellenius, B. H. 1984. Ignorance is bliss—And that is why we sleep so well. *Geot. News, Can. Geotech. Soc. and the U. S. Nat. Soc. of the Int. Soc. of Soil Mech. and Found. Eng.* 2(4): 14–15.

Fellenius, B. H. 1989. Tangent modulus of piles determined from strain data. *Proc. 1989 Found. Cong., Geotech. Div., ASCE.* 1: 500–510.

Goble, G. G., Rausche, F., and Likins, G. 1980. The analysis of pile driving—a state-of-the-art. *Proc. 1st Int. Sem. of the Appl. Stress-wave Theory to Piles, Stockholm.* A. A. Balkema, Rotterdam, pp. 131–161.

Goudreault, P. A. and Fellenius, B. H. 1990. *Unipile Version 1.02 Users Manual.* Unisoft Ltd., Ottawa, 76 p.

Goudreault, P. A. and Fellenius, B. H. 1993. *Unisettle Version 1.1 Users Manual.* Unisoft Ltd., Ottawa, 58 p.

GRL, 1993. *Background and Manual on GRLWEAP Wave Equation Analysis of Pile Driving*. Goble, Rausche, Likins, Cleveland, OH.

Hannigan, P. J. 1990. *Dynamic Monitoring and Analysis of Pile Foundation Installations*. Deep Foundation Institute, Sparta, NJ, 69 p.

Holtz, R. D. and Kovacs, W. D. 1981. *An Introduction to Geotechnical Engineering*. Prentice Hall, New York, 780 p.

Ismael, N. F. 1985. Allowable bearing pressure from loading tests on Kuwaiti soils. *Can. Geotech. J.* 22(2):151–157.

Janbu, N. 1963. Soil compressibility as determined by oedometer and triaxial tests. *Proc. Eur. Conf. Soil Mech. Found. Eng., Wiesbaden.* 1:19–25, 2:17–21.

Janbu, N. 1965. Consolidation of clay layers based on non-linear stress-strain. *Proc. 6th Int. Conf. on Soil Mech. Found. Eng., Montreal,* 2:83–87.

Janbu, N. 1967. Settlement calculations based on the tangent modulus concept. *University of Trondheim, Nor. Inst. of Tech., Geotech. Inst. Bull. 2.* 57 p.

Kany M. 1959. *Beitrag zur berechnung von flachengrundungen*. Wilhelm Ernst und Zohn, Berlin, 201 p.

Ladd, C. C. 1991. Stability evaluation during staged construction. The twenty-second Terzaghi Lecture. *J. of Geotech. Eng., ASCE.* 117(4): 540–615.

Lambe, T. W. and Whitman, R. V. 1979. *Soil Mechanics*. John Wiley & Sons, New York.

Meyerhof, G. G. 1976. Bearing capacity and settlement of pile foundations. The eleventh Terzaghi lecture. *J. of Geotech. Eng., ASCE.* 102(GT3): 195–198.

Mitchell, J. K. 1976. *Fundamentals of Soil Behavior.* John Wiley & Sons, New York.

Newmark, N. M. 1935. Simplified computation of vertical stress below foundations. *Univ. Illinois, Eng. Expnt. Stat. Circ.,* 24.

Newmark, N. M. 1942. Influence chart for computation of stresses in elastic foundations. *Univ. Illinois Eng. Expnt. Stat. Bull. 338,* 61(92).

Nordlund, R. L. 1963. Bearing capacity of piles in cohesionless soils. *J. Geotech. Eng., ASCE.* 89(SM3): 1–35.

Rausche, F., Moses, F., and Goble. G. G. 1972. Soil resistance predictions from pile dynamics. *J. Geotech. Eng., ASCE.* 98(SM9): 917–937.

Rausche, F., Goble, G. G., and Likins, G. E. 1985. Dynamic determination of pile capacity. *J. of the Geotech. Eng. ASCE.* 111(3): 367–383.

Schmertmann, J. H. 1978. *Guidelines for Cone Penetration Test, Performance, and Design.* U.S. Federal Highway Administration, Washington, Report FHWA-TS-78-209, 145 p.

Taylor, D. W. 1948. *Fundamentals of Soil Mechanics.* Wiley & Sons, New York.

Terzaghi, K. 1955. Evaluation of coefficients of subgrade reaction. *Geotechnique.* 5(4): 297–326.

Vesic, A. S. 1967. *A Study of Bearing Capacity of Deep Foundations.* Final Report Project B-189, Geor. Inst. of Tech., Engineering Experiment Station, Atlanta, GA, 270 p.

Vijayvergiya, V. N. and Focht, J. A., Jr. 1972. A new way to predict the capacity of piles in clay. *Proc. 4th Ann. Offshore Tech. Conf.* 2: 865–874.

Winterkorn, H. F. and Fang, H.-Y. 1975. *Foundation Engineering* (1st ed.). Van Nostrand Reinhold.

For Further Information

Foundation Engineering Handbook. (1st ed.). 1975. Edited by Winterkorn, H. F. and Fang, H.-Y. Van Nostrand Reinhold, New York.

Foundation Engineering Handbook. (2nd ed.). 1991. Edited by Fang, H.-Y., Van Nostrand Reinhold, New York.

American Society of Civil Engineers, Geotechnical Engineering Division, *Congress on Foundation Engineering.* F. H. Kulhawy, editor. Evanston, IL, June 1989, Vols. 1 and 2.

American Society of Civil Engineers, *Geotechnical Engineering Congress.* F. G. McLean, editor. Boulder, CO, June 1991, Vols. 1 and 2.

American Society of Civil Engineers, Geotechnical Engineering Division, *Speciality Conference on Vertical and Horizontal Deformations for Foundations and Embankments,* Houston, June 1994, Vols. 1 and 2.

Lambe, T. W. and Whitman, R. V. 1979. *Soil Mechanics.* Series in Soil Engineering. John Wiley & Sons, New York.

Mitchell, J. K. 1976. *Fundamentals of Soil Behavior.* Series in Soil Engineering. John Wiley & Sons, New York.

Taylor, D. W. 1948. *Fundamentals of Soil Mechanics.* John Wiley & Sons, New York.

23

Geosynthetics

R. D. Holtz
University of Washington, Seattle

23.1 Introduction

In only a very few years, geosynthetics (geotextiles, geogrids, and geomembranes) have joined the list of traditional civil engineering construction materials. Often the use of a geosynthetic can significantly increase the safety factor, improve performance, and reduce costs in comparison with conventional construction alternatives. In the case of embankments on extremely soft foundations, geosynthetics can permit construction to take place at sites where conventional construction alternatives would be either impossible or prohibitively expensive.

Two recent conferences dealt specifically with geosynthetics and soil improvement, and their proceedings deserve special mention. See Holtz [1988a] and Borden *et al.* [1992].

After an introduction to geosynthetic types and properties, recent developments in the use of geosynthetics for the improvement of soils in foundations and slopes will be summarized. Primary applications are to drainage and erosion control systems, temporary and permanent roadways and railroads, soil reinforcement, and waste containment systems.

There are a number of **geotextile**-related materials, such as webs, mats, nets, grids, and sheets, which may be made of plastic, metal, bamboo, or wood, but so far there is no ASTM (American Society for Testing and Materials) definition for these materials. They are "geotextile-related" because they are used in a similar manner, especially in reinforcement and stabilization situations, to geotextiles.

Geotextiles and related products such as nets and grids can be combined with **geomembranes** and other synthetics to take advantage of the best attributes of each component. These products are called *geocomposites*, and they may be composites of geotextile–geonets, geotextile–geogrids, geotextile–geomembranes, geomembrane–geonets, geotextile–polymeric cores, and even three-dimensional polymeric cell structures. There is almost no limit to the variety of geocomposites that are possible and useful. The general generic term encompassing all these materials is *geosynthetic*.

0-8493-8953-4/95/$0.00 + $.50

Types and Manufacture

A convenient classification for geosynthetics is given in Fig. 23.1. For details on the composition, materials, and manufacture of geotextiles and related materials, see Koerner and Welsh [1980], Rankilor [1981], Giroud and Carroll [1983], Christopher and Holtz [1985], Veldhuijzen van Zanten [1986], Ingold and Miller [1988], and Koerner [1990a]. Most geosynthetics are made from synthetic polymers such as polypropylene, polyester, polyethylene, polyamide, and PVC. These materials are highly resistant to biological and chemical degradation. Natural fibers such as cotton, jute, and bamboo can be used as geotextiles and geogrids, especially for temporary applications, but they have not been promoted or researched as widely as geotextiles made from synthetic polymeric materials.

Applications and Functions

More than 150 separate applications of geosynthetics have been identified [Koerner, 1990a]. Table 23.1 lists those for geotextiles and related products, and the primary application areas are filtration and drainage, erosion protection and control, roadways and asphalt pavement overlays, and reinforced walls, slopes, and embankments. The primary geotextile functions are filtration, drainage, separation, and reinforcement. In virtually every application, the geotextile also provides one or more secondary functions of filtration, drainage, separation, and reinforcement (see Table 23.1). Geogrids, nets, webs, fascines, and so on are primarily used for roadway stabilization and all

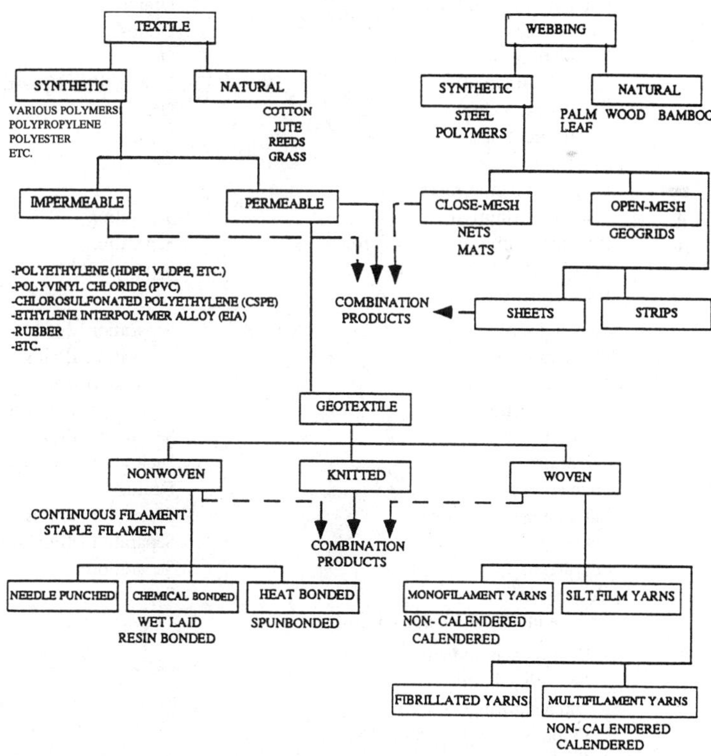

FIGURE 23.1 Classification of geosynthetics. [Adapted (from Rankilor, P. R. 1981. *Membranes in Ground Engineering*. John Wiley & Sons, New York) by Christopher, B. R. and Holtz, R. D. 1989. *Geotextile Design and Construction Guidelines*. U.S. Federal Highway Administration, National Highway Institute, Report No. FHWA-HI-90-001.]

Table 23.1 Representative Applications and Controlling Functions of Geotextiles

Primary Function	Application	Secondary Functions
Separation	Unpaved roads (temporary and permanent)	Filter, drains, reinforcement
	Paved roads (secondary and primary)	Filter, drains
	Construction access roads	Filter, drains, reinforcement
	Working platforms	Filter, drains, reinforcement
	Railroads (new construction)	Filter, drains, reinforcement
	Railroads (rehabilitation)	Filter, drains, reinforcement
	Landfill covers	Drains, reinforcement
	Preloading (stabilization)	Drains, reinforcement
	Marine causeways	Filter, drains, reinforcement
	General fill areas	Filter, drains, reinforcement
	Paved and unpaved parking facilities	Filter, drains, reinforcement
	Cattle corrals	Filter, drains, reinforcement
	Coastal and river protection	Filter, drains, reinforcement
	Sports fields	Filter, drains
Drainage-transmission	Retaining walls	Separation, filter
	Vertical drains	Separation, filter
	Horizontal drains	Reinforcement
	Below membranes (drainage of gas and water)	Reinforcement
	Earth dams	Filter
	Below concrete (decking and slabs)	—
Reinforcement	Pavement overlays	—
	Concrete overlays	—
	Subbase reinforcement in roadways and railways	Filter
	Retaining structures	Drains
	Membrane support	Separation, drains, filter
	Embankment reinforcement	Drains
	Fill reinforcement	Drains
	Foundation support	Drains
	Soil encapsulation	Drains, filter separation
	Net against rockfalls	Drains
	Fabric retention systems	Drains
	Sandbags	—
	Reinforcement of membranes	—
	Load redistribution	Separation
	Bridging nonuniformity soft soil areas	Separation
	Encapsulated hydraulic fills	Separation
	Bridge piles for fill placement	—
Filter	Trench drains	Separation, drains
	Pipe wrapping	Separation, drains
	Base course drains	Separation, drains
	Frost protection	Separation, drainage, reinforcement
	Structural drains	Separation, drains
	Toe drains in dams	Separation, drains
	High embankments	Drains
	Filter below fabric-form	Separation, drains
	Silt fences	Separation, drains
	Silt screens	Separation
	Culvert outlets	Separation
	Reverse filters for erosion control:	
	Seeding and mulching	
	Beneath gabions	
	Ditch amoring	
	Embankment protection, coastal	
	Embankment protection, rivers and streams	
	Embankment protection, lakes	
	Vertical drains (wicks)	Separation

Source: Christopher, B. R. and Holtz, R. D. 1989. *Geotextile Design and Construction Guidelines.* U.S. Federal Highway Administration, National Highway Institute, Report No. FHWA-HI-90-001.

types of earth reinforcement. They also have significant applications in waste containment systems when used together with geotextile filters and geomembranes (geocomposites).

Historical and Recent Developments

Giroud [1986] and Fluet [1988] give interesting accounts of the history of geosynthetics. Few developments have had such a rapid growth and strong influence on so many aspects of civil engineering practice. In 1970, there were only five or six geotextiles available. Today more than 250 different materials are sold in the U.S. as geotextiles and geogrids; worldwide, the number probably exceeds 400. The size of the geotextile market, both in terms of square meters produced and their dollar value, is indicative of their influence. For example, in 1991, more than 300 million square meters of geotextiles and related materials such as geogrids were sold in North America. The worldwide consumption was probably more than twice this amount. The value of these materials is probably close to $500 million. Since the total cost of the construction is at least four or five times the cost of the geosynthetic itself, the impact of these materials on civil engineering construction is very large indeed.

Another important recent development is the rapid growth of the literature on geosynthetics. This literature includes books, conference proceedings, and technical journals devoted to various aspects of geosynthetics. A listing of geosynthetic literature can be found in Holtz and Paulson [1988] and Cazzuffi and Anzani [1992].

Design and Selection

In the early days of geotextiles, selection and specification were primarily by type or brand name. This was satisfactory as long as there were only a few geotextiles on the market and the choices available to the designer were limited. Today, however, with such a wide variety of geosynthetics available, this approach is inappropriate. The recommended approach for designing, selecting, and specifying geosynthetics is no different than what is commonly practiced in any geotechnical engineering design. First, the design should be made *without* geosynthetics to see if they really are needed. If conventional solutions are impractical or uneconomical, design calculations using reasonable engineering estimates of the required geosynthetic properties are carried out. Next, generic or performance type specifications are written so that the most appropriate and economical geosynthetic is selected, consistent with the properties required for its function, constructibility, and endurance. In addition to conventional soils and materials testing, geosynthetic testing will very likely be required. Finally, as with any other construction, design with geosynthetics is not complete until construction has been satisfactorily carried out. Therefore, careful field inspection during construction is essential for a successful project.

Properties and Tests

Because of the wide variety of geosynthetics available (Fig. 23.1), along with their different polymers, filaments, bonding mechanisms, thicknesses, masses, and so on, they have a wide range of physical and mechanical properties. A further complicating factor is the variability of some properties, even within the same manufactured lot or roll. Differences may sometimes be due to the test procedures themselves.

Many of our current geosynthetic tests were developed by the textile and polymer industries, often for quality control of the manufacturing process. Consequently the test values from these tests may not relate well to the civil engineering conditions of a particular application. Furthermore, soil confinement or interaction is not accounted for in most geosynthetics testing. Research is now underway to provide test procedures and soil–geosynthetic interaction properties which are more appropriate for design. Geotextile testing is discussed in detail by Christopher and Holtz [1985] and Koerner [1990a].

Specifications

Good specifications are essential for the success of any civil engineering project, and this is even more critical for projects in which geosynthetics are to be used. Christopher and DiMaggio [1984] provide some guidance on writing generic and performance-based geotextile specifications.

23.2 Filtration, Drainage, and Erosion Control

One of the most important uses for geotextiles is as a filter in drainage and erosion control applications. Drainage examples include trench and French drains, interceptor drains, blanket drains, pavement edge drains, and structural drains, to name just a few. Permanent erosion control applications include coastal and lakeshore revetments, stream and canal banks, cut and fill slope protection, and scour protection. In all these applications, geotextiles are used to replace graded granular filters used in conjunction with the drainage aggregate, perforated pipe, rip rap, and so on. When properly designed, geotextiles can provide comparable performance at less cost, provide consistent filtration characteristics, and they are easier and therefore cheaper to install. Although erosion control technically does not improve the soil, prevention of both external and internal erosion in residual and structured soils is an important design consideration.

Geotextiles can also be used to temporarily control and minimize erosion or transport of sediment from unprotected construction sites. In some cases, geotextiles provide temporary protection after seeding and mulching but before vegetative ground cover can be established. Geotextiles may also be used as armor materials in diversion ditches and at the ends of culverts to prevent erosion. Probably the most common application is for silt fences, which are a substitute for hay bales or brush piles, to remove suspended particles from sediment-laden runoff water.

Filtration Design Concepts

For a geotextile to satisfactorily replace a graded granular filter, it must perform the same functions as a graded granular filter: (1) prevent soil particles from going into suspension; (2) allow soil particles already in suspension to pass the filter (to prevent clogging or blinding); and (3) have a sufficiently high permeability and flow rate so that no back pressure develops in the soil being protected.

How a geotextile filter functions is discussed in detail by Bell *et al.* [1980, 1982], Rankilor [1981], Christopher and Holtz [1985], and Koerner [1990a]. The factors that control the design and performance of a geotextile filter are (1) physical properties of the geotextile, (2) soil characteristics, (3) hydraulic conditions, and (4) external stress conditions.

After a detailed study of research carried out here and in Europe on both conventional and geotextile filters, Christopher and Holtz [1985] developed a widely used design procedure for geotextile filters for drainage and permanent erosion-control applications. The level of design required depends on the critical nature of the project and the severity of the hydraulic and soil conditions. Especially for critical projects, consideration of the risks involved and the consequences of possible failure of the geotextile filter require great care in selecting the appropriate geotextile. For such projects and for severe hydraulic conditions, very conservative designs are recommended. As the cost of the geotextile is usually a minor part of the total project or system cost, geotextile selection should not be based on the lowest material cost. Also, expenses should not be reduced by eliminating laboratory soil–geotextile performance testing when such testing is recommended by the design procedure.

The three design criteria which must be satisfied are (1) soil retention (piping resistance), (2) permeability, and (3) clogging criteria. For both permeability and clogging, different approaches are recommended for critical/severe applications. Furthermore, laboratory filtration tests must be performed to determine clogging resistance. It is not sufficient to simply rely on retention and

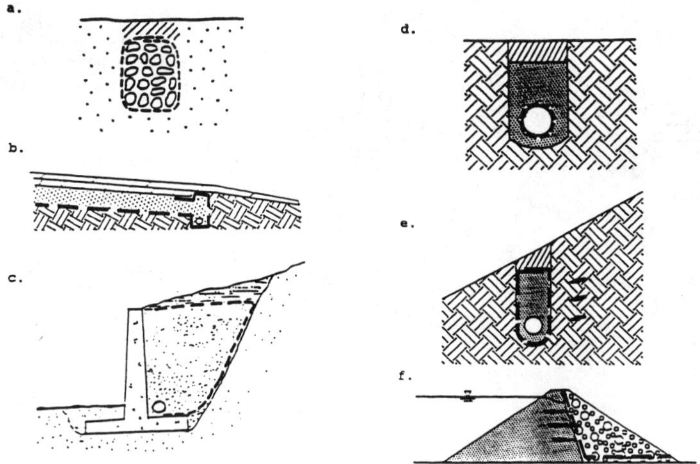

FIGURE 23.2 Drainage applications: (a) trench drains, (b) blanket and pavement edge drains, (c) structural drains, (d) pipe wraps, (e) interceptor drains, and (f) drains in dams. (*Source:* Christopher, B. R. and Holtz, R. D. 1989. *Geotextile Design and Construction Guidelines.* U.S. Federal Highway Administration, National Highway Institute, Report No. FHWA-HI-90-001.)

permeability to control clogging potential. Finally, mechanical and index property requirements for durability and constructibility are given. Constructibility is sometimes called *survivability*, and it depends on the installation conditions. The best geotextile filter design in the world is useless if the geotextile does not survive the construction operations.

Fischer *et al.* [1990] have proposed a design procedure based on the pore size distribution of the geotextile filter.

Applications

The more common applications of geotextiles in drainage and erosion control are shown in Figs. 23.2, 23.3, and 23.4. Construction of geotextile filters in these applications is described in detail by Christopher and Holtz [1985, 1989].

Prefabricated Drains

In the last few years, prefabricated geocomposite drainage materials have become available as a substitute for conventional drains with and without geotextiles. Geocomposites are probably most practical for lateral drainage situations [Figs. 23.2(b), (c), and (e)] and in waste containment systems in conjunction with clay or geomembrane liners [Koerner, 1990a]. Another important soil improvement application of geocomposites is the use of prefabricated vertical ("wick") drains to accelerate the consolidation of soft compressible cohesive soil layers. Because they are much less expensive to install, geocomposite drains have made conventional sand drains obsolete. Design principles and installation techniques are given by Holtz *et al.* [1991].

23.3 Geosynthetics in Temporary and Permanent Roadways and Railroads

A very important use of geosynthetics is to stabilize roadways and railroads. The primary function of the geosynthetic in these applications is that of separation; secondary functions such as

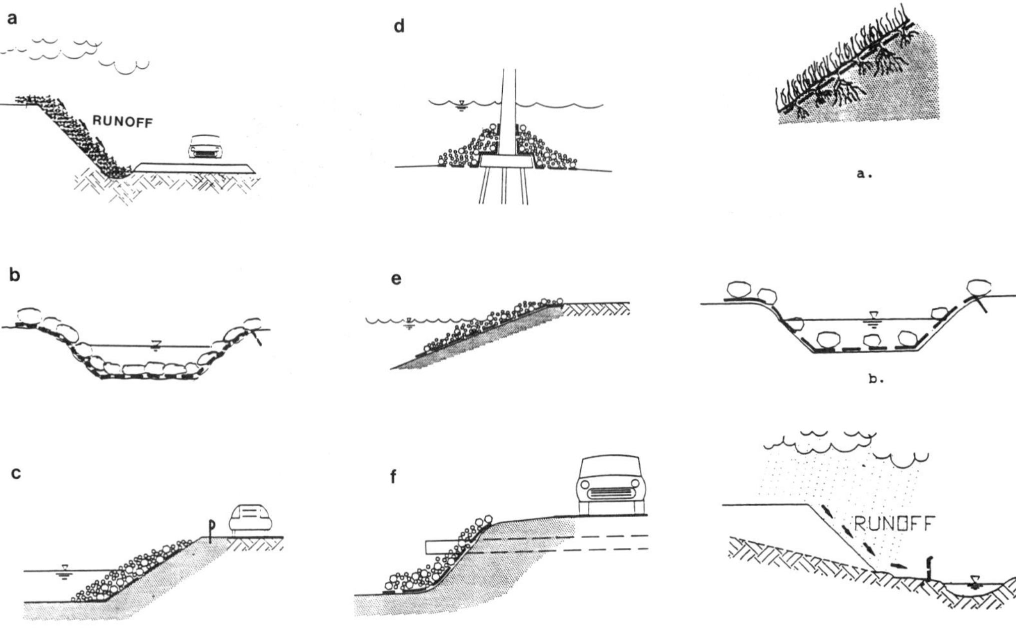

FIGURE 23.3 Permanent erosion control applications: (a) slope protection, (b) diversion ditches, (c) stream and canal banks, (d) scour protection, (e) beaches, and (f) culvert outlets. (*Source:* Christopher, B. R. and Holtz, R. D. 1989. *Geotextile Design and Construction Guidelines.* U.S. Federal Highway Administration, National Highway Institute, Report No. FHWA-HI-90-001.)

FIGURE 23.4 Temporary erosion control applications: (a) vegetative cover, (b) diversion ditches, and (c) silt fences. (*Source:* Christopher, B. R. and Holtz, R. D. 1989. *Geotextile Design and Construction Guidelines.* U.S. Federal Highway Administration, National Highway Institute, Report No. FHWA-HI-90-001.)

filtration, drainage, and possibly reinforcement may also be more or less present. Recommended references on geotextiles in various aspects of roads and railroads are Steward *et al.* [1977], Rankilor [1981], Christopher and Holtz [1985, 1989], Fluet [1986], and Koerner [1990a].

In the early days of geotextiles, a number of design procedures were developed to use geotextiles in temporary facilities such as haul roads, contractor staging yards, mine and timber storage and haulage facilities, temporary construction platforms, and so on. Virtually all methods provided design charts which indicated that the use of a geotextile could reduce the thickness of aggregate required for a given subgrade and traffic condition. Unfortunately, most of these design methods permitted some rutting to occur in the subgrade even with the geotextile in place; obviously, rutting is not desirable for permanent facilities. Although the geotextile functions are similar for both temporary and permanent roadways, because of the difference in performance requirements, design methods for temporary roads should *not* be used to design permanent roads. However, as noted by Christopher and Holtz [1985], there still are a number of significant advantages to using geotextiles in permanent roadways.

Geosynthetics in roadways are most cost effective on soft subgrades (CBR values less than 2) with sensitive silty or clayey soils (CH, ML, MH, A-6, A-7, etc.), and at sites with the water table near the ground surface and where there is poor equipment mobility. If such conditions are present in residual and structured soil subgrades, geotextiles should also be effective stabilizers of subgrades on these soils. One important consequence of using geotextiles in roadway construction is that the contractors are required to be more careful during construction to avoid damaging the geotextile. Such care usually results in reduced stress and damage to the subgrade soils, which is especially appropriate for residual and structured soil subgrades.

Although there is strong evidence that the primary function of geotextiles in roadways is separation [Steward *et al.*, 1977], research has suggested a number of possible geosynthetic reinforcement mechanisms [Christopher and Holtz, 1985]. For example, a geogrid placed in the aggregate base course will provide a significant increase in load-carrying capacity [Haas *et al.*, 1988].

Design Approaches

Design of geosynthetics in both temporary and permanent roadways is discussed at length by Christopher and Holtz [1985, 1989] and Koerner [1990a]. Recommended temporary road design methods are discussed by Steward, *et al.* [1977], Bender and Barenberg [1978], Giroud and Noiray [1981], Haliburton and Barron [1983], Houlsby *et al.* [1989], and Milligan *et al.* [1989]. The last two papers are summarized by Jewell [1990].

Because most of the available design methods for permanent roadways lack field or controlled experimental verification, Christopher and Holtz [1991] proposed a simple procedure which assumes that the first aggregate stabilization layer often required to permit construction at very soft, wet sites acts as an unpaved road subjected to only a few passes of construction equipment. The geotextile acts primarily as a separator, and geotextile survivability must be taken into account [Christopher and Holtz, 1985, 1989]. It should be noted that no structural support is attributed to the geotextile in this design procedure. The method does not change the design thickness required for traffic or other design considerations. It only allows aggregate to be saved which would otherwise be consumed in constructing the first stabilization lift.

Detailed guidelines for the construction of roads using geotextiles are given by Christopher and Holtz [1985, 1989].

23.4 Geosynthetics for Reinforcement

One of the most important applications of geosynthetics in geotechnical engineering is soil reinforcement, an important aspect of soil improvement. Applications include reinforcing the base of embankments constructed on very soft foundations, increasing the stability and steepness of slopes, and reducing the earth pressures behind retaining walls and abutments. In the first two applications, geosynthetics permit construction that otherwise would be cost-prohibitive or in some cases impossible. In the case of retaining walls, significant cost savings are possible in comparison with conventional retaining wall construction.

Other reinforcement and stabilization applications in which geosynthetics have also proven to be very effective include large area stabilization and natural slope reinforcement.

Reinforced Embankments

In only a few years, geosynthetic reinforcement has joined the list of the more traditional soil improvement methods for increasing the stability of embankments on very soft foundations. Concepts for using geosynthetics for reinforcement are indicated in Fig. 23.5. Discussion of these concepts as well as detailed design procedures are given by Christopher and Holtz [1985, 1989], Bonaparte *et al.* [1987], Bonaparte and Christopher [1987], Holtz [1989a, 1989b, 1990], and Humphrey and Rowe [1991]. As our approach is to design against failure, types of unsatisfactory behavior (Fig. 23.6) which are likely to require reinforcement must be assumed so that an appropriate stability analysis may be carried out.

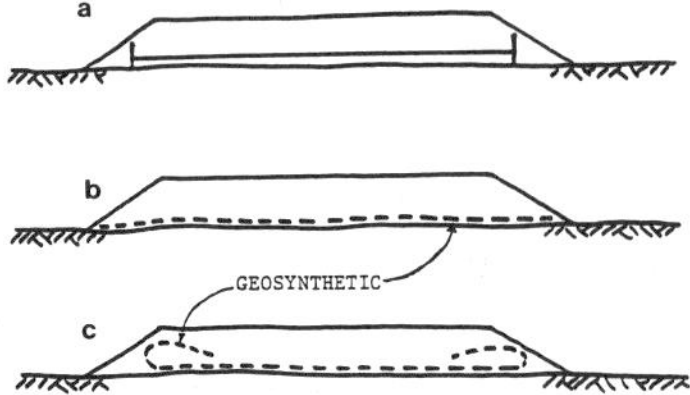

FIGURE 23.5 Concepts for using geosynthetics to reinforce embankments. (*Source:* Holtz, R. D. 1990. Design and construction of geosynthetically reinforced embankments on very soft soils. In *Performance of Reinforced Soil Structures, Proc. Int. Reinforced Soil Conf.* Glasgow, British Geotechnical Society, pp. 391–402.)

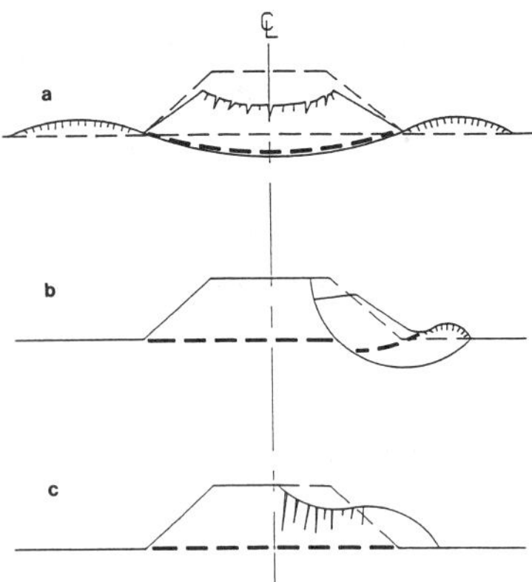

FIGURE 23.6 Unsatisfactory behavior that may occur in reinforced embankments. (*Source:* Christopher, B. R. and Holtz, R. D. 1989. *Geotextile Design and Construction Guidelines.* U.S. Federal Highway Administration, National Highway Institute, Report No. FHWA-HI-90-001.)

The design steps are as follows:

1. Check overall bearing capacity.
2. Check edge bearing capacity or edge slope stability.
3. Conduct a sliding wedge analysis for embankment spreading.
4. Perform an analysis to limit geosynthetic deformations.
5. Determine geosynthetic strength requirements in the longitudinal direction.

Based on these stability calculations, the minimum geosynthetic strengths required for stability at an appropriate factor of safety can be determined. A detailed discussion of geosynthetic properties and specifications is given in the references cited above.

The importance of proper construction procedures for geosynthetic reinforced embankments cannot be overemphasized. A specific construction sequence is usually required in order to avoid failures during construction. Appropriate site preparation, low ground pressure equipment, small initial lift thicknesses, and partially loaded hauling vehicles may be required. Fill placement, spreading, and compaction procedures are also very important. A detailed discussion of construction procedures for reinforced embankments on very soft foundations is given by Christopher and Holtz [1985, 1989].

It should be noted that all geosynthetic seams must be positively joined. For geotextiles, this means sewing; for geogrids, some type of positive clamping arrangement must be used. Careful inspection is essential, as the seams are the "weak link" in the system and seam failures are common in embankments which are improperly constructed.

Slope Stability

Geosynthetics have been very effectively used many times both here and abroad to stabilize failed slopes. Cost savings result because the slide debris is reused together with geosynthetic reinforcement in the reinstatement of the slope. It is also possible that even though foundation conditions are satisfactory, slopes of a compacted embankment fill may be unstable at the desired slope angle. Costs of fill and right-of-way plus other considerations may require a steeper slope

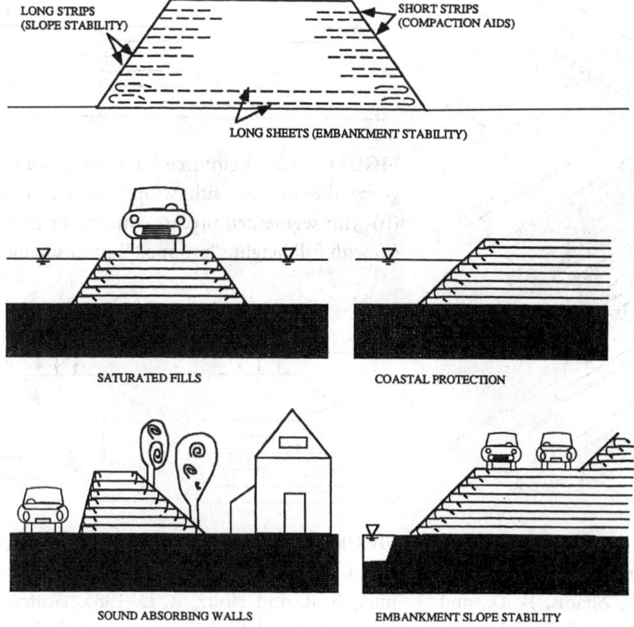

FIGURE 23.7 Examples of multilayer geosynthetic slope reinforcement. (*Source:* Christopher, B. R. and Holtz, R. D. 1985. *Geotextile Engineering Manual.* STS Consultants Ltd., Northbrook, IL; U.S. Federal Highway Administration, Report No. FHWA-TS-86/203.)

than is stable in compacted soils. Typically, embankment slope reinforcement is placed in layers as the embankment is constructed in lifts (see Fig. 23.7).

Embankment slope stability analyses have been developed by Murray [1982], Jewell *et al.* [1984], Schneider and Holtz [1986], Schmertmann *et al.* [1987], Verduin and Holtz [1989], Christopher *et al.* [1989], and Christopher and Leshchinsky [1991].

Reinforced Retaining Walls and Abutments

Retaining walls are used where a slope is uneconomical or not technically feasible. Retaining walls with reinforced backfills are very cost-effective, especially for higher walls; also, they are more flexible and thus more suitable for poor foundation conditions. The concept was developed in France by H. Vidal in the mid-1960s. His system, called *reinforced earth,* is shown in Fig. 23.8. Steel strips or ties are used to reduce the earth pressure against the wall face ("skin"). The design and construction of Vidal-type reinforced earth walls are now well established, and many thousands have been successfully built throughout the world in the last 25 years.

The use of geotextiles as reinforcing elements started in the early 1970s because of questions about possible corrosion of the metallic strips in reinforced earth. Systems using sheets of geosynthetics rather than steel strips are shown in Fig. 23.9. The most commonly used system is shown in Fig. 23.9(a), in which the geosynthetic provides the facing as well as the reinforcing elements in the wall.

Most designs of geotextile reinforced retaining walls utilize classical earth pressure theory combined with tensile-resistance *tiebacks,* in which the reinforcement extends back beyond the assumed failure plane (see Fig. 23.10). This approach is discussed by Christopher and Holtz [1985, 1989], Bonaparte *et al.* [1987], Christopher *et al.* [1989], and Allen and Holtz [1991]. Both internal and external stability must be considered. Failure modes and geosynthetic properties required for

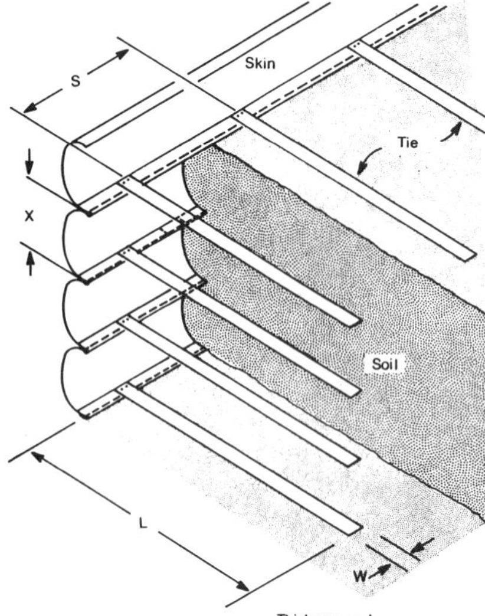

FIGURE 23.8 Components of reinforced earth. (*Source:* Lee, K. L., Adams, B. D., and Vagneron, J. M. J. 1973. Reinforced earth retaining walls. *J. Soil Mech. Found., ASCE.* 99(SM10):745–764.

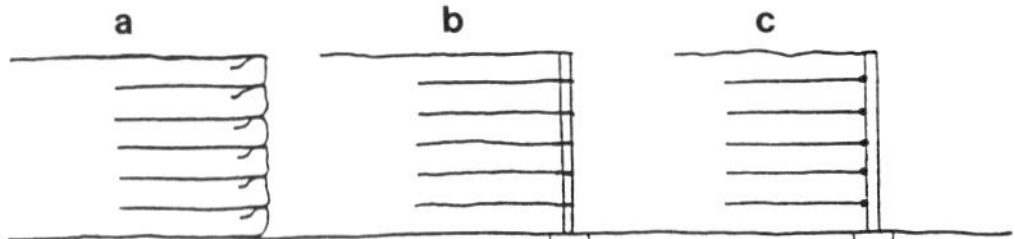

FIGURE 23.9 Reinforced retaining wall systems using geosynthetics: (a) with wraparound geosynthetic facing, (b) with segmented precast concrete or timber panels, and (c) with full height ("propped") precast panels.

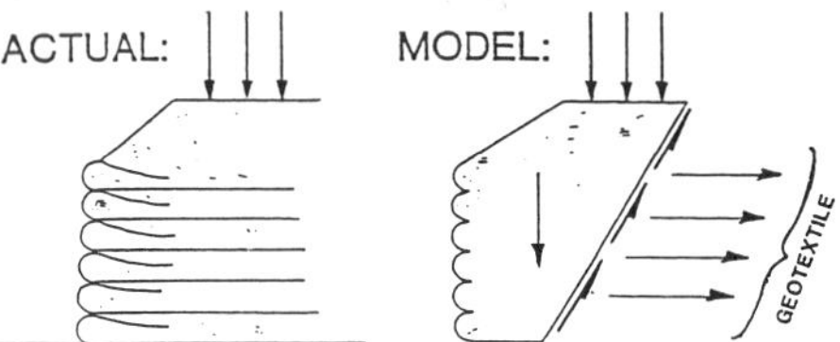

FIGURE 23.10 Actual geosynthetic-reinforced wall compared to its analytical model. (*Source:* Christopher, B. R. and Holtz, R. D. 1985. *Geotextile Engineering Manual.* STS Consultants Ltd., Northbrook, IL; U.S. Federal Highway Administration, Report No. FHWA-TS-86/203.)

design are given in Table 23.2; these are patterned after the failure modes for reinforced earth. Christopher and Holtz [1985, 1989], Bonaparte *et al.* [1987], and Christopher *et al.* [1989] discuss how these properties may be obtained by laboratory and other tests.

Typically, sliding of the entire reinforced mass controls the length of the reinforcing elements. Thus Christopher and Holtz [1989] and Christopher *et al.* [1989] recommend starting with a sliding analysis first (considering surcharges, etc.) and then designing for internal stability before proceeding to check the other components of external stability (bearing capacity, overturning, and slope stability). Surcharges may be considered in the usual manner, and both stiff and flexible facings are possible. For design, K_0 and K_A may be assumed, depending on the rigidity of the facing and the amount of yielding likely to occur during construction.

Backfill for geosynthetically reinforced walls should be free draining. This is important for stability considerations and because an impervious permanent facing is often used; thus, drainage outward through the wall face may not be possible. For permanent construction, some type

Table 23.2 Failure Modes in Geosynthetically Reinforced Retaining Walls

Geosynthetic Failure Mode	Corresponding Reinforced Earth Failure Mode	Property Required
Geosynthetic rupture	Ties break	Geosynthetic tensile strength
Geosynthetic pullout	Ties pull out	Soil-geosynthetic friction
Creep	Creep	Creep resistance

Source: Christopher, B. R. and Holtz, R. D. 1985. *Geotextile Engineering Manual.* STS Consultants Ltd., Northbrook, IL; U.S. Federal Highway Administration, Report No. FHWA-TS-86/203.

of permanent facing is required because of possible deterioration of the geosynthetic due to ultraviolet radiation. This is the only application in which the geosynthetic is not entirely buried, and consequently, some ultraviolet deterioration and loss of stability, at least locally, is possible. Permanent facings which have been successfully used include shotcrete, sprayed asphalt emulsion, precast concrete elements hung on the wall, segmented precast concrete blocks, and separate timber or precast concrete facades.

Construction procedures for geosynthetic reinforced walls and abutments are quite straightforward. Details are given by Steward *et al.* [1977] and Christopher and Holtz [1985, 1989].

23.5 Geosynthetics in Waste Containment Systems

Geomembranes and geocomposite drainage systems are commonly used today in the construction and remediation of containment and disposal systems for hazardous, industrial, and domestic wastes. Although not directly part of soil improvement, their importance in environmental geotechnology deserves special mention. Tremendous advances have been made in recent years in geomembrane technology, yet the seams and joints between sheets of geomembrane are still the weak link in these systems. Inspection during construction is crucial; improper installation or damage to the membrane during construction operations may compromise the integrity of the entire containment system. Helpful references on geomembrane technology and related systems include Koerner [1990a, 1990b], Bonaparte [1990], and Rollin and Rigo [1991].

Defining Terms

Geomembrane: Continuous membrane-type liners and barriers composed of asphaltic, polymeric, or a combination of asphaltic and polymeric materials with sufficiently low permeability so as to control fluid migration in a geotechnical engineering-related man-made project, structure, or system.

Geotextile: The American Society for Testing and Materials (ASTM) has defined a *geotextile* as any permeable textile material used with foundation, soil, rock, earth, or any other geotechnical engineering-related material as an integral part of a man-made project, structure, or system.

References

Allen, T. M. and Holtz, R. D. 1991. Design of retaining walls reinforced with geosynthetics. *Geotech. Eng. Congr. 1991, ASCE.* Geotechnical Special Publication No. 27, II:970–987.

Bell, J. R., Hicks, R. G., *et al.* 1980; 1982. Evaluation of test methods and use criteria for geotechnical fabrics in highway application. Oregon State University, Corvallis, Interim Report No. FHWA/RD-80/021; Final Report No. FHWA/RD-82.

Bender, D. A. and Barenberg, E. J. 1978. Design and behavior of soil–fabric–aggregate systems. *Transp. Res. Rec.* 671:64–75.

Bonaparte, R., ed. 1990. *Waste Containment Systems: Construction, Regulation, and Performance. Proc. Symp.* San Francisco, Geotechnical Special Publication No. 26, ASCE.

Bonaparte, R. and Christopher, B. R. 1987. Design and construction of reinforced embankments over weak foundations. *Transp. Res. Rec.* 1153:26–39.

Bonaparte, R., Holtz, R. D., and Giroud, J. P. 1987. Soil reinforcement design using geotextiles and geogrids, *in* Geotextile Testing and the Design Engineer, *Am. Soc. Test. Mater., Spec. Test. Publ. 952*:69–116.

Borden, R. H., Holtz, R. D., and Juran, I., eds. 1992. *Grouting, Soil Improvement, and Geosynthetics, Proc. Conf.* New Orleans, Geotechnical Special Publication No. 30, ASCE.

Cazzuffi, D. and Anzani, A. 1992. List of reference documents. IGS Education Committee, *IGS News*, 8(1):2-page supplement.

Christopher, B. R. and Holtz, R. D. 1989. *Geotextile Design and Construction Guidelines.* U.S. Federal Highway Administration, National Highway Institute, Report No. FHWA-HI-90-001.

Christopher, B. R. and DiMaggio, J. A. 1984. Specifying geotextiles. *Geotech. Fabrics Rep.* 2(2): 21–25.

Christopher, B. R., Gill, S. A., Giroud, J. P., Juran, I., Mitchell, J. K., Schlosser, F., and Dunnicliff, J. 1989. *Reinforced Soil Structures,* Vol. I, *Design and Construction Guidelines,* Vol. II, *Summary of Research and Systems Information.* FHWA, Report No. FHWA-RD-89-043.

Christopher, B. R. and Holtz, R. D. 1985. *Geotextile Engineering Manual.* STS Consultants Ltd., Northbrook, IL; U.S. Federal Highway Administration, Report No. FHWA-TS-86/203.

Christopher, B. R., and Holtz, R. D. 1991. Geotextiles for subgrade stabilization in permanent roads and highways. In *Proc. Geosynth.* Atlanta, Vol. 2, pp. 701–713.

Christopher, B. R. and Leshchinsky, D. 1991. Design of geosynthetically reinforced slopes. *Proc. Geotech. Eng. Congr.* Boulder, Geotechnical Special Publication No. 27, ASCE, Vol. II, pp. 988–1005.

Fischer, G. R., Christopher, B. R., and Holtz, R. D. 1990. Filter criteria based on pore size distribution. In *Proc. 4th Int. Conf. Geotext., Geomembr. Relat. Prod. Vol. 1.* The Hague, pp. 289–294.

Fluet, J. E., Jr., ed. 1986. Special issue on railroads. *Geotext. Geomembr.* 3(2, 3):89–219.

Fluet, J. E., Jr. 1988. Geosynthetics for soil improvement: A general report and keynote address. *Proc. Symp. Geosynth. Soil Improvement,* ed. R. D. Holtz. Geotechnical Special Publication No. 18, ASCE, pp. 1–21.

Giroud, J. P. and Noiray, L. 1981. Design of geotextile-reinforced, unpaved roads. *J. Geotech. Eng., ASCE.* 107(G79):1233–1254.

Giroud, J. P. and Carroll, R. G., Jr. 1983. Geotextile products. *Geotech. Fabrics Rep.* 1(1):12–15.

Giroud, J. P. 1986. From geotextiles to geosynthetics: A revolution in geotechnical engineering. *Proc. 3rd Int. Conf. Geotext. Vol. I.* Vienna, pp. 1–18.

Haas, R., Walls, J., and Carroll, R. G. 1988. Geogrid reinforcement of granular bases in flexible pavements. *Transp. Res. Rec.* 1188:19–27.

Haliburton, T. A. and Barron, J. V. 1983. Optimum depth method for design of fabric reinforced unpaved roads. *Transp. Res. Rec.* 916:26–32.

Holtz, R. D., ed. 1988a. *Geosynthetics for Soil Improvement, Proc. Symp. ASCE Spring Convention.* Nashville, Special Geotechnical Publication No. 18.

Holtz, R. D. 1988b. Geosynthetics in civil engineering construction. In *New Horizons in Construction Materials,* ed. S. P. Shah and D. Y. Lee, pp. 38–57.

Holtz, R. D. 1989a. Design and construction of embankments on very soft soils. In *Proc. 31st Annu. Minn. Geotech. Conf.* St. Paul, pp. 1–35.

Holtz, R. D. 1989b. Treatment of problem foundation for highway embankments. In *Synthesis of Highway Practice 147,* National Cooperative Highway Research Program, Transportation Research Board.

Holtz, R. D. 1990. Design and construction of geosynthetically reinforced embankments on very soft soils. In *Performance of Reinforced Soil Structures, Proc. Int. Reinforced Soil Conf.* Glasgow, British Geotechnical Society, pp. 391–402.

Holtz, R. D. 1991. Geosynthetics in civil engineering. In *Use of Geosynthetics in Dams, Eleventh Annual USCOLD Lecture Series,* White Plains, New York, United States Commission on Large Dams, pp. 1–19.

Holtz, R. D. and Paulson, J. N. 1988. Geosynthetic literature. *Geotech. News.* 6(1):13–15.

Holtz, R. D., Jamiolkowski, M. B., Lancellotta, R., and Pedroni, R. 1991. *Prefabricated Vertical Drains: Design and Performance.* Butterworth/CIRIA copublication series, Butterworths-Heinemann, London.

Houlsby, G. T., Milligan, G. W. E., Jewell, R. A., and Burd, H. J. 1989. A new approach to the design of unpaved roads—Part I. *Ground Eng.* 22(3):25–29.

Humphrey, D. N. and Rowe, R. K. 1991. Design of reinforced embankments—Recent developments in the state-of-the-art. *Geotech. Eng. Congr. 1991, ASCE.* Geotechnical Special Publication No. 27, II:1006–1020.

Ingold, T. S. and Miller, K. S. 1988. *Geotextiles Handbook.* Telford, London.

Jewell, R. A. 1990. Strength and deformation in reinforced soil design. In *Proc. 4th Int. Conf. Geotext. Geomembr. Relat. Prod. Vol. 3.* The Hague, pp. 913–946.

Jewell, R. A., Paine, N., and Woods, R. I. 1984. Design methods for steep reinforced embankments. *Polymer Grid Reinforcement., Proc. Conf.* London, pp. 70–81.

Koerner, R. M. 1990a. *Designing with Geosynthetics*, 2nd ed. Prentice Hall, New York.

Koerner, R. M. 1990b. Preservation of the environment via geosynthetic containment systems. *Proc. 4th Int. Conf. Geotext. Geomembr. Relat. Prod. Vol. 3.* The Hague, pp. 975–988.

Koerner, R. M. and Welsh, J. P. 1980. *Construction and Geotechnical Engineering Using Synthetic Fibers.* John Wiley & Sons, New York.

Lee, K. L., Adams, B. D., and Vagneron, J. M. J. 1973. Reinforced earth retaining walls. *J. Soil Mech. Found., ASCE.* 99(SM10):745–764.

Milligan, G. W. E., Jewell, R. A., Houlsby, G. T., and Burd, H. J. 1989. A new approach to the design of unpaved roads–Part II. *Ground Eng.* 22(8):37–43.

Murray, R. T. 1982. Fabric reinforcement of embankments and cuttings. In *Proc. 2nd Int. Conf. Geotext. Vol. III.* Las Vegas, pp. 707–713.

Rankilor, P. R. 1981. *Membranes in Ground Engineering.* John Wiley & Sons, New York.

Rollin, A. and Rigo, J. M., eds. 1991. *Geomembranes: Identification and Performance Testing*, RILEM Report 4, Chapman and Hall, London.

Schmertmann, G. R., Bonaparte, R., Chouery-Curtis, V. E., and Johnson, R. D. 1987. Design charts for geogrid-reinforced soil slopes. In *Proc. Geosynth. 87, Vol. 1.* New Orleans, pp. 108–120.

Schneider, H. R. and Holtz, R. D. 1986. Design of slopes reinforced with geotextiles and geogrids. *Geotext. Geomembr.* 3(1):29–51.

Steward J., Williamson, R., and Mohney, J. 1977. *Guidelines for Use of Fabrics in Construction and Maintenance of Low-Volume Roads.* USDA, Forest Service, Portland, Oregon (also published as FHWA Report No. FHWA-TS-78-205.

Veldhuijzen van Zanten, R., ed. 1986. *Geotextiles and Geomembranes in Civil Engineering.* John Wiley & Sons, New York.

Verduin, J. R. and Holtz, R. D. 1989. Geosynthetically reinforced slopes: A new design procedure. In *Proc. Geosynth. '89, Vol. 1.* San Diego, pp. 279–290.

24

Geotechnical Earthquake Engineering

Jonathan D. Bray
University of California at Berkeley

24.1 Introduction

Geotechnical factors often exert a major influence on damage patterns and loss of life in earthquake events. For example, the localized patterns of heavy damage during the 1985 Mexico City and 1989 Loma Prieta, California, earthquakes provide grave illustrations of the importance of understanding the seismic response of deep clay deposits and loose, saturated sand deposits. The near failure of the Lower San Fernando dam in 1971 due to liquefaction of the upstream shell materials is another grave reminder that we must strive to understand the seismic response of critical earth structures. The characteristics and distribution of earth materials at a project site significantly influence the characteristics of the earthquake ground motions, and hence significantly influence the seismic response of the constructed facilities at a site. Moreover, the composition and geometry of earth structures, such as earth dams and solid waste landfills, significantly affect their seismic response. Geotechnical considerations therefore play an integral role in the development of sound earthquake-resistant designs. In this chapter, geotechnical earthquake engineering phenomena such as site-specific amplification, soil liquefaction, and seismic slope stability are discussed. Case histories are used to illustrate how earthquakes affect engineered systems, and established, simplified empirical procedures that assist engineers in assessing the effects of these phenomena are presented. The field of earthquake engineering is quite complex, so the need for exercising engineering judgment based on appropriate experience is emphasized.

24.2 Earthquake Strong Shaking

The development and transmission of earthquake energy through the underlying geology is quite complex, and a site-specific seismic response study requires an assessment of the primary factors influencing the ground motion characteristics at a site. They are

- Earthquake source mechanism
- Travel path geology

0-8493-8953-4/95/$0.00 + $.50

- Topographic effects
- Earthquake magnitude
- Distance from zone of energy release
- Local soil conditions

Earthquakes are produced in a particular geologic setting due to specific physical processes. A midplate earthquake (e.g., New Madrid) will differ from a plate margin earthquake (e.g., San Andreas) [see Nuttli, 1982]. The principal descriptive qualities of the earthquake source are the type of fault displacement (strike-slip, normal, or reverse), depth of the rupture, length of the rupture, and duration of the rupturing. The characteristics of the rock which the seismic waves travel through influence the frequency content of the seismic energy. Significant topographic features (e.g., basins) can focus and hence amplify earthquake motions. The **magnitude** of an earthquake is related to the amount of energy released during the event. The difference between earthquakes of different magnitudes is significant; for example, a magnitude 7 earthquake event releases nearly a thousand times more energy than a magnitude 5 event. The potential for seismic damage will typically increase with earthquake events of greater magnitude. Seismic energy attenuates as it travels away from the zone of energy release and spreads out over a greater volume of material. Hence, the intensity of the bedrock motion will typically decrease as the distance of a particular site from the zone of energy release increases. A number of **attenuation relationships** based on earthquake magnitude and distance from the earthquake fault rupture are available [e.g., Nuttli and Herrmann, 1984; Joyner and Boore, 1988; Idriss, 1991]. Local soil conditions may significantly amplify ground shaking, and some soil deposits may undergo severe strength loss resulting in ground failure during earthquake shaking. The last three factors listed above (magnitude, distance, local soil conditions) are usually the most important factors, and most seismic studies focus on these factors.

There are several earthquake magnitude scales, so it is important to use these scales consistently. The earliest magnitude scale, local magnitude (M_L), was developed by Richter [1935] and is defined as the logarithm of the maximum amplitude on a Wood-Anderson torsion seismogram located at a distance of 100 km from the earthquake source [Richter, 1958]. Other related magnitude scales include surface wave magnitude, M_s, and body wave magnitudes, m_b and m_B [Gutenberg and Richter, 1956]. These magnitude scales are based on measurement of the amplitude of the seismic wave at different periods (M_L at 0.8 s, m_b and m_B between 1 s and 5 s, and M_s at 20 s), and hence they are not equivalent. The moment magnitude, M_w, is different from these other magnitude scales because it is directly related to the dimensions and characteristics of the fault rupture. Moment magnitude is defined as

$$M_w = \tfrac{2}{3}\log M_0 - 10.7 \tag{24.1}$$

where M_0 is the seismic moment in dyne-cm, with $M_0 = \mu \cdot A_f \cdot D$; μ = shear modulus of material along the fault plane (typically 3×10^{11} dyne/cm^2), A_f = area of fault rupture in cm^2, and D = average slip over the fault rupture in cm [Hanks and Kanamori, 1979]. Heaton *et al.* [1982] has shown that these magnitude scales are roughly equivalent up to $M_w = 6$, but that magnitude scales other than M_w reach limiting values for higher moment magnitude earthquake events (i.e., $\max m_b \approx 6$, $\max M_L \approx 7$, $\max m_B \approx 7.5$, and $\max M_s \approx 8$). Thus, the use of moment magnitude is preferred, but the engineer must use the appropriate magnitude scale in available correlations between engineering parameters and earthquake magnitude.

The earthquake motion characteristics of engineering importance are

- Intensity
- Frequency content
- Duration

Geologic and Seismologic Evaluation	Geotechnical Evaluation	Structural Design
• Identify Seismic Sources	• Site Response	• Dynamic Analysis
• Potential for Surface Rupture	• Liquefaction Potential	- Pseudo-static
• Size and Frequency of Events	• Seismic Stability	- Time History
• Develop Rock Motions	• Soil-Structure Interaction	• Design Considerations

FIGURE 24.1 Seismic hazard assessment.

The intensity of ground shaking is usually portrayed by the **maximum horizontal ground acceleration (MHA)**, but since velocity is a better indicator of the earthquake energy that must be dissipated by an engineered system, it should be used as well. The MHA developed from a site-specific seismicity study should be compared to that presented in the U.S. Geological Survey maps prepared by Algermissen *et al.* [1991] and the seismic zone factor (Z) given in the Uniform Building Code. The frequency content of the ground motions is typically characterized by its **predominant period** (T_p). The predominant period of the ground motions at a site tends to increase with higher magnitude events and with greater distances from the zone of energy release [see Idriss, 1991]. Earthquake rock motions with a concentration of energy near the fundamental periods of the overlying soil deposit and structure have a greater potential for producing amplified shaking and seismic damage. Lastly, the **duration of strong shaking** is related to the earthquake's magnitude and is typically described by the duration of the earthquake record in which the intensity is sufficiently high to be of engineering importance (i.e., MHA exceeding around $0.05g$) or the equivalent number of cycles of strong shaking [see Seed and Idriss, 1982].

A seismic hazard assessment generally involves those items listed in Fig. 24.1. Earthquake engineering is a multidisciplinary field that requires a coordinated effort. The success of the geotechnical evaluation depends greatly on the results of the geological and seismological evaluation, and the results of the geotechnical evaluation must be compatible with the requirements of the structural design.

24.3 Site-Specific Amplification

The localized patterns of heavy damage during the 1989 Loma Prieta earthquake in northern California demonstrate the importance of understanding the seismic response of deep soil deposits. Well over half of the economic damage and more than 80% of the loss of life occurred on considerably less than 1% of the land within 80 km of the fault rupture zone largely as a result of site-specific effects [Seed *et al.,* 1990]. For example, in the Oakland area, which is 70 km away from the rupture zone, maximum horizontal ground accelerations were amplified by a factor of 2 to 4 and spectral accelerations at some frequencies were amplified by a factor of 3 to 8 [Bray *et al.,* 1992]. The dramatic collapse of the elevated highway I-880 structure, in which 38 people died, is attributed in part to these amplified strong motions [Hough *et al.,* 1990]. Hundreds of buildings in the San Francisco Bay area sustained significant damage because of earthquake strong shaking. These observations are critical to many cities as deep soil deposits exist in many earthquake-prone areas around the world. For example, records of the January 31, 1986 northeastern Ohio earthquake suggest that similar site-specific amplification effects could occur in the central U.S. and produce heavy damage during a major event in the New Madrid seismic zone [Nuttli, 1987].

Response spectra are typically used to portray the characteristics of the earthquake shaking at a site. The **response spectrum** shows the maximum response induced by the ground motions in damped single-degree-of-freedom structures of different **fundamental periods**. Each structure has a unique fundamental period at which the structure tends to vibrate when it is allowed to vibrate freely without any external excitation. The response spectrum indicates how a particular

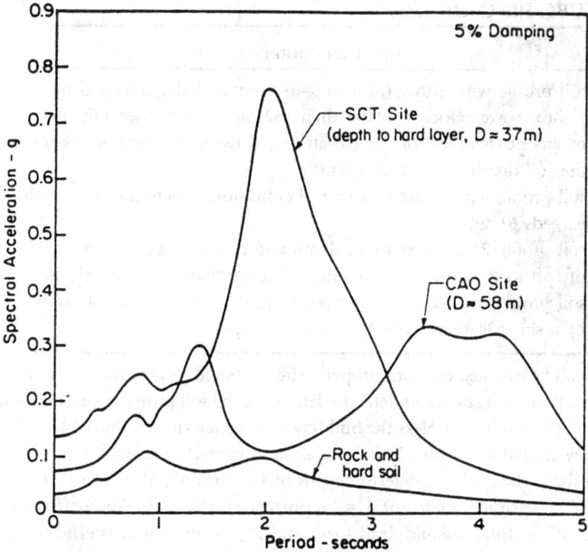

FIGURE 24.2 Acceleration response spectra for motions recorded in Mexico City during the 1985 Mexico City earthquake (after Seed, H. B., Romo, M. P., Sun, J., Jaime, A., and Lysmer, J. 1987. Relationships between Soil Conditions and Earthquake Ground Motions in Mexico City in the Earthquake of Sept. 19, 1985. Earthquake Engineering Research Center, Report No. UCB/EERC-87/15, University of California, Berkeley).

structure with its inherent fundamental period would respond to the earthquake ground motion. For example, referring to Fig. 24.2, a low-period structure (say, $T = 0.1$ s) at the SCT building site would experience a maximum acceleration of 0.14g, whereas a higher-period structure (say, $T = 2.0$ s) at the SCT site would experience a maximum acceleration of 0.74g for the same ground motions.

The response spectra shown in Fig. 24.2 illustrate the pronounced influence of local soil conditions on the characteristics of the observed earthquake ground motions. Since Mexico City was located approximately 400 km away from the earthquake's epicenter, the observed response at rock and hard soil sites was fairly low (i.e., the spectral accelerations were less than 0.1g at all periods). Damage was correspondingly negligible at these sites. At the Central Market site (CAO), spectral accelerations were significantly amplified at periods of around 1.3 s and within the range of 3.5 s to 4.5 s. Since buildings at the CAO site did not generally have fundamental periods within these ranges, damage was fairly minor. The motion recorded at the SCT building site, however, indicated significant amplification of the underlying bedrock motions with a maximum horizontal ground acceleration (the spectral acceleration at a period of zero) over four times that of the rock and hard soil sites and with a spectral acceleration at $T = 2.0$ s over seven times that of the rock and hard soil sites. Major damage, including collapse, occurred to structures with fundamental periods ranging from about 1 s to 2 s near the SCT building site and in areas with similar subsurface conditions. At these locations, the soil deposit's fundamental period *matched* that of the overlying structures, creating a resonance condition that amplified strong shaking and caused heavy damage.

The 1991 Uniform Building Code (UBC) utilizes site coefficients in its pseudostatic design base shear procedure to limit damage due to local soil conditions (see Table 24.1). For example, the site coefficient for soil characteristics (S factor) is increased to 2.0 for the soft soil profile S_4, and an S factor of 1.0 is used at rock sites where no soil-induced amplification occurs. However, a deposit of stiff clay greater than 61 m thick, such as those that underlay Oakland, would be categorized as soil

Table 24.1 1991 UBC Site Coefficients[1,2]

Type	Description	S factor
S_1	A soil profile with either (a) a rock-like material characterized by a shear wave velocity greater than 762 mps or by other suitable means of classification, or (b) stiff or dense soil condition where the soil depth is less than 61 m.	1.0
S_2	A soil profile with dense or stiff soil conditions where the soil depth exceeds 61 m.	1.2
S_3	A soil profile 21 m or more in depth and containing more than 6 m of soft to medium stiff clay but not more than 12 m of soft clay.	1.5
S_4	A soil profile containing more than 12 m of soft clay characterized by a shear wave velocity less than 152 mps.	2.0

[1]The site factor shall be established from properly substantiated geotechnical data. In locations where the soil properties are not known in sufficient detail to determine the soil profile type, soil profile S_3 shall be used. Soil profile S_4 need not be assumed unless the building official determines that soil profile S_4 may be present at the site, or in the event that soil profile S_4 is established by geotechnical data.

[2]The total design base shear (V) is determined from the formula $V = Z \cdot I \cdot C \cdot W/R_w$, where $C = 1.25S/T^{2/3} \leq 2.75$, Z = seismic zone factor, I = importance factor, S = site coefficient, T = fundamental period of structure, W = total seismic dead load, and R_w = reduction coefficient based on the lateral load-resisting system (see UBC). Hence, $V \propto S$ if $C < 2.75$.

profile S_2 with an S factor of only 1.2. The seismic response of the deep stiff clay sites during the 1989 Loma Prieta earthquake with spectral amplification factors on the order of 3 to 8 suggest that we may be currently underestimating the seismic hazard at these sites. Earthquake engineering is a relatively young field of study, and additional research is required to support the evolution of safer building codes.

The UBC also allows dynamic analyses of structural systems and provides the normalized response spectra shown in Fig 24.3. The spectral acceleration of a structure can be estimated from this figure given an estimate of the system's fundamental period (T), the peak ground acceleration (MHA) of the design event, and the classification of the subsurface soil conditions. At longer periods ($T > 0.5$ s), the spectral accelerations for deep soil sites (soil type 2) and soft soil sites (soil type 3) are significantly higher than that for rock and stiff soils (soil type 1). The engineer can also use wave propagation analyses [e.g., SHAKE91; see Idriss and Sun, 1992] to develop a site-specific design response spectrum based on the geologic, seismologic, and soil characteristics associated with the project site. The seismic response of earth materials is dictated primarily by geometric considerations and by the soil's dynamic properties (e.g., shear modulus and damping characteristics). The shear modulus gives an indication of the stiffness of the soil system, whereas the **damping ratio** provides a measure of the soil system's ability to dissipate energy under cyclic loading. Soils exhibit strain-dependent dynamic properties so that as earthquake strong shaking increases and the strain induced in the soil increases, the material's damping ratio increases and its shear modulus decreases [see Seed *et al.*, 1984; Vucetic and Dobry, 1991]. As material damping increases, the soil-induced amplification tends to decrease. As the material stiffness decreases, however, the fundamental period of the soil system increases, and this may affect the amplification of higher-period motions.

The amplification of higher-period ground motions that may match the fundamental period of the building located at the site is one of the most critical concerns in seismic site response studies. If the building's fundamental period is close to that of the site, an earthquake with a concentration of energy around this period would have the potential to produce heavy damage to this structure. The matching of the building and site fundamental periods creates a resonant condition that can amplify shaking. Consideration of a different structural system whose fundamental period does not match that of the underlying soil deposit might be prudent. Otherwise, the design criteria should be more stringent to limit damage to the building during an earthquake event.

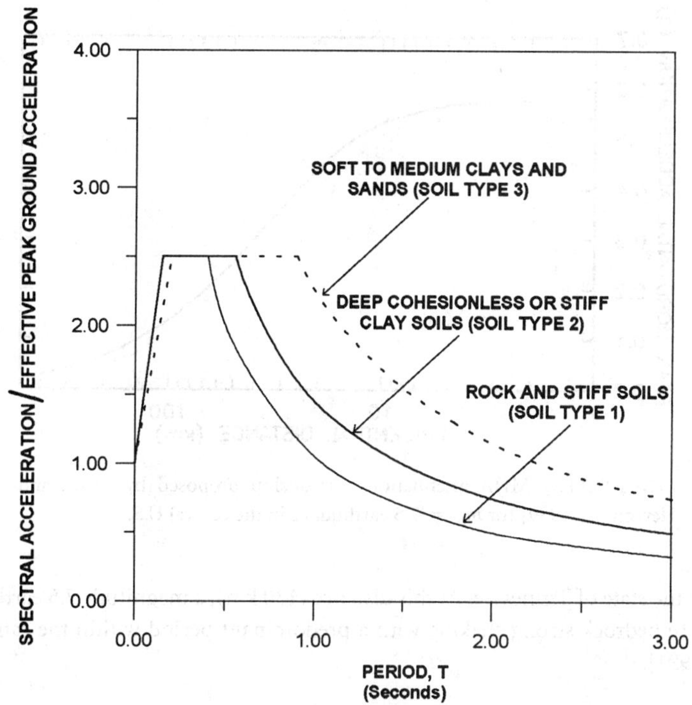

FIGURE 24.3 1991 UBC normalized acceleration response spectra.

Illustrative Site Response Problem

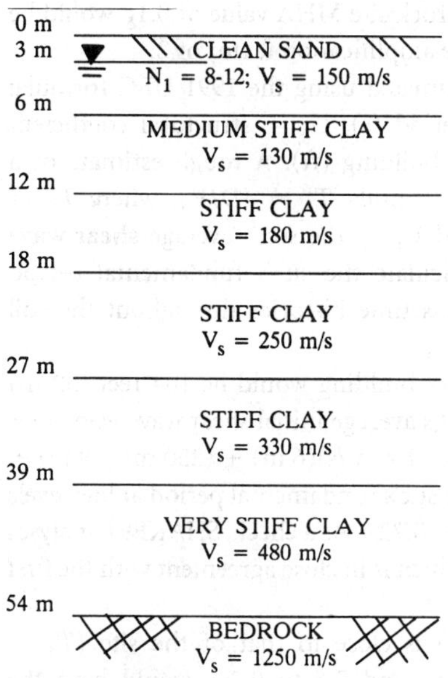

FIGURE 24.4 Subsurface conditions of project site discussed in problem 24.1.

Problem 24.1

An eight-story building will be constructed at a deep soil site in Tennessee. As shown in Fig. 24.4, the deep soil site contains surficial deposits of loose, saturated sand overlying a thick clay deposit. The design earthquake is a magnitude (m_B) 7.5 event occurring at a distance of 130 km from the site. Develop a preliminary estimate of the maximum horizontal acceleration (MHA) at the site and evaluate the potential for soil-induced amplification of earthquake shaking near the building's fundamental period.

Earthquake Strong Shaking

Limited data is available to develop attenuation curves for the deeper-focus eastern-U.S. earthquakes, and hence this is an area of ongoing research. Nuttli and Herrmann [1984] proposed the attenuation curve shown in Fig. 24.5 for earthquakes likely to occur in the eastern and central U.S. For a site located 130 km from the zone of energy release for an $m_B = 7.5$ event, the bedrock MHA would be on the order of 0.1g. This magnitude of MHA is comparable with what established building codes (e.g., BOCA, SBCCI, and UBC) would recommend for the

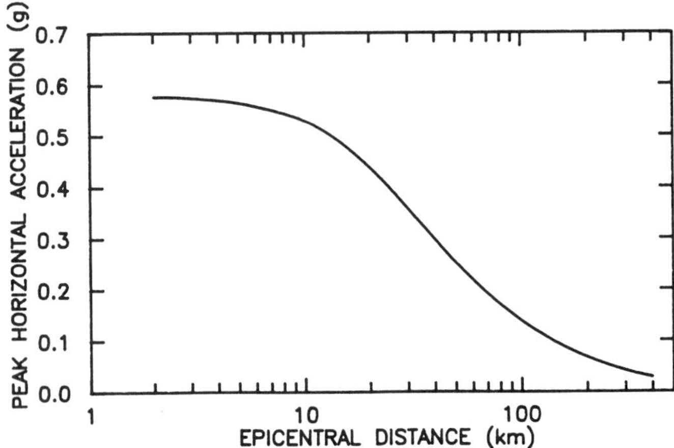

FIGURE 24.5 MHA attenuation relationship proposed by Nuttli and Herrmann [1989] for $m_B = 7.5$ earthquake in the central U.S.

central part of the state of Tennessee. At this distance (130 km), a magnitude 7.5 earthquake would tend to produce bedrock strong shaking with a predominant period within the range of 0.4 s to 0.7 s [Idriss, 1991].

Site-Specific Amplification

Comparisons between MHA recorded at deep clay soil sites and those recorded at rock sites indicate that MHA at deep clay soil sites can be 1 to 3 times greater than those at rock sites when $MHA_{rock} \approx 0.1g$ [Bray *et al.*, 1992; Idriss, 1991]. The results of one-dimensional (columnar) dynamic response analyses also suggest MHA amplification factors on the order of 1 to 3 for deep clay sites at low values of MHA_{rock}. A proposed relationship between the MHA at soft clay sites and the MHA at rock sites is shown in Fig. 24.6. The estimated rock site MHA value of $0.1g$ would be increased to $0.2g$ for this deep clay site using the average site amplification factor of 2.

The fundamental period of typical buildings can be estimated using the 1991 UBC formula: $T = C_t \cdot (h_n)^{3/4}$, where T = building's fundamental period (s), C_t = structural coefficient = 0.020 for a typical building, and h_n = height of the building (ft). A rough estimate of a level site's fundamental period can be calculated by the formula $T_s = 4D/V_s$, where T_s = the site's fundamental period, D = the soil thickness, and V_s = the soil's average **shear wave velocity.** Computer programs such as SHAKE91 can calculate the site's fundamental period as well as calculate horizontal acceleration and shear stress time histories throughout the soil profile.

Assuming a story height of 13 feet (4 m), the eight-story building would be 104 feet (32 m) high, and $T = 0.020 \cdot (104 \text{ ft})^{3/4} = 0.65$ s. The soil deposit's average initial shear wave velocity is estimated to be $V_s = [(150 \text{ m/s})(6 \text{ m}) + (130 \text{ m/s})(6 \text{ m}) + (180 \text{ m/s})(6 \text{ m}) + (250 \text{ m/s})(9 \text{ m}) + (330 \text{ m/s})(12 \text{ m}) + (480 \text{ m/s})(15 \text{ m})]/54 \text{ m} = 300$ m/s. The site's fundamental period at low levels of shaking is then approximately $T_s = 4 \cdot (54 \text{ m})/300 \text{ m/s} = 0.72$ s. As a check, SHAKE91 analyses calculate the site's fundamental period to be 0.71 seconds, which is in close agreement with the first estimate ($T_s = 0.72$ s).

Since the building's fundamental period ($T = 0.65$ s) is close to that of the site ($T_s = 0.72$ s), an earthquake with a concentration of energy around 0.6 to 0.7 s would have the potential to produce heavy damage to this structure. In fact, SHAKE91 results indicate a maximum spectral acceleration amplification factor of almost 15 at a period around 0.7 s. In comparison, this site would be classified as an S_3 site with a site coefficient (or site amplification factor) of 1.5, and the

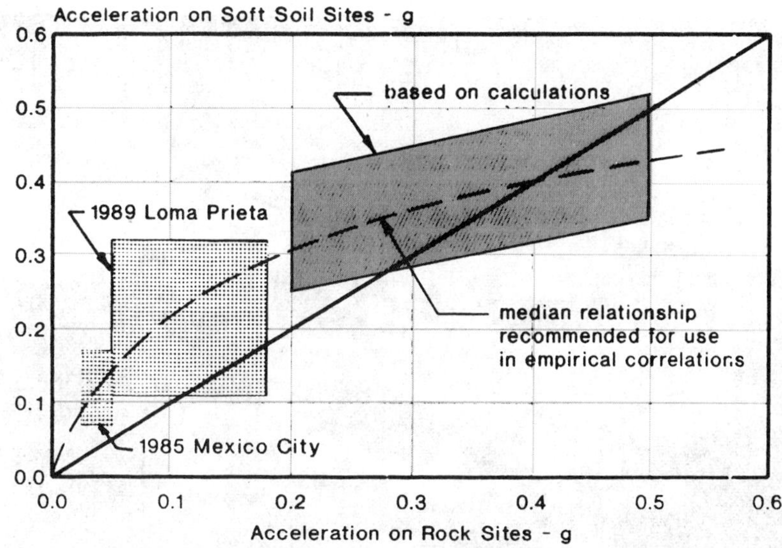

FIGURE 24.6 Variations of peak horizontal accelerations (MHA) at soft soil sites with accelerations at rock sites (after Idriss, I. M. 1991. Earthquake Ground Motions at Soft Soil Sites. *Proceedings, Second International Conference on Recent Advances in Geotechnical Engineering and Soil Dynamics,* March 11–15, St. Louis, pp. 2265–2272).

design base shear would be multiplied by an amplification factor of $C = 2.5$ for this site (see Table 24.1). Alternatively, using the normalized response spectrum for deep stiff clay soils in the UBC (Fig. 24.3), the spectral response at $T = 0.7$ s would be twice the MHA value of $0.2g$.

24.4 Soil Liquefaction

Unlike many other construction materials, cohesionless soils such as sand possess negligible strength without effective confinement. The overall strength of an uncemented sand depends on particle interaction (interparticle friction and particle rearrangement), and these particle interaction forces depend on the effective confining stresses. Soil **liquefaction** is a phenomenon resulting when the pore water pressure (the water pressure in the pores between the soil grains) increases, thereby reducing the effective confining stress and hence the strength of the soil. Seed and Idriss [1982] present this qualitative explanation of soil liquefaction:

> If a saturated sand is subjected to ground vibrations, it tends to compact and decrease in volume; if drainage is unable to occur, the tendency to decrease in volume results in an increase in pore water pressure, and if the pore water pressure builds up to the point at which it is equal to the overburden pressure, the effective stress becomes zero, the sand loses its strength completely, and it develops a liquefied state.

If the strength of the ground underlying a structure reduces below that required to support the overlying structure, excessive structural movements can occur and damage the structure. The pore water pressure in the liquefied soil may be sufficient to cause the liquefied soil to flow up through the overlying material to the ground surface producing sand boils, lateral spreading, and ground breakage. This dramatic seismic response of a saturated, loose sand deposit can pose obvious hazards to constructed facilities, and the potential for soil liquefaction should be assessed in seismic regions where such soil deposits exist.

FIGURE 24.7 Liquefaction-induced damage of the Marina Research Facility at Moss Landing, California, caused by the 1989 Loma Prieta earthquake (after Seed, R. B., Dickenson, S. E., Riemer, M. F., Bray, J. D., Sitar, N., Mitchell, J. K., Idriss, I. M., Kayen, R. E., Kropp, A., Harder, L. F., and Power, M. S. 1990. Preliminary Report on the Principal Geotechnical Aspects of the October 17, 1989 Loma Prieta Earthquake. Earthquake Engineering Research Center, Report No. UCB/EERC-90/05, University of California).

An example of the structural damage that can result from soil liquefaction is shown in Fig. 24.7. The Marine Research Facility at Moss Landing, California was a group of modern one- and two-story structures founded on concrete slabs. This facility was destroyed beyond repair by foundation displacements as a result of liquefaction of the foundation soils during the 1989 Loma Prieta earthquake [Seed *et al.*, 1990]. The facility settled a meter or two and lateral spreading of the structure "floating" on the liquefied soil below the slab foundation stretched the facility by 2 m, literally pulling it apart.

Engineering procedures for evaluating liquefaction potential have developed rapidly in the past twenty years, and a well-accepted approach is the Seed and Idriss [1982] simplified procedure for evaluating soil liquefaction potential. The average cyclic shear stress imparted by the earthquake in the upper 12 m of a soil deposit can be estimated using the equation developed by Seed and Idriss [1982]:

$$(\tau/\sigma_0')_d \approx 0.65 \cdot \text{MHA}/g \cdot \sigma_0/\sigma_0' \cdot r_d \qquad (24.2)$$

where

$(\tau/\sigma_0')_d$ = average cyclic stress ratio developed during the earthquake
MHA = maximum horizontal acceleration at the ground surface
g = acceleration of gravity
σ_0 = total stress at depth of interest
σ_0' = effective stress (total stress minus pore water pressure) at depth
r_d = reduction in acceleration with depth ($r_d \approx 1 - 0.008 \cdot \text{depth, m}$)

For a magnitude (M_s) 7.5 event at a level ground site, the cyclic stress ratio required to induce liquefaction, $(\tau/\sigma_0')_l$, of the saturated sand deposit can be estimated using the empirically based

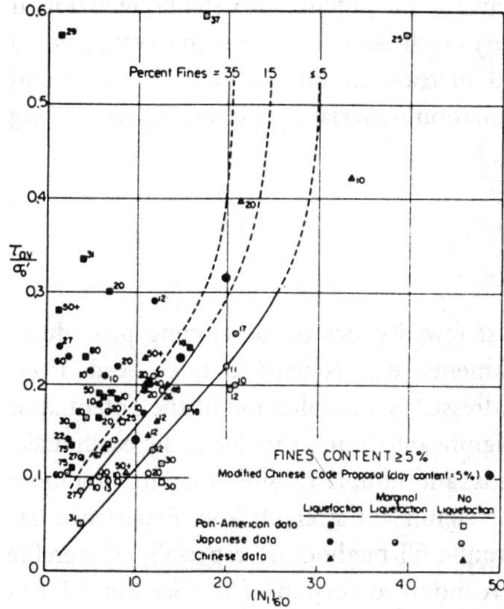

FIGURE 24.8 Relationship between stress ratios causing liquefaction and $(N_1)_{60}$-values for silty sands for $M_s = 7\frac{1}{2}$ earthquakes (after Seed, H. B., Tokimatsu, K., Harder, L. F., and Chung, R. M. 1985. Influence of SPT Procedures in Soil Liquefaction Resistance Evaluations. *Journal of the Geotechnical Engineering Division,* ASCE. (111) 12:1425–1445).

standard penetration test (SPT) correlations developed by Seed *et al.* [1985]. The SPT blowcount (the number of hammer blows required to drive a standard sampling device 1 foot into the soil deposit) provides an index of the in situ state of a sand deposit, and especially of its relative density. Field measured SPT blowcount numbers are corrected to account for overburden pressure, hammer efficiency, saturated silt, and other factors. A loose sand deposit will generally have low SPT blowcount numbers (4–10) and a dense sand deposit will generally have high SPT blowcount numbers (30–50). Moreover, changes in factors that tend to increase the cyclic loading resistance of a deposit similarly increase the SPT blowcount. Hence, the cyclic stress required to induce soil liquefaction can be related to the soil deposit's penetration resistance. Well-documented sites where soil deposits did or did not liquefy during earthquake strong shaking were used to develop the correlation shown in Fig. 24.8. Correction factors can be applied to the $(\tau/\sigma_0')_l$ versus SPT correlation presented in Fig. 24.8 to account for earthquake magnitude, greater depths, and sloping ground [see Seed and Harder, 1990].

Finally, the liquefaction susceptibility of a saturated sand deposit can be assessed by comparing the cyclic stress ratio required to induce liquefaction, $(\tau/\sigma_0')_l$, with the average cyclic stress ratio developed during the earthquake, $(\tau/\sigma_0')_d$. A reasonably conservative factor of safety should be employed ($\approx$1.5) because of the severe consequences of soil liquefaction.

Illustrative Soil Liquefaction Problem

Problem 24.2

Evaluate the liquefaction susceptibility of the soil deposit shown in Fig. 24.4.

Soil Liquefaction

At this site, MHA $\approx 0.2g$ (see Problem 24.1). At a depth of 5 m, $\sigma_0 = \rho_t \cdot g \cdot z = (2.0\ \text{Mg/m}^3)(9.81\ \text{m/s}^2)(5\ \text{m}) = 98$ kPa; $\sigma_0' = \sigma_0 - \gamma_w \cdot z_w = 98\ \text{kPa} - (1\ \text{Mg/m}^3)(9.81\ \text{m/s}^2)(3\ \text{m}) = 68$ kPa; and $r_d \approx 1.0 - 0.008(5m) = 0.96$. Hence, using Eq. (24.2), the average cyclic stress ratio developed during the earthquake is

$$(\tau/\sigma_0')_d \approx 0.65 \cdot (0.2g/g) \cdot \frac{98\ \text{kPa}}{68\ \text{kPa}} \cdot 0.96 = 0.18 \tag{24.3}$$

The corrected penetration resistance of the sand at a depth of 5 m is around 10, and using Fig. 24.8, for a clean sand with less than 5% fines, $(\tau/\sigma_0')_l \approx 0.11$.

The factor of safety (FS) against liquefaction is then

$$\text{FS} = (\tau/\sigma_0')_l/(\tau/\sigma_0')_d = 0.11/0.18 = 0.6 \tag{24.4}$$

Since the factor of safety against liquefaction is less than 1.0, the potential for soil liquefaction at the site is judged to be high. Modern ground improvement techniques (e.g., dynamic compaction) may be used to densify a particular soil deposit to increase its liquefaction resistance and obtain satisfactory performance of the building's foundation material during earthquake strong shaking.

24.5 Seismic Slope Stability

Considerable attention has been focused over the last few decades on developing procedures to analyze the seismic performance of earth embankments [e.g., Newmark, 1965; Seed, 1979; Marcuson *et al.,* 1992]. The first issue that must be addressed is an evaluation of the potential of the materials comprising the earth structure to lose significant strength under cyclic earthquake loading. Saturated cohesionless materials (gravels, sands, and nonplastic silts) that are in a loose state are prime candidates for liquefaction and hence significant strength loss. Experience has shown that cohesionless materials placed by the hydraulic fill method are especially vulnerable to severe strength loss as a result of strong shaking. A modified version of the Seed and Idriss [1982] simplified method has been employed to evaluate the liquefaction potential of cohesionless soils in earth slopes and dams [see Seed and Harder, 1990]. Certain types of clayey materials have also been shown to lose significant strength as a result of cyclic loading. If clayey materials have a small percentage of clay-sized particles, low liquid limits, and high water contents, the material's cyclic loading characteristics should be determined by cyclic laboratory testing [Seed and Idriss, 1982].

The potentially catastrophic consequences of an earth embankment material that undergoes severe strength loss during earthquake shaking is demonstrated by the near failure of the Lower San Fernando dam as a result of the 1971 San Fernando earthquake [Seed *et al.,* 1975]. The center and upstream sections of the dam slid into the reservoir because a large section within the dam liquefied. The slide movement left only 1.5 m of earth fill above the reservoir level. Fortunately, the reservoir level was 11 m below the original crest at the time of the earthquake. Still, because of the precarious condition of the dam after the main shock, 80,000 people living downstream of the dam were ordered to evacuate until the reservoir could be lowered to a safe elevation.

Surveys of earth dam performance during earthquakes suggest that embankments constructed of materials that are not vulnerable to severe strength loss as a result of earthquake shaking (most well-compacted clayey materials, unsaturated cohesionless materials, and some dense saturated sands, gravels, and silts) generally perform well during earthquakes [Seed *et al.,* 1978]. The embankment, however, may undergo some level of permanent deformation as a result of the earthquake shaking. With well-built earth embankments experiencing moderate earthquakes, the magnitude of permanent seismic deformations should be small, but marginally stable earth embankments experiencing major earthquakes may undergo large deformations that may jeopardize the structure's integrity. Simplified procedures have been developed to evaluate the potential for seismic instability and seismically induced permanent deformations [e.g., Seed, 1979; Makdisi and Seed, 1978], and these procedures can be used to evaluate the seismic performance of earthen structures and natural slopes.

In pseudostatic slope stability analyses, a factor of safety against failure is computed using a static limit equilibrium stability procedure in which a pseudostatic, horizontal inertial force, which represents the destabilizing effects of the earthquake, is applied to the potential sliding mass. The horizontal inertial force is expressed as the product of a seismic coefficient, k, and the weight, W, of the potential sliding mass (Fig. 24.9). If the factor of safety approaches unity, the embankment is considered unsafe. Since the seismic coefficient designates the horizontal force to be used in the stability analysis, its selection is crucial. The selection of the seismic coefficient must be coordinated with the selection of the dynamic material strengths and minimum factor of safety, however, as

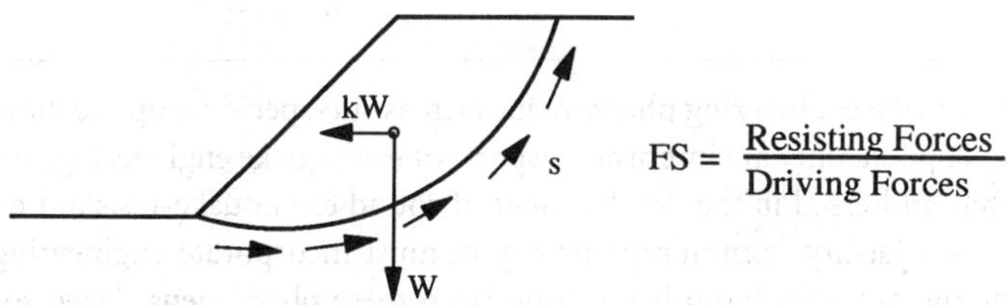

FIGURE 24.9 Pseudostatic slope stability analysis.

these parameters work together to achieve a satisfactory design. For earth embankments, case histories are available which have guided the selection of these parameters. For example, Seed [1979] recommends using appropriate dynamic material strengths, a seismic coefficient of 0.15, and a minimum factor of safety of 1.15 to ensure that an embankment composed of materials that do not undergo severe strength loss performs satisfactorily during a major earthquake.

A significant limitation of the pseudostatic approach is that the horizontal force, representing the effects of an earthquake, is constant and acts in only one direction. With dynamically applied loads, the force may be applied for only a few tenths of a second before the direction of motion is reversed. The result of these transient forces will be a series of displacement pulses rather than complete failure of the slope. Normally, a certain amount of limiting displacement during an earthquake event is considered tolerable. Hence, if conservative strength properties are selected and the seismic coefficient represents the maximum disturbing force (i.e., maximum shear stress induced by the earthquake on the potential sliding surface; see Seed and Martin, 1966), a factor of safety of one is likely to ensure adequate seismic performance. Other conservative combinations of these parameters could be developed, but their use must be evaluated in the context that their use in analysis ensures a design that performs well during earthquakes (i.e., tolerable deformations).

Seismically induced permanent deformations are generally calculated using a Newmark-type procedure. The Newmark [1965] analysis assumes that relative slope movements would be initiated when the inertial force on a potential sliding mass was large enough to overcome the yield resistance along the slip surface, and these movements would stop when the inertial force decreased below the yield resistance and the velocities of the ground and sliding mass coincided. The yield acceleration is defined as the average acceleration producing a horizontal inertial force on a potential sliding mass which gives a factor of safety of one and can be calculated as the seismic coefficient in pseudostatic slope stability analyses that produces a safety factor of one. By integrating the equivalent average acceleration [see Bray *et al.*, 1993] acting on the sliding surface in excess of this yield acceleration as a function of time, the displacement of the slide mass can be estimated. A commonly used procedure for calculating seismically induced permanent deformations has been developed by Franklin and Chang [1977] and computer programs are available [e.g., Pyke and Beikae, 1991]. Simplified chart solutions have been developed by Makdisi and Seed [1978] for earth embankments.

An emerging area of concern in many regions of the world is the seismic performance of waste fills. For example, recent U.S. federal regulations (40 CFR 258-USEPA 1991: Subtitle D) require that municipal solid waste landfills located in seismic impact zones be designed to resist earthquake hazards. Since these designated seismic impact zones encompass nearly half of the continental U.S., these regulations are having a pronounced impact on the design of new landfills and the lateral expansion of existing landfills. The results of a comprehensive study of the effects of the characteristics of the waste fill, subsurface soils, and earthquake ground motions are presented in Bray *et al.* [1993]. The investigators found that the seismic loading strongly depends on the dynamic properties and height of the waste fill. General design considerations regarding the seismic stability of solid waste landfills are discussed in Repetto *et al.* [1993].

24.6 Summary

Geotechnical earthquake engineering phenomena such as site-specific amplification, soil liquefaction, and seismic slope stability are important aspects of earthquake engineering, and these aspects must be adequately addressed in the development of sound earthquake-resistant designs. Seismic risk assessments of a facility, community, or region must incorporate engineering analyses that properly evaluate the potential hazards resulting from these phenomena. Deep soil deposits can amplify the underlying bedrock ground motions and produce intense levels of shaking at significant distances from the earthquake's epicenter. Under sufficient cyclic loading, loose, saturated sand deposits may suddenly liquefy, undergo severe strength loss, and fail as a foundation or dam material. Seismically induced permanent deformations of a landfill's liner system can jeopardize the integrity of the system and potentially release pollutants into the environment. Simplified empirical procedures employed to evaluate these hazards have been presented and they provide a starting point. The field of earthquake engineering is quite complex, however, and there are many opportunities for future research.

Defining Terms

Attenuation relationship: Provides the value of an engineering parameter versus distance from the zone of energy release of an earthquake.

Damping ratio: An indication of the ability of a material to dissipate vibrational energy.

Duration of strong shaking: Duration of the earthquake record in which the intensity is sufficiently high to be of engineering importance (i.e., MHA $\geq 0.05g$).

Fundamental period: Period at which a structure tends to vibrate when allowed to vibrate freely without any external excitation.

Liquefaction: Phenomenon resulting when the pore-water pressure within saturated particulate material increases dramatically, resulting in a severe loss of strength.

Magnitude: Measure of the amount of energy released during an earthquake. Several magnitude scales exist (e.g., local magnitude, M_L, and moment magnitude, M_w).

Maximum horizontal ground acceleration (MHA): Highest horizontal ground acceleration recorded at a free-field site (i.e., not in a structure) during an earthquake.

Predominant period (T_p): Period at which most of the seismic energy is concentrated, often defined as the period at which the maximum spectral acceleration occurs.

Response spectrum: Displays maximum response induced by ground motions in damped single-degree-of-freedom structures of different fundamental periods.

Shear wave velocity (V_s): Speed that shear waves travel through a medium. An indication of the dynamic stiffness of a material. Note that $G = \rho V_s^2$, where G = shear modulus and ρ = mass density.

References

Algermissen, S. T., Perkins, D. M., Thenhous, P. C., Hanson, S. L., and Bender, B. L. 1991. Probabilistic earthquake acceleration and velocity maps for the United States and Puerto Rico. *U.S.G.S. Misc. Field Stud.* Map MF2120.

Allen and Hoshall. 1985. *An Assessment of Damage and Casualties.* Federal Emergency Management Agency Report, Sections V and XI.

Bray, J. D., Chameau, J. L., and Guha, S. 1992. Seismic response of deep stiff clay deposits. In *Proc. 1st Can. Symp. Geo. Nat. Hazards,* Vancouver, BC, May 6–9, pp. 167–174.

Bray, J. D., Repetto, P. C., Angello, A. J., and Byrne, R. J. 1993. An overview of seismic design issues for solid waste landfills. In *Proc. Geosynth. Res. Inst. Conf.,* Drexel Univ., Philadelphia, Penn., Dec.

Building Officials & Code Administrators International, Inc. 1992. *National Building Code.* Country Club Hills, Illinois.

Franklin, A. G. and Chang, F. K. 1977. Earthquake resistance of earth and rockfill dams. Misc. Pap. S-71-17, U. S. Army Waterways Experiment Station, Vicksburg, MS, November, 1977.

Gutenberg, B. and Richter, C. F. 1956. Earthquake magnitude, intensity, energy and acceleration. *B. Seism. Soc. Am.* 46(2): 143–145.

Hanks, T. C. and Kanamori, H. 1979. A moment magnitude scale. *J. Geophys. Res.* 82: 2981–2987.

Heaton, T. H., Tajima, F., and Mori, A. W. 1982. *Estimating Ground Motions Using Recorded Accelerograms.* Report by Dames & Moore to Exxon Production Res. Co., Houston.

Hopper, M. G., ed. 1985. *Estimation of Earthquake Effects Associated with Large Earthquakes in the New Madrid Seismic Zone.* U.S. Geological Survey Open-File Report 85-457.

Hough, S. E., Friberg, P. A., Busby, R., Field, E. F., Jacob, K. H., and Borcherdt, R. D. 1990. Sediment-induced amplification and the collapse of the Nimitz Freeway, *Nature* 344:853–855.

Hynes, M. E. and Franklin, A. G. 1984. Rationalizing the seismic coefficient method. Misc. Pap. GL-84-13. U.S. Army Engineer WES, Vicksburg, MS.

Idriss, I. M. 1985. Evaluating seismic risk in engineering practice. In *Proc. 11th Int. Conf. Soil Mech. Found. Eng.* August 12–16, 1985, San Francisco, California, Vol. 1, pp. 255–320.

Idriss, I. M. 1991. Earthquake ground motions at soft soil sites. In *Proc. 2nd Int. Conf. Recent Adv. Geo. Eng. Soil Dyn.* March 11–15, St. Louis, pp. 2265–2272.

Idriss, I. M. and Sun, J. I. 1992. *User's Manual for SHAKE91—A Computer Program for Conducting Equivalent Linear Seismic Response Analyses of Horizontally Layered Soil Deposits.* Center for Geotechnical Modeling, Dept. of Civil and Environ. Engrg., Univ. of California, Davis, Nov.

International Conference of Building Officials. 1991. *Uniform Building Code.* Whittier, Calif.

Joyner, W. B. and Boore, D. M. 1988. Measurement characteristics and prediction of strong ground motion: State-of-the-art reports. In *Proc. Spec. Conf. Earthquake Eng. Soil Dyn. II,* ASCE, pp. 43–102.

Makdisi, F. I and Seed, H. B. 1978. Simplified procedure for estimating dam and embankment earthquake-induced deformation. *J. Geotech. Eng., ASCE.* 104(7): 849–867.

Marcuson, W. F., III, Hynes, M. E., and Franklin, A. G. 1992. Seismic stability and permanent deformation analyses: The last twenty-five years. In *Proc. ASCE Spec. Conf. Stability and Performance of Slopes and Embankments—II,* Berkeley, California, June 28–July 1, pp. 552–592.

Newmark, N. M. 1965. Effects of earthquakes on dams and embankments. *Geotechnique.* 15(2): 139–160.

Nuttli, O. W. 1982. Advances in seismicity and tectonics. In *Proc. 3rd Int. Earthquake Microzonation Conf.* Vol. I, pp. 3–24.

Nuttli, O. W. and Herrmann, R. B. 1984. Ground motion of Mississippi Valley earthquakes. *J. Tech. Top. Civ. Eng., ASCE.* 110:(54–69).

Nuttli, O.W. 1987. The Current and Projected State-of-Knowledge on Earthquake Hazards. Unpublished report presented in St. Louis, Missouri.

Pyke, R. and Beikae, M. 1991. TNMN. Taga Engineering Software Services, Lafayette, CA.

Repetto, P. C., Bray, J. D., Byrne, R. J., and Augello, A. J. 1993. Seismic design of landfills. In *Proc. 13th Cent. Pa. Geotech. Semin.,* Penn. DOT & ASCE, April 12–14.

Richter, C. F. 1935. An instrumental earthquake scale. *B. Seism. Soc. Am.* 25(1):1–32.

Richter, C. F. 1958. *Elementary Seismology.* W. H. Freeman, San Francisco.

Seed, H. B. and Martin, G. R. 1966. The seismic coefficient in earth dam design. *J. Soil Mech. Found., ASCE.* 92(3): 25–58.

Seed, H. B. 1979. Considerations in the earthquake-resistant design of earth and rockfill dams. *Geotechnique.* 29(3): 215–263.

Seed, H. B., Lee, K. L., Idriss, I. M., and Makdisi, F. I. 1975. The slides in the San Fernando dams during the earthquake of February 9, 1971. *J. Geotech. Eng., ASCE.* 101(7):889–911.

Seed, H. B., Makdisi, F. I., and DeAlba, P. 1978. Performance of earth dams during earthquakes. *J. Geotech. Eng., ASCE.* 104(7):967–994.

Seed, H. B. and Idriss, I. M. 1982. Ground motions and soil liquefaction during earthquakes. Monograph, Earthquake Engineering Research Institute, Berkeley, Calif.

Seed, H. B. 1983. Earthquake-resistant design of earth dams. Paper presented at ASCE National Convention, Philadelphia, Penn., May 16–20, pp. 41–64.

Seed, H. B., Wong, R. T., Idriss, I. M., and Tokimatsu, K. 1984. Moduli and damping factors for dynamic analyses of cohesionless soils. Earthquake Engineering Research Center, Report No. UCB/EERC-84/14, University of California, Berkeley, CA, October.

Seed, H. B., Tokimatsu, K., Harder, L. F., and Chung, R. M. 1985. Influence of SPT procedures in soil liquefaction resistance evaluations. *J. Geotech. Eng., ASCE.* (111)12:1425–1445.

Seed, H. B., Romo, M. P., Sun, J., Jaime, A., and Lysmer, J. 1987. Relationships between soil conditions and earthquake ground motions in Mexico City in the earthquake of Sept. 19, 1985. Earthquake Engineering Research Center, Report No. UCB/EERC- 87/15, University of California, Berkeley.

Seed, R. B. and Harder, Jr., L. F. 1990. SPT-based analysis of cyclic pore pressure generation and undrained residual strength. *H. Bolton Seed Memorial Symp.*, Vol. II, May, pp. 351–376.

Seed, R. B. and Bonaparte, R. 1992. Seismic analysis and design of lined waste fills: Current practice. In *Proc. ASCE Spec. Conf. Stability and Performance of Slopes and Embankments—II.* Berkeley, Calif., June 28–July 1, pp. 1521–1545.

Seed, R. B., Dickenson, S. E., Riemer, M. F., Bray, J. D., Sitar, N., Mitchell, J. K., Idriss, I. M., Kayen, R. E., Kropp, A., Harder, L. F., and Power, M. S. 1990. Preliminary report on the principal geotechnical aspects of the October 17, 1989 Loma Prieta earthquake. Earthquake Engineering Research Center, Report No. UCB/EERC-90/05, University of California.

Southern Building Code Congress International. 1992. *Standard Building Code.* Birmingham, Alabama.

United States Code of Federal Regulations, Title 40. 1991. Protection of the Environment, Part 258, *Solid Waste Disposal Facility Criteria.*

Vucetic, M. and Dobry, R. 1991. Effect of soil plasticity on cyclic response. *J. Geotech. Eng., ASCE.* 117(1):89–107.

For Further Information

"Ground Motions and Soil Liquefaction during Earthquakes" by Seed and Idriss [1982] provides an excellent overview of site-specific amplification and soil liquefaction. Seed and Harder [1990] present an up-to-date discussion of soil liquefaction. "Evaluating Seismic Risk in Engineering Practice" by Idriss [1985] provides an excellent discussion of seismicity and geotechnical earthquake engineering. Seismic stability considerations in earth dam design are presented in "Considerations in the Earthquake-Resistant Design of Earth and Rockfill Dams" by Seed [1979]. Seismic design issues concerning solid waste landfills are presented in Bray *et al.* [1993].

25

Geo-Environment

Pedro C. Repetto
Woodward-Clyde Consultants

25.1 Introduction

The objective of this chapter is to present the main applications of geotechnical engineering to environmental projects. There are numerous examples of these applications, among which the most frequent are waste disposal projects, environmental compliance projects, and environmental remediation projects.

Although some projects could be classified under more than one category, in this introduction they are separated depending on whether the project is (1) to be designed prior to the disposal of waste; (2) the objective of the project is to control **releases** to the environment caused by current activities or to bring a facility into compliance with current environmental or health and safety regulations; or (3) the objective is to remediate a site impacted by previous releases and/or control releases from contaminants or wastes accumulated at a site. Although there is not uniformity in these definitions, generally the term *environmental compliance* is applied to active facilities, whereas *environmental remediation* is applied to closed facilities.

The main objective of the three categories of projects mentioned above is to reduce or minimize impacts to the environment. Releases that impact the environment may occur through the disposal or spreading of **solid wastes** or **liquid wastes,** the infiltration or runoff of liquids that have been in contact with wastes, or through gaseous emissions. These releases may, in turn, impact different media. They may carry contaminants into soils located in upland areas or into sediments within water bodies, transport liquids or dissolved contaminants through groundwater or surface water bodies, or be released to the atmosphere.

The geo-environmental structures used to reduce or minimize the propagation of contaminants through the different media are called containment systems. The most commonly used containment systems are **liner** systems, cover systems (or caps), and slurry wall cut-offs. The use of liner systems is generally limited to new projects or expansions, while cover systems and slurry walls are used with all categories of projects.

This chapter presents a brief description of the main categories of environmental projects mentioned above, followed by a detailed discussion of liner and cover systems, including materials used for their construction and some construction-related aspects.

0-8493-8953-4/95/$0.00 + $.50

Waste Disposal Projects

As indicated above, the first category of environmental projects comprises those projects designed and constructed to contain waste materials prior to the disposal of waste. This includes projects such as landfills, **lagoons,** and tailings impoundments. Although tailings impoundments are basically landfills, their design and operation, requiring simultaneous management of liquids and solids, differs substantially from typical landfills, in which mostly solids are managed, and therefore it is justified to consider them as a different type of project.

A brief discussion of landfill projects is presented below as an introduction to the use of containment systems in environmental projects. For brevity, a discussion of tailings impoundments and lagoons is omitted. However, the reader should keep in mind that several of the elements of landfill projects have similar application for tailings impoundments and lagoons.

Modern landfills are complex structures that include a number of systems to prevent undue releases to the environment. The main systems are:

- Bottom liner system
- Final cover system
- Surface water management system
- **Leachate** management system (leachate collection, removal, and disposal systems)
- Gas management system
- Environmental **monitoring** system

Although all these systems are important to prevent undue releases to the environment, only the bottom liner and the final cover systems will be discussed, since the other systems are beyond the scope of this chapter. The function of the liner system is to contain leachate and prevent it from migrating downward into the subsoil or groundwater. The function of the final cover is to control infiltration of precipitation into the waste, to prevent contact of runoff with the waste, to prevent the displacement or **washout** of wastes to surrounding areas, to reduce the potential for disease **vectors** (birds, insects, rodents, etc.), and to control the emission of gases.

Landfills may be classified in accordance with several criteria. The most common criteria used to classify landfills are type of waste, type of liner system, and geometrical configuration. Based on current federal regulations, landfills are classified with respect to the type of waste as **municipal solid waste** (MSW), **hazardous waste,** and other types of solid waste. Federal regulations regarding these three types of landfills are published in Sections 258, 264, and 257, respectively, of the Code of Federal Regulations. State regulations are, however, frequently more stringent than federal regulations. State regulations frequently include specific regulations for landfills where other types of waste will be disposed, such as **industrial waste, construction demolition debris,** and residual waste.

With respect to the liner system, landfills are classified as single or double lined. Liner systems comprising only one (primary) liner are called single-liner systems. Liner systems comprising a primary and a secondary liner, with an intermediate leachate detection system, are called double-liner systems. Each of the liners (primary or secondary) may consist of only one layer [clay, geomembrane, or geosynthetic clay liner (GCL)] or adjacent layers of two of these materials, in which case it is called a composite liner. A detailed discussion of liner systems is presented later in this chapter.

The geometrical configuration of a landfill is the result of a number of factors that have to be considered during the design. It is generally intended to design the subgrade excavation as deep as possible, so that airspace is maximized and soils are made available for daily, intermediate, and final covers, structural fill, soil liners, perimeter and intercell berms, sedimentation basin berms, and access roads. However, the footprint shape and extent, cell layout, excavation depth, and subbase grading are strongly influenced by the following factors:

- Depth to the water table or uppermost aquifer. Some regulations require a minimum separation from the water table. Even if regulations do not require separation from the water table, excavation dewatering has a significant cost impact.

- Depth to bedrock. The cost of hard rock excavation negatively impacts the economic feasibility of a project. Additionally, excavated rock generally has little use in landfill projects.
- Stability of the foundation. Weak foundation soils, as well as areas previously mined or susceptible to sinkhole development, may cause instability problems.
- Site topography and stability of natural and cut slopes. In flat areas only fill embankments and cut slopes need to be considered, whereas excavations close to natural slopes may affect their stability and need to be considered in selecting the landfill footprint.
- Stability of cut slopes. The stability of cut slopes is a function of their height, inclination, groundwater conditions, and the strength and unit weight of the in situ soils. The inclination and height of the cut slopes affects the landfill airspace and the volume of soils to be excavated.
- Soils balance. The soils balance refers to the types and volumes of soils available and needed. An adequate soils balance allows a sufficient volume of soils to be available from on-site excavations and, at the same time, avoids excessive excavation that would have to be stockpiled on-site or disposed off-site.
- Permeability of natural soils. In some cases a natural clay liner may be acceptable, which would eliminate the need for low-permeability borrow for an engineered liner.
- Required airspace and available area. A minimum airspace may be required to make the landfill project economically feasible.
- Waste stream. The waste rate is directly related to the life of each cell. Normally cells are dimensioned so that they do not remain empty for an extended period of time.
- Filling sequence. One of the design goals is to avoid excessive stockpiling of excavated soils and to minimize leachate production by placing the final cover as soon as possible.
- Grading of the landfill floor. A minimum slope is required for the leachate collection system.

Finally, from the point of view of geometrical configuration, the most common landfill types are:

- Area fill. Landfilling progresses with little or no excavation (Fig. 25.1). Normally used in relatively flat areas with shallow water table.
- Aboveground and below-ground fill. The landfill consists of a two-dimensional arrangement of large cells that are excavated one at a time. Once two adjacent cells are filled, the area between them is also filled (Fig. 25.2). Normally used in relatively flat areas without a shallow water table.
- Valley fill. The area to be filled is located between natural slopes (Fig. 25.3). It may include some below-ground excavation.
- Trench fill. This method is similar to the aboveground and below-ground configuration, except that the cells are narrow and parallel (Fig. 25.4). It is generally used only for small waste streams.

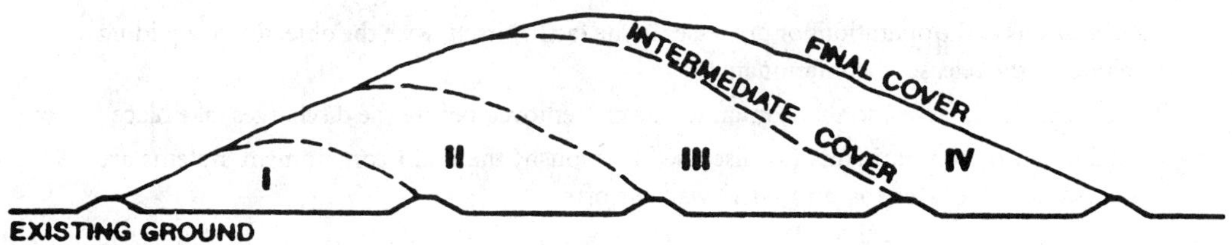

FIGURE 25.1 Area fill.

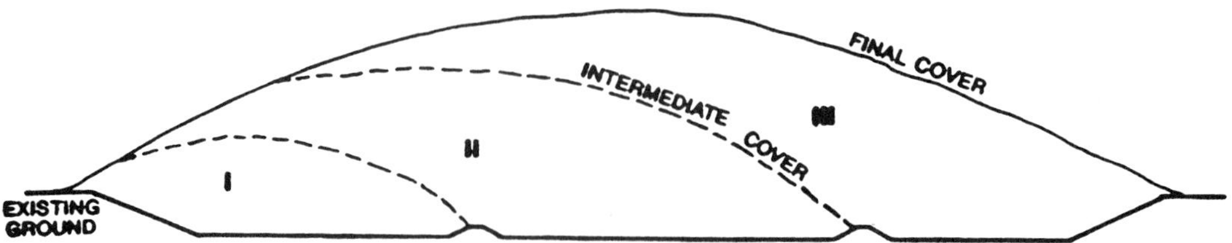

FIGURE 25.2 Aboveground and below-ground fill.

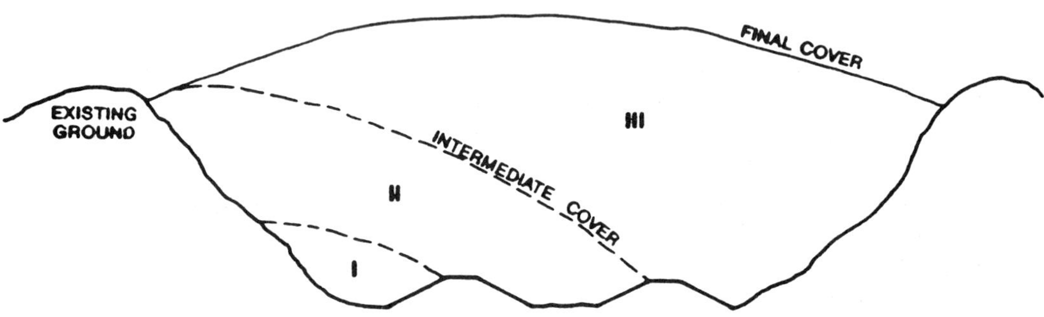

FIGURE 25.3 Valley fill.

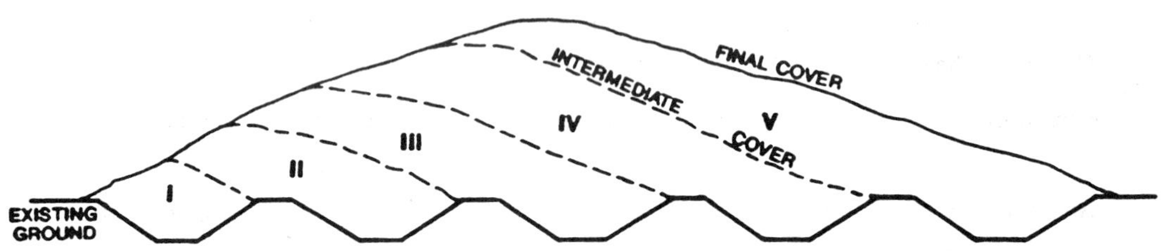

FIGURE 25.4 Trench fill.

Environmental Compliance Projects

The second category comprises a very broad range of projects, in which the aim is to bring a facility into compliance with current regulations (including health and safety) and/or to control releases to the environment caused from current site activities. As mentioned above, these releases may occur through solid, liquid, or gaseous discharges, and may affect soils, sediments, water bodies, or the atmosphere. The type of facility involved may be of any kind: landfill, tailings impoundment, lagoon, or any industrial facility.

The actions for controlling releases from active facilities are multiple and can be classified into three major groups:

- Modifications to the operations or processes in the facility itself, with the objective of avoiding or reducing the release of contaminants
- Treatment of discharges, so that contaminants are removed before the discharges take place
- Containment of contaminants (as discussed previously, the main containment systems are liner systems, cover systems, and slurry wall cut-offs)

The first two groups of actions listed above are beyond the scope of this chapter. Containment systems are discussed later in this chapter.

Environmental Remediation Projects

This category comprises projects in which the aim is to control releases to the environment from wastes or contaminants accumulated at a site and/or to remediate areas whose environmental functions have been impaired by previous site activities. As discussed previously, this category of projects generally applies to sites where activities have ceased. As in the case of environmental compliance projects, the types of activities that took place at the site can be of any kind.

The actions used to control releases from closed facilities can be classified into four major groups:

- Removal of the wastes or contaminated materials, and disposal at an appropriate facility
- Fixation or removal of the contaminants in place, so that releases are prevented
- Interception and treatment of contaminated releases
- Isolation of the wastes or contaminated materials by means of containment systems, so that releases are prevented

Of these four groups, only the use of containment systems is within the scope of this chapter and is discussed next.

25.2 Geo-Environmental Containment Systems

Liners

Liner systems are containment elements constructed under the waste to control infiltration of contaminated liquids into the subsoil or groundwater. The contaminated liquid, or leachate, may be part of the waste itself or may originate from water that has infiltrated into the waste.

Liner systems consist of multiple layers which fulfill specific functions. The description presented below refers specifically to landfill liner systems. However, the main characteristics of liner systems are similar for other applications. Landfill liner systems may consist, from top to bottom, of the following functional layers:

- *Protective layer.* This is a layer of soil, or other appropriate material, that separates the **refuse** from the rest of the liner to prevent damage from large objects.
- *Leachate collection layer.* This is a high-permeability layer, whose function is to collect leachate from the refuse and to convey it to sumps from where it is removed. Frequently the functions of the protective layer and the leachate collection layer are integrated in one single layer of coarse granular soil.
- *Primary liner.* This is a low-permeability layer (or layers of two different low-permeability materials in direct contact with each other). Its function is to control the movement of leachate into the subsoil.
- *Secondary leachate collection layer, or leakage detection layer.* This is a high-permeability (or high-transmissivity, if geosynthetic) layer designed to detect and collect any leachate seeping through the primary liner. This layer is used only in conjunction with a secondary liner.
- *Secondary liner.* This is a second (or backup) low-permeability layer (or layers of two different low-permeability materials in direct contact with each other). Not all liner systems include a secondary liner.
- *Drainage layer.* In cases where the liner system is close or below the water table, a high-permeability (or high-transmissivity, if geosynthetic) blanket drainage layer is generally placed under the liner system to control migration of moisture from the foundation to the liner system.
- *Subbase.* This layer is generally of intermediate permeability. Its function is to separate the liner system from the natural subgrade or structural fill.

These layers are normally separated by geotextiles to prevent migration of particles between layers, or to provide cushioning or protection of geomembranes.

As indicated above, liner systems may have a primary liner only or may include primary and secondary liners. In the first case it is called a *single-liner* system, and if it has a primary and a secondary liner it is called a *double-liner* system. Further, each of the liners (primary or secondary) may consist of one layer only (low-permeability soil, geomembrane, or GCL) or adjacent layers of two of these materials, in which case it is called a *composite* liner. There are multiple combinations of these names, some of which are given below as examples (obviously there are many more combinations):

- Single synthetic liner: primary liner only, consisting of a geomembrane.
- Single soil liner: primary liner only, consisting of a low-permeability soil layer.
- Double synthetic liner: primary and secondary liners, each consisting of a geomembrane, separated by a secondary leachate collection layer.
- Single composite liner: primary liner only, consisting of a geomembrane in direct contact with a low-permeability soil layer. (This is the design standard specified in Subtitle D regulations of the Resource Conservation and Recovery Act for **municipal solid waste landfill units**).
- Double composite liner: primary and secondary liners, each consisting of a geomembrane in direct contact with a low-permeability soil layer, separated by a secondary leachate collection layer (see Fig. 25.5).

Low-Permeability Covers

A low-permeability cover is a containment system constructed on top of the waste, primarily to control the infiltration of precipitation. Cover systems control infiltration in two ways: (1) by providing a low-permeability barrier and (2) by promoting runoff with adequate grading of the

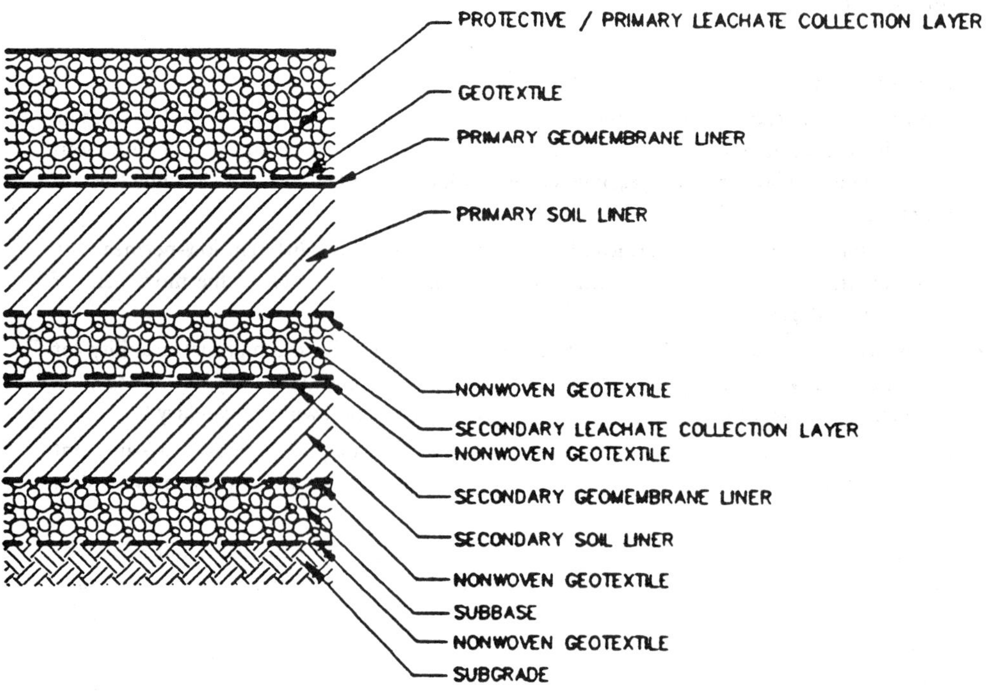

FIGURE 25.5 Double composite liner system.

final surface. Other functions of cover systems are to prevent contact of runoff with waste, to prevent spreading or washout of wastes, to reduce disease vectors, and to control the emission of gases.

Cover systems also consist of multiple layers with specific functions related to the management of surface water. Of the precipitation that falls on a cover, part becomes surface water runoff, a portion infiltrates into the cover, and the remainder is lost through evapotranspiration. A low-permeability cover may include, from top to bottom, the following functional layers (see Fig. 25.6):

- *Erosion (vegetative) layer.* This is a layer of soil capable of supporting vegetation (typically grass) and with good resistance to erosion due to surface water runoff.
- *Infiltration layer.* The functions of this layer, also frequently called **cover material,** are to protect the barrier layer from frost penetration and to separate the precipitation that does not evaporate into runoff and infiltration. Each of these two components of the flow has to be controlled separately: the runoff through adequate erosion resistance of the erosion layer and the infiltration through adequate hydraulic capacity of the underlying drainage layer.
- *Drainage layer.* This is a permeable layer whose function is to convey the water that infiltrates into the cover system. A cover without a drainage layer is susceptible to damage by pop-outs.

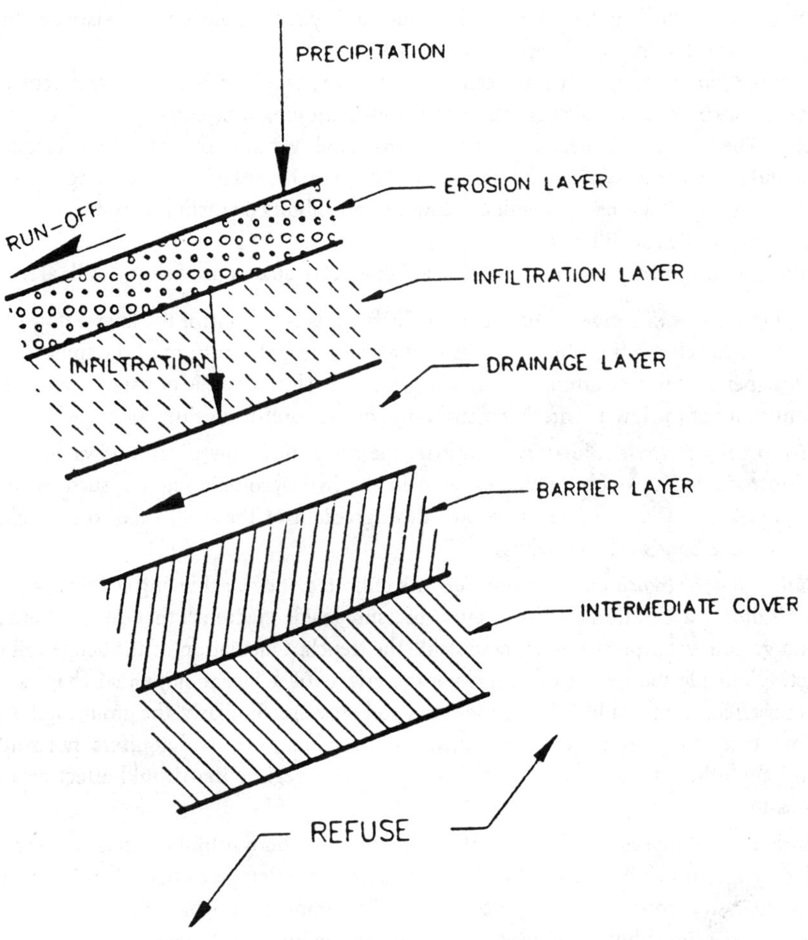

FIGURE 25.6 Cover functional layers.

- *Barrier layer.* This is a low-permeability layer (or layers of two different low-permeability materials in direct contact with each other) whose function is to control infiltration into the waste.

The erosion and the infiltration layers are normally constructed of soils, since they must support vegetation and provide frost protection. The drainage layer can be constructed of a permeable soil, a geonet, or a geocomposite. The barrier layer can be constructed of a low-permeability soil, a geomembrane, a GCL, or adjacent layers of two of these materials, in which case it is called a *composite* barrier layer. Some of these layers are also generally separated by geotextiles.

Slurry Walls

Slurry walls are a type of cut-off wall whose function in environmental projects is to control the horizontal movement of contaminated liquids. The single most important characteristic of slurry walls is that the stability of the trench walls during excavation is maintained without any bracing or shoring by keeping the trench filled with a slurry.

During excavation, stability of the trench walls is achieved by maintaining a pressure differential between the slurry inside the trench and the active soil and groundwater pressures acting from outside of the trench. In permeable soils there is initially some loss of slurry, due to the infiltration of slurry from the trench into the surrounding soils. After a short period of time, a low-permeability crust, called *cake,* forms on the trench walls; it provides a hydraulic barrier between the slurry and the groundwater. If the length of trench that is open at any given time is short relative to the height of the trench, the stability is improved by arching effect.

After some length of trench has been excavated, the slurry is displaced and replaced by a permanent backfill mix, consisting either of a soil–bentonite admixture or a bentonite–cement admixture. The backfill mix needs to have low permeability, but at the same time adequate shear strength and low compressibility. The low permeability is achieved by incorporating bentonite into the backfill mix, and the shear strength and low compressibility by including some granular soil or cement as part of the backfill mix.

The most important aspects related to the design and construction of slurry walls are:

- *Keying of the wall bottom.* The intent of slurry walls is to control the lateral flow of liquids. In order to achieve this objective, it is necessary to have the bottom of the trench keyed into a low-permeability stratum, so that flow under the slurry wall is not significant. Therefore, continuity of the low-permeability stratum requires careful consideration.
- *Hydrogeologic effects.* Slurry walls restrict the horizontal flow of groundwater. The effects of introducing this containment system in the hydrogeologic regime, such as mounding upgradient of the wall and drawdown downgradient of the wall, need to be evaluated by means of groundwater modeling.
- *Stability of the trench during excavation.* As indicated above, achieving stability of the trench walls during excavation requires a slurry pressure inside the trench that exceeds the active soil and groundwater pressures acting outside the trench. Calculation of the active soil pressure acting outside the trench must include the effects of adjacent slopes, surcharges, and any other effects that would increase the active soil pressure. Similarly, the groundwater pressure must take into account the effects of upward gradient, confined aquifers, horizontal flow, and any other characteristic of the hydrogeological regime that would affect groundwater pressures.
- *Backfill mix composition.* The backfill mix requires low permeability, adequate shear strength, and low compressibility. Ideally, this is achieved by selecting a backfill mix consisting of granular soil, cohesive soil, and bentonite. The granular soil provides shear strength and reduces compressibility, whereas the cohesive soil reduces the amount of bentonite required to achieve low permeability.

25.3 Liners and Covers

Although the functions of liners and covers are quite different, they use similar materials and are subjected to similar construction considerations. Therefore, for simplicity, the materials used for liners and covers are presented together in this section.

Materials used for the construction of liners and covers can be classified with respect to their origin and function. With respect to their origin they are classified into two major groups: natural soils and geosynthetic materials. With respect to their function the classification is more complex and includes the following:

- *Barrier layers.* These are low-permeability layers of natural soils (clay, soil–bentonite admixtures) or geosynthetic materials (geomembranes, GCLs).
- *Drainage layers.* These are high-permeability (if soil) or high-transmissivity (if geosynthetic) layers. In the first case they consist of granular soils and in the latter of geonets or geocomposites.
- *Fill.* The uses of fill include applications such as grading, intercell berms, perimeter berms, and sedimentation basin berms. Obviously this category is limited to natural soils.
- *Vegetative layer.* Vegetative layers are used as the uppermost layer of final covers, or as erosion protection for permanent or temporary slopes. This category is also limited to natural soils.
- *Filtration/protection/cushion layers.* Liners and covers consist of functional layers of soils of very different gradations, or geosynthetic materials that require special protection to prevent damage during installation or operation. Separation and protection are generally provided by geotextiles placed between functional layers. Geotextiles are placed as filters between soil layers of different gradations to prevent migration of soil particles, as cushion below geomembranes to be installed on granular soils, or as protection on top of geomembranes placed below granular soils.
- *Tensile reinforcement.* In cases where large deformations can be expected in liners or covers, geogrids or geotextiles are used as tensile reinforcement to avoid excessive tensile strains in other elements of the liner or cover system. Additionally, sliding is a potential failure mode in steep cover slopes. In this case geogrids or geotextiles are installed embedded within the layer susceptible to sliding to provide a stabilizing tensile force.

Soils

Barrier Layers

Barrier layers constructed of natural soils may consist of clay or a soil–bentonite admixture. In the case of admixtures, the required bentonite content is selected based on laboratory permeability tests performed with different bentonite percentages.

The typical requirement of most regulations for barrier layers is to have a permeability not exceeding 10^{-7} cm/s. However, since most regulations do not specify under what conditions this value is to be determined, the project specifications must address those conditions. The main factors that influence the permeability of a given soil or soil–bentonite admixture are discussed below:

- *Compaction dry density.* All other factors being constant, the higher the compaction dry density of a soil, the lower its permeability. However, at the same compaction dry density, the permeability also varies as a function of the compaction moisture content and the compaction procedure. The compaction dry density is normally specified as a percentage of the standard or modified Proctor test maximum dry density.
- *Compaction moisture content.* All other factors maintained constant, the permeability of a given soil increases with increasing compaction moisture content. It should be noted, however, that as the moisture content approaches saturation, it becomes more difficult to achieve a

target dry density, and other properties of the compacted soil, such as compressibility or shear strength, may become critical.

- *Compaction procedure.* It has been observed that specimens compacted to identical dry densities and moisture contents may exhibit different permeabilities if compacted following different procedures. Therefore, if laboratory test results are close to the required permeability, the compaction procedure used for preparation of laboratory specimens should be similar to the field compaction procedures.
- *Consolidation pressure.* In geo-environmental applications, barrier layers may be subjected to high permanent loads while in service, as is the case in landfill liner systems. In these cases there is an increase in dry density due to consolidation under the service loads. On the other hand, permanent loads are minimal on cover systems and an increase in dry density with time should not be expected. Therefore, permeability tests should be performed using a consolidation pressure consistent with the expected loads.

The combination of compaction dry densities and moisture contents that yields a permeability not exceeding a specified value can be determined by means of a permeability window. A permeability window is constructed by performing permeability tests on a series of specimens compacted at different dry densities and moisture contents, covering the ranges of dry density and moisture content of interest. The specimens should be prepared using a compaction procedure consistent with the field procedure (generally Proctor compaction) and the tests should be performed at a consolidation pressure similar to those to be produced by the expected loads.

A practical way of determining a permeability window is described below:

- A large sample of the soil, sufficient to perform three Proctor tests and to prepare nine permeability specimens, should be carefully homogenized, so that differences due to individual specimens are avoided.
- Specific gravity of the soil is determined and the saturation line is plotted in a dry density–moisture graph (see Fig. 25.7).

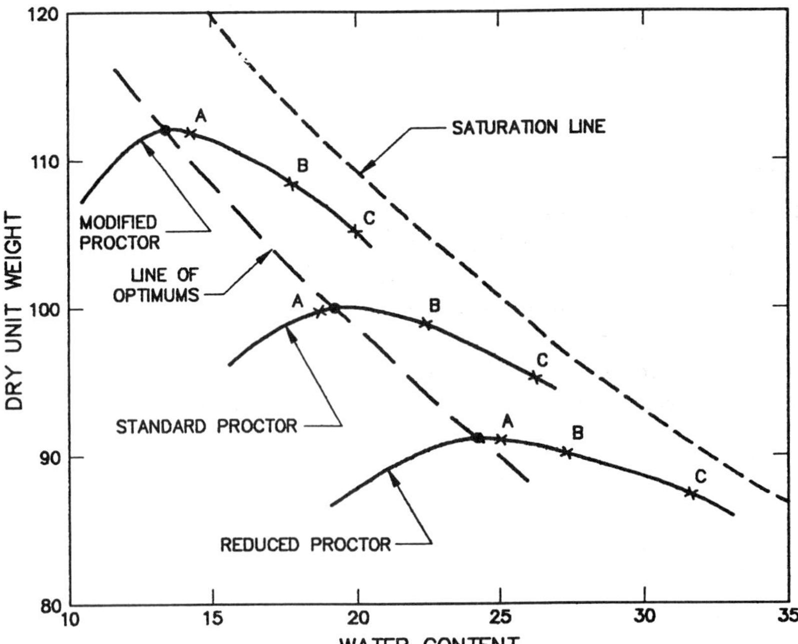

FIGURE 25.7 Dry density–moisture graph.

- Three Proctor tests are performed (modified, standard, and reduced) and plotted in the dry density–moisture graph (Fig. 25.7). The reduced Proctor is a test performed with the same hammer as the standard Proctor, but using a reduced number of blows per layer. Assuming a four-inch mold is used, generally the reduced Proctor is performed using 10 or 15 blows per layer.
- For each of the three Proctor tests, the maximum density and the optimum moisture content are determined, and a line of optimums connecting the three curves is drawn (Fig. 25.7).
- For each of the Proctor curves, three moisture contents are selected corresponding to points on the Proctor curve. The moisture contents typically range from the optimum moisture content to about 95% saturation (points *A*, *B*, and *C* in Fig. 25.7).
- Specimens for permeability testing are prepared at each of the three moisture contents selected above, compacted with the same energy and procedure of the Proctor curve used to select the moisture contents.
- The same procedure is followed for the other two Proctor curves, thus obtaining a total of nine permeability specimens.
- Flexible wall permeability tests are performed on the nine specimens, using a consolidation pressure consistent with the expected loads. Typically the consolidation pressure is selected equal to K_0 times the vertical stress.
- The results of the permeability tests are plotted on the dry density–moisture graph and isopermeability lines are drawn by interpolating between the nine test points (Fig. 25.8). If needed, additional test points may be selected to improve accuracy or to verify questionable points.
- Finally, the permeability window is selected as the area bound by the isopermeability line corresponding to the required permeability and the saturation line. Additionally, the permeability window is generally truncated by moisture content lower and upper bounds. The line of optimums is generally selected as the moisture content lower bound, because cohesive soils compacted dry of optimum are too rigid and may easily crack. The moisture content upper bound is selected based on unconfined strength and compressibility, depending on the specific application.

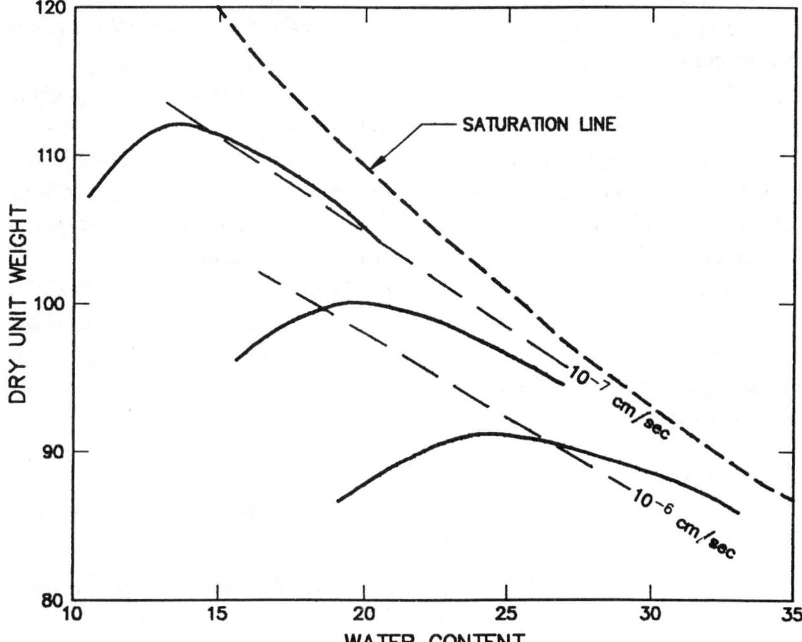

FIGURE 25.8 Permeability window.

It should be noted that selecting moisture contents on the Proctor curves and compacting the specimens following the Proctor procedure does not provide points on a grid. However, this procedure is preferred because the compaction procedure is uniform and considered to be more representative of the field procedures. Some laboratories prefer to prepare the specimens following a rectangular grid on the dry density–moisture graph, compacting the specimens within a mold using a jack. In these cases the procedure is less representative of the field conditions and some differences can be expected with respect to tests performed on specimens prepared using Proctor procedures.

The discussion presented above refers, in general, to tests in which the permeant is clean water. However, in geo-environmental applications, the permeant is leachate in the case of liners. The flow of leachate through cohesive soils may produce changes in the cations, which in turn may affect the permeability of the soil. In order to evaluate these changes, leachate compatibility permeability tests are performed as described below.

Leachate compatibility permeability tests are long-duration permeability tests, in which distilled water is used initially as the permeant and then leachate from the facility (or a similar leachate) is used as the permeant. These tests are continued until several pore volumes (typically three or four) of leachate have circulated through the specimen. During this period changes in the permeability of the soil are monitored. A leachate compatibility permeability test of a low-permeability soil generally lasts several months.

Drainage Layers

Drainage layers constructed of natural soils consist of sands or gravels, and may be used in liners or covers. Drainage layers are designed to have a hydraulic capacity adequate to convey the design flow rate without significant head buildup. The hydraulic capacity of a drainage layer is a function of its permeability, thickness, and slope. Frequently the transmissivity, defined as the product of the permeability by the thickness, is used. The main considerations regarding drainage layers are listed below:

- The gradation of the granular soil to be used for a drainage layer is selected based on the required permeability. In general, there are no strict gradation requirements, since the requirements for filtration between adjacent soil layers are satisfied by means of a geotextile acting as a filter. Frequently, gradation requirements are driven mostly by availability (i.e., aggregates commercially available).
- Drainage layers in liner systems must resist the attack of leachate, which is generally acidic. Therefore calcareous granular soils must be avoided.
- Drainage layers are frequently adjacent to geomembranes, which are susceptible to puncture. Geotextiles are used to protect geomembranes from granular soils. However, if gravel-size soil is used, geotextiles may not provide sufficient protection from angular gravel. Therefore, the maximum angularity of gravels is generally limited to subangular.
- The minimum drainage layer thickness that can be practically constructed is approximately six inches. However, if the drainage layer has to be placed on a geomembrane, driving construction equipment on the drainage layer may seriously damage the geomembrane. In these cases placement considerations control the minimum layer thickness and must be carefully evaluated.

Fill

The requirements for fills are, in general, similar to those for other types of engineering projects. The following types of fills are frequent in geo-environmental projects:

- *Grading fill.* There are no strict requirements for grading fills in which fill slopes are not constructed. Depending on the thickness of the fill and the loads to be applied on them, compressibility may be an important consideration. When grading fill is used to form fill slopes, coarse granular soil is used to provide adequate shear strength.

- *Structural fill.* This category comprises fill used for elements such as intercell berms and perimeter berms in landfills, or perimeter berms for leachate ponds. When these elements will be subjected to significant lateral pressures, these berms are constructed of coarse granular soils. These berms are generally lined, so permeability is not an issue.
- *Water-containment berms fill.* Berms containing water-retaining structures, such as sedimentation and detention basin berms, require a combination of low permeability and high shear strength. These two conditions are difficult to satisfy simultaneously, since soils of low permeability are weak, and vice versa. In these cases the type of fill generally used consists of a granular soil with significant fines content, which provides intermediate permeability and shear strength. Alternatively, lined berms may be constructed.

Vegetative Layer

As explained previously, the vegetative layer is the uppermost layer of a cover system. Vegetative layers must be adequate to support vegetation, generally grass, and must have adequate resistance to erosion.

In order to support vegetation, the soil must contain sufficient nutrients. Nutrients can also be supplied by adding limestone or other fertilizers. For information about this topic, consult the state erosion and sediment control manual or the county Soil Conservation District, since the requirements vary as a function of climate.

The erosion that a vegetative layer may suffer is a function of the soil type and the slope inclination and length. The soil loss is estimated by means of the Universal Soil Loss Equation, published by the U.S. Department of Agriculture. The maximum soil loss recommended by the U.S. Environmental Protection Agency for landfill covers is 2 tons/acre/year.

It should be noted that specifications frequently refer to vegetative layers as "topsoil." The term *topsoil* has a specific meaning from an agricultural point of view and is generally more expensive than other soils that can also support vegetation with adequate fertilization. For these reasons, it is recommended to use the term *topsoil* only when that type of soil is specifically required.

Geosynthetics

A large number of geosynthetic materials are used in environmental applications and there are many different tests to characterize their properties. This prevents a detailed presentation within the space allocated in this section. Therefore this section presents a brief summary of the most common types of geosynthetic materials generally used in environmental applications and their properties. For a detailed discussion on the applications and testing of geosynthetic materials, product data, and manufacturers, Koerner [1994], GRI and ASTM test methods and standards, and the Specifier's Guide of the Geotechnical Fabrics Report [Industrial Fabrics Association International, 1993] are recommended.

In general, two types of test are included in geosynthetic specifications: conformance tests and performance tests. Conformance tests are performed prior to installing the geosynthetics, to demonstrate compliance of the materials with the project specifications; some of the conformance tests are frequently provided by the manufacturer. Performance testing is done during the construction activities, to ensure compliance of the installed materials and the installation procedures with the project specifications.

Barrier Layers

Geosynthetic barrier layers may consist of geomembranes, also called flexible membrane liners (FMLs), or geosynthetic clay liners (GCLs).

As discussed previously, geomembranes are used as barrier layers for landfill liner and cover systems. They are also used as canal liner, surface impoundment liner and cover, tunnel liner, dam liner, and leach pad liner. In general, geomembranes are classified with respect to the polymer that they are made of and their surficial roughness. These two classifications are discussed below.

The most common polymers used for geomembranes are high density polyethylene (HDPE), very low density polyethylene (VLDPE), polypropylene, and PVC. Selection of the polymer is based primarily on chemical resistance to the substances to be contained. The polymer most widely used for landfill liner systems is HDPE, since it has been shown to adequately resist most landfill leachates. In the case of cover system barrier layers, flexibility is frequently an important selection factor, since landfill covers are subjected to significant settlements. VLDPE geomembranes are generally more flexible than HDPE and therefore are frequently used for cover systems.

The chemical resistance of geomembranes and other geosynthetic materials is evaluated by means of the EPA 9090 Compatibility Test. In this test, the initial physical and mechanical properties of a geosynthetic are determined (baseline testing) prior to any contact with the chemicals to be contained (leachate in the case of landfills). Then, geosynthetic specimens are immersed in tanks containing those chemicals, at 23 and 50°C. Specimens are removed from the tanks after 30, 60, 90, and 120 days of immersion and tested to determine their physical and mechanical properties. Comparison of these properties with the results of the baseline testing serves as an indicator of the effect of the chemicals on that specific geosynthetic material.

With respect to surface roughness, geomembranes are classified as smooth or textured. Smooth geomembranes are less expensive and easier to install than textured geomembranes, but exhibit a low interface friction angle (as low as 6 to 8 degrees) with other geosynthetics and with cohesive soils. Textured geomembranes provide a higher interface friction angle. The reader is warned that interface friction angles are not fixed values and must be evaluated on a case-specific basis, since they vary with parameters such as relative displacement (peak versus residual strength), normal stress, moisture conditions, backing used in the test (soil or rigid plates), and so on.

The main physical and mechanical properties generally used to characterize geomembranes and the test procedures to measure those properties are as follows:

Thickness	ASTM D751 and D1593
Density	ASTM D1505
Tensile properties	ASTM D638
Yield strength	
Break strength	
Elongation at yield	
Elongation at break	
Puncture resistance	FTMS 101C method 2065
Tear resistance	ASTM D1004
Low-temperature brittleness	ASTM D746
Carbon black content	ASTM D1603
Environmental stress crack resistance	ASTM D5397

Seaming of geomembranes is performed by bonding together two sheets. The main methods for seaming geomembranes are:

- *Extrusion welding.* A ribbon of molten polymer is extruded on the edge of one of the sheets or in between the two sheets. This method is applicable only to polyethylene and polypropylene geomembranes.
- *Thermal fusion.* Portions of the two sheets are melted using a hot wedge or hot air. This method is applicable to all types of geomembranes.
- *Solvent or adhesive processes.* These methods are not applicable to polyethylene and polypropylene geomembranes.

Seam strength (shear and peel) of geomembranes is controlled during installation by means of performance (destructive) testing performed in accordance with ASTM D4437 procedures.

GCLs consist of a thin layer of bentonite sandwiched between two geotextiles or bonded to a geomembrane. GCLs are currently available under the following registered names: Gundseal, Claymax,

Shear-Pro, Bentofix, Bentomat, and NaBento. Gundseal is manufactured with a geomembrane on one side only, while the other products have geotextiles on both sides. The geomembrane and geotextiles are fixed to the bentonite layer by means of adhesives, needle-punched fibers, or stitches. The types of geotextiles used include several combinations of woven and nonwoven geotextiles.

The main physical and mechanical properties generally used to characterize GCLs and the test procedures to perform those tests are as follows:

Bentonite mass per unit area	ASTM D3776
GCL permeability	ASTM D5084
Base bentonite properties	
Moisture content	ASTM D4643
Swell index	USP-NF-XVII
Fluid loss	API 13B
Geotextiles	
Weight	ASTM D3776
Thickness	ASTM D1593
GCL tensile strength	ASTM D4632
Percent elongation	ASTM D463

Evaluation of the strength of a GCL must take into account its internal shear strength and the interface strength between the GCL and adjacent materials. The reader is warned that both the internal and interface shear strengths are not fixed values and must be evaluated on a case-specific basis, since they vary with parameters such as relative displacement (peak versus residual strength), normal stress, hydration of the bentonite, backing used in the test, and so on. Special attention must be given to the effects of long-term shear (creep) on needle-punched fibers and stitches, and to the squeezing of hydrated bentonite through the geotextiles. Seaming of GCLs is performed by overlapping adjacent sheets.

Drainage Layers

Geosynthetic drainage layers used as part of liner and cover systems may consist of geonets or geocomposites. Geocomposites consist of a geonet with factory-welded geotextile on one side or on both sides.

Polymers used for geonets are polyethylene (PE), HDPE, and medium density polyethylene (MDPE). The EPA 9090 Compatibility Test is also used to evaluate the chemical resistance of geonets and geocomposites. Geotextiles of various types and weights are attached to geonets to manufacture geocomposites, the most common type being nonwoven.

The main physical, mechanical, and hydraulic properties generally used to characterize geonets and geocomposites, and the test procedures to measure those properties, are thickness (ASTM D5199), compressive strength at yield (ASTM D1621), and in-plane flow rate [transmissivity (ASTM D4716)]. The most important property of geonets and geocomposites, in relation to their performance as drainage layers, is the transmissivity. The reader is warned that the results of laboratory transmissivity tests on geonets and geocomposites vary with several parameters, including normal stress, gradient, the type and weight of geotextiles attached, and the backing used in the tests. Furthermore, transmissivity values determined in short-term laboratory tests must be decreased, applying several correction factors to calculate long-term performance values. These correction factors account for elastic deformation of the adjacent geotextiles into the geocomposite core space, creep under normal load, chemical clogging and/or precipitation of chemicals, and biological clogging. A complete discussion of the correction factors recommended for various applications is presented by Koerner [1990].

In addition to the properties of the geonet or geocomposite, the properties of the geotextiles attached to the geonet are generally specified separately as discussed below. Seaming of geonets and geocomposites is generally performed using plastic ties, which are not intended to transfer stresses.

Geotextiles

Geotextiles may be used to perform several different functions. The most important are:

- *Separation.* Consists of providing separation between two different soils to prevent mixing. A typical application is placement of a granular fill on a soft subgrade.
- *Reinforcement.* Applications of geotextiles for reinforcement are identical to those of geogrids, discussed in the next section.
- *Filtration.* The geotextile is designed as a filter to prevent migration of soil particles across its plane. Several aspects need to be considered associated with this function: filtration (opening size relative to the soil particles), permittivity (flow rate perpendicular to the geotextile), and clogging potential.
- *Drainage.* In-plane capacity to convey flow or transmissivity.
- *Cushioning/protection.* The geotextile serves to separate a geomembrane from a granular soil, to protect the geomembrane from damage.

Geotextiles are classified with respect to the polymer that they are made of and their structure. The polymers most commonly used to manufacture geotextiles are polypropylene and polyester. With respect to their structure geotextiles are classified primarily as woven or nonwoven. Each of these types of structure is in turn subdivided, depending on the manufacturing process, as follows: woven (monofilament, multifilament, slit-film monofilament, slit-film multifilament); or nonwoven (continuous filament heat bonded, continuous filament needle punched, staple needle punched, spun bonded, or resin bonded).

The main physical, mechanical, and hydraulic properties generally used to characterize geotextiles and the test procedures to measure those properties are:

Specific gravity	ASTM D792 or D1505
Mass per unit area	ASTM D5261
Percent open area (wovens only)	CWO-22125
Apparent opening size	ASTM D4751
Permittivity	ASTM D4491
Transmissivity	ASTM D4716
Puncture strength	ASTM D4833
Burst strength	ASTM D3786
Trapezoid tear strength	ASTM D4533
Grab tensile/elongation	ASTM D4632
Wide width tensile/elongation	ASTM D4595
UV resistance	ASTM D4355

Seaming of geotextiles is performed by sewing. Seam strength is tested following ASTM D4884.

Most geotextiles are susceptible to degradation under ultraviolet light. Therefore, appropriate protection is required during transport, storage, and immediately after installation.

Geogrids

Geogrids are used to provide tensile reinforcement. Typical applications include:

- *Reinforcement of slopes and embankments.* Potential failure surfaces would have to cut across layers of geogrid. The resisting forces on the potential failure surface would be comprised of the shear strength of the soil and the tensile strength of the geogrid.
- *Reinforcement of retaining walls.* In this application, some of the soil pressure that would act against the retaining wall is transferred by friction to the part of the reinforcing geogrid layer adjacent to the wall, while the rest of the geogrid (away from the wall) provides passive anchorage.

- *Unpaved roads.* The stiffness of the geogrid allows distribution of loads on a larger area and prevents excessive rutting.
- *Reinforcement of cover systems.* A potential mode of failure of relative steep covers is sliding of a soil layer on the underlying layer (veneer-type sliding). To control this type of failure, a geogrid layer is embedded within the unstable soil layer to provide a stabilizing tensile force. The opposite side of the geogrid must be anchored or must develop sufficient passive resistance to restrain its displacement.
- *Bridging of potential voids under liner systems.* When liner systems are constructed on existing waste (vertical expansions) or in areas where sinkholes may develop, geogrids are used within the liner system to allow bridging of voids.

Geogrids are made of polyester, polyethylene, and polypropylene. If exposed to waste or leachate, selection of the polymer is based on chemical resistance, as in the case of geomembranes. Depending on the direction of greater strength and stiffness, geogrids are classified as uniaxial or biaxial.

The main physical and mechanical properties generally used to characterize geogrids and the test procedures to measure those properties are mass per unit area (ASTM D5261), aperture size, wide width tensile strength (ASTM D4595), and long-term design strength (GRI GG4).

Defining Terms

The definitions presented in this section have been extracted from the Code of Federal Regulations (CFR), 40 CFR 258, "EPA Criteria for Municipal Solid Waste Landfills," and 40 CFR 261, "Identification and Listing of Hazardous Waste." Definitions not available in the federal regulations were obtained from the Virginia Solid Waste Management Regulations.

Agricultural waste: All solid waste produced from farming operations, or related commercial preparation of farm products for marketing.

Commercial solid waste: All types of solid waste generated by stores, offices, restaurants, warehouses, and other nonmanufacturing activities, excluding residential and industrial wastes.

Construction/demolition/debris landfill: A land burial facility engineered, constructed, and operated to contain and isolate construction waste, demolition waste, debris waste, inert waste, or combinations of the above solid wastes.

Construction waste: Solid waste which is produced or generated during construction, remodeling, or repair of pavements, houses, commercial buildings, and other structures. Construction wastes include, but are not limited to, lumber, wire, sheetrock, broken brick, shingles, glass, pipes, concrete, paving materials, and metal and plastics if the metal or plastics are part of the materials of construction or empty containers for such materials. Paints, coatings, solvents, asbestos, any liquid, compressed gases or semiliquids, and garbage are not construction wastes.

Cover material: Compactable soil or other approved material which is used to blanket solid waste in a landfill.

Debris waste: Wastes resulting from land-clearing operations. Debris wastes include, but are not limited to, stumps, wood, brush, leaves, soil, and road spoils.

Demolition waste: That solid waste which is produced by the destruction of structures and their foundations and includes the same materials as construction wastes.

Garbage: Readily putrescible discarded materials composed of animal, vegetal, or other organic matter.

Hazardous waste: The definition of hazardous waste is fairly complex and is provided in 40 CFR Part 261, "Identification and Listing of Hazardous Waste." The reader is referred to this regulation for a complete definition of hazardous waste. A solid waste is classified as hazardous waste if it is not excluded from regulations as a hazardous waste; it exhibits characteristics of ignitability, corrosivity, reactivity, or toxicity as specified in the regulations; or it is listed in the regulations (the regulations include two types of hazardous wastes: from nonspecific sources and from specific sources).

Household waste: Any solid waste (including garbage, trash, and sanitary waste in septic tanks) derived from households (including single and multiple residences, hotels and motels, bunkhouses, ranger stations, crew quarters, campgrounds, picnic grounds, and day-use recreation areas).

Industrial solid waste: Solid waste generated by a manufacturing or industrial process that is not a hazardous waste regulated under Subtitle C of RCRA. Such waste may include, but is not limited to, waste resulting from the following manufacturing processes: electric power generation; fertilizer/agricultural chemicals; food and related products/by-products; inorganic chemicals; iron and steel manufacturing; leather and leather products; nonferrous metals manufacturing/foundries; organic chemicals; plastics and resins manufacturing; pulp and paper industry; rubber and miscellaneous plastic products; stone, glass, clay, and concrete products; textile manufacturing; transportation equipment; and water treatment. This term does not include mining waste or oil and gas waste.

Industrial waste landfill: A solid waste landfill used primarily for the disposal of a specific industrial waste or a waste which is a by-product of a production process.

Inert waste: Solid waste which is physically, chemically, and biologically stable from further degradation and considered to be nonreactive. Inert wastes include rubble, concrete, broken bricks, bricks, and blocks.

Infectious waste: Solid wastes defined to be infectious by the appropriate regulations.

Institutional waste: All solid waste emanating from institutions such as, but not limited to, hospitals, nursing homes, orphanages, and public or private schools. It can include infectious waste from health care facilities and research facilities that must be managed as an infectious waste.

Lagoon: A body of water or surface impoundment designed to manage or treat wastewater.

Leachate: A liquid that has passed through or emerged from solid waste and contains soluble, suspended, or miscible materials removed from such waste.

Liquid waste: Any waste material that is determined to contain free liquids.

Liner: A continuous layer of natural or synthetic materials beneath or on the sides of a storage or treatment device, surface impoundment, landfill, or landfill cell that severely restricts or prevents the downward or lateral escape of hazardous waste, hazardous waste constituents, or leachate.

Litter: Any solid waste that is discarded or scattered about a solid waste management facility outside the immediate working area.

Monitoring: All methods, procedures, and techniques used to systematically analyze, inspect, and collect data on operational parameters of the facility or on the quality of air, groundwater, surface water, and soils.

Municipal solid waste: Waste which is normally composed of residential, commercial, and institutional solid waste.

Municipal solid waste landfill unit: A discrete area of land or an excavation that receives household waste, and that is not a land application unit, surface impoundment, injection well, or waste pile. A municipal solid waste landfill unit also may receive other types of RCRA Subtitle D wastes, such as commercial solid waste, nonhazardous sludge, small quantity generator waste, and industrial solid waste. Such a landfill may be publicly or privately owned.

Putrescible waste: Solid waste which contains organic material capable of being decomposed by microorganisms and causing odors.

Refuse: All solid waste products having the character of solids rather than liquids and which are composed wholly or partially of materials such as garbage, trash, rubbish, litter, residues from cleanup of spills or contamination, or other discarded materials.

Release: Any spilling, leaking, pumping, pouring, emitting, emptying, discharging, injection, escaping, leaching, dumping, or disposing into the environment solid wastes or hazardous constituents of solid wastes (including the abandonment or discarding of barrels, containers, and other closed receptacles containing solid waste). This definition does not include any release which results in exposure to persons solely within a workplace; release of source, by-product, or special nuclear material from a nuclear incident, as those terms are defined by the Atomic Energy Act of 1954; and the normal application of fertilizer.

Rubbish: Combustible or slowly putrescible discarded materials which include but are not limited to trees, wood, leaves, trimmings from shrubs or trees, printed matter, plastic and paper products, grass, rags, and other combustible or slowly putrescible materials not included under the term *garbage.*

Sanitary landfill: An engineered land burial facility for the disposal of household waste which is located, designed, constructed, and operated to contain and isolate the waste so that it does not pose a substantial present or potential hazard to human health or the environment. A sanitary landfill also may receive other types of solid wastes, such as commercial solid waste, nonhazardous sludge, hazardous waste from conditionally exempt small-quantity generators, and nonhazardous industrial solid waste.

Sludge: Any solid, semisolid, or liquid waste generated from a municipal, commercial, or industrial wastewater treatment plant, water supply treatment plant, or air pollution control facility exclusive of treated effluent from a wastewater treatment plant.

Solid waste: Any garbage (refuse), sludge from a wastewater treatment plant, water supply treatment plant, or air pollution control facility, and other discarded material, including solid, liquid, semisolid, or contained gaseous material resulting from industrial, commercial, mining, and agricultural operations and from community activities. Does not include solid or dissolved materials in domestic sewage, or solid or dissolved materials in irrigation return flows or industrial discharges that are point sources subject to permit under 33 U.S.C. 1342, or source, special nuclear, or by-product material as defined by the Atomic Energy Act of 1954, as amended.

Special wastes: Solid wastes that are difficult to handle, require special precautions because of hazardous properties, or the nature of the waste creates management problems in normal operations.

Trash: Combustible and noncombustible discarded materials. Used interchangeably with the term *rubbish.*

Vector: A living animal, insect, or other arthropod which transmits an infectious disease from one organism to another.

Washout: Carrying away of solid waste by waters of the base flood.

Yard waste: That fraction of municipal solid waste that consists of grass clippings, leaves, brush, and tree prunings arising from general landscape maintenance.

References

Commonwealth of Virginia, Department of Waste Management. 1993. Solid Waste Management Regulations VR 672-20-10.

Geosynthetic Research Institute, Drexel University. 1991. GRI Test Methods and Standards.

Industrial Fabrics Association International. 1993. Geotechnical fabrics report: Specifiers guide. 10:(9).

Koerner, R. M. 1994. *Designing with Geosynthetics,* 3rd ed. Prentice Hall, New York.

United States Code of Federal Regulations, Title 40 Protection of the Environment, Part 258. 1991. EPA Criteria for Municipal Solid Waste Landfills.

United States Code of Federal Regulations, Title 40 Protection of the Environment, Part 261. 1990. Identification and Listing of Hazardous Waste.

For Further Information

D'Appolonia, D. J. 1980. Soil–bentonite slurry trench cutoffs. *J. Geotech. Div., ASCE.* April 1980: 399–417.

Geosynthetic Research Institute, Drexel University. 1990. Landfill Closures: Geosynthetics, Interface Friction and New Developments.

Geosynthetic Research Institute, Drexel University. 1992. MQC/MQA and CQC/CQA of geosynthetics. *Proc. 6th GRI Seminar.*

Geosynthetic Research Institute, Drexel University. 1993. Geosynthetic liner systems: innovations, concerns and designs. *Proc. 7th GRI Seminar.*

Millet, R. A. and Perez, J-Y. 1981. Current USA practice: Slurry wall specifications. *J. Geotech. Div., ASCE.* August 1981:1041–1056.

Millet, R. A., Perez, J-Y., and Davidson, R. R. 1992. USA Practice Slurry Wall Specifications 10 Years Later, Slurry Walls: Design, Construction and Quality Control, ASTM STP 1129.

Morgenstern, N. and Amis-Tahmasseb, I. 1965. The stability of a slurry trench in cohesionless soils. *Geotechnique.* December 1965:387–395.

NSF. 1993. Flexible Membrane Liners—NSF International Standard, NSF54-1993.

Repetto, P. C. and Foster, V. E. 1993. Basic considerations for the design of landfills. *Proc. 1st Annual Great Lakes Geotechnical/Geoenvironmental Conf.* The University of Toledo, Ohio.

USEPA, Office of Emergency and Remedial Response. 1984. Slurry Trench Construction for Pollution Migration Control, EPA-540/2-84-001.

USEPA, Office of Solid Waste and Emergency Response. 1986. Technical Guidance Document: Construction Quality Assurance for Hazardous Waste Land Disposal Facilities, EPA/530-SW-86-031.

USEPA. 1993. Office of Research and Development. Quality Assurance and Quality Control for Waste Containment Facilities, Technical Guidance Document EPA/600/R-93/182.

USEPA. 1993. Office of Research and Development. Report of Workshop on Geosynthetic Clay Liners, EPA/600/R-93/171.

USEPA. 1993. Office of Research and Development. Proceedings of the Workshop on Geomembrane Seaming, EPA/600/R-93/112.

USEPA, Office of Solid Waste and Emergency Response. 1982. Evaluating Cover Systems for Solid and Hazardous Waste.

USEPA, Risk Reduction Engineering Laboratory. 1988. Guide to Technical Resources for the Design of Land Disposal Facilities, EPA/625/6-88/018.

26

In Situ Subsurface Characterization

J. David Frost
Georgia Institute of Technology

Susan E. Burns
Georgia Institute of Technology

26.1 Introduction

The in situ **subsurface** characterization section of a civil engineering handbook published 20 years ago would have been dominated by details of the standard penetration test with perhaps no more than a passing reference to some other test methods. As a result of significant technological advances in the past two decades and, perhaps equally important, increased recognition that there is a direct relationship between the efficiency of a design and the quality of the parameters on which this design is based, discussion of a much broader range of test methods is now appropriate in a text such as this. **Invasive** and **noninvasive** test methods using a variety of penetrometers and wave propagation techniques (e.g., cone **penetration testing**, seismic reflection/refraction testing, dilatometer testing, and pressuremeter testing) are now routinely used in many instances in preference to, or at least as a complement to, the standard penetration test. A listing of the more common techniques is given in Table 26.1.

26.2 Subsurface Characterization Methodology

The process of characterizing a site should begin long before the first boring or sounding is advanced. In most cases, there will be information available either at the immediate site or at least in the general vicinity such that some initial impressions can be synthesized with respect to the subsurface conditions and the types of potential problems which may be encountered during the proposed development at the site. Example sources and types of information which may be available are summarized in Table 26.2.

When this available data has been synthesized, the engineer can then develop a site investigation strategy to supplement/complement the existing information and help achieve the objectives of the exploration program, including:

- Determine the subsurface stratigraphy (geologic profile), including the interface between fill and natural materials and the depth to bearing strata (e.g., bedrock) if appropriate

0-8493-8953-4/95/$0.00 + $.50

Table 26.1 Summary of Common In Situ Subsurface Characterization Techniques

Test	Invasive/ Noninvasive	Sample Recovered	Usage
Standard penetration test	Invasive	Yes	Extensive
Cone penetration test	Invasive	No	Extensive
Pressuremeter test	Invasive	No	Moderate
Dilatometer	Invasive	No	Moderate
Vane shear test	Invasive	No	Moderate
Becker density test	Invasive	Yes	Limited
Borehole seismic test	Invasive	No	Extensive
Surface seismic test	Noninvasive	No	Extensive

Table 26.2 Sources and Types of Background Information

Data Source	Information Available
Topographic maps	Maps published by the U.S. Geological Survey showing site terrain, dams, surface water conditions, rock quarries
Previous geologic studies	Soil types, current and previous river and lake locations, floodplains, groundwater conditions, rock profiles
Soil survey data	Maps published by the Department of Agriculture profiling the upper 6 to 10 feet of soil
Previous engineering reports	Site geological description, record of fills or cuts, groundwater information, floodplains, wetlands, previous construction activity
Aerial photogrammetry	Macroscopic identification of topography, surface water drainage/erosion patterns, vegetation
State/municipal well logs	Groundwater table information, pumping rates, water table drawdown
Seismic potential	Maps published by the U.S. Geological Survey delineating seismicity zones in the U.S.
Personal reconnaissance	Identification of geological features through the examination of road cuts, vegetation, slopes, rivers, rivers, previously constructed buildings

- Investigate the groundwater conditions, including the location of water-bearing seams as well as perched aquifer and permanent groundwater table elevations
- Obtain samples of subsurface materials for additional laboratory testing as appropriate
- Install any instrumentation as required to permit additional assessment of the subsurface environment at subsequent time intervals (e.g., piezometers, inclinometers, thermistors).

26.3 Subsurface Characterization Techniques

As noted above, the range of test methods available today for subsurface characterization programs has increased significantly over the past few decades. For discussion purposes, they are considered herein under the following broad categories:

- Test pits
- Conventional drilling and sampling
- Penetration testing
- **Geophysical testing**
- Other testing techniques

Additional details of these categories are given below.

Test Pits

Test pits are a valuable technique for investigating near-surface conditions under a variety of scenarios. Typical depths of 15 to 20 feet are readily excavated with backhoe equipment of the type generally available on most construction sites. Excavations to greater depths are possible with long-boom equipment or if a multiple-layer excavation is made. The method becomes less efficient with deeper test pits since the area of the excavation typically increases for deeper holes as the sides are sloped to facilitate excavation and personnel access and safety. Among the advantages of test pits are that the engineer can clearly document and photograph the subsurface stratigraphy, and the recovery of bulk samples for laboratory compaction and other tests requiring large samples is easy. Near-surface groundwater and cohesionless soils can combine to make excavation difficult as soil caving undermines the edges of the test pit. Although, unfortunately, less frequently used nowadays than the authors consider appropriate, block sampling techniques are easily conducted in the base or side of a test pit.

Conventional Drilling and Sampling

Depending on the anticipated subsurface conditions and the specific objectives of the investigation program, a number of conventional drilling and sampling techniques are available. An example field borehole log is shown in Fig. 26.1. Typical boring techniques used include auger drilling, rotary drilling, cable tool drilling, and percussion drilling. Factors ranging from the anticipated stratigraphy (sequence and soil type) to depth requirements can influence the method chosen. A summary of the main advantages and disadvantages for the various methods is given in Table 26.3.

Samples of soil and rock for subsequent analysis and testing can be obtained using a variety of techniques. These may range from chunk samples (taken from flights of augers) to split spoon samples (disturbed samples), which are typically obtained by driving a split barrel sampler as in the standard penetration test [ASTM D1586], to thin-walled tube samples (**undisturbed samples**), which can be obtained using one of a variety of mechanical or hydraulic insertion devices [ASTM D1587]. A summary of the factors pertinent to the selection of a specific sampling technique is listed in Table 26.4.

Penetration Testing

The term *penetration testing* is being used herein to describe a variety of test procedures which involve the performance of a controlled application and recording of loads and/or deformations as a tool is being advanced into the subsurface. For the purposes of this text, this includes pressuremeter tests [ASTM D4719] performed in predrilled holes (although obviously this is strictly not a penetration-type test as defined above). In some cases, the loads and/or deformations are recorded continuously as the device is being inserted into the ground, while in other cases measurements are made when the insertion process is halted at predetermined intervals. An assessment of in situ testing is given in Table 26.5. Brief descriptions of the most common methods follow.

Standard Penetration Testing

Standard penetration testing refers to a test procedure wherein a split tube sampler is driven into the ground with a known force and the number of blows required to drive the sampler 12 inches is recorded as an *N* value [de Mello, 1971]. The standard test procedure [ASTM D1586] refers to sampler devices which have an outside diameter of 5.1 cm, an inside diameter of 3.5 cm, and a length somewhere in the range of 50 to 80 cm to retain the soil sample. The sampler is driven into the ground with a drive weight of 63.5 kg dropping 76 cm. A variety of different hammer types are available. These range from donut and safety hammers, which are manually operated through the use of a rope and cathead, to automatic trip hammers. There is little question that this is still

FIELD BOREHOLE LOG

Boring Number	EW-39	**Depth**	60 ft
Project	ASW-2578	**Sheet**	1 of 1
Drill Rig	CME-77	**Date**	12/4/92
Elevation	500 ft above MSL	**Driller**	J. A. Smith

Elev	Stratum Depth	Visual Soil Description		D (ft)	SR (in)	N (blows/ft)	Remarks
500		Topsoil, grass, roots					
	4.7						
		Firm dark brown silty fine to medium sand with trace gravel (SM)		6.5	7	19 (8_10_9)*	
490							G.W. table at 10' at time of drilling
				22.5	10	17 (7_9_8)	
480							
470	30.2						
		Soft black silty clay with trace of fine sand (OL-OH)		32.3	10	4 (1_2_2)	
	38.6						
460		Firm brown silty medium sand with trace gravel (SM)		39.0	9	20 (9_10_10)	
450							
	54.2						
		Dense brown silty fine to medium sand with trace gravel (SM)		56.0	8	82 (35_40_42)	
440	60.0						
		Boring terminated at 60.0'					
430							

D	Sample Depth (ft)
SR	Sample Recovery (in)
N	Penetration in blows per foot *(Blows per 6" increment)

FIGURE 26.1 Typical field boring log.

probably the most widely used penetration test device in the U.S. although there is clearly more widespread recognition of the many limitations of the test device resulting from equipment and operator error sources. The principal advantages and disadvantages of standard penetration testing are summarized in Table 26.6.

Table 26.3 Comparison of Various Drilling Methods

Drilling Method		Advantages	Disadvantages
Auger drilling	Hollow stem	• Rapid • Inexpensive • Visual recognition of changes in strata • Hole easily cased to prevent caving • Soil/water samples easily recovered, although disturbed • No drilling fluid required	• Depth limited to approximately 80–100 ft • Cannot drill through rock • Can have heave in sands • Limited casing diameter
	Solid stem	• Rapid • Inexpensive • Small borehole required	• Sampling difficulty • Borehole collapse on removal
Rotary drilling	Direct	• Rapid • Used in soil or rock • Casing not required • Wells easily constructed • Soil disturbance below borehole minimal • Easily advances borehole through dense layers	• Drilling fluid required • No water table information during drilling • Difficult to identify particular strata • Sampling not possible during boring • Slow in coarse gravels
	Air	• Rapid • Used in soil or rock • Capable of deep drilling • No water-based drilling fluid required	• Casing required in soft heaving soils • Relatively expensive
Cable tool		• Inexpensive • Small quantities of drilling fluid required • Used in soil or rock • Water levels easily determined	• Minimum casing diameter 4 in. • Steel casing required • Slow • Screen required to take water sample • Depth limited to approximately 50–60 ft • Difficult to detect thin layers
Percussion drilling (Becker density test)		• Measure penetration resistance of gravelly soils • Relatively operator independent • Estimate pile drivability • Continuous profiling • Designed for gravels and cobbles	• Equipment strongly influences test results • Based on empirical correlations

Table 26.4 Selection of Sampling Technique

Sample Type	Sample Quality	Suitability for Testing
Block sample	Excellent	Classification, water content, density, consolidation, shear strength
Thin-walled tube, piston	Very good	Classification, water content, density, consolidation, shear strength
Thin-walled tube	Good	Classification, water content, density, consolidation, shear strength
Split spoon	Poor	Classification, water content
Auger/wash cuttings	Very poor	Soil identification

Table 26.5 Assessment of In Situ Testing

Advantages	Disadvantages
Rapid	No sample recovered (except SPT)
Inexpensive	Indirect measurement related through calibration
Difficult deposits can be tested	Complex data reduction
In situ stress, pore fluid, temperature conditions	Relies heavily on empirical correlations
Real-time measurements	Unknown boundary conditions
Reproducible results	Unknown drainage conditions
Large volume of soil tested	Strain-rate effects
Continuous or semicontinuous profiling	Nonuniform strains applied
	Specialized equipment and skilled operators often required

Table 26.6 Assessment of Standard Penetration Testing

Advantages	Disadvantages
Commonly available	Based on empirical correlations
Applicable to most soils	Significant operator/equipment influences (See Navfac DM7.1)
Sample (disturbed) recovered	Not useful in gravels, cobbles
Rapid/inexpensive	Not useful in sensitive clays

Cone Penetration Testing

Cone penetration testing refers to a test procedure wherein a conical-shaped probe is pushed into the ground and the penetration resistance is recorded [Robertson and Campanella, 1983]. The standard test procedure [ASTM D3441] refers to test devices which have a cone with a 60° point angle and a base diameter of 3.57 cm that results in a projected cross-sectional area of 10 cm^2. While original cones operated with an incremental mechanical system, most new cones are electronic and are pushed continuously at a rate of 2 cm/sec. Other frequent additions to a penetrometer include a friction sleeve with an area of 150 cm^2 and a porous element which permits the pore water pressure to be recorded by a pressure transducer. A typical cone penetration test system along with details of an electronic piezo-friction cone are illustrated in Fig. 26.2. Simultaneous continuous measurements of tip resistance, q_c, side friction, f_s, and pore pressure, u, are recorded. Appropriate corrections are required to account for unequal end areas behind the tip of the penetrometer. An example cone sounding record is shown in Fig. 26.3. The principal advantages and disadvantages of cone penetration testing are summarized in Table 26.7. Cone penetrometers are being used for an increasing number of applications as new sensors are being developed and incorporated into penetration devices for a variety of geotechnical and geo-environmental applications, as summarized in Table 26.8.

Dilatometer Testing

The flat plate dilatometer test [Marchetti, 1980; Schmertmann, 1986] was originally introduced to provide an easy method for determining the horizontal soil pressures acting on laterally loaded piles. The present design of the dilatometer blade consists of a flat blade 1.5 cm thick by 9.6 cm wide with a 6.0 cm diameter membrane on one face, as shown in Fig. 26.4. The test is performed by advancing the blade by quasi-static push at a rate of 2 cm/s. At regular intervals, typically every 20 cm, two or three pressure readings are obtained. The A pressure reading is a membrane liftoff pressure and is obtained just as the membrane begins to move. The B pressure reading is the pressure required to cause the center of the membrane to move 1.1 mm into the soil mass. If desired, a C pressure reading may be obtained by controlling the rate of deflation of the membrane and finding the pressure at which the membrane once again comes in contact with its seat. The A

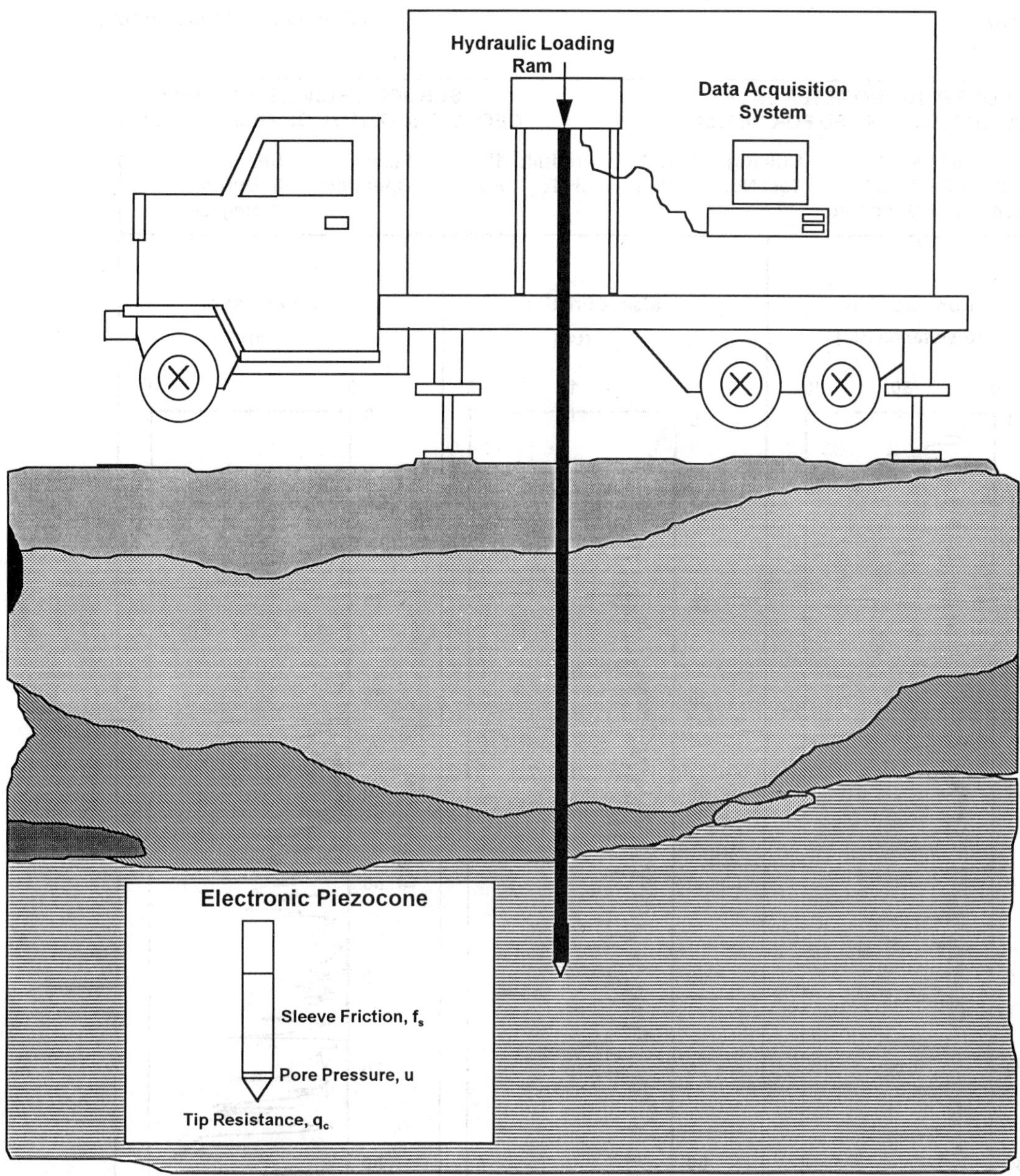

FIGURE 26.2 Cone penetration test system.

Table 26.7 Assessment of Cone Penetration Testing

Advantages	Disadvantages
Rapid/inexpensive	No sample recovered
Reproducible results	Penetration depth limited to 150–200 ft
Continuous tip resistance, sleeve friction, and pore pressure (piezocone) profile	Normally cannot push through gravel
Accurate, detailed subsurface stratigraphy/identification of problem soils	Requires special equipment and skilled operators
Real-time measurements	Most analysis based on correlations
Pore pressure dissipation tests allow prediction of permeability and c_h	
Models available to predict strength, stress history, compressibility	

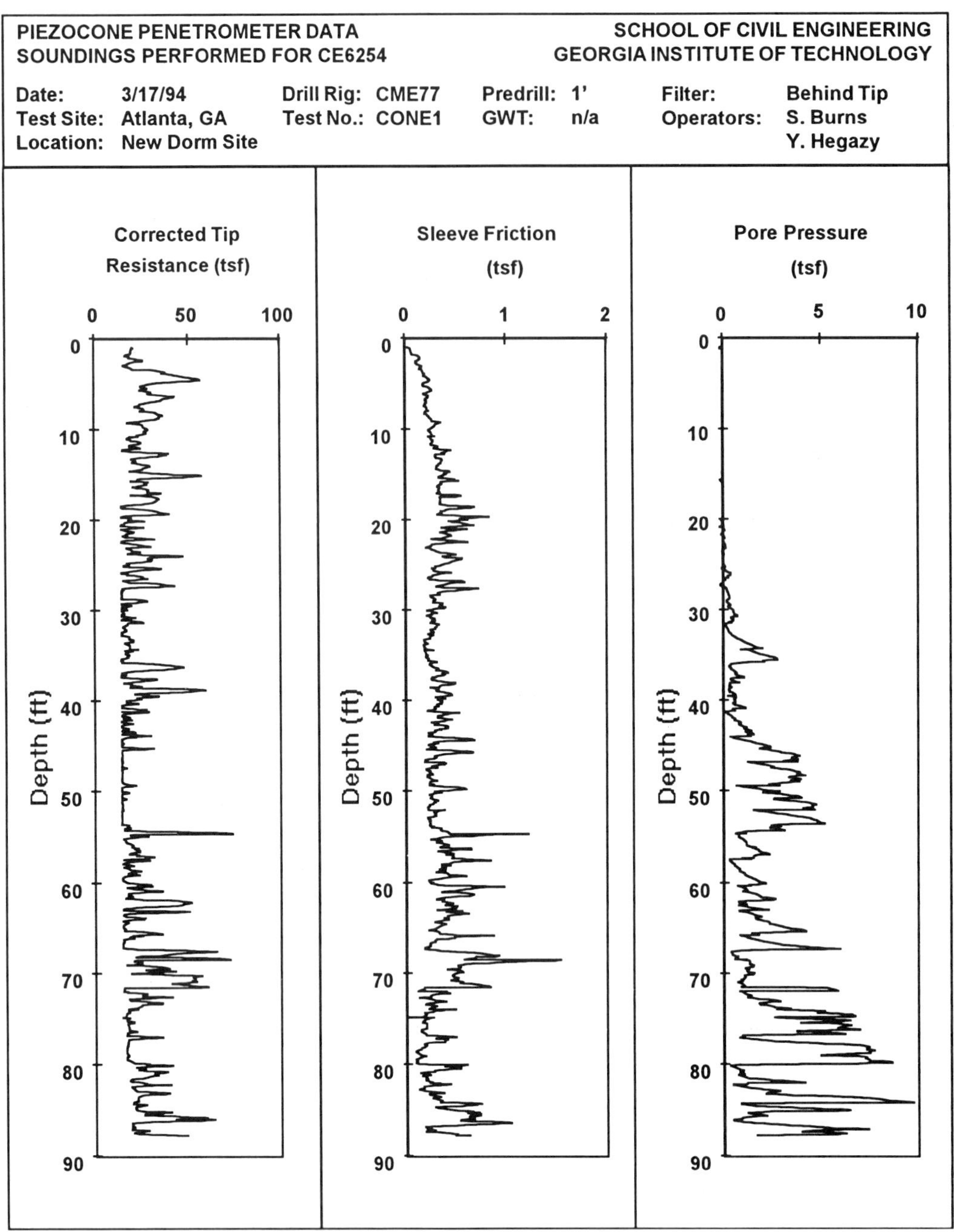

FIGURE 26.3 Typical cone penetrometer record.

Table 26.8 Specialized Cone Penetrometers

Sensor	Application
Accelerometer	Measurement of seismic wave velocity
Nculear moisture content sensors	Measurement of soil moisture content
Resistivity electrodes	Identification of pore characteristics and fluids
Laser-induced fluorescence	Hydrocarbon detection
Temperature	Measurement of cone body temperature
Hydrocarbon sensors	Detection of BTEX chemicals in pore fluid and vadose zone

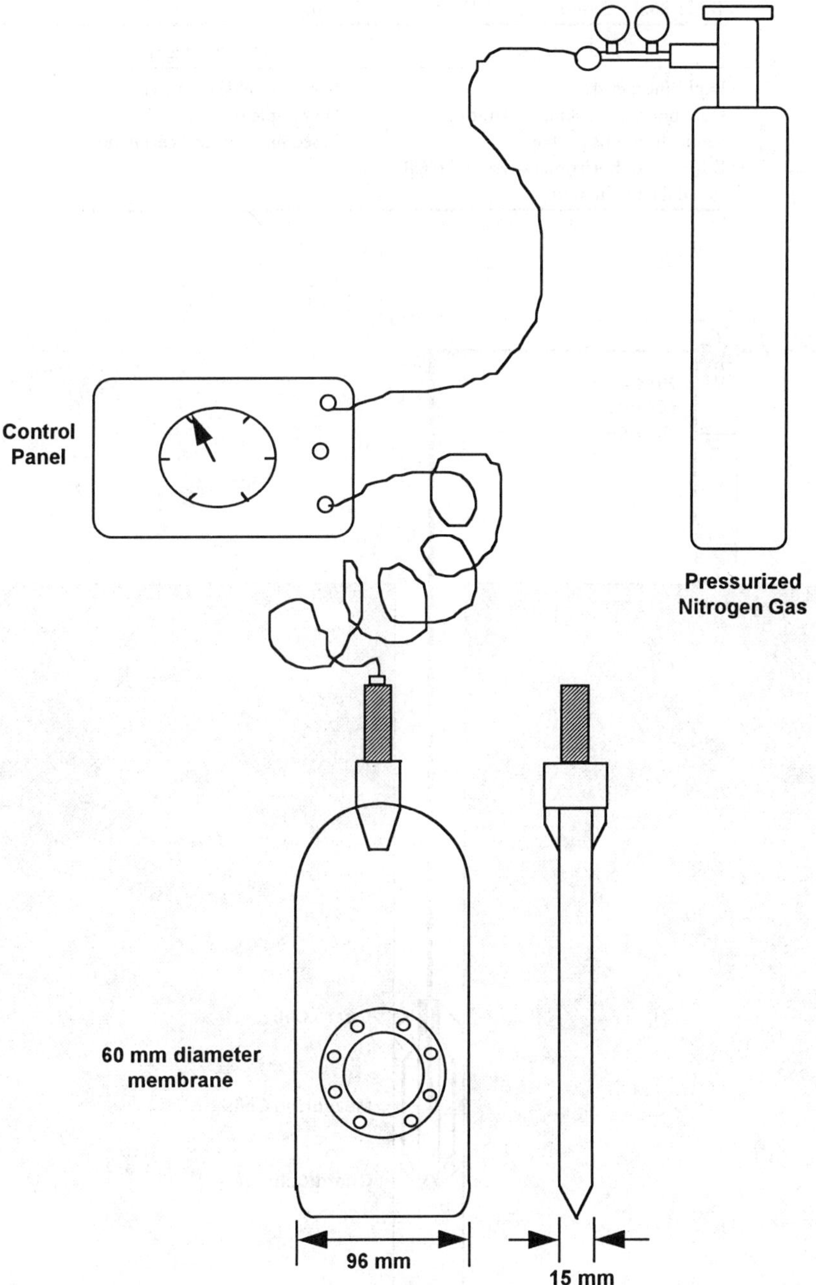

FIGURE 26.4 Dilatometer test system.

and B pressure readings, corrected for membrane stiffness to P_1 and P_0, respectively, are used to define a number of dilatometer indices:

$$\text{Dilatometer index, } E_D = 34.7(P_1 - P_0)$$
$$\text{Horizontal stress index, } K_D = (P_1 - U_0)/(\sigma'_{vo})$$
$$\text{Material index, } I_D = (P_1 - P_0)/(P_0 - U_0)$$

The C pressure reading, corrected for membrane stiffness, is thought to provide an upper bound to the induced pore pressures.

Table 26.9 Assessment of Dilatometer Testing

Advantages	Disadvantages
Rapid/inexpensive	Not applicable in gravels
Does not require skilled operators	No sample recovered
Semicontinuous profile	Based on empirical correlations
Estimates of horizontal stress and OCR	
Rapid data reduction	

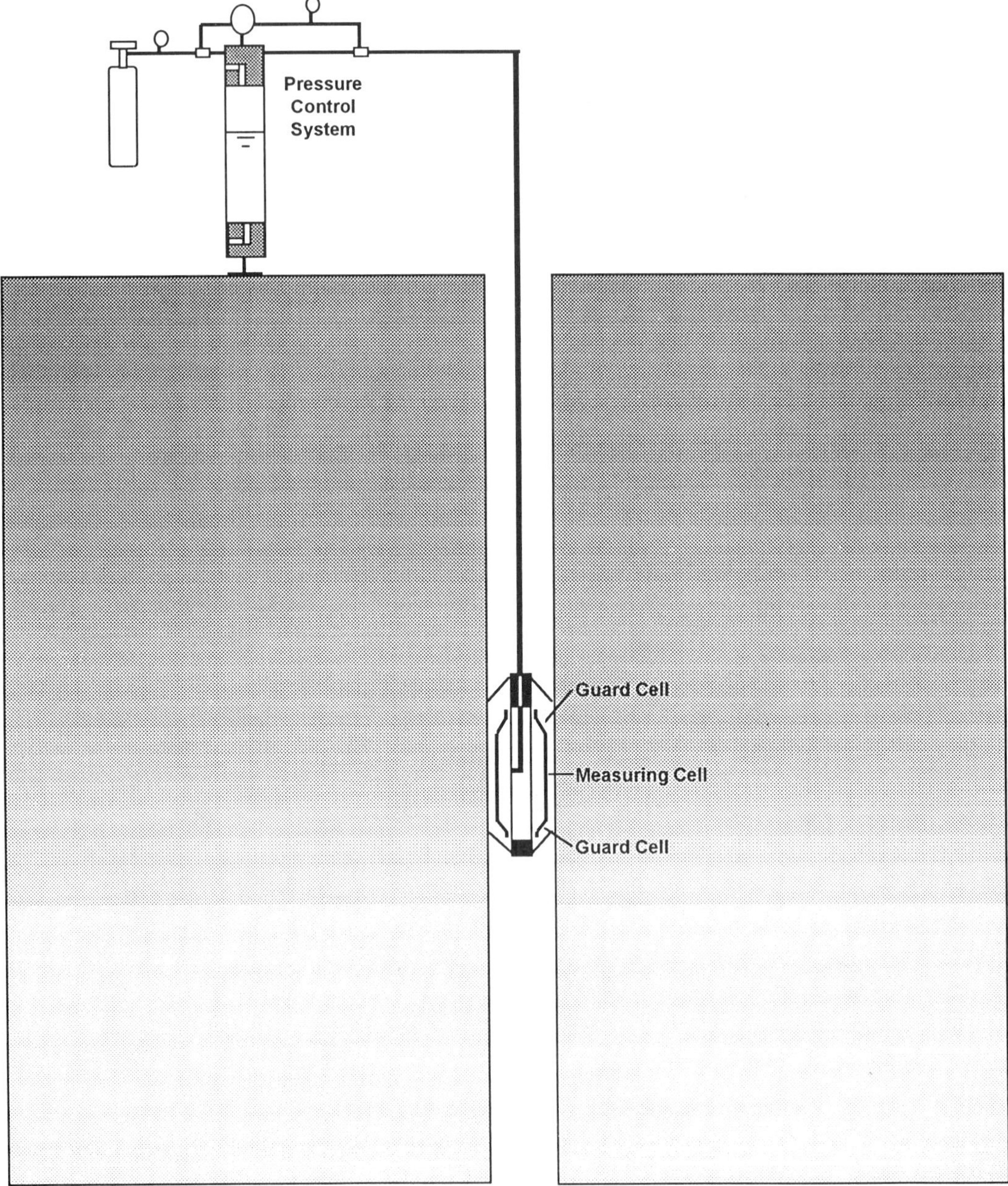

FIGURE 26.5 Pressuremeter test system.

Using these dilatometer indices and numerous correlations which have been developed, a large number of soil parameters can be estimated. The principal advantages and disadvantages of dilatometer testing are summarized in Table 26.9.

Pressuremeter Testing

The pressuremeter test [Baguelin *et al.*, 1978] typically consists of placing an inflatable cylindrical probe in a predrilled borehole and recording the changes in pressure and volume as the probe is inflated. The standard test procedure (ASTM D4719) uses probes with typical diameters ranging between 4.4 and 7.4 cm while the length of the inflatable portion of the probe on which the soil response is based varies between about 30 and 60 cm depending on whether the unit is a single-cell type or has guard cells at either end of the measuring cell. The probe can be expanded using equal pressure increments or equal volume increments. A schematic of a typical test arrangement is shown in Fig. 26.5. Pressuremeter soundings consist of tests performed at 1 m intervals, although clearly this is a function of the site geology and the purpose of the investigation. The test results, appropriately corrected for membrane stiffness and hydrostatic pressure between the control unit and the probe, are plotted as shown in Fig. 26.6, from which the pressuremeter modulus, E_{PM}, and the limit pressure, P_L, are determined. Using these pressuremeter indices and numerous correlations which have been developed, a large number of soil parameters can be estimated. The principal advantages and disadvantages of pressuremeter testing are summarized in Table 26.10.

One of the key factors which affects the results of the pressuremeter test is the amount of stress relief which occurs before the probe is expanded. To minimize this problem, guidelines for borehole sizes and the test sequence are given in ASTM D4719 for probes requiring a predrilled borehole. Alternatively, self-boring devices can be used to reduce the impact of stress relief.

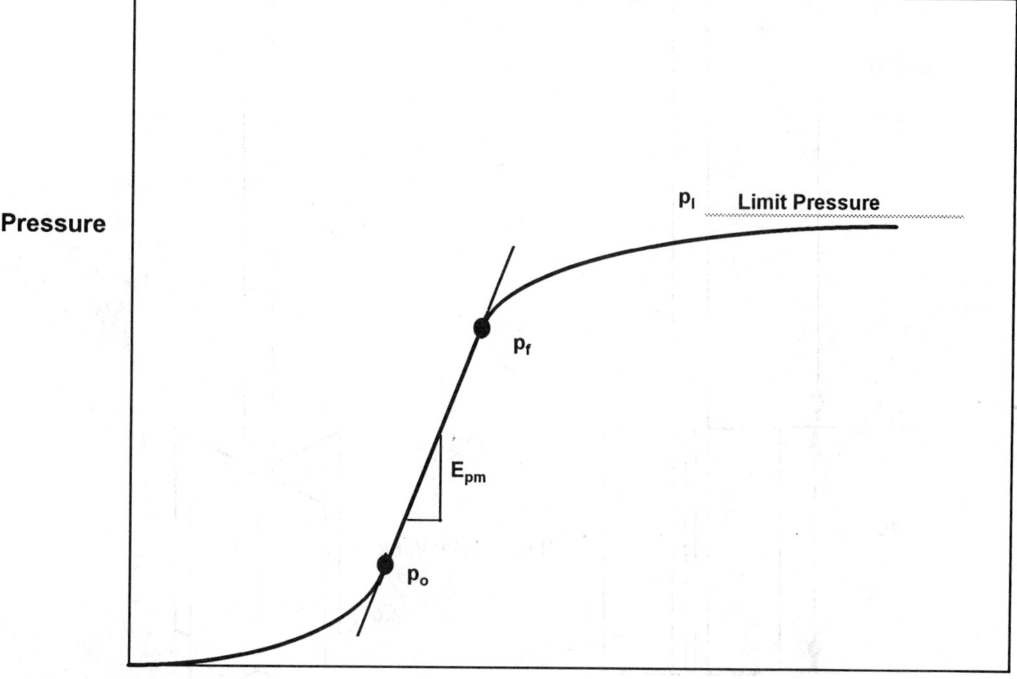

FIGURE 26.6 Typical pressuremeter test result.

Table 26.10 Assessment of Pressuremeter Testing

Advantages	Disadvantages
Applicable in most soils	Expensive
In situ measurements of horizontal stresses, deformability, strength	Specialized equipment and skilled operators required
Prediction of modulus	Delicate equipment
	Independent soil characterization required
	Prebored hole may be required

Vane Shear Test

The vane shear test consists of placing a four-bladed vane in the undisturbed soil at the bottom of a boring and determining the torsional force required to cause a cylindrical surface to be sheared by the vane [Becker *et al.*, 1987]. The test is applicable for cohesive soils. The standard test procedure [ASTM D2573] uses vanes with typical diameters ranging between 3.8 and 9.2 cm and lengths of 7.6 to 18.4 cm, as shown in Fig. 26.7. Selection of the vane size depends on the soil type with larger vanes used in softer clays so as to provide measured torque values of a reasonable magnitude. The torque is applied at a relatively slow rate of the order of 0.1°/s which results in times to failure of

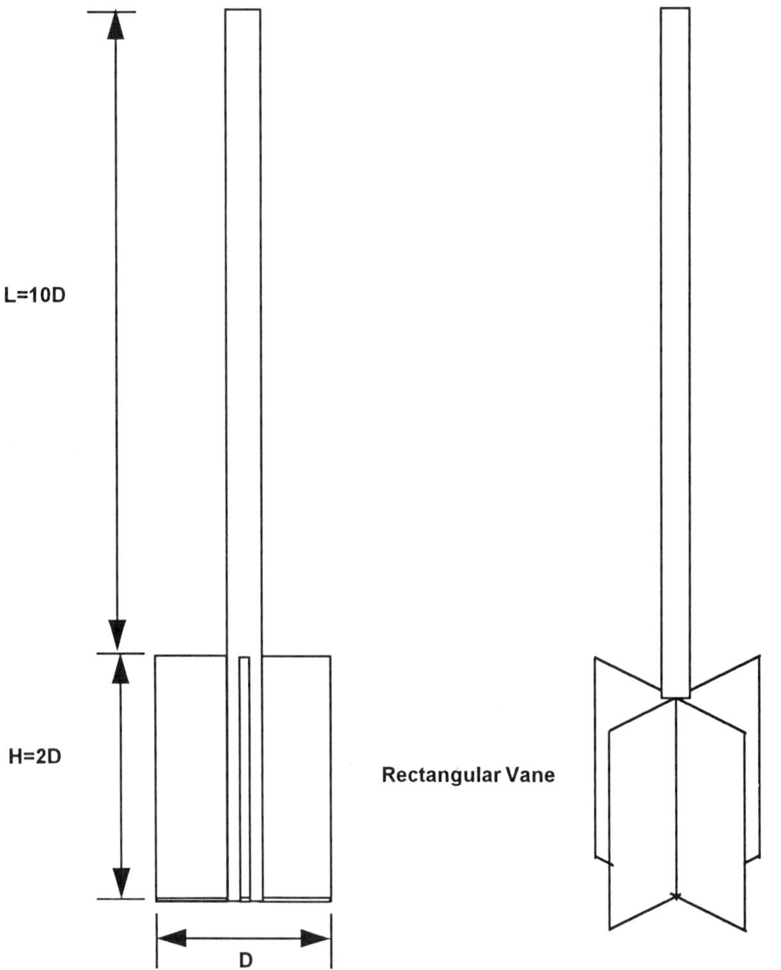

FIGURE 26.7 Vane shear test system.

Table 26.11 Assessment of Vane Shear Testing

Advantages	Disadvantages
Rapid/inexpensive	Only applicable in soft clays
Applicable to sensitive clays	Point measurement
Theoretical basis	Generally only undrained shear strength measurements
Measurement of shear strength, remolded shear strength, and sensitivity	No sample recovered
	Prebored hole may be required
	Independent soil characterization required

2 to 10 minutes depending on soil type. The shear strength of the soil is calculated as the product of the torque applied and a constant depending on the geometry of the vane. The principal advantages and disadvantages of vane shear testing are summarized in Table 26.11.

Geophysical Testing

Geophysical testing techniques [Woods, 1978] for investigating subsurface conditions have become a frequently used tool by engineers. They offer a number of advantages over other investigation techniques, including the noninvasive nature of the methods and the volume of soil for which properties are determined. The most common methods are seismic reflection and seismic refraction. The basis of these methods is that the time for seismic waves to travel between a source and receiver can be used to interpret information about the material through which it travels. Depending on the arrangement of the source and receivers, the subsurface environment can thus be characterized. In general, the methods require a subsurface profile where the layer stiffnesses and hence wave velocities increase with depth. Advantages and disadvantages of geophysical test methods are given in Table 26.12.

Seismic Reflection

Seismic reflection is used to describe methods where the time for the reflection of a seismic wave induced at the surface is recorded. A typical test configuration is shown in Fig. 26.8. This method involves study of complete wave trains from multiple receivers to characterize the subsurface; thus, interpretation of the test results can be subjective.

Table 26.12 Assessment of Geophysical Testing

Method	Advantages	Disadvantages
Downhole	Only one borehole required	Attenuation with depth
	Relatively inexpensive	Invasive
	Measurement of seismic soil properties	No sample recovered
		Limited by depth of borehole
Crosshole	Minimum of two boreholes required	Expensive
	No attenuation with depth	Invasive
	Measurement of seismic soil properties	Possible refraction interference
		No sample recovered
		Limited by depth of borehole
Surface	Noninvasive	Complex data analysis
	Inexpensive	Special equipment and skilled operators required
	Measurement of seismic soil properties	No sample recovered
	No boreholes required	Attenuation with depth
	Environmental applications due to limited contaminant exposure	Refraction method applicable only when velocities increase with depth
		Possible refraction interference

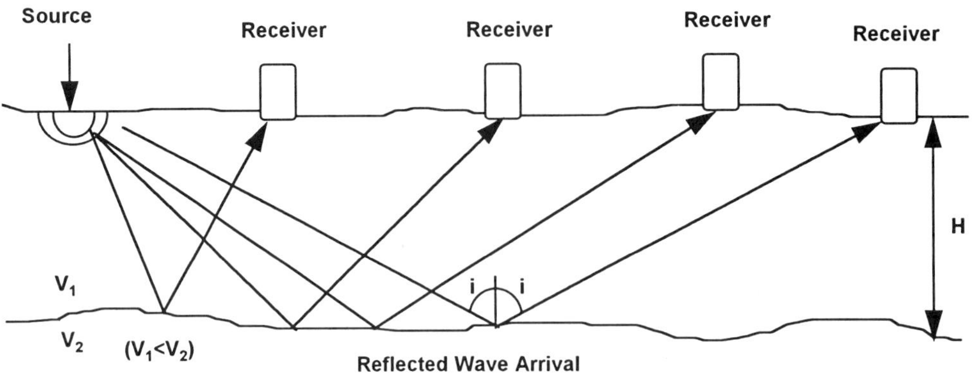

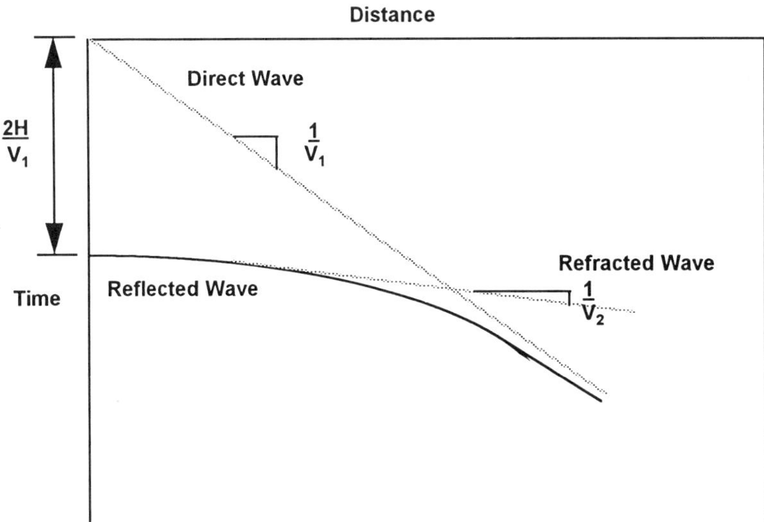

FIGURE 26.8 Seismic reflection test configuration.

Seismic Refraction

Seismic refraction is used to describe methods where the time for seismic waves which are refracted when they encounter a stiffer material in the subsurface are recorded. A typical test configuration is shown in Fig. 26.9. Unlike reflection methods, refraction methods only rely on the time for first arrivals; thus, interpretation of the results can be more straightforward.

Crosshole Testing

Crosshole seismic testing differs from the methods described above in that the source and receiver are located at the same depth in adjacent boreholes and the time for seismic waves to travel between these instruments is recorded. The standard test procedure for crosshole testing [ASTM D4428] involves drilling a minimum of three boreholes in line spaced about 3 m apart. A PVC casing is then grouted in place to ensure a good couple between the source/receiver and the PVC casing and between the PVC casing and the surrounding soil. A typical configuration is shown in Fig. 26.10.

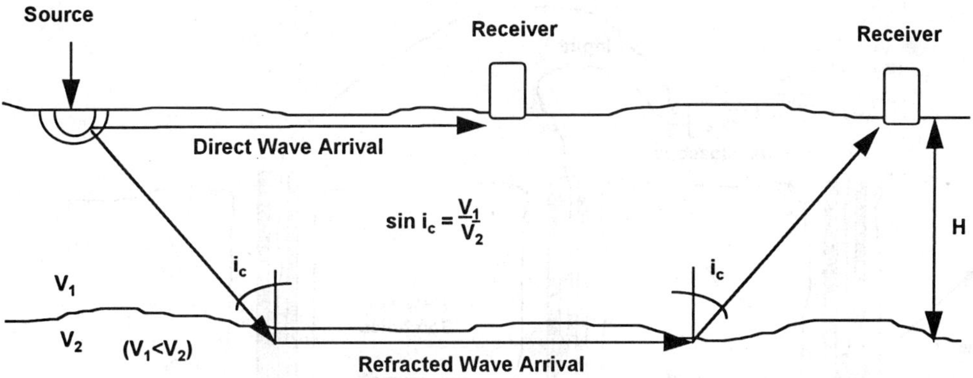

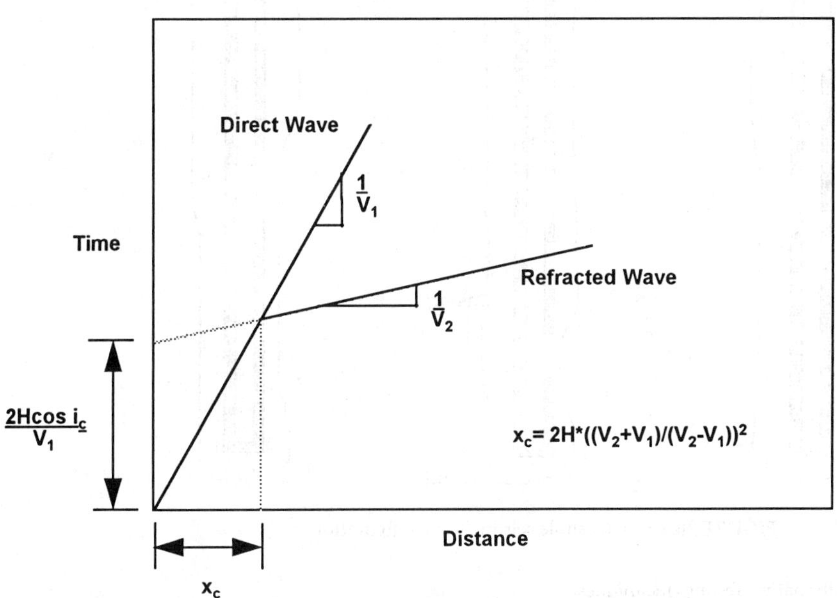

FIGURE 26.9 Seismic refraction test configuration.

Other Testing Techniques

While the specific test methods described above represent those that are most frequently used, there are a large number of other devices and methods that are available and should be considered by the engineer designing a site investigation program. A number of these methods are used extensively in geo-environmental site characterization programs while others are still in development or are available only for use on a limited basis. Nevertheless, since the efficiency and quality of any foundation design is directly dependent on the quality of the subsurface information available, the engineer should be aware of all possible investigation tools available and select those which can best suit the project at hand. Recognition of the simple fact that the expenditure of an additional few thousand dollars at the site investigation stage could result in the savings of many thousands or even millions of dollars as a result of an inefficient design or, worse, a failed foundation system is important. Accordingly, Table 26.13 contains a listing of several other testing techniques which should be considered.

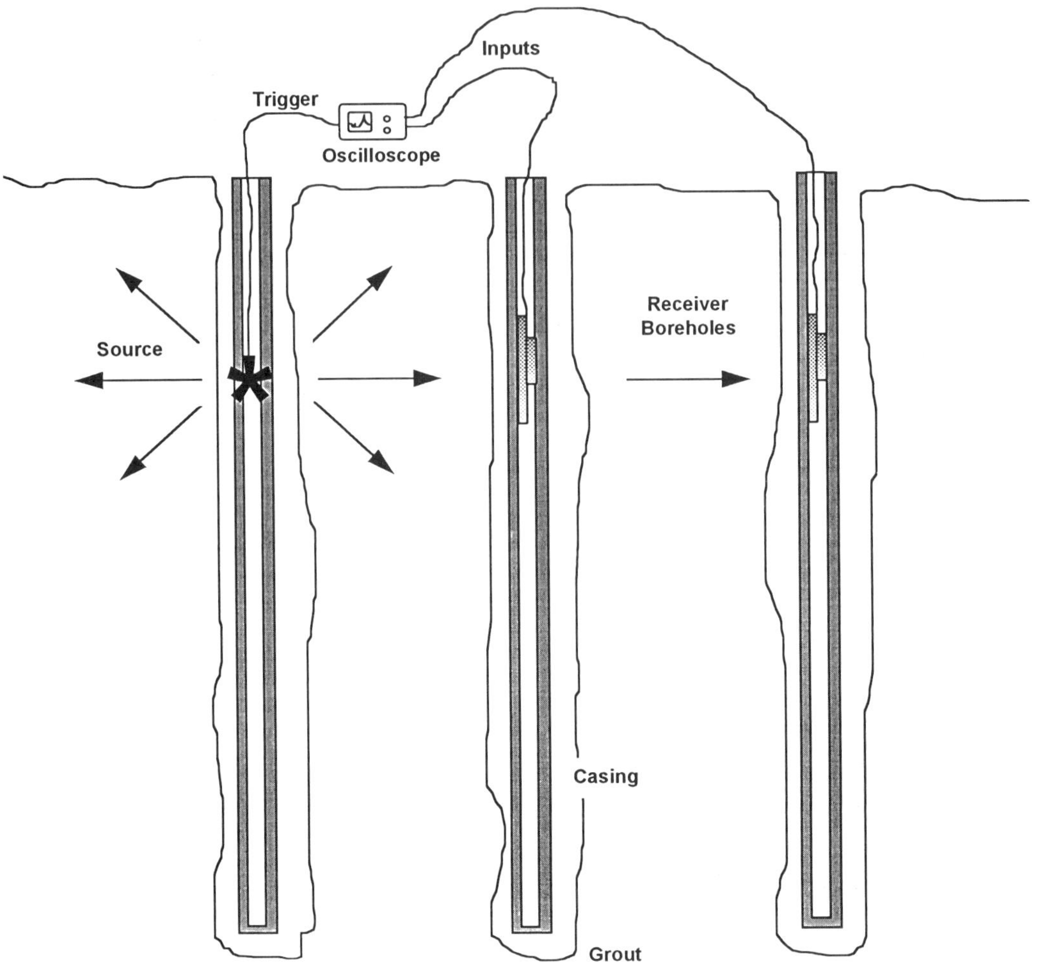

FIGURE 26.10 Crosshole seismic test configuration.

Table 26.13 Alternative Testing Techniques

Test	Usage	Reference
Iowa stepped blade	Lateral stress measurement	Handy *et al.*, 1982
Borehole shear test	Shear strength measurement	Handy *et al.*, 1967
Screwplate	In situ determination of modulus	Schmertmann, 1970
Plate load test	Incremental loading of a plate model of a foundation to predict ultimate bearing capacity	Marsland, 1972
Field direct shear	Strength of fissured soils	Marsland, 1971
Field hydraulic conductivity test	In situ measurement of hydraulic conductivity	Daniel, 1989

26.4 Shipping and Storage of Samples

Use of the best available techniques for drilling and sampling can be negated if appropriate procedures are not used for shipping and storing samples. Accordingly, an integral part of the planning of any site investigation program should be the identification of procedures required for shipping samples to a laboratory and for their subsequent storage prior to testing. Typical details of procedures and containers appropriate for maintaining subsurface samples in a condition as close as possible to their undisturbed state are available [for example, ASTM D3213, ASTM D4220, ASTM D5079].

Defining Terms

Geophysical testing: Test procedures which involve the application and recording of the travel of relatively low frequency, high amplitude waves in the subsurface.

Invasive: Test procedure which involves physical insertion of a test instrument into the subsurface.

Noninvasive: Test procedure which does not involve physical insertion of a test instrument into the subsurface.

Penetration testing: Test procedures which involve the performance of a controlled application and recording of loads and/or deformations as a device is being advanced into the subsurface.

Subsurface: Matrix of soil, rock, groundwater, and pores from which earth structures will be made and on which buildings will be supported.

Undisturbed sampling: Retrieval of samples from subsurface for subsequent laboratory evaluation and testing with minimum of disturbance.

References

ASTM 1194, Standard Test Method for Bearing Capacity of Soil for Static Load and Spread Footings, Vol. 04.08.

ASTM D1586, Standard Test Method for Penetration Test and Split Barrel Sampling of Soils, Vol. 04.08.

ASTM D1587, Standard Practice for Thin-Walled Sampling of Soils, Vol. 04.08.

ASTM D2573, Standard Test Method for Field Vane Shear Test in Cohesive Soil, Vol. 04.08.

ASTM D3213, Standard Practices for Handling, Storing and Preparing Soft Undisturbed Marine Soil, Vol. 04.08.

ASTM D3441, Standard Test Method for Deep, Quasi-Static, Cone and Friction Cone Penetration Tests of Soil, Vol. 04.08.

ASTM D4220, Standard Practices for Preserving and Transporting Soil Samples, Vol. 04.08.

ASTM D4428, Standard Test Methods for Crosshole Seismic Testing, Vol. 04.08.

ASTM D4719, Standard Test Method for Pressuremeter Testing in Soils, Vol. 04.08.

ASTM D5079, Standard Practice for Preserving and Transporting Rock Samples, Vol. 04.08.

Baguelin, F., Jezequel, J. F., and Shields, D. H. 1978. The pressuremeter and foundation engineering. *Trans. Tech.*, 617 pp.

Becker, D. E., Crooks, J. H. A., and Been, K. 1987. Interpretation of the field vane test in terms of in situ and yield stresses. In *ASTM Symp. Lab. Field Vane Shear Strength Test.* Tampa.

Daniel, D. E. 1989. In situ hydraulic conductivity tests for compacted clay. *J. Geotech. Eng., ASCE.* 115(9):1205–1226.

de Mello, V. F. B. 1971. The standard penetration test. In *Pro. 4th Pan Am Conf. Soil Mech. Found. Eng.*, Puerto Rico. 1:1–86.

Demartinecourt, J. P. and Bauer, G. E. 1983. The modified borehole shear device. *Geotech. Test. J., ASTM.* 6(1):24–29.

Handy, R. L. and Fox, N. S. 1967. A soil borehole direct shear test device. *Highway Res. News* 27:42–51.

Handy, R. L., Remmes, B., Moldt, S., Lutenegger, A. J., and Trott, G. 1982. In situ stress determination by Iowa stepped blade. *J. Geotech. Eng., ASCE.* 108(GT11):1405–1422.

Harder, L. F. and Seed, H. B. 1986. *Determination of Penetration Resistance for Coarse Grained Soils Using the Becker Hammer Drill*, Report No. UCB/EERC-86-06, University of California–Berkeley.

Janbu, N. and Senneset, K. 1973. Field compressometer—Principles and applications. *Proc. 8th Int. Conf. Soil Mech. Found. Eng.* Moscow, 1.1:191–198.

Marchetti, S. 1980. In situ tests by flat dilatometer. *J. Geotech. Eng., ASCE.* 106(GT3):299–321.

Marsland, A. 1971. Large in situ tests to measure the properties of stiff fissured clays. *Proc. Aust.—N. Z. Conf. Geomech.*, Melbourne, 1:180–189.

Marsland, A. 1972. Clays subjected to in situ plate tests. *Ground Eng.* 5, No. 5(6):24–31.
Robertson, P. K. and Campanella, R. G. 1983. Interpretation of cone penetration tests. Part I: Sand, Part II: Clay. *Can. Geotech. J.* 20(4):718–745.
Schmertmann, J. H. 1970. Suggested Method for Screwplate Load Test. *Am. Soc. Test. Mater., Spec. Tech. Publ. 479:* 81–85.
Schmertmann, J. H. 1986. Suggested method for performing flat dilatometer test. ASTM *Geotech. Test. J., ASTM.* 9(2):93–101.
Woods, R. D. 1978. Measurement of dynamic soil properties. *Proc. ASCE Spec. Conf. Earthquake Eng. Soil Dynamics*, Pasadena, 1:91–178.

For Further Information

There is a very extensive bibliography describing the numerous test devices and methods introduced in this chapter. There have been a number of important conferences, symposia, and workshops over the past two decades. The interested reader is encouraged to review the proceedings of such meetings for additional information. The principal proceedings include the following:

- Proceedings of ASCE Specialty Conference on In Situ Measurement of Soil Properties, Raleigh, USA, 1975.
- Proceedings of ASCE Specialty Session on Cone Penetration Testing and Experience, St. Louis, USA, 1981.
- Proceedings of First European Symposium on Penetration Testing, ESOPT I, Stockholm, Sweden, 1974.
- Proceedings of Second European Symposium on Penetration Testing, ESOPT II, Amsterdam, Holland, 1982.
- Proceedings of ASCE Specialty Conference on Use of In Situ Tests in Geotechnical Engineering, (IN SITU '86), Blacksburg, USA, 1986.
- Proceedings of First International Symposium on Penetration Testing, (ISOPT I), Orlando, USA, 1988.

In addition to the above proceedings, a number of substantive reports have been written by various researchers/practitioners about specific test devices. Some of the more notable ones include the following:

- Mitchell, J. K., Guzikowski, F., and Villet, W.C.B., *The Measurement of Soil Properties In Situ,* Geotechnical Engineering Report # LBL-636, University of California, Berkeley, 1978.
- Robertson, P. K., and Campanella, R. G., *Guidelines for Geotechnical Design Using CPT and CPTU*, Soil Mechanics Report # 120, University of British Columbia, 1989.
- Miran, J., and Briaud, J. L., *The Cone Penetrometer Test*, Geotechnical Report, Texas A&M University, 1990.
- Davidson, J. L., Bloomquist, D. G., and Basnett, C. R., *Dilatometer Guidelines and the Effects of Dynamic Insertion,* Report # FL/DOT/MO/345-89, University of Florida, 1988.
- Whittle, A. J., Aubeny, C. P., Rafalovich, A., Ladd, C. C., and Baligh, M. M., *Prediction and Interpretation of In Situ Penetration Tests in Cohesive Soils*, Report # R91-01, Massachusetts Institute of Technology, 1991.
- Schmertmann, J. H., *Guidelines for Using the Marchetti DMT for Geotechnical Design*, Volumes 3 and 4, Report # FHWA-PA-024+84-24 and Report # FHWA-PA-025+84-24, NTIS, 1988.
- Kulhawy, F. H., and Mayne, P. W., *Manual on Estimating Soil Properties for Foundation Design*, Report # EPRI EL-6800, Electric Power Research Institute, 1990.

27

In Situ Testing and Field Instrumentation

Rodrigo Salgado
Purdue University

27.1 Introduction

Geotechnical design requires assigning values to quantities such as unit weights, stresses, forces, displacements, and strains inside soil or rock masses, as well as at the interface of these masses with structures made of steel, concrete, or other materials. **Instrumentation** is used (1) to determine the values of soil or rock properties, as well as stresses or water pressures, before the design and construction stages of a geotechnical project, and (2) to measure, during either the construction or the useful life of the structure, quantities whose values were assumed, computed, or specified during design.

In situ tests provide one way to determine soil parameters before the design and construction phases of a project are started. Such tests are performed by subjecting a volume of soil in situ, in its natural condition, to a certain excitation and measuring its response. Examples of excitations and measured responses include pushing a penetrometer into the soil and measuring the force resisting penetration, or generating a shear wave at one point and measuring the time it takes to reach another point.

Some instruments used for measuring initial variables do not subject the soil to any excitation, but attempt to measure quantities as if they were not present. **Piezometers**, for example, may be used to get information about the groundwater pattern at a site before any construction takes place. In order for that to be done effectively, the piezometers have to be installed with a minimum of disturbance and they must operate in such a way that they interfere as little as possible with the original groundwater pattern.

It is important to determine initial soil or rock parameters because subsequent behavior will be determined to a great extent by the values of the initial parameters. A clear illustration of this is the different values of stresses and strains at failure in a triaxial compression test reached by initially dense and initially loose samples under the same initial confining stresses.

0-8493-8953-4/95/$0.00 + $.50

If the determination of the initial values of soil parameters is limited by our ability to measure them in situ, the determination of values for the same quantities during construction or the service life of the structure is further complicated by the imperfection of our theories and predictive methods. If limiting the difference between predicted and actual values will permit changing construction procedures to reduce costs or avoid a catastrophic failure of the facility during its service life, for example, then the use of instruments to measure a number of quantities at selected locations may be in order.

So geotechnical instrumentation, generally speaking, consists in selecting the following:

1. The type and number of instruments or devices to be used
2. The specific locations where measurements are to be made
3. Installation procedures
4. The data acquisition system
5. The methods for analyzing and interpreting the data

Instrumentation technology and methods of interpretation of the results from in situ tests and field instrumentation have reached a high degree of sophistication. Successful planning, execution, and use of the results require the following technical background:

1. Knowledge of soil and rock behavior to select which quantities to measure and where to measure them, as well as to interpret the results.
2. Knowledge of the devices available in the market to make a selection that will measure reliably what is needed with the accuracy required, for the duration necessary, at the least possible cost.

It is beyond the scope of this chapter to elaborate on the first requirement above, which is addressed by other chapters in this handbook. Focus will be placed on discussing devices, their use, and interpretation of measurements.

Deep foundations, underground excavations, braced excavations, mining excavations, natural or excavated slopes, bridges, embankments on compressible ground, and dams are some structures where geotechnical instrumentation to monitor the performance of the structure during construction or operation of the facility may be used. The quantities whose measurement may become necessary include pore water pressures, total stresses in soils, stresses in rock masses, loads or stresses in structural elements, deformation and displacements, and temperature. These quantities are measured and recorded by instruments and data acquisition systems using mechanical, hydraulic, pneumatic, or electrical means of sensing, transmitting, reading, and storing data. We will start with a discussion of the equipment available for in situ testing and geotechnical instrumentation and how they measure different quantities. That will be followed by a discussion of interpretation procedures, and by examples of applications.

27.2 In Situ Testing Devices and Other Instruments to Assess Initial Conditions at a Site

Initial State of the Soil at the Site

Geotechnical design requires an assessment of the properties of the soil or rocks at the site where construction activity will take place. Instrumentation for monitoring the performance of a structure requires that initial values be known for the quantities to be monitored, or else it will be difficult to assess how much change has taken place and how far the facility is from a state at which it no longer functions adequately. The nature of the project will ultimately determine the quantities whose values must be assessed. Some information should always be gathered:

1. The groundwater regime at the site (in most projects the water table level and its seasonal variation is all that is needed).
2. The nature of soil and rock found at different depths as far down as is thought to be of consequence; for soils, the grain-size distribution and some assessment of soil mineralogy and geologic origin, plastic and liquid limit of clays, minimum and maximum void ratios of sands; and for rocks, geologic origin, degree of weathering, frequency and dimensions of discontinuities would be some determinations to be made.
3. Any particular geologic characteristic, such as an underground cavity or a fault.

The following are important quantities in projects where a substantial change in stress state will take place as a result of construction: the stress state at points of interest in the soil mass; distribution of void ratio, or, for sands, relative density; and interparticle cementation, if any. Projects where underground water flows are important include dams, tunnels, and waste-containment structures. In such projects the values and seasonal variations of water pressures at the site must be quantified in greater detail.

Devices for In Situ Testing

In situ testing devices can be used for determining initial values for relevant variables at a site. Their use differs philosophically from the use of laboratory tests performed on samples recovered from the ground. When samples are recovered, one expects to make all the measurements he or she needs on those samples, such as mass density, strength, compressibility, permeability, and so on. How reliable the measurements are depends on how badly the samples were disturbed during sampling, how representative the samples are of the soil conditions in situ, and on the quality and soundness of the laboratory testing. If it could be assumed that there was no disturbance, that the tests did measure what they are supposed to measure, that the soil was homogeneous throughout the site (so that the recovered samples do not miss any important characteristics of the deposit), then there would be no need for in situ tests. It is precisely because these assumptions are not reasonable that in situ tests are used.

In situ tests do not provide a way to measure directly the fundamental properties of the soil. The measurements that are made must be related to the quantities of interest by means of theoretical relationships or empirical correlations. The fundamental concept underlying in situ test interpretation is that the number of independent measurements made must be equal to the number of soil parameters to be determined. In addition, there must be the same number of relationships linking the unknown parameters to the independent measurements. So, conceptually, in situ test interpretation is very much like solving a system of equations with the same number of equations as there are unknowns.

During the site investigation phase of a project at least one form of in situ test is always performed. The most common tests in the U.S. are the standard penetration test (SPT) and the cone penetration test (**CPT**). Other in situ tests include the plate load test (PLT), the dilatometer test (DMT), the vane shear test (VST), and the pressuremeter test (PMT).

The SPT is performed inside boreholes by advancing a spoon sampler into the base of the borehole by blows from a hammer with a standard weight of 140 pounds falling from a height of 30 inches. The number of blows required to advance the sampler a distance of 1 foot into the soil is recorded and considered to be indicative of the soil density or consistency, the stress state, and the nature and size distribution of particles and soil structure. The approach to SPT interpretation is usually based on the use of correlations of several different properties with the blow count. Unfortunately, SPT blow counts also depend on the equipment and procedure used to perform the test. The following were all found to affect the test results: procedure used to raise and drop the hammer (number of turns of the rope around the pulley if the cathead system, which is the most common in the U.S., is used), hammer type, string length, presence of a liner inside the sampler,

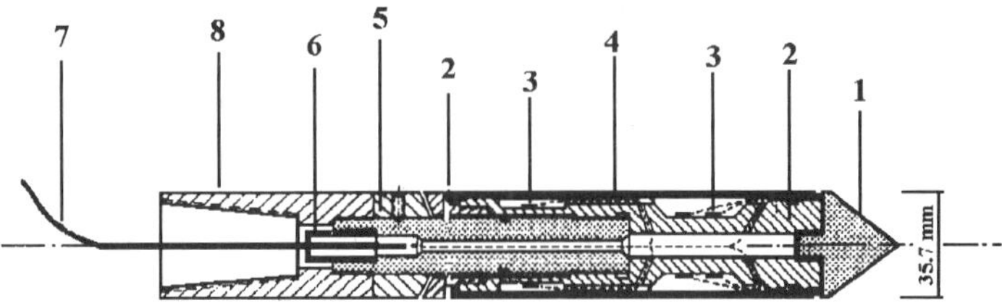

FIGURE 27.1 The standard electrical cone penetrometer. [After De Ruiter, J. 1971. Electric penetrometer for site investigation. *J. Soil. Mech. Found., ASCE.* 97(SM2). Reproduced with permission of the publisher.]

sampler condition, borehole condition, and drilling method. Attempts to standardize the test have been made [Seed *et al.*, 1986; Skempton, 1986]. That, associated with the familiarity engineers have with the test, is responsible for the continuing importance and use of the SPT in geotechnical field testing.

In the CPT a number of quantities are measured as a cylindrical penetrometer (Fig. 27.1) with a conical tip is pushed vertically down into the soil at a rate of two centimeters per second. The most important quantity measured in the CPT is the vertical force acting on the tip during penetration. The ratio of this force to the projected or plane area of the tip is the cone penetration resistance q_c. The **cone penetrometer** is a very versatile device. Many different types of sensors can be incorporated into the cone, enabling it to measure a number of additional quantities: the frictional resistance f_s along a lateral sleeve, the shear wave velocity V_s along the penetration path, the pore pressure u generated during penetration at the face of the conical tip or behind it, as well as a number of quantities of interest in geoenvironmental projects, such as temperatures, electrical resistivities, organic content in the pore fluid, and other pore fluid chemistry parameters [Mitchell, 1988].

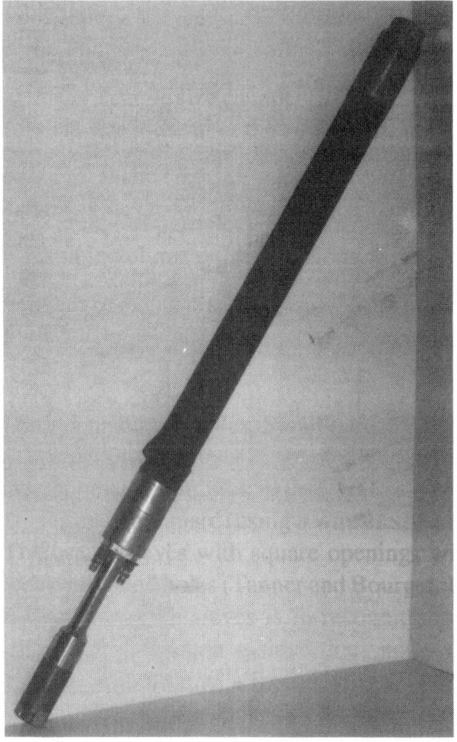

FIGURE 27.2 A pressuremeter. (Photograph by A. Bernal and A. Karim, Bechtel Geotechnical Laboratory, Purdue University.)

The advancement of the cone penetrometer through soil can be mathematically modeled [Salgado, 1993; Salgado and Mitchell, 1994], and the influence of the operator and hardware on the values of the measured quantities is virtually nonexistent. The advantages of the device over the SPT, the only other in situ test that is more frequently used, are so numerous that it will likely replace it as the test of choice in the future. For interpretation of the CPT refer to ASTM D 3441-86, Salgado [1993], Salgado and Mitchell [1994], Robertson and Campanella [1986], and Jamiolkowski *et al.* [1985].

The PMT is performed by expanding a cylindrical membrane (Fig. 27.2) in the soil while simultaneously measuring the pressure inside the membrane and the displacement at the center of the membrane. There are essentially two installation methods: in one a borehole is prebored and the pressuremeter is lowered into it (this is known as the Menard pressuremeter), and in the other the pressuremeter itself has a drilling tool at its lower end that is used to install it at the desired depth (this is the so-called self-boring pressuremeter). Apart from any disturbance due to drilling (and unloading in the case of the Menard pressuremeter), the pressure-strain curve obtained in a PMT, which may have unloading-reloading loops, is related to the stress-strain relationship of soils (which itself is a function of the initial soil state), and interpretation of a pressuremeter test is essentially obtaining the stress-strain curve from the pressure-strain PMT curve. Attempts to extrapolate the pressuremeter curve so that a limit pressure can be obtained, if successful, provide an indication as to the ultimate strength of the material.

The problems with the pressuremeter are assessing the degree of disturbance due to installation and the effect of the length-to-diameter (L/D) ratio of the membrane. If the L/D is small, only for very small displacements can the expansion be assumed to approximate a cylindrical cavity expansion, as is usually assumed in the interpretation methods. For higher levels of strain, the effects of L/D must be taken into account.

The **cone pressuremeter** (or **pressurecone**) is a cone penetrometer with an expandable cylindrical membrane, as in a pressuremeter, with a large L/D value [Schnaid and Houlsby, 1991]. It is a relatively new in situ device that can be advantageous over a traditional pressuremeter because it permits extrapolation of limit pressure that is superior to that which is possible with the pressuremeter. It has the advantage that cone penetrometer data also result from the test. Installation, however, affects the soil in a completely different way from that of a Menard or self-boring pressuremeter, and so parameters derived from the small-strain part of the expansion curve may be very different for the different types of devices.

References on pressuremeter test performance and interpretation include ASTM D 4719-87, Yu [1994], Manassero [1992], Yu [1990], Mair and Wood [1987], and Hughes *et al.* [1977].

The DMT is performed by pushing down into the soil the blade shown in Fig. 27.3 and simultaneously measuring the resistance to penetration, in the same way as in a CPT, and stopping penetration at preselected depths for loading the soil in a different way. At such depths the circular

FIGURE 27.3 A dilatometer. (Photograph by A. Bernal and A. Karim, Bechtel Geotechnical Laboratory, Purdue University.)

membrane on one of the sides of the dilatometer is expanded and the pressures at deflections of 0 and 1.1 millimeters are recorded and corrected for membrane stiffness. Based on these two fundamental parameters a number of empirical relationships have been proposed for soil parameters of interest in design [Marchetti, 1980; Schmertmann, 1986; Fretti *et al.*, 1993].

For further information on in situ testing devices refer to Jamiolkowski *et al.* [1985].

27.3 Instrumentation for Monitoring Performance during or after Construction

Planning of an Instrumentation Program

The planning of an instrumentation program starts as soon as the need for one is established. There are multiple reasons why one might want to use instrumentation. It may be that the design has enough flexibility to be changed if measurements made during early stages of construction indicate changes to be desirable. The facility may be so important and its loss may lead to so much damage or danger that instrumentation can be used to detect any deviations from the assumed or predicted behavior of the facility, in which case previously planned actions can be taken to either correct the deviation or minimize losses. Sometimes a new concept is used, and instrumentation is used to show or ascertain that it indeed works. Whatever the reason, detailed planning is essential for a successful instrumentation project. Figure 27.4 illustrates the main steps in the planning of an instrumentation program.

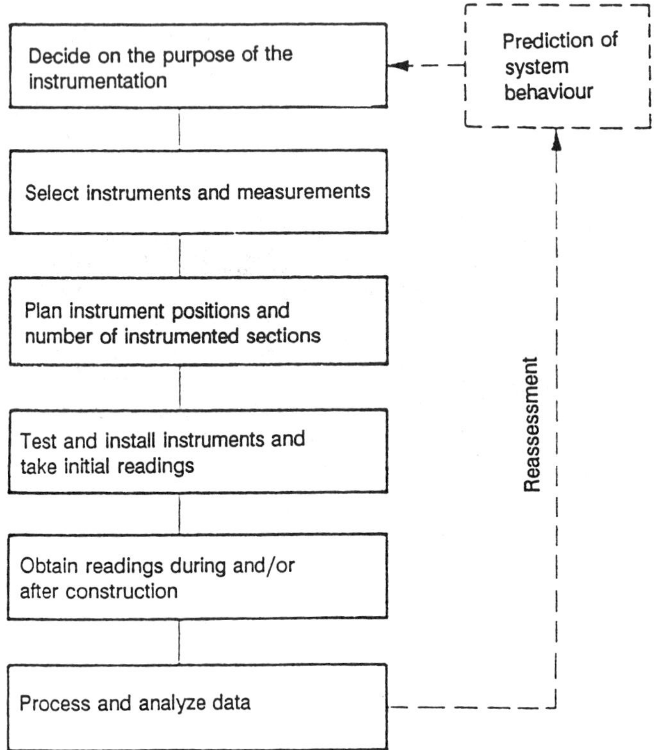

FIGURE 27.4 Planning of an instrumentation program. [After Ortigão, J. A. R. 1986. *Instrumentation of Embankments on Soft Soils: A Review of Instruments and Techniques* (in Portuguese). Unpublished notes.]

The definition of the objective of the instrumentation program usually follows the realization that something about the project is not known well enough, and that measurements of a number of quantities at certain locations would be desirable for the sake of economy or safety. A person or team of specialists is then charged with planning the program. After first acquainting itself with all the details of the project, the team may visualize all possible ways the construction might behave so as to choose which quantities to measure, where to measure them, and select adequate instruments to do so.

Selecting the instruments requires estimating the magnitudes of changes in the quantities to be measured resulting from construction or operation of the facility. The maximum expected change of the parameter to be measured will allow the definition of the range for the instrument selected to measure it. The minimum expected change will permit defining the sensitivity or resolution of the instrument. Such predictions are also useful in defining criteria for remedial action, if the purpose of the instrumentation is to ensure proper and safe operation of the facility.

Simplicity should be strived for. The simpler the instrumentation, the less the risk of malfunctioning. Simplicity is directly related to the working principle of the device. It decreases in the following order [Dunnicliff, 1988]: optical systems→ mechanical systems→ hydraulic systems→ pneumatic systems→ electrical systems.

The following are properties an instrument should have [after Dunnicliff, 1988]:

1. *Conformance.* The instrument should conform to the surrounding soil or rock in such a way that it does not alter the values of the properties it is intended to measure.
2. *Accuracy.* The measure should be as close as possible to the true value.
3. *Repeatability.* Under identical conditions, an instrument must measure the same value for the quantity it is intended to measure whenever a measurement is taken.
4. *Resolution.* The smallest division of the instrument readout scale must be compatible with the minimum expected change in the quantity being measured.
5. *Sensitivity.* Change in output per unit change in the input. In an electrical transducer, for example, the smaller the sensitivity, the smaller the voltage generated by a given change in the input, which may lead to an excessively large noise-to-signal ratio.
6. *Linearity.* The relationship between input and output should be as linear as possible for the range in which the input is to be measured.
7. *Hysteresis.* It is undesirable that a different value for the quantity of interest be measured depending on whether the quantity is increasing or decreasing, a behavior known as hysteresis. This difference should be kept to a minimum, particularly if cyclic measurements are needed.
8. *Noise.* Random variations in the measurement should be limited.

There are two approaches for selecting instrument location. Instruments can be installed in trouble spots, such as points where it is expected that there will be large stress concentrations or large pore pressure increases, or where the material is particularly weak. Alternatively, they can be placed at a number of representative points or zones in the deposit.

Testing and installation of the instrumentation and the taking of initial readings is the next major step. This will typically be carried out by joint efforts of the owner, the design consultant, the instrumentation team, and the construction contractor. The specific division of labor among the parties will be very much dependent on the circumstances of the project.

The taking of readings and their processing and analysis must be carried out in a systematic, organized way. Careful record keeping is crucial. It should include taking note of any significant events that may later be correlated with unanticipated or significant changes in the quantities measured.

According to Ortigão and Almeida [1988], the approximate costs of an instrumentation program are as follows. Instrumentation acquisition: 10%; installation: 20%; data acquisition: 20%; analysis: 50%. It is important to allocate enough funds for the analysis and interpretation of the results from the beginning. An instrumentation program will fail, regardless of how well the instruments, their locations, and everything else was selected, if there are no resources left when the time comes to analyze the numbers.

In the following sections we discuss the measurement of quantities of importance in geotechnical engineering and some of the devices used.

Instruments

A **sensor** is a device that detects the change of a certain variable. A **transducer** is a device that converts one type of signal into another type. A transducer may, for example, transform a mechanical input, such as a load, into an electrical output, such as a voltage differential. If remote reading is required, the signal of a transducer must be conveyed to the reading unit. The precise way in which data is transmitted and read depends on the data acquisition system employed.

The two sensors used to measure liquid pressure in geotechnical instrumentation are the **manometer** and the **Bourdon pressure gage.**

The manometer is a U-tube filled with a liquid, whose two ends may be connected to different pressure sources. At any elevation through the manometer, the pressure on one side of the tube must equal the pressure on the other side. The pressure is measured at the interface between two fluids, such as water and air or mercury and water. It is possible to determine the unknown pressure by recognizing that the unknown pressure must balance the weight of the column of the heavier fluid above the elevation of the interface on the other branch of the manometer, which can be read. The use of a heavy liquid, such as mercury, allows large pressures to be effectively read as they will translate into small liquid columns in the manometer.

The Bourdon gage is a C-curved tube that uncoils upon pressurization. Appropriate calibration permits associating the uncoiling motion of the tube to the change in pressure, which can be read by means of a dial (Fig. 27.5).

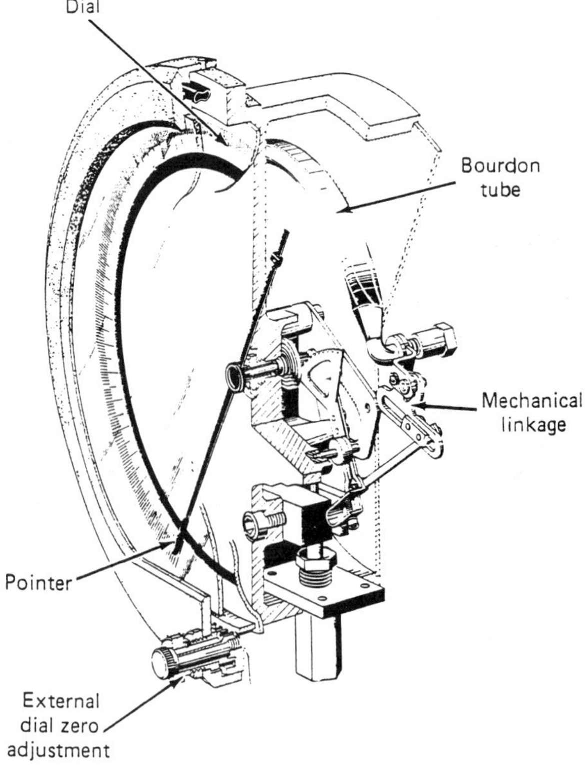

FIGURE 27.5 Bourdon pressure gage. (Courtesy of Dresser Industries. Used by permission.)

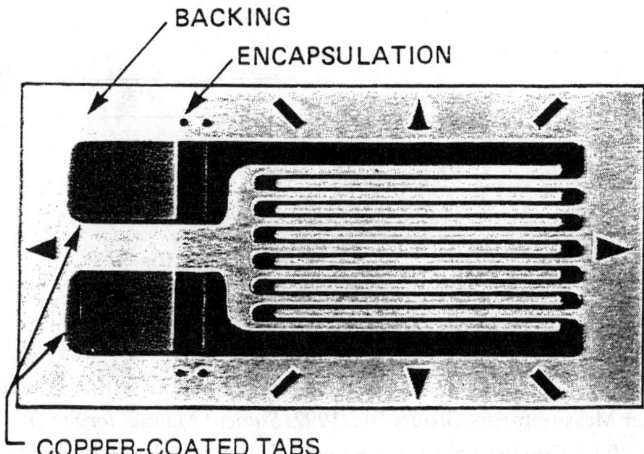

FIGURE 27.6 An electrical resistance strain gage. (After Measurements Group, Inc. 1992. *Student Manual for Strain Gage Technology.* Bulletin 309D. Courtesy of Measurements Group, Inc., Raleigh, NC, USA.)

In case remote reading is desired, a pressure transducer can be used in place of the manometer or the Bourdon tube to convert the liquid pressure to an electrical signal.

Some electricity-based devices used in geotechnical instrumentation include different types of **strain gage**, the linear variable differential transducer **(LVDT), vibrating wire** transducers, and hardware to transmit, receive, and process data.

An **electrical resistance strain gage** is a conductor with the property that electrical resistance changes with the change in length. The two types of electrical resistance strain gages most often used in geotechnical applications are the bonded foil and the weldable strain gage.

The **bonded foil strain gage** consists of a thin foil of a strain-sensitive alloy such as constantan, annealed constantan, or nickel-chromium alloy bonded to a thin plastic film (Fig. 27.6), which can be attached to the surface of the structure to be monitored [Measurements Group Inc., 1992]. The **weldable strain gage** is made by attaching a resistance element to a thin stainless steel mounting. The resistance element may be a bonded foil transducer or something else. The steel mounting can be welded to a steel structural element. The typical length of a strain gage is between one-sixteenth of an inch and six inches. A typical range is 2 to 5% strain, and the resolution can be as low as 10^{-6}. The output of electrical resistance strain gages is measured using a **Wheatstone bridge** (Fig. 27.7). The Wheatstone bridge is used to determine the value of an unknown resistance R_3. A voltage is

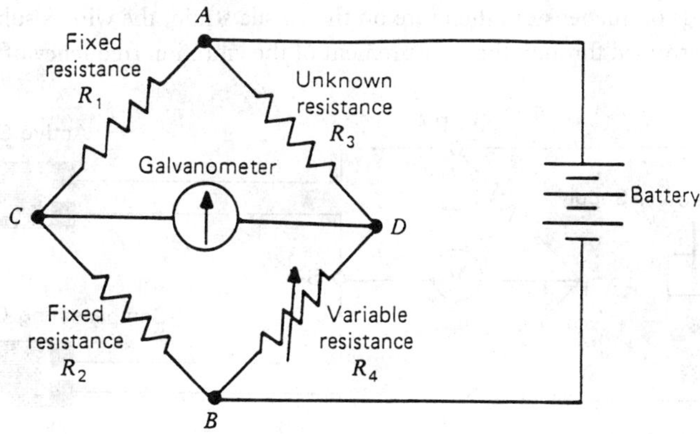

FIGURE 27.7 A Wheatstone bridge. (After Dunnicliff, J. 1988. *Geotechnical Instrumentation for Monitory Field Performance.* John Wiley & Sons, New York.)

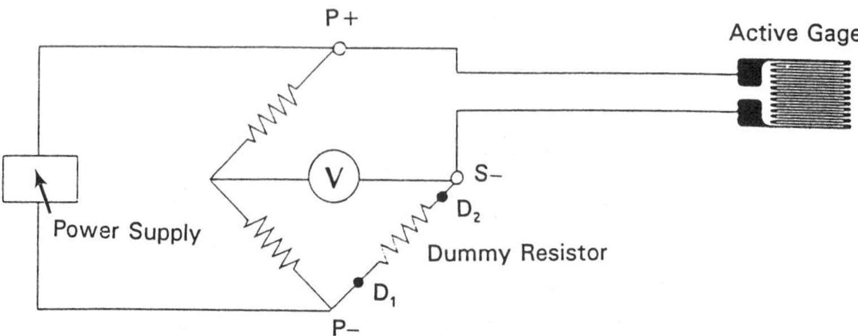

FIGURE 27.8 The simplest scheme for strain measurement with a Wheatstone bridge circuit. (After Measurements Group, Inc. 1992. *Student Manual for Strain Gage Technology.* Bulletin 309D. Courtesy of Measurements Group, Inc., Raleigh, NC, USA.)

applied between A and B and the current is monitored between C and D. Resistances R_1 and R_2 are known, and resistance R_4 is a potentiometer than can be adjusted until no current flows between C and D. When that happens,

$$R_3 = R_4 \frac{R_1}{R_2}$$

The simplest possible arrangement is that in Fig. 27.8, where a single strain gage appears as one of the resistances in the Wheatstone bridge. Changes in strain will reflect in changes in resistance that are measured by the bridge. The disadvantage of this scheme is that changes in resistance in the wires leading to the gage due to temperature variations will read as if strain had taken place. This effect can be reduced by using short, thick lead wires, and can be corrected using the scheme of Fig. 27.9, in which two exactly identical gages are connected as adjacent arms of the bridge. Effects due to temperature will appear on both sides and will cancel out from the above expression. Elimination of temperature effects in single-gage installations can be accomplished using the three-wire scheme of Fig. 27.10.

An LVDT is composed of a magnetic core oriented along the axes of two coils: a primary coil, to which AC voltage is applied, and a secondary coil, in which AC voltage appears as a result of the application of voltage in the primary coil (Fig. 27.11). The AC voltage can be converted to DC by a phase-sensitive device. The difference between the two voltages is exactly zero when the magnetic core is at the central position, and becomes either negative or positive depending on the direction in which the core moves off center.

In vibrating wire transducers a tensioned wire is maintained in resonance. The vibration frequency depends on the tension, therefore on the tensile strain, the wire is subjected to. So the strain can be determined through the measurement of the vibration frequency of the wire.

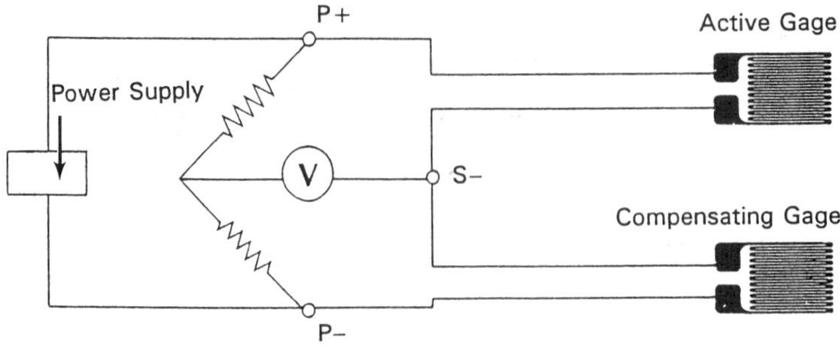

FIGURE 27.9 Double-gage scheme for measurement of strain. (After Measurements Group, Inc. 1992. *Student Manual for Strain Gage Technology.* Bulletin 309D. Courtesy of Measurements Group, Inc., Raleigh, NC, USA.)

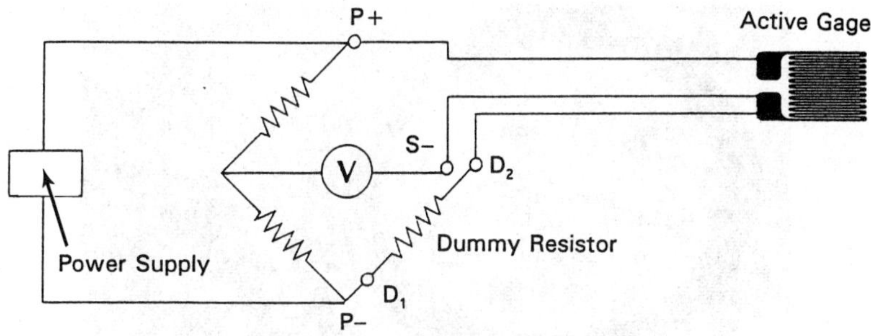

FIGURE 27.10 Double-wire scheme for measurement of strain. (After Measurements Group, Inc. 1992. *Student Manual for Strain Gage Technology.* Bulletin 309D. Courtesy of Measurements Group, Inc., Raleigh, NC, USA.)

Measurement of Load and Stresses

Loads are measured by **load cells.** The measurement of loads is needed for load testing of deep foundations, tiebacks, and rock bolts, to name a few applications.

The measurement of strains can also indicate the magnitude of stresses or loads in a structure. Strain gages can be placed almost anywhere in a structure. They are called surface-mounted strain gages if applied to an exposed surface, and embedment strain gages if installed in the interior of a mass.

There are a variety of load cells: mechanical, hydraulic, electrical resistance, vibrating wire, photoelastic, and others. A calibrated hydraulic jack is also sometimes used for measuring loads when accuracy is not a great concern.

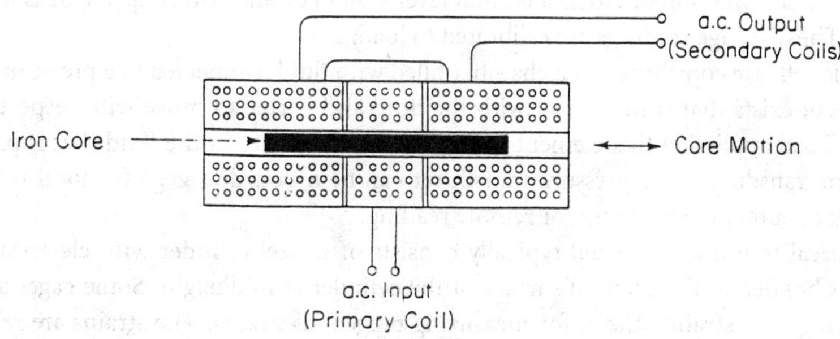

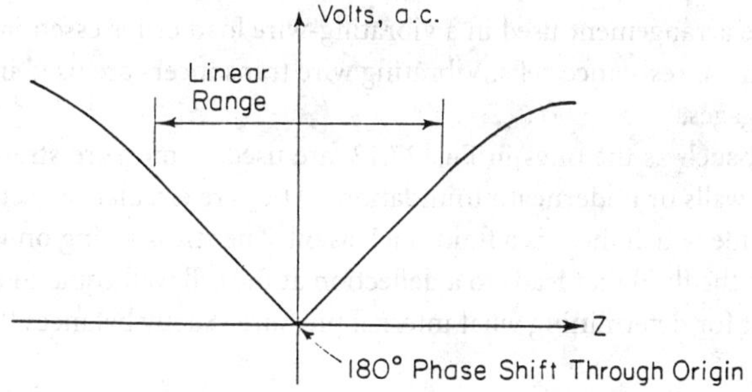

FIGURE 27.11 Linear variable differential transformer (LVDT). (After Richart, F. E., Jr., Woods, R. D., and Hall, J. R., Jr. 1970. *Vibrations of Soils and Foundations.* Prentice Hall, New York, ©1970, p. 282, Fig. 9.5. Reprinted by permission of Prentice-Hall, Englewood Cliffs, N.J.)

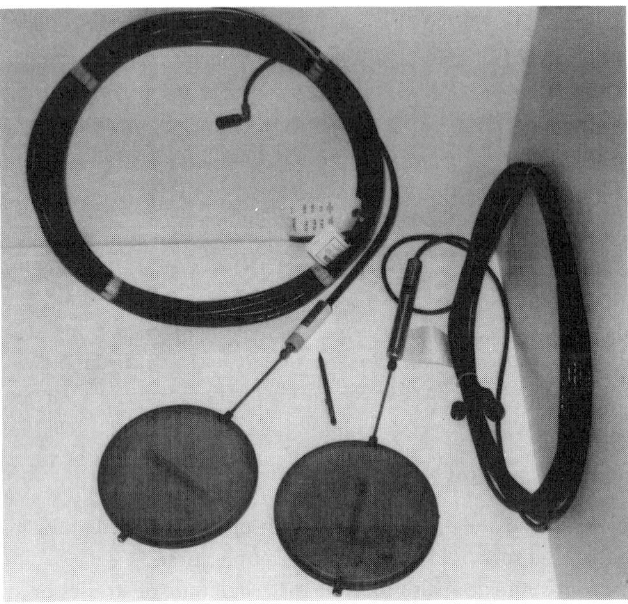

FIGURE 27.12 Pressure cells for measurement of stresses in soil masses or contact stresses between soil masses and concrete or steel surfaces. (Photograph by A. Bernal and A. Karim, Bechtel Geotechnical Laboratory, Purdue University.)

Mechanical load cells are based on the existence of a direct relationship between deformation and load. The deformation of either a torsion lever system or an elastic cup spring is measured by dial gages. The dial gage readings are calibrated to load.

Hydraulic cells are constituted of a chamber filled with fluid, connected to a pressure transducer. A flat element exists that is in contact with the fluid and is free to move with respect to the rest of the cell. Load applied at this element will reflect in a pressure in the fluid that is picked up by the pressure transducer. The pressure transducer can be a Bourdon gage for local reading or an electrical or pneumatic transducer for remote reading.

An electrical resistance load cell typically consists of a steel cylinder with electrical resistance strain gages bonded to the external surface of the cylinder at midheight. Some gages are oriented for measuring axial strain, others, for measuring tangential strains. The strains are related to the applied load. The gages are arranged and connected so as to minimize errors associated with load eccentricity.

The strain gage arrangement used in a vibrating-wire load cell is essentially the same as the one adopted for electrical resistance cells. Vibrating wire transducers are used in place of the electrical resistance strain gages.

Pressure cells, such as the ones in Fig. 27.12, are used to measure stresses. They may be used behind retaining walls or underneath foundations. They are circular or rectangular pads with thin metallic walls inside which there is a fluid, such as oil. The stress acting on walls of the cell is equal to the pressure in the fluid that leads to a deflection at the cell wall equal to zero. Different internal mechanisms exist for determining what internal pressure exactly balances the outside stress.

Measurement of Deformation

A reliable way of measuring displacements of the ground surface or parts of structures that can be observed above the ground surface is by surveying techniques.

The measurement of absolute motions requires that reference points be established. The reference for the measurement of a vertical displacement is called a **benchmark**, while for horizontal displacement measurements a **reference monument** is used. Reference stations should be motionless. If no permanent structure, known to be stable and free of deformation, exists, the reference station (benchmark or monument) must be built. A benchmark may consist of a rod or pipe extending all the way down to a relatively incompressible layer, isolated from the surrounding soils so as not to undergo any deformation, which can be accomplished by placing the rod inside a pipe without allowing any contact between the rod and the pipe.

Points where deformations are to be determined are called **measuring points** [Dunnicliff, 1988]. These should satisfy three requirements: first, they must be well marked, so that accurate measurements can be made; second, they must be placed at the actual seat of movement; and third, they must last for the time during which measurements will be needed. For measurements on structures the first requirement can be satisfied by the use of bolts embedded into wall surfaces. Sometimes a grid, scale, or bull's-eye may be used for improved readings.

If ground surface displacement measurements are desired and a pavement exists, particularly if it is a structural pavement, it is important to place the measuring point below the pavement, in the ground, to get an accurate measurement of the displacement. It is also important to go beyond the freeze-and-thaw depth, and any layer of expansive or collapsible soil, unless the motions related to these phenomena are to be determined.

Surface **extensometers** are used to measure the change in distance between two points located on a surface, which may be the ground surface. There are a variety of such devices. A mechanical extensometer, for example, is an instrument that can be fixed to two points on a surface, with any variation in the distance between them being read from a dial gage.

A **tiltmeter** (Fig. 27.13) is used for measuring the rotational component of a motion. Tiltmeters may be useful to monitor landslides where rotation is significant, the tilt of retaining walls, and other similar applications. Again, there are mechanical, vibrating-wire, and other types of tiltmeters, and the discussion of each specific type is beyond the scope of this chapter.

An **inclinometer** is a device used to measure the inclination with respect to the vertical. A borehole is drilled and fitted with a casing that will allow the inclinometer to be periodically lowered down the casing all the way to a depth where no motion is expected. This allows the evolution

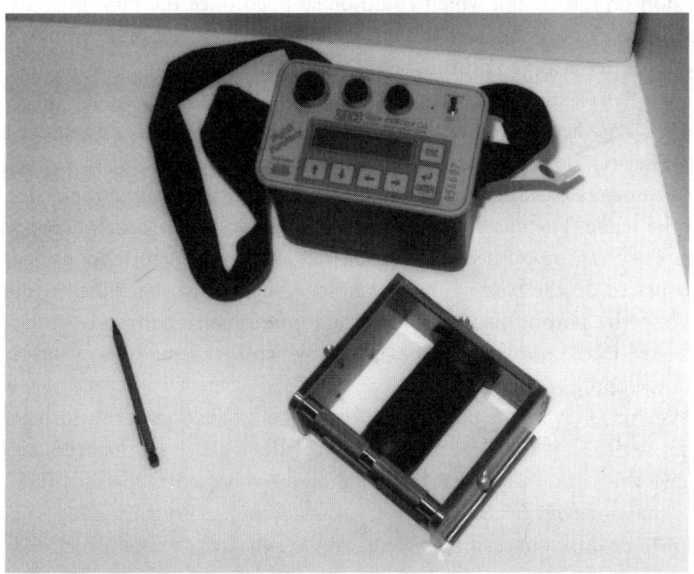

FIGURE 27.13 Tiltmeter and readout unit. (Photograph by A. Bernal and A. Karim, Bechtel Geotechnical Laboratory, Purdue University.)

of the borehole profile with time to be assessed. The obvious applications of such devices are monitoring of landslides, measurement of horizontal displacements in embankments in soft ground and of motion undergone by retaining walls, and assessment of the location of a developing failure surface.

Settlement plates are used to monitor the deformations underneath embankments, for example. The plates, usually square steel plates at least 3 ft in size, are embedded underneath the embankment. They are connected to one-inch rods that extend all the way to the surface and are isolated from the soil by means of a casing. The rods are used for measuring the displacements. The principle is essentially the same for **telltales**. Telltales are rods extending from a point inside a structural element, such as a drilled shaft, and isolated from the surrounding concrete (or other material) by means of a casing. The displacement of the tip of the rod will then be the same as that of the point where its end has been placed.

In dynamic or vibration problems the velocity and acceleration may also be needed in addition to displacement. The transducer is attached to a point on a surface, whose motion is to be determined. Because of the internal stiffness, inertia, and damping of the transducer, what is measured may have very little to do with the motion to be described. By making appropriate selection of the stiffness and damping of the transducer it is possible to measure either acceleration or velocity accurately [Richart *et al.*, 1970]. The transducers are accordingly referred to as accelerometers and velocity transducers.

Measurement of Pore Water Pressure

Piezometers are used to measure groundwater pressure. The knowledge of the pressure in the groundwater permits computing effective stresses or forces exerted by the groundwater on buried structures. The pore pressure may be easy to calculate if the depth to the groundwater table is known and hydrostatic conditions are known to exist, but there are several situations where there is enough uncertainty about pore pressure values to warrant the use of piezometers to measure them. One typical example is the placement of piezometers in earth dams to monitor long-term pore pressure changes. Ineffective drainage could lead to pore pressure buildups that would be impossible to predict if perfect drainage were assumed.

Additional uses of piezometers could include the monitoring of pore pressures during embankment construction on a soft soil, which could be stopped once the pore pressures increased so much as to reduce the factor of safety to a value close to 1, and the monitoring of pore pressure dissipation around a pile after it has been driven to its final depth, which would allow the selection of the right time to start a **static load test.**

Piezometers are classified according to their working principle as hydraulic, pneumatic, and vibrating-wire. They can be further classified as simple or remote-reading.

Hydraulic piezometers consist of a porous tip installed at the point where the pore pressure readings are to be made. They are usually installed at the base of a borehole with a diameter of 50 to 100 mm. They can be pushed into the base, so they are surrounded by natural material, or alternatively placed on the base of the borehole, which is then backfilled with sand that will completely involve the porous tip. The Casagrande piezometer consists of a single tube [Fig. 27.14(a)] connected to the porous tip. The level to which the water rises in the tube reflects the pore pressure at the porous tip.

A different system is shown in Fig. 27.14(b), where two tubes connect the porous tip to a remote pressure-reading device. The double-tube system provides a check on the pressure reading: if the tubes do not give the same reading, at least one of them has entrapped air. The readout units can be mercury manometers, electrical pressure transducers, or Bourdon gauges. Advantages and disadvantages of these different reading systems are summarized in Table 27.1.

In vibrating-wire piezometers, as well as in pneumatic piezometers, the pore pressure acts on one side of a diaphragm. The difference between them resides in the mechanism, on the other side of the diaphragm, used to measure the pore pressure.

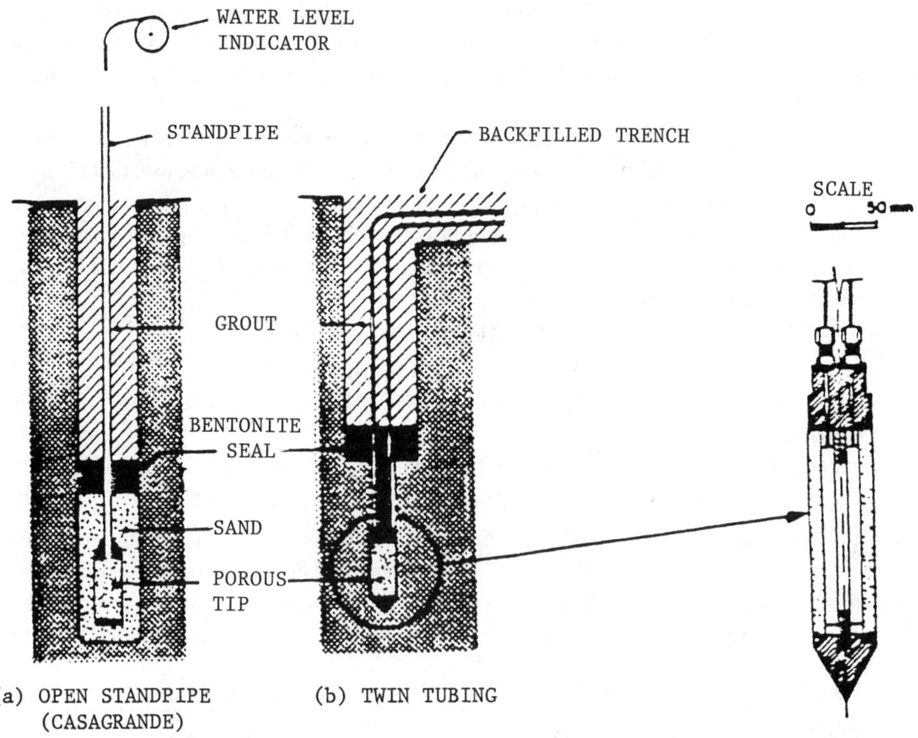

FIGURE 27.14 Piezometers: (a) Casagrande; (b) double-tube. [After Ortigão, J. A. R. 1986. *Instrumentation of Embankments on Soft Soils: A Review of Instruments and Techniques* (in Portuguese). Unpublished notes.]

Table 27.1 Characteristics of Readout Units for Hydraulic Twin Tubing Piezometers

Readout unit	Disadvantages	Advantages
Individual mercury manometers	• Large instrument housing needed to accommodate readout panels • Tedious to obtain readings • Frequent deairing of the readout unit may be necessary when negative pressures are read at readout elevation • Limited pressure range (due to panel height limitation)	• Simple • Fault-proof • Appropriate to low pore pressure range (embankments on soft soil) • Very reliable
Pressure transducer gages	• Relatively high cost of electric transducer and readout unit • Electronic instruments not as robust as manometers or Bourdon gages	• Easy to read • Readings directly in engineering units • Transducer unit can be portable or exchangeable • Any pressure range • Lower cost for increasing number of piezometers • Small instrument housing
Individual Bourdon gages	• Individual gage necessary • High cost of accurate gages • Not recommended for low pore pressure range (as in the case of embankments on soft soils)	• Usually appropriate for high pressure range (embankment dams) • Simple • Fault-proof

Source: Ortigão, J. A. R. 1986. *Instrumentation of Embankments on Soft Soils: A Review of Instruments and Techniques* (in Portuguese). Unpublished notes.

In a vibrating-wire piezometer the diaphragm is connected to a tensioned wire that undergoes vibration and is connected to a frequency measuring device at the other end. As the pore pressure increases or decreases, the stress in the wire changes, leading to a change in the frequency of vibration of the wire.

In a pneumatic piezometer, inner tubes containing a fluid, often an oil, circulate past the diaphragm. If the pressure in the circulating oil is smaller than the pore pressure, there is no flow. This is true of the piezometer in its inactive state. Flow starts if the oil pressure is increased until it becomes greater than the pore pressure. After flow is established, balancing of the pressures will cause the flow to stop in the trailing tube of the piezometer. The pressure reading at this instant is indicative of the pore pressure value.

The following are desirable with respect to piezometers [Drnevich, 1992]:

1. Accuracy—the correct pore pressure should be measured.
2. Lag time should be small, so that it does not take too long for the correct pore pressure to be measured.
3. Installation should not disturb the soil so much that accuracy or lag time are affected.
4. The instrument should be sufficiently durable for the project at hand.

The time lag is a direct result of the working principle of the piezometer: in order for a pore pressure change to be recorded, a given volume of water must flow in or out the porous tip, and that may require a long time in fine soils. The lag time of the piezometer must be compatible with the speed with which pore pressures are expected to change. Ortigão and Almeida [1988] show

Table 27.2 Advantages and Disadvantages of Different Types of Piezometers

Piezometer	Advantages	Disadvantages
Open Stand pipe	• Simple, low cost, easy to install • High reliability • Self-deairing • Allows field permeability tests	• Long response time in soils of low permeability • For embankment on soft soils, difficulties interfacing vertical standpipe with embankment construction
Hydraulic twin tubing	• Porous tip deairing possible • Very reliable and simple (no calibration necessary) • Appropriate for long-term installations • Suitable for field permeability tests • Easy installation • Low cost of porous tip	• Time-consuming system deairing • Limitation on pressure measuring elevation (maximum of six meters above piezometric head) to avoid cavitation
Pneumatic	• Small and portable readout unit • Pore pressure value read directly (no correction necessary for differences in level) • No limitation imposed by cavitation on piezometric level measurement difference • Very low response time	• Deairing of the porous tip usually not possible • Soil permeability tests usually not possible • Very careful installation required to ensure tip saturation • Possible failure of diaphragm in long-term installations • Medium to high cost of porous tip
Electrical	• Automatic data acquisition easily implemented • Difference in elevation between tip and readout unit not a factor in installation • Very short response time	• High cost • Dealring of the porous tip usually not possible • Soil permeability tests through porous tip usually not possible • Possibility of electric failures (mainly in long-term installations)

Source: Ortigão, J. A. R. 1986. *Instrumentation of Embankments on Soft Soils: A Review of Instruments and Techniques* (in Portuguese). Unpublished notes.

that the lag time for a 95% pressure equalization for a typical Casagrande piezometer in a soil with a hydraulic conductivity of 10^{-10} m/s is about 1000 hours, while a vibrating-wire or pneumatic piezometer would require only about 10 seconds for a measurement to be obtained.

Table 27.2 summarizes advantages and disadvantages of different types of piezometers.

27.4 Examples of Uses of Instrumentation

Static Load Tests

The great challenge in the design of deep foundations is to ascertain, for a given displacement at the pile head, what fraction of the load applied at the pile head is carried by side friction, and what fraction by base or tip resistance.

The main purpose of a static load test on a pile or drilled shaft is to obtain a curve describing the relationship between the load and the displacement at the head of the pile. So-called **instrumented static load tests** have a more general objective. In such tests it is also desired to establish the distribution of the load along the pile or drilled shaft (the normal force at a cross section of the pile decreases with depth, because an increasing proportion of the load is carried by friction between the pile and the surrounding soil).

A typical arrangement for load-testing a pile is shown in Fig. 27.15 (ASTM D 1143). Reaction piles, usually two, are installed on each side of the test pile. They should be sufficiently distant from the test pile not to interfere. That usually means at least 5 to 8 pile diameters [Poulos and Davis, 1990]. A load cell should be used to measure the load applied at the pile head. The dial gages are used to measure the displacement at the pile head. They are affixed to a reference beam that should be protected from sunlight and temperature changes, and whose supports must be placed sufficiently far from the test and anchor piles not to be moved during the test. The load distribution along the pile shaft is obtained by measuring the deformation at different depths. This can be accomplished by using strain gages, telltales (Fig. 27.16), or **Mustran cells** (Fig. 27.17) [Crowther, 1988]. A telltale is a steel bar whose tip is placed at a depth where the value of the displacement is desired. Telltales are particularly convenient for use in drilled shafts, in which they are installed inside PVC tubes that isolate them from the surrounding concrete. This ensures that they will move freely and the displacement measured at the pile head will be equal to the displacement

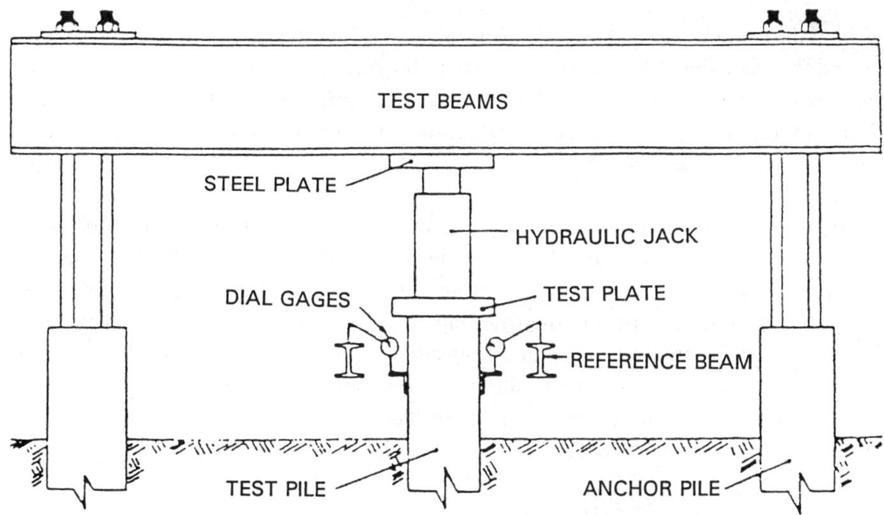

FIGURE 27.15 Setup for pile static load test. [After ASTM D 1143-81 (reapproved 1987). *Standard Method for Piles under Static Load.*]

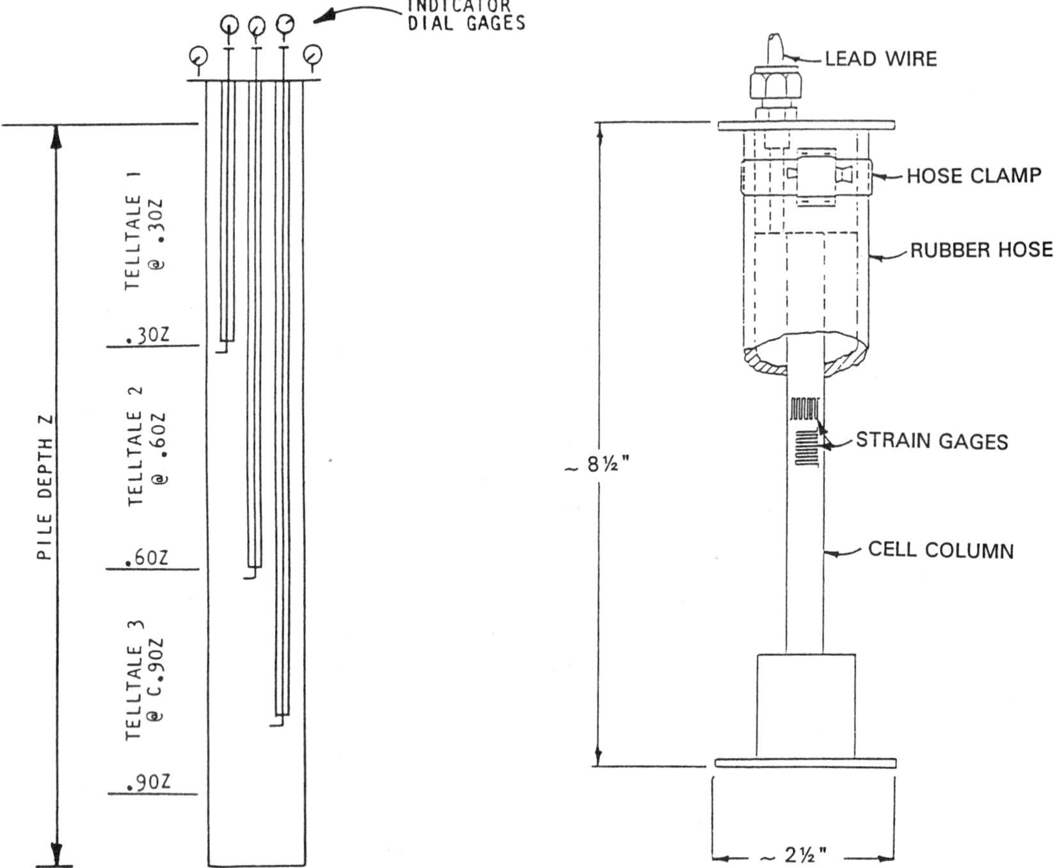

FIGURE 27.16 Possible arrangement of telltales to determine load distribution along a drilled shaft. (After Crowther, C. L. 1988. *Load Testing of Deep Foundations.* John Wiley & Sons, New York. Reprinted by permission of John Wiley & Sons, Inc.)

FIGURE 27.17 Mustran cell. (After Crowther, C. L. 1988. *Load Testing of Deep Foundations.* John Wiley & Sons, New York. Reprinted by permission of John Wiley & Sons, Inc.)

at the tip of the telltale. If several telltales are installed in a single pile at different depths, the average strain, and therefore the average stress and load, in the section of the pile between two telltale tips can be easily computed as the ratio of the difference in displacements to the difference in depths.

A Mustran (multiple strain transducer) cell consists of a short bar to which a number of electrical resistance strain gages are attached protected by a plastic sheath. The protective sheath is filled with an inert gas.

Mustran cells are sometimes used in instrumented tests of drilled shafts. The Mustran cell or telltale can be connected to a rebar before concrete placement, as illustrated in Fig. 27.18.

If the strain is known at any pile section, from a strain gage or Mustran cell reading, the stress and force there can be computed as the strain times the modulus of concrete or steel, depending on the pile, and as the stress times the area, respectively. The load carried by side friction between the surface and any depth is equal to the difference between the total load at the pile head and the normal load at the section of the pile at that depth. The larger the number of sections where gages have been placed, the better the definition of the load distribution along the pile shaft.

Embankments on Soft Ground

Although there are a number of alternatives to the construction of embankments on soft ground (relocation to a site with soils with better characteristics, ground improvement, and others) it is

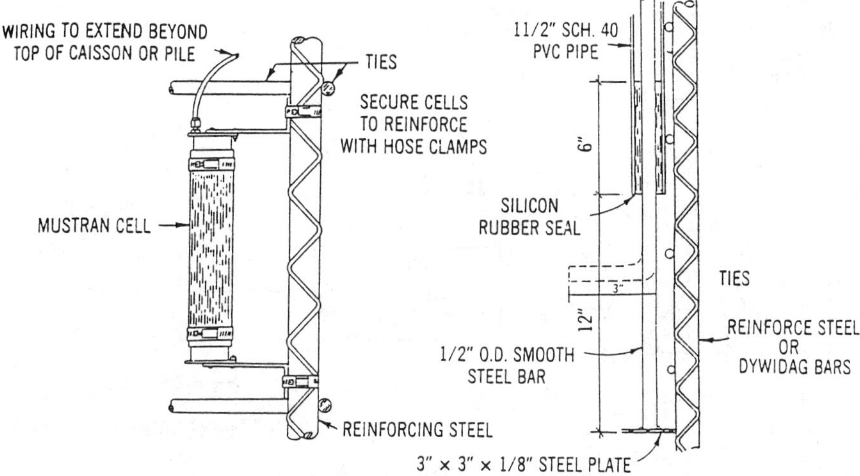

FIGURE 27.18 Attachment of Mustran cell or telltale to rebars in a drilled shaft. (After Crowther, C. L. 1988. *Load Testing of Deep Foundations.* John Wiley & Sons, New York. Reprinted by permission of John Wiley & Sons, Inc.)

frequently necessary to adopt such a solution for technical or economical reasons. When that is the case, the designer faces the following questions:

1. How much will the soft soil compress under the embankment loading?
2. How fast will compression take place?
3. How should construction proceed so as to ensure that there will be no bearing capacity failure of the soft soil?
4. What techniques can be effectively used to accelerate construction and open the structure for use by the public or owner?

A first attempt to answer these questions will be based on assessing the initial soil state (density, stress state, pore pressure) at all the points at the site and using soil mechanics theories to predict how the soil state will change with time, and whether such changes will cause only limited deformation or may lead to failure. Initial state, as previously discussed, is determined using in situ tests and other site investigation techniques, as well as piezometers.

The structure may be sufficiently important or costly for the uncertainties involved in such an approach to warrant the use of instrumentation to help answer the questions outlined above. An instrumentation program for an embankment on soft soil will typically consist of a means to measure the settlement at the base of the embankment (using settlement plates, for example), the pore pressures at different points in the foundation soil (using piezometers), and devices that may indicate the development of a slip surface (such as inclinometers).

Driven Piles

A designer of driven piles is in most cases ultimately interested in determining the capacity of a pile to sustain static loads. Still, there is an obvious relationship (although it is not necessarily easy to establish) between how hard it is to drive a pile into the ground and the static capacity it will ultimately have. Instrumentation is often used to estimate static capacity from measurements made at the pile head during driving.

A pile is driven into the ground by blows from a hammer onto the pile head. Each of these blows propagates down the pile as compressive stress and displacement waves. At each section of the pile the displacement wave mobilizes a soil resistance, which will itself generate a compression

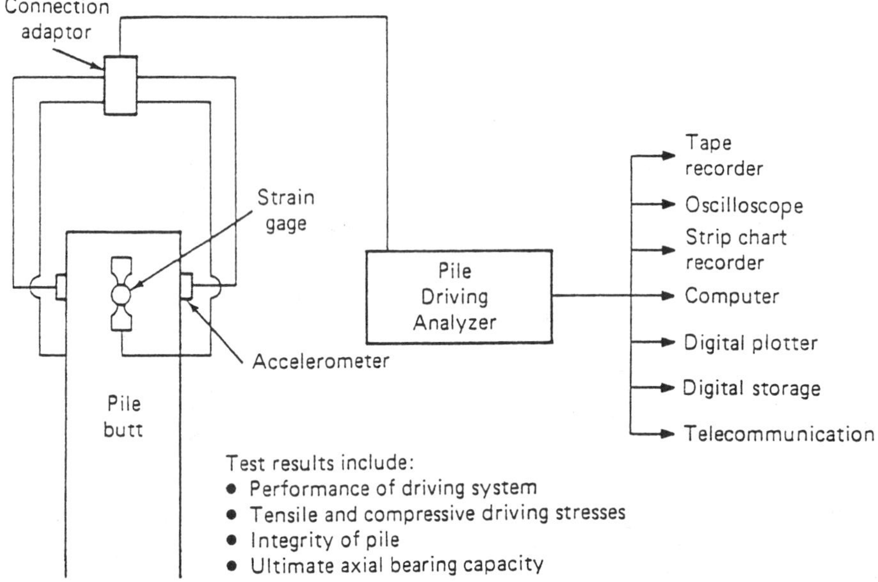

FIGURE 27.19 Setup for static capacity determination based on measurements at the pile head during driving. (After Dunnicliff, J. 1988. *Geotechnical Instrumentation for Monitory Field Performance.* John Wiley & Sons, New York. Reprinted by permission of John Wiley & Sons, Inc.)

wave in the upward direction and a tension wave in the downward direction. When any of these waves reaches the base of the pile, a base resistance may be mobilized, generating a new wave. In addition, any changes in the cross section of the pile (such as one due to damage during driving) will also cause waves to reflect partially or totally. Instrumentation, coupled with complex analyses of wave propagation in the pile and some simplifying assumptions, aims at determining static soil resistances along the shaft and at the base from the dynamic measurements at the pile head.

The measurements made at the pile head are essentially two: strain (from which stress and force can be obtained) and acceleration. An accelerometer and a strain gage are therefore needed at the pile head. A system that uses this scheme is shown in Fig. 27.19. The measurements are combined, in the pile driver analyzer (PDA), with a simplified analysis to give field estimates of static capacity. The same measurements can be combined with a more sophisticated analysis, such as the one contained in the program CAPWAP, to give better estimates of static capacity. PDA and CAPWAP are trademarks of Pile Dynamics, Inc. and Goble Rausche Likins and Associates, respectively, both of Cleveland, Ohio.

Defining Terms

Benchmark: Reference for measurement of vertical displacements using surveying methods.

Bonded foil strain gage: A gage for measure of strain consisting of a thin foil of one of a number of alloys sensitive to straining bonded to a thin plastic film.

Bourdon pressure gage: A gage for measuring hydraulic pressure, consisting of a curved tube that uncoils when pressurized by an amount directly related to the pressure.

Cone penetrometer: A cylindrical, instrumented device with a conical point that permits the measurement of many quantities, the most important of which is the resistance offered by the soil to vertical penetration measured at the conical tip.

Cone pressuremeter: A cone penetrometer with an expandable cylindrical membrane above the conical tip; the membrane can be expanded and measurements taken as if a pressuremeter test were being performed.

CPT: Cone penetration test. An in situ test performed by pushing a cone penetrometer down into the ground and simultaneously measuring several quantities.

Electrical resistance strain gage: A conductor whose electrical resistance changes as it is deformed.

Extensometer: A device that gives the distance between two points on a surface.

Inclinometer: Device used for measuring lateral displacement that runs down inside a cased borehole.

In situ tests: Tests performed in situ with devices that subject the soil to a certain type of loading and measure the soil response (e.g., the cone penetration test and the pressuremeter test).

Instrumentation: The collection of devices and the way they are arranged and linked to measure quantities deemed of interest in a particular geotechnical project.

Instrumented static load tests: When said of a deep foundation, the measurement of displacements or strains at different levels in the deep foundation element as load is applied at the head of the element.

Load cells: Devices used to measure load.

LVDT: The linear variable differential transformer, a device used to measure displacements, consisting of a central magnetic core and two coils.

Manometer: A U-tube filled with two fluids, used to measure the pressure at their interface.

Measuring points: Points where measurements are to be taken using surveying techniques.

Mustran cells: Multiple strain transducers cells, devices consisting of a bar with strain transducers attached, used to obtain strain measurements inside drilled shafts or precast concrete piles.

Piezometer: Device used to measure hydraulic pressure.

Pressure cells: Cells used to measure stresses inside soil masses or at soil-concrete or soil-steel interfaces.

Pressurecone: Cone pressuremeter.

Reference monument: (also **Horizontal control station**) A reference for horizontal displacement measurements.

Seismic cone: Cone penetrometer that permits the determination of a shear wave velocity profile with depth.

Sensor: A device that allows the measurement of a certain quantity at a point.

Settlement plates: Plates that are embedded in a soil mass at a depth where the vertical displacements are to be determined.

Static load test: When said of a deep foundation, the measurement of the displacements at the pile head corresponding to a sequence of loads applied also at the pile head; different rates of load application are possible, as well as different measurements and locations where measurements are taken.

Strain gage: A gage used to measure strain.

Telltale: A steel rod or bar placed inside a concrete pile, but separated from the concrete by a PVC casing that is not in contact with the telltale, whose purpose is to determine the displacement at the concrete section where its tip is located.

Tiltmeter: A device to measure the rotational component of a displacement.

Transducer: A device that transforms one energy type (such as mechanical) into another (such as electrical or pneumatic).

Vibrating wire: When said of a transducer, a principle by which the strain or stress in a tensioned wire can be related to its frequency of vibration.

Weldable strain gage: A bonded foil or other type of strain gage attached to a steel plate that can be welded to a steel structure.

Wheatstone bridge: An arrangement of two known electrical resistances with one variable resistance and one unknown resistance; adjustment of the variable resistance until no current passes between a certain pair of points in the circuit allows the determination of the unknown resistance; useful because many strain transducers are based on the dependence of the resistance of certain materials on deformation.

References

Crowther, C. L. 1988. *Load Testing of Deep Foundations.* John Wiley & Sons, New York.

De Ruiter, J. 1971. Electric penetrometer for site investigation. *J. Soil Mech. Found., ASCE.* 97(SM2): 457–472.

Drnevich, V. P. *Notes on In-Situ Testing and Field Instrumentation* (unpublished).

Dunnicliff, J. 1988. *Geotechnical Instrumentation for Monitory Field Performance.* John Wiley & Sons, New York.

Fretti, C., Lo Presti, D., and Salgado, R. 1992. The research dilatometer: In-situ and calibration test results. *Revista Italiana di Geotecnica,* XXVI(4).

Hanna, T. H. 1985. *Field Instrumentation in Geotechnical Engineering.* Trans Tech Publications, Clausthal-Zellerfeld, Germany.

Hansmire, W. H., Russell, H. A., and Rawnsley, R. P. 1984. *Instrumentation and Evaluation of Slurry Wall Construction, Vol. 1, Interpretation of Field Measurements.* U.S. Department of Transportation, Federal Highway Administration, Rep. No. FHWA/RD-84-053.

Harrell, A. S. and Stokoe II, K. H. 1984. *Integrity Evaluation of Drilled Piers by Stress Waves.* Res. Rep. 257-IF, Center for Transportation Research, University of Texas, Austin.

Haws, E. T., Lippard, D. C., Tabb, R., and Burland, J. B. 1974. Foundation instrumentation for the National Westminster Bank tower. In *Proc. Symp. Field Instrum. Geotech. Eng.* British Geotechnical Society, Butterworths, London, pp. 180–193.

Herceg, E. G. 1972. *Handbook of Measurement and Control.* Schaevitz Engineering, Pennsauken, NJ.

Hirsch, T. J., Coyle, H. M., Lowery, L. L., and Samson, C. H. 1970. Instruments, performance and method of installation. In *Proc. Conf. Design Installation Pile Found. Cell. Struct.* Lehigh University, Bethlehem, PA. Envo Publishing, Bethlehem, PA, pp. 173–190.

Hughes, J. M. O., Wroth, C. P., and Windle, D. 1977. Pressuremeter tests in sands. *Geotechnique.* 27(4):455–477.

Hunt, R. E. 1984. *Geotechnical Engineering Investigation Manual.* McGraw Hill, New York.

Hvorslev, M. J. 1951. *Time Lag and Soil Permeability in Ground-water Observations,* U.S. Army Corps of Engineers, Waterways Experiment Station, Vicksburg, MS, Bull. No. 36.

Hvorslev, M. J. 1976. *The Changeable Interaction Between Soils and Pressure Cells: Tests and Review at the Waterways Experiment Station,* U.S. Army Corps of Engineers, Waterways Experiment Station, Vicksburg, MS, Tech. Rep. No. S-76-7.

Jamiolkowski, M., Ladd, C., Germaine, J., and Lancellotta, R. 1985. New Developments in Field and Laboratory Testing of Soils. *Proc. XI ICSMFE,* San Francisco, Balkema.

Mair, R. J. and Wood, D. M. 1987. *Pressuremeter Testing—Methods and Interpretation.* Butterworths, London.

Manassero, M. 1992. Finite cavity expansion in dilatant soils: Loading analysis, discussion. *Geotechnique.* 42(4):649–654.

Marchetti, S. 1980. In situ test by flat dilatometer. *J. Geotech. Eng., ASCE.* 106(GT3):299–321.

Measurements Group Inc. 1992. *Student Manual for Strain Gage Technology.* Bulletin 309D.

Mitchell, J. K. 1988. New developments in penetration tests and equipment. In *Penetration Testing 1988, ISOPT-1,* Vol. 1, ed. J. De Ruiter, pp. 245–261. A. A. Balkema, Rotterdam.

Ortigão, J. A. R. 1986. *Instrumentation of Embankments on Soft Soils: A Review of Instruments and Techniques* (in Portuguese). Unpublished notes.

Ortigão, J. A. R. and Almeida, M. S. S. 1988. Stability and deformation of embankments on soft clay. In *Handbook of Civil Engineering Practice,* Vol. III, ed. P. N. Chereminisoff, N. P. Chereminisoff, and S. L. Cheng, pp. 267–336. Technomic Publishing, New Jersey.

Poulos, H. G. and Davis, E. H. 1990. *Pile Foundations Analysis and Design.* Robert F. Krieger Publishing, Malabor, FL.

Richart, F. E., Jr., Woods, R. D., and Hall, J. R., Jr. 1970. *Vibrations of Soils and Foundations.* Prentice Hall, New York.

Robertson, P. K. and Campanella, R. G. 1986. *Guidelines for Use, Interpretation, and Application of the CPT and CPTU.* Soil Mechanics Series No. 105, Department of Civil Engineering, University of British Columbia.

Salgado, R. 1993. *Analysis of Penetration Resistance in Sands.* Unpublished doctoral dissertation. University of California, Berkeley.

Salgado, R. and Mitchell, J. K. 1994. Extra-terrestrial soil property determination by CPT. In *Proc. 8th Int. Conf. Int. Assoc. Comput. Methods Adv. Geomechanics*, Vol. 2, ed. H. Siriwardane and M. M. Zaman, pp. 1781–1788, Morgantown, May.

Schmertmann, J. H. 1986. Suggested method for performing the flat dilatometer test. *Geotech. Test. J., ASTM.* 9(2):93–101.

Schnaid, F. and Houlsby, G. 1991. An assessment of chamber size effects in the calibration of in situ tests in sand. *Geotechnique.* 41(3):437–445.

Seed, H. B., Tokimatsu, K., Harder, L. F., and Chung, R. 1985. Influence of SPT procedures in soil liquefaction resistance evaluations. *J. Geotech. Eng., ASCE.* 109(GT3):458–482.

Skempton, A. W. 1986. Standard penetration test procedures and the effects in sands of overburden pressure, relative density, particle size, ageing and overconsolidation. *Geotechnique.* 36(3):425–447.

Yu, H. S. 1990. *Cavity Expansion Theory and its Application to the Analysis of Pressuremeters.* Unpublished doctoral dissertation. University of Oxford.

Yu, H. S. 1994. The state parameter from self-boring pressuremeter tests in sand. *J. Geotech. Eng., ASCE.* In press.

For Further Information

The books by Dunnicliff, Hanna, and Hunt (see references) are excellent references to devices and procedures of field instrumentation. Pile load tests are better addressed in more specific publications, such as Crowther's book. For instrumentation of dynamic or vibratory problems, a good starting point is Richart *et al.* There has been much recent research on the interpretation of in situ tests, so it is best to review recent journal papers if that is the specific interest.

ASTM has standards for the use of many of the devices discussed in this chapter. A volume is published every year with standards for geotechnical engineering, including the following: ASTM D 1143-81 (reapproved 1987). *Standard Method for Piles under Static Compressive Load;* ASTM D 4719-87. *Standard Method for Pressuremeter Testing in Soils;* and ASTM D 3441-86. *Standard Method for Deep, Quasi-Static, Cone and Friction-Cone Penetration Tests of Soils.*

Geotechnical engineering journals are a good source for new improvements in the instruments and techniques, as well as for case histories where instrumentation has been used. Journals in English include *Journal of Geotechnical Engineering* of the ASCE, *Geotechnique* of the Institute of Civil Engineers of Great Britain, *Canadian Geotechnical Journal, Journal of Soil Testing* of the ASTM, and *Soils and Foundations* of Japan. Some foreign journals publish papers both in the native language of the country where the journal is published and in English. *Rivista Italiana di Geotecnica* of Italy and *Solos e Rochas* of Brazil often carry interesting papers on in situ tests and field instrumentation.

Catalogs and other publicity material of companies manufacturing instruments, as well as reference manuals for the instruments, are a good source of information, both general and specific. Some companies manufacturing instrumentation devices include the Slope Indicator Company (Sinco) of Seattle, Washington, Geokon of Lebanon, New Hampshire, Geotest of Evanston, Illinois, Omega of Stamford, Connecticut, and Dresser Industries of Newtown, Connecticut.

Also of interest are periodicals such as *Experimental Stress Analysis Notebook of the Measurements Group, Inc.*, of Raleigh, North Carolina. They carry current information on transducers and other devices, as well as articles on the general subject of instrumentation and materials testing.

The Hoover Dam, straddling the mighty Colorado River between Nevada and Arizona, was built to harness the river and bring life to the vast desert regions of the American Southwest. Designed and administered by the U.S. Bureau of Reclamation, this project took several thousand workers five years to build in hazardous conditions and scorching heat. They overcame unprecedented engineering challenges to finish more than two years ahead of schedule in 1935.

Built in the steep and narrow Black Canyon, the structure is an ingenious hybrid of an arch dam and a gravity dam. It can withstand the crushing force of the river by arch action and through its massive weight. The dam complex contains enough concrete to pave a two-lane highway from San Francisco to New York. This would have required more than a century to cool after setting, were it not for an innovative cooling process that circulated cold water through pipes embedded in successive blocks of concrete.

Hoover Dam has bestowed abundant benefits for millions of people on its surrounding regions: drinking water for more than 15 million people; irrigation of 1.5 million acres of farm land; enough energy to serve 500,000 homes; flood control; and recreation through 110-mile-long Lake Mead, the nation's largest artificial reservoir. Tourists also benefit, for few sights can equal the grandeur of the 726-foot-tall monolith flanked by the dramatic canyon walls and backed by the sparkling clear-blue waters of Lake Mead. (Photo courtesy of the American Society of Civil Engineers.)

IV

Hydraulic Engineering

Jacques W. Delleur
Purdue University

GLOBAL FRESHWATER RESOURCES are of the order of one million cubic miles. Most of this water is in groundwater, less than one-half mile deep within the earth. Only 30,300 cubic miles of this resource reside in freshwater lakes and streams. However, all of this water is in a continuous movement known as the "hydrologic cycle." The dynamic nature of this movement is quite variable. The response time of urban runoff is of the order of minutes to hours, the average residence time of atmospheric moisture is only a little over nine days, while the global average residence time of freshwater in streams is about 10 days, and that of groundwater varies between 2 weeks to 10,000 years.

Because of the diversity of the amounts of water involved, the variability of the response times and the myriad of water uses, civil engineers must deal with a multitude of physical and management water problems. Some of these problems are concerned with water supply for cities, industries, and agriculture; drainage of urban areas; and the collection of used water. Other problems deal with flows in rivers, channels, and estuaries and flood protection; others are concerned with oceans and lakes, hydropower generation, or water transportation. Although the emphasis of this section is on water quantity, many problems are also concerned with water quality.

Because of the multitude of different types of problems, hydraulic engineering is subdivided into a number of specialties, many of which are the object of separate chapters in this section. These specialties all have fluid mechanics as a common basis. Since the concern is with water, there is little interest in gases; the fundamental science is hydraulics, the science of motion of incompressible fluids.

Chapter 28, "Fundamentals of Hydraulics," deals with properties of fluids, hydrostatics, kinematics, and dynamics of liquids. A separate chapter, Chapter 29, is devoted to open channel hydraulics because of the importance of free surface flows in civil engineering applications. Erosion, deposition, and transport of sediments are important in the design of stable channels and stable structures. Sediment resuspension has important implications for water quality. These problems are treated in Chapter 33, "Sediment Transport in Open Channels."

The flow of water in the natural environment, such as rainfall and subsequent infiltration, evaporation, and flow in rills and streams, is the purview of Chapter 30, "Surface Water Hydrology." Because of the uncertainty of natural events the analysis of hydrologic data requires the use of statistics such as frequency analysis. Surface water runoff in the urban environment is discussed separately in Chapter 31, "Urban Drainage," because of the short time of concentration, the specific techniques, the possible need for detention basins, and the potential for urban runoff quality problems. Hydrology is generally separated into surface water hydrology and subsurface hydrology depending upon whether the emphasis is on surface water or on groundwater. Chapter 32, "Groundwater Engineering," is concerned with hydraulics of wells, land subsidence due to excessive pumping, contaminant transport, site remediation, and landfills.

Many hydraulic structures (Chapter 35) have been developed for the storage, conveyance, and control of natural flows. These structures include dams, spillways, pipes, open channels, outlet works, energy-dissipating structures, turbines, and pumps. The interface between land and ocean and lakes is part of Chapter 34, "Coastal Engineering." Chapter 34 contains a discussion of the mechanics of ocean waves and their transformation in shallow water and resulting coastal circulation. It also includes a discussion of coastal processes and their influence on coastal structures.

Various forms of software and software packages are available to facilitate design and analysis. Many of these were originally developed by government agencies and later improved by private companies, which add preprocessors and postprocessors that greatly facilitate the input of data and the plotting of output results. Models such as SWMM and the HEC series have well-defined graphical and expert system interfaces associated with model building, calibration, and presentation of results. Several of the public domain packages are listed in Chapter 36, "Simulation in Hydraulics and Hydrology," and detailed examples of application to urban drainage are also given.

It is not sufficient for civil engineers to deal only with the physical problems associated with water resources. They are also concerned with planning and management of these resources. Conceptualization and implementation of strategies for delivering water of sufficient quantity and quality to meet societal needs in a cost-effective manner are dealt with in Chapter 37, "Water Resources Planning and Management."

Hydraulic engineering is a rapidly changing field. The electronic encapsulation of knowledge and information in the form of software, databases, expert systems, and geographical information systems has produced a "Copernican revolution in hydraulics" [Abbott, 1994]. The Europeans have coined the term *hydroinformatics* to designate the association of computational hydraulic modeling and information systems. Thus hydraulic and hydrologic models become part of larger computer-based systems that generate information for the different interests of the water resources managers. HYDRONAUT [Swartzenbroekx, 1993] is an example of a linkage of the Wallingford storm sewer design and analysis package [WALLRUS, Price, 1981] with databases, geographical information systems, and computer-aided design systems.

The 1993 floods in the Mississippi and Missouri basins indicated the importance of hydrologic forecasting, the design of flood protection structures, and water management at the basin scale while taking into account the environmental, ecological, and economic impacts. According to Starosolski [1991], the old approach of first designing the engineering project and then considering the ecological effects should be replaced by a systems approach in which the hydraulic, environmental, and ecological aspects are all included in the planning, execution, and operation of water resources projects.

Progress in a selected topic of hydraulic engineering is presented in the Hunter Rouse lecture at the annual meeting of the Water Resources Engineering Division of the American Society of Civil Engineers and is subsequently published in the *Journal of Water Resources Engineering* (formerly *Journal of Hydraulic Engineering*). The proceedings of the congresses of the International Association of Hydraulic Research, held every four years, summarize the advances in the field, primarily in the keynote papers.

Both in surface water and groundwater hydrology recent researchers have been inspired by the improved ability to observe and model the many heterogeneities of surface and material properties and of transport processes [van Genuchten, 1991]. Remote sensing now makes it possible to model land surface hydrologic processes at the global scale. Then these processes can be included in general circulation models of the atmosphere [Wood, 1991]. The research accomplishments in surface water and groundwater hydrology are summarized every four years in reports from several countries to the International Union of Geodesy and Geophysics. The U.S. quadrennial report is prepared by the American Geophysical Union and is published in *Reviews of Geophysics.*

Similarly, in coastal engineering recent mathematical models make it possible to simulate large-scale coastal behavior, that is, scales larger than tens of kilometers and time scales of decades. These models include waves, currents, and sediment transport. Models capable of describing the interaction with the bottom topography are under development for short-term coastal behavior [De Vriend, 1991].

References

Abbott, M. B. 1994. Hydroinformatics: A Copernican revolution in hydraulics. *Journal of Hydraulic Research,* Vol. 32, pp. 1–13, and other papers in this special issue.

De Vriend, H. B. 1991. Mathematical modelling and large-scale coastal behavior. *Journal of Hydraulic Research,* Vol. 29, pp. 727–755.

Price, R. K. 1981. Wallingford storm sewer design and analysis package. In *Proc. 2nd. Int. Conf. Urban Storm Drainage,* Urbana-Champaign, Illinois.

Starosolski, O. 1991. Hydraulics and the environmental partnership in sustainable development. *Journal of Hydraulic Research,* Vol. 29, pp. 5–7, and other papers in this extra issue.

Swartzenbroekx, P. 1993. HYDRONAUT: A new, standardized tool towards the global hydraulic design of large sewerage systems in Flanders. In *Proc. 6th Int. Conf. Urban Storm Drainage*, Niagara Falls, Ontario, Canada, vol. II, pp. 1478–1483.

van Genuchten, M. T. 1991. Progress and opportunities in hydrologic research, 1987–1990. *Review of Geophysics*, Supplement, April 1991, pp. 189–192. U.S. National Report to International Union of Geodesy and Geophysics, 1987–1990.

Wood, E. F. 1991. Global scale hydrology: Advances in land surface modelling. *Review of Geophysics*, Supplement, pp. 193–201. U.S. National Report to International Union of Geodesy and Geophysics, 1987–1990.

28

Fundamentals of Hydraulics

D. A. Lyn
Purdue University

Engineering hydraulics is concerned broadly with civil engineering problems in which the flow or management of fluids, primarily water, plays a role. Solutions to this wide range of problems require an understanding of the fundamental principles of fluid mechanics in general and hydraulics in particular, and these are summarized in this chapter.

28.1 Properties of Fluids

The material properties of a fluid, which may vary, sometimes sensitively, with temperature, pressure, and composition (if the fluid is a mixture), determine its mechanical behavior. The physical properties of water and other common fluids are tabulated in Tables 28.1 through 28.8 in Section 28.9.

0-8493-8953-4/95/$0.00 + $.50

Fluid Density and Related Quantities

The density of a fluid, denoted as ρ, is defined as its mass per unit volume, with units kg/m^3 or slug/ft^3. The specific gravity, denoted as s, refers to the dimensionless ratio of the density of a given material to the density of pure water, ρ_{ref}, at a reference temperature and pressure (often taken to be 4°C and 1 standard atmosphere): $s = \rho/\rho_{\text{ref}}$. The specific weight, denoted as γ, is the weight per unit volume, or the product of ρ and g, the gravitational acceleration: $\gamma = \rho g$.

Fluid Viscosity and Related Concepts

Fluids are *Newtonian* if their strain rate is linearly proportional to the applied shear stress and is zero when the latter is zero. The strain rate can be related to gradients of fluid velocity. The proportionality constant is the *dynamic viscosity*, denoted by μ, with units N s/m^2 or lb s/ft^2. Frequently μ arises in combination with ρ, so that a *kinematic viscosity*, defined as $\nu \equiv \mu/\rho$, with units m^2/s or ft^2/s, is defined. Some materials, e.g., mud, may, under certain circumstances, behave as fluids but may not exhibit such a relationship between shear stress and the strain rate. These are termed *non-Newtonian fluids*. The most common fluids, air and water, are Newtonian. The concept of an ideal fluid, for which $\mu = 0$, is useful; effects of fluid friction (shear) are thereby neglected. This approximation is not valid in the vicinity of a solid boundary, where the *no-slip* condition must be satisfied; that is, at the fluid-solid interface the velocity of the fluid must be equal to the velocity of the solid surface.

Vapor Pressure

The vapor pressure of a pure liquid, denoted as p_v, with units Pa abs or lb/ft^2 abs ("abs" refers to the absolute pressure scale, see Section 28.2), is the pressure exerted by its vapor in a state of vapor-liquid equilibrium at a given temperature. It derives its importance from the phenomenon of *cavitation*, the term applied to the genesis, growth, and eventual collapse of vapor bubbles (cavities) in the interior of a flowing liquid when the fluid pressure at some point in the flow is reduced below the vapor pressure. This leads to reduced performance and possibly damage to pumping and piping systems and spillways.

Bulk Modulus and Speed of Sound

The bulk modulus of elasticity, denoted by E_v, is the ratio of the change in pressure, p (see Section 28.2), to the relative change in density: $E_v \equiv \rho(\partial p/\partial \rho)$, with units Pa or lb/ft^2, and measures the compressibility of a material. Common fluids such as water and air may generally be treated as *incompressible*; that is, $E_v \to \infty$. Compressibility effects become important, however, where large changes in pressure occur suddenly (as in *waterhammer* problems resulting from fast-closing valves) or where high speeds are involved (as in supersonic flow). The kinematic quantity, the *speed of sound*, defined by $c \equiv \sqrt{E_v/\rho}$, may be more convenient than E_v to use.

Surface Tension and Capillary Effects

The interface between immiscible fluids acts like an infinitely thin membrane that supports a tensile force. The magnitude of this force per unit length of a line on this surface is termed the *coefficient of surface tension*, denoted as σ, with units N/m or lb/ft. Effects of surface tension are usually important only for problems involving highly curved interfaces and small length scales (e.g., in small-scale laboratory models). *Capillary effects* refer to the rise or depression of fluid in small-diameter tubes or in porous media due to surface tension. The value of σ varies with the fluids forming the interface and is affected by chemically active agents (surfactants) at the interface.

28.2 Fluid Pressure and Hydrostatics

Pressure in a fluid is a normal compressive stress. It is considered positive even though compressive, and its magnitude does not depend on the orientation of the surface on which it acts. In most practical applications, pressure differences rather than absolute pressures are important, since only the former induce net forces and drive flows. Absolute pressures are relevant, however, in problems involving cavitation, since the vapor pressure (see Section 28.1) must be considered. A convenient pressure scale, the *gage pressure* scale, can be defined with zero corresponding to atmospheric pressure, p_{atm}, rather than an absolute pressure scale with zero corresponding to the pressure in an ideal vacuum. The two scales are related by $p_{gage} = p_{abs} - p_{atm}$. At the free surface of a water body (Fig. 28.1) that is exposed to the atmosphere, $p_{gage} = 0$, since $p_{abs} = p_{atm}$. Unless otherwise specified, the gage pressure scale is always used in the following.

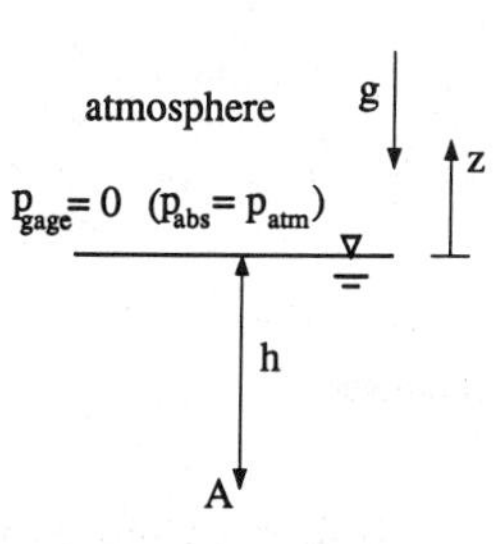

FIGURE 28.1 Pressure scales and coordinate system for Eq. (28.1).

The Hydrostatic Equation

Hydrostatics is concerned with fluids that are stagnant or in uniform motion. Relative motion (shear forces) and acceleration (inertial forces) are excluded; only pressure and gravitational forces are assumed to act. A balance of these forces yields

$$\frac{dp}{dz} = -\rho(z)g = -\gamma(z) \tag{28.1}$$

which describes the rate of change of p with elevation z, and where the chosen coordinate system is such that gravity acts in a direction opposite to that of increasing z (Fig. 28.1). In Eq. (28.1), ρ or γ may vary in the z direction, as is often the case in the thermally stratified atmosphere or lake. Where ρ or γ is constant over a certain region, Eq. (28.1) may be integrated to give

$$p(z) - p(z_0) = -\gamma(z - z_0). \tag{28.2}$$

Only differences in pressure and differences in elevation are of consequence, so that a pressure datum or elevation datum can be arbitrarily chosen. In Fig. 28.1, if the elevation datum is chosen to coincide with the free surface, and the gage pressure scale is used, then the pressure at a point A located at a depth h below the free surface (at $z = -h$) is given by $p_A = \gamma h$. If $\gamma \approx 0$, as in the case of gases, and elevation differences are not large, then $p \approx$ constant. Eq. (28.2) can also be rearranged to give

$$\frac{p(z)}{\gamma} + z = \frac{p(z_0)}{\gamma} + z_0 \tag{28.3}$$

The *piezometric* **head**, with dimensions of length, is defined as the sum of the *pressure head*, p/γ, and the *elevation head*, z. Equation (28.3) states that the piezometric head is constant everywhere in a constant-density fluid where hydrostatic conditions prevail.

Forces on Plane Surfaces

The force on a surface S due to hydrostatic pressure is obtained by integration over the surface. If the surface is plane, a single resultant point force can be found that is equivalent to the distributed pressure load (see Fig. 28.2). For the case of constant-density fluid, the magnitude of this resultant force F may be determined from

$$F = \int_S p\,dS = p_c A \tag{28.4}$$

where $p_c = p_0 + \gamma h_c$ is the pressure at the centroid of the surface, situated at a depth h_c, and A is the area of the surface. Since $p_c = F/A$, p_c is interpreted as the average pressure on the surface.

This resultant point force acts compressively, normal to the surface, through a point termed the *center of pressure.* If the surface is inclined at an angle θ to the horizontal, the coordinates of the center of pressure (x_{cp}, y_{cp}) in a coordinate system in the plane of the surface with origin at the centroid of the surface are

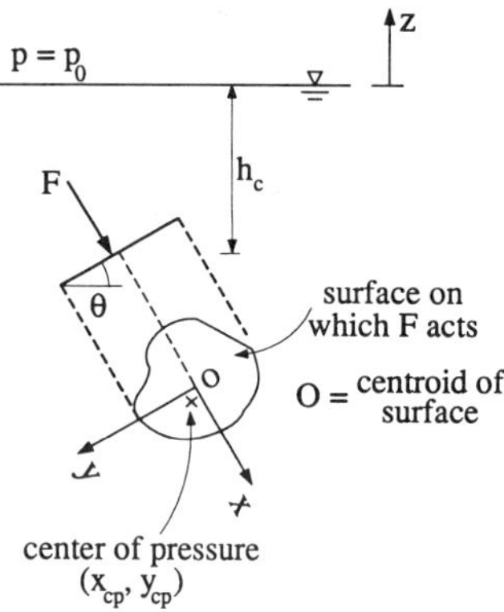

FIGURE 28.2 Hydrostatic force on a plane surface.

$$(x_{cp}, y_{cp}) = \left[\frac{(\gamma \sin\theta) I_{xy}}{F}, \frac{(\gamma \sin\theta) I_{xx}}{F}\right] \tag{28.5}$$

where I_{xx} is the area moment of inertia; I_{xy} the product of inertia of the plane surface, both with respect to the centroid of the surface; and y is positive in the direction below the centroid. The properties of common plane surfaces, such as centroids and moments of inertia, are discussed in the section on mechanics of materials. The surface is often symmetrically loaded, so that $I_{xy} = 0$ and hence, $x_{cp} = 0$; in other words, the center of pressure is located directly below the centroid on the line of symmetry. If the surface is horizontal, the center of pressure coincides with the centroid. Further, as the surface becomes more deeply submerged, the center of pressure approaches the centroid, $(x_{cp}, y_{cp}) \rightarrow (0, 0)$, because the numerators of Eq. (28.5) remain constant while the denominator increases (p_c increases). For plane surfaces that can be decomposed into simpler elementary surfaces, the magnitude of the resultant force can be computed as the vector sum of the forces on the elementary surfaces. The coordinates of the center of pressure of the entire surface are then determined by requiring a balance of moments.

Forces on Curved Surfaces

For general curved surfaces it may no longer be possible to determine a single resultant force equivalent to the hydrostatic load. Three mutually orthogonal forces equivalent to the hydrostatic load can, however, be found. The horizontal forces are treated differently from the vertical force. The horizontal forces acting on the plane projected surfaces are equal in magnitude and have the same line of action as the horizontal forces acting on the curved surface. For the curved surface ABC in Fig. 28.3, the plane projected surface is represented as $A'C'$. The results of the previous section can thus be applied to find the horizontal forces on $A'C'$, and hence on ABC.

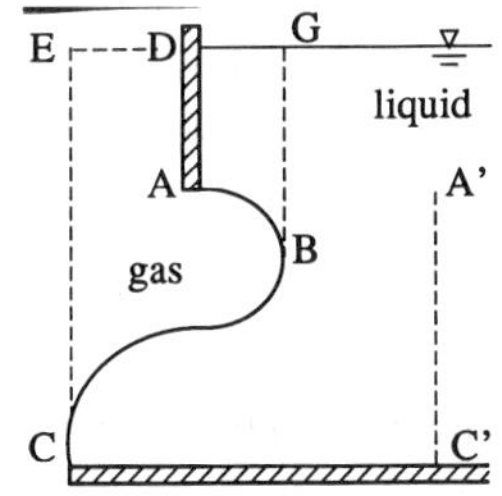

FIGURE 28.3 Hydrostatic forces on a curved surface.

A systematic procedure to deal with the vertical forces distinguishes between those surfaces that are exposed to the hydrostatic load from above, like the surface AB in Fig. 28.3, and those surfaces exposed to a hydrostatic load from below, like the surface BC. The vertical force on each of these subsurfaces is equal in magnitude to the weight of the volume of (possibly imaginary) fluid lying above the curved surface to a level where the pressure is zero, usually to a water surface level. It acts through the center of gravity of that fluid volume. The vertical force acting on *AB* equals in magnitude the weight of fluid in the volume, *ABGDA*, while the vertical force acting on *BC* equals in magnitude the weight of the imaginary fluid in the volume *BGECB*. If the load acts from above, as on *AB*, the direction of the force is downward; if the load acts from below, as on *BC*, the direction of the force is upward. The net vertical force on a surface is the algebraic sum of upward and downward components. If the net vertical force is upward, it is often termed the *buoyant force.* The line of action is again found by a balance of moments.

In the special case of a curved surface that is a segment of a circle or a sphere, a single resultant force can be obtained, because hydrostatic pressure acts *normal* to the surface, and all normals intersect at the center. The magnitudes and direction of the components in the vertical and horizontal directions can be determined according to the procedure outlined in the previous paragraph, but these components must act through the center. The analysis for curved surfaces can also be applied to plane surfaces. In some problems, it may even be simpler to deal with horizontal and vertical components, rather than the seemingly more direct formulas of the previous section.

Example 28.1: Force on a Vertical Dam Face. What are the magnitude and direction of the force on the vertical rectangular dam (Fig. 28.4), of height H and width W, due to hydrostatic loads, and where is the center of pressure?

Solution. Equation (28.4) is applied and, using Eq. (28.2), $p_c = \gamma h_c$, where $h_c = H/2$ is the depth at which the centroid of the dam is located. Since the area is $H \times W$, the magnitude of the force $F = \gamma W H^2/2$. The center of pressure is found from Eq. (28.5), using $\theta = 90°$, $I_{xx} = WH^3/12$, and $I_{xy} = 0$. Thus, $x_{cp} = 0$, and $y_{cp} = [\gamma(1)(WH^3/12)]/[\gamma W H^2/2] = H/6$. The center of pressure is located at a distance $H/6$ directly below the centroid of the dam, or a distance of $2H/3$ below the water surface. The direction of the force is normal and compressive to the dam face, as shown.

Example 28.2: Force on an Arch Dam. Consider a constant-radius arch dam with a vertical upstream face (Fig. 28.5). What is the net horizontal force acting on the dam face due to hydrostatic forces?

Solution. The projected area is $A_p = (2R \sin\theta) \times H$, while the pressure at the centroid of the projected surface is $p_c = \gamma H/2$. The magnitude of the horizontal force is thus $\gamma R H^2 \sin\theta$, and the center of pressure lies (as in Example 1) $2H/3$ below the water surface on the line of symmetry of the dam face. Because the face is assumed vertical, the vertical force on the dam is zero.

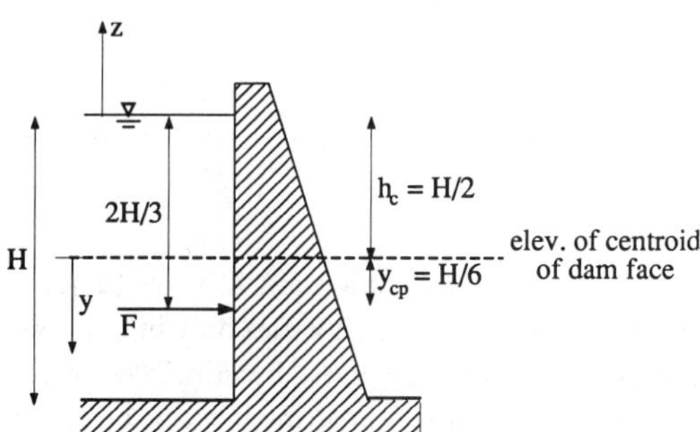

FIGURE 28.4 Hydrostatic forces on a vertical dam face.

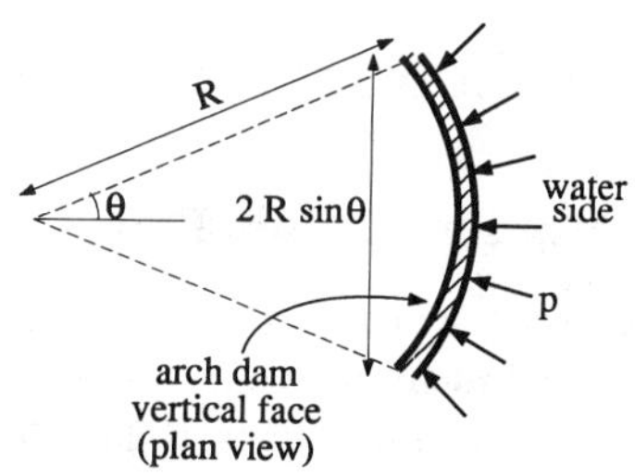

FIGURE 28.5 Hydrostatic forces on the face of a vertical arch dam.

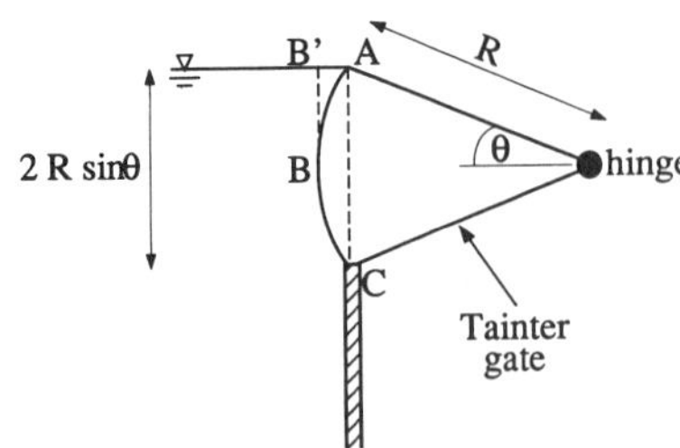

FIGURE 28.6 Hydrostatic forces on a radial (Tainter) gate.

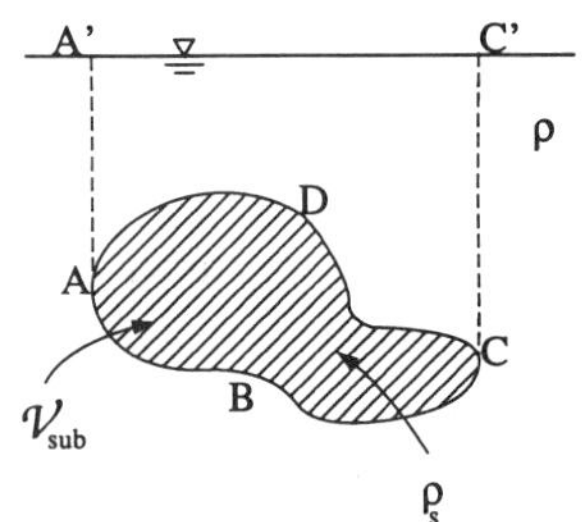

FIGURE 28.7 Hydrostatic forces on a general submerged body.

Example 28.3: Force on a Tainter Gate. Consider a radial (Tainter) gate of radius R with angle θ and width W (Fig. 28.6). What are the horizontal and vertical forces acting on the gate?

Solution. For the horizontal force computation, the area of the projected surface is $A_p = (2R\sin\theta)W$, and the pressure at the centroid of the projected surface is $p_c = \gamma(2R\sin\theta)/2$, since the centroid is located at a depth of $(2R\sin\theta)/2$. Hence, the magnitude of the horizontal force is $2\gamma(R\sin\theta)^2 W$, and it acts at a distance $2(2R\sin\theta)/3$ below the water surface. The vertical force is the sum of a downward force, equal in magnitude to the weight of the fluid in the volume $AB'BA$, and an upward force, equal in magnitude to the weight of the fluid in the volume $AB'BCA$. This equals in magnitude the weight of the fluid in the volume $ABCA$, namely, $\gamma WR^2(\theta - \sin\theta\cos\theta)$, and acts upwards through the center of gravity of this volume. Alternatively, because the gate is an arc of a circle, the horizontal and vertical forces act through the center O, and no net moment is created by the fluid forces.

Example 28.4: Archimedes' Law of Buoyancy. Consider an arbitrarily shaped body of density ρ_s, submerged in a fluid of density ρ (Fig. 28.7). What is the net vertical force on the body due to hydrostatic forces?

Solution. Note that the net horizontal force is necessarily zero, since the horizontal forces on projected surfaces are equal and opposite. The surface in contact with the fluid, and therefore exposed to the hydrostatic load, is divided into an upper surface, marked by curve ADC, and a lower surface, marked by curve ABC. The vertical force on ADC equals in magnitude the weight of the fluid in the volume $ADCC'A'A$ and acts downward, since the vertical component of the hydrostatic forces on this surface acts downward. The vertical force on ABC equals in magnitude the weight of the fluid in the volume $ABCC'A'A$, but this acts upward, since the vertical component of the hydrostatic forces on this surface acts upward. The vector sum F_b of these vertical forces acts upward through the center of gravity of the submerged volume and equals in magnitude the weight of the fluid volume displaced by the body, $F_b = \rho g \mathcal{V}_{sub}$, where $\mathcal{V}_{sub}$ is the submerged volume of the body. This result is known as *Archimedes' principle*. The effective weight of the body in the fluid is $W_{eff} = g(\rho_s - \rho)\mathcal{V}_{sub}$.

28.3 Fluids in Nonuniform Motion

Description of Fluid Flow

A flow is described by the velocity vector $\mathbf{u}(x, y, z, t) = (u, v, w)$ at a point in space (x, y, z) and at a given instant in time t. It is one-, two-, or three-dimensional if it varies only in one, two, or three coordinate directions. If flow characteristics do not vary in a given direction, flow is said to be *uniform* in that direction. If a flow does not vary with time, it is termed *steady*; otherwise, it is *unsteady*. Although a flow may strictly speaking be unsteady and three-dimensional, it may for some practical purposes be approximated as a steady one-dimensional flow.

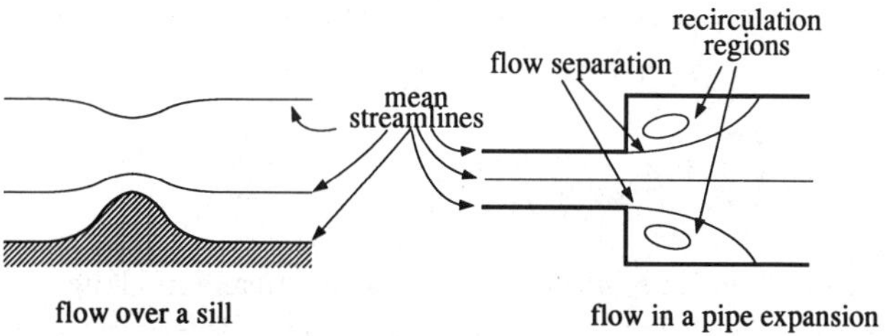

FIGURE 28.8 Examples of mean streamlines and flow separation.

A **streamline** is a line to which the velocity vectors are tangent; thus, the streamline shows the instantaneous direction of flow at each of its points (see Fig. 28.8 for examples of streamlines). As such, there can be no flow across streamlines; flow boundaries, such as a solid impervious boundary or the free water surface, must therefore coincide with streamlines (actually stream surfaces). If an unsteady flow is approximated as being steady, the concept of time-averaged streamlines, corresponding to the time-averaged velocity field, is useful. A collection of streamlines can be used to give a picture of the overall flow pattern. Streamlines should, however, be distinguished from *streaklines*, which are formed when dye or particles are injected at a point into the flow, as is often done in flow visualization. Streaklines and streamlines coincide in a strictly steady flow but differ in an unsteady flow.

Flow Classification

Broad classes of qualitatively similar flow phenomena may be distinguished. *Laminar* flows are characterized by gradual and regular variations over time and space, with little mixing occurring between individual fluid elements. In contrast, **turbulent** flows are unsteady, with rapid and seemingly random instantaneous variations in flow variables, such as velocity or pressure, over time and space. A high degree of bulk mixing accompanies these fluctuations, with implications for transport of momentum (fluid friction) and contaminants. In dealing with predominantly turbulent flows in practice, the hydraulic engineer is concerned primarily with time-averaged characteristics.

Flow **separation** occurs when a streamline begins at a solid boundary (at the separation point) and enters the region of flow (see Fig. 28.8). This may happen when the solid boundary diverges sufficiently sharply in the streamwise direction, and it is associated with downstream *recirculating regions* (regions with closed streamlines). The latter are sometimes termed *dead-water zones*, and their presence may have important implications for mixing efficiency. In flows around bodies, flow separation creates a low-pressure region immediately downstream of the body, termed the *wake*. The difference in pressure upstream and downstream of the body may therefore contribute significantly to flow resistance (see Section 28.7).

The Bernoulli Theorem

In the case of an ideal constant-density fluid moving in a gravitational field, a balance along a streamline among inertial (acceleration), pressure, and gravitational forces yields the *Bernoulli theorem*. This states that, on a streamline,

$$\int_s \frac{\partial u_s}{\partial t}\, ds + \left(\frac{p}{\rho} + gz + \frac{u_s^2}{2}\right) = \text{constant} \tag{28.6}$$

where the integration is performed along the streamline, and u_s is the *magnitude* of the velocity at any point on the streamline. Equation (28.6) applies on a given streamline, and the constant will in general vary for different streamlines. The Bernoulli equation for steady flows ($\partial u_s/\partial t = 0$) is usually expressed as

$$\left(\frac{p}{\gamma} + z + \frac{u_s^2}{2g}\right)_A = \left(\frac{p}{\gamma} + z + \frac{u_s^2}{2g}\right)_B \tag{28.7}$$

where the subscripts A and B indicate two points on the same streamline along which the Bernoulli theorem is applied (see Fig. 28.9). The quantity $p/\gamma + z + u_s^2/2g$ is termed the *Bernoulli constant* and is the sum of the piezometric head and the *velocity head* $u_s^2/2g$. For steady uniform flows, $(u_s^2/2g)_B = (u_s^2/2g)_A$, and Eq. (28.7) reduces to the equality of piezometric heads, as found before for hydrostatic conditions. The Bernoulli theorem thus generalizes the hydrostatic result to account for the nonuniformity or acceleration of the flow field by including a possibly changing velocity head. Although obtained from a momentum balance, Eq. (28.7) is often interpreted in terms of the conservation of mechanical energy. The terms gz and $u_s^2/2$ are identified as the gravitational potential and the kinetic energy of the fluid per unit mass. The pressure-work or flow-work term, p/ρ, is interpreted as the work per unit mass that could be done by the fluid at that pressure.

The Bernoulli theorem provides information about variations along the streamline direction. In the direction normal to the streamline, a similar force balance shows that the piezometric head is constant in the direction normal to the streamline, provided the flow is steady and parallel or nearly parallel. In other words, hydrostatic conditions prevail at a flow cross section in a direction normal to the nearly parallel streamlines. This is implicitly assumed in much of hydraulics.

Example 28.5: Orifice Flow. An *orifice* is a closed-contour opening in a wall of a tank or in a plate at a pipe cross section. A simple example of flow through an orifice from a large tank discharging into the atmosphere is shown in Fig. 28.10. What is the exit velocity at (or near) the orifice?

Solution. The Bernoulli theorem is applied on a streamline between a point A on the free surface and a point B at the *vena contracta* of the orifice—the section at which the jet area is a minimum. At the vena contracta the streamlines are straight, such that hydrostatic conditions prevail, and the pressure in the fluid must be the same throughout that section. Since the pressure on the surface of the jet is zero, the pressure $p_B = 0$. Since the tank is open to the atmosphere, $p_A = 0$, and since the tank is large, the velocity head in the tank, $(u_s^2/2g)_A$, is negligible. Hence, $(u_s^2/2g)_B = z_A - z_B = H$, where H is the elevation difference between the tank free surface and the vena contracta. The exit velocity at the vena contracta is given by $u_s = \sqrt{2gH}$, a result also known as *Torricelli's theorem*. The related result that the discharge (see the definition in Section 28.4), $Q \propto A\sqrt{2gH}$, where A is the area of the orifice, arises also in discussions of culverts flowing full and of spillways.

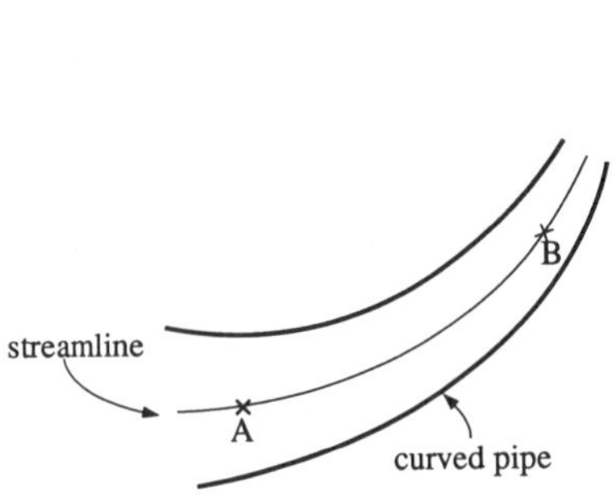

FIGURE 28.9 Flow streamline along which the Bernoulli theorem is applied.

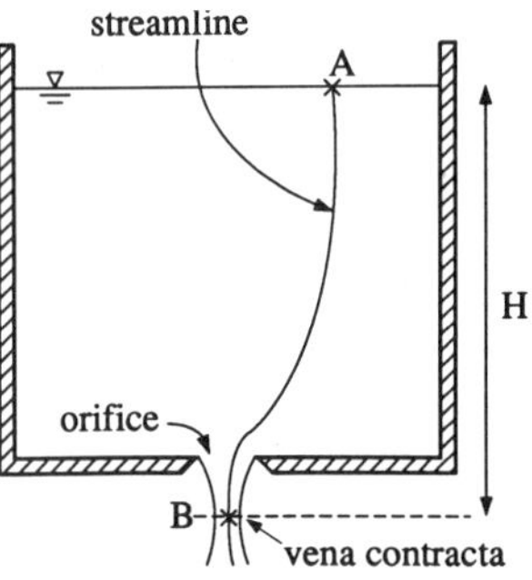

FIGURE 28.10 Flow through an orifice in the bottom of a large tank.

28.4 Fundamental Conservation Laws

The analysis of flow problems is based on three main conservation laws, namely, the conservations of mass, momentum, and energy. For most hydraulic problems it suffices to formulate these laws in integral form for one-dimensional flows, to which the following discussion is restricted. A systematic approach is based on the analysis of a *control volume*, which is an imaginary volume bounded by *control surfaces* through which mass, momentum, and energy may pass. The control volume plays the role of the free body in statics as a device to organize and systematize the accounting of fluxes and forces. In the following, the fluid is assumed to be of constant density ρ, and the control volume is assumed fixed in space and nondeforming.

Fluxes

The *flux* of a quantity, such as mass or momentum, across a control surface S is defined as the amount that is transported across S per unit time. The mass flux $\dot{m}$ is defined as

$$\dot{m} = \int_S \rho u \, dS = \rho Q = \rho V A \tag{28.8}$$

where the velocity u may be nonuniform over the control surface; the discharge or volume flow rate is $Q = \int_S u \, dS$; and V is the average velocity over the area A. The momentum flux can similarly be expressed as

$$\dot{m}\beta \mathbf{V} = \int_S \rho \mathbf{u} \, u \, dS \tag{28.9}$$

The *momentum correction factor* β accounts for nonuniformity of the velocity profile across S and is defined by Eq. (28.9). If u is uniform over S, then $\beta = 1$. The momentum flux, unlike the mass flux, is a vector quantity and is therefore associated with a direction as well as a magnitude. The mechanical energy flux can be expressed as

$$\dot{m}gH \equiv \dot{m}g\left(\frac{p}{\gamma} + z + \alpha\frac{V^2}{2g}\right) = \int_S \gamma\left(\frac{p}{\gamma} + z + \frac{u^2}{2g}\right) u \, dS \tag{28.10}$$

where $H = p/\gamma + z + \alpha V^2/2g$ is called the *total head* (if $\alpha = 1$, H is essentially the Bernoulli constant). Hydrostatic conditions across the control surface have been assumed in obtaining Eq. (28.10). The *kinetic energy correction factor* α accounts for nonuniformities in the velocity profile across the control surface and is defined as

$$\alpha \equiv \frac{1}{A}\int_S \left(\frac{u}{V}\right)^3 dS \tag{28.11}$$

For fully developed turbulent flows in pipes and rectangular channels, β and α are generally close to unity, but they may deviate significantly from unity in a channel with a complicated cross-sectional geometry or in separated flows.

The Conservation Equations

The law of conservation of mass, also termed *mass balance* or *continuity*, states that the change in time of fluid mass in a control volume cv must be balanced by the sum of all mass fluxes crossing all control surfaces:

$$\frac{d}{dt}\int_{cv} \rho \, d\mathcal{V} = \sum \dot{m}_{\text{in}} - \sum \dot{m}_{\text{out}} \tag{28.12}$$

where the subscripts "in" and "out" refer to fluxes into and out of the control volume. For steady flow with one inlet and one outlet, Eq. (28.12) simplifies to $\dot{m}_{\text{in}} = \dot{m}_{\text{out}}$ and

$$Q = V_{\text{in}}A_{\text{in}} = V_{\text{out}}A_{\text{out}} \tag{28.13}$$

The law of conservation of momentum, based on Newton's second law, states that the change in time of fluid momentum in a control volume is equal to the sum of all momentum fluxes across all control surfaces plus all forces **F** acting on the control volume:

$$\frac{d}{dt}\int_{cv} \rho \mathbf{u}\, d\mathcal{V} = \sum(\dot{m}\beta\mathbf{V})_{\text{in}} - \sum(\dot{m}\beta\mathbf{V})_{\text{out}} + \sum\mathbf{F} \tag{28.14}$$

For steady flow with one inlet and one outlet, Eq. (28.14) simplifies to

$$\dot{m}\left[(\beta\mathbf{V})_{\text{out}} - (\beta\mathbf{V})_{\text{in}}\right] = \sum\mathbf{F} \tag{28.15}$$

since $\dot{m} = \dot{m}_{\text{in}} = \dot{m}_{\text{out}}$ because of mass conservation. Unlike the mass conservation equation, the momentum balance equation is a vector equation. Care must therefore be taken in considering different components and accounting for the directions of the individual terms. The analysis is identical to free-body analysis in statics, except that fluxes of momentum must also be considered in the force balance.

The law of conservation of energy, based on the first law of thermodynamics, states that the change in time of the total energy of a system is equal to the rate of heat input minus the rate at which work is being done by the system. For problems with negligible heat transfer, this can be usefully expressed in terms of fluxes of mechanical energy as

$$\frac{d}{dt}\int_{cv} \rho\left(\frac{u^2}{2} + gz\right) d\mathcal{V} = \sum(\dot{m}gH)_{\text{in}} - \sum(\dot{m}gH)_{\text{out}} - \sum\dot{W}_s - \sum\dot{m}gh_L \tag{28.16}$$

where $\dot{W}_s$ represents the *shaft work* done by the system, as in pumps and turbines. The head loss, $h_L \geq 0$, represents the conversion of useful mechanical energy per unit weight of fluid into unrecoverable internal or thermal energy. For the frequent case of a steady flow with a single inlet and a single outlet, Eq. (28.16) becomes

$$H_{\text{in}} + H_p = H_{\text{out}} + H_t + h_L \tag{28.17a}$$

where the shaft work term $\dot{W}_s = -\dot{W}_p + \dot{W}_t$ (where the rate of work done by the system on the pump is $-\dot{W}_p = -\dot{m}gH_p$, and the rate of work done by the system on the turbine is $\dot{W}_t = \dot{m}gH_t$). H_p and H_t represent respectively the head per unit weight of liquid delivered by a pump or lost to a turbine. In expanded form, this is often expressed as

$$\left(\frac{p}{\gamma} + z + \alpha\frac{V^2}{2g}\right)_{\text{in}} + H_p = \left(\frac{p}{\gamma} + z + \alpha\frac{V^2}{2g}\right)_{\text{out}} + H_t + h_L \tag{28.17b}$$

Because of its similarity to Eq. (28.7), the energy Eq. (28.17a) is often also termed loosely the (generalized) Bernoulli equation.

Energy and Hydraulic Grade Lines

For the typical steady one-dimensional nearly horizontal flows, *hydraulic* and *energy grade lines* (HGL and EGL respectively) are useful as graphical representation of the piezometric and the total head, respectively. For flows in which frictional effects are neglected, the EGL is simply a horizontal line, since the total head must remain constant. If frictional effects are considered, the EGL slopes downward in the direction of flow, since the total head is reduced by frictional losses. The slope is termed the *friction slope*, often denoted by $S_f \equiv h_f/L$, where h_f is the frictional head loss over a conduit of length L. In pipe flows S_f is *not* related to the pipe slope. The EGL rises only in the case of energy input, such as by a pump. For flows that are uniform in the streamwise direction, the HGL runs parallel to the EGL, since the velocity head is constant. The HGL excludes the velocity head and so lies below the EGL; it coincides with the EGL only where the velocity head is negligible, such as in a reservoir or large tank. Even without energy input or output, the HGL may rise or fall due to a decrease or increase in flow area, leading to an increase or decrease in velocity head. The elevation of the HGL above the pipe centerline is equal to the pressure head; if the HGL crosses or lies below the pipe centerline, this implies that the pressure head is zero or negative; that is, the static pressure is equal to or below atmospheric pressure, which may have implications for cavitation. Since the pressure at the free surface of an open-channel flow is necessarily zero, the HGL for an open-channel flow coincides with the free surface.

Example 28.6: The Venturi Tube. Many devices for measuring discharges depend on reducing the flow area, thus increasing the velocity, and measuring the resulting difference in piezometric head or pressure across the device. An example is the *Venturi tube* (Fig. 28.11), which consists of a short contraction section, a throat section of constant diameter, and a long gradual diffuser (expansion) section. Static pressure taps are located upstream of the contraction and at the throat, since the streamlines can be considered straight at these sections, thus justifying the use of the hydrostatic assumption. The analysis begins with the choice of control volume as shown, with inlet and outlet control surfaces at the pressure tap locations. The Bernoulli theorem is applied on a streamline between points A and B with $V = Q/A$ to give

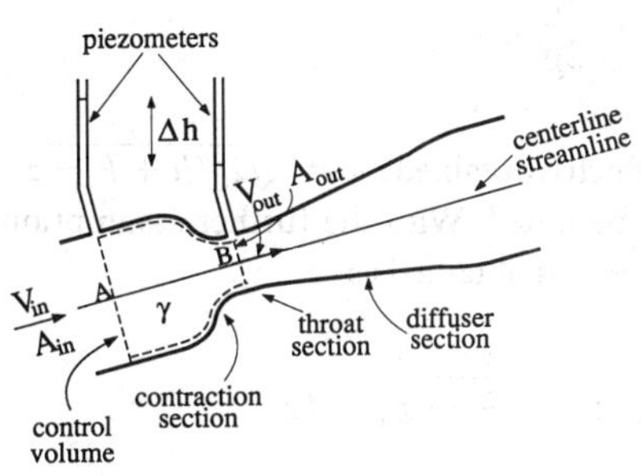

FIGURE 28.11 Flow through a Venturi tube.

$$\Delta h = \left(\frac{p_A}{\gamma} + z_A\right) - \left(\frac{p_B}{\gamma} + z_B\right) = \frac{(Q/A)_B^2}{2g} - \frac{(Q/A)_A^2}{2g} = \frac{Q^2}{2g}\left(\frac{1}{A_A^2} - \frac{1}{A_B^2}\right).$$

Thus, the flow rate is directly related to the change in piezometric head, Δh, which can be simply measured by means of *piezometers*—namely, open tubes in which liquid can rise freely without overflowing. The level to which the liquid rises in a piezometer coincides with the HGL, since it is a free surface. The discharge Q can therefore be expressed as

$$Q = \frac{C_d A_{\text{out}}}{\sqrt{1 - (A_{out}/A_{\text{in}})^2}} \sqrt{2g\Delta h}$$

where the discharge coefficient C_d is an empirical coefficient introduced to account for various approximations in the analysis (see Section 28.8 for further discussion of C_d for Venturi tubes). In typical applications, the pressure leads are brought to the same elevation, and hence the difference in piezometric heads reduces to a pressure difference.

Example 28.7: The Rectangular Sharp-Crested Weir. The sharp-crested weir (Fig. 28.12) is commonly used to measure discharges in open channels by a simple measurement of water level upstream of the weir. It consists of a thin plate at the end of an open channel, over which the flow discharges freely into the atmosphere. The crest of the weir is the top of the plate. The jet flow, or *nappe*, just beyond the crest should be completely *aerated*—that is, at atmospheric pressure. The discharge Q is to be related to the weir head h, the elevation of the upstream free surface above the *weir crest*.

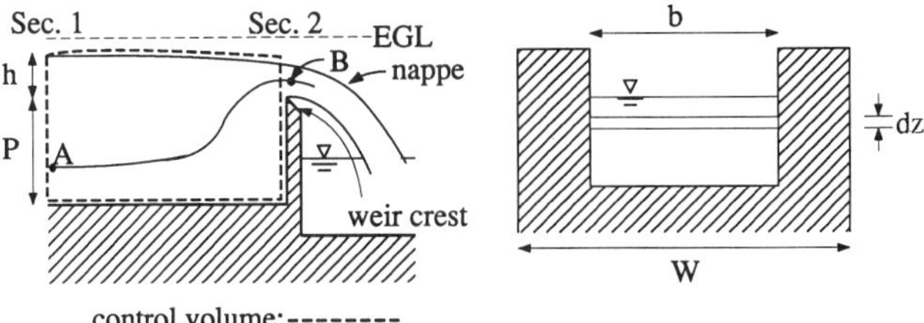

FIGURE 28.12 Flow over a sharp-crested rectangular weir.

Solution. With the control volume as shown, mass conservation implies $Q = V_1A_1 = V_2A_2$. Cross section 1 is chosen so that the flow is nearly parallel, and hence hydrostatic conditions prevail. As such, the piezometric head at cross section 1 is constant, $(p/\gamma + z) = h + P$, with the channel bottom as datum. The Bernoulli equation is applied on the streamline shown between points A and B at cross section 1 and at cross section 2, with the result that

$$h + P = \left(\frac{p}{\gamma} + z\right)_{\mathrm{B}} + \frac{u_B^2}{2g}$$

For an aerated nappe, $p \approx 0$ at any *point* at section 2, from which is obtained $u_B = \sqrt{2g(h + P - z)}$. The similarity between this and the orifice result should be noted. With the further assumption that the velocity is uniform across the weir crest, the discharge is obtained as

$$Q = \int_P^{h+P} u_{\mathrm{B}}(z)\, dA = \int_P^{h+P} \sqrt{2g(h + P - z)}\, b\, dz$$

The upper limit of integration assumes that there is no *drawdown* at the weir; that is, no depression of the free surface below the upstream level. The final result is that

$$Q = C_d \left(\tfrac{2}{3}\, b\, \sqrt{2gh^3}\right)$$

where a discharge coefficient C_d has been introduced to account for any approximations that have been made. A more complete discussion of C_d for weirs is given in Section 28.8. The "weir" discharge relation $Q \sim bh^{3/2}$ also arises in discussions of spillways.

Example 28.8: Forces on a Pipe Bend Anchor Block. A vertical pipe bend is to be anchored by a block (Fig. 28.13). The pressures at the inlet and outlet are p_{in} and p_{out}, the steady discharge is

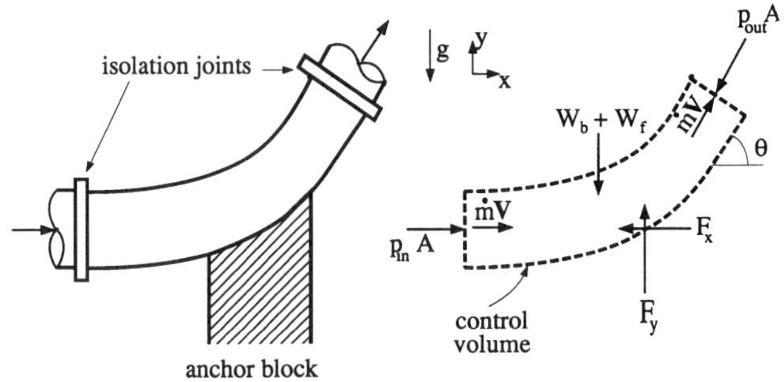

FIGURE 28.13 Forces on an anchor block supporting a pipe bend.

$Q = VA$, and the pipe cross-sectional area is A. The weight of the bend is W_{bend}, and the weight of the fluid in the bend is W_f. The entire load is assumed to be carried by the anchor block, so a control volume is considered as shown with force components on the pipe bend due to the anchor block, F_x and F_y. The coordinate system is chosen so that velocities and forces are positive upwards or to the right. The velocity vectors are $\mathbf{V}_{\text{in}} = (V, 0)$ and $\mathbf{V}_{\text{out}} = (V\cos\theta, V\sin\theta)$; the pressure force at the inlet is $(p_{\text{in}}A, 0)$ and at the outlet $(-p_{\text{out}}A\cos\theta, -p_{\text{out}}A\sin\theta)$. Here p_{in} and p_{out} refer to the pressure at the centroid of the control surface (the centerline) if a hydrostatic pressure distribution is assumed. If $\beta_{\text{in}} \approx \beta_{\text{out}} \approx 1$, the two components of the momentum conservation equation can be written as

$$\dot{m}(V\cos\theta - V) = p_{\text{in}}A - (p_{\text{out}}A)\cos\theta - F_x \qquad \text{(x momentum)}$$

$$\dot{m}V\sin\theta = -(p_{\text{out}}A)\sin\theta + F_y - W_{\text{bend}} - W_f \qquad \text{(y momentum)}$$

If the other quantities are known, then the forces F_x and F_y can be computed. The forces on the block due to the pipe must be equal in magnitude and opposite in direction.

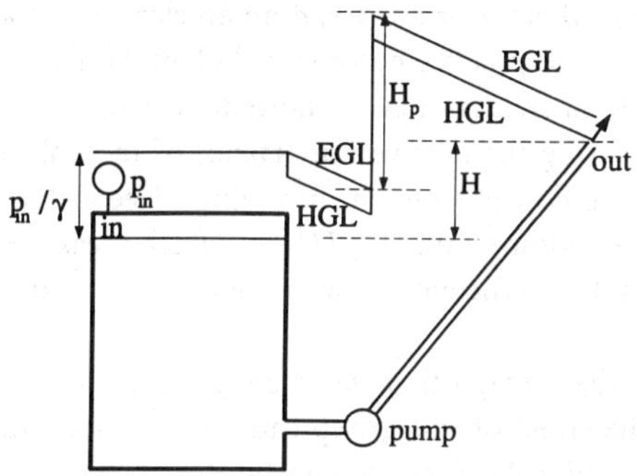

FIGURE 28.14 Energy analysis and associated hydraulic and energy grade lines of a pipe–pump system.

Example 28.9: Energy Equation in a Pipe System with Pump. Fluid is pumped from a large pressurized tank or reservoir at pressure p_{in} through a pipe of uniform diameter discharging into the atmosphere (Fig. 28.14). The difference in elevation between the fluid level in the tank and the pipe end is H. If the head loss in the pipe is known to be h_L, and the head delivered by the pump is H_p, what is the discharge Q?

Solution. Equation (28.17b) is applied to the control volume with inlet and outlet as shown and yields

$$\left(\alpha\frac{V^2}{2g}\right)_{\text{out}} = \left(\frac{p}{\gamma}\right)_{\text{in}} + H_p - h_L - H$$

where the velocity head in the tank has been neglected, and the pressure at the outlet is set to zero since the pipe discharges into the atmosphere. The discharge is calculated as $Q = VA$, where A is the pipe cross-sectional area. The EGL and HGL begin at a level p_{in}/γ above the fluid level in the tank because the tank is pressurized, and they coincide because the velocity head in the tank is negligible. In the pipe flow region, they both slope downward in the direction of flow, the HGL always running parallel but below the EGL, because the constant pipe diameter implies a constant difference due to the constant velocity head. At the pump, energy is added to the system, so both the EGL and the HGL rise abruptly, with the magnitude of the rise equalling the head delivered by the pump, H_p. At the outlet, the HGL intersects the pipe centerline because the pressure head there is zero (discharge into the atmosphere).

28.5 Dimensional Analysis and Similitude

Analysis based purely on conservation equations (including the Bernoulli theorem) is generally not sufficient for the solution of engineering problems. It must be complemented by empirical correlations or results from scale model tests. Dimensional analysis guides the organization of empirical data and the design of scale models.

The Buckingham–Pi Theorem and Dimensionless Groups

Dimensions (as distinct from units) are basic physical quantities. In hydraulics the basic dimensions are those of mass $[M]$, length $[L]$, and time $[T]$, though that of force $[F]$ may sometimes be more conveniently substituted for $[M]$. In this section, square brackets indicate dimensions. A physically sound equation describing a physical phenomenon must be dimensionally homogeneous in that all terms in the equation must have the same dimensions. The basic theoretical result is the Buckingham–Pi theorem, which states that, for a problem involving N independent physical variables with M basic dimensions, $(N - M)$ independent *dimensionless groups* (of variables) can be formed. The design of empirical correlations and scale models, therefore, needs only to consider the $(N - M)$ dimensionless groups rather than the original N variables in order to describe the flow phenomena completely. Further, a relationship among the dimensionless groups relevant to a problem is automatically dimensionally homogeneous and as such satisfies a requirement for a physically sound description.

The two most useful dimensionless groups in hydraulics are the Reynolds number, $\text{Re} \equiv \rho UL/\mu$, and the Froude number, $\text{Fr} \equiv U/\sqrt{gL}$, where U and L are velocity and length scales characteristic of a given problem. The former is interpreted as measuring the relative importance of inertial forces ($ma \sim \rho U^2 L^2$) to viscous forces ($\tau A \sim \mu UL$), where m is a mass, a an acceleration, τ a shear stress, and A an area. At sufficiently high Re, flows in practice become turbulent. Similarly, for given boundary geometry, high-Re flows are more likely than low-Re flows to separate. The Froude number may be similarly interpreted as measuring the relative importance of inertial to gravitational forces ($\sim \rho g L^3$). It plays an essential role in flow phenomena involving a free surface in a gravitational field. In small-scale models the Weber number, $\text{We} = \rho L U^2/\sigma$, which measures the relative importance of surface tension effects (σ is the coefficient of surface tension), may also arise.

An argument that can often be applied considers the asymptotic case where a dimensionless group becomes very large or very small, such that the effect of this group can be neglected. An example of this argument is that used in the case of high-Re flows, where flow characteristics become essentially independent of Re (see the discussion of the Moody diagram, Section 28.6).

Similitude and Hydraulic Modeling

Similitude between hydraulic scale model and prototype is required if predictions based on the former are to be applicable to the latter. Three levels of similarity are geometric, kinematic, and dynamic. Geometric similarity implies that all length-scale ratios in both model and prototype are the same. Kinematic similarity requires, in addition to geometric similarity, that all time-scale ratios be the same. This implies that streamline patterns in model and prototype must be geometrically similar. Finally, dynamic similarity requires, in addition to kinematic similarity, that all mass-scale ratios be the same. This implies that all force-scale ratios at corresponding points in model and prototype flows must be the same. Equivalently, similitude requires that all but one relevant independent dimensionless group be the same in model and prototype flows. Typically, dynamic similarity is formulated in terms of dimensionless groups representing force ratios; for example, $\text{Re}_p = \text{Re}_m$, $\text{Fr}_p = \text{Fr}_m$, where the subscripts p and m refer to prototype and model quantities, respectively.

Practical hydraulic scale modeling is complicated, because strict similitude is generally not feasible, and it must be decided which dimensionless groups can be neglected, with the possible need to correct results *a posteriori*. In many hydraulic models the effects of Re are neglected, based on an implicit assumption of high-Re similarity, and only Fr scaling is satisfied, since it is argued that free-surface gravitational effects are more important than viscous effects. Flow resistance, which may still depend on viscous effects, may therefore be incorrectly modeled, so empirical corrections to the model results for flow resistance may be necessary before they can be applied to the prototype situation. Similarly, geometric similarity is often not achieved in large-scale models of river systems or tidal basins, because this would imply excessively small flow depths, with extraneous viscous and surface tension effects playing an erroneously important role. *Distorted* modeling, with different vertical and horizontal length scales, is therefore often applied. These deviations from strict similitude are discussed in more detail by Yalin [1971] and Sharpe [1981] specifically for problems arising in hydraulic modeling.

Example 28.10: Pump Performance Parameters. The power required by a pump, $\dot{W}_p$ $[ML^2/T^3]$, varies with the impeller diameter D $[L]$, the pump rotation speed n $[1/T]$, the discharge Q $[L^3/T]$, and the fluid density ρ $[M/L^3]$. How can this relationship be expressed in terms of dimensionless groups?

Solution. It follows from the Buckingham–Pi theorem that only two dimensionless groups may be formed, since five variables ($\dot{W}_p$, D, n, Q, and ρ) and three dimensions ($[M]$, $[L]$, $[T]$) are involved. The dimensionless groups are not unique, and different groups may be appropriate for different problems. Three basic variables involving the basic dimensions are chosen, such as n, D, and ρ. Mass (m), length (l), and time (t) scales are formed from these basic variables, such as $m = \rho D^3$; $l = D$; $t = 1/n$. The remaining variables are then made dimensionless by these scales, as in $\dot{W}_p/(ml^2/t^3) = \dot{W}_p/[(\rho D^3)D^2n^3]$, and $Q/(l^3/t) = Q/nD^3$. These are the power and the flow-rate (or discharge) coefficients, respectively, of a pump. A relationship between these dimensionless groups can be written as $\dot{W}_p/[(\rho D^3)D^2n^3] = \mathcal{F}(Q/nD^3)$, which can be used to characterize the performance of a series of similar pumps.

Example 28.11: Spillway Model. A dam spillway is to be modeled in the laboratory. Strict similitude requires $\mathrm{Re}_p = \mathrm{Re}_m$ and $\mathrm{Fr}_p = \mathrm{Fr}_m$, or equivalently, $U_pL_p/\nu_p = U_mL_m/\nu_m$ and $U_p/\sqrt{g_pL_p} = U_m/\sqrt{g_mL_m}$. Since $g_p = g_m$, this implies that $\nu_m/\nu_p = (L_m/L_p)^{3/2}$. For practical scale ratios this result is not feasible, since no common fluid has such a small ν_m. The test is therefore conducted using Froude scaling $\mathrm{Fr}_p = \mathrm{Fr}_m$, with the only condition on Re_m being that Re_m must be sufficiently high (say, $\mathrm{Re}_m > 5000$) such that Reynolds number effects can be assumed negligible. $\mathrm{Fr}_m = \mathrm{Fr}_p$ requires that $U_m/U_p = (L_m/L_p)^{1/2}$, which in turn implies that $Q_m/Q_p = (U_mL_m^2)/(U_pL_p^2) = (L_m/L_p)^{5/2}$.

28.6 Velocity Profiles and Flow Resistance in Pipes and Open Channels

Flow Resistance in Fully Developed Flows

A *fully developed* steady flow in a conduit (pipe or open channel) is defined as a flow in which velocity characteristics do not change in the streamwise direction. This occurs in straight pipe or channel sections of constant geometry far from any transitions such as entrances or exits. Under these conditions, application of the momentum and energy conservation equations yields a balance between shear forces on the conduit boundary and gravitational or pressure forces, or

$$\tau_w = \gamma \frac{A}{P}\frac{h_f}{L} \tag{28.18}$$

where τ_w is the average shear stress on the conduit boundary, A is the cross-sectional flow area, P is the wetted perimeter, and h_f is the head loss due to boundary friction over a conduit section of length L. The *wetted perimeter* is the length of perimeter of the conduit that is in contact with the fluid. For a circular pipe flowing full (Fig. 28.15), the wetted perimeter is the pipe circumference, or $P = \pi D$.

Equation (28.18) is also frequently written as $\tau_w = \gamma R_h S_f$, where $R_h \equiv A/P$ is called the *hydraulic radius* and $S_f \equiv h_f/L$ is the *energy* or *friction slope*. For a circular pipe flowing full,

$R_h = A/P = D/4$. A *shear velocity* u_* can be defined such that $u_*^2 \equiv \tau_w/\rho$. It follows that

$$u_*^2 = gR_hS_f \tag{28.19}$$

These expressions hold for both laminar and turbulent flows. The frictional head loss h_f increases linearly with the length L of the conduit and so may be conveniently expressed in terms of the Darcy–Weisbach friction factor f as

$$h_f = f\frac{L}{4R_h}\frac{V^2}{2g} \tag{28.20}$$

or $(u_*/V)^2 = f/8$.

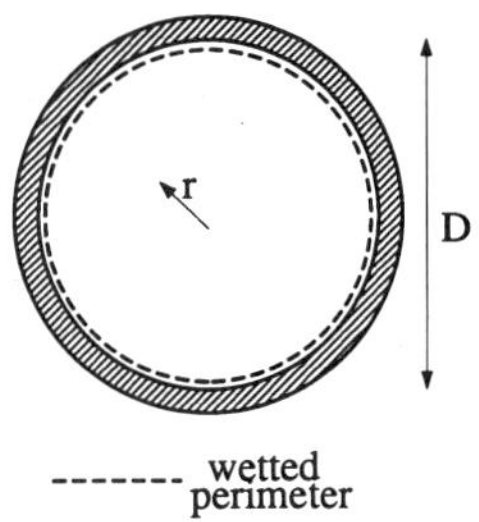

FIGURE 28.15 Coordinate system for pipe flow velocity profile, and the wetted perimeter for a pipe flowing full.

Laminar Velocity Profiles

Velocity profiles for steady fully developed laminar flows in a circular pipe and in an infinitely wide open channel can be theoretically obtained. For a circular pipe, it can be shown that

$$\frac{u(r)}{u_*} = \frac{1}{4}\frac{u_*}{V}\text{Re}\left[1 - \left(\frac{2r}{D}\right)^2\right] \tag{28.21}$$

where $u(r)$ is the velocity at a radial distance r measured from the centerline (Fig. 28.15), and the pipe Reynolds number $\text{Re} = VD/\nu$. For an infinitely wide open channel,

$$\frac{u(y)}{u_*} = \frac{1}{8}\frac{u_*}{V}\text{Re}\left(\frac{y}{h}\right)\left(2 - \frac{y}{h}\right) \tag{28.22}$$

where $u(y)$ is the velocity at a vertical distance y measured from the channel bottom (Fig. 28.16), h is the flow depth, and the channel Reynolds number $\text{Re} = V(4R_h)/\nu$. Note that $R_h = h$ for an infinitely wide channel. Both profiles show a quadratic variation, characteristic of uniform laminar flows.

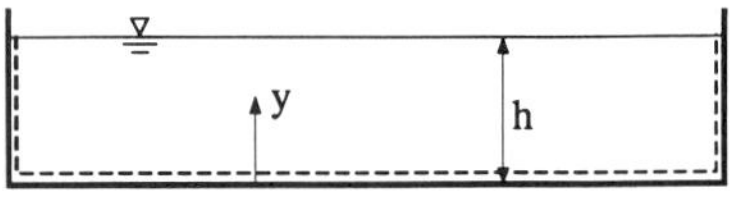

FIGURE 28.16 Coordinate system for open-channel flow velocity profile, and the corresponding wetted perimeter.

Friction Relationships for Laminar Flows

A simple relation between f and Re can be obtained by integrating Eq. (28.21) or (28.22) over the cross section of the flow. For a circular pipe,

$$f = \frac{64}{\text{Re}} \quad \text{and} \quad h_f = \frac{32\nu L}{gD^2}V \tag{28.23}$$

while for an infinitely wide, open channel,

$$f = \frac{96}{\text{Re}} \quad \text{and} \quad h_f = \frac{3\nu L}{gh^2}V \tag{28.24}$$

where the appropriate definition of the Reynolds number must be used. A characteristic of laminar flows is a linear variation of h_f with V, as seen in both Eqs. (28.23) and (28.24).

Turbulent Velocity Profiles

Because a complete theory for turbulent pipe or channel flows is lacking, a greater reliance on empirical results is necessary in discussing turbulent velocity profiles. Two types of profiles are the *log-law* profile and the *power-law* profile. The log-law profile is more physically sound, but a detailed discussion becomes complicated. A useful approximate form may be given as

$$\frac{u(\xi)}{u_*} = \frac{V}{u_*} + \frac{1}{\kappa}\ln\frac{\xi}{\Delta} + B \tag{28.25}$$

where ξ is the coordinate measured from the wall, $\kappa \approx 0.4$ is the von Kármán constant, Δ is the depth in the case of open-channel flow and the radius in the case of pipe flow, and B is a constant with value ≈ 2.5 for flow in a wide open channel and ≈ 3.7 for pipe flows. This profile is not valid very near the boundary ($\xi \to 0$) or near the centerline or free surface ($\xi \to \Delta$). In some problems, the power-law profile may be more convenient; it is expressed as

$$\frac{u(\xi)}{u_{\max}} = \left(\frac{\xi}{\Delta}\right)^{1/m} \tag{28.26}$$

where $u_{\max}$ is the maximum velocity attained in the flow at $\xi = \Delta$; the wall is located at $\xi = 0$; and m increases slowly with increasing Re from $m = 6$ at Re $= 5000$ to $m = 10$ for Re $> 2 \times 10^6$. In real open-channel flows, the maximum velocity may not occur at the free surface, due to the effects of secondary currents, and Eq. (28.26) must be accordingly interpreted.

Effects of Roughness

The effects of small-scale roughness of solid boundaries are negligible for laminar flows but become important for turbulent flows. Wall roughness for a given conduit material is characterized by a typical height k_s of roughness elements. The wall is said to be *hydrodynamically smooth* provided $k_s < \delta_\nu$, where the thickness of the viscous sublayer $\delta_\nu \approx 5\nu/u_*$. Similarly, the wall is said to be *fully rough* if $k_s \gg \delta_\nu$. Precise information regarding k_s is usually available only for new pipes, and with age k_s is likely to increase. A range of values of k_s is given on the Moody diagram, Fig. 28.22 in Section 28.9.

Friction Relationship for Turbulent Flows

The turbulent velocity profiles can be integrated to give friction relationships for steady fully developed turbulent flows. The well-known Colebrook–White formula,

$$\frac{1}{\sqrt{f}} = -0.86\ln\left(\frac{k_s}{3.7D} + \frac{2.51}{\mathrm{Re}\sqrt{f}}\right) \tag{28.27}$$

is based on a log-law profile. Given Re and k_s/D, f can be determined. Because this formula is implicit and transcendental for f, its graphical form (the Moody diagram, Fig. 28.22 in Section 28.9) may be more convenient to use. On log-log coordinates, curves of f vs. Re at constant k_s/D are plotted. If Re and k_s/D are known and f desired, the curve corresponding to the given k_s/D is found, and the value of f corresponding to the given Re on that curve is read. For given k_s/D, the curves of f vs. Re level off for sufficiently high Re, which is an example of the high-Re similarity mentioned in Section 28.5. In this "fully rough" regime, f is independent of Re and depends only on k_s/D, so that for given k_s/D the head loss $h_f \propto Q^2$, which is characteristic of high-Re turbulent flows.

For noncircular conduits or open-channel flows, the Moody diagram can be used as an approximation by substituting an effective diameter $D_{\text{eff}} = 4R_h$ for D in determining the relative roughness and the Reynolds number. In some problems, k_s/D or Re may not be directly available, because the discharge Q or the pipe diameter D is to be determined. For such problems, an iterative solution is required.

Explicit Friction Relationships for Turbulent Flows

In problems involving predominantly friction losses, the iterative solution can be avoided through the use of explicit friction formulas, which are also useful in computer-aided solutions of flow problems. The following dimensionless formulas, adapted from Swamee and Jain [1976], are also based on the Colebrook–White formula.

- Given k_s/D and Re, to find h_f or f use

$$\frac{h_f}{(L/D)(V^2/2g)} = f = \frac{0.25}{\{\ln\left[(k_s/3.7D) + (5.74/\text{Re}^{0.9})\right]\}^2}, \tag{28.28}$$

- Given k_s/D and $S_f = h_f/L$, to find Q use

$$\frac{Q}{\sqrt{gS_fD^5}} = -0.965 \ln\left[\frac{k_s}{3.7D} + \left(\frac{3.17\nu^2}{gS_fD^3}\right)^{1/2}\right], \tag{28.29}$$

- Given k_s, Q, and S_f, to find D use

$$\frac{gS_fD^5}{Q^2} = 0.125\left[\left(\frac{gS_fk_s^5}{Q^2}\right)^{1/4} + \left(\frac{\nu^5}{gS_fQ^3}\right)^{1/5}\right]^{1/5}. \tag{28.30}$$

These formulas are valid for $10^{-6} < k_s/D < 10^{-2}$ and $5000 < \text{Re} < 10^8$.

Other traditional explicit formulas, intended for use with water only, are encountered in practice. The Chezy–Manning equation is used frequently in open-channel flows, whereas the Hazen–Williams equation is frequently used in the waterworks industry. These equations, while convenient, approximate the more generally applicable Colebrook–White-type formulas only over a restricted range of Reynolds numbers and relative roughness. In practice this may be compensated by judiciously adjusting the relevant coefficients to particular situations.

Minor Losses

Head losses h_M that are caused by localized disturbances to a flow, such as by valves or pipe fittings or in open-channel transitions, are traditionally termed *minor losses*. Since they are not necessarily negligible, a more precise term would be *transition losses*, because they typically occur at transitions from one fully developed flow to another. Being due to localized disturbances, they do not vary with the conduit length but are instead described by an overall head loss coefficient K_m. The greater the flow disturbance, the larger the value of K_m. For example, as a valve is gradually closed, the associated value of K_m for the valve increases. Similarly, K_m for different pipe fittings will vary somewhat with pipe size, decreasing with increasing pipe size, since fittings are generally not geometrically similar for different pipe sizes. Values of K_m for various types of transitions are given in Table 28.9 in Section 28.9. An alternative treatment of minor losses replaces the sources of the minor losses by equivalent lengths of pipes, which are often tabulated. In this way, the problem is converted to one with entirely friction losses, to which the explicit equations can be directly

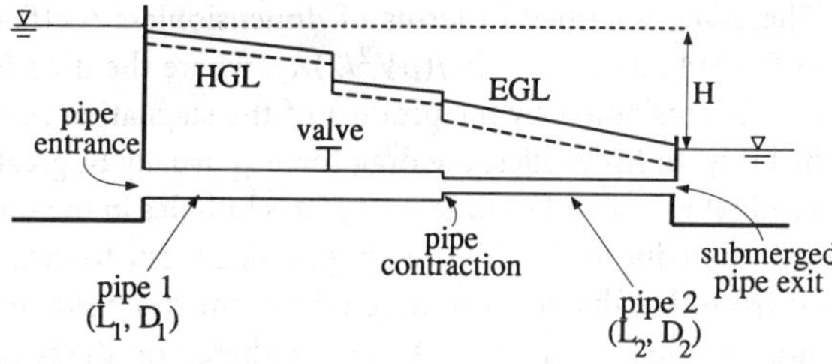

FIGURE 28.17 A pipe system with minor (transitional) losses.

applied. The total head loss h_L of Eq. (28.17) consists of contributions from both friction and transition losses:

$$h_L = h_f + h_m = \sum_j \left(\frac{fL}{D}\right)_j \frac{V_j^2}{2g} + \sum_i (K_m)_i \frac{V_i^2}{2g} \tag{28.31}$$

In some problems with long pipe sections and few instances of minor losses, minor losses may be negligible.

Example 28.12: A Piping System with Minor Losses. Consider a flow between two reservoirs through the pipe system shown in Fig. 28.17. If the difference in elevation between the two reservoirs is H, what is the discharge Q?

Solution. The energy equation between the two free surfaces yields $H = h_L$, since pressure and velocity heads are zero. The total head loss consists of friction losses in pipe 1 and pipe 2 as well as minor losses at the entrance, the valve, the contraction, and the submerged exit:

$$h_L = H = \frac{Q^2}{2g}\left[\frac{f_1L_1}{D_1}\frac{1}{A_1^2} + \frac{f_2L_2}{D_2}\frac{1}{A_2^2} + \left\{(K_m)_{\text{ent}} + (K_m)_{\text{valve}}\right\}\frac{1}{A_1^2} + \left\{(K_m)_{\text{contr}} + (K_m)_{\text{exit}}\right\}\frac{1}{A_2^2}\right]$$

where the subscripts refer to pipes 1 and 2. The minor loss coefficients are obtained from tabulated values; for example, from Table 28.9, $(K_m)_{\text{ent}} = 0.5$. The friction factors f_1 and f_2 may differ, since relative roughness and Reynolds number may differ in the two pipes. Further, since Q is not known, neither Re_1 nor Re_2 can be computed, and hence neither f_1 nor f_2 can be directly determined; an iterative solution is necessary if the Moody diagram is used.

If an explicit formula is used, the presence of minor-losses terms would still require a trial and error solution of a nonlinear equation. The iterative solution would begin by making initial guesses of f_1 and f_2, computing the resulting Q, from which Re_1 and Re_2 could be estimated; and finally checking that the guesses were indeed consistent. The HGL and EGL run parallel to each other in the two pipes; the distance between them is, however, larger in pipe 2 because of the larger velocity head (due to the smaller pipe diameter). The energy slope is constant for each pipe section but will generally differ in the two pipe sections. The losses at transitions, shown as abrupt drops in the grade lines, are exaggerated for visual purposes.

28.7 Hydrodynamic Forces on Submerged Bodies

Bodies moving in fluids or stationary bodies in a moving fluid experience *hydrodynamic* forces in addition to the hydrostatic forces discussed in Section 28.2. A net force in the direction of mean flow is termed a *drag force,* whereas a net force in the direction normal to the mean flow is termed a *lift force*. These are described in terms of dimensionless coefficients of drag or lift, defined as $C_D \equiv F_D/(\rho V^2/2)A_p$ and $C_L \equiv F_L/(\rho V^2/2)A_p$, where the drag force F_D and the lift force F_L have been made dimensionless by the product of the stagnation pressure $\rho V^2/2$ and the projected area A_p of the body. In hydraulics, the drag force is usually of greater interest, as in the determination of the terminal velocity of sediment or of gas bubbles in the water column, or loads on structures. The contribution due to skin friction drag, which is due to shear stresses on the body surface, is distinguished from that due to form drag (also termed *pressure drag*), which is due to normal stresses (pressure) on the body surface. At low Re (based on the relative velocity of body and fluid) or in flows around streamlined bodies, skin friction drag will dominate, while in high-Re flows around bluff bodies flow separation will occur, and with the formation of the low-pressure wake region, form drag will dominate.

Theoretical results are available only for low-Re flows around bodies with simple geometrical shapes. For steady flow around solid spheres in a fluid of infinite extent and Re $\rightarrow 0$, $C_D = 24/\text{Re}$, where Re $= UD/\nu$, U being the relative velocity of the body in a fluid with kinematic viscosity ν, and D being the diameter of the sphere. In this so-called Stokes flow regime (Re < 1), the terminal velocity w_T of a solid sphere (i.e., the steady velocity it attains when falling under gravity in a stagnant fluid of infinite extent) is given by $w_T = g(s_s - s_f)D^2/18\nu$, where s_s and s_f are the specific gravities of the body and the fluid. For larger Re and for more complicated geometries, empirical correlations for C_D as a function of Re and other factors such as relative roughness must be used. A standard drag curve that plots C_D vs. Re is available for common geometries, such as smooth spheres or infinitely long circular cylinders. In these cases, C_D exhibits a form of high-Re similarity (see Section 28.5 and 28.6) in the range $10^3 < \text{Re} < 10^5$, where it attains an approximately constant value: $C_D \approx 0.5$ for smooth spheres, and $C_D \approx 1$ for smooth, infinitely long circular cylinders. For high-Re flows around bluff bodies without sharp edges to fix the separation point, C_D exhibits an abrupt decrease at a critical value of Re, because the separation point on the body moves downstream, resulting in a narrower low-pressure wake region. For a smooth sphere or a smooth, infinitely long cylinder, the critical value is observed to be $\approx 3 \times 10^5$ (smooth sphere) and $\approx 5 \times 10^5$ (smooth cylinder), but surface roughness will decrease the critical value.

Example 28.13: Terminal Velocity of a Solid Sphere. Consider a metal sphere (density ρ_s, diameter D), falling under gravity in fluid (density ρ, kinematic viscosity ν). When the sphere attains its terminal velocity w_T, its effective weight (including buoyancy effects), $g(\rho_s - \rho)\pi D^3/6$, is balanced by the drag force $F_D = C_D(\rho w_T^2/2)A_p$, where $A_p = \pi D^2/4$, so that $w_T = \sqrt{4g[(\rho_s/\rho) - 1]D/(3C_D)}$. Since C_D varies with Re $= w_T D/\nu$, and w_T is unknown, the standard drag curve cannot be used directly. The iterative solution might begin with an initial guess for C_D, from which a w_T and a Re can be computed. A solution is obtained when the computed Re is consistent with the guessed C_D in agreeing with the standard drag curve.

28.8 Discharge Measurements

Discharge measurements are made for monitoring and control purposes. Measurement methods may be divided into those for pipe and those for open-channel flows. The following summarizes some results for both types.

Pipe Flow Measurements

The most popular methods for measurements of discharge Q in a pipe of diameter D and cross-sectional area A have depended on the production of a pressure difference Δp across a device that constricts the flow. Foremost among such devices are various types of orifice and Venturi meters, the basic theory of which has been outlined in Examples 28.5 and 28.6. In general, Q is related to

Δh, the difference in piezometric head across the orifice, by

$$Q = \frac{C_d}{\sqrt{1-\beta^4}} A_0 \sqrt{2g(\Delta h)} \tag{28.32}$$

where A_0 is cross-sectional area of the contraction and $\beta = d/D$, d being the diameter of the contracted section. The discharge coefficient C_d may vary according to the device, the location of the pressure taps, β, and Re. The head loss across the device, h_m, is given as a fraction of the measured differential pressure head. The common thin-plate orifice meter, Fig. 28.18(a), is square-edged and concentric with the pipe. It is used for clean fluids and is inexpensive, but it is associated with relatively high head loss. Miller [1989] gives a correlation for C_d for an orifice with corner taps:

$$C_d = 0.5959 + 0.0312\beta^{2.1} - 0.184\beta^8 + \frac{91.71\beta^{2.5}}{\mathrm{Re}^{0.75}} \tag{28.33}$$

and a correlation for h_m as

$$\frac{h_m}{\Delta h} = 1 - 0.24\beta - 0.52\beta^2 - 0.16\beta^3 \tag{28.34}$$

These correlations are valid for $0.2 < \beta < 0.75$ and $10^4 < \mathrm{Re} < 10^7$.

In comparison to orifice meters, the Venturi, Fig. 28.18(b), incurs low head losses, is suitable for flows with suspended solids, and exhibits less variation in performance characteristics. Disadvantages include higher initial cost and the length and weight of the device. C_d for a Venturi meter ranges from 0.975 to 0.995 (for $\mathrm{Re} > 5 \times 10^4$), while $h_m/\Delta h$ varies with the exit cone angle as [Miller, 1989]

$$\frac{h_m}{\Delta h} = 0.436 - 0.83\beta + 0.59\beta^2, \qquad \theta = 15^\circ \tag{28.35a}$$

$$= 0.218 - 0.42\beta + 0.38\beta^2, \qquad \theta = 7^\circ \tag{28.35b}$$

A variety of other flow meters are in use, and only a few are noted here. *Elbow meters* are based on the pressure differential between the inner and the outer radius of the elbow. Attractive because of their low cost, they tend to be less accurate than orifice or Venturi meters because of a relatively small pressure differential. *Rotameters* or *variable-area* meters are based on the balance between the upward fluid drag on a float, located in an upwardly diverging tube, and the weight of the float. By the choice of float and tube divergence, the equilibrium position of the float can be made linearly proportional to the flow rate. More recently developed nonmechanical devices include the

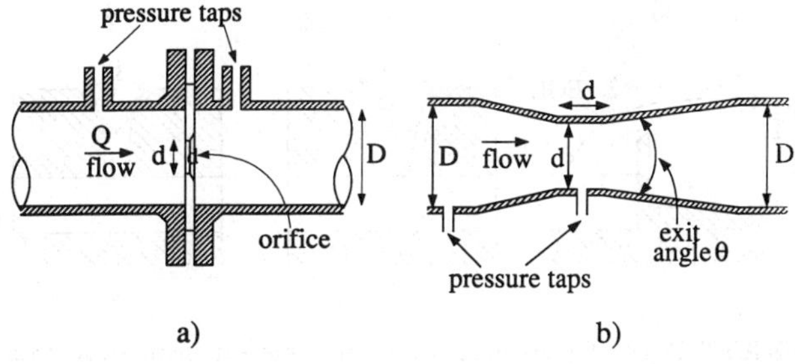

FIGURE 28.18 Pipe flow-measuring devices (a) Thin-plate orifice with corner pressure taps. (b) Venturi tube.

electromagnetic flow meter and the *ultrasonic* flow meter. In the former a voltage is induced between two electrodes, which are located in the pipe walls but in contact with the fluid. The fluid must be conductive, but it can be multiphase. The output is linearly proportional to Q, independent of Re, and insensitive to velocity profile variations. Ultrasonic flow meters are based on the transmission and reception of acoustic signals, which can be related to the flow velocity. An attractive feature of some ultrasonic meters is that they can be clamped onto a pipe rather than being installed in place as is the case of most other devices.

Open-Channel Flow Measurements

Open-channel flow measurements are typically based on measurements of flow depth, which is then correlated with discharge in head–discharge curves. The most common measurement structures may be divided into *weirs* and critical-depth (Venturi) *flumes*. Weirs are raised obstructions in a channel, with the top of the obstruction being termed the *crest*. The streamwise extent of the weir may be short or long, corresponding to sharp-crested and broad-crested weirs, respectively. A critical-depth flume is a constriction built in an open channel, which causes the flow to become *critical* (Fr $=$ 1) in, or near, the throat of the flume. As shown in Fig. 28.19(a), the width of the channel will be denoted by B, the height of the weir crest above the channel bottom is P, and the upstream elevation of the free surface relative to the weir crest is h. The fluid is assumed to be water, and viscous and surface tension effects are assumed negligible.

For a rectangular sharp-crested weir, Fig. 28.19(b) (see also Example 7), the head–discharge formula may be expressed as

$$Q = C_d \tfrac{2}{3} \sqrt{2g}\, b h^{3/2} \tag{28.36}$$

where b is the width of the weir opening and C_d is a discharge coefficient that varies with b/B and h/P as [Bos, 1989]

$$C_d = 0.602 + 0.075(h/P), \qquad b/B = 1 \tag{28.37a}$$

$$= 0.593 + 0.018(h/P), \qquad b/B = 0.6 \tag{28.37b}$$

$$= 0.588 - 0.002(h/P), \qquad b/B = 0.1 \tag{28.37c}$$

Equation (28.37) should give good results if $h > 0.03$ m, $h/P < 2$, $P > 0.1$ m, $b > 0.15$ m, and the tailwater level should be at least 0.05 m below the crest. The triangular or V-notch weir, Fig. 28.19(c), is often chosen when a wide range of discharges is expected, since it is able to remain fully aerated at low discharges. The head–discharge formula for a triangular opening with an angle θ may be expressed as

$$Q = C_d \tfrac{8}{15} \sqrt{2g}\, \tan(\theta/2)\, h^{5/2} \tag{28.38}$$

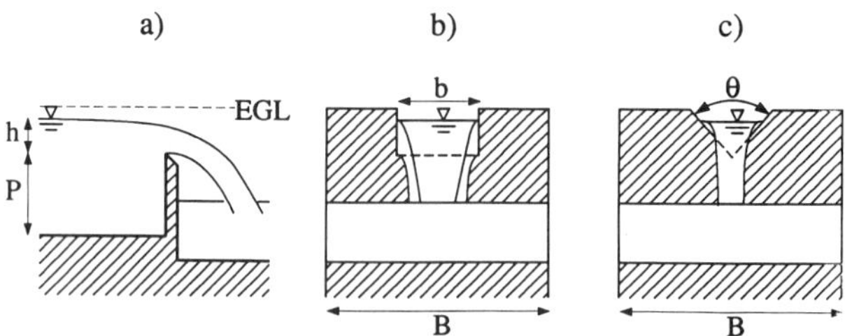

FIGURE 28.19 Open-channel discharge measurement structures. (a) Side view of a sharp-crested weir. (b) Front view of a rectangular weir. (c) Front view of a triangular (V-notch) weir.

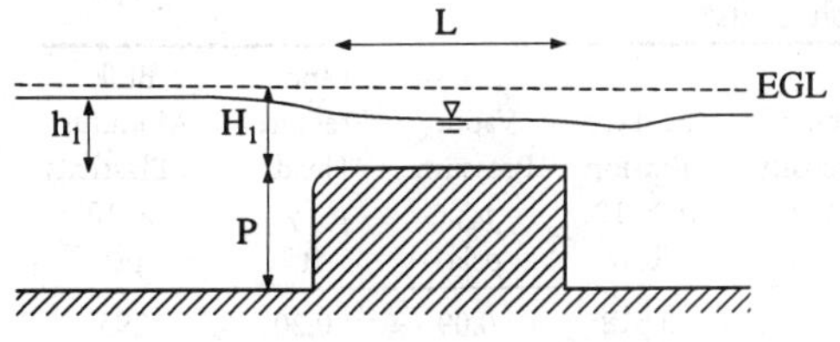

FIGURE 28.20 Broad-crested weir.

If $h/P \leq 0.4$ and $\tan(\theta/2)h/B \leq 0.2$, then C_d varies only slightly with θ: $C_d = 0.58 \pm 0.01$ for $25° < \theta < 90°$ [Bos, 1989]. It is also recommended that $h > 0.05$ m, $P > 0.45$ m, and $B > 0.9$ m.

Under field conditions, where sharp-crested weirs may present maintenance problems, the broad-crested weir (Fig. 28.20) provides a more robust gaging structure. It relies on the establishment of critical flow over the weir crest.

In order to justify the neglect of frictional effects and the assumption of hydrostatic conditions over the weir, the length L of the weir should satisfy $0.07 \leq H_1/L \leq 0.5$, where H_1 is the upstream total head, referred to the elevation of the weir crest. The head–discharge formula may be expressed as

$$Q = C_d C_v \tfrac{2}{3} \sqrt{2g/3}\, b h^{3/2} \tag{28.39}$$

where C_d is a discharge coefficient; Bos [1989] recommends that $C_d = 0.93 + 0.10H_1/L$. C_v is a coefficient accounting for a nonzero approach velocity ($C_v \geq 1$, and $C_v = 1$ if the approach velocity is neglected, in which case $H_1 = h$).

In situations where deposition of silt or other debris may pose a problem, a critical-depth flume may be more appropriate than a weir. Since it too is based on the establishment of critical flow in the throat of the flume, its approximate theoretical analysis is essentially the same. The most common example is the Parshall flume (Fig. 28.21), which resembles a Venturi tube in having a converging section with a level bottom, a throat section with a downward-sloping bottom, and a diverging section with an upward-sloping bottom. The dimensioning of the Parshall flume is standardized; as such, the calibration curve for each size (e.g., defined by the throat width) is different, since the different sizes are *not* geometrically similar. The head–discharge relation is typically expressed in terms of a piezometric head h_c at a prescribed location in the converging section, which is different for each size. This relationship is given by $Q = Kh_c^m$, where the exponent m ranges from 1.52 to 1.6, and the dimensional coefficient K varies with the size of the throat width b_c with K ranging from 0.060 for $b_c = 1$ in. to 35.4 for $b_c = 50$ ft. More detailed information concerning K can be found in Bos [1989].

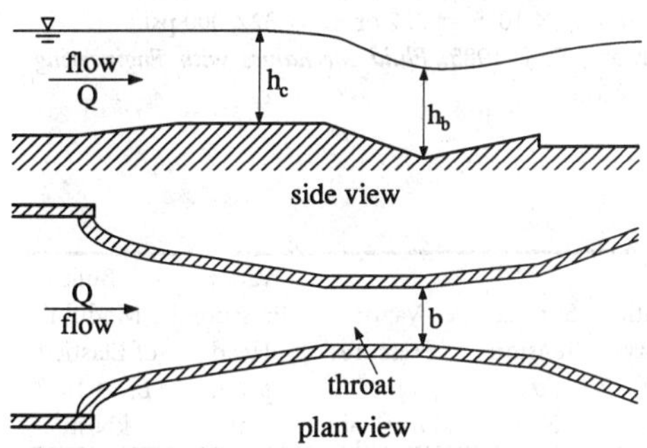

FIGURE 28.21 Critical-depth or Venturi flume (Parshall type).

28.9 Additional Tables and Figures

Tables 28.1 and 28.2 list the physical properties of water, whereas Tables 28.3 and 28.4 list the physical properties of air at standard atmospheric pressure. Tables 28.5 and 28.6 give the standard properties of the earth's atmosphere at several altitudes. Tables 28.7 and 28.8 list the properties of various other common liquids.

Figure 28.22 displays the Moody diagram, which yields the friction factor f depending on roughness and Reynolds number.

Table 28.9 gives values of the transition loss coefficient K_m for various types of transitions.

Table 28.1 Physical Properties of Water in English Units*

Temperature, °F	Specific Weight γ, lb/ft^3	Density ρ, slugs/ft^3	Viscosity $\mu \times 10^5$, lb·s/ft^2	Kinematic Viscosity $\nu \times 10^5$, ft^2/s	Surface Tension $\sigma \times 10^2$ lb/ft	Vapor Pressure p_v, psia	Vapor Pressure Head p_v/γ, ft	Bulk Modulus of Elasticity $E_v \times 10^{-3}$. psi
32	62.42	1.940	3.746	1.931	0.518	0.09	0.20	293
40	62.43	1.940	3.229	1.664	0.514	0.12	0.28	294
50	62.41	1.940	2.735	1.410	0.509	0.18	0.41	305
60	62.37	1.938	2.359	1.217	0.504	0.26	0.59	311
70	62.30	1.936	2.050	1.059	0.500	0.36	0.84	320
80	62.22	1.934	1.799	0.930	0.492	0.51	1.17	322
90	62.11	1.931	1.595	0.826	0.486	0.70	1.61	323
100	62.00	1.927	1.424	0.739	0.480	0.95	2.19	327
110	61.86	1.923	1.284	0.667	0.473	1.27	2.95	331
120	61.71	1.918	1.168	0.609	0.465	1.69	3.91	333
130	61.55	1.913	1.069	0.558	0.460	2.22	5.13	334
140	61.38	1.908	0.981	0.514	0.454	2.89	6.67	330
150	61.20	1.902	0.905	0.476	0.447	3.72	8.58	328
160	61.00	1.896	0.838	0.442	0.441	4.74	10.95	326
170	60.80	1.890	0.780	0.413	0.433	5.99	13.83	322
180	60.58	1.883	0.726	0.385	0.426	7.51	17.33	318
190	60.36	1.876	0.678	0.362	0.419	9.34	21.55	313
200	60.12	1.868	0.637	0.341	0.412	11.52	26.59	308
212	59.83	1.860	0.593	0.319	0.404	14.70	33.90	300

*In this table and in the others to follow, if $\mu \times 10^5 = 3.746$, then $\mu = 3.746 \times 10^{-5}$lb · s/ft^2, etc. For example, at 80°F, $\sigma \times 10^2 = 0.492$ or $\sigma = 0.00492$ lb/ft and $E_c \times 10^{-3} = 322$ or $E_r = 322,000$ psi.

Source: Daugherty, R. L., Franzini, J. B., and Finnemore, E. J. 1985. *Fluid Mechanics with Engineering Applications*, 8th ed., McGraw-Hill, New York. With permission.

Table 28.2 Physical Properties of Water in SI Units

Temperature, °C	Specific Weight γ, kN/m^3	Density ρ, kg/m^3	Viscosity $\mu \times 10^3$, N·s/m^2	Kinematic Viscosity $\nu \times 10^6$, m^2/s	Surface Tension σ, N/m	Vapor Pressure p_v, kN/m^2, abs	Vapor Pressure Head p_v/γ, m	Bulk Modulus of Elasticity $E_v \times 10^{-6}$. kN/m^2
0	9.805	999.8	1.781	1.785	0.0756	0.61	0.06	2.02
5	9.807	1000.0	1.518	1.519	0.0749	0.87	0.09	2.06
10	9.804	999.7	1.307	1.306	0.0742	1.23	0.12	2.10
15	9.798	999.1	1.139	1.139	0.0735	1.70	0.17	2.14
20	9.789	998.2	1.002	1.003	0.0728	2.34	0.25	2.18
25	9.777	997.0	0.890	0.893	0.0720	3.17	0.33	2.22
30	9.764	995.7	0.798	0.800	0.0712	4.24	0.44	2.25
40	9.730	992.2	0.653	0.658	0.0696	7.38	0.76	2.28
50	9.689	988.0	0.547	0.553	0.0679	12.33	1.26	2.29
60	9.642	983.2	0.466	0.474	0.0662	19.92	2.03	2.28
70	9.589	977.8	0.404	0.413	0.0644	31.16	3.20	2.25
80	9.530	971.8	0.354	0.364	0.0626	47.34	4.96	2.20
90	9.466	965.3	0.315	0.326	0.0608	70.10	7.18	2.14
100	9.399	958.4	0.282	0.294	0.0589	101.33	10.33	2.07

Source: Daugherty, R. L., Franzini, J. B., and Finnemore, E. J. 1985. *Fluid Mechanics with Engineering Applications*, 8th ed., McGraw-Hill, New York. With permission.

Table 28.3 Physical Properties of Air at Standard Atmospheric Pressure in English Units

Temperature T, °F	Temperature T, °C	Density $\rho \times 10^3$, slugs/ft³	Specific Weight $\gamma \times 10^2$, lb/ft³	Viscosity $\mu \times 10^7$, lb·s/ft²	Kinematic Viscosity $\nu \times 10^4$, ft²/s
−40	−40.0	2.94	9.46	3.12	1.06
−20	−28.9	2.80	9.03	3.25	1.16
0	−17.8	2.68	8.62	3.38	1.26
10	−12.2	2.63	8.46	3.45	1.31
20	−6.7	2.57	8.27	3.50	1.36
30	−1.1	2.52	8.11	3.58	1.42
40	4.4	2.47	7.94	3.62	1.46
50	10.0	2.42	7.79	3.68	1.52
60	15.6	2.37	7.63	3.74	1.58
70	21.1	2.33	7.50	3.82	1.64
80	26.7	2.28	7.35	3.85	1.69
90	32.2	2.24	7.23	3.90	1.74
100	37.8	2.20	7.09	3.96	1.80
120	48.9	2.15	6.84	4.07	1.89
140	60.0	2.06	6.63	4.14	2.01
160	71.1	1.99	6.41	4.22	2.12
180	82.2	1.93	6.21	4.34	2.25
200	93.3	1.87	6.02	4.49	2.40
250	121.1	1.74	5.60	4.87	2.80

Source: Daugherty, R. L., Franzini, J. B., and Finnemore, E. J. 1985. *Fluid Mechanics with Engineering Applications*, 8th ed., McGraw-Hill, New York. With permission.

Table 28.4 Physical Properties of Air at Standard Atmospheric Pressure in SI Units

Temperature T, °C	Temperature T, °F	Density ρ, kg/m³	Specific Weight γ, N/m³	Viscosity $\mu \times 10^5$, N·s/m²	Kinematic Viscosity $\nu \times 10^5$, m²/s
−40	−40	1.515	14.86	1.49	0.98
−20	−4	1.395	13.68	1.61	1.15
0	32	1.293	12.68	1.71	1.32
10	50	1.248	12.24	1.76	1.41
20	68	1.205	11.82	1.81	1.50
30	86	1.165	11.43	1.86	1.60
40	104	1.128	11.06	1.90	1.68
60	140	1.060	10.40	2.00	1.87
80	176	1.000	9.81	2.09	2.09
100	212	0.946	9.28	2.18	2.31
200	392	0.747	7.33	2.58	3.45

Source: Daugherty, R. L., Franzini, J. B., and Finnemore, E. J. 1985. *Fluid Mechanics with Engineering Applications*, 8th ed., McGraw-Hill, New York. With permission.

Table 28.5 The ICAO Standard Atmosphere in English Units

Elevation above Sea Level, ft	Temp, °F	Absolute Pressure, psia	Specific Weight γ, lb/ft^3	Density ρ, slugs/ft^3	Viscosity $\mu \times 10^7$, lb·s/ft^2
0	59.0	14.70	0.07648	0.002377	3.737
5,000	41.2	12.24	0.06587	0.002048	3.637
10,000	23.4	10.11	0.05643	0.001756	3.534
15,000	5.6	8.30	0.04807	0.001496	3.430
20,000	−12.3	6.76	0.04070	0.001267	3.325
25,000	−30.1	5.46	0.03422	0.001066	3.217
30,000	−47.8	4.37	0.02858	0.000891	3.107
35,000	−65.6	3.47	0.02367	0.000738	2.995
40,000	−69.7	2.73	0.01882	0.000587	2.969
45,000	−69.7	2.15	0.01481	0.000462	2.969
50,000	−69.7	1.69	0.01165	0.000364	2.969
60,000	−69.7	1.05	0.00722	0.000226	2.969
70,000	−69.7	0.65	0.00447	0.000140	2.969
80,000	−69.7	0.40	0.00277	0.000087	2.969
90,000	−57.2	0.25	0.00168	0.000053	3.048
100,000	−40.9	0.16	0.00102	0.000032	3.150

Source: Daugherty, R. L., Franzini, J. B., and Finnemore, E. J. 1985. *Fluid Mechanics with Engineering Applications*, 8th ed., McGraw-Hill, New York. With permission.

Table 28.6 The ICAO Standard Atmosphere in SI Units

Elevation above Sea Level, km	Temp, °C	Absolute Pressure, kN/m^2, abs	Specific Weight γ, N/m^3	Density ρ, kg/m^3	Viscosity $\mu \times 10^5$, N·s/m^2
0	15.0	101.33	12.01	1.225	1.79
2	2.0	79.50	9.86	1.007	1.73
4	− 4.5	60.12	8.02	0.909	1.66
6	−24.0	47.22	6.46	0.660	1.60
8	−36.9	35.65	5.14	0.526	1.53
10	−49.9	26.50	4.04	0.414	1.46
12	−56.6	19.40	3.05	0.312	1.42
14	−56.5	14.20	2.22	0.228	1.42
16	−56.5	10.35	1.62	0.166	1.42
18	−56.5	7.57	1.19	0.122	1.42
20	−56.5	5.53	0.87	0.089	1.42
25	−51.6	2.64	0.41	0.042	1.45
30	−40.2	1.20	0.18	0.018	1.51

Source: Daugherty, R. L., Franzini, J. B., and Finnemore, E. J. 1985. *Fluid Mechanics with Engineering Applications*, 8th ed., McGraw-Hill, New York. With permission.

Table 28.7 Physical Properties of Common Liquids at Standard Atmospheric Pressure in English Units

Liquid	Temperature T, °F	Density ρ, slug/ft^3	Specific Gravity, s	Viscosity $\mu \times 10^5$, lb·s/ft^2	Surface Tension σ, lb/ft	Vapor Pressure p_v, psia	Modulus of Elasticity E_v, psi
Benzene	68	1.74	0.90	1.4	0.002	1.48	150,000
Carbon tetrachloride	68	3.08	1.594	2.0	0.0018	1.76	160,000
Crude oil	68	1.66	0.86	15	0.002		
Gasoline	68	1.32	0.68	0.62	. . .	8.0	
Glycerin	68	2.44	1.26	3100	0.004	0.000002	630,000
Hydrogen	−430	0.14	0.072	0.043	0.0002	3.1	
Kerosene	68	1.57	0.81	4.0	0.0017	0.46	
Mercury	68	26.3	13.56	3.3	0.032	0.000025	3,800,000
Oxygen	−320	2.34	1.21	0.58	0.001	3.1	
SAE 10 oil	68	1.78	0.92	170	0.0025		
SAE 30 oil	68	1.78	0.92	920	0.0024		
Water	68	1.936	1.00	2.1	0.005	0.34	300,000

Source: Daugherty, R. L., Franzini, J. B., and Finnemore, E. J. 1985. *Fluid Mechanics with Engineering Applications*, 8th ed., McGraw-Hill, New York. With permission.

Table 28.8 Physical Properties of Common Liquids at Standard Atmospheric Pressure in SI Units

Liquid	Temperature T, °C	Density ρ, kg/m^3	Specific Gravity, s	Viscosity $\mu \times 10^4$, N·s/m^2	Surface Tension σ, N/m	Vapor Pressure p_v, kN/m^2,abs	Modulus of Elasticity $E_v \times 10^{-6}$, N/m^2
Benzene	20	895	0.90	6.5	0.029	10.0	1,030
Carbon tetrachloride	20	1,588	1.59	9.7	0.026	12.1	1,100
Crude oil	20	856	0.86	72	0.03		
Gasoline	20	678	0.68	2.9	. . .	55	
Glycerin	20	1,258	1.26	14,900	0.063	0.000014	4,350
Hydrogen	−257	72	0.072	0.21	0.003	21.4	
Kerosene	20	808	0.81	19.2	0.025	3.20	
Mercury	20	13,550	13.56	15.6	0.51	0.00017	26,200
Oxygen	−195	1,206	1.21	2.8	0.015	21.4	
SAE 10 oil	20	918	0.92	820	0.037		
SAE 30 oil	20	918	0.92	4,400	0.036		
Water	20	998	1.00	10.1	0.073	2.34	2,070

Source: Daugherty, R. L., Franzini, J. B., and Finnemore, E. J. 1985. *Fluid Mechanics with Engineering Applications*, 8th ed., McGraw-Hill, New York. With permission.

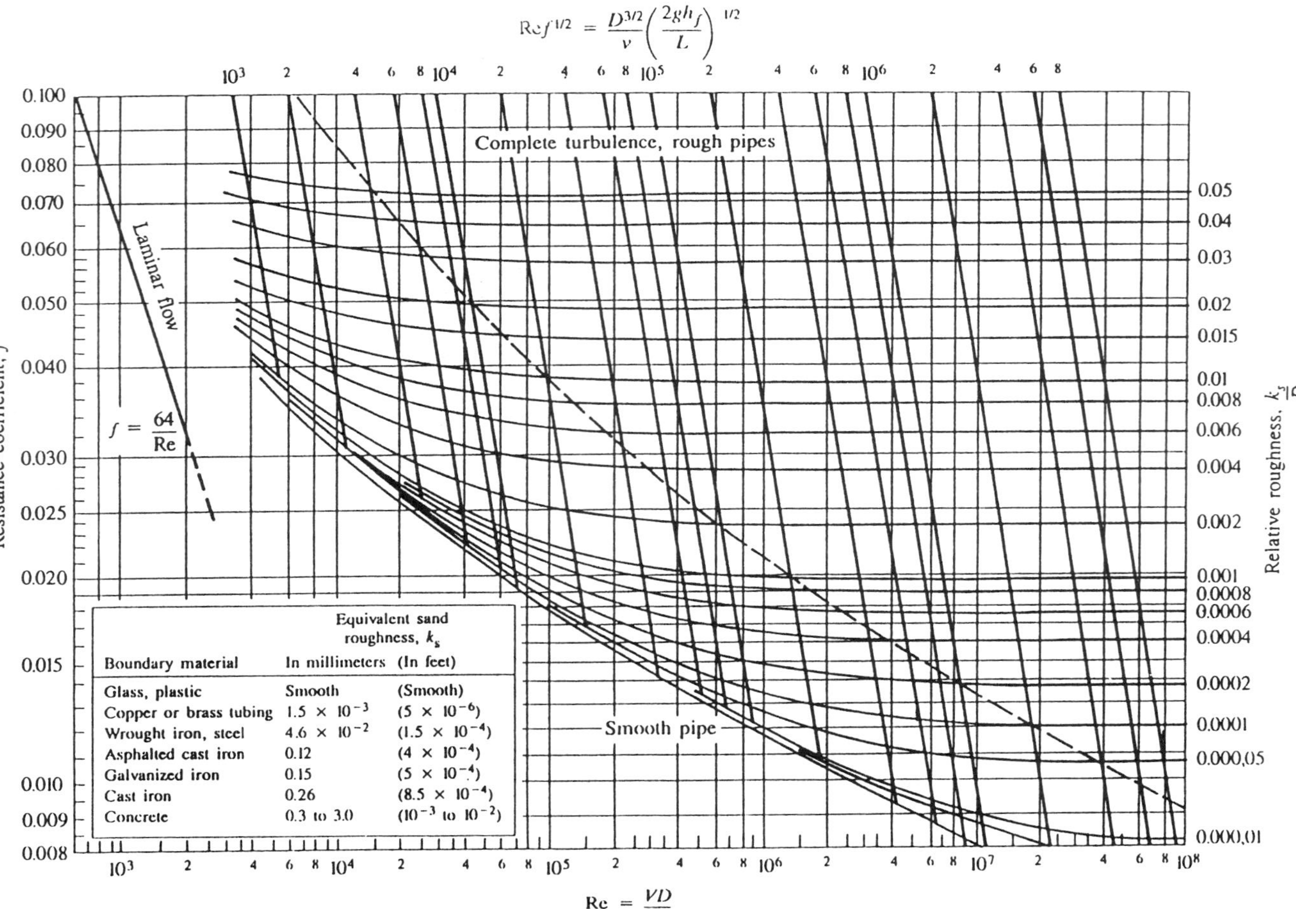

FIGURE 28.22 The Moody diagram. *Source*: Roberson, J. A. and Crowe, C. T. 1993. *Engineering Fluid Mechanics*, 5th ed. Houghton Mifflin, Boston, MA.

Table 28.9 Table of Transition Loss Coefficients

Type of Fitting	Screwed			Flanged		
Diameter	1 in.	2 in.	4 in.	2 in.	4 in.	8 in.
Globe valve (fully open)	8.2	6.9	5.7	8.5	6.0	5.8
(half open)	20	17	14	21	15	14
(one-quarter open)	57	48	40	60	42	41
Angle valve (fully open)	4.7	2.0	1.0	2.4	2.0	2.0
Swing check valve (fully open)	2.9	2.1	2.0	2.0	2.0	2.0
Gate valve (fully open)	0.24	0.16	0.11	0.35	0.16	0.07
Return bend	1.5	.95	.64	0.35	0.30	0.25
Tee (branch)	1.8	1.4	1.1	0.80	0.64	0.58
Tee (line)	0.9	0.9	0.9	0.19	0.14	0.10
Standard elbow	1.5	0.95	0.64	0.39	0.30	0.26
Long sweep elbow	0.72	0.41	0.23	0.30	0.19	0.15
45° elbow	0.32	0.30	0.29			
Square-edged entrance 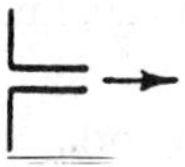			0.5			
Reentrant entrance 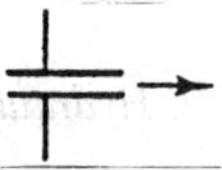			0.8			
Well-rounded entrance 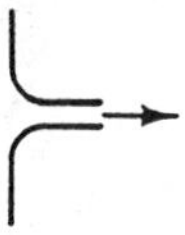			0.03			
Pipe exit			1.0			
	Area ratio					
Sudden contraction[b] 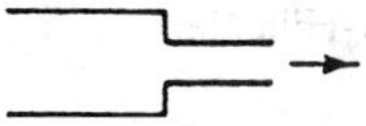	2:1		0.25			
	5:1		0.41			
	10:1		0.46			
	Area ratio A/A_o					
Orifice plate	1.5:1		0.85			
	2:1		3.4			
	4:1		29			
	$\geq 6:1$		$2.78\,[(A/A_0) - 0.6]^2$			
Sudden enlargement[c] 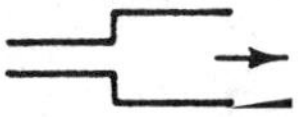			$[1 - (A_1/A_2)]^2$			
90° miter bend (without vanes) 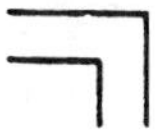			1.1			
(with vanes) 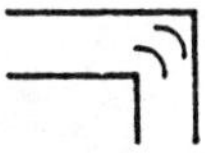			0.2			
General contraction (30° included angle)			0.02			
(70° included angle) 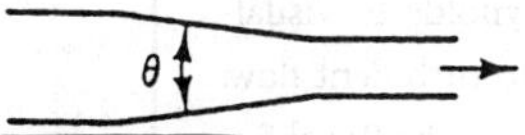			0.07			

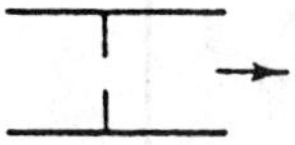

Source: Potter, M. C. and Wiggert, D. C. 1991. *Mechanics of Fluids,* Prentice-Hall, Englewood Cliffs, NJ. With permission.

[a] Values for other geometries can be found in *Technical Paper 410*, The Crane Company, 1957.

[b] Based on exit velocity V_2.

[c] Based on entrance velocity V_1.

Defining Terms

Head: A measure of mechanical energy per unit weight of fluid in terms of the height of a column of stagnant fluid.

Separation: A flow phenomenon in which streamlines that, initially, closely follow a flow boundary break abruptly away from the boundary and typically enclose a downstream region of recirculating flow.

Streamline: A line in a flow field, each point of which is tangent to the instantaneous velocity vector at that point.

Turbulence: A state of flow that is characterized by apparently random changes in time and space of flow variables and is associated with a high degree of mixing.

References

Bos, M. G., ed. 1989. *Discharge Measurement Structures*, 3rd rev. ed., Publication 20, Int. Inst. Land Reclamation and Improvement, Wageningen, The Netherlands.

Miller, R. W. 1989. *Flow Measurement Engineering*, 2nd ed., McGraw-Hill, New York.

Sharpe, J. J. 1981. *Hydraulic Modelling*, Butterworths, London.

Swamee, P. K. and Jain, A. K. 1976. Explicit equations for pipe flow problems. *J. Hydraulics Div., ASCE.* **102** (HY5): 657–664.

Yalin, M. S. 1971. *Theory of Hydraulic Models*. Macmillan, London.

For Further Information

Most of the topics treated are discussed in greater detail in standard elementary texts on fluid mechanics or hydraulics, including the following:

Daugherty, R. L., Franzini, J. B., and Finnemore, E. J. 1985. *Fluid Mechanics*, 8th ed., McGraw-Hill, New York.

Roberson, J. A. and Crowe, C. T. 1993. *Engineering Fluid Mechanics*, 5th ed., Houghton Mifflin, Boston, MA.

Fox, R. W. and McDonald, A. T. 1992. *Introduction to Fluid Mechanics*, 4th ed., McGraw-Hill, New York.

The original Reynolds apparatus. This apparatus was utilized by Osborne Reynolds to visualize the transition from laminar to turbulent flow. (From Philosophical Transactions of the Royal Society of London, 1883, Vol. 174, Part III, p. 935, plate 73.)

29

Open Channel Hydraulics

†Aldo Giorgini
Purdue University

Open channel hydraulics is a subject of great importance to civil engineers. It deals with flow problems involving a free surface: manufactured channels for water supply, irrigation, drainage, or power; ditches, rivers, and estuaries; conduits flowing partly full, such as pipes, sewers, and tunnels. Open channel hydraulics includes permanent flows that are steady in time, varied flows that have changes in depth and velocity along the channel, and transient flows that are time dependent. This chapter deals only with rigid-boundary channels without sediment deposition or erosion. The laminar regime is unimportant in open channel flow and only turbulent flows are considered. Only the one-dimensional treatment of uniform, nonuniform, and unsteady flows, which are so important in civil engineering practice, is discussed. Design aspects of structures involving free surface flows are discussed in Chapter 35. Sediment transport in open channels is covered in Chapter 33.

†Due to the untimely death of Professor Giorgini, this chapter was completed by his colleague, Dr. Jacques W. Delleur. Dr. Giorgini's class notes were used as a source of material, particularly for the figures, and every effort was made to retain the spirit of his manuscript. This chapter is dedicated to him.

0-8493-8953-4/95/$0.00 + $.50

29.1 Definitions and Principles

Open channel flow is the flow of a liquid with a free surface. This type of flow may occur either in an open channel, that is, in a convex vessel without upper closure (as in rivers, canals, and ditches) or in closed conduits, when only part of the cross section is occupied by the liquid (as in sewers).

The flow is assumed to be streamlined quasi-parallel, or one-dimensional. For any given cross section the following terminology and notation are used:

- The region occupied by the liquid is called the *flow area, A*.
- The part of the cross section perimeter which is below the water surface profile is called the *wetted perimeter, P*. (See Chapter 28, Section 28.6.)
- The length of the free surface is called the *topwidth, T*.
- The width of a rectangular channel is its *breadth, b*.
- The *elevation of the bottom* with respect to an arbitrary reference level is z.
- For prismatic channels, the vertical distance to the free surface from the lowest point of the cross section is called *depth of flow, y*, or *water depth* (whenever the liquid is water). Referring to Fig. 29.1, y_1 is the water depth at point A and d_1 is the thickness of the stream [Chow, 1959]. For small slopes normally encountered in rivers and canals $d \approx y$. The pressure head on the channel bottom is $y \cos^2 \theta = d \cos \theta$. For $\theta < 6°$ the error in approximating pressure head by y is less than 1%. For nonprismatic channels, like the natural channels, the term *water surface elevation, h*, is more meaningful.
- The *hydraulic depth* is $D = A/T$.
- The *hydraulic radius* is $R = A/P$.
- The *density* of the liquid, ρ, is the mass per unit volume and the *specific weight*, γ, is the weight per unit volume, so that $\gamma = \rho g$, where g is the gravitational acceleration.

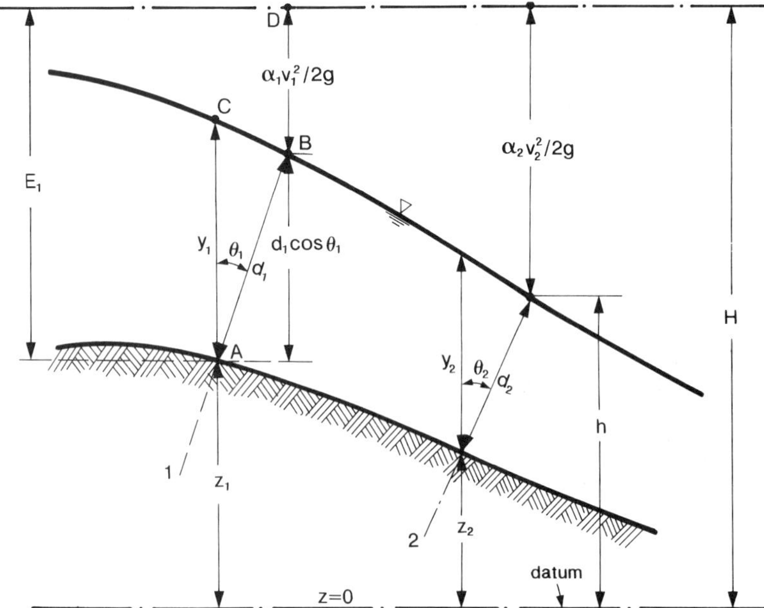

FIGURE 29.1 Definitions: y_1, depth of stream at A; d_1, thickness of stream at A; E_1, specific energy at A; $\alpha_1 V_1^2/2g$, velocity head at A; h, piezometric head or water surface elevation; H, total head.

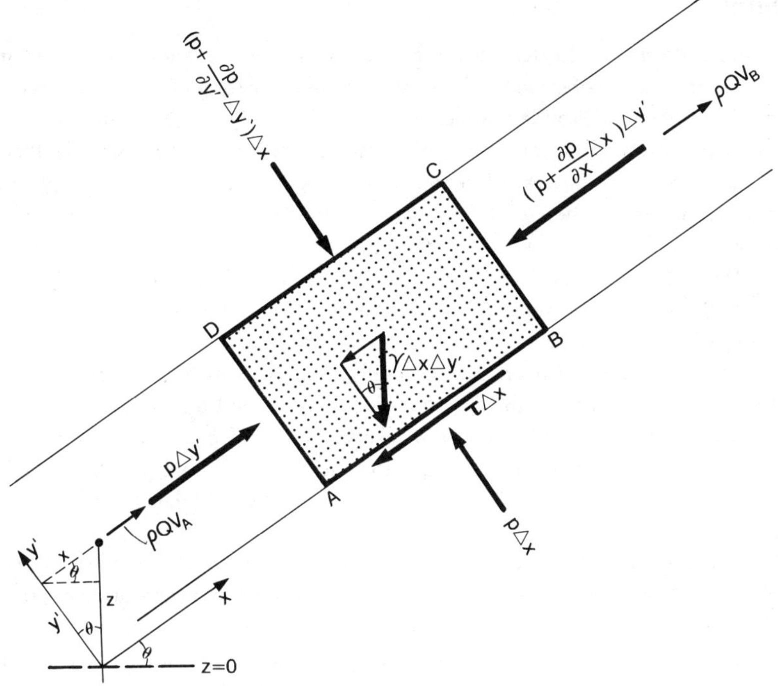

FIGURE 29.2 Forces on a fluid element and momentum flux.

The direction of flow is indicated by the space variables x, and the two coordinates orthogonal to each other and to x are called y' and z' (Fig. 29.2). Since the flow considered is parallel, the total volume of water flowing per unit time across an orthogonal cross section is the *flow rate* or *discharge* at location x and at time t, $Q(x, t)$, given by

$$Q(x,t) = \int_A\!\!\int v(x, y', z', t)\, dy' dz' = V(x,t)A(x,t) \tag{29.1}$$

where $v(x, y', z', t)$ is the local velocity at coordinates x, y', z' at time t; the integral extends across the whole flow area $A(x, t)$ at station x at time t, and $V(x, t)$ is the mean velocity or average discharge per unit area at station x and at time t.

Classification of Flows

Flows are usually classified according to the characteristics of the free surface. If the flow rate changes along the direction of flow (due to addition or subtraction of flow), it is a *spatially varied flow.* The most common case is the *spatially invariant flow. Steady flows* are time-invariant and *unsteady flows* are time-varying. If the flow area varies very gradually along the direction of flow it is *gradually varied.* If the variation of cross-sectional area is rather abrupt, so that the streamlines (see Chapter 28, Section 28.3) converge or diverge rapidly, then the flow is *rapidly varied.* In prismatic channels of constant slope, if the free surface has the same slope as the channel bottom, then all cross sections have the same flow area and therefore have the same kinematic characteristics. This type of flow is called *uniform flow* or *normal flow.* Uniform flow seldom occurs in natural channels, because the cross-sectional profiles usually vary significantly along the flow direction.

Flow Regimes

Since free surface flows are affected by the gravitational, viscous, and surface tension forces, different flow regimes can be observed in different regions of the three-dimensional space spanned by the Reynolds number, the Froude number, and the Weber number. The Weber number (which is a measure of the effects of surface tension) is unimportant for channels larger than narrow fissures, and therefore most analyses of free surface flow regimes can be done in the plane spanned by the Reynolds number and the Froude number. The Reynolds number $\text{Re} = \rho VR/\mu$ (where ρ is the density and μ is the dynamic viscosity of the liquid) increases as inertial effects predominate over viscous effects. The flow may be laminar, transitional, or turbulent, with a transitional range that spans from $\text{Re} = 500$ to an upper limit that may vary according to other flow characteristics, but which is usually less than $\text{Re} = 12{,}500$. In most natural and manufactured open channels Re is larger than 12,500 and so hydraulics generally deals with turbulent flows, albeit laminar and transitional flow may occur in the so-called sheet flow, found on a large area under rain or immediately after a rain event and in the flow of a viscous fluid in a small trough such as that used to drain the oil from a car [see Eq. (28.22)]. In general, the Reynolds number is high enough that the friction effects are independent of the Reynolds number and depend only on the relative roughness.

The only parameter left is the Froude number, which is a measure of the effects of gravity. It is defined as $\text{Fr} = V/C$, where $C = (gy)^{1/2}$ is the celerity or velocity of propagation of gravity waves generated by small free surface perturbations such as the celerity of a small wavelet created by a stone dropped on the surface of a still lake of depth y. The flow is said to be *subcritical* if $\text{Fr} < 1$, *supercritical* where $\text{Fr} > 1$, and *critical* where $\text{Fr} = 1$. When the flow is supercritical the velocity of the stream is larger than the celerity of the perturbation. Any change in free surface caused by an observer at any cross section cannot propagate upstream, and, therefore, the observer cannot modify or control the characteristics of the stream at that cross section. *Supercritical streams can be controlled only by upstream devices*, in the sense that, with a supercritical stream, the controlling device is upstream of the observer. When the flow is subcritical, the stream velocity is less than the celerity of perturbation. Any change in free surface caused by an observer at any cross section can propagate upstream. Therefore, *subcritical streams can be controlled by downstream devices.*

29.2 Balance and Conservation Principles

The laws of nature are usually written in a "balance" form. These have been discussed in Chapter 28.

Conservation of Mass

The principle of conservation of mass states that the time rate of change of mass inside the control volume is equal to the balance between the inflowing and outflowing mass through the control surfaces [see Eq. (28.12)]. For a homogeneous liquid, the density in open channel flow is usually assumed constant. The conservation of mass becomes a conservation of volume.

In the case of steady flow in an open channel, the volume inside the control volume or channel reach is constant and $\sum Q_{\text{in}} = \sum Q_{\text{out}}$ or

$$A_1V_1 = A_2V_2 \tag{29.2}$$

where the indices 1 and 2 refer to the entrance and the exit cross sections of the reach. For a rectangular channel

$$q = V_1d_1 = V_2d_2 = \text{Constant} \tag{29.3}$$

where the specific flow rate, $q = Q/b$, is the flow rate per unit width of the channel.

Conservation of Momentum

The conservation of momentum states that the time rate of change of the momentum inside the control volume is equal to the sum of all forces acting on the control volume plus the difference between the incoming and outgoing momentum fluxes [see Eq. (28.14)]. Note that it is a vector equation. The momentum equation in the x direction for steady flow and with constant density ρ and discharge Q is

$$\sum \overline{F} = \rho Q(\beta_2 V_2 - \beta_1 V_1) \tag{29.4}$$

where $\sum \overline{F}$ is the sum of the forces (with proper sign convention) along the direction of flux acting on the control region. These forces typically include the hydrostatic pressure forces on the end sections, the weight of liquid between the end sections, and the shear force on the wetted surface of the channel. The subscripts 1 and 2 refer to the entrance and exit sections of the control volume, respectively. β is a dimensionless number called the *momentum correction factor* which reflects the fact that the velocity is not constant through the section:

$$\beta = \frac{1}{V^2 A} \int_A v^2 dA = \frac{1}{A} \int_A \left(\frac{v}{V}\right)^2 dA \tag{29.5}$$

The value of β is often taken as 1.0 but its value increases beyond 1.0 as the degree of irregularity of the cross section increases.

With the assumption of $\beta = 1$, the momentum conservation equation (29.4) reduces to

$$\sum \overline{F} = \rho Q(V_2 - V_1) \tag{29.6}$$

Piezometric Head

As a first application of the momentum equation, consider a control region delimited by the channel, the water surface, and two cross sections at x and $x + \Delta x$ as shown in Fig. 29.2 and apply the momentum equation *normal* to the flow. Since the momentum fluxes and eventual shears along the face AB have no orthogonal component, the net pressure force in the y' direction, $\partial p/\partial y' \Delta x \Delta y'$, must be balanced by the orthogonal component of the weight of the control region: $\gamma \Delta x \Delta y' \cos\theta = \gamma \Delta x \Delta y' \partial z/\partial y'$; thus, $\partial/\partial y'(z + p/\gamma) = 0$, or

$$\left(z + \frac{p}{\gamma}\right) = h \tag{29.7}$$

This result is fundamental to the field of free surface flow hydraulics based on the parallel flow approximation. All the points of the same cross section of a parallel flow have the same piezometric head, which coincides with the elevation of the free surface for that cross section if the slope is not too steep, say, $\theta < 6°$. This result suggests two corollaries: the pressure distribution within a given cross section of a parallel flow is hydrostatic, and the free surface profile of a parallel flow is the piezometric line of the free surface flow (see Fig. 29.1).

Boundary Shear

A second application of the momentum equation, this time along the direction of motion, considers as control region a full slice of channel as presented in Fig. 29.2. In this case too the momentum fluxes do not directly enter the picture because the momentum flux entering the control region equals the momentum flux exiting [i.e., $V_2 = V_1$ in Eq. (29.4)]. Furthermore, if the depth is constant, the difference between the pressure forces acting on the entry and exit faces of the control region must be zero. Therefore the streamwise component of the weight of the fluid in the control region must be balanced by the shear force acting on the wetted perimeter. Thus, if τ is the average shear stress on the channel wall, $P\Delta x\tau = \gamma A\,\Delta x \sin\theta$, or, since $R = A/P$ has been defined as the hydraulic radius,

$$\tau = \gamma RS \tag{29.8}$$

where $S = -dh/dx$, the loss of piezometric head per unit channel length, is misnamed "slope" of the hydraulic grade line. Equation (29.8) will be reintroduced when the effects of viscosity and of wall roughness are introduced.

Total Thrust and Specific Force

Another application of the momentum Eq. (29.4) concerns a small channel reach defined by two parallel cross sections AD and BC similar to Fig. 29.2 but with varying depth. The forces in the flow direction are the hydrostatic forces on the end surfaces and the component of the weight of the liquid in the flow direction. Friction forces are neglected. The momentum equation in the flow direction is thus

$$\gamma\bar{d}_1A_1 - \gamma_2\bar{d}_2A_2 - W\sin\theta = \beta_2\rho\frac{Q^2}{A_2} - \beta_1\rho\frac{Q^2}{A_1} \tag{29.9}$$

where $\bar{d}$ represents the depth of the center of gravity and W is the weight of the liquid in the control volume. This can be written as

$$\left(\gamma\bar{d}_1A_1 + \beta_1\rho\frac{Q^2}{A_1}\right) = \left(\gamma\bar{d}_2A_2 + \beta_2\rho\frac{Q^2}{A_2}\right) + W\sin\theta \tag{29.10}$$

This equation contains the term

$$F = \gamma\bar{d}A + \beta\rho\left(Q^2/A\right) \tag{29.11}$$

which is the **total thrust** acting on a cross section; it is the sum of the *hydrostatic thrust* $\gamma\bar{d}A$ and the *momentum flux* $\beta\rho\left(Q^2/A\right)$. The function F varies with d as illustrated in Fig. 29.3. When all terms of Eq. (29.11) are divided by the specific weight, γ, the term F/γ is known as the **specific force** [Chow, 1959].

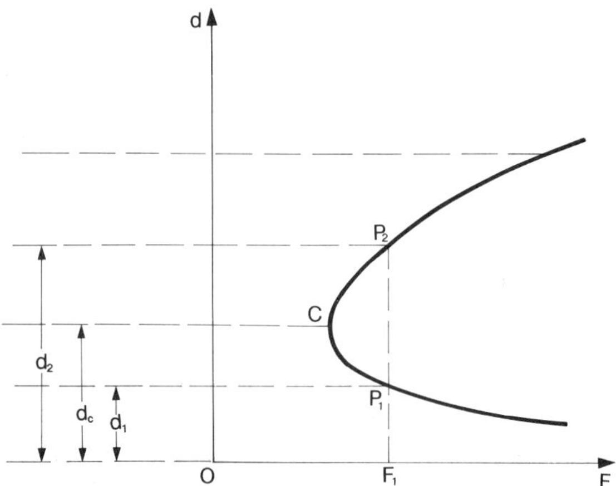

FIGURE 29.3 Total thrust curve.

The value of d for which the minimum of F occurs is called the *critical stream thickness*, d_c, and it can be found by zeroing the derivative of Eq. (29.11), that is, by solving the equation

$$A\frac{\mathrm{d}\bar{d}}{\mathrm{d}d} + \bar{d}T = \beta\frac{Q^2T}{gA^2} \tag{29.12}$$

since $T\mathrm{d}d = \mathrm{d}A$ and Q is constant. This equation can be solved analytically only for a few channel geometries.

Balance of Mechanical Energy

The mechanical energy of a body of mass m is the sum of its potential energy mgz and its kinetic energy $\frac{1}{2}mv^2$, where z is the elevation of the mass m above a reference datum. The conservation of energy states that the time rate of change in mechanical energy in the control volume is equal to the flux of mechanical energy at the inlet and outlet sections, plus the work done by the pressure forces at the inlet and at the outlet of the control region and plus the loss of mechanical energy in the control region [see Eq. (28.16)]. For steady flows the energy equation becomes

$$z_1 + \frac{p_1}{\gamma} + \alpha_1\frac{V_1^2}{2g} = z_2 + \frac{p_2}{\gamma} + \alpha_2\frac{V_2^2}{2g} + h_L \tag{29.13}$$

where V is the average velocity, h_L is the head loss, and α is the kinetic energy correction factor that accounts for the nonuniformity of the velocity across the section:

$$\alpha = \frac{1}{A}\int_A \left(\frac{v}{V}\right)^3 dA \tag{29.14}$$

α takes values close to 1, but increases as the channel cross section becomes more irregular. Referring to Fig. 29.1, Eq. (29.13) is rewritten for prismatic open channels as

$$z_1 + d_1\cos\theta_1 + \alpha_1 V_1^2/(2g) = z_2 + d_2\cos\theta_2 + \alpha_2 V_2^2/(2g) + h_L \tag{29.15}$$

Specific Energy

The **specific energy,** E, is defined as

$$E = d\cos\theta + \alpha V^2/(2g) = d\cos\theta + \alpha Q^2/(2gA^2) \tag{29.16}$$

and so Eq. (29.15) becomes

$$E_2 = E_1 + z_1 - z_2 - h_L \tag{29.17}$$

For a rectangular channel, $q = Q/b$, Eq. (29.16) can be written as

$$E = d\cos\theta + \alpha q^2/(2gd^2) \tag{29.18}$$

The specific energy E for q constant is shown in Fig. 29.4(a). The depth d_c, corresponding to the minimum specific energy, is the critical depth. For one value of $E > E_{\min}$ there are two possible depths: $d_1 < d_c$ corresponding to supercritical flow and $d_2 > d_c$ for subcritical flow. The two depths d_1 and d_2 are called **alternate depths**. This equation can also be presented in dimensionless form, as shown in Fig. 29.4(b) and Table 29.1. Assuming that α does not vary with d, the critical depth is found by taking the derivative of E with respect to d and equating to zero. This yields the following expression for the critical depth in a rectangular channel:

$$d_c = [(\alpha q^2)/(g\cos\theta)]^{1/3} \tag{29.19}$$

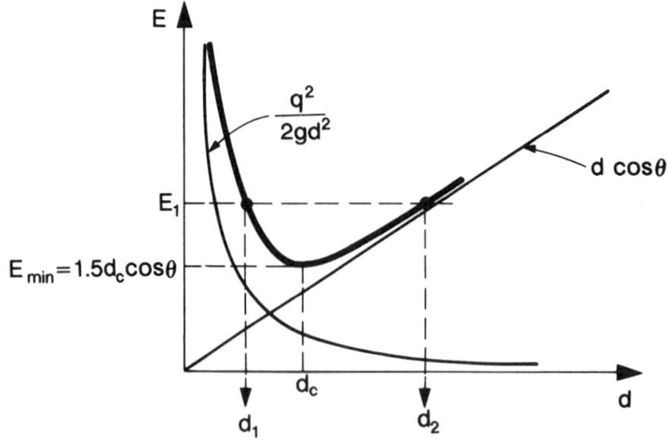

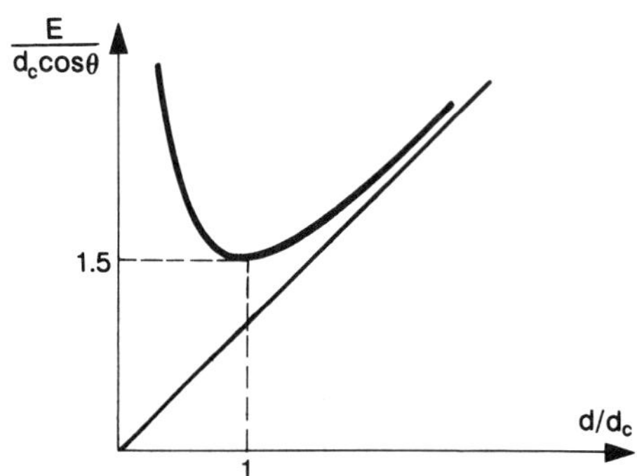

FIGURE 29.4 (a)Specific energy diagram; (b) nondimensional specific energy diagram.

and the corresponding minimum of specific energy is

$$E_{\min} = \frac{3}{2} d_c \cos\theta \tag{29.20}$$

Equations (29.16) and (29.17) are the foundations for the calculation of water surface profiles. In fact given a flow rate Q and the geometry of cross section 1, with the water thickness there d_1, Eq. (29.16) allows us to find E_1. Then Eq. (29.17) allows us to find E_2, if an estimate of the head losses h_L is possible. Once E_2 is known, Eq. (29.16), used as an equation in d, permits the calculation of the two possible stream thicknesses d_1 and d_2.

Critical Flow

Critical flow occurs when the given discharge Q flows with minimal specific energy or minimal total thrust. In principle the two requirements seem different. Minimizing the specific energy, that is, minimizing E given by Eq. (29.16), one obtains

$$\cos\theta A^3/T = \alpha Q^2/g \tag{29.21}$$

(if α is not considered a function of d), and, minimizing the total thrust, one obtains Eq. (29.12) rewritten here to facilitate comparison with Eq. (29.21):

Table 29.1 Specific Energy Table

$E/d_c \cos\theta$ E/y_c	d_1/d_c y_1/y_c	d_2/d_c y_2/y_c	$E/d_c \cos\theta$ E/y_c	d_1/d_c y_1/y_c	d_2/d_c y_2/y_c
1.500	1.000	1.000	2.500	0.500	2.414
1.505	0.944	1.060	2.600	0.500	2.521
1.510	0.923	1.086	2.800	0.462	2.733
1.515	0.906	1.107	3.000	0.442	2.942
1.520	0.893	1.125	3.500	0.402	3.458
1.525	0.881	1.141	4.000	0.371	3.963
1.530	0.871	1.156	4.500	0.347	4.475
1.540	0.853	1.182	5.000	0.327	4.980
1.550	0.838	1.207	5.500	0.310	5.483
1.560	0.825	1.229	6.000	0.296	5.986
1.570	0.812	1.250	7.000	0.273	6.990
1.580	0.801	1.270	8.000	0.254	7.992
1.590	0.791	1.289	9.000	0.239	8.994
1.600	0.782	1.308	10.000	0.226	9.995
1.625	0.761	1.351	11.000	0.215	10.996
1.650	0.742	1.392	12.000	0.206	11.997
1.675	0.726	1.431	14.000	0.190	13.997
1.700	0.711	1.468	16.000	0.178	15.998
1.750	0.685	1.539	18.000	0.167	17.998
1.800	0.663	1.606	20.000	0.159	19.999
1.850	0.644	1.671	25.000	0.142	24.999
1.900	0.627	1.734	30.000	0.129	29.999
2.000	0.597	1.855	35.000	0.120	35.000
2.100	0.572	1.971	40.000	0.112	40.000
2.200	0.551	2.085	45.000	0.106	45.000
2.300	0.532	2.196	50.000	0.100	50.000
2.400	0.515	2.306			

$$\frac{A^3}{T}\frac{\mathrm{d}\bar{d}}{\mathrm{d}d} + A^2\bar{d} = \frac{\beta Q^2}{g} \tag{29.22}$$

In the above equations, both A and T are functions of d. Solving Eqs. (29.21) and (29.22) for Q/A, the following two expressions for the critical velocity for small slopes are obtained:

$$V_c = \sqrt{(gD)/\alpha} \tag{29.23}$$

and

$$V_c = \left[(g/\beta)\left(D\frac{\mathrm{d}\bar{d}}{\mathrm{d}d} + \bar{d}\right)\right]^{1/2} \tag{29.24}$$

where $D = A/T$ is the hydraulic depth. The two equations coincide, in the case of rectangular channels, when $\alpha = \beta$. Equations (29.23) and (29.24) have been derived by assuming that α and β are not dependent on d, which is usually not the case. The calculation of d_c is usually done by solving Eq. (29.21) numerically.

Hydraulic Jump

A hydraulic jump is a sudden increase in depth that occurs whenever the flow changes from supercritical to subcritical as a result of the deceleration of the flow. Momentum balance requires that, for a rectangular channel on a horizontal surface, the depth before the jump, y_1, and the depth after the jump, y_2, are related by the expression

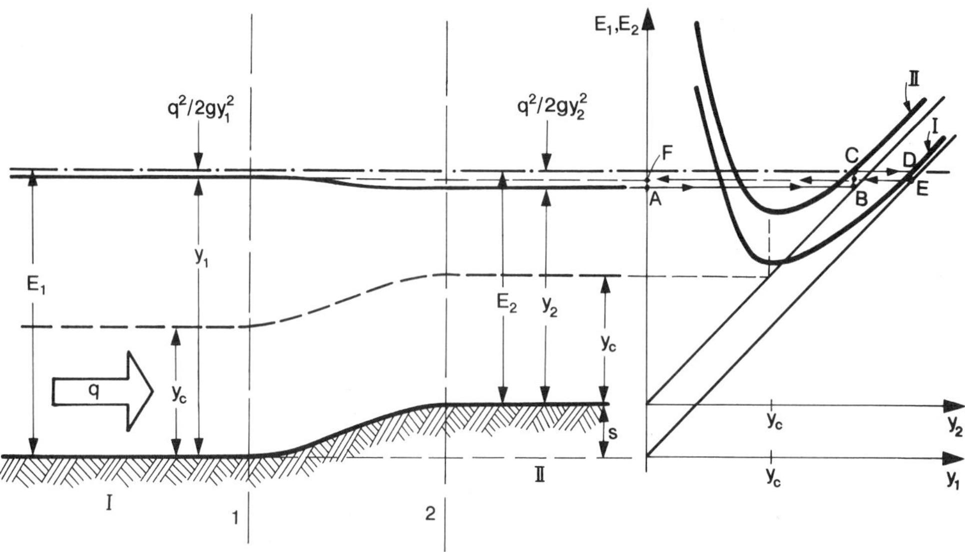

FIGURE 29.5 Subcritical stream on upward step.

$$y_2/y_1 = \tfrac{1}{2}[(1 + 8\,\mathrm{Fr}_1^2)^{1/2} - 1] \tag{29.25}$$

where $\mathrm{Fr}_1 = V_1/(gy_1)^{1/2}$ is the Froude number of the flow before the jump. The depths y_1 and y_2 are called **conjugate depths**. The head loss in the jump is given by

$$h_j = [y_1 + V_1^2/(2g)] - [y_2 + V_2^2/(2g)] = (y_2 - y_1)^3/(4y_1y_2). \tag{29.26}$$

The length of the jump is approximately $6y_2$ for Froude numbers $\mathrm{Fr} > 4.5$. The hydraulic jump is an effective means of dissipating excess kinetic energy of supercritical flows.

Example 29.1: Subcritical Flow on a Step. Consider the subcritical flow in the rectangular horizontal channel shown in Fig. 29.5 which presents an upward step of height s with respect to the direction of the current. The specific energy upstream of the step is larger than the minimum $3y_c/2$. The water surface elevation at section 1 is higher than at section 2. Given $q = 10\ \mathrm{m^2/s}$, $y_2 = 3.92$ m, and $s = 0.5$ m, find y_1, assuming no losses.

1. Calculate $y_c = (q^2/g)^{1/3} = 2.17$ m.
2. Find $E_2 = y_2 + q^2/(2gy_2^2) = 4.25$ m.
3. Find $E_1 = E_2 + s$ (step height) $= 4.75$ m.
4. Enter Table 29.1 with $E_1/y_c = 2.19$.
5. Find $y_1/y_c = 2.07$. Thus $y_1 = 4.5$ m. This result can also be obtained by solving Eq. (29.18). If the bottom of the upward step increases very gradually, one can find intermediate points along the step, as shown in Fig. 29.5. It is useful to draw the critical depth line in order to show the relative distance of the free surface profile from it. Figure 29.5 also shows a graphical technique based on the curve $E(y)$. Follow the path A, B, C, D, E, F. The limit case occurs when the specific energy on the upper part of the step is a minimum, that is, when the depth is critical.

Example 29.2: Supercritical Flow on a Step. Consider the supercritical flow in the rectangular horizontal channel of Fig. 29.6, which presents an upward step with respect to the direction of the current. The specific energy upstream of the step is larger than the minimum specific energy $3y_c/2$ and the flow is rapid. The water depth at cross section 2 is larger than at section 1 due to the

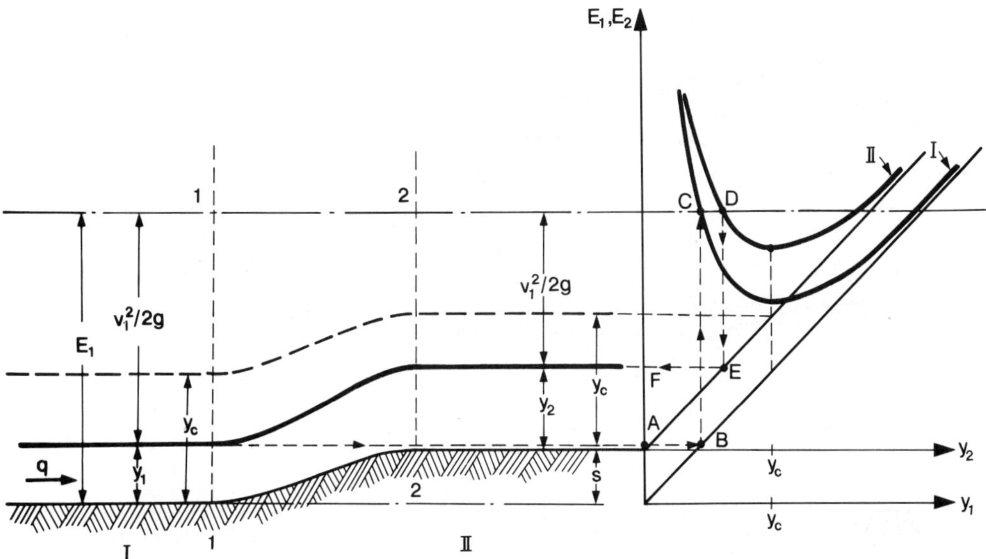

FIGURE 29.6 Supercritical stream on upward step.

decreased kinetic energy of the stream. Given $q = 10\ \text{m}^2/\text{s}, y_1 = 1.2$ m, and $s = 0.5$ m, find y_2 assuming no losses.

1. Find $y_c = (q^2/g)^{1/3} = 2.17$ m.
2. Find $E_1 = y_1 + q^2/(2g\,y_1^2) = 4.74$ m.
3. Find $E_2 = E_1 - s = 4.24$ m.
4. Enter Table 29.1 with $E_2/y_c = 1.95$.
5. Find $y_2/y_c = 0.612$, thus $y_2 = 1.32$ m. This result can also be obtained from Eq. (29.18). Figure 29.6 shows a graphical technique based on the curve $E(y)$. Follow the path A, B, C, D, E, F. The limit case occurs when the specific energy on the upper part of the step is a minimum, that is, when the depth is critical.

Example 29.3: Constriction in a Subcritical Flow. Consider the horizontal rectangular channel of Fig. 29.7 with a constriction in the direction of the current. The specific energy upstream of the constriction is larger than $E_{\text{min}} = 3y_c/2$. The water depth at section 2 is shallower than at section 1 as the kinetic energy of the stream increases. Given $q_1 = 10\ \text{m}^2/\text{s}, y_2 = 5.46$ m, find y_1, assuming no losses.

1. Find $y_c = (q_1^2/g)^{1/3} = 2.17$ m and $y_c' = [(2q_1)^2/g]^{1/3} = 3.44$ m.
2. Find $E_2 = y_2 + q_2^2/2gy_2^2 = 6.14$ m with $q_2 = 2q_1$.
3. $E_1 = 6.14$ m.
4. Enter Table 29.1 with $E_1/y_c = 2.83$.
5. Find $y_1/y_c = 2.80$ m, thus $y_1 = 6.0$ m. It is seen that $E_2 > 1.53\ y_c'/2$, thus the flow is subcritical in the constriction. The limiting case occurs when $y_2 = y_c'$. Further information on sills, contractions, and expansions for subcritical and supercritical flows can be found in Ippen [1950].

29.3 Uniform or Normal Flow

When a flow rate Q is discharged in a constant slope prismatic channel, a constant-depth steady state will be reached somewhere in the channel. Such flow is called *uniform flow* or *normal flow*.

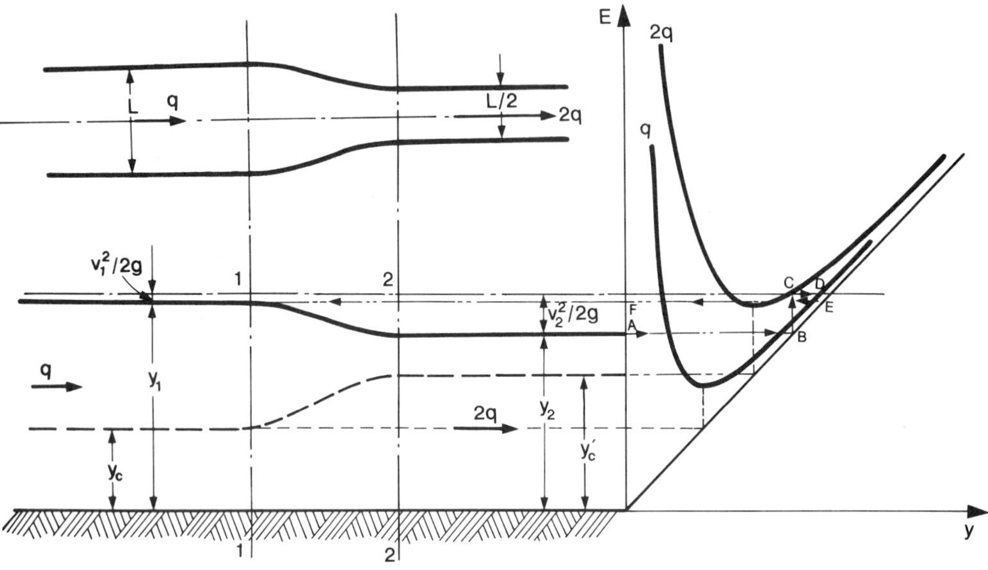

FIGURE 29.7 Subcritical stream in a channel contraction.

The shear stress τ can be written as

$$\tau = \rho V^2 f \text{ (relative roughness)} \tag{29.27}$$

where f is a nondimensional function. Manning has proposed for f the function

$$f = gn^2/(K_M^2 R^{1/3}) \tag{29.28}$$

where n is a nondimensional number characterizing the roughness of a surface and K_M is a constant with dimension $L^{1/3}T^{-1}$ with value 1 $m^{1/3}$/s or 1.486 $ft^{1/3}$/s. For other considerations concerning the dimensionality of n, see Chow [1959, p. 98]. Typical values of n are given in Table 29.2. More detailed information on n can be found in Chow [1959]. Using Eq. (29.28) in Eq. (29.27) and eliminating τ between it and Eq. (29.8) one obtains *Manning's equation*:

$$V = \frac{K_M}{n} R^{1/6}\sqrt{RS} = \frac{K_M}{n} R^{2/3}\sqrt{S} \tag{29.29}$$

Table 29.2 Manning's Roughness Coefficient

Nature of Surface	$n_{\min}$	$n_{\max}$
Neat cement surface	0.010	0.013
Concrete, precast	0.011	0.013
Cement mortar surfaces	0.011	0.015
Concrete, monolithic	0.012	0.016
Cement rubble surfaces	0.017	0.030
Canals and ditches, smooth earth	0.017	0.025
Canals		
Dredged in earth, smooth	0.025	0.033
In rock cuts, smooth	0.025	0.035
Rough beds and weeds on sides	0.025	0.040
Rock cuts, jagged and irregular	0.035	0.045
Natural streams:		
Smoothest	0.025	0.033
Roughest	0.045	0.060
Very weedy	0.075	0.150

or, multiplying this by the flow area A,

$$Q = \frac{K_M}{n} AR^{2/3} \sqrt{S} = \frac{K_M}{n} \frac{A^{5/3}}{P^{2/3}} \sqrt{S} \tag{29.29a}$$

where $K_M = 1$ for standard metric units and $K_M = 1.486$ for standard English units. The multiplier of $\sqrt{S}$ in Eq. (29.29a) is called the conveyance K of the cross section. Other commonly used uniform flow relationships include the Chézy equation,

$$V = C\sqrt{RS} \tag{29.30}$$

where the Chézy C and Manning's n are related by $C = K_M R^{1/6}/n$. Strickler [1923] and others related n to the diameter of sand grains of the stream bed. Subramanya [1982] gives a form of Strickler's equation as

$$n = 0.047\, d_{50}^{1/6} \tag{29.30a}$$

where d_{50} is the diameter of the bed material in meters such that 50% is smaller by weight.

The solution of Eq. (29.29) for d yields the stream thickness d_0 for normal flow. Where the channel slope is negligible it is called the *normal depth* y_0 and can be found by solving numerically the following rearrangement of Eq. (29.29a):

$$\frac{A^{5/3}}{P^{2/3}} = \frac{nQ}{K_M \sqrt{S}} \tag{29.31}$$

In the case of a very wide rectangular channel Eq. (29.31) reduces to

$$y_0 = [nq/(K_M \sqrt{S})]^{3/5} \tag{29.32}$$

where q is the discharge per unit width, since $R = A/P \approx y$.

Comparison of d_0 with d_c, given by Eq. (29.21), determines whether the normal flow is supercritical, critical, or subcritical. Critical conditions for a wide rectangular channel can be found by equating (29.32) to (29.19) to obtain

$$\frac{\alpha^5 q}{g^5 (\cos\theta)^5} = \frac{n^9}{K_M^9 S^{9/2}} \tag{29.33}$$

where $S = \sin\theta \approx \tan\theta$. For the usual conditions where S is small (of the order of 0.01) the above equation yields

$$\frac{\alpha^5 q}{g^5} = \frac{n^9}{K_M^9 S^{9/2}} \tag{29.34}$$

This equation yields the **critical slope** S_c of a given wide rectangular channel when the values of q, n, and g are given. A channel slope less than S_c is called **mild** and a slope larger than S_c is called **steep**. The same equation can be used to find the *critical discharge* q_c, when S, n, and g are fixed, or the *critical roughness* n_c, when S, g, and q are fixed.

29.4 Composite Cross Sections

This subsection is concerned with the determination of the discharge, global roughness, normal depth and critical depth of a composite section, as shown in Fig. 29.8(a). The three parts of the global channel behave as three different channels in parallel with the same slope and the same water surface elevation. The wetted perimeters of the subsections include only the portions of the

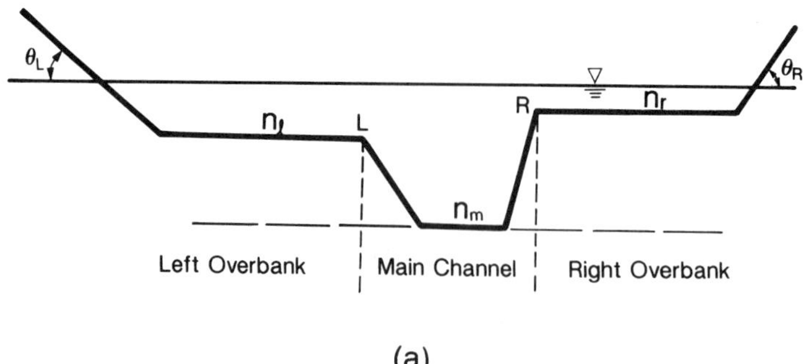

(a)

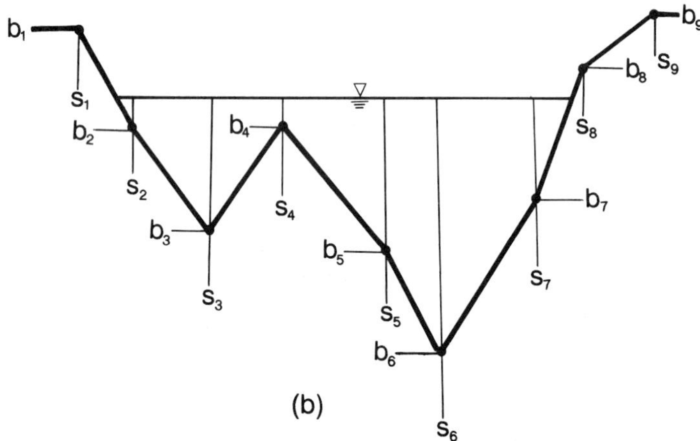

(b)

FIGURE 29.8 Composite cross sections.

solid boundary belonging to that section. The total discharge is

$$Q = K_M \left(\frac{1}{n_l} \frac{A_l^{5/3}}{P_l^{2/3}} + \frac{1}{n_m} \frac{A_m^{5/3}}{P_m^{2/3}} + \frac{1}{n_r} \frac{A_r^{5/3}}{P_r^{2/3}} \right) \sqrt{S} \tag{29.35}$$

where the subscripts l, m, r refer to the left overbank, the main channel, and the right overbank. From Eq. (29.35) with A representing the total cross-sectional area and P the total wetted perimeter, the global value of n given the individual values n_l, n_m, n_r is

$$\frac{1}{n} = \sum_l^r \frac{1}{n_i} \frac{(A_i/A)^{5/3}}{(P_i/P)^{2/3}} \tag{29.36}$$

Given the quantities Q, S, n_l, n_m, and n_r, and given the functions $A_l(y)$, $A_m(y)$, $A_r(y)$, $P_l(y)$, $P_m(y)$, and $P_r(y)$, the *normal depth* is the root y_0 of Eq. (29.35). Then y_0 can be substituted to find the actual values of the areas and wetted perimeters of the subsections to obtain Q_l, Q_c, and Q_r.

To find the critical depth it is necessary to obtain an estimate of the kinetic energy correction factor. For an irregular cross section, α defined in Eq. (29.14) becomes

$$\alpha = \frac{1}{A} \sum_l^r \left(\frac{V_i}{V} \right)^3 A_i = \sum_l^r \left(\frac{Q_i}{Q} \right)^3 \left(\frac{A}{A_i} \right)^2 = K_M^3 \frac{A^2}{Q^3} S^{3/2} \sum_l^r \frac{1}{n_i^3} \frac{A_i^3}{P_i^2} \tag{29.37}$$

Substituting Eq.(29.37) into Eq. (29.21), assuming $\cos\theta \approx 1$, yields

$$\frac{gQ}{K_M^3 S^{3/2}} = \frac{T}{A}\sum_{l}^{r}\frac{1}{n_i^3}\frac{A_i^3}{P_i^2} \tag{29.38}$$

The **critical depth** is obtained by solving Eq. (29.38) for $y = y_c$. Numerical techniques are generally needed to obtain the root, y_c, of Eq. (29.38).

Usually, river cross sections are given as sequences of station abscissas, s, and elevations, b, and a general cross section is approximated in polygonal form, as shown in Fig. 29.8(b). For each trapezoidal element, with z being the water surface elevation,

$$Q_i = \frac{K_M}{n_i}\frac{A_i^{5/3}}{P_i^{2/3}}\sqrt{S} = \frac{K_M}{n_i}\frac{(s_{i+1}-s_i)^{5/3}\left(z-\dfrac{b_{i+1}+b_i}{2}\right)^{5/3}}{\left((s_{i+1}-s_i)^2+(b_{i+1}-b_i)^2\right)^{1/3}}\sqrt{S} \tag{29.39}$$

29.5 Steady, Gradually Varied Flow Equation

In gradually varied flows the depth changes gradually in the flow direction. The slope of the channel bottom, S_0, the slope of the water surface, and the slope of the energy grade line S_f are all different. This section is concerned with the development of an expression for the rate of change of the depth along the channel, called the *gradually varied flow equation.*

From Fig. 29.1, it is seen that the total head H is

$$H = z + d\cos\theta + \alpha V^2/2g \tag{29.40}$$

Taking the derivative of both sides of Eq. (29.40) with respect to x, recognizing that $dH/dx = -S_f$ and $dz/dx = -S_0$, and assuming θ small,

$$\frac{dy}{dx} = \frac{S_0 - S_f}{1+\alpha\dfrac{d}{dy}\left(\dfrac{V^2}{2g}\right)} = S_0\frac{1-S_f/S_0}{1-\dfrac{\alpha Q^2}{g}\dfrac{T}{A^3}} \tag{29.41}$$

since $V = Q/A$ and $dA/dy = T$, where T is the free surface width, assuming Q constant.

When Q varies along the channel,

$$\frac{dy}{dx} = \frac{S_0 - S_f - \lambda(Q/gA^2)\,dQ/dx}{1-\mathrm{Fr}^2} \tag{29.41a}$$

with $\lambda = 1$ or 2 for decreasing or increasing discharge, respectively.

By substituting for S_f and S_0 the expressions $S_f = Q^2/K^2$ and $S_0 = Q^2/K_0^2$, where K is the conveyance at any depth and K_0 is the conveyance at normal depth, and substituting $\alpha\,Q^2/g = A_c^3/T_c$, Eq. (29.41) becomes

$$\frac{dy}{dx} = S_0\frac{1-(K_0/K)^2}{1-\dfrac{A_c^3/T_c}{A^3/T}} = S_0\frac{N(y)}{D(y)} \tag{29.42}$$

In Eq. (29.42) the subscript 0 refers to normal depth and the subscript c refers to critical depth. Equation (29.42) is the *differential equation for the water surface profile $y(x)$ in a prismatic channel.* It is a first-order differential equation, which can be solved, at least in principle, by direct integration:

$$\int_{y_1}^{y}\frac{D(y')}{N(y')}\,dy' = S_0(x-x_1) \tag{29.43}$$

where y_1 is the boundary condition at x_1. The boundary condition for a given reach is usually given at the downstream boundary if the type of flow in the reach is subcritical, or at the upstream boundary if the flow in the reach is supercritical.

As Eq. (29.43) is analytically integrable only under very particular conditions of channel geometry and resistance law, some general observations will be made here, valid for any prismatic channel geometry and for Manning's resistance law. The numerator of Eq. (29.42) becomes zero where the depth of the water approaches y_0. The derivative of y with respect to x is or approaches zero when either the curve coincides with the y_0 line (that is, with normal flow) or as the curve approaches the y_0 line asymptotically.

The denominator becomes zero where $y = y_c$, implying that if a free surface profile approaches the y_c line, it must do so with infinite slope, that is, vertically. In the neighborhood of $y \approx y_c$, however, the model of Eq. (29.42) is inadequate because of the large curvature of the streamlines. Also for very small values of the water depth y the derivative dy/dx tends to become infinite. For any reach of given constant slope S_0, the lines $y = y_0$ and $y = y_c$, together with the bottom line $y = 0$, divide the x, y plane into three significant regions (when $y_0 \neq y_c$) for the analysis of the qualitative behavior of the surface water profiles. With these observations in mind, a brief presentation of all possible types of water surface profiles is made in the next section.

29.6 Water Surface Profiles Analysis

Different classes of profiles can be distinguished, according to the relative magnitude of the bottom slope, S_0, and the critical slope S_c calculated by Eq. (29.34).

The Steep-Slope Class

In the steep-slope class $(S_0 > S_c)(y_0 < y_c)$ (Figs. 29.9, 29.10) the lines $y = 0, y = y_0, y = y_c$ divide the channel considered into three regions, $0 < y < y_0, y_0 < y < y_c, y > y_c$. The free-surface profiles belonging to the three regions are called S_3, S_2, and S_1, respectively (Figs. 29.9 and 29.10).

The S_1 Profile $(y > y_c)$

Since in this case in Eq. (29.42) $N > 0$ and $D > 0$, dy/dx is always positive and y has the tendency to grow as x increases from a starting datum. If this datum coincides with the critical depth y_c, the profile grows from it with infinite slope. The slope decreases very rapidly to become S_0 for large values of x (where the values of y are large) because the ratio N/D tends to 1. This means that the S_1 profile has a horizontal asymptote for $x \to \infty$. This is illustrated in Fig. 29.9. Since S_1 is a subcritical profile, it is drawn from downstream to upstream always below a horizontal line through the control point P_1 (through which the profile is known to pass) and cuts the y_c line vertically at Q. In practice, before point Q can be reached a **hydraulic jump** occurs.

The S_2 Profile $(y_0 < y < y_c)$

Since in this case $N > 0$ and $D < 0$, dy/dx is always negative and y has the tendency to decrease as x increases from the starting datum. If this datum coincides with the critical depth y_c, the slope of the profile decreases from an initial infinite slope to reach asymptotically zero because the numerator tends to be zero for $x \to \infty$. This means that the S_2 profile has the y_0 line as an asymptote for $x \to \infty$. Since S_2 is a supercritical profile, it is drawn from upstream to downstream through the control point P_2, and approaches asymptotically the y_0 line as shown in Fig. 29.9.

The S_3 Profile $(0 < y < y_c)$

In this case $N < 0$ and $D < 0$; therefore, dy/dx is always positive, and the water surface profile tends to approach the y_0 line asymptotically downstream. Upstream the depth tends to decrease

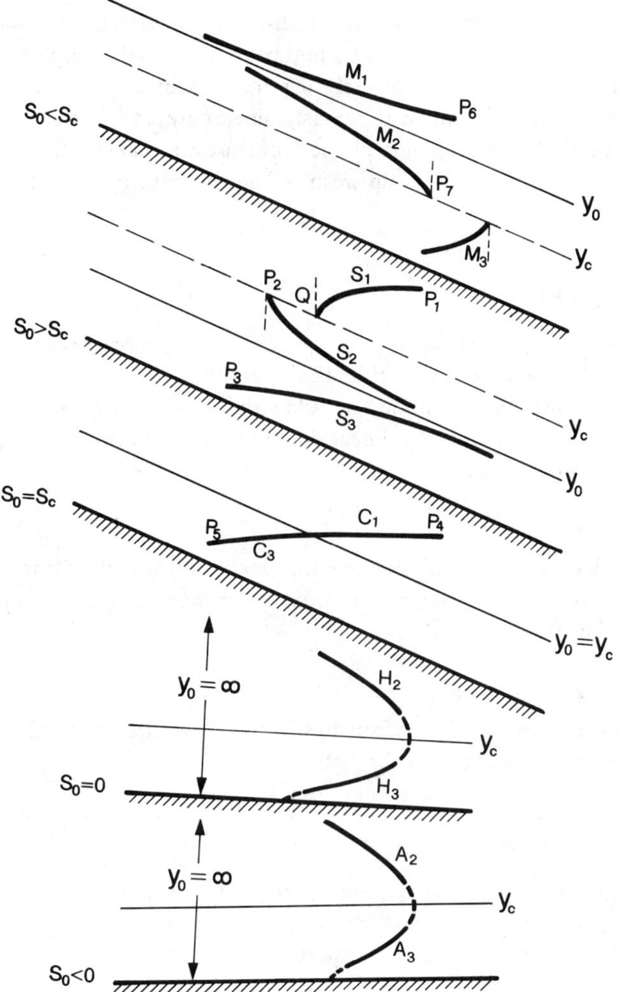

FIGURE 29.9 Gradually varied flow surface curves.

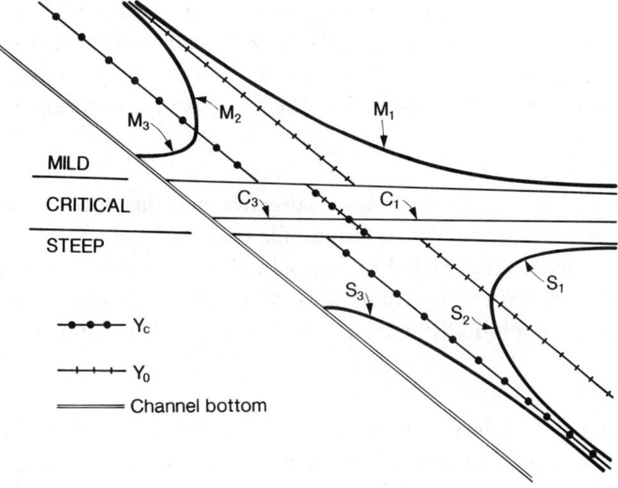

FIGURE 29.10 Surface curves synoptic diagram.

approaching zero. The water surface profile will cut the bottom vertically, but the mathematical continuation of the profile below the channel bottom has a horizontal asymptote at the upstream infinity. The S_3 profile starts therefore from zero depth and infinite slope and rapidly curves to a maximum elevation; the depth grows continuously, albeit only asymptotically as x approaches infinity. The small depth behavior is not physical because the parallel flow hypothesis does not hold. The S_3 profile is drawn from upstream to downstream because it is a supercritical stream.

The Critical-Slope Profiles

The critical-slope profiles $(S_0 = S_c)(y_0 = y_c)$ (Figs. 29.9, 29.10) rarely occur in nature because they are usually unstable. Theoretically for this class of profiles, since $y_0 = y_c$, there are only two regions to consider, the one above and the one below the $y_0 = y_c$ line. The slope of the water surface profile as it approaches the $y_0 = y_c$ line is slightly above the horizontal, roughly $S_0/9$ above horizontal [Giorgini, 1987].

The C_1 Profile $(y > y_c = y_0)$

The profile is therefore only slightly off the horizontal line throughout its extent. Since the stream is subcritical, it is drawn from downstream to upstream, starting from the datum point P_4, always below the horizontal line through P_4.

The C_3 Profile $(0 < y < y_c = y_0)$

Since the stream is supercritical, it is drawn from upstream to downstream, starting from the datum point P_5, always above the horizontal line through P_5.

The Mild-Slope Profiles

For mild-slope profiles $(S_0 < S_c)(y_0 > y_c)$ (Figs. 29.9, 29.10), the lines $y = 0, y = L_c$ divide the channel into three regions $0 < y < y_c, y_c < y < y_0, y > y_c$. The free-surface profiles belonging to the three regions are called M_3, M_2, and M_1, respectively.

The M_1 Profile $(y > y_0)$

In this case $N > 0$ and $D > 0, dy/dx$ is always positive, and therefore y tends to grow as the stream proceeds downstream, where the free-surface profile slope tends to S_0. Since the depth becomes closer to y_0 upstream, the water surface profile has two asymptotes: the line y_0 upstream and a horizontal line downstream, as shown in Fig. 29.9. Since the stream is subcritical it is drawn from downstream to upstream, starting from the datum point P_6, always above the horizontal line through P_6, tending asymptotically to the y_0 line. The M_1 profile is a *backwater profile.*

The M_2 Profile $(y_c < y < y_0)$

In this case $N < 0$ and $D > 0, dy/dx$ is always negative, and therefore the water depth tends to decrease from upstream (where the surface profile tends to be asymptotic to the y_0 line) to downstream, where it approaches the y_c line vertically. Since the stream is subcritical it is drawn from downstream to upstream, starting from the datum P_7, always above a line through P_7 parallel to the bottom, tending asymptotically to the y_0 line. The M_2 profile is a *drawdown profile.*

The M_3 Profile $(0 < y < y_c)$

In this case $N < 0$ and $D < 0$; therefore, $dy/dx > 0$. Since dy/dx tends to be infinite as y approaches either zero or y_c, the profile has an inverted-S shape, as in Fig. 29.9. It can be shown that the tangent to the profile at the inflection point is only slightly larger than S_0, which means that the profile always has a slope above the horizontal line.

The Horizontal Channel Profile, $S_0 = 0$

Equation (29.41) can be integrated when $S_0 = 0$ [Chow, 1959, p. 260]. For an infinitely wide channel and using Manning's formula, the distance x from the brink (where the flow is assumed critical) to the point where the depth is y is given by

$$\frac{n^2 g}{3K_M^2 y_c^{4/3}} x = -\frac{1}{13}\left(\frac{y}{y_c}\right)^{13/3} + \frac{1}{4}\left(\frac{y}{y_c}\right)^{4/3} - \frac{9}{52} \tag{29.44}$$

which is represented in Fig. 29.9. Region 1 does not exist because y_0 tends to infinity as S_0 tends to zero. The two profiles H_2 and H_3 are very similar to the profiles M_2 and M_3, with the exception that the drawdown profile H_2 tends to grow indefinitely proportionally to $|x|^{3/13}$ as x tends to $-\infty$.

The Adverse-Slope Profile, $S_0 < 0$

In this case the differential Eq. (29.41) becomes

$$\frac{dy}{dx} = -|S_0| \frac{1 + S_f\,|S_0|}{1 - \dfrac{\alpha Q^2}{g}\dfrac{T}{A^3}} = -\,|S_0|\,\frac{N(y)}{D(y)} \tag{29.45}$$

(Fig. 29.9). Since N is always positive and D is positive for $y > y_c$ and negative for $y < y_c$, the water depth decreases streamwise for $y > y_c$ and increases streamwise for $y < y_c$. The profile A_3, below the y_c line, is similar to the M_3 and H_3 profiles. The profile A_2 has a horizontal asymptote. As the water depth increases counterstreamwise, Eq. (29.45) yields $dy/dx = -\,|S_0|$ for $x \to -\infty$, which means that the A_2 profile has a horizontal asymptote in that direction. The A_1 profile does not exist as the normal depth is not a real number.

29.7 Qualitative Solution of Open Channel Flow Profiles

The differential equation for the free-surface profiles is a first-order differential equation, and as such it needs only one boundary condition to define the water-surface profile in a given reach. The location of this boundary condition is usually called the **control section** in open channel hydraulics. The first step in the calculation of a flow profile is the determination of the control sections and the profile types. In simple cases, such as long channels with an abrupt change of slope, as illustrated in Fig. 29.11, it is obvious that the control section is at the change of slope and the curve types are as shown in the figure.

The stream flowing into a generic open channel can be considered as a sequence of gradually varied flow reaches possibly separated by one or more rapidly varied flow sections. Some of these

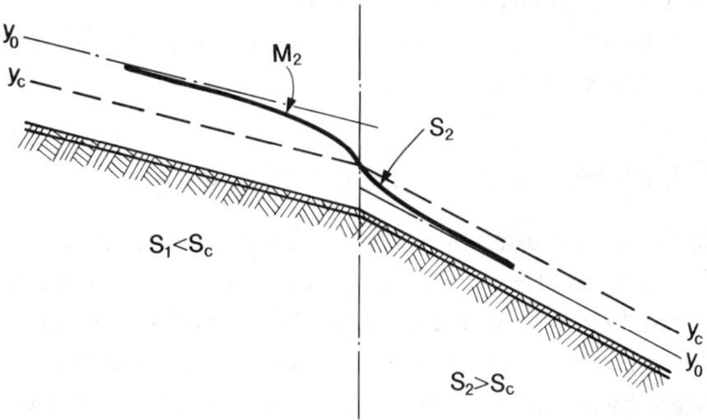

FIGURE 29.11 Channel with abrupt change of slope.

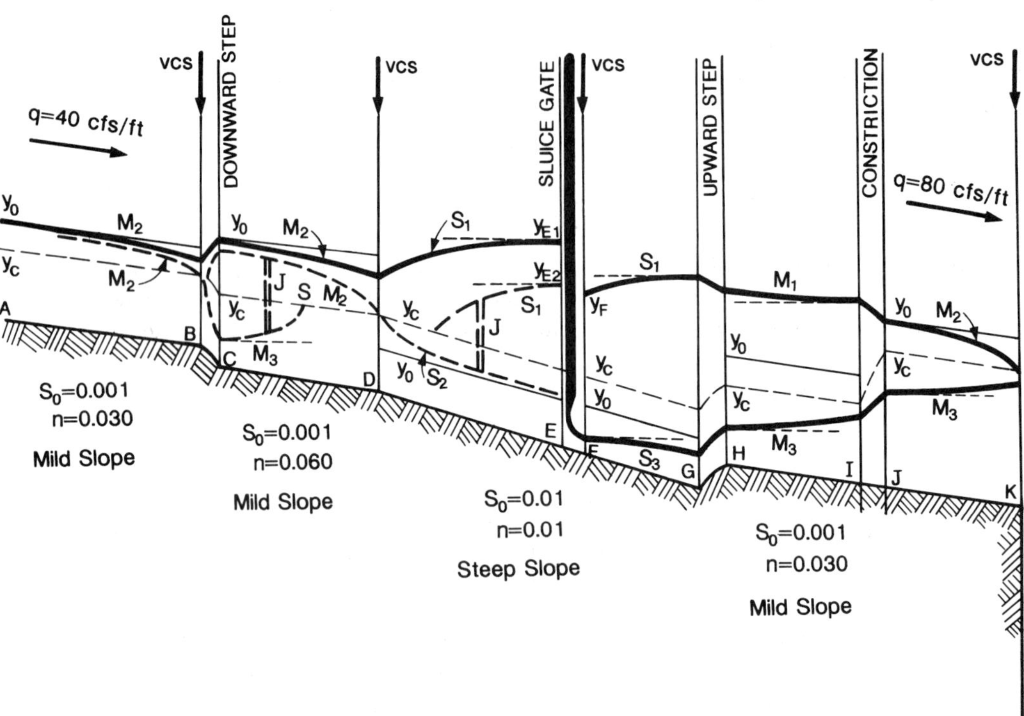

FIGURE 29.12 Qualitative solution of open channel flow profiles.

(upward steps, downward steps, channel contractions, channel expansions without energy losses, and hydraulic jumps) have been considered earlier. Others include weirs and gates, spillways and culverts, and so on. The conceptual solution is illustrated in Fig. 29.12.

The gradually varied flow reaches AB, CD, DE, FG, HI, JK are interrupted by the rapidly varied flow sections: BC, a downward step, EF, a sluice gate, GH, an upward step, IJ, a contraction. Discharges, slopes, and roughness coefficients are shown in Fig. 29.12. The solution process can be divided into the following three phases.

Phase I—Determination of Critical Depths and Normal Depths

Calculate the values of y_c and y_0 throughout the rectangular channel length. From Eq. (29.19), $y_c = (q^2/g)^{1/3} = 3.68$ ft from A to I. Similarly, $y_c = 5.84$ ft from J to K. From Eq. (29.32), $y_0 = [0.03 \times 40/(1.49 \times \sqrt{0.001})]^{3/5} = 6.97$ ft on AB. Similarly, $y_0 = 10.57$ ft on CD; $y_0 = 1.81$ ft on DE and FG; $y_0 = 6.97$ ft on HI; $y_0 = 10.57$ ft on JK.

The y_c line and of the y_0 lines are then sketched as shown in Fig. 29.12. Observe those cross sections where the water surface elevation is known.

Phase II—Virtual Control Section Determination

The adjective virtual indicates the uncertainty that these cross sections are actually "controlling" the water surface profile. Only at the end of the solution process will it be determined whether a virtual control section actually does control the flow. Recall that subcritical streams are controlled from downstream and supercritical streams are controlled from upstream.

Start the search process by inquiring whether the cross section A is a virtual control section (VCS). No indication of this is given by Fig. 29.12. Thus presume that, if there is a control section upstream, it is so far away as to have no effect on A.

Next consider the rapidly varying section BC as a possible site for a control section. It is obvious that if any cross section between B and C is a control section, it should be at B, because if no loss of energy occurs between B and C (as assumed), B is the point of lowest specific energy. This would make cross section B a *natural control*, that is, a section that controls both upstream and downstream.

As seen in Fig. 29.12, there is the possibility of another natural control at D. Moving downstream consider the segment EF, where a sluice gate is situated. If the efflux from the sluice gate is not submerged downstream, then the water surface elevation at F, the vena contracta (see Chapter 28) of the efflux, is known: it is actually $c_C a$ where a is the gate opening and c_C is a contraction coefficient. The water depth at the cross section through E can be found as the alternate of the depth at F, since negligible loss of energy occurs between E and F.

Go now to the rapidly varying section GH. If any cross section between G and H is a control section, this is H because, with no energy loss, H has the lowest specific energy. It could possibly be a natural control; that is, the water surface profile would cut the cross section through H at the critical depth entering region 3 of the reach HI. However, it is easy to see that this is not possible, because GH is followed by a mild slope whose M_3 profile is incompatible with the above geometric description.

The same argument is valid for the rapidly varying section IJ: neither I nor J nor any cross section in between can be a control. The only section left is K, which is a virtual control. In this case, since no water level is offered in the pond after K, it must be assumed that, if K is a control (a natural control), the water depth is critical there.

Phase III—Profile Sketching

K: Critical depth
JK: M_2 profile
IJ: Hydraulic drop due to subcritical flow in constriction
HI: M_1 profile
GH: Hydraulic drop due subcritical flow on upward step
FG: S_1 profile interrupted by sluice gate with submerged efflux if conjugate depth to vena contracta less than y_F; otherwise possible hydraulic jump downstream of F

Otherwise
F: Supercritical stream control, jet at F
FG: S_3 profile
GH: Upward step in supercritical stream; assume y_c line not cut, otherwise jump occurs between F and this section
HI: M_3 profile, supercritical stream
IJ: Supercritical flow in contraction, hydraulic rise, assume y_c line not cut, flow jets out at K; otherwise hydraulic jump between F and this section
* With the profiles illustrated in Fig. 29.12 it is still possible to have a hydraulic jump between F and K in some circumstances, if a pair of conjugate depths exist between the two profiles drawn between F and K, and depending on the circumstances that created the discharge in the channel. Assume that a hydraulic jump does not occur or that it submerges the sluice gate efflux in order to emphasize the concept of *virtual* control section. In the first case F is a control and K is not, in the second F is not and K is.

Upstream of sluice gate with submerged efflux
F: $E_F = y_F + q^2/(2g\ c_c^2 a^2)$, assume $E_E = E_F$; find y_{E1} as alternate subcritical depth corresponding to E_F
DE: S_1 profile, assume y_c line not cut, subcritical flow
CD: M_2 profile

BC: Rise due to subcritical flow in hydraulic drop, assume y_c line not cut
AB: M_2 profile

Otherwise with free sluice gate efflux
E: y_{E2} is alternate depth to $y = c_c a$; note $y_{E2} < y_{E1}$
ED: S_1 profile, assume y_c line is cut, thus control upstream
D: Natural control is assumed; downstream of D: S_2 profile, jump between D and E; upstream of D: M_2 profile
BC: Hydraulic rise, assume y_c line is cut
B: Control overdrop in BC, followed by M_3 and jump between C and S
AB: M_2 profile

29.8 Methods of Calculation of Flow Profiles

The integration of the gradually varied flow Eq. (29.42) can be performed by direct or numerical integration. The *direct integration* method is limited to prismatic channels. It requires the definitions of the *hydraulic exponent for critical flow* [Chow, 1959, p. 66] and the *hydraulic exponent for uniform flow* [Chow, 1959, p. 131] and involves the use of the *varied flow function* developed and tabulated by Bakhmeteff [1932] and extended by Chow [1959, p. 254, tables pp. 641–655]. Closed-form solutions are not available except for some simplified cases such as an infinitely wide channel on a horizontal bottom.

The *numerical integration* of the varied flow equation is not limited to prismatic channels; it is easily programmed and can also be performed with the help of spreadsheet software. Equation (29.41) can be written in finite difference form as

$$\Delta x = \Delta E/(S_0 - \overline{S}_f) \tag{29.46}$$

where $\Delta E = E_2 - E_1$ is the change in specific energy between the end points of the reach of length Δx and $\overline{S}_f$ is the average slope of the energy grade line over the reach. The latter value can be taken as the average of S_{f1} and S_{f2} where

$$S_{fi} = n_i^2 V_i^2/(K_M^2 R_i^{4/3}), \qquad i = 1, 2 \tag{29.47}$$

where the subscripts 1 and 2 represent the ends of the reach and K_M is 1 for metric units and 1.486 for English units. Numerical solutions are discussed briefly for the following problem. Given the flow rate, Q, the cross-sectional geometry, the bottom slope, S_0, and the roughness coefficient, n, calculate the elevation of the water surface profile.

Consider first the case of a *prismatic channel.* The curve type and the control point are identified as discussed in the preceding section. Then successive arbitrary values of the depth y consistent with the profile type are selected from the control point in the upstream direction for mild slopes or in the downstream direction for steep slopes. For each value of y, the area, A, the hydraulic radius, R, the velocity, $V = Q/A$, the velocity head, $\alpha V^2/2g$, the specific energy, $E = y + \alpha V^2/2g$, and the slope of the energy grade line, S_{fi}, are calculated. The average slope of the energy grade line between consecutive depths, $\overline{S}_f$, is calculated. The distances between consecutive depths are calculated by successive application of Eq. (29.46). Since the depths y and the distances Δx have been computed, the surface profile can be plotted. This procedure is known as the *direct step method.* It does not involve any trial and error. The accuracy depends on the selected changes in the consecutive values of the depth.

When the channel is *not prismatic,* the *standard step method* is used. At fixed stations where the cross sections are known, water surface elevations, h, are assumed consistent with the profile type. The depth y is the difference between the water surface and channel bottom elevations. The area, A, the velocity, V, the velocity head, $\alpha V^2/2g$, and the elevation of the energy grade line,

$H = h + \alpha V^2/2g$, are calculated. Then note the value H_1 so obtained at point 1 at the end of the reach with the assumed water elevation. Then $\overline{S}_f \Delta x = h_f$ is the friction head loss in the reach. Eddy losses due to converging or diverging reaches are calculated as $h_e = k(\Delta \alpha V^2/2g)$ with $k = 0$ to 0.1 and 0.2 for converging and diverging reaches, respectively, and $k \approx 0.5$ for abrupt contraction or expansion [Chow, 1959, p. 267]. The elevation of the energy line at point 1 is recalculated as $H_1' = H_2 + h_f + h_e$ where H_2 is the elevation of the energy line at point 2, at the other end of the reach, known from the previous iteration. If this value of H_1' is sufficiently close to the value of H_1 previously noted, the assumed elevation of the water surface at the section 1 is taken as correct. Otherwise a new elevation is assumed and the calculations are repeated until convergence is obtained. Detailed examples of the direct step and the standard step methods can be found in Chow [1959] and French [1985]. Simple illustrations of the direct and standard step methods follow. Other numerical methods of integration of the gradually varied flow equation include the predictor-corrector method [Prasad, 1970].

Example 29.4: Direct Step Method. A trapezoidal channel having a bottom width $b = 10$ ft, side slopes two horizontal on one vertical, $n = 0.02$, and $S_0 = 0.0016$ carries a discharge of 160 cfs. Assume $\alpha = 1.0$. A dam creates a depth of 4.5 ft. Calculate the backwater profile.

1. Find $y_0 = 3.0$ ft from Eq. (29.29a):

$$160 = (1.486/0.02)(10y_0 + 2y_0^2)[(10y_0 + 2y_0^2)/(10 + 2\sqrt{5}y_0)]^{2/3}(0.0016)^{1/2}$$

2. Find $y_c = 1.76$ ft from Eq. (29.21):

$$[(10y_c + 2y_c^2)^3/(10 + 4y_c)] = 160^2/32.2$$

3. Find the type of curve: since $4.5 > y_0 > y_c$, it is an M_1 curve.
4. Direct step method: Table 29.3. Only two segments shown, y arbitrarily selected between $y = 4.5$ and $y_0 = 3.0$. Calculation proceeds in upstream direction. Better accuracy is obtained by taking smaller increments.

Example 29.5: Standard Step Method. Same data as in Example 29.4. Elevation of channel bottom at downstream end is 500.00 ft. The calculations are shown in Table 29.4. Initial water elevation is $h = 504.5$ ft. Only two successful trial values of h are shown. Calculation proceeds in upstream direction. Better accuracy is obtained by taking smaller increments.

Software

Computer programs are available to estimate water surface profiles in natural streams, including the effect of constrictions due to bridges or culverts. One such software package, developed by the Hydrologic Engineering Center of the U.S. Army Corps of Engineers, is called HEC-2. Another program is WSPRO, part of the HYDRAIN package developed for the Federal Highway Administration. (See Chapter 36 for sources and availability.)

Table 29.3 Example 29.4—Direct Step Method

y (ft)	A (ft)	R (ft)	$R^{2/3}$ ($ft^{2/3}$)	$V = Q/A$ (ft)	$\alpha V^2/2g$ (ft)	E^a (ft)	ΔE (ft)	S_f^b	$\bar{S}_f$	$S_o - \bar{S}_f$	Δx^c (ft)	x (ft)
4.50	85.500	2.838	2.005	1.871	0.0544	4.5544		0.0001578				
4.25	78.625	2.711	1.944	2.035	0.0643	4.3143	0.2401	0.0001985	0.0001781	0.0014219	169	169
4.00	72.000	2.582	1.882	2.222	0.0767	4.0767	0.2376	0.0002525	0.0002255	0.0013745	173	342

[a] $E = y + \alpha V^2/2g$.
[b] Equation (29.47).
[c] Equation (29.46).

Table 29.4 Example 29.5—Standard Step Method

x (ft)	Trial h (ft)	y (ft)	A (ft^2)	R (ft)	$V = Q/A$ (ft/s)	$\alpha V^2/2g$ (ft)	H_1 (ft)	S_f	$\bar{S}_f$	Δx_1^g (ft)	$h_e = \bar{S}_f \Delta x$ (ft)	H_2 (ft)
0	504.50[a]	4.50	85.500	2.838	1.871	0.0544	504.5544[e]	0.0001578[f]				504.5544
170[b]	504.52[c]	4.25[d]	78.625	2.711	2.035	0.0643	504.5843	0.0001985	0.0001781	170	0.03028	504.5847[h]
340	504.54	4.00	72.000	2.582	2.222	0.0767	504.6167	0.0002525	0.0002255	170	0.03833	504.6230

[a] Initial condition.
[b] Station at which cross section is known.
[c] Trial value.
[d] 504.52 − (500 + 170 × 0.0016).
[e] 504.50 + 0.0544.
[f] Equation (29.47).
[g] From column 1.
[h] $H_2 + h_e$ = 504.5544 + 0.03028. If this value is close enough to H_1, the value of h is accepted; otherwise, a new value of h is tried.

29.9 Unsteady Flows

Application of the conservation of mass and conservation of momentum to an elementary reach of open channel, as depicted in Fig. 29.2, yields the *Saint Venant equations:*

$$T\frac{\partial y}{\partial t} + V\frac{\partial A}{\partial x} + A\frac{\partial V}{\partial x} = 0 \tag{29.48}$$

$$\frac{\partial V}{\partial t} + V\frac{\partial V}{\partial x} + g\frac{\partial y}{\partial x} - gS_0 + gS_f = 0 \tag{29.49}$$

where T is the free surface width, y is the depth, A is the cross-sectional area, and S_0 and S_f are the bottom and energy grade line slopes, respectively. In Eqs. (29.48) and (29.49) the depth y and the velocity V are the dependent variables and x and t are the independent variables. These equations can also be written using y and the discharge Q or the water surface elevation and V, or Q and h as dependent variables [Lai, 1986].

Initial and boundary conditions need to be specified. Typical initial conditions include the depth and velocity at time $t = 0$ along the whole length of the stream. Boundary conditions must be specified both upstream and downstream. Typical upstream boundary conditions are the flow hydrograph or the stage hydrograph at the upstream end of the channel. The downstream boundary condition is often a rating curve (water surface elevation vs. discharge relationship).

The Saint Venant equations are nonlinear hyperbolic partial differential equations. They can be solved by the *method of characteristics.* The dependent variables V and y are calculated at the intersections of a network of characteristic curves in the *x-t* plane (of the independent variables) whose slopes are given by

$$dx/dt = V + (gy)^{1/2} \qquad dx/dt = V - (gy)^{1/2} \tag{29.50}$$

As $(gy)^{1/2}$ is the speed of propagation of an infinitesimal wave in still water, the characteristic curves are the paths of propagation of infinitesimal disturbances in the upstream and downstream directions (in the case of subcritical flow) [Chaudhry, 1993]. Closed-form solutions are not available except for some simplified cases. Numerical methods based on the method of characteristics have been developed. One of them is the *method of specified intervals* [Wylie and Streeter, 1993]. Solutions based on the method of characteristics are often used to check the validity of other types of numerical solutions.

Finite difference solutions of the Saint Venant equations have been formulated using explicit schemes and implicit schemes. Explicit schemes are easy to program but tend to be unstable, whereas implicit schemes are cumbersome to program but are generally stable [Chaudhry, 1993; Fread, 1993]. Linearized versions of the Saint Venant equations, obtained by assuming small deviations from a steady base flow and where the energy grade line is approximated by $S_f \propto Q\,|Q_0|$ (where Q_0 is the steady base flow), may be solved analytically for prismatic channels and may be solved efficiently in the general case [Dronkers, 1964].

Defining Terms

Alternate depths: The subcritical or tranquil depth, y_T, and the supercritical or rapid depth, y_R, for a given level of specific energy.

Conjugate depths: The depths before and after a hydraulic jump corresponding to a given level of the total thrust or the specific force.

Control section: A section in a nonuniform flow at which the depth is known a priori and which serves as a boundary condition for the calculation of a water surface profile. Often the flow goes through critical at the control section.

Critical depth: The depth of flow corresponding to the minimum specific energy.

Critical slope: The slope of the bottom of a channel in which the normal depth and the critical depth coincide.

Critical velocity: The velocity occurring at critical depth. It is equal to the celerity of propagation of an infinitesimal wave in still water.

Hydraulic jump: The sudden increase in depth that occurs whenever the flow passes from supercritical to subcritical; accompanied by turbulence and large energy loss.

Mild slope: A slope less than the critical slope.

Specific energy: The sum of the depth of flow plus the velocity head.

Specific force: The total thrust per unit weight of liquid.

Steep slope: A slope larger than the critical slope.

Subcritical flow: The flow that occurs when the velocity is smaller than the critical velocity and the depth is larger than the critical depth.

Supercritical flow: The flow that occurs when the velocity is larger than the critical velocity and the depth is smaller than the critical depth.

Total thrust: The sum of the hydrostatic force on a channel cross section plus the momentum flux through that section.

References

Bakhmeteff, B. A. 1932. *Hydraulics of Open Channels,* Engineering Societies Monographs, McGraw-Hill, New York.

Chaudhry, M. H. 1993. *Open-Channel Flow*, Prentice Hall, Englewood Cliffs, NJ.

Chow, Ven Te. 1959. *Open-Channel Hydraulics*, McGraw-Hill, New York.

Dronkers, J. J. 1964. *Tidal Computations in Rivers and Coastal Waters*, North Holland, Amsterdam, Netherlands and Intersciences, New York.

Fread, D. L. 1993. Flood routing. In Maidment, D. R., ed., *Handbook of Hydrology*, McGraw-Hill, New York.

French, R. H. 1985. *Open-Channel Hydraulics*, McGraw-Hill, New York.

Giorgini, A. 1987. Open channel hydraulics. Unpublished class notes, School of Civil Engineering, Purdue University, West Lafayette, IN.

Henderson, F. M. 1966. *Open Channel Flow*, Macmillan, New York.

Ippen, A. T. 1950. Channel transitions and controls. In Rouse, H., ed. *Engineering Hydraulics*, John Wiley & Sons, New York.

Lai, C. 1986. *Numerical Modeling of Unsteady Open-Channel Flow*, Advances in Hydroscience, Vol. 14, pp. 162–323, Academic Press.

Nezu, I. and Nakagawa, H. 1993. *Turbulence in Open-Channel Flows* (IAHR Monograph Series), A.A. Balkema, Rotterdam, Netherlands.
Prasad, R. 1970. Numerical method of computing flow profiles. *J. Hydraul. Eng., ASCE.* 96(1):75–86.
Strickler, A. 1923. Beiträge zur Frage der Geschwindigkeitsformel und der Rauhigkeitszahlen für Ströme, Kanäle und geschlossene Leitungen (Some contributions to the problem of the velocity formula and roughness factors for rivers, canals, and closed conduits), *Mitteilungen des eidgenössischen Amtes für Wasserwirtschaft,* Bern, Switzerland, no. 16.
Subramanya, K. 1982. *Flow in Open Channels,* Vol. 1, Tata McGraw-Hill, New Delhi, India.
Wylie, E. B. and Streeter, V. L. 1993. *Fluid Transients in Systems,* Prentice Hall, Englewood Cliffs, NJ.
Yen, B. C., ed. 1992. *Channel Flow Resistance: Centennial of Manning's Formula,* Water Resources, Littleton, CO.

For Further Information

Bakhmeteff [1932] provided the first extensive American text in open channel hydraulics and gives the expansions and tabulation of the varied flow function.

Chow [1959] wrote an extensive text that is the classic reference on the subject. Chow extended Bakhmeteff's varied flow function tables. Extensive tabular solutions of problems are given and they can be adapted to spreadsheet software.

Henderson [1966] provides a classic presentation of open channel flow and includes a detailed discussion of unsteady flows, flood routing, and kinematic waves. Also covered are sediment transport and similitude and models.

French [1985], in addition to the classical open channel hydraulics, provides chapters on turbulent diffusion, buoyant surface jets, unsteady gradually, and rapidly varied flow and hydraulic models.

Yen [1992] provides a series of articles on the history of Manning's formula and on the concepts of roughness and energy losses in open channel flow.

Chaudhry [1993] emphasizes unsteady flow problems and their solution by numerical methods. Also included are two-dimensional flows and finite elements applied to both one- and two-dimensional flows. Several FORTRAN computer programs and a diskette are provided.

Nezu and Nakagawa [1993] deal with open channel turbulence, boundary layers, and turbulent transport processes in rivers and estuaries.

30

Surface Water Hydrology

A. R. Rao
Purdue University

30.1 Introduction

The Hydrologic Cycle

Hydrologic cycle is the name given to the intricate continuum of water movement in the atmosphere, hydrosphere, and lithosphere. A schematic diagram of the hydrologic cycle is shown in Fig. 30.1.

Water evaporates from the oceans and land and becomes a part of the atmosphere. The water vapor either is carried in the atmosphere or returns to earth in the form of **precipitation**. A portion of the precipitation falling on land may be intercepted by vegetation and returned directly to the atmosphere by evaporation. Precipitation that reaches the earth may evaporate or be transpired by plants; it may flow over the ground surface and reach streams as surface water; or it may infiltrate the soil. The infiltrated water may flow over the upper soil regions and reach surface water, or it may percolate into deeper zones and become groundwater. Groundwater, in turn, may reach the streams naturally, or it may be pumped, used, and discarded to become a part of the surface water system.

0-8493-8953-4/95/$0.00 + $.50

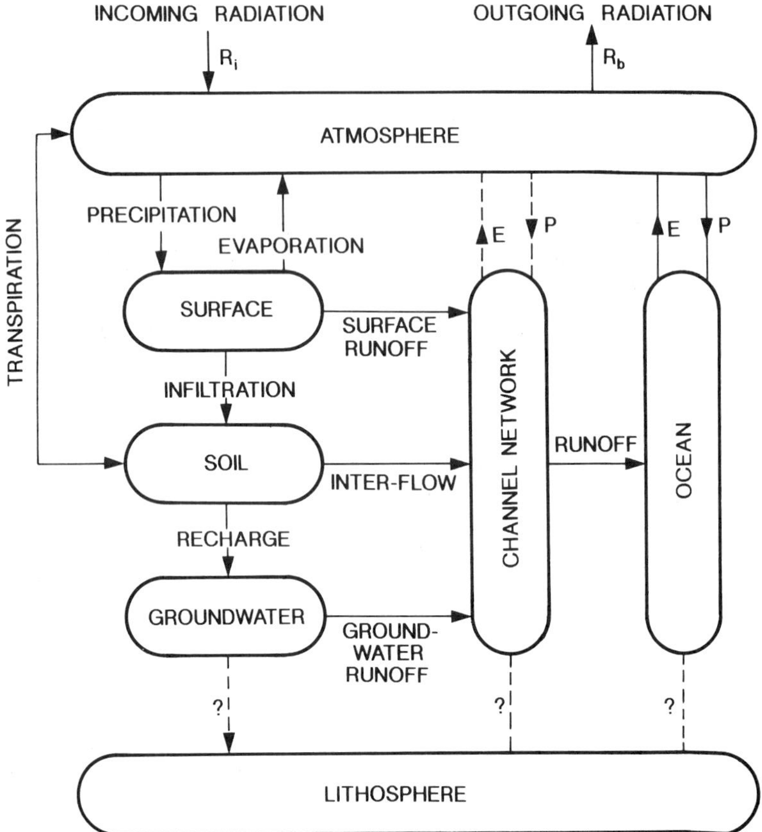

FIGURE 30.1 A systems representation of the hydrologic cycle (after Dooge, 1973).

The general volume balance equation for the hydrologic cycle may be written as

$$\frac{dS}{dt} = I(t) - O(t) \tag{30.1}$$

where S is the storage, including the surface, soil moisture, groundwater, and interception storage; t is the time; $I(t)$ is the input, which includes precipitation in all its forms; and $O(t)$ is the output, which includes **surface** and **subsurface runoff,** *evaporation*, *transpiration*, and **infiltration**.

According to world water balance studies [UNESCO, 1978], about 50 in. of precipitation falls on oceans of area 361.3 million km^2. The corresponding values for land areas are 31 in. and 148.8 million km^2. The evaporation rates are 55 and 19 in. respectively from the oceans and land areas. Of the 12 in. of water that remains on land, 0.2 in. becomes groundwater and 11.8 in. reaches the oceans as surface water. Because a large part of the precipitation falling on the earth remains as ice, the amount of surface water that is available for use is quite limited. In this chapter several aspects of surface water hydrology are discussed.

Historical Development

Chow [1964] has classified the development of hydrology into eight periods. The first of these is called the *period of speculation* (from ancient times to A.D. 1400), during which there were many speculations, but no formal science, about the concept of the hydrologic cycle. However, during this period, practical aspects of hydrologic knowledge were studied and used to build civil works. During the period from A.D. 1400 to 1600 (*the period of observation*), hydrological variables were simply observed, and Leonardo da Vinci and Bernard Palissy understood the hydrological cycle. Hydrological measurements started during the *period of measurement* (1600 to 1700); the *science* of hydrology may be said to have begun during this period and continued through the *period of experimentation* (1700 to 1800). The foundations of modern hydrology were laid during the *period of modernization* (1800 to 1900). The *period of empiricism* (1900 to 1930) gradually gave way to the period of rationalization (1930 to 1960), during which theoretical developments in hydrography, infiltration, and groundwater processes took place. Accelerating theoretical developments have made the period from 1950 to the present a period of theorization. The recent development of computers has made it possible to construct and verify theories of increasing complexity, an endeavor which has become the mainstay of modern hydrologic research.

30.2 Precipitation

Atmospheric Processes

For precipitation to occur, air must be lifted and cooled whether by the passage of fronts, when a warm air mass is lifted over cooler air (*frontal* cooling); by the passage of warm air over mountain ranges (*orographic* cooling); or by the rising of locally heated air, such as in a thunderstorm cell (*convective* cooling).

As the air is cooled, water condenses on microscopic particles called *nuclei*, and this process is called *nucleation*. Dust and salt particles are common condensation nuclei. The water particles that result from nucleation grow by condensation and by coming into contact with neighboring particles. They start descending as they become heavier; they may coalesce with other water drops, or they may decrease in size because of evaporation during descent. If conditions are favorable, these water drops reach the ground as rain, snow, or sleet. The particular form taken by precipitation is dictated by the atmospheric conditions encountered during the descent of water drops.

Measurement of Precipitation

Rainfall and snowfall are commonly measured. Both nonrecording and recording gauges are used for rainfall measurement. Recording rain gauges are used to measure rainfall depth at predetermined time intervals that can be as small as a minute. Nonrecording rain gauges are read at larger time intervals. Common recording rain gauges are of the *weighing*, the *tipping-bucket*, or the *float* type. In each one of these, a record of rainfall depth against time is obtained. Depth and density of snow packs, in addition to the water equivalent of snow, are also commonly measured, as these are useful in estimating the water yield from snow packs. Measurement of snow depth is complicated, because of the strong effect of wind on snow [Garstka, 1964].

Temporal Variation of Precipitation

Rainfall is measured as depth, in inches or mm. The rates are usually expressed as in./hr or mm/hr, although longer durations such as days, months, and years are also used. The rainfall rate at a particular moment, as during a storm, is called the **rainfall intensity**, and the time unit is usually an hour. In general, rainfall intensity is highly variable with time. A plot of intensity or depth against duration is called a *hyetograph* of rainfall, whereas a plot of the sum of the rainfall depth against time is called a *mass curve*. A mass curve whose abscissa and ordinate are dimensionless is called a *dimensionless mass curve*. A typical hyetograph, mass curve, and dimensionless mass curve are shown in Fig. 30.2.

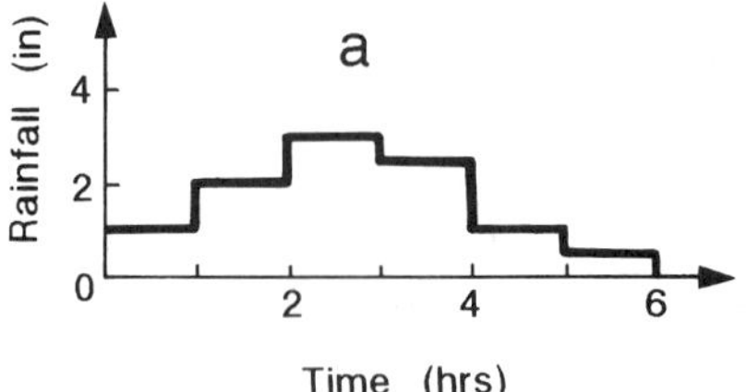

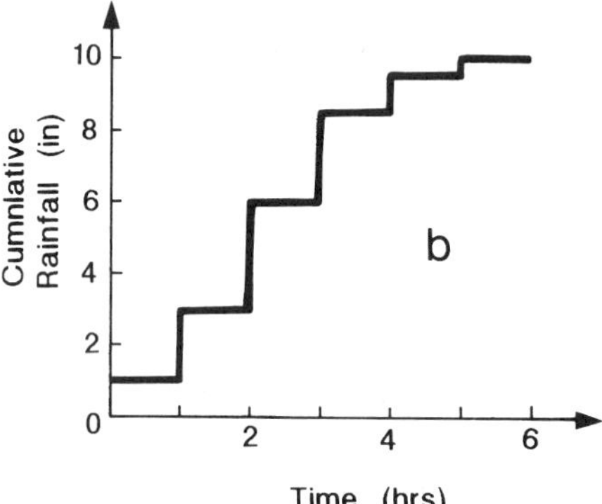

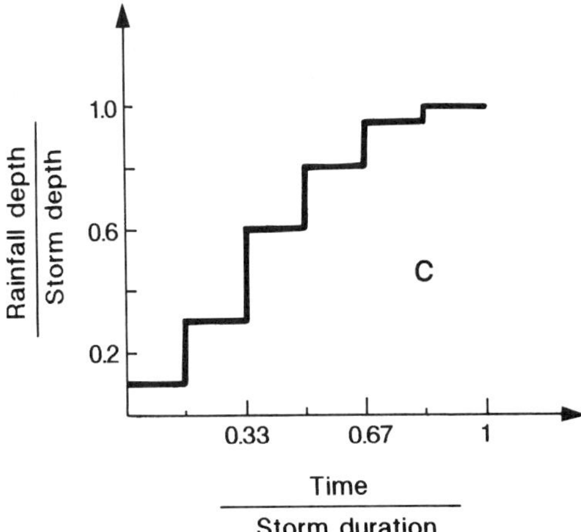

FIGURE 30.2 (a) Hyetograph. (b) Mass curve. (c) Dimensionless mass curve.

Spatial Variation of Precipitation

Rainfall measurements are taken at different points in an area. The spatial structure of storms and their internal variation cannot be adequately represented by a point measurement or even by many point measurements made over a region. Consequently, there have been attempts to relate point rainfall measurements to spatial average rainfall. As the area represented by a point measurement increases, the reliability of the data from a point as a representative of the average over a region decreases. As drainage areas become larger than a few square miles, point data must be adjusted to estimate areal data [Hershfield, 1961]. If the drainage area is larger than 8 mi^2, an area reduction factor obtained from the rainfall atlas [Hershfield, 1961] is applied to the point rainfall depth values.

Average Rainfall over an Area

The *arithmetic average* method, the *Thiessen polygon* method, and the **isohyetal** methods are commonly used to compute average rainfall over an area. These and other methods have been investigated by Singh and Chowdhury [1986], who concluded that they give comparable results, especially for longer storm durations.

Let the *average rainfall* over an area be $\overline{P}$. Let the rainfall measured by N rain gauges over an area A be $P_1, P_2, \ldots P_N$. The general expression for $\overline{P}$ is given in Eq.(30.2), where W_i are the weights:

$$\overline{P} = \sum_{j=1}^{N} W_j P_j, \qquad W_j \in (0, 1) \tag{30.2}$$

Different methods of estimation of rainfall give different W_j values. For the arithmetic average, the weights W_j are the same and are equal to $(1/N)$. To compute the *Thiessen* average rainfall, the locations of rain gauges are joined by straight lines on a map of the area. These are bisected to develop a Thiessen polygon such that each rain gauge with rainfall P_j is located in a part of a watershed of area A_j. The sum of areas A_j equal the watershed area A. The Thiessen weights W_j are given by A_i/A and add up to unity. The Thiessen average does not consider the spatial distribution of rainfall, but it takes into account the spatial distribution of rain gauges.

In the *isohyetal* method, lines of equal rainfall depth, or *isohyetal lines*, are first estimated. The variation in rainfall between rain gauges may be assumed to be linear for interpolation purposes. The areas A_j between isohyetal lines are measured, and these add up to A. The number of areas enclosed within isohyets is usually not equal to the number of rain gauges. The weights W_j are equal to A_j/A. The rainfall values P_j in Eq. (30.2) are the average rainfall values between the isohyets. In the isohyetal method the spatial distribution of rainfall variation is explicitly considered.

Intensity–Duration–Frequency (i–d–f) Curves

The largest rainfall depth measured over a specified duration, during a year, is an extreme value. Because the durations are fixed, we will have a series of rainfall extreme values. For example, we will have extreme rainfall depths corresponding to 30 min, 1 hr, 6 hr, and so forth. These extreme values—which are also the annual maximum values in this case—are random variables, denoted by x. The probability that x is larger than a value x_T is called the *exceedance probability* and is indicated by $P(x \geq x_T)$ or p. The recurrence interval T is the average time elapsed between occurrences of $x \geq x_T$. The *return period* or *recurrence interval* between events that exceed or equal x_T is the *average time* T between exceedances of the event. The exceedance probability and the return period are inversely related; the inverse of the return period is also called the **frequency**.

By analyzing annual maximum rainfall intensities corresponding to a duration—such as one hr or 6 hrs—the rainfall intensity–frequency relationships are obtained. A set of rainfall intensity (i)–frequency (f) relationships corresponding to different rainfall durations (d) are called the intensity (i)–duration (d)–frequency (f) curves. The i–d–f curves for Indianapolis, presented in Fig. 30.3, are typical of these.

It is difficult to use the i–d–f curves in computer analysis. Consequently, they are represented as empirical equations such as Eq. (30.3), where i is the intensity in in./hr; t is the duration in hours; T is the frequency in years; C and d are constants corresponding to a location, and m and n are exponents. For Indianapolis, $C = 1.5899$, $d = 0.725$, $m = 0.2271$ and $n = 0.8797$, and the resulting equation is valid for durations between 1 and 36 hr.

$$i = \frac{CT^m}{(t + d)^n} \tag{30.3}$$

These empirical equations give results that are more accurate than those obtained by interpolation of graphs.

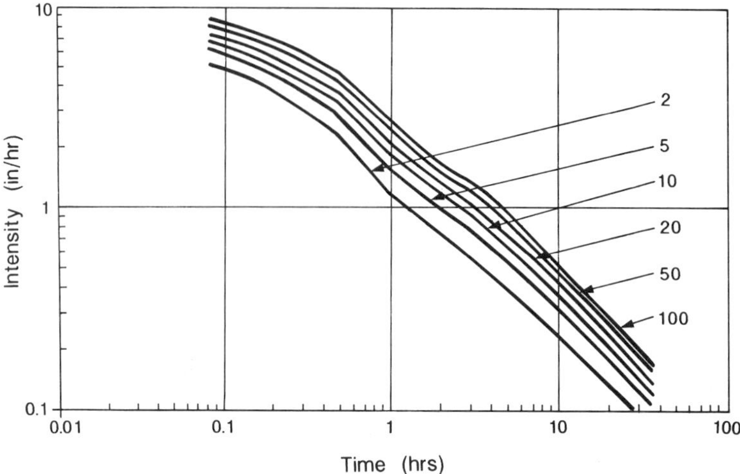

FIGURE 30.3 Intensity–duration–frequency curves for Indianapolis. *Source:* Purdue, A. M., Jeong, G. D., and Rao, A. R. 1992. *Statistical Characteristics of Short Time Increment Rainfall,* p. 64. Tech. Rept. CE-EHE-92-09, School of Civil Engineering, Purdue University, West Lafayette, IN.

Dimensionless Mass Curves

In order to estimate hyetographs from rainfall depth–duration data, dimensionless mass curves are used. These dimensionless mass curves are developed by classifying observed rainfall data into different quartiles. For example, if the largest rainfall depth occurs during the first quartile of a storm duration, then it is called a *first-quartile storm.* The cumulative rainfall in these storms is made dimensionless by dividing the rainfall depths by the total rainfall depth and corresponding times by the storm duration. These dimensionless mass curves of rainfall are analyzed to establish their frequencies of occurrence. These are published in graphical form as shown in Fig. 30.4 and Table 30.1.

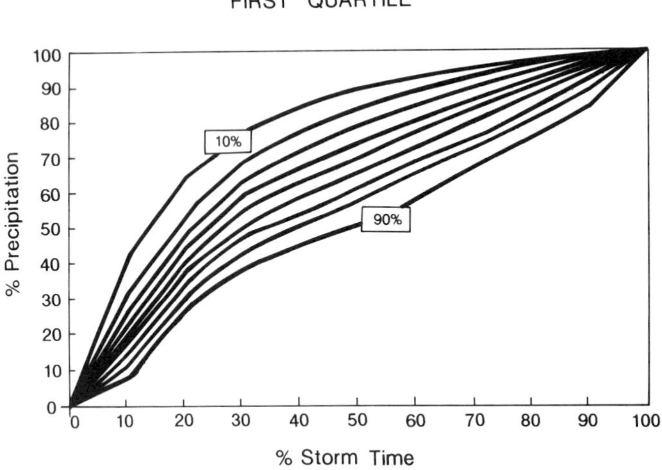

FIGURE 30.4 Dimensionless first-quartile cumulative rainfall curves for Indianapolis.

Table 30.1 Dimensionless Mass Curves (10% Level) for Four Quartiles

% Storm Time	I	II	III	IV
0	0.00	0.00	0.00	0.00
10	42.00	16.36	14.36	19.35
20	64.35	32.73	25.26	30.00
30	76.36	58.10	33.33	36.36
40	83.84	76.36	41.82	43.53
50	89.63	87.50	54.72	50.00
60	92.50	92.54	78.18	54.55
70	95.00	95.69	92.35	63.64
80	97.00	97.00	96.43	82.50
90	98.57	98.67	98.73	96.73
100	100.00	100.00	100.00	100.00

Given a rainfall depth and duration, the dimensionless mass curve information is used to generate hyetographs, which may be used as inputs to rainfall–runoff models or with unit hydrographs to generate runoff hydrographs. The dimensionless hyetographs thus provide an easy method for generating rainfall hyetographs.

Chen [1983] has developed a method of generating intensity–duration curves for different frequencies or recurrence intervals. The method is based on 10 yr–1 hr (R_1^{10}), 10 yr–24 hr (R_{24}^{10}), and 100 yr–1 hr (R_1^{100}) rainfall *depths* for the location of interest. These are available from the National Weather Service (NWS) publications, such as TP-40, NWS HYDRO-35, or NOAA Atlas 2 in the United States [Hershfield, 1961; Frederick *et al.*, 1977; Miller *et al.*, 1973]. The intensity–duration–frequency relationship used in this study is of the form of Eq. (30.3).

In this method, the ratios R_1^{10}/R_{24}^{10} and R_1^{100}/R_1^{10} are formed. The R_1^{10}/R_{24}^{10} ratio is the x axis of Fig. 30.5, and by using this value, a_1, d, and n are read off from Fig. 30.5. The rainfall intensity corresponding to duration t and recurrence interval T is estimated by using Eq. (30.4) where r_1^{10} is the 10 year–1 hour rainfall intensity, which is the same as R_1^{10} because the duration is 1 hour.

$$r_t^T = \frac{a_1(r_1^{10}) \log(10^{2-x} T^{x-1})}{(t+d)^n} \tag{30.4}$$

Chen's [1983] method has been demonstrated to give very good estimates of rainfall intensity–duration–frequency relationships for rainfall in the U.S.

Other Probabilistic Aspects

Numerous probabilistic models have been developed and used to characterize various rainfall properties. For analyzing annual maximum rainfall data, log normal and extreme-value (III) distributions have been successfully used. The Weibull distribution has been used to characterize the time between precipitation events. Two-parameter gamma distribution and bivariate exponential distribution have been used to characterize the storm depths and durations. A discussion of these models is found in Eagleson [1970] and Bras [1990].

30.3 Evaporation and Transpiration

Evaporation

Evaporation is the process by which water is removed from an open water surface. The rate of evaporation depends on two factors: the energy available to provide the latent heat of vaporization, and the rate of transport of water vapor from the water surface. Evaporation from ponds, rivers, and lakes depends on solar radiation and wind velocity. The gradient of **specific humidity**, which is the ratio of the water vapor pressure e to the atmospheric pressure p, also affects the evaporation rate.

Evaporation from a body of water can be estimated by using the law of conservation of energy. By using a control volume and estimating the energy inputs to and outputs from it, the evaporation

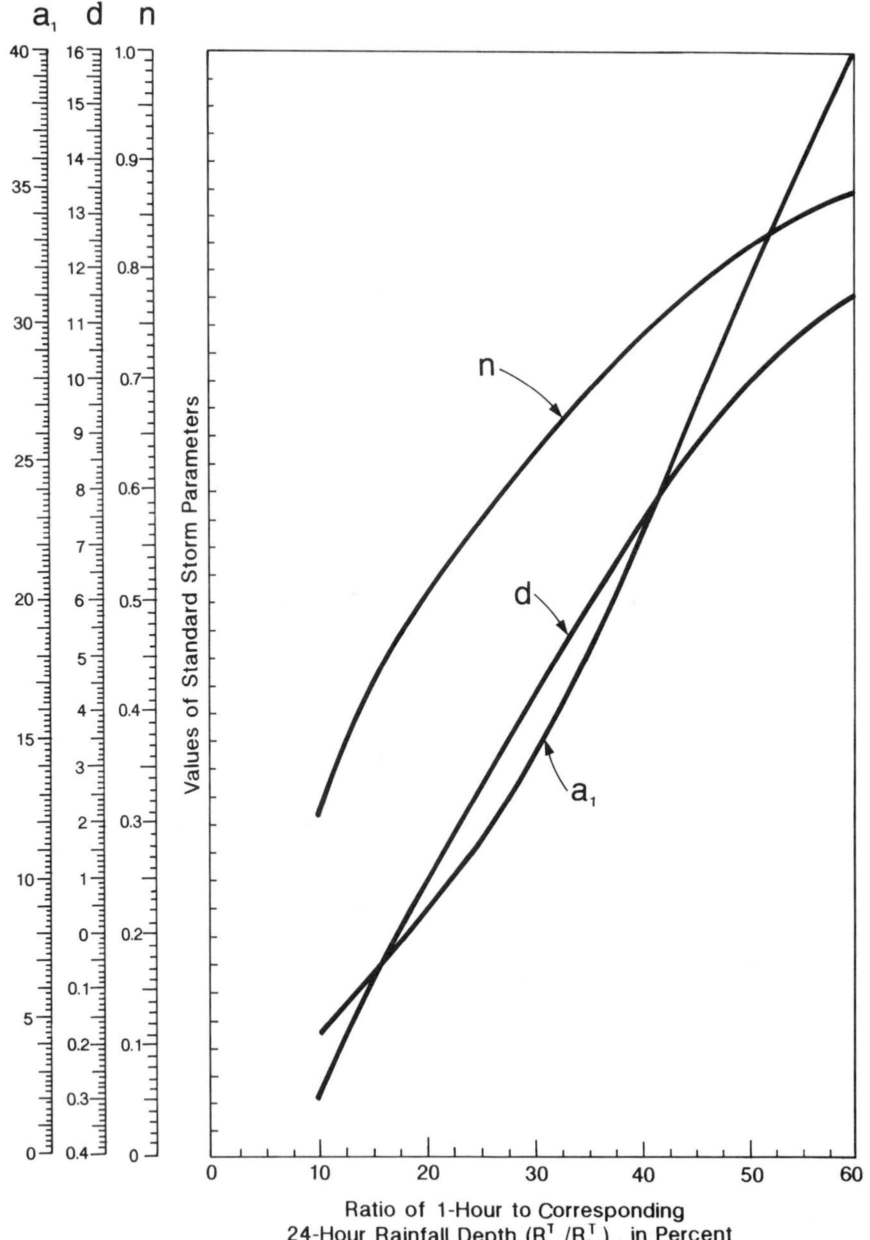

FIGURE 30.5 Coefficients and exponents for use with Chen's [1983] method.

rate E_1 (mm/day) can be shown to be as in Eq. (30.5), where R_w is the net short-wave radiation received in W/m^2:

$$E_1 = \frac{R_w}{l_v \rho_w} \tag{30.5}$$

where l_v is the latent heat of vaporization (kJ/kg) and ρ_w is the density of water in kg/m^3.

In deriving Eq. (30.5), the sensible heat flux and the heat transfer from the water to the ground are neglected. These assumptions are not realistic. Furthermore, there is no provision in Eq. (30.5) for removal of water vapor from the water surface. In order to eliminate this strong drawback,

Thornthwaite and Holzman [1939] derived an equation that includes the wind velocity to estimate evaporation from open water surfaces. This equation has been simplified for operational application and is given as Eq. (30.6) [Chow *et al.*, 1988], where k is the von Kármán constant, usually assumed to be 0.4; ρ_a is the density of air (kg/m^3); u_2 is the wind speed (m/sec) at height z_2; ρ_w is the density of water in kg/m^3; z_0 is a roughness height; p is the atmospheric pressure (Pa); e_{sat} is the saturated vapor pressure (Pa); e_2 is the actual vapor pressure (Pa); and E_2 is in mm/day.

$$E_2 = \frac{0.622\, k^2\, \rho_a\, u_2}{p\, \rho_w\, [\ln(z_2/z_0)]^2}(e_{\text{sat}} - e_a) \tag{30.6}$$

Penman [1948] developed a comprehensive theory of evaporation, and his equation involves both the estimates E_1 and E_2. The Penman evaporation estimate E is given in Eq. (30.7) where γ is the **psychrometric constant** and Δ is the slope of the saturated vapor pressure curve at air temperature T_a.

$$E = \frac{\Delta}{\Delta + \gamma} E_1 + \frac{\gamma}{\Delta + \gamma} E_2 \tag{30.7}$$

$$\gamma = \frac{C_p\, K_h\, p}{0.622 l_v\, K_w} \tag{30.8}$$

$$D = \frac{4098 e_{\text{sat}}}{(237.3 + T)^2} \tag{30.9}$$

In Eqs. (30.8) and (30.9), C_p is the specific heat at constant pressure; K_h is the heat diffusivity; K_w is the vapor eddy diffusivity; the ratio (K_h/K_w) is usually assumed to be unity; and T is the temperature (°C). Priestley and Taylor [1972] analyzed Eq. (30.7) and found that for evaporation over large areas, the second term in Eq. (30.7) is about 30% of the first one. Based on this observation they developed Eq. (30.10) the Priestley–Taylor equation, which reduces the computations involved.

$$E = 1.3 \frac{\Delta}{\Delta + \gamma} E_1 \tag{30.10}$$

The Penman equation is the most accurate of all equations used to estimate evaporation from open water surfaces. However, the data required to use it, such as solar radiation, humidity, and wind speed, may not be easily available. In such cases, simpler evaporation equations [ASCE, 1973; Doorenbos and Pruitt, 1977] are used.

Most commonly, evaporation is measured by using *evaporation pans*. The *class A pan* used in the U.S. is 4 ft (120.67 cm) in diameter and 10 in.(25.4 m) deep. The pan is made of Monel metal or of unpainted galvanized iron and is placed on a wooden support to facilitate air circulation beneath it. In addition to the pan, an *anemometer* to measure the wind speed, a precipitation gauge, and thermometers to measure water and air temperatures are also used. Water level in the pan is measured and reset to a fixed level each day. The change in water level, adjusted to account for precipitation, is the evaporation that has occurred during that day. Further details about evaporation pans, those used in both the U.S. and in other countries, are found in WMO [1981].

The rates measured in evaporation pans are usually larger than those measured in larger water bodies. The ratio of lake evaporation to pan evaporation is called the **pan coefficient**. In order to estimate evaporation from a larger water body, the pan evaporation is multiplied by the pan coefficient. An average pan coefficient in the U.S. is about 0.7, which varies with locations and with seasons at a location.

Evapotranspiration

The process by which water in soil, vegetation, and land surface is converted into water vapor is called **evapotranspiration**. Both transpiration of water by vegetation and evaporation of water from soil, vegetation, and water surfaces are included in evapotranspiration. Because it includes water vapor generated by all the mechanisms, evapotranspiration plays a major role in water balance computations. *Consumptive use* includes evapotranspiration and the water used by plant tissue. In practice, *evapotranspiration* and *consumptive use* are used interchangeably.

The process by which plants transfer water from roots to leaf surfaces, from which it evaporates, is called *transpiration*. The rate of transpiration greatly depends on sunshine and on seasons and moisture availability. Transpiration rates of different plant types vary. As transpiration ends up in evaporation, transpiration rates are affected by the same meteorological variables as evaporation. Therefore, it is common practice to combine transpiration and evaporation and express the total as evapotranspiration.

Potential evapotranspiration (PET) [Thornthwaite *et al.*, 1944] is a concept in common usage in evapotranspiration computations. PET is the evapotranspiration rate that occurs when the moisture supply is unlimited and is a good indicator of optimum crop water requirements. The *reference crop evapotranspiration* (ET_0), introduced by Doorenbos and Pruitt (1977), and similar to PET, is the rate of evapotranspiration from an extended surface of 8 to 15 cm tall green grass cover of uniform height, actively growing, completely shading, and not short of water. Thus, ET_0 is the PET of short green grass, which is the reference crop.

The PET is equivalent to evaporation from free water surface of extended proportions. However, the heat storage capacity of this water body is assumed to be negligibly small. Consequently, methods used to estimate PET and evaporation are similar. Evapotranspiration and PET are based on (1) temperature, (2) radiation, (3) combination or Penman, and (4) pan evaporation methods. These are considered next.

The Blaney–Criddle formula, Eq. (30.11), which is widely used to estimate crop water requirements, is typical of evapotranspiration formulas based only on temperature,

$$F = PT \tag{30.11}$$

where F is the evapotranspiration for a given month in inches; P is the ratio of total daytime hours in a given month to the total daytime hours in a year; and T is the monthly mean temperature in degrees Fahrenheit. Doorenbos and Pruitt [1977] modified the Blaney–Criddle formula to include actual insolation time, minimum relative humidity, and daytime wind speed.

Another well known temperature-based evapotranspiration estimation method is that proposed by Thornthwaite et al. [1944]. In this method, the heat index I_j is defined in terms of the mean monthly temperature T_j (° C). The annual temperature efficiency index J is the sum of 12 monthly heat indices I_j.

$$I_j = \left(\frac{T_j}{5}\right)^{1.514}; \qquad J = \sum_{j=1}^{12} I_j \tag{30.12}$$

The potential evapotranspiration is computed by

$$\text{PET} = K\,\text{PET}_0 \tag{30.13}$$

$$\text{PET}_0 = 1.6\left(\frac{10T_j}{J}\right)^{c} \tag{30.14}$$

$$c = 0.000000675J^3 - 0.0000771J^2 + 0.49239$$

where PET_0 is the PET at 0° latitude in centimeters per month. K in Eq. (30.13) varies from month to month and is given in Thornthwaite et al. [1944] and in Ponce [1989].

Priestley and Taylor [1972] developed a formula to compute PET based only on the radiation part of the Penman equation, Eq. (30.7):

$$\mathrm{PET} = \frac{1.260\Delta(R_w/l_v\rho_w)}{\Delta + \gamma} \tag{30.15}$$

The Penman equation is also used to calculate PET. Penman [1952] suggested the use of crop coefficients (0.6 in winter and 0.8 in summer) to compute the evapotranspiration rates.

In pan evaporation models, the PET is given by Eq. (30.16) where K_p is the pan coefficient and E_p is the pan evaporation.

$$\mathrm{PET} = K_p E_p \tag{30.16}$$

Pan evaporation is widely used to estimate PET. Guidelines to choose appropriate pan coefficients are found in Doorenbos and Pruitt [1977].

30.4 Infiltration

Process and Variability

Water on the soil surface enters the soil by infiltration. **Percolation** is the process by which water moves through the soil under the influence of gravity. The soil exposed to atmosphere is not usually saturated, so flow near the ground surface is through an unsaturated medium. The percolated water may reach groundwater storage, or it may transpire back to the surface.

As water travels from the surface to the groundwater, two forces act on it. Gravity attracts the flow towards groundwater, and capillary forces attract it to capillary spaces. Consequently, rate of percolation decreases with the passage of time and leads to decreasing rates of infiltration. The *infiltration capacity*, f_p, is the maximum rate at which infiltration can occur and is affected by conditions such as the soil moisture; the actual rate of infiltration is f_i. The infiltration rate and capacity are the same when the rate of supply of water i_s is equal to or greater than f_p. Infiltration theories assume that i_s is equal to or greater than f_p. Under these conditions, the maximum infiltration rate f_0 occurs at the beginning of a storm, and the rate approaches a constant rate f_c as the soil becomes saturated. The rate at which f_i approaches f_c, and the final values of f_i, depend on the characteristics of the soil and initial soil moisture (Fig. 30.6).

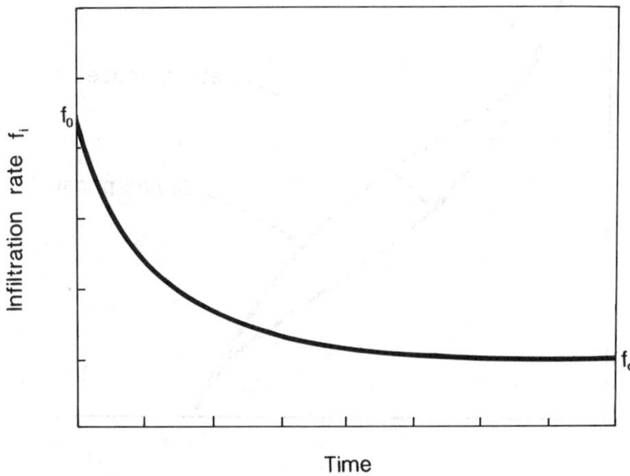

FIGURE 30.6 Infiltration curves.

The energy possessed by the fluid due to suction forces $\psi(\theta)$ of a soil is a function of *volumetric soil moisture* θ, which is the ratio of the volume of water in soil to the volume of the soil. The capillary potential $\psi(\theta)$ is related to the piezometric head h as in Eq. (30.17) where z is the elevation head, h is the piezometric head, and p_c/γ is the pressure head.

$$h = \frac{p_c}{\gamma} + z = \psi(\theta) + z \tag{30.17}$$

The head p_c is negative; hence, p_c/γ is called the *suction head.* The capillary potential, for the same soil moisture θ, depends on whether the soil is in the wetting or drying phase. For the same θ, $\psi(\theta)$ is higher when the soil is drying than when it is wetting (Fig. 30.7).

Infiltration Models

Considering only the flow in the vertical direction z, the infiltration of water is governed by Eq. (30.18), where t is the time, $D_z(\theta)$ is the diffusivity, and $K_z(\theta)$ is the hydraulic conductivity in the z direction [Bras, 1990].

$$\frac{\partial \theta}{\partial t} = \frac{\partial}{\partial z}\left[D_z(\theta)\frac{\partial \theta}{\partial z} + K_z(\theta)\right] \tag{30.18}$$

The diffusivity $D_z(\theta)$ is given by Eq. (30.19), where the hysteresis effects of $\psi(\theta)$ (Fig. 30.7) are ignored and $\psi(\theta)$ is considered as a single-valued function of θ.

$$D_z(\theta) = K_z(\theta)\frac{\partial \psi(\theta)}{\partial \theta} \tag{30.19}$$

To simplify Eq. (30.18), $K_z(\theta)$ is considered to be a constant or small in comparison with $D_z(\theta)(\partial\theta/\partial z)$. If it is further assumed that $D_z(\theta)$ is constant and equal to D, Eq. (30.18) reduces to the diffusion equation Eq. (30.20). Assuming a semi-infinite soil system and the boundary conditions in Eq. (30.21), Eagleson [1970] shows that the solution of Eq. (30.20) is given by Eq. (30.22).

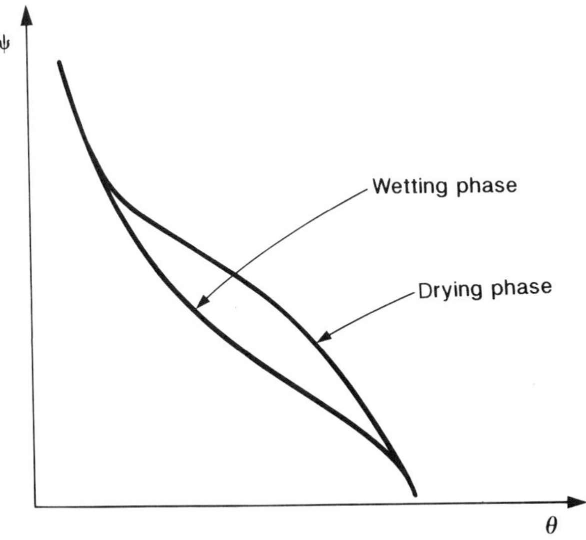

FIGURE 30.7 Capillary potential vs. soil moisture.

$$\frac{\partial \theta}{\partial t} = D\frac{\partial^2 \theta}{\partial z^2} \tag{30.20}$$

$$\theta = \begin{cases} \theta_i & z \leq 0; \quad t = 0 \\ \theta_o & z = 0; \quad t > 0 \end{cases} \tag{30.21}$$

$$f_p = f_c + (f_0 - f_c)e^{-Dl^2 t} \tag{30.22}$$

where

$$f_p = f_c \qquad z = 0, \qquad t = \infty$$

$$f_p = f_0 \qquad z = 0, \qquad t_0 = 0$$

The solution in Eq. (30.22) is applicable at the soil surface ($z = 0$). The variable l in Eq. (30.22) is a characteristic length dependent on z. Equation (30.22) is a form of the infiltration equation derived by Horton [1939, 1940]. Philip [1960] presented an analytical solution to the infiltration equation when the initial and boundary conditions correspond to those in Eq. (30.21), with θ_o equal to porosity n. An approximate form of Philip's solution is given in Eq. (30.23), where S, called the *sorptivity,* and K are parameters related to diffusivity, hydraulic conductivity, and initial soil moisture.

$$f_p(t) = \tfrac{1}{2}St^{-1/2} + K \tag{30.23}$$

Empirical Infiltration Equations

Although there is a strong theoretical basis from which infiltration equations such as Horton's and Philip's are derived, numerous empirical equations are in use. Most of these are based on the observations that there is an initial infiltration rate f_0, that depends on the soil type and antecedent moisture conditions and that the rate decreases to infiltration capacity f_c under a supply rate i_s that is higher than f_0. The third parameter besides f_0 and f_c is the *rate of decay.*

Some of the more popular empirical infiltration equations are considered here. Singh [1992] has derived these models by using a systems approach. Quite often the supply rate is smaller than the infiltration capacity. In such a case all the rainfall supply is assumed to infiltrate into the soil until the total rainfall and the infiltration depths are the same. The time at which the infiltration and rainfall depths are the same is called the *ponding time.*

Horton's Equation

Horton's equation, Eq. (30.24), is the same as Eq. (30.22) with Dl^2 replaced by k.

$$f_p(t) = f_c + (f_0 - f_c)e^{-kt} \tag{30.24}$$

The cumulative infiltration $F_p(t)$ corresponding to Eq. (30.24) is given by Eq. (30.25). If a constant rainfall rate i_s is assumed, the ponding time t_p is given by Eq. (30.26).

$$F_p = f_c t + \frac{(f_0 - f_c)}{k} \tag{30.25}$$

$$t_p = \frac{1}{i_s k}\left[f_0 - i_s + f_c \ln\left(\frac{f_0 - f_c}{i_s - f_c}\right)\right] \tag{30.26}$$

Philip's Infiltration Equation

Philip's equation is not an empirical equation. However, the cumulative infiltration and the time of ponding for rainfall of constant intensity i_s for this equation are given by Eqs. (30.27) and (30.28).

$$F_p = St^{1/2} + At \tag{30.27}$$

$$t_p = S^2(i_s - A/2)/2i_s(i_s - A)^2, \qquad i_s > A \tag{30.28}$$

The Green and Ampt Model

Green and Ampt [1911] proposed an infiltration model that is given by Eq. (30.29), where the cumulative infiltration is given by Eq. (30.30).

$$f_p = K\left(\frac{\psi\Delta\theta}{F(t)} + 1\right) \tag{30.29}$$

$$F_p = Kt + \psi\Delta\theta \ln\left(1 + \frac{F_p}{\psi\Delta\theta}\right) \tag{30.30}$$

In Eqs. (30.29) and (30.30), K is the hydraulic conductivity; $\Delta\theta$ is the difference between the soil porosity n and moisture content θ; and ψ is the soil suction head. In practice, the hydraulic conductivity K, the soil suction head ψ, and the porosity are obtained from published sources such as Rawls et al. [1983]. Instead of the difference $\Delta\theta$, that factor is expressed in terms of effective porosity θ_e and effective saturation s_e as in Eq. (30.31).

$$\Delta\theta = (1 - s_e)\theta_e \tag{30.31}$$

Effective porosity θ_e is the difference between soil porosity and the residual moisture content θ_r after it has been thoroughly drained, Eq. (30.32).

$$\theta_e = n - \theta_r \tag{30.32}$$

The effective saturation s_e is given by Eq. (30.33).

$$s_e = \frac{\theta - \theta_r}{n - \theta_r} \tag{30.33}$$

Values of θ_e are listed for different soils by Rawls *et al.* [1983]. Consequently, if the effective saturation of the soil, s_e, and the other parameters listed by Rawls *et al.* [1983] are known, the Green and Ampt model may be used to estimate infiltration.

30.5 Surface Runoff

Process and Measurement

An area that drains into a stream at a given location via a network of streams is called a *watershed.* Rain that falls on a watershed fills the *depression storage,* which consists of storage provided by natural depressions in the landscape; it is temporarily stored on vegetation as *interception*; and it infiltrates into the soil. After these demands are satisfied, water starts flowing over the land, and this process is called *overland flow.* Water that is stored in the upper soil layer may emerge from the soil and join the overland flow. The overland flow lasts only for short distances, after which it is collected in small channels called *rills.* Flows from these rills reach *channels.* Flow in channels reaches the mainstream.

When rainfall is of low intensity, the overland–rill–channel flow sequence may not occur. In such cases only the land near the streams contributes to the flow; these areas are called *variable source* or *partial areas.* Only a small area of watersheds contribute to stream flows in a humid region.

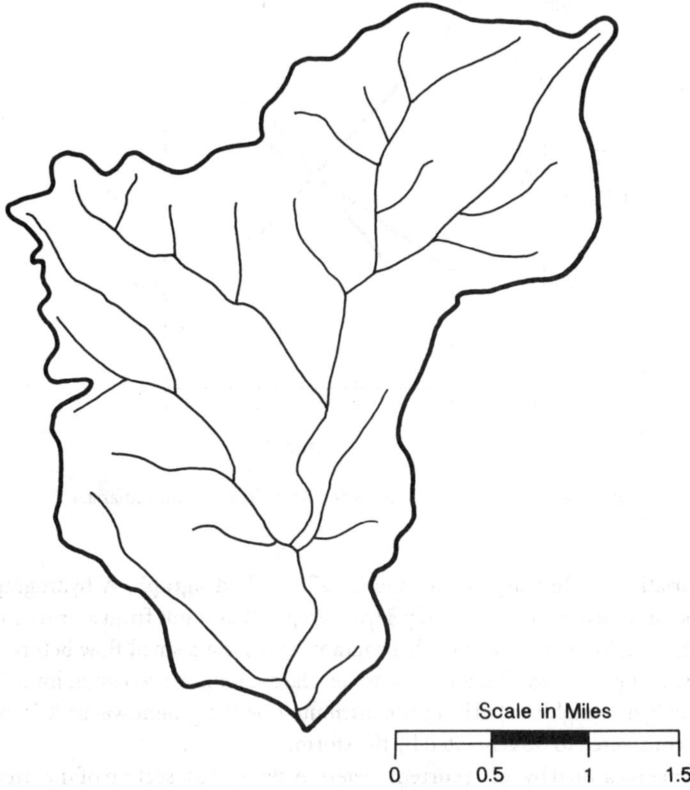

FIGURE 30.8 Drainage map of Bear Creek Basin, Indiana.

The transformation of rainfall to runoff is affected by the stream network, by precipitation, by soil, and land use. A watershed consists of a network of streams, as shown in Fig. 30.8. Channels that start from upland areas are called the *first-order* channels. Horton [1945] developed a stream order system in which, when two streams of order i join together, the resulting stream is of order $i + 1$. Several laws of stream orders are developed by Horton [1945].

If a watershed has N_i streams of order i and N_{i+1} of order $i + 1$, the ratio of N_i/N_{i+1} is called the bifurcation ratio R_B. The ratio of average stream lengths $\overline{L}_{i+1}$ and $\overline{L}_i$ belonging to orders $i + 1$ and i is denoted by R_L, and the ratio R_A of average areas $\overline{A}_i$ and $\overline{A}_{i+1}$ is called the area ratio. These ratios vary over a small range for each watershed. The drainage density D of a watershed is the ratio of total stream length to the area of the watershed. Higher values of D represent a highly developed stream network, and vice versa. Plots of L_i, A_i, and N_i against the order i for an Indiana watershed are shown in Fig. 30.9.

The second factor that significantly affects runoff is rainfall. The spatial and temporal rainfall distribution and the history of rainfall preceding a storm affect runoff from watersheds. Rainfall is usually treated as a lumped variable, because spatial rainfall data are not commonly available.

The third factor that affects runoff characteristics is the land use. As watersheds are changed from rural to urban or from forested to clear-cut condition, runoff from these watersheds changes drastically. For example, when a rural watershed is urbanized, the peak discharges from the urban watershed may be more than 100% higher than runoff from the rural watershed for the same rainfall. The time to peak discharge would also be considerably shorter, and the runoff volume much larger, in urban watersheds than in rural watersheds.

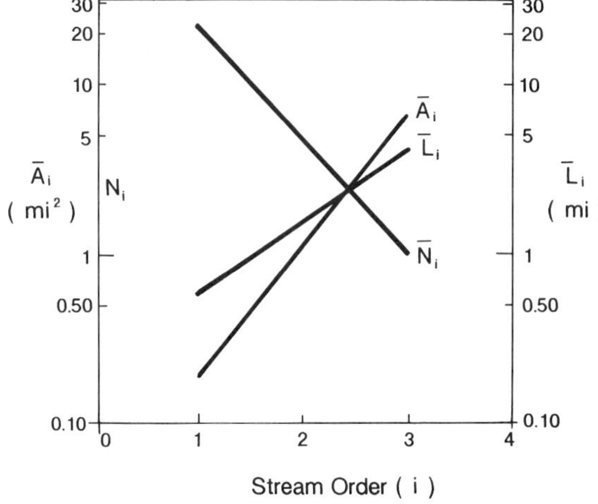

FIGURE 30.9 Horton's ratios for Bear Creek Basin, Indiana.

A plot of variation of discharge with time is called a **hydrograph**. A hydrograph may have different time scales, such as hourly or daily. Hydrographs that result from storms are called *storm hydrographs* (Fig. 30.10). A typical storm hydrograph may have a small flow before the discharge increases on the rising limb, reaches a peak, and decreases along the recession limb. The flow that exists before the hydrograph starts rising is contributed by the groundwater; it is called the *base flow* and is not considered to be generated by the storm.

Stream flows are measured by using current meters. A stable cross section of the stream is selected and divided into a number of sections. Velocities in each section are measured and averaged. The product of the average velocity and the area of the section gives the discharge in that section. The sum of discharges measured in different sections gives the discharge at the stream at that cross section.

Discharges are uniquely related to the water levels in stable stream channels. A plot of discharges against water level elevations, called *river stages*, is called a *rating curve* of a stream at a gauging station (Fig. 30.11). Once a rating curve is established for a river cross section, only the stages are measured, and discharges are computed by using the rating curve. Discharges are recorded

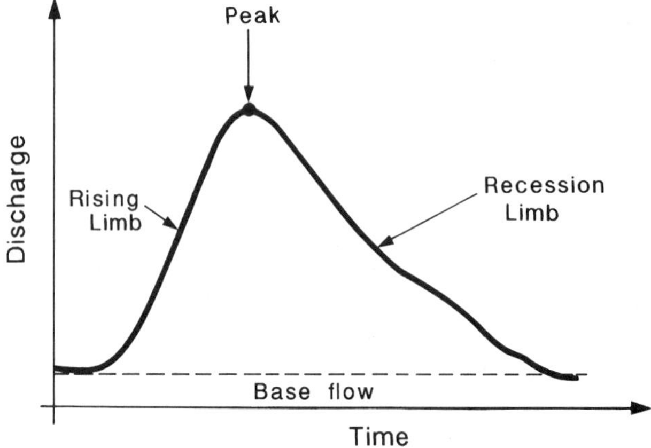

FIGURE 30.10 Single-peaked hydrograph.

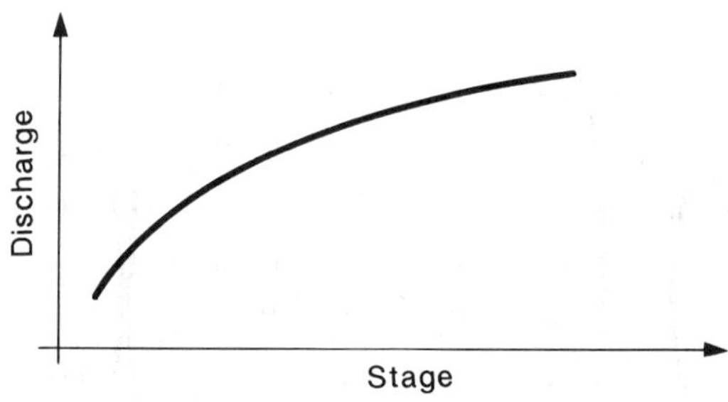

FIGURE 30.11 Rating curve.

continuously or at finite time intervals. They are also transmitted electronically to a central location, where they are recorded for dissemination. Details about river gauging are found in Rantz *et al.* [1982].

Unit Hydrographs

One of the common and important problems in hydrology is the estimation of runoff hydrographs that result from rainfall. These hydrographs are needed for various purposes, such as design of drainage and hydraulic structures and flood flow forecasting. **Unit hydrograph** theory is commonly used to estimate runoff hydrographs.

Although surface runoff has several components such as base flow, interflow, and overland flow, two components of surface runoff are commonly recognized. These are the base flow and direct runoff. There are numerous methods for separating base flow from the observed hydrograph to obtain the direct runoff hydrograph. The simplest among these assumes a constant base flow value. The volume under the direct runoff hydrograph may be expressed in units of depth (cm or in.) by dividing it by the watershed area.

The rainfall hyetograph associated with the surface runoff hydrograph may also be separated into two components: the *effective rainfall hyetograph* and *losses*. The ϕ-index method is commonly used to derive the effective rainfall hyetograph. The rainfall depth under an effective rainfall hyetograph is the same as the depth of direct runoff.

The *unit hydrograph* is the direct runoff hydrograph resulting from one unit (1 in. or 1 cm) of effective rainfall occurring uniformly in space and time over a unit period of time. The duration of effective rainfall is the "unit" for which the unit hydrograph is estimated. A unit hydrograph is derived by dividing the direct runoff hydrograph ordinates by the direct runoff depth.

If a unit hydrograph of duration D_1 is available for a watershed, a unit hydrograph of duration D_2 for the same watershed may be developed by using the S-curve method. In order to develop an S-curve, unit hydrographs are displaced in time (or lagged) by a time interval D_1. Unit hydrograph ordinates at a given time are summed to obtain the S-curve. The S-curve is displaced by duration D_2, and the difference between the two S-curve ordinates is multiplied by D_1/D_2 to get the unit hydrograph of duration D_2.

The dependence of unit hydrographs on effective rainfall duration can be eliminated by assuming the time interval between the ordinates of the runoff hydrograph to be the duration of the unit hydrograph and also of the effective rainfall pulses. Under these assumptions, the direct runoff **Q** is related to the effective rainfall **P** and unit hydrograph **U**, as in Eq. (30.34), where **Q** and **U** are vectors, and **P** is a matrix.

$$\mathbf{Q} = \mathbf{PU} \tag{30.34}$$

In Eq. (30.34), the direct runoff ordinates are $Q_1, Q_2, \ldots, Q_i$; the effective rainfall ordinates are $P_1, P_2, \ldots, P_j$, and the unit hydrograph ordinates are $U_1, U_2, \ldots, U_k$. Equation (30.34) may be expressly written as

$$\begin{pmatrix} Q_1 \\ Q_2 \\ \cdot \\ \cdot \\ \cdot \\ Q_i \end{pmatrix} = \begin{pmatrix} P_1 & 0 & \cdots & 0 \\ P_2 & P_1 & \cdots & 0 \\ \vdots & \vdots & \ddots & \vdots \\ P_j & P_{j-1} & \cdots & P_{j-k+1} \\ 0 & P_j & \cdots & P_{j-k+2} \\ \vdots & \vdots & \ddots & \vdots \\ 0 & 0 & \cdots & P_{j-1} \\ 0 & 0 & \cdots & P_j \end{pmatrix} \begin{pmatrix} U_1 \\ U_2 \\ \cdot \\ \cdot \\ \cdot \\ U_{k-1} \\ U_k \end{pmatrix} \tag{30.35}$$

The relationship between i, j, and k is given as

$$i = j + k - 1 \tag{30.36}$$

Expanding Eq. (30.35), we get Eq. (30.37). These equations are used to compute direct runoff given effective precipitation **P** and unit hydrograph ordinates **U**. They may also be used to estimate unit hydrograph ordinates by forward substitution:

$$\begin{aligned} Q_1 &= P_1U_1 \\ Q_2 &= P_2U_1 + P_1U_2 \\ Q_3 &= P_3U_1 + P_2U_2 + P_1U_3 \\ &\vdots \\ Q_{i-1} &= P_jU_{k-1} + P_{j-1}U_k \\ &\vdots \\ U_1 &= Q_1/P_1 \\ U_2 &= (Q_2 - P_2U_1)/P_1 \\ U_3 &= (Q_3 - P_3U_1 - P_2U_2)/P_1 \\ &\vdots \end{aligned} \tag{30.37}$$

The major problem with the forward substitution method of computing unit hydrograph ordinates is that the errors in the estimated unit hydrograph ordinates propagate and magnify.

In order to avoid the amplification of errors and to get stable unit hydrograph estimates, several other methods have been developed [Singh, 1988]. One of the commonly used methods is the least squares method. The least squares estimate of the unit hydrograph is given by Eq. (30.38), where superscript T indicates the vector transpose.

$$\mathbf{U} = [\mathbf{P}^{\mathrm{T}}\mathbf{P}]^{-1}\mathbf{P}^{\mathrm{T}}\mathbf{Q} \tag{30.38}$$

Synthetic Unit Hydrographs

For many watersheds, especially small ones, rainfall–runoff data may not be available to develop unit hydrographs and use them to estimate runoff. In such cases, relationships developed between unit hydrograph characteristics derived by using observed rainfall data and watershed and effective rainfall characteristics are used to generate unit hydrographs. These hydrographs are called *synthetic* unit hydrographs.

Snyder [1938] developed synthetic unit hydrographs by using data from Appalachian highland watersheds that varied in size from about 10 to 10,000 mi^2. A number of studies followed Snyder's study, many of which were designed to develop unit hydrographs from urban watersheds. A few, representative synthetic unit hydrograph methods are discussed in the following paragraphs.

Sarma *et al.* [1969] provided a set of equations to estimate the watershed *time lag* (defined as the time interval between the centroids of effective rainfall and direct runoff), the time to peak T_p, and peak discharge Q_p of observed runoff. In Eqs. (30.39) through (30.41), A is the area (mi^2); U is the fraction of imperviousness of the watershed; P_E is the effective rainfall depth (in.); and T_R is the duration of effective rainfall (hr).

$$t_L = 0.831\,A^{0.458}\,(1+U)^{-1.662}\,P_E^{-0.267}\,T_R^{0.371} \tag{30.39}$$

$$Q_p = 484.1\,A^{0.723}\,(1+U)^{1.516}\,P_E^{1.113}\,T_R^{-0.403} \tag{30.40}$$

$$T_p = 0.775\,A^{0.323}\,(1+U)^{-1.285}\,P_E^{-0.195}\,T_R^{0.634} \tag{30.41}$$

The time lag t_L is the parameter k in the instantaneous unit hydrograph of the single linear reservoir model in Eq. (30.42). The unit hydrograph of an urban watershed can thus be estimated by Eq. (30.42), and the direct runoff can be computed by using Eq. (30.34).

$$u(t) = \frac{1}{k}e^{-t/k} \tag{30.42}$$

Espey *et al.* [1977] developed another set of equations to estimate synthetic unit hydrographs along the lines of Snyder [1939]. In Eqs. (30.43) through (30.47) L is the length along the main channel (ft), S is the main channel slope, determined by $H/0.8L$, where H is the difference in elevation between two points on the main channel bottom, of which the first point is at a distance of $0.22L$ downstream from the watershed boundary and the second point is at the downstream point. I (percent) is the watershed impervious area (equal to 5% for undeveloped watershed); ϕ is a dimensionless conveyance factor; A is the watershed area (mi^2); t_p (min) is the time to peak; U_p (cfs) is the peak flow of the unit hydrograph; t_b (min) is the unit hydrograph base time; and W_{75} and W_{50} are the times of 75% and 50% of unit hydrograph peak discharge.

$$t_p = 3.1L^{0.23}\,S^{-0.25}\,I^{0.18}\,\phi^{1.57} \tag{30.43}$$

$$U_p = 31.62 \times 10^3\,A^{0.96}\,T_p^{-1.07} \tag{30.44}$$

$$t_b = 125.89 \times 10^3\,A\,Q_p^{-0.95} \tag{30.45}$$

$$W_{50} = 16.22 \times 10^3\,A^{0.93}\,Q_p^{-0.92} \tag{30.46}$$

$$W_{75} = 3.24 \times 10^3\,A^{0.79}\,U_p^{-0.78} \tag{30.47}$$

The watershed conveyance factor ϕ is estimated from Fig. 30.12.

SCS Method

The **SCS** (U.S. Department of Agriculture Soil Conversion Service) method is based on the time of concentration t_c (hrs) and the watershed area A (mi^2). The duration of the unit hydrograph ΔD (hours) is given by Eq. (30.48); the time to peak t_p (hours) and the peak discharge q_p (cfs) of the unit hydrograph are given by Eqs. (30.49) and (30.50); and the base time unit is given by Eq. (30.51).

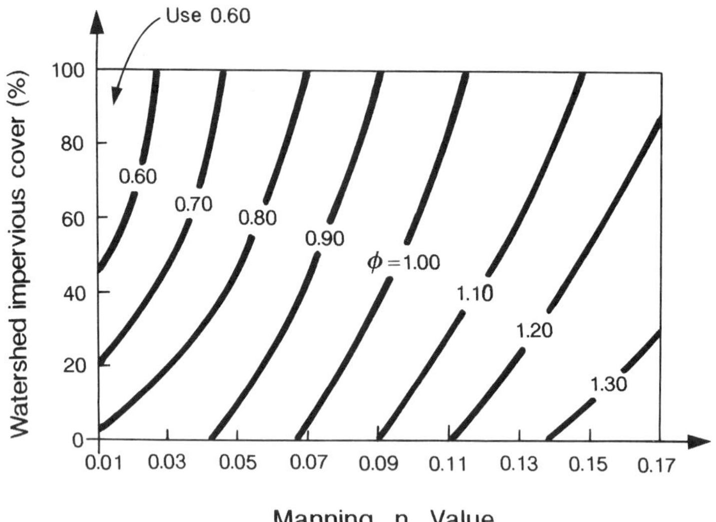

FIGURE 30.12 Watershed conveyance factor ϕ. *Source:* After Espey, W. H., Jr., Altman, D. G., and Groves, C. B. 1977. *Nomographs for Ten-Minute Unit Hydrographs for Small Urban Watersheds.* Tech. Memo 32, Urban Water Resources, Res. Prog., ASCE, New York, NY.

$$\Delta D = 0.133 t_c \tag{30.48}$$

$$t_p = 0.67 t_c \tag{30.49}$$

$$q_p = \frac{484A}{t_p} \tag{30.50}$$

$$t_b = 2.67 t_p \tag{30.51}$$

A triangular unit hydrograph of duration ΔD can be constructed by using Eqs. (30.49) through (30.51). A unit hydrograph can also be generated by the dimensionless unit hydrograph given in SCS [1972].

30.6 Flood Routing through Channels and Reservoirs

As runoff from land enters into a channel, the volume of water temporarily stored in the channel increases. After the end of precipitation, water moves down the channel and the discharge decreases. At a cross section of a channel, this increase in stage and its decrease at the end of a storm is analogous to the passage of a wave, and hence these are called *flood waves.*

Whether a flood wave moves down a channel or through a reservoir, water is temporarily stored in the channel or in the reservoir and is naturally drained out or released. **Flood routing** is the name given to a set of techniques that have been developed to analyze the passage of a flood wave through the system. *Hydraulic routing* is flood routing in which the equations that govern the motion of a flood wave (St. Venant's equations) are used. In *hydrologic routing,* a one-dimensional, lumped, continuity equation is solved to estimate the passage of a flood wave. *Reservoir routing* is similar to hydrologic routing of flood waves.

Hydraulic Routing

The St. Venant equations, which are the basic equations of motion describing the passage of a flood wave down a channel, are as follows. Equation (30.52) is the continuity equation, and Eq. (30.53) the momentum equation.

$$\frac{\partial Q}{\partial x} + \frac{\partial A}{\partial t} = 0 \tag{30.52}$$

$$\frac{1}{A}\frac{\partial Q}{\partial t} + \frac{1}{A}\frac{\partial}{\partial x}\left(\frac{Q^2}{A}\right) + g\frac{\partial y}{\partial x} - g(s_o - s_f) = 0 \tag{30.53}$$

In Eqs. (30.52) and (30.53), Q is the discharge; A is the cross-sectional area of flow; t is the time; x is the distance along the channel; y is the depth; g is the acceleration due to gravity; and s_o and s_f are respectively the slopes of the channel bottom and of the energy grade line. These equations are first-order, nonlinear, hyperbolic partial differential equations. Given a set of boundary and initial conditions, they can be solved for discharge and depth [Lai, 1986]. Another input is the roughness coefficient, Manning's n. The St. Venant equations are often solved as simplifications, because approximate solutions are sufficient in many cases. Two of these simplified solutions are the kinematic wave and diffusion approximations.

Kinematic Wave Approximation

In this case the momentum equation is reduced to Eq. (30.54), and the continuity equation is retained:

$$\frac{\partial Q}{\partial x} + \frac{\partial A}{\partial t} = 0 \tag{30.52}$$

$$s_o = s_f \tag{30.54}$$

In this approximation the dynamic terms in the momentum equation are ignored. Consequently the discharge and stage variation, both in time and space, must be small for the kinematic wave approximation to be valid.

Diffusion Approximation

In this case, in addition to the slope terms, the term involving $\partial y/\partial x$ in Eq. (30.53) is retained to give the system of Eq. (30.52) and (30.55). When the stage variation with distance is significant, the diffusion wave approximation gives better results than the kinematic wave approximation.

$$\frac{\partial Q}{\partial x} + \frac{\partial A}{\partial t} = 0 \tag{30.52}$$

$$\frac{\partial y}{\partial x} - s_o + s_f = 0 \tag{30.55}$$

The kinematic wave approximation has been used in urban hydrology and for modeling flows from small areas. The diffusion wave approximation is used to route floods through streams. Although a few closed-form solutions are available for simple cases of kinematic wave routing, numerical methods are used to solve even these simpler equations [Singh, 1988]. Hydrologic routing is often used instead because of its simplicity.

Hydrologic Channel Routing

The storage S in a channel reach is a function of both the inflow I and outflow O from a reach. The continuity equation for this system is given as

$$\frac{dS}{dt} = I - O \tag{30.56}$$

The routing problem is to estimate the outflow given the inflow and an initial outflow. In the *Muskingum method,* the storage function is assumed to be a linear function of inflow and outflow, as in

$$S = K[xI + (1 - x)O] \tag{30.57}$$

Writing Eq. (30.57) in a discrete form in terms of storage at times j and $j + 1$, we get Eq. (30.58), where Δt is the time interval between j and $j + 1$.

$$S_{j+1} - S_j = \frac{[I_{j+1} + I_j]}{2}\Delta t - \frac{[O_{j+1} + O_j]}{2}\Delta t \tag{30.58}$$

Substituting Eqs. (30.59) and (30.60) into Eq. (30.58), we get Eq. (30.61), where C_1, C_2 and C_3 are defined in Eq. (30.62).

$$S_{j+1} = K[xI_{j+1} + (1 - x)O_{j+1}] \tag{30.59}$$

$$S_j = K[xI_j + (1 - x)O_j] \tag{30.60}$$

$$Q_{j+1} = C_1 I_{j+1} + C_2 I_j + C_3 O_j \tag{30.61}$$

$$C_1 = \frac{\Delta t - 2Kx}{N}, \qquad C_2 = \frac{\Delta t + 2Kx}{N}, \qquad C_3 = \frac{2K(1 - x) - \Delta t}{N} \tag{30.62}$$

where

$$N = 2K - 2Kx + \Delta t \tag{30.63}$$

$$C_1 + C_2 + C_3 = 1 \tag{30.64}$$

In order to use Eq. (30.61), the coefficients C_1, C_2, and C_3, which depend on K, Δt, and x, must be known. To estimate K, by using a set of inflow and outflow hydrographs, different values of x are assumed and $xI_j + (1 - x)O_j$ values are plotted against computed storage values. The value of x that gives a linear relationship between observed and computed storages is selected, and the slope of the fitted line is the best estimate of the K value. K and x values are also estimated by using the method of moments, the method of least squares, and optimization methods. If several sets of inflow and outflow hydrographs are used, average K and x values are used as the best estimates for the reach. These K and x values, the inflow hydrograph, and the initial outflow value are used to compute the outflow hydrograph from the reach, thus completing the stream flow routing through the channel reach. A detailed discussion of these methods is found in Singh [1988].

Reservoir Routing

Reservoir routing is the procedure by which the outflow hydrograph from a reservoir is computed given the inflow hydrograph, the initial outflow or reservoir level, and the storage characteristics of the reservoir. A linear relationship between storage and inflow and outflow, such as that assumed in channel routing, cannot be assumed in this case.

The continuity equation, Eq. (30.56), forms the basis of the routing method in this case also. The discrete form of the continuity equation, Eq. (30.58), is rewritten as

$$\left(\frac{2S_{j+1}}{\Delta t} + O_{j+1}\right) = (I_{j+1} + I_j) + \left(\frac{2S_j}{\Delta t} - O_j\right) \tag{30.65}$$

By using the relationship between the storage and outflow relationships that are unique for each reservoir system, Eq. (30.65) is solved iteratively for discharge. Various methods have been developed to solve Eq. (30.65); one of these methods—the storage indication method—is as follows.

By using the reservoir elevation–discharge and elevation–storage data, a curve relating $(2S/\Delta t + O)$ and discharge O is developed. By using the initial storage–outflow information, the $2S_0/\Delta t - O_0$ value is estimated. The $2S_1/\Delta t + O_1$ is then estimated by Eq. (30.65), and by using the $2S/\Delta t + O_1$ vs. O relationship, the discharge O_1, and hence $2S_1/\Delta t - O_1$, is computed. By using O_1, I_2, and I_1 in Eq. (30.65), $2S_2/\Delta t + O_2$ is computed, and O_2 is evaluated by the curve of $2s/\Delta t + O$ vs. O. The procedure is repeated until the end of the inflow hydrograph.

30.7 Statistical Analysis of Hydrologic Data

Important hydrologic processes such as floods or droughts, which are extreme events, are treated as random events. The theory of probability is used to estimate the probabilities of occurrence of these events. The emphasis in statistical analysis is on the events rather than on the physical processes that generate them. In the **frequency analysis** of floods, the emphasis is on the frequency of occurrence of these events.

Probability Distributions and Parameter Estimates

The common probability distributions used in hydrologic analysis and their parameters are listed in Table 30.2. There are various methods of parameter estimation, the simplest of which is

Table 30.2 Probability Distributions for Fitting Hydrologic Data

Distribution	Probability Density Function	Range	Parameters
Normal	$F(x) = \dfrac{1}{\sigma\sqrt{2\pi}} \exp\left(-\dfrac{(x-\mu)^2}{2\sigma^2}\right)$	$-\infty \le x \le \infty$	μ, σ $\mu = \bar{x}, \quad \sigma = s_x$
Log normal	$f(x) = \dfrac{1}{x\sigma\sqrt{2\pi}} \exp\left(-\dfrac{(y-\mu_y)^2}{2\sigma_y^2}\right)$ where $y = \log x$	$x > 0$	μ_y, σ_y $\mu_y = \bar{y}, \quad \sigma_y = s_y$
Gamma	$f(x) = \dfrac{\gamma^\beta x^{\beta-1} e^{-\gamma x}}{\Gamma(\beta)}$ where $\Gamma = \gamma$ function	$x \ge 0$	λ, β $\beta = \left(\dfrac{\bar{x}}{s_x}\right)^2; \quad \lambda = \dfrac{\bar{x}}{s_x^2}$
Pearson Type III (3-parameter gamma)	$f(x) = \dfrac{\gamma^\beta (x-\epsilon)^{\beta-1} e^{-\gamma(x-\epsilon)}}{\Gamma(\beta)}$	$x \ge \epsilon$	λ, β, ϵ $\lambda = \dfrac{S_y}{\sqrt{P}}; \quad \beta = \left(\dfrac{2}{C_{s2}}\right)^2$ $\epsilon = \bar{x} - s_x\sqrt{\beta}$
Log Pearson Type III	$f(x) = \dfrac{\gamma^\beta (y-\epsilon)^{\beta-1} e^{-\gamma(y-\epsilon)}}{x\Gamma(\beta)}$ where $y = \log x$	$\log x \ge \epsilon$	λ, β, ϵ $\lambda = s_y/\sqrt{\beta};$ $\beta = \left[\dfrac{2}{C_{sy}}\right]^2$ $\epsilon = \bar{y} - s_y\sqrt{\beta}$
Extreme Value Type I	$f(x) = \dfrac{1}{\alpha} \exp\left[-\dfrac{x-u}{\alpha} - \exp\left(-\dfrac{x-u}{\alpha}\right)\right]$	$-\infty < x < \infty$	α, u $\alpha = \dfrac{\sqrt{6}\, s_x}{\pi}$ $u = \bar{x} - 0.5772\alpha$

the *method of moments*. In this method, moments estimated from the data are equated to the expressions of moments of distributions, and the resulting equations are solved for the distribution parameters. The mean, standard deviation, and skewness coefficient are the three moments commonly computed from the data.

The mean $\bar{x}$, the standard deviation s_x, and the skewness coefficient CS_x computed by using the data $x_1, x_2, \ldots, x_N$ are given in Eqs. (30.66) through (30.68).

$$\bar{x} = \frac{\sum_{i=1}^{N} x_i}{N} \tag{30.66}$$

$$s_x = \sqrt{\frac{1}{N-1}\sum_{i=1}^{N}(x_i - \bar{x})^2} \tag{30.67}$$

$$CS_x = \frac{N\sum_{i=1}^{N}(x_i - \bar{x})^3}{(N-1)(N-2)S_x^3} \tag{30.68}$$

The coefficient of variation CV is the ratio of the standard deviation to the mean of the data. If CS_x is zero, then the distribution of x_i is symmetric; otherwise it is positively or negatively distributed depending on the sign of CS_x. Expressions relating the moments in Eqs. (30.66) through (30.68) and the parameters of the distributions are shown in Table 30.2.

Frequency Analysis of Hydrologic Data

Frequency analysis of hydrologic data is conducted to estimate the magnitude of the variate corresponding to a recurrence interval T. The recurrence interval and the exceedance probability p, or $P(x \geq x_T)$, are inversely related.

$$p = P(x \geq x_T) = \frac{1}{T} \tag{30.69}$$

The probability of occurrence of an event x_T, $F_x(x_T)$, is the probability that the random variable x is smaller than x_T, $p(x < x_T)$. Therefore, the recurrence interval T and the probability of occurrence of an event are related to each other as

$$P(x < x_T) = F_x(x_T) = 1 - \frac{1}{T} = \frac{T-1}{T} \tag{30.70}$$

The relationships in Eqs. (30.69) and (30.70) are used to derive the relationships between x_T and the corresponding T. For example, the probability of type I extreme value distribution, or the EV(I) distribution, is given in Eq. (30.71) [Gumbel, 1958]. The function $(x_T - u)/\alpha$ is called the reduced variate y_T.

$$F_x(x_T) = \exp\left[-\exp\left(-\frac{x_T - u}{\alpha}\right)\right] \tag{30.71}$$

Solving Eq. (30.71) for y_T and substituting Eq. (30.70) for $F_x(x_T)$ into the resulting equation, Eq. (30.72) is obtained.

$$y_T = -\ln\left[\ln\left(\frac{T}{T-1}\right)\right] \tag{30.72}$$

$$x_T = u + \alpha y_T \tag{30.73}$$

Therefore, y_T is computed for specific values of T, u, and α by Eq. (30.72), and the corresponding value of the variate is computed by Eq. (30.73). If annual maximum flood data are analyzed, the corresponding resulting flood magnitude x_T is called the T-year flood.

Chow [1951] generalized relationships such as those in Eq. (30.73) for use in hydrologic frequency analysis. If observed data x_i are used, then the relationship between the magnitude of the variable x_T and the corresponding recurrence interval T is given by Eq. (30.74), where k_T is the hydrologic frequency factor, which is related to T and $F_x(x_T)$ as in Eq. (30.70).

$$x_T = \bar{x} + k_T s_x \tag{30.74}$$

If log-transformed data are used in the frequency analysis, $y_i = \log x_i$, the relationship corresponding to Eq. (30.74) is given in Eq. (30.75), where $\bar{y}$ and s_y are the mean and standard deviation of log-transformed data.

$$y_T = \bar{y} + k_T S_y \tag{30.75}$$

The skewness coefficient of log-transformed data is denoted by C_{sy}. The relationships between k_T and T for normal, EV(1), and log Pearson type (III) distribution are shown in Table 30.3.

Hydrologic data are plotted on probability paper, on which the ordinates represent x_T and the abscissas represent either the exceedance probability or the cumulative probability of a distribution. These exceedance probabilities are estimated by using plotting-position formulas. A general representation of the plotting-position formulas is given in Eq. (30.76), where i is the rank of the variate, with $i = 1$ for the largest value. Several commonly used plotting position formulas may be derived from Eq. (30.76). For example, C is equal to 0.5 for Hazen's, 0 for Weibull's, 0.3 for Chegodayev's, 0.375 for Blom's, 0.33 for Tukey's, and 0.44 for Gringorten's formulas.

$$P(x \geq x_T) = \frac{1}{T} = \frac{i - C}{N + 1 - 2C} \tag{30.76}$$

If the data obey the distribution whose paper they are plotted on, they fall approximately on a straight line. Straight lines may be fitted to these data and extrapolated to estimate the x_T values corresponding to recurrence interval T.

Table 30.3 Frequency Factors for Commonly Used Distributions

Distribution	k_T
Normal	For a given T, compute $p = 1/T$ and $a = \sqrt{\ln(1/p^2)}$
	$k_{TN} = a - \left(\dfrac{2.525517 + 0.802853a + 0.010328a^2}{1 + 1.432788a + 0.189269a^2 + 0.001308a^3}\right)$
	If $p > 0.5$, $a = \sqrt{\ln(1/(1-p)^2)}$ and $x_T = -x_T$.
EV(I)	$k_{TE} = -\dfrac{\sqrt{6}}{\pi}0.5772 + \ln\left\{\ln\left(\dfrac{T}{T-1}\right)\right\}$
LP (III)	For a given T, compute k_{TN}. $b = c_{sy}/6$
	$k_{TL} = k_{TN} + (k_{TN}^2 - 1)b + \frac{1}{3}(k_{TN}^3 - 6k_{TN})b^2 - (k_{TN}^2 - 1)b^3 + bk_{TN}^4 + \frac{1}{3}k_{TN}^5$

The Water Resources Council Method

The U.S. Water Resources Council [Benson, 1968] has recommended that the LP (III) distribution be used for flood frequency analysis. In this method, the annual maximum flood data are log-transformed and the statistics $\bar{y}$, s_y, and C_{sy} are calculated. The variance of the "station skew" C_{sy}, denoted by $V(C_{sy})$, is given by Eq. (30.77), where A and B are defined in Eq. (30.78a through d) [Wallis *et al.*, 1976].

$$V(C_{sy}) = 10^{A-B\log_{10}(N/10)} \tag{30.77}$$

$$A = -0.33 + 0.08\,|\,C_{sy}\,| \qquad \text{for } |\,C_{sy}\,| \le 0.9 \tag{30.78a}$$

$$A = -0.52 + 0.3\,|\,C_{sy}\,| \qquad \text{for } |\,C_{sy}\,| > 0.9 \tag{30.78b}$$

$$B = 0.94 - 0.26\,|\,C_{sy}\,| \qquad \text{for } |\,C_{sy}\,| \le 1.50 \tag{30.78c}$$

$$B = 0.55 \qquad \text{for } |\,C_{sy}\,| > 1.5 \tag{30.78d}$$

The map skewness (C_m) is interpolated from Fig. 30.13. The variance of the map skewness $V(C_m)$ for the U.S. is 0.3025. The weighted skewness coefficient C_{sw} is given by Eq. (30.79).

$$C_{sw} = \frac{V(C_m)C_{sy} + V(C_{sy})C_m}{V(C_m) + V(C_{sy})} \tag{30.79}$$

This skewness coefficient C_{sw} is used to compute b in Table 30.3 and k_{TL} values corresponding to recurrence interval T. The logarithm of the flow is computed by Eq. (30.75), which is used to compute the value of the flow x_T. A computer program, HECWRC [U.S. Army Corps of Engineers, 1982], and its derivatives exist to perform these computations.

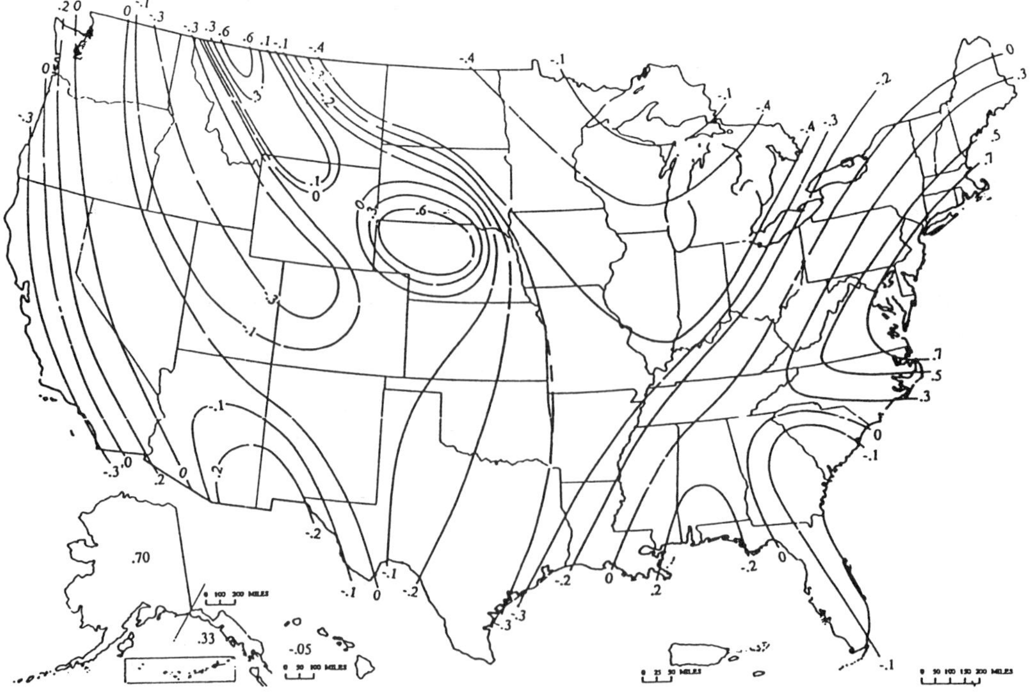

FIGURE 30.13 Map of skewness coefficients of annual maximum flows [U.S. Water Resources Council, 1981].

Defining Terms

Evapotranspiration: The process by which water in soil, vegetation, and land surface is converted to water vapor.

Flood routing: A technique used to analyze the passage of a flood wave through a channel or a reservoir.

Frequency analysis: A probabilistic analysis of hydrologic data conducted to estimate the recurrence intervals of specified precipitation events.

Hydrograph: A plot of variation of discharge with time.

Hydrologic cycle: The continuum of the movement of water in the atmosphere, hydrosphere, and lithosphere.

Infiltration: The process by which water on the soil surface enters the soil.

Isohyetal: Equal values of precipitation measured or observed.

Pan coefficient: Ratio of lake evaporation to pan evaporation.

Percolation: Process by which water moves through the soil under the influence of gravity.

Potential evapotranspiration: Evapotranspiration rate that would occur if the moisture supply were unlimited.

Precipitation: Water removed in its solid or liquid phase from the atmosphere and deposited on the surface.

Psychrometric constant: A constant involving specific heat, latent heat of vaporization, and other variables, which is used in evaporation estimation.

Rainfall frequency: Time interval between occurrence of a rainfall depth over a specific duration.

Rainfall intensity: Rate at which rain is falling at a particular time.

SCS: Soil Conservation Service, a division of the U.S. Department of Agriculture.

Specific humidity: Ratio of water vapor pressure to the atmospheric pressure.

Surface runoff, subsurface runoff: Movement of water over land surface.

Unit hydrograph: The direct runoff hydrograph that would result from one unit of effective rainfall occurring uniformly in space and time over a unit period of time.

References

ASCE. 1973. *Consumptive Use of Water and Irrigation Water Requirements,* ed. M. E. Jensen. ASCE, New York.

Benson, M. A. 1968. Uniform flood-frequency estimating methods for federal agencies. *Water Resources Res.* 4(5):891–908.

Bobée, B. and Ashkar, F. 1991. *The Gamma Family and Derived Distributions Applied in Hydrology.* Water Resources Publications, Colorado.

Bras, R. L. 1990. *Hydrology.* Addison-Wesley, Reading, MA.

Chen, C. L. 1983. Rainfall intensity–duration–frequency formulas. *J. Hydraulic Eng., ASCE.* 109(12):1603–1621.

Chow, V. T. 1951. A general formula for hydrologic frequency analysis. *Trans. AGU.* 32(2):231–237.

Chow, V. T., Maidment, D. R., and Mays, L. W. 1988. *Applied Hydrology.* McGraw-Hill, New York.

Chow, V. T. 1964. *Handbook of Applied Hydrology.* McGraw-Hill, New York.

Cunnane, C. 1989. *Statistical Distributions for Flood Frequency Analysis.* WMO Operational Hydrology Report No. 33, WMO-No. 718.

Doorenbos, J. and Pruitt, W. O. 1977. *Crop Water Requirements.* Irrigation and Drainage Paper 24, FAO, Rome, Italy.

Eagleson, P. S. 1970. *Dynamic Hydrology.* McGraw-Hill, New York.

Espey, W. H., Jr., Altman, D. G., and Graves, C. B. 1977. *Nomographs for Ten-Minute Unit Hydrographs for Small Urban Watersheds.* Tech. Memo 32, Urban Water Resources, Res. Prog., ASCE, New York.

Frederick, R. H., Myers, V. A., and Auciello, E. P. 1977. *Five to 60 Minute Precipitation Frequency for the Eastern and Central United States.* NOAA Technical Memo NWS HYDRO-35, National Weather Service, Silver Spring, MD.

Garstka, W. U. 1964. Snow and snow survey. Chap. 10 in *Handbook of Applied Hydrology,* ed. V. T. Chow. McGraw-Hill, Inc., New York.

Green, W. H. and Ampt, G. 1911. Studies of soil physics part I—the flow of air and water through soils. *J. Agric. Sci.* 4:1–24.

Gumbel, E. 1958. *Statistics of Extremes.* Columbia University Press, New York.

Hershfield, D. M. 1961. *Rainfall Frequency Atlas of the United States for Durations from 30 Minutes to 24 Hours and Return Periods from 1 to 100 Years.* Tech. Paper No. 40, U.S. Dept. of Commerce, U.S. Weather Bureau, Washington, D.C.

Horton, R. E. 1939. Analysis of runoff plot experiments with varying infiltration capacity. *Trans. AGU,* Part IV:693–711.

Horton, R. E. 1940. An approach toward a physical interpretation of infiltration capacity. *Soil Sci. Soc. Am. J.* 5:399–417.

Horton, R. E. 1945. Erosional development of streams and their drainage basins; hydrophysical approach to quantitative morphology. *Bull. Geol. Soc. Am.* 56:275–370.

Lai, C. 1986. *Numerical Modeling of Unsteady Open Channel Flow.* Advances in Hydroscience, 14, Academic Press, Orlando, FL.

Miller, J. F., Frederick, R. H., and Tracey, R. J. 1973. *Precipitation-Frequency Atlas of the Conterminous Western United States (by States).* NOAA Atlas 2, 11 vols. National Weather Service, Silver Spring, MD.

Penman, H. L. 1948. Natural evaporation from open water, bare soil and grass. *Proc. R. Soc.,* London, Ser. A., 193:120–146.

Penman, H. L. 1952. The physical basis of irrigation control. *Proc. 13th Int. Hortic. Congr.,* London.

Philip, J. R., 1960. General method of exact solution of the concentration dependent diffusion equation. *Aust. J. Phys.* 13(1):1–12.

Ponce, V. M. 1989. *Engineering Hydrology.* Prentice Hall, Englewood Cliffs, NJ.

Priestley, C. H. B. and Taylor, R. J. 1972. On the assessment of surface heat flux and evaporation using large-scale parameters. *Monthly Weather Rev.* 100:81–92.

Purdue, A. M., Jeong, G. D., and Rao, A. R. 1992. *Statistical Characteristics of Short Time Increment Rainfall,* Tech. Rept. CE-EHE-92-09, School of Civil Engineering, Purdue University, W. Lafayette, IN, p. 64.

Rantz, S. E. *et al.*, 1982. *Measurement and Computation of Streamflow, Vol. 1: Measurement of Stage and Discharge.* Water Supply Paper 2175, U.S. Geological Survey, Washington, D.C.

Rawls, W. J., Brakensiek, D. L., and Miller, N. 1983. Green-Ampt infiltration parameters from soils data. *J. of Hydraul. Eng., ASCE.* 109(1):62–70.

Sarma, P. B. S., Delleur, J. W., and Rao, A. R. 1969. *A Program in Urban Hydrology Part II. An Evaluation of Rainfall-Runoff Models for Small Urbanized Watersheds and the Effect of Urbanization on Runoff.* Tech. Rept. No. 9, Water Resources Research Center, Purdue University, West Lafayette, IN, p. 240.

SCS (Soil Conservation Service) 1972. *Hydrology.* Sec. 4 of National Soil Conservation Service, USDA, Washington, DC.

Singh, V. P. 1988. *Hydrologic Systems Vol. 1, Rainfall-Runoff Modelling.* Prentice Hall, Englewood Cliffs, NJ.

Singh, V. P. and Chowdhury, P. K. 1986.Comparing some methods of estimating mean areal rainfall. *Water Resources Bull.* 22(2):275–82.

Snyder, F. F. 1938. Synthetic unit-graphs. *Trans. AGU* 19:447–454.

Thornthwaite, C.W., *et al.*, 1944. Report of the Committee on Transpiration and Evaporation, 1943–1944. *Trans. AGU* 25, pt. V:683–693.

Thornthwaite, C. W. and Holzman, B. 1939. The determination of evaporation from land and water surfaces. *Monthly Weather Rev.* 67: 4–11.

UNESCO. 1978. *World Water Balance and Water Resources of the Earth.* Paris, France.

U.S. Army Corps of Engineers. 1982. *Flood Flow Frequency Analysis, Computer Program 723-X6-L7550, User's Manual.* Davis, CA.

U.S. Water Resources Council. 1981. *Guidelines for Determining Flood Flow Frequency.* Bulletin 17B, Washington, D.C.

Wallis, J. R., Matalas, N. C., and Slack, J. R. 1976. Just a moment. *Water Resources Res.* 10(2):211–219.

WMO (World Meteorological Organization). 1981. *Guide to Hydrological Practices, Vol. 1: Data Acquisition and Processing.* Report No. 168, Geneva, Switzerland.

For Further Information

In addition to Bras [1990], Chow [1964], and Eagleson [1970], the following handbooks and textbooks in hydrology contain further details of the topics discussed in this section.

Maidment, D. R. (ed.) 1993. *Handbook of Hydrology.* McGraw-Hill, New York.

Gupta, Ram S. 1989. *Hydrology and Hydraulic Systems.* Prentice Hall, Englewood Cliffs, NJ.

Linsley, R. K. Jr., Kohler, M.A., and Paulhus, J. L. H. *Hydrology for Engineers.* McGraw-Hill, New York.

McCuen, R. H. 1989. *Hydrologic Analysis and Design.* Prentice Hall, Englewood Cliffs, NJ.

Bedient, P. B. and Huber, W. C. 1988. *Hydrology and Flood Plain Analysis.* Addison Wesley, Reading, MA.

Veissman, W., Jr., Lewis, G. L., and Knapp, J. W. 1989. *Introduction to Hydrology,* 3rd ed. Harper and Row, New York.

Singh, V. P. 1992. *Elementary Hydrology.* Prentice Hall, Englewood Cliffs, NJ.

Shaw, E. M. 1988. *Hydrology in Practice,* 2nd ed. VNR International, London, UK.

31

Urban Drainage

A. R. Rao
Purdue University

C. B. Burke
Christopher B. Burke Engineering, Ltd.

T. T. Burke, Jr.
Purdue University

31.1 Introduction

Urbanization drastically alters the hydrologic and meteorological characteristics of watersheds. Because of the changes in surface and heat retention characteristics brought about by buildings and roads, heat islands develop in urban areas. Increases in nucleation and photoelectric gases due to urbanization result in higher smog, precipitation and related activities, and lower radiation in urban areas compared to the surrounding rural areas. Some of these meteorologic effects of urbanization are discussed by Lowry [1967] and Landsburg [1981].

When an area is urbanized, trees and vegetation are removed, the drainage pattern is altered, conveyance is accelerated, and the imperviousness of the area is increased because of the construction of residential or commercial structures and roads. Increased imperviousness decreases infiltration with a consequent increase in the volume of runoff. Improvements in a drainage system cause runoff to leave the urbanized area faster than from a similar undeveloped area. Consequently, the time for runoff to reach its peak is shorter for an urban watershed than for an undeveloped watershed. The peak runoff from urbanized watersheds, on the other hand, is larger than from similar undeveloped watersheds. The effects of urbanization on runoff are summarized in Table 31.1.

Urban stormwater drainage collection and conveyance systems are designed to remove runoff from urbanized areas so that flooding is avoided and transportation is not adversely affected. A schematic diagram of a typical urban stormwater drainage collection and conveyance system is shown in Fig. 31.1. The cost of this and similar systems is directly dependent on the recurrence interval of rainfall used in the design. Rainfall with 5- to 10-year recurrence intervals is most often used in the sizing and design of the urban stormwater drainage collection and conveyance systems. To accommodate areas that encounter frequent floods or high losses due to flooding and to reduce the potential for downstream flooding, stormwater storage facilities are developed to temporarily store the stormwater and to release it after a storm has passed over the area. Examples of large-scale facilities are the Tunnel and Reservoir Plan (TARP) of the Metropolitan Water Reclamation District of Greater Chicago and the deep tunnel project in Milwaukee. Smaller-scale facilities to temporarily

0-8493-8953-4/95/$0.00 + $.50

Table 31.1 Potential Hydrologic Effects of Urbanization

Urbanizing Influence	Potential Hydrologic Response
Removal of trees and vegetation	Decrease in evapotranspiration and interception; increase in stream sedimentation
Initial construction of houses, streets, and culverts	Decrease infiltration and lowered groundwater table; increased storm flows and decreased base flows during dry periods
Complete development of residential, commercial, and industrial areas	Increased imperviousness reduces time of runoff concentration thereby increasing peak discharges and compressing the time distribution of flow; volume of runoff and flood damage potential greatly increased
Construction of storm drains and channel improvements	Local relief from flooding; concentration of floodwaters may aggravate flood problems downstream

Source: American Society of Civil Engineers Tech. Memo 24, 1974.

detain and release stormwater to the storm sewer system after the passage of storms (detention facilities) or to retain it and let the water infiltrate or evaporate (retention facilities) are commonly used in suburban flood control projects and are often required for a new development. Detention basins are commonly used to prevent downstream flooding and are often designed such that they increase the aesthetic appeal of areas in which they are placed.

For many watersheds, models such as TR-20 [SCS, 1982], developed by the Soil Conservation Service of the U.S. Department of Agriculture, and HEC-1 [HEC, 1985], developed by the Hydrologic Engineering Center of the U.S. Army Corps of Engineers, are commonly used by designers. These models are used to size collection and conveyance systems as well as stormwater storage facilities, investigate alternative scenarios, and evaluate existing drainage systems.

In this chapter commonly used formulas and methods to compute peak discharge and runoff hydrographs, such as the rational method formula and the SCS method, are discussed first. Methods

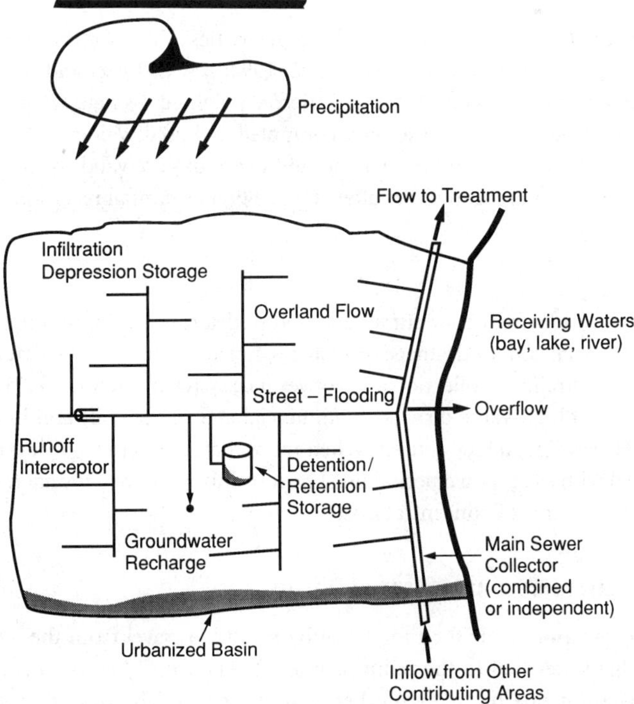

FIGURE 31.1 Schematic diagram of an urban storm-drainage system.

for sizing of detention basins are discussed next, followed by a discussion of water quality aspects of urban runoff.

31.2 The Rational Method

The **rational method** is widely used to determine peak discharges from a given watershed. It was introduced into the U.S. by Kuichling (1889); a large majority of the engineering offices which deal with storm drainage work in the U.S. use the rational method. This popularity is due to its simplicity and perhaps to tradition.

In the rational method, the peak rate of surface flow from a given watershed is assumed to be proportional to the watershed area and the average rainfall intensity over a period of time just sufficient for all parts of the watershed to contribute to the outflow. The rational formula is shown in Eq. (31.1),

$$Q = CiA, \tag{31.1}$$

where Q is the peak discharge (cfs), C is the ratio of peak runoff rate to average rainfall rate over the watershed during the time of concentration (runoff coefficient), i is the rainfall intensity (inches/hour) and A is the contributing area of the watershed (acres).

The rational method is usually applied to drainage basins less than 200 acres in area, but should be used carefully for basins greater than 5 acres. The basic assumptions used in the rational formula are as follows: (1) The rainfall is uniform over the watershed. (2) The storm duration associated with the peak discharge is equal to the time of concentration for the drainage area. (3) The runoff coefficient C depends on the rainfall return period, and is independent of storm duration and reflects infiltration rate, soil type, and antecedent moisture condition. The coefficient C, the rainfall intensity, i, and the area of the watershed, A, are estimated in order to use the rational method.

Runoff Coefficient

The runoff coefficient C reflects the watershed characteristics. Values of the runoff coefficient C are found in drainage design manuals [Burke, *et al.*, 1994; ASCE, 1992] and in textbooks [Chow *et al.*, 1988]. If a watershed has different land uses, a weighted average C based on the actual percentage of lawns, streets, roofs, and so on is computed and used. Values of C must be carefully selected. The C values usually found in manuals and textbooks are valid for recurrence intervals up to 10 years. These values are sometimes altered when higher rainfall return periods are used.

Rainfall Intensity

Rainfall intensity-duration-frequency curves are used to determine rainfall intensities used in the rational method. Local custom or drainage ordinances dictate the use of a particular return period. In the design of urban drainage collection and conveyance systems, a return period of 5 to 10 years is generally selected. In high-value districts (commercial and residential) and in flood protection works, a 50- or 100-year frequency is used. When more than one return period is used, the costs and risks associated with each return period must be scrutinized. In the rational method, the storm duration is equal to the time of concentration.

Time of Concentration and Travel Time

The **time of concentration,** t_c, is the time taken by runoff to travel from the hydraulically most distant point on the watershed to the point of interest. The *time of travel,* T_t, is the time taken by water to travel from one point to another in a watershed. The time of concentration may be visualized as the sum of the travel times in components of a drainage system. The different components include overland flow, shallow concentrated flow, and channel flow. As an area is

Table 31.2 Equations for Overland Flow Travel Time

Name	Equation for t_t	Notes
Regan [1972]	$t_t = \dfrac{L^{0.6} n^{0.6} K}{i^{0.4} S^{0.3}}$	n is Manning's roughness coefficient
Kerby [1959]	$t_t = 0.827 \left[\dfrac{NL}{\sqrt{S}}\right]^{0.467}$	$L < 1200$ ft
Federal Aviation Agency [1956]	$t_t = 1.8(1.1 - C)\dfrac{L}{\sqrt{100S}}$	airport areas C = runoff coefficient
Izzard [1946]	$t_t = \dfrac{2}{60}\dfrac{.0007i + c}{S^{1/3}} L \left[\dfrac{iL}{43,200}\right]^{-2/3}$ or $t_t = \dfrac{41cL^{1/3}}{(Ci)^{2/3} S^{1/3}}$	$iL < 500$,
Overton and Meadows [1970]	$t_t = \dfrac{0.007(nL)^{0.8}}{(P_2)^{0.5} S^{0.4}}$	

where t_t is the overland flow time (min), L is the basin length (ft),
S is the basin slope (ft/ft), i is the rainfall intensity (in./h),
c is the retardance coefficient and C is the runoff coefficient,
n is Manning's n,
P_2 is the 2-year 24-hour rainfall (in.)
$K = 56$
N is a roughness coefficient

urbanized, the quality of flow surface and conveyance facilities are improved, and the times of travel and concentration generally decrease. On the other hand, ponding and reduction of land slopes which may accompany urbanization increase times of travel and concentration.

Overland flows are assumed to have maximum flow lengths of about 300 ft. From about 300 ft to the point where the flow reaches well-defined channels, the flow is assumed to be of the shallow concentrated type. After the flow reaches open channels it is characterized by Manning's formula.

Some commonly used formulas employed in the determination of the overland flow travel time are shown in Table 31.2. Most of these equations relate the overland time of travel to the basin length, slope, and surface roughness. Two equations, by Izzard and Regan, include rainfall intensity as a factor, which necessitates an iterative solution.

The average velocities for shallow concentrated flow are estimated by Eqs. (31.2) and (31.3) for unpaved and paved areas respectively, where V is the average velocity in ft/s and S is the slope of the land surface in ft/ft [SCS, 1986]:

$$\text{Unpaved:} \quad V = 16.13(S)^{0.5} \tag{31.2}$$

$$\text{Paved:} \quad V = 20.33(S)^{0.5} \tag{31.3}$$

Flows in open channels are characterized by Manning's formula. In sewered watersheds, the time of concentration is calculated by estimating the overland flow *travel time,* which is called the **inlet time,** the gutter flow time, and the *time of travel in sewers.* Often, the inlet time, which is the time taken by water to reach inlets, is assumed to be between 5 and 30 min. In flat areas with widely spaced street inlets, an inlet time of 20 to 30 minutes is assumed [ASCE, 1992]. These inlet times are added to the flow time in the sewer or channel to determine the travel time at a downstream location.

The flow time in sewers is usually calculated by choosing a pipe or channel configuration and calculating the velocity. The time is then found by

$$t = \frac{L}{60V} \tag{31.4}$$

where t is the travel time in the sewer (min), L is the length of the pipe or channel (ft), and V is the velocity in the sewer (ft/s).

Table 31.3 Typical Velocities in Natural Waterways

	Velocity in		
Average Slope of Waterway (%)	Natural Channel (Not Well Defined) (ft/s)	Shallow Channel (ft/s)	Main Drainage Channel (ft/s)
1–2	1.5	2–3	3–6
2–4	3.0	3–5	5–9
4–6	4.0	4–7	7–10
6–10	5.0	6–8	—

Source: AASHTO, 1991.

Manning's formula is commonly used to estimate the travel times in sewers. The sewer is assumed to flow full and a velocity is computed. Manning's n values commonly used for this purpose are found in Chow [1959]. Sometimes, natural channels are used to convey storm runoff. In these cases the velocities given in Table 31.3 (AASHTO, 1991) may be used in Eq. (31.4). The drainage area, A, used in the rational formula is determined from topographic maps and field surveys.

Application of the Rational Method

The choice of parameters in the rational method is subjective. Consequently, variations occur in designs. Since rainfall intensity values are derived from statistical analyses and may not represent actual storm events, it is impossible to have a storm of a specified design intensity and duration associated with the results from the rational method. The procedure for the application of the rational method is as follows: (1) The contributing basin area A (acres) is determined by using maps or plans made specifically for the basin. (2) By using the land use information, appropriate C values are determined. If the land has multiple uses, a composite C value is estimated by Eq. (31.5):

$$C_{\text{comp}} = \frac{(C_1 A_1 + C_2 A_2 + \cdots + C_n A_n)}{A} \tag{31.5}$$

where $C_1, C_2, \ldots, C_n$ are the runoff coefficients associated with the $A_1, A_2, \ldots, A_n$ respectively and A is the sum of $A_1, A_2, \ldots, A_n$.(3) The time of concentration is estimated by summing the travel time components. (4) The rainfall intensity is determined by using an intensity-duration-frequency diagram and the time of concentration as the storm duration. (5) The peak runoff (cfs) is computed by multiplying C, i, and A. (6) If there is another basin downstream, the time of concentration from the upstream basin is added to the travel time in the channel. This time of concentration is compared to the time of concentration of the second basin and the larger of the two is used as the new time of concentration. Examples of the application of rational method are found in Burke *et al.* [1994]. Because of the assumptions and the simplistic approach on which rational method is based, its application is not recommended for watersheds larger than five acres.

31.3 The Soil Conservation Service (SCS) Methods

The Soil Conservation Service has developed a method to estimate rainfall excess, P_e, from total rainfall, P, based on the total ultimate abstraction, S. In this method, P_e is given by Eq. (31.6), where P must be greater than $0.2S$, which is the initial storage capacity I_a.

$$P_e = \frac{(P - 0.2S)^2}{P + 0.8S} \tag{31.6}$$

If P is less than $0.2S$, P_e is assumed to be zero. The rainfall excess, P_e, has units of inches. In this method, the abstraction S is related to the "curve number" CN as in Eq. (31.7).

Table 31.4 Soil Classification Table

Name	Class	Name	Class	Name	Class
Abscota	A	Bewleyville	B	Check to Waga	D
Digby	B	Door	B	Cincinnatic	C
Ade	A	Birds	(C/D)	Chetwynd	B

Source: SCS, 1972.

$$S = \frac{1000}{CN} - 10 \tag{31.7}$$

The curve number CN is also related to soil types and land uses. Soils are divided into four classes A through D, based on infiltration characteristics. Type A soils have the maximum and D soils the minimum infiltration capacity, with B and C soils falling in between. Tables containing soil names and types are available in SCS [1972] and a portion of this table is shown in Table 31.4. Curve numbers for selected land use and AMC (II) are shown in Table 31.5.

Application of the SCS Method

In order to use this method, the area A is subdivided into subareas $A_1, A_2, \ldots, A_n$ so that each of the subareas has a uniform land use. By identifying the soil types and land uses in each of these

Table 31.5 Runoff Curve Numbers for Selected Land Uses (Antecedent Moisture Condition II, $I_a = 0.2S$)

Land Use Description		Hydrologic Soil Group			
		A	B	C	D
Cultivated land: without conservation treatment		72	81	88	91
with conservation treatment		62	71	78	81
Pasture or range land: poor condition		68	79	86	89
good condition		39	61	74	80
Meadow: good condition		30	58	71	78
Wood or forest land: thin stand, poor cover, no mulch		45	66	77	83
good cover		25	55	70	77
Open spaces, lawns, parks, golf courses, cemeteries, etc.					
good condition: grass cover on 75% or more of the area		39	61	74	80
fair condition: grass cover on 50% to 75% of the area		49	69	79	84
Commercial and business areas (85% impervious)		89	92	94	95
Industrial districts (72% impervious)		81	88	91	93
Residential					
Average lot size	Average % impervious	77	85	90	92
1/8 acre or less	65	61	75	83	87
1/4 acre	38	57	72	81	86
1/3 acre	30	54	70	80	85
1/2 acre	25	51	68	79	84
acre	20				
Paved parking lots, roofs, driveways, etc.		98	98	98	98
Streets and roads:					
paved with curbs and storm sewers		98	98	98	98
gravel		76	85	89	91
dirt		72	82	87	89

Source: SCS, 1972.

subareas, the CN values for the subareas are estimated from Table 31.4. The rainfall excess P_e is computed by Eq. (31.6). A rainfall excess hyetograph may then be computed by using the Huff curves. The rainfall excess hyetograph thus generated may be used with the SCS-unit hydrograph method to compute a direct runoff hydrograph. The direct runoff hydrograph can be used to size collection and conveyance systems and to estimate detention storage volumes.

31.4 Detention Storage Design

The increased runoff volume produced by the urbanization of watersheds can result in downstream flooding. Storage facilities are designed to receive runoff from developed upstream watersheds and release it downstream at a reduced rate. This reduced rate is determined by using parameters fixed by local ordinances or by calculating the available capacity of the downstream storm sewer network. Some of the methods used to compute the required storage volumes of detention storages are discussed in this section.

Types of Storage Facilities

Storage facilities can be divided into two general categories as **detention** and **retention.** These are also called dry or wet detention ponds. Detention storage is the temporary storage of the runoff which is in excess of that released. After a storm ends, the facility is emptied and resumes its normal function. Ponds, parking lots, rooftops, and parks are common detention facilities, which may be designed to temporarily store a limited amount of runoff.

Retention facilities are designed to retain runoff for an indefinite period of time. Ponds and lakes are examples of retention facilities that are used in subdivisions to enhance urban drainage projects.

Computation of Detention Storage Volumes

The primary goal in the design of detention facilities is to provide the necessary storage volume. If infiltration and evaporation are neglected during the runoff period, the continuity equation for a detention pond may be written as in Eq. (31.8).

$$I(t) - O(t) = \frac{\Delta S}{\Delta t} \tag{31.8}$$

where

$I(t)$ = inflow to the pond from the sewer network at time (t) (cfs)
$O(t)$ = outflow from the pond into the downstream drainage network at time (t) (cfs)
ΔS = change in storage (ft^3) in time interval Δt (seconds)
Δt = time interval

Equation (31.8) may also be written as

$$(I_1 + I_2)\frac{\Delta t}{2} - (O_1 + O_2)\frac{\Delta t}{2} = S_2 - S_1 \tag{31.9}$$

where subscripts 1 and 2 denote the flows and storages at times t_1 and t_2.

When inflow and outflow hydrographs are known, the largest accumulated values of $S_2 - S_1$ found in Eq. (31.9) are the required storage. The following is a discussion of some of the methods used to estimate detention storage volumes. For retention facilities the outflow rate is equal to the sum of the evaporation and infiltration rates. These are negligible during a storm and the required volume is therefore equal to the runoff volume.

Storage Determination by Using the Rational Method

The rational method discussed previously is extended to compute detention storage volumes by multiplying the peak flow rate by the storm duration. The allowable peak flow rate (release rate) leaving the detention pond, $O(t)$, is calculated by using the contributing undeveloped area, A_U, the runoff coefficient applicable for undeveloped condition, C_U, and rainfall intensity, i_U, associated with the time of concentration of the undeveloped basin. The return period for the intensity i_U is normally fixed by local ordinances or is based on the design parameters of the larger downstream drainage network. The allowable outflow rate is assumed to remain constant for all storm durations, t_d. Therefore the volume corresponding to t_d, $V(t_d)$, is the product of $O(t)$ and t_d. This is illustrated in Fig. 31.2, where the lines $V_I(t_d)$ and $V_O(t_d)$ are the inflow and outflow volumes at times t_d.

The rate of inflow to the detention basin, $I(t)$, is calculated by using the contributing developed area, A_D, the developed runoff coefficient, C_D, and a rainfall intensity, i_D, corresponding to storm duration, t_d, and the return period. Thus, for various durations, the peak flow and the volume of runoff are computed. The maximum difference between the inflow and outflow volumes is the required detention pond storage. This is shown in Fig. 31.2 as $S_{\max}$. The method may also be expressed as in Eq. (31.10)

$$S(t_d) = [C_D i_D A_D - C_U i_U A_U] \frac{t_d}{12} \tag{31.10}$$

where $S(t_d)$ is the required storage (acre $-$ ft), and t_d is the storm duration in hours. Various storm durations, t_d, are selected and the largest value of $S(t_d)$ is selected as the required volume of the detention pond.

The procedure to size a detention storage facility by using the rational method is as follows:(1) The area, A_U, runoff coefficient, C_U, and time of concentration for the undeveloped site are determined. By using the appropriate intensity-duration-frequency curve, the intensity, i_U, corresponding to the return period for the allowable outflow rate is estimated. (2) The runoff (O) from the undeveloped site ($O = C_U i_U A_U$) is computed. (3) The runoff coefficient corresponding to the developed conditions, C_D, is estimated. (4) The rainfall intensities (i_d) for various durations (t_d) are obtained for different return periods. Recommended durations are 10, 20, 30, 40, and 50 minutes and 1, 1.5, 2, 3, 4, 5, 6, 7, 8, 9, and 10 hours. (5) The inflow rate to the detention pond, $I(t_d) = C_D i_D A_D$, is computed. (6) The required storage for each duration, $S(t_d) = [I(t_d) - 0]t_d$ is calculated. (7) The largest volume $S(t_d)$ is selected as the design volume.

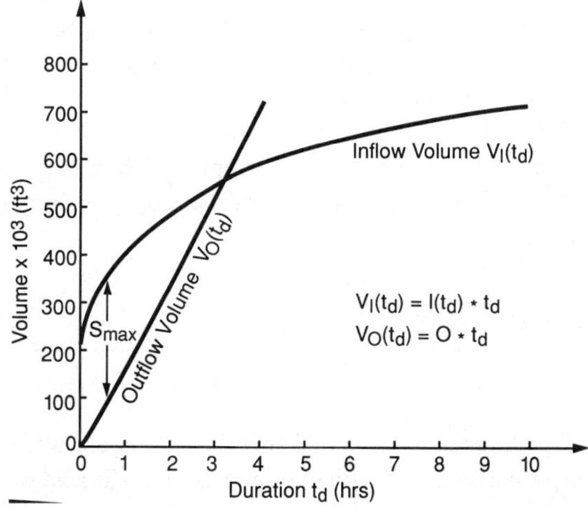

FIGURE 31.2 Graphical representation of storage volumes as determined by the rational method.

Various agencies have set guidelines for selection of i_U, i_D, C_U, and C_D. For example, the Metropolitan Water Reclamation District of Greater Chicago (MWRDGC) uses the criteria that i_U should be based on a 3-year return period, i_D should be based on a 100-year return period, C_U should be less than or equal to 0.15, and C_D should be 0.45 for pervious areas and 0.9 for impervious areas.

It may be impossible to collect and convey all of the runoff from a given watershed under certain conditions. The result is that some runoff is discharged directly into the downstream drainage network without being detained. To compensate for this unrestricted release, the allowable release rate, O, is reduced by that amount.

Soil Conservation Service Hydrograph Method

Methods were discussed earlier by which the stormwater runoff hydrographs can be estimated by the SCS method. These hydrographs are used to compute detention storage volume.

As previously described, the difference between the inflow from the developed watershed and the allowable outflow from a detention pond is the required storage volume. The outflow is determined by using characteristics of the undeveloped watershed and a rainfall frequency equal to or less than that which a receiving system can handle, or is prescribed by local ordinance.

Figure 31.3 shows inflow and outflow hydrographs. The difference between the hydrographs, which is the required storage, is shown as the shaded region. At point C, the detention pond inflow rate is equal to the outflow rate when it will start to empty. In Fig. 31.3 the outflow rate is assumed to remain constant. The outflow rate will depend upon the depth of water in the pond and the type of outlet structure (i.e., weir, orifice, or pipe). It should also be noted that, for a detention pond, the inflow and outflow volumes are equal.

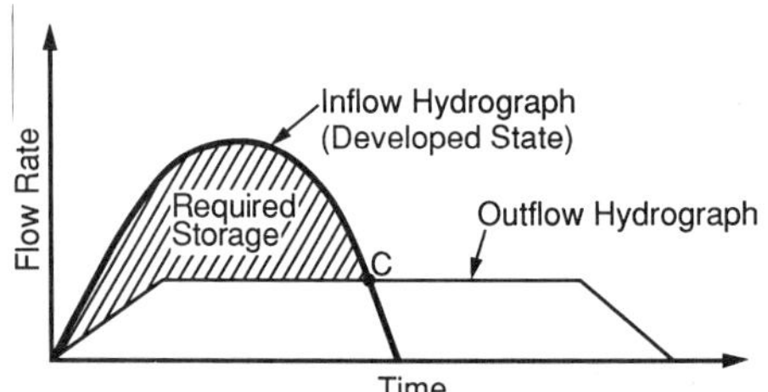

FIGURE 31.3 Inflow and outflow hydrographs for a hypothetical detention pond.

The following is an outline of the procedure used to determine the required storage volume by the SCS hydrograph method. (1) Calculate the curve numbers for the basin in developed and undeveloped condition. (2) Find the time of concentration, t_c, for the basin in undeveloped and developed condition. (3) From t_c, calculate the duration of the unit hydrograph ΔD, T_p, and q_p for the developed and undeveloped basins. (4) Determine the coordinates of the inflow and outflow unit hydrographs. (5) From the design storm duration, depth, time distribution, and frequency, calculate the cumulative rainfall at ΔD intervals for both the undeveloped and developed states. (6) Using the basin curve number and ultimate abstraction S, calculate the cumulative runoff $P_e(t_d)$ at each ΔD interval by using the rainfall data in Step 5. (7) Calculate the storm hydrographs by using the effective rainfall and direct runoff data. (8) Using the peak flow from the undeveloped state as the peak outflow, calculate the outflow hydrograph as determined by the type of outflow structure. (9) Calculate the required storage by using the developed hydrograph and outflow data or by using routing methods.

31.5 Quality of Urban Runoff

J. W. Delleur

The principal sources of pollutants accumulated in urban areas are wet and dry atmospheric deposition, litter accumulation, and traffic. Urban rainfall is generally acid with pH values below 5. This acidity damages pavements, sewers, and buildings. Rain contains nitrates, ammonia, sulfates and sulfites, phosphorus, lead, mercury, and so on. Dry atmospheric deposition includes fine particles carried by wind from some distance and particles from local sources such as traffic, construction, and industrial activities. Dusts originating from urban litter and from soil are present in urban air. Traffic generates dust as road surfaces are abraded by vehicles; traffic resuspends dust and shifts it to the curb or the median, contributing pollutants such as soil, grease, lead, rubber, and neoprene. Most of the street refuse occurs within three feet of the curb. For this reason it is often expressed per unit of curb length. The fraction of street refuse passing a 1/8-inch sieve is referred to as "dust and dirt." Many of the particles are picked up and transported by the runoff on the impervious surfaces. Additional information on pollutant deposition and washoff can be found in Novotny and Chesters [1981].

The largest and most comprehensive investigation of urban storm runoff quality was the *Nationwide Urban Runoff Program* (NURP) undertaken by the U.S. Environmental Protection Agency (EPA) [1983] and the U.S. Geological Survey (USGS). It included 2300 storm events at 81 sites in 22 different cities geographically distributed across the U.S. The principal conclusions quoted from the executive summary are:

1. Heavy metals (especially copper, lead, and zinc) are by far the most prevalent priority pollutant constituents found in urban runoff.
2. The organic priority pollutants are detected less frequently and at a lower concentration than heavy metals.
3. Coliform bacteria are present at high levels in urban runoff and can be expected to exceed EPA water quality criteria during and immediately after storm events in many surface waters, even those providing a high degree of dilution.
4. Nutrients are generally present in urban runoff.
5. Oxygen-demanding substances are present in urban runoff at concentrations approximating those in secondary treatment plant discharges.
6. Total suspended solids concentrations in urban runoff are fairly high in comparison to treatment plant discharges.

For the Nationwide Urban Runoff Program, EPA adopted the following constituents as standard pollutants characterizing urban runoff:

TSS	Total suspended solids
BOD	Biochemical oxygen demand
COD	Chemical oxygen demand
TP	Total phosphorus (as P)
SP	Soluble phosphorus (as P)
TKN	Total Kjeldahl nitrogen (as N)
NO_{2+3}-N	Nitrite and nitrate (as N)
Cu	Total copper
Pb	Total lead
Zn	Total zinc

These standard pollutants can be considered as representatives of others. These pollutants are examined both in point and nonpoint source studies and include representatives of solids, oxygen-consuming constituents, nutrients, and heavy metals. Most of the data in the EPA report consist of flow-weighted average concentrations—that is, the **event mean concentration**— for each pollutant for each runoff event.

The median and coefficient of variation of the event mean concentrations (EMC) of several constituents found in urban runoff are given in Table 31.6. These data can be used for planning purposes. Driscoll [1986] showed that for most sites and most pollutants, the EMCs are lognormally distributed and are weakly correlated with runoff volumes. The corresponding EMC mean concentrations are given in Table 31.7.

The mean annual load can be estimated for urban runoff constituents by choosing the appropriate rainfall and runoff coefficient values and selecting the EMC value from Table 31.7. Estimates of annual pollutant loads in kg/Ha/year for different types of urban developments are listed in

Table 31.6 Water Quality Characteristics of Urban Runoff

Constituent	Event-to-Event Variability in EMCs (Coef Var)	Site Median EMC For Median Urban Site	For 90th Percentile Urban Site
TSS (mg/1)	1–2	100	300
BOD (mg/1)	0.5–1.0	9	15
COD (mg/1)	0.5–1.0	65	140
Tot. P (mg/1)	0.5–1.0	0.33	0.70
Sol. P (mg/1)	0.5–1.0	0.12	0.21
TKN (mg/1)	0.5–1.0	1.50	3.30
NO_{2+3}-N (mg/1)	0.5–1.0	0.68	1.75
Tot. Cu (μg/1)	0.5–1.0	34	93
Tot. Pb (μg/1)	0.5–1.0	144	350
Tot. Zn (μg/1)	0.5–1.0	160	500

Source: U.S. EPA, 1983, Vol. 1, Table 6.17, pp. 6–43.

Table 31.7 EMC Mean Values Used in Load Comparison

Constituent	Median Urban Site	Site Median EMC 90th Percentile Urban Site	Values Used in Load Comparison
TSS (mg/1)	141–224	424–671	180–548
BOD (mg/1)	10–13	17–21	12–19
COD (mg/1)	73–92	157–198	82–178
Tot. P (mg/1)	0.37–0.47	0.78–0.99	0.42–0.88
Sol. P (mg/1)	0.13–0.17	0.23–0.30	0.15–0.28
TKN (mg/1)	1.68–2.12	3.69–4.67	1.90–4.18
NO_{2+3}-N (mg/1)	0.76–0.96	1.96–2.47	0.86–2.21
Tot. Cu (μg/1)	38–48	104–132	43–118
Tot. Pb (μg/1)	161–204	391–495	182–443
Tot. Zn (μg/1)	179–226	559–707	202–633

Source: U.S. EPA 1983, Vol. 1, Table 6.24, pp. 6–60.

Table 31.8 Annual Urban Runoff Loads kg/Ha/Year

Constituent	Site Mean Con. mg/1	Residential	Commercial	All Urban
Assumed *Rv*		0.3	0.8	0.35
TSS	180	550	1460	640
BOD	12	36	98	43
COD	82	250	666	292
Total P	0.42	1.3	3.4	1.5
Sol. P	0.15	0.5	1.2	0.5
TKN	1.90	5.8	15.4	6.6
NO_{2+3}-N	0.86	2.6	7.0	3.6
Tot. Cu	0.043	0.13	0.35	0.15
Tot. Pb	0.182	0.55	1.48	0.65
Tot. Zn	0.202	0.62	1.64	0.72

Note: Assumes 40 inches/year rainfall as a long-term average.
Source: U.S. EPA, 1983, Vol. 1, Table 6–25, pp. 6–64.

Table 31.8 on the basis of a 40-inch annual rainfall and for the runoff coefficients, *Rv*, also shown in Table 31.8.

Rainfall events produce bursts of runoff from the highly impervious urban areas. These slugs of runoff are conveyed by the drainage system to the receiving streams. Thus rivers and streams that receive urban runoff are exposed to pulses of contaminated urban runoff. EPA [1983] has made a probabilistic analysis of the concentration characteristics of the instream storm pulses. The recurrence intervals of instream concentrations during storm events are shown in Fig. 31.4 for copper, assuming upstream concentration of zero and conditions corresponding to the 50th percentile of all urban sites. The concentrations are shown for end-of-pipe discharge, untreated urban runoff, and treated urban runoff with 60% removal efficiency. The concentrations are compared to stream target concentrations associated with different degrees of adverse impact (significant mortality, threshold effect, and EPA maximum) on more sensitive biological species. This type of presentation is useful to indicate how the frequency of concentrations may be interpreted to infer the presence or the degree of severity of a pollution problem. The quality of receiving streams can be simulated by models such as QUAL2E [EPA, 1987].

Stormwater quality can be enhanced by on-site detention, inlet controls, appropriate planning, real-time controls, in-system controls, and treatment. These techniques are discussed in Torno [1989]. Detention ponds with a permanent water volume during inter-storm periods, called *wet detention ponds,* are effective pollutant removal devices. Principles of sizing these ponds are given by Hvitved-Jacobsen [1989]. Other methods of enhancement of urban runoff quality include street cleaning, on-site grass area infiltration, porous pavement parking lots, and sediment retention. Rehabilitation of inadequate drainage systems can be planned to enhance stormwater quality. Delleur [1994] developed a matrix that summarizes the types of structural, hydraulic, and environmental failures of urban drainage infrastructure and lists methods of diagnosis and rehabilitation. Details on planning and modeling, programs, methods, and case studies of urban drainage rehabilitation can be found in Macaitis [1994].

Modeling the pollutant load draining from urban watersheds typically includes the estimation of the pollutant deposition between storm events and the pollutant washoff during storm events. This is the procedure used in the Storm Water Management Model (SWMM) developed for EPA [Huber and Dickinson, 1988]. SWMM is the most complete simulation model of the hydrologic, hydraulic, and environmental aspects of urban drainage systems developed in the U.S. The model includes several computation modules or blocks. The principal blocks are the RUNOFF block,

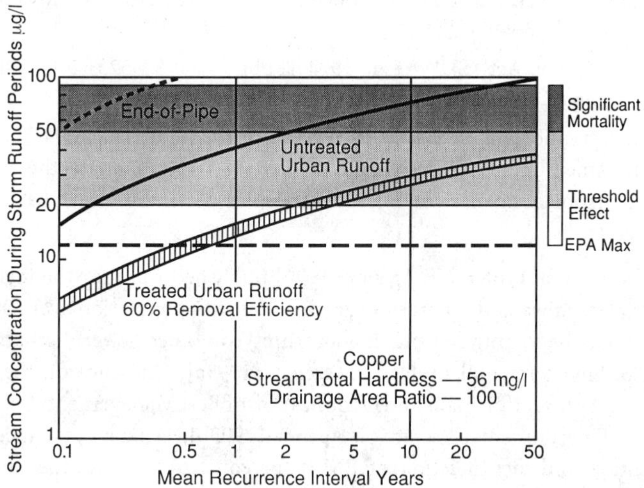

FIGURE 31.4 Recurrence intervals for pollutant concentrations. (*Source:* U.S. EPA 1983, Vol. 1, Fig. 7–5, pp. 7–11.)

which simulates the quantity and quality of the runoff, the TRANSPORT block, which routes the quantity and quality through the drainage system, and the STORAGE/TREATMENT block, which simulates the routing of flows and pollutant through a storage/treatment facility. The model is described in more detail in Chapter 36.

Three *pollutant deposition* models are included in the RUNOFF block:

1. A linear deposition model: $L = rate * time$ where L is the surface load in mass per unit area, *rate* is the deposition rate, and *time* is the number of dry days. This model correctly simulates yearly loads but is not adequate for short time periods.
2. An exponential model obtained by the combination of a linear accumulation and a removal model:

$$L = limit[1 - \exp(-k\ time)] \tag{31.11}$$

where *limit* is the maximum surface load in mass per area, *time* is the number of dry days, and k is the rate of particle removal.

3. A Michaelis-Menton model

$$L = limit * time/(a + time) \tag{31.12}$$

where a is the number of days after which half the maximum buildup has been deposited.

The *washoff* model is

$$d(Pshed)/dt = -rcoeff * Pshed * r^{washpow} \tag{31.13}$$

where $Pshed$ is the pollutant surface load, r is the runoff rate, and the coefficient $rcoeff$ and the exponent $washpow$ are quantities to be determined during the calibration procedure. Alley *et al.* [1980] proposed an alternative washoff model, based on sediment transport theory, called the particulate transport model (PTM). The RUNOFF block can also simulate runoff and pollutant loads generated by snowmelt.

In the U.S. many federal regulations impact on the disposal of urban runoff. Some of these are:

- Federal Water Pollution Control Act of 1972 (Public Law 92-500) and the Clean Water Act (CWA) Amendments of 1977 (Public Law 95-217). Section 208 is concerned with discharges from urban-industrial areas. Pollutant discharges from a point source into navigable waters are prohibited unless authorized by a National Pollutant Discharge Elimination System (NPDES) permit. EPA has developed National Strategies to control urban runoff and combined sewer overflows (CSO).
- Safe Drinking Water Act (SDWA) of 1974 (Public Law 93-523) and 1986 Amendments regulate injection of wastewater into aquifers.
- Resource Conservation and Recovery Act (RCRA) of 1976 (Public Law 94-580) and Hazardous Solid Waste Amendments of 1984 (Public Law 98-616) mandate the protection of the environment from accidental or unregulated spills of hazardous substances and leaking underground tanks.

The U.S. Environmental Protection Agency [1990] published rules for the regulation of municipal and industrial stormwater discharges. Municipalities with population over 100,000 must apply for a permit to discharge stormwater. Rules concerning combined sewer discharges were recently issued by the U.S. Environmental Protection Agency [1993a]. Guidance on the development of stormwater pollution prevention plans and identification of Best Management Practices (BMP) have been given by the Environmental Protection Agency [1993b]. Detailed guidance for stormwater pollution prevention and identification of BMPs for construction activities and for industrial activities have also been published by EPA [1992a,b]. Overviews of the EPA regulations have been given by Elder [1989] and by Roesner [1993]. Guidance for stormwater sampling has been given by Smoley [1993].

Defining Terms

Detention storage: Storage used to temporarily store stormwater and to release it gradually after the storm is over.

Event mean concentration (EMC): The average pollutant concentration during the runoff caused by a storm event.

Inlet time: Time taken by water to reach inlets.

Rational method: A method to estimate peak runoff from small watersheds.

Retention storage: A facility to store stormwater. The stored water is allowed to infiltrate and evaporate.

Time of concentration: Time taken by runoff to travel from the hydraulically most distant point on the watershed to the point of interest.

Washoff: amount of pollutant which is entrained by runoff from urban surfaces.

References

AASHTO (American Association of State Highway and Transportation Officials). 1991. *Model Drainage Manual,* Suite 225, 444 N. Capitol St. N.W., Washington D.C., 20001.

Alley, W. M., Ellis F. W., and Sutherland, R. C. 1980. Toward a more deterministic urban runoff quality model. *Proc. Int. Symp. Urban Runoff.* University of Kentucky, Lexington, KY, pp. 171–182.

American Society of Civil Engineers. 1974. *Urban Water Resources Research Program, Management of Urban Storm Runoff.* Tech-Memo 24. New York.

American Society of Civil Engineers. 1992. *Design and Construction of Urban Stormwater Management Systems.* ASCE, New York.

Bedient, P. B. and Huber, W. C. 1988. *Hydrology and Floodplain Analysis.* Addison-Wesley, Reading, MA.

Burke, C. B., Burke, T. T., Jr., and McCormick, D. L. 1994. *Stormwater Drainage Manual.* Tech. Rept. H-94-4, Highway Extension and Research Project for Indiana Counties and Cities, Purdue University, West Lafayette, IN.

Chow, V. T. 1959. *Open Channel Hydraulics.* McGraw-Hill, New York.

Chow, V. T., Maidment, D. R., and Mays, L. W. 1988. *Applied Hydrology.* McGraw-Hill, New York.

Delleur, J. W. 1994. Sewerage failure, diagnosis and rehabilitation. In Macaitis, 1994.

Driscoll, E. D. 1986. Lognormality of point and non-point source pollutant concentrations. In Urbonas and Roesner, 1986.

Elder, J. R. 1989. The regulation of CSO's and storm water in the United States. In Torno, 1989.

Environmental Protection Agency. 1983. *Results of the Nationwide Urban Runoff Program.* Final Report. NTIS Accession No. PB 84-185552.

Environmental Protection Agency. 1987. *The Enhanced Stream Water Quality Models QUAL2E and QUAL2E-UNCAS, Documentation and User Model.* Report EPA/600/3-87/007. Environmental Research Laboratory, Athens, GA.

Environmental Protection Agency. 1990. *National Pollutant Discharge Elimination System Permit Application Regulations for Stormwater Discharges, Final Rule.* Federal Register, Vol. 55, No. 222. U.S. Govt. Print. Off., Washington, D.C., Nov. 16.

Environmental Protection Agency. 1992a. *Stormwater Management Activities for Construction Activities, Developing Pollution Prevention Plans and Best Management Practices.* Report EPA-832-R-92-005. Office of Wastewater Enforcement and Compliance, Washington, D.C.

Environmental Protection Agency. 1992b. *Stormwater Management for Industrial Activities, Developing Pollution Prevention Plans and Best Management Practices.* Report EPA-832-R-92-006. Office of Wastewater Enforcement and Compliance, Washington, D.C.

Environmental Protection Agency. 1993a. *Combined Sewer Overflow Policy: Draft Guidance Availability.* Federal Register, Vol. 58, No. 11. U.S. Govt. Print. Off., Washington, D.C., Jan. 19.

Environmental Protection Agency. 1993b. *Investigation of Inappropriate Pollutant Entries into Drainage Systems.* Report EPA/600/R-92/238. Office of Research and Development, Washington, D.C.

Hall, M. J. 1984. *Urban Hydrology.* Elsevier, New York.

Huber, W. C. and Dickinson, R. E. 1988. *Storm Water Management Model, Version 4: User's Manual.* U.S. EPA, Environmental Research Laboratory, Office of Research and Development, Athens, GA.

Hydrologic Engineering Center. 1985. *HEC-1, Flood Hydrograph Package, Users Manual.* U.S. Army Corps of Engineers, Davis, CA.

Hvitved-Jacobsen, T. 1989. Design criteria for detention pond quality. In Torno, 1989.

Kuichling, E. 1889. The relation between the rainfall and the discharge of sewers in populous Districts. *Trans. ASCE.* 20: 1–56.

Landsburg, H. E. 1981. *The Urban Climate.* Academic Press, New York.

Lowry, W. P. 1967. The climate of cities. *Sci. Am.* 217(2): 15–23.

Macaitis, W., ed. 1994. *Programs and Methods of Urban Drainage Infrastructure Rehabilitation.* ASCE, New York.

Novotny, V. and Chesters, G. 1981. *Handbook of Nonpoint Pollution, Sources and Management.* Van Nostrand Reinhold, New York.

Novotny, V., Imhoff, K. R., Olthof, M., and Krenkel, P. A. 1989. *Karl Imhoff's Handbook of Urban Drainage and Waste Disposal.* Wiley & Sons, New York.

Roesner, L. A. 1993. Overview of federal law and USEPA regulations for urban runoff. *Proc. 6th Intl. Conf. on Urban Storm Drainage, Vol. 2, 1958–1963.* Niagara Falls, Ontario.

Sheaffer, J. R., Wright, K. R., Taggart, W. E., and Wright, R. M. 1982. *Urban Storm Drainage Management.* Dekker, New York.

Smoley, C. K. 1993. *NPDES Storm Water Sampling, Guidance Manual, U.S. EPA, Office of Water.* CRC Press, Boca Raton, FL.

Soil Conservation Service. 1972. *National Engineering Handbook.* Washington, D.C.

Soil Conservation Service. 1982. *TR-20 Computer Program for Project Formulation Hydrology.* Tech. Release No. 20. Washington, D.C., May.

Soil Conservation Service. 1986. *Urban Hydrology for Small Watersheds.* Tech. Release No. 55. Washington, D.C.

Torno, H. C., ed. 1989. *Urban Stormwater Quality Enhancement — Source Control, Retrofitting and Combined Sewer Technology. Proc. Eng. Foundation Conference.* ASCE, New York.

Urbonas, B. and Roesner, L. A., eds. 1986. *Concentrations in Urban Runoff Quality Impact and Quality Enhancement Technology.* ASCE, New York.

Urbonas, B. and Stahre, P. 1993. *Stormwater: Best Management Practices and Detention for Water Quality, Drainage and CSO Management.* Prentice Hall, Englewood Cliffs, NJ.

Walesh, S. G. 1989. *Urban Surface Water Management.* Wiley & Sons, New York.

Wanielista, M. P. 1993. *Stormwater Management.* Wiley & Sons, New York.

For Further Information

Kibler, D. F., ed. 1982. *Urban Stormwater Hydrology.* American Geophysical Union, Washington, D.C.

Wanielista [1993] covers the quantity and quality aspects of urban drainage.

Sheaffer *et al.* [1982] consider the desirable goals and legal aspects of drainage and flood control as well as the strategic drainage planning process.

Hall [1984] provides a general text on urban hydrology including water quality and stormwater management.

Urbonas and Stahre [1993] discuss urban runoff best management practices, detention for water quality, and combined sewer overflow management.

Novotny *et al.* [1989] discuss urban drainage, combined sewer overflow management, sewage and wastewater treatment, and protection of receiving waters.

Walesh [1989], in addition to hydrology and hydraulics of storm sewers, also considers pollution loads, detention/retention facilities, and computer modeling.

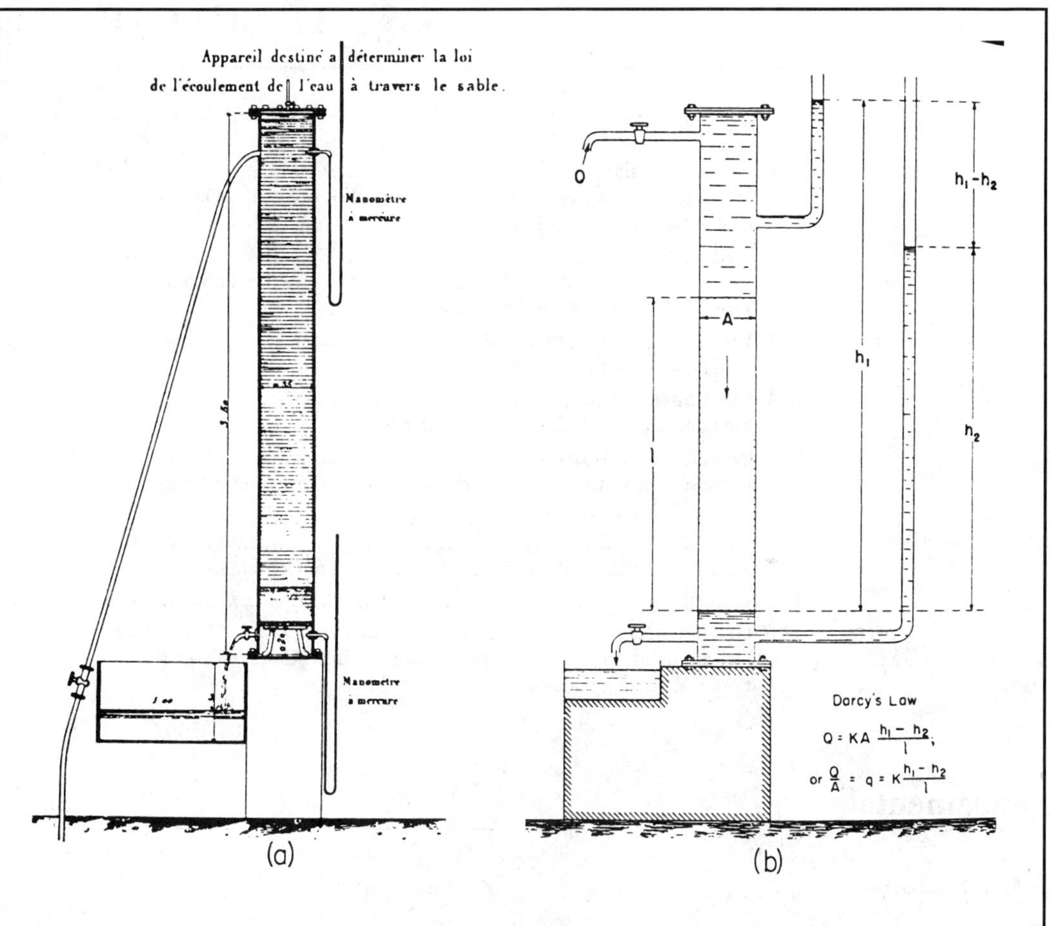

The two apparatuses illustrate Henry Darcy's experiment on the flow of water through sand. Darcy's original apparatus on the left, used in 1885 and 1886, is with mercury manometers. On the right is the equivalent apparatus with water manometers and Darcy's own statement of the law expressed in the bottom right corner. (From Hubbert, M. King, *History of Geophysics*, Vol. 3, 1968, 1987. With permission.)

32

Groundwater Engineering

Jacques W. Delleur
Purdue University

32.1 Fundamentals

Introduction

This section on groundwater engineering is concerned with the occurrence, movement, use, and quality of water below ground. The subsection on fundamentals deals with the definitions, the properties of the unsaturated and saturated zones, and the physics of the movement of subsurface water. Specific engineering applications such as well hydraulics, well construction, contaminant transport, containment of contaminants, landfills, and geostatistics are discussed in the following subsections.

Subsurface Water

The **water table** is the level at which the groundwater is at atmospheric pressure. The zone between the ground surface and the water table is called the **vadose zone**. It contains some water that is held between the soil particles by capillary forces. Immediately above the water table is the

0-8493-8953-4/95/$0.00 + $.50

capillary fringe where the water fills the pores. The zone above the capillary fringe is often called the *unsaturated zone.* Below the capillary fringe is the *saturated zone.* (The **saturation ratio** is the fraction of the volume of voids occupied by water.) The water above the water table is below atmospheric pressure, and the water below the water table is above atmospheric pressure. Only the water below the water table, the *groundwater*, is available to supply wells and springs. Recharge of the groundwater occurs primarily by percolation through the unsaturated zone. The geologic formations that yield water in usable quantities, to a well or a spring, are called **aquifers**. If the upper surface of the saturated zone in the aquifer is free to rise or to decline the aquifer is said to be an **unconfined aquifer**. The upper boundary at atmospheric pressure is the water table, also called the **phreatic surface.** If the water completely fills the formation, the aquifer is **confined** and the saturated zone is the thickness of the aquifer. If the confining material is impermeable it is called an **aquiclude.** If the confining layer is somewhat permeable in the vertical direction, thus permitting recharge, it is called an **aquitard.** When a layer restricts downward infiltration toward the main water table a *perched* aquifer with a separate **perched water** table may be formed. A perched aquifer is, in general, of limited areal extent, and if used as a water supply, extreme caution should be exerted because of its ephemeral nature. If the water in a well in a confined aquifer rises above the top of the aquifer, the well is an **artesian well,** and the aquifer is in artesian condition.

Physical Properties

The *porosity*, n, is the ratio of the volume of voids, V_v, to the total volume, V_t, of the rock or soil:

$$n = V_v/V_t = [V_t - V_s]/V_t$$

where V_s is the volume of solids. The *void ratio*, e, used in soil mechanics is defined as $e = V_v/V_s$ so that $1/n = 1 + 1/e$. The fraction of void space between grains of soil or of unconsolidated rock is referred to as *primary porosity*. Porosity due to fracturing of the rock or chemical dissolution is called *secondary porosity*. Typical values of the porosity are given in Table 32.1. The *effective porosity*, n_e, is the pore fraction that actually contributes to the flow; isolated and dead-end pores are excluded. In unconsolidated sediments coarser than 50 μm, n_e is of the order of $0.95n$ to $0.98n$. When all the voids are occupied by water the soil is saturated. Otherwise the fraction of the voids occupied by water is the volumetric *water content* designated by θ, which is dimensionless. When the soil is saturated, the soil moisture is $\theta_s = n$. After the soil has been drained the remaining soil moisture is the *residual moisture* θ_r. In unsaturated soils the *effective porosity* is $\theta_e = n - \theta_r$ and the *effective saturation* is defined as $s_e = (\theta - \theta_r)/(n - \theta_r)$.

Table 32.1 Values of Porosity, Permeability, and Hydraulic Conductivity

Material	Porosity, n (%)	Permeability, k (cm^2)	Hydraulic Conductivity, K (cm/s)
Unconsolidated deposit			
Gravel	25–40	10^{-3}–10^{-6}	10^{2}–10^{-1}
Sand	25–50	10^{-5}–10^{-9}	1–10^{-4}
Silt	35–50	10^{-8}–10^{-12}	10^{-3}–10^{-7}
Clay	40–70	10^{-12}–10^{-15}	10^{-7}–10^{-10}
Rocks			
Fractured basalt	5–50	10^{7}–10^{-11}	10^{-2}–10^{-6}
Karst limestone	5–50	10^{-5}–10^{-9}	1–10^{-4}
Sandstone	5–30	10^{-9}–10^{-13}	10^{-4}–10^{-8}
Limestone, dolomite	0–20	10^{-9}–10^{-12}	10^{-4}–10^{-7}
Shale	0–10	10^{-12}–10^{-16}	10^{-7}–10^{-11}
Fractured crystalline rock	0–10	10^{-7}–10^{-11}	10^{-2}–10^{-6}
Dense crystalline rock	0–5	10^{-13}–10^{-17}	10^{-8}–10^{-12}

The *hydraulic conductivity*, K, is a measure of the ability of a fluid to flow through a porous medium. It is the volume rate of flow, Q, per unit gross area, A, of soil or rock under a hydraulic gradient $\partial h/\partial s$:

$$K = -(Q/A)(\partial h/\partial s)^{-1}$$

For *saturated flows* the hydraulic conductivity, K, depends on the porous medium through the *permeability*, k, and on the fluid properties through the density, ρ, and the viscosity, μ. These properties are related by the following equation:

$$K = k\rho g/\mu$$

so that a usual expression for k is

$$k = -(Q/A)(\mu/\rho g)(\partial h/\partial s)^{-1}$$

For spheres $k = Cd^2$, where C is a constant, k has the dimension of L^2 and K has the units of LT^{-1}. Ranges of values of the permeability and hydraulic conductivity are given in Table 32.1. Several formulas exist in the literature that estimate the hydraulic conductivity of granular noncohesive materials. Most are of the form

$$K = (g/\nu)\, C\, \phi(n)\, d_e^2$$

where g is the acceleration of gravity, ν is the kinematic viscosity, C is a coefficient, $\phi(n)$ is a function of the porosity, and d_e is the effective grain diameter, with the variables in a consistent set of units. Vukovic and Soro [1992] list 10 formulas of this type. Two of the simplest formulas are the *Hazen formula* with $C = 6 \cdot 10^{-4}$, $\phi(n) = [1 + 10(n - 0.26)]$, $d_e = d_{10}$ which is applicable for 0.1 mm $< d_e < 3$ mm and $d_{60}/d_{10} < 5$ and the *USBR formula* with $C = 4.8 \cdot 10^{-4} d_{20}{}^{0.3}$, $\phi(n) = 1$, $d_e = d_{20}$ and is applicable to medium sand grains with $d_{60}/d_{10} < 5$, where d_{10} is the particle size such that 10% are finer.

When the flow occurs horizontally through a series of n equally thick layers in parallel, of hydraulic conductivities $K_1, K_2, \ldots, K_n$, the equivalent hydraulic conductivity of the system of layers is the arithmetic average of the conductivities. When the flow occurs vertically through a stack of n equal layers in series, the equivalent hydraulic conductivity of the system of layers is the harmonic mean of the conductivities.

For an *unsaturated condition* the hydraulic conductivity is also a function of the moisture content of the soil and is designated by $K(\theta)$. When the soil is saturated, the *saturated hydraulic conductivity* is designated by K_s. The ratio of the hydraulic conductivity for a given moisture content to the saturated conductivity is the *relative conductivity*, K_r. Brooks and Corey [1964] gave the following formula for the hydraulic conductivity of unsaturated porous materials:

$$K(\theta) = K_s[(\theta - \theta_r)/(n - \theta_r)]^\lambda$$

where λ is an experimentally obtained coefficient. Other formulas have been given by Campbell [1974] and van Genuchten [1980].

The *transmissivity*, T, is the product of the hydraulic conductivity and the thickness, b, of the aquifer: $T = Kb$. It has the units of L^2T^{-1}. The *storage coefficient*, or *storativity*, S, is the volume of water yielded per unit area per unit drop of the piezometric surface. For unconfined aquifers the drop of the water table corresponds to a drainage of the pore space and the storage coefficient is also called the **specific yield**. In an unconfined aquifer the amount of water that can be stored per unit rise of the water table per unit area is called the *fillable porosity*, f, where $f = \theta_s - \theta$. In a confined aquifer, when the water pressure decreases the fluid expands and the fraction of the weight to be carried by the solid matrix increases, resulting in a decrease of the pore space. Since the compressibility of the water is very small its decompression contributes only a small fraction to the storage coefficient. Leakage of an overlying unconfined aquifer through aquitard can also contribute to the yield of a semiconfined aquifer. Values of S typically vary between $5 \cdot 10^{-2}$ and 10^{-5} for confined aquifers.

Darcy's Law

The *volumetric flow rate* $Q\,[\mathrm{L^3T^{-1}}]$ across a gross area A of a formation with a hydraulic conductivity $K\,[\mathrm{LT^{-1}}]$ under a hydraulic gradient $\partial h/\partial s$ in the s direction is given by Darcy's law:

$$Q = -KA\,\partial h/\partial s = qA$$

where q is the *specific discharge* or flow rate per unit area $[\mathrm{LT^{-1}}]$, also called *Darcy velocity*. The *hydraulic head* h is the sum of the elevation head z and the pressure head p/γ. The minus sign in the above equation indicates that the flow takes place from high to low head, namely in the direction of the decreasing head. The *pore velocity*, $v = q/n_e$, is the average flow velocity in the pores or the average velocity of transport of solutes that are nonreactive. The one-dimensional form of Darcy's law is

$$q = K[(p_1/\gamma + z_1) - (p_2/\gamma + z_2)]/L$$

where the subscripts 1 and 2 refer to the two points at which the pressure head and the elevation head are considered and L is the distance between these points. Darcy's law implies that the flow is laminar as is generally the case. However, in some cases, as in karstic limestone and in rocks with large fractures, the flow may be turbulent. In such cases the flow rate is not proportional to the hydraulic gradient but to a power of the hydraulic gradient. Darcy's law as given above applies to *isotropic* media, that is, where the hydraulic conductivity is independent of direction. It also applies to flows where the direction of the hydraulic conductivity corresponds to the direction of the hydraulic gradient. In *nonisotropic* media the hydraulic conductivity depends upon the direction. Then a hydraulic conductivity *tensor* is used and Darcy's law is expressed as a *tensor equation* [de Marsily, 1986].

Dupuit Assumption

The one-dimensional form of Darcy's law applies to simple flow problems in the vertical or horizontal direction. In some cases with both horizontal and vertical components, the horizontal component dominates and the vertical component can be neglected. The flow can then be approximated as a horizontal flow uniform across the depth. This is the Dupuit assumption.

32.2 Hydraulics of Wells

Steady Flow to a Well

The steady flow to a well of radius r_w fully penetrating a *confined* aquifer (Fig. 32.1) with a transmissivity T is given by the *Thiem equation*:

$$Q = 2\pi T[h - h_w]/\ln(r/r_w)$$

where h is the hydraulic head at a distance r and h_w is the hydraulic head at the well. For a well fully penetrating an unconfined aquifer (Fig. 32.2) the equation for the flow rate obtained using the Dupuit assumption is

$$Q = \pi K[h^2 - h_w^2]/\ln(r/r_w)$$

When solved for the head at the well, this equation does not yield accurate results because of the neglect of the vertical flow component.

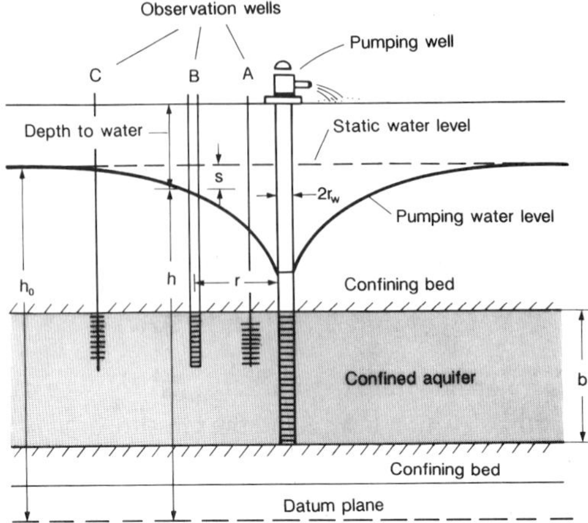

FIGURE 32.1 Well in a confined aquifer. (*Source:* Heath, R. C. 1984. *Basic Gound-Water Hydrology*. U.S. Geological Survey Water Supply Paper 2220, U.S. Government Printing Office.)

Transient Flow to a Well

Pumping a well causes a *cone of depression*, or *drawdown*, of the water table of an unconfined aquifer or of the piezometric surface for a confined aquifer. The drawdown $s(r, t)$ at a distance r from a fully penetrating well at time t after the beginning of pumping at a constant rate Q from a confined aquifer with transmissivity T and storage constant S is given by the *Theis equation*:

$$s(r,t) = [Q/(4\pi T)] \int_u^{\infty} (e^{-z}/z)\, dz = [Q/(4\pi T)] W(u)$$

where $u = r^2 S/(4Tt)$. The integral in this equation is the exponential integral, also known as the *well function* $W(u)$. This function can be expanded as

$$W(u) = -0.577216 - \ln u + u - u^2/2\ 2! + u^3/3\ 3! - u^4/4\ 4! + \cdots$$

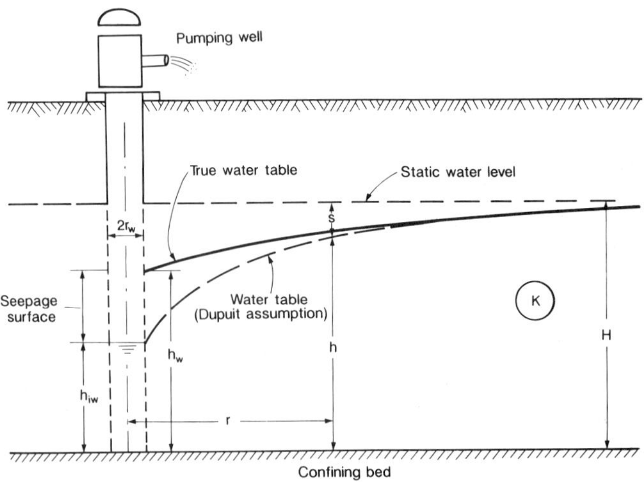

FIGURE 32.2 Well in an unconfined aquifer. (*Source:* Bouwer, H. 1978. *Ground Water Hydrology*. McGraw-Hill, New York.)

For $u < 0.01$ only the first two terms need to be considered and the drawdown is approximated by *Jacob's equation*:

$$s(r,t) = [2.30\,Q/(4\pi T)]\log_{10}[(2.25\,Tt)/(r^2S)]$$

The drawdown from an unconfined aquifer with horizontal and vertical hydraulic conductivities K_r and K_z, respectively, has been given by Neuman [1975]. A simplified form, given by Freeze and Cherry [1979], is

$$s(r,t) = [Q/(4\pi T)]W(u_a, u_b, \eta)$$

where

$$u_a = r^2S/(4Tt) \quad \text{and} \quad u_b = r^2S_y/(4Tt) \qquad \eta = r^2K_z/(b^2K_r)$$

b is the initial saturated thickness, S_y is the specific yield, and S is the elastic storage coefficient. This solution is valid for $S_y \gg S$. Freeze and Cherry [1979] give a plot of the well function $W(u_a, u_b, \eta)$.

For a pumped *leaky confined aquifer* with constants T and S, separated from an unpumped upper aquifer by an aquitard of thickness b' and constants K' and S', Hantush and Jacob [1955] obtained a relationship for the drawdown which can be written as

$$s(r,t) = [Q/(4T)]W(u, r/B)$$

where

$$u = (r^2S)/(4Tt) \quad \text{and} \quad r/B = r[K'/(Tb')]^{1/2}$$

Values of $W(u, r/B)$ can be found in Bouwer [1978], Freeze and Cherry [1979], and Fetter [1988].

Pumping Tests

The hydraulic properties of aquifers can be determined by pumping a well at constant discharge and observing the drawdown at an observation well for a period of time. For confined aquifers the Thiem *steady state* equation yields only the transmissivity

$$T = [Q\ln(r_2/r_1)]/[2\pi(s_1 - s_2)]$$

from the observed drawdowns s_1 and s_2 at distances r_1 and r_2 from the pumped well.

For confined aquifers the transient state Jacobs equation yields both the transmissivity and the storage constant based on a semilog straight-line plot (Fig. 32.3) of the observed drawdown (arithmetic scale) versus the time since pumping began (logarithmic scale) as

$$T = 2.3Q/(4\pi\Delta s) \qquad S = 2.25Tt_0/r^2$$

where Δs is the increase in drawdown per log cycle of t and t_0 is the time intercept of the straight-line fitted through the drawdowns at the several times. Only the observations corresponding to very small times violate Jacobs assumption that $u = r^2S/(4Tt) < 0.01$ and do not fall on the straight line. This approach is known as *time-drawdown* analysis. If simultaneous drawdown observations are taken at different distances, then the *distance drawdown* analysis can be used. In this latter approach a semilogarithmic plot of the drawdown (arithmetic scale) versus the distance r from the well (logarithmic scale) is used (Fig. 32.4) and the aquifer constants are given by

$$T = 2.3Q/(2\pi\Delta s) \qquad S = 2.25Tt/r_0^2.$$

where Δs is the drawdown across one log cycle of r, t is the time at which the drawdowns were measured, and r_0 is the distance intercept of the straight line fitted through the drawdowns at several distances.

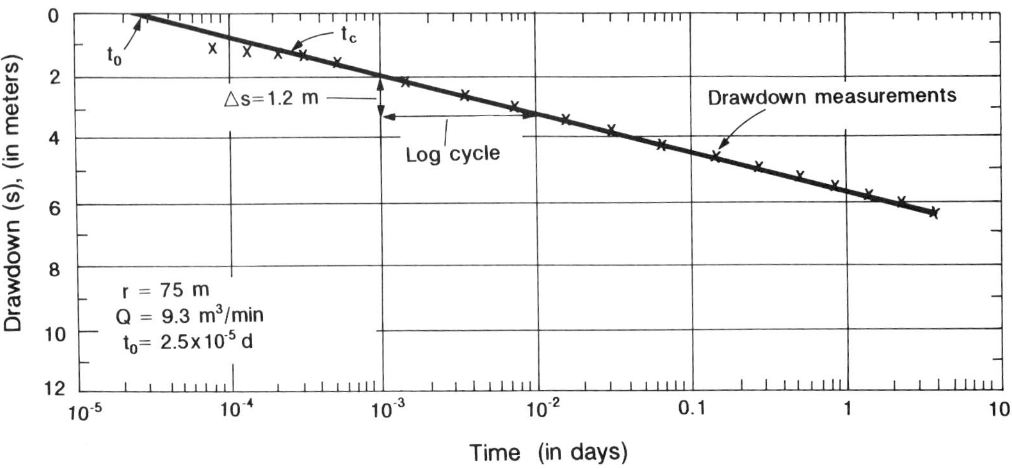

FIGURE 32.3 Time–drawdown analysis. (*Source:* Heath, R. C. 1984. *Basic Ground-Water Hydrology*. U.S. Geological Survey Water Supply Paper 2220, U.S. Government Printing Office.)

Application of the Theis equation and solutions for unconfined aquifers require more elaborate graphic solutions [Bouwer, 1978; Fetter, 1988; Freeze and Cherry, 1979] or computer solutions [Boonstra, 1989].

A well test can also be performed using only the drawdown measurements at the pumped well without observation wells. This type of test is called the *single well test.* The previous equations assume laminar flow and a linear relationship between drawdown through the geologic formation and discharge. As the flow reaches the gravel pack around the well screen the velocity increases and the flow becomes turbulent except for very small pumping rates. The total drawdown at the well s_t is thus the sum of the aquifer drawdown s_a and the *well loss* s_w [Walton, 1962]:

$$s_t = s_a + s_w = BQ + CQ^2$$

where the constant C is related to well characteristics. If the well is pumped at different rates for the same length of time, a plot can be prepared of the total drawdown versus discharge. A tangent

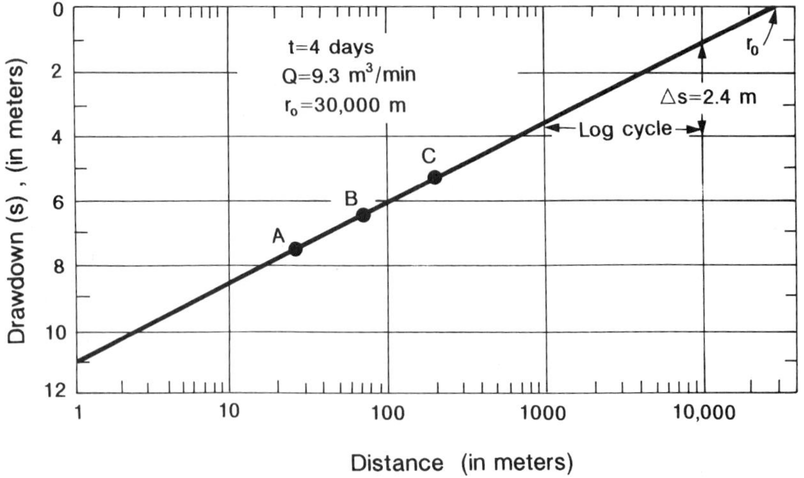

FIGURE 32.4 Distance–drawdown analysis. (*Source*: Heath, R. C. 1984. *Basic Ground-Water Hydrology*. U.S. Geological Survey Water Supply Paper 2220, U.S. Government Printing Office.)

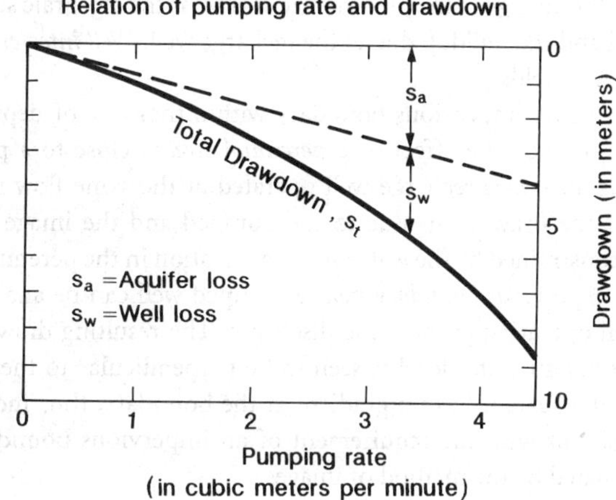

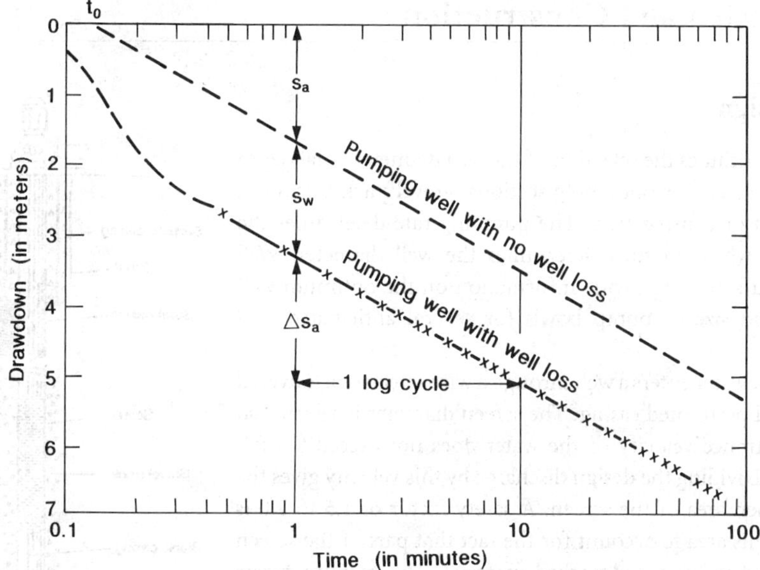

FIGURE 32.5 Single-well test. (*Source*: Heath, R. C. 1984. *Basic Ground-Water Hydrology*. U.S. Geological Survey Water Supply Paper 2220, U.S. Government Printing Office.).

at the origin will separate the aquifer drawdown and the well loss. The time–drawdown plot is then performed for a constant discharge and the transmissivity is determined as before. A line is plotted parallel to the straight-line portion of the time–drawdown observations at a distance s_w above the observations. This line determines the time intercept t_0 used in the relationship for the storage constant S (Fig. 32.5).

Multiple Wells and Boundaries

For multiple wells the total drawdown s_t is the sum of the drawdowns due to the individual wells. For the case of a confined aquifer the Theis formula yields

$$s_t = (4\pi T)^{-1}[Q_1 W(u_1) + Q_2 W(u_2) + \cdots]$$

where $u_i = r_i^2 S/(4Tt_i)$ in which r_i is the distance from the ith pumping well to the observation point and t_i is the time since pumping began at well i with a discharge Q_i. In the case of a pumping well and a *recharge well*, the change in the piezometric surface is the algebraic sum of the drawdown due to the pumped well and the buildup due to the recharge well. *Well interference* is an important matter in the design of well fields.

A recharge boundary or an impervious boundary within the cone of depression modifies the shape of the drawdown curve. The effect of a *perennial stream* close to a pumped well can be analyzed by considering an *image* recharge well operated at the same flow rate. The drawdown is the algebraic sum of the drawdowns due to the pumped and the image recharge wells. The resulting water level is constrained by the water surface elevation in the perennial stream. Similarly, the effect of a vertical *impervious boundary* near a pumped well can be analyzed by considering an image pumping well operating at the same discharge. The resulting drawdown is the sum of the drawdowns. The resultant water level is seen to be perpendicular to the vertical impervious boundary. This horizontal surface has no gradient at the boundary, thus indicating that there is no flow, which is consistent with the requirement of an impervious boundary. More elaborate boundaries can be simulated by the method of images.

32.3 Well Design and Construction

Well Design

Well design includes the selection of the well diameter, total depth of the well, screen or open hole sections, gravel pack thickness, and method of construction. The pumping rate determines the pump size, which in turn determines the well diameter. Well pump manufacturers provide information on the optimum well diameter and size of pump bowls for several anticipated well yields.

Generally, water enters a well through a wire screen or a louvered or shuttered perforated casing. The screen diameter is selected so that the entrance velocity of the water does not exceed 0.1 ft/s (0.03 m/s). Dividing the design discharge by this velocity gives the required open area of the screen. A safety factor of 1.5 to 2.0 is applied to this area to account for the fact that part of the screen may be blocked by gravel-packed material. The manufacturers supply the open areas of screen per lineal foot for different slot sizes and screen diameters. The required length of the screen is obtained by dividing the required area by the open area per lineal foot. Screens are normally installed in the middle 70 to 80% of confined aquifers and the lower 30 to 40% of unconfined aquifers.

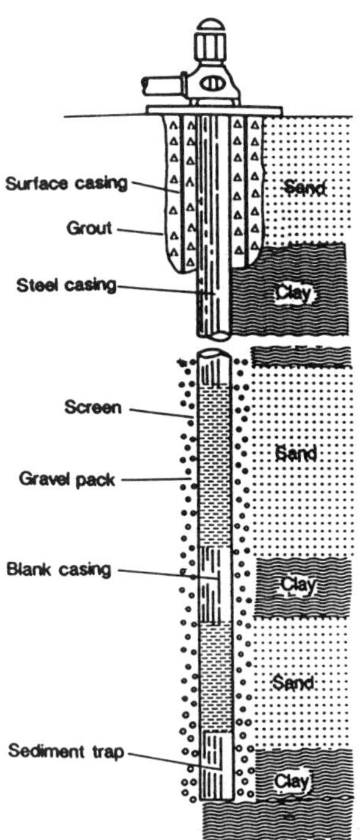

FIGURE 32.6 Supply well with multiple screen and gravel pack. (*Source*: Heath, R. C. 1984. *Basic Ground-Water Hydrology*. U.S. Geological Survey Water Supply Paper 2220, U.S. Government Printing Office.)

A gravel envelope, or *gravel pack* (Fig. 32.6), is used around the screen to prevent fine material from entering the well. Gravel packs make it possible to use larger screen slots, thus reducing the well loss. They also increase the effective radius of the well. Gravel packs are used in fine-textured aquifers in which D_{90} (the sieve size retaining 90% of the material) is less than 0.25 mm; they have a coefficient of uniformity $D_{40}/D_{90} < 3$. Wells dug through multiple layers of sand and clay are generally constructed with a gravel pack (Fig. 32.6). A sand bridge is usually provided at the top of the gravel pack to separate it from the impermeable grout that extends to the surface. The well casing should extend somewhat higher than the ground level and a concrete slab sloping away from the well is provided to prevent surface runoff from entering into the well.

Table 32.2 Suitability of Different Well-Construction Methods to Geologic Conditions

					Drilled		
						Rotary	
Characteristics	Dug	Bored	Driven	Jetted	Percussion (Cable Tool)	Hydraulic	Air
Maximum practical depth, in m (ft)	15 (50)	30 (100)	15 (50)	30 (100)	300 (1000)	300 (1000)	250 (800)
Range in diameter, in cm (in.)	1–6 m (3–20 ft)	5–75 (2–30)	3–6 (1–2)	5–30 (2–12)	10–46 (4–18)	10–61 (4–24)	10–25 (4–10)
Unconsolidated material:							
Silt	X	X	X	X	X	X	
Sand	X	X	X	X	X	X	
Gravel	X	X			X	X	
Glacial till	X	X			X	X	
Shell and limestone	X	X		X	X	X	
Consolidated material:							
Cemented gravel	X				X	X	X
Sandstone					X	X	X
Limestone					X	X	X
Shale					X	X	X
Igneous and metamorphic rocks					X	X	X

Source: Heath, R. C. 1984. *Basic Ground-Water Hydrology.* U.S. Geological Survey Water Supply Paper 2220, U.S. Government Printing Office.

Construction Methods

The principal methods of well construction include digging, boring, driving, jetting, percussion drilling, hydraulic rotary drilling, and air rotary drilling. Table 32.2 indicates the suitability of the several well construction methods according to geologic conditions.

After the construction is completed the well is developed, stimulated, and sterilized. The removal of fine sand and construction mud is called *well development.* This can be accomplished by water or air surging. *Stimulation* is a technique to increase the well production by loosening consolidated material around the well. For example, high-pressure liquid can be injected into the well to increase the size of the fractures in the rock surrounding the well. Finally, water supply wells should be *sterilized.* This can be done with chlorine or some other disinfectant.

Centrifugal pumps are normally used for water supply wells. If the water level in the well is *below* the centerline of the pump it is necessary to check that the available positive suction head exceeds the required positive suction head specified by the manufacturer in order to avoid cavitation. (See the section on pumps in "Hydraulic Structures," Chapter 35.) For high heads, *multiple stage pumps* are used. *Submersible pumps* avoid the need of long shafts and are used for very deep wells. Drillers keep **well logs** or records of the geologic formation encountered.

For further details about well-drilling methods, well design, well screens, well pumps, and their maintenance, the reader is referred to the work of Driscoll [1986].

32.4 Land Subsidence

Introduction

Groundwater pumping causes a downward movement of the water table or of the piezometric surface, which in turn can cause a downward movement of the land surface called *subsidence,* or *consolidation.* This movement can be a few centimeters to several meters. If the subsidence is not uniform, the differential settlement can produce severe damage to structures.

Calculation of Subsidence

Consider a unit area of a horizontal plane at a depth Z below the ground surface. The total downward pressure due to the weight of the overburden on the plane is resisted partly by the upward hydrostatic pressure P_h and partly by the *intergranular pressure* P_i exerted between the grains of the material:

$$P_t = P_h + P_i \quad \text{or} \quad P_i = P_t - P_h$$

A lowering of the water table results in a decrease of the hydrostatic pressure and a corresponding increase in the intergranular pressure. If P_{i1} and P_{i2} denote the intergranular pressures before and after a drop in the water table or piezometric surface, the subsidence S_u can be calculated as

$$S_u = Z[P_{i1} - P_{i2}]/E$$

where E is the modulus of elasticity of the soil. If there are layers of different soil types, the subsidences are calculated for each layer and added to yield the total subsidence. As the modulus of elasticity of clayey materials is much less than that of sands and gravel, most of the settlement takes place in the clayey layers.

The previous equation can also be used to calculate the rebound when the intergranular pressure decreases. Caution must be exercised because the modulus of elasticity usually is not the same for decompression as for compression. This is particularly the case for clays. For Boston blue clay the rebound modulus of elasticity is only about 50% of that for compression [Bouwer, 1978, p. 323]. If subsidence has occurred for a long time, complete rebound is unlikely to occur.

If there is an upward vertical flow, the head loss due to friction as the water flows in the pores results in an increase in the hydrostatic pressure. This in turn results in a decrease in the intergranular pressure. A condition known as *quicksand* is reached when the intergranular pressure vanishes and the sand loses its bearing capacity. Horizontal flow of the ground water can cause a lateral displacement that can result in damage to wells.

32.5 Contaminant Transport

Introduction

This section describes the transport and fate of constituents in groundwater. *Constituent* is a general term that does not necessarily imply a polluting substance. The *fate* of a constituent depends on its transport through the groundwater and includes possible decay or reactions that may occur. Transport can occur by **advection**, **diffusion**, and dispersion. Transport of solutes can be accompanied by chemical processes such as precipitation, dissolution, **sorption**, radioactive decay, and biochemical processes such as biodegradation.

Advection

Advection is the movement of a constituent as a result of the flow of groundwater. It is the most important mechanism of solute transport. The average flow velocity in the pores v is obtained by dividing the Darcy or gross velocity by the effective porosity n_e or $v_x = -[K/n_e]/(\partial h/\partial x)$, where K is the hydraulic conductivity and $\partial h/\partial x$ is the hydraulic gradient. The one-dimensional differential equation for the advection transport in the x direction is given by

$$\partial C/\partial t = -v_x\, \partial C/\partial x$$

where C is the solute concentration (M/L^3).

Diffusion and Dispersion

Diffusion is the spreading of the solute due to molecular activity. The mass flux of solute per unit area per unit time F is given by Fick's law: $F = -D_d(\partial C/\partial x)$, where D_d is the diffusion coefficient (L^2/T). For solutes in an infinite medium this coefficient is of the order of 10^{-9} to $2 \cdot 10^{-9}$ m^2/s at 20 °C and varies slightly with temperature. The diffusion coefficient in a porous medium is reduced by a factor of 0.1 to 0.7 for clays and sands, respectively [de Marsily, 1986, p. 233] because of the *tortuosity* of the flow paths, and is designated by D^*. The variability of the pore sizes, the multiplicity of flow paths of different lengths, and the variation of the velocity distribution in the pores of different sizes result in a mechanical spreading known as dispersion. The *longitudinal* dispersion in the flow direction is larger than the *transverse* dispersion perpendicular to the flow direction. When advection and dispersion are the dominant transport mechanisms diffusion is a second-order effect.

Molecular diffusion and **mechanical dispersion** are grouped under the term **hydrodynamic dispersion**. The longitudinal and transverse dispersion coefficients D_L and D_T, respectively, are given by

$$D_L = \alpha_L v_x + D^* \quad \text{and} \quad D_T = \alpha_T v_x + D^*$$

where α_L and α_T are the longitudinal and transverse dynamic dispersivity and D^* is the molecular diffusion coefficient in the porous medium. The relative importance of the dynamic dispersivity and the molecular diffusion can be determined from the value of the *Peclet* number. It is defined as $P_e = v_x L/D_d$, where L is a characteristic length of the porous medium, generally taken as the mean diameter of the grains or the pores. The longitudinal advective dispersion dominates over the molecular diffusion when $P_e > 10$, and the transverse advective dispersion dominates when $P_e > 100$. The dispersion coefficients α_L and α_T are known to vary with the scale at which they are measured. Fetter [1993, pp. 71–73] suggested a regression equation $\alpha_L = 0.1x$ where x is the flow length (both α_L and x are in meters). Neuman [1990] has proposed a fractal scaling of the hydraulic conductivity and dispersivity.

The two-dimensional diffusion-dispersion in a uniform flow in the x direction in a homogeneous aquifer is governed by the following equation:

$$\partial C/\partial t = D_L\, \partial^2 C/\partial x^2 + D_T\, \partial^2 C/\partial y^2 - v_x\, \partial C/\partial x$$

Sorption

This discussion is limited to the cases of *adsorption* when the solute in the groundwater becomes attached to the surface of the porous medium and *cation exchange* when positively charged ions in the solute are attracted by negatively charged clay particles. The relationships relating the solute concentration C of a substance to the mass of that substance adsorbed per unit mass in the solid phase, F, are called *isotherms* because they are determined at constant temperature. The simplest is the linear isotherm $F = K_d C$, where K_d is the distribution coefficient. Nonlinear isotherms have been proposed. The effect of the adsorption is to retard the transport of the substance. The resulting one-dimensional advection-diffusion-dispersion equation for a uniform flow in the x direction in a homogeneous aquifer is

$$R\, \partial C/t = D_L\, \partial^2 C/\partial x^2 - v_x\, \partial C/\partial x$$

where $R = 1 + K_d \rho_b / n_e$ is the *retardation factor* in which n_e is the effective porosity and ρ_b is the bulk density of the porous medium.

In the case of organic compounds the partition coefficient is $K_d = K_{oc} f_{oc}$ where K_{oc} is the partition coefficient with respect to organic carbon and f_{oc} is the fraction of organic carbon in the soil. A number of regression equations have been obtained that relate K_{oc} to the octanol–water partition coefficient and to the aqueous solubility [de Marsily, 1986; Fetter, 1993].

Multiphase Flow

Liquids that are not miscible with water are called **non-aqueous-phase liquids (NAPL)**. In the unsaturated zone four phases may be present: soil, water, air, and NAPL. Many contaminant problems are associated with the movement of NAPL. The NAPL can have densities that are less than that of water and are called *light non-aqueous-phase liquids* **(LNAPL)** or they can have densities that are larger than that of water and are called *dense non-aqueous-phase liquids* **(DNAPL)**. In an unconfined aquifer an LNAPL will tend to float near the water table whereas a DNAPL will tend to sink to the bottom of the aquifer.

When two liquids compete for the pore space one will preferentially spread over the grain surface and wet it. The **wettability** depends upon the interfacial tension between the two fluids. In the case of oil-and-water systems, water is the wetting fluid in the saturated zone but in the unsaturated zone oil is the wetting fluid if the soil is very dry. The *relative permeability* is the ratio of the permeability of a fluid at a given saturation to the intrinsic permeability. The relative permeability of the wetting fluid is designated by k_{rw} and that of the nonwetting fluid by k_{rnw}. For two-phase flow, Darcy's laws for the wetting and nonwetting liquids are, respectively,

$$
\begin{aligned}
Q_w &= -[k_{rw} k \rho_w / \mu_w] A \, \partial h_w / \partial s \\
Q_{nw} &= -[k_{rnw} k \rho_{nw} / \rho_{nw}] A \, \partial h_{nw} / \partial s
\end{aligned}
$$

where the subscripts w and nw refer to the wetting and nonwetting fluids, respectively. If an LNAPL (e.g., oil) is spilled on the ground, it will infiltrate, move vertically in the vadose zone, and, if sufficient quantity is available, will eventually reach the top of the capillary fringe. Some *residual* NAPL remains in the vadose zone. As the NAPL (oil) accumulates over the capillary zone an *oil table* (oil surface at atmospheric pressure) forms, and the water capillary fringe becomes thinner and eventually completely disappears. The oil table then rests on the water table [Abdul, 1988]. The mobile oil below the oil table moves downward along the slope of the water capillary fringe. Soluble constituents of the LNAPL are dissolved in the ground water and are transported by advection and diffusion close to the water table. The residual NAPL left behind in the vadose zone partitions into vapor and liquid phases depending upon the degree of volatility and of water solubility. The thickness of LNAPL in a monitoring well is larger than that of free LNAPL in the subsurface.

If a DNAPL (e.g., chlorinated hydrocarbon) spills on the ground surface, under the force of gravity, it migrates through the vadose zone and through the saturated zone, eventually reaching an impervious layer. A layer of DNAPL then accumulates over the impervious layer. The mobile DNAPL then migrates along the slope of the impervious surface, which does not necessarily coincide with the slope of the water table and the direction of the ground flow. Monitoring wells placed just at the top of the impervious layer will show the presence of the DNAPL at the bottom of the well. If the well extends into the impervious layer the DNAPL will also fill that portion of the monitoring well below the impervious surface that acts as a sump.

32.6 Remediation

Monitoring Wells

Before any site remediation work is undertaken it is necessary to explore the aquifer and the extent of the groundwater contamination. Monitoring wells are used principally to measure the elevation of the water table or the piezometric level, collect water samples for chemical analysis and eventually observe the presence of non-aqueous-phase liquids (lighter or denser than water), and collect samples of these non-aqueous-phase liquids. The equipment and supplies must be decontaminated before they are used in a water quality monitoring well.

If the purpose of the well is for observation of the water elevation only, a one-inch casing is adequate. If water samples are required, a two-inch casing is necessary. Screens are used to allow the water into the well. In unconfined aquifers the screens must be placed so that they extend approximately from 5 ft above the expected high water table to 5 ft below the expected low water table level. **Piezometer** screens for confined aquifers are shorter and generally have a length of 2 to 5 ft. The screen is surrounded by a *filter pack* consisting of medium to coarse silica sand. The filter pack extends about 2 ft above the screen. A seal is placed on top of the filter pack. It consists of a 2 ft layer of fine sand and an optional 2 ft layer of granular bentonite for further sealing. If there is a leachate plume, several wells with different depths and screen lengths may be necessary to intercept the plume. Multilevel sampling devices that are installed in a single casing have been developed.

Monitoring of the water quality in the vadose zone can be accomplished with *lysimeters*, which are installed in a bore hole above the water table. The lysimeter consists principally of a porous cup mounted at the lower end of a tube with a stopper at the upper end. Because the soil water pressure is below atmospheric, a suction must be applied so that the water penetrates into the porous cup. The water accumulated in the porous cup is then pumped into a flask at ground level.

Removal and Containment of Contaminants

Control of the source will prevent the continued addition of pollutant. The three principal methods of source control are removal, containment, and hydrodynamic isolation. Removal of the source will require transportation of the waste and its final disposition in an environmentally acceptable manner. Containment of the waste can generally be accomplished by a cutoff wall made of soil-bentonite slurry or concrete. The waste can also be isolated hydrodynamically by installing a pumping well immediately downstream of the contaminant plume so that the flow through the contaminated zone is captured by the well. The shape of the capture zone with a single well at the origin of the coordinate axes has been given by Javandel and Tsang [1986] for a confined aquifer as

$$y = \pm Q/(2BU) - Q/(2\pi BU)\tan^{-1}(y/x)$$

where Q is the pumping rate, B is the thickness of the aquifer, and U is the regional Darcy velocity. Javandel and Tsang [1986] also give equations for the capture zone formed by several wells. Figure 32.7 shows the capture zones for several values of Q/BU. The curve that fully encloses the plume

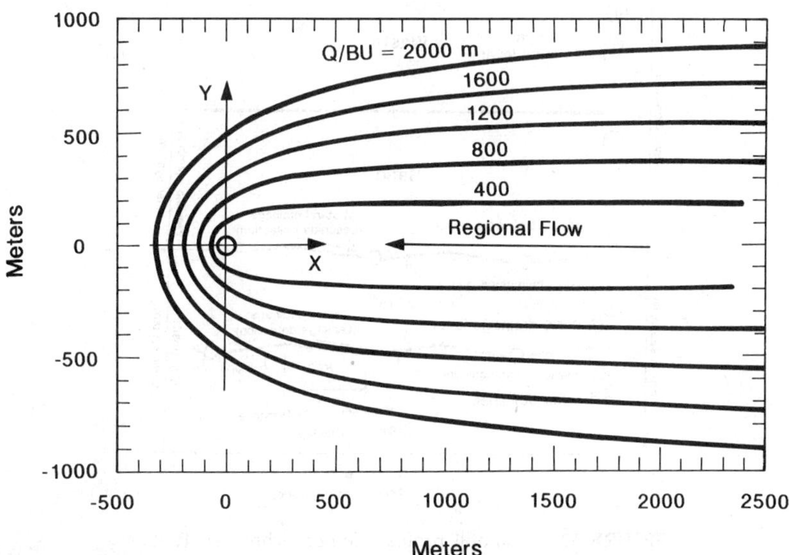

FIGURE 32.7 Capture zone for a single well in a confined aquifer. (*Source:* Javandel, I. and Tsang, C. F. 1986. Capture zone type curves: A tool for aquifer cleanup. *Ground Water* 24(5):616–625.)

is selected. The required pumping rate is obtained by dividing the value of the parameter Q/BU by the product of the aquifer thickness, B, and the regional flow velocity, U.

32.7 Landfills

A typical landfill consists of three major layers: a top, a middle, and a bottom subprofile (Fig. 32.8). The purpose of the top subprofile is to cover the waste, minimize rainfall infiltration into the waste, and provide an exterior surface that is resistant to erosion and deterioration. The middle subprofile includes the waste layer, a lateral drainage layer with the leachate collection system underlain by a flexible membrane liner. The bottom subprofile includes an additional drainage layer, a leakage detection system, and a barrier soil liner.

The design and operation of landfills are controlled by federal and local regulations. The federal regulations include Subtitle C landfill regulations of the Resource Conservation and Recovery Act (RCRA) as amended by the Hazardous and Solid Waste Amendments (HSWA) of 1984 and the Minimum Technology Guidance [USEPA, 1988]. The RCRA regulations mandate that below the waste layer there must be double liners with a leak detection system. According to the guidance the double liner system includes a synthetic liner, a secondary leachate collection system, and a composite liner consisting of a synthetic liner over a low-permeability soil or a thick low-permeability soil liner. The soil liner should have an in-place hydraulic conductivity not exceeding

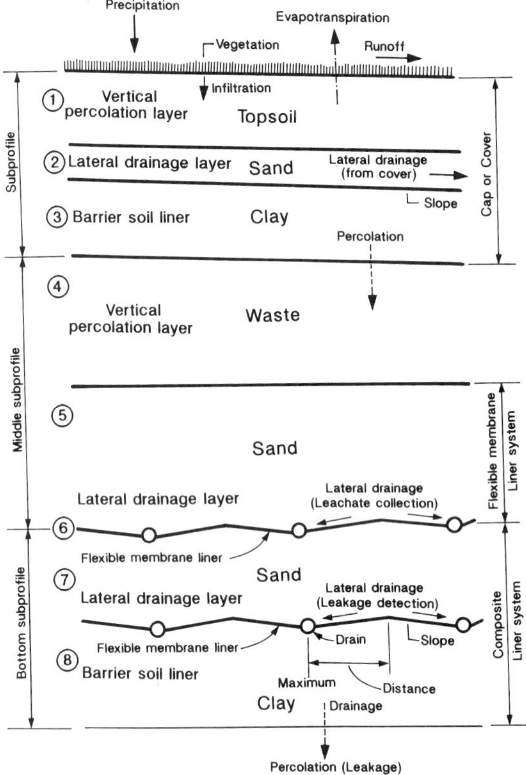

FIGURE 32.8 Landfill profile. (*Source:* Schroeder, P. R., Peyton, R. L., McEnroe, B. M., and Sjostrom, J. W. 1992a *Hydrologic Evaluation of Landfill Performance (HELP) Model, Vol. III: User's Guide for Version 2.* Department of the Army.)

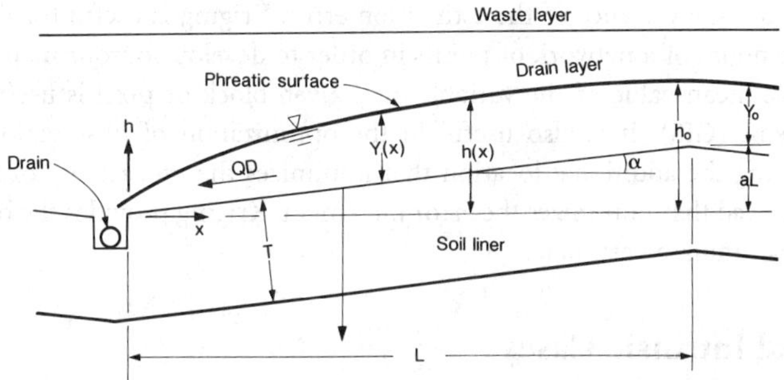

FIGURE 32.9 Landfill drainage layer. (*Source:* Peyton, R. L. and Schroeder, P. R. 1990. Evaluation of landfill liner designs. *J. Environ. Eng., ASCE.* 116(3):421–437.)

$1 \cdot 10^{-7}$ cm/s and a thickness of at least 3 ft. The primary and secondary leachate collection systems should include a drainage layer with a thickness of at least 1 ft with a saturated hydraulic conductivity of at least $1 \cdot 10^{-2}$ cm/s and a minimum bottom slope of 2%. The leachate depth cannot exceed 1 ft.

The simplified steady state equations governing the moisture flow through the landfill as given by Peyton and Schroeder [1990] are as follows. The lateral drainage per unit area Q_D is given by

$$Q_D = 2C_1 K_D Y h_0 / L^2$$

where K_D is the saturated hydraulic conductivity of the lateral drainage layer (Fig. 32.9), Y is the average saturated depth over the liner (in.), h_0 is the head above the drain at the crest of the drainage layer (in.), L is the drainage length (in.), and $C_1 = 0.510 + 0.00205\alpha L$, where α is the dimensionless slope of the drainage layer (ft/ft). The saturated depth at the crest of the drainage layer, y_0 (in.), is

$$y_0 = Y^{1.16} / [\alpha L]^{0.16}$$

and the vertical percolation rate through the soil liner Q_P is given by

$$Q_P = L_F K_P [Y + T] / T$$

where L_F is the synthetic liner leakage factor, K_P is the saturated hydraulic conductivity of the soil liner, and T is the thickness of the soil liner. Graphs for the estimation of the synthetic liner leakage factor can be found in Peyton and Schroeder [1990] and in Schroeder *et al.* [1992a, b]. The Hydrologic Evaluation of Landfill Performance model [HELP, Schroeder *et al.*, 1992a, b] solves extended forms of the above equations for Q_P, Q_D, and Y.

32.8 Geostatistics

Definition of Kriging

Hydrologic and hydrogeologic variables such as hydraulic conductivity, hydraulic head, storage coefficient, transmissivity, rainfall, and solute concentration are functions of space. These quantities, although very variable, are not completely random and often exhibit a spatial correlation or structure. The study of such variables was developed by Matheron [1971] under the name of *regionalized variables.* **Kriging** is a method of optimal estimation of the magnitude of a regionalized variable at a point or over an area, given the observations of this variable at a number of locations.

Kriging also provides the variance of the estimation error. Kriging is useful for the estimation of the variable at the nodes of a network of points in order to develop contour maps of the variable. An estimate of the mean value of the variable on a given block or pixel is useful in geographic information systems (GIS). It is also useful in the optimization of observation networks, for example by choosing the additional location that minimizes the uncertainty or by choosing the location to be removed that minimizes the error increment. Kriging provides the best (in the mean square sense) linear unbiased estimate.

Stationary and Intrinsic Cases

In the stationary case the mean m of the observations $Z(x)$ at $x = (x, y, z)$ is the same everywhere and the correlation between two observations $Z(x_1)$ and $Z(x_2)$ depends only on their relative distance $\mathbf{h} = \mathbf{x}_1 - \mathbf{x}_2$ or

$$E[Z(\mathbf{x})] = m, \qquad E\{[Z(\mathbf{x}_1) - m][z(\mathbf{x}_2) - m]\} = C(\mathbf{h})$$

where $C(\mathbf{h})$ is called the *covariance* function. A useful generalization is the intrinsic case, in which the increments of the variables are stationary. In the intrinsic case,

$$E[Z(\mathbf{x}_1) - Z(\mathbf{x}_2)] = 0, \qquad E\{[Z(\mathbf{x}_1) - Z(\mathbf{x}_2)]^2\} = 2\gamma(\mathbf{h})$$

where $\gamma(\mathbf{h})$ is the *semivariogram* and $2\gamma(\mathbf{h})$ is the *variogram*. Minus γ must be a conditionally positive definite function. Examples of acceptable semivariogram functions include the following [de Marsily, 1986, p. 303; Kitinadis, 1993, p. 20.6]:

The power model	ωh^{λ}	$\lambda < 2$
The spherical model	$\omega[\frac{3}{2}(h/a) - \frac{1}{2}(h/a)^3]$	$h < a$
	ω	$h > a$
The exponential model	$\omega[1 - \exp(-h/a)]$	
The gaussian model	$\omega\{1 - \exp[-(h/a)^2]\}$	

where h is the length of the vector $\mathbf{h}$ and the parameters a and λ are selected to fit the empirical semivariogram.

Estimation

The pairs of observation points are classified according to distances, d_1, d_2, $d_3, \ldots$; for example, $0 < d_1 < 1$ km, $1 < d_2 < 2$ km, $2 < d_3 < 4$ km. For each class the average distance and the average of $\frac{1}{2}(Z_i - Z_j)^2$ are calculated. The plot of these quantities (the mean distances against the half mean squares of the observation differences) is the *empirical semivariogram*. An acceptable semivariogram is fitted to the empirical one. The estimate Z_0^* of the variable Z at the desired point $\mathbf{x}_0$ based on the observations Z_i at the points $\mathbf{x}_i$, $i = 1, \ldots, n$ is

$$Z_0^* = \Sigma_i \lambda_0^i Z_i$$

The kriging coefficients λ_0^i are obtained from the kriging equations [de Marsily, 1986, p. 296]:

$$\Sigma_j \lambda_0^j \gamma(\mathbf{x}_i - \mathbf{x}_j) + \mu = \gamma(\mathbf{x}_i - \mathbf{x}_0), \qquad i = 1, \ldots, n$$

and the unbiasedness constraint is

$$\Sigma_i \lambda_0^i = 1$$

where λ_0^j and μ are unknowns, $\gamma(\mathbf{x}_i - \mathbf{x}_i)$ are fitted semivariogram values for the distance between the observation points $\mathbf{x}_i$ and $\mathbf{x}_j$ and $\gamma(\mathbf{x}_i - \mathbf{x}_0)$ are fitted semivariogram values for the distances between the observation points $\mathbf{x}_i$ and the point $\mathbf{x}_0$ at which the estimate is being obtained. The variance of the error of estimation is

$$\sigma^2 = \Sigma_i \lambda_0^i \gamma(\mathbf{x}_i - \mathbf{x}_0) + \mu$$

and the 95% confidence interval of the estimate is approximately

$$Z_0^* = \Sigma_i \lambda_0^i Z_i \pm 2\sigma$$

Extension and Software

The methodology can be extended to the cases of two or more observation variables that are correlated. This extended technique is known as *cokriging* and its estimate is superior to that obtained by kriging each variable independently without considering their correlation.

Two public domain software packages have been developed by EPA: GEOEAS [Englund and Sparks, 1988] and GEOPACK [Yates and Yates, 1989].

Defining Terms

Advection: Transport of a solute due to mass movement of ground water and not dispersion or diffusion.

Artesian well: A well in which the water rises above top of the upper confining layer of an aquifer under pressure condition. If the piezometric level is above the ground the well is free flowing. Named after Artois, a former province of northern France.

Aquiclude: A geologic formation that is essentially impermeable.

Aquifer: A geologic formation that is water saturated and sufficiently permeable to yield economically important amounts of water to wells and springs.

Aquifuge: A geologic formation that does not contain or transmit water.

Aquitard: A geologic formation that is less permeable than an aquifer and that restricts the flow of water.

Confined aquifer: An aquifer which is confined between two layers of impervious material.

Diffusion: Spreading of a solute in groundwater due to molecular diffusion.

DNAPL: Denser than water non-aqueous-phase liquid, such as chlorinated solvents.

Hydrodynamic dispersion: Spreading of a solute in groundwater due to the combined effect of diffusion and mechanical dispersion.

Kriging: A statistical method to obtain the best (in the mean square sense) linear unbiased estimate of a hydrologic variable (such as hydraulic conductivity) at a point or over an area given values of the same variable at other locations.

LNAPL: Lighter than water non-aqueous-phase liquid, such as gasoline.

Mechanical dispersion: Spreading of a solute in groundwater due to heterogeneous permeability in the porous medium.

Non-aqueous-phase liquids (NAPL): Liquids that are not miscible with water.

Perched water: A saturated zone of a limited extent located above the water table due to a local impermeable layer.

Phreatic surface: The water table or free surface in an unconfined aquifer.

Piezometer: A tube with a small opening penetrating an aquifer for the purpose of observing the hydraulic head.

Saturation ratio: Fraction of the volume of voids occupied by water.

Sorption: Chemical reaction between a solute in groundwater and solid particles, which results in a bonding of part of the solute and the porous medium.

Specific yield: Volume of water drained per unit area of an unconfined aquifer due to a unit drop of the water table.

Unconfined aquifer: An aquifer without a covering confining layer in which the water surface or water table is free to move up or down.

Vadose zone: The zone between the ground surface and the top of the capillary fringe immediately over the water table.

Water table: The water free surface at atmospheric pressure at the top of an unconfined aquifer.

Well log: A description of the types and depths of the geologic materials encountered during the drilling of a well.

Wettability: Preferential spreading of one liquid over a solid surface in a two-liquid flow. In the saturated zone water is the wetting fluid. In the vadose zone the non-aqueous-phase liquid usually is the wetting fluid.

References

Abdul, A. S. 1988. Migration of petroleum products through sandy hydrogeologic systems. *Monitor. Rev.* 8(4):73–81.

Bear, J. 1979. *Hydraulics of Groundwater*. McGraw-Hill, New York.

Bear, J. and Verruijt. 1987. *Modeling Groundwater Flow and Pollution*. Reidel, Boston.

Boonstra, J. 1989. *SATEM: Selected Aquifer Test Evaluation Methods*, Int. Inst. for Land Reclamation and Improvement, Wageningen, The Netherlands.

Bouwer, H. 1978. *Ground Water Hydrology*. McGraw-Hill, New York.

Brooks, R. H. and Corey, A. T. 1964. Hydraulic properties of porous media, *Hydrology Paper 3*, Colorado State University, Fort Collins, CO.

Campbell, G. S. 1974. A simple method for determining the unsaturated conductivity from moisture retention data. *Soil Sci.* 117:311–314.

Davis, S. N. 1969. Porosity and Permeability of Natural Materials. In *Flow through Porous Media*, ed. R. J. M. De Wiest, pp. 54–89. Academic Press, New York.

Deutsch, C. V. and Journel, A. G. 1992. *GSLIB: Geostatistical Software Library and User's Guide*. Oxford University Press, New York.

de Marsily, G. 1986. *Quantitative Hydrogeology, Groundwater Hydrology for Engineers*. Academic Press, Orlando, FL.

Driscoll, F. G. 1986. *Groundwater and Wells*, 2nd ed. Johnson Division, St. Paul, MN.

Englund, E. and Sparks, A. 1988. *GEOEAS (Geostatistical Environmental Assessment Software) User's Guide*. USEPA 600/4-88/033a, Las Vegas.

Fetter, C. W. 1988. *Applied Hydrogeology*. Macmillan, New York.

Fetter, C. W. 1993. *Contaminant Hydrogeology*. Macmillan, New York.

Freeze, R. A. and Cherry, J. A. 1979. *Groundwater*. Prentice-Hall, Englewood Cliffs, NJ.

Hantush, M. S. and Jacob, C. E. 1955. Nonsteady radial flow in an infinite leaky aquifer. *Trans. Am. Geophy. Union.* 36:95–100.

Heath, R. C. 1984. *Basic Ground-Water Hydrology*. U.S. Geological Survey Water Supply Paper 2220, U.S. Government Printing Office.

Javandel, I. and Tsang, C. F. 1986. Capture zone type curves: A tool for aquifer cleanup. *Ground Water* 24(5):616–625.

Kitanidis, P. K. 1993. Geostatistics. In *Handbook of Hydrology*, ed. D. R. Maidment. McGraw-Hill, New York.

Matheron, G. 1971. The theory of regionalized variables and its applications. Paris School of Mines, *Cah. Cent. Morphologie Math.* 5. Fontainebleau, France.

Neuman, S. P. 1975. Analysis of pumping test data from anisotropic unconfined aquifers considering delayed gravity response. *Water Resour. Res.* 11(2):329–342.

Neuman, S. P. 1990. Universal scaling of hydraulic conductivity and dispersivities in geologic media. *Water Resour. Res.* 26(8):1749–1758.

National Research Council 1984. *Groundwater Contamination.* National Academy Press, Washington, D.C.

Peyton, R. L. and Schroeder, P. R. 1990. Evaluation of landfill liner designs. *J. Environ. Eng., ASCE.* 116(3):421–437.

Schroeder, P. R., Peyton, R. L., McEnroe, B. M., and Sjostrom, J. W. 1992a. *Hydrologic Evaluation of Landfill Performance (HELP) Model, Vol III: User's Guide for Version 2.* Department of the Army. Waterways Experiment Station, Vicksburg, MI.

Schroeder, P. R., McEnroe, B. M., Peyton, R. L., and Sjostrom, J. W. 1992b. *Hydrologic Evaluation of Landfill Performance (HELP) Model, Vol. IV: Documentation for Version 2.* Department of the Army. Waterways Experiment Station, Vicksburg, MI.

U.S. Department of Agriculture. 1951. *Soil Survey Manual,* Handbook No. 18. U.S. Government Printing Office, Washington, D.C.

U.S. Environmental Protection Agency. 1988. *Guide to Technical Resources for the Design of Land Disposal Facilities,* Report EPA/625/6-88/018, USEPA Risk Reduction Engineering Laboratory and Center for Environmental Resource Information, Cincinnati, Ohio.

van Genuchten, M. T. 1980. A closed-form equation for predicting the hydraulic conductivity of unsaturated soils. *Soil Sci. Soc. Am. J.* 32:329–334.

Vukovic, M. and Soro, A. 1992. *Determination of Hydraulic Conductivity of Porous Media from Grain-Size Composition.* Water Resources Publications, Littleton, CO.

Walton, W. C. 1962. *Selected Analytical Methods for Well and Aquifer Evaluations.* Illinois State Water Survey Bulletin 49.

Yates, S. R. and Yates, M. V. 1989. *Geostatistics for Waste Management: User's Manual for GEOPACK.* Kerr Environmental Research Laboratory, Office of Research and Development, USEPA, Ada, OK.

Young, R. N., Mohamed, A. M. O., and Warkentin, B. P. 1992. *Principles of Contaminant Transport in Soils.* Elsevier, Amsterdam.

For Further Information

Heath [1984] provides an excellent, well-illustrated introduction to groundwater flow, wells, and pollution.

National Research Council [1984] provides a well-documented, nonmathematical introduction to groundwater contamination, including case studies.

Driscoll [1986] provides a wealth of practical information on well hydraulics, well drilling, well design, well pumps, well maintenance and rehabilitation, and groundwater monitoring.

Freeze and Cherry [1979] provide a textbook with a detailed treatment of groundwater flow and transport processes.

Bear [1979] and Bear and Verruijt [1987] provide an in-depth study of groundwater flow and contaminant transport from a mathematical perspective.

Young *et al.* [1992] discuss the basic principles of contaminant transport in the unsaturated zone.

Fetter [1993] provides a general treatment of contaminant transport in groundwater and a discussion of groundwater monitoring and site remediation.

Deutsch and Journel [1992] provide a didactic review of kriging as well as a collection of geostatistical routines and Fortran source code for PC computers.

33

Sediment Transport in Open Channels

D. A. Lyn
Purdue University

33.1 Introduction

The erosion, deposition, and transport of sediment by water arise in a variety of situations with engineering implications. Erosion must be considered in the design of stable channels or the design for local scour around bridge piers. Resuspension of possibly contaminated bottom sediments have consequences for water quality. Deposition is often undesirable since it may hinder the operation or shorten the working life of hydraulic structures and navigational channels. Sediment traps are specifically designed to promote the deposition of suspended material in order to minimize their downstream impact, for example, on cooling water inlet works or in water treatment plants. A large literature exists on approaches to problems involving sediment transport; this chapter can only introduce the basic concepts in summary fashion. It is oriented primarily to applications in steady uniform flows in a sand-bed channel; problems involving flow nonuniformity, unsteadiness, and gravel beds are only briefly mentioned. Coastal processes are treated in Chapter 34. Cohesive sediments—for which physicochemical attractive forces may lead to the aggregation of particles—are not considered at all. The finer fractions (clays and silts) that are susceptible to aggregation are found more in estuarial and coastal shelf regions than in streams. A recent review of problems in dealing with cohesive sediments is given by Mehta *et al.* [1989a,b].

0-8493-8953-4/95/$0.00 + $.50

33.2 The Characteristics of Sediment

Density, Size, and Shape

The density of sediment depends on its composition. Typical sediments in alluvial water bodies consist mainly of quartz, the specific gravity of which can be taken as $s = 2.65$. The specific weight is therefore $\gamma_s = 165.4$ lb/ft^3, or 26.0 kN/m^3. In many formulas the effective specific weight, which includes the effect of buoyancy, is used: $(s - 1)\gamma$, where γ is the specific weight of water.

The exact shape of a sediment particle is not spherical, and so a compact specification of its geometry or size is not feasible. Two practical measures of grain size are:

1. *The sedimentation or aerodynamic diameter.* The diameter of the sphere of the same material with the same fall velocity, w, (see below for definition) under the same conditions.
2. *The sieve diameter.* The length of a side of the square sieve opening through which the particle will just pass.

Because size determination is most often performed with sieves, the available data for sediment size usually refer to the sieve diameter, which is taken to be the geometric mean of the adjacent sieve meshes—that is, the mesh size through which the particle has passed and the mesh size at which the particle is retained. The sedimentation diameter is related empirically to the sieve diameter by means of a shape factor, SF, which increases from 0 to 1 as the particle becomes more spherical (for a well-worn sand, SF $\approx$ 0.7).

Size Distribution

Naturally occurring sediment samples exhibit a range of grain diameters. A characteristic diameter, d_a, may be defined in terms of the percent, a, by weight of the sample that is smaller than d_a. Thus, for a sample with $d_{84} = 0.35$ mm, 84% by weight of the sample is less than 0.35 mm in diameter. The median size is denoted as d_{50}. Frequently, the grain size distribution is assumed to be lognormally distributed, and a geometric mean diameter and standard deviation are defined as $d_g = \sqrt{d_{16}d_{84}}$, and $\sigma_g = \sqrt{d_{84}/d_{16}}$. For a lognormal distribution, $d_{50} = d_g$ and the arithmetic mean diameter $d_m = d_g e^{(0.5 \ln^2 \sigma_g)}$. Similarly, d_a can be determined from d_g and σ_g from the relation $d_a = d_g \sigma_g^{Z_a}$, where Z_a is the standard normal variate corresponding to the value of a. For example, if $a = 65\%$, $d_g = 0.35$ mm, and $\sigma_g = 1.7$, then $Z_a = 0.39$ and so $d_{65} = (0.35 \text{ mm})(1.7)^{0.39} = 0.43$ mm. In natural sand-bed streams σ_g typically ranges between 1.4 and 2, but in gravel-bed streams it may attain values greater than 4. Qualitative discussions of sediment size may be based on a standard sediment grade scale established by the American Geophysical Union. A simplified grade scale divides the size range into cobbles and boulders ($d > 64$ mm), gravels (2 mm $< d <$ 64 mm), sands (0.06 mm $< d <$ 2 mm), silts (0.004 mm $<$ $d <$ 0.06 mm), and clays ($d <$ 0.004 mm).

Fall (or Settling) Velocity

The terminal velocity of a particle falling alone through a stagnant fluid of infinite extent is called its *fall* or *settling velocity, w.* The standard drag curve for a spherical particle provides a relationship between d and w. For nonspherical sand particles in water, the fall velocity at various temperatures can be determined from Fig. 33.1 if the sieve diameter and SF are known or can be assumed (note the different fall velocity scales). As an example, for a geometric sieve diameter of 0.3 mm and SF = 0.7, the fall velocity in water at 10°C is determined as $\approx$3.6 cm/s. In a horizontally flowing turbulent suspension the actual mean fall velocity of a given particle may be influenced by neighboring particles (hindered settling) and by turbulent fluctuations.

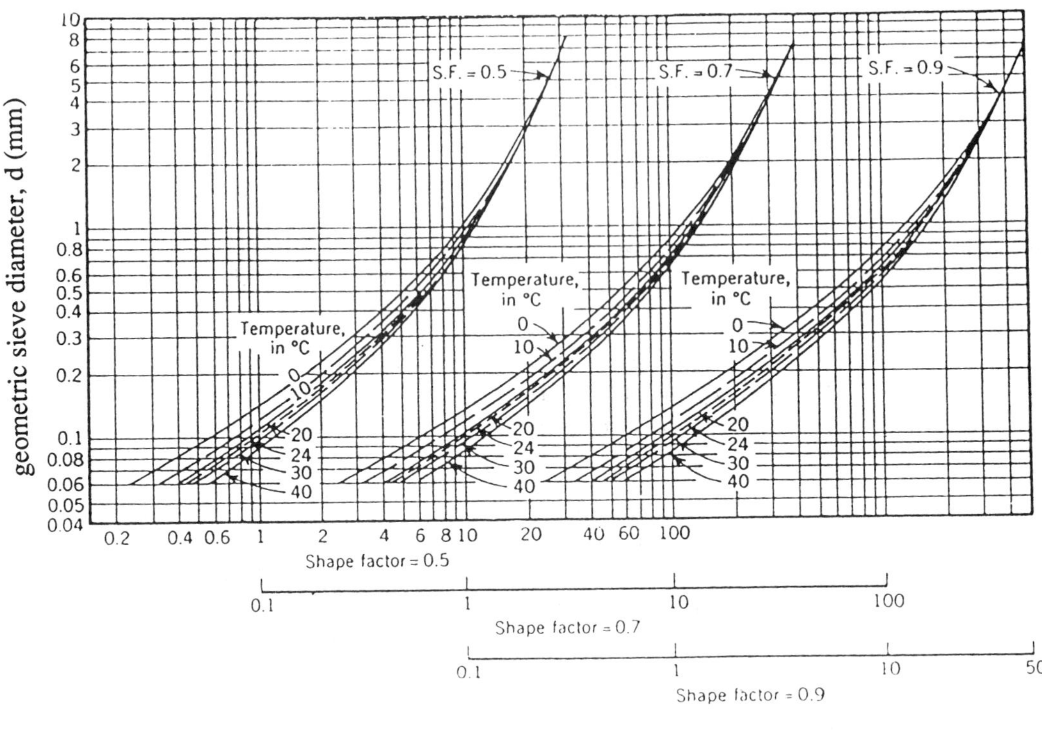

FIGURE 33.1 Relationship between fall velocity, sand-grain diameter, and shape factor in water. (*Source:* Vanoni, 1975.)

Angle of Repose

The angle of repose of a sediment particle is important in describing the initiation of its motion and hence sediment erosion of an inclined surface, such as a stream bank. It is defined as the angle, θ, at which the particle is just in equilibrium with respect to sliding due to gravitational forces. It will vary with particle size, shape, and density; empirical curves for some of these variations are given in Fig. 33.2.

33.3 Flow Characteristics and Dimensionless Parameters; Notation

The important flow characteristics are those associated with open-channel or more generally free-surface flows (see Chapter 28 or 29 for more details). These are the total water discharge, Q (or the discharge per unit width, $q = Q/B$, where B is the average width of the channel); the mean velocity, $U = Q/A$, where A is the channel cross-sectional area; the hydraulic radius, R_h, (or for a very wide channel the flow depth, $R_h \approx H$); and the energy or friction slope, S_f. The total bed shear stress, τ_b, and the related quantity, the shear velocity, $u_* = \sqrt{\tau_b/\rho} = \sqrt{gR_h S_f}$, are also important.

Much of sediment transport engineering remains highly empirical, and so the organization of information in terms of dimensionless parameters becomes important (see the discussion of dimensional analysis in Chapter 28). Sediment and flow characteristics may be combined in several dimensionless parameters that arise repeatedly in sediment transport. A dimensionless bed shear stress, also termed the *Shields parameter* (described later), can be defined as

$$\Theta \equiv \frac{\tau_b}{\gamma(s-1)d} = \frac{u_*^2}{g(s-1)d} = \frac{R_h S_f}{(s-1)d} \tag{33.1}$$

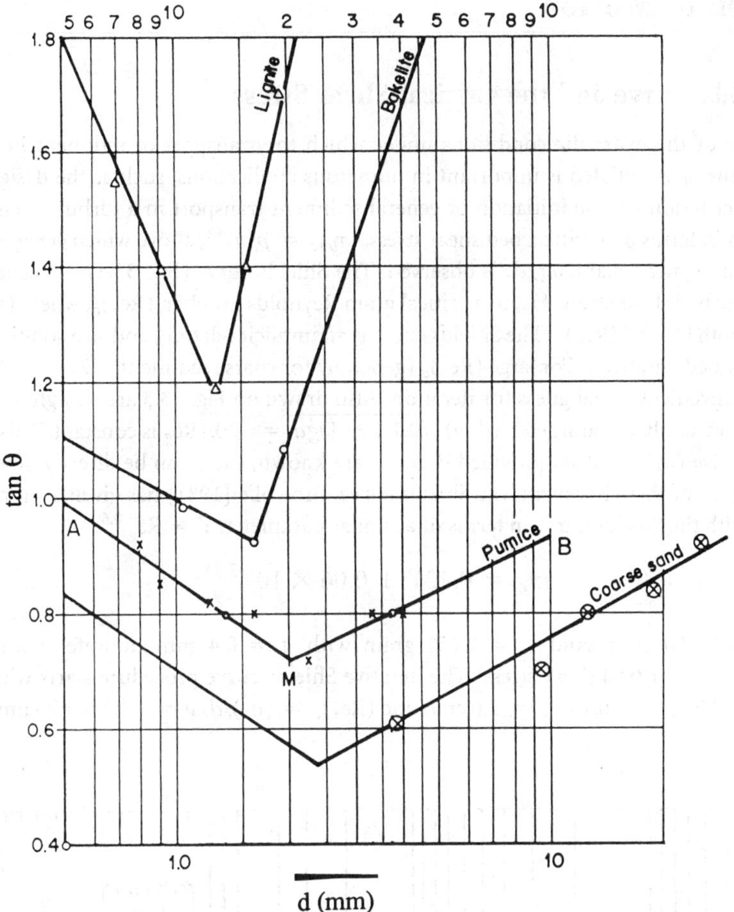

FIGURE 33.2 Angle of repose as a function of size and material. (*Source:* Simons and Sentürk, 1992.)

Two grain Reynolds numbers can be usefully defined as

$$\text{Re}_g \equiv \frac{\sqrt{g(s-1)d^3}}{\nu} \quad \text{and} \quad \text{Re}_* \equiv \frac{u_* d}{\nu} \tag{33.2}$$

A grain Froude number can be defined as

$$\text{Fr}_g \equiv \frac{U}{\sqrt{g(s-1)d}} = \left(\frac{U}{u_*}\right)\sqrt{\Theta} \tag{33.3}$$

A dimensionless **sediment discharge,** Φ, per unit width may be defined as

$$\Phi \equiv \frac{g_s/\gamma_s}{\sqrt{g(s-1)d^3}} \tag{33.4}$$

where $g_s = \gamma q \overline{C}$ is the weight flux of sediment per unit width and $\overline{C}$ is the flux-weighted mass or weight concentration of sediment (see the section on "Sediment Transport"). In the above definitions various characteristic grain diameters and shear velocities may be used according to the context.

33.4 Initiation of Motion

The Shields Curve and the Critical Shear Stress

A knowledge of the hydraulic conditions under which the transport of sediment in an alluvial channel begins or is initiated is important in numerous applications, such as the design of stable channels. A criterion for the initiation of general sediment transport in a turbulent channel flow may be given in terms of a critical bed shear stress, $(\tau_b)_c = \rho(u_*)_c^2$, above which *general* motion of bed sediment of mean diameter, d, is observed. The Shields curve (Fig. 33.3) correlates a critical dimensionless bed shear stress, Θ_c, to a critical grain Reynolds number, $(\mathrm{Re}_*)_c$, where $(u_*)_c$ is used in defining both Θ_c and $(\mathrm{Re}_*)_c$. The Shields curve is an implicit relation, and so a solution for $(u_*)_c$ must be obtained iteratively. For large $(\mathrm{Re}_*)_c$ (generally for coarse sediment), $\Theta_c \rightarrow \approx 0.06$, which provides a convenient initial guess for iteration. Also drawn on Fig. 33.3 are straight oblique lines along which an auxiliary parameter, $(d/\nu)\sqrt{0.1(s-1)gd} = \sqrt{0.1\mathrm{Re}_g}$ is constant. This parameter does not involve $(u_*)_c$, and so, provided d and ν are known, $(u_*)_c$ can be directly determined by the intersection of these lines with the Shields curve. Brownlie [1981] has given a simple formula consistent with the Shields curve in terms of a similar parameter, $Y \equiv \mathrm{Re}_g^{-0.6}$:

$$\Theta_c = 0.22Y + 0.06 \times 10^{-7.7Y} \tag{33.5}$$

Example 33.1 Given a sand ($s = 2.65$) grain with $d = 0.4$ mm in water with $\nu = 0.01$ cm^2/s, what is the critical shear stress? The iterative Shields curve procedure starts with an initial guess, $\Theta_c = 0.06$ implying $(u_*)_c = 2.0$ cm/s and $(\mathrm{Re}_*)_c = (u_*)_c d/\nu = 8$. This is inconsistent with

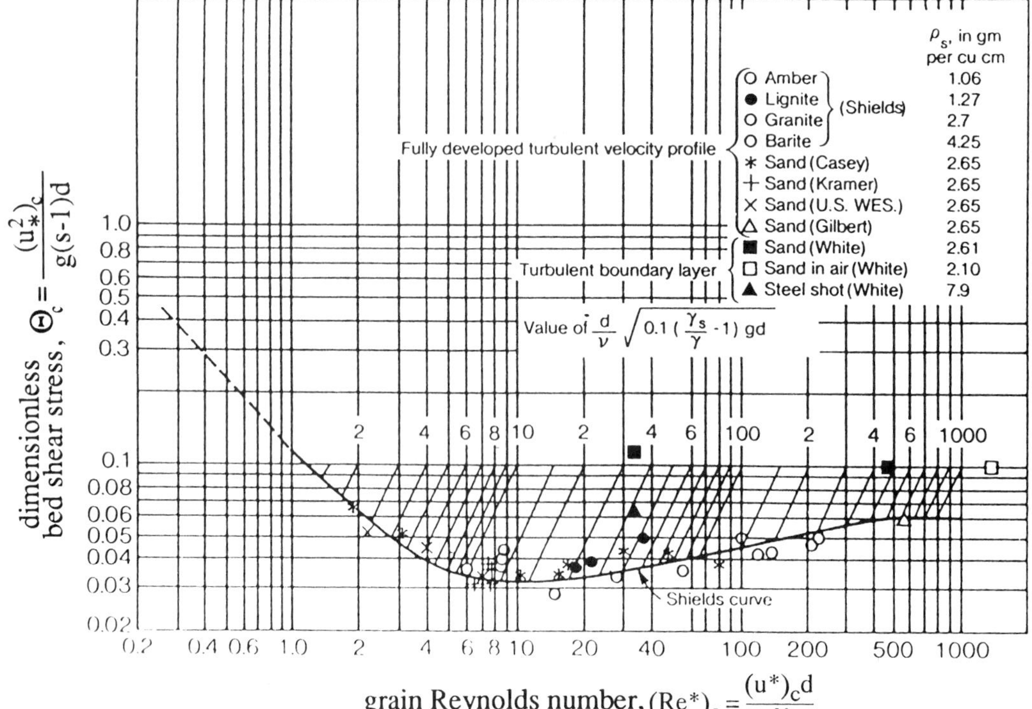

FIGURE 33.3 The Shields diagram relating critical shear stress to hydraulic and particle characteristics. (*Source:* Chang, 1988.)

the Shields curve, which indicates $\Theta_c = 0.032$ for $(\text{Re}_*)_c = 8$. The procedure is iterated by making another guess, $\Theta_c = 0.032$, which yields $\tau_c = 0.21$ kPa corresponding to $(\text{Re}_*)_c = 5.8$. This result is sufficiently consistent with the Shields curve, and so the iteration can be stopped. More directly, the auxiliary parameter, $(d/\nu)\sqrt{0.1(s-1)gd} = 10$, can be computed, and the line corresponding to this value intersects the Shields curve at $\Theta_c = 0.034$. The use of the Brownlie empirical formula, Eq. (33.5), gives more directly, with $\text{Re}_g = 32.2$ and $Y = 0.125$, $\Theta_c = 0.034$.

Instead of using $(\tau_b)_c$, traditional procedures for the design of stable channels have often been formulated in terms of a critical average velocity, U_c, or critical unit-width discharge, q_c, above which sediment transport begins. The specification of U_c or q_c is complicated by the necessity for a friction law—that is, a relationship between U and τ_b—which is discussed in later.

The Effect of Slope

The above criterion is applicable to grains on a surface with negligible slope, as is usually the case for grains on the channel bed. Where the slope of the surface on which grains are located is appreciable—for example, on a river bank—its effect cannot be neglected. With the inclusion of the additional gravitational forces, a force balance reveals that $(\tau_b)_c$ is reduced by a fraction involving the angle of repose of the grain. The ratio between the value of $(\tau_b)_c$ including slope effects and its value for a horizontal surface is given by:

$$\frac{[(\tau_b)_c]_{\text{slope}}}{[(\tau_b)_c]_{\text{no slope}}} = \left(1 - \frac{\sin^2\phi}{\sin^2\theta}\right)^{1/2} \tag{33.6}$$

where ϕ is the angle of the sloping surface and θ is the angle of repose of the grain. On a horizontal surface $\phi = 0$ and the ratio is unity, whereas if $\phi = \theta$, no shear is required to initiate sediment motion (consistent with the definition of the angle of repose).

Summary

Although the Shields curve is widely accepted as a reference, controversy remains concerning its details and interpretation—for example, its behavior for small Re_* [Raudkivi, 1990] and the effect of fluid temperature [Taylor and Vanoni, 1972]. The random nature of turbulent flow and random magnitudes of the instantaneous bed shear stresses motivate a probabilistic approach to the initiation of sediment motion. The critical shear stress given by the Shields curve can be accordingly interpreted as being associated with a probability that sediment particle of given size will begin to move. It should not be interpreted as a criterion for zero sediment transport, and design relations for zero transport, if based on the Shields curve, should include a significant factor of safety [Vanoni, 1975].

33.5 Flow Resistance and Stage-Discharge Predictors

The stage-discharge relationship or rating curve for a channel relating the uniform-flow water level (stage) or hydraulic radius, R_h, to the discharge, Q, is determined by channel flow resistance. For flow conditions above the threshold of motion, the erodible sand bed is continually subject to scour and deposition, so that the bed acts as a deformable or "movable" free surface. The flat or plane bed—that is, one in which large-scale features are absent—is often unstable, and **bed forms** (Fig. 33.4) such as dunes, ripples, and antidunes develop. Dunes, which exhibit a mild upstream slope and a sharper downstream slope, are the most commonly occurring of bed forms in sand-bed channels. Ripples share the same shape as dunes, but are smaller in dimensions. They may be found in combination with dunes, but are generally thought to be unimportant except in streams at small

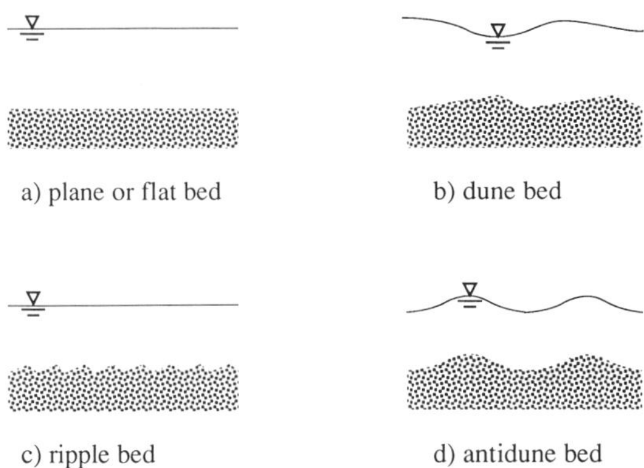

FIGURE 33.4 Sketch of various bed forms.

depths and low velocities. Antidunes assume a smoother, more symmetric sinusoidal shape, which results in less flow resistance, and are associated with steeper streams. Antidunes differ from dunes in moving upstream rather than downstream and in being associated with water surface variations that are in phase rather than out of phase with bed surface variations.

In fixed-bed, open-channel flows, resistance is characterized by a Darcy–Weisbach friction factor (see Chapter 28) or a Manning's n (see Chapter 29), which are assumed to vary only slowly or not at all with discharge. For movable or erodible beds, substantial changes in flow resistance may occur as the bed forms develop or are washed out. Very loosely, as transport intensity (as measured, e.g., by the Shields parameter, Θ) increases, ripples evolve into dunes, which in turn become plane or transition beds, to be followed by antidunes. Multiple depths may be consistent with the same discharge, and the rating curve may exhibit discontinuities (Fig. 33.5). These discontinuities are attributed to a short-term transition from low-velocity high-resistance flow over ripples and dunes,

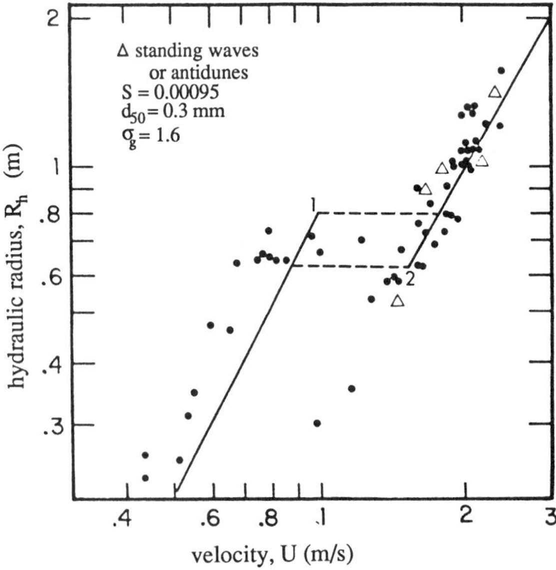

FIGURE 33.5 Stage-discharge data reported by Dawdy [1961] for the Rio Grande River near Bernalillo, New Mexico. The solid lines are predictions using the Brownlie formulas, Eqs. (33.9–11). (*Source:* Brownlie, 1981.)

termed *lower-regime* flow, to high-velocity low-resistance flow over plane, transition, or antidune bed, termed *upper-regime flow*, or vice versa. Because of these two possible regimes, movable-bed friction formulas (unlike fixed-bed friction formulas) must include a method to determine the flow regime.

Form and Grain Resistance Approach

In flows over dunes and ripples, form resistance due to flow separation from dune tops provides the dominant contribution to overall resistance. Yet the processes involved in determining bedform characteristics are more directly related to the actual bed shear stress (as in the problem of initiation of motion). Much of sediment transport modeling has distinguished between form and grain (skin) resistance (see section on hydrodynamic forces in Chapter 28 for the distinction between the two types of flow resistance). A *fictitious* overall bed shear stress, $\tau_o \equiv \gamma R_h S$, is taken as the sum of a contribution from grain resistance, τ', and a contribution from form resistance, τ''. Since τ_o is usually correlated empirically with τ', it remains only to determine τ' from given hydraulic parameters. The traditional approach estimates τ' from fixed-bed friction formulas for plane beds. A simple effective example of this approach to stage-discharge prediction is from Engelund and Hansen [1967][with extension by Brownlie (1983)] and correlates a dimensionless total shear stress, Θ, with a dimensionless grain shear stress, Θ'. The Engelund–Hansen formula is as follows:

$$\Theta = 1.581\sqrt{\Theta' - 0.06}, \qquad 0.06 < \Theta' < 0.55 \quad \text{lower regime} \tag{33.7a}$$

$$= \Theta', \qquad 0.55 < \Theta' < 1 \quad \text{upper regime} \tag{33.7b}$$

$$= \frac{1}{[1.425(\Theta')^{-1.8} - 0.425]^{1/1.8}}, \qquad \Theta' > 1 \quad \text{upper regime} \tag{33.7c}$$

$\Theta'(\equiv R_h' S_f/(s-1)d_{50})$ is related to U by a friction formula for a plane fixed bed:

$$\frac{U}{\sqrt{gR_h' S_f}} = 5.76 \log_{10} \frac{5.51R_h'}{d_{65}} \tag{33.8}$$

The lower regime corresponds to flows over dune-covered beds, such that $\Theta > \Theta'$ for values of Θ' not too close to 0.06, whereas in the transition or upper regime ($\Theta < 1$) corresponding to plane beds or beds with antidunes, $\Theta = \Theta'$ since form resistance is expected to be small. The Engelund–Hansen formula was originally developed based on large-flume laboratory data with d_{50} in the range 0.19 mm–0.93 mm and σ_g of 1.3 for the finest sediment and 1.6 for the others.

Overall Resistance Approach

The distinction between grain (skin) and form resistance is physically sound, but the use of a plane fixed-bed friction formula such as Eq. (33.8) cannot be justified rigorously for beds with dunes and ripples, and the need for a further correlation between Θ and Θ' is inconvenient. A simpler approach relating Θ (or R_h) directly to Q or other dimensionless parameters may therefore be more attractive. Guided by dimensional analysis, Brownlie [1983] performed regression analyses on a large data set of laboratory and field measurements and proposed the following stage-discharge formulas, known as the Brownlie formulas:

$$\frac{R_h}{d_{50}} = 0.0576(s-1)^{0.945}\mathrm{Fr}_g^{1.889} S_f^{-0.735} \sigma_g^{0.303}, \qquad \text{lower regime} \tag{33.9a}$$

$$= 0.0348(s-1)^{0.833}\mathrm{Fr}_g^{1.665} S_f^{-0.767} \sigma_g^{0.214}, \qquad \text{upper regime} \tag{33.9b}$$

where σ_g is the geometric standard deviation and Fr_g is the grain Froude number. To determine whether the flow is in lower or upper regime, the following criteria are applied.

For $S_f > 0.006$, only upper regime flow is observed.

For $S_f < 0.006$, additional criteria are formulated in terms of a modified grain Froude number, $\mathcal{F} \equiv \mathrm{Fr}_g/[1.74S_f^{-1/3}]$, and a modified grain Reynolds number, $\Delta \equiv u'_* d_{50}/11.6\nu$, where u'_* is the shear velocity corresponding to the upper regime flow, that is, due primarily to grain resistance.

1. The *lower* limit of the *upper regime* is given as

$$\log_{10} \mathcal{F}_{\mathrm{up}} = -0.0247 + 0.152 \log_{10} \Delta + 0.838(\log_{10} \Delta)^2 \quad \Delta < 2 \tag{33.10a}$$

$$= \log_{10} 1.25 \quad \Delta \geq 2 \tag{33.10b}$$

2. The *upper* limit of the *lower regime* is given as

$$\log_{10} \mathcal{F}^{\mathrm{low}} = -0.203 + 0.0703 \log_{10} \Delta + 0.933(\log_{10} \Delta)^2 \quad \Delta < 2 \tag{33.11a}$$

$$= \log_{10} 0.8 \quad \Delta \geq 2 \tag{33.11b}$$

The range of conditions covered by the data set was: $3 \times 10^{-6} < S_f < 3.7 \times 10^{-2}$, 0.088 mm $< d_{50} <$ 2.8 mm, 0.012 m^3/s/m $< q <$ 40 m^3/s/m, $0.025 < R_h < 17$ m, and temperatures between 0 and 63°C.

Example 33.2 Given a wide alluvial channel with unit width discharge, $q = 1.6$ m^3/s/m, and bed slope, $S = 0.00025$, and sand characteristics, $d_{50} = 0.35$ mm, $\sigma_g = 1.7$, estimate the normal flow depth. The approximation is made that, for a wide channel, $R_h \approx H$ and $U = q/H$, where H is the required flow depth. With the Brownlie formulas, substitution of the given values yields for the lower regime [Eq. (33.9a)], $H_{\mathrm{low}} = 1.80$ m, and for the upper regime [Eq. (33.9b)], $H_{\mathrm{up}} = 1.30$ m. To determine which of these depths is possible, the criteria for the two regimes are examined. Since $u'_* = \sqrt{gH_{\mathrm{up}}S_f} = 0.056$ m/s, $\Delta = u'_* d_{50}/11.6\nu = 1.70$, assuming $\nu = 10^{-6}$ m^2/s. From Eqs. (33.10–11) the limits of the flow regimes for $\Delta = 1.70$ are determined as $\mathcal{F}_{\mathrm{up}} = 1.13$ and $\mathcal{F}^{\mathrm{low}} = 0.73$. The parameter $\mathcal{F} = (q/H)/[1.74S_f^{-1/3}\sqrt{g(s-1)d_{50}}]$ is evaluated using $H = H_{\mathrm{up}}$ to be $\mathcal{F} = 0.59 < \mathcal{F}_{\mathrm{up}}$—that is, below the lower limit of the upper flow regime—and using H_{low} to be $\mathcal{F} = 0.43 < \mathcal{F}^{\mathrm{low}}$, below the upper limit of the lower flow regime. The flow must then be in the lower regime, and so the required flow depth $H = H_{\mathrm{low}} = 1.80$ m. In some cases the above criteria may still not be sufficient to give a unique solution.

The Engelund–Hansen formula involves two unknowns—H and H' (the "fictitious" depth related to grain resistance)—and so requires iteration. H' may be initially guessed, for example, at 1 m, from which $\Theta' = 0.43$. According to Eq. (33.7) this falls in the lower regime, and from Eq. (33.7a) $\Theta = 0.96$, so that $H = 2.22$ m. From Eq. (33.8), the mean velocity is computed (assuming a lognormal size distribution, $d_{65} \approx 0.43$ mm for $\sigma_g = 1.7$, see the discussion on "Size Distribution") as $U = 1.17$ m/s, and hence $q = UH = 2.60$ m^2/s. Since this is not consistent with the given $q = 1.6$ m^2/s, the iteration is continued. A final iteration yields $\Theta = 0.76$ and $H = 1.75$ m, which agrees well with the result using the Brownlie formulas.

Critical Velocity

Given a stage-discharge predictor, a formula for the critical velocity U_c (see "The Shields Curve and the Critical Shear Stress" for definition) can be obtained. The Brownlie lower regime equation [Eq. (33.9a)] can be expressed in terms of a critical grain Froude number, $(\mathrm{Fr}_g)_c$, as

$$(\mathrm{Fr}_g)_c \equiv \frac{U_c}{\sqrt{g(s-1)d_{50}}} = 4.60 \frac{\Theta_c^{0.529}}{S_f^{0.141}\sigma_g^{0.161}} \tag{33.12}$$

where Θ_c is obtained from the Shields curve [e.g., Eq. (33.5)]. Based on experiments, Neill [1967] gave a simpler conservative design formula for coarse particles:

$$(\mathrm{Fr}_g)_c^2 = 2.5\left(\frac{H}{d}\right)^{0.2} \tag{33.13}$$

This gives a more conservative result than Eq. (33.12).

Example 33.3 A channel is to be designed to carry a discharge with no sediment of 5 m^3/s on a slope of 0.001. The bed material has a median diameter $d_{50} = 8$ mm and a geometric standard deviation $\sigma_g = 3$. Determine the width, B, and depth, H, at which the channel will not erode. Assume for simplicity a rectangular channel cross section and rigid banks. For $d_{50} = 8$ mm it is found from Eq. (33.5) that $\Theta_c = 0.054$, so from Eq. (33.12), $(F_g)_c = 2.18$ and $U_c = 0.78$ m/s. This is substituted into the lower regime friction equation, Eq. (33.9a), to give $R_h = BH/(B + 2H) = 0.72$ m. Since $Q = U_c BH = 5$ m^3/s, H is determined as 0.9 m, and so $B = 7.1$ m. If Eq. (33.13) is used with a Manning–Strickler friction law, then H is found to be 0.46 m and B to be 12.4 m, with $U_c = 0.86$ m/s. Equation (33.12), being based on the Shields curve, is not intended to be used for design for zero transport (see summary of previous section), whereas Eq. (33.13) was intended as a design equation for zero transport and so gives a more conservative value (smaller H and hence smaller bed shear stress).

Summary

The prediction of flow depth in alluvial channels remains an uncertain art with much room for judgment. Estimates using various predictors should be considered and the use of field data specific to the problem should be exploited where feasible to arrive at a range of predictions. If the regime is correctly predicted, the better stage-discharge relations are generally reliable to within 10–15% in predicting depth.

33.6 Sediment Transport

Three modes of sediment transport are distinguished: wash load, suspended load, and bed load. *Wash load* refers to very fine suspended material that—because of its very small fall velocities—interacts little with the bed. It will not be further considered since it is determined by upstream supply conditions rather than by local hydraulic parameters. *Suspended load* refers to material that is transported downstream primarily in suspension far from the bed, but that because of sedimentation and turbulent mixing still interacts significantly with the bed. Finally, *bed load* refers to material that remains generally close to the bed in the bed load region, being transported mainly through rolling or in short hops (termed *saltation*). The relative importance of the two modes of sediment transport may be roughly inferred from the ratio of settling velocity to shear velocity, w/u_*. For $w/u_* < 0.5$ suspended load transport is likely dominant, while for $w/u_* > 1.5$ bed load transport is likely dominant. Suspended and bed loads are termed *bed-material load* to distinguish them from the wash load, which may only be very weakly, if at all, related to material found in bed samples.

The total sediment load or discharge, G_T, is considered here as the sum of only the suspended load discharge, G_S, and the bed load discharge, G_B, and is defined as the mass or more usually the weight flux of sediment material passing a given cross section (SI units of kg/s or N/s; English units of slugs/s or lb/s). A total sediment discharge (by weight) per unit width is often used:

$$g_T = \gamma q\overline{C} = \gamma \int_0^H uc\,dy \tag{33.14}$$

where u and c are the mean velocity and mass (or weight) concentration at a point in the water column and $\overline{C}$ is a mean flux-weighted mass (or weight) concentration defined by Eq. (33.14). Because of a nonuniform velocity profile, $\overline{C}$ is not equal to the depth-averaged concentration, $\langle C \rangle \equiv (1/H) \int_0^H c\, dy$.

Suspended Load Models

The prediction of g_T given appropriate sediment characteristics and hydraulic parameters has been attempted by treating bed load and suspended load separately, but this approach is fraught with difficulties. The traditional approach derives a differential equation for conservation of sediment assuming uniform conditions in the streamwise direction:

$$\epsilon_s \frac{dc}{dy} + wc = 0 \tag{33.15}$$

where ϵ_s is a turbulent diffusion or mixing coefficient for sediment. The first term represents a net upward sediment flux due to turbulent mixing, while the second term is interpreted as the net downward flux due to sedimentation (w is the settling velocity). A solution for the vertical distribution of sediment concentration, $c(y)$, depends on a model for ϵ_s and a boundary condition at or near the bed. The well-known Rouse solution,

$$\frac{c(y)}{c_a} = \left(\frac{H-y}{y}\frac{a}{H-a}\right)^{Z_R} \tag{33.16}$$

with the Rouse exponent $Z_R \equiv w/\beta\kappa u_*$, assumes an eddy viscosity mixing model with $\epsilon_s = \beta u_* y(1 - y/H)$, where β is a coefficient relating momentum to sediment diffusion and κ stems from the assumption of a log-law velocity profile. It avoids a precise specification of the bottom boundary condition by introducing a reference concentration, c_a, at a reference level $y = a$ taken close to the bed. Here $u_* = \sqrt{gR_h S_f}$ refers to the overall shear velocity rather than to that component attributed to grain resistance.

Although Eq. (33.16) can usually be made to fit measured concentration profiles approximately with an appropriate choice of Z_R, its predictive use is limited by the lack of information concerning β, κ, and particularly c_a, which may vary with hydraulic and sediment characteristics. In the simplest models $\beta = 1$ and $\kappa = 0.4$, which assumes that sediment diffusion is identical to momentum diffusion and the velocity profile follows the log-law (see section on turbulent flows in Chapter 28) profile exactly as in plane fixed-bed flows. A more complicated recent model [van Rijn, 1984a] proposed that β is correlated with w/u_* and that κ varies with suspended sediment concentration. The suspended load discharge per unit width is computed using Eq. (33.16) as

$$g_S = \gamma \int_a^H uc\, dy \tag{33.17}$$

with u typically taken as a log-law profile. To determine g_T, a formula for predicting g_B—the transport per unit width in the bed load region $y < a$—must be coupled with Eq. (33.17), and the reference level $y = a$ must be chosen at the limit of the bed load region. In flows with bed forms neither Eq. (33.15) nor Eq. (33.16) can be rigorously justified, since bed conditions are not uniform in the streamwise direction and the log-law velocity profile is inadequate to describe velocity and stress profiles near the bed [Lyn, 1993].

Bed Load Models and Formulas

Bed load models are used either in cases where bed load transport is dominant or to complement suspended load models in total load computations. Most available formulas can be written in terms of the dimensionless bed load transport, Φ_b, and a dimensionless grain shear stress, Θ' (see section 33.2 for definitions). Only two such models, one traditional and one more recent, are described. The Meyer–Peter–Muller bed load formula was based on laboratory experiments with coarse sediments (mean diameter range: 0.4 mm–30 mm) with very little suspended load.

Meyer–Peter–Muller Bed Load Formula

$$\Phi_b = 0.08\,[(\Theta'/\Theta_c) - 1]^{3/2}, \qquad (\Theta'/\Theta_c) > 1 \tag{33.18}$$

where the dimensionless critical shear stress $\Theta_c = 0.047$ and Θ' is a fraction of the dimensionless total shear stress $\Theta' = (k/k')^{3/2}\Theta$. Based on a plane fully rough, fixed-bed friction law of Strickler type, k/k' is computed from

$$\frac{k}{k'} = 0.12\left(\frac{d_{90}}{R_h}\right)^{1/6}\frac{U}{\sqrt{gR_hS_f}} \tag{33.19}$$

In the Meyer–Peter–Muller formula, the characteristic grain size used in defining Φ_b and Θ' is the mean diameter, d_m (which can be related to d_g if necessary, see "Size Distribution").

A more recent bed load model from van Rijn [1984b]—intended for predicting bed load dominated transport as well as for complementing a suspended load model—is similar in form:

van Rijn Bed Load Formula

$$\Phi_b = 0.053\,\frac{[(\Theta'/\Theta_c) - 1]^{2.1}}{Re_g^{0.2}} \qquad (\Theta'/\Theta_c) > 1 \tag{33.20}$$

Θ_c is determined from a Shields curve relation, and Θ' is computed from a fully rough plane-bed friction formula of log-law form [see Eq. (33.8)],

$$\frac{U}{u'_*} = 5.75\,\log_{10}\frac{12R_h}{k_s} \tag{33.21}$$

where the roughness height, $k_s = 3d_{90}$. The median grain diameter, d_{50}, is used in defining Φ_b, Θ', and Re_g. In tests with laboratory and field data Eq. (33.20) performed on average as well as other well-known bed models, including the Meyer–Peter–Muller formula. Equating q_B to a sediment flux based on a reference mass concentration, c_a, at a reference level $y = a$, van Rijn [1984b] obtained a semiempirical relation for c_a to be used with a suspended load model:

$$c_a = 0.015\,s\left(\frac{d_{50}}{a}\right)\frac{[(\Theta'/\Theta_c) - 1]^{1.5}}{Re_g^{0.2}} \qquad (\Theta'/\Theta_c) > 1 \tag{33.22}$$

where a is chosen to be one-half of a bed form height for lower regime flows, or the roughness height for upper regime flows with a minimum value chosen arbitrarily to be $0.01H$.

Example 33.4 Given quartz ($s = 2.65$) sediment with $d_{50} = 1.44$ mm, $\sigma_g = 2.2$, in a uniform flow of hydraulic radius $R_h = 0.62$ m, in a wide channel of slope $S = 0.00153$, and average velocity $U = 0.8$ m/s, what is the sediment discharge per unit width? A bed load–dominated sediment discharge is indicated by $w/u_* \approx$ (16 cm/s)/(9.6 cm/s) $= 1.6$, based on d_{50}, and $u_* = \sqrt{gR_hS_f} = 0.096$ m/s. In the Meyer–Peter–Muller formula, $k/k' = 0.43$, where $d_{90} = 3.9$ mm. Hence, since $d_m = 1.96$ mm, it is found that $\Theta' = 0.082$. This gives $\Phi_b = 0.052$, from which $g_B = \gamma_s\Phi_b\sqrt{g(s-1)d_m^3} = 0.47$ N/s/m or $\overline{C} = g_B/\gamma q = 97$ ppm by mass. If the van Rijn formula is used $u'_* = 0.050$ m/s, so that $\Theta' = 0.106$. From the Shields curve $\Theta_c = 0.039$ for $\mathrm{Re}_g = 219$, so that $\Phi_b = 0.056$ or $g_B = 0.32$ N/s/m or $\overline{C} = 66$ ppm. The given parameter values correspond to field measurements in the Hii River in Japan where the reported $\overline{C}$ was 191 ppm [from the data compiled by Brownlie, 1981], which may have included some suspended load as well as wash load.

Total Load Models

The distinction between suspended load and bed load is conceptually useful, but, as has been noted previously in other contexts, this does not necessarily yield any predictive advantages since neither component can as yet be treated satisfactorily for most practical problems. As such, simpler empirical formulas that directly relate g_T (or equivalently, $\overline{C}$) to sediment and hydraulic parameters remain attractive and have often performed as well or better than more complicated formulas in practical predictions. Only two of the many such formulas will be discussed.

Engelund–Hansen Total Load Formula

The formula of Engelund and Hansen [1967] was developed along with their stage-discharge formula (see section 33.5 for the range of experimental parameters). The dimensionless total transport per unit width, Φ_T, is related to the grain Froude number, Fr_g, and $\Theta = R_h S_f/(s-1)d_{50}$, with characteristic grain size, d_g, by

$$\Phi_T = 0.05\,\mathrm{Fr}_g^2\,\Theta^{3/2} \tag{33.23}$$

Brownlie Total Load Formula

The Brownlie formula was originally stated in terms of the mean sediment transport (mass or weight) concentration, $\overline{C}$, as:

$$\overline{C} = 0.00712\,c_f\,[\mathrm{Fr}_g - (\mathrm{Fr}_g)_c]^{1.98} S_f^{0.66}\left(\frac{d_{50}}{R_h}\right)^{0.33} \tag{33.24}$$

where $c_f = 1$ for laboratory data and $c_f = 1.27$ for field data. $(\mathrm{Fr}_g)_c$ is the critical grain Froude number corresponding to the initiation of sediment motion given by Eq. (33.12). In terms of Φ_T and Θ, Eq. (33.24) can be expressed (assuming $R_h \approx H$) with rounding as:

$$\Phi_T = (0.00712/s)c_f\,[\mathrm{Fr}_g - (\mathrm{Fr}_g)_c]^2 \mathrm{Fr}_g[(s-1)\Theta]^{2/3} \tag{33.25}$$

Example 33.5 The total load formulas should also be applicable to bed load–dominated transport as in Example 33.4. In that case the Engelund–Hansen formula, with $\mathrm{Fr}_g^2 = 27.5$ and $\Theta = 0.40$, predicts $\Phi_T = 0.35$, corresponding to $g_T = 2.0$ N/s/m and $\overline{C} = 411$ ppm by weight. This is approximately twice the measured value. The Brownlie formula, with $(\mathrm{Fr}_g)_c = 1.8$ from Eq. (33.12) and $c_f = 1.27$, yields $\Phi_T = 0.16$, corresponding to $g_T = 0.91$ N/s/m and $\overline{C} = 189$ ppm by weight, which agrees well with the measured value of 191 ppm. This exceptionally close agreement is fortuitous, and predictions for other cases are likely to be substantially worse.

Measurement of Sediment Transport

In addition to, and contributing to, the difficulties in describing and predicting accurately sediment transport, total load measurements, particularly in the field, are associated with much uncertainty. Natural alluvial channels may exhibit a high degree of spatial and temporal nonuniformities that are not specifically considered in the "averaged" models discussed above. Standard methods of suspended load measurements in streams include the use of depth-integrating samplers that collect a continuous sample as they are lowered at a constant rate (depending on stream velocity) into the stream, and the use of point-integrating samplers that incorporate a valve mechanism to restrict sampling, if desired, to selected points or intervals in the water column. Such sampling assumes that the sampler is aligned with a dominant flow direction and that the velocity at the sampler intake is equal to the stream velocity. In the vicinity of a dune-covered bed these conditions cannot be fulfilled. The finite size of the suspended load samplers implies that they cannot measure the bed load discharge, which must therefore be measured with a different sampler or estimated with a bed load model. A bed load sampler, such as the U.S.G.S. Helley–Smith sampler, will necessarily interact with and possibly change the erodible bed. Calibration is therefore necessary—for example, in the laboratory using a sediment trap—but this may vary with several parameters, including the particular type of sampler used, the transport rate, and grain size [Hubbell, 1987]. Unless full-scale tests are performed, questions of model-prototype similitude also arise.

Summary: Expected Accuracy of Transport Formulas

The reliability of sediment transport formulas is relatively poor. Some of this poor performance may be attributed to measurement uncertainties. The best general sediment discharge formulas available have been found to predict values of g_T that are within one-half to twice the observed value for only about 75% of cases [Brownlie, 1981; van Rijn, 1984a,b; Chang, 1988]. Circumspection is advised in basing engineering decisions on such formulas, especially when they are embedded in sophisticated computer models of long-term deposition or erosion. Where feasible, site-specific field data should be exploited and used to complement model predictions.

33.7 Special Topics

The preceding sections have been limited to the simplest sediment-transport problems involving steady uniform flow. The following discussion deals briefly with more specialized and complex problems.

Local Scour

Hydraulic structures, such as bridge piers or abutments, that obstruct or otherwise change the flow pattern in the vicinity of the structure may cause localized erosion or scour. Changes in flow characteristics lead to changes in sediment transport capacity and hence to a local disequilibrium between actual sediment load and the capacity of the flow to transport sediment. A new equilibrium may eventually be restored as hydraulic conditions are adjusted through scour. *Clear-water scour* occurs when there is effectively zero sediment transport upstream of the obstruction, that is, $\mathrm{Fr}_g < (\mathrm{Fr}_g)_c$ upstream. *Live-bed scour* occurs when there would be general sediment transport even in the absence of the local obstruction, that is, $\mathrm{Fr}_g > (\mathrm{Fr}_g)_c$ upstream. Additional difficulties in treating local scour are flow nonuniformity and unsteadiness. The many different types and geometries of hydraulic structures lead to a wide variety of scour problems, which precludes any detailed unified treatment. Design for local scour requires many considerations and the results that follow should be considered only as a part of the design process.

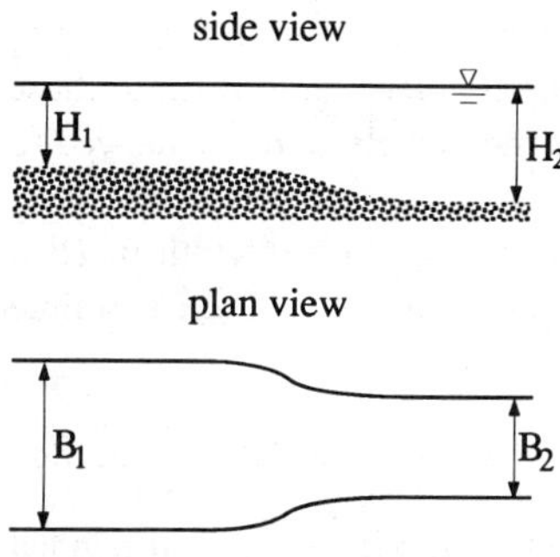

FIGURE 33.6 Definition sketch of channel constriction causing local scour.

Empirical formulas have been developed for special scour problems; only two are presented here. Consider a channel contraction sufficiently long that uniform flow is established in the contracted section, which is uniformly scoured (Fig. 33.6). The entire discharge is assumed to flow through the main and the contracted channels. Application of conservation of water and sediment (assuming a simple transport formula of power-law form, $g_T \sim U^m$) results in

$$\frac{H_1}{H_2} = \left(\frac{B_2}{B_1}\right)^{\alpha} \qquad (33.26)$$

where the subscripts 1 and 2 indicate the contracted (2) or the main (1) channels, H the flow depth, and B the channel width. The exponent α varies from 0.64 to 0.86, increasing with τ_c/τ_1, where τ_c is the critical shear stress for the bed material and τ_1 is total bed shear stress in the main channel. A value of $\alpha = 0.64$—corresponding to $\tau_c/\tau_1 \ll 1$, that is, significant transport in the main channel—is often used.

Scour around bridge piers has been much studied in the laboratory but field studies have been hampered by inadequate instrumentation and measurement procedures. A recent design formula for live-bed scour based on regression analysis of field measurements [Froehlich, 1988] typifies such empirical relationships:

$$\frac{y_s}{b} = 0.32\,K\left(\frac{b'}{b}\right)^{0.62}\left(\frac{H_1}{b}\right)^{0.46} Fr_1^{0.20}\left(\frac{b}{d_{50}}\right)^{0.08} + 1 \qquad (33.27)$$

where K is a coefficient depending on pier geometry with a value ranging from 0.7 for a sharp-nosed pier to 1 for a circular pier to 1.3 for a square-nosed pier and $Fr_1 = U_1/\sqrt{gH_1}$ is the upstream Froude number. The quantities b and b' are the actual pier width and the projected pier width (relative to the main flow direction), and H_1 is the upstream flow depth, as shown in Fig. 33.7. The conditions on which Eq. (33.27) is based ranged from $d_{50} = 0.008$ mm to 90 mm, angle of attack of $\theta = 0$–$35°$, pier width/pier length of 1 to 12, and upstream Froude numbers $Fr_1 = .07$–0.8. Eq. (33.27) gives a conservative prediction of the scour depth, but does not account for nonuniform sediment size distribution or for bed forms because of lack of data.

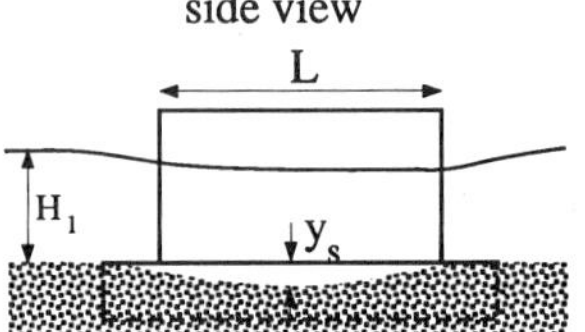

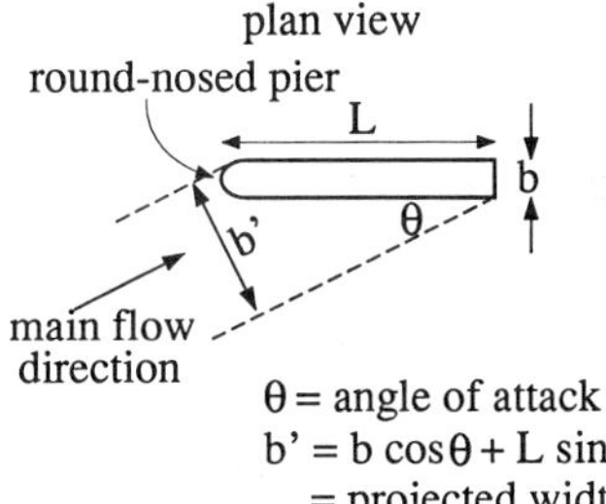

θ = angle of attack
b' = b cosθ + L sinθ
= projected width

FIGURE 33.7 Definition sketch of bridge pier causing local scour.

Unsteady Aspects

Many problems in channels involve nonuniform flows and slow long-term changes, such as **aggradation** (an increase in bed elevation due to net deposition) or degradation (a decrease in bed elevation due to net erosion). The problem is formulated generally in terms of three (differential) balance equations: conservation of mass of water, of momentum (or energy), and of sediment. For gradually varied flows the first two equations are identical in form to those encountered in fixed-bed problems (see Chapter 29), except that the bed elevation is allowed to change with time.

A control volume analysis of a channel reach of cross-sectional area A and bed width B_b (Fig. 33.8) shows that conservation (continuity) of sediment over a small reach of length Δx in a time interval Δt requires

$$\Delta[(1-p)z]B_b\Delta x + \Delta[\langle C\rangle]A\Delta x + (G_T|_{x+\Delta x} - G_T|_x)\Delta t = S_s\Delta t\Delta x \qquad (33.28)$$

where p is the porosity of the bed, z the bed elevation, t the time variable, $\langle C\rangle$ the suspended sediment concentration averaged over the cross-sectional area, G_T the total sediment discharge, and S_s is included as a possible external sediment source strength per unit length. The first term represents the change over time in the bulk volume of sediment in the bed due to net deposition or

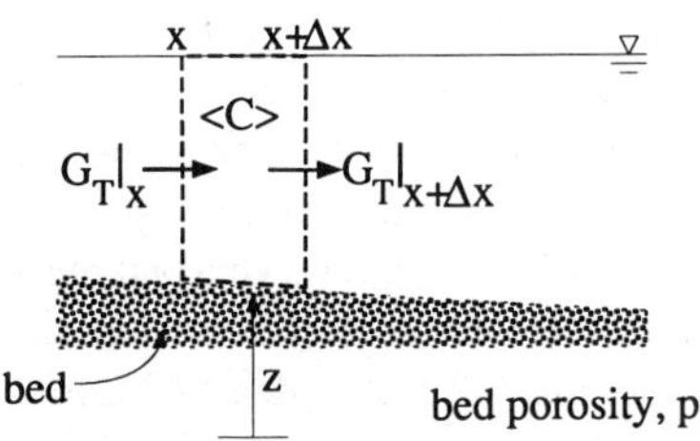

FIGURE 33.8 Sketch of control volume used in unsteady analysis.

erosion (bed storage); the second term represents the change over time in total volume of suspended sediment in the water column (water column storage); the third term stems from differences in sediment discharge between the channel cross sections bounding the control volume; and the fourth term allows for distributed sediment sources. The second term is often assumed negligible, so that in its differential form (dividing through by $\Delta x \Delta t$ and taking the limit as $\Delta x \to 0$, $\Delta t \to 0$) Eq. (33.28) is written as

$$\frac{\partial[(1-p)zB_b]}{\partial t} + \frac{\partial G_T}{\partial x} = S_s \tag{33.29}$$

which is sometimes referred to as the Exner equation. A total load computation as in section 33.6 is performed to determine G_T. This assumes implicitly that a quasi-equilibrium has been established in which the sediment discharge at any section is equal to the sediment transport capacity as specified by conventional total load computations. Thus, the quality of the unsteady model depends not only on the quasi-equilibrium assumption but also on the quality of the predictions of available total load formulas.

Numerical methods are used to solve Eq. (33.29) simultaneously with the flow equations (water continuity and the momentum/energy equations). In practice numerical models often solve the flow equations first and then the sediment continuity equation, under the implicit assumption that changes in bed elevations occur much more slowly than changes in water-surface elevation. Of the many unsteady alluvial-river models described in the literature, HEC-6 for scour and deposition in rivers and reservoirs may be mentioned as a member of the well-known HEC series of channel models [Hydrologic Engineering Center, 1991]. An evaluation [Committee on Hydrodynamic Models for Flood Insurance Studies, 1983] of several models, including HEC-6, noted the following general deficiencies: unreliable formulation and/or inadequate understanding of sediment-transport capacity, of flow resistance, of **armoring** (see below), and of bank erosion. In spite of the intervening years this evaluation may still be taken as a cautionary note in using such models.

Effects of Nonuniform Size Distribution

Natural sediments exhibit a size distribution (also termed *gradation*), and, since the grain diameter profoundly influences transport, the effects of size distribution are likely substantial. The crudest models of such effects incorporate distribution parameters such as the geometric standard deviation, σ_g, in empirical formulas, for example, in the Brownlie formulas. An alternative approach more appropriate for computer modeling divides the distribution into a finite number of discrete size classes. Each size class is characterized by a single grain diameter, and results such as the Shields curve or the Rouse equation are applied to each separate size class, where they are presumably more valid. Total transport is then determined by a summation of the transport in each size class.

The heterogeneous bed material—which constitutes a source or sink of grains of different size classes—must be taken into account. Conventional bed load or total load transport equations or even initiation of motion criteria may not necessarily apply to individual size classes in a mixture. The transport or entrainment into suspension of one size class may influence transport or entrainment of other size classes, so individual size classes may not be treated independently of each other. This is often handled by the use of empirical "hiding" coefficients. Finer bed material may under erosive conditions be preferentially entrained into the flow, with the result that the remaining bed material becomes coarser. This will reduce the rate of erosion relative to the case in which the bed consists of uniformly sized fine material. If the available fine material is eventually depleted, suspended load transport will be reduced or in the limit entirely suppressed. Eventually, a coarse layer termed the *armor layer* may develop, which consists of material that is not erodible under the given flow condition and protects or "armors" the finer material below it from erosion, thereby substantially reducing sediment transport and local scour. Armoring may also have consequences for flow resistance, since size distribution characteristics of the bed will vary with varying bed shear stress and hence affect bed roughness and bed forms. In this way episodic high-transport events such as floods may have an enduring impact on sediment transport as well as flow depths. Various detailed numerical models of the armoring process have been developed, and the reader is directed to the literature for further information [Borah *et al.*, 1982; Sutherland, 1987; Andrews and Parker, 1987; Holly and Rahuel, 1990a,b; Hydrological Engineering Center, 1991].

Gravel-Bed Streams

Channels in which the bed material consists primarily of coarse material in the gravel and larger range are typically situated in upland mountain regions with high bed slopes ($S > 0.005$), in contrast to sand-bed channels, which are found on flatter slopes of lower-lying regions. The same basic concepts summarized in previous sections apply also to gravel-bed streams, but the possibly very wide range of grain sizes introduces particular difficulties. Bed forms such as dunes play less of a role, and so grain resistance can often be assumed dominant; hence an upper-regime stage-discharge relationship can be applied. The effects of large-scale roughness elements such as cobbles and boulders that may even protrude through the water surface may however not be well described by formulas based primarily on data from sand-bed channels. Instead of a gradually varying bed elevation, riffle-pool (or step-pool) sequences of alternating shallow and deep flow regions may occur. The wide size range results in transport events that may be highly nonuniform across the stream and highly unsteady in the sense of being dominated by episodic events. Armoring may also need to be considered. The coarse grain sizes increase the relative importance of bed load transport. The highly nonuniform and unsteady nature of the transport hinders reliable field measurements. Much debate has surrounded the topic of appropriate sampling of the bed surface material to characterize the grain-size distribution. The traditional grid method of Wolman [1954] draws a regular grid over the bed of the chosen reach, with the gravel (cobble or boulder) found at each gridpoint being included in the sample.

A friction law proposed specifically for mountain streams is that of Bathurst [1985] based on data from English streams (60 mm $< d_{50} <$ 343 mm, $0.0045 < S < 0.037$, 0.3 m^3/s $< Q <$ 195 m^3/s):

$$\sqrt{\frac{8}{f}} = 5.62 \log_{10} \frac{H}{d_{84}} + 4 \tag{33.30}$$

with a reported uncertainty of $\pm 30\%$. An earlier formula from Limerinos [1970] is identical in form except that R_h is used instead of the depth H and Manning's n is sought rather than f:

$$\frac{K}{\sqrt{g}} \frac{R_h^{1/6}}{n} = \sqrt{\frac{8}{f}} = 5.7 \log_{10} \frac{R_h}{d_{84}} + 3.4 \tag{33.31}$$

K is the dimensional constant associated with Manning's equation ($K = 1$ for metric units, and $K = 1.49$ for English units). Using laboratory and field data, Bathurst [1987] assessed various criteria for the initiation of motion and bed load discharge formulas [including the Meyer–Peter–Muller formula, Eq. (33.18)]. He recommended a modified Schoklisch formula for larger rivers ($Q > 50$ m^3/s) where sediment supply is not a constraint:

$$q_{sb} = \frac{2.5}{s} S^{3/2}(q - q_c) \tag{33.32}$$

where the the critical unit-width discharge q_c is given by

$$q_c = 0.21 \frac{\sqrt{g d_{16}^3}}{S^{1.12}} \tag{33.33}$$

Here, q_{sb} is the volumetric unit-width bed load discharge, and the units are metric in both equations. These gravel-bed formulas, although representative, are not necessarily the best for all problems; they should be applied with caution and a dose of skepticism.

Defining Terms

Aggradation: Long-term increase in bed level over an extended reach due to sediment deposition.

Armoring: A phenomenon in which a layer of coarser particles that are nonerodible under the given flow condition protects the underlying layer of finer erodible particles.

Bed forms: Features on an erodible channel bed that depart from a plane bed, such as dunes or ripples.

Bed load: The part of the total sediment discharge that is transported primarily very close to the bed.

Critical shear stress: The bed shear stress above which general sediment transport is said to begin.

Critical velocity: The mean velocity above which general sediment transport is said to begin.

Local scour: Erosion occurring over a region of limited extent due to local flow conditions, such as may be caused by the presence of hydraulic structures.

Sediment discharge: The downstream mass or weight flux of sediment.

Suspended load: The part of the total sediment discharge that is transported primarily in suspension.

References

Andrews, E. D. and Parker, G. 1987. Formation of a coarse surface layer as the response to gravel mobility. In *Sediment Transport in Gravel-Bed Rivers,* eds. C. R. Thorne, J. C. Bathurst, and R. D. Hey. Wiley-Interscience, Chichester, UK.

Bathurst, J. C. 1985. Flow resistance estimation in mountain rivers. *J. Hydraulic Eng.* 111(4):625–643.

Bathurst, J. C. 1987. Bed load discharge equations for steep mountain rivers. In *Sediment Transport in Gravel-Bed Rivers,* eds. C. R. Thorne, J. C. Bathurst, and R. D. Hey. Wiley-Interscience, Chichester, UK.

Borah, D. K., Alonso, C. V., and Prasad, S. N. 1982. Routing graded sediment in streams: Formations. *J. Hydraulics Div., ASCE.* 108(HY12):1486–1503.

Brownlie, W. R. 1981. *Prediction of Flow Depth and Sediment Discharge in Open Channels.* Rept. KH-R-43A, W. M. Keck Lab. Hydraulics and Water Resources, Calif. Inst. Tech., Pasadena, CA.

Brownlie, W. R. 1983. Flow depth in sand-bed channels. *J. Hydraulic Eng.* 109(7):959–990.

Chang, H. H. 1988. *Fluvial Processes in River Engineering.* John Wiley & Sons, New York.

Committee on Hydrodynamic Models for Flood Insurance Studies. 1983. *An Evaluation of Flood-Level Prediction Using Alluvial-River Models.* National Academy Press, Washington, D.C.

Dawdy, D. R. 1961. Depth-discharge relations of alluvial streams. *Water-Supply Paper 1498-C.* U.S. Geological Survey, Washington, D.C.

Engelund, F. and Hansen, E. 1967. *A Monograph on Sediment Transport in Alluvial Streams.* Tekniske Vorlag, Copenhagen, Denmark.

Froehlich, D. C. 1988. Analysis of On-Site Measurements of Scour at Piers. *Proc. 1988 National Conf. Hydraulic Engineering.* Colorado Springs, CO. Aug. 8-12, p. 534–539.

Holly, F. M., Jr., and Rahuel, J.-L. 1990a. New numerical/physical framework for mobile-bed modeling, part 1: Numerical and physical principles. *J. Hydraulic Research.* 28(4):401–416.

Holly, F. M., Jr., and Rahuel, J.-L. 1990b. New numerical/physical framework for mobile-bed modeling, part 1: Test applications. *J. Hydraulic Research.* 28(5):545–563.

Hubbell, D. W. 1987. Bed load sampling and analysis. In *Sediment Transport in Gravel-Bed Rivers,* eds. C. R. Thorne, J. C. Bathurst, and R. D. Hey. Wiley-Interscience, Chichester, UK.

Hydrologic Engineering Center. 1991. *HEC-6: Scour and Deposition in Rivers and Reservoir.* U.S. Army Corps of Engineers, Davis, CA.

Limerinos, J. T. 1970. *Determination of the Manning coefficient from Measured Bed Roughness in Natural Channels.* Water-Supply Paper 1989-B. U.S. Geological Survey, Washington, D.C.

Lyn, D. A. 1993. Turbulence measurements in open-channel flows over artificial bed forms. *J. Hydraulic Eng.* 119(3):306–326.

Mehta, A. J., Hayter, E. J., Parker, W. R., Krone, R. B., and Teeter, A. M. 1989. Cohesive sediment transport. I: Process description. *J. Hydraulic Eng.* 115(8):1076–1093.

Mehta, A. J., McAnally, W. H., Hayter, E. J., Teeter, A. M., Schoellhammer, D., Heltzel, S. B., and Carey, W. P. 1989. Cohesive sediment transport. II: Application. *J. Hydraulic Eng.* 115(8): 1094–1112.

Neill, C. R. 1967. Mean velocity criterion for scour of coarse uniform bed material. *Proc. 12th Congress Int. Assoc. Hydraulic Research.* Fort Collins, CO, p. 46–54.

Raudkivi, A. J. 1990. *Loose Boundary Hydraulics,* 3rd ed. Pergamon Press, Tarrytown, NY.

Sutherland, A. J. 1987. Static armour layers by selective erosion. In *Sediment Transport in Gravel-Bed Rivers,* eds. C. R. Thorne, J. C. Bathurst, and R. D. Hey. Wiley-Interscience, Chichester, UK.

Taylor, B. D. and Vanoni, V. A. 1972. Temperature effects in low-transport, flat-bed flows. *J. Hydraulics Div., ASCE.* 97(HY8):1427–1445.

Vanoni, V. A., ed. 1975. *Sedimentation Engineering.* ASCE Manual No. 54. ASCE, New York.

van Rijn, L. 1984a. Sediment transport, part 1: Bed load transport. *J. Hydraulic Eng.* 110(10): 1431–1456.

van Rijn, L. 1984b. Sediment transport, part 2: Suspended load transport. *J. Hydraulic Eng.* 110(11):1613–1641.

Wolman, M. G. 1954. *The Natural Channel of Brandywine Creek, Pennsylvania.* Prof. Paper 271. U.S. Geological Survey, Washington, D.C.

For Further Information

A standard comprehensive reference dealing in detail with many aspects of sediment transport engineering is *Sedimentation Engineering,* ed. V. A. Vanoni [1975], ASCE Manual No. 54. More recent books include *Fluvial Processes in River Engineering,* H. H. Chang [1988], Prentice-Hall; *Loose Boundary Hydraulics,* A. J. Raudkivi [1990], 3rd ed., Pergamon Press; and *Sediment Transport Technology,* D. B. Simons and F. Sentürk [1992], Water Resources Publications. The topic of scour is discussed extensively in *Scouring,* H. N. C. Breusers and A. J. Raudkivi [1991], Balkema. Hydraulic considerations in the design and siting of bridges, including scour, are dealt with in *Guide to Bridge Hydraulics,* C. R. Neill, ed. [1967], Univ. Toronto Press. Specific problems associated with gravel-bed streams including armoring are addressed in the proceedings of a 1985 workshop, *Sediment Transport in Gravel-Bed Rivers,* eds. C. R. Thorne, J. C. Bathurst, and R. D. Hey [1987], John Wiley & Sons. Information on settling basins and sediment traps can be found in *Sedimentation: Exclusion and Removal of Sediment from Diverted Water,* A. J. Raudkivi [1993], Balkema. Practical advice regarding field measurements of sediment transport, including site selection and sampling methods, is given in *Field Methods for Measurement of Fluvial Sediment,* H. P. Guy and V. W. Norman [1970] in the series *Techniques of Water-Resources Investigations of the United States Geological Survey,* book 3, chapter C2.

34

Coastal Engineering

William L. Wood
Purdue University

Guy A. Meadows
University of Michigan

34.1 Wave Mechanics

Waves on the surface of a natural body of open water are the result of disturbing forces that create a deformation which is restored to equilibrium by gravitational and surface tension forces. Surface waves are characterized by their height, length, and the water depth over which they are traveling. Figure 34.1 shows a two-dimensional sketch of a sinusoidal surface wave propagating in the x-direction. The **wave height,** H, is the vertical distance between its crest and leading trough. Wavelength, L, is the horizontal distance between any two corresponding points on successive waves, and wave period is the time required for two successive crests or troughs to pass a given point. The **celerity** of a wave, C, is the speed of propagation of the wave form (phase speed), defined as $C = L/T$. Most ocean waves are progressive, that is, their wave form appears to travel at celerity C relative to a background. Standing waves remain stationary relative to a background, occur from the interaction of progressive waves traveling in opposite directions, and are often observed near reflective coastal features. Progressive deep ocean waves are oscillatory, meaning that the water particles making up the wave do not exhibit a net motion in the direction of wave propagation. However, waves entering shallow water begin to show a net displacement of water in the direction of propagation and are classified as translational. The equilibrium position used to reference surface wave motion, still water level (SWL), is $z = 0$, and the bottom is located at $z = -d$ (Fig. 34.1).

The free surface water elevation, η, for a natural water wave propagating over an irregular, permeable bottom may appear quite complex. However, by assuming that viscous effects are negligible (concentrated near the bottom), flow is irrotational and incompressible, and wave height

0-8493-8953-4/95/$0.00 + $.50

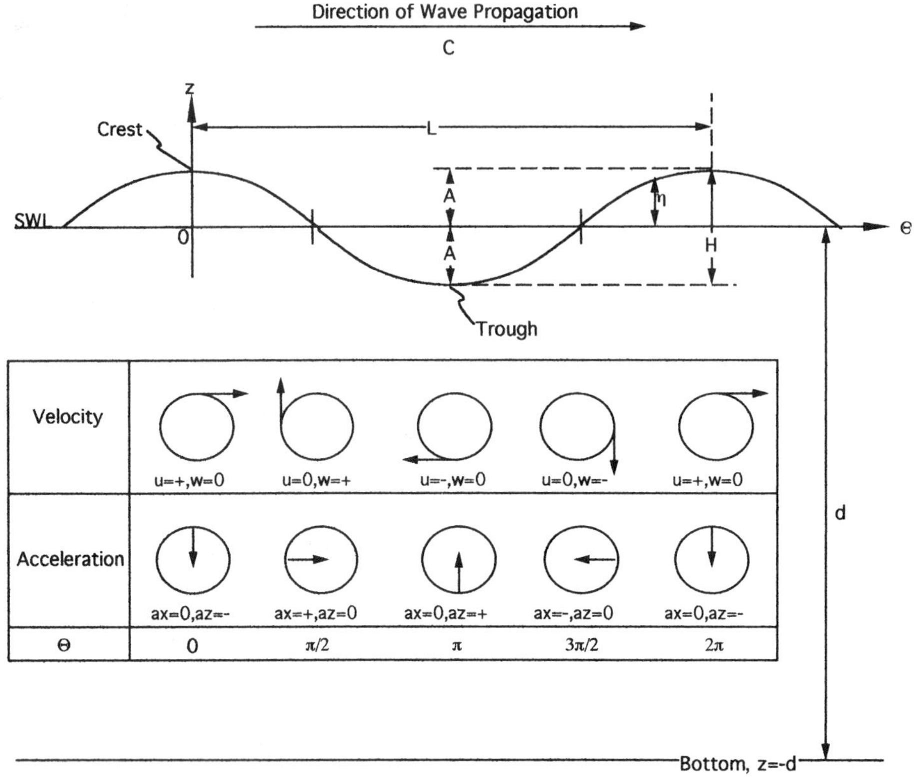

FIGURE 34.1 Definition sketch of free surface wave parameters for a linear progressive wave. Shown also are the fluid particle velocities and accelerations associated with each portion of the wave.

is small compared to wavelength, a remarkably simple solution can be obtained for the surface wave boundary value problem. This simplification, referred to as linear, small-amplitude wave theory, is extremely accurate and easy to use in many coastal engineering applications. Furthermore, the linear nature of this formulation allows for the free surface to be represented by superposition of sinusoids of different amplitudes and frequencies, which facilitates the application of Fourier decomposition and associated analysis techniques.

Progressive, Small-Amplitude Waves—Properties

The equation for the free surface displacement of a progressive wave is

$$\eta = A\cos(kx - \sigma t) \tag{34.1}$$

where amplitude is $A = H/2$, wave number is $k = 2\pi/L$, and wave frequency is $\sigma = 2\pi/T$. The expression relating individual wave properties and wave depth, d, to the propagation behavior of these waves is the dispersion relation,

$$\sigma^2 = gk\tanh kd \tag{34.2}$$

where g is the acceleration of gravity.

From Eq. (34.2) and the definition of celerity (C) it can be shown that

$$C = \frac{\sigma}{k} = \frac{gT}{2\pi} \tanh kd \tag{34.3}$$

and

$$L = \frac{gT^2}{2\pi} \tanh kd \tag{34.4}$$

The hyperbolic function $\tanh kd$ approaches useful simplifying limits of 1 for large values of kd (deep water) and kd for small values of kd (shallow water). Applying these limits to (34.3) and (34.4) results in expressions for deep water of

$$C_o = \frac{gT}{2\pi} = 5.12T \qquad \text{(English units, ft/s)}$$

or

$$C_o = 1.56T \qquad \text{(SI units, m/s)}$$

and

$$L_o = \frac{gT^2}{2\pi} = 5.12T^2 \qquad \text{(English units, ft)} \tag{34.5}$$

or

$$L_o = 1.56T^2 \qquad \text{(SI units, m)}$$

A similar application for shallow water results in

$$C = \sqrt{gd} \tag{34.6}$$

which shows that wave speed in shallow water is dependent only on water depth. The normal limits for deep and shallow water are $kd > \pi$ and $kd < \pi/10$ ($d/L > \frac{1}{2}$ and $d/L < \frac{1}{20}$), respectively, although modification of these limits may be justified for specific applications. The region between these two limits ($\pi/10 < kd < \pi$) is defined as intermediate depth water and requires use of the full Eqs. (34.3) and (34.4).

Some useful functions for calculating wave properties at any water depth, from deep water wave properties, are

$$\frac{C}{C_o} = \frac{L}{L_o} = \tanh \frac{2\pi d}{L} \tag{34.7}$$

Values of d/L can be calculated as a function of d/L_o by successive approximations using

$$\frac{d}{L} \tanh \frac{2\pi d}{L} = \frac{d}{L_o} \tag{34.8}$$

The term d/L has been tabulated as a function of d/L_o by Wiegel [1954] and is presented, along with many other useful functions of d/L, in [U.S. Army Corps of Engineers, 1984, Appendix C].

Particle Motions

The horizontal component of particle velocity beneath a wave is

$$u = \frac{H}{2}\sigma\frac{\cosh k(d+z)}{\sinh kd}\cos(kx - \sigma t) \tag{34.9}$$

The corresponding acceleration is

$$a_x = \frac{\partial u}{\partial t} = \frac{H}{2}\sigma^2\frac{\cosh k(d+z)}{\sinh kd}\sin(kx - \sigma t) \tag{34.10}$$

The vertical particle velocity and acceleration are, respectively,

$$w = \frac{H}{2}\sigma\frac{\sinh k(d+z)}{\sinh kd}\sin(kx - \sigma t) \tag{34.11}$$

and

$$a_z = \frac{\partial w}{\partial t} = -\frac{H}{2}\sigma^2\frac{\sinh k(d+z)}{\sinh kd}\cos(kx - \sigma t) \tag{34.12}$$

It can be seen from Eqs. (34.9) and (34.11) that the horizontal and vertical particle velocities are 90° out of phase at any position along the wave profile. Extreme values of horizontal velocity occur in the crest (+, in the direction of wave propagation) and trough (−, in the direction opposite to the direction of wave propagation) while extreme vertical velocities occur midway between the crest and trough, where water displacement is zero. The u and w velocity components are each at a minimum at the bottom and both increase as distance upward in the water column increases. Maximum vertical acceleration corresponds to maximum horizontal velocity and maximum horizontal acceleration corresponds to maximum vertical velocity. Figure 34.1 provides a graphic summary of these relationships.

The particle displacements can be obtained by integrating the velocity with respect to time and simplified by using the dispersion relationship [Eq. (34.2)] to give a horizontal displacement

$$\xi = -\frac{H}{2}\frac{\cosh k(d+z_o)}{\sinh kd}\sin(kx_o - \sigma t) \tag{34.13}$$

and vertical displacement

$$\zeta = \frac{H}{2}\frac{\sinh k(d+z_o)}{\sinh kd}\cos(kx_o - \sigma t) \tag{34.14}$$

where (x_o, z_o) is the mean position of an individual particle. It can be shown by squaring and adding the horizontal and vertical displacements that the general form of a water particle trajectory beneath a wave is elliptical. In deep water particle paths are circular, and in shallow water they are highly elliptical, as shown in Fig. 34.2.

Pressure Field

The pressure distribution beneath a progressive water wave is given by the following form of the Bernoulli equation:

$$p = -\rho g z + \rho g \eta K_p(z) \tag{34.15}$$

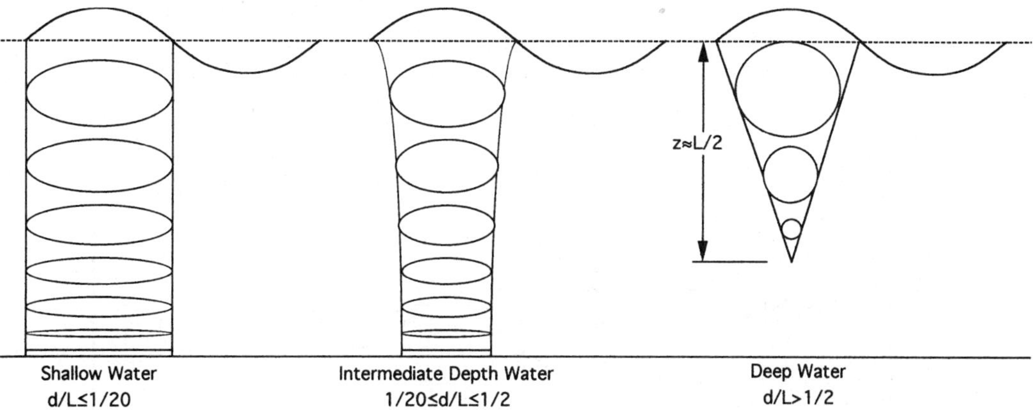

FIGURE 34.2 Fluid particle paths for linear progressive waves in different relative water depths.

where ρ is fluid density and K_p, the pressure response coefficient, is

$$K_p = \frac{\cosh k(d+z)}{\cosh kd} \tag{34.16}$$

which will always be less than 1, below mean still water level. The first term in Eq. (34.15) is the *hydrostatic pressure* and the second is the *dynamic pressure* term. This dynamic pressure term accounts for two factors that influence pressure: the free surface displacement η and the vertical component of acceleration.

A frequently used method for measuring waves at the coast is to record pressure fluctuations from a bottom-mounted pressure gage. Isolating the dynamic pressure (P_D) from the recorded signal by subtracting out the hydrostatic pressure gives the relative free surface displacement:

$$\eta = \frac{P_D}{\rho g K_p(-d)} \tag{34.17}$$

where $K_p(-d) = 1/\cosh kd$. It is necessary, therefore, when determining wave height from pressure records to apply the dispersion relationship [Eq. (34.2)] to obtain K_p from the frequency of the measured waves. It is important to note that K_p for short-period waves is very small at the bottom $(-d)$, which means that very-short-period waves may not be measured by a pressure gage.

A summary of the formulations for calculating linear wave theory wave characteristics in deep, intermediate, and shallow water is presented in Table 34.1. Comprehensive presentations of linear, small-amplitude theory can be found in Wiegel [1954], U.S. Army Corps of Engineers [1984], and Dean and Dalrymple [1991].

Wave Energy

Progressive surface water waves possess potential energy from the free surface displacement and kinetic energy from the water particle motions. From linear wave theory it can be shown that the average potential energy per unit surface area for a free surface sinusoidal displacement, restored by gravity, is

$$\overline{E}_p = \frac{\rho g H^2}{16} \tag{34.18}$$

Table 34.1 Summary of Linear Wave Theory, Wave Characteristics, $\theta = (kx - \sigma t)$

Relative Depth	Shallow Water, $\frac{d}{L} < \frac{1}{20}$	Transitional Water, $\frac{1}{20} < \frac{d}{L} < \frac{1}{2}$	Deep Water, $\frac{d}{L} > \frac{1}{2}$
Wave profile	Same as for transitional water	$\eta = \frac{H}{2}\cos[kx - \sigma t] = \frac{H}{2}\cos\theta$	Same as for transitional water
Wave celerity	$C = \frac{L}{T} = \sqrt{gd}$	$C = \frac{L}{T} = \frac{gT}{2\pi}\tanh(kd)$	$C = C_o = \frac{L}{T} = \frac{gT}{2\pi}$
Wavelength	$L = T\sqrt{gd} = CT$	$L = \frac{gT^2}{2\pi}\tanh(kd)$	$L = L_o = \frac{gT^2}{2\pi} = C_o T$
Group velocity	$C_g = C = \sqrt{gd}$	$C_g = nC = \frac{1}{2}\left[1 + \frac{2kd}{\sinh(2kd)}\right]\cdot C$	$C_g = \frac{1}{2}C = \frac{gT}{4\pi}$
Water particle velocity			
Horizontal	$u = \frac{H}{2}\sqrt{\frac{g}{d}}\cos\theta$	$u = \frac{H}{2}\sigma\frac{\cosh k(z+d)}{\sinh kd}\cos\theta$	$u = \frac{\pi H}{T}e^{kz}\cos\theta$
Vertical	$w = \frac{H\pi}{T}\left(1 + \frac{z}{d}\right)\sin\theta$	$w = \frac{H}{2}\sigma\frac{\sinh k(z+d)}{\sinh kd}\sin\theta$	$w = \frac{\pi H}{T}e^{kz}\sin\theta$
Water particle accelerations			
Horizontal	$a_x = \frac{H\pi}{T}\sqrt{\frac{g}{d}}\sin\theta$	$a_x = \frac{H}{2}\sigma^2\frac{\cosh k(z+d)}{\sinh kd}\sin\theta$	$a_x = 2H\left(\frac{\pi}{T}\right)^2 e^{kz}\sin\theta$
Vertical	$a_z = -2H\left(\frac{\pi}{T}\right)^2\left(1 + \frac{z}{d}\right)\cos\theta$	$a_z = -\frac{H}{2}\sigma^2\frac{\sinh k(z+d)}{\sinh kd}\cos\theta$	$a_z = -2H\left(\frac{\pi}{T}\right)^2 e^{kz}\cos\theta$
Water particle displacements			
Horizontal	$\xi = -\frac{HT}{4\pi}\sqrt{\frac{g}{d}}\sin\theta$	$\xi = -\frac{H}{2}\frac{\cosh k(z+d)}{\sinh kd}\sin\theta$	$\xi = -\frac{H}{2}e^{kz}\sin\theta$
Vertical	$\zeta = \frac{H}{T}\left(1 + \frac{z}{d}\right)\cos\theta$	$\zeta = \frac{H}{2}\frac{\sinh k(z+d)}{\sinh kd}\cos\theta$	$\zeta = \frac{H}{2}e^{kz}\cos\theta$
Subsurface pressure	$p = \rho g(\eta - z)$	$p = \rho g\eta\frac{\cosh k(z+d)}{\cosh kd} - \rho gz$	$p = \rho g\eta e^{kz} - \rho gz$

Likewise, the average kinetic energy per unit surface area is

$$\overline{E}_k = \frac{\rho g H^2}{16} \tag{34.19}$$

and the total average energy per unit surface area is

$$\overline{E} = \overline{E}_p + \overline{E}_k = \frac{\rho g H^2}{8} \tag{34.20}$$

The unit surface area considered is a unit width times the wavelength L so that the total energy per unit width is

$$E_T = \frac{1}{8}\rho g H^2 L \tag{34.21}$$

The total energy per unit surface area in a linear progressive wave is always equipartitioned as one-half potential energy and one-half kinetic energy.

Energy flux is the rate of energy transfer across the sea surface in the direction of wave propagation. The average energy flux per wave is

$$F_E = ECn \tag{34.22}$$

where

$$n = \frac{C_g}{C} = \frac{1}{2}\left(1 + \frac{2kd}{\sinh 2kd}\right) \tag{34.23}$$

and C_g is the **group speed,** defined as the speed of energy propagation. In deep water $n = \frac{1}{2}$ and in shallow water $n = 1$, indicating that energy in deep water travels at half the speed of the wave while in shallow water energy propagates at the same speed as the wave.

Wave Shoaling

Waves entering shallow water conserve period and, with the exception of minor losses, up to breaking, conserve energy. However, wave celerity decreases as a function of depth and wave length shortens correspondingly. Therefore, the easiest conservative quantity to follow is the energy flux [given in (Eq. 34.22)], which remains constant as a wave shoals. Equating energy flux in deep water (H_o, C_o) to energy flux at any shallow water location (H_x, C_x) results in the general shoaling relation

$$\frac{H_x}{H_o} = \left(\frac{1}{2n}\frac{C_o}{C_x}\right)^{1/2} \tag{34.24}$$

where n is calculated from Eq. (34.23) and C_o/C_x can be obtained from Eq. (34.7). Therefore, by knowing the deep water wave height and period (H_o, T_o) and the bathymetry of a coastal region, the shoaling wave characteristics (H_x, C_x, L_x) can be calculated at any point, x, prior to breaking. A limitation to Eq. (34.24) is that it does not directly incorporate the effect of deep water angle of approach to the coast.

Wave Refraction

It can be shown that a deep water wave approaching a coast at an angle α_o and passing over a coastal bathymetry characterized by straight, parallel contours refracts according to Snell's law:

$$\frac{\sin\alpha_o}{C_o} = \frac{\sin\alpha}{C} \tag{34.25}$$

Since waves in shallow water slow down as depth decreases, application of Snell's law to a plane parallel bathymetry indicates that wave crests tend to turn to align with the bathymetric contours. Unfortunately, most offshore bathymetry is both irregular and variable along a coast, and the applicable **refraction** techniques involve a nonlinear partial differential equation, which can be solved approximately by various computer techniques [Noda, 1974; RCPWAVE; and others]. However, there are more easily applied ray tracing methods that use Snell's law applied to idealized bathymetry (bathymetry that has been "smoothed" to eliminate abrupt turns and steep gradients). The U.S. Army Corps of Engineers distributes an easy-to-use set of PC computational programs, the Automated Coastal Engineering System (ACES), including a Snell's law ray tracing program.

Considering two or more **wave rays** propagating shoreward over plane parallel bathymetry (Fig. 34.3), it is possible to have the rays either converge or diverge. Under these conditions, the energy per unit area may increase (convergence) or decrease (divergence) as a function of the perpendicular distance of separation between wave rays b_o and b_x. Using the geometric relationships shown in Fig. 34.3, Eq. (34.24) is modified to account for convergence and divergence of wave rays as

$$\frac{H_x}{H_o} = \left(\frac{1}{2n}\frac{C_o}{C_x}\right)^{1/2}\left(\frac{b_o}{b_x}\right)^{1/2} \tag{34.26}$$

also written as

$$H_x = H_o K_s K_R \tag{34.27}$$

where K_s is the shoaling coefficient and K_R is the refraction coefficient. This expression is equally valid between any two points along a wave ray in shallow water.

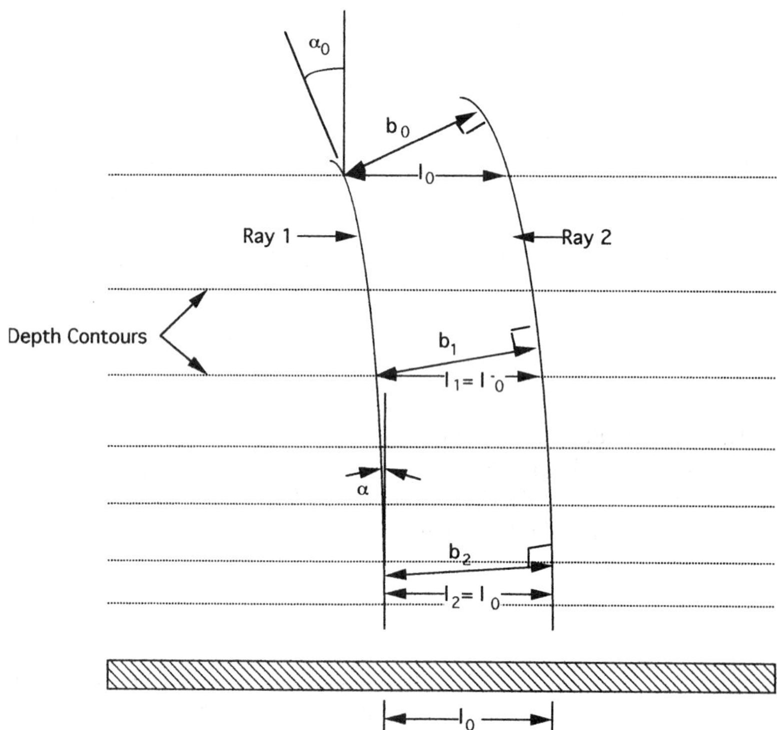

FIGURE 34.3 Definition sketch for wave rays refracting over idealized plane parallel bathymetry.

Wave Diffraction

Wave diffraction is a process by which energy is transferred along the crest of a wave from an area of high energy density to an area of low energy density. There are two important coastal engineering applications of diffraction. First, as wave rays converge and diverge in response to natural changes in bathymetry, the K_R term in Eq. (34.27) will increase and decrease, respectively. As a result, energy will move along the wave crest from areas of convergence to areas of divergence. It is, therefore, necessary to consider the effects of both refraction and diffraction when calculating wave height transformation due to shoaling. The U.S. Army Corps of Engineers RCPWAVE is a PC-compatible program capable of doing these calculations for "smoothed" bathymetry.

The fundamental equations used to carry out diffraction calculations are based on the classical Sommerfield relation:

$$\eta = \frac{AkC}{g} \cosh kd |F(r, \Psi)| e^{ikCt} \tag{34.28}$$

where

$$K' = \frac{H_D}{H_i} = |F(r, \Psi)| \tag{34.29}$$

The second, and perhaps most important, application of **wave diffraction** is that due to wave-structure interaction. For this class of problems, wave diffraction calculations are essential for obtaining the distribution of wave height in harbors or behind engineered structures. Three primary types of wave-structure diffraction are important to coastal engineering (Fig. 34.4): (a) diffraction at the end of a single breakwater (semi-infinite); (b) diffraction through a harbor entrance (gap diffraction); and (c) diffraction around an offshore breakwater.

The methods of solution for all three of these wave-structure interactions are similar, but are restricted by some important assumptions. For each case there is a *geometric shadow zone* on the sheltered side of the structure, a *reflected wave zone* on the front or incident wave side of the structure, and an *"illuminated" zone* in the area of direct wave propagation [see Fig. 34.4(a)]. The solution to $F(r, \Psi)$ is complicated; U.S Army Corps of Engineers [1984] provides a series of templates for determining diffraction coefficients K', defined as the ratio of wave height in the zone affected by diffraction to the unaffected incident wave height, for semi-infinite breakwaters and for breakwater gaps between 1 and 5 wavelengths (L) wide. For breakwater gaps greater than $5L$ the semi-infinite templates are used independently, and for gaps $1L$ or less, a separate set of templates is provided [U.S. Army Corps of Engineers, 1984]. A basic diffraction-reflection calculation program is also provided in ACES [1992].

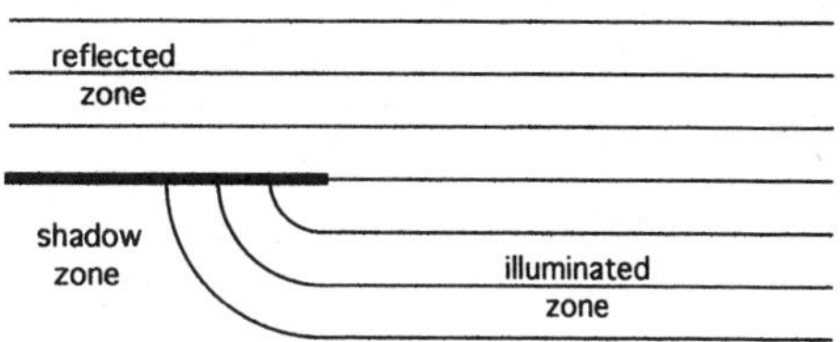

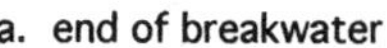
a. end of breakwater

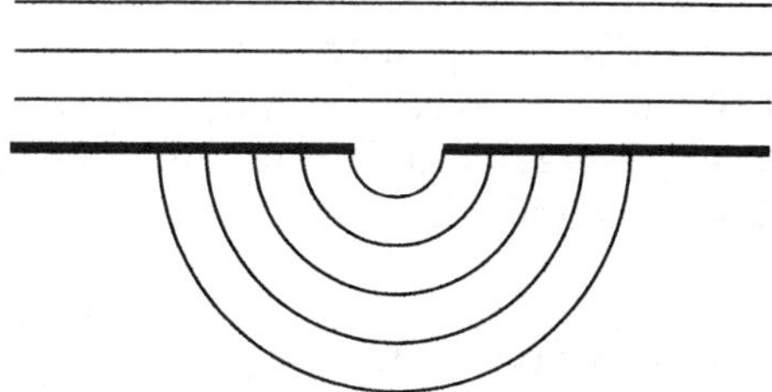
b. through breakwaters

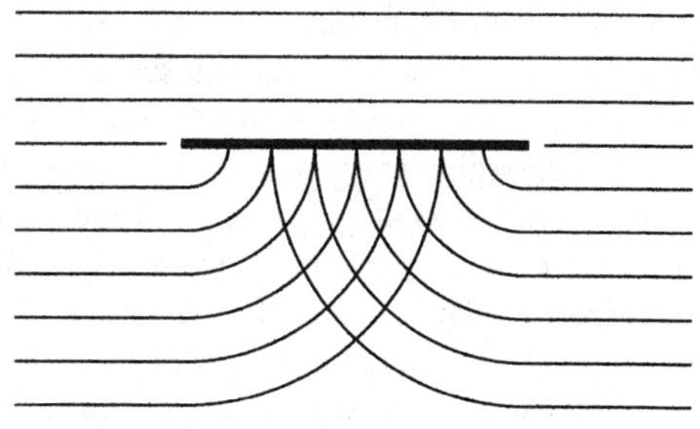
c. around breakwater

FIGURE 34.4 Wave diffraction patterns around breakwaters.

Wave Breaking

Waves propagating into shallow water tend to experience an increase in wave height to a point of instability at which the wave breaks, dissipating energy in the form of turbulence and work done on the bottom. Breaking waves are classified as **spilling breakers,** generally associated with low-sloping bottoms and a gradual dissipation of energy; **plunging breakers,** generally associated with steeper-sloping bottoms and a rapid, often spectacular, "explosive" dissipation of energy; and **surging breakers**, generally associated with very steep bottoms and a rapid, narrow region of energy dissipation. A widely used classic criterion [McCowan, 1894] applied to shoaling waves relates breaker height H_b to depth of breaking d_b through the relation

$$H_b = 0.78 d_b \tag{34.30}$$

However, this useful estimate neglects important shoaling parameters such as bottom slope (m) and deep water wave angle of approach (α_o). Dean and Dalrymple [1991] used Eq. (34.26) and McCowan's breaking criterion to solve for breaker depth (d_b), distance from the shoreline to the breaker line (x_b), and breaker height (H_b) as

$$d_b = \frac{1}{g^{1/5}\kappa^{4/5}}\left(\frac{H_o^2 C_o \cos\alpha_o}{2}\right)^{2/5} \tag{34.31}$$

$$x_b = \frac{d_b}{m} \tag{34.32}$$

and

$$H_b = \kappa d_b = \kappa m x_b = \left(\frac{\kappa}{g}\right)^{1/5}\left(\frac{H_o^2 C_o \cos\alpha_o}{2}\right)^{2/5} \tag{34.33}$$

where m is beach slope and $\kappa = H_b/d_b$. Dalrymple *et al.* [1977] compared the results of a number of laboratory experiments with Eq. (34.33) and found that it underpredicts breaker height by approximately 12% (with $\kappa = 0.8$). Wave breaking is still not well understood and caution is urged when dealing with engineering design in the active breaker zone.

34.2 Ocean Wave Climate

The Nature of the Sea Surface

As the wind blows across the surface of the sea, a large lake, or a bay, momentum is imparted from the wind to the sea surface. Of this momentum, approximately 97% is used in generating the general circulation (currents) of the water body and the remainder supplies the development of the surface wave field. Although this surface wave momentum represents a small percentage of the total momentum, it results in an enormous quantity of wave energy.

Within the region of active wind-wave generation, the sea surface becomes very irregular in size, shape, and direction of propagation of individual waves. This disorderly surface is referred to as **sea.** As waves propagate from their site of generation, they tend to sort themselves out into a more orderly pattern. This phenomenon, known as *dispersion,* is due to the fact that longer-period waves tend to travel faster, while short-period waves lag behind [see Eq. (34.2)]. Therefore, **swell** is a term applied to waves which have propagated outside the region of active wind-wave generation and are characterized by a narrow distribution of periods, a regular shape, and a narrow direction of travel.

Given these distinctions between sea and swell it is reasonable to expect that the statistical description of the sea surface would be very different for each case. The wave spectrum is a plot of the

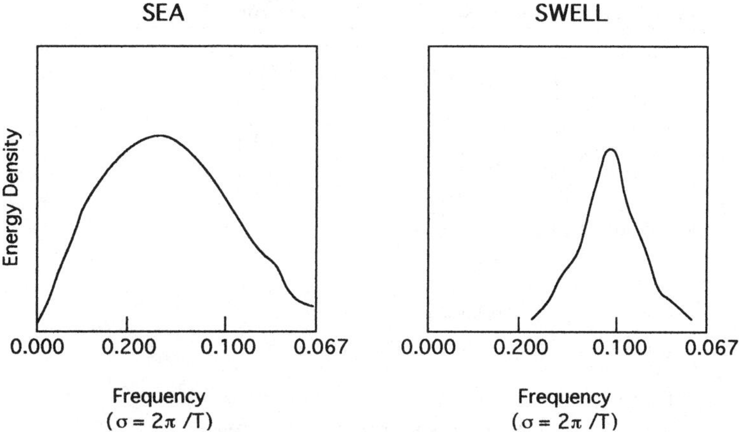

FIGURE 34.5 Characteristic energy diagrams showing the difference between sea and swell.

energy associated with each frequency ($f = 1/T$) component of the sea surface. The difference between sea and swell spectra is shown schematically in Fig. 34.5. Note that the sea spectrum is typically broadly distributed in frequency while the swell spectrum is narrowly distributed in frequency (tending toward monochromatic waves).

The directional distribution of wave energy propagation is given by the two-dimensional directional wave energy spectrum. Just as in the case of the one-dimensional wave energy spectrum, the two-dimensional spectrum provides a plot of wave energy versus frequency; however, the spectrum is further defined by the direction of wave propagation.

Wave Prediction

The wave height and associated energy contained in the sea surface is generally dependent on three parameters: the speed of the wind, U, measured at 10 m above the sea surface; the open water distance over which the wind blows, **fetch** length, x; and the length of time the wind does work on the sea surface, duration, t. The growth of a wind-driven sea surface may be limited by either the fetch or duration, producing a sea state less than "fully arisen" (maximum energy) for a given wind speed.

One-dimensional wave prediction models generally consist of equations that estimate wave height and wave period at a particular location and time as a function of fetch length and wind speed. Three examples of one-dimensional wave prediction formulas are provided in Table 34.2. It should be noted that the wind speed utilized in these wave prediction models must be obtained from, or corrected to, a height of 10 m above the water surface. A widely used approximation for correcting wind speed to 10 m, from a wind measured at height z over the open ocean, is

$$U_{10} = U_z \left(\frac{10}{z}\right)^{1/7} \tag{34.34}$$

If the wind speed is measured near the coast, the exponent used for this correction is 2/7. In the event that overwater winds are not available, overland winds may be utilized, but need to be corrected for frictional resistance. This is due to the fact that the increased roughness typically present over land sites serves to modify the wind field. A concise description of this methodology is presented in the U.S. Army Corps of Engineers [1984].

Since the natural sea surface is statistically complex, the wave height is usually expressed in terms of the average of the one-third largest waves or the **significant wave height,** H_s. The significant

Table 34.2 One-dimensional Wave Prediction Formulas

SMB	$H_s = 0.283g^{-1}U^2 \tanh\left[0.0125\left(\frac{gx}{U_{10}^2}\right)^{0.42}\right]$
	$T = 7.54g^{-1}U \tanh\left[0.077\left(\frac{gx}{U_{10}^2}\right)^{0.25}\right]$
JONSWAP	$H_s = 0.0016g^{0.5}U_x^{0.5}$
	$T = 0.286g^{0.62}U_{10}^{0.33}x^{0.33}$
Donelan	$H_s = 0.00366g^{-0.62}U_{10}^{1.24}x^{0.38}(\cos\phi)^{1.24}$
	$T = 0.54g^{-0.77}U_{10}^{0.54}x^{0.23}(\cos\phi)^{0.54}$
Definitions	H_s = significant wave height (in meters)
	T = peak energy wave period (in seconds)
	U_{10} = wind speed at 10 m height (in meters per second)
	x = fetch length (in wave direction for Donelan formulas)
	ϕ = angle between wind and waves
	g = 9.8 m/s

wave period corresponds to the energy peak in the predicted wave spectrum. Other expressions for wave height which are commonly used in design computations are H_{max}, the maximum wave height, H_{rms}, the root mean square wave height, $\overline{H}$, the average wave height, H_{10}, the average of the highest 10% of all waves, H_1, the average of the highest 1% of all waves (see Fig. 34.6). The energy-based parameter commonly used to represent wave height is H_{10}, which is an estimate of the significant wave height fundamentally related to the energy distribution of a wave train. Table 34.3 summarizes the relationships among these various wave height parameters.

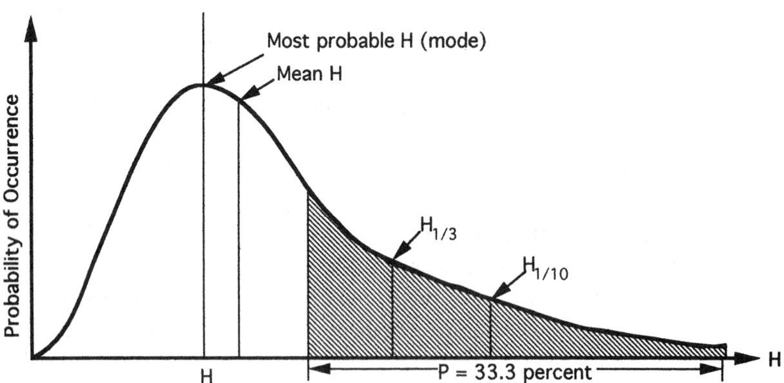

FIGURE 34.6 Statistical distribution of wave heights.

Table 34.3 Summary of Approximate Statistical Wave Height Relations

	$\bar{H}$	H_{rms}	H_s	H_{10}	H_1	H_{max}
$\bar{H}$		0.89	0.63	0.49	0.38	0.33
H_{rms}	1.13		0.71	0.56	0.42	0.38
H_s	1.60	1.42		0.79	0.60	0.53
H_{10}	2.03	1.80	1.27		0.76	0.68
H_1	2.67	2.37	1.67	1.31		0.89
H_{max}	2.99	2.65	1.87	1.47	1.12	

Note: See text for explanations of various wave height designations.

When predicting wave generation by hurricanes, the determination of fetch and duration is much more difficult due to large changes in wind speed and direction over short time frames and distances. Typically, the wave field associated with the onset of a hurricane or large storm will consist of a locally generated sea superimposed on swell components from other regions of the storm [U.S. Army Corps of Engineers, 1984].

Wave Data Information Sources

Several sources of wave data exist in both statistical and time series form. The National Climatic Data Center (NCDC) has compiled summaries of ship observations, over many open water areas, into the series *Summary of Synoptic Meteorological Observations* (SSMO). These publications present statistical summaries of numerous years of shipboard observations of wind, wave, and other environmental conditions. These publications may be purchased through NCDC/NOAA, Asheville, NC.

The National Data Buoy Center (NDBC) is responsible for the archiving of wave and weather data collected by their network of moored, satellite-reporting buoys. These buoys report hourly conditions of wave height and period, as well as wind speed and direction, air and sea temperature, and other meteorological data. This information can be obtained in time series form (usually hourly observations) from the NDBC office.

Another statistical summary of wind and wave data is available from the U.S. Army Corps of Engineers, Waterways Experiment Station, Coastal Engineering Research Center (CERC). The primary purpose of the CERC Wave Information Study (WIS) is to provide an accurate and comprehensive database of information of the long-term wave climate. The WIS generally uses a complete series of yearly wind records that vary in length from 20 to 32 years. The study considers the effects of ice cover where applicable and reflects advances in the understanding of the physics involved in wave generation, propagation, and dissipation, employing currently developed techniques to model these processes. The summary tables generated from the WIS hindcast include percent occurrence of wave height and period by direction, a wave "rose" diagram, the mean significant wave height by month and year, the largest significant wave height by month and year, and total summary statistics for all of the years at each station. In addition, the study also provides return period tables for the 2-, 5-, 10-, 20-, and 50-year design waves. WIS reports for ocean coastal areas of the U.S. and the Great Lakes can be obtained from CERC.

34.3 Water Level Fluctuations

Long-period variations of water level occur over a broad range of time scales, greater than those of sea waves and swell. These types of fluctuations include astronomical tides, **seiches, tsunamis,** wave setup, and storm surge, as well as very-long-period (months to years) variations related to climatologic and eustatic processes.

Tides

Tides are periodic variations in mean sea level caused by gravitational forces among the earth, moon, and sun and by the centrifugal force balance of the three-body earth-moon-sun system. Although complicated, the resultant upward or downward variation in mean sea level at a point on the earth's surface can be predicted quite accurately. Complete discussions of tidal dynamics are given in Defant [1961], Neumann and Pierson [1966], and Apel [1990]. The specific computational approach currently being used for official tide prediction in the U.S. is described in Pore and Cummings [1967]. Tide tables for the coastlines of the U.S. can be obtained from the U.S. Department of Commerce, National Ocean Service, Rockville, MD.

Tides tend to follow a lunar (moon) cycle and thus show a recurrence pattern of approximately one month. During this one-month cycle there will be two periods of maximum high and low

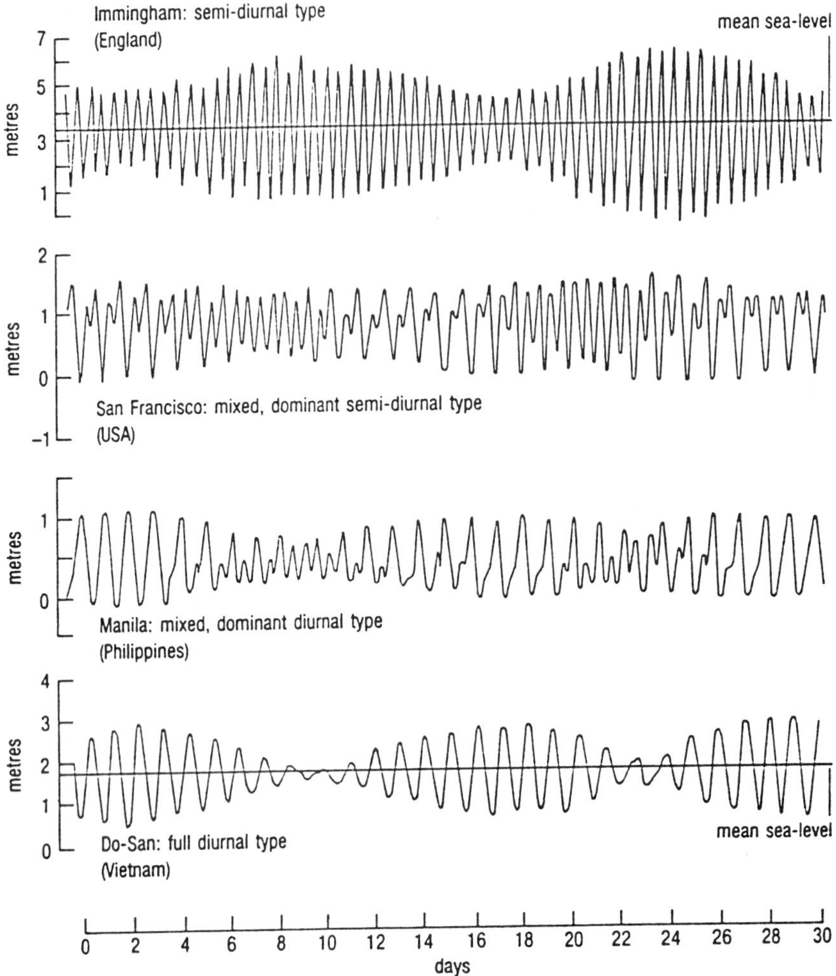

FIGURE 34.7 Different types of tides that may occur at a coast.

water level variation, called **spring tides,** and two periods of minimum high and low water level variation, called **neap tides** (Fig. 34.7). Figure 34.7 also illustrates the different types of tides that may occur at the coast. These tides may be *diurnal,* high and low tides that occur once daily; *semidiurnal,* high and low tides that occur twice daily; or *mixed,* two highly unequal high and low tides that occur daily. Diurnal tides occur on a lunar period of 24.84 hours and semidiurnal tides occur on a half-lunar period of 12.42 hours. Therefore, the time of occurrence of each successive high or low tide advances approximately 50 (diurnal) or 25 (semidiurnal) minutes.

Seiches

Seiches are long-period standing waves formed in enclosed or semienclosed basins such as lakes, bays, and harbors. Seiches are usually generated by abrupt rapid changes in pressure or wind stress. The natural free oscillation period, T_n, of a seiche in an *enclosed* rectangular basin of constant depth is

$$T_n = \frac{2l_b}{n\sqrt{gd}} \tag{34.35}$$

where l_b is the basin length in the direction of travel and n is the number of nodes along the basin length. The maximum period occurs at the fundamental, where $n = 1$. The natural free oscillation period for an *open* rectangular basin (analogous to a bay or harbor) is

$$T'_{n'} = \frac{4l_{b'}}{(1 + 2n')\sqrt{gd}} \tag{34.36}$$

where n' is the number of nodes between the node at the opening and the antinode at the opposite end. A complete discussion of seiches is presented in Huthinson [1957].

Tsunami

Tsunamis are long-period progressive gravity waves generated by sudden violent disturbances such as earthquakes, volcanic eruptions, or massive landslides. These long waves usually travel across the open seas at shallow water wave speeds, $c = \sqrt{gd}$, and thus can obtain speeds of hundreds of kilometers per hour. These relatively low waves (tens of centimeters) on the open ocean are greatly amplified at the coast and have been recorded at heights greater than 30 meters.

Wave Setup

Wave setup is defined as the superelevation of mean water level caused by wave action at the coast [U.S. Army Corps of Engineers, 1984]. This increase in mean water level occurs between the breaking point and the shore. U.S. Army Corps of Engineers [1984] gives the following formula for calculating wave setup:

$$S_w = 0.15d_b - \frac{g^{0.5}(H'_o)^2 T}{64\pi d_b^{0.66}} \tag{34.37}$$

where d_b is the depth of breaking and H'_o is the unrefracted deep water wave height. Wave setup is typically of the order of centimeters.

Storm Surge

Storm conditions often produce major changes in water level as a result of the interaction of wind and atmospheric pressure on the water surface. Severe storms and hurricanes have produced surge heights greater than 8 meters on the open coast and can produce even higher surges in bays and estuaries. Although prediction of storm surge heights depends on many factors [U.S. Army Corps of Engineers, 1984], an estimate of sea surface slope at the coast, caused by wind stress, τ_s, effects, can be calculated as

$$\frac{dz}{dx} = \lambda \frac{\tau_s}{\gamma d} \tag{34.38}$$

where λ, an experimentally determined variable, ranges from 0.7 to 1.8 (average value of 1.27) and

$$\tau_s = 0.58U_{10}^{2.22} \tag{34.39}$$

where τ_s is kg-m^2 and U_{10} is the wind speed in m/s measured at 10 meters above the sea surface. A similar simple estimate of pressure setup (P_s) in meters at the storm center for a stationary storm can be made using the relation

$$P_s = 0.0136\Delta P \tag{34.40}$$

where ΔP is the difference between the normal pressure and the central pressure of the storm measured in millimeters of mercury.

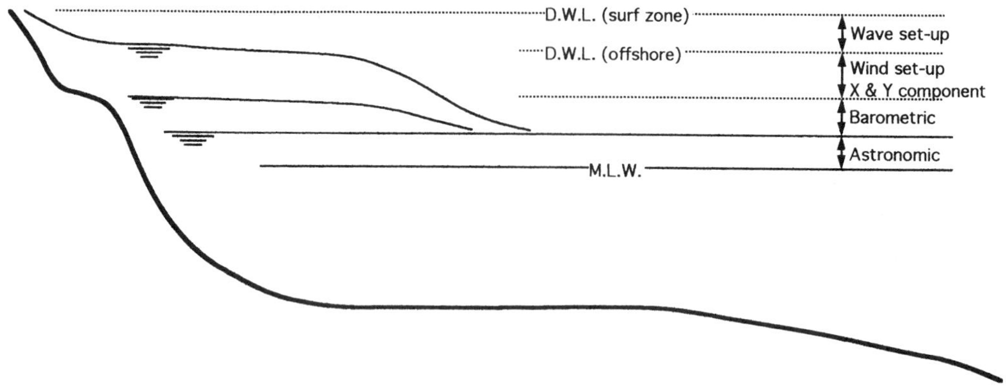

FIGURE 34.8 Schematic of the design water level components.

Climatologic Effects

For most coastal engineering on ocean coasts, climatologic effects on sea level are small and may be neglected. However, on large lakes such as the Great Lakes, water levels may vary by tens of centimeters per year and meters per decade. Great Lakes water level information is provided monthly by the National Oceanic and Atmospheric Administration and local large lake water level information is usually available from state agencies. Accurate knowledge of sea and lake level change is essential to successful coastal engineering design.

Design Water Level

Design of coastal structures usually requires the determination of a maximum and minimum design water level (DWL). Design water level is computed as an addition of the various water level fluctuation components as follows:

$$\mathrm{DWL} = d + A_s + S_w + P_s + W_s \tag{34.41}$$

where d is chart depth referenced to mean low water; A_s is the astronomical tide; S_w is wave setup, defined in Eq. (34.37); P_s is the pressure setup, defined in Eq. (34.40); and W_s is the wind setup calculated using Eq. (34.38). It is important to recognize that all of these components may not occur in phase and, therefore, Eq. (34.41) will tend to result in extreme maximum and minimum DWL values. It should also be recognized that some of these component effects are amplified at the shore. Figure 34.8 shows a schematic illustration of the design water level components and their relative magnitude.

34.4 Coastal Processes

Coastal regions can take a wide variety of forms. Of the total U.S. shoreline, 41% is exposed to open water wave activity, and the remainder is protected. Outside of Alaska, about 30% is rocky and approximately 14% of the shoreline has beaches. Approximately 24% of the U.S. shoreline is eroding.

Beach Profiles

Nearshore profiles oriented perpendicular to the shoreline have characteristic features which emulate the influence of local littoral processes. A typical beach profile possesses a sloping near-shore bottom, one or more sand bars, one or more flat beach berms, and a bluff or escarpment. As waves move toward shore, they first encounter the beach profile in the form of a sloping nearshore bottom. When a wave reaches a water depth of approximately 1.3 times the wave height, the wave will break. Breaking results in the dissipation of wave energy by the generation of turbulence and the transport of sediment. As a result, waves suspend sediment and transport it to regions of lower energy, where it is deposited. In many cases, this region is a sand bar. Therefore, the beach profile is constantly adjusting to the incident wave conditions.

The Equilibrium Beach

Although beaches seldom reach a steady state profile, it is convenient to consider them as responding to variable incident wave conditions by approaching an ideal long-term configuration that depends on steady incident conditions and sediment characteristics. Bruun [1954] and Dean [1977], in examining natural beach profiles, found that the typical profile was well defined by the relationship

$$d(x) = Ax^q \tag{34.42}$$

where A is a scale coefficient that depends on the sediment characteristics, q is a shape factor, found both theoretically and experimentally to be approximately $\frac{2}{3}$, x is the distance offshore, and d is the water depth.

This concept of the equilibrium profile serves as a basis for many models of coastal sediment movement and bathymetric change. When a profile is disturbed from equilibrium by a short-term disturbance, such as storm-induced erosion, the profile is then hypothesized to adjust accordingly toward a new equilibrium state defined by the long-term incident wave conditions and sediment characteristics of the site.

Beach Sediments

Beach sediments range from fine sands to cobbles. The size and character of sediments and the slope of the beach are related to the forces to which the beach is exposed and the type of material available on the coast. The origin of coastal materials can be from inland or offshore sources or from erosion of coastal features. Coastal transport processes and riverine transport bring these materials to the beach and nearshore zone for disbursement through wave and current activity. Beaches arising from erosion of coastal features tend to be composed of inorganic materials while beaches in tropical latitudes can be composed of shell and coral reef fragments. Finer particles are typically kept in suspension in the nearshore zone and transported away from beaches to more quiescent waters such as lagoons and estuaries or deeper offshore regions.

Longshore Currents

As waves approach the shoreline at an oblique angle, a proportion of the wave orbital motion is directed in the longshore direction. This movement gives rise to longshore currents. These currents flow parallel to the shoreline and are restricted primarily between the breaker zone and the shoreline. Longshore current velocities vary considerably across the surf zone, but a typical mean value is approximately 0.3 m/s. Expressions for calculating longshore current velocity range from theoretically based to completely empirical expressions. However, all of the expressions are calibrated using measured field data. Two accepted formulations for calculating mean longshore current velocities ($\overline{V}$) are the theoretically based formulation of Longuet-Higgins [1970],

$$\overline{V} = 20.7m(gH_b)^{1/2} \sin 2\alpha_b \tag{34.43}$$

where m is beach slope, H_b is wave height at breaking, α_b is wave crest angle at breaking, and g is gravitational acceleration; and the empirical formulation of Komar and Inman [1970],

$$\overline{V} = 2.7u_{\text{max}} \sin \alpha_b \cos \alpha_b \tag{34.44}$$

where u_{max} is the maximum particle velocity at breaking.

Cross-Shore Currents

Cross-shore currents and resultant transport can be caused by mass transport in shoaling waves, wind-induced surface drift, wave-induced setup, irregularities on the bottom, and density gradients. These factors can produce cross-shore currents ranging in intensity from diffuse return flows, visible as turbid water seaward of the surf zone, to strong, highly organized rip currents. Rip currents are concentrated flows which carry water seaward through the breaker zone. There is very little information on the calculation of cross-shore current velocities in the breaker zone.

Sediment Transport

A schematic drawing of the sediment transport (littoral transport) along a coast is shown in Fig. 34.9. This diagram depicts the distribution of sediments for an uninterrupted length (no shoreline structures) of natural coastline. When there is sufficient wave energy, a system of sediment erosion, deposition, and transport is established as shown in Fig. 34.9.

A finite amount of sediment S_{Tin} is transported by longshore currents into a section of the coast (nearshore zone). Wave action at the shore erodes the beach and dune-bluff sediment (S_B) and carries it into the longshore current. This same wave action lifts sediment from the bottom (S_w) for potential transport by the longshore current. Finally, wave action and cross-shore currents at the offshore boundary move sediment onshore or offshore (S_{on} and S_{off}) depending on wave and bottom conditions. This transport at the outer limit of the nearshore zone can usually be assumed negligible with respect to the other transports. The summation of these various transports over time provide a measure of the net sediment budget for a section of coast.

An estimate of the longshore sediment transport rate can be obtained by the energy flux method using deep water wave characteristics applied to calculate the longshore energy flux factor entering the surf zone (P_{ls})

$$P_{\text{ls}} = 0.05\rho g^{3/2} H_{so}^{5/2} (\cos\alpha_o)^{1/4} \sin 2\alpha_o \tag{34.45}$$

where ρ is the density of water, H_{so} is the deep water significant wave height, and α_o is the deep water wave angle of approach. The actual quantity of sediment transported can be calculated

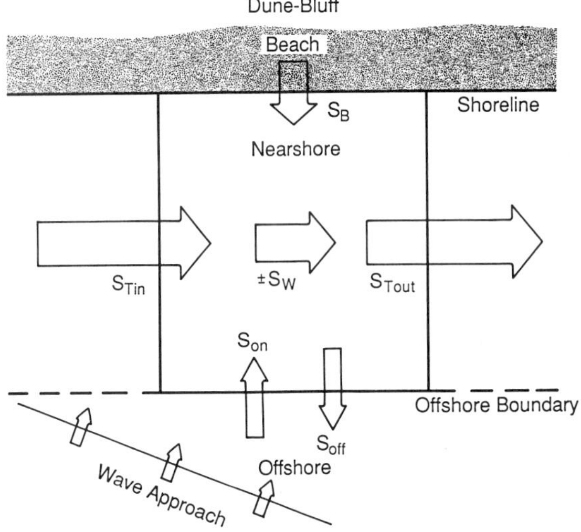

FIGURE 34.9 Box diagram of sediment transport components at the coast.

directly from P_{ls} as

$$Q\left(\frac{\text{m}^3}{\text{yr}}\right) = 1290P_{ls}\left(\frac{\text{J}}{\text{m}-\text{s}}\right) \tag{34.46}$$

or

$$Q\left(\frac{\text{yd}^3}{\text{yr}}\right) = 7500P_{ls}\left(\frac{\text{ft}-\text{lb}}{\text{ft}-\text{s}}\right) \tag{34.47}$$

These calculations should be carried out using an offshore wave climatology distributed in wave height and angle. A complete description of this methodology is given in U.S. Army Corps of Engineers [1984, Chapter 4].

34.5 Coastal Structures and Design

There are four general categories of coastal engineering problems which may require structural solutions: shoreline stabilization, backshore (dune-bluff) protection, inlet stabilization, and harbor protection [U.S. Army Corps of Engineers, 1984]. Figure 34.10 shows the types of structures or protective works in each of these four coastal engineering problem areas. A listing of factors that should be considered in evaluating each of these problem areas is also given in Fig. 34.10. Hydraulic considerations include wind, waves, currents, storm surge or wind setup, water level variation, and bathymetry. Sedimentation considerations include sediment classification, distribution properties, and characteristics; direction and rate of littoral transport; net versus gross littoral transport; and shoreline trend and alignment. Control structure considerations include selection of the protective

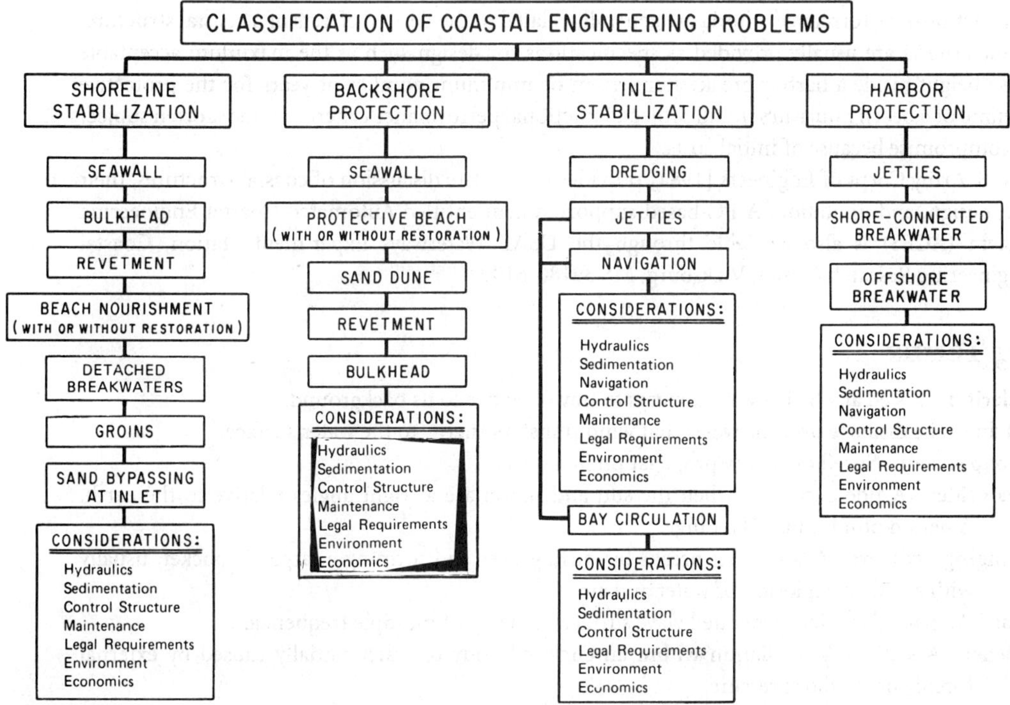

FIGURE 34.10 Classification of coastal engineering problems at the coast. (*Source:* U.S. Army Corps of Engineers. 1984. *Shore Protection Manual,* CERC/WES, Vicksburg, MS.)

works with respect to type, use, effectiveness, cost, and environmental impact [U.S. Army Corps of Engineers, 1984]. The other factors listed in Fig. 34.10 are more generally understood and will not be elaborated on further. It is important to remember that a "no action" alternative should also be considered as a possible solution for any one of these categories of coastal problems.

Structural Selection Criteria

A diverse set of criteria needs to be considered in the selection and design of coastal structures. Structural stability criteria and functional performance criteria encompass two areas of primary concern for selection and evaluation of coastal structures.

Structural stability criteria are usually associated with extreme environmental conditions which may cause severe damage to or failure of a coastal structure. These stability criteria are, therefore, related to episodic events in the environment (severe storms, hurricanes, earthquakes) and are often evaluated on the basis of risk of encounter probabilities. A simple method for evaluating the likelihood of encountering an extreme environmental event is to calculate the encounter probability (E_p) as

$$E_p = 1 - \left(1 - \frac{1}{T_R}\right)^{\Lambda} \tag{34.48}$$

where T_R is the return period and Λ is the design life of the structure [Borgman, 1963]. The greatest limitation to structural stability criteria selection is the need for a long-term database on critical environmental variables sufficient to determine reasonable return periods for extreme events. For example, coastal wave data for U.S. coasts is geographically sparse, and in most locations where it exists the period of collection is ten years. Since most coastal structures have a design life well in excess of ten years, stability criteria selection often relies on extrapolation of time-limited data or statistical modeling of environmental processes.

Functional performance criteria are generally related to the desired effects of a coastal structure. These criteria are usually provided as specifications for design such as the maximum acceptable wave height inside a harbor breakwater system or minimum number of years for the protective lifetime of a beach nourishment fill project. Functional performance criteria are most often subject to compromise because of initial costs.

U.S. Army Corps of Engineers [1984] provides a complete discussion of coastal structures, their use, design, and limitation. A PC-based support system entitled Automated Coastal Engineering System (ACES) is also available through the USAE Waterways Experiment Station, Coastal Engineering Research Center, Vicksburg, MS 39180-6199.

Defining Terms

Celerity: Speed at which a wave appears to travel relative to its background.

Fetch: The distance over the water that wind transfers energy to the water surface.

Group speed: Speed of energy propagation in a wave.

Neap tide: A tide occurring when the sun and moon are at right angles relative to the earth. A period of minimal tidal range.

Plunging breaker: A wave that breaks by curling over and forming a large air pocket, usually with a violent explosion of water and air.

Sea: An irregular, wind-generated wave surface made up of multiple frequencies.

Seiche: A stationary oscillation within an enclosed body of water initially caused by external forcing of wind or pressure.

Significant wave height (H_s): The average height of the one-third highest waves in a wave height distribution.

Spilling breaker: A wave that breaks by gradually entraining air along the leading face and that slowly decays as it moves shoreward.

Spring tide: A tide occurring when the sun and moon are in line relative to the earth. A period of maximum tidal range.

Swell: A very regular series of wind-generated surface waves made up of a single dominant frequency.

Tsunami: A long-period, freely traveling wave usually caused by a violent disturbance such as an underwater earthquake, volcanic eruption, or landslide.

Wave diffraction: The spread of energy laterally along a wave crest usually due to the interaction of a wave with a barrier.

Wave height: The vertical distance from a wave crest to the trough of the preceding wave.

Wave refraction: The bending of a wave crest as it enters shallow water caused by a differential in speed between the deeper and shallower portions of the crest.

Wave ray: A line drawn perpendicular to a wave crest indicating the direction of propagation of the wave.

References

Apel, J. R. 1990. *Principles of Ocean Physics,* Academic Press, San Diego.

Borgman, L. 1963. Risk criteria, *J. Waterways and Harbors, ASCE,* 3(89):1–36.

Bruun, P. 1954. *Coast Erosion and the Development of Beach Profiles,* Technical Memorandum No. 44, Beach Erosion Board.

Dalrymple, R. A., Eubanks, R. A., and Biekemeier, W. A. 1977. Wave-induced circulation in shallow basins, *J. Waterways, Ports, Coastal and Ocean Division, ASCE,* 103:117–135.

Dean, R. G. 1977. *Equilibrium Beach Profiles: U.S. Atlantic and Gulf Coasts,* Ocean Engineering Technical Report No. 12, University of Delaware, Newark.

Dean, R. G. and Dalrymple, R. A. 1991. *Water Wave Mechanics for Engineers and Scientists,* World Scientific, Singapore.

Defant, A. 1961. *Physical Oceanography,* Macmillan, New York.

Huthinson, E. G. 1957. *A Treatise on Limnology,* vol. 1, John Wiley & Sons, New York.

Komar, P. D. and Inman, D. L. 1970. Longshore sand transport on beaches, *J. Geophysical Research,* 75(30):5914–5927.

Longuet-Higgins, M. S. 1970. Longshore currents generated by obliquely incident sea waves, *J. Geophysical Research,* 75(33):6778–6789.

McCowan, J. 1894. On the highest wave of permanent type, *Phil. Mag.,* Series 5(38):351–357.

Neumann, G. and Pierson, W. J. 1966. *Principles of Physical Oceanography,* Prentice Hall, Englewood Cliffs, NJ.

Noda, E. K. 1974. Wave-induced nearshore circulation, *J. Geophysical Research,* 79(27):4097–4106.

Pore, N. A. and Cummings, R. A. 1967. *A Fortran Program for the Calculation of Hourly Values of Astronomical Tide and Time and Height of High and Low Water,* Technical Memorandum WBTM TDL-6, U.S. Dept. of Commerce, Washington, D.C.

U.S. Army Corps of Engineers. 1984. *Shore Protection Manual,* CERC/WES, Vicksburg, MS.

U.S. Army Corps of Engineers. 1992a. *Automated Coastal Engineering System (ACES), User's Guide,* Coastal Engineering Research Center, Vicksburg, MS.

U.S. Army Corps of Engineers. 1992b. *Coastal Modeling System (CMS), User's Guide,* Coastal Engineering Research Center, Vicksburg, MS.

Wiegel, R. L. 1954. *Gravity Waves, Tables of Functions,* Engineering Foundation Council on Wave Research, Berkeley, CA.

35

Hydraulic Structures

Jacques W. Delleur
Purdue University

35.1 Introduction

This section covers the principles of the hydraulic design of the more usual hydraulic structures found in civil engineering practice. These include reservoirs, dams, spillways and outlet works, energy dissipation structures, open channel transitions, culverts, bridge constrictions, pipe systems, and pumps.

35.2 Reservoirs

Classification According to Use

Reservoirs are used for the storage of water, for the purpose of conservation for later use, for mitigation of flood damages, for balancing time-varying supply and demand (for water supply or generation of hydropower), for storing excess runoff in urbanized areas, and so on. If a reservoir is built for one purpose only, it is said to be a *single-objective* reservoir; if it is built to satisfy multiple purposes, it is called a *multiobjective* reservoir.

0-8493-8953-4/95/$0.00 + $.50

For example, a single-purpose reservoir could be built for water supply and a multipurpose reservoir could be built for flood control and hydropower. In the latter example, the two objectives yield contradictory requirements. For flood control it is desired to have the reservoir as empty as possible in order to have as much free capacity as possible for a forthcoming flood, whereas the hydropower generation requires a reservoir nearly full in order to have the maximum head over the turbines and thus generate the maximum amount of power possible. These conflicting objectives are reconciled by assigning certain portions of the reservoir to the respective objectives. These proportions can vary throughout the year. For example, there are rainy seasons when the danger of flood is high and the reservoir portion assigned to flood control should be larger than during drought seasons. These time-varying storage assignments and the constraints on the amounts of water that can be released downstream at a certain time form the operating rules of the reservoir.

Reservoir Characteristics

The capacity-elevation and the area-elevation relationships are fundamental. Figure 35.1 shows the area and capacity curves for Lake Mead obtained from surveys in 1935 and 1963–64. The decrease in capacity between the two surveys is due to the accumulation of sediment. The minimum pool level corresponds to the elevation of the invert of the lowest outlet or sluiceway. The volume below this level is called **dead storage** and can be used for accumulation of sediments. The level corresponding to the crest of the spillway is the normal pool level and the volume difference between the normal pool level and the minimum pool level is the useful storage. When there is overflow over the spillway the additional storage above the normal pool is the **surcharge storage.** The discharge that can be guaranteed during the most critical dry period of record is the safe yield; any additional release is the secondary yield.

For flood routing in reservoirs (see Chapter 30, "Surface Hydrology," for details) the water surface in reservoirs is generally assumed to be horizontal. However, this is not the case when the reservoir is narrow and long, similar to a wide river. In such cases the water surface has a gradient that increases with increasing discharge. Techniques for the calculation of water surface profiles under steady and time-varying flows are discussed in Chapter 29, "Open Channel Hydraulics."

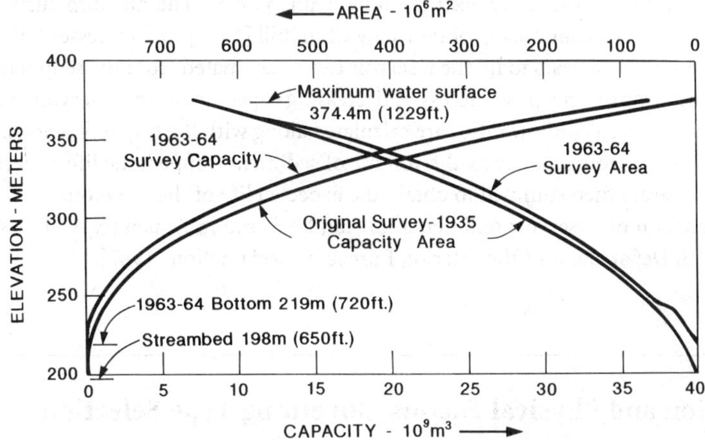

FIGURE 35.1 Area and capacity curves for Lake Mead. (*Source:* U.S. Department of the Interior, Bureau of Reclamation. 1987. *Design of Small Dams,* p. 531. U.S. Government Printing Office, Denver, CO.)

Capacity of a Reservoir

The Rippl diagram, or mass curve, is a plot of the cumulative inflow in the reservoir as a function of time. This curve exhibits segments with steep slopes during periods of large inflow in the reservoir and segments with relatively flat slopes during periods of low inflows or droughts. This inflow is generally adjusted to account for the evaporation losses and required releases downstream. When the losses plus the downstream requirements exceed the inflow, the curve shows periods of decreasing cumulative adjusted inflow. After a sharp rise in the curve it may exhibit a sharp decrease in slope. When such changes occur the reservoir is usually full. At these several bends in the adjusted cumulative inflow, lines are traced with a common slope equal to the demand in volume per unit of time. These accumulated demand lines will be above the accumulated adjusted inflow curve for a period of time. For such periods the largest departure between the accumulated demand line and the cumulative adjusted inflow curve represents the required volume of the reservoir.

When long data series are analyzed, the sequent peak algorithm of Thomas and Fiering [1963] is preferred. The accumulated differences between inflows and demand are plotted as a function of time. The curve exhibits a sequence of peaks and troughs. The first peak and the next higher peak, the sequent peak, are located. The difference between the first peak and the lowest subsequent valley represents the storage for the period. The next sequent peak is identified and the required storage after the first sequent peak is found. The process is repeated for the whole study period and the maximum storage is identified.

Reservoir Sedimentation

Methods of determining the suspended sediment carried by streams have been discussed in Chapter 33, "Sediment Transport." The results of sediment samplings can often be summarized by rating curves of the type

$$Q_s = aQ_w^b \tag{35.1}$$

in which Q_s is the suspended transport (tons per day), Q_w is the flow (cfs or cms), and a and b are coefficients. The larger particles of sediments carried by the streams leading to reservoirs are deposited at the head of the reservoirs, forming deltas. The smaller particles are carried in density currents and eventually are deposited in the lower parts of the reservoir near the dam.

The accumulation of sediments limits the useful life of a reservoir. The trap efficiency of reservoirs is a function of the ratio of reservoir capacity to annual inflow volume. Envelopes and medium trap efficiency curves have been given by Brune [1953]. The medium curve is shown in Fig. 35.2. Also shown is a relationship obtained by Churchill [1949] for Tennessee Valley Authority reservoirs. The number of years to fill the reservoir can be estimated from the sediment inflow and the trap efficiency. For this purpose successive decreasing capacities of the reservoir are considered and the respective capacity-inflow ratios are calculated along with the trap efficiencies. By dividing the increments in reservoir capacity by the volume of sediment trapped, the life of each increment is calculated. These are then summed to obtain the expected life of the reservoir.

Further discussion on reservoir sedimentation can be found in Appendix A of *Design of Small Dams* by the U.S. Department of the Interior, Bureau of Reclamation [1987].

35.3 Dams

Classification and Physical Factors Governing Type Selection

Dams are generally classified according to the material used in the structure and the basic type of design. Concrete gravity dams depend on their weight for their stability, concrete arch dams transfer the hydrostatic forces to their abutments by arch action, and in concrete buttress dams the

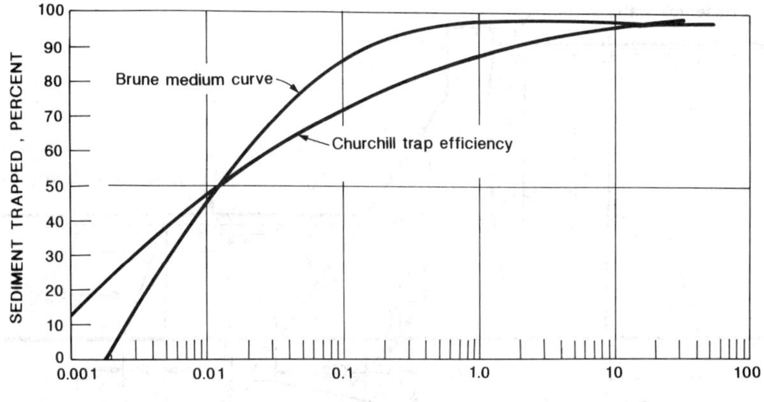

FIGURE 35.2 Trap efficiency curves. (Adapted from U.S. Department of the Interior, Bureau of Reclamation. 1987. *Design of Small Dams*, p. 542. U.S. Government Printing Office, Denver, CO.)

hydrostatic force is resisted by a slab that transmits the load to buttresses perpendicular to the dam axis. Embankment dams can be subdivided into earth-fill dams and rock-fill dams.

The principal physical factors governing the choice of dam type are the topography, the geology, the availability of materials, and the hydrology. The topography usually governs the basic choice of dam. Low, rolling plains would suggest an earth-fill dam, whereas a narrow valley with high rock walls would suggest a concrete structure or a rock-fill dam. Saddles in the periphery of the reservoir may provide ideal locations for spillways, especially emergency spillways. The foundation geology is also of major significance in dam selection. Good rock foundations are excellent for all types of dams, while gravel foundations are suitable for earth-fill and rock-fill dams. Silt or fine sand foundations are not suitable for rock-fill dams but can be used for earth-fill dams with flat slopes. The availability of materials (soils and rock for the embankment, riprap and concrete aggregate near the dam site) weigh heavily in economic considerations. The hydrologic condition of stream flow characteristics and rainfall will influence the method of diversion of the flows during construction and the construction time. Finally, if the dam is located in a seismic-prone area the horizontal and the vertical components of the earthquake acceleration on the dam structure and on the impounded water must be considered in the analysis of dam stability.

Stability of Gravity Dams

The principal forces to be considered are the weight of the dam, the hydrostatic force, the uplift force, the earthquake forces, the ice force, and the silt force. The gravity force is equal to the weight of concrete, V_C, plus the weight of such appurtenances as gates and bridges. This force passes through the center of gravity of the dam. The horizontal component of the hydrostatic force per unit width of the dam (see Chapter 28) is

$$H_w = \gamma h^2/2 \tag{35.2}$$

where γ is the specific weight of the water, and h is the depth of water at the vertical projection of the upstream face of the dam. This force acts at a distance $h/3$ above the base of the dam (Fig. 35.3). The vertical component of the hydrostatic force, V_w, is equal to the weight of water vertically above the upstream face of the dam. This force passes through the center of gravity of this wedge of water. There may also be horizontal and vertical hydrostatic forces, H'_w and V'_w, respectively, on the downstream face of the dam. Water may eventually seep between the dam masonry and its foundation, creating an **uplift pressure.** The most conservative design assumes that the uplift

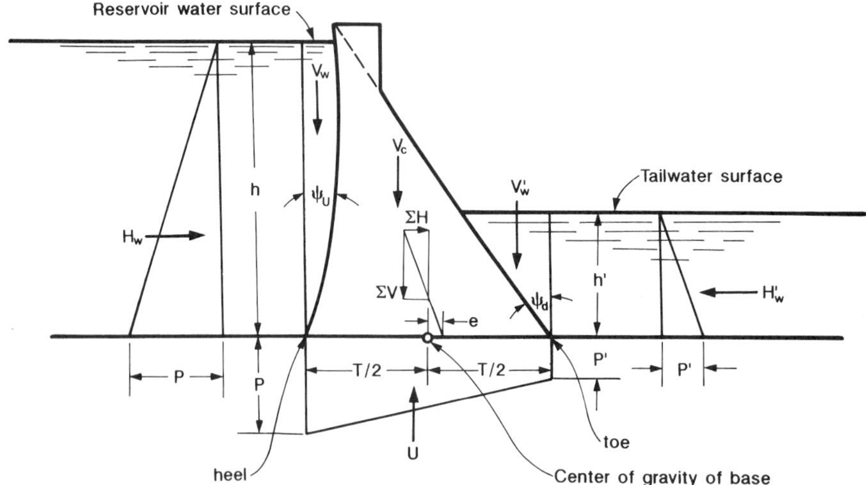

FIGURE 35.3 Forces acting on a concrete gravity dam. (Adapted from U.S. Department of the Interior, Bureau of Reclamation. 1987. *Design of Small Dams*, p. 317. U.S. Government Printing Office, Denver, CO.)

pressure varies linearly from a full hydrostatic pressure at the upstream face to the full tailwater pressure at the downstream face. The uplift force is then

$$U = \gamma[(h + h')/2]T \tag{35.3}$$

where h and h' are the water depths at the upstream and downstream faces, respectively, and T is the width of the dam base. The uplift force acts vertically upward through the center of gravity of the trapezoidal uplift pressure diagram. If there is a drain located at about 25% of the base width from the heel of the dam, the uplift pressure can be assumed to decrease to 2/3 of the full upstream hydrostatic pressure at the drain and then to zero at the toe, assuming no downstream hydrostatic pressure. Field measurements of uplift pressures are quoted in Yeh and Abdel-Malek [1993]. Suggested values of the earthquake accelerations in terms of the distance from the source of energy release and of the Richter scale magnitude as well as information on the increase in water pressure can be found in U.S. Department of the Interior, Bureau of Reclamation [1987]. According to the same source, an acceptable criterion for ice force is 10,000 lb/ft of contact between the dam and the ice for an assumed ice depth of 2 ft or more. An acceptable criterion for saturated silt pressure is to use an equivalent fluid with a specific weight of 85 lb/ft^3 for the horizontal component and 120 lb/ft^3 for the vertical component.

The stability analysis includes the following steps:

- Calculate the resultant vertical force above base of section, $\sum V$.
- Calculate the resultant horizontal force above base of section, $\sum H$.
- Calculate the overturning moment (clockwise in Fig. 35.3) about the toe of the section, $\sum M_0$.
- Calculate the stabilizing moment (counterclockwise in Fig. 35.3) about the toe, $\sum M_s$.
- Calculate the factor of safety against overturning, $\text{FS}_0 = \sum M_s / \sum M_0$.
- Calculate the resultant horizontal or sliding force above the base, $\sum H$.
- Calculate the available friction force $F_f = \mu \sum V$, where μ is the friction coefficient.
- Calculate the factor of safety against sliding $\text{FS}_s = F_f / \sum H$.
- Calculate the distance from the toe to the point where the resultant of the vertical forces cuts the base $x = [\sum M_s - \sum M_0] / \sum V$.
- Calculate the eccentricity of the load $e = T/2 - x$.

- Calculate the normal stresses $\sigma = \sum V/A \pm M\,c/I$, where c is the distance from the center of the section to its edge, I is the moment of inertia of the section about its centroidal axis, and $M = e\sum V$ is the moment due to the eccentricity of the load.
- Calculate the stress parallel to the face of the dam, $\sigma' = \sigma/\cos^2\psi_u$.
- Verify that $FS_0 > 2$.
- Verify that $FS_s > 1.0$.
- Verify that the stresses are within specifications.

Arch Dams

Arch dams are curved in plan so that they transmit part of the water pressure to the canyon walls of the valleys in which they are built. The arch action requires a unified monolithic concrete structure. Arch dams are classified as thin if the ratio of their base thickness to their structural height is less than 0.2 and thick if that ratio is larger than 0.3. The upstream side of an arch dam is called the **extrados** and the downstream side is the **intrados** (Fig. 35.4).

The structural analysis of arch dams assumes that it can be considered as a series of horizontal arch ribs and a series of vertical cantilevers. The load is distributed among these two actions in such a way that the arch and cantilever deflections are equal. This analysis is a specialized subject of structural engineering. The U.S. Department of the Interior, Bureau of Reclamation [1977] has published an extensive book on the subject. Only the simplified cylinder theory is summarized below.

The forces acting on an arch dam are the same as on a gravity dam, but their relative importance is not the same. The uplift is less important because of the comparatively narrow base and the ice pressure is more important because of the large cantilever action. If the arch has a radius r and a central angle θ, the horizontal hydrostatic force due to a head h on a rib of unit thickness is $H_h = \gamma h 2r \sin(\theta/2)$. (See Chapter 28.) This force is balanced by the abutment reaction in the upstream direction, $R_y = 2R\sin(\theta/2)$. By equating the two forces, the abutment reaction (Fig. 35.4) becomes $R = \gamma hr$. If the working stress of the concrete is σ_w, the thickness of the arch rib is $t = \gamma hr/\sigma_w$. It is seen that with this simplified theory the thickness increases linearly with the depth. The volume of a single arch rib with a cross section A is $V = rA\theta$, where θ is in radians. Because of the relationship between the thickness t and the radius, the angle that minimizes this volume of concrete can be shown to be $\theta = 133^\circ\,34'$. For this angle the radius r of the arch in a valley of width B is $r = (B/2)/\sin(66^\circ\,47') = 0.544B$. One could select an arch of constant radius with an average angle around $133^\circ\,34'$. The angle would be larger at the top and smaller at the base. Alternatively, one could select a fixed angle and determine the valley width B at various depths, calculate the radius r required, and then the necessary thickness t. A compromise between these two cases consists in keeping the radius fixed for a few sections and varied in others.

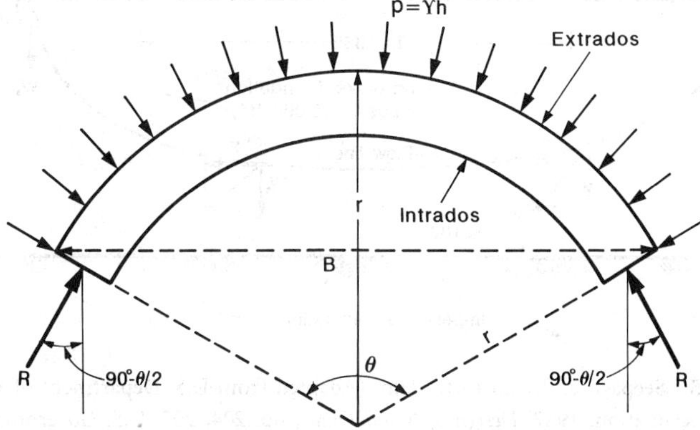

FIGURE 35.4 Hydrostatic pressure and resulting thrust on an arch rib.

Earth Dams

Early earth dams were homogeneous simple embankments as are many levees today. Most earth dams are zoned embankments. They may have an impermeable core made out of clay or a combination of clay, sand, and fine gravel. This core may be flanked on the upstream and downstream sides by more pervious zones or shells. These zones support and protect the impervious core. The upstream zone provides stability against rapid drawdown while the downstream zone controls the seepage and the position of the lower phreatic surface. In addition, there can be filters between the impervious zone and the downstream shell and a drainage layer below the downstream shell. *Diaphragm-type* dams are generally built on pervious foundations. Their impervious core is extended downward by a cutoff wall generally made of concrete. This wall is often accompanied by a horizontal impervious clay blanket under the base of the upstream face and extending further in the upstream direction. This lengthens the seepage path and reduces its quantity. The Bureau of Reclamation does not recommend this type of design for small dams because of potential cracking of the concrete wall due to differential movement induced by consolidation of the embankment.

The amount of seepage under a dam can be approximated roughly using Darcy's formula (see Chapter 32, "Groundwater Engineering"), $Q = KAh/L$, where K is the hydraulic conductivity of the foundation material, A is the gross cross-sectional area of the foundation through which the flow takes place, and h/L is the hydraulic gradient, the difference of head divided by the length of the flow path. For the conditions shown in Fig. 35.5 with a hydraulic conductivity K of 25,000 ft/yr $= 0.00079$ ft/s, a head differential $h = 210 - 175 = 35$ ft over a distance $L = 165$ ft resulting in a hydraulic gradient $h/L = 35/165 = 0.212$ ft/ft, and a flow cross-sectional area of $(170 - 100) \times 1 = 70$ ft^2 per ft of width, the discharge per unit width is $Q = (0.00079)(0.212)(70) = 0.012$ ft^3/s.

The flow through the foundation produces seepage forces due to the friction between the water and the pores of the foundation material. These forces on soil segments are labeled F_1, F_2 in Fig. 35.5 and W_s is the submerged weight of the soil segment. As the flow section is restricted, the flow velocity and the friction forces increase so that F_2 and F_3 are larger than F_1 and F_4. As the water percolates upward at the toe of the dam the seepage force tends to lift the soil. If F_4 is larger than W_s the soil could be "piped out," which is referred to as a piping failure.

The amount of seepage through an earth dam and its foundation, if the latter is pervious, can be estimated from a flow net, a network of streamlines (see Chapter 28, Description of Fluid Flow)

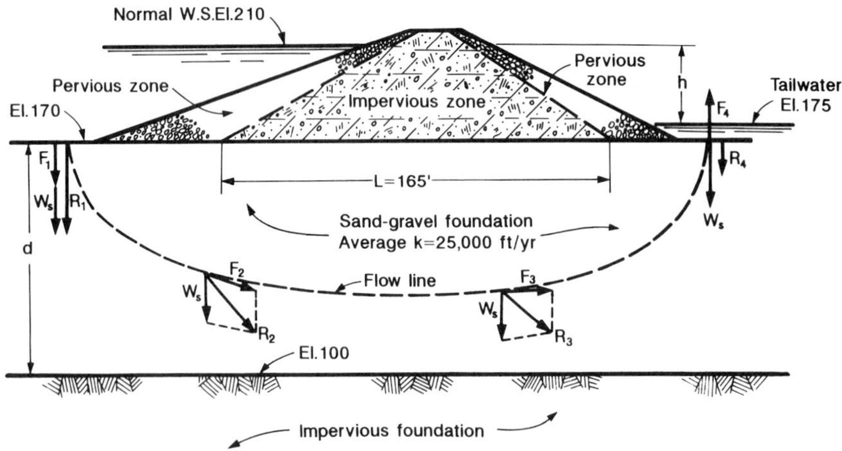

FIGURE 35.5 Seepage under an earth dam. (Adapted from U.S. Department of the Interior, Bureau of Reclamation. 1987. *Design of Small Dams*, pp. 204, 205. U.S. Government Printing Office, Denver, CO.)

tangent to the flow velocity vector, and the equipotential lines, which are normal to the streamlines and are lines of equal pressure head. The network of lines forms figures that tend to be squares when the number of lines becomes large. The streamlines are drawn in such a way that the amount of flow between consecutive streamlines is the same. The energy or head drop between consecutive equipotential lines is also the same. The exterior streamline along which the pressure is atmospheric is the phreatic line. It can be approximated by a parabola [Morris and Wiggert, 1972, p. 243]. The amount of flow through a unit width of dam is $q = N'kh/N$, where N is the number of equipotential drops, N' is the number of flow paths (i.e., the number of spaces between streamlines), and K is the hydraulic conductivity of the material.

A simple method of stability analysis for small earth dams assumes that the surface of failure is cylindrical. The slide mass above an arbitrary slip-circle is divided into a number of vertical slices. The factor of safety is equal to the ratio of the sum of the stabilizing moments to the sum of the destabilizing moments of the several slices about the center of the slip-circle (for details see the subsection on slope stability in Chapter 24, "Geotechnical Earthquake Engineering").

The upstream face of earth dams must be protected against erosion and wave action. This can be achieved by covering the upstream face with riprap or a concrete slab. Both should be placed over a filter of graded material and the slab should have drainage weep holes. The downstream face should be protected against erosion using grass or soil cement. Geotextiles are porous synthetic fabrics that do not degrade. They can be used for separation of materials in a zoned embankment to prevent the migration of fines and relieve pore pressure. Geomembranes are impervious and are used to reduce seepage (see Chapter 24).

35.4 Spillways

Spillway Design Flood

Spillways are structures that release the excess flood water that cannot be contained in the allotted storage. In contrast, *outlet works* regulate the release of water impounded by dams. As earth-fill and rock-fill dams are likely to be destroyed if overtopped, it is imperative that the spillways designed for these dams have adequate capacity to prevent overtopping of the embankment. For dams in the high hazard category (i.e., those higher than 40 ft. with an impoundment of more than 10,000 acre-feet and whose failure would involve loss of life or damages of disastrous proportions), the design flow is based on the probable maximum precipitation (PMP). The U.S. National Weather Service has developed generalized PMP charts for the region east of the 105th meridian [Schreiner and Reidel, 1978] and the National Academy of Sciences [1983] has published a map that indicates the appropriate NWS hydrometeorological reports (HMR) for the region west of the 105th meridian (HMR 38 for California, HMR 43 for Northwest States, HMR 49 for Colorado River and Great Basin Drainage). The U.S. Army Corps of Engineers [1982] has issued hydrologic evaluation guidelines with recommended spillway design floods for different sizes of dams and hazard categories. For minor structures, inflow design floods with return periods of 50 to 200 years may be used if permitted by the responsible control agency. The flood hydrograph is estimated and routed through the reservoir assumed to be full to obtain the spillway design flood. (See Chapter 30 for details on hydrograph estimation and flood routing.) It is often economical to have two spillways: a *service spillway* designed for frequently occurring outflows and an *emergency spillway* for extreme event floods. Modifications of dams to accommodate major floods have been reviewed by USCOLD (1992).

Overflow Spillways

The overflow spillway has an ogee-shaped profile that closely conforms to the lower **nappe** or sheet of water falling from a ventilated sharp crested weir (see Chapter 28). Thus for flow at the design

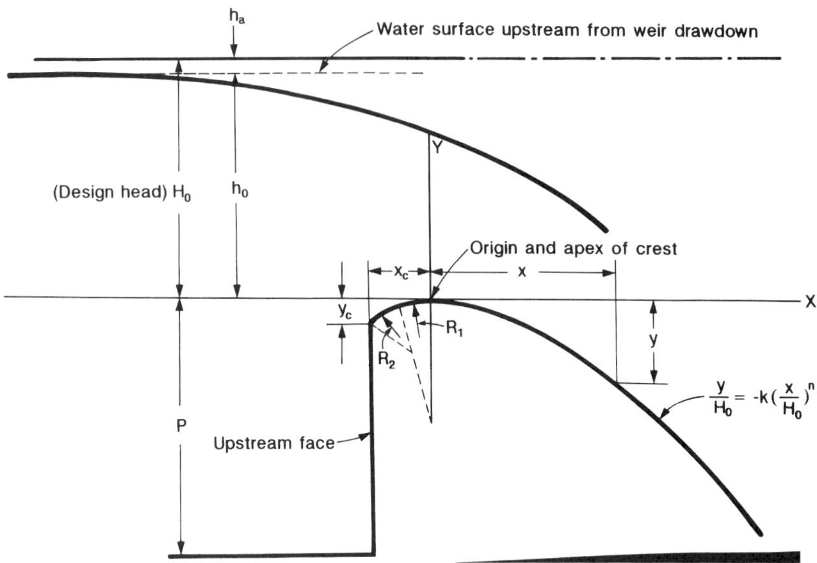

FIGURE 35.6 Ogee-shaped overflow spillway profile. (Adapted from U.S. Department of the Interior, Bureau of Reclamation. 1987. *Design of Small Dams*, p. 366. U.S. Government Printing Office, Denver, CO.)

discharge the water glides over the spillway crest with almost no interference from the boundary. Below the ogee curve the profile follows a tangent with the slope required for structural stability (see "Stability of Gravity Dams" earlier in this chapter). At the bottom of this tangent there is a reverse curve that turns the flow onto the apron of a stilling basin or into a discharge channel. The shape of the ogee is shown in Fig. 35.6 and can be expressed as

$$y/H_0 = -K(x/H_0)^n \tag{35.4}$$

where x and y are the horizontal and vertical distances from the crest, H_0 is the design head including the velocity head of approach h_a, and K and n are coefficients that depend on the slope of the upstream face of the spillway and h_a. For a vertical face and a negligible velocity of approach $K = 0.5$ and $n = 1.85$. Other values can be found in U.S. Department of the Interior, Bureau of Reclamation [1987]. The discharge over an overflow spillway is given by

$$Q = CLH_e^{3/2} \tag{35.5}$$

where Q is the discharge, C is the discharge coefficient (units $L^{1/2}T^{-1}$), H_e is the actual head over the weir including the velocity head of approach, h_a, and L is the effective length of the crest. The basic discharge coefficient for a vertical-faced ogee crest is designated by C_0 and is shown in Fig. 35.7 as a function of the design head H_0. For a head H_e other than the design head, H_0, for an ogee with sloping upstream face and for the effect of the downstream flow conditions, U.S. Department of the Interior, Bureau of Reclamation [1987] gives correction factors to be applied to C_0 to obtain the discharge coefficient C. When the flow over the crest is contracted by abutments or piers the effective length of the crest is

$$L = L' - 2(NK_p + K_a)H_e \tag{35.6}$$

where L' is the net crest length; N is the number of piers; and K_p is the pier contraction coefficient, with a value of 0.02 for square-nosed piers with rounded corners (radius = 0.1 pier thickness), 0.01 for rounded-nose piers, and 0.0 for pointed-nose piers. K_a is the abutment contraction coefficient with a value of 0.20 for square abutments with headwall perpendicular to the flow and 0.10 for rounded abutment ($0.5\,H_0 \leq r \leq 0.15\,H_0$) with headwalls perpendicular to the flow.

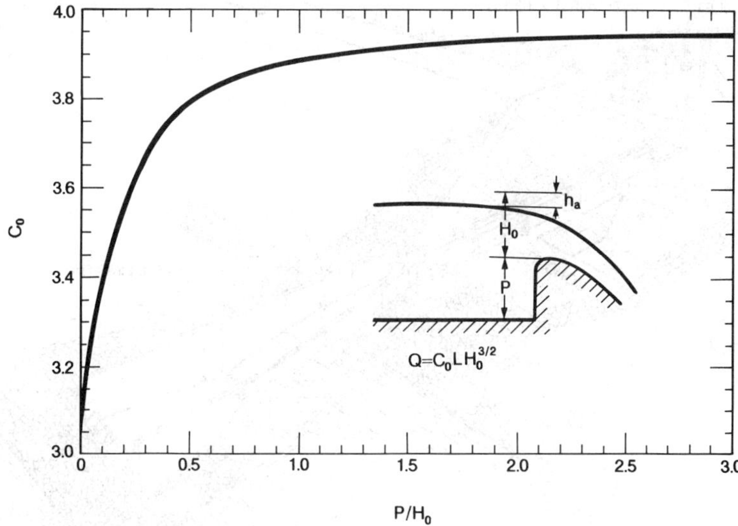

FIGURE 35.7 Discharge coefficient for Q in ft^3/s, with L and H_0 in feet for vertical-faced ogee crest. (Adapted from U.S. Department of the Interior, Bureau of Reclamation. 1987. *Design of Small Dams*, p. 370. U.S. Government Printing Office, Denver, CO.)

Other Types of Spillways

In straight drop or free overfall spillways the flow drops freely from the crest, which has a nearly vertical face. The underside of the nappe must be ventilated. Scour will occur at the base of the overfall if no protection is provided. A hydraulic jump will form if the overfall jet impinges on a flat apron with sufficient tailwater depth.

Chute spillways convey the discharge through an open channel placed along dam abutments or through saddles in the reservoir peripheries. They are often used in conjunction with earth dams. Generally, upstream of the crest the flows are subcritical, passing through critical at the control section and accelerating at supercritical velocity until the terminal structure is reached. Because of the high flow velocity all vertical curves and changes in alignment must be very gradual. Concrete floor slabs are provided with expansion joints that must be kept watertight and drains are provided at intervals under the slab to prevent piping.

Side channel spillways are placed parallel to and along the upper reaches of the discharge channels. Thus flows pass over the crest and then turn approximately 90° to run into the parallel discharge channel. This layout is advantageous for narrow canyons where there is not sufficient space to accommodate an overflow or a chute spillway. Flows from the side channel can be directed to an open channel or to a closed conduit or to a tunnel leading to the terminal structure. Discharge in the side channel increases in the downstream direction. Details of the analysis of such spatially varying flows are treated in Chow [1959] and U.S. Department of the Interior, Bureau of Reclamation [1987].

In shaft or **morning glory spillways** the water first passes over a circular weir discharging into a vertical or sloping shaft followed by a horizontal or nearly horizontal tunnel leading to the terminal structure. The overflow crest is often provided with antivortex piers, as shown in Fig. 35.8. At low flows the discharge over the spillway is $Q = CLh^{3/2}$, where h is the head over the crest and L is the crest length. When the shaft is full and the inlet is submerged the spillway functions as a pipe flowing full. The discharge is then proportional to $H^{1/2}$, where H is the difference between the reservoir elevation and the elevation of the pipe outlet. Trash racks may be desirable to avoid clogging or damage by debris. In high shaft spillways there is the possibility of **cavitation** in the bend between the vertical shaft and the conduit. This situation must be avoided.

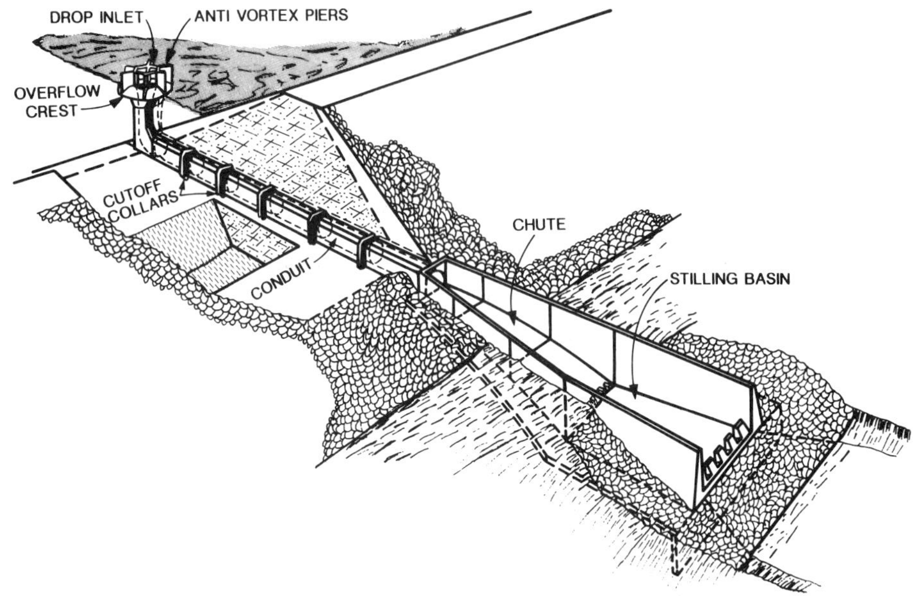

FIGURE 35.8 Shaft spillway. (*Source:* U.S. Department of the Interior, Bureau of Reclamation. 1987. *Design of Small Dams*, p. 358. U.S. Government Printing Office, Denver, CO.)

Siphon spillways can be used when large capacities are not required, space is limited, and fluctuations of reservoir level must be maintained within close limits. When the siphon flows full the discharge is given by an orifice equation (see Chapter 28).

$$Q = C_d A[2gH]^{1/2} \tag{35.7}$$

where the coefficient of discharge C_d is approximately 0.9 and H is the difference in elevation between the reservoir and the siphon outlet. In order to avoid cavitation the maximum average velocity in the siphon must be limited to (in English units)

$$V = [90.5 r_i \log(r_o/r_i)]/(r_o - r_i) \tag{35.8}$$

where r_o and r_i are the outside and inside radii at the crown of the siphon, assuming a free vortex at the crown [Morris and Wiggert, 1972]. Without contraction at the outlet, the maximum permissible head, H, associated with this velocity is

$$H = (1 + k_e + k_f + k_b)V^2/(2g) \tag{35.9}$$

where k_e, k_f, k_b are the coefficients for entrance, friction, and bend losses, respectively.

Spillway Crest Gates

Several types of gates can be installed on spillway crests in order to obtain additional storage. By opening the gates, partial or full spillway discharge capacity can be obtained. The radial or **Tainter gate** has an upstream surface which is a sector of a cylinder. Thus the hydrostatic force goes through the pivot or trunnion located at the center of the circular arc (see Chapter 28). Other types of crest gates include vertical lift gates, drum gates, and roller gates. Horizontal boards or stop logs laid between grooved piers can be used in small installations. Flashboards or wooden panels held by vertical pins anchored on the crest of the spillway are sometimes used in small installations to temporarily raise the water surface.

35.5 Outlet Works

Components and Layout

The purpose of the outlet works is to regulate the operational outflows from the reservoir. The intake structure forms the entrance to the outlet works. It may also include trash racks, fish screens, and gates. The conduit entrance may be vertical, inclined, or horizontal. The conduit may be free flowing or under pressure. For low dams the outlet may be a gated open channel. For higher earth dams the outlet may be a cut-and-cover conduit or a tunnel through an abutment. For concrete dams the outlet is generally a pipe embedded in the masonry or the outlet is formed through the spillway using a common stilling basin to dissipate the excess energy of both the spillway and outlet works outflows. Diversion tunnels used for the construction can in some cases be converted into outlet works. Examples of layout of outlet works may be found in U.S. Department of the Interior, Bureau of Reclamation [1987].

Hydraulics of Outlet Works

When the outlet is under pressure it performs as a system of pipes and fittings in series, as shown in Fig 35.9. It typically includes trash racks, an inlet, conduits, expansions, contractions, bends, guard gates (usually fully open or fully closed for the purpose of isolating a segment of the system), and control valves for the regulation of the flow. The total head H_T, the difference in elevation between the reservoir and the centerline of the outlet, is used in overcoming the losses and producing the velocity head at the exit:

$$H_T = \sum h_l + h_v \tag{35.10}$$

where $\sum h_l$ is the sum of the applicable losses due to trash racks, entrance, bends, friction, expansion, contraction, and gate valves and h_v is the exit velocity head. These losses are expressed as $h_l = KV^2/(2g)$, except for the contraction and expansion losses, which are expressed as $h_l = K(V_1^2 - V_2^2)/(2g)$. Appropriate values of the coefficient K may be found in Chapter 28 as well as in U.S. Department of the Interior, Bureau of Reclamation [1987] and in Brater [1982]. Additional information on design of trashracks may be found in ASCE [1993].

When the outlet works flow as open channels the flows are usually controlled by head gates. As the channel may be nonprismatic, the flow profile is calculated by the procedure described in Chapter 29. An example of design can be found in U.S. Department of the Interior, Bureau of Reclamation [1987].

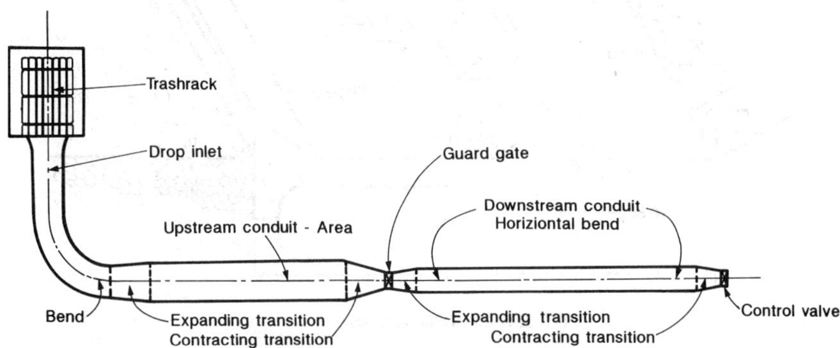

FIGURE 35.9 Outlet works pressure conduit. (Adapted from U.S. Department of the Interior, Bureau of Reclamation. 1987. *Design of Small Dams,* p. 457. U.S. Government Printing Office, Denver, CO.)

35.6 Energy Dissipation Structures

When spillways or outlet works flows reach the downstream river a large portion of the static head has been converted into kinetic energy. Energy dissipation structures are therefore needed to prevent scour at the toe of the dam or erosion in the receiving stream or damage to the adjacent structures. As the flow from the spillway or the outlet works is usually supercritical, the hydraulic jump provides an efficient way of dissipating energy as the flow goes from supercritical to subcritical. The ratio of the depth d_1 before the jump to the conjugate depth d_2 after the jump is

$$d_2/d_1 = \frac{1}{2}[(1 + 8F^2)^{1/2} - 1] \tag{35.11}$$

where $F = V_1/(gd_1)^{1/2}$ is the Froude number of the incoming flow. The head dissipated in the jump, h_j, is

$$h_j = (d_1 + V_1^2/2g) - (d_2 + V_2^2/2g) \tag{35.12}$$

where V_1 and V_2 are the velocities before and after the jump. The U.S. Bureau of Reclamation has developed several types of **stilling basins** to stabilize the position of the jump and improve energy dissipation. Figure 35.10 shows a type III stilling basin for $F > 4.5$ and $V_1 < 60$ ft/s. The tailwater depth in the downstream channel is taken equal to the conjugate depth d_2. Similar information for other ranges of Froude numbers can be found in U.S. Department of the Interior, Bureau of Reclamation [1987].

Usually the conjugate depth d_2 and the tailwater depth TW in the discharging stream cannot be matched for all discharges. The elevation of the floor of the stilling pool can be set such that d_2 and TW match at the maximum discharge. If $TW > d_2$ at lower discharges the conjugate depth d_2 can be raised by widening the stilling basin, and a closer fit between the two rating curves can be

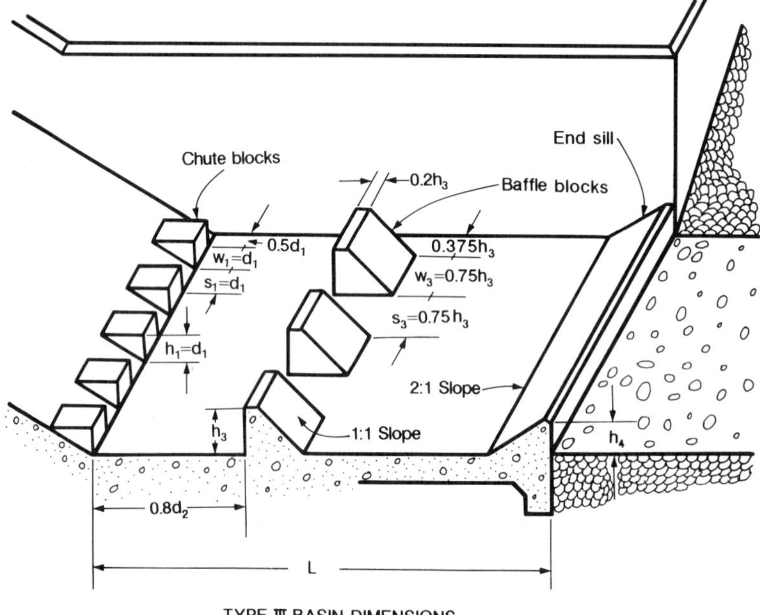

FIGURE 35.10 USBR stilling basin type III for Froude numbers larger than 4.5 and incoming flow velocity less than 60 ft/s. (*Source:* U.S. Department of the Interior, Bureau of Reclamation. 1987. *Design of Small Dams*, p. 391. U.S. Government Printing Office, Denver, CO.)

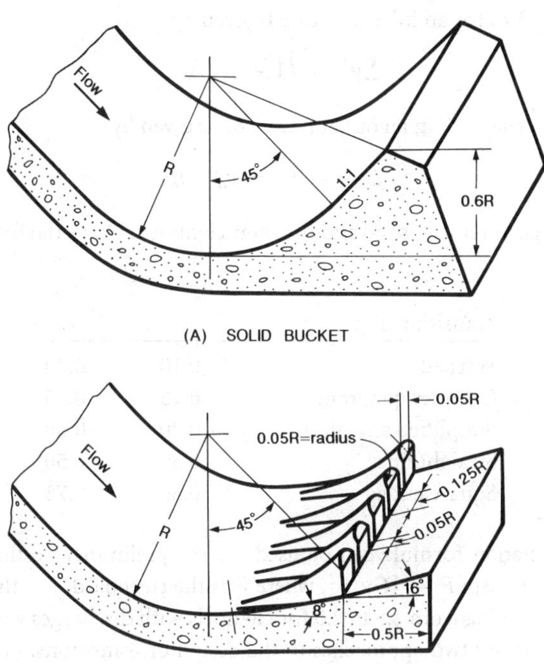

FIGURE 35.11 Submerged bucket dissipators. (*Source:* U.S. Department of the Interior, Bureau of Reclamation. 1987. *Design of Small Dams*, p. 398. U.S. Government Printing Office, Denver, CO.)

obtained. For some rating curves the tailwater and the conjugate depth rating curves can be matched for an intermediate discharge (see U.S. Department of the Interior, Bureau of Reclamation, 1987, p. 397). Whether the basin is widened depends on hydraulic and economic considerations.

The submerged bucket dissipator can be used when the tailwater is too deep for the formation of a hydraulic jump. There are two types: the solid bucket and the slotted bucket. Figure 35.11 shows the geometry of these dissipators. The dissipation is due to the formation of two rollers rotating in opposite directions. Curves for their design can be found in U.S. Department of the Interior, Bureau of Reclamation [1987]. The slotted bucket is the preferred design although the range of acceptable discharges is more limited. Other types of smaller dissipation structures are the straight drop spillway, the slotted grating dissipator, and the impact-type stilling basin. The latter type can be used with an open chute or a closed conduit and its performance does not depend on the tailwater.

35.7 Open Channel Transitions

Subcritical Transitions

Transitions are needed to connect channels of different cross sections—for example, to connect a trapezoidal channel to a rectangular flume or to a circular conduit to cross over a valley on an aqueduct or under a valley with an inverted siphon, respectively. These typically are contracting transitions. Likewise the transition from a rectangular flume or a circular conduit to a trapezoidal channel usually is through an expanding transition. Chow [1959] recommends an optimum maximum angle of 12.5° between the channel axis and a line connecting the channel sides. The drop of water surface, $\Delta y'$, for an inlet structure is given by

$$\Delta y' = (1 + c_i)\Delta h_v \tag{35.13}$$

and the rise in water surface, y', in an outlet transition is given by

$$\Delta y' = (1 - c_o)\Delta h_v \tag{35.14}$$

where Δh_v is the change in velocity head and the coefficients c_i, c_o have the following values [Chow, 1959]:

Transition Type	c_i	c_o
Warped	0.10	0.20
Cylinder quadrant	0.15	0.25
Simplified straight line	0.20	0.30
Straight line	0.30	0.50
Square ended	0.30	0.75

The Bureau of Reclamation formula can be used for the preliminary estimates of freeboard in channels less than 12 ft deep: $F = [Cy]^{1/2}$, where F is the freeboard, y is the depth, both in feet, and C is a coefficient varying from 1.5 to 2.5 for channels with discharges varying from 20 to 3000 ft^3/s, respectively. There are two approaches to the design of transitions: (1) a free water surface is assumed (for example, two reversed parabolas), and the depth is calculated for assumed width and side slope [Chow, 1959, pp. 310–317; French, 1985]; or (2) the boundaries are set first and the surface is calculated [Vittal and Chiranjeevi, 1983; French, 1985].

Supercritical Contractions

Contractions designed for subcritical flows will not function properly for supercritical flows. Generally, with supercritical flow, wave patterns are formed in the contraction and propagate in the downstream channel. Supercritical flow contractions are best designed for rectangular channels. The converging angles on each side produce two oblique hydraulic jumps that make an angle with the original flow direction. A second pair of oblique jumps is created by the diverging angle at the downstream end of the contraction. Ippen and Dawson [1951] devised a design such that the disturbances caused by the converging angles are canceled by the disturbances caused by the diverging angles, so there is no wave pattern in the channel downstream of the contraction. This design will function properly only for the specified Froude number in the upstream channel. Additional details on the design of supercritical transitions can be found in Ippen [1950], Chow [1959], Henderson [1966], French [1985] and Sturm [1985].

35.8 Culverts

Flow Types

Culverts are short conduits that convey flows under a roadway or other embankment. They are generally constructed of concrete or corrugated metal. Common shapes include circular, rectangular, elliptical, and arch. Culverts may flow full or partly full. When the culvert flows full it functions as a pipe under pressure. When it flows partly full it functions as an open channel and the flow can be subcritical, critical, or supercritical. (See Chapter 29, "Open Channel Hydraulics.") The flow conditions that can occur in culverts are shown in Fig. 35.12, where H_1 is the headwater height and H_2 is the vertical distance from the water surface at the outlet to the culvert invert at the inlet.

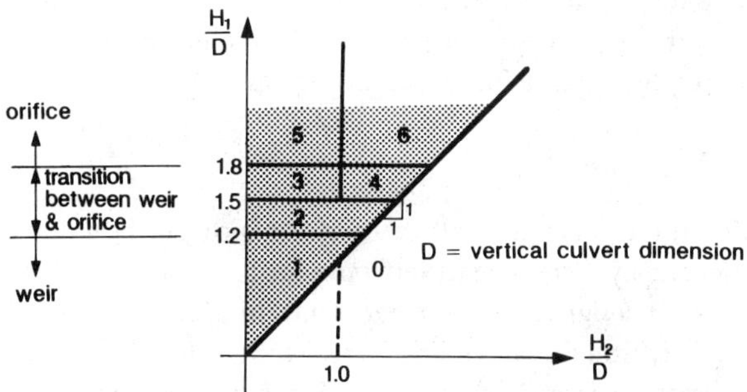

FIGURE 35.12 Culvert flow domains. 0—zero flow; 1—submerged weir flow; 2—submerged transition flow; 3—transition flow; 4—minimum (transition or outlet control flow); 5—orifice flow; 6—minimum (orifice or outlet control flow). (*Source:* GKY and Associates, Inc. 1992. *HYDRAIN—Integrated Drainage Design Computer System, Volume IV: HYCLV—Culverts.* Report to the Federal Highway Administration.)

Partly full flow can occur with inlet control or with outlet control. When operating under inlet control the flow becomes critical just inside the entrance and the flow is supercritical through the length of the culvert if the outlet is unsubmerged; if the outlet is submerged a hydraulic jump forms in the barrel. For low unsubmerged headwater the entrance of the culvert operates as a weir (Eq. 35.5). When the headwaters submerge the entrance it performs as an orifice (Eq. 35.7). From tests by the National Bureau of Standards performed for the Bureau of Public Roads (now the Federal Highway Administration), equations have been obtained to calculate the headwater above the inlet invert, for unsubmerged and submerged inlet control [Norman *et al.*, 1985, Appendix A].

For a full flow condition the total loss, H_L, through the conduit is (in English units)

$$H_L = [1 + k_e + 29\,n^2 L/R^{1.33}]V^2/2g \tag{35.15}$$

where k_e is an entrance loss coefficient, L is the length of the culvert in feet, R is the hydraulic radius in feet, n is Manning's roughness coefficient, and V is the flow velocity in feet per second. For a full flow condition the energy and hydraulic grade line are shown in Fig. 35.13, and the relation between points at the free surface upstream and downstream of the culvert is

$$HW + V_u^2/(2g) = TW + V_d^2/(2g) + H_L \tag{35.16}$$

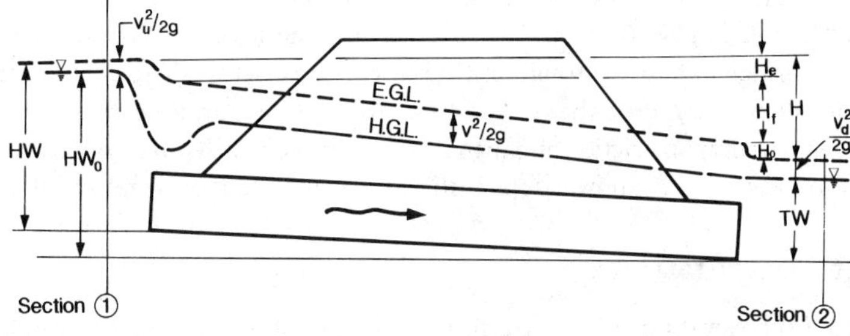

FIGURE 35.13 Energy and hydraulic grade line for full flow condition. (*Source:* Norman, J. M., Houghtalen, R. J., and Johnston, W. J. 1985. *Hydraulic Design of Highway Culverts.* Federal Highway Administration Report No. FHWA-IP-85-15.)

where HW is the headwater depth above the outlet invert, V_u is the upstream velocity of approach, TW is the tailwater depth above the outlet invert, V_d is the downstream velocity, and H_L is the total loss through the conduit given in the preceding equation.

Inlets

For culverts operating under inlet control, the performance can be improved by reducing the contraction at the inlet and by increasing the effective head. These objectives can be achieved with tapered inlets. The simplest design is the side tapered inlet in which the side walls are flared between the throat and the face resulting in an enlarged face section. In addition, the effective head can be increased by depressing the inlet. This can be achieved by an upstream depression between the wingwalls or a sump upstream of the face section. A more elaborate design is the slope tapered inlet, which has flared sidewalls. A fall is incorporated between the throat and the face sections.

Software

The principal public domain computer programs for the hydraulic design and analysis of culverts are the command driven program HYCLV and the interactive program HY8, part of the HYDRAIN [GKY, 1992] package prepared for the Federal Highway Administration, and HEC-2, developed by the Hydrologic Engineering Center of the U.S. Army Corps of Engineers. These programs are discussed in more detail in Chapter 36, "Computational Hydraulics and Hydrology."

35.9 Bridge Constrictions

Backwater and Discharge Approaches

The hydraulics of bridges can be approached from the point of view of the highway engineer or from that of the hydrologist. The highway engineer is concerned with the amount of backwater created by a bridge constriction. Approach embankments are often extended in the flood plain to reduce the span of the bridge proper and thus decrease the cost of bridge crossings. The flow is thus forced to pass through a constriction that may produce a backwater. In contrast the hydrologist is concerned with the indirect determination of the discharge of a large flood from high-water-mark observations and from the geometry of the constriction. Indirect methods of discharge determination are needed when the flow rate is beyond the range of the rating curves at gaging stations.

Flow Types

Three different types of flow can occur at a bridge constriction. The first occurs when the flow is subcritical both in the stream and through the constriction. This is the flow type most generally encountered, illustrated in Fig. 35.14 and discussed below. In the second type the flow is subcritical upstream of the bridge but goes through critical in the constriction. The flow can then return to subcritical immediately as it exits the constriction, or the water surface can dip below the critical depth downstream of the constriction and then return to its normal depth through a hydraulic jump. The third type occurs when the flow is supercritical through both the stream and the constriction.

Backwater Computation

The backwater superelevation, h_1^*, upstream of a constriction with subcritical flow is given by Bradley [1978]:

$$h_1^* = K^* \alpha_2 V_{n2}^2/(2g) + \alpha_1 [(A_{n2}/A_4)^2 - (A_{n2}/A_1)^2] V_{n2}^2/(2g) \qquad (35.17)$$

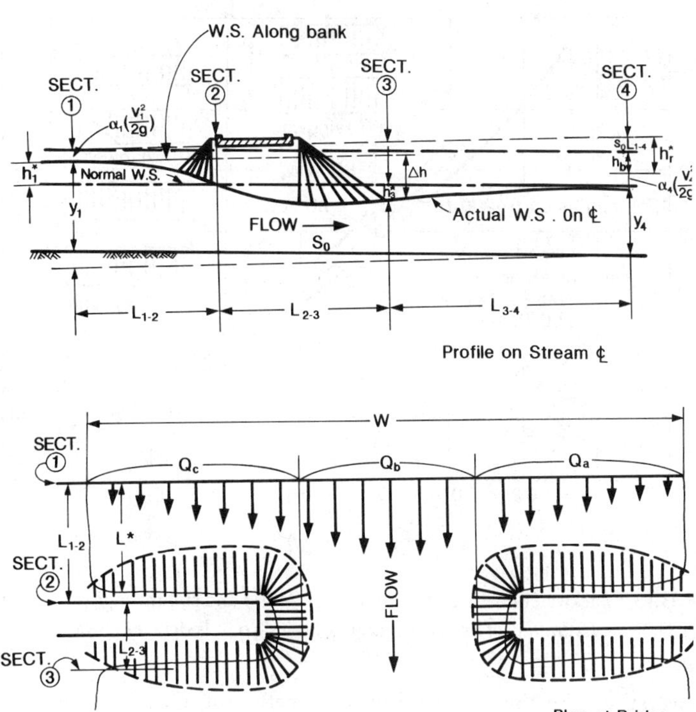

FIGURE 35.14 Normal bridge crossing with spillthrough abutments. (*Source:* Bradley, J. N. 1978. *Hydraulics of Bridges.* Hydraulic Series No. 1, Federal Highway Administration.)

where K^* is the backwater coefficient, A_{n2} is the gross water area in constriction measured below normal stage, $V_{n2} = Q/A_{n2}$ is the average velocity in the constriction, A_4 is the area at section 4 where the normal depth is reestablished, A_1 is the water area at section 1 including that produced by the backwater, and α_1 is the kinetic energy correction factor $\alpha_1 = \sum qv^2/(QV_1^2)$. In Fig. 35.15 the basic backwater coefficient K_b is given as a function of the bridge opening ratio, $M = Q_b/Q$, where Q_b is the flow in the portion of channel within projected length of bridge (Fig. 35.14) and Q is the total discharge. Incremental coefficients to account for the effects of the piers, opening eccentricity, skewed crossing, dual bridges, bridge girder submergence, and backwater in the stream may be found in Bradley [1978]. The sum of K_b and the incremental coefficients yields K^*.

Discharge Estimation

The discharge, Q, through a bridge constriction, under subcritical flow condition, can be calculated from Chow [1959], Kindsvater *et al.* [1953]:

$$Q = CA_3[2g(\Delta h - h_f + \alpha_1 V_1^2/(2g))]^{1/2} \tag{35.18}$$

where C is a discharge coefficient, A_3 is the flow area at section 3, Δh is the drop in water surface between section 1 and section 3, V_1 is the mean velocity at section 1, and h_f is the head loss between sections 1 and 3. This loss is calculated from

$$h_f = L_a[Q/(K_1K_3)^{1/2}]^2 + L[Q/K_3]^2 \tag{35.19}$$

where L_a is the approach length from section 1 to the upstream face of the abutment, L is the length of the contraction, and $K_1 = (1.486/n)A_1R_1^{2/3}$ is the conveyance at section 1. Similarly,

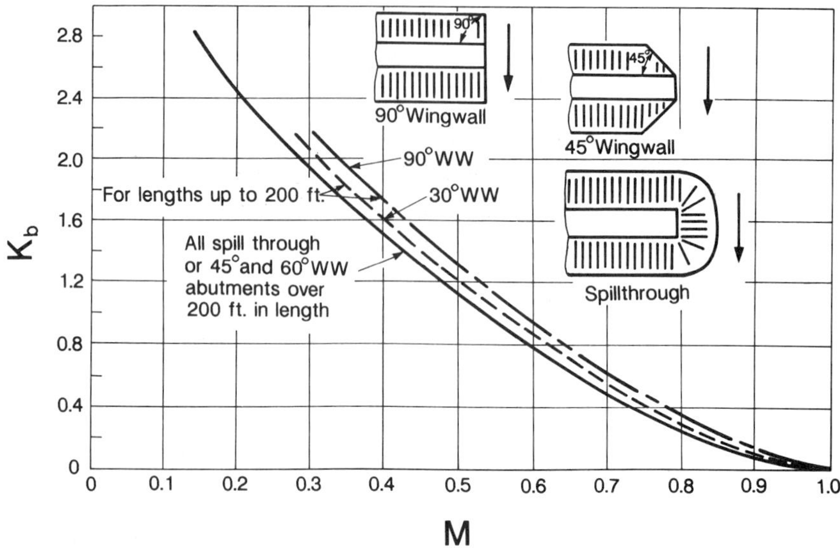

FIGURE 35.15 Backwater coefficient for wingwall (WW) and spillthrough abutments. (*Source:* Bradley, J. N. 1978. *Hydraulics of Bridges.* Hydraulic Series No. 1, Federal Highway Administration.)

K_3 is the conveyance at section 3. The discharge coefficient, C, depends on the shape of the abutment, the ratio of the contraction length to the contraction width, L/b, and the contraction ratio $m = 1 - K_b/K_1$, where b is the constriction width, K_b is the conveyance of the contracted section, and K_1 is the conveyance of the approach section 1. Figure 35.16 gives the base discharge coefficient for the bridge opening with spillthrough abutments. Chow [1959] and Kindsvater *et al.* [1953] give curves for determination of the discharge coefficient for different abutment types and multiplicative correction factors to account for the effects of the Froude number, the rounding and chamfering of the abutment, the skewness and eccentricity of the bridge, the possible submergence of the bridge girders, and the piers and piles.

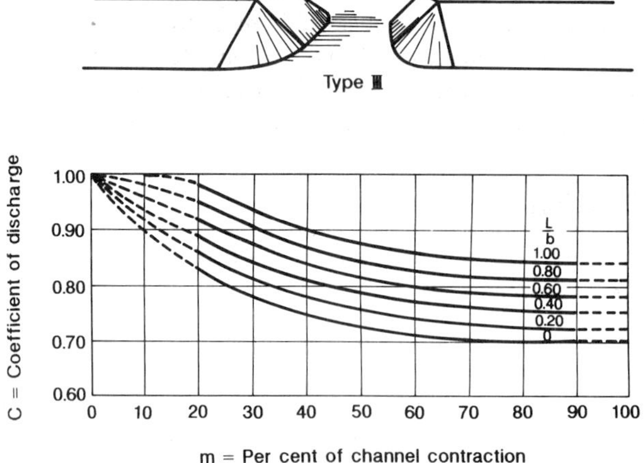

FIGURE 35.16 Discharge coefficient for spillthrough abutment bridge opening. (*Source:* Kindsvater, C. E., Carter, R. W., and Tracy, H. J. 1953. *Computation of the Peak Discharge at Contractions.* U.S. Geological Survey, Circular No. 284.)

Software

The principal public domain computer programs for hydraulics of bridges are WSPRO—Step Backwater and Bridge Hydraulics, part of the HYDRAIN (GKY, 1992) package prepared for the Federal Highway Administration, and HEC-2, developed by the Hydrologic Engineering Center of the U.S. Army Corps of Engineers. These programs are discussed in some detail in Chapter 36, "Computational Hydraulics and Hydrology."

35.10 Pipes

Networks

For pipe network calculations it is convenient to express the friction loss in a pipe by an equation of the form

$$h_L = KQ^x \tag{35.20}$$

where K includes the effects of the pipe diameter, length, and roughness as well as the fluid viscosity. Typical of this form is the Hazen–Williams formula, which can be written as

$$h_f = [4.73C^{-1.852}LD^{-4.87}]Q^{1.852} \tag{35.21}$$

The Darcy–Weisbach formula can be written as

$$h_f = [8fL\pi^{-2}g^{-1}D^{-5}]Q^2 \tag{35.22}$$

In the above formulas h_f is the friction loss, L is the pipe length, D is the diameter, all in feet, Q is the discharge in ft^3/s, f is the friction factor (obtained from the Moody diagram—see Chapter 28), and C is the Hazen–Williams coefficient, typical values of which are given in Table 35.1.

Two basic relationships must be satisfied in a network: the continuity or conservation of mass at each junction and the energy relationship around any closed loop. The continuity relationship requires that the sum of the flows entering a node be equal to the sum of the flows leaving it. The energy relationship requires that the algebraic sum of the head losses in any loop be zero using an appropriate sign convention (for example, flows are positive in the counterclockwise direction). The same sign convention applies to all loops of the network.

A numerical procedure for the calculation of the flows of liquids in pipe networks has been proposed by Cross [1936]. It includes the following steps: (1) assume a discharge in each pipe, Q_g, so that the continuity requirement is satisfied at each node; (2) for each pipe loop calculate a first order correction to the discharge

$$\Delta Q = -\sum KQ_g^x / \left[\sum |xKQ_g^{x-1}|\right] \tag{35.23}$$

taking into account the sign convention in the numerator of the right-hand side; (3) in each pipe loop add the corrections algebraically to flow in each pipe (note that a pipe that is common to two loops has different signs depending on which loop is considered, and this pipe will receive two corrections); (4) with Q_g being the new flows, calculate a new correction ΔQ for each loop; (5) in each loop add the correction algebraically to each pipe; and (6) repeat steps 4 and 5 until the correction becomes sufficiently small. For large networks computer programs can be used (see the section on software later in this chapter).

Table 35.1 Hazen–Williams Coefficients

Pipe Material	Condition	Size	C
Cast iron	New	All	130
	5 years old	$\geq$ 12″	120
		8″	119
		4″	118
	10 years old	$\geq$ 24″	113
		12″	111
		4″	107
	20 years old	$\geq$ 24″	100
		12″	96
		4″	89
	30 years old	$\geq$ 30″	90
		16″	87
		4″	75
	40 years old	$\geq$ 30″	83
		16″	80
		4″	64
	50 years old	$\geq$ 40″	77
		24″	74
		4″	55
Welded steel	Same as cast iron 5 years older		
Riveted steel	Same as cast iron 10 years older		
Wood stave	Average value regardless of age		120
Concrete or concrete lined	Large sizes, good workmanship, steel forms		140
	Large sizes, good workmanship, wooden forms		120
	Centrifugally spun		135
Vitrified	In good condition		110
Plastic or drawn tubing			150

Source: Wood, D. J. 1980. *Computer Analysis of Flow in Pipe Networks Including Extended Period of Simulation.* User's Manual. Office of Continuing Education and Extension of the College of Engineering, University of Kentucky, Lexington, KY.

Water Hammer

When the flow of a liquid in a pipe is stopped abruptly (due to a rapid valve closure, for example), the kinetic energy is transformed into elastic energy and a train of positive and negative pressure waves travels up and down the pipe until the energy is dissipated by friction. When the liquid is water this is known as a **water hammer** (Fig. 35.17). The elasticity of the liquid and the pipe material needs to be taken into account. Consider the elastic properties of the water and the pipe: the bulk modulus of the liquid, E (about 3×10^5 psi or 2 GN/m^2 for water) and the modulus of elasticity of the pipe material E_p (about 30×10^6 psi or 200 GN/m^2 for steel). The velocity or *celerity*, c, of the pressure wave is given by

$$c^2 = (E/\rho)[1 + ED/(E_p t_p)]^{-1} \tag{35.24}$$

where ρ is the fluid density, t_p is the thickness of the pipe wall, and D is the pipe diameter. For an initial flow velocity V the rise in pressure head due to the sudden valve closure is obtained from the momentum principle as

$$\Delta H = \Delta p/\gamma = Vc/g \tag{35.25}$$

This is the pressure head obtained when the time of closure of the valve, t_c, is less than the round trip travel time of the pressure wave $2L/c$. For a longer closing time, t, the pressure head

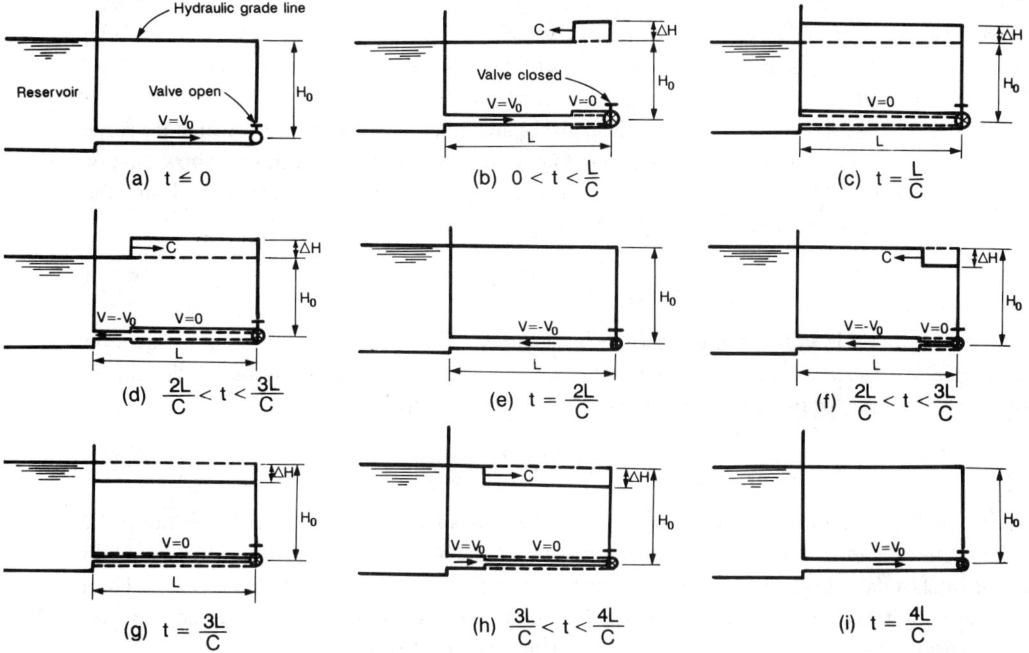

FIGURE 35.17 Water hammer cycle due to instantaneous valve closure.

can be approximated as $(t/t_c)\Delta H$. However, more accurate results can be obtained by numerical integration of the transient flow equations [Morris and Wiggert, 1972; Wylie and Streeter, 1993; Borg 1993]. For an in-depth treatment of water hammer see, for example, Wylie and Streeter [1993], Borg [1993], Rich [1963], Parmakian [1963], Chaudhry [1987], and Jaeger [1977]. There are computer programs for the analysis of water hammer (see the section on software later in this chapter).

Surge Tanks

Surge tanks are standpipes that are installed in large piping systems to relieve the water hammer pressure when a valve is suddenly closed and to provide a reserve of liquid when a valve is suddenly opened. In hydropower installations they are located close to the turbine gates. In pumping installations they are located on the discharge side of the pumps to protect against low pressures during stoppage of the pumps. A simple surge tank is connected directly to the **penstock** (Fig. 35.18).

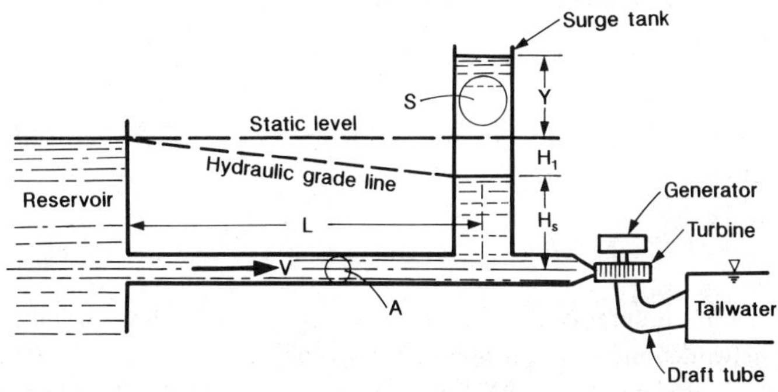

FIGURE 35.18 Simple surge tank.

An orifice surge tank has an orifice in the connection between the tank and the pipe, often with a larger coefficient of discharge for flow out of the tank. A differential surge tank consists of two concentric surge tanks; the inner one is usually a simple surge tank that provides a rapid response but has a small volume. The outer and larger tank is usually an orifice tank.

Consider a horizontal pipe of cross-sectional area A and length L between a reservoir and a surge tank of cross section S, in which the instantaneous water level is at an elevation y above that of the reservoir (Fig. 35.18). V_0 is the steady state flow velocity in the pipe. At the time of closure the surge water elevation obtained neglecting friction is

$$y_{\max} = V_0[(A/S)(L/g)]^{1/2} \tag{35.26}$$

If the pipe is fairly long the friction should be included. This results in a differential equation that requires numerical solution [Morris and Wiggert, 1972]. The minimum cross-sectional area required for stability of the surge tank derived by Thoma and cited by Rich [1963] is

$$S = (AL)/(2gkH_s) \tag{35.27}$$

where $k = H_l/V^2$ and H_s is the steady state head in the surge tank. This area is multiplied by a minimum safety factor of 1.5 for simple surge tanks and 1.25 for orifice surge tanks to obtain reasonably fast damping of the water oscillations according to Borg [1993] and Rich [1963]. For a more detailed treatment of surge tanks and other surge suppressing devices see Wylie and Streeter [1993], Borg [1993], Rich [1963], Chaudry [1987], and Jaeger [1977].

Anchors and Bends

Forces on pipe bends and anchor blocks are obtained by the momentum equation as illustrated in Chapter 28.

External Forces on Pipes and Temperature Stresses

Buried pipes must support the loads due to gravity earth forces and live loads. Load and supporting strength depend on installation conditions. Design details and specifications can be found in ACI, ASTM, AASHTO, or FHWA specifications and industry manuals. ASCE [1992] Manual of Practice 77 (Chapter 14 on structural requirements) gives a good state-of-the-art review. For concentrated and distributed loads superimposed on buried pipes the reader is referred to the AASHTO code, the Portland Cement Association [1951], and the American Concrete Pipe Association [1988] for wheel loads, as well as to ASCE [1992] for a discussion of the Boussinesq theory for concentrated loads.

A temperature change ΔT on a pipe of modulus of elasticity E and coefficient of thermal expansion α will induce a longitudinal stress, assuming fixed ends, given by

$$\sigma = \alpha E \Delta T \tag{35.28}$$

For steel, approximate values of the physical constants are $E = 30 \times 10^6$ psi and $\alpha = 6.5 \times 10^{-6}\ ^\circ\text{F}^{-1}$.

Software

A collection of BASIC programs covering most aspects of hydraulics, including flow in pipe manifolds, pipe networks, and unsteady flow in pipes, can be found in Casey [1992]. For large pipe networks computer software such as KY PIPES [Wood, 1980] is used. These programs typically analyze any network configuration, including storage tanks, pumps, valves, meters, fittings, pressure regulating valves, and check valves. (See Chapter 36, "Computational Hydraulics and Hydrology".) FORTRAN computer programs for water hammer analysis can be found in Wylie and Streeter [1993] and in Chaudhry [1987]. A FORTRAN program for water level oscillations in a simple surge tank is given in Chaudhry [1987]. Tabular presentations of the numerical integration of the unsteady flow equations that can be adapted to spreadsheet software such as EXCEL, LOTUS, and QUATTRO PRO can be found in Morris and Wiggert [1972, p. 335] for water hammer and in Rich [1963] for water hammer, surge tanks, and stability analysis.

35.11 Pumps

Centrifugal Pumps

Centrifugal pumps are those most commonly used in civil engineering applications. The rotating part of the centrifugal pump is the impeller. It consists of blades or vanes attached to the hub. If the blades are enclosed by plates or shrouds on the top and bottom sides, the impeller is closed. Impellers without shrouds (i.e., open impellers) are less prone to become clogged when the liquids contain suspended matter, but closed impellers are more efficient. In radial flow impellers the fluid is forced outward in the radial direction, which is perpendicular to the axis (see Fig. 35.19), while in axial flow impellers the fluid exits along the axis. In the mixed flow pumps the impeller imparts velocities that have both radial and axial components. The flow exits from the impeller into a casing called the volute. Centrifugal pumps can be single stage or multistage. Deep well pumps often are multistage, which is equivalent to several stages or impellers in series so that the total head generated by the pump is the sum of the heads imparted to the fluid at each stage.

The impeller exerts a torque on the fluid. This torque can be calculated as the change in angular momentum, which is obtained by multiplying the lever arm r by the terms of the momentum equation

$$T = \rho Q(V_{t2}r_2 - V_{t1}r_1) \tag{35.29}$$

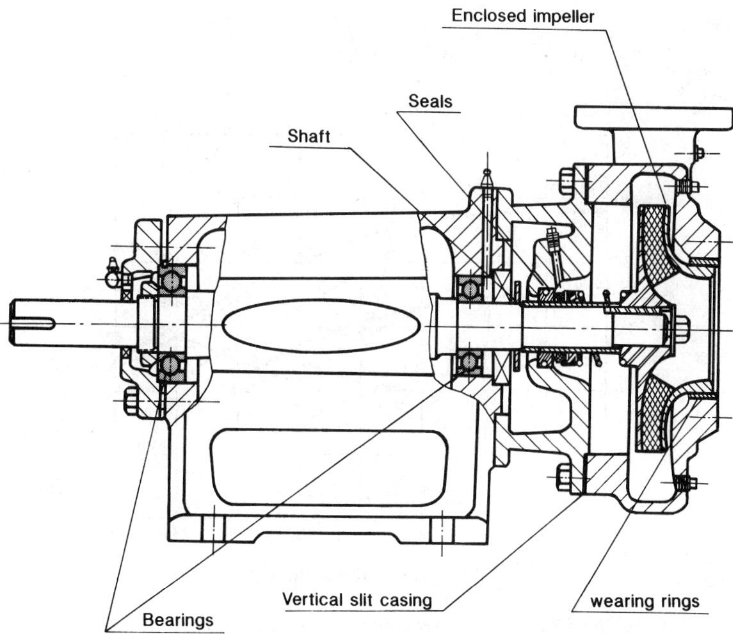

FIGURE 35.19 Radial pump cross section. (Adapted from Peerless Pump Co.)

where V_{t1} and V_{t2} are the tangential components of the flow velocities at the entrance and at the exit of the impeller, respectively, and r_1 and r_2 are the radii at the entrance and exit of the vane. If e is the pump efficiency, then the power to be supplied to the pump shaft by the motor is (HP in English units and kW in SI units)

$$\text{HP} = (T\omega)/(550e) = (\gamma Q H_p)/(550e), \quad \text{kW} = T\omega/e = \gamma Q H_p/e \tag{35.30}$$

where H_p is the head imparted by the pump to the fluid. H_p can be obtained from the energy equation written between the suction side (subscript s) and the discharge side (subscript d) of the pump:

$$z_s + p_s/\gamma + V_s^2/2g + H_p = z_d + p_d/\gamma + V_d^2/2g \tag{35.31}$$

Pump Characteristics

Figure 35.20 is an example of a pump characteristic curve. Pump manufacturers supply characteristic curves for their several designs and operating speeds. The head at zero discharge is called the shutoff head. The head and capacity corresponding to the maximum efficiency are the nominal head and discharge of the pump. Also shown is the net positive suction head (NPSH) curve. The NPSH is the difference between the total head on the suction side of the pump above the atmospheric head, p_a/γ, and the vapor pressure head, p_v/γ:

$$\text{NPSH} = p_a/\gamma + p_{s/\gamma} + V_s^2/(2g) - p_v/\gamma \tag{35.32}$$

If the liquid surface elevation in the supply reservoir is below the axis of the impeller by a distance z_s, there is a suction lift equal to $z_s + h_{fs}$ (where h_{fs} is the friction head loss in the suction pipe) and the pressure head on the suction side of the pump will be negative, that is, below atmospheric:

$$p_s/\gamma = -z_s - h_{fs} - V_s^2/(2g)$$

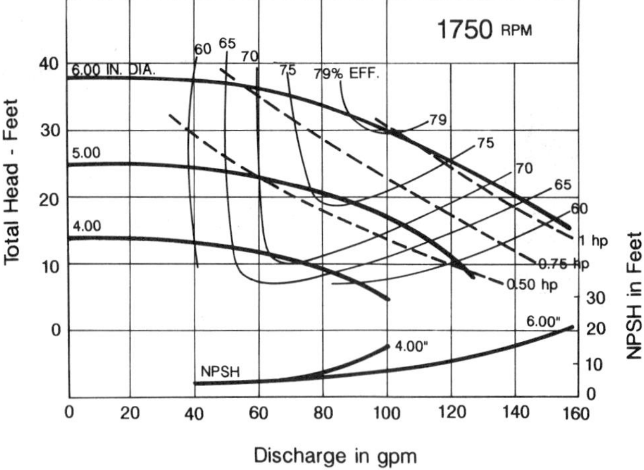

FIGURE 35.20 Characteristic curves of a centrifugal pump. (Adapted from Peerless Pump Co.)

To avoid cavitation the suction pressure head, p_s/γ, must be such that the available NPSH calculated by the above equation is larger than the required NPSH determined from the manufacturer curves or specifications. Equivalently, the calculated value of the cavitation parameter defined as $\sigma = \text{NPHS}/H_p$ must be below the critical value supplied by the manufacturer.

The basic similarity parameter for pumps is the specific speed, defined as

$$N_s = NQ^{1/2}/H_p^{3/4} \tag{35.33}$$

in which N_s is the specific speed, N is the rotational speed in rpm, Q is the discharge in gpm, and H_p is the head in feet. The specific speed can be interpreted as the rpm of a homologous pump operating at a discharge of 1 gpm under a head of 1 ft. The specific speed can be derived from the similarity parameter $W_p/(\rho D^5 n^3)$ obtained in Chapter 28 by noting that the power W_p is proportional to $\gamma Q H_p$ and that the discharge Q is proportional to $D^2 H_p^{1/2}$ or equivalently that the diameter D is proportional to $Q^{1/2}/H_p^{1/4}$. Figure 35.21 shows the optimum efficiency of water pumps as a function of the specific speed.

Pump Systems

Pumps generally operate as part of a system including one or more pipes and perhaps other pumps in series or in parallel. When pumps operate in series the discharge is the same through each pump but the head imparted to the fluid is the sum of the heads imparted by each pump. When the pumps operate in parallel the head is the same but the system discharge is the sum of the discharges of the individual pumps. The operating head and discharge of a pump–pipe system is at the intersection of the pump and pipe discharge-head curves as shown in Fig. 35.22, where Δz is the static lift and $\sum h_f$ is the sum of the head losses in the suction and discharge pipes.

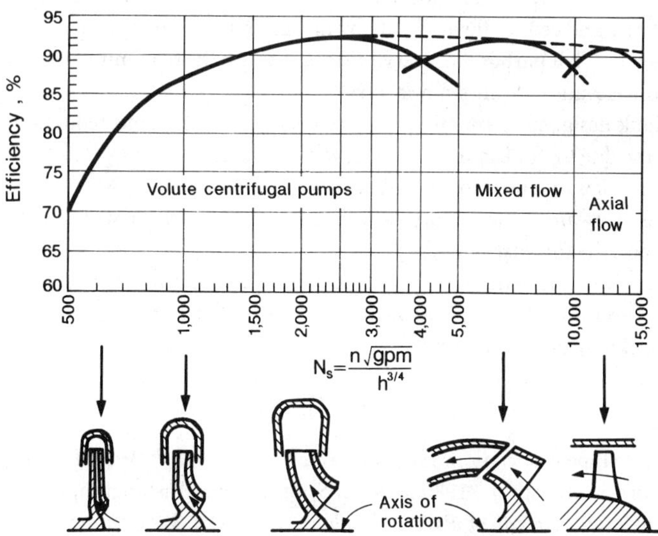

FIGURE 35.21 Optimum efficiency of water pumps as a function of specific speed. (*Source:* Daugherty, R. L., Franzini, J. B., and Finnemore, E. J. 1985. *Fluid Mechanics with Engineering Applications*, 8th ed. McGraw-Hill, New York.)

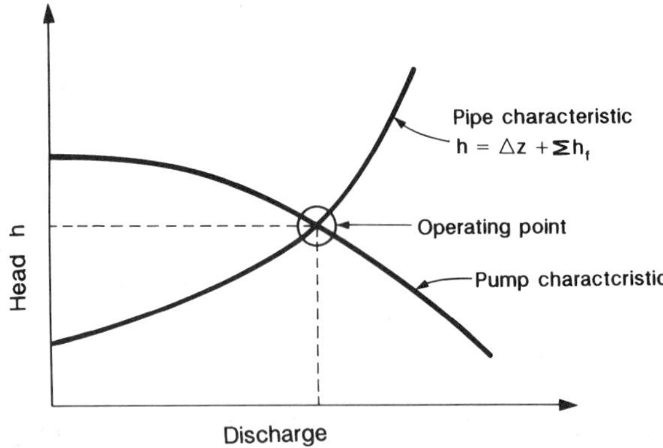

FIGURE 35.22 Pump and pipe system.

Defining Terms

Cavitation: The formation of water vapor cavities due to reduction of local pressure in the water. The cavitation bubbles collapse as they are swept into regions of higher pressure. Continuous implosion of cavitation bubbles severely damages concrete and metal surfaces.

Dead storage: That part of the reservoir storage below the elevation of the lowest outlet.

Extrados: Upstream side of an arch dam.

Intrados: Downstream side of an arch dam.

Morning glory spillway: A spillway consisting of a circular overflow section leading into a vertical shaft connected to a horizontal tunnel. The vertical cross section resembles that of a morning glory flower.

Nappe: The upper and lower nappes are the upper and lower water surface profiles that water takes as it flows over a ventilated sharp-crested weir.

Penstock: A large pipe to carry water to the turbines of a hydroelectric power plant.

Stilling basin: A structure, usually located at the foot of a spillway, to dissipate energy over a short distance and reduce flow velocity by means of a hydraulic jump.

Surcharge storage: That part of the reservoir storage above the maximum operating level, usually taken as the elevation of the spillway crest.

Surge tank: Tank designed to contain pressure upsurges due to rapid reduction of flow velocity in a pipeline due to, for example, a quick valve closure.

Tainter gate: A radial gate used to control flow over a spillway.

Uplift pressure: Upward vertical hydrostatic pressure at an interface, such as between the bottom of a dam and the foundation.

Water hammer: A pressure transient in a pipe system due to a rapid reduction of flow velocity caused, for example, by an adjustment of the setting of a control valve or a change in the operation of a turbine or a pump.

References

American Concrete Pipe Association. 1988. *Concrete Pipe Handbook*, 3rd. ed. ACPA, Arlington, VA.

American Iron and Steel Institute. 1980. *Modern Sewer Design.* Washington, D.C.

ASCE. 1993. *Hydraulic Design of Reversible Flow Trashracks.* Task Committee on Design and Performance of Reversible Flow Trashracks. New York.

ASCE. 1992. *Design and Construction of Urban Stormwater Management Systems.* ASCE Manuals and Reports of Engineering Practice No. 77, WEF Manual of Practice FD-20. New York.

Borg, J. E. 1993. Hydraulic transients. In *Davis's Handbook of Applied Hydraulics*, 4th ed., ed. V. J. Zipparro and H. Hasen. McGraw-Hill, New York.

Bradley, J. N. 1978. *Hydraulics of Bridges.* Hydraulic Series No. 1, Federal Highway Administration. Washington, D.C.

Brater, E. F. 1982. *Handbook of Hydraulics*, 6th ed. McGraw-Hill, New York.

Brune, G. M. 1953. Trap efficiency of reservoirs. *Trans. Am. Geophys. Union.* 34(407). Washington, D.C.

Casey, T. J. 1992. *Water and Wastewater Engineering Hydraulics.* Oxford University Press, Oxford.

Chaudhry, M. H. 1987. *Applied Hydraulic Transients*, 2nd ed. Van Nostrand Reinhold, New York.

Chow, V. T. 1959. *Open Channel Hydraulics.* McGraw-Hill, New York.

Churchill, M. A. 1949. Discussion of "Analysis and use of sediment data" by L. C. Gottschalk. In *Proc. Fed. Interagency Sedimentation Conf.* Denver, CO, p. 139.

Concrete Pipe Association of Indiana. 1974. *Concrete Pipe Design Manual.* Arlington, VA.

Cross, H. 1936. *Analysis of Flow in Networks of Conduits or Conductors.* University of Illinois Bulletin 286. Urbana, IL.

Daugherty, R. L., Franzini, J. B., and Finnemore, E. J. 1985. *Fluid Mechanics with Engineering Applications*, 8th ed. McGraw-Hill, New York.

French, R. H. 1985. *Open Channel Hydraulics.* McGraw-Hill, New York.

GKY and Associates, Inc. 1992. *HYDRAIN—Integrated Drainage Design Computer System, Volume IV: HYCLV—Culverts.* Report to the Federal Highway Administration.

Henderson, F. M. 1966. *Open Channel Flow.* Macmillan, New York.

Hydrologic Engineering Center, 1990. *HEC 2, Water Surface Profiles.* User's Manual, Version 4.5, U.S. Army Corps of Engineers, Davis, CA.

ICOLD. 1992. *Dams and Environment, Socio-economic Impacts.* Bulletin 86, International Commission on Large Dams. Paris.

ICOLD. 1980. *Dams and the Environment.* Bulletin 35, International Commission on Large Dams (and following reports), Commission Internationale des Grands Barrages, Paris.

Ippen, A. T. 1950. Channel transitions and controls. In *Engineering Hydraulics*, ed. H. Rouse. John Wiley & Sons, New York.

Ippen, A. T. and Dawson, J. H. 1951. Design of channel contractions. *Trans. ASCE.* 116:326–346.

Jaeger, C. 1977. *Fluid Transients in Hydro-Electric Practice.* Blackie, Glasgow.

Kindsvater, C. E., Carter, R. W., and Tracy, H. J. 1953. *Computation of the Peak Discharge at Contractions.* U.S. Geological Survey, Circular No. 284.

Linsley, R. K, Franzini, J. B., Freyberg, D. L., and Tchobanoglous, G. 1992. *Water Resources Engineering*, 4th ed. McGraw-Hill, New York.

Mays, L. W. 1979. Optimal design of culverts under uncertainties. *J. Hydraul., ASCE.* 105:443–460.

Morris, H. M. and Wiggert, J. M. 1972. *Applied Hydraulics in Engineering.* Ronald Press, New York.

National Academy of Sciences. 1983. *Safety of Existing Dams: Evaluation and Improvement.* National Academy Press, Washington, D.C.

National Clay Pipe Institute. 1972. *Clay Pipe Engineering Manual.* Crystal Lake, IL.

Norman, J. M., Houghtalen, R. J., and Johnston, W. J. 1985. *Hydraulic Design of Highway Culverts.* Federal Highway Administration Report No. FHWA-IP-85-15.

Parmakian, J. 1963. *Water Hammer Analysis.* Dover, New York.

Portland Cement Association. 1951. *Vertical Pressure on Culverts under Wheel Loads on Concrete Pavement Slabs*, Publication No. ST-65. PCA, Skokie, IL.

Rich, G. R. 1963. *Hydraulic Transients*, 2nd ed. Dover, New York.

Schreiner, L. C. and Reidel, J. T. 1978. *Probable Maximum Precipitation Estimates, United States East of the 105th Meridian.* NOAA Hydrometeorological Report 51, National Weather Service, Washington, D.C.

Sturm, T.W. 1985. Simplified design of contractions in supercritical flow. *J. Hydraul. Eng.* 111(5):871–875.

Thomas, H. A., Jr. and Fiering, M. B. 1963. The nature of the storage yield function. In *Operation Research in Water Quality Management.* Harvard University Water Program. Cambridge, MA.

Tung, Y. K. and Mays, L. W. 1980. Risk analysis for hydraulic design. *J. Hydraul., ASCE.* 106(5): 893–913.

USCOLD. 1991. *Key References for Hydraulic Design.* United States Committee on Large Dams, Denver, CO.

USCOLD. 1992. *Modification of Dams to Accomodate Major Floods.* 12th Annual USCOLD Lecture Series. Denver, CO.

U.S. Army Corps of Engineers. 1982. *National Program of Inspection of Nonfederal Dams.* Final report to Congress.

U.S. Department of the Interior, Bureau of Reclamation. 1987. *Design of Small Dams.* U.S. Government Printing Office, Denver, CO.

U.S. Department of the Interior, Bureau of Reclamation. 1977. *Design of Arch Dams.* U.S. Government Printing Office, Denver, CO.

Vittal, N. and Chiranjeevi, V. V. 1983. Open channel transitions: Rational method of design. *J. Hydraul. Eng., ASCE.* 109(1):99–115.

Wood, D. J. 1980. *Computer Analysis of Flow in Pipe Networks Including Extended Period of Simulation.* User's Manual. Office of Continuing Education and Extension of the College of Engineering, University of Kentucky, Lexington, KY.

Wylie, E. B. and Streeter, V. L. 1993. *Fluid Transients in Systems.* Prentice Hall, Englewood Cliffs, NJ.

Yeh, C. H. and Abdel-Maalek, R. 1993. Concrete dams. In *Davis's Handbook of Applied Hydraulics*, 4th ed., ed. V. J. Zipparro and H. Hasen. McGraw-Hill, New York.

Young, G., Childrey, M., and Trent, R. 1974. Optimal design for highway drainage culverts. *J. Hydraul., ASCE.* 100(HY7):971–994.

Young, G. and Krolak, J. S. 1992. *HYDRAIN Integrated Drainage Design Computer System, Version 4.* GKY and Associates, Springfield, VA.

Zipparro, V. J. and Hasen, H., eds. 1993. *Davis's Handbook of Applied Hydraulics*, 4th ed. McGraw-Hill, New York.

For Further Information

Davis's Handbook of Applied Hydraulics [Zipparro and Hasen, 1993] provides a detailed treatment of reservoirs, natural channels, canals and conduits, dams, spillways, fish facilities, hydroelectric power, navigation locks, irrigation, drainage, water distribution, and wastewater conveyance.

The U.S. Army Corps of Engineers' *Hydraulics Design Criteria* includes practical formulas, nomographs, and design standards on most hydraulic structures.

USCOLD [1991] lists key references for hydraulic design on the following subjects: cavitation in hydraulic structures, increasing discharge capacity of existing projects, spillway design floods, fish passage, gas transfer at hydraulic structures, trash rack vibrations, energy dissipation, and terminal structures for spillways, outlet works, thermal stratification, and instream thermal simulation.

ICOLD [1980, 1992] has published several bulletins (35, 37, 50, 65, 66, and 86) on the environmental effects of dams and reservoirs. Bulletin 35 is technical while the other reports include case histories in several countries.

Design of Small Dams (U.S. Dept. of the Interior, 1987) and *Design of Arch Dams* (U.S. Dept. of the Interior, 1977) provide a wealth of information on dam design and the associated appurtenances.

36

Simulation in Hydraulics and Hydrology

A. R. Rao
Purdue University

C. B. Burke
Christopher B. Burke Engineering, Ltd.

T. T. Burke, Jr.
Purdue University

36.1 Introduction

During the last two decades computer models have become ubiquitous in hydraulics and hydrology. These models have been of invaluable help in hydrologic analysis and design. In this chapter several models are briefly introduced and two models, HEC-1 and TR-20, are illustrated in depth.

Many of these models are quite complex. Books such as that by Hoggan [1989] have been written to explain and illustrate them. It is not possible to discuss these models in the depth required to illustrate their full capabilities. Consequently the approach taken in this section is to give synoptic descriptions of more important uses of these models. The detailed illustrations of HEC-1 and TR-20 are given to demonstrate their characteristics and utility.

36.2 Some Commonly Used Models

Name: Storm Water Management Model (SWMM)
PC and Main Frame versions

Source: Center for Exposure Assessment Modeling (CEAM)
U.S. Environmental Protection Agency
Office of Research and Development
Environmental Research Laboratory
College Station Road
Athens, GA 30613-0801
(706) 546-3549

0-8493-8953-4/95/$0.00 + $.50

User's manual:

Huber, W. C. and Dickinson, R. E. 1988. *Storm Water Management Model (SWMM), Version 4, User's Manual.*
EPA/600/3-88/001a
NTIS accession No. PB88 236 641

Roesner, R. A., Aldrich, J. A., and Dickinson, R. E. *Storm Water Management Model (SWMM), Version 4, User's Manual Part B, Extran Addendum.* 1989.
EPA/600/3-88/0001b
NTIS accession no. PB88 236 658

Availability of user's manuals:

National Technical Information Service (NTIS)
5825 Port Royal Road
Springfield, Virginia 22161
(703) 487-4650

or:

Dr. Wayne Huber
Oregon State University
Department of Civil Engineering
Apperson Hall 202
Corvallis, OR 97331-2302
(503) 737-6150

SWMM is a comprehensive simulation model of urban runoff quantity and quality including surface and subsurface runoff, snow melt, routing through the drainage network, storage, and treatment. Single event and continuous simulation can be performed. Basins may have storm sewers, combined sewers, and natural drainage. The model can be used for planning and design purposes. In the planning mode continuous simulation is used on a coarse schematization of the catchment, and statistical analyses of the hydrographs and pollutographs are produced. In the design mode simulation is performed at shorter time steps using a more detailed schematization of the catchment and specific rainfall events.

The model consists of an executive block, four computational blocks (Runoff, Transport, EXTRAN, and Storage/Treatment), and five service blocks (Statistics, Graph, Combine, Rain, and Temp). The EXECUTIVE block performs a number of control tasks such as assignment of logical units and files, sequencing of computational blocks and error messages. The RUNOFF block generates runoff given the hydrologic characteristics of the catchment, the rainfall, and/or the snow melt hyetographs and the antecedent conditions of the basin. The runoff quality simulation is based on conceptual buildup and washoff relationships (see Chapter 31 on quality of runoff). The TRANSPORT block routes the flows and pollutants through the drainage system. It also generates dry-weather flow and infiltration into the sewer system. As a kinematic wave routing is used (see Chapter 30), backwater effects are not modeled. The EXTENDED TRANSPORT block (EXTRAN) simulates gradually varied unsteady flow using an explicit finite difference form of the Saint Venant Equations (see Chapter 29). EXTRAN can handle backwater effects, conduit pressurization, and closed and looped networks. EXTRAN is limited to quantity and consequently does not perform quality simulation. The STORAGE/TREATMENT block routes flows and up to three different pollutants through storage and a treatment plant with up to five units or processes. The STATISTICS block calculates certain statistics such as moments of event data and produces tables of magnitude, return period, and frequency. The COMBINE block combines the output of SWMM runs from the same or different blocks to model larger areas. The RAIN block reads National Weather Service rainfall data and therefore can be used to read long precipitation records and can perform storm event analysis. The TEMP block reads temperature, evaporation, and wind speed data from the National Weather Service.

Name: Water Surface Profiles (HEC-2)
PC and Main Frame versions

Source: Hydrologic Engineering Center
609 Second Street
Davis, CA 95616
(916) 440-2105

User's manuals:
U.S. Army Corps of Engineers
Water Resources Support Center
The Hydrologic Engineering Center
609 Second Street
Davis, CA 95616

Source code and processing programs:
Available from many vendors who advertise in *Civil Engineering* magazine. HEC also maintains and distributes a list of sources.

The program is intended for calculating water surface profiles for steady gradually varied flow in natural or artificial channels. Both subcritical and supercritical flow profiles can be calculated. The effects of various obstructions such as bridges, culverts, weirs, and structures in the flood plain may be considered in the computations. The computational procedure is based on the solution of the one-dimensional energy equation with energy loss due to friction evaluated with Manning's equation. The computational procedure is generally known as the standard step method. The program is also designed for application in flood plain management and flood insurance studies to evaluate floodway encroachments and to designate flood hazard zones. Also, capabilities are available for assessing the effects of channel improvements and levees on water surface profiles. Input and output units may be either English or metric.

Supplementary Programs

A data edit program (EDIT-2) which checks the data cards for various input errors is available.

A Fortran graphics program (Hydraulics Graphics Package) which produces HEC-2 cross section and profile plots in interactive or batch modes is also available. Documentation for the Hydraulics Graphics Package may be obtained from the Hydrologic Engineering Center.

Two versions of HEC-2 have been developed by the Hydrologic Engineering Center (HEC). A standard version of HEC-2 has been developed on the CDC 6600 computer. It may be used with minor modifications on most medium to large computers. A microcomputer version of HEC-2 has been developed for use on the IBM PC/XT microcomputer. FORTRAN source code for both versions and an executable module for the micro version are available.

Name: Modular Three-Dimensional Finite-Difference
Groundwater Flow Model (MODFLOW)
PC and Main Frame Versions

Source: U.S. Geological Survey
437 National Center
12201 Sunrise Valley Drive
Reston, VA 22092
(703) 648-5695

User's manual:
McDonald, M. G. and Harbaugh, A. V.
U.S. Geological Survey Open-File Report

Books and Open-File Reports Section
U.S. Geological Survey
Federal Center
Box 25425
Denver, CO 80225
(303) 236-7476

Source code and processing programs:

International Groundwater Modelling Center (IGWMC)
Institute for Groundwater Research and Education
Colorado School of Mines
Golden, CO 80401-1887
Phone (303) 273-3103
Fax (303) 273-3278

or:

Scientific Software Group
P.O. Box 23041
Washington, D.C. 20026-3041
Phone (703) 620-9214
Fax (703) 620-6793

or:

Geraghty & Miller
Software Modeling Group
10700 Parkridge Boulevard, Suite 600
Reston, VA 20091
Phone (703) 758-1200
Fax (703) 758-1201

MODFLOW is a finite-difference groundwater model that simulates flow in three dimensions. The modular structure of the program consists of a main program and a series of highly independent subroutines called *modules*. The modules are grouped into *packages*. Each package deals with a specific feature of the hydrologic system which is to be simulated, such as flow from rivers or flow into drains, or with a specific method of solving linear equations which describe the flow system, such as strongly implicit procedure or slice-successive overrelaxation.

Groundwater flow within the aquifer is simulated using a block-centered finite-difference approach. Layers can be simulated as confined, unconfined, or a combination of confined and unconfined. Flow associated with external entities, such as wells, areal recharge, evapotranspiration, drains, and streams, can also be simulated. The finite-difference equations can be solved using either the strongly implicit procedure or slice-successive overrelaxation.

Application of the computer program to solve groundwater flow problems requires knowledge of the following hydrogeologic conditions: (1) hydraulic properties of the aquifer, (2) the shape and physical boundaries of the aquifer system, (3) flow conditions at the boundaries, and (4) initial conditions of groundwater flow and water levels.

The accuracy of the calibrated mathematical model is dependent on the assumptions and approximations in the finite-difference numerical solutions and the distribution and quality of data. Hydraulic properties of the aquifer deposits (estimated by model calibration) can be used to define the flow system and evaluate impacts that would be produced by changes in stress, such as pumping. However, three main limitations that constrain the validity of the model are:

1. The inability of the numerical model to simulate all the complexities of the natural flow system. The assumptions used for construction of the model affect the output and are simplified relative to the natural conditions.

2. The distribution of field data; for example, water level or lithologic data may not be areally or vertically extensive enough to define the system adequately.
3. The model is probably not unique. Many combinations of aquifer properties and recharge-discharge distributions can produce the same results, particularly because the model was calibrated for a predevelopment (steady state) condition. For example, a proportionate change in total sources and sinks of water with respect to transmissivity would result in the same steady state model solution.

Name: HEC-6 Scour and Deposition in Rivers and Reservoirs
PC and Main Frame versions

Source: Hydrologic Engineering Center
U.S. Army Corps of Engineers
609 Second Street
Davis, CA 95616-4687
(916) 756-1104

User's manual:

U.S. Army Corps of Engineers. 1993. *HEC-6: Scour and Deposition in Rivers and Reservoirs—User's Manual.*

Hydrologic Engineering Center. 1987. *COED: Corps of Engineers Editor User's Manual.*

Thomas, W. A., Gee, D. M., and MacArthur, R. C. 1981. *Guidelines for the Calibration and Application of Computer Program HEC-6.*

Availability of user's manuals:

National Technical Information Service
5825 Port Royal Road
Springfield, Virginia 22161
(703) 487-4650
(703) 321-8547

HEC-6 is a member in the series of numerical models developed by the U.S. Army Corps of Engineers Hydrologic Engineering Center for studies of hydraulic or hydrologic problems. It is a one-dimensional model of a river or reservoir where entrainment, deposition, or transport of sediment occur. It is intended for use in the analysis of long-term river or reservoir response to changes in flow or sediment conditions. HEC-6 models changes as a sequence of steady states; the hydraulic and sediment parameters of each may vary during the sequence. It can be applied to analyze problems arising in or from stream networks, channel dredging, levee design, and reservoir deposition.

The hydraulic model is essentially identical to that used in HEC-2 (see the description of HEC-2 for more details), with water-surface profiles being computed using the standard step method and with flow resistance being modeled with Manning's equation. Although flow resistance due to bed forms is not separately considered, Manning's n can be specified as a function of discharge. In a "fixed bed" mode, the sediment transport model can be turned off, and HEC-6 can be used as a limited form of HEC-2 without capabilities for special problems such as bridges or islands. The input formats of both HEC-2 and HEC-6 are very similar, so those familiar with HEC-2 should have little difficulty in setting up the input data for HEC-6.

The sediment transport model is based on the Exner equation (see Chapter 33). It treats graded sediments, ranging from clays and silts to boulders up to 2048 mm in diameter, by dividing both inflow and bed sediments into up to 20 size classes. Several standard transport formulas for sand and gravel sizes are available, and provision for a user-developed formula is included. Additional features include the modeling of armoring, clay and silt transport, and cohesive sediment scour. Because it is a one-dimension model, it does not model bank erosion, meanders, or non-uniform transport at a given cross section.

36.3 TR-20 Program

The TR-20 computer program was developed by the Soil Conservation Service [1982] to assist in the hydrologic evaluation of storm events for water resource projects. TR-20 is a single-event model that computes direct runoff resulting from synthetic or natural rainfall events. There is no provision for recovery of initial abstraction or infiltration during periods without rainfall. The program develops runoff hydrographs from excess precipitation and routes the flow through stream channels and reservoirs. It combines the routed hydrograph with those from tributaries and computes the peak discharges, their times of occurrence, and the water elevations at any desired reach or structure. Up to nine different rainstorm distributions over a watershed under various combinations of land treatment, flood control structures, diversions, and channel modifications may be used in the analyses. Such analyses can be performed on as many as 200 reaches and 99 structures in any one continuous run [SCS, 1982]. The program can be obtained from: *National Technical Information Service, U.S. Department of Commerce, 5285 Port Royal Road, Springfield, VA 22161, (703) 487-4600.*

Summary of TR-20 Input Structure

The input requirements of TR-20 are few. If data from actual rainfall events are not used, the depth of precipitation is the only required meteorological input. For each subarea, the drainage area, runoff curve number, and time of concentration are required; the antecedent soil moisture (AMC) condition (i.e., I, II, or III) can be specified, although the SCS now recommends only AMC II, the so-called average runoff condition [SCS, 1986]. For each channel reach, the length is defined; the channel cross section is defined by the discharge and end area data at different elevations; a routing coefficient, which is optional, may also be specified. If the channel-routing coefficient is not given as input, it will be computed by using a modified Att-Kin (attenuation-kinematic) procedure.

The modified Att-Kin routing procedure is described in TR-66 [SCS, 1982]. It uses storage and kinematic models to attenuate and translate natural flood waves. For each channel reach, the Att-Kin procedure routes an inflow hydrograph through the reach using a storage model. With the storage and kinematic models, flows are routed as a linear combination so that the outflow hydrograph satisfies the conservation of mass at the time to peak of the outflow hydrograph. Used separately, the storage routing provides only attenuation without translation; the kinematic routing provides only translation and distortion without attenuation of the peak [SCS, 1982].

For each structure it is necessary to describe the outflow characteristics with an elevation-discharge-storage relationship. The time increment for all computations and any baseflow in a channel reach must be specified.

Input Structure

The input for a TR-20 run consists of the following four general statement types: Job Control, tabular data, Standard Control statements, and Executive Control statements.

Job Control

The Job Control statements are the JOB statement, two TITLE statements, ENDATA statement, the ENDCMP statement, and ENDJOB statement. The JOB and two TITLE statements *must* appear first and second, respectively, in a run and the ENDJOB statement must appear last. The ENDATA statement separates the standard control and the tabular data from the executive control statements. The ENDCMP statement is used to signify the end of a pass through a watershed or a sub-watershed thereby allowing conditions to be changed by the user before further processing.

Input data are entered according to the specifications in a blank form. Each line of data on the form is a record entered into the computer file. The 80 columns across the top of the form represent the 80 positions on a record. Each statement is used to perform a specific operation. The order of records determines the sequence in which the operations are performed.

A name appears in columns 4 through 9 of all lines except those containing tabular data. These names identify the type of operation performed or type of data entered. A number corresponding to the name in columns 4 through 9 is placed in columns 2 and 11. The digit in column 11 is the operation number and identifies the type of operation, which must be entered unless it is a blank.

Tabular Data

The tabular data serve as support data for the problem description. These data precede both the Standard and Executive Control and have six possible tables:

1. *Routing coefficient table:* the relationship between the streamflow routing coefficient and velocity.
2. *Dimensionless hydrograph table:* the dimensionless curvilinear unit hydrograph ordinates as a function of dimensionless time.
3. *Cumulative rainfall tables.*
4. *Stream cross-section data tables:* a tabular data summary of the water surface elevation–discharge–cross sectional end area relationship.
5. *Structure data table:* a tabular summary of the water surface elevation-discharge-reservoir storage relationship.
6. *Read-discharge hydrograph data table:* a hydrograph that is entered directly into the program as data.

For example, the data needed to support a cumulative rainfall table include a rainfall table number, time increment, runoff coefficient (if desired), and the cumulative rainfall values for the modeled rainfall event. The cumulative rainfall values used in the following examples are the SCS Type II distribution.

Standard Control Operations

Standard Control consists of the following six subroutine operations. There can be up to 600 Standard Control statements for each TR-20 job.

1. RUNOFF: an instruction to develop a subbasin runoff hydrograph
2. RESVOR: an instruction to route a hydrograph through a structure or reservoir
3. REACH: an instruction to route a hydrograph through a channel reach
4. ADDHYD: an instruction to combine two hydrographs
5. SAVMOV: an instruction to move a hydrograph from one computer memory storage location to another
6. DIVERT: an instruction for a hydrograph to be separated into two parts

As an example, the data needed to support the standard RUNOFF control operation are the area, curve number, and time of concentration. The operations are indicated on the input lines in columns 2, 4–9, and 11. The number 6 is placed in column 2 to indicate standard control. The name of the subroutine operation is placed in columns 4–9. The operation number, which is placed after the operation name (e.g., 1 for RUNOFF, 6 for DIVERT), is placed in column 11.

Column 61, 63, 65, 67, 69, and 71 of standard control records are used to specify the output. The individual output options are selected by placing a 1 in the appropriate column. If either the column is left blank or a zero is inserted, the corresponding option will not be selected. Table 36.1 summarizes the output options. If only a summary table is desired, it is convenient to place SUMMARY in columns 51–57 in the JOB statement and leave all the output options on the standard control records blank.

Executive Control

The Executive Control has two functions: (1) to execute the Standard Control statements, and (2) to provide additional data necessary for processing. The Executive Control consists of six types

Table 36.1 Output Options for TR-20

Output Option	A "1" in Column	Produces the Following Printout
PEAK	61	Peak discharge and corresponding time of peak and elevation (maximum storage elevation for a structure).
HYD	63	Discharge hydrograph ordinates.
ELEV	65	Stage hydrograph ordinates (reach elevations for a cross section and water surface elevation for structures).
VOL	67	Volume of water under the hydrograph in inches depth, acre-feet, and ft^3/sec-hours.
SUM	71	Requests the results of the subroutine be inserted in the summary tables at the end of the job.

of statements. They are LIST, BASFLO, INCREM, COMPUT, ENDCMP, and ENDJOB. The Executive Control statements are placed after the Standard Control statements and tabular data to which they pertain.

The Standard Control is used to describe the physical characteristics of the watershed. The Executive Control is used to prescribe the hydrologic conditions of the watershed including the baseflow. The performance for the Executive Control is directed by the COMPUT statement. Its purpose is to describe the rainfall and the part of the watershed over which that rainfall is to occur [SCS, 1982].

For the examples discussed in this section, the INCREM, COMPUT, ENDCMP, and ENDJOB will be the only executive control statements required to fulfill the operations requested.

Preparation of Input Data

Preparation of input data can be divided into the following requirements and functions:

1. Prepare a schematic drawing that conveniently identifies the locations, drainage areas, curve numbers (CN), times of concentration (t_c), and reach lengths for the watershed. It should display all alternate structural systems together with the routing and evaluation reaches through which they are to be analyzed.
2. Establish a Standard Control list for the watershed.
3. List the tabular data to support the requirements of the Standard Control list. This may consist of structure data, stream cross section data, cumulative rainfall data, and the dimensionless hydrograph table.
4. Establish the Executive Control statements that describe each storm and alternative situation that is to be analyzed through the Standard Control list.

Calculations

For a large watershed it may be necessary to divide the watershed into subbasins. Each subbasin is determined by finding the different outlet points or design points within the watershed, then finding the area contributing to those points.

1. *Area.* The area of each subbasin in square miles (mi^2).
2. *Curve number.* The curve number (CN) for each subbasin.
3. *Time of concentration.* Following the curve number, the time of concentration (t_c) is specified for each subbasin.

Forms for Input Data

Blank forms for the TR-20 input data are found in SCS [1982]. The example problems therein demonstrate how the forms are completed. The process is straightforward because of the instructions given and the preselected columns for the data. The output options are shown in Table 36.1.

Examples

A 188.5-acre basin is modeled to determine the discharge and required storage as a result of development. The first example models a subbasin with existing conditions. The second example models the same subbasin with proposed conditions and a detention basin. Lastly, the third example incorporates the entire basin to determine the peak discharge at the outlet. The area is located in Indianapolis and is shown in Fig. 36.1.

Example 36.1. Referring to Fig. 36.1, drainage area 1 is modeled to determine the runoff for present conditions. This area is a small part of the total drainage area of the watershed in Fig. 36.1.

Hydrological Input Data.

The cumulative rainfall data used are the SCS Type II rainfall distribution. The intensity-duration-frequency tables for Indianapolis, Indiana, are used with the 6-hour rainfall depth for the 10-year event to generate the surface-runoff hydrograph for the subbasin. The SCS 24-hour distribution was scaled to give a 6-hour hyetograph. The point rainfall depth is 3.23 inches/hour.

Calculations.

Area. The area of subbasin 1 is 13.5 acres or 0.021 mi^2.
Curve number. There are two different land uses for this subbasin; therefore both areas must be calculated and a composite curve number determined for the respective area. An open area of 7.6 acres has a CN of 74. The other land use is commercial with 5.9 acres and a CN of 94. The product of curve number and area is 1117. The sum of the product of the curve numbers and areas is then divided by the total area of the drainage area 1 to find an overall composite CN of 83.
Time of concentration. The time of concentration is computed by assuming 300 ft of sheet flow, 350 ft of shallow concentrated flow over unpaved surface, and channel flow of 1150 ft. The channel flow is computed to the lake, not the watershed boundary. The time is 0.24 h.

Computer Input.

The following section demonstrates where, for this example, the information is input to the computer. The following discussion is based on the input file shown in Fig. 36.2.

1. The first input is the JOB and TITLE. For this example NOPLOTS is entered in columns 61–67, which indicates that cross section discharge-area plots are not desired. Two TITLE statements are used to describe the problem and the rainfall information.
2. Next the cumulative rainfall table is used to describe the rainfall event. The statement describing this operation is RAINFL. There are six standard rainfall distributions which may be used and these are preloaded into TR-20. The user need not input these tables, but use only the proper rain table identification number on the COMPUT statement, in column 11. The user may override any of these tables by entering a new RAINFL table and entering 7, 8, or 9 in column 11. For this example a 7 is placed in column 11 indicating that a rainfall table will be specified by the user. Rainfall depths in inches can also be used. If these are used, the data code in column 2 needs to be changed to 8. The rainfall table of SCS Type II distribution is placed in the proper columns below the RAINFL statement as shown on Fig. 36.2. The time increment used for this example is 0.05 h, determined by the number of cumulative rainfall points and the storm duration. The number of rainfall increments is 20, so that each increment is 0.05 of the total. This number 0.05 is entered into columns 25–36. The 6-hour storm duration is entered on the COMPUT statement.

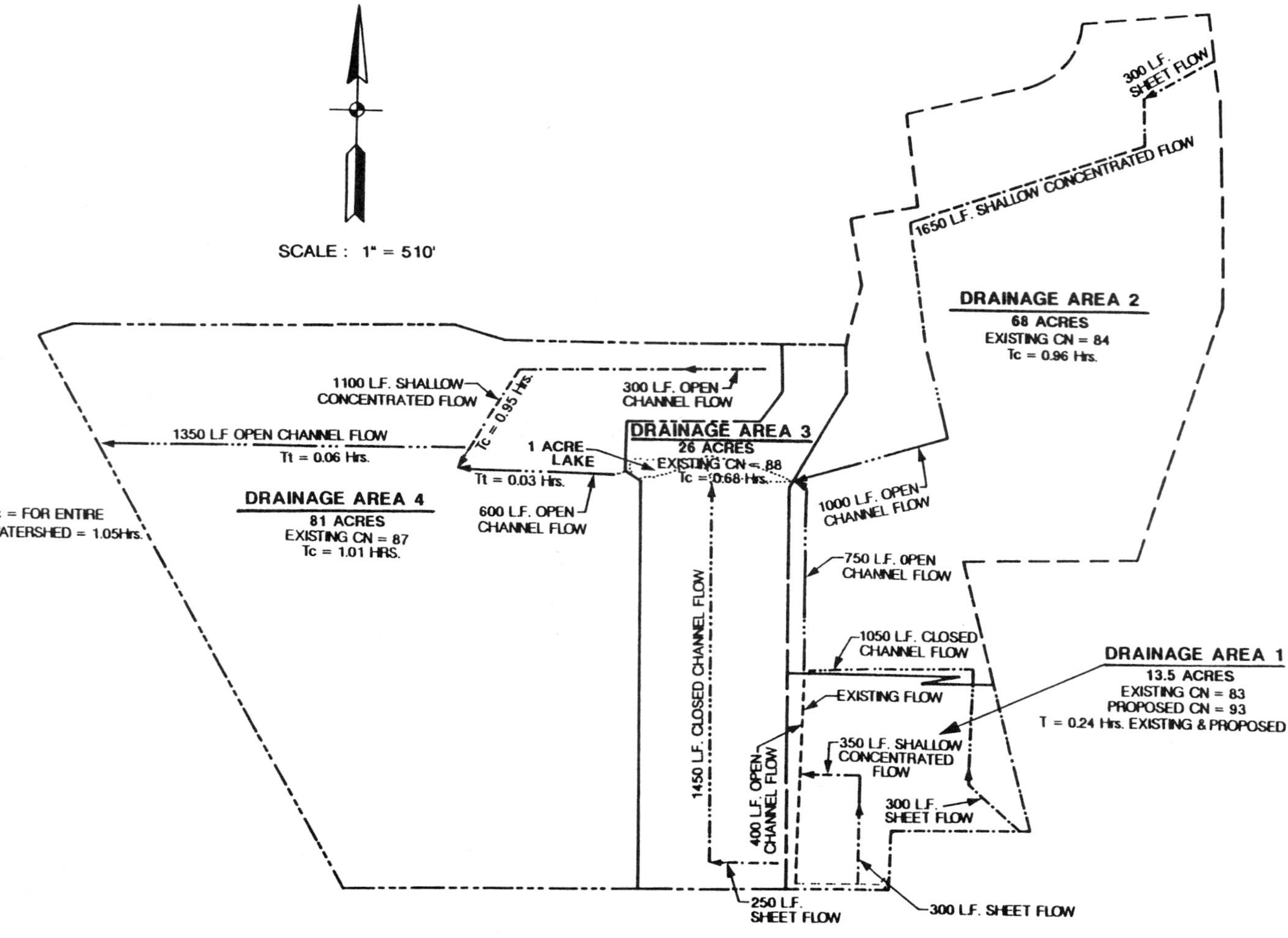

FIGURE 36.1 Location map.

```
JOB TR-20                                                  NOPLOTS
TITLE      EXAMPLE 1 - DRAINAGE AREA 1
TITLE      EXISTING CONDITIONS
5 RAINFL 7              0.05
8            0.         .0125       .025        .04         .06
8            .08        .10         .13         .165        .22
8            .64        .78         .835        .87         .895
8            .92        .94         .96         .98         .99
8            1.0        1.0         1.0         1.0         1.0
9 ENDTBL
6 RUNOFF 1   1        1 0.021       83.           0.24      1 1 0 1 0 1 AREA 1
  ENDATA
7 INCREM 6              0.1
7 COMPUT 7   1     1    0.00        3.23        6.          7 2   1   1 10-YR
  ENDCMP 1
  ENDJOB 2
```

FIGURE 36.2 Input for Example 36.1.

3. For this example, the runoff from area 1 is computed with the RUNOFF statement with the proper codes, described previously. This is the only standard control statement needed for this example. ENDATA indicates that the input information for the runoff is complete. Comments in columns 73 through 80 are helpful reminders.
4. The Executive Control statement is then used to indicate which computations are desired. This runoff from area 1 is computed for the 10-year rainfall. The INCREM statement causes all the hydrographs generated within the TR-20 program to have a time increment of 0.1 hours, as specified in columns 25–36, unless the main time increment is changed by a subsequent INCREM statement. The 7 in column 61 refers the COMPUT to rainfall table 8, from the cumulative rainfall table. The AMC number is given in column 63 for each COMPUT (computation). For this example the AMC is 2. The starting time, rainfall depth, and rainfall duration are given for each recurrence interval. In this example, the starting time is 0.0, the rainfall depth is 3.23 in., and the duration is 6 h. These values are placed in their corresponding columns. A COMPUT and an ENDCMP statement are necessary for each recurrence interval. ENDJOB is then used to signify that all the information has been given and the calculations are to begin. The Executive Control data are shown on Fig. 36.2.

Output.

The type and amount of output are controlled by input options on the JOB record and by the output options on the Standard Control records. For this example, by declaring NOPLOTS in columns 61–67 of the JOB record, plots are suppressed. The RUNOFF statement requested to have the peak discharge, volume of water under the hydrograph, and a summary table of the results, because a "1" was placed in columns 61, 67, and 71 in this statement. The output for this example is shown in Fig. 36.3.

Summary/Explanation.

The table in Fig. 36.3 summarizes all of the information obtained. From this table it can be seen that, for a rainfall event with a 10-year recurrence interval, the amount of runoff from Area 1 is 1.64 inches for a 6-hour duration. The peak discharge occurs after 3.08 hours and has a flow rate of 30.72 cfs.

Example 36.2

Description.

The runoff from area 1 is computed for developed conditions and a pond is added inside area 1. A stage-storage-discharge relationship is required for the structure, and the attenuation effects are evaluated. The hydrologic input data are the same as that which was used in Example 36.1.

```
EXECUTIVE CONTROL OPERATION INCREM                                                          RECORD ID
+                                         MAIN TIME INCREMENT =   .10 HOURS

EXECUTIVE CONTROL OPERATION COMPUT                                                          RECORD ID 10-YR
+                                         FROM XSECTION   1
+                                                          TO XSECTION   1
       STARTING TIME =   .00    RAIN DEPTH = 3.23    RAIN DURATION=  6.00    RAIN TABLE NO.= 7    ANT. MOIST. COND= 2
       ALTERNATE NO.= 1         STORM NO.= 1    MAIN TIME INCREMENT =   .10 HOURS


OPERATION RUNOFF    CROSS SECTION   1

          PEAK TIME(HRS)               PEAK DISCHARGE(CFS)              PEAK ELEVATION(FEET)
             3.08                            30.72                           (RUNOFF)
             5.35                             2.40                           (RUNOFF)

TIME(HRS)          FIRST HYDROGRAPH POINT =    .00 HOURS      TIME INCREMENT =  .10 HOURS      DRAINAGE AREA =     .02 SQ.MI.
   2.00   DISCHG          .00       .00       .02       .12       .26       .49       .88      1.32      4.93     16.04
   3.00   DISCHG        27.38     30.47     22.64     17.67     14.47      9.95      7.49      6.23      4.98      4.34
   4.00   DISCHG         3.92      3.35      3.05      2.95      2.92      2.91      2.81      2.56      2.42      2.38
   5.00   DISCHG         2.37      2.36      2.37      2.37      2.37      2.16      1.65      1.36      1.25      1.22
   6.00   DISCHG         1.20       .98       .46       .16       .06       .02       .01       .00

  RUNOFF VOLUME ABOVE BASEFLOW =  1.64 WATERSHED INCHES,     22.18 CFS-HRS,      1.83 ACRE-FEET;   BASEFLOW =     .00 CFS


EXECUTIVE CONTROL OPERATION ENDCMP                                                          RECORD ID
+                                         COMPUTATIONS COMPLETED FOR PASS   1

EXECUTIVE CONTROL OPERATION ENDJOB                                                          RECORD ID


TR20 XEQ                          EXAMPLE 1 - DRAINAGE AREA 1                               JOB  1    SUMMARY
    REV PC 09/83(.2)              EXISTING CONDITIONS                                                  PAGE   1

SUMMARY TABLE 1 - SELECTED RESULTS OF STANDARD AND EXECUTIVE CONTROL INSTRUCTIONS IN THE ORDER PERFORMED
                  (A STAR(*) AFTER THE PEAK DISCHARGE TIME AND RATE (CFS) VALUES INDICATES A FLAT TOP HYDROGRAPH
                   A QUESTION MARK(?) INDICATES A HYDROGRAPH WITH PEAK AS LAST POINT.)

SECTION/   STANDARD               RAIN  ANTEC  MAIN        PRECIPITATION                          PEAK DISCHARGE
STRUCTURE   CONTROL   DRAINAGE   TABLE  MOIST  TIME   ------------------------  RUNOFF  -------------------------------------
   ID      OPERATION    AREA       #    COND  INCREM  BEGIN   AMOUNT  DURATION  AMOUNT  ELEVATION   TIME      RATE      RATE
                       (SQ MI)                 (HR)   (HR)    (IN)     (HR)      (IN)     (FT)      (HR)     (CFS)     (CSM)


  ALTERNATE    1    STORM    1
+ ____________________________
 XSECTION   1  RUNOFF     .02      7     2     .10     .0     3.23     6.00      1.64      ---      3.08      30.72    1463.1


SUMMARY TABLE 3 - DISCHARGE (CFS) AT XSECTIONS AND STRUCTURES FOR ALL STORMS AND ALTERNATES

XSECTION/           DRAINAGE
STRUCTURE             AREA       STORM NUMBERS..........
  ID                (SQ MI)           1

0 XSECTION    1          .02
+____________________________
   ALTERNATE     1                30.72
END OF  1 JOBS IN THIS RUN
```

FIGURE 36.3 Output from Example 36.1.

Calculations.

Once the proposed pond is designed, the discharge and storage is computed at different elevations by the user. The discharge from the pond is controlled by the outlet structure. Table 36.2 shows the stage-storage-discharge relationship for the pond. For example, at elevation 783.0 feet, the discharge is 15.0 cfs and the storage is 0.88 acre-ft. This is an ungated control structure with headwater control.

The CN for the developed condition is 93 and the time of concentration is 0.24 hours.

Table 36.2 Stage-Storage-Discharge Relationship for Pond in Example 36.2

Stage	Discharge	Storage
780.5	0	0.0
781.0	2	0.004
781.5	4	0.014
782.0	8	0.019
782.5	12	0.046
783.0	15	0.88
783.5	20	1.33
784.0	23	1.67
784.5	28	2.07
785.0	30	2.46
785.5	32.5	2.85
786.0	35	3.23
786.5	36	3.49
787.0	40	4.52

Computer Inputs (Fig. 36.4).

1. The TITLE statement is adjusted to show the proper example problem number and description.
2. As mentioned previously, the same rainfall table will be used.
3. After the ENDTBL statement of the rainfall table, the structure table for the pond is added. The initial values for the stage-storage-discharge relationship must have the elevation of the invert of the outlet structure and 0.0 values for the discharge and storage. All data entered in this table must have decimal points and must increase between successive lines of data. The STRUCT table must have a number, placed in columns 16 and 17 of the first row of the STRUCT statement. This number, 01, should be the same as the structure number on the RESVOR standard control statement.
4. The RUNOFF Standard Control statement for area 1 is used to obtain the runoff from the 0.021 square miles. The CN and t_c for the developed conditions replace the numbers used in Example 36.1.
5. A RESVOR Standard Control statement is used to have the hydrograph calculated from area 1 routed through the pond structure. This is done by placing "01" (the pond structure

```
JOB TR-20                                                    NOPLOTS
TITLE      EXAMPLE 2 - DRAINAGE AREA 1
TITLE      PROPOSED CONDITIONS
 5 RAINFL 7             0.05
 8          0.          .0125        .025        .04         .06
 8          .08         .10          .13         .165        .22
 8          .64         .78          .835        .87         .895
 8          .92         .94          .96         .98         .99
 8          1.0         1.0          1.0         1.0         1.0
 9 ENDTBL
 3 STRUCT       01
 8                      780.5        0.0         0.0
 8                      781.0        2.0         0.004
 8                      781.5        4.0         0.014
 8                      782.0        8.0         0.019
 8                      782.5        12.0        0.046
 8                      783.0        15.0        0.88
 8                      783.5        20.0        1.33
 8                      784.0        23.0        1.67
 8                      784.5        28.0        2.07
 8                      785.0        30.0        2.46
 8                      785.5        32.5        2.85
 8                      786.0        35.0        3.23
 8                      786.5        36.0        3.49
 8                      787.0        40.0        4.52
 9 ENDTBL
 6 RUNOFF 1   1       1 0.021        93.          0.24       1 1 0 1 0 1
 6 RESVOR 2     01 1  2                                      1 0 0 1 0 1
   ENDATA
 7 INCREM 6             0.1
 7 COMPUT 7   1      01 0.00         3.23        6.          7 2  1  1
   ENDCMP 1
   ENDJOB 2
```

FIGURE 36.4 Input for Example 36.2.

```
SUMMARY TABLE 1 - SELECTED RESULTS OF STANDARD AND EXECUTIVE CONTROL INSTRUCTIONS IN THE ORDER PERFORMED
              (A STAR(*) AFTER THE PEAK DISCHARGE TIME AND RATE (CFS) VALUES INDICATES A FLAT TOP HYDROGRAPH
               A QUESTION MARK(?) INDICATES A HYDROGRAPH WITH PEAK AS LAST POINT.)

SECTION/   STANDARD              RAIN  ANTEC  MAIN           PRECIPITATION                             PEAK DISCHARGE
STRUCTURE   CONTROL  DRAINAGE  TABLE  MOIST  TIME  ------------------------- RUNOFF  --------------------------------------
  ID       OPERATION   AREA      #    COND  INCREM  BEGIN   AMOUNT DURATION  AMOUNT   ELEVATION   TIME        RATE        RATE
                      (SQ MI)                (HR)   (HR)    (IN)    (HR)     (IN)      (FT)      (HR)       (CFS)       (CSM)


  ALTERNATE    1   STORM    1
+
XSECTION   1  RUNOFF     .02     7     2     .10     .0     3.23    6.00     2.48      ---       3.05       46.63      2220.4
STRUCTURE  1  RESVOR     .02     7     2     .10     .0     3.23    6.00     2.47    783.07      3.45       15.72       748.8
```

FIGURE 36.5 Output from Example 36.2.

Table 36.3 Summary of Example 36.2

Location	Peak Discharge (cfs)	Runoff Amount (inches)	Water Surface Elevation (feet)	Time to Peak (hours)
Area 1	46.63	2.48	—	3.05
Pond	15.72	2.47	783.07	3.45

number) in columns 19 and 20. The routed hydrograph is placed in a computer memory location by placing a number 1 through 7, not already used, in column 23. Therefore, a 2 is placed in column 23.

6. The INCREM and COMPUT Executive Control statements are the same as in Example 36.1. With these statements, the peak discharge rate and time will be computed, as well as the peak elevation in the pond. In order for the flow in pond to be analyzed, the last cross section must be specified as the pond structure. This is done by placing the cross section number of the culvert in columns 20 and 21 of the COMPUT statement. Again, ENDJOB is the last statement.

Output.

The output format is the same as Example 36.1 because the output options in columns 61–70 are the same, with the addition of the options chosen for the RESVOR statement. A condensed version of the output is shown on Fig. 36.5 and Table 36.3.

Summary.

As in Example 36.1, the summary table on the last page of the output summarizes all of the information obtained. This table contains the developed runoff information for area 1 and the peak flow and elevation through the pond. The pond used to control the runoff from area 1 is checked to see if it is sufficient. The peak discharge elevation is 783.07 feet for the 10-year storm. This indicates that the pond does not overtop for the storm event modeled.

Example 36.3.

Description.

The entire watershed is analyzed to determine the amount of runoff from the 0.3 square miles. This example incorporates the same area as Example 36.1, but also includes the runoff from the three other subbasins under existing conditions from the entire watershed. The watershed is subdivided into 4 subbasins, as shown on Fig. 36.1. Area 1 has been analyzed with and without the pond in the two previous examples. There is a 1-acre lake in the middle of the watershed which takes in the runoff from Areas 1, 2, and 3. The discharge from this lake is routed through a channel to the

Table 36.4 Stage-Storage-Discharge Relationship for the 1-Acre Lake in Example 36.3

Stage	Discharge	Storage
773.0	0.0	0.0
773.5	3.0	0.34
774.0	5.0	0.86
774.5	7.0	1.33
775.0	9.0	1.87
775.5	11.0	2.41
776.0	13.0	3.03
776.5	19.0	3.56
777.0	116.0	4.37
777.5	325.0	5.17
778.0	656.0	5.96

outlet of the watershed. Table 36.4 shows the stage-storage-discharge relationship for the 1-acre lake.

The hydrologic input data is the same as in the previous examples. This refers to the rainfall data for the SCS Type II distribution and the rainfall amount for a 10-year recurrence interval in Indianapolis. Another storm, the 100-year 12-hour event, was used to demonstrate how easy it is to add additional storms.

Calculations.

Table 36.5 contains the information from the computation of the CN and t_c values. The longest t_c is for drainage area 4. The total time of concentration for the entire watershed is 1.05 hours.

Computer Input.

The input file from Example 36.1 is edited to include the information for the additional subbasins. A reach is added by using the tabular control XSECTN and the corresponding codes. The STRUCT table is changed to reflect the 1-acre table. The Standard Control statements, RUNOFF and ADDHYD, are needed for the additional subbasins. The area, CN, and t_c are described on each RUNOFF statement. COMPUT and ENDCMP statements are used to add the 100-year storm event values. The ADDHYD statement is used to combine two hydrographs computed from the respective RUNOFF statement. The ADDHYD is used because there are only two input hydrographs entries for each RUNOFF statement. The input is shown in Fig. 36.6.

For example, the hydrographs from areas 1, 2, and 3 are combined and routed through the 1-acre lake before adding runoff from area 4. Therefore, two ADDHYD statements are used to combine the hydrographs computed from the pond in area 1 and the runoff from area 2. Another ADDHYD statement is used to combine the previously obtained hydrographs with the hydrograph computed from area 3. The 1-acre pond in area 3 is then described by a RESVOR statement. This statement refers to STRUCT 02 for the stage-storage-discharge relationship. The runoff from all

Table 36.5 Curve Numbers and Times of Concentration for Example 36.3

Drainage Area	Area (acres)	Land Use	CN	Area * CN
		Drainage Area 2		
	36	Residential	86	3096
2	11	Commercial	94	1034
	21	Open Area	74	1554
Total	68			5684
		Composite curve number		84
		Drainage Area 3		
	19	Residential	86	1634
3	7	Commercial	94	658
	N/A	Open Area		
Total	26			2292
		Composite curve number		88
		Drainage Area 4		
	28	Residential	86	2408
4	37	Commercial	94	3478
	16	Open Area	74	1184
Total	81			7070
		Composite curve number		87

```
JOB TR-20                                           SUMMARY  NOPLOTS
TITLE     EXAMPLE 3 - ENTIRE WATERSHED
TITLE     PROPOSED CONDITIONS
 5 RAINFL 7              0.05
 8          0.          .0125        .025        .04         .06
 8          .08         .10          .13         .165        .22
 8          .64         .78          .835        .87         .895
 8          .92         .94          .96         .98         .99
 8          1.0         1.0          1.0         1.0         1.0
 9 ENDTBL
 2 XSECTN   006         1.0          770.0
 8                      759.0        0.0         0.0
 8                      760.0        15.09       5.0
 8                      761.0        51.53       12.0
 8                      762.0        110.34      21.0
 8                      763.0        194.31      32.0
 8                      764.0        306.33      45.0
 8                      765.0        372.72      71.0
 8                      766.0        670.69      121.0
 8                      767.0        1208.35     195.0
 8                      768.0        2034.72     293.0
 8                      769.0        2751.61     455.0
 8                      770.0        4579.98     723.0
 9 ENDTBL
 3 STRUCT     01
 8                      773.0        0.0         0.0
 8                      773.5        3.0         0.34
 8                      774.0        5.0         0.86
 8                      774.5        7.0         1.33
 8                      775.0        9.0         1.87
 8                      775.5        11.0        2.41
 8                      776.0        13.0        3.03
 8                      776.5        19.0        3.56
 8                      777.0        116.0       4.37
 8                      777.5        325.0       5.17
 8                      778.0        656.0       5.96
 9 ENDTBL
 6 RUNOFF 1   1       1 0.021        93.         0.24                    AREA 1
 6 RUNOFF 1   2       2 0.106        84.         0.96                    AREA 2
 6 RUNOFF 1   3       3 0.041        88.         0.68                    AREA 3
 6 ADDHYD 4   4   1 2 4
 6 ADDHYD 4   5   3 4 1
 6 RESVOR 2    01 1   2                                                  LAKE
 6 REACH  3   6   2   3 2050.                                            REACH
 6 RUNOFF 1   7       4 0.127        87.         1.01                    AREA 4
 6 ADDHYD 4   8   3 4 5
   ENDATA
 7 INCREM 6              0.1
 7 COMPUT 7   1     8   0.00         3.23        6.          7 2   1   1 10-YR
   ENDCMP 1
 7 COMPUT 7   1     8   0.00         5.24        12.         7 2   1   2 100-YR
   ENDCMP 1
   ENDJOB 2
```

FIGURE 36.6 Input for Example 36.3.

three subbasins was routed through the pond by placing a "1," the combined hydrograph number, in column 19 of the RESVOR statement. The discharge from the pond is then routed through a channel to the outlet of the watershed. The routing is described by a REACH statement which refers to XSECT 006. The XSECT is the stage-discharge and area relationship for the 2050-foot channel. The relationship is computed by using the channel cross section shown in Fig. 36.7. The RUNOFF statement is then used to describe the runoff from area 4. Finally, the hydrograph from the REACH and area 4 runoff are added together with the ADDHYD statement.

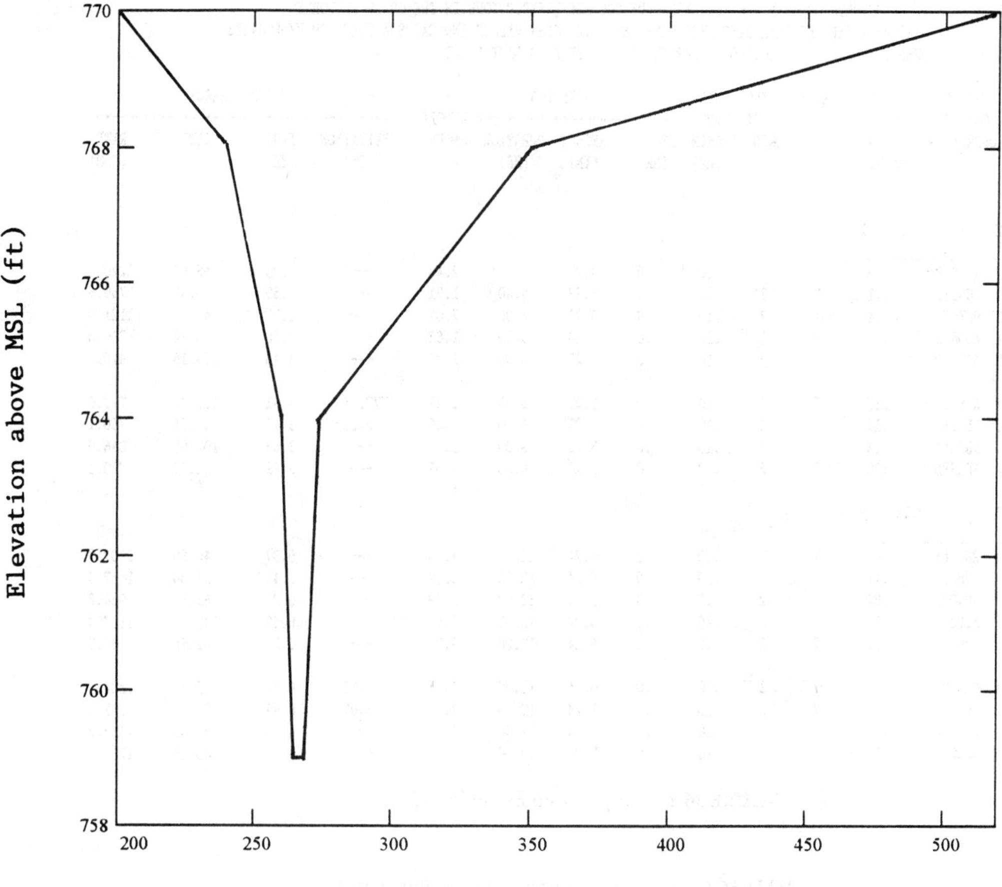

FIGURE 36.7 Cross section of reach through drainage area 4.

The Executive Control statements are the same as those used in the previous two example problems, except the last cross section number used must be placed in columns 20 and 21 and the additional COMPUT and ENDCMP. The COMPUT statements require the input of the initial and final cross sections that are to be analyzed. Because the ADDHYD was the last Standard Control statement, its cross section number (columns 14 and 15) is placed in columns 20 and 21 of the COMPUT statement.

Output.

The type and amount of output can be controlled by input options on the JOB record and by the output options on the Standard Control records. All of the output options on the Standard Control statements were left blank because only a summary table is desired. To accomplish this, SUMMARY is placed in columns 51–57 of the JOB statement. The output is in Fig. 36.8.

Summary.

From the summary tables of the output also shown in Fig. 36.8, it is determined that, for a 10-year storm, the peak discharge is 233 cfs, which occurs after 3.71 hours. For the 100-year 12-hour storm, the peak discharge is 361 cfs with a time to peak at 6.46 hours. Output from each subbasin can be inspected to determine the amount it contributes to the overall runoff from the watershed. The outflow from drainage areas 1, 2, and 3 are routed through the 1-acre pond and then through

SUMMARY TABLE 1 - SELECTED RESULTS OF STANDARD AND EXECUTIVE CONTROL INSTRUCTIONS IN THE ORDER PERFORMED
(A STAR(*) AFTER THE PEAK DISCHARGE TIME AND RATE (CFS) VALUES INDICATES A FLAT TOP HYDROGRAPH
A QUESTION MARK(?) INDICATES A HYDROGRAPH WITH PEAK AS LAST POINT.)

SECTION/ STRUCTURE ID	STANDARD CONTROL OPERATION	DRAINAGE AREA (SQ MI)	RAIN TABLE #	ANTEC MOIST COND	MAIN TIME INCREM (HR)	PRECIPITATION BEGIN (HR)	PRECIPITATION AMOUNT (IN)	PRECIPITATION DURATION (HR)	RUNOFF AMOUNT (IN)	PEAK DISCHARGE ELEVATION (FT)	PEAK DISCHARGE TIME (HR)	PEAK DISCHARGE RATE (CFS)	PEAK DISCHARGE RATE (CSM)
ALTERNATE 1 STORM 1													
XSECTION 1	RUNOFF	.02	7	2	.10	.0	3.23	6.00	2.48	---	3.05	46.63	2220.4
XSECTION 2	RUNOFF	.11	7	2	.10	.0	3.23	6.00	1.71	---	3.60	80.62	760.6
XSECTION 3	RUNOFF	.04	7	2	.10	.0	3.23	6.00	2.02	---	3.37	46.00	1121.9
XSECTION 4	ADDHYD	.13	7	2	.10	.0	3.23	6.00	1.83	---	3.50	90.94	716.0
XSECTION 5	ADDHYD	.17	7	2	.10	.0	3.23	6.00	1.88	---	3.43	136.16	810.5
STRUCTURE 1	RESVOR	.17	7	2	.10	.0	3.23	6.00	1.87	777.04	3.61	130.97	779.6
XSECTION 6	REACH	.17	7	2	.10	.0	3.23	6.00	1.88	762.21	3.73	127.71	760.1
XSECTION 7	RUNOFF	.13	7	2	.10	.0	3.23	6.00	1.94	---	3.63	107.55	846.9
XSECTION 8	ADDHYD	.30	7	2	.10	.0	3.23	6.00	1.91	---	3.71	232.79	789.1
ALTERNATE 1 STORM 2													
XSECTION 1	RUNOFF	.02	7	2	.10	.0	5.24	12.00	4.43	---	5.97	46.89	2232.7
XSECTION 2	RUNOFF	.11	7	2	.10	.0	5.24	12.00	3.49	---	6.43	125.86	1187.4
XSECTION 3	RUNOFF	.04	7	2	.10	.0	5.24	12.00	3.89	---	6.22	63.91	1558.8
XSECTION 4	ADDHYD	.13	7	2	.10	.0	5.24	12.00	3.65	---	6.40	143.16	1127.3
XSECTION 5	ADDHYD	.17	7	2	.10	.0	5.24	12.00	3.71	---	6.30	202.98	1208.2
STRUCTURE 1	RESVOR	.17	7	2	.10	.0	5.24	12.00	3.70	777.21	6.34	202.33	1204.3
XSECTION 6	REACH	.17	7	2	.10	.0	5.24	12.00	3.71	763.07	6.47	201.82	1201.3
XSECTION 7	RUNOFF	.13	7	2	.10	.0	5.24	12.00	3.79	---	6.46	159.03	1252.2
XSECTION 8	ADDHYD	.30	7	2	.10	.0	5.24	12.00	3.74	---	6.46	360.85	1223.2

FIGURE 36.8 Output from Example 36.3.

Table 36.6 Summary of Results for Example 36.3

	6-Hour Storn Event		12-Hour Storm Event	
Drainage Area	Time of Peak (hours)	Peak Discharge (cfs)	Time of Peak (hours)	Peak Discharge (cfs)
1	3.05	46.64	5.97	46.93
2	3.60	80.62	6.43	125.86
3	3.37	46.00	6.22	63.91
Reach	3.73	127.71	6.47	201.82
4	3.63	107.55	6.46	159.03
Outlet	3.71	232.79	6.46	360.85

a channel. The reach had very little effect on the time of peak or the discharge. The channel would have to be much wider and longer for any attenuation to be realized. The results are also summarized below in Table 36.6.

36.4 The HEC-1 Model

Uses of HEC-1

HEC-1 is a flood hydrograph package developed by the U.S. Army Corps of Engineers Hydrologic Engineering Center (HEC) that is used: (1) to estimate unit hydrographs, loss rates, and streamflow routing parameters from measured data, (2) to simulate streamflow from historical or design rainfall data, (3) to compute frequency curves of flood damage and expected annual damages, and (4) to simulate reservoir outflows for dam safety analysis [Bedient and Huber, 1988].

In addition, HEC-1 program has several other features that may be used for snowmelt simulation, dam safety analysis, pumping and diversion schemes, multiple-flood and multiple-plan analyses, and simulation of precipitation depth-area relationships. Information about the manual and software of HEC-1 can be obtained by writing to: *The Hydrologic Engineering Center, U.S. Army Corps of Engineers, Attention: Training Division, 609 Second Street, Davis, CA 95616.* Part of the software that accompanies HEC-1 is a full-screen text-editing program called COED, which provides on-line help screens and input variable documentation to prepare input files.

The HEC Data Storage System (DSS) [HEC, 1983] allows the transfer of data between HEC programs. The data are identified by unique labels called PATHNAMES, which are specified when the data are created or retrieved. Thus, a hydrograph computed by HEC-1 can be labeled and stored in DSS for later retrieval as input to HEC-2, for example. The DSS program has several utility programs for manipulating data.

Summary of HEC-1 Input Structure

When developing a precipitation-runoff model using HEC-1, boundaries of the basin are initially identified. Most often, the basin is subdivided into smaller subbasins depending on the study objectives, drainage pattern, and other factors. Points where runoff information is needed are identified. The model can be structured to produce hydrographs at any desired location. As different areas of a large basin may have different hydrologic response characteristics, it is important to select an appropriate computational time interval and subdivide the watershed so that lumped parameters provide a reasonable depiction of the subbasins.

There are several methods which may be used in HEC-1 to compute surface runoff. These are based on: (1) the unit hydrograph computed from observed data, (2) the Clark or the Snyder synthetic unit hydrographs, (3) the SCS methods (curve number method and SCS unit hydrograph), and (4) the kinematic wave method for overland hydrograph [Bedient and Huber, 1988]. Losses may be computed by using one of the four methods, shown in Table 36.7.

Three methods are available in HEC-1 for flood routing: (1) Muskingum-storage coefficient plus travel time through each reach, (2) normal-depth channel routing in which Manning's equation is used to compute a table of storage versus outflow values for use in modified pulse routing, and (3) kinematic wave-outflow from a reach based on continuity and Manning's equations. There are other methods not discussed here that deviate slightly from the three methods mentioned above.

Table 36.7 Estimation of Losses in HEC-1

Method	Description
Initial and constant	Initial loss volume is satisfied and then constant loss rate begins.
HEC exponential	Loss function is related to antecedent moisture condition and is a continuous function of soil wetness.
SCS curve number	Initial loss is satisfied before calculating cumulative runoff as a function of cumulative rainfall.
Holtan method	Infiltration rate is computed as an exponential function of available soil moisture storage from Holtan's equation.
Grren-Ampt method	The Green-Ampt infiltration function is combined with an initial abstraction to compute rainfall losses.

Input Data Structure

In the HEC-1 model, "records" of input that indicate the format for the river basin data and determine output formats are used. A "record" refers to an input line of 80 columns in either fixed or free format. Fixed format consists of ten 8-column fields, and free format allows data values to be given continuously, separated by commas.

In these cards there are two identification characters in columns 1 and 2 that form a unique two-character classification system. The first character identifies a data category, and the second indicates a type of data or option within the category. For example, an "R" in the first column indicates routing method, which has seven subcategories as indicated by the second character in the line identifiers: RC, RK, RL, RM, RN, RS, and RT. For example, RK is used to indicate the kinematic wave channel routing method. A list of the basic data categories is shown in Table 36.8. For detailed information on different lines of input and their methods, one should refer to Appendix A of the HEC-1 user's manual. Line input descriptions and variable definitions are also provided in help files of the text editor COED.

Table 36.8 Basic Data Categories

Job Control	
I_	Job initialization
V_	Variable Output summary
O_	Optimization
J_	Job type
Hydrology and Hydraulics	
K_	Job step control
H_	Hydrograph transformation
Q_	Hydrograph data
B_	Basin data
P_	Precipitation data
L_	Loss (infiltration) data
U_	Unitgraph data
M_	Melt data
R_	Routing data
S_	Storage data
D_	Diversion data
W_	Pump withdrawal data
Economics and End of Job	
E_	Economics data
ZZ	End of job
Retrieve and Store Data in DSS	
Z_	Specify appropriate pathnames
BZ	Retrieval of runoff parameters

The sequence of lines for an input file follows a specific order that describes the flow and routing. The job control data come first, followed by the hydrology and hydraulics information, and finally the economic information. The input file always begins with an ID in the first record. Other job control data which follow describe the job type, output control, time interval, time span, and the type of units selected—where the default is English units. The hydrology and hydraulics data include all the information necessary to simulate various physical processes in the river basin. The KK record is used at the beginning of each hydrograph calculation, followed by other lines that provide data to compute the hydrograph. For example, subbasin runoff, hydrograph combination, and a routing step begin with a KK record, giving the input a "block" structure. The economic data begins with an EC line and is used to compute the expected annual damages. A ZZ record is the last line in an input file, signaling the end of the job.

Time data are entered by using IT and IN records. A time interval and a time span are used to calculate runoff hydrographs which are defined with the variables on the IT record. Time series data are input with different time parameters than those used for simulation. For example, precipitation data can have different starting times, dates, and time intervals. These are specified on the IN record which is located just before the data. Precipitation data are interpolated to fit the simulation time conditions which have been specified on the IT record. The precipitation data may be input as basin average rainfall, gauge data, synthetic storm data, incremental or cumulative rainfall as a time series, probable maximum storm, and so on [Hoggan, 1989]. If precipitation does not change from one subbasin to another, then the data may be input only once, with the first KK group. The different precipitation input options are shown in Table 36.9.

After a storm hydrograph has been computed, channel and reservoir routing are performed using methods indicated by R records. Storage-discharge data is input on the SV/SQ records and the SQ/SE lines are used to input a rating curve of discharge versus elevation. A number of records are used to model the outflow structures for reservoir routing. The SA, SV, and SE records describe reservoir area-volume-elevation and SS or SL records specify spillway and low-level outlet data.

Table 36.9 Precipitation Input Options for HEC-1

Line Identification	Type of Storm
PI or PC	Incremental or cumulative rainfall as a time series
PB	Basin average precipitation
PG	Total rainfall for a nonrecording gage or to indicate that recorded data follow on PI or PC lines
PR, PW and PT, PW	Specify which gages and corresponding weights are to be used to calculate a subbasin's average precipitation
PH	Hypothetical storm from depth-duration data
PM	Probable maximum storm (may not be useful for all cases)
PS	Standard project storm

Two or more hydrographs are summed by the HC combination record. The HC step uses the most recently calculated hydrographs in the system. The progression of hydrologic calculation follows the order of the KK blocks. If a gauge exists at any point to measure the flows, the QO record can be used to compare computed and observed flows. An example of the input data organization is shown in Table 36.10.

The previous description covers the basics ideas for modeling a watershed. There are other options and features. The following case studies are used to demonstrate some of the features of the model described above.

Example 36.4. The runoff from area 1 in Fig. 36.1 is to be estimated using HEC- 1. The area of the subbasin is 13.5 acres. The curve number and the time of concentration for this area are determined to be 83 and 0.24 hours, respectively. The precipitation data used for the subbasin are for a 10-year event. The 10-year 6-hour rainfall depth for this location is 3.23 inches.

Table 36.10 Example of Input Data Organization

Segment	Code	Description of Data
Job control	ID	Title and description of model
	IT/IN	Time interval and beginning time
	IO	Output control for entire job
Runoff from first subbasin	KK	Station name
	KM	Comment card (optional)
	BA	Basin area
	P_	Precpitation method
	L_	Loss rate method
	U_	Rainfall transformation method
Route hydrograph from subbasin to next point	KK	Station name
	KM	Comment card (optional)
	R_	Routing method
Runoff from next subbasin	K	Station name
	KM	Comment card (optional)
Combine routed hydrographs	KK	Station name
	KM	Comment card (optional)
	HC	No. of hydrographs to be combined
Process continues until mouth of the watershed is reached		
	ZZ	Indicates end of job

```
ID   EXAMPLE 1
ID   DRAINAGE AREA 1
ID   EXISTING CONDITIONS - 10-YEAR EVENT
IT    18       0       0      25
IO     5       0
KK  SUB1
KM HYDROGRAPH FOR SUB1
BA 0.021
PB  3.23
PC   0.0   .0125    .025     .04     .06     .08     .10     .13    .165     .22
PC   .64     .78    .835     .87    .895     .92     .94     .96     .98     .99
PC   1.0
LS     0      83
UD  .144
ZZ
```

FIGURE 36.9 Input data for Example 36.4.

The input data are shown in Fig. 36.9. The details of information shown in Fig. 36.9 are discussed below. The first three lines are used to describe the location and conditions, as well as for other identification. These comments are written in columns 3 through 80, after an ID is placed as the identifier.

IT: The IT line specifies the time. The number of minutes in the tabulation interval is 18, which is placed in field 1 of the IT line. According to the HEC-1 manual, for this problem, Δt should be about 2 min. However, Δt of 2 min would generate too many runoff values. Therefore an 18-min value is used for illustration. Since there are 20 values for the precipitation data and each interval is 18 minutes, the total storm duration is 6 hours. Field 4 is used to give the number of hydrograph ordinates to be computed, specified here as 50 out of a maximum of 300.

IO: The IO line is used to control the output. The "5" in the first field enables printing of the job specification and master summary. A "0" in the second field signals that printer plots will be suppressed. The KO line is used to change output control.

KK: A hydrograph will be calculated following the data after KK "block" of lines. The station location identification is placed in field 1 in the KK line. It must be a unique identifier for each block.

KM: The message in the KM line will be printed at the beginning of the output for each station or plan.

BA: The drainage area specified by BA in square miles is placed in field 1. Area 1 is 13.5 acres or 0.021 square miles.

PB: This is the basin average precipitation line. The total basin average precipitation in inches is 3.23 for a 10-year 6-hour event, which is placed in field 1.

PC: The cumulative precipitation distribution is given on the PC lines. If the time interval and starting time are different for this subbasin than what was specified on the IT line, then an IN line must be put immediately before this PI line to specify the times for this hydrograph. There are 20 cumulative precipitation values for this time distribution. The values are the same as those used in the TR-20 examples. Because there are only ten fields available per line and 0.0 is the starting value, three PC lines are used.

LS: The SCS curve number loss rate is specified here. Since a "0" was placed in the first field, the initial abstraction is 0.2S. Field 2 is used to specify the SCS curve number, which is 83. The third field is used to give the percent of the drained basin that is impervious.

UD: Used to specify the SCS lag in hours. The time of lag is 0.6 times the time of concentration.

ZZ: Indicates the end of a job.

```
                         EXAMPLE 1
                         DRAINAGE AREA 1
                         EXISTING CONDITIONS - 10-YEAR EVENT

5 IO         OUTPUT CONTROL VARIABLES
                   IPRNT          5  PRINT CONTROL
                   IPLOT          0  PLOT CONTROL
                   QSCAL         0.  HYDROGRAPH PLOT SCALE

  IT         HYDROGRAPH TIME DATA
                    NMIN         18  MINUTES IN COMPUTATION INTERVAL
                   IDATE      1   0  STARTING DATE
                   ITIME       0000  STARTING TIME
                      NQ         25  NUMBER OF HYDROGRAPH ORDINATES
                  NDDATE      1   0  ENDING DATE
                  NDTIME       0712  ENDING TIME
                   ICENT         19  CENTURY MARK

              COMPUTATION INTERVAL    .30 HOURS
                   TOTAL TIME BASE   7.20 HOURS

   ENGLISH UNITS
       DRAINAGE AREA           SQUARE MILES
       PRECIPITATION DEPTH     INCHES
       LENGTH, ELEVATION       FEET
       FLOW                    CUBIC FEET PER SECOND
       STORAGE VOLUME          ACRE-FEET
       SURFACE AREA            ACRES
       TEMPERATURE             DEGREES FAHRENHEIT


                                        RUNOFF SUMMARY
                                  FLOW IN CUBIC FEET PER SECOND
                               TIME IN HOURS,  AREA IN SQUARE MILES

                            PEAK   TIME OF    AVERAGE FLOW FOR MAXIMUM PERIOD     BASIN    MAXIMUM   TIME OF
     OPERATION    STATION   FLOW    PEAK                                           AREA     STAGE    MAX STAGE
+                                             6-HOUR     24-HOUR     72-HOUR

     HYDROGRAPH AT
+                  SUB1      24.    3.00        4.          3.          3.          .02


*** NORMAL END OF HEC-1 ***
```

FIGURE 36.10 Output from Example 36.4.

The results from this example in Fig. 36.10 shows that the peak flow for the present conditions from drainage area 1 is 24 cfs. The time to peak is 3.0 hours. These values are slightly less than the values obtained from TR-20, which may be due to the different Δt value used in this example.

Example 36.5. The flow from area 1 with developed conditions is estimated first, as in the previous example. Flow is then routed through the pond. A stage-storage-discharge relationship is calculated for the pond, as shown in Table 36.2. The developed curve number and the time of concentration for this example are 93 and 0.24 hours, respectively.

Using the input as in the previous case, CN is changed from 83 to 93 on the LS record. As another hydrograph is computed for the pond a KK record is used. The pond elevation-storage-discharge relationship is input with the SV (for storage), SE (for elevation), and SQ (for discharge). The RS record is required for storage routing. The ELEV and 780.5 in the RS record indicates the starting elevation of the pond. The "1" indicates the number of steps used in storage routing. These are the only additions needed. The input is shown on Fig. 36.11.

```
ID   EXAMPLE 8.2.2
ID   DRAINAGE AREA 1 AND PROPOSED POND
ID   PROPOSED CONDITIONS - 10-YEAR EVENT
IT    18       0       0      50
IO     5       0
KK  SUB1
KM HYDROGRAPH FOR SUB1
BA 0.021
PB  3.23
PC   0.0   .0125    .025     .04     .06     .08     .10     .13    .165     .22
PC   .64     .78    .835     .87    .895     .92     .94     .96     .98     .99
PC   1.0
LS     0      93       0
UD  .144
KK  POND
KM ROUTE SUB1 HYDROGRAPH THROUGH DETENTION POND
RS     1    ELEV   780.5
SV     0   0.004   0.014   0.019   0.046    0.88    1.33    1.67    2.07    2.46
SV  2.85    3.23    3.49    4.52
SE 780.5   781.0   781.5   782.0   782.5   783.0   783.5   784.0   784.5   785.0
SE 785.5   786.0   786.5   787.0
SQ     0     2.0     4.0     8.0    12.0    15.0    20.0    23.0    28.0    30.0
SQ  32.5    35.0    36.0    40.0
ZZ
```

FIGURE 36.11 Input for Example 36.5.

```
                              RUNOFF SUMMARY
                        FLOW IN CUBIC FEET PER SECOND
                     TIME IN HOURS,  AREA IN SQUARE MILES

                            PEAK   TIME OF     AVERAGE FLOW FOR MAXIMUM PERIOD     BASIN    MAXIMUM     TIME OF
      OPERATION    STATION   FLOW     PEAK                                           AREA     STAGE    MAX STAGE
+                                               6-HOUR      24-HOUR     72-HOUR

      HYDROGRAPH AT
+                     SUB1     39.    3.00        6.           2.          2.        .02

      ROUTED TO
+                     POND     15.    3.60        6.           2.          2.        .02
+                                                                                           783.01        3.60


*** NORMAL END OF HEC-1 ***
```

FIGURE 36.12 Output from Example 36.5.

The output is in the same format as in the previous case since the IO record is not changed. The output is given in Fig. 36.12.

The runoff from this area has increased from the previous example problem due to the development to nearly 40 cfs. The flowrate from the area has decreased due to the detention pond. The peak outflow from the pond is 15 cfs at a time to peak of 3.60 hours. The pond is large enough to detain the flow from development as the peak elevation reached for the 6-hour 10-year event is 4 feet below the top elevation of the pond.

Example 36.6. The entire watershed of 0.30 square miles is analyzed to compute the runoff hydrograph. Runoff from the watershed, which is subdivided into a total of 4 subbasins shown in Fig. 36.1, is computed for a 12-hour 100-year event. The proposed pond considered earlier will be neglected for this case. There is a 1-acre lake in the middle of the watershed which takes in the runoff from areas 1, 2, and 3. The discharge from the pond is then routed through a channel to the outlet of the watershed. The stage-storage-discharge relationship for the 1 acre lake is given in Table 36.4.

The curve numbers and times of concentration are shown in Table 36.5. The area given in acres on Fig. 36.1 is converted to square miles and a stage-storage-discharge relationship is calculated by the user for the 1-acre lake in the same manner as for TR-20 Example 36.3.

Four additional hydrographs are input for the areas 2, 3, and 4, one routing hydrograph and two inputs for adding hydrographs together. The KK record is used before each station computation, subbasin runoff, routing, and so on.

```
ID   EXAMPLE 3
ID   ENTIRE WATERSHED
ID   PROPOSED CONDITIONS - 100-YEAR EVENT
IT    36       0        0       40
IO     5       0
KK  SUB1
KM HYDROGRAPH FOR SUB1
BA 0.021
PB  5.24
PC   0.0   .0125     .02     .04     .06     .08     .10     .13    .165     .22
PC   .64     .78    .835     .87    .895     .92     .94     .96     .98     .99
PC   1.0
LS     0      93       0
UD  .144
KK  SUB2
KM HYDROGRAPH FOR SUB2
BA  .106
LS     0      84       0
UD  .576
KK  SUB3
KM HYDROGRAPH FOR SUB3
BA  .041
LS     0      88       0
UD  .408
KK COMBN
KM COMBINE HYDROGRAPHS FOR ROUTING
HC     3
KK  LAKE (1 ACRE LAKE)
KM ROUTE COMBINED HYDROGRAPH THROUGH LAKE
RS     1    ELEV     773
SV     0     .34     .86    1.33    1.87    2.41    3.03    3.56    4.37    5.17
SV  5.96
SE   773   773.5     774   774.5     775   775.5     776   776.5     777   777.5
SE   778
SQ     0       3       5       7       9      11      13      19     116     325
SQ   656
KK  SUB4
KM ROUTE POND OUTFLOW TO SUB4
RD
RC   .04     .04     .04    2050     .01
RX   200     240     260     265     269     274     350     520
RY   770     768     764     759     759     764     768     770
KK  SUB4
KM HYDROGRAPH FOR SUB4
BA  .127
LS     0      87       0
UD  .606
KK  ROAD
KM COMBINE POND OUTFLOW WITH SUB4
KO     0       2
HC     2
ZZ
```

FIGURE 36.13 Input for Example 36.6.

The routing input is straightforward. There are three records (RC, RX, and RY) for describing the channel. In the RC record the values for the channel's Manning's n, length, and slope are input. The RX and RY records define the cross section of the channel as shown in Fig. 36.7. HEC-1 computes the stage-storage-discharge relation internally from this information. The HC record is used for combing the subbasins. The number in the first field indicates the number of hydrographs to be combined. A KO record is added in the KK ROAD "block" to indicate a change in the desired output. A "2" in the second field calls for a hydrograph to be printed in the output. The problem input file is shown in Fig. 36.13.

The input, hydrograph at the outlet, and summary table are shown in Figs. 36.14 and 36.15 for the output of Example 36.6.

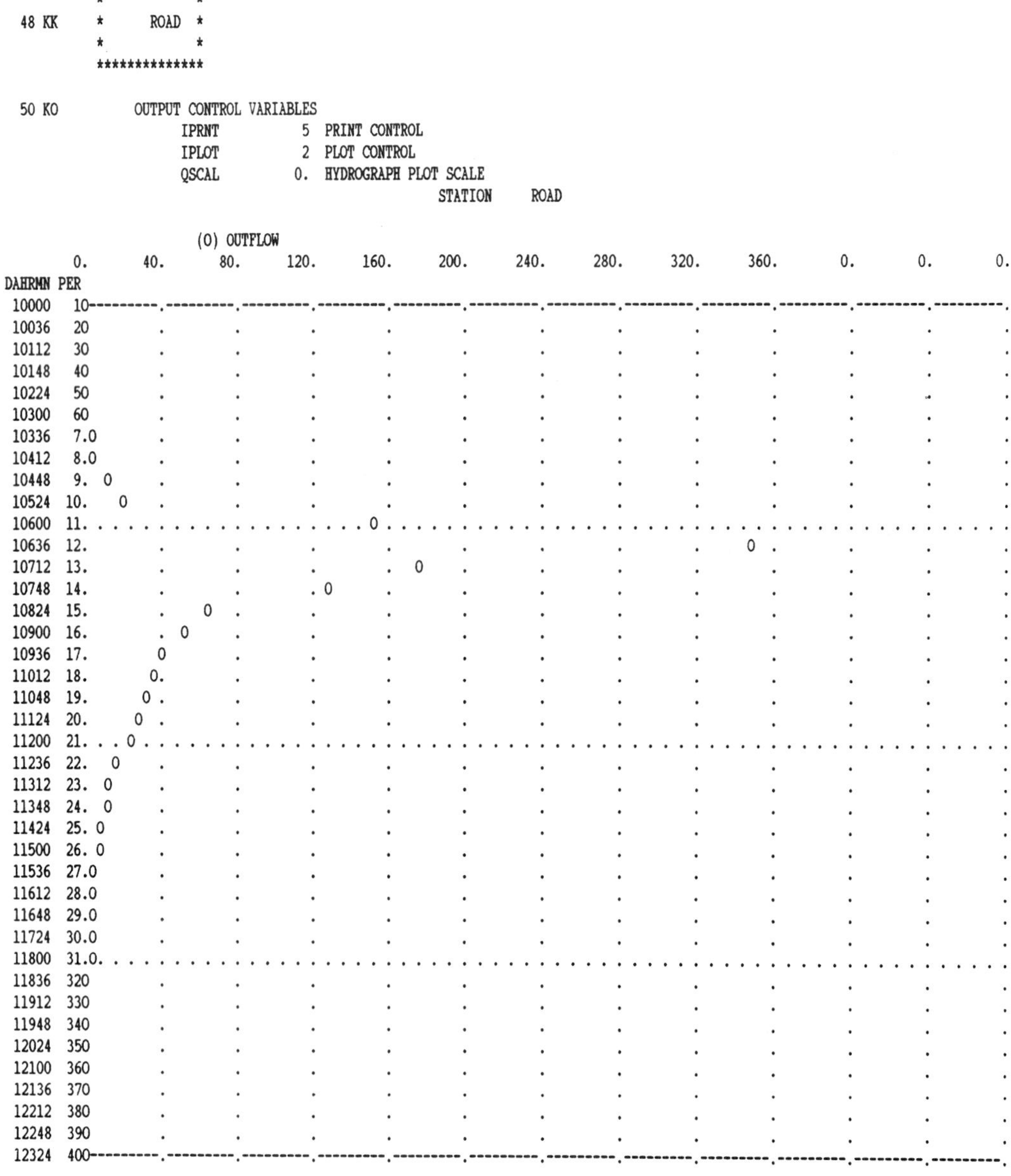

```
             **************
             *            *
 48 KK       *     ROAD   *
             *            *
             **************

 50 KO          OUTPUT CONTROL VARIABLES
                     IPRNT          5  PRINT CONTROL
                     IPLOT          2  PLOT CONTROL
                     QSCAL          0. HYDROGRAPH PLOT SCALE
                                                    STATION     ROAD

                         (0) OUTFLOW
          0.      40.       80.      120.      160.      200.      240.      280.      320.      360.       0.        0.        0.
DAHRMN PER
 10000  10---------.---------.---------.---------.---------.---------.---------.---------.---------.---------.---------.---------.
 10036  20         .         .         .         .         .         .         .         .         .         .         .         .
 10112  30         .         .         .         .         .         .         .         .         .         .         .         .
 10148  40         .         .         .         .         .         .         .         .         .         .         .         .
 10224  50         .         .         .         .         .         .         .         .         .         .         .         .
 10300  60         .         .         .         .         .         .         .         .         .         .         .         .
 10336  7.0        .         .         .         .         .         .         .         .         .         .         .         .
 10412  8.0        .         .         .         .         .         .         .         .         .         .         .         .
 10448  9.  0      .         .         .         .         .         .         .         .         .         .         .         .
 10524  10.   0    .         .         .         .         .         .         .         .         .         .         .         .
 10600  11. . . . . . . . . . . . . . . . . . . . 0 . . . . . . . . . . . . . . . . . . . . . . . . . . . . . . . . . . . . . . . .
 10636  12.        .         .         .         .         .         .         .         .       0 .         .         .         .
 10712  13.        .         .         .         .    0    .         .         .         .         .         .         .         .
 10748  14.        .         .         .  0      .         .         .         .         .         .         .         .         .
 10824  15.        .      0  .         .         .         .         .         .         .         .         .         .         .
 10900  16.        .  0      .         .         .         .         .         .         .         .         .         .         .
 10936  17.        0         .         .         .         .         .         .         .         .         .         .         .
 11012  18.       0.         .         .         .         .         .         .         .         .         .         .         .
 11048  19.      0 .         .         .         .         .         .         .         .         .         .         .         .
 11124  20.     0  .         .         .         .         .         .         .         .         .         .         .         .
 11200  21. . . 0 . . . . . . . . . . . . . . . . . . . . . . . . . . . . . . . . . . . . . . . . . . . . . . . . . . . . . . . . .
 11236  22.  0     .         .         .         .         .         .         .         .         .         .         .         .
 11312  23.  0     .         .         .         .         .         .         .         .         .         .         .         .
 11348  24.  0     .         .         .         .         .         .         .         .         .         .         .         .
 11424  25. 0      .         .         .         .         .         .         .         .         .         .         .         .
 11500  26. 0      .         .         .         .         .         .         .         .         .         .         .         .
 11536  27.0       .         .         .         .         .         .         .         .         .         .         .         .
 11612  28.0       .         .         .         .         .         .         .         .         .         .         .         .
 11648  29.0       .         .         .         .         .         .         .         .         .         .         .         .
 11724  30.0       .         .         .         .         .         .         .         .         .         .         .         .
 11800  31.0. . . . . . . . . . . . . . . . . . . . . . . . . . . . . . . . . . . . . . . . . . . . . . . . . . . . . . . . . . . .
 11836  320        .         .         .         .         .         .         .         .         .         .         .         .
 11912  330        .         .         .         .         .         .         .         .         .         .         .         .
 11948  340        .         .         .         .         .         .         .         .         .         .         .         .
 12024  350        .         .         .         .         .         .         .         .         .         .         .         .
 12100  360        .         .         .         .         .         .         .         .         .         .         .         .
 12136  370        .         .         .         .         .         .         .         .         .         .         .         .
 12212  380        .         .         .         .         .         .         .         .         .         .         .         .
 12248  390        .         .         .         .         .         .         .         .         .         .         .         .
 12324  400---------.---------.---------.---------.---------.---------.---------.---------.---------.---------.---------.---------.
```

FIGURE 36.14 Output from Example 36.6.

RUNOFF SUMMARY
FLOW IN CUBIC FEET PER SECOND
TIME IN HOURS, AREA IN SQUARE MILES

	OPERATION	STATION	PEAK FLOW	TIME OF PEAK	AVERAGE FLOW FOR MAXIMUM PERIOD			BASIN AREA	MAXIMUM STAGE	TIME OF MAX STAGE
+					6-HOUR	24-HOUR	72-HOUR			
	HYDROGRAPH AT									
+		SUB1	35.	6.00	9.	3.	3.	.02		
	HYDROGRAPH AT									
+		SUB2	105.	6.60	37.	10.	10.	.11		
	HYDROGRAPH AT									
+		SUB3	50.	6.00	16.	4.	4.	.04		
	3 COMBINED AT									
+		COMBN	168.	6.60	62.	17.	17.	.17		
	ROUTED TO									
+		LAKE	236.	6.60	57.	17.	17.	.17		
+									777.29	6.60
	ROUTED TO									
+		SUB4	213.	6.60	57.	17.	17.	.17		
	HYDROGRAPH AT									
+		SUB4	136.	6.60	48.	13.	13.	.13		
	2 COMBINED AT									
+		ROAD	349.	6.60	105.	31.	31.	.30		

SUMMARY OF KINEMATIC WAVE - MUSKINGUM-CUNGE ROUTING
(FLOW IS DIRECT RUNOFF WITHOUT BASE FLOW)

ISTAQ	ELEMENT	DT	PEAK	TIME TO PEAK	VOLUME	INTERPOLATED TO COMPUTATION INTERVAL			
						DT	PEAK	TIME TO PEAK	VOLUME
		(MIN)	(CFS)	(MIN)	(IN)	(MIN)	(CFS)	(MIN)	(IN)
SUB4	MANE	4.97	226.26	402.17	3.71	36.00	213.24	396.00	3.75

FIGURE 36.15 Output from Example 36.6.

References

Bedient, P. B. and Huber W. C. 1988. *Hydrology and Floodplain Analysis.* Addison-Wesley, Reading, MA.

Burke, C. B., and Burke, T. T., Jr. 1994. *Stormwater Drainage Manual.* Tech Rept. H-94-6. Highway Extension and Research Project for Indiana Counties and Cities, Purdue University, West Lafayette, IN.

Hoggan, D. H. 1989. *Computer Assisted Flood Plain Hydrology and Hydraulics.* McGraw-Hill, New York.

Hydrologic Engineering Center. 1983. *HECDSS User's Guide and Utility Program Manual,* U.S. Army Corps of Engineers, Davis, CA.

Hydrologic Engineering Center, 1990. *HEC-1 Flood Hydrograph Package, User's Manual,* U.S. Army Corps of Engineers, Davis, CA.

Soil Conservation Service. 1982. *TR-20 Computer Program for Project Formulation Hydrology.* Tech. Release No. 20. Washington, DC.

37

Water Resources Planning and Management

J. R. Wright
Purdue University

M. H. Houck
George Mason University

37.1 Introduction

Water resources engineering is concerned with the conceptualizing, design, and implementation of strategies for delivering water of sufficient quality and quantity to meet societal needs in a cost-effective manner. Alternatives that can be engineered to accomplish this essential function include development of new water supplies, regulation of natural sources of water, transfer of water over large distances, and treatment of degraded water so that it can be reused. The challenges for water resources engineers are (1) to identify the essential characteristics of a given water resources problem, (2) to recognize feasible alternatives for resolving the problem, (3) to systematically evaluate all feasible alternatives in terms of the goals and objectives of the decision maker, and (4) to present a clear and concise representation of the trade-offs that exist between various alternatives.

Water Resources Decision Making

Because large-scale planning, development, and management of water resources systems generally take place in the public sector, the individual or individuals responsible for making decisions about, or selecting from, a set of development alternatives are usually not the engineers who perform the technical analyses related to a given problem domain. The decision-making topology is rather more like that presented in Fig. 37.1. At the top of the topology is the decision maker, usually an elected official or his or her appointee. This individual assumes the responsibility for selecting a course of action that best achieves the goals and objectives of his or her constituency. During the course of the decision-making process, the decision maker interacts with other interested parties, such as local, state and federal government agencies; independent organizations and groups; industry; and individuals. At the bottom of the hierarchy are all data that pertain to the problem domain, including hydrologic data, economic and other cost data, demographic and historic data, and information about relevant structural and management technologies. The water resources engineer selects, from

0-8493-8953-4/95/$0.00 + $.50

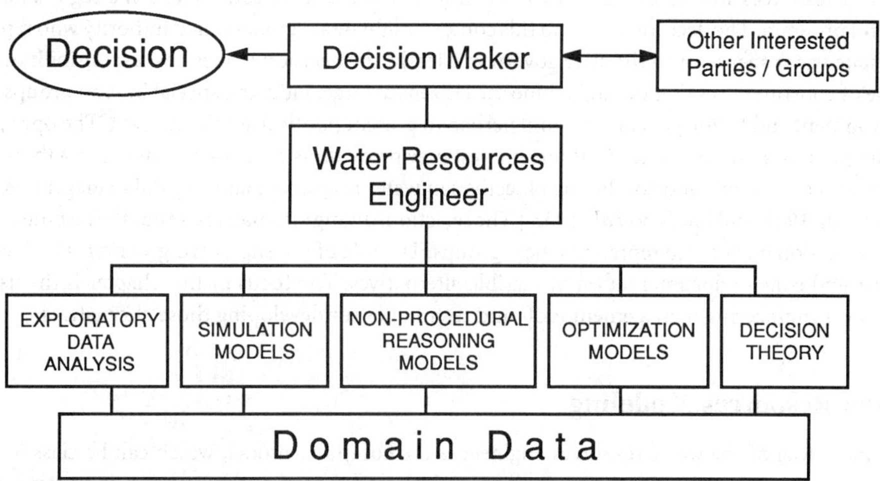

FIGURE 37.1 The water resources engineer uses a wide range of analytical tools and techniques to identify and evaluate alternative development plans and management strategies. The selection of a particular project or plan from a set of alternatives identified by the engineer is the responsibility of a decision maker.

a wide range of modeling and analysis technologies, those that can best evaluate these data and provides the decision maker with information about the trade-offs that exist among and between multiple and conflicting management objectives. Consequently, the water resources systems analyst must be skilled in problem identification, proficient in the use of different modeling methods and technologies, and willing and able to interact with nontechnical managers and decision makers.

Comprehensive water resources planning and management is generally conducted in several separate but related phases requiring input from a wide range of specialists, including urban and regional planners, economists and financial planners, government agency personnel, citizens groups, architects, sociologists, real estate agents, civil engineers, hydrologists, and environmental specialists. The National Environmental Policy Act of 1969 **(NEPA)** requires the preparation of an environmental impact statement for every major federal action (program, project, or licensing action) "significantly" affecting the quality of the human environment. Most water projects fall under this legislation and most states have prepared regional guidelines for complying with NEPA. Special-purpose developments often require the assistance of specialists from other disciplines, such as soil scientists, agricultural specialists, crop experts, computer specialists, and legal experts. These individuals are involved in one or more of several planning and management phases:

- Establishment of project goals and objectives
- Collection of relevant data
- Identification of feasible best-compromise alternative solutions
- Preliminary impact assessment
- Formulation of recommendation(s)
- Implementation (detailed structural design, construction, and/or policy implementation)
- Operation, management, and sustainment

Because most large-scale water resources projects involve many different constituencies having different goals and objectives, a multiobjective perspective through this process is essential. Consider, for example, the problem of developing an operating strategy for a large multipurpose reservoir designed to provide water for irrigated agriculture, municipal and industrial water supply, water-based recreation, and power generation (see Chapter 35 for a comprehensive discussion of water resources structures). The reservoir may also be a critical component in a regional flood control program. The decision maker in this context might be a regional water authority who reports to a state water agency or to the state governor. Clearly, the decision maker has responsibilities to a range of constituencies that might include the citizens at large, industry, special interest groups, the environment and future generations, and perhaps a present political administration. The operating strategy that best meets the needs of one interest may prove disastrous for another. [For a thorough discussion of multiple and conflicting objectives in water resources planning and management, see Goodman, 1984; and Lindsey *et al.*, 1992.] The selection among alternatives is thus the responsibility of the decision maker who represents these groups. The role of the engineering analyst is to develop a clear and concise documentation of feasible alternatives. The focus of this chapter is the use of analytical engineering management tools and techniques for developing those alternatives.

Water Resources Modeling

The main tool of the water resources engineer is the computer model, which can be classified by (1) structure and function (**optimization** or **simulation**), (2) degree of uncertainty in system inputs (**deterministic** or **stochastic**), (3) level of fluctuation in economic or environmental conditions being modeled (**static** or **dynamic**), (4) distribution of model data (**lumped** or **distributed**), and (5) type of decision to be made (investment or operations/management). Each model configuration has inherent strengths and weaknesses, and each has its proper role in water resources planning. The challenge for the water resources engineer is not to determine which is better, but which is most appropriate for a particular situation given available resources including time, money, computer resources, and data.

Optimization vs. Simulation Models

A variety of water resources management and planning models can be formulated and solved as **mathematical programs.** Mathematical programs are optimization models having the following general structure:

$$\text{Optimize } Z = f(x_1, x_2, \ldots, x_n) \tag{37.1}$$

subject to:

$$g_1(x_1, x_2, \ldots, x_n) \leq b_1 \tag{37.2}$$

$$g_2(x_1, x_2, \ldots, x_n) \leq b_2 \tag{37.3}$$

$$\vdots$$

$$g_m(x_1, x_2, \ldots, x_n) \leq b_m \tag{37.4}$$

$$x_j \geq 0 \qquad j = 1, 2, \ldots, J \tag{37.5}$$

Equation 37.1 is called the **objective function** and is specified as a mathematical criterion for measuring the "goodness" of any given solution. The objective function is some function of a set of non-negative variables known as **decision variables** $(x_1, x_2, \ldots x_n)$, whose values we are interested in finding such that the value of the objective function Z is optimized—*maximized* or *minimized.* The optimal solution to the mathematical program is the set of values for the decision variables that provides the "best" value for the objective function.

The quality of the solution as measured by the objective function is constrained by a set of equations, appropriately called constraint equations or **constraints**—Eqs. (37.2–4) above. The functions $g_1, g_2, \ldots, g_m$ also depend on the value of the decision variables, and are each restricted by a set of constants $b_1, b_2, \ldots, b_m$, commonly referred to as the right-hand-side vector. While the typical presentation of the mathematical program formulation indicates that the constraints $g_i | i = 1, 2, \ldots, m$ are *less than or equal* to their respective right-hand-side constant, *equality* and *greater than or equal* to relations may also be specified.

Finally, the mathematical programming model formulation includes the requirement that all decision variables are nonnegative—Eq. (37.5). From an engineering management standpoint, decision variables represent those factors of a problem over which the engineer has control, such as the amount of resource to allocate to a particular activity, the appropriate size of a component of a structure, the time at which something should begin, or cost that should be charged for a service. Clearly, negative levels for such things have no physical meaning. If a water resources planning or management model can be constructed to adhere to the rigid structure of the mathematical program, a variety of solution methodologies are available to solve these models (Hillier and Lieberman, 1990).

Undoubtedly the most widely used analytical procedure employed in the area of water resources systems engineering is simulation (or descriptive) modeling. The main characteristics of this modeling methodology are (1) problem complexities can be incorporated into the model at virtually any level of abstraction deemed appropriate by the model designer or user (in contrast to the rigid structure required by optimization models), and (2) the model results do not inherently represent good solutions to engineering problems; these models reflect the structure and function of the system being modeled and do not attempt to suggest changes in design or configuration towards improving a given scenario.

Simulation models may be time or event sequenced. In time-sequenced simulation, time is represented as a series of discrete time steps $t = 0, 1, 2, \ldots, N$ of an appropriate length—perhaps hours, days, weeks, or months depending on the system being modeled. At the end of each time period t all model parameters would be updated (recomputed) resulting in a new system state at the beginning of time step $t + 1$. The relationships among and between model parameters may be deterministic or stochastic through this updating process, again, depending on the design of the system and the level of abstraction assumed by the model. Model inputs, both initially and throughout the simulation, may follow parameter distributions as discussed in the previous section or may be input from external sources such as monitoring instrumentation or databases.

Models that simulate physical or economic water resource systems can also be event sequenced, wherein the model is designed to simulate specific events or their impacts whenever they occur. These events might be input as deterministic or stochastic events or they might be triggered by the conditions of the system. In any case, the model responds to these events as they occur regardless of their timing relative to simulated real time. Regardless of the treatment of time through the simulation process, these models can be either deterministic or stochastic.

With increasingly powerful computer technology, extremely complicated simulation models can be developed that emulate reality to increasingly high levels of accuracy. Very complicated systems can be modeled through many time steps and these models can be "exercised" heavily (run many times with different parameter settings and/or data inputs) to understand the system being modeled better. A number of commercial vendors market simulation systems that can be used to design and develop simulation models.

Historically, optimization models have been used as screening models in water resources planning and management analyses. The gross level of abstraction required to "fit" a particular problem to this rigid structure, coupled with the heavy computational burden required to solve these models, precluded the construction of large and accurate systems representation. Once a general solution strategy or set of alternatives was identified, simulation models could be constructed for purposes of more detailed analysis and "what-if"-type analyses. With the advent of increasingly powerful computing capability, we are seeing a much tighter integration of these and other modeling technologies.

Deterministic vs. Stochastic Models

Water resources models can also be classified by the level of uncertainty that is present in model parameters and hydrologic inputs. A model is said to be deterministic if all input parameters and expected future unregulated streamflows and other time series are assumed to be known with certainty and defined specifically within model constraint equations. If, on the other hand, only the probability distributions of these streamflows are assumed to be known within the model, the model is said to be stochastic (see Chapter 30 for a more complete discussion of statistical hydrologic analysis).

Both optimization and simulation models can be either deterministic or stochastic. Stochastic models are generally more complex than deterministic models, having more variables and constraints or limiting conditions. But deterministic models, having parameters and inputs based on average values over potentially long time periods, are usually optimistic; system benefits are overestimated, while costs and system losses are generally underestimated. If sufficient information and computational resources are available, stochastic models (either optimization or simulation) are generally superior (Loucks *et al.*, 1981).

Static vs. Dynamic Models

Models vary in the manner in which changes in model parameters occur over time. In a particular watershed, for example, while actual or predicted streamflows might vary over time, the probability distribution for those flows may not change appreciably from one year to the next and may thus be assumed static within the corresponding model. Dynamic models, on the other hand, assume changing conditions over time and attempt to incorporate such changes into the analyses being conducted. Dynamic models tend to be more complex and require more computational effort to solve, but usually provide more accurate results, assuming that adequate data are available to calibrate the models appropriately; this is particularly true for investment models. Static models can be significantly larger in scope. Models may be static in terms of some factors (physical characteristics) while dynamic in terms of others (economies).

Investment vs. Operations/Management Models

Models may also be classified in terms of the time frame within which the analysis is being performed or for which the resulting decision will be made. Long-term decisions dealing with selecting investment strategies including things like physical changes to facilities (reservoir capacity expansion or hydroelectric facility development, for example) are characteristically different from short-term decisions, such as determining the appropriate reservoir release at a particular point in time. Models used to develop operating or management strategies for a water resources system can generally be more detailed than those designed to recommend longer-term investment decisions, which frequently consider actions taken over multiple time periods.

Lumped vs. Distributed Data Models

Lumped data models are those that assume single values (average values, for example) for parameters that are, in fact, distributed spatially within the region being modeled. Distributed data models use spatially distributed data as input to the corresponding model. Clearly, distributed data models require considerably more data and probably computational effort while providing a much more realistic representation of the physical system being studied. Lumped parameter models are generally much more efficient to solve and may be appropriate in cases where insufficient or incomplete data sources are available. With the explosive growth in the use of geographic information systems and corresponding availability of spatial data, distributed data models are becoming much more popular, at least within the research arena (Maidment, 1993).

37.2 Evaluation of Management Alternatives

For any given water resources problem, an alternative may be represented as a set of investment decisions, each having a specific time stream of costs and benefits. The level of each investment is a variable, the best value of which depends on the values of other variables and the goals and objectives of the decision maker. The questions that must be considered systematically by the water resources engineer in providing meaningful guidance to the decision maker are (1) how should each variable be evaluated economically? (2) what is the set of values for these variables such that the resulting alternative best satisfies a given objective? and (3) what is the best set of alternatives, and how can one be assured that there are no better alternatives? A number of proven modeling techniques are available to address these questions.

Consider a set of alternate water resources projects P consisting of individual projects $p \in P$. Each project may be specified as a set of values for a discrete number of decision variables x_j^p, $j = 1, 2, \ldots, n_p$. Each plan is fully specified by a vector of these decision variables and their values X_p. A common goal of water managers is to identify that plan which maximizes **net benefits** (NB):

$$\text{Maximize} \sum_{p \in P} \text{NB}(X_p) \tag{37.6}$$

When the benefits (or costs) of a particular project alternative are most properly evaluated in economic terms, the value of a particular investment component of any given project depends at least in part on the timing of that particular investment. Because different water resources investment alternatives may have different useful lives, it is important that they be compared using a common framework. While a comprehensive treatment of engineering economic analysis is beyond the scope of this handbook, a brief outline of an approach to evaluating alternatives is offered. Basic understanding of the time value of money, as well as finance principles, is important in an overall analysis of complex investment strategies. The interested reader is referred to Blanchard and Fabrycky [1990]; Fabrycky and Blanchard [1991]; Grant *et al.* [1990]; Thuesen and Fabrycky [1989]; and White *et al.* [1989] for additional information on performing comprehensive engineering economic analysis.

Discount factors are used to determine the value of a particular investment over time.

Let **PV** = the **present value** of an amount of money (principal),
FV = the **future value** of an amount (or value) of money,
$\boldsymbol{i}$ = the **interest rate** each period, and
$\boldsymbol{n}$ = the life (in periods) of the investment.

Given an investment at the present time, the future value of that investment n time periods into the future is given by the *single-payment, compound amount factor:*

$$\text{FV} = \text{PV}(1 + i)^n \tag{37.7}$$

The present value of costs (or benefits) resulting from some future payment is computed using the reciprocal of this factor, which is referred to as the *single-payment, present worth factor:*

$$\text{PV} = \text{FV}(1 + i)^{-n} \tag{37.8}$$

Suppose that a particular investment (water resources project, for example) is anticipated to return a stream of future benefits NB_t over T discrete time periods, $t = 1, 2, \ldots, T$. The present value of this stream of benefits is computed:

$$\text{PV} = \sum_{t=1}^{T} \text{NB}_t (1 + i)^{-t} \tag{37.9}$$

The benefits resulting from water resources project investments may be justified as a continuous stream with an infinite time horizon. The present value of benefits over time can also be annualized using the *capital recovery factor* (CR), which determines the value of a recurring benefit stream generated by an initial investment (present value at the beginning of a project's life, for example):

$$\text{CR} = \left[\frac{i(1+i)^T}{(1+i)^T - 1}\right] \tag{37.10}$$

such that

$$\text{NB} = \left[\frac{i(1+i)^T}{(1+i)^T - 1}\right]\text{PV} \tag{37.11}$$

As a supplement to using present value of net benefits as a criterion to evaluate projects, the computation of a benefit-cost ratio (present value of benefits divided by present value of costs) is often used to perform preliminary screening of alternatives:

$$\frac{B}{C} = \frac{\text{PVB}}{\text{PVC}} \tag{37.12}$$

Alternatives having a benefit-cost ratio less than 1 might be removed from further consideration. However, because costs are typically easier to identify and estimate, care should be used in itemizing and valuing project benefits for this purpose.

For public investments, the appropriate interest rate to be used in comparing projects is a matter of public record and is based on the average yield on federal government bonds having approximately the same maturity period as the economically useful life of the project being evaluated. [The Water Resources Development Act of 1974 (P.L. 9302511)]. The assumption of a constant interest rate is standard practice for these types of investments. A comprehensive discussion of inflationary considerations in project evaluation is presented in Hanke *et al.* [1975].

37.3 Water Quantity Management Modeling

Among the largest public investments are those designed to stabilize the flow of water in rivers and streams. A stream that may carry little or no water during a significant portion of the year may experience extremely large (perhaps damaging) flows during peak periods. A storage reservoir may be employed to retain excess water from these peak flow periods for **conservation** use during low-flow periods (water supply, low-flow augmentation for environmental protection, irrigation, power production, navigation, recreation, etc.) or to contain peak flows for purposes of reducing downstream flood damage **(flood control).** In this section we discuss methods for managing surface-water quantity pursuant to the development of comprehensive management alternatives.

The management of a reservoir or system of reservoirs is achieved through a set of **operating rules** that govern releases from the reservoir as a function of such things as inflows into the impoundment, demand for water, storage volumes, and reservoir elevations. Design of the reservoir storage volume, the spillway, and other reservoir components depend on these rules.

Deterministic Reservoir Models

While there is always a significant amount of uncertainty in any hydrologic system, preliminary analysis of alternatives can be accomplished using **deterministic models:** models that do not explicitly consider uncertainty in hydrologic variables. Deterministic methods are particularly useful as screening models, which can identify alternatives for further analyses using more complete and thorough system representations.

Reservoir Storage Requirements

The factors that determine the extent to which streamflows can be stored for future use are (1) the capacity of the **impoundment**, and (2) the manner in which the reservoir is operated. It

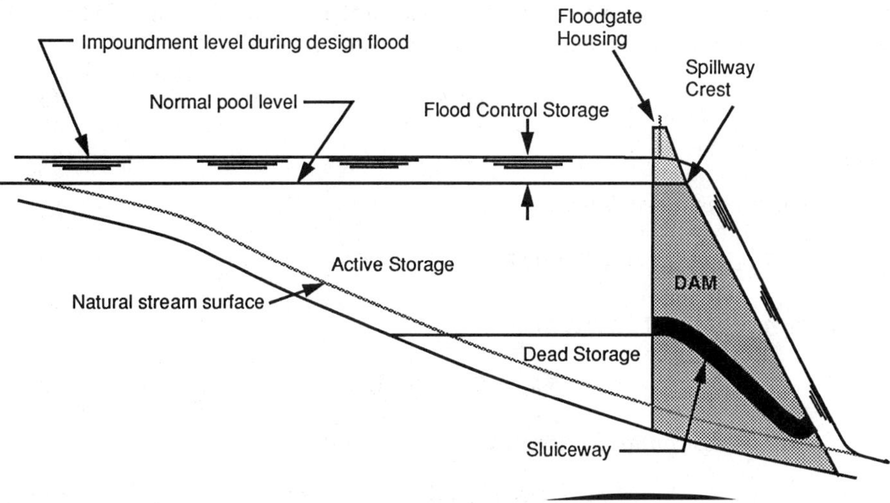

FIGURE 37.2 Schematic cross section of a typical reservoir system showing the relative location of different storage types.

is important to realize that the use of a reservoir to manage streamflows will necessarily result in a net loss of total water due to evaporation and seepage. Only if the benefits of regulation will be greater than these losses should a reservoir be utilized. Thus, the main goal of storage-reservoir design is a thorough analysis of the relationship between **firm yield**—the maximum quantity of flow that can be guaranteed with high reliability at a given site along a stream at all times—and reservoir capacity.

The impoundment behind a reservoir consists of three components (see Fig. 37.2): (1) **dead storage**—that volume used for sediment control and retention of a permanent pool; (2) **active storage**—that volume used to meet water demands such as for water supply, irrigation, conservation, recreation, and power production (head); and (3) **flood control storage**—that volume of the reservoir reserved to contain flood flows, thereby protecting downstream assets. The determination of reservoir capacity and operating strategy should consider all three storage components with a goal of achieving a least-cost design strategy (total storage volume) that satisfies a predetermined set of operational constraints.

A variety of methods are available for performing these analyses including the *mass diagram analysis* [Rippl, 1883; Fair *et al.*, 1966], which finds the maximum positive cumulative difference between a sequence of specified reservoir releases and known historical (or simulated) inflows; the *sequent peak procedure* [Thomas and Fiering, 1963], which is a bit less cumbersome than the mass diagram method; and a variety of *optimization methods* [see, for example, Yeh, 1982], which can consider multiple reservoir systems.

The optimization methods incorporate mass balance constraints of the following form:

$$ST_{t+1} = ST_t + PP_t + QF_t - R_t - EV_t \tag{37.13}$$

where

ST_t = storage in the reservoir at the beginning of time period t

PP_t = precipitation amount during time period t

QF_t = reservoir inflow during time period t

R_t = reservoir release during time period t

EV_t = evaporation from reservoir during time period t

Defining K_a as the minimum active storage capacity of the reservoir for a specified firm yield release level R^*, the mathematical program can be formulated as follows:

$$\min K_a \tag{37.14}$$

subject to

conservation of mass in each period:

$$ST_t - ST_{t+1} - \hat{R}_t = R^* - QF_t - PP_t + EV_t \quad t = 1, 2, \ldots, T \tag{37.15}$$

reservoir capacity cannot be exceeded:

$$ST_t - K_a \leq 0 \quad t = 1, 2, \ldots, T \tag{37.16}$$

storages and releases cannot be negative:

$$ST_t, \hat{R}_t \geq 0 \quad t = 1, 2, \ldots, T \tag{37.17}$$

where $\hat{R}_t$ is volume released in excess of the firm release R^*. The model can be solved to find optimal values for K_a, ST_t, and $\hat{R}_t$, and can be solved repeatedly with different values of R^* to find the relationship (trade-off) between minimum reservoir storage and firm yield. This is a deterministic model that uses a particular sequence of streamflows, and the estimates of required storage volume are dependent on the adequacy of this streamflow sequence to represent the actual stochastic nature of the streamflows.

Flood Control Planning

Prevention (or mitigation) of damage due to flooding can be achieved by (1) containing excess streamflow in upstream reservoirs, or (2) confining the excess flow within a channel using levees, flood walls, or closed conduits. Flood flows generally occur during certain periods of the year and last for relatively short periods of time. The likelihood of a flood event of a given magnitude is usually described by its **return period;** the average interval in years between the occurrence of a flood of a specified magnitude and an equal or larger flood.

The probability that a T-year flood will be exceeded in any given year is $1/T$. Let PQ be a random annual peak flood flow and PQ_T a particular peak flood flow having a return period of T years. Then the probability of PQ equaling or exceeding PQ_T is given by:

$$Pr[PQ \geq PQ_T] = \frac{1}{T} \tag{37.18}$$

For continuous distributions:

$$Pr[PQ \geq PQ_T] = 1 - F_{PQ}(PQ_T) \tag{37.19}$$

or

$$F_{PQ}(PQ_T) = 1 - \frac{1}{T} \tag{37.20}$$

where $F_{PQ}(PQ_T)$ is the cumulative distribution function of annual peak flows. The peak flow at any potential damage site resulting from a flood of return period T will be some function of the upstream reservoir flood storage capacity K_f and the reservoir operating policy:

$$PQ_T = f_T(K_f) \tag{37.21}$$

By assuming a reservoir operating policy for flood flow releases and using an appropriate flood-routing simulation model, the function $f_T(\cdot)$ can be defined by routing different floods through the reservoir using different storage capacities K_f [Lawler, 1964]. Damage functions can then be derived using field surveys at potential damage sites [James and Lee, 1971].

As an alternative to containing peak flows in upstream reservoirs in order to protect downstream sites, channel modifications at the potential damage site can be implemented. In addition, a number of nonstructural measures have been adopted to reduce the impact of flooding [Johnson, 1977]. Frequently, combinations of upstream containment and at-site improvements are considered. If cost functions can be assumed for various improvement alternatives, optimization models can be constructed to identify least-cost options for improvements or for determining optimal strategies using benefit-cost analyses [Loucks *et al.*, 1981].

Among the most widely used reservoir simulation models is the U.S. Army Corps of Engineers, Hydrologic Engineering Center, **HEC-5** program [U.S. Army COE, 1982]. This model was developed to assist in the analysis of multipurpose, multireservoir systems, and contains routines to model flood control operation (including the computation of expected annual damages), determination of firm yield, hydropower systems simulation, multiple-purpose, multiple-reservoir system operation and analysis, and simulation of pumped storage projects. The model incorporates an operating priority as set forth in Table 37.1.

Water Supply Objectives

The models discussed above may explicitly consider the net benefits that would result from the provision of water for domestic, commercial, and industrial water supplies, as well as supplies to irrigation and other agricultural uses. Nonagricultural water users generally base their investments on the availability of firm water yields because of the importance of the water input. A benefit function can be incorporated directly into the model (usually as an objective function in an optimization model), provided that the expected net benefit of some specified target allocation can be assumed. Constraints are included to ensure that the allocation to meet demand in any period not exceed the available yield at that time and location. For those water uses for which economic benefit/loss data are available, constraints specifying minimum acceptable allocations can be used.

Analysis of agricultural investments are generally more activity-specific than for nonagricultural users of water and should thus consider additional input factors such as the availability of capital, labor, and seasonal availability of water. Frequently, a detailed model is used for each irrigation

Table 37.1 Reservoir Operation Priorities in HEC-5

Condition	Normal Priority	Optional Priority
During flooding at downstream location	No release for power requirements	Release for primary power
If primary power releases can be made without increasing flooding downstream	Release down to top of buffered pool	Release down to top of inactive pool (level 1)
During flooding at downstream location	No releases for minimum flow	Release minimum desired flow
If minimum desired flows can be made without increasing flooding downstream	Release minimum flow between top of conservation and top of buffered pool	Same as normal
If minimum required flows can be made without increasing flooding downstream	Release minimum flow between top of conservation and top of inactive pool	Same as normal
Diversions from reservoirs (except when diversion is a function of storage)	Divert down to top of buffered pool	Divert down to top of inactive pool (level 1)

Source: U.S. Army Corps of Engineers, 1982.

site within a region or watershed and may contain detailed information about soil types, cropping patterns, rainfall and runoff profiles, variation in crop yield as a function of water allocation, and perhaps market crop pricing. In cases where detailed irrigation analyses are not possible, a general benefit function may be assumed and incorporated into the planning model. A more complete discussion of the factors to be included in the analysis of water use for irrigation may be found in Chapter 14 of Linsley *et al.* [1992].

Power Production Objectives

The potential for power production at a reservoir site depends on the flow rate of water that can pass through generation turbines and the potential head available. Power **plant capacity** is the maximum power that can be generated under normal head at full flow, while **firm power** is the amount of power that can be sustained, available 100% of the time. **Firm energy** is the energy produced with the plant operating at the level of firm power. Power that is generated in excess of firm power is called **secondary** or **interruptible power.** The problem of determining reservoir storage sufficient to produce a specified firm energy (or some surrogate benefit function) is similar to that of determining firm yield, but, because firm energy, flow rate, **gross head,** and storage are nonlinearly related, it is more difficult to estimate. Nonlinear optimization models, and in some cases mass curve analysis, may be used to estimate the minimum storage necessary to provide a given firm yield. More common linear optimization models may also be used to analyze hydropower potential: Typically, a value of head is assumed, thereby linearizing the power equations; the resulting optimization model is solved; and the actual heads are compared to the assumed values. Adjustment of the assumed values can be made if the discrepancy is too large, and the process is repeated. Simulation models may also be used to evaluate power potential, but they require an explicit specification of the operating rules for the reservoir. A more complete discussion of the use of optimization models to size and control hydropower reservoirs can be found in Major and Lenton [1979]; Loucks *et al.* [1981]; and Mays and Tung [1992].

Flow Augmentation and Navigation

It is often desirable to use water stored in a reservoir to augment downstream flows for instream uses such as natural habitat protection, recreation, navigation, and general water quality considerations. Not only is the volume of flow important, but the regulation of water temperature may also be of concern. Dilution of wastewater or runoff such as from agricultural sources is another potential objective of reservoir management. Assuming that appropriate target values can be established for flow augmentation during different times of the year, these considerations can readily be incorporated into simulation and optimization models used for developing reservoir operating strategies. By constraining the appropriate streamflow yields at a specific time and location to be no less than some minimal acceptable value, it is possible to estimate the degradation in the resulting value of the objective function (or quantifiable net benefits) that would result from such a policy.

Real-Time Operations

Long-term operations and planning models attempt to address questions such as: how large should the storage volume (reservoir) be? what types and capacities of turbines would be best for hydroelectric energy production? and how much freeboard should be allocated for flood control? Many long-term operations and planning models are based on a time scale of months or seasons (and sometimes weeks), so that any operating rules from these models serve as guides to real-time operations but do not well define the operating policy to be followed in the short term.

Real-time operations models attempt to answer the question of how best to operate a reservoir or water control system in the short term (perhaps hours or days), using the existing physical system. Real-time operations models can have all of the model characteristics already described, but they typically differ because (1) they have a short time horizon (days or weeks compared to years); and (2) they are used repeatedly. For example, at the beginning of a particular day, the system state—current storages, anticipated flows for the next week, etc.—is input to the model; the the model is solved to determine the optimal releases to be made during each of the next seven days; recommended release for today is actually made; and at the beginning of tomorrow, the whole process is repeated with the actual new system state and forecasts.

Systems Expansion

The water resources management models discussed thus far have assumed a single (static) planning horizon with constant parameters (demand, release targets, etc.). More typically, investments of this magnitude span many years, during which change is continuous. While these models provide reasonable solutions for a particular future time period, they are not well suited for analyses over multiple time periods or when multiple stages of development are required. When longer-term planning is required, **dynamic expansion models** should be considered. Two types of models are used most frequently: (1) integer programs to select an investment sequence, and (2) dynamic programming models.

The capacity expansion integer program assumes a finite set of expansion investments for each project site over a finite number of time periods, t. If the present value of net benefits of each investment is known as a function of when that investment is undertaken (NB_{st} = net present value of benefits if investment at site s is undertaken in time period t), then an objective function can be written to maximize total benefits across all projects during the planning period:

$$\max \sum_{t} \sum_{s} \mathrm{NB}_{st} X_{st} \tag{37.22}$$

where $X_{st} = 1$, if investment s is undertaken in period t, and 0, otherwise. Constraints ensuring that each investment can be undertaken only once (or that each must be undertaken exactly once), constraints on total expenditures, and constraints that enforce requirements for dependencies among investments can also be included as necessary in this capacity expansion model. This model can also be used to determine the optimal magnitude of a particular investment, such as determining the optimal net increase in storage to add to a particular reservoir [Morin, 1973; Loucks *et al.*, 1981].

Integer programs are relatively easy to develop but computationally expensive to solve. In contrast, dynamic programming is an alternate approach for solving capacity expansion problems, and one that has become extremely popular among water resources professionals over the past decade. Dynamic programming is particularly useful for determining strategies for making decisions about a sequence of interrelated activities (investments) where nonlinearities in the objective function of constraints are present. These models tend to be much more efficient to solve, but, because there is no standard mathematical formulation available for dynamic programs, they may require more time and care to develop [Loucks *et al.*, 1981; Mays and Tung, 1992]. Dynamic programming has been used to model to a wide range of water resources planning and management activities [Esogbue, 1989].

Stochastic Reservoir Modeling

The deterministic modeling approach discussed above was based on an assumed profile of system inputs (e.g., streamflows). However, the historical record of streamflows upon which these analyses are typically based may not be sufficiently representative of long-term conditions. In stochastic reservoir modeling, available records are considered to be only a sampling of long-term hydrologic processes and are thus used to estimate the statistical properties of the underlying stochastic process. Two approaches have been developed to use the resulting stochastic model: (1) incorporate the stochastic model (e.g., a Markov model of streamflow) directly in an optimization model; and (2) use the stochastic model to generate **synthetic records**, which are then used as inputs for a set of deterministic models.

Consider the second option—generating synthetic records that reflect the main statistical properties of the observed historic flows (mean, standard deviations, first-order correlation coefficients, etc.). Multiple sets of these streamflow records may then be used in a variety of river basin models. (Synthetic sequences of other processes are frequently used in river basin simulation modeling, including rainfall, evaporation, temperature, and economic factors.) There are two basic approaches to synthetic streamflow generation. If a representative historic record is available, and if streamflow is believed to be a stationary stochastic process (wherein the main statistical parameters of the process do not change over time), then a statistical model can be used to generate sequences that reproduce characteristics of that historic record [Matalas and Wallis, 1976]. For river basins or watersheds that have experienced significant changes in runoff due to such factors as modified cropping practices, changing land use, urbanization, or major changes in groundwater resources (see Chapter 32), the assumption of stationarity may not be valid. In these instances rainfall runoff models may be useful in generating streamflow sequences that can be used effectively for water resources engineering [Chow *et al.*, 1988].

The use of synthetic sequences in simulation models is an acceptable way of dealing with uncertainty in the analysis of water resources planning and management alternatives. Recall, however, that simulation models are unable to generate optimal alternatives. The optimization modeling methodologies discussed previously can also incorporate hydrologic variability (and processes uncertainties) for solution by the appropriate optimization algorithms.

The direct incorporation of the stochastic model in an optimization model permits explicit probabilistic constraints or objectives to be included. One example of this approach to the incorporation of hydrologic uncertainty into surface-water optimization models is through the use of chance constraints, which can suggest optimal reservoir size and operating strategy while enforcing prespecified levels of reliability for release and storage requirements [ReVelle *et al.*, 1969; Sniedovich, 1980]. This technique involves the transformation of probabilistic constraints into their deterministic equivalents,

$$P[S_{\min,t} \leq ST_t] \geq \alpha_{ST_t} \qquad t = 1, 2, \ldots, T \tag{37.23}$$

which ensures that the probability of the storage at the beginning of time period t being greater than or equal to a minimum storage value ($S_{\min,t}$) will be greater than or equal to some reliability level for storage in time period t, α_{ST_t}. Similarly, for reservoir release levels:

$$P[R_{\min,t} \leq R_t \leq R_{\max,t}] \geq \alpha_{R_t} \qquad t = 1, 2, \ldots, T \tag{37.24}$$

The result is a set of linear decision rules that relate releases from the reservoir to storage, inflows, and other model decision variables [Loucks and Dorfman, 1975; Stedinger *et al.*, 1983]. This methodology can be extended to help determine strategies for developing and implementing multiple reservoir systems having multiple water uses.

Dynamic programming models can also be used to develop reservoir operation strategies under assumptions of hydrologic uncertainty (Esogbue, 1989). Flows are treated as random variables discretized into a number of different possible values. The total flow sequence is treated as a stochastic process—frequently a first-order Markov process—in which the inflow during period t is dependent on the inflow during period $t - 1$. Transition probabilities represent the likelihood of observing two specific successive flow levels, which in turn allow the computation of conditional probabilities that become the transition probability matrix for a dynamic program.

Groundwater Modeling

Comparable modeling approaches have been developed and used to solve quantity and quality problems associated with groundwater. The modeling approaches include optimization and simulation, lumped and distributed, deterministic and stochastic, and static and dynamic modeling. For a complete description of these problems and the modeling methods used to solve them, see Freeze and Cherry [1979] and Willis and Yeh [1987].

37.4 Data Considerations

For at least four reasons there is presently a great deal of interest in the use of spatial data as a framework for conducting water resources engineering: (1) enormous resources are being expended to collect and maintain water resources and physical attribute data within a spatial context [Hastings *et al.*, 1991]; (2) there is a clear relationship between physical land features, which are spatially distributed, and hydrologic surface and subsurface processes [Maidment, 1993]; (3) **geographic information systems (GIS)** have matured to the point that they can be used to simulate hydrologic processes [Engel *et al.*, 1993; Englund, 1993]; and (4) the integration of water resources models, spatial databases, and sophisticated user interfaces has resulted in the development of powerful (spatial) decision support systems that can be understood and used by water professionals [Fedra, 1993].

Geographic information systems is a technology for storing, manipulating, and displaying geo-referenced spatial data. While the data manipulation capabilities of GIS are still rather crude and simplistic by modeling standards, the computational sophistication of these systems and the efficiency with which they can store and display data are impressive [Star and Estes, 1990]. A framework for integration of these spatial databases, traditional lumped parameter data and contextual information, with specialized water resources models is presented in Fig. 37.3. Physical, hydrologic, and possibly biological and economic data are represented as attribute maps or "layers" (A_1 = slope, A_2 = soils type, A_3 = vegetation, A_4 = depth to aquifer, etc., for example). Lumped parameter information and model parameters might be stored in a database or provided by monitoring instrumentation. The models base might include optimization screening models, or more detailed simulation models. The user would interact with the models through a graphical user interface that would allow such options as display of systems status, modification of systems parameters and data sets, and display of model results.

In addition to advances in spatial data management and conventional (simulation and optimization) modeling technologies, the use of other nonprocedural modeling technologies is also being explored to improve our understanding of water resources systems. Though simulation models

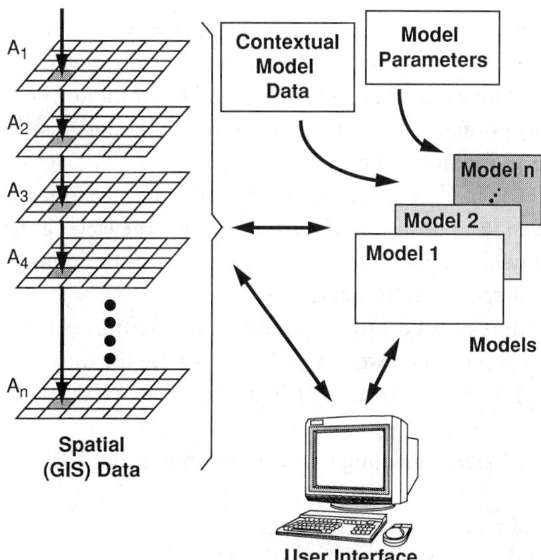

FIGURE 37.3 Functional representation of a spatial decision support system for water resources planning and management that integrates geo-referenced spatial and non-spatial data, one or more specialized models, and a custom user interface.

have become an important part of the technology of water resources systems engineering, many well-known and -accepted programs have become extremely large and complex, precluding use by inexperienced professionals. In addition, some system relationships are too poorly understood to be simulated, but may be well understood (or at least expertly managed or operated) by individuals who possess special knowledge and expertise. Expert systems and other techniques from the field of artificial intelligence, such as artificial neural networks, may be used to model water resources systems when such is the case [see, for example, Bender *et al.*, 1993; Buch *et al.*, 1993; Burde *et al.*, 1994; Davis *et al.*, 1990; McKinney *et al.*, 1992, 1993; VanBlargan *et al.*, 1990].

Defining Terms

Active storage: The portion of water in a reservoir that is used to meet societal needs, such as for water supply, irrigation, etc.

Conservation: The confinement of excess flow for future societal use.

Constraints: Functional relationships that limit collective values of decision variables in a management model.

Dead storage: The portion of water in a reservoir that is used for sediment collection, recreation, hydropower generation, etc.

Decision variable: An operational or design parameter, the value of which is to be determined through formal analysis (for example, reservoir capacity).

Deterministic models: Models whose inputs and parameters are assumed to be known with certainty.

Dynamic models: Models of processes the main characteristics of which are believed to change significantly over time.

Expansion models: Models that can analyze water resources investments over multiple time or planning horizons.

Firm energy: The energy produced with the plant operating at the level of firm power.

Firm power: The amount of power that can be sustained; available 100 percent of the time.

Firm yield: The largest flow of water that can be provided continuously from a stream or reservoir.

Flood control storage: Portion of volume in a reservoir that is reserved for storing excess flow during flooding in order to protect downstream assets.

Future value, FV: The future value of an amount (or value) of money.

Geographic information systems, GIS: Spatial database technology used increasingly for water resources planning and management.

Gross head: The difference in elevation between the water surface immediately upstream of the reservoir structure and the elevation of the point where the water enters the turbine.

HEC-5: A simulation model developed by the U.S. Army Corps of Engineers designed to be used in evaluating multipurpose, multireservoir systems.

Impoundment: The volume of water stored behind a water resources reservoir.

Interest rate, i: The fraction of borrowed capital paid as a fee for the use of that resource.

Interruptible (secondary) power: Power that is generated in excess of firm power; cannot be sustained.

Mathematical program: A rigid modeling structure commonly used for water resources planning and management.

***n*:** The life (in periods) of an economic investment.

NEPA: The National Environmental Policy Act of 1969.

Net benefits, NB: The total value of benefits expected to result from some investment minus total costs.

Objective function: The function used to evaluate feasible management alternatives.

Operating rules: The policy for operation of a water resource system expressed as a set of management rules.

Optimization: A modeling procedure that determines the optimal values for the decision variables of a model.

Plant capacity: The maximum power that can be generated under normal head at full flow.

Present value, PV: The present value of an amount of money (principal).

Return period, T: The average interval in years between the occurrence of a flood of a specified magnitude and an equal or larger flood.

Rule curve: A mapping of the boundaries between different storage zones for a multipurpose reservoir over time.

Simulation: A modeling methodology that is used to describe the structure and function of a system. The model may be used to find good solution strategies for a variety of management problems.

Static models: Models of processes the major characteristics of which are not assumed to change significantly over time.

Stochastic models: Models whose inputs and parameters are assumed to follow a specified probability distribution.

Synthetic record: A streamflow sequence that is generated based on statistical properties from a historic record of streamflows.

References

Bender, M. J., Simonovic, S. P, Burn, D. H., and Mohammed, W. 1993. Interactive computer graphics for expert-system verification. *ASCE Water Res. Plan. Manage.* 19(5): 518–530.

Buch, A. M., Mazumdar, H. S., and Pandey, P. C. 1993. Application of artificial neural networks in hydrological modeling: A case study of runoff simulation of a Himalayan glacier basin. *Proc. 1993 Int. Joint Conf. Neural Netw.: Part 1 of 3.* Nagoya, Japan. Oct. 25–29.

Burde, M., Jackel, T., Hemker, H., and Dieckmann, R. 1994. Expert system shell SAFRAN and its use to estimate the contaminant load of groundwater and soil caused by deposition of air pollutants: System use for decision making in a practical planning problem. *European Water Pollution Control.* 4(1): 20–24.

Davis, R., Cuddy, S., Laut, P., Goodspeed, J., and Whigham, P. 1990. Integrated GIS and models for assisting the managers of an army training area. *Proc. ASCE Symp. Watershed Plan. Anal. in Action.* Durango, CO. July 9–11.

Chow, V. T., Maidment, D. R., and Mays, L. W. 1988. *Handbook of Applied Hydrology.* McGraw-Hill, New York.

Engel, B. A., Raghavan, S., and Rewerts, C. 1993. A spatial decision support system for modeling and managing non-point source pollution. In Goodchild, M. F., Parks, B. O., and Steyaert, L. T. (eds.) *Environmental Modeling with GIS.* Oxford University Press, New York.

Englund, E. J. 1993. Spatial simulation: Environmental applications. Goodchild, M. F., Parks, B. O., and Steyaert, L. T. (eds.) *Environmental Modeling with GIS.* Oxford University Press, New York.

Esogbue, A. O. (ed.) 1989. *Dynamic Programming for Optimal Water Resources Systems Analysis.* Prentice Hall, Engelwood Cliffs, N.J.

Fair, G. M., Geyer, J .C., and Okun, D. A. 1966. *Water and Wastewater Engineering, Vol. 1: Water Supply and Wastewater Removal.* John Wiley & Sons, New York.

Fang, S-C. and Puthenpura, S. 1993. *Linear Optimization and Extensions: Theory and Algorithms.* Prentice Hall, Engelwood Cliffs, NJ.

Fedra, K. 1993. GIS and environmental modeling. In Goodchild, M. F., Parks, B. O., and Steyaert, L. T. (eds.) *Environmental Modeling with GIS.* Oxford University Press, New York.

Freeze, R. A. and Cherry, J. A. 1979. *Groundwater.* Prentice Hall, Englewood Cliffs, NJ.

Goodman, A. S. 1984. *Principles of Water Resources Planning.* Prentice Hall, Englewood Cliffs, NJ.

Hanke, S. H., Carver, P. H., and Bugg, P. 1975. Project evaluation during inflation. *Water Resources Research.* 11(4):511–514.

Hastings, D. A., Kineman, J. J., and Clark, D. M. 1991. Development and applications of global databases: Considerable progress, but more collaboration needed. *Int. J. Geograph. Inform. Syst.* 5(1):137–146.

Hillier, F. S. and Lieberman G. J. 1990. *Introduction to Operations Research,* 5th ed. McGraw-Hill, New York.

James, L. D. and Lee, R. R. 1971. *Economics of Water Resources Planning.* McGraw-Hill, New York.

Johnson, W. K. 1977. *Physical And Economic Feasibility Of Non-structural Flood Plain Management Measures.* Hydrologic Engineering Center, U.S. Army Corps of Engineers, Davis, CA.

Lawler, E. A. 1964. Flood routing. In *Handbook of Applied Hydrology,* V. T. Chow, ed. McGraw-Hill, New York.

Linsley, R. K., Franzini, J. B., Freyberg, D. L., and Tchobanoglous, G. 1992. *Water-Resources Engineering,* 4th ed. McGraw-Hill, New York.

Loucks, D. P. and Dorfman, P. 1975. An evaluation of some linear decision rules in chance-constrained models for reservoir planning and operations. *Water Resources Research.* 11(6): 777–782.

Loucks, D. P., Stedinger, J. R., and Haith, D. A. 1981. *Water Resource Systems Planning and Analysis.* Prentice Hall, Englewood Cliffs, NJ.

McKinney, D. C., Maidment, D. R., and Tanriverdi, M. 1992. Water planning using an expert GIS. *Proc. ASCE 1992 Nat. Conf. Water Resources Plan. Manage.—Water Forum '92.* Baltimore, MD. August.

McKinney, D. C., Maidment, D. R., and Tanriverdi, M. 1993. Expert geographic information systems for Texas water planning. *ASCE Water Resources Plan. Manage.* March–April, pp. 170–183.

Maidment, D. R. 1993. GIS and hydrological modeling. In Goodchild, M. F., Parks, B. O., and Steyaert, L. T. (eds.)*Environmental Modeling with GIS.* Oxford University Press, New York.

Major, D. C. and Lenton, R. L. 1979. *Applied Water Resources Systems Planning.* Prentice Hall, Englewood Cliffs, NJ.

Matalas, N. C. and Wallis, J. R. 1976. Generation of synthetic flow sequences. In Biswas, A. K. (ed.) *Systems Approach to Water Management.* McGraw-Hill, New York.

Mays, L. W. and Tung, Y.-K. 1992. *Hydrosystems Engineering and Management.* McGraw-Hill, New York.

Moore, J. W. 1989. *Balancing the Needs of Water Use.* Springer-Verlag, New York.

Morin, T. L. 1973. Optimal sequencing of capacity expansion projects. *J. Hydraul. Div. ASCE.* Vol. 99, No. HY9.

Rehak, D. 1983. Expert systems in water resource management. In James, W. and Torno, H. (eds.)*Proc. ASCE Conf. Emerg. Techniques in Storm Water Flood Manage.* Niagara on the Lake, Ontario, Canada. October.

ReVelle, C. S., Jores, E., and Kirby, W. 1969. The linear decision rule in reservoir management and design: Development of the stochastic model. *Water Resources Research.* 5(4): 767–777.

Rippl, W. 1883. The capacity of storage reservoirs for water supply. *Proc. Inst. Civ. Eng.* (Brit.) Vol. 71.

Sniedovich, M. 1980. Analysis of a chance-constrained reservoir control model. *Water Resources Research.* 16(4): 849–854.

Star, J. and Estes, J. 1990. *Geographic Information Systems: An Introduction.* Prentice Hall, Engelwood Cliffs, NJ.

Stedinger, J. R., Sule, R. B., and Pei, D. 1983. Multiple reservoir system screening models. *Water Resources Research.* 16(6): 1383–1393.

Thomas, H. A., Jr., and Fiering, M. B. 1963. The nature of the storage yield function. *Operations Research in Water Quality Management.* Harvard Water Resources Group, Cambridge, MA.

U.S. Army Corps of Engineers, Hydrologic Engineering Center HEC-5. 1982. *Simulation of Flood Control and Conservation Systems Users Manual.* Davis, CA..

VanBlargan, E. J., Ragan, R. M., and Schaake, J. C. 1990. Hydrologic geographic information systems. *Transportation Research Record.* 1261:44–51.

Willis, R. and Yeh, W. W.-G. 1987. *Groundwater Systems Planning and Management.* Prentice Hall, Engelwood Cliffs, NJ.

Yeh, W. W.-G. 1982. *State of the Art Review: Theories and Applications of Systems Analysis Techniques to the Optimal Management and Operation of a Reservoir System.* UCLA-ENG-82-52. University of California, Los Angeles. June.

For Further Reading

American Society of Civil Engineers. 1994. Water policy and management: Solving the problems. *Proc. 21st Ann. Conf. Water Resources Plan. Manage.* New York.

Ang, A.H.-S. and Tang, W. H. 1984. *Probability Concepts in Engineering Planning and Design, Volume 1: Decision, Risk and Reliability.* John Wiley & Sons, New York.

Blanchard, B. S. and Fabrycky, W. J. 1990. *Systems Engineering and Analysis.* Prentice Hall, Englewood Cliffs, NJ.

Chavatal, V. *Linear Programming.* 1980. W.H. Freeman, New York.

Emmert, B. 1993. *Expert Systems, Decision Support Systems, and Computer-Assisted Instruction for Water Resource Management: January 1985–June 1993.* Quick Bibliography Series. National Agricultural Library, Beltsville, Md.

Fabrycky, W. J. and Blanchard, B. S. 1991. *Life-Cycle Cost and Economic Analysis.* Prentice Hall, Englewood Cliffs, NJ.

Grant, E. L., Ireson, W. G., and Leavenworth, R. S. 1990. *Principles of Engineering Economy,* 8th ed. Ronald Press, New York.

Mays, L.W. 1992. *Hydrosystems Engineering and Management.* McGraw-Hill, New York.

Thuesen, G. J. and Fabrycky, W. J. 1989. *Engineering Economy,* 7th ed. Prentice Hall, Englewood Cliffs, NJ.

White, J. A., Agee, M. H., and Case, K. E. 1989. *Principles of Engineering Economic Analysis.* 3rd ed., John Wiley & Sons, New York.

The World Trade Center in New York City was built by the Port Authority of New York and New Jersey to serve as the region's headquarters for international trade. The seven-building center is dominated by its famed Twin Towers, rising 1350 feet as the second-tallest buildings in the world.

The statistics associated with the construction from 1966 to 1972 include the following: more than 1.2 million cubic yards of excavated earth were placed in the Hudson River to create the 23.5-acre Battery Park City site; more than 200,000 tons of steel were used in construction; the 425,000 cubic yards of concrete used in building the World Trade Center is enough to build a sidewalk from New York City to Washington, D.C.; and 43,600 windows in the Twin Towers contain 600,000 square feet of glass cleaned by automatic window-washing machines. The World Trade Center provides office space for 50,000 workers and hosts 1.8 million visitors per year on its popular observation decks.

The Twin Towers were designed to withstand once-in-a-century wind blasts of 150 mph and the head-on impact of a Boeing 707—the largest plane in the skies when the towers were erected. The strength of its unique steel curtain wall design was proven in February 1993 when the Center withstood a tragic terrorist bombing in its underground parking garage. (Photo courtesy of the Port Authority of New York and New Jersey.)

V

Materials Engineering

W. L. Dolch
Purdue University (Emeritus)

THE STUDY OF CONSTRUCTION MATERIALS is important to most areas of civil engineering. Civil engineers are concerned, in one way or another, with building. And if one is going to build something, one has to build it out *of* something, so civil engineers must be able to choose wisely among a variety of possible materials. Some are better for one use and others for another. The factors that influence this choice are complicated; it is not a simple, cookbook (or even handbook) matter. If this choice is to be more than empirical it must be illuminated with an understanding of why a given material behaves as it does. Only then can the many factors that determine a material's suitability be properly considered. How much time, money, and effort have been wasted because someone made a conventional, but wrong, choice of material?

Civil engineers are interested in construction materials, which are designed to bear loads, so the civil engineer is interested primarily in the mechanical properties of materials. But the study and understanding of civil engineering materials covers a wide spectrum, from purely mechanical properties to purely chemical ones. Involved are the many aspects that determine a substance's composition, structure, surface properties, and reactivity, as well as its defects, elasticity, plasticity, viscoelasticity, and fracture. Effective use of materials demands some degree of understanding of all these aspects.

A further consideration is the great complexity of construction materials. Most are polyphase substances of variable and uncertain composition, finely divided, and with poorly documented physical and chemical properties. As an example, after a hundred years of research, we still do not understand the nature of hardened cement paste, the "glue" that holds concrete together. Progress in the understanding of such substances requires the full range of knowledge and investigative techniques available to materials research.

It is not possible to be a renaissance person in materials, to be really expert in all the disciplines that are important. But it is necessary, for the field engineer as well as the research specialist, to have a reasonable familiarity with the principles and vocabulary of this wide spectrum. Hence the need for education in chemistry, physics, and materials science in addition to mechanics. The modern civil engineering materials curriculum is designed to foster such education.

The chapters in this section cover the conventional materials: metals, wood, concrete, and bituminous materials and mixtures. Soils, although a civil engineering material, are covered in the geotechnical section, owing to the traditional subdivision of the field. The coverage of concrete is somewhat greater than for the other materials.

These chapters were designed to be as informative as possible in a short offering to provide a rapid means for the civil engineer to review and become informed about recent developments. The chapters are descriptive of the important aspects of each material. They do not contain extensive data compilations, as are found in some handbooks. The data needed for design and use of construction materials vary so greatly with the circumstances that to include all detail is not practical. Each of the materials is the subject of a vast literature. References are given to guide the reader to more detailed treatments and data.

38

Nature and Properties of Concrete

W. L. Dolch
Purdue University (Emeritus)

38.1 Introduction

Concrete is a mixture of a discontinuous aggregate phase and a continuous binder phase that, after hardening, holds the aggregate together into a compact mass capable of bearing significant loads. There are many binders—asphalt, sulfur, epoxy resin, and so on—but the unqualified term *concrete* means that the binder is principally **hardened cement paste,** which is the product of the chemical reaction of portland cement and water. So the term *concrete* is short for *portland cement concrete.*

Concrete is the most widely used material; water is the only substance used by humans in larger quantities. Concrete is a material that has been used in some form since ancient times. But the era of modern concrete dates from the work of Smeaton in the mid-18th century, and more importantly that of Aspdin in the first part of the nineteenth, which was the advent of the first truly "portland" cement [Stanley, 1980].

The important properties of any concrete are **workability, durability,** strength, and economy. All aspects of the formulation and use of concrete are designed to provide these properties. Matters of durability are covered in another chapter and will not be considered here.

Quality concrete is obtained not only by using the proper ingredients in the proper amounts but also by employing correct handling procedures during the placement and early life of the concrete. But these latter are procedural rather than materials problems, and are frequently under the control of persons other than the engineer. Therefore, such matters as mixing, transportation, placing, and finishing, although important, are not treated in detail in this chapter. Excellent information is available in several publications of the Portland Cement Association.

In this chapter reference will be made to various organizations, and the abbreviations by which they are known in the trade will usually be used. They are as follows: American Concrete Institute

0-8493-8953-4/95/$0.00 + $.50

(ACI), the worldwide technical society dealing with concrete; American Society for Testing and Materials (now officially ASTM), the most important standards-writing organization in the world; and Portland Cement Association (PCA), the trade association for the U.S. cement industry.

With a subject of this breadth a detailed treatment is impossible in so limited a space. Cement and concrete are the subjects of many texts and references. Recommended ones are listed under "For Further Information" at the end of this chapter. Deserving of special mention is the excellent PCA manual, *Design and Control of Concrete Mixtures* [Kosmatka and Panarese, 1988]. The reader is referred to these, to the references listed, and to the pertinent ASTM standards for essential details.

38.2 Properties of Concrete

This chapter is concerned with the nature and properties of concrete, and with how one ensures desirable properties and avoids undesirable ones. The engineer is concerned with concrete in two states—fresh and hardened. When concrete is freshly mixed it can be manipulated and reshaped. These are the operations of transporting, delivery, placement, consolidation, and finishing. Concrete during this stage is termed **fresh** or *plastic.* After a comparatively short time, usually a few hours, the concrete becomes so firm that it can no longer be reshaped, and then it is said to have *set.* Newly set concrete has little strength and is termed *green.* As time goes by, especially if proper access to moisture, or *curing,* is available, it gains strength and is referred to as *hardened.*

The essential properties of fresh concrete are workability and a proper setting time. The essential properties of hardened concrete are strength and durability. Further, the material must be economical, and there are many circumstances when beauty is also an important property.

38.3 Constituents of Concrete

The constituents of concrete are one or more cementing materials, water, and aggregates. Additionally, there may be chemical or mineral **admixtures.** Reinforcing steel, although obviously incorporated in some concrete, is not part of it.

Portland Cement

In the usual or *ordinary* concrete the cementing material is portland cement. If other components that contribute to the cementing action are used, the concrete is no longer ordinary and is usually termed a *special* concrete. Such materials are discussed in another chapter and will not be considered here.

Portland cement is manufactured by heating in a kiln a mixture of a calcareous and a silicious, usually argillaceous, material. This process is straightforward, although chemically complex. Details are seldom of concern to the civil engineer; they can be found in various texts [Bye, 1983]. The name *portland* originated early on because of the visual similarity of hardened portland cement concrete to a well-known building stone that was (and is) quarried on the Bill of Portland on the south coast of England.

The product of the kiln operation is small nodules termed *clinker.* Clinker is interground with a small amount of an **addition** (usually gypsum) to become the portland cement of commerce. It is marketed in sacks of 94 lb, or (usually) in bulk. Portland is a **hydraulic** cement, which means it can set and harden under water, in contrast to substances (such as common plaster) that depend on drying for their cohesion.

Chemically, portland cement is a mixture of four major constituents and several minor ones. The major constituents are impure varieties of the compounds tricalcium silicate (Ca_3SiO_5), dicalcium silicate (Ca_2SiO_4), tricalcium aluminate ($Ca_3Al_2O_6$), and tetracalcium aluminoferrite ($Ca_4Al_2Fe_2O_{10}$).

Cement chemists have adopted a shorthand notation in which molecular entities are given a single-letter symbol. A partial list is as follows: C = CaO; S = SiO_2; A = Al_2O_3; F = Fe_2O_3; and H = H_2O. For historical reasons, and because of their impure state, the major compounds in cement are usually referred to by historic names, the shorthand notations, or more generic terms. The tricalcium silicate phase is called "alite" or C_3S (pronounced "C three S"); the dicalcium silicate phase is called C_2S, or sometimes "belite"; the tricalcium aluminate phase is called C_3A, or just the "aluminate phase"; and the tetracalcium aluminoferrite is called C_4AF, or just the "ferrite (or iron) phase."

The ASTM specification for portland cement is Designation C 150. This and all other ASTM standards are found in the annually issued ASTM Book of Standards [ASTM, 1994]. Specification C 150 sets forth chemical and physical standards for portland cement; the requisite accompanying test methods are also ASTM standards.

The specification defines five types of portland cement, designated by Roman numerals: I, II, III, IV, V. Type I is for general use and is what is delivered if no special properties are required. Type II is for moderate **heat of hydration** and moderate **sulfate resistance.** These two types make up the great majority of the cement on the market. Type III is so-called high-early-strength cement, used in situations where a more rapid set and strength gain are desired. Type IV has a low heat of hydration and is available only on special order for mass concrete. Type V has higher sulfate resistance and is obtainable in places where sulfate attack is a major problem, such as desert regions.

The cement specifications of other countries resemble those of the U.S. in terms of physical and chemical requirements. Some have a wider variety of types, for example, high early strength and super-high early strength, in Japan. Germany specifies several classes of strength development for each type.

Some cement has interground with it an air-entraining addition, so that the concrete will be resistant to the destructive action of freezing. Such cements are termed *air-entraining,* and are designated by an A following the number (e.g., Type IA). **Air entrainment** is covered in the chapter on durability, and is better obtained with admixtures, so little cement nowadays is air-entraining, and it may be a vanishing species.

C 150 sets forth the chemical compositions that define the five ASTM types. The physical properties specified in C 150 include **fineness,** strength potential, and time of set. Fineness is a measure of the average particle size, or the fineness of grind, of the cement; it has units of surface area per unit weight, and a larger value means a smaller particle. Modern cements are much more finely ground than ones in former days, because the larger area exposed to the mixing water results in a more rapid hydration of the cement and strength gain of the concrete.

The strength potential of a cement is determined by making a **mortar** with standard sand and a given **water-to-cement ratio (w/c),** since that is the primary factor that determines the strength of a cementitious mixture. Cubes are cured for a specified time, and then tested in compression. The minimum strength values specified by C 150 are conservative, lower than one usually finds, and designed mostly to ensure against defective or "dead" cement.

The setting time is determined by penetration tests of cement pastes (cement plus water) or mortars made with standard sand. The specified times are arbitrary and have no close correspondence to the setting time that a concrete made with the cement would have.

The other property of portland cement of most interest to civil engineers is its specific gravity, since this value is needed for some proportioning procedures. Methods are available for its determination, but frequently a value of 3.15 can be assumed with sufficient accuracy.

There are hydraulic cements other than portland that are used in construction. Blended cements are mixtures of portland and a **pozzolan** or ground **granulated blast furnace slag.** They are treated in the chapter on special cements and concretes. The other hydraulic cement of greatest importance to civil engineers is high-alumina cement, which is used mostly as a refractory material but also in construction, especially in Europe.

Aggregates

By far the largest amount of aggregate used in concrete is mineral aggregate—gravel and crushed stone. Good concrete can be made with either. A local supply is almost always used, since the shipment of large amounts of heavy material over long distances is prohibitively expensive. Aggregate differs in quality, and in some locations high-quality aggregate is now in short supply. Sometimes nonmineral aggregates are used. These are manufactured, sometimes waste, materials. Chief among them is crushed blast furnace slag that has been slowly air-cooled.

Concrete aggregates are divided into fine and coarse, based on retention on the 4.75 mm (No. 4) sieve. Fine aggregate obtained from a natural supply is called sand. A mortar is a cementitious mixture without coarse aggregate. Aggregates are also categorized into normal-weight, lightweight, and heavyweight aggregates. Lightweight aggregates are used to make lightweight concrete, where the dead weight of the structure is an important design parameter. Heavyweight aggregates are used for radiation-shielding concrete.

The civil engineer is usually concerned with the physical, rather than the chemical or mineralogical, properties of aggregates. The main exception is the alkali–aggregate reaction, which is treated in the chapter on durability of concrete. The physical properties of most interest are the specific gravity, **porosity** and voids, and particle size distribution, or grading.

Concrete aggregates are almost always porous materials, and an aggregation of such particles has two kinds of spaces. The intraparticle spaces are called the pores, and the interparticle spaces are called the voids, although sometimes the terms are, unfortunately, used interchangeably. The pores, along with the nature of the solids, determine the various specific gravities of the individual pieces; they are functions of the kind of substance and not of the nature of its aggregation or packing. The voids determine the specific gravity, or unit weight, of a packing. Both properties are important to mix design.

The weight of a packing is termed the *dry-rodded unit weight* and is determined according to ASTM C 29. The value is the weight of unit volume of the dry packing that has been rodded into a container. Such a value includes the voids as part of the volume in question.

When only the pores, and not the voids, are considered, the total volume of pores and solids (i.e., that inside the "skin" of the individual piece of aggregate) is called the bulk volume. The mass of unit dry bulk volume is the bulk density. If the pores contain water, the mass of unit wet bulk volume is the bulk-saturated surface-dry density. These two parameters, calculated as the respective specific gravities, are determined according to ASTM C 127.

The weight of water absorbed into the pores by unit weight of dry solid is its **absorption,** also determined by C 127. The volume of water divided by that of the pores (i.e., the degree to which they are full of water) is called the degree of saturation, or simply the saturation. It and the absorption are frequently quoted as a percentage.

The particle size distribution of the pieces of an aggregate is termed the **grading.** Gradings are continuous if no sizes are missing between the largest and smallest; if they are, it is a *gap* grading. If the particles are comparatively large, the grading is *coarse*; if small, it is *fine*. If the size range between largest and smallest is comparatively large, the grading is *long;* if not, it is *short.* The shortest grading is *single-size,* when everything is the same.

With aggregates, the size is determined by the sieve openings that the piece will, and will not, pass. So a group of particles is said to pass such-and-such a sieve and be retained on its smaller neighbor in the nest. Sieve analysis is performed according to ASTM C 136. If the sample is a fine aggregate, the sizes of the openings of the sieves in the nest differ from each other by a factor of two. The standard set goes from 4.75 mm (No. 4) to 150 μm (No. 100) and thus establishes five sizes of material. For coarse aggregates, the two factor is too large, and intermediate sizes are used up to the largest, the so-called D_{max}.

Sieve analysis is usually shown as a cumulative (integral) distribution curve, plotting the percentage passing a given sieve against the logarithm of its size. The spaces between the increments on the abscissa are then equal, owing to the two factor between sieve sizes.

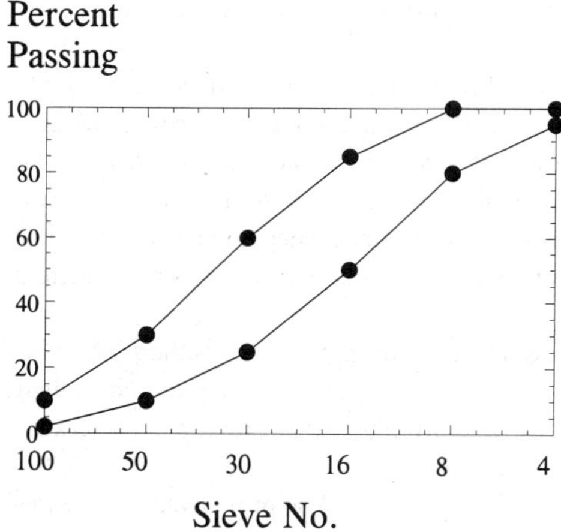

FIGURE 38.1 Grading curves for fine aggregate limits, according to ASTM C 33.

The specification for concrete aggregates is ASTM C 33, and Fig. 38.1 is a plot of its recommendations for fine aggregate, showing the permitted envelope between the coarsest and finest gradings. It should be noted that the official sieve designation is now the size of its openings (ASTM E 11), but many practitioners still refer to the sieve number, the openings per inch, as is done in the figure.

A concise alternative to the grading curve is the **fineness modulus** (FM). This number is obtained by summing the cumulative fractions (percentages divided by 100) *retained* on the sieves (i.e., the value retained if each were the only sieve). For example, the FM of the finer of the two gradings in Fig. 38.1 is 2.15. The parameter is curiously misnamed, since a coarser grading has a larger fineness modulus. The fineness modulus is proportional to the logarithmic average of the particle size, so many gradings can have the same FM. Its usefulness is that it is sensitive to changes in the grading over time. Unfortunately, the FM is affected most by changes in the coarser fractions, which are comparatively less important to the workability of the concrete.

Water

The usual requirement is that the mixing water be potable, although some that is not would be suitable for concrete. It depends on what the impurities are. A good list is given in the PCA manual [Kosmatka and Panarese, 1988]. Seawater can usually be used, but not if the concrete is prestressed or reinforced, since the chloride promotes steel corrosion.

Admixtures

An admixture is a substance, other than the main ingredients, that is added at the mixer. If such a substance is interground with the cement, it is termed an *addition.* Admixtures are of two kinds: mineral and chemical.

Mineral admixtures, such as fly ash and silica fume, are added in comparatively large amounts and are covered in the chapter on special cements and concretes. Chemical admixtures are added in small amounts in order to impart desired properties to the concrete. Reviews of the details of chemical admixtures can be found in the compilation edited by Ramachandran [1984] and in the book by Dodson [1990].

The ASTM specification for chemical admixtures is C 494. It recognizes seven types, defined by their effects—on the setting time and strength gain and on the water content for normal workability. The requirements of C 494 are designed to ensure that the material will do what it is supposed to do without having a harmful effect on strength.

Air-Entraining Agents

These admixtures are used to produce air-entrained concrete. Air entrainment is necessary to protect concrete that is exposed to freezing and thawing when it is highly saturated with water. Most exterior concrete, particularly flatwork (pavements, sidewalks, etc.), in a freezing environment is susceptible to deterioration if it is not air entrained [Dolch, 1984].

Air entrainment also improves the workability of concrete and reduces **segregation** and **bleeding.** Its only drawback is the strength reduction it brings about, roughly 4% for each 1% of entrained air. Sometimes this strength reduction can be offset by proportioning changes permitted by air entrainment.

The air entraining agent is a surface-active, or soaplike, substance that is added in amounts (solids/solids) of a few hundredths of a percent of the weight of the cement. It serves to stabilize the air bubbles that are incorporated into the concrete by the mixing action. The molecules of the agent are adsorbed at the bubble surface and prevent coalescence and loss from the mixture. The result is the presence of very many small bubbles throughout the hardened cement paste. Their total volume is about 4–8% of the volume of the concrete, and their size is from a few micrometers to about a millimeter.

The air content of fresh concrete is determined in several ways, but the pressure method (ASTM C 231) and the volumetric method (ASTM C 173) are by far the most common. They give equivalent results, except that the pressure method is not applicable to lightweight-aggregate concrete, because the large porosity of the aggregate introduces too great an error.

These methods determine the total air content of the fresh concrete. The proper value decreases as the maximum size of the coarse aggregate increases. The reason is that a larger maximum size corresponds to a smaller amount of cement paste in the concrete, and the air affects only this phase.

More important, the air content is not the determinative factor in the resistance a concrete exhibits to freezing. The factor that matters is a geometric aspect of the air-bubble system in the hardened paste. This is the **spacing factor,** a length roughly half the average distance between bubbles. This spacing factor should be smaller than about 200 μm (0.008 in.). This and other geometric factors of the air-bubble system are determined microscopically on the hardened concrete, according to the recommendations of ASTM C 457.

Accelerators

These admixtures accelerate the hardening and strength development of concrete. They are used primarily in cold-weather concreting, when the chemical reactions and strength development are retarded by the lower ambient temperature.

By far the most common accelerator is calcium chloride, which is dissolved in the mixing water and added in about one percent of the weight of the cement. Too much results in a quick set, which is undesirable. Calcium chloride is not used in prestressed concrete because of the corrosive effect of the chloride ion, although it is not considered to harm the steel in ordinary reinforced concrete that will be dry in service. Chloride-free accelerators have been developed to try to solve the steel corrosion problem; chief among these are calcium formate and sodium thiocyanate. Others, such as calcium nitrite, have accelerating action and also a protective effect on the steel.

Although the early strength of concrete is increased by the use of an accelerator, the ultimate strength developed will be somewhat less—a few percent. This fact is part of a generalization: anything that increases the early strength of concrete—accelerators, higher temperature—will result in a slightly decreased ultimate strength. And the opposite is also true: anything that decreases the early strength—retarders, lower temperature (obviously not including freezing)—will generate a slightly higher ultimate strength. The ultimate strength differences, in either direction, are not great enough to be a design consideration.

Water Reducers and Retarders

A water-reducing admixture is one that permits the removal of some of the mixing water, up to about 10%, without the sacrifice of workability. High-range water reducers permit even greater water reduction; they will be considered later.

If the water content is reduced, the strength will be increased, at the same cement content. Alternatively, the cement content, and consequently the cost, can be reduced to maintain the same w/c and strength that the concrete would have had without the admixture. Still another possibility is to add the water reducer without changing the mix proportions. This leads to greater **slump,** and the effect is that of a plasticizer, useful in placing concrete around concentrated reinforcing steel.

A retarding admixture delays setting, hardening, and strength development of the concrete. The usual use is in hot-weather concreting, when the setting would be so quick as to interfere with proper placing and finishing.

Some inorganic substances are retarders, but they are usually an accidental problem, as when equipment is contaminated with residual lead or zinc salts, or phosphates.

Most modern retarders are organic compounds that are also water reducers. The chief materials are of three classes: lignosulfonate salts, hydroxycarboxylic acids, and "sugars," usually short-chain glucose polymers. These substances are added at rates of up to a few tenths of a percent of the weight of the cement. To obtain only water reduction, an accelerator is added to overcome the retarding effect.

Water reduction is accomplished by the admixture molecule adsorbing on the surface of the cement particle and rendering it electrostatically repelling to its fellows. So the cement is more dispersed and less likely to form flocs that entrap water that does not contribute to plasticity. The result is a lower water requirement for a given workability of the concrete.

The mechanism of set retardation is complex and not well understood. Adsorption of the admixture molecules on the solids results in a blocking action that decreases the reactivity of the cement with the water.

The whole picture, both the effects on workability and on hydration, is complicated by the processes of retardation that the cement normally undergoes. The reason for the interground gypsum in portland cement is to retard what might in its absence be a rapid, or *flash,* set. The gypsum dissolves and reacts with the aluminate phase to produce a complex compound that goes by its mineral name of ettringite. This process retards the aluminate reactions and allows normal placement and finishing of the concrete.

Sometimes difficulty is experienced when a water-reducing retarder is used with a given cement. These problems are collectively called *admixture incompatibility.* They usually take the form of a shortened set time and a more rapid loss of slump than the usual gradual change that is normally experienced. If this happens, placement becomes difficult or impossible. Sometimes the difficulty is in the opposite direction, and setting is greatly retarded. Such a delay disrupts schedules and increases costs, but if one waits long enough the concrete will usually become hardened.

These problems are complex, and they have no single or simple cause. Frequently an increase in the sulfate content of the cement is beneficial. Alternatively, if the admixture is added a few minutes after initial mixing of the concrete, this *delayed addition* technique is sometimes helpful.

Superplasticizers

During the 1970s in the U.S., and earlier elsewhere, admixtures came into use that are about twice as effective as the conventional water reducers. These are called superplasticizers or, more accurately but less conveniently, high-range water reducing admixtures (HRWRA).

Chemically, these materials are low-molecular-weight polymers, originally of three classes: sulfonated polynaphthalene-formaldehyde, sulfonated polymelamine-formaldehyde, and purified lignosulfonates. They are added at about 1% of the weight of cement, roughly ten times the concentration of conventional water reducers, which makes them expensive, and this cost was, and is, a limiting factor on their use.

These substances are used in two ways. They bring about a water reduction of around 20% or more at constant slump. So they can be used to produce a constant w/c at a lower cement content, as with conventional admixtures. But the cost of superplasticizers makes this a less attractive option. More commonly, none or only some cement is removed to yield a workable concrete with a considerably lower w/c and much higher strength. All ultra-high-strength concretes contain a superplasticizer, since this is the only way to preserve workability at very low water contents.

The other way these admixtures are used is to produce "flowing" concrete, by not removing any water. Such concrete has a high slump, greater than 7.5 in. by definition, and yet exhibits cohesion and no segregation. The concrete is self-leveling in slabs and can be placed more easily in regions of high reinforcement, thus saving labor costs.

One problem with the use of superplasticizers has been the serious loss of slump that frequently occurs in the concrete. For this reason, superplasticizers are added to the mixture just before discharge of the fresh concrete. Even so, slump loss can be a problem that requires various measures to control. A "second generation" of admixtures has been advocated as superior in this regard [Whiting and Dziedzic, 1990].

38.4 Nature of Cement Paste

When cement and water interact chemically, the reactions are collectively termed *hydration.* The anhydrous compounds are chemically transformed into new substances, hydrates, that make up the hydrated cement. About a fourth of this material is lime, calcium hydroxide, $Ca(OH)_2$. The rest is a complex mixture, which has a composition that varies with the type of cement and the duration of hydration. Most of it is a silicate hydrate, which is termed C-S-H so as to avoid the details of a complicated composition. So hydrated cement is mostly C-S-H and lime, with small amounts of aluminum and iron phases.

The mixture of cement, water, and hydration products is termed cement paste, first fresh, and later hardened. The physical properties of hardened cement paste are more important to the engineer than is its chemical nature. Paste is a mixture of two primary morphologies. The lime is present as comparatively large crystals. The rest is primarily a very finely divided, porous solid called C-S-H gel, because its ultimate particle size is colloidal, and it possesses the properties of other solid gels.

The small particle size of hardened paste is reflected in a large **specific surface,** the surface area of the solids per unit mass. A fully hydrated cement paste will have a specific surface of about 200 m^2/g, when determined by the water vapor sorption technique. If the large lime crystals are ignored, the finely divided remainder has a specific surface of about 270 m^2/g.

Hardened paste is an inherently porous material; it cannot be produced without pore space. The minimum is about one-quarter of the total volume, excluding any unhydrated cement. The larger the original water content, the larger the pore volume. Ordinary concrete, of a w/c of about 0.5, will have a paste half of which is pores. In a mature paste these spaces range in size from that of a few water molecules, about 1 nm (10 Å), up to about 0.1 μm. The small pore size is responsible for the extremely small **permeability** of paste and, in general, for the watertightness of a properly made concrete.

The fact that additional mixing water generates additional pore space, which results in lower strength and higher permeability, is the central insight of good concrete practice. The mixing water should be kept to the absolute minimum that is consistent with proper workability. This is the golden rule of concrete.

38.5 Mix Design

Mixture proportioning, commonly called *mix design,* is the process of selecting the proportions of a concrete mixture in order to secure the desired properties in the final product. This is a necessary, although not a sufficient, step in the process; a proper mix design can result in a poor concrete due to mistakes in handling, curing, and so on. But these are operational, rather than materials, problems, and their details, although important, are not within the scope of this chapter.

Approximate Methods

For small jobs simple methods may be suitable. One old rule of thumb is proportioning by loose volume; for example, a 1:2:3 mix means one bucketful of cement, two of sand, and three of coarse aggregate. Indeed, the size of a sack of cement was originally determined by its being one loose cubic foot.

ACI Method

The best and most popular method of rational mix design is sponsored and monitored by ACI Committee 211 [ACI, 1994]. It is also known as the method of absolute volumes, because the rationale is that a unit volume is filled with the constituents of the concrete, leaving no empty space except the internal porosity of the (coarse) aggregate.

This method is based on three principles. First, the water requirement depends only on the slump and the maximum size of the coarse aggregate (D_{max}), and on whether or not the concrete is air entrained. This is called the Lyse rule. It implies that the richness of the mix (i.e., the cement content) and the grading of the aggregate are not variables, and is true only to a point, that is, only for normal concretes. Slump, ASTM C 143, a rough measure of workability, is determined by the kind of section being placed (slabs, footings, beams, etc.). The larger the slump, the higher the water requirement. D_{max} is determined by the size of section and the space between reinforcing. The larger the D_{max}, the lower the water requirement, because of a smaller content of fine material that needs to be lubricated by the fluid paste. Air entrainment is determined by the necessity to protect from freezing-and-thawing damage. But in these terms, its effect is to increase the fluidity and workability and so lower the water demand of the concrete.

Once these parameters have been decided, the Lyse rule permits the water content to be decided. Figure 38.2 shows the relationship among these variables, for a non-air-entrained concrete.

The water content must be adjusted for any free water present on the fine aggregate and for any absorption of mix water that may occur into the coarse aggregate during or shortly after mixing. This latter amount is somewhat problematical; a value of half the 24-hour absorption is sometimes assumed for the uptake by an initially dry coarse aggregate.

Second, the strength of concrete is determined only (or at least primarily) by the water-to-cement ratio (w/c). This is sometimes called the Abrams rule, after the man who first enunciated this important relationship. The values depend on whether the concrete is air entrained, since the presence of air lowers the strength. Figure 38.3 shows the influence of w/c on compressive strength, as given in the ACI [1994] reference. So the cement content is determined by the desired strength and the previously determined water content.

If the concrete will be exposed to freezing-and-thawing or sulfate attack, durability considerations rather than strength determine the maximum permissible w/c, and hence the cement content.

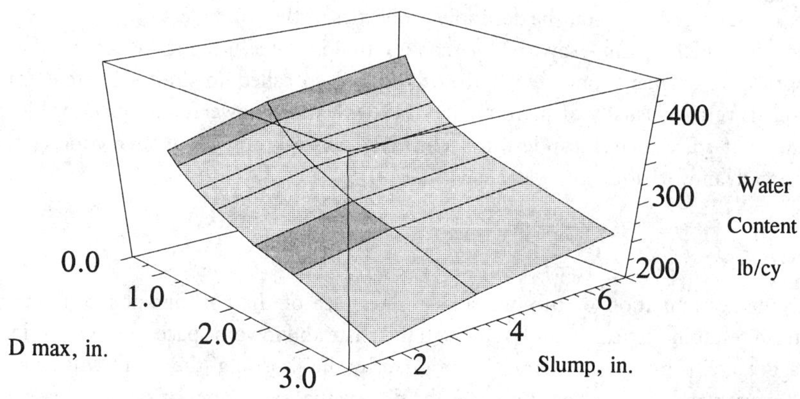

FIGURE 38.2 Influence of maximum aggregate size and slump on the water requirement of non-air-entrained concrete. (Source of data: ACI [1994].)

Third, the amount of coarse aggregate required depends on its maximum size and on the fineness (modulus) of the fine aggregate. This rule arises from the desirability of a continuous, long gradation for the combined aggregate. A larger D_{max} means more coarse aggregate, ipso facto. A larger fineness modulus of the fine aggregate means that it is coarser, so less coarse aggregate is required for the combination. Of these two, the influence of D_{max} is considerably greater than that of the FM of the sand.

These factors determine the required volume of dry-rodded coarse aggregate. Since the ingredients are batched by weight rather than by volume, the dry-rodded unit weight of the coarse aggregate must be known.

Once the amounts of these three ingredients—water, cement, and coarse aggregate—are known, the remainder of the space is filled with the fine aggregate. This step formerly required a knowledge of the specific gravity of the sand, since it is also batched by weight. But now the method gives a table of trial weights of the total concrete, and by difference the fine aggregate, based on its assumed density. This variant is easier and recognizes the fact that the procedure only gives the proportions of a trial batch anyway. The final proportions are determined by testing trail batches and making any necessary adjustments.

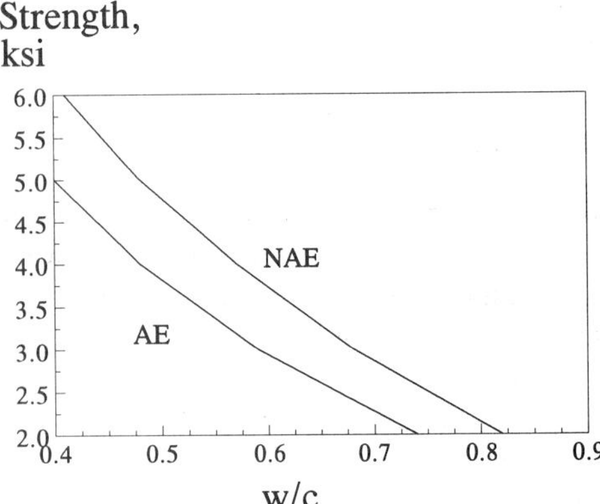

FIGURE 38.3 Relationship between compressive strength of concrete and its water-to-cement ratio. NAE is non-air-entrained and AE is air-entrained concrete. (Source of data: ACI [1994].)

38.6 Properties of Plastic Concrete

Plastic concrete must have certain properties that permit it to be placed into its final form so that the hardened concrete will fulfill its engineering design.

Slump

The slump test (ASTM C 143) is the most frequently performed test on fresh concrete; often it is the only one performed. It is considered by many to be a measure of workability, but it really is a measure of consistency, or wetness. Its main function is to give a ready indication if the concrete tested has a different water content than intended. Its importance derives from the critical significance of an excessive w/c on the desirable properties of the concrete.

Concrete with a high slump—say, greater than about 6 in.—has a higher water content than is ever appropriate for ordinary concrete. On the other hand, so-called no-slump concrete, for which the test result is zero, is usually of high quality and, with more powerful compaction devices, is coming into greater use. The slump test, by definition, tells nothing about such concretes, which may differ greatly in workability.

Workability

Bleeding is the accumulation of mix water at the surface of the consolidated concrete. When excessive it can result in vertical channels through the paste and in void spaces, usually under pieces of coarse aggregate. If bleed water is reincorporated by the finishing, the result will be a surface region that is weak and porous. Segregation is the differential settlement of the aggregate, causing it to accumulate in the bottom of the section and to leave a mortar-rich region in the top.

Workability is the most important property of plastic concrete. It is defined as the ease with which a concrete can be transported, placed, and consolidated without excessive bleeding or segregation. So defined, workability is essentially subjective and cannot be measured. Further, workability depends on the means of working. A concrete that is easily worked with vibration may be completely unworkable when the time-honored method of boots and shovels is used.

Even so, several test methods have been devised to measure, if not workability, something useful and akin to it. The "Kelly ball" test (C 360) measures the penetration of a hemispherical weight into the fresh concrete. It has the advantage of speed and can be used on any flat surface. It makes a convenient substitute for the slump test; the ball penetration value is about half the slump.

Two tests that have been developed for no-slump and similar concretes are the Vebe test and the compacting factor test. Both are used more elsewhere than in the U.S. They are described in the standard texts; neither is yet an ASTM method.

The compacting factor is the ratio of the densities of a concrete that has been allowed to fall a standard distance into a container to that of the fully compacted concrete. The Vebe test, the more useful of the two, measures the time to reform a slump-shaped sample into a new, cylindrical shape. This time, in seconds, is called the Vebe time, and is a useful parameter, especially for stiff concrete that will be placed by vibration.

Rheological Properties

None of the test methods described above, subjective definitions aside, measures workability in any fundamental sense. Rheological parameters are the logical descriptors of such a system. Several devices, basically rotational viscosimeters, have been devised for use with concrete. This work shows concrete to be a modified Bingham body, the flow diagram of which is shown in Fig. 38.4. The intercept is the yield point, the shear stress needed for flow to begin, and the slope of the curve is the plastic viscosity. Both would seem, intuitively, to be significant to the workability of concrete, which must first be put into motion and then kept moving, whatever the details of the placement operation. These parameters have been measured for concrete, along with the influence thereon of mixture proportions, air entrainment, and chemical and mineral admixtures [Tattersall and Banfill, 1983].

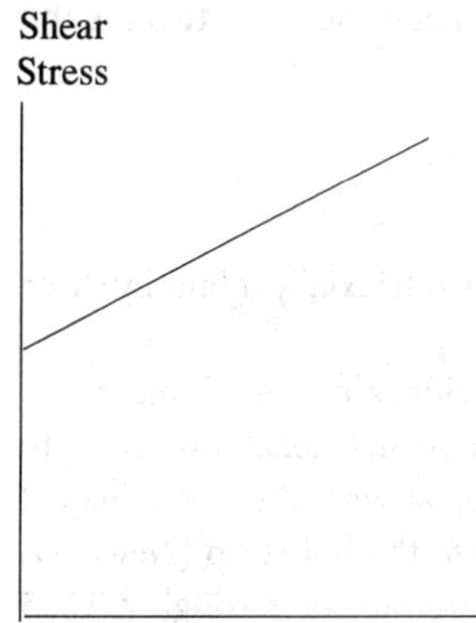

FIGURE 38.4 Flow diagram for a Bingham body. Concrete obeys such a pattern closely.

Setting and Hardening

The other property that fresh concrete must have is a proper setting time—neither so short that handling and finishing are impossible nor so long that the construction schedule is impaired. Of course, if the concrete is site-mixed and the finishing is minimal, the time need not be long, but if the concrete is ready-mixed and the schedule is interrupted so that the truck must wait, then a longer time is required. Sometimes it is so long that more water must be added to recover lost slump. This practice, always undesirable but sometimes unavoidable, is called **retempering.**

Concrete has two important setting times—initial and final. They are sometimes defined operationally. Initial set is the time when the concrete can no longer be reworked into a new shape. Final set is the time when the concrete will bear small loads, such as a person's weight, without noticeable indentation.

The test method for determining the setting behavior of concrete is ASTM C 403. In this test a (Proctor-type) penetrometer needle is pushed into a mortar sample sieved from the concrete, and the penetration resistance is determined as a function of time after mixing. A penetration resistance of 500 psi (3.5 Mpa) is arbitrarily defined as initial set, and one of 4000 psi (28 Mpa) as final set.

Assuming that the cement meets specification requirements, the most important parameter affecting setting time is temperature. As with other chemical reactions, a higher temperature promotes an exponentially more rapid (early) hydration rate in cement. Admixtures are used to control this problem, as already discussed. Cold-weather and hot-weather concreting also sometimes require heating or cooling of the ingredients, and of the placed concrete. Details are discussed in the ACI Manual [1994].

38.7 Properties of Hardened Concrete

The main function of concrete structures is to bear loads, so the most important properties of concrete are mechanical. If the loads are the deadweight of the material or are imposed externally, we are concerned with the strength and rigidity of the concrete. If the stresses and strains are the result of internal processes, we are concerned with the durability of the concrete. Such strains can arise from several sources. Thermal strains arise from temperature changes. Shrinkage strains arise from loss or gain of water. Freezing-and-thawing strains arise from the conversion of water to ice in the pores of the system. Strains from chemical attack are the result of chemical reactions in the concrete. The first two kinds of durability problems will be discussed here; the latter two are the subject of a separate chapter.

Failure of Concrete

Concrete is a brittle material that usually fails by fracture. If it is restrained triaxially it fails by shear in much the same manner as a soil.

The failure (strength) theory obeyed by concrete is, on the tension side, a modified maximum tensile strain theory that changes to the maximum distortional energy theory in triaxial compression. The failure envelope in principal-stress space looks like a bullet, coaxial with the line of equal principal stresses, and with the origin (zero stress point) located close to the bullet tip [Newman, 1968]. For ordinary concrete, the critical tensile strain at which failure occurs is roughly 0.015 percent (150 microstrains). Any stress state, derived from whatever loadings, that develops this critical strain will bring about failure by fracture.

Thermal Properties

The most important thermal properties of concrete are the **coefficient of thermal expansion** and the **thermal conductivity.** The expansion coefficient is the more important to matters of durability. The thermal conductivity is important to the fire resistance of concrete. Damage due to fire is a subtopic of concrete technology that is not usually considered part of durability, and will not be treated here.

Any temperature change of concrete generates consequent strains. If these strains are nonuniform, owing to the presence of temperature gradients or to physical restraint of the member, cracking may occur. Such cracking is properly called *thermal cracking.* Unfortunately, sometimes this phenomenon is called *shrinkage cracking,* thus confusing it with the effects of drying. Thermal strains arise from two kinds of temperature changes—those brought about by external heat or cold, and those from internal heat generated by the exothermic hydration reactions of the cement.

The magnitude of any thermal expansion or contraction of concrete is proportional to its coefficient of expansion. This value is founded in a mixture rule, as with most composite materials; that is, the contribution of each component depends on the value of its coefficient and on how much of it is present. Since the aggregate is the component present in the largest amount, its coefficient largely determines that of the concrete. Carbonate rocks—limestones and dolomites—

have a considerably lower value than do silicious rocks—quartzites, sandstones, granites. Therefore, concrete made with carbonate rocks will have a lower expansion coefficient, roughly two-thirds that of concrete made with silicious aggregates.

Externally generated temperature changes depend on the climate in which the concrete finds itself. Leaving out fire, jet blast, and the like, they are not ordinarily large enough to bring about disruptive strains. The only exception is the problem of blowups in pavements. A blowup is an upward buckling of the pavement, caused by its thermal expansion when heated by the sun [Kerr and Shade, 1983]. The problem frequently occurs at contraction joints that have not been properly maintained and cleaned of debris accumulated during the winter when they are opened by the thermal contraction of the slabs.

Another thermal problem occurs early in the life of the concrete. The hydration reactions of the cement liberate heat, and the temperature of the concrete increases. Such increase is greater for a thicker section. The adiabatic temperature rise can be about 40°C (70°F), but is, of course, smaller when heat is lost to the surroundings. For flatwork of ordinary thickness, the temperature begins to fall not long after final set. The decline is greater at the surface than in the interior, so differential strains are set up that, if large enough, will cause cracking. Along with shrinkage strains that act in concert, these temperature effects are the cause of the early contraction cracks that occur in flatwork. Control (contraction) joints are sawed or formed so as to confine this cracking to predetermined locations.

Thermal strains are a major problem in mass concrete, such as locks and dams, in which the loss of hydration heat is slow because of the thickness of the section or lift. Various means, including low cement content, use of pozzolans, and cooling of the ingredients and the concrete, are needed to control the thermal strains and cracks that would otherwise occur. Details are given in the report of ACI Committee 207 [ACI, 1994].

Shrinkage

The other cause of early cracking of concrete is the shrinkage that occurs when the material dries. As with other finely divided, porous materials (clay, wood), hardened cement paste shrinks and swells when it loses and gains water. If the water loss is nonuniform, or if the volume change is restrained, differential tensile strains will occur that may cause cracking of the concrete. As with thermal stresses, shrinkage stresses are especially a problem in the earliest stages of hardening, when the green concrete is weakest and least able to resist tensile strains. This is an important reason for proper curing; the worst thing to do, as happens not infrequently, is to allow the concrete to dry after finishing, with no curing at all or with it delayed too long.

Various mechanisms have been identified as responsible for the development of shrinkage stresses and strains. When the paste is first drying the responsible mechanism is primarily the generation of **capillary pressure** across the menisci that form in the pores. The result is a tension in the water phase that is transmitted hydraulically to the solids and generates the volume decrease, as also happens in drying soils. After the greater part of the water has been lost, and the remainder exists primarily as films on the finely divided cement paste solids, the mechanism becomes one based on changes in the **surface energy** of those solids.

If a hardened cement paste is dried from a saturated to a dry state, its (bulk) volumetric change will be about 3%. Since the paste is mechanically isotropic, this change corresponds to a linear (shrinkage) strain of about 1%, more than enough to crack the paste and render it useless as a structural material. A primary function of the aggregate in concrete is to restrain this shrinkage to acceptable limits. The quantitative relationship was due to Pickett [1956] and is discussed in various texts. It postulates that the shrinkage is parabolically related to the volumetric proportion of the paste (plus air voids), that is, to the nonaggregate volume of the concrete. This is the reason to use as large a maximum size of coarse aggregate as possible, because the larger it is, the less the amount of paste (and cost, incidentally).

Shrinkage is also greater for higher water contents, for the same reason; the paste volume is increased. If the w/c also increases, the result is a paste with a higher porosity and lower elastic modulus that deforms more under a given shrinkage stress.

Strength and Elasticity

Strength is ordinarily considered the most important property of concrete, although one can always get enough strength, practically speaking, if one is willing to pay for it by buying more cement; durability problems are sometimes more intractable. Mechanical properties of concrete have been reviewed frequently [Neville, 1971].

Influence of Porosity

The strength of any porous, brittle solid is related to its porosity, that is, the volumetric proportion of pores. The two most common relationships state that the relative strength, S_r, its ratio to that of a zero-porosity material, is related to the porosity, ϵ, either exponentially,

$$S_r = e^{-b\epsilon} \tag{38.1}$$

or parabolically,

$$S_r = (1 - \epsilon)^m \tag{38.2}$$

where b and m are constants. If the porosity is not large, these two relationships predict much the same value.

This dependency of strength, and other important properties such as elastic modulus and permeability, on the porosity is the basis of the water-to-cement ratio (Abrams) rule. Any increase in porosity that is brought about by excess water will result in lower strength and elastic modulus, and higher permeability.

For hardened cement paste the term *porosity,* used in this sense, requires a special definition. It does not mean the total porosity, since paste cannot be produced with a total porosity smaller than a minimum value, about 0.25, as described earlier. So this becomes the *zero porosity* state, and the porosity that is generated by excess mixing water and contributes to weakness is that in excess of this minimum amount.

Lowered strength is also caused by porosity of another sort, the entrained air bubbles, each volume percent of which removes about 4% of strength. This is, of course, a linear law, but if the porosity is small, a few percent for the entrained air, this rule is practically as good as the equations given above.

The sizes of these kinds of pores are greatly different; the largest pore in the gel is about 0.1 μm diameter, and the smallest entrained air bubble is about 4 μm. But they all reduce the strength according to the same principal. Powers [1958] was the first to show that the strength of these materials follows a cubic law, that is, with $m = 3$ in Eq. (38.2). The influence of w/c and air content on the strength of concrete has been extensively reviewed by Popovics [1985].

Elastic Properties

The Young's modulus of concrete develops roughly with the strength. A rule of thumb for ordinary concrete is that the Young's modulus is 1000 times the compressive strength. A better relation is that of the ACI [1994], which states that the modulus, E (in psi), is

$$E = 33W^{1.5}\sigma^{0.5} \tag{38.3}$$

where W is unit weight, lb/yd^3, and σ is compressive strength, psi. The influence of porosity on the elastic modulus parallels that of its influence on strength.

The elastic modulus can be determined statically, by the simultaneous measurement of load and strain in a compression test. The stress-strain curve for concrete is slightly curved, concave to the strain axis. The ASTM standard, C 469, specifies the calculation of a chord modulus; that is, the slope of a line joining the points on the curve corresponding to a strain of fifty microstrains and a stress of 40% of ultimate. This value is only a little smaller than the secant modulus.

The other way to measure the modulus is dynamically. ASTM C 215 describes the determination of the fundamental resonant frequency of a beam sample. An input transducer, activated by a tuned audio oscillator, excites the vibrations of the beam, and the signal from a pickup indicates the frequency at which the vibration is a maximum. This fundamental frequency is related to the density and the elastic modulus of the concrete. Other equipment is on the market that is basically a frequency meter that measures the vibrational frequency when a beam sample is struck a light blow.

The other dynamic method is the measurement of the speed of traverse through the concrete of an ultrasonic pulse between two transducers. The pulse velocity is related to the density and the modulus, but also the Poisson ratio must be known to calculate an accurate value. If a crack interrupts the path of the elastic wave, the transit time will be greatly increased. A field device called the Soniscope uses this principle to investigate the integrity of structures. The dynamic moduli are larger than the static values. This difference is smaller for higher strength concrete.

The Poisson ratio of concrete varies from about 0.15 to 0.20. Naik and Malhotra [1991] state that a value of 0.2–0.3 is appropriate for the pulse velocity calculation.

Compressive Strength

By far the most common method of determining the strength of concrete, and of mortars and other cementitious mixtures, is to test in simple (unconfined) compression. The strength is the maximum stress sustained. The ASTM method is C 39. The samples are molded cylinders with a height-to-diameter ratio of two, or drilled cores that are trimmed as close to this ratio as practical. The conventional cylinder size has been 6 by 12 inches, but smaller ones, depending on the coarse aggregate size, are increasingly used. The samples must be kept dripping wet until tested, because if they dry, even a little, capillary pressures are generated that amount to a confining stress that will increase the strength. The ends of the specimens are capped to avoid stress concentrations; reusable caps are available.

Most of the world other than the U.S. uses cubes as strength test specimens. Because of an influence of the height/diameter ratio, the strength of cubes will be somewhat higher than that of 2/1 cylinders.

The strength of properly cured concrete increases parabolically with time, as more cement hydration product is formed and the porosity of the paste decreases. The conventional test age is 28 days, but owing to the exigencies of practice, testing is frequently done earlier, at three or seven days. Accelerated methods, such as heating, have been devised so the concrete can be tested when it is only hours or a day or two old. These are covered by ASTM C 684.

Ordinary concrete will have compressive strengths from about 2000 psi (14 Mpa) to 6000 psi (42 Mpa). So-called controlled-low-strength-material, with even lower strengths, is used for backfill, bedding, and so on. High-strength concrete was thought not long ago to be any stronger than 6000 psi (42 Mpa), but now the boundary has risen to around 10,000 psi (70 Mpa). Concrete of considerably higher strength than this is made routinely, and the record goes up constantly. Such concrete is now termed *high-performance* concrete, to emphasize that it has many desirable properties, such as low permeability and high durability, in addition to high strength. See Zia *et al.* [1991] for a state-of-the-art review.

Other than w/c and curing time, the strength of ordinary concrete is mostly dependent on the hydration temperature, because of its influence on the reaction rate of the cement. A system that incorporates both time and temperature into a combined index of the degree of hydration is called the **maturity** method [Carino, 1991]. This method is used for in situ testing to determine, for example, when concrete is strong enough for form removal. The ASTM standard is C 1074.

Flexural and Tensile Strength

Since concrete fails primarily in tension, a tensile test has been thought more logical than compression, by some. The problem has always been how to hold the specimen, since one cannot cut threads in a brittle material, and wedge grips would shatter it. Several indirect methods have therefore been devised. Chief among these is the flexural test in which a beam is bent, and the failure is from the tensile stress so generated. The failure tensile stress is called the flexural strength or the modulus of rupture. Its value is about one-seventh of the compressive strength, for ordinary concrete. The ASTM methods are C 78 and C 293, for third-point and center-point loading, respectively. The third-point loading, which generates a constant bending moment, and stress, over the center third of the beam, is more common; the slightly lower strength it gives, compared with center-point loading, illustrates a general principle.

All brittle materials fail by the propagation of flaws, or cracks. The larger the crack, the lower the stress needed to propagate it; this is the Griffith principle. If the flaws are of random occurrence, a larger sample will have a higher probability of having a large flaw that is critically oriented with respect to the stress field. Therefore, large test specimens have lower strengths than do small ones. More of the third-point flexure specimen is exposed to the maximum tensile stress, compared with the center-point, so its average flexural strength is lower. The same principle applies to compression testing also, because this is really an indirect tension test, and the specimen fails by the development of a maximum tensile strain.

The ends of a beam previously broken in flexure can be used for compression test specimens; this is method C 116.

The other widely used indirect tensile test is the splitting, or *Brazilian* test. A cylinder is laid on its side in the testing machine and loaded in compression, through two thin strips of plywood to prevent shattering from stress concentrations. This loading develops uniform tension across most of the diametral section, and the sample splits in two. The failure tensile stress is calculated and called the splitting tensile strength. It is smaller than the flexural strength, because a larger region is tested; in one test series the differences were about 250 psi [Narrow and Ullberg, 1963].

Nondestructive Test Methods

A number of other more or less nondestructive test methods are used to indicate the quality of concrete. They are suitable for testing the structure itself; some are also useful for laboratory specimens. These methods are reviewed in detail by Malhotra and Carino [1991].

Perhaps the most used of these methods is the rebound hammer, also called the Schmidt or Swiss, after the inventor and his nationality; it is described in C 805. A spring-loaded plunger is allowed to fall on the surface of the concrete. It will bounce back, and the extent of this rebound is measured and called the rebound number. The harder (and stronger) the concrete, the larger the rebound number. The device is easily portable, which is one of its attractions.

Correlations exist between the rebound number and the compressive strength, but they depend on various factors, and a naive assumption of the values given with the instrument is usually a mistake. Further, the indication is only of the quality of the surface region. The Schmidt hammer is an inexpensive device, mostly useful for a comparison among concretes at various locations in a structure.

Another penetration method is called the Windsor probe, ASTM C 803. A "pistol" is used to fire a 6.3-mm-diameter steel pin into the concrete. The penetration is determined (inversely) by measuring the exposed length of the probe, which is correlated with the strength of the concrete. The test, while more complicated than the rebound test, is still comparatively simple and quick. The strength correlation is not especially good, and the test is not appropriate for high-strength concrete. It also leaves a hole that must usually be patched, for appearance' sake. The greatest usefulness of this method is, again, as a comparative rather than an absolute strength determination.

The traditional way of determining the actual strength of concrete in a structure has been to test cores drilled from it. The cores are soaked in lime water and then tested in compression. The ASTM standard is C 42.

Two alternatives have been developed that are quasi-nondestructive: the pullout test, C 900, and the break-off test, C 1150. In the pullout test an embedded metallic insert, cast into the concrete, is pulled from the surface. The force needed to do this is measured and can be converted to a stress that is related to the strength of the concrete.

In the break-off test a cylindrical segment is formed at the surface, either by drilling a core or by casting in a sleeve that is later removed. The resulting cantilever segment, 55 mm in diameter and 70 mm deep, is then broken off by a hydraulic device that applies a concentrated load at the outer end. The break-off number is the pressure of the actuating pump that is needed to fracture the specimen.

Both methods are relatively simple to perform. They obviate the need for a core drill and a compression tester, both bulky items, and for the time-consuming soaking, capping, and testing of cores. Whether these methods give better results than the testing of conventional cores may be a matter of opinion. The holes generated by all three methods must usually be patched.

Failure Process

Various techniques, including observation of cracks by x-ray or dye absorption, acoustic emission, and longitudinal and transverse strain measurement, have been used to show that the failure process of concrete is complex and occurs over much of the compressive loading [Shah and Slate, 1968; Mindess, 1983a,b]. Direct observation shows cracks, caused by shrinkage or other volume changes, to exist at coarse aggregate boundaries even before loading starts. During about the first half of the loading, due to stress concentrations at the coarse aggregate interfaces, these bond cracks increase in size, and new ones form. At loads of about two-thirds ultimate, the bond cracks extend into the mortar matrix and many new ones form at voids and other points of stress concentration. This is also the point at which the accumulating crack volume causes the volumetric strain to reverse from being increasingly compressive to being less so, until finally it becomes tensile. In the final stage of loading the cracks link up into large failure regions, and the sample can bear no further load and fails. This final failure region can be extended to considerably larger strains, at lower stresses (strain softening), thus generating the complete stress-strain curve, if a very hard machine or its experimental equivalent is used. So even greatly fractured concrete still has some cohesion and can bear appreciable loads.

Fracture Mechanics of Concrete

Traditional concrete design presupposes no tensile strength for the material. So concrete was useful only when subjected to purely compressive loads or when reinforced or prestressed. The brittle nature of concrete and its type of failure, which proceeds by the extension of cracks, logically invited the application of fracture mechanics, which had a head start in metal design owing to the brittle failure of supposedly ductile structures and to the advent of higher-strength, more-brittle steels.

The original work, on glass, was done by Griffith. He balanced the elastic strain energy released when a crack enlarges with the surface energy absorbed by the creation of new crack surfaces. The resulting equation stated that the failure tensile stress normal to a crack, σ, is related to the size of the crack, c, by

$$\sigma^2 c = (2/\pi) E \gamma \tag{38.4}$$

where E is Young's modulus and γ is surface energy. The elastic modulus and the surface energy are both structure-insensitive properties of the material, much as the density or molecular weight.

Further work showed that processes other than new surface creation also consumed energy during fracture. For metals that process is plastic deformation at the crack tip due to the high stress concentrations present there. This *plastic work* is usually much larger than the surface energy.

Development of the field gave rise to two new parameters. The symbol G was given to the strain energy release rate, its change with crack size. It is equal to twice the sum of the surface energy and plastic work. The other parameter was called the stress intensity factor, K, which arose from the stress analysis of the region near the crack tip. K is defined as

$$K = \sigma(\pi c)^{1/2}(f) \tag{38.5}$$

where f is a geometric factor and the other symbols are as previously defined. These two, G and K, are related to each other, $K^2 = GE$, and are alternate ways of combining all the energy-consuming factors in a fracture failure. The critical value of K is a material property called the fracture toughness; it has become the design parameter for many alloys. It relates the combination of stress and crack size a material can tolerate without catastrophic fracture failure, as indicated by Eq. (38.5).

These parameters were first measured for concrete in about 1970. The main technique has been a flexural test of a beam with a sawed notch on the tension side. A review of early work is by Mindess [1983a,b]. This work showed that hardened cement paste is a fairly brittle material, although its fracture energy is several times the value of its surface energy, so some other process is involved. The value of G (and K) for concrete is about ten times that of cement paste, because of the presence of the aggregate, which interferes with crack propagation.

Since little plastic deformation occurs with concrete, the "extra" energy absorbed has been postulated to be that occurring in a fracture process zone of extensive microcracking ahead of the main crack tip, the brittle analog of the plastic deformation region of metals. Direct observation of cracks in the electron microscope has shown a multiple and tortuous cracking, but not the zone proposed by some models.

The simple design application of linear elastic fracture mechanics, that outlined above, to concrete structures is complicated by several factors. Among these is a slow crack growth that occurs and continuously changes the size in question. This long-term instability of the cracks is thought to be responsible for the **static fatigue** experienced by concrete. Another complication is a size effect, the net result of which is an increase in brittleness and fracture susceptibility as the size of the member increases. This effect calls into question all the fracture parameters measured with ordinary-size laboratory specimens, as well as the applicability of these models to really large structures.

These complications have led to debate about whether fracture mechanics is applicable at all to concrete. The result has been a very large amount of important research. Recent work has concerned the development of better models, some of which involve nonlinear effects. These developments and the state of the art with respect to the applicability of fracture mechanics to design are reviewed in Elfgren [1989] and the reports of ACI Committees 446 and 224 [ACI, 1994].

Permeability

Permeability is a measure of the ease of fluid flow through porous media. Its significance with concrete concerns both the performance of the material when it serves to contain fluids—dams, tanks, barriers, and so on—and its durability, which is usually related to its ability to keep water and aggressive solutions from entering into its interior.

Permeability is proportional to the square of the size of the channels through which the fluid flows. The permeability of concrete is small, owing to the very small size of the pores of the cement paste. It increases with increasing w/c, as more porosity is generated by the extra water. But almost any intact concrete will be relatively impermeable. If, however, the concrete is cracked, these comparatively wide channels permit a greatly increased flow of fluid. So any concrete that must be highly impermeable must be maintained free of cracking.

The term "permeability" has recently been used in another sense, which refers to the ease of passage of ions in solution through the pores. The chief concern is the attack of chloride ions on embedded steel. The corrosion of reinforcing steel, especially in bridge decks, is a most important materials problem. A test method has been devised, ASTM C 1202, which basically is used to determine the electrical conductance of concrete to ionic transport, and has been correlated with its resistance to chloride ion penetration.

Defining Terms

The report of ACI Committee 116 [ACI, 1994] is an extensive list of definitions of terms relating to cement and concrete.

Absorption: The weight of water absorbed by a porous solid, per unit weight of dry solid.

Addition: A substance that is interground with the clinker at the time of production of the cement. Calcium sulfate, usually gypsum, is the most common addition.

Admixture: A substance other than cement, water, and aggregates that is added at the mixer to impart desirable properties to the concrete. Certain substances, such as fibers, that are added to special concretes are not normally considered admixtures.

Air entrainment: The process of incorporating many small air bubbles in the cement paste. These bubbles protect the concrete from damage from freezing-and-thawing exposure.

Bleeding: The separation of mixing water onto the surface of a plastic concrete.

Capillary pressure: The pressure difference across an interface. For a meniscus in a small tube of radius r, the capillary pressure is $2\gamma/r$, where γ is the surface tension of the liquid.

Coefficient of thermal expansion: The relative volume change, or length change, of a substance caused by unit temperature change.

Durability: The quality of permanence under deleterious influences other than design loads.

Fineness: A measure of the surface area of the cement solids, and thereby, of its average particle size. A high fineness means a small particle size. Units are area/unit weight of cement.

Fineness modulus: A numerical factor obtained from the grading curve and related to the average logarithmic size of the particles.

Fresh: Concrete from the time it is mixed until the time it is placed in the forms.

Grading: The distribution, according to particle size, of particulate material.

Granulated blast furnace slag: The by-product of iron production in a blast furnace. Essentially calcium silicates and aluminosilicates, the molten material is granulated by being cooled rapidly in water.

Hardened cement paste: The hard product formed from the cement and water. It consists of hydrated cement, pore space, and any unhydrated cement. Air voids are not part of the paste. The hardened paste's volume is (almost) the sum of the volumes of cement and water.

Heat of hydration: The heat released when the cement and water interact chemically. The value depends on the type of cement, but is about 100 cal/g (418 kJ/kg).

Hydraulic: When applied to a cement, one that is capable of setting and hardening under water.

Maturity: A function that combines the effects of time and temperature of curing on the strength of concrete.

Mortar: A cementitious mixture without coarse aggregate; the portion of a concrete other than the coarse aggregate is referred to as the mortar fraction or portion.

Permeability: The volumetric flow rate of a fluid of unit viscosity through unit cross-sectional area of a porous material under a unit pressure gradient. Permeability is proportional to the porosity of the material and to the square of the pore size.

Porosity: The ratio of the volume of pores to the total (bulk) volume of a porous solid.

Pozzolan: A silicious substance that will react with the lime of a hydrated cement to form cementitious substances. The term is derived from Pozzuoli, a small village near Naples, where the Romans obtained a pozzolan of high quality.

Retempering: The practice of adding additional mixing water to concrete, to recover lost slump or merely to make placement easier.

Segregation: The partial separation of the constituents of a plastic concrete, resulting in a higher concentration of aggregate at the bottom.

Slump: A measure of consistency. The slump is the distance a 12-in. truncated cone of concrete will slump down when the molding cone is removed.

Spacing factor: A geometric parameter of the air-bubble system in a hardened concrete. It is roughly half the average distance between bubbles.

Specific surface (area): The surface area of unit weight of a substance. The fineness of cement is a specific surface.

Static fatigue: The failure of concrete under long-time loading at a lower stress than it could sustain in the usual brief test.

Sulfate resistance: Resistance to chemical attack by sulfate-containing solutions. The usual circumstances involve desert soils or seawater.

Surface energy: The energy needed to create unit area of new surface.

Thermal conductivity: The rate of heat flow through unit cross section of a material when subjected to unit temperature gradient.

Water-to-cement ratio (w/c): The weight ratio of the amount of water to that of cement.

Workability: The ease with which a concrete can be transported, placed, and consolidated without undesirable bleeding or segregation.

References

ACI. 1994. *ACI Manual of Concrete Practice,* American Concrete Institute, Detroit.

ASTM. 1994. *Annual Book of ASTM Standards,* ASTM, Philadelphia. The volume on cement is 04.01; that on concrete and aggregates is 04.02.

Bye, G. C. 1983. *Portland Cement. Composition, Production and Properties,* Pergamon Press, New York.

Carino, N. J. 1991. The maturity method. In Malhotra and Carino, 1991, pp. 101–146.

Dodson, V. H. 1990. *Concrete Admixtures,* Van Nostrand Reinhold, New York.

Dolch, W. L. 1984. Air entraining admixtures. In Ramachandran, 1984, pp. 269–302.

Elfgren, L., ed. 1989. *Fracture Mechanics of Concrete Structures,* Chapman and Hall, London.

Kerr, A. D. and Shade, P. J. 1983. Analytic approach to concrete pavement blowups. *Transp. Res. Rec.* 930:78–83.

Kosmatka, S. H. and Panarese, W. C. 1988. *Design and Control of Concrete Mixtures,* 13th ed., Portland Cement Association, Skokie, IL.

Malhotra, V. M. and Carino, N. J., eds. 1991. *CRC Handbook on Nondestructive Testing of Concrete,* CRC Press, Boca Raton, Florida.

Mindess, S. 1983a. Mechanical Performance of Cementitious Systems. In Barnes, P., ed., *Structure and Performance of Cements,* pp. 319–364, Applied Science Publishers, London.

Mindess, S. 1983b. The application of fracture mechanics to cement and concrete. In Wittmann, F., ed., *Fracture Mechanics of Concrete,* pp. 1–30, Elsevier, Amsterdam.

Naik, T. R. and Malhotra, V. M. 1991. The ultrasonic pulse velocity method. In Malhotra and Carino, 1991, pp. 169–188.

Naik, T. R. 1992. Maturity of concrete: Its application and limitations. In Malhotra, V. M., ed., *Advances in Concrete Technology,* pp. 329–360, Energy, Mines and Resources, Ottawa, Canada.

Narrow, I. and Ullberg, E. 1963. Correlation between tensile splitting strength and flexural strength of concrete. *Proc. ACI* 60(1):27–38.

Neville, A. M. 1971. *Hardened Concrete: Physical and Mechanical Aspects,* ACI Monograph No. 6, Detroit.

Newman, K. 1968. Criteria for the behavior of plain concrete under complex states of stress. In *The Structure of Concrete,* pp. 255–275, Cement and Concrete Association, London.

Pickett, G. 1956. Effect of aggregate on shrinkage of concrete and hypothesis concerning shrinkage. *Proc. ACI* 52:581–590.

Popovics, S. 1985. New formulas for the prediction of the effect of porosity on concrete strength. *Proc. ACI* 82(2):136–146.

Powers, T. C. 1958. Structure and physical properties of hardened cement paste. *J. Am. Ceram. Soc.* 41(1):1–6.

Ramachandran, V. S., ed. 1984. *Concrete Admixtures Handbook,* Noyes Publications, Park Ridge, NJ.

Shah, S. P. and Slate, F. O. 1968. Internal microcracking, mortar-aggregate bond and the stress-strain curve of concrete. In *The Structure of Concrete,* pp. 82–92, Cement and Concrete Association, London.

Stanley, C. C. 1980. *Highlights in the History of Concrete,* Cement and Concrete Association, Wexam Springs, Slough, England.

Tattersall, G. H. and Banfill, P. F. G. 1983. *The Rheology of Fresh Concrete,* Pitman Books, London.

Whiting, D. and Dziedzic, W. 1990. Effects of "second-generation" high range water-reducers on durability and other properties of hardened concretes. *ACI SP-122:* 81–105.

Zia, P., Leming, M. L., and Ahmed, S. H. 1991. *High Performance Concretes,* Strategic Highway Research Program, National Research Council, Washington.

For Further Information

The best reference text on concrete is *Properties of Concrete* by A. M. Neville, 3rd ed. (Pitman, London, 1981).

Other excellent texts are *Concrete* by S. Mindess and J. F. Young (Prentice-Hall, Englewood Cliffs, NJ, 1981) and *Concrete* by P. K. Mehta and P. J. M. Monteiro, 2nd ed. (Prentice-Hall, Englewood Cliffs, NJ, 1993).

Significance of Tests and Properties of Concrete and Concrete-Making Materials, Klieger, P. and Lamond, J., eds., ASTM STP 169C (1994) is an excellent reference to many topics.

The standard on chemical aspects is *Cement Chemistry* by H. F. W. Taylor (Academic Press, San Diego, 1990).

The materials of concrete are covered extensively by S. Popovics in *Concrete-Making Materials* (Hemisphere Publishing, Washington, 1979).

A review of the modern situation is *Advances in Concrete Technology,* V. M. Malhotra, ed. (Canadian Communications Group, Ottawa, 1992).

Many practical details, on concrete as well as other construction materials, are given in handbook format in *Construction Materials* by C. Hornbostel (Wiley, New York, 1991).

39

Durability of Concrete

W. L. Dolch
Purdue University (Emeritus)

Sidney Diamond
Purdue University

39.1 Freezing-and-Thawing Durability

Durability of concrete refers to its permanence under destructive influences other than structural loads. Many ambient or environmental conditions are responsible for a lack of durability of concrete. The avoidance of these difficulties involves choices among the materials that make up the concrete as well as those of good practice in its preparation and use. Concrete cannot be made so strong as to be immune from these effects, and to ignore them is fatal.

Of the several possible causes of nondurable concrete, only the two most important will be discussed in this chapter. These are difficulty caused by freezing and thawing and difficulty caused by chemical attack.

The problems of freezing and thawing can arise in either the paste or the coarse-aggregate components of the concrete. That in the paste is, for practical purposes, completely solved by using air-entrained concrete. Aggregate problems are more difficult; the only real solution is to use a better aggregate.

Chemical problems are of two large categories: alkali-aggregate reaction and sulfate attack. Each of these can be broken down into several subcategories. The solution to alkali-aggregate problems is to use a low-alkali cement or a nonsusceptible aggregate, or sometimes, a pozzolan. The solution to sulfate attack problems is to use a concrete that is properly proportioned, placed, and cured and a cement with a low tricalcium aluminate content.

A chemical problem that is not, strictly speaking, one of concrete is corrosion of the embedded steel. This problem is also treated in this chapter.

Freezing and Thawing, Field Appearance

Difficulty from freezing-and-thawing exposure requires that the concrete become highly saturated with water. Therefore any concrete not so exposed will be, ipso facto, free of such problems. Flatwork, such as pavements and sidewalks, is usually the most affected, for obvious reasons.

0-8493-8953-4/95/$0.00 + $.50

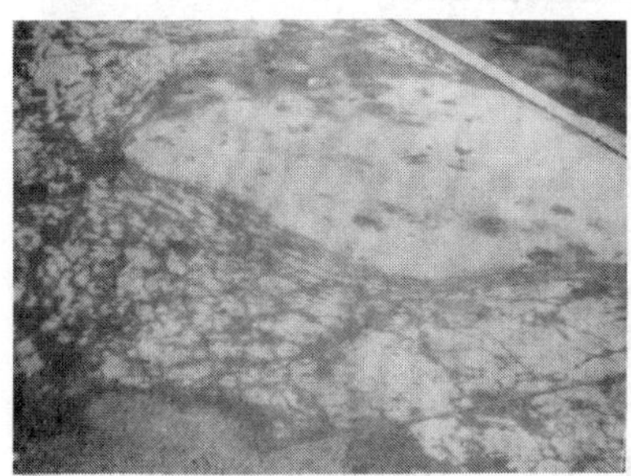

FIGURE 39.1 Freezing-and-thawing paste failure of sidewalk.

Vertical sections are comparatively immune except where they make an edge or corner with a horizontal surface. The trouble always starts where the water has easiest access: at edges, corners, cracks, joints, and surfaces. Freezing-and-thawing failure of concrete has a characteristic appearance, and usually can be diagnosed visually.

Paste Failure

The cracking caused by paste failure is always a pattern of approximately parallel cracks that proceed inward from the places where the concrete has first become highly saturated with water. The usual crack pattern is shown in Fig. 39.1, in which the susceptibility of the region of

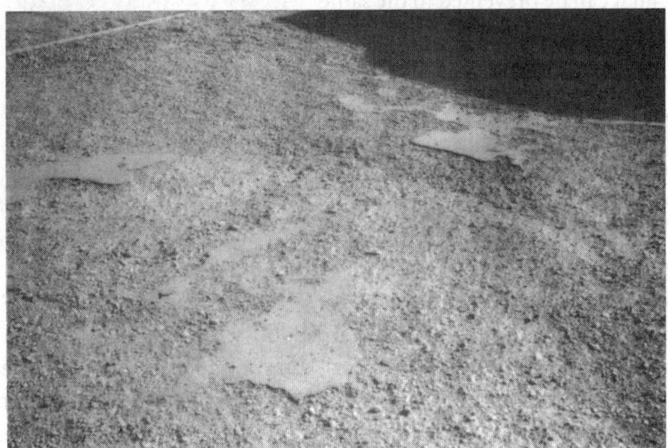

FIGURE 39.2 Scaling of sidewalk.

FIGURE 39.3 Salt scaling of sidewalk.

high water saturation is obvious. The other kind of paste difficulty originates at the upper surface, usually of flatwork, and is termed scaling. Figure 39.2 illustrates a typical example. Here most or all of the upper surface scales away, but usually the lower regions are not destroyed. Scaling is exacerbated by the application of deicing salt to the concrete; this is sometimes termed *salt scaling*. An example is shown in Fig. 39.3. The salt from the tires has scaled the sidewalk, but the air-entrained driveway is immune. The preventive measure is the same for any kind of paste failure—air entrainment.

Aggregate Failure, D-cracking

One of the crack patterns caused by coarse aggregate failure in freezing and thawing looks much the same as paste failure—parallel cracks proceeding inward from regions of high saturation. This pattern is known as D-cracking. It can be distinguished from paste failure, if at all, by having a somewhat greater spacing between cracks. A typical example is shown in Fig. 39.4.

Aggregate Failure, Popouts

The other kind of freezing-and-thawing failure caused by aggregates is a surface phenomenon called a popout. The difficulty occurs in coarse aggregate pieces close to the surface. The result is an expelled flake of mortar and half of a piece of aggregate, with the other half remaining at the bottom of the depression. Figure 39.5 shows an example. Popouts are unsightly but do not constitute a structural problem.

Sometimes a flake of mortar will scale off the top of a piece of coarse aggregate that is close to the surface of the concrete. Such a failure is called a popoff. It is basically a paste failure, a sort of localized scaling, and it can be prevented in the same way as any other paste failure.

FIGURE 39.4 D-cracking of pavement due to freeze-thaw failure of aggregate.

FIGURE 39.5 Popout of piece of gravel in pavement.

Freezing-and-Thawing Mechanism, Paste Failure

Paste failures have been attributed to several mechanisms. The two most significant are the hydraulic pressure hypothesis [Powers, 1945] and the gel water diffusion hypothesis [Powers and Helmuth, 1953]. Both depend on ice forming in paste pores that are greater than critically saturated with water. Critical saturation is about 90%, since water expands about 10% on freezing. But a significant factor is that the water in the pores of a cement paste does not all freeze at the same temperature. The freezing point is depressed in small pores, so only about half of the water in an ordinary paste freezes close to the normal freezing point, and perhaps a third does not freeze until about −40°C (−40°F) [Fontenay and Sellevold, 1980].

Hydraulic Pressure Theory

The hydraulic pressure theory postulates the following scenario. The water freezes first in the largest pores, and the growing ice body forces unfrozen water through the rest of the paste. The hydraulic pressure associated with this flow results in the development of tensile stresses in the concrete. The farther this water must flow, the greater the developed pressure and stress. If the flow path is larger than a critical distance, the result will be a fracture of the matrix. For the paste in the usual concrete, this critical distance is of the order of 0.2 mm (0.008 in.), and is so small because of the low permeability of the paste.

Gel Water Diffusion Theory

The gel water diffusion mechanism is sometimes called the osmotic hypothesis, because it is thermodynamically similar to osmosis. The driving force is a free energy difference between the unfrozen water in smaller pores and adjacent ice in larger pores. The gel water will then migrate to the ice, hence the name of the process. If the larger pore is full of ice, a higher pressure will develop, which results in tensile stress in the concrete. This mechanism is similar to that responsible for the growth of ice lenses in soils, except on a micro rather than a macro scale. This mechanism does not require expansion of the liquid when it freezes. Its validity is supported by the destruction of porous materials when nonaqueous liquids, which contract upon solidification, are frozen in the pores.

Opinion differs about the comparative importance for concrete of the osmotic mechanism and the hydraulic pressure concept. A time factor is involved, in that expansion has been shown to increase as the material is held at the subfreezing temperature. The fact that less field difficulty is experienced in those climates where the concrete remains frozen for a long time, and more where cycles of freezing and thawing occur, may speak against the significance of this theory.

Salt Scaling

The further problem arising in the paste and caused by the application of deicing salts to the surface of the concrete is explained in two ways. The "melting" of the surface ice that occurs causes a temperature shock that lowers the temperature of the surface regions and brings about the freezing of still more water inside the paste. Further, the dissolved salt concentrates in the pores of the paste that are partly frozen and results in a true osmotic pressure that acts in concert with the other internal pressures. These effects exacerbate the scaling that would otherwise occur. The result is salt scaling. This difficulty is not a chemical effect; any other soluble substance shows the same results as sodium chloride or calcium chloride, which are the two most common deicers.

Some substances that have been used as deicers do cause a failure that may look much like salt scaling, but is a chemical effect. Notable compounds are sulfates and ammonium salts, which are components of many fertilizers. Such substances should not be used to deice concrete.

Influence of Air Entrainment

Air entrainment is the most important advance in the technology of concrete in the last fifty years or so [Dolch, 1984]. It was discovered accidentally, and since that time has become firmly established as the only way to be certain of protection from freezing-and-thawing problems. A properly air-entrained concrete with a sufficiently low water-to-cement ratio will, for all practical purposes, be immune from such difficulty arising in the paste. Trouble from bad aggregate, on the other hand, is not influenced to an important degree by air entrainment.

Air-Entraining Agents

Air entrainment is achieved by incorporating an air-entraining agent in the concrete, usually as an admixture dissolved in the mixing water. Sometimes the agent is interground with the cement as an addition, and the result is an air-entraining cement, designated by a capital A after the ASTM type (e.g., type IA). But the use of such cements is decreasing, and the time will probably soon come when they will no longer be manufactured.

Air-entraining agents are surfactants, and they operate by stabilizing the air bubbles that are incorporated into the concrete by the mixing action. The result is the presence of many (about a million per cubic inch) very small (average size about 0.05 mm) air bubbles that make up several percent of the volume of the concrete.

Factors That Influence Air Entrainment

A variety of factors can influence the amount of air that a given dose of agent will cause to be entrained in plastic concrete. Factors that make for a low air content are the presence of mineral admixtures, a rich mix, a cement of higher fineness, an elevated temperature, and a low water content. The obverse of these conditions increases the air content. These matters are considered by Dodson [1990]. Usually such problems can be solved by a change in the dosage of the air-entraining agent, which is a reason to prefer admixtures to the use of air-entraining cements.

Sometimes an air-entraining agent is "sensitive" to other mix components, and the desired air content is difficult to obtain. In such an event consideration should be given to using a different fine aggregate or a different cement.

The protection against freezing that is afforded by air entrainment comes at a price; the strength of the concrete is decreased by air entrainment, about 4% for each 1% of air. Frequently this strength decrease can be offset by a lowered water/cement (w/c) ratio that is made possible by the increased workability imparted to the concrete by air entrainment. The many beneficial changes to both the plastic and hardened concrete make a case for air entraining all concrete, except where very high strength is important.

Protective Mechanism of Entrained Air

The air bubbles remain empty unless the concrete is long exposed to water, especially under head. They protect the concrete from freezing distress by serving as closely spaced reservoirs into which the unfrozen water can migrate without generating disruptive stresses. The amount of air required depends on the maximum size of the coarse aggregate. The range is from about 4 to 7% of the volume of the concrete, with the larger amount being necessary for the smaller coarse aggregate. In addition to proper air entrainment, a durable concrete should have a comparatively low w/c ratio (below about 0.45), so that the paste will be impermeable and will have a limited capacity for freezable water. See ACI [1993] and PCA [1988] for specific recommendations. Air-entraining agents should meet ASTM C 260. This and all other ASTM designations are found in ASTM [1993].

The total amount of entrained air, that is, the air content mentioned above, is not the determinative factor with respect to protection of the concrete from frost effects. The important parameter is a geometric one, called the spacing factor [Powers, 1949; Dolch, 1984]. This factor is roughly half the average distance between adjacent air bubbles in the paste. If this factor is smaller than about 200 μm (0.008 in.), experience has shown that the concrete will be durable in freezing. Sometimes this critical spacing factor will be larger, for example in concrete of low w/c ratio, but exact values have yet to be firmly established. The spacing factor, the air content, and other parameters of the bubble system in hardened concrete are determined microscopically using the methods of ASTM C 457.

For field concerns, if an air-entraining agent meets ASTM C 260, it is presumed to give a bubble system with a proper spacing factor.

High-Strength Concrete and Air Entrainment

For high-strength concrete, the need for air entrainment to provide durability has been questioned. Such concrete usually contains silica fume and superplasticizers and has a very low w/c ratio. The resulting paste may have a pore structure in which none of the water can freeze at temperatures ordinarily met in the field. If this is the case, no entrained air is necessary. The matter is reviewed by Philleo [1986]. A prudent conclusion, nevertheless, is to use air entrainment in high-strength concrete that will become highly saturated with water and frozen.

Salt Scaling

Scaling, even when it is made worse by deicing salts, is prevented in ordinary concrete by proper air entrainment. But when mineral admixtures—fly ash, blast furnace slag, silica fume, and so on—are used, the concrete is more sensitive to salt scaling, and some restrictions are usually considered necessary. These take the form of limits on the amount of mineral admixture that may be used.

In addition to materials considerations, scaling is greatly influenced by the finishing and curing operations. Overfinishing brings fines and water to the surface, even with air-entrained concrete, which normally bleeds little. The result is a weak surface region of high w/c ratio that is nondurable and scales easily. Proper curing, preferably with access to free water, is also extremely important to freedom from scaling. However, the concrete needs to dry for some days before the first freeze, so that the pores are no longer completely saturated with water. Good practice dictates that deicing salt not be used on concrete during its first winter of life, if it was placed late in the fall.

Aggregate Failure

Poor freezing-and-thawing durability caused by the aggregate component of concrete arises in the coarse, rather than the fine, aggregate. Fine aggregate does not cause such difficulty owing to a size effect that makes the larger sizes most susceptible. The problem is a hydraulic pressure phenomenon that has several aspects. In every instance, critical saturation is necessary (i.e., the aggregate must have a degree of saturation in its pores greater than about 90%). This high saturation can occur in the aggregate, since its pores are comparatively large, only if the adjacent paste is completely saturated. In such a circumstance three possibilities exist [Verbeck and Landgren, 1960].

If the porosity (and, therefore, the absorption) of the aggregate is small enough, no disruptive pressures can result even if all the water in the aggregate freezes, because the resulting strains are too small. This circumstance is called elastic accommodation. The critical absorption is a few tenths of a percent, and there are some aggregates, usually igneous or metamorphic rocks, that are therefore immune from freezing-and-thawing difficulty.

Aggregates with larger porosity and absorption are susceptible to deterioration if they become critically saturated. Two kinds of distress are possible: D-cracking and popouts. These have characteristic appearances that were discussed above.

Mechanism of Popouts

Popouts result from internal failure in aggregate pieces that have comparatively small pores and low permeability and are located close to the surface of the concrete, where they become easily saturated. Cherts and shales, which are common components of many gravels, are the most important examples.

The difficulty is worse in large pieces of the aggregate, so a coarse aggregate with a comparatively large maximum size may give problems while a smaller size of the same material will be comparatively immune [Stark and Klieger, 1973].

Sometimes the problem components of gravel supplies are so prevalent that the only recourse is to beneficiate the supply by removing the offending materials. This can be done in several ways; flotation with high-density liquids is the most common.

In order for the aggregate piece to experience popout difficulty it must become critically saturated. High saturation is less likely if the covering mortar is made with a low w/c paste that is well cured and relatively impermeable, and if the piece is set a little below the surface. So a concrete that is properly proportioned, finished, and cured and that has a small maximum aggregate size may experience little difficulty even if the aggregate components are inherently problematical. However, the only way to be completely free of popouts is to use a coarse aggregate that does not contain susceptible components.

Mechanism of D-Cracking

D-cracking is caused by coarse aggregates that have comparatively large pores and high permeabilities. The most important kinds of rocks are many limestones, dolomites, and sandstones. If these pieces become critically saturated and frozen, the excess water is expelled from them, causing a failure in the adjacent paste rather than in the aggregate piece itself. The trouble is experienced throughout the concrete, not just in the surface. The difficulty begins, as it does in all other cases, where the concrete becomes most easily saturated—at the edges, joints, cracks, and corners.

Air entrainment does little good, because its capacity for relief close to the aggregate piece is overwhelmed by the large demand. Again, larger maximum sizes are worse, and the use of smaller-sized aggregate can delay, but usually not eliminate, difficulty. The only practical way to avoid D-cracking is to use an aggregate that is not susceptible. In many locations, such aggregates are difficult to find. D-cracking is a serious problem that is not yet completely understood or solved.

Test Methods for Freezing-and-Thawing Durability

Tests for freezing-and-thawing durability fall into two classes–those done on concrete and those done on components of the concrete, usually coarse aggregates. If one is concerned with aggregate quality, it is obviously easier to test the aggregate itself, without going to the trouble of making concrete. Frequently, however, concrete testing is required in order to get the most complete information on which to base recommendations.

Tests on Concrete

Tests on concrete usually involve the freezing of the sample and some measurement of property changes that are indicative of durability performance.

The Rapid Freezing-and-Thawing Test, ASTM C 666. The most frequently used ASTM standard is C 666. In this test method concrete prisms or cylinders are frozen and thawed between 4.4°C (40°F) and −17.8°C (0°F) during a period of between 2 and 5 hours. Two regimes are permitted. Method A requires freezing and thawing while constantly immersed in water. This method obviously necessitates a container for each specimen, and is the more complex of the two. Method B specifies freezing in air and thawing in water.

No specification is made of the materials from which the concrete is made or of any pretreatment of them or the concrete. Such matters depend on what one is trying to find out. Sometimes a given mix design is being tested, so its requirements determine the makeup and conditioning of the concrete. If one is testing the durability of the paste, as when qualifying an air-entraining agent, one obviously uses a durable aggregate, so it will cause no problem. If one is testing an aggregate's durability, a properly air-entrained paste must be used.

The performance of an aggregate in this test will depend a great deal on its moisture content at the time of freezing, which in turn will be determined by its exposure and its moisture content as mixed in the concrete. The most severe test is of aggregate that has been vacuum saturated prior to mixing.

When to use method A or B is also left to the tester's choice. Method A is obviously more severe, because there is no opportunity for the concrete to dry and become less saturated, as there is with method B.

As stated in the method, it is not intended to provide an indication of the time a concrete will remain durable in the field. The method is purely comparative, and, especially when its results are considered in conjunction with field experience, it provides valuable assistance in the selection of aggregates and other components.

The properties of the concrete that are required to be measured to monitor its performance in this test method are the dynamic modulus of elasticity and the weight. Measurement of expansion (i.e., length change) is optional.

Weight change is the oldest of these methods and is now little used, because it usually indicates only superficial changes and not those of the mass of the concrete.

Length change of the concrete is determined according to ASTM designation C 157. A length comparator is used, and the indicator, typically a dial micrometer, is readable to 2.5 μm (0.0001 in.). There is no universally accepted definition of the length change that constitutes "failure" in C 666. Frequently an expansion of 0.1% is considered failure, but sometimes a smaller value is adopted. Stark and Klieger [1973] used 0.033%. The reason for a permanent length increase is obviously the many microcracks that result from the freezing-and-thawing failure.

The presence of these same cracks results in a decrease in the elastic modulus. Method C 666 prescribes that the modulus be determined dynamically, using the resonant frequency technique of ASTM C 215. The dynamic elastic modulus can also be determined by measuring the velocity of an elastic wave through the sample. The two techniques do not generally yield the same result, so one must use the prescribed technique.

The results of the modulus determination are translated into a parameter called the durability factor (DF), defined as

$$\mathrm{DF} = PN/M \tag{39.1}$$

where P is the relative dynamic elastic modulus (the measured value divided by the original value), in percent, of the sample after it has experienced N freeze-thaw cycles, and M is the predetermined number of cycles at which the test will be terminated if P has not fallen to some predetermined minimum value. The method specifies a minimum P of 60% and an M value of 300 cycles. A sample whose modulus remains unchanged will have a DF of 100, and the lower the value, the greater the damage the sample will have sustained. No DF can be selected a priori to differentiate "durable" from "nondurable" concrete. A value of 60 is suggested in some of the standard texts, but a higher value would certainly be desirable.

Sometimes an alternate parameter, such as the number of cycles to reduce P to 80%, has been used as a numerical indicator of the freezing-and-thawing test results.

Method C 666 is not a highly precise or reproducible test, especially at intermediate DF values, which are those for which the decision about the use of the material in question is difficult.

The Critical-Dilation Test, ASTM C 671. Test method C 666 is unrealistically severe. The freezing rate is higher than that normally experienced in the field, and the concrete has no opportunity for seasonal drying, which almost always occurs. Powers [1955] proposed a method to take cognizance of these factors, and this test has become ASTM C 671. This method is essentially a direct way of discovering how long a sample must be soaked in water before it becomes so saturated that it will expand by a destructive amount when subjected to a single freeze.

The preconditioning of the sample is not specified; it depends on the purpose of the test. Usually, the sample is dried somewhat, so as to approximate the condition of the concrete prior to the winter exposure expected. Then the sample is soaked in water, and every two weeks it is subjected to a single freeze cycle. The freezing is done with the sample in a strain frame, the indicator of which is a linear variable differential transformer capable of indicating a deformation of 0.25 μm. This process is continued until the expansion on freezing begins to increase sharply, by a factor of 2 or more times the previous change. This is called critical expansion, and the soaking period prior to it is considered the time of immunity from frost damage for the concrete.

This method has been used little, in spite of its logical basis, perhaps owing to its complexity. It has been advocated for use with high-strength concrete [Philleo, 1986].

Method C 671 can be used to evaluate the frost susceptibility of aggregates in concrete. The details are given in ASTM C 682. The aggregate is used to make an air-entrained concrete that is then tested by the critical dilation procedure. Usually the aggregate and the resulting concrete are preconditioned to be similar in moisture content to that anticipated for the field condition. The time to critical dilation is then considered to be an indication of the period during which the concrete will be safe from freeze-thaw difficulty. For further information on both these methods, see the bibliography at the end of ASTM C 682.

The Scaling Test, ASTM C 672. The resistance of a concrete to scaling when deicing salts are applied to it can be determined by means of method C 672. A test slab is cast and finished. Then a dike, usually made of portland-cement mortar or epoxy mortar, is placed around the perimeter of the top. After curing and any surface treatment that is being evaluated (e.g., a protective coating), a calcium chloride solution is ponded on the top, and the assembly is frozen and thawed at a rate of one cycle per day. The scaling behavior of the slab is assessed using a visual rating on a 0 (best) to 5 (worst) scale.

Other properties of concrete, such as its absorption, sorptivity, permeability, and depth of carbonation, are indirect indicators of its durability, because the same pore structure of the material determines them all.

Tests on Unconfined Aggregate

Tests on the aggregate alone are to be preferred, for the sake of simplicity, to those that require the making of concrete. Whether any such test gives adequate information on the durability to be expected of the aggregate is a matter of dispute. The cautious assessment is that the safest thing to do is to make concrete and subject it to the tests described above. Nevertheless, tests on the aggregate are usually performed and frequently give important, even decisive, information.

Early in this chapter it was pointed out that the pore structure of the paste, best described by its pore size distribution, is determinative of the important properties that govern its durability.

Among these are the porosity, absorption, degree of saturation, permeability, sorptivity, and the freezing point of the pore water. The same is true of aggregate materials. Many aggregate tests involve these properties.

The pore structure determines the ability of the aggregate to acquire a critical degree of saturation, so it is the property that is fundamental to an aggregate's frost durability.

Tests for Specific Gravity and Absorption, ASTM C 127. The specific gravity and absorption of an aggregate are determined by the nature and volume of its solid portion, the volume of its pores, and the extent to which they are filled with water.

The pores are the intraparticle empty (when dry) spaces. The interparticle spaces in a collection of pieces are called voids. Unfortunately, sometimes the words are used interchangeably. The voids are an important property of a packing of particles, such as gravel or crushed stone. They influence the dry rodded unit weight (ASTM C 29), which is a property that is needed for the calculations of concrete proportioning, or mix design. But they have no bearing on the durability of the material. The parameters determined by C 127, the several specific gravities and the absorption, are properties of the individual pieces and not of their packing, so they are influenced by the pores, not the voids.

Test method C 127 involves weighing a sample of the aggregate in air in both the oven-dry and the saturated-surface-dry condition, and weighing it when immersed in water. These three weights permit the calculation of the three specific gravities: the bulk, the bulk-saturated-surface-dry (BSSD), and the apparent, and also the absorption.

Since all these parameters, except the bulk specific gravity, depend on the amount of water in the pores, it is necessary to specify how the sample was wetted. There are two main procedures. The standard prescribed in C 127 is to immerse the oven-dry material in water for 24 hours, and this, the so-called 24-hour condition, is assumed if no other description is given. The "vacuum-saturated" state is reached by evacuating the air from the pores of a sample, then admitting water and allowing the sample to remain immersed for a time. Naturally, the vacuum-saturated values are larger than the 24-hour ones.

Specific Gravity. The specific gravity values themselves are not used as indicators of aggregate durability, except in an exclusionary sense. Of course, for a given solid component, a larger porosity will result in a smaller specific gravity. And generally, those rocks with a high porosity and relatively low specific gravity are likely to be comparatively nondurable. But the size of the pores is also important. Sweet [1948] found that small pores are worse than large ones. Indeed, many aggregates with a comparatively large porosity, but also with large pores, are durable to freeze-thaw exposure in concrete, because they do not stay critically saturated. Most lightweight aggregates are examples.

However, for substances with small pores, such as cherts and shales, a low specific gravity is an indicator of poor performance. ASTM C 33, the aggregate specification standard, places limits on the amount of chert with a BSSD specific gravity smaller than 2.40.

Absorption. The absorption has long been used as an indicator of an aggregate's frost durability, and many specifications contain limits, such as 3% or so. But the usefulness of such limits is doubtful. As outlined above, if a rock's porosity is so small that it can elastically accommodate the freezing of pore water, it should be durable. But such a porosity corresponds to an absorption of less than 1%. Such aggregates are rare, especially in sedimentary rock regions. Similarly, a large porosity and absorption frequently means a nondurable material, but sometimes such aggregates are durable, because their pores are large. So an absorption specification is in difficulty on both ends. Many pavements have developed D-cracking with aggregates that were thought to be safe because of their reasonable absorption values. Data on aggregate absorption as related to frost durability of concrete are given by Gaynor and Meininger [1967].

Degree of Saturation. The accepted mechanisms for frost action in aggregates require the development of a critically high degree of saturation in the pores. So if the aggregate cannot develop such a high saturation, it should be safe from freezing. Durable aggregates have indeed shown low degrees of saturation when vacuum saturated. The only trouble with such a test is its difficulty. To determine the degree of saturation requires an independent determination of either the porosity or the true specific gravity (i.e., that of the solids themselves). Methods exist for such determinations, but they are tedious [Dolch, 1978]. For this reason, little work has been done on this aspect.

The Sulfate Soundness Test, ASTM C 88. The rationale behind test method C 88 is the supposed similarity between ice growth in a pore and the crystallization of a salt therein. A dry sample that is retained on a given sieve is soaked in a solution of either sodium sulfate or magnesium sulfate. It is then oven dried. This constitutes one cycle. ASTM C 33 specifies a five-cycle test. At the end of the test the sample is resieved, and the result reported is the amount of the original that now passes the given sieve. This is called the soundness loss. C 33 requires a loss of less than 18% for magnesium sulfate or 12% for sodium sulfate, since it is less severe.

The process that occurs in the rock pores is a crystallization of the anhydrous salt during the oven-drying portion of the cycle, and then a rehydration and conversion to the hydrated salt when the sample is soaked. The growth of the hydrated crystals exerts pressure on the pore walls and breaks down the piece so that it passes the sieve on which it was formerly retained.

The results of this test do not correlate well with freezing-and-thawing tests of the aggregate in concrete. If the aggregate has a large soundness loss it probably will be nondurable in concrete, although there are exceptions. The converse is not at all true. Many aggregates with a low soundness loss are nondurable, although of course some show the expected relationship. Typical data are given by Gaynor and Meininger [1967]. These data show that about twice as many samples with a low loss are nondurable than are durable. The interpretation of the test is considered in detail by Dolar-Mantuani [1978, 1983].

The Expected Durability Factor. Kaneuji *et al.* [1980] attempted to relate freeze-thaw durability of an aggregate to its pore structure. Aggregates were batched vacuum saturated, and the concrete was tested according to C 666. The aggregate pore size distributions were determined by the mercury intrusion method. Durability factors were correlated with various aspects of the pore size distribution curves. The best correlation gave rise to an equation that defines a so-called expected durability factor (EDF) as follows:

$$\mathrm{EDF} = 0.6/(V) + 6d_{\mathrm{av}} + 3 \tag{39.2}$$

where the coefficients are approximate, and V is the total intruded pore volume (cm^3/g) and d_{av} is the average pore diameter (μm), that is, the diameter corresponding to half of the total volume intruded.

Correlation of field performance with the EDF of the aggregate showed nondurable aggregates to have an EDF of less than about 50 [Winslow *et al.*, 1982]. If the test is done on an inhomogeneous material, such as gravel, a statistical selection must be made that reflects the lithologic composition, and testing need not be done on the portion with low absorption values.

Equation (39.2) shows the EDF to depend on the pore volume and size, with a larger volume or a smaller size being worse, from the durability standpoint, as was long known. As an example, since V is roughly the same as the fractional absorption value, if V is 1%, or 0.01, the EDF will be greater than 60, and the aggregate will be durable, no matter what the average pore size. On the other hand, if d_{av} is greater than 10 μm, the EDF will be greater than 60, again indicating durability no matter what the absorption.

This method has been little used so far, perhaps because the mercury porosimeter, although commercially available, is expensive.

39.2 Chemically Related Durability

This section provides an overview of the major classes of chemically related durability problems that can affect concrete in service, along with details of the mechanisms responsible, practical aspects of the deterioration and degradation that may be induced, and indications of appropriate preventive and ameliorative measures.

Most civil engineers are familiar with portland-cement concrete as a construction material designed to bear loads and resist the range of applied stresses that may be met in service. However, familiarity with concrete as a physical-chemical system subject to damage and deterioration by chemical effects arising from exposure to normal service environments is not widespread.

In this section we first describe porosity and pore solution characteristics of concretes that may influence chemically induced durability problems. We then discuss certain significant inhomogeneities found in concrete structures that influence their resistance to these problems. Finally we provide separate descriptions of the causes, mechanisms, and preventive measures of the important individual classes of chemically induced problems.

Concrete Porosity and Pore Solutions

As has been indicated previously, portland-cement concrete is a porous material containing several classes of pores in the cement paste component, and frequently containing aggregate of appreciable porosity. The degree to which the pores in a given concrete mass are interconnected and support fluid flow and the diffusion of dissolved substances varies significantly with water/cement ratio, the maturity of the concrete, and various processing and environmental factors. Generally speaking, the greater the interconnection of the pore system that persists in the concrete, the greater the potential concern with most chemically related durability problems [Mindess and Young, 1981].

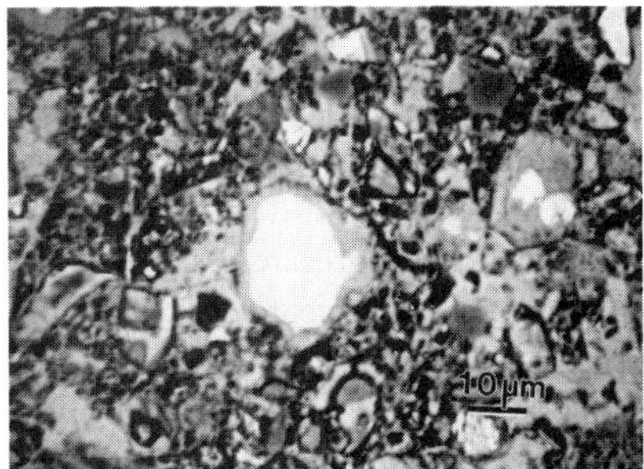

FIGURE 39.6 Scanning electron micrograph of cement paste (water/cement ratio 0.4, hydrated for 7 days). Capillary pore spaces are black areas. White area in center of figure is unhydrated cement grain; gray areas are cement hydration products.

Figure 39.6 shows a backscatter-mode scanning electron micrograph of a hardened cement paste providing a good representation of a capillary pore system in a relatively young portland-cement paste.

It is often not appreciated that seemingly dry concrete may not be dry internally. Almost always, even in dry climates, concrete retains a significant content of pore fluid within the internal pore system. Except in very arid climates concrete rarely dries out more than several inches back from the exposed surfaces [Stark, 1992].

The fluid contained within the pores of concrete is an aqueous solution. In recent years it has become possible to express this pore solution from concrete cores in specially constructed high pressure devices, and to recover a small amount for chemical analysis. Figure 39.7 shows such an apparatus, used for some years in concrete research at Purdue University.

It turns out that concrete pore fluid is not a solution of calcium hydroxide, as had been supposed for many years. Rather, the primary components are potassium and sodium hydroxide, at combined concentrations of the order of 0.4 to 1.0 Normal. Thus the internal pore solution in concrete is extremely alkaline. The range of pH values is between about 13.2 to almost 14. Concrete that is partially dry may have still higher concentrations of alkali hydroxide, but contain smaller amounts of fluid, which may be restricted to films and isolated local menisci at interstices within the pore system.

The potassium and sodium hydroxide found in concrete pore solutions are normally derived from alkali impurities within the cement. Most portland cements contain greater or lesser contents

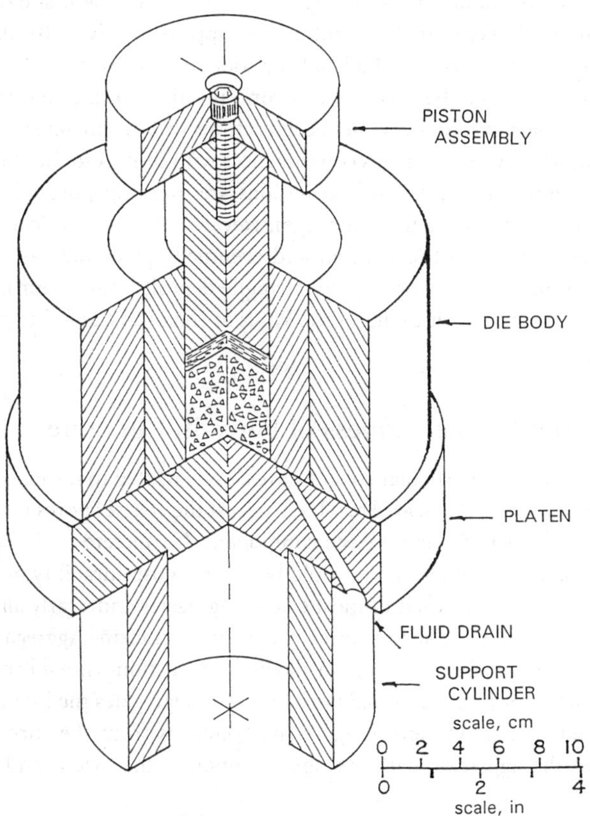

FIGURE 39.7 High-pressure pore solution device in use at Purdue University. Specimen is confined by a rigid die body. Up to 250,000 pounds of force is applied to the piston, resulting in stresses sufficient to remold the concrete and expel the pore solution.

of alkalis in the form of soluble alkali or alkali-calcium sulfates, and also as alkali ions in solid solution in the major portland-cement clinker components.

The combined content of sodium and potassium in a given cement is ordinarily determined by the manufacturer and stated as a "Na_2O equivalent." This parameter is the sum of the weight percent of Na_2O and 0.658 times the weight percent of K_2O. It has been established that the concentration of alkali hydroxide eventually developed in saturated concrete is about 0.7 N per percent Na_2O equivalent of the cement, assuming a water/cement ratio of approximately 0.5 [Diamond, 1989]. Leaching of alkalis from concrete that is permeated by water flow can reduce this value. Conversely, inclusion of alkali-bearing chemical admixtures, incorporation of dissolved salts from the outside environment, or sometimes internal chemical reaction with certain alkali-containing aggregate minerals can increase the alkali hydroxide concentration.

The alkali hydroxide concentration present in the pore solution of a given concrete can have a significant effect on its durability in service. As indicated subsequently, several varieties of alkali aggregate reaction can take place if high alkali hydroxide concentrations are developed in concretes that also contain reactive aggregate components. Conversely, high alkali hydroxide concentrations tend to protect reinforcing and prestressing steel from corrosion, although the effectiveness of such protection varies with other circumstances.

Concrete pore solutions also contain much smaller concentrations of other dissolved constituents. Appreciable concentrations of sulfate ions are present during the first half day or so after placement of the concrete, but dissolved sulfate almost disappears by the end of the first day under normal circumstances. Calcium ions in very small concentrations (approximately 0.001–0.005 N) may be retained, although the high concentration of alkali hydroxide ordinarily precludes the presence of more than token amounts. Despite their usually low concentrations in the solution, both sulfate and calcium ions may take part in secondary chemical reactions within concrete (especially delayed ettringite formation), which may have serious consequences with respect to durability.

Ordinarily, silicon-, aluminum-, and iron-bearing ions are found in pore solutions in almost undetectable concentrations (parts per million range). Nevertheless, both silicon and aluminum may be readily transported through the solution and eventually precipitate as a component of either ASR gel (silicon) or of secondary or delayed ettringite deposits (aluminum). Such transport of ions through the solution can be critical in several kinds of chemical durability problems, as will be discussed later.

Size Distribution and Connectedness of Pores in Concrete

The durability characteristics of a particular concrete are closely related to the extent of void space, the distribution of pore sizes, and the degree to which at least the coarser pores form a continuous pore structure within the hardened cement paste component. If concrete could be produced in a fully dense condition, that is, without pores, it would be free of most durability problems.

The pore volumes within the rock itself of most coarse aggregates and nearly all sands are small enough to be unimportant except in special cases. So-called D-cracking aggregates have already been discussed. Some of the aggregates involved in alkali-silica reactions (notably opals and cherts) contain appreciable contents of interconnected pores. In these aggregates the interconnected pores provide pathways for internal movement of aggressive solutions from the surrounding cement paste to the interior of the aggregate particles, thus promoting the extent and rapidity of the alkali-silica reactions.

A highly significant content of pores is always found within the cement paste component of concrete. This paste pore system arises from several distinct effects. The largest pores are deliberately or accidentally entrained air voids; these usually remain empty for long periods of time. Pores of a smaller size range, so-called capillary pores, represent remnants of spaces between cement grains that were originally occupied by water at the time the concrete set, and which were not wholly filled in by later deposition of cement hydration products. The finest pores, the so-called gel pores, are typically only of the order of nanometers in size, and usually are considered to be pore spaces developed within, rather than between, individual particles of calcium silicate hydrate, the primary product of cement hydration. These do not ordinarily influence durability.

The extent, size distribution, and degree of interconnectedness of the capillary pore system may be markedly reduced by the inclusion in the concrete of supplementary cementing materials, or "mineral admixtures," notably silica fume, blast furnace slag, and fly ash. Such inclusion may have extremely beneficial effects on resistance to certain classes of durability problems.

Generally speaking, the characteristics of the capillary pore system that are potentially harmful in concrete are associated with higher water/cement ratios. Accordingly, in standardized mix design procedures such as that recommended by the American Concrete Institute [ACI Committee 201, 1991], restrictions are placed on the maximum water/cement ratio that can be used for concrete exposed to conditions where durability problems may become a factor. Part of the benefit associated with supplementary cementing materials derives from the reduction in water content of the concrete mix that may often be possible when such materials are used.

This is especially true when these materials are used in combination with high doses of chemical admixtures that promote dispersion of the normally flocculated agglomerates of cement particles that are present in the fresh cement paste in concrete. In particular, high-range water-reducing admixtures (also called superplasticizers) permit concrete to be batched at much lower water contents than usual. They are particularly effective in this regard when used in combination with silica fume, or to some extent, with fly ash. The resulting concretes usually have much finer and less interconnected capillary pore systems and may consequently be much less susceptible to chemical durability problems [Malhotra *et al.*, 1987].

One of the important features of ordinary concrete is the existence of so-called interfacial transition zones surrounding aggregate particles. Such zones, typically 40 to 50 μm in width, are more porous and have a higher degree of interconnectedness of pores than does the cement paste more distant from aggregate particles. The existence of interconnected pathways for flow and diffusion in close proximity to the aggregates can constitute a special factor in many concrete durability problems. Part of the beneficial effects arising from use of silica fume and superplasticizers appears to be due to the virtual elimination of this porous interfacial transition zone when this admixture combination is used [Bentur *et al.*, 1988].

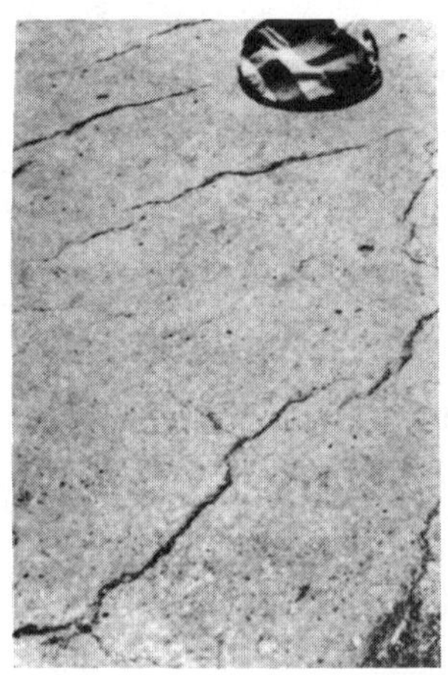

FIGURE 39.8 Plastic shrinkage cracks in a concrete. (Used by permission.)

Local Effects at the Outer Zones of Concretes

To some extent, the outer surface of a concrete slab or structural member can be regarded as a sort of skin through which the external chemical and physical environment acts on the concrete. The degree to which this skin is dense, impermeable, and free of cracks considerably influences the progress of chemical durability problems.

Curing is important in this regard. To the extent that near-surface water is allowed to escape from concrete members that are improperly cured, hydration in the near-surface zone is slowed or halted. Poor curing may thus produce a surface zone of incompletely hydrated cement paste, with a correspondingly coarse and interconnected pore structure. While hydration may resume if the concrete surface is subsequently rewetted, the degree to which the effects of poor curing on the near-surface pore structure can be reversed is limited.

Another factor influencing the degree to which the concrete outer layer can isolate the interior of the concrete from its surroundings is the occurrence or nonoccurrence of plastic shrinkage cracks. These can develop during the first day of hydration as a result of excessive evaporation before adequate strength to resist the tensile forces arising from the evaporation has been developed. An illustration is provided in Fig. 39.8. Such cracks provide direct pathways for deleterious substances to enter the interior of the concrete. To some extent thermal cracks may be harmful in the same fashion.

Inhomogeneities within Concrete

The extent to which the local water content can vary in different parts of a fresh concrete member is not generally appreciated. Regions of higher local water content can result in interconnected local pore systems that may tend to seriously compromise resistance to durability problems.

One of the important variations is at the upper surface of the concrete, especially in flatwork. A certain degree of sedimentation takes place within a fresh concrete mass before it sets. The sedimentation brings water (so-called bleed water) to the upper surface. In consequence the upper portion of the concrete may have a substantially higher local water content than the bulk of the concrete. Similar (but usually smaller) increases in local water content may also occur at the side surfaces of the member as well. Even within the heart of a concrete member, significant local inhomogeneities almost always occur in the form of vertical bleeding channels and horizontal pockets of water-rich paste resulting from bleed water trapped under coarse aggregate particles.

The existence of pore-rich paste transition zones at aggregate surfaces has previously been mentioned as regions of high local permeability. Additionally, microcracks tend to develop at these interfaces in consequence of even modest stresses induced by loading or drying. The microcracks further increase local permeability of these zones.

In consequence of these effects, much conventionally proportioned and well-placed concrete is in fact ill equipped to withstand various chemical processes of deterioration arising through exposure to the outdoor environment. When the significant amount of concrete that is improperly proportioned or placed is added, the widespread existence of concrete durability problems is not surprising.

Specific Causes of Chemically Induced Concrete Deterioration

Concrete in service is susceptible to a variety of durability problems, usually induced by environmental exposure but occasionally internally caused, which can lead to premature deterioration and sometimes to destruction of the concrete. Of these, the major problems involve alkali-aggregate reactions, steel corrosion, and externally caused or internally generated sulfate attack. Each of these involves a separate complex of chemical and physical factors, and may be prevented or ameliorated by specific measures that need to be separately considered.

Alkali-Aggregate Reactions

There are at least two categories of these reactions: the more common alkali-silica reaction and the rarer alkali-carbonate reaction. Some authorities subdivide alkali-silica reactions into two types, considering the so-called late, slow reactions that occur with some silicate rocks as a separate class of alkali-silicate reactions.

In all cases, the concrete problems are caused by long-term chemical reactions between certain aggregate components and the alkali hydroxide pore solutions found in all portland-cement concretes. In all these reactions, the higher the alkali hydroxide concentration developed in the pore solution, the more complete the reaction processes tend to be in a given time. However, the extent of concrete distress produced can be influenced by many other factors.

Alkali-Silica Reactions

Alkali-silica reaction (ASR) occurrences were first recognized in California in the 1940s, but have subsequently been found in many countries. In the U.S., reactive aggregates are more common in the western and southwestern states, in certain parts of the southeast, and in some of the central states. For many years the northeastern states were thought to be free of these difficulties, but that apparent immunity has not held up, and a number of incidents of ASR have been reported in the northeast in recent years.

Field evidence for the occurrence of ASR includes expansion and development of polygonal (or "map") cracking on the surface of affected members, especially when accompanied by reaction product gel exuding from the cracks. An illustration is shown in Fig. 39.9.

FIGURE 39.9 Typical map-cracking pattern developed by alkali-silica reaction in a bridge parapet wall.

Other kinds of distress may sometimes show features similar to those of ASR. To assist in recognition of the problem, a field test has recently been developed to identify ASR reaction gel in a concrete [Natesaiyer and Hover, 1989]. Nevertheless, even if the presence of ASR is confirmed, a petrographic examination is generally necessary to confirm that the distress observed is actually due to the ASR.

Many instances are known, especially with the late, slow-reactive aggregates, where distress is manifest after some years, but only small amounts of gel are produced and may be hard to find.

ASR gel is produced by the reaction of alkali hydroxide in pore solutions with any of a wide variety of siliceous components in aggregate. The usual list includes amorphous silica, especially volcanic glass, opal, and amorphous components in cherts; some crystalline forms of silica, including tridymite, cristobalite, and highly strained quartz; silica or siliceous components in such sedimentary and metamorphic rocks as greywackes, argillites, phyllites, and siltstones; and disseminated silica (usually as skeletal remains of small organisms) in limestone. Synthetic glasses (window glass, soft-drink bottles, etc.) are highly reactive and should never be incorporated in concrete.

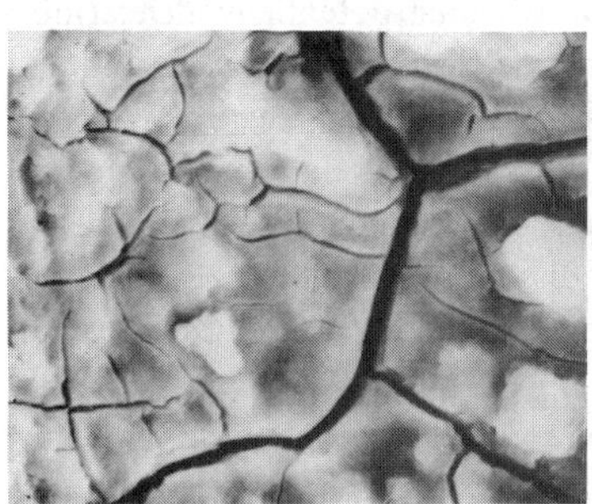

FIGURE 39.10 Scanning electron micrograph of alkali-silica reaction product gel.

The usual explanation for the distress involves (1) the formation of ASR gel by the chemical reaction between the alkali hydroxide and the reactive aggregate, and (2) expansion and cracking of the concrete resulting from swelling induced by the uptake of additional water or pore solution into the gel after its formation. The uptake of water is usually attributed to osmotic effects [Diamond, 1989].

An illustration of the characteristic scanning electron microscopic appearance of ASR gel is shown in Fig. 39.10. It should be appreciated that the fine-scale cracking shown in the figure is not itself the result of expansion, but rather represents the effect of shrinkage due to drying and evacuating the specimen so that it can be observed in the microscope.

Studies of ASR gel compositions reveal wide variations in the relative proportions of chemical elements present, but indicate that ASR gel is composed of alkali (Na and K), calcium, silicon, and often small amounts of aluminum. Hydroxide ions and water molecules are inevitably present. The calcium appears to be a necessary component; laboratory studies of ASR in the absence or near absence of calcium hydroxide in the concrete suggest that in such unusual circumstances the ASR reaction merely tends to dissolve the affected aggregate component without producing gel, and generates little or no distress.

Some studies have detected crystalline reaction products in some concretes, presumably as the results of secondary recrystallization from primary gel [Regourd-Moranville, 1989]. It is not thought that the crystalline reaction products themselves contribute to the distress.

Much effort has gone into the development of laboratory test methods to determine whether or not a given aggregate will be reactive. Some of these methods have been standardized by ASTM, but recent developments in other countries have produced an additional broad spectrum of test methods.

The premier ASTM test (ASTM C 227) involves measurement of the progressive expansion of mortar bars containing the suspected reactive aggregate during prolonged exposure at 38°C and 100% RH. The aggregate in question is ground to a specific fine-aggregate gradation, and the mortar is made with either a known high-alkali cement or the potential job cement for a given construction project. Expansion in excess of 0.10% in one year (or 0.05% in six months) is considered sufficient to treat the aggregate as potentially deleteriously expansive.

In what amounts to a radical modification of this test, ASTM method P 214 has recently been provisionally adopted after a period of development in several countries, most notably South Africa. In this test method mortar bars are immersed in a 1 N sodium hydroxide solution at 80°C for two weeks, with expansions measured at intervals. Expansions in excess of 0.2% during this two-week period of immersion are considered evidence that the aggregate is potentially deleteriously expansive. The driving force for the adoption of this form of testing is obviously the fact that results can be obtained in a few weeks rather than the many months required for the "traditional" C 227 mortar bar test.

ASTM C 289, designated as the test for "Potential Reactivity of Aggregates (Chemical Method)," or, more informally, the "quick chemical test," is now rarely used and is considered by most authorities to be unreliable. The test involves placing finely ground aggregate into 1 N sodium hydroxide solution at 80°C in a sealed container for 24 hours. The extent of reaction is monitored by a determination of the amount of dissolved silica and of the degree of reduction in the OH ion concentration of the solution.

ASTM also provides method C 295, "Petrographic Examination of Aggregates for Concrete." This is not a test method per se, but is rather a series of guidelines for petrographic (microscopic) examination of the aggregate, including a determination of whether potentially deleterious components are present. The services of a skilled and experienced petrographer are required.

A wide spectrum of non-ASTM test methods to distinguish ASR-prone aggregates have been developed in various countries (China, Japan, France) in recent years. These include measurements of expansion in mortars tested under various autoclaved conditions, concrete prism expansion tests, and a major modification of ASTM C 289 in which rates of dissolution of silica are considered, and the reduction of the alkalinity is monitored in terms of change in Na^+ ion concentration rather than in OH^- ion concentration [Sorrentino *et al.*, 1992].

Aggregates that are potentially deleteriously reactive should be excluded from use in concrete for obvious reasons. Occasionally economic or other factors compel their use, in which case suitable precautions to prevent or limit the distress that would otherwise occur must be taken. These include the following.

Use of low alkali cements. The concentration of alkali hydroxide in the pore solution depends most closely on the alkali content of the cement used. Selecting a cement with minimal alkali content (traditionally, one with equivalent Na_2O less than 0.60%) can serve to minimize the degree of ASR and mitigate its consequences. European practice takes into account the cement content of the concrete as well as the alkali content of the cement; most agencies suggest limiting the alkali content (as equivalent Na_2O) to less than about 3 kg/m^3 of concrete.

It should be appreciated that limiting the alkali content is not always effective. Sometimes relatively dilute alkali hydroxide solutions become concentrated in certain portions of the concrete structure despite the low overall level, and cause local distress. Alkali compounds from external sources, notably salt or salt spray, can penetrate the concrete, and the alkali salt can be converted to alkali hydroxide. Occasionally, internally derived alkali hydroxide is developed from alkali released to the pore solution from alkali-bearing aggregate components by reaction with calcium hydroxide. These noncement alkali sources may generate sufficiently augmented alkali hydroxide concentrations to cause distress in concrete despite the use of low alkali cement.

Incorporation of slag, silica fume, fly ash, or natural or artificial pozzolans. Use of ground granulated blast furnace slag, or the other silica-bearing concrete amendments in reasonable proportions, can serve to mitigate or prevent distress due to ASR, despite the fact that most of these components themselves have appreciable alkali contents. Slag is usually used in substantial proportion, replacing as much as 50% of the portland cement. Other amendments, notably silica fume, are normally used in smaller proportion. ASTM method C 441 prescribes measurement of the reduction in expansion of mortar bars containing standard reactive aggregates to determine the effectiveness of a particular amendment. However, there are some indications that the results of such testing are unreliable [Chen *et al.*, 1992].

Decreasing the supply of water. Concretes batched at low water/cement ratios have only a limited supply of internal water that might cause swelling of ASR gel, and the permeability of such concretes to outside water is much reduced. Thus the deleterious consequences of ASR may be slowed significantly by lack of water. Experience generally has indicated that when external moisture is not readily available to the interior of concrete, and if the concrete self-desiccates sufficiently that the relative humidity in the pores falls to less than about 80%, no adverse expansion due to

ASR occurs. However, the chemical reaction is not precluded, and subsequent wetting may produce rapid and significant expansions.

Use of chemical admixtures. Chemical admixtures for the prevention of ASR have been developed by several companies. At least one commercial admixture supplier is now marketing such a product containing a lithium salt as the active ingredient. There is a considerable body of technical literature suggesting that lithium salts may be effective in such use [Diamond and Ong, 1992]. Other compounds for the prevention of ASR distress may also be effective.

Once distress due to ASR has become evident, few options for treatment exist. Modifying the environment of the structure affected to eliminate any external sources of water and drying out the concrete have been helpful. However, attempts to fill and stabilize existing cracks by injecting epoxy or other resins have not generally been successful where ongoing ASR continues to take place. Use of polymer coatings have had some success, especially silane-based coatings that permit diffusion of water vapor out of the concrete but prevent further wetting from the outside.

Alkali-Carbonate Reactions

Alkali-carbonate reactions (ACRs) have been much less studied than alkali-silica reactions, presumably because their recognized occurrences are much less common. The carbonate-bearing aggregates that cause difficulties have traditionally been held to constitute only a small subset of carbonate rocks. Specifically, ACR-prone aggregates have been characterized as impure, clay-bearing dolomitic rocks in which the dolomite occurs as separate rhombs within the rock structure. The rhombs are disseminated in a finer-grained matrix of illite clay and calcite. The reaction that has usually been considered responsible for the distress is dedolomitization—the conversion of dolomite [$CaMg(CO_3)_2$] to calcite [$CaCO_3$] and brucite [$Mg(OH)_2$] by action of the alkali hydroxide in the concrete pore solution. Expansion and cracking accompanying this conversion have been attributed to the formation of the brucite and calcite in situ within the piece of aggregate. The clay also plays a role. The details of the mechanism are still unclear.

Recently, these ideas have been modified by several authors, who suggest that impure carbonate rocks of other characteristics may also react. Katayama [1992] suggested that deleterious expansion may occur with carbonate rock of various lithologies, as long as the impurity (insoluble residue) content is of the order of 10% or more, and Prince and Perami [1992] suggested that, based on purely chemical grounds, any dolomite would undergo dedolomitization to produce brucite at a sufficiently high concentration of dissolved alkali.

On the other hand, it is known that some carbonate aggregates that are not necessarily reactive in the dedolomitization mode may contain disseminated silica as an impurity. If the aggregates are sufficiently porous that alkali hydroxide can reach and react with the silica in the interior of the aggregate, significant deterioration due fundamentally to ASR-type reactions can occur. Thus limestone aggregates in general (i.e., calcitic as well as dolomitic) may need to be tested for potential ASR reactivity.

Steel Corrosion in Concrete

A common and widespread cause of deterioration in reinforced concrete (especially in bridge decks and parking garages) is steel corrosion. Steel corrosion can also occur in prestressed members, in which case its effects on the structure may be even more damaging than when it occurs in conventional reinforced concrete.

Both conventional reinforcing bars and the high-strength steel used for prestressing cables are ordinarily immune from corrosion when embedded in concrete. This is due to a passivation that arises from the high alkali concentration of the pore solution. Passivation requires the establishment and maintenance of a thin oxide film around the exposed surface of the steel. Such a passivating film is ordinarily produced and maintained without difficulty around steel in the interior of concrete.

Unfortunately, the presence of chloride ions in the concrete pore solution (from chloride-bearing components batched with the original concrete or from subsequent penetration of dissolved chloride salt from the outside) interferes with passivation and renders the steel susceptible to corrosion if oxygen can reach it [Cook and McCoy, 1977].

A similar result may derive from an entirely different mechanism. The outer layer of any concrete exposed to the atmosphere is subject to chemical reaction with carbon dioxide present in the air. This carbonation reaction has certain chemical effects on the solid components of the affected concrete, but more important, it greatly reduces the OH^- ion concentration of the solution in the pores of the affected zone. If this zone penetrates to the depth of the reinforcing steel, the passivating film cannot be maintained and the steel may begin active corrosion. Relationships between the progress of carbonation and important concrete factors such as density, cement content, and water/cement ratio have been explored by Treadaway *et al.* [1983].

It is usual to consider that the time response of corrosion effects in concrete can be subdivided into a so-called induction period prior to the onset of active corrosion, and a period of active corrosion. Efforts may be concentrated on lengthening the induction period as much as possible, minimizing the rate at which active corrosion occurs once passivation is destroyed, or both.

The deterioration of the concrete is not usually associated with the corrosion process per se, but with the accumulation of its products in the form of voluminous rust in the restricted space surrounding the steel in the concrete. As rust accumulates it generates expansive stresses that eventually lead to cracking and spalling of the concrete cover. Figure 39.11 shows the severity of concrete deterioration that can result from corrosion.

For both carbonation and salt-induced corrosion, the depth of the concrete cover over the steel is critical. Recommended practice is usually to provide at least a 50 mm thick layer of concrete over the steel, but because of poor field control, actual placement of the steel within the section may be highly variable. The quality of the cover concrete (especially its permeability) is also of critical importance.

Steel cables contained in grout tubes in the usual posttensioned prestressing practice may be subject to corrosion if the grout does not completely fill the tube and thus keep all of the steel passivated.

Anchors and other critical points in prestressed concrete systems are also susceptible to *stress corrosion*. Here the response is to stress as well as to the maintenance of passivation. The deleterious effects of stress corrosion in prestressed concrete structures may be much more important than the spalling effect produced with reinforcing bars, and in critical cases may result in sudden structural

FIGURE 39.11 Corrosion damage to a bridge deck pavement due to salt application. (Used by permission.)

failure and consequent tragedy. A well-documented example, that of the collapse of the Berlin Congress Hall, was described by Isecke [1983].

Prevention of steel-corrosion effects in concrete may involve a number of elements. These include the following:

1. A rigorous effort to preclude more than trace amounts of chloride from being batched with the concrete.
2. Careful attention to steel placement and to maintenance of the depth and the quality of the concrete cover.
3. Use of nearly impermeable, high-performance concretes containing silica fume or slag, and superplasticizers, batched at extremely low water contents. Such concretes delay the onset of active corrosion and may cause the rate of corrosion, once started, to be so slow as to make corrosion unimportant.
4. Use of corrosion-preventing admixtures. Admixtures based on calcium nitrite, and others based on more complicated organic systems, are being marketed.
5. Use of epoxy coatings around the steel. Epoxy-coated steel has been mandated by the Federal Highway Administration for use in concrete bridge decks where salt exposure is expected, but recent occurrences have cast some doubt as to the complete effectiveness of such coatings in preventing corrosion.

In some structures, notably concrete bridges, it is possible to prevent corrosion by installing cathodic protection systems. In such systems a continuous source of DC voltage must be applied to the steel to render it cathodic, and thus not subject to metal loss. Maintenance of such systems requires a continuous effort, but when conscientiously undertaken can provide protection indefinitely. A review of the systems used and the results obtained in cathodic protection of bridge decks in the Province of Ontario has been supplied by Schell *et al.* [1987].

External Sulfate Attack

This form of concrete deterioration stems from internal reactions that occur in some concretes when dissolved sulfate ions penetrate into the body of the concrete from its surroundings. The classic form of sulfate attack occurs with concrete exposed to groundwater bearing appreciable sulfate concentrations. Such groundwaters are common in many parts of the western U.S., but are found occasionally elsewhere in the country. The distress is marked by expansion followed by progressive cracking that can, in severe cases, lead to complete destruction of the concrete.

FIGURE 39.12 Scanning electron micrograph showing rod-shaped ettringite crystals developed as a result of severe sulfate attack on a masonry-mortar wall.

Sulfate attack occurs primarily with concrete made from cements that have high contents of tricalcium aluminate, one of the normal components. Sulfate ions diffusing into such concrete from the groundwater react with aluminate compounds in the unhydrated cement or its hydration products to produce ettringite. Ettringite (calcium aluminate trisulfate hydrate) is a normal product of tricalcium aluminate hydration, and is deposited early in the hydration process. The amount formed in normal hydration is usually constrained by the limited sulfate content of the cement used. If additional ettringite is formed by penetration of sulfate at later ages, a local volume increase can occur and lead to expansive stresses and cracking of the concrete. An illustration of the effects of such attack on the internal structure of the hydrated cement paste is provided in Fig. 39.12. The local breakdown of the paste and the rod-shaped secondary ettringite particles are apparent.

The problem can be avoided by using so-called sulfate-resisting portland cements (ASTM types II for moderate sulfate resistance and

V for sulfate resistance) in which the content of tricalcium aluminate is limited. On the other hand, the problem may be exacerbated by the use of certain high-calcium (ASTM class C) fly ashes in concrete, which may contain additional tricalcium aluminate. The problem of predicting the sulfate resistance of concretes containing fly ashes or other mineral admixtures has been discussed by Mehta [1988].

A form of distress, also usually classed as sulfate attack, may occur in concretes immersed in seawater. The details of seawater attack on concrete are more complicated. To begin with, dissolved magnesium sulfate is a significant constituent of seawater, as is dissolved calcium carbonate. Both influence the effects of seawater on concrete.

To some extent seawater is protective of concrete, in that a layer of aragonite (a form of calcium carbonate) and brucite (magnesium hydroxide) tends to precipitate on the outer exposed surfaces of the concrete. However, intrusion of dissolved salts is not completely prevented, especially in the case of permeable concrete of high water/cement ratio and relatively low cement content. A series of complex reactions can occur within the concrete involving the decomposition of certain paste constituents followed by precipitation of secondary gypsum (calcium sulfate dihydrate) and of brucite (magnesium hydroxide). These processes often result in a general softening of the cement paste with little or no expansion, and eventually a breakdown of the concrete may occur. It appears that ettringite formation does not play a major role in this form of sulfate attack.

To minimize the deterioration from either groundwater sulfate or seawater, it is usually recommended that the concrete be of relatively high cement content and low water/cement ratio so as to produce as dense and impermeable a material as possible. The use of superplasticizers appears to be indicated, and the possibility of incorporating slag or silica fume should be investigated. Slag cement concrete has had a generally good history in seawater exposure.

Reinforced or prestressed concrete exposed to seawater is of course also highly subject to corrosion damage. Such damage complicates efforts to minimize the chemical attack on the cement paste in the concrete, since cracking or spalling due to corrosion may serve to provide entry of the dissolved seawater constituents into the bulk of the concrete.

Internal Sulfate Attack (Delayed Ettringite Formation)

In recent years durability problems related to the expansive form of sulfate attack have been reported to occur in circumstances where no contact with external sulfate took place. Most, but not all, of these occurrences involve steam-cured concrete, often railroad ties. In many of these occurrences cracks were found around aggregates and through the paste that contained late-deposited or "secondary" ettringite.

In steam-cured concrete, most or all of the original ettringite is destroyed during the hot part of the steam-curing cycle. Additional ettringite is not usually produced for some weeks, but if the concrete is exposed to moist conditions, ettringite is slowly deposited. This delayed ettringite deposition is implicated as having induced the cracking and expansion.

However, in many of these occurrences, simultaneous indications of ASR in the concrete were also noted. It appears from laboratory experiments that significant ASR expansion and cracking may actually occur during the brief period of steam curing. In these circumstances the delayed ettringite tends to deposit in the cracks induced by ASR effects, or in some cases, in cracks thought to have been started by ASR but enlarged by repeated mechanical loading [Shayan and Quick, 1992]. The degree to which a given concrete damage is due to ASR and the degree to which it is due to delayed ettringite deposition are difficult to establish when both are present, and probably varies from occurrence to occurrence.

It appears that the use of clearly non-ASR-prone aggregates and of low-alkali cements would be prudent to avoid ASR in steam-cured concrete, and cements low in C_3A and sulfate should be used to minimize delayed ettringite deposition. In addition, the maximum temperature of the steam-curing cycle should be limited; most authorities recommend temperatures not higher than 80°C.

References

ACI (American Concrete Institute) Committee 201. 1991. Guide to durable concrete. *ACI Mater. J.* 88(5):553.

ACI (American Concrete Institute). 1993. *ACI Manual of Concrete Practice 1993*, Part 1. American Concrete Institute, Detroit.

ASTM (American Society for Testing and Materials). 1993. *Annual Book of ASTM Standards*, Vol. 04.01, Cement Lime; Gypsum, and Vol. 04.02, Concrete and Aggregates. ASTM, Philadelphia.

Bentur, A., Goldman, A., and Cohen, M. D. 1988. The contribution of the transition zone to the strength of high quality silica fume concrete. *Mater. Res. Soc. Symp. Proc.* 114:97–103.

Chen, H., Soles, J. A., and Malhotra, V. M. 1992. Investigations of supplementary cementing materials for reducing alkali-aggregate reaction. *Int. Workshop Alkali-Aggregate React. Concrete: Occurrences, Test. Control.* CANMET—Energy, Mines, and Resources Canada, Ottawa, Canada.

Cook, H. K. and McCoy, W. J. 1977. Influence of chloride in reinforced concrete, *in* Chloride Corrosion of Steel in Concrete, *Am. Soc. Test. Mater., Spec. Tech. Publ. 629:* 20–29.

Diamond, S. 1989. ASR—Another look at mechanisms. In *Proc. 8th Int. Conf. Alkali-Aggregate React.* Kyoto, Japan, pp. 83–94.

Diamond, S. and Ong, S. 1992. The mechanisms of lithium effects on ASR. In *Proc. 9th Int. Conf. Alkali-Aggregate React.* London, pp. 269–278.

Dodson, V. H. 1990. *Concrete Admixtures.* Van Nostrand Reinhold, New York.

Dolar-Mantuani, L. 1983. *Handbook of Concrete Aggregates.* Noyes, Park Ridge, NJ.

Dolar-Mantuani, L. 1978. Soundness and deleterious substances. In *Significance of Tests and Properties of Concrete and Concrete-Making Materials,* Spec. Tech. Pub. 169B, p. 744. ASTM, Philadelphia.

Dolch, W. L. 1984. Air entraining admixtures. In *Concrete Admixtures Handbook*, ed. V. S. Ramachandran, p. 269. Noyes, Park Ridge, NJ.

Dolch, W. L. 1978. Porosity. In *Significance of Tests and Properties of Concrete and Concrete-Making Materials,* Spec. Tech. Pub. 169B, p. 646. ASTM, Philadelphia.

Fontenay, C. and Sellevold, E. J. 1980. Ice formation in hardened cement paste. I. Mature water-saturated pastes; *in* Durability of Building Materials and Components, *Am. Soc. Test. Mater., Spec. Tech. Publ.* 691:425.

Gaynor, R. D. and Meininger, R. C. 1967. Investigation of aggregate durability in concrete. *Highway Res. Rec.* 196:25.

Isecke, B. 1983. Failure analysis of the collapse of the Berlin Congress Hall. In *Corrosion of Reinforcement in Concrete Construction*, ed. A. P. Crane. Ellis Horwood, Chichester, England.

Kaneuji, M., Winslow, D. W., and Dolch, W. L. 1980. The relationship between an aggregate's pore size distribution and its freeze-thaw durability in concrete. *Cem. Concrete Res.* 10:433.

Katayama, T. 1992. A critical review of alkali-carbonate reactions—Is their reactivity useful or harmful? In *Proc. 9th Int. Conf. Alkali-Aggregate React.* London, pp. 508–518.

Malhotra, V. M., Ramachandran, V. S., Feldman, R. F., and Aitcin, P. C. 1987. *Condensed Silica Fume in Concrete.* CRC Press, Boca Raton, FL.

Mehta., P. K. 1988. Scientific basis for determining the sulfate resistance of blended cements. In *Fly Ash and Coal Conversion By-Products: Characterization, Utilization and Disposal V, Mater. Res. Society Symp. Proc.* 113:145–152.

Mindess, S. and Young, J. F. 1981. *Concrete.* Prentice Hall, Englewood Cliffs, NJ.

Natesaiyer, K. and Hover, K. C. 1989. Some field studies of the new in situ method for identification of alkali silica reaction products. In *Proc. 8th Int. Conf. Alkali-Aggregate React.* Kyoto, Japan, pp. 555–560.

PCA (Portland Cement Association). 1988. *Design and Control of Concrete Mixtures,* 13th ed. Portland Cement Association, Skokie, IL.

Philleo, R. E. 1986. Frost susceptibility of high-strength concrete. *ACI Spec. Publ. SP-100:* 819.

Powers, T. C. and Helmuth, R. A. 1953. Theory of volume changes in hardened portland cement pastes during freezing. *Proc. Highway Res. Board* 32:285.

Powers, T. C. 1955. Basis considerations pertaining to freezing and thawing tests. *Proc. Am. Soc. Test. Mater.* 55:403.

Powers, T. C. 1945. A working hypothesis for further studies of frost resistance of concrete. *Proc. Am. Concrete Inst.* 41:245.

Powers, T. C. 1949. The air requirements of frost-resistant concrete. *Proc. Highway Res. Board* 29:184.

Prince, W. and Perami, R. 1992. Mechanism of alkali-dolomite reaction. In *Proc. 9th Int. Conf. Alkali-Aggregate React.* London, pp. 799–806.

Regourd-Moranville, M. 1989. Products of reaction and petrographic examination. In *Proc. 8th Int. Conf. Alkali-Aggregate React.* Kyoto, Japan, pp. 445–456.

Schell, H. C., Manning, D. G., and Pianca, F. 1987. A decade of bridge deck cathodic protection in Ontario. In *Proc. Corros. 87, Symp. Corros. Met. Concrete.* National Association of Corrosion Engineers, Houston, TX, pp. 65–78.

Shayan, A. and Quick, G. W. 1992. Microscopic features of cracked and uncracked concrete railway sleepers. *ACI Mater. J.* 89(4):348–361.

Sorrentino, D., Clement, J. Y., and Golberg, J. M. 1992. A new approach to characterize the chemical reactivity of the aggregates. In *Proc. 9th Int. Conf. Alkali-Aggregate React.* London, pp. 1109–1116.

Stark, D. and Klieger, P. 1973. Effect of maximum size of coarse aggregate on D-cracking in concrete pavements. *Highway Res. Rec.* 441:33.

Stark, D. 1992. The moisture condition of field concrete exhibiting alkali-silica reactivity. *Int. Workshop Alkali-Aggregate React. Concrete: Occurrences, Test. Control.* CANMET—Energy, Mines, and Resources Canada, Ottawa, Canada.

Sweet, H. S. 1948. Research on Concrete Durability as Affected by Coarse Aggregate. *Proc. Highway Res. Board* 48:988.

Treadaway, K. W. J., MacMillan, G., Hawkins, P., and Fontenay, C. 1983. The influence of concrete quality on carbonation in Middle Eastern conditions—A preliminary study. In *Corrosion of Reinforcement in Concrete Construction*, ed. Allan P. Crane. Ellis Horwood, Chichester, England.

Verbeck, G. J. and Landgren, R. 1960. Influence of physical characteristics of aggregates on frost resistance of concrete. *Proc. Am. Soc. Test. Mater.* 60:1063.

Winslow, D. N., Lindgren, M. K., and Dolch, W. L. 1982. Relation between pavement D-cracking and coarse-aggregate pore structure. *Transp. Res. Rec.* 853:17.

For Further Information

The best reference text on concrete is by A. M. Neville, *Properties of Concrete,* 3rd ed. Pitman, London, 1981.

An excellent introduction to all practical aspects of concrete is the latest edition of the PCA manual, *Design and Control of Concrete Mixtures,* Portland Cement Association, Skokie, IL.

State-of-the-art reviews of all aspects of concrete are in the annually updated *ACI Manual of Concrete Practice,* 5 volumes, American Concrete Institute, Detroit. Durability and other materials aspects are covered in part 1.

The *ASTM Book of Standards,* ASTM, Philadelphia, is issued annually and contains all ASTM specifications and test methods. The volumes on cement and concrete/aggregates are 04.01 and 04.02, respectively.

40

Special Cements and Concretes

Menashi D. Cohen
Purdue University

This chapter discusses the characteristics of special cements and concretes, focusing primarily on the mineral admixtures fly ash, silica fume, and ground granulated blast-furnace (GGBF) slag when used in portland cement concrete.

40.1 Blended Cements and Mineral Admixtures

Blended cements generally contain a combination of portland cement and a mineral admixture (or mineral additive) in large quantity. Blended cements are hydraulic, and they are often referred to as special cements. The term *special concrete* refers to a concrete containing a special cement. Blended cements occasionally impart superior strength and durability in concrete compared to plain portland cement concrete.

Other terms synonymous with the term *mineral admixture* are *pozzolan* and *supplementary cementing material.* There are many mineral admixtures that are used in practice to replace some,

0-8493-8953-4/95/$0.00 + $.50

or occasionally all, of the portland cement in concrete. Most of them are industrial by-products. The principal ones used in concrete, and their typically recommended amounts when used as a partial replacement of portland cement in concrete, are:

1. Fly ash: (high-lime = 20 to 35%, low-lime = 12 to 25%)
2. Silica fume: 5 to 11%
3. GGBF slag: 30 to 65%

Fly ash and silica fume are collected for the purpose of meeting environmental air-quality standards. The above materials have historically represented a disposal burden to the original producer.

Pozzolans react chemically with the calcium hydroxide of the hydrated portland cement to produce a calcium-silicate-hydrate gel that is cementitious. This process is called the *pozzolanic reaction*. The gel is essentially the same principal source of strength as is developed in hydrating portland cement. The pozzolanic reaction rate, however, is slower than that seen when portland cement hydrates, and the full benefit for strength enhancement is probably not realized for several weeks or months after mixing, placing, and curing.

Uses of Mineral Admixtures

For several years organizations connected to the construction industry have been involved in research about energy conservation as it relates to the production of portland cement concrete. These organizations have encouraged the use of less energy-intensive materials, such as the previously mentioned mineral admixtures, as addition or partial replacement for the relatively more energy-intensive portland cement.

The decision to use mineral admixtures depends on several factors and is not always governed by strength and durability considerations. Some of the factors involved in consideration for their use follow.

Special or High Performance

The terms *special* and *high-performance concrete* are often used in conjunction with the use of special cements. Some special cements can impart desirable properties such as high strength, reduction in water demand, lowered permeability, improved durability, and reduced heat liberation (during setting and hydration). Mineral admixtures are also used to correct for deficiencies in concrete. For instance, fly ash is used to provide fines naturally missing from fine aggregate. This avoids workability problems.

Cost and Energy Savings

The production of portland cement is more energy intensive than the production of mineral admixtures. The energy required to produce portland cement is approximately 10 MJ/kg (wet process) or 8 MJ/kg (dry process). For fly ash and GGBF slag the values are 0.05 to 0.49 MJ/kg and 1.65 to 2.23 MJ/kg, respectively [Berry, 1980].

Depending on the degree of reinforcement and type of concrete, portland cement can account for 30 to 50% of the energy spent in making reinforced concrete. Any energy savings realized during the production of the cement portion of concrete sill contribute substantially to total energy savings associated with concrete production, which would be reflected in reduced construction cost [Berry, 1980]. Therefore, replacing some part of the portland cement used in concrete with a mineral admixture leads to an overall energy savings—hence cost savings (in most instances)—per cubic meter of concrete placed.

Mineral admixtures can also help reduce costs typically associated with concrete. This is often attributed to improved workability, finishing qualities, pumping, injection in post-tension tendon ducts, intrusion into preplaced aggregate, and placement of concrete. Curing costs with mineral admixtures in concrete usually go up—or at least, so do the consequences of not curing adequately. Furthermore, most concretes containing silica fume are harder to finish unless broom texture (or some texture) during the initial finishing pass (or thereabouts) is employed. Some (the minority of) fly ash concretes finish worse than straight portland cement concrete.

Environmental Protection

The manufacture and production of portland cement involve processes that are destructive to earth and its environment. The use of mineral admixtures in concrete not only helps to reduce the amount of portland cement needed, but provides a beneficial way to use by-products, especially relative to disposing of them in the air or water, or in landfills.

Supply

The use of mineral admixtures in concrete offers an attractive feature in developing countries where a low supply of portland cement sometimes exists.

Comparison of the Mineral Admixtures

Considerable differences exist between the chemical and physical properties of fly ash, silica fume, and GGBF slag. Table 40.1 [from Philleo, 1989] provides typical compositions of the mineral admixtures. Although they are all described as siliceous or glassy, there are significant differences between their silica contents, ranging from weight percent of 30s for slag and fly ash (high-lime) to 90s for silica fume.

GGBF slag has essentially no free lime—it is bound as Ca within the glass. Glass predominates in GGBF slag. Other types of slag and other processing of slag may be associated with lime in the materials, but not GGBF slag. For silica fume, generally, the lower the Si content of the ferro-alloy, the lower its SiO_2 content.

The three major mineral admixtures significantly vary in their non-SiO_2 components. Major non-SiO_2 mass is carbon for both fly ash and silica fume, and iron oxides for fly ash. GGBF slag is essentially all glass with a small amount of iron oxide. Some fly ashes have significant calcium content too.

In Europe, according to RILEM, the mineral admixtures are classified as cementitious, highly pozzolanic, normally pozzolanic, or cementitious and pozzolanic. Table 40.2 [from Mehta and Monteiro, 1993, with modification] provides the mineralogical composition and particle characteristics of the high- and low-lime fly ashes, GGBF slag, and silica fume grouped according to the RILEM classification.

Methods of Use

There are four ways in which these mineral admixtures can be used in concrete [Philleo, 1989]:

1. As raw materials in making portland cement.
2. As a blending ingredient in making blended cement.

Table 40.1 Typical Chemical Composition of Supplementary Cementing Materials

% by Mass	Portland Cement	Slag	Low-Lime Fly Ash	High-Lime Fly Ash	Silica Fume
SiO_2	22	35	50	35	90
Al_2O_3	6	8	25	20	2
Fe_2O_3	3	0	10	5	2
CaO	63	40	1	20	0

Source: Philleo, 1989.

Table 40.2 Mineralogical Composition and Particle Characteristics

Classification	Mineralogical Composition	Particle Characteristics
	Cementitious	
Slag	• Mostly silicate glass (containing Ca, Mg, Al, silica) • Crystalline compounds of melilite group may be present in small quantity	• Unprocessed material is of sand size and contains 10–15% moisture • Before use it is dried and ground to $<$45 μm (~ 500 m^2/kg Blaine) • Particles have rough texture and sharp or angular shape
	Cementitious and Pozzolanic	
High lime ash	• Mostly silicate glass (containing Ca, Mg, Al, and alkalis) • Some crystalline matters (C_3A, quartz, free lime, periclase, also some $C\bar{S}$ and $C_4A_3\bar{S}$ present in case of high sulfur coals • Little unburned coal, $\leq$ 2% • Light in color • Sometimes generates high heat of hydration due to free CaO present • Sometimes contains high content of MgO (could be periclase) • Could be used as raw material for portland cement (requires less heat during kiln production)	• 10–15% particles $>$ 45 μm (finer than low-lime fly ash) • High fineness: 300–400 m^2/kg Blaine • Most particles are solid spheres, less than 20 μm in diameter. • Particle surface generally smooth, but not as clean as in low-lime fly ash
	Highly Active Pozzolan	
Silica fume	• Essentially pure noncrystalline silica	• Extremely fine powder, ~20,000 m^2/kg surface area by BET N_2 • 0.1 μm average diameter
Rice husk ash (burned under controlled condition)	• Particle size $<$ 45 μm, but highly cellular with extremely large surface area, ~60,000 m^2/kg by BET N_2	
	Normal Pozzolan	
Low-lime fly ash	• Mostly silicate glass (containing Al, Fe, alkalis) • Some crystalline (quartz, mullite, sillimanite, hematite, and magnetite) • Unburned carbon, usually $<$ 5%, but may be as high as 10% • Note: aluminosilicates: sillimanite ($Al_2O_3.SiO_2$), mullite ($3Al_2O_3.2SiO_2$) are slender needles and exist in interior of glassy spheres)	• 15–30% of particles $>$ 45 μm • 200–300 m^2/kg surface area by Blaine • Mosts particles are solid spheres averaging 20 μm in diameter • Cenospheres (hollow spheres fly ash particle) and plerospheres (spheres of fly ash particles that contain other fine fly ash particles in their cavity) may be present • Particles have smooth surfaces
	Weak Pozzolan	
Slowly cooled slag, bottom ash, boiler slag, field-burnt rice husk ash	• Essentially crystalline silicate minerals and only a small amount of noncrystalline matter may be present	• The materials must be pulverized to very fine particle size in order to develop some pozzolanic activity • Ground particles are rough in texture

Source: Mehta and Monteiro, 1992.

3. As a dry ingredient added at a concrete batching plant or on-site.
4. As an ingredient of slurry added at the concrete batch plant.

GBF slag is the only material that has been used in all four ways. Throughout most of the world it is used primarily as an ingredient of blended cement, and most of the world's blended cements are portland blast-furnace slag cements (25 to 70%). In North America, GGBF slag is the most commonly used form of slag (as a dry ingredient added at the concrete batch plant).

Fly ash is usually batched as a dry ingredient added at the batch plant, although a significant amount of portland-pozzolan cement is manufactured. Silica fume is usually added at the concrete batch plant. In the U.S. the use of fly ash in blended cements constitutes only a small portion of the cement market. Approximately 2% of all cements are blended with fly ash. In Iceland and Canada it is an ingredient of blended cement. Because of the difficulty in handling this extremely fine material, it is frequently stored and batched in the slurry or dry-densified form.

Specifications

There are many specifications for mineral admixtures and blended cements. Historically, the specifications were developed independently, so there are separate specifications for each material. Critics say that they contain many irrelevant requirements and that if only pertinent requirements were included they could easily be combined into a common specification [Philleo, 1989].

In the ASTM standards blended cements are grouped and specified in ASTM C595 (prescription) and in the new (1993) performance document, ASTM C1157. Fly ash and slag specifications are included separately in ASTM C618 and ASTM C989, respectively. A new ASTM specification, C1240, for use of silica fume in mortar, has recently been issued. The Canadian specification CAN/CSA-A23.5-M86 is the first to combine all materials into a single document.

ASTM C595 gives the chemical and physical requirements for blended hydraulic cements. The standard is highlighted in Table 40.3.

In Europe slag is used in different quantities in blended cements. In England, BS 146 allows a maximum of 65%, while BS 4246 (low heat) allows 50 to 90%. Blended cements with such a large amount of slag are not appropriate for cold weather because of low heat and low rate of strength development of the concrete. In Germany, Eisenportland blended cement allows 30% slag and Hochofen allows 31 to 85% slag replacement. In France, *ciment de haut fourneau* allows 65 to 75% slag. The Trief process in Belgium involves a GGBF slag in the form of a slurry which is fed directly into the concrete mixer together with portland cement and aggregate. Here, the cost of drying the slag is avoided, and grinding in the wet state results in a greater fineness than would be obtained in dry grinding for the same power input [Neville, 1981].

Production and Use of Blended Cements

ASTM C595 allows preparation of blended cement by any of these methods:

1. Intergrinding the two components.
2. Intergrinding individual components separately and subsequently blending in special cyclones and machines.
3. By both methods 1 and 2 [Berry, 1980].

Intergrinding

With blended cements the cement maker often prefers to intergrind the cement clinker and the mineral admixture rather than to interblend. With fly ash, however, there is some evidence to suggest that intergrinding breaks down the spherical particles and reduces one of the major advantages of fly ash, which is the reduction of water demand. On the other hand, surface area of the pozzolan is increased during intergrinding, and this leads to an increase in the pozzolanic activity of the fly ash.

Table 40.3 Highlight of ASTM C595 for Blended Cements

Blended Cement	Type	Options[1]	Composition	Wt. %
1. Slag-modified portland cement (modified-p.c.)[2]	One type, probably I(SM)	MS A MH	p.c. + slag[5]	(wt. slag/total wt.) < 25%
2. Portland blast-furnace slag cement[5]	IS	MS A MH	p.c. + slag	(wt. slag/total wt.) = 25–70%
3. Slag cement[3]	S	A	Slag + p.c. (or) Slag + hydrated lime Type N or S[4]	(wt. slag/total wt.) ≥ 70%
4. Pozzolan-modified p.c.[6]	I(PM)	MS A MH	p.c. + pozzolan[7] (or) portland blast-furnace slag cement (item #2 above) + pozzolan	(wt. pozzolan/total wt.) ≤ 15%
5. Portland-pozzolan cement[8]	IP[9]	MS A MH		(wt. pozzolan/total wt.) = 15–40%
	P[9]	MS A MH		

[1] MS = moderate sulfate resistant, A = air-entrainment, MH = moderate heat of hydration

[2] Items # 1, 2, and 3 are similar except that their slag content is different. Further, item #3 allows hydrated lime to be used. Use item #1, which requires less slag than item #3, when special characteristics attributed to larger quantities of slag in portland blast-furnace slag cements are not desired.

[3] The term *slag* is used interchangeably with *finely granulated blast-furnace slag.* Type S is used in combination with portland cement in making concrete and with hydrated lime in making masonry mortar.

[4] Hydrated lime Types N and S are generally used for coats, plasters, addition to portland cement, and stucco, among other applications. More information is available in ASTM C207. Type N is normal hydrated lime used for masonry purposes. Types S and SA differ from Types N and NA principally by their ability to develop high, early plasticity and higher water retention, and by a limitation on unhydrated oxide content. Note that Types NS and SA indicate presence of intentional air-entrainment.

[5] Type IS is similar to ASTM Type IV (low heat) cement. BS 146 allows up to 65% slag. British Standard BS4246: low heat portland–blast furnace slag cement allows 50 to 90% slag.

[6] Similar in characteristics to ASTM Type II cement. Type I(PM) allows less pozzolan (<15%) compared to item #5, which is 15 to 40%. The cement should not be used when special characteristics attributed to larger pozzolan quantities are desired.

[7] Finely divided pozzolan.

[8] Similar in characteristics to ASTM Type IV cement.

[9] Type IP is used in general concrete construction. Type P is used where high early strengths at early ages are not required (see strength requirements in ASTM 595).

This explains why concrete producers are so interested in the cement manufacturer's production process and how this may affect the concrete-making characteristics of cement [Newman, 1986].

There is no evidence to suggest that intergrinding with natural pozzolans has any effect on the portland cement [Cook, 1986]. In some instances, GBF slag intergrinding can produce a blend in which the portland cement clinker is overground and the slag remains coarser than optimal.

Separate Grinding and Adding at the Mixer

This method gives control over fineness. One can also control the cement composition to suit a variety of types of concrete. This is not a blended cement [Newman, 1986].

Separate Grinding and Interblending

Although this method is ideal, it is difficult to interblend two finely ground materials [Newman, 1986].

With the exception of a few studies, research and field data relate to cementing combinations that were separately batched [Berry, 1980; Newman, 1986]. The major properties probably are not influenced by the method chosen to introduce the components, and all methods are thought to produce the same concrete properties, provided other conditions are similar. However, it is important to ensure that when ingredients are added separately at the mixer, adequate time is given for mixing and dispersing of the cements. The time for measuring the start of mixing should also begin after all ingredients are put in the mixer [Berry, 1980].

Use of Mineral Admixtures as Slurry

Fly ash, silica fume, and GGBF slag have been provided as a slurry for the production of concrete. There are several advantages in using these materials in the form of a slurry, including absence of dusting during loading and unloading, especially in the case of silica fume. There are disadvantages too. The cost of transportation may increase, considering that up to 50% of the slurry's mass is the water. Other disadvantages are associated with the mix proportion because the exact water content of the slurry may change due to evaporation of part of the water. Problems may also be associated with freezing of water during freezing weather, and gelling and flocculation of particles in the slurry.

GGBF slag was used as a slurry in the construction of a hydroelectric dam in Glenmoriston, Scotland, in 1953. In this process the slag was wet ground and stored as a slurry until it was mixed with portland cement and aggregate during batching of the concrete. The advantages claimed were a saving in fuel for drying the slag and greater efficiency of grinding in the wet state [Berry, 1980].

A mixture of 50-50 fly ash–water slurry was used in Okutadami Dam in Japan. The problem of pack setting and maintaining dry-stored fly ash uniformity was avoided [Berry, 1980].

Selecting Manufacturer Blend or Blending at Concrete Mixer

In the U.S. only 1 to 2% of cements are blended by the manufacturer. Blended cement is the responsibility of the cement manufacturer, while a concrete mixer blend is the responsibility of the batching plant.

Several important factors could be considered when engineers face the decision of using a mineral admixture as a blend prepared by the cement manufacturer or blending mineral admixture with other ingredients of concrete at the concrete mixer. The latter choice, blending at the mixer, provides certain advantages over the former [Newman, 1986]. These include control of the amount of cementing materials that can be added independently to the concrete, ultimately controlling engineering properties and cost.

There are some disadvantages associated with separately adding materials at the concrete mixer [Newman, 1986]. The batching plant may require (1) greater management; (2) modification of the plant (i.e., improved air-filters, silo venting, aeration, transfer systems, and batching controls); (3) construction of separate and bigger silos for mineral admixtures because of smaller density of mineral admixtures; (4) that additions go to the correct silos and that the correct materials are batched; (5) increased quality control, including materials testing, laboratory trial mixtures, mixture proportioning, and concrete testing; and (6) construction of new tanks (i.e., for silica fume slurries) or silos (dry products).

Users are cautioned against use of mineral admixtures that are contaminated, especially fly ash. Fly ash can be contaminated at the power plant by the addition of additives used by the producer to meet stack emission standards. Some have no harmful effect on either the fly ash or the concrete in which it is used, while others, notably soda ash, can cause expansion and extreme efflorescence of the concrete. Such efflorescence, besides being unsightly, can cause loss of bond of paint on the concrete and thus cause serious maintenance problems. Because of this, the user should be certain that the producer and marketer of fly ash understand the seriousness of the problem and take steps to ensure against it [ACI Comm. 212 Report, 1989].

40.2 Pozzolans

Fly ash and silica fume, along with other mineral admixtures such as natural pozzolan and rice husk ash, are pozzolans. Although GGBF slag has been mentioned in some publications as a pozzolan, it should be viewed as a cementitious material with a pozzolanic component. ASTM C219 (Terminology) defines a pozzolan as

> a siliceous or siliceous and aluminous material, which in itself possesses little or no cementitious value but which will, in finely divided form and in the presence of moisture, chemically react with calcium hydroxide at ordinary temperatures to form compounds possessing cementitious properties.

There are two types of pozzolans—natural and industrial. The natural pozzolans are classified as Type N according to ASTM specification C595. They include volcanic ash, tuff, diatomaceous earth, and some calcined clays. The industrial pozzolans include fly ash, silica fume, and rice husk ash.

Until recently fly ash was regarded only as a pozzolan. However, high-lime fly ash has become widely used and does not need an external source to produce some cementitious properties. Therefore, it is both cementitious and pozzolanic. The term *mineral admixture* has been suggested to describe all classes of GGBF slag, fly ash, pozzolans, and other cement supplements with a distinction being drawn on the basis of their self-cementing capabilities. But this terminology has been criticized by some as reflecting the use of these materials in small quantities, as is the case for chemical admixtures. In Canada the term *supplementary cementing material* has been adopted in specifications. Although it is lengthy, it does describe the role of the material [Philleo, 1989].

Pozzolan–Portland Cement Reaction

The nature of the pozzolan-portland cement reaction is complex, and details are not well established. The reaction process can be presented as simple equations—(40.1) through (40.3) [Cook, 1986; Mindess and Young, 1981].

In the first step, Eq. (40.1), C_3S of portland cement reacts with water to produce calcium silicate hydrate gel (C-S-H) and calcium hydroxide (CH).

$$C_3S + \text{aq.} \rightarrow \text{C-S-H} + \text{CH} \tag{40.1}$$

where $C = CaO$, $S = SiO_2$, $A = Al_2O_3$, and $H = H_2O$.

The hydration of the C_2S of the portland cement follows a similar pattern as its counterpart, C_3S. If pozzolan is present in the system, the calcium hydroxide produced from Eq. (40.1) reacts with the amorphous and reactive silica (S) or alumina (A) supplied by the pozzolan. The reaction is called *pozzolanic* and is expressed ideally according to Eqs. (40.2) and (40.3) below:

$$\text{CH} + \text{S} + \text{H} \xrightarrow{\text{aq.}} \text{C-S-H (calcium silicate hydrates)} \tag{40.2}$$

$$\text{CH} + \text{A} + \text{H} \xrightarrow{\text{aq.}} \text{C-A-H (calcium aluminate hydrates)} \tag{40.3}$$

Normally in the literature, pozzolanic reaction refers to Eq. (40.2), although both Eqs. (40.2) and (40.3) describe a pozzolanic reaction.

The C-S-H gel is known to be the source of strength in portland cement concrete. The calcium hydroxide constitutes more than 25 volume percent of the matrix. It grows in the solution as relatively large and brittle crystals, and due to their morphology they do not contribute to strength. Cracks can propagate more easily through regions populated by calcium hydroxide crystals than through C-S-H gel, especially at the aggregate–cement paste interface where there are large pockets of calcium hydroxide crystals and voids. The calcium hydroxide crystals are also vulnerable to attack by carbon dioxide and other chemicals such as sulfates. Their vulnerability poses durability problems for the concrete. According to Eq. (40.2), pozzolans increase the proportions of C-S-H in the portland cement paste at the expense of calcium hydroxide. The rate of reaction between CH and S is slow and resembles the hydration of C_2S with water. It can be deduced that pozzolan addition is like adding extra C_2S to the cement.

The reaction according to Eq. (40.3) produces calcium aluminate hydrates (C-A-H). The hydrates can cause a durability problem in concrete. If they react with sulfate they could generate hydrates, such as ettringite, that are expansive in nature and can lead to cracking of the concrete. Therefore, if a pozzolan is used to improve the sulfate resistance of concrete, it must have a low amount of alumina.

It has been stated that pozzolans accelerate the hydration of portland cement, but this finding is by no means consistent and depends on the characteristics of both the portland cement and pozzolan [Cook, 1986].

The hydration of pozzolan follows a topochemical process—that is, a surface reaction. The reaction of portland cement with water can be accelerated by the presence of pozzolan particles, because they can act as nucleation sites for the reaction products. Future dissolution of C_3S is stimulated by the combination of calcium ions (Ca^{2+}) with pozzolan, which lowers the concentration of Ca^{2+}. During the hydration of pozzolan–portland cement, hydrates assemble and show zonal structures surrounding the C_3S grains. The composition of each layer is constant within the layer, but the C/S ratio in each layer decreases toward the pozzolan grain, and it becomes a porous structure. There is a clearance between the pozzolan grain and the surrounding layer of hydrates, and the clearance has a relationship to the alkalis dissolved from the pozzolan. Hydration of SiO_2 occurs by the formation of an amorphous layer and continued swelling and bursting of the layer [Ogawa *et al.*, 1980].

Some pozzolans have high amounts of alkalis, and they can create problems if the aggregates used are reactive. However, some Italian cement blends containing high levels of alkalis did not develop problems associated with alkali-silica reaction when reactive aggregates were used, but in fact improved performance. Use of GGBF slag with alkali content exceeding 1.0% has demonstrated significant reductions in expansion when highly reactive aggregates were used [KOCH Minerals Co. tests, Bulletin No. 10004; ACI 226, 1987b]. Therefore, the role of alkali in portland cement–pozzolan reaction is not well established [Cook, 1986].

Evaluation of Pozzolanic Activity

Laboratory test methods are available for measuring the activity of a pozzolan or its suitability in concrete. The surest way to determine the suitability of a given pozzolan is to test it in mortar or concrete for the property of interest (for instance, strength, water requirement, or durability). But this is not always practical, mostly owing to the time-consuming nature of the tests, especially those for strength and durability. One method for assessing the suitability of a pozzolan can be described by its pozzolanic activity index, which can be measured by chemical, physical, or mechanical methods.

The pozzolanic activity index of a pozzolan describes its ability to react with calcium hydroxide to produce C-S-H gel and to improve strength. The chemical composition of a pozzolan is not a clear indication of its pozzolanic activity. The crystal structure (i.e., degree of crystallinity, or amorphous nature) is a better indicator. For instance, quartz that is mainly crystalline SiO_2 is nonreactive with calcium hydroxide, while a volcanic ash that contains amorphous or glassy SiO_2 is reactive even if it contains less than half of the silicon as SiO_2. Particle size and shape and texture of the reactive silica are also important in determining the pozzolanic activity.

Three groups of tests are available to determine the pozzolanic activity—chemical, physical, and mechanical—although correlation among the groups is poor [Cook, 1986]. One type of chemical activity test involves measuring the amount of $SiO_2 + Al_2O_3 + Fe_2O_3$ rendered soluble as a result of the pozzolanic reaction in either alkalis or acids. Solubility in cold hydrochloric acid is the basis of the Feret-modified Florentin method [Cook, 1986]. Another category of chemical testing measures the decrease in Ca^{2+} concentration when pozzolan is added to a saturated lime solution. This method was developed by Fratini [Cook, 1986] and has been adopted in many countries. The problem, however, is that the test takes 7 days to complete and thus it is not suitable for rapid evaluation of a pozzolan. Further, there is no correlation between the results obtained in the chemical tests with strength of mortars or concretes.

The most promising method involves the use of x-ray diffraction to monitor the progress of lime uptake. Thermogravimetric analyses and DTA are also used in the physical test. Good correlations with 6-month and 1-year strengths have been indicated [Cook, 1986].

Compressive strength is the most meaningful mechanical parameter that describes the suitability and activity of a pozzolan. The disadvantage of strength tests is that the rate of strength development in portland cement–pozzolan and pozzolan-lime mixtures is relatively low. However, one can accelerate the rate of strength development. This is done by curing the specimen for 7 days at 50°C and comparing the strength values with those of specimens cured for 7 days at 18°C. The differences in the strength values are related to pozzolanic activity. For acceptable pozzolans the minimum difference should be approximately 2500 psi for 40% cement replacement and 1500 psi for 20% cement replacement with the pozzolan. Most standards use high temperature to accelerate curing [Cook, 1986].

The pozzolanic activity index can be measured according to ASTM C311. The test gives a measure of lime fixation. The test works for some pozzolans, but not all—that is, good correlation between the pozzolanic activity index and the concrete strength is obtained for only some pozzolans. This is probably because the microstructures of two concretes made with two pozzolans of similar lime fixation properties may not be the same owing to the different water-to-cement ratio used in the plain mortar and in mortar containing a pozzolan.

The pozzolanic activity index test was not proposed to evaluate the pozzolanic activity, but only to ensure whether the amount of pozzolan added to portland cement is enough to cause the cement-wetting solution to be unsaturated with respect to calcium hydroxide. Unfortunately, this method of evaluating the index is widespread in some countries [Cook, 1986].

The suitability of slag is determined by the slag activity index test, according to ASTM C989. The test is similar to the pozzolanic activity index test, but the amount of slag in the mixture is higher (a mass ratio of 50:50 slag to reference portland cement combination is used) and there are restrictions on the alkali content of the reference portland cement used. Another test for slags is based on ASTM E1085. This test measures the hydraulic activity of slag and is mainly used for quality control of slag production and some indication of its fineness. Another test is used for gauging slag activity—the ASTM C1073 test, which accelerates 100% GGBF slag mortar cubes with both sodium hydroxide and heat in a wet environment. This is a 1-day test.

Characteristics of Pozzolanic Reaction

The pozzolanic reaction has a considerable influence on the behavior and properties of concrete. This is because the reaction alters the chemical and microstructural nature of the concrete. The characteristics of the pozzolanic reaction include the following:

1. *The reaction is lime consuming instead of lime producing.* This influences the durability of paste in an acidic environment.
2. *The reaction is a pore refiner.* The reaction product, C-S-H gel, fills up the capillary spaces and transforms the large capillary pores into smaller ones. Figure 40.1 shows a typical behavior that is obtained by mercury intrusion porosimetry for a mature plain portland cement paste and a paste containing portland cement and a pozzolan (silica fume). Generally, adding a pozzolan to portland cement will not cause a major change in the total porosity, but it tends to refine or modify the pore structure of the paste. This is accomplished by (a) shifting

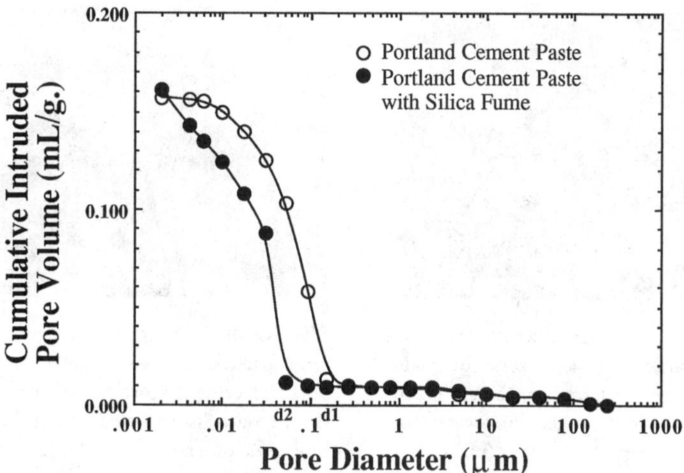

FIGURE 40.1 Typical mercury intrusion porosimetry curve for a mature portland cement paste and a paste containing portland cement and a pozzolan (silica fume). (Note: $d1$ = threshold diameter for portland cement paste, and $d2$ = threshold diameter for portland cement paste with silica fume.)

the threshold diameter (corresponding to the diameter at which the pore volume increases sharply) to a size smaller than before, that is, from $d1$ to $d2$ where $d2 < d1$, and (b) increasing the slope of the cumulative volume-diameter curve as the curve approaches the vertical axis. The increase in the slope of the curve indicates an increase in the required pressure (or degree of difficulty) by which mercury is intruded in the smaller and smaller pores.

3. *The reaction is a grain-size refiner.* The nucleation of calcium hydroxide around smaller pozzolan particles will have the effect of replacing large and oriented calcium hydroxide crystals with numerous smaller and less-oriented crystals of calcium hydroxide and other reaction products. In effect, the microstructure becomes more homogeneous.
4. *The reaction is a transition-zone refiner.* The pozzolanic reaction at the interface of cement paste and aggregate—that is, the transition zone—can significantly alter the microstructure of this zone and, hence, the mechanical and durability properties of the concrete. The influence of pozzolan on the transition zone can extend to distances in the bulk paste as far as 50 microns from the aggregate surface. Considering that typical distances between sand particles in the mortar portion of concrete are somewhere between 150 microns and 200 microns, the influence of pozzolans on the microstructure and, hence, on important properties of the concrete is significant.

Illustrations of the transition zone refinement are provided in Figs. 40.2 and 40.3 for plain portland cement mortar [Bentur and Cohen, 1987] and Fig. 40.4 for portland cement with silica fume mortar [Bentur and Cohen, 1987]. Figure 40.2 shows the microstructure of the paste–sand particle interfacial zone of a plain portland cement mortar specimen cured for one day in saturated lime-water as observed by SEM (magnification 2000×). The top portion of the micrograph is a socket formerly occupied by a piece of aggregate. There is a continuous thin layer (< 1 micron in thickness), presumably the duplex film, discussed by Barnes *et al.* [1978]. This layer is composed of amorphous calcium hydroxide and is in direct contact with the sand particle surface. Beyond the duplex film, and at a distance up to about 10 to 20 microns (and sometimes up to 50 microns), there is the interfacial zone where relatively little hydration product is present, and big gaps are observed. These gaps are partially filled with porous hydration products (not shown here) or with large calcium hydroxide crystals (shown), the latter being more typical. Calcium hydroxide crystals

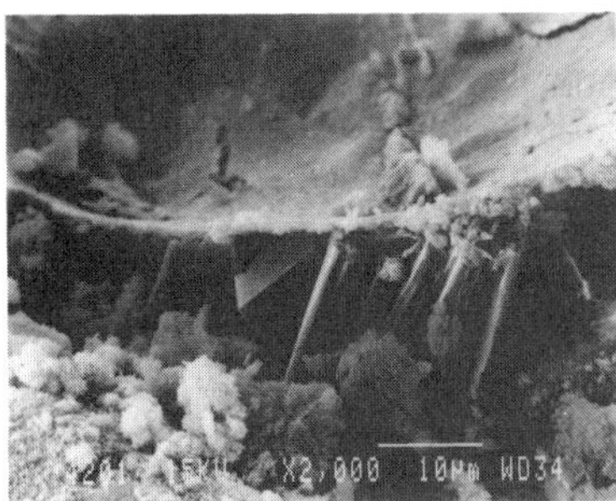

FIGURE 40.2 Microstructure of paste-sand particle interfacial zone of plain portland cement mortar specimen cured for one day in saturated lime-water solution as observed by SEM (magnification 2000×). (*Source:* Bentur and Cohen, 1987.)

FIGURE 40.3 Microstructure of paste-sand particle interfacial zone of plain portland cement mortar specimen cured for 28 days in saturated lime-water solution as observed by SEM (magnification 5000×). (*Source:* Bentur and Cohen, 1987.)

do not have a cementitious characteristic and occupy as much as one quarter of the volume of the hydration products, and in plain concrete they tend to congregate around aggregate particles. The calcium hydroxide crystals at the interface prevent good bond between the paste (lower left corner) and the sand particle.

Figure 40.3 shows the microstructure of the paste–sand particle interfacial zone of a plain portland cement mortar specimen cured for 28 days in saturated lime-water solution, as observed by SEM (magnification 5000×). The top portion of the micrograph is a socket formerly occupied by a sand particle. The duplex film is clearly evident. The interface microstructure is relatively more dense than the 1-day-old mortar specimen in Fig. 40.2. Some gaps can be clearly seen, running parallel to the interface, at a distance of about 10 to 20 microns from the sand particle surface. These gaps are remnants of the bigger gaps observed at earlier hydration that resulted from the mixing water being trapped at the paste–sand particle interface.

Figure 40.4 shows the microstructure of the paste–sand particle interfacial zone of a portland cement with silica fume mortar specimen cured for 28 days in saturated lime-water solution as observed by SEM (magnification 2000×). The top portion shows a sand particle socket with no calcium hydroxide crystals beneath the interface. The paste matrix was extremely dense and extended usually to the particle's surface. Unlike in plain mortar, there are no massive calcium hydroxide particles or gaps present. Even the presence of a duplex film cannot be resolved at this age. The paste forms an excellent bond to the aggregate, producing strong and impermeable mortar.

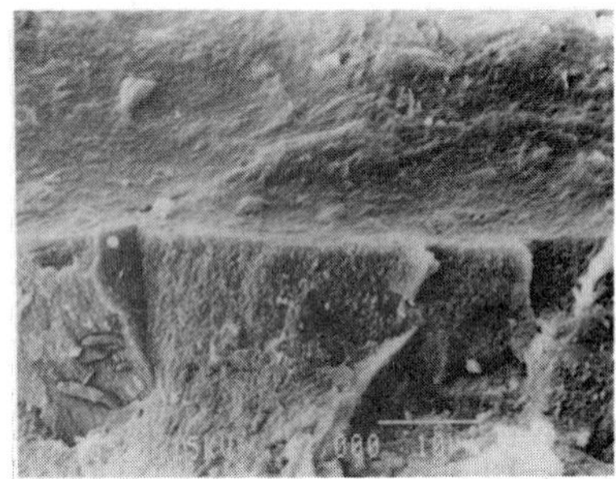

FIGURE 40.4 Microstructure of paste-sand particle interfacial zone of portland cement with silica fume mortar specimen cured for 28 days in saturated lime-water solution as observed by SEM (magnification 2000×). (*Source:* Bentur and Cohen, 1987.)

The above characteristics associated with the pozzolanic reaction can lead to vast improvements in strength and durability properties of concrete. Attempting to obtain compressive strength values of 8000 to 10,000 psi or higher in plain concrete can prove to be technologically difficult and uneconomical. Higher strengths can be achieved, but the use of a large cement content in the mix may be required. High strength can be achieved in a plain concrete mix containing about 1000 lb/yd^3 of cement (11 bags), low water-to-cement ratio, and a high-range water-reducing admixture (HRWRA).

However, such a mix is usually undesirable due to (a) high cost and high heat of hydration generated during setting (because of high cement content), (b) susceptibility to large amounts of drying shrinkage, creep, and cracking, also due to the large cement content, and (c) the presence of a large amount of calcium hydroxide in the bulk paste and the transition zone limiting the strength of the concrete.

Using a mineral admixture, high-strength concrete can be obtained with less difficulty and concern. Less cement can be used if a pozzolan is used as partial replacement. A concrete mix containing 800 to 850 lb/yd^3 cement (8.5 to 9 bags), 100 lb/yd^3 low-lime fly ash, and water–to–cement+fly ash ratio of 0.30:0.33, plus a HRWRA can provide a strength of 10,000 psi after about 7 weeks. The same strength can be obtained with even less cement, 650 to 670 lb/yd^3 (7 bags) and the same water–to–cement+fly ash ratio if high-lime fly ash, 160 to 170 lb/yd^3, is used instead of low-lime.

Pozzolanic reaction can lead to significant reduction in concrete permeability due to the modification of the pore structure of the paste, even without a change in total porosity. Generally, a pozzolan can reduce the permeability by one to three orders of magnitude. A lower amount of calcium hydroxide in portland cement–pozzolan concrete compared to plain concrete can lead to improvement in durability in some cases, for instance, resistance to acid attack and carbonation shrinkage. However, a lower amount of calcium hydroxide is not always desirable. Calcium hydroxide can provide protection for the reinforcing steel against corrosion by maintaining a high degree of alkalinity in the pore water system of the concrete. However, some researchers, including Diamond [1975], have reported that the resistance of concrete to corrosion is not related to its pore solution alkalinity (a chemical factor) but to its permeability (a physical factor).

One major use of pozzolans, especially fly ash, is in massive structures such as dams, owing to their ability to lower the heat of hydration released during the setting period. This is illustrated in Fig. 40.5 [Compton and MacInnis, 1952] where the reduction in temperature rise in fly ash concrete is evident. This reduction is due to the role of pozzolans in reducing the heat generated

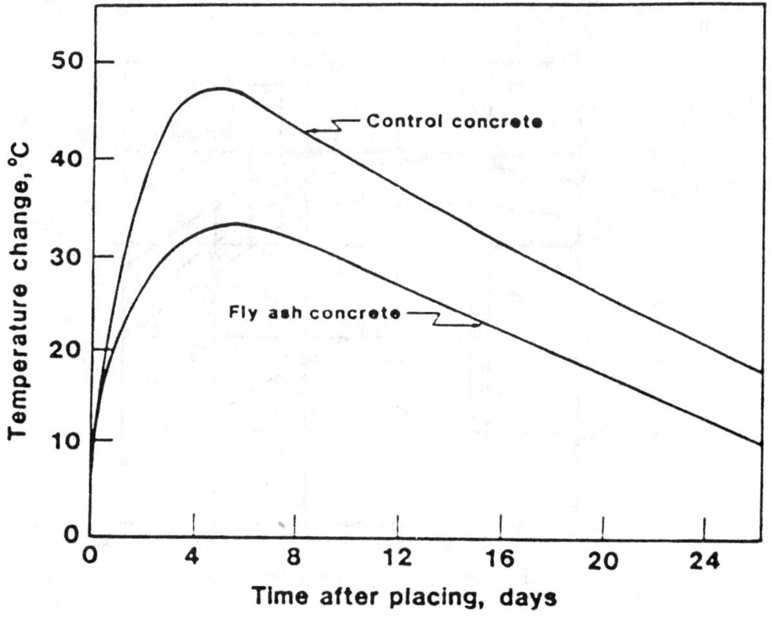

FIGURE 40.5 Rate of heat liberated as measured by calorimetry. (*Source:* Compton and MacInnis, 1952.)

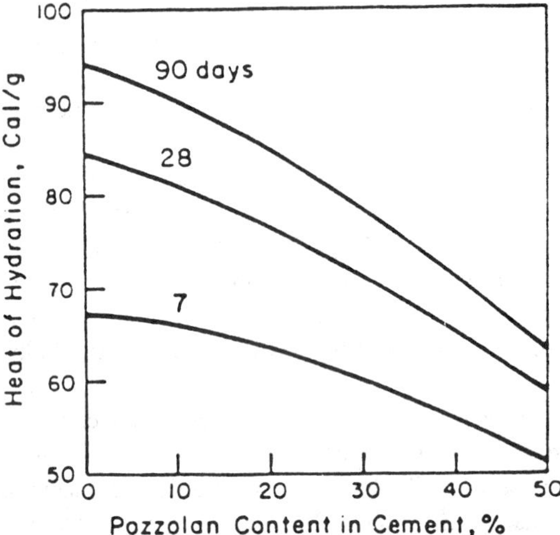

FIGURE 40.6 Effect of pozzolan content in cement on the amount of heat of hydration. (*Source:* Massazza and Costa, 1979.)

during the setting and hardening of concrete. Figure 40.6 [Massazza and Costa, 1979] shows the general trends for the total heat liberated during hydration for plain portland cement paste and portland cement paste including a natural pozzolan. Figure 40.6 shows that the heat generated decreases as the amount of pozzolan used increases. However, only a maximum limit in the amount of pozzolan should be established; otherwise the strength of the concrete, especially at the early ages, might be compromised, as shown in Fig. 40.7 [Massazza and Costa, 1979]. Figure 40.7 shows the relative

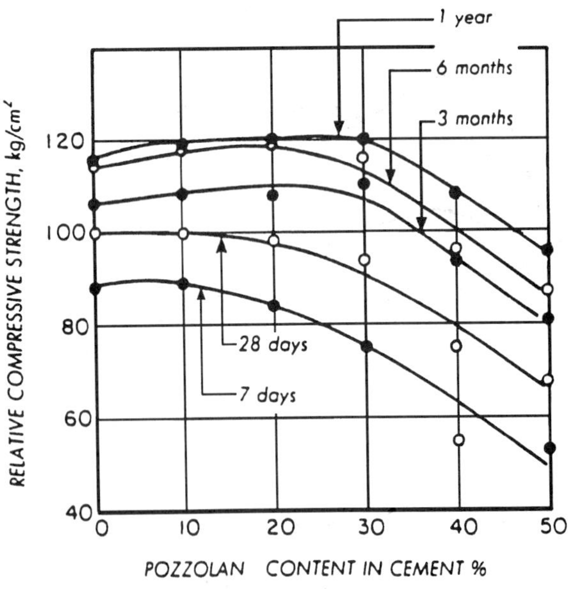

FIGURE 40.7 Effect of pozzolan content in cement on the relative compressive strength of concrete. (*Source:* Massazza and Costa, 1979.)

compressive strength values (relative to portland cement concrete without pozzolan) as a function of the pozzolan content.

The period at which concrete containing a pozzolan starts to show higher strength than the concrete without the pozzolan is a function of the mixture proportions including the amount of pozzolan and the type and quality of the ingredients. For concrete that has been properly proportioned, the effective period varies generally in the order of 1 to 2 days for silica fume, 3 to 7 days for high-lime fly ash and GGBF slag, and 14 days to 4 weeks for low-lime fly ash and natural pozzolans.

The observation of strength must be made relative to the rate of strength development at early ages, and also to the ultimate strength. The figure shows that at early ages an increase in the amount of the pozzolan can lead to lower strengths. However, at later ages (i.e., 3 to 6 months) there appears to be an optimum amount of pozzolan that leads to a maximum strength. For this series the optimum content of Santorin earth was 20%, but the optimum varies for different pozzolans and mixture proportions. For GGBF slags (grade 120), it appears that a 28-day compressive strength optimum is achieved around a 35 to 40% replacement of portland cement [ACI 226, 1987b]. At this level, a 20 to 35% strength increase occurs.

The presence of a pozzolan leads to an increase in the magnitude of drying shrinkage of concrete. This is probably due to the presence of larger amounts of C-S-H gel. However, the cracking tendency becomes less, owing to stronger transition zones and a better aggregate bond.

Pozzolans do not impair the resistance of concrete to freezing-and-thawing cycles if the concrete is adequately air entrained. Finely ground pozzolans require higher-than-normal dosage of air-entraining admixture to obtain the desired air content or the desired spacing factor. The influence of the specific mineral admixtures such as fly ash, silica fume, and slag on the chemical stability of air-entraining admixtures will be discussed in the following sections.

There are other possible disadvantages to using pozzolans [Price, 1975]. Inferior quality concrete can be obtained if the quality of the pozzolan is poor or if excessive amounts are used. This can lead to:

1. Reduced rate of hardening and development of strength.
2. Increased drying shrinkage.
3. Increased water requirement (which can lead to increased drying shrinkage).
4. Low resistance to freezing-and-thawing cycles.

40.3 Fly Ash Concrete

Production and Uses

Fly ash is produced by the combustion of finely ground coal, injected with a stream of hot air into the furnace at electricity-generating stations. On entry into the furnace, where the temperature is about 1500°C, the carbonaceous content of the coal is burnt quickly. The remaining inorganic matter present, such as shale and clay (essentially consisting of silica, alumina, and iron oxide) melt while in suspension . Upon subsequent rapid cooling, as it is carried out of the furnace with the flue gases, these materials form fine spherical particles [Dhir, 1986]. Although most of the fly ash particles are solid spheres of glass, sometimes a small number are spheres which are completely hollow (referred to as cenospheres) or hollow spheres containing many smaller spheres (referred to as plerospheres).

About 80% of the coal ash is eventually carried out of the furnace with the flue gases and must be removed before they are discharged to the atmosphere [Dhir, 1986]. This collected material is called *fly ash* in the U.S., a term coined in the 1930s [Berry and Malhotra, 1986]. In the U.K. fly ash is known as *pulverized fuel ash* (pfa), in France as *cendres volantes*, in Germany as *flugashe*, and in the Spanish language as *cenizas volantes*. The remainder of the coal ash falls to the bottom of the furnace where it sinters to form a coarse material called *furnace bottom ash* or *bottom ash*.

Fly ash was first used in concrete construction in the early 1930s when a number of utilities began to investigate means of profitably utilizing it. The early successful performance of fly ash in concrete led to its use in Hungry Horse Dam, Montana, which was started in 1948. Fly ash was used for the purpose of heat reduction in that project. Up to 32.4% of the portland cement was replaced with fly ash, which was shipped from the Chicago area. Fly ash has also been used in a number of other dams and important works since then [Dhir, 1986]. Fly ash is being used in a number of civil engineering applications besides dams. These include pavements, highway structures, and buildings.

Fly ash is also used in a variety of concrete materials including high-strength concrete, roller-compacted concrete, and controlled low-strength material, among others.

Methods of Collecting Fly Ash

Fly ash is collected from flue gas in one of two ways:

1. Baghouse filter
2. Electrostatic precipitator

In a baghouse precipitator, fly ash is filtered by tubular fabric bags through which the flue gases pass on their way to the smoke stack. The bags are made of Teflon-coated glass fibers, polyester, or acrylic cloth. Bags are used for high-lime coal to reduce stack emissions [Tikalsky, 1989]. In a baghouse, the collected layers of fly ash act as the filter medium—not the cloth itself. The baghouse filter method is used mainly in industrialized nations because of stringent air-pollution requirements. The fly ash collected in a baghouse is more homogeneous than that collected in an electrostatic precipitator. Baghouse filtering is a good method to remove fine particles (0.3 to 1 micron in size), whereas electrostatic precipitators tend to act as classifiers, which create inhomogeneity [Tikalsky, 1989].

Baghouse filtering is the more expensive and more efficient of the two processes. Bags need to be replaced with wear, and this also occasionally creates a disposal problem for the worn bags. In the U.S. approximately one-third of the fly ash is collected in baghouse filters, and the remaining two-thirds are collected by electrostatic precipitators.

Fly ash collected via electrostatic precipitators is often subsequently *scrubbed* from the system with water, which forms a slurry. Fly ash recovered this way has, to date, only rarely been used to make portland cement concrete. Usually, this fly ash is stabilized with lime, GGBF slag, or another binder.

Collected dry fly ash is stored in silos. It is often later transported to landfills or empty coal mineshafts. A portion is used as backfill, soil stabilizer, or mineral admixture for concrete. Fly ash is composed of the same bulk materials as portland cement with one exception. Fly ash, and especially the low-lime type, is deficient in the lime that is required to make the C-S-H and CAH. However, this lime can be introduced as slaked lime from an external source, and then the pozzolanic reaction can occur [Tikalsky, 1989].

Marketing Dilemma

Both portland cement and fly ash are produced from finely divided ground rock by burning at about the same temperature. Also, both consist essentially of the same oxides (SiO_2, Al_2O_3, CaO, and Fe_2O_3) although these are present in differing proportions [Dhir, 1986].

However, portland cement is marketed as a hydraulic cement, while fly ash has been considered a waste material. This has helped create a deeply entrenched bias against the use of fly ash as a cementitious component in concrete. This could be avoided by using a more appropriate name such as pozzolan, mineral admixture, or supplementary cementing material [Dhir, 1986].

Physical Properties of Fly Ash

The type of fly ash produced depends on the type of coal that is combusted. The different kinds of coal result from differences in the coal formation processes over geological time periods [Helmuth, 1987]. Generally, fly ash collected from the combustion of subbituminous low-sulfur lignite coals contains more lime (or calcium) and less iron than fly ash collected from the combustion of bituminous coal. The former also contains little unburned carbon. Fly ash is divided into two classes: high-lime fly ash and low-lime fly ash.

Low-lime fly ash contains less than 10% lime. It is usually produced from anthracite and bituminous coals. It roughly corresponds to ASTM class F fly ash [Dhir, 1986]. It possesses a true pozzolanic property and needs an activator to undergo pozzolanic reaction. Early uses of fly ash concrete were mainly with low-lime fly ash, when use of low sulfur coal was not required for environmental problems. Most of the data in the early literature concerns low-lime fly ash [Price, 1975].

High-lime fly ash contains more than 10% lime, and sometimes a considerable amount of it—as much as 50%; however, a typical amount is about 20%. It roughly corresponds to ASTM class C fly ash [Dhir, 1986]. This fly ash possesses some cementitious (self-hardening) properties in addition to its pozzolanic properties. This fly ash is usually produced from subbituminous and lignite coals. A great amount of low sulfur coal is available in the U.S. (approximately 1.45 trillion tons or 1.32×10^{15} kg). This kind of coal exists mainly in the eastern part of Montana, Wyoming, and the western Dakotas. The coal from these areas is being shipped to power plants in the midwestern states [Price, 1975].

High-lime fly ash is low in carbon, has a high fineness, and because of the smooth and spherical shape of its particles it can contribute to a desirable workability of a concrete mixture. It is light in color, which sometimes is a desirable feature. Unfortunately, occasionally some of the lime is free lime and slacks readily with generation of considerable heat. Furthermore, some high-lime fly ashes have MgO contents greater than 7%; though probably not in the form of periclase, it should be examined to be sure [Price, 1975]. A fly ash containing a high amount of lime should be investigated from the standpoint of adding to the raw feed at the cement plant, as such ash should reduce the amount of limestone required in the kiln and also lower the heat requirements to produce clinker. This is because a portion of lime would already be free of CO_2 [Price, 1975].

The particle size distributions of low-lime and high-lime fly ashes are given in Fig. 40.8 [from Mehta, 1983], alongside those of portland cement and silica fume. Note that high-lime fly ash contains more fines and less coarse particles than low-lime fly ash, probably because there is less carbon and less quartz in low-lime fly ash.

Due to the rapid rate of cooling of the flue gas in the furnace, fly ash particles are glassy in nature, although there are often some crystalline compounds. The composition of the glasses is dependent upon the composition of the coal and the temperature at which it is burnt. The major differences in fly ash glass composition lie in the amount of calcium present in the ash [Tikalsky, 1989]. However, the nature of the glass in the two fly ashes is different. Generally, the x-ray diffraction pattern (Cu $K\alpha$ radiation) shows a halo peak or diffuse band between 30° to 34° 2θ for high-lime fly ash. Low-lime fly ash shows the diffuse band between 21° to 25° 2θ. Figure 40.9 [Mehta, 1983] shows typical x-ray diffraction patterns of high- and low-lime fly ash, and for comparison, granulated blast-furnace slag. Note that the diffuse bands of slag and high-lime fly ash appear at similar locations. This is probably due to the high lime content of both materials. As shown in

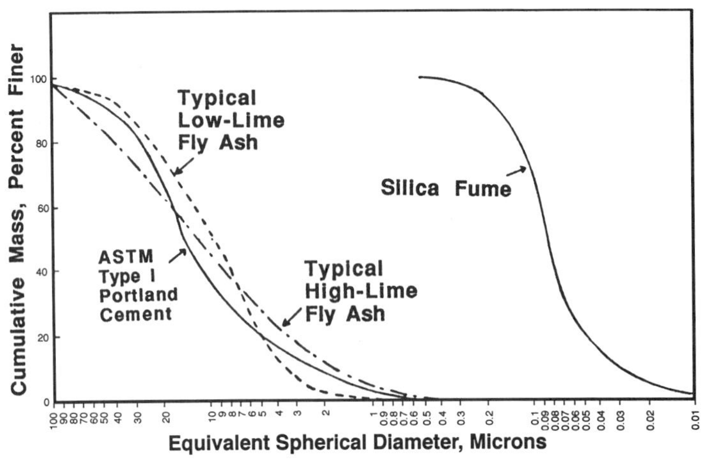

FIGURE 40.8 Particle size distribution of low-lime fly ash, high-lime fly ash, portland cement, and silica fume. (*Source:* Mehta, 1983.)

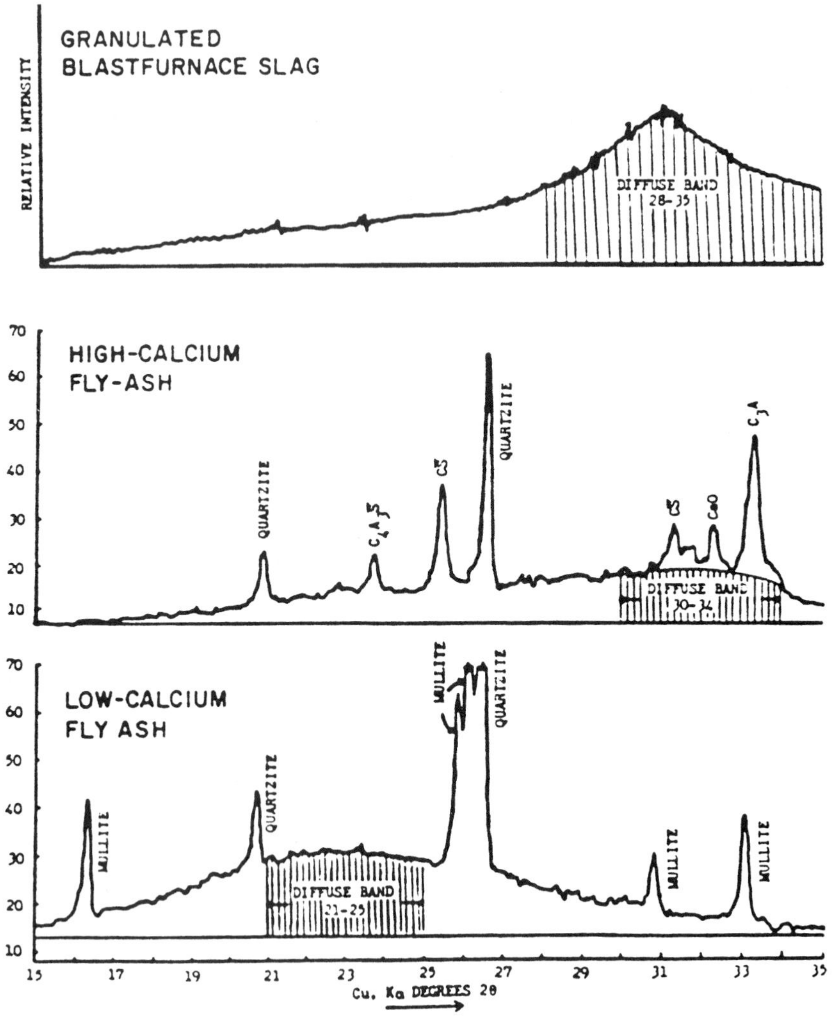

FIGURE 40.9 Typical x-ray diffraction patterns of granulated blast-furnace slag and fly ashes. (*Source:* Mehta, 1983.)

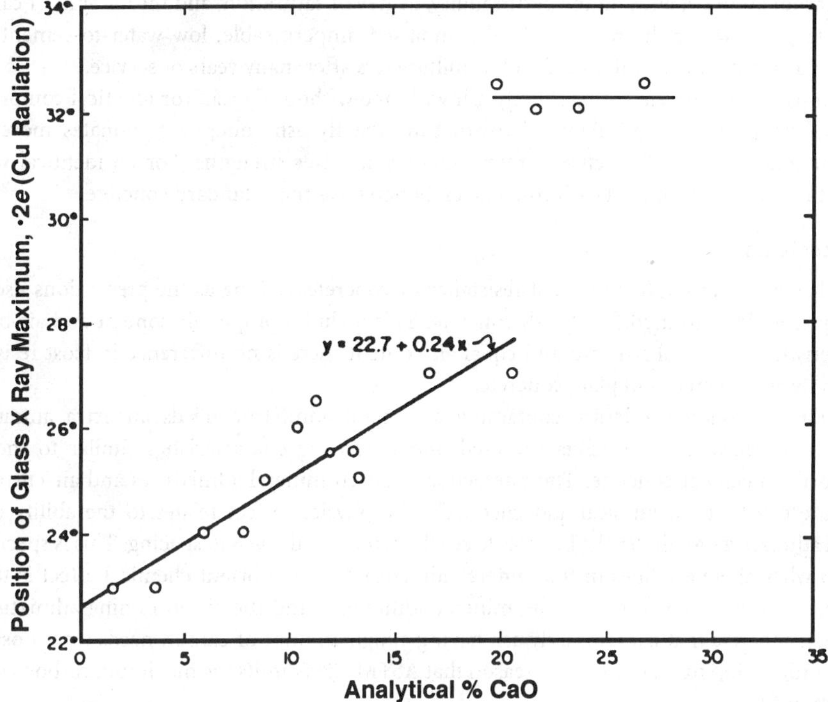

FIGURE 40.10 Position of the diffuse band and CaO content. (*Source:* Diamond, 1983.)

Fig. 40.10, [Diamond, 1983a], the position of the diffuse band of a glass varies approximately linearly with its CaO content until about 21% CaO content. Then the position becomes stable and remains constant at 32° 2θ.

Specifications

ASTM C618 provides the specification for high-lime (Class C) and low-lime (Class F) fly ashes. Classification of fly ash is based mainly on a minimum amount of the sum of the oxides $SiO_2 + Fe_2O_3 + Al_2O_3$. The minimum amount is 70.0% for low-lime fly ash and 50.0% for high-lime fly ash. The selection of these limits is arbitrary and has been questioned because many high-lime fly ash sources meet the requirements of low-lime, while all low-lime fly ashes meet the requirements of high-lime fly ash.

In practice, there does not appear to be a relationship between the sum of oxides in a given fly ash and the engineering performance of a concrete containing it. For instance, a fly ash containing a large amount of the oxides $SiO_2 + Fe_2O_3 + Al_2O_3$ does not necessarily perform better in concrete compared to a fly ash that contains a smaller amount [Mehta, 1987]. Suggestions have been made to introduce a parameter in the standard that defines the glass content of the fly ash.

Durability Properties

Fly ash has a significant effect on the durability characteristics of concrete. Some of these effects are highlighted below [Berry and Malhotra, 1986].

Depth of Carbonation

Carbonation attack occurs in moist conditions when carbon dioxide from the air attacks the calcium hydroxide of the concrete, and to a lesser extent the C-S-H hydrates, to form calcium carbonate. Carbonation attack can lead to an increase in permeability, shrinkage, and steel to corrosion. The rate of carbonation depends on the permeability, degree of saturation, and the amount of calcium hydroxide available for the reaction. Well-compacted, impermeable, low water-to-cement ratio concrete carbonates to a depth of only a few millimeters after many years of service.

In comparing a concrete containing fly ash with one without fly ash, for identical compressive strength, and probably equivalent microstructure, the fly ash concrete carbonates more than the plain concrete. The difference is more obvious at lower strengths. For an identical water–cementitious materials ratio, fly ash concrete carbonates less than standard concrete.

Frost Resistance

Fly ash has no adverse effect on frost resistance of concrete, as long as the precautions used for plain concrete are also used for fly ash concrete. This includes proper air content and air-bubble characteristics. For equal strength and equal air content there is no difference in frost resistance between fly ash concrete and plain concrete.

Generally, a concrete mixture containing a mineral admixture needs an extra amount of air-entraining admixture to achieve desired bubble-spacing characteristics similar to those of plain portland cement concrete. The interaction between mineral admixtures and air-entraining admixture can be both physical and chemical. The physical effect relates to the ability of the mineral admixtures to hinder the bubbles to reach their desired size and spacing. This is specifically more so when there are fines in the mineral admixture. An important chemical effect is that of the interaction between carbon in the mineral admixtures and the air-entraining admixture. In particular, concrete that contains a fly ash having a high amount of carbon needs extra dosage of air-entraining admixture. It is for this reason that ASTM C595 limits the maximum carbon content in fly ash to 6%.

The amount of air-entraining admixture necessary to achieve the desired spacing is a function of the concrete mixture proportions, in general, and the type and amount of mineral admixture, in particular.

There is no existing formula as to what the exact dosage of air-entraining admixture should be. The practical approach to establish the correct dosage is to do trial batches.

Sulfate Attack

The influence of fly ash on the resistance of concrete to sulfate attack is variable and depends on the type of fly ash. Dunstan's work in the 1970s and 1980s [1976; 1980; 1981] resulted in a method that quantifies the influence of fly ash on the resistance of concrete to sulfate attack. The parameter R, referred to as Dunstan resistance factor, was defined as:

$$R = (\mathrm{C} - 5)/\mathrm{F} \tag{40.4}$$

where C = percent CaO in fly ash, and F = percent Fe_2O_3 in fly ash. According to Equation (40.4), as the value of R increases for a fly ash, the resistance of the fly ash concrete to sulfate attack decreases.

Figures 40.11 and 40.12 [Dunstan, 1980] show the effects of the R-value of fly ash on the expansion of concrete subjected to sulfate attack. Generally, a lower R-value for a fly ash (i.e., low-lime fly ash) is associated with a lower concrete expansion, as shown in Fig. 40.11. However, this is not a rule and there are exceptions. A higher R-value (i.e., high-lime fly ash) is also generally associated with a higher expansion of concrete containing that fly ash, Fig. 40.12, but exceptions are to be expected as well. In general, high-lime fly ash, by virtue of its higher R-value (compared to low-lime fly ash), results in a higher concrete expansion.

The findings of Dunstan are summarized in Table 40.4. The suggested limits of R-values (at 25% cement replacement) and the expected performance of the fly ash concrete under sulfate attack (for ASTM Type II cement at 0.45 water-to-cement ratio) are listed. The Bureau of Reclamation uses a more detailed table for the selection of appropriate fly ash [Pierce, 1982].

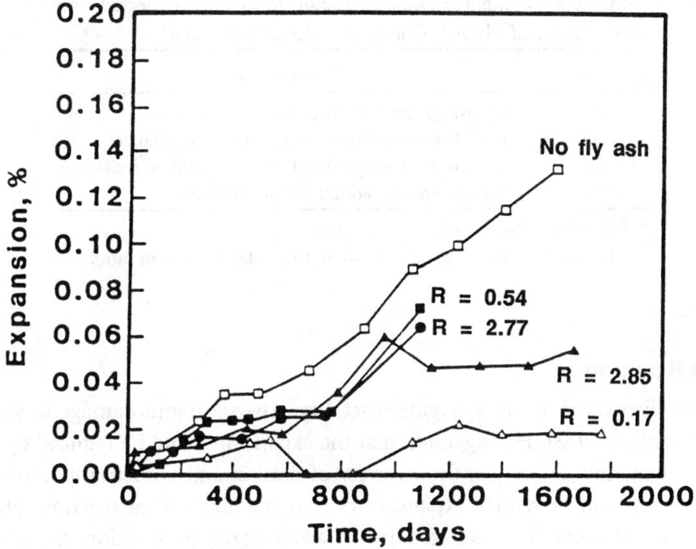

FIGURE 40.11 Expansion due to sulfate attack on a concrete containing low-lime fly ashes with varying R-values. (*Source:* Dunstan, 1980.)

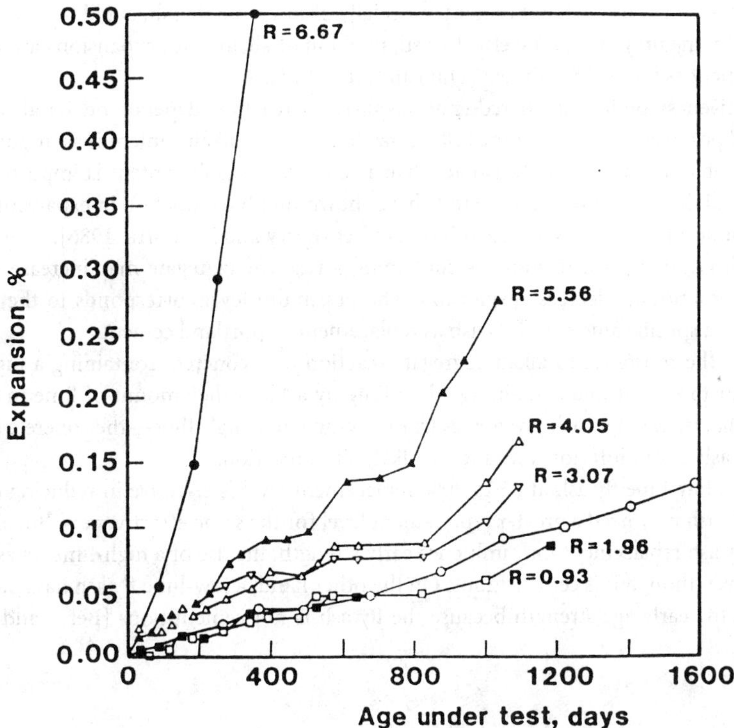

FIGURE 40.12 Expansion due to sulfate attack on a concrete containing high-lime fly ashes with varying R-values. (*Source:* Dunstan, 1980.)

Table 40.4 Limits for Dunstan Factor, R, and the Expected Performance of Fly Ash Concrete under Sulfate Attack

R Limits[1]	Sulfate Resistance[2]
<0.75	Fly ash greatly improves sulfate resistance
0.75–1.5	Fly ash moderately improves sulfate resistance
1.5–3.0	Fly ash has no significant effect on sulfate resistance
>3.0	Fly ash greatly reduces sulfate resistance

[1] At 25% replacement level for cement.
[2] Relative to ASTM Type II cement at 0.45 water-to-cement ratio.

Alkali-Silica Reaction

Stanton in 1940 discovered alkali-aggregate reactivity expansion and damage to some concretes in California [Stanton, 1942]. He suggested that the expansion could be reduced by adding finely ground reactive materials. Subsequently, a variety of natural and industrial pozzolans, including fly ash, have been found to reduce expansion due to the alkali-silica reaction. However, these admixtures are not effective in a second form of alkali-aggregate reaction, the alkali-carbonate reaction [Berry and Malhotra, 1986].

Pepper and Mather [1959] showed that the minimum content of fly ash (as percent by weight of cement) that is needed to reduce expansion due to alkali-silica reaction by 75% is about 30 to 40%. Questions have been raised as to whether the early strength losses caused by such high amounts of fly ash replacement of portland cement, especially of low-lime fly ash, could be tolerable. It has been shown in one project that the effective suppression of alkali-silica expansion can be achieved by a replacement as low as 25% fly ash [Duncan *et al.*, 1973].

The effectiveness of fly ash in reducing alkali-silica reaction depends on its alkali content, fineness, and pozzolanic activity. Some tests show that the total alkali content is as important as the available (soluble) alkalis. Since fly ash acts like a cement, its alkali content is important. Fly ash acts as an alkali diluter. However, some tests have shown that the reduction in expansion is greater than could be accounted for by simple dilution effect [Berry and Malhotra, 1986].

Small addition of fly ash to mortars containing a reactive aggregate may increase expansion, whereas larger amounts decrease expansion. The pessimum level corresponds to the maximum expansion for a specific amount of fly ash as replacement of portland cement.

In general, the resistance to alkali-aggregate reaction of a concrete containing a high-lime fly ash is inferior to that of one containing a low-lime fly ash. As the amount of lime in the fly ash increases—that is, as the fly ash progresses from low-lime to a high-lime—the concrete containing the given fly ash has an inferior resistance to alkali-silica reaction.

In practice, low-lime fly ash at 25 to 30% replacement level is effective in reducing expansion. High-lime fly ash may need a greater replacement level for the same effectiveness. But since a high volume of fly ash replacement may influence early strength, the use of a high-lime fly ash may not be suitable even though it is cementitious. On the other hand, a low-lime fly ash has a larger effect on reducing the early age strength because the fly ash is not cementitious [Berry and Malhotra, 1986].

Corrosion

Fly ash improves concrete's resistance to corrosion in three ways [Berry and Malhotra, 1986]: (1) It provides an alkaline medium in the immediate vicinity of the steel surface; (2) it offers a physical and chemical barrier to the ingress of moisture, oxygen, carbon dioxide, chlorides, and other aggressive agents; and (3) it provides a relatively electrically resistive medium around the steel members.

Fresh Properties

Fly ash particles have complex interaction within fresh concrete thereby affecting the rheological properties as follows [Dhir, 1986]:

Fly ash improves the workability of fresh concrete. This is due to a number of causes. Fly ash, or any other ultra-fine powder, can physically disperse cement flocs, thereby freeing more paste to lubricate aggregates and to improve workability. This is analogous to adding a chemical plasticizing agent. There is an increase in packing efficiency of the solids.

Fly ash particles also act as ball bearings because of their spherical shape. This leads to improvement of workability. In addition, each spherical particle of fly ash will carry with it an associated surface film of water because of its geometry and hydrophilic nature. Thus, there is a water entrainment that ensures uniform dispersion of water, and at fine loci where water is needed. Also, bleeding is reduced because of this mechanism.

Since fly ash has a lower density than cement, there is an increased paste volume for a given weight of composite. However, concrete designed for equal "mass" of paste is generally less workable than concrete designed for equal "volume" of paste, negating the above theory.

With the increase in packing efficiency and dispersion of water, there is increased tortuosity of flow channels. This, therefore, restricts movement of free water in plastic concrete and reduces bleeding.

The presence of fly ash in fresh concrete often leads to improved mobility and pumping characteristics of the concrete; to more homogeneous, dense, and stable microstructure; and to reduced rate of evaporation of mixed water (i.e., smaller water channels make the evaporation of water more difficult). Last, but not least, for equal slump, fly ash concrete can be handled more easily than concrete without fly ash.

Mechanical Properties

Elastic Properties

For practical purposes, the Young's moduli and Poisson's ratio of plain and fly ash concretes are similar [Berry and Malhotra, 1986]. For the same compressive strength, however, the Young's modulus of fly ash concrete is higher than the Young's modulus of plain concrete. For the same Young's modulus, the compressive strength of fly ash concrete is less than that of plain concrete, as expressed by the well-used ACI formula for plain concrete shown in Fig. 40.13 [Ghosh and Timusk, 1981].

Load-Independent Volume Changes

The use of fly ash in normal proportions does not significantly influence the drying shrinkage of concrete.

Thermal Expansion

The thermal expansion coefficient of concrete is affected by the presence of fly ash. A 40% fly ash level reduces the coefficient by 4%. The values reported for paste are 11.0 to 20.0 me/C, and for concrete containing limestone and quartzite 4.0 to 11.7 me/C [Gifford and Ward, 1982].

Mixture Proportioning

The proportioning procedures for concretes incorporating fly ash are similar to those used for other concretes, ACI 211. The method described by ACI 211 is based on a simple replacement method.

There exist several proportioning alternatives when fly ash is used as a mineral admixture and not a manufacturer-prepared cement blend. These are listed and discussed below using mainly these sources: Olek [1987], Olek and Diamond [1989], and Berry and Malhotra [1986].

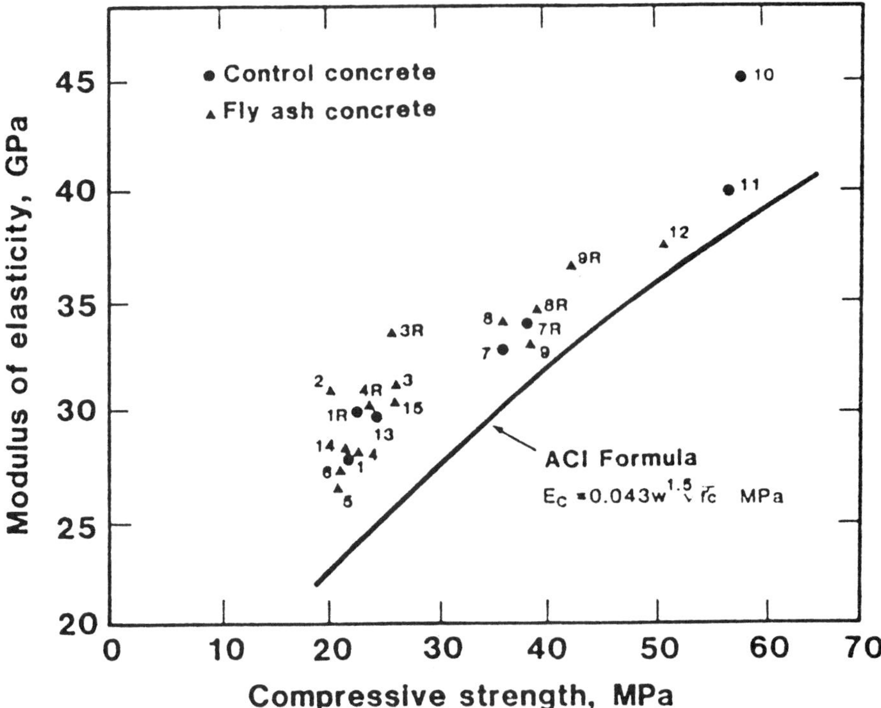

FIGURE 40.13 Modulus of elasticity versus compressive strength of fly ash concrete as compared to plain portland cement concrete according to the ACI formula. (*Source:* Ghosh and Timusky, 1981.)

The simple replacement method involves partial replacement of cement by fly ash. It can be done by the absolute volume basis or by the weight basis. In the absolute volume basis the portland cement concrete mix is designed for a desired compressive strength and slump. The mix is then altered by replacing a fixed amount of cement by an equivalent volume of fly ash, on an absolute basis. Also water/cement ratio changes to water/cement + fly ash ratio. Note that the absolute volume of cement + fly ash remains the same as that of the original cement, and adjustment is made in calculating water/cement + fly ash by weight. This method reduces the total weight of cementitious materials since the specific gravity of fly ash is normally less than that of the cement. Water content must be adjusted.

Figure 40.14 shows a comparison of the relative rates of strength increase of plain concrete and fly ash concrete proportioned by the simple replacement of portland cement. Generally, for replacements of 30% (or less) fly ash, the later-age strength (usually after 28 days curing) is greater than that of reference concrete.

For some concrete containing high amounts of fly ash, even after ten years the strength of fly ash concrete does not equal that of plain concrete. Therefore, the main difficulty of this proportioning method is the low early strength and the low amount of fly ash that can be tolerated. In the literature most of the data are for low-lime fly ash concrete. For high-lime fly ash concrete the situation is different, because the fly ash is cementitious as well as pozzolanic.

In the weight basis method the composition of plain concrete is adjusted by replacing a given weight of cement by an equal weight of fly ash. Since the specific gravity of fly ash is lower than that of portland cement, the absolute volume of cement + fly ash is greater than the absolute volume of the cement alone. In this method the increased volume (or yield) is compensated by reduction in volume of aggregate used. Water must also be adjusted.

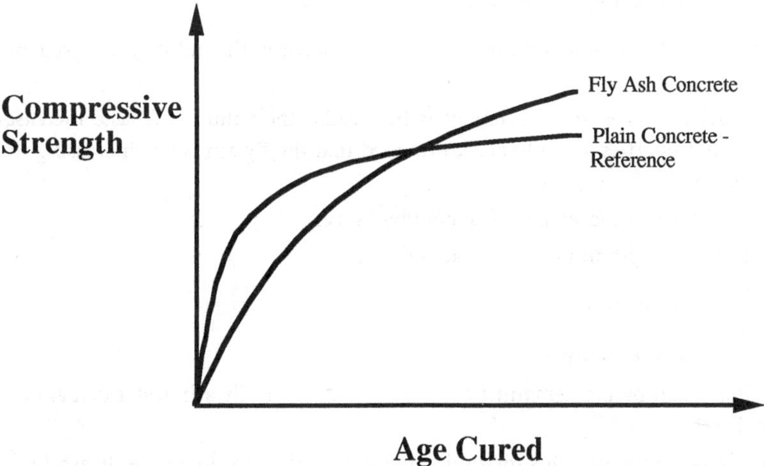

FIGURE 40.14 Comparison of the relative rates of strength increase of plain concrete (reference) and fly ash concrete proportioned by simple replacement of portland cement. Generally, for replacements of 30% (or less) fly ash, the later-age strength (usually after 28 days curing) is greater than that of reference concrete.

In most cases reported for weight replacement (for low-lime fly ash), the early age strength was higher but still not as high as that of plain concrete. But by adding water reducers, the mix design can be modified further to obtain equal strengths even at 1 day of curing with 30% replacement of cement by weight with low-lime fly ash.

The cement replacements method seems to work better with high-lime fly ash than with low-lime fly ash, since high-lime fly ash is cementitious and can contribute to strength at an earlier time. The general trend is illustrated in Fig. 40.15.

A simple replacement method was used for production of high-strength concrete, meeting target strength at 28 days and significant increases in strength at 56 and 180 days. There is a recommendation to move the specifications to accept criteria to 56 and 90 days.

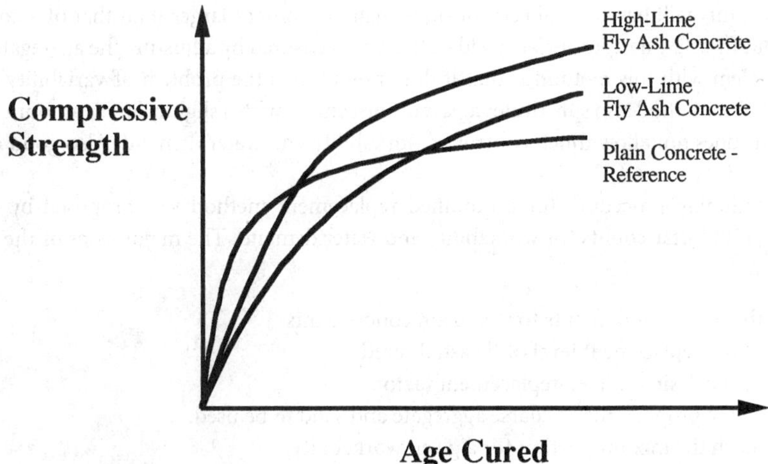

FIGURE 40.15 Comparison of the relative rates of strength increase of plain concrete (reference), high-lime fly ash concrete, and low-lime fly ash concrete by simple replacement of portland cement.

The advantages of the simple replacement methods are:

- Straightforwardness—quantitative relationships between the individual ingredients stay the same.
- When partial replacement of cement is by weight, the volume of paste increases and the rheological properties are enhanced, provided that the fly ash is not increasing the demand for water.
- Lower cost due to replacement of cement by fly ash.
- Potential for strength improvement at later age.

The drawbacks of this method are:

- There is a large variability.
- Many important parameters must be considered: percent fly ash, water demand, and nature of the fly ash.
- Low early-age strength—this method is not appropriate for low-lime fly ash (although it is used frequently).

The addition method requires addition of fly ash to a concrete mixture without removing cement. This method results in a substantial increase in the cementitious content of the mix, because the adjustment is made in the aggregate content. The main advantages of this method are:

- Improved workability.
- Reduced bleeding in lean concrete.
- Useful for low strength or superplasticized concretes.
- Useful for making concrete with good resistance to sulfate attack.
- Improvement in the early-age strength, when compared with the simple replacement method.

The modified replacement method considers the slow rate of early-age strength gains. The method requires that the weight of fly ash put into a mix be greater than the weight of cement removed. Adjustment in the fine aggregate content is made, just as in simple replacement methods. This method requires that the fly ash be used as a cement replacement and a sand replacement.

The procedures developed for this method overcame the problem of slow early-age strength gain of fly ash concrete, a problem associated with the simple replacement methods. In all cases, the fly ash concrete mix will have a total cementitious material content larger than that of a comparable mix without fly ash. The appropriate yield of the mix is ensured by adjusting the aggregate content.

The problem with this method is that it does not address the problem of variability of fly ash properties. The method also substitutes a part of the cement with a slightly larger amount of a given fly ash. This does not allow much control of workability and water demand. Also good durability is not ensured.

A proportioning procedure for a modified replacement method was proposed by Olek and Diamond [1989]. It accounts for workability and water demand. The major steps of the proposed procedure are:

1. Set the basic requirements for the plain concrete mix.
2. Select the replacement level of fly ash desired.
3. Select the desired under-replacement factor.
4. Select the proportions of coarse aggregate and sand to be used.
5. Establish the mix proportion for a given workability
6. Prepare a trial batch (measure unit weight and calculate yield).

This rational method considers the potential advantages and versatility of fly ash, which the other methods do not. This method allows a more efficient use of the fly ash. The method quantifies the influence of fly ash on various concrete properties and utilizes that quantification at the design stage. One such technique is the cement efficiency factor, which allows treating fly ash as a low-grade cement and permits its influence on strength to be quantified. Unfortunately, the problem with this method is the difficulty in determining the value of the factor.

40.4 Silica Fume Concrete

Production

During the past decade significant attention has been invested in another pozzolan—silica fume (SF). The advantages of concrete containing silica fume over a plain portland cement concrete are mainly in its higher strengths and lower permeabilities. Under practical conditions it can attain strengths of approximately 14,000 psi (96 MPa)—about 2 to 3 times the strength of portland cement concrete. This means a significant reduction in weight and size of a structure can be achieved. The lower permeability can also make it a very durable material.

Silica fume has been deemed one of a new generation of construction materials. Structures ranging from the Tjorn Bridge in Sweden to the Kinzua Dam in the U.S. have been built or repaired using this material. Silica fume concrete will continue to play a significant and important role in the civil engineering profession.

It is reported [Sellevold and Nilsen, 1987] that the first tests of silica fume in concrete were made at the laboratories of the Norwegian Institute of Technology in 1950. In 1952, silica fume was used in the field in the Blindtarmen concrete tunnel in Oslo. This tunnel is exposed to acidic and high-sulfate water in the Oslo alum-shale region [Sellevold and Nilsen, 1987; Buck and Burkes, 1981]. The first documented use of silica fume in structural concrete took place in Norway in 1971 [Sellevold and Nilsen, 1987].

With the start of large-scale use of filters in the mid-1970s for environmental regulation and protection purposes, the use of silica fume in concrete spread to other countries such as Denmark, Sweden, and Iceland. Reports of its use appeared in the United States and Canada in the late 1970s and early 1980s [Sellevold and Nilsen, 1987; Mehta and Gjorv, 1982; Malhotra and Carette, 1983; Sheetz *et al.*, 1981]. By then, in Canada, silica fume was being used in ready-mixed concrete for commercial applications.

Numerous other names for silica fume can be found in the literature. These include: condensed silica fume, microsilica, ferrosilicon dust, arc-furnace silica, silica flue dust, silica dust, silica flour, dust, amorphous silica, volatized silica, and very-fine-grained siliceous fly ash [Sellevold and Nilsen, 1987; Elkem Chemicals, 1985; Luther and Smith, 1991]. Silica fume, condensed silica fume, and microsilica seem to be the most commonly used names. The majority refer to this material simply as silica fume.

Silica fume is a by-product of the ferrosilicon industry, where alloys are made in a high-temperature (approximately 2000°C) submerged electric-arc furnace. Silica fume is the dust collected from these furnace exhaust systems.

Ferrosilicon alloys commonly are produced with nominal silicon contents of 50, 75, and 90%. When the silicon content reaches 98% in the alloy, the alloy is called *silicon metal.* As the silicon content increases in the alloy, the SiO_2 content will increase in the silica fume. Silica fume from grades 75% and higher has been sold in the United States for use in concrete.

During the production of silica fume, the ferrosilicon is drawn off as a liquid from the bottom of the furnace. A vapor rises from the furnace and quickly oxidizes. As it cools, particles are condensed as spheres and trapped in bag filters. The silica fume is usually further processed to remove impurities [Sellevold and Nilsen, 1987; Mehta, 1983; Luther, 1989a]. Processing the initially collected silica fume by densifying (compacting), slurrifying, or bagging it makes the material more easily handled and shipped.

Silica fume is also collected as a by-product in the production of other silicon alloys, such as ferrochrom-silicon, silicon-manganese, ferro-silicon-magnesium, and calcium-silicon. The amount of silica fume produced from one ton of silicon or ferro-silicon alloy varies with the alloy type. There are few published data available on the properties of these silica fumes except that silica fume from ferrochromium has properties somewhat similar to those obtained from ferrosilicon [Aitcin *et al.*, 1980]. In contrast, ferromanganese fume, with a reported SiO_2 content of only 0.04%, not surprisingly contributes no strength-enhancing properties [Luther and Kojundic, 1991], although its high specific gravity, chemistry, and color make it valuable in grouts and fluxes. The use of a specific silica fume should be avoided unless data on its performance in concrete are available [ACI Comm. 234, 1994].

Often, the production of one ton of silicon generates about 0.60 tons of silica fume, whereas one ton of FeSi 75% alloy generates from 0.20 to 0.45 tons [Malhotra *et al.*, 1987; Killin, 1973].

Physical and Chemical Properties

Silica fume is a mineral admixture composed mainly of very fine amorphous silicon dioxide (SiO_2), which is present as individual spheres with a number of agglomerates. Most of the silica fume particles are smaller than 1 micron in diameter, and the majority range in size from 0.01 to 0.3 micron with an average diameter of about 0.1 micron. Silica fume particles are 50 to 100 times smaller than cement or fly ash particles. Surface areas range from 1700 m^2/kg to 2500 m^2/kg although 2000 m^2/kg seems to be the most common figure associated with silica fume used in concrete. The surface area of silica fume is much larger than that of plain portland cement and fly ash (typically 300 to 600 m^2/kg). The specific gravity of silica fume is usually 2.2, lower than portland cement (3.10–3.23) and fly ash (2.4–2.7).

Typical oxide analyses of commercial-grade silica fume, portland cement, and low and high-lime fly ashes are presented in Table 40.5.

SiO_2 by itself is clear. The color of silica fume is determined by the non-silica components, which typically include carbon and iron oxide. Usually the color of silica fume is light gray to dark gray. The more the carbon content, the darker is the gray color. The carbon comes from the burning of coal and wood, and its amount depends on the burning conditions and type of alloy being produced. Depending on the heat recovery system, some silica fume can be white in color. The color can be brown if the alloy contains manganese [Luther and Kojundic, 1991].

Table 40.5 Oxide Composition (Mass Percent) of a Commercial-Grade Silica Fume, Portland Cement, and High-Lime and Low-Lime Fly Ash[1]

Oxide Composition	Silica Fume[2]	Portland Cement[3]	Fly Ash[4]	
			Low Lime	High Lime
SiO_2	93.7	20.80	42.9	32.6
Al_2O_3	0.3	5.28	26.1	15.2
Fe_2O_3	0.8	2.13	21.1	10.3
CaO	0.2	64.42	1.72	31.0
SO_3		3.14	1.00	5.54
MgO	0.2	1.50	5.23	3.63
Na_2O	0.2	0.08	0.27	0.74
K_2O	0.5	0.64	2.60	0.95
Carbon	2.6		1.4	9.30

[1] Values are for specific products. Results are generally variable, specifically for fly ashes. (Compare with general values in Table 40.1)

[2] North American silica fume [Malhotra and Carette, 1983].

[3] ASTM Type I p.c., Lone Star Industries, Inc. Greencastle, IN.

[4] Indiana fly ashes [Kilgour, 1988].

Mechanism of Hydration and Strength Development

Silica fume can operate in concrete in two ways: as a pozzolan and as a filler (other terms used instead of filler are *packing* and *microfiller*).

As a pozzolan, silica fume reacts with calcium hydroxide that is liberated during cement hydration. The two-step pozzolanic reaction is expressed in Eqs. (40.1) and (40.2), presented earlier. The pozzolanic reaction between silica fume and calcium hydroxide improves the strength and reduces the permeability of the concrete.

As a filler, silica fume transforms large concrete pores into finer ones, both in the matrix and at the interface of cement paste and inclusion—that is, aggregate—fiber (Figs. 40.2 and 40.3). Silica fume does not, as commonly perceived by users, increase the density of the concrete. Hence the term *densify*, which is sometimes used in the literature, may be misleading. Regarding the effect on total porosity, silica fume has no significant effect on the total porosity of the matrix, but it does alter the pore size distribution by shifting the pores to a range of finer sizes. A reduction in the size of threshold diameter is also a characteristic of the influence of silica fume (see Fig. 40.1 for illustration).

There are questions as to whether the pozzolanic reaction or the filler effect makes the greater contribution to strength and durability improvements. It is now generally accepted that at early ages (up to 3 days or so) the filler effect is more important to strength contribution than the pozzolanic reaction. At later ages the effect of the pozzolanic reaction becomes important and contributes to strength [Cohen and Klitsikas, 1986; Goldman and Bentur, 1993; Diamond, 1984; Sellevold, 1987]. The exact details of strength improvement are still not understood, and more research is needed.

Transition Zone vs. Bulk Paste Modification

There is also controversy as to whether strength increase is due to (a) the ability of silica fume to modify the microstructure of the cement paste-aggregate interface (transition zone)—transition zone modification, or (b) the ability of silica fume to enhance the strength of the bulk paste, that is, bulk paste modification [Darwin and Slate, 1970; Cong *et al.*, 1992].

In a recent paper [Cohen *et al.*, 1994], an experimental program was introduced which was aimed at separating these two mechanisms. This was attempted by testing the dynamic Young's moduli of mortar specimens with various aggregate size (or aggregate surface area). Other parameters such as water-to-cementitious materials ratio and paste and aggregate content were kept constant. By varying the aggregate surface area one could isolate and quantify the influence of the transition zone on the mechanical properties of the mortar. Consequently, one could separate the effect of silica fume on the transition zone from that on the bulk paste.

Figure 40.16 [Cohen *et al.*, 1994] shows the relationship between dynamic E and aggregate surface area. In the portland cement mortars, as the aggregate surface area increases, E decreases. This indicates that an increase in the volume of the transition zone leads to a reduction in E. The drop in E is monotonic; that is, initially the drop occurs at a high rate, and subsequently the rate of drop slows down. In the silica fume mortars there is no major decrease in E with increase in aggregate surface area.

The lack of sensitivity of silica fume mortar to aggregate surface area is due to the ability of silica fume to modify the cement paste-aggregate interfaces or transition zones. This observation, plus the sensitivity of portland cement mortar to aggregate surface area confirms the significance of the transition and its weakening effect in portland cement mortar. This implies that the effects of silica fume on the transition zone, (i.e., transition zone modification) are important. The improved performance due to silica fume addition is evident as early as one day and continues at later ages.

The effect of silica fume on the E values of silica fume mortars as early as one day can be mainly due to the filler effect [Goldman and Bentur, 1993] of the silica fume. This effect can remain significant at later ages.

In the series containing high-surface-area aggregate, silica fume mortar (mix 3) had a higher E than the corresponding portland cement mortar (mix 8). This is an important observation.

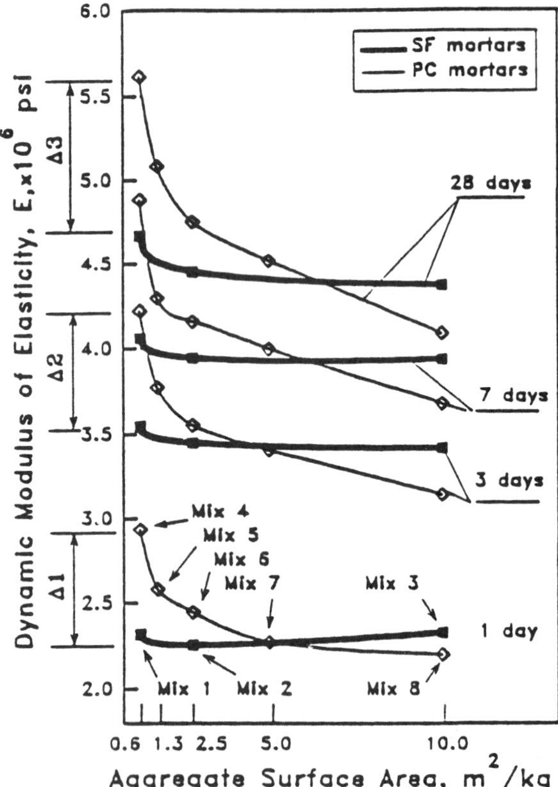

FIGURE 40.16 Relationship between dynamic Young's modulus and the aggregate surface area (exposed) for plain portland cement mortar and portland cement mortar containing silica fume cured at different ages. (*Source:* Cohen *et al.*, 1994.)

The presence of high-surface-area aggregate, such as in sandy concretes (typical of ready-mixed concretes in the U.S.), indicates the overwhelming significance of transition zones in such concretes. In fact, Diamond *et al.* [1982] have shown that there is little amount, if any, of bulk paste in ready-mixed concrete and that most of the paste is transition zone. Because of the large amount of transition zone paste in the mortar portion of ready-mixed concrete, the advantages of adding silica fume to it are clearly demonstrated.

Specifications

ASTM C1240, "Silica Fume for Use in Hydraulic-Cement Concrete and Mortar," was adopted in 1993. The mandatory chemical requirements include a minimum SiO_2 content of 85.0%, a maximum moisture content of 3.0%, and a maximum loss-on-ignition of 6.0%. The Canadian specification, CAN/CAS—A23.5-M86, has the same chemical requirements as ASTM C1240 in its limitations for SiO_2 content (minimum 85.0%) and loss-on-ignition (maximum 6.0%). However, the Canadian standard does not set a limit on the moisture content, but instead, it sets a 1.0% maximum limit on the SO_3 content. ASTM C1240 addresses the SO_3 content indirectly by establishing a maximum sulfate expansion as part of its optional physical requirements. The maximum expansion limits in ASTM C1240 are: 0.10% in six months for moderate sulfate resistance, 0.05% in six months for high sulfate resistance, and 0.05% in one year for very high sulfate resistance. The other physical requirements of ASTM C1240 and CAN/CAS—A23.5-M86 are nearly identical.

Fresh Properties

Used as an admixture, silica fume can improve the properties of both fresh and hardened concrete. Used as a partial replacement for cement, it can provide significant savings without degrading the concrete quality, provided it costs less than portland cement.

Due to its very high surface area, silica fume concrete requires an increased amount of water for a given slump. For concrete of normal cement content, HRWRA should always be used with silica fume to achieve normal consistency. Otherwise, the mixture will have a poor workability. Silica fume usually benefits concrete only when an HRWRA is used.

Considering fresh concrete, the mix is usually cohesive and there is no risk of segregation during handling of the concrete, and a desired finish can usually be achieved. There is little or no bleeding, so under hot, arid, and windy conditions, protective measures must be taken to guard against plastic shrinkage cracking in floor and pavement slabs. It is, therefore essential to finish the concrete as soon as possible after it has been placed and to protect the surface from drying [*Indian Concrete Journal*, 1985; Luther and Smith, 1991].

For reasons that are not clear, silica fume concrete can experience slump loss. No clear explanation for this phenomenon has been found. Continuous addition of HRWRA may not be a good way to solve this problem. However, other studies aimed at quantifying the slump retention of silica fume concrete and portland cement concrete showed similar slump loss [Luther and Smith, 1991]. More research is needed in this area.

Since there is already an abundance of fines supplied by the silica fume, a sand with a coarse grading may be preferred [Bentur *et al.*, 1987; *Indian Concrete Journal*, 1985].

For high silica fume dosage mixtures, slump is a less reliable indicator of consistency than with portland cement concrete. Cohesion due to silica fume particles can have an influence on slump. For the majority of silica fume concrete, however, slump may usually be 1 to 1.5 inches [Luther and Smith, 1991]. These effects may be attributed to the cohesion of silica fume concrete.

Mechanical Properties

Generally, the available research data point to an improvement of strength and durability properties when silica fume is added to the mix. The beneficial influence becomes apparent from about 3 to 28 days after mixing, casting, and curing, at normal temperatures [ACI Comm. 226, 1987a]. The improvements due to silica fume addition are increased compressive strength, reduced permeability, and improved resistance to alkali-silica reaction.

Compressive strengths of about 7200 psi at 28 days have been reported for silica fume concrete [Malhotra and Carette, 1983; ACI Comm. 226, 1987a]. This can be compared to about 5500 psi for normal concrete. A compressive strength of 31,000 psi was reported as early as 1981 [Bache, 1981; ACI Comm. 226, 1987a; Wolsiefer, 1984]. Laboratory strengths of silica fume concrete made with portland cement reached 38,000 psi by 1990, and the strengths of calcium aluminate–silica fume concretes were approaching 70,000 psi [Luther, 1990a]. At this time it is unlikely that such formulations could be used in the field and in ready-mixed concrete owing to economic and technological limitations. Today's main range for strength is 13,000 to 14,000 psi. Young's modulus at 28 days is in the order of 6×10^6 psi, and Poisson's ratio of about 0.21.

Results on creep and drying shrinkage vary. Generally, for small amounts of silica fume replacement (say 10%) creep and shrinkage of the concrete are basically the same as those of plain concrete. Although, theoretically, it is to be expected that for two concretes containing equal amounts of paste, the one containing the silica fume would creep and shrink more, results do not clearly indicate this [Luther and Hansen, 1989]. One reason creep and shrinkage in high-strength silica fume concrete may be lower than anticipated may be that full hydration of portland cement does not occur. This means there is less overall hydration product (although what has hydrated is more efficiently utilized), and the unhydrated interior portions of the cement grains act as aggregate and restrain both creep and shrinkage.

According to ACI Committee 234 draft report [1994], due to the limited published data on the subject and because of the different nature of the creep tests used in various investigations, no specific conclusions can be drawn as to the creep of silica fume concrete. However, it is stated that creep of silica fume concrete is no higher than that of concrete without silica fume and of equal strength. The report also indicates that the drying shrinkage of silica fume concrete (after 28 days of moist curing) is generally comparable to that of plain concrete, regardless of the water–to–cementitious materials ratio.

Durability Properties

Permeability

Reduced permeability is attributed to the filler effect (especially at early ages) that silica fume particles have in concrete and to the pozzolanic reaction (especially at later ages). This is evident in reduced rates of water evaporation in oven drying [Bentur *et al.*, 1987] and of chloride impermeability [Cohen and Olek, 1989]. Mercury intrusion porosimetry (MIP) tests provide further confirmation, as illustrated in Fig. 40.1. The reduction in the value of the threshold diameter has a significant contribution to the reduction in the permeability. Another general observation in the MIP curves of portland cement paste and silica fume paste is that at the low end of the pore size regime the curve associated with silica fume has a slope that is higher than that of portland cement paste. This perhaps indicates the presence of finer pores in the silica fume paste. Mehta and Gjorv [1982] showed that there is a significant reduction of about one order of magnitude in the volume of pores larger than 1000 when silica fume is added. Such large pores made up 6% of the total pore volume in the silica fume pastes and 60% of the volume of pastes without silica fume.

Furthermore, tests conducted at Purdue University showed a significant reduction in chloride permeability using the AASHTO T277 (American Association of State Highway and Transportation Officials) 6-hour rapid chloride permeability test [Cohen and Olek, 1989]. For example, a normal concrete specimen (that is, without silica fume), containing 625 lb/yd^3 cement and 220 lb/yd^3 water for a water–to–cementitious material ratio of 0.35, gave a reading of approximately 1800 coulombs, while the silica fume concrete containing 583 lb/yd^3 portland cement, 65 lb/yd^3 silica fume, and 228 lb/yd^3 water for a water–to–portland cement+silica fume ratio of 0.35 and a silica fume–to–silica fume+portland cement ratio of 0.10 showed only 300 coulombs. As the amount of silica fume increased the charge passed decreased. Table 40.6 [from Wolsiefer, unpublished] shows chloride ion permeability values to be expected for specific types of concrete. Research reported by Luther [1990a] showed strong chloride screening improvements (salt ponding-type test) attributed

Table 40.6 Rapid Chloride Ion Permeability Table

Charge Passed (Coulombs)	Chloride Permeability	Typical of:
4,000 (greater than)	High	High water-cement ratio (0.6) conventional PCC
2,000 to 4,000	Moderate	Moderate water-cement ratio (0.4–0.5) conventional PCC
1,000 to 2,000	Low	Low water-cement ratio (0.4) conventional PCC
100 to 1,000	Very low	Latex-modified concrete, silica fume concrete (5 to 15%), and internally sealed concrete
100 (less than)	Negligible	Polymer-impregnated concrete, polymer concrete, and high–silica fume percentage concrete (15 to 20%)

Source: Wolsiefer, unpublished.

to silica fume dosage as low as 4.5%. Increasing both silica fume dose and concrete maturity were also shown to reduce the electrical charge passed. Further data is needed to compare results between the rapid-chloride-ion permeability test and the ponding test.

Steel Corrosion, Concrete Carbonation, and Electrical Resistivity

Silica fume reduces the rate of carbonation of concrete. Silica fume also increases the electrical resistivity of concrete significantly. For example, for a plain concrete the electrical resistivity was 4200 ohms-cm, while for a concrete containing 11% silica fume in the cement the resistivity was 70,000 ohms-cm, and for 20% silica fume it was 110,000 ohms-cm. Higher values of electrical resistivity indicate higher resistance to corrosion. Since the process of corrosion and the rate of corrosion of steel depend on the rate of oxygen transport through the concrete and on its electrical resistance, silica fume can be incorporated in concrete without concern for the corrosion of the steel [ACI Comm. 234, 1994]. Field observation has confirmed this conclusion [Luther, 1991].

Alkali-Silica Reaction

Silica fume, like most fly ashes and natural pozzolans, can be used to reduce or prevent the deleterious expansion caused by alkali-silica reaction in concrete [ACI Comm. 226, 1987a; Idorn, 1983]. Silica fume binds the sodium and potassium ions in portland cement and helps reduce or prevent the detrimental effect of the alkali-silica reaction. It can also help change the characteristic of the alkali-silica gel and make it less expansive [Diamond, 1983b]. Silica fume is being found a very desirable material for combating alkali-silica reactions in Iceland [Gjorv, 1983; Olafsson, 1982], a country with a history of such problems.

As first reported in Scandinavia by Idorn Associates, some densified (compacted) silica fume that has failed to completely disperse in concrete has been linked to causing alkali-silica reaction.

Sulfate Attack

Most of the literature deals with one type of sulfate attack—sodium sulfate. Information on magnesium sulfate attack on silica fume concrete is incomplete [Cohen and Bentur, 1988; Cohen and Mather, 1991]. Most of the data point to the benefits of silica fume in reducing sulfate attack. Mather [1982] tested three high-C_3A-content portland cements with various pozzolans and noted that the greatest benefit was obtained with silica fume.

Silica fume concrete has shown good resistance in sodium sulfate exposure. It is possible to use a blend of ASTM Type I portland cement with silica fume to replace the ASTM Type V sulfate resistance portland cement [Cohen and Bentur, 1988].

Work dealing with magnesium sulfate action on pastes of portland cement + silica fume showed that the presence of silica fume can be deleterious [Cohen and Bentur, 1988; Bonen and Cohen 1992a, 1992b]. The pastes experienced strength drop, mass loss, and expansion. These results were confirmed on tests with silica fume concrete [Goldman, 1987].

Freezing and Thawing

Information regarding freeze-thaw durability of silica-fume concretes is not conclusive and, in fact, some results are contradictory. This is due to increased sensitivity to numerous parameters involved in the test.

Air-entrained specimens containing up to 20% silica fume showed good performance when tested according to ASTM C666, Procedure A. With relatively low water–to–cement + silica fume ratio (0.4) and more (20 to 30%) silica fume, freeze-thaw durability was unsatisfactory [ACI Comm. 226, 1987a; Sellevold and Nilsen, 1987]. No clear explanation can be provided for the poor durability. It would be expected that, if anything, durability would improve with increased content of silica fume. Perhaps the adverse effect of reduced permeability on moisture movement is the reason for the lower durability. Contradictory results have been reported by Wolsiefer [1984], who showed good freeze-thaw durability for dense, 0.3 water–to–cement+silica fume ratio, silica fume concrete using ASTM C666, Procedure A.

The critical bubble spacing for silica fume concrete is less than that for portland cement concrete, which is 200 micron (0.008 in.). This is attributed to the higher impermeability of the silica fume paste and thus the greater development of hydraulic pressure compared to portland cement paste.

Care must be taken in applying standard test methods, developed for ordinary concrete, to problems of silica-fume concrete. The available data on frost resistance are inconclusive, and even contradictory. There is, however, a general trend which indicates a potential problem with concretes containing a high silica fume content (20 to 30%). There is also concern that the general standard test methods (i.e., ASTM C666) applied to normal concrete may not be applicable to silica fume concrete and may need to be modified [Cohen *et al.*, 1992]. Use of ASTM C671 in the critical dilation method for assessing resistance to freezing-and-thaw of high-strength silica fume concrete should be investigated as a possible replacement for the ASTM C666 procedure.

The influence of air-entraining admixtures on frost resistance and the plausibility of obtaining a non-air-entrained frost-resistant high-strength silica fume concrete has been a subject of interest and was reviewed by Cohen *et al.* [1992]. Current practice suggests that air-entrainment should be used when concrete containing silica fume is subjected to freezing-and-thawing action, even for concrete with a low water-to-cementitious materials ratio [Luther and Smith, 1991; Luther, 1989b].

Chemical Attack

Mehta [1985] has studied the resistance of silica fume concrete to a variety of aggressive chemicals. These included 1% hydrochloric acid, 1% sulfuric acid, 1% lactic acid, 5% acetic acid, 5% ammonium sulfate, and 5% sodium sulfate. The results indicated that silica fume concrete had very high resistance to all the chemicals except ammonium sulfate. Improved resistance to chemical attack may be due to the lowered permeability of silica fume concrete. When silica fume concrete was exposed to the ammonium sulfate the C-S-H gel decomposed. These concentrations are high, and usually they are not encountered in practice. However, if the silica fume concrete is in the splash zone, the continuous wetting and drying can lead to damage even if the concentration of the solution is low.

Abrasion Erosion Resistance

An important and well-known benefit of silica fume to concrete is the improvements to abrasion-erosion resistance of the concrete. Silica fume concrete is an excellent material that can be used in water structures that are subjected to forces of abrasion and erosion [Holland *et al.*, 1986; Luther, 1989b, 1990b]. Silica fume concrete has also demonstrated cavitation resistance about two orders of magnitude greater than that of portland cement concrete [Luther, 1990a].

Thermal Expansion

As with portland cement concrete, the coefficient of thermal expansion of silica fume concrete depends primarily on the thermal properties of the aggregate [ACI Comm. 234, 1994]. Therefore, the parameters that influence the thermal expansion properties of plain concrete should influence in the same manner those of silica fume concrete.

Field and Novel Applications

Silica fume has been used successfully in civil engineering concrete structures because of its ability to achieve high strength and high resistance to abrasion/erosion, and low permeability. Silica fume in concrete has been used in a variety of field applications ranging from repairs and rehabilitations to construction of tunnels, high-rise structures, bridge decks, overlays, and underwater castings. The first large-scale application of silica fume in the U.S. was in the Kinzua Dam Stilling basin located in western Pennsylvania. The purpose of using silica fume was to increase the abrasion resistance of the concrete [Luther, 1989b].

The first advertisement in an engineering magazine was in *Civil Engineering* [Anon., 1983]. The emphasis was on strength. Since there was a limited market for high-strength concrete, suppliers began to emphasize durability and began to promote reduced permeability for bridge decks. Since there was not enough data available to satisfy certain state transportation agencies, the material was not readily accepted, and the focus therefore shifted to parking structures. Today, in the U.S., the primary applications of silica fume concrete are for (a) impermeability, and (b) high strength, mainly in parking structures. The use of silica fume in bridge decks does not occupy a major market because of the small volume of concrete involved [Holland, 1989], but it is a growing application [Luther, 1988; Luther, 1993].

Initially, in the U.S., silica fume was used as a cement replacement. While this practice is still continuing in some other countries, because of the high cost of silica fume (about \$500 to \$700 per ton delivered) the focus now is on using it for performance enhancement [Holland, 1989]. Nevertheless, even with this high price the use of silica fume only doubles the cost of the cementitious material in the concrete, assuming a 10% or so silica fume level. The increase in the cost of the material due to use of silica fume in most cases is insignificant compared to the overall cost of the project and the potential benefits that can be gained.

A few of the projects using silica fume concrete are listed in Table 40.7 along with the 28-day strengths. The table shows the wide applicability of silica fume concrete and its usefulness in both small and large projects.

Table 40.7 Some Project Listings in the U.S. and Abroad

Location	Application	28-Day Compressive Strength psi (MPa)
U.S. [Courtesy of Elkem Materials Inc.]		
California	Los Angeles River Channel	9,000 (62)
Pennsylvania	stabilization of rocks	9,000 (62)
Pennsylvania	Kinzua Dam Stilling Basin	13,000 (90)
Massachusetts	industrial floor	7,000 (48)
Ohio	industrial floor	7,000 (48)
Kentucky	bridge overlay	10,000 (69)
Ohio	bridge overlay	8,500 (59)
Massachusetts	bridge overlay	13,000 (90)
Ohio	parking lot	7,000 (48)
Massachusetts	footings	11,000 (76)
West Virginia	bridge replacement (full depth)	7,000 (48)
Ohio	parking garage	7,000 (48)
Missouri	parking garage	6,500 (45)
Massachusetts	floating dock	13,000 (90)
Vermont	hockey rink overlay	11,000 (76)
Maryland	columns	11,000 (76)
Pennsylvania	parapet walls/bridge	8,000 (55)
International:		
Sweden [Elkem Materials Inc., Magne and Sellevold, 1987]	Tjorn Bridge, wharves warehouse	9,000 (62)
Denmark [Justesen, 1981]	sea floor, sea walls, tank walls sewage treatments, prestressing tendons, bridge piers	
Norway [Elkem Materials Inc., Magne and Sellevold, 1987]	grain silos, tunnels, housing projects, deck slabs, loading buoys, bridge structures, lightweight blocks, prestress/precast elements	$5{,}800^{+}$ (40^{+})

Silica fume has also been found to be beneficial in other applications that are novel in one way or another. Specifically, it is being used in machine and structural components that undergo severe wear. They are used for making concrete paddles for concrete and asphalt mixers. These replace the conventional cast iron and steel-reinforced rubber subject to heavy abrasive wear. Silica fume is used also for numerous other components and structures subjected to wear, including pipes and pipe bends which transport abrasive materials (i.e, fly ash, glass, etc.), scoop feeders for cement mills, and bolt-on wear panels for wear protection of larger components of industrial transport systems and processing equipment (courtesy of Aalborg Portland, Denmark).

Silica fume is also used for special components requiring ultra-high strength, such as tooling and molding that can be used by the sheet metal (courtesy of Aalborg Portland, Denmark) and aerospace [Wise *et al.*, 1985] industries. It has also been used for situations requiring high temperature and high impact resistance such as refractory concrete, concrete security vaults, and explosive-resistant structures [Luther, 1990a; Luther and Smith, 1991].

In the high-technology ceramics field, Diamond [1985] states that there is yet no mention made of cement or cementitious components. It seems fair to conclude that the successful performance of silica fume concrete as cutting tools and as other abrasion-resistant materials might lead to its acceptance as a high-performance ceramic material.

Important Practical Considerations

The potential benefits of using silica fume in concrete are now recognized and well established. There are a number of important practical considerations that must be accounted for when using silica fume in concrete. These are as follows.

Storing, Packing, and Transporting

The ultra-fine size of the particles and its low bulk density make silica fume extremely hard to store, pack, and transport. Silica fume requires a large volume to store and transport; for example, four truck loads of silica fume are required to equal in weight one truck load of cement. On the other hand, the normally small amount of silica fume in mixes converts into two to three truckloads of cement per truckload of silica fume [ACI Comm. 226, 1987a].

Dusting is a major problem when the silica fume powder is poured into the mix. The solution to this problem was introduced by the Norwegians who proposed to transform the dry powder into a slurry. This transformation was achieved by mixing the powder with water on an equal mass basis. Small amounts of additives were added to prevent flocculation.

There are, however, several difficulties associated with the use of a slurry. First, special equipment needs to be manufactured to store the slurry. Second, the gelling tendency of the silica fume particles in water, even in the presence of anti-gelling additives, requires periodic agitation of the slurry. Third, use of the slurry in cold climates creates difficulties because of freezing. Fourth, but not least, the higher cost of transportation due to the large amount of water present in the slurry increases the cost of silica-fume concrete and makes it uneconomical in certain situations.

Another alternative form of delivery is the dry-densified (compacted) powder. A number of methods have been developed for densifying silica fume powder [Aitcin *et al.*, 1980]. Normally a continuous dry mixing of the silica fume powder can lead to densification [Cohen *et al.*, 1989, 1990]. Ordinarily a 50 to 60% bulk densification is achieved—that is, the original and fine silica fume particles agglomerate into larger sizes. In a microscope they appear as spheres, mostly larger than 100 micron [Cohen *et al.*, 1989, 1990]. The main advantage of densification is that silica fume can be used satisfactorily in this form, so avoiding problems associated with the slurry and undensified forms.

Mixing, Slump, and Curing

Because of the high fineness and generally lowered water-to-cementitious materials ratios used, silica fume makes concrete sticky, cohesive, and susceptible to plastic shrinkage cracking. Since there is little or no bleeding, under hot and windy conditions protective measures must be taken to guard against plastic shrinkage cracking in floor and pavement slabs. It is therefore essential to finish the concrete as soon as possible and protect the surface from drying out [Cohen *et al.*, 1989, 1990]. Although bleeding is diminished when using an HRWRA, the concrete still remains cohesive and its slump may become more difficult to control.

These qualities are not always negative. High cohesivity of silica fume concrete has been used advantageously for underwater concreting in Denmark [Aitcin *et al.*, 1981; Luther, 1989b, 1990a].

Other Considerations

A detailed discussion on the practical aspects of using silica fume in the field, including problems associated with handling, mixing, casting, and curing silica fume concrete is presented in a paper by Holland [1989] and in another by Holland and Luther [1987]. Some of the highlights of the conclusions of the papers are the following:

1. Silica fume is not a cure for bad practice.
2. Silica fume is less forgiving than plain portland cement concrete if there is any attempt to cut corners.
3. To add silica fume to get higher strength and impermeability requires strict attention to details and careful adherence to good practice.

The above comments also apply to the general class of pozzolans.

Effects of Form of Silica Fume

Effects of the form of silica fume (water-slurry, dry-densified powder, dry-undensified powder) on properties of high-quality plain concrete have been studied [Cohen and Olek, 1989; Fidjestol *et al.*, 1989; Luther, 1990c; Luther *et al.*, 1992]. Fidjestol *et al.* [1989] and Luther *et al.* [1992] indicate that at the age of 28 days, water-slurry and dry-undensified forms of silica fume provided somewhat greater contribution to strength when compared with the dry-densified form. The data from Cohen and Olek [1989], however, indicate that the 28-day strength of dry-densified silica fume concrete, Fig. 40.17, compared well with the dry-undensified and slurry silica fume concretes.

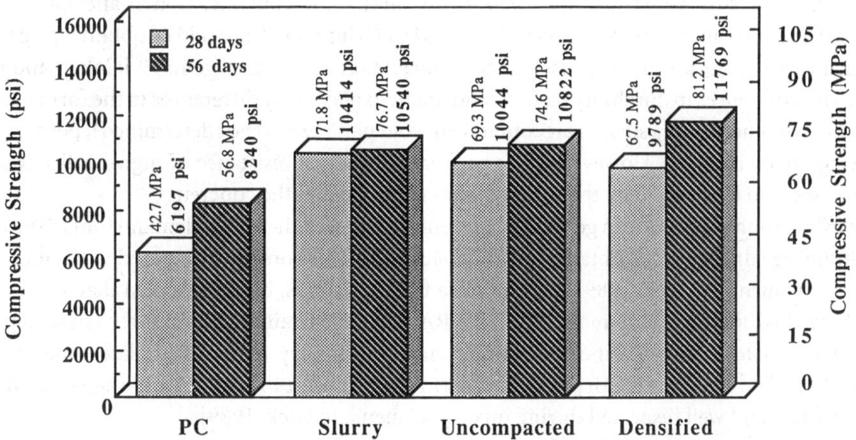

FIGURE 40.17 Compressive strengths at 28 and 56 days of plain concrete, water-slurry silica fume concrete, dry-undensified silica fume concrete, and dry-densified silica fume concrete. (*Source:* Cohen and Olek, 1989.)

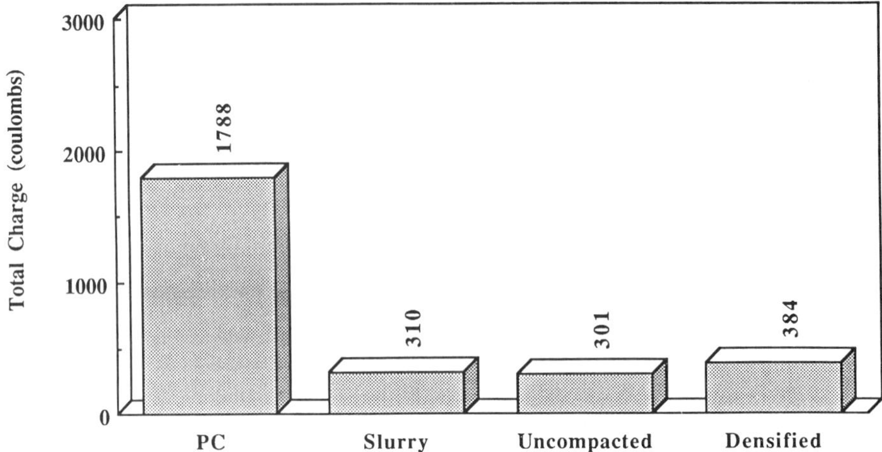

FIGURE 40.18 Total charge passed (coulombs) by rapid chloride ion permeability test of plain concrete, water-slurry silica fume concrete, dry-undensified silica fume concrete, and dry-densified silica fume concrete. (*Source:* Cohen and Olek, 1989.)

Cohen and Olek [1989] also report that at a later age, 56 days, the dry-densified silica fume provided the largest contribution to strength. It is suggested that some of the initially undispersed or partially crushed silica fume agglomerates can be available for pozzolanic reaction at later ages. This time-capsule-like behavior [Cohen and Olek, 1989] can be a positive characteristic, as delayed pozzolanic reactions can perhaps be more effective in satisfying long-term strength and durability requirements. However, this behavior demands that concrete be protected from potentially aggressive exposure while waiting for the increased pozzolanic reaction. Fidjestol *et al.* [1989] also observed some undispersed agglomerates after mixing of concrete, but they did not report post-28-day strength values.

The behavior of the dry-undensified form of silica fume is more similar to that of the water-slurry form, as evident from Fig. 40.17.

Results from the rapid chloride permeability measurements indicate that all forms of silica fume tend to increase resistance to chloride permeability to about the same level [Cohen and Olek, 1989; Fidjestol *et al.*, 1989; Luther, 1990c], as shown in Fig. 40.18 [from Cohen and Olek, 1989].

Regarding the frost resistance of air-entrained silica fume concrete, Cohen and Olek [1989] showed that a durability factor well above the ASTM C666 limit of 60% could be obtained, regardless of which form of silica fume was used in the concrete, as shown in Fig. 40.19 [Cohen and Olek, 1989]. The differences in durability factor could not be explained by differences in the forms of silica fume used, as other factors such as air content, air-void parameters (not determined), permeability, and strength are involved. However, it was suggested that frost resistance of high-quality concrete does not seem to be affected by the form of silica fume used in the concrete.

Regarding fresh properties, in general, plain concrete showed the lowest demand for HRWRA and air-entraining admixture for obtaining similar slump and air content, compared with silica fume concretes [Cohen and Olek, 1989]. For the silica fume mixtures, it was observed that water-slurry silica fume had the highest demand for HRWRA and air-entraining admixture. These demands were intermediate for dry-densified silica fume and lowest for dry-undensified silica fume concrete. The intermediate levels for the dry-densified form again suggest that some of the agglomerates are being crushed and well dispersed during mixing [Cohen and Olek, 1989].

It was pointed out as well that, for the laboratory-scale specimens used in that investigation [Cohen and Olek, 1989], the dry-densified form of silica fume was the easiest to handle during the mixing. The concrete containing this form also showed the best finishability.

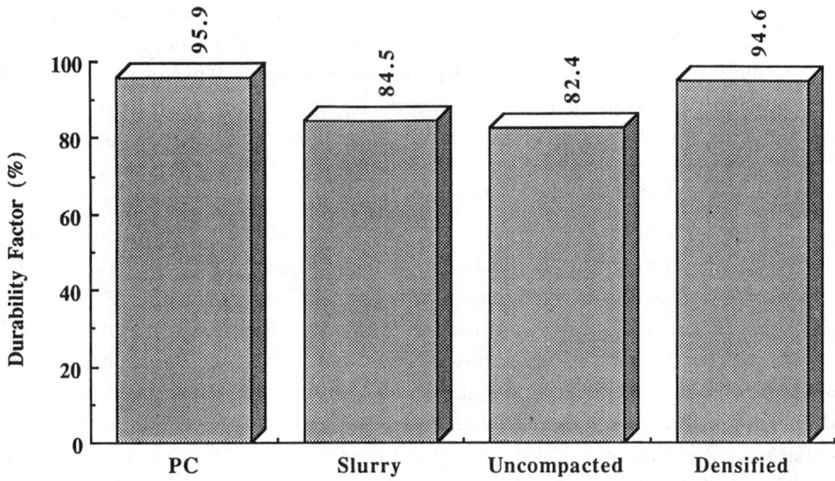

FIGURE 40.19 Durability factors at 300 cycles of plain concrete, water-slurry silica fume concrete, dry-undensified silica fume concrete, and dry-densified silica fume concrete. (*Source:* Cohen and Olek, 1989.)

Mixture Proportioning Concepts

Currently, neither a standard nor a formal recommendation exists for mixture proportioning of silica fume concrete. However, different groups and organizations have developed their own procedures, and in one way or more, have similarities with each other. Silica fume concrete mixtures have been proportioned based on several concepts and considerations as described below.

Replacement and Addition Methods

Silica fume can be used as (1) partial replacement for the cement (from 5 to 20% of the weight of cement, although 40% dosage has been used), or, (2) addition to cement or concrete. Item (1) is normally done for economic reasons. Item (2) is done to improve concrete properties, both in the fresh and hardened states. Silica fume can be either blended with cement or added directly to the concrete.

Water-Reducing and Inherent Effects

Silica fume concrete mixtures can also be proportioned in terms of two effects [Bentur *et al.*, 1987]:

1. *Water reducing.* Associated with the reduction in water-to-cementitious materials ratio obtained when silica fume is added in combination with superplasticizer while maintaining equal slump with the concrete without silica fume.
2. *Inherent.* Associated with the increase in strength in the silica fume concrete compared with a similar water-to-cementitious materials ratio concrete without silica fume but with equal slump.

At 28 days after casting and curing, the inherent effect showed greater compressive strength compared to water-reducing effect.

Efficiency Factor *K*

Silica fume concrete mixtures can be proportioned based on an efficiency factor, K. For example, an efficiency factor of 3 indicates that 1 kg of silica fume can replace 3 kg of portland cement in concrete and still result in concrete having the same compressive strength. K values in the literature range from 2 to 5. The value for K may be determined graphically or with the aid of an equation, as presented in the reference by Sellevold and Radjy [1983].

Aitcin and Vezina [1984] have also provided an equation for an efficiency factor, which is based on a similar principle. The criticism of the method has been that the efficiency factor for durability (carbonation, chloride permeability) is not as high as that for strength. This means that silica fume designed for strength may exhibit poor durability property, especially if the strength is low and the cement content is permitted to fall to values that are too low [Rickne and Svensson, 1982].

Particle Packing and Cement Chemistry

The concept of using particle packing and cement chemistry as means of obtaining ultra-high strength products was used first by H. Bache, Z. Fordos, L. Hjorth, and co-workers at Aalborg Portland in Denmark. The highly modified cement paste had a large dosage of silica fume in addition to portland cement and a minimum amount of water. A sufficient amount of an HRWRA or, as sometimes referred to by its trade name, superplasticizer, was used to deflocculate the silica fume and portland cement particles. This system is given the acronym *DSP* (densified with small particles) referring in general to densified systems containing homogeneously arranged ultrafine particles [Bache, 1981; Diamond, 1985]. The objective is for silica fume particles to fill up the empty spaces normally occupied by water or air. A special mixing technique must be used to ensure consolidation and homogeneity [Diamond, 1985].

Other investigators have used this concept in developing other products. For the development of ultra-high strength DASH 47 cementitious tooling and molding material, CEMCOM research group has looked at two ways of proportioning silica fume [CEMCOM Internal Report; Wise *et al.*, 1985; Sheetz *et al.*, 1981]: particle packing and cement chemistry. The particle-packing criteria relates to the optimum particle packing of the powder constituents based on the size of particles: an ideal 7:1 radius ratio for optimum packing, and ratios of bulk density to true density of the particles [Milewski, 1978; Nilsen and Young, 1977]. Here the coarse aggregates were stainless steel 0.2 to 2.0 mm, the matrix was cement approximately 30 microns in size, and the fines were silica fume with particles smaller than 5 microns.

The optimum weight percentage of silica fume calculated on this basis is about 7%, which is typical in concrete applications. The reason for the deliberate raising of the silica content is based on consideration of cement chemistry, which is the second criteria.

Chemical consideration indicates that the highest strength is obtained with 30 to 40 weight % of silica fume. However, in practice the silica fume content is less than this range owing to difficulties associated with mixing and cohesivity.

40.5 Ground Granulated Blast-Furnace Slag Concrete

Terminology

Granulated blast-furnace (GBF) slag that is ground to a fineness approaching that of portland cement, or finer, becomes strongly cementitious. There are several products containing GBF slag–sourced material used in concrete, and these are described more extensively under *Specifications and Standards*. Here, GBF slag that is ground and added to a concrete mixer at a concrete plant is referred to as *ground granulated blast-furnace* (GGBF) slag. Whenever GBF slag is used in a blended cement, the blended cement will have one of three different names if it is an ASTM C595 blended cement. One of these blended cements is called *slag cement* (Type S), which contains at least 70% by mass of GBF slag–sourced material. (Many refer to GGBF slag as *slag cement*, regardless of correct terminology. It is suggested that whenever one hears the term *slag cement* being used, a series of questions should be asked to learn whether the material being discussed is GGBF slag or truly slag cement.)

The use of ground granulated iron blast-furnace slag in concrete offers a great potential for utilizing an industrial by-product and saving energy in cement production. Only 1.67 gigajoules of energy is expended for producing 1 ton of slag cement containing 65% slag. In comparison 3.53 gigajoules/ton is necessary for portland cement production. Blast-furnace slag is used in concrete in many countries. In France and China more than 40% of the total cement production is GBF slag–sourced material, considerably higher than in the U.S., where it is only 1%.

History

The first use of ground granulated blast-furnace slag as a cementitious material dates back to 1774 when Loriot made a mortar using slag in combination with slaked lime. In 1892 Langen studied and proposed a granulation process. Others looked at air-granulation processes and started to recognize the importance of glass content and slag composition. The first commercial use of slag-lime cements was in 1895 in Germany. In 1889 the Paris underground Metro System was built using slag. The first production of portland blast-furnace slag cement was in Germany in 1892. Now in Europe blended cements have typically 20% by weight of GGBF slag [ACI Committee 226, 1987b].

The first U.S. production of portland blast-furnace slag cement (a blended cement) was in 1896. Until the 1950s slag was used as raw material for making portland cement and also as a cementitious material blended with portland cement, hydrated lime, gypsum, or anhydrite. After the 1950s it was used as a separate cementitious material added at the concrete mixer with portland cement. This method is popular in the U.S., South Africa, United Kingdom, Japan, Canada, and Australia for two reasons: (1) slag can be ground separately from portland cement, so that both portland cement and slag can be ground to their optimum fineness, and (2) proportions can be controlled better [ACI Committee 226 report, 1987b].

Production

Ground granulated blast-furnace slag used in civil engineering practice is a by-product of the production of iron in a blast furnace. Slags come also from the production of other metals, including steel, copper, lead, and nickel. Slag from these metal processes, however, is not blast-furnace slag. These other slags (other than iron-sourced) have varying suitability for use in concrete with current technology.

Iron production is the starting point for the production of steel. The production of iron occurs in a blast furnace, Fig. 40.20 [from Regourd, 1985]. In producing iron, ore is added into the furnace from the top. Coke and limestone or dolomite fluxing material are also added with the iron ore, and sometimes scrap iron is added too. The coke is used as fuel to produce heat. It also acts as a reducing agent separating the oxygen from the iron. Flux is added to facilitate the flow of the melted material.

At the top of the furnace the flux loses a molecule of CO_2 and turns into lime. The temperature in the furnace is sufficiently high (1300°C to 1600°C) to cause chemical reactions and melting to occur. Fe_2O_3 and Fe_3O_4 are reduced to Fe [Regourd, 1986]. The molten iron is thus obtained. It is then used for production of steel. The molten material is periodically tapped from the blast furnace. Because of the differences in densities, the molten iron and the slag separate, with the slag floating on top of the iron [Lee, 1974]. After the iron is tapped, the slag is then tapped from the furnace.

Composition

The main components of BF slag are silica and alumina (from the iron ore) combined with calcium and magnesium oxides (from the fluxing sources). The ternary C-S-A phase diagram for constant amount of MgO is shown in Fig. 40.21 [Regourd, 1985]. The diagram depicts typical ranges of

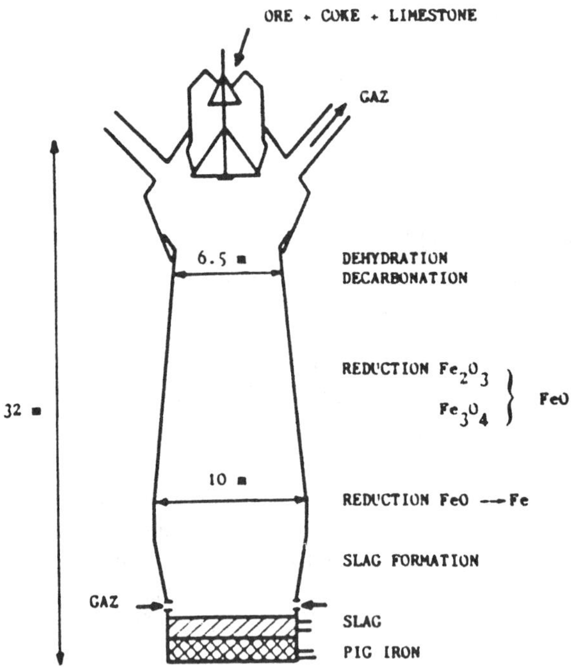

FIGURE 40.20 Production process of slag in a blast furnace. (*Source:* Regourd, 1985.)

composition for cements and mineral admixtures including BF slag. The BF slag with the lower CaO and higher SiO_2 content is referred to as *acid slag*, while that with higher CaO and lower SiO_2 content is *basic slag*. The similarity of BF slag and portland cement compositions make BF slag a good raw material for the manufacture of portland cement, since no energy is expended driving off the CO_2, as must be done if limestone is used as a raw material. Table 40.8 provides typical ranges of chemical composition of BF slags.

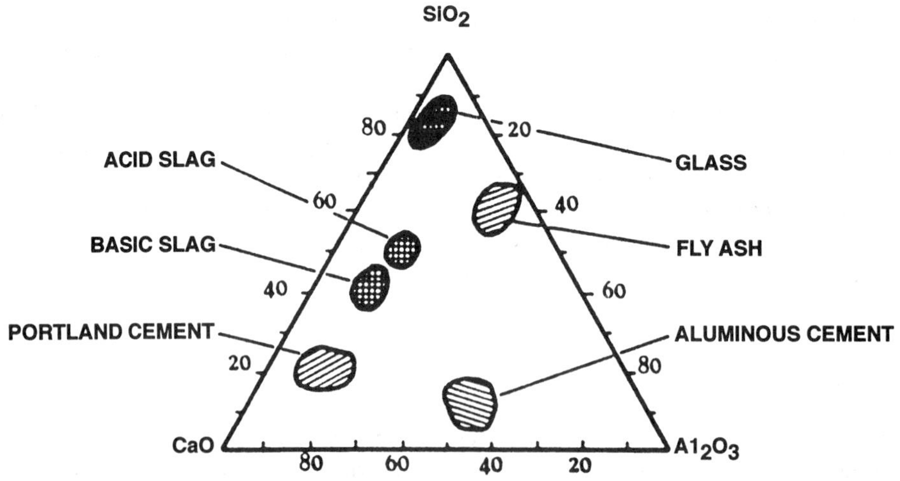

FIGURE 40.21 Ternary phase diagram for cements and mineral admixtures. (*Source:* Regourd,1985.)

Types of Slag, Their Reactivity, and Uses

Table 40.8 Range of Chemical Composition of Slag

Compound	Lee[1]	NSA[2]
Cao	36–43%	32–45%
SiO_2	28–36%	32–45%
Al_2O_3	12–22%	7–16%
MgO	4–11%	5–15%
Sulfur	1–2%	1–2%
Fe_2O_3	0.3–1.7%	0.1–1.5%
MnO	—	0.2–1.0%

[1] From Lee, 1974.

[2] From the National Slag Association, Pub. NSA 188-1, Alexandria, VA.

The type of slag obtained depends on how it is cooled and handled. There are four types: (1) slow- or air-cooled slag, (2) expanded or foamed, (3) pelletized, and (4) granulated. In 1986, in the U.S., of the 15.4 million short tons of slag produced, 13.5 million short tons were air cooled, 1.2 million short tons were granulated or pelletized, and 0.7 million short tons were expanded or foamed. (The Bureau of Mines reports pelletized as granulated. This may lead to some confusion, in part, because the industry uses both terms for the same slag. Part of a pelletized product can be used as a granulated BF slag. The remainder [the majority] is a lightweight aggregate. Some people continue to refer to pelletized BF slag as granulated [Luther, M.D., private communication].) No data was made available for the pelletized form [data from Bureau of Mines Minerals Yearbook, 1987].

The oldest type is the slow- or air-cooled slag. This is produced if the slag is allowed to solidify and cool slowly in the air, which leads to the formation of crystalline and stable solids of C-A-S-M, especially melilite (C-A-M), which is a solid solution of gehlenite (C_2AS) and akermanite (C_3MS_2) [Regourd, 1986]. Normally, the slag is poured into a ditch and left there until solidification at atmospheric temperature. Sometimes cooling is accelerated by applying water to the solidified surface of the air-cooled slag. This type of slag is strong and durable. Its mechanical properties are similar to those of basalt. It is used as aggregate in concrete and bituminous pavements, base course in highways and airport projects, railroad ballast, trickling filtering medium, and septic tanks absorption beds [Josephson *et al.*, 1949]. Also, because of its lighter unit weight than normal aggregates, it is used in roofing.

The expanded or foamed slag is obtained by rapid cooling of the hot slag. Rapid cooling is done by controlled quenching often in conjunction with steam formation. The process can lead to partial or complete formation of glassy materials, which have a higher energy than the crystalline counterparts. Rapid cooling occurs by introducing a limited or controlled amount of water, or water with steam or compressed air. The gases of steam produce a cellular structure with a low unit weight, which depending on gradation can be 50 to 65 lb/ft^3. This material is used as lightweight aggregate in concrete construction for structural purposes and floor fills, other concrete products, masonry units, and lightweight fills and embankments [Lee, 1974; National Slag Association]. This slag varies from crystalline to glassy in nature [Lewis, 1986].

Pelletized slag is sometimes considered an expanded slag and part of it may be considered a granulated material. This slag is the result of air granulation. The process was developed by the National Slag Limited in Canada and allows for better control of the emission of gases. The slag is poured into a rotating drum while water is sprayed onto it. The drum spins and breaks the slag, letting it fly for about 15 meters, which is sufficient to let the particles form pellets. This slag is highly glassy and cementitious [Lewis, 1986; and Regourd, 1986]. It contains less moisture than the water-granulated slag and does not need to be pre-dried before grinding. This translates into savings of energy. Further, because of its high porosity, this slag might require less energy to grind than water-granulated slag. In Canada, pelletized slag is mostly ground and used as a GGBF slag, which is regionally referred to as *slag cement*. It also has been used as lightweight aggregate in concrete floor slabs and block products. The properties of the pelletized slag can be controlled by variations in the production process. Therefore, it is possible to produce pellets with more crystalline material to be used as aggregates. or a more glassy pellet for cementitious applications [Emery, 1989]. There is less capital investment (by about 25%) required to pelletize slag, compared to the granulated material [Thomas, 1979; Emery, 1989].

Granulated blast-furnace (GBF) slag is produced by quenching BF slag with a large amount of water. This quenching method is commonly referred to as the *jet process*. A material high in glass content is produced in the form of sand-sized granules. These granules are then ground to the specific surface area desired (ground to a fineness usually similar to or finer than portland cement—and types of blended cements). Of all the slags, GBF slag is the primary source used as a mineral admixture in concrete. When the GBF slag (the granules) are ground to a fineness that imparts useful cementitious properties, the material is referred to as *ground granulated blast-furnace (GGBF) slag*.

Slag is a versatile material and there are other uses for it in civil engineering. These include uses in bases for pavements, runways, parking areas, aggregate in concrete block or pavers, for agricultural purposes to neutralize acid soils and induce growth of plants (as the elements present in slag stimulate plant growth), as fill materials including backfill for abutments and for preventing cast iron pipes from corroding because of the high alkalinity in the presence of moisture, and for immobilizing certain toxic and radioactive wastes. It also can be used as a fine aggregate in concrete or asphaltic cement pavement.

Pollution Control

Jet process GBF slag production requires the use of large quantities of water to cool molten slag. Concern for water pollution has been expressed as a disincentive for investment in granulation facilities. However, in practice much of the water leaves the system as steam, and the minerals in the water remain with the slag. Also, air-emissions concerns have recently caused greater consideration of installing jet process granulators, because this process reduces air emissions relative to other BF slag processes.

Specifications and Standards

ASTM C989 describes three grades of GGBF slag: 120, 100, and 80. The grades correspond more or less to the 28-day strength ratio of mortar cubes containing a slag–reference portland cement mixture to those containing a reference portland cement mixture.

More finely ground versions of GGBF slag exist, and these are specialty products. Sometimes GGBF slag is blended with small amounts of other materials to adjust the concrete properties.

ASTM C595 describes three types of blended cements that contain granulated blast-furnace slag. These are: (1) type 1 (SM) slag-modified portland cement, (2) type 1S, portland blast-furnace slag cement, and (3) type S, slag cement (see Table 40.3).

The GBF blast-furnace slag used in an ASTM C595 blended cement must (at the fineness at which it is ground in the blended cement) demonstrate a slag activity (strength relative to a reference) at least 75% that of the associated portland cement.

ASTM C1157 is a relatively new performance specification for blended cement. Essentially, any slag material may be used to make blended cement under the ASTM C1157 specification, as long as the performance requirements are met.

The British Standard, BS 6699 (1986), considers the amount of glass in quantitative terms (glass content $>$ 66% by mass). The glass content has been a controversial issue because of the difficulty involved in measuring it. Different techniques give scattered results, some going as far as to suggest that there is no relationship between glass content and slag activity index [Douglas and Zerbino, 1986]. The Austrian Standards require that the slag be predominantly glassy. BS 6699 also provides a requirement for the hydraulic modulus $[(CaO + MgO + Al_2O_3)/SiO_2)]$, which must be larger than 1. No such modulus is used in the ASTM standards. It is thought that hydraulic moduli fail to provide adequate criteria, due to complexity of the system. The Slag Activity Index for mortar in ASTM C989 appears to provide a satisfactory evaluation of the role of slag in mortar and concrete [ACI Committee 226, 1987b].

Since the fineness of slag is similar to that of portland cement, the Blaine method is practical for specifying the fineness.

The slag portion of the Canadian specification is similar to ASTM C989, a nearly complete performance standard that accepts slag on the basis of an activity test in which 50% of a reference portland cement is replaced by GGBF slag. The reference portland cement must have a sodium alkali equivalent content between 0.60 and 0.90%. In general, a single performance specification is recommended with the only physical requirements being fineness and an activity test in which there is a 10% volume replacement of cement by silica fume, 30% replacement by fly ash, and 50% replacement by slag.

Chemical Activation of Slag

The hydration of portland cement with GGBF slag is complicated and not well understood. However, portland cement hydrates in a normal manner and the calcium hydroxide that is released acts as a starter or activator of the slag. Further hydration of the slag is direct and does not depend on combination with lime.

Generally, the factors that determine the cementitious properties of GGBF slag are:

1. Chemical composition of slag
2. Alkali concentration of reacting system
3. Glass content of slag
4. Fineness of slag and portland cement
5. Temperature during early phase of hydration [ACI Committee 226, 1987b]

GGBF slag reacts more slowly with water than does portland cement, but it can be activated chemically. No matter what activator is used, C-S-H gel is always present. Basically there are three types of activators [Regourd *et al.*, 1980]: (1) alkaline activators, (2) sulfate activators, and (3) mixed activators. Alkaline activators include soda or lime. If the activator is soda the products of the reaction are C-S-H, C_4AH_{13}, and C_2ASH_8, which is gehlenite. C_4AH_{13} and C_2ASH_8 are hexagonal plates that play the role of crystalline bridges between the slag grains. The C-S-H gel appears to cover the slag grains. The C/S ratio produced in slag hydration is about 1.0, which is smaller than the typical ratio of 1.5 found in mature C-S-H gel produced in portland cement hydration. Soda can be in the form of NaOH solution. It can be replaced with sodium carbonate or sodium silicate [Regourd *et al.*, 1980]. Calcium hydroxide provides the lime activation. The lime can be supplied as such or from the hydration of portland cement. The products of reaction are C-S-H gel and C_4AH_{13}.C_2ASH_8 does not form in the presence of calcium hydroxide.

Sulfate activators are gypsum, hemihydrate, anhydrite, and phosphogypsum. Sulfate activation from gypsum leads to formation of C-S-H gel, ettringite, and $Al(OH)_3$, which is aluminum hydroxide.

Mixed activation occurs in the presence of gypsum and lime (or gypsum and soda). $Al(OH)_3$ converts to ettringite. Sodium is found in the form of Na_2SO_4. Soda plays the role of activator, while gypsum and lime enter into the hydration reaction [Regourd *et al.*, 1980]. Other activators include calcium chloride, portland cement, and cement kiln dusts.

Hydration of GGBF Slag

Portland cement provides two slag activants: lime and gypsum. Lime is liberated by hydration of C_3S and C_2S of the portland cement. Gypsum is readily available as an addition. The first to hydrate in a portland cement–slag blend is portland cement. The main products produced from its reaction are C-S-H, calcium hydroxide, and ettringite. The composition of the C-S-H gel is different from that produced from the hydration of C_2S and C_3S of portland cement. The time when the slag hydration starts is not easily detectable. Times as high as 28 days and as low as 2 days have been suggested. In the x-ray diffraction pattern slag hydration manifests itself by a reduction in the calcium hydroxide content in the paste. At high slag content, nearly 80% of the calcium hydroxide disappears.

Similar to other mineral admixtures, portland cement blended with GGBF slag has a lower hydration rate and cumulative heat liberation than does portland cement by itself. This is of greatest effect in blends containing at least 60% GGBF slag. Mixtures containing slag dosages optimized for strength (around 35% to 45% GGBF slag) may not show less heat, though. This is an important requirement for mass concrete.

Fresh Properties

Like portland cement, GGBF slag is a crushed material and thus angular in shape. However, its surface texture is much smoother than that of portland cement, and this results in an improvement in workability. Further, its specific gravity is lower than portland cement's—2.9 compared to 3.15. The replacement of cement with GGBF slag on an equal mass-to-mass basis will result in a larger paste volume in concrete. This leads to improvement in workability. For the same water–to–cementitious materials ratio, increasing the level of GGBF slag replacement increases slump [Taylor, 1974].

In practice, the use of GGBF slag is often accompanied by reduced water content for equivalent cohesion, flow, and compaction characteristics, particularly when using a pump or a mechanical vibrator [Reeves, 1980].

The fineness of the GGBF slag has an influence on slump. While increasing fineness leads to improvement in hydraulic activity, it does not usually lead to either a decrease in slump or an increase in the water demand. The variations in water demand and slump should not vary significantly within normal parameters of expected production fineness, about $5000 \pm 500\ cm^2/g$ [Meusel and Rose, 1983].

The initial rate of reaction between GGBF slag and water, being slower than that of portland cement and water, causes an increase in time of setting. Generally, the setting times of concrete increase with increasing percentage of GGBF slag used [Reeves, 1980; Vladimir *et al.*, 1986]. The degree to which the time of setting is affected is dependent on the initial curing temperature of the concrete, the proportions of the blend used, the water–to–cementitious materials ratio, and the characteristics of the powders used [Fulton, 1974].

The extended time of setting of GGBF slag concrete at low temperatures may be beneficial in minimizing the occurrence of cold joints, but it may also result in increased pressure on vertical formwork. The increase in setting time is perhaps an indicator that the plasticity or working time of the mixture can be extended, but this is not necessarily related to the speed at which workability loss may occur. Heat generated will be reduced as the percentage of GGBF slag is increased, and this could have a significant effect on workability loss. Although no information is available on the workability retention, the contractor's general feeling is that slag cement concretes retain their workability for longer periods than do portland cement concretes.

The finishability of GGBF slag concretes is generally viewed as being improved relative to portland cement concrete. This is particularly noted up to the 50% GGBF slag dose. GGBF slag concretes employing ASTM C989 grade 120 material generally show less bleeding than portland cement concretes. Coarser GGBF slag materials usually may show the same or more bleeding than portland cement concrete.

Color

GGBF slag affects the color of hardened concrete. Usually, the higher the GGBF slag dose, the lighter the concrete color (or shade of gray), given other factors being constant. Also, because GGBF slag use in concrete lightens the color, colored architectural concrete may be achieved with less coloring additive, and brighter colors may be achieved too [Luther, M.D., personal communication].

Curing and Greening

Occasionally a GGBF slag concrete that has been adequately cured (having been provided continuous access to water) may be greenish during the first few days. This is more prevalent at higher GGBF slag dosages and with some portland cements. This young-age color, due to ferrous iron from the slag, is an indication that curing was probably adequate. The color usually changes to light gray or white within a few days of air drying and/or exposure to sunlight [Luther, M.D., personal communication].

Mechanical Properties

GGBF slag cements provide about the same strength in concrete mixtures as do portland cements. However, the rate of strength development is different. For portland cement and GGBF slag concretes of equal 28-day strength, portland cement has higher early-age strength, but GGBF slag concrete has a higher strength at later ages [Regourd, 1986]. If GGBF slag concrete comes under stress before the reaction is underway, it may not be able to perform as designed.

The influence of slag on the mechanical properties of concrete follows patterns similar to those observed with high-lime fly ash. However, the amount of replacement of portland cement with GGBF slag is considerably higher than that for fly ash or silica fume. In fact, with a proper alkali activator, a concrete mixture can be designed with 100% GGBF slag as the cement.

A comparison of the strength development of GGBF slag mortar in terms of the slag activity index is shown in Fig. 40.22 [from ACI Committee 226, 1987b]. For high-slag replacement levels, it is a good practice to use a higher ASTM grade slag because the strength of the concrete is expected to be lowered.

An approach adopted in some European countries and in Japan for evaluating the performance of GGBF slag concrete focuses on these parameters of slag [Kobayashi *et al.*, 1979]:

1. Basicity index expressed as: $(CaO + MgO + Al_2O_3)/SiO_2$
2. Amount of glassy material
3. Fineness
4. Amount of gypsum addition

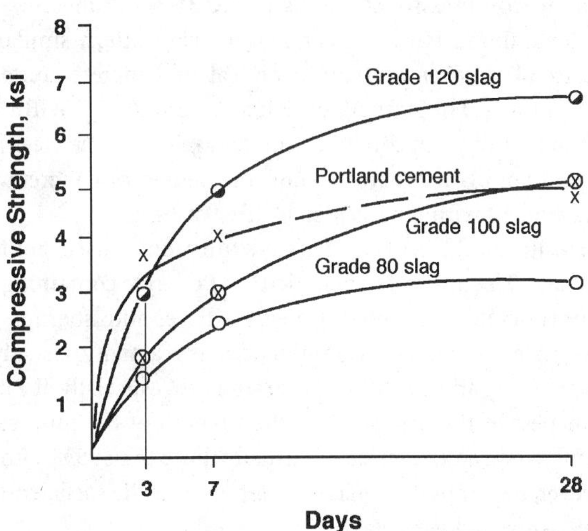

FIGURE 40.22 Strength relationship in terms of slag activity index of mortar containing a typical GGBF slag meeting ASTM C989 requirements, compared to portland cement mortar. (*Source:* ACI Committee 226, 1987b.)

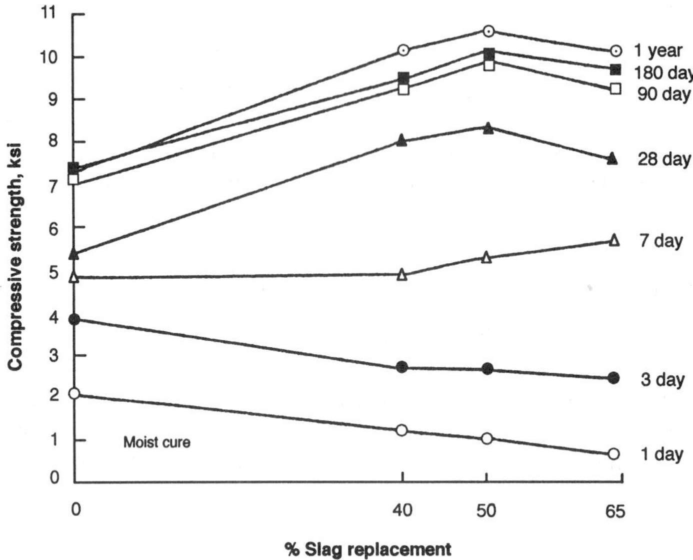

FIGURE 40.23 Influence of GGBF slag replacement on mortar cube strength. (*Source:* Hogan and Meusel, 1981.)

The basicity index provides indirect information on the chemical composition of the slag. It is a parameter that defines the amount of cementitious materials. A slag having a higher basicity gives a higher early-age strength, one with a lower basicity gives a higher later-age strength [Endo *et al.*, 1987].

Figure 40.23 [Hogan and Meusel, 1981] shows the relationship between the replacement level of a given slag and the compressive strength of mortar. For later-age strength (28 days and later), there appears to be an optimum replacement content of 50%. Not all slags show such optimum replacement level. When such a level exists, it varies with the characteristics and type of the slag used.

Other data pertaining to compressive strengths of GGBF slag concrete and portland cement concrete indicate that flexural and tensile strengths follow a pattern similar to the compressive strength, and that the trend of strength gain is typical of concrete containing other mineral admixtures, for example, fly ash. The strength of concrete containing GGBF slag at early age (one week or less) is lower than that of plain concrete. At later age (after one week) concrete containing slag has a higher strength than plain concrete. Thus, the benefit of concrete containing a normal amount of slag becomes evident within a week or so after casting.

For GGBF slag concrete, the modulus of elasticity as a function of age is similar to the compressive strength as a function of age. The modulus of elasticity of concrete containing slag is essentially the same as portland cement concrete of an equal strength [Klieger and Isberner, 1967].

The amount of glassy material in a slag also influences its reactivity. Usually, the more the glass, the more the hydraulic reactivity and contribution to strength, especially at early ages. Generally, as the glass content and fineness of the slag increase, the strength of the concrete containing the slag increases. The amount of glassy material in slag is usually more than 95%, although lower amounts can be acceptable. However, the amount of glassy materials must be sufficiently high to provide the advantageous properties associated with glass.

The fineness of GGBF slag also influences the strength of concrete. A finer slag provides enhancement in strength. For a GGBF slag concrete containing 40 to 65% slag replacement level, the 7-day compressive strength of an ASTM C109 mortar can increase from 4000 psi to 5500 psi by increasing the Blaine fineness from 4000 to 6500 cm^2/g [Hogan and Meusel, 1981].

Gypsum can be added to a GGBF slag in addition to the portland cement. An addition of up to 7% of gypsum can result in slight increase in the early-age compressive strength, but a decrease in the long-term strength [Kobayashi *et al.*, 1979; Kobayashi, 1979].

GGBF slag addition can lead to a finer microstructure and a smaller porosity of the matrix. As a result, drying shrinkage is expected to be higher owing to larger capillary forces, but field observations do not suggest such differences from portland cement concrete. However, a finer microstructure can also give a higher strength and modulus of elasticity. Thus, the magnitudes of drying shrinkage of GGBF slag concrete and plain concrete cannot be compared simply on the basis of microstructure or strength and elasticity alone. Concrete containing GGBF slag creeps less than plain concrete.

There are few data regarding thermal properties. The replacement of cement by GGBF slag has only a small influence. Results obtained so far have been contradictory. While one report showed a small increase in the thermal coefficient with a replacement of 75% cement with slag [Bamforth, 1980], another indicated a 5% reduction with a 70% slag replacement [Wainwright, 1987]. One report indicated that the coefficient for an air-stored 1:6 gravel concrete made with slag cement was 7.6×10^{-6}/F while that for plain concrete of the same mixture proportion was 7.3×10^{-6}/F. For water-stored concrete, the coefficients were 6.9×10^{-6}/F for slag concrete and 6.8×10^{-6}/F for plain concrete [Bonnell and Harper, 1951].

Durability Properties

GGBF slag has basically the same effect on concrete durability as do other mineral admixtures. The lower permeability of GGBF slag concrete is a principal reason for its superior durability properties. The coefficients of permeability of GGBF slag cement concrete for 30% and 60% replacement levels and of plain concrete are provided in Table 40.9 [Virtanen, 1983]. Its lower calcium hydroxide content makes it less susceptible to attack by chemicals.

Table 40.9 Coefficient of Permeability in 10^{11} cm/sec

Age	Portland Cement	30% Slag	60% Slag
28 days	30.0	9.9	2.0
90 days	3.0	<1.0	<1.0
1 year	1.0	<1.0	<1.0

Source: Virtanen, 1983.

GGBF slag concrete may carbonate faster than portland cement concrete if inadequately cured. Such concrete could be expected to carbonate at a slower rate owing to its decreased permeability and the presence of less calcium hydroxide. Recent work by the Forschungs-gemeinschaft Eisenhüllen-schlacken (German Slag Institute) on cores from tidal zones of 40-year old marine structures in northern France showed carbonation depths of less than 8 mm [Geisler and Lang, 1994].

It has been suggested that the less basic C-S-H gel carbonates faster owing to its higher surface area [Roy and Li, 1986]. The advantages of low permeability of GGBF slag concrete are lost on carbonation. The reduction of calcium hydroxide by carbonation can cause concern from the point of view of corrosion of steel. Other results indicate no difference in the carbonation of GGBF slag concrete compared to portland cement concrete [Duda, 1987].

Theoretically, the susceptibility of steel to corrosion should increase in GGBF slag concrete since both the pozzolanic reaction and the carbonation will reduce the alkalinity of the concrete pore solution. However, several factors in GGBF slag prevent this action. The relatively high lime content of industrial slags normally prevents the alkalinity of the concrete dropping to critical levels [Nakamura *et al.*, 1986] . However, lower chloride ion diffusion rates, rather than high alkalinity, are considered to be the main advantage of GGBF slag in preventing the corrosion of steel.

The low rapid chloride permeability results associated with GGBF slag concrete [Luther *et al.*, 1994] also suggest that the electrical resistivity of these concretes is high. This suggests that the concrete corrosion current may be lowered, and the rate of corrosion of embedded steel, likewise, may be slowed.

Using slag as a partial replacement for a Type I portland cement can provide the concrete good resistance to sulfate attack, with properties exceeding those of a Type II and a Type V sulfate resisting concrete. Improvements in sulfate resistance are due to (a) a dilution effect and the presence of less calcium hydroxide, which suppresses the formation of ettringite; (b) SO_4^{-2} ions have less tendency to diffuse in the concrete, since the pH of the pore solution of GGBF slag concrete is lower compared to plain concrete (the tendency of SO_4 ions to diffuse in GGBF slag concrete is less than that for Type V portland cement concrete); and (c) it has been suggested that SO_4^{-2} ions are not only not harmful but their presence may be beneficial, because they can act as an activator for the hydration of the slag. However, in using slag to combat sulfate attack one must be careful regarding the alumina content of the slag. Slags containing a higher alumina content (18% or higher) may impart lowered sulfate resistance to concrete. A 7.8% alumina slag when used as 50:50 replacement with Type I portland cement is equivalent to a Type V sulfate resistant portland cement concrete (ASTM C989). The presence of alumina in slag is not necessarily harmful. It can be rendered harmless if it forms solid solution with C-S-H gel.

GGBF slag concrete is also durable in freezing-and-thawing provided it is adequately air entrained. Studies [Litvan and Meyer, 1986; Canham *et al.*, 1987] indicate that slag cement concrete may need a higher dosage of air-entraining admixture, especially at low water–to–cementitious materials ratios, than portland cement concrete to obtain similar air content. The different chemistry of the hydrates and pore solution change the effectiveness of air-entraining admixtures and the air-bubble stability [Canham *et al.*, 1987], which suggests different air-void specifications for slag cements concrete may need to be specified. However, in field applications there have been no particular difficulties reported regarding achieving a traditionally acceptable air-void system.

Recent work indicated GGBF slag concretes can provide improved deicer scaling resistance if the appropriate GGBF slag dose is selected [Luther *et al.*, 1994]. Here, for ASTM C989 grade 100 or grade 120 materials, a 25% GGBF slag dose showed improved resistance, and a 35% GGBF slag dose achieved similar performance. GGBF slag dosages at 50% or 65% were more susceptible to deficiencies in wet curing, apparently, than the reference portland cements, unless a sealer was employed. In field testing, the use of sealers improved deicer scaling regardless of the concrete type.

It is known that slag has a potential benefit in concrete in reducing the expansion due to alkali-silica reaction. However, the mechanism is not well understood. The reduction in expansion has been associated with a number of factors, including: (a) the dilution effect related to addition of slag and the reduction of calcium hydroxide, (b) the presence of more C-S-H gel and the resulting ability to tie up more of the calcium hydroxide and alkalis, and last, but not least, (c) the lower permeability and diffusivity of GGBF slag concrete.

Mixture Proportioning

The proportioning procedure for concretes incorporating slag is similar to those used in proportioning other concrete as described in the ACI Committee 211 report. Considerable research has been done on mix proportioning of concrete with fly ash, but relatively little has been reported on proportioning of GGBF slag concrete. Normally, the same methods used for fly ash concrete can be used for slag, whether it is used as blended cement or as an admixture added at the time of batching. Any number of proportioning methods can be used when slag is added separately: simple replacement methods, rational methods, and modified or rational replacement methods.

In mixture proportioning, one must account for differences in the characteristics of fly ash and slag particles. The fly ash particles are mostly spherical with a smooth surface. In contrast, slag particles are angular and have a rough surface texture. Even though these particle shape characteristics are important considerations, they have no practical bearing on proportioning. Like fly ash, most GGBF slags somewhat reduce water demand at low to moderate dose (up to

~50%). Fineness of the portland cement and its alkali and C_3A content play a greater role than shape differences. For portland cement concretes, even these effects (alkali, C_3A) are often still neglected.

Acknowledgments

The reviews, comments, and constructive criticisms provided by Mr. Mark D. Luther, KOCH Minerals Company, and Professor Jan Olek, Purdue University, are thankfully acknowledged. Some of the information in the section dealing with slag have been obtained from term papers written by present and former graduate students enrolled in the author's course, CE 597—*Special Cements and Concretes*. Thanks are specifically due to Mr. D. Constantiner (Ph.D. candidate), Dr. J.-D. Lin, Dr. C. Lobo, Dr. K. Matsukawa, and Dr. S. Ong for their hard work and contributions.

References

ACI Committee 212. 1989. Chemical admixtures for concrete. *Amer. Conc. Inst. Matls. J.* 86(3): 297–327.

ACI Committee 226. 1987a. Ground-granulated-blastfurnace slag as a cementitious constituent in concrete. *ACI Manual of Concrete Practice*, Part 1. American Concrete Institute, Detroit, MI.

ACI Committee 226. 1987b. Silica fume in concrete. *Amer. Conc. Inst. Matls. J.* 84(2):158–166.

ACI Committee 234. 1994. Silica fume in concrete. Draft Report. American Concrete Inst., Detroit, MI.

Aitcin, P. C. and Vezina, D. 1984. Resistance to freezing and thawing of silica fume concrete. *Cement, Concrete, and Aggregates*. 6(1):38–42.

Aitcin, P. C., Pinsonneault, P., and Fortin, R. 1980. Vaporization of silica fumes. *Seminar on Industrial Fillers.* CANMET, Energy, Mines and Resources, Ottawa, Canada.

Aitcin, P. C., Pinsonneault, P., and Rau, G. 1981. The use of condensed silica fume in concrete. *Effects of Fly Ash in Concrete, Matls Res. Soc. Sympos. Proc.*, pp. 316–325.

Anon. 1983. *Civil Eng.* 53(12):7.

Bache, H. H. 1981. Densified cement/ultra-fine particle-based materials. Presented at the 2nd International Conference on Superplasticizers in Concrete, Ottawa, Canada.

Bamforth, P. B. 1980. In situ measurements of the effect of partial portland cement replacement using either fly ash or ground-granulated-blastfurnace slag on the performance of mass concrete. *Proc. Inst. Civil Engrs.* 69(2):777–800.

Barnes, B. D., Diamond, S., and Dolch, W. L. 1978. Contact zone between portland cement paste and glass aggregate surface. *Cement Conc. Res.* 8:233–243.

Bentur, A. and Cohen, M. D. 1987. The effect of condensed silica fume on the microstructure of the interfacial zone in portland-cement mortars. *J. Amer. Ceram. Soc.* 70:738–743.

Bentur, A., Goldman, A., and Cohen, M. D. 1987. The contribution of the transition zone to the strength of high quality silica-fume concretes. *Bonding in Cementitous Composites. Matls. Res. Soc. Sympos. Proc.* 114:97–103.

Berry, E. E. 1980. Concrete made with supplementary cementing materials. In Malhotra, V. M., ed, *Progress in Concrete Technology*, pp. 333–365. CANMET, Energy, Mines, and Resources, Ottawa, Canada.

Berry, E. E. and Malhotra, V. M. 1986. *Fly Ash In Concrete.* Spec. Pub. 85-3. American Concrete Inst., Detroit, MI.

Berry, E. E. and Malhotra, V. M. 1987. Fly Ash Concrete. In Malhotra, V. M., ed., *Supplementary Cementing Materials for Concrete*, Chap. 2. CANMET, Energy, Mines, and Resources, Ottawa, Canada.

Bonen, D. and Cohen, M. D. 1992a. The course of magnesium sulfate attack on portland-cement and portland-cement-with-silica-fume pastes. I. Microstructural analysis. *Cement Conc. Res.* 22(1): 169–180.

Bonen, D. and Cohen, M. D. 1992b. The course of magnesium sulfate attack on portland-cement and portland-cement-with-silica-fume pastes. II. Chemical and mineralogical analyses. *Cement Conc. Res.* 22(4):697–708.

Bonnell, D. G. R. and Harper, F. C. 1951. The thermal expansion of concrete. *Natl. Bld. Studies.* Tech. Paper No. 7.

Buck, A. and Burkes, J. P. 1981. Characterization and reactivity of silica fume. *Proc. 3rd Intnl. Conf. Cement Microscopy,* pp. 279–285

Canham, L., Page, C. L., and Nixon, P. J. 1987. Aspects of the pore solution chemistry of blended cements related to the control of alkali silica reaction. *Cement Conc. Res.* 17:839–844.

CEMCOM Internal report. DASH 47. The Development of a High-Strength Cementitious Tooling/Molding Material. CEMCOM Research Associates, Laanham, MD.

Cohen, M. D. and Bentur, A. 1988. Durability of portland-cement silica-fume pastes in magnesium sulfate and sodium sulfate solutions. *J. Amer. Conc. Inst.* 85(3):148–157.

Cohen, M. D. and Klitsikas, M. 1986. Mechanisms of hydration and strength development in portland cement composites containing silica fume particles. *Indian Conc. J.* 60(11):296–300.

Cohen, M. D. and Mather, B. 1991. Sulfate attack on concrete—research needs. *Amer. Conc. Inst. Matls. J.* 88(1):62–69.

Cohen, M. D. and Olek, J. 1989. Effects of form of silica fume on engineering performance of portland-cement concrete. *Conc. Intnl.: Design and Construction* 11(11):43–47.

Cohen, M. D., Goldman, A., Chen, W. F. 1994. The role of silica fume in mortar. Transition zone versus bulk paste modification. *Cement Conc. Res.* 24(1).

Cohen, M. D., Olek, J., and Dolch, W. L. 1989. Effects of the form of silica fume on plastic-shrinkage cracking of portland-cement paste. *Proc. 3rd Intnl. Conf. on Fly Ash, Silica Fume, and Natural Pozzolans in Concrete.* Spec. Pub. 114, pp. 431–443. American Concrete Inst., Detroit, MI.

Cohen, M. D., Olek, J., and Dolch, W. L. 1990. Mechanisms of plastic shrinkage cracking in portland-cement-silica-fume paste and mortar. *Cement Conc. Res.* 20(1):103–119.

Cohen, M. D., Zhou, Y., and Dolch, W. L. 1992. Non-air-entrained high-strength concrete—Is it frost resistant? *Amer. Conc. Inst. Matls. J.* 89(4):406–415.

Compton, F. R. and MacInnis, C. 1952. Field trial of fly-ash concrete. *Ontario Hydro Res. News,* pp. 18–21.

Cong, X., Garg, S., Darwin, D., and McCabe, S. L. 1992. Role of silica fume in compressive strength of cement paste, mortar, and concrete. *Amer. Conc. Inst. Matls. J.* 89(4):375–387.

Cook, D. J. 1986. Natural pozzolans in concrete. In Swamy, R. N., ed., *Concrete Technology and Design. Vol 3. Cement Replacement Materials,* pp. 1–39. Surrey Univ. Press, Black & Sons, London.

Darwin, D. and Slate, F. O. 1970. Effect of paste-aggregate bond strength on behavior of concrete. *J. Matls.* 5(1):86–98.

Dhir, R. K. 1986. Pulverized fuel ash. In Swamy, R. N., ed., *Concrete Technology and Design. Vol. 3, Cement Replacement Materials,* pp. 197–255. Surrey Univ. Press, Black & Sons, London.

Diamond, S. 1975. Long-term status of calcium hydroxide saturation of pore solution in hardened cements. *Cement Conc. Res.* 5(6):607–616.

Diamond, S. 1983a. On the glass present in low-calcium and in high-calcium flyashes. *Cement Conc. Res.* 13(4):459–464.

Diamond, S. 1983b. Effects of microsilica (silica fume) on pore solution chemistry of cement pastes. *J. Amer. Ceram. Soc.* 66(5):C 82–84.

Diamond, S. 1984. *Scientific Basis for the Use of Microsilica in Concrete.* Elkem Chemicals, Kristiansand, Norway.

Diamond, S. 1985. Very high-strength cement-based materials—A prospective. *Very High-Strength Cement-Based Materials. Matls. Res. Soc. Sympos. Proc.* 42:233–243.

Diamond, S., Mindess. S., and Lovell, J. 1982. On the spacing between aggregate grains in concrete and the dimensions of the *aureole de transition. Proc. Liaisons Pates de Ciment Materiaux Associes, RILEM, Toulouse, France:* C 42–46.

Douglas, E. and Zerbino, R. 1986. Characterization of granulated and pelletized-blastfurnace slag. *Cement Conc. Res.* 16(5):662.

Duda, A. 1987. Aspects of sulfate resistance of steelwork-slag cements. *Cement Conc. Res.* 17:373–384.

Duncan, M. A., Swenson, E. G., Gillott, J. E., and Foran, M. 1973. Akali-aggregate reaction in Nova Scotia. I. Summary of five-year study. *Cement Conc. Res.* 3:55–59.

Dunstan, E. R. 1976. *Performance of Lignite and Sub-Bituminous Fly Ash in Concrete—a Progress Report.* Report REC- ERC-76-1, U.A. Bureau of Reclamation, Denver, CO.

Dunstan, E. R. 1980. A possible method for identifying fly ashes that will improve the sulfate resistance of concretes. *Cement Conc. and Aggregates.* 2:20–30.

Dunstan, E. R. 1981. The effect of fly ash on concrete alkali-aggregate reaction. *Cement Conc. and Aggregates.* 3:101–104.

Elkem Chemicals. 1985. How microsilica improves concrete. *Concr. Construction Pub.* 30(4):327–332.

Emery, J. J. 1989. *The Blast Furnace and the Manufacture of Pig Iron.* National Slag Assoc. Bull. MF 182-2. David Williams Co., New York.

Endo, H., Kodama, K., Nakagawa, O., and Takada, M. 1987. Influences of ground-granulated-blast-furnace slag on mixture proportion and strength of concrete. (In Japanese). *Japan Soc. Civil Engrs., Sympos. Applic. of Ground Granulated Blastfurnace Slag in Concrete.*

Fidjestol, P., Luther, M. D., Danielssen, T., Obuchowicz, M., and Tutokey, S. 1989. Silica fume—efficiency versus form of delivery. *Amer. Conc. Inst., Spec. Pub. 114,* pp. 568–584.

Fulton, F. S. 1974. *The Properties of Portland Cements Containing Milled Granulated Blastfurnace Slag.* South Africa Portland Cement Inst., Johannesburg.

Geisler, J. and Lang, E. 1994. Long-term durability of non-air-entrained concrete structures exposed to marine environments and freezing-and-thawing cycles. *Third CANMET/ACI Intnl. Conf. Durability of Conc.* Nice, France.

Ghosh, R. S. and Timusky, J. 1981. Creep of fly-ash concrete. *J. Amer. Conc. Inst.* 78(5):351–357.

Gifford, P. M. and Ward, M. A. 1982. Results of laboratory tests on lean mass concrete utilizing PFA to a high level of cement replacement. *Proc. Intnl. Sympos. Use of PGA in Concrete, Leeds, England,* pp. 221–230.

Gjorv, O. E. 1983. Durability of concrete containing condensed silica fume. In Malhotra, V. M., ed., *Fly Ash, Silica Fume, Slag, and Other Mineral By-Products in Concrete.* Spec. Pub. 79, pp. 695–703. American Concrete Inst., Detroit, MI.

Goldman, A. 1987. Properties of concretes and pastes with microsilica. *Thesis.* Israel Inst. of Technology, Haifa.

Goldman, A. and Bentur, A. 1993. The influence of microfillers on enhancement of concrete strength. *Cement Conc. Res.* 23(4):962–972.

Helmuth, R. A. 1987. Fly Ash in Cement and Concrete. Portland Cement Assoc., Skokie, IL.

Hogan, F. J. and Meusel, J. W. 1981. Evaluation for durability and strength development of a ground-granulated-blastfurnace slag. *Cement Conc. and Aggregates* 3:40–52.

Holland, T. C. 1989. Working with silica fume in ready-mixed concrete—USA experience. *Proc. 3rd Intnl. Conf. on Fly Ash, Silica Fume, and Natural Pozzolans in Concrete.* Spec. Pub. 114, pp. 763–781. American Concrete Inst., Detroit, MI.

Holland, T. C. and Luther, M. D. 1987. Improving concrete quality with silica fume. *Lewis H. Tuthill Intnl. Symp. on Concrete and Concrete Construction.* Spec. Pub. 104, pp. 107–112. American Concrete Inst., Detroit, MI.

Holland, T. C., Krysa, A., Luther, M. D., and Liu, T. C. 1986. Use of silica-fume concrete to repair abrasion-erosion damage in the Kinzua Dam stilling basin. *Proc. 2nd Intnl. Conf. on Fly Ash, Silica Fume, and Natural Pozzolans in Concrete.* Spec. Pub. 91, pp. 841–864. American Concrete Inst., Detroit.

Idorn, G. M. 1983. Thirty years with alkalis in concrete. *Proc. 6th Intnl. Conf. on Alkalis in Concrete.* Danish Concrete Assoc., Copenhagen, pp. 21–38.

Indian Conc. J. 1985. Silica fume and concrete (editorial comment). 59(8):197–198, 203.

Josephson, G. W., Sillers, F., Jr., and Runner, D. G. 1949. *Iron Blastfurnace Slag Production, Processing, Properties, and Uses.* Bull. 479. U.S. Depart. of the Interior, GPO, Washington, DC.

Justesen, C. F. 1981. *Performance of DENSIT Injection Grout for Prestressing Tendons.* Project Reprt. CFJ/VAS/HSQ/EJ/ul. Aalborg Portland Cement, Aalborg, Denmark.

Kilgour, C. L. 1988. Composition and properties of Indiana fly ashes. *Thesis.* Purdue University, West Lafayette, IN.

Killin, A. M. 1973. Progress report air pollution control study of the ferroalloy industry. *Electric Furnace Proc.* 31:66.

Klieger, P. and Isberner, A. W. 1967. *Laboratory Studies of Blended Cements—Portland-Blastfurnace-Slag Cements.* R & D Bull. 218. Portland Cement Association, Skokie, IL.

Kobayashi, K. 1979. *Partial Replacement of Portland Cement by Ground-Granulated-Blastfurnace Slag.* Japan-U.S. Science Seminar, San Francisco.

Kobayashi, K., Uomoto, T., and Shima, F. 1979. Influences of ground-granulated-blastfurnace slag as an admixture for concrete in its compressive strength and drying shrinkage (in Japanese). *Conc. J. of the Japan Conc. Inst.* 17(5):87–95.

KOCH Minerals Co. 1992. *GranCem Cement for Alkali-Silica Reaction.* KOCH Minerals Co., Wichita, KS.

KOCH Minerals Co. 1992. *GranCem Cement Brochure.* KOCH Materials Co., Wichita, KS.

Lee, A. R. 1974. *Blastfurnace and Steel Slag—Production, Properties, and Uses.* Wiley & Sons, New York.

Lewis, D. W. 1986. *Cementitious Applications of Ground-Iron-Blastfurnace Slag.* Technical Seminar MF 184-5. National Slag Association, Washington, DC.

Litvan, G. G. and Meyer, A. 1986. Carbonation of granulated- blastfurnace-slag-cement concrete during twenty years of field exposure. *Proc. 2nd Intnl. Conf. on Fly Ash, Silica Fume, and Natural Pozzolans in Concrete.* Spec. Pub. 91, pp. 1445–1462. American Concrete Inst., Detroit.

Luther, M. D. 1988. Silica-fume (microsilica)-concrete in bridges in the United States. *Trans. Res. Record* 1204:11–20.

Luther, M. D. 1989a. Silica-fume materials and action in concrete. *Proc. of the Seminar on Recent Advances in Concrete Technology.* Michigan State Univ.:13.2–13.3.

Luther, M. D. 1989b. Microsilica (silica fume) concrete durability in severe environments. *Amer. Soc. Civil Engrs., Proc. 7th Annual Struc. Cong., Matls. Processing Sec.* pp. 95–105.

Luther, M. D. 1990a. High-performance-silica-fume (microsilica)-modified-cementitious repair materials. *Trans. Res. Record* 1284:88–94.

Luther, M. D. 1990b. Case studies of microsilica (silica fume) concrete repair projects. *Amer. Soc. Civil Engrs. Structures Congress Abstracts*, pp. 533–534.

Luther, M. D. 1990c. Silica fume (microsilica): Production, materials, and action in concrete. *How To Effectively Use the Newest Admixtures.* Bradley University.

Luther, M. D. 1991. What revolution? Letter to the editor. *Concrete Producer News,* March, p. 2.

Luther, M. D. 1993. Review of silica-fume (microsilica)-concrete use in bridges. *Conc. International* 15(4):29–33.

Luther, M. D. and Hansen, W. 1989. Comparison of creep and shrinkage of high-strength-silica-fume concretes with fly-ash concretes of similar strength. *Proc. 3rd Intnl. Conf. on Fly Ash, Silica Fume, and Natural Pozzolans in Concrete.* Spec Pub. 114, pp. 573–591. American Concrete Inst., Detroit, MI.

Luther, M. D. and Kojundic, A. N. 1991. Rapid chloride permeability test and microsilica (silica fume) concrete for new construction and for rehabilitation of deteriorated concrete. *Proc. CANMET/ACI 3rd Intnl. Silica Fume Workshop.* Washington, DC, pp. 207–239.

Luther, M. D. and Smith, F. A. 1991. Silica-fume (microsilica) fundamentals for use in concrete. *Utilization of Cement-Based Materials.* Engineering Foundation, New York.

Luther, M. D., Mmikols, W. J., De Maio, A. J., and Whitlinger, J. E. 1994. Scaling resistance of ground-granulated-blastfurnace-slag concretes. *Proc. 3rd Intnl. Conf. on Durability of Concrete.* Nice, France.

Luther, M. D., Prisby, R. D., and Kojundic, D. N. 1992. Long-term strength development of silica-fume (microsilica) concrete: Influence of mixing procedure and silica fume product form. *Proc. 4th CANMET/ACI Intnl. Conf. on Fly Ash, Silica Fume, Slag, and Natural Pozzolans in Concrete.* Istanbul, pp. 427–449.

Magne, M. and Sellevold, E. J. 1987. Effect of microsilica on the durability of concrete structures. *Conc. International* 9(12):39–43.

Malhotra V. M. and Carette, G. G. 1983. Silica-fume concrete—properties, applications, and limitations. *Conc. International* 5(5):40–46.

Malhotra, V. M., Ramachandran, V. S., and Feldman, R. F. 1987. *Condensed Silica Fume in Concrete.* CRC Press, Boca Raton, FL.

Massazza, F. and Costa, U. 1979. Aspects of the pozzolanic activity and properties of pozzolanic cements. *Il Cemento* 76:3–13.

Mather, K. 1982. Current research in sulfate resistance at the Waterways Experiment Station. *George Verbeck Symp. on Sulfate Resistance of Concrete.* Spec. Pub. 77, pp. 63–74. American Concrete Inst., Detroit, MI.

Mehta, P. K. 1983. Pozzolanic and cementitious by-products as mineral admixtures for concrete—A critical review. In Malhotra, V. M., ed., *Fly Ash, Silica Fume, Slag, and Other Mineral By-Products in Concrete.* Spec. Pub. 79, pp. 695–703. American Concrete Inst., Detroit, MI.

Mehta, P. K. 1985. Studies on chemical resistance of low-water-cement ratio concretes. *Cement Conc. Res.* 15:969–978.

Mehta, P. K. and Gjorv, O. E. 1982. Properties of portland-cement concrete containing fly ash and condensed silica fume. *Cement Conc. Res.* 12(5):587–595.

Mehta, P. K. and Monteiro, P. J. M. 1993. *Concrete*, 2nd ed. Prentice Hall, Englewood Cliffs, NJ.

Mehta, P. K. 1987. Natural pozzolans. In Malhotra, V. M., ed., *Supplementary Cementing Materials,* Chap. 1. CANMET, Energy, Mines, and Resources, Ottawa, Canada.

Meusel, J. W. and Rose, J. H. 1983. Production of granulated-blastfurnace slag at Sparrow's Point, and workability and strength potential of concrete incorporating the slag. In Malhotra, V. M., ed., *Fly Ash, Silica Fume and Other Mineral By-Products in Concrete.* Spec. Pub. 79, pp. 867–890. American Concrete Inst., Detroit, MI.

Milewski, J. V. 1978. *Handbook of Fillers and Reinforcements for Plastics*, Chap. 4. Van Nostrand Reinhold, New York.

Mindess, S. and Young, J. F. 1981. *Concrete.* Prentice Hall, Englewood Cliffs, NJ.

Nakamura, N., Sakai, M., Koibuchi, K., and Iijima, Y. 1986. Properties of high-strength concrete incorporating very finely-ground-granulated-blastfurnace slag. *Proc. 2nd Intnl. Conf. on Fly Ash, Silica Fume, and Natural Pozzolans in Concrete.* Spec. Pub. 91, pp. 1361–1380. American Concrete Inst., Detroit, MI.

National Slag Association. *Slag.* Bull. 188-1. NSA, Washington, DC.

Neville, A. M. 1981. *Properties of Concrete,*. 3rd ed. Pitman, London.

Newman, K. 1986. Blended cements or concrete mixer blends—which does the concrete producer prefer? *Proc. 2nd Intnl. Conf. on Fly Ash, Silica Fume, and Natural Pozzolans in Concrete.* Spec. Pub. 91, pp. 1590–1601. American Concrete Inst., Detroit, MI.

Nilsen, J. A. and Young, J. F. 1977. Additions of colloidal silicas and silicates to portland-cement pastes. *Cement Conc. Res.* 7:277–282.

Ogawa, K., Uchikawa, H., Takemoto, K., and Yassui, I. 1980. The mechanisms of the hydration in the system C_3S-pozzolans. *Cement Conc. Res.* 10(5):683–696.

Olafsson, H. 1982. *Effect of Silica Fume on the Properties of Cement and Concrete.* Report BML82.610. Norwegian Institute of Technology, Trondheim.

Olek, J. 1987. Properties of fly-ash concrete. *Thesis.* Purdue University, West Lafayette, IN.

Olek, J. and Diamond, S. 1989. Proportioning of constant paste composition fly-ash concrete mixes. *Amer. Conc. Inst. Matls. J.* 86(2):159–166.

Owens, J. 1987. Slag-iron and steel. *Minerals Yearbook,* 738. U.S. Bureau of Mines, Washington, DC.

Pepper, L. and Mather, B. 1959. Effectiveness of mineral admixtures in preventing excessive expansion of concrete due to alkali-aggregate reaction. *Proc. Amer. Soc. Testing Matls.* 59:1178–1202.

Philleo, R. E. 1989. Slag or other supplementary materials. *Proc. 3rd International Conf. on Fly Ash, Silica Fume, and Natural Pozzolans in Concrete.* Spec. Pub. 114, pp. 1197–1207. American Concrete Inst., Detroit, MI.

Pierce, J. S. 1982. Use of fly ash in combating sulfate attack in concrete. *Proc. 6th Intnl. Sympos. on Fly Ash Utilization.* Reno, NV, pp. 208–231.

Price, W. H. 1975. Pozzolans—A review. *Proc. Amer. Conc. Inst.* 72:225–232.

Reeves, C. M. 1980. *The Production, Properties and Applications of Blastfurnace Slag with Particular Reference to Portland-Blastfurnace-Cement Concretes, Mortars, and Grouts Made With Cemsave.* Cement and Concrete Association, Slough, Bucks, England.

Regourd, M. 1985. Slags and slag cements. In Roy, D. M., ed., *Modules in Cement Science,* pp. 93–111. Materials Education Council, Pennsylvania State University, University Park, PA.

Regourd, M. 1986. Slags and slag cements. In Swamy, R. N., ed., *Concrete Technology and Design. Vol. 3. Cement Replacement Materials,* pp. 73–99. Surrey Univ. Press, Black & Sons, London.

Regourd, M., Mortureux, B., Gautier, B., Hornnin, H., and Volant, J. 1980. Characterization et activation thermique des ciments au laitier. *7th Int. Congr. Chem. Cement,* Paris, 1980, Vol. II, III, pp. 105–111, Edition Septima, Paris.

Rickne, S. and Svensson, C. 1982. The new Tjorn bridge. *Nordisk Betong* 2–4:213–217.

Roy, D. M. and Li, S. 1986. Investigation of relations between porosity, pore structure, and chloride ion diffusion of fly-ash and blended-cement pastes. *Cement Conc. Res.* 16(5):749–759.

Sellevold, E. J. 1987. The function of condensed silica fume in high strength concrete. *Proc. Sympos. on Utilization of High-Strength Concrete.* Stavanger, Norway, pp. 39–49.

Sellevold, E. J. and Nilsen, T. 1987. Condensed silica fume in concrete: A world review. In Malhotra, V. M., ed., *Supplementary Cementing Materials for Concrete.* Chap. 3. CANMET, Energy, Mines, and Resources, Ottawa, Canada.

Sellevold, E. J. and Radjy, F. F. 1983. Condensed silica fume (microsilica) in concrete: Water demand and strength development. In Malhotra, V. M., ed., *Fly Ash, Silica Fume, Slag and Other Mineral By-Products in Concrete.* Spec. Pub. 79, pp. 677–689. American Concrete Inst., Detroit, MI.

Sheetz, B. E., Grutzeck, M., Strickler, D. W., and Roy, D. M. 1981. Effect of composition of additives upon microstructures of hydrated cement composites. *Proc. 3rd Intnl. Conf. on Cement Microscopy.* Houston, TX.

Stanton, T. E. 1942. Expansion of concrete through reaction between cement and aggregate. *Trans. Amer. Soc. Civil Engrs.* 2:68–85; 1940. *Proc. Amer. Soc. Civil Engrs.* 66:1781–1811.

Taylor, I. F. 1974. Cutting costs with slag cement. *Civil Eng.*, pp. 45–49.

Thomas, A. 1979. *Metallurgical and Slag Cements—the Indispensable Energy Savers.* Inst. Elec., Electronic Engrs., Cement Industry Technical Conf. The Engineering Foundation, New York.

Tikalsky, P. J. 1989. The effect of fly ash on the sulfate resistance of concrete. Thesis. University of Texas, Austin.

Transportation Research Circular. 1990. *Admixtures and Ground Slag for Concrete.* Circ. No. 365. Transportation Research Board, Washington.

Virtanen, J. 1983. Freeze-thaw resistance of concrete containing blastfurnace slag, fly ash and condensed silica fume. In Malhotra, V. M., ed., *Fly Ash, Silica Fume, Slag, and Other Mineral By-Products in Concrete.* Spec. Pub. 79, pp. 695–703. American Concrete Inst., Detroit, MI.

Vladimir, S., Dubovoy, S. D., Steven, H. G., and Paul, K. 1986. Effects of ground-granulated-blastfurnace slag on some properties of pastes, mortars, and concretes. In Frohnsdorff, G., ed., *Blended Cements,* STP 897. American Society for Testing and Materials, Philadelphia, PA.

Wainwright, P. J. 1987. Properties of fresh and hardened concrete. In Swamy, R. N., ed., *Concrete Technology and Design. Vol. 3. Cement Replacement Materials.* Chap. 4. Surrey Univ. Press, Black & Sons, London.

Wise, S., Satkowski, J. A., Scheetz, B., Rizer, J. M., MacKenzie, M. L., and Double, D. D. 1985. The development of a high-strength cementitious tooling/molding material. *Very High-Strength Cement-Based Materials. Matls. Res. Soc. Sympos. Proc.* 42:253–263.

Wolsiefer, J. 1984. Ultra high-strength field-placeable concrete with silica fume admixture. *Conc. Intnl.* 6(4):25–31.

For Further Information

Lea, F. M. 1971. *The Chemistry of Cement and Concrete,* 3rd ed., Chemical Publishing, New York.

Malhotra, V. M. 1987. *Supplementary Cementing Materials for Concrete.* CANMET, Energy, Mines, and Resources, Ottawa, Canada.

Skalny, J., ed. 1989, 1990. *Materials Science of Concrete,* Parts I and II. American Ceramic Soc., Westerville, OH.

41

Fracture and Fatigue in Steel Structures

Mark D. Bowman
Purdue University

41.1 Brittle Fracture of Steel Structures

Brittle fracture is defined by Barsom and Rolfe [1987] as "a type of catastrophic failure in structural materials that usually occurs without prior plastic deformation and at extremely high speeds—as high as 2130 meters/sec (7000 feet per second)." Often, a structure that fractures in a brittle fashion exhibits little prior ductility or deformation. Moreover, because the fracture occurs at stresses well below the yield strength and at a rapid rate with virtually no forewarning, the consequences can be catastrophic.

There have been many examples of brittle fractures during the past century. An early example occurred in October 1886 on Long Island, New York when a 76-meter- (250-foot-) tall water standpipe failed during acceptance testing. Another notable fracture was the failure of a molasses tank in Boston, Massachusetts in January 1919, which resulted in a number of deaths and injuries when the tank ruptured. The energy release upon fracture of the tank was so great that a portion of the Boston elevated railroad was knocked over.

During the Second World War, a number of Liberty-class ships were manufactured by the U.S. to ship materials needed for the war effort. Many of these vessels developed very large cracks (up to 3 m) in the deck near the middle of the ship at sharp hatch corners and at ladder openings and were subsequently classified as dangerous. Moreover, a number of ships completely fractured, resulting in the loss of many sailors and materials.

Highway and railway bridge structures are also susceptible to brittle fracture. One of the most serious bridge failures in the U.S. was at the Silver Bridge in Point Pleasant, West Virginia, which collapsed on 15 December 1967. The fracture was particularly ominous since it occurred with little warning while many were commuting from work during the afternoon rush hour. A total of 46 people were killed and many others were injured. The bridge was a suspension-type structure, built in 1928, which utilized a pair of eyebar members for the suspension elements. The collapse occurred when one eyebar link fractured after a small, 3-mm (1/8-in.) crack developed on the inside of the eyebar.

Several other bridge fractures also have been reported. For example, a structural bent on the Dan Ryan Rapid Transit Elevated Line developed a large crack that was discovered by a commuter [Koncza, 1979]. The structure was immediately closed to traffic and the Chicago Transit Authority

0-8493-8953-4/95/$0.00 + $.50

quickly repaired the structure so that the busy line could be returned to service. Also, a brittle fracture completely severed the bottom flange and web of a highway bridge fascia girder on the I-79 Neville Island Bridge over the Ohio River, west of Pittsburgh [FHWA, 1986]. The fracture initiated from cracks that developed in a repair weld in the bottom flange. A repair weld was also responsible for the brittle fracture of a significant portion of the tension flange and web of a steel box girder railway bridge [Bruestle, 1990]. The 2.44-m (8-ft) crack was traced to a flaw in a repair weld that experienced cold cracking as a result of improper welding procedures.

Significant Factors in Brittle Fracture

The appearance of a brittle fracture surface is often granular or crystalline. The fracture surface is usually flat with small shear lips on the edges. Very little reduction in cross section occurs during the fracture, which corroborates the lack of ductility exhibited during a brittle fracture. Another characteristic common in brittle fracture is the presence of a herringbone pattern on the fracture surface that points to the site of fracture initiation.

The following factors are known to play a significant role in the fracture resistance of a steel structure:

- Geometrical configuration
- Material characteristics
- Environment

If all three factors increase the brittle fracture susceptibility of a steel, then the likelihood of a brittle fracture is significantly enhanced. Likewise, if the factors all exhibit a low brittle fracture susceptibility, then a brittle fracture is unlikely. Consequently, the engineer can influence brittle fracture susceptibility during the design stage by controlling these critical factors.

Geometrical Configuration

The geometry of a structure plays a major role in the fracture resistance of a structure. Particularly important are sudden changes in the geometry, which occur whenever two members are joined together by riveting, bolting, or welding. Sudden changes in geometry also occur when it is necessary to add openings in a structure, introduce changes in cross section thickness or width, or when poor fabrication procedures introduce unwanted notches, grooves, or cracks. Sudden changes in geometry cause a stress concentration that elevates the nominal stress at the geometric anomaly. The stress concentration effect can be minimized by using rounded corners and smooth fillets at thickness and width transitions.

Several tests, each with a different geometry, have been devised to measure the fracture toughness of steel structures. For example, a *Charpy V-notch test* uses a small square beam section with a V-notch machined in the middle of one side, while the *NDT drop weight test* uses a notch cut into a weld bead that is placed at the middle of a plate section. A comparison of the nil ductility transition temperature measured by several different fracture toughness tests reveals a wide range of results for various thicknesses of materials and indicates that the geometry of the test specimen plays a significant role in the fracture toughness measured.

Material Characteristics

It is well known that the composition of a structural steel influences the yield strength, tensile strength, ductility, and weldability. Steel composition also influences fracture toughness. For example, low Charpy V-notch transition temperatures generally correspond to low carbon steels, and vice versa [Parker, 1957].

Manganese is also known to increase the Charpy energy transition temperature as the content increases. The influence of other elements is not conclusive, since some elements can cause different effects at different levels, some elements interact with others, and some elements exhibit different effects on unhardened steels than on hardened steels.

Another important factor is ferrite grain size in pearlite steels. Fine-grain steels generally have lower transition temperatures than coarse-grain steels.

Environment

Environment is a broad term involving external influences to which the structure is subjected. Included are factors such as temperature, corrosive elements, radiation, and structural loading. The geometry and the material characteristics of a structure do not change after the structure is fabricated, unless the structure is physically altered through retrofitting or heat treatment. The environment, on the other hand, can and often does continuously change throughout the service life of the structure.

Ambient temperature can significantly influence the fracture toughness of a steel structure. This influence is clearly demonstrated in a Charpy V-notch test, in which the amount of energy absorbed in fracturing the V-notch coupon is directly related to temperature, as shown in Fig. 41.1. At low temperatures, the energy absorption of Steel A is quite low and falls in the lower shelf region of the fracture toughness curve. As the temperature increases, however, a transition occurs and the fracture toughness reaches an upper shelf or plateau. Each heat of steel has a unique Charpy V-notch characteristic curve. A decrease in the fracture toughness of the steel would cause the characteristic curve to shift to the right, as indicated by Steel B in Fig. 41.1.

The stress in a structure also significantly influences the propensity for brittle fracture. For example, tests conducted using 1.8-m- (6-ft-) wide plate specimens demonstrated that stress levels as low as one-fourth of the tensile strength can propagate a brittle fracture, but that stresses near the yield strength are needed to initiate the fracture [Mosborg *et al.*, 1957]. As the state of stress that a structure experiences increases, the structure becomes increasingly sensitive to internal cracks and discontinuities.

In addition to the level of stress, the rate at which a structure experiences a loading also plays a role in the fracture resistance. The fracture toughness of a steel that is loaded at various rates is shown in Fig. 41.2. The figure illustrates that the fracture toughness of the steel when loaded slowly

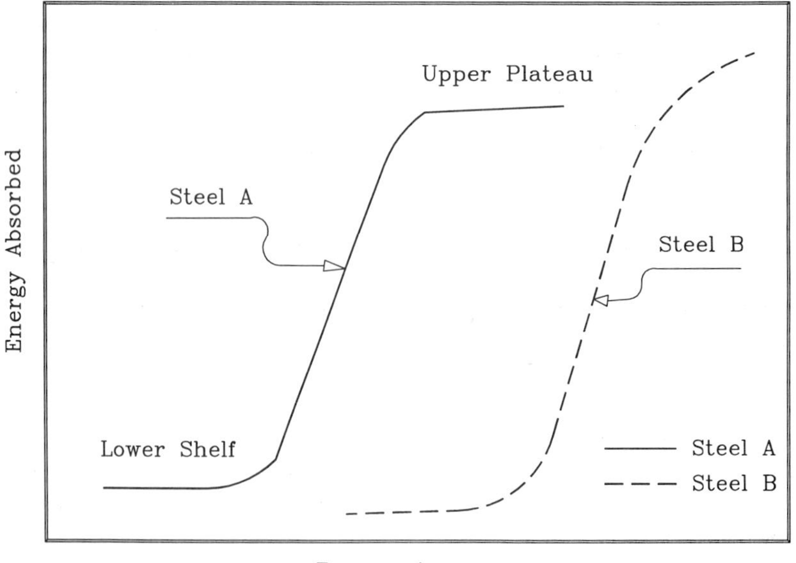

FIGURE 41.1 Absorbed energy in Charpy V-notch test versus temperature for two steels.

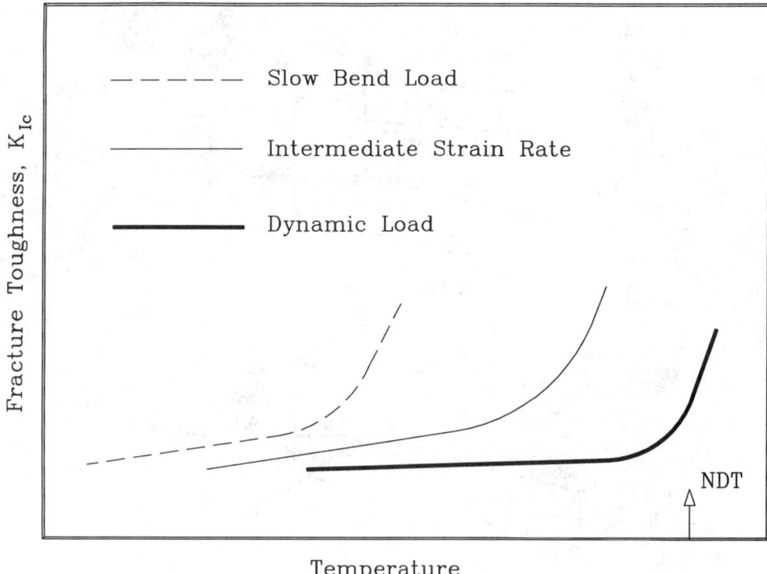

FIGURE 41.2 Fracture toughness K_{Ic} versus temperature for various loading rates.

is significantly greater than when loaded dynamically. This characteristic is important in bridge and crane structures, which are often subjected to loadings rates somewhere between static and impact.

Another factor that can influence brittle fracture resistance is the presence of residual stresses. Residual stresses are internal, self-equating stresses that can be generated as a result of differential cooling or mechanical straining of a structure. Welding, for example, causes differential heating of a structure and often results in residual stresses that approach the yield strength of the base metal as the welds cool. Residual stresses alone, however, are generally not responsible for brittle fracture. Instead, residual stresses are detrimental when they act together with an applied stress and geometrical discontinuities, resulting in a significant stress concentration.

Fracture Strength Evaluation

The fracture behavior of structural members has been widely studied. Griffith [1920] proposed one of the first models to evaluate the fracture behavior of brittle materials with sharp discontinuities. The Griffith model is an energy method that is based upon the elastic energy supplied at the crack tip during an incremental increase in crack length exceeding the elastic energy available at the crack tip associated with the incremental increase in crack length.

The energy-balance approach is not suitable for conditions of stable crack extension, such as fatigue and stress-corrosion crack growth. The stress-intensity parameter, however, is suitable for stable crack growth conditions. Consequently, linear-elastic fracture mechanics is more commonly used to evaluate crack growth behavior and fracture resistance of structural members.

Fracture Mechanics

Fracture mechanics is used to relate stresses in the vicinity of a crack tip to the nominal stresses applied in a structure, as well as geometry of the structure and characteristics of the crack. In linear-elastic fracture mechanics, the stress field in the crack tip vicinity can be described by a single variable, K, called the stress-intensity factor. To conduct the stress analysis needed to describe K, one of three modes of crack surface displacement must be defined, as shown in Fig. 41.3 [Paris and Sih, 1965]. Mode I is used to describe crack opening, mode II for crack sliding, and mode III for crack tearing.

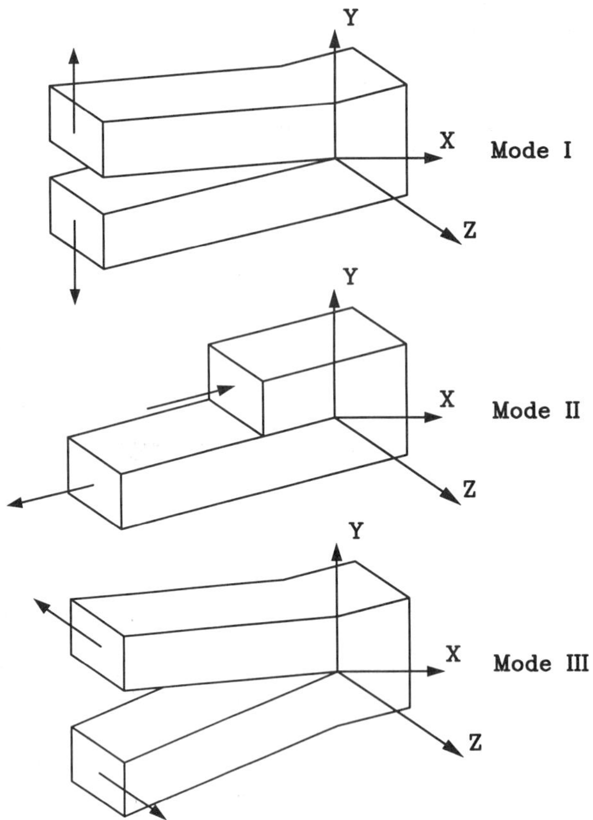

FIGURE 41.3 Modes of crack surface displacement.

The general form of the stress-intensity factor for crack opening (mode I) is given by

$$K = \sigma \sqrt{a} f(g) \tag{41.1}$$

where σ is the applied nominal stress in the structure, a is the crack length, and $f(g)$ is a parameter that depends upon the member and crack geometry.

The stress-intensity factor for an infinite plate [Fig. 41.4(a)] that is subjected to a remote stress σ that is oriented perpendicular to a through-thickness crack of length $2a$ is given by

$$K = \sigma \sqrt{\pi a} \tag{41.2}$$

This expression can be modified to obtain the approximate solution for a crack in the center of a plate of finite width W [Fig. 41.4(b)] by adding a finite-width correction term to the above equation:

$$K_I = \sigma \sqrt{\pi a} \sqrt{\frac{W}{\pi a} \tan \frac{\pi a}{W}} \tag{41.3}$$

Stress-intensity factor solutions for many common structural configurations can be found in convenient summary tables [Paris and Sih, 1965; Tada *et al.*, 1973]. Also, approximate solutions can be used for unique problems having geometries that cannot be modelled using a handbook solution. For example, the following general expression [Albrecht and Yamada, 1977] can be used to obtain an approximate solution for K_I:

$$K_I = \sigma \sqrt{\pi a} F_E F_S F_W F_G \tag{41.4}$$

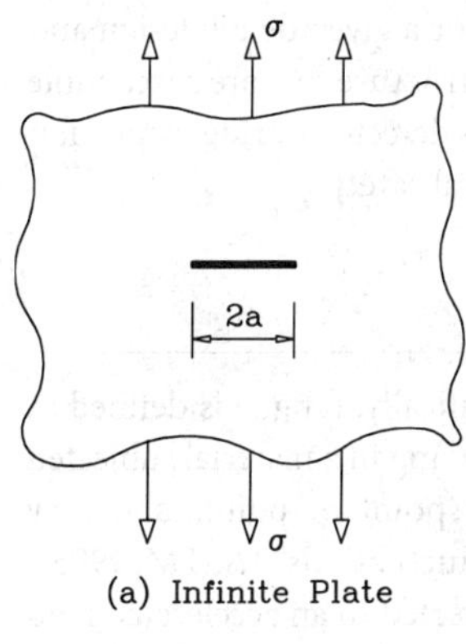

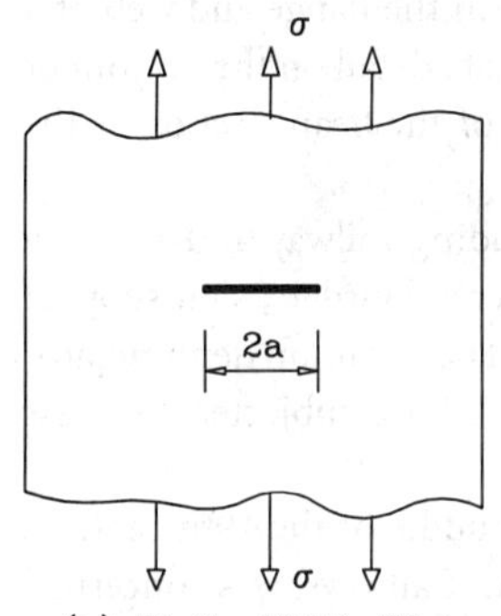

FIGURE 41.4 Plate configurations for stress-intensity factor.

where σ is the remote stress, and F_E, F_S, F_W, and F_G are, respectively, the correction factors for elliptical crack shape, free surface, finite width, and geometrical variations in stress distribution. Only those correction factors that apply are utilized in the analysis, while the other factors are taken as unity.

Fracture Toughness

The limiting characteristics for brittle fracture can be examined from the variables in Eq. (41.2). Knowledge of any two variables would allow one to solve for the remaining variable. For example, if the value of the stress σ and a limiting value for the stress-intensity factor K_I were known, then the maximum crack size at fracture could be determined. Limiting values for the stress-intensity do exist and are described by K_c, K_{Ic}, and K_{Id}. The critical stress-intensity factor when unstable crack growth occurs at a particular temperature for static loading, K_c, depends on the material thickness and the constraint. The limiting value of K_c for a slow loading rate is known as K_{Ic} for plane strain, or maximum constraint, conditions. For dynamic loadings the limiting value is defined as K_{Id}. A thorough description of determination of the limiting K_I values can be found in Barsom and Rolfe [1987].

Many tests besides K_I tests have been used to assess fracture toughness of a steel. Other tests are often used for economic reasons, since the K_I fracture toughness test can be quite expensive. The Charpy V-notch test, the dynamic tear test, and the crack tip opening displacement test are a few of the more common tests found in use today. The Charpy V-notch test, for example, is used to establish toughness limits for highway [AASHTO, 1992] and railway [AREA, 1994] bridge construction.

Preventing Brittle Fracture

Brittle fracture can be prevented or at least minimized by controlling critical variables that influence the fracture phenomenon. Although many factors and influences can be identified, they primarily involve one or more of the three primary variables: material, stress, and crack geometry.

The material can play a significant role in the fracture resistance of a structure. A steel should be selected that has adequate toughness for the loading and lowest anticipated service temperature. Consideration may need to be given also to the fracture toughness of the weld metal and the adjacent heat affected zones (HAZ) of welded structures. Welding procedures should be utilized that will not embrittle the steel in the vicinity of the weld. Proper control of preheat and interpass temperatures during placement of the welds should minimize underbead cracking and limit the hardenability of the HAZ.

While fracture toughness of a material is important, controlling the fracture susceptibility of a steel structure through selection of materials with outstanding fracture toughness may not be economical. The stresses developed in a structure are primarily influenced by the geometrical section properties of the structure. During the design stage of a project, the engineer has a great deal of flexibility to produce a structure with desired stress levels. Care should be taken during fabrication and construction of steel members to utilize procedures that prevent the introduction of unwanted stresses. Notches, fillets, holes, and sudden or sharp changes in cross section often can elevate stresses and encourage crack initiation. Also, unwanted flaws or discontinuities in the welds should be limited to acceptable specification sizes; flaws larger than the specification must be corrected.

Crack size often determines when rapid crack extension will occur for a given detail design and stress level. Regular inspection intervals to monitor those regions of a structure that are susceptible to fracture will help in efforts to assess the status of the structure. Cracks detected during inspection can be evaluated, monitored, and corrected without impairing structural safety.

41.2 Fatigue of Steel Structures

Fatigue is a problem that has plagued structures for many years. Specifically, fatigue is defined as "the process of progressive localized permanent structural change occurring in a material subjected to conditions that produce fluctuating stresses and strains at some point or points and may culminate in cracks or complete fracture after a significant number of fluctuations" [ASTM, 1993]. Crack formation is readily evident when structural prototypes are subjected to an accelerated load test in a laboratory setting. For example, a large fatigue crack is visible in the flange and web of a rolled beam section in Fig. 41.5. The crack initiated at a welded cover plate detail on the bottom of the flange and slowly propagated through the flange and into the web of the beam section as the beam specimen was subjected to several thousands of loading cycles.

Many different kinds of structures develop fatigue problems, including railway and highway bridges, vehicle suspension systems, offshore structures, drilling rigs, crane girders, sign support structures, aircraft, ships, and earthmoving equipment, to name a few. The common denominator in the development of fatigue cracks in these structures is that they all are subjected to stress fluctuations at localized stress concentrations.

Some of the earliest fatigue tests were conducted by Wohler in the middle of the 19th century. Wohler, who was chief locomotive engineer for the Royal Lower Silesian Railways, was concerned about the fatigue strength of axles. He built a rotating fatigue machine to experimentally evaluate the fatigue strength of railway axles. Based on extensive testing, Wohler advanced two fundamental laws of fatigue [Cazaud, 1953]:

- "Iron and steel may fracture under a unit stress not merely less than the static rupture stress, but also less than the elastic limit, if the stress is repeated a sufficient number of times."
- "However many times the stress-cycle is repeated, rupture will not take place if the range of stress between the maximum and minimum stress is less than a certain limiting value."

Since Wohler's early tests, numerous additional studies have been conducted to understand and evaluate fatigue for particular structural applications. One example of these efforts is research conducted to address the many different types of fatigue problems reported in highway and railway structures. For example, the development of fatigue cracks at the end of welded, partial-length

FIGURE 41.5 Fatigue crack in tension flange and web of beam section with a welded cover plate.

cover plates in the Yellow Mill Pond Bridge in 1970 [Bowers, 1973] was typical of fatigue problems developed in many similar bridge structures. Numerous fatigue tests [Munse and Stallmeyer, 1962; Fisher *et al.*, 1970] were conducted using various cover plate geometries and welding configurations to better understand the cyclic behavior of members with welded cover plates and to develop design guidelines that would minimize the type of fatigue cracking exhibited at Yellow Mill Pond Bridge.

A second type of fatigue cracking that has developed in many bridge structures is known as *distortion-induced cracking.* The Belle Fourche Bridge, for example, developed fatigue cracks in the web regions of plate girders at the end of vertical stiffeners that were used to connect transverse diaphragm members [Fisher, 1984]. These cracks initiated as a result of secondary bending stresses that occurred in the web between the end of the vertical stiffener and the flange. The secondary bending stress was caused by diaphragms transferring load generated by differential displacement of adjacent bridge girders. Although relevant research [Fisher *et al.*, 1990] has been conducted to evaluate displacement-induced fatigue, work is continuing due to the numerous cases of this type of cracking.

Significant Factors in Fatigue

The fatigue life of a structure can be divided into two broad categories. The number of loading cycles required to initiate a crack is known as fatigue crack initiation life, N_I, while the number of cycles needed to propagate a given crack to failure is known as fatigue crack propagation life, N_P. The total fatigue life is the sum of the initiation and propagation fatigue lives.

Stress versus Cyclic Life

The total fatigue life, initiation plus propagation, is used to design bridge and building components that are subjected to cyclic loadings. The total fatigue life can be represented by a simple diagram known as an *S*-*N* curve. As shown in Fig. 41.6, the logarithm of the stress range is plotted versus the logarithm of the number of loading cycles to failure. Data for *S*-*N* curves are obtained by subjecting several specimens with the same structural detail to loadings that produce a constant-amplitude stress range until failure occurs. A best-fit line through all of the fatigue data can be used to represent the expected cyclic life. For the fatigue strength of a given structural detail, design specifications often use a lower bound that is roughly two standard deviations below the mean strength.

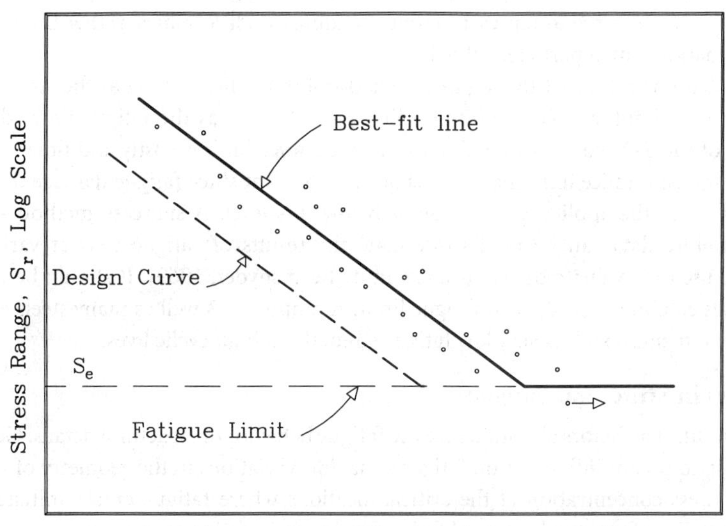

FIGURE 41.6 Stress range versus number of cycles to failure.

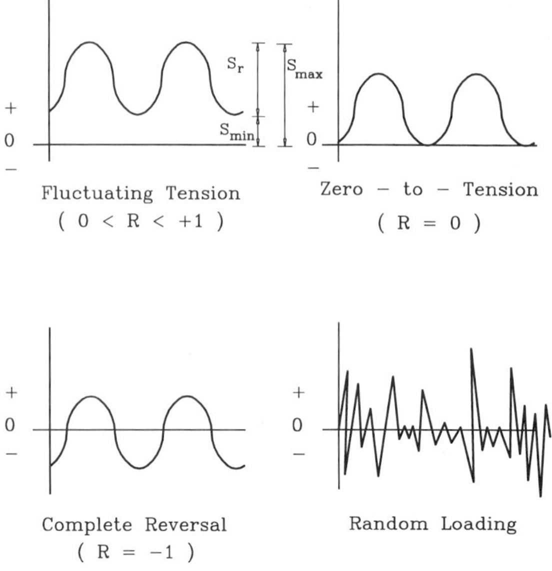

FIGURE 41.7 Various stress cycles caused by loading of a structural detail.

The stress cycle that a structure experiences is shown in Fig. 41.7. The loading may cause the stress to fluctuate between a maximum and minimum stress value. The difference between the maximum and minimum stress is known as the stress range, S_r. As shown in Fig. 41.7, when the minimum and maximum stress change but the stress range is unchanged, the mean stress varies. The influence of the mean stress developed during a loading cycle can be represented by the stress ratio, R, which is the ratio of the minimum to maximum stress [Munse, 1964]. Zero-to-tension loadings provide an R value of zero, while fully reversed loadings produce a value of -1. Earlier editions of the highway and railway specifications utilized diagrams known as *modified Goodman diagrams,* which considered the influence of mean load (R ratio) on the cyclic life. The influence of mean stress is usually not a primary consideration, however, and current specifications utilize stress range in the S-N curve to represent fatigue loadings at all R values, rather than maximum stress in combination with a particular R value.

Figure 41.6 demonstrates that the log linear relationship applies as long as the stress is greater than the endurance limit, S_e. The endurance limit, also known as the fatigue limit, defines the lower portion of the S-N curve. Evaluation of the endurance limit is costly and time consuming. Knowledge of the endurance limit, however, is quite important since fatigue damage theoretically will not occur when the applied stress range is below this level. A staircase method—in which survival and failure data can be used to evaluate the results of fatigue tests at various stress levels—can be used to evaluate the endurance limit [Reemsnyder, 1969]. It should be noted that not all materials exhibit a well-defined fatigue limit. Aluminum, as well as many steels exposed to corrosive environments, exhibits very low fatigue strengths at high cyclic lives.

Other Factors in Structural Fatigue

A number of additional factors also influence the fatigue behavior of structural details. Geometry is one of the most dominant influences on fatigue behavior. Variations in the geometry of a structure influence the stress concentration at the critical locations where fatigue cracks initiate. Because a significant portion of the total fatigue life is often spent in fatigue crack initiation, even small changes in the geometry—such as an increase in the weld reinforcement in a welded joint—can alter the stress concentration and significantly change the cyclic life.

Material strength is not generally believed to be a significant factor in fatigue. Cyclic tests have shown that the fatigue limit can be directly related to the tensile strength in polished specimens. When notches are introduced, however, the influence of a material on fatigue strength is diminished. Tests conducted by Fisher *et al.* [1970], using three distinct steels for both longitudinally welded girders and rolled beams with partial-length cover plate details, demonstrated that the fatigue strength was influenced by detail type more than by material strength.

The rate of loading for mild carbon steel has little influence for cyclic frequencies between 6 to 3000 cycles per minute [Barsom and Rolfe, 1987]. A decrease in the fatigue strength can occur for lower frequencies when plastic deformation occurs or when corrosive environments are present.

Temperature can also influence the fatigue behavior of steels. The effect of temperature on fatigue, however, is often ignored in building and bridge construction because fatigue strength is lowest near room temperature. At lower temperatures the fatigue strength increases dramatically, although a significant decrease in fracture toughness also occurs. Also, between room temperature and 315°C (600°F) fatigue strength increases slightly [Frost *et al,* 1974].

Loading History

Most fatigue tests conducted in the laboratory are constant-amplitude tests in which the stress cycle is repeated until the specimen fails or the test is stopped due to the development of a significant fatigue crack. For some structures, such as a support shaft for a rotating biological aeration unit, the same stress cycle is repeated on each rotation. For most structures, however, the stress cycle for each loading event is different from the others.

The fatigue damage caused by each loading cycle can be accumulated to determine the fatigue life when the loading history is variable. A damage model known as the Palmgren-Miner rule [Miner, 1945] is used to sum up cycle damage ratios until a summation of unity is obtained.

$$\sum\left(\frac{n_i}{N_i}\right) = 1.0 \tag{41.5}$$

In the above equation, n_i is the number of cycles of loading applied at the ith stress range, and N_i is the number of cycles of loading required to cause failure for continuous loading at the ith stress range. Constant amplitude S-N curves are used to determine the cyclic life at a given stress level.

When using the Palmgren–Miner rule, the loading sequence must be divided into a number of discrete stress range loading events that represent the load history. Another approach, which allows more flexibility for a particular type of structure, is to use a probability density function to model a particular loading history. Schilling *et al.* [1978] demonstrated that a Rayleigh probability curve could be skewed to accurately model field load data.

The fatigue damage caused by a variable loading event can be determined from the root-mean-square (RMS) approach, which is used to correlate constant-amplitude fatigue data and variable-amplitude load spectra. The RMS is obtained from the square root of the mean of the squares of the individual load cycles in a spectrum.

Fatigue Crack Initiation and Propagation

The total fatigue life of a structural detail involves the nucleation of a crack, stable crack growth, and rapid crack extension. As previously noted, the S-N curve is often used to represent the entire range of behavior—from the formation of a crack to fracture. Unfortunately, no unified method has been developed to compute the entire fatigue life of a structural component. Instead, analytical methods are available to determine the portion of the total cyclic life devoted to fatigue crack initiation and fatigue crack propagation.

The cyclic life necessary to initiate a small fatigue crack is known as the fatigue crack initiation life, N_I. To calculate the crack initiation life, the critical location where a fatigue crack will form must be known, and the behavior of the material at that location is assumed to be represented by the cyclic behavior of small, smooth specimens. The general requirements for analysis of the crack initiation life are [Topper and Morrow, 1970]:

- A mechanics analysis to relate nominal stress and strains to local stress and strains at the notch root
- Knowledge of the cyclic stress-strain properties
- Knowledge of the fatigue properties of the metal
- A cumulative damage procedure

The damage incurred on a given loading reversal is computed on the basis of the local strain and stress developed at the critical location (notch root). Applications of the approach are reported by Martin [1973] and Bowman and Munse [1983].

Fatigue crack propagation life, N_P, is the number of loading cycles necessary to propagate a crack from an initial to a final size. The stable growth of a fatigue crack, as shown in Fig. 41.8, is given by the following expression for crack growth rate [Paris and Erdogan, 1963]:

$$\frac{da}{dN} = C(\Delta K)^m \tag{41.6}$$

where da/dN is the crack growth rate, ΔK is the range in stress-intensity factor, and C and m are material constants. Equation (41.6) applies for values of ΔK between a lower threshold limit, ΔK_{th}, associated with no crack growth and an upper limit associated with rapid crack extension.

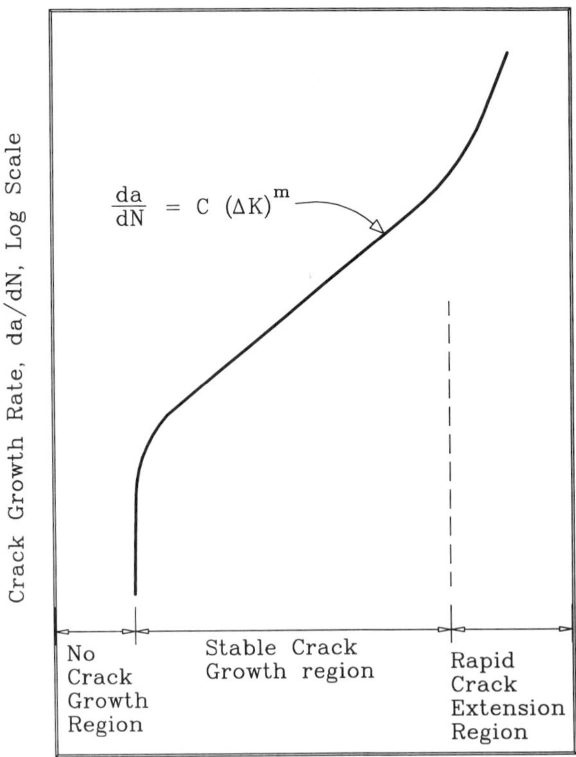

FIGURE 41.8 Crack growth rate versus range in stress-intensity factor.

The values of C and m can be determined experimentally by carefully monitoring the growth of a fatigue crack versus number of loading cycles; typical values of the material constants can be found in a number of sources [Barsom and Rolfe, 1987]. Computation of the crack propagation life is performed by first estimating ΔK and then solving Eq. (41.6) by numerical integration between an initial and final crack size. An example of the approach is reported by Lawrence and Munse [1973], and a discussion of the influence of mean stresses and crack closure is provided by Elber [1971].

Preventing Fatigue

Fatigue can be controlled by careful attention to the factors that influence fatigue crack initiation and growth. The use of structural details that are known to have a low fatigue resistance should be avoided whenever possible. If information on a particular detail is not available, such as a fatigue category classification in a design specification, then data must be collected to properly evaluate the resistance of the detail.

The fabrication quality of a structure is also important. Care should be taken to ensure that temporary connections used during erection—such as temporary tack welds or other unspecified welded or bolted details—do not deleteriously influence the fatigue resistance of the structure. Grinding of the fabricated structure, when necessary, should be performed parallel to the direction of the loading, so that unwanted stress raisers at the grinding marks are not introduced perpendicular to the applied stress.

The environment of the structural detail must be known if the fatigue resistance is to be controlled. The load history will influence the performance of a detail and must be known to realistically predict expected performance. The presence of corrosive agents, such as salt or moisture, should be avoided if possible. However, if exposure to a corrosive environment cannot be avoided, the influence of the agent on fatigue crack initiation and propagation must be considered.

Finally, a comprehensive inspection program should be utilized to detect fatigue cracks as early as possible. A number of nondestructive inspection tools can be used to detect the presence of surface or subsurface cracks or discontinuities. Nondestructive testing, however, should be used only by qualified inspectors and engineers. By detecting fatigue cracks at an early stage, a catastrophic failure can be avoided and the cracks can be monitored and corrected when adequate resources are available.

References

Albrecht, P. and Yamada, K. 1977. Rapid calculation of stress intensity factors. *Amer. Soc. Civ. Eng., J. Struc. Div.* 103:377-389.

AASHTO. 1992. *Standard Specifications for Highway Bridges,* 15th ed. American Assoc. of State Highway and Transportation Officials, Washington, D.C.

AREA. 1994. Specifications for steel railway bridges. *Manual for Railway Engineering,* Chap. 15. American Railway Engineering Association, Chicago.

ASTM. 1993. Standard definitions of terms relating to fatigue, E1150. *Annual Book of ASTM Standards,* Vol. 03.01. ASTM, Philadelphia, PA.

Barsom, J. M. and Rolfe, S. T. 1987. *Fracture and Fatigue Control in Structures—Applications of Fracture Mechanics.* Prentice Hall, Englewood Cliffs, NJ.

Bowers, D. G. 1973. Loading history of span 10 on Yellow Mill Pond viaduct. *Highway Res. Record* 428:64–71.

Bowman, M. D. and Munse, W. H. 1983. Fatigue behavior of welded steel butt joints containing artificial dicontinuities. *Welding J., Res. Suppl.* 42(2):36s–44s.

Bruestle, K. 1990. Repairs to cracks on welded girders on Burlington Northern's Latah Creek bridge. *Proc. Amer. Railway Eng. Assoc.* 91:226–235.

Cazaud, R. 1953. *Fatigue of Metals.* Chapman & Hall, London.

Elber, W. 1971. The significance of fatigue crack closure. *Amer. Soc. Test. Matls.* Spec. Tech. Pub. 486:230–242.

FHWA. 1986. *Training Course—Inspection of Fracture-Critical Bridge Members.* Federal Highway Administration, GPO, Washington, D.C.

Fisher, J. W. 1984. *Fatigue and Fracture in Steel Bridges: Case Studies.* John Wiley & Sons, New York.

Fisher, J. W., Frank, E. H., Hirt, M. A., and McNamee, B. M. 1970. *Effect of Weldments on the Fatigue Strength of Steel Beams.* NCHRP Rept. 102. Highway Research Board, Washington, D.C.

Fisher, J. W., Jin, J., Wagner, D. C., and Yen, B. T. 1990. *Distortion-Induced Fatigue Cracking in Steel Bridges.* NCHRP Rept. 336. Transportation Research Board, Washington, D.C.

Frost, H. E., Marsh, E. J., and Pook, L. P. 1974. *Metal Fatigue.* Oxford University Press, London.

Griffith, A. A. 1920. The phenomena of rupture and flow in solids. *Phil. Trans. Roy. Soc.* (London) A221:163–189.

Koncza, L. 1979. Bridge failure on the Chicago Transit Authority's Dan Ryan line. *Proc. Amer. Railway Eng. Assoc.* 80:392–402.

Lawrence, F. V. and Munse, W. H. 1973. Fatigue crack propagation in butt welds containing joint penetration defects. *Welding J., Res Supply.* 52:221s–225s.

Martin, J. F. 1973. Fatigue damage analysis for irregular-shaped structures subjected to representative loads. *Thesis.* University of Illinois, Urbana.

Miner, M. 1945. Cumulative damage in fatigue. *J. Appl. Mech.* 67:159–164.

Mosborg, R. J., Hall, W. J., and Munse, W. H. 1957. Arrest of brittle fractures in wide steel plates. *Welding J.* 36(9).

Munse, W. H. (L. Grover, ed.). 1964. *Fatigue of Welded Steel Structures.* Welding Research Council, New York.

Munse, W. H. and Stallmeyer, J. E. 1962. Fatigue in welded beams and girders. *Highway Res. Board Bull.* 315.

Paris, P. C. and Erdogan, F. 1963. A critical analysis of crack-propagation laws. *Amer. Soc. Mech. Engrs., J. Basic Eng.* 85(D)(4):528–534.

Paris, P. C. and Sih, G. C. 1965. Stress analysis of cracks. *Amer. Soc. Test. Matls.* Spec. Tech. Pub. 381.

Parker, E. R. 1957. *Brittle Behavior of Engineering Structures.* John Wiley & Sons, New York.

Reemsnyder, H. S. 1969. Procurement and analysis of structural fatigue data. *Amer. Soc. Civ. Eng., J. Struc. Div.* 95:1533–1551.

Schilling, C. G., Klippstein, K. H., Barson, J. M., and Blake, G. T. 1978. *Fatigue of Welded Steel Bridge Members Under Variable-Amplitude Loadings.* NCHRP Rept. 188. Transportation Research Board, Washington, D.C.

Tada, H., Paris, P.C., and Irwin, G. R. 1973. *Stress Analysis of Cracks Handbook.* Del Research, Hellertown, PA.

Topper, T. H. and Morrow, J. 1970. *Simulation of the Fatigue Behavior at the Notch Root in Spectrum-Loaded Members.* TAM Rept. 333. University of Illinois, Urbana.

For Further Information

Early fracture toughness testing techniques are described in *Fracture Toughness Testing and Its Applications,* Spec. Tech. Pub. 381, ASTM, Philadelphia, PA.

A description of the fracture control requirements for highway bridge structures is given in *Guide Specifications for Fracture Critical Non-Redundant Steel Bridge Members,* American Association of State Highway and Transportation Officials, Washington, D.C.

Fatigue analysis techniques, including crack initiation models, are described in *Fatigue Design Handbook,* Society for Automotive Engineers.

A summary of fatigue S-N curves for many different strengths and detail configurations is given in *Fatigue Data Bank and Data Analysis Investigations,* SRS Report 405, University of Illinois, Urbana.

Case histories of bridge structures that have experienced fatigue and fracture problems are described in *Fatigue and Fracture in Steel Bridges: Case Studies,* John Wiley & Sons, New York.

42

Wood as a Construction Material

John F. Senft
Purdue University

42.1 Introduction

This brief introduction to wood as a material is written primarily to inform the practicing civil engineer about what wood is; its cellular makeup; and, therefore, how it may be expected to react under various loading conditions. Space limitations preclude much detail; instead, references are given to lead the reader to detailed cause-effect relationships. Emphasis is placed on those items and relationships that most often lead to wood misuse or problems of proper wood use in structural applications and that may provide useful, practical guidelines for successful wood use.

What Is Wood?

Next to stone, wood is perhaps the building material earliest used by humans. Despite its complex chemical nature, wood has excellent properties which lend themselves to human use. It is readily and economically available; easily machinable; amenable to fabrication into an infinite variety of sizes and shapes using simple on-site building techniques; exceptionally strong relative to its weight; a good heat and electrical insulator; and—of increasing importance—it is a renewable and

0-8493-8953-4/95/$0.00 + $.50

biodegradable resource. However, it also has some drawbacks that the user must be aware of. It is a "natural" material and, as such, it comes with an array of defects (**knots**, irregular grain, etc.); it is subject to decay if not kept dry; it is flammable; and it is anisotropic.

Definitions

In order to understand how best to use wood as a structural material, a few terms must be understood. A tree is a marvel of nature; it comes in a variety of species, sizes, shapes, and utilization potentials. However, all trees have some basic characteristics in common:

Growth ring: The portion of wood of a tree produced during one growing season. In the temperate zones this is also called an *annual ring*.

Earlywood: The portion of a growth ring that is formed early in the growing season. It normally contains larger cells with thinner walls. Earlywood is relatively low in **density** and is followed by latewood as the growing season progresses.

Latewood: The portion of a growth ring that is formed later in the growing season. Cells tend to be smaller in size and have thicker, denser cell walls.

Heartwood: The innermost growth rings of a tree; may be darker in color than the outermost growth rings (called **sapwood**). Contains phenolic compounds that in some species impart decay resistance to the heartwood.

Sapwood: The outermost growth rings of a tree; always light brown to cream-colored in all species; never decay- or insect-resistant. Sapwood and heartwood together make up the "wood" of commercial use.

Bark: The outside covering of a tree, which protects the tree from invasion by insects, disease, and decay. The bark is separated from the wood of a tree by a thin layer of cells, the *cambium*, which is able to produce new cells annually to increase a tree in diameter as an annual ring is added.

Hardwood: Trees that are deciduous, i.e., trees whose leaves are broad and are generally shed each year in the temperate zones. Typical hardwoods include oaks, maples, and poplar. It is important to realize that not all hardwoods are "hard"; balsa is a hardwood, for example. The major use of hardwoods is in furniture and cabinet manufacture.

Softwood: Typically "evergreen" trees with needle-like leaves. Includes Douglas fir, pines, spruces, cedars, and hemlock. Traditionally softwoods have been used primarily for structural timbers and are graded specifically for this purpose. The wood of softwoods ranges from soft to quite hard.

Wood cells: Long, thin, hollow units that make up wood. Most cells are oriented with their long axis roughly parallel to the axis of the tree. However, cell orientation may vary, as around a tree branch to form a knot in a board cut from a tree, and some cells are in groups called **wood rays**, which are oriented horizontally and radiate outward from the center of the tree.

Wood rays: A band of cells radiating outward from the center of the tree toward the bark. The long axis of these ray cells is horizontal; ray cells are used for food storage and horizontal translocation of fluids in a tree.

Wood Chemistry and Anatomy

Chemically, wood of all species is composed of five basic components: **cellulose**, in the form of long-chain molecules in large groups that make up threadlike structures called **microfibrils**; **hemicellulose**; **lignin**; *extractives*; and *ash*. Cellulose gives the wood its strength, particularly along the microfibrillar direction, and constitutes 40 to 50% of the wood by volume, depending upon species. Hemicellulose is about 20 to 35% of the wood of a tree by volume and is a more readily soluble form of cellulose; it is a polysaccharide often referred to as "fungi food." Lignin is the natural adhesive that glues the cellulose molecules and wood cells together to give the wood its rigidity and its viscoelastic and thermoplastic properties. Extractives typically constitute about 1 to 5% of the wood, and while they have little or no effect on wood strength, they impart resistance to decay and insects to those species that are termed durable. Extractives may also impart color to the heartwood. It is important to note that while all species probably contain some amount of extractives in the heartwood portion of a tree, they do not necessarily create a coloration different from that of the sapwood, nor do they necessarily impart any degree of durability or toxicity to insects and fungi to the heartwood. Ash normally is about 1% of the volume of wood.

The microfibrils are oriented at 5 to 30° to the cell axis; it is their orientation within a cell, as well as their very small shrinkage along their length as wood dries compared to their relatively large shrinkage between adjacent microfibrils, that is directly responsible for two major characteristics of wood: disparate strength properties and shrinkage properties along and across the grain, i. e., wood's anisotropic nature. Wood's cellular orientation, with most of the cells oriented longitudinally (approximately parallel to the axis of the tree) and bands of ray cells oriented radially, produces wood properties that are generally taken as *orthotropic*: longitudinal, radial, or tangential (tangent to the growth rings and perpendicular to the wood rays). From a practical point of view, radial and tangential properties are of a similar order of magnitude; thus, wood is usually viewed as having properties "along the grain" and "across the grain." A study of compression strength values in Table 42.1 will emphasize the fact that wood is widely variable in its properties and is generally much stronger along the grain than it is across the grain.

Wood species differ one from another, despite the fact that all wood is made up of basically the same chemical components. Inter- and intraspecies differences may be accounted for by several factors:

1. *Different cell types.* Softwoods have primarily one cell type: an all-purpose cell called a *tracheid*, which is responsible for both wood strength and vertical translocation of fluids. Hardwoods, on the other hand, have a number of different cell types with more specialized functions. Wood strength reflects those different cell types. Likewise, particularly in hardwoods, the proportions, or mix, of cell types also affects wood properties.
2. *Proportion of wood ray cells and size of wood rays.* In general, the softwood species tend to have small, narrow wood rays. Hardwoods, on the other hand, have rays that range in size from too small to be easily seen with the eye (buckeye, willow, cottonwood) to large (oak species). Ray size and appearance, along with more distinctive heartwood coloration, have led to the preference for hardwoods in furniture and panel manufacture as well as to species strengths and use differences.
3. *Site.* This may broadly include numerous aspects of tree growth: wet versus dry site, low versus high elevation (differences in water and temperature), weather cycles, shaded versus sunny site, fertile versus less fertile site, etc. These factors in turn affect the length of a tree's growing season and, hence, the width of the growth rings. The width of an annual growth ring tends to affect overall wood density and, thereby, species and individual tree properties. As a rough rule of thumb, softwoods with wider-than-normal growth rings (say, less than four rings per inch as seen on a tree cross section) tend to be low in density and have lower strength properties. Hardwoods in general tend to have normal or higher-than-normal strength properties as growth ring width increases.

Wood strength properties also vary from the center (pith) of the tree outward toward the bark. The innermost growth rings for most species studied (particularly for softwood species) tend to be lower in density, weaker, and more prone to **warp** in product form than the outer rings. Although this characteristic varies between species, it is generally limited to the first 10 to 20 growth rings from the pith, with those nearest the pith being generally weakest and gradually increasing in

Table 42.1 Clear Wood Strength Values (Metric Units) Unadjusted for End Use and Measures of Variation for Commercial Species of Wood in the Unseasoned Condition[a]

	Modulus of Rupture[b]		Modulus of Elasticity[c]		Compression Parallel to Grain, Max. Crushing Strength		Shear Strength		Compression Perpendicular to Grain: Fiber Stress at Proportional Limit		Compression Perpendicular to Grain: Mean Stress at 0.04 in. Deformation[d,e]	Specific Gravity	
	Avg. MPa	Standard Deviation MPa	Avg. MPa	Standard Deviation MPa	Avg. MPa	Standard Deviation MPa	Avg. MPa	Standard Deviation MPa	Avg. MPa	Standard Deviation MPa	MPa	Avg.	Standard Deviation
Cedar													
Western red	35.74	5.25	6,474	1.54	19.13	3.40	5.32	0.79	1.68	0.45	2.96	0.31	0.027
Douglas fir[f]													
Coast	52.85	9.08	10,756	2.17	26.09	5.06	6.23	0.90	2.63	0.74	4.83	0.45	0.057
Interior West	53.18	9.11	10,432	2.23	26.70	5.51	6.45	0.94	2.88	0.81	4.87	0.46	0.058
Interior North	51.28	8.02	9,715	1.89	23.92	4.15	6.53	0.87	2.45	0.69	4.61	0.45	0.049
Interior South	46.77	6.26	8,012	1.38	21.46	3.37	6.57	1.05	2.32	0.65	3.99	0.43	0.045
Fir													
Balsam fir	38.04	3.81	8,625	0.99	18.14	1.95	4.56	0.57	1.29	0.21	2.34	0.32	0.025
Subalpine fir	33.78	4.58	7,253	1.25	16.49	2.50	4.80	0.71	1.32	0.30	2.40	0.31	0.032
Pine													
Eastern White	33.99	5.44	6,853	1.51	16.82	3.03	4.67	0.66	1.50	0.42	2.68	0.35	0.035
Lodgepole	37.85	6.05	7,419	1.63	18.00	3.24	4.72	0.66	1.74	0.49	3.05	0.39	0.039
Ponderosa	35.37	5.66	6,874	1.51	16.89	3.04	4.85	0.68	1.94	0.54	3.39	0.39	0.039
Sugar	33.74	4.57	7,115	1.33	16.95	2.66	4.95	0.72	1.48	0.30	2.63	0.34	0.027
Western White	32.32	4.78	8,225	1.77	16.78	2.80	4.67	0.68	1.32	0.32	2.40	0.35	0.034

Redwood													
Old Growth	51.71	8.29	8,115	1.79	29.03	5.23	5.54	0.77	2.92	0.82	4.94	0.39	0.039
Second-Growth	40.82	6.53	6,584	1.45	21.44	3.86	6.16	0.86	1.85	0.52	3.24	0.34	0.034
Spruce													
Englemann	32.44	4.77	7,095	1.43	15.03	2.94	4.39	0.44	1.36	0.34	2.47	0.033	0.033
Sitka	39.02	6.25	8,481	1.87	18.41	3.32	5.22	0.73	1.92	0.54	3.35	0.38	0.038
Hickory													
Shagbark	75.98	12.16	10,797	2.37	31.58	5.68	10.48	1.47	5.81	1.63	9.53	0.64	0.064
Maple													
Sugar	64.95	10.39	10,659	2.34	27.72	4.99	10.10	1.41	4.45	1.25	7.36	0.57	0.057
Oak, Red													
Northern	57.23	9.16	9,329	2.05	23.72	4.27	8.37	1.17	1.13	1.19	6.81	0.56	0.056
Southern	47.71	7.63	7,867	1.73	20.89	3.76	6.44	0.90	3.77	1.05	6.29	0.53	0.053
Oak, White													
Live	82.25	13.16	10,859	2.39	37.44	6.74	15.24	2.13	14.06	3.94	22.63	0.81	0.081
White	57.23	9.16	8,591	1.89	24.55	4.42	8.61	1.21	4.63	1.30	7.65	0.60	0.060
Swamp	67.98	10.88	10,983	2.41	30.06	5.41	8.94	1.25	5.27	1.48	8.66	0.64	0.064

[a] *Source*: Adapted from Tables 1 and 2, ASTM. 1992. *Standard Practice for Establishing Clear Wood Strength Values, ASTM Designation D2555-88*. American Society for Testing and Materials, Philadelphia, PA.

[b] Modulus of rupture values are applicable to material 51 mm (2 in.) in depth.

[c] Modulus of elasticity values are applicable at a ratio of shear span to depth of 14.

[d] Based on a 51 mm (2 in.) wide steel plate bearing on the center of a 51 mm (2 in.) wide by 51 mm (2 in.) thick by 152 mm (6 in.) long specimen oriented with growth rings parallel to load.

[e] A coefficient of variation of 28% can be used as an approximate measure of variability of individual values about the stresses tabulated.

[f] The regional description of Douglas fir is that given on pp. 54–55 of U.S. Forest Service Research Paper FPL 27, "Western Wood Density Survey Report No. 1."

strength as rings are added. Since these innermost rings contain the most knots, they also tend to become relegated to the lower grades of lumber and do not usually end up in structurally critical members. Their warpage characteristics also tend to relegate them to nonstructural uses. (One exception to this, however, is sometimes found in the use of pith-centered, nominal 4-by-4s used as concrete formwork.)

42.2 Wood Defects as They Affect Wood Strength

The major problems which arise in wood use may be attributed either to the effects of grain distortions (cell orientation or alignment), to the effects of excess moisture, or to defects that occur as a result of the drying process. The specific defects taken into account in the grading of lumber products include

Knots: The result of cutting across a branch in lumber manufacture. If the branch is cut perpendicular to its axis, the knot is round or oblong and presents a miniature aspect of a tree with visible growth rings. Knots may be *live* (cut through a living branch with intact tissue) or *dead* (cut through a dead branch stub with loose bark, usually resulting in a knothole). If the saw is oriented so as to cut along the length of a branch, the knot is greatly elongated and is termed a *spike knot*. Due to the obvious grain distortion around knots, they are areas of severe strength reduction. The lumber grading process takes this into account by classifying lumber grade by knot size, number, type, and location within the member. Knots located along the edge of a piece are, for example, restricted in size more than are knots located along the centerline of the member.

Slope of grain: A deviation of cell orientation from the longitudinal axis of the member. Slope of grain may be a natural phenomenon wherein the grain is at some angle to the tree axis (termed *spiral grain*), or it may be the result of sawing the member nonparallel to the tree axis. Slope of grain has a negative effect upon wood strength properties. A slope of 1:20 has minimal effect, but a slope of 1:6 reduces strength to about 40% in bending and to about 55% in compression parallel to the grain. Tensile strength is even more adversely affected.

Wane: Lack of wood. Wane occurs whenever a board is sawn so as to intersect the periphery of the tree, resulting in one edge or portion of an edge of a board being rounded or including bark. Limited amounts of wane are permitted, depending upon lumber grade. The effect of wane on wood strength or nailing surface is obvious.

Shake: A lengthwise separation of the wood, which usually occurs between or through the annual growth rings. Shakes are limited in grading since they present a plane of greatly reduced shear strength. Shake may occur as a result of severe wind that bends a tree to produce an internal shear failure, or as a result of subsequent rough handling of the tree or its products.

Splits and ***cracks:*** Separations of the wood cells along the grain, most often the result of drying stresses as the wood shrinks. Cracks are small, whereas splits extend completely through the thickness of a piece. Splits at the ends of the member, particularly along the central portion of a beam, are limited in grading.

Insect attack: Insect attack may range from small blemishes that do not affect strength to large voids or extensive damage in the wood as the result of termite or other insect infestation. Insect attack is usually treated as equivalent to the effect of similarly sized knotholes.

Decay: Decay, caused by wood-destroying fungi, is precluded from wood use except for certain species in lower grades because the strength-reducing effects of fungal attack are quite significant even before visible evidence (wood discoloration, punkiness) appears. It is important to note that decay organisms require moisture to live and grow; hence, the presence of active decay or mold implies access to a source of moisture. Moist wood will *always* decay, unless the wood is **preservative**-treated or is of a very durable species.

42.3 Physical Properties of Wood

The practicing civil engineer should be knowledgeable about several important physical properties of wood:

Specific Gravity (SG)

As a general rule, specific gravity (SG) and the major strength properties of wood are directly related. SG for the major native structural species ranges from roughly 0.30 to 0.90. The southern pines and Douglas fir are widely used structural species that are known to exhibit wide variation in SG; this is taken into account in the lumber grading process by assigning higher allowable design values to those pieces having narrower growth rings (more rings per inch) or more dense latewood per growth ring and, hence, higher SG.

Moisture Content (MC) and Shrinkage

Undoubtedly, wood's reaction to moisture provides more problems than any other factor in its use. Wood is *hygroscopic*; that is, it picks up or gives off moisture to equalize with the relative humidity and temperature in the atmosphere. As it does so, it changes in strength; bending strength can increase by about 50% in going from green to a **moisture content** (MC) found in wood members in a residential structure, for example. Elasticity values can also increase, but only about 20% , over a similar moisture change range. Wood also shrinks as it dries, or swells as it picks up moisture, with concomitant warpage potential. Critical in this process is the **fiber saturation point (fsp),** the point (about 25% moisture content, on **oven-dry** basis) below which the hollow center of the cell has lost its fluid contents, the cell walls begin to dry and shrink, and wood strength begins to increase. The swelling and shrinkage processes are reversible and approximately linear between fiber saturation point and 0% MC. Due to its chemical and cellular makeup, wood shrinks along the grain only about 0.1 to 0.2%. Shrinkage across the grain may range from about 3 to 12% in going from *green* (above fiber saturation point) to 0% MC, depending not only on species but also on grain orientation (radial versus tangential). Wood decay or fungal stain do not occur when the MC is below 20%.

There is no practical way to prevent moisture change in wood; most wood finishes and coatings only slow the process down. Thus, vapor barriers, adequate ventilation, exclusion of water from wood, or preservative treatment are absolutely essential in wood construction. The *National Design Specification for Wood Construction*® (NDS®) (American Forest and Paper Association, 1991) contains explicit guidelines for the treatment of wood moisture content in regard to various structural design modes, including fastener design.

Moisture content is defined by the equation

$$\text{MC (\%)} = \frac{\text{Wet weight} - \text{Oven-dry weight}}{\text{Oven-dry weight}} \times 100 \tag{42.1}$$

The MC of wood may be in excess of 200 percent as it is cut from a log, particularly in species that are low in SG and contain thin-walled cells. It is important to note that a wood member dries most rapidly from the ends and that as it dries below fsp, shrinkage stresses may result in cracks, checks, splits, and warpage. These defects are limited in stress-grades of lumber.

Thermal Properties/Temperature Effects

Although wood is an excellent heat insulator, its strength and other properties are affected adversely by exposure for extended periods to temperatures above about 100°F. Refer to NDS (American

Forest and Paper Association, 1991). The combination of high relative humidity or MC and high temperatures, as in unventilated attic areas, can have serious effects on roof sheathing materials and structural elements over and above the potential for attack by decay organisms. Simple remedies and caution usually prevent any problems. At temperatures above 220°F, wood takes on a *thermoplastic* behavior. This characteristic, which is rarely encountered in normal construction, is an advantage in the manufacture of some reconstituted board products, where high temperatures and pressures are utilized.

Durability

Although design texts classify various species as nondurable, moderately resistant to decay, or resistant to decay, it is best to note that only the heartwood of a species may be resistant to decay and that of the readily available species, only a few are effectively resistant in their natural state (redwood, cypress, western red cedar, and black locust). On the other hand, many structural softwood species are made very durable against fungal and insect attack when properly preservative-treated with creosote, pentachlorophenol, chromated copper arsenate (CCA), or ammoniacal copper arsenate (ACA).

Chemical Effects

Wood is relatively chemically inert. Although, obviously, wood will deteriorate when in contact with some acids and bases, some species have proven very useful for food containers (berry boxes and crates) because they are nontoxic and impart no taste to the foods contained therein. Wood structures have also found widespread use as storage facilities for salt and fertilizer chemicals.

42.4 Mechanical Properties of Selected Species

Major Engineering Properties

As stated before, wood strength depends on several factors unique to wood as a material; these factors include species (and associated inherent property variability), wood (properties parallel or perpendicular to the grain), MC at time of use, **duration of load**, and lumber grade (reflective of type and degree of defects present). Tables 42.1 and 42.2 present average strength property values for selected structural species. It is important to note that these values are for wood that is straight-grained, defect-free, and at a green MC (i.e., above fsp). The data are best used for property comparisons between species. These basic properties are modified to arrive at allowable design properties (refer to Section 42.7 and to [ASTM, 1992] and [ASTM, 1988]) following a general format of

$$F = \frac{(\bar{X} - 1.645\ \mathrm{SD})}{F_{\mathrm{ADJ}}}(F_{\mathrm{MC}})(F_{\mathrm{SIZE}})(F_{\mathrm{DENS}})(F_{\mathrm{SR}})$$

where

F = allowable design stress

$\bar{X}$ = average property value, green MC

SD = standard deviation for the property; this reduction of 1.645 standard deviations from the mean establishes a 95% lower exclusion value on the mean. Design values for E and compression perpendicular to the grain, considered to be non–life-threatening properties, are derived directly from mean values, however, with no adjustment for property variability.

F_{MC} = a factor to correct for an increase in property value if the product is dried. The factor applies only to members of nominal 4 inches or less in least dimension. (Larger members are not dried prior to installation, and subsequent drying defects in large sizes may act to nullify much of the strength increase due to drying.)

F_{SIZE} = a correction for member size. Test observation shows that larger-depth beams fail at lower stress levels than smaller beams, and allowable bending stress values are corrected by a factor of $(2/d)^{1/9}$, where d = member depth in inches.

F_{DENS} = a density correction factor for southern pine or Douglas fir members that are slow-grown and of a high density.

F_{SR} = a "strength ratio" factor for lumber grade (defects). Current grades for visually graded structural lumber are Select Structural (SR 65), No. 1 (SR 55), No. 2 (SR 45), and No. 3 (SR 26). The SR value, as a percentage, expresses the ratio of the strength of a member with its permitted defects to the strength it would have if defect-free.

F_{ADJ} = an adjustment factor to convert table values from short-time duration of load to "normal" duration (assumed to be an accumulated 10-year period of full design load) plus a factor of safety.

The reader is referred to the ASTM documents (ASTM, 1988, 1992) and the lumber grade rules manuals for stress grading details and for the factors which apply to related products, such as scaffold plank, etc.

Note that while this format is still in use and indicates major factors considered in lumber grade-strength assessment, it has been largely replaced for some properties by "in-grade testing" data, described in Section 42.7.

Strengths and Weaknesses

Wood in bending is amazingly strong for its weight; however, in many applications beam size is limited more by deflection criteria than by strength. Young's modulus (E) values for native species will range from about 3450 MPa (0.5 million psi) to about 17,250 MPa (2.5 million psi). Wood in tension parallel to the grain is exceedingly strong; however, it is readily affected by wood defects, particularly by knots and slope of grain. For this reason tensile allowable design properties are taken as $0.55 \times$ bending values. Also, tension perpendicular to the grain properties are very weak, and fracture in this mode is abrupt; design for this mode of possible failure is not acceptable (a singular exception to this rule is made for laminated arches, haunched frames, and similar members where shear stress is unavoidable; allowable design values in these cases are in the range of only 0.10 MPa to 0.20 MPa (15 to 30 psi).

Wood strength in compression necessarily must consider grain angle since compressive strength varies inversely with grain angle from 0° (parallel to the grain) to 90° (perpendicular to the grain). This relationship is described by Hankinson's formula:

$$N = \frac{C_{\parallel} \times C_{\perp}}{(C_{\parallel})(\sin^2 \phi) + C_{\perp}(\cos^2 \phi)} \tag{42.2}$$

where

N = normal stress on a surface
$C_{\parallel}$ = compressive strength parallel to the grain
$C_{\perp}$ = compressive strength perpendicular to the grain
ϕ = grain angle measured in degrees from parallel to the member long axis

This relationship is also used in the design of fasteners in a joint where members converge from different angles or where eccentric bearing occurs.

Table 42.2 Clear Wood Strength Values (English Units) Unadjusted for End Use and Measures of Variation for Commercial Species of Wood in the Unseasoned Condition[a]

	Modulus of Rupture[b]		Modulus of Elasticity[c]		Compression Parallel to Grain, Max. Crushing Strength		Shear Strength		Compression Perpendicular to Grain: Fiber Stress at Proportional Limit		Compression Perpendicular to Grain: Mean Stress at 0.04 in. Deformation[d,e]	Specific Gravity	
	Avg. (psi)	Standard Deviation (psi)	Avg. (ksi)	Standard Deviation (ksi)	Avg. (psi)	Standard Deviation (psi)	Avg. (psi)	Standard Deviation (psi)	Avg. (psi)	Standard Deviation (psi)	(psi)	Avg.	Standard Deviation
Cedar													
Western red	5,184	761	939	223	2774	493	771	115	244	65	430	0.31	0.027
Douglas fir[f]													
Coast	7,665	1317	1560	315	3784	734	904	131	382	107	700	0.45	0.057
Interior West	7,713	1322	1513	324	3872	799	936	137	418	117	707	0.46	0.058
Interior North	7,438	1163	1409	274	3469	602	947	126	356	100	669	0.45	0.049
Interior South	6,784	908	1162	200	3113	489	953	153	337	94	578	0.43	0.045
Fir													
Balsam fir	5,517	552	1251	143	2631	283	662	83	187	31	340	0.32	0.025
Subalpine fir	4,900	664	1052	182	2391	363	696	103	192	44	348	0.31	0.032
Pine													
Eastern White	4,930	789	994	219	2440	439	678	95	218	61	389	0.35	0.035
Lodgepole	5,490	878	1076	237	2610	470	685	96	252	71	443	0.39	0.039
Ponderosa	5,130	821	997	219	2450	441	704	99	282	79	491	0.39	0.039
Sugar	4,893	663	1032	193	2459	386	718	105	214	43	382	0.34	0.027
Western White	4,688	693	1193	257	2434	406	677	98	192	46	348	0.35	0.034

Old Growth	7,500	1202	1177	259	4210	758	803	112	424	119	716	0.39	0.039
Second-Growth	5,920	947	955	210	3110	560	894	125	269	75	470	0.34	0.034
Spruce													
Englemann	4,705	692	1029	207	2180	427	637	64	197	50	358	0.33	0.033
Sitka	5,660	906	1230	271	2670	481	757	106	279	78	486	0.38	0.038
Hickory													
Shagbark	11,020	1763	1566	344	4580	824	1520	213	843	236	1382	0.64	0.064
Maple													
Sugar	9,420	1507	1546	340	4020	724	1465	205	645	181	1067	0.57	0.057
Oak, Red													
Northern	8,300	1328	1353	298	3440	619	1214	170	164	172	987	0.56	0.056
Southern	6,920	1107	1141	251	3030	545	934	131	547	153	912	0.53	0.053
Oak, White													
Live	11,930	1909	1575	346	5430	977	2210	309	2039	571	3282	0.81	0.081
White	8,300	1328	1246	274	3560	641	1249	175	671	188	1109	0.60	0.060
Swamp	9,860	1578	1593	350	4360	785	1296	181	764	214	1256	0.64	0.064

[a] *Source*: Adapted from Tables 1 and 2, ASTM. 1992. *Standard Practice for Establishing Clear Wood Strength Values, ASTM Designation D2555-88.* American Society for Testing and Materials, Philadelphia, PA.

[b] Modulus of rupture values are applicable to material 51 mm (2 in.) in depth.

[c] Modulus of elasticity values are applicable at a ratio of shear span to depth of 14.

[d] Based on a 51 mm (2 in.) wide steel plate bearing on the center of a 51 mm (2 in.) wide by 51 mm (2 in.) thick by 152 mm (6 in.) long specimen oriented with growth rings parallel to load.

[e] A coefficient of variation of 28% can be used as an approximate measure of variability of individual values about the stresses tabulated.

[f] The regional description of Douglas fir is that given on pp. 54–55 of U.S. Forest Service Research Paper FPL 27, "Western Wood Density Survey Report No. 1."

Stability must be taken into account for columns with a ratio of column length to least cross-sectional dimension greater than around 10. Column stability is a function of compressive strength, E, end fixation, and manufacture (sawn lumber versus round piles versus **glu-lam**) (American Forest and Paper Association, 1991). Wood is very strong in shear across the grain, but very weak and variable in shear along the grain; for this reason horizontal shear stress over supports in bending beams is often a critical design consideration, particularly for short, deep beams. The designer of wood structures must at all times be aware of wood species, grade, grain angle in localized areas, and MC effects.

Recent design criteria for bending beams (American Forest and Paper Association, 1991) take beam stability into account. As in column design, beam strength, elasticity, and other factors enter in. In cases where combined stresses occur (tension or compression combined with bending, as in roof or ceiling loads applied to timber trusses), possible two-dimensional bending is also taken into account.

In general, structures made of stress-graded material do not fail; their performance record is excellent. Several reasons may account for this. Wood tends to be resilient; members also tend to "share" their loads. Load-sharing design factors (called *repetitive member factors*) may be applied, but they are considered to be conservative. Three or more bending members fastened together and stabilized, as by flooring over joists, in sizes limited to nominal 2 to 4 inches thick may be termed **repetitive member use**, and allowable bending design stresses may be increased by a conservative 15%. In addition, allowable design stresses are, in general, also conservative; there are several built-in safety factors in the derivation process. The visual grading process for structural lumber is predicated upon setting an upper limit on permissible defect sizes and number; most pieces within a grade have defects somewhat less in size or degree than is permitted in the grade, resulting in an additional margin of safety. Wood structural failure is most often attributed to poor design or failure of the fasteners.

Duration of Load Effects

Wood is one of those materials that exhibits duration of load (DOL) effects; it will withstand higher loads at failure if the load is applied over a shorter period of time as opposed to a longer time to create failure. All strength properties show this effect except E and compression perpendicular to the grain. For design purposes **normal duration of load** for full design load is taken as ten years; other load durations are accounted for by multiplier factors, given in Table 42.3. There is some evidence indicating, however, that for lower grades of material the limiting defect (a large knot, for example) is the major factor in determining failure and that for low-grade lumber the DOL factors for short DOLs should be ignored or applied with caution.

Table 42.3 Frequently Used Load Duration Factors for Wood Construction[a]

Load Duration	Factor	Typical Design Load
Permanent	0.9	Dead load
Ten years	1.0	Occupancy live load
Two months	1.15	Snow load
Seven days	1.25	Construction load
Ten minutes	1.6[b]	Wind/earthquake load
Impact	2.0	Impact load

[a] *Source*: Taken from Table 2.3.2, NDS. 1991. *National Design Specification for Wood Construction.* American Forest and Paper Association, Washington, DC.

[b] The 1.6 factor was 1.33 in previous editions of NDS; some code agencies may not accept the new value.

Wood, due to its plastic nature, creeps under load. A safety factor is recommended in designing where deformation is important. In beam design a general rule of thumb is that instantaneous bending deflection under load is about half that which will result over the normal life of the structure. For long columns subject to buckling, a safety factor of 3 for E is recommended. An increasing rate of creep deformation is indicative of impending failure; measurement of beam deflection over time, for example, may be used to indicate potential problems (as in flat roof ponding) and required reinforcement.

Table 42.4 Coefficient of Variation (COV) Values for Wood Properties[a]

Property	COV
SG	0.10
Modulus of Rupture	0.16
E	0.22
Compression, parallel	0.18
Compression, perpendicular	0.28
Shear	0.14

[a] *Source:* Adapted from Table 4-5, Forest Products Laboratory. 1987. *Wood Handbook: Wood as an Engineering Material.* Agricultural Handbook 72, USDA, Washington, D.C.

Strength Variability

As a biological material, wood is variable in all its properties. Variability is measured by coefficient of variation values, Table 42.4, where the values listed are broad species values.

A property variability effect is also imposed by the presence of minor defects within the wood (small grain deviations near knots; rapid growth in some annual rings of conifers, which tend to lower density; very slow growth in any species; or wood in close proximity to the pith, which is inherently low in density and has a propensity to shrink excessively). Generally this is taken into account mathematically in the design stress determination and grading processes; however, caution must be used whenever load situations require the loading of single members, as in scaffolding.

Age Effects

If wood in use is kept dry and free from mechanical and insect damage, it will remain nearly unchanged in its properties over time. Timbers removed from old structures may be reused. The only cautionary action is to have any structural members regraded to account for any increase in cracks or splits due to the continued drying of a piece in use. Most wood members will be in the range of 12 to 20% MC at time of installation. They may, over time, dry to as low as 5 to 7% MC, depending on their ambient conditions. Although wood strength may be expected to increase as the MC decreases, defects that occur due to the drying process tend to offset or nullify strength increases. This is particularly evident in large, heart-centered members, which may develop large cracks from the outer surface of the piece to the center (pith) of the piece. Shear strength may be affected in beams, and column strength/stability may also be affected.

42.5 Structural Products and Their Uses

Beginning with small trees and limbs, then lumber and plywood, structural uses of wood have evolved slowly into "artificial" products, such as wood composite beams and laminated veneer lumber, which have greatly expanded the architectural design capabilities of wood structures as well as conserving a valuable natural resource through more complete utilization. Thin-kerf saws, improved veneer production techniques, better adhesives, and extensive research and development on modern timber products have led not only to new products but also to the efficient utilization of more of the tree and of a much wider range of species.

The simplest form of wood for use is the *pole* or *piling*. This is merely a delimbed and debarked tree. Common species used for this purpose are the southern pines and Douglas fir, both of which must be preservative-treated prior to use. Western red cedar is also used extensively but does not need to be treated. Poles are often **incised** (surface perforated to permit deeper, more uniform preservative penetration) before treatment. Poles have wide usage in farm structures, in the utility field, and to obtain a rustic appearance in restaurants and residential construction. Piling is, of course, used in marine structures or as foundation supports when driven into the ground. Piling has been known to have been used for several decades, removed, inspected, and reused; the exclusion of oxygen underground essentially eliminates the danger of decay except at ground level. For this reason poles and piling should be inspected regularly for signs of deterioration at any point where wood, moisture, and oxygen meet for any significant period of time and where the ambient temperature lies between 15° and 35°C. In dealing with poles and piling (long columns) it is essential that lateral stability and adequate bracing against buckling be carefully considered.

Lumber is the most common wood construction material. Allowable design stresses for the many softwood species and grades of lumber are available; refer to NDS (American Forest and Paper Association, 1991) or to lumber grade rules books. Some hardwood species have assigned design stresses; however, stress-graded hardwoods are virtually nonexistent in most areas. Paradoxically, hardwoods have long been used in the East and Midwest regions for farm structures, but as general construction lumber, not with allowable stress ratings. The reason for this is because most hardwoods are valued for their esthetic appearance and, therefore, command considerably higher prices in the furniture trade than most softwoods do in the structural markets.

Structural lumber comes in several grades and is manufactured in two-foot increments of length. It is important to note that structural lumber (synonymous with the term *stress-graded lumber*) is graded to be used as single, unaltered members. Cutting the piece along its length or across its width essentially nullifies any grade marking.

Since lumber is purchased in discrete lengths by grade and by width and thickness (2-by-6, etc.), the stocking and marketing of a vast multitude of different sizes, lengths, and grades has necessitated species grouping. Species of similar properties and uses have been grouped (via specific procedures outlined in ASTM D-2555 [ASTM, 1988]) for marketing ease. For example, "southern pine" consists of a mixture of as many as eight distinct hard pines; "hem-fir" is made up of western hemlock combined with several fir species. The allowable design values are derived to reflect the mix of species, with limits imposed by the species with the lowest average property values. Lumber is also categorized by size and use classes as

1. Dimension lumber (often referred to as "joists and planks"): nominal 2 in. or 4 in. thick and nominal 2 in. or more in width; graded primarily for edgewise or flatwise bending
2. Beams and stringers: nominal 5 in. and thicker with a width at least 5 cm (2 in.) greater than nominal thickness; graded for strength in bending when loaded on the narrow face
3. Posts and timbers: pieces of square or nearly square cross section, 5 in. by 5 in. nominal thickness or larger; graded primarily for use as posts or columns
4. Stress-rated boards: lumber less that 2 in. in thickness and 2 in. or wider

These classes do not preclude use for other purposes; e.g., post and timber grades also have allowable bending stresses assigned to them.

Lumber is purchased by its **nominal size**, e.g., 2-by-6; its actual size is somewhat less. The nominal size represents the member as it would be before reduction in size due to the removal of saw kerf, shrinkage due to drying, and reduction in size due to planing smooth after drying. Nominal and actual green and dry sizes are given in Table 42.5.

Plywood was the first of a large number of wood panel products produced for structural purposes. It is made of an odd number of layers of veneer; each alternate layer is laid with the grain at right angles to adjacent layers. The odd number of layers ensures that the panel is "balanced" in terms of strength and shrinkage about the panel neutral axis. Obviously, this necessitates that the grain on the face plies be placed so as to utilize the panel's strength along the grain. Plywood is manufactured in several standard thicknesses; 0.64 cm to 3.8 cm ($\frac{1}{4}$ to $1\frac{1}{2}$ in.) are common thicknesses. The most common species used on the outermost plies for structural purposes are southern pine and Douglas fir; however, dozens of different species may be used for the interior, or core, plies. Only the face plies are required to be of the designated species. Plywood is typically manufactured in 122 cm $\times$ 244 cm (4 ft by 8 ft) panels, but special sizes are available. Plywood is typically used as floor underlayment and roof sheathing, but it has been largely replaced by other panel products for these uses. It is also widely used for concrete forming where special, surface-treated panels are available. Plywood is produced in several grades in regard to glueline durability. Panels may be rated as *interior* (for interior use only; the glueline is not to be exposed to moisture) or *exterior*. Exterior panels are classed as Exposure 1, which has a fully waterproof bond and is designed for applications where long construction delays may be expected or where high moisture conditions may be encountered in service; or Exposure 2, which is intended for protected

Table 42.5 American Standard Lumber Sizes for Structural Lumber[a]

	Thickness (in.)			Face Width (in.)		
Item	Nominal	Minimum Dry	Dressed Green	Nominal	Minimum Dry	Dressed Green
Dimension	2	1 1/2	1 9/16	2	1 1/2	1 9/16
	2 1/2	2	2 1/16	3	2 1/2	2 9/16
	3	2 1/2	2 9/16	4	3 1/2	3 9/16
	3 1/2	3	3 1/16	6	5 1/2	5 5/8
	4	3 1/2	3 9/16	8	7 1/4	7 1/2
	4 1/2	4	4 1/16	10	9 1/4	9 1/2
				12	11 1/4	11 1/2
				14	13 1/4	13 1/2
Timbers	5 and larger	—	1/2 in. less than nominal	5 and larger	—	1/2 in. less than nominal

[a] Adapted from Table 5-6, Forest Products Laboratory. 1987. *Wood Handbook: Wood as an Engineering Material.* Agricultural Handbook 72, USDA, Washington, DC.

construction applications where only moderate delays in providing protection from moisture may be expected. Structural plywood panels are grade-stamped with glueline type and recommended span rating. A special grade, *marine plywood*, is also available; it has improved durability and a higher grade of veneer in the inner plies. Consult the American Plywood Association for technical advice on plywood design and use (*American Plywood Association*, 1992a,b).

Whereas lumber is sawn from a log with a considerable waste factor for slabs and sawdust, and plywood is made from thin sheets of veneer peeled from higher-quality logs, other panel products have been developed from technology that permits the conversion of entire logs into particles, chips, or carefully sized flakes with insignificant waste factors, and subsequently into panels with predictable engineering properties and a wide array of marketable sizes and thicknesses. **Particleboard**, the oldest of these products, is made from particles bonded under pressure with, usually, a urea-formaldehyde adhesive. Its intended use is as a substrate for overlaid sheet materials or as underlayment. Strength properties are relatively low. A majority of the particleboard produced is utilized by the furniture and related industries. Like all panel products, its properties depend upon the material input and process: species, size and shape of the particles or flakes, orientation of the flakes, adhesive used, density of the finished product, thickness of the panel, and pressure and temperature at which the panels are formed. Particleboards are not generally considered as "structural" materials because of lower levels of strength and use of a nondurable adhesive.

However, as a subset of generic particleboard, composites made from large flakes and exterior-grade/waterproof adhesives have evolved into structural-use panels. They may be "engineered" for a specific purpose by varying the pressing temperature, pressure, adhesive amount, and flake orientation. Board strength may be designed into the product. However, unlike lumber and most materials where bending strength and E tend to be properties of primary importance, panel products need to be evaluated also for their internal bond strength, shear properties, and fastener strength. This has resulted in a wide variety of panel products with properties patterned toward specific end uses. **Oriented-strand board** (OSB) is one of the most recent such products; its flakes are mechanically oriented to align them to be closely parallel to the long axis of the panel on the outer faces of the panel in order to attain properties more closely resembling those of plywood in bending, while still maintaining relatively low density and the economic advantage of maximum resource use and low cost. Panel weight can be varied by producing a layered product with the inner portion of the panel made up to have different properties than the faces. Typical uses of OSB panels are roof sheathing, floor sheathing, and web material in composite wood I-beams or in shear walls.

Table 42.6 Ranges of Physical and Mechanical for Commercially Available Flakeboard[a]

	Density		Modulus of Rupture		Modulus of Elasticity	
Type of Flakeboard	kg/m^3	pcf	MPa	psi	MPa	ksi
Waferboard	608.70–720.83	38–45	13.79–20.68	2000–3000	3.10–4.48	450–650
Oriented strand board						
Parallel to alignment	608.70–800.92	38–50	27.58–48.26	4000–7000	5.17–8.96	750–1300
Perpendicular to alignment	—	—	10.34–24.13	1500–3500	2.07–3.45	300–500

	In-Plane Shear Strength		Internal Bond	
	MPa	psi	MPa	psi
Waferboard	8.27–12.41	1200–1800	0.34–0.69	50–100
Oriented strand board				
Parallel to alignment	6.89–10.34	1000–1500	0.48–0.69	70–100
Perpendicular to alignment	—	—	—	—

[a] Excerpted from Table 22-5, Forest Products Laboratory. 1987. *Wood Handbook: Wood as an Engineering Material.* Agricultural Handbook 72, USDA, Washington, DC.

Physical and mechanical property values for various flakeboard types are given in Table 42.6; note that the tabled values are *not* design allowable values, and manufacturers' specifications should be referred to.

Glued-laminated beams (glu-lam) have been in use for decades and have the distinct advantage of being able to be produced in nearly any size or shape desired. The only practical limitation is the difficulty in transporting large structural beams or arches to the building site. Glu-lam is made by gluing thin boards, usually 1 or 2 in. (2.5 or 5.1 cm) thick lumber, together over forms to achieve the desired size and shape of a solid member. All the boards used are dried prior to lay-up of the member to greatly reduce drying defects associated with large, sawn timbers, and by using thin layers, natural defects are evenly distributed throughout the beam. Current technology in this field is centered on methods to combine accurate grade separation of high-quality lumber with computerized design placement of high-strength outer laminations balanced against lower-strength and lower-grade material in the inner laminations to achieve maximum material utilization at economical cost. The boards used in glu-lam, particularly for outer laminations, are a specialty grade with special grading criteria. Laminating grades of L1, L2, and L3 are graded with additional restrictions on permitted defects and growth rates over the normal visual grades. "Standard" beam lay-up configuration and sizes, as well as design criteria, are available from the American Institute of Timber Construction. Lumber used in glu-lam manufacture is generally of a softwood species, although glu-lam from hardwoods is available. Manufacture is in conformance with ANSI/AITC A190.1-1983. Finished beams come in three appearance grades: Industrial, Architectural, and Premium. Beams may be preservative-treated if necessary. Sizes and shapes will vary, but "standard" widths are 6.3 cm (2 ½ in.), 7.9 cm (3 ⅛ in.), 13.0 cm (5 ⅛ in.), 17.1 cm (6 ¾ in.), 22.2 cm (8 ¾ in.), and 27.3 cm (10 ¾ in.); number of laminations and depth of member can be as many as 50 or more laminations and several feet in depth. The American Institute of Timber Construction has a technical staff to aid in glu-lam design.

A new breed of product, **structural composite lumber** (SCL), has appeared recently on the market; it is a panel product made of thin layers of veneer or of wood strands mixed with a waterproof adhesive. The veneered product, termed **laminated-veneer-lumber** (LVL), is a miniature glu-lam. The strand products are known as *parallel-strand-lumber* (PSL) and *laminated-strand-lumber* (LSL); the latter, being the newest, evolved directly from the oriented-strand board-manufacturing process. SCL products are generally produced in flat panels or *billets* (of selected species) 122 cm to 244 cm (4 to 8 ft) wide, 3.8 cm (1 ½) in. or more thick, and up to 12.3 m (40 ft) long. These products are available in larger sizes than sawn lumber and tend to be significantly stronger than lumber of equal size (due to redistribution and minimization of defects), but, due to the stringent

Table 42.7 Design Stress Values for Code-Approved LVL Products[a]

Design Stress	Range MPa	psi
Flexure	15.17–28.96	2200–4200
Tension parallel-to-grain	11.03–19.31	1600–2800
Compression parallel-to-grain	16.55–22.06	2400–3200
Compression perpendicular to grain:		
Perpendicular to glueline	2.76–4.14	400–600
Parallel to glueline	2.76–5.52	400–800
Horizontal shear:		
Perpendicular to glueline	1.38–2.07	200–300
Parallel to glueline	0.69–1.38	100–200
Modulus of elasticity	12,410–19,310	1.8×10^6–2.8×10^6

[a] *Source:* Forest Products Laboratory. 1987. *Wood Handbook: Wood as an Engineering Material.* Agricultural Handbook 72, USDA, Washington, DC. pp. 10–13.

manufacturing process, they also tend to be somewhat more expensive. They are normally used for purposes requiring high strength or stiffness in both residential and light industrial/commercial construction. Table 42.7 lists design stress values for building code-approved LVL products.

With the plethora of engineered specialty products, it is no surprise that a vast array of structural composite products has also become available. Solid lumber structural members, while still preferred for many traditional uses, particularly in residential construction, are being replaced by wood I-beams with LVL flanges and OSB webs. Metal bar webs and lumber or LVL flanges are common. Toothed, stamped metal plate connectors are used as fasteners for a myriad of structural frames, trusses, and components. Design software has kept pace with these trends and is able to factor into a design the various decisions on temperature effects, moisture effects, duration of load, and combined stress situations as well (Triche and Suddarth, 1993).

42.6 Preservatives

For all practical purposes only a few native species are truly immune to fungal deterioration, and then, as stated earlier, only the heartwood portion of the wood is decay-resistant. Availability and economy usually dictate that where decay resistance is required, preservative treatment is a must. Any structural component which is in contact with the ground, subject to periodic wetting (leakage or rain), or in a high-relative-humidity atmosphere for extended time periods, may be expected to decay.

There are several preservatives available; degree of exposure and the use of the member will indicate which specific preservative to use. In all cases a pressure treatment is required; dip treating, soaking, or painting the surface with a preservative solution are only temporary deterrents at best and are not recommended where structural integrity is required. Creosote, one of the oldest and most effective treatments, is used primarily for treating utility poles and marine piling. It is an oilborne preservative of high toxicity and is not recommended where human contact is anticipated. A number of arsenic-containing treatments are commonly used. CCA (chromated copper arsenate) is used with dimension lumber, particularly with southern pine, and ACA (ammoniacal copper arsenate) is also commonly used. Both CCA and ACA are waterborne preservatives that are pressure-impregnated into dry (below fsp) lumber; the chemicals become permanently bonded to the wood as the wood becomes redried after treatment. It is very important to know that until the wood has become dry again after treatment, it is dangerous to handle. Resawn wood that is wet on the inside of the piece, even if it appears dry on the outside, can produce arsenic poisoning. It is also important to know that even under high impregnation pressures the depth of penetration of the preservative into the wood may be incomplete. Resawing may expose untreated wood to decay;

treatment after cutting or boring members to final size is recommended. CCA and ACA treatments are commonly used for foundations, decks, and greenhouses. Dry CCA- and ACA-treated lumber is approved for human contact use. Under no circumstances are wood scraps of CCA- or ACA-treated wood to be burned in the open air; this will ultimately release poisonous arsenic and chromium compounds into the air.

Borate compounds are effective wood preservatives and are economical and nontoxic to humans and animals. Unfortunately, they also leach out of the wood rather readily when subjected to rain or wet conditions. Research on these and other compounds may result in a new family of leach-resistant, nontoxic-to-humans preservatives for wood in the future.

Preservative-treated structural lumber is available in several grades, depending upon intended use and retention level. Table 42.8 lists desired use-retention levels; retention levels are part of the information given on the grade stamp of treated lumber.

Table 42.8 Preservative Retention Level–Use Recommendations

Retention Level (lbs/cu. ft)	Use/Exposure
0.25	Above ground
0.40	Ground contact/fresh water
0.60	Wood foundation
2.50	Salt water

42.7 Grades and Grading of Wood Products

Lumber stress-grading procedures for structural purposes are under the jurisdiction of the American Lumber Standards Committee and follow guidelines given in several ASTM documents. Although several rules-writing agencies publish grading rules with grade descriptions, they all conform to ALSC guidelines and restrictions and, therefore, the common grades are identical for American and Canadian producers. There are currently two grading methodologies: visual grading and machine stress rating (MSR). Visual grading is accomplished by skilled graders who visually assess the size and location of various defects and other characteristics on all four faces of a board. The main defects assessed include slope of grain, knots (size, number, and location relative to the edges of the piece), wane, **checks** and splits, decay (not permitted except for "white speck" in some grades), and low density for the species. Strength reductions for various defects are termed **strength ratio** (SR) values and are applied to strength values representing a statistical 95 percent lower confidence limit on mean strength for the property and species. Strength ratio factors delimit the grades as Select Structural (SR = .65), No. 1 (SR = .55), No. 2 (SR = .45), and No. 3 (SR = .26). Because E and compression perpendicular to the grain are not considered to be life-threatening properties, their use is treated differently. The SR value for all grades of lumber for compression perpendicular to the grain is 1.00. E values do not use an SR term; instead "quality factors" are used. Quality factors, dictated by ASTM standard (ASTM, 1992), are less severe than SR factors. Special dense grades (Dense Select Structural, etc.) are assigned to slow-growing, dense pieces of southern pine and Douglas fir.

Structural lumber is produced in three MC categories: S-GRN (surfaced in the green MC condition, above 19 percent MC); S-DRY (surfaced in the dry condition, maximum MC of any piece is 19 percent); and MC-15 (surfaced at a maximum MC of 15 percent). Southern pine grade rules have additional designations of KD for kiln-dried material and AD for air-dried material; drying lumber in a kiln is accomplished at temperatures high enough to effectively kill insects and to dry areas of accumulated pitch. Southern pine may be labeled MC-15AD, MC-15KD, MC-19AD, etc. Pieces over nominal 4 inches in thickness are normally sold S-GRN.

The various strength characteristics, as outlined in ASTM documents, including an appropriate safety factor, are applied to each board by the grader to arrive at a relatively conservative assessment of a grade. Every stress-graded piece is required to have a grade stamp on it; the grade stamp contains five pieces of information: producing mill number; grading association under which the grade rules have been issued; species or species group; moisture content at time of grading (e. g., S-DRY); and lumber grade.

For the most common softwood species and a few hardwood species used in construction, the recently completed, and very extensive, in-grade testing program has brought about some shifting of the allowable design stress values. These tests of numerous grades of lumber in various species and across most nominal 2-by and 4-by sizes were conducted to ascertain whether the design stresses accurately represented what was in the marketplace. Up-to-date knowledge of within-grade variability in property values and information on the presence or absence of basic forest resource quality shifts were also obtained. After careful analysis, many of the design values were reassigned; some species-size-grade categories warranted an increase while others were reduced. Fewer species groups are now marketed. More realistic values have resulted across the board, leading to improved reliability in wood structures.

The machine stress rating (MSR) grading process relies on a statistical relationship between a nondestructively determined E value and the bending strength of the piece. Thus, each piece is rapidly flexed flatwise in a machine to obtain an average E for the piece and a low-point E; the average E is then statistically, by species, used to assign a bending stress value, with the low-point E serving as a limiting factor in the process of assigning a grade. The MSR process lends itself to more accurate strength/E evaluation and also to closer quality control programs, because it makes an actual piece-by-piece test for one property. In general, this process has been limited to identifying higher-quality material for use in specialized industries, e.g., truss and wood I-beam manufacture. The MSR grading process has an added advantage in that lumber graded by this process is grade-stamped with a combination bending stress and E value; e.g., $2100F_b$-$1.8E$. Different combinations may easily be selected to meet special market demands. Currently over a dozen grade categories are available, ranging from $900F_b$-$1.0E$ to $2850F_b$-$2.3E$. Although the MSR grading process is somewhat capital-intensive, it has definite advantages in terms of grading accuracy and reliability.

Panel products are produced and graded in close relationship to quality control (QC) program results. That is, panels are produced according to strict manufacturing parameters, and quality is monitored via regular production line QC test procedures. The reader is referred to APA documents or manufacturer's specifications for the many various grades and uses of panel products such as siding, sheathing, structural plywood, OSB products, etc. (American Plywood Association, 1992a, 1992b).

42.8 Wood Fasteners and Adhesives

Fasteners come in a wide variety of sizes, shapes, and types. Nails are the most common fastener used in construction. Design loads for nails depend upon type of nail (common wire, threaded hardened steel, spike, coated, etc.), wood species or density, thickness of the members being fastened together, nail diameter and length, depth of penetration of the point into the main member, and failure mode. Various failure mode criteria have been incorporated into fastener design in the 1991 edition of NDS. Fastener design for bolts, lag screws, and shear connectors have similar design considerations. Most wood fasteners used in construction tend to have rather large safety factors incorporated into their design and tend to form tenacious joints; however, small deformations of fasteners at a joint also tend to result in serious deformations and structural deficiency over time. Inadequate or inappropriate joint design is a common cause of building problems that are often mistakenly attributed to "wood failure." For all fastener designs the following aspects must be kept in mind: DOL (shorter term loads allow higher design values per fastener); MC factors for dry, partially seasoned, or wet conditions at time of fabrication or in subsequent service; service temperature; group action (a reduction of design load for a series of fasteners in a row); the effect of having a metal side plate in lieu of a wood side plate; and whether the fastener is loaded in lateral or withdrawal mode. In general, placing fasteners into the end grain of wood is to be avoided, certainly so in a withdrawal mode.

Metal plate connectors are also commonly used; they are almost universally used in truss fabrication for residential and light frame construction. Although accurate plate placement is critical, their performance has proven their utility for decades, and numerous computer software packages are available to design structural frames or components that integrate metal plate fasteners into the design. Various types of joist hangers and heavier metal fixtures are also readily available and tend to speed construction of larger structures, particularly where modular components can be fabricated on site. Various fastener manufacturers provide engineering specifications for use of their specific products.

There are several wood-to-wood adhesives which produce joints stronger than the wood itself; however, their use is generally restricted to controlled factory conditions where temperature, adhesive age and formulation, press time, pressure, and adequate QC can be carefully monitored. Wood-to-wood bonds that are well made perform satisfactorily, but they require adequate, uniform pressure on smooth, clean, well-mated surfaces with an even glue spread. A waterproof adhesive is strongly recommended if adhesives are used as a fastener for wood in construction; phenol-resorcinol-formaldehyde is one that is waterproof (it is also widely used for plywood and panel product manufacture). Field gluing is generally to be avoided; however, the use of epoxy resins, particularly for repair work, is practiced successfully.

42.9 Where Do Designers Go Wrong? Typical Problems in Wood Construction

Wood and wood products are relatively simple engineering materials, but the conception, design, and construction process is fraught with problems and places to err. In using wood in its many forms and with its unique inherent characteristics, there are problem areas which seem to present easily overlooked pitfalls. As gentle reminders for caution, some of these areas are discussed below.

Wood and water do not mix well. Wood is hygroscopic and, unless preservative-treated, rots when its MC rises above 20 percent. It must be protected in some way. Minor roof leakage often leads to pockets of decay, which may not be noticed until severe decay or actual failure has occurred. Stained areas on wood siding or at joints may indicate metal fastener rust associated with a wet spot or decay in adjoining, supporting members. In many cases what appears to be a minor problem ends up as major and sometimes extensive repair. Improper installation or lack of an adequate vapor barrier can result in serious decay in studs within a wall as well as paint peel on exterior surfaces. Ground contact of wood members can lead to decay as well as providing ready access to wood-deteriorating termites. Placement of preservative-treated members between the ground and the rest of the structure (as a bottom sill in a residence) is usually a code requirement. Timber arches for churches, office buildings, and restaurants are usually affixed to a foundation by steel supports; if the supports are not properly installed, they may merely form a receptacle for rain or condensation to collect, enter the wood through capillary action, and initiate decay. Once decay is discovered, major repair is indicated; preservative treatment to a decayed area may prevent further decay, but it will not restore the strength of the material. Elimination of the causal agent (moisture) is paramount. Visible decay usually means that significant fungal deterioration has progressed for 1 to 2 feet along the grain of a member beyond where it is readily identifiable.

Pay attention to detail. In an area that has high relative humidity, special precautions should be taken. A structure that is surrounded by trees or other vegetation or that prevents wind and sun from drying action, is prone to high humidity nearly every day, particularly on a north side. Likewise, if the structure is near a stream or other source of moisture, it may have moisture problems. Home siding in this type of atmosphere may warp or exhibit heavy mildew or fungal stain. Buildings with small (or nonexistent) roof overhangs are susceptible to similar siding problems if the siding is improperly installed, allowing water or condensation to enter and accumulate behind the siding. Inadequate sealing and painting of a surface can add to the problem. In a classic example, a three-story home on a tree-shaded area next to a small stream and with no roof overhang had poorly installed siding, which subsequently warped so badly that numerous pieces fell off of the home. Poor architecture, poor site, poor construction practice, and poor judgment combined to create a disaster. This type of problem becomes magnified in commercial structures, where large surfaces are covered with wood panel products that tend to swell in thickness at their joints if they are not properly sealed and protected from unusual moisture environments. If properly installed, these materials provide economical, long-term, excellent service.

Wood is viscoelastic and will creep under load. This has created widespread problems in combination with clogged or inadequate drains on flat roofs. Ponding, with increasing roof joist deflection, can lead to ultimate roof failure. In situations where floor or ceiling deflection is important, a rule of thumb to follow is that increased deflection due to long-term creep may be assumed to be about equal to initial deflection under the design loading. In some cases the occupants of a building will report that they can hear wood members creaking, particularly under a snow load or ponding action. This is a good indication that the structure is overstressed and failure, or increasing creep deformation with impending failure, is imminent. Deflection measurements over a several-week period can often isolate the problem and lead to suitable reinforcement.

Repair structural members correctly. Epoxy resin impregnation and other techniques are often used to repair structural members. These methods are said to be particularly effective in repairing decayed areas in beams and columns. Removal of decayed spots and replacement by epoxy resin is acceptable only if the afflicted members are also shielded from the original causal agent—excess moisture or insect attack. Likewise, if a wood adhesive must be used as a fastener in an exposed area, use a waterproof adhesive; "water-resistant" or carpenter's glue won't do. Although several wood adhesives will produce a wood-to-wood bond stronger than the wood itself, most of these adhesives are formulated for, and used in, furniture manufacture, where the wood is dry (about 6 to 7% MC) at time of fabrication and is presumed to be kept that way. Structural-use adhesives (unless they are specially formulated epoxy or similar types) may be used where the wood is not above about 20% MC. Structural-use adhesives must also be gap-fillers; i.e., they must be able to form a strong joint between two pieces of wood that are not always perfectly flat, close-fitting surfaces. In addition, the adhesive should be waterproof. The most common and readily available adhesive that meets these criteria is a phenol-resorcinol-formaldehyde adhesive, a catalyzed, dark purple-colored adhesive which is admirably suited to the task.

Protect materials at the job site. Failure to do so has caused plywood and other panel products to become wet through exposure to rain so that they delaminate, warp severely, or swell in thickness to the point of needing to be discarded. Lumber piled on the ground for several days or more, particularly in hot, humid weather, will pick up moisture and warp or acquire surface fungi and stain. This doesn't harm the wood if it is subsequently dried again, but it does render it esthetically unfit for exposed use. To repeat, wood and water don't mix.

Take time to know what species and grades of lumber you require, and then inspect it. Engineers and architects tend to order the lumber grade indicated by mathematical calculations; carpenters use what is provided to them. Unlike times past, no one seems to be ultimately responsible for appropriate quality until a problem arises and expensive rework is needed. Case in point: a No. 2 grade 2-by, which is tacitly presumed to be used in conjunction with other structural members to form an integrated structure, is not satisfactory for use as scaffolding plank or to serve a similar, critical function on the job site where it is subjected to large loads independent of neighboring planks.

Inspect the job site. Make sure that panel products, such as plywood, OSB, or flakeboard, are kept under roof prior to installation. Stacked on the ground or subjected to several weeks of rainy weather, not only will these panels warp, but they may lose their structural integrity over time. "An ounce of prevention," etc.

Be aware of wood and within-grade variability due to the uniqueness of tree growth and wood defects. It is often wise to screen lumber to cull out pieces that have unusually wide growth rings or wood that is from an area including the pith (center) of the tree. This material often tends to shrink along its length as much as ten times the normal amount due to an inherently high microfibrillar angle in growth rings close to the pith. In truss manufacture this has resulted in the lower chords of some trusses in a home (lower chords in winter being warmer and drier) to shorten as they dry, while the top chords don't change MC as much. The result is that the truss will bow upward, separating by as much as an inch from interior partitions—very disconcerting to the inhabitants and very difficult to cure. A good component fabricator is aware of this phenomenon and will buy higher-quality material to at least minimize the potential problem. Conversely, avoid the expensive, "cover all the bases" approach of ordering only the top grade of the strongest species available.

Inspect all timber connections during erection. Check on proper plate fasteners on trusses and parallel chord beams after installation; plates should have sufficient teeth fully embedded into each adjoining member. Occasionally in a very dense piece the metal teeth will bend over rather than penetrate into the wood properly. A somewhat similar problem arises if wood frames or trusses are not handled properly during erection; avoid undue out-of-plane bending in a truss during transport or erection since this will not only highly stress the lumber but may also partially remove the plates holding the members together. Bolted connections must be retightened at regular intervals for about a year after erection to take up any slack due to subsequent lumber drying and shrinkage.

Perhaps one of the major causes of disaster is the lack of adequate bracing during frame erection. This is a particularly familiar scenario on do-it-yourself projects, such as by church groups or unskilled erection crews. Thin, 2-by lumber is inherently unstable in long lengths; design manuals and warning labels on lumber or product shipments testify to this, yet the warnings are continually disregarded. Unfortunately, the engineer, designer, or architect and materials supplier often are made to share the resulting financial responsibility.

Be aware of wood's orthotropicity. A large slope of grain around a knot or a knot strategically poorly placed can seriously alter bending or compressive strength and are even more limiting in tension members. Allowable design values for tension parallel to the grain are dictated by an ASTM standard (ASTM, 1992) as being 55% of allowable bending values because test results have indicated that slope of grain or other defects greatly reduce tensile properties. Different orthotropic shrinkage values, due to grain deviations or improper fastening of dissimilar wood planes, can lead to warpage and subsequent shifts in load-induced stresses. Care must be taken when using multiple fasteners (bolts, split rings, etc.) to avoid end splits as wood changes MC, particularly if the members are large and only partially dried at time of installation. When installing a deep beam that is end-supported by a heavy steel strap hanger, it is often best to fasten the beam to the hanger by a single bolt, installed near the lower edge of the beam. This will provide the necessary restraint against lateral movement, whereas multiple bolts placed in a vertical row will prevent the beam from normal shrinkage in place and often induce splits in the ends of the beam as the beam tries to shrink and swell with changes in relative humidity. Not only are the end splits unsightly, but they also reduce the horizontal shear strength of the beam at a critical point. In addition, if the beam has several vertically aligned bolts and subsequently shrinks, the bolts will become the sole support of the beam independent of the strap hanger, as shrinkage lifts the beam free of the supporting strap hanger.

Use metal joist hangers and other fastening devices; they add strength and efficiency in construction to a job. Toe-nailing the end of a joist may restrain it from lateral movement, but it does little to prevent it from overturning if there is no stabilizing decking. Erection stresses caused by carpenters and erection crews standing or working on partially completed framework are a leading cause of member failure and job site injury.

In renovating old structures, as long as decay is not present, the old members can be reused. However, because large sawn timbers tend to crack as they dry in place over a period of time, the members must be regraded by a qualified grader. The dried wood (usually well below 19% MC) has increased considerably in strength, perhaps counterbalancing the decrease in strength due to deep checking and/or splitting. End splits over supports should be carefully checked for potential shear failure.

Wood and fire pose a unique situation. Wood burns, but in larger sizes—15 cm (6 in.) and larger—the outer shell of wood burns slowly and, as the wood turns to charcoal, the wood becomes insulated and ceases to support combustion. Once the fire has been extinguished, the wood members can be removed, planed free of char, and reused, but at a reduced section modulus. Smaller members can also be fire retardant–treated to the degree that they will not support combustion. However, treating companies should be consulted in regard to any possible strength-reducing effects due to the treatment, particularly where such members are to be subjected to poorly ventilated areas of high temperature and high relative humidity, as in attic spaces. In recent years newly developed fire retardant treatments have reacted with wood when in a high temperature–high relative humidity environment to seriously deteriorate the wood in treated plywood or truss members. These chemicals, presumably withdrawn from the marketplace, act slowly over time, but have contributed to structural failure in the attics of numerous condominium-type buildings. Preventive measures where such problems may be anticipated include the addition of thermostatically controlled forced-air venting (the easiest and probably most effective measure); the addition of an insulation layer to the underside of the roof to reduce the amount of heat accumulation in the attic due to radiant heat absorption from the sun; and the installation of a vapor barrier on the floor of the attic to reduce the amount of water vapor from the underlying living units.

In using preservative-treated wood it is always best—certainly so when dealing with larger members—to make all cuts to length, bore holes, cut notches, etc., prior to treatment. Depth of preservative treatment in larger members is usually not complete, and exposure of untreated material through cutting may invite decay. Determination of the depth of penetration of a preservative by noting a color change in the wood is hazardous; penetration may be more or less than is apparent to the eye. Deep checking as a large member dries will often expose untreated wood to fungal organisms or insects. Periodic treatment by brushing preservative into exposed cracks is highly recommended. This is particularly true for log home–type construction. Modern log home construction utilizes partially seasoned materials with shaped sections, which not only increase the insulative quality of the homes but also tend to balance, or relieve, shrinkage forces to reduce cracking. Treated or raised nonwood foundations are recommended.

Wood is an excellent construction material, tested and used effectively over the years for a myriad of structural applications—provided one takes the time to understand its strengths and weaknesses and to pay appropriate attention to detail. Knowing species and lumber grade characteristics and how a member is to be used, not only in a structure but also during erection, can go a long way toward trouble-free construction.

42.10 Wood and the Environment

Trees are nature's only renewable resource for building materials. Trees use energy from the sun and carbon dioxide to create cellulose while cleansing the atmosphere and giving off oxygen. Wood is a significant "storehouse" for carbon, and it does all this with little or no input from people. Converting trees into useful products requires much less energy than is needed for other construction materials. Considering any other structural material in terms of production costs to the environment and use through recycling and ultimate disposal, wood is certainly the most environmentally benign material in use today. It is renewable, available, easily converted into products, recyclable, and biodegradable with no toxic residues.

What is the current status of the forest resource? What are the factors affecting the resource and its use? The answers seem to depend upon whom you ask, because there are no explicit, easy answers. Environmental concerns and the "green movement" have resulted in national policy shifts regarding forest use. Large acreages have been set aside as wilderness areas to remain totally unavailable for timber management and harvest. Harvesting on national timberlands has been drastically curtailed. This will necessarily shift harvesting pressure more to industrial and private commercial forestland. While commercial forestland, which is held by numerous wood products companies, is quite productive and provides more product per acre than other sources, privately held timberlands tend to be the least managed of the nation's forested acres. Thus, even though a valid argument can be made for sequestering national timberlands or reducing their production of timber products, shifting the nation's demand for wood to the private, noncommercial sector may not be wise in the long run. On a brighter note, in many traditional timber states, regulations require that cut areas be properly and promptly restocked and waste greatly reduced. It has been estimated that in several western states six trees are planted for each one that is harvested (but under normal forest growth patterns only one or two of the six survive to reach maturity). In most regions of the U.S. increases in tree growth significantly exceed harvest and mortality due to fire, old age, and disease. In other words, there is more timber being produced annually than is being harvested by a significant margin. But that's not the whole story. Average tree diameter has been steadily declining as we have harvested the biggest, best, and most economically available trees. Likewise, individual tree quality has been decreasing, and many forested areas are inaccessible or uneconomical for harvesting operations. Balanced against these factors is the impact of advancing technology. Such products as structural composite lumber, LVL, OSB, glu-lam, and other products, as well as improved grading and strength assessment techniques, have stretched the resource remarkably. We use nearly 100% of the tree for useful products; waste has been significantly reduced. Lesser-known species are being utilized; for example, aspen, a "junk tree" species not long ago, is now the mainstay for many OSB plants in the upper Midwest because of its low density, availability in large quantities, and desirable panel properties. Thin-kerf saws, improved kiln-drying technology, environmentally friendly preservatives, waste conversion to fuel energy (modern integrated paper mills may be over 90% energy self-sufficient, paper industry average energy self-sufficiency value is 56%), and more reliable product grading and QC programs are just a few successful resource-stretching innovations.

New technologies have made tremendous strides and old technologies are being updated. In some areas it is predicted that by the turn of the century wood will be a significant fuel source. Fast-grown tree plantations of hardwoods represent an economical fuel source that doesn't require mining and is there when the sun isn't shining or the wind isn't blowing. Wood chips added to coal significantly reduce sulfur and carbon emissions. Over 1000 wood-burning plants are in operation, and their combined output is reputed to be the equivalent of three large nuclear reactors (Anonymous, 1993). Wood waste from manufacturing operations and demolition refuse may become a valuable, environmentally acceptable fuel source.

Although controversy regarding just how we are to allocate the nation's timber resource to provide for endangered species, increased demand for "wild" areas, and increasing numbers of products made from wood will certainly continue, it is possible to retain many of the "natural" aspects of forests and still obtain products from this remarkable resource on a sustainable basis if attention is paid to proper management and skillful utilization. Current (1990s) controversy over environmental policies will lead to acceptable compromise in time. However, the nature of trees and all the wood products derived from them will assure wood a prominent place as a highly preferred, environmentally desirable, and economically competitive building material.

Defining Terms

Bark: The outside covering of a tree which protects the tree from invasion by insects, disease, and decay.

Cellulose: A long-chain molecule that constitutes the major building block of plant material. Cellulose is the element responsible for wood's strength along the grain.

Check: Often referred to as a "seasoning check"; a separation of the wood cells as a result of shrinkage during drying. Checks tend to reduce shear strength.

Decay: Results from fungal attack of wood. Fungi may attack either the wood cellulose and hemicellulose or the lignin; in either case even the early stages of decay seriously reduce wood strength and make it unsuitable for any structural purpose.

Density: Weight per unit volume of wood. The weight is always in the oven-dry condition (i.e., zero percent MC); the volume, however, may be measured at any stated MC (most often green, 12%, or oven-dry). Since wood shrinks as it dries, density values vary, and the MC base must be stated as, for example, "volume measured at 12% MC." Similarly, specific gravity values will vary depending on the MC at which volumetric measurements are made.

Duration of load (DOL): The ability of wood members to sustain larger loads for shorter periods of time without failure than they can sustain without failure for longer periods of time. Structures that will have design loads imposed upon them for shorter than "normal" DOL are permitted to have an increase in allowable design stresses in bending, compression parallel to the grain, tension parallel to the grain, and shear. Normal DOL is defined as a 10-year cumulative loading duration.

Earlywood: The portion of a growth ring that is formed early in the growing season. It normally contains larger cells with thinner walls. Earlywood is relatively low in density and is followed by latewood as the growing season progresses. See *Latewood.*

Fiber saturation point (fsp): A point (about 25% MC, depending on species) in the drying of wood at which the hollow centers of the wood cells have lost their moisture, leaving the cell walls fully saturated. Wood does not begin to shrink until its MC has dropped below fsp.

Flat-sawn: Lumber that has its wide face in a tangential plane (i.e., cut approximately tangent to the annual growth rings); also called plain-sawn. See *Quarter-sawn.*

Glu-lam: A structural product made by gluing structural boards together; the grain of all pieces is oriented along the long axis of the member. Glu-lam may be manufactured into a variety of curved beams, arches, or irregular shapes. See also *Laminated-veneer-lumber.*

Growth ring: The portion of wood of a tree produced during one growing season. In the temperate zones this is also called an annual ring.

Hardwood: Trees that are deciduous; that is, trees whose leaves are broad and are generally shed each year in the temperate zones. Typical hardwoods include oaks, maples, and poplar.

Heartwood: The innermost growth rings of a tree; may be darker in color than the outermost growth rings (called *sapwood*). Contains phenolic compounds that in some species impart decay resistance to the heartwood.

Hemicellulose: A long-chain molecule similar to cellulose, but more easily soluble in dilute acid or basic solutions.

Incising: Perforation of the surface of a pole or wood member by a series of chisel-like knives prior to preservative treatment to permit deeper and more uniform infusion of the preservative into the wood.

Knots: The result of a branch having been cut in the manufacture of a board. If the branch is cut at a right angle to its length, the knot appears as a branch cross section; if the branch is cut along its length, appearing to go across the face of a board, it is termed a spike knot.

Laminated-veneer-lumber (LVL): A board product made by gluing pieces of thin lumber or veneer together to make a larger member, usually formed into long flat panels suitable for remanufacture into common lumber sizes. The grain of all pieces is oriented along the long axis of the panel. See also *Glu-lam* and *Plywood.*

Latewood: The portion of a growth ring that is formed later in the growing season. Cells tend to be smaller in size and have thicker, denser cell walls than in earlywood. See *Earlywood.*

Lignin: A highly complex molecule that acts as an adhesive to bond wood cellulose units together. Although plastic in nature, especially at elevated temperatures, lignin gives wood its rigidity.

Microfibrils: Groups of cellulose or hemicellulose molecules that form long, threadlike macromolecules. Layers of microfibrils form the wood cell walls and are ultimately responsible for wood's anisotropic properties.

Moisture content (MC): The amount (weight) of water in a piece of wood, expressed as a percent of the weight of an oven-dry (0% MC) piece.

Nominal size: A convenient size nomenclature that approximates the size of a member in a log prior to being sawn into lumber; a nominal 2-by-4, when sawn, dried, and surfaced, has an actual, usable size of $1\frac{1}{2} \times 3\frac{1}{2}$ in., for example.

Normal duration of load: Ten years, cumulative, of the load for which a member or structure has been designed. Allowable design stresses may be modified for shorter or longer load periods.

Oriented-strand board (OSB): A flat board product made from wood flakes that are long and narrow and are oriented along the length of the panel, and a waterproof adhesive, so that the properties and characteristics resemble those of solid lumber.

Oven-dry weight: The weight of a piece of wood at 0% MC; normally determined by drying a small block of wood at 103°C until repeated weighings indicate that all moisture has been removed from the block.

Particleboard: A flat board product made from wood particles or flakes mixed with an adhesive and formed under pressure and elevated temperature. Not sold for structural use. See also *Oriented-strand board.*

Preservative: A chemical, usually in solution form, that is forced into wood (usually under pressure) to preserve the wood from attack by insects and decay organisms.

Plywood: A flat glued panel made from thin sheets of veneer with alternate plies having grain directions oriented 90 degrees to adjacent plies.

Quarter-sawn: Lumber that has its wide face in a radial plane (i.e., cut approximately parallel to the wood rays). Lumber is quarter-sawn to accentuate the wood figure due to the large wood rays in some species. See *Flat-sawn.*

Repetitive member use: A structural situation wherein three or more bending members in sizes of 2 to 4 inches in nominal thickness (as for floor joists) are fastened together and stabilized, as by flooring, to form an interactive unit. Repetitive members are permitted a 15% increase in allowable bending stress.

Sapwood: The outermost growth rings of a tree; always light brown to cream-colored in all species; never decay- or insect-resistant. Sapwood and heartwood together make up the "wood" of commercial use.

Shake: A lengthwise separation of the wood that may occur between or through annual growth rings.

Slope of grain: A deviation of cell orientation from the longitudinal axis of a member; measured as rise/run (as in 1 in 10). Slope of grain may be a growth phenomenon (as when grain deviates around a knot) or a manufacturing defect caused by failure to cut a member "along the grain."

Softwood: Typically "evergreen" trees with needle-like leaves. Includes Douglas fir, pines, spruces, cedars, and hemlock.

Specific gravity (SG): The ratio of the density of a material to that of an equal volume of water. See *Density.*

Split: A separation of wood cells along the grain; splits are deeper and more serious than cracks or checks in that splits penetrate completely through the thickness of a member. Splits most often occur at the ends of a member and reduce shear strength.

Strength ratio (SR): A factor applied in lumber stress grading to account for wood defects. It represents a ratio of the strength of a piece with all its defects to the strength the piece would have if no defects were present.

Structural composite lumber (SCL): A panel product made of thin layers of veneer or wood strands mixed with a waterproof adhesive. See LVL and PSL.

Wane: A lack of wood that occurs in lumber manufacture when the edge of a member intersects the periphery (bark) of a tree so that the edges are not "square."

Warp: Any deviation from a true or plane surface, including bow, crook, cup, and twist. *Bow* is a deviation flatwise from a straight line drawn from end to end of a piece. *Crook* is a deviation edgewise from a straight line drawn from end to end of a piece. *Cup* is a deviation in the face of a piece from a line drawn from edge to edge of a piece. *Twist* is a deviation flatwise, or a combination flatwise and edgewise, of a piece, resulting in the raising of one corner of a piece while the other three corners remain in contact with a flat surface.

Wood cells: Long, thin, hollow units that make up wood. Most cells are oriented with their long axis roughly parallel to the axis of the tree.

Wood rays: Groups of cells whose long axes are oriented horizontally and that radiate outward from the center of a tree. Wood rays are responsible for horizontal translocation of fluids in a tree and are partially responsible for the difference in radial versus tangential shrinkage and swellage in wood.

References

American Forest and Paper Association. 1991. *National Design Specification for Wood Construction* and *Supplement.* American Forest and Paper Association, Washington, D.C.

American Plywood Association. 1992. *APA Product Guide, Grades and Specifications.* American Plywood Association, Tacoma, WA.

American Plywood Association. 1992. *Guide to Wood Design Information.* American Plywood Association, Tacoma, WA.

Anonymous. 1993. Electric Utilities Study an Old, New Source of Fuel: Firewood. *Wall Street J.*, Dec. 2, 1993, p. A1.

ASTM. 1992. *Standard Practice for Establishing Structural Grades and Related Allowable Properties for Visually Graded Lumber, ASTM Designation D245-92.* American Society for Testing and Materials, Philadelphia, PA.

ASTM. 1988. *Standard Test Methods for Establishing Clear Wood Strength Values, ASTM Designation D2555-88.* American Society for Testing and Materials, Philadelphia, PA.

Forest Products Laboratory. 1987. *Wood Handbook: Wood as an Engineering Material.* Agricultural Handbook 72, USDA, Washington, D.C.

Triche, M. H. and Suddarth, S. K. 1993. *Purdue Plane Structures Analyzer 4.* Purdue Research Foundation, Purdue University, West Lafayette, IN.

For Further Information

Dietz, A. G. H., Schaffer, E. L., and Gromala, D. J., eds. 1982. *Wood as a Structural Material.* Vol. II of the *Clark C. Heritage Memorial Series on Wood.* Education Modules for Materials Science and Engineering Project. The Pennsylvania State University, University Park, PA.

Freas, A. D., Moody, R. C., and Soltis, L. A. 1986. *Wood: Engineering Design Concepts.* Vol. IV of the *Clark C. Heritage Memorial Series on Wood.* Materials Education Council, The Pennsylvania State University, University Park, PA.

Hoadley, R. B. 1980. *Understanding Wood: A Craftsman's Guide to Wood Technology.* The Taunton Press, Newtown, CT.

Hoyle, R. J. and Woeste, F. E. 1989. *Wood Technology in the Design of Structures*, 5th ed. Iowa State University Press, Ames, IA.

Wangaard, F. F. 1981. *Wood: Its Structure and Properties.* Vol. I of the *Clark C. Heritage Memorial Series on Wood.* Educational Modules for Materials Science and Engineering Project, The Pennsylvania State University, University Park, PA.

American Forest and Paper Association
1250 Connecticut Avenue NW
Washington DC 20036

The American Institute of Timber Construction
11818 SE Mill Plain Blvd.
Suite 415
Vancouver WA 98684-5092

American Plywood Association
PO Box 11700
Tacoma WA 98411

American Wood Preserver Association
PO Box 849
Stevensville MD 21666

Forest Products Laboratory
U.S. Department of Agriculture
One Gifford Pinchot Drive
Madison WI 53705-2398

National Particleboard Association
18928 Premiere Court
Gaithersburg MD 20879

Wood Truss Council of America
111 E Wacker Drive
Chicago IL 60601

43

Bituminous Materials and Mixtures

Jesus Larralde
California State University, Fresno

Mang Tia
University of Florida

43.1 Introduction

The term *bituminous materials* is generally used to denote substances in which bitumen is present or from which it can be derived [Goetz and Wood, 1960]. Bitumen is defined as an amorphous, black or dark-colored (solid, semisolid, or viscous) cementitious substance, composed principally of high molecular weight hydrocarbons, and soluble in carbon disulfide [ASTM, 1994]. For civil engineering applications, bituminous materials include primarily asphalts and tars. Asphalts may occur in nature (natural asphalts) or may be obtained from petroleum processing (petroleum asphalts). Tars do not occur in nature and are obtained as condensates in the processing of coal, petroleum, oil-shale, wood, or other organic materials [ASTM, 1994]. Pitch is formed when a tar is partially distilled so that the volatile constituents have evaporated off from it. The term *bituminous mixtures* is generally used to denote the combinations of bituminous materials (as binders), aggregates, and additives.

This chapter presents the basic principles and practices of the usage of bituminous materials and mixtures in pavement construction. In recent years the use of tars in highway construction has been very limited due to concern about the possible emission of hazardous fumes when tars are heated. Thus, this chapter deals primarily with asphalts and asphalt mixtures.

43.2 Bituminous Materials

Types of Bituminous Materials Used in Pavement Construction

Asphalt cement is an asphalt that has been specially refined as to quality and consistency for direct use in the manufacture of asphalt pavements, and has a penetration at 25°C (see the discussion

0-8493-8953-4/95/$0.00 + $.50

titled "Commonly Used Tests on Asphalt Cements and Their Significance") of between 5 and 300 [ASTM, 1994]. An asphalt cement has to be heated to an appropriate high temperature in order to be fluid enough to be mixed and placed.

Cutback asphalt is a liquid asphalt which is a blend of asphalt and petroleum solvents (such as gasoline and kerosene). A cutback asphalt can be mixed and placed with little or no application of heat. After a cutback asphalt is applied and exposed to the atmosphere, the solvent will gradually evaporate, leaving the asphalt cement to perform its function as a binder.

Emulsified asphalt (or asphalt emulsion) is an emulsion of asphalt cement and water which contains a small amount of emulsifying agent. In a normal emulsified asphalt, the asphalt cement is in the form of minute globules in suspension in water. An emulsified asphalt can be mixed and applied without any application of heat. After an asphalt emulsion is applied, sufficient time is required for the emulsion to break and the water to evaporate to leave the asphalt cement to perform its function as a binder. In an *inverted emulsified asphalt,* minute globules of water are in suspension in a liquid asphalt, which is usually a cutback asphalt. Inverted asphalt emulsions are seldom used in pavement applications.

Commonly Used Tests on Asphalt Cements and Their Significance

In this section, the purpose and significance of the commonly used tests on asphalt cements are described. Readers may refer to the appropriate standard test methods for detailed description of the test procedures.

Penetration Test

The *penetration test* is one of the oldest and most commonly used tests on asphalt cements or residues from distillation of asphalt cutbacks or emulsions. The standardized procedure for this test can be found in ASTM D5 [ASTM, 1994]. It is an empirical test that measures the consistency (hardness) of an asphalt at a specified test condition. In the standard test condition, a standard needle of a total load of 100 g is applied to the surface of an asphalt sample at a temperature of 25°C for 5 seconds. The amount of penetration of the needle at the end of 5 seconds is measured in units of 0.1 mm (or penetration unit). A softer asphalt will have a higher penetration, while a harder asphalt will have a lower penetration. The standard test condition is implied unless other test conditions are specified. Other test conditions that have been used include (1) 0°C, 200 g, 60 seconds, and (2) 46°C, 50 g, 5 seconds.

The penetration test can be used to designate grades of asphalt cement and to measure changes in hardness due to age hardening or changes in temperature.

Flash Point Test

The *flash point test* determines the temperature to which an asphalt can be safely heated in the presence of an open flame. The test is performed by heating an asphalt sample in an open cup at a specified rate and determining the temperature at which a small flame passing over the surface of the cup will cause the vapors from the asphalt sample temporarily to ignite or flash. The commonly used flash point test methods include (1) the Cleveland open cup (ASTM D92), and (2) the Tag open cup (ASTM D1310). The Cleveland open cup method is used on asphalt cements or asphalts with relatively higher flash points, while the Tag open cup method is used on cutback asphalts or asphalts with flash points of less than 79°C.

Minimum flash point requirements are included in the specifications for asphalt cements for safety reasons. Flash point tests can also be used to detect contaminating materials such as gasoline or kerosene in an asphalt cement. Contamination of an asphalt cement by such materials can be indicated by a drop in flash point. When the flash point test is used to detect contaminating materials, the Pensky-Martens closed tester method (ASTM D93), which tends to give more indicative results, is normally used. In recent years, the flash point test results have been related to the hardening potential of an asphalt. An asphalt with a high flash point is more likely to have a lower hardening potential in the field.

Solubility Test

An asphalt consists primarily of bitumens, which are high-molecular-weight hydrocarbons soluble in carbon disulfide. The solubility test measures the amount of bitumen in an asphalt by means of its solubility in carbon disulfide. In the standard test (ASTM D4), a small sample of about 2 g of the asphalt is dissolved in 100 ml of carbon disulfide and the solution is filtered through a filtering mat in a filtering crucible. The amount of material retained on the filter is then determined and expressed as a percentage of the weight of the original asphalt. Due to the extreme flammability of carbon disulfide, trichloroethylene may be substituted for carbon disulfide in routine testing. However, carbon disulfide should be used in referee testing.

The solubility test is used to detect contamination in an asphalt cement. Specifications for asphalt cements normally require a minimum bitumen content of 99%.

Ductility Test

The ductility test (ASTM D113) measures the distance a standard asphalt sample will stretch without breaking under a standard testing condition (5 cm/min at 25°C). Specifications for asphalt cements normally contain requirements for minimum ductility. However, the significance of this test is disputed. While a low ductility may indicate poor adhesive properties of an asphalt, a high ductility is usually related to high temperature susceptibility of an asphalt.

Viscosity Test

The viscosity test measures the viscosity of an asphalt. Both the viscosity test and the penetration test measure the consistency of an asphalt at some specified temperatures and are used to designate grades of asphalts. The advantage of using the viscosity test as compared with the penetration test is that the viscosity test measures a fundamental physical property rather than an empirical value.

Viscosity is defined as the ratio between the applied shear stress and induced shear rate of a fluid. The relationship between shear stress, shear rate, and viscosity can be expressed as:

$$\text{Shear rate} = \text{Shear stress / Viscosity} \tag{43.1}$$

When shear rate is expressed in units of 1/seconds and shear stress in units of pascal, viscosity will be in units of pascal-seconds. One pascal-second is equal to 10 poises. The lower the viscosity of an asphalt, the faster the asphalt will flow under the same stress.

For a Newtonian fluid, the relationship between shear stress and shear rate is linear, and thus the viscosity is constant at different shear rates or shear stresses. However, for a non-Newtonian fluid, the relationship between shear stress and shear rate is not linear, and thus the apparent viscosity will change as the shear rate or shear stress changes. Asphalts tend to behave as slightly non-Newtonian fluids, especially at lower temperatures. When different methods are used to measure the viscosity of an asphalt, the test results might be significantly different, since the different methods might be measuring the viscosity at different shear rates. It is thus very important to indicate the test method used when viscosity results are presented.

The most commonly used viscosity test on asphalt cements is the *absolute viscosity test by vacuum capillary viscometer* (ASTM D2171). The standard test temperature is 60°C. The absolute viscosity test measures the viscosity in units of poises. The viscosity at 60°C represents the viscosity of the asphalt at the maximum temperature a pavement is likely to experience in most parts of the U.S.

When the viscosity of an asphalt at a higher temperature (such as 135°C) is to be determined, the most commonly used test is the *kinematic viscosity test* (ASTM D2170), which measures the kinematic viscosity in units of stokes or centistokes. Kinematic viscosity is defined as:

$$\text{Kinematic viscosity} = \text{Viscosity / Density} \tag{43.2}$$

When viscosity is in units of poises and density in units of g/cm^3, the kinematic viscosity will be in units of stokes. Theoretically, to convert from kinematic viscosity (in units of stokes) to absolute viscosity (in units of poises), one simply multiplies the number of stokes by the density in units of g/cm^3. However, due to the fact that an asphalt might be non-Newtonian and that the kinematic viscosity test and the absolute viscosity test are run at different shear rates, conversion by this method will not produce accurate results and can only serve as a rough estimation. The standard temperature for the kinematic test on asphalt cement is 135°C. The viscosity at 135°C approximately represents the viscosity of the asphalt during mixing and placement of a hot mix.

Thin Film Oven and Rolling Thin Film Oven Tests

When an asphalt cement is used in the production of asphalt concrete, it has to be heated to an elevated temperature and mixed with a heated aggregate. The hot asphalt mixture is then hauled to the job site, placed, and compacted. By the time the compacted asphalt concrete cools down to the normal pavement temperature, significant hardening of the asphalt binder has already taken place. The properties of the asphalt in service are significantly different from those of the original asphalt.

Since the performance of the asphalt concrete in service depends on the properties of the hardened asphalt binder in service rather than the properties of the original asphalt, the properties of the hardened asphalt in service need to be determined and controlled.

The *thin film oven test* (TFOT) procedure (ASTM D1754) was developed to simulate the effects of heating in a hot-mix plant operation on an asphalt cement. In the standard TFOT procedure, the asphalt cement sample is poured into a flat-bottomed pan to a depth of about ⅛ inch. The pan with the asphalt sample in it is then placed on a rotating shelf in an oven and kept at a temperature of 163°C for five hours. The properties of the asphalt before and after the TFOT procedure are measured to determine the change in properties that might be expected after a hot-mix plant operation.

The *rolling thin film oven test* (RTFOT) procedure (ASTM D2872) was developed for the same purpose as the TFOT and designed to produce essentially the same effect as the TFOT procedure on asphalt cement. The advantages of the RTFOT over the TFOT are that (1) a larger number of samples can be tested at the same time, and (2) less time is required to perform the test. In the standard RTFOT procedure, the asphalt cement sample is placed in a specially designed bottle, which is then placed on its side on a rotating shelf, in an oven kept at 163°C, and rolled continuously for 75 minutes. Once during each rotation, the opening of the bottle passes an air jet that provides fresh air to the asphalt in the bottle for increased oxidation rate.

Ring and Ball Softening Point Test

The ring and ball softening point test (ASTM D36) measures the temperature at which an asphalt reaches a certain softness. When an asphalt is at its softening point temperature, it has approximately a penetration of 800 or an absolute viscosity of 13,000 poises. This conversion is only approximate and can vary from one asphalt to another, due to the non-Newtonian nature of asphalts and the different shear rates used by these different methods.

The softening point temperature can be used along with the penetration to determine the temperature susceptibility of an asphalt. Temperature susceptibility of an asphalt is often expressed as:

$$M = [\log(p_2) - \log(p_1)]/(t_2 - t_1) \tag{43.3}$$

where

M = temperature susceptibility
t_1, t_2 = temperatures in °C
p_1 = penetration at t_1
p_2 = penetration at t_2

Since an asphalt has approximately a penetration of 800 at the softening point temperature, the softening point temperature (SPT) can be used along with the penetration at 25°C to determine the temperature susceptibility as:

$$M = [\log(\text{pen at } 25°\text{C}) - \log(800)]/(25 - \text{SPT}) \qquad (43.4)$$

The M computed in this manner can then be used to compute a *penetration index* (PI) as follows:

$$\text{PI} = (20 - 500M)/(1 + 50M) \qquad (43.5)$$

The penetration index is an indicator of the temperature susceptibility of an asphalt. A high PI indicates a low temperature susceptibility. Normal asphalt cements have a PI between −2 and +2. Asphalt cements with a PI of more than +2 are of low temperature susceptibility, while those with a PI of less than −2 are of excessively high temperature susceptibility.

Conventional Methods of Grading and Specifications of Asphalt Cements

There are three commonly used methods of grading asphalt cements. These three methods are (1) grading by penetration at 25°C, (2) grading by absolute viscosity at 60°C, and (3) grading by absolute viscosity of aged asphalt residue after the rolling thin film oven test (RTFOT) procedure. These three methods of grading and the associated ASTM specifications of asphalt cements are presented and discussed in this section.

The method of grading asphalt cements by standard penetration at 25°C was the first systematic method developed and is still used by a few highway agencies in the world. The standard grades by this method include 40/50, 60/70, 85/100, 120/150, and 200/300 asphalts, which have penetrations of 40 to 50, 60 to 70, 85 to 100, 120 to 150, and 200 to 300, respectively. The Asphalt Institute recommends the use of a 120/150 or 85/100 pen. asphalt in the asphalt concrete for cold climatic conditions, with a mean annual temperature of 7°C or lower. For warm climatic conditions, with a mean air temperature between 7 and 24°C, a 85/100 or 60/70 pen. asphalt is recommended. For hot climatic conditions, with a mean annual air temperature of 24°C or greater, the use of a 40/50 or 60/70 pen. asphalt is recommended [Asphalt Institute, 1991].

ASTM D946 [ASTM, 1994] provides a specification for penetration-graded asphalt cements. Table 43.1 shows the specification for 60/70 and 85/100 pen. asphalts as examples. According to this specification, the only requirement on the consistency of the asphalt cements is the penetration at 25°C. There is no requirement on the consistency at either a higher or lower temperature, and thus no requirement on the temperature susceptibility of the asphalt cements. Two asphalts may be of the same penetration grade and yet have substantially different viscosities at 60°C. This problem is illustrated in Fig. 43.1. Thus, it is clear that specifying the penetration grade alone does not ensure that the asphalt used will have the appropriate viscosities at the expected service temperatures.

Table 43.1 Requirements for 60/70 and 85/100 Penetration Asphalt Cements

	Penetration Grade			
	60/70		85/100	
Test	Min	Max	Min	Max
Penetration at 25°C, 0.1 mm	60	70	85	100
Flash point (Cleveland open cup), °C	232	—	232	—
Ductility at 25°C, cm	100	—	100	—
Solubility in trichloroethylene, %	99.0	—	99.0	—
Retained penetration after TFOT, %	52	—	47	—
Ductility at 25°C after TFOT, cm	50	—	75	—

Source: ASTM. 1994. ASTM D946 Standard Specification for Penetration-Graded Asphalt Cement for Use in Pavement Construction. *Annual Book of ASTM Standards.* Volume 04.03, pp. 91–92. ASTM, Philadelphia, PA.

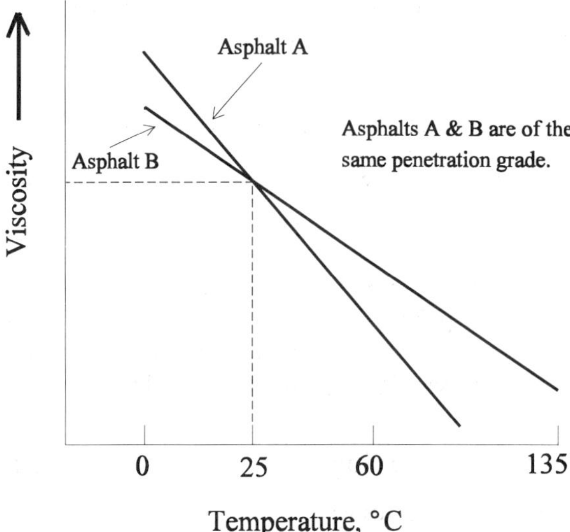

FIGURE 43.1 Variation in viscosity of two penetration-graded asphalts at different temperatures.

Other requirements in the specification are (1) minimum flash point temperature, (2) minimum ductility at 25°C, (3) minimum bitumen content by means of solubility in trichloroethylene, and (4) penetration and ductility at 25°C of the asphalt after aging by the TFOT procedure.

Since penetration is an empirical test, grading by penetration was thought to be unscientific. Considerable efforts were made in the 1960s to grade asphalts using fundamental units. Early attempts were made to grade asphalts by viscosity at 25°C and 20°C. However, problems were encountered in measuring viscosity at such low temperatures. With some reluctance, the temperature for grading asphalt by viscosity was moved to 60°C, which represents approximately the highest temperature pavements experienced in most parts of the U.S. When an asphalt is graded by this system, it is designated as AC, followed by a number that represents its absolute viscosity at 60°C in units of 100 poises. For example, an AC-20 would have an absolute viscosity of around 2000 poises at 60°C. An AC-20 roughly corresponds to a 60/70 pen. asphalt. However, due to the possible effects of different temperature susceptibility and non-Newtonian behavior, the conversion from a viscosity grade to a penetration grade may be different for different asphalts. Figure 43.2 shows the effects of different temperature susceptibility on the viscosity variation of two asphalts that have the same viscosity grade. In an effort to control this variation, the requirements for a minimum penetration at 25°C and a minimum viscosity at 135°C were added to the specification.

ASTM D3381 [ASTM, 1994] provides two different specifications for asphalt cements graded by absolute viscosity of the original asphalt at 60°C. Table 43.2 shows the requirements for AC-10 and AC-20 grade asphalts in the two specifications as examples. The main difference between these two specifications is that one requires a lower temperature susceptibility than the other. Limits on temperature susceptibility are specified through a minimum required penetration at 25°C and a minimum required kinematic viscosity at 135°C. The other requirements are similar to those in the specification of penetration-graded asphalts, namely: (1) minimum flash point temperature, (2) minimum ductility at 25°C, (3) minimum bitumen content by means of solubility in trichloroethylene, and (4) required properties of the asphalt after aging by the TFOT procedure (by means of maximum viscosity at 60°C and ductility at 25°C).

The third asphalt grading system involves grading asphalts according to their viscosity when placed on the road (after aging due to the heating and mixing process). This grading system has been adopted by several western states in the U.S. Grading is based on the absolute viscosity at 60°C of the asphalt residue after the rolling thin film oven test (RTFOT), which simulates the effects of

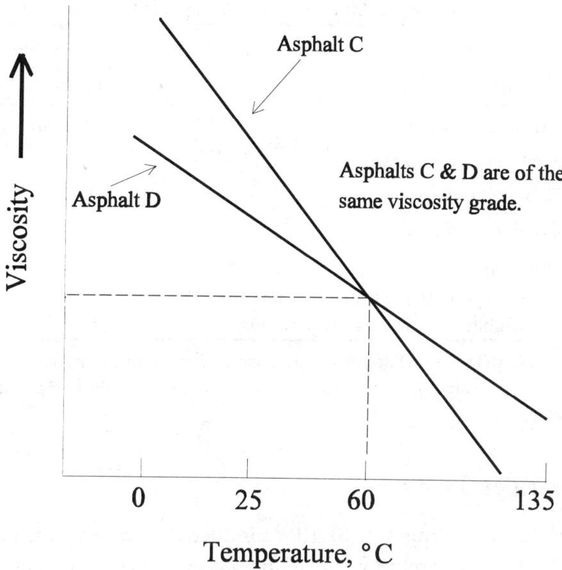

FIGURE 43.2 Variation in viscosity of two viscosity-graded asphalts at different temperatures.

Table 43.2 Requirements for AC-10 and AC-20 Asphalt Cements

		Viscosity Grade			
		AC-10		AC-20	
Test on Original Asphalt		Spec. 1	Spec. 2	Spec. 1	Spec. 2
Absolute viscosity at 60°C, poises		1000±200	1000±200	2000±400	2000±400
Kinematic viscosity at 135°C, cSt		150	250	210	300
Penetration at 25°C, 0.1 mm		70	80	40	60
Flash point (Cleveland open cup), °C		219	219	232	232
Solubility in trichloroethylene, min., %		99.0	99.0	99.0	99.0
Tests on residue from TFOT	Viscosity at 60°C, max., poises	5000	5000	10000	10000
	Ductility at 25°C, min., cm	50	75	20	50

Source: ASTM. 1994. ASTM D3381 Standard Specification for Viscosity-Graded Asphalt Cement for Use in Pavement Construction. *Annual Book of ASTM Standards.* Volume 04.03, pp. 297–298. ASTM, Philadelphia, PA.

the hot-mix plant operation. An asphalt graded by this system is designated as AR, followed by a number that represents the viscosity of the aged residue at 60°C in units of poises. For example, an AR-8000 would have an aged residue with an absolute viscosity of around 8000 poises. An AR-8000 would roughly correspond to an AC-20 or a 60/70 pen. asphalt. However, it should be recognized that the conversion from an AR grade to an AC grade depends on the hardening characteristics of the asphalt.

ASTM D3381 [ASTM, 1994] provides a specification for asphalt cements graded by viscosity of aged residue after the RTFOT process. Table 43.3 shows the specification for AR-4000 and AR-8000 grade asphalts as examples. According to this specification, temperature susceptibility is specified through requiring a minimum penetration at 25°C and a minimum kinematic viscosity at 135°C of the residue after the RTFOT. Similar to the requirements in the specifications for the other two grading systems, there are requirements on (1) ductility at 25°C of the aged residue, (2) minimum flash point of the original asphalt, and (3) minimum bitumen content of the original asphalt. Another requirement in this specification is a minimum percent of retained penetration after the RTFOT, which can serve as a check on the composition and aging characteristics of the asphalt.

Table 43.3 Requirements for AR-4000 and AR-8000 Asphalt Cements

		Viscosity Grade	
Test on Residue from RTFOT		AR-4000	AR-8000
Absolute viscosity at 60°C, poises		4000±1000	8000±2000
Kinematic viscosity at 135°C, min., cSt		275	400
Penetration at 25°C, min., 0.1 mm		25	20
% of original penetration, min.		45	50
Ductility at 25°C, min., cm		75	75
Test on original asphalt	Flash point (Cleveland open cup), min., °C	227	232
	Solubility in trichloroethylene, min., %	99.0	99.0

Source: ASTM. 1994. ASTM D3381 Standard Specification for Viscosity-Graded Asphalt Cement for Use in Pavement Construction. *Annual Book of ASTM Standards.* Volume 04.03, pp. 297–298. ASTM, Philadelphia, PA.

SHRP-Proposed Binder Tests

The Strategic Highway Research Program (SHRP) conducted a $50 million research effort from October 1987 to March 1993 to develop performance-based test methods and specifications for asphalts and asphalt mixtures. The proposed new tests and specifications have been standardized by the American Association of State Highway and Transportation Officials (AASHTO). The significance of the SHRP-proposed binder tests is described in this section. The detailed procedures can be found in the AASHTO publications for these tests [AASHTO, 1993a].

Pressure-Aging Vessel

The SHRP *pressure-aging vessel* (PAV) procedure is proposed to be used for simulation of long-term aging of asphalt binders in service. According to the proposed method (AASHTO Designation PP1), the asphalt samples are first aged in the standard RTFOT. Pans containing 50 grams of RTFOT residue are then placed in the PAV, which is pressurized with air at 2.1 ± 0.1 MPa, and aged for 20 hours. As many as 10 pans can be placed in the PAV. The proposed range of PAV temperature to be used is between 90 and 110°C. The PAV temperature to be used will depend on the climatic condition of the region where the binders will be used. A higher PAV temperature could be used for a warmer climatic condition, while a lower temperature could be used for a colder climatic condition.

Dynamic Shear Rheometer Test

The dynamic shear rheometer test measures the viscoelastic properties of an asphalt binder by testing it in an oscillatory mode. The general method had been used by researchers long before the SHRP researchers adopted and standardized the method for the purpose of asphalt specification. Typically, in a dynamic shear rheometer test, a sample of asphalt binder is placed between two parallel steel plates. The top plate is oscillated by a precision motor with a controlled angular velocity, w, while the bottom plate remains fixed. From the measured torque and angle of rotation, the shear stress and shear strain can be calculated. The oscillatory strain, γ, can be expressed as:

$$\gamma = \gamma_0 \sin wt \tag{43.6}$$

where γ_0 = peak shear strain and w = angular velocity in radian/second. The shear stress, τ, can be expressed as

$$\tau = \tau_0 \sin(wt + \delta) \tag{43.7}$$

where τ_0 = peak shear stress and δ = phase shift angle. The following parameters are usually computed from the test data:

1. Complex shear modulus. $G^* = \tau_0/\gamma_0$ (43.8)
2. Dynamic viscosity. $\eta^* = G^*/w$ (43.9)
3. Storage modulus. $G' = G^* \cos\delta$ (43.10)
4. Loss modulus. $G'' = G^* \sin\delta$ (43.11)
5. Loss tangent. $\tan\delta = G''/G'$ (43.12)

How are the results of a dynamic rheometer test related to the basic rheologic properties of the tested binder? This question can be answered by analyzing how a viscoelastic material would behave in such a test. For simplicity, the test binder is modeled by a Maxwell model with a shear modulus of G and a viscosity of η. When the test binder is modeled in this manner, it can be shown analytically that the complex shear modulus G^* is equal to:

$$G^* = \tau_0/\gamma_0 = w\eta/(1 + \eta^2 w^2/G^2)^{1/2} \tag{43.13}$$

The dynamic viscosity, η^*, can be derived to be:

$$\eta^* = G^*/w = \eta/(1 + \eta^2 w^2/G^2)^{1/2} \tag{43.14}$$

The loss tangent $\tan\delta$, can be derived to be:

$$\tan\delta = G/w\eta \tag{43.15}$$

It can be seen from the above equations that the complex shear modulus, G^*, will approximate the shear modulus, G, when the angular velocity, w, is extremely high. The dynamic viscosity, η^*, will approach the true viscosity, η, when w is extremely low.

The storage modulus, G', and loss modulus, G'', which are computed from the results of the dynamic shear rheometer test, can be derived to be the following:

$$G' = G^* \cos\delta = w\eta/(1 + \eta^2 w^2/G^2)^{1/2}(1 + G^2/\eta^2 w^2)^{1/2} \tag{43.16}$$

$$G'' = G^* \sin\delta = G/(1 + \eta^2 w^2/G^2)^{1/2}(1 + G^2/\eta^2 w^2)^{1/2} \tag{43.17}$$

It can be seen that at high w, G' would approach the true shear modulus, G. At low w, G'' would approach the true viscosity, η.

SHRP standardized the dynamic shear rheometer test (AASHTO Designation TP5) for use in measuring the asphalt properties at high and intermediate service temperatures for specification purposes. In the standardized test method, the oscillation speed is specified as 10 radians/second. The amplitude of shear strain to be used depends on the stiffness of the binder, and varies from 1% for hard materials tested at low temperatures to 13% for relatively softer materials tested at high temperatures. There are two standard sample sizes. For relatively softer materials, a sample thickness (gap) of 1 mm and a sample diameter (spindle diameter) of 25 mm are used. For harder materials, a sample thickness of 2 mm and a sample diameter of 8 mm are used. The two values to be measured from each test are the complex shear modulus, G^*, and the phase angle, δ. These two test values are then used to compute $G^*/\sin\delta$ and $G^* \sin\delta$. In the SHRP-proposed asphalt specification, permanent deformation is controlled by requiring the $G^*/\sin\delta$ of the binder at the highest anticipated pavement temperature to be greater than 1.0 kPa before aging and 2.2 kPa after the RTFOT process. Fatigue cracking is controlled by requiring that the binder after PAV aging should have a $G^*/\sin\delta$ value of less than 5000 kPa at a specified intermediate pavement temperature.

Bending Beam Rheometer Test

The bending beam rheometer test (AASHTO Designation TP1) was proposed by SHRP to measure the stiffness of asphalts at low service temperatures. The standard asphalt test specimen is a rectangular prism with a width of 12.5 mm, a height of 6.25 mm and a length of 125 mm. The test specimen is submerged in a temperature-controlled fluid bath and simply supported with a distance between supports of 102 mm. For specification testing, the test samples are fabricated from PAV-aged asphalt binders that simulate the field-aged binders. In the standard testing procedure, after the beam sample has been properly preconditioned, a vertical load of 100 gram-force is applied to the middle of the beam for a total of 240 seconds. The deflection of the beam at the point of load is recorded during this period and used to compute for the creep stiffness of the asphalt by the following equation:

$$S(t) = PL^3/4bh^3\delta(t) \tag{43.18}$$

where

$S(t)$ = creep stiffness at time t
P = applied load, 100 g
L = distance between beam supports, 102 mm
b = beam width, 12.5 mm
h = beam height, 6.25 mm
$\delta(t)$ = deflection at time t

The above equation is similar to the equation that relates the deflection at the center of the beam to the elastic modulus of an elastic beam according to the classical beam theory. The instantaneous deflection in the original equation is replaced by the time-dependent deflection $\delta(t)$, while the elastic modulus in the original equation is replaced by the time-dependent creep stiffness $S(t)$.

For SHRP specification purposes, the bending beam rheometer test is to be run at 10°C above the expected minimum pavement temperature, $T_{\min}$. SHRP researchers [Anderson and Kennedy, 1993] claimed that the stiffness of an asphalt after 60 seconds at $T_{\min} + 10°\text{C}$ is approximately equal to its stiffness at $T_{\min}$ after 2 hours loading time, which is related to low-temperature cracking potential. The SHRP binder specification requires the stiffness at the test temperature after 60 seconds to be less than 300 MPa to control low-temperature cracking.

The second parameter obtained from the bending beam rheometer test result is the m-value. The m-value is the slope of the log stiffness versus log time curve at a specified time. A higher m-value would mean that the asphalt would creep at a faster rate to reduce the thermal stress and would be more desirable to reduce low-temperature cracking. SHRP specification requires the m-value at 60 seconds to be greater than or equal to 0.30.

Direct Tension Test

The SHRP direct tension test (AASHTO Designation TP3) measures the failure strain of an asphalt binder. In this test, a small "dog bone"–shaped asphalt specimen is pulled at a constant, slow rate until it breaks. The amount of elongation at failure is used to compute the failure strain. The test specimen is 30 mm long and has a cross section of 6 mm by 6 mm at the middle portion. For SHRP specification purposes, the direct tension test is to be run on PAV-aged binders at the same test temperature as for the bending beam rheometer test; 10°C above the minimum expected pavement temperature. SHRP specifications require that the failure strain at this condition should not be less than 1% in order to control low-temperature cracking.

Brookfield Rotational Viscometer Test

SHRP selected the Brookfield rotational viscometer test as specified by ASTM D4402 for use in measuring the viscosity of binders at elevated temperatures to ensure that the binders are sufficiently fluid when being pumped and mixed at the hot-mix plants. In the Brookfield rotational viscometer test, the test binder sample is held in a temperature-controlled cylindrical sample chamber, and a cylindrical spindle, which is submerged in the sample, is rotated at a specified constant speed. The torque required to maintain the constant rotational speed is measured and used to calculate the shear stress according to the dimensions of the sample chamber and spindle. Similarly, the rotational speed is used to calculate the shear rate of the test. Viscosity is then calculated by dividing the computed shear stress by the computed shear rate.

Compared to the capillary tube viscometers, the rotational viscometer provides larger clearances between the components. Therefore, it can be used to test modified asphalts containing larger particles, which could plug up a capillary viscometer tube. Another advantage of the rotational viscometer is that the shear stress–versus–shear rate characteristics of a test binder can be characterized over a wide range of stress or strain levels.

For SHRP specification purposes, the rotational viscosity test is run on the original binder at 135°C. The maximum allowable viscosity at this condition is 3 Pa · s.

SHRP-Proposed Binder Specification

The SHRP-proposed performance-graded asphalt specification (AASHTO MP1) uses grading designations that correspond to the maximum and minimum pavement temperatures of the specified region. The designation starts with "PG" and is followed by the maximum and the minimum anticipated service temperature in °C. For example, a PG-64-22 grade asphalt is intended for use in a region where the maximum pavement temperature (based on average 7-day maximum) is 64°C and the minimum pavement temperature is −22°C. A PG-52-46 grade asphalt is for use where the maximum pavement temperature is 52°C and the minimum pavement temperature is −46°C.

Table 43.4 shows the SHRP-proposed specification for three performance grades of asphalts (PG-52-16, PG-52-46, and PG-64-22) as examples. The specified properties are constant for all grades, but the temperatures at which these properties must be achieved vary according to the climate in which the binder is to be used. It is possible that an asphalt can meet the requirements for several different grades.

All grades are required to have a flash point temperature of at least 230°C for safety purposes, and to have a viscosity no greater than 3 Pa · s at 135°C to ensure proper workability during mixing and placement.

Dynamic shear rheometer tests are run on the original and RTFOT-aged binders at the maximum pavement design temperature. The minimum required values of $G^*/\sin\delta$ at this temperature are 1.0 kPa and 2.2 kPa for the original and RTFOT-aged binders, respectively. These requirements are intended to control pavement **rutting**.

Dynamic shear rheometer tests are also run on PAV-aged binders at an intermediate temperature, which is equal to 4°C plus the mean of the maximum and minimum pavement design temperatures. For example, for a PG-52-46 grade, the intermediate temperature is 7°C. The maximum allowable value of $G^* \sin\delta$ at this condition is 5000 kPa. This requirement is intended to control pavement fatigue cracking.

Bending beam rheometer tests and direct tension tests are run on PAV-aged binders at a temperature 10°C above the minimum pavement design temperature. For example, for a PG-52-46, the test temperature is −36°C. At a loading time of 60 seconds, the stiffness is required to be no greater than 300 MPa, and the m-value is required to be no less than 0.3. The failure strain is required to be at least 1%.

Effects of Properties of Asphalt Binders on the Performance of Asphalt Pavements

Effects of Viscoelastic Properties of Asphalt

When an asphalt concrete surface is cooled in winter time, stresses are induced in the asphalt concrete. These stresses can be relieved by the flowing of the asphalt binder within the asphalt mixture. However, if the viscosity of the asphalt binder is too high at this low temperature, the flow

Table 43.4 Examples of SHRP Performance-Graded Binder Specification

Performance Grade	PG-52-16	PG-52-40	PG-64-22
Average 7-day maximum pavement design temperature, °C	52	52	64
Minimum pavement design temperature, °C	−16	−40	−22
Original Binder			
Flash point temperature, minimum, °C		230	
Viscosity: maximum, 3 Pa·s Test temperature, °C		135	
Dynamic shear @ 10 rad/s: G*/sin δ, minimum, 1.00 kPa Test temperature, °C	52	52	64
Rolling Thin Film Oven Residue			
Mass loss, maximum, %		1.00	
Dynamic shear @ 10 rad/s: G*/sin δ, minimum, 2.20 kPa Test temperature, °C	52	52	64
Pressure-Aging Vessel Residue			
PAV aging temperature, °C	90	100	100
Dynamic shear @ 10 rad/s: G*/sin δ, maximum, 5000 kPa Test temperature, °C	22	7	25
Creep stiffness @ 60 s: *S*, maximum, 300 MPa *m*-value, minimum, 0.30 Test temperature, °C	−6	−36	−12
Direct tension @ 1.0 mm/min: Failure strain, minimum, 1.0% Test temperature, °C	−6	−36	−12

Source: AASHTO. 1993. AASHTO Designation MP1 Standard Specification for Performance-Graded Asphalt Binder. *AASHTO Provisional Standard.* AASHTO, Washington, D.C.

of the asphalt binder may not be fast enough to relieve the high induced stresses. Consequently, low-temperature cracking may occur. The viscosity of asphalt at which low-temperature cracking can occur is dependent on the cooling rate of the pavement as well as the characteristics of the asphalt concrete. However, as a rough prediction of low-temperature cracking, a limiting viscosity of 2×10^{10} poises can be used [Davis, 1987]. If the viscosity of the asphalt binder at the lowest anticipated temperature is kept lower than this limiting value, low-temperature cracking is unlikely to occur.

The effects of the elastic property of asphalt on low-temperature cracking can be understood by analyzing how a viscoelastic material as modeled by a Maxwell model with a shear modulus of G and a viscosity of η would release its stress after it is subjected to a forced strain γ_0 (which could be caused by a sudden drop in pavement temperature). If the material is subjected to a forced strain of γ_0 at $t = 0$, the instantaneous induced stress is equal to $\gamma_0 G$, but the stress decreases with time according to the following expression:

$$\tau = \gamma_0 G e^{-Gt/\eta} \tag{43.19}$$

It can be seen that the rate of stress release is proportional to G/η. The reciprocal of this parameter, η/G, is commonly known as the *relaxation time.* To maximize the rate of relaxation, it is desirable to have a low relaxation time, η/G, or a higher G/η. As presented earlier, the parameter tan δ—as obtained from the dynamic shear rheometer test—is directly proportional to G/η. Thus, a high tan δ value is desirable to reduce the potential for low-temperature pavement cracking. Experimental data show that tan δ of an asphalt always decreases with decreasing temperature. Goodrich [1991] stated that when testing is done at an angular velocity, w, of 0.1 radian/second, the temperature at which tan δ of the binder is equal to 0.4 corresponds approximately to the temperature at which the asphalt mixture reaches its limiting stiffness.

Another critical condition of an asphalt concrete is at the highest pavement temperature, at which the asphalt mixture is the weakest and most susceptible to plastic flow when stressed. When the other factors are kept constant, an increase in the viscosity of the asphalt binder will increase the shear strength and subsequently the resistance to plastic flow of the asphalt concrete. With respect to resistance to plastic flow of the asphalt concrete, it is preferable to have a high asphalt viscosity at the highest anticipated pavement temperature. Results by Goodrich [1988] indicate that a low tan δ value of the binder (as obtained from the dynamic rheometer test) tends to correlate with a high creep compliance of the asphalt mixture, which indicates high rutting resistance. Thus, a low tan δ value of the binder is desirable to reduce rutting potential.

The effectiveness of the mixing of asphalt cement and aggregate and the effectiveness of the placement and compaction of the hot asphalt mix are affected greatly by the viscosity of the asphalt. The Asphalt Institute recommends that the mixing of asphalt cement and aggregate should be done at a temperature where the viscosity of the asphalt is 1.7 ± 0.2 poises. Compaction should be performed at a temperature where the viscosity of the asphalt cement is 2.8 ± 0.3 poises [Epps *et al.*, 1983]. These viscosity ranges are only offered as guidelines. The actual optimum mixing and compaction temperatures will depend on the characteristics of the mixture as well as the construction environment.

In the selection of a suitable asphalt cement to be used in a certain asphalt paving project, the main concerns are (1) whether the viscosity of the asphalt at the lowest anticipated service temperature

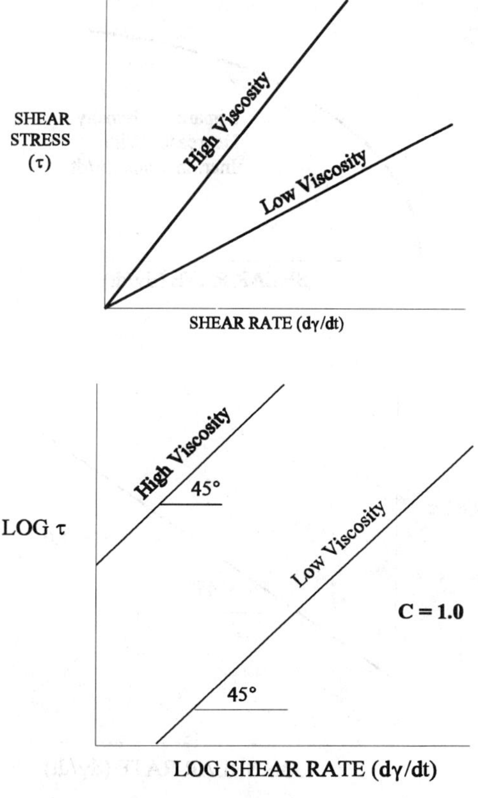

FIGURE 43.3 Newtonian flow characteristics.

would be low enough to avoid low-temperature cracking of the asphalt concrete, (2) whether the viscosity of the asphalt at the highest anticipated temperature would be high enough to resist rutting, and (3) whether the required temperatures for proper mixing and placement would be too high. The present specifications of asphalt cements are quite general in nature. Two asphalt cements can belong to the same grade and yet may have different viscosity-temperature characteristics and give completely different performances. It is thus very important that the suitability of an asphalt cement for a particular paving project be evaluated rationally by considering the viscosity at the anticipated service temperatures.

Effects of Newtonian and Non-Newtonian Flow Properties of Asphalt

The flow behavior of asphalt cements can be classified into four main categories, namely Newtonian, pseudoplastic, Bingham-plastic, and dilatant. Asphalt cements usually exhibit **Newtonian** or near-Newtonian flow behavior, especially at temperatures in excess of 25°C. A Newtonian flow behavior is characterized by a linear shear stress–shear rate relationship, as shown in Fig. 43.3. The **shear susceptibility,** C, is defined as the slope of the plot of log (shear stress) versus log (shear rate). For a Newtonian flow behavior, C is equal to 1.00.

The type of flow behavior where a reduction in viscosity is experienced with increased stress is termed **pseudoplastic.** The shear stress–shear rate relationship for a pseudoplastic fluid is shown in Fig. 43.4. It can be seen that the shear rate increases more rapidly at higher stresses. The shear susceptibility, C, is less than 1.0 in this case.

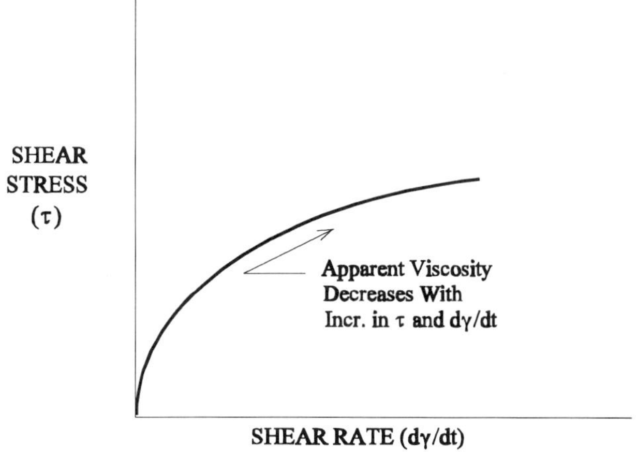

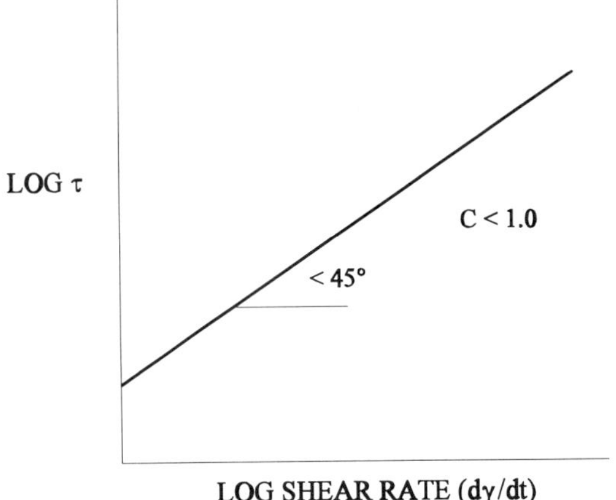

FIGURE 43.4 Pseudoplastic flow characteristics.

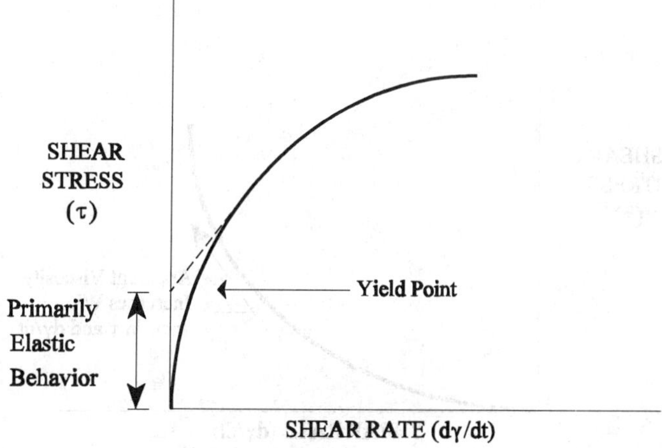

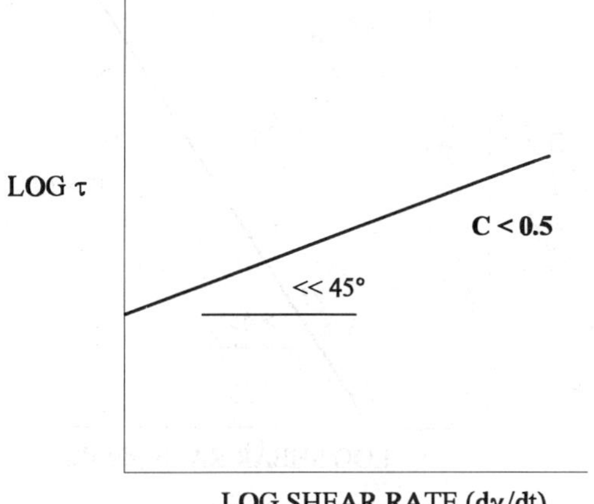

FIGURE 43.5 Bingham-plastic flow characteristics.

The shear stress–shear rate relationship for a **Bingham-plastic material** is illustrated in Fig. 43.5. When the stress is below a certain stress level, there is no flow. When the stress is above the yield point, the flow characteristic is likely to be highly pseudoplastic, with a C of less than 0.5. Highly air-blown asphalts usually exhibit Bingham-plastic behavior at low temperatures.

The type of flow behavior in which the apparent viscosity increases with increased stress is referred to as **dilatant.** The shear stress–shear rate characteristics of dilatant behavior are shown in Fig. 43.6. For this type of flow behavior, C is greater than 1.

What are the effects of the flow behavior of the asphalt cement on the performance of the asphalt pavement? The answer to this question is still not definitive at this point. However, some research results have indicated that asphalts with high shear susceptibility (C) have been related to tender mixes [Epps *et al.*, 1983], as well as high temperature susceptibility and high aging indices [Kandhal *et al.*, 1973].

The effect of non-Newtonian flow behavior on the measured viscosity is clear. When an asphalt exhibits a non-Newtonian flow behavior, the measured viscosity changes as the shear stress or shear rate used for the test changes. This effect must be properly accounted for. When the viscosity of the asphalt is used to predict the behavior of the asphalt concrete in service, the viscosity at a stress level close to the anticipated stress level in service should be used.

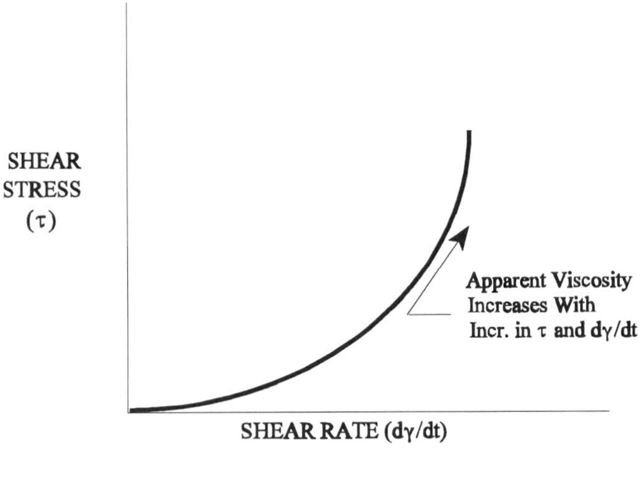

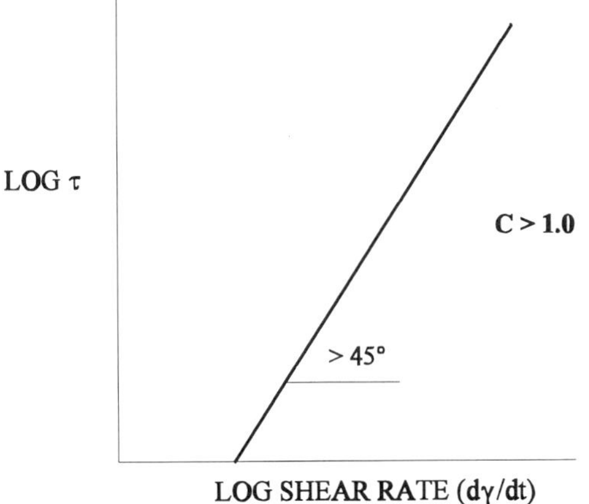

FIGURE 43.6 Dilatant flow characteristics.

Effects of Hardening Characteristics of Asphalt

An important factor that affects the durability of an asphalt concrete is the rate of hardening of the asphalt binder. The causes of hardening of asphalt have been attributed to oxidation, loss of volatile oils, and polymerization (changes in structure). Among all these possible factors, oxidation is generally considered to be the prime cause of asphalt hardening.

The most severe hardening of asphalt occurs during the mixing process. The viscosity of the asphalt binder immediately after the asphalt concrete is placed on the road is usually 2 to 4 times the viscosity of the original asphalt cement. The asphalt binder continues to harden through service; its viscosity could reach as high as 10 to 20 times the viscosity of the original asphalt cement. The rate of asphalt hardening is dependent on asphalt composition, mixing temperature, air voids content, and climatic conditions. It usually increases with increased mixing temperature, increased air voids content in the asphalt mix, and increased service (air) temperature.

Excessive hardening of the asphalt binder will cause the asphalt concrete to be too brittle and low-temperature cracking to occur. It may also cause the asphalt binder to partially lose its adhesion and cohesion, and subsequently it may cause raveling (progressive disintegration of pavement material and separation of aggregates from it) in the asphalt concrete.

A certain amount of hardening of the asphalt binder during the mixing process is usually expected and designed for. If an asphalt binder has not hardened sufficiently during the mixing process (due to low mixing temperature or the peculiar nature of the asphalt), the asphalt binder may be too soft at placement. This may cause the asphalt mix to be difficult to compact (tender mix) and to have a low resistance to rutting in service. If the tenderness of an asphalt concrete disappears within a few weeks after construction, the problem is most likely caused by a slow-setting asphalt. These types of asphalts require an excessive amount of time to "set up" after they are heated up and returned to normal ambient temperature. Asphalts containing less than 10 percent asphaltenes appear to have a greater probability of producing slow-setting asphalt mixtures. Asphaltenes are the high molecular weight fraction of an asphalt which can be separated from the other asphalt fractions by dissolving an asphalt in a specified solvent (such as n-heptane as used in the ASTM D4124 Methods for Separation of Asphalt into Four Fractions) [ASTM, 1994]. The asphaltenes, which are insoluble in the solvent, would be precipitated out in this method.

Types and Grades of Cutback Asphalts

Cutback asphalts are classified into three main types on the basis of the relative speed of evaporation of the solvents in them. A *rapid-curing* (RC) cutback asphalt is composed of an asphalt cement and a solvent of a volatility similar to that of naphtha or gasoline, which evaporates at a fast speed. A *medium-curing* (MC) cutback asphalt contains a solvent of a volatility similar to that of kerosene, which evaporates at a medium speed. A *slow-curing* (SC) cutback asphalt contains an oil of relatively low volatility.

Within each type, cutback asphalts are graded by kinematic viscosity at 60°C. An asphalt is designated by the type followed by the lower limit of the kinematic viscosity at 60°C in units of centistokes (cSt). The upper limit for the viscosity is twice its lower limit. For example, an "RC-70" designates a rapid-curing cutback asphalt with a kinematic viscosity at 60°C ranging between 70 and 140 cSt, while an "SC-800" designates a slow-curing cutback asphalt with a viscosity ranging between 800 and 1600 cSt. The standard specifications for SC, MC, and RC cutback asphalts can be found in ASTM Designation D2026, D2027, and D2028, respectively [ASTM, 1994].

The standard practice for selection of cutback asphalts for pavement construction and maintenance can be found in ASTM D2399 [ASTM, 1994].

Types and Grades of Emulsified Asphalts

Emulsified asphalts (or asphalt emulsions) are divided into three major kinds, namely anionic, cationic, and nonionic, on the basis of the electrical charges of the asphalt particles in the emulsion. An anionic asphalt emulsion has negatively charged asphalt particles and is usually more suitable for use with calcareous aggregates, which tend to have positive surface charges. A cationic asphalt emulsion has positively charged asphalt particles and is usually more suitable for use with siliceous aggregates, which tend to have negative surface charges. A nonionic asphalt emulsion contains asphalt particles that are electrically neutral. Nonionic asphalt emulsions are not used in pavement applications.

Asphalt emulsions are further classified into three main types on the basis of how quickly the suspended asphalt particles revert to asphalt cement. The three types are *rapid-setting* (RS), *medium-setting* (MS), and *slow-setting* (SS). An RS emulsion is designed to demulsify (to break away from the emulsion form such that asphalt particles are no longer in suspension) upon contact with an aggregate, and thus has little or no ability to mix with an aggregate. It is best used in spraying applications where mixing is not required but fast setting is desirable. An MS emulsion is designed to have good mixing characteristics with coarse aggregates and to demulsify after proper mixing. It is suitable for applications where mixing with coarse aggregate is required. An SS emulsion is designed to be very stable in the emulsion form, and is suitable for use where good flowing characteristics are desired or where mixing with fine aggregates is required. The three types of cationic asphalt emulsions are denoted as CRS, CMS, and CSS. The absence of the letter "C" in front of the emulsion type denotes an anionic type.

Two other standard types of anionic asphalt emulsions available are *high-float rapid-setting* (HFRS) and *high-float medium-setting* (HFMS). This type of asphalt emulsion contains an asphalt cement that has a Bingham-plastic characteristic (resistant to flow at low stress level). This flow property of the asphalt permits a thicker film coating on an aggregate without danger of runoff.

Within each type, asphalt emulsions are graded by the viscosity of the emulsion and the hardness of the asphalt cement. The lower viscosity grade is designated by a number "1" and the higher viscosity grade is designated by a number "2," which is placed after the emulsion type. A letter "h" that follows the number "1" or "2" designates that a harder asphalt cement is used. For example, an "RS-1" designates a rapid-setting anionic type with a relatively low viscosity. An "HFMS-2h" designates a high-float medium-setting anionic type having a relatively higher viscosity and containing a hard base asphalt. A "CSS-1h" designates a slow-setting cationic type having a relatively lower viscosity and containing a hard base asphalt. Standard specifications for anionic and cationic emulsified asphalts can be found in ASTM D977 and D2397, respectively.

The standard practice for selection and use of emulsified asphalts in pavement construction and maintenance can be found in ASTM D3628 [ASTM, 1994].

43.3 Bituminous Mixtures

Types of Bituminous Mixtures

The term *bituminous mixture* can be generically applied to the combination of bituminous materials, properly graded aggregates, and other material additives. Bituminous mixtures are utilized in a wide range of applications including roofing, dam facing, reservoir lining, etc., although their main use is as paving material [Asphalt Institute, 1977]. Thus, the emphasis of this section is on the utilization of bituminous mixtures in pavement construction.

Bituminous mixtures for diverse applications are obtained by mixing different types of asphalt binders with aggregates of various gradations. In the following paragraphs the mixtures most commonly used in paving construction are discussed.

Hot-mix asphalt (HMA), also known as asphalt cement concrete, asphalt concrete, bituminous concrete, asphalt paving mix, bituminous mix, bituminous paving mix, etc., is typically produced in a stationary mixing plant, transported to the site, and placed and compacted to a specified density. In the production of HMA mixtures, asphalt cement is liquefied by heating to a temperature high enough to allow proper mixing with the aggregates but low enough to prevent premature aging of the asphalt cement. The aggregates for HMA mixtures are usually selected with care and controlled to comply with stringent specifications for particle size distribution and other properties. HMA mixtures are used in construction of full-depth asphalt pavements, wearing and surface courses, and asphalt concrete overlays for flexible and rigid pavements.

Depending on the gradation characteristics of the aggregates, HMA mixtures can be either dense graded, open graded, or gap graded. Dense-graded mixtures consist of aggregates with a continuous grading and asphalt binder. A dense-graded mixture with nominal size of aggregate greater than 25.4 mm (1 in.) is called a *large-stone mix*. By contrast, a sand mix is a dense-graded mix without coarse aggregates, with 100% of the aggregate particles passing the 9.5 mm (3/8 in.) sieve. Open-graded mixtures consist of an asphalt binder and primarily coarse aggregate particles with a minimal amount of fine aggregate particles. These types of mixtures exhibit a very open structure with high permeability that allows water to drain through. Also, open-graded mixtures exhibit a rough surface texture that enhances contact with vehicle tires, increasing the skid resistance. Gap-graded mixtures consist of asphalt binder and a gap-graded aggregate, that is, an aggregate in which the particles between the No. 4 (4.75 mm) and No. 40 (425 μm) are missing or are present in a very small amount.

Cold-laid mixes are produced by mixing (generally at ambient temperature) aggregates with an asphalt emulsion or with an asphalt cutback. This type of bituminous mixture is generally used in construction of pavement bases or for pavement patching and repair.

Mixed-in-place or road mixes are produced by mixing mineral aggregates and asphalt cutback or asphalt emulsion. Road mixes are produced near the construction site in transportable plants or by mixing the asphalt and aggregate materials on the road surface by means of graders or special road mixing equipment. These types of mixtures are used in base and subbase construction and in pavements for low traffic volume.

Stone matrix asphalt (SMA) mixes are produced by combining gap-graded aggregates with an asphalt-rich binder mastic. These types of mixtures have successfully been used in pavement construction in Europe to prevent excessive rutting.

Slurry seal mixes are utilized in crack filling and sealing of medium and highly textured pavement surfaces. Slurry seals are mixtures of emulsified asphalt, water, well-graded fine aggregate, and mineral filler. Different slurry seal mixes can be obtained by combining asphalt emulsions of varying setting times with aggregates of 3 mm (1/8 in.) to 9.5 mm (3/8 in.) maximum particle size.

Performance Requirements of HMA Mixtures

The ideal HMA mix is one that is able to maintain the serviceability of the pavement in which it is used, without maintenance cost. *Pavement serviceability* (AASHTO, 1993) represents the ability of a pavement to provide a comfortable ride to the users. Serviceability can deteriorate due to the action of traffic and weather, which continuously subject the pavement to changes in stress, temperature, and moisture. These changes produce pavement distress which can occur in different forms, traditionally grouped into four categories: distortion, disintegration, cracking, and skid hazard (AI MS-16). Pavement distress can be the result of deficiencies in materials and mixtures, faulty construction, substandard pavement design, inadequate drainage, or combinations thereof. Table 43.5 shows a list of different pavement distresses and their possible causes.

In order to minimize the material-related distress in pavements with HMA mixtures, careful mix design and quality control have to be exercised to obtain the required properties of the mix

Table 43.5 Types of Pavement Distress and Their Causes

	Possible Causes			
Distress Type	Structural Failure	Mix Composition	Temperature or Moisture Changes	Faulty Construction
Fatigue cracking	X			
Bleeding		X		X
Block cracking		X	X	
Corrugations	X	X		X
Depressions	X			X
Reflection cracking			X	
Lane/shoulder dropoff or heave	X		X	X
Lane/shoulder separation	X		X	X
Longitudinal and transverse cracking			X	X
Patch deterioration	X	X	X	X
Polished aggregate		X	X	X
Potholes	X		X	X
Pumping and water bleeding	X		X	X
Raveling and weathering		X		X
Rutting	X	X		X
Slippage cracks				X
Swell			X	

Source: Federal Highway Administration. 1988. *FHWA Pavement Rehabilitation Manual.* Report No. FHWA-ED-88-025, March.

in service. The mix has to be stable, durable, skid resistant, and yet workable and economic (AI MS-2). Mix stability—that is, the resistance to permanent deformation—is especially critical during hot weather when the viscosity of the asphalt cement is low and the traffic loads are primarily supported by the particle-to-particle contact of the aggregate. Stability is increased by using aggregates with rough surface texture, dense gradations, relatively low asphalt content, harder asphalts, and well-compacted mixtures [Monismith *et al.*, 1985]. HMA mixtures have to be resistant to cracking produced by repeated traffic loading. The resistance of the mix to this type of cracking, called *fatigue cracking,* is dependent primarily on the amount and grade of asphalt, the aggregate gradation, and the compaction and amount of voids in the mixture [Monismith *et al.*, 1985].

At low temperature, HMA mixes are vulnerable to thermal cracking. This susceptibility to cracking results from the reduction of the tensile failure strain of the asphalt cement at low temperature. Asphalt cements of different grades exhibit different values of ultimate strain; in general, cements with low penetration grade (high viscosity) have lower ultimate tensile strains. Nevertheless, asphalt cements of the same grade but from different sources may have different tensile failure strains which lead to varying resistance to low-temperature cracking. Thus, low-temperature cracking can be minimized by proper selection of asphalt cement binder [Leung and Anderson, 1987; Roberts *et al.*, 1991].

HMA mixtures must have adequate durability, that is, adequate resistance to disintegration produced by weather and traffic. Mixture durability can be reduced by the aging or hardening of the asphalt cement, which can be accidentally induced by excessive heating during production. Premature aging can also occur in the mixture in service when oxidation of the asphalt cement takes place in mixtures with high permeability or low cement content. Some HMA mixes are susceptible to loss of adhesion between aggregates and asphalt cement. This loss of adhesion, or aggregate stripping, is the physical separation of the asphalt cement and aggregate, primarily due to the action of water or water vapor. Stripping is primarily related to the surface texture of the aggregate and to its chemical and mineralogical nature which defines the aggregate-asphalt compatibility [Kennedy *et al.*, 1982]. Mixture durability can also be reduced by aggregate degradation resulting from freeze and thaw cycles.

HMA mixtures used in surface courses must be able to provide and maintain sufficient skid resistance to permit normal turning and braking for vehicles. Aggregate characteristics, such as texture, shape, and resistance to polish, are the main factors controlling the skid resistance of mixes. The amount of asphalt cement, especially excessive asphalt cement that causes flushing or bleeding, can also reduce the skid resistance.

The desired properties in the HMA mixture result from the careful selection of the amounts and types of asphalt cement and aggregates. The asphalt cement primarily provides viscous strength, cohesion, and impermeability to the mix. The aggregates primarily provide resistance to load and deformation. Mix durability is primarily dependent on the amount of asphalt relative to the inter-particle volume and particle surface of the aggregates. Ideally, the amount of asphalt cement binder in the mix has to be low enough to promote particle-to-particle interaction in the aggregates but high enough to reduce air and water permeability, which accelerate binder hardening and aging.

Aggregates for HMA Mixtures

Aggregates for bituminous mixtures may be of natural, processed, or synthetic origin. Natural aggregates such as sand and gravel result from the weathering of rocks. Rock weathering is produced by physical agents causing rock disintegration or by chemical agents causing rock decomposition. The soil resulting from the weathering of rocks may remain directly over the rock of origin (residual soil) or it may be moved from the rock of origin by the action of water, wind, or ice (transported soil). Natural aggregates for bituminous mixtures are mostly obtained from transported soils, in river deposits, alluvial fans, and glacial outwash. Processed aggregates may be obtained by crushing rock or crushing oversized gravel and boulders. Crushing reduces the size of the rock particles to make them suitable for use in bituminous mixtures. Crushing is followed by screening to adjust the particle size to that required in the aggregate, and especially to eliminate the very fine or the very large particles. Crushing also changes the texture and shape of the particles, which are influential in the performance of the bituminous mixture.

Synthetic aggregates may be obtained as a by-product of some industrial processes or from the processing of raw materials for ultimate use as aggregates. Blast-furnace slag, for instance, which is a by-product of the smelting of iron in blast furnaces, is commonly used as aggregate. Other synthetic aggregates are manufactured by high-temperature processing of clay, shale, slate, and other natural materials. These aggregates are typically light and may have high resistance to abrasion. Materials obtained from the recycling of waste products such as glass and tires have also been studied as potential sources of aggregates for bituminous mixtures, especially because of the increasing awareness of the need for protection of the environment.

Physical Properties of Aggregates

The suitability of aggregates to be used in HMA mixtures depends on their physical and mineralogical properties and to a lesser extent on their chemical composition. The physical properties of aggregates that primarily control the performance of HMA mixtures include: gradation, cleanliness, toughness, durability, surface texture, particle shape, and absorption.

Aggregate Gradation

Aggregate sizes are typically divided into coarse [particle size greater than No. 4 sieve (4.75 mm), although some agencies use the No. 8 (2.36 mm) or the No. 10 (2.00 mm)], fine [particle size between No. 4 (4.75 mm) and No. 200 (0.075 mm) sieves], and mineral filler [at least 70% by weight passing No. 200 (0.075 mm) sieve]. Aggregate particle size is defined in terms of the maximum particle size and the particle size distribution. The maximum particle size (which is defined as the smallest sieve through which 100 percent of the aggregate particles pass, or as the largest sieve that retains about 10 percent of the particles) affects the economy and workability of the mix. Large particle sizes reduce the consumption of binder per unit volume of mix. However, larger particle sizes make it more difficult to obtain proper compaction in the mix, especially if the maximum particle size exceeds one-half the thickness of the compacted pavement layer [U.S. Army Corps of Engineers, 1991]. The use of large aggregate particles to produce large-stone mixes has recently gained considerable attention as a potential technique to reduce rutting. These mixtures require special attention to obtain proper compaction and to avoid segregation.

The particle size distribution is most commonly expressed as the weight percents of particle sizes mechanically screened with sieves of square openings. Other techniques are also used to separate the particles of different sizes, especially the fine particles, including sedimentation, light scattering, light blocking, etc. The most common way to define the particle size distribution, though, is in terms of the aggregate gradation, which is expressed in terms of weight percentages of particles retained (or passing) through a set of sieves with successively decreasing opening (ASTM C136). The gradation curve is the graphical representation of the particle size distribution with the ordinate defining the percent by weight passing a given size on an arithmetic scale, while the abscissa is the particle size plotted to a logarithmic scale (Fig. 43.7). The sieves typically used for aggregates for HMA mixtures range in opening from 50 mm (2 in.) to 0.075 mm (No. 200) sieve.

Several aggregate gradations have been proposed as optimum for the performance of HMA mixes. The gradation that results in the best packing of particles, with minimum voids between them, produces maximum resistance to loads because of the large number of points with particle-to-particle contact. However, such an agglomerate of particles in an HMA mixture has a minimum volume of voids to be filled with the asphalt binder, which is required to ensure impermeability and cohesion in the mixture. The mixture with low voids in the mineral aggregate will tend to bleed, exuding the asphalt binder to the surface under the action of traffic. Also, a tightly packed aggregate results in a mixture that is very sensitive to small variations in asphalt content (Roberts *et al.*, 1991).

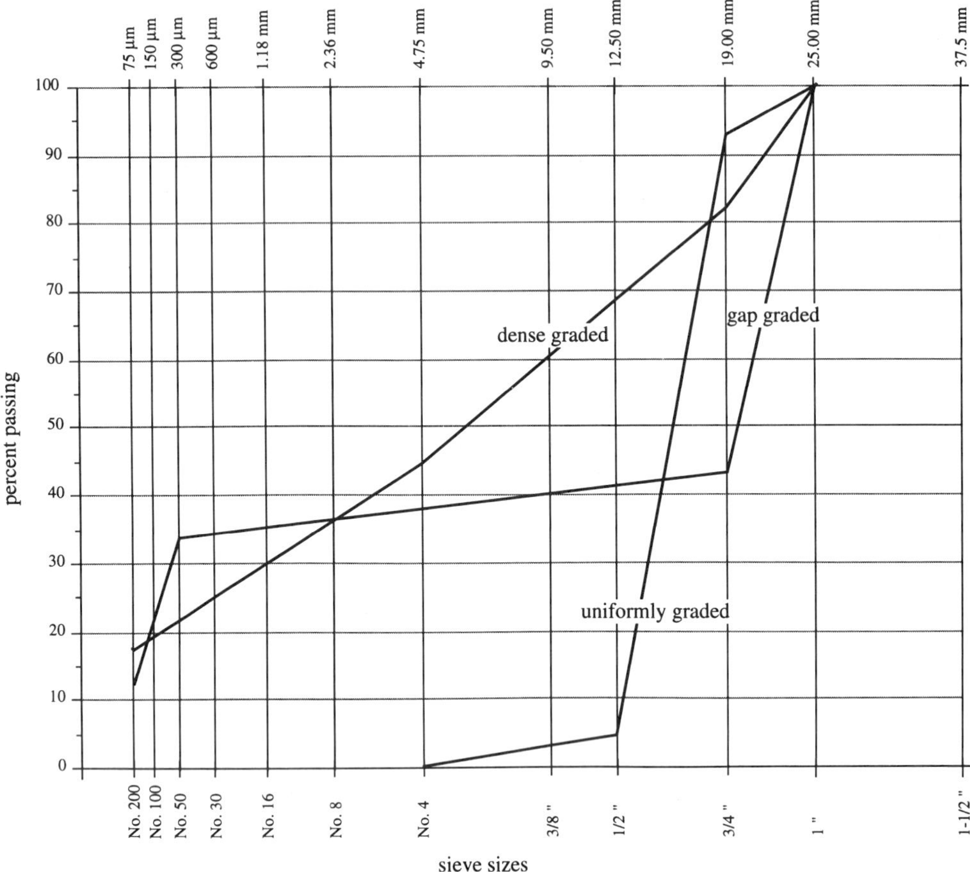

FIGURE 43.7 Various gradations of aggregates plotted on a 0.45 power chart.

The gradation proposed by Fuller and Thompson (Fuller and Thompson, 1907), commonly known as *Fuller's gradation,* is defined by:

$$P_i = \left(\frac{d_i}{D}\right)^n \tag{43.20}$$

where P_i is the total percent passing through ith, d_i is the opening of sieve ith, and D is the maximum particle size of the aggregate. The value of n was initially proposed as 0.5. Later the Federal Highway Administration recommended a value of 0.45. Aggregate gradations of maximum density obtained from Fuller's equations are shown in Figs. 43.8 and 43.9 for n values of 0.5 and 0.45, respectively. The theoretical gradation of maximum density does not necessarily result in minimum voids between the aggregate particles. Aggregate particles with an angular shape or with a rough surface texture will result in higher voids than aggregates with the same gradation but with round or smooth particles. Also, the nominal maximum particle size is defined by the size of the sieve through which 100% of the particles pass, but the actual size of the largest particles may be different from the sieve size. Thus the maximum density line should be drawn from the largest sieve on which some aggregate is retained intersecting the horizontal line corresponding to the actual amount of material passing such a sieve. Since the amount of material passing the No. 200 sieve affects considerably the performance of HMA mixes and is usually strictly controlled, the lower end of the maximum density line should pass the line of the No. 200 (0.075 mm) sieve at the desired amount of material passing that sieve.

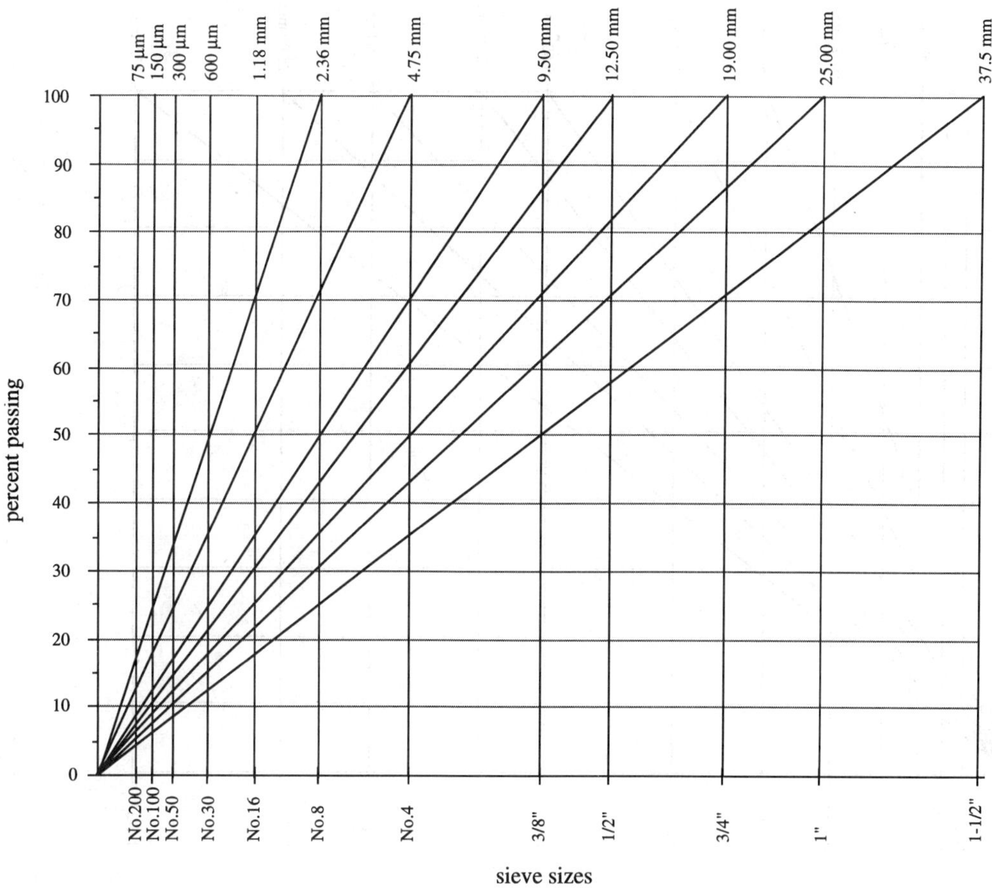

FIGURE 43.8 Gradations of maximum density on 0.50 power chart.

It is usually recommended, though, to use a gradation curve that is approximately parallel to the maximum density curve but two to four points above (fine) or below (coarse) the maximum density curve. Finely graded aggregates are more workable and easier to compact. Coarser aggregates are more difficult to compact, but they produce mixtures that are more resistant to permanent deformation (rutting). Aggregate gradation affects the voids in the mineral aggregate (VMA)—that is, the volume between particles of the compacted mixture, expressed as a percent of the total volume of the mixture. A mixture with low VMA is very sensitive to the asphalt content and will tend to check and shove during placement and compaction. Under traffic, a mixture with low VMA will tend to rut and bleed if the asphalt content is too high or will tend to ravel if the asphalt content is too low. An aggregate gradation that makes an *S* across the maximum density line will result in an HMA mixture with a tendency to segregate. Aggregate gradations that exhibit a hump as compared with the maximum density line will be more tender; that is, they will be more difficult to compact and it will be more difficult to achieve in the field the density obtained in laboratory samples. The amount of material finer than the No. 200 (0.075 mm) sieve (mineral filler), as well as the volume ratio of filler to asphalt binder, will influence the compactability of the mixture. The compactability of the HMA mixture increases with increasing amounts of filler up to a certain point, after which the mixture becomes stiffer. The optimum filler-asphalt ratio has been reported to be approximately 0.17.

Specifications for aggregate gradation are usually expressed by a gradation band that indicates the upper and lower limits of material passing (or retained) through a series of sieves. Gradation specifications for HMA mixes are recommended in various ASTM specifications; in practice,

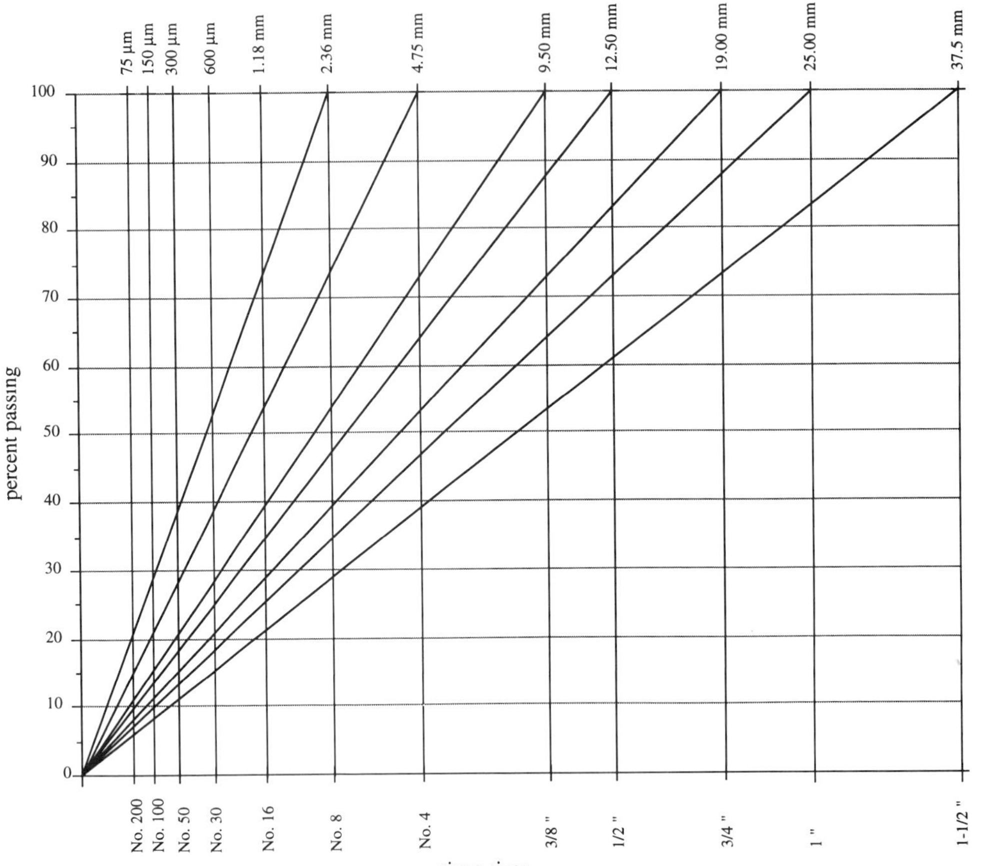

FIGURE 43.9 Gradations of maximum density on 0.45 power chart.

however, most gradation specifications have been developed by trial and error and local experience by different state agencies and are slightly different or more stringent than the ASTM-recommended gradations. The aggregate gradation requirements in the new SHRP mix design method are shown in Table 43.6.

Aggregate Blends

Commonly, two or more aggregates have to be combined to obtain a blend that meets the gradation specifications. Several techniques have been developed over the years to calculate the relative amounts of different aggregates to be mixed to obtain a blend with a specified gradation. The most common method to obtain the blend proportions is by trial and error. This method consists of calculating the gradation of the aggregate blend that is obtained from a proposed mixture of aggregates with assumed relative amounts. The gradation of the resulting blend is compared to the required gradation band. The gradation of the blend is calculated by a simple linear mixing rule. That is, if aggregates A, B, C, etc., of known gradations are to be blended in proportions a, b, c, etc., respectively, the gradation of the blend is calculated as:

$$P_i = aA_i + bB_i + cC_i + \cdots \tag{43.21}$$

where P_i = percent of material in aggregate blend passing through the ith sieve

$a, b, c, \ldots$ = weight proportions of aggregates $A, B, C, \ldots$

$A_i, B_i, C_i, \ldots$ = percent of material in aggregates $A, B, C, \ldots$ passing through ith sieve.

Table 43.6 Aggregate Gradation Requirements of New SHRP Mix Design Method *(continues)*

Sieve Size		Control Points		Restricted Zone	
(mm)		Minimum	Maximum	Minimum	Maximum
		37.5 mm Nominal Size			
37.50	1-1/2″	90.00	100.00		
25.40	1″				
19.00	3/4″				
12.50	1/2″				
9.50	3/8″				
4.75	No. 4			34.70	34.70
2.36	No. 8	15.00	41.00	23.30	27.30
1.18	No. 16			15.50	21.50
0.60	No. 30			11.70	15.70
0.30	No. 50			10.00	10.00
0.15	No. 100				
0.075	No. 200	0.00	6.00		
		25 mm Nominal Size			
37.50	1-1/2″				
25.40	1″	90.00	100.00		
19.00	3/4″				
12.50	1/2″				
9.50	3/8″				
4.75	No. 4			39.50	39.50
2.36	No. 8			26.80	30.80
1.18	No. 16			18.10	24.10
0.60	No. 30			13.60	17.60
0.30	No. 50			11.40	11.40
0.15	No. 100				
0.075	No. 200	1.00	7.00		
		19 mm Nominal Size			
37.50	1-1/2″				
25.40	1″				
19.00	3/4″	90.00	100.00		
12.50	1/2″				
9.50	3/8″				
4.75	No. 4				
2.36	No. 8	23.00	49.00	34.60	34.60
1.18	No. 16			22.30	28.30
0.60	No. 30			16.70	20.70
0.30	No. 50			13.70	13.70
0.15	No. 100				
0.075	No. 200	2.00	8.00		
		12.5 mm Nominal Size			
37.50	1-1/2″				
25.40	1″				
19.00	3/4″				
12.50	1/2″	90.00	100.00		
9.50	3/8″				
4.75	No. 4				
2.36	No. 8			39.10	39.10
1.18	No. 16			25.60	31.60
0.60	No. 30			19.10	23.11
0.30	No. 50			15.50	15.50
0.15	No. 100				
0.075	No. 200	2.00	10.00		

Table 43.6 (continued) Aggregate Gradation Requirements of New SHRP Mix Design Method

Sieve Size		Control Points		Restricted Zone	
(mm)		Minimum	Maximum	Minimum	Maximum
		9.5 mm Nominal Size			
37.50	1-1/2″				
25.40	1″				
19.00	3/4″				
12.50	1/2″				
9.50	3/8″	90.00	100.00		
4.75	No. 4				
2.36	No. 8	32.00	67.00	47.20	47.20
1.18	No. 16			31.60	37.60
0.60	No. 30			23.50	27.50
0.30	No. 50			18.70	18.70
0.15	No. 100				
0.075	No. 200	2.00	10.00		

To facilitate the calculation of the correct proportions, the gradations of the aggregates to be blended can be plotted together with the specification band and the midrange of the band. The aggregate(s) with the gradation(s) closest to the specification band is to be used in the highest proportion in the blend. For instance, three aggregates whose gradations are shown in Table 43.7 are to be blended to obtain a gradation with the specifications shown in the table. Various blend proportions are shown in Table 43.8.

Other methods to determine blend proportions include graphical techniques as well as computer algorithms [Asphalt Institute, 1977; Neumann, 1964; Popovics, 1973; Roberts *et al.*, 1991]. The selection of the optimum blend includes the consideration of the gradation specifications and

Table 43.7 Illustration of Three Aggregate Gradations to Be Blended to Obtain a Specified Gradation

	Agg. A	Agg. B	Agg. C	Gradation
Sieve Size	Total Percent Passing			Specifications
3/8 in.	100	100	100	100
No. 4	100	100	82	95–100
No. 8	100	90	65	—
No. 16	100	80	47	45–80
No. 30	90	72	12	—
No. 50	84	66	3	10–30
No. 100	68	10	1	2–10

Table 43.8 Gradation Results of Various Trial Blend Proportions

						Trial 1	Trial 2	Trial 3
	Agg. A	Agg. B	Agg. C	Specifications		A = 20% B = 60%	A = 0% B = 50%	A = 0% B = 60%
Sieve Size	(Total Percent Passing)			Range	Midrange	C = 20%	C = 50%	C = 40%
3/8 in.	100	100	100	100	100	100.00	100.00	100.00
No. 4	100	100	90	95–100	97.5	98.00	95.00	96.00
No. 8	100	90	75	—	—	89.00	82.50	84.00
No. 16	100	80	47	45–80	62.5	77.40	63.50	66.80
No. 30	90	68	12	—	—	61.20	40.00	45.60
No. 50	78	40	3	10–30	20	40.20	21.50	25.20
No. 100	61	10	1	2–10	6	18.40	5.50	6.40

potential performance of the HMA mix as well as the economy of blending aggregates that may be from different sources.

Aggregate Cleanliness

The presence of dust and foreign matter such as vegetation, soft particles, clay lumps, etc., in the aggregates is undesirable. Aggregate cleanliness can be determined by visual inspection, although it is better to obtain an objective measure of such cleanliness. Standard tests to obtain an index of cleanliness include the sand equivalent test (ASTM D2419) and the test for presence of clay lumps and friable particles (ASTM C142). In the sand equivalent test, the proportion of clay and silt particles relative to the amount of fine aggregate is determined. Clean aggregates, with a small proportion of clay and dust particles, will have a high sand equivalent value, which is usually specified at a minimum of about 25 to 35 for aggregates to be used in HMA mixtures. The presence of friable particles, which can break down as a result of freeze/thaw or wet/dry cycles, can reduce the durability of HMA mixes. Some of the friable particles can break down during aggregate processing and can be removed by washing. More resistant friable particles will break down during mixing or, worse yet, during service under the action of traffic. Specifications usually limit the amount of friable particles (particles that can be broken by the fingers after a 24-hour period of soaking of the previously washed aggregate) to a maximum of 1% by weight of the total aggregate.

Aggregate Toughness

Aggregates are subjected to abrasion and polishing during stockpiling, mixing, placing, and compaction, and during service under the action of traffic. Not only potentially friable particles have to be removed from the aggregates but the aggregate particles themselves have to be hard and tough. An index of the toughness and hardness of aggregates is usually obtained with the *Los Angeles abrasion test* (ASTM C131 and C535), which consists of placing the aggregates with a known gradation into a steel drum with a charge of steel balls. The drum is rotated at a speed of approximately 30 rpm for 500 revolutions, and the aggregate gradation is determined afterwards. The change of weight percent of material passing a designated sieve defines the Los Angeles abrasion value. The wear of coarse aggregates for HMA mixes is typically limited to a maximum L.A. abrasion value of 40 to 60%.

Aggregate Soundness

Aggregates in paving mixtures must be resistant to weathering action, that is, freezing/thawing and wetting/drying cycles. Although many engineers believe that aggregates in HMA mixes are not susceptible to freeze damage, mostly because they are first dried and then coated with asphalt during production (thus preventing the penetration of moisture into the aggregate), still several tests have been devised to evaluate their potential susceptibility to frost damage. For instance, the AASHTO T103 test recommends three ways to determine aggregate resistance to disintegration by freezing and thawing. In the test, aggregate samples with a defined gradation are first fully or partially immersed in water or in an ethyl alcohol–water solution. After immersion, the samples are subjected to alternate cycles of freezing and thawing to determine their gradation after a specified number of cycles, which can be from 16 to 50 cycles, depending on the immersion procedure. The resistance to freezing and thawing is measured by the relative amount of material that is not disintegrated and that is retained on the original sieves. The ASTM C88 test provides an index of weathering resistance of aggregates by determining the resistance to disintegration by saturated solutions of sodium sulfate or magnesium sulfate. Aggregate samples are prepared with a specified gradation and are subjected to alternate cycles of saturation by immersion in sulfate solution followed by drying. After a predetermined number of cycles, the aggregates are cleaned from the sulfate by washing and then dried. Their gradation is then determined and the relative amounts of material that has been disintegrated are calculated from the amounts of material passing a set of reference sieves. The average of the weight losses of each size fraction is then calculated.

Aggregate Specific Gravity

Specific gravity of any material is defined as the ratio of the mass of the material to the mass of an equal volume of gas-free, distilled water at a specified temperature. The mass ratio is numerically equal to the weight ratio; thus, the latter is typically used in the calculation of the specific gravity. Aggregates are porous materials; therefore, a distinction has to be made between the apparent volume of the aggregate particles and the volume of the solid portion of the aggregate particles, excluding the volume of the porous that can be saturated with water or with asphalt cement. Furthermore, the weight of the aggregate particles varies with the degree of saturation or the presence of moisture inside the particles. The specific gravity of aggregate particles is determined by weighing the aggregate particles in air and calculating their volume from the weight of water displaced when they are immersed in it as described in ASTM C127. Several definitions of specific gravity are commonly distinguished depending on the saturation conditions of the aggregate particles and on how their volume is measured, as illustrated in Fig. 43.10. The effective specific

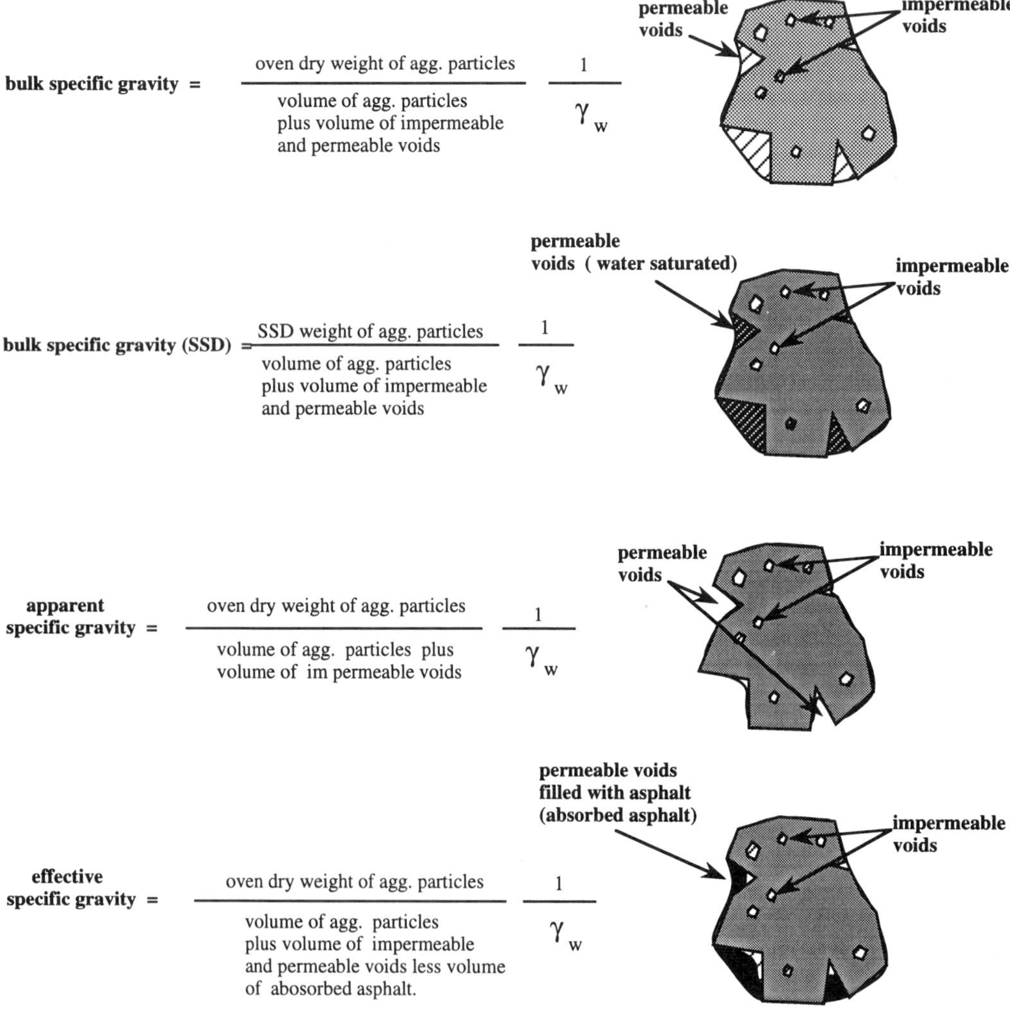

FIGURE 43.10 Illustration of various definitions of aggregate specific gravity.

gravity, commonly used in bituminous mixtures, is the specific gravity of the aggregates immersed in the matrix of asphalt binder and accounts for the absorption of asphalt in the pores of the particles (Fig. 43.10).

Mineralogical and Chemical Composition

Although the suitability of aggregates for HMA mixtures is primarily determined from their physical properties, mineralogical and chemical composition of the aggregates provide additional insight about the potential performance of the aggregates in the mixture. The mineralogical composition of the aggregates depends primarily on the rock of origin. Most common minerals in aggregates are listed in Table 43.9 (ASTM C294).

Aggregate mineralogy influences various aspects of the performance of bituminous mixtures. For instance, the adhesion of asphalt to the aggregate surface is higher in carbonate aggregates than in siliceous aggregates. The presence of certain minerals as coatings on the surface of the aggregate particles affects the bond with the asphalt cement and the propensity to absorb water or moisture. Clay, gypsum, iron oxides, silt, and other minerals may have either poor adhesion with the asphalt binder or a propensity to absorb moisture and break the bond between the aggregate and the asphalt. Certain minerals such as quartz and feldspars are hard and resistant to polish, thus enabling the asphalt mixture to maintain its skid resistance under the abrasive effect of traffic. Aggregates from sedimentary rocks such as limestone and dolomite, in contrast, can have a tendency to be polished under the action of traffic.

The chemical composition of aggregates, which is generally given in terms of oxides, is not as informative of their potential performance in paving mixtures as is the mineralogical composition. Nonetheless, the presence of certain substances can lead to performance problems. For instance, the presence of water-soluble and moisture-absorbing substances can produce mixtures that are susceptible to moisture damage in the form of aggregate stripping (loss of adhesion between asphalt and aggregate), raveling, or loss of stability. Other substances may be susceptible to oxidation, hydration, or carbonation.

Table 43.9 Rocks of Origin and Minerals in Aggregates

Rock	Minerals
Igneous	Silica
Granite	Quartz
Syenite	Opal
Diorite	Chalcedony
Gabbro	Tridymite
Peridotite	Cristobalite
Pegmatite	Silicates
Volcanic Glass	Feldspars
Obsidian	Ferromagnesian
Pumice	Hornblende
Tuff	Augite
Scoria	Clay
Perlite	Illites
Pitchstone	Kaolins
Felsite	Chlorites
Basalt	Montmorillonites
Sedimentary	Mica
Conglomerate	Zeolite
Sandstone	Carbonate
Quartzite	Calcite
Graywacke	Dolomite
Subgraywake	Sulfate
Arkose	Gypsum
Claystone	Anhydrite
Siltstone	Iron Sulfide
Arguillite	Pyrite
Shale	Marcasite
Carbonates	Pyrrhotite
Limestone	Iron Oxide
Dolomite	Magnetite
Marl	Hematite
Chalk	Geothite
Chert	Ilmenite
Metamorphic	Limonite
Marble	
Metaquartzite	
Slate	
Phyllite	
Schist	
Amphibolite	
Hornfels	
Gneiss	
Serpentinite	

Voids in HMA Mixtures

The compacted HMA mixture consists of aggregates, asphalt binder, mineral filler, and voids. After mixing, placing, and compaction, the mixture exhibits a relatively continuous matrix of asphalt binder (with or without fillers or additives) and trapped air voids, and the discontinuous phase

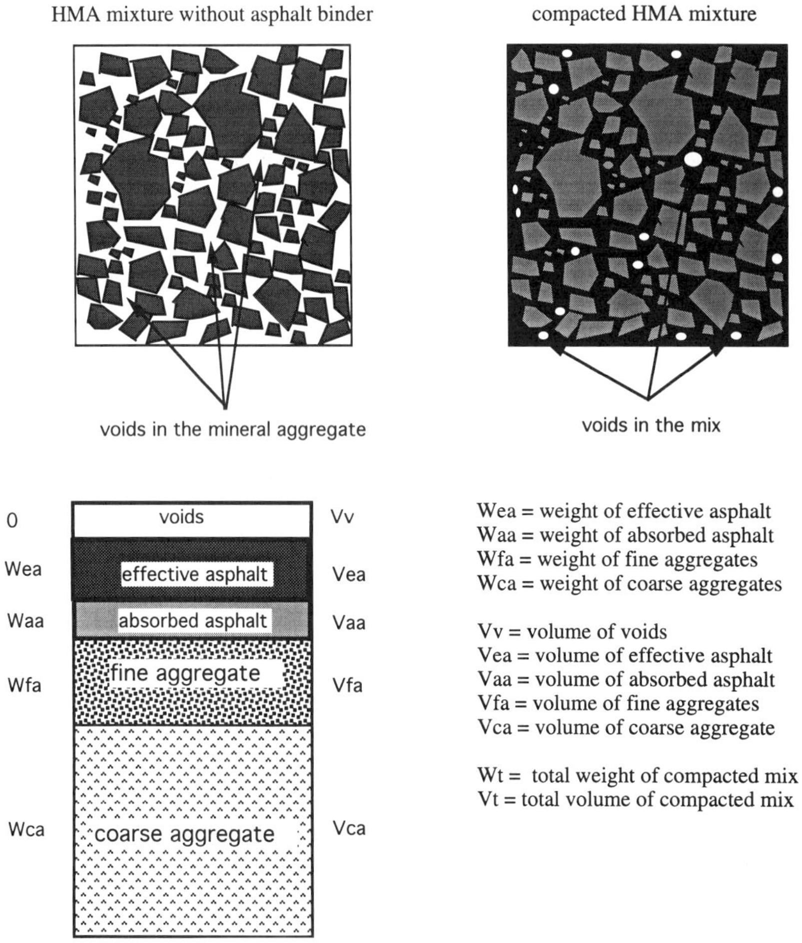

HMA mixture without asphalt binder

compacted HMA mixture

voids in the mineral aggregate

voids in the mix

0	voids	Vv
Wea	effective asphalt	Vea
Waa	absorbed asphalt	Vaa
Wfa	fine aggregate	Vfa
Wca	coarse aggregate	Vca

Wea = weight of effective asphalt
Waa = weight of absorbed asphalt
Wfa = weight of fine aggregates
Wca = weight of coarse aggregates

Vv = volume of voids
Vea = volume of effective asphalt
Vaa = volume of absorbed asphalt
Vfa = volume of fine aggregates
Vca = volume of coarse aggregate

Wt = total weight of compacted mix
Vt = total volume of compacted mix

FIGURE 43.11 Schematic representation of mixture components and voids.

of fine and coarse aggregates (Fig. 43.11). Of all the asphalt binder used in the mixture, a part is absorbed in the surface porosity of the aggregates, and the remaining asphalt (effective asphalt) fills the spaces between the aggregate particles. The bulk specific gravity of the compacted HMA mixture is the ratio between the weight in air of the compacted HMA mix and the weight of an equal volume of water at a stated temperature (ASTM D1189, AASHTO T275).

$$G_{cm} = \frac{W_t}{V_t}\frac{1}{\gamma_w} \tag{43.22}$$

where

G_{cm} = bulk specific gravity of compacted HMA mix
W_t = dry weight of compacted HMA mix $= W_{ea} + W_{aa} + W_{fa} + W_{ca}$
W_{ea} = weight of effective asphalt
W_{aa} = weight of absorbed asphalt
W_{fa} = weight of fine aggregate
W_{ca} = weight of coarse aggregate
V_t = volume of compacted HMA mix including permeable voids
$= V_v + V_{ea} + V_{fa} + V_{ca}$ (see Fig. 43.11)

V_v = volume of voids
V_{ea} = volume of effective asphalt
V_{fa} = volume of fine aggregate
V_{ca} = volume of coarse aggregate

The *theoretical maximum specific gravity* or *theoretical maximum density* (TMD; Rice specific gravity) of a bituminous paving mixture is the ratio between the dry weight in air of the compacted bituminous mixture and the weight of a volume of gas-free, distilled water equal to the void-free volume of the bituminous mixture (ASTM D2041). The void-free volume of the bituminous mixture is obtained from the weight of water displaced by the mixture after it is immersed in water and subjected to partial vacuum to force the air out of the mix:

$$G_{mt} = \frac{W_t}{V_t - V_v} \frac{1}{\gamma_w} \tag{43.23}$$

where

G_{mt} = maximum theoretical specific gravity of bituminous mixture
W_t = dry weight of compacted HMA mix
V_t = volume of compacted HMA mix including permeable voids
V_v = volume of voids in the mix
γ_w = unit weight of water

The *voids in the mineral aggregate* (VMA) is defined as the volume of the space between the aggregate particles in the compacted mix, which includes the volume of voids and the volume of effective asphalt, as a percent of the total volume of the mix:

$$\text{VMA} = \frac{V_t - V_{fa} - V_{ca}}{V_t}(100) \tag{43.24}$$

or

$$\text{VMA} = \frac{V_v + V_{ea}}{V_t}(100) \tag{43.25}$$

where

VMA = voids in mineral aggregate (percent)
V_t = total volume of mix
V_{fa} = volume of fine aggregate
V_{ca} = volume of coarse aggregate
V_v = volume of voids
V_{ea} = volume of effective asphalt

The *voids in total mix* (VTM) is the effective volume of voids in the mix that is not occupied by aggregates or asphalt, expressed as a percent of the total volume of the mix:

$$\text{VTM} = \frac{V_v}{V_t}(100) \tag{43.26}$$

or

$$\text{VTM} = \frac{V_t - V_{ea} - V_{fa} - V_{ca}}{V_t}(100) \tag{43.27}$$

where

VTM = voids in total mix (percent)
V_t = volume of compacted HMA mix including permeable voids
V_v = volume of voids
V_{ea} = volume of effective asphalt
V_{fa} = volume of fine aggregate
V_{ca} = volume of coarse aggregate

Voids filled with asphalt (VFA) is the percent of the voids in the mineral aggregate that is filled with asphalt:

$$\text{VFA} = \frac{V_{ea}}{V_v + V_{ea}}(100) \tag{43.28}$$

or

$$\text{VFA} = \frac{V_{ea}}{\text{VMA} \times V_t/100}(100) \tag{43.29}$$

where

V_t = volume of compacted HMA mix including permeable voids
V_v = volume of voids
V_{ea} = volume of effective asphalt

Asphalt content is the percent ratio between the weight of the asphalt cement and the total weight of the HMA mix. In some agencies the percent of asphalt content is expressed with respect to the weight of the aggregates. For asphalt content with respect to total weight of mix,

$$\text{AC} = \frac{W_{ea} + W_{aa}}{W_t}(100) \tag{43.30}$$

For asphalt content with respect to weight of aggregates,

$$\text{AC} = \frac{W_{ea} + W_{aa}}{W_{fa} + W_{ca}}(100) \tag{43.31}$$

where

W_{ea} = weight of effective asphalt
W_{aa} = weight of absorbed asphalt
W_{fa} = weight of fine aggregate
W_{ca} = weight of coarse aggregate

Asphalt absorption (*Paa*) is the weight percent of asphalt absorbed in the aggregate with respect to the weight of the aggregate:

$$P_{aa} = \frac{W_{aa}}{W_{fa} + W_{ca}}(100) \tag{43.32}$$

Mechanical Properties of HMA Mixtures

Strength

The two most common ways to obtain a measure of the strength of HMA mixtures are by means of the Marshall (ASTM D1559) and Hveem (ASTM D1560) tests. These tests are empirical, and they do not measure a fundamental property in the mix. However, they have been used for approximately 50 years, and a considerable amount of performance experience has been accumulated throughout the world. In the Marshall procedure, the strength, or Marshall stability, is measured in terms of the maximum load that a cylindrical specimen can support while maintained at a temperature of 60°C (140°F). The load is applied through semicircular testing heads moving at a rate of 50 mm/min (2 in./min). The strength measured by the Hveem stability actually reflects the resistance of the mix to lateral deformation while subjected to an axial load. The Hveem stability thus depends on the combined effects of cohesion and internal friction in the mix. As in the Marshall test, the Hveem stability is determined while the specimen is maintained at 60°C (140°F), which is intended to simulate the most critical temperature expected in the field. Another way of measuring the strength in HMA mixtures is with the indirect tensile test (ASTM D4123). A cylindrical specimen is loaded with a single load (or repeated load to determine the fatigue strength) applied along the vertical diametrical plane, which produces a relatively uniform tensile stress. The indirect tensile strength test is typically conducted while the specimen is maintained at 25°C (77°F), although conducted at lower temperatures it can provide information about the thermal cracking potential of the mix.

Stiffness

The stiffness of HMA mixtures is quantified in terms of the modulus of elasticity measured under either dynamic loading (dynamic complex modulus, ASTM D3497) or repeated loading (resilient modulus, ASTM D4123). The dynamic complex modulus is determined by subjecting a cylindrical specimen—typically 152 mm (6 in.) in diameter by 300 mm (12 in.) in length—to uniaxial, sine-wave loading and then measuring the resulting deformations. The modulus is then calculated from the stress produced by the load divided by the total strain (elastic and inelastic strain). Several specimens from the same mix are typically tested at various temperatures from 5°C (41°F) to 40°C (104°F) and at load frequencies of 1 hertz to 16 hertz. The resilient modulus is determined by measuring the indirect tensile stiffness modulus of a specimen under repeated loading. A cylindrical specimen is subjected to repeated, square-wave loading applied along the diametral plane causing repeated, indirect tensile stresses. The lateral deformations (perpendicular to the diametral plane) in the specimen are measured while under loading. The resilient modulus is then quantified as the applied indirect tensile stress divided by the recoverable (resilient) strain. The applied repeated stress is typically between 5 and 20% of the indirect tensile strength and is applied at intervals of 1 second with 0.1 seconds of loading and 0.9 seconds of rest period.

Another method to determine the stiffness of HMA mixtures, developed by Van der Poel [1964] and modified by Heukelom and Klomp [1964] and McLeod [1976], is based on the composition of the mix and the stiffness of the asphalt cement. The composition of the mix is defined in terms of the volume proportion of aggregate, and the stiffness of the asphalt cement is derived from its penetration and viscosity determined at specific temperatures. The value of the stiffness modulus of HMA mixtures obtained with this method is only an approximation, however.

Fatigue Strength

Pavement performance life, which is quantified in terms of the cumulative number of traffic load repetitions to reach failure, depends on pavement thickness and on pavement fatigue strength. The fatigue strength of the pavement in turn depends on the fatigue strength of the constituent materials, including the asphalt-bound materials. The fatigue strength of bituminous mixtures is determined in the laboratory by subjecting either beam specimens or cylindrical specimens to repeated flexural loading or to repeated indirect tensile loading, respectively. In the flexural fatigue test, beam specimens fixed at both ends are subjected to haversine loading applied at the third points of the span, producing a repeated uniform bending moment over the center portion of the beam. The deflection caused by the load is measured at the center of the beam. The loading heads are attached to the upper and lower sides of the beam specimen to force the beam to follow a haversine deflection pattern. The load is applied at a rate of 1 to 2 hertz. The test can be conducted maintaining the repeated load at a constant amplitude until the specimen fails (constant stress test), or it can be conducted maintaining the repeated deformation in the specimen at a constant amplitude until the load reaches a value 25 to 50% of the initial load (constant strain test). The fatigue strength test results obtained from the constant stress test can be shown graphically by plotting on the vertical axis the applied stress and on the horizontal axis the logarithm of the number of load cycles to failure (S-N curve). Also the log of the initial strains can be plotted versus the log of the number of cycles to failure. The results of the constant strain test are plotted in a graph with the vertical axis representing the log of applied strain versus the log of the number of cycles to failure. Also, the log of the applied stress can be plotted versus the log of the number of cycles to failure.

Another method to determine the fatigue strength of bituminous mixtures is through the indirect tensile strength test described earlier. To determine the fatigue strength, though, the splitting load is applied in a repeated manner until the specimen fails. The applied tensile stress and resulting tensile strain are calculated assuming homogeneous, isotropic, and elastic behavior. The test results can be represented in terms of log of applied stress versus log of number of cycles to failure, or log of applied strain versus log of cycles to failure.

Creep

Creep defines the time-dependent deformation that results from sustained loading. Part of the deformation due to creep is recovered after the load is removed, but the other portion remains as permanent deformation. The creep characteristics of bituminous mixtures define their potential to allow permanent deformation or rutting to occur in the pavement. The creep characteristics of HMA mixtures are generally determined following one of three testing procedures: uniaxial static unconfined loading, uniaxial static confined loading, or indirect tensile loading. The uniaxial static unconfined creep test is the most common because of its relative simplicity. In the test a specimen, at a specified temperature, is maintained under a static uniaxial load sustained for a specified period of time. The deformation of the specimen is measured during the loading period and the creep recovery is measured after the load is removed. However, although the test is simple to perform, the test conditions are not representative of the creep that occurs in the field under repeated traffic loading. Also, the magnitude of the applied pressure is limited, as well as the test temperature, since in general HMA mixtures have relatively low strength under sustained loads and unconfined conditions. A more representative creep behavior can be obtained by conducting the test with the specimen confined with a pressure of approximately 135 kPa (20 psi). With the confining pressure, the test can be performed at higher temperatures (representative of the hottest conditions expected in the field) and at higher sustained loads (more representative of current high tire pressures). The test results can be expressed in terms of total deformation (or strain), permanent deformation (or permanent strain), or in terms of *creep compliance,* which is the value calculated by dividing the strain in the specimen at a specified time by the applied stress.

HMA Mix Design Methods

The selection of types and amounts of constituents in the HMA mixture is carried out in several stages. First the type of asphalt binder and the aggregate gradation are selected based on the job specifications and the expected traffic and climatic conditions of the area where the mix will be used. Test specimens are then fabricated with various mix proportions to determine the properties of the mixes. The test results on the specimens are analyzed and the optimum proportion of each of the mix components is then selected.

Several procedures to determine the mix proportions have been developed over the years, including the *Hubbard-Field* method, the *Marshall* method, the *Hveem* method, the *Texas* method, and others. Recently, a new methodology for mix design has been developed as part of the Strategic Highway Research Program (SHRP) of the National Research Council. The Federal Highway Administration (FHWA) is currently implementing the application of this method. The Marshall and Hveem methods are described herein. Also, the SHRP mix design is briefly presented, although at this time the method is still in the development and implementation stages.

Marshall Mix Design

This design method, described in ASTM D1559, is the most commonly used method to design HMA mixtures throughout the world. The method is primarily used for mixes with a maximum size of aggregate of up to 25.4 mm (1 in.), although it can be used with some modifications to design mixes with larger aggregate. The method includes acceptance tests for aggregates and asphalt cement to comply with applicable specifications and traffic and climatic conditions. The mix proportions are then selected by analyzing the stability, flow, voids, and density characteristics of specimens prepared with various trial proportions. The method is illustrated in Fig. 43.12 and outlined as follows (for detailed description of test procedure see ASTM D1559).

Aggregate Selection. The target aggregates to be used in the mixture are tested to verify compliance with all applicable specifications and acceptance tests for abrasion resistance, soundness, absorption, gradation, etc. Blend proportions are obtained if aggregates from various sources are to be used.

Binder Selection. Select asphalt cement grade for geographical location (climatic conditions) and expected traffic. Verify compliance with applicable specifications (viscosity, penetration, solubility, flash point, softening point, etc.) and determine mixing temperature (to obtain viscosity of 170 ± 20 cSt) and compaction temperature (to obtain 280 ± 30 cSt). Also, determine the specific gravity of the asphalt cement for the entire range of expected temperatures.

Specimen Preparation. Prepare materials to make three specimens [101.6 mm (4 in.) in diameter by 63.5 mm (2.5 in.) in height] for each asphalt content to be tested: approximately 23 kg ($\approx$50 lb) of aggregates and 4 liters ($\approx$1 gal) of asphalt cement to prepare 18 specimens (5 different asphalt contents and 1 trial mix; three specimens each). Aggregates and asphalt cement are heated to the mixing temperature to prepare trial mixtures. Prepare the mixtures using 5 different asphalt contents at 0.5% increments above and below the expected optimum asphalt content. An additional mix for three specimens is prepared with an asphalt content near the optimum to determine the Rice specific gravity (TMD) and the effective specific gravity of the aggregates. The previously weighed aggregates and asphalt cement are thoroughly mixed until all aggregate particles are coated with asphalt cement. Place the mixture at the compaction temperature in preheated molds with a disc of filter paper at the bottom and compact it, with the mold resting on a special compaction pedestal and the compaction hammer—preheated at 93 to 149°C (200 to 300°F)—applying the specified number of compaction blows. Rotate the mold upside-down, place another filter paper on the top of the specimen, and apply the same number of compaction blows on the new top. Allow the specimens to cool down to room temperature and extract them from the molds with a hydraulic jack. After curing overnight at room temperature, determine bulk specific gravity of each specimen (AASHTO T166). Determine the Rice specific gravity (TMD) and the effective specific gravity of the aggregates on the loose mixture (ASTM D2041).

Stability and Flow Tests. Place the previously cured specimens in a water bath at 60°C (140°F) for 30 to 40 minutes. Remove the specimens one by one from the water bath, remove the excess water from their surface, and place them in the head of the Marshall press. Load each specimen to failure and record maximum load in pounds (*Marshall stability*) and deformation at failure in 0.254 mm (0.01 in.) (*Marshall flow*).

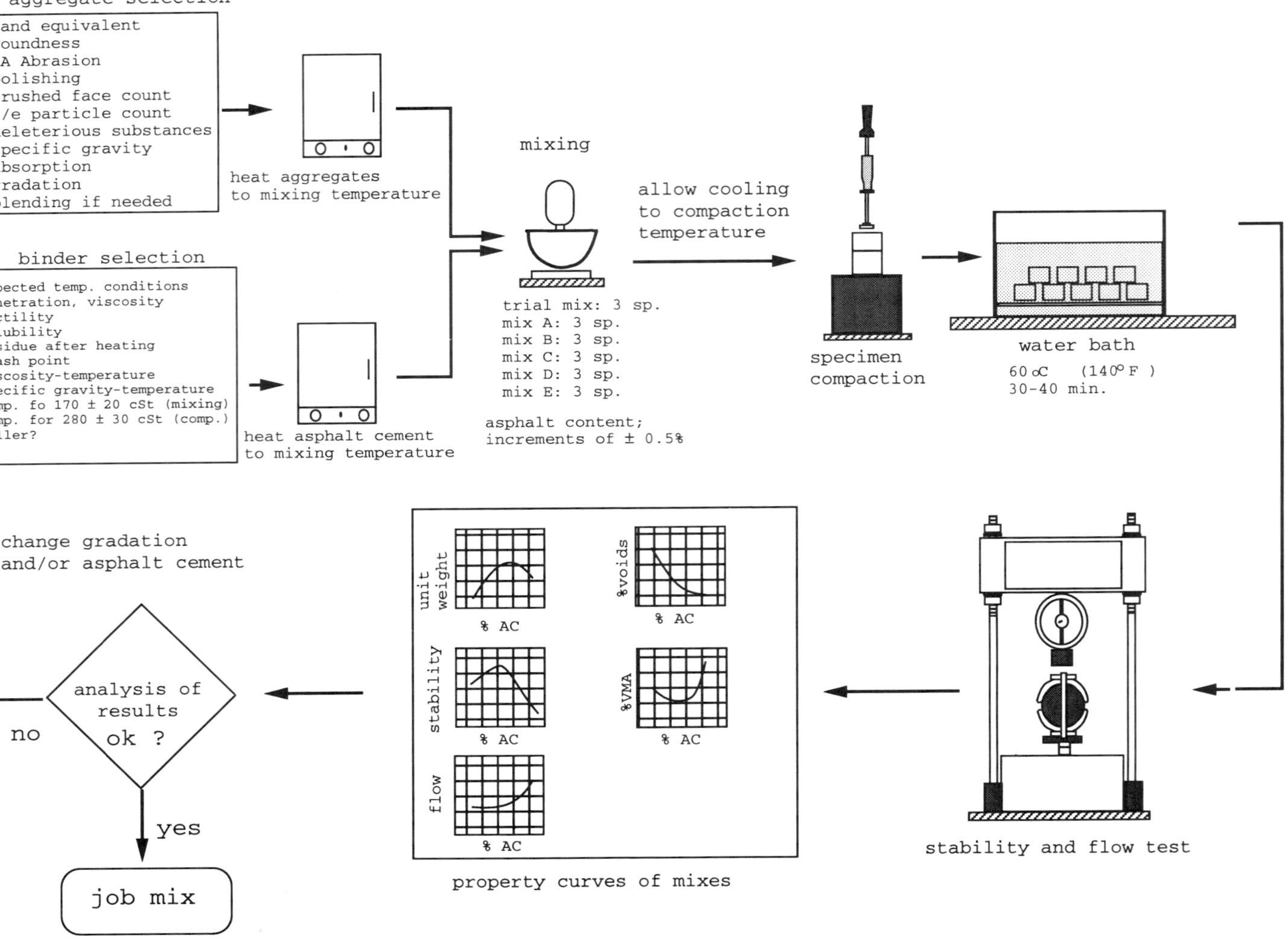
aggregate selection
sand equivalent
soundness
LA Abrasion
polishing
crushed face count
f/e particle count
deleterious substances
specific gravity
absorption
gradation
blending if needed
heat aggregates
to mixing temperature
binder selection
expected temp. conditions
penetration, viscosity
ductility
solubility
residue after heating
flash point
viscosity-temperature
specific gravity-temperature
temp. fo 170 ± 20 cSt (mixing)
temp. for 280 ± 30 cSt (comp.)
filler?
heat asphalt cement
to mixing temperature
mixing
trial mix: 3 sp.
mix A: 3 sp.
mix B: 3 sp.
mix C: 3 sp.
mix D: 3 sp.
mix E: 3 sp.
asphalt content;
increments of ± 0.5%
allow cooling
to compaction
temperature
specimen
compaction
water bath
60 °C (140° F)
30-40 min.
stability and flow test
unit
weight
% AC
%voids
% AC
stability
% AC
%VMA
% AC
flow
% AC
property curves of mixes
change gradation
and/or asphalt cement
analysis of
results
ok ?
no
yes
job mix

FIGURE 43.12 Schematic diagram of Marshall mix design procedure.

Density/Voids Calculations and Test Results. Calculate the density/void properties of each specimen, including:

Bulk specific gravity
Unit weight = Bulk specific gravity × 1 (g/cm^3) or [Bulk specific gravity × 62.4 (lb/ft^3)]
Voids in mineral aggregate (VMA)
Voids in total mix (VTM)
Voids filled with asphalt (VFA)

Correct the Marshall stability values for each specimen to compensate for variable specimen height by multiplying the measured Marshall stability times the correction factor given in ASTM D1559. Calculate the average values of each mixture of three specimens and tabulate the results. Prepare the following plots:

Asphalt content versus unit weight
Asphalt content versus Marshall stability
Asphalt content versus Marshall flow
Asphalt content versus VTM
Asphalt content versus VMA
Asphalt content versus VFA

Expected Trends in Plots. The following effects have to be observed in the plots, with increasing asphalt content:

Marshall stability increases, reaches a maximum value, and then decreases.
Marshall flow increases.
Unit weight increases, reaches a peak, and then decreases.
VTM increases.
VMA decreases, reaches a minimum, then increases.
VFA increases.

Selection of Optimum Asphalt Content. The methodology for selecting the optimum asphalt content from the Marshall test results varies among agencies. Two methods are illustrated as follows. The Asphalt Institute procedure (AI MS-13):

1. Obtain the average of the asphalt contents required for maximum stability, maximum density, and midpoint of specified range of VMA.
2. Obtain from the test plots the values of stability, flow, VTM, and VMA corresponding to the average asphalt content calculated in 1.
3. Verify that values determined in 2 comply with acceptability criteria.

The National Asphalt Paving Association procedure (NAPA TAS-14):

1. Determine asphalt content corresponding to the midpoint of the specified VTA.
2. Determine the stability, flow, VMA, and VFA corresponding to the asphalt content selected in 1.
3. Compare each of the values from 2 to the specified values. Redesign the mixture if any of the values is not within specifications.

Hveem Mix Design

The Hveem design method has traditionally been used in Indiana, Maine, and most of the western states of the U.S. The method (ASTM D1560) is similar to the Marshall procedure in selecting the optimum asphalt content from the test results of trial mixes prepared with various amounts of asphalt cement. However, the specimens prepared with various contents of asphalt cement are tested in a triaxial testing device that registers the later pressure produced by the specimens when they are subject to axial load. The aim of the Hveem method is to select a mixture with well-graded aggregates and with as much asphalt binder as the mixture tolerates without losing stability. Also, a minimum of 3% of VTM is required in the mixture. The Hveem design method is illustrated in Fig. 43.13 and outlined in the following paragraphs.

As in the Marshall method, the Hveem method for mix design first requires that the asphalt binder and aggregates be selected based on the applicable specifications, expected weather and traffic conditions, economic considerations, etc. The procedure is then slightly different from the Marshall method, as follows.

Estimation of Amount of Binder. An estimate of the amount of asphalt binder is made from the centrifuge kerosene equivalent (CKE) for fine aggregates and from the percent oil retained (POR) for the coarse aggregates. For the fine portion of the aggregate, the relative weight amount of kerosene that remains absorbed by the particles after being saturated by submersion and then dried by centrifuge action is called the *centrifuge kerosene equivalent.* The CKE defines the surface constant of the fine aggregate K_f. The surface constant of the coarse portion of the aggregate, K_c, is determined from the weight percent of SAE No. 10 oil (POR) that is retained on the surface of the particles after submersion in the oil followed by draining by gravity during a specified amount of time. The values of K_f and K_c are combined and used to determine the tentative percent of asphalt cement to be used in the mix (AI MS-2).

Specimen Preparation. After determining the tentative amount of asphalt cement in the mixture, several mixes are prepared with various percents of asphalt cement content—2% above and 1% below the value previously determined by the CKE and POR procedures and in increments of 0.5%. The specimens are prepared by compacting the mixtures using a kneading compactor (ASTM D1561). The specimens are compacted by first applying 10 to 50 blows with the compacting foot under a pressure of 1.69 MPa (250 psi) followed by 150 blows under a foot pressure of 3.39 MPa (500 psi). The specimens are then heated to 140°F or 23°F and compacted under a static pressure of 6.79 MPa (1000 psi).

Specimen Testing. The compacted specimens are then tested to determine the Hveem stability, density/void content, and amount of swell. The Hveem stability is a measure of the specimen's resistance to lateral deformation while under confinement. In the test, specimens are first preheated to 140°F for an hour and then placed inside the stabilometer. The specimens are subjected to triaxial stresses consisting of a lateral confining pressure, produced by a manual hydraulic pump, and a vertical stress, produced by the head of a testing machine. Under constant axial load, the confining pressure is increased from 5 psi to 100 psi by manually displacing the piston in the pump to force the pump's fluid inside the stabilometer to replace the reduction in volume of the specimen under increasing confining pressure. The Hveem stability is calculated as:

$$S = \frac{22.2}{P_h \times \dfrac{D}{P_v - P_h} + 0.222} \tag{43.33}$$

where

S = Hveem stability
P_v = vertical pressure on specimen
P_h = horizontal pressure on specimen
D = lateral deformation of specimen (0.1 in. = 1 turn of handle of the manual pump)

The unit weight and percent of voids (VTM) are determined for all tested specimens in a manner similar to that used in the Marshall test. In addition, the trial mixtures are tested to determine the

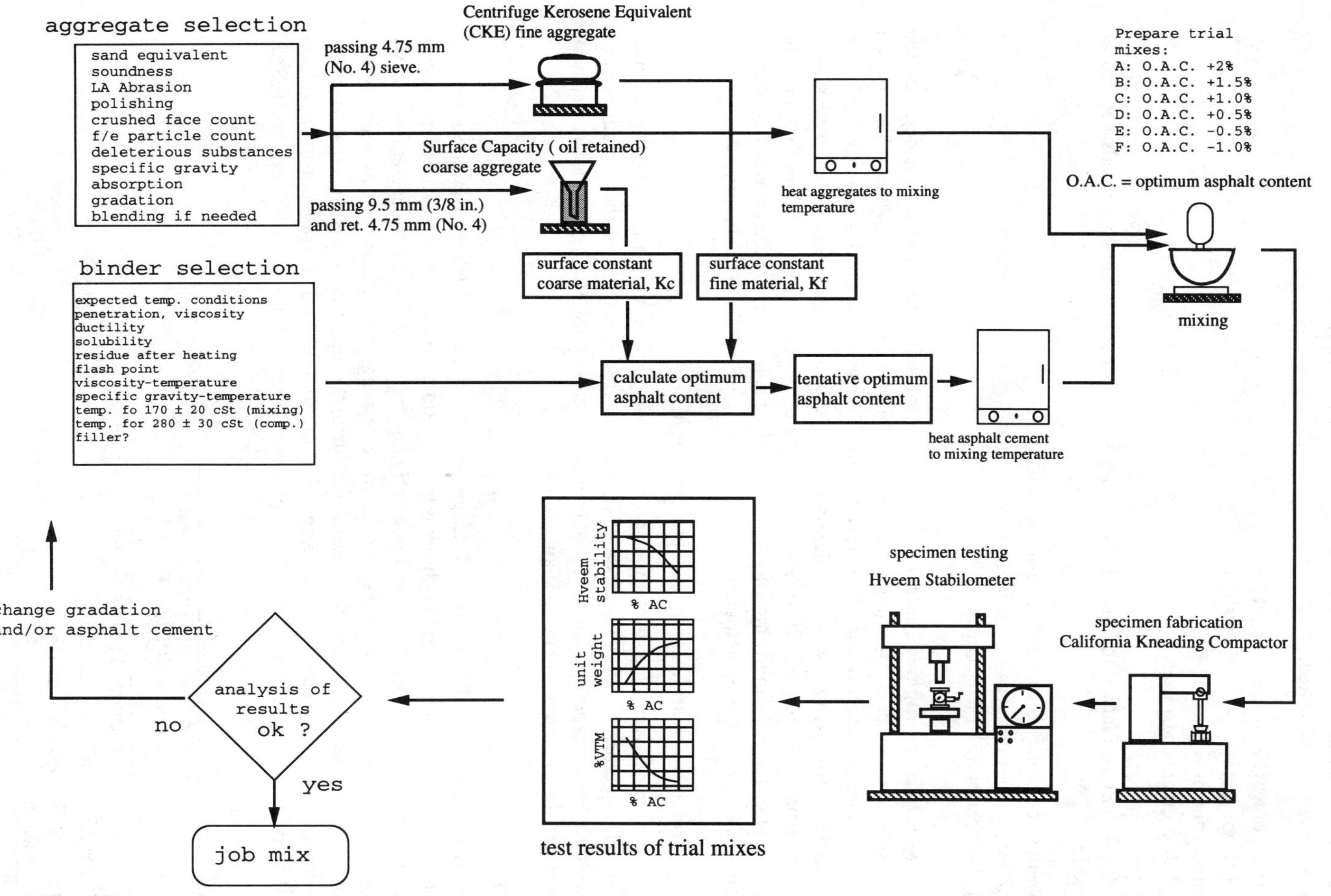

aggregate selection
sand equivalent
soundness
LA Abrasion
polishing
crushed face count
f/e particle count
deleterious substances
specific gravity
absorption
gradation
blending if needed
passing 4.75 mm (No. 4) sieve.
passing 9.5 mm (3/8 in.) and ret. 4.75 mm (No. 4)
Centrifuge Kerosene Equivalent (CKE) fine aggregate
Surface Capacity (oil retained) coarse aggregate
heat aggregates to mixing temperature
Prepare trial mixes:
A: O.A.C. +2%
B: O.A.C. +1.5%
C: O.A.C. +1.0%
D: O.A.C. +0.5%
E: O.A.C. -0.5%
F: O.A.C. -1.0%
O.A.C. = optimum asphalt content
mixing
binder selection
expected temp. conditions
penetration, viscosity
ductility
solubility
residue after heating
flash point
viscosity-temperature
specific gravity-temperature
temp. fo 170 ± 20 cSt (mixing)
temp. for 280 ± 30 cSt (comp.)
filler?
surface constant coarse material, Kc
surface constant fine material, Kf
calculate optimum asphalt content
tentative optimum asphalt content
heat asphalt cement to mixing temperature
Hveem stability
unit weight
%VTM
% AC
% AC
% AC
test results of trial mixes
specimen testing
Hveem Stabilometer
specimen fabrication
California Kneading Compactor
change gradation and/or asphalt cement
analysis of results ok ?
no
yes
job mix

FIGURE 43.13 Schematic diagram of Hveem mix design procedure.

amount of swell resulting from water absorption. Compacted specimens are maintained at room temperature for at least one hour prior to the swell test and then placed in a mold with a perforated plate on top. Five hundred ml of water are added to the specimens in the molds and after 24 hours the amount of vertical swell and water absorption are recorded. After the tests are completed, the following results are tabulated:

Asphalt content versus unit weight
Asphalt content versus Hveem stability
Asphalt content versus VTM
Asphalt content versus swell

Expected Trends in Plots. To consider the test results valid, the following trends have to be observed:

Hveem stability decreases with increasing asphalt content.
Unit weight increases at a diminishing rate with increasing asphalt content.
VTM decreases with increasing asphalt content.

Selection of Optimum Asphalt Content. Selection of optimum asphalt content is based on the prevention of excess asphalt (flushing), maximum stability, a minimum of 4% of VTM, and an amount of swell of not more than 0.03 in. (760μm). The selection process is illustrated in Fig. 43.14. First the two mixtures with asphalt content that produces excess flushing or exudation of asphalt cement are eliminated. Of the remaining three mixtures, the two mixtures that meet the requirements of Hveem stability are selected. The final asphalt content is then selected to obtain a VTM of 4% or greater and an amount of swell of not more than 0.03 in. (760μm).

Limitations of the Marshall and Hveem Design Methods

The Marshall and Hveem design methods do not provide procedures to measure fundamental mechanical properties of HMA mixtures. The results are test-specific, and their validity resides primarily on the past experience accumulated over many years of use and the empirical correlations between mix design results and performance results.

SHRP Mix Design

A new system for material selection and design of HMA mixtures has been developed through the program SHRP of the Federal Highway Administration. The new system, known as *superior asphalt pavements* (Superpave), includes the selection of the asphalt binder, the aggregates, and the amounts of mixed components. Unlike previous mixture design procedures, the new design system combines the material properties with pavement structural properties to predict actual pavement performance in terms of rutting and cracking. The mixture design is carried out in accordance with three different levels of expected traffic, expressed in terms of *equivalent single axle load* (ESAL)

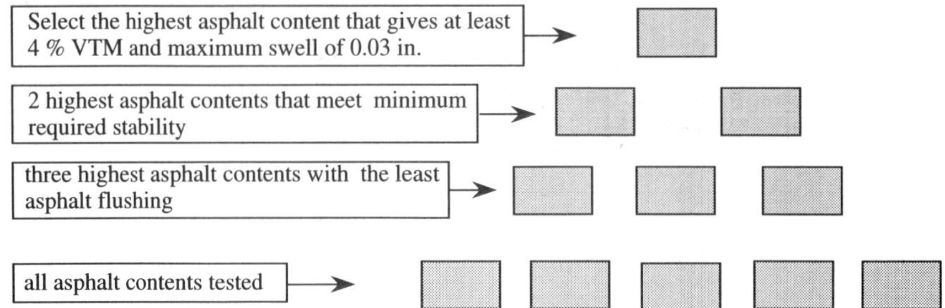

FIGURE 43.14 Selection of asphalt content in Hveem mix design procedure.

repetitions. The degree of refinement and complexity of the design procedure depends on the expected traffic.

Asphalt Binder Selection. Asphalt binders are selected on basis of the climate in which the mixture will serve according to the SHRP-proposed binder specifications discussed earlier. The physical property requirements for asphalt binder selection are constant among all grades of asphalts, but for the different asphalt grades the requirements must be met at different temperatures. For instance, two asphalt binders classified as PG 58-28 and PG 64-34 must meet all the physical requirements (dynamic shear, creep stiffness, etc.) at the pavement design temperatures of 58 and 28°C and 64 and 34°C, respectively. Asphalt binder grading is based on the assumption that the mixture will be subjected to fast-moving loads. For those cases in which loads are expected to be applied slowly (e.g., at road intersections), the required grade is increased one or two grades in the high-temperature limit.

Aggregate Selection. The selection of aggregates is based on two groups of physical property requirements and on gradation. The first group of physical requirements—which includes angularity (coarse and fine aggregates), amount of flat and elongated particles, and clay content—are the most critical in achieving high performance of HMA mixtures. The values of the physical requirements vary depending on the expected traffic and the location of the mixture in the pavement. For example, coarse aggregate angularity, which is defined by the percent of particles with one and two fractured surfaces, is required to be 95/90 for mixtures to be placed near the pavement surface and for an expected traffic level between 10^7 ESALs and 3×10^7 ESALs [where 95/90 denotes the percent of particles with one fractured face (95%) and with two fractured faces (90%)]; angularity is required to be 100/100 for a mixture to be used near the pavement surface and with an expected traffic greater than 3×10^7 ESALs. The percentage of flat and elongated particles (coarse aggregate particles with an aspect ratio of 1:5 or more) is specified at a maximum of 10% for traffic levels greater than 3×10^6 ESALs. The clay content is controlled by specifying a minimum of sand equivalent value. For example, the minimum sand equivalent value for a traffic level of less than 3×10^6 ESALs is 40.

The second group of physical requirements includes toughness, soundness, and deleterious materials. The required values specified in this second group may be determined by local conditions, however. For instance, the toughness, expressed as the percent loss of materials in the Los Angeles abrasion test, is limited to approximately 35 to 45%. The soundness, measured as the percent loss of material in the sodium or magnesium sulfate soundness test, is limited to approximately 10 to 20%. The weight percentage of deleterious materials may be limited to as low as 0.2% or as high as 10%, depending on the nature of the deleterious materials present.

Aggregate gradations are specified in 0.45 power gradation charts for five different aggregate designations (Table 43.5), which include the nominal sizes of 9.5, 12.5, 19, 25, and 37.5 mm. The nominal maximum size is defined as the nominal opening of the sieve one sieve size larger than the first sieve to retain more than 10% by weight of the aggregate. The specified gradation charts consist of two sets of points. One set of points (control points) limits the upper and lower values of percent passing within which aggregate gradation must pass. The other set of points delimits a restricted zone—between an intermediate sieve and the 300 μm sieve—that should not be crossed by the line of the aggregate gradation. The aggregate gradation specified with the mentioned control points typically exhibit a characteristic *S* shape. It is recommended, although not required, that the aggregate gradation passes below the restricted zone—that is, that the aggregate is coarser than the aggregate of maximum density in the range of intermediate particle sizes.

Selection of Mix Proportions. Selection of mix proportions is based on the test results of specimens fabricated with various trial mixtures. The loose mixtures are oven aged for four hours at 135°C prior to fabrication of the specimens in a new gyratory compactor (Superpave gyratory compactor, SGC). This new device used in the compaction of the test specimens is intended to reflect more realistically the compactability of the trial mixes to achieve densities similar to those achieved in actual pavements and under actual climatic conditions. In the compactor the height and diameter of the specimen are determined while the specimen is being compacted so that a direct relationship between specimen density and number of gyrations is obtained. Thus, mix compactability and potential tender mixture behavior can be identified and avoided during the mix design stage.

The mix design methodology is divided into three design procedures of increasing complexity. Design level I (volumetric design) consists of the fabrication of test specimens using the SGC and selection of the asphalt content on the basis of voids (VTM, VMA, and VFA) and the ratio of dust to effective asphalt content. After an asphalt content is selected based on the voids and compaction properties of the mix, a moisture sensitivity test is performed to verify if the mix will be susceptible to moisture damage. The moisture sensitivity of the mixtures is quantified as the ratio between (1) the indirect tensile strength of specimens subjected to moisture conditioning, and (2) the indirect tensile strength of control specimens tested directly after compaction.

Design levels II and III use the results of level I (volumetric design) as a starting point but, in addition, evaluate trial mixtures for their potential performance in the pavement. The potential pavement performance is expressed in terms of permanent deformation, fatigue cracking, and low-temperature cracking. The potential pavement performance is derived from the results of specimens tested with two new devices: the *Superpave shear tester* (SST) and the *indirect shear tester* (IDT). Various tests are performed in the SST, including volumetric test, uniaxial strain test, repeated shear test, simple shear test, and frequency sweep test. Two tests are performed in the IDT: (1) creep compliance and strength at low temperature (to predict low-temperature cracking), and (2) strength tests at intermediate temperatures (to predict fatigue cracking). Three models are utilized to predict the mixture performance: permanent deformation (rut depth vs. ESALs), fatigue cracking (percent fatigue-cracked area vs. ESALs), and low-temperature cracking (crack spacing vs. pavement life).

The SHRP method for mixture design is only briefly presented in the previous paragraphs. More detailed information can be found in the sources listed under "For Further Information."

Defining Terms

Bingham-plastic material: A material that behaves as a solid (which would not flow) when the shear stress is below its yield strength, but behaves as a fluid (which would flow under stress) when the shear stress is above this yield point.

Dilatant fluid: A non-Newtonian fluid whose viscosity increases as the shear rate or shear stress increases.

Newtonian fluid: A fluid whose viscosity remains constant with changes in shear rate or shear stress. The relationship between shear stress and shear rate is linear for this type of fluid.

Pseudoplastic fluid: A non-Newtonian fluid whose viscosity decreases as the shear rate or shear stress increases.

Rutting: Permanent vertical depression of pavement surface along the wheel paths.

Shear susceptibility: The slope of the plot of log (shear stress) versus log (shear rate). It is usually denoted as C. For a Newtonian fluid, C is equal to 1.

References

AASHTO. 1993a. *AASHTO Provisional Standards.* American Assoc. of State Highway and Transportation Officials, Washington, D.C.

AASHTO. 1993b. *Guide for Design of Pavement Structures.* American Assoc. of State Highway and Transportation Officials, Washington, D.C.

Anderson, D. A. and Kennedy, T. W. 1993. Development of SHRP binder specifications. *Proc. Assoc. Asphalt Paving Technol.* 62:481–528.

Asphalt Institute. 1977. *A Brief Introduction to Asphalt and Some of Its Uses,* 7th ed. Asphalt Institute, Lexington, KY.

Asphalt Institute. 1989. *Mix Design Methods for Asphalt Concrete and Other Hot-Mix Types.* Asphalt Institute, Lexington, KY.

Asphalt Institute. 1991. *Thickness Design—Asphalt Pavements for Highways and Streets.* Manual Series No.1. Asphalt Institute, Lexington, KY.

ASTM. 1994. *Annual Book of ASTM Standards.* ASTM, Philadelphia, PA.

Brock, J. D. 1988. *Hot Mix Asphalt Segregation: Causes and Cures.* Asphalt Pavement Association, Riverdale, MD.

Davis, R. L. 1987. Relationship between the rheological properties of asphalt and the rheological properties of mixtures and pavements. *Amer. Soc. Testing Matls. Spec. Tech. Pub.* 941: 28–50.

Epps, J. A., Button, J. W., and Gallaway, B. M. 1983. *Paving with Asphalt Cements Produced in the 1980's.* NCHRP Rept. 269. Transportation Research Board, Washington, D.C.

Fuller, W. B. and Thompson, S. E. 1907. The laws of proportioning concrete. *Trans. Amer. Soc. Civil Engrs.* 59:67–143.

Goetz, W. H. and Wood, L. E. 1960. Bituminous materials and mixtures. In Woods, K. B., ed., *Highway Engineering Handbook,* Sec. 18. McGraw-Hill, New York.

Goodrich, J. L. 1988. Asphalt and polymer-modified asphalt properties related to the performance of asphalt-concrete mixes. *Proc. Assoc. of Asphalt Paving Technol.* 57:116–175.

Goodrich, J. L. 1991. Asphalt binder rheology, asphalt concrete rheology, and asphalt-concrete mix properties. *Proc. Assoc. Asphalt Paving Technol.* 60:80–120.

Heukelom, W. and Klomp, A.J.G. 1964. Road design and dynamic loading. *Proc. Assoc. Asphalt Paving Technol.* 33:92–125.

Kandhal, P. S., Sandvig, L. D., and Wenger, M. E. 1973. Shear susceptibility of asphalts in relation to pavement performance. *Proc. Assoc. Asphalt Paving Technol.* 42:99–125.

Kennedy, T. W., Roberts, F. L., and Lee, K. W. 1982. Evaluation of moisture susceptibility of asphalt mixtures using the Texas freeze-thaw pedestal test. *Proc. Assoc. Asphalt Paving Technol.* 51:327–341.

Lee, D. 1973. Review of aggregate blending techniques. *Highway Res. Record.* 441:111–127.

Leung, S. C. and Anderson, K. O. 1987. Evaluation of asphalt cements for low-temperature performance. *Trans. Res. Record* 1115:23–32.

McLeod, N. W. 1976. Asphalt cements: Pen-vis number and its application to moduli stiffness. *J. Testing Eval.* 4(4).

Monismith, C. L., Epps, J. A., and Finn, F. N. 1985. Improved asphalt mix design. *Proc. Assoc. Asphalt Paving Technol.* 54:347–392.

Neumann, D. L. 1964. Mathematical method for blending aggregates. *Amer. Soc. Civil Engrs., J. Construc. Div.* 2:1–13.

Popovics, S. 1973. Methods for the determination of required blending proportions for aggregates. *Highway Res. Record.* 441:65–75.

Roberts, F., Kandhal, P. S., Brown, E. R., Lee, D. Y., and Kennedy, T. W. 1991. *Hot Mix Asphalt Materials, Mixture Design, and Construction.* Natl. Asphalt Paving Association, Education Foundation, Lanham, MA.

U.S. Army Corps of Engineers. 1991. *Hot Mix Asphalt Paving Handbook.* Pub. UN-13. U.S. Army Corps of Engineers, GPO, Washington, D.C.

Vallerga, B. A. and Lovering, W. R. 1985. Evolution of the Hveem stabilometer method of designing asphalt paving mixtures. *Proc. Assoc. Asphalt Paving Technol.* 54:243–265.

Van der Poel, C. 1964. A general system describing the viscoelastic properties of bitumen and its relation to routine test data. *J. Appl. Chem.*

Von Quintus, H. L., Scherocman, J. A., Hughes, C. S., and Kennedy, T. W. 1991. *NCHRP Report 338: Asphalt-Aggregate Mixture Analysis System: AAMAS*. Transportation Research Board, National Research Council, Washington, D.C.

Von Quintus, H. L., Hughes, C. S., and Scherocman, J. A. 1992. *NCHRP Asphalt-Aggregate Mixture Analysis System*. TRR 1353, pp. 90–99. Transportation Research Board, National Research Council, Washington, D.C.

White, T. D. 1985. Marshall procedures for design and quality control of asphalt mixtures. *Proc. Assoc. Asphalt Paving Technol.* 54:265–284.

For Further Information

This is a list of books that give comprehensive coverage :

Abraham, H. 1960. *Asphalt and Allied Substances*, 6th ed. D. Van Nostrand, Princeton, NJ.

Barth, E. J. 1962. *Asphalt Science and Technology*, 5 vols. Gordon and Breach, New York.

Hoiberg, A. J., ed. 1979. *Bituminous Materials: Asphalts, Tars, and Pitches*, 3 vols. R.E. Krieger, Huntington, NY.

Pfeiffer, J. P. 1950. *The Properties of Asphaltic Bitumen*. Elsevier, New York.

Road Research Laboratory. 1963. *Bituminous Materials in Road Construction*. Dept. of Scientific and Industrial Research, HMSO, London.

Traxler, R. N. 1961. *Asphalt—Its Composition, Properties, and Uses*. Reinhold, New York.

Wallace, H. A. and Martin, J. R. 1967. *Asphalt Pavement Engineering*. McGraw-Hill, New York.

In addition, many manuals dealing with all aspects of asphalt pavement design, construction, and testing are published and regularly updated by the Asphalt Institute, P.O. Box 14052, Lexington, KY 40512, and by the National Asphalt Paving Association, 6811 Kenilworth Ave., Riverdale, MD 20737.

In 1994 the Sears Tower celebrated its 20th anniversary as the world's tallest skyscraper. This 1454-foot-tall building contains a gross area of 4.5 million square feet, the equivalent of 80 football fields, and is designed for a daily population of 16,500 people. The Sears Tower has 110 stories, 103 passenger elevators, 18 escalators, and 181 drinking fountains. The great height provides for ample open space at ground level; only 40% of the site is occupied by the building. The structural system is steel frame and bundled tube. The cladding is black aluminum with bronze-tinted glass.

The architect and structural engineer was Skidmore, Owings & Merrill in Chicago. The contractor was Diesel Construction, which is now named Morse Diesel Inc. The developer was Sears, Roebuck & Company. The electrical and mechanical engineer was Jaros, Baum & Bolles from New York. (Photo by Hedrich-Blessing and courtesy of Skidmore, Owings & Merrill.)

VI

Structural Engineering

W. F. Chen
Purdue University

STRUCTURAL ENGINEERING IS CONCERNED with the application of structural theory, theoretical and applied mechanics, and optimization to the design, analysis, and evaluation of building structures, bridges, and cable and shell structures. The science of structural engineering includes the understanding of the physical properties of engineering materials, the development of methods of analysis, the study of the relative merits of various types of structures and method of fabrication and construction, and the evaluation of their safety, reliability, economy, and performance.

The study of structural engineering includes such typical topics as strength of materials, structural analysis in both classical and numerical computer methods, structural design in both steel and concrete as well as wood and masonry, solid mechanics, and probabilistic methods. The types of structures involved in a typical structural engineering work include bridges, buildings, offshore platforms, containment vessels, reactor vessels, and dams. Research in structural engineering can include such topics as high-performance computing, computer graphics, computer-aided analysis and design, stress analysis, structural dynamics and earthquake engineering, structural fatigue, structural mechanics, structural models and experimental methods, structural safety and reliability, and structural stability.

The scope of this section is indicated by the outline of the contents. It sets out initially to examine the basic properties and strength of materials (Chapter 44), goes on to show how these properties affect the analysis (Chapter 45) and design process of these structures made of either steel (Chapters 46 and 47) or concrete (Chapter 48), and finally outlines some of the mathematical techniques by which the safety and reliability issues of these structures so designed may be evaluated and their performance assessed (Chapter 49).

Recent demands for improvements of the nation's infrastructure, which includes, among other public facilities, the highway system and over 550,000 bridges, have increased the number of structural engineers employed by highway departments and consulting firms. Graduates with advanced degrees in structural engineering in the areas of experimental works, large-scale computer applications, computer-aided design and engineering, interactive graphics, and knowledge-based expert systems are in great demand by consulting firms, private industry, government and national laboratories, and educational institutions. The rapid advancement in computer hardware, particularly in the computing and graphics performance of personal computers and workstations, is making future structural engineering more and more oriented toward computer-aided engineering. Increased computational power will also make hitherto unrealized approaches feasible. For example, this will make the rigorous consideration of the life-cycle analysis and performance assessment of large structural systems feasible and practical. High-performance computing in structural engineering is now a subject of intense research interest. Good progress has been made, but much more remains to be done.

44

Mechanics of Materials

Austin D. Pan
Purdue University

Egor P. Popov
University of California at Berkeley

44.1 Introduction

The subject of **mechanics of materials** involves analytical methods for determining the *strength, stiffness* (deformation characteristics), and **stability** of the various members in a structural system. Alternatively, the subject may be called the strength of materials, mechanics of solid deformable

0-8493-8953-4/95/$0.00 + $.50

bodies, or simply mechanics of solids. The behavior of a member depends not only on the fundamental laws that govern the equilibrium of forces but also on the mechanical characteristics of the material. These mechanical characteristics come from the laboratory, where materials are tested under accurately known forces and their properties are carefully observed and measured. For this reason, it is seen that mechanics of materials is a blended science of experiment and Newtonian postulates of analytical mechanics.

The advent of computer technology has made possible remarkable advances in analytical methods, notably the finite element method, for solving problems of mechanics of materials. Prior to the computer, practical solutions were largely restricted to simple and idealized problems. Today, the technology is capable of analyzing complex three-dimensional structural systems with nonlinear material properties. Although this chapter will be limited to presenting the classical topics, the relatively simple methods employed are unusually useful as they apply to a vast number of technically important and practical problems of structural engineering.

44.2 Stress

Method of Sections

Engineering mechanics is in large part the study of the nature of forces within a body. To study these internal forces a uniform approach—the method of sections—is applied. In this approach an arbitrary plane cuts the original solid body into two distinct parts (see Fig. 44.1).[1] If the body as a whole is in equilibrium, then each part must also be in equilibrium, which leads to the fundamental conclusion that the external forces are balanced by the internal forces.

Definition of Stress

Stress is defined as the intensity of forces acting on infinitesimal areas of a cut section (Fig. 44.1). The mathematical definition of stress τ is

$$\tau = \lim_{\Delta A \to 0} \frac{\Delta P}{\Delta A} \tag{44.1}$$

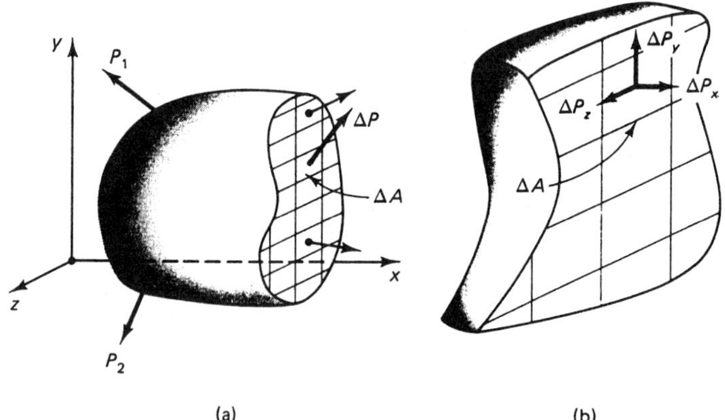

FIGURE 44.1 Sectioned body: (a) free body with some internal forces; (b) enlarged view with components of ΔP.

[1]Figures 44.1 through 44.42 and Tables 44.1 through 44.3 are reprinted from Popov, E. P., 1990. *Engineering Mechanics of Solids,* ©1990, reprinted by permission of Prentice-Hall, Englewood Cliffs, NJ.

It is advantageous to resolve these intensities perpendicular and parallel to the section (see Fig. 44.1). The perpendicular component is the normal stress. The parallel components are the shearing stresses. Normal stresses that pull away from the section are tensile stresses while those that push against it are compressive stresses. Normal stresses are alternatively designated by the letter σ.

Stress Tensor

A cube of infinitesimal dimensions is isolated from a body as shown in Fig. 44.2. All stresses acting on this cube are identified on the figure. The first subscript of τ gives reference to the axis that is perpendicular to the plane on which the stress is acting; the second subscript gives the direction of the stress. Using this cube, the state of stress at any point can be defined by three components on each of the three mutually perpendicular axes. In mathematical terminology this is called a tensor, and the matrix representation of the stress tensor is

$$\begin{bmatrix} \tau_{xx} & \tau_{xy} & \tau_{xz} \\ \tau_{yx} & \tau_{yy} & \tau_{yz} \\ \tau_{zx} & \tau_{zy} & \tau_{zz} \end{bmatrix} \tag{44.2}$$

This is a second-rank tensor requiring two indices to identify its components. There are three normal stresses: $\tau_{xx} \equiv \sigma_x$, $\tau_{yy} \equiv \sigma_y$, $\tau_{zz} \equiv \sigma_z$; and six shearing stresses: τ_{xy}, τ_{yx}, τ_{yz}, τ_{zy}, τ_{zx}, τ_{xz}. For brevity, a stress tensor can be written in indicial notation as τ_{ij} where it is understood that i and j can assume designations x, y, and z. To satisfy the requirement of moment equilibrium, a stress tensor is symmetric, i.e., $\tau_{ij} = \tau_{ji}$. Thus subscripts of τ are commutative and shear stresses on mutually perpendicular planes are numerically equal. Moment equilibrium cannot be satisfied by a single pair of shear stresses; two pairs are required with their arrowheads meeting at diametrically opposite corners of an element.

When stresses on the z plane do not exist, the third column and third row of Eq. (44.2) are zeros, and this two-dimensional state is referred to as plane stress. If all shear stresses are absent, only the diagonal terms remain and the stresses are said to be triaxial; for plane stress $\tau_{zz} = 0$ and the state of stress is biaxial. If τ_{yy} is further eliminated the state of stress is referred to as uniaxial.

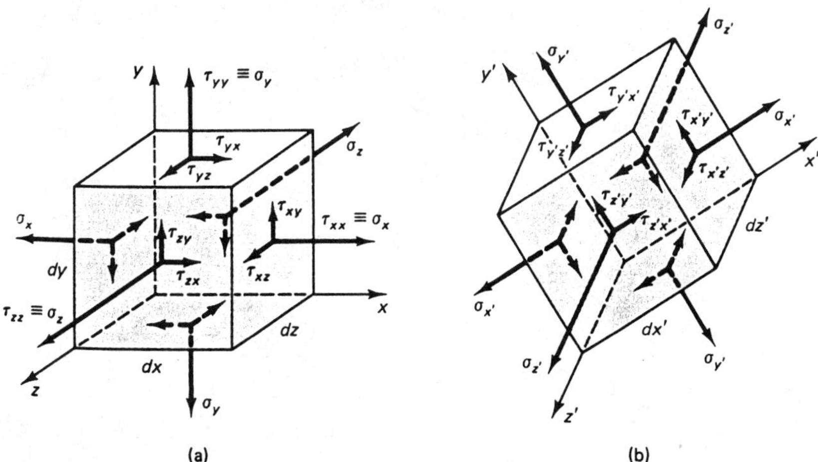

FIGURE 44.2 (a) General state of stress acting on an infinitesimal element in the initial coordinate system. (b) General state of stress acting on an infinitesimal element defined in a rotated system of coordinate axes. All stresses have positive sense.

Differential Equations for Equilibrium

For an infinitesimal element to be in equilibrium the following equation must hold in the x direction:

$$\frac{\partial \sigma_x}{\partial x} + \frac{\partial \tau_{yx}}{\partial y} + \frac{\partial \tau_{zx}}{\partial z} + X = 0 \tag{44.3}$$

where X is the inertial or body forces. Similar equations hold for the y and z directions. These equations are applicable whether a material is elastic, plastic, or viscoelastic. Note that there are not enough equations of equilibrium to solve for the unknown stresses. Thus all problems in stress analysis are internally intractable or indeterminate and can only be solved when supplemented by other equations given by kinematic requirements and the mechanical properties of the material.

Stress Analysis of Axially Loaded Bars

Figure 44.3 shows a bar loaded axially by a force P and a free-body diagram isolated by the section *aa*, which is at an angle θ with the vertical. The equilibrium force on the inclined section is equal to P and is resolved into the two components: the normal force component, $P \cos \theta$, and the shear component, $P \sin \theta$. The area of the inclined section is $A/\cos \theta$. Therefore, given the definition of

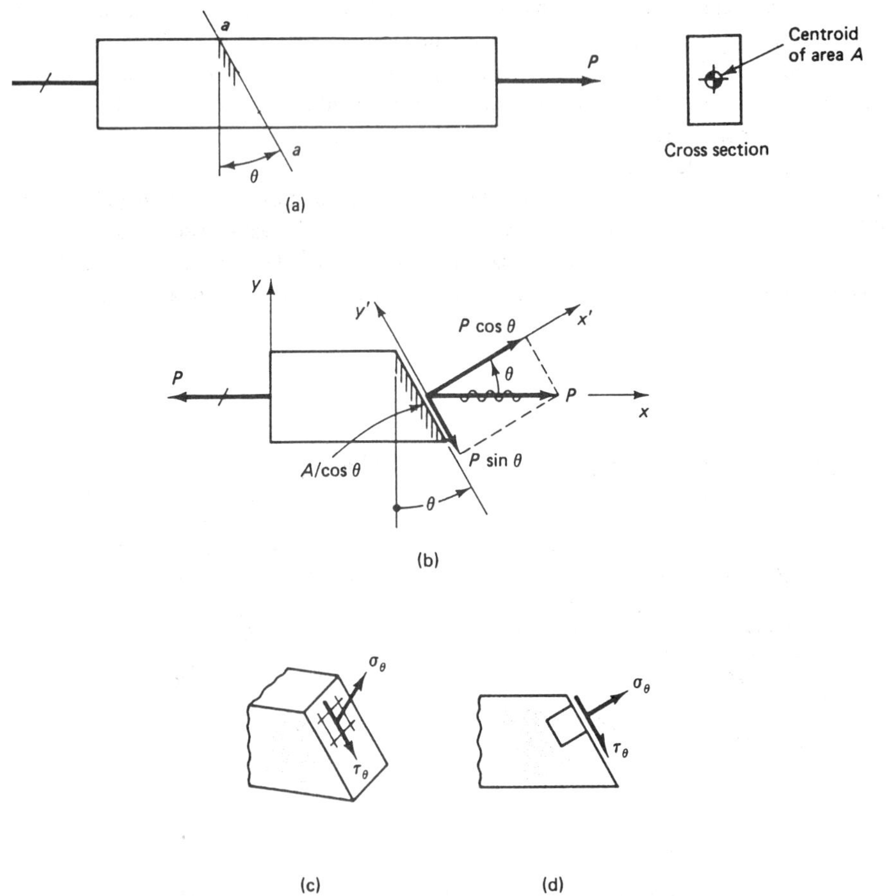

FIGURE 44.3 Sectioning of a prismatic bar loaded axially.

stress as force per unit area, the normal stress σ_θ and the shear stress τ_θ are given by the following equations:

$$\sigma_\theta = \frac{P}{A}\cos^2\theta \tag{44.4}$$

and

$$\tau_\theta = \frac{P}{A}\sin\theta\cos\theta \tag{44.5}$$

These equations show that the normal and shear stresses vary with the angle θ. Therefore the maximum normal stress is reached when $\theta = 0$ or when the section is perpendicular to the x axis, and $\sigma_{\max} = \sigma_x = P/A$. By differentiating Eq. (44.5) the maximum shear stress occurs on planes either $\pm 45°$ with the axis of the bar, and $\tau_{\max} = P/2A = \sigma_x/2$. It is important to note that the basic procedure of engineering mechanics of solids used here gives the average or mean stress at a section.

44.3 Strain

Normal Strain

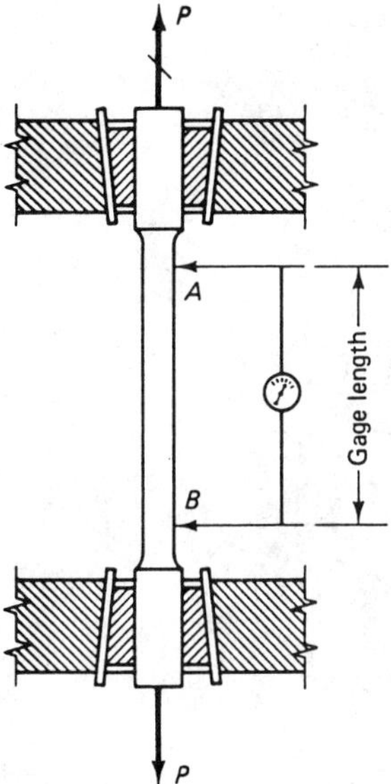

FIGURE 44.4 Diagram of a tension specimen in a testing machine.

A solid body deforms when subjected to an external load or a change of temperature. When an axial rod shown in Fig. 44.4 is subjected to an increasing force P, a change in length occurs between any two points, such as A and B. The gage length L_0 is the initial distance between A and B. If after loading the observed length is L, the gage elongation $\Delta L = L - L_0$. The elongation ε per unit of initial gage length is then given as

$$\varepsilon = \frac{L - L_0}{L_0} = \frac{\Delta L}{L_0} \tag{44.6}$$

This expression defines the *extensional strain*, which is a dimensionless quantity. Since this strain is associated with the normal stress, it is usually called the *normal strain*.

In some applications, where strains are large, one defines the *natural* or *true strain* $\bar{\varepsilon}$. The strain increment $d\varepsilon$ for this strain is defined as dL/L, where L is the instantaneous length of the specimen and dL is the incremental change in length. Analytically,

$$\bar{\varepsilon} = \int_{L_0}^{L} dL/L = \ln L/L_0 = \ln(1+\varepsilon) \tag{44.7}$$

Stress-Strain Relationships

The mechanical behavior of real materials under loads is of primary importance. Information on this behavior is generally provided by plotting stress against strain from experiments as shown in Fig. 44.5 for a variety of engineering materials. Each material has its own characteristics. Typical

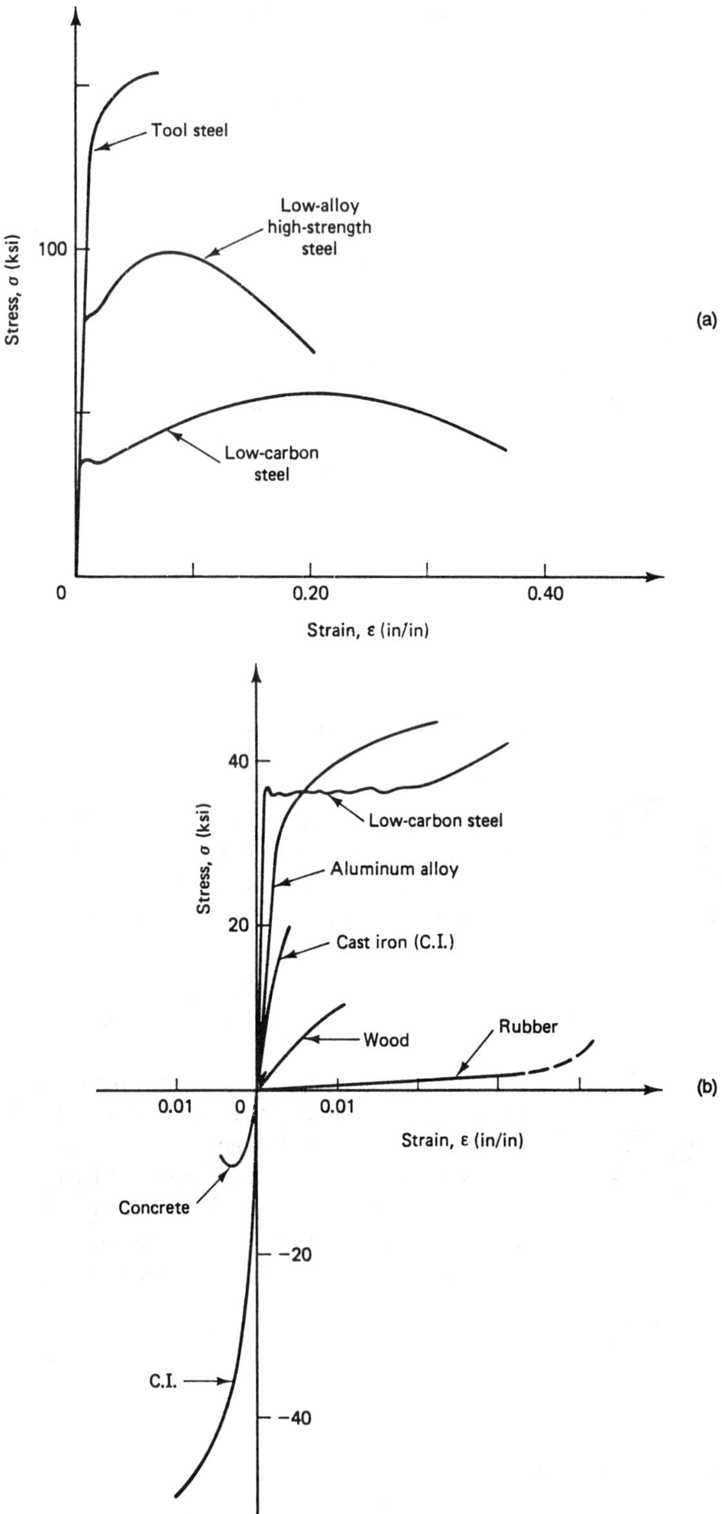

FIGURE 44.5 (a) Typical stress-strain diagrams for different steels. (b) Typical stress-strain diagrams for different materials.

physical properties of some common materials are given in Table 44.1. It should be noted that even for the same material the stress-strain diagram will differ depending on the temperature at which the test was conducted, the speed of the test, and a number of other variables. Materials capable of withstanding large strains without a significant increase in stress are referred to as ductile (see Fig. 44.6). The converse applies to brittle materials. Stresses computed on the basis of the original area of a specimen are referred to as conventional or engineering stresses. For some materials, such as mild steel and aluminum, significant transverse contraction or expansion takes place near the breaking point, referred to as necking. Dividing the applied load by the corresponding actual area of a specimen gives the so-called true stress; see Fig. 44.6(a). The vast majority of engineering materials are generally assumed to be completely homogeneous (the same from point to point) and isotropic (having essentially the same physical properties in different directions).

Hooke's Law

As illustrated by Fig. 44.6(a), the experimental values of stress versus strain obtained for mild steel lie essentially on a straight line for a limited range from the origin. This idealization is known as Hooke's Law and can be expressed by the equation

$$\sigma = E\varepsilon \tag{44.8}$$

which simply means that stress is directly proportional to strain. The constant of proportionality E is called the *elastic modulus, modulus of elasticity,* or *Young's modulus.* Graphically, E is the slope of the straight line from the origin to point A, the proportional or elastic limit of the material. Physically, the elastic modulus represents the stiffness of the material to an imposed load and it is a definite property of a material (Table 44.1). Almost all materials have an initial range where the relationship between stress and strain is linear and Hooke's law is applicable.

In Fig. 44.6(a), the highest point B corresponds to the ultimate strength of the material. Stress associated with the plateau *ab* in Fig. 44.6(a) is the yield strength of the material. The yield strength is often so near the proportional limit that the two may be taken to be the same. For materials with no well-defined yield strength, the *offset method* is usually used, in which a line offset an arbitrary amount—generally, 0.2% of strain—is drawn parallel to the straight-line portion of the material; the yield strength is taken where the line offset intersects the stress-strain curve.

Constitutive Relations

The relationships between stress and strain are frequently referred to as **constitutive relations**, or laws. A linear elastic material implies that stress is directly proportional to strain and Hooke's law is applicable. A nonlinear elastic material responds in a nonproportional manner, yet when unloaded returns back along the loading path to its initial stress-free state. If in stressing an inelastic or plastic material its elastic limit is exceeded, on unloading it usually responds approximately in a linearly elastic manner, but a permanent deformation, or set, develops at no external load. Figure 44.7 shows three other types of idealized stress-strain diagrams. A rigid-perfectly plastic diagram [Fig. 44.7(a)] is applicable to problems in which the elastic strains can be neglected in relation to the plastic ones. If both the elastic and perfectly plastic strains have to be included, Fig. 44.7(b) is more appropriate. Beyond the elastic range, many materials resist additional stress, a phenomenon referred to as strain hardening [Fig. 44.7(c)]. For a more accurate idealization, equations capable of representing a wide range of stress-strain curves have been developed, for example, by Ramberg and Osgood [Ramberg and Osgood, 1943]. This equation is

$$\frac{\varepsilon}{\varepsilon_0} = \frac{\sigma}{\sigma_0} + \frac{3}{7}\left(\frac{\sigma}{\sigma_0}\right)^n \tag{44.9}$$

Table 44.1 Typical Physical Properties of and Allowable Stresses for Some Common Materials (in U.S. Customary System of Units)

Material	Unit Weight, lb/in.^{3}	Ultimate Strength, ksi			Yield Strength, ksi		Allow Stresses, psi		Elastic Moduli $\times 10^6$ psi		Coef. of Thermal Expans. $\times 10^{-6}$ per °F
		Tens.	Comp.	Shear	Tens.	Shear	Tens. or Comp.	Shear	Tens. or Comp.	Shear	
Aluminum alloy (extruded): 2024-T4		60	—	32	44	25			10.6	4.00	12.9
Aluminum alloy (extruded): 6061-T6	0.100	38	—	24	35	20			10.0	3.75	13.0
Cast iron: gray		30	120	—	—	—			13	6	5.8
Cast iron: malleable	0.276	54	—	48	36	24			25	12	6.7
Concrete: 8 gal/sack		—	3	—	—	—	−1,350	66	3	—	6.0
Concrete: 6 gal/sack	0.087	—	5	—	—	—	−2,250	86	5	—	
Magnesium alloy, AM100A	0.065	40	—	21	22	—			6.5	2.4	14.0
Steel: 0.2% carbon (hot-rolled)		65	—	48	36	24	±24,000	14,500			
Steel: 0.6% carbon (hot-rolled)	0.283	100	—	80	60	36			30	12	6.5
Steel: 0.6% carbon (quenched)		120	—	100	75	45					
Steel: 3.5% Ni, 0.4% C		200	—	150	150	90					
Wood: Douglas fir (coast)	0.018	—	7.4	1.1	—	—	±1,900	120	1.76	—	—
Wood: Southern pine (longleaf)	0.021	—	8.4	1.5	—	—	±2,250	135	1.76	—	—

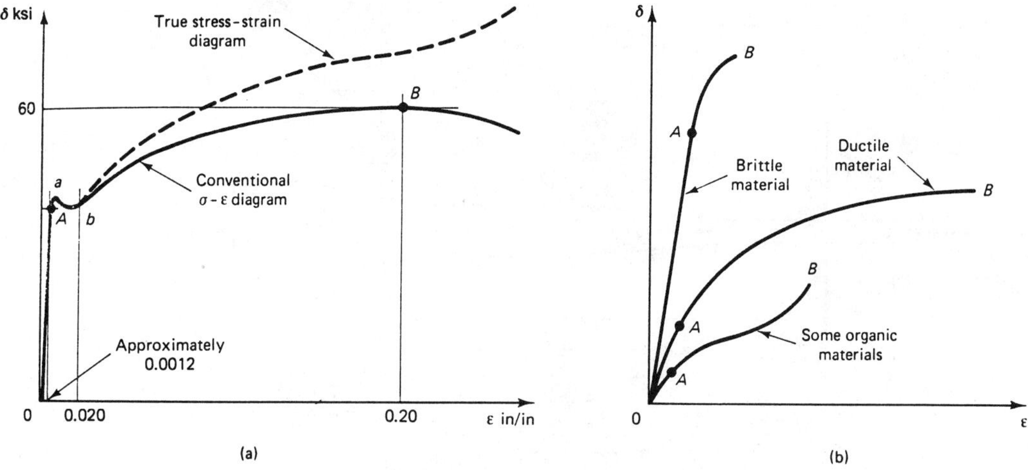

FIGURE 44.6 Stress-strain diagrams. (a) Mild steel. (b) Typical materials.

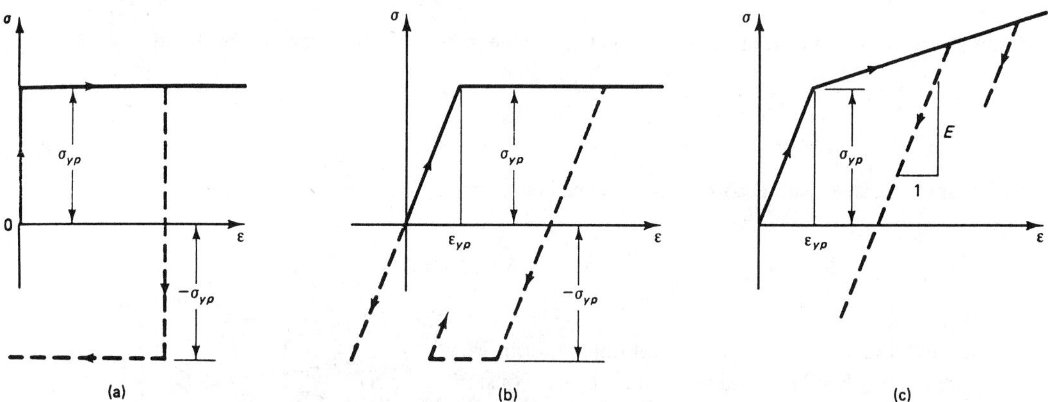

FIGURE 44.7 Idealized stress-strain diagrams: (a) rigid-perfectly plastic material, (b) elastic-perfectly plastic material, and (c) elastic-linearly hardening material.

where the constants ε_0 and σ_0 correspond to the yield point and n is a characteristic constant of the material. This equation has the important advantage of being a continuous mathematical function by which an instantaneous or tangent modulus E_t defined as

$$E_t = \frac{d\sigma}{d\varepsilon} \tag{44.10}$$

can be uniquely determined.

Deformation of Axially Loaded Bars

Consider the axially loaded bar shown in Fig. 44.8 and recast Eq. (44.6) for a differential element dx. Let du be the change in length. The normal strain ε_x in the x direction is

$$\varepsilon_x = \frac{du}{dx} \tag{44.11}$$

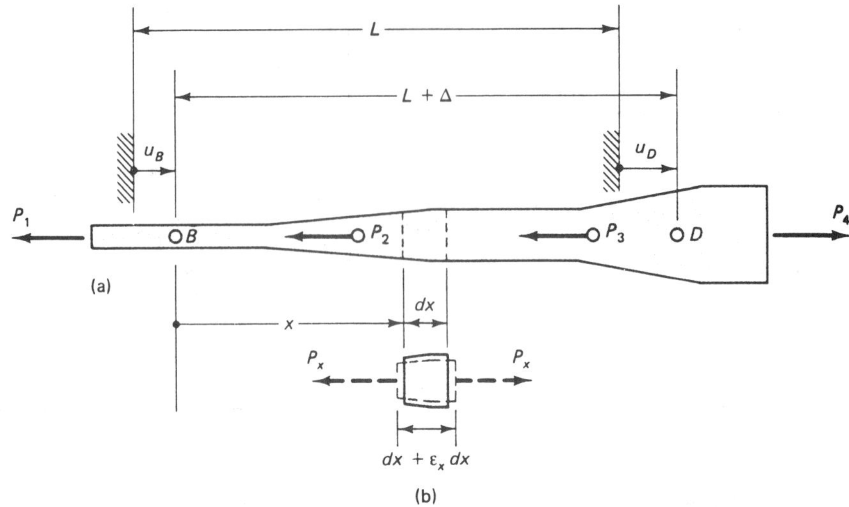

FIGURE 44.8 An axially loaded bar.

Assuming the origin of x at B, and integrating, the change in length Δ between points D and B is

$$\Delta = \int_0^L \varepsilon_x \, dx \tag{44.12}$$

For linear elastic materials, according to Hooke's law, $\varepsilon_x = \sigma_x / E$, where $\sigma_x = P_x / A_x$. Thus,

$$\Delta = \int_0^L \frac{P_x \, dx}{A_x E_x} \tag{44.13}$$

where the force P_x, area A_x, and elastic modulus E_x can vary along the length of a bar. For a beam of length L with constant cross-sectional area A and elastic modulus E, $\Delta = PL/AE$.

Poisson's Ratio

As illustrated by Fig. 44.9, when a solid body is subjected to an axial tension, it contracts laterally; when it is compressed, the material expands laterally. These lateral deformations on a relative basis are termed *lateral strains* and bear a constant relationship to the axial strains. This constant is a definite property of a material and is called Poisson's ratio. It is denoted by ν and is defined as follows:

$$\nu = \left| \frac{\text{Lateral strain}}{\text{Axial strain}} \right| \tag{44.14}$$

The value of ν fluctuates for different materials over a relatively narrow range. Generally, it is on the order of 0.25 to 0.35. The largest possible value is 0.5 and is normally attained during plastic flow and signifies constancy of volume.

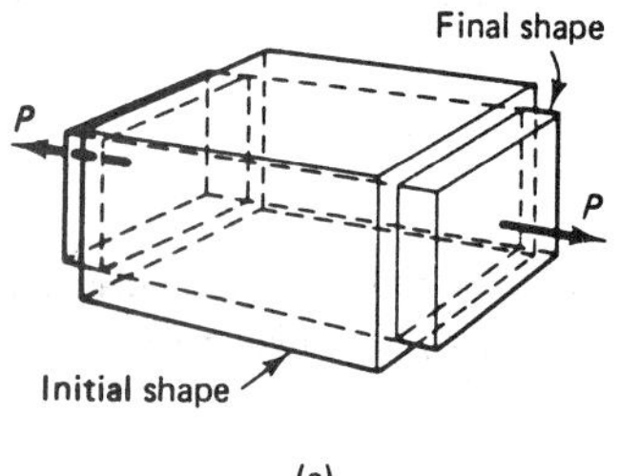

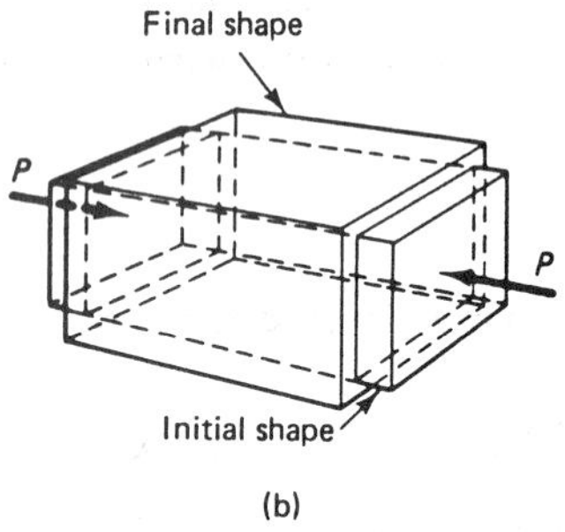

FIGURE 44.9 (a) Lateral contraction and (b) lateral expansion of solid bodies subjected to axial forces (Poisson's effect).

Thermal Strain and Deformation

Solid bodies expand as temperature increases and contract as it decreases. The thermal strain ε_T caused by a change in temperature from T_0 to T can be expressed as

$$\varepsilon_T = \alpha(T - T_0) \tag{44.15}$$

where α is an experimentally determined coefficient of linear thermal expansion (see Table 44.1).

For a body of length L subjected to a uniform temperature, the extensional deformation Δ_T due to a change in temperature of $\delta T = T - T_0$ is

$$\Delta_T = \alpha(\delta T)L \tag{44.16}$$

Saint-Venant's Principle and Stress Concentrations

Saint-Venant's principle states that the manner of force application on stresses is important only in the vicinity of the region where the force is applied. Figure 44.10 illustrates how normal stresses at a distance equal to the width of the member are essentially uniform. Only near the location where the concentrated force is applied is the stress nonuniform. Saint-Venant's principle also applies to changes in cross section, as shown by Fig. 44.11. The ratio of the maximum stress to the average stress is called the stress concentration factor K, which depends on the geometrical proportions of the members. The maximum normal stress can then be expressed as

$$\sigma_{\max} = K\sigma_{av} = K\frac{P}{A} \tag{44.17}$$

Many stress concentration factors K can be found tabulated in the literature [Roark and Young, 1975].

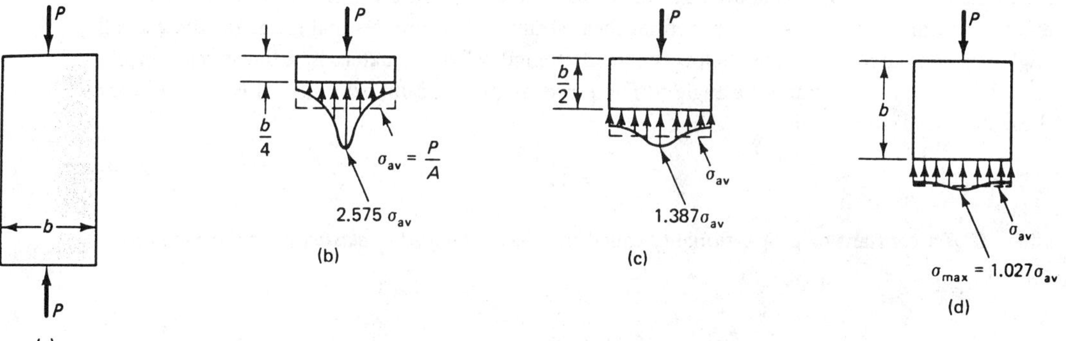

FIGURE 44.10 Stress distribution near a concentrated force in a rectangular elastic plate.

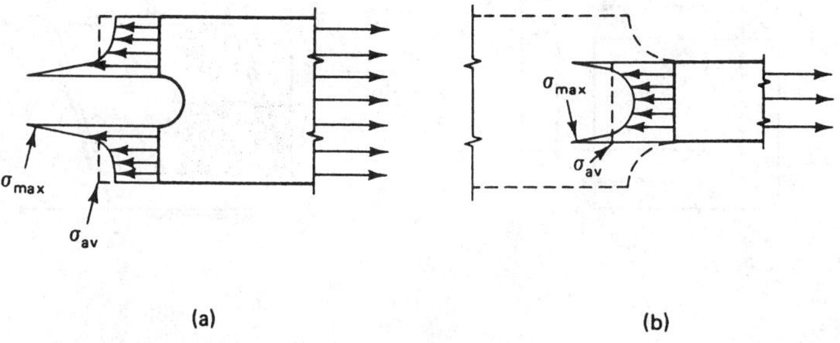

FIGURE 44.11 Meaning of the stress-concentration factor K.

Elastic Strain Energy for Uniaxial Stress

The product of force (stress multiplied by area) and deformation is the internal work done in a body by the externally applied forces. This internal work is stored in an elastic body as the internal elastic energy of deformation or the elastic strain energy. The internal elastic strain energy U for an infinitesimal element subjected to uniaxial stress is

$$dU = \tfrac{1}{2}\sigma_x \varepsilon_x \, dV \tag{44.18}$$

where dV is the volume of the element.

In the elastic range, Hooke's law applies, $\sigma_x = E\varepsilon_x$. Then

$$U = \int_{\text{vol}} \frac{\sigma_x^2}{2E} \, dV \tag{44.19}$$

The strain energy stored in an elastic body per unit volume of the material (its strain-energy density) U_0 is equal to $\sigma_x^2/2E$. Substituting the value of the stress at the proportional limit gives the modulus of resilience, an index of a material's ability to store or absorb energy without permanent deformation. Analogously, the area under a complete stress-strain diagram gives a measure of a material's ability to absorb energy up to fracture and is called its toughness.

44.4 Generalized Hooke's Law

Stress-Strain Relationships for Shear

An element in pure shear is shown in Fig. 44.12. The change in the initial right angle between any two imaginary planes in a body defines shear strain γ. For infinitesimal elements, these small angles are measured in radians. Below the yield strength of most materials, a linear relationship exists between pure shear and the angle γ. Therefore, mathematically, extension of Hooke's law for shear stress and strain reads

$$\tau = G\gamma \tag{44.20}$$

where G is a constant of proportionality called the *shear modulus of elasticity*, or the *modulus of rigidity*.

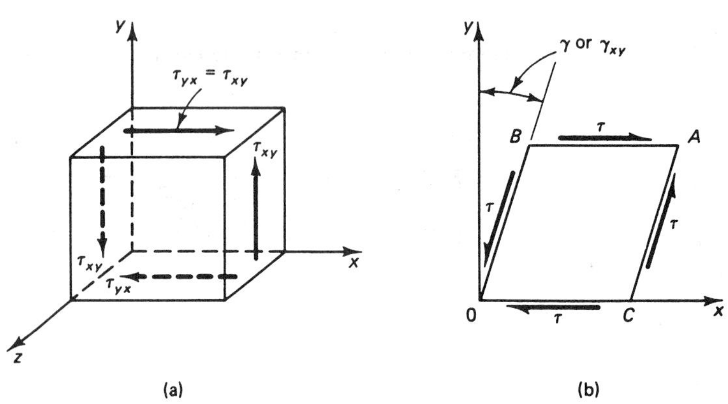

FIGURE 44.12 Element in pure shear.

Elastic Strain Energy for Shear Stress

An expression for the elastic strain energy for shear stresses may be established in a manner analogous to that for the elastic strain energy in uniaxial stress. Thus,

$$U_{\text{shear}} = \int_{\text{vol}} \frac{\tau^2}{2G} \, dV \tag{44.21}$$

Mathematical Definition of Strain

Since strains generally vary from point to point, the definitions of strain must relate to an infinitesimal element. As shown in Fig. 44.13(a), consider an extensional strain taking place in one direction. Points A and B move to A' and B', respectively. During straining, point A experiences a displacement u. Point B experiences a displacement $u + \Delta u$ since in addition to the rigid-body displacement u, common to the whole element Δx, a stretch Δu takes place within the element. On this basis, the definition of the extensional or normal strain is

$$\varepsilon = \lim_{\Delta x \to 0} \frac{\Delta u}{\Delta x} = \frac{du}{dx} \tag{44.22}$$

For the two-dimensional case shown in Fig. 44.13(b), a body is strained in orthogonal directions and subscripts must be attached to ε to differentiate between the directions of the strains, and it is also necessary to change the ordinary derivatives to partial ones. Therefore, if at a point of a body, u, v, and w are the three displacement components occurring, respectively, in the x, y, and z directions of the coordinate axes, the basic definitions of normal strain become

$$\varepsilon_x = \frac{\partial u}{\partial x} \qquad \varepsilon_y = \frac{\partial v}{\partial y} \qquad \varepsilon_z = \frac{\partial w}{\partial z} \tag{44.23}$$

In addition to normal strain, an element can also experience shear strain, as shown in Fig. 44.13(c) for the xy plane. This inclines the sides of the deformed element in relation to the x and the y axes. Since v is the displacement in the y direction, as one moves in the x direction, $\partial v/\partial x$ is the slope of the initially horizontal side of the infinitesimal element. Similarly, the vertical side

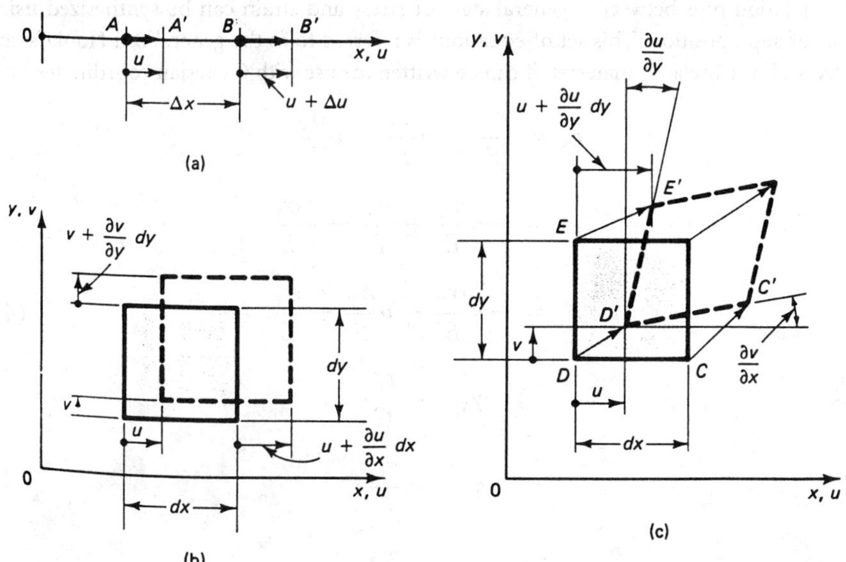

tilts through an angle $\partial u/\partial y$. On this basis, the initially right angle *CDE* is reduced by the amount $\partial v/\partial x + \partial u/\partial y$. Analogous descriptions can be used for shear strains in the *xz* and *yz* planes. Therefore, for small angle changes, the definitions of the shear strain become

$$\begin{aligned} \gamma_{xy} &= \gamma_{yx} = \frac{\partial v}{\partial x} + \frac{\partial u}{\partial y} \\ \gamma_{xz} &= \gamma_{zx} = \frac{\partial w}{\partial x} + \frac{\partial u}{\partial z} \\ \gamma_{yz} &= \gamma_{zy} = \frac{\partial w}{\partial y} + \frac{\partial v}{\partial z} \end{aligned} \tag{44.24}$$

It is assumed that tangents of small angles are equal to angles measured in radians, and positive sign applies for the shear strain when the element is deformed as depicted in Fig. 44.13(c).

Strain Tensor

In order to obtain the strain tensor, an entity which must obey certain laws of transformation, it is necessary to redefine the shear strain $\varepsilon_{xy} \equiv \varepsilon_{yx}$ as one-half of γ_{xy}. The strain tensor in matrix representation can then be assembled as follows:

$$\begin{bmatrix} \varepsilon_x & \dfrac{\gamma_{xy}}{2} & \dfrac{\gamma_{xz}}{2} \\ \dfrac{\gamma_{yx}}{2} & \varepsilon_y & \dfrac{\gamma_{yz}}{2} \\ \dfrac{\gamma_{zx}}{2} & \dfrac{\gamma_{zy}}{2} & \varepsilon_z \end{bmatrix} \equiv \begin{bmatrix} \varepsilon_{xx} & \varepsilon_{xy} & \varepsilon_{xz} \\ \varepsilon_{yx} & \varepsilon_{yy} & \varepsilon_{yz} \\ \varepsilon_{zx} & \varepsilon_{zy} & \varepsilon_{zz} \end{bmatrix} \tag{44.25}$$

The strain tensor is symmetric. Just as for the stress tensor, using indicial notation, one can write ε_{ij} for the strain tensor. For a two-dimensional problem, the third row and column are eliminated and one has the case of plane strain.

Generalized Hooke's Law for Isotropic Materials

Six basic relationships between a general state of stress and strain can be synthesized using the principle of superposition. This set of equations is referred to as the generalized Hooke's law and for isotropic linearly elastic materials it can be written for use with Cartesian coordinates as

$$\begin{aligned} \varepsilon_x &= \frac{\sigma_x}{E} - \nu\frac{\sigma_y}{E} - \nu\frac{\sigma_z}{E} \\ \varepsilon_y &= -\nu\frac{\sigma_x}{E} + \frac{\sigma_y}{E} - \nu\frac{\sigma_z}{E} \\ \varepsilon_z &= -\nu\frac{\sigma_x}{E} - \nu\frac{\sigma_y}{E} + \frac{\sigma_z}{E} \\ \gamma_{xy} &= \frac{\tau_{xy}}{G} \\ \gamma_{yz} &= \frac{\tau_{yz}}{G} \\ \gamma_{zx} &= \frac{\tau_{zx}}{G} \end{aligned} \tag{44.26}$$

E, G, and ν Relationship

Using the relationship between shear and extensional strains and the fact that a pure shear stress at a point can be alternatively represented by the normal stresses at 45° with the directions of the shear stresses, one can obtain the following relationship among E, G, and ν:

$$G = \frac{E}{2(1+\nu)} \tag{44.27}$$

Dilatation and Bulk Modulus

For volumetric changes in elastic materials subjected to stress, change in volume per unit volume, often referred to as *dilatation*, is defined as

$$e = \varepsilon_x + \varepsilon_y + \varepsilon_z \tag{44.28}$$

The shear strains cause no change in volume.

If an elastic body is subjected to hydrostatic pressure of uniform intensity p, then, based on the generalized Hooke's law, it can be shown that

$$\frac{-p}{e} = k = \frac{E}{3(1-2\nu)} \tag{44.29}$$

The quantity k represents the ratio of the hydrostatic compressive stress to the decrease in volume and is called the modulus of compression, or bulk modulus.

The six equations of Eq. (44.26) have an inverse which can be solved to express stresses in terms of strain and may be written as

$$\begin{aligned}
\sigma_x &= \lambda e + 2G\varepsilon_x \\
\sigma_y &= \lambda e + 2G\varepsilon_y \\
\sigma_z &= \lambda e + 2G\varepsilon_z \\
\tau_{xy} &= G\gamma_{xy} \\
\tau_{yz} &= G\gamma_{yz} \\
\tau_{zx} &= G\gamma_{zx}
\end{aligned} \tag{44.30}$$

where

$$\lambda = \frac{\nu E}{(1+\nu)(1-2\nu)} \tag{44.31}$$

44.5 Torsion

Torsion of Circular Elastic Bars

To establish a relation between the internal torque and the stresses it sets up in members with circular solid and tubular cross sections, it is necessary to make two assumptions (Fig. 44.14): (1) a plane section of material perpendicular to the axis of a circular member remains plane after the torque is applied, that is, no warpage or distortion of parallel planes normal to the axis of a member takes place; (2) shear strains γ vary linearly from the central axis reaching γ_{max} at the periphery. This means that in Fig. 44.14 an imaginary plane such as DO_1O_3C moves to $D'O_1O_3C$ when the torque is applied, and the radii O_1D and O_2B remain straight. In the elastic case, shear stresses vary linearly from the central axis of a circular member, as shown in Fig. 44.15. Thus the maximum $\tau_{\max}$ occurs at the radius c from the center, and at any distance ρ from O, the shear stress is $(\rho/c)\tau_{\max}$. For equilibrium, the internal resisting torque must equal the externally applied torque T. Hence,

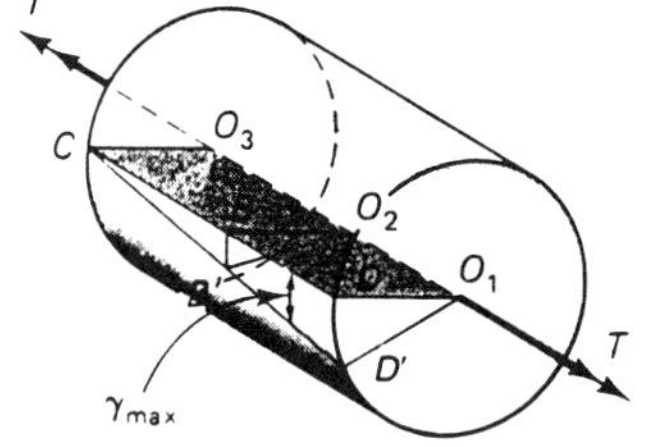

FIGURE 44.14 Variation of strain in circular member subjected to torque.

$$\frac{\tau_{\max}}{c}\int_A \rho^2\,dA = T \tag{44.32}$$

However, $\int_A \rho^2\,dA$ is the polar moment of inertia of a cross-sectional area and is a constant. By using the symbol J for the polar moment of inertia of a circular area, that is, $J = \pi c^4/2$, Eq.(44.32) may be written more compactly as

$$\tau_{\max} = \frac{Tc}{J} \tag{44.33}$$

For a circular tube with inner radius b, $J = (\pi c^4/2) - (\pi b^4/2)$.

Angle-of-Twist of Circular Members

The governing differential equation for the angle-of-twist for solid and tubular circular elastic shafts subjected to torsional loading can be determined by referring to Fig. 44.16, which shows a differential shaft of length dx under a differential twist $d\phi$. Arc DD' can be expressed as $\gamma_{\max}\,dx = d\phi\,c$. Then substituting Eq. (44.33) and assuming Hooke's law is applicable, the governing differential equation for the angle-of-twist is obtained:

$$\frac{d\phi}{dx} = \frac{T}{JG} \quad \text{or} \quad d\phi = \frac{T\,dx}{JG} \tag{44.34}$$

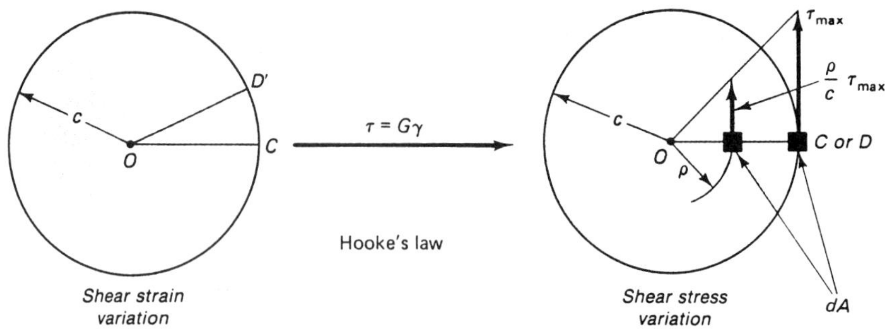

FIGURE 44.15 Shear strain assumption leading to elastic shear stress distribution in a circular member.

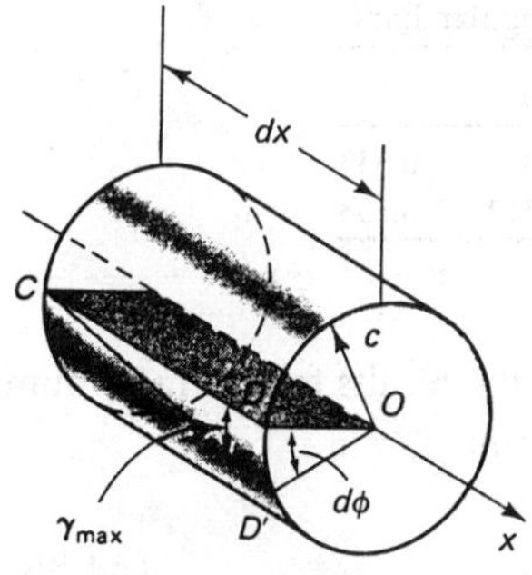

FIGURE 44.16 Deformation of a circular bar element due to torque.

Hence, a general expression for the angle-of-twist between any two sections A and B on a shaft is

$$\phi = \phi_B - \phi_A = \int_A^B d\phi = \int_A^B \frac{T_x\, dx}{J_x G} \tag{44.35}$$

where ϕ_B and ϕ_A are, respectively, the global shaft rotations at ends B and A. In this equation, the internal torque T_x and the polar moment of inertia J_x may vary along the length of a shaft. The direction of the angle of twist ϕ coincides with the direction of the applied torque. For an elastic shaft of length L and constant cross section, $\phi = TL/JG$.

Torsion of Solid Noncircular Members

The two assumptions made for circular members do not apply for noncircular members. Sections perpendicular to the axis of a member warp when a torque is applied. The nature of the distortions that take place in a rectangular section can be surmised from Fig. 44.17. For a rectangular member, the corner elements do not distort at all. Therefore shear stresses at the corners are zero; they are maximum at the midpoints of the long sides. Analytical solutions for torsion of rectangular, elastic members have been obtain [Timoshenko and Goodier, 1970]. The methods used are mathematically

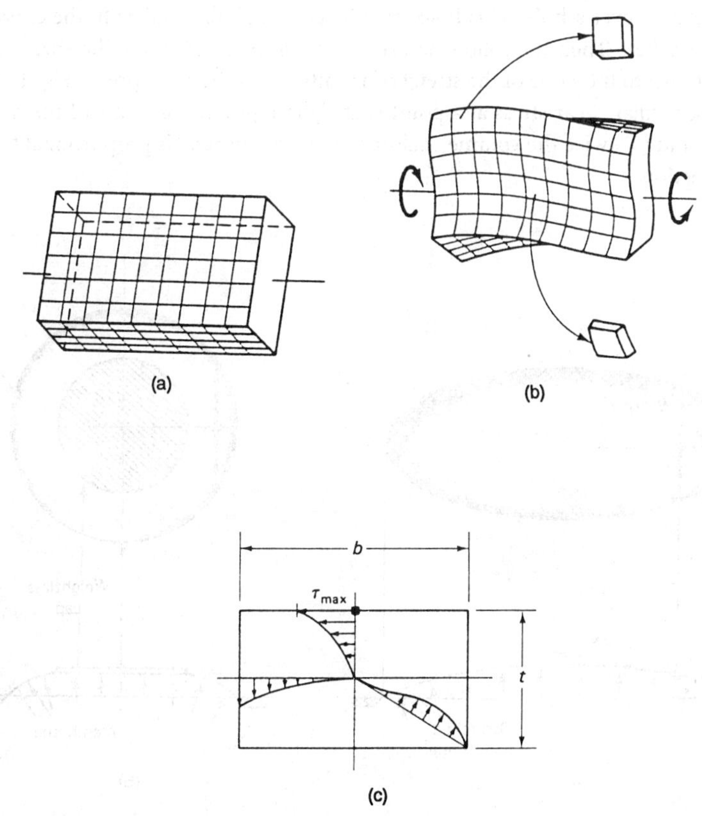

FIGURE 44.17 Rectangular bar (a) before and (b) after a torque is applied. (c) Shear stress distribution in a rectangular shaft subjected to a torque.

Table 44.2 Table of Torsional Coefficients for Rectangular Bars

b/t	1.00	1.50	2.00	3.00	6.00	10.0	∞
α	0.208	0.231	0.246	0.267	0.299	0.312	0.333
β	0.141	0.196	0.229	0.263	0.299	0.312	0.333

complex and beyond the scope of the present discussion. However, the results for the maximum shear stresses and the angle-of-twist can be put into the following form:

$$\tau_{\max} = \frac{T}{\alpha b t^2} \quad \text{and} \quad \phi = \frac{TL}{\beta b t^3 G} \tag{44.36}$$

where T is the applied torque, b is the length of the long side, and t is the thickness or width of the short side of a rectangular section. The values of parameters α and β depend on the ratio b/t given in Table 44.2. For thin sections, where b is much greater than t, the values of α and β approach 1/3. It is useful to recast the second Eq. (44.36) to express the torsional stiffness k_t for a rectangular section, giving

$$k_t = \frac{T}{\phi} = \beta b t^3 \frac{G}{L} \tag{44.37}$$

For cases that cannot be conveniently solved mathematically, the membrane analogy has been devised. It happens that the solution of the partial differential equation that must be solved in the elastic torsion problem is mathematically identical to that for a thin membrane, such as a soap film, lightly stretched over a hole. This hole must be geometrically similar to the cross section of the shaft being studied. Then the following can be shown to be true: (1) the shear stress at any point is proportional to the slope of the stretched membrane at the same point (Fig. 44.18); (2) the direction of a particular shear stress at a point is at right angles to the slope of the membrane at the same point; and (3) twice the volume enclosed by the membrane is proportional to the torque carried by the section.

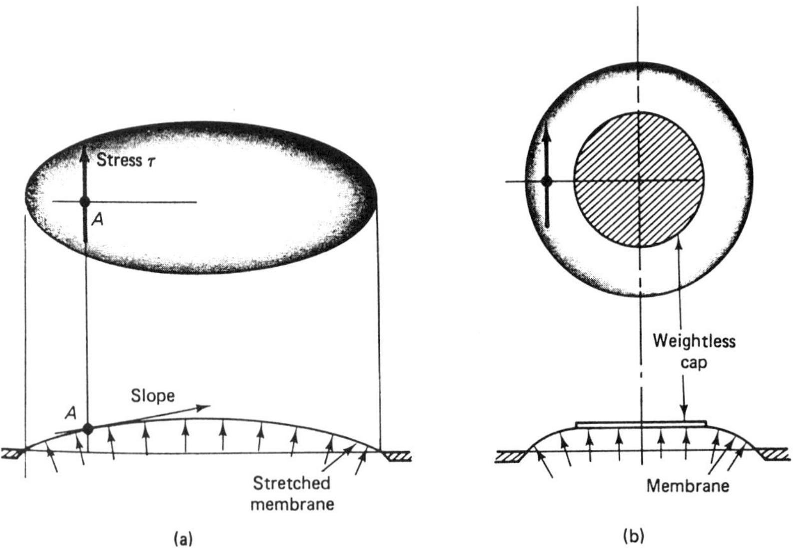

FIGURE 44.18 Membrane analogy: (a) simply connected region, (b) multiply connected (tubular) region.

For plastic torsion, the sand-heap analogy has been developed, and it has similar interpretations as those in the membrane analogy [Nadai, 1950]. Dry sand is poured onto a raised flat surface having the shape of the cross section of the member. The surface of the sand heap so formed assumes a constant slope. The volume of the sand heap, and hence its weight, is proportional to the fully plastic torque carried by a section.

44.6 Bending

The Basic Kinematic Assumption

Consider a typical element of the beam between two planes perpendicular to the beam axis, as shown in Fig. 44.19. In side view, such an element is identified in the figure as $abcd$. When such a beam is subjected to equal end moments M_z acting around the z axis, Fig. 44.19(b), this beam bends in the plane of symmetry, and the planes initially perpendicular to the beam axis slightly tilt. Nevertheless, the lines such as ad and bc becoming $a'd'$ and $b'c'$ remain straight. This observation forms the basis for the fundamental hypothesis of the flexure theory. It may be stated thus: plane sections through a beam taken normal to its axis remain plane after the beam is subjected to bending.

In pure bending of a prismatic beam, the beam axis deforms into a part of a circle of radius ρ, as shown by Fig. 44.19(b). For an element defined by an infinitesimal angle $d\theta$, the fiber length ef of the beam axis is given as $ds = \rho\, d\theta$. Hence,

$$\frac{d\theta}{ds} = \frac{1}{\rho} = \kappa \tag{44.38}$$

where the reciprocal of ρ defines the axis curvature κ.

The fiber length gh located on a radius $(\rho - y)$ can be found similarly and the difference between fiber lengths gh and ef can be expressed as $(-y\, d\theta)$, which is equal to du since the deflection and

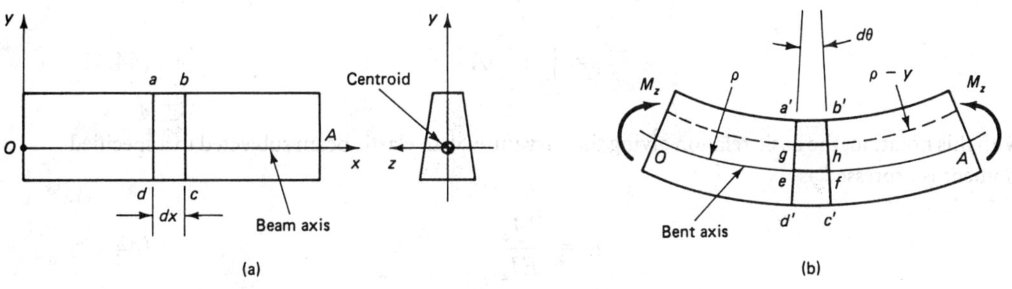

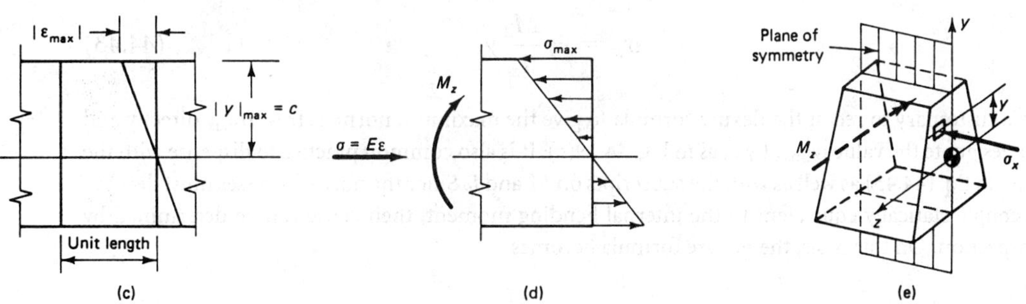

FIGURE 44.19 Assumed behavior of elastic beam in bending.

rotations of the beam axis are very small. Then one obtains the normal strain, $\varepsilon_x = du/dx$, as

$$\varepsilon_x = -\kappa y \tag{44.39}$$

This equation establishes the expression for the basic kinematic hypothesis for the flexure theory: the strain in a bent beam varies along the beam depth linearly with y.

The Elastic Flexure Formula

By using Hooke's law, the expression for the normal strain given by Eq. (44.39) can be recast into a relation for the normal longitudinal stress σ_x:

$$\sigma_x = E\,\varepsilon_x = -E\,\kappa y \tag{44.40}$$

To satisfy equilibrium, the sum of all forces at a section in pure bending must vanish, $\int_A \sigma_x\, dA = 0$, which can be rewritten as $-E\kappa \int_A y\, dA = 0$. By definition, the integral $\int_A y\, dA = \bar{y}A$, where $\bar{y}$ is the distance from the origin to the centroid of an area A. Since the integral equals zero here and area A is not zero, distance $\bar{y}$ must be set equal to zero. Therefore the z axis must pass through the centroid of a section. In bending theory, this axis is also referred to as the neutral axis of a beam. On this axis both the normal strain ε_x and the normal stress σ_x are zero. Based on this result, linear variation in strain is schematically shown in Fig. 44.19(c). The corresponding elastic stress distribution in accordance with Eq. (44.40) is shown in Fig. 44.19(d). Both the absolute maximum strain $\varepsilon_{\max}$ and the absolute maximum stress $\sigma_{\max}$ occur at the largest value of y.

Equilibrium requires the additional condition that the sum of the externally applied and the internal resisting moments must vanish. For the beam segment in Fig. 44.19(d), this yields $M_z = E\kappa \int_A y^2\, dA$. In mechanics, the last integral, depending only on the geometrical properties of a cross-sectional area, is called the rectangular moment of inertia or second moment of inertia of the area A and is designated by I. Since I must always be determined with respect to a particular axis, it is often meaningful to identify it with a subscript corresponding to such an axis. For the case considered, this subscript is z, that is,

$$I_z = \int_A y^2\, dA \tag{44.41}$$

With this notation, the basic relation giving the curvature of an elastic beam subjected to a specified moment is expressed as

$$\kappa = \frac{M_z}{EI_z} \tag{44.42}$$

By substituting Eq. (44.42) into Eq. (44.40), the elastic flexure formula for beams is obtained:

$$\sigma_x = -\frac{M_z}{I_z} y \tag{44.43}$$

It is customary to recast the flexure formula to give the maximum normal stress $\sigma_{\max}$ directly and to designate the value $|y|_{\max}$ by c, as in Fig. 44.19(c). It is also common practice to dispense with the sign in Eq. (44.43) as well as with the subscripts on M and I. Since the normal stresses must develop a couple statically equivalent to the internal bending moment, their sense can be determined by inspection. On this basis, the flexure formula becomes

$$\sigma_{\max} = \frac{Mc}{I} \tag{44.44}$$

Elastic Strain Energy in Pure Bending

Using the section "Elastic Strain Energy for Uniaxial Stress" as the basis, the elastic strain energy for beams in pure bending can be found. By substituting the flexure formula into Eq. (44.18) and integrating over the volume V of the beam, the expression for the elastic strain energy U in a beam in pure bending is obtained.

$$U = \int_0^L \frac{M^2\,dx}{2EI} \tag{44.45}$$

Unsymmetric Bending and Bending with Axial Loads

Consider the rectangular beam shown in Fig. 44.20, where the applied moments M act in the plane $abcd$. By using the vector representation for **M** shown in Fig. 44.20(b), this vector forms an angle α with the z axis and can be resolved into the two components, M_y and M_z. Since the cross section for this beam has symmetry about both axes, Eqs. (44.40)–(44.44) are directly applicable. By assuming elastic behavior of the material, a superposition of the stresses caused by M_y and M_z is the solution to the problem. Hence, using Eq. (44.43),

$$\sigma_x = -\frac{M_z y}{I_z} + \frac{M_y z}{I_y} \tag{44.46}$$

A line of zero stress, i.e., a neutral axis, forms at an angle β with the z axis and can be determined from the following equation:

$$\tan\beta = \frac{I_z}{I_y}\tan\alpha \tag{44.47}$$

In general, the neutral axis does not coincide with the normal of the plane in which the applied moment acts.

Superposition can again be employed to include the effect of axial loads, leading Eq. (44.46) to be generalized into

$$\sigma_x = \frac{P}{A} - \frac{M_z y}{I_z} + \frac{M_y z}{I_y} \tag{44.48}$$

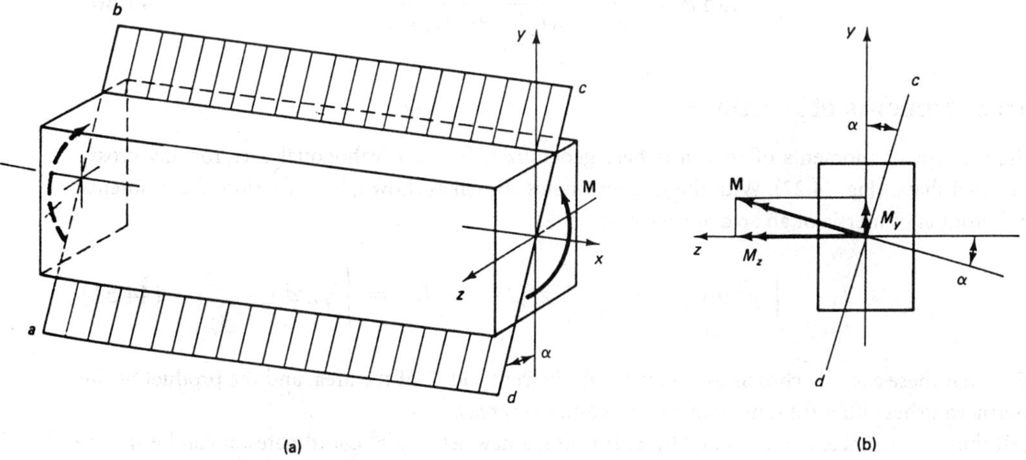

FIGURE 44.20 Unsymmetrical bending of a beam with doubly symmetric cross section.

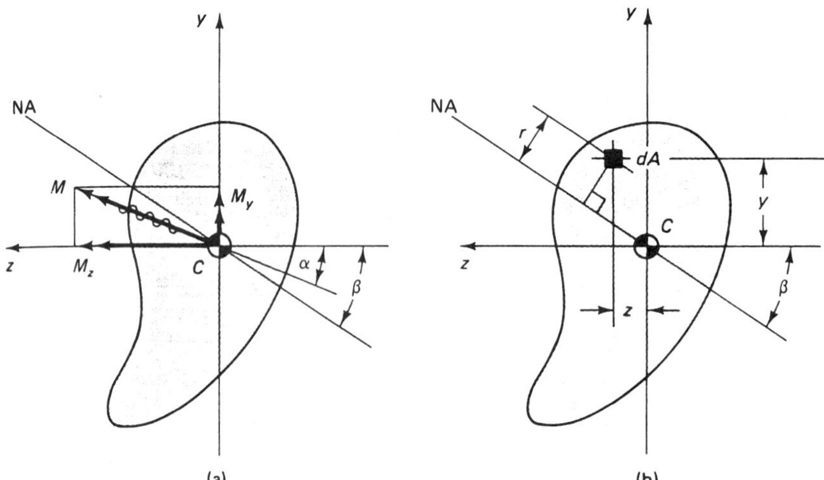

FIGURE 44.21 Bending of unsymmetric cross section.

where P is taken positive for axial tensile forces, and bending takes place around the two principal y and z axes. Further, if an applied axial force causes compression, a member must be stocky, or a buckling problem of the type considered in Section 44.9 may arise.

Bending of Beams with Unsymmetric Cross Section

A general equation for pure bending of elastic members of arbitrary cross section (Fig. 44.21), whose reference axes are not the principal axes can be formulated using the same approach as for the symmetrical cross sections. This generalized flexure formula is

$$\sigma_x = -\frac{M_z I_y + M_y I_{yz}}{I_y I_z - I_{yz}^2} y + \frac{M_y I_z + M_z I_{yz}}{I_y I_z - I_{yz}^2} z \tag{44.49}$$

By setting this equation equal to zero, the angle β for locating the neutral axis in the arbitrary coordinate system is obtained, giving

$$\tan\beta = \frac{y}{z} = \frac{M_y I_z + M_z I_{yz}}{M_z I_y + M_y I_{yz}} \tag{44.50}$$

Area Moments of Inertia

The concept of moments of inertia is here generalized for two orthogonal axes for any cross-sectional shape (Fig. 44.22). With the yz coordinates chosen as shown, by definition, the moments and product of inertia of an area are given as

$$I_z = \int y^2\, dA \qquad I_y = \int z^2\, dA \qquad I_{yz} = \int yz\, dA \tag{44.51}$$

Note that these axes are chosen to pass through the centroid C of the area, and the product of the inertia vanishes either for doubly or singly symmetric areas.

If the orthogonal axes are rotated by θ, forming a new set of $y'z'$ coordinates, it can be shown that the moments and product of inertia are transformed to the following quantities:

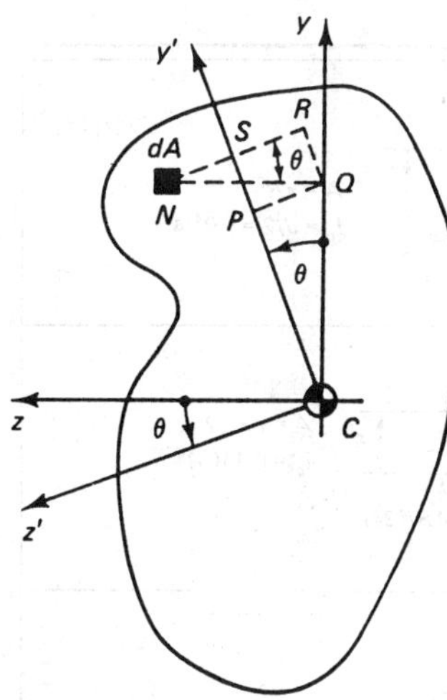

FIGURE 44.22 Rotation of coordinate axes.

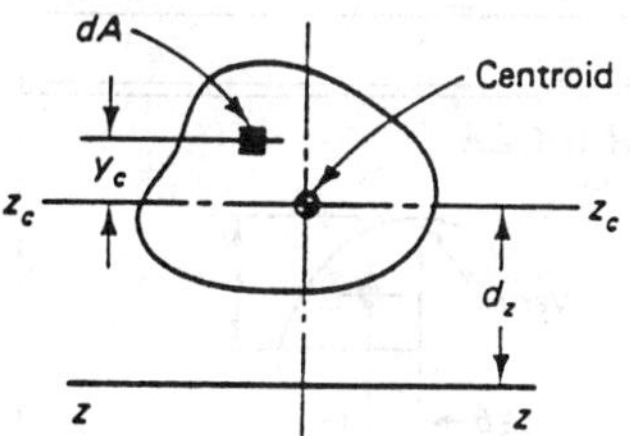

FIGURE 44.23 Area for deriving the parallel-axis theorem.

$$I_{z'} = \frac{I_z + I_y}{2} + \frac{I_z - I_y}{2} \cos 2\theta + I_{yz} \sin 2\theta$$

$$I_{y'} = \frac{I_z + I_y}{2} - \frac{I_z - I_y}{2} \cos 2\theta - I_{yz} \sin 2\theta$$

$$I_{y'z'} = -\frac{I_z - I_y}{2} \sin 2\theta + I_{yz} \cos 2\theta \tag{44.52}$$

Note that the sum of the moments of inertia around two mutually perpendicular axes is invariant, that is, $I_{y'} + I_{z'} = I_y + I_z$.

Table 44.3 provides formulas for the areas, centroids, and moments of inertia of some simple shapes. Most cross-sectional areas used may be divided into a combination of these simple shapes. To find I for an area composed of several simple shapes, the parallel-axis theorem (sometimes called the transfer formula) is necessary. It can be stated as follows: the moment of inertia of an area around any axis is equal to the moment of inertia of the same area around a parallel axis passing through the area's centroid, plus the product of the same area and the square of the distance between the two axes. Hence,

$$I_z = I_{zc} + Ad_z^2 \tag{44. 53a}$$

where d_z is the distance from the centroid of the subarea to the centroid of the whole area, as shown in Fig. 44.23. In calculation, Eq. (44.53a) must be applied to each subarea into which a cross-sectional area has been divided and the results summed to obtain I_z for the whole section:

$$I_z(\text{whole section}) = \sum (I_{zc} + Ad_z^2) \tag{44.53b}$$

44.7 Shear Stresses in Beams

Shear Flow

Consider an elastic beam made from several continuous longitudinal planks whose cross section is shown in Fig. 44.24. To make this beam act as an integral member, it is assumed that the planks are fastened at intervals by vertical bolts. If an element of this beam, Fig. 44.24(b), is subjected to bending moments $+M_A$ at end A and $+M_B$ at end B, bending stresses that act normal to the sections are developed. These bending stresses vary linearly from the neutral axis in accordance with the flexure formula My/I. The top plank of the beam element is isolated as shown in Fig. 44.24(c). The forces acting perpendicular to ends A and B of this plank may be determined by summing the bending stresses over their respective areas. Denoting the total force acting normal to the area $fghj$ by F_B and remembering that, at section B, M_B and I are constants, one obtains the following relation:

$$F_B = -\frac{M_B}{I} \int_{\text{area}=fghj} y \, dA = -\frac{M_B Q}{I} \tag{44.54}$$

Table 44.3 Useful Properties of Areas

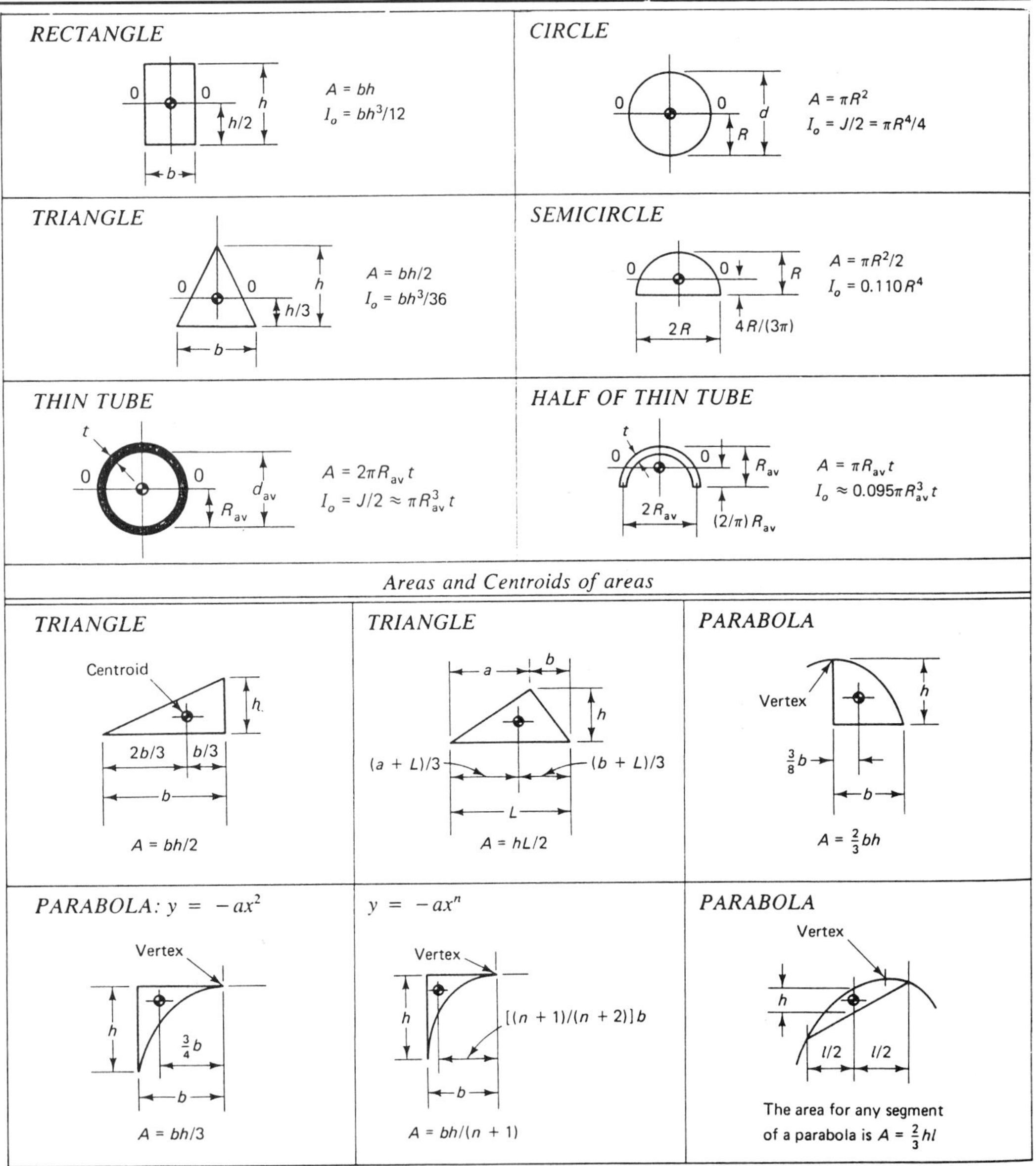

where

$$Q = \int_{\text{area}=fghj} y\, dA = A_{fghj}\bar{y} \tag{44.55}$$

The integral defining Q is the first, or static, moment of area $fghj$ around the neutral axis. By definition, $\bar{y}$ is the distance from the neutral axis to the centroid of A_{fghj}. Similarly, one can express the total force acting normal to the area $abcd$ as $F_A = -M_A Q/I$.

If M_A is not equal to M_B, which is always the case when shears are present at the adjoining sections, F_A is not equal to F_B. Equilibrium of the horizontal forces in Fig. 44.24(c) may be attained

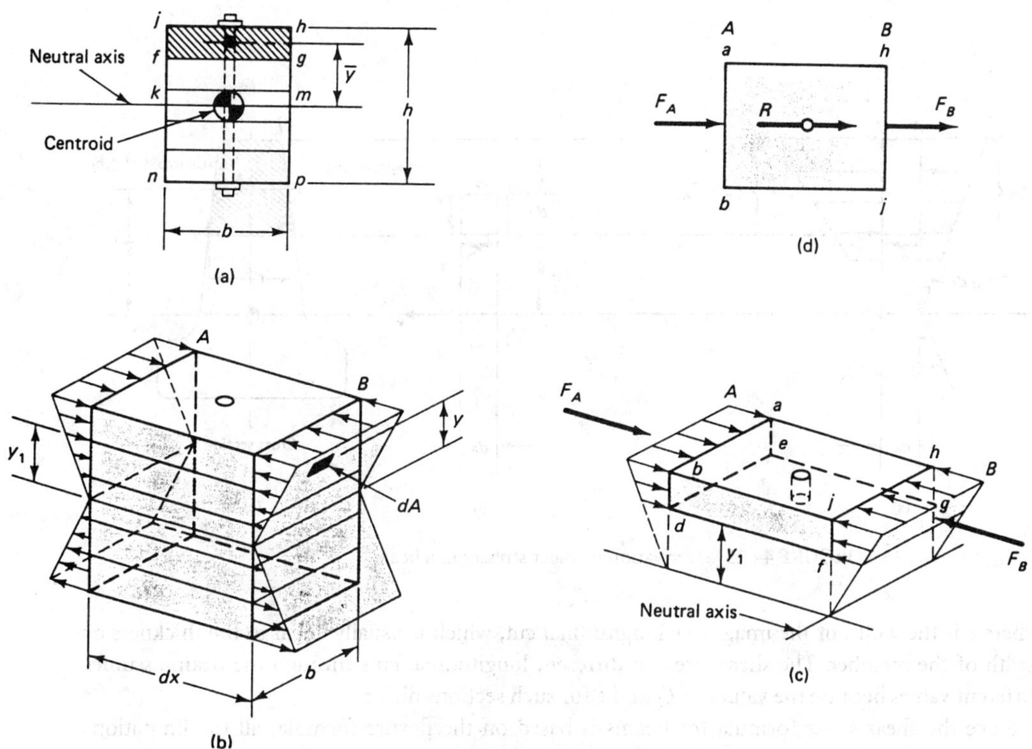

FIGURE 44.24 Elements for deriving shear flow in a beam.

only by developing a horizontal resisting force in the bolt R, as in Fig. 44.24(d). Taking a differential beam element of length dx, $M_B = M_A + dM$ and $dF = |F_B| - |F_A|$, and substituting these relations into the expression for F_A and F_B found above, one obtains $dF = dM\,Q/I$. It is more significant to obtain the force per unit length of beam length, dF/dx, which will be designated by q and will be referred to as the shear flow. Then, noting that $dM/dx = V$, one obtains the following expression for the shear flow in beams:

$$q = \frac{VQ}{I} \tag{44.56}$$

In this equation, I is the moment of inertia of the entire cross-sectional area around the neutral axis and Q extends only over the cross-sectional area of the beam to one side at which q is investigated.

Shear-Stress Formula for Beams

The shear-stress formula for beams may be obtained by modifying the shear flow formula. In a solid beam, the force resisting dF may be developed only in the plane of the longitudinal cut taken parallel to the axis of the beam, as shown in Fig. 44.25. Therefore, assuming that the shear stress τ is uniformly distributed across the section of width t, the shear stress in the longitudinal plane may be obtained by dividing dF by the area $t\,dx$. This yields the horizontal shear stress τ, which for an infinitesimal element is numerically equal to the shear stress acting on the vertical plane [see Fig. 44.25(b)]. Since $q = dF/dx$, one obtains

$$\tau = \frac{VQ}{It} \tag{44.57}$$

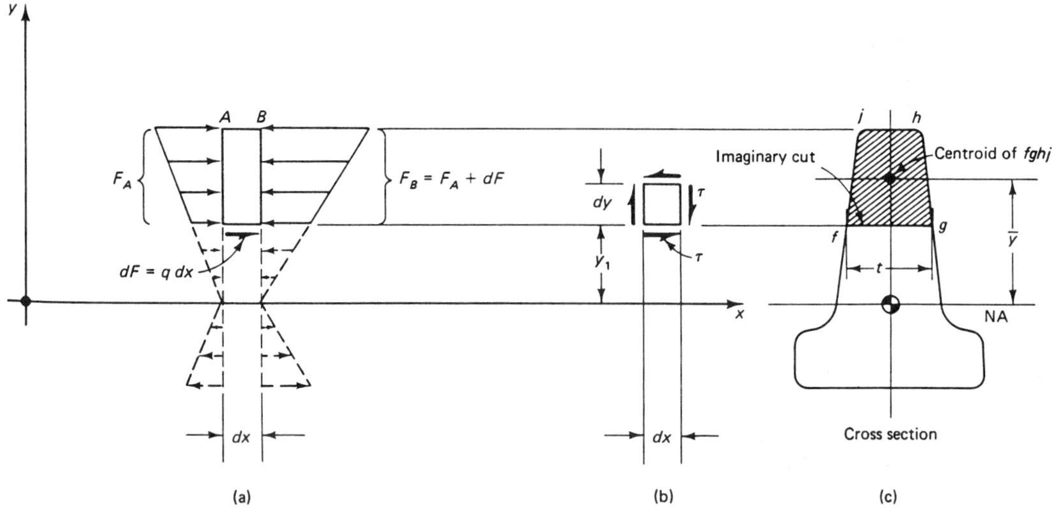

FIGURE 44.25 Derivation of shear stresses in a beam.

where t is the width of the imaginary longitudinal cut, which is usually equal to the thickness or width of the member. The shear stress at different longitudinal cuts through the beam assumes different values because the values of Q and t for such sections differ.

Since the shear-stress formula for beams is based on the flexure formula, all the limitations imposed on the flexure formula apply. The material is assumed to be elastic with the same elastic modulus in tension as in compression. The theory applies only to straight beams. Moreover, in certain cases, such as a wide flange section, the shear-stress formula may not satisfy the requirement of a stress-free boundary condition. However, no appreciable error is involved by using Eq. (44.57) for thin-walled members, and the majority of beams belong to this group.

Shear Center

The channel section shown in Fig. 44.26 does not have a vertical axis of symmetry. Thus with bending around the horizontal axis, there is a tendency for the channel to twist around some longitudinal axis. To prevent twisting and thus maintain the applicability of the flexure formula, the externally applied force P shown in Fig. 44.26(c) must be applied in such a manner as to balance the internal couple $F_1 h$. This location is called the shear center or center of twist and is designated by the letter S. The shear center for any cross section lies on a longitudinal line parallel

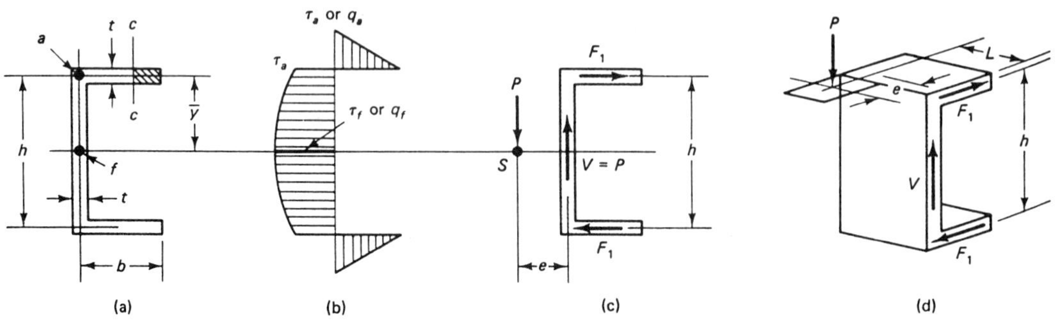

FIGURE 44.26 Deriving location of shear center for a channel.

to the axis of the beam. Any transverse force applied through the shear center causes no torsion of the beam. For a channel section, the shear center location measured by e from the center of the web is equal to $b^2h^2t/4I$. For a symmetrical angle, the shear center is located at the intersection of the centerlines of its legs.

44.8 Transformation of Stress and Strain

Transformation of Stress

In stress analysis, a more general problem often arises, as illustrated in Fig. 44.27, in which element A is subjected to a normal stress σ_x due to axial pull and bending, and simultaneously experiences a direct shear stress τ_{xy}. The combined normal stress with the shear stress requires a consideration of stresses on an inclined plane, such as shown by Fig. 44.27(b). Since an inclined plane may be chosen arbitrarily, the state of stress at a point can be described in an infinite number of ways, all of which are equivalent. The planes on which the normal or the shear stresses reach their maximum intensity have a particularly significant effect on materials.

An infinitesimal element of unit thickness, as shown in Fig. 44.28(a), is used to describe the state of two-dimensional stress. To determine stresses on any inclined plane, the fundamental procedure involves isolating a wedge and using the equations of the equilibrium of forces. By multiplying the stresses shown in Fig. 44.28(b) by their respective areas, a diagram with the forces acting on the wedge is constructed, as in Fig. 44.28(c). Then, by applying the equations of static

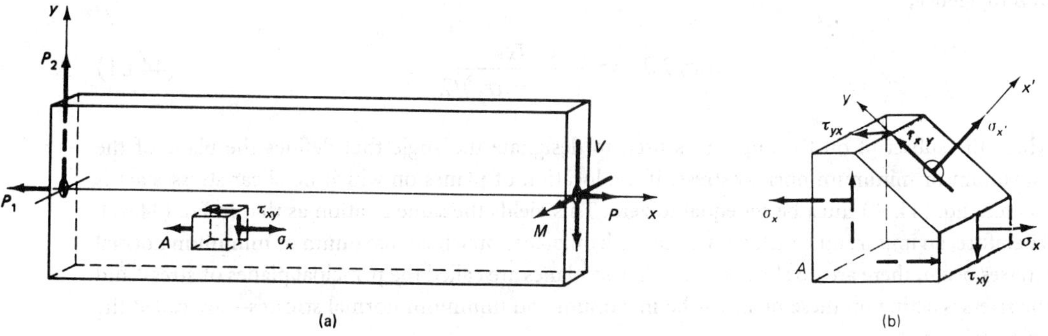

FIGURE 44.27 State of stress at a point on different planes.

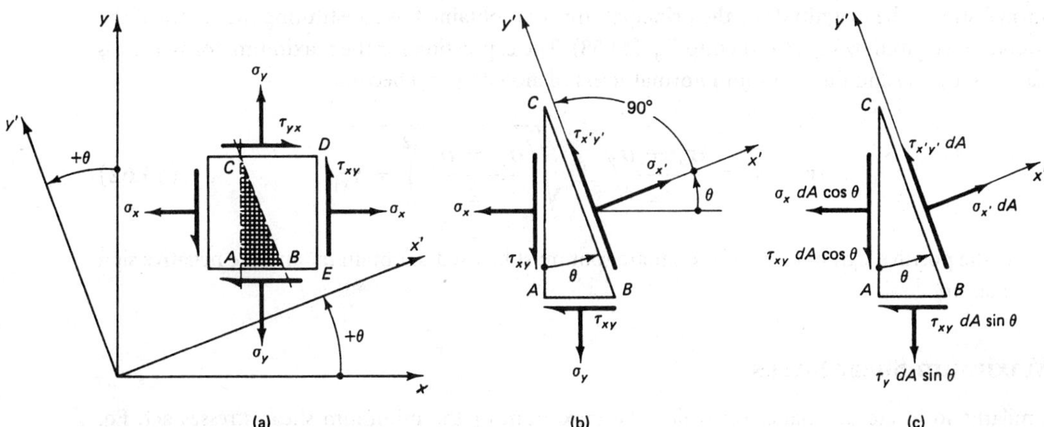

FIGURE 44.28 Derivation of stress transformation on an inclined plane.

equilibrium to the forces acting on the wedge, the stresses $\sigma_{x'}$, $\sigma_{y'}$, and $\tau_{x'y'}$ are obtained:

$$\sigma_{x'} = \frac{\sigma_x + \sigma_y}{2} + \frac{\sigma_x - \sigma_y}{2} \cos 2\theta + \tau_{xy} \sin 2\theta \tag{44.58}$$

$$\sigma_{y'} = \frac{\sigma_x + \sigma_y}{2} - \frac{\sigma_x - \sigma_y}{2} \cos 2\theta - \tau_{xy} \sin 2\theta \tag{44.59}$$

$$\tau_{x'y'} = -\frac{\sigma_x - \sigma_y}{2} \sin 2\theta + \tau_{xy} \cos 2\theta \tag{44.60}$$

These equations are the equations of transformation of stress from one set of coordinates to another. The sign convention assumed for positive stress and positive angle θ are depicted in Fig. 44.28(a).

By adding Eqs. (44.58) and (44.59), $\sigma_{x'} + \sigma_{y'} = \sigma_x + \sigma_y$, meaning that the sum of the normal stresses on any two mutually perpendicular planes remain the same, that is, invariant, regardless of the angle θ.

Principal Stresses

Interest often centers on the determination of the largest possible stress, as given by Eqs. (44.58)–(44.60), and the planes on which such stresses occur. To find the plane for a maximum or a minimum normal stress, Eq. (44.58) is differential with respect to θ and the derivative is set equal to zero. Hence,

$$\tan 2\theta_1 = \frac{\tau_{xy}}{(\sigma_x - \sigma_y)/2} \tag{44.61}$$

where the subscript on the angle θ is used to designate the angle that defines the plane of the maximum or minimum normal stress. If the location of planes on which no shear stresses act is wanted, Eq. (44.60) must be set equal to zero. This yields the same relation as that in Eq. (44.61). Therefore, an important conclusion is reached: on planes on which maximum or minimum normal stresses occur, there are no shear stresses. These planes are called the principal planes of stress, and the stresses acting on these planes—the maximum and minimum normal stresses—are called the principal stresses.

Equation (44.61) has two roots that are 90° apart. One of these roots locates a plane on which the maximum normal stress acts; the other plane locates the corresponding plane for the minimum normal stress. The magnitude of the principal stresses is obtained by substituting the values of the double angle given by Eq. (44.61) into Eq. (44.58). The expression for the maximum normal stress (denoted by σ_1) and the minimum normal stress (denoted by σ_2) becomes

$$\sigma_{1\text{ or }2} = \frac{\sigma_x + \sigma_y}{2} \pm \sqrt{\left(\frac{\sigma_x - \sigma_y}{2}\right)^2 + \tau_{xy}^2} \tag{44.62}$$

where the positive sign in front of the square root must be used to obtain σ_1 and the negative sign to obtain σ_2.

Maximum Shear Stress

Similarly, to locate the planes on which the maximum or the minimum shear stresses act, Eq. (44.60) must be differentiated with respect to θ and the derivative set equal to zero. Hence,

$$\tan 2\theta_2 = -\frac{(\sigma_x - \sigma_y)/2}{\tau_{xy}} \tag{44.63}$$

where the subscript 2 designates the plane on which the shear stress is a maximum or minimum. Again, the two planes defined by this equation are mutually perpendicular. Moreover, Eq. (44.63) is a negative reciprocal of Eq. (44.61). This means that the angles that locate the planes of maximum or minimum shear stress form angles of 45° with the planes of principal stresses. A substitution of the results of Eq. (44.63) into Eq. (44.60) gives the maximum and the minimum values of the shear stresses.

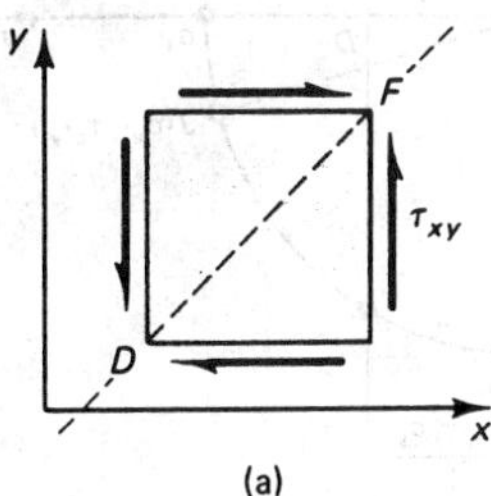

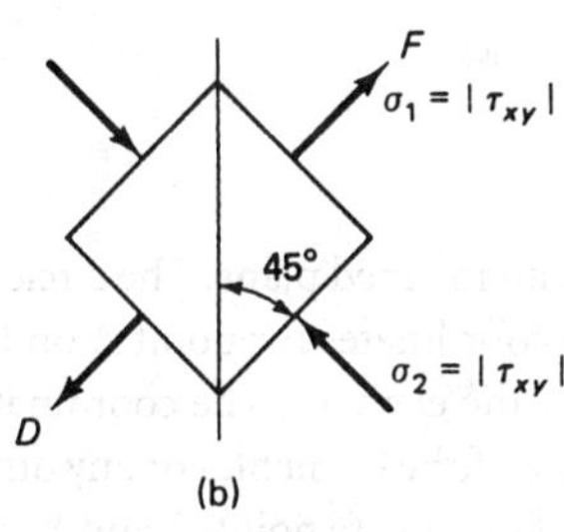

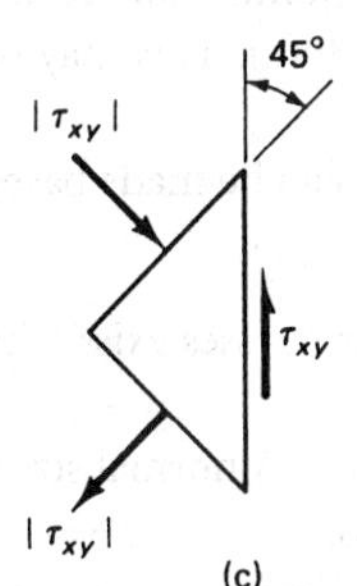

FIGURE 44.29 Equivalent representations for pure shear stress.

$$\tau_{\text{max or min}} = \pm\sqrt{\left(\frac{\sigma_x - \sigma_y}{2}\right)^2 + \tau_{xy}^2} \tag{44.64}$$

Thus, the maximum shear stress differs from the minimum shear stress only in sign. From the physical point of view, these signs have no meaning, and for this reason, the largest shear stress regardless of sign will often be called the maximum shear stress. By substituting θ_2 into Eq. (44.58), the normal stresses σ' that act on the planes of the maximum shear stresses are

$$\sigma' = \frac{\sigma_x + \sigma_y}{2} \tag{44.65}$$

If σ_x and σ_y in Eq. (44.65) are the principal stress, τ_{xy} is zero and Eq. (44.64) simplifies to

$$\tau_{\text{max}} = \frac{\sigma_1 - \sigma_2}{2} \tag{44.66}$$

For pure shear, with the absence of normal stresses, the principal stresses are numerically equal to the shear stress, as displayed by Fig. 44.29.

Mohr's Circle of Stress

The equations of stress transformation given by Eqs. (44.58), (44.59), and (44.60) may be presented graphically. They can be shown to represent a circle written in parametric form,

$$(\sigma_{x'} - a)^2 + \tau_{x'y'}^2 = b^2 \tag{44.67}$$

where the center of the circle C is at $(a, 0)$, and the circle radius is equal to b:

$$a = \frac{\sigma_x + \sigma_y}{2} \tag{44.68}$$

$$b = \sqrt{\left(\frac{\sigma_x - \sigma_y}{2}\right)^2 + \tau_{xy}^2} \tag{44.69}$$

In a given problem, σ_x, σ_y, and τ_{xy} are the three known stresses of the element. Hence, if a circle satisfying Eq. (44.67) is plotted, as shown in Fig. 44.30, the simultaneous values of a point (x, y) on

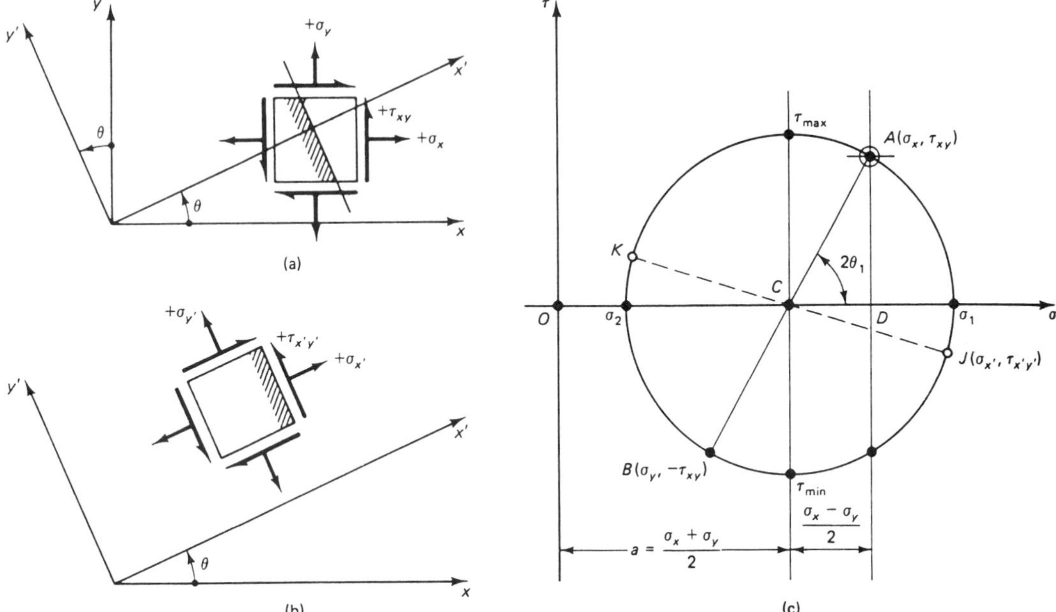

FIGURE 44.30 Mohr's circle of stress.

this circle correspond to $\sigma_{x'}$ and $\tau_{x'y'}$ for a particular orientation of an inclined plane. The circle so constructed is called the *circle of stress* or *Mohr's circle of stress.* The coordinates for point A on the circle correspond to the stresses in Fig. 44.30(a) on the right face of the element. The coordinates for the conjugate point B correspond to the stresses on the upper face of the element. For any other orientation θ of an element, such as shown in Fig. 44.30(b), a pair of conjugate points J and K can always be found on the circle to give the corresponding stresses, as in Fig. 44.30(c). This may be easily accomplished by rotating the AB axis by a corresponding 2θ.

The following important observations regarding the state of stress at a point can be made based on the Mohr's circle.

1. The largest possible normal stress is σ_1 and the smallest is σ_2. No shear stresses exist with these principal stresses.
2. The largest shear stress τ_{max} is numerically equal to the radius of the circle. A normal stress equal to $(\sigma_1 + \sigma_2)/2$ acts on each of the planes of maximum shear stress.
3. If $\sigma_1 = \sigma_2$, Mohr's circle degenerates into a point, and no shear stresses develop in the xy plane.
4. If $\sigma_x + \sigma_y = 0$, the center of Mohr's circle coincides with the origin of the $\sigma - \tau$ coordinates, and the state of pure shear exists.
5. The sum of the normal stresses on any two mutually perpendicular planes is invariant, that is,

$$\sigma_x + \sigma_y = \sigma_1 + \sigma_2 = \sigma_{x'} + \sigma_{y'} = \text{constant}$$

Principal Stresses for a General State of Stress

Consider a general state of stress and define an infinitesimal tetrahedron, as shown in Fig. 44.31. The unknown stresses are sought on an arbitrary oblique plane ABC in the three-dimensional xyz coordinate system. A set of known stresses on the other three faces of the mutually perpendicular planes of the tetrahedron is given. These stresses are the same as those shown in Fig. 44.2. A unit

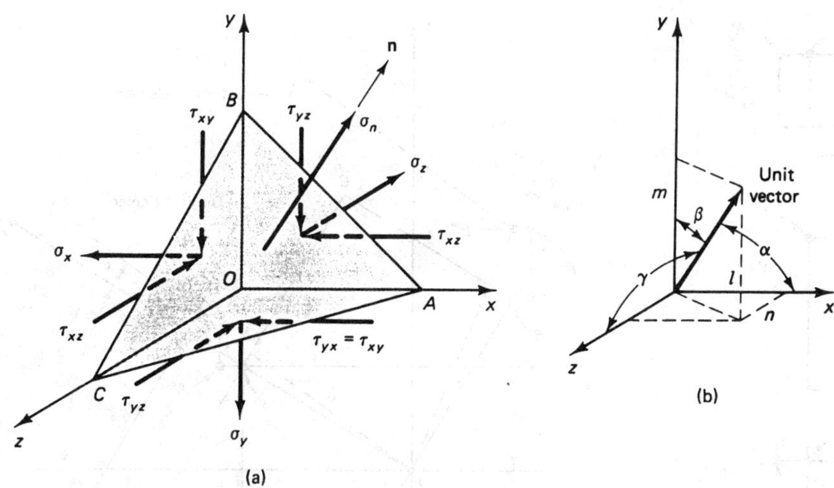

FIGURE 44.31 Tetrahedron for deriving a principal stress on an oblique plane.

normal **n** to the oblique plane defines its orientation. This unit vector is identified by its direction cosines l, m, and n as illustrated by Fig. 44.31(b). Further, if the infinitesimal area ABC is defined as dA, then the three areas of the tetrahedron along the coordinate axes are $dA\ l$, $dA\ m$, and $dA\ n$. Force equilibrium for the tetrahedron can now be written by multiplying the stresses given in Fig. 44.31 by the respective areas established. It will be assumed that only a normal stress σ_n (i.e., a principal stress) is acting on face ABC; a system of linear homogeneous equations is then obtained that has nontrivial solution if and only if the determinant of the coefficients of l, m, and n vanishes. Expansion of this determinant gives

$$\sigma_n^3 - I_\sigma \sigma_n^2 + II_\sigma \sigma_n - III_\sigma = 0 \tag{44.70}$$

where

$$\begin{aligned} I_\sigma &= \sigma_x + \sigma_y + \sigma_z \\ II_\sigma &= (\sigma_x\sigma_y + \sigma_y\sigma_z + \sigma_z\sigma_x) - (\tau_{xy}^2 + \tau_{yz}^2 + \tau_{zx}^2) \\ III_\sigma &= \sigma_x\sigma_y\sigma_z + 2\tau_{xy}\tau_{yz}\tau_{xz} - (\sigma_x\tau_{yz}^2 + \sigma_y\tau_{xz}^2 + \sigma_z\tau_{xy}^2) \end{aligned}$$

The constants I_σ, II_σ, and III_σ are invariant since if the initial coordinate system is changed, thereby changing the three mutually perpendicular planes of the tetrahedron, the σ_n on the inclined plane must remain the same. In general, Eq. (44.70) has three real roots. These roots are the eigenvalues of the determinant and are the principal stresses of the problem.

Transformation of Strain

The transformation of normal and shear strain from one set of rotated axes to another (Fig. 44.32) is completely analogous to the transformation of normal and shear stresses presented earlier. Fundamentally, this is because both stresses and strains are second-rank tensors and mathematically obey the same laws of transformation. One may then obtain equations of strain transformation from the equations of stress transformation by simply substituting the normal stress σ with the normal strain ε, and the shear stress τ with shear strain γ. Hence, the basic expressions for strain transformation in a plane in an arbitrary direction defined by the x' axis are

$$\varepsilon_{x'} = \frac{\varepsilon_x + \varepsilon_y}{2} + \frac{\varepsilon_x - \varepsilon_y}{2}\cos 2\theta + \frac{\gamma_{xy}}{2}\sin 2\theta \tag{44.71}$$

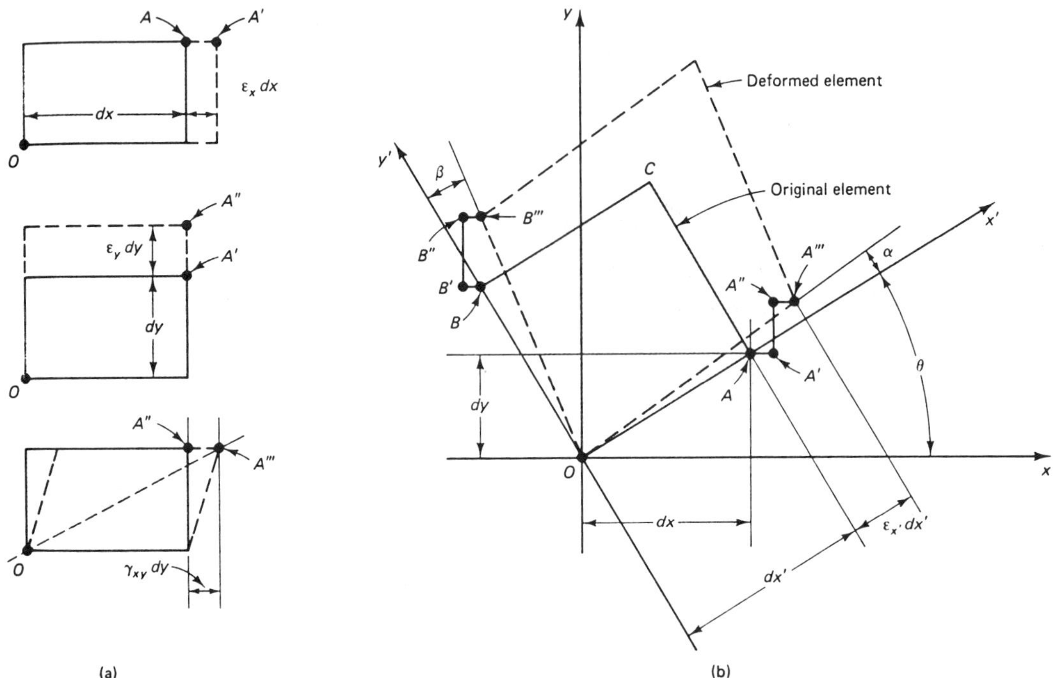

FIGURE 44.32 Exaggerated deformations of elements for deriving strains along new axes.

$$\gamma_{x'y'} = -(\varepsilon_x - \varepsilon_y)\sin 2\theta + \gamma_{xy}\cos 2\theta \tag{44.72}$$

The sign convention adopted corresponds to the element distortions shown in Fig. 44.32(a) for positive strain. Likewise, Mohr's circle of strain can be constructed, where every point on the circle gives two values: one for the normal strain, the other for the shear strain divided by 2.

Yield and Fracture Criteria

Thus far the idealized mathematical procedures for determining the states of stress and strain, as well as their transformation to different coordinates, have been presented. However, it must be pointed out that the precise response of real materials to such stresses and strains defies accurate formulations. As yet no comprehensive theory can provide accurate predictions of material behavior under the multitude of static, dynamic, impact, and cyclic loading, as well as temperature effects. Only the classical idealizations of yield and fracture criteria for materials are discussed in this section.

The maximum shear-stress theory, or simply the maximum shear theory, results from the observation that in a ductile material slip occurs during yielding along critically oriented planes. This suggests that the maximum shear stress plays the key role, and it is assumed that yielding of the material depends only on the maximum shear stress attained within an element. Therefore, whenever a certain critical value τ_{cr} is reached, yielding in an element commences. For a given material, this value usually is set equal to the shear stress at yield σ_{yp} in simple tension or compression. Hence, according to Eq. (44.64), $\tau_{max} \equiv \tau_{cr} = \sigma_{yp}/2$.

In applying this criterion to a biaxial plane stress problem, two different cases arise. In one case, if the signs of the principal stresses σ_1 and σ_2 are the same, the maximum shear stress is of the same magnitude as would occur in a simple uniaxial stress. Therefore, the criteria corresponding to this case are

$$|\sigma_1| \leq \sigma_{yp} \quad \text{and} \quad |\sigma_2| \leq \sigma_{yp} \tag{44.73}$$

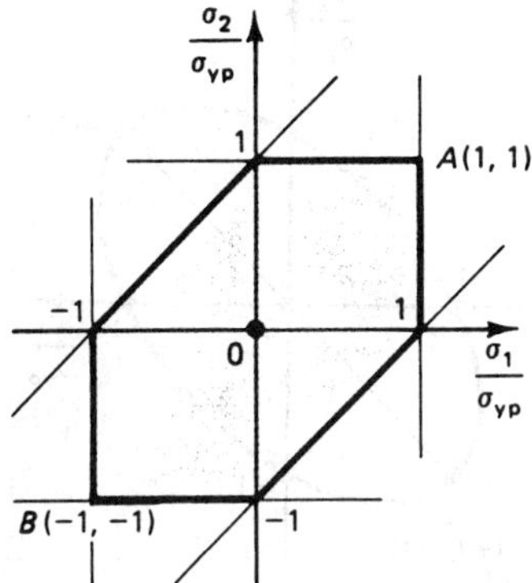

FIGURE 44.33 Yield criterion based on maximum shear stress.

In the second case, the signs of σ_1 and σ_2 are opposite, and the maximum shear stress $\tau_{max} = (|\sigma_1| + |\sigma_2|)/2$. Hence,

$$\left| \pm \frac{\sigma_1 - \sigma_2}{2} \right| \leq \frac{\sigma_{yp}}{2} \tag{44.74}$$

A plot of Eqs. (44.73) and (44.74) is shown in Fig. 44.33. If a point defined by σ_1/σ_{yp} and σ_2/σ_{yp} falls on the hexagon shown, a material begins and continues to yield. No such stress points can lie outside the hexagon because one of the three yield criteria equations would be violated. The stress points falling within the hexagon indicate that a material behaves elastically. The derived yield criterion for perfectly plastic material is often referred to as the *Tresca yield condition* and is one of the widely used laws of plasticity.

Another widely accepted criterion of yielding for ductile isotropic materials is based on energy concepts. In this approach, the total elastic energy is divided into two parts: one associated with the volumetric changes of the material, and the other causing shear distortions. By equating the shear distortion energy at yield point in simple tension to that under combined stress, the yield criterion for combined stress is established.

Based on the concept of superposition, it is possible to consider the stress tensor of the three principal stresses—σ_1, σ_2, and σ_3—to consist of two additive component tensors:

$$\begin{bmatrix} \sigma_1 & 0 & 0 \\ 0 & \sigma_2 & 0 \\ 0 & 0 & \sigma_3 \end{bmatrix} = \begin{bmatrix} \bar{\sigma} & 0 & 0 \\ 0 & \bar{\sigma} & 0 \\ 0 & 0 & \bar{\sigma} \end{bmatrix} + \begin{bmatrix} \sigma_1 - \bar{\sigma} & 0 & 0 \\ 0 & \sigma_2 - \bar{\sigma} & 0 \\ 0 & 0 & \sigma_3 - \bar{\sigma} \end{bmatrix} \tag{44.75}$$

where $\bar{\sigma} = (\sigma_1 + \sigma_2 + \sigma_3)/3$ is the hydrostatic stress. The stresses associated with the first tensor component are the same in every possible direction and this tensor is called the spherical stress tensor, or dilatational stress tensor. The second tensor component causes no volumetric changes but distorts or deviates the element from its initial cubic shape. It is called the deviatoric, or distortional stress tensor.

Extending the results of elastic strain energy for uniaxial stress, generalizing for three dimensions, and expressing in terms of the principal stresses, the total strain energy per unit volume (i.e., strain density) can be found:

$$U_{\text{total}} = \frac{1}{2E}(\sigma_1^2 + \sigma_2^2 + \sigma_3^2) - \frac{\nu}{E}(\sigma_1\sigma_2 + \sigma_2\sigma_3 + \sigma_3\sigma_1) \tag{44.76}$$

The strain energy per unit volume due to the dilatational stress can be determined from this equation and expressed in terms of the principal stress. Thus,

$$U_{\text{dilatation}} = \frac{1 - 2\nu}{6E}(\sigma_1 + \sigma_2 + \sigma_3)^2 \tag{44.77}$$

By subtracting Eq. (44.77) from Eq. (44.76), simplifying, and noting that $G = E/2(1 + \nu)$, one finds the distortion strain energy for combined stress:

$$U_{\text{distortion}} = \frac{1}{12G}[(\sigma_1 - \sigma_2)^2 + (\sigma_2 - \sigma_3)^2 + (\sigma_3 - \sigma_1)^2] \tag{44.78}$$

According to the basic assumption of the distortion-energy theory, the expression of Eq. (44.78) must be equal to the maximum elastic distortion energy in simple tension, which is equal to $2\sigma_{yp}^2/12G$. After minor simplifications, one obtains the basic law for yielding of an ideally plastic material:

$$(\sigma_1 - \sigma_2)^2 + (\sigma_2 - \sigma_3)^2 + (\sigma_3 - \sigma_1)^2 = 2\sigma_{yp}^2 \tag{44.79}$$

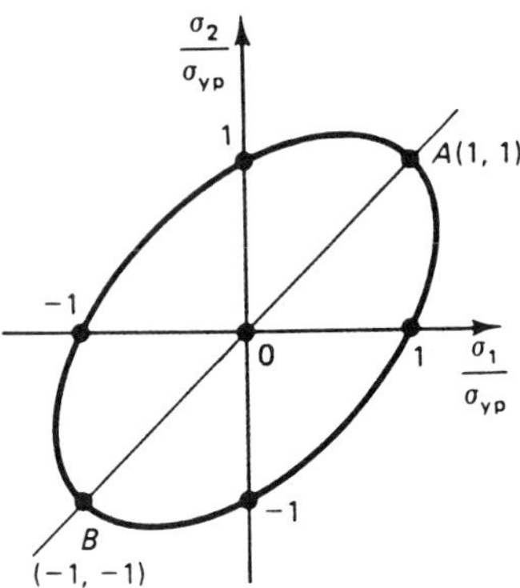

FIGURE 44.34 Yield criterion based on maximum distortion energy.

For plane stress, $\sigma_3 = 0$, Eq. (44.79) is an equation of an ellipse, a plot of which is shown in Fig. 44.34. Any stress falling within the ellipse indicates that the material behaves elastically. Points on the ellipse indicate that the material is yielding.

In three-dimensional stress space, the yield surface becomes a cylinder with an axis having all three direction cosines equal to $1/\sqrt{3}$. Such a cylinder is shown in Fig. 44.35. The ellipse in Fig. 44.34 is simply the intersection of this cylinder with the $\sigma_1 - \sigma_2$ plane. It can be shown that the yield surface for the maximum shear stress criterion is a hexagon that fits into a tube, as in Fig. 44.35(b). The fundamental relation given by Eq. (44.79) is widely used for perfectly plastic materials and is often referred to as Huber-Hencky-Mises yield condition, or simply the von Mises yield condition.

The maximum normal stress theory, or maximum stress theory, asserts that failure or fracture of a material occurs when the maximum normal stress at a point reaches a critical value regardless of the other stresses. Only the largest principal stress must be determined to apply this criterion. The

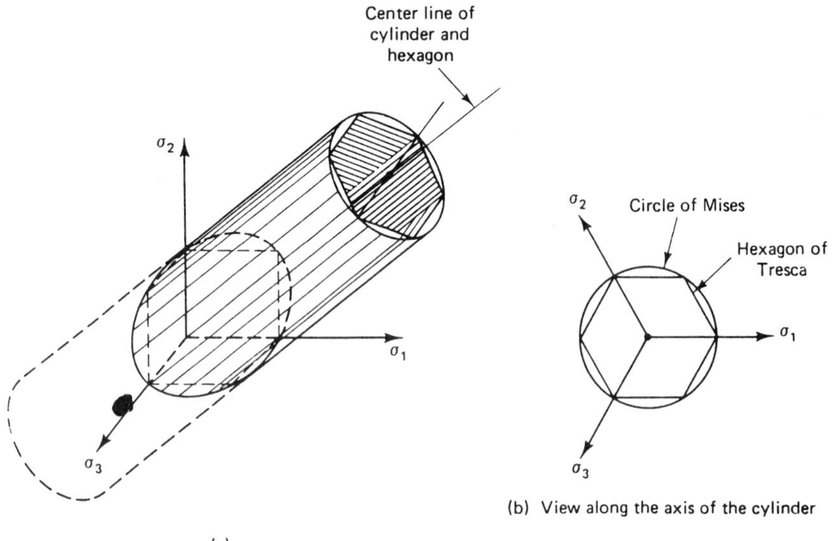

FIGURE 44.35 Yield surfaces for triaxial state of stress.

critical value of stress σ_{ult} is usually determined in a tensile experiment. Experimental evidence indicates that this theory applies well to brittle materials in all range of stresses, providing a tensile principal stress exists. Failure is characterized by the separation, or cleavage, fracture. The maximum stress theory can be interpreted on graphs, as shown in Fig. 44.36. Unlike the previous theories, the stress criterion gives a bounded surface of the stress space.

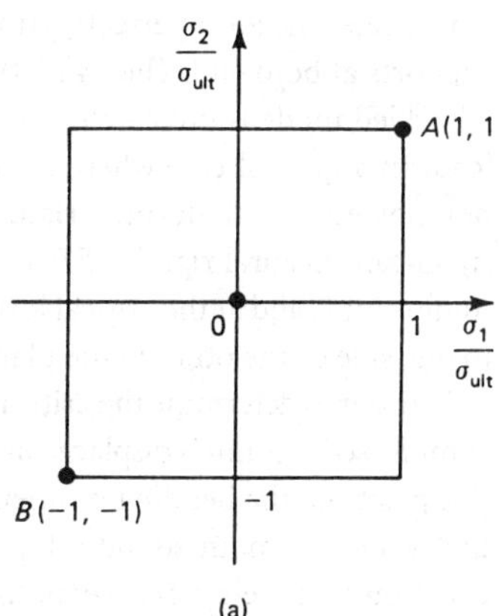

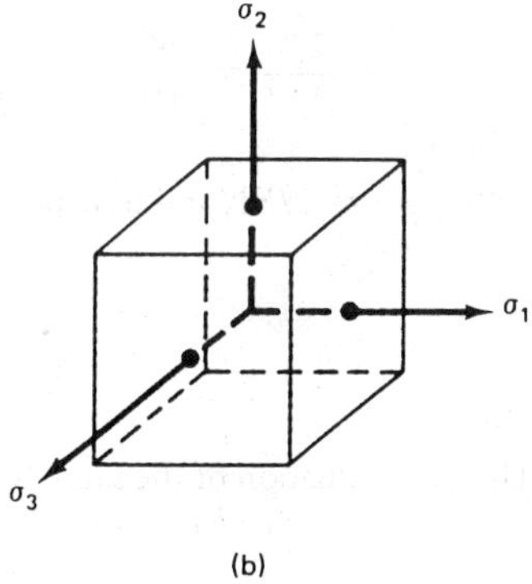

FIGURE 44.36 Fracture envelope based on maximum stress criterion.

44.9 Stability of Equilibrium: Columns

Governing Differential Equation for Deflection

Figure 44.37 shows a deflected beam segment with point A' directly above its initial position A by a displacement v_A. The tangent to the elastic curve at the same point and a plane section with the centroid at A' rotate through an angle dv/dx. Therefore, assuming small angles, the displacement u of a material point at a distance y from the elastic curve is

$$u = -y\frac{dv}{dx} \tag{44.80}$$

Next, recall Eq. (44.22), which states that $\varepsilon_x = du/dx$. Therefore, from Eq. (44.80), $\varepsilon_x = -yd^2v/dx^2$. The same normal strain also can be found from Eqs. (44.26) and (44.43) yielding $\varepsilon_x = -My/EI$. On equating the two alternative expressions for ε_x, one obtains the governing differential equation for small deflections of elastic beams as

$$\frac{d^2v}{dx^2} = \frac{M}{EI} \tag{44.81}$$

Buckling Theory for Columns

The procedures of stress and deformation analysis in a state of stable equilibrium were discussed in the preceding sections. But not all structural systems are necessarily stable. The phenomenon of structural instability occurs in numerous situations where compressive stresses are present. The consideration of material strength alone is not sufficient to predict the behavior of such members. Stability considerations are primary in some structural systems.

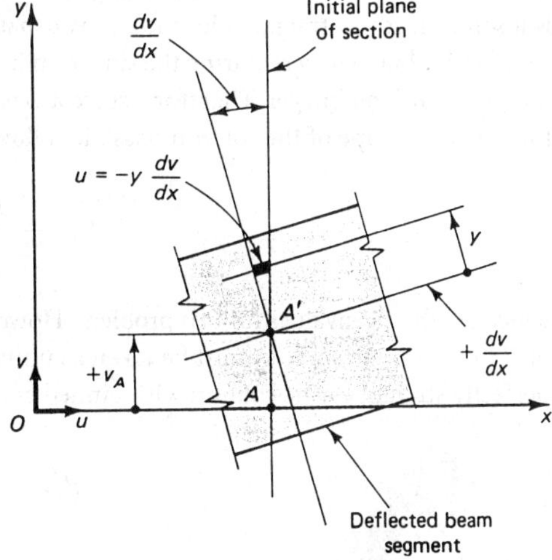

FIGURE 44.37 Longitudinal displacement in a beam due to rotation of a plane section.

Consider the ideal perfectly straight column with pinned supports at both ends (Fig. 44.38). The least force at which a buckled mode is possible is the critical or Euler buckling load. In a general case where a compression member does not possess equal flexural rigidity in all directions, the significant flexural rigidity EI of a column depends on the minimum I, and at the critical load a column buckles either to one side or the other in the plane of the major axis.

In order to determine the critical load for this column, the compressed column is displaced as shown in Fig. 44.38(b). In this position, the bending moment is $-Pv$. By substituting this value of moment into Eq. (44.81), the differential equation for the elastic curve for the initially straight column becomes

$$\frac{d^2v}{dx^2} = \frac{M}{EI} = -\frac{P}{EI}v \tag{44.82}$$

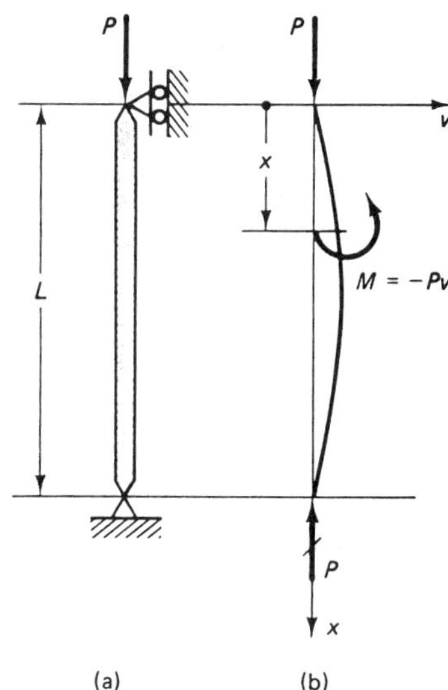

FIGURE 44.38 Column pinned at both ends.

Letting $\lambda^2 = P/EI$, and transposing,

$$\frac{d^2v}{dx^2} + \lambda^2 v = 0 \tag{44.83}$$

This is an equation of the same form as the one for simple harmonic motion, and its solution is

$$v = A\sin\lambda x + B\cos\lambda x \tag{44.84}$$

where A and B are arbitrary constants that must be determined from the boundary conditions. These conditions are $v(0) = 0$ and $v(L) = 0$. Hence, $B = 0$ and

$$A\sin\lambda L = 0 \tag{44.85}$$

This equation can be satisfied by taking $A = 0$. However, with A and B each equal to zero, this is a solution for a straight column, and is usually referred to as a trivial solution. An alternative solution is obtained by requiring the sine term in Eq. (44.85) to vanish. This occurs when λL equals $n\pi$, where n is an integer. Therefore, since λ was defined as $\sqrt{P/EI}$, the nth critical P_n that makes the deflected shape of the column possible follows from solving $\sqrt{P/EI}L = n\pi$. Hence,

$$P_n = \frac{n^2\pi^2 EI}{L^2} \tag{44.86}$$

which are the eigenvalues for this problem. However, since in stability problems only the least value of P_n is of importance, n must be taken as unity, and the critical or Euler load P_{cr} for an initially perfectly straight elastic column with pinned ends becomes

$$P_{cr} = \frac{\pi^2 EI}{L^2} \tag{44.87}$$

where E is the elastic modulus of the material, I is the least moment of inertia of the constant cross-sectional area of a column, and L is its length. This case of a column pinned at both ends is often referred to as the fundamental case.

According to Eq. (44.84), at the critical load, since $B = 0$, the equation of the buckled elastic curve is

$$v = A \sin \lambda x \tag{44.88}$$

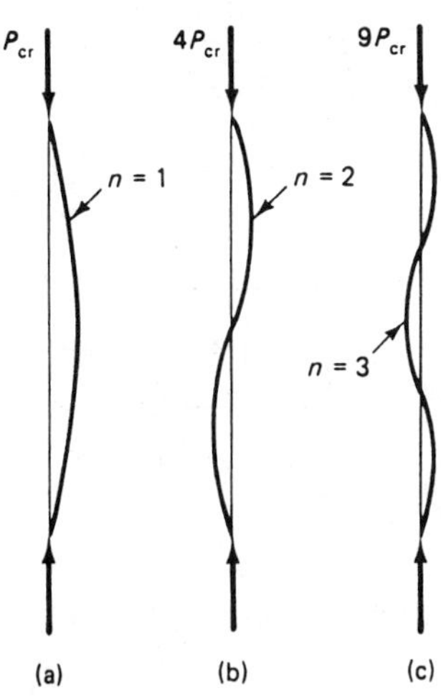

FIGURE 44.39 First three buckling modes for a column pinned at both ends.

This is the characteristic, or eigenfunction, of this problem, and, since $\lambda = n\pi/L$, n can assume any integer value. There is an infinite number of such functions. In this linearized solution, amplitude A of the buckling mode remains indeterminate. For the fundamental case $n = 1$, the elastic curve is a half-wave sine curve. This shape and the modes corresponding to $n = 2$ and 3 are shown in Fig. 44.39. The higher modes have no physical significance in buckling problems, since the least critical buckling load occurs at $n = 1$.

It should be noted that the elastic modulus E was used for the derivation of the Euler formulas for columns; therefore, the formulas are applicable while the material behavior remains linearly elastic. To bring out this significant limitation, Eq. (44.87) is rewritten in a different form. By definition, $I = Ar^2$, where A is the cross-sectional area and r is the radius of gyration. Substitution of this relation into Eq. (44.87) gives

$$\sigma_{cr} = \frac{P_{cr}}{A} = \frac{\pi^2 E}{(L/r)^2} \tag{44.89}$$

where the critical stress σ_{cr} for a column is the average stress over the cross-sectional area A of a column at the critical load P_{cr} The ratio L/r of the column length to the least radius of gyration is the column slenderness ratio. Note that σ_{cr} always decreases with increasing ratios of L/r. Since Eq. (44.89) is based on elastic behavior, σ_{cr} determined by this equation cannot exceed the proportional limit.

Euler Loads for Columns with Different End Restraints

The same procedure as that discussed before can be used to determine the critical axial loads for columns with different boundary conditions. The solutions of buckling problems are very sensitive to the end restraints. Some of these different cases of end restraint combinations are shown in Fig. 44.40. It can be shown that the buckling formulas for all these cases can be made to resemble the fundamental case, Eq. (44.87), provided the effective column lengths are used instead of the actual column length. The effective column length L_e for the fundamental case is L, but for a free-standing column it is $2L$. For a column fixed at both ends, the effective length is $L/2$, and for a column fixed at one end and pinned at the other it is $0.7L$. For a general case, $L_e = KL$, where K is the effective length factor, which depends on the end restraints.

Generalized Euler Buckling-Load Formulas

Beyond the proportional limit, it may be said that a column of different material has been created, since the stiffness of the material is no longer represented by the elastic modulus. At this point the material stiffness is given instantaneously by the tangent to the stress-strain curve, that is, by the tangent modulus E_t; see the section on constitutive relations. The column remains stable if its new flexural rigidity E_tI is sufficiently large, and it can carry a higher load. Substitution of the tangent modulus E_t for the elastic modulus E is the only modification necessary to make the elastic

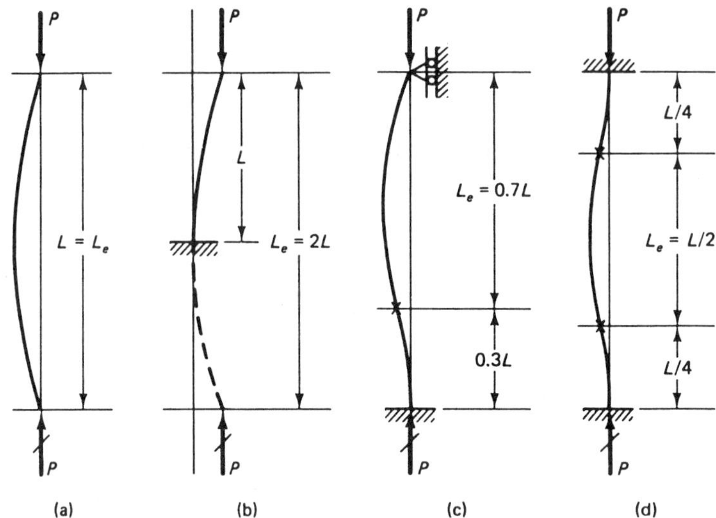

FIGURE 44.40 Effective lengths of columns with different restraints.

buckling formulas applicable in the inelastic range. Hence, the generalized Euler buckling-load formula, or the tangent modulus formula, becomes

$$\sigma_{cr} = \frac{\pi^2 E_t}{(L/r)^2} \tag{44.90}$$

Eccentric Loads and the Secant Formula

Since no column is perfectly straight and the applied forces are not perfectly concentric, the behavior of real columns may be studied with some statistically determined imperfections or possible misalignments of the applied loads. To analyze the behavior of an eccentrically loaded column, consider the column shown in Fig. 44.41. The differential equation for the elastic curve is the same as for a concentrically loaded column, derived in the section on buckling theory for columns. However, the boundary conditions are now different. At the upper end, v is equal to the eccentricity e of the applied load. And given that the elastic curve has a vertical tangent at the midheight of the column, the equation for the elastic curve can be obtained:

$$v = e\left(\frac{\sin \lambda L/2}{\cos \lambda L/2} \sin \lambda x + \cos \lambda x\right) \tag{44.91}$$

No indeterminacy of any constants appears in this equation, and the maximum deflection v_{max} can be found from it. This maximum occurs at $L/2$. Hence, $v_{max} = e \sec(\lambda L/2)$.

For the column shown in Fig. 44.41, the largest bending moment M is developed at the point of maximum deflection and numerically is equal to Pv_{max}. Therefore, the maximum compressive stress occurring in the column can be computed by superposition of the axial and bending stresses, that is, $(P/A) + (Mc/I)$, to give

$$\sigma_{max} = \frac{P}{A}\left(1 + \frac{ec}{r^2} \sec \frac{L}{r}\sqrt{\frac{P}{4EA}}\right) \tag{44.92}$$

This equation, because of the secant term, is known as the secant formula for columns, and it applies to columns of any length, provided the maximum stress does not exceed the elastic limit.

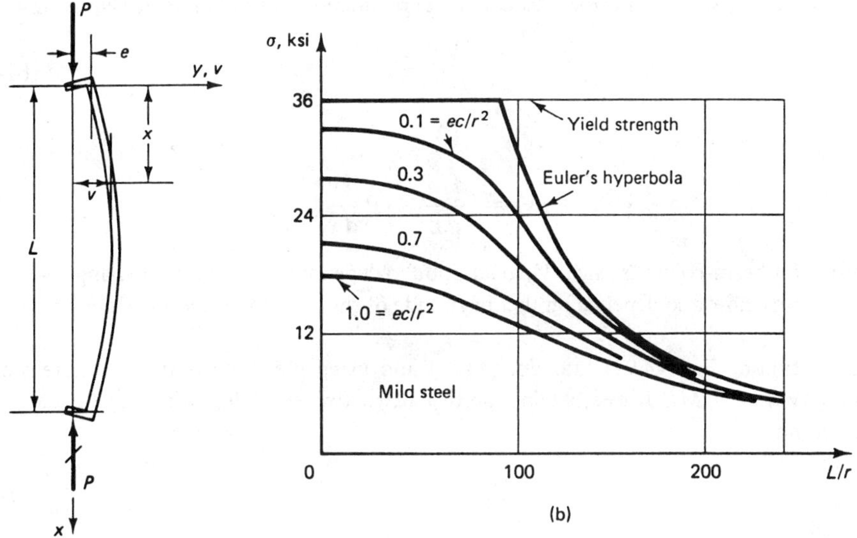

FIGURE 44.41 Results of analyses for different columns by the secant formula.

A condition of equal eccentricities of the applied forces in the same direction causes the largest deflection. Note that in Eq. (44.92), the relation between $\sigma_{\max}$ and P is not linear; $\sigma_{\max}$ increases faster than P. Therefore, the solutions for maximum stresses in columns caused by different axial forces cannot be superposed; instead, the forces must be superposed first, and then the stresses can be calculated. A plot of Eq. (44.92) is shown in Fig. 44.41. Note the large effect load eccentricity has on short columns and the negligible one on very slender columns.

Differential Equations for Beam-Columns

As was the case in the preceding section, a member acted on simultaneously by an axial force and transverse forces or moments causing bending is referred to as a beam-column. To obtain the governing differential equation consider the beam-column element shown in Fig. 44.42. Applying

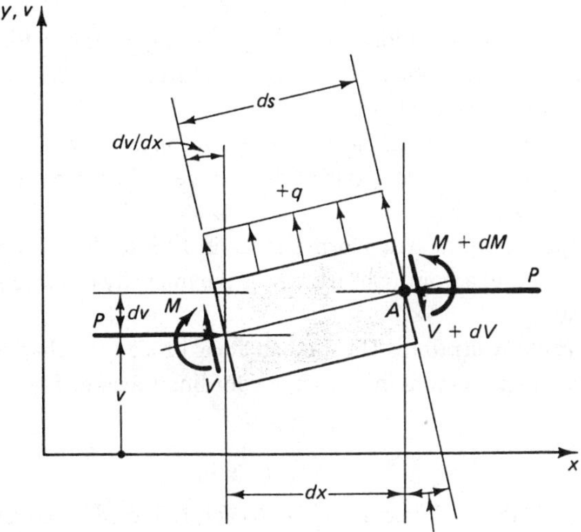

FIGURE 44.42 Beam-column element.

the equilibrium equations and small deflection approximations, one obtains two equations:

$$\frac{dV}{dx} = q \tag{44.93}$$

and

$$V = \frac{dM}{dx} + P\frac{dv}{dx} \tag{44.94}$$

Therefore, for beam-columns, shear V, in addition to depending on the rate of change of moment M as in beams, now also depends on the magnitude of the axial force and the slope of the elastic curve.

On substituting Eq. (44.94) into Eq. (44.93) and using the usual beam curvature-moment relation $d^2v/dx^2 = M/EI$, one obtains the two alternative governing differential equations for beam columns:

$$\frac{d^2M}{dx^2} + \lambda^2 M = q \tag{44.95}$$

or

$$\frac{d^4v}{dx^4} + \lambda^2\frac{d^2v}{dx^2} = \frac{q}{EI} \tag{44.96}$$

where, as before, $\lambda^2 = P/EI$.

Defining Terms

Constitutive relation: The relationship between stress and strain. A linear elastic material implies that stress is directly proportional to strain and Hooke's law is applicable. A nonlinear elastic material responds in a nonproportional manner, yet when unloaded returns back along the loading path to its initial stress-free state. For an inelastic or plastic material, when the elastic limit is exceeded, upon unloading a permanent deformation, or set, develops at no external load.

Mechanics of materials: A branch of applied mechanics that deals with the behavior of various load-carrying members. It involves analytical methods for determining strength, stiffness, and stability.

Stability: Stability refers to the ability of a load-carrying member to resist buckling under compressive loads. For an elastic material, the least force at which a buckled mode is possible is the critical or Euler buckling load.

Strain: Deformation per unit length. Normal or extensional strain is the change in length per unit of initial gage length. Shear strain is the change in initial right angle between any two imaginary planes in a body.

Stress: Intensity of force per unit area. Normal stress is the intensity of force perpendicular to or normal to the section at a point. Shear stress is the intensity of force parallel to the plane of the elementary area.

Transformation of stress and strain: The mathematical process for changing the components of the state of stress or strain given in one set of coordinate axes to any other set of rotated axes.

References

Nadai, A. 1950. *Theory of Flow and Fracture of Solids,* Vol. 1, 2nd ed., McGraw-Hill, New York.

Ramberg, W. and Osgood, W. R. 1943. *Description of Stress-Strain Curves by Three Parameters.* National Advisory Committee on Aeronautics, TN 902.

Roark, R. J. and Young, W. C. 1975. *Formulas for Stress and Strain,* 5th ed. McGraw-Hill, New York.
Timoshenko, S. and Goodier, J. N. 1970. *Theory of Elasticity,* 3rd ed., McGraw-Hill, New York.

For Further Information

For further information the reader may consult *Engineering Mechanics of Solids* by Egor P. Popov, which served as the main source of information for this chapter. Permission from Prentice-Hall to use the figures from that textbook is gratefully acknowledged.

For a more advanced treatment of the subject matter, the series of books written by S. P. Timoshenko and coauthors can be consulted: *Theory of Elasticity, Theory of Elastic Stability, Theory of Plates and Shells,* and *Strength of Materials.* Other good sources include *Theory of Flow and Fracture of Solids* by A. Nadai; *Introduction to the Mechanics of Continuous Medium* by L. E. Malven; and *Mathematical Theory of Elasticity* by A. E. H. Love. *Mechanics of Elastic Structures* by J. T. Oden and E. A. Ripperger includes a good presentation of the problem of torsion and warpage.

For further information regarding the mechanical characteristics of materials, there are these references: *Materials Science and Engineering* by W. D. Callister; *Introduction to Material Science for Engineers* by J. F. Shackelford; and *Material Science for Engineers* by L. H. Van Vlack.

On the finite element method, one may refer to *The Finite Element Method* by O. C. Zienkiewicz and R. L. Taylor and *Finite Element Fundamentals* by R. H. Gallagher.

Two early books on the topic of stability are *Buckling Strength of Metal Structures* by F. Bleich and *Principles of Structural Stability* by H. Ziegler.

The American Society of Civil Engineers prints the *Journal of Engineering Mechanics,* which covers activities and developments in the field of applied mechanics as they relate to civil engineering.

45

Structural Analysis

N. E. Shanmugam
National University of Singapore

J. Y. Richard Liew
National University of Singapore

0-8493-8953-4/95/$0.00 + $.50

45.1 Introduction to Structural Analysis

A *structure* is an assembly of **members** interconnected at joints. *Structural analysis* is the determination of forces and deformations of the structure due to applied loads. *Structural design* involves the arrangement and proportioning of structures and their parts such that the assembled structure is capable of supporting the loads to which it may be subjected within the design limit states. Full-scale tests are sometimes carried out on mass-produced structures in order to gather data to be used in assessing the analytical model, which is an idealization of the actual structure. The structural idealization should relate the actual behavior to material properties, structural details, loading, and boundary conditions as accurately as is practicable.

All structures that occur in practice are three-dimensional, but in many cases it is possible to idealize them as two-dimensional structures in orthogonal directions. *Joints* in a structure are those points at which two or more members of the structure are connected. A **truss** is a structural system consisting of members which are designed to resist only axial forces and the members are assumed to be connected at their ends. A structural system in which joints are capable of transferring end moments is called a *frame*. Members in this system are assumed to be capable of resisting bending moment, axial force, and shear force. A structure is said to be two-dimensional or planar if all the members lie in the same plane. **Beams** are those members that are subjected to bending or flexure. They are usually thought of as being in horizontal positions and loaded with gravitational or vertical loads. *Ties* are members that are subjected to axial tension only, and *struts* (columns or posts) are members subjected to axial compression only.

45.2 Boundary Conditions

A *hinge* [Fig. 45.1(a)] represents a pin connection to a structural assembly; it does not allow translational movements. It is assumed to be frictionless and to allow complete rotation of a member with respect to the others. A *roller* [Fig. 45.1(b)] represents a kind of support that permits the attached structural part to rotate freely with respect to a surface and to translate freely in the direction parallel to the surface. No translational movement in any other direction is allowed. A *fixed support* [Fig. 45.1(c)] does not allow rotation or translation in any direction. A *rotational spring* [Fig. 45.1(d)] represents a support that provides some rotational restraint but does not provide any translational restraint. A *translational spring* [Fig. 45.1(e)] provides partial restraints along the direction of deformation.

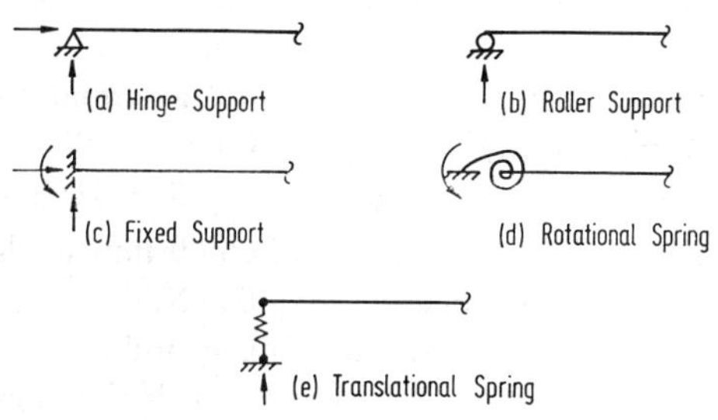

FIGURE 45.1 Various boundary conditions.

45.3 Loads

It is important to estimate the type and magnitude of the loads that may be applied to a structure during its life. The worst possible combinations of the estimated load should also be assessed accurately.

Loads may be broadly classified as *dead loads*, which are of constant magnitude and remain in one position, and *live loads*, which may change in position and magnitude. Dead loads may include a structure's own weight and other loads such as frame walls, floors, roof, plumbing, and fixtures that are permanently attached to the frame. Live loads may include those caused by construction operations, wind, rain, earthquakes, blasts, and temperature changes in addition to those which are movable, such as furniture, warehouse materials, and snow. *Snow loads* are due to the accumulation of snow on roof surfaces. Historical data for snow loads in various areas are available in several specifications. *Ponding* results when water on a flat roof accumulates faster than it runs off. *Wind loads* act as pressures on vertical windward surfaces, pressures, or suction on sloping windward surfaces. ASCE Task Committee on Wind Forces [*Wind Forces on Structures,* 1961] has prepared a report in which information concerning wind-pressure coefficients for various types of structures and maximum wind velocity for particular geographical areas are available.

Impact loads are caused by suddenly applied loads or by the vibration of moving or movable loads and are usually taken as a fraction of the live loads. *Earthquake loads* are those forces caused by the acceleration of the ground surface during an earthquake. **Static loads** are noncyclic loads that produce no dynamic effects.

45.4 Reactions

A structure that is initially at rest and remains at rest when acted upon by applied loads is said to be in a state of *equilibrium*. The resultant of the external loads on the body and the supporting forces or reactions is zero. If a structure or part thereof is to be in equilibrium under the action of a system of loads, it must satisfy the six static equilibrium equations:

$$\begin{aligned} \sum F_x &= 0, \quad \sum F_y = 0, \quad \sum F_z = 0 \\ \sum M_x &= 0, \quad \sum M_y = 0, \quad \sum M_z = 0 \end{aligned} \tag{45.1}$$

The summation in these equations is for all the components of the forces and of the moments about each of the three axes. If a body is subjected to forces that lie in one plane—say, *xy*—the above equations are reduced to

$$\sum F_x = 0, \quad \sum F_y = 0, \quad \sum M_z = 0 \tag{45.2}$$

Consider, for example, the beam shown in Fig. 45.2 under the action of the loads shown. The **reaction** at B must act perpendicular to the surface on which the rollers are constrained to roll upon. The support reactions and the applied loads, which are resolved in vertical and horizontal directions, are shown in Fig. 45.2(b).

We know from geometry that $B_y = \sqrt{3}B_x$. Equation (45.2) can be used to determine the magnitude of the support reactions. Taking the moment about B gives $10A_y - 346.4 \times 5 = 0$, from which $A_y = 173.2$ kN. Equating the sum of vertical forces $\sum F_y$ to zero gives $173.2 + B_y - 346.4 = 0$, and hence we get $B_y = 173.2$ kN. Therefore $B_x = B_y/\sqrt{3} = 100$ kN. Equilibrium in the horizontal direction, $\sum F_x = 0$, gives $A_x - 200 - 100 = 0$, and hence $A_x = 300$ kN.

There are three unknown reaction components at a fixed end, two at a hinge, and one at a roller. If, for a particular structure, the total number of reaction components equals the number of equations available, the unknowns may be calculated from the equilibrium equations, and the

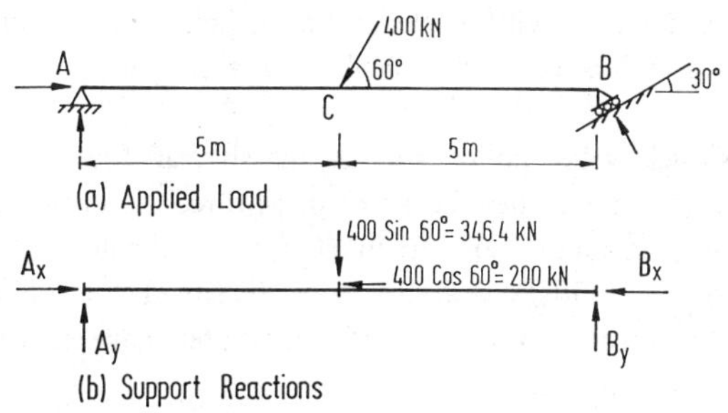

FIGURE 45.2 Beam in equilibrium.

structure is then said to be **statically determinate** externally. Should the number of unknowns be greater than the number of equations available, the structure is **statically indeterminate** externally; if less, it is unstable externally. The ability of a structure to adequately support the loads applied to it depends not only on the number of reaction components but also on the arrangement of those components. It is possible for a structure to have as many (or more) reaction components as there are equations available and yet be unstable. This condition is referred to as *geometric instability*.

45.5 Principle of Superposition

Much of structural analysis is based on the principle of superposition, which is applicable only when the displacements are linear functions of applied loads. The principle may be stated as follows: if the structural behavior is linearly elastic the forces acting on a structure may be separated or divided in any convenient fashion and the structure analyzed for the separate case. The final results can then be obtained by adding up the individual results. This is applicable to the computation of structural responses such as moment, shear, deflection, and so on. There are two situations in which the principle of superposition cannot be applied. The first case is associated with instances where the geometry of the structure is appreciably altered under load. The second case in which superposition is not valid is in situations where the structure is composed of a material in which the stress is not linearly related to the strain.

45.6 Idealized Analytical Models

Structural analyzes are carried out on an analytical model that is an idealization of the actual structure. Any complex structure can be considered to be built up of simpler units or components called members or **elements**. Engineering judgment must be used in defining the idealized structure such that it represents the actual structural behavior as accurately as possible. All structures are three-dimensional, but in many cases it is possible to consider the structure as being two-dimensional in two mutually perpendicular directions.

Broadly speaking, structural elements can be classified into three categories. Structures consisting of line elements, such as a bar, beam, or column, for which the length is much larger than breadth and height are termed *skeletal structures*. A variety of skeletal structures can be obtained by connecting line elements together using hinged, rigid, or semirigid joints. Depending on whether the axes of these members lie in one plane or in different planes these structures are termed *plane structures* or *space structures*. The line elements in these structures under load may be subjected to a single force such as the axial force or a combination of forces such as shear, moment, torsion, and axial. In the first case the structures are referred to as truss-type; in the second, frame-type.

Structures in which members have two dimensions (length and breadth) of the same order and much larger than the thickness are termed *plated structures*. These structural elements may be plane or curved in plane, in which case they are called plates or shells, respectively. These units are generally used in combination with beams and bars. Reinforced concrete slabs supported on beams, box-girders, plate-girders, cylindrical shells, and water tanks are typical examples of plate and shell structures.

In structures such as thick-walled hollow spheres, massive raft foundations, and dams, all three dimensions are of the same order. The analysis of these three-dimensional structures is complex even when several simplifying assumptions are made. It should be noted that a structural engineer is concerned, for the most part, with skeletal structures. Recent advancements in matrix methods of structural analysis and the advent of more powerful computers have enabled the economic design of skeletal as well as plated structures.

45.7 Flexural Members

One of the most common structural elements is a beam; it is subjected to bending or flexure due to loads acting transversely to its centroidal axis or sometimes by loads acting both transversely and parallel to this axis. The following discussions are limited to straight beams, that is, beams in which the centroidal axis is a straight line with shear center coinciding with the centroid of the cross section. It is also assumed that all the loads and reactions lie in a simple plane that also contains the centroidal axis of the flexural member and a principal axis of every cross section. If these conditions are satisfied, the beam will simply bend in the plane of loading without twisting.

Axial Force, Shear Force, and Bending Moment

The *axial force* at any transverse cross section of a straight beam is the algebraic sum of the components acting parallel to the axis of the beam of all loads and reactions applied to the portion of the beam on either side of that cross section. The **shear force** at any transverse cross section of a straight beam is the algebraic sum of the components acting transverse to the axis of the beam of all the loads and reactions applied to the portion of the beam on either side of the cross section. The **bending moment** at any transverse cross section of a straight beam is the algebraic sum of the moments, taken about an axis passing through the centroid of the cross section. The axis about which the moments are taken is, of course, normal to the plane of loading.

Relation between Load, Shear, and Bending Moment

When a beam is subjected to transverse loads, it should be recognized that there exist certain relationships between load, shear, and bending moment. Let us consider, for example, the beam shown in Fig. 45.3 subjected to some arbitrary loading, p.

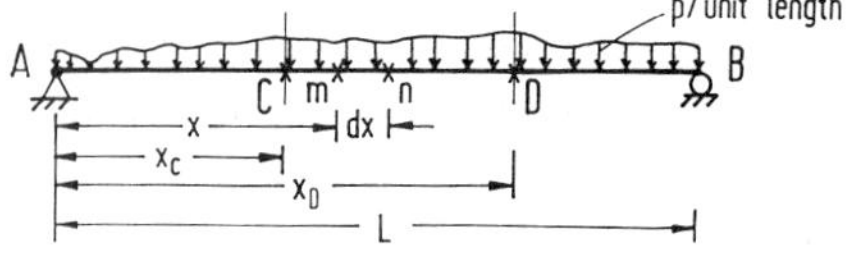

FIGURE 45.3 A beam under arbitrary loading.

Let S and M be the shear and bending moment, respectively, for any point m at a distance x, which is measured from A, being positive when measured to the right. Corresponding values of the shear and bending moment at point n at a differential distance dx to the right of m are $S + dS$ and $M + dM$, respectively. It can be shown, neglecting the second-order quantities, that

$$p = \frac{dS}{dx} \tag{45.3}$$

and

$$S = \frac{dM}{dx} \tag{45.4}$$

The relationship stated mathematically in Eq. (45.3) shows that the rate of change of shear at any point is equal to the intensity of load applied to the beam at that point. Therefore, the difference in shear at two cross sections C and D is

$$S_D - S_C = \int_{x_C}^{x_D} p\,dx$$

We can write this in the same way for the moment as

$$M_D - M_C = \int_{x_C}^{x_D} S\,dx$$

Shear and Bending Moment Diagrams

When a beam is being analyzed or designed for a system of static loads, it is useful to have diagrams from which the value of the shear and bending moment at any cross section can readily be obtained. In order to plot the shear force and bending moment diagrams it is necessary to adopt a sign convention for these responses. A shear force is considered to be positive if it produces a clockwise moment about a point in the free body on which it acts. A negative shear force produces a counterclockwise moment about the point. The bending moment is positive if it causes compression in the upper fibers of the beam and tension in the lower fibers. In other words, the **sagging moment** is positive and the **hogging moment** is negative. The construction of these diagrams is explained better with an example, given in Fig. 45.4.

The section at E of the beam is in equilibrium under the action of applied loads and internal forces acting at E, as shown in Fig. 45.5. There must be an internal vertical force and an internal bending moment to maintain equilibrium at section E. The vertical force or the moment can be obtained as the algebraic sum of all forces or the algebraic sum of the moment of all forces that lie on either side of section E.

The construction of shear force and bending moment diagrams is straightforward but needs some explanation. The shear on a cross section an infinitesimal distance to the right of point A is +55 k, and therefore the shear diagram rises abruptly from zero to +55 at this point. In the portion

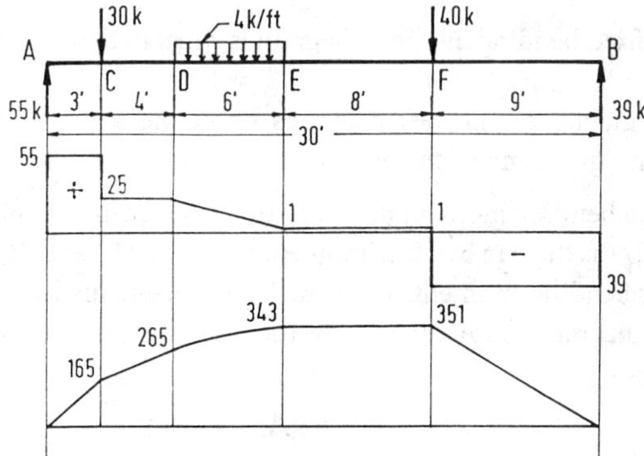

FIGURE 45.4 Bending moment and shear force diagrams.

FIGURE 45.5 Internal forces.

AC, since there is no additional load, the shear remains +55 on any cross section throughout this interval, and the diagram is horizontal as shown (Fig. 45.4). An infinitesimal distance to the left of *C* the shear is +55, but an infinitesimal distance to the right of this point the 30 load has caused the shear to be reduced to +25. Therefore, at point C, there is an abrupt change in the shear diagram from +55 to +25. In the same manner, the shear force diagram for the portion *CD* of the beam remains a rectangle. In the portion *DE*, the shear on any cross section a distance x from point D is

$$S = 55 - 30 - 4x = 25 - 4x$$

which indicates that the shear diagram in this portion is a straight line decreasing from an ordinate of +25 at D to +1 at E. The remainder of the shear force diagram can easily be verified in the same way. It should be noted that, in effect, a concentrated load is assumed to be applied at a point, and hence at such a point the ordinate to the shear diagram changes abruptly by an amount equal to the load.

In the portion AC, the bending moment at a cross section a distance x from point A is $M = 55x$. Therefore, the bending moment diagram starts at zero at A and increases along a straight line to an ordinate of +165 at point C. In the portion CD, the bending moment at any point a distance x from C is $M = 55(x + 3) - 30x$. Hence, the bending moment diagram in this portion is a straight line increasing from 165 at C to 265 at D. In the portion DE, the bending moment at any point a distance x from D is $M = 55(x + 7) - 30(x + 4) - 4(x^2/2)$. Hence, the bending moment diagram in this portion is a curve with an ordinate of 265 at D and 343 at E. In an analogous manner, the remainder of the bending moment diagram can easily be constructed.

Bending moment and shear force diagrams for some simple loading and boundary conditions are given in Fig. 45.6.

Built-In Beams

When the ends of a beam are held so firmly that they are not free to rotate under the action of applied loads, the beam is known as a **built-in beam** or fixed beam and it is indeterminate. The bending moment diagram for such a beam can be considered to consist of two parts: the free bending moment diagram obtained by treating the beam as if the ends are simply supported, and the fixed moment diagram resulting from the restraints imposed on the ends of the beam. The solution of a fixed beam is greatly simplified by considering Mohr's principles, which state the following:

1. The area of the fixed bending moment diagram is equal to that of the free bending moment diagram.
2. The centers of gravity of the two diagrams lie in the same vertical line (i.e., they are equidistant from a given end of the beam).

The construction of a bending moment diagram for a fixed beam is explained with the example shown in Fig. 45.7. *PQUT* is the free bending moment diagram, M_s, and *PQRS* is the fixed moment diagram, M_i. The net bending moment diagram, M, is shown shaded. If A_s is the area of the free bending moment diagram and A_i the area of the fixed moment diagram, then, from the first Mohr's principle, we have

$$A_s = A_i$$

that is,

$$\frac{1}{2} \times \frac{Wab}{L} \times L = \frac{1}{2}(M_A + M_B) \times L$$

$$M_A + M_B = \frac{Wab}{L} \tag{45.5}$$

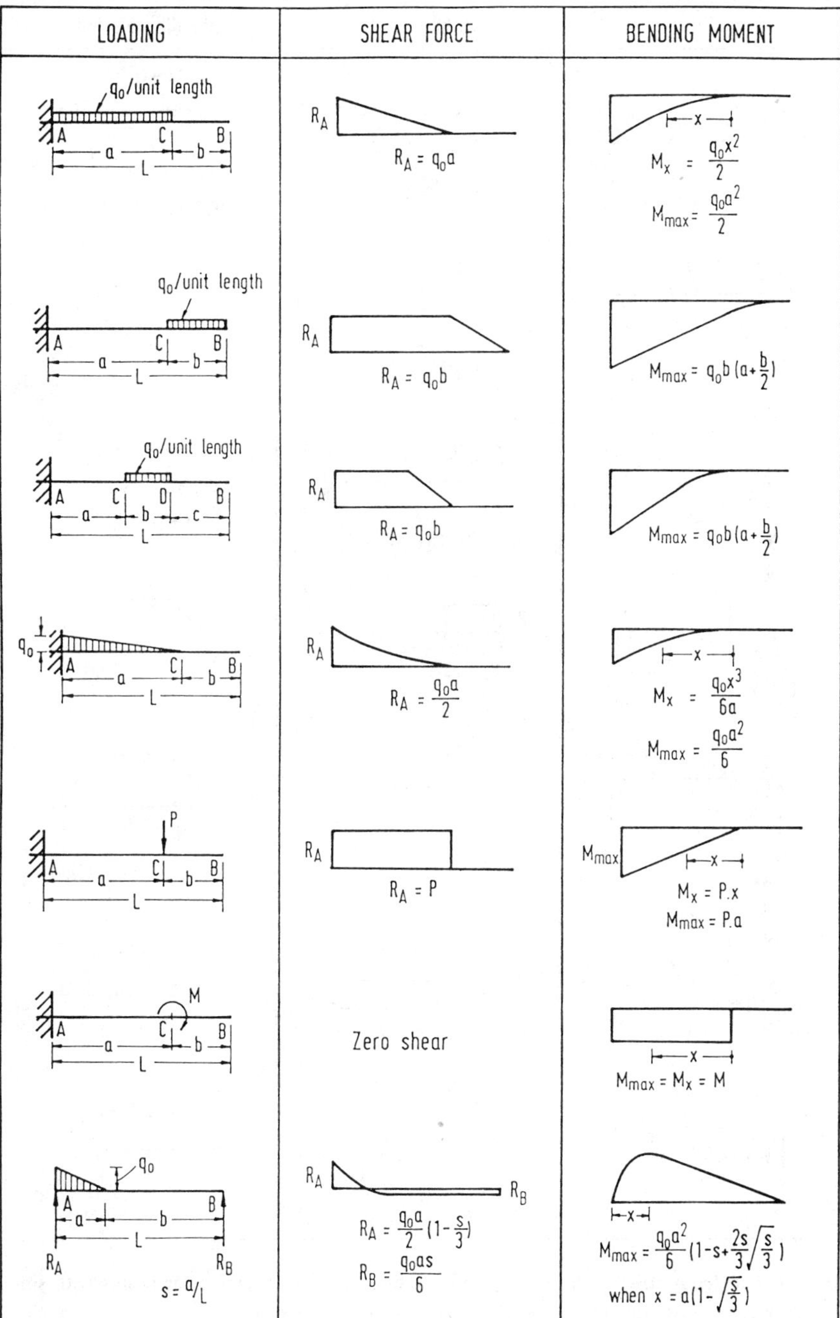

FIGURE 45.6 Shear force and bending moment diagrams for beams with simple boundary conditions subjected to selected loading cases. *(continues)*

From the second principle, equating the moment about A of A_s and A_i, we have

$$M_A + 2M_B = \frac{Wab}{L^3}(2a^2 + 3ab + b^2) \tag{45.6}$$

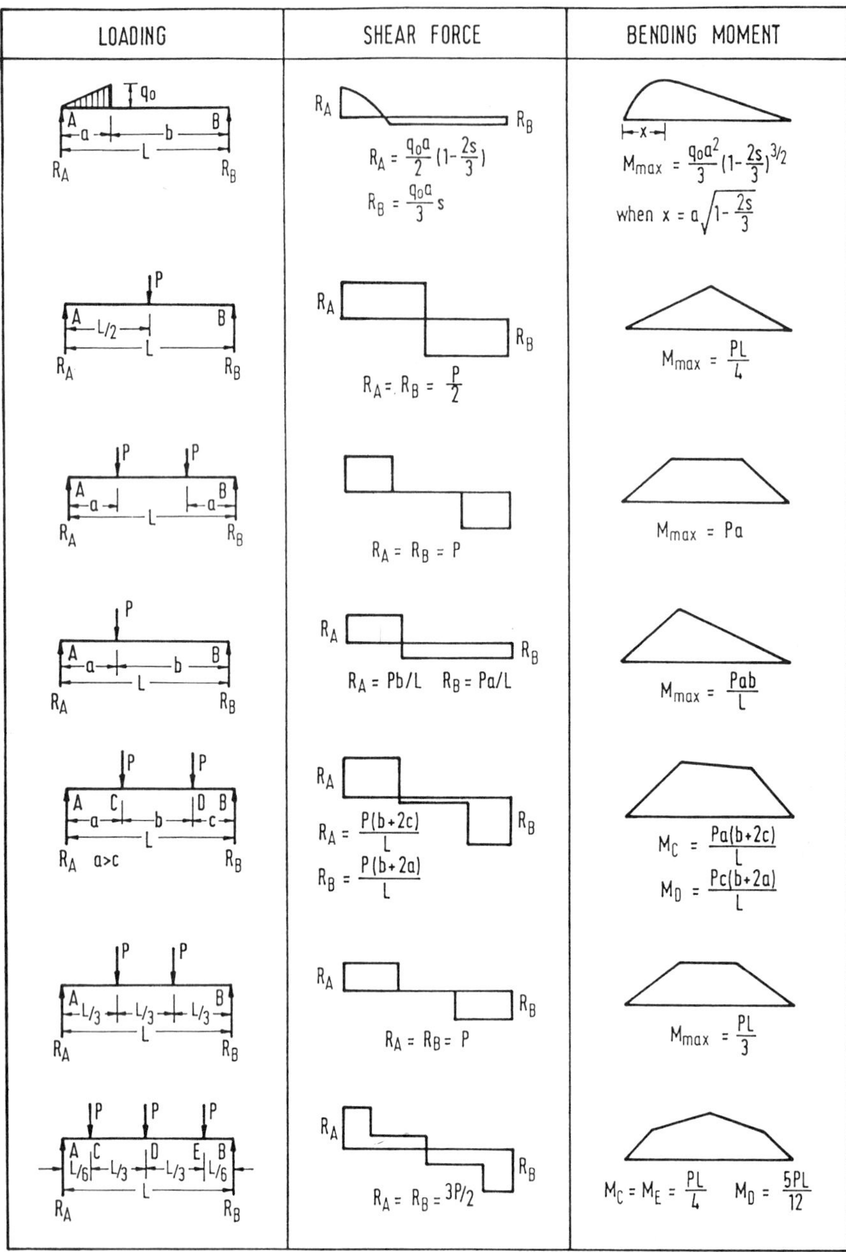

FIGURE 45.6 (continued) Shear force and bending moment diagrams for beams with simple boundary conditions subjected to selected loading cases. *(continues)*

Solving Eq. (45.5) and (45.6) for M_A and M_B we get

$$M_A = \frac{Wab^2}{L^2}$$

$$M_B = \frac{Wa^2b}{L^2}$$

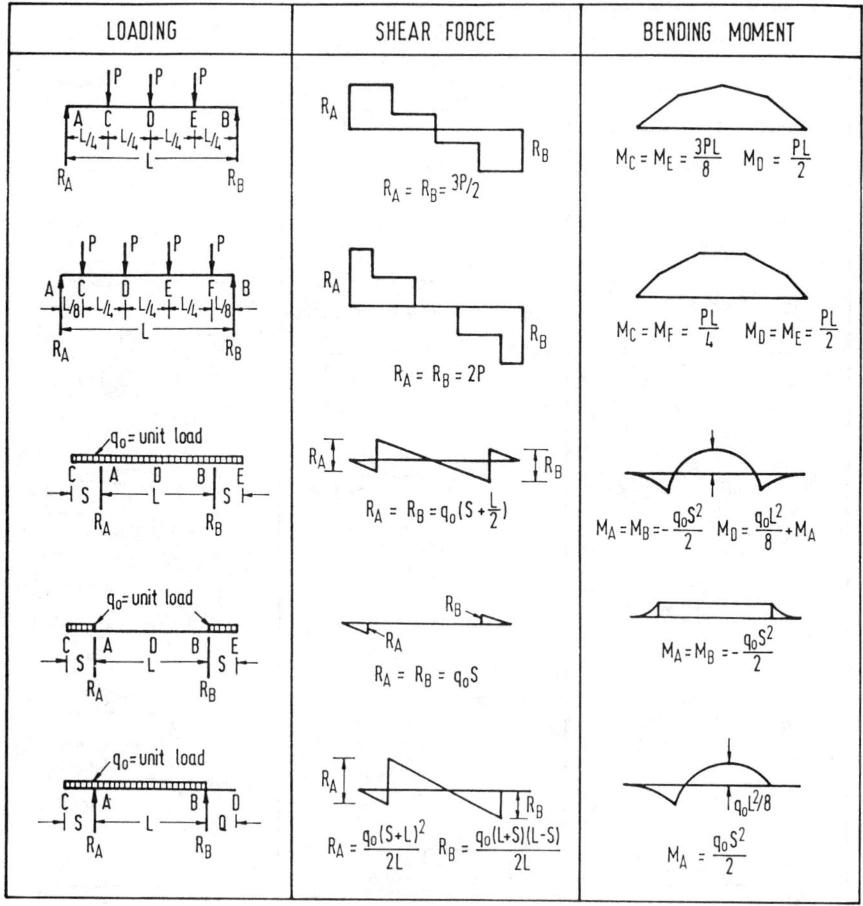

FIGURE 45.6 (continued) Shear force and bending moment diagrams for beams with simple boundary conditions subjected to selected loading cases.

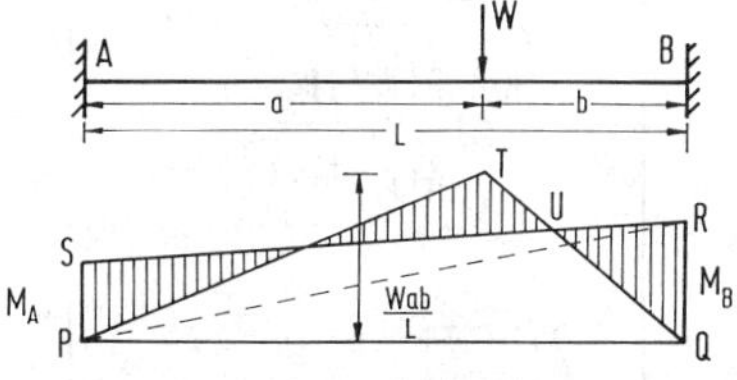

FIGURE 45.7 Fixed-end beam.

Shear force can be determined once the bending moment is known. The shear forces at the ends of the beam (i.e., at A and B) are

$$S_A = \frac{M_A - M_B}{L} + \frac{Wb}{L}$$

$$S_B = \frac{M_B - M_A}{L} + \frac{Wa}{L}$$

Bending moment and shear force diagrams for some typical loading cases are shown in Fig. 45.8.

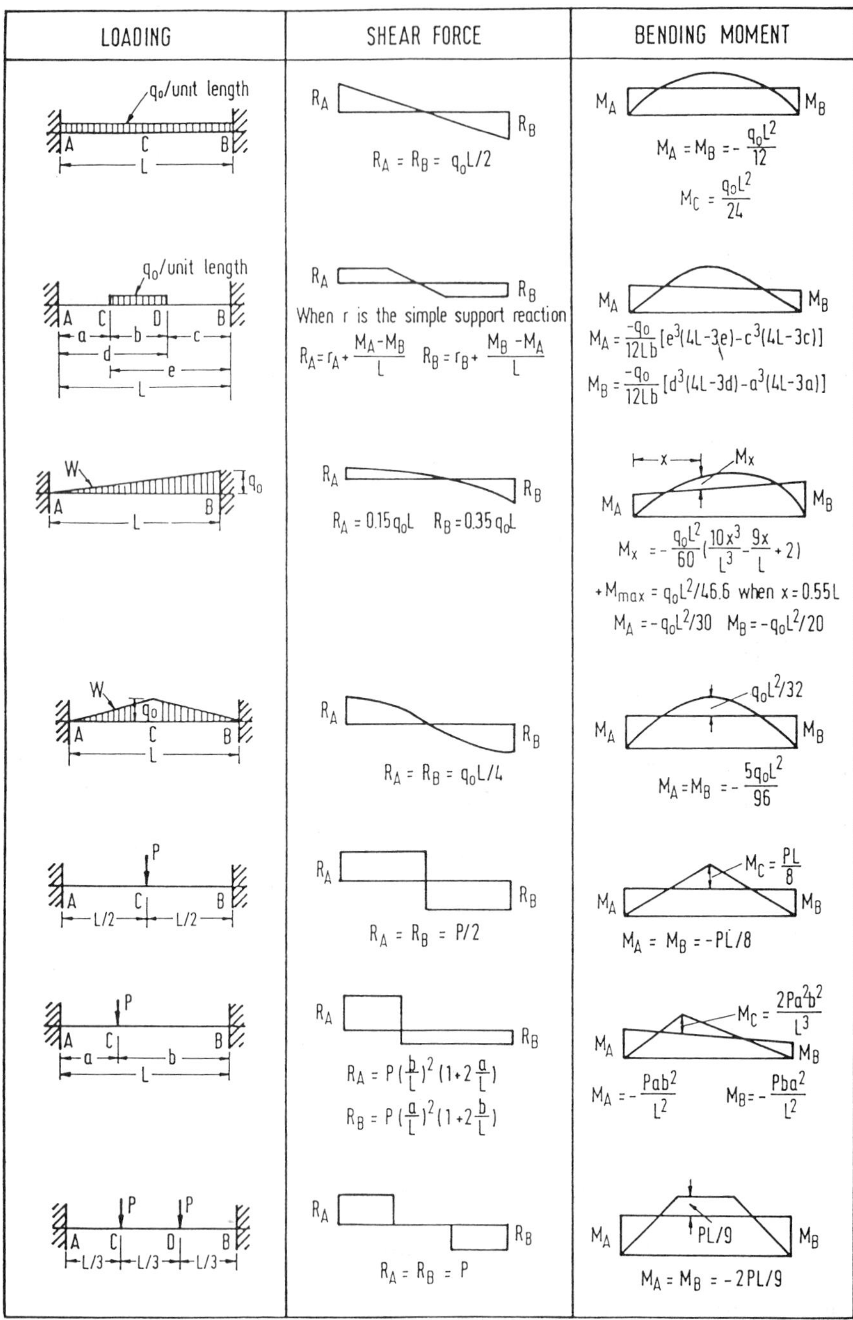

FIGURE 45.8 Shear force and bending moment diagrams for built-up beams subjected to typical loading cases. *(continues)*

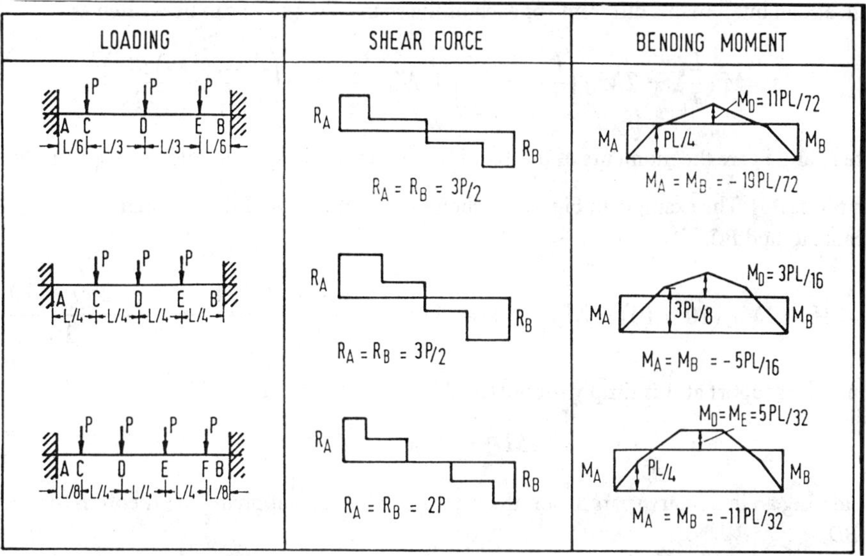

FIGURE 45.8 (continued) Shear force and bending moment diagrams for built-up beams subjected to typical loading cases.

Continuous Beams

Continuous beams, like fixed beams, are statically indeterminate. Bending moments in these beams are functions of the geometry, moments of inertia, and modulus of elasticity of individual members besides the load and span. They can be determined by Clapeyron's theorem of three moments, the moment distribution method, or the slope deflection method. In this section an example of a two-span continuous beam is solved by Clapeyron's theorem of three moments. The theorem is applied to two adjacent spans at a time and the resulting equations in terms of unknown support moments are solved. The theorem states that

$$M_A L_1 + 2M_B(L_1 + L_2) + M_C L_2 = 6\left(\frac{A_1 x_1}{L_1} + \frac{A_2 x_2}{L_2}\right) \tag{45.7}$$

where M_A, M_B, and M_C are the hogging moments at the supports A, B, and C, respectively, of two adjacent spans of length L_1 and L_2 (Fig. 45.9); A_1 and A_2 are the area of bending moment diagrams produced by the vertical loads on the simple spans AB and BC, respectively; x_1 is the centroid of A_1 from A; and x_2 is the distance of the centroid of A_2 from C. If the beam section is constant

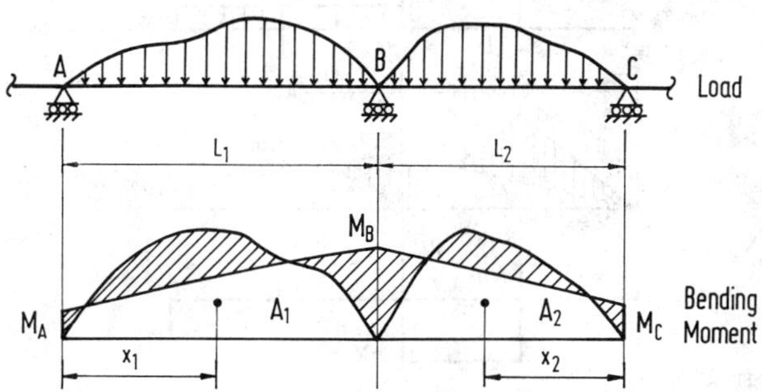

FIGURE 45.9 Continuous beams.

within a span but remains different for each of the spans, Eq. (45.7) can be written as

$$M_A \frac{L_1}{I_1} + 2M_B \left(\frac{L_1}{I_1} + \frac{L_2}{I_2} \right) + M_C \frac{L_2}{I_2} = 6 \left(\frac{A_1 x_1}{L_1 I_1} + \frac{A_2 x_2}{L_2 I_2} \right) \tag{45.8}$$

where I_1 and I_2 are the moments of inertia of beam sections in span L_1 and L_2, respectively.

Example 45.1. The example in Fig. 45.10 shows the application of this theorem.

Spans *AC* and *BC*:

$$M_A \times 10 + 2M_C(10+10) + M_B \times 10 = 6 \left[\frac{\frac{1}{2} \times 500 \times 10 \times 5}{10} + \frac{\frac{2}{3} \times 250 \times 10 \times 5}{10} \right]$$

Since the support at *A* is simply supported, $M_A = 0$. Therefore,

$$4M_C + M_B = 1250 \tag{45.9}$$

Considering an imaginary span *BD* on the right side of *B*, and applying the theorem for spans *CB* and *BD*,

$$M_c \times 10 + 2M_B(10) + M_D \times 10 = 6 \times \frac{\frac{2}{3} \times 10 \times 5}{10} \times 2$$

$$M_C + 2M_B = 500 \qquad (\because M_C = M_D) \tag{45.10}$$

Solving Eqs. (45.9) and (45.10), we get

$$M_B = 107.2 \text{ kNm}$$

$$M_C = 285.7 \text{ kNm}$$

Shear force, S, at *A* is

$$\frac{M_A - M_C}{L} + 100 = -28.6 + 100 = 71.4 \text{ kN}$$

Shear force, S, at *C* is

$$\left(\frac{M_C - M_A}{L} + 100 \right) + \left(\frac{M_C - M_B}{L} + 100 \right) = (28.6 + 100) + (17.9 + 100)$$

$$= 246.5 \text{ kN}$$

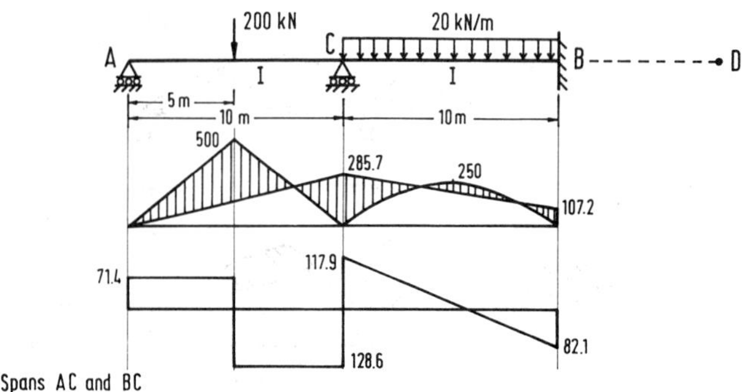

FIGURE 45.10 Example 45.1.

Shear force, S, at B is

$$\left(\frac{M_B - M_C}{L} + 100\right) = -17.9 + 100 = 82.1 \text{ kN}$$

The bending moment and shear force diagrams are shown in Fig. 45.10.

Deflection of Beams

Beams, when loaded, deform and result in damage to attached materials if allowable values of deformations are exceeded. It becomes, therefore, necessary to assess the deflection values accurately. There are several methods available for determining deflections. They are the moment-area method, the conjugate-beam method, the virtual work method and Castigliano's second theorem, among others.

The elastic curve of a member is the shape the neutral axis takes when the member deflects under load. The inverse of the radius of curvature at any point of this curve is obtained as

$$\frac{1}{R} = \frac{M}{EI} \tag{45.11}$$

where M is the bending moment at the point and EI is the flexural rigidity of the beam section. Since the deflection is small, $1/R$ is approximately taken as d^2y/dx^2, so one can write

$$M = EI\frac{d^2y}{dx^2} \tag{45.12}$$

where y is the deflection of the beam at distance x measured from the origin of the coordinate. The change in slope in a distance dx can be expressed as $M(dx/EI)$, and hence the slope in a beam is obtained as

$$\theta_B - \theta_A = \int_A^B \frac{M}{EI}\, dx \tag{45.13}$$

Eq. (45.13) may be stated as follows: the change in slope between the tangents to the elastic curve at two points is equal to the area of the M/EI diagram between the two points.

Once we determine the changes in slope between tangents to the elastic curve, the deflection can be obtained by integrating further the slope equation thus obtained. In a distance dx the neutral axis changes in direction by an amount $d\theta$. The deflection of one point on the beam with respect to the tangent at another point due to this angle change is equal to x (the distance from the point at which deflection is desired to the particular differential distance) times $d\theta$:

$$d\delta = x\, d\theta$$

To determine the total deflection from the tangent at a point A to the tangent at another point B on the beam, it is necessary to obtain a summation of the products of each $d\theta$ angle (from A to B) times the distance to the point where deflection is desired, or

$$\delta_B - \delta_A = \int_A^B \frac{Mx\, dx}{EI} \tag{45.14}$$

The deflection of a tangent to the elastic curve of a beam with respect to a tangent at another point is equal to the moment of M/EI diagram between the two points, taken about the point at which deflection is desired.

Moment Area Method

The moment area method is most conveniently used for determining slopes and deflections for beams in which the direction of the tangent to the elastic curve at one or more points is known, such as contilever beams, where the tangent at the fixed end does not change in slope. The method is applied easily to beams loaded with concentrated loads because the moment diagrams consist of straight lines. These diagrams can be broken down into single triangles and rectangles. Beams supporting uniform loads or uniformly varying loads may be handled by integration. Properties of some of the shapes of M/EI diagrams that designers usually come across are given in Fig. 45.11.

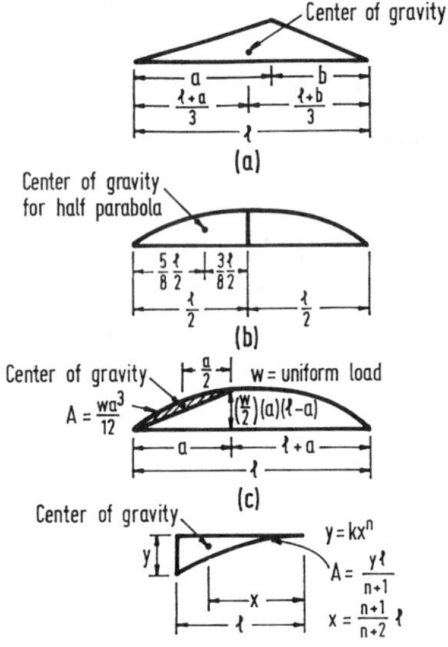

FIGURE 45.11 Typical M/EI diagram.

It should be understood that the slopes and deflections obtained using the area moment theorems are with respect to tangents to the elastic curve at the points being considered. The theorems do not directly give the slope or deflection at a point in the beam as compared to the horizontal axis (except in one or two special cases); they give the change in slope of the elastic curve from one point to another or the deflection of the tangent at one point with respect to the tangent at another point. There are some special cases in which beams are subjected to several concentrated loads or the combined action of concentrated and uniformly distributed loads. In such cases it is advisable to separate the concentrated loads and uniformly distributed loads, and the moment-area method can be applied separately to each of these loads. The final responses are obtained by the principle of superposition.

Let us, for example, consider a simply supported beam subjected to a uniformly distributed load, as shown in Fig. 45.12. We will use the moment area method to determine the expression for deflection at the center.

The tangents to the elastic curve at each end of the beam are inclined. The deflection δ_1 of the tangent at the left end from the tangent at the right end is found as $ql^4/24EI$. The distance from

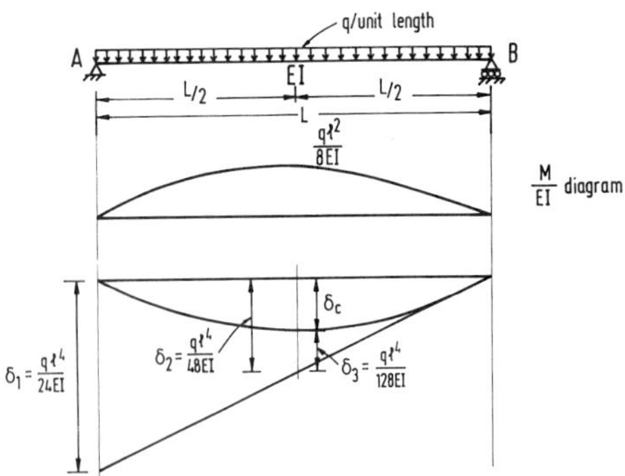

FIGURE 45.12 Deflection—simply supported beam under a uniformly distributed load.

the original chord between the supports and the tangent at the right end, δ_2, can be computed as $ql^4/48EI$. The deflection of a tangent at the center from a tangent at the right end, δ_3, is determined in this step as $ql^4/128EI$. The difference between δ_2 and δ_3 gives the centerline deflection as $(5/384)(ql^4/EI)$.

Curved Flexural Members

The flexural formula is based on the assumption that the beam to which the bending moment is applied is initially straight. Many members that are subjected to bending, however, are curved before a bending moment is applied to them. Such members are called *curved beams*. It is important to determine the effect of initial curvature of a beam on the stresses and deflections caused by loads applied to the beam in the plane of initial curvature. In the following discussion all the conditions applicable to the straight-beam formula are assumed valid, except that the beam is initially curved.

Let the curved beam *DOE* shown in Fig. 45.13 be subjected to the loads Q. The surface on which the fibers do not change in length is called the *neutral surface*. The total deformations of the fibers between two normal sections such as *AB* and A_1B_1 are assumed to vary proportionally with the distances of the fibers from the neutral surface. The top fibers are compressed while those at the bottom are stretched (i.e., the plane section before bending remains plane after bending).

In Fig. 45.13 the two lines *AB* and A_1B_1 are two normal sections of the beam before the loads are applied. The change in the length of any fiber between these two normal sections after bending is represented by the distance along the fiber between the lines A_1B_1 and $A'B'$; the neutral surface is represented by NN_1, and the stretch of fiber PP_1 is $P_1P'_1$, and so on. For convenience it will be assumed that the line *AB* is a line of symmetry and does not change direction.

The total deformations of the fibers in the curved beam are proportional to the distances of the fibers from the neutral surface. However, the strains of the fibers are not proportional to these distances because the fibers are not of equal length. Within the elastic limit the stress on any fiber in the beam is proportional to the strain of the fiber, and hence the elastic stresses in the fibers of a curved beam are not proportional to the distances of the fibers from the neutral surface. The resisting moment in a curved beam, therefore, is not given by the expression $\sigma I/c$. Hence the neutral axis in a curved beam does not pass through the centroid of the section. The distribution of stress over the section and the relative position of the neutral axis are shown in Fig. 45.13(b); if the beam were straight, the stress would be zero at the centroidal axis and would vary proportionally with the distance from the centroidal axis as indicated by the dot-dash line in the figure. The stress on a normal section such as *AB* is called the *circumferential stress*.

Signs

The bending moment M is positive when it decreases the radius of curvature and negative when it increases the radius of curvature; y is positive when measured toward the convex side of the beam

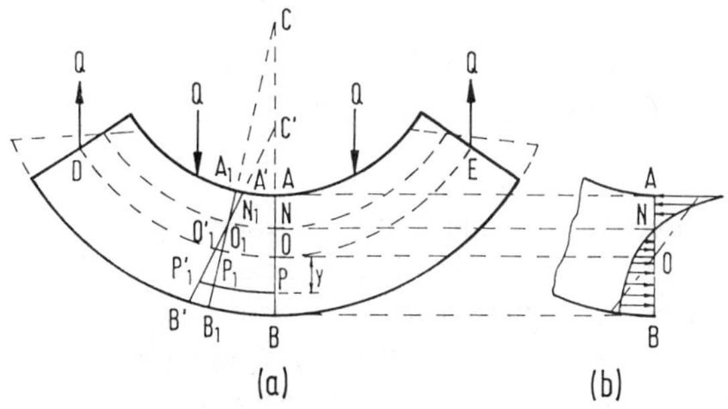

FIGURE 45.13 Bending of curved beams.

and negative when measured toward the concave side, that is, toward the center of curvature. With these sign conventions, when σ is positive it is a tensile stress.

Circumferential Stress at Any Point in a Curved Beam

Figure 45.14 shows a free-body diagram of the portion of the body on one side of the section; the equations of equilibrium are applied to the forces acting on this portion. The equations obtained are

$$\sum F_z = 0 \quad \text{or} \quad \int \sigma\, da = 0 \qquad (45.15)$$

$$\sum M_z = 0 \quad \text{or} \quad \mathrm{M} = \int y\sigma da \qquad (45.16)$$

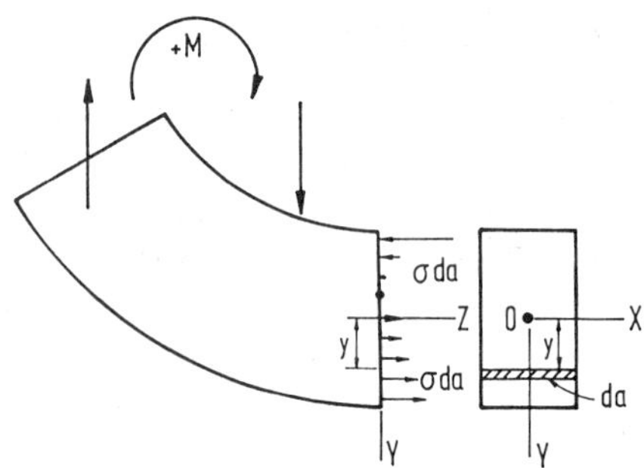

FIGURE 45.14 Free-body diagram of a curved beam segment.

Figure 45.15 represents the part ABB_1A_1 of Fig. 45.13(a) enlarged; the angle between the two sections AB and A_1B_1 is $d\theta$. The bending moment causes the plane A_1B_1 to rotate through an angle $\Delta\, d\theta$, thereby changing the angle this plane makes with the plane BAC from $d\theta$ to $(d\theta + \Delta\, d\theta)$; the center of curvature is changed from C to C', and the distance of the centroidal axis from the center of curvature is changed from R to ρ. It should be noted that y, R, and ρ at any section are measured from the centroidal axis and not from the neutral axis.

It can be shown that the bending stress σ is given by the relation

$$\sigma = \frac{M}{aR}\left(1 + \frac{1}{Z}\frac{y}{R+y}\right) \qquad (45.17)$$

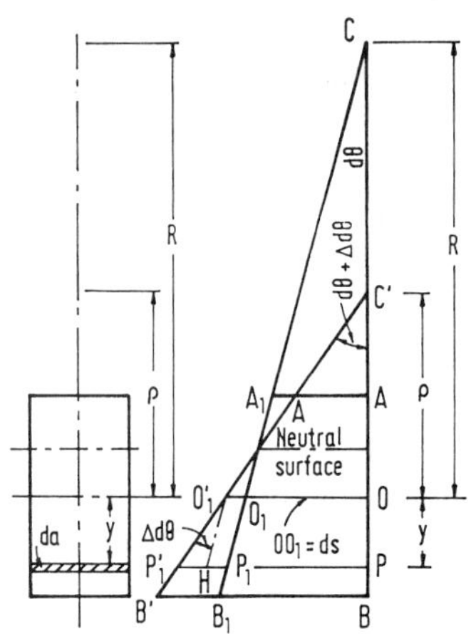

FIGURE 45.15 Curvature in a curved beam.

where

$$Z = -\frac{1}{a}\int \frac{y}{R+y}\, da$$

σ is the tensile or compressive (circumferential) stress at a point at the distance y from the centroidal axis of a transverse section at which the bending moment is M; R is the distance from the centroidal axis of the section to the center of curvature of the central axis of the unstressed beam; a is the area of the cross section; Z is a property of the cross section, the values of which can be obtained from the expressions for various areas given in Table 45.1. [Detailed information can be obtained from Seely and Smith, 1952.]

Table 45.1 Analytical Expressions for Z *(continues)*

Section	Z
h, c, R	$Z = \frac{1}{4}\left(\frac{c}{R}\right)^2 + \frac{1}{8}\left(\frac{c}{R}\right)^4 + \frac{5}{64}\left(\frac{c}{R}\right)^6 + \frac{7}{128}\left(\frac{c}{R}\right)^8 + \ldots$ $Z = -1 + 2\left(\frac{R}{c}\right)^2 - 2\left(\frac{R}{c}\right)\sqrt{\left(\frac{R}{c}\right)^2 - 1}$
h, c, R	$Z = \frac{1}{3}\left(\frac{c}{R}\right)^2 + \frac{1}{5}\left(\frac{c}{R}\right)^4 + \frac{1}{7}\left(\frac{c}{R}\right)^6 + \ldots$ $Z = -1 + \frac{R}{h}\left[\log_e\left(\frac{R+c}{R-c}\right)\right]$
h, b_1, b, c_1, c_2, R	$Z = -1 + \frac{R}{ah}\left\{[b_1 h + (R+c_1)(b-b_1)]\log_e\left(\frac{R+c_1}{R-c_2}\right) - (b-b_1)h\right\}$ $Z = -1 + \frac{2R}{(b+b_1)h}\left\{\left[b_1 + \frac{b-b_1}{h}(R+c_1)\right]\log_e\left(\frac{R+c_1}{R-c_2}\right) - (b-b_1)\right\}$
h, c_1, c_2, R	$Z = -1 + 2\frac{R}{h^2}\left[(R+c_1)\log_e\left(\frac{R+c_1}{R-c_2}\right) - h\right]$
h, c, R	$Z = \frac{1}{4}\left(\frac{c}{R}\right)^2 + \frac{1}{8}\left(\frac{c}{R}\right)^4 + \frac{5}{64}\left(\frac{c}{R}\right)^6 + \frac{7}{128}\left(\frac{c}{R}\right)^8 + \ldots$ $Z = -1 + 2\left(\frac{R}{c}\right)^2 - 2\left(\frac{R}{c}\right)\sqrt{\left(\frac{R}{c}\right)^2 - 1}$

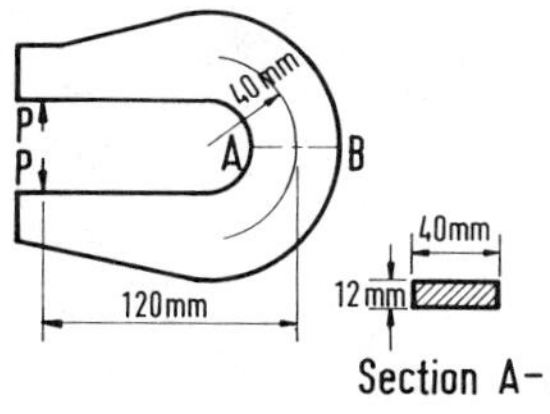

FIGURE 45.16 Example 45.2.

Example 45.2. The frame shown in Fig. 45.16 is subjected to a load $P = 1780$ N. Calculate the circumferential stress at A and B assuming that the elastic strength of the material is not exceeded.

We know from Eq. (45.17)

$$\sigma = \frac{P}{a} + \frac{M}{aR}\left(1 + \frac{1}{Z}\frac{y}{R+y}\right)$$

Table 45.1 (continued) Analytical Expressions for *Z* *(continues)*

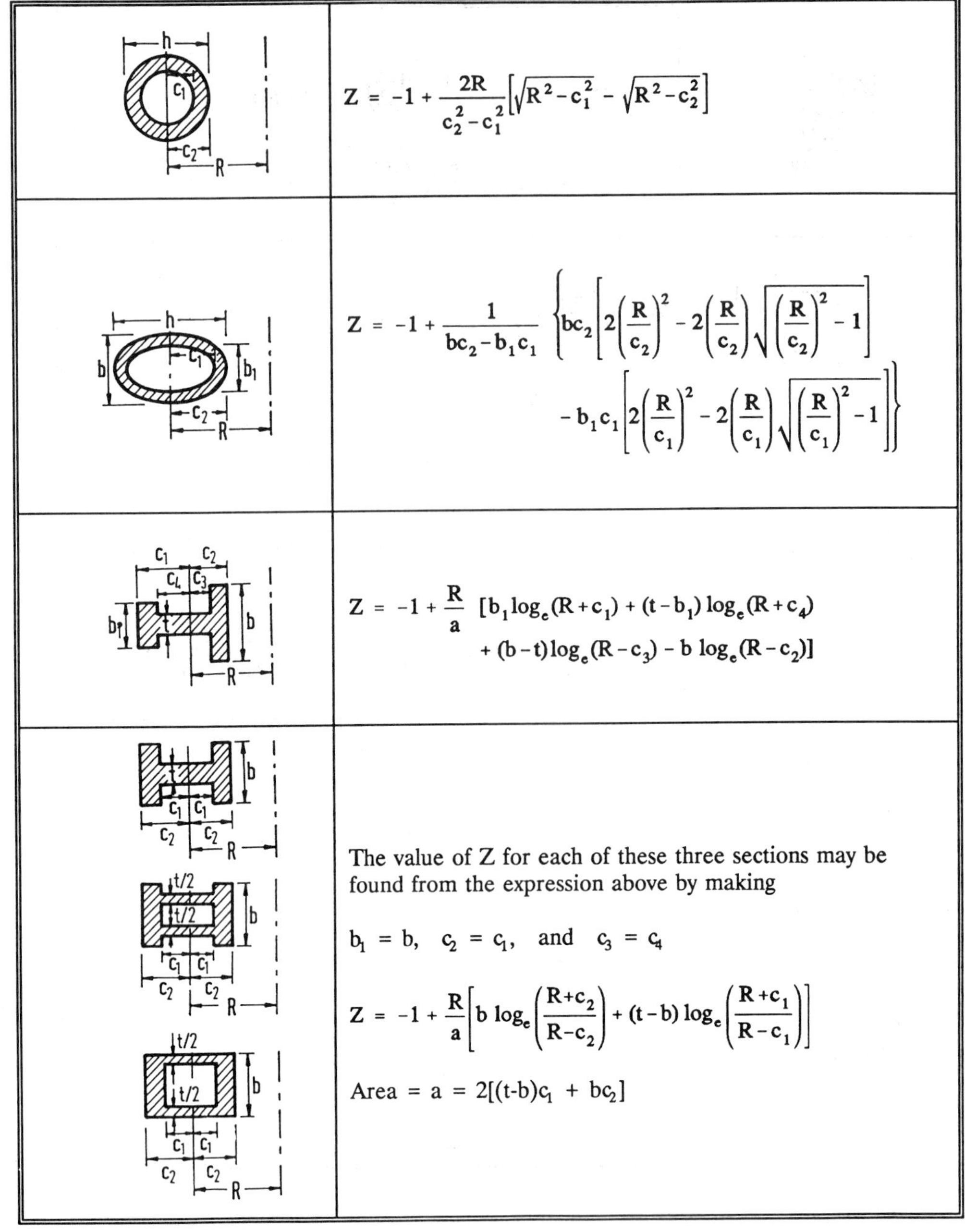

	$Z = -1 + \frac{2R}{c_2^2 - c_1^2}\left[\sqrt{R^2 - c_1^2} - \sqrt{R^2 - c_2^2}\right]$
	$Z = -1 + \frac{1}{bc_2 - b_1c_1}\left\{bc_2\left[2\left(\frac{R}{c_2}\right)^2 - 2\left(\frac{R}{c_2}\right)\sqrt{\left(\frac{R}{c_2}\right)^2 - 1}\right] - b_1c_1\left[2\left(\frac{R}{c_1}\right)^2 - 2\left(\frac{R}{c_1}\right)\sqrt{\left(\frac{R}{c_1}\right)^2 - 1}\right]\right\}$
	$Z = -1 + \frac{R}{a}\left[b_1\log_e(R+c_1) + (t-b_1)\log_e(R+c_4) + (b-t)\log_e(R-c_3) - b\log_e(R-c_2)\right]$
	The value of Z for each of these three sections may be found from the expression above by making $b_1 = b, \quad c_2 = c_1, \quad \text{and} \quad c_3 = c_4$ $Z = -1 + \frac{R}{a}\left[b\log_e\left(\frac{R+c_2}{R-c_2}\right) + (t-b)\log_e\left(\frac{R+c_1}{R-c_1}\right)\right]$ Area $= a = 2[(t-b)c_1 + bc_2]$

Table 45.1 (continued) Analytical Expressions for *Z*

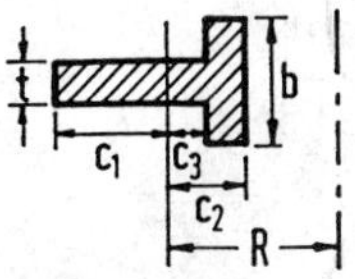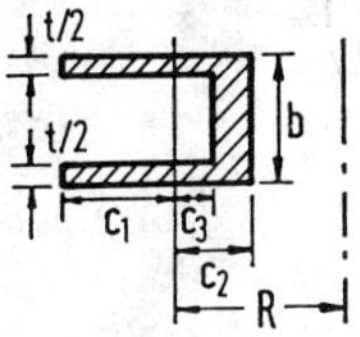	In the expression for the unequal I given above make $c_4 = c_1$ and $b_1 = t$, then $Z = -1 + \frac{R}{a}[t \log_e(R+c_1) + (b-t)\log_e(R-c_3) - b\log_e(R-c_2)]$ Area = a = $tc_1 - (b-t)c_3 + bc_2$
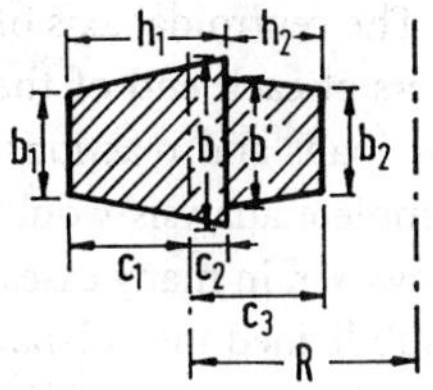	$Z = -1 + \frac{R}{a}\left\{\left[b_1 + \frac{b-b_1}{h_1}(R+c_1)\right]\log\frac{R+c_1}{R-c_2} + \left[b_2 - \frac{b'-b_2}{h_2}(R-c_3)\right]\log\frac{R-c_2}{R-c_3} + (b'-b_2) - (b-b_1)\right\}$

Source: Seely, F. B. and Smith, J. O. 1952.

where

a = area of rectangular section = $40 \times 12 = 480\ \text{mm}^2$

R = 40 mm

y_A = −20 mm

y_B = +20 mm

P = 1780 N

$M = -1780 \times 120 = -213{,}600$ N mm

From Table 45.1, for a rectangular section

$$Z = -1 + \frac{R}{h}\left[\log_e \frac{R+c}{R-c}\right]$$

$$h = 40\ \text{mm}$$

$$c = 20\ \text{mm}$$

Hence,

$$Z = -1 + \frac{40}{40}\left[\log_e \frac{40+20}{40-20}\right] = 0.0986$$

Therefore,

$$\sigma_A = \frac{1780}{480} + \frac{-213{,}600}{480 \times 40}\left(1 + \frac{1}{0.0986}\frac{-20}{40-20}\right)$$

$$= 105.4\ \mathrm{N/mm^2} \quad \text{(Tensile)}$$

$$\sigma_B = \frac{1780}{480} + \frac{-213{,}600}{480 \times 40}\left(1 + \frac{1}{0.0986}\frac{20}{40+20}\right)$$

$$= -45\ \mathrm{N/mm^2} \quad \text{(Compressive)}$$

45.8 Truss Structures

A structure that is composed of a number of bars pin-connected at their ends to form a stable framework is called a *truss*. If all the bars lie in a plane, the structure is a *planar truss*. It is generally assumed that loads and reactions are applied to the truss only at the joints. The centroidal axis of each member is straight, coincides with the line connecting the joint centers at each end of the member, and lies in a plane that also contains the lines of action of all the loads and reactions. In reality, most truss structures are three-dimensional in nature and a complete analysis would require consideration of the full spatial interconnection of the members. However, in many cases, such as bridge structures and simple roof systems, the framework can be subdivided into planar components for analysis as planar trusses without seriously compromising the accuracy of the results. Figure 45.17 shows some typical idealized planar truss structures.

There exists a relation between the number of members, m, number of joints, j, and reaction components, r. The expression

$$m = 2j - r$$

must be satisfied if it is to be statically determinate internally. r is the least number of reaction components required for external stability. If m exceeds $(2j - r)$, the excess members are called redundant members and the truss is said to be statically indeterminate.

Truss analysis gives the bar forces in a truss; for a statically determinate truss, these bar forces can be found by employing the laws of statics to ensure internal equilibrium of the structure.

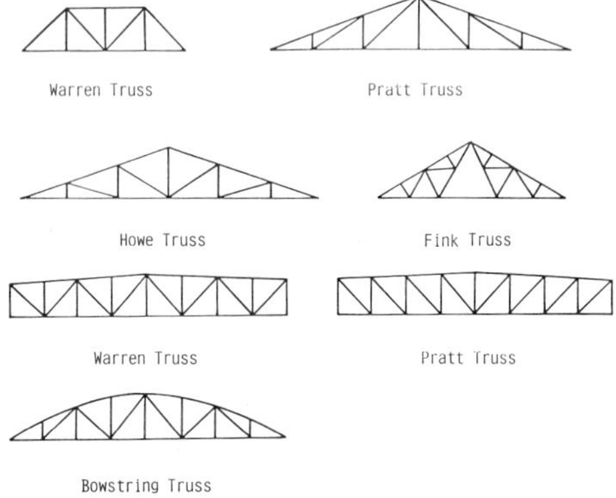

FIGURE 45.17 Typical planar trusses.

The process requires repeated use of free-body diagrams, from which individual bar forces are determined. The *method of joints* is a technique of truss analysis in which the bar forces are determined by the sequential isolation of joints—the unknown bar forces at one joint are solved and become known bar forces at subsequent joints. The other method is known as the *method of sections*, in which equilibrium of a part of the truss is considered.

Method of Joints

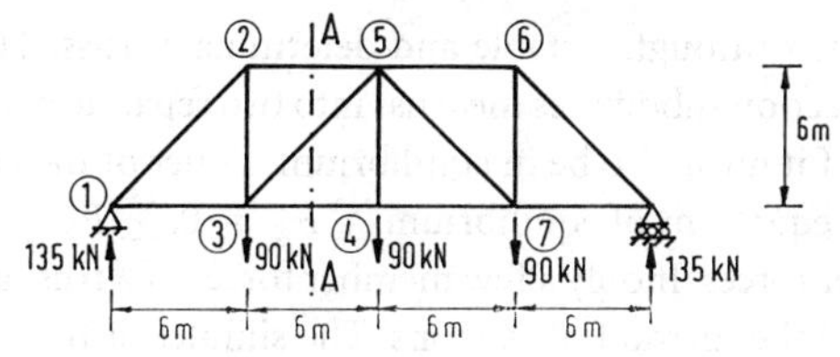

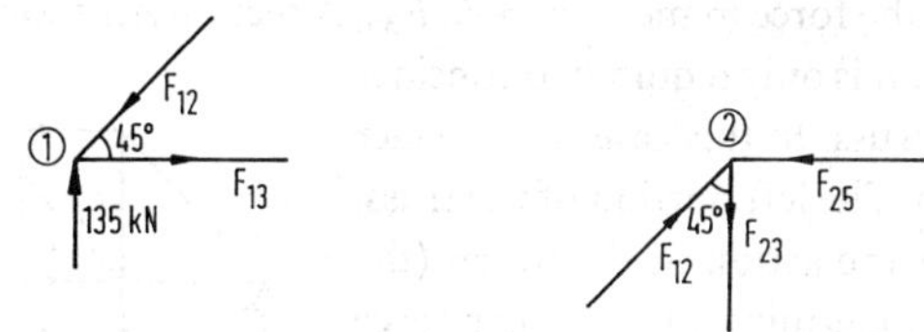

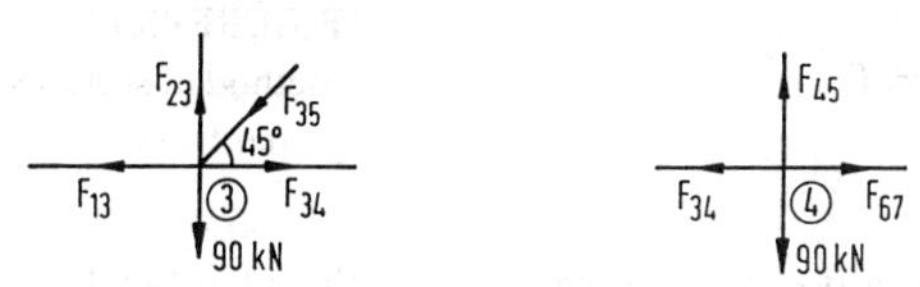

FIGURE 45.18 Method of joints, planar truss.

An imaginary section may be completely passed around a joint in a truss. The joint has become a free body in equilibrium under the forces applied to it. The equations $\sum H = 0$ and $\sum V = 0$ may be applied to the joint to determine the unknown forces in members meeting there. It is evident that no more than two unknowns can be determined at a joint with these two equations. The member forces in a truss shown in Fig. 45.18 may be obtained by using the method of joints as follows.

The truss is symmetrically loaded, and it is sufficient to solve half the truss by considering joints 1 through 5. At joint 1 we see that there are two unknown forces. Summation of the vertical components of all forces at joint 1 gives

$$135 - F_{12} \sin 45^\circ = 0$$

which in turn gives the force in member 1–2, $F_{12} = 190.0$ kN (compressive). Similarly, summation of the horizontal components gives

$$F_{13} - F_{12} \cos 45^\circ = 0$$

Substituting for F_{12} we get the force in member 1–3, $F_{13} = 135$ kN (tensile). Now, joint 2 is cut completely and it is found that there are two unknown forces F_{25} and F_{23}. Summation of the vertical components give $F_{12} \cos 45^\circ - F_{23} = 0$. Therefore $F_{23} = 135$ kN (tensile). Summation of the horizontal components gives $F_{12} \sin 45^\circ - F_{25} = 0$ and hence $F_{25} = 135$ kN (compressive). After solving for joints 1 and 2 one proceeds to take a section around joint 3, at which there are now two unknown forces F_{34} and F_{35}. Summation of the vertical components at joint 3 gives

$$F_{23} - F_{35} \sin 45^\circ - 90 = 0$$

Substituting for F_{23}, one gets $F_{35} = 63.6$ kN (compressive). Summing the horizontal components and substituting for F_{13}, one gets

$$-135 - 45 + F_{34} = 0$$

Therefore, $F_{34} = 180$ kN (tensile). The next joint involving two unknowns is joint 4. When we consider a section around it, the summation of the vertical components at joint 4 gives $F_{45} = 90$ kN (tensile). Now, the forces in all the members on the left half of the truss are known, and by symmetry the forces in the remaining members can be determined. The forces in all the members of a truss can also be determined by making use of the method of sections.

Method of Sections

In this method, an imaginary cutting line is drawn through a stable and determinate truss. This imaginary cutting line is called a section. Thus, a section subdivides the truss into two separate parts. Since the entire truss is in equilibrium, any part of it must also be in equilibrium. Either of the two parts of the truss can be considered and the three equations of equilibrium $\sum F_x = 0, \sum F_y = 0$, and $\sum M = 0$ can be applied to solve for member forces. If only a few member forces of a truss are needed, the quickest way to find these forces is by the method of sections. The simpler of the two parts is chosen so that the time taken for the solution can be minimized. The example considered in the preceding section (Fig. 45.18) is once again considered now.

Let us assume that it is required to determine the force in member 3–5, F_{35}. A section *AA* should be run to cut member 3–5 as shown in Fig. 45.18. It is only required to consider the equilibrium of one of the two parts of the truss. In this case we consider the portion of the truss on the left of the section. The left portion of the truss, as shown in Fig. 45.19, is in equilibrium under the action of the forces (the external and internal forces). It is required to determine the member force F_{35}. Considering the equilibrium of forces in the vertical direction one can obtain

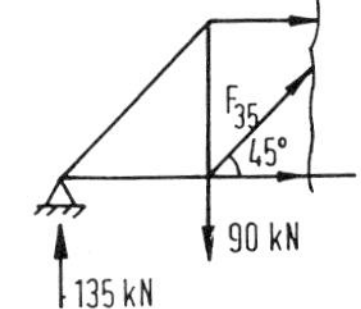

FIGURE 45.19 Method of sections, planar truss.

$$135 - 90 + F_{35} \sin 45^\circ = 0$$

Therefore, F_{35} is obtained as $-45\sqrt{2}$ kN, the negative sign indicating that the member force is compressive. This result is the same as the one obtained by the method of joints. The other member forces cut by the section can be obtained by considering the other equalibrium equations, $\sum M = 0$. More sections can be taken in the same way so as to solve for other member forces in the truss. It should be noted that the most important advantage of this method is that one can obtain the required member force without solving for the other member forces.

Compound Trusses

A compound truss is formed by interconnecting two or more simple trusses. It may be possible to ship the fabricated simple trusses individually and to interconnect the simple trusses on the construction site to form a long-span compound truss. Figure 45.20 shows some examples of compound trusses. A typical roof truss is shown in Fig. 45.20(a), in which two simple trusses are interconnected by means of a single member and a common joint. The compound truss shown in Fig. 45.20(b) is commonly used in bridge construction; in this case, three members are used to interconnect

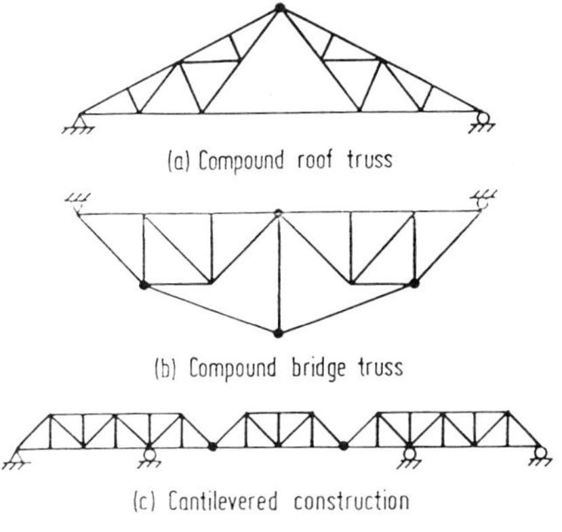

(a) Compound roof truss
(b) Compound bridge truss
(c) Cantilevered construction

FIGURE 45.20 Compound trusses.

two simple trusses at a common joint. Three simple trusses interconnected at their common joints are shown in Fig. 45.20(c).

Basically the method of sections is used to determine the member forces in the interconnecting members of compound trusses similar to those shown in Figs. 45.20(a) and (b). However, in the case of a cantilevered truss the middle simple truss is isolated as a free-body diagram to find its reactions; these reactions are reversed and applied to the interconnecting joints of the other two simple trusses. After the interconnecting forces between the simple trusses are found, the simple trusses are analyzed by the method of joints or the method of sections.

Stability and Determinacy

The simplest stable and determinate plane truss has three members, three joints, and three reaction components. To form a stable and determinate plane truss of n joints, the three members of the original triangle plus two additional members for each of the remaining $(n - 3)$ joints are required. Thus, the minimum total number of members, m, required to form an internally stable plane truss is $m = 2n - 3$. If a stable, simple plane truss of n joints and $(2n - 3)$ members is supported by three independent reaction components, the structure is stable and determinate when subjected to general loading. If the stable, simple plane truss has more than three reaction components, the structure is externally indeterminate, which means all of the reaction components cannot be determined from the three available equations of statics. If the stable, simple plane truss has more than $(2n - 3)$ members, the structure is internally indeterminate and hence all of the member forces cannot be determined from the $2n$ available equations of statics in the method of joints. The analyst must examine the arrangement of the truss members and the reaction components to determine if the simple plane truss is stable. Some simple plane trusses having $(2n - 3)$ members are not necessarily stable.

Braced Barrel Vaults

During recent years, the advantages of barrel vaults as viable and often highly suitable forms for covering not only low-cost industrial buildings, warehouses, large-span hangars, indoor sports stadia, and gymnasia but also large cultural and leisure centers have been realized. The impact of industrialization on prefabricated braced barrel vaults has proved to be the most significant factor leading to lower costs for these structures. The response of those barrel vaults which work as if they were thin isotropic or anisotropic shell structures can be classified into three types.

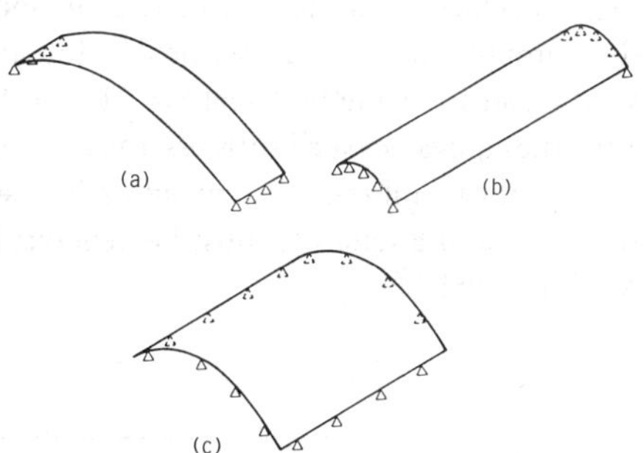

FIGURE 45.21 Different mode responses of braced barrel vaults: (a) arch mode, (b) beam mode, (c) shell mode.

Arch Mode

If the structure is supported only along its sides as shown in Fig. 45.21(a) then it works principally as an **arch** structure. Displacement in the radial direction is resisted by circumferential bending stiffness.

Beam Mode

If a barrel vault is supported only at its ends as shown in Fig. 45.21(b) then it works principally as a **simple beam**. This class of barrel vault has to resist longitudinal compressive stresses near the crown of the vault, longitudinal tensile stresses toward the free edges, and shear stresses toward the supports. The end support could be provided by means of a thin diaphragm. If the diaphragm is rigid then the structure will behave as a clamped beam.

Shell Mode

If a barrel vault is supported along both sides and both ends as shown in Fig. 45.21(c) then the structure responds as a **thin shell**. This structure looks as if it is partly an arch and partly a beam. The loads acting on the structure will be resisted by a complex system of forces acting tangentially to the surface of the barrel vault.

The arch-type shell barrel vaults work mainly by a system of shell bending stresses acting so as to deform the shell in a radial direction while the beam and shell vaults work mainly by in-plane shell forces. For these latter structures bending stiffness is required only to maintain the stability of those parts of the structure which work in compression.

Any barrel vault which is properly triangulated and also any with a stressed skin will behave essentially like the shell structures described above. Those that are not fully triangulated (and without a stressed skin) will have their own individual force distribution, which will depend on the particular bracing configuration employed and will certainly include members working in flexure.

Structural Idealization for Analysis

Two different approaches are in use for physical idealization of braced barrel vaults. In the first approach the structure is represented by an *equivalent shell*, and in the second approach the structure is analyzed directly as a general assembly of discrete simple members. This second approach is called the **space frame** approach if the joints are assumed rigid and the **space truss** approach if the joints are assumed pinned.

With the equivalent shell approach the members are assumed to be smeared over the surface of the barrel vault so as to produce a shell structure with the same stiffness as the original framework. The equivalent shell is then analyzed algebraically or numerically. This approach is restricted to frameworks which are regular in both geometry and cross-sectional properties. The stiffness method is usually employed when the structure is analyzed as a general assembly of discrete members.

Double-layer structures can be analyzed in several ways. An equivalent shell can be produced and analyzed, a full double-layer space frame or truss can be analyzed, and some double-layer structures can be replaced by an equivalent single-layer structure which is then analyzed as a single-layer space frame or truss.

The equivalent shell method can be applied only to regular structures while the stiffness method can handle an arbitrary structural configuration. The computation time is reasonable for an equivalent shell analysis whereas it may be enormous with a stiffness method analysis. The work involved in an equivalent shell analysis includes calculating the equivalent shell stiffness, calculating the forces in the equivalent shell, and calculating the forces in the members. To carry out a stiffness method analysis one needs computer facilities and software, and a data file must be generated describing the structure. [For further studies refer to Mullord, 1985.]

45.9 Influence Lines

Structures such as bridges, industrial buildings with traveling cranes, and frames supporting conveyer belts are subjected to moving loads. Each member of these structures must be designed for the most severe conditions that can possibly be developed in that member. In design, live loads should be placed at the positions where they will produce these severe conditions. The critical positions for placing live loads will not be the same for every member. On some occasions it is possible by inspection to determine where to place the loads to give the most critical forces, but on many other occasions it is necessary to resort to certain criteria to find the locations. The most useful of these methods is the **influence lines** method.

An influence line for a particular response such as reaction, shear force, bending moment, or deflection is defined as a diagram the ordinate to which at any point equals the value of that response attributable to a unit load acting at that point on the structure. Influence lines provide a systematic procedure for determining how the force in a given part of a structure varies as the applied load moves about on the structure. They are primarily used to determine where to place live loads to cause maximum force and to compute the magnitude of those forces. Knowledge of influence lines helps to study the structural response under different moving load conditions.

Influence Lines for Shear in Simple Beams

Figure 45.22 shows influence lines for shear at two sections of a simply supported beam. It is assumed that positive shear occurs when the sum of the transverse forces to the left of a section is in the upward direction or when the sum of the forces to the right of the section is downward. A unit force is placed at various locations and the shear forces at sections 1-1 and 2-2 are obtained for each position of the unit load. These values give the ordinate of the influence line, with which the influence line diagrams for shear force at sections 1-1 and 2-2 can be constructed. Note that the slope of the influence line for shear on the left of the section is equal to the slope of the influence line on the right of the section. This information is useful in drawing shear force influence lines in other cases.

Influence Lines for Bending Moments in Simple Beams

Influence lines for bending moments at sections 1-1 and 2-2 of the simple beam considered in Fig. 45.22 are plotted as shown in Fig. 45.23. For a section, when the sum of the moments of all the

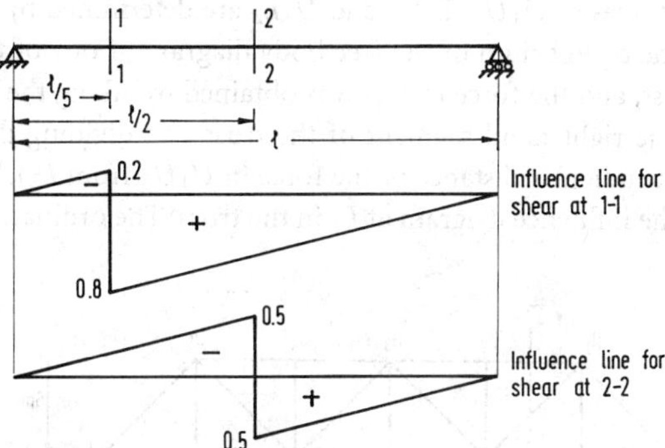

FIGURE 45.22 Influence line for shear force.

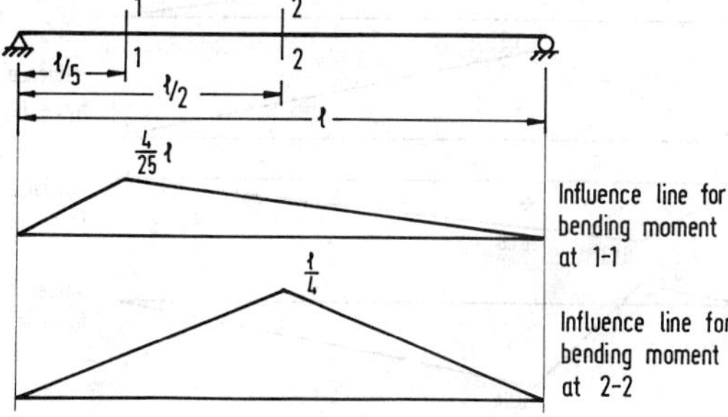

FIGURE 45.23 Influence line for bending moment.

forces to the left is clockwise or when the sum to the right is counterclockwise, the moment is positive. The bending moment values at sections 1-1 and 2-2 are obtained for various positions of unit load and plotted as shown in the figure.

It should be understood that a shear or bending moment diagram shows the variation of shear or moment across an entire structure for loads fixed in one position. On the other hand, an influence line for shear or moment shows the variation of that response at one particular section in the structure caused by the movement of a unit load from one end of the structure to the other.

Influence lines can be used to obtain the value of a particular response when the beam is subjected to any particular type of loading. If, for example, a uniform load of intensity q_0 per unit length is acting over the entire length of the simple beam shown in Fig. 45.22 the shear force at section 1-1 is given by the product of the load intensity q_0 and the net area under the influence line diagram. The net area is equal to $0.3l$ and the shear force at section 1-1 is, therefore, equal to $0.3q_0l$. In the same way, the bending moment at the section can be found as the area of the corresponding influence line diagram times the intensity of loading, q_0. The bending moment at the section is, therefore, $0.08l^2 \times q_0$ or $0.08\, q_0l^2$.

Influence Lines for Trusses

Influence lines for support reactions and member forces may be constructed in the same manner as those for various beam functions. They are useful to determine the maximum load that can be applied to the truss. The unit load moves across the truss, and the ordinates for the responses under consideration may be computed for the load at each panel point. Member force, in most cases, need not be calculated for every panel point, because certain portions of influence lines can readily be seen to consist of straight lines for several panels. One method used for calculating the forces in a chord member of a truss is by the method of sections discussed earlier.

The truss shown in Fig. 45.24 is considered for illustrating the construction of influence lines for trusses. The member forces in U_1U_2, L_1L_2, and U_1L_2 are determined by cutting through section 1-1 and considering the equilibrium of the free body diagram of one of the truss segments. Unit load is placed at L_1 first, and the force in U_1U_2 is obtained by taking the moment about L_2 of all the forces acting on the right-hand segment of the truss and dividing the resulting moment by the lever arm (the perpendicular distance of the force in U_1U_2 from L_2). The value thus obtained gives the ordinate of the influence diagram at L_1 in the truss. The ordinate at L_2 obtained similarly

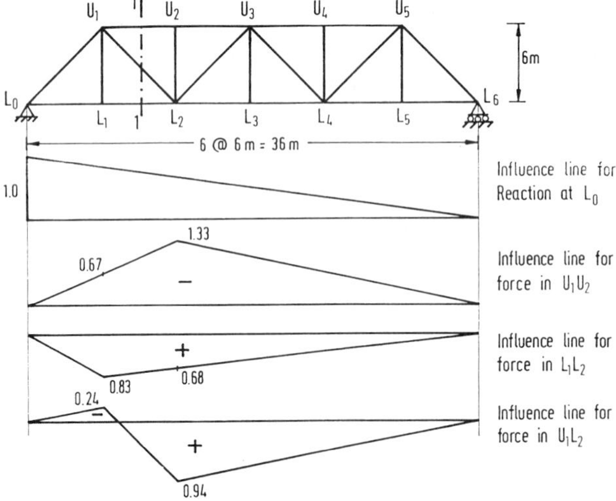

FIGURE 45.24 Influence line for truss.

represents the force in U_1U_2 for unit load placed at L_2. The influence line can be completed with two other points, one at each of the supports. The force in the member L_1L_2 due to unit load placed at L_1 and L_2 can be obtained in the same manner, and the corresponding influence line diagram can be completed. By considering the vertical component of force in the diagonal of the panel the influence line for force in U_1L_2 can be constructed. Figure 45.24 shows the respective influence diagram for member forces in U_1U_2, L_1L_2, and U_1L_2. Influence line ordinates for the force in a chord member of a curved-chord truss may be determined by passing a vertical section through the panel and taking moments at the intersection of the diagonal and the other chord.

45.10 Analysis of Frames

Rigid frames are statically indeterminate in general; special methods are required for their analysis. Slope **deflection** and moment distribution methods are two such methods commonly employed. Slope deflection is a method that takes into account the flexural deformations such as rotations and displacements and involves solutions of simultaneous equations. Moment distribution involves successive cycles of computation, each cycle drawing closer to the "exact" answers. The method is more labor intensive but yields accuracy equivalent to that obtained from exact methods. This method remains the most important hand-calculation method for the analysis of frames.

Slope Deflection Method

This method is a special case of the stiffness method of analysis, and it is convenient for hand analysis of small structures. Moments at the ends of frame members are expressed in terms of the rotations and deflections of the joints. Members are assumed to be of constant section between each pair of supports. It is further assumed that the joints in a structure may rotate or deflect, but the angles between the members meeting at a joint remain unchanged.

The member force-displacement equations that are needed for the slope deflection method are written for a member *AB* in a frame. This member, which has its undeformed position along the *x* axis, is deformed into the configuration shown in Fig. 45.25. The positive axes, along with the positive member-end force components and displacement components, are shown in the figure.

The equations for end moments are written as

$$M_{AB} = \frac{2EI}{l}(2\theta_A + \theta_B - 3\psi_{AB}) + M_{FAB}$$

$$M_{BA} = \frac{2EI}{l}(2\theta_B + \theta_A - 3\psi_{AB}) + M_{FBA} \qquad (45.18)$$

where M_{FAB} and M_{FBA} are fixed-end moments at supports A and B, respectively, due to the applied load. ψ_{AB} is the rotation as a result of the relative displacement between the member ends *A* and *B*

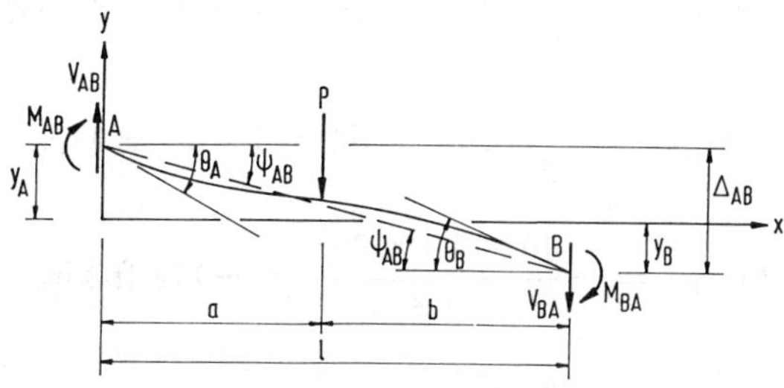

FIGURE 45.25 Deformed configuration of a beam.

given as

$$\psi_{AB} = \frac{\Delta_{AB}}{l} = \frac{y_A + y_B}{l} \tag{45.19}$$

Fixed-end moments for some loading cases may be obtained from Fig. 45.8. The slope deflection equations in Eq. (45.18) show that the moment at the end of a member depends on member properties EI, dimension l, and displacement quantity. The last term in the equations reflect the transverse loading on the member.

Application of the Slope Deflection Method to Frames

The slope deflection equations may be applied to statically indeterminate frames with or without sidesway. A frame may deflect to one side if the loads, **moments of inertia**, and dimensions of the frame are not symmetrical about the centerline. Application of the slope deflection method can be illustrated with the following example, shown in Fig. 45.26.

Let Δ be the sidesway to the right of the frame. Equation (45.18) can be applied to each of the members of the frame as follows. Member AB:

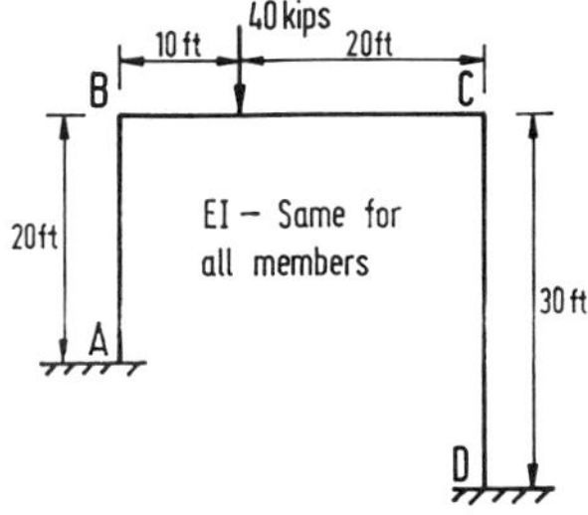

FIGURE 45.26 Slope deflection method.

$$M_{AB} = \frac{2EI}{20}\left(2\theta_A + \theta_B - \frac{3\Delta}{20}\right) + M_{FAB}$$

$$M_{BA} = \frac{2EI}{20}\left(2\theta_B + \theta_A - \frac{3\Delta}{20}\right) + M_{FBA}$$

$$\theta_A = 0, \qquad M_{FAB} = M_{FBA} = 0$$

Hence,

$$M_{AB} = \frac{2EI}{20}(\theta_B - 3\psi) \tag{45.20}$$

$$M_{BA} = \frac{2EI}{20}(2\theta_B - 3\psi) \tag{45.21}$$

where $\psi = \Delta/20$. Member BC:

$$M_{BC} = \frac{2EI}{30}(2\theta_B + \theta_C - 3 \times 0) + M_{FBC}$$

$$M_{CB} = \frac{2EI}{30}(2\theta_C + \theta_B - 3 \times 0) + M_{FCB}$$

$$M_{FBC} = -\frac{40 \times 10 \times 20^2}{30^2} = -178 \text{ ft-kips}$$

$$M_{FCB} = -\frac{40 \times 10^2 \times 20}{30^2} = 89 \text{ ft-kips}$$

Hence

$$M_{BC} = \frac{2EI}{30}(2\theta_B + \theta_C) - 178 \tag{45.22}$$

$$M_{CB} = \frac{2EI}{30}(2\theta_C + \theta_B) + 89 \tag{45.23}$$

Member *CD*:

$$M_{CD} = \frac{2EI}{30}\left(2\theta_C + \theta_D - \frac{3\Delta}{30}\right) + M_{FCD}$$

$$M_{DC} = \frac{2EI}{30}\left(2\theta_D + \theta_C - \frac{3\Delta}{30}\right) + M_{FDC}$$

$$M_{FCD} = M_{FDC} = 0$$

Hence

$$M_{CD} = \frac{2EI}{30}\left(2\theta_C - 3 \times \frac{2}{3}\psi\right)$$

$$= \frac{2EI}{30}(2\theta_C - 2\psi) \tag{45.24}$$

$$M_{DC} = \frac{2EI}{30}\left(\theta_C - 3 \times \frac{2}{3}\psi\right)$$

$$= \frac{2EI}{30}(\theta_C - 2\psi) \tag{45.25}$$

$$\sum M_B = M_{BA} + M_{BC} = 0$$

Substituting for M_{BA} and M_{BC} one obtains

$$\frac{EI}{30}(10\theta_B + 2\theta_C - 9\psi) = 178$$

or

$$10\theta_B + 2\theta_C - 9\psi = 267/K \qquad \text{where } K = EI/20 \tag{45.26}$$

$$\sum M_C = M_{CB} + M_{CD} = 0$$

Substituting for M_{CB} and M_{CD} we get

$$\frac{2EI}{30}(4\theta_C + \theta_B - 2\psi) = -89$$

or

$$\theta_B + 4\theta_C - 2\psi = -\frac{66.75}{K} \tag{45.27}$$

$$\sum H = H_A + H_D = 0$$

or

$$\frac{M_{AB} + M_{BA}}{l_{AB}} + \frac{M_{CD} + M_{DC}}{l_{CD}} = 0$$

Substituting for M_{AB}, M_{BA}, M_{CD}, and M_{DC}, and simplifying,

$$2\theta_B + 12\theta_C - 70\psi = 0 \tag{45.28}$$

Solution of Eqs. (45.26) to (45.28) results in

$$\theta_B = \frac{45.45}{K}$$

$$\theta_C = \frac{20.9}{K}$$

and

$$\psi = \frac{12.8}{K} \tag{45.29}$$

Substituting for θ_B, θ_C, and ψ from Eqs. (45.29) into Eqs. (45.20) to (45.25), we get $M_{AB} = 10.10$ ft-kips, $M_{BA} = 93$ ft-kips, $M_{BC} = -93$ ft-kips, $M_{CB} = 90$ ft-kips, $M_{CD} = -90$ ft-kips, and $M_{DC} = -62$ ft-kips.

Moment Distribution Method

The slope deflection method of frame analysis involves the solution of inconvenient simultaneous equations. One does not have to solve such equations in the case of the moment distribution method, which involves successive cycles of computation, each cycle drawing closer to the exact answers. The calculations may be stopped after two or three cycles, giving a very good approximate analysis, or they may be carried on to whatever degree of accuracy is desired. Moment distribution remains the most important hand-calculation method for the analysis of continuous beams and frames, and it may be solely used for the analysis of small structures.

The theory and application of the moment distribution method is simple, and readers will be able to grasp quickly the principles involved. Some of the terms constantly used in moment distribution are fixed-end moments, unbalanced moment, distributed moments, and carryover moments. When all of the joints of a structure are clamped to prevent any joint rotation, the external loads produce certain moments at the ends of the members to which they are applied. These moments are referred to as *fixed-end moments*. Initially, the joints in a structure are considered to be clamped. When a joint is released, it rotates if the sum of the fixed-end moments at the joint is not zero. The difference between zero and the actual sum of the end moments is the *unbalanced moment*. After the clamp at a joint is released, the unbalanced moment causes the joint to rotate. The rotation twists the ends of the members at the joint and changes their moments. In other words, rotation of the joint is resisted by the members and resisting moments are built up in the members as they are twisted. Rotation continues until equilibrium (the point at which the resisting moments equal the unbalanced moment) is reached, at which time the sum of the moments at the joint is equal to zero. The moments developed in the members resisting rotation are the *distributed moments*. The distributed moments in the ends of the member cause moments in the other ends, which are assumed fixed, and these are the *carryover moments*.

Sign Convention

The moments at the end of a member are assumed to be positive when they tend to rotate the member clockwise about the joint. This implies that the resisting moment of the joint would be counterclockwise. Accordingly, under gravity-loading conditions the fixed-end moment at the left end is counterclockwise ($-ve$) and at the right end is clockwise ($+ve$).

Fixed-End Moments

Fixed-end moments for several cases of loading may be found in Fig. 45.8. Application of moment distribution may be explained with reference to an example as follows.

The continuous beam shown in Fig. 45.27 is considered in this example. Fixed-end moments are computed for each of the three spans. At joint B the unbalanced moment is obtained and the clamp is removed. The joint rotates, thus distributing the unbalanced moment to the B ends of spans BA and BC in proportion to their distribution factors. The values of these distributed moments are carried over at one-half rate to the other ends of the members. When equilibrium is reached, joint B is clamped in its new rotated position and joint C is released afterwards. Joint C rotates under its unbalanced moment until it reaches equilibrium, the rotation causing distributed moments in the C ends of members CB and CD and their resulting carryover moments. Joint C is now clamped and joint B is released. This procedure is repeated for joints B and C until the release of a joint causes negligible rotation. This process is called *moment distribution.*

The stiffness factors and distribution factors are computed as follows:

$$DF_{BA} = \frac{K_{BA}}{\sum K} = \frac{I/20}{I/20 + I/30} = 0.6$$

$$DF_{BC} = \frac{K_{BC}}{\sum K} = \frac{I/30}{I/20 + I/30} = 0.4$$

$$DF_{CB} = \frac{K_{CB}}{\sum K} = \frac{I/30}{I/30 + I/25} = 0.45$$

$$DF_{CD} = \frac{K_{CD}}{\sum K} = \frac{I/25}{I/30 + I/25} = 0.55$$

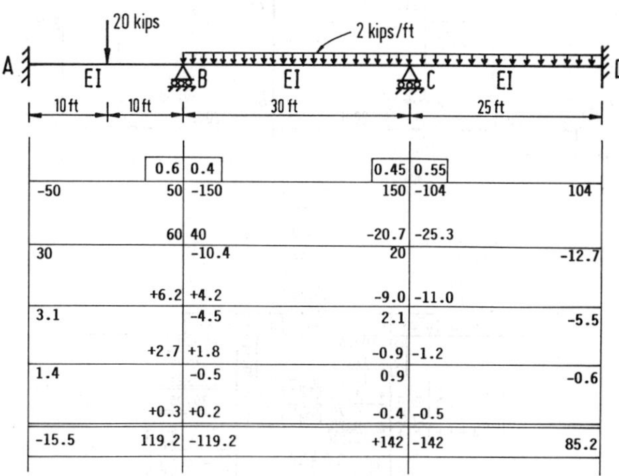

	0.6	0.4	0.45	0.55	
-50	50	-150	150	-104	104
	60	40	-20.7	-25.3	
30		-10.4	20		-12.7
	+6.2	+4.2	-9.0	-11.0	
3.1		-4.5	2.1		-5.5
	+2.7	+1.8	-0.9	-1.2	
1.4		-0.5	0.9		-0.6
	+0.3	+0.2	-0.4	-0.5	
-15.5	119.2	-119.2	+142	-142	85.2

FIGURE 45.27 Continuous beam by moment distribution.

Fixed-end moments: $M_{FAB} = -50$ ft-kips; $M_{FBA} = 50$ ft-kips; $M_{FBC} = -150$ ft-kips; $M_{FCB} = 150$ ft-kips; $M_{FCD} = -104$ ft-kips; and $M_{FDC} = 104$ ft-kips.

When a clockwise couple is applied at the near end of a beam, a clockwise couple of half the magnitude is set up at the far end of the beam. The ratio of the moments at the far and near ends is defined as the *carryover factor*, and it is $\frac{1}{2}$ in the case of a straight prismatic member. The carryover factor was developed for carrying over to fixed ends, but it is applicable to simply supported ends, which must have final moments of zero. It can be shown that the beam simply supported at the far end is only three-fourths as stiff as the one that is fixed. If the stiffness factors for end spans that are simply supported are modified by three-fourths and the simple end is initially balanced to zero, no carryovers are made to the end afterward. This simplifies the moment distribution process significantly.

Moment Distribution for Frames

Moment distribution for frames without sidesway is similar to that for continuous beams. Where sidesway is possible, however, it must be taken into account, because the movements or deflections cause twisting and affect the moments in members. The example shown in Fig. 45.28 illustrates the applications of moment distribution for a frame without sidesway.

$$DF_{BA} = \frac{EI/20}{(EI/20) + (EI/20) + (2EI/20)} = 0.25 \qquad (45.30)$$

Similarly, $DF_{BE} = 0.50$; $DF_{BC} = 0.25$; $M_{FBC} = -100$ ft-kips; $M_{FCB} = 100$ ft-kips; $M_{FBE} = 50$ ft-kips; and $M_{FEB} = -50$ ft-kips.

Structural frames that occur in practice are usually subjected to sway in one direction or another due to structural asymmetry and eccentricity of loading. The sway deflections affect the moments, resulting in an unbalanced moment. These moments could be obtained for the computed deflections and added to the originally distributed fixed-end moments. The sway moments are distributed to columns. Should a frame have columns all of the same length and stiffness, the sidesway moments will be the same for each column. However, should the columns have

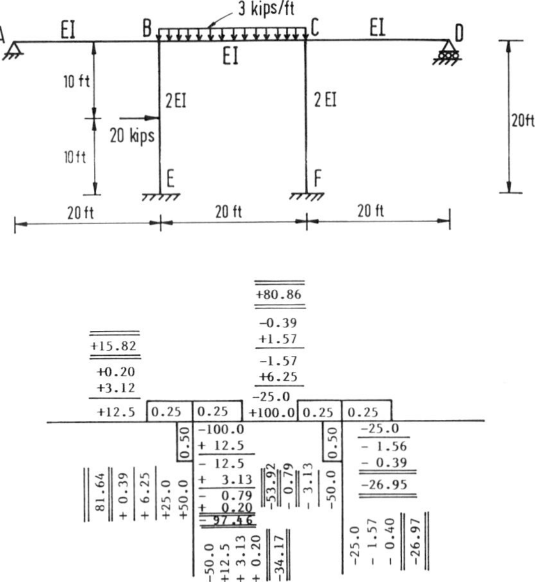

FIGURE 45.28 Nonsway frame by moment distribution.

differing lengths and/or stiffnesses, this will not be the case. The sidesway moments should vary from column to column in proportion to their I/l^2 values.

The frame in Fig. 45.29 shows a frame subjected to sway, and the process of obtaining the final moments is illustrated for this frame.

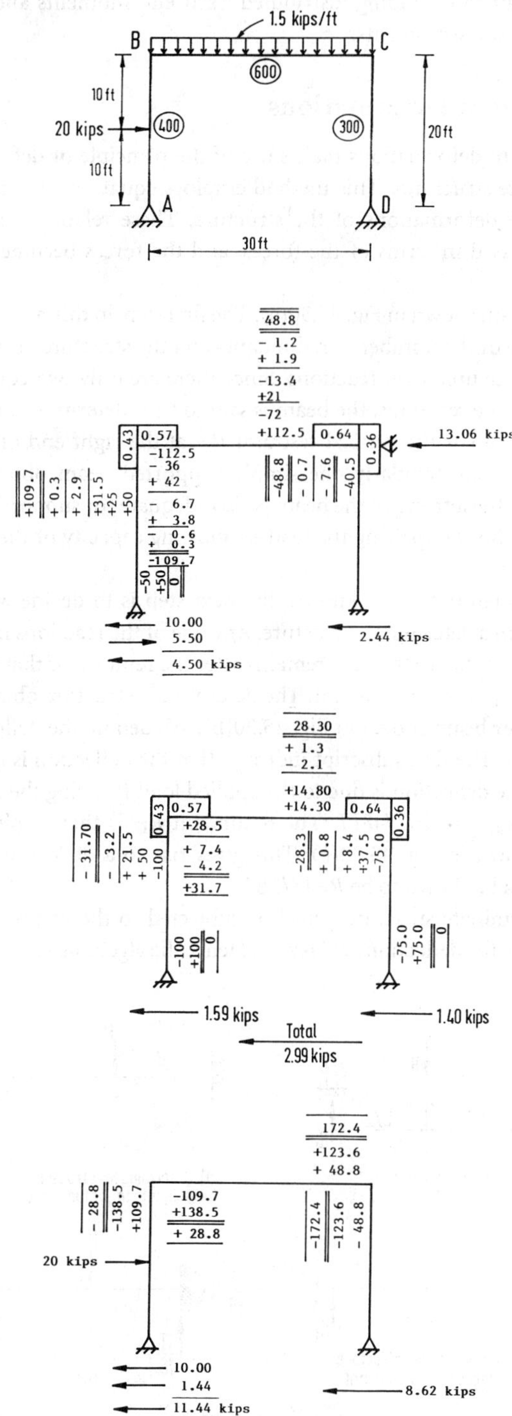

FIGURE 45.29 Sway frame by moment distribution.

The frame sways to the right. The sidesway moment can be assumed in the ratio

$$\frac{400}{20^2} : \frac{300}{20^2} \quad \text{or} \quad 1:0.75$$

Final moments are obtained by adding distributed fixed-end moments and 13.06/2.99 times the distributed assumed sidesway moments.

Method of Consistent Deformations

The method of consistent deformations makes use of the principle of deformation compatibility to analyze indeterminate structures. This method employs equations that relate the forces acting on the structure to the deformations of the structure. These relations are formed so that the deformations are expressed in terms of the forces, and the forces become the unknowns in the analysis.

Let us consider the beam shown in Fig. 45.30(a). The first step, in this method, is to determine the degree of indeterminacy or the number of redundants that the structure possesses. As shown in the figure, the beam has three unknown reactions. Since there are only two equations of equilibrium available for calculating the reactions, the beam is said to be indeterminate to the first degree. The cantilever beam that results if the vertical restraint R_C at the right end of the beam is removed suffices to support the load. Similarly, the simply supported beam obtained by removing the moment restraint M_A at the left end of the beam is also adequate for supporting the load. Restraints that can be removed without impairing the load-supporting capacity of the structure are referred to as *redundants.*

Once the number of redundants is known, the next step is to decide which reaction is to be removed in order to form a determinate structure. Any one of the reactions may be chosen to be the redundant provided that a stable structure remains after the removal of that reaction. For example, let us take the reaction R_C as the redundant. The determinate structure obtained by removing this restraint is the **cantilever** beam shown in Fig. 45.30(b). We denote the deflection at end C of this beam, due to P, by Δ_{CP}. The first subscript indicates that the deflection is measured at C and the second subscript that the deflection is due to the applied load P. Using the moment-area method, it can be shown that $\Delta_{CP} = 5PL^3/48EI$. The redundant R_C is then applied to the determinate cantilever beam, as shown in Fig. 45.30(c). This gives rise to a deflection Δ_{CR} at point C, the magnitude of which can be shown to be $R_C L^3/3EI$.

In the actual indeterminate structure, which is subjected to the combined effects of the load P and the redundant R_C, the deflection at C is zero. Hence the algebraic sum of the deflection Δ_{CP} in

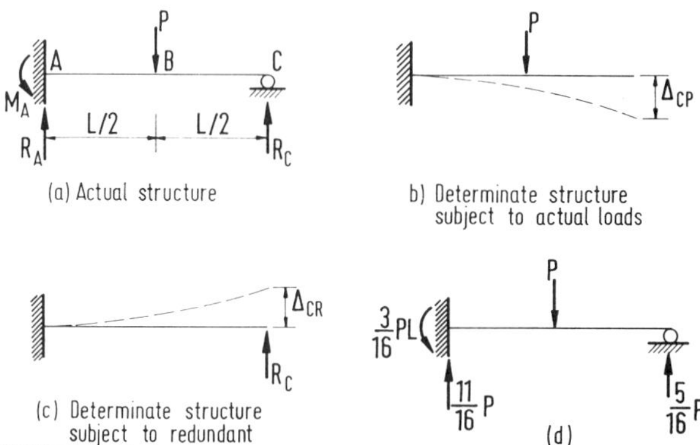

FIGURE 45.30 Solving a beam problem by the method of consistent deformations.

Fig. 45.30(b) and the deflection Δ_{CR} in Fig. 45.30(c) must vanish. Assuming downward deflections to be positive, we write

$$\Delta_{CP} - \Delta_{CR} = 0 \tag{45.31}$$

or

$$\frac{5PL^3}{48EI} - \frac{R_C L^3}{3EI} = 0$$

from which

$$R_C = \frac{5}{16}P$$

Equation (45.31), which is used to solve for the redundant, is referred to as an equation of consistent deformations.

Once the redundant R_C has been evaluated, one can determine the remaining reactions by applying the equations of equilibrium to the structure in Fig. 45.30(a). Thus $\sum F_y = 0$ leads to

$$R_A = P - \frac{5}{16}P = \frac{11}{16}P$$

and $\sum M_A = 0$ gives

$$M_A = \frac{PL}{2} - \frac{5}{16}PL = \frac{3}{16}PL$$

A free-body diagram of the beam, showing all the forces acting on it, is shown in Fig. 45.30(d).

The steps involved in the method of consistent deformations are as follows: (1) The number of redundants in the structure are determined. (2) Enough redundants to form a determinate structure are removed. (3) The displacements that the applied loads cause in the determinate structure at the points where the redundants have been removed are then calculated. (4) The displacements at these points in the determinate structure due to the redundants are obtained. (5) At each point where a redundant has been removed, the sum of the displacements calculated in steps (3) and (4) must be equal to the displacement that exists at that point in the actual indeterminate structure. The redundants are evaluated using these relationships. (6) Once the redundants are known, the remaining reactions are determined using the equations of equilibrium.

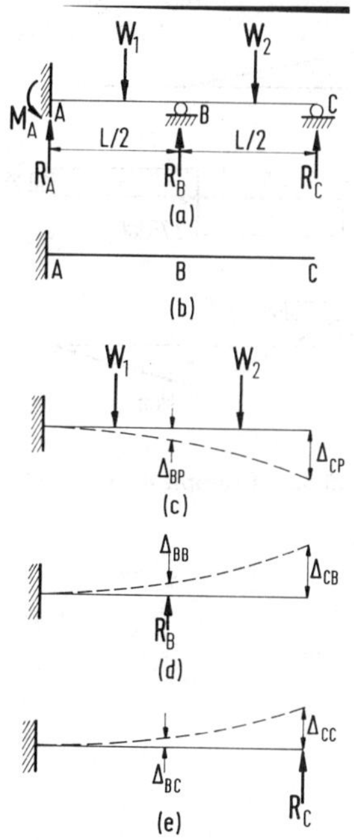

FIGURE 45.31 Beam with two redundant reactions.

Structures with Several Redundants

The method of consistent deformations can be applied to structures with two or more redundants. For example, the beam in Fig. 45.31(a) is indeterminate to the second degree and has two redundant reactions. If we let the reactions at B and C be the redundants, then the determinate structure obtained by removing these supports is the cantilever beam shown in Fig. 45.31(b). To this determinate structure we apply separately the given load [Fig. 45.31(c)] and the redundants R_B and R_C one at a time [Figs. 45.31(d) and (e)].

Since the deflections at B and C in the original beam are zero, the algebraic sum of the deflections in Fig. 45.31(c), (d), and (e) at these same points must also vanish. Thus

$$\Delta_{BP} - \Delta_{BB} - \Delta_{BC} = 0$$

$$\Delta_{CP} - \Delta_{CB} - \Delta_{CC} = 0 \tag{45.32}$$

It is useful in the case of complex structures to write the equations of consistent deformations in the form

$$\Delta_{BP} - \delta_{BB}R_B - \delta_{BC}R_C = 0$$

$$\Delta_{CP} - \delta_{CB}R_B - \delta_{CC}R_C = 0 \tag{45.33}$$

where δ_{BC}, for example, denotes the deflection at B due to a unit load at C in the direction of R_C. Solution of Eq. (45.33) gives the redundant reactions R_B and R_C.

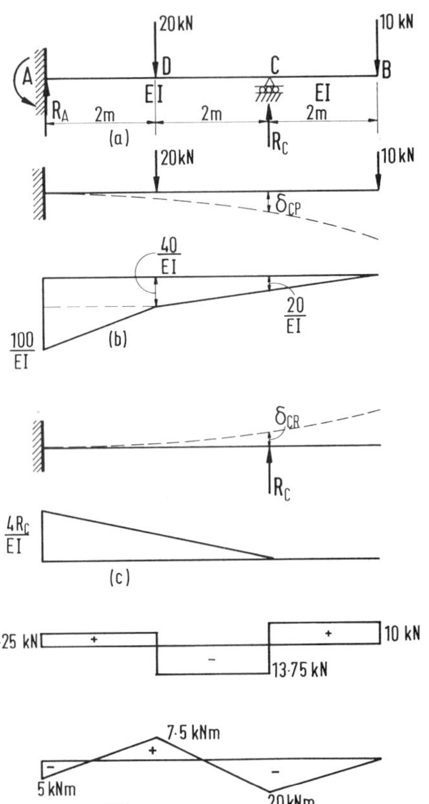

FIGURE 45.32 Example 45.3.

Example 45.3. Determine the reactions for the beam shown in Fig. 45.32 and draw its shear force and bending moment diagrams.

It can be seen from the figure that there are three reactions, M_A, R_A, and R_C, one more than that required for a stable structure. The reaction R_C can be removed to make the structure determinate. We know that the deflection at support C of the beam is zero. One can determine the deflection δ_{CP} at C due to the applied load on the cantilever in Fig. 45.32(b). In the same way the deflection δ_{CR} at C due to the redundant reaction on the cantilever [Fig. 45.32(c)] can be determined. The compatibility equation gives

$$\delta_{CP} - \delta_{CR} = 0$$

By the moment area method,

$$\begin{aligned}\delta_{CP} &= \frac{20}{EI} \times 2 \times 1 + \frac{1}{2} \times \frac{20}{EI} \times 2 \times \frac{2}{3} \times 2 \\ &\quad + \frac{40}{EI} \times 2 \times 3 + \frac{1}{2} \times \frac{60}{EI} \times 2 \times \left(\frac{2}{3} \times 2 + 2\right) \\ &= \frac{1520}{3EI}\end{aligned}$$

$$\delta_{CR} = \frac{1}{2} \times \frac{4R_C}{EI} \times 4 \times \frac{2}{3} \times 4 = \frac{64R_C}{3EI}$$

Substituting for δ_{CP} and δ_{CR} in the compatibility equation one obtains

$$\frac{1520}{3EI} - \frac{64R_C}{3EI} = 0$$

from which

$$R_C = 23.75 \text{ kN} \uparrow$$

By using statical equilibrium equations we get

$$R_A = 6.25 \text{ kN} \uparrow$$

and $M_A = 5$ kNm.

The shear force and bending moment diagrams are shown in Fig. 45.32(c) and (d).

Nonprismatic Members

In the above discussion of framed-structure analysis, structures composed of only uniform elements were considered. However, the members forming a structure are nonprismatic in many instances, or have variable rigidities such as those shown in Fig. 45.33. The expression for member constants, including the fixed-end actions, the flexibility and stiffness coefficients, and the stiffness and carryover factors necessary for a moment distribution, derived specifically for prismatic members, are not valid for nonprismatic members. These member constants are to be determined in order to carry out the analysis of structures made up of nonprismatic members either by slope-deflection, moment distribution, or matrix methods. Various integral formulas expressing these constants are given in the following sections.

Fixed-End Actions

Consider a member of varying fluxural rigidity with both ends fixed and subjected to the bending action caused by member loads, as shown in Fig. 45.34. For such members, fixed-end moment M_1 and support reaction V_1 can be given, respectively, as

$$M_1 = \left(\frac{w}{2}\right) \frac{\left(\int_0^l (x^2\,dx)/I\right)^2 - \int_0^l (x\,dx)/I \int_0^l (x^3\,dx)/I}{\int_0^l dx/I \int_0^l (x^2\,dx)/I - \left(\int_0^l (x\,dx)/I\right)^2} \tag{45.34}$$

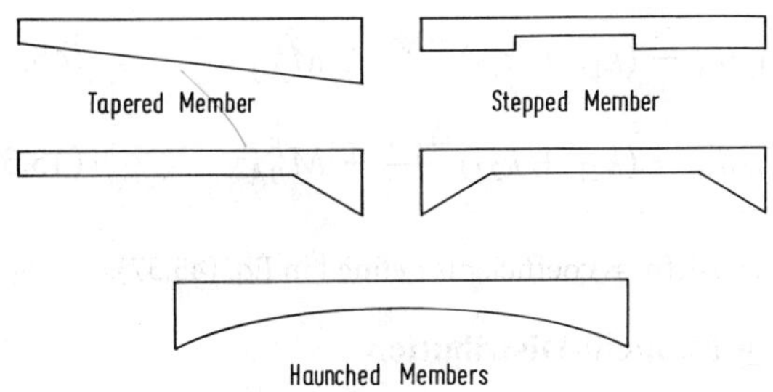

FIGURE 45.33 Beams with varying rigidity.

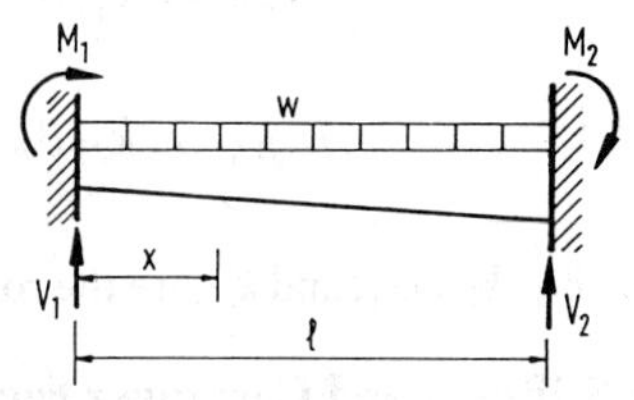

FIGURE 45.34 Fixed-end beam with varying rigidity.

$$V_1 = \frac{w}{2}\frac{\int_0^l dx/I \int_0^l (x^3\,dx)/I - \int_0^l (x\,dx)/I \int_0^l (x^2\,dx)/I}{\left(\int_0^l (x\,dx)/I\right)^2 - \int_0^l dx/I \int_0^l (x^2\,dx)/I} \tag{45.35}$$

where I is the moment of inertia of the beam cross section and is a function of x. The fixed-end moment and support reaction at 2 can also be obtained from Eqs. (45.34) and (45.35) by taking the integral origin at 2 and using reverse sign. For a member of varying I, M_1 and M_2 are not equal except in a symmetrical system.

Flexibility Matrix of a Beam Element

For the beam element with a variable cross section subjected to end moments R_1 and R_2 with the corresponding rotations r_1 and r_2, as in Fig. 45.35, the flexibility matrix is given as

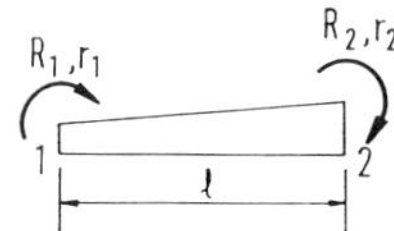

FIGURE 45.35 Beam element with varying rigidity.

$$[f] = \begin{bmatrix} f_{11} & f_{12} \\ f_{21} & f_{22} \end{bmatrix} = \frac{1}{El^2}\begin{bmatrix} \int_0^l \frac{(l-x)^2\,dx}{I} & -\int_0^l \frac{(l-x)(x)\,dx}{I} \\ -\int_0^l \frac{(l-x)(x)\,dx}{I} & \int_0^l \frac{x^2\,dx}{I} \end{bmatrix} \tag{45.36}$$

Stiffness Matrix of a Beam Element

For the same beam element in Fig. 45.35, the stiffness matrix is given as

$$[k] = \begin{bmatrix} k_{11} & k_{12} \\ k_{21} & k_{22} \end{bmatrix} = \frac{E}{\int_0^l \frac{dx}{I}\int_0^l \frac{x^2dx}{I} - \left(\int_0^l \frac{x\,dx}{I}\right)^2}\begin{bmatrix} \int_0^l \frac{x^2\,dx}{I} & \int_0^l \frac{(l-x)(x)\,dx}{I} \\ \int_0^l \frac{(l-x)(x)\,dx}{I} & \int_0^l \frac{(l-x)^2\,dx}{I} \end{bmatrix} \tag{45.37}$$

The Generalized Slope-Deflection Equations

The generalized slope-deflection equations applicable to cases involving variable moments of inertia are given with reference to Fig. 45.25 as

$$M_{AB} = k_{11}\theta_A + k_{12}\theta_B - (k_{11} + k_{12})\frac{\Delta_{AB}}{l} + M_{AB}^F \tag{45.38}$$

$$M_{BA} = k_{21}\theta_A + k_{22}\theta_B - (k_{21} + k_{22})\frac{\Delta_{AB}}{l} + M_{BA}^F \tag{45.39}$$

where k_{11}, k_{12}, k_{21}, and k_{22} are the rotational stiffness coefficients defined in Eq. (45.37).

The Stiffness and Carryover Factor for Moment Distribution

We know that rotational stiffness for an end of a member is the end moment required to produce a unit rotation at this end (simple end) while the other end is fixed. The carryover factor is the ratio of induced moment at the other end (fixed) to the applied moment at this end. For a member with varying I we can write, with reference to Fig. 45.36, the rotational stiffness at end 1 as

$$S_{12} = \frac{E\int_0^l (x^2\,dx)/I}{\int_0^l dx/I \int_0^l (x^2\,dx)/I - \left(\int_0^l (x\,dx)/I\right)^2} \tag{45.40}$$

The rotational stiffness of end 2 of the member 1–2 is

$$S_{21} = \frac{E\int_0^l (l-x)^2\,dx/I}{\int_0^l dx/I \int_0^l (x^2\,dx)/I - \left(\int_0^l (x\,dx)/I\right)^2} \tag{45.41}$$

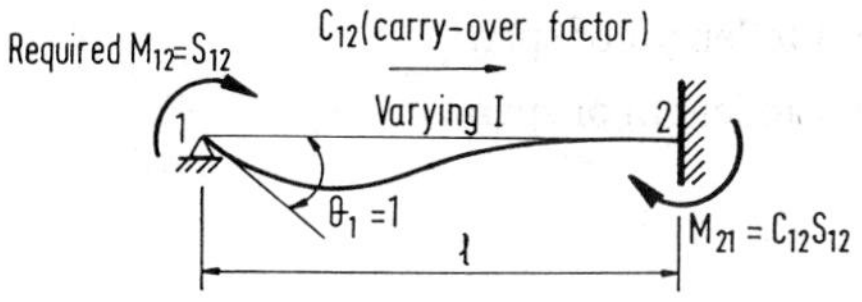

FIGURE 45.36 Carryover factor.

From Fig. 45.36, the induced moment M_{21} at end 2 is the carryover moment from end 1 and is therefore equal to $C_{12}S_{12}$. Thus one can write

$$C_{12} = \frac{\int_0^l (l-x)x\,dx/I}{\int_0^l x^2\,dx/I} \tag{45.42}$$

and

$$C_{12} = \frac{\int_0^l (l-x)x\,dx/I}{\int_0^l (l-x)^2\,dx/I} \tag{45.43}$$

Fixed-End Moment Due to Joint Translation

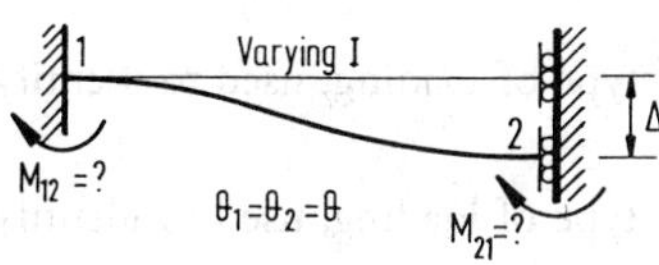

FIGURE 45.37 Beam of varying I subjected to end translation.

For a member having varying I subjected to a pure relative end translation Δ, as shown in Fig. 45.37, the moment restraints at end 1 and 2 can readily be obtained as

$$\begin{aligned} M_{12} &= -(S_{12} + C_{21}S_{21})\frac{\Delta}{l} \\ &= -S_{12}(1 + C_{12})\frac{\Delta}{l} \end{aligned} \tag{45.44}$$

$$\begin{aligned} M_{21} &= -(S_{21} + C_{12}S_{12})\frac{\Delta}{l} \\ &= -S_{21}(1 + C_{21})\frac{\Delta}{l} \end{aligned} \tag{45.45}$$

If end 2 is hinged, the modified fixed-end moment at 1, called M_{12}, resulting from the relative end translation Δ can be found by first assuming both ends fixed and subsequently restoring end 2 to its original hinged condition. Thus,

$$M'_{12} = M_{12} - C_{21}M_{21}$$

Using Eqs. (45.44) and (45.45), we obtain

$$M'_{12} = -S_{12}(1 - C_{12}C_{21})\frac{\Delta}{l} \tag{45.46}$$

Beam Factors and Moment Coefficients for Nonprismatic Members

Determination of fixed-end moments and stiffness and carryover factors for a nonprismatic member involves considerable labor. Graphs and tables have, therefore, been made available to determine the above data for shapes that are commonly used. Tables 45.2 and 45.3 are some of these tables taken from the *Handbook of Frame Constants* published by the Portland Cement Association (PCA). More of these tables and the relevant derivations of formulas may be obtained from the PCA handbook. The nomenclature in Tables 45.2 and 45.3 is follows:

a_A = ratio of length of haunch at end A to length of span
a_B = ratio of length of haunch at end B to length of span
b = ratio of distance from loading point to end A to length of span
b_1 = ratio of distance from loading point to end B to length of span
C_{AB} = carryover factor of member AB at end A
C_{BA} = carryover factor of member AB at end B
E = modulus of elasticity
h_A = depth of member at end A
h_B = depth of member at end B
h_C = depth of member at minimum section
I_C = moment of inertia of section of minimum depth
I_x = moment of inertia of any section distance x from A or x_1 from B
k_{AB} = stiffness factor of member AB at end A; stiffness = $(k_{AB}EI_c)/L$
k_{BA} = stiffness factor of member AB at end B
L = length of member
M_{AB} = fixed-end moment at end A of member AB for any type of loading; used to identify column of fixed-end member coefficients in tables
M_{BA} = fixed-end moment at end B of member AB for any type of loading; used to identify column of fixed-end member coefficients in tables
P = concentrated load
r_A = $(h_A - h_c)/h_c$ for rectangular cross section at end A
r_B = $(h_B - h_c)/h_c$ for rectangular cross section at end B
w = uniform load
W_A = weight of haunch per linear foot at end A
W_B = weight of haunch per linear foot at end B
x = distance from variable point to end A
x_1 = distance from variable point to end B

More information on this topic may be obtained from Yuan [1992] and Hibbeler [1990].

45.11 Energy Methods in Structural Analysis

Energy methods are a powerful tool in obtaining numerical solutions of statically indeterminate problems. The basic quantity required is the *strain energy*, or work stored due to deformations, of the structure.

Table 45.2 Straight Haunches—Constant Width

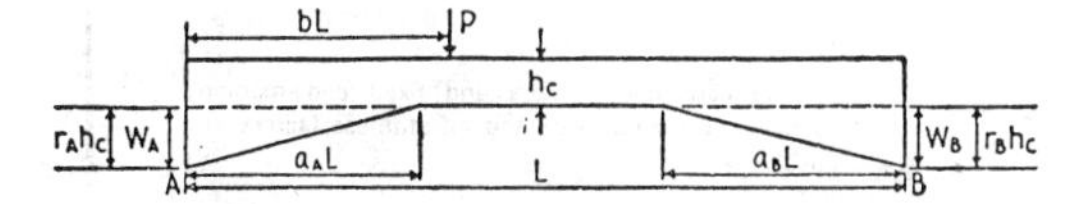

Note:

All carry-over factors and fixed end moment coefficients are negative and all stiffness factors are positive.

Right Haunch a_B	r_B	Carry-over Factors C_{AB}	C_{BA}	Stiffness Factors k_{AB}	k_{BA}	Unif. Load F.E.M. Coef. $\times wL^2$ M_{AB}	M_{BA}	Concentrated Load F.E.M.—Coef. $\times PL$, b = 0.1 M_{AB}	M_{BA}	0.3 M_{AB}	M_{BA}	0.5 M_{AB}	M_{BA}	0.7 M_{AB}	M_{BA}	0.9 M_{AB}	M_{BA}	Haunch Load at Left F.E.M. Coef. $\times W_A L^2$ M_{AB}	M_{BA}	Right F.E.M. Coef. $\times W_B L^2$ M_{AB}	M_{BA}
$a_A = 0.1$				a_B = variable						TABLE 27				$r_A = 0.4$				r_B = variable			
0.1	0.4	0.552	0.552	4.83	4.83	0.0889	0.0889	0.0884	0.0060	0.1629	0.0617	0.1340	0.1340	0.0617	0.1629	0.0060	0.0884	0.0016	0.0000	0.0000	0.0016
	0.6	0.568	0.550	4.90	5.06	0.0871	0.0923	0.0883	0.0063	0.1615	0.0643	0.1310	0.1395	0.0583	0.1693	0.0048	0.0908	0.0016	0.0000	0.0000	0.0016
	1.0	0.591	0.548	4.98	5.37	0.0847	0.0969	0.0881	0.0067	0.1596	0.0679	0.1271	0.1472	0.0538	0.1780	0.0032	0.0938	0.0016	0.0000	0.0000	0.0016
	1.5	0.608	0.547	5.04	5.61	0.0828	0.1005	0.0880	0.0070	0.1582	0.0708	0.1241	0.1531	0.0504	0.1846	0.0021	0.0959	0.0016	0.0001	0.0000	0.0017
	2.0	0.619	0.546	5.08	5.77	0.0817	0.1027	0.0879	0.0071	0.1573	0.0725	0.1221	0.1569	0.0483	0.1888	0.0015	0.0971	0.0016	0.0001	0.0000	0.0017
0.2	0.4	0.501	0.540	4.99	5.55	0.0854	0.0963	0.0881	0.0070	0.1595	0.0689	0.1273	0.1483	0.0552	0.1766	0.0058	0.0889	0.0016	0.0000	0.0004	0.0059
	0.6	0.637	0.535	5.12	6.09	0.0821	0.1029	0.0878	0.0074	0.1567	0.0746	0.1215	0.1600	0.0492	0.1887	0.0044	0.0914	0.0016	0.0000	0.0003	0.0061
	1.0	0.689	0.528	5.31	6.92	0.0774	0.1126	0.0874	0.0083	0.1525	0.0833	0.1130	0.1776	0.0405	0.2066	0.0028	0.0946	0.0016	0.0000	0.0002	0.0063
	1.5	0.730	0.523	5.46	7.63	0.0736	0.1205	0.0870	0.0090	0.1491	0.0906	0.1060	0.1924	0.0335	0.2212	0.0016	0.0967	0.0016	0.0001	0.0001	0.0064
	2.0	0.757	0.519	5.57	8.12	0.0709	0.1258	0.0868	0.0095	0.1468	0.0956	0.1014	0.2022	0.0290	0.2307	0.0011	0.0979	0.0016	0.0001	0.0001	0.0065
0.3	0.4	0.642	0.523	5.10	6.26	0.0836	0.1003	0.0879	0.0075	0.1572	0.0745	0.1232	0.1580	0.0524	0.1820	0.0060	0.0883	0.0016	0.0000	0.0013	0.0122
	0.6	0.698	0.513	5.29	7.19	0.0794	0.1093	0.0875	0.0084	0.1533	0.0834	0.1153	0.1754	0.0449	0.1974	0.0047	0.0907	0.0016	0.0000	0.0011	0.0127
	1.0	0.784	0.499	5.60	8.79	0.0728	0.1237	0.0867	0.0100	0.1466	0.0981	0.1025	0.2036	0.0337	0.2216	0.0030	0.0940	0.0016	0.0001	0.0007	0.0134
	1.5	0.858	0.489	5.88	10.34	0.0671	0.1365	0.0861	0.0114	0.1407	0.1119	0.0910	0.2299	0.0246	0.2426	0.0019	0.0962	0.0016	0.0001	0.0005	0.0139
	2.0	0.909	0.481	6.09	11.50	0.0632	0.1455	0.0856	0.0125	0.1363	0.1222	0.0831	0.2487	0.0180	0.2567	0.0013	0.0976	0.0016	0.0001	0.0003	0.0143
0.4	0.4	0.673	0.503	5.17	6.91	0.0829	0.1016	0.0877	0.0078	0.1559	0.0780	0.1212	0.1626	0.0523	0.1802	0.0062	0.0872	0.0016	0.0000	0.0032	0.0196
	0.6	0.747	0.488	5.41	8.29	0.0782	0.1118	0.0872	0.0097	0.1510	0.0895	0.1118	0.1837	0.0449	0.1955	0.0050	0.0895	0.0016	0.0000	0.0027	0.0206
	1.0	0.870	0.466	5.83	10.90	0.0705	0.1295	0.0862	0.0113	0.1423	0.1104	0.0961	0.2207	0.0335	0.2198	0.0033	0.0927	0.0016	0.0001	0.0019	0.0222
	1.5	0.986	0.448	6.26	13.77	0.0633	0.1468	0.0853	0.0138	0.1337	0.1322	0.0808	0.2583	0.0238	0.2416	0.0022	0.0951	0.0016	0.0001	0.0013	0.0235
	2.0	1.071	0.437	6.60	16.17	0.0581	0.1599	0.0845	0.0157	0.1269	0.1497	0.0693	0.2876	0.0174	0.2566	0.0015	0.0966	0.0016	0.0001	0.0010	0.0243
0.5	0.4	0.691	0.483	5.22	7.46	0.0824	0.1009	0.0876	0.0081	0.1551	0.0795	0.1198	0.1623	0.0522	0.1764	0.0064	0.0863	0.0016	0.0000	0.0060	0.0274
	0.6	0.780	0.461	5.49	9.28	0.0776	0.1113	0.0870	0.0096	0.1496	0.0926	0.1100	0.1841	0.0450	0.1901	0.0052	0.0884	0.0016	0.0000	0.0052	0.0290
	1.0	0.938	0.430	5.99	13.08	0.0695	0.1302	0.0859	0.0125	0.1397	0.1181	0.0932	0.2243	0.0341	0.2127	0.0036	0.0914	0.0016	0.0001	0.0039	0.0315
	1.5	1.101	0.406	6.56	17.79	0.0616	0.1502	0.0846	0.0159	0.1289	0.1476	0.0762	0.2680	0.0249	0.2338	0.0025	0.0939	0.0016	0.0001	0.0029	0.0339
	2.0	1.230	0.390	7.05	22.23	0.0555	0.1667	0.0836	0.0190	0.1199	0.1737	0.0628	0.3046	0.0186	0.2491	0.0017	0.0954	0.0016	0.0001	0.0022	0.0357
$a_A = 0.2$				a_B = variable						TABLE 28				$r_A = 0.4$				r_B = variable			
0.1	0.4	0.540	0.601	5.55	4.99	0.0963	0.0854	0.0889	0.0058	0.1766	0.0552	0.1483	0.1273	0.0689	0.1595	0.0070	0.0881	0.0059	0.0004	0.0000	0.0016
	0.6	0.556	0.600	5.63	5.23	0.0944	0.0887	0.0887	0.0060	0.1752	0.0576	0.1452	0.1327	0.0652	0.1660	0.0054	0.0905	0.0059	0.0004	0.0000	0.0016
	1.0	0.579	0.597	5.74	5.56	0.0918	0.0934	0.0885	0.0064	0.1733	0.0610	0.1409	0.1403	0.0602	0.1748	0.0036	0.0936	0.0059	0.0004	0.0000	0.0016
	1.5	0.596	0.596	5.81	5.81	0.0898	0.0968	0.0883	0.0067	0.1718	0.0636	0.1377	0.1461	0.0564	0.1815	0.0024	0.0957	0.0059	0.0004	0.0000	0.0017
	2.0	0.606	0.595	5.87	5.98	0.0886	0.0990	0.0682	0.0068	0.1709	0.0653	0.1356	0.1497	0.0541	0.1857	0.0017	0.0970	0.0059	0.0004	0.0000	0.0017
0.2	0.4	0.588	0.588	5.75	5.75	0.0926	0.0926	0.0885	0.0065	0.1732	0.0618	0.1412	0.1412	0.0618	0.1732	0.0065	0.0885	0.0059	0.0004	0.0004	0.0059
	0.6	0.623	0.583	5.91	6.32	0.0891	0.0991	0.0882	0.0071	0.1704	0.0671	0.1350	0.1527	0.0552	0.1855	0.0050	0.0911	0.0058	0.0005	0.0003	0.0061
	1.0	0.674	0.575	6.15	7.20	0.0840	0.1087	0.0877	0.0079	0.1661	0.0752	0.1259	0.1701	0.0455	0.2037	0.0031	0.0944	0.0058	0.0005	0.0002	0.0063
	1.5	0.714	0.569	6.35	7.97	0.0799	0.1166	0.0873	0.0088	0.1626	0.0821	0.1183	0.1847	0.0377	0.2186	0.0019	0.0966	0.0058	0.0006	0.0001	0.0064
	2.0	0.741	0.565	6.49	8.50	0.0772	0.1218	0.0871	0.0092	0.1602	0.0867	0.1133	0.1945	0.0327	0.2283	0.0012	0.0978	0.0058	0.0006	0.0001	0.0065
0.3	0.4	0.628	0.570	5.89	6.49	0.0908	0.0964	0.0883	0.0071	0.1709	0.0669	0.1369	0.1506	0.0589	0.1785	0.0068	0.0878	0.0059	0.0005	0.0015	0.0121
	0.6	0.682	0.559	6.12	7.48	0.0862	0.1052	0.0878	0.0080	0.1670	0.0751	0.1284	0.1676	0.0507	0.1941	0.0053	0.0902	0.0058	0.0005	0.0012	0.0126
	1.0	0.766	0.543	6.52	9.19	0.0791	0.1194	0.0870	0.0096	0.1601	0.0889	0.1147	0.1955	0.0384	0.2185	0.0034	0.0935	0.0058	0.0006	0.0008	0.0133
	1.5	0.838	0.532	6.89	10.86	0.0729	0.1322	0.0863	0.0111	0.1540	0.1021	0.1022	0.2217	0.0282	0.2399	0.0022	0.0958	0.0057	0.0007	0.0006	0.0138
	2.0	0.888	0.524	7.15	12.13	0.0687	0.1412	0.0858	0.0122	0.1495	0.1117	0.0934	0.2406	0.0214	0.2543	0.0014	0.0972	0.0057	0.0008	0.0004	0.0142
0.4	0.4	0.657	0.549	5.97	7.16	0.0900	0.0975	0.0881	0.0075	0.1696	0.0701	0.1348	0.1549	0.0587	0.1766	0.0070	0.0867	0.0058	0.0005	0.0036	0.0194
	0.6	0.729	0.531	6.28	8.62	0.0850	0.1075	0.0875	0.0087	0.1646	0.0808	0.1248	0.1755	0.0505	0.1919	0.0056	0.0890	0.0058	0.0006	0.0030	0.0204
	1.0	0.849	0.507	6.81	11.41	0.0767	0.1248	0.0865	0.0110	0.1558	0.1004	0.1078	0.2121	0.0379	0.2165	0.0038	0.0923	0.0058	0.0007	0.0022	0.0221
	1.5	0.961	0.488	7.37	14.52	0.0690	0.1420	0.0854	0.0134	0.1468	0.1211	0.0912	0.2496	0.0271	0.2388	0.0025	0.0948	0.0057	0.0009	0.0015	0.0234
	2.0	1.044	0.475	7.81	17.15	0.0632	0.1551	0.0846	0.0154	0.1398	0.1379	0.0785	0.2792	0.0199	0.2544	0.0017	0.0964	0.0057	0.0010	0.0011	0.0242
0.5	0.4	0.674	0.526	6.03	7.72	0.0895	0.0968	0.0880	0.0077	0.1688	0.0715	0.1333	0.1546	0.0585	0.1728	0.0072	0.0858	0.0058	0.0005	0.0067	0.0270
	0.6	0.760	0.502	6.37	9.64	0.0844	0.1069	0.0873	0.0092	0.1633	0.0837	0.1228	0.1758	0.0506	0.1865	0.0059	0.0880	0.0058	0.0006	0.0058	0.0286
	1.0	0.913	0.468	7.01	13.67	0.0757	0.1253	0.0861	0.0121	0.1531	0.1075	0.1047	0.2154	0.0387	0.2092	0.0041	0.0910	0.0057	0.0008	0.0044	0.0311
	1.5	1.070	0.442	7.74	18.75	0.0672	0.1450	0.0848	0.0155	0.1420	0.1355	0.0863	0.2589	0.0284	0.2306	0.0028	0.0935	0.0056	0.0010	0.0033	0.0335
	2.0	1.194	0.424	8.37	23.57	0.0606	0.1614	0.0836	0.0187	0.1325	0.1606	0.0716	0.2957	0.0213	0.2463	0.0020	0.0952	0.0056	0.0013	0.0025	0.0353
$a_A = 0.3$				a_B = variable						TABLE 29				$r_A = 0.4$				r_B = variable			
0.1	0.4	0.523	0.642	6.26	5.10	0.1003	0.0836	0.0883	0.0060	0.1820	0.0524	0.1580	0.1232	0.0745	0.1572	0.0075	0.0379	0.0122	0.0013	0.0000	0.0016
	0.6	0.539	0.641	6.35	5.35	0.0983	0.0870	0.0881	0.0062	0.1803	0.0549	0.1546	0.1287	0.0704	0.1637	0.0059	0.0903	0.0122	0.0014	0.0000	0.0016
	1.0	0.561	0.638	6.48	5.69	0.0955	0.0916	0.0878	0.0066	0.1785	0.0582	0.1502	0.1361	0.0651	0.1727	0.0039	0.0934	0.0121	0.0015	0.0000	0.0016
	1.5	0.577	0.636	6.57	5.96	0.0934	0.0951	0.0876	0.0069	0.1771	0.0607	0.1468	0.1419	0.0610	0.1794	0.0026	0.0957	0.0121	0.0016	0.0000	0.0017
	2.0	0.587	0.635	6.63	6.14	0.0921	0.0973	0.0876	0.0070	0.1760	0.0624	0.1446	0.1454	0.0585	0.1837	0.0019	0.0969	0.0121	0.0016	0.0000	0.0017
0.2	0.4	0.570	0.628	6.49	5.89	0.0964	0.0908	0.0878	0.0068	0.1785	0.0589	0.1506	0.1369	0.0669	0.1709	0.0071	0.0883	0.0121	0.0015	0.0005	0.0059
	0.6	0.604	0.622	6.68	6.48	0.0927	0.0972	0.0875	0.0073	0.1754	0.0642	0.1439	0.1483	0.0598	0.1833	0.0055	0.0909	0.0120	0.0017	0.0004	0.0060
	1.0	0.653	0.614	6.96	7.41	0.0873	0.1068	0.0870	0.0081	0.1711	0.0722	0.1344	0.1656	0.0493	0.2018	0.0034	0.0942	0.0119	0.0019	0.0003	0.0062
	1.5	0.692	0.607	7.21	8.21	0.0829	0.1148	0.0866	0.0091	0.1674	0.0789	0.1263	0.1801	0.0409	0.2168	0.0021	0.0965	0.0118	0.0020	0.0002	0.0064
	2.0	0.717	0.603	7.37	8.77	0.0800	0.1200	0.0862	0.0096	0.1649	0.0835	0.1210	0.1900	0.0355	0.2267	0.0013	0.0976	0.0118	0.0021	0.0001	0.0065
0.3	0.4	0.608	0.608	6.65	6.65	0.0945	0.0945	0.0875	0.0073	0.1762	0.0640	0.1461	0.1461	0.0640	0.1762	0.0073	0.0875	0.0121	0.0016	0.0016	0.0121
	0.6	0.660	0.596	6.93	7.68	0.0897	0.1033	0.0870	0.0083	0.1719	0.0720	0.1371	0.1630	0.0553	0.1918	0.0058	0.0900	0.0120	0.0018	0.0013	0.0126
	1.0	0.741	0.579	7.40	9.47	0.0822	0.1175	0.0861	0.0100	0.1649	0.0856	0.1225	0.1909	0.0419	0.2164	0.0037	0.0934	0.0118	0.0022	0.0009	0.0133
	1.5	0.811	0.566	7.83	11.23	0.0755	0.1303	0.0854	0.0115	0.1584	0.0985	0.1093	0.2170	0.0306	0.2380	0.0024	0.0958	0.0116	0.0026	0.0006	0.0138
	2.0	0.859	0.557	8.15	12.57	0.0709	0.1394	0.0848	0.0127	0.1537	0.1080	0.0998	0.2360	0.0232	0.2527	0.0016	0.0972	0.0115	0.0028	0.0004	0.0142
0.4	0.4	0.635	0.585	6.75	7.33	0.0937	0.0956	0.0873	0.0078	0.1748	0.0671	0.1439	0.1504	0.0636	0.1742	0.0077	0.0864	0.0120	0.0017	0.0039	0.0193
	0.6	0.705	0.565	7.10	8.85	0.0884	0.1055	0.0868	0.0090	0.1696	0.0775	0.1333	0.1707	0.0547	0.1895	0.0061	0.0887	0.0118	0.0021	0.0033	0.0203
	1.0	0.820	0.539	7.73	11.76	0.0796	0.1227	0.0856	0.0113	0.1605	0.0967	0.1154	0.2071	0.0412	0.2144	0.0042	0.0921	0.0116	0.0025	0.0024	0.0219
	1.5	0.928	0.518	8.38	15.02	0.0714	0.1400	0.0843	0.0139	0.1510	0.1170	0.0978	0.2446	0.0295	0.2370	0.0027	0.0947	0.0114	0.0031	0.0017	0.0233
	2.0	1.007	0.504	8.91	17.77	0.0652	0.1533	0.0835	0.0160	0.1437	0.1338	0.0842	0.2745	0.0218	0.2529	0.0019	0.0963	0.0112	0.0035	0.0012	0.0241
0.5	0.4	0.650	0.561	6.81	7.90	0.0932	0.0949	0.0871	0.0080	0.1739	0.0684	0.1423	0.1501	0.0632	0.1704	0.0079	0.0855	0.0120	0.0018	0.0073	0.0267
	0.6	0.732	0.534	7.20	9.88	0.0877	0.1049	0.0866	0.0094	0.1683	0.0802	0.1310	0.1710	0.0547	0.1842	0.0064	0.0877	0.0118	0.0022	0.0063	0.0283
	1.0	0.878	0.497	7.95	14.05	0.0785	0.1232	0.0853	0.0124	0.1577	0.1034	0.1121	0.2104	0.0420	0.2071	0.0045	0.0908	0.0115	0.0028	0.0048	0.0309
	1.5	1.028	0.468	8.80	19.33	0.0695	0.1429	0.0836	0.0160	0.1462	0.1307	0.0926	0.2539	0.0308	0.2287	0.0030	0.0934	0.0112	0.0035	0.0036	0.0333
	2.0	1.147	0.449	9.54	24.37	0.0625	0.1594	0.0825	0.0193	0.1362	0.1561	0.0769	0.2908	0.0232	0.2447	0.0022	0.0950	0.0110	0.0042	0.0027	0.0352

Source: Portland Cement Association, *Handbook of Frame Constants.*

Table 45.3 Parabolic Haunches–Constant Width

Note:

All carry-over factors and fixed end moment coefficients are negative and all stiffness factors are positive.

Right Haunch		Carry-over Factors		Stiffness Factors		Unif. Load F.E.M. Coef. × wL^2		Concentrated Load F.E.M.—Coef. × PL										Haunch Load at			
								b										Left		Right	
								0.1		0.3		0.5		0.7		0.9		F.E.M. Coef. × $W_A L^2$		F.E.M. Coef. × $W_B L^2$	
a_B	r_B	C_{AB}	C_{BA}	k_{AB}	k_{BA}	M_{AB}	M_{BA}	M_{AB}	M_{BA}	M_{AB}	M_{BA}	M_{AB}	M_{BA}	M_{AB}	M_{BA}	M_{AB}	M_{BA}	M_{AB}	M_{BA}	M_{AB}	M_{BA}
$a_A = 0.1$				a_B = variable						TABLE 1				$r_A = 0.4$				r_B = variable			
0.1	0.4	0.537	0.537	4.56	4.56	0.0873	0.0873	0.0869	0.0066	0.1583	0.0621	0.1313	0.1313	0.0621	0.1583	0.0066	0.0869	0.0008	0.0000	0.0000	0.0008
	0.6	0.549	0.536	4.60	4.72	0.0860	0.0898	0.0868	0.0069	0.1573	0.0640	0.1293	0.1354	0.0597	0.1630	0.0056	0.0888	0.0008	0.0000	0.0000	0.0008
	1.0	0.566	0.535	4.66	4.94	0.0841	0.0933	0.0866	0.0071	0.1557	0.0668	0.1262	0.1410	0.0563	0.1694	0.0043	0.0913	0.0008	0.0000	0.0000	0.0008
	1.5	0.580	0.534	4.71	5.12	0.0827	0.0964	0.0865	0.0074	0.1547	0.0691	0.1239	0.1459	0.0536	0.1750	0.0033	0.0934	0.0008	0.0000	0.0000	0.0008
	2.0	0.590	0.533	4.74	5.25	0.0817	0.0983	0.0864	0.0076	0.1539	0.0707	0.1223	0.1492	0.0517	0.1786	0.0026	0.0947	0.0008	0.0000	0.0000	0.0008
0.2	0.4	0.572	0.530	4.67	5.04	0.0845	0.0933	0.0866	0.0072	0.1558	0.0674	0.1263	0.1420	0.0568	0.1693	0.0058	0.0885	0.0008	0.0000	0.0002	0.0030
	0.6	0.598	0.528	4.75	5.38	0.0821	0.0981	0.0864	0.0077	0.1538	0.0715	0.1222	0.1503	0.0525	0.1782	0.0046	0.0909	0.0008	0.0000	0.0001	0.0031
	1.0	0.635	0.523	4.87	5.92	0.0786	0.1053	0.0861	0.0084	0.1508	0.0778	0.1162	0.1628	0.0460	0.1914	0.0030	0.0940	0.0008	0.0000	0.0001	0.0032
	1.5	0.667	0.519	4.98	6.39	0.0754	0.1119	0.0857	0.0090	0.1482	0.0828	0.1110	0.1737	0.0406	0.2027	0.0019	0.0961	0.0008	0.0000	0.0001	0.0032
	2.0	0.689	0.516	5.06	6.75	0.0737	0.1157	0.0856	0.0094	0.1464	0.0872	0.1074	0.1815	0.0369	0.2106	0.0013	0.0974	0.0008	0.0000	0.0000	0.0033
0.3	0.4	0.603	0.521	4.75	5.50	0.0827	0.0974	0.0864	0.0078	0.1538	0.0720	0.1226	0.1505	0.0537	0.1763	0.0057	0.0885	0.0008	0.0000	0.0005	0.0064
	0.6	0.642	0.515	4.87	6.08	0.0795	0.1041	0.0861	0.0085	0.1509	0.0782	0.1169	0.1627	0.0479	0.1883	0.0045	0.0910	0.0008	0.0000	0.0004	0.0066
	1.0	0.702	0.506	5.07	7.03	0.0746	0.1147	0.0856	0.0096	0.1463	0.0884	0.1079	0.1824	0.0392	0.2069	0.0029	0.0941	0.0008	0.0000	0.0003	0.0069
	1.5	0.755	0.498	5.25	7.95	0.0702	0.1242	0.0851	0.0107	0.1421	0.0978	0.0998	0.2005	0.0317	0.2233	0.0018	0.0963	0.0008	0.0000	0.0002	0.0071
	2.0	0.793	0.492	5.38	8.67	0.0673	0.1310	0.0847	0.0115	0.1390	0.1049	0.0939	0.2140	0.0265	0.2350	0.0012	0.0975	0.0008	0.0000	0.0001	0.0072
0.4	0.4	0.629	0.510	4.82	5.95	0.0816	0.0998	0.0862	0.0082	0.1524	0.0755	0.1201	0.1564	0.0519	0.1787	0.0058	0.0882	0.0008	0.0000	0.0012	0.0105
	0.6	0.681	0.500	4.97	6.78	0.0778	0.1080	0.0858	0.0091	0.1487	0.0838	0.1131	0.1720	0.0452	0.1921	0.0046	0.0905	0.0008	0.0000	0.0010	0.0111
	1.0	0.764	0.485	5.23	8.25	0.0718	0.1213	0.0851	0.0107	0.1426	0.0979	0.1016	0.1982	0.0358	0.2136	0.0030	0.0937	0.0008	0.0000	0.0007	0.0118
	1.5	0.842	0.472	5.49	9.79	0.0664	0.1339	0.0844	0.0124	0.1368	0.1121	0.0908	0.2238	0.0273	0.2330	0.0019	0.0959	0.0008	0.0000	0.0005	0.0121
	2.0	0.901	0.463	5.69	11.07	0.0625	0.1434	0.0839	0.0137	0.1322	0.1234	0.0827	0.2437	0.0217	0.2468	0.0013	0.0972	0.0008	0.0000	0.0004	0.0124
0.5	0.4	0.648	0.497	4.90	6.39	0.0809	0.1008	0.0860	0.0089	0.1519	0.0783	0.1188	0.1598	0.0511	0.1794	0.0060	0.0876	0.0008	0.0000	0.0023	0.0153
	0.6	0.711	0.481	5.05	7.46	0.0763	0.1099	0.0854	0.0100	0.1468	0.0881	0.1103	0.1779	0.0440	0.1938	0.0047	0.0900	0.0008	0.0000	0.0020	0.0160
	1.0	0.818	0.458	5.36	9.56	0.0698	0.1253	0.0846	0.0120	0.1398	0.1060	0.0971	0.2085	0.0340	0.2152	0.0031	0.0932	0.0008	0.0000	0.0017	0.0171
	1.5	0.922	0.443	5.73	11.92	0.0639	0.1402	0.0838	0.0141	0.1323	0.1251	0.0844	0.2408	0.0253	0.2351	0.0020	0.0955	0.0008	0.0000	0.0011	0.0181
	2.0	1.003	0.429	6.05	14.05	0.0590	0.1524	0.0831	0.0159	0.1263	0.1411	0.0745	0.2667	0.0195	0.2493	0.0014	0.0967	0.0008	0.0000	0.0008	0.0187
$a_A = 0.2$				a_B = variable						TABLE 2				$r_A = 0.4$				r_B = variable			
0.1	0.4	0.530	0.572	5.04	4.67	0.0933	0.0845	0.0885	0.0058	0.1693	0.0568	0.1420	0.1263	0.0674	0.1558	0.0072	0.0866	0.0030	0.0002	0.0000	0.0008
	0.6	0.543	0.571	5.09	4.83	0.0919	0.0870	0.0884	0.0060	0.1683	0.0587	0.1398	0.1303	0.0649	0.1605	0.0062	0.0886	0.0030	0.0002	0.0000	0.0008
	1.0	0.560	0.570	5.15	5.06	0.0900	0.0905	0.0883	0.0063	0.1669	0.0613	0.1367	0.1359	0.0613	0.1671	0.0047	0.0912	0.0030	0.0002	0.0000	0.0008
	1.5	0.574	0.569	5.21	5.25	0.0884	0.0934	0.0882	0.0065	0.1657	0.0634	0.1342	0.1406	0.0583	0.1725	0.0036	0.0933	0.0030	0.0002	0.0000	0.0008
	2.0	0.583	0.568	5.25	5.39	0.0874	0.0953	0.0881	0.0066	0.1649	0.0649	0.1325	0.1438	0.0563	0.1762	0.0029	0.0945	0.0030	0.0002	0.0000	0.0008
0.2	0.4	0.565	0.565	5.16	5.16	0.0903	0.0903	0.0883	0.0063	0.1668	0.0618	0.1367	0.1367	0.0618	0.1668	0.0063	0.0883	0.0030	0.0002	0.0002	0.0030
	0.6	0.591	0.562	5.26	5.52	0.0878	0.0951	0.0881	0.0067	0.1648	0.0657	0.1325	0.1449	0.0571	0.1757	0.0050	0.0907	0.0030	0.0002	0.0001	0.0031
	1.0	0.627	0.558	5.40	6.08	0.0841	0.1022	0.0877	0.0073	0.1618	0.0715	0.1261	0.1572	0.0502	0.1891	0.0033	0.0938	0.0030	0.0002	0.0001	0.0032
	1.5	0.659	0.553	5.53	6.59	0.0810	0.1083	0.0875	0.0079	0.1592	0.0767	0.1207	0.1680	0.0444	0.2005	0.0021	0.0960	0.0030	0.0002	0.0001	0.0032
	2.0	0.681	0.550	5.62	6.96	0.0789	0.1125	0.0873	0.0083	0.1573	0.0805	0.1168	0.1759	0.0404	0.2085	0.0014	0.0973	0.0030	0.0002	0.0000	0.0033
0.3	0.4	0.596	0.555	5.26	5.64	0.0881	0.0943	0.0881	0.0068	0.1648	0.0661	0.1329	0.1450	0.0585	0.1737	0.0062	0.0883	0.0030	0.0002	0.0006	0.0064
	0.6	0.634	0.549	5.39	6.24	0.0851	0.1009	0.0877	0.0074	0.1619	0.0719	0.1269	0.1571	0.0523	0.1858	0.0049	0.0907	0.0030	0.0002	0.0005	0.0066
	1.0	0.693	0.539	5.64	7.25	0.0799	0.1113	0.0873	0.0084	0.1573	0.0815	0.1174	0.1765	0.0429	0.2046	0.0032	0.0939	0.0030	0.0002	0.0003	0.0068
	1.5	0.745	0.531	5.85	8.22	0.0753	0.1208	0.0868	0.0094	0.1530	0.0905	0.1088	0.1944	0.0348	0.2213	0.0020	0.0961	0.0030	0.0002	0.0002	0.0071
	2.0	0.783	0.525	6.01	8.98	0.0722	0.1275	0.0865	0.0101	0.1499	0.0973	0.1025	0.2079	0.0292	0.2331	0.0013	0.0974	0.0030	0.0003	0.0002	0.0072
0.4	0.4	0.622	0.543	5.34	6.11	0.0873	0.0967	0.0879	0.0072	0.1634	0.0694	0.1303	0.1507	0.0566	0.1762	0.0063	0.0879	0.0030	0.0002	0.0014	0.0106
	0.6	0.673	0.533	5.52	6.97	0.0833	0.1047	0.0875	0.0080	0.1597	0.0771	0.1229	0.1661	0.0494	0.1900	0.0050	0.0903	0.0030	0.0002	0.0011	0.0110
	1.0	0.755	0.517	5.84	8.52	0.0771	0.1178	0.0868	0.0095	0.1536	0.0905	0.1108	0.1920	0.0393	0.2116	0.0033	0.0935	0.0030	0.0003	0.0007	0.0118
	1.5	0.831	0.503	6.15	10.15	0.0713	0.1303	0.0862	0.0110	0.1476	0.1039	0.0994	0.2174	0.0300	0.2311	0.0021	0.0958	0.0030	0.0003	0.0006	0.0121
	2.0	0.889	0.493	6.39	11.51	0.0672	0.1396	0.0857	0.0122	0.1430	0.1147	0.0906	0.2373	0.0237	0.2450	0.0014	0.0971	0.0030	0.0003	0.0004	0.0124
0.5	0.4	0.640	0.527	5.40	6.58	0.0870	0.0978	0.0877	0.0077	0.1627	0.0718	0.1286	0.1541	0.0556	0.1768	0.0065	0.0874	0.0030	0.0002	0.0026	0.0153
	0.6	0.703	0.511	5.60	7.67	0.0823	0.1066	0.0871	0.0089	0.1579	0.0810	0.1200	0.1718	0.0480	0.1915	0.0051	0.0897	0.0030	0.0002	0.0022	0.0160
	1.0	0.807	0.488	5.97	9.65	0.0753	0.1214	0.0863	0.0105	0.1506	0.0979	0.1061	0.2021	0.0371	0.2131	0.0034	0.0929	0.0030	0.0003	0.0017	0.0171
	1.5	0.910	0.470	6.40	12.38	0.0686	0.1361	0.0855	0.0125	0.1430	0.1160	0.0928	0.2341	0.0279	0.2326	0.0022	0.0953	0.0030	0.0003	0.0012	0.0180
	2.0	0.989	0.456	6.74	14.65	0.0632	0.1483	0.0849	0.0144	0.1369	0.1314	0.0819	0.2601	0.0213	0.2472	0.0015	0.0967	0.0029	0.0003	0.0009	0.0186
$a_A = 0.3$				a_B = variable						TABLE 3				$r_A = 0.4$				r_B = variable			
0.1	0.4	0.521	0.603	5.50	4.75	0.0974	0.0827	0.0885	0.0057	0.1763	0.0537	0.1505	0.1226	0.0720	0.1538	0.0078	0.0864	0.0064	0.0005	0.0000	0.0008
	0.6	0.533	0.602	5.56	4.92	0.0959	0.0851	0.0884	0.0059	0.1752	0.0554	0.1482	0.1265	0.0693	0.1585	0.0066	0.0884	0.0064	0.0005	0.0000	0.0008
	1.0	0.550	0.601	5.64	5.16	0.0939	0.0886	0.0883	0.0062	0.1738	0.0579	0.1450	0.1321	0.0654	0.1652	0.0050	0.0911	0.0064	0.0006	0.0000	0.0008
	1.5	0.564	0.600	5.70	5.36	0.0923	0.0915	0.0882	0.0064	0.1726	0.0600	0.1423	0.1368	0.0623	0.1707	0.0039	0.0931	0.0064	0.0006	0.0000	0.0008
	2.0	0.573	0.599	5.74	5.49	0.0912	0.0935	0.0881	0.0066	0.1718	0.0615	0.1405	0.1400	0.0602	0.1744	0.0031	0.0944	0.0064	0.0006	0.0000	0.0008
0.2	0.4	0.555	0.596	5.64	5.26	0.0943	0.0881	0.0883	0.0062	0.1737	0.0585	0.1450	0.1329	0.0661	0.1648	0.0068	0.0881	0.0064	0.0006	0.0002	0.0030
	0.6	0.580	0.593	5.76	5.63	0.0917	0.0931	0.0880	0.0066	0.1717	0.0622	0.1405	0.1409	0.0611	0.1738	0.0054	0.0905	0.0064	0.0006	0.0001	0.0031
	1.0	0.616	0.588	5.93	6.22	0.0877	0.1002	0.0877	0.0073	0.1686	0.0678	0.1339	0.1532	0.0537	0.1873	0.0035	0.0937	0.0063	0.0007	0.0001	0.0032
	1.5	0.647	0.583	6.07	6.74	0.0845	0.1064	0.0874	0.0078	0.1659	0.0728	0.1282	0.1639	0.0476	0.1988	0.0023	0.0959	0.0063	0.0007	0.0001	0.0032
	2.0	0.669	0.580	6.18	7.12	0.0823	0.1105	0.0872	0.0082	0.1640	0.0764	0.1241	0.1716	0.0433	0.2069	0.0015	0.0973	0.0063	0.0007	0.0000	0.0033
0.3	0.4	0.585	0.585	5.76	5.76	0.0923	0.0923	0.0880	0.0067	0.1717	0.0625	0.1410	0.1410	0.0625	0.1717	0.0067	0.0880	0.0064	0.0006	0.0006	0.0064
	0.6	0.623	0.578	5.92	6.38	0.0888	0.0989	0.0877	0.0073	0.1688	0.0682	0.1347	0.1529	0.0560	0.1839	0.0053	0.0905	0.0063	0.0007	0.0005	0.0066
	1.0	0.681	0.568	6.19	7.42	0.0834	0.1093	0.0872	0.0084	0.1640	0.0774	0.1248	0.1722	0.0460	0.2029	0.0034	0.0938	0.0063	0.0008	0.0003	0.0068
	1.5	0.732	0.559	6.44	8.43	0.0786	0.1187	0.0867	0.0093	0.1597	0.0861	0.1158	0.1901	0.0374	0.2197	0.0021	0.0960	0.0062	0.0009	0.0002	0.0071
	2.0	0.769	0.552	6.62	9.22	0.0753	0.1254	0.0863	0.0101	0.1565	0.0928	0.1092	0.2035	0.0313	0.2317	0.0014	0.0973	0.0062	0.0009	0.0002	0.0072
0.4	0.4	0.610	0.572	5.84	6.23	0.0911	0.0946	0.0879	0.0071	0.1703	0.0657	0.1383	0.1466	0.0607	0.1741	0.0068	0.0876	0.0063	0.0007	0.0015	0.0105
	0.6	0.660	0.561	6.06	7.13	0.0870	0.1026	0.0875	0.0079	0.1666	0.0732	0.1306	0.1618	0.0534	0.1882	0.0054	0.0901	0.0063	0.0007	0.0012	0.0110
	1.0	0.740	0.544	6.42	8.73	0.0805	0.1156	0.0867	0.0094	0.1603	0.0861	0.1180	0.1874	0.0422	0.2100	0.0035	0.0933	0.0062	0.0009	0.0008	0.0118
	1.5	0.815	0.530	6.78	10.43	0.0744	0.1280	0.0861	0.0109	0.1542	0.0992	0.1059	0.2129	0.0323	0.2296	0.0022	0.0957	0.0062	0.0010	0.0006	0.0121
	2.0	0.872	0.519	7.06	11.85	0.0701	0.1374	0.0855	0.0121	0.1494	0.1098	0.0969	0.2328	0.0255	0.2436	0.0015	0.0970	0.0061	0.0012	0.0005	0.0124
0.5	0.4	0.628	0.556	5.93	6.70	0.0904	0.0958	0.0877	0.0076	0.1696	0.0680	0.1366	0.1500	0.0594	0.1743	0.0070	0.0872	0.0063	0.0007	0.0027	0.0153
	0.6	0.690	0.540	6.15	7.83	0.0860	0.1046	0.0871	0.0088	0.1649	0.0769	0.1277	0.1674	0.0515	0.1896	0.0056	0.0895	0.0063	0.0008	0.0023	0.0160
	1.0	0.793	0.513	6.55	10.08	0.0786	0.1191	0.0863	0.0103	0.1576	0.0932	0.1132	0.1974	0.0400	0.2111	0.0037	0.0928	0.0062	0.0009	0.0018	0.0170
	1.5	0.894	0.496	7.06	12.70	0.0713	0.1339	0.0854	0.0125	0.1496	0.1108	0.0989	0.2294	0.0300	0.2312	0.0023	0.0951	0.0061	0.0011	0.0012	0.0180
	2.0	0.970	0.479	7.43	15.08	0.0660	0.1457	0.0847	0.0143	0.1433	0.1258	0.0879	0.2554	0.0231	0.2460	0.0016	0.0966	0.0060	0.0014	0.0009	0.0186

Source: Portland Cement Association, *Handbook of Frame Constants.*

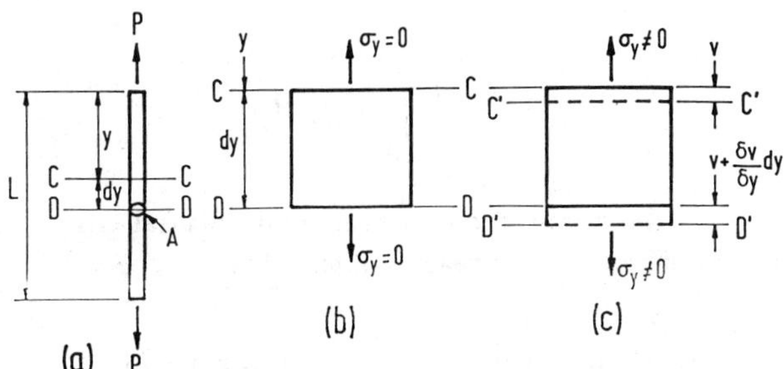

FIGURE 45.38 Axially loaded bar.

Strain Energy Due to Normal Stresses

In an axially loaded bar with constant section the applied load causes normal stress σ_y, as shown in Fig. 45.38. The tensile stress σ_y increases from zero to a value σ_y as the load is gradually applied. The original, unstrained position of any section such as C–C will be displaced by an amount v. A section D–D located a differential length below C–C will have been displaced by an amount $v + (\delta v/\delta y)dy$. As σ_y varies with the applied load, from zero to σ_y, the work done by the forces external to the element can be shown to be

$$dV = \frac{1}{2E}\sigma_y^2 A\,dy = \frac{1}{2}\sigma_y \varepsilon_y A\,dy \tag{45.47}$$

where A is the area of cross section of the bar and ε_y is the strain in the direction of σ_y.

Strain Energy in Bending

It can be shown that the strain energy of a differential volume $dx\,dy\,dz$ stressed in tension or compression in the x direction only by a normal stress σ_x will be

$$dV = \frac{1}{2E}\sigma_x^2\,dx\,dy\,dz = \frac{1}{2}\sigma_x \varepsilon_x\,dx\,dy\,dz \tag{45.48}$$

When σ_x is the bending stress given by $\sigma_x = My/I$ (see Fig. 45.39), then

$$dV = \frac{1}{2E}\frac{M^2 y^2}{I^2}dx\,dy\,dz$$

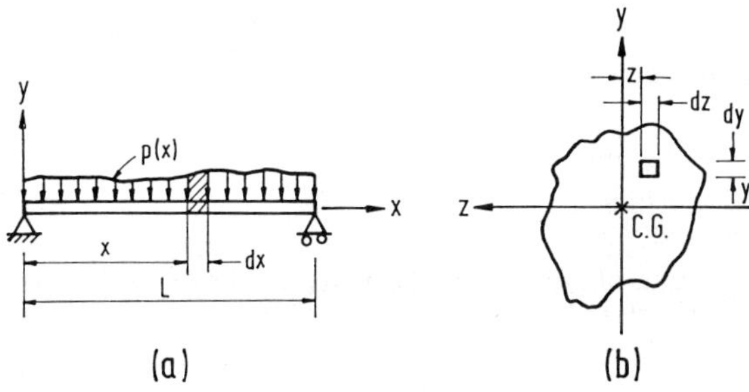

FIGURE 45.39 Beam under arbitrary bending load.

where

$$I = \int\int_A y^2 dz\, dy$$

is the moment of inertia of the cross-sectional area about the neutral axis.

The total strain energy of bending of a beam is obtained as

$$V = \int\int\int_{\text{volume}} \frac{1}{2E}\frac{M^2}{I^2} y^2 dz\, dy\, dx$$

where

$$I = \int\int_{\text{area}} y^2 dz\, dy$$

Therefore

$$V = \int_{\text{length}} \frac{M^2}{2EI} dx \tag{45.49}$$

Strain Energy in Shear

Figure 45.40 shows an element of volume $dx\, dy\, dz$ subjected to shear stress τ_{xy} and τ_{yx}. For static equilibrium, it can readily be shown that

$$\tau_{xy} = \tau_{yx}$$

The shear strain is defined as AB/AC. For small deformations, it follows that

$$\gamma_{xy} = \frac{AB}{AC}$$

Hence, the angle of deformation γ_{xy} is a measure of the shear strain. The strain energy for this differential volume is obtained as

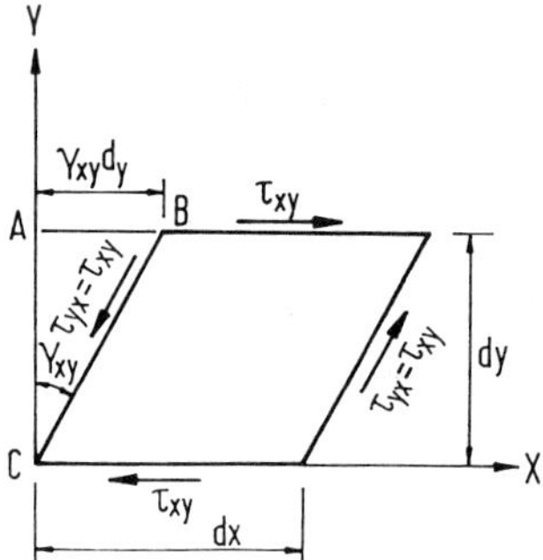

FIGURE 45.40 Shear loading.

$$dV = \frac{1}{2}(\tau_{xy}\, dz\, dx)\gamma_{xy}\, dy = \frac{1}{2}\tau_{xy}\gamma_{xy}\, dx\, dy\, dz \tag{45.50}$$

Hooke's law for shear stresses and strains is

$$\gamma_{xy} = \frac{\tau_{xy}}{G} \tag{45.51}$$

where G is the shear modulus of elasticity of the material. The expression for strain energy in shear reduces to

$$dV = \frac{1}{2G}\tau_{xy}{}^2\, dx\, dy\, dz \tag{45.52}$$

The Energy Relations in Structural Analysis

The energy relations (the law of conservation of energy, the theorem of virtual work, the theorem of minimum potential energy, and Castigliano's theorem) are of fundamental importance in structural engineering and are used in various ways in structural analysis.

The Law of Conservation of Energy

There are many ways of stating this law. For the purpose of structural analysis it will be sufficient to state it in the following form:

If a structure and the external loads acting on it are isolated so that these neither receive nor give out energy, then the total energy of this system remains constant.

A typical application of the law of conservation of energy can be made by referring to Fig. 45.41, which shows a cantilever beam of constant cross section subjected to a concentrated load at its end. If bending strain energy only is considered, the external work equals the internal work:

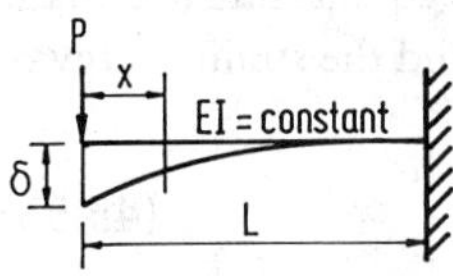

FIGURE 45.41 Cantilever beam.

$$\frac{P\delta}{2} = \int_0^L \frac{M^2\,dx}{2EI}$$

Substituting $M = -Px$ and integrating along the length gives

$$\delta = \frac{PL^3}{3EI} \tag{45.53}$$

The Theorem of Virtual Work

The theorem of virtual work can be derived by considering the beam shown in Fig. 45.42. The full curved line represents the equilibrium position of the beam under the given loads. Assume the beam to be given an additional small deformation consistent with the boundary conditions. This is called a virtual deformation and corresponds to increments of deflection $\Delta_{y1}, \Delta_{y2}, \ldots, \Delta_{yn}$ at loads $P_1, P_2, \ldots, P_n$ as shown by the broken line.

The change in potential energy (PE) of the loads is given by

$$\Delta(\text{PE}) = \sum_{i=1}^{n} P_i \Delta y_i \tag{45.54}$$

By the law of conservation of energy this must be equal to the internal strain energy stored in the beam. Hence, we may state the theorem of virtual work in the following form:

If a body in equilibrium under the action of a system of external loads is given any small (virtual) deformation, then the work done by the external loads during this deformation is equal to the increase in internal strain energy stored in the body.

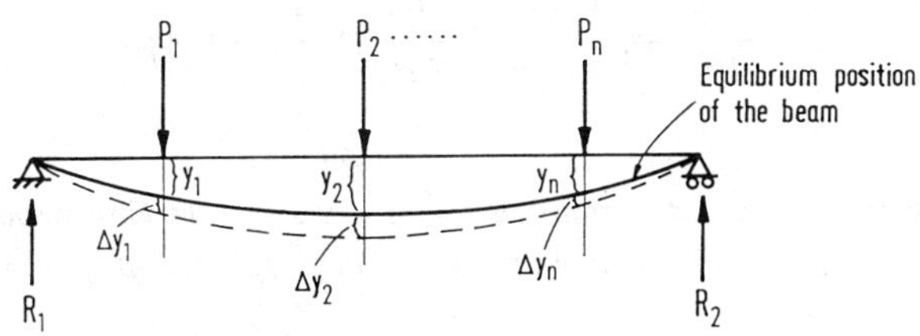

FIGURE 45.42 Equilibrium of a simply supported beam under loading.

The Theorem of Minimum Potential Energy

Let us consider the beam shown in Fig. 45.43. The beam is in equilibrium under the action of loads $P_1, P_2, P_3, \ldots, P_i, \ldots, P_n$. The curve ACB defines the equilibrium positions of the loads and reactions. Now apply by some means an additional small displacement to the curve so that it is defined by $AC'B$. Let y_i be the original equilibrium displacement of the curve beneath a particular load P_i. The additional small displacement is called δ_{yi}. The potential energy of the system while it is in the equilibrium configuration is found by comparing the potential energy of the beam and loads in equilibrium and in the undeflected position. If the change in potential energy of the loads is W and the strain energy of the beam is V, the total energy of the system is

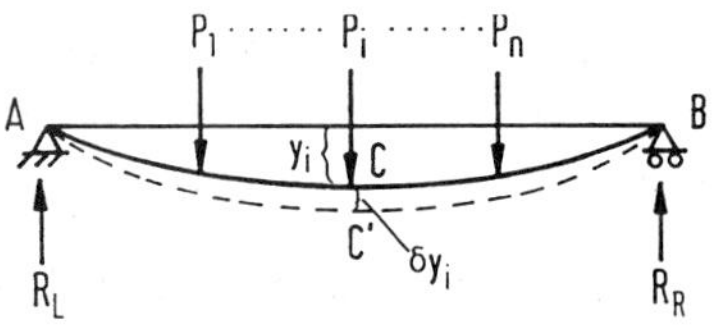

FIGURE 45.43 Simply supported beam under point loading.

$$U = W + V \tag{45.55}$$

If we neglect the second-order terms, then

$$\delta U = \delta(W + V) = 0 \tag{45.56}$$

The above is expressed as the theorem of minimum potential energy, which can be stated as follows:

Of all displacements satisfying given boundary conditions, those that satisfy the equilibrium conditions make the potential energy a minimum.

Castigliano's Theorem

An example of application of energy methods to the field of structural engineering is Castigliano's theorem. The theorem applies only to structures stressed within the elastic limit, and all deformations must be linear homogeneous functions of the loads. Castigliano's theorem can be derived using the expression for total potential energy as follows. For a beam in equilibrium loaded as in Fig. 45.42, the total energy is

$$U = -[P_1 y_1 + P_2 y_2 + \cdots + P_j y_j + \cdots + P_n y_n] + V \tag{45.57}$$

For an elastic system, the strain energy, V, turns out to be one-half the change in the potential energy of the loads:

$$V = \frac{1}{2}\sum_{i=1}^{i=n} P_i y_i \tag{45.58}$$

Castigliano's theorem results from studying the variation in the strain energy, V, produced by a differential change in one of the loads, say P_j.

If the load P_j is changed by a differential amount δP_j and if the deflections y are linear functions of the loads, then

$$\frac{\partial V}{\partial P_j} = \frac{1}{2}\sum_{i=1}^{i=n} P_i \frac{\partial y_i}{\partial P_j} + \frac{1}{2} y_j = y_j \tag{45.59}$$

Castigliano's theorem is stated as follows:

The partial derivatives of the total strain energy of any structure with respect to any one of the applied forces is equal to the displacement of the point of application of the force in the direction of the force.

To find the deflection of a point in a beam that is not the point of application of a concentrated load, one should apply a load $P = 0$ at that point and carry the term P into the strain energy equation. Finally, introduce the true value of $P = 0$ into the expression for the answer.

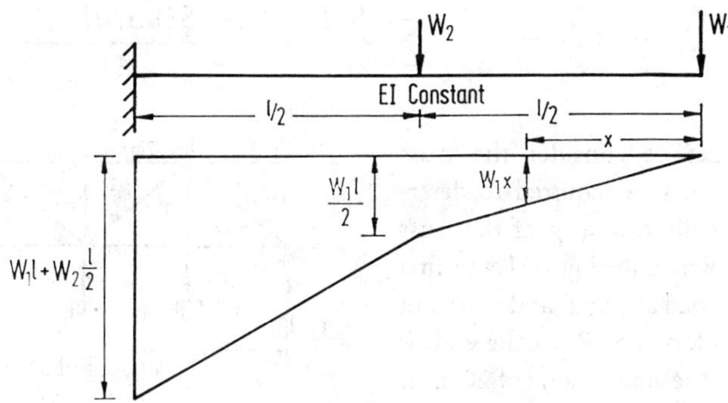

FIGURE 45.44 Example 45.4.

Example 45.4. Determine the bending deflection at the free end of a cantilever loaded as shown in Fig. 45.44.
Solution.

$$V = \int_0^L \frac{M^2}{2EI} dx$$

$$\Delta = \frac{\partial V}{\partial W_1} = \int_0^L \frac{M}{EI}\frac{\partial M}{\partial w_1} dx$$

$$M = W_1 x \qquad 0 < x < \frac{L}{2}$$

$$= W_1 x + W_2(x - \frac{l}{2}) \qquad \frac{L}{2} < x < L$$

$$\Delta = \frac{1}{EI}\int_0^{l/2} W_1 x \times x\, dx + \frac{1}{EI}\int_{l/2}^{l} [W_1 x + W_2(x - \frac{l}{2})]x\, dx$$

$$= \frac{W_1 l^3}{24EI} + \frac{7W_1 l^3}{24EI} + \frac{5W_2 l^3}{48EI}$$

$$= \frac{W_1 l^3}{3EI} + \frac{5W_2 l^3}{48EI}$$

Castigliano's theorem can be applied to determine deflection of trusses as follows. We know that the increment of strain energy for an axially loaded bar is given as

$$dV = \frac{1}{2E}\sigma_y^2 A\, dy$$

Substituting $\sigma_y = S/A$, where S is the axial load in the bar, and integrating over the length of the bar, the total strain energy of the bar is given as

$$V = \frac{S^2 L}{2AE} \tag{45.60}$$

The deflection component Δ_i of the point of application of a load P_i in the direction of P_i is given as

$$\Delta_i = \frac{\partial V}{\partial P_i} = \frac{\partial}{\partial P_i}\sum \frac{S^2 L}{2AE} = \sum \frac{S(\partial S/\partial P_i)L}{AE}$$

Example 45.5. Let us consider the truss shown in Fig. 45.45. It is required to determine the vertical deflection at g of the truss when loaded as shown in the figure. Let us first replace the 20 kips load at g by P and carry out the calculations in terms of P. At the end, P will be replaced by the actual value of 20 kips.

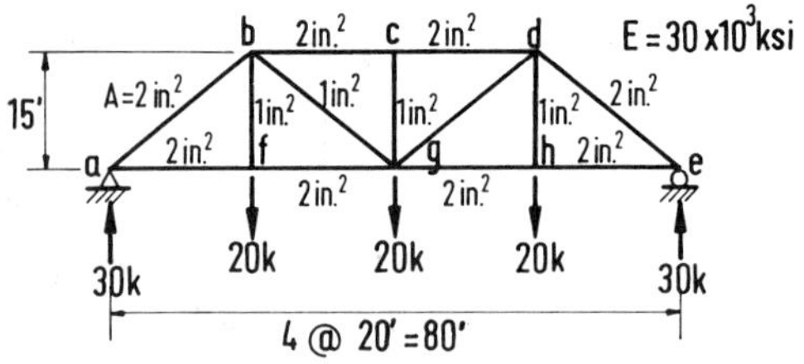

FIGURE 45.45 Example 45.5.

Member	A (in.^{2})	L (ft)	S	$\frac{\delta S}{\delta P}$	n	$nS\frac{\delta S}{\delta P}\frac{L}{A}$
ab	2	25	$-(33.3 + 0.83P)$	-0.83	2	$(691 + 17.2P)$
af	2	20	$(26.7 + 0.67P)$	0.67	2	$(358 + 9P)$
fg	2	20	$(26.7 + 0.67P)$	0.67	2	$(358 + 9P)$
bf	1	15	20	0	2	0
bg	1	25	$0.83P$	0.83	2	$34.4P$
bc	2	20	$-26.7 - 1.33P$	-1.33	2	$(710 + 35.4P)$
cg	1	15	0	0	1	0

n indicates the number of similar members

$$\sum \frac{S(\delta S/\delta P)L}{A} \qquad 2117 + 105P$$

With

$$P = 20 \text{ kips}$$

$$\Delta_g = \sum \frac{S(\delta S/\delta P)L}{AE} = \frac{(2117 + 105 \times 20) \times 12}{30 \times 10^3}$$

$$= 1.69 \text{ in.}$$

Unit Load Method

The unit load method is a versatile tool in the solution of deflections of both trusses and beams. Consider an elastic body in equilibrium under loads $P_1, P_2, P_3, P_4, \ldots, P_n$ and a load p applied at point O, as shown in Fig. 45.46. By Castigliano's theorem, the component of the deflection of point O in the direction of the applied force p is

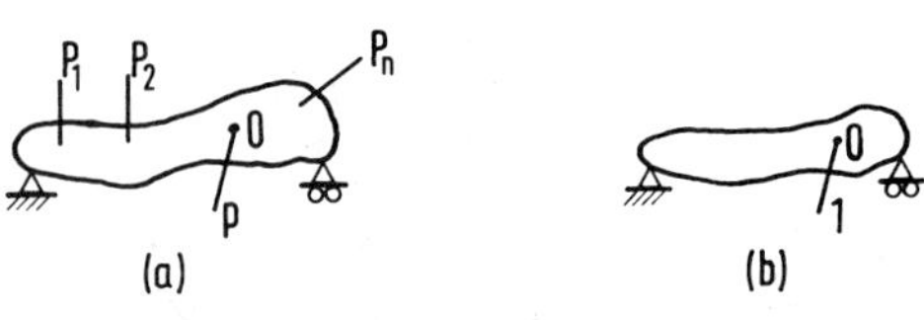

FIGURE 45.46

$$\delta_{O_p} = \frac{\partial V}{\partial p} \tag{45.61}$$

where V is the strain energy of the body. It has been shown in Eq. (45.49) that the strain energy of a beam, neglecting shear effects, is given by

$$V = \int_O^L \frac{M^2}{2EI}dx$$

Also it was shown that if the elastic body is a truss, from Eq. (45.60),

$$V = \sum \frac{S^2 L}{2AE}$$

For a beam, therefore, from Eq. (45.61),

$$\delta_{O_p} = \int_L \frac{M(\partial M/\partial p)dx}{EI} \tag{45.62}$$

and for a truss,

$$\delta_{O_p} = \sum \frac{S(\partial S/\partial p)L}{AE} \tag{45.63}$$

The bending moments M and the stresses S are functions of the load p as well as of the loads P_1, $P_2, \ldots, P_n$. Let a unit load be applied at O on the elastic body and the corresponding moment be m if the body is a beam, and the stresses in the members of the body be u if the body is a truss. For the body in Fig. 45.46 the moments M and the stresses S due to the system of forces $P_1, P_2, \ldots,$ P_n and p at O applied separately can be obtained by superposition as

$$M = M_p + pm \tag{45.64}$$

$$S = S_p + pu \tag{45.65}$$

where M_P and S_P are, respectively, moments and stresses produced by $P_1, P_2, \ldots, P_n$. Then

$$\frac{\partial M}{\partial p} = m = \text{Moments produced by a unit load at } O \tag{45.66}$$

$$\frac{\partial S}{\partial p} = u = \text{Stresses produced by a unit load at } O \tag{45.67}$$

Using Eqs. (45.66) and (45.67) in Eqs. (45.62) and (45.63), respectively,

$$\delta_{O_p} = \int_L \frac{Mm\,dx}{EI} \tag{45.68}$$

$$\delta_{O_p} = \sum \frac{SuL}{AE} \tag{45.69}$$

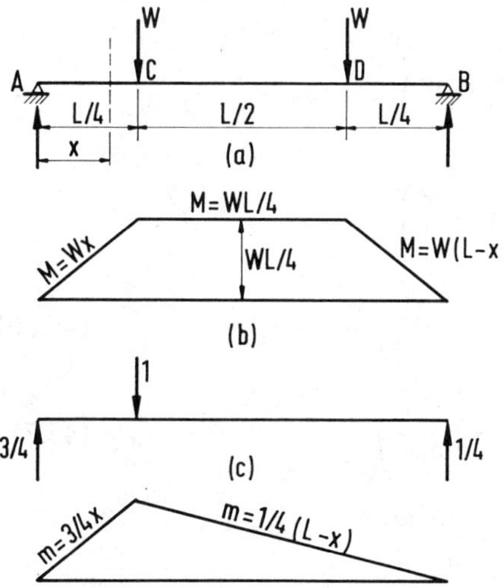

FIGURE 45.47 Example 45.6.

Example 45.6. Determine, using the unit load method, the deflection at C of a simple beam of constant cross section loaded as shown in Fig. 45.47(a).

Solution. The bending moment diagram for the beam due to the applied loading is shown in Fig. 45.47(b). A unit load is applied at C, where it is required to determine the deflection as shown in Fig. 45.47(c), and the corresponding bending moment diagram is shown in Fig. 45.47(d). Using

Eq. 45.68, we now have

$$\begin{aligned}\delta_C &= \int_O^L \frac{Mm\,dx}{EI}\\ &= \frac{1}{EI}\int_O^{L/4}(Wx)\left(\frac{3}{4}x\right)dx + \frac{1}{EI}\int_{L/4}^{3L/4}\left(\frac{WL}{4}\right)\frac{1}{4}(L-x)\,dx\\ &\quad + \frac{1}{EI}\int_{3L/4}^{L} W(L-x)\frac{1}{4}(L-x)\,dx\\ &= \frac{WL^3}{48EI}\end{aligned}$$

[Further details on energy methods in structural analysis may be found in Borg and Gennaro, 1959.]

45.12 Bending of Thin Plates

When the thickness of an elastic body is small compared to its other dimensions, we call it a **thin plate**. The plane parallel to the faces of the plate and bisecting the thickness of the plate, in the undeformed state, is called the *middle plane* of the plate. When the deflection of the middle plane is small compared with the thickness h, it can be assumed that (1) there is no deformation in the middle plane; (2) the normals of the middle plane before bending are deformed into the normals of the middle plane after bending; and (3) the normal stresses in the direction transverse to the plate can be neglected.

Based on these assumptions, all stress components can be expressed by the deflection w of the plate, which is a function of the two coordinates in the plane of the plate. This function has to satisfy a linear partial differential equation, which, together with the boundary conditions, completely defines w.

Figure 45.48(a) shows a plate element cut from a plate whose middle plane coincides with the xy plane. The middle plane of the plate is subjected to a lateral load of intensity q. It can be shown [Fig. 45.48(b)] that by considering the equilibrium of the plate element, the stress resultants are given as

$$\begin{aligned}M_x &= -D\left(\frac{\partial^2 w}{\partial x^2} + \nu\frac{\partial^2 w}{\partial y^2}\right)\\ M_y &= -D\left(\frac{\partial^2 w}{\partial y^2} + \nu\frac{\partial^2 w}{\partial x^2}\right)\\ M_{xy} &= -M_{yx} = D(1-\nu)\frac{\partial^2 w}{\partial x\,\partial y}\end{aligned} \tag{45.70}$$

$$Q_y = -D\frac{\partial}{\partial y}\left(\frac{\partial^2 w}{\partial x^2} + \frac{\partial^2 w}{\partial y^2}\right) \tag{45.71}$$

$$\begin{aligned}V_x &= \frac{\partial^3 w}{\partial x^3} + (2-\nu)\frac{\partial^3 w}{\partial x\,\partial y^2}\\ V_y &= \frac{\partial^3 w}{\partial y^3} + (2-\nu)\frac{\partial^3 w}{\partial y\,\partial x^2}\end{aligned} \tag{45.72}$$

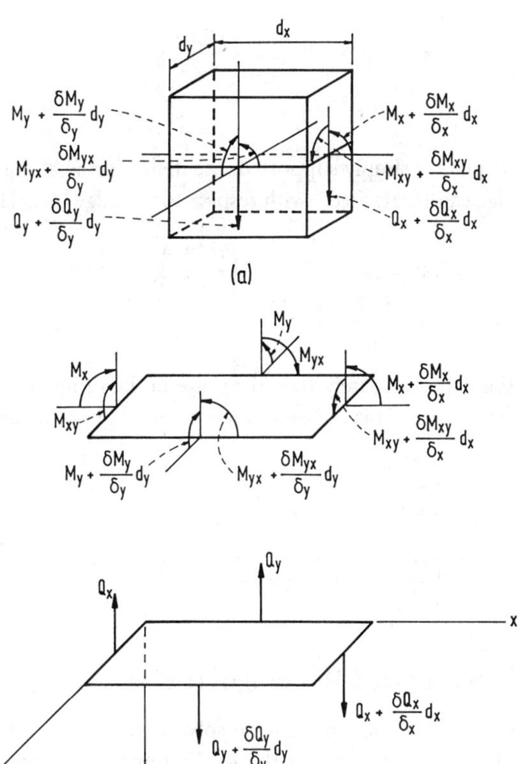

FIGURE 45.48 (a) Plate element, (b) stress resultants.

$$R = 2D(1 - \nu)\frac{\partial^2 w}{\partial x \, \partial y} \tag{45.73}$$

where M_x and M_y are bendings moments per unit length in the x and y directions, M_{xy} and M_{yx} are twisting moments per unit length, Q_x and Q_y are shearing forces per unit length, V_x and V_y are supplementary shear forces, and R is the corner force. D represents the flexural rigidity of the plate per unit length and is given by $[Eh^3/12(1 - \nu^2)]$, where E is the modulus of elasticity and ν is Poisson's ratio. The governing equation for the plate is obtained as

$$\frac{\partial^4 w}{\partial x^4} + 2\frac{\partial^4 w}{\partial x^2 \partial y^2} + \frac{\partial^4 w}{\partial y^4} = \frac{q}{D} \tag{45.74}$$

Any plate problem should satisfy the governing Eq. (45.74) and boundary conditions of the plate.

Boundary Conditions

There are three basic boundary conditions for plate problems: the clamped edge, the simply supported edge, and the free edge.

Clamped Edge

For this boundary condition, the edge is restrained such that the deflection and slope are zero along the edge. If we consider the edge $x = a$ to be clamped we have

$$(w)_{x=a} = 0 \qquad \left(\frac{\partial^2 w}{\partial x^2}\right)_{x=a} = 0 \tag{45.75}$$

Simply Supported Edge

If the edge $x = a$ of the plate is simply supported, the deflection w along this edge must be zero. At the same time this edge can rotate freely with respect to the edge line. This means that

$$(w)_{x=a} = 0 \qquad \left(\frac{\partial^2 w}{\partial x^2}\right)_{x=a} = 0 \tag{45.76}$$

Free Edge

If the edge $x = a$ of the plate is entirely free, there are no bending and twisting moments, and there is no vertical shearing force. This can be written in terms of w, the deflection, as

$$\left(\frac{\partial^2 w}{\partial x^2} + \nu\frac{\partial^2 w}{\partial y^2}\right)_{x=a} = 0$$

$$\left(\frac{\partial^3 w}{\partial x^3} + (2-\nu)\frac{\partial^3 w}{\partial x\,\partial y^2}\right)_{x=a} = 0 \tag{45.77}$$

Bending of Simply Supported Rectangular Plates

A number of the plate bending problems may be solved directly by solving the differential Eq. (45.74). The solution, however, depends on the loading and boundary conditions. Consider a simply supported plate subjected to a sinusoidal loading, as shown in Fig. 45.49.

The differential Eq. (45.74) in this case becomes

$$\frac{\partial^4 w}{\partial x^4} + 2\frac{\partial^4 w}{\partial x^2 \partial y^2} + \frac{\partial^4 w}{\partial y^4} = \frac{q_0}{D}\sin\frac{\pi x}{a}\sin\frac{\pi y}{b} \tag{45.78}$$

The boundary conditions for the simply supported edges are

$$w = 0, \qquad \frac{\partial^2 w}{\partial x^2} = 0 \quad \text{for } x = 0 \text{ and } x = a$$

$$w = 0, \qquad \frac{\partial^2 w}{\partial y^2} = 0 \quad \text{for } y = 0 \text{ and } y = b \tag{45.79}$$

The deflection function

$$w = w_0 \sin\frac{\pi x}{a}\sin\frac{\pi y}{b} \tag{45.80}$$

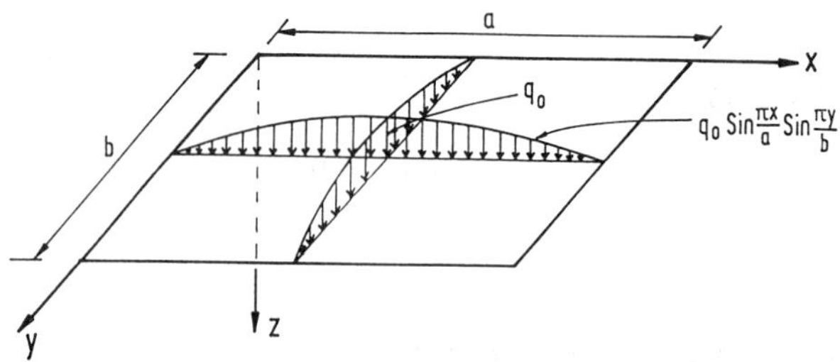

FIGURE 45.49 Rectangular plate under sinusoidal loading.

can be seen to satisfy all the boundary conditions in Eq. (45.79). w_0 must be chosen to satisfy Eq. (45.78). Substitution of Eq. (45.80) into Eq. (45.78) gives

$$\pi^4\left(\frac{1}{a^2}+\frac{1}{b^2}\right)^2 w_0 = \frac{q_0}{D}$$

The deflection surface for the plate can, therefore, be found as

$$w = \frac{q_0}{\pi^4 D\left(\frac{1}{a^2}+\frac{1}{b^2}\right)^2}\sin\frac{\pi x}{a}\sin\frac{\pi x}{b} \tag{45.81}$$

Using Eqs. (45.81) and (45.70) we find the expressions for moments as

$$M_x = \frac{q_0}{\pi^2\left(\frac{1}{a^2}+\frac{1}{b^2}\right)^2}\left(\frac{1}{a^2}+\frac{\nu}{b^2}\right)\sin\frac{\pi x}{a}\sin\frac{\pi y}{b}$$

$$M_y = \frac{q_0}{\pi^2\left(\frac{1}{a^2}+\frac{1}{b^2}\right)^2}\left(\frac{\nu}{a^2}+\frac{1}{b^2}\right)\sin\frac{\pi x}{a}\sin\frac{\pi y}{b}$$

$$M_{xy} = \frac{q_0(1-\nu)}{\pi^2\left(\frac{1}{a^2}+\frac{1}{b^2}\right)^2 ab}\cos\frac{\pi x}{a}\cos\frac{\pi y}{b} \tag{45.82}$$

Maximum deflection and maximum bending moments, which occur at the center of the plate, can be written by substituting $x = a/2$ and $y = b/2$ in Eqs. (45.81) and (45.82) as

$$w_{\max} = \frac{q_0}{\pi^4 D\left(\frac{1}{a^2}+\frac{1}{b^2}\right)^2}$$

$$(M_x)_{\max} = \frac{q_0}{\pi^2\left(\frac{1}{a^2}+\frac{1}{b^2}\right)^2}\left(\frac{1}{a^2}+\frac{\nu}{b^2}\right)$$

$$(M_y)_{\max} = \frac{q_0}{\pi^2\left(\frac{1}{a^2}+\frac{1}{b^2}\right)^2}\left(\frac{\nu}{a^2}+\frac{1}{b^2}\right) \tag{45.83}$$

If the plate is square, then $a = b$ and Eq. (45.83) becomes

$$w_{\max} = \frac{q_0 a^4}{4\pi^4 D}$$

$$(M_x)_{\max} = (M_y)_{\max} = \frac{(1+\nu)}{4\pi^2}q_0 a^2 \tag{45.84}$$

If the simply supported rectangular plate is subjected to any kind of loading given by

$$q = q(x, y) \tag{45.85}$$

the function $q(x, y)$ should be represented in the form of a double trigonometric series as

$$q(x,y) = \sum_{m=1}^{\infty}\sum_{n=1}^{\infty} q_{mn} \sin\frac{m\pi x}{a} \sin\frac{n\pi y}{b} \tag{45.86}$$

where q_{mn} is given by

$$q_{mn} = \frac{4}{ab}\int_0^a\int_0^b q(x,y) \sin\frac{m\pi x}{a} \sin\frac{n\pi y}{b} dx\, dy \tag{45.87}$$

From Eqs. (45.78), (45.85), (45.86) and (45.87) we can obtain the expression for deflection as

$$w = \frac{1}{\pi^4 D}\sum_{m=1}^{\infty}\sum_{n=1}^{\infty} \frac{q_{mn}}{\left(\dfrac{m^2}{a^2}+\dfrac{n^2}{b^2}\right)^2} \sin\frac{m\pi x}{a} \sin\frac{n\pi y}{b} \tag{45.88}$$

If the applied load is uniformly distributed of intensity q_0, we have

$$q(x,y) = q_0$$

and from Eq. (45.87) we obtain

$$\begin{aligned} q_{mn} &= \frac{4q_0}{ab}\int_0^a\int_0^b \sin\frac{m\pi x}{a} \sin\frac{n\pi y}{b} dx\, dy \\ &= \frac{16q_0}{\pi^2 mn} \end{aligned} \tag{45.89}$$

where m and n are odd integers. $q_{mn} = 0$ if m or n or both of them are even numbers. We can, therefore, write the expression for deflection of a simply supported plate subjected to uniformly distributed load as

$$w = \frac{16q_0}{\pi^6 D}\sum_{m=1}^{\infty}\sum_{n=1}^{\infty} \frac{\sin\dfrac{m\pi x}{a} \sin\dfrac{n\pi y}{b}}{mn\left(\dfrac{m^2}{a^2}+\dfrac{n^2}{b^2}\right)^2} \tag{45.90}$$

where $m = 1, 3, 5, \ldots$ and $n = 1, 3, 5, \ldots$

The maximum deflection occurs at the center and it can be written by substituting $x = a/2$ and $y = b/2$ in Eq. (45.90) as

$$w_{\max} = \frac{16q_0}{\pi^6 D}\sum_{m=1}^{\infty}\sum_{n=1}^{\infty} \frac{(-1)^{[(m+n)/2]-1}}{mn\left(\dfrac{m^2}{a^2}+\dfrac{n^2}{b^2}\right)^2} \tag{45.91}$$

Eq. (45.91) is a rapidly converging series and a satisfactory approximation can be obtained by taking only the first term of the series; for example, in the case of a square plate,

$$w_{\max} = \frac{4q_0 a^4}{\pi^6 D} = 0.00416\frac{q_0 a^4}{D}$$

Assuming $\nu = 0.3$, we get, for the maximum deflection,

$$w_{\max} = 0.0454\frac{q_0 a^4}{Eh^3}$$

The expressions for bending and twisting moments can be obtained by substituting Eq. (45.90) into Eq. (45.70). Figure 45.50 shows some loading cases and the corresponding loading functions.

The above solution for uniformly loaded cases is known as the Navier solution. If two opposite sides, say, $x = 0$ and $x = a$, of a rectangular plate are simply supported, the solution taking the deflection function as

$$w = \sum_{m=1}^{\infty} Y_m \sin \frac{m\pi x}{a} \tag{45.92}$$

No.	Load $q(x,y) = \sum_m \sum_n q_{mn} \sin\frac{m\pi x}{a} \sin\frac{n\pi y}{b}$	Expansion Coefficients q_{mn}
1	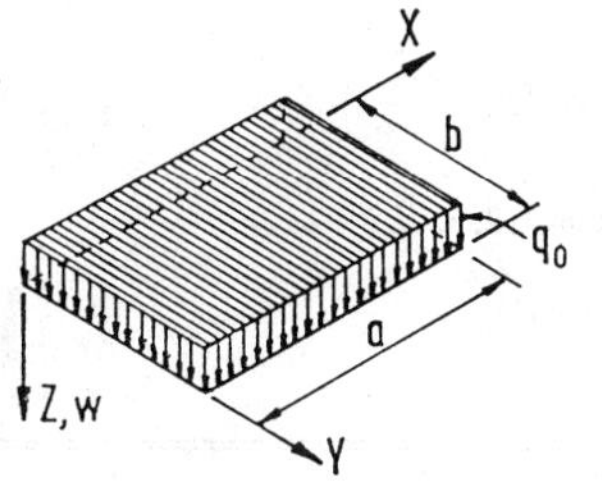	$q_{mn} = \dfrac{16q_0}{\pi^2 mn}$ (m, n = 1, 3, 5, ...)
2	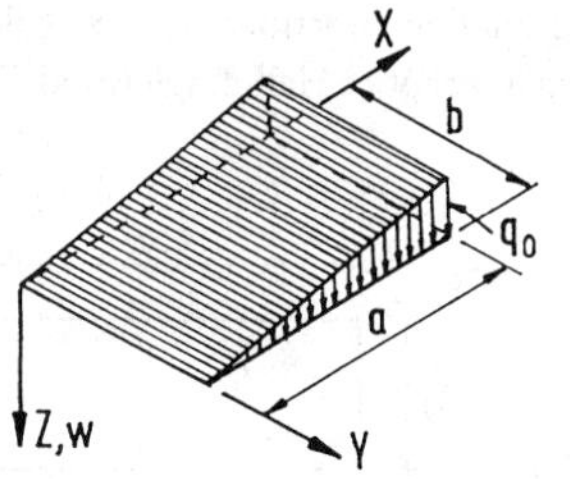	$q_{mn} = \dfrac{-8q_0 \cos m\pi}{\pi^2 mn}$ (m, n = 1, 3, 5, ...)
3	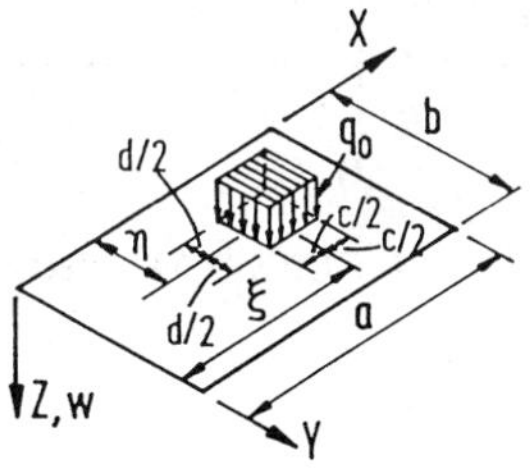	$p_{mn} = \dfrac{16q_0}{\pi^2 mn} \sin\dfrac{m\pi\xi}{a} \sin\dfrac{n\pi\eta}{b} \times \sin\dfrac{m\pi c}{2a} \sin\dfrac{n\pi d}{2b}$ (m, n = 1, 3, 5, ...)
4	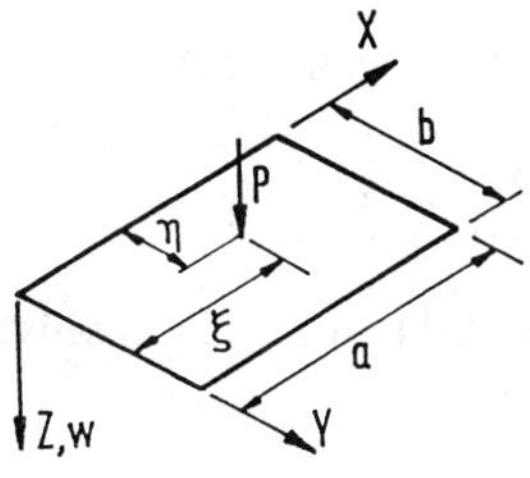	$q_{mn} = \dfrac{4q_0}{ab} \sin\dfrac{m\pi\xi}{a} \sin\dfrac{n\pi\eta}{b}$ (m, n = 1, 2, 3, ...)

FIGURE 45.50 Typical loading on plates and loading functions. (From Szilard, R. 1974. *Theory and Analysis of Plates—Classified and Numerical Methods.* Prentice Hall, Englewood Cliffs, NJ.) *(continues)*

No.	Load $q(x,y) = \sum_M \sum_n q_{mn} \sin\frac{m\pi x}{a} \sin\frac{n\pi y}{b}$	Expansion coefficients q_{mn}
5	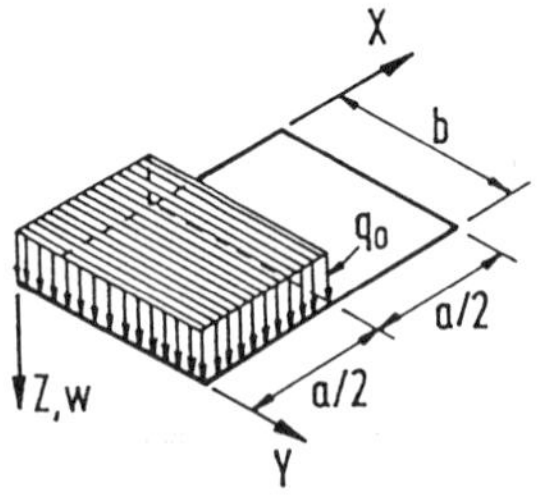 X, b, q_0, a/2, a/2, Z,w, Y	$q_{mn} = \frac{8q_0}{\pi^2 mn}$ for m,n = 1, 3, 5, ... $q_{mn} = \frac{16q_0}{\pi^2 mn}$ for $\begin{cases} m = 2, 6, 10, ... \\ n = 1, 3, 5, ... \end{cases}$
6	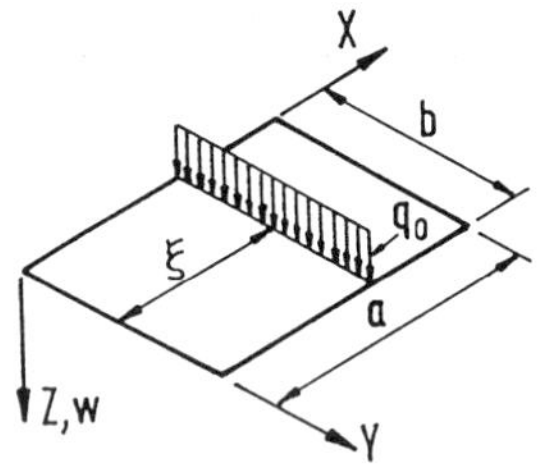 X, b, q_0, ξ, a, Z,w, Y	$q_{mn} = \frac{4q_0}{\pi a n} \sin\frac{m\pi \xi}{a}$ (m, n = 1, 2, 3, ...)

FIGURE 45.50 (continued) Typical loading on plates and loading functions. (From Szilard, R. 1974. *Theory and Analysis of Plates—Classified and Numerical Methods.* Prentice Hall, Englewood Cliffs, NJ.)

can be adopted. This solution was proposed by Levy. Equation (45.92) satisfies the boundary conditions $w = 0$ and $\partial^2 w/\partial x^2 = 0$ on the two simply supported edges. Y_m should be determined such that it satisfies the boundary conditions along the edges $y = \pm(b/2)$ of the plate shown in Fig. 45.51 and also the equation of the deflection surface

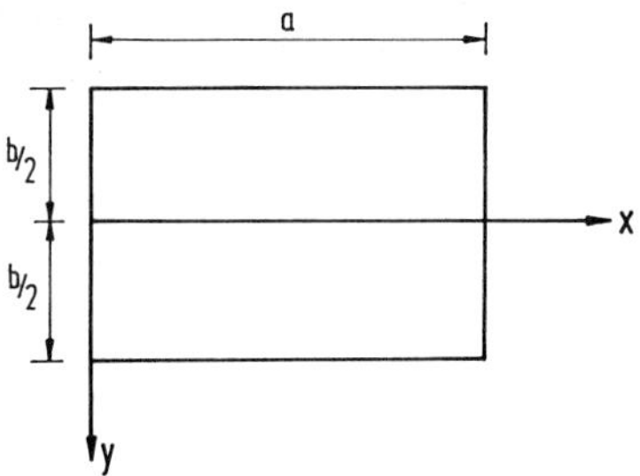

FIGURE 45.51 Rectangular plate.

$$\frac{\partial^4 w}{\partial x^4} + 2\frac{\partial^4 w}{\partial x^2 \partial y^2} + \frac{\partial^4 w}{\partial y^4} = \frac{q_0}{D} \qquad (45.93)$$

where q_0 is the intensity of uniformly distributed load.

The solution for Eq. (45.93) can be taken in the form

$$w = w_1 + w_2 \qquad (45.94)$$

for a uniformly loaded, simply supported plate. w_1 can be taken in the form

$$w_1 = \frac{q_0}{24D}\left(x^4 - 2ax^3 + a^3 x\right) \qquad (45.95)$$

representing the deflection of a uniformly loaded strip parallel to the x axis. It satisfies Eq. (45.93) and also the boundary conditions along $x = 0$ and $x = a$.

The expression w_2 has to satisfy the equation

$$\frac{\partial^4 w_2}{\partial x^4} + 2\frac{\partial^4 w_2}{\partial x^2 \partial y^2} + \frac{\partial^4 w_2}{\partial y^4} = 0 \qquad (45.96)$$

and must be chosen such that Eq. (45.94) satisfies all boundary conditions of the plate. Taking w_2 in the form of series given in Eq. (45.92) it can be shown that the deflection surface takes the form

$$w = \frac{q_0}{24D}(x^4 - 2ax^3 + a^3x) + \frac{q_0 a^4}{24D}\sum_{m=1}^{\infty}\left(A_m \cosh\frac{m\pi y}{a} + B_m \frac{m\pi y}{a}\sinh\frac{m\pi y}{a} + C_m \sinh\frac{m\pi y}{a} + D_m \frac{m\pi y}{a}\cosh\frac{m\pi y}{a}\right)\sin\frac{m\pi x}{a} \tag{45.97}$$

Observing that the deflection surface of the plate is symmetrical with respect to the x axis we keep in Eq. (45.97) only the even function of y; therefore, $C_m = D_m = 0$. The deflection surface takes the form

$$w = \frac{q_o}{24D}(x^4 - 2ax^3 + a^3x) + \frac{q_0 a^4}{24D}\sum_{m=1}^{\infty}\left(A_m \cosh\frac{m\pi y}{a} + B_m \frac{m\pi y}{a}\sinh\frac{m\pi y}{a}\right)\sin\frac{m\pi x}{a} \tag{45.98}$$

Developing the expression in Eq. (45.95) into a trigonometric series, the deflection surface in Eq. (45.98) is written as

$$w = \frac{q_0 a^4}{D}\sum_{m=1}^{\infty}\left(\frac{4}{\pi^5 m^5} + A_m \cosh\frac{m\pi y}{a} + B_m \frac{m\pi y}{a}\sin\frac{m\pi y}{a}\right)\sin\frac{m\pi x}{a} \tag{45.99}$$

Substituting Eq. (45.99) in the boundary conditions

$$w = 0 \qquad \frac{\partial^2 w}{\partial y^2} = 0 \tag{45.100}$$

one obtains the constants of integration A_m and B_m, and the expression for deflection may be written as

$$w = \frac{4q_0 a^4}{\pi^5 D}\sum_{m=1,3,5,\ldots}^{\infty}\frac{1}{m^5}\left(1 - \frac{\alpha_m \tanh\alpha_m + 2}{2\cosh\alpha_m}\cosh\frac{2\alpha_m y}{b} + \frac{\alpha_m}{2\cosh\alpha_m}\frac{2y}{b}\sinh\frac{2\alpha_m y}{b}\right)\sin\frac{m\pi x}{a} \tag{45.101}$$

where $\alpha_m = m\pi b/2a$.

Maximum deflection occurs at the middle of the plate, $x = a/2, y = 0$ and is given by

$$w_{\max} = \frac{4q_0 a^4}{\pi^5 D}\sum_{m=1,3,5,\ldots}^{\infty}\frac{(-1)^{(m-1)/2}}{m^5}\left(1 - \frac{\alpha_m \tanh\alpha_m + 2}{2\cosh\alpha_m}\right) \tag{45.102}$$

Solution of plates with arbitrary boundary conditions are complicated. It is possible to make some simplifying assumptions for plates with the same boundary conditions along two parallel edges in order to obtain the desired solution. Alternatively, energy methods can be applied more efficiently to solve plates with complex boundary conditions. However, it should be noted that the accuracy of results depends on the deflection function chosen. These functions must be so chosen that they satisfy at least the kinematic boundary conditions.

Figure 45.52 gives formulas for deflection and bending moments of rectangular plates with typical boundary and loading conditions.

Case No.	Structural System and Static Loading	Deflection and Internal Forces
1	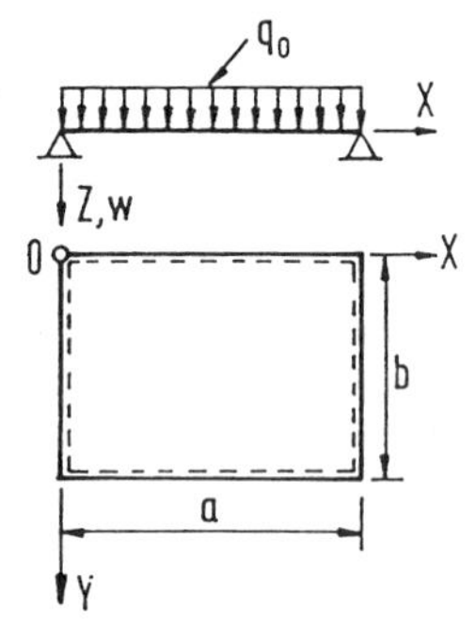 q_0, X, Z,w, 0, X, b, a, Y	$w = \dfrac{16q_0}{\pi^6 D}\sum_m\sum_n \dfrac{\sin\dfrac{m\pi x}{a}\sin\dfrac{n\pi y}{b}}{mn\left(\dfrac{m^2}{a^2}+\dfrac{n^2}{b^2}\right)^2}$ $m_x = \dfrac{16q_0a^2}{\pi^4}\sum_m\sum_n \dfrac{\left(m^2+\nu\dfrac{n^2}{\epsilon^2}\right)\sin\dfrac{m\pi x}{a}\sin\dfrac{n\pi y}{b}}{mn\left(m^2+\dfrac{n^2}{\epsilon^2}\right)^2}$ $m_y = \dfrac{16q_0a^2}{\pi^4}\sum_m\sum_n \dfrac{\left(\dfrac{n^2}{\epsilon^2}+\nu m^2\right)\sin\dfrac{m\pi x}{a}\sin\dfrac{n\pi y}{b}}{mn\left(m^2+\dfrac{n^2}{\epsilon^2}\right)^2}$ $\epsilon = \dfrac{b}{a},\quad m = 1, 3, 5, \ldots, \infty;\qquad n = 1, 3, 5, \ldots, \infty$
2	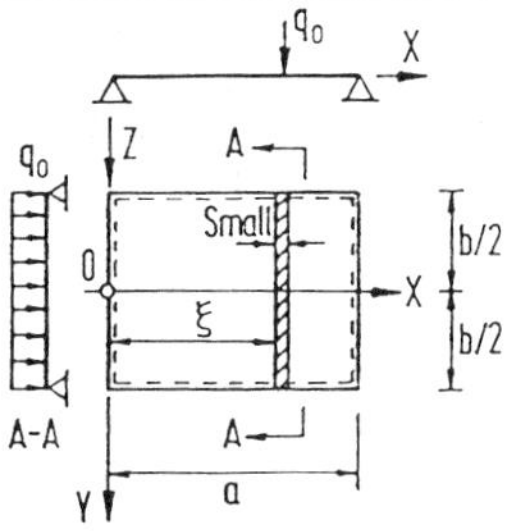 q_0, X, q_0, Z, A, Small, 0, X, b/2, b/2, ξ, A-A, A, Y, a	$w = \dfrac{a^4}{D\pi^4}\sum_{m=1}^{\infty}\dfrac{P_m}{m^4}\left(1-\dfrac{2+\alpha_m\tanh\alpha_m}{2\cosh\alpha_m}\cos\lambda_m y + \dfrac{\lambda_m y\sinh\lambda_m y}{2\cosh\alpha_m}\right)\sin\lambda_m x$ where $P_m = \dfrac{2q_0}{a}\sin\dfrac{m\pi\xi}{a}\qquad \lambda_m = \dfrac{m\pi}{a}$ $m = 1, 2, 3, \ldots\qquad \alpha_m = \dfrac{m\pi b}{2a}$
3	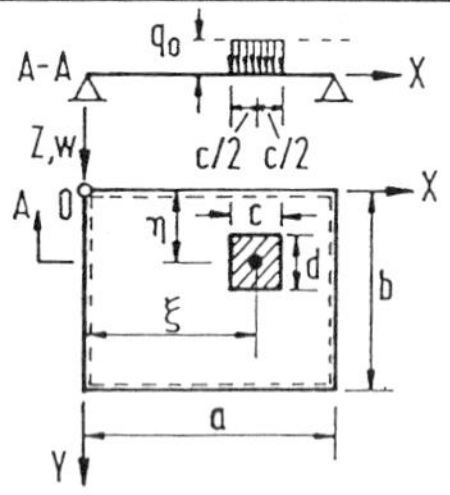 A-A, q_0, X, Z,w, c/2 c/2, A, 0, X, η, c, d, b, ξ, a, Y	$w = \dfrac{16q_0}{D\pi^6}\sum_m\sum_n \dfrac{\sin\dfrac{m\pi\xi}{a}\sin\dfrac{n\pi\eta}{b}\sin\dfrac{m\pi c}{b}\sin\dfrac{n\pi d}{2b}}{mn\left(\dfrac{m^2}{a^2}+\dfrac{n^2}{b^2}\right)^2}$ $\times \sin\dfrac{m\pi x}{a}\sin\dfrac{n\pi y}{b}$ $m = 1, 2, 3, \ldots$ $n = 1, 2, 3, \ldots$
4	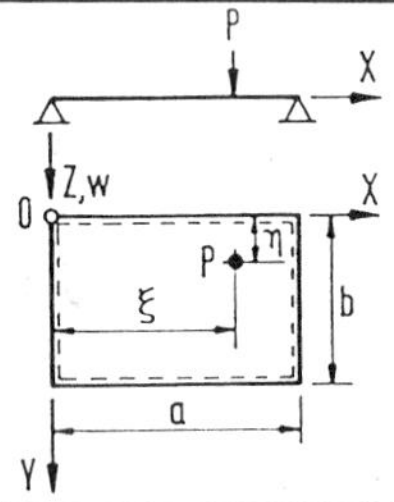 P, X, Z,w, 0, X, P, η, ξ, b, a, Y	$w = \dfrac{4P}{D\pi^4 ab}\sum_m\sum_n \dfrac{\sin\dfrac{m\pi\xi}{a}\sin\dfrac{n\pi\eta}{b}\sin\dfrac{m\pi x}{a}\sin\dfrac{n\pi y}{b}}{\left(\dfrac{m^2}{a^2}+\dfrac{n^2}{b^2}\right)^2}$ $m = 1, 2, 3, \ldots$ $n = 1, 2, 3, \ldots$

FIGURE 45.52 Typical loading and boundary conditions for rectangular plates. (From Szilard, R. 1974. *Theory and Analysis of Plates—Classified and Numerical Methods.* Prentice Hall, Englewood Cliffs, NJ.)

Bending of Circular Plates

In the case of a symmetrically loaded circular plate, the loading is distributed symmetrically about the axis perpendicular to the plate through its center. In such cases, the deflection surface to which the middle plane of the plate is bent will also be symmetrical. The solution of circular plates can be conveniently carried out by using polar coordinates.

Stress resultants in a circular plate element are shown in Fig. 45.53. The governing differential equation is expressed in polar coordinates as

$$\frac{1}{r}\frac{d}{dr}\left\{r\frac{d}{dr}\left[\frac{1}{r}\frac{d}{dr}\left(r\frac{dw}{dr}\right)\right]\right\} = \frac{q}{D} \tag{45.103}$$

where q is the intensity of loading.

In the case of a uniformly loaded circular plate, Eq. (45.103) can be integrated successively, and the deflection at any point at a distance r from the center can be expressed as

$$w = \frac{q_0 r^4}{64D} + \frac{C_1 r^2}{4} + C_2 \log\frac{r}{a} + C_3 \tag{45.104}$$

where q_0 is the intensity of loading and a is the radius of the plate. C_1, C_2, and C_3 are constants of integration to be determined using the boundary conditions.

For a plate with clamped edges under uniformly distributed load q_0, the deflection surface reduces to

$$w = \frac{q_0}{64D}(a^2 - r^2)^2 \tag{45.105}$$

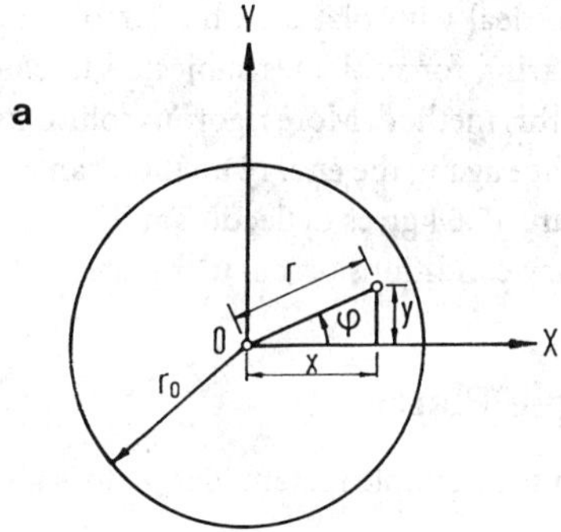

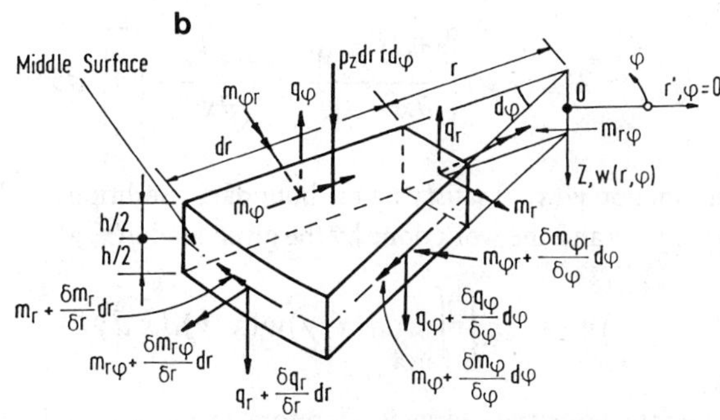

FIGURE 45.53 (a) Circular plate, (b) stress resultants.

The maximum deflection occurs at the center, where $r = 0$, and is given by

$$w = \frac{q_0 a^4}{64D} \tag{45.106}$$

Bending moments in the radial and tangential directions, respectively, are given by

$$M_r = \frac{q_0}{16}[a^2(1+\nu) - r^2(3+\nu)]$$

$$M_t = \frac{q_0}{16}[a^2(1+\nu) - r^2(1+3\nu)] \tag{45.107}$$

The method of superposition can be applied in calculating the deflections for circular plates with simply supported edges. The expressions for deflection and bending moment are given as follows:

$$w = \frac{q_0(a^2 - r^2)}{64D}\left(\frac{5+\nu}{1+\nu}a^2 - r^2\right)$$

$$w_{\max} = \frac{5+\nu}{64(1+\nu)}\frac{q_0 a^4}{D} \tag{45.108}$$

$$M_r = \frac{q_0}{16}(3+\nu)(a^2 - r^2)$$

$$M_t = \frac{q_0}{16}[a^2(3+\nu) - r^2(1+3\nu)] \tag{45.109}$$

This solution can be used to deal with plates with a circular hole at the center and subjected to concentric moment and shearing forces. Plates subjected to concentric loading and concentrated loading also can be solved by this method. More rigorous solutions are available to deal with irregular loading on circular plates. Once again, the energy method can be employed advantageously to solve circular plate problems. Figure 45.54 gives deflection and bending moment expressions for typical cases of loading and boundary conditions on circular plates.

Strain Energy of Simple Plates

The strain energy expression for a simple rectangular plate is given by

$$U = \frac{D}{2}\int\int_{\text{area}}\left\{\left(\frac{\partial^2 w}{\partial x^2} + \frac{\partial^2 w}{\partial y^2}\right)^2 - 2(1-\nu)\left[\frac{\partial^2 w}{\partial x^2}\frac{\partial^2 w}{\partial y^2} - \left(\frac{\partial^2 w}{\partial x \partial y}\right)^2\right]\right\} dx\,dy \tag{45.110}$$

A suitable deflection function $w(x, y)$ satisfying the boundary conditions of the given plate may be chosen. The strain energy U and the work done by the given load $q(x, y)$

$$W = -\int\int_{\text{area}} q(x,y)w(x,y)dx\,dy \tag{45.111}$$

can be calculated. The total potential energy is, therefore, given as $V = U + W$. Minimizing the total potential energy, the plate problem can be solved.

Case No.	Structural System and Static Loading	Deflection and Internal Forces	
1	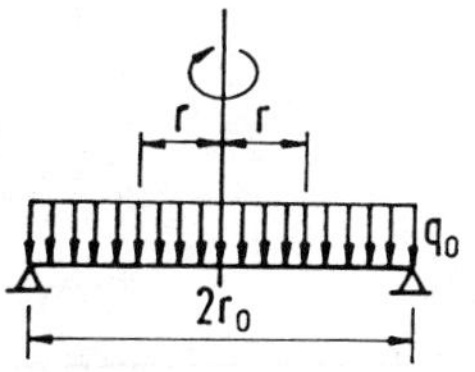 r, r, q_0, $2r_0$	$w = \frac{q_0 r_0^4}{64D(1+\nu)}[2(3+\nu)C_1 - (1+\nu)C_0]$ $m_r = \frac{q_0 r_0^2}{16}(3+\nu)C_1$ $m_\theta = \frac{q_0 r_0^2}{16}[2(1-\nu) - (1+3\nu)C_1]$ $q_r = \frac{q_0 r_0}{2}\rho$	$\rho = \frac{r}{r_0}$ $C_0 = 1 - \rho^4$ $C_1 = 1 - \rho^2$
2	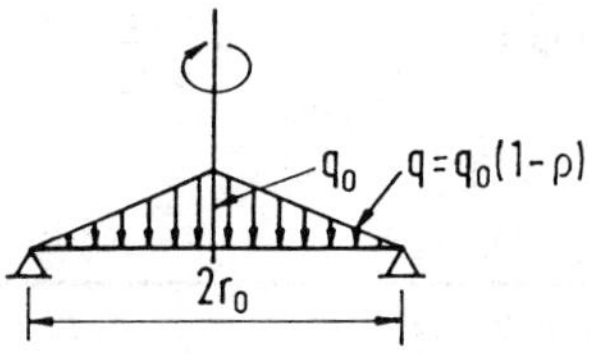 q_0, $q = q_0(1-\rho)$, $2r_0$	$w = \frac{q_0 r_0^4}{14400D}\left[\frac{3(183+43\nu)}{1+\nu} - \frac{10(71+29\nu)}{1+\nu}\rho^2 + 225\rho^4 - 64\rho^5\right]$ $(m_r)_{\rho=0} = (m_\varphi)_{\rho=0} = \frac{q_0 r_0^4}{720}(71+29\nu);$ $(q_r)_{\rho=1} = -\frac{q_0 r_0}{6}$	$\rho = \frac{r}{r_0}$
3	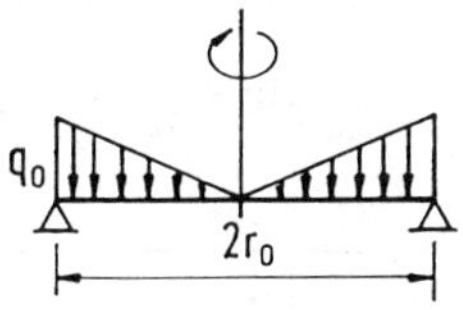 q_0, $2r_0$	$w = \frac{q_0 r_0^4}{450D}\left[\frac{3(6+\nu)}{1+\nu} - \frac{5(4+\nu)}{1+\nu}\rho^2 + 2\rho^5\right]$ $(m_r)_{\rho=0} = (m_\varphi)_{\rho=0} = \frac{q_0 r_0^2}{45}(4+\nu);$ $(q_r)_{\rho=1} = -\frac{q_0 r_0}{3}$	$\rho = \frac{r}{r_0}$
4	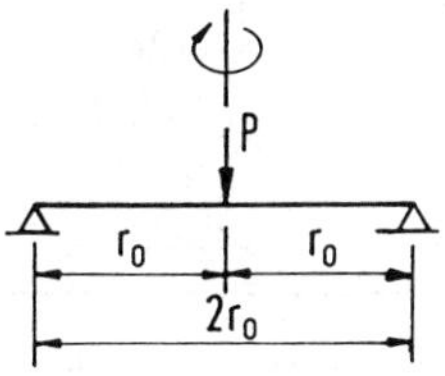 P, r_0, r_0, $2r_0$	$w = \frac{P r_0^2}{16\pi D}\left[\frac{3+\nu}{1+\nu}C_1 + 2C_2\right]$ $m_r = \frac{P}{4\pi}(1+\nu)C_3$ $m_\varphi = \frac{P}{4\pi}[(1-\nu) - (1+\nu)C_3]$ $q_r = \frac{P}{2\pi r_0 \rho}$	$C_1 = 1-\rho^2$ $C_2 = \rho^2 \ln \rho$ $C_3 = \ln \rho$ $\rho = \frac{r}{r_0}$

FIGURE 45.54 Typical loading and boundary conditions for circular plates. (From Szilard, R. 1974. *Theory and Analysis of Plates—Classified and Numerical Methods.* Prentice Hall, Englewood Cliffs, NJ.) *(continues)*

The term

$$\left[\frac{\partial^2 w}{\partial x^2}\frac{\partial^2 w}{\partial y^2} - \left(\frac{\partial^2 w}{\partial x \partial y}\right)^2\right]$$

is known as the Gaussian curvature. If the function $w(x, y) = f(x) \cdot \phi(y)$ (product of a function of x only and a function of y only) and $w = 0$ at the boundary are assumed, then the integral of the Gaussian curvature over the entire plate equals zero. Under these conditions

$$U = \frac{D}{2}\int\int_{\text{area}}\left(\frac{\partial^2 w}{\partial x^2} + \frac{\partial^2 w}{\partial y^2}\right)^2 dx\, dy$$

Figure 6.2.12.7 (Cont'd)

Case No.	Structural System and Static Loading	Deflection and Internal Forces
5	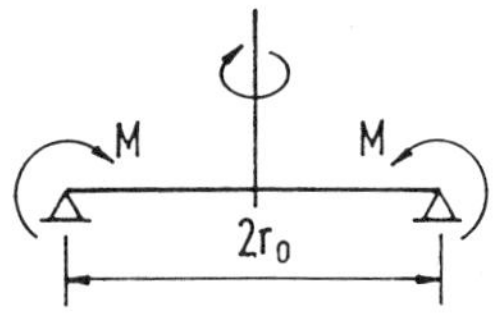 M, M, $2r_0$	$w = \dfrac{Mr_0^2}{2D(1+\nu)C_1}$ $m_r = m_\varphi = M$ $q_r = 0$ $C_1 = 1 - \rho^2, \quad \rho = \dfrac{r}{r_0}$
6	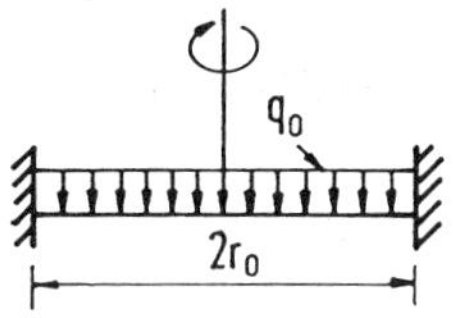 q_0, $2r_0$	$w = \dfrac{q_0 r_0^4}{64D}(1-\rho^2)^2 \qquad q_r = -\dfrac{q_0 r_o}{2}\rho$ $m_r = \dfrac{q_0 r_0^2}{16}[1+\nu-(3+\nu)\rho^2] \qquad \rho = \dfrac{r}{r_o}$ $m_\varphi = \dfrac{q_0 r_0^2}{16}[1+\nu-(1+3\nu)\rho^2]$
7	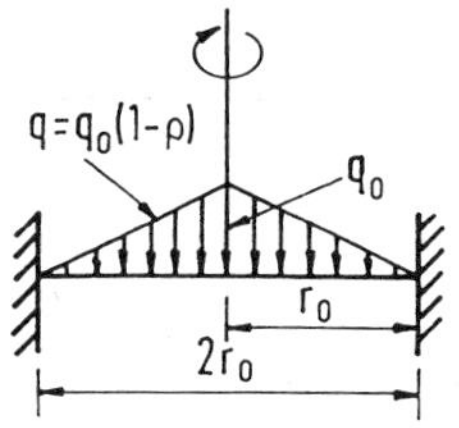 $q = q_0(1-\rho)$, q_0, r_0, $2r_0$	$w = \dfrac{q_0 r_0^4}{14400D}(129-290\rho^2+225\rho^4-64\rho^5)$ $(m_r)_{\rho=0} = (m_\varphi)_{\rho=0} = \dfrac{29 q_0 r_0^2}{720}(1+\nu) \qquad (q_r)_{\rho=1} = -\dfrac{q_0 r_0}{6}$ $(m_r)_{\rho=1} = (m_\varphi)_{\rho=1} = -\dfrac{7 q_0 r_0^2}{120} \qquad \rho = \dfrac{r}{r_0}$
8	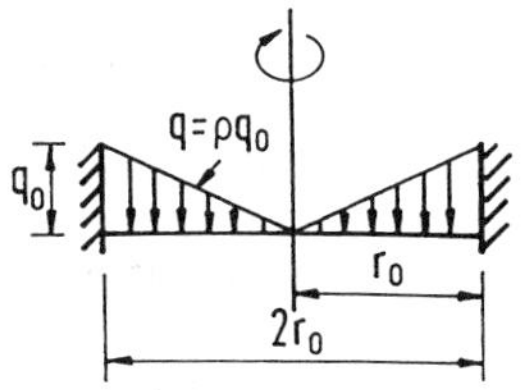 $q = \rho q_0$, q_0, r_0, $2r_0$	$w = \dfrac{q_0 r_0^4}{450D}(3-5\rho^2+2\rho^5) \qquad q_r = -\dfrac{q_0 r_0}{3}\rho^2$ $m_r = \dfrac{q_0 r_0^2}{45}[1+\nu-(4+\nu)\rho^3] \qquad \rho = \dfrac{r}{r_0}$ $m_\varphi = \dfrac{q_0 r_0^2}{45}[1+\nu-(1+4\nu)\rho^3]$
9	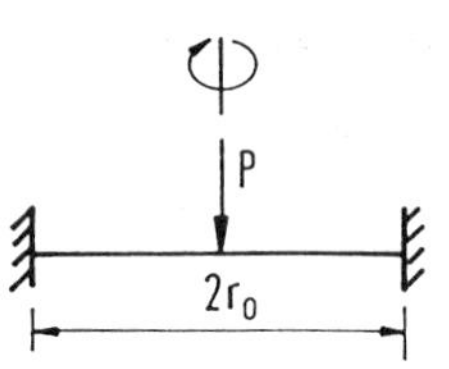 P, $2r_0$	$w = \dfrac{P r_0^2}{16\pi D}(1-\rho^2+2\rho^2 \ln \rho) \qquad q_r = -\dfrac{P}{2\pi r_0 \rho}$ $m_r = -\dfrac{P}{4\pi}[1+(1+\nu)\ln \rho] \qquad \rho = \dfrac{r}{r_0}$ $m_\varphi = -\dfrac{P}{4\pi}[\nu+(1+\nu)\ln \rho]$

FIGURE 45.54 (continued) Typical loading and boundary conditions for circular plates. (From Szilard, R. 1974. *Theory and Analysis of Plates—Classified and Numerical Methods.* Prentice Hall, Englewood Cliffs, NJ.)

If polar coordinates instead of rectangular coordinates are used and axial symmetry of loading and deformation is assumed, the equation for strain energy U takes the form

$$U = \frac{D}{2}\int\int_{\text{area}} \left\{ \left(\frac{\partial^2 w}{\partial r^2} + \frac{1}{4}\frac{\partial w}{\partial r} \right)^2 - \frac{2(1-\nu)}{r}\frac{\partial w}{\partial r}\frac{\partial^2 w}{\partial r^2} \right] r\, dr\, d\theta \tag{45.112}$$

and the work done, W, is written as

$$W = -\int\int_{\text{area}} qwr\,dr\,d\theta \tag{45.113}$$

Detailed treatment of plate theory can be found in Timoshenko and Woinowsky-Krieger [1959] and Gould [1988].

45.13 Membrane Action and Bending of Thin Shells

As in the case of plates, a thin shell is defined as a shell with a thickness relatively small compared with its other dimensions. Deformations should not be large compared with the thickness. The primary difference between a shell structure and a plate structure is that the former has a curvature in the unstressed state, whereas the latter is assumed to be initially flat. The presence of initial curvature is of little consequence as far as flexural behavior is concerned. The membrane behavior, however, is affected significantly by the curvature. Membrane action in a surface is caused by in-plane forces. These forces may be primary forces caused by applied edge loads or edge deformations, or they may be secondary forces resulting from flexural deformations.

In the case of the flat plates, secondary in-plane forces do not give rise to appreciable membrane action unless the bending deformations are large. Membrane action due to secondary forces is, therefore, neglected in small deflection theory. If the surface, as in the case of shell structures, has an initial curvature, membrane action caused by secondary in-plane forces will be significant regardless of the magnitude of the bending deformations.

We can draw the following analogies to illustrate the difference in plate and shell behavior. A beam resists transverse loads by development of bending and shear stress; a **cable** can resist the same load through tension alone. A plate is likened to a two-dimensional beam and resists transverse loads by two-dimensional bending and shear: a membrane is the two-dimensional equivalent of the cable and resists loads through tensile stresses. Imagine a membrane with large deflections [Fig. 45.55(a)]; reverse the load and the membrane and we have the structural shell [Fig. 45.55(b)], provided the shell is stable for the type of load shown. The membrane resists the load through tensile stresses, but the ideal thin shell must be capable of developing both tension and compression.

Stress Resultants in Shell Element

Let us consider an infinitely small shell element formed by two pairs of adjacent planes which are normal to the middle surface of the shell and which contain its principal curvatures, as shown in Fig. 45.56(a). The thickness of the shell is denoted as h. Coordinate axes x and y are taken tangent at O to the lines of principal curvature and the axis z normal to the middle surface. r_x and r_y are the principal radii of curvature lying in the xz and yz planes, respectively. The resultant forces per

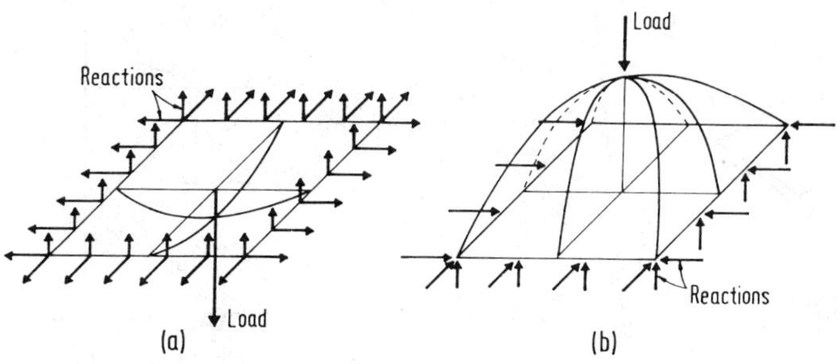

FIGURE 45.55

unit length of the normal sections are given as

$$N_x = \int_{-h/2}^{h/2} \sigma_x \left(1 - \frac{z}{r_y}\right) dz \qquad N_y = \int_{-h/2}^{h/2} \sigma_y \left(1 - \frac{z}{r_x}\right) dz$$

$$N_{xy} = \int_{-h/2}^{h/2} \tau_{xy} \left(1 - \frac{z}{r_y}\right) dz \qquad N_{yx} = \int_{-h/2}^{h/2} \tau_{yx} \left(1 - \frac{z}{r_x}\right) dz$$

$$Q_x = \int_{-h/2}^{h/2} \tau_{xz} \left(1 - \frac{z}{r_y}\right) dz \qquad Q_y = \int_{-h/2}^{h/2} \tau_{yz} \left(1 - \frac{z}{r_x}\right) dz \qquad (45.114)$$

The bending and twisting moments per unit length of the normal sections are given by

$$M_x = \int_{-h/2}^{h/2} \sigma_x z \left(1 - \frac{z}{r_y}\right) dz \qquad M_y = \int_{-h/2}^{h/2} \sigma_y z \left(1 - \frac{z}{r_x}\right) dz$$

$$M_{xy} = \int_{-h/2}^{h/2} \tau_{xy} z \left(1 - \frac{z}{r_y}\right) dz \qquad M_{yx} = \int_{-h/2}^{-h/2} \tau_{yx} z \left(1 - \frac{z}{r_x}\right) dz \qquad (45.115)$$

It is assumed, in bending the shell, that linear elements AD and BC (Fig. 45.56), which are normal to the middle surface of the shell, remain straight and become normal to the deformed middle surface of the shell. If the conditions of a shell are such that bending can be neglected, the problem of stress analysis is greatly simplified since the resultant moments [Eq. (45.115)] vanish along with shearing forces Q_x and Q_y in Eq. (45.114). Thus the only unknowns are N_x, N_y, and $N_{xy} = N_{yx}$, and these are called membrane forces.

Membrane Theory of Shells of Revolution

Shells having the form of surfaces of revolution find extensive application in various kinds of containers, tanks, and domes. Consider an element of a shell cut by two adjacent meridians and two parallel circles, as shown in Fig. 45.57. There will be no shearing forces on the sides of the

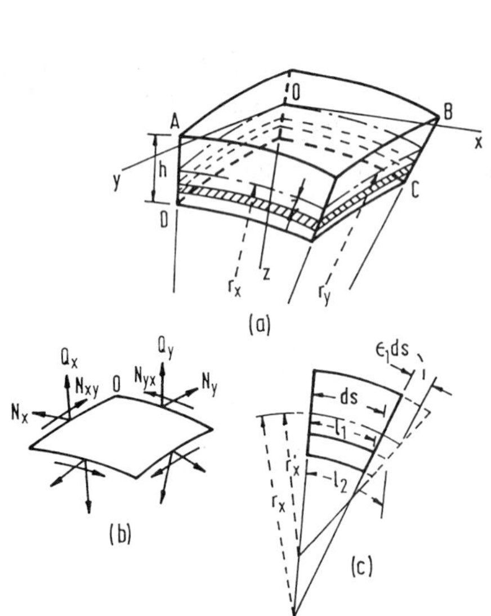

FIGURE 45.56 A shell element.

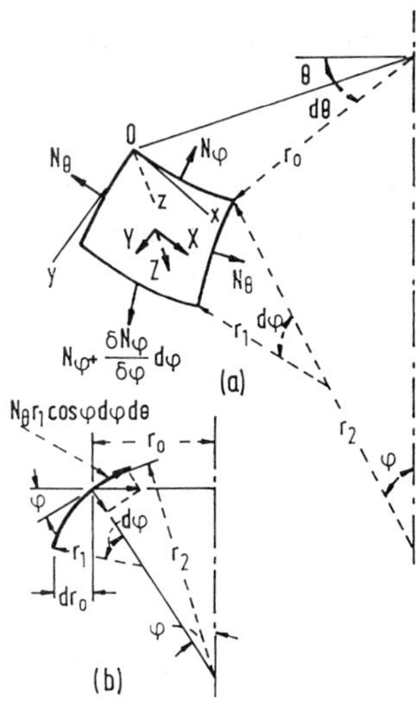

FIGURE 45.57 An element from shells of revolution—symmetrical loading.

element because of the symmetry of loading. By considering the equilibrium in the direction of the tangent to the meridian and z, two equations of equilibrium are written, respectively, as

$$\frac{d}{d\phi}(N_\phi r_0) - N_\theta r_1 \cos\phi + Y r_1 r_0 = 0$$

$$N_\phi r_0 + N_\theta r_1 \sin\phi + Z r_1 r_0 = 0 \tag{45.116}$$

The forces N_θ and N_ϕ can be calculated from Eq. (45.116) if the radii r_0 and r_1 and the components Y and Z of the intensity of the external load are given.

Spherical Dome

The spherical shell shown in Fig. 45.58 is assumed to be subjected to its own weight; the intensity of the self-weight is assumed as a constant value q_0 per unit area. Considering an element of the shell at an angle ϕ, the self-weight of the portion of the shell above this element is obtained as

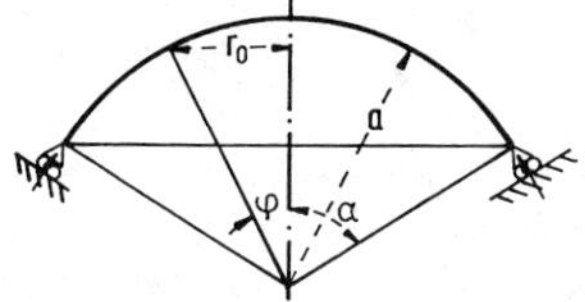

FIGURE 45.58 Spherical dome.

$$r = 2\pi \int_0^\phi a^2 q_0 \sin\phi \, d\phi$$

$$= 2\pi a^2 q_0 (1 - \cos\phi)$$

Considering the equilibrium of the portion of the shell above the parallel circle defined by the angle ϕ, we can write

$$2\pi r_0 N_\phi \sin\phi + R = 0 \tag{45.117}$$

Therefore,

$$N_\phi = -\frac{aq(1-\cos\phi)}{\sin^2\phi} = -\frac{aq}{1+\cos\phi}$$

We can write from Eq. (45.116)

$$\frac{N_\phi}{r_1} + \frac{N_\theta}{r_2} = -Z \tag{45.118}$$

Substituting for N_ϕ and R into Eq. (45.118),

$$N_\theta = -aq\left(\frac{1}{1+\cos\phi} - \cos\phi\right)$$

It is seen that the forces N_ϕ are always negative. There is thus a compression along the meridians that increases as the angle ϕ increases. The forces N_θ are also negative for small ϕ. The stresses as calculated above will represent the actual stresses in the shell with great accuracy if the supports are of such a type that the reactions are tangent to meridians as shown in the figure.

Conical Shells

If a force P is applied in the direction of the axis of the cone, as shown in Fig. 45.59, the stress distribution is symmetrical and we obtain

$$N_\phi = -\frac{P}{2\pi r_0 \cos\alpha}$$

By Eq. (45.118), one obtains $N_\theta = 0$.

In the case of a conical surface in which the lateral forces are symmetrically distributed, the membrane stresses can be obtained by using Eqs. (45.117) and (45.118). The curvature of the meridian in the case of a cone is zero and hence $r_1 = \infty$; Eqs. (45.117) and (45.118) can, therefore, be written as

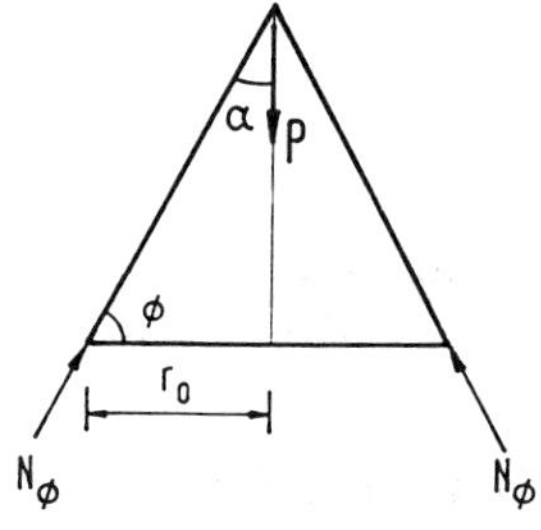

FIGURE 45.59 Conical shell.

$$N_\phi = -\frac{R}{2\pi r_0 \sin\phi}$$

and

$$N_\theta = -r_2 Z = -\frac{Z r_0}{\sin\phi}$$

If the load distribution is given, N_ϕ and N_θ can be calculated independently.

For example, a conical tank filled with a liquid of specific weight γ is considered as shown in Fig. 45.60. The pressure at any parallel circle mn is

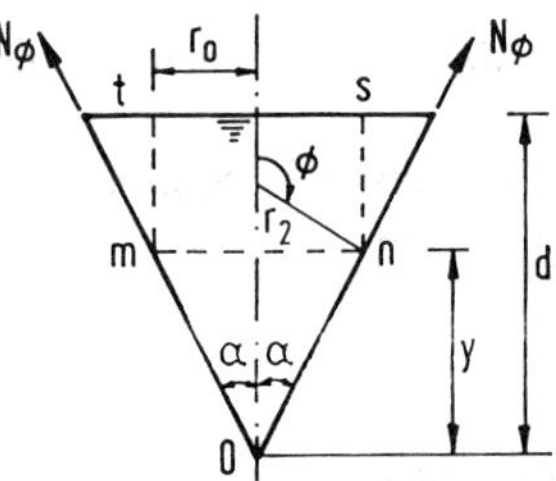

FIGURE 45.60 Inverted conical tank.

$$p = -Z = \gamma(d - y)$$

For the tank, $\phi = \alpha + (\pi/2)$ and $r_0 = y \tan\alpha$. Therefore,

$$N_\theta = \frac{\gamma(d - y) y \tan\alpha}{\cos\alpha}$$

N_θ is maximum when $y = d/2$ and hence

$$(N_\theta)_{\max} = \frac{\gamma d^2 \tan\alpha}{4\cos\alpha}$$

The term R in the expression for N_ϕ is equal to the weight of the liquid in the conical part mno and the cylindrical part must be as shown in Fig. 45.59. Therefore,

$$\begin{aligned} R &= -\left[\tfrac{1}{3}\pi y^3 \tan^2\alpha + \pi y^2 \tan^2\alpha (d - y)\right]\gamma \\ &= -\pi\gamma y^2 (d - \tfrac{2}{3}y) \tan^2\alpha \end{aligned}$$

Hence,

$$N_\phi = \frac{\gamma y (d - \frac{2}{3}y) \tan\alpha}{2\cos\alpha}$$

N_ϕ is maximum when $y = \frac{3}{4}d$ and

$$(N_\phi)_{\max} = \frac{3}{16}\frac{d^2\gamma \tan\alpha}{\cos\alpha}$$

The horizontal component of N_ϕ is taken by the reinforcing ring provided along the upper edge of the tank. The vertical components constitute the reactions supporting the tank.

Shells in the Form of a Surface of Revolution Subjected to Unsymmetrical Loading

Consider an element cut from a shell by two adjacent meridians and two parallel circles (Fig. 45.61). In the general case shear forces $N_{\varphi\theta} = N_{\theta\varphi}$ in addition to normal forces N_φ and N_θ will act on the sides of the element. Projecting the forces on the element in the y direction we obtain the equation

$$\frac{\partial}{\partial\varphi}(N_\varphi r_0) + \frac{\partial N_{\theta\varphi}}{\partial\theta} r_1 - N_\theta r_1 \cos\varphi + Y r_1 r_0 = 0 \tag{45.119}$$

Similarly, the forces in the x direction can be summed to give

$$\frac{\partial}{\partial\varphi}(r_0 N_{\varphi\theta}) + \frac{\partial N_\theta}{\partial\theta} r_1 + N_{\theta\varphi} r_1 \cos\varphi + X r_0 r_1 = 0 \tag{45.120}$$

Since the projection of shearing forces on the z axis vanishes, the third equation is the same as Eq. (45.118). The problem of determining membrane stresses under unsymmetrical loading reduces to the solution of Eqs. (45.118), (45.119), and (45.120) for given values of the components X, Y, and Z of the intensity of the external load.

Membrane Theory of Cylindrical Shells

It is assumed that the generator of the shell is horizontal and parallel to the x axis. An element is cut from the shell by two adjacent generators and two cross sections perpendicular to the x axis, and its position is defined by the coordinate x and the angle φ. The forces acting on the sides of the element are shown in Fig. 45.62(b).

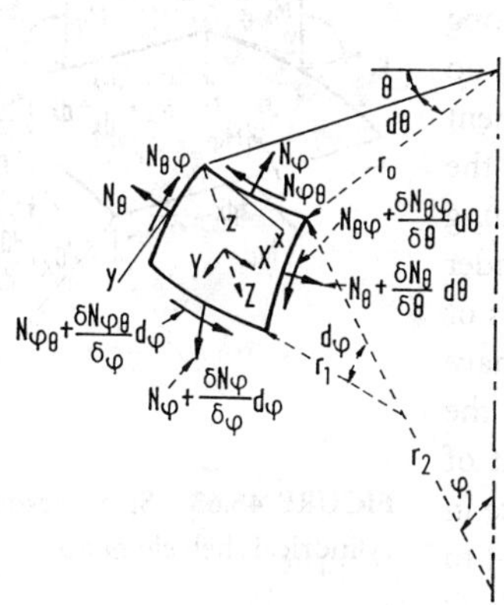

FIGURE 45.61 An element from shells of revolution—unsymmetrical loading.

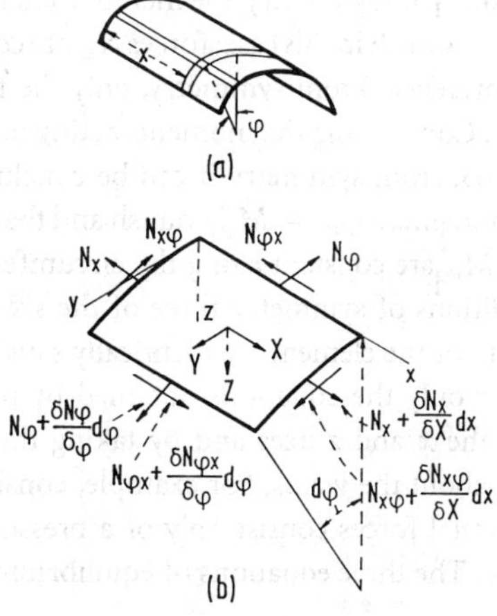

FIGURE 45.62 Membrane forces on a cylindrical shell element.

The components of the distributed load over the surface of the element are denoted as X, Y, and Z. Considering the equilibrium of the element and summing the forces in the x direction, we obtain

$$\frac{\partial N_x}{\partial x} r\, d\varphi\, dx + \frac{\partial N_{\varphi x}}{\partial \varphi} d\varphi\, dx + Xr\, d\varphi\, dx = 0$$

The corresponding equations of equilibrium in the y and z directions are given, respectively, as

$$\frac{\partial N_{x\varphi}}{\partial x} r\, d\varphi\, dx + \frac{\partial N_{\varphi}}{\partial \varphi} d\varphi\, dx + Yr\, d\varphi\, dx = 0$$

$$N_{\varphi}\, d\varphi\, dx + Zr\, d\varphi\, dx = 0$$

The three equations of equilibrium can be simplified and represented in the following form:

$$\frac{\partial N_x}{\partial x} + \frac{1}{r}\frac{\partial N_{x\varphi}}{\partial \varphi} = -X$$

$$\frac{\partial N_{x\varphi}}{\partial x} + \frac{1}{r}\frac{\partial N_{\varphi}}{\partial \varphi} = -Y$$

$$N_{\varphi} = -Zr \tag{45.121}$$

In each particular case we readily find the value of N_φ. Substituting this value in the second equation, we then obtain $N_{x\varphi}$ by integration. Using the value of $N_{x\varphi}$ thus obtained we find N_x by integrating the first equation.

Circular Cylindrical Shell Loaded Symmetrically with Respect to Its Axis

In practical applications, problems in which a circular shell is subjected to the action of forces distributed symmetrically with respect to the axis of the cylinder are common. To establish the equations required for the solution of these problems, we consider an element, as shown in Figs. 45.62(a) and 45.63, and consider the equations of equilibrium. From symmetry, the membrane shearing forces $N_{x\varphi} = N_{\varphi x}$ vanish in this case; forces N_φ are constant along the circumference. From symmetry, only the forces Q_z do not vanish. Considering the moments acting on the element in Fig. 45.63, from symmetry it can be concluded that the twisting moments $M_{x\varphi} = M_{\varphi x}$ vanish and that the bending moments M_φ are constant along the circumference. Under such conditions of symmetry three of the six equations of equilibrium of the element are identically satisfied. We have to consider only the equations obtained by projecting the forces on the x and z axes and by taking the moment of the forces about the y axis. For example, consider a case in which external forces consist only of a pressure normal to the surface. The three equations of equilibrium are

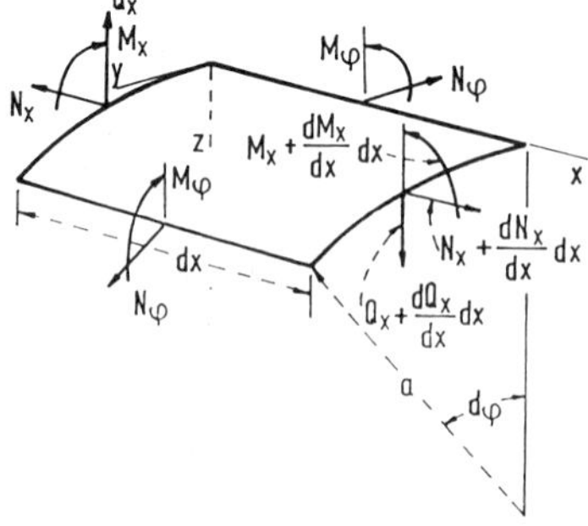

FIGURE 45.63 Stress resultants in a cylindrical shell element.

$$\frac{dN}{dx} a\, dx\, d\varphi = 0$$

$$\frac{dQ_x}{dx} a\, dx\, d\varphi + N_\varphi\, dx\, d\varphi + Za\, dx\, d\varphi = 0$$

$$\frac{dM_x}{dx} a\, dx\, d\varphi - Q_x a\, dx\, d\varphi = 0 \tag{45.122}$$

The first one indicates that the forces N_x are constant, and they are taken equal to zero in the further discussion. If they are different from zero, the deformation and stress corresponding to such constant forces can be easily calculated and superposed on stresses and deformations produced by lateral load. The remaining two equations are written in the following simplified form:

$$\frac{dQ_x}{dx} + \frac{1}{a} N_\varphi = -Z$$

$$\frac{dM_x}{dx} - Q_x = 0 \tag{45.123}$$

These two equations contain three unknown quantities: N_φ, Q_x, and M_x. We need, therefore, to consider the displacements of points in the middle surface of the shell.

The component v of the displacement in the circumferential direction vanishes because of symmetry. Only the components u and w in the x and z directions, respectively, are to be considered. The expressions for the strain components then become

$$\varepsilon_x = \frac{du}{dx} \qquad \varepsilon_\varphi = -\frac{w}{a} \tag{45.124}$$

By Hooke's law, we obtain

$$N_x = \frac{Eh}{1-\nu^2}(\varepsilon_x + \nu\varepsilon_\varphi) = \frac{Eh}{1-\nu^2}\left(\frac{du}{dx} - \nu\frac{w}{a}\right) = 0$$

$$N_\varphi = \frac{Eh}{1-\nu^2}(\varepsilon_\varphi + \nu\varepsilon_x) = \frac{Eh}{1-\nu^2}\left(-\frac{w}{a} + \nu\frac{du}{dx}\right) = 0 \tag{45.125}$$

From the first of these equation it follows that

$$\frac{du}{dx} = \nu\frac{w}{a}$$

and the second equation gives

$$N_\varphi = -\frac{Ehw}{a} \tag{45.126}$$

Considering the bending moments, we conclude from symmetry that there is no change in curvature in the circumferential direction. The curvature in the x direction is equal to $-d^2w/dx^2$. Using the same equations as for plates, we then obtain

$$M_\varphi = \nu M_x$$

$$M_x = -D\frac{d^2w}{dx^2} \tag{45.127}$$

where

$$D = \frac{Eh^3}{12(1 - \nu^3)}$$

is the flexural rigidity per unit length of the shell.

Eliminating Q_x from Eq. (45.123) we obtain

$$\frac{d^2 M_x}{dx^2} + \frac{1}{a} N_\varphi = -Z$$

from which, by using Eqs. (45.126) and (45.127), we obtain

$$\frac{d^2}{dx^2}\left(D\frac{d^2 w}{dx^2}\right) + \frac{Eh}{a^2} w = Z \qquad (45.128)$$

All problems of symmetrical deformation of circular cylindrical shells thus reduce to the integration of Eq. (45.128).

The simplest application of this equation is obtained when the thickness of the shell is constant. Under such conditions Eq. (45.128) becomes

$$D\frac{d^4 w}{dx^4} + \frac{Eh}{a^2} w = Z$$

Using the notation

$$\beta^4 = \frac{Eh}{4a^2 D} = \frac{3(1 - \nu^2)}{a^2 h^2} \qquad (45.129)$$

Equation (45.129) can be represented in the simplified form

$$\frac{d^4 w}{dx^4} + 4\beta^4 w = \frac{Z}{D} \qquad (45.130)$$

The general solution of this equation is

$$w = e^{\beta x}(C_1 \cos \beta x + C_2 \sin \beta x) + e^{-\beta x}(C_3 \cos \beta x + C_4 \sin \beta x) + f(x) \qquad (45.131)$$

Detailed treatment of shell theory can be obtained from Timoshenko and Woinowsky-Krieger [1959] and Gould [1988].

45.14 Computer Methods of Structural Analysis

In this method of structural analysis, a set of simultaneous equations that describe the load-deformation characteristics of the structure under consideration are formed. These equations are solved using matrix algebra to obtain the load-deformation characteristics of discrete or finite elements into which the structure has been subdivided. Matrix algebra is ideally suited for setting up and solving equations on the computer. Matrix structural analysis has two methods of approach: the flexibility method, in which forces are used as independent variables; and the stiffness method, which employs deformations as the independent variables. The two methods are also called the force method and the displacement method, respectively.

Flexibility Method

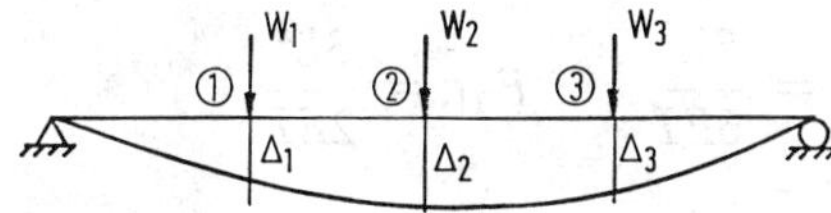

FIGURE 45.64 Simple beam under concentrated loads.

In a structure, the forces and displacements are related to one another by using stiffness influence coefficients. Let us consider, for example, a simple beam in which three concentrated loads W_1, W_2, and W_3 are applied at sections 1, 2, and 3, respectively, as shown in Fig. 45.64. The deflection at section 1, Δ_1, can be expressed as

$$\Delta_1 = F_{11}W_1 + F_{12}W_2 + F_{13}W_3$$

where F_{11}, F_{12}, and F_{13} are called flexibility coefficients and are, respectively, defined as the deflection at section 1 due to unit loads applied at sections 1, 2, and 3. Deflections at sections 2 and 3 are similarly given as

$$\Delta_2 = F_{21}W_1 + F_{22}W_2 + F_{23}W_3$$

and

$$\Delta_3 = F_{31}W_1 + F_{32}W_2 + F_{33}W_3 \tag{45.132}$$

These expressions are written in matrix form as

$$\begin{Bmatrix} \Delta_1 \\ \Delta_2 \\ \Delta_3 \end{Bmatrix} = \begin{bmatrix} F_{11} & F_{12} & F_{13} \\ F_{21} & F_{22} & F_{23} \\ F_{31} & F_{32} & F_{33} \end{bmatrix} \begin{Bmatrix} W_1 \\ W_2 \\ W_3 \end{Bmatrix} \quad \text{or} \quad \{\Delta\} = [F]\{W\} \tag{45.133}$$

The matrix $[F]$ is called the flexibility matrix. It can be shown, by applying Maxwell's reciprocal theorem [Borg and Gennaro, 1959], that the matrix $\{F\}$ is a symmetric matrix.

Flexibility matrix for a cantilever beam loaded, as shown in Fig. 45.65, can be constructed as follows. The first column in the flexibility matrix can be generated by applying a unit vertical load at the free end of the cantilever as shown in Figure 45.65(b) and making use of the moment area

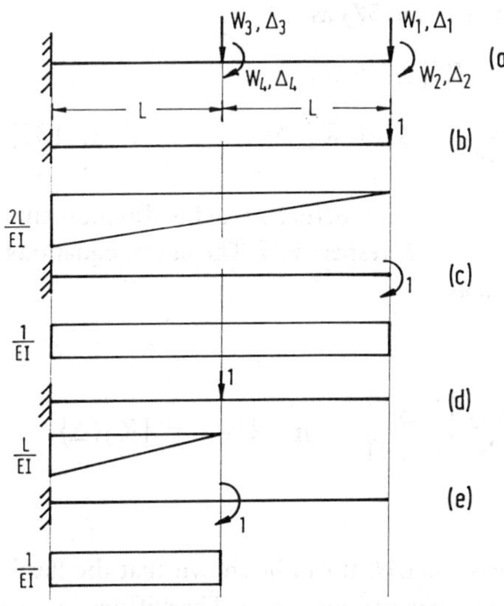

FIGURE 45.65 Cantilever beam.

method. We get

$$F_{11} = \frac{8L^3}{3EI} \qquad F_{21} = \frac{2L^2}{EI} \qquad F_{31} = \frac{5L^3}{6EI} \qquad F_{41} = \frac{3L^2}{2EI}$$

Columns 2, 3, and 4 are, similarly, generated by applying the unit moment at the free end and the unit force and unit moment at midspan as shown in Fig. 45.65(c), (d), and (e), respectively. Combining the results thus obtained, one gets the flexibility matrix as

$$\begin{Bmatrix} \Delta_1 \\ \Delta_2 \\ \Delta_3 \\ \Delta_4 \end{Bmatrix} = \frac{1}{EI} \begin{bmatrix} \frac{8L^3}{3} & 2L^2 & \frac{5L^3}{6} & \frac{3L^2}{2} \\ 2L^2 & 2L & \frac{L^2}{2} & L \\ \frac{5L^3}{6} & \frac{L^2}{2} & \frac{L^3}{3} & \frac{L^2}{2} \\ \frac{3L^2}{2} & L & \frac{L^2}{2} & L \end{bmatrix} \begin{Bmatrix} W_1 \\ W_2 \\ W_3 \\ W_4 \end{Bmatrix} \tag{45.134}$$

The above method to generate the flexibility matrix for a given structure is extremely impractical. It is therefore recommended to subdivide a given structure into several elements and form the flexibility matrix for each of the elements. The flexibility matrix for the entire structure is then obtained by combining the flexibility matrices of the individual elements.

Force transformation matrix relates what occurs in these elements to the behavior of the entire structure. Using the conditions of equilibrium, it relates the element forces to the structure forces. The principle of conservation of energy may be used to generate transformation matrices.

Stiffness Method

Forces and deformations in a structure are related to one another by means of stiffness influence coefficients. Let us consider, for example, a simply supported beam subjected to end moments W_1 and W_2 applied at supports 1 and 2, and let the respective rotations be denoted as Δ_1 and Δ_2 as shown in Fig. 45.66. We can now write the expressions for end moments W_1 and W_2 as

$$\begin{aligned} W_1 &= K_{11}\Delta_1 + K_{12}\Delta_2 \\ W_2 &= K_{21}\Delta_1 + K_{22}\Delta_2 \end{aligned} \tag{45.135}$$

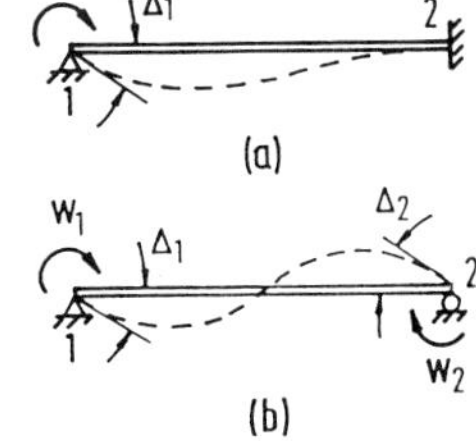

FIGURE 45.66 Simply supported beam.

where K_{11} and K_{12} are stiffness influence coefficients defined as moments at 1 due to unit rotation at 1 and 2, respectively. The above equations can be written in matrix form as

$$\begin{Bmatrix} W_1 \\ W_2 \end{Bmatrix} = \begin{bmatrix} K_{11} & K_{12} \\ K_{21} & K_{22} \end{bmatrix} \begin{Bmatrix} \Delta_2 \\ \Delta_2 \end{Bmatrix} \quad \text{or} \quad \{W\} = [K]\{\Delta\} \tag{45.136}$$

The matrix $[K]$ is the stiffness matrix. It can be shown that the flexibility matrix of a structure is the inverse of the stiffness matrix and vice versa. The stiffness matrix of the whole structure is formed out of the stiffness matrices of the individual elements that make up the structure.

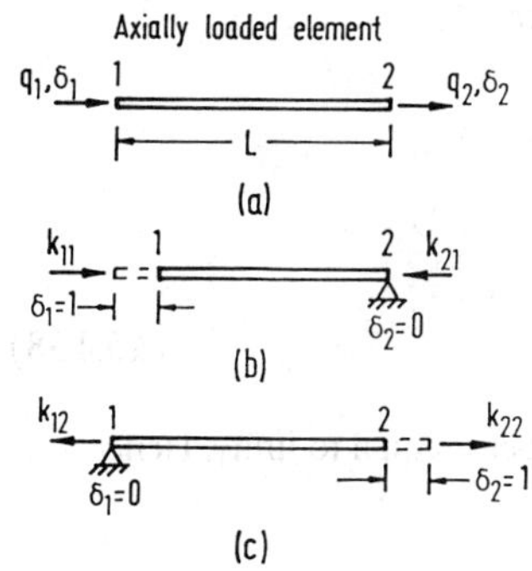

Element Stiffness Matrix of an Axially Loaded Member

Figure 45.67 shows an axially loaded member of constant cross-sectioned area with element forces q_1 and q_2 and displacements δ_1 and δ_2. They are shown in their respective positive directions. With unit displacement $\delta_1 = 1$ at node 1, as shown in Fig. 45.67, axial forces at nodes 1 and 2 are

$$k_{11} = \frac{EA}{L} \qquad k_{21} = -\frac{EA}{L}$$

In the same way, by setting $\delta_2 = 1$, as shown in Fig. 45.67, the corresponding forces are

FIGURE 45.67 Axially loaded member.

$$k_{12} = -\frac{EA}{L} \qquad k_{22} = \frac{EA}{L}$$

The stiffness matrix is

$$\begin{Bmatrix} q_1 \\ q_2 \end{Bmatrix} = \begin{bmatrix} k_{11} & k_{12} \\ k_{21} & k_{22} \end{bmatrix} \begin{Bmatrix} \delta_1 \\ \delta_2 \end{Bmatrix} \quad \text{or} \quad \begin{Bmatrix} q_1 \\ q_2 \end{Bmatrix} = \frac{EA}{L} \begin{bmatrix} 1 & -1 \\ -1 & 1 \end{bmatrix} \begin{Bmatrix} \delta_1 \\ \delta_2 \end{Bmatrix} \tag{45.137}$$

Element Stiffness Matrix of a Flexural Member

The stiffness matrix for the flexural element shown in Fig. 45.68 can be constructed as follows. The forces and the corresponding displacements (the moments, the shears, and the corresponding rotations and translations at the ends of the member) are defined in the figure. The matrix equation that relates these forces and displacements can be written in the form

$$\begin{bmatrix} q_1 \\ q_2 \\ q_3 \\ q_4 \end{bmatrix} = \begin{bmatrix} k_{11} & k_{12} & k_{13} & k_{14} \\ k_{21} & k_{22} & k_{23} & k_{24} \\ k_{31} & k_{32} & k_{33} & k_{34} \\ k_{41} & k_{42} & k_{43} & k_{44} \end{bmatrix} \begin{bmatrix} \delta_1 \\ \delta_2 \\ \delta_3 \\ \delta_4 \end{bmatrix}$$

The terms in the first column consist of the element forces q_1 through q_4 that result from displacement $\delta_1 = 1$ when $\delta_2 = \delta_3 = \delta_4 = 0$. This means that a unit vertical displacement is imposed at the left end of the member while translation at the right end and rotation at both ends are prevented, as shown in Fig. 45.68. The four member forces corresponding to this deformation can be obtained using the moment-area method.

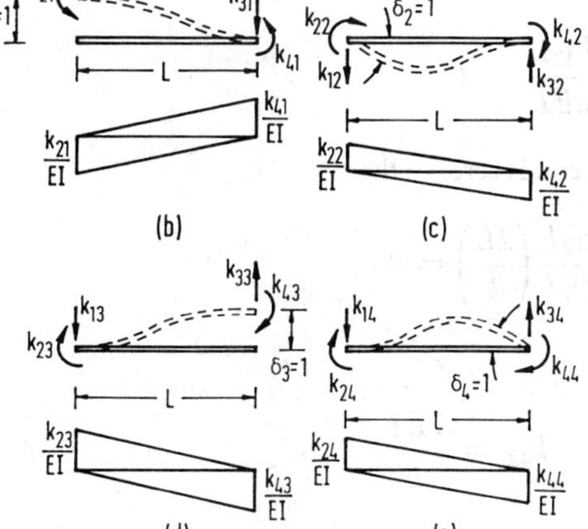

FIGURE 45.68 Beam element—stiffness matrix.

The change in slope between the two ends of the member is zero and the area of the

M/EI diagram between these points must, therefore, vanish. Hence

$$\frac{k_{41}L}{2EI} - \frac{k_{21}L}{2EI} = 0$$

and

$$k_{21} = k_{41} \tag{45.138}$$

The moment of the M/EI diagram about the left end of the member is equal to unity. Hence

$$\frac{k_{41}L}{2EI}\left(\frac{2L}{3}\right) - \frac{k_{21}L}{2EI}\left(\frac{L}{3}\right) = 1$$

and in view of Eq. (45.138),

$$k_{41} = k_{21} = \frac{6EI}{L^2}$$

Finally, the moment equilibrium of the member about the right end leads to

$$k_{11} = \frac{k_{21} + k_{41}}{L} = \frac{12EI}{L^3}$$

and from equilibrium in the vertical direction we obtain

$$k_{31} = k_{11} = \frac{12EI}{L^3}$$

The forces act in the directions indicated in Fig. 45.68(b). To obtain the correct signs, one must compare the forces with the positive directions defined in Fig. 45.68(a). Thus

$$k_{11} = \frac{12EI}{L^3} \qquad k_{21} = -\frac{6EI}{L^2} \qquad k_{31} = -\frac{12EI}{L^3} \qquad k_{41} = \frac{6EI}{L^2}$$

The second column of the stiffness matrix is obtained by letting $\delta_2 = 1$ and setting the remaining three displacements equal to zero, as indicated in Fig. 45.68(c). The area of the M/EI diagram between the ends of the member for this case is equal to unity, and hence,

$$\frac{k_{22}L}{2EI} - \frac{k_{42}L}{2EI} = 1$$

The moment of the M/EI diagram about the left end is zero, so that

$$\frac{k_{22}L}{2EI}\left(\frac{L}{3}\right) - \frac{k_{42}L}{2EI}\left(\frac{2L}{3}\right) = 0$$

Therefore, one obtains

$$k_{22} = \frac{4EI}{L} \qquad k_{42} = \frac{2EI}{L}$$

From the vertical equilibrium of the member,

$$k_{12} = k_{32}$$

and the moment equilibrium about the right end of the member leads to

$$k_{12} = \frac{k_{22} + k_{42}}{L} = \frac{6EI}{L^2}$$

Comparison of the forces in Fig. 45.68(c) with the positive directions defined in Fig. 45.68(a) indicates that all the influence coefficients except k_{12} are positive. Thus

$$k_{12} = -\frac{6EI}{L^2} \qquad k_{22} = \frac{4EI}{L} \qquad k_{32} = \frac{6EI}{L^2} \qquad k_{42} = \frac{2EI}{L}$$

Using Figures 45.68(d) and (e), the influence coefficients for the third and fourth columns can be obtained. The results of these calculations lead to the following element-stiffness matrix:

$$\begin{bmatrix} q_1 \\ q_2 \\ q_3 \\ q_4 \end{bmatrix} = \begin{bmatrix} \frac{12EI}{L^3} & -\frac{6EI}{L^2} & -\frac{12EI}{L^3} & -\frac{6EI}{L^2} \\ -\frac{6EI}{L^2} & \frac{4EI}{L} & \frac{6EI}{L^2} & \frac{2EI}{L} \\ -\frac{12EI}{L^3} & \frac{6EI}{L^2} & \frac{12EI}{L^3} & \frac{6EI}{L^2} \\ -\frac{6EI}{L^2} & \frac{2EI}{L} & \frac{6EI}{L^2} & \frac{4EI}{L} \end{bmatrix} \begin{bmatrix} \delta_1 \\ \delta_2 \\ \delta_3 \\ \delta_4 \end{bmatrix} \tag{45.139}$$

Note that Eq. (45.139) defines the element-stiffness matrix for a flexural member with constant flexural rigidity EI.

If axial load in a frame member is also considered the general form of an element-stiffness matrix for an element shown in Fig. 45.69 becomes

$$\begin{bmatrix} q_1 \\ q_2 \\ q_3 \\ q_4 \\ q_5 \\ q_6 \end{bmatrix} = \begin{bmatrix} \frac{EA}{L} & 0 & 0 & -\frac{EA}{L} & 0 & 0 \\ 0 & \frac{12EI}{L^3} & -\frac{6EI}{L^2} & 0 & -\frac{12EI}{L^3} & -\frac{6EI}{L^2} \\ 0 & -\frac{6EI}{L^2} & \frac{4EI}{L} & 0 & \frac{6EI}{L^2} & \frac{2EI}{L} \\ -\frac{EA}{L} & 0 & 0 & \frac{EA}{L} & 0 & 0 \\ 0 & -\frac{12EI}{L^3} & \frac{6EI}{L^2} & 0 & \frac{12EI}{L^3} & \frac{6EI}{L^2} \\ 0 & -\frac{6EI}{L^2} & \frac{2EI}{L} & 0 & \frac{6EI}{L^2} & \frac{4EI}{L} \end{bmatrix} \begin{bmatrix} \delta_1 \\ \delta_2 \\ \delta_3 \\ \delta_4 \\ \delta_5 \\ \delta_6 \end{bmatrix} \tag{45.140}$$

or

$$[q] = [k_c][\delta] \tag{45.141}$$

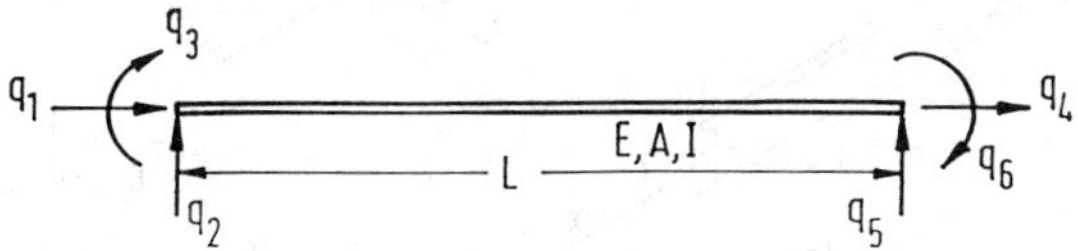

FIGURE 45.69 Beam element with axial force.

Grillages

Structures in which the members all lie in one plane with loads being applied in the direction normal to this plane are called **grillages.** This type of structure is commonly adopted in building floor systems, bridge decks, ship decks, and so on. The grid floor, for purpose of analysis by using the matrix method, can be treated as a space frame. However, the solution can be simplified by considering the grid floor as a planar grid. A typical grid floor is shown in Fig. 45.70(a). The significant member forces in a member and the corresponding deformations are, as shown in Fig. 45.70(b),

$$\begin{bmatrix} m_x \\ p_y \\ m_z \end{bmatrix} \quad \text{and} \quad \begin{bmatrix} \theta_x \\ \delta_y \\ \theta_z \end{bmatrix}$$

The member stiffness matrix can be written as

$$[K] = \begin{bmatrix} \frac{GJ}{L} & 0 & 0 & -\frac{GJ}{L} & 0 & 0 \\ 0 & \frac{12EI_z}{L^3} & \frac{6EI_z}{L^2} & 0 & -\frac{12EI_z}{L^3} & \frac{6EI_z}{L^2} \\ 0 & \frac{6EI_z}{L^2} & \frac{4EI_z}{L} & 0 & -\frac{6EI_z}{L^2} & \frac{2EI_z}{L} \\ -\frac{GJ}{L} & 0 & 0 & \frac{GJ}{L} & 0 & 0 \\ 0 & -\frac{12EI_z}{L^2} & -\frac{6EI_z}{L^2} & 0 & \frac{12EI_z}{L^3} & -\frac{6EI_z}{L^2} \\ 0 & \frac{6EI_z}{L^2} & \frac{2EI_z}{L} & 0 & -\frac{6EI_z}{L^2} & \frac{4EI_z}{L} \end{bmatrix} \tag{45.142}$$

Structure Stiffness Matrix

Equation (45.141) has been expressed in terms of the coordinate system of the individual members. In a structure consisting of many members there would be as many systems of coordinates as there are members. Before the internal actions in the members of the structure can be related, all forces and deflections must be stated in terms of one single system of axes common to all—the structure axes. The transformation from element to structure coordinates is carried out separately for each element and the resulting matrices are then combined to form the structure-stiffness matrix. A separate transformation matrix $[T]$ is written for each element and a relation of the form

$$[\delta]_n = [T]_n[\Delta]_n \tag{45.143}$$

is written in which $[T]_n$ defines the matrix relating the element deformations of element n to the structure deformations at the ends of that particular element. The element and structure forces are

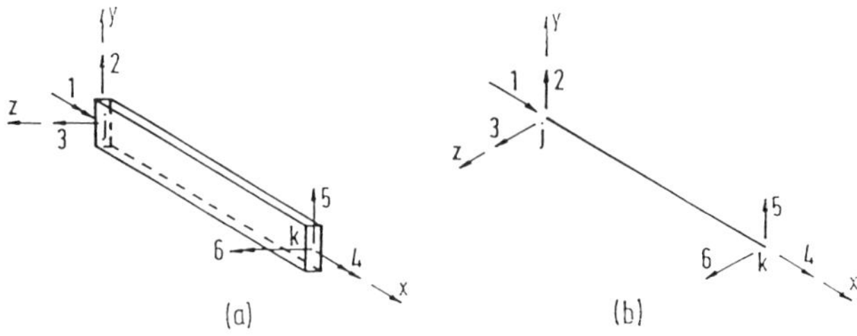

FIGURE 45.70 A truss member.

related in the same way as the corresponding deformations as

$$[q]_n = [T]_n[W]_n \tag{45.144}$$

where $[q]_n$ contains the element forces for element n and $[W]_n$ contains the structure forces at the extremities of the element. The transformation matrix $[T]_n$ can be used to transform element n from its local coordinates to structure coordinates. We know that for an element n the force-deformation relation is given as

$$[q]_n = [k]_n[\delta]_n$$

Substituting for $[q]_n$ and $[\delta]_n$ from Eqs. (45.142) and (45.143) one obtains

$$[T]_n[W]_n = [k]_n[T]_n[\Delta]_n$$

or

$$\begin{aligned}[W]_n &= [T]_n^{-1}[k]_n[T]_n[\Delta]_n \\ &= [T]_n^{\mathrm{T}}[k]_n[T]_n[\Delta]_n \\ &= [K]_n[\Delta]_n\end{aligned}$$

$$[K]_n = [T]_n^{\mathrm{T}}[k]_n[T]_n \tag{45.145}$$

$[K]_n$ is the stiffness matrix which transforms any element n from its local coordinate to structure coordinates. In this, each element is transformed individually from element coordinate to structure coordinate and the resulting matrices are combined to form the stiffness matrix for the entire structure.

Member stiffness matrix $[K]_n$ in structure coordinates for the truss member shown in Fig. 45.71, for example, is given as

$$[K]_n = \frac{AE}{L}\begin{bmatrix} \lambda^2 & \lambda\mu & -\lambda^2 & -\lambda\mu \\ \lambda\mu & \mu^2 & -\lambda\mu & -\mu^2 \\ -\lambda^2 & -\lambda\mu & \lambda^2 & \lambda\mu \\ -\lambda\mu & -\mu^2 & \lambda\mu & \mu^2 \end{bmatrix}\begin{matrix} i \\ j \\ k \\ l \end{matrix} \tag{45.146}$$

where $\lambda = \cos\phi$ and $\mu = \sin\phi$.

To construct $[K]_n$ for a given member it is necessary to have the values of λ and μ for the member. In addition, the structure coordinates i, j, k, and l at the extremities of the member must be known.

Member stiffness matrix $[K]_n$ in structural coordinates for a flexural member shown in Fig. 45.72 can be written as

$$[K]_n =$$

$$\begin{bmatrix} \left(\lambda^2\frac{AE}{L} + \mu^2\frac{12EI}{L^3}\right) & & & \text{(Symmetric)} & & \\ \mu\lambda\left(\frac{AE}{L} - \frac{12EI}{L^3}\right) & \left(\mu^2\frac{AE}{L} + \lambda^2\frac{12EI}{L^3}\right) & & & & \\ -\mu\left(\frac{6EI}{L^2}\right) & \lambda\left(\frac{6EI}{L^2}\right) & \frac{4EI}{L} & & & \\ \left(-\lambda^2\frac{AE}{L} - \mu^2\frac{12EI}{L^3}\right) & \mu\lambda\left(\frac{AE}{L} - \frac{12EI}{L^3}\right) & \mu\left(\frac{6EI}{L^2}\right) & \left(\lambda^2\frac{AE}{L} + \mu^2\frac{12EI}{L^3}\right) & & \\ -\mu\lambda\left(\frac{AE}{L} - \frac{12EI}{L^3}\right) & \left(-\mu^2\frac{AE}{L} - \lambda^2\frac{12EI}{L^3}\right) & -\lambda\left(\frac{6EI}{L^2}\right) & \mu\lambda\left(\frac{AE}{L} - \frac{12EI}{L^3}\right) & \left(\mu^2\frac{AE}{L} + \lambda^2\frac{12EI}{L^3}\right) & \\ -\mu\left(\frac{6EI}{L^2}\right) & \lambda\left(\frac{6EI}{L^2}\right) & \frac{2EI}{L} & \mu\left(\frac{6EI}{L^2}\right) & -\lambda\left(\frac{6EI}{L^2}\right) & \frac{4EI}{L} \end{bmatrix} \tag{45.147}$$

where $\lambda = \cos\phi$ and $\mu = \sin\phi$.

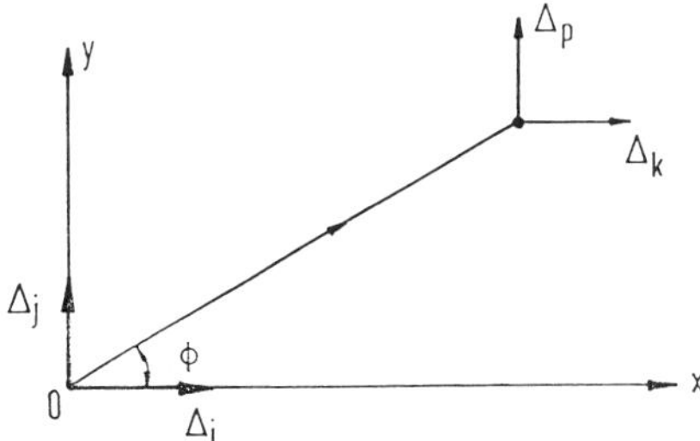

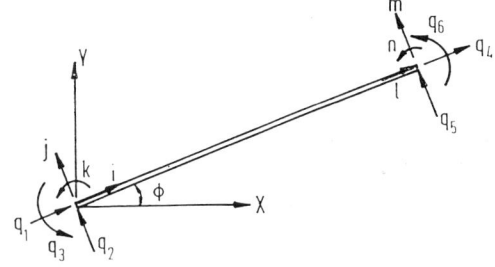

FIGURE 45.71 A flexural member in global coordinates.

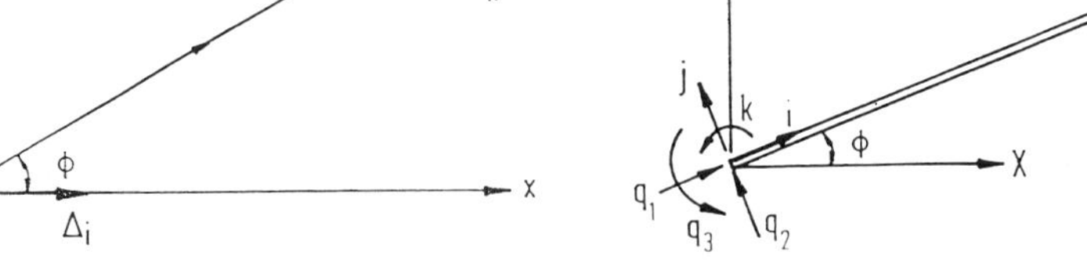

FIGURE 45.72 A grid member.

Example 45.7. Determine the displacement at the loaded point of the truss shown in Fig. 45.73. Both members have same area of cross section $A = 3$ in.^{2} and $E = 30 \times 10^3$ ksi.

The details required to form the element stiffness matrix with reference to structure coordinates axes are listed as follows:

Member	Length	ϕ	λ	μ	i	j	k	l
1	10 ft	90°	0	1	1	2	3	4
2	18.9 ft	32°	0.85	0.53	1	2	5	6

We now use these data in Eq. (45.146) to form $[K]_n$ for the two elements. For member 1,

$$\frac{AE}{L} = \frac{3 \times 30 \times 10^3}{120} = 750$$

$$[K]_1 = \begin{array}{c} \begin{matrix} 1 & 2 & 3 & 4 \end{matrix} \\ \begin{bmatrix} 0 & 0 & 0 & 0 \\ 0 & 750 & 0 & -750 \\ 0 & 0 & 0 & 0 \\ 0 & -750 & 0 & 750 \end{bmatrix} \end{array} \begin{matrix} 1 \\ 2 \\ 3 \\ 4 \end{matrix}$$

For member 2,

$$\frac{AE}{L} = \frac{3 \times 30 \times 10^3}{18.9 \times 12} = 397$$

$$[K]_2 = \begin{array}{c} \begin{matrix} 1 & 2 & 5 & 6 \end{matrix} \\ \begin{bmatrix} 286 & 179 & -286 & -179 \\ 179 & 111 & -179 & -111 \\ -286 & -179 & 286 & 179 \\ -179 & -111 & 179 & 111 \end{bmatrix} \end{array} \begin{matrix} 1 \\ 2 \\ 5 \\ 6 \end{matrix}$$

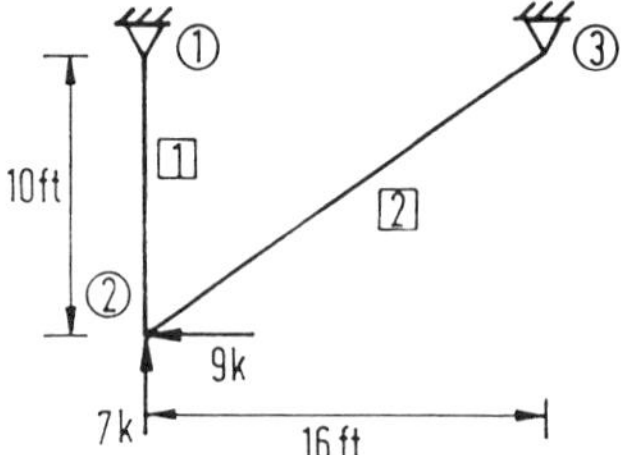

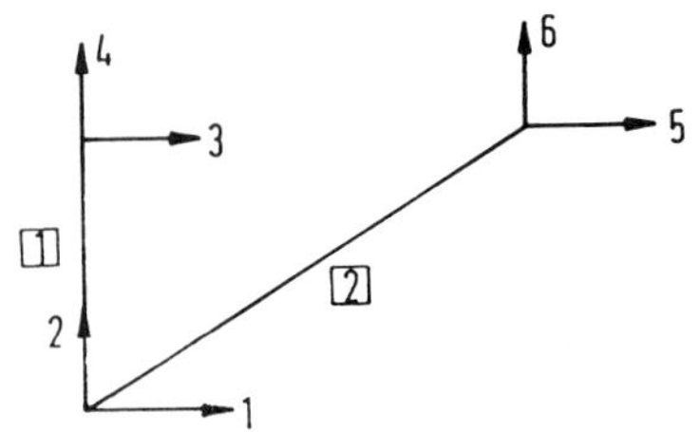

FIGURE 45.73 Example 45.7.

Combining the element stiffness matrices $[K]_1$ and $[K]_2$ one obtains the structure stiffness matrix as follows.

$$\begin{bmatrix} W_1 \\ W_2 \\ W_3 \\ W_4 \\ W_5 \\ W_6 \end{bmatrix} = \begin{bmatrix} 286 & 179 & 0 & 0 & -286 & -179 \\ 179 & 861 & 0 & -750 & -179 & -111 \\ 0 & 0 & 0 & 0 & 0 & 0 \\ 0 & -750 & 0 & 750 & 0 & 0 \\ -286 & -179 & 0 & 0 & 286 & 179 \\ -179 & -111 & 0 & 0 & 179 & 111 \end{bmatrix} \begin{bmatrix} \Delta_1 \\ \Delta_2 \\ \Delta_3 \\ \Delta_4 \\ \Delta_5 \\ \Delta_6 \end{bmatrix}$$

The stiffness matrix can now be subdivided to determine the unknowns. Let us consider Δ_1 and Δ_2, the deflections at joint 2, which can be determined in view of $\Delta_3 = \Delta_4 = \Delta_5 = \Delta_6 = 0$ as follows:

$$\begin{bmatrix} \Delta_1 \\ \Delta_2 \end{bmatrix} = \begin{bmatrix} 286 & 179 \\ 179 & 861 \end{bmatrix}^{-1} \begin{bmatrix} -9 \\ 7 \end{bmatrix}$$

or

$$\Delta_1 = 0.042 \text{ in. to the left}$$
$$\Delta_2 = 0.0169 \text{ in. upward}$$

Example 45.8. A simple triangular frame is loaded at the tip by 20 kips, as shown in Fig. 45.74. Assemble the structure stiffness matrix and determine the displacements at the loaded node.

Member	Length (in.)	A (in.^{2})	I (in.^{4})	ϕ	λ	μ
1	72	2.4	1037	0	1	0
2	101.8	3.4	2933	45°	0.707	0.707

For members 1 and 2 the stiffness matrices in structure coordinates can be written by making use of Eq. (45.147):

$$[K]_1 = 10^3 \times \begin{array}{c} \begin{array}{cccccc} 1 & 2 & 3 & 4 & 5 & 6 \end{array} \\ \begin{bmatrix} 1 & 0 & 0 & -1 & 0 & 0 \\ 0 & 1 & 36 & 0 & -1 & 36 \\ 0 & 36 & 1728 & 0 & -36 & 864 \\ -1 & 0 & 0 & 1 & 0 & 0 \\ 0 & -1 & -36 & 0 & 1 & -36 \\ 0 & 36 & 864 & 0 & -36 & 1728 \end{bmatrix} \end{array} \begin{array}{c} 1 \\ 2 \\ 3 \\ 4 \\ 5 \\ 6 \end{array}$$

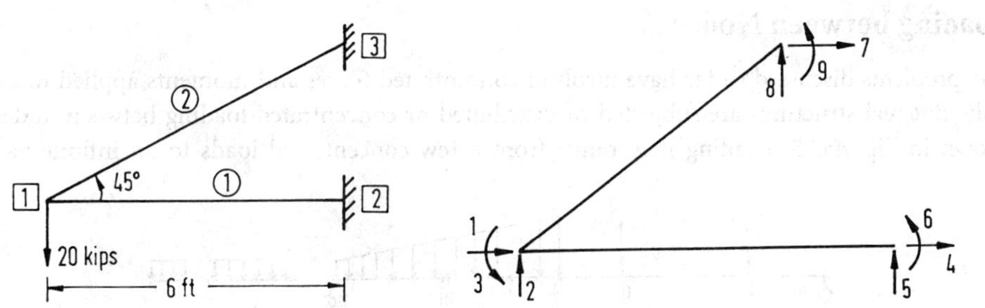

FIGURE 45.74 Example 45.8.

and

$$[K]_1 = 10^3 \times \begin{array}{c} \begin{array}{cccccc} \;\;1 & \;\;\;2 & \;\;\;3 & \;\;\;\;7 & \;\;\;\;8 & \;\;\;\;9 \end{array} \\ \left[\begin{array}{cccccc} 1 & 0 & -36 & -1 & 0 & -36 \\ 0 & 1 & 36 & 0 & 1 & 36 \\ -36 & 36 & 3457 & 36 & -36 & 1728 \\ -1 & 0 & 36 & 1 & 0 & 36 \\ 0 & 1 & -36 & 0 & 1 & -36 \\ -36 & 36 & 1728 & 36 & -36 & 3457 \end{array}\right] \end{array} \begin{array}{c} \\ 1 \\ 2 \\ 3 \\ 7 \\ 8 \\ 9 \end{array}$$

Combining the element stiffness matrices $[K]_1$ and $[K]_2$ one obtains the structure stiffness matrix as follows:

$$[K] = 10^3 \times \left[\begin{array}{ccccccccc} 2 & 0 & -36 & -1 & 0 & 0 & -1 & 0 & -36 \\ 0 & 2 & 72 & 0 & -1 & 36 & 0 & 1 & 36 \\ -36 & 72 & 5185 & 0 & -36 & 864 & 36 & -36 & 1728 \\ -1 & 0 & 0 & 1 & 0 & 0 & 0 & 0 & 0 \\ 0 & -1 & -36 & 0 & 1 & -36 & 0 & 0 & 0 \\ 0 & 36 & 864 & 0 & -36 & 1728 & 0 & 0 & 0 \\ -1 & 0 & 36 & 0 & 0 & 0 & 1000 & 0 & 36 \\ 0 & 1 & -36 & 0 & 0 & 0 & 0 & 1 & -36 \\ -36 & 36 & 1728 & 0 & 0 & 0 & 36 & 36 & 3457 \end{array}\right] \begin{array}{c} 1 \\ 2 \\ 3 \\ 4 \\ 5 \\ 6 \\ 7 \\ 8 \\ 9 \end{array}$$

The deformations at joints 2 and 3 corresponding to Δ_5 to Δ_9 are zero since the joints 2 and 4 are restrained in all directions. Canceling the rows and columns corresponding to zero deformations in the structure stiffness matrix one obtains the force deformation relation for the structure:

$$\begin{bmatrix} F_1 \\ F_2 \\ F_3 \end{bmatrix} = \begin{bmatrix} 2 & 0 & -36 \\ 0 & 2 & 72 \\ -36 & 72 & 5185 \end{bmatrix} \times 10^3 \begin{bmatrix} \Delta_1 \\ \Delta_2 \\ \Delta_3 \end{bmatrix}$$

Substituting for the applied load $F_2 = -20$ kips the deformations are given as

$$\begin{bmatrix} \Delta_1 \\ \Delta_2 \\ \Delta_3 \end{bmatrix} = \begin{bmatrix} 2 & 0 & -36 \\ 0 & 2 & 72 \\ -36 & 72 & 5185 \end{bmatrix}^{-1} \times 10^{-3} \begin{bmatrix} 0 \\ -20 \\ 0 \end{bmatrix}$$

or

$$\begin{bmatrix} \Delta_1 \\ \Delta_2 \\ \Delta_3 \end{bmatrix} = \begin{bmatrix} 6.66 \text{ in.} \\ -23.334 \text{ in.} \\ 0.370 \text{ rad} \end{bmatrix} \times 10^{-3}$$

Loading between Nodes

The problems discussed so far have involved concentrated forces and moments applied to nodes only. But real structures are subjected to distributed or concentrated loading between nodes, as shown in Fig. 45.75. Loading may range from a few concentrated loads to an infinite variety

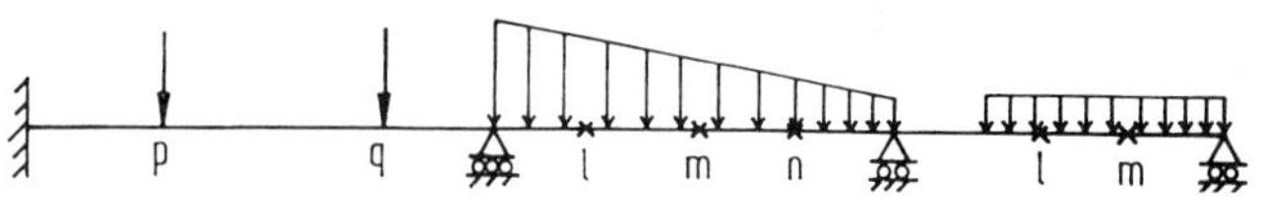

FIGURE 45.75 Loading between nodes.

of uniform or nonuniformly distributed loads. The solution method of matrix analysis must be modified to account for such load cases.

One way to treat such loads in the matrix analysis is to insert artificial nodes, such as p and q as shown in Fig. 45.75. The degrees of freedom corresponding to the additional nodes are added to the total structure and the necessary additional equations are written by considering the requirements of equilibrium at these nodes. The internal member forces on each side of nodes p and q must equilibrate the external loads applied at these points. In the case of distributed loads, suitable nodes such as l, m, and n (Fig. 45.75) are selected arbitrarily, and the distributed loads are lumped as concentrated loads at these nodes. The degrees of freedom corresponding to the arbitrary and real nodes are treated as unknowns. There are different ways of obtaining equivalence between the lumped and the loading. In all cases the lumped loads must be statically equivalent to the distributed loads they replace.

The method of introducing arbitrary nodes is not a very elegant procedure because the number of unknown degrees of freedom make the solution procedure laborious. The approach that is of most general use with the displacement method is one employing the related concepts of artificial joint restraint, fixed-end forces, and equivalent nodal loads.

The Finite Element Method

Many problems that confront the design analyst, in practice, cannot be solved by analytical solutions. This is particularly true for problems involving complex material properties and boundary conditions. Numerical methods, in such cases, provide approximate but acceptable solutions. Of the many numerical methods developed before and after the advent of computers, the finite element method has proved to be a powerful tool. This method can be regarded as a natural extension of the matrix methods of structural analysis. It can accommodate complex and difficult problems such as nonhomogenity, nonlinear stress-strain behavior, and complicated boundary conditions. The finite element method is applicable to a wide range of boundary value problems in engineering and dates back to the mid-1950s with the pioneering work by Argyris [1955], Clough [1960], and others. The method was first applied to the solution of plane stress problems and extended subsequently to the solution of plates, shells, and axisymmetric solids.

Concept of the Finite Element Method

The finite element method is based on the representation of a body or a structure by an assemblage of subdivisions called finite elements, as shown in Fig. 45.76. These elements are considered to be connected at nodes. Displacement functions are chosen to approximate the variation of displacements over each finite element. Polynomials are commonly employed to express these functions. Equilibrium equations for each element are obtained by means of the principle of minimum potential energy. These equations are formulated for the entire body by combining the equations for the individual elements so that the continuity of displacements is preserved at the

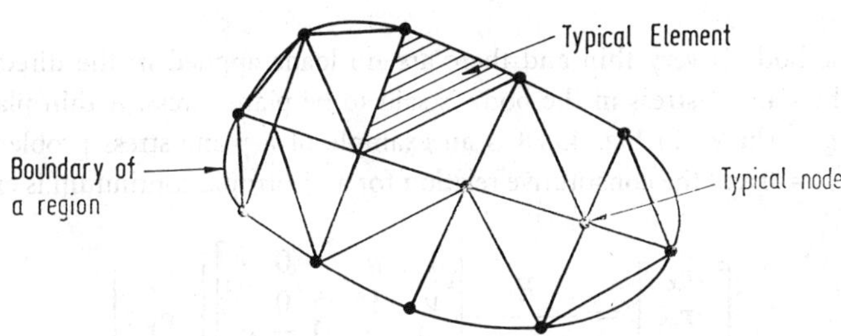

FIGURE 45.76 Assemblage of subdivisions.

nodes. The resulting equations are solved, satisfying the boundary conditions in order to obtain the unknown displacements.

The entire procedure of the finite element method involves the following steps: (1) the given body is subdivided into an equivalent system of finite elements; (2) a suitable displacement function is chosen; (3) the element stiffness matrix is derived using a variational principle of mechanics, such as the principle of minimum potential energy; (4) the global stiffness matrix for the entire body is formulated; (5) the algebraic equations thus obtained are solved to determine unknown displacements; and (6) element strains and stresses are computed from the nodal displacements.

Basic Equations from Theory of Elasticity

Figure 45.77 shows the state of stress in an elemental volume of a body under load. It is defined in terms of three normal stress components σ_x, σ_y, and σ_z and three shear stress components τ_{xy}, τ_{yz}, and τ_{zx}. The corresponding strain components are three normal strains ε_x, ε_y, and ε_z and three shear strains γ_{xy}, γ_{yz}, and γ_{zx}. These strain components are related to the displacement components u, v, and w at a point as follows:

$$\begin{aligned}
\varepsilon_x &= \frac{\partial u}{\partial x} & \gamma_{xy} &= \frac{\partial v}{\partial x} + \frac{\partial u}{\partial y} \\
\varepsilon_y &= \frac{\partial v}{\partial y} & \gamma_{yz} &= \frac{\partial w}{\partial y} + \frac{\partial v}{\partial z} \\
\varepsilon_z &= \frac{\partial w}{\partial z} & \gamma_{zx} &= \frac{\partial u}{\partial z} + \frac{\partial w}{\partial x}
\end{aligned} \tag{45.148}$$

The relations given in Eq. (45.148) are valid in the case of the body experiencing small deformations. If the body undergoes large or finite deformations, higher-order terms must be retained.

The stress-strain equations for isotropic materials may be written in terms of Young's modulus and Poisson's ratio as

$$\begin{aligned}
\sigma_x &= \frac{E}{1-\nu^2}[\varepsilon_x + \nu(\varepsilon_y + \varepsilon_z)] \\
\sigma_y &= \frac{E}{1-\nu^2}[\varepsilon_y + \nu(\varepsilon_z + \varepsilon_x)] \\
\sigma_z &= \frac{E}{1-\nu^2}[\varepsilon_z + \nu(\varepsilon_x + \varepsilon_y)] \\
\tau_{xy} &= G\gamma_{xy} \qquad \tau_{yz} = G\gamma_{yz} \qquad \tau_{zx} = G\gamma_{zx}
\end{aligned} \tag{45.149}$$

Plane stress

When the elastic body is very thin and there are no loads applied in the direction parallel to the thickness, the state of stress in the body is said to be plane stress. A thin plate subjected to in-plane loading as shown in Fig. 45.78 is an example of a plane stress problem. In this case, $\sigma_z = \tau_{yz} = \tau_{zx} = 0$ and the constitutive relation for an isotropic continuum is expressed as

$$\begin{bmatrix} \sigma_x \\ \sigma_y \\ \sigma_{xy} \end{bmatrix} = \frac{E}{1-\nu^2} \begin{bmatrix} 1 & \nu & 0 \\ \nu & 1 & 0 \\ 0 & 0 & \dfrac{1-\nu}{2} \end{bmatrix} \begin{bmatrix} \varepsilon_x \\ \varepsilon_y \\ \gamma_{xy} \end{bmatrix} \tag{45.150}$$

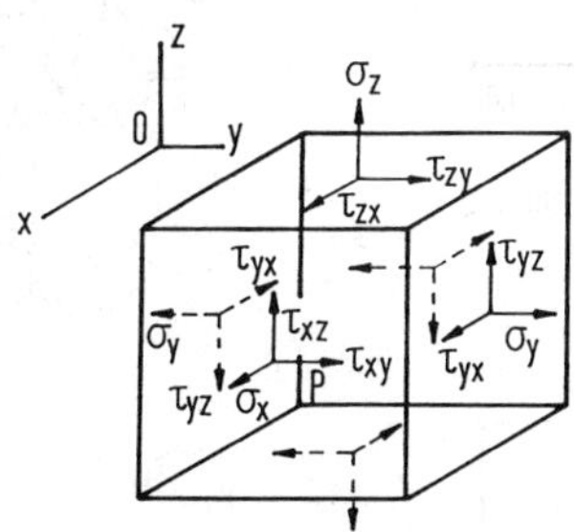

FIGURE 45.77 State of stress in an elemental volume.

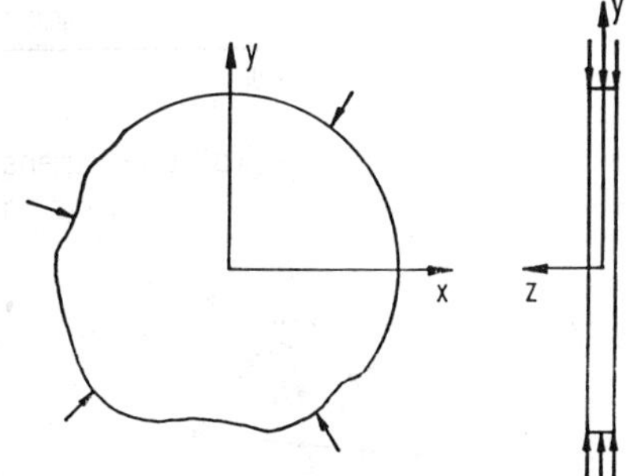

FIGURE 45.78 Plane-stress problem.

Plane Strain

The state of plane strain occurs in members that are not free to expand in the direction perpendicular to the plane of the applied loads. Examples of some plane strain problems are retaining walls, dams, long cylinders, tunnels, and so on, as shown in Fig. 45.79. In these problems ε_z, γ_{yz}, and γ_{zx} will vanish, and hence,

$$\sigma_z = \nu(\sigma_x + \sigma_y)$$

The constitutive relations for an isotropic material is written as

$$\begin{bmatrix} \sigma_x \\ \sigma_y \\ \tau_{xy} \end{bmatrix} = \frac{E}{(1+\nu)(1-2\nu)} \begin{bmatrix} (1-\nu) & \nu & 0 \\ \nu & (1-\nu) & 0 \\ 0 & 0 & \dfrac{1-2\nu}{2} \end{bmatrix} \begin{bmatrix} \varepsilon_x \\ \varepsilon_y \\ \gamma_{xy} \end{bmatrix} \tag{45.151}$$

Element Shapes and Discretization

The process of subdividing a continuum is an exercise of engineering judgment. The choice by an analyst depends on the geometry of the body. A finite element generally has a simple one-, two-, or three-dimensional configuration. The boundaries of elements are often straight lines, and the elements can be one-dimensional, two-dimensional, or three-dimensional, as shown in Fig. 45.80. While subdividing the continuum one has to decide the number, shape, size, and configuration of the elements in such a way that the original body is simulated as closely as possible. Nodes must be located in locations where abrupt changes in geometry, loading, and material properties occur. A node must be placed at the point of application of a concentrated load because all loads are converted into equivalent nodal-point loads.

It is easy to subdivide a continuum into a completely regular one having the same shape and size. But problems encountered in practice do not involve regular shapes; they may have regions of steep gradients of stresses. A finer subdivision may be necessary in regions where stress concentrations

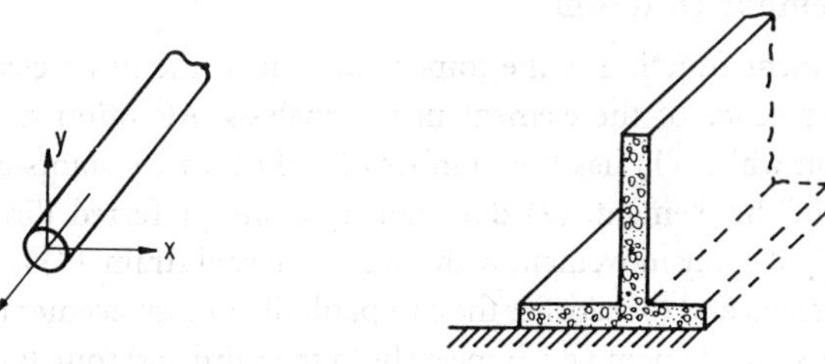

FIGURE 45.79 Practical examples of plane-strain problems.

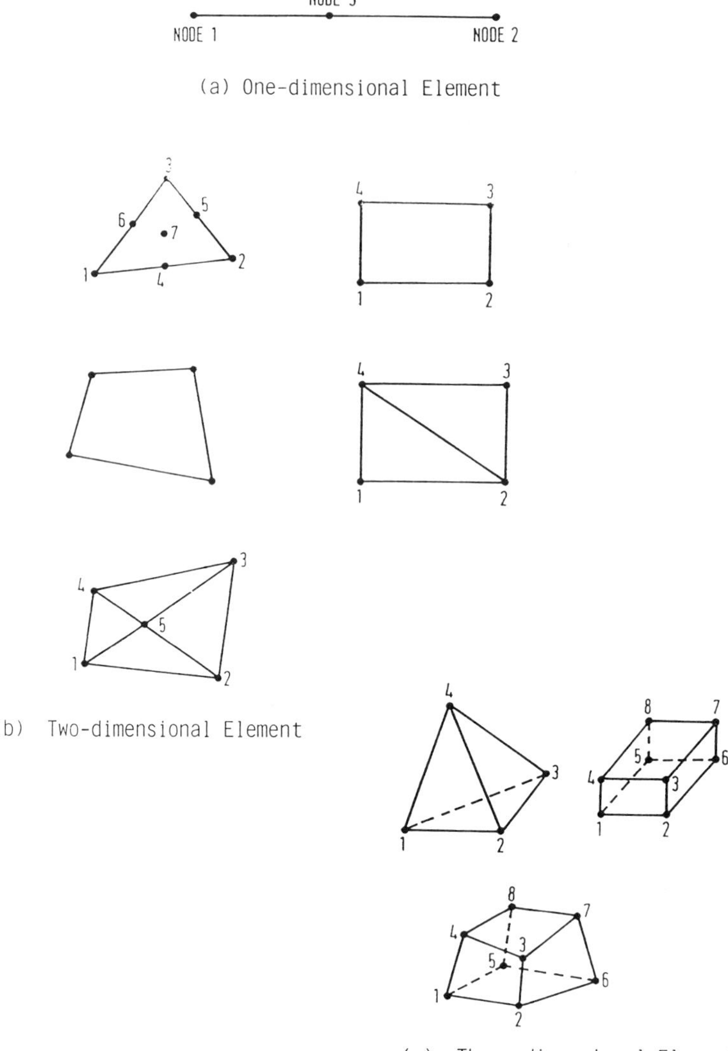

FIGURE 45.80 (a) One-dimensional element, (b) two-dimensional elements, (c) three-dimensional elements.

are expected in order to obtain a useful approximate solution. Typical examples of mesh selection are shown in Fig. 45.81.

Choice of Displacement Function

Selection of displacement function is the important step in the finite element analysis, since it determines the performance of the element in the analysis. Attention must be paid to select a displacement function which (1) has the number of unknown constants as the total number of degrees of freedom of the element, (2) does not have any preferred directions, (3) allows the element to undergo rigid-body movement without any internal strain, (4) is able to represent states of constant stress or strain, and (5) satisfies the compatibility of displacements along the boundaries with adjacent elements. An element which meets both the third and fourth requirements is known as a *complete element.*

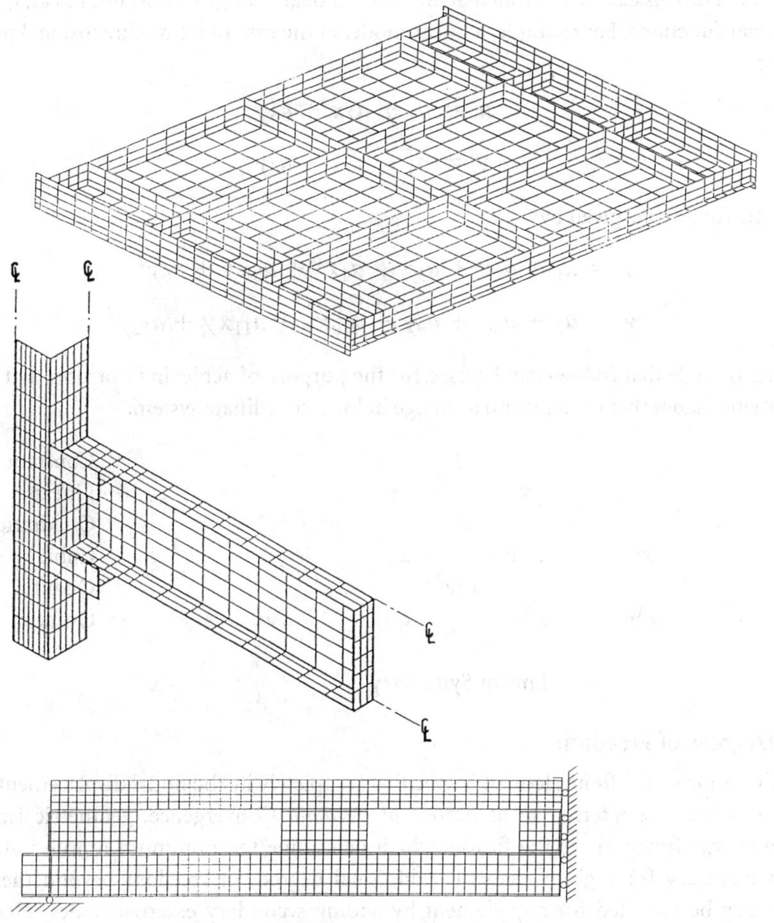

FIGURE 45.81 Typical examples of finite element mesh.

A polynomial is the most common form of displacement function. Mathematics of polynomials are easy to handle in formulating the desired equations for various elements and are convenient in digital computation. The degree of approximation is governed by the stage at which the function is truncated. Solutions closer to exact solutions can be obtained by including more terms. The polynomials are of the general form

$$w(x) = a_1 + a_2x + a_3x^2 + \cdots + a_{n+1}x^n \tag{45.152}$$

where a is the generalized displacement amplitude. The general polynomial form for a two-dimensional problem can be given as

$$\begin{aligned} u(x,y) &= a_1 + a_2x + a_3y + a_4x^2 + a_5xy + a_6y^2 + \cdots + a_my^n \\ v(x,y) &= a_{m+1} + a_{m+2}x + a_{m+3}y + a_{m+4}x^2 + a_{m+5}xy \\ &\quad + a_{m+6}y^2 + \cdots + a_{2m}y^n \end{aligned}$$

where

$$m = \sum_{i=1}^{n+1} i \tag{45.153}$$

These polynomials can be truncated at any desired degree to give constant, linear, quadratic, or higher-order functions. For example, a linear model in the case of a two-dimensional problem can be given as

$$\begin{aligned} u &= a_1 + a_2x + a_3y \\ v &= a_4 + a_5x + a_6y \end{aligned} \tag{45.154}$$

A quadratic function is given by

$$\begin{aligned} u &= a_1 + a_2x + a_3y + a_4x^2 + a_5xy + a_6y^2 \\ v &= a_7 + a_8x + a_9y + a_{10}x^2 + a_{11}xy + a_{12}y^2 \end{aligned} \tag{45.155}$$

The Pascal triangle that follows can be used for the purpose of achieving isotropy, that is, to avoid displacement shapes that change with a change in local coordinate system:

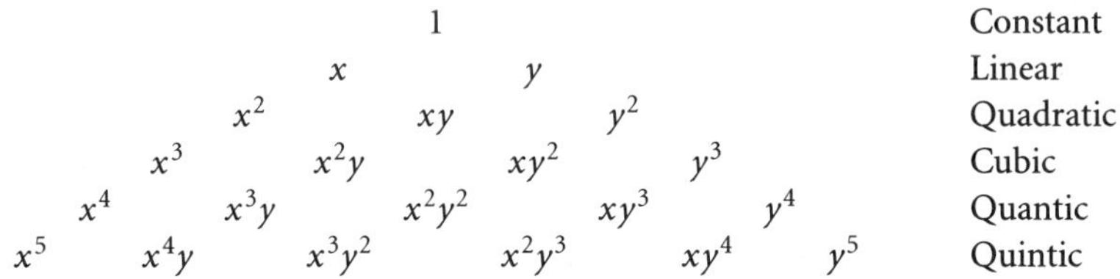

Nodal Degrees of Freedom

The deformation of the finite element is specified completely by the nodal displacement, rotations, or strains, which are referred to as *degrees of freedom*. Convergence, geometric isotropy, and potential energy function are the factors which determine the minimum number of degrees of freedom necessary for a given element. Additional degrees of freedom beyond the minimum number may be included for any element by adding secondary external nodes. Elements with additional degrees of freedom are called higher-order elements. Elements with more additional degrees of freedom are more flexible.

Isoparametric Elements

The scope of finite element analysis is also measured by the variety of element geometries that can be constructed. Formulation of element stiffness equations requires the selection of displacement expressions with as many parameters as there are node-point displacements. In practice, for planar conditions, only the four-sided (quadrilateral) element finds as wide an application as the triangular element. The simplest form of quadrilateral, the rectangle, has four node points and involves two displacement components at each point for a total of eight degrees of freedom. In this case one would choose four-term expressions for both u and v displacement fields. If the description of the element is expanded to include nodes at the midpoints of the sides, an eight-term expression would be chosen for each displacement component.

The triangle and rectangle can approximate the curved boundaries only as a series of straight-line segments. A closer approximation can be achieved by means of *isoparametric* coordinates. These are nondimensionalized curvilinear coordinates whose description is given by the same coefficients as are employed in the displacement expressions. The displacement expressions are chosen to ensure continuity across element interfaces and along supported boundaries, so geometric continuity is ensured when the same forms of expressions are used as the basis of description of the element boundaries. Elements in which the geometry and displacements are described in terms of the same parameters and are of the same order are called *isoparametric elements*. The isoparametric concept enables one to formulate elements of any order which satisfy the completeness and compatibility requirements and which have isotropic displacement functions.

Isoparametric Families of Elements—Definitions and Justifications

For example, let u_i represent nodal displacements and x_i represent nodal x coordinates. The interpolation formulas are

$$u = \sum_{i=1}^{m} N_i u_i \qquad x = \sum_{i=1}^{n} N_i' x_i$$

where N_i and N_i' are shape functions written in terms of the intrinsic coordinates. The value of u and the value of x at a point within the element are obtained in terms of nodal values of u_i and x_i from the above equations when the (intrinsic) coordinates of the internal point are given. Displacement components v and w in the y and z directions are treated in a similar manner.

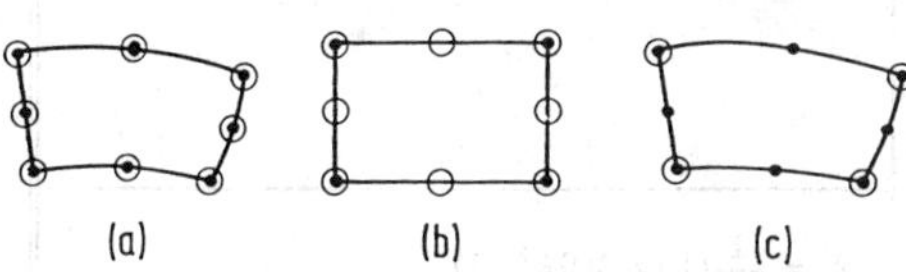

FIGURE 45.82 (a) Isoparametric element, (b) subparametric element, (c) superparametric element.

The element is isoparametric if $m = n, N_i = N_i'$, and the same nodal points are used to define both element geometry and element displacement [Fig. 45.82(a)]; the element is subparametric if $m > n$, the order of N_i larger than N_i' [Fig. 45.82(b)]; the element is superparametric if $m < n$, the order of N_i smaller than N_i' [Fig. 45.82(c)]. The isoparametric elements can correctly display rigid-body and constant-strain modes.

Element Shape Functions

The finite element method is not restricted to the use of linear elements. Most commercially available finite element codes allow the user to select elements with linear or quadratic interpolation functions. In the case of quadratic elements fewer elements are needed to obtain the same degree of accuracy in the nodal values. Also, the two-dimensional quadratic elements can be shaped to model a curved boundary. Shape functions can be developed based on the following properties: (1) each shape function has a value of 1 at its own node and 0 at each of the other nodes, (2) the shape functions for two-dimensional elements are 0 along each side that the node does not touch, and (3) each shape function is a polynomial of the same degree as the interpolation equation. Shape functions for typical elements are given in Fig. 45.83.

Formulation of Stiffness Matrix

It is possible to obtain all the strains and stresses within the element and to formulate the stiffness matrix and a consistent load matrix once the displacement function has been determined. This consistent load matrix represents the equivalent nodal forces which replace the action of external distributed loads.

As an example, let us consider a linearly elastic element of any of the types shown in Fig. 45.84. The displacement function may be written in the form

$$\{f\} = [P]\{A\} \tag{45.156}$$

where $\{f\}$ may have two components $\{u, v\}$ or may simply be equal to w, $[P]$ is a function of x and y only, and $\{A\}$ is the vector of undetermined constants. If Eq. (45.156) is applied repeatedly to the nodes of the element one after the other, we obtain a set of equations of the form

$$\{D^*\} = [C]\{A\} \tag{45.157}$$

Element name	Configuration	DOF	Shape functions
Two-node linear element	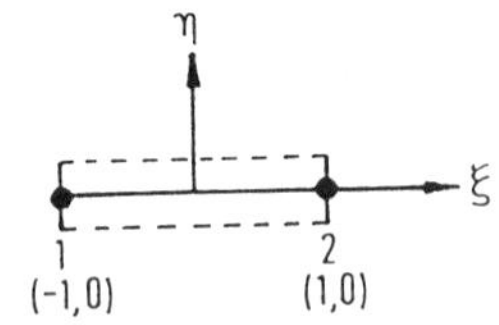 η, ξ; 1 (-1,0), 2 (1,0)	+	$N_i = \frac{1}{2}(1+\xi_0)$; $i = 1, 2$
Three-node parabolic element	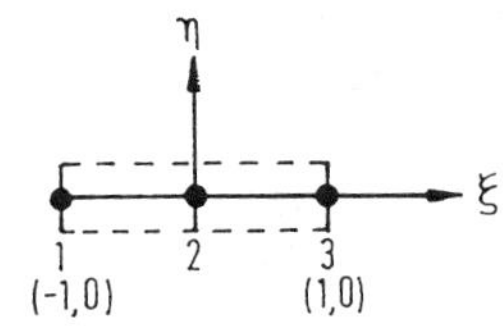 η, ξ; 1 (-1,0), 2, 3 (1,0)	+	$N_i = \frac{1}{2}\xi_0(1+\xi_0)$; $i = 1, 3$ $N_i = (1-\xi^2)$; $i = 2$
Four-node cubic element	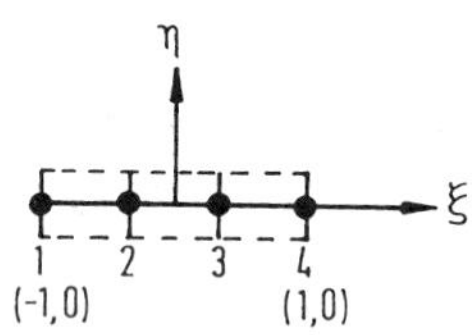 η, ξ; 1 (-1,0), 2, 3, 4 (1,0)	+	$N_i = \frac{1}{16}(1+\xi_0)(9\xi^2-1)$ $i = 1, 4$ $N_i = \frac{9}{16}(1+9\xi_0)(1-\xi^2)$ $i = 2, 3$
Five-node quartic element	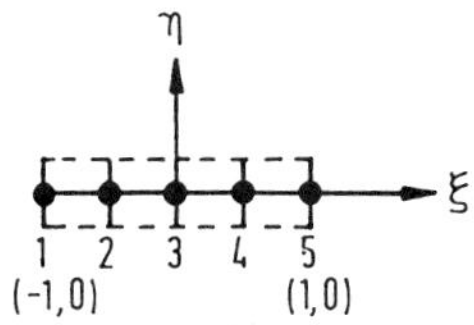 η, ξ; 1 (-1,0), 2, 3, 4, 5 (1,0)	+	$N_i = \frac{1}{6}(1+\xi_0)\{4\xi_0(1-\xi^2)$ $+ 3\xi_0\}$ $i = 1, 5$ $N_i = 4\xi_0(1-\xi^2)(1+4\xi_0)$ $i = 2, 4$ $N_3 = (1-4\xi^2)(1-\xi^2)$

FIGURE 45.83 Shape functions for typical elements. (*Source:* Kardestuncer, H. and Norrie, D. H., eds. 1988. *Finite Element Handbook.* McGraw-Hill, New York.) *(continues)*

where $\{D^*\}$ is the nodal parameter and $[C]$ is the relevant nodal coordinate. The undetermined constant $\{A\}$ can be expressed in terms of the nodal parameter $\{D^*\}$ as

$$\{A\} = [C]^{-1}\{D^*\} \tag{45.158}$$

Substituting Eq. (45.158) into (45.156),

$$\{f\} = [P][C]^{-1}\{D^*\} \tag{45.159}$$

Constructing the displacement function directly in terms of the nodal parameters, one obtains

$$\{f\} = [L]\{D^*\} \tag{45.160}$$

Element name	Configuration	DOF	Shape functions
Four-node plane quadrilateral	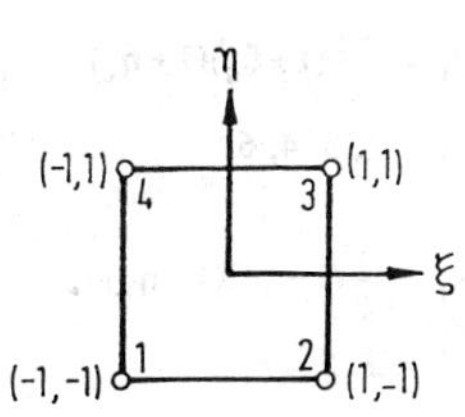	u, v	$N_i = \frac{1}{4}(1+\xi_0)(1+\eta_0)$; $i = 1, 2, 3, 4$
Eight-node plane quadrilateral	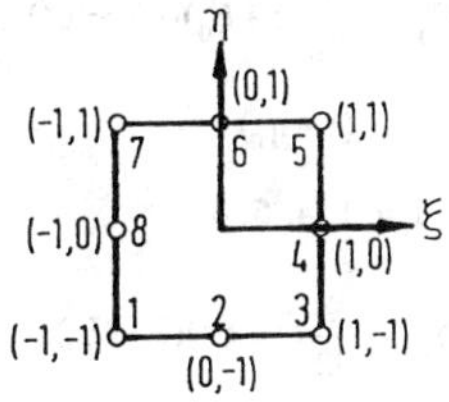	u, v	$N_i = \frac{1}{4}(1+\xi_0)(1+\eta_0)$ $(\xi_0+\eta_0-1)$; $i = 1, 3, 5, 7$ $N_i = \frac{1}{2}(1-\xi^2)(1+\eta_0)$ $i = 2, 6$ $N_i = \frac{1}{2}(1-\eta^2)(1+\xi_0)$ $i = 4, 8$
Twelve-node plane quadrilateral	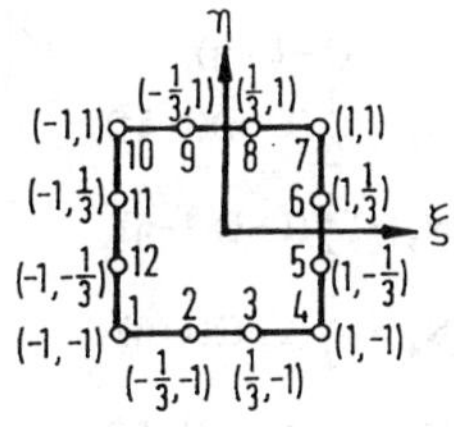	u, v	$N_i = \frac{1}{32}(1+\xi_0)(1+\eta_0)$ $(-10+9(\xi^2+\eta^2))$ $i = 1, 4, 7, 10$ $N_i = \frac{9}{32}(1+\xi_0)(1+\eta^2)$ $(1+9\eta_0)$ $i = 5, 6, 11, 12$ $N_i = \frac{9}{32}(1+\eta_0)(1-\xi^2)$ $(1+9\xi_0)$ $i = 2, 3, 8, 9$

FIGURE 45.83 (continued) Shape functions for typical elements. (*Source:* Kardestuncer, H. and Norrie, D. H., eds. 1988. *Finite Element Handbook.* McGraw-Hill, New York.) *(continues)*

where $[L]$ is a function of both (x, y) and $(x, y)_{i,j,m}$ given by

$$[L] = [P][C]^{-1} \tag{45.161}$$

The various components of strain can be obtained by appropriate differentiation of the displacement function. Thus,

$$\{\varepsilon\} = [B]\{D^*\} \tag{45.162}$$

Serial no.	Element name	Configuration	DOF	Shape functions
4	Six-node linear Quadrilateral	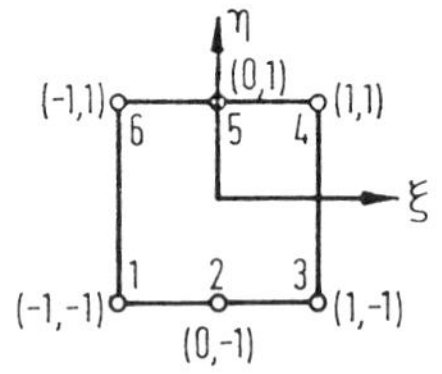 η; ξ; (-1,1) 6; (0,1) 5; (1,1) 4; (-1,-1) 1; (0,-1) 2; (1,-1) 3	u, v	$N_i = \frac{\xi_0}{4}(1+\xi_0)(1+\eta_0)$ $i = 1, 3, 4, 6$ $N_i = \frac{1}{2}(1-\xi^2)(1+\eta_0)$ $i = 2, 5$
5	Eight-node plane quadrilateral	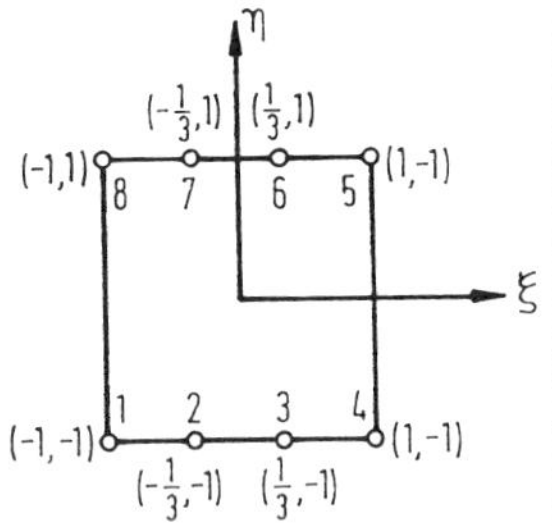 η; ξ; (-1,1) 8; $(-\frac{1}{3},1)$ 7; $(\frac{1}{3},1)$ 6; (1,-1) 5; (-1,-1) 1; $(-\frac{1}{3},-1)$ 2; $(\frac{1}{3},-1)$ 3; (1,-1) 4	u, v	$N_1 = \frac{1}{32}(1+\xi_0)(-1+9\xi^2)$ $(1+\eta_0)$ $i = 1, 4, 5, 8$ $N_i = \frac{9}{32}(1-\xi^2)(1+9\xi_0)$ $(1+\eta_0)$ $i = 2, 3, 6, 7$
6	Seven-node plane quadrilateral	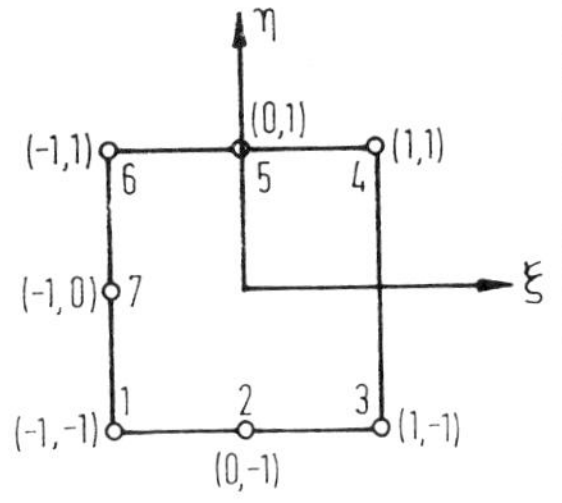 η; ξ; (-1,1) 6; (0,1) 5; (1,1) 4; (-1,0) 7; (-1,-1) 1; (0,-1) 2; (1,-1) 3	u, v	$N_1 = \frac{1}{4}(1-\xi)(1-\eta)$ $(1+\xi+\eta)$ $N_2 = \frac{1}{2}(1-\eta)(1-\xi^2)$ $N_3 = \frac{\xi}{4}(1+\xi)(1-\eta)$ $N_4 = \frac{\xi}{4}(1+\xi)(1+\eta)$ $N_5 = \frac{1}{2}(1+\eta)(1-\xi^2)$ $N_6 = -\frac{1}{4}(1-\xi)(1+\eta)$ $(1+\xi-\eta)$ $N_7 = \frac{1}{2}(1-\xi)(1-\eta^2)$

FIGURE 45.83 (continued) Shape functions for typical elements. (*Source:* Kardestuncer, H. and Norrie, D. H., eds. 1988. *Finite Element Handbook.* McGraw-Hill, New York.)

$[B]$ is derived by differentiating appropriately the elements of $[L]$ with respect to x and y. The stresses $\{\sigma\}$ in a linearly elastic element are given by the product of the strain and a symmetrical elasticity matrix $[E]$. Thus

$$\{\sigma\} = [E]\{\varepsilon\}$$

or

$$\{\sigma\} = [E][B]\{D^*\} \tag{45.163}$$

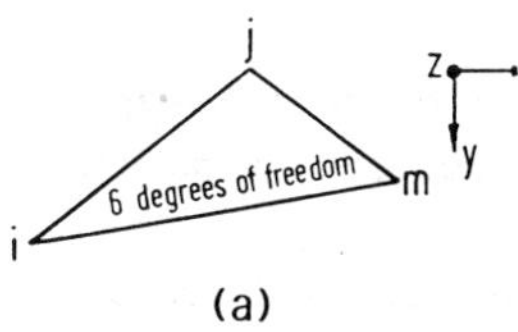

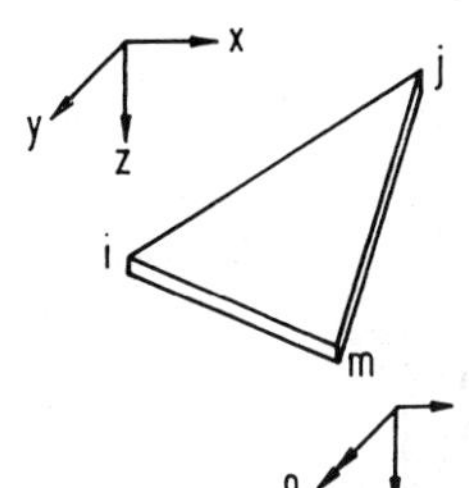

FIGURE 45.84 Degrees of freedom: (a) triangular plane-stress element, (b) triangular bending element.

The stiffness and the consistent load matrices of an element can be obtained using the principle of minimum total potential energy. The potential energy of the external load in the deformed configuration of the element is

$$W = -\{D^*\}^{\mathrm{T}}\{Q^*\} - \int_a \{f\}^{\mathrm{T}}\{q\}\,da \qquad (45.164)$$

where $\{Q^*\}$ represents concentrated loads at nodes and $\{q\}$ is the distributed loads per unit area. Substituting for $\{f\}^T$ from Eq. (45.160) one obtains

$$W = -\{D^*\}^{\mathrm{T}}\{Q^*\} - \{D^*\}^{\mathrm{T}} \int_a [L]^{\mathrm{T}}\{q\}\,da \qquad (45.165)$$

Note that the integral is taken over the area a of the element. The strain energy of the element integrated over the entire volume, v, is

$$U = \frac{1}{2}\int_v \{\varepsilon\}^{\mathrm{T}}\{\sigma\}\,dv$$

Substituting $\{\varepsilon\}$ and $\{\sigma\}$ from Eqs. (45.162) and (45.163), respectively,

$$U = \frac{1}{2}\{D^*\}^{\mathrm{T}}\left(\int_v [B]^{\mathrm{T}}[E][B]\,dv\right)\{D^*\} \qquad (45.166)$$

The total potential energy of the element is

$$V = U + W$$

or

$$V = \frac{1}{2}\{D^*\}^{\mathrm{T}}\left(\int_v [B]^{\mathrm{T}}[E][B]\,dv\right)\{D^*\} - \{D^*\}^{\mathrm{T}}\{Q^*\} - \{D^*\}^{\mathrm{T}}\int_a [L]^{\mathrm{T}}\{q\}\,da \qquad (45.167)$$

Using the principle of minimum total potential energy, we obtain

$$\left(\int_v [B]^{\mathrm{T}}[E][B]\,dv\right)\{D^*\} = \{Q^*\} + \int_a [L]^{\mathrm{T}}\{q\}\,da$$

or

$$[K]\{D^*\} = \{F^*\} \qquad (45.168)$$

where

$$[K] = \int_v [B]^{\mathrm{T}}[E][B]\,dv \qquad (45.169)$$

and

$$\{F^*\} = \{Q^*\} + \int_a [L]^{\mathrm{T}}\{q\}\,da \qquad (45.170)$$

Plates Subjected to In-Plane Forces

The simplest element available in two-dimensional stress analysis is the triangular element. The stiffness and consistent load matrices of such an element will now be obtained by applying the equation derived in the previous section.

Consider the triangular element shown in Fig. 45.84(a). There are two degrees of freedom per node and a total of six degrees of freedom for the entire element. We can write

$$u = A_1 + A_2 x + A_3 y$$

and

$$v = A_4 + A_5 x + A_6 y$$

expressed as

$$\{f\} = \begin{Bmatrix} u \\ v \end{Bmatrix} = \begin{bmatrix} 1 & x & y & 0 & 0 & 0 \\ 0 & 0 & 0 & 1 & x & y \end{bmatrix} \begin{Bmatrix} A_1 \\ A_2 \\ A_3 \\ A_4 \\ A_5 \\ A_6 \end{Bmatrix} \tag{45.171}$$

or

$$\{f\} = [P]\{A\} \tag{45.172}$$

Once the displacement function is available, the strains for a plane problem are obtained from

$$\varepsilon_x = \frac{\partial u}{\partial x} \qquad \varepsilon_y = \frac{\partial v}{\partial y}$$

and

$$\gamma_{xy} = \frac{\partial u}{\partial y} \frac{\partial v}{\partial x}$$

The matrix $[B]$ relating the strains to the nodal displacement $\{D^*\}$ is thus given as

$$[B] = \frac{1}{2\Delta} \begin{bmatrix} b_i & 0 & b_j & 0 & b_m & 0 \\ 0 & c_i & 0 & c_j & 0 & c_m \\ c_i & b_i & c_j & b_j & c_m & b_m \end{bmatrix} \tag{45.173}$$

where b_i, c_i, and so on are constants related to the nodal coordinates only. The strains inside the element must all be constant (hence the name of the element).

For derivation of strain matrix, only isotropic material is considered. The plane stress and plane strain cases can be combined to give the following elasticity matrix which relates the stresses to the strains:

$$[E] = \begin{bmatrix} C_1 & \text{Symmetrical} & \\ C_1 C_2 & C_1 & \\ 0 & 0 & C_{12} \end{bmatrix} \tag{45.174}$$

where

$$C_1 = \bar{E}(1 - \nu) \quad \text{and} \quad C_2 = \nu \qquad \text{for plane stress}$$

and

$$C_1 = \frac{\bar{E}(1-\nu)}{(1+\nu)(1-2\nu)} \quad \text{and} \quad C_2 = \frac{\nu}{(1-\nu)} \qquad \text{for plane strain}$$

and for both cases,

$$C_{12} = C_1(1 - C_2)/2$$

and $\bar{E}$ is the modulus of elasticity.

The stiffness matrix can now be formulated according to Eq. (45.169):

$$[E][B] = \frac{1}{2\Delta}\begin{bmatrix} C_1 & C_1C_2 & 0 \\ C_1C_2 & C_1 & 0 \\ 0 & 0 & C_{12} \end{bmatrix}\begin{bmatrix} b_i & 0 & b_j & 0 & b_m & 0 \\ 0 & c_i & 0 & c_j & 0 & c_m \\ c_i & b_i & c_j & b_j & c_m & b_m \end{bmatrix}$$

where Δ is the area of the element.

The stiffness matrix is given by Eq. (45.169) as

$$[K] = \int_v [B]^{\mathrm{T}}[E][B]\,dv$$

The stiffness matrix has been worked out algebraically to be

$$[K] = \frac{h}{4\Delta}\begin{bmatrix} \begin{array}{l} C_1b_i^2 \\ +C_{12}c_i^2 \end{array} & & & & & \\ \begin{array}{l} C_1C_2b_ic_i \\ +C_{12}b_ic_i \end{array} & \begin{array}{l} C_1c_i^2 \\ +C_{12}b_i^2 \end{array} & & \text{Symmetrical} & & \\ \begin{array}{l} C_1b_ib_j \\ +C_{12}c_ic_j \end{array} & \begin{array}{l} C_1C_2b_jc_i \\ +C_{12}b_ic_j \end{array} & \begin{array}{l} C_1b_j^2 \\ +C_{12}c_j^2 \end{array} & & & \\ \begin{array}{l} C_1C_2b_ic_j \\ +C_{12}b_jc_i \end{array} & \begin{array}{l} C_1c_ic_j \\ +C_1b_ib_j \end{array} & \begin{array}{l} C_1C_2b_jc_j \\ +C_{12}b_jc_j \end{array} & \begin{array}{l} C_1c_j^2 \\ +C_{12}b_j^2 \end{array} & & \\ \begin{array}{l} C_1b_ib_m \\ +C_{12}c_ic_m \end{array} & \begin{array}{l} C_1C_2b_mc_i \\ +C_{12}b_ic_m \end{array} & \begin{array}{l} C_1b_jb_m \\ +C_{12}c_jc_m \end{array} & \begin{array}{l} C_1C_2b_mc_j \\ +C_{12}b_jc_m \end{array} & \begin{array}{l} C_1b_m^2 \\ +C_{12}c_m^2 \end{array} & \\ \begin{array}{l} C_1C_2b_ic_m \\ +C_{12}b_mc_i \end{array} & \begin{array}{l} C_1c_ic_m \\ +C_{12}b_ib_m \end{array} & \begin{array}{l} C_1C_2b_jc_m \\ +C_{12}b_mc_j \end{array} & \begin{array}{l} C_1c_jc_m \\ +C_{12}b_jb_m \end{array} & \begin{array}{l} C_1C_2b_mc_m \\ +C_{12}b_mc_m \end{array} & \begin{array}{l} C_1c_m^2 \\ +C_{12}b_m^2 \end{array} \end{bmatrix} \tag{45.175}$$

Beam element

The stiffness matrix for a beam element with two degrees of freedom (one deflection and one rotation) can be derived in the same manner as for other finite elements using Eq. (45.169).

The beam element has two nodes, one at each end, and two degrees of freedom at each node, giving it a total of four degrees of freedom. The displacement function can be assumed as

$$f = w = A_1 + A_2x + A_3x^2 + A_4x^3$$

that is,

$$f = [1\,x\,x^2\,x^3]\begin{Bmatrix} A_1 \\ A_2 \\ A_3 \\ A_4 \end{Bmatrix}$$

or

$$f = [P]\{A\}$$

With the origin of the x and y axis at the left-hand end of the beam, we can express the nodal-displacement parameters as

$$\begin{aligned}
D_1^* &= (w)_{x=0} &&= A_1 + A_2(0) + A_3(0)^2 + A_4(0)^3 \\
D_2^* &= \left(\frac{dw}{dx}\right)_{x=0} &&= A_2 + 2A_3(0) + 3A_4(0)^2 \\
D_3^* &= (w)_{x=l} &&= A_1 + A_2(l) + A_3(l)^2 + A_4(l)^3 \\
D_4^* &= \left(\frac{dw}{dx}\right)_{x=l} &&= A_2 + 2A_3(l) + 3A_4(l)^2
\end{aligned}$$

or

$$\{D^*\} = [C]\{A\}$$

where

$$\{A\} = [C]^{-1}\{D^*\}$$

and

$$[C]^{-1} = \begin{bmatrix} 1 & 0 & 0 & 0 \\ 0 & 1 & 0 & 0 \\ \dfrac{-3}{l^2} & \dfrac{-2}{l} & \dfrac{3}{l^2} & \dfrac{-1}{l} \\ \dfrac{2}{l^3} & \dfrac{1}{l^2} & \dfrac{-2}{l^3} & \dfrac{1}{l^2} \end{bmatrix}$$

Using Eq. (45.161), we obtain

$$[L] = [P][C]^{-1}$$

or

$$[L] = \left[\left(1 - \frac{3x^2}{l^2} + \frac{2x^3}{l^3}\right)\middle|\left(x - \frac{2x^2}{l} + \frac{x^3}{l^2}\right)\middle|\left(\frac{3x^2}{l^2} - \frac{2x^3}{l^3}\right)\middle|\left(-\frac{x^2}{l} + \frac{x^3}{l^2}\right)\right] \quad (45.176)$$

Neglecting shear deformation,

$$\{\varepsilon\} = -\frac{d^2y}{dx^2} \quad (45.177)$$

Substituting Eq. (45.176) into Eq. (45.160) and the result into Eq. (45.177),

$$\{\varepsilon\} = \left[\frac{6}{l^2} - \frac{12x}{l^3}\middle|\frac{4}{l} - \frac{6x}{l^2}\middle| - \frac{6}{l^2} + \frac{12x}{l^3}\middle|\frac{2}{l} - \frac{6x}{l^2}\right]\{D^*\}$$

or

$$\{\varepsilon\} = [B]\{D^*\}$$

The moment-curvature relationship is given by

$$M = \bar{E}I\left(-\frac{d^2y}{dx^2}\right)$$

where $\bar{E}$ is the modulus of elasticity.

We know that $\{\sigma\} = [E]\{\varepsilon\}$, so we have for the beam element

$$[E] = \bar{E}I$$

The stiffness matrix can now be obtained from Eq. (45.169) written in the form

$$[k] = \int_0^1 [B]^{\mathrm{T}}[d]][B]\,dx$$

with the integration over the length of the beam. Substituting for $[B]$ and $[E]$, we obtain

$$[k] = \bar{E}I\int_0^1 \begin{bmatrix} \frac{36}{l^4}-\frac{144x}{l^5}+\frac{144x^2}{l^6} & & \text{Symmetrical} & \\ \frac{24}{l^3}-\frac{84x}{l^4}+\frac{72x^2}{l^5} & \frac{16}{l^2}-\frac{48x}{l^3}+\frac{36x^2}{l^4} & & \\ \frac{-36}{l^4}+\frac{144x}{l^5}-\frac{144x^2}{l^6} & \frac{-24}{l^3}+\frac{84x}{l^4}-\frac{72x^2}{l^5} & \frac{36}{l^4}-\frac{144x}{l^5}+\frac{144x^2}{l^6} & \\ \frac{12}{l^3}-\frac{60x}{l^4}+\frac{72x^2}{l^5} & \frac{8}{l^2}-\frac{36x}{l^3}+\frac{36x^2}{l^4} & \frac{-12}{l^3}+\frac{60x}{l^4}-\frac{72x^2}{l^5} & \frac{4}{l^2}-\frac{24x}{l^3}+\frac{36x^2}{l^4} \end{bmatrix} dx$$

or

$$[k] = \bar{E}I \begin{bmatrix} \frac{12}{l^3} & & \text{Symmetrical} & \\ \frac{6}{l^2} & \frac{4}{l} & & \\ \frac{-12}{l^3} & \frac{-6}{l^2} & \frac{12}{l^3} & \\ \frac{6}{l^2} & \frac{2}{l} & \frac{-6}{l^2} & \frac{4}{l} \end{bmatrix} \tag{45.178}$$

Plates in Bendings—Rectangular Element

For the rectangular bending element shown in Fig. 45.85 with three degrees of freedom (one deflection and two rotations) at each node, the displacement function can be chosen as a

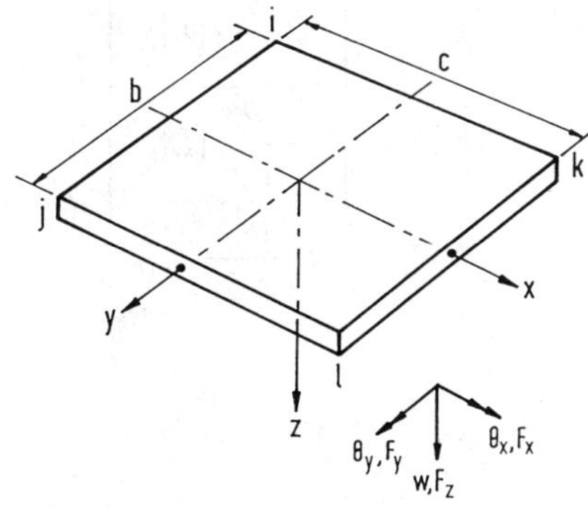

FIGURE 45.85 Rectangular bending element.

polynomial with twelve undetermined constants:

$$\begin{aligned}\{f\} = w =& A_1 + A_2x + A_3y + A_4x^2 + A_5xy + A_6y^2 + A_7x^3 \\ &+ A_8x^2y + A_9xy^2 + A_{10}y^3 + A_{11}x^3y + A_{12}xy^3\end{aligned} \tag{45.179}$$

or

$$\{f\} = \{P\}\{A\}$$

The displacement parameter vector is defined as

$$\{D^*\} = \{w_i, \theta_{xi}, \theta_{yi} | w_j, \theta_{xj}, \theta_{yj} | w_k, \theta_{xk}, \theta_{yk} | w_l, \theta_{xl}, \theta_{yl}\}$$

where

$$\theta_x = \frac{\partial w}{\partial y} \quad \text{and} \quad \theta_y = -\frac{\partial w}{\partial x}$$

As in the case of a beam it is possible to derive from Eq. (45.179) a system of 12 equations relating $\{D^*\}$ to constants $\{A\}$. The last equation is

$$w = \left[[L]_i | [L]_j \left| [L]_k \right| [L]_l \right] \{D^*\} \tag{45.180}$$

The curvatures of the plate element at any point (x, y) are given by

$$\{\varepsilon\} = \left\{ \begin{array}{c} \dfrac{-\partial^2 w}{\partial x^2} \\ \dfrac{-\partial^2 w}{\partial y^2} \\ \dfrac{2\partial^2 w}{\partial x \, \partial y} \end{array} \right\}$$

By differentiating Eq. (45.180), we obtain

$$\{\varepsilon\} = \left[[B]_i | [B]_j \left| [B]_k \right| [B]_l \right] \{D^*\} \tag{45.181}$$

or

$$\{\varepsilon\} = \sum_{r=i,j,k,l} [B]_r \{D^*\}_r \tag{45.182}$$

where

$$[B]_r = \begin{bmatrix} -\dfrac{\partial^2}{\partial x^2}[L]_r \\ \hdashline -\dfrac{\partial^2}{\partial y^2}[L]_r \\ \hdashline 2\dfrac{\partial^2}{\partial x \, \partial y}[L]_r \end{bmatrix} \tag{45.183}$$

and

$$\{D^*\}_r = \{w_r, \theta_{xr}, \theta_{yr}\} \tag{45.184}$$

For an isotropic plate, the moment-curvature relationship is given by

$$\{\sigma\} = \{M_x \quad M_y \quad M_{xy}\} \tag{45.185}$$

$$[E] = N \begin{bmatrix} 1 & \nu & 0 \\ \nu & 1 & 0 \\ 0 & 0 & \frac{1-\nu}{2} \end{bmatrix} \tag{45.186}$$

and

$$N = \frac{\overline{E}h^3}{12(1-\nu^2)} \tag{45.187}$$

For orthotropic plates with the principal directions of orthotropy coinciding with the x and y axes, no additional difficulty is experienced. In this case we have

$$[E] = \begin{bmatrix} D_x & D_1 & 0 \\ D_1 & D_y & 0 \\ 0 & 0 & D_{xy} \end{bmatrix} \tag{45.188}$$

where D_x, D_1, D_y, and D_{xy} are the orthotropic constants used by Timoshenko and Woinowsky-Krieger [1959],

$$\left.\begin{aligned} D_x &= \frac{E_x h^3}{12(1-\nu_x\nu_y)} \\ D_y &= \frac{E_y h^3}{12(1-\nu_x\nu_y)} \\ D_1 &= \frac{\nu_x E_y h^3}{12(1-\nu_x\nu_y)} = \frac{\nu_y E_x h^3}{12(1-\nu_x\nu_y)} \\ D_{xy} &= \frac{Gh^3}{12} \end{aligned}\right\} \tag{45.189}$$

where E_x, E_y, ν_x, ν_y, and G are the orthotropic material constants and h is the plate thickness.

Unlike the strain matrix for the plane stress triangle [see Eq. (45.173)], the stress and strain in the present element vary with x and y. In general, we calculate the stresses (moments) at the four corners. These can be expressed in terms of the nodal displacements by Eq. (45.163), which, for an isotropic element, takes the form shown in Table 45.4.

The stiffness matrix corresponding to the 12 nodal coordinates can be calculated by

$$\{k^*\} = \int_{-b/2}^{b/2}\int_{-c/2}^{c/2} [B]^{\mathrm{T}}[E][B]\,dx\,dy \tag{45.190}$$

For an isotropic element, this gives

$$\{k^*\} = \frac{N}{15cb}[T][\overline{k}][T] \tag{45.191}$$

where

$$[T] = \begin{bmatrix} [T_s] & & & \text{Submatrices not} \\ & [T_s] & & \text{shown are} \\ & & [T_s] & \text{zero} \\ & & & [T_s] \end{bmatrix} \tag{45.192}$$

$$[T_s] = \begin{bmatrix} 1 & 0 & 0 \\ 0 & b & 0 \\ 0 & 0 & c \end{bmatrix} \tag{45.193}$$

and $[\overline{K}]$ is given by the matrix shown in Table 45.5.

Table 45.4 Stresses Calculated for an Isotropic Element

$$
\begin{Bmatrix} \{\sigma\}_i \\ \{\sigma\}_j \\ \{\sigma\}_k \\ \{\sigma\}_l \end{Bmatrix} = \frac{N}{cb}
\left[\begin{array}{ccc|ccc|ccc|ccc}
6p^{-1}+6\nu p & 4\nu c & -4b & -6\nu p & 2\nu c & 0 & -6p^{-1} & 0 & -2b & 0 & 0 & 0 \\
6p+6\nu p^{-1} & 4c & -4\nu b & -6p & 2c & 0 & -6\nu p^{-1} & 0 & -2\nu b & 0 & 0 & 0 \\
-(1-\nu) & -(1-\nu)b & (1-\nu)c & (1-\nu) & 0 & -(1-\nu)c & (1-\nu) & (1-\nu)b & 0 & -(1-\nu) & 0 & 0 \\
\hline
-6\nu p & -2\nu c & 0 & 6p^{-1}+6\nu p & -4\nu c & -4b & 0 & 0 & 0 & -6p^{-1} & 0 & -2b \\
-6p & -2c & 0 & 6p+6\nu p^{-1} & -4c & -4\nu b & 0 & 0 & 0 & -6\nu p^{-1} & 0 & -2\nu b \\
-(1-\nu) & 0 & (1-\nu)c & (1-\nu) & -(1-\nu)b & -(1-\nu)c & (1-\nu) & 0 & 0 & -(1-\nu) & (1-\nu)b & 0 \\
\hline
-6p^{-1} & 0 & 2b & 0 & 0 & 0 & 6p^{-1}+6\nu p & 4\nu c & 4b & -6\nu p & 2\nu c & 0 \\
-6\nu p^{-1} & 0 & 2b & 0 & 0 & 0 & 6p+6\nu p^{-1} & 4c & 4\nu b & -6p & 2c & 0 \\
-(1-\nu) & -(1-\nu)b & 0 & (1-\nu) & 0 & 0 & (1-\nu) & (1-\nu)b & (1-\nu)c & -(1-\nu) & 0 & -(1-\nu)c \\
\hline
0 & 0 & 0 & -6p^{-1} & 0 & 2b & -6\nu p & -2\nu c & 0 & 6p^{-1}+6\nu p & -4\nu c & 4b \\
0 & 0 & 0 & -6\nu p^{-1} & 0 & 2\nu b & -6p & -2c & 0 & 6p+6\nu p^{-1} & -4c & 4\nu b \\
-(1-\nu) & 0 & 0 & (1-\nu) & -(1-\nu)b & 0 & (1-\nu) & 0 & (1-\nu)c & -(1-\nu) & (1-\nu)b & -(1-\nu)c
\end{array}\right]
\begin{Bmatrix} \{D^*\}_i \\ \{D^*\}_j \\ \{D^*\}_k \\ \{D^*\}_l \end{Bmatrix}
$$

Note: $p = c/b$.

Table 45.5 $[\bar{K}]$ Matrix

$$
[\bar{K}] = \begin{bmatrix}
60p^{-2} + 60p^{2} - 12\nu + 42 & & & & & & & & & & & \\
30p^{2} + 12\nu + 3 & 20p^{2} - 4\nu + 4 & & & & \text{Symmetrical} & & & & & & \\
-(30p^{-2} + 12\nu + 3) & -15\nu & 20p^{-2} - 4\nu + 4 & & & & & & & & & \\
30^{-2} - 60p^{2} + 12\nu - 42 & -30p^{2} + 3\nu - 3 & -15p^{-2} + 12\nu + 3 & 60p^{-2} + 60p^{2} - 12\nu + 42 & & & & & & & & \\
30p^{2} - 3\nu + 3 & 10p^{2} + \nu - 1 & 0 & -(30p^{2} + 12\nu + 3) & 20p^{2} - 4\nu + 4 & & & & & & & \\
-15p^{-2} + 12\nu + 3 & 0 & 10p^{-2} + 4\nu - 4 & -(30p^{-2} + 12\nu + 3) & 15\nu & 20p^{-2} - 4\nu + 4 & & & & & & \\
-60p^{-2} + 30p^{2} + 12\nu - 42 & 15p^{2} - 12\nu - 3 & 30p^{-2} - 3\nu + 3 & -30p^{-2} - 30p^{2} - 12\nu + 42 & 15p^{2} + 3\nu - 3 & 15p^{-2} + 3\nu - 3 & 60p^{-2} + 60p^{2} - 12\nu + 42 & & & & & \\
15p^{2} - 12\nu - 3 & 10p^{2} + 4\nu - 4 & 0 & -15p^{2} - 3\nu + 3 & 5p^{2} - \nu + 1 & 0 & 30p^{2} + 12\nu + 3 & 20p^{2} - 4\nu + 4 & & & & \\
-30p^{-2} + 3\nu - 3 & 0 & 10p^{-2} + \nu - 1 & -15p^{-2} - 3\nu + 3 & 0 & 5p^{-2} - \nu + 1 & 30p^{-2} + 12\nu + 3 & 15\nu & 20p^{-2} - 4\nu + 4 & & & \\
-30p^{-2} - 30p^{2} - 12\nu + 42 & -15p^{2} - 3\nu + 3 & 15p^{-2} + 3\nu - 3 & -60p^{-2} + 30p^{2} + 12\nu - 42 & -15p^{2} + 12\nu + 3 & 30p^{-2} - 3\nu + 3 & 30p^{-2} - 60p^{2} + 12\nu - 42 & -30p^{2} + 3\nu - 3 & 15^{-2} - 12\nu - 3 & 60p^{-2} + 60p^{2} - 12\nu + 42 & & \\
15p^{2} - 3\nu - 3 & 5p^{2} - \nu + 1 & 0 & -15p^{2} + 12\nu + 3 & 10p^{2} + 4\nu - 4 & 0 & 30p^{2} - 3\nu + 3 & 10p^{2} + \nu - 1 & 0 & -(30p^{2} + 12\nu + 3) & 20p^{2} - 4\nu + 4 & \\
-15p^{-2} - 3\nu - 3 & 0 & 5p^{-2} - \nu + 1 & -30p^{-2} + 3\nu - 3 & 0 & 10p^{-2} + \nu - 1 & 15p^{-2} - 12\nu - 3 & 0 & 10p^{-2} + 4\nu - 4 & 30p^{-2} + 12\nu + 3 & -15\nu & 20p^{-2} - 4\nu + 4
\end{bmatrix}
$$

If the element is subjected to a uniform load in the z direction of intensity q, the consistent load vector becomes

$$[Q_q^*] = q \int_{-b/2}^{b/2} \int_{-c/2}^{c/2} [L]^{\mathrm{T}}\, dx\, dy \tag{45.194}$$

where $\{Q_q^*\}$ are 12 forces corresponding to the nodal displacement parameters. Evaluating the integrals in this equation gives

$$\{Q_q^*\} = qcb \begin{Bmatrix} 1/4 \\ b/24 \\ -c/24 \\ ---- \\ 1/4 \\ -b/24 \\ -c/24 \\ ---- \\ 1/4 \\ b/24 \\ c/24 \\ ---- \\ 1/4 \\ -b/24 \\ c/24 \end{Bmatrix} \tag{45.195}$$

More details on the finite element method can be found in Desai and Abel [1972] and Ghali and Neville [1978].

45.15 Cables and Arches

Cables are used in many important structures, such as bridges and electrical transmission structures. The cable is a flexible structure with no moment-carrying capacity. Every section of the cable is in pure tension. When a cable is subjected to external loads it deforms in such a way that there is no bending moment at any section of the cable. The deformed shape of cable changes with the position of applied loads. Cables thus obtained are called *funicular.* If the deformed shape of a cable is inverted the resulting configuration is called a *funicular arch.* The internal forces in this case are compressive. Arch shapes that do not follow the deformed configuration under certain loading conditions will be subjected to a bending moment, and the design of such arches should account for bending stresses.

Cables

Cables have to resist more than one loading condition (e.g., dead load; dead load and live load). Since the cable shape changes with the type of loading it is necessary to consider the analysis of such cables subjected to concentrated loading and uniformly distributed loading separately.

Cables Subjected to Concentrated Vertical Loads

A cable supported at two points A and B is subjected to concentrated vertical loads $W_1, W_2, \ldots, W_n$ at distances $x_1, x_2, \ldots, x_n$ from the support A, as shown in Fig. 45.86.

Since all the external loads are vertical, the horizontal components of the two support reactions at A and B, namely R_{Ax} and R_{Bx}, must be equal and opposite to each other; that is,

$$R_{Ax} = R_{Ay} = H$$

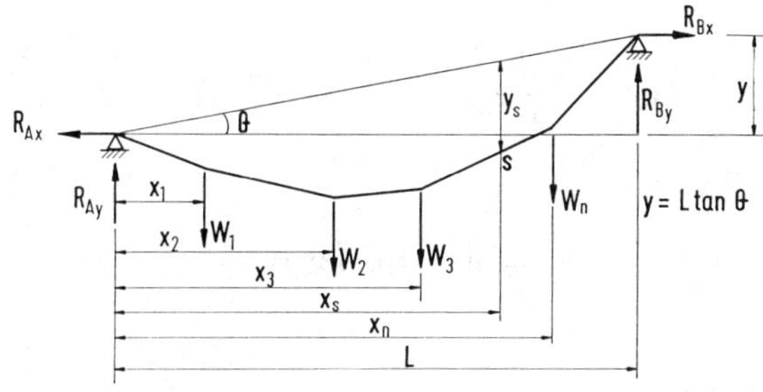

FIGURE 45.86 A cable supported at two points.

At any point along the cable the horizontal component of cable tension is constant and is equal to H. The support reaction can be computed as follows. We know that

$$\sum M_B = 0$$

Assuming clockwise moment positive,

$$\sum M_B = R_{Ay}L + H(L\tan\theta) - \sum m_B = 0$$

or

$$R_{Ay} = \frac{\sum m_B}{L} - H\tan\theta \tag{45.196}$$

The cable is assumed to be perfectly flexible and the bending moment is, therefore, zero at all points. Taking the moment about any point S and equating to zero,

$$R_{Ay}x_s - H(y_s - x_s\tan\theta) - \sum m_s = 0 \tag{45.197}$$

where $\sum m_s$ represents the sum of moments about δ of those loads acting to the left of S.

Substituting for R_{Ay} from Eqs. (45.196) into (45.197) and simplifying, one obtains

$$H = \frac{x_s}{y_s}\frac{1}{L}\sum m_B - \frac{1}{y_s}\sum m_s \tag{45.198}$$

All the support reactions can be computed from the above equations. Tension at any point on the cable can also be determined since the value of H is now known.

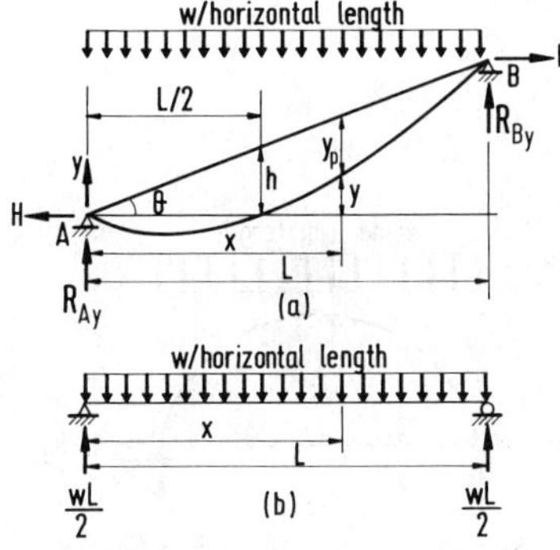

FIGURE 45.87 A cable subjected to uniformly distributed load.

Cables Subjected to Uniformly Distributed Vertical Loads

Let us consider a cable supported at A and B and subjected to a uniformly distributed load of q_0 per unit length along the horizontal span. If the sag is small the self-weight of the cable can be treated as the load per unit of horizontal length. The cable and the loading are shown in Fig. 45.87.

Equation (45.198) can be made use of to determine the horizontal component, H, of the support reactions.

Therefore,

$$Hy_s = \frac{q_0 L}{2}x - q_0 x\frac{x}{2} = \frac{q_0 x}{2}(L - x) \tag{45.199}$$

If the sag at midspan is h, then

$$y_s = h \quad \text{when}\, x = \frac{L}{2}$$

Hence

$$H = \frac{q_0 L^2}{8h} \tag{45.200}$$

Other reaction components can be obtained by making use of equations of equilibrium. Substituting for H from Eq. (45.200) into (45.199), one obtains

$$y_s = \frac{4hx}{L^2}(L - x) \tag{45.201}$$

Equation (45.201) defines the cable profile with x measured horizontally from the left support and y_s measured vertically from the chord AB.

Arches

An arch is essentially a compressive structure, and it can be visualized as an inverted cable. An arch is sufficiently rigid to resist bending moments. There are three types of arches: three-hinged arches, two-hinged arches, and fixed arches, as shown in Fig. 45.88. The two end supports for an arch can be either at the same level or at different levels. The three-hinged arch is statically determinate whereas the other two are not.

Three-Hinged Arch

Let us consider a three-hinged arch subjected to uniformly distributed load, as shown in Fig. 45.89. From symmetry, the vertical reactions at the two end supports are each equal to one-half the total load: $R_{Ay} = R_{By} = wL/2$. The horizontal components at the supports must be equal and opposite to each other. Taking the moment (clockwise position) about the hinge at c and equating to zero,

$$M_c = 0 = R_{Ay}\frac{L}{2} - Hy_c - w\frac{L}{2}\frac{L}{4}$$

or

$$H = \frac{wL^2}{8y_c}$$

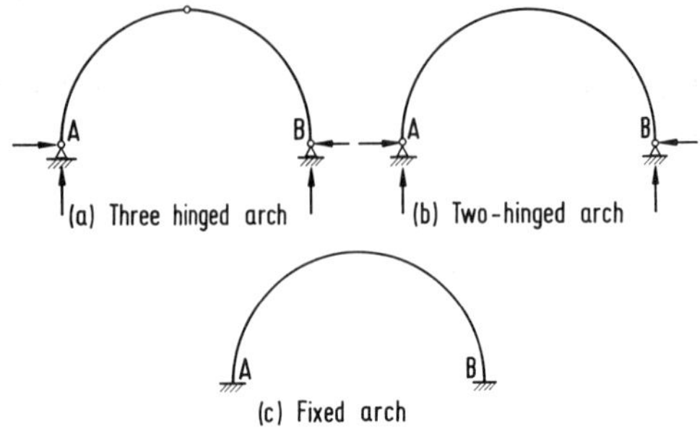

FIGURE 45.88 Three types of arches.

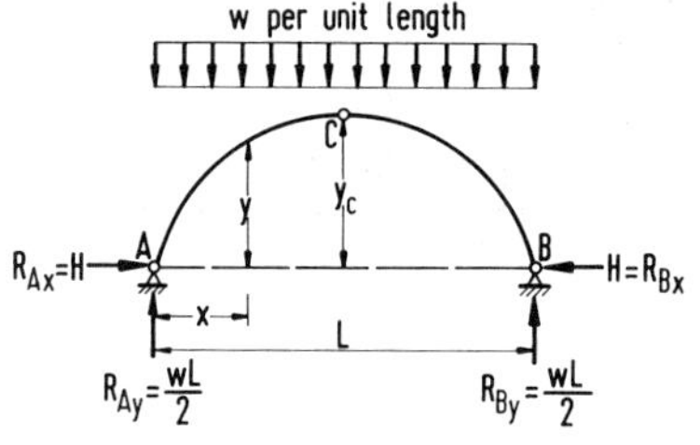

FIGURE 45.89 A three-hinged arch subjected to uniformly distributed load.

We see that the horizontal component of the support reactions is a function of y_c, the rise of the arch at the intermediate hinge. The bending moment (clockwise position) at any section at distance x from the support A is given by

$$M_x = \frac{wL}{2}x - Hy - \frac{wx^2}{2}$$

In certain cases where the bending moment is zero at all sections along the arch, the arch is in pure compression. For this case $Hy = (wL/2)x - (wx^2/2)$ and the shape of the arch is funicular.

Two-Hinged Arch

When an arch is supported at the two ends by means of hinges, as shown in Fig. 45.88(b), it is called a two-hinged arch. In this case the unknown reaction components are in excess of the three equations of equilibrium and the arch is, therefore, statically indeterminate to the first degree. The principle of superposition can be adopted for the analysis of two-hinged arches. The arch is made statically determinate by introducing a roller on one of the supports, and the analysis is carried out to determine the reactions and the movement in the direction of the roller. Further analysis is carried out to determine the force required in the direction of the roller to force it to its original position. By superposition of the results from the two analyses the support reactions of the two-hinged arch are obtained. The horizontal thrust at the left support, A, is given by

$$H_A = \frac{\int My\frac{ds}{EI}}{\int y^2\frac{ds}{EI} + \int \frac{\cos^2\alpha\, ds}{AE}} \tag{45.202}$$

where

M = moment at any section due to the given loads, with redundants removed
E = modulus of elasticity of the material
I = moment of inertia of any section of the arch
A = area of any cross section
α = the angle of inclination of the tangent to the arch axis at the horizontal section
y = rise of the arch at the section
ds = elemental length of arch at the section.

It is usual to resort to an approximate numerical solution since, in most practical cases, the integration in Eq. (45.202) is not practical. The arch axis is divided into a convenient number of segments of length ds and the integration is carried out as the summation of results obtained for elemental length, ds.

45.16 Inelastic Analysis

Practical Inelastic Analysis Methods

Inelastic analysis models, in general, may be classified into two main approaches. The first approach is known as **plastic-zone analysis.** The analysis follows explicitly the gradual spread of yielding throughout the volume of the structure. Material yielding in the member is modeled by discretization of members into several beam-column elements, and subdivision of the cross sections into many fibers. The effects of residual stresses, initial geometric imperfections, and material strain hardening can all be included in the analysis. Because of the refined discretization of the members and their cross sections, the plastic-zone analysis can predict accurately the inelastic response of the structure.

The second approach is known as *plastic hinge analysis.* The analysis assumes that structural elements remain elastic except at critical regions where zero-length **plastic hinges** are allowed to form. This type of analysis is computationally more efficient and more economical than the plastic-zone analysis.

Plastic hinge analyses can be further classified into two categories: first-order analysis and **second-order analysis.** For the first-order plastic hinge analysis, equilibrium is formulated based on undeformed structural geometry. Thus, only the inelasticity effects that influence the strength of the structure are included. The geometric nonlinear effects on the equilibrium of the structure are not considered. First-order plastic hinge analysis predicts the maximum load of the structure corresponding to the formation of a plastic collapse mechanism [Chen and Sohal, 1994]. The analysis essentially predicts the same maximum load as the rigid-plastic analysis approaches.

If equilibrium is formulated based on deformed structural geometry, the analysis is second-order. The need for second-order analyses of steel frames is increasing in view of the American Institute of Steel Construction specifications [AISC, 1986], which give explicit permission for the engineer to compute load effects from a direct second-order analysis. Second-order elastic analysis is also the preferred method in several modern limit states specifications [Eurocode 3, 1990; CSA, 1989].

The application of plastic hinge analysis for building design started as early as 1914 [Kazinczy, 1914]. Basic tests verifying the concepts of plastic hinge theory were reported in 1927 [Maier-Leibnitz, 1927]. Since then significant developments have been made to the plastic hinge theory of structures both in the U.S. and Great Britain [Horne, 1954; Johnson *et al.*, 1953; Beedle, 1958; Discroll, 1965]. Plastic hinge analysis is finding considerable application in continuous beams and low-rise building frames where members are loaded primarily in flexure. For tall building frames and for frames with slender columns subjected to sidesway, failure occurs normally due to geometric nonlinear effects and progressive yielding prior to the formation of a plastic mechanism [SSRC, 1988].

This section presents the virtual work principle to explain the fundamental theorems of plastic hinge analysis. Simple and approximate techniques of practical plastic analysis methods are then introduced. The concept of hinge-by-hinge analysis is presented. Finally, more advanced topics such as second-order elastic, second-order elastic-plastic, and advanced inelastic analysis techniques are introduced.

Ductility of Steel

Plastic analysis is strictly applicable for material that can undergo large deformation without fracture. Steel is clearly a material that resembles such behavior when subjected to tensile force. The actual stress-strain curves for steel of various grades are shown in Fig. 45.90. They may be idealized and represented by a bilinear curve as shown in Fig. 45.91. The assumption in this idealization is that before the yield stress is reached, the material is assumed to behave linearly elastic, and after the yield stress is reached, the strain increases indefinitely without further increase in stress. This material idealization is generally known as *elastic-perfectly plastic* behavior. For a compact section, the attainment of initial yielding does not result in failure. The compact section will have reserved plastic strength that depends on the shape of the cross section.

Inelastic Force Redistribution

The benefit of using steel as a ductile material can be demonstrated by the three-bar system shown in Fig. 45.92(a). From the equilibrium condition of the system,

$$2T_1 + T_2 = P \qquad (45.203)$$

Since the system is statically indeterminate, a compatibility condition is required to determine the force distribution in the system. A compatibility condition requires that the longitudinal

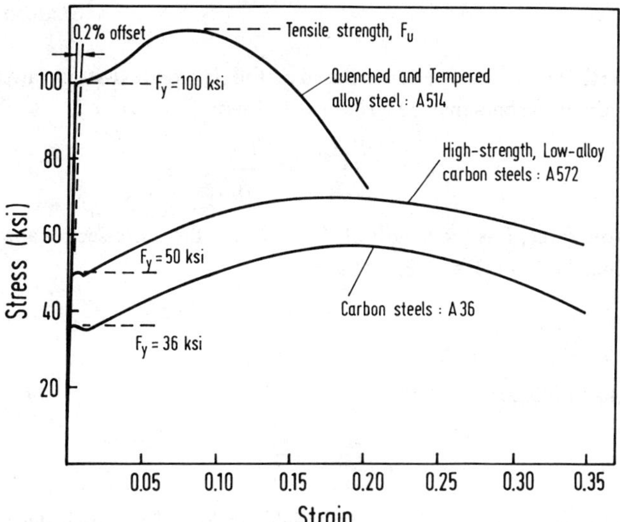

FIGURE 45.90

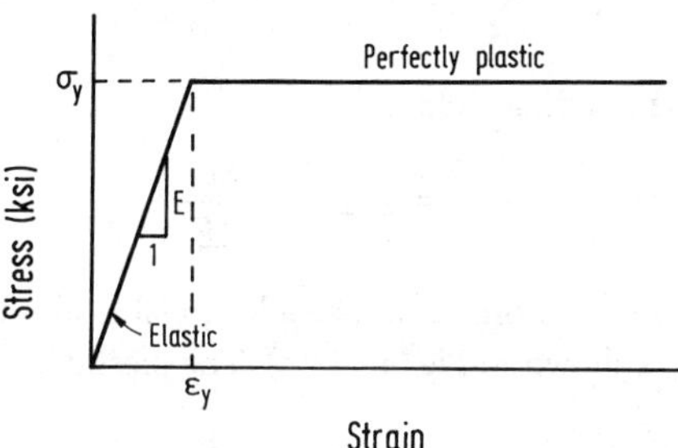

FIGURE 45.91

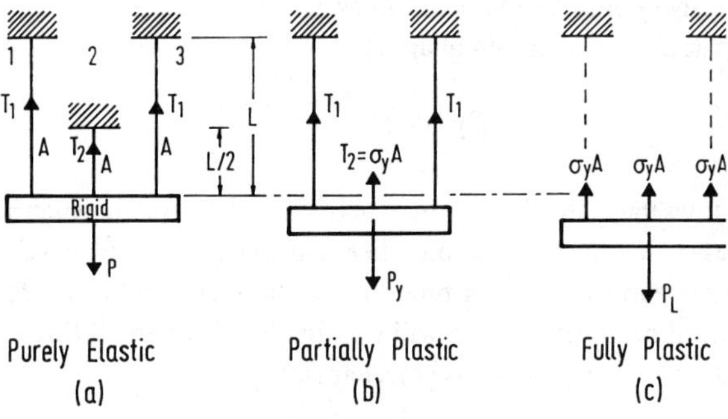

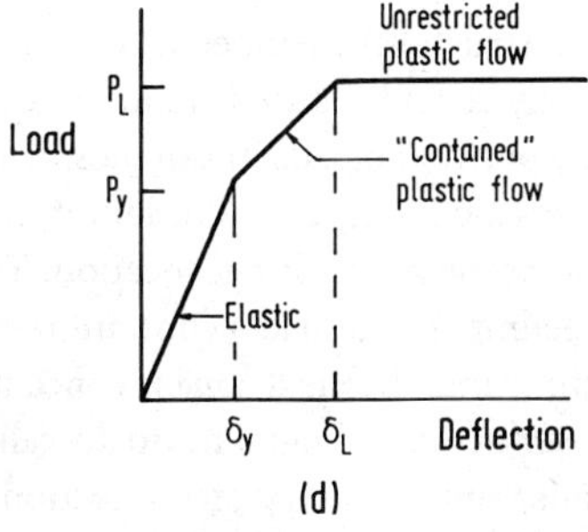

FIGURE 45.92

displacements of the three bars must be equal. Based on the elastic stress-strain law, the displacement and force relationship of the bars may be written as follows:

$$\delta = \frac{T_1 L_1}{AE} = \frac{T_2 L_2}{AE} \tag{45.204}$$

where T is the tensile force, L is the length of the rod, A is the cross-sectional area, and E is the elastic modulus. Since $L_2 = L_1/2 = L/2$, we have

$$T_1 = \frac{T_2}{2} \tag{45.205}$$

From Eqs. (45.204) and (45.205), we have

$$T_2 = \frac{P}{2} \tag{45.206}$$

The load at which the structure reaches the first yield [in Fig. 45.92(b)] is determined by letting $T_2 = \sigma_y A$ in Eq. (45.206):

$$P_y = 2T_2 = 2\sigma_y A \tag{45.207}$$

the corresponding displacement at first yield is

$$\delta_y = \varepsilon_y L = \frac{\sigma_y L}{2E} \tag{45.208}$$

After bar 2 is yielded, the system continues to take additional load until all three bars reach their maximum strength of $\sigma_y A$, as shown in Fig. 45.92(c). The plastic limit load of the system is thus written as

$$P_L = 3\sigma_y A \tag{45.209}$$

The process of successive yielding of bars in this system is known as *force redistribution.*

The displacement at the incipient of collapse is

$$\delta_L = \varepsilon_y L = \frac{\sigma_y L}{E} \tag{45.210}$$

Figure 45.92(d) shows the load-displacement behavior of the system when subjected to increasing force. As load increases, the most critical member (bar 2) will reach its maximum strength first. As bar 2 is yielded, the force in the member remains constant, and additional loads on the system are taken by the less critical bars. The system will eventually fail when all three bars are fully yielded, thus transforming the system into a failure mechanism.

Plastic Hinge

A plastic hinge is said to form in a structural member when the cross section is fully yielded. However, plastic hinges do not usually form because it requires a large change in slope to occur over a small length of member at the position of maximum moment.

When a cross section develops a plastic hinge, any addition of moment will cause the beam to rotate with little increase in moment at the plastic hinge location. The fully yielded zone acts as if it were a hinge that can undergo indefinite rotation at a constant restraining moment M_p.

Most plastic analysis theories assume that the yield zone is concentrated at zero length plasticity. In fact, the yield zone is developed over a certain length, normally called the *hinge length,* depending on the loading, boundary conditions, and geometry of the section. The hinge lengths of beams (ΔL) with different support and loading conditions are shown in Fig. 45.93.

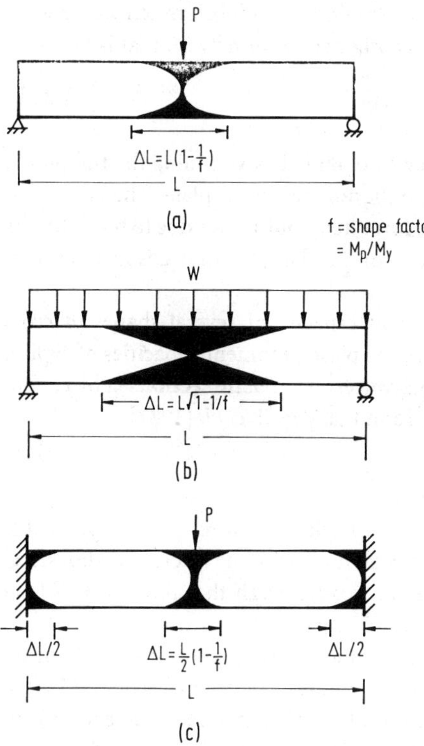

FIGURE 45.93

Plastic hinges are developed first at the sections subjected to the greatest curvature deformation. The possible locations for plastic hinges to develop are at the points of concentrated loads, the intersections of members involving a change in geometry, and the point of zero shear for member under uniform distributed load.

Plastic Moment

A knowledge of full **plastic moment** capacity of a section is important in plastic analysis. If the plastic moments of various members of a structural system are known, then the plastic limit load of the system can be determined. The full plastic moment is the maximum moment of resistance of a fully yielded cross section. The cross section must be fully compact in order to develop full plastic moment. In other words, the component plates of a section must not buckle prior to the attainment of full moment capacity.

The plastic moment capacity, M_p, of a cross section depends on the material yield stress and the section geometry. The procedure for the calculation of M_p may be summarized in the following two steps. First, locate the plastic neutral axis. This is done by considering equilibrium of forces normal to the cross section. Figure 45.94(a) shows a cross section of arbitrary shape subjected to increasing moment. The plastic neutral axis is determined by equating the force in compression (C) to that in tension (T). If the entire cross section is made of the same material, the plastic neutral axis can be determined by dividing the cross-sectional area into two equal parts. If the cross section is made of more than one type of material, the plastic neutral axis must be determined by summing the normal forces and setting them equal to zero. Second, the plastic moment capacity is determined by obtaining the moment generated by the tensile and compressive forces.

Consider an arbitrary section with area $2A$ and with one axis of symmetry. The section is strengthened by a cover plate of area a, as shown in Fig. 45.94(b). Assume that the yield strengths of the original section and the cover plate are σ_{yo} and σ_{yc}, respectively. At the full plastic state, the

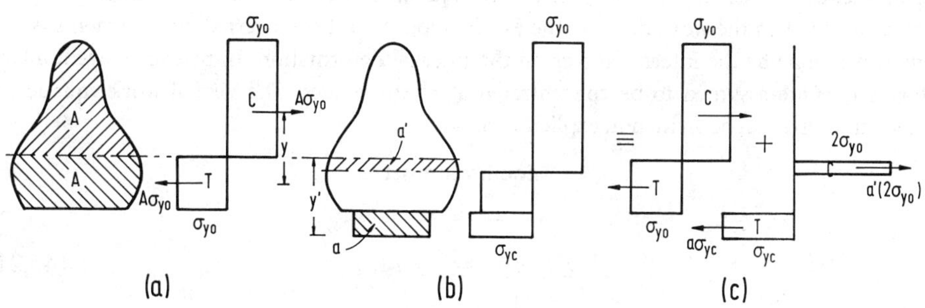

FIGURE 45.94

total axial force acting on the cover plate is $a\sigma_{yc}$. In order to maintain equilibrium of force in axial direction, the plastic neutral axis must shift down from its original position by a', that is,

$$a' = \frac{a\sigma_{yc}}{2\sigma_{yo}} \tag{45.211}$$

The resulting plastic capacity of the built-up section may be obtained by summing the full plastic moment of the original section and the moment contribution by the cover plate. The additional capacity is equal to the moment caused by the cover plate force $a\sigma_{yc}$ and a force due to the fictitious stress $2\sigma_{yo}$ acting on the area a' resulting from the shifting of the plastic neutral axis from tension zone to compression zone, as shown in Fig. 45.94(c).

Figure 45.95 shows the computation of plastic moment capacity of several shapes of cross section. Based on the principle developed in this section, the plastic moment capacities of typical cross sections may be generated. Additional information concerning **plastic sections** subjected to combined bending, torsion, shear, and axial load can be found in Mrazik *et al.* [1987].

Theory of Plastic Analysis

The main assumptions for plastic analysis are as follows: (1) the structure is made of ductile material that can undergo large deformations beyond elastic limit without fracture or buckling, and (2) the deflections of the structure under loading are small enough that the effect of load acting on the deformed geometry (second-order forces) can be ignored.

An exact plastic hinge solution must satisfy three basic conditions: equilibrium, mechanism, and plastic moment. The plastic analysis disregards the continuity condition as required by the elastic analysis of indeterminate structures. The formation of plastic hinge in members leads to discontinuity of slope. A *mechanism condition* occurs when sufficient plastic hinges are formed to allow the structure to deform into a mechanism. Since plastic analysis utilizes the limit of resistance of a member's plastic strength, the plastic moment condition is required to ensure that the resistance of the cross sections is not violated anywhere in the structure. The equilibrium condition, which is the same condition to be satisfied in elastic analysis, requires that the sum of all applied forces and reactions be equal to zero, and it requires that all internal forces be self-balanced.

When all three conditions are satisfied, the resulting plastic analysis for limiting load is the "correct" limit load. The collapse loads for simple structures such as beams and **portal frames** can be solved easily using a direct approach or through visualization of the formation of a correct collapse mechanism. However, for more complex structures, the exact solution satisfying all the three conditions may be difficult to predict. Thus simple techniques using approximate methods of analysis are often used to assess these solutions. These techniques are presented in the following sections.

Principle of Virtual Work

The virtual work principle relates a system of forces in equilibrium to a system of compatible displacements. For example, if a structure in equilibrium is given a set of small compatible displacements, then the work done by the external loads on these external displacements is equal to the work done by the internal forces on the internal deformation. In plastic analysis, internal deformations are assumed to be concentrated at plastic hinges. The virtual work equation for hinged structures can be written in explicit form as

Equilibrium set

$$\sum P_i \delta_j = \sum M_i \theta j \tag{45.212}$$

Displacement set

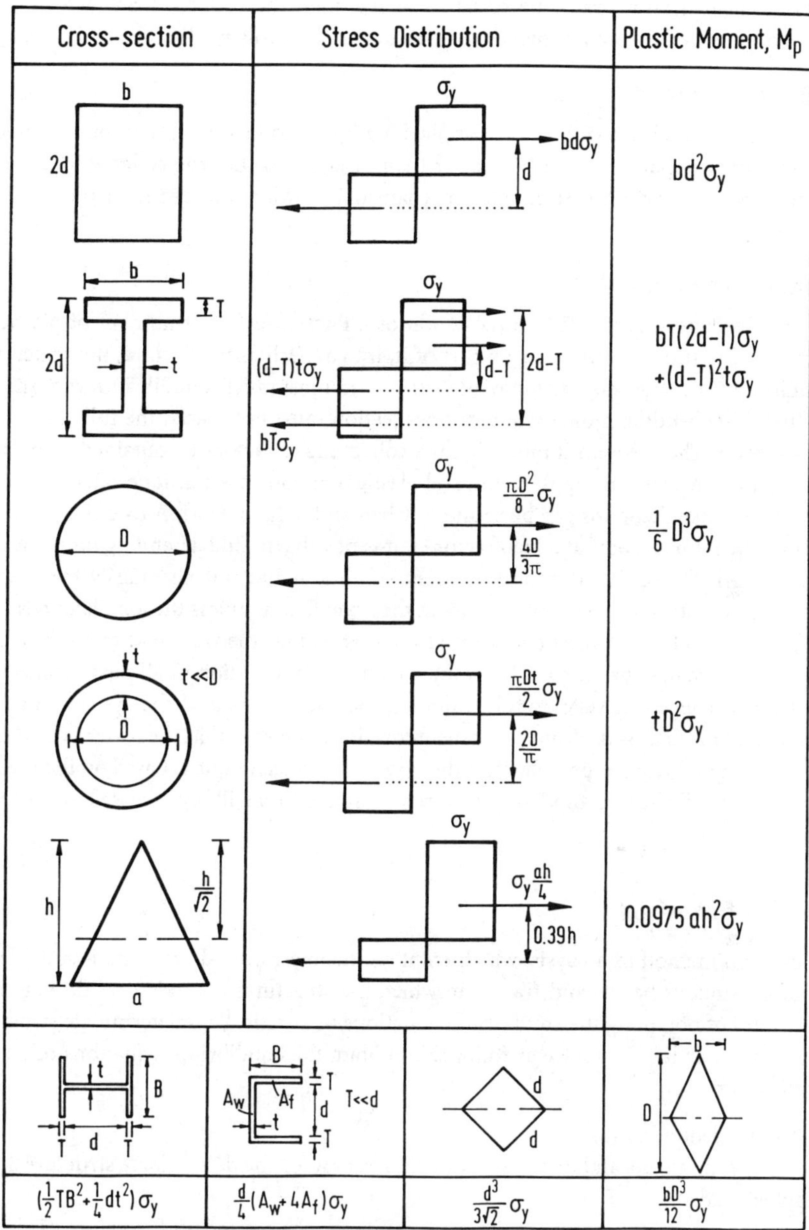

FIGURE 45.95

where P_i is an external load and M_i is the internal moment at a hinge location. Both P_i and M_i constitute an equilibrium set and so must be in equilibrium. δ_j are the displacements under the point loads P_i in the direction of the loads. θ_j are the plastic hinge rotations under the moment M_i. Both δ_j and θ_j constitute a displacement set, and they must be compatible with each other.

Lower Bound Theorem

For a given structure, if there exists any distribution of bending moments in the structure that satisfies both the equilibrium and plastic moment conditions, then the load factor, λ_L, computed from this moment diagram must be equal to or less than the collapse load factor, λ_c, of the structure. Lower bound theorem provides a safe estimate of the collapse limit load; that is, $\lambda_L \leq \lambda_c$.

Upper Bound Theorem

For a given structure subjected to a set of applied loads, a load factor, λ_u, computed based on an assumed collapse mechanism must be greater than or equal to the true collapse load factor, λ_c. Upper bound theorem, which uses only the mechanism condition, overestimates the collapse limit load; that is, $\lambda_u \geq \lambda_c$.

Uniqueness Theorem

A structure at collapse has to satisfy three conditions. First, a sufficient number of plastic hinges must be formed to turn the structure, or part of it, into a mechanism. Second, the structure must be in equilibrium (i.e., the bending moment distribution must satisfy equilibrium with the applied loads). Finally, the bending moment at any cross section must not exceed the full plastic value of that cross section. The theorem implies that the collapse load factor, λ_c, obtained from the three basic conditions (mechanism, equilibrium, and plastic moment) has a unique value.

Proofs of the three theorems can be found in Chen and Sohal [1994]. A useful corollary of the lower bound theorem is that if at a load factor, λ, it is possible to find a bending moment diagram that satisfies both the equilibrium and moment conditions but not necessarily the mechanism condition, then the structure will not collapse at that load factor, unless the load happens to be the collapse load. A corollary of the upper bound theorem is that the true load factor at collapse is the smallest possible one that can be determined from a consideration of all possible mechanisms of collapse. This concept is very useful in finding the collapse load of the system from various combinations of mechanisms. From these theorems, it can be seen that the lower bound theorem is based on the equilibrium approach while the upper bound technique is based on the mechanism approach. The equilibrium method and the mechanism method will be discussed in the following sections.

Equilibrium Method

The equilibrium method of analysis, which employs the lower bound theorem, is suitable for the analysis of continuous beams and frames in which the structural redundancies do not exceed 2. The procedures of obtaining the equilibrium equations of a statically indeterminate structure and evaluating its plastic limit load are as follows. To obtain the equilibrium equations of a statically indeterminate structure:

1. Select the redundant(s).
2. Free the redundants and draw the moment diagram for the determinate structure under the applied loads.
3. Draw the moment diagram for the structure due to the redundant forces.
4. Superimpose the moment diagrams in steps 2 and 3.
5. Obtain maximum moments at critical sections of the structure utilizing the moment diagram in step 4.

To evaluate the plastic limit load of the structure:

1. Select value(s) of redundant(s) such that the plastic moment condition is not violated at any section in the structure.
2. Determine the load corresponding to the selected redundant(s).
3. Check for the formation of a mechanism. If a collapse mechanism condition is met, then the computed load is the exact plastic limit load. Otherwise, it is a lower bound solution.
4. Adjust the redundant(s) and repeat steps 1 to 3 until the exact plastic limit load is obtained.

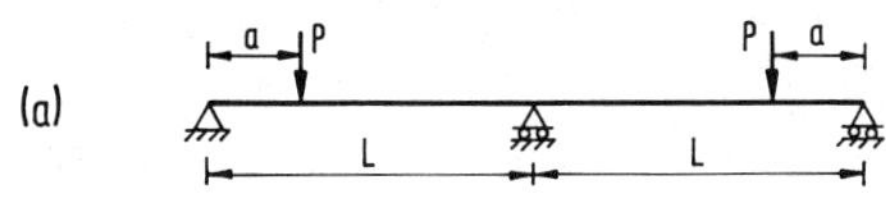

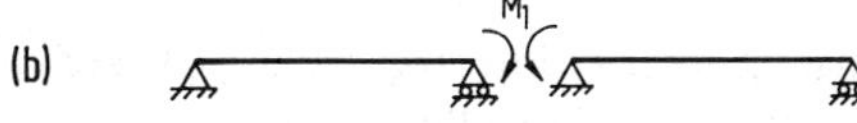

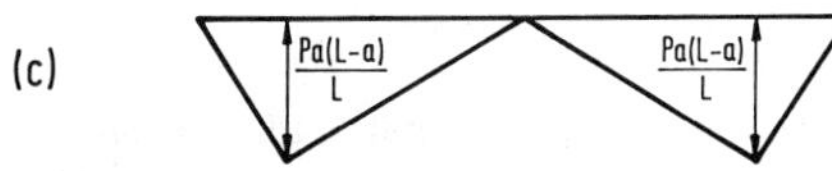

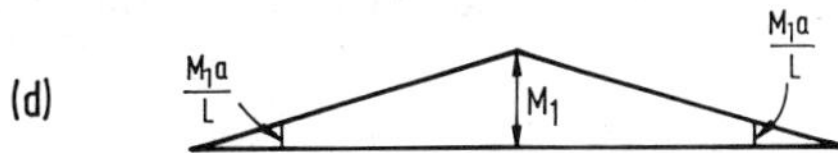

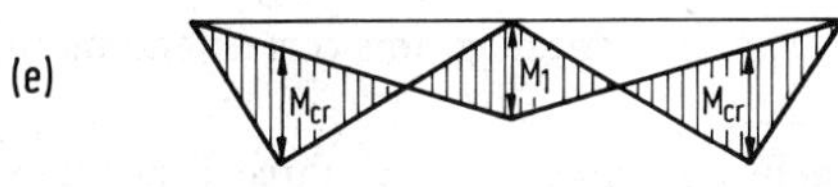

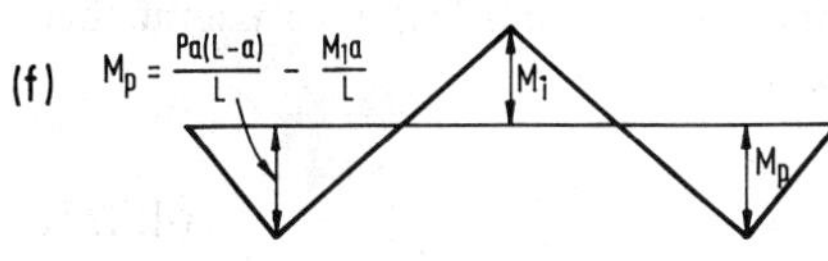

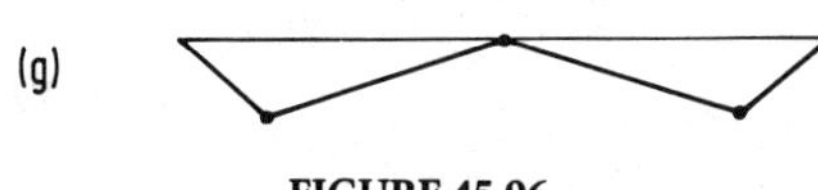

FIGURE 45.96

Analysis of Continuous Beams

Figure 45.96(a) shows a two-span continuous beam analyzed using the equilibrium method. The plastic limit load of the beam is calculated based on the step-by-step procedure described in the previous section as follows:

1. Select the redundant force M_1, which is the bending moment at the intermediate support, as shown in Fig. 45.96(b).
2. Free the redundants and draw the moment diagram for the determinate structure under the applied loads, as shown in Fig. 45.96(c).
3. Draw the moment diagram for the structure due to the redundant moment M_1, as shown in Fig. 45.96(d).
4. Superimpose the moment diagrams in Figs. 45.96(c) and (d). The results are shown in Fig. 45.96(e).
5. The moment diagram in (e) is redrawn on a single straight baseline. The critical moment in the beam is

$$M_{cr} = \frac{Pa(L-a)}{L} - \frac{M_1 a}{L} \tag{45.213}$$

The maximum moment at critical sections of the structure utilizing the moment diagram in (e) is obtained. By letting $M_{cr} = M_p$, the resulting moment distribution is shown in Fig. 45.96(f).

6. A lower bound solution may be obtained by selecting a value of redundant moment M_1. For example, if $M_1 = 0$ is selected, the moment diagram is reduced to that shown in Fig. 45.96(c). By equating the maximum moment in the diagram to the plastic moment, M_p, we have

$$M_{cr} = \frac{Pa(L-a)}{L} = M_p \tag{45.214}$$

which gives $P = P_1$ as

$$P_1 = \frac{M_p L}{a(L-a)} \tag{45.215}$$

The moment diagram in Fig. 45.96(c) shows a plastic hinge formed at each span. Since two plastic hinges in each span are required to form a plastic mechanism, the load P_1 is a lower bound solution. However, if the redundant moment M_1 is set equal to the plastic moment M_p, then, letting the maximum moment in Fig. 45.96(f) equal the plastic moment, we have

$$M_{cr} = \frac{Pa(L-a)}{L} - \frac{M_p a}{L} = M_p \tag{45.216}$$

which gives $P = P_2$ as

$$P_2 = \frac{M_p(L + a)}{a(L - a)} \tag{45.217}$$

7. Since a sufficient number of plastic hinges has formed in the beams [Fig. 45.96(g)] to arrive at a collapse mechanism, the computed load, P_2, is the exact plastic limit load.

Analysis of Portal Frames

A pinned-base rectangular frame subjected to vertical load V and horizontal load H is shown in Fig. 45.97(a). All the members of the frame, AB, BD, and DE, are made of the same section with moment capacity M_p. The objective is to determine the limit value of H if the frame's width-to-height ratio L/h is 1.0.

Procedure. The frame has one degree of redundancy. The redundant for this structure can be chosen as the horizontal reaction at E. Figures 45.97(b) and (c) show the resulting determinate frame loaded by the applied loads and redundant force. The moment diagrams corresponding to these two loading conditions are shown in Fig. 45.97(d) and (e).

The horizontal reaction S should be chosen in such a manner that all three conditions (equilibrium, plastic moment, and mechanism) are satisfied. Formation of two plastic hinges is necessary to form a mechanism. The hinges may be formed at B, C, and D. Let us assume that a plastic hinge forms at D, as shown in Fig. 45.97(e). We have

$$S = \frac{M_p}{h} \tag{45.218}$$

Corresponding to this value of S, the moments at B and C can be expressed as

$$M_B = Hh - M_p \tag{45.219}$$

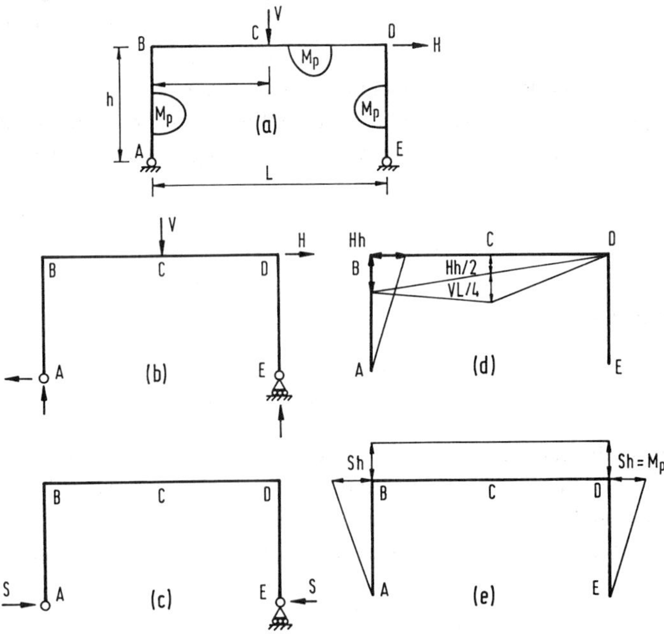

FIGURE 45.97

$$M_C = \frac{Hh}{2} + \frac{VL}{4} - M_p \tag{45.220}$$

The condition for the second plastic hinge to form at B is $|M_B| > |M_C|$. From Eqs. (45.219) and (45.220) we have

$$Hh - M_p > \frac{Hh}{2} + \frac{VL}{4} - M_p \tag{45.221}$$

and

$$\frac{V}{H} < \frac{h}{L} \tag{45.222}$$

The condition for the second plastic hinge to form at C is $|M_C| > |M_B|$. From Eqs. (45.219) and (45.220) we have

$$Hh - M_p < \frac{Hh}{2} + \frac{VL}{4} - M_p \tag{45.223}$$

and

$$\frac{V}{H} > \frac{h}{L} \tag{45.224}$$

For a particular combination of V, H, L, and h, H is calculated as follows. When $L/h = 1$ and $V/H = \frac{1}{3}$, we have

$$M_B = Hh - M_p \tag{45.225}$$

$$M_C = \frac{Hh}{2} + \frac{Hh}{12} - M_p = \frac{7}{12}Hh - M_p \tag{45.226}$$

Since $|M_B| > |M_C|$, the second plastic hinge will form at B and the corresponding value for H is

$$H = \frac{2M_p}{h} \tag{45.227}$$

When $L/h = 1$ and $V/H = 3$, we have

$$M_B = Hh - M_p \tag{45.228}$$

$$M_C = \frac{Hh}{2} + \frac{3}{4}Hh - M_p = \frac{5}{4}Hh - M_p \tag{45.229}$$

Since $|M_C| > |M_B|$, the second plastic hinge will form at C and the corresponding value for H is

$$H = \frac{1.6M_p}{h} \tag{45.230}$$

Mechanism Method

This method is based on the upper bound theorem, which states that the load computed on the basis of an assumed failure mechanism is never less than the exact plastic limit load of a structure. Thus, it always predicts the upper bound solution of the collapse limit load. It can also be shown that the minimum upper bound is the limit load itself. The procedure of using the mechanism method has the following two steps:

1. Assume a failure mechanism and form the corresponding work equation from which an upper bound value of the plastic limit load can be estimated.

2. Write the equilibrium equations for the assumed mechanism and check the moments to see whether the plastic moment condition is met everywhere in the structure.

To obtain the true limit load using the mechanism method, it is imperative to determine every possible collapse mechanism, some of which are combinations of certain independent mechanisms. Once the independent mechanisms have been identified, a work equation may be established for each combination and the corresponding collapse load determined. The lowest load among those obtained by considering all possible combinations of independent mechanisms is the correct plastic limit load.

Number of Independent Mechanisms

The number of possible independent mechanisms, n, for a structure can be determined from

$$n = N - R \tag{45.231}$$

where N is the number of critical sections at which plastic hinges might form and R is the degree of redundancy of the structure.

Critical sections generally occur at the points of concentrated loads, at joints where two or more members are meeting at different angles, and at sections where there is an abrupt change in section geometries or properties. To determine the number of redundants R of a structure, it is necessary to free sufficient supports or restraining forces in structural members so that the structure becomes an assembly of several determinate substructures.

Figure 45.98 shows two examples. The cuts made in each structure reduce the structural members to either cantilevers or simply supported beams. The fixed-end beam requires a shear force and a moment to restore continuity at the cut section, and thus $R = 2$. For the two-story frame, an axial force, shear, and moment are required at each cut section for full continuity, and thus $R = 12$.

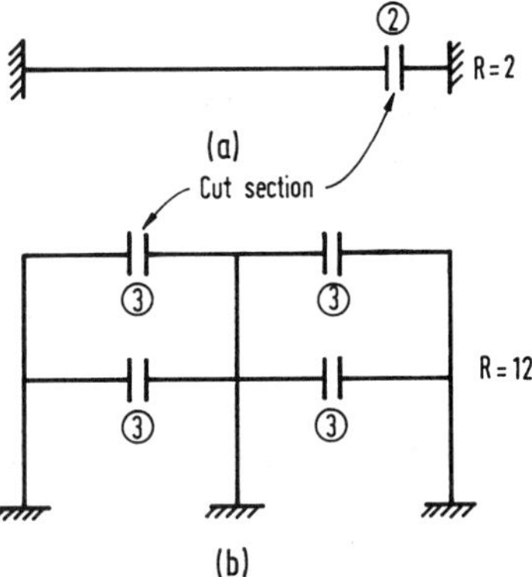

FIGURE 45.98

Types of Mechanism

Figure 45.99(a) shows a frame structure subjected to a set of loading. The frame may fail by different types of collapse mechanisms depending on the magnitude of loading and the frame's configuration. The collapse mechanisms are as follows.

1. *Beam mechanism.* Possible mechanisms of this type are shown in Fig. 45.99(b).
2. *Panel mechanism.* This collapse mode is associated with sidesway, as shown in Fig. 45.99(c).
3. *Gable mechanism.* This collapse mode is associated with the spreading of column tops with respect to the column bases, as shown in Fig. 45.99(d).
4. *Joint mechanism.* This collapse mode is associated with the rotation of joints whose adjoining members developed plastic hinges and deformed under an applied moment, as shown in Fig. 45.99(e).
5. *Combined mechanism.* This can be a partial collapse mechanism, as shown in Fig. 45.99(f), or a complete collapse mechanism, as shown in Fig. 45.99(g).

The principal rule for combining independent mechanisms is to obtain a lower value of collapse load. The combinations are selected such that the external work becomes a maximum and the internal work becomes a minimum. Thus the work equation would require that the mechanism involve as many applied loads as possible and at the same time eliminate as many plastic hinges as possible. This procedure will be illustrated in the following example.

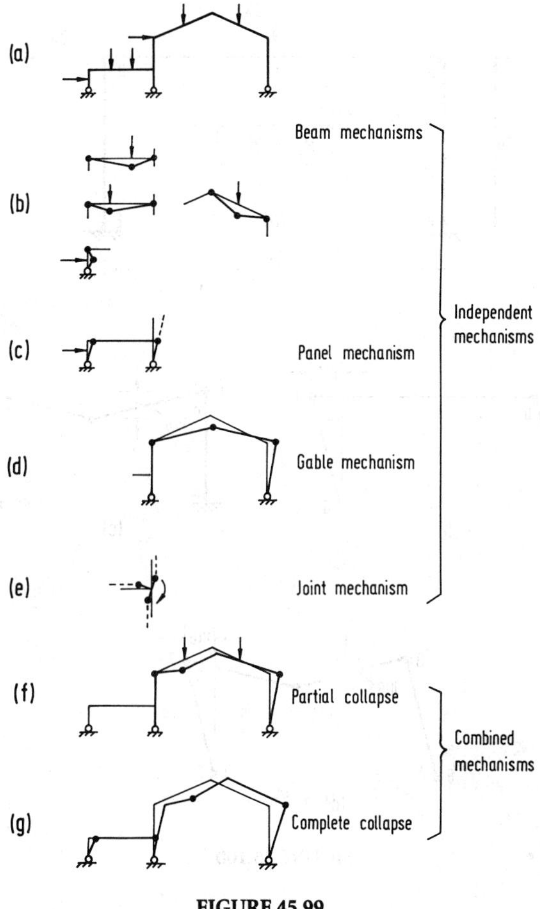

FIGURE 45.99

Example 45.9. A fixed-end rectangular frame has a uniform section with $M_p = 20$ and carries the load shown in Fig. 45.100. Determine the value of the load ratio λ at collapse.
Solution.

Number of possible plastic hinges: $N = 5$
Number of redundancies: $R = 3$
Number of independent mechanisms: $N - R = 2$

The two independent mechanisms are shown in Figs. 45.100(b) and (c), and the corresponding work equations are as follows. For the panel mechanism:

$$20\lambda = 4(20) = 80 \quad \Rightarrow \lambda = 4$$

For the beam mechanism:

$$30\lambda = 4(20) = 80 \quad \Rightarrow \lambda = 2.67$$

The combined mechanisms are now examined to see whether they will produce a lower λ value. It is observed that only one combined mechanism is possible. The mechanism is shown in Fig. 45.100(d) and involves cancellation of the plastic hinge at B. The calculation of the limit load is

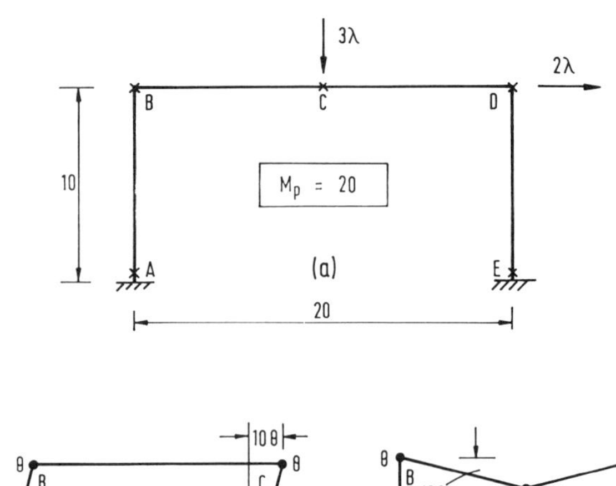

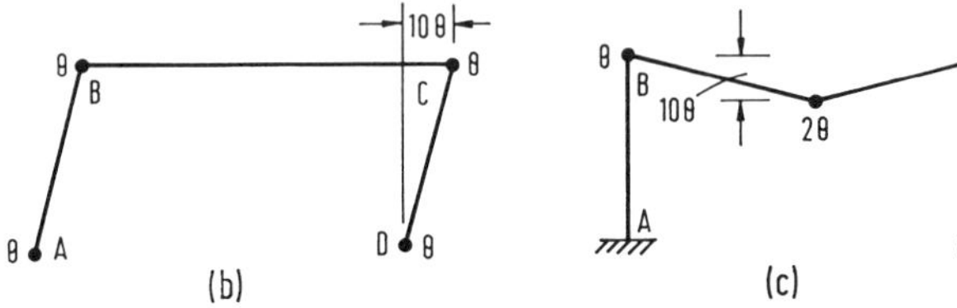

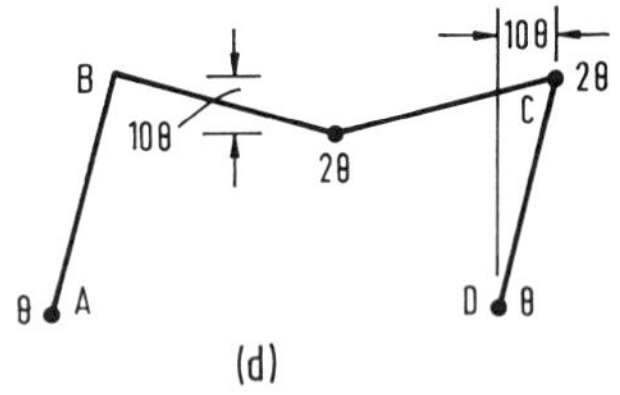

FIGURE 45.100

described as follows:

Panel mechanism:	$20\lambda = 4(20)$
Beam mechanism:	$30\lambda = 4(20)$
	$50\lambda = 8(20)$
Cancel plastic hinge:	$-2(20)$
Combined mechanism:	$50\lambda = 6(20) \quad \rightarrow \lambda = 2.4$

The combined mechanism results in a smaller value for λ, and no other possible mechanism can produce a lower load. Thus, $\lambda = 2.4$ is the collapse load.

Frame Subjected to Distributed Load

When a frame is subjected to distributed loads, the maximum moment and hence the plastic hinge location is not known in advance. The exact location of the plastic hinge may be determined by writing the work equation in terms of the unknown distance and then maximizing the plastic moment by formal differentiation.

Consider the frame shown in Fig. 45.101(a). The sidesway collapse mode in Fig. 45.101(b) leads to the following work equation:

$$4M_p = 24(10\theta) \tag{45.232}$$

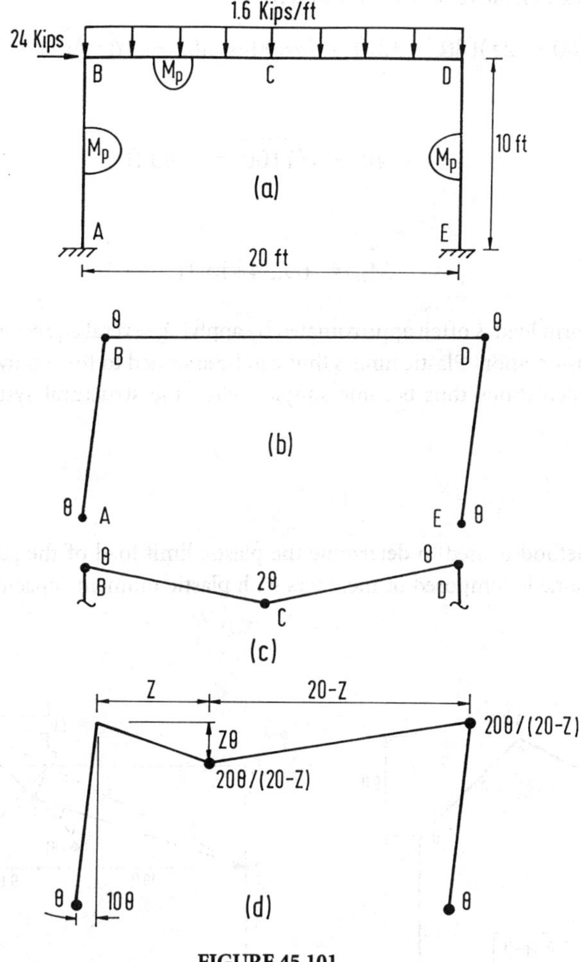

FIGURE 45.101

which gives

$$M_p = 60 \text{ kip-ft}$$

The beam mechanism of Fig. 45.101(c) gives

$$4M_p\theta = \tfrac{1}{2}(10\theta)32$$

which gives

$$M_p = 40 \text{ kip-ft}$$

The correct mechanism is shown in Fig. 45.101(d), where the distance Z from the plastic hinge location is unknown. The work equation is

$$24(10\theta) + \frac{1}{2}(1.6)(20)(z\theta) = M_p\left\{2 + 2\left(\frac{20}{20-z}\right)\right\}\theta$$

which gives

$$M_p = \frac{(240 + 16z)(20 - z)}{80 - 2z}$$

To maximize M_p, the derivative of M_p is set to zero:

$$(80 - 2z)(80 - 32z) + (4800 + 80z - 16z^2)(2) = 0$$

which gives

$$z = 40 - \sqrt{1100} = 6.83 \text{ ft}$$

and

$$M_p = 69.34 \text{ kip-ft}$$

In practice, uniform load is often approximated by applying several equivalent point loads to the member under consideration. Plastic hinges thus can be assumed to form only at the concentrated load points; the calculations thus become simpler when the structural system is getting more complex.

Gable Frames

The mechanism method is used to determine the plastic limit load of the gable frame shown in Fig. 45.102. The frame is composed of members with plastic moment capacity of 270 kip-in. The

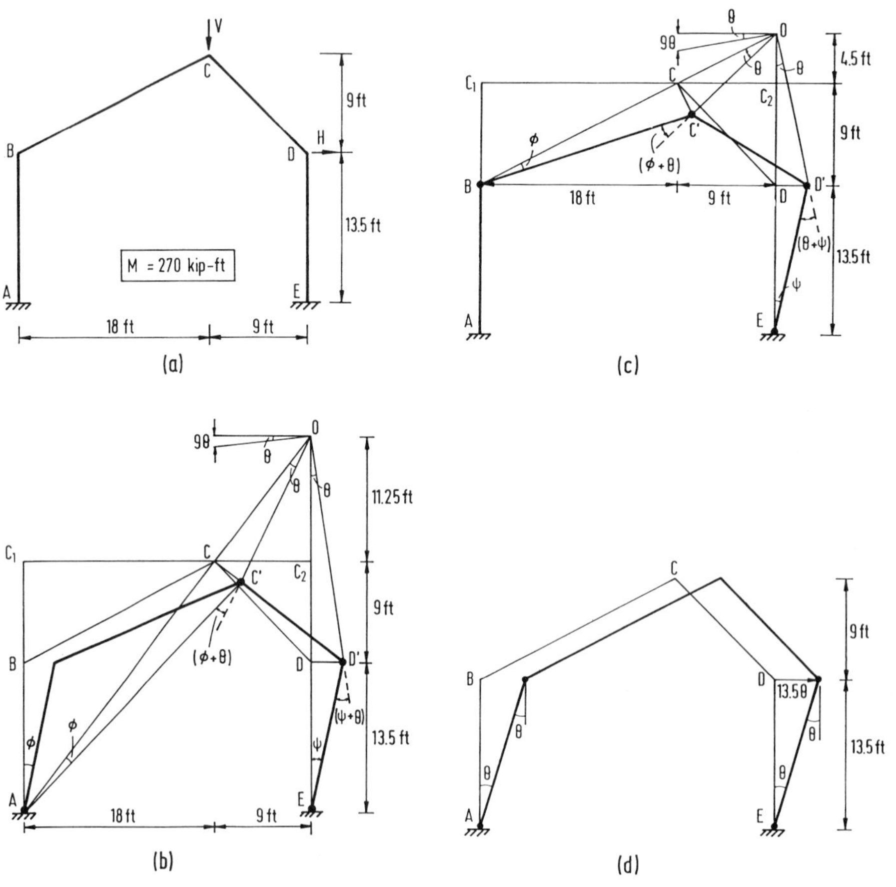

FIGURE 45.102

column bases are fixed. The frame is loaded by a horizontal load H and vertical concentrated load V. A graph of V and H causing the collapse of the frame can be produced.

Consider the three modes of collapse as follows.

Mechanism with Plastic Hinges at *A*, *C*, *D*, and *E*

The mechanism is shown in Fig. 45.102(b). The instantaneous center O for member CD is located at the intersection of AC and ED extended. From similar triangles ACC_1 and OCC_2, we have

$$\frac{OC_2}{CC_2} = \frac{C_1A}{C_1C}$$

which gives

$$OC_2 = \frac{C_1A}{C_1C}CC_2 = \frac{22.5(9)}{18} = 11.25 \text{ ft}$$

From triangles ACC' and $CC'O$, we have

$$AC(\phi) = OC(\theta)$$

which gives

$$\phi = \frac{OC}{AC}\theta = \frac{CC_2}{C_1C}\theta = \frac{9}{8}\theta = \frac{1}{2}\theta$$

Similarly, from triangles ODD' and EDD', the rotation at E is given as

$$DE(\psi) = OD(\theta)$$

which gives

$$\psi = \frac{OD}{DE}\theta = 1.5\theta$$

From the hinge rotations and displacements, the work equation for this mechanism can be written as

$$V(90) + H(13.5\psi) = M_p[\phi + (\phi + \theta) + (\theta + \psi) + \psi]$$

Substituting values for ψ and ϕ and simplifying, we have

$$V + 2.25H = 180$$

Mechanism with Plastic Hinges at *B*, *C*, *D*, and *E*

Figure 45.102(c) shows the mechanism in which the plastic hinge rotations and displacements at the load points can be expressed in terms of the rotation of member CD about the instantaneous center O.

From similar triangles BCC_1 and OCC_2, we have

$$\frac{OC_2}{CC_2} = \frac{BC_1}{C_1C}$$

which gives

$$OC_2 = \frac{BC_1}{C_1C}CC_2 = \frac{9}{18}(9) = 4.5 \text{ ft}$$

From triangles BCC' and $CC'O$, we have

$$BC(\phi) = OC(\theta)$$

which gives

$$\phi = \frac{OC}{BC}\theta = \frac{OC_2}{BC_1}\theta = \frac{4.5}{9}\theta = \frac{1}{2}\theta$$

Similarly, from triangles ODD' and EDD', the rotation at E is given as

$$DE(\psi) = OD(\theta)$$

which gives

$$\psi = \frac{OD}{DE}\theta = \theta$$

The work equation for this mechanism can be written as

$$V(9\theta) + H(13.5\psi) = M_p[\phi + (\phi + \theta) + (\theta + \psi) + \psi]$$

Substituting values of ψ and ϕ and simplifying, we have

$$V + 1.5H = 150$$

Mechanism with Plastic Hinges at *A*, *B*, *D*, and *E*

The hinge rotations and displacements corresponding to this mechanism are shown in Fig. 45.102(d). The rotation of all hinges is θ. The horizontal load moves by 13.5θ but the horizontal load has no vertical displacement. The work equation becomes

$$H(13.5\theta) = M_p(\theta + \theta + \theta + \theta)$$

or

$$H = 80 \text{ kips}$$

The interaction equations corresponding to these three mechanisms are plotted in Fig. 45.103. By carrying out moment checks, it can be shown that mechanism 1 is valid for portion AB of the curve, mechanism 2 is valid for portion BC, and mechanism 3 is valid only when $V = 0$.

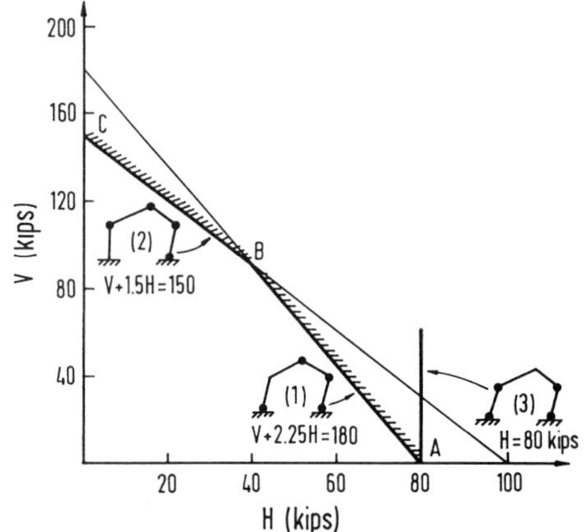

FIGURE 45.103

Analysis Charts for Gable Frames

Pinned-Base Gable Frames

Figure 45.104(a) shows a pinned-end gable frame subjected to uniform gravity load and horizontal load at the column top. The collapse mechanism is shown in Fig. 45.104(b). The work equation may be used to determine the plastic limit load. First, the instantaneous center of rotation O is

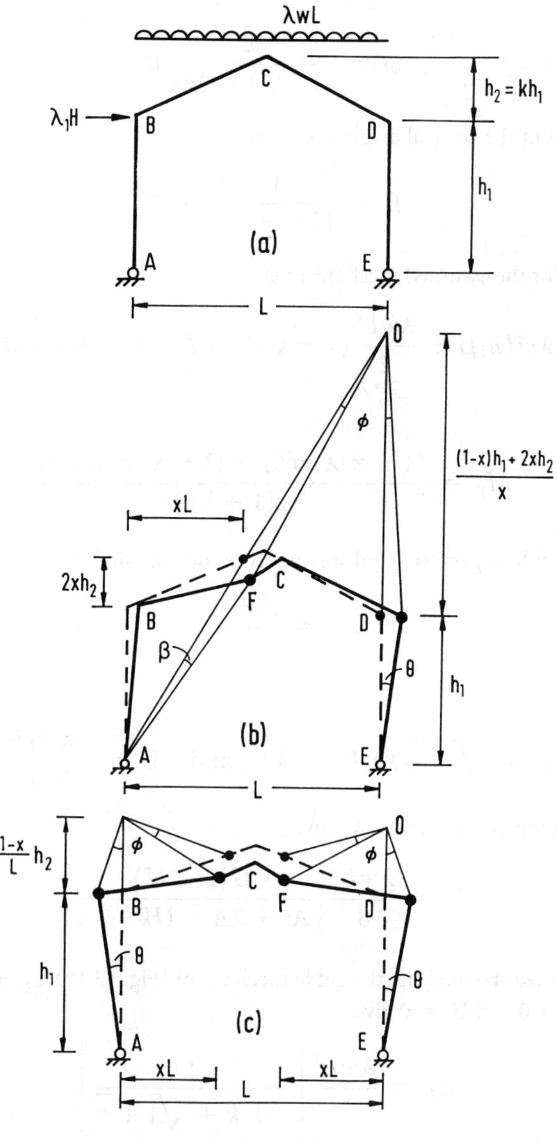

FIGURE 45.104

determined by considering similar triangles:

$$\frac{OE}{CF} = \frac{L}{xL} \quad \text{and} \quad \frac{OE}{CF} = \frac{OE}{h_1 + 2xh_2} \tag{45.233}$$

and

$$OD = OE - h_1 = \frac{(1-x)h_1 + 2xh_2}{x} \tag{45.234}$$

From the horizontal displacement of D,

$$\theta h_1 = \phi OD \tag{45.235}$$

of which

$$\phi = \frac{x}{(1-x)+2xk}\theta \tag{45.236}$$

where $k = h_2/h_1$. From the vertical displacement at C,

$$\beta = \frac{1-x}{(1-x)+2xk}\theta \tag{45.237}$$

The work equation for the assumed mechanism is

$$\lambda_1 H h_1 \beta + \frac{\lambda w L^2}{2}(1-x)\phi = M_p(\beta + 2\phi + \theta) \tag{45.238}$$

which gives

$$M_p = \frac{(1-x)\lambda_1 H h_1 + (1-x)x\lambda w L^2/2}{2(1+kx)} \tag{45.239}$$

Differentiating M_p with respect to x and solving for x, one obtains the solution

$$x = \frac{A-1}{k} \tag{45.240}$$

where

$$A = \sqrt{(1+k)(1-Uk)} \quad \text{and} \quad U = \frac{2\lambda_1 H h_1}{\lambda w L^2} \tag{45.241}$$

Substituting for x in the expression for M_p gives

$$M_p = \frac{\lambda w L^2}{8}\left[\frac{U(2+U)}{A^2 + 2A - Uk^2 + 1}\right] \tag{45.242}$$

In the absence of horizontal loading, the gable mechanism [Fig. 45.104(c)] is the failure mode. In this case, letting $H = 0$ and $U = 0$ gives

$$M_p = \frac{\lambda w L^2}{8}\left[\frac{1}{1+k+\sqrt{1+k}}\right] \tag{45.243}$$

which is the formula derived by Horne [1964].

The equations just derived can be used to produce the chart shown in Fig. 45.105 by which the value of M_p can be determined rapidly knowing the values of

$$k = \frac{h_2}{h_1} \quad \text{and} \quad U = \frac{2\lambda_1 H h_1}{\lambda w L^2} \tag{45.244}$$

Fixed-Base Gable Frames

A similar chart can be generated for a fixed-base gable frame, as shown in Fig. 45.106. Thus, if the values of loading, λw and $\lambda_1 H$, and frame geometry, h_1, h_2, and L, are known, the parameters k and U can be evaluated and the corresponding value of $M_p/(\lambda w L^2)$ can be read directly from the appropriate chart. The required value of M_p is obtained by multiplying the value of $M_p/(\lambda w L^2)$ by $\lambda w L^2$.

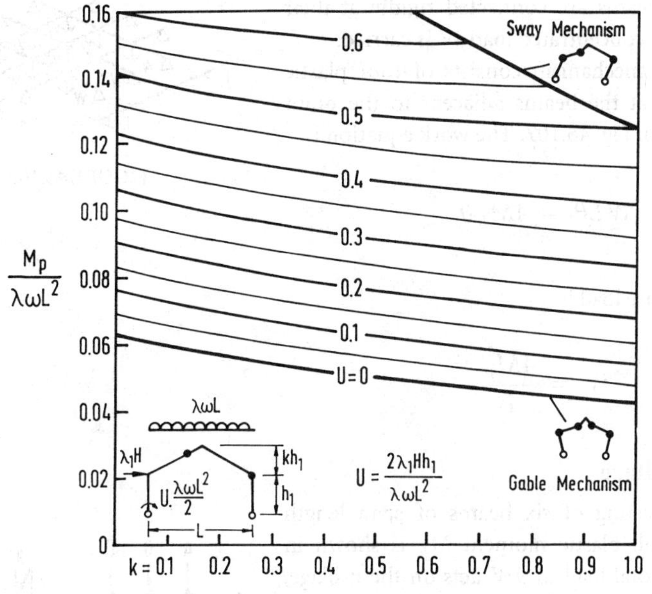

FIGURE 45.105

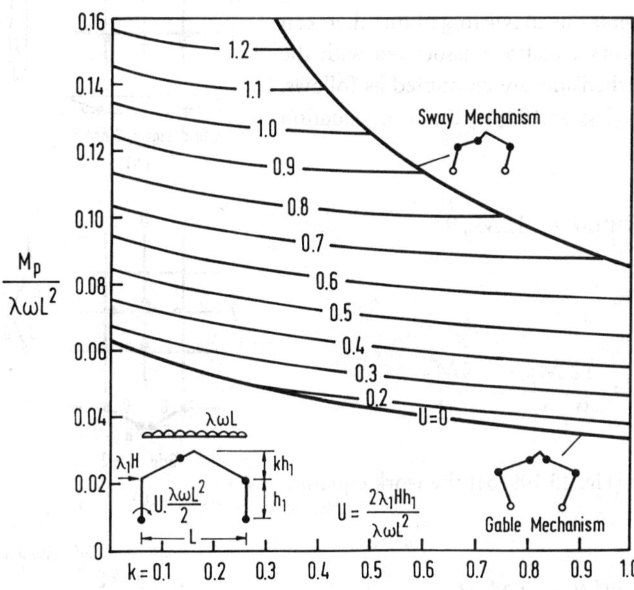

FIGURE 45.106

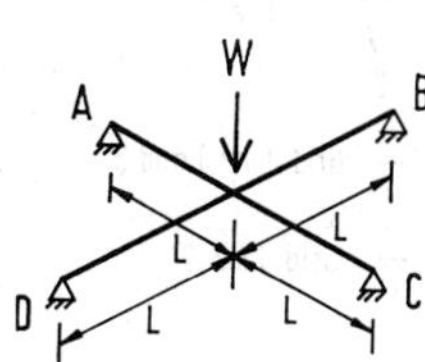

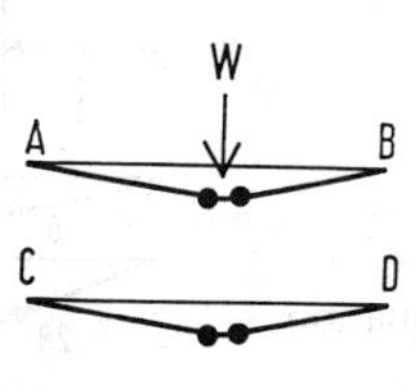

FIGURE 45.107

Grillages

A grillage is a type of structure consisting of straight beams lying on the same plane, subjected to loads acting perpendicular to the plane. An example of such a structure is shown in Fig. 45.107. The grillage consists of two equal simply supported beams of **span** length $2L$ and full plastic moment

M_p. The two beams are connected rigidly at their centers, where a concentrated load W is carried.

The collapse mechanism consists of four plastic hinges formed at the beams adjacent to the point load, as shown in Fig. 45.107. The work equation is

$$WL\theta = 4M_p\theta$$

where the collapse load is

$$W = \frac{4M_p}{L}$$

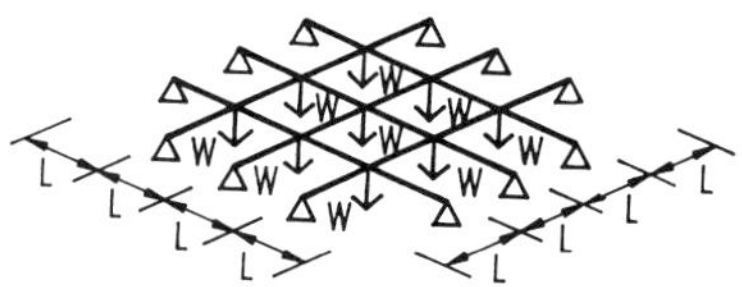

FIGURE 45.108

Six-Beam Grillage

A grillage consisting of six beams of span length $4L$ each and full plastic moment M_p is shown in Fig. 45.108. A total load of $9W$ acts on the grillage, split into concentrated loads W at the nine nodes. Three collapse mechanisms are possible. Assuming the twisting moments due to twisting of members can be ignored, the work equations associated with the three collapse mechanisms are computed as follows. For mechanism 1 [Fig. 45.109(a)] the work equation is

$$9wL\theta = 12M_p\theta$$

where

$$w = \frac{12}{9}\frac{M_p}{L} = \frac{4M_p}{3L}$$

For mechanism 2 [Fig. 45.109(b)] the work equation is

$$wL\theta = 8M_p\theta$$

where

$$w = \frac{8M_p}{L}$$

For mechanism 3 [Fig. 45.109(c)] the work equation is

$$w2L2\theta + 4 \times w2L\theta = M_p(4\theta + 8\theta)$$

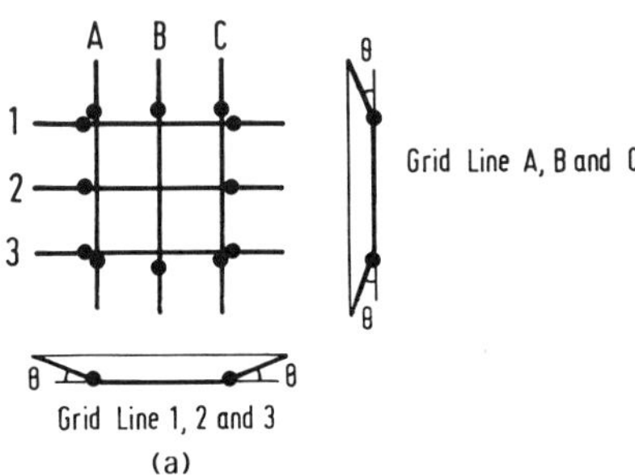

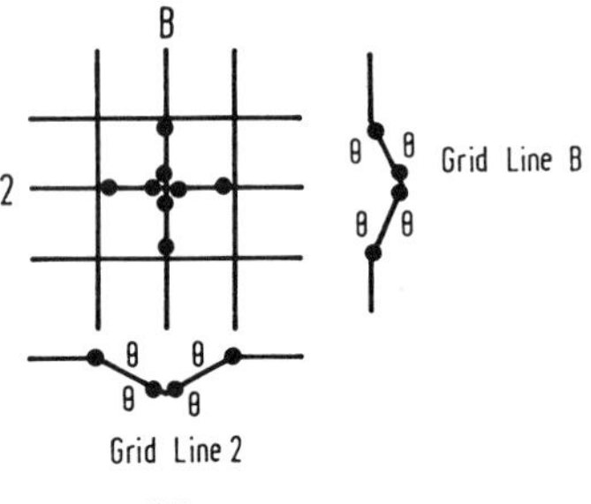

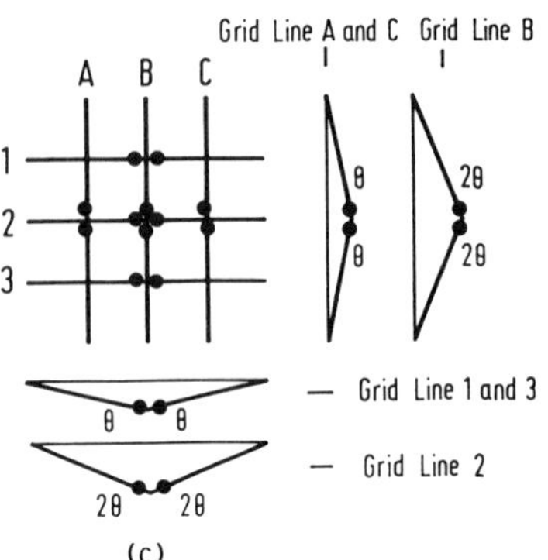

FIGURE 45.109

where

$$w = \frac{M_p}{L}$$

The lowest upper bound load corresponds to mechanism 3. This can be confirmed by conducting a moment check to ensure that no bending moments are violating the plastic moment condition. Additional discussion of plastic analysis of grillages can be found in Baker and Heyman [1969] and Heyman [1971].

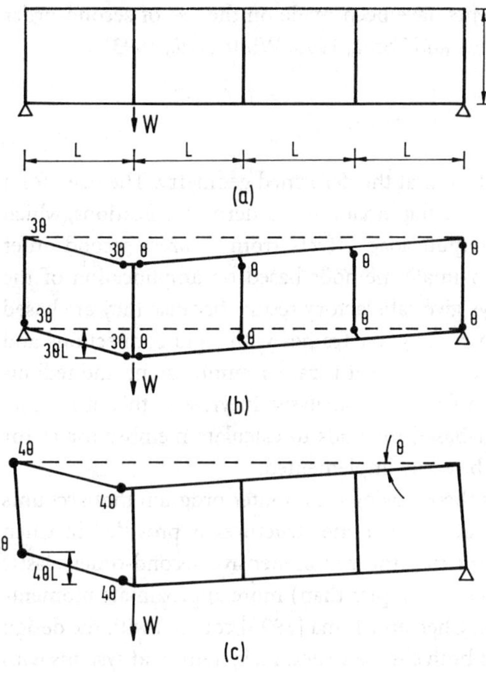

FIGURE 45.110

Vierendeel Girders

Vierendeel girders resist loading mainly by bending of members, which makes them suitable for plastic analysis. Figure 45.110 shows a simply supported girder in which all members are rigidly jointed and have the same plastic moment M_p. It is assumed that axial loads in the members do not cause instability. Two possible collapse mechanisms are considered, as shown in Fig. 45.110(b) and (c). The work equation for mechanism 1 is

$$W3\theta L = 20M_p\theta$$

so

$$W = \frac{20M_p}{3L}$$

The work equation for mechanism 2 is

$$W3\theta L = 16M_p\theta$$

or

$$W = \frac{16M_p}{3L}$$

It can be easily proved that the collapse load associated with mechanism 2 is the correct limit load. This is done by constructing an equilibrium set of bending moments and checking that they are not violating the plastic moment condition.

First-Order Hinge-By-Hinge Analysis

Instead of finding the collapse load of the frame, it may sometimes be useful to obtain information about the distribution and redistribution of forces prior to reaching the collapse load. Elastic-plastic hinge analysis (also known as hinge-by-hinge analysis) determines the order of plastic hinge formation, the load factor associated with each plastic-hinge formation, and member forces in the frame between each hinge formation. Thus the state of the frame can be defined at any load factor rather than only at the state of collapse. This allows a more accurate determination of member forces at the design load level.

Educational and commercial software packages are available for elastic-plastic hinge analysis [Chen and Sohal, 1994]. The computations of deflections for simple beams and multistory frames can be done using the virtual work method [Chen and Sohal, 1994; ASCE, 1971; Beedle, 1958; Knudsen *et al.,* 1953]. The basic assumption of first-order elastic plastic hinge analysis is that the deformations of the structure are insufficient to radically alter the equilibrium equations. This assumption ceases to be true for slender members and structures, and the method gives unsafe predictions of limit loads. Significant developments have been made on the use of second-order elastic plastic hinge analysis for frame design [Chen and Toma, 1993; White *et al.,* 1993].

Second-Order Elastic Analysis

Second-order analysis considers structural equilibrium at the deformed geometry. The need for a second-order elastic analysis of steel frames is increasing in view of modern specifications, which give explicit permission for the engineer to compute load effects from a direct second-order elastic analysis. Although there are several approximate methods based on amplification of the first-order moment, these methods do not always give satisfactory results because they are based on simplified assumptions and are applicable to rectangular frames with rigid connections and smaller displacements. Furthermore, for complex structures it is rather cumbersome and tedious to calculate the amplification factors applied to a first-order analysis. In view of this, it is more convenient and straightforward to use computer-based methods to calculate member forces for beam-column design as long as the methods can be easily implemented.

A comprehensive second-order elastic analysis theory using a computer program that accounts for both P-δ and P-Δ effects in general types of plane frame structures is provided in Chen and Toma [1993]. It is not surprising to discover that the comprehensive second-order elastic analysis can be as efficient as (and in certain respects, simpler than) more approximate moment-amplification methods. The computer program in Chen and Toma [1993] coupled with the design procedure described may be used in the design of both simple structural systems and systems with irregular geometry and leaned columns.

Second-Order Elastic-Plastic Hinge Analysis

A number of recent researches have focused on the development of second-order elastic-plastic hinge methods based on large-displacement small-strain theory and the assumption that inelasticity is concentrated at "zero-length" plastic hinges. This research has demonstrated that the elastic-plastic hinge analysis is capable of predicting the strength and load-displacement characteristics of large-scale frameworks [Orbison, 1982; Powell and Chen, 1986; Chen and Lui, 1991; Liew, 1992; Chen and Toma, 1993; White *et al.,* 1991, 1993].

This section presents a stiffness matrix approach relating the element end forces and end displacements of a beam-column element. Attention is focused on elastic formulation based on large-displacement small strain theory. The extension of the theory to include plastic-hinge formation is included. Finite element formulation based on the force-space plasticity approach may also be found in White *et al.* [1993] and Orbison [1982], among others.

Second-Order Formulation

The basis of the formulation is that the beam-column element is prismatic and initially straight. An update-Lagrangian approach [Bathe, 1982] is assumed. There are two methods to incorporate second-order effects: the stability function approach and the geometric stiffness (or finite element) approach. The stability function approach is based directly on the governing differentiation equations of the problem, whereas the stiffness approach is based on an assumed cubic polynomial variation of the transverse displacement along the element length. Therefore, the stability function approach is more exact in terms of representing the member stability behavior.

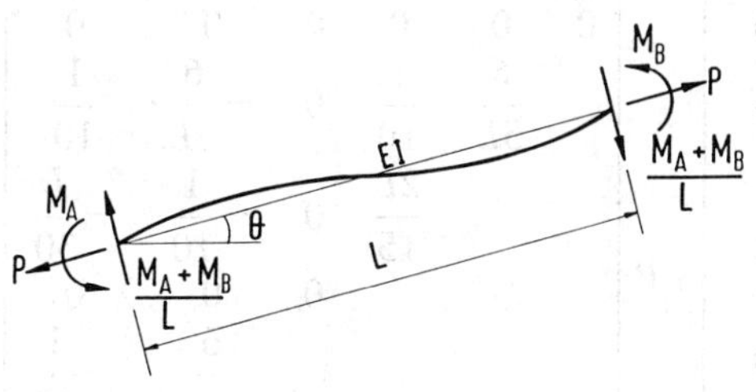

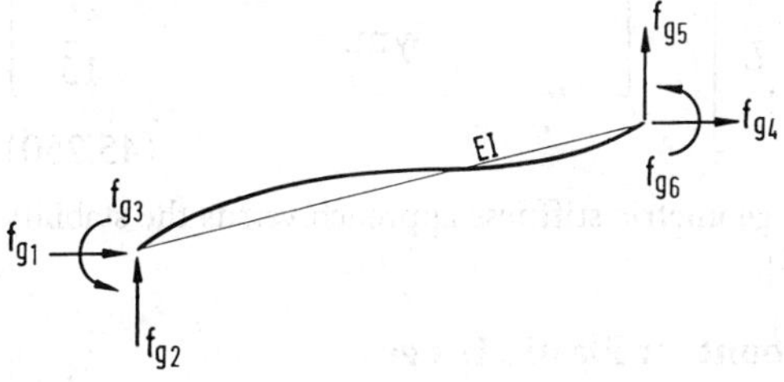

FIGURE 45.111 Frame element subjected to end forces.

For either approach, the linearized element stiffness equations may be expressed in either incremental or total force and displacement forms as

$$[K]\{d\} + \{F_f\} = \{F\} \tag{45.245}$$

where $[K]$ is the element stiffness matrix, $\{d\}$ is the element nodal displacement vector, $\{F_f\}$ is the element fixed-end force vector, and $\{F\}$ is the nodal force vector. If the stability function approach is employed, the stiffness matrix of a two-dimensional beam-column element such as the one shown in Fig. 45.111 may be written as

$$[K] = \begin{bmatrix} \frac{EA}{L} & 0 & 0 & -\frac{EA}{L} & 0 & 0 \\ & \frac{2(S_{ii}+S_{ij})}{L^2} - \frac{P}{L} & \frac{S_{ii}+S_{ij}}{L} & 0 & \frac{-2(S_{ii}+S_{ij})}{L^2} + \frac{P}{L} & \frac{S_{ii}+S_{ij}}{L} \\ & & & \frac{EA}{L} & 0 & 0 \\ & \text{Symmetric} & & & \frac{2(S_{ii}+S_{ij})}{L^2} - \frac{P}{L} & \frac{-(S_{ii}+S_{ij})}{L} \\ & & & & & S_{ii} \end{bmatrix} \tag{45.246}$$

where S_{ii} and S_{ij} are the member stiffness coefficients obtained from the elastic beam-column stability functions [Livesley and Chandler, 1956] and P is positive in compression. These coefficients may be expressed as

$$S_{ii} = \begin{cases} \frac{EI}{L} \quad \frac{\rho\sin(\rho) - \rho^2\cos(\rho)}{2 - 2\cos(\rho) - \rho\sin(\rho)} & \text{for } P < 0 \\ \frac{EI}{L} \quad \frac{\rho^2\cosh(\rho) - \rho\sinh(\rho)}{2 - 2\cosh(\rho) + \rho\sinh(\rho)} & \text{for } P > 0 \end{cases} \tag{45.247}$$

$$S_{ij} = \begin{cases} \frac{EI}{L} \quad \frac{\rho^2 - \rho\sin(\rho)}{2 - 2\cos(\rho) - \rho\sin(\rho)} & \text{for } P < 0 \\ \frac{EI}{L} \quad \frac{\rho\sin(\rho) - \rho^2}{2 - 2\cosh(\rho) + \rho\sinh(\rho)} & \text{for } P > 0 \end{cases} \tag{45.248}$$

where the parameter ρ is defined as

$$\rho = \pi\sqrt{\frac{|P|}{P_e}} = \pi\sqrt{\frac{|P|}{\pi^2 EI/L^2}} = L\sqrt{\frac{|P|}{EI}} \tag{45.249}$$

If the geometric stiffness approach is employed, the resulting element stiffness may be written as

$$[K] = \frac{EI}{L}\begin{bmatrix} \frac{A}{I} & 0 & 0 & -\frac{A}{I} & 0 & 0 \\ & \frac{12}{L^2} & \frac{6}{L} & 0 & -\frac{12}{L^2} & \frac{6}{L} \\ & & 4 & 0 & -\frac{6}{L} & 2 \\ & & & \frac{A}{I} & 0 & 0 \\ & & & & \frac{12}{L^2} & -\frac{6}{L} \\ & \text{Sym.} & & & & 4 \end{bmatrix} + P\begin{bmatrix} 0 & 0 & 0 & 0 & 0 & 0 \\ & \frac{6}{5L} & \frac{1}{10} & 0 & -\frac{6}{5L} & \frac{1}{10} \\ & & \frac{2L}{15} & 0 & -\frac{1}{10} & -\frac{L}{30} \\ & & & 0 & 0 & 0 \\ & & & & \frac{6}{5L} & -\frac{1}{10} \\ & & \text{Sym.} & & & \frac{2L}{15} \end{bmatrix} \tag{45.250}$$

Detailed discussions on the limitations of the geometric stiffness approach versus the stability function approach are given in White *et al.* [1993].

Modification of Element Equations to Account for Plastic Hinges

There are two commonly used approaches for representing plastic hinge behavior in the elastic-plastic hinge formulation. The most basic approach is to model the plastic hinge behavior as a real hinge for the purpose of calculating the element stiffness. The change in moment capacity due to the change in axial force can be accommodated directly in the numerical formulation. The change in moment is determined in the force recovery at each solution step such that, for continued plastic loading, the new force point is positioned at the strength surface at the current value of axial force. The detailed description of these procedures is given by Chen and Lui [1991], Liew [1992], and Lee and Basu [1989], among others.

Alternatively, the elastic-plastic hinge model may be formulated based on the extending and contracting plastic hinge model. The plastic hinge can rotate and extend or contract for plastic loading and axial force. The formulation can follow the force-space plasticity concept using the normality flow rule relative to the cross-section surface strength [Chen and Han, 1988]. Formal derivations of the beam-column element based on this approach have been presented by Porter and Powell [1971] and Orbison [1982].

Refined Plastic Hinge Approach

The main limitation of the conventional elastic-plastic hinge approach is that it overestimates the strength of columns that fail by inelastic flexural buckling. The key reason for this limitation is the modeling of a member by a perfect elastic element between the plastic hinge locations. Furthermore, the elastic-plastic hinge model assumes that material behavior changes abruptly from the elastic state to the fully yielded state. The element under consideration exhibits a sudden stiffness reduction upon the formation of a plastic hinge. This approach, therefore, overestimates the stiffness of a member loaded into the inelastic range [Liew and Chen, 1993; Liew *et al.*, 1994; White *et al.*, 1993]. This has led to further research and development of an alternative method called the *refined plastic hinge approach.* This approach is based on the following improvements to the elastic-plastic hinge model:

1. A column tangent-modulus model E_t is used in place of the elastic modulus E to represent the distributed plasticity along the length of a member due to axial force effects. The member inelastic stiffness, represented by the member axial and bending rigidities E_tA and E_tI, is assumed to be the function of axial load only. In other words, E_tA and E_tI can be thought of as the properties of an effective core of the section, considering column action

only. The tangent modulus concept was earlier explored by Beedle and Tall [1960] and Tall [1974] because of the effect of early yielding in the cross section due to residual stresses, which was believed to be the cause for the low strength of inelastic column buckling. The tangent modulus approach also has been utilized in previous work by Orbison *et al.* [1982], Liew [1992] and White *et al.* [1993] to improve the accuracy of the elastic-plastic hinge approach for structures in which members are subjected to large axial forces.

2. Distributed plasticity effects associated with flexure are captured by gradually degrading the member stiffness at the plastic hinge locations as yielding progresses under increasing load as the cross-section strength is approached. Several models of this type have been proposed in recent literature based on extensions to the elastic-plastic hinge approach [Powell and Chen, 1986] as well as the tangent modulus inelastic hinge approach [Liew *et al.*, 1993; White *et al.*, 1993]. The rationale of modeling stiffness degradation associated with both axial and flexural actions is that the tangent modulus model represents the column strength behavior in the limit of pure axial compression, and the plastic hinge stiffness degradation model represents the beam behavior in pure bending. Thus the combined effects of these two approaches should also satisfy the cases in which the member is subjected to combined axial compression and bending.

It has been shown that with the above two improvements, the refined plastic hinge model can be used with sufficient accuracy to provide a quantitative assessment of a member's performance up to failure. Detailed descriptions of the method and discussion of results generated by the method are given in Liew et al. [1993], White *et al.* [1993], and Chen and Toma [1993].

Plastic Zone Analysis

Plastic zone analyses can be classified into two main types: 3-D shell element and 2-D beam-column approaches. In the 3-D plastic zone analysis, the structure is modeled using a large number of finite 3-D shell elements, and the elastic constitutive matrix, in the usual incremental stress-strain relations, is replaced by an elastic-plastic constitutive matrix once yielding is detected. This analysis approach typically requires numerical integration for the evaluation of the stiffness matrix. Based on a deformation theory of plasticity, the combined effects of normal and shear stresses may be accounted for. The 3-D spread-of-plasticity analysis is computationally intensive and best suited for analyzing small-scale structures.

The second approach for plastic zone analysis is based on the use of beam-column theory, in which the member is discretized into many beam-column segments, and the cross section of each segment is further subdivided into a number of fibers. Inelasticity is typically modeled by the consideration of normal stress only. When the computed stresses at the centroid of any fibers reach the uniaxial normal strength of the material, the fiber is considered as yielded. Compatibility is treated by assuming that full continuity is retained throughout the volume of the structure in the same manner as for elastic range calculations. Most of the plastic zone analysis methods developed are meant for planar (2-D) analysis [Chen and Toma, 1993; White, 1985; Vogel, 1985]. Three-dimensional plastic zone techniques are also available involving various degrees of refinements [White, 1988; Wang and Nethercot, 1988].

A plastic zone analysis, which includes the spread of plasticity, residual stresses, initial geometric imperfections, and any other significant second-order behavioral effects, is often considered to be an exact analysis method. Therefore, when this type of analysis is employed, the checking of member interaction equations is not required. However, in reality, some significant behavioral effects (such as performance of joints and connections) tend to defy precise numerical and analytical modeling. In such cases, a simpler method of analysis that adequately captures the inelastic behavior would be sufficient for engineering applications. Second-order plastic hinge analysis is still the preferred method for advanced analysis of large-scale steel frames.

45.17 Structural Dynamic

Equation of Motion

The essential physical properties of a linearly elastic structural system subjected to external dynamic loading are its mass, stiffness, and energy absorption or damping. The principle of dynamic analysis may be illustrated by considering a simple single-story structure, as shown in Fig. 45.112. The structure is subjected to a time-varying force $f(t)$. k is the spring constant that relates the lateral story deflection x to the story shear force, and the dashpot relates the damping force to the velocity by a damping coefficient c. If the mass, m, is assumed to concentrate at the beam, the structure becomes a single-degree-of-freedom (SDOF) system. The equation of motion of the system may be written as follows:

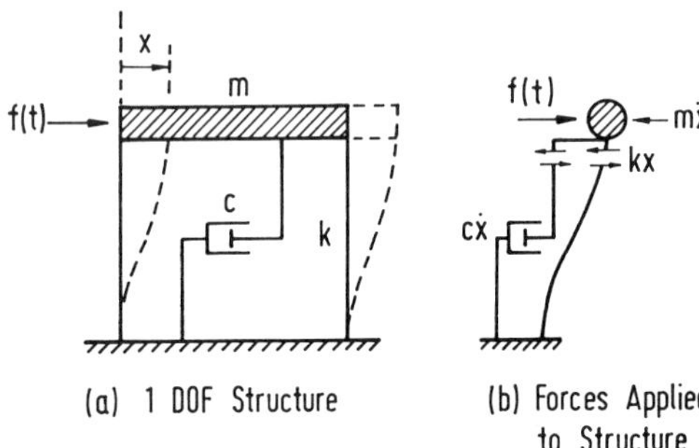

FIGURE 45.112

$$m\ddot{x} + c\dot{x} + kx = f(t) \tag{45.251}$$

Various solutions to Eq. (45.251) can give an insight into the behavior of the structure under dynamic situations.

Free Vibration

In this case the system is set into motion and allowed to vibrate in the absence of applied force $f(t)$. Letting $f(t) = 0$, Eq. (45.251) becomes

$$m\ddot{x} + c\dot{x} + kx = 0 \tag{45.252}$$

Dividing Eq. (45.252) by the mass, m, we have

$$\ddot{x} + 2\xi\omega\dot{x} + \omega^2 x = 0 \tag{45.253}$$

where

$$2\xi\omega = \frac{c}{m} \quad \text{and} \quad \omega^2 = \frac{k}{m} \tag{45.254}$$

The solution to Eq. (45.253) depends on whether the vibration is damped or undamped.

Case 1: Undamped Free Vibration

In this case, $c = 0$, and the solution to the equation of motion may be written as follows:

$$x = A \sin \omega t + B \cos \omega t \tag{45.255}$$

where $\omega = \sqrt{k/m}$ is the circular frequency. A and B are constants that can be determined by the initial boundary conditions. In the absence of external forces and damping the system will vibrate indefinitely in a repeated cycle of vibration with an amplitude of

$$X = \sqrt{A^2 + B^2} \tag{45.256}$$

and a natural frequency of

$$f = \frac{\omega}{2\pi} \tag{45.257}$$

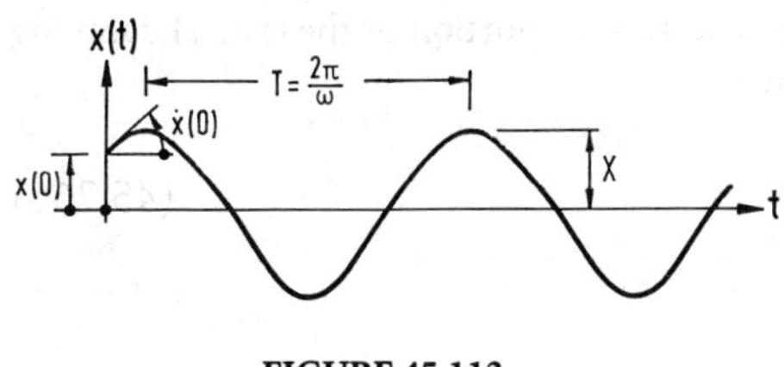

FIGURE 45.113

The natural period is

$$T = \frac{2\pi}{\omega} = \frac{1}{f} \tag{45.258}$$

The undamped free vibration motion as described by Eq. (45.255) is shown in Fig. 45.113.

Case 2: Damped Free Vibration

If the system is not subjected to applied force and damping is present, the corresponding solution becomes

$$x = A\exp(\lambda_1 t) + B\exp(\lambda_2 t) \tag{45.259}$$

where

$$\lambda_1 = \omega\left[-\xi + \sqrt{\xi^2 - 1}\right] \tag{45.260}$$

$$\lambda_2 = \omega\left[-\xi - \sqrt{\xi^2 - 1}\right] \tag{45.261}$$

The solution of Eq. (45.259) changes its form with the value of ξ defined as

$$\xi = \frac{c}{2\sqrt{mk}} \tag{45.262}$$

If $\xi^2 < 1$, the equation of motion becomes

$$x = \exp(-\xi\omega t)(A\cos\omega_d t + B\sin\omega_d t) \tag{45.263}$$

where ω_d is the damped angular frequency, defined as

$$\omega_d = \sqrt{(1 - \xi^2)\omega} \tag{45.264}$$

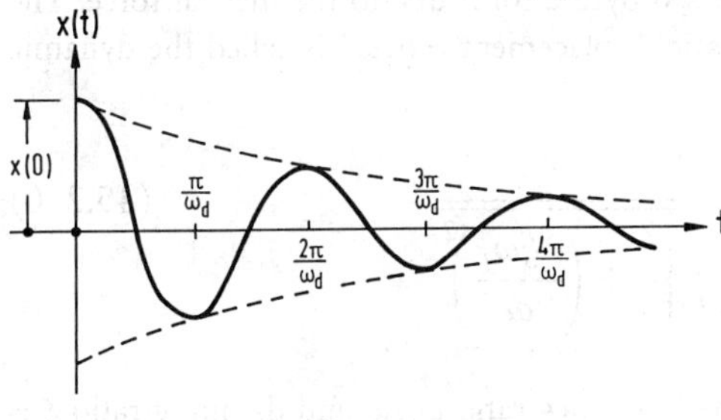

FIGURE 45.114

For most structures ξ is very small (about 0.01) and therefore $\omega_d \approx \omega$. The system oscillates about the neutral position as the amplitude decays with time t. Figure 45.114 illustrates an example of such motion. The rate of decay is governed by the amount of damping present.

If the damping is large, oscillation will be prevented. This happens when $\xi^2 > 1$ and the behavior is generally referred to as overdamped. The motion of such behavior is shown in Fig. 45.115.

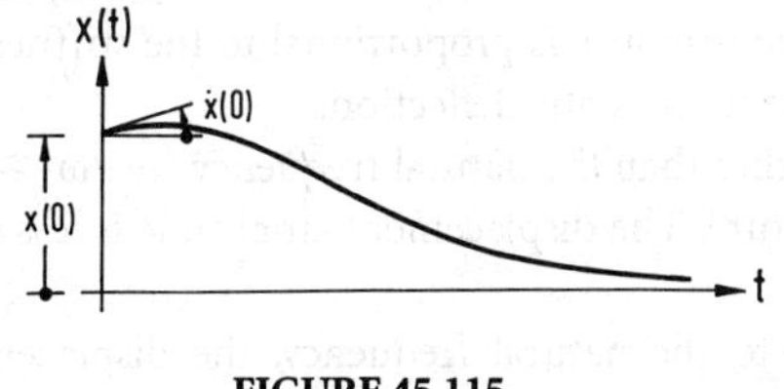

FIGURE 45.115

Damping with $\xi^2 = 1$ is called critical damping. This is the case where minimum damping is required to prevent oscillation and the critical damping coefficient is given as

$$c_{cr} = 2\sqrt{km} \tag{45.265}$$

The degree of damping in the structure is often expressed as a proportion of the critical damping value. Referring to Eqs. (45.262) and (45.265), we have

$$\xi = \frac{c}{c_{cr}} \tag{45.266}$$

where ξ is the damping ratio.

Forced Vibration

If a structure is subjected to a sinusoidal motion such as a ground acceleration of $\ddot{x} = F \sin \omega_f t$, it will oscillate, and after some time the motion of the structure will reach a steady state. For example, the equation of motion due to the ground acceleration [from Eq. (45.253)] is

$$\ddot{x} + 2\xi\omega\dot{x} + \omega^2 x = -F \sin \omega_f t \tag{45.267}$$

The solution to the above equation consists of two parts: the complimentary solution given by Eq. (45.255) and the particular solution. If the system is damped, oscillation corresponding to the complementary solution will decay with time. After some time the motion will reach a steady state, and the system will vibrate at a constant amplitude and frequency. This motion, which is called forced vibration, is described by the particular solution expressed as

$$x = C_1 \sin \omega_f t + C_2 \cos \omega_f t \tag{45.268}$$

It can be observed that the steady forced vibration occurs at the frequency of the excited force, ω_f, not the natural frequency of the structure, ω.

Substituting Eq. (45.268) into (45.267), the displacement amplitude can be shown to be:

$$X = -\frac{F}{\omega^2} \frac{1}{\sqrt{\left\{1 - \left(\frac{\omega_f}{\omega}\right)^2\right\}^2 + \left(\frac{2\xi\omega_f}{\omega}\right)^2}} \tag{45.269}$$

The term $-F/\omega^2$ is the static displacement caused by the force due to the inertial force. The ratio of the response amplitude relative to the static displacement $-F/\omega^2$ is called the dynamic displacement amplification factor, D, given as

$$D = \frac{1}{\sqrt{\left\{1 - \left(\frac{\omega_t}{\omega}\right)^2\right\}^2 + \left(\frac{2\xi\omega_f}{\omega}\right)^2}} \tag{45.270}$$

The variation of the magnification factor with the frequency ratio ω_f/ω and damping ratio ξ is shown in Fig. 45.116.

When the dynamic force is applied at a frequency much lower than the natural frequency of the system ($\omega_f/\omega \ll 1$), the response is quasi-static. The response is proportional to the stiffness of the structure, and the displacement amplitude is close to the static deflection.

When the force is applied at a frequency much higher than the natural frequency ($\omega_f/\omega \gg 1$), the response is proportional to the mass of the structure. The displacement amplitude is less than the static deflection ($D < 1$).

When the force is applied at a frequency close to the natural frequency, the displacement amplitude increases significantly. The condition at which $\omega_f/\omega = 1$ is called resonance.

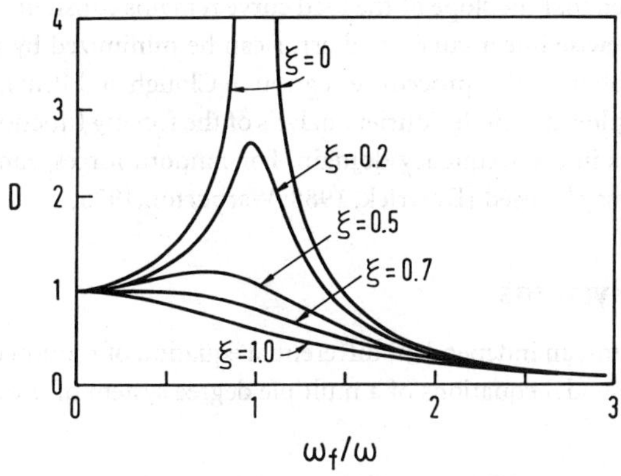

FIGURE 45.116

Similarly, the ratio of the acceleration response relative to the ground acceleration may be expressed as

$$D_a = \left|\frac{\ddot{x} + \ddot{x}_g}{\ddot{x}_g}\right| = \sqrt{\frac{1 + \left(\frac{2\xi\omega_f}{\omega}\right)^2}{\left\{1 - \left(\frac{\omega_f}{\omega}\right)^2\right\}^2 + \left(\frac{2\xi\omega_f}{\omega}\right)^2}} \tag{45.271}$$

where D_a is the dynamic acceleration magnification factor.

Response to Suddenly Applied Load

Consider a spring-mass damper system onto which a load P_0 is applied suddenly. The differential equation is given by

$$M\ddot{x} + c\dot{x} + kx = P_0 \tag{45.272}$$

If the system is started at rest, the equation of motion is

$$x = \frac{P_0}{k}\left[1 - \exp(-\xi\omega t)\left\{\cos\omega_d t + \frac{\xi\omega}{\omega_d}\sin\omega_d t\right\}\right] \tag{45.273}$$

If the system is undamped, then $\xi = 0$ and $\omega_d = \omega$, and we have

$$x = \frac{P_0}{k}\left[1 - \cos\omega_d t\right] \tag{45.274}$$

The maximum displacement is $2(P_0/k)$ corresponding to $\cos\omega_d t = -1$. Since P_0/k is the maximum static displacement, the dynamic amplification factor is equal to 2. The presence of damping would naturally reduce the dynamic amplification factor and the force in the system.

Response to Time-Varying Loads

Some forces and ground motions that are encountered in practice are rather complex in nature. In general, numerical analysis is required to predict the response of such effects, and the finite element method is one of the most common techniques employed in solving such problems.

The evaluation of responses due to time-varying loads can be carried out using the piecewise exact method. In using this method, the loading history is divided into small time intervals. Between these points, it is assumed that the slope of the load curve remains constant. The entire load history is represented by a piecewise linear curve, and error can be minimized by reducing the length of the time steps. A description of this procedure is given in Clough and Penzien [1993].

Other techniques employed include Fourier analysis of the forcing function followed by solution for Fourier components in the frequency domain. For random forces, random vibration theory and spectrum analysis may be used [Dowrick, 1988; Warburton,1976].

Multiple Degree Systems

In multiple degree systems, an independent differential equation of motion can be written for each degree of freedom. The nodal equations of a multiple degree system of n degrees of freedom may be written as

$$[m]\{\ddot{x}\} + [c]\{\dot{x}\} + [k]\{x\} = \{F(t)\} \tag{45.275}$$

where $[m]$ is a symmetrical $n \times n$ matrix of mass, $[c]$ is a symmetrical $n \times n$ matrix of damping coefficient, and $\{F(t)\}$ is the force vector, which is zero in the case of free vibration.

Consider a system under free vibration without damping. The general solution of Eq. (45.275) is assumed in the form of

$$\begin{Bmatrix} x_1 \\ x_2 \\ \vdots \\ x_n \end{Bmatrix} = \begin{bmatrix} \cos(\omega t - \phi) & 0 & 0 & 0 \\ 0 & \cos(\omega t - \phi) & 0 & 0 \\ \vdots & \vdots & \vdots & \vdots \\ 0 & 0 & 0 & \cos(\omega t - \phi) \end{bmatrix} \begin{Bmatrix} C_1 \\ C_2 \\ \vdots \\ C_n \end{Bmatrix} \tag{45.276}$$

where angular frequency ω and phase angle ϕ are common to every x. In this assumed solution, ϕ and $C_1, C_2, \ldots,$ and C_n are the constants to be determined from the initial boundary conditions of the motion and ω is a characteristic value (eigenvalue) of the system.

Substituting Eq. (45.276) into Eq. (45.275) yields

$$\begin{bmatrix} k_{11} - m_{11}\omega^2 & k_{12} - m_{12}\omega^2 & \cdots & k_{1n} - m_{1n}\omega^2 \\ k_{21} - m_{21}\omega^2 & k_{22} - m_{22}\omega^2 & \cdots & k_{2n} - m_{2n}\omega^2 \\ \vdots & \vdots & \vdots & \vdots \\ k_{n1} - m_{n1}\omega^2 & k_{n2} - m_{n2}\omega^2 & \cdots & k_{nn} - m_{nn}\omega^2 \end{bmatrix} \begin{Bmatrix} C_1 \\ C_2 \\ \vdots \\ C_n \end{Bmatrix} \cos(\omega t - \phi) = \begin{Bmatrix} 0 \\ 0 \\ \vdots \\ 0 \end{Bmatrix} \tag{45.277}$$

or

$$\left[[k] - \omega^2[m]\right]\{C\} = \{0\} \tag{45.278}$$

where $[k]$ and $[m]$ are the $n \times n$ matrices, ω^2 and $\cos(\omega t - \phi)$ are scalars, and $\{C\}$ is the amplitude vector. For the nontrivial solution, $\cos(\omega t - \phi) \neq 0$; thus the solution to Eq. (45.278) requires the determinant of $\left|[k] - \omega^2[m]\right| = 0$. The expansion of the determinant yields a polynomial of n degree as a function of ω^2, the n roots of which are the eigenvalues $\omega_1, \omega_2, \omega_n$.

If the eigenvalue ω for a normal mode is substituted in Eq. (45.278), the amplitude vector $\{C\}$ for that mode can be obtained. $\{C_1\}, \{C_2\}, \{C_3\}, \{C_n\}$ are therefore called the eigenvectors, the absolute values of which must be determined through initial boundary conditions. The resulting motion is a sum of n harmonic motions, each governed by the respective natural frequency ω_i, written as

$$\{x\} = \sum_{i=1}^{n} \{C_i\} \cos(\omega_i t - \phi_i) \tag{45.279}$$

Distributed Mass Systems

Although many structures may be approximated by lumped mass systems, in practice all structures are distributed mass systems consisting of an infinite number of particles. Consequently, if the motion is repetitive, the structure has an infinite number of natural frequency and mode shapes. The analysis of a distributed-parameter system is entirely equivalent to that of a discrete system once the mode shapes and frequencies have been determined, because in both cases the amplitudes of the modal response components are used as generalized coordinates in defining the response of the structure.

In principle an infinite number of these coordinates are available for a distributed-parameter system, but in practice only a few modes, usually those of lower frequencies, will provide a significant contribution to the overall response. Thus the distributed-parameter system can be converted to a discrete system form in which only a limited number of modal coordinates is used to describe the response.

Flexural Vibration of Beams

The motion of the distributed mass system is best illustrated by a classical example of a uniform beam of span length L and flexural rigidity EI and a self-weight of m per unit length, as shown in Fig. 45.117(a). The beam is free to vibrate under its self-weight. From Fig. 45.117(b), dynamic equilibrium of a small beam segment of length dx requires

$$\frac{\partial V}{\partial x} dx = m\, dx \frac{\partial^2 y}{\partial t^2} \tag{45.280}$$

where

$$\frac{\partial^2 y}{\partial x^2} = \frac{M}{EI} \tag{45.281}$$

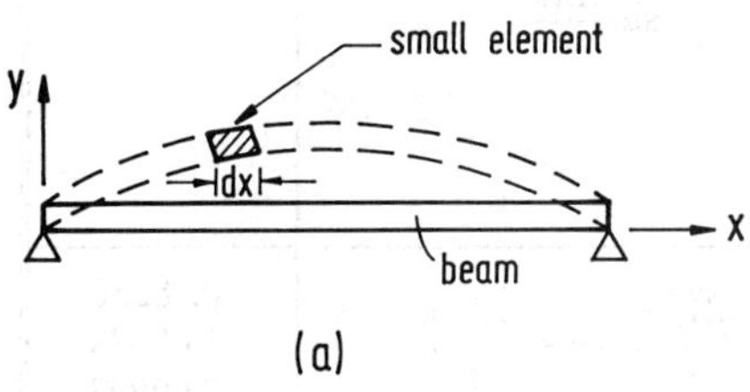

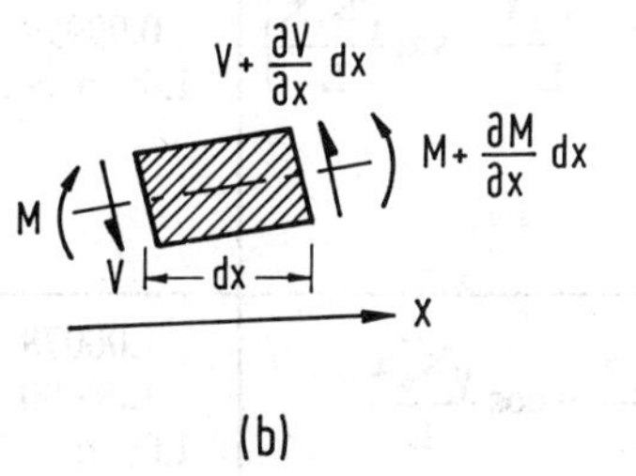

FIGURE 45.117

and

$$V = -\frac{\partial M}{\partial x} \qquad \frac{\partial V}{\partial x} = -\frac{\partial^2 M}{\partial x^2} \tag{45.282}$$

Substituting these equations into Eq. (45.280) leads to the equation of motion of the flexural beam:

$$\frac{\partial^4 y}{\partial x^4} + \frac{m}{EI}\frac{\partial^2 y}{\partial t^2} = 0 \tag{45.283}$$

The equation just derived can be solved for beams with given sets of boundary conditions. The solution consists of a family of vibration mode with corresponding natural frequencies. Standard results are available in Table 45.6 to compute the natural frequencies of uniform flexural beams of different supporting conditions. Methods are also available for dynamic analysis of continuous beams [Clough and Penzien, 1993].

Table 45.6 Frequencies and Mode Shapes of Beams in Flexural Vibration

$$f_n = \frac{k_n}{2\pi}\sqrt{\frac{EI}{mL^4}} \text{ HZ}$$

n = 1, 2, 3...

L = Length (m)

EI = Flexural Rigidity (Nm^2)

M = Mass per unit length (kg/m)

Boundary Conditions	K_n; n = 1,2,3	Mode Shape $y_n\left(\frac{x}{L}\right)$	A_n; n = 1,2,3...
Pinned - Pinned 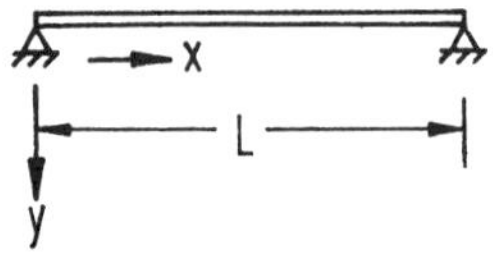	$(n\pi)^2$	$\sin\frac{n\pi x}{L}$	-
Fixed - Fixed 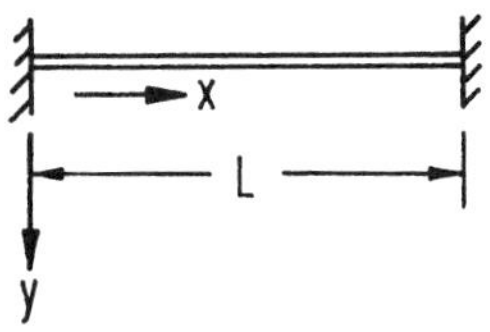	22.37 61.67 120.90 199.86 298.55 $(2n+1)\frac{\pi^2}{4}$; n > 5	$\cosh\frac{\sqrt{K_n}x}{L} - \cos\frac{\sqrt{K_n}x}{L} - A_n\left(\sinh\frac{\sqrt{K_n}x}{L} - \sin\frac{\sqrt{K_n}x}{L}\right)$	0.98250 1.00078 0.99997 1.00000 0.99999 1.0; n > 5
Fixed - Pinned 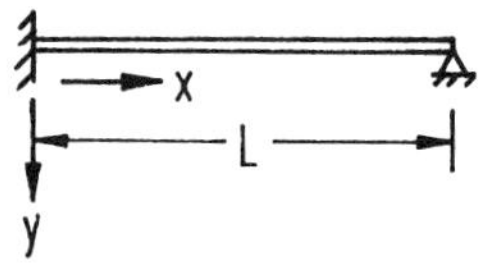	15.42 49.96 104.25 178.27 272.03 $(4n+1)^2\frac{\pi^2}{4}$; n > 5	$\cosh\frac{\sqrt{K_n}x}{L} - \cos\frac{\sqrt{K_n}x}{L} - A_n\left(\sinh\frac{\sqrt{K_n}x}{L} - \sin\frac{\sqrt{K_n}x}{L}\right)$	1.00078 1.00000 1.0; n > 3
Cantilever 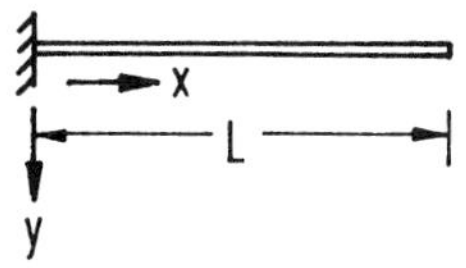	3.52 22.03 61.69 120.90 199.86 $(2n-1)^2\frac{\pi^2}{4}$; n > 5	$\cosh\frac{\sqrt{K_n}x}{L} - \cos\frac{\sqrt{K_n}x}{L} - A_n\left(\sinh\frac{\sqrt{K_n}x}{L} - \sin\frac{\sqrt{K_n}x}{L}\right)$	0.73410 1.01847 0.99922 1.00003 1.0; n > 4

Shear Vibration of Beams

Beams can deform by flexure or shear. Flexural deformation normally dominates the deformation of slender beams. Shear deformation is important for short beams and in higher modes of slender beams. Table 45.7 gives the natural frequencies of uniform beams in shear, neglecting flexural deformation. The natural frequencies of these beams are inversely proportional to the beam length L rather than L^2, and the frequencies increase linearly with the mode number.

Table 45.7 Frequencies and Mode Shapes of Beams in Shear Vibration

$$f_n = \frac{K_n}{2\pi}\sqrt{\frac{KG}{\rho L^2}} \quad \text{HZ}$$

L = Length
K = Shear Coefficient (Cowper, 1966)
G = Shear Modulus = E/[2(1+υ)]
ρ = Mass Density

Boundary Condition	K_n; n = 1,2,3...	Mode Shape $y_n\left(\frac{x}{L}\right)$
Fixed - Free 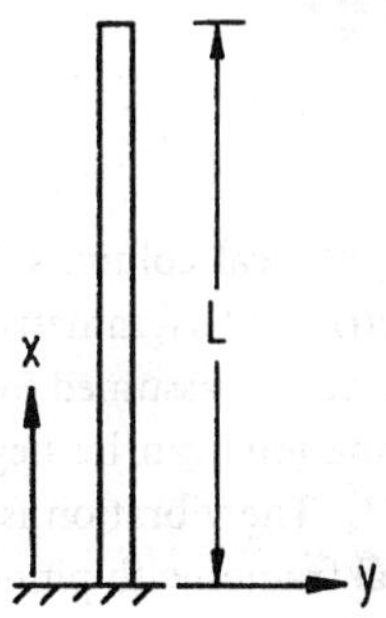	$n\pi$; n = 1,2,3...	$\cos\frac{n\pi x}{L}$; n = 1,2,3...
Fixed - Fixed 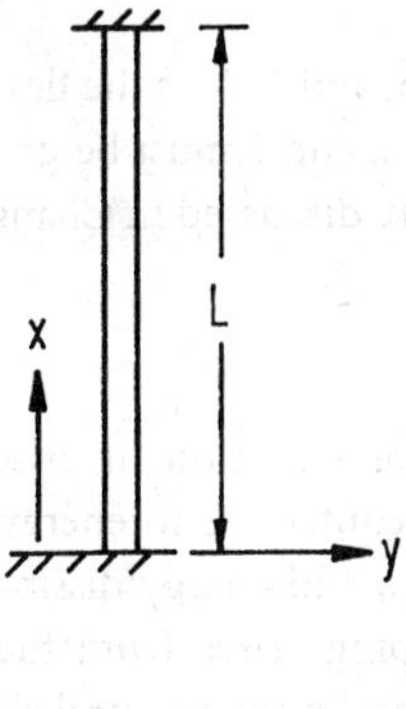	$n\pi$; n = 1,2,3...	$\sin\frac{n\pi x}{L}$; n = 1,2,3...

Combined Shear and Flexure

The transverse deformation of real beams is the sum of flexure and shear deformations. In general, numerical solutions are required to incorporate both the shear and flexural deformation in the prediction of the natural frequency of beams. For beams with comparable shear and flexural deformations, the following simplified formula may be used to estimate the beam's frequency:

$$\frac{1}{f^2} = \frac{1}{f_f^2} + \frac{1}{f_s^2} \tag{45.284}$$

where f is the fundamental frequency of the beam, and f_f and f_s are the fundamental frequencies predicted by the flexure and shear beam theory [Rutenberg, 1975].

Natural Frequency of Multistory Building Frames

Tall building frames often deform more in the shear mode than in flexure. The fundamental frequencies of many multistory building frameworks can be approximated [Housner, 1963; Rinne, 1952] by

$$f = \alpha \frac{\sqrt{B}}{H} \tag{45.285}$$

where α is approximately equal to 11 $\sqrt{\text{m/s}}$, B is the building width in the direction of vibration, and H is the building height. This empirical formula suggests that a shear beam model with f inversely proportional to H is more appropriate than a flexural beam model for predicting natural frequencies of buildings.

Portal Frames

A portal frame consists of a cap beam rigidly connected to two vertical columns. The natural frequencies of portal frames vibrating in the fundamental symmetric and asymmetric modes are shown in Tables 45.8 and 45.9, respectively. The beams in these frames are assumed to be uniform and sufficiently slender so that shear, axial, and torsional deformations can be neglected. The method of analysis of these frames is given in Yang and Sun [1973]. The vibration is assumed to be in the plane of the frame, and the results are presented for portal frames with pinned and fixed bases.

If the beam is rigid and the columns are slender and uniform, but not necessarily identical, then the natural fundamental frequency of the frame can be approximated using the following formula [Robert, 1979]:

$$f = \frac{1}{2\pi} \left[\frac{12 \sum E_i I_i}{L^3 (M + 0.37 \sum M_i)} \right]^{1/2} \text{Hz} \tag{45.286}$$

where M is the mass of the beam, M_i is the mass of the i column, and $E_i I_i$ is the flexural rigidity of the i column. The summation refers to the sum of all columns, and i must be greater than or equal to 2. Additional results for frames with inclined members are discussed in Chang [1978].

Damping

Damping is found to increase with the increasing amplitude of vibration. It arises from the dissipation of energy during vibration. The mechanisms which contribute to energy dissipation are material damping, friction at interfaces between components, and energy dissipation due to foundation interaction with soil, among others. Material damping arises from the friction at bolted connections and frictional interaction between structural and nonstructural elements such as partitions and cladding.

Table 45.8 Fundamental Frequencies of Portal Frames in Asymmetrical Mode of Vibration

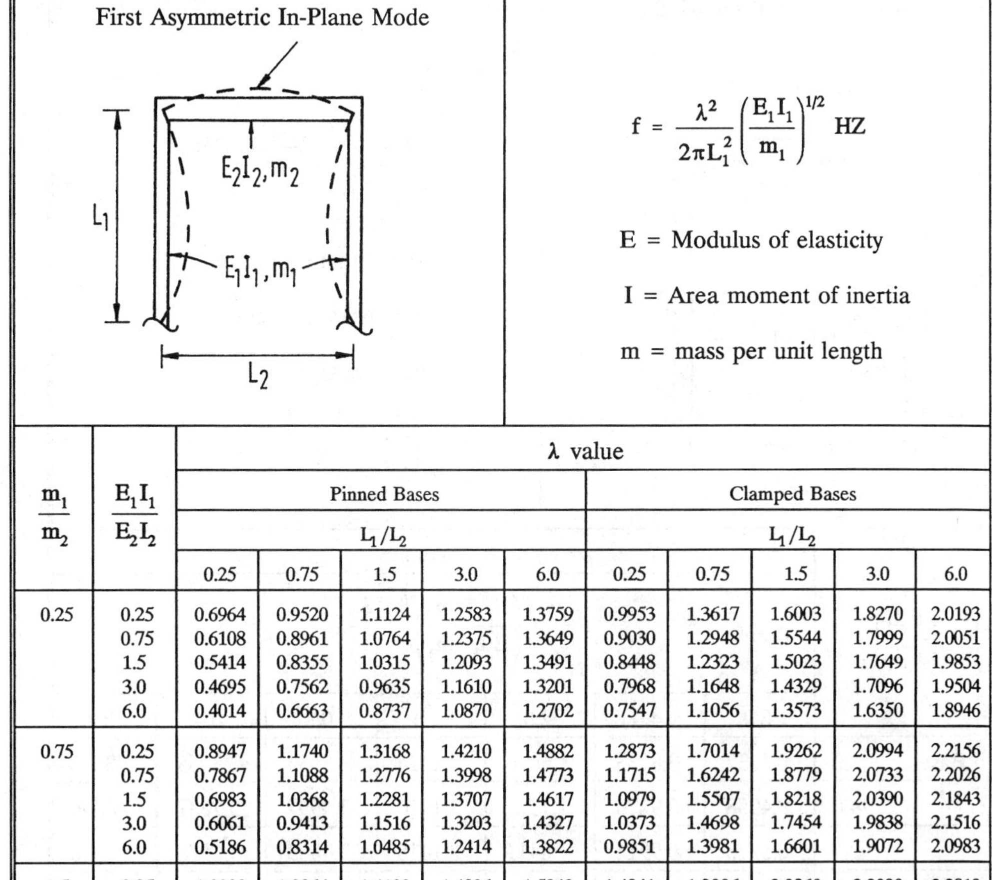

$\frac{m_1}{m_2}$	$\frac{E_1I_1}{E_2I_2}$	λ value									
		Pinned Bases					Clamped Bases				
		L_1/L_2					L_1/L_2				
		0.25	0.75	1.5	3.0	6.0	0.25	0.75	1.5	3.0	6.0
0.25	0.25	0.6964	0.9520	1.1124	1.2583	1.3759	0.9953	1.3617	1.6003	1.8270	2.0193
	0.75	0.6108	0.8961	1.0764	1.2375	1.3649	0.9030	1.2948	1.5544	1.7999	2.0051
	1.5	0.5414	0.8355	1.0315	1.2093	1.3491	0.8448	1.2323	1.5023	1.7649	1.9853
	3.0	0.4695	0.7562	0.9635	1.1610	1.3201	0.7968	1.1648	1.4329	1.7096	1.9504
	6.0	0.4014	0.6663	0.8737	1.0870	1.2702	0.7547	1.1056	1.3573	1.6350	1.8946
0.75	0.25	0.8947	1.1740	1.3168	1.4210	1.4882	1.2873	1.7014	1.9262	2.0994	2.2156
	0.75	0.7867	1.1088	1.2776	1.3998	1.4773	1.1715	1.6242	1.8779	2.0733	2.2026
	1.5	0.6983	1.0368	1.2281	1.3707	1.4617	1.0979	1.5507	1.8218	2.0390	2.1843
	3.0	0.6061	0.9413	1.1516	1.3203	1.4327	1.0373	1.4698	1.7454	1.9838	2.1516
	6.0	0.5186	0.8314	1.0485	1.2414	1.3822	0.9851	1.3981	1.6601	1.9072	2.0983
1.5	0.25	1.0300	1.2964	1.4103	1.4826	1.5243	1.4941	1.9006	2.0860	2.2090	2.2819
	0.75	0.9085	1.2280	1.3707	1.4616	1.5136	1.3652	1.8214	2.0390	2.1842	2.2695
	1.5	0.8079	1.1514	1.3203	1.4326	1.4982	1.2823	1.7444	1.9837	2.1515	2.2521
	3.0	0.7021	1.0482	1.2414	1.3821	1.4694	1.2141	1.6583	1.9070	2.0983	2.2206
	6.0	0.6011	0.9279	1.1335	1.3024	1.4191	1.1570	1.5808	1.8198	2.0234	2.1693
3.0	0.25	1.1597	1.3898	1.4719	1.5189	1.5442	1.7022	2.0612	2.1963	2.2756	2.3190
	0.75	1.0275	1.3202	1.4326	1.4981	1.5336	1.5649	1.9834	2.1515	2.2520	2.3070
	1.5	0.9161	1.2412	1.3821	1.4694	1.5182	1.4752	1.9063	2.0982	2.2206	2.2899
	3.0	0.7977	1.1333	1.3024	1.4191	1.4896	1.4015	1.8185	2.0233	2.1693	2.2595
	6.0	0.6838	1.0058	1.1921	1.3391	1.4395	1.3425	1.7382	1.9366	2.0964	2.2094
6.0	0.25	1.2691	1.4516	1.5083	1.5388	1.5545	1.8889	2.1727	2.2635	2.3228	2.3385
	0.75	1.1304	1.3821	1.4694	1.5181	1.5440	1.7501	2.0980	2.2206	2.2899	2.3268
	1.5	1.0112	1.3023	1.4191	1.4896	1.5287	1.6576	2.0228	2.1693	2.2595	2.3101
	3.0	0.8827	1.1919	1.3391	1.4395	1.5002	1.5817	1.9358	2.0963	2.2095	2.2802
	6.0	0.7578	1.0601	1.2277	1.3595	1.4502	1.5244	1.8550	2.0110	2.1380	2.2309

The amount of damping in a building can never be predicted precisely, and design values are generally derived based on dynamic measurements of structures of a corresponding type. Damping can be measured based on the rate of decay of free vibration following an impact, by spectral methods based on analysis of response to wind loading, or by force excitation by mechanical vibrator at varying frequencies to establish the shape of the steady state resonance curve. However,

Table 45.9 Fundamental Frequencies of Portal Frames in Symmetrical Mode of Vibration

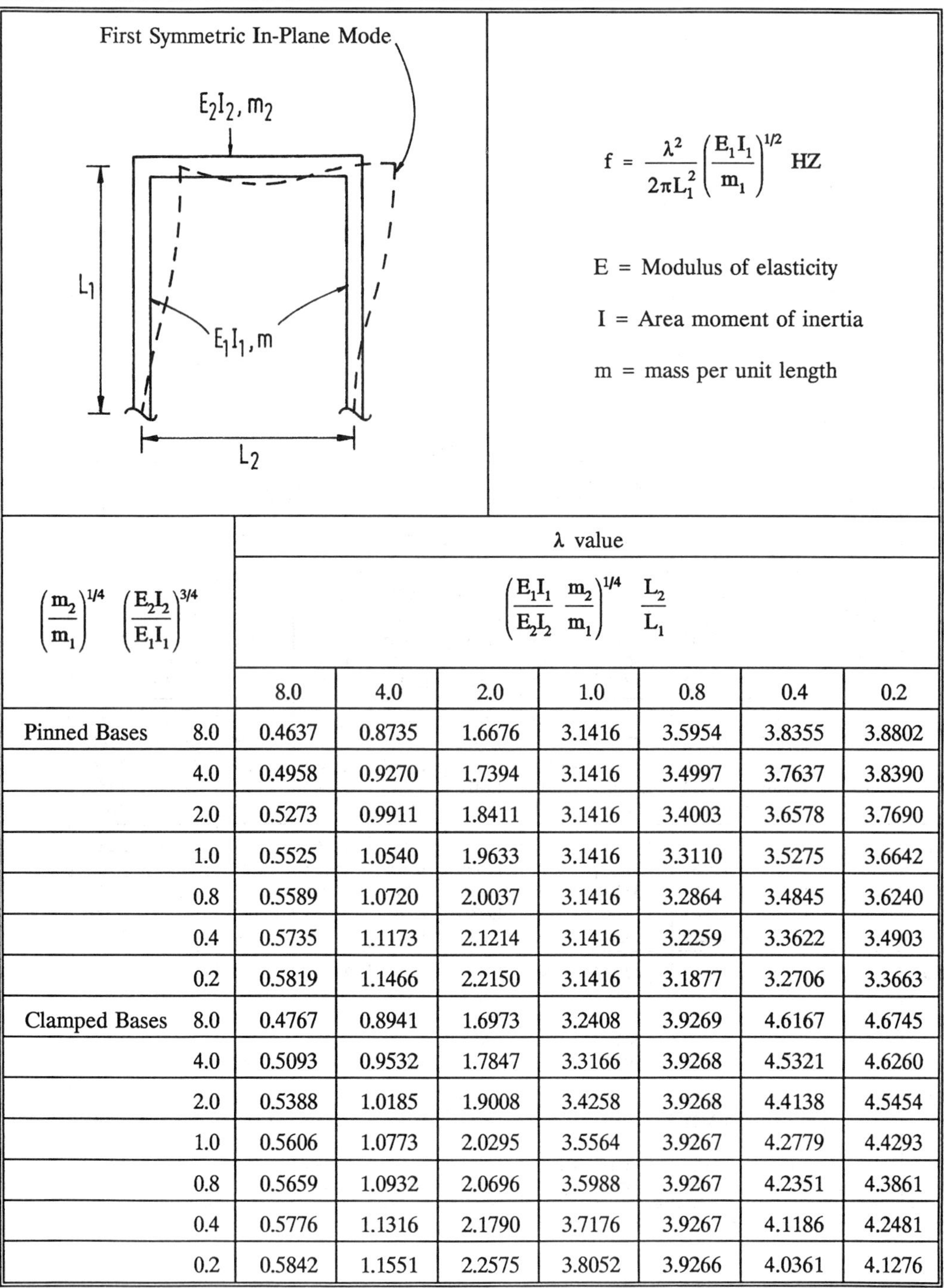

$$f = \frac{\lambda^2}{2\pi L_1^2}\left(\frac{E_1 I_1}{m_1}\right)^{1/2} \text{ HZ}$$

E = Modulus of elasticity

I = Area moment of inertia

m = mass per unit length

$\left(\frac{m_2}{m_1}\right)^{1/4}\left(\frac{E_2 I_2}{E_1 I_1}\right)^{3/4}$		λ value: $\left(\frac{E_1 I_1}{E_2 I_2}\frac{m_2}{m_1}\right)^{1/4}\frac{L_2}{L_1}$						
		8.0	4.0	2.0	1.0	0.8	0.4	0.2
Pinned Bases	8.0	0.4637	0.8735	1.6676	3.1416	3.5954	3.8355	3.8802
	4.0	0.4958	0.9270	1.7394	3.1416	3.4997	3.7637	3.8390
	2.0	0.5273	0.9911	1.8411	3.1416	3.4003	3.6578	3.7690
	1.0	0.5525	1.0540	1.9633	3.1416	3.3110	3.5275	3.6642
	0.8	0.5589	1.0720	2.0037	3.1416	3.2864	3.4845	3.6240
	0.4	0.5735	1.1173	2.1214	3.1416	3.2259	3.3622	3.4903
	0.2	0.5819	1.1466	2.2150	3.1416	3.1877	3.2706	3.3663
Clamped Bases	8.0	0.4767	0.8941	1.6973	3.2408	3.9269	4.6167	4.6745
	4.0	0.5093	0.9532	1.7847	3.3166	3.9268	4.5321	4.6260
	2.0	0.5388	1.0185	1.9008	3.4258	3.9268	4.4138	4.5454
	1.0	0.5606	1.0773	2.0295	3.5564	3.9267	4.2779	4.4293
	0.8	0.5659	1.0932	2.0696	3.5988	3.9267	4.2351	4.3861
	0.4	0.5776	1.1316	2.1790	3.7176	3.9267	4.1186	4.2481
	0.2	0.5842	1.1551	2.2575	3.8052	3.9266	4.0361	4.1276

these methods may not be easily carried out if several modes of vibration close in frequency are presented.

Table 45.9 gives values of modal damping that are appropriate for use when amplitudes are low. Higher values are appropriate at larger amplitudes where local yielding may develop (e.g., in seismic analysis).

Numerical Analysis

Many less complex dynamic problems can be solved without much difficulty by hand methods. For more complex problems, such as determination of natural frequencies of complex structures, calculation of response due to time-varying loads and response spectrum analysis to determine seismic forces may require numerical analysis. The finite element method has been shown to be a versatile technique for this purpose.

The global equations of an undamped motion, in matrix form, may be written as

$$[M]\{\ddot{x}\} + [K]\{\dot{x}\} = \{F(t)\} \tag{45.287}$$

where

$$[K] = \sum_{i=1}^{n}[k_i] \qquad [M] = \sum_{i=1}^{n}[M_i] \qquad [F] = \sum_{i=1}^{n}[f_i] \tag{45.288}$$

are the global stiffness, mass, and force matrices, respectively. $[k_i]$, $[m_i]$, and $\{f_i\}$ are the stiffness, mass, and force of the ith element, respectively. The elements are assembled using the direct stiffness method to obtain the global equations such that intermediate continuity of displacements is satisfied at common nodes and, in addition, interelement continuity of acceleration is also satisfied.

Equation (45.287) contains the matrix equations discretized in space. To solve the equation, discretization in time is also necessary. The general method used is called direct integration. There are two methods for direct integration: implicit and explicit. The first, and simplest, is the explicit method known as the central difference method [Biggs, 1964]. The second, more sophisticated but more versatile, is the implicit method known as the Newmark method [Newmark, 1959]. Other integration methods are also available in Bathe [1982].

The natural frequencies are determined by solving Eq. (45.287) in the absence of force $F(t)$ as

$$[M]\{\ddot{x}\} + [K]\{\dot{x}\} = 0 \tag{45.289}$$

The standard solution for $x(t)$ is given by the harmonic equation in time,

$$\{x(t)\} = \{X\}e^{i\omega t} \tag{45.290}$$

where $\{X\}$ is the part of the nodal displacement matrix (natural modes) assumed to be independent of time, i is the imaginary number, and ω is the natural frequency.

Differentiating Eq. (45.290) twice with respect to time, we have

$$\ddot{x}(t) = \{X\}(-\omega^2)e^{i\omega t} \tag{45.291}$$

Substitution of Eqs. (45.290) and (45.291) into (45.289) yields

$$e^{-i\omega t}([K] - \omega^2[M])\{X\} = 0 \tag{45.292}$$

Since $e^{i\omega t}$ is not zero, we obtain

$$([K] - \omega^2[M])\{X\} = 0 \tag{45.293}$$

Equation (45.293) is a set of linear homogeneous equations in terms of displacement mode $\{X\}$. It has a nontrivial solution if the determinant of the coefficient matrix $\{X\}$ is nonzero; that is,

$$[K] - \omega^2[M] = 0 \tag{45.294}$$

In general, Eq. (45.294) is a set of n algebraic equations, where n is the number of degrees of freedom associated with the problem.

Defining Terms

Arch: Principal load-carrying member curved in elevation; resistance to applied loading developed by axial thrust and bending.

Beam: A straight or curved structural member, primarily supporting loads applied at right angles to the longitudinal axis.

Bending moment: Bending moment due to a force or a system of forces at a cross section is computed as the algebraic sum of all moments to one side of the section.

Built-in beam: A beam restrained at its ends against vertical movement and rotation.

Cables: Flexible structures with no moment-carrying capacity.

Cantilever: A beam restrained against movement and rotation at one end and free to deflect at the other end.

Continuous beam: A beam that extends over several supports.

Deflection: Movement of a structure or parts of a structure under applied loads.

Element: Part of a cross section forming a distinct part of the whole.

Grillage: Structures in which the members all lie in one plane with loads being applied in the direction normal to this plane.

Hogging moment: Bending moment causing upward deflection in a beam.

Influence line: An influence line indicates the effect at a given section of a unit load placed at any point on the structure.

Member: Any individual component of a structural frame.

Moment of inertia: The second moment of area of a section about the elastic neutral axis.

Plastic analysis: Analysis assuming redistribution of moments within the structure in a continuous construction.

Plastic hinge: Position at which a member has developed its plastic moment of resistance.

Plastic moment: Moment capacity allowing for redistribution of stress within a cross section.

Plastic section: A cross section which can develop a plastic hinge with sufficient rotational capacity to allow redistribution of bending moments within the section.

Portal frame: A single-story continuous plane frame deriving its strength from bending resistance and arch action.

Reaction: The load carried by each support.

Rigid frame: An indeterminate plane frame consisting of members with fixed end connections.

Sagging moment: An applied bending moment causing a sagging deflection in the beam.

Second-order analysis: Analysis considering the equilibrium formulated based on deformed structural geometry.

Shear force: An internal force acting normal to the longitudinal axis; given by the algebraic sum of all forces to one side of the section chosen.

Simple beam: A beam restrained at its end only against vertical movement.

Space frame: A three-dimensional structure.

Span: The distance between the supports of a beam or a truss.

Static load: A noncyclic load that produces no dynamic effects.

Statically determinate structure: A structure in which support reactions may be found from the equations of equilibrium.

Statically indeterminate structure: A structure in which equations of equilibrium are not sufficient to determine the reactions.

Thin plate: A flat surface structure in which the thickness is small compared to the other dimensions.

Thin shell: A curved surface structure with a thickness relatively small compared to its other dimensions.

Truss: A coplanar system of structural members joined at their ends to form a stable framework.

References

AISC (American Institute of Steel Construction). 1986. *Load and Resistance Factor Design Specification for Structural Steel Buildings,* 1st. ed. AISC, Chicago, IL.

Argyris, J. H. 1960. *Energy Theorems and Structural Analysis.* Butterworth, London. (Reprinted from *Aircraft Engineering,* Oct. 1954–May 1955.)

Armenakas, A. E. 1991. *Modern Structural Analysis.* McGraw-Hill, New York.

ASCE. 1971. *Plastic Design in Steel—A Guide and Commentary, Manual 41.* American Society of Civil Engineers, New York.

Baker, L. and Heyman, J. 1969. *Plastic Design of Frames: 1. Fundamentals.* Cambridge University Press, Cambridge.

Bathe, K. J. 1982. *Finite Element Procedures in Engineering Analysis.* Prentice Hall, Englewood Cliffs, NJ.

Beedle, L. S. 1958. *Plastic Design of Steel Frames.* John Wiley & Sons, New York.

Biggs, J. M. 1964. *Introduction to Structural Dynamic.* McGraw-Hill, New York.

Borg, S. F. and Gennaro, J. J. 1959. *Advanced Structural Analysis.* Van Nostrand, Princeton, NJ.

CSA (Canadian Standards Association). 1989. *Limit States Design of Steel Structures.* CAN/CSA-S16.1-M89.

Chajes, A. 1983. *Structural Analysis.* Prentice Hall, Englewood Cliffs, NJ.

Chang, C. H. 1978. Vibration of frames with inclined members. *J. Sound Vib.* 56:201–214.

Chen, W. F. and Lui, E. M. 1991. *Stability Design of Steel Frames.* CRC Press, Boca Raton, FL.

Chen, W. F. and Toma, S. 1993. *Advanced Analysis in Steel Frames.* CRC Press, Boca Raton, FL.

Chen, W. F. and Han, D. J. 1988. *Plasticity for Structural Engineers.* Spring-Verlag, New York.

Clough, R. W. 1960. The finite element method in plane stress analysis. *Proc. 2nd Conf. Electronic Computation, ASCE,* Pittsburgh, Pa., Sept. 8–9.

Clough, R. W. and Penzien, J. 1993. *Dynamics of Structures,* 2nd ed. McGraw-Hill, New York.

Coats, R. C., Coutie, M. G., and Kong, F. K. 1980. *Structural Analysis,* 2nd ed. Thomas Nelson and Sons, Surrey.

Cowper, G. R. 1966. The shear coefficient in Timoshenko's beam theory. *J. Appl. Mech.* 33:335–340.

Desai, C. S. and Abel, J. F. 1972. *Introduction to the Finite Element Method.* Van Nostrand Reinhold, New York.

Discroll, G. C., Plastic design of multi-story frames. Lecture Notes and Design Aids, Report No. 273.20 and 273.24, Fritz Engineering Laboratory, Lehigh University, Bethlehem, PA.

Dowrick, D. J. 1988. *Earthquake Resistant Design for Engineers and Architects,* 2nd ed. John Wiley & Sons, New York.

Eurocode 3. 1990. *Design of Steel Structures: Part I—General Rules and Rules for Buildings,* vol. 1. Eurocode Edited Draft, Issue 3.

Ghali, A. and Neville, A. M. 1978. *Structural Analysis.* Chapman and Hall, London.

Gould, P. L. 1988. *Analysis of Shells and Plates.* Springer-Verlag, New York.

Gutkowski, R. M. 1990. *Structures: Fundamental Theory and Behaviour.* Van Nostrand Reinhold, New York.

Heyman, J. 1971. *Plastic Design of Frames: 2. Applications.* Cambridge University Press, Cambridge.

Hibbeler, R. C. 1990. *Structural Analysis.* Macmillan, New York.

Hodge, Jr., P. G. 1963. *Limit Analysis of Rotationally Symmetric Plates and Shells.* Prentice Hall, Englewood Cliffs, NJ.

Horne, M. R. 1954. A moment distribution method for the analysis and design of structures. *Proc. Inst. Civ. Eng.,* April.

Horne, M. R. 1964. *The Plastic Design of Columns.* The British Constructional Steelwork Association, Publication No. 23.

Housner, G. W. and Brody, A. G. 1963. Natural periods of vibration of buildings. *J. Eng. Mech. Div., ASCE.* 89:31–65.

Jenkins, W. M. 1969. *Matrix and Digital Computer Methods in Structural Analysis.* McGraw-Hill, London.

Johnston, B. G., Beedle, L. S., and Yang, C. H. 1953. An evaluation of plastic analysis as applied to structural design. *Weld. J.* 32(5):P224-s.

Kardestuncer, H. and Norrie, D. H., 1988. *Finite Element Handbook.* McGraw-Hill, New York.

Kazinczy, G. 1914. Experiments with clamped girders. *Betonszemle,* vol. 2, no. 4, p. 68; no. 5, p. 83; no. 6, p. 101.

Kennedy, J. B. and Madugula, M. K. S. 1990. *Elastic Analysis of Structures.* Harper & Row, New York.

Knudsen, K. E., Yang, C. H., Johnson, B. G., and Beedle, L. S. 1953. Plastic strength and deflections of continuous beams. *Weld. J.* 32(5):240–245.

Liew, J. Y. R. 1992. *Advanced Analysis for Frame Design.* Ph.D. dissertation, School of Civil Engineering, Purdue University, West Lafayette, Indiana.

Liew, J. Y. R. and Chen, W. F. 1993. A trend toward advanced analysis. In *Advanced Analysis in Steel Frames: Theory, Software and Applications,* ed. W. F. Chen and S. Toma. CRC Press, Boca Raton, FL.

Liew, J. Y. R., White, D. W., and Chen, W. F. 1993. Second-order refined plastic hinge analysis for frame design. *J. Struct. Eng., ASCE.* 119(11):3196–3237.

Liew, J. Y. R., White, D. W., and Chen, W. F. National load plastic hinge method for frame design. *J. Struct. Eng., ASCE.* 120(5):1434–1454.

Livesley, R. K. 1983. *Finite Element: An Introduction for Engineers.* Cambridge University Press, Cambridge.

Livesley, R. K. and Chandler, D. B. 1956. *Stability Functions for Structural Frameworks.* Manchester University Press, Manchester.

Maier-Leibnitz, H. 1927. Contributions to the problem of ultimate carrying capacity of simple and continuous beams of structural steel and timber. *Die Bautechnik.* 1(6).

Makowski, Z. S., ed. 1985. *Analysis, Design and Construction of Braced Barrel Vaults.* Elsevier, London.

McCormac, J. and Elling, R. E. 1988. *Structural Analysis.* Harper & Row, New York.

McGuire, W. and Gallagher, R. H. 1979. *Matrix Structural Analysis.* John Wiley & Sons, New York.

Meek, J. L. 1971. *Matrix Structural Analysis.* McGraw-Hill, International Students Edition, Tokyo.

Melosh, R. J. 1990. *Structural Engineering Analysis by Finite Elements.* Prentice Hall, Englewood Cliffs, NJ.

Mrazik, A., Skaloud, M., and Tochacek, M. 1987. *Plastic Design of Steel Structures.* Halsted Press, New York.

Mullord, P. 1985. Introduction to the analysis of braced barrel vaults. In *Analysis, Design and Construction of Braced Barrel Vaults,* ed. Z. S. Makowski, pp. 36–40. Elsevier, London.

Newmark, N. M. 1959. A method of computation for structural dynamic. *J. Eng. Mech., ASCE.* 85(EM3):67–94.

Orbison, J. G. 1982. *Nonlinear Static Analysis of Three-Dimensional Steel Frames.* Department of Structural Engineering, Report No. 82-6, Cornell University.

Owens, G and Knowles, P. R., eds. 1993. *Steel Designers' Manual,* 5th ed. The Steel Construction Institute, Blackwell Scientific.

Porter, F. L. and Powell, G. M. 1971. *Static and Dynamic Analysis of Inelastic Frame Structures.* Report No. EERC 71-3, Earthquake Engineering Research Center, University of California, Berkeley.

Powell, G. H. and Chen, P. F. 1986. 3D beam-column element with generalized plastic hinges. *J. Eng. Mech., ASCE.* 112(7):627–641.

Rinne, J. E. 1952. Building code provisions for seismic design. In *Proc. Symp. Earthquake Blast Effects Struct.* Los Angeles, CA, pp. 291–305.

Robert, D. B. 1979. *Formulas for Natural Frequency and Mode Shapes.* Van Nostrand Reinhold, New York.

Rutenberg, A. 1975. Approximate natural frequencies for coupled shear walls. *Earthquake Eng. Struct. Dynam.* 4:95–100.

Seely, F. B. and Smith, J. O. 1952. *Advanced Mechanics of Materials,* pp. 137–157. John Wiley & Sons, New York.

Smith, J. C. 1988. *Structural Analysis.* Harper & Row, New York.

SSRC (Structural Stability Research Council). 1988. *Guide to Stability Design Criteria for Metal Structures,* ed. T. V. Galambos, 4th ed. John Wiley & Sons, New York.

Szilard, R. 1974. *Theory and Analysis of Plates—Classified and Numerical Methods.* Prentice Hall, Englewood Cliffs, NJ.

Tall, L., ed. 1974. *Structural Steel Design,* 2nd. ed. Ronald Press, New York.

Timoshenko, S. P. and Goodier, J. N. 1970. *Theory of Elasticity.* McGraw-Hill, New York.

Timoshenko, S. P. and Woinowsky-Krieger, S. 1959. *Theory of Plates and Shells.* McGraw-Hill, New York.

Utku, S., Norris, C. H., and Wilbur, J. B. 1991. *Elementary Structural Analysis.* McGraw-Hill, New York.

Vogel, U. 1985. Calibrating frames. *Stahlbau.* 10:295–301.

Wang, Y. C. and Nethercot, D. A. 1988. Ultimate strength analysis of three-dimensional column subassemblages with flexible connections. *J. Constr. Steel Res.* 9:2235–2264.

Wang, C. K. 1953. *Statically Indeterminate Structures.* McGraw-Hill, New York.

Wang, C. K. 1973. *Introductory Structural Analysis with Matrix Methods.* Prentice Hall, Englewood Cliffs, NJ.

Warburton, G. B. 1976. *The Dynamical Behaviour of Structures,* 2nd ed. Pergamon Press, New York.

West, H. H. 1980. *Analysis of Structures.* John Wiley & Sons, New York.

White, D. W. 1985. *Material and Geometric Nonlinear Analysis of Local Planar Behavior in Steel Frames Using Iterative Computer Graphics.* M.S. Thesis., Cornell University, Ithaca, NY.

White, D. W., Liew, J. Y. R., and Chen, W. F. 1993. Toward advanced analysis in LRFD. in *Plastic Hinge Based Methods for Advanced Analysis and Design of Steel Frames—An Assessment of the State-of-the-Art,* ed. D. W. White and W. F. Chen, Structural Stability Research Council, Lehigh University, Bethlehem, PA., March, pp. 95–173.

White, D. W. 1988. *Analysis of Monotonic and Cyclic Stability of Steel Frame Subassemblages.* Ph.D. thesis, Cornell University, Ithaca, NY.

Wind Forces on Structures. 1961. Task Committee on Wind Forces, Committee on Loads and Stresses, Structural Division, ASCE, Final Report. *Trans. ASCE.* 126(II):1124–1125.

Yang, Y. T. and Sun, C. T. 1973. Axial-flexural vibration of frameworks using finite element approach. *J. Acoust. Soc. Am.* 53:137–146.

Yuan, Y. H. 1992. *Elementary Theory of Structures.* Prentice-Hall, Singapore.

Zienkiewiez, O. C. 1971. *the Finite Element Method in Engineering Science.* McGraw-Hill, London.

For Further Information

The *Structural Engineering Handbook* by E. H. Gaylord and C. N. Gaylord provides a reference work on structural engineering and deals with planning, design, and construction of a variety of engineering structures.

The *Finite Element Handbook* by H. Kardestuncer and D. H. Norrie presents the underlying mathematical principles, the fundamental formulations, and both commonly used and specialized applications of the finite element method.

46

Structural Steel Design

E. M. Lui
Syracuse University

46.1 Materials

Stress-Strain Behavior of Structural Steel

A schematic diagram of an engineering stress-strain curve of steel obtained from a simple tension test is shown in Fig. 46.1. Three regions are identified:

0-8493-8953-4/95/$0.00 + $.50

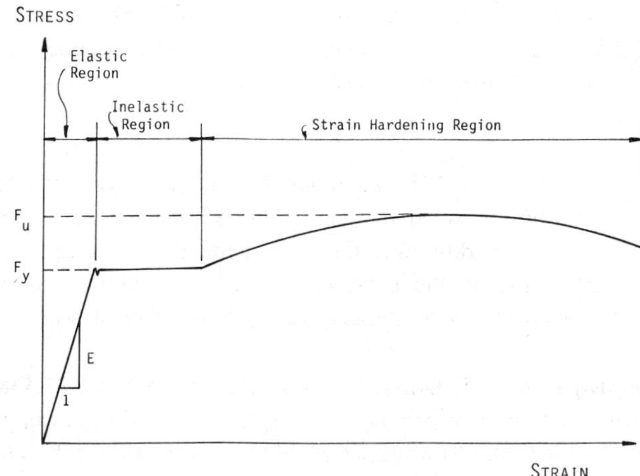

FIGURE 46.1 Schematic stress-strain curve of structural steel.

Elastic Region. In this region the stress is proportional to the strain, and Hooke's law applies. The constant of proportionality is the modulus of elasticity, or Young's modulus, *E*. The modulus of elasticity for steel has values ranging from 28,000–30,000 ksi (190–210 GPa). The modulus of elasticity does not vary appreciably for the different grades of steel used in construction, and a value of 29,000 ksi (200 GPa) is often used for design. The elastic region ends when the stress reaches F_y, the yield stress. For stress below F_y no plastic, or permanent, deformation will occur in the steel section.

Inelastic Region. In this region the steel section deforms plastically under a constant stress, F_y. The extent of this deformation differs for different steel grades. Generally, the ductility (the ability of a material to undergo plastic deformation prior to fracture) decreases with increasing steel strength. Ductility is a very important attribute of steel. The ability of structural steel to deform considerably before failure by fracture allows the structure to undergo force redistribution when yielding occurs, and it enhances the energy absorption characteristic of the structure.

Strain-Hardening Region. In this region deformation is accompanied by an increase in stress. The peak point of the engineering stress-strain curve is the ultimate stress, F_u. F_u is the largest stress the material can attain under uniaxial condition. In a uniaxial tension test, the specimen experiences nonuniform plastic deformation (necking) once the stress reaches F_u. Beyond F_u deformation proceeds at a rapid rate, and equilibrium can be maintained only by a reduction in the applied load. For design purposes, F_u is often regarded as the stress at which failure is imminent.

Types of Steel

Structural steels used for construction purposes are generally grouped into several major American Society for Testing and Materials (ASTM) classifications:

Carbon Steels (ASTM A36, ASTM A529, ASTM 709). In addition to iron, the main ingredients of this category of steels are carbon (maximum content 1.7%) and manganese (maximum content 1.65%), with a small amount ($< 0.6\%$) of silicon and copper. Depending on the amount of carbon content, different types of carbon steels can be identified:

Low-carbon steel: carbon content $< 0.15\%$
Mild carbon steel: carbon content varies from 0.15 to 0.29%
Medium-carbon steel: carbon content 0.30–0.59%
High-carbon steel: carbon content 0.60–1.70%

The most commonly used structural carbon steel has a mild carbon content. It is extremely ductile and is suitable for both bolting and welding. ASTM A36 is used mainly for buildings. ASTM A529 is occasionally used for bolted and welded building frames and trusses. ASTM 709 is used primarily for bridges.

High-Strength, Low-Alloy Steels (ASTM A441, ASTM A572). These steels possess enhanced strength as a result of the presence of one or more alloying agents, such as chromium, copper, nickel, silicon, and vanadium, in addition to the basic elements of iron, carbon, and manganese. Normally, the total quantity of all the alloying elements is below 5% of the total composition. These steels generally have higher corrosion-resistant capability than carbon steels.

Corrosion-Resistant, High-Strength, Low-Alloy Steels (ASTM A242, ASTM A588). These steels have enhanced corrosion-resistant capability because of the addition of copper as an alloying element. Corrosion is severely retarded when a layer of patina (an oxidized metallic film) is formed on the steel surfaces. The process of oxidation normally takes place within one to three years and is signified by a distinct appearance of a deep reddish-brown to black coloration of the steel. For the process to take place, the steel must be subjected to a series of wetting-drying cycles. These steels, especially ASTM 588, are used primarily for bridges and transmission towers (in lieu of galvanized steel), where members are difficult to access for periodic painting.

Quenched and Tempered Alloy Steels (ASTM A852, ASTM A514). The quantities of alloying elements used in these steels are in excess of those used in carbon and low-alloy steels. In addition, they are heat-treated by quenching and tempering to enhance their strengths. These steels do not exhibit well-defined yield points. Their yield stresses are determined by the 0.2% offset strain method. These steels, despite their enhanced strength, have reduced ductility (Fig. 46.2), and care must be exercised in their usage, as the design limit state for the structure or structural elements may be governed by serviceability considerations (e.g., deflection, vibration) or local buckling (under compression).

Table 46.1 gives a summary of the specified minimum yield stresses (F_y), the specified minimum tensile strengths (F_u), and general usages for these various categories of steels.

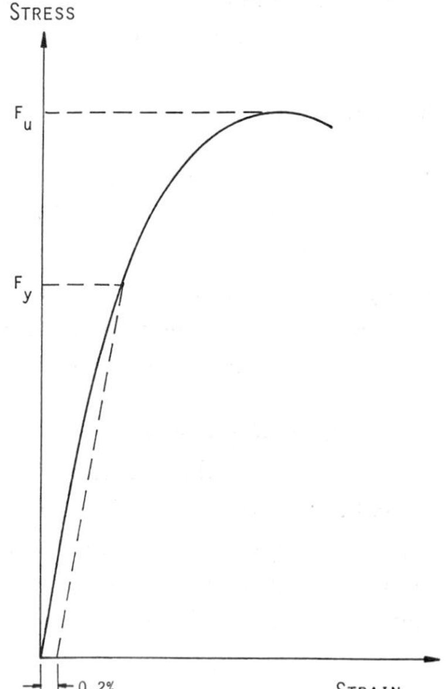

FIGURE 46.2 Schematic stress-strain curve of high-strength steel.

Fireproofing of Steel

Although steel is an incombustible material, its strength (F_y, F_u) and stiffness (E) reduce quite noticeably at temperatures normally reached in fires when other materials in a building burn. Exposed steel members, which will be subjected to high temperature when a fire occurs, should be fireproofed to conform to the fire ratings set forth in city codes. Fire ratings are expressed in units of time (usually hours) beyond which the structural members under a standard ASTM Specification (E119) fire test will fail under a specific set of criteria.

Various approaches are available for fireproofing steel members. Steel members can be fireproofed by encasement in concrete if a minimum cover of 2 inches of concrete is provided. If the use of concrete is undesirable (because it adds weight to the structure), a lath and plaster (gypsum) ceiling placed underneath the structural members supporting the floor deck of an upper story can

Table 46.1 Types of Steels

ASTM Designation	F_y (ksi)[a]	F_u (ksi)[a]	Availability	General Usages
A36	36	58–80	For all shapes, plates, and bars up to 8 inches	Riveted, bolted, and welded buildings and bridges
A529	42	60–85	For Group 1 shapes[b], plates, and bars up to ½ inch	Similar to A36. The higher yield stress for A529 steel allows for savings in weight
	50	70–100	Groups 1 and 2 shapes, plates and bars up to 1 ½ inch	
A441	40 42 46 50	60 63 67 70	For all shapes, plates, and bars up to 8 inches[b,c]	Similar to A36. Used when weight savings and enhanced durability are desired
A572 Grade 42	42	60	All shapes, plates, and bars up to 6 inches	Similar to A441. Grades 60 and 65 not suitable for welded bridges
Grade 50	50	65	All shapes, plates, and bars up to 4 inches	
Grade 60	60	75	Groups 1 and 2 shapes, plates, and bars up to 1 ¼ inch	
Grade 65	65	80	Group 1 shapes, plates, and bars up to 1 ¼ inch	
A242	42 46 50	63 67 70	For all shapes, plates, and bars up to 4 inches[b,c]	Riveted, bolted and welded buildings and bridges. Used when weight savings and enhanced atmospheric corrosion resistance are desired. Specific instructions must be provided for welding.
A588	42 46 50	63 67 70	For all shapes (F_y = 50 ksi only), plates, and bars up to 8 inches[c]	Similar to A242. Atmospheric corrosion resistance is about four times that of A36 steel.
A852	70	90–110	For plates only, up to 4 inches	
A514	90 100	100–130	For plates only, up to 6 inches[c]	Primarily for welded bridges. Avoid usage if ductility is important.

[a] 1 ksi = 6.895 MPa.
[b] See Table 46.2 for structural shape size groupings.
[c] Availability depends on specific steel grade and F_y.

be used. In lieu of such a ceiling, spray-on materials, such as mineral fibers, perlite, vermiculite, or gypsum, can also be used for fireproofing. Other means of fireproofing include placing steel members away from the source of heat, circulating liquid coolant inside box or tubular members, and using insulative paints. These special paints foam and expand when heated, thus forming a shield for the members [Rains, 1976].

For a more detailed discussion of structural steel design for fire protection, refer to the latest edition of AISI (American Iron and Steel Institute) Publication No. FS3, *Fire-Safe Structural Steel—A Design Guide.* Additional information on fire-resistant standards and fire protection can be found in the AISI booklets on *Fire Resistant Steel Frame Construction*; *Designing Fire Protection for Steel Columns*; and *Designing Fire Protection for Steel Trusses,* as well as in Part VII of the *Uniform Building Code.*

Corrosion Protection of Steel

Atmospheric corrosion occurs when steel is exposed to a continuous supply of water and oxygen. The rate of corrosion can be reduced if a barrier is used to keep water and oxygen from contact with the surface of bare steel. Painting is a practical and cost-effective way to protect steel from corrosion. The Steel Structures Painting Council issues specifications for the surface preparation and the painting of steel structures for corrosion protection of steel. In lieu of painting, the use of corrosion-resistant steel, such as ASTM A242 and A588 steel or galvanized steel, may be considered.

Structural Steel Shapes

Steel sections used for construction are available in a variety of shapes and sizes. In general, there are three procedures by which steel shapes can be formed: hot-rolled, cold-formed, and welded. Regardless of the manner by which the steel shape is formed, it must be manufactured to meet ASTM standards. Commonly used steel shapes include the familiar I or H sections, consisting of a *web* forming the crossbar of the H and *flanges* forming the uprights of the H, further classified as wide flange (W) sections or American Standard beam (S) sections; bearing pile (HP) sections; American Standard channel (C) sections; angle (L) sections; and tee (WT) sections as well as bars, plates, pipes, and tubular sections. H sections that, by dimensions, cannot be classified as W or S shapes are designated as miscellaneous (M) sections, and C sections that, by dimensions, cannot be classified as American Standard channels are designated as miscellaneous channel (MC) sections.

Hot-rolled shapes are classified in accordance with their tensile property into five size groups by the American Institute of Steel Construction (AISC). The groupings are shown in Table 46.2. Groups 4 and 5 shapes and group 3 shapes with flange thickness exceeding 1 ½ in. are generally used for application as compression members. When weldings are used, care must be exercised to minimize the possibility of cracking in regions at the vicinity of the welds by carefully reviewing the material specification and fabrication procedures of the pieces to be joined.

Structural Fasteners

Steel sections can be fastened together by rivets, bolts, or welds. While rivets were used quite extensively in the past, their use in modern steel construction has become almost obsolete. Bolts have essentially replaced rivets as the primary means to connect nonwelded structural components.

Bolts

Four basic types of bolts are commonly in use; they are designated by ASTM as A307, A325, A490, and A449.

A307 bolts, called *unfinished* or *ordinary* bolts, are made from low-carbon steel. Two grades (A and B) are available in diameters from ¼ in. to 4 in. in ⅛ in. increments. They are used primarily for low-stress connections and for secondary members.

A325 and A490 bolts are called *high-strength* bolts. A325 bolts are made from a heat-treated medium-carbon steel. They are available in three types:

Type 1: Bolts made of medium-carbon steel
Type 2: Bolts made of low-carbon martensite steel

Type 3: Bolts having atmospheric-corrosion resistance and weathering characteristics comparable to A242 and A588 steel

A490 bolts are made from quenched and tempered alloy steel and thus have higher strength than A325 bolts. As with A325 bolts, three types (Types 1 to 3) are available. Both A325 and A490 bolts are available in diameters from ½ in. to 1½ in. in ⅛ in. increments. They are used for general construction purposes.

Table 46.2 Structural Shape Size Groupings

Structural Shape	Group 1	Group 2	Group 3	Group 4	Group 5
W shapes		W44 × 230 to 262	W44 × 290 to 335		
		W40 × 149 to 268	W40 × 277 to 328	W40 × 362 to 655	
		W36 × 135 to 256	W36 × 230 to 359	W36 × 393 to 798	W36 × 848
		W33 × 118 to 152	W33 × 201 to 354	W33 × 387 to 619	
		W30 × 90 to 211	W30 × 261	W30 × 292 to 581	
		W27 × 84 to 194	W27 × 235 to 258	W27 × 281 to 539	
	W24 × 55.62	W24 × 68 to 176	W24 × 192 to 229	W24 × 250 to 492	
	W21 × 44 to 57	W21 × 62 to 166	W21 × 182 to 223	W21 × 248 to 402	
	W18 × 35 to 71	W18 × 76 to 143	W18 × 158 to 192	W18 × 211 to 311	
	W16 × 26 to 57	W16 × 67 to 100			
	W14 × 22 to 53	W14 × 61 to 132	W14 × 145 to 211	W14 × 233 to 550	W14 × 605 to 730
	W12 × 14 to 58	W12 × 65 to 106	W12 × 120 to 190	W12 × 210 to 336	
	W10 × 12 to 45	W10 × 49 to 112			
	W8 × 10 to 48	W8 × 58.67			
	W6 × 9 to 25				
	W5 × 16.19				
	W4 × 13				
M shapes	To 20 lb/ft				
S shapes	To 35 lb/ft	Over 35 lb/ft			
HP shapes		To 102 lb/ft	Over 102 lb/ft		
American standard channels (C shapes)	To 20.7 lb/ft	Over 20.7 lb/ft			
Miscellaneous channels (MC shapes)	To 28.5 lb/ft	Over 28.5 lb/ft			
Angles (L shapes) and structural bar size	To ½ in.	½ to ¾ in.	Over ¾ in.		

(1 lb = 4.45 N, 1 ft = 0.305 m, 1 in. = 25.4 mm)
Structural tees from W, M, and S shapes fall in the same group designation as the structural shapes from which they are cut.

Table 46.3 Proof Stresses

ASTM Designation	Bolt Diameter (in.)	Proof Stress[a] (ksi)[b]
A307	1/4 to 4	Not applicable
A325	1/2 to 1	85
	1 1/8 to 1 1/2	74
A490	1/2 to 1 1/2	120
A449	1/4 to 1	85
	1 1/8 to 1 1/2	74
	1 3/4 to 3	55

[a] Proof load (in kips) can be obtained by multiplying the given proof stress by A_s (tensile stress area of the bolt in in.^{2}), where $A_s = 0.7854[D - (0.9743/n)]^2$, D = nominal diameter of the bolt (in in.), n = number of threads per inch.
[b] 1 kip = 4.45 kN, 1 in. = 25.4 mm.

A449 bolts are made from quenched and tempered steel and are available in diameters from 1/4 in. to 3 in. A449 bolts are used when diameters over 1 1/2 in. are needed. They are also used for anchor bolts and threaded rods.

High-strength bolts are required to be tightened to specific stresses (called *proof stresses*) so that a clamping force will be developed on the joint. The values for these proof stresses are given in Table 46.3.

Bolts used in *slip-critical* conditions (i.e., conditions for which the integrity of the connected parts is dependent on the frictional force developed between the interfaces of the joint), and in conditions where the bolts are subjected to direct tension, are required to be tightened to develop a pretension force equal to about 70% of the minimum tensile stress F_u of the material from which the bolts are made. This can be accomplished by using the turn-of-the-nut method, the calibrated wrench method, alternate-design fasteners, or a direct tension indicator.

Turn-of-the-Nut Method. One-third to one full turn (depending on the bolt length-to-diameter ratio and slope of the surfaces with respect to the bolt axis [RCSC, 1986], see the following table) is applied to the bolt after the bolt has been brought to a snug-tight condition. *Snug-tight* is defined as

	Disposition of Outer Faces of Bolted Parts		
Bolt length (underside of head to end of bolt)	Both faces normal to bolt axis	One face normal to bolt axis and the other face sloped not more than 1:20 (beveled washer not used)	Both faces sloped not more than 1:20 from normal to bolt axis (beveled washers not used)
$l_b \leq 4d_b$	1/3 turn	1/2 turn	2/3 turn
$4d_b < l_b \leq 8d_b$	1/2 turn	2/3 turn	5/6 turn
$8d_b < l_b \leq 12d_b$	2/3 turn	5/6 turn	1 turn

d_b = bolt diameter
l_b = bolt length

Note 1: Nut rotation is relative to bolt regardless of the element (nut or bolt) being turned. For bolts installed by 1/2 turn and less, the tolerance should be plus or minus 30°; for bolts installed by 2/3 turn and more, the tolerance should be plus or minus 45°.
Note 2: For bolt length exceeding $12d_b$, the required rotation must be determined by testing in a suitable tension-measuring device which simulates conditions of solidly fitted steel.

the condition that exists when all plies of the connecting parts are in firm contact with one another. It is the tightness obtained by the full effort of a construction worker using an ordinary spud wrench, or the tightness obtained after a few impacts of an impact wrench.

Calibrated Wrench Method. A calibrated wrench is a wrench which can be adjusted to stall at a specified torque that corresponds to the pretension force required to be developed in the bolts. To account for unavoidable variations in inducing the pretension force, the wrench should be calibrated to produce a bolt tension at least 5% in excess of the required pretension force. In addition, it is necessary that the wrenches be calibrated daily and that hardened washers be used.

Alternate-Design Fasteners. In this installation technique, special fasteners designed to indicate bolt tension indirectly or provide the required pretension force automatically are used. An example is the use of bolts with splined ends that extend beyond the threaded portion of the bolts. The desired bolt tension is attained when this splined end is sheared off using a special wrench during tightening.

Direct Tension Indicator. In this procedure a special hardened washer, containing a series of protrusions on one face in the form of arches, is placed between the bolt head and the gripped material. The arches will be flattened as the bolt is tightened. The magnitude of the gap is a measure of the bolt tension. Properly tensioned bolts should have a gap that measures 0.0015 in. or less.

Bolts neither used for slip-critical conditions nor subjected to direct tension need only be tightened to a snug-tight condition.

Welds

Welding is a very effective means to connect two or more pieces of materials together. Welding can be done with or without filler materials, although most weldings used for construction utilize filler materials. The filler materials used in modern welding processes are electrodes. Table 46.4

Table 46.4 Electrode Designations

Welding Processes	Electrode Designations	Remarks
Shielded Metal Arc Welding (SMAW)	E60XX E70XX E80XX E100XX E110XX	The E denotes electrode. The first two digits indicate tensile strength in ksi.[a] The two Xs represent numbers indicating the usage of the electrode.
Submerged Arc Welding (SAW)	F6X-EXXX F7X-EXXX F8X-EXXX F10X-EXXX F11X-EXXX	The F designates a granular flux material. The digit(s) following the F indicate the tensile strength in ksi (6 means 60 ksi, 10 means 100 ksi, etc.). The digit before the hyphen gives the Charpy V-notched impact strength. The E and the Xs that follow represent numbers relating to the use of the electrode.
Gas Metal Arc Welding (GMAW)	ER70S-X ER80S ER100S ER110S	The digits following the letters ER represent the tensile strength of the electrode in ksi.
Flux-Core Arc Welding (FCAW)	E6XT-X E7XT-X E8XT E10XT E11XT	The digit(s) following the letter E represent the tensile strength of the electrode in ksi (6 means 60 ksi, 10 means 100 ksi, etc.).

[a] 1 ksi = 6.895 MPa.

summarizes the electrode designations used for the four most commonly used welding processes [AWS, 1988].

The four most commonly used welding processes are briefly described as follows:

Shielded Metal Arc Welding. SMAW is one of the simplest welding processes for welding structural steel. It is limited primarily to manual application. In this process, a consumable electrode is melted from a coated electrode wire and transferred by an electric arc to the base metal. The coating of the electrode is converted partly into a shielding gas and partly into a slag. The shield serves to exclude air (thus preventing oxides from forming) and stabilize the weld during the welding process. After cooling, the slag, which has floated to the surface of the weld, may be removed by chipping.

Submerged Arc Welding. SAW can be performed manually, semiautomatically, or automatically. In SAW the arc is not visible because it is shielded by a granular, fusible material called *flux*, which covers the metal surfaces to be welded. The filler material is provided by either a consumable electrode or a supplementary welding rod. Submerged arc welds have been found to have uniformly high quality. The process is used quite often in shop fabrication operations.

Gas Metal Arc Welding. GMAW utilizes an electrode which is in the form of a continuous coil of wire fed through a gun-shaped device. The shielding is provided from an externally supplied gas or gas mixture or from a combination of a gas and a flux. GMAW allows the filler material to be transferred at a high rate. GMAW using CO_2 as gas shield works well for low-carbon and low-alloy steel used in buildings and bridges.

Flux-Core Arc Welding. FCAW is similar to GMAW except that the continuously fed electrode is tubular and contains the flux material with its core. The core material provides the shield during welding. Frequently, additional gas (CO_2) shielding is used. FCAW is a useful welding process in cold-weather application.

To ensure their quality, finished welds should be inspected by qualified welding inspectors. Methods available for weld inspections include visual methods, liquid penetrants, magnetic particles, ultrasonic equipment, and radiographic methods. Discussion of these and other welding inspection techniques can be found in the Welding Handbook [AWS, 1987].

Weldability of Steel

Most ASTM specification construction steels are weldable. In general, the strength of the electrode used should equal or exceed the strength of the steel being welded [AWS, 1987]. For various chemical elements in steel, the following table gives percentage ranges within which good weldability is ensured [Blodgett, undated].

Element	Percentage Range for Good Weldability	Percent Requiring Special Care
Carbon	0.06–0.25	0.35
Manganese	0.35–0.80	1.40
Silicon	0.10 max.	0.30
Sulfur	0.035 max.	0.050
Phosphorus	0.030 max.	0.040

46.2 Design Philosophy and Design Formats

Design Philosophy

Structural design should be performed to satisfy three criteria: strength, serviceability, and economy. *Strength* pertains to the general integrity and safety of the structure under extreme load conditions. The structure is expected to withstand occasional overloads without severe distress and damage during its lifetime. *Serviceability* refers to the proper functioning of the structure as related to its appearance, maintainability, and durability under normal, or **service load,** conditions. Deflection, vibration, permanent deformation, cracking, and corrosion are some design considerations associated with serviceability. *Economy* concerns the overall material and labor costs required for the design, fabrication, erection, and maintenance processes of the structure.

Design Formats

At present, steel design can be performed in accordance with one of the following three formats:

Allowable Stress Design (ASD)

This design methodology has been in use for decades for steel design of buildings and bridges. It continues to enjoy popularity among structural engineers engaged in steel building design. In **allowable stress** (or *working stress*) **design**, member stresses computed under the action of service (or *working*) loads are compared to some predesignated stresses, called *allowable* stresses. The allowable stresses are usually expressed as a function of the yield stress (F_y) or tensile stress (F_u) of the material. To account for overload, understrength, and approximations used in structural analysis, a factor of safety is applied to reduce the nominal resistance of the structural member to a fraction of its tangible capacity. The general format for an allowable stress design has the form

$$\frac{R_n}{\mathrm{FS}} \geq \sum_{i=1}^{m} Q_{ni} \tag{46.1}$$

where

R_n = nominal resistance of the structural component expressed in units of stress
Q_{ni} = service or working stress computed from the applied working load of type i
FS = factor of safety
i = load type (dead, live, wind, etc.)
m = number of load types considered in the design

R_n/FS represents the allowable stress of the structural component.

Plastic Design (PD)

Plastic design makes use of the fact that steel sections have reserved strength beyond the first yield condition. When a section is under flexure, yielding of the cross section occurs in a progressive manner, commencing with the fibers farthest away from the neutral axis and ending with the fibers nearest the neutral axis. This phenomenon of progressive yielding, referred to as *plastification,* means that the cross section does not fail at first yield. The additional moment that a cross section can carry in excess of the moment that corresponds to first yield varies depending on the shape of the cross section. To quantify such reserved capacity, a quantity called *shape factor,* defined as the ratio of the *plastic moment* (moment that causes the entire cross section to yield, resulting in the formation of a **plastic hinge**) to the *yield moment* (moment that causes yielding of the extreme fibers only) is used. The shape factor for hot-rolled I-shaped sections bent about the strong axes has a value of about 1.15. The value is about 1.50 when these sections are bent about their weak axes.

For an indeterminate structure, failure of the structure will not occur after the formation of a plastic hinge. After complete yielding of a cross section, force (or, more precisely, moment) redistribution will occur, in which the unfailed portion of the structure continues to carry any additional loadings. Failure will occur only when enough cross sections have yielded to render the structure unstable, resulting in the formation of a *plastic collapse mechanism.*

In plastic design the factor of safety is applied to the applied loads to obtain **factored loads.** A design is said to have satisfied the strength criterion if the load effects (i.e., forces, shears, and moments) computed using these factored loads do not exceed the nominal plastic strength of the structural component. Plastic design has the form

$$R_n \geq \gamma \sum_{i=1}^{m} Q_{ni} \tag{46.2}$$

where

R_n = nominal plastic strength of the member
Q_{ni} = nominal load effect from loads of type i
γ = load factor
i = load type
m = number of load types

In steel building design the load factor is given by the AISC Specification as 1.7, if Q_n consists of dead and live gravity loads only, and as 1.3, if Q_n consists of dead and live gravity loads acting in conjunction with wind or earthquake loads.

Load and Resistance Factor Design (LRFD)

LRFD is a probability-based limit state design procedure. In its development, both load effects and resistance were treated as random variables. Their variabilities and uncertainties were represented by frequency distribution curves. A design is considered satisfactory according to the strength criterion if the resistance exceeds the load effects by a comfortable margin. The concept of safety is represented schematically in Fig. 46.3. Theoretically, the structure will not fail unless R is less than

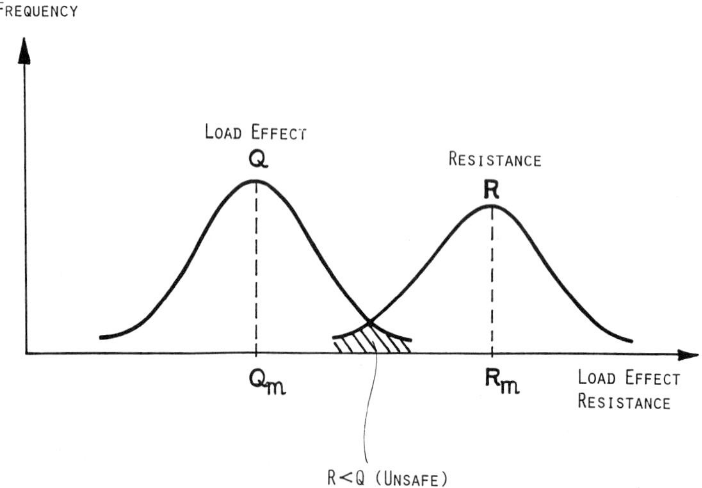

FIGURE 46.3 Frequency distribution of load effect and resistance.

Q, as shown by the shaded portion in the figure where the R and Q curves overlap. The smaller this shaded area, the less likely that the structure will fail. In actual design, a **resistance factor** ϕ is applied to the nominal resistance of the structural component to account for any uncertainties associated with the determination of its strength, and a **load factor** γ is applied to each load type to account for the uncertainties and difficulties associated with determining its actual load magnitude. Different load factors are used for different load types to reflect the varying degree of uncertainty associated with the determination of load magnitudes. In general, a lower load factor is used for a load that is more predictable and a higher load factor is used for a load that is less predictable. Mathematically, the LRFD format takes the form

$$\phi R_n \geq \sum_{i=1}^{m} \gamma_i Q_{ni} \tag{46.3}$$

where ϕR_n represents the **design** (or usable) **strength** and $\sum \gamma_i Q_{ni}$ represents the required strength or load effect for a given load combination. For steel building design, the values of ϕ are given by the AISC Specification and are summarized in Table 46.5. Table 46.6 shows the load combinations to be used on the right-hand side of Eq. (46.3). For a safe design, all load combinations should be investigated, and the design is based on the worst-case scenario.

LRFD is based on the limit state design concept. A **limit state** is defined as a condition in which a structure or structural component becomes unsafe (that is, a violation of the strength limit state) or unsuitable for its intended function (that is, a violation of the serviceability limit state). In a limit state design, the structure or structural component is designed in accordance to its limits of usefulness, which may be strength-related or serviceability-related.

The American Association of State Highway and Transportation Officials (AASHTO) *Standard Specifications for Highway Bridges* [AASHTO, 1992] provides a different set of load combinations for the design of highway bridges. The resistance factor is taken as unity, while the load factor γ varies from values of 1.2 to 1.3, depending on the load group. The γ factor is to be used in conjunction with additional load factors (the β factors) for each load group to obtain the factored load for design. Since the resistance factor assumes no value other than unity, the limit state design in AASHTO is termed *load factor design* rather than *load and resistance factor design.*

Allowable stress design has been the norm of the design profession for decades. Nevertheless, the worldwide trend is toward the limit state approach to design. This change in design philosophy is evidenced by the adoption and general usage of various limit state specifications in Canada, Europe, and Japan. In view of this trend and in cognizance of the likelihood that LRFD will be the mainstream design method in the 21st century, only LRFD provisions will be discussed in this chapter. Information on allowable stress design or plastic design methodologies can be found in the list of publications at the end of this chapter.

46.3 Tension Members

Limit State Design of Tension Member Components

According to the AISC-LRFD Specification [AISC, 1993], tension members designed to resist a factored axial force of P_u, calculated using the load combinations shown in Table 46.6, must satisfy the condition

$$\phi_t P_n \geq P_u \tag{46.4}$$

The design strength $\phi_t P_n$ is evaluated based on the limit states of (1) yielding in gross section; (2) fracture in effective net section; and (3) block shear.

Table 46.5 AISC Resistance Factors

Structural Component	Limit State	Value of ϕ
Tension members	Yielding	0.90
	Fracture	0.75
Pin-connected members	Tension	0.75
	Shear	0.75
	Bearing	0.75
Columns	Buckling	0.85
Beams, plate girders	Flexure	0.90
	Shear	0.90
Beam flanges	Bending	0.90
Beam webs	Local web yielding	1.00
	Web crippling	0.75
	Sidesway web buckling	0.85
	Compression buckling	0.90
A307 bolts	Tension	0.75
	Shear in bearing-type connections	0.75
High-strength bolts	Tension	0.75
	Shear in bearing-type connections	0.75
	Shear in slip-critical joints	1.00 (0.85 for long-slotted holes when the load is in the direction of the slot)
Bolt holes	Bearing	0.75
	Shear rupture	0.75
Complete-penetration groove welds	Tension or compression normal to effective area	0.90 (base material)
	Tension or compression parallel to axis of welds	0.90 (base material)
	Shear on effective area	0.90 (base material) 0.80 (weld electrode)
Partial-penetration groove welds	Tension normal to effective area	0.90 (base material) 0.80 (weld electrode)
	Compression normal to effective area	0.90 (base material)
	Tension or compression parallel to axis of weld	0.90 (base material)
	Shear parallel to axis of welds	0.75 (base material or weld electrode)
Fillet welds	Tension or compression parallel to axis of welds	0.90 (base material)
	Shear on effective area	0.75 (base material or weld electrode)
Plug or slot welds	Shear parallel to faying surfaces (on effective area)	0.75 (base material or weld electrode)
Connecting elements (splice, gusset plates)	Yielding	0.90
	Fracture	0.75
	Block shear	0.75
	Shear	0.90
Concrete pedestals	Bearing	0.60

Table 46.6 Load Factors and Load Combinations

$1.4D$

$1.2D + 1.6L + 0.5(L_r \text{ or } S \text{ or } R)$

$1.2D + 1.6(L_r \text{ or } S \text{ or } R) + (0.5L \text{ or } 0.8W)$

$1.2D + 1.3W + 0.5L + 0.5(L_r \text{ or } S \text{ or } R)$

$1.2D \pm 1.0E + 0.5L + 0.2S$

$0.9D \pm (1.3W \text{ or } 1.0E)$

where

D = dead load
L = live load
L_r = roof live load
W = wind load
S = snow load
E = earthquake load
R = nominal load due to initial rainwater or ice exclusive of the ponding contribution

The load factor on L in the third, fourth, and fifth load combinations shown above must equal 1.0 for garages, areas occupied as places of public assembly, and all areas where the live load is greater than 100 psf.

Yielding in Gross Section

Yielding in the cross section away from the joint should be avoided to prevent excessive deformation that results when steel yields. The design strength for this limit state is evaluated from the equation

$$\phi_t P_n = 0.90[F_y A_g] \tag{46.5}$$

where

0.90 = resistance factor for tension
F_y = specified minimum yield stress of the material
A_g = gross cross-sectional area of the member

Fracture in Effective Net Section

Fracture of the cross section at the joint should be avoided, to prevent the loss of load-carrying capacity of the member. The design strength for this limit state is evaluated from the equation

$$\phi_t P_n = 0.75[F_u A_e] \tag{46.6}$$

where

0.75 = resistance factor for fracture in tension
F_u = specified minimum tensile strength
A_e = effective net cross-sectional area of the member

The effective net area is calculated from the equation

$$\begin{aligned} A_e &= U A_n \\ &= U\left[A_g - \sum_{i=1}^{m} d_{ni} t_i + \sum_{j=1}^{k}\left(\frac{s^2}{4g}\right)_j t_j\right] \end{aligned} \tag{46.7}$$

where U is a reduction coefficient given by Munse and Chesson [1963]:

$$U = 1 - \frac{\bar{x}}{l} \leq 0.90 \tag{46.8}$$

in which l is the length of the connection (for bolted connections, l is the out-to-out distance of the extreme bolts; for welded connections, l is the length of the welds) and $\bar{x}$ is the connection eccentricity as shown in Fig. 46.4. U should be calculated using the maximum value of $\bar{x}$. This reduction coefficient is introduced to account for the shear lag effect that arises when some elements of the cross section in a joint are not connected, rendering the connection less effective

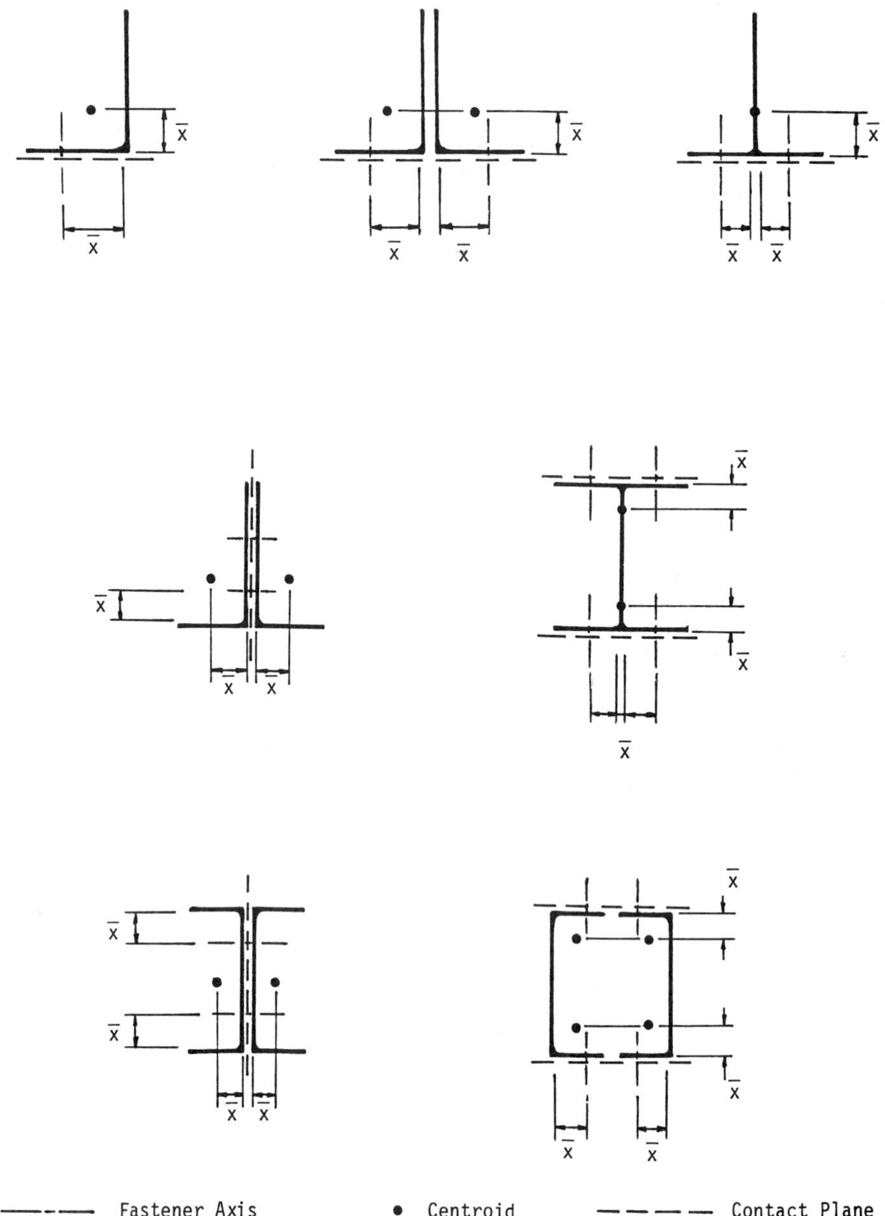

FIGURE 46.4 Definition of $\bar{x}$ for selected cross sections.

in transmitting the applied load. The terms in brackets in Eq. (46.7) constitute the so-called *net section*, A_n. The various terms are defined as follows.

A_g = gross cross-sectional area

d_n = nominal diameter of the hole (bolt cutout). It is taken as the nominal bolt diameter plus 1/8 in. (3.2 mm) for tension fracture (i.e., the fracture section is normal to the direction of the applied load), or the nominal bolt diameter plus 1/16 in. (1.6 mm) for shear fracture (i.e., the fracture section is parallel to the direction of the applied load).

t = thickness of the component element

s = longitudinal center-to-center spacing (pitch) of any two consecutive fasteners in a chain of staggered holes

g = transverse center-to-center spacing (gage) between two adjacent fastener gage lines in a chain of staggered holes

The second term inside the brackets of Eq. (46.7) accounts for loss of material due to bolt cutouts; the summation is carried for all bolt cutouts lying on the failure line. The last term inside the brackets of Eq. (46.7) indirectly accounts for the effect of the existence of a combined stress state (tensile and shear) along an inclined failure path associated with staggered holes. The summation is carried for all staggered paths along the failure line. This term vanishes if the holes are not staggered. Normally, it is necessary to investigate different failure paths that may occur in a connection. The critical failure path is the one giving the smallest value for A_e.

Block Shear

Block shear failure occurs when a segment of the connecting member is torn out as a result of the combined effects of tension and shear. Block shear must be checked if the load is transmitted by some but not all of the component elements of the cross section. The design strength for block shear is determined from the following two conditions:

Tension Fracture—Shear Yield.

$$\phi_t P_n = 0.75[0.60 F_y A_{gv} + F_u A_{nt}] \quad (46.9)$$

Shear Fracture—Tension Yield.

$$\phi_t P_n = 0.75[0.60 F_u A_{nv} + F_y A_{gt}] \quad (46.10)$$

where

0.75 = resistance factor for block shear

F_y, F_u = specified minimum yield stress and tensile strengths, respectively

A_{gv} = gross area of the torn-out segment under shear

A_{nt} = net area of the torn-out segment under tension

A_{nv} = net area of the torn-out segment under shear

A_{gt} = gross area of the torn-out segment under tension

Normally, it is necessary to investigate both the tension fracture—shear yield and the shear fracture—tension yield criteria. The *larger* of the two values calculated is to be used for $\phi_t P_n$.

Example. Select a double-channel tension member, as shown in Fig. 46.5(a), to carry a dead load D of 40 kips and a live load L of 100 kips. The member is 15 feet long. Six 1 in. diameter A325 bolts in standard-size holes are used to connect the member to a 3/8 in. gusset plate. Use A36 steel (F_y = 36 ksi, F_u = 58 ksi) for all the connected parts.

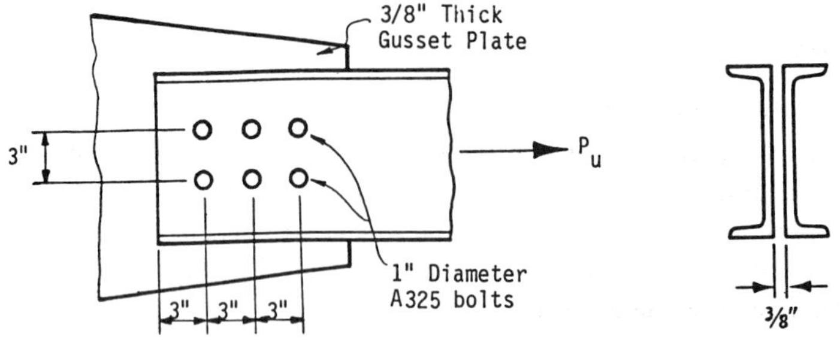

(a) A Double Channel Tension Member

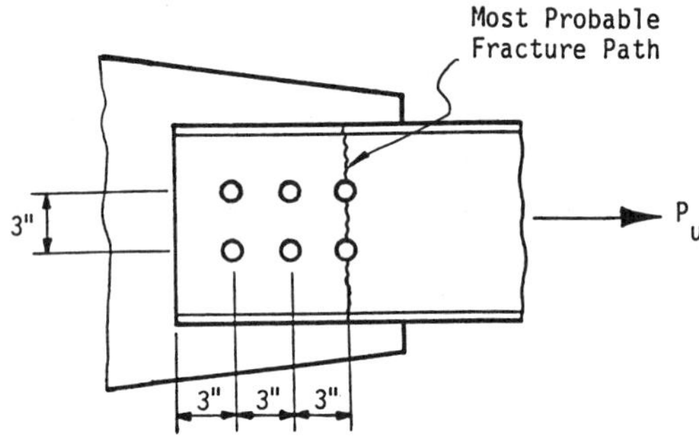

(b) Fracture Failure

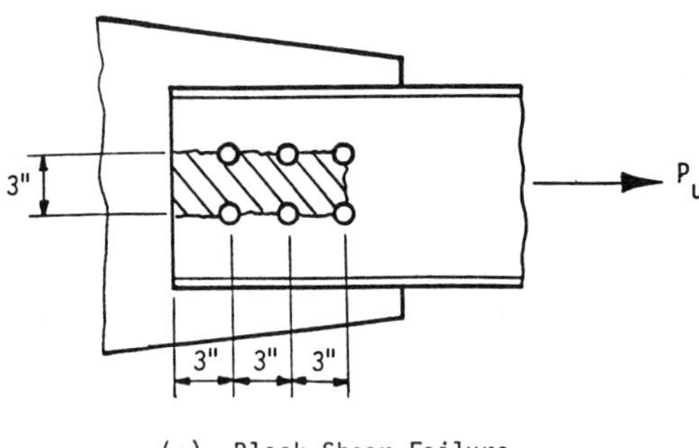

(c) Block Shear Failure

FIGURE 46.5 Design of a double-channel tension member. (1 in. = 25.4 mm.)

Load Combinations. From Table 46.6, the applicable load combinations are

$$1.4D = 1.4(40) = 56 \text{ kips}$$

$$1.2D = 1.6L = 1.2(40) + 1.6(100) = 208 \text{ kips}$$

The design of the tension member is to be based on the larger of the two, i.e., 208 kips, and so *each* channel is expected to carry 104 kips.

Yielding in Gross Section. Using Eqs. (46.4) and (46.5), the gross area required to prevent cross section yielding is

$$0.90[F_y A_g] \geq P_u$$

$$0.09[(36)(A_g)] \geq 104$$

$$(A_g)_{\text{req'd}} \geq 3.21 \text{ in.}^2$$

From the section properties table contained in the AISC LRFD Manual [AISC, 1994] one can select the following trial sections: C8 × 11.5 ($A_g = 3.38$ in.^{2}), C9 × 13.4 ($A_g = 3.94$ in.^{2}), C8 × 13.75 ($A_g = 4.04$ in.^{2}).

Check for the Limit State of Fracture in Effective Net Area. The above sections are checked for the limiting state of fracture in the following table.

Section	A_g (in.^{2})	t_w (in.)	$\bar{x}$ (in.)	U^a	$A_e{}^b$ (in.^{2})	$\phi_t P_n$ (kips)
C8 × 11.5	3.38	0.220	0.571	0.90	2.6	113.1
C9 × 13.4	3.94	0.233	0.601	0.90	3.07	133.5
C8 × 13.75	4.04	0.303	0.553	0.90	3.02	131.4

[a] Eq. (46.8).
[b] Eq. (46.7), Fig. 46.5 (b).

From the last column of the above table, it can be seen that fracture is not a problem for any of the trial sections.

Check for the Limit State of Block Shear. Figure 46.5(c) shows a possible block shear failure mode. To avoid block shear failure, the larger value obtained from Eq. (46.9) and Eq. (46.10) should exceed $P_u = 104$ kips.

For the C8 × 11.5 section:

$$\begin{aligned} A_{gv} &= 2(9)(0.220) = 3.96 \text{ in.}^2 \\ A_{nv} &= A_{gv} - 5(1 + 1/16)(0.220) = 2.79 \text{ in.}^2 \\ A_{gt} &= (3)(0.220) = 0.66 \text{ in.}^2 \\ A_{nt} &= A_{gt} - 1(1 + 1/8)(0.220) = 0.41 \text{ in.}^2 \end{aligned}$$

Substituting these results into Eqs. (46.9) and (46.10), we obtain $\phi_v P_n = 82$ and 90.6 kips, respectively. Since the larger of the two values, i.e., 90.6 kips, does not exceed $P_u = 104$ kips, the C8 × 11.5 section is not adequate. Significant increase in block shear strength is not expected from the C9 × 13.4 section, because its web thickness t_w is just slightly over that of the C8 × 11.5 section. As a result, we shall check the adequacy of the C8 × 13.75 section instead.

For the C8 × 13.75 section:

$$\begin{aligned} A_{gv} &= 2(9)(0.303) = 5.45 \text{ in.}^2 \\ A_{nv} &= A_{gv} - 5(1 + 1/16)(0.303) = 3.84 \text{ in.}^2 \\ A_{gt} &= (3)(0.303) = 0.91 \text{ in.}^2 \\ A_{nt} &= A_{gt} - 1(1 + 1/8)(0.303) = 0.57 \text{ in.}^2 \end{aligned}$$

Substituting these results into Eqs. (46.9) and (46.10), we obtain $\phi_v P_n = 113$ and 125 kips, respectively. The larger of these two values, i.e., 125 kips, exceeds the required strength P_u of 104 kips. Therefore, block shear will not be a problem for the C8 × 13.75 section.

Check for the Limiting Slenderness Ratio. Using the parallel axis theorem, the least radius of gyration of the double-channel cross section is calculated to be 0.96 in. Therefore, $L/r = (15)(12)/0.96 = 187.5$, which is less than the recommended maximum value of 300.

Check for the Adequacy of the Connection. The calculations are shown in an example in the section 46.12.

Longitudinal Spacing of Connectors. The spacing of connectors in built-up tension members consist of elements in continuous contact shall conform to the spacing requirements for fasteners discussed in section 46.12. If tie plates are used, the longitudinal spacing of fasteners shall not exceed 6 inches.

Use 2C8 × 13.75 connected intermittently at 6-in. interval.

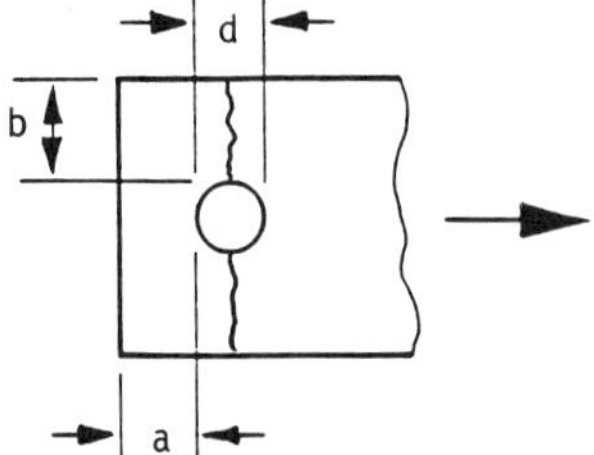

TENSION FRACTURE

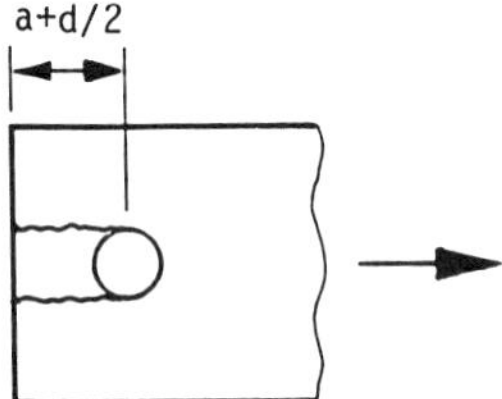

LONGITUDINAL SHEAR

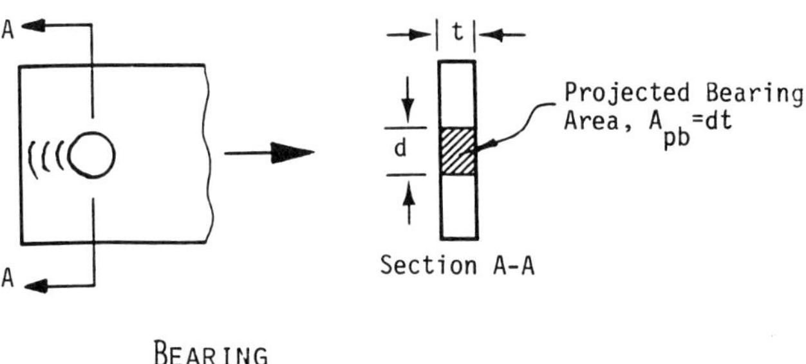

BEARING

FIGURE 46.6 Failure modes of pin-connected members.

Pin-Connected Members

Pin-connected members shall be designed to satisfy the limit states of (1) tension on the effective net area; (2) shear on the effective area; and (3) bearing on the projected pin area (Fig. 46.6).

Tension on Effective Net Area

To prevent fracture of the net section, the design strength for this limit state, calculated from the following equation, should not exceed P_u:

$$\phi_t P_n = 0.75[2\, t b_{\text{eff}} F_u] \tag{46.11}$$

Shear on Effective Area

To prevent failure by longitudinal shear, the design strength calculated from the following equation should not exceed P_u:

$$\phi_{sf} P_n = 0.75[0.60 A_{sf} F_u] \tag{46.12}$$

Bearing on Projected Pin Area

To prevent bearing failure on milled surfaces, the design strength calculated from the following equation should not exceed P_u:

$$\phi P_n = 0.75[1.8 A_{pb} F_y] \tag{46.13}$$

The terms in the foregoing equations are given as follows:

a = shortest distance from edge of the pin hole to the edge of the member measured in the direction of the force

A_{pb} = projected bearing area = dt

A_{sf} = $2t(a + d/2)$

b_{eff} = $2t + 0.63$, but not more than the actual distance from the edge of the hole to the edge of the part measured in the direction normal to the applied force

d = pin diameter

t = plate thickness

Threaded Rods

According to the AISC LRFD Specification [AISC, 1993], threaded rods designed as tension members shall have a gross area A_b given by

$$A_b \geq \frac{P_u}{\phi 0.75 F_u} \tag{46.14}$$

where

A_b = gross area of the rod computed using a diameter measured to the outer extremity of the thread

P_u = factored tensile load

ϕ = resistance factor, given as 0.75

F_u = specified minimum tensile strength

Limiting Slenderness Ratio

Except for rods in tension, members designed on the basis of tensile force preferably should have a slenderness ratio L/r not exceeding 300, where L is the member length and r is the least radius of gyration of the cross section.

46.4 Compression Members

Compression members can fail by yielding, inelastic buckling, or elastic buckling depending on the slenderness ratio of the members. Members with low slenderness ratios tend to fail by yielding, whereas members with high slenderness ratios tend to fail by elastic buckling. Most compression members used in construction have intermediate slenderness ratios, and so the predominant mode of failure is inelastic buckling. Member buckling can occur in one of three different modes: flexural, torsional, and flexural-torsional. *Flexural buckling* occurs in members with doubly symmetric or doubly antisymmetric cross sections (e.g., I, Z) and in members with singly symmetric sections (e.g., C, T, equal-legged L, double L) when such sections are buckled about an axis that is perpendicular to the axis of symmetry. *Torsional buckling* occurs in members with doubly symmetric sections such as cruciform or built-up shapes with very thin walls. *Flexural-torsional buckling* occurs in members with singly symmetric cross sections (e.g., C, T, equal-legged L, double L) when such sections are buckled about the axis of symmetry and in members with unsymmetric cross sections (e.g., unequal-legged L). Normally, torsional buckling of symmetric shapes and flexural-torsional buckling of unsymmetric shapes are not important in the design of hot-rolled compression members: either they do not govern or their buckling strengths do not differ significantly from the corresponding weak-axis flexural buckling strengths. However, torsional and flexural-torsional buckling modes may govern for sections that have relatively thin component plates.

In addition to slenderness ratio and cross-sectional shape, the behavior of compression members is affected by the relative thickness of the component elements that constitute the cross section. The relative thickness of a component element is quantified by the width-thickness ratio (b/t) of the element. The width-thickness ratios of some selected steel shapes are shown in Fig. 46.7. If the width-thickness ratio falls within a limiting value denoted by the AISC LRFD Specification [AISC, 1993] as λ_r (Table 46.7), local buckling of the component element will not occur. However, if the width-thickness ratio exceeds λ_r, consideration of local buckling in the design of the compression member is required.

Design Criterion for Compression Members

Compression members are to be designed so that the following criterion is satisfied:

$$\phi_c P_n \geq P_u \tag{46.15}$$

where $\phi_c P_n$ is the design strength for compression and P_u is the factored axial force. $\phi_c P_n$ is to be calculated as follows:

Flexural Buckling (Width-Thickness Ratio $\leq \lambda_r$)

For $\lambda_c \leq 1.5$ (inelastic buckling without local buckling),

$$\phi_c P_n = 0.85[A_g(0.658^{\lambda_c^2})F_y] \tag{46.16a}$$

and, for $\lambda_c > 1.5$ (elastic buckling without local buckling),

$$\phi_c P_n = 0.85\left[A_g\left(\frac{0.877}{\lambda_c^2}\right)F_y\right] \tag{46.16b}$$

where

$\lambda_c = (KL/r\pi)\sqrt{(F_y/E)}$
A_g = gross cross-sectional area
F_y = specified minimum yield stress
E = modulus of elasticity
K = effective length factor
L = unbraced member length
r = radius of gyration about axis of buckling

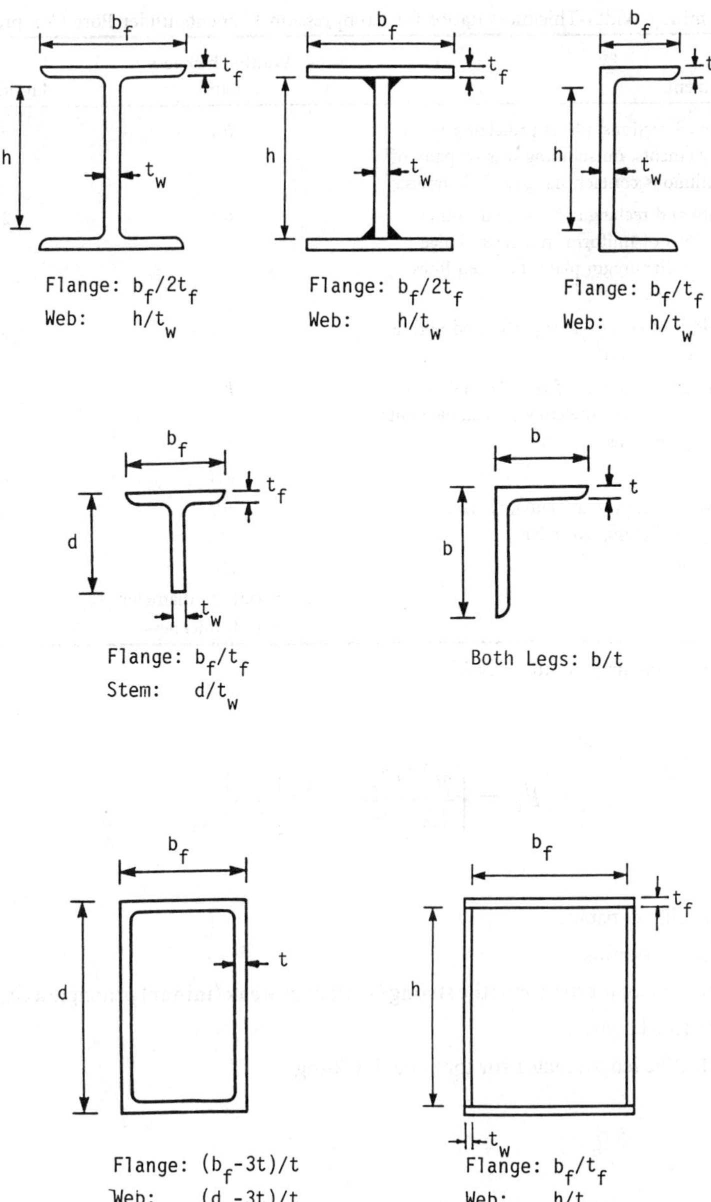

FIGURE 46.7 Definition of width-thickness ratio for selected cross sections.

A schematic plot of Eqs. (46.16a) and (46.16b) is shown in Fig. 46.8. The design is considered satisfactory if this strength curve envelops the factored load for a given value of λ_c.

Torsional Buckling (Width-Thickness Ratio $< \lambda_r$)

$\phi_c P_n$ is calculated from Eq. (46.16a) or Eq. (46.16b), whichever applies, but with λ_c replaced by λ_e, where

$$\lambda_e = \sqrt{F_y/F_e} \tag{46.17}$$

Table 46.7 Limiting Width-Thickness Ratios for Compression Elements under Pure Compression

Component Element	Width-Thickness Ratio	Limiting Value, λ_r
Flanges of I-shaped sections; plates projecting from compression elements; outstanding legs of pairs of angles in continuous contact; flanges of channels.	b/t	$95/\sqrt{F_y}$
Flanges of square and rectangular box and hollow structural sections of uniform thickness; flange cover plates and diaphragm plates between lines of fasteners or welds.	b/t	$238/\sqrt{F_y}$
Unsupported width of cover plates perforated with a succession of access holes	b/t	$317/\sqrt{F_y}$
Legs of single angle struts; legs of double angle struts with separators; unstiffened elements (i.e., elements supported along one edge)	b/t	$76/\sqrt{F_y}$
Stems of tees	d/t	$127/\sqrt{F_y}$
All other uniformly compressed elements (i.e., elements supported along two edges)	b/t h/t_w	$253/\sqrt{F_y}$
Circular hollow sections	D/t D = outside diameter t = wall thickness	$3300/F_y$

F_y = specified minimum yield stress in ksi.

with

$$F_e = \left[\frac{\pi^2 E C_w}{(K_z L)^2} + GJ\right]\frac{1}{I_x + I_y} \tag{46.18}$$

in which

C_w = warping constant
G = shear modulus
I_x, I_y = moment of inertia about the **strong** (**major**) or **weak** (**minor**) principal **axis,** respectively
J = torsional constant
K_z = effective length factor for torsional buckling

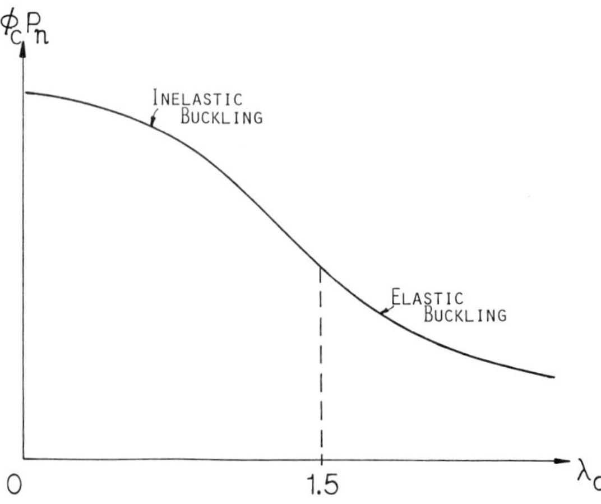

FIGURE 46.8 Column strength curve.

Table 46.8 Approximate Equations for C_w and J

Structural Shape	Warping Constant, C_w	Torsional Constant, J
I	$h'^2 I_c I_t/(I_c + I_t)$	$\sum C_i (b_i t_i^3/3)$
C	$(b' - 3E_0)h'^2 b'^2 t_f/6 + E_0^2 I_x$ where $E_0 = b'^2 t_f/(2b' t_f + h' t_w/3)$	where b_i = Width of component element i t_i = Thickness of component element i C_i = Correction factor for component element i (see figure below)
T	$(b_f^3 t_f^3/4 + h''^3 t_w^3)/36$ (≈ 0 for small t)	
L	$(l_1^3 t_1^3 + l_2^3 t_2^3)/36$ (≈ 0 for small t)	

b' = distance measured from toe of flange to centerline of web
h' = distance between centerline lines of flanges
h'' = distance from centerline of flange to tip of stem
l_1, l_2 = length of the legs of the angle
t_1, t_2 = thickness of the legs of the angle
b_f = flange width
t_f = average thickness of flange
t_w = thickness of web
I_c = moment of inertia of compression flange taken about the axis of the web
I_t = moment of inertia of tension flange taken about the axis of the web
I_x = moment of inertia of the cross section taken about the major principal axis

The warping constant C_w and the torsional constant J are tabulated for various steel shapes in the AISC LRFD Manual [AISC, 1994]. Equations for calculating approximate values for these constants for some commonly used steel shapes are shown in Table 46.8.

Flexural-Torsional Buckling (Width-Thickness Ratio $\leq \lambda_r$)

The design calculation is the same as for torsional buckling, except that F_e is now given by the following formulas:

For Singly Symmetric Sections

$$F_e = \frac{F_{es} + F_{ez}}{2H}\left[1 - \sqrt{\frac{1 - 4F_{es}F_{ez}H}{(F_{es} + F_{ez})^2}}\right] \tag{46.19}$$

where

F_{es} = F_{ex} if x is the axis of symmetry of the cross section, or F_{ey} if y is the axis of symmetry of the cross section

$F_{ex} = \pi^2 E/(K_x L/r_x)^2$

$F_{ey} = \pi^2 E/(K_y L/r_x)^2$

$H = 1 - (x_0^2 + y_0^2)/r_0^2$

in which

K_x, K_y = effective length factors for buckling about the x and y axes, respectively
L = unbraced member length
r_x, r_y = radii of gyration about the x and y axes, respectively
x_0, y_0 = the shear center coordinates with respect to the centroid (Fig. 46.9)
$r_0^2 = x_0^2 + y_0^2 + r_x^2 + r_y^2$

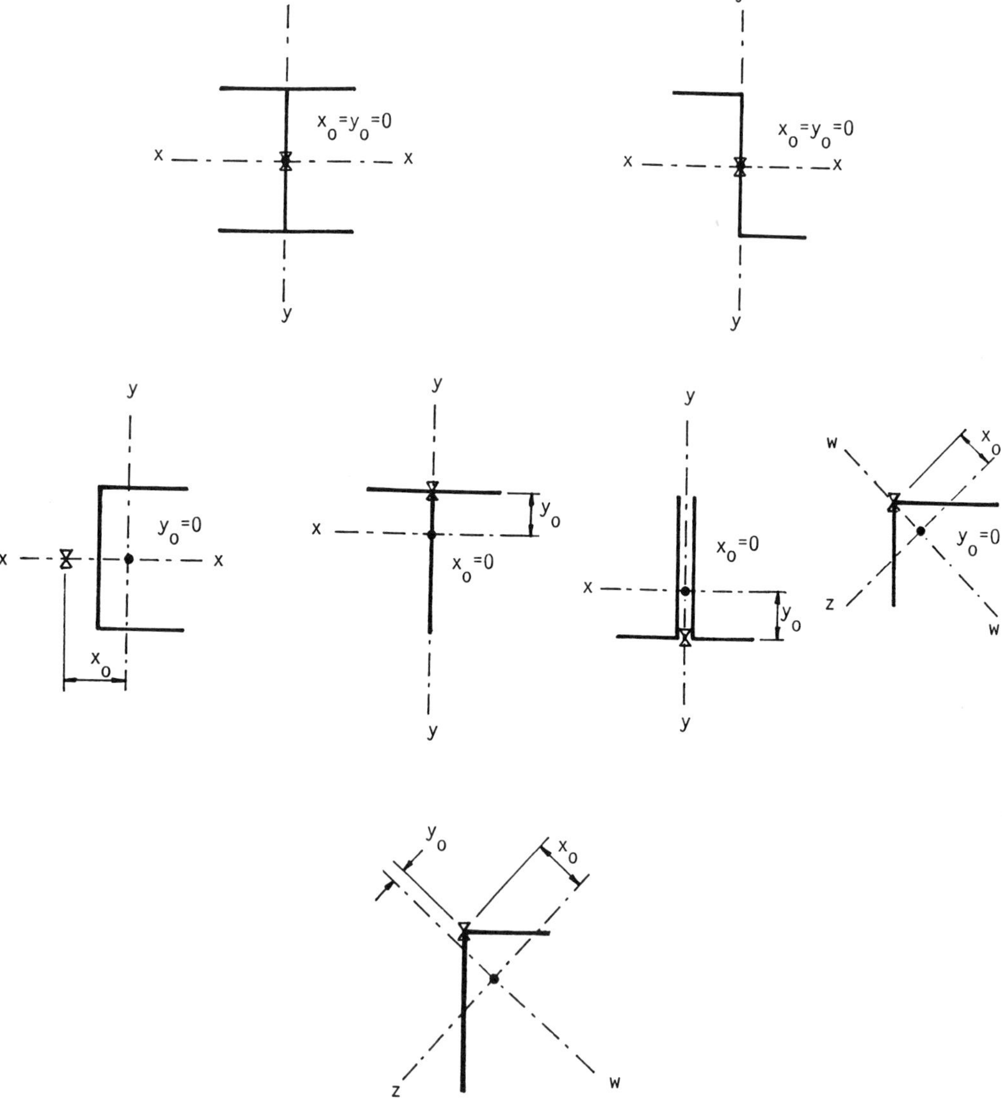

FIGURE 46.9 Location of shear center for selected cross sections.

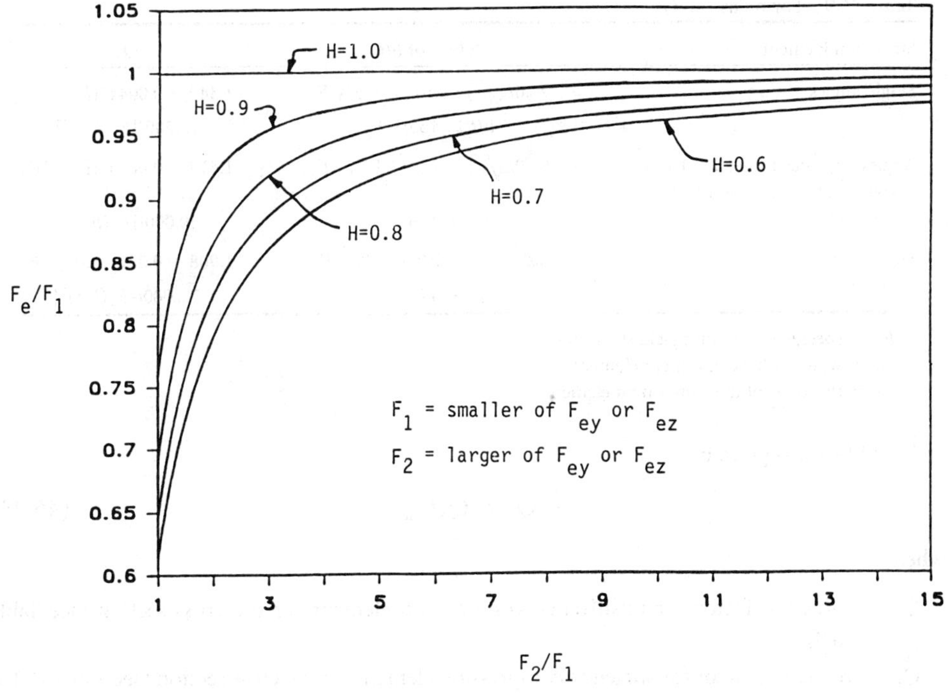

FIGURE 46.10 Chart for F_e.

Numerical values for r_0 and H are given for hot-rolled I, C, T and L and double-L sections in the AISC LRFD Manual [AISC, 1994].

In lieu of Eq. (46.19), F_e can be determined graphically from Fig. 46.10 [Zahn and Iwankiw, 1989].

For Unsymmetric Sections. F_e is to be obtained by solving the following cubic equation:

$$(F_e - F_{ex})(F_e - F_{ey})(F_e - F_{ez}) - F_e^2(F_e - F_{ey})\left(\frac{x_0}{r_0}\right)^2 - F_e^2(F_e - F_{ex})\left(\frac{y_0}{r_0}\right)^2 = 0 \tag{46.20}$$

The terms in the above equation are defined the same as in Eq. (46.19).

Local Buckling (Width-Thickness Ratio $\geq \lambda_r$)

Local buckling in a component element of the cross section is accounted for in design by introducing a reduction factor Q in the column equations, Eqs. (46.16a), (46.16b), as follows:

Inelastic Buckling with Local Buckling. For $\lambda_e\sqrt{Q} \leq 1.5$

$$\phi_c P_n = 0.85[A_g Q(0.658^{Q\lambda_e^2})F_y] \tag{46.21}$$

Elastic Buckling with Local Buckling. For $\lambda_e\sqrt{Q} > 1.5$

$$\phi_c P_n = 0.85\left[A_g\left(\frac{0.877}{\lambda_e^2}\right)F_y\right] \tag{46.22}$$

Table 46.9 Formulas for Q_s

Structural Element	Range of b/t	Q_s
Single angles	$76.0/\sqrt{F_y} < b/t < 155/\sqrt{F_y}$	$1.340 - 0.00447(b/t)\sqrt{F_y}$
	$b/t \geq 155/\sqrt{F_y}$	$15{,}500/[F_y(b/t)^2]$
Angles or plates projecting from rolled columns or other compression members	$95.0/\sqrt{F_y} < b/t < 176/\sqrt{F_y}$	$1.415 - 0.00437(b/t)\sqrt{F_y}$
	$b/t \geq 176/\sqrt{F_y}$	$20{,}000/[F_y(b/t)^2]$
Stems of tees	$127/\sqrt{F_y} < b/t < 176/\sqrt{F_y}$	$1.908 - 0.00715(b/t)\sqrt{F_y}$
	$b/t \geq 176/\sqrt{F_y}$	$20,000/[F_y(b/t)^2]$

F_y = specified minimum yield stress in ksi
b = width of the component element
t = thickness of the component element

The Q factor is given by

$$Q = Q_s Q_a \tag{46.23}$$

where

Q_s = reduction factor for unstiffened compression elements of the cross section (see Table 46.9)

Q_a = reduction factor for stiffened compression elements of the cross section (see Table 46.10)

Built-Up Compression Members

Built-up members are members made by bolting or welding together two or more standard structural shapes. For a built-up member to be fully effective (i.e., if all component structural shapes are to act as one unit rather than as individual units), the following conditions must be satisfied:

1. The ends of the built-up member must be prevented from slippage during buckling.
2. Adequate fasteners must be provided along the length of the member.
3. The fasteners must be able to provide sufficient gripping force on all the component shapes being connected.

Table 46.10 Formula for Q_a

The effective area is equal to the summation of the effective areas of the stiffened elements of the cross section. The effective area of a stiffened element is equal to the product of its thickness t and its effective width b_e given by the following formulas:

For flanges of square and rectangular sections of uniform thickness:

$$b_e = \frac{326t}{\sqrt{f}}\left[1 - \frac{64.9}{(b/t)\sqrt{f}}\right] \leq b$$

For other uniformly compressed elements:

$$b_e = \frac{326t}{\sqrt{f}}\left[1 - \frac{57.2}{(b/t)\sqrt{f}}\right] \leq b$$

where

b = actual width of the stiffened element
f = computed elastic compressive stress in the stiffened elements, in ksi

Condition 1 is satisfied if continuous welds are used throughout the length of the built-up compression member. all component shapes in contact at the ends of the member are connected by a weld having a length not less than the maximum width of the member, or by fully tightened bolts spaced longitudinally not more than four diameters apart for a distance equal to $1\frac{1}{2}$ times the maximum width of the member.

Condition 2 is satisfied if continuous welds are used throughout the length of the built-up compression member.

Condition 3 is satisfied if either welds or fully tightened bolts are used as the fasteners.

While condition 1 is mandatory, conditions 2 and 3 can be violated in design. If condition 2 or condition 3 is violated, the built-up member is not fully effective, and slight slippage among component shapes may occur. To account for the decrease in capacity due to slippage, a modified slenderness ratio is used for the computation of the design compressive strength when buckling of the built-up member is about an axis coincident with or parallel to at least one plane of contact for the component shapes. The modified slenderness ratio $(KL/r)_m$ is given as follows:

Condition 2 Violated. If intermittent welds or fully tightened bolts are used

$$\left(\frac{KL}{r}\right)_m = \sqrt{\left(\frac{KL}{r}\right)_0^2 + 0.82\frac{(h/2r_{ib})^2}{1+(h/2r_{ib})^2}\left(\frac{a}{r_{ib}}\right)^2} \tag{46.24}$$

Condition 3 Violated. If snug-tight bolts are used,

$$\left(\frac{KL}{r}\right)_m = \sqrt{\left(\frac{KL}{r}\right)_0^2 + \left(\frac{a}{r_i}\right)^2} \tag{46.25}$$

In the foregoing equations, $(KL/r)_0 = (KL/r)_x$ if the buckling axis is the x axis and at least one plane of contact between component shapes is parallel to that axis; $(KL/r)_o = (KL/r)_y$ if the buckling axis is the y axis and at least one plane of contact is parallel to that axis. a is the distance between fasteners measured along the longitudinal axis of the member. r_i is the minimum radius of gyration of the component element. r_{ib} is the radius of gyration of the component element relative to its centroidal axis parallel to the member axis of buckling. h is the distance between centroids of component elements perpendicular to the member buckling axis.

No modification to (KL/r) is necessary if the buckling axis is perpendicular to the planes of contact of the component shapes. Modifications to both $(KL/r)_x$ and $(KL/r)_y$ are required if the built-up member is so constructed that planes of contact exist in both the x and y directions of the cross section.

Once the slenderness ratio is computed, the design compression strength is to be calculated from Eqs. (46.16a,b) to (46.22), whichever applies depending on the cross section geometry (doubly symmetric, singly symmetric, unsymmetric) and component element width-thickness ratio (possibility of local buckling) of the built-up shape.

An additional requirement for the design of built-up members is that a/r_i does not exceed $\frac{3}{4}$ of the governing slenderness ratio of the built-up member. This provision is provided to prevent component shape buckling from occurring between adjacent fasteners before the built-up member buckles overall.

Example. Determine the size of a pair of cover plates to be bolted, using snug-tight bolts, to the flanges of a W24 × 229 section, as shown in Fig. 46.11, to increase its design strength $\phi_c P_n$ by 15%. Also, determine the spacing of the bolts in the longitudinal direction of the built-up column. The effective lengths of the section about the major $(KL)_x$ and minor $(KL)_y$ axes are both equal to 20 feet. A36 steel is to be used.

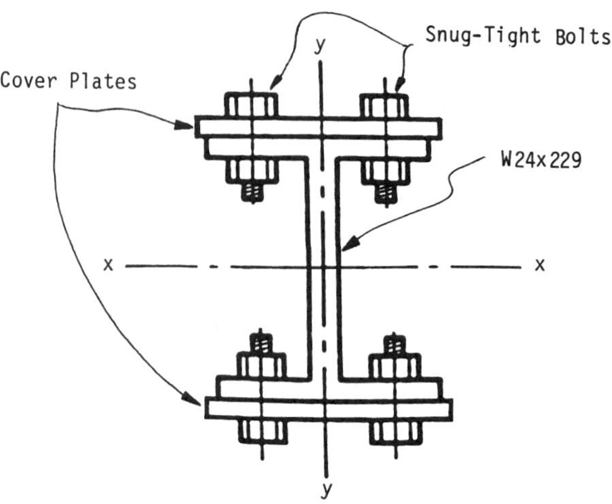

FIGURE 46.11 Design of cover plates for a compression member.

Determine Design Strength for the W24 × 229 Section. Since $(KL)_x = (KL)_y$ and $r_x > r_y$, $(KL/r)_y$ will be greater than $(KL/r)_x$ and the design strength will be controlled by flexural buckling about the minor axis. Using section properties, $r_y = 3.11$ in. and $A = 67.2$ in.^{2}, obtained from Part 1 of the AISC LRFD Manual (AISC, 1994), the slenderness parameter λ_c about the minor axis can be calculated as follows:

$$(\lambda_c)_y = \frac{1}{\pi}\left(\frac{KL}{r}\right)_y \sqrt{\frac{F_y}{E}} = \frac{1}{3.142}\left(\frac{20 \times 12}{3.11}\right)\sqrt{\frac{36}{29,000}} = 0.865$$

Since λ_c is less than 1.5, Eq. (46.16a) is to be used to calculate the design strength of the section:

$$\phi_c P_n = 0.85\left[67.2(0.658^{0.865^2})36\right] = 1503 \text{ kips}$$

Alternatively, the above value of $\phi_c P_n$ can be obtained directly from the column tables contained in Part 3 of the AISC LRFD Manual.

Determine Design Strength for the Built-Up Section. The built-up section is expected to possess a design strength that is 15% in excess of the design strength of the $W24 \times 229$ section, so

$$(\phi_c P_n)_{\text{req'd}} = (1.15)(1503) = 1728 \text{ kips}$$

Determine the Size of the Cover Plates. After cover plates are added, the resulting section is still doubly symmetric. Therefore, the overall failure mode is still flexural buckling. For flexural buckling about the minor axis (y-y), no modification to KL/r is required, since the buckling axis is perpendicular to the plane of contact of the component shapes and no relative movement between the adjoining parts is expected. However, for flexural buckling about the major (x-x) axis, modification to KL/r is required since the buckling axis is parallel to the plane of contact of the adjoining structural shapes, and slippage between the component pieces will occur. We shall design the cover plates assuming that flexural buckling about the minor axis will control, and check for flexural buckling about the major axis later.

A W24 × 229 section has a flange width of 13.11 inches; so, as a trial, use cover plates with widths of 13 inches, as shown in Figure 46.11(a). Denoting t as the thickness of the plates, we have

$$(r_y)_{\text{built-up}} = \sqrt{\frac{(I_y)_{\text{W-shape}} + (I_y)_{\text{plates}}}{A_{\text{W-shape}} + A_{\text{plates}}}} = \sqrt{\frac{651 + 183.1t}{67.2 + 26t}}$$

and

$$(\lambda_c)_{y,\text{ built-up}} = \frac{1}{\pi}\left(\frac{KL}{r}\right)_{y,\text{ built-up}}\sqrt{\frac{F_y}{E}} = 2.69\sqrt{\frac{67.2 + 26t}{651 + 183.1t}}$$

Assuming $(\lambda)_{y,\text{ built-up}}$ is less than 1.5, one can substitute the above expression for λ_c in Eq. (46.16a). With $\phi_c P_n$ equal to 1728, we can solve for t. The result is $t \approx$ ½ in. Backsubstituting $t =$ ½ into the above expression, we obtain $(\lambda)_{c,\text{built-up}} = 0.884$, which is indeed < 1.5. ***So try 13″ × ½″ cover plates.***

Check for Local Buckling. For the I section:

$$\text{Flange:} \quad \left[\frac{b_f}{2t_f} = 3.8\right] < \left[\frac{95}{\sqrt{F_y}} = 15.8\right]$$

$$\text{Web:} \quad \left[\frac{h_c}{t_w} = 22.5\right] < \left[\frac{253}{\sqrt{F_y}} = 42.2\right]$$

For the cover plates, if ¾-in. diameter bolts are used and assuming an edge distance of 1 ¼ in., the width of the plate between fasteners will be 13 − 2.5 = 10.5 in. Therefore, we have

$$\left[\frac{b}{t} = \frac{10.5}{1/2} = 21\right] < \left[\frac{238}{\sqrt{F_y}} = \frac{238}{\sqrt{36}} = 39.7\right]$$

Since the width-thickness ratios of all component shapes do not exceed the limiting width-thickness ratio for local buckling, local buckling is not a concern.

Check for Flexural Buckling about the Major (x-x) Axis. Since the built-up section is doubly symmetric, the governing buckling mode will be flexural buckling regardless of the axes. Flexural buckling will occur about the major axis if the modified slenderness ratio $(KL/r)_m$ about the major axis calculated using Eq. (46.25) exceeds $(KL/r)_y$. Otherwise, buckling will occur about the minor axis. In order to arrive at an optimal design, we shall determine the longitudinal fastener spacing a such that the modified slenderness ratio $(KL/r)_m$ about the major axis will be equal to $(KL/r)_y$. That is, we shall solve for a from the equation

$$\left[\left(\frac{KL}{r}\right)_m = \sqrt{\left(\frac{KL}{r}\right)_x^2 + \left(\frac{a}{r_i}\right)^2}\right] = \left[\left(\frac{KL}{r}\right)_y = 78.9\right]$$

In the foregoing equation $(KL/r)_x$ is the slenderness ratio about the major axis of the built-up section; r_i is the least radius of gyration of the component shapes, which in this case is the cover plate.

Substituting $(KL/r)_x = 21.56$, $r_i = r_{\text{cover-plate}} = \sqrt{(I/A)_{\text{cover-plate}}} = \sqrt{[(1/2)^2/112]} = 0.144$ into the foregoing equation, we obtain $a = 10.9$ in. Since $(KL) = 20$ ft., we shall use $a = 10$ in. for the longitudinal spacing of the fasteners.

Check for Component Shape Buckling between Adjacent Fasteners.

$$\left[\frac{a}{r_i} = \frac{10}{0.144} = 69.44\right] > \left[\frac{3}{4}\left(\frac{KL}{r}\right)_y = \frac{3}{4}(78.9) = 59.2\right]$$

Since the component shape buckling criterion is violated, we need to decrease the longitudinal spacing from 10 in. to 8 in.

Use 13″ × ½″ cover plates bolted to the flanges of the W24 × 229 section by ¾ in. diameter fully tightened bolts spaced 8 in. longitudinally.

Tapered Compression Members

For doubly symmetric web-tapered compression members having a linearly varied web depth with $d_L \leq$ smaller of $(d_0 + 0.268L, 7d_0)$, where d_L is the depth at the larger end of the member, d_0 is the depth at the smaller end of the member, and L is the length of the member, the design compression strength is given by Eqs. (46.16a) and (46.16b), provided that (1) A_g is calculated using the smallest area of the tapered member and (2) λ_c is replaced by λ_{eff}, where

$$\lambda_{\text{eff}} = \frac{S}{\pi}\sqrt{\frac{QF_y}{E}} \tag{46.26}$$

in which

S = $K_\lambda L/r_{0x}$ for strong axis buckling and KL/r_{0y} for weak axis buckling

K_λ = effective length factor about the x axis, obtained from a rational analysis; see, for example, charts by Lee, Morrell, and Ketter (1972)

K = effective length factor about the x axis (assuming the member is prismatic)

r_{0x}, r_{0y} = radii of gyration about the x and y axes, respectively, evaluated based on the smaller end of the member

F_y = specified minimum yield stress

E = modulus of elasticity

Q = $Q_s Q_a$ (see Tables 46.9 and 46.10)

Effective Length Factor

The effective length factor K is a factor which, when multiplied by the actual unbraced length L of an end-restrained compression member, will yield an equivalent pinned-ended member whose buckling strength is the same as that of the original end-restrained member. For a prismatic member, the effective length factor can be determined from Fig. 46.12 [AISC, 1994] or Fig. 46.13.

	(a)	(b)	(c)	(d)	(e)	(f)
Buckled shape of column is shown by dashed line	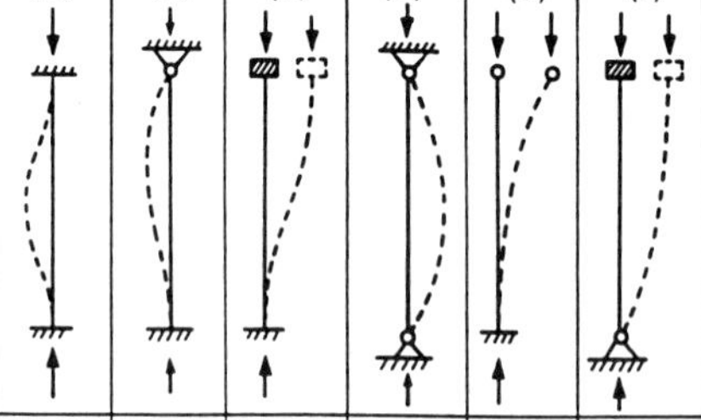					
Theoretical K value	0.5	0.7	1.0	1.0	2.0	2.0
Recommended design value when ideal conditions are approximated	0.65	0.80	1.2	1.0	2.10	2.0
End condition code	 Rotation fixed and translation fixed Rotation free and translation fixed Rotation fixed and translation free Rotation free and translation free					

FIGURE 46.12 *K* factor table.

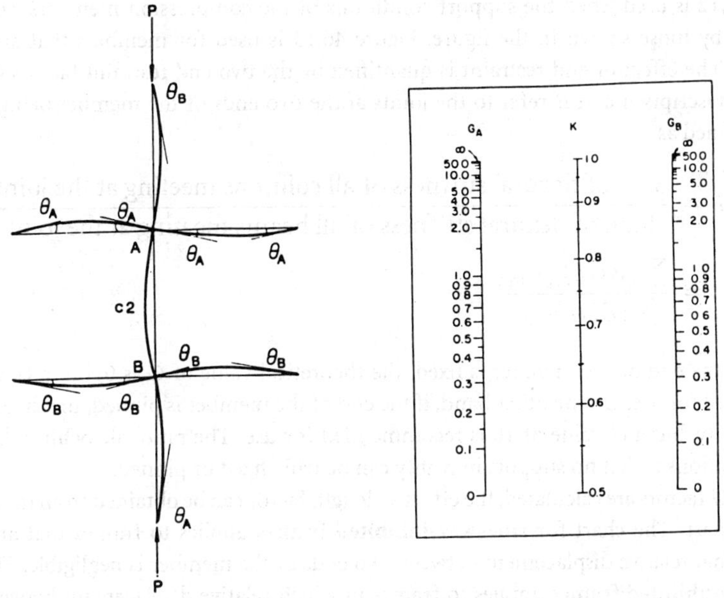

(A) SWAY INHIBITED CASE

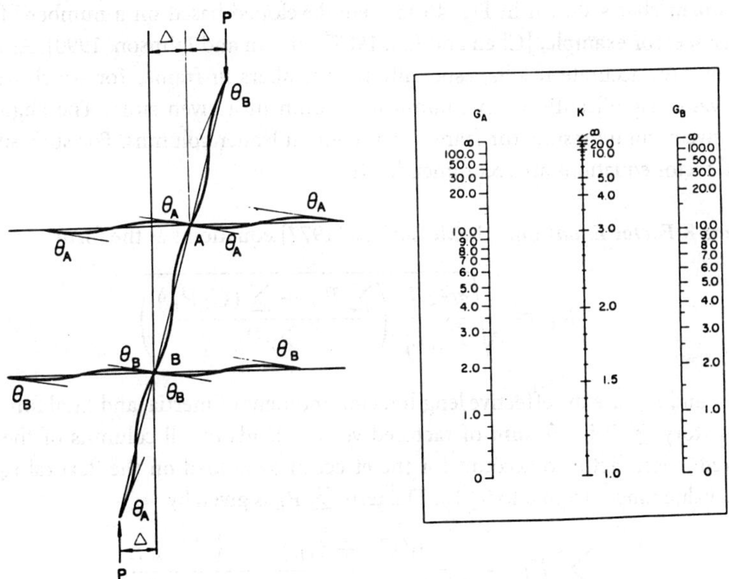

(B) SWAY UNINHIBITED CASE

FIGURE 46.13 *K* factor alignment charts. (a) Sidesway-inhibited frames. (b) Sidesway-uninhibited frames.

Figure 46.12 is used when the support conditions of the compression members can be closely represented by those shown in the figure. Figure 46.13 is used for members that are parts of a framework. The effect of end restraint is quantified by the two end restraint factors G_A and G_B, where the subscripts A and B refer to the joints at the two ends of the member being considered and G is defined as

$$G = \frac{\text{Sum of flexural stiffness of all columns meeting at the joint}}{\text{Sum of flexural stiffness of all beams meeting at the joint}}$$

$$= \frac{\sum(EI/L)_{\text{column}}}{\sum(EI/L)_{\text{beam}}} \tag{46.27}$$

Note that if the end of the member is fixed, the theoretical value of G is 0, but a G value of 1 is recommended for use. On the other hand, if the end of the member is pinned, the theoretical value of G is infinity, but a G value of 10 is recommended for use. The rationale behind the foregoing recommendations is that no support in reality can be truly fixed or pinned.

Once the G factors are calculated, the effective length factor can be obtained from the appropriate alignment chart. The chart for **sidesway-inhibited frames** applies to frames that are braced in such a way that relative displacement between two ends of the member is negligible. The chart for **sidesway-uninhibited frames** applies to frames in which relative displacement between member ends is not negligible. Although the charts were developed assuming elastic behavior for all members, inelasticity in the columns can be accounted for by multiplying the end-restrained factors by the quantity E_t/E, where E_t is the tangent modulus (Yura, 1971; Disque, 1973).

The alignment charts shown in Fig. 46.13 were developed based on a number of simplifying assumptions; see, for example, [Chen and Lui, 1987; Salmon and Johnson, 1990]. As a result, they do not always give accurate results, especially for members in frames for which the parameter $L\sqrt{(P/EI)}$ varies significantly from column to column in a given story. The alignment charts also fail to give accurate results for frames that contain leaner columns. For such situations, the following K factor equations are recommended for use:

LeMessurier's K Factor Equation. LeMessurier's [1977] equation has the form

$$K_i = \sqrt{\frac{\pi^2 EI_i}{P_{ui}L_i^2}\left(\frac{\sum P_u + \sum(C_L P_u)}{\sum P_L}\right)} \tag{46.28}$$

where K_i, I_i, and P_{ui} are the effective length factor, moment of inertia, and axial force of member i in a given story. $\sum P$ is the sum of factored vertical loads on all columns of the story. C_L is a stiffness reduction factor to account for the effect of axial load on the flexural rigidity of the member; its value ranges from 0 to 0.216. The term $\sum P_L$ is given by

$$\sum P_L = \frac{6(G_A + G_B) + 36}{2(G_A + G_B) + G_A G_B + 3}\frac{EL}{L^2} \tag{46.29}$$

Although charts are provided for the determination of C_L and P_L, it was noted by LeMessurier that in a multistory building the values of G_A and G_B do not vary significantly. Letting $G_A = G_B = G_S$, the following equations can be used to calculate approximate values for C_L and $\sum P_L$:

$$C_L = \frac{0.216}{(1 + G_s)^2} \tag{46.30}$$

$$\sum P_L = \frac{12}{1 + G_s}\frac{EI}{L^2} \tag{46.31}$$

If we assume that (1) the total gravity load associated with the sway buckling of a story is equal to the sum of the buckling loads of all columns in the story that provide it sidesway resistance, and (2) each individual column buckling load is obtained from the equation $P_{ek} = \pi^2 EI/(KL)^2$ where K is obtained from the alignment chart for sidesway-uninhibited case, a simplified form of Eq. (46.28) can be written as

$$K_i = \sqrt{\frac{\pi^2 E I_i}{P_i L_i^2}\left(\frac{\sum P_u}{\sum P_{ek}}\right)} \tag{46.32}$$

The advantage of using Eq. (46.32) over Eq. (46.30) is that there is no need to determine C_L or $\sum P_L$.

Lui's K Factor Equation. Lui's [1992] equation has the form

$$K_i = \sqrt{\frac{\pi^2 E I_i}{P_{ui} L_I^2}\left[\left(\sum \frac{P_u}{L}\right)\left(\frac{1}{5\sum_\eta} + \frac{\Delta_I}{\sum_H}\right)\right]} \tag{46.33}$$

where $\sum(P_u/L)$ is the sum of the factored axial force-to-length ratios of all members in a story, and $\sum \eta$ is the sum of the member stiffness indexes of all members in the story. The member stiffness index η is given by

$$\eta = \left[3 + 4.8\left(\frac{M_A}{M_B}\right) + 4.2\left(\frac{M_A}{M_B}\right)^2\right]\frac{EI}{L^3} \tag{46.34}$$

In the foregoing equation (M_A/M_B) is the ratio of the smaller to larger end moments of the member; it is taken as positive if the member bends in reverse curvature and negative if the member bends in single curvature. Values for M_A and M_B are to be obtained from a first-order analysis of the frame subjected to a set of fictitious lateral forces applied at each story in proportion to the story factored gravity loads. Δ_I in Eq. (46.33) is the interstory deflection produced by these fictitious lateral forces, and $\sum H$ is the sum of the fictitious lateral forces at and above the story under consideration.

In the event that both M_A and M_B are zero, as in the case for a pinned–pinned leaner column, the ratio (θ_A/θ_B), where θ is the member end rotation with respect to its chord, should be used in place of (M_A/M_B) in Eq. (46.34). For instance, if the leaner column bends in reverse curvature, η should be taken as $12EI/L^3$, but if the leaner column bends in single curvature, η should be taken as $2.4EI/L^3$.

Limiting Slenderness Ratio

The governing slenderness ratio (KL/r) of compression members preferably should not exceed 200.

Example. Determine the effective length factors for the two columns of the portal frame shown in Fig. 46.14(a).

Alignment Chart Approach. Using Eq. (46.27), for column AB we have

$$G_A = 1 \text{ (fixed support)}, \qquad G_B = \frac{(30{,}000)(288)/(20)(12)}{(30{,}000)(768)/(30)(12)} = 0.56$$

and from the sway-uninhibited alignment chart of Fig. 46.13(b) we obtain $K = 1.24$.

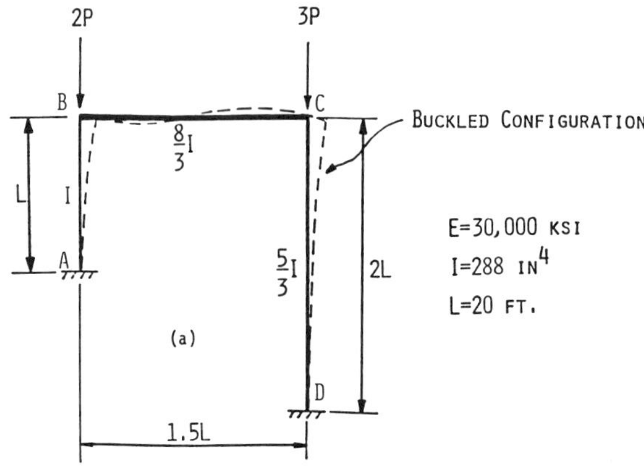

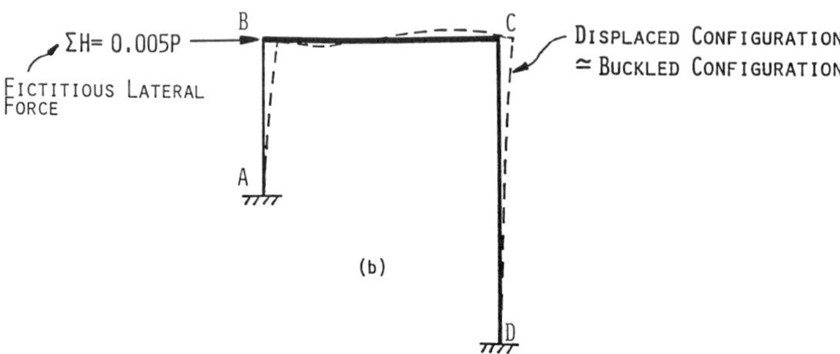

FIGURE 46.14 K factor determination. 1 kip = 4.45 kN, 1 ft. = 0.305 m, 1 inch = 25.4 mm.

Similarly, for column CD we have

$$G_c = \frac{(30{,}000)(480)/(40)(12)}{(30{,}000)(768)/(30)(12)} = 0.47, \qquad G_D = 1 \text{ (fixed support)}$$

so from the alignment chart we obtain $K = 1.22$.

LeMessurier's Approach. To use Eq. (46.38) it is necessary to determine C_L and P_L. The determination of these two terms is facilitated by the use of a special chart developed by LeMessurier [1977]. From the chart, we obtain for column AB, $C_L = 0.075$ and $P_L = 945$ kips; and for column CD, $C_L = 0.08$ and $P_L = 425$ kips. Now, using Eq. (46.28), we have for column AB,

$$K = \sqrt{\frac{(\pi^2)(30{,}000)(288)}{(2P)(20 \times 12)^2}\left[\frac{5P + (0.075)(2P) + (0.08)(3P)}{(945 + 425)}\right]} = 1.7$$

and for column CD,

$$K = \sqrt{\frac{(\pi^2)(30{,}000)(480)}{(3P)(40 \times 12)^2}\left[\frac{5P + (0.075)(2P) + (0.08)(3P)}{(945 + 425)}\right]} = 0.9$$

Lui's Approach. To use Lui's approach, we must first subject the frame to a fictitious lateral force and perform a first-order analysis on the frame to determine the member end moments and story deflection. The magnitude of this fictitious lateral force is quite arbitrary as long as it is proportional to the applied story gravity loads. In this example, the magnitude of the fictitious lateral force is taken as 0.1% of the applied story gravity loads. Therefore, a lateral force of $0.005P$ is used. This is shown in Fig. 46.14(b). The object of subjecting the frame to a fictitious force is to create a deflected configuration of the frame that somewhat resembles the actual buckled configuration.

Performing a first-order analysis on the frame shown in Fig. 46.14(b), we obtain for column AB, $M_A = 0.530P$ and $M_B = 0.419P$, and for column CD, $M_C = 0.245P$ and $M_D = 0.256P$. From Eq. (46.34), η for column AB is calculated to be 5.89, and η for column CD is calculated to be 1.49. So $\sum \eta = 7.38$. Also from the analysis, we obtain $\Delta_I = 0.000713P$. So $\Delta_I / \sum H = 0.143$. Now, using Eq. (46.33), we have for column AB,

$$K = \sqrt{\frac{(\pi^2)(30{,}000)(288)}{(2P)(20 \times 12)^2} \left[\left(\frac{2P}{20 \times 12} + \frac{3P}{40 \times 12}\right)\left(\frac{1}{5 \times 7.38} + 0.143\right)\right]} = 1.35$$

and for column CD,

$$K = \sqrt{\frac{(\pi^2)(30{,}000)(480)}{(3P)(40 \times 12)^2} \left[\left(\frac{2P}{20 \times 12} + \frac{3P}{40 \times 12}\right)\left(\frac{1}{5 \times 7.38} + 0.143\right)\right]} = 0.71$$

Theoretical K Factors. The theoretical K factors obtained using an eigenvalue analysis [Halldorsson and Wang, 1968] are 1.347 and 0.710 for column AB and column CD, respectively. Thus, it can be seen that the alignment chart approach may give erroneous results when used for members in frames for which the parameter $L\sqrt{(P/EI)}$ varies significantly from column to column.

46.5 Flexural Members

According to the width-thickness ratios of the component elements, steel sections used for **flexural members** are classified as compact, noncompact, and slender-element sections. *Compact sections* are sections that can develop the cross-section plastic moment (M_p) under flexure and sustain that moment through a large hinge rotation without fracture. *Noncompact sections* are sections that either cannot develop the cross-section full plastic strength or cannot sustain a large hinge rotation at M_p, probably due to local buckling of the flanges or web. *Slender-element sections* are sections that fail by local buckling of component elements long before M_p is reached. A section is considered compact if all its component elements have width-thickness ratios less than λ_p. A section is considered noncompact if one or more of its component elements have width-thickness ratios that fall between λ_p and λ_r. A section is considered to be a slender element if one or more of its component elements have width-thickness ratios that exceed λ_r. Expressions for λ_p and λ_r are given in Table 46.11.

In addition to the compactness of the steel section, another important consideration for beam design is the lateral unsupported (unbraced) length of the member. For beams bent about their strong axes, the limit states vary depending on the number and spacing of lateral supports provided to brace the compression flange of the beam. The compression flange of a beam behaves somewhat like a compression member. It buckles if adequate lateral supports are not provided, in a phenomenon called *lateral torsional buckling*. Depending on the lateral unsupported length of the beam, lateral torsional buckling may or may not be accompanied by yielding. Thus, lateral torsional buckling can be inelastic or elastic. If the lateral unsupported length is large, the limit state is elastic lateral torsional buckling. If the lateral unsupported length is smaller, the limit state is inelastic lateral torsional buckling. For compact-section beams with adequate lateral supports, the limit state is full yielding of the cross section (i.e., plastic hinge formation). For noncompact-section beams with adequate lateral supports, the limit state is flange or web local buckling.

Table 46.11 λ_p and λ_r for Members under Flexural Compression

Component Element	Width-Thickness Ratio	λ_p	λ_r
Flanges of I-shaped rolled beams and channels	b/t	$65/\sqrt{F_y}$	$141/\sqrt{(F_y - 10)}$
Flanges of I-shaped hybrid or welded beams	b/t	$65/\sqrt{F_{yf}}$ (nonseismic) $52/\sqrt{F_{yf}}$ (seismic) F_{yf} = yield stress of flange	$162/\sqrt{(F_{yw} - 16.5)/k_c}$ F_{yw} = yield stress of web $k_c = 4/\sqrt{h/t_w}, 0.35 \leq k_c \leq 0.763$
Flanges of square and rectangular box and hollow structural sections of uniform thickness; flange cover plates and diaphragm plates between lines of fasteners or welds	b/t	$190/\sqrt{F_y}$	$238/\sqrt{F_y}$
Unsupported width of cover plates perforated with a succession of access holes	b/t	NA	$317/\sqrt{F_y}$
Legs of single angle struts; legs of double angle struts with separators; unstiffened elements	b/t	NA	$76/\sqrt{F_y}$
Stems of tees	d/t	NA	$127/\sqrt{F_y}$
Webs in flexural compression	h/t_w	$640/\sqrt{F_y}$ (nonseismic) $520/\sqrt{F_y}$ (seismic)	$970/\sqrt{F_y}$
Webs in combined flexural and axial compression	h/t_w	For $P_u/\phi_b P_y \leq 0.125$: $640(1 - 2.75P_u/\phi_b P_y)/\sqrt{F_y}$ (nonseismic) $520(1 - 1.54P_u/\phi_b P_y)/\sqrt{F_y}$ (seismic) For $P_u/\phi_b P_y > 0.125$: $191(2.33 - P_u/\phi_b P_y)/\sqrt{F_y} \geq 253/\sqrt{F_y}$ (nonseismic) $152(2.89 - P_u/\phi_b P_y)/\sqrt{F_y}$ (seismic) $\phi_b = 0.90$ P_u = factored axial force $P_y = A_g F_y$	$970/\sqrt{F_y}$
Circular hollow sections	D/t D = outside diameter; t = wall thickness	$2070/F_y$ $1300/F_y$ for plastic design	$8970/F_y$

All stresses have units of ksi.
See Fig. 46.7 for definition of b, h, and t.

For beams bent about their weak axes, lateral torsional buckling will not occur, so the lateral unsupported length has no bearing on the design. The limit states for such beams will be the formation of a plastic hinge, if the section is compact, or flange or web local buckling, if the section is noncompact.

The design of flexural members should satisfy the following criteria: (1) flexural strength criterion; (2) shear strength criterion; (3) criteria for concentrated loads; and (4) deflection criterion.

Flexural Strength Criterion

Flexural members must be designed to satisfy the flexural strength criterion of

$$\phi_b M_n \geq M_u \tag{46.35}$$

where $\phi_b M_n$ is the flexural design strength and M_u is the factored moment. The flexural design strength is determined as follows:

Compact-Section Members Bent about Their Major Axes

Plastic Hinge Formation. For $L_b \leq L_p$,

$$\phi_b M_n = 0.90 M_p \tag{46.36}$$

Inelastic Lateral Torsional Buckling. For $L_p < L_b \leq L_r$,

$$\phi_b M_n = 0.90 C_b \left[M_p - (M_p - M_r) \left(\frac{L_b - L_p}{L_r - L_p} \right) \right] \leq 0.90 M_p \tag{46.37}$$

Elastic Lateral Torsional Buckling. For $L_b > L_r$, for I-shaped members and channels,

$$\phi_b M_n = 0.90 C_b \left[\frac{\pi}{L_b} \sqrt{E I_y G J + \left(\frac{\pi E}{L_b} \right)^2 I_y C_w} \right] \leq 0.90 M_p \tag{46.38}$$

For solid rectangular bars and symmetric box sections,

$$\phi_b M_n = 0.90 C_b \frac{57{,}000 \sqrt{JA}}{L_b / r_y} \leq 0.90 M_p \tag{46.39}$$

The variables used in the foregoing equations are defined as follows:

L_b = lateral unsupported length of the member
L_p, L_r = limiting lateral unsupported lengths given in the table on the next page
M_p = $F_y Z_x$
M_r = $F_L S_x$ for I-shaped sections and channels, $F_y S_x$ for solid rectangular bars, $F_{yf} S_{\text{eff}}$ for box sections
F_L = smaller of $(F_{yf} - F_r)$ or F_{yw}
F_{yf} = flange yield stress, ksi
F_{yw} = web yield stress
Fr = 10 ksi for rolled sections, 16.5 ksi for welded sections
F_y = specified minimum yield stress

L_p	L_r
I-shaped sections, channels:	
$300r_y/\sqrt{F_{yf}}$	$[r_y X_1/F_L]\{\sqrt{[1+\sqrt{(1+X_2F_L^2)}]}\}$
where	where
r_y = radius of gyration about minor axis, in. F_{yf} = flange yield stress, ksi	r_y = radius of gyration about minor axis, in. $X_1 = (\pi/S_x)\sqrt{(EGJA/2)}$ $X_2 = (4C_W/I_y)(S_x/GJ)^2$ F_L = smaller of $(F_{yf} - F_r)$ or F_{yw} F_{yf} = flange yield stress, ksi F_{yw} = web yield stress, ksi F_r = 10 ksi for rolled shapes, 16.5 ksi for welded shapes S_x = elastic section modulus about the major axis, in.^{3} (use S_{xc}, the elastic section modulus about the major axis with respect to the compression flange if the compression flange is larger than the tension flange); I_y = moment of inertia about the minor axis in.^{4} J = torsional constant, in.^{4} C_W = warping constant, in.^{6} E = modulus of elasticity, ksi G = shear modulus, ksi
Solid rectangular bars, symmetric box sections:	
$[3750r_y\sqrt{(JA)}]/M_p$	$[57{,}000r_y\sqrt{(JA)}]/M_p$
where	where
r_y = radius of gyration about minor axis, in. J = torsional constant, in.^{4} A = cross-sectional area, in.^{2} M_p = plastic moment capacity = F_yZ_x F_y = yield stress, ksi Z_x = plastic section modulus about the major axis, in.^{3}	r_y = radius of gyration about minor axis, in. J = torsional constant, in.^{4} A = cross-sectional area, in.^{2} $M_r = F_{yf}S_x$ for solid rectangular bar, $F_{yf}S_{\text{eff}}$ for box sections F_y = yield stress, ksi F_{yf} = flange yield stress, ksi S_x = elastic section modulus about the major axis, in.^{3}

Note: L_p given in this table are valid only if the bending coefficient C_b is equal to unity. If $C_b > 1$, the value of L_p can be increased. However, using the L_p expressions given above for $C_b > 1$ will give conservative value for the flexural design strength.

S_x = elastic section modulus about the major axis
Z_x = plastic section modulus about the major axis
I_y = moment of inertia about the minor axis
J = torsional constant
C_w = warping constant
E = modulus of elasticity
G = shear modulus
$C_b = 12.5M_{max}/(2.5M_{max} + 3M_A + 4M_B + 3M_C)$
M_{max}, M_A, M_B, M_C = maximum moment, quarter-point moment, midpoint moment, and three-quarter point moment along the unbraced length of the member, respectively

Compact-Section Members Bent about Their Minor Axes

Plastic Hinge Formation.

$$\phi_b M_n = 0.90 M_{py} = 0.90 F_y Z_y \tag{46.40}$$

Noncompact-Section Members Bent about Their Major Axes

Flange or Web Local Buckling. For $L_b \leq L'_p$,

$$\phi_b M_n = \phi_b M'_n = 0.90\left[M_p - (M_p - M_r)\left(\frac{\lambda - \lambda_p}{\lambda_r - \lambda_p}\right)\right] \tag{46.41}$$

where L_p, L_r, M_p, M_r are defined as before for compact-section members,

$$L'_p = L_p + (L_r - L_p)\left(\frac{M_p - M'_n}{M_p - M_r}\right) \tag{46.42}$$

and, for flange local buckling,

$\lambda = b_f/2t_f$ for I-shaped members, b_f/t_f for channels
$\lambda_p = 65/\sqrt{F_y}$
$\lambda_r = 141/\sqrt{(F_y - 10)}$

For web local buckling,

$\lambda = h_c/t_w$
$\lambda_p = 640/\sqrt{F_y}$
$\lambda_r = 970/\sqrt{F_y}$

in which

b_f = flange width
t_f = flange thickness
h_c = twice the distance from the neutral axis to the inside face of the compression flange less the fillet or corner radius
t_w = web thickness

Inelastic Lateral Torsional Buckling. For $L'_p < L_b \leq L_r$, $\phi_b M_n$ is given by Eq. (46.37), except that the limit $0.90M_p$ is to be replaced by the limit $0.90M'_n$.

Elastic Lateral Torsional Buckling. For $L_b > L_r$, $\phi_b M_n$ is the same as for compact-section members as given in Eq. (46.38) or Eq. (46.39).

Noncompact-Section Members Bent about Their Minor Axes

Flange or Web Local Buckling. $\phi_b M_n$ is given by Eq. (46.41), regardless of the value of L_b.

Schematic plots of the flexural design strength discussed in the foregoing sections are shown in Fig. 46.15. A design is considered satisfactory for flexure if the strength curve envelops the factored moment for a given unbraced beam length.

Slender-Element Sections

Refer to the section on "Plate Girders."

Tees and Double Angles Bent about Their Major Axes

The design flexural strength for tees and double-angle beams with flange and web slenderness ratios less than the corresponding limiting slenderness ratios λ_r shown in Table 46.11 is given by

$$\phi_b M_n = 0.90 \left[\frac{\pi \sqrt{EI_y GJ}}{L_b} (B + \sqrt{1 + B^2}) \right] \leq 0.90(1.5M_y) \tag{46.43}$$

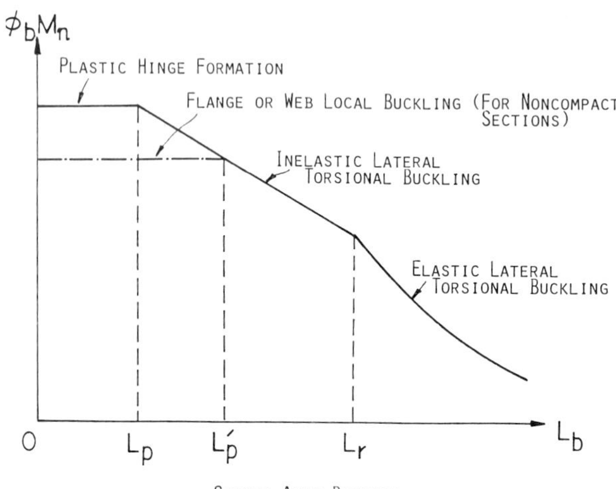

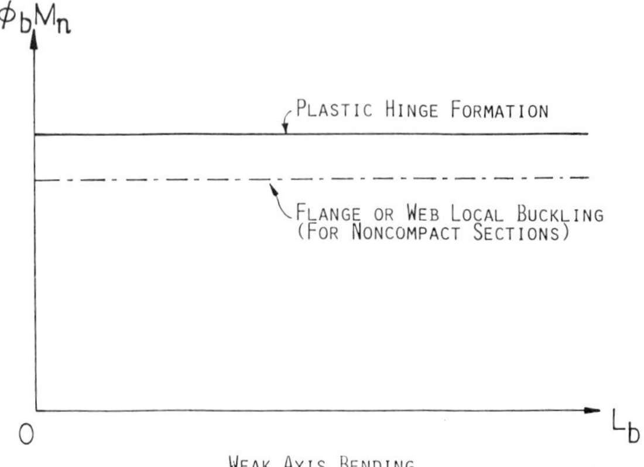

FIGURE 46.15 Beam flexural strength curves.

where

$$B = \pm 2.3\left(\frac{d}{L_b}\right)\sqrt{\frac{I_y}{J}} \tag{46.44}$$

Use the plus sign for B if the entire length of the stem along the unbraced length of the member is in tension. Otherwise, use the minus sign. The other variables in Eq. (46.43) are defined as before in Eq. (46.38).

Shear Strength Criterion

For practically all structural shapes commonly used for construction purposes, the shear strength contribution from the flanges is small compared to that from the webs. As a result, the design shear strength for flexural members is determined based on the webs only. For a satisfactory design, the design web shear strength must exceed the factored shear acting on the cross section, i.e.,

$$\phi_v V_n \geq V_u \tag{46.45}$$

Depending on the slenderness ratios of the webs, three limit states can be identified: shear yielding, inelastic shear buckling, and elastic shear buckling. The design shear strength that corresponds to each of these limit states are given as follows:

Shear Yielding of Web. For $h/t_w \leq 187\sqrt{(k_v/F_{yw})}$,

$$\phi_v V_n = 0.90[0.60 F_{yw} A_w] \tag{46.46}$$

Inelastic Shear Buckling of Web. For $187\sqrt{(k_v/F_{yw})} \leq h/t_w \leq 234\sqrt{(k_v/F_{yw})}$,

$$\phi_v V_n = 0.90\left[0.60 F_{yw} A_w \frac{187\sqrt{k_v/F_{yw}}}{h/t_w}\right] \tag{46.47}$$

Elastic Shear Buckling of Web. For $234\sqrt{(k_v/F_{yw})} < h/t_w \leq 260$,

$$\phi_v V_n = 0.90\left[A_w \frac{26{,}400 k_v}{(h/t_w)^2}\right] \tag{46.48}$$

The variables used in the foregoing equations are defined as follows:

h = clear distance between flanges less the fillet or corner radius
t_w = web thickness
F_{yw} = yield stress of web
A_w = $d t_w$
d = overall depth of section
k_v = $5 + 5/(a/h)^2$
a = distance between adjacent transverse stiffeners

The web plate coefficient k_v is taken as 5 if a/h exceeds the smaller of 3 or $[260/(h/t_w)]^2$, or if no stiffeners are present. Transverse stiffeners are not required if Eq. (46.45) is satisfied or when $h/t_w \leq 418/\sqrt{F_{yw}}$.

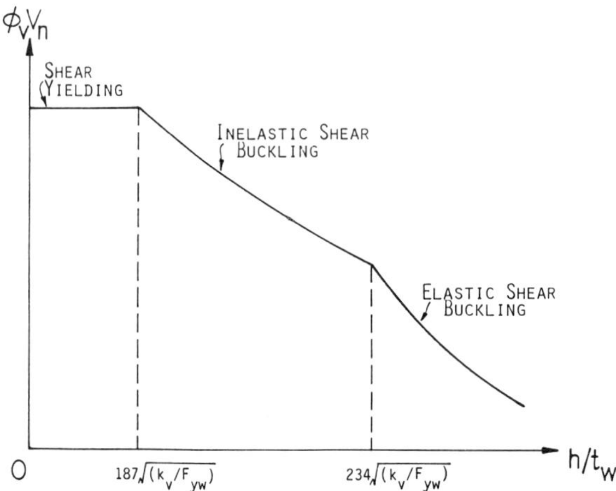

FIGURE 46.16 Beam shear strength curve.

A schematic plot of the shear strength curve described by Eqs. (46.46) to (46.48) is shown in Fig. 46.16. A design is considered satisfactory for shear if the curve envelops the factored shear for a given web slenderness.

Criteria for Concentrated Loads

When concentrated loads are applied normal to the flanges in planes parallel to the webs of flexural members, the flange(s) and web(s) must be checked to ensure that they have sufficient strengths ϕR_n to withstand the concentrated forces R_u, i.e.,

$$\phi R_n \geq R_u \tag{46.49}$$

The design strengths for a variety of limit states are as follows:

Local Flange Bending. If the concentrated load that acts on the beam flange is tensile, the tips of the flange, by virtue of its flexibility, will be pulled away from the web, leading ultimately to failure by fracture. To prevent this type of failure, the loaded flange should possess sufficient thickness and strength. The design strength is given by

$$\phi R_n \geq 0.90[6.25 t_f^2 F_{fy}] \tag{46.50}$$

where

t_f = flange thickness of the loaded flange, in.
F_{yf} = flange yield stress, ksi

Local Web Yielding. The design strengths for yielding of the beam web at the toe of the fillet, under tensile or compressive loads acting on one or both flanges, are as follows. If the load acts at a distance from the beam end that exceeds the depth of the member,

$$\phi R_n = 1.00\left[(5k + N)F_{yw}t_w\right] \tag{46.51}$$

If the load acts at a distance from the beam end that does not exceed the depth of the member,

$$\phi R_n = 1.00[(2.5k + N)F_{yw}t_w] \tag{46.52}$$

where

k = distance from outer face of flange to web toe of fillet

N = length of bearing on the beam flange

F_{yw} = web yield stress

t_w = web thickness

Web Crippling. The design strengths for crippling of a beam web under compressive loads acting on one or both flanges are as follows. If the load acts at a distance from the beam end that exceeds half the depth of the beam,

$$\phi R_n = 0.75\left\{1.35t_w^2\left[1+3\left(\frac{N}{d}\right)\left(\frac{t_w}{t_f}\right)^{3/2}\right]\sqrt{F_{yw}t_f/t_w}\right\} \tag{46.53}$$

If the load acts at a distance from the beam end that does not exceed half the depth of the beam,

$$\phi R_n = 0.75\left\{68t_w^2\left[1+3\left(\frac{N}{d}\right)\left(\frac{t_w}{t_f}\right)^{3/2}\right]\sqrt{F_{yw}t_f/t_w}\right\} \tag{46.54}$$

where

d = overall depth of the section, in.

t_f = flange thickness, in.

the other variables are the same as those defined in Eqs. (46.50) and (46.51).

Sidesway Web Buckling. Sidesway web buckling may occur in the web of a member if a compressive concentrated load is applied to a flange that is not restrained against relative movement by stiffeners or lateral bracings. If the compression flange is restrained against rotation about the longitudinal member axis and $(h/t_w)(l/b_f)$ does not exceed 2.3, the sidesway web buckling design strength for the member is

$$\phi R_n = 0.85\left\{\frac{C_r t_w^3 t_f}{h^2}\left[1+0.4\left(\frac{h/t_w}{l/b_f}\right)^3\right]\right\} \tag{46.55}$$

If the compression flange is not restrained against rotation about the longitudinal member axis and $(h/t_w)(l/b_f)$ does not exceed 1.7,

$$\phi R_n = 0.85\left\{\frac{C_r t_w^3 t_f}{h^2}\left[0.4\left(\frac{h/t_w}{l/b_f}\right)^3\right]\right\} \tag{46.56}$$

where

C_r = 960,000 when $M_u < M_y$ at the location of the forces, ksi

= 480,000 when $M_u \geq M_y$ at the location of the forces, ksi

t_w = web thickness, in.

h = clear distance between flanges less the fillet or corner radius, in.

b_f = flange width, in.

l = largest laterally unbraced length along either flange at the point of load, in.

Compression Buckling of the Web. This limit state may occur in members with unstiffened webs when both flanges are subjected to compressive forces. The design strength for this limit state is

$$\phi R_n = 0.90\left[\frac{4{,}100t_w^3\sqrt{F_{yw}}}{d_c}\right] \tag{46.57}$$

The variables in Eq. (46.57) are the same as those defined in Eqs. (46.55) and (46.56).

Use of Stiffeners

Stiffeners must be provided in pairs if any one of the foregoing strength criteria is violated. If the local flange bending or the local web yielding criterion is violated, the stiffener pair to be provided to carry the excess R_u need not extend more than one-half the web depth. The stiffeners must be welded to the loaded flange if the applied force is tensile; they either bear on or are welded to the loaded flange if the applied force is compressive.

If the web crippling or the compression web buckling criterion is violated, the stiffener pair to be provided must extend the full height of the web. They shall be designed as axially loaded compression members (see section 46.4) with an effective length factor $K = 0.75$ and a cross section A_g composed of the cross-sectional areas of the stiffeners plus $25t_w{}^2$ for interior stiffeners and $12t_w{}^2$ for stiffeners at member ends.

Deflection Criterion

Deflection is a serviceability consideration. As a result, service loads (not factored loads) are used in calculating beam deflections. Since most beams are fabricated with a camber, which somewhat offsets the dead load deflection, consideration is often given to deflection due to live load only. For beams supporting plastered ceilings, the service live load deflection preferably should not exceed $L/360$, where L is the beam span. A larger deflection limit can be used if due considerations are given to ensure the proper functioning of the structure.

46.6 Continuous Beams

Continuous beams shall be designed in accordance with the criteria for flexural members given in the preceding section. However, a 10% reduction in negative moments due to gravity loads is allowed at the supports, provided that:

1. The maximum positive moment between supports is increased by $\frac{1}{10}$ the average of the negative moments at the supports.
2. The section is compact.
3. The lateral unbraced length does not exceed L_{pd}, computed as shown below.
4. The beam is not a hybrid member.
5. The beam is not made of high-strength steel.
6. The beam is continuous over the supports (i.e., not cantilevered).

For I-shaped members, L_{pd} is given by

$$L_{pd} = \frac{3{,}600 + 2{,}200(M_1/M_2)}{F_y} r_y \tag{46.58}$$

For solid rectangular bars and symmetric box beams,

$$L_{pd} = \frac{5{,}000 + 3{,}000(M_1/M_2)}{F_y} r_y \geq 3{,}000 r_y/F_y \tag{46.59}$$

in which

F_y = yield stress of compression flange, ksi

M_1/M_2 = ratio of smaller to larger moment within the unbraced length, taken as positive if the moments cause reverse curvature and negative if the moments cause single curvature

r_y = radius of gyration about the minor axis, in.

Example. Select the lightest W section for the three-span continuous beam shown in Fig. 46.17(a) to support a uniformly distributed dead load of 1.5 kip/ft and a uniformly distributed live load of 3 kip/ft. The beam is laterally braced at the supports A, B, C, and D. Use A36 steel.

Load Combinations. The beam is to be designed based on the worst load combination of Table 46.6. By inspection, the load combination $1.2D + 1.6L$ will control the design. Thus, the beam will be designed to support a factored uniformly distributed dead load of $1.2 \times 1.5 = 1.8$ kip/ft and a factored uniformly distributed live load of $1.6 \times 3 = 4.8$ kip/ft.

Placement of Loads. The uniform dead load is to be applied over the entire length of the beam, as shown in Fig. 46.17(b). The uniform live load is to be applied to spans AB and CD, as shown in Fig. 46.17(c), to obtain the maximum positive moment, and it is to be applied to spans AB and BC, as shown in Fig. 46.17, to obtain the maximum negative moment.

Reduction of Negative Moment at Supports. Assuming the beam is compact and $L_b < L_{pd}$ (these assumptions will be checked later), a 10% reduction in support moment due to gravity load is allowed, provided that the maximum moment is increased by $\frac{1}{10}$ the average of the negative support moments. This reduction is shown in the moment diagrams as solid lines in Figs. 46.17(b) and 46.17(d). (The dotted lines in these figures represent the unadjusted moment diagrams.) This provision for support moment reduction takes into consideration the beneficial effect of moment redistribution in continuous beams, and it allows for the selection of a lighter section if the design is governed by negative moments. Note that no reduction in negative moments is made to the case when only spans AB and CD are loaded. This is because for this load case, the negative support moments are less than the positive in-span moments.

Determination of the Required Flexural Strength, M_u. Combining load case 1 and load case 2, the maximum positive moment is found to be 256 kip-ft. Combining load case 1 and load case 3, the maximum negative moment is found to be 266 kip-ft. Thus, the design will be controlled by the negative moment, so $M_u = 266$ kip-ft.

Beam Selection. A beam section is to be selected based on Eq. (46.35). The critical segment of the beam is span BC. For this span, the lateral unsupported length, L_b, is equal to 20 ft. The bending coefficient, C_b, is conservatively taken as 1. The selection of a beam section is facilitated by the use of a series of beam charts contained in Part 4 of the AISC LRFD Manual (AISC, 1994). Beam charts are plots of flexural design strength $\phi_b M_n$ of beams as a function of the lateral unsupported length L_b based on Eqs. (46.36) to (46.38). A beam is considered satisfactory for the limit state of flexure if the beam strength curve envelops the required flexural strength for a given L_b.

For the present example, $L_b = 20$ ft and $M_u = 266$ kip-ft, the lightest section (the first solid curve that envelops $M_u = 266$ kip-ft for $L_b = 20$ ft) obtained from the chart is a W16 × 67 section. Upon adding the factored dead weight of this W16 × 67 section to the specified loads, the required strength increases from 266 kip-ft to 269 kip-ft. Nevertheless, the beam strength curve still envelops this required strength for $L_b = 20$ ft; therefore, the section is adequate.

Check for Compactness. For the W16 × 67 section,

$$\text{Flange:} \quad \left[\frac{b_f}{2t_f} = 7.7\right] < \left[\frac{65}{\sqrt{F_y}} = 10.8\right]$$

$$\text{Web:} \quad \left[\frac{h_c}{t_w} = 35.9\right] < \left[\frac{640}{\sqrt{F_y}} = 106.7\right]$$

Therefore, the section is compact.

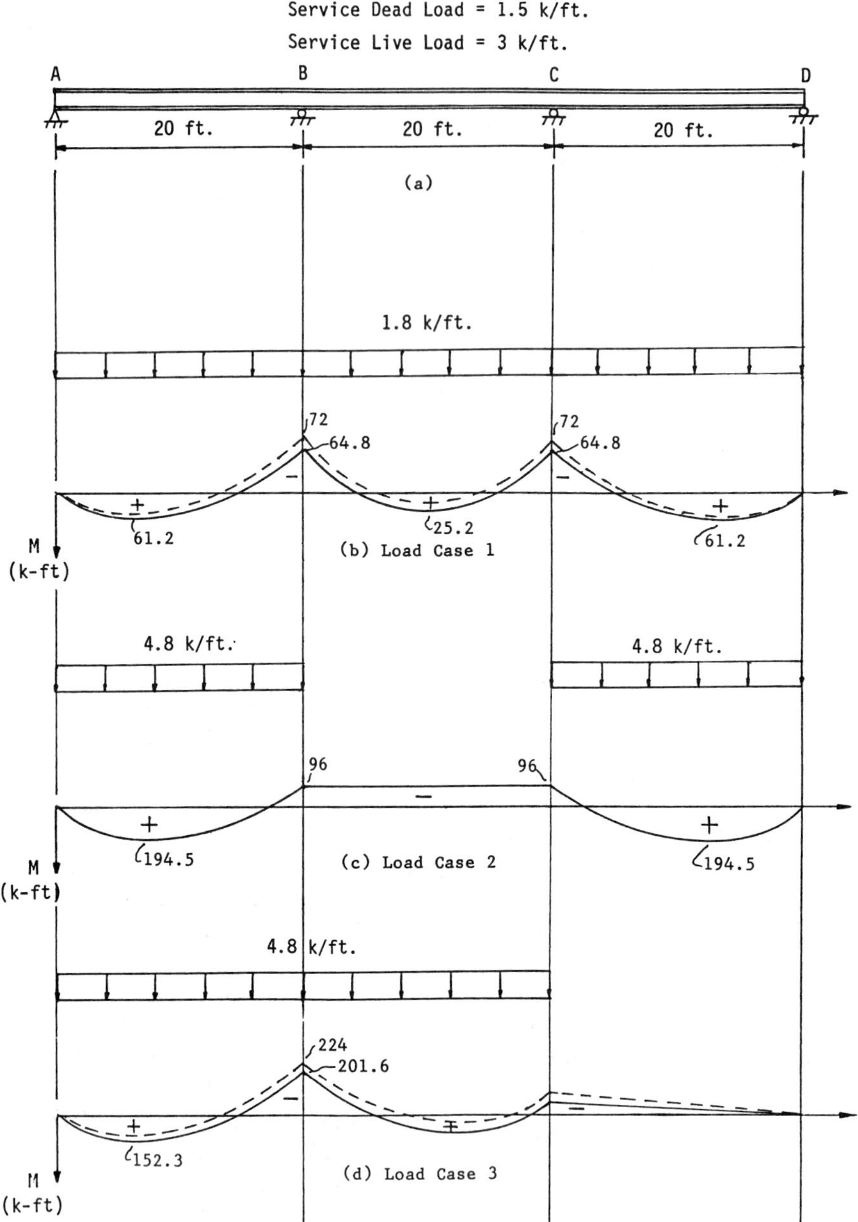

FIGURE 46.17 Design of a three-span continuous beam. 1 kip = 4.45 kN, 1 ft. = 0.305 m.

Check Whether $L_b < L_{pd}$. Using Eq. (46.58), with $M_1/M_2 = 0$, $r_y = 2.46$ in., and $F_y = 36$ ksi, we have $L_{pd} = 246$ in. (or 20.5 ft). Since $L_b = 20$ ft is less than $L_{pd} = 20.5$ ft, the assumption made earlier is validated.

Check for the Limit State of Shear. The selected section must satisfy the shear strength criterion of Eq. (46.45). From structural analysis it can be shown that maximum shear occurs just to the left of support B under load case 1 (for dead load) and load case 3 (for live load). It has a magnitude of 81.8 kips. For the W16×67 section, $h/t_w = 35.9$, which is less than $187\sqrt{(k_v/F_{yw})} = 69.7$, so the design shear strength is given by Eq. (46.46). We have, for $F_{yw} = 36$ ksi and $A_w = dt_w = (16.33)(0.395)$,

$$[\phi_v V_n = 0.90(0.60 F_{yw} A_w) = 125 \text{ kips}] > [V_u = 81.8 \text{ kips}]$$

Therefore, shear is not a concern. Normally, the limit state of shear will not control unless for short beams subjected to heavy loads.

Check for Limit State of Deflection. Deflection is a serviceability limit state. As a result, a designer should use service (not factored) loads for deflection calculations. In addition, most beams are cambered to offset deflection caused by dead loads, so only live loads are considered in deflection calculations. From structural analysis, it can be shown that maximum deflection occurs in spans AB and CD when (service) live loads are placed on those two spans. The magnitude of the deflection is 0.297 in. Assuming the maximum allowable deflection is $L/360$, where L is the span length between supports, we have an allowable deflection of $20 \times 12/360 = 0.667$ in. Since the calculated deflection is less than the allowable deflection, deflection is not a problem.

Check for the Limit States of Web Yielding and Web Crippling at Points of Concentrated Loads. From structural analysis it can be shown that maximum support reaction occurs at support B when the beam is subjected to loads shown as load case 1 (for dead load) and load case 3 (for live load). The magnitude of the reaction R_u is 157 kips. Assuming point bearing, i.e., $N = 0$, we have, for $d = 16.33$ in., $k = 1.375$ in., $t_f = 0.665$ in., and $t_w = 0.395$ in.,

$$\text{Web yielding:} \quad [\phi R_n = \text{Eq. (46.51)} = 97.8 \text{ kips}] < [R_u = 157 \text{ kips}]$$

$$\text{Web crippling:} \quad [\phi R_n = \text{Eq. (46.53)} = 123 \text{ kips}] < [R_u = 157 \text{ kips}]$$

Thus, both the web yielding and web crippling criteria are violated. As a result, we need to provide web stiffeners or a bearing plate at support B. Suppose we choose the latter. The size of the bearing plate is to be determined by solving Eq. (46.51) and Eq. (46.53) for N, given $R_u = 157$ kips. Solving Eq. (46.51) and Eq. (46.53) for N, we obtain $N = 4.2$ in. and 3.3 in., respectively. So use $N = 4.25$ in. The width of the plate, B, should conform with the flange width, b_f, of the W-section. The W16 × 67 section has a flange width of 10.235 in., so use $B = 10.5$ in. For uniformity, use the same size plate at all the supports. The bearing plates are to be welded to the supporting flange of the W section.

Use a W16 × 67 section. Provide bearing plates of size 10.5″ × 4″ at all supports.

46.7 Combined Flexure and Axial Force

Doubly or singly symmetric members subject to combined flexure and axial force are to be designed in accordance with the following interaction equations. For $P_u/\phi P_n \geq 0.2$,

$$\frac{P_u}{\phi P_n} + \frac{8}{9}\left(\frac{M_{ux}}{\phi_b M_{nx}} + \frac{M_{uy}}{\phi_b M_{ny}}\right) \leq 1.0 \tag{46.60}$$

For $P_u/\phi P_n < 0.2$,

$$\frac{P_u}{2\phi P_n} + \left(\frac{M_{ux}}{\phi_b M_{nx}} + \frac{M_{uy}}{\phi_b M_{ny}}\right) \leq 1.0 \tag{46.61}$$

where, if P is tensile,

P_u = factored tensile axial force
P_n = design tensile strength (see section on "Tension Members")
M_u = factored moment (preferably obtained from a second-order analysis)
M_n = design flexural strength (see section on "Flexural Members")
$\phi = \phi_t$ = resistance factor for tension = 0.90
ϕ_b = resistance factor for flexure = 0.90

and, if P is compressive,

$$P_u = \text{factored compressive axial force}$$
$$P_n = \text{design compressive strength (see section 46.4)}$$
$$M_u = \text{factored moment (see following discussion)}$$
$$M_n = \text{design flexural strength (see section 46.5)}$$
$$\phi = \phi_c = \text{resistance factor for compression} = 0.85$$
$$\phi_b = \text{resistance factor for flexure} = 0.90$$

The factored moment M_u should be determined from a second-order elastic analysis. In lieu of such an analysis, the following equation may be used:

$$M_u = B_1 M_{nt} + B_2 M_{lt} \tag{46.62}$$

where

M_{nt} = factored moment in member, assuming the frame does not undergo lateral translation [see Fig. 46.18(b)]
M_{lt} = factored moment in member as a result of lateral translation [see Fig. 46.18(c)]
$B_1 = C_m/(1 - P_u/P_e) \geq 1.0$
$P_e = \pi^2 EI/(KL)^2$, with $K \leq 1.0$ in the plane of bending
C_m = a coefficient to be determined from the following discussion
$B_2 = 1/[1 - (\sum P_u \Delta_{oh}/\sum HL)$ or $B_2 = 1/[1 - (\sum P_u/\sum P_e)$
$\sum P_u$ = sum of all factored loads acting on and above the story under consideration
Δ_{oh} = first-order interstory translation
$\sum H$ = sum of all lateral loads acting on and above the story under consideration
L = story height
$P_e = \pi^2 EI/(KL)^2$

For end-restrained members that do not undergo relative joint translation and are not subject to transverse loading between their supports in the plane of bending, C_m is given by

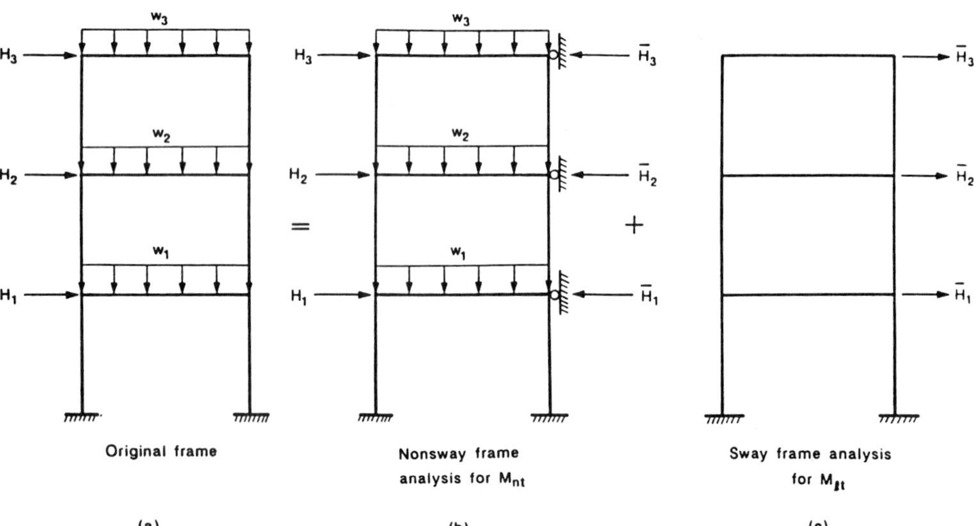

FIGURE 46.18 Calculation of M_{nt} and M_{lt}.

$$C_m = 0.6 - 0.4(M_1/M_2)$$

where M_1/M_2 is the ratio of the smaller to larger member end moments. The ratio is positive if the member bends in reverse curvature and negative if the member bends in single curvature. For end-restrained members that do not undergo relative joint translation and are subject to transverse loading between their supports in the plane of bending,

$$C_m = 0.85$$

For unrestrained members that do not undergo relative joint translation and are subject to transverse loading between their supports in the plane of bending,

$$C_m = 1.00$$

The selection of trial sections for use as **beam-columns** is facilitated by rewriting the interaction equations, Eq. (46.60) and Eq. (46.61), into the so-called *equivalent axial load* form. For $P_u/\phi_c P_n > 0.2$,

$$P_u + m_x M_{ux} + m_y U M_{uy} \leq \phi_c P_n \tag{46.63}$$

For $P_u/\phi_c P_n \leq 0.2$,

$$\frac{P_u}{2} + \frac{9}{8} m_x M_{ux} + \frac{9}{8} m_y U M_{uy} \leq \phi_c P_n \tag{46.64}$$

where

$$m_x = \tfrac{8}{9}(\phi_c P_n / \phi_b M_{nx})$$
$$m_y U = \tfrac{8}{9}(\phi_c P_n / \phi_b M_{ny})$$

Numerical values for m and U are provided in the AISC *Manual of Steel Construction* [AISC, 1994]. The advantage of using Eq. (46.63) and Eq. (46.64) for preliminary design is that the terms on the left-hand side of the inequality can be regarded as an equivalent axial load, $(P_u)_{\text{eff}}$. The similarity in form between the two equations and Eq. (46.15) allows the designer to take advantage of the column tables provided in the manual for selecting trial sections.

46.8 Biaxial Bending

Members subjected to bending about both principal axes (e.g., purlins on an inclined roof) should be designed for **biaxial bending**. Since both the moment about the major axis, M_{ux}, and the moment about the minor axis, M_{uy}, create flexural stress over the cross section of the member, to avoid yielding at the most severely stressed point the following equation for the yielding limit state must be satisfied:

$$f_{un} \leq \phi_b F_y \tag{46.65}$$

where

$f_{un} = M_{ux}/S_x + M_{uy}/S_y$ = the flexural stress under factored loads
S_x, S_y = elastic section moduli about the major and minor axes, respectively
$\phi_b = 0.90$
F_y = specified minimum yield stress

In addition, the limit state for lateral torsional buckling about the major axis should also be checked, i.e.,

$$\phi_b M_{nx} \geq M_{ux} \tag{46.66}$$

where $\phi_b M_{nx}$ is the design flexural strength about the major axis (see section 46.5). Note that lateral torsional buckling will not occur about the minor axis.

Equation (46.65) can be rearranged to give

$$S_x \geq \frac{M_{ux}}{\phi_b F_y} + \frac{M_{uy}}{\phi_b F_y}\left(\frac{S_x}{S_y}\right) \approx \frac{M_{ux}}{\phi_b F_y} + \frac{M_{uy}}{\phi_b F_y}\left(3.5\frac{d}{b_f}\right) \tag{46.67}$$

The approximation $(S_x/S_y) \approx (3.5d/b_f)$, where d is the overall depth and b_f is the flange width, was suggested by Gaylord et al. [Gaylord, Gaylord, and Stallmeyer, 1992] for doubly symmetric I-shaped sections. The use of Eq. (46.67) greatly facilitates the selection of trial sections for use in biaxial bending problems.

46.9 Combined Bending, Torsion, and Axial Force

Members subjected to the combined effect of bending, torsion, and axial force should be designed to satisfy the following limit states:

Yielding under Normal Stress.

$$\phi F_y \geq f_{un} \tag{46.68}$$

where

ϕ = 0.90
F_y = specified minimum yield stress
f_{un} = maximum shear stress determined from an elastic analysis under factored loads

Yielding under Shear Stress.

$$\phi(0.6F_y) \geq f_{uv} \tag{46.69}$$

where

ϕ = 0.90
F_y = specified minimum yield stress
f_{uv} = maximum shear stress determined from an elastic analysis under factored loads

Buckling.

$$\phi_c F_{cr} \geq f_{un} \quad \text{or} \quad f_{uv}, \quad \text{whichever is applicable} \tag{46.70}$$

where

$\phi_c F_{cr}$ = $\phi_c P_n/A_g$, in which $\phi_c P_n$ is the design compressive strength of the member (see section 46.4), and A_g is the gross cross-section area
f_{un}, f_{uv} = normal and shear stresses as defined in Eq. (46.68) and Eq. (46.69)

46.10 Frames

Frames are designed as a collection of structural components such as beams, beam-columns (columns), and connections. According to the restraint characteristics of the connections used in the construction, frames can be designed as fully restrained (rigid), partially restrained (semi-rigid), and unrestrained (simple) framings. The design of fully restrained frames necessitates the use of connections capable of transmitting all or a significant portion of the moment developed between the connecting members. The rigidity of the connections must be such that the angles between intersecting members should remain virtually unchanged under factored loads. The design of partially restrained frames is permitted upon evidence that the connections deliver a predictable amount of moment restraint. The design of unrestrained frames is based on the assumption that the connections provide no moment restraint to the beam insofar as gravity loads are concerned, but these connections should have adequate capacity to resist **drift** due to wind moments. Partially restrained and unrestrained constructions often incur inelastic deformation in the connections. The connections used in these constructions must be proportioned to possess sufficient ductility to avoid overstress of the fasteners or welds.

Regardless of the types of constructions used, due consideration must be given to account for the frame instability ($P - \Delta$) effect, by the use of either second-order analysis or Eq. (46.62). The end-restrained effect on members should also be accounted for by the use of the effective length factor (see the discussion of "Effective Length Factor" in section 46.4).

Example. Select the lightest W section for the simple portal frame shown in Fig. 46.19(a). The frame is braced at points A, B, C, and D for out-of-plane deflection. Use A36 steel. For simplicity, use the same size section for the beam and the columns.

Load Combinations. Refer to Table 46.6. By inspection the two load combinations that need to be investigated are

$$1.2D + 1.6S + 0.8W$$

$$1.2D + 1.3W + 0.5S$$

The analysis results for M_{nt} and M_{lt} (see Fig. 46.18) based on the above two load combinations are shown in Figs. 46.19(b) and 46.19(c), respectively. From the analysis results, it can seen that the load combination $1.2D + 1.6S + 0.8W$ will control the design.

Section Selection. From Fig. 46.19(b), it can be seen that the maximum first-order moment occurs at C and has a magnitude of $37.7 + 25.85 = 63.55$ kip-ft. Although the design should take into consideration second-order effects, we shall first select a trial section based on this first-order moment of 63.55 kip-ft and then check the adequacy of the trial section later for second-order effects.

With the aid of the beam charts contained in Part 4 of the AISC-LRFD Manual, the lightest section that satisfies the flexural strength criterion of Eq. (46.35) is found to be a W12 × 30 section. Section properties for this section useful for later calculations are given as follows: $A = 8.79$ in.^{2}, $I_x = 238$ in.^{4}, $I_y = 20.3$ in.^{4}, $r_x = 5.21$ in., $r_y = 1.52$ in., $J = 0.46$ in.^{4}, $C_w = 720$ in.^{6}, $L_p = 6.3$ ft, and $L_r = 19.1$ ft. (The above section properties are obtained from the Section Properties Tables, the Flexural-Torsion Properties Tables, and the Uniform Load Constants Tables in Part 1 and Part 4 of the AISC-LRFD Manual [AISC, 1994].)

Calculate M_u. M_u is to be calculated using Eq. (46.62). The second-order effects are accounted for by the term B_1 ($P - \delta$ moment magnification factor) and B_2 ($P - \Delta$ moment magnification factor) in the equation. To calculate B_1, we use the equation

$$B_1 = \frac{C_m}{1 - \dfrac{P_u}{P_e}} = \frac{0.6}{1 - \dfrac{19.88}{\pi^2(29{,}000)(238)/(1)(180)^2}} = 0.606$$

Since 0.606 is less than 1, therefore use $B_1 = 1$. Note that the effective length factor K used for calculating P_e in the above equation is for a "nonsway" frame; see Fig. 46.18(b). It was conservatively taken as unity in the foregoing calculation. To calculate B_2, we use the equation

$$B_2 = \frac{1}{1 - \dfrac{\sum P_u}{\sum P_e}} = \frac{1}{1 - \dfrac{35.28}{\pi^2(29{,}000)(238)/(1.33)(180)}} = 1.03$$

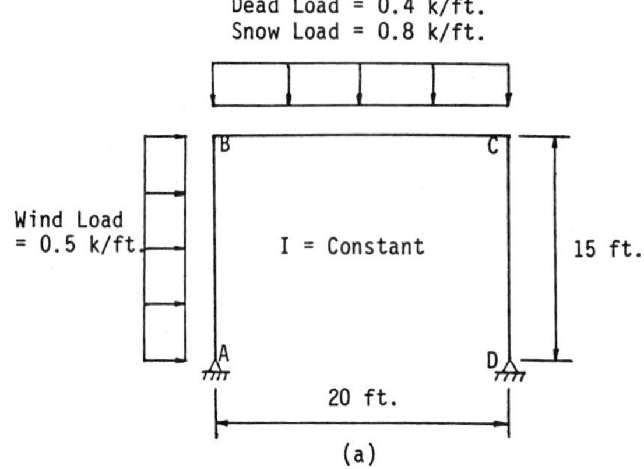

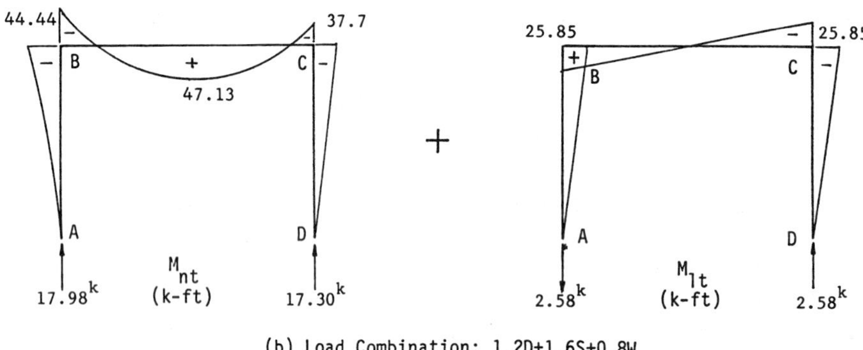

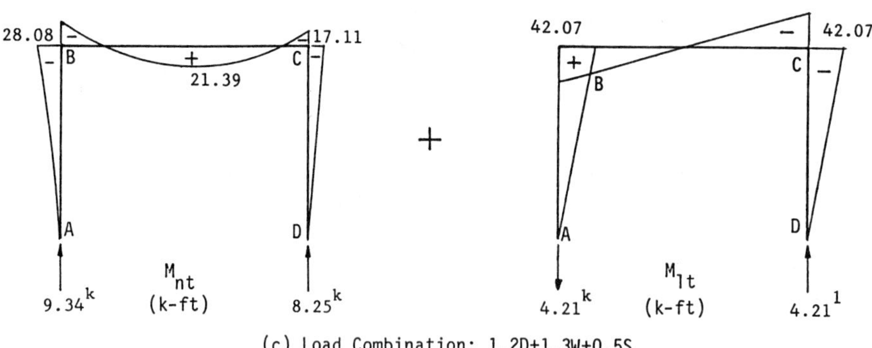

FIGURE 46.19 Design of a simple portal frame. 1 kip = 4.45 kN, 1 ft. = 0.305 m.

Note that the effective length factor K used for calculating P_e in the foregoing equation is for a "sway" frame; see Fig. 46.18(c). The value $K = 1.95$ was obtained from the alignment chart of Fig. 46.13(b) with $G_A = 10$ (pinned support) and $G_B = \sum(EI/L)_{\text{column}}/\sum(EI/L)_{\text{beam}} = 1.33$. Thus, we have

$$M_u = B_1 M_{nt} + B_2 M_{lt} = (1)(37.7) + (1.03)(25.85) = 64.3 \text{ kip-ft}$$

Check Adequacy of the* W12 × 30 *Section Used as a Beam. For member BC, $L_b = 20$ ft, $C_b = 1$. Since $[L_b = 20 \text{ ft}] > [L_r = 19.1 \text{ ft}]$, the design flexural strength $\phi_b M_n$ of the section is to be calculated from Eq. (46.38). Substituting $L_b = 20$ ft (240 in.), $C_b = 1$, $E = 29{,}000$ ksi, $G = 11{,}200$ ksi, $I_y = 20.3$ in.^{4}, $J = 0.46$ in.^{4}, and $C_w = 720$ in.^{6} into Eq. (46.38), we obtain $\phi_b M_n = 70.4$ kip-ft, which is larger than $M_u = 64.3$ kip-ft. Therefore, the section is adequate.

Check Adequacy of the* W12 × 30 *Section Used as a Beam-Column. For members AB and BC, the adequacy of the W12 × 30 section must be checked in accordance with Eq. (46.60) or Eq. (46.61), whichever is appropriate depending on the value of $P_u/\phi_c P_n$. From Fig. 46.19(b), we know $P_u = 19.88$ kips. To calculate $\phi_c P_n$, we must first calculate the slenderness parameter λ_c about the major and minor axes:

$$(\lambda_c)_x = \frac{(KL/r)_x}{\pi}\sqrt{\frac{F_y}{E}} = \frac{(1.95)(180)/5.21}{3.142}\sqrt{\frac{36}{29{,}000}} = 0.756$$

$$(\lambda_c)_y = \frac{(KL/r)_y}{\pi}\sqrt{\frac{F_y}{E}} = \frac{(1)(180)/1.52}{3.142}\sqrt{\frac{36}{29{,}000}} = 1.33$$

Since $(\lambda_c)_y > (\lambda_c)_x$, flexural buckling about the minor axis will control. The design compressive strength $\phi_c P_n$ of the section will be evaluated based on minor-axis buckling. Substituting $(\lambda_c)_y = 1.33$ and $A = 8.79$ in.^{2} into Eq. (46.16a), we obtain $\phi_c P_n = 128.3$ kips. The ratio $P_u/\phi_c P_n = 0.155$ is less than 0.2, so Eq. (46.61) is to be used.

Before we can check for the adequacy of the section using Eq. (46.61), we need to determine the design flexural strength $\phi_b M_n$ of the member. For members AB and CD, $L_b = 15$ ft, which falls between $L_p = 6.3$ ft and $L_r = 19.1$ ft, so $\phi_b M_n$ is to be determined from Eq. (46.37). The bending coefficient C_b in Eq. (46.37) for member CD is calculated from the equation

$$\begin{aligned} C_b &= \frac{12.5M_{\text{max}}}{2.5M_{\text{max}} + 3M_{\text{quarter-point}} + 4M_{\text{midpoint}} + 3M_{\text{three-quarterpoint}}} \\ &= \frac{12.5(63.55)}{2.5(63.55) + 3(47.66) + 4(31.77) + 3(15.89)} \\ &= 1.67 \end{aligned}$$

Substituting $C_b = 1.67$, $M_p = F_y Z_x = (36)(43.1) = 1551.6$ kip-in. or 129 kip-ft, $M_r = (F_{yf} - F_r)S_x = (36 - 10)(38.6) = 1003.6$ kip-in. $= 83.6$ kip-ft, $L_p = 6.3$ ft, $L_b = 15$ ft, and $L_r = 19.1$ ft into Eq. (46.37), we obtain $\phi_b M_n = 147$ kip-ft, which exceeds $0.90M_p = 116$ kip-ft. Therefore, use $\phi_b M_n = \phi_b M_p = 116$ kip-ft.

Now, using Eq. (46.61), we have

$$\frac{P_u}{2\phi_c P_n} + \left(\frac{M_{ux}}{\phi_b M_{nx}} + \frac{M_{uy}}{\phi_b M_{ny}}\right) = \frac{19.88}{(2)(128.3)} + \left(\frac{64.3}{116} + 0\right) = 0.63$$

which is less than 1; therefore, the section is adequate. (In fact, the section is overdesigned when used as beam-column for members AB and CD. For this example, it is clear that the beam, i.e., member BC, controls the design.)

Check Frame Drift. From structural analysis it can be easily shown that the lateral deflection experienced by the frame under service wind load is 0.88 in. This is considered acceptable, so frame drift is not a problem.

Use W12 × 30 section for the frame.

46.11 Plate Girders

Plate girders are built-up beams. They are used as flexural members to carry extremely large lateral loads. According to the LRFD Specification, a flexure member is considered a beam if the

Table 46.12 Web Stiffener Requirements

Range of Web Slenderness	Stiffener Requirements
$\frac{h}{t_w} \leq 260$	Plate girder can be designed without web stiffeners.
$260 < \frac{h}{t_w} \leq \frac{14{,}000}{\sqrt{F_{yf}(F_{yf}+16.5)}}$	Plate girder must be designed with web stiffeners. The spacing of stiffeners, a, can exceed $1.5h$. The actual spacing is determined by the shear criterion.
$\frac{14{,}000}{\sqrt{F_{yf}(F_{yf}+16.5)}} < \frac{h}{t_w} \leq \frac{2000}{\sqrt{F_{yf}}}$	Plate girder must be designed with web stiffeners. The spacing of stiffeners, a, cannot exceed $1.5h$.

a = clear distance between stiffeners.
h = clear distance between flanges when welds are used or the distance between adjacent lines of fasteners when bolts are used.
t_w = web thickness.
F_{yf} = compression flange yield stress, ksi.

slenderness of the web h/t_w does not exceed $970/\sqrt{F_y}$. If $h/t_w > 970/\sqrt{F_y}$, the member is considered a plate girder. Because of the large web slenderness, plate girders are often designed with web stiffeners to reinforce the flexural buckling strength of the web and to allow for postbuckling (shear) strength (i.e., tension field action) to develop. Table 46.12 summarizes the requirements for web stiffeners for plate girders based on the web slenderness ratio h/t_w.

Normally, the depths of plate girder sections are so large that simple beam theory, which postulates that plane sections before bending remain plane after bending, does not apply. As a result, a different set of strength criteria is required for plate girder design.

Flexural Strength Criterion

Doubly or singly symmetric single-web plate girders loaded in the plane of the web should satisfy the flexural strength criterion of Eq. (46.35). For the limit state of yielding of tension flange, the plate girder design flexural strength is given by

$$\phi_b M_n = 0.90[S_{xt} R_e F_{yt}] \tag{46.71}$$

For the limit state of buckling of compression flange,

$$\phi_b M_n = 0.90\,[S_{xc} R_{PG} R_e F_{\text{cr}}] \tag{46.72}$$

where

S_{xt} = section modulus referred to the tension flange
$= I_x/c_t$
S_{xc} = section modulus referred to the compression flange
$= I_x/c_c$
I_x = moment of inertia about the major axis
c_t = distance from neutral axis to extreme fiber of the tension flange
c_c = distance from neutral axis to extreme fiber of the compression flange
R_{PG} = plate girder bending strength reduction factor
$= 1 - [a_r/(1{,}200 + 300a_r)](h/t_w - 970/\sqrt{F_{\text{cr}}}) \leq 1.0$
R_e = hybrid girder factor
$= [12 + a_r(3m - m^3)/(12 + 2a_r)] \leq 1.0$
(R_e = 1 for nonhybrid girder)
a_r = ratio of web area to compression flange area ≤ 10

Range of Slenderness	F_{cr}
Flange local buckling	
$\dfrac{b_f}{2t_f} \le \dfrac{65}{\sqrt{F_{yf}}}$	F_{yf}
$\dfrac{65}{\sqrt{F_{yf}}} < \dfrac{b_f}{2t_f} \le \dfrac{150}{\sqrt{F_{yf}}}$	$F_{yf}\left[1 - \dfrac{1}{2}\left(\dfrac{(b_f/2t_f) - (65/\sqrt{F_{yf}})}{(150/\sqrt{F_{yf}}) - (65/\sqrt{F_{yf}})}\right)\right] \le F_{yf}$
$\dfrac{b_f}{2t_f} > \dfrac{150}{\sqrt{F_{yf}}}$	$\dfrac{26{,}200\left(4/\sqrt{h/t_w}\right)}{\left(b_f/2t_f\right)^2}, 0.35 \le 4/\sqrt{h/t_w} \le 0.763$
Lateral torsional buckling	
$\dfrac{L_b}{r_T} \le \dfrac{300}{\sqrt{F_{yf}}}$	F_{yf}
$\dfrac{300}{\sqrt{F_{yf}}} < \dfrac{L_b}{r_T} \le \dfrac{756}{\sqrt{F_{yf}}}$	$C_bF_{yf}\left[1 - \dfrac{1}{2}\left(\dfrac{(L_b/r_T) - (300/F_{yf})}{(756/\sqrt{F_{yf}}) - (300/\sqrt{F_{yf}})}\right)\right] \le F_{yf}$
$\dfrac{L_b}{r_T} > \dfrac{756}{\sqrt{F_{yf}}}$	$\dfrac{286{,}000C_b}{(L_b/r_T)^2}$

m = ratio of web yield stress to flange yield stress or to F_{cr}
F_{yt} = tension flange yield stress
F_{cr} = critical compression flange stress (see following table)

and

b_f = compression flange width
t_f = compression flange thickness
L_b = lateral unbraced length of the girder
r_T = $\sqrt{[(t_f b_f^{\,3}/12 + h_c t_w^{\,3}/72)/(b_f t_f + h_c t_w/6)]}$
h_c = twice the distance from the neutral axis to the inside face of the compression flange less the fillet
t_w = web thickness
F_{yf} = yield stress of compression flange, ksi
C_b = Bending coefficient (see discussion of "Flexural Strength Criterion" in section 46.5)

F_{cr} must be calculated for both flange local buckling and lateral torsional buckling. The smaller value of F_{cr} is used in Eq. (46.72).

The plate girder bending strength reduction factor R_{PG} is a factor to account for the reduced ability of the section to carry bending moments as a result of web buckling due to flexure. Because of the high web slenderness, web buckling usually occurs prior to reaching the full nominal moment strength.

The hybrid girder factor is a reduction factor to account for the lower yield strength of the web when the nominal moment capacity is computed, assuming a homogeneous section made entirely of the higher yield stress of the flange.

Shear Strength Criterion

Plate girders can be designed with or without consideration for tension field action. *Tension field action* refers to the postbuckling (shear) strength of the girder that develops when sufficient web stiffeners are provided for the girder to carry the applied loads by truss-type action. If tension field action is considered, intermediate web stiffeners must be provided and spaced at a distance, a, such that a/h is smaller than 3 or $[260/(h/t_w)]^2$, whichever is smaller. Also, one must check the flexure-shear interaction of Eq. (46.76), if appropriate. Consideration of tension field action is not allowed if (1) the panel is an end panel, (2) the plate girder is a hybrid girder, (3) the plate girder is a web-tapered girder, or (4) a/h is larger than 3 or $[260/(h/t_w)]^2$, whichever is smaller.

The design shear strength of a plate girder should exceed the factored shear [Eq. (46.45)]. If tension field action is not considered, $\phi_v V_n$ is the same as for a beam and is given in Eqs. (46.46) to (46.48). If tension field action is considered and $h/t_w \leq 187\sqrt{k_v/F_{yw}}$, $\phi_v V_n$ is given by

$$\phi_v V_n = 0.90[0.60 A_w F_{yw}] \tag{46.73}$$

If $h/t_w > 187\sqrt{k_v/F_{yw}}$,

$$\phi_v V_n = 0.90\left[0.60 A_w F_{yw}\left(C_v + \frac{1 - C_v}{1.15\sqrt{1 + (a/h)^2}}\right)\right] \tag{46.74}$$

where

k_v = $5 + 5/(a/h)^2$ (k shall be taken as 5.0 if a/h exceeds 3.0 or $[260/(h/t_w)]^2$, whichever is smaller)

A_w = $d t_w$

F_{yw} = web yield stress, ksi

C_v = shear coefficient, calculated according to the following table:

Range of h/t_w	C_v
$187\sqrt{\dfrac{k_v}{F_{yw}}} \leq \dfrac{h}{t_w} \leq 234\sqrt{\dfrac{k_v}{F_{yw}}}$	$\dfrac{187\sqrt{k_v/F_{yw}}}{h/t_w}$
$\dfrac{h}{t_w} > 234\sqrt{\dfrac{k_v}{F_{yw}}}$	$\dfrac{44{,}000 k_v}{(h/t_w)^2 F_{yw}}$

Flexure-Shear Interaction

Plate girders designed for tension field action must satisfy the flexure-shear interaction criterion in regions of high moment and shear. In regions where

$$0.6\phi V_n \leq V_u \leq \phi V_n \quad \text{and} \quad 0.75\phi M_n \leq M_u \leq \phi M_n \tag{46.75}$$

the following flexure-shear interaction equation must be checked:

$$\frac{M_u}{\phi M_n} + 0.625\frac{V_u}{\phi V_n} \leq 1.375 \tag{46.76}$$

where $\phi = 0.90$.

Bearing Stiffeners

Bearing stiffeners must be provided for a plate girder at unframed girder ends and at points of concentrated loads where the web yielding or the web crippling criterion is violated (see discussion of "Criteria for Concentrated Loads" in section 46.5). Bearing stiffeners must be provided in pairs and extend from the upper flange to the lower flange of the girder. Denoting b_{st} as the width of one stiffener and t_{st} as its thickness, bearing stiffeners must be proportioned to satisfy the following limit states:

Local Buckling.

$$\frac{b_{st}}{t_{st}} \leq \frac{95}{\sqrt{F_y}} \tag{46.77}$$

Compression. Equation (46.15) must be satisfied. The design compressive strength $\phi_c P_n$ is to be determined based on an effective length factor K of 0.75 and an effective area , A_{eff}, equal to the area of the bearing stiffeners plus a portion of the web. For end bearing, this effective area is equal to $2(b_{st}t_{st}) + 12t_w^2$; and for interior bearing, this effective area is equal to $2(b_{st}t_{st}) + 25t_w^2$, where t_w is the web thickness. The slenderness parameter, λ_c, is to be calculated using a radius of gyration, $r = \sqrt{(I_{st}/A_{eff})}$, where $I_{st} = t_{st}(2b_{st} + t_w)^3/12$.

Bearing. Equation (46.49) must be satisfied. The bearing strength ϕR_n on milled surfaces is given by

$$\phi R_n = 0.75[1.8F_y A_{pb}] \tag{46.78}$$

where F_y is the yield stress and A_{pb} is the bearing area.

Intermediate Stiffeners

Intermediate stiffeners must be provided if (1) the shear strength capacity is calculated based on tension field action, (2) the shear criterion is violated (i.e., when V_u exceeds $\phi_v V_n$), or (3) the web slenderness h/t_w exceeds $418/\sqrt{F_{yw}}$. Intermediate stiffeners can be provided in pairs, or on one side of the web only, in the form of plates or angles. They should be welded to the compression flange and the web, but they may be stopped short of the tension flange. The following requirements apply to the design of intermediate stiffeners.

Local Buckling. The width-thickness ratio of the stiffener must be proportioned so that Eq. (46.77) is satisfied to prevent failure by local buckling.

Stiffener Area. The cross-sectional area of the stiffener must satisfy the criterion

$$A_{st} \geq \frac{F_{yw}}{F_y}\left[0.15Dht_w(1 - C_v)\frac{V_u}{\phi_v V_n} - 18t_w^2\right] \geq 0 \tag{46.79}$$

where

F_y = yield stress of stiffeners
D = 1.0 for stiffeners in pairs
= 1.8 for single angle stiffeners
= 2.4 for single plate stiffeners

The other terms in Eq. (46.79) are defined as before in Eq. (46.73) and Eq. (46.74).

Stiffener Moment of Inertia. The moment of inertia for stiffener pairs taken about an axis in the web center, or for single stiffeners taken the face of contact with the web plate, must satisfy the criterion

$$I_{st} \geq a t_w^3\left[\frac{2.5}{(a/h)^2} - 2\right] \geq 0.5a t_w^3 \tag{46.80}$$

Stiffener Length. The length of the stiffeners, l_{st}, should fall within the range

$$h - 6t_w < l_{st} < h - 4t_w \qquad (46.81)$$

where h is the clear distance between the flanges less the widths of the flange-to-web welds, and t_w is the web thickness.

If intermittent welds are used to connect the stiffeners to the girder web, the clear distance between welds must not exceed $16t_w$ or 10 in. If bolts are used, their spacing shall not exceed 12 in.

Stiffener Spacing. The spacing of the stiffeners, a, shall be determined from the shear criterion $\phi_v V_n \geq V_u$. This spacing shall not exceed the smaller of 3 and $[260/(h/t_w)]^2 h$.

Example. Design the cross section of an I-shaped plate girder, shown in Fig. 46.20(a), to support a factored moment M_u of 4600 kip-ft (dead weight of the girder is included). The girder is a 60-ft long, simply supported girder. It is laterally supported at every 20-ft interval. Use A36 steel.

Proportion of the Girder Web. Ordinarily, the overall depth-to-span ratio d/L of a building girder is in the range 1:12 to 1:10. Accordingly, begin with $h = 70$ in. Also, knowing that h/t_w of a plate girder is in the range $970/\sqrt{F_{yf}}$ and $2000/\sqrt{F_{fy}}$, begin with $t_w = 5/16$ in.

Proportion of the Girder Flanges. For a preliminary design, the required area of the flange can be determined using the flange area method:

$$A_f \approx \frac{M_u}{F_y h} = \frac{4600 \text{ kip-ft} \times 12 \text{ in./ft}}{(36 \text{ ksi})(70 \text{ in.})} = 21.7 \text{ in.}^2$$

So, let $b_f = 20$ in. and $t_f = 1\,1/8$ in., giving $A_f = 22.5$ in.^{2}

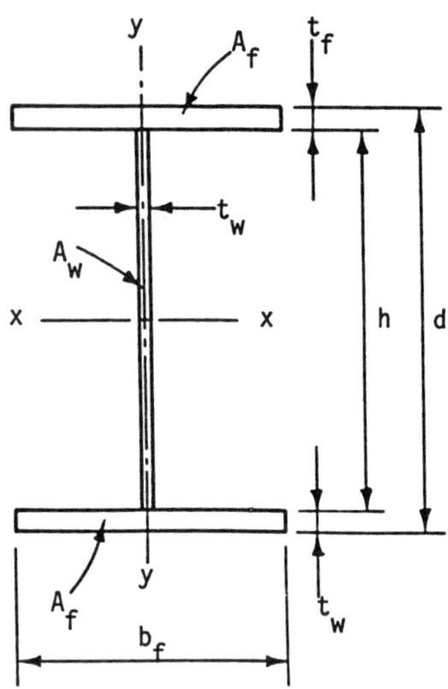

(a) Plate Girder Nomenclature

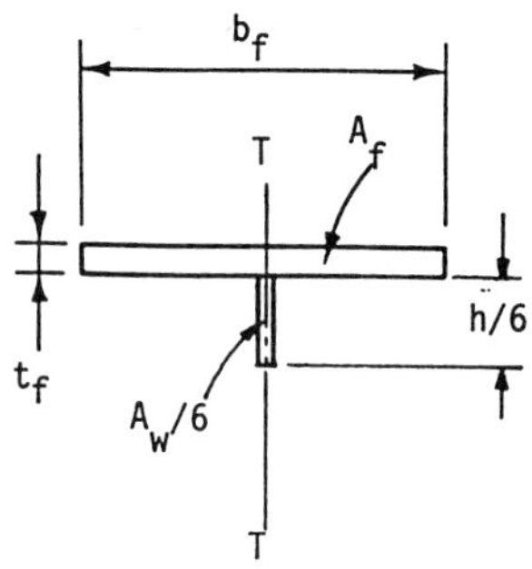

(b) Calculation of r_T

FIGURE 46.20 Design of a plate girder cross section.

Determine the Design Flexural Strength $\phi_b M_n$ of the Girder I_x.

$$\begin{aligned} I_x &= \sum \left[I_i + A_i Y_i^2\right] \\ &= [8932 + (21.88)(0)^2] + 2[2.37 + (22.5)(35.56)^2] \\ &= 65{,}840 \text{ in.}^4 \end{aligned}$$

Determine S_{xt}, S_{xc}.

$$S_{xt} = S_{xc} = \frac{I_x}{c_t} = \frac{I_x}{c_c} = \frac{65{,}840}{35 + 1.125} = 1823 \text{ in.}^3$$

Determine r_T. Refer to Fig. 46.20(b):

$$r_T = \sqrt{\frac{I_T}{A_f + \frac{1}{6}A_w}} = \sqrt{\frac{(1.125)(20)^3/12 + (11.667)(5/16)^3/12}{22.5 + \frac{1}{6}(21.88)}} = 5.36 \text{ in.}$$

Determine F_{cr}. For flange local buckling (FLB),

$$\left[\frac{b_f}{2t_f} = \frac{20}{2(1.125)} = 8.89\right] < \left[\frac{65}{\sqrt{F_{yf}}} = \frac{65}{\sqrt{36}} = 10.8\right] \quad \text{so} \quad F_{\text{cr}} = F_{yf} = 36\,\text{ksi}$$

For lateral torsional buckling (LTB),

$$\left[\frac{L_b}{r_T} = \frac{20 \times 12}{5.36} = 44.8\right] < \left[\frac{300}{\sqrt{F_{yf}}} = \frac{300}{\sqrt{36}} = 50\right] \quad \text{so} \quad F_{\text{cr}} = F_{yf} = 36 \text{ ksi}$$

Determine R_{PG}.

$$R_{PG} = 1 - \left[\frac{a_r}{1{,}200 + 300a_r}\right]\left[\frac{h}{t_w} - \frac{970}{\sqrt{F_{\text{cr}}}}\right] \leq 1.0$$

$$= 1 - 0.00065\left[\frac{70}{(5/16)} - \frac{970}{\sqrt{36}}\right] = 0.96$$

Determine $\phi_b M_n$.

$$\phi_b M_n = \text{smaller of} \begin{cases} 0.90\, S_{xt} R_e F_{yt} = (0.90)(1823)(1)(36) = 59{,}065 \\ 0.90\, S_{xc} R_{PG} R_e F_{\text{cr}} = (0.90)(1823)(0.96)(1)(36) = 56{,}700 \end{cases}$$

$$= 56{,}700 \text{ kip-in.}$$

$$= 4725 \text{ kip-ft}$$

Since $[\phi_b M_n = 4775 \text{ kip-ft}] > [M_u = 4600 \text{ kip-ft}]$, the cross section is acceptable.

Use web plate 5/16″ × 70″ and two flange plates 1⅛″ × 20″ for the girder cross section.

Example. Design bearing stiffeners for the plate girder of the preceding example for a factored end reaction of 260 kips.

Since the girder end is unframed, bearing stiffeners are required at the supports. The size of the stiffeners must be selected to ensure that the limit states of local buckling, compression, and bearing are not violated.

Limit State of Local Buckling. Refer to Fig. 46.21; try $b_{\text{st}} = 8$ in. To avoid a problem with local buckling, $b_{\text{st}}/2t_{\text{st}}$ must not exceed $95/\sqrt{F_y} = 15.8$. Therefore, try $t_{\text{st}} = ½$ in. So $b_{\text{st}}/2t_{\text{st}} = 8$, which is less than 15.8.

Limit State of Compression.

$$A_{\text{eff}} = 2(b_{\text{st}}t_{\text{st}}) + 12t_w^2 = 2(8)(0.5) + 12(5/16)^2 = 9.17 \text{ in.}^2$$

$$I_{\text{st}} = t_{\text{st}}(2b_{\text{st}} + t_w)^3/12 = 0.5[2(8) + 5/16]^3/12 = 181 \text{ in.}^4$$

$$r_{\text{st}} = \sqrt{(I_{\text{st}}/A_{\text{eff}})} = \sqrt{(181/9.17)} = 4.44 \text{ in.}$$

$$Kh/r_{\text{st}} = 0.75(70)/4.44 = 11.8$$

$$\lambda_c = (Kh/\pi r_{\text{st}})\sqrt{(F_y/E)} = (11.8/3.142)\sqrt{(36/29{,}000)} = 0.132$$

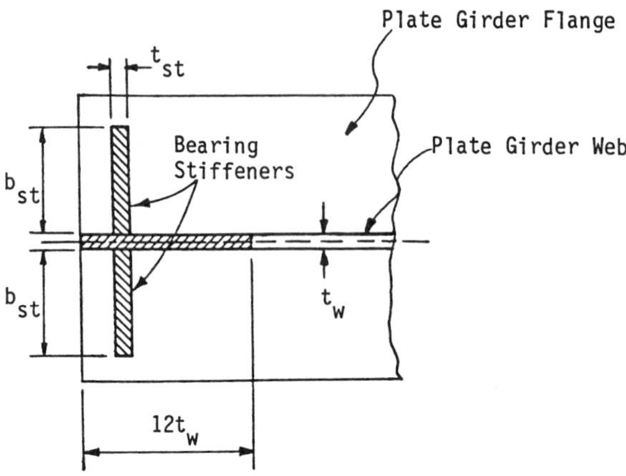

FIGURE 46.21 Design of bearing stiffeners.

and, from Eq. (46.16a),

$$\phi_c P_n = 0.85(0.658^{\lambda_c^2})F_y A_{st} = 0.85(0.658)^{0.132^2}(36)(9.17) = 279 \text{ kips}$$

Since $\phi_c P_n > 260$ kips, the design is satisfactory for compression.

Limit State of Bearing. Assuming there is a ¼-in. weld cutout at the corners of the bearing stiffeners at the junction of the stiffeners and the girder flanges, the bearing area for the stiffener pairs is

$$A_{pb} = (8 - 0.25)(0.5)(2) = 7.75 \text{ in.}^2$$

Substituting this into Eq. (46.78), we have $\phi R_n = 0.75(1.8)(36)(7.75) = 377$ kips, which exceeds the factored reaction of 260 kips. So bearing is not a problem.

Use two ½″ × 8″ plates for bearing stiffeners.

46.12 Connections

Connections are structural elements used for joining different members of a framework. Connections can be classified according to

- The type of connecting medium used: bolted connections, welded connections, bolted-welded connections, riveted connections
- The type of internal forces the connections are expected to transmit: shear (Type PR, semirigid, simple) connections, moment (Type FR, rigid) connections
- The type of structural elements that made up the connections: single-plate angle connections, double web angle connections, top and seated angle connections, seated beam connections, etc.
- The type of members the connections are joining: beam-to-beam connections (beam splices), column-to-column connections (column splices), beam-to-column connections, hanger connections

To design a connection properly, a designer must have a thorough understanding of the behavior of the joint under loads. Different modes of failure can occur depending on the geometry of the connection and the relative strengths and stiffnesses of the various components of the connection. To ensure that the connection can carry the applied loads, a designer must check for all limit states pertinent to each component of the connection and for the connection as a whole.

Bolted Connections

Bolted connections are connections whose components are fastened together primarily by bolts. The four basic types of bolts commonly used for steel construction are discussed as "Structural Fasteners" in Section 46.1. Depending on the direction and line of action of the loads relative to the orientation and location of the bolts, the bolts may be loaded in tension, shear, or a combination of tension and shear. For bolts subjected to shear forces, the design shear strength of the bolts also depends on whether or not the threads of the bolts are excluded from the shear planes. A letter X or N is placed at the end of the ASTM designation of the bolts to indicate whether the threads are excluded or not excluded from the shear planes, respectively. Thus, A325X denotes A325 bolts whose threads are excluded from the shear planes and A490N denotes A490 bolts whose threads are not excluded from the shear planes. Because of the reduced shear areas for bolts whose threads are not excluded from the shear planes, these bolts have lower design shear strengths than their counterparts whose threads are excluded from the shear planes.

Bolts can be used in both bearing-type connections and slip-critical connections. *Bearing-type* connections rely on bearing between the bolt shanks and the connecting parts to transmit forces. Some slippage between the connected parts is expected to occur for this type of connections. *Slip-critical* connections rely on the frictional force developed between the connecting parts to transmit forces. No slippage between connecting elements is expected for this type of connections. Slip-critical connections are used for structures designed for vibratory or dynamic loads such as bridges, industrial buildings, and buildings in regions of high seismicity. Bolts used in slip-critical connections are denoted by the letter F after their ASTM designation, e.g., A325F, A490F.

Bolt Holes. Holes made in the connected parts for bolts may be standard-sized, oversized, short-slotted, or long-slotted. Table 46.13 gives the maximum hole dimension for ordinary construction usage.

Standard holes can be used for both bearing-type and slip-critical connections. Oversized holes are to be used only for slip-critical connections. Short-slotted and long-slotted holes can be used for both bearing-type and slip-critical connections, provided that, when such holes are used for bearing, the direction of the slot is transverse to the direction of loading.

Bolts Loaded in Tension. If a tensile force is applied to the connection such that the direction of load is parallel to the longitudinal axes of the bolts, the bolts will be subjected to tension. The following condition must be satisfied for bolts under tensile stresses:

$$\phi_t F_t \geq f_t \tag{46.82}$$

Table 46.13 Nominal Hole Dimensions (in.) (1 in. = 25.4 mm)

Bolt Diameter (in.)	Hole Dimensions: Standard (dia.)	Oversized (dia.)	Short-Slotted (width × length)	Long-Slotted (width × length)
$\frac{1}{2}$	$\frac{9}{16}$	$\frac{5}{8}$	$\frac{9}{16} \times \frac{11}{16}$	$\frac{9}{16} \times 1\frac{1}{4}$
$\frac{5}{8}$	$\frac{11}{16}$	$\frac{13}{16}$	$\frac{11}{16} \times \frac{7}{8}$	$\frac{11}{16} \times 1\frac{9}{16}$
$\frac{3}{4}$	$\frac{13}{16}$	$\frac{15}{16}$	$\frac{13}{16} \times 1$	$\frac{13}{16} \times 1\frac{7}{8}$
$\frac{7}{8}$	$\frac{15}{16}$	$1\frac{1}{16}$	$\frac{15}{16} \times 1\frac{1}{8}$	$\frac{15}{16} \times 2\frac{3}{16}$
1	$1\frac{1}{16}$	$1\frac{1}{4}$	$1\frac{1}{16} \times 1\frac{5}{16}$	$1\frac{1}{16} \times 2\frac{1}{2}$
$\geq 1\frac{1}{8}$	$d + \frac{1}{16}$	$d + \frac{5}{16}$	$(d + \frac{1}{16}) \times (d + \frac{3}{8})$	$(d + \frac{1}{16}) \times (2.5d)$

Table 46.14 Nominal Tensile Strength of Bolts

Bolt Type	F_t, ksi (Static Loading)	F_t, ksi (Fatigue Loading)
A307	45.0	Not allowed
A325	90.0	If number of cycles is less than 20,000, F_t = same as for static loading. If number of cycles is from 20,000 to 500,000, $F_t = 0.30F_u$.
A490	113	If number of cycles is more than 500,000, $F_t = 0.25F_u$ where F_u = minimum specified tensile strength, ksi.

1 ksi = 6.895 MPa.

where

ϕ_t = 0.75
f_t = tensile stress produced by the factored loads, ksi
F_t = nominal tensile strength, given in Table 46.14

Bolts Loaded in Shear. When the direction of load is perpendicular to the longitudinal axes of the bolts, the bolts will be subjected to shear. The condition that needs to be satisfied for bolts under shear stresses is

$$\phi_v F_v \geq f_v \tag{46.83}$$

where

ϕ_v = 0.60 (for A307 bolts)
= 0.65 (for A325 and A490 bolts in bearing-type connections)
= 1.00 (for A325 and A490 bolts in slip-critical connections when standard, oversized, short-slotted, or long-slotted holes with load perpendicular to the slots are used)
= 0.85 (for A325 and A490 bolts in long-slotted holes when long-slotted holes with load in the direction of the slots are used)
f_v = shear stress produced by the factored loads, ksi
F_v = nominal shear strength given in Table 46.15.

Bolts Loaded in Combined Tension and Shear. If a tensile force is applied to a connection such that its line of action is at an angle with the longitudinal axes of the bolts, the bolts will be subjected to combined tension and shear. The conditions that need to be satisfied are

$$\phi_v F_v \geq f_v \quad \text{and} \quad \phi_t F_t \geq f_t \tag{46.84}$$

where ϕ_v, F_v, f_v are defined as in Eq. (46.83); $\phi_t = 1.0$; f_t = tensile stress due to factored loads, ksi; and F_t = nominal tension stress limit for combined tension and shear, given in Table 46.16.

Bearing Strength at Fastener Holes. Connections designed on the basis of bearing rely on the bearing force developed between the fasteners and the holes to transmit forces and moments. The limit state for bearing must therefore be checked to ensure that bearing failure will not occur. Bearing strength is independent of the type of fastener. This is because the bearing stress is more critical on the parts being connected than on the fastener itself. The AISC specification provi-

Table 46.15 Nominal Shear Strength of Bolts

Bolt Type	F_v, ksi[a]
A307	24 (regardless of whether or not threads are excluded from shear planes)
A325N	48[b]
A325X	60[b]
A325F[c]	17.0 (for standard-sized holes) 15.0 (for oversized and short-slotted holes) 12.0 (for long-slotted holes when direction of load is transverse to the slots)
A490N	60[b]
A490X	75[b]
A490F[c]	21.0 (for standard-sized holes) 18.0 (for oversized and short-slotted holes) 15.0 (for long-slotted holes when direction of load is transverse to the slots)

[a] 1 ksi = 6.895 MPa.

[b] Tabulated values shall be reduced by 20% if the bolts are used to split tension members having a fastener pattern whose length, measured parallel to the line of action of the force, exceeds 50 inches.

[c] Tabulated values are applicable only to class A surface, i.e., clean mill surface and blast-cleaned surface with class A coatings (with slip coefficient = 0.33). For design strengths with other coatings, see RCSC "Load and Resistance Factor Design Specification to Structural Joints Using ASTM A325 or A490 Bolts" [RCSC, 1986].

Table 46.16 Tension Stress Limit for Bolts under Combined Tension and Shear

Bolt Type	Bearing-Type Connections: Threads Not Excluded from the Shear Plane	Bearing-Type Connections: Threads Excluded from the Shear Plane
A307	$F_t = 59 - 1.9f_v \leq 45$	
A325	$F_t = 117 - 1.9f_v \leq 90$	$F_t = 117 - 1.5f_v \leq 90$
A490	$F_t = 147 - 1.9f_v \leq 113$	$F_t = 147 - 1.5f_v \leq 113$
Slip-Critical Connections		
$F_t = [1 - (T/T_b)]$ × (values given in Table 46.15 for A325F and A490F bolts)		

T = service tensile force
T_b = pretension load = $0.70A_b$ × (proof stress given in Table 46.3)
A_b = nominal cross-sectional area of bolt, in.^{2}

sions for bearing strength are based on preventing excessive hole deformation. As a result, the design bearing strength ϕR_n is expressed as a function of the type of holes (standard, oversized, slotted); bearing area (bolt diameter times the thickness of the connected parts); the bolt spacing (s); the edge distance (L_e); the strength of the connected parts (F_u); and the number of fasteners in the direction of the bearing force. Table 46.17 summarizes the expressions used for calculating the bearing strength and the conditions under which each expression is valid.

Minimum Fastener Spacing. To ensure safety and efficiency, and to maintain clearances between bolt nuts as well as to provide room for wrench sockets, AISC provides the following guidelines for fastener spacing:

Table 46.17 Design Bearing Strength

Conditions	ϕR_n, ksi
1. For standard or short-slotted holes with $L_e \geq 1.5d$, $s \geq 3d$, and number of fasteners in the direction of bearing ≥ 2	$0.75[2.4dtF_u]$
2. For long-slotted holes with direction of slot transverse to the direction of bearing and $L_e \geq 1.5d$, $s \geq 3d$, and the number of fasteners in the direction of bearing ≥ 2	$0.75[2.0dtF_u]$
3. For fasteners closest to the edge of the connected parts or if neither condition 1 nor 2 above is satisfied with $L_e < 1.5d$	$0.75[L_e tF_u] \leq 0.75[2.4dtF_u]$
4. If hole deformation is not a design consideration and adequate spacing and edge distance is provided (see sections on Minimum Fastener Spacing and Minimum Edge Distance)	$0.75[3.0dtF_u]$

L_e = edge distance, measured from the edge of the connected part to the center of the hole. For long-slotted hole, the edge distance is measured from the edge of the connected part to the center of the end of the slot nearest to the edge.
s = hole spacing (i.e., center-to-center distance between adjacent holes measured in the direction of bearing)
d = nominal bolt diameter, in.
t = thickness of the connected part, in.
F_u = specified minimum tensile strength of the connected part, ksi

If bearing strength is determined based on condition 1 or 2 in Table 46.17,

$$s \geq 3d \tag{46.85}$$

where s is the fastener spacing and d is the nominal fastener diameter.

If bearing strength is determined based on other conditions,

$$s \geq \frac{P}{\phi F_u t} + \frac{d_h}{2} + C_1 \tag{46.86}$$

where

P = force transmitted by *one* fastener to the critical connected part, kips
ϕ = 0.75
F_u = specified minimum tensile strength, ksi
t = thickness of the critical connected part, in.
d_h = nominal diameter of standard size hole, in.
C_1 = a constant obtained from Table 46.18

For oversized and slotted holes, the spacing evaluated in accordance with Eq. (46.85) or (46.86) shall not be less than one bolt diameter.

Table 46.18 Values for C_1 (in.)

			Slotted Holes		
				Parallel to Line of Force	
Nominal Diameter of Fastener (in.)	Standard Holes	Oversized Holes	Transverse to Line of Force	Short Slots	Long Slots[a]
$\leq 7/8$	0	1/8	0	3/16	$3d/2 - 1/16$
1	0	3/16	0	1/4	1 7/16
$\geq 1\,1/8$	0	1/4	0	5/16	$3d/2 - 1/16$

[a] When length of slot is less than the value shown in Table 46.13, C_1 may be reduced by the difference between the value shown and the actual slot length.

Table 46.19 Minimum Edge Distance

Conditions	Minimum Edge Distance, in.
If the bearing strength is calculated based on Condition 1 or 2 in Table 46.17	1.5d
If the bearing strength is calculated based on the other conditions	Larger of $P/\phi F_u t$ or values given in Table 46.20

d, ϕ, F_u, t are as defined in Eqs. (46.85) and (46.86).

Minimum Edge Distance. To prevent excessive deformation and shear rupture at the edge of the connected part, a minimum edge distance L must be provided in accordance with values given in Tables 46.19 and 46.20. For oversized or slotted holes, the minimum edge distance shall be that shown in Table 46.19 plus an increment C_2, whose values are given in Table 46.21.

Maximum Fastener Spacing and Edge Distance. A limit is placed on the maximum value for the spacing between adjacent fasteners, to prevent the possibility of gaps forming or buckling from occurring in between fasteners when the load to be transmitted by the connection is compressive. The edge distance is also limited, to prevent prying action from occurring. The maximum fastener spacing and edge distance shall not exceed the smaller of $12t$ (where t is the thickness of the connected part) or 6 in.

Example. Check the adequacy of the connection shown in Fig. 46.5. The bolts are 1-in. diameter A325N bolts in standard holes.

Check Bolt Capacity. All bolts are subjected to double shear. Therefore, the design shear strength of the bolts will be twice that shown in Table 46.15. Assuming each bolt carries an equal share of the factored applied load, we have, in regard to Eq. (46.83),

$$[\phi_v F_v = 0.65(2 \times 54) = 70.2 \text{ ksi}] > \left[f_v = \frac{208}{(6)(\pi 1^2/4)} = 44.1 \text{ ksi}\right]$$

So the shear capacity of the bolt is adequate.

Check Bearing Capacity of the Connected Parts. With reference to Table 46.17, it can be easily determined that condition 1 applies for the present problem. Therefore, we have

$$[\phi R_n = 0.75(2.4)(1)\left(\frac{3}{4}\right)(58) = 39.2 \text{ kips}] > \left[R_u = \frac{208}{6} = 34.7 \text{ kips}\right]$$

Table 46.20 Minimum Edge Distance for Standard Holes (in.)

Nominal Fastener Diameter (in.)	At Sheared Edges	At Rolled Edges of Plates, Shapes and Bars or Gas-Cut Edges
1/2	7/8	3/4
5/8	1 1/8	7/8
3/4	1 1/4	1
7/8	1 1/2	1 1/8
1	1 3/4	1 1/4
1 1/8	2	1 1/2
1 1/4	2 1/4	1 5/8
Over 1 1/4	1 3/4 × diameter	1 1/4 × diameter

1 in. = 25.4 mm.

Table 46.21 Edge Distance Increment for Oversized or Slotted Holes (in.)

Nominal Diameter of Fastener (in.)[a]	Oversized Holes	Slotted Holes: Slot Transverse to Edge: Short Slot	Slotted Holes: Slot Transverse to Edge: Long Slot[b]	Slotted Holes: Slot Parallel to Edge
≤ 7/8	1/16	1/8	$3d/4$	0
1	1/8	1/8		
≤ 1 1/8	1/8	3/16		

[a] 1 in. = 25.4 mm.
[b] If the length of the slot is less than the maximum shown in Table 46.13, the value shown may be reduced by one-half the difference between the maximum and the actual slot lengths.

so bearing is not a problem. Note that bearing on the gusset plate is more critical than bearing on the webs of the channels, because the thickness of the gusset plate is less than the combined thickness of the double channels.

Check Bolt Spacing. From Eq. (46.85), the minimum bolt spacing is $3d = 3(1) = 3$ in. The maximum bolt spacing is the smaller of $12t = 12(.303) = 3.64$ in. or 6 in. The actual spacing is 3 in., which falls within the range of 3 to 3.64 in., so bolt spacing is adequate.

Check Edge Distance. From Tables 46.19 and 46.20, it can be determined that the minimum edge distance is $1.5d = 1.5$ inches. The maximum edge distance allowed is the smaller of $12t = 12(0.303) = 3.64$ in. or 6 in. The actual edge distance is 3 in., which falls within the range of 1.5 to 3.64 in., so edge distance is adequate.

The connection is adequate.

Bolted Hanger-Type Connections

A typical hanger connection is shown in Fig. 46.22. In the design of such connections, the designer must take into account the effect of *prying action,* which results when flexural deformation occurs in the tee flange or angle leg of the connection (Fig. 46.23). Prying action tends to increase the tensile force, called *prying force,* in the bolts. To minimize the effect of prying, the fasteners should be placed as close to the tee stem or outstanding angle leg as the wrench clearance will permit (see "Table on Threaded Fasteners—Assembling Clearances" in Volume II, "Connections," of the AISC-LRFD Manual [AISC, 1994]). In addition, the flange and angle thickness should be proportioned so that the full tensile capacities of the bolts can be developed.

Two failure modes can be identified for hanger-type connections: formation of plastic hinges in the tee flange or angle leg at cross sections 1 and 2 (see Fig. 46.23), and tensile failure of the bolts when the tensile force, including prying action B_c, exceeds the tensile capacity of the bolt B. Since the determination of the actual prying force is rather complex, the design equations for the

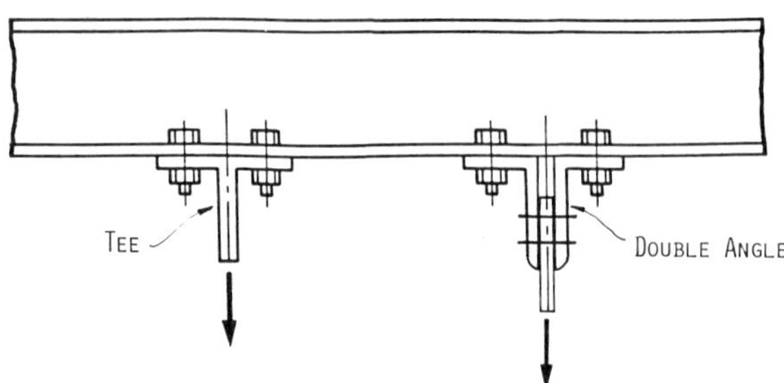

FIGURE 46.22 Hanger connections.

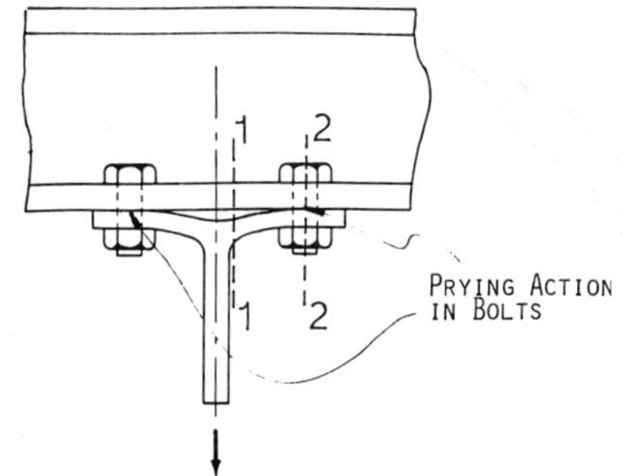

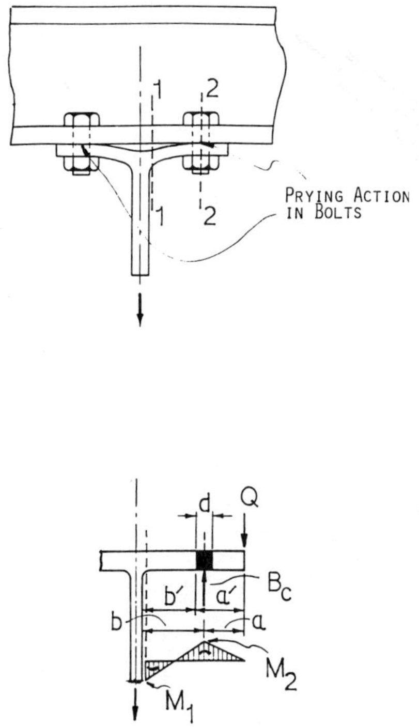

FIGURE 46.23 Mechanism of prying action.

required thickness for the tee flange or angle leg and B_c are semiempirical in nature. They are given as follows:

$$t_{\text{req'd}} = \sqrt{\frac{4T_u b'}{\phi_b p F_y (1 + \delta\alpha)}} \tag{46.87}$$

$$B_c = T_u \left[1 + \frac{\delta\alpha}{(1 + \delta\alpha)} \frac{b'}{a'}\right] \tag{46.88}$$

where

$$\alpha = \frac{\left(\frac{B}{T_u} - 1\right)\frac{a'}{b'}}{\delta\left[1 - \left(\frac{B}{T_u} - 1\right)\frac{a'}{b'}\right]} \tag{46.89}$$

α should be set to zero if it is < 0, and it should be set to 1 if it is > 1.

$$\delta = \frac{p - d'}{p} = \frac{\text{Net area at bolt line}}{\text{Gross area at web face}} \tag{46.90}$$

and

ϕ_b = 0.90
B = design tensile strength of one bolt = $\phi F_t A_b$, kips (ϕF_t is given in Table 46.14, and A_b is the nominal diameter of the bolt)
T_u = factored applied tension per bolt, kips
$a' = a + d/2$
$b' = b - d/2$
a = distance from bolt centerline to edge of tee flange or angle leg but not more than $1.25b$, in.
b = distance from bolt centerline to face of tee stem or outstanding leg, in.
d' = diameter of bolt hole = bolt diameter + ⅛″, in.
p = length of flange tributary to each bolt measured along the longitudinal axis of the tee or double angle section, in.

A design is considered satisfactory if the thickness of the tee flange or angle leg t_f exceeds $t_{\text{req'd}}$ and if the design tensile strength of the bolt B exceeds B_c.

Note that if t_f is different from $t_{\text{req'd}}$ (which is often the case), α should be recomputed using the following equation

$$\alpha = \frac{1}{\delta}\left[\frac{4T_u b'}{\phi_b p t_f^2 f_y} - 1\right] \tag{46.91}$$

As before, the value of α should be limited to the range $0 \leq \alpha \leq 1$. This new value of α is to be used in Eq. (46.88) to recalculate B_c. The design is satisfactory if the new value of B_c does not exceed the tensile capacity of the bolt B.

Bolted Bracket-Type Connections

Figure 46.24 shows three commonly used bracket-type connections. The bracing connection shown in Fig. 46.24(a) should be designed so that the line of action of the force passes through the centroid

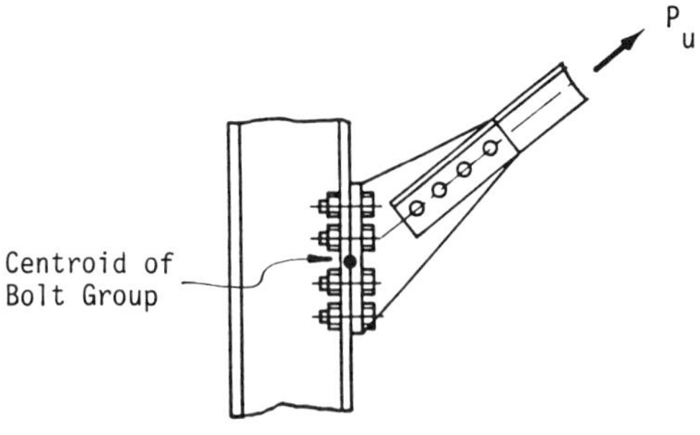

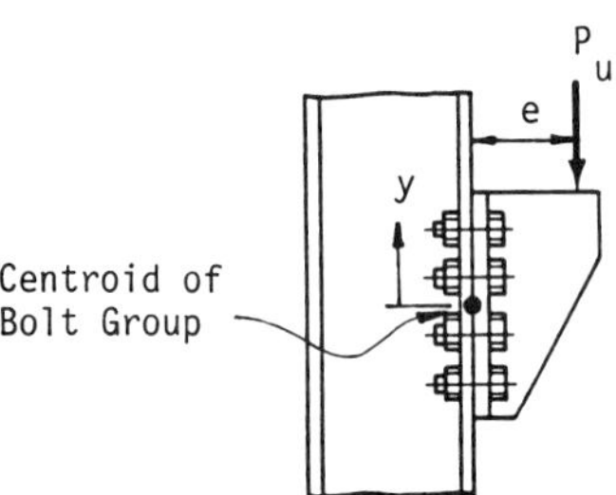

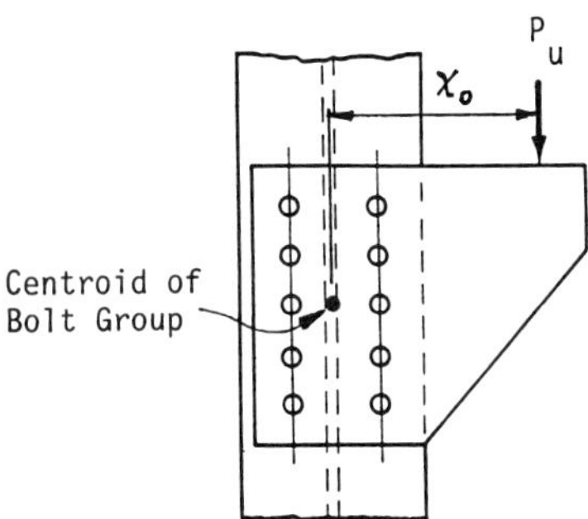

FIGURE 46.24 Bolted bracket-type connections.

of the bolt group. It is apparent that the bolts connecting the bracket to the column flange are subjected to combined tension and shear. As a result, the capacity of the connection is limited to the combined tensile-shear capacities of the bolts in accordance with Eq. (46.84). For simplicity, f_v and f_t are to be computed assuming that both the tensile and shear components of the force are distributed evenly to all bolts. In addition to checking for the bolt capacities, the bearing capacities (Table 46.17) of the column flange and the bracket should also be checked. If the axial component of the force is significant, the effect of prying should also be considered.

In the design of the eccentrically loaded connections shown in Fig. 46.24(b), it is assumed that the neutral axis of the connection lies at the center of gravity of the bolt group. As a result, the bolts above the neutral axis will be subjected to combined tension and shear, so Eq. (46.84) needs to be checked. The bolts below the neutral axis are subjected to shear only, so Eq. (46.83) applies. In calculating f_v, one can assume that all bolts in the bolt group carry an equal share of the shear force. In calculating f_t, one can assume that the tensile force varies linearly from a value of zero at the neutral axis to a maximum value at the bolt farthest away from the neutral axis. Using this assumption, f_t can be calculated from the expression $P_u ey/I$, where y is the distance from the neutral axis to the location of the bolt above the neutral axis, and $I = \sum A_b y^2$ is the moment of inertia of the bolt areas, with A_b = cross-sectional area of each bolt. The capacity of the connection is determined by the capacities of the bolts and the bearing capacity of the connected parts.

The capacity of the eccentrically loaded bracket connection shown in Fig. 46.24 is given by AISC [1994] as

$$P_u = C(\phi F_v) \tag{46.92}$$

where

ϕr_v = $\phi F_v A_b$ is the design load of one fastener (see the discussion of bolts loaded in shear in this section for ϕF_v)

C = a coefficient tabulated in the AISC-LRFD Manual [AISC, 1994]

C is a function of the number of vertical lines of bolts; the horizontal and vertical bolt spacings; the horizontal distance from the centroid of the bolt group to P_u (labeled χ_0 in the figure) and the angle P_u makes with the vertical.

To prevent failure of the bracket plate by shear rupture, the shear rupture strength of the plate must also be investigated. The shear rupture criterion is given by

$$\phi(0.60F_u)A_{nv} \geq P_u \tag{46.93}$$

where

ϕ = 0.75

F_u = specified minimum tensile stress

A_{nv} = net area subject to shear

Bolted Shear Connections

Shear connections are connections designed to resist shear force only. These connections are not expected to provide appreciable moment restraint to the connection members. Examples of these connections are shown in Fig. 46.25. The framed beam connection shown in Fig. 46.25(a) consists of two web angles, which are often shop-bolted to the beam web and then field-bolted to the column flange. The seated beam connection shown in Fig. 46.25(b) consists of two flange angles, often shop-bolted to the beam flange and field-bolted to the column flange. To enhance the strength and stiffness of the seated beam connection, a stiffened seated beam connection, shown in Fig. 46.25(c), is sometimes used to resist large shear force. Shear connections must be designed to sustain appreciable deformation, and yielding of the connections is expected. The need for ductility often limits the thickness of the angles that can be used. Most of these connections are designed with angle thickness not exceeding ⅝ in.

The design of the connections shown in Fig. 46.25 is facilitated by the use of a set of tables contained in the AISC-LRFD Manual. These tables give design loads for the connections with specific dimensions based on the limit states of bolt shear, bearing strength of the connection, bolt bearing with different edge distances, and block shear (for coped beams).

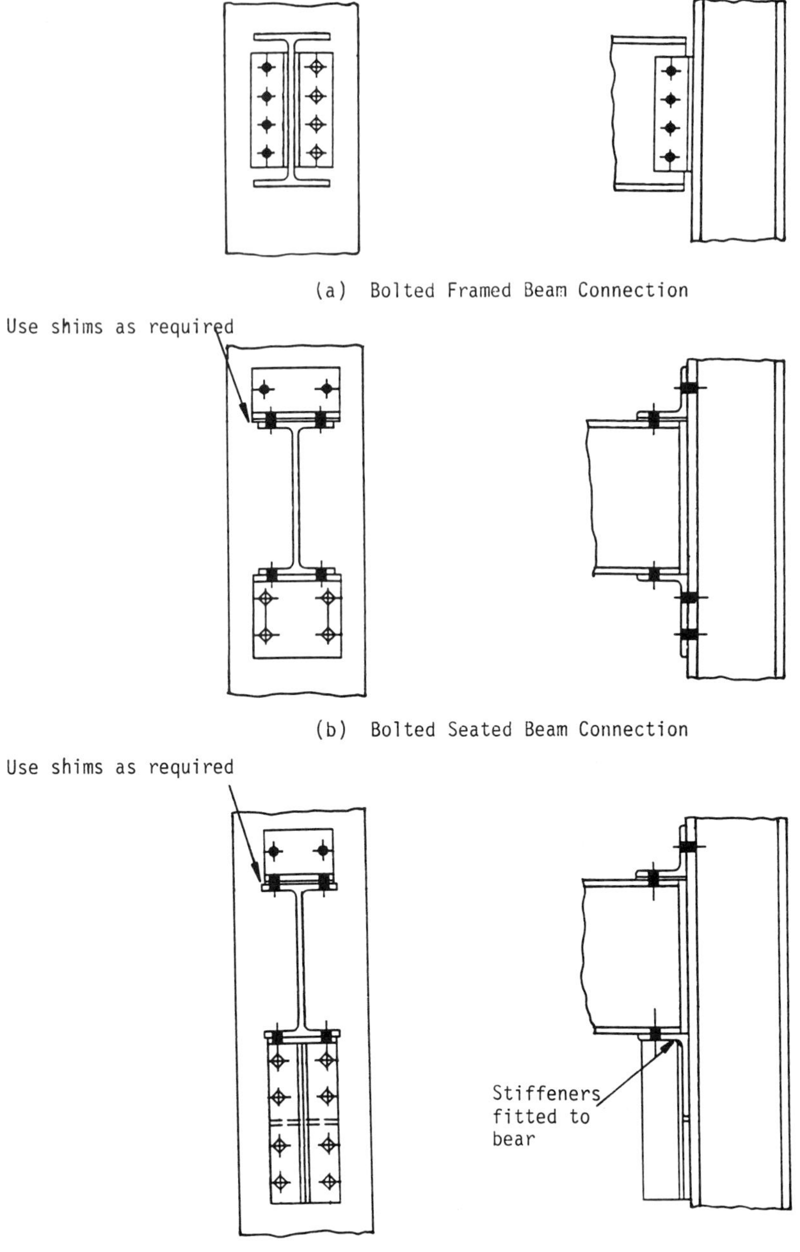

FIGURE 46.25 Examples of bolted shear connections. (a) Bolted framed beam connection. (b) Bolted seated beam connection. (c) Bolted stiffened seated beam connection.

Bolted Moment-Resisting Connections

Moment-resisting connections are connections designed to resist both moment and shear. These connections are often referred to as *rigid* or *fully restrained* connections, as they provide full continuity between the connected members and are designed to carry the full factored moments. Figure 46.26 shows some examples of moment-resisting connections. Additional examples can be found in the AISC-LRFD Manual and Chapter 4 of the AISC Manual on Connections [AISC, 1992].

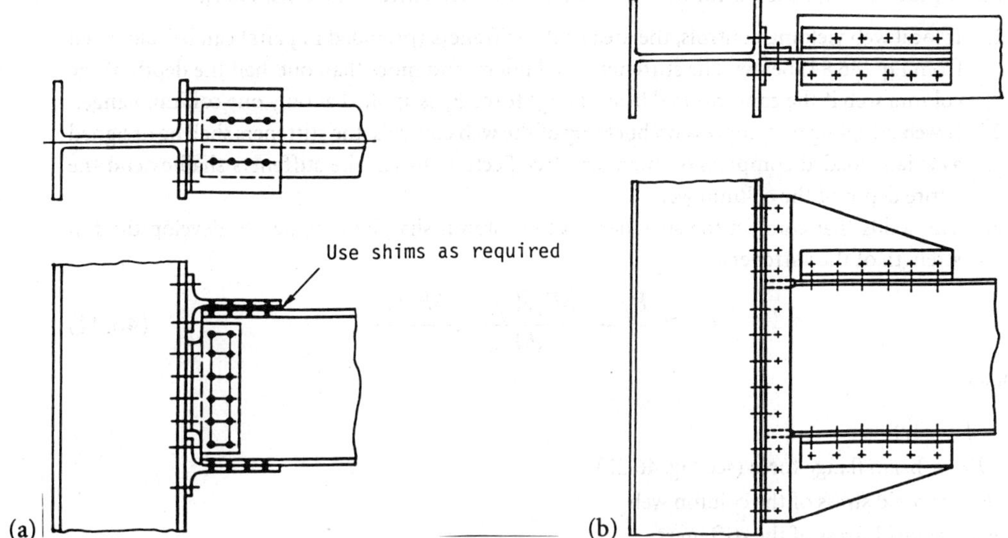

FIGURE 46.26 Examples of bolted moment connections.

Design of Moment-Resisting Connections

An assumption used quite often in the design of moment connections is that the factored moment is carried solely by the flanges of the beam. The moment is converted to a couple F_f, given by $F_f = M_u/(d - t_f)$, acting on the beam flanges, as shown in Fig. 46.27.

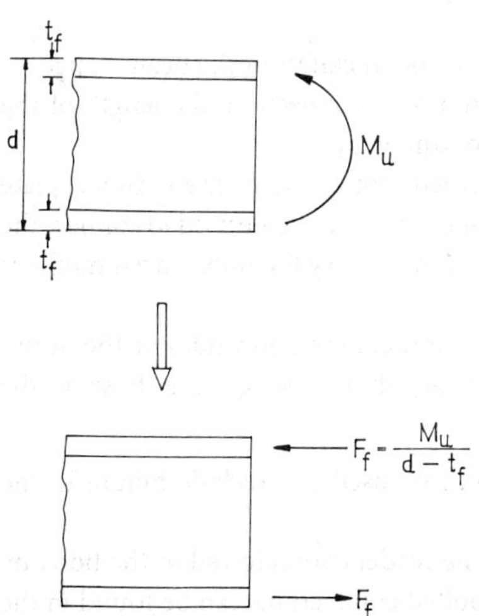

FIGURE 46.27 Flange forces in moment connections.

The design of the connection for moment is considered satisfactory if the capacities of the bolts and connecting plates or structural elements are adequate to carry the flange force F_f. Depending on the geometry of the bolted connection, this may involve checking (a) the shear and tensile capacities of the bolts; (b) the yield and fracture strength of the moment plate; (c) the bearing strength of the connected parts; and (d) bolt spacing and edge distance, as discussed in the foregoing sections.

As for shear, it is common practice to assume that all the shear resistance is provided by the shear plates or angles. The design of the shear plates or angles is governed by the limit states of bolt shear, bearing of the connected parts, and *shear rupture:* failure by fracture due to high shear. To avoid shear rupture, Eq. (46.93) with P_u replaced by V_u (where V_u is the factored shear force), must be satisfied.

If the moment to be resisted is large, the flange force may cause bending of the column flange or local yielding, crippling, or buckling of the column web. To prevent failure due to bending of the column flange or local yielding of the column web (for a tensile F_f) as well as local yielding, crippling, or buckling of the column web (for a compressive F_f), column stiffeners should be provided:

- If F_f is tensile and its magnitude exceeds ϕR_n given by Eq. (46.50) for column flange bending or by Eq. (46.51) for column web yielding
- If F_f is compressive and its magnitude exceeds ϕR_n given by Eq. (46.51) for column web yielding, by Eq. (46.53) for column web crippling, by Eq. (46.57) for column web buckling

Following is a set of guidelines for the design of column web stiffeners [AISC, 1994]:

1. If local web yielding controls, the area of the stiffeners (provided in pairs) can be calculated from Eq. (46.94) below. The stiffeners need not extend more than one-half the depth of the column web if the concentrated beam flange force F_f is applied at only one column flange.
2. If web crippling or compression buckling of the web controls, the stiffeners shall be designed as axially loaded compression members (see Section 46.4). The stiffeners shall extend the entire depth of the column web.
3. The welds that connect the stiffeners to the column shall be designed to develop the full strength of the stiffeners.

$$A_{\text{st}} \geq \frac{F_f - \phi F_{yc}(t_{fb} + 5k)t_{wc}}{\phi F_{yst}} \tag{46.94}$$

where

ϕ = 1.0
F_f = beam flange force (see Fig. 46.27)
F_{yc} = yield stress of the column web
F_{yst} = yield stress of the stiffeners
k = distance from outer face of flange to web toe of fillet
t_{fb} = beam flange thickness
t_{wc} = column web thickness

In addition, the following recommendations are given:

1. The width of the stiffener plus one-half of the column web thickness should not be less than one-half the width of the beam flange or of the moment connection plate that delivers the beam flange force F_f.
2. The stiffener thickness should not be less than one-half the thickness of the beam flange.
3. If only one flange of the column is connected by a moment connection, the length of the stiffener plate does not have to exceed one-half the column depth.
4. If both flanges of the column are connected by moment connections, the stiffener plate should extend through the depth of the column web, and welds should be used to connect the stiffener plate to the column web, with sufficient strength to carry the unbalanced moment on opposite sides of the column.
5. If column stiffeners are required on both the tension and compression sides of the beam, the sizes of the stiffeners on the tension side of the beam should be equal to those on the compression side for ease of construction.

In lieu of stiffener plates, a stronger column section should be used to preclude failure in the column flange and web.

For a more thorough discussion of bolted connections, the readers are referred to the book by Kulak et al. [1987]. Examples on the design of a variety of bolted connections can be found in the AISC LRFD Manual [AISC, 1994] and the AISC Manual on Connections [AISC, 1992].

Welded Connections

Welded connections are connections whose components are joined together primarily by welds. The four most commonly used welding processes are discussed under "Structural Fasteners" in section 46.1. Welds can be classified according to

- Form of weld (groove, fillet, plug, and slot)
- Position of the weld (horizontal, vertical, overhead, and flat)
- Types of joint (butt, lap, corner, edge, and tee)

Although fillet welds are weaker than groove welds, they are used much more often, because they allow for larger tolerances during erection than groove welds. Plug and slot welds are expensive to make, and they do not provide much reliability in transmitting tensile forces perpendicular to the *faying surfaces* (the surfaces to be joined). Furthermore, quality control of such welds is difficult, because inspection of the welds is rather arduous. As a result, plug and slot welds are normally used only for stitching different parts of the members together.

Welding Symbols. A shorthand notation giving important information on the location, size, length, etc., for the various types of welds has been developed by the American Welding Society [AWS, 1987] to facilitate the detailing of welds. This system of notation is reproduced in Fig. 46.28.

Design Strength of Welds. The design strength of welds is taken as the lower value of the design strength of the base material ϕF_{BM} and the design strength of the weld electrode ϕF_w. These design strengths are summarized in Table 46.22 [AISC, 1993].

The design strength of welds shown in the table should exceed the required strength obtained by dividing the load to be transmitted by the effective area of the welds.

Effective Area of Welds. The effective area of groove welds is equal to the product of the width of the part joined and the *effective throat thickness.* The effective throat thickness of a full-penetration groove weld is taken as the thickness of the thinner part joined. The effective throat thickness of a partial-penetration groove weld is taken as the depth of the chamfer for J, U, bevel, or V (with bevel $\geq 60°$) joints, and it is taken as the depth of the chamfer minus ⅛ in. for bevel or V joints if the bevel is between 45° and 60°. For flare bevel groove welds the effective throat thickness is taken as $5R/16$, and for flare V-groove the effective throat thickness is taken as $R/2$ (or $3R/8$ for GMAW process when $R \geq 1$ in.). R is the radius of the bar or bend.

The effective area of fillet welds is equal to the product of the length of the fillets, including returns, and the effective throat thickness. The effective throat thickness of a fillet weld is the shortest distance from the root of the joint to the face of the diagrammatic weld, as shown in Fig. 46.29. Thus, for an equal-leg fillet weld, the effective throat is given by 0.707 times the leg dimension. For a fillet weld made by the submerged-arc welding process (SAW), the effective throat thickness is taken as the leg size (for ⅜-in. and smaller fillet welds) or as the theoretical throat plus 0.11 in. (for fillet weld over ⅜ in.). The larger value for the effective throat thickness for welds made by the SAW process is due to the inherently superior quality of such welds.

The effective area of plug and slot welds is taken as the nominal cross-sectional area of the hole or slot in the plane of the faying surface.

Size and Length Limitations of Welds. To ensure the effectiveness of the welds, certain size and length limitations are imposed by the AISC LRFD specification. For partial-penetration groove welds, minimum values for the effective throat thickness are shown in Table 46.23.

For fillet welds the following size and length limitations apply:

Minimum size of leg: The minimum leg size is given in Table 46.24.
Maximum size of leg: Along the edge of a connected part less than ¼ in. thick, the maximum leg size is equal to the thickness of the connected part. For thicker parts, the maximum leg size is $t - 1/16$ in. where t is the thickness of the part.
Minimum length of weld: The minimum length of a fillet weld is 4 times its leg size. If a shorter length is used, the leg size of the weld must be taken as ¼ of its length for purposes of stress computation. The length of fillet welds used for flat bar tension members shall not be less than the width of the bar if the welds are provided in the longitudinal direction only. The transverse

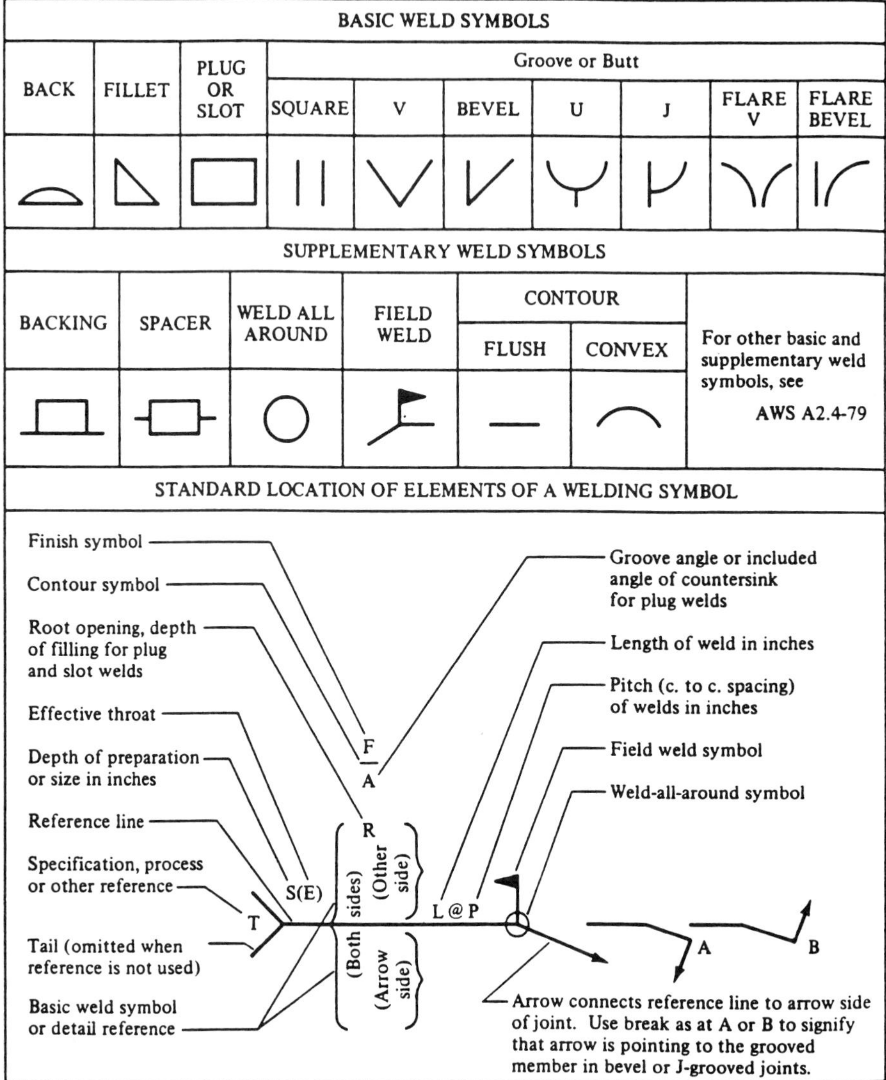

Note:

Size, weld symbol, length of weld and spacing must read in that order from left to right along the reference line. Neither orientation of reference line nor location of the arrow alters this rule.

The perpendicular leg of ◺, ⊬, ⊬, |⌈ weld symbols must be at left.

Arrow and Other Side welds are of the same size unless otherwise shown. Dimensions of fillet welds must be shown on both the Arrow Side and the Other Side Symbol.

The point of the field weld symbol must point toward the tail.

Symbols apply between abrupt changes in direction of welding unless governed by the "all around" symbol or otherwise dimensioned.

These symbols do not explicitly provide for the case that frequently occurs in structural work, where duplicate material (such as stiffeners) occurs on the far side of a web or gusset plate. The fabricating industry has adopted this convention: that when the billing of the detail material discloses the existence of a member on the far side as well as on the near side, the welding shown for the near side shall be duplicated on the far side.

FIGURE 46.28 Basic weld symbols.

Table 46.22 Design Strength of Welds

Types of Weld and Stress[a]	Material	ϕF_{BM} or ϕF_w	Required Weld Strength Level[b,c]
	Full-Penetration Groove Welds		
Tension normal to effective area	Base	$0.90F_y$	"Matching" weld must be used
Compression normal to effective area	Base	$0.90F_y$	Weld metal with a strength level equal to or less than "matching" must be used
Tension of compression parallel to axis of weld			
Shear on effective area	Base Weld electrode	$0.90[0.60F_y]$ $0.80[0.60F_{EXX}]$	
	Partial-Penetration Groove Welds		
Compression normal to effective area	Base	$0.90F_y$	Weld metal with a strength level equal to or less than "matching" weld metal may be used
Tension or compression parallel to axis of weld[d]			
Shear parallel to axis of weld	Base Weld electrode	$0.75[0.60F_{EXX}]$	
Tension normal to effective area	Base Weld electrode	$0.90F_y$ $0.80[0.60F_{EXX}]$	
	Fillet Welds		
Stress on effective area	Base Weld electrode	$0.75[0.60F_{EXX}]$ $0.90F_y$	Weld metal with a strength level equal to or less than "matching" weld metal may be used
Tension or compression parallel to axis of weld[d]	Base	$0.90F_y$	
	Plug or Slot Welds		
Shear parallel to faying surfaces (on effective area)	Base Weld electrode	$0.75[0.60F_{EXX}]$	Weld metal with a strength level equal to or less than "matching" weld metal may be used

[a] See discussion in text for effective area.
[b] See Table 4.1.1, AWS D1.1 for "matching" weld material.
[c] Weld metal one strength level stronger than "matching" weld metal will be permitted.
[d] Fillet welds and partial-penetration groove welds joining component elements of built-up members, such as flange-to-web connections, may be designed without regard to the tensile or compressive stress in these elements parallel to the axis of the welds.

distance between longitudinal welds should not exceed 8 in. unless the effect of shear lag is accounted for by the use of an effective net area.

End returns: End returns must be continued around the corner and must have a length of at least 2 times the weld (leg) size.

Welded Connections for Tension Members

Figure 46.30 shows a tension angle member connected to a gusset plate by fillet welds. The factored axial force P_u is assumed to act along the center of gravity of the angle. To avoid eccentricity, the lengths of the two fillet welds must be proportioned so that their resultant will also act along the center of gravity of the angle. From equilibrium considerations, the following equations can be written. Summing forces along the axis of the angle,

$$(\phi F_M)t_{\text{eff}}L_1 + (\phi F_M)t_{\text{eff}}L_2 = P_u \tag{46.95}$$

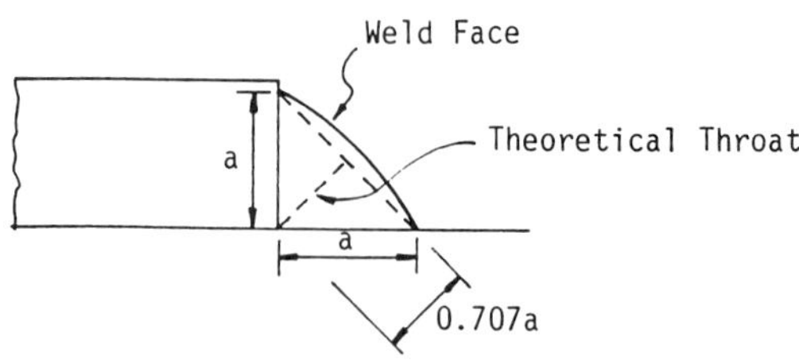

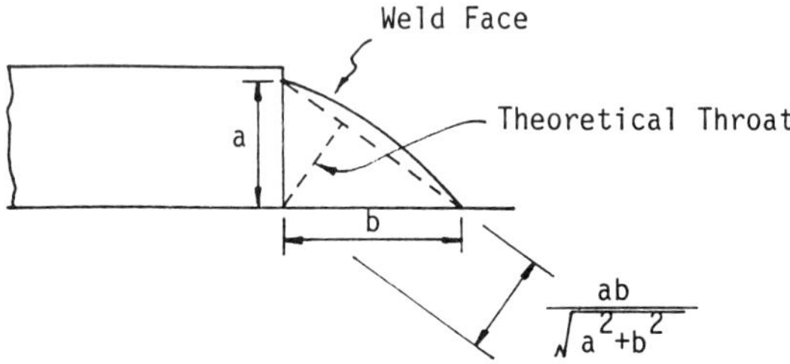

FIGURE 46.29 Effective throat of fillet welds.

Table 46.23 Minimum Effective Throat Thickness of Partial-Penetration Groove Welds

Thickness of the Thicker Part Joined, t (in.)	Minimum Effective Throat Thickness (in.)
$t \leq 1/4$	1/8
$1/4 < t \leq 1/2$	3/16
$1/2 < t \leq 3/4$	1/4
$3/4 < t \leq 1\,1/2$	5/16
$1\,1/2 < t \leq 2\,1/4$	3/8
$2\,1/4 < t \leq 6$	1/2
$t > 6$	5/8

1 in. = 25.4 mm.

Table 46.24 Minimum Leg Size of Fillet Welds

Thickness of Thicker Part Joined, t (in.)	Minimum Leg Size (in.)
$t \leq 1/4$	1/8
$1/4 < t \leq 1/2$	3/16
$1/2 < t \leq 3/4$	1/4
$t > 3/4$	5/16

1 in. = 25.4 mm.

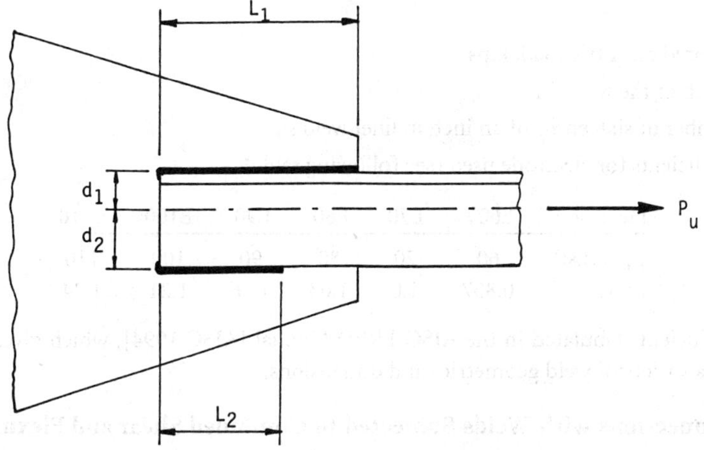

FIGURE 46.30 An eccentrically loaded welded tension connection.

Summing moments about the center of gravity of the angle,

$$(\phi F_M)t_{\text{eff}}L_1d_1 = (\phi F_M)t_{\text{eff}}L_2d_2 \tag{46.96}$$

where ϕF_M is the design strength of the welds, ϕF_{BM} or ϕF_w, as given in Table 46.22; t_{eff} is the effective throat thickness; L_1, L_2 are the lengths of the welds; and d_1, d_2 are the transverse distances from the center of gravity of the angle to the welds. The two equations can be used to solve for L_1 and L_2. If end returns are used, the added strength of the end returns should also be included in the calculations.

Welded Bracket-Type Connections

A typical welded bracket connection is shown in Fig. 46.31. The load capacity for the connection is given by AISC [1994] as

$$P_u = CC_1DL \tag{46.97}$$

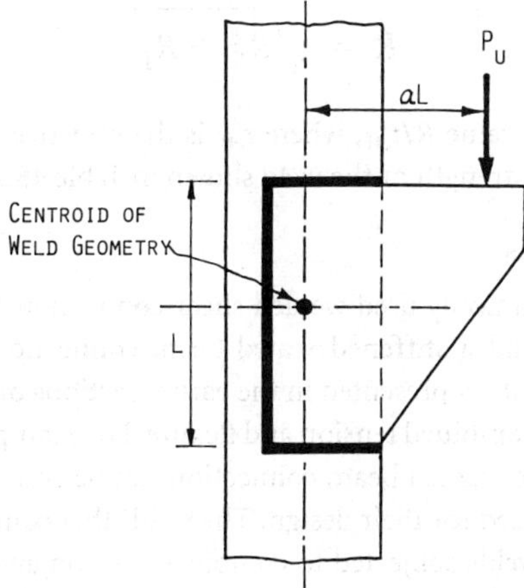

FIGURE 46.31 An eccentrically loaded welded bracket connection.

where

P_u = factored eccentric load, kips
L = length of the weld, in.
D = number of sixteenths of an inch in fillet weld size
C_1 = coefficients for electrode used (see following table)

Electrode	E60	E70	E80	E90	E100	E110
F_{EXX}(ksi)	60	70	80	90	100	110
C_1	0.857	1.0	1.03	1.16	1.21	1.34

C = coefficients tabulated in the AISC-LRFD Manual [AISC 1994], which gives values of C for a variety of weld geometries and dimensions.

Welded Connections with Welds Subjected to Combined Shear and Flexure

Figure 46.32 shows a welded framed connection and a welded seated connection. The welds for these connections are subjected to combined shear and flexure. For purpose of design, it is common practice to assume that the shear force per unit length acting on the welds, R_S, is a constant and is given by

$$R_S = \frac{P_u}{2L} \tag{46.98}$$

where P_u is the factored eccentric load and L is the length of the weld.

In addition to shear, the welds are subjected to flexure as a result of load eccentricity. There is no general agreement on how the flexure stress should be distributed on the welds. One approach is to assume that the stress distribution is linear with half the weld subjected to tensile flexure stress and half subjected to compressive flexure stress. Based on this stress distribution and ignoring the returns, the flexure tension force per unit length of weld R_F acting at the top of the weld can be written as

$$R_F = \frac{Mc}{I} = \frac{P_u e(L/2)}{2L^3/12} = \frac{3P_u e}{L^2} \tag{46.99}$$

where e is the load eccentricity.

The resultant force per unit length acting on the weld R is then

$$R = \sqrt{R_S^2 + R_F^2} \tag{46.100}$$

For a satisfactory design, the value R/t_{eff}, where t_{eff} is the effective throat thickness of the weld, should not exceed the design strength of the weld shown in Table 46.22.

Welded Shear Connections

Figure 46.33 shows three commonly used welded shear connections: a framed beam connection, a seated beam connection, and a stiffened seated beam connection. These connections can be designed by using the information presented in the earlier sections on welds subjected to eccentric shear and welds subjected to combined tension and flexure. For example, the welds that connect the angles to the beam web in the framed beam connection can be considered as eccentrically loaded welds, so Eq. (46.97) can be used for their design. The welds that connect the angles to the column flange can be considered as welds subjected to combined tension and flexure, so Eq. (46.100) can be used for their design. Like bolted shear connections, welded shear connections are expected to exhibit appreciable ductility, so the use of angles with thickness in excess of ⅝ in. should be avoided. To prevent shear rupture failure, the shear rupture strength [Eq. (46.93)] of the critically loaded connected parts should be checked (with P_u replaced by the factored shear force V_u).

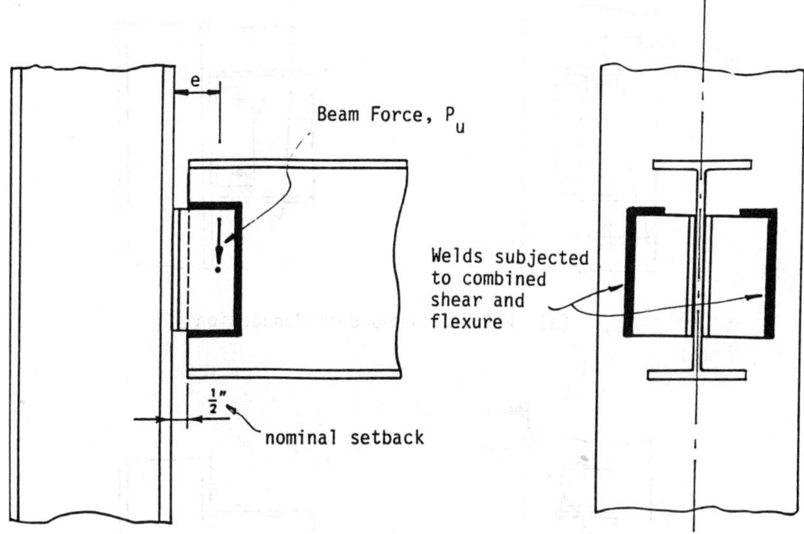

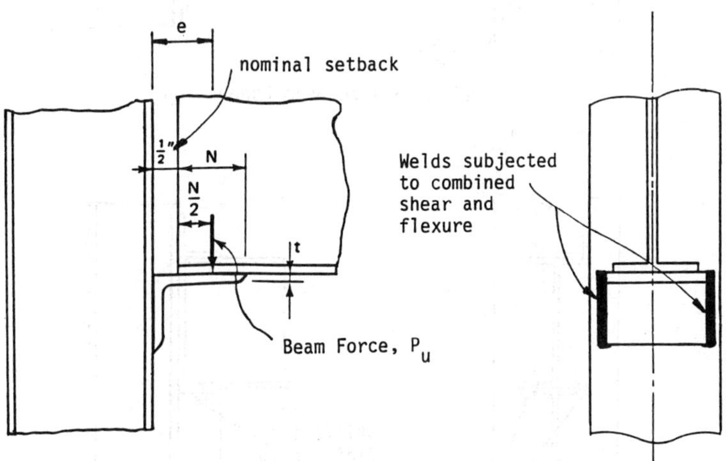

FIGURE 46.32 Examples of welds subjected to combined shear and flexure.

To facilitate the design of these connections, the AISC LRFD Manual [AISC, 1994], provides a set of tables by which the weld capacities and shear rupture strengths for different connection dimensions can be checked readily.

Welded Moment-Resisting Connections

Welded moment-resisting connections (Fig. 46.34), like bolted moment-resisting connections, must be designed to carry the full factored moment M_u and the factored shear V_u. To simplify the design procedure, it is customary to assume that the factored moment, to be represented by a couple F_f as shown in Fig. 46.27, is to be carried by the beam flanges and that the factored shear is to be carried by the beam web. The connected parts (moment plates, welds, etc.) are then designed

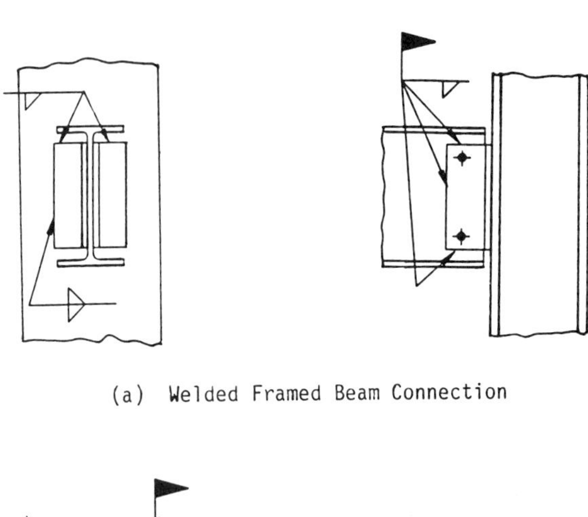

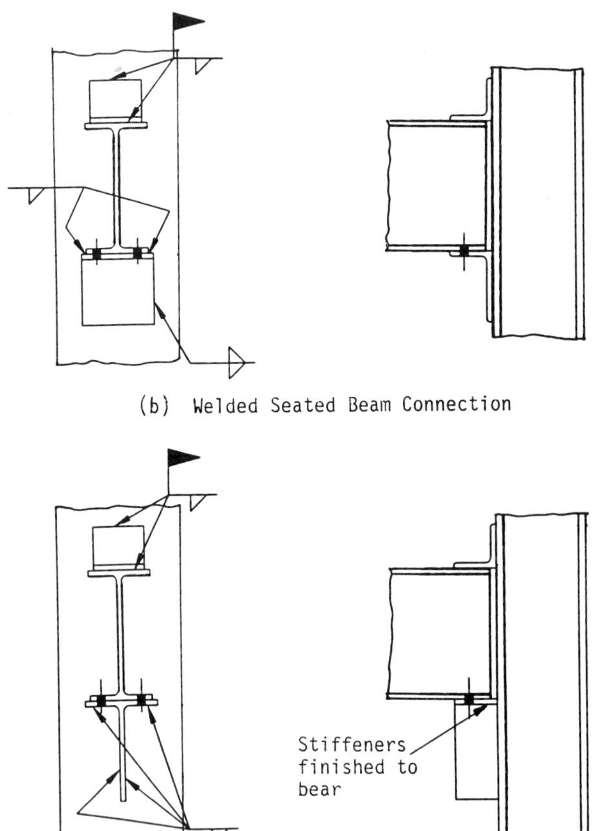

FIGURE 46.33 Examples of welded shear connections. (a) Framed beam connection. (b) Seated beam connection. (c) Stiffened seated beam connection.

to resist the forces F_f and V_u. Depending on the geometry of the welded connection, this may require checking (a) the yield and fracture strength of the moment plate; (b) the shear and tensile capacity of the welds; (c) the shear rupture strength of the shear plate.

If the column to which the connection is attached is weak, the designer should consider the use of column stiffeners to prevent failure of the column flange and web due to bending, yielding,

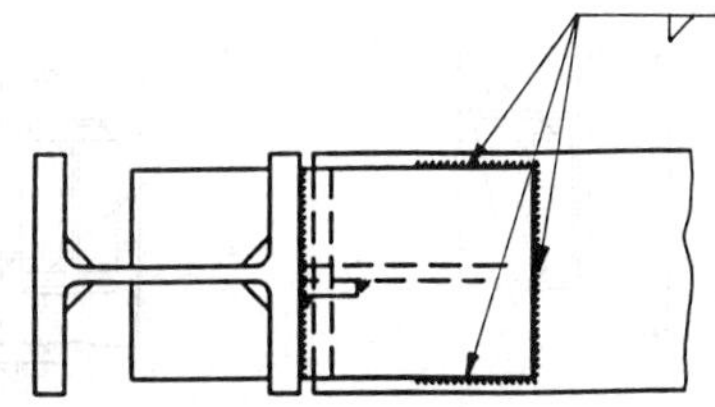

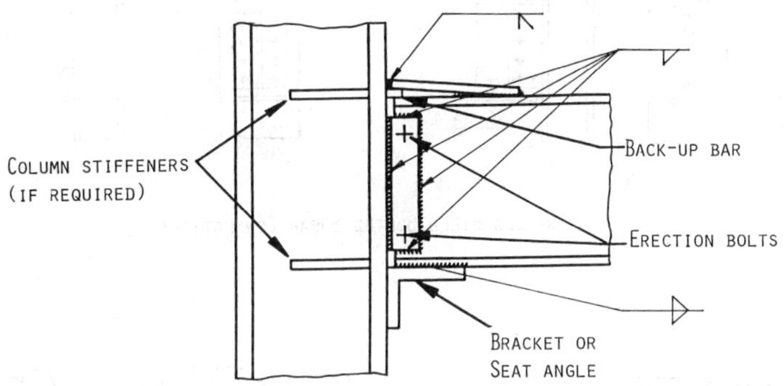

FIGURE 46.34 Examples of welded moment connections.

crippling, or buckling. See the discussion on "Design of Moment-Resisting Connections" under "Bolted Connections," this section.

Examples on the design of a variety of welded shear and moment-resisting connections can be found in the AISC-LRFD Manual [AISC, 1994] and the AISC Manual on Connections [AISC, 1992].

Shop-Welded, Field-Bolted Connections

A large percentage of connections used for construction are shop-welded and field-bolted. These connections are usually more cost-effective than fully welded connections, and their strength and ductility characteristics often rival those of fully welded connections. Figure 46.35 shows some of these connections. The design of shop-welded, field-bolted connections is also covered in Volume II the AISC-LRFD Manual and the AISC Manual on Connections. In general, the following should be checked:

1. Shear/tensile capacities of the bolts and welds
2. Bearing strength of the connected parts
3. Yield and fracture strength of the moment plate
4. Shear rupture strength of the shear plate

Also, as for any other types of moment connections, column stiffeners must be provided if any one of the following criteria is violated: column flange bending, local web yielding, and crippling and compression buckling of the column web.

Beam and Column Splices

Beam and column splices (Fig. 46.36) are used to connect beam or column sections of different sizes. They are also used to connect beams or columns of the same size if the design calls for an extraordinarily long span. Splices should be designed for both moment and shear, unless it is the intention of the designer to utilize the splices as internal hinges. If splices are used for internal

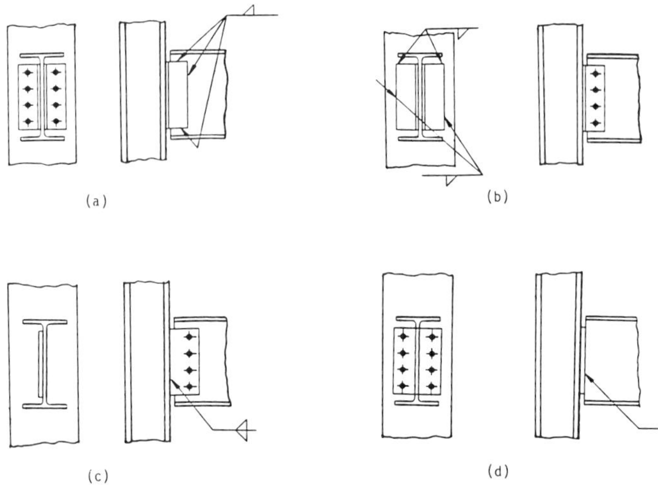

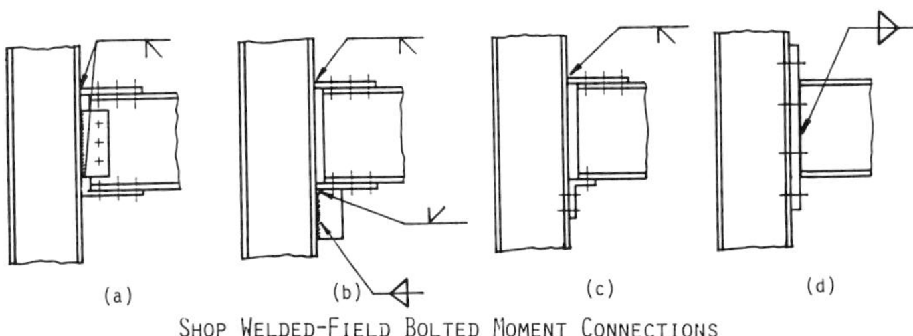

FIGURE 46.35 Examples of shop-welded, field-bolted connections.

hinges, provisions must be made to ensure that the connections possess adequate ductility to allow for large hinge rotation.

Splice plates are designed according to their intended functions. Moment splices should be designed to resist the flange force $F_f = M_u/(d - t_f)$ (Fig. 46.27) at the splice location. In particular, the following limit states need to be checked: yielding of gross area of the plate; fracture of net area of the plate (for bolted splices); bearing strengths of connected parts (for bolted splices); shear capacity of bolts (for bolted splices); and weld capacity (for welded splices). Shear splices should be designed to resist the factored shear V_u. The limit states that need to be checked include shear rupture of the splice plates; shear capacity of bolts under an eccentric load (for bolted splices); bearing capacity of the connected parts (for bolted splices); shear capacity of bolts (for bolted splices); and weld capacity under an eccentric load (for welded splices). Design examples of beam and column splices can be found in the AISC Manual [AISC, 1994] and the AISC Manual of Connections [AISC, 1992].

46.13 Column Base Plates and Beam Bearing Plates

Column Base Plates

Column base plates are steel plates, placed at the bottom of columns, whose function is to transmit the column loads to the concrete pedestal. The design of a column base plate involves two major

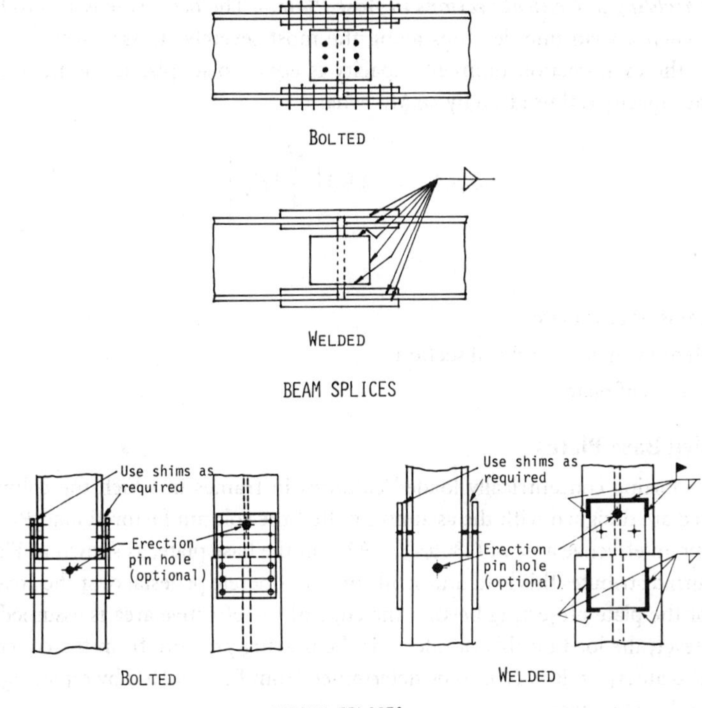

FIGURE 46.36 Examples of bolted and welded beam and column splices.

steps: (1) determining the size $B \times N$ of the plate and (2) determining the thickness t_p of the plate. Generally, the size of the plate is determined based on the limit state capacity of concrete in bearing, and the thickness of the plate is determined based on the limit state of plastic bending of critical sections in the plate. Depending on the types of forces (axial force, bending moment, shear force) the plate will be subjected to, the design procedures differ slightly. In all cases, a layer of grout should be placed between the base plate and its support, for the purpose of leveling, and anchor bolts should be provided to stabilize the column during erection or to prevent uplift for cases involving large bending moment.

Limit State Capacity of Concrete in Bearing. The design bearing strength of concrete is given by the equation

$$\phi_c P_p = 0.60\left[0.85 f_c' A_1 \sqrt{\frac{A_2}{A_1}}\right] \le 0.60\left[1.7 f_c' A_1\right] \tag{46.101}$$

where

$f_{c'}$ = compressive strength of concrete
A_1 = area of base plate
A_2 = area of concrete pedestal

From Eq. (46.101) it can be seen that the bearing capacity increases when the concrete area is greater than the plate area. This is due to the beneficial effect of confinement. The upper limit of the bearing strength is obtained when $A_2 = 4A_1$. Presumably, the concrete area in excess of $4A_1$ is not effective in resisting the load transferred through the base plate.

Limit State of Yielding of Critical Sections in Base Plate. The base plate is said to have reached its limit state when a yield line develops along the most severely stressed sections. A yield line develops when the cross-section moment capacity is equal to its plastic moment capacity. The design moment capacity is thus given by $\phi_b M_p$, where

$$\phi_b M_p = 0.90\left[\frac{t_p^2}{4} l F_y\right] \tag{46.102}$$

in which

t_p = thickness of base plate
l = length of yield line at critical section
F_y = yield stress of plate

Axially Loaded Base Plates

Base plates supporting concentrically loaded columns in frames in which the column bases are assumed pinned are designed with the assumption that the column factored load P_u is uniformly distributed over an effective area of $0.80b_f \times 0.90d$ on the base plate, as shown in Fig. 46.37. The base plate in turn distributes this load uniformly on the concrete pedestal over the area of the plate. The portion of the plate projecting beyond the edge of the effective area is assumed to act as an inverted cantilever; the load on this cantilever is the bearing pressure from the concrete pedestal. The area of the plate $A_1 = B \times N$ is to be determined from Eq. (46.101) by replacing $\phi_c P_n$ by P_u and solving for A_1. This gives

$$A_1 = \frac{1}{A_2}\left[\frac{P_u}{(0.60)(0.85f_c')}\right]^2 > \frac{P_u}{(0.60)(1.7f_c')} \tag{46.103}$$

For an efficient design, the value of m and n (Fig. 46.37) should not differ considerably. This condition can be attained if

$$N \approx \sqrt{A_1} + 0.50(0.95d - 0.80b_f) \tag{46.104}$$

Once N is calculated, B can be determined from

$$B = \frac{A_1}{N} \tag{46.105}$$

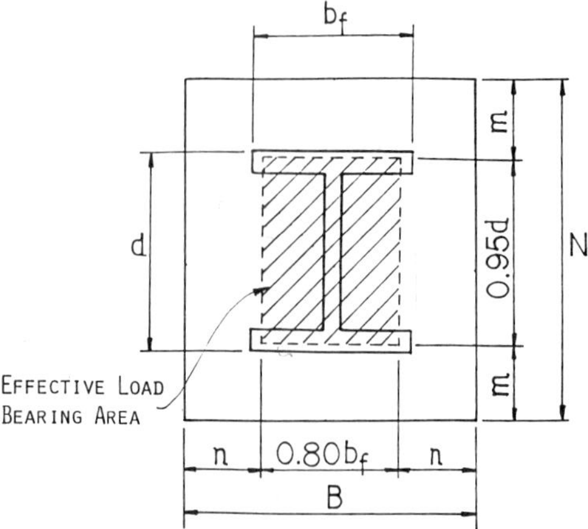

FIGURE 46.37 Centrally loaded column base plates.

The thickness of the plate is to be determined from Eq. (46.102) by replacing $\phi_b M_p$ by M_u, where M_u is the factored moment developed along the critical section for bending of the plate. This critical section is along the width of the effective area, with $l = B$, if $m > n$, and along the length of the effective area, with $l = N$, if $m < n$. Thus, substituting $M_u = (Bm^2/2)(P_u BN)$ if $m > n$, or $M_u = (Nn^2/2)(P_u/BN)$ if $m < n$, into Eq. (46.102) and solving for t_p gives

$$t_p = (\text{larger of } m \text{ or } n)\sqrt{\frac{2P_u}{0.90F_y BN}} \tag{46.106}$$

Lightly Loaded Base Plates

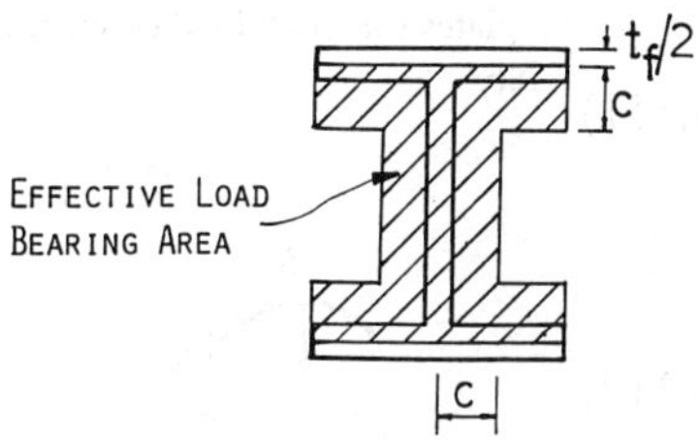

FIGURE 46.38 Centrally loaded column base plates (for lightly loaded columns).

These are base plates whose sizes are equal to or slightly larger than the column dimensions. For these plates, the values of m and n are approximately zero, so the critical sections for bending are along the inside edge of the column flanges and along the sides of the web. The area of the plate is to be determined from Eqs. (46.103) to (46.105), with an additional requirement that $A_1 > b_f d$. The thickness of the plate is to be determined by assuming that a portion of the factored column load $P_0 = (b_f d/BN)P_u$ is uniformly distributed over an H-shaped area A_H (Fig. 46.38), given by

$$A_H = \frac{P_0}{0.6(0.85f_c'\sqrt{A_2/b_f d})} \geq \frac{P_0}{0.60(1.7f_c')} \tag{46.107}$$

From geometry, the value c in Fig. 46.38 can be calculated to be

$$c = \frac{1}{4}\left[(d + b_f - t_f) - \sqrt{(d + b_f - t_f)^2 - 4(A_H - t_f b_f)}\right] \tag{46.108}$$

from which

$$t_p = C\sqrt{\frac{2P_0}{0.90F_y A_H}} \tag{46.109}$$

Base Plates for Tubular and Pipe Columns

The design concepts for base plates just discussed for I-shaped sections can be applied to the design of base plates for rectangular tubes and circular pipes. The critical section used to determine the plate thickness should be based on 0.95 times the outside column dimension for rectangular tubes and 0.80 times the outside dimension for circular pipes [Dewolf and Ricker, 1990].

Base Plates with Moments

For columns in frames designed to carry moments at the base, base plates must be designed to support both axial forces and bending moments. If the moment is small compared to the axial force, the base plate can be designed without consideration of the tensile force that may develop in the anchor bolts. However, if the moment is large, this effect should be considered. To quantify the relative magnitude of this moment, an eccentricity $e = M_u/P_u$ is used. The general procedures for the design of base plates for different values of e will be given in the following [Dewolf and Ricker, 1990].

Small Eccentricity, $e \leq N/6$. If e is small, the bearing stress is assumed to distribute linearly over the entire area of the base plate (Fig. 46.39). The maximum bearing stress is given by

$$f_{\max} = \frac{P_u}{BN} + \frac{M_u c}{I} \tag{46.110}$$

where $c = N/2$ and $I = BN^3/12$.

The size of the plate is to be determined by a trial-and-error process. The size of the base plate should be such that the bearing stress calculated using Eq. (46.110) does not exceed $\phi_c P_p/A_1$, given by

$$0.60\left[0.85 f_c' \sqrt{\frac{A_2}{A_1}}\right] \leq 0.60\left[1.7 f_c'\right] \tag{46.111}$$

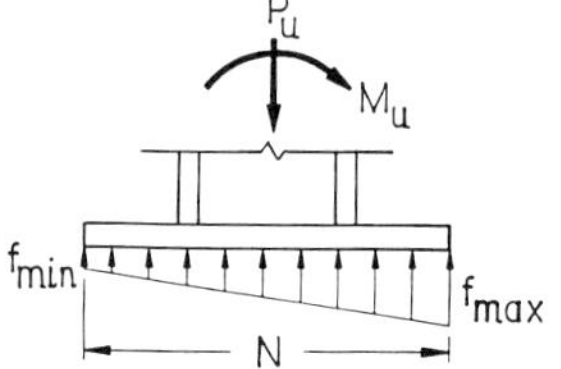

FIGURE 46.39 Column base plates for small load eccentricity.

The thickness of the plate is to be determined from

$$t_p = \sqrt{\frac{4M_{plu}}{0.9F_y}} \tag{46.112}$$

where M_{plu} is the moment per unit width of critical section in the plate. M_{plu} is to be determined by assuming that the portion of the plate projecting beyond the critical section acts as an inverted cantilever loaded by the bearing pressure. The moment calculated at the critical section divided by the length of the critical section (i.e., B) gives M_{plu}.

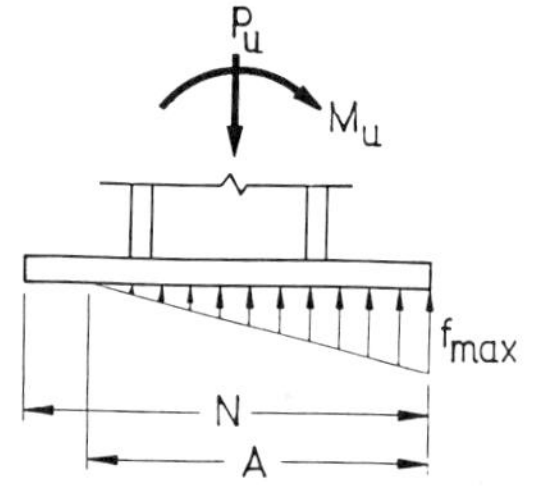

FIGURE 46.40 Column base plates for moderate load eccentricity.

Moderate Eccentricity, $N/6 < e \leq N/2$. For plates subjected to moderate moments, only a portion of the plate will be subjected to bearing stress (Fig. 46.40). Ignoring the tensile force in the anchor bolt in the region of the plate where no bearing occurs, and denoting A as the length of the plate in bearing, the maximum bearing stress can be calculated from force equilibrium considerations as

$$f_{\max} = \frac{2P_u}{AB} \tag{46.113}$$

where $A = 3(N/2 - e)$ is determined from moment equilibrium. The plate should be proportioned such that $f_{\max}$ does not exceed the right-hand side of Eq. (46.111). t_p is to be determined from Eq. (46.112).

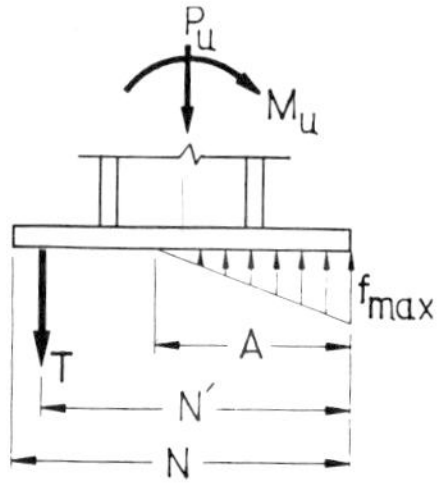

FIGURE 46.41 Column base plates for large load eccentricity.

Large Eccentricity, $e > N/2$. For plates subjected to large bending moments so that $e > N/2$, one needs to take into consideration the tensile force that develops in the anchor bolts (Fig. 46.41). Denoting T as the resultant force in the anchor bolts, force equilibrium requires that

$$T + P_u = \frac{f_{\max}AB}{2} \tag{46.116}$$

and moment equilibrium requires that

$$P_u\left(N' - \frac{N}{2}\right) + M_u = \frac{f_{\max}AB}{2}\left(N' - \frac{A}{3}\right) \tag{46.115}$$

The foregoing equations can be used to solve for A and T. The size of the plate is to be determined using a trial-and-error process. The size should be chosen such that f_{max} does not exceed the left-hand side of Eq. (46.111); A should be smaller than N'; and T should not exceed the tensile capacity of the bolts.

Once the size of the plate is determined, the plate thickness t_p is to be calculated using Eq. (46.112). Note that there are two critical sections on the plate: one on the compression side of the plate and the other on the tension side of the plate. Two values of M_{plu} are to be calculated, and the larger value should be used to calculate t_p.

Base Plates with Shear

Under normal circumstances, the factored column base shear is adequately resisted by the frictional force developed between the plate and its support. Additional shear capacity is also provided by the anchor bolts. For cases in which exceptionally high shear force is expected, such as in a bracing connection or a connection in which uplift reduces the frictional resistance, the use of shear lugs may be necessary. Shear lugs can be designed based on the limit states of bearing on concrete and bending of the lugs. The size of the lug should be proportioned such that the bearing stress on concrete does not exceed $0.60(0.85f_c')$. The thickness of the lug can be determined from Eq. (46.112). M_{plu} is the moment per unit width at the critical section of the lug. The critical section is taken to be at the junction of the lug and the plate (Fig. 46.42).

Anchor Bolts

Anchor bolts are provided to stabilize the column during erection and to prevent uplift for cases involving large moments. Anchor bolts can be *cast-in-place* bolts or *drilled-in* bolts. The latter are placed after the concrete is set and are not often used; their design is governed by the manufacturer's specifications. Cast-in-place bolts are hooked bars, bolts, or threaded rods with nuts (Fig. 46.43) placed before the concrete is set. Of the three types of cast-in-place anchors shown in the figure, the hooked bars are recommended for use only in axially loaded base plates. They are not normally relied on to carry significant tensile force. Bolts and threaded rods with nuts can be used for both axially loaded base plates and base plates with moments. Threaded rods with nuts are used when the length and size required for the specific design exceed those of standard-size bolts. Failure of bolts or threaded rods with nuts occurs when their tensile capacities are reached. Failure is also considered to occur when a cone of concrete is pulled out from the pedestal. This cone pullout type of failure is depicted schematically in Fig. 46.44. The failure cone is assumed to radiate out from the bolt head or nut at an angle of 45° with tensile failure occurring along the surface of the cone at an average stress of $4\sqrt{f_{c'}}$ where $f_{c'}$ is the compressive strength of concrete in psi. The load

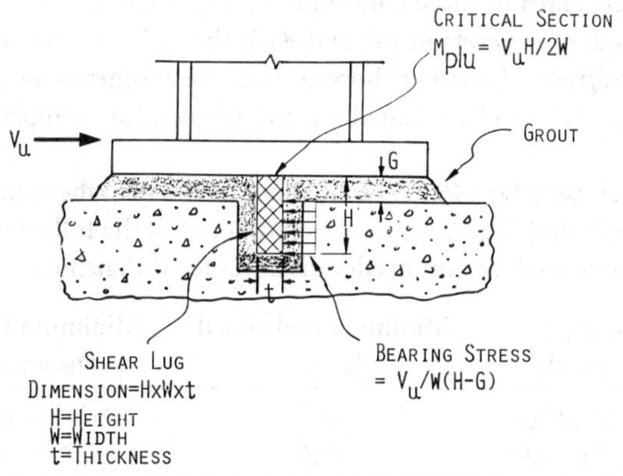

FIGURE 46.42 Column base plates for shear.

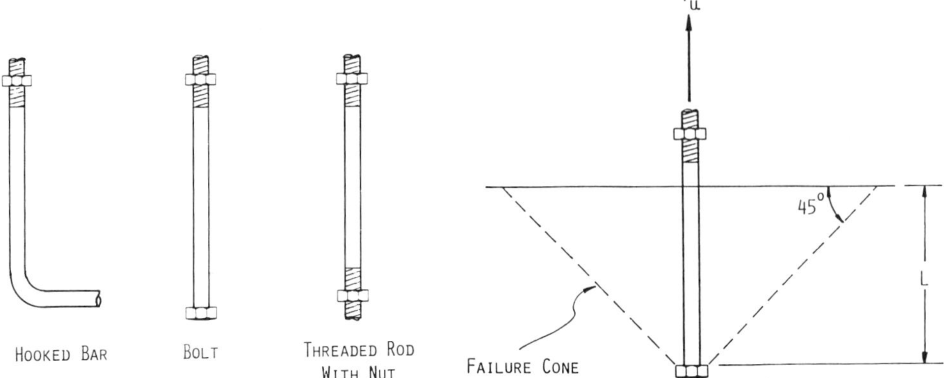

FIGURE 46.43 Types of base plate anchors.

FIGURE 46.44 Cone pullout failure.

that will cause this cone pullout failure is given by the product of this average stress and the projected area of the cone A_p [Marsh and Burdette, 1985b]. The design of anchor bolts is thus governed by the limit states of tensile fracture of the anchors and cone pullout.

Limit State of Tensile Fracture. The area of the anchor should be such that

$$A_g \geq \frac{T_u}{\phi_t 0.75 F_u} \tag{46.116}$$

where A_g is the required gross area of the anchor, F_u is the minimum specified tensile strength, and ϕ_t is the resistance factor for tensile fracture, which is equal to 0.75.

Limit State of Cone Pullout. From Fig. 46.44, it is clear that the size of the cone is a function of the length of the anchor. Provided that there is sufficient edge distance and spacing between adjacent anchors, the amount of tensile force required to cause cone pullout failure increases with the embedded length of the anchor. This concept can be used to determine the required embedded length of the anchor. Assuming that the failure cone does not intersect another failure cone or the edge of the pedestal, the required embedded length can be calculated from the equation

$$L \geq \sqrt{\frac{A_p}{\pi}} = \sqrt{\frac{(T_u / \phi_t 4 \sqrt{f_{c'}})}{\pi}} \tag{46.117}$$

where A_p is the projected area of the failure cone; T_u is the required bolt force in pounds; $f_{c'}$ is the compressive strength of concrete in psi; and ϕ_t is the resistance factor, assumed to be equal to 0.75. If failure cones from adjacent anchors overlap one another or intersect with the pedestal edge, the projected area A_p must be adjusted accordingly; see, for example, [Marsh and Burdette, 1985a].

The length calculated using Eq. (46.117) should not be less than the recommended values given by Shipp and Haninge [1983]. These values are reproduced in the following table. Also shown in the table are the recommended minimum edge distances for the anchors.

Bolt Type (Material)	Minimum Embedded Length	Minimum Edge Distance
A307 (A36)	$12d$	$5d > 4$ in.
A325 (A449)	$17d$	$7d > 4$ in.

d = nominal diameter of the anchor.

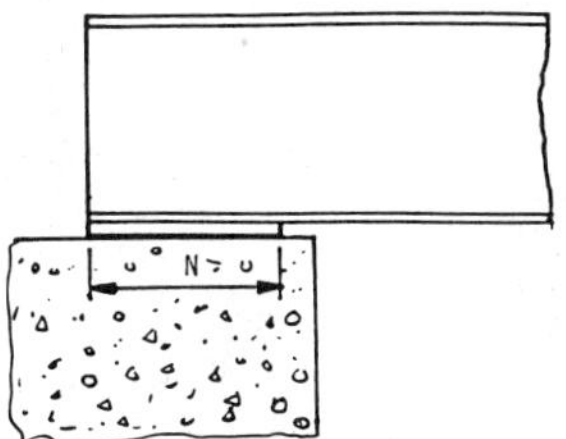

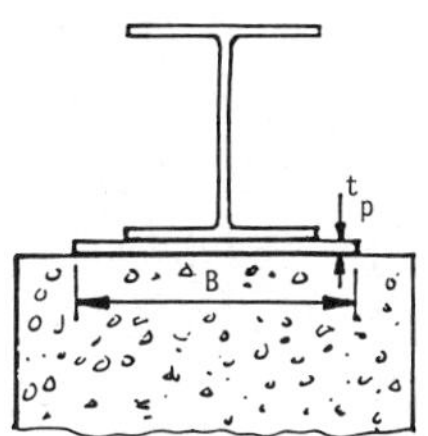

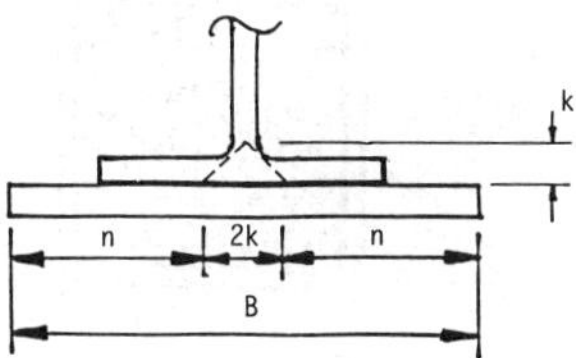

FIGURE 46.45 Beam bearing plates.

Beam Bearing Plates

Beam bearing plates are provided between main girders and concrete pedestals to distribute the girder reactions to the concrete supports (Fig. 46.45). Beam bearing plates may also be provided between cross beams and girders if the cross beams are designed to sit on the girders.

Beam bearing plates are designed based on the limit states of web yielding, web crippling, bearing on concrete, and plastic bending of the plate. The dimension of the plate along the beam axis, i.e., N, is determined from the web yielding [Eq. (46.51), Eq. (46.52)] or web crippling [Eq. (46.53), Eq. (46.54)] criterion, whichever is more critical. The dimension B of the plate is determined from Eq. (46.105) with A_1 calculated using Eq. (46.103). P_u in Eq. (46.103) is to be replaced by R_u, the factored reaction at the girder support.

Once the size $B \times N$ is determined, the plate thickness t_p can be calculated using the equation

$$t_p = \sqrt{\frac{2R_u n^2}{0.90F_y BN}} \tag{46.118}$$

where R_u is the factored girder reaction, F_y is the yield stress of the plate, and $n = (B - 2k)/2$, in which k is the distance from toe of web fillet to the outer surface of the flange. This equation was developed based on the assumption that the critical sections for plastic bending in the plate occur at a distance k from the centerline of the web.

46.14 Composite Members

Composite members are structural members made from two or more materials. The majority of composite sections used for construction are made from steel and concrete. Steel provides strength, and concrete provides rigidity. The combination of the two materials often results in efficient load-carrying members. Composite members may be *concrete-encased* or *concrete-filled*. For concrete-encased members, Fig. 46.46(a), concrete is cast around steel shapes. In addition to

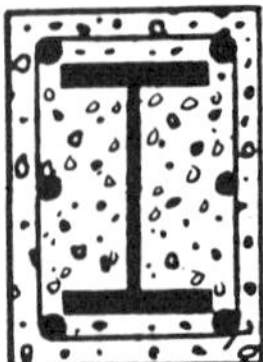

(a) Concrete Encased Composite Section

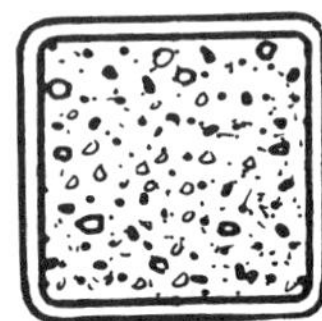

(b) Concrete Filled Composite Sections

FIGURE 46.46 Composite columns. (a) Concrete-encased. (b) Concrete-filled.

enhancing strength and providing rigidity to the steel shapes, the concrete acts as a fireproofing material to the steel shapes. It also serves as a corrosion barrier, shielding the steel from corroding under adverse environmental conditions. For concrete-filled members, Fig. 46.46(b), structural steel tubes are filled with concrete. In both concrete-encased and concrete-filled sections the rigidity of the concrete often eliminates the problem of local buckling experienced by some slender elements of the steel sections.

One disadvantage associated with composite sections is that concrete creeps and shrinks. Furthermore, uncertainties with regard to the mechanical bond developed between the steel shape and the concrete often complicate the design of beam–column joints.

Composite Columns

According to the LRFD Specification [AISC, 1993], a member is qualified to be a composite column if:

1. The cross-sectional area of the steel shape is at least 4% of the total composite area. If this condition is not satisfied, the member should be designed as a reinforced concrete column.
2. Longitudinal reinforcements and lateral ties are provided for concrete-encased members. The cross-sectional area of the reinforcing bars must be 0.007 in.^{2} per in. of bar spacing. To avoid spalling, lateral ties shall be placed at a spacing not greater than $\frac{2}{3}$ the least dimension of the composite cross section. For fire and corrosion resistance, a minimum clear cover of 1.5 in. must be provided.
3. The compressive strength of concrete f_c' used for the composite section falls within the range 3 ksi to 8 ksi for normal-weight concrete and not less than 4 ksi for lightweight concrete. These limits are set because they represent the range of test data available for the development of the design equations.
4. The specified minimum yield stress for the steel shapes and reinforcing bars used in calculating the strength of the composite column does not exceed 55 ksi. This limit is set

because this stress corresponds to a strain below which the concrete remains unspalled and stable.

5. The minimum wall thickness of the steel shapes for concrete-filled members is equal to $b\sqrt{(F_y/3E)}$ for rectangular sections of width b and $D\sqrt{(F_y/8E)}$ for circular sections of outside diameter D.

Design Compressive Strength. The design compressive strength must exceed the factored compressive force in accordance with Eq. (46.15). For $\lambda_c \leq 1.5$, the design compressive strength is given as

$$\phi_c P_n = 0.85\left[(0.658\lambda_c^2)A_s F_{my}\right] \tag{46.119}$$

For $\lambda_c < 1.5$,

$$\phi_c P_n = 0.85\left[\left(\frac{0.877}{\lambda_c^2}\right)A_s F_{my}\right] \tag{46.120}$$

where

$$\lambda_c = \frac{KL}{r_m \pi}\sqrt{\frac{F_{my}}{E_m}} \tag{46.121}$$

$$F_{my} = F_y + c_1 F_{yr}\left(\frac{A_r}{A_s}\right) + c_2 f_c'\left(\frac{A_c}{A_s}\right) \tag{46.122}$$

$$E_m = E + c_3 E_c\left(\frac{A_c}{A_s}\right) \tag{46.123}$$

A_c = area of concrete, in.^{2}
A_r = area of longitudinal reinforcing bars, in.^{2}
A_s = area of steel shape, in.^{2}
E = modulus of elasticity of steel, ksi
E_c = modulus of elasticity of concrete, ksi
F_y = specified minimum yield stress of steel shape, ksi
F_{yr} = specified minimum yield stress of longitudinal reinforcing bars, ksi
f_c' = specified compressive strength of concrete, ksi
r_m = radius of gyration of the steel shape, pipe, or tubing; for steel shapes, r_m shall not be less than 0.3 times the overall thickness of the composite cross section in the plane of buckling, in.
c_1, c_2, c_3 = coefficients given in the following table

Type of Composite Section	c_1	c_2	c_3
Concrete-encased shapes	0.7	0.6	0.2
Concrete-filled pipes and tubings	1.0	0.85	0.4

In addition to satisfying Eq. (46.15) with $\phi_c P_n$ calculated using Eq. (46.119) or Eq. (46.120), the design must also satisfy the following bearing condition for concrete. Denoting $\phi_c P_{nc}$

($= \phi_c P_{n,\text{composite section}} - \phi_c P_{n,\text{steel shape alone}}$) as the portion of compressive strength resisted by the concrete and A_B as the loaded area, then if supporting concrete area > loaded area,

$$\phi_c P_{nc} \leq 0.60[1.7 f_c' A_B] \tag{46.124}$$

Composite Beams

For steel beams fully encased in concrete, the design flexural strength $\phi_b M_n$ can be computed using either an elastic analysis or a plastic analysis.

If an elastic analysis is used, ϕ_b is to be taken as 0.90. A linear strain distribution is assumed for the cross section, with zero strain at the neutral axis and maximum strains at the extreme fibers. The stresses are then computed by multiplying the strains by E (for steel) or E_c (for concrete). Maximum stress in steel shall be limited to F_y, and maximum stress in concrete shall be limited to $0.85 f_{c'}$. Tensile strength of concrete shall be neglected. M_n is to be calculated by integrating the resulting stress block about the neutral axis.

If a plastic analysis is used, ϕ_c is to be taken as 0.90, and M_n is assumed to be equal to M_p, the plastic moment capacity of the steel section alone.

Composite Beam-Columns

Composite beam-columns shall be designed to satisfy the interaction equation of Eq. (46.60) or Eq. (46.61), whichever is applicable, with $\phi_c P_n$ calculated based on Eqs. (46.119) to (46.123), P_e calculated using the equation $P_e A_s F_{my} / \lambda^2_{c'}$ and $\phi_b M_n$ calculated using the following equation [Galambos and Chapuis, 1980]:

$$\phi_b M_n = 0.90 \left[Z F_y + \frac{1}{3}(h_2 - 2c_r) A_r F_{yr} + \left(\frac{h_2}{2} - \frac{A_w F_y}{1.7 f_c' h_1} \right) A_w F_y \right] \tag{46.125}$$

in which

Z = plastic section modulus of the steel section, in.^{3}

c_r = average of the distance measured from the compression face to the longitudinal reinforcement in that face and the distance measured from the tension face to the longitudinal reinforcement in that face, in.

h_1 = width of the composite section perpendicular to the plane of bending, in.

h_2 = width of the composite section parallel to the plane of bending, in.

A_r = cross-sectional area of longitudinal reinforcing bars, in.^{2}

A_w = web area of the encased steel shape (= 0 for concrete-filled tubes)

$F_y r$ = yield stress of longitudinal reinforcement, ksi

F_y = yield stress of steel shape, ksi

If $0 < (P_u / \phi_c P_n) \leq 0.3$, a linear interpolation of M_n, calculated using the above equation assuming $P_u / \phi_c P_n = 0.3$ and that for beams with $P_u / \phi_c P_n = 0$ (see discussion on "Composite Beams" above) should be used.

Composite Floor Slab

A composite floor slab (Fig. 46.47) can be designed as *shored* or *unshored*. In shored construction, temporary shores are used during construction to support the dead and accidental live loads until the concrete cures, and the supporting beam is designed on the basis of its ability to develop composite action to support all factored loads after the concrete cures. In unshored construction, temporary shores are not used; as a result, the steel beam alone must be designed to support

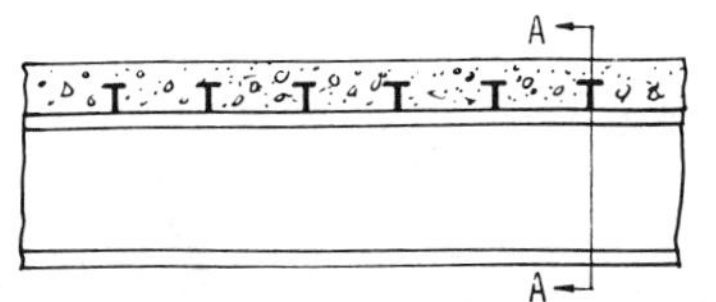

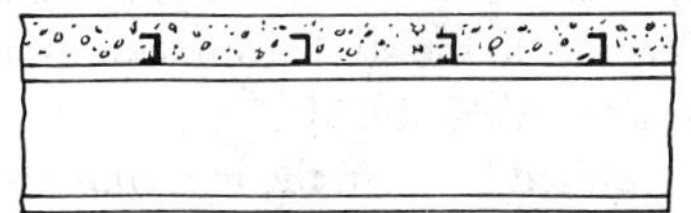

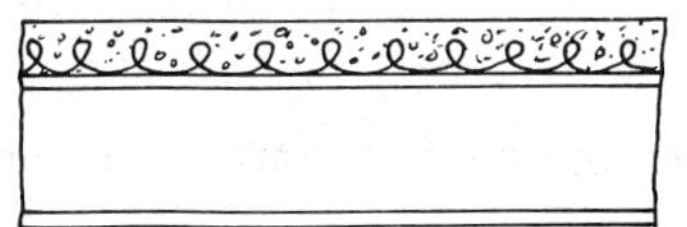

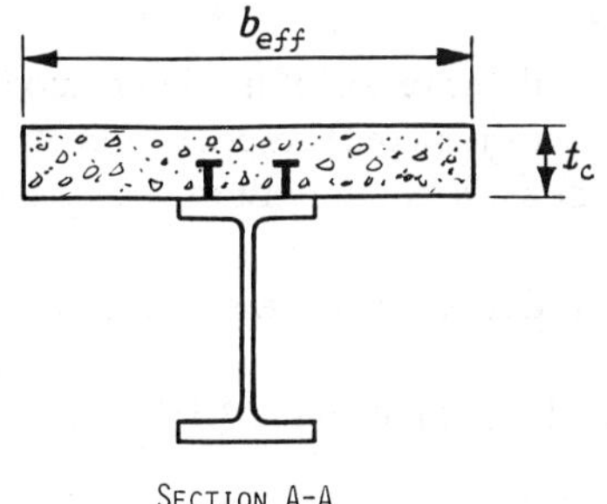

FIGURE 46.47 Composite floor slabs.

the dead and accidental live loads before the concrete has attained 75% of its specified strength. After the concrete is cured, the composite section should have adequate strength to support all factored loads.

Composite action for the composite floor slab shown in Fig. 46.47 is developed as a result of the presence of shear connectors. The horizontal shear force V_h, which should be designed for at the interface of the steel beam and the concrete slab, is given in regions of positive moment by

$$V_h = \min\,(0.85f_c'A_c, A_sF_y, \textstyle\sum Q_n) \quad (46.126)$$

In regions of negative moment,

$$V_h = \min\,(A_rF_{yr}, \textstyle\sum Q_n) \quad (46.127)$$

where

f_c' = compressive strength of concrete, ksi
A_c = effective area of the concrete slab = $t_c b_{\text{eff}'}$, in.^{2}
t_c = thickness of the concrete slab, in.
b_{eff} = effective width of the concrete slab, in.
= min $(L/4, s)$, for an interior beam
= min $(L/8+$ distance from beam centerline to edge of slab, $s/2+$ distance from beam centerline to edge of slab), for an exterior beam
L = beam span measured from center to center of supports, in.
s = spacing between centerline of adjacent beams, in.
A_s = cross-sectional area of the steel beam, in.^{2}
F_y = yield stress of the steel beam, ksi
A_r = area of reinforcing steel within the effective area of the concrete slab, in.^{2}
F_{yr} = yield stress of the reinforcing steel, ksi
$\sum Q_n$ = sum of nominal shear strengths of the shear connectors, kips

The nominal shear strength of a shear connector (used without a formed steel deck) is given for a stud shear connector by

$$Q_n = 0.5A_{sc}\sqrt{f_c'E_c} \leq A_{sc}F_u \quad (46.128)$$

For a channel shear connector,

$$Q_n = 0.3(t_f + 0.5t_w)L_c\sqrt{f_c'E_c} \quad (46.129)$$

where

A_{sc} = cross-sectional area of the shear stud, in.^{2}
f_c' = compressive strength of concrete, ksi

E_c = modulus of elasticity of concrete, ksi
F_u = minimum specified tensile strength of the shear stud, ksi
t_f = flange thickness of the channel, in.
t_w = web thickness of the channel, in.
L_c = length of the channel, in.

If a formed steel deck is used, Q_n must be reduced by a reduction factor. The reduction factor depends on whether the deck ribs are perpendicular or parallel to the steel beam. Expressions for the reduction factor are given in the AISC LRFD Specification [AISC, 1993].

For full composite action, the number of connectors required between the *maximum* moment point and the *zero* moment point of the beam is given by

$$N = \frac{V_h}{Q_n} \tag{46.130}$$

For partial composite action, the number of connectors required is governed by the condition $\phi_b M_n \geq M_u$.

The placement and spacing of the shear connectors should comply with the following guidelines:

1. The shear connectors shall be uniformly spaced with the region of maximum moment and zero moment. However, the number of shear connectors placed between a concentrated load point and the nearest zero moment point must be sufficient to resist the factored moment M_u.
2. Except for connectors installed in the ribs of formed steel decks, shear connectors shall have at least 1 in. of lateral concrete cover.
3. Unless located over the web, diameter of shear studs must not exceed 2.5 times the thickness of the beam flange.
4. The longitudinal spacing of the studs should fall in the range from 6 times the stud diameter to 8 times the slab thickness, if a solid slab is used, or from 4 times the stud diameter to 8 times the slab thickness, if a formed steel deck is used.

The design flexural strength $\phi_b M_n$ of the composite beam with shear connectors is determined as follows:

Regions of Positive Moment. For $h_c/t_w \leq 640/\sqrt{F_{yf}}$, $\phi_b = 0.85$, and M_n = the moment capacity determined using a plastic stress distribution, assuming concrete crushes at a stress of $0.85 f_c'$ and steel yields at a stress of F_y. If a portion of the concrete slab is in tension, the strength contribution of that portion of concrete is ignored. The determination of M_n using this method is very similar to the technique used for computing the moment capacity of a reinforced concrete beam according to the ultimate strength method.

For $h_c/t_w > 640/\sqrt{F_{yf}}$, $\phi_b = 0.90$, and M_n = the moment capacity determined using superposition of elastic stress, considering the effect of shoring. The determination of M_n using this method is similar to the technique used for computing the moment capacity of a reinforced concrete beam according to the working stress method.

Regions of Negative Moment. $\phi_b M_n$ is to be determined for the steel section alone, in accordance with the requirements discussed in section 46.5.

To facilitate design, numerical values of $\phi_b M_n$ for composite beams with shear studs in solid slabs are given in tabulated form by the AISC LRFD Manual [AISC, 1994]. Values of $\phi_b M_n$ for composite beams with formed steel decks are given in a publication by the Steel Deck Institute [1989].

46.15 Plastic Design

Plastic analysis and design is permitted only for steels with yield stress not exceeding 65 ksi. The reason for this is that steels with high yield stress lack the ductility required for inelastic deformation at hinge locations. Without adequate inelastic deformation, moment redistribution, which is an important characteristic for plastic design, cannot take place.

In plastic design the predominant limit state is the formation of plastic hinges. Failure occurs when sufficient plastic hinges have formed for a collapse mechanism to develop. To ensure that plastic hinges can form and can undergo large inelastic rotation, the following conditions must be satisfied:

- Sections must be compact. That is, the width-thickness ratios of flanges in compression and webs must not exceed λ_p in Table 46.11.
- For columns, the slenderness parameter λ_c (Section 46.4) shall not exceed $1.5K$, where K is the effective length factor, and P_u from gravity and horizontal loads shall not exceed $0.75 A_g F_y$.
- For beams, the lateral unbraced length L_b shall not exceed L_{pd} where, for doubly and singly symmetric I-shaped members loaded in the plane of the web,

$$L_{pd} = \frac{3{,}600 + 2{,}200(M_1/M_2)}{F_y} I_y \tag{46.131}$$

and, for solid rectangular bars and symmetric box beams,

$$L_{pd} = \frac{5{,}000 + 3{,}000(M_1/M_2)}{F_y} r_y \geq \frac{3{,}000 r_y}{F_y} \tag{46.132}$$

In the above equations, M_1 is the smaller end moment and M_2 is the larger end moment (M_1/M_2 is positive when the member bends in reverse curvature and negative when the member bends in single curvature) within the unbraced length of the beam; r_y is the radius of gyration about the minor axis, in in.; and F_y is the specified minimum yield stress, in ksi.

L_{pd} is not defined for beams bent about their minor axes nor for beams with circular and square cross sections because these beams do not experience lateral torsional buckling when loaded.

Plastic Design of Columns and Beams

Provided that the aforementioned limitations are satisfied, the design of columns must meet the requirement of Eq. (46.15), and the design of beams must meet the requirements of Eq. (46.35), with $\phi_b M_n = \phi_b M_p$, and Eq. (46.45). For beams subjected to concentrated loads, the requirements of Eq. (46.49) should also be checked.

Plastic Design of Beam-Columns

Beam-columns designed on the basis of plastic analysis shall satisfy the interaction Eq. (46.60) or (46.61), whichever is applicable. M_u in these equations shall be determined from a second-order plastic analysis.

Defining Terms

ASD: Acronym for Allowable Stress Design.

Beam-columns: Structural members whose primary function is to carry loads both along and transverse to their longitudinal axes.

Biaxial bending: Simultaneous bending of a member about two orthogonal axes of the cross section.

Built-up members: Structural members made of structural elements joined together by bolts, welds, or rivets.

Composite members: Structural members made of both steel and concrete.

Compression members: Structural members whose primary function is to carry compressive loads along their longitudinal axes.

Design strength: Resistance provided by the structural member, obtained by multiplying the nominal strength of the member by a resistance factor.

Drift: Lateral deflection of a building.

Factored load: The product of the nominal load and a load factor.

Flexural members: Structural members whose primary function is to carry loads transverse to their longitudinal axes.

Limit state: A condition in which a structural or structural component becomes unsafe (strength limit state) or unfit for its intended function (serviceability limit state).

Load factor: A factor to account for the unavoidable deviations of the actual load from its nominal value and for uncertainties in structural analysis in transforming the applied load into a load effect (axial force, shear, moment, etc.).

LRFD: Acronym for Load and Resistance Factor Design.

Major axis (or **strong axis**): Axis of a cross section about which the moment of inertia is a maximum.

Minor axis (or **weak axis**): Axis of a cross section about which the moment of inertia is a minimum.

PD: Acronym for Plastic Design.

Plastic hinge: A yielded zone of a structural member in which the internal moment is equal to the plastic moment of the cross section.

Resistance factor: A factor to account for the unavoidable deviations of the actual resistance of a member from its nominal value.

Service load: Nominal load expected to be supported by the structure or structural component under normal usage.

Sidesway-inhibited frames: Frames in which lateral deflections are prevented by a system of bracing.

Sidesway-uninhibited frames: Frames in which lateral deflections are not prevented by a system of bracing.

References

AASHTO. 1992. *Standard Specification for Highway Bridges,* 15th ed. American Association of State Highway and Transportation Officials, Washington, D.C.

AISC. 1992. *Manual of Steel Construction—Volume II: Connections,* ASD 1st ed./LRFD 1st ed., American Institute of Steel Construction, Chicago, IL.

AISC. 1993. *Load and Resistance Factor Design Specification for Structural Steel Buildings.* American Institute of Steel Construction, Chicago, IL.

AISC. 1994. *Manual of Steel Construction—Load and Resistance Factor Design, Volume I: Structural Members, Specifications & Codes, Volume II: Connections,* 2nd ed. American Institute of Steel Construction, Chicago, IL.

ASTM. 1985. *Specification for Heat-Treated Steel Structural Bolts, 150 ksi Minimum Tensile Strength (A490-85).* American Society for Testing and Materials, Philadelphia, PA.

ASTM. 1986a. *Specification for High Strength Bolts for Structural Steel Joints (A325-86).* American Society for Testing and Materials, Philadelphia, PA.

ASTM. 1986b. *Specification for Quenched and Tempered Steel Bolts and Studs (A449-86).* American Society for Testing and Materials, Philadelphia, PA.

ASTM. 1988. *Specification for Carbon Steel Bolts and Studs, 60000 psi Tensile Strength (A307-88a).* American Society for Testing and Materials, Philadelphia, PA.

AWS. 1987. *1, Welding Technology,* 8th ed. *Welding Handbook.* American Welding Society, Miami, FL.

AWS. 1988. *Structural Welding Code—Steel,* 11th ed. American Welding Society, Miami, FL.

Blodgett, O. W. Undated. Distortion ... How to Minimize It with Sound Design Practices and Controlled Welding Procedures Plus Proven Methods for Straightening Distorted Members. *Bulletin G261.* The Lincoln Electric Company, Cleveland, OH.

Chen, W. F. and Lui, E. M. 1987. *Structural Stability—Theory and Implementation.* Elsevier, New York.

Dewolf, J. T. and Ricker, D. T. 1990. *Column Base Plates. Steel Design Guide Series 1.* American Institute of Steel Construction, Chicago, IL.

Disque, R. O. 1973. Inelastic K-factor in column design. *AISC Eng. J.* 10(2):33–35.

Galambos, T. V. and Chapuis, J. 1980. *LRFD Criteria for Composite Columns and Beam Columns.* Washington University, Department of Civil Engineering, St. Louis, MO.

Gaylord, E. H., Gaylord, C. N., and Stallmeyer, J. E. 1992. *Design of Steel Structures,* 3rd ed. McGraw-Hill, New York.

Halldorsson, O. P. and Wang, C.-K. 1968. Stability analysis of frameworks by matrix methods. *J. Struct., ASCE,* 94(7):1745–1760.

Kulak, G. L., Fisher, J. W. and Struik, J.H.A. 1987. *Guide to Design Criteria for Bolted and Riveted Joints,* 2nd ed. John Wiley & Sons, New York.

Lee, G. C., Morrel, M. L., and Ketter, R. L. 1972. Design of Tapered Members. *WRC Bulletin No.* 173.

LeMessurier, W. J. 1977. A practical method of second order analysis, part 2—Rigid frames. *AISC Eng. J.* 14(2):49–67.

Lui, E. M. 1992. A novel approach for K factor determination. *AISC Eng. J.* 29(4):150–159.

Marsh, M. L. and Burdette, E. G. 1985a. Multiple bolt anchorages: Method for determining the effective projected area of overlapping stress cones. *AISC Eng. J.* 22(1):29–32.

Marsh, M. L. and Burdette, E. G. 1985b. Anchorage of steel building components to concrete. *AISC Eng. J.* 22(1):33–39.

Munse, W. H. and Chesson E., Jr. 1963. Riveted and bolted joints: Net section design. *J. Struct., ASCE.* 89(1):107–126.

Rains, W. A. 1976. A new era in fire protective coatings for steel. *Civ. Eng., ASCE,* September:80–83.

RCSC. 1986. *Load and Resistance Factor Design Specification for Structural Joints Using ASTM A325 or A490 Bolts.* American Institute of Steel Construction, Chicago, IL.

Salmon, C. G. and Johnson, J. E. 1990. *Steel Structures—Design and Behavior,* 3rd ed. Harper & Row, New York.

Shipp, J. G. and Haninge, E. R. 1983. Design of headed anchor bolts. *AISC Eng. J.* 20(2):58–69.

Steel Deck Institute. 1989. *LRFD Design Manual for Composite Beams and Girders with Steel Deck.* Steel Deck Institute, Canton, OH.

Yura, J. A. 1971. The effective length of columns in unbraced frames. *AISC Eng. J.* 8(2):37–42.

Zahn, C. J. and Iwankiw, N. R. 1989. Flexural-torsional buckling and its implications for steel compression member design. *AISC Eng. J.* 26(4):143–154.

For Further Information

The following publications provide additional sources of information for the design of steel structures:

General Information

Beedle, L. S., ed. 1991. *Stability of Metal Structures—A World View.* 2nd ed. Structural Stability Research Council, Lehigh University, Bethlehem, PA.

Chen, W. F. and Lui, E. M. 1991. *Stability Design of Steel Frames.* CRC Press, Boca Raton, FL.

Galambos, T. V., ed. 1988. *Guide to Stability Design Criteria for Metal Structures.* 4th ed. Wiley, New York.

Trahair, N. S. 1993. *Flexural-Torsional Buckling of Structures.* CRC Press, Boca Raton, FL.

Allowable Stress Design

Adeli, H. 1988. *Interactive Microcomputer-Aided Structural Steel Design.* Prentice Hall, Englewood Cliffs, NJ.

Cooper, S. E. and Chen, A. C. 1985. *Designing Steel Structures—Methods and Cases.* Prentice Hall, Englewood Cliffs, NJ.

Crawley, S. W. and Dillon, R. M. 1984. *Steel Buildings Analysis and Design.* 3rd ed. Wiley, New York.

Fanella, D. A., Amon, R., Knobloch, B., and Mazumder, A. 1992. *Steel Design for Engineers and Architects.* 2nd ed. Van Nostrand Reinhold, New York.

Kuzmanovic, B. O. and Willems, N. 1983. *Steel Design for Structural Engineers.* 2nd ed. Prentice Hall, Englewood Cliffs, NJ.

McCormac, J. C. 1981. *Structural Steel Design,* 3rd ed. Harper & Row, New York.

Segui, W. T. 1989. *Fundamentals of Structural Steel Design.* PWS-KENT, Boston, MA.

Spiegel, L. and Limbrunner, G. F. 1986. *Applied Structural Steel Design.* Prentice Hall, Englewood Cliffs, NJ.

Plastic Design

ASCE. 1971. *Plastic Design in Steel—A Guide and Commentary,* 2nd ed. ASCE Manual No. 41, ASCE-WRC, New York.

Horne, M. R. and Morris, L. J. 1981. *Plastic Design of Low-Rise Frames.* Constrado Monographs, Collins, London, England.

Load and Resistance Factor Design

Geschwindner, L. F., Disque, R. O., and Bjorhovde, R. 1994. *Load and Resistance Factor Design of Steel Structures.* Prentice Hall, Englewood Cliffs, NJ.

McCormac, J. C. 1989. *Structural Steel Design—LRFD Method.* Harper & Row, New York.

Segui, W. T. 1994. *LRFD Steel Design.* PWS, Boston, MA.

Smith, J. C. 1991. *Structural Steel Design—LRFD Approach.* John Wiley & Sons, New York.

47

Cold-Formed Steel Structures

J. Rhodes
University of Strathclyde, Glasgow

N. E. Shanmugam
National University of Singapore

47.1 Introduction to Cold-Formed Steel Sections

General Introduction

Cold-formed steel products find extensive application in modern construction, both low-rise and high-rise steel buildings. Primary as well as secondary framing members in low-rise construction are fabricated using cold-formed steel sections, while roof and floor decks, steel joists, wall panels, door and window frames, and sandwich panel partitions built out of cold-formed steel sections have been successfully used in tall buildings. In addition, these products are used in car bodies, railway coaches, storage racks, grain bins, highway products and transmission towers. Although the uses of these products are many and varied, and many widely different products with tremendous diversity of shapes, sizes, and applications are produced in steel using the cold-forming process, this chapter is primarily concerned with the design of **cold-formed steel structural members** for use in building construction. However, the general design philosophies developed in the chapter are

0-8493-8953-4/95/$0.00 + $.50

applicable in many cases over a wide range of other uses. More detailed information on cold-formed steel structures are available in books by Yu [1991], Rhodes [1991] and Hancock [1988].

Manufacturing Methods

Cold-forming is the term used to describe the manufacture of products by forming material in the cold state from a strip or sheet of uniform **thickness**. A variety of methods of forming are used for cold-formed products in general, but in the case of structural sections the main methods used are folding, press-braking and rolling.

Folding is the simplest process, in which specimens of short length and simple geometry are produced from a sheet of material by folding a series of bends. This process has very limited application.

Press-braking is more widely used, and a greater variety of cross-sectional forms can be produced by this process. Here a section is formed from a length of strip by pressing the strip between shaped dies to form the profile shape. Usually each bend is formed separately. This process also has limitations on the profile geometry that can be formed and, often more importantly, on the lengths of sections that can be produced.

The major cold-forming process used for large-volume production is *cold-rolling*. In this process strip material is formed into the desired profile shape by feeding it continuously through successive pairs of rolls. Each pair of rolls brings the form of the strip progressively closer to the final profile shape, as illustrated in Fig. 47.1. The number of pairs of rolls, or "stages," required depends on the thickness of the material and the complexity of the profile to be formed.

The cold-rolling process can be used to produce prismatic sections of virtually any profile, from a wide range of materials, with a high degree of consistency and accuracy to any desired length. Sections are rolled at speeds varying from about 10 m/min up to about 100 m/min, depending on the

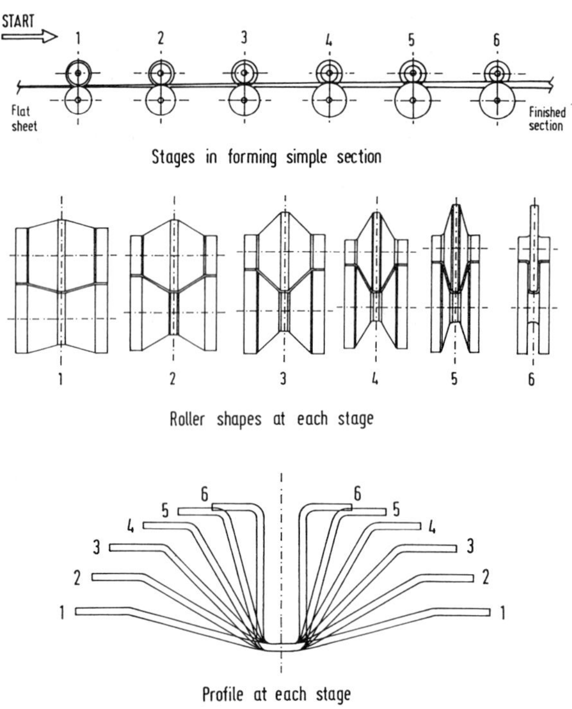

FIGURE 47.1 Stages in roll-forming a simple section.

FIGURE 47.2 Typical shapes produced by roll-forming.

complexity of the profile, material being formed, equipment used, etc. Holes, notches, and cutouts can be produced in a member during the rolling process, and pregalvanized or precoated steel strip is often used to eliminate corrosion and to produce aesthetically pleasing finished products. Typical profiles produced are shown in Fig. 47.2.

Applications of Cold-Formed Steel

Trapezoidal profiles and the many variations on the trapezoidal shape are now widely used for industrial and commercial buildings, sports arenas, hotels, restaurants, and many other types of building construction. A variety of advances have been made in the field of profiled **sheeting** in relatively recent times. The use of attachment systems that ensure watertightness of roofs has been an area of development, as has the incorporation of stiffeners in the profiles. The use of foam-filled sandwich panels, in which the roof or wall covering is combined with the insulation to give superior structural performance in addition to other advantages, is an area of rapid growth. There has also been rapid growth in the steel-concrete **composite flooring slab** area in the last few years. Steel decks acting compositely with concrete have long been used in the U.K. and U.S.A.

Roof purlins, which are ideally suited for production as cold-rolled sections, account for a substantial proportion of cold-formed steel usage in buildings. Two basic shapes are used for purlins in the U.K.: the Z shape [Fig. 47.3(a)], which was introduced from the U.S., and the Sigma shape, [Fig. 47.3(b)]. Both of these shapes are very efficient in acting in conjunction with the sheeting to produce a high structural performance. Recent research and development effort has led to refinements in the Z such as the Zeta shape [Fig. 47.3(c)] and the UltraZED shape [Fig. 47.3(d)]. Purlin thicknesses in use range from about 0.047 in. (1.2 mm) to about 0.126 in. (3.2 mm), and material of yield strength of 50 ksi (350 N/mm^2) is becoming widely used in the production of purlins.

Storage platforms and mezzanine floor systems form another area of growing use of cold-formed steel members. In these systems the columns are often hot-rolled sections such as square hollow sections or I sections, and the **beams** are cold-formed sections. In lattice-beam construction the boom members are generally cold-formed sections of hat or similar shape, and the lattice members may be tubular or made from round bar or other cold-formed shapes. The concept of preengineered buildings, made largely from cold-formed steel sections in the factory and erected on-site, has been a constantly recurring theme in the development of the cold-rolled sections industry, and the use of steel stud wall systems is a further step in this direction. Storage racking is another area that forms a significant outlet for cold-formed steel products. This accounts for perhaps 20% of all

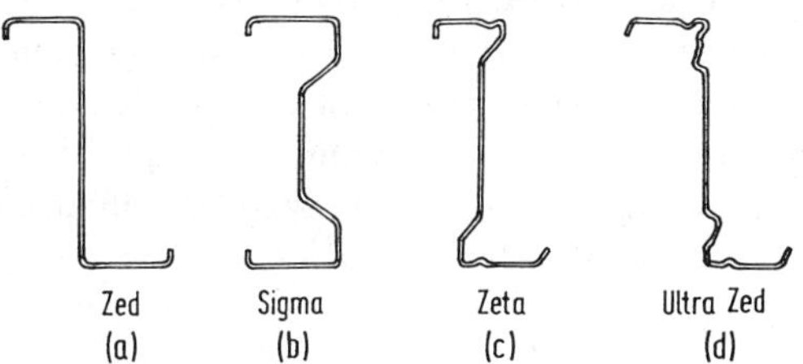

FIGURE 47.3 Types of roof purlin presently in common use.

the constructional use of cold-formed sections and utilizes a substantial proportion of perforated members. Storage installations range from relatively small shelving systems to extremely large and sophisticated pallet racking systems.

Advantages of Cold-Formed Steel

Cold-formed steel products have several advantages over hot-rolled steel sections. The main attractions of cold-formed steel sections are

- Lightness
- High strength and stiffness
- Ease of fabrication and mass production
- Fast and easy erection and installation
- Substantial elimination of delays due to weather
- More accurate detailing
- Nonshrinking and noncreeping at ambient temperatures
- Absence of formwork
- Termite-proof and rotproof
- Uniform quality
- Economy in transportation and handling
- Noncombustibility

The combination of these advantages can result in cost saving during construction [Yu, 1991].

Design Codes and Specifications

Since the late 1970s cold-formed steel has taken on a new importance in Europe, and there has been a period of substantial activity in research and in the development of new design codes. This began with the publication of a new Swedish design specification in 1982 [Swedish Code for Light-Gauge Metal Structures, 1982], followed by European Recommendations at various stages. Insofar as the design method is concerned, some specifications use the allowable stress design approach, whereas others are based on **limit state** design. The American Iron and Steel Institute (AISI) includes both allowable stress design (ASD) [AISI, 1986] and limit state design (LRFD) [AISI, 1991]. Eurocode 3 [1989], Part 1.3, and the British standard BS5950: Part 5 [British Standards Institution, 1987] deal with design of cold-formed steel members. Both Canada [Canadian Standards Association, 1989] and Australia [Hancock, 1988] have developed their own codes.

Range of Thicknesses

The provisions of codes apply primarily to steel sections with thickness not more than 0.33 in. (8 mm), although the use of thicker material is not precluded. Minimum thicknesses for specific applications are set by practical considerations, such as damage tolerance during handling, and of course, by the economics of the particular applications. With regard to the maximum thickness, 0.33 in. (8 mm) is about the limiting thickness normally rolled, although sections of up to about 0.8 in. (20 mm) can be rolled for specific applications.

Properties of Steel

The design strength of the steel used should be taken as the **yield strength** of the material, provided that the steel has an ultimate tensile strength about 20% or more greater than the yield strength. If this is not the case, the design strength is reduced accordingly in some codes. In the case of steels that have no clearly defined yield strength, either the 0.2% proof stress or the stress at 0.5% total elongation in a tensile test may be taken as the design strength. The **yield points** of steels listed in the AISI Specification range from 25 to 70 ksi (172 to 483 MPa).

The strength of members that fail by **buckling** is also a function of the modulus of elasticity E, the value of which is recommended as 29,500 ksi (203 kN/mm^2) by AISI in its specification for design purposes. Poisson's ratio is taken as 0.3.

Effects of Cold-Forming

Cold-forming increases the yield and ultimate tensile strengths of the material being formed in the vicinity of the bend areas. Experiments suggest that the average increase in yield strength of a section is dependent on the number of bends, the area of the section, the material thickness, and the difference between the ultimate and yield strengths of the material. Significant enhancement of the yield strength due to cold-forming is only found for sections composed largely of corners with small radii and having small width-to-thickness ratios. For more slender cross sections the benefits of cold-forming on enhancement of yield strength become substantially reduced. If a member is subjected to tension only, it is quite permissible to use the increased yield strength for the complete cross section. If, however, the member is subjected to compression or combinations of compression and bending, then each element should be considered separately after the initial determination of the average yield strength of a formed section. In determination of the compression yield strength of an individual element, the width-to-thickness ratio of the element is important. There is no conclusive evidence to support the design use of enhanced yield strength in the presence of **local buckling**, which occurs for slender elements; thus, for slender elements subject to local buckling behavior, enhancement of the yield strength due to cold-forming should not be taken into account.

Since any operation on the formed material that introduces heat, such as welding, annealing, or galvanizing, will affect the material properties, use of the enhanced yield strength is prohibited if any such operation is carried out.

Calculation of Section Properties

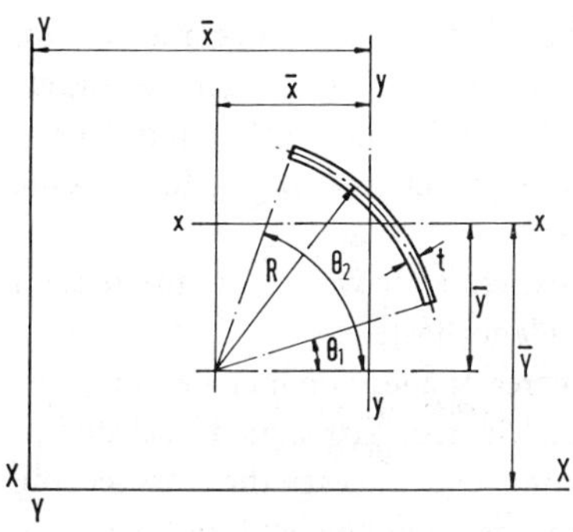

FIGURE 47.4 Geometry of a corner.

Since many cold-formed steel sections have thin walls and small radii, in many cases the determination of section properties can be simplified by assuming that the material is concentrated at the centerline of the section and replacing the area of elements by straight or curved "line elements." The thickness dimension t is introduced after the linear computations have been completed.

Properties of Corners

For a corner element having the geometry shown in Fig. 47.4 the properties are as follows, with the angles given in radian measure.

$$A = Rt(\theta_2 - \theta_1), \quad \bar{y} = \frac{R(\cos\theta_1 - \cos\theta_2)}{(\theta_2 - \theta_1)}, \quad \bar{x} = \frac{R(\sin\theta_2 - \sin\theta_1)}{(\theta_2 - \theta_1)} \tag{47.1}$$

$$I_{xx} = R^3t\left\{\frac{1}{2}(\theta_2 - \theta_1) - \frac{1}{4}(\sin 2\theta_2 - \sin 2\theta_1) - \frac{(\cos\theta_1 - \cos\theta_2)^2}{(\theta_2 - \theta_1)}\right\} \quad (47.2)$$

$$I_{yy} = R^3t\left\{\frac{1}{2}(\theta_2 - \theta_1) + \frac{1}{4}(\sin 2\theta_2 - \sin 2\theta_1) - \frac{(\sin\theta_2 - \sin\theta_1)^2}{(\theta_2 - \theta_1)}\right\} \quad (47.3)$$

$$I_{xy} = R^3t\left\{\frac{1}{4}(\cos 2\theta_1 - \cos 2\theta_2) - \frac{\sin(\theta_1 + \theta_2) - \frac{1}{2}(\sin 2\theta_1 + \sin 2\theta_2)}{(\theta_2 - \theta_1)}\right\} \quad (47.4)$$

In the particular case of a right-angled corner with $\theta_1 = 0$, $\theta_2 = \pi/4$, the properties may be written as follows:

$$A = \frac{\pi R t}{2} = 1.57Rt, \quad \bar{y} = \frac{2R}{\pi} = 0.637R, \quad \bar{x} = \frac{2R}{\pi} = 0.637R \quad (47.5)$$

$$I_{xx} = I_{yy} = R^3t\left(\frac{\pi}{4} - \frac{2}{\pi}\right) = 0.149R^3t \quad (47.6)$$

$$I_{xy} = R^3t\left(\frac{1}{2} - \frac{2}{\pi}\right) = -0.137R^3t \quad (47.7)$$

Formulas for the bending properties of a number of common cross sections, based on centerline dimensions, are given in Table 47.1.

Effects of Holes

In evaluation of the section properties of members in bending or compression, holes made specifically for fasteners such as screws or bolts may be neglected on the basis that the hole is filled with materials in any case. However, for any other openings or holes, the reduction in cross-sectional area and cross-sectional properties caused by these holes or openings should be taken into account. If the section properties are to be evaluated analytically, they should be calculated considering the net cross section that has the most detrimental arrangement of the holes that are not specifically for fasteners. This is not necessarily the same cross section for bending analysis and compression analysis. This is illustrated in Fig. 47.5, where for the channel section shown the net cross section *A–A* has a smaller area than cross section *B–B* and is therefore critical with regard to purely compressional behavior. The second moment of area about *x–x* and minimum section modulus of cross section *B–B* with regard to axis, however, are less than those of section *A–A*, and for bending strength section *B–B* is critical.

In the case of tension members, fasteners do not in themselves effectively resist the tension loading, which tends to open the fastener holes; and holes made for fasteners must also be taken into consideration for tension loading. In determining the net area of a tension member, the cross section that has the largest area of holes should be considered. The area that should be deducted from the gross cross-sectional area is the total cross-sectional areas of all holes in the cross section. In deducting the area of fastener holes, the nominal hole diameter should be used. In the case of countersunk holes, the countersunk area should also be deducted. In a tension member that has staggered holes, the weakening effects of holes that are not in the same cross section, but close enough to interact with the holes in a given cross section, should be taken into account. If two lines of holes are far apart, one line of holes does not have any effect on the strength of the section at the position of the other line of holes. If the lines are close, however, then each line of holes affects the other.

Table 47.1 Formulas for Bending Properties of Typical Sections *(continues)*

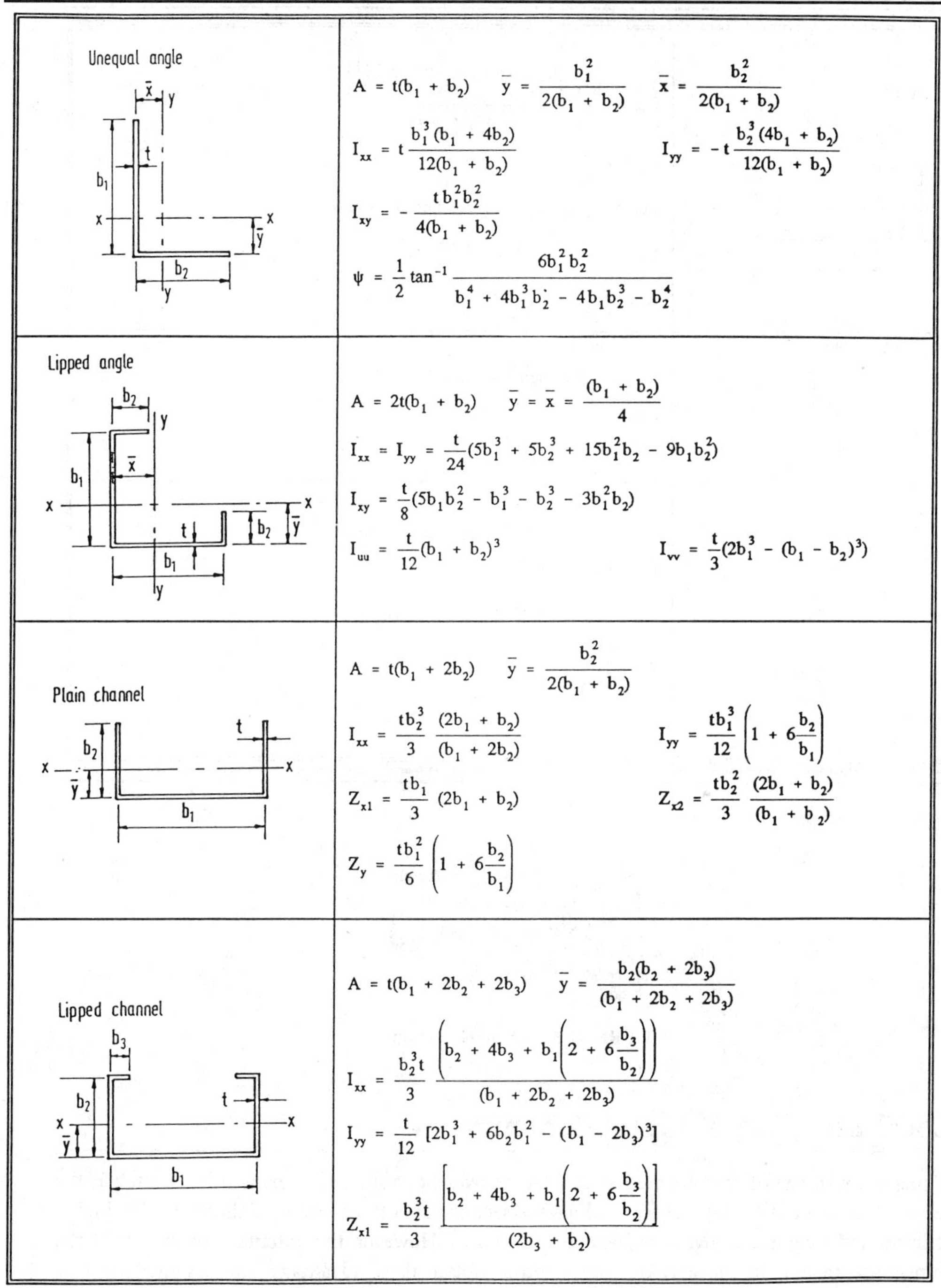

Unequal angle	$A = t(b_1 + b_2)$ $\bar{y} = \frac{b_1^2}{2(b_1 + b_2)}$ $\bar{x} = \frac{b_2^2}{2(b_1 + b_2)}$ $I_{xx} = t\frac{b_1^3(b_1 + 4b_2)}{12(b_1 + b_2)}$ $I_{yy} = -t\frac{b_2^3(4b_1 + b_2)}{12(b_1 + b_2)}$ $I_{xy} = -\frac{t b_1^2 b_2^2}{4(b_1 + b_2)}$ $\psi = \frac{1}{2}\tan^{-1}\frac{6b_1^2 b_2^2}{b_1^4 + 4b_1^3 b_2 - 4b_1 b_2^3 - b_2^4}$
Lipped angle	$A = 2t(b_1 + b_2)$ $\bar{y} = \bar{x} = \frac{(b_1 + b_2)}{4}$ $I_{xx} = I_{yy} = \frac{t}{24}(5b_1^3 + 5b_2^3 + 15b_1^2 b_2 - 9b_1 b_2^2)$ $I_{xy} = \frac{t}{8}(5b_1 b_2^2 - b_1^3 - b_2^3 - 3b_1^2 b_2)$ $I_{uu} = \frac{t}{12}(b_1 + b_2)^3$ $I_{vv} = \frac{t}{3}(2b_1^3 - (b_1 - b_2)^3)$
Plain channel	$A = t(b_1 + 2b_2)$ $\bar{y} = \frac{b_2^2}{2(b_1 + b_2)}$ $I_{xx} = \frac{tb_2^3}{3}\frac{(2b_1 + b_2)}{(b_1 + 2b_2)}$ $I_{yy} = \frac{tb_1^3}{12}\left(1 + 6\frac{b_2}{b_1}\right)$ $Z_{x1} = \frac{tb_1}{3}(2b_1 + b_2)$ $Z_{x2} = \frac{tb_2^2}{3}\frac{(2b_1 + b_2)}{(b_1 + b_2)}$ $Z_y = \frac{tb_1^2}{6}\left(1 + 6\frac{b_2}{b_1}\right)$
Lipped channel	$A = t(b_1 + 2b_2 + 2b_3)$ $\bar{y} = \frac{b_2(b_2 + 2b_3)}{(b_1 + 2b_2 + 2b_3)}$ $I_{xx} = \frac{b_2^3 t}{3}\left(\frac{b_2 + 4b_3 + b_1\left(2 + 6\frac{b_3}{b_2}\right)}{(b_1 + 2b_2 + 2b_3)}\right)$ $I_{yy} = \frac{t}{12}[2b_1^3 + 6b_2 b_1^2 - (b_1 - 2b_3)^3]$ $Z_{x1} = \frac{b_2^3 t}{3}\left[\frac{b_2 + 4b_3 + b_1\left(2 + 6\frac{b_3}{b_2}\right)}{(2b_3 + b_2)}\right]$

Table 47.1 (continued) Formulas for Bending Properties of Typical Sections

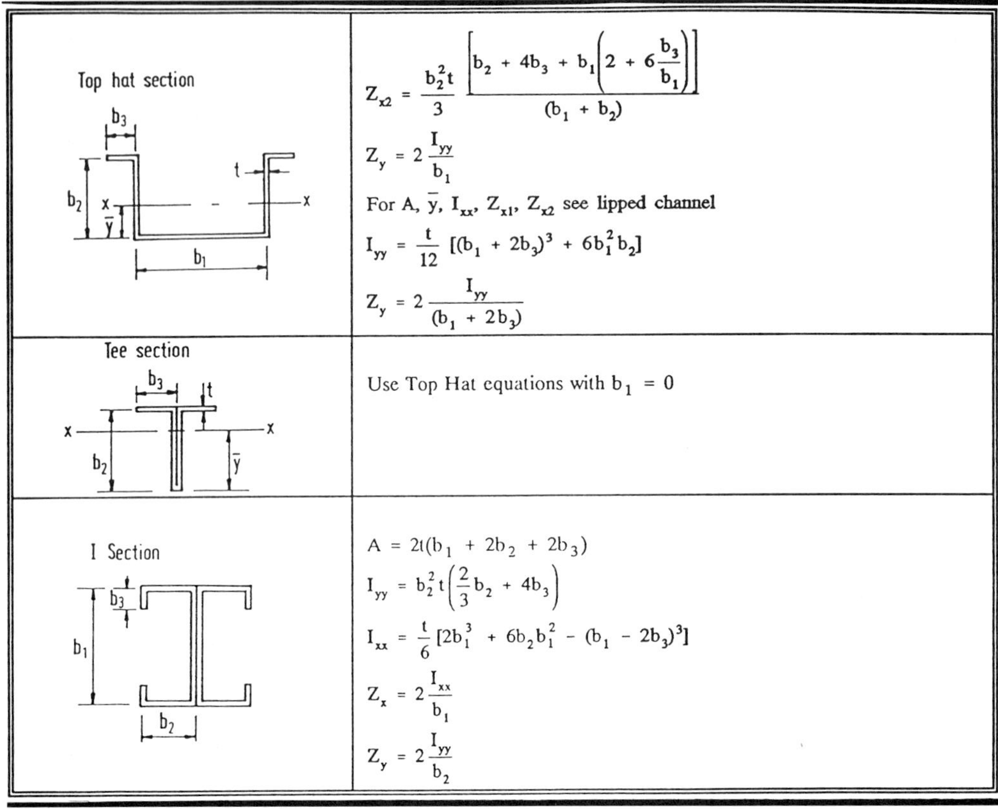

Top hat section	$Z_{x2} = \frac{b_2^2 t}{3}\left[\frac{b_2 + 4b_3 + b_1\left(2 + 6\frac{b_3}{b_1}\right)}{(b_1 + b_2)}\right]$ $Z_y = 2\frac{I_{yy}}{b_1}$ For A, $\bar{y}$, I_{xx}, Z_{x1}, Z_{x2} see lipped channel $I_{yy} = \frac{t}{12}[(b_1 + 2b_3)^3 + 6b_1^2 b_2]$ $Z_y = 2\frac{I_{yy}}{(b_1 + 2b_3)}$
Tee section	Use Top Hat equations with $b_1 = 0$
I Section	$A = 2t(b_1 + 2b_2 + 2b_3)$ $I_{yy} = b_2^2 t\left(\frac{2}{3}b_2 + 4b_3\right)$ $I_{xx} = \frac{t}{6}[2b_1^3 + 6b_2 b_1^2 - (b_1 - 2b_3)^3]$ $Z_x = 2\frac{I_{xx}}{b_1}$ $Z_y = 2\frac{I_{yy}}{b_2}$

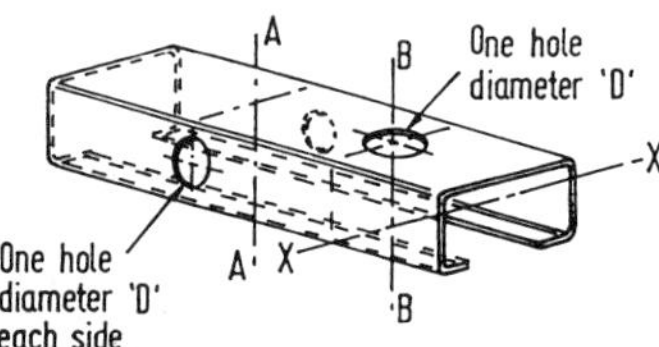

FIGURE 47.5 Channel section with holes.

47.2 Local Buckling of Plate Elements

A major advantage of cold-formed steel sections over hot-rolled sections is to be found in the relative thinness of the material from which the sections are often formed. This can lead to highly efficient and weight-effective members and structures. However, the potential advantages of the thin walls can only be partially obtained, and to obtain these advantages the designer must be aware of the phenomena associated with thin-walled members and their effects on design analysis. Perhaps the most important of these phenomena is local buckling.

Local Buckling of Plates

When a thin plate is loaded in compression, the possibility of local buckling arises. This type of buckling is so called because the length of the buckles that form is similar to the dimensions of

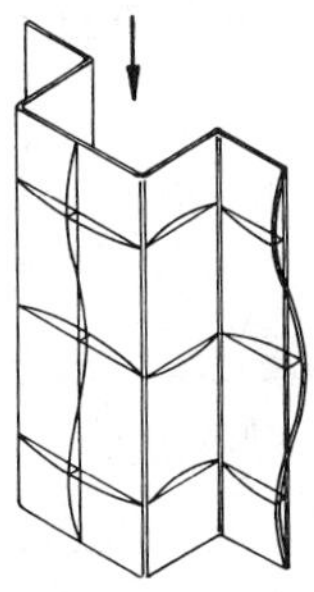

FIGURE 47.6 Local buckling in a thin-walled section.

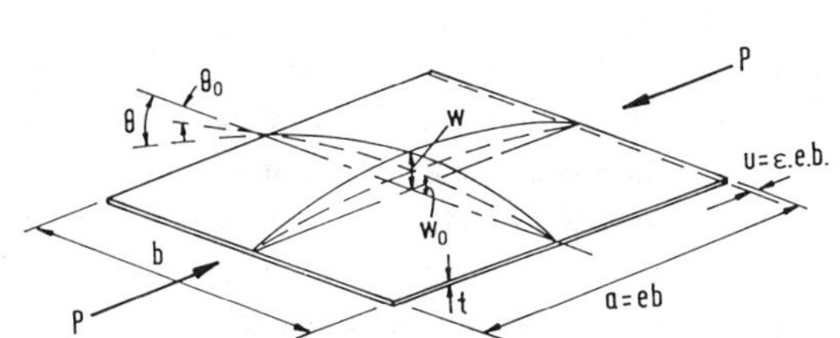

FIGURE 47.7 Uniformly compressed plates.

the cross section, rather than to the length of the structure as is normally the case with other types of buckling. Elastic local buckling in a member is characterized by a number of ripples or buckles becoming evident in the component plates, as illustrated in Fig. 47.6. This local buckling has substantial bearing on the stiffness and strength of the member, and to gain some insight into the local buckling phenomenon we shall now examine local buckling of plate elements.

Local Buckling Analysis

Consider the plate shown in Fig. 47.7, supported on all four edges and compressed uniformly on its longitudinal edges to produce a displacement u as shown in the figure. Due to the loading, we shall assume that out-of-plane deflections w occur as shown. We can examine the local buckling situation from a consideration of the strain energy in the plate. The strain energy due to bending U_B can be written in terms of the deflections as given in Eq. 45.110 in Chapter 45.

The strain energy in the plate due to the membrane actions is given by

$$U_D = \frac{1}{2}\int_{\text{vol}} p_x \varepsilon_x \, d(\text{vol}) = \frac{Eta}{2}\int\left(\bar{\varepsilon} - \frac{1}{2a}\int\left(\frac{\partial w}{\partial x}\right)^2\right)^2 dy \tag{47.8}$$

where $\bar{\varepsilon}$ is the "nominal" applied strain, equal to u/a, and E is the modulus of elasticity.

The total strain energy stored in the plate is given by the sum of bending and membrane energies. Since the displacement of the plate ends is prescribed, the principle of minimum potential energy requires that the strain energy is a minimum. The simplest case of local buckling, very often considered in design, is that of a plate simply supported on all four edges. In this case the deflections w at buckling are given by the expression

$$w = w_c \sin\frac{n\pi x}{a}\sin\frac{\pi y}{b} \tag{47.9}$$

in which n indicates the number of half–sine waves into which the plate buckles in the x direction. Substituting for w in $U(= U_D + U_B)$ and performing the integrations gives the strain energy in terms of the deflection magnitude coefficient, w_c. Using the principle of minimum potential energy, it can be shown that the critical stress (p_{cr}) to cause local buckling is given as

$$p_{cr} = E\bar{\varepsilon}_{\text{cr}} = \frac{\pi^2 D}{b^2 t}\left[\frac{nb}{a} + \frac{a}{nb}\right]^2 = \frac{K\pi^2 E}{12(1-\nu^2)}\left(\frac{t}{b}\right)^2 \tag{47.10}$$

the coefficient K is called the *buckling coefficient,* and for the case in question is given by

$$K = \left[\frac{nb}{a} + \frac{a}{nb}\right]^2 \tag{47.11}$$

The variation of K with variation in plate length-to-width ratio a/b is shown in Fig. 47.8 for various numbers of buckles n. As may be observed from this figure, the minimum value of K is 4, and this occurs when the length of the plate is equal to n times the plate width. For long plates, the number of buckle half-waves that occur is approximately the same as the ratio of plate length to its width, and the buckling coefficient is very close to 4. For plates with different support conditions the value of the buckling coefficient becomes different from 4.

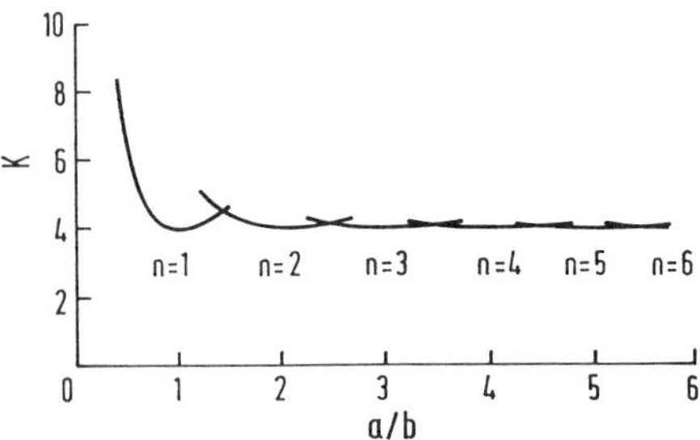

FIGURE 47.8 Variation of buckling coefficient with length for a simply supported plate.

Thus, the value of the stress required theoretically to produce local buckling varies inversely as the square of the plate width-to-thickness ratio. Plates with lower width-to-thickness ratios will theoretically yield before local buckling, and plates with higher width-to-thickness ratios will buckle before yielding. This statement holds true only for perfect plates. In a practical situation imperfections are always present, and the effects of local buckling are generally to be observed at stresses less than the theoretical buckling stress. Furthermore, local buckling does not necessarily signify the attainment of the full load capacity of a plate. For very thin plates, local buckling occurs at low stress levels, and such plates can sustain loads greatly in excess of the **buckling load**. This capacity to carry loads beyond the local buckling load provides the means for advantageous use of thin plates, but it also requires knowledge of the adverse effects of local buckling to ensure safe design.

Postbuckling Analysis

If the plate has buckled, the magnitude of the local buckles is related to the compression magnitude. The variation of stresses with variation in strains after buckling can be shown to be

$$p_x = E\left[\bar{\varepsilon} - \frac{4}{3}(\bar{\varepsilon} - \bar{\varepsilon}_{cr}) \sin^2 \frac{\pi y}{b}\right] \tag{47.12}$$

Thus, the average stress varies across the plate as indicated in Fig. 47.9. Strips of plate near the supports are relatively unaffected by local buckling and carry increased loading as further compression is applied, while strips of plate near the center shed load and offer very little resistance to further compression. The plate may be thought of as consisting of a series of slender columns linked together, with those near the edges largely prevented from buckling and those near the center buckling relatively freely. The plate does not behave completely like a column and lose all its compression resistance after local buckling, but its resistance after buckling is confined to portions of the plate near the supported edges.

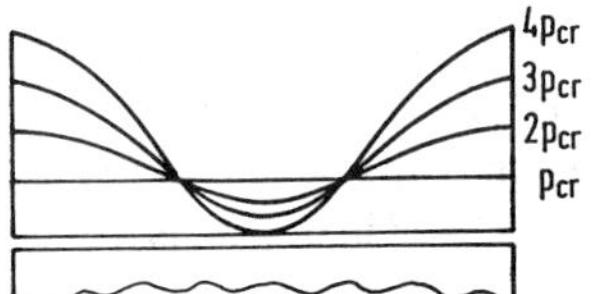

FIGURE 47.9 Variation of stress across a plate after local buckling.

The load on the plate at any end compression is obtained by summing the stresses across the plate, i.e.

$$p = t\int p_x dy = Etb\left[\bar{\varepsilon} - \frac{2}{3}(\bar{\varepsilon} - \bar{\varepsilon}_{cr})\right] = \frac{Etb}{3}\left[\bar{\varepsilon} + 2\bar{\varepsilon}_{cr}\right] \tag{47.13}$$

The plate load grows after buckling with increasing compression, but the rate of growth is substantially reduced relative to that before buckling.

The analysis shows that after buckling the plate stiffness decreases, the edge stresses increase more quickly with load than before buckling, and the plate center becomes inefficient at resisting load, as shown by the load shedding in this area. The plating could be considered effective in resisting compression only over a short width adjacent to the supports. As a rule of thumb, in ship design it was considered that a plate of width equal to 25 times the thickness could effectively resist compression adjacent to each support. Thus, for a plate with supports on each edge, a total width of 50t was considered to be effective in resisting compression. This was the origin of the **effective width** concept, now used widely in design analysis.

The Effective Width Concept

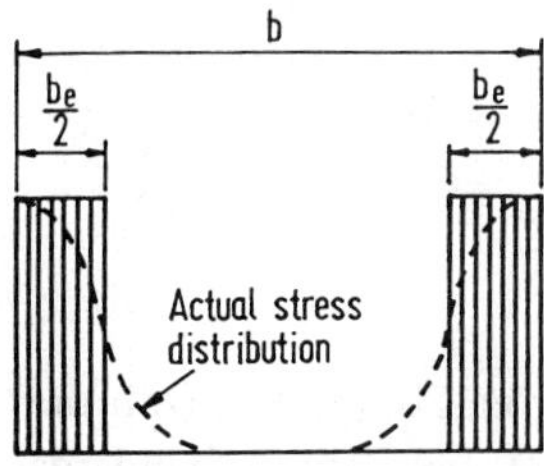

FIGURE 47.10 Effective width idealization.

The effective width concept, as generally used in design, assumes that the portions of a plate element near the supports are fully effective in resisting load and the remainder of the element is completely ineffective. This is illustrated in Fig. 47.10, in which the varying stress distribution across an element is idealized into a constant stress acting over the two effective portions, and the center part of the plate is considered completely ineffective and stress-free.

In 1932 von Kármán *et al.* produced the first theoretical explanation of the effective width concept, and in the years that followed, many investigators produced further insight into this field. At present there is a variety of methods available for rigorous analysis of plates under many conditions of loading and support, ranging from approaches that have been set up for relatively easy use by the reader to the numerical approaches using finite elements or finite differences.

Research into cold-formed steel began largely in the United States, and much of the original work was carried out at Cornell University by George Winter. In dealing with local buckling, Winter [1947] produced an empirical variation of the von Kármán effective width expression that, with minor modifications, has been accepted in the U.S. and in many other countries for analysis of local buckling. This expression, in the form used widely at present, is as follows:

$$\frac{b_{\text{eff}}}{b} = \sqrt{\frac{p_{\text{cr}}}{p_{\max}}}\left(1 - 0.22\sqrt{\frac{p_{\text{cr}}}{p_{\max}}}\right) \tag{47.14}$$

where

$p_{\max}$ = maximum edge stress in the plate
b_{eff} = effective width of the compression element

As per AISI recommendation, $b_{\text{eff}} = b$ for $\sqrt{p_{\max}/p_{\text{cr}}} \leq 0.673$, and the b_{eff} is calculated from Eq. (47.14) for $\sqrt{p_{\max}/p_{\text{cr}}} > 0.673$.

The basic effective width expression developed for BS 5950: Part 5 [British Standards Institution, 1987] is as follows:

$$\frac{b_{\text{eff}}}{b} = \left[1 + 14\left(\sqrt{\frac{f_c}{p_{\text{cr}}}} - 0.35\right)^4\right]^{-0.2} \tag{47.15}$$

in which f_c is the edge stress corresponding to the yield stress when failure is said to occur. This expression is a little more cumbersome than that of Eq. (47.14) but still perfectly usable on a calculator or a small microcomputer.

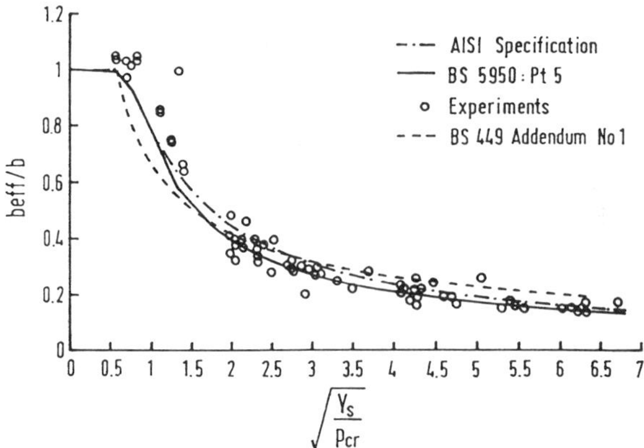

FIGURE 47.11 Effective widths from design codes and tests from various sources.

Figure 47.11 shows comparison of the basic effective width expression, Eq. (47.15), with the AISI expression, Eq. (47.14); the *CL* factors of Addendum No. 1 to BS 449 [British Standards Institution, 1975]; and experiments from various sources. The experiments shown here are all on simply supported plates; even so, a significant degree of scatter is noticeable.

Classification of Elements

When the effective width approach is used to deal with elements of sections, the different types of elements used in such sections must taken into consideration. These may be classified into four groups: **stiffened elements, unstiffened elements,** *edge-stiffened elements,* and *intermediately stiffened elements.* Examples of each type of element are shown in Fig. 47.12.

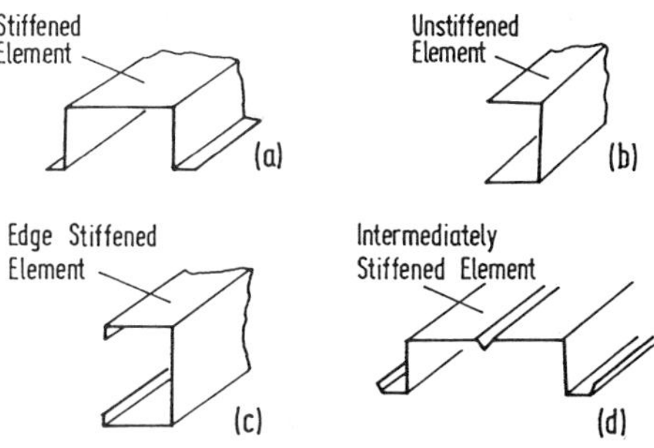

FIGURE 47.12 Types of elements found in cold-formed sections.

Stiffened elements are elements that are supported by a substantial element on both longitudinal edges. If the supporting elements are themselves stiffened or edge-stiffened elements, this type of element can have a width-to-thickness ratio up to 500. For such an element the minimum *K* factor applied is 4.

Unstiffened elements are elements that are supported along only one longitudinal edge. In this case local buckling arises much more quickly than for stiffened elements; therefore, the width-to-thickness ratio that can be covered by the code is considerably reduced. The relevant maximum width-to-thickness ratio is 60, and the minimum *K* factor is 0.425.

Since unstiffened elements are severely affected by local buckling, it is common practice to convert these into stiffened elements by folding the free edge of these elements to produce a lip or a similar edge stiffener, to support this edge and prevent local buckling of the edge. If the edge stiffener satisfies the requirements of the code, such an element may be treated as a stiffened element. For such an *edge-stiffened element,* however, the maximum width-to-thickness ratios are severely curtailed by the codes. In the case of an element stiffened by a simple lip, the maximum allowable width-to-thickness ratio is 60, while if any other type of edge stiffener is used, the width-to-thickness ratio can be increased to 90.

In order to improve the behavior of stiffened elements, intermediate stiffeners are often employed. This type of stiffener is usually formed during the rolling process and has the effect of transforming a slender, high–b/t ratio element into two or more relatively compact subelements. This can substantially increase the effectiveness of the element. The total width of *intermediately stiffened elements* is limited to 500 times the material thickness.

Stiffened Elements

Buckling Coefficients

Stiffened elements under uniform compression are considered to be governed by effective width expression. The minimum K factor used for such elements is 4, but if higher values can be justified, they may be used. The K factors applicable to compact elements of low b/t ratio can be very much less than for slender elements in the same section. In general, the elements in a section that are restrained from buckling at their natural stress induce premature buckling in the elements that restrain them, and if one element has a buckling coefficient greater than 4, others will have buckling coefficients less than 4. This mathematically correct but rather pessimistic view has been incorporated in some design codes, and it invariably leads to lower design loads than would occur if all elements were considered to be simply supported.

However, it has been observed [Rhodes, 1987] that for restraining elements the effects of premature local buckling are negligible. Only when the applied loading is sufficient to attain the natural (simply supported) buckling stress of such elements do these elements suffer the effects of local buckling to any substantial degree.

Stiffened Elements under Eccentric Compression

For some stiffened elements, such as webs of beams, the loading is not pure compression but some combination of axial loading and in-plane bending. In cases where this type of loading occurs, the codes treat the situation in either of two ways, depending on the degree of bending involved. For beam webs in which the stress changes from compression to tension across the element, the effective width approach is replaced by a limiting stress approach.

For cases in which an element is subjected to a combination of compression and in-plane bending in which the stress is compressive on both unloaded edges, sufficiently accurate analysis may be obtained as follows. If the stress varies from f_{c1} at one edge to f_{c2} at the other edge, as illustrated in Fig. 47.13, then the mean value of these stresses should be taken as the stress on the element. Using this stress, together with the critical stress based on a K factor of 4, the effective width may be found as for uniformly compressed elements. The effective portions of the element are considered to be equally distributed adjacent to each supported edge, as for uniformly compressed elements, and the ultimate load on the element may be assumed to occur when the maximum stress reaches yield. This rather rough and ready method has some theoretical backing and gives reasonable and conservative results when compared with tests.

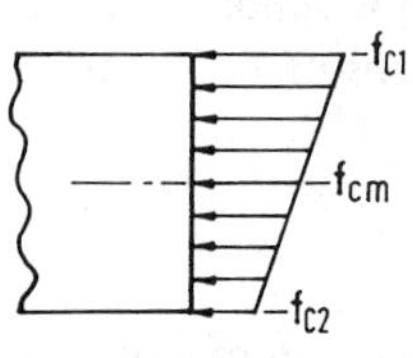

FIGURE 47.13 Element with varying stress distribution.

Unstiffened Elements

In open sections there are normally unstiffened elements. As these elements buckle much more quickly than stiffened elements, they constitute a rather unsatisfactory type of element with regard to load capacity, and because of this the cold-formed counterparts of such common hot-rolled sections as channels, angles, and hat sections are often lipped to increase the resistance to local buckling.

Unstiffened elements can also be analyzed using the approach employed earlier for stiffened elements. If we assume that the out-of-plane deflections of an unstiffened element under load, as shown in Fig. 47.14, are given by the expression

$$w = w_{\max}\frac{y}{b}\sin\frac{\pi x}{a} \qquad (47.16)$$

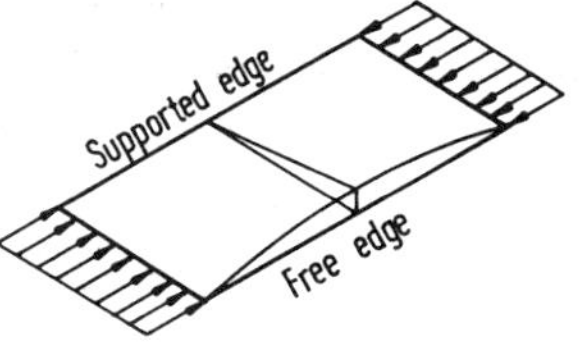

FIGURE 47.14 Unstiffened element under compression.

then the same type of analysis gives the following results:

$$K = \left(\frac{b}{a}\right)^2 + 0.425 \qquad (47.17)$$

and the relative stiffness before and after buckling is

$$\frac{E^*}{E} = \frac{4}{9} \qquad (47.18)$$

The variation of K with variation in element length-to-width ratio is shown in Fig. 47.15. In this case the element buckles into a single half-wave regardless of its length, according to the analysis used. This is only true in the cases where the supported edge is simply supported, such as elements of angle sections. For unstiffened elements that have some restraint on their supported edge, several buckles are found if the element is long. However, the simply supported–free condition gives a lower bound to the strength and stiffness of more restrained elements and leads to safe design. The buckling coefficients applicable to a fully fixed supported edge are also shown in Fig. 47.15 for comparison purposes.

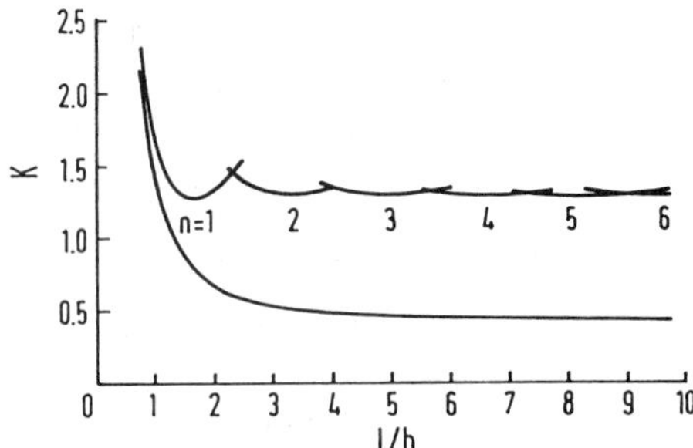

FIGURE 47.15 Variation of buckling coefficients for unstiffened elements with element length.

The minimum value of K in this case is 0.425 for an element with simple support on the supported edge, based on a Poisson ratio of 0.3. Thus, the buckling load of an unstiffened element is reduced by a factor of $4/0.425 = 9.4$ from that of a stiffened element.

Effective Widths

Returning to our approximate analysis, if we examine the load-compression behavior of a perfect plate after buckling and plot the relevant curves for stiffened and unstiffened elements, normalized with respect to the buckling point, we find that the postbuckling stiffness of the unstiffened element is greater than that of a stiffened element at a given multiple of the buckling stress. As the effective width expressions contain the critical stress, it follows that the same expression will be rather conservative for unstiffened elements.

In the British Code the effective widths so obtained are increased for unstiffened elements. The method used is to initially determine the effective width, b_{eff}, as for a stiffened element but using the buckling coefficient applicable to the unstiffened element under examination. This is then converted to an enhanced effective width for an unstiffened element, using the following equation:

$$b_{\text{eu}} = 0.89b_{\text{eff}} + 0.11b \qquad (47.19)$$

Stiffeners

Edge Stiffeners

Because of the low buckling resistance of unstiffened elements and the rather unfortunate consequences that can arise, the benefits of incorporating edge stiffeners are plain to see. Adequately edge-stiffened elements can be substantially stronger than the unstiffened counterparts. In general an edge stiffener is required to eliminate, or at least minimize, any tendency for the otherwise unsupported edge of an element to displace out of plane. If a stiffener is adequate, the stiffened element will not incur deflections at the stiffened edge, and the element can be treated as a stiffened element. If, however, the stiffener does not have sufficient flexural rigidity to prevent out-of-plane deflections of the edge, the stiffener is said to be inadequate.

The precise requirements for adequacy are even now undergoing change. In various cold-formed steel design codes in the past, it was accepted that for adequacy a stiffener should prevent buckling of the element edge until the element buckled as a stiffened plate with K factor of 4. This was shown [Desmond *et al.*, 1981] to be an insufficient requirement, as the edge stiffener must also be able to prevent the edge of the element from buckling even after local buckling has occurred—indeed, until it fails as a stiffened element—if it is to do its job correctly. The requirements for adequacy in the latest AISI code and in the European Recommendations have been based on this premise.

However, this is not the only difficulty in assessing stiffener adequacy. Rigorous analysis shows that the required rigidity of an edge stiffener depends not only on the geometry of the element to be stiffened but also on the geometry of the section as a whole, and indeed on the geometry of the stiffener itself [Rhodes, 1983]. Thus, obtaining a single formula that covers all these variables with accuracy is a daunting task.

Intermediate Stiffeners

Intermediate stiffeners are becoming more widely used in cold-formed steel members. The advantages of replacing slender elements by more effective subelements of relatively compact proportions, at the expense of a little extra material for the stiffener are apparent. As with edge stiffeners, the intermediate stiffeners used must have adequate rigidity to prevent deflection in the element in the region of the stiffener. Adequate and inadequate stiffeners of this type are illustrated in Fig. 47.16. The geometry of an element with a single intermediate stiffener is shown in Fig. 47.17.

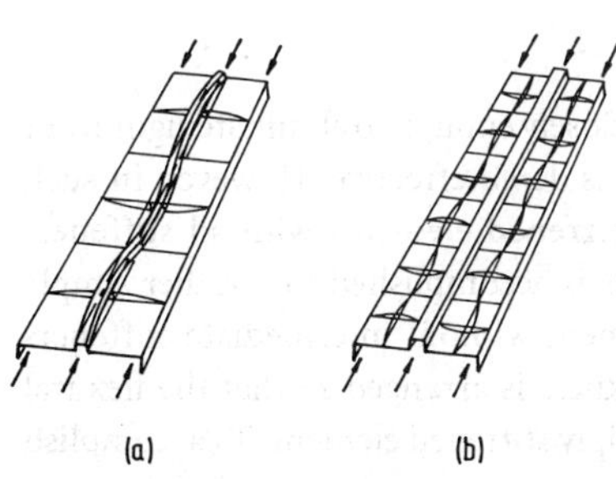

FIGURE 47.16 Intermediate stiffeners. (a) Inadequate stiffeners. (b) Adequate stiffeners.

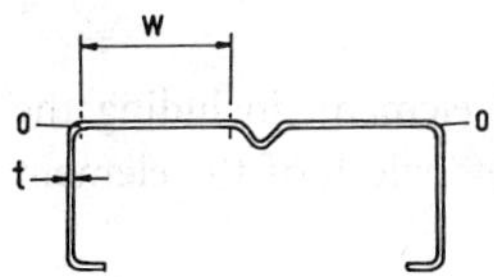

FIGURE 47.17 Geometry of an element with a single intermediate stiffener.

If the rigidity requirement is attained by an intermediate stiffener, the effective area of the stiffened element may, under certain conditions, be evaluated from the sum of the effective areas of each individual subelement, analyzed as stiffened elements, and the stiffener area. The conditions under which this is allowable are that the width-to-thickness ratios of the subelements are less than 60. Since in this case the out-of-plane deflections are eliminated near the stiffener, the hypothetical situation is that under uniform compression here there is no shedding of load near the stiffener, and this area of the element becomes fully stressed. Each subelement is therefore stressed in the same manner as a stiffened element of width w. The stiffener is also fully stressed and plays its full part in resisting load.

Reductions in Capacity for High Width-to-Thickness Ratios

If the subelement width-to-thickness ratios are greater than 60, reductions in the subelement effective area and in the effective stiffener area must be introduced. The main reasons for this are to be found in the examination of **beam** behavior. Beams that have very wide, slender flanges, either in tension or compression, suffer from the tendency of these flanges to move toward the neutral axis of the section under loading.

BS 5950: Part 5 [British Standards Institution, 1987] takes the adverse effects into account in a rather simple way, based on the AISI specification prior to the current version. If the subelement width-to-thickness ratio w/t is greater than 60, the effective width of the subelement, b_{eff}, is replaced by a reduced effective width, b_{cr}, determined from the expression

$$\frac{b_{\text{cr}}}{t} = \frac{b_{\text{eff}}}{t} - 0.1\left(\frac{w}{t} - 60\right) \tag{47.20}$$

The effective stiffener area is also reduced. For w/t less than 60, the stiffener is taken as fully effective. For w/t greater than 90, the ratio of effective stiffener area A_{eff} to full stiffener area A_{st} is taken as the same as the ratio of effective subelement area to full subelement area:

$$A_{\text{eff}} = A_{\text{st}}\frac{b_{\text{cr}}}{w} \tag{47.21}$$

For w/t values between 60 and 90 a linear interpolation formula is used to obtain the effective stiffener area. This is

$$A_{\text{eff}} = A_{\text{st}}\left(3 - 2\frac{b_{\text{cr}}}{w} + \frac{1}{30}\left(1 - \frac{b_{\text{cr}}}{w}\right)\frac{w}{t}\right) \tag{47.22}$$

This expression gives stiffener effective area varying linearly from A_{st} at $w/t = 60$ to $A_{\text{st}}b_{\text{cr}}/w$ at $w/t = 90$.

Multiple Intermediate Stiffeners

If an element has many intermediate stiffeners that are spaced closely enough to eliminate significant local buckling (i.e., $w/t < 30$), then all stiffeners may be considered effective. However, in such a case local buckling that involves the complete **multiply stiffened element,** with all stiffeners participating in the buckling, has to be guarded against. This is accomplished in a rather simple way, by considering the complete element as a stiffened element without intermediate stiffeners but with a fictitious equivalent thickness. The fictitious thickness is arranged so that the flexural rigidity of the stiffened element is the same as that of the multiply stiffened element. To accomplish this the equivalent thickness t_s is taken as

$$t_s = \left(\frac{12I_s}{w_s}\right)^{1/3} \tag{47.23}$$

where I_s is the second moment of area of the full multiply stiffened element, including the intermediate stiffeners, about its own neutral axis, and w_s is the complete width of the element between two webs.

Example 47.1. Compute the effective width of the compression (top) flange of the beam shown in Fig. 47.18. Assume that the compressive stress in the flange is 25 ksi. $E = 29{,}500$ ksi, $\nu = 0.3$.
Solution. The following solution is based on AISI design code.

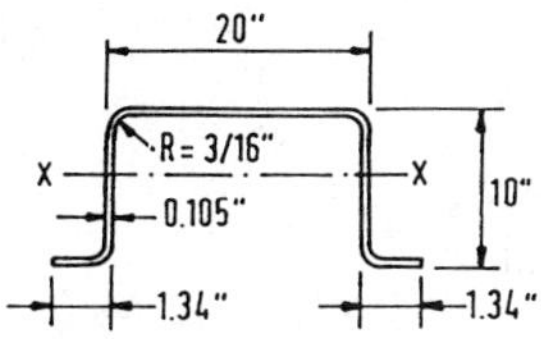

FIGURE 47.18 Beam cross section for Example 47.1.

As the first step, compute p_{max}/p_{cr}

$$K = 4.0$$

$$\begin{aligned} b &= 20 - 2(R + t) \\ &= 20 - 2(0.1875 + 0.105) = 19.42 \text{ in.} \end{aligned}$$

$$\frac{b}{t} = \frac{19.42}{0.105} = 185$$

$$\begin{aligned} p_{cr} &= \frac{4\pi^2 E}{12(1-\nu^2)} \frac{1}{(b/t)^2} \\ &= 3.12 \text{ ksi} \end{aligned}$$

$$p_{max} = 25 \text{ ksi}$$

$$\sqrt{\frac{p_{max}}{p_{cr}}} = 2.83 > 0.673$$

Therefore,

$$\begin{aligned} b_{eff} &= b\sqrt{\frac{p_{cr}}{p_{max}}}\left(1 - 0.22\sqrt{\frac{p_{cr}}{p_{max}}}\right) \\ &= 6.52 \text{ in.} \end{aligned}$$

Example 47.2. Calculate the effective width of the compression flange of the box section (Fig. 47.19) to be used as a beam bending about the x axis. Use $p_{max} = 33$ ksi. Assume that the beam webs are fully effective and that the bending moment is based on initiation of yielding. Take $E = 29{,}500$ ksi and $\nu = 0.3$.

Solution. The solution is in accordance with the AISI Code.

Because the compression flange of the given section is a uniformly compressed stiffened element, that is supported by a web on each longitudinal edge, the effective width of the flange can be computed by using Eq. (47.14) with $K = 4.0$.

Given that the bending strength of the section is based on initiation of yielding,

$$\bar{y} \geq 3 \text{ in.}$$

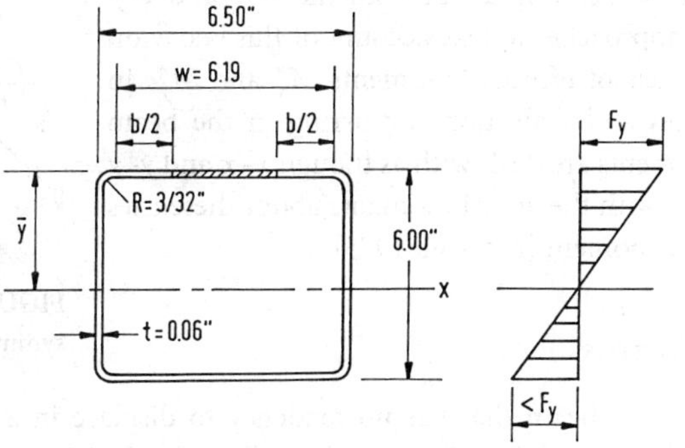

FIGURE 47.19 Box section for Example 47.2.

$$\frac{b}{t} = \frac{6.19}{0.06} = 103.2$$

$$p_{cr} = \frac{4\pi^2 E}{12(1-\nu^2)}\frac{1}{(b/t)^2} = 10.01 \text{ ksi}$$

$$\sqrt{\frac{p_{max}}{p_{cr}}} = \sqrt{\frac{33}{10.01}} = 1.816 > 0.673$$

Therefore,

$$b_{eff} = b\sqrt{\frac{p_{cr}}{p_{max}}}\left[1 - 0.22\sqrt{\frac{p_{cr}}{p_{max}}}\right]$$

$$= 2.997 \text{ in.}$$

$$\approx 3 \text{ in.}$$

47.3 Members Subject to Bending

Because of the thin-walled nature of cold-formed steel sections, the effects of local buckling must, in most cases, be taken into account in determining the moment capacity and flexibility of beams. In the design of beam webs, the capacity of webs to withstand concentrated loads or support reactions must be ensured, and the interaction of different effects must be taken into account. In the case of unbraced beams the possibility of lateral **torsional buckling** arises, and the designer must be able to guard against this phenomenon. In certain circumstances the use of simple bending theory cannot give realistic estimates of beam behavior. The most obvious of these circumstances arises in the design analysis of beams having unsymmetrical cross sections. If such beams are not restrained continuously along their lengths, these must be analyzed, taking into account the unsymmetrical nature of the behavior.

Bending of Unsymmetrical Cross Sections

Consider a thin-walled beam having a general nonsymmetrical cross section as shown in Fig. 47.20. If the x–x and y–y axes shown in the figure are not the principal axes of the cross section, the application of a moment about either one of the axes will cause the beam to bend about both axes; for example, a moment M applied about axis x–x will cause bending about both the x–x and y–y axes. There are several approaches to take account of this behavior, one of these being the use of effective moments M_x^* and M_y^*. In this method the stresses and deflections occurring in the beam under the action of moments are dealt with as though x–x and y–y were principal axes, but with the actual moments about these axes replaced by the effective moment [Megson, 1975].

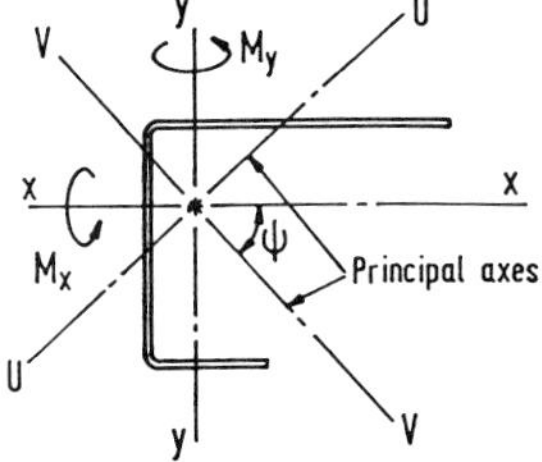

FIGURE 47.20 Beam of unsymmetrical cross section.

Laterally Stable Beams

A laterally stable beam is a beam that has no tendency to displace in a direction perpendicular to the direction of loading. A beam may be laterally stable by virtue of its shape, or it may be considered to be laterally stable if it is braced sufficiently to prevent potential displacements out of the plane of loading. For laterally stable beams, local buckling is the major weakening effect. In the analysis of laterally stable beams according to BS 5950: Part 5, limiting web stress is used to take into account the possibility of local buckling in the webs, and the effective width approach is used to take account of local buckling in the compression elements.

Limiting Web Stress

The effects of local buckling resulting from varying bending stresses in thin webs of beams can be quite substantial. The buckling stress in a web is generally very much greater than in a compression element of the same geometry, but if the web depth-to-thickness ratio is large, local buckling of the web can still have significant influence on the beam strength. For webs under pure bending, the minimum buckling coefficient is approximately 23.9 as compared with 4 for a uniformly compressed plate.

The effects of local buckling on web strength are not so easily taken into consideration as in the case of elements under uniform compression. Effective width approaches have been investigated to take local buckling of webs into account, and have been found to accomplish this task very well, but at the expense of adding further complexity to the analysis. This is mainly due to the necessity to position the effective and ineffective portions correctly. A number of design codes, including the AISI Code, use the effective width approach. If a web has intermediate stiffeners, these will assist the web in resisting local buckling. There has been substantial research into this topic.

Effective Width of Compression Elements

In the case of beams the elements under compression are considered to have effective widths less than their actual widths, while all other elements are considered to have their actual dimensions. A typical effective cross section is shown in Fig. 47.21. The effective width of the compression element or elements is evaluated using effective width expression. The buckling coefficients in the case of beams are in most circumstances greater than the minimum values, since buckling of the compression elements is generally (but not always) restrained by the adjacent webs.

Moment Capacity

The moment capacity of the cross section is determined on the basis that the maximum compressive stress on the section is p_0. This leads to two possible types of failure analysis, depending on the situation on the tension side of the cross section, as indicated in Fig. 47.22. If the geometry of the effective cross section is such that the compressive stress reaches p_0 before the maximum tensile

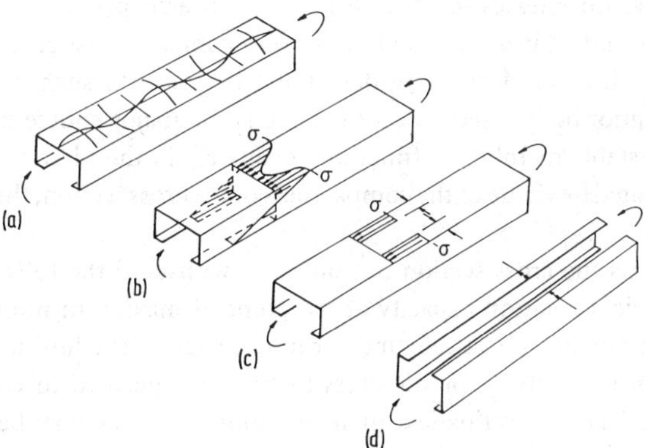

FIGURE 47.21 Effective width concept applied to laterally stable beams. (a.) Local buckles. (b) Stress distribution. (c) Idealized stress distribution on compression element. (d) Effective section.

stress reaches the yield stress, as in Fig. 47.22, then the moment capacity is evaluated using the product of compression section modulus and p_0, That is,

$$M_c = p_0 \times Z_c \qquad (47.24)$$

where Z_c is the compression section modulus of the effective cross section. This situation will occur in the case of members that have wide tension elements or that have substantially ineffective compression elements.

If, on the other hand, the tensile stresses reach yield before p_0 is attained on the compression side, as in Fig. 47.22(b), the designer is allowed to take advantage of the plastic redistribution of tension stresses and thus obtain higher predictions of moment capacity than would be the case if first yield were taken as the criterion. This necessitates an increase in the complexity of the analysis if the added capacity is to be obtained. If simplicity of analysis is more important than the requirement to obtain the most beneficial estimate of capacity, the moment capacity can also be obtained using the product of the yield stress and tension modulus of the effective cross section. This may well, however, lead to significant underestimates of the member capacity.

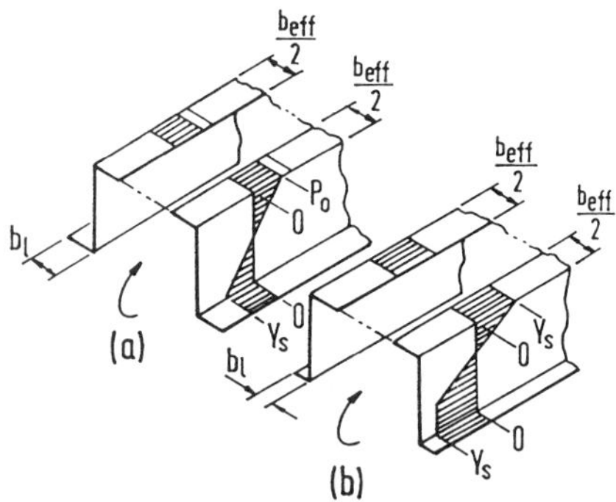

FIGURE 47.22 Failure criteria for laterally stable beams. (a) Failure by compression yield; tensile stress elastic. (b) Tensile stresses reach yield before failure; elastoplastic stress distribution.

Determination of Deflections

It is often required to satisfy deflection limitations at the working load, and in such a case the use of the fully reduced properties will often give overconservative results. This occurs because the effective section properties reduce progressively, and for thin-walled cross sections these will be greater at working load than at ultimate load. In general the use of the fully reduced section properties overestimates deflections at loads below ultimate, whereas the use of the full section properties underestimates these deflections.

Plastic Bending Capacity

The potential for local buckling and its adverse effects is not always present for cold-formed steel sections. Since material thicknesses of up to 8 mm are covered primarily by codes and greater thicknesses are not precluded, it is possible to have very compact cross sections, even in the case of large members, in which local buckling does not take place. In such cases the potential for fully plastic design cannot be ignored. Local buckling is the major source of impairment of the capacity of a laterally stable member to function adequately in the plastic range, and when this phenomenon is eliminated by virtue of the compactness of the cross section, the member can behave plastically.

For compact elements the cross section not only can withstand the fully plastic moment but also can provide sufficient rotation capacity at the point of maximum moment to allow plastic redistribution of the moments in statically indeterminate beams. The limiting width-to-thickness ratios for compression elements of plastic cross section are specified in codes. Sections whose compression elements have b/t ratios less than the limiting values may be designed using the principle of plastic analysis, providing various other qualifying features are complied with.

The other qualifications that must be complied with are as follows:

1. The member is laterally stable.
2. The virgin yield strength of the material is used, and enhanced yield due to cold-forming effects is neglected.

3. The depth-to-thickness ratio of the compression portion of the web is less than the value specified in the codes.
4. The maximum shear force is less than the value limited by the code.
5. The angle between any web and the loading plane does not exceed 20°.
6. The ratio of ultimate to yield strength is at least 1.08, and the total elongation at failure in a tensile test is not less than 10% over a 2 in. gauge length.

These qualifications are imposed largely on the basis of engineering judgment to avoid any possibility of underdesign through the use of plastic analysis.

Web Crushing

An important effect that must be avoided in the use of cold-formed steel beams is local crushing at support points or points of concentrated load. The thinness of the web material makes cold-formed sections susceptible to such behavior if they are supported directly on the bottom elements over a short support length. Web crushing is characterized by localized buckling in the immediate vicinity of the concentrated load or support point as illustrated in Fig. 47.23. This type of buckling signifies the limit of the load capacity of a beam and must be avoided.

In the most commonly used cold-formed beams—roof purlins—web crushing is avoided by the use of cleats, which support the beam using bolts fixed through the web, thus eliminating the high compressive stresses that would be incurred if the beam were supported through its bottom flange. The use of cleats is illustrated in Fig. 47.24 and is a most effective way of overcoming the problem of web crushing.

If cleats are not to be used, the main method of ensuring that web crushing does not occur is to make the length of support sufficiently large to avoid the possibility. The capacity of a beam web to withstand concentrated loading is dependent on the web D/t ratio, the material yield strength, the length over which the load or support takes place, the corner radius of the supported flange, the web angle, the general geometry of the cross section, and the position of the load or support point on the member.

If concentrated loads are applied close to the ends of a member, the capacity of the web to resist these loads is less than it would be for loads applied far from the ends, since it is easier for the web to buckle out of plane if it has material only on one side of the support to resist buckling. In BS 5950: Part 5 [British Standards Institution, 1987] the rules governing web crushing were adapted from the 1980 AISI Specification, which is based largely on tests carried out at Cornell University [Winter and Pian, 1946; Zetlin, 1955], with refinements produced by more recent testing at the University of Missouri—Rolla [Hetrakul and Yu, 1980]. A more detailed consideration of the web crushing problem and the setup of the AISI design rules is given in Yu [1991].

Shear in Webs

The primary functions of webs are to keep the flanges apart and to carry the shear loadings. It is necessary for safe design to ensure that the shear stresses in the webs do not become unacceptably

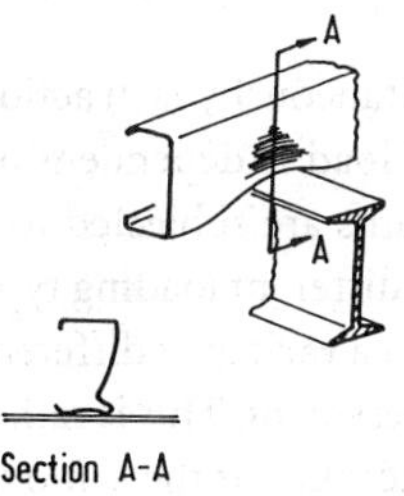

FIGURE 47.23 Web crushing at support.

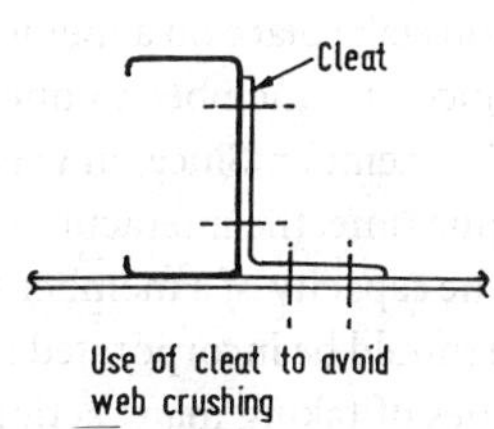

FIGURE 47.24 Use of cleat to avoid web crushing.

large. In thin webs there are two potential sources of danger regarding the behavior of the web in shear. The first is the possibility of shear stresses inducing yield of the material, and the second is the possibility of shear buckling in the web.

Material Yielding in Shear

With regard to material yielding, the von Mises yield criterion predicts that in the case of a material under pure shear, yielding will occur when the shear stress reaches 0.577 times the yield stress in simple tension. In AISI, with a **safety factor** equal to 1.44, shear stresses equal to 0.4 times the yield stress are allowed at the strength limit state. In determining the shear stress limitations with regard to yield resistance, two different provisions are given, one applying to the maximum shear stress in the web and the other applying to the average shear stress in the web. The average shear stress in the web is obtained by dividing the shear force by the web area. The use of the average shear stress in design calculations is simple and expedient, but it should be borne in mind that the shear stress is not normally constant across the web, and the maximum shear stress should also be checked.

Web Buckling Due to Shear

In short deep beams with thin webs, as illustrated in Fig. 47.25, local buckling due to shear becomes a potential problem. Under shear loading the form of the local buckles is rather different from that produced by direct stresses. The main difference lies in the fact that shear buckles are oriented at some angle to the axis of the web, as indicated in the figure. The degree of orientation depends on the relative magnitudes of the direct stresses and shear stresses in the webs, and becomes a maximum of 45° when only shear is present. This type of buckling has been investigated by many researchers [Rockey, 1967; Allen and Bulson, 1980]. After buckling, the web can withstand further loading due to tensile stresses that arise to resist shear deformation. This resistance is known as **tension-field action**, and for hot-rolled sections tension field action may be used in design to improve the design capacity.

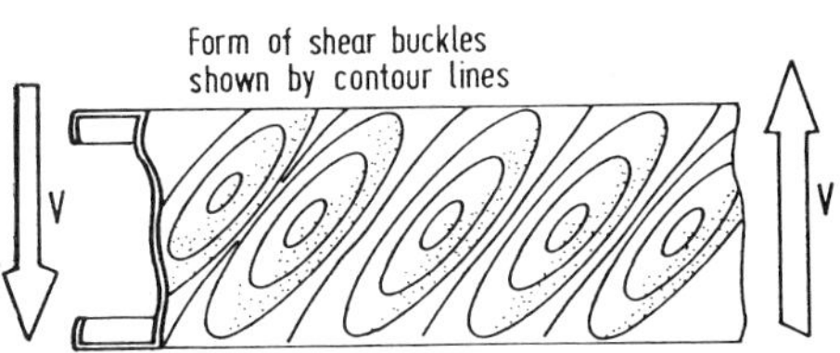

FIGURE 47.25 Shear buckling of thin web.

In the case of cold-formed steel sections the variety of possible sections that must be covered by the design rules precludes the use of tension-field action in the general case in the light of present knowledge, and shear buckling is taken as the limiting factor. However, in view of the underlying sources of increased safety, the reductions in buckling resistance (such as those due to imperfections) that are taken into account for other forms of buckling are disregarded in the case of web buckling due to shear. In determining the shear buckling resistance, the worst case of shear on a web is considered.

Combined Effects

When different load actions take place on a member simultaneously, each action affects the general behavior, and the resistance of a member to one type of load is dependent on the magnitude of all the load actions on the member. Since, in general, beams are subjected to shear, bending, and support loading at the same time, the interactions of each different loading type should be checked out. Ideally, in assessing the capacity of a member subject to a variety of different actions, all actions that contribute to failure should be incorporated in the assessment. This is rather difficult, however, because of the complexities of taking many actions into account at the same time, and in practice only the main actions are included in the interaction equations. Design codes give interaction equations dealing with the combinations of bending and web crushing and web crushing and shear.

Combined Bending and Web Crushing

Since the web crushing provisions were adapted from the AISI Specification, the rules in BS 5950: Part 5 governing the load capacity of beams under combined bending and web crushing were naturally taken from the same source. The AISI rules were based on a series of tests carried out at the University of Missouri—Rolla [Hetrakul and Yu, 1980]. Two different interaction formulas are given in BS 5950: Part 5: one for single-thickness webs and the other for I beams made from channels connected back to back.

For single-thickness webs the relevant interaction equation is

$$1.2\left(\frac{F_w}{P_w}\right) + \left(\frac{M}{M_c}\right) \leq 1.5 \tag{47.25}$$

For I beams, or for any section in which the web is provided with a high degree of rotational restraint at its junction with the flange,

$$1.1\left(\frac{F_w}{P_w}\right) + \left(\frac{M}{M_c}\right) = 1.5 \tag{47.26}$$

In both of these equations F_w is the concentrated web load or reaction; M is the applied bending moment at the point of application of the web load; P_w and M_c are the web crushing capacity and moment capacity of the member. These equations are, of course, subject to the overriding conditions that P cannot be greater than P_w and M cannot be greater than M_c. The interaction diagrams are shown in Fig. 47.26(a), which indicates that single-thickness webs are considered to be affected to a somewhat greater extent by the combination of effects than are I beam webs.

Combined Bending and Shear

The interaction of shear force and bending is covered in BS 5950: Part 5 by the equation

$$\left(\frac{F_v}{P_v}\right)^2 + \left(\frac{M}{M_c}\right)^2 \leq 1 \tag{47.27}$$

where F_v and P_v are, respectively the shear force and shear capacity. This equation is illustrated in Figure 47.26(b) and is again the same as that used in the AISI Specification. The AISI Specification also has further provisions for webs fitted with transverse stiffeners at the load points.

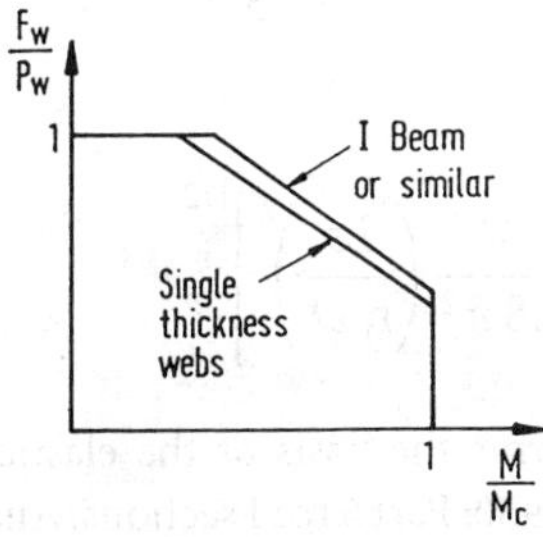

FIGURE 47.26 (a) Interaction diagram for combined web crushing and shear.

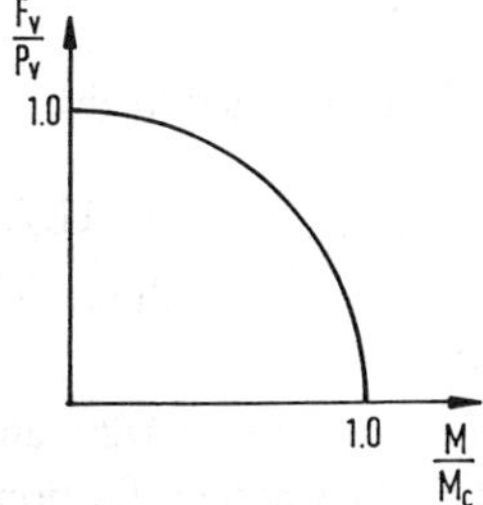

FIGURE 47.26 (b) Interaction diagram for combined bending and shear.

Lateral Buckling

Lateral buckling, sometimes called lateral-torsional buckling, generally occurs when a beam that is bent about its major axis develops a tendency to displace laterally (i.e., perpendicularly to the direction of loading) and twist. Many, if not most, beams used in cold-formed construction are restrained against lateral movement—in many cases continuously restrained by roof or wall **cladding**. In other cases restraint is afforded by other members connected to the beam in question or by bracing, such as antisag bars. Such restraints reduce the potentiality of lateral buckling but do not necessarily eliminate the problem. For example, roof purlins are generally restrained against lateral displacement by the cladding, but under wind uplift, which induces compression in the unrestrained flange, lateral buckling is still a common cause of failure. This occurs because of the flexibility of the restraining cladding and the distortional flexibility of the purlin itself, which permits lateral movement to occur in the compression flange even if the other flange is supported.

A further point that should also be noted is that, contrary to the statement made in BS 5950: Part 5, it is not a necessary condition for lateral torsional buckling that bending take place about the major axis. In some cases beams that are bent about the minor axis may undergo this type of buckling behavior. In general, lateral torsional buckling is closely related to torsional **flexural buckling** in columns. Any cross section that is susceptible to torsional flexural buckling may also have lateral buckling tendencies.

Elastic Lateral Buckling Resistance Moment

In the case of an I beam, theoretical analysis [Allen and Bulson, 1980] shows that the elastic critical moment M_E, for such a beam of length L bent in the plane of the web is given by the expression

$$M_E = \frac{\pi}{L}\left[\frac{EI_1}{\gamma}\left(GJ + EC_w\frac{\pi^2}{L^2}\right)\right]^{1/2} \tag{47.28}$$

where I_1 is the second moment of area about an axis through the web, C_w is the warping constant, G is the shear modulus, J is the torsion constant, and $\gamma = 1 - I_1/I_2$, I_2 being the second moment of area about the neutral axis perpendicular to the web.

Using the relationships

$$I_1 \approx \frac{B^3t}{6} = Ar_y^2 \qquad C_2 \approx I_1 \times \frac{D^2}{4}$$

where B is the flange width, D the beam depth, A the area of cross section, r_y the radius of gyration about y axis, and

$$G = \frac{E}{2(1+\nu)} = \frac{E}{2.6} \qquad J = \frac{At^2}{3}$$

then Eq. (47.28) can be rearranged to give

$$M_E = \frac{\pi^2 AED}{2(L/r_y)^2}\left[1 + \frac{4}{7.8\pi^2}\left(\frac{Lt}{r_yD}\right)^2\right]^{1/2} \tag{47.29}$$

The term $4/7.8\pi^2$ is very close to 1/20, and this forms the basis of the elastic lateral buckling resistance moment used in AISI Specification and BS 5950: Part 5 for I sections. Analysis of channels gives very similar results, and these can be dealt with using the same equation. In this equation the **effective length** L_E is used instead of L, and a coefficient C_b, which accounts for the variation in moment along a beam, is also incorporated.

In the case of Z section beams, it is rather difficult to envisage such beams being used completely unrestrained against lateral movement, as the unsymmetrical behavior would make them highly flexible. The vast majority of Z sections are used as purlins, with a high degree of lateral restraint due to roof cladding; even types of cladding that are classed as nonrestraining offer sufficient restraint to enable these purlins to function more or less as laterally braced members if lateral buckling is not considered. In the case of Z sections that are not restrained laterally or have very light restraint, the lateral buckling resistance is taken in AISI Specification and BS 5950: Part 5 as half of that calculated for a channel or I section. This recommendation is based on tests in the U.S. by Winter [1947].

Variation in Moment along a Beam

The coefficient C_b is used to take account of the variation in moment along a beam. Without this coefficient the buckling resistance is calculated on the basis of a uniform moment acting all along the beam, which is a most severe condition. If the moment varies along the beam, then the maximum moment to cause lateral buckling will be greater than that analyzed on the basis of pure moment, and this is taken into account by the C_b factor.

C_b acts as a multiplying factor, and if the elastic lateral buckling moment derived for pure bending is multiplied by this factor, the resulting values of M_E become good approximations to the elastic lateral buckling moments for the case of linearly varying bending moment along a beam. These coefficients were derived on the basis of a linearly varying bending moment distribution, but within the limitations they may be used also in the case of nonlinearly varying moments.

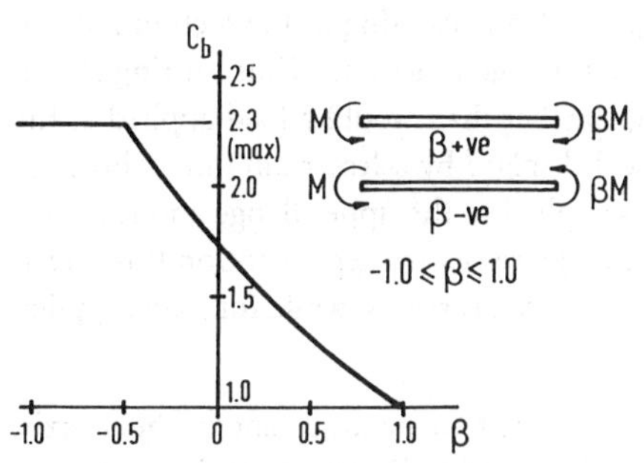

FIGURE 47.27 Variation of C_b factor with distribution of moment over the span.

The C_b factors used in AISI Specification and BS 5950: Part 5 are, with reference to Fig. 47.27,

$$C_b = 1.75 - 1.05\beta + 0.3\beta^2 \leq 2.3 \quad (47.30)$$

in which β is the ratio of the end moments.

If the maximum moment within the beam span between supports is less than the larger of the end moments, Eq. (47.30) can be used to determine C_b. If the maximum moment within the span is greater than the larger end moment, C_b must be taken as unity.

Effective Lengths

The restraints afforded by the supports can have a substantial effect on the lateral buckling resistance of beams. The expressions given for M_E assume that no resistance to warping is afforded by the support. If the supports can resist torsion, however, increases in buckling resistance can be obtained. These increases in buckling resistance are derived using the well-known effective length concept, in which it is assumed that the beam has an effective length different from its actual length for the purpose of determining buckling resistance.

In estimating the effective length with regard to lateral buckling, the engineer is required to exercise a degree of judgment. The effective lengths are directly affected by the degree of restraint on rotation of the beam at the supports or bracing points, and the rotations that require examination occur about three perpendicular axes, as shown in Fig. 47.28. Some assessment must be made regarding the degree of restraint afforded by the support about each axis. If it is considered that no restraint is provided against rotation about any axis, then for safe design the effective length should be taken as 1.1 times the actual span between supports or bracing members.

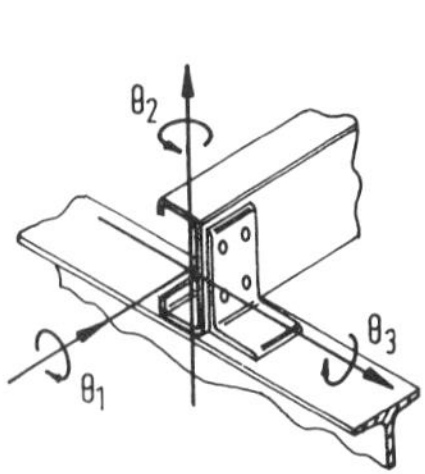

FIGURE 47.28 Support conditions with regard to effective length determination.

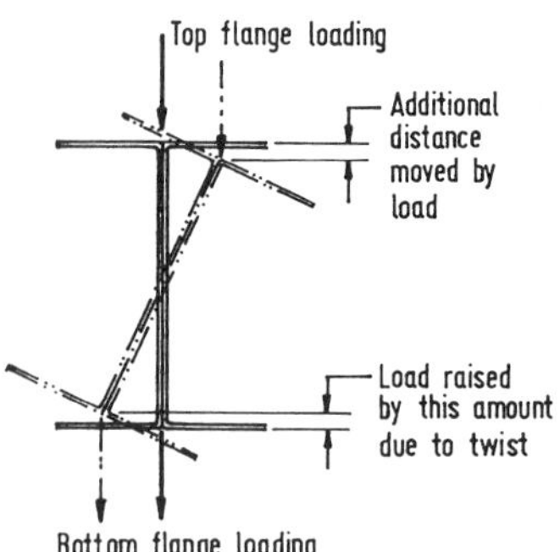

FIGURE 47.29 Stabilizing and destabilizing loads.

Destabilizing Loads

The elastic buckling resistance moment was determined initially on the basis of pure moment loading on simply supported beams. This was then modified to take account of moment variation via the C_b factors and to take account of the support restraints via the use of effective lengths. One further factor that must be taken into account concerns the position of the loading on the cross section. If we consider the I section shown in Fig. 47.29, any twisting of the section reduces the vertical distance between the shear center and the web flange junctions. Thus, during lateral buckling a load applied to the upper flange will displace farther than would a load applied at the shear center while a load applied to the lower flange will displace by a lesser amount. Thus, the work done by the load during buckling is greatest if it is applied to the upper flange and least if it is applied to the lower flange, and the values of the buckling stresses are dependent on this effect. For loads applied above the shear center the buckling resistance decreases, while for loads applied below the shear center the buckling resistance increases.

Example 47.3. Figure 47.30(a) shows a hat section subjected to bending about the x axis. Assuming the yield point of steel is 50 ksi, determine the allowable bending moment in accordance with AISI Specifications.

Solution. *Calculation of sectional properties:* Midline dimensions shown in Fig. 47.30(b) are used for calculation.

$$R' = R + t/2 = 0.240 \text{ in.}$$

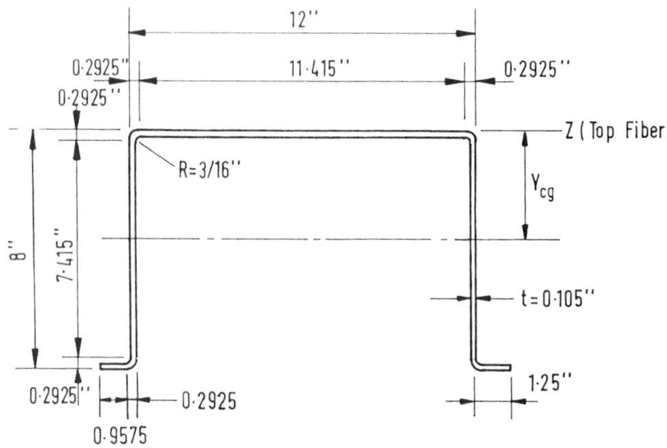

FIGURE 47.30 Example 47.3. (a) Hat section.

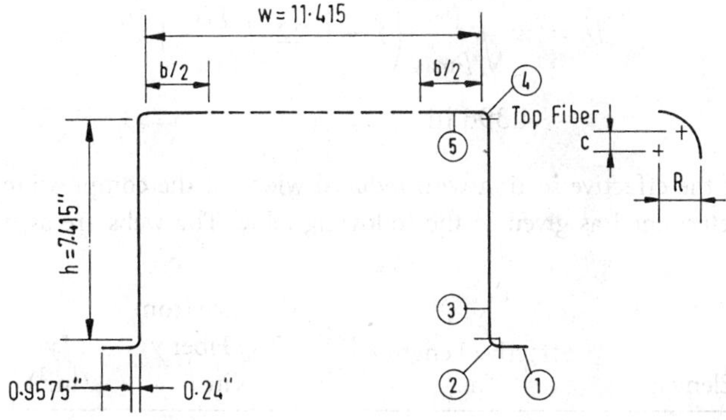

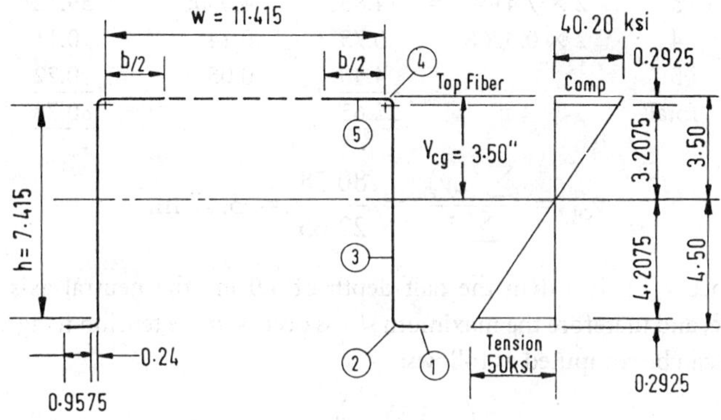

FIGURE 47.30 (continued) (b) Line elements. (c) Effective width and stress distribution using fully effective web.

Arc length of the corner element:

$$L = 1.57R' = 0.3768 \text{ in.}$$

$$c = 0.637R' = 0.1529 \text{ in.}$$

Location of neutral axis, first approximation: For the compression flange,

$$w = 12 - 2(R + t) = 11.415 \text{ in.}$$

$$\frac{w}{t} = 108.71$$

$$p_{cr} = \frac{4\pi^2 E}{12(1 - \nu^2)} \frac{1}{(w/t)^2}$$

$$= \frac{4 \times \pi^2 \times 29,500}{12 \times (1 - 0.3^2)} \frac{1}{(108.71)^2} = 9.02 \text{ ksi}$$

$$p_{max} = 50 \text{ ksi}$$

$$\sqrt{\frac{p_{max}}{p_{cr}}} = \sqrt{\frac{50}{9.02}} = 2.35 > 0.673$$

$$b = w\sqrt{\frac{p_{cr}}{p_{max}}}\left(1 - 0.22\sqrt{\frac{p_{cr}}{p_{max}}}\right)$$
$$= 4.396 \text{ in.}$$

The neutral axis of the effective section with reduced width of the compression flange equal to 4.396 in. can be determined as given in the following table. The webs are assumed to be fully effective.

Element	Effective Length L (in.)			Distance from Top Fiber y (in.)	Ly (in.^{2})
1	2 × 0.9575	=	1.92	7.95	15.23
2	2 × 0.3768	=	0.75	7.86	5.90
3	2 × 7.4150	=	14.83	4.0000	59.32
4	2 × 0.3768	=	0.75	0.14	0.11
5			4.40	0.05	0.22
Total			22.65		80.78

$$y_{cg} = \frac{\sum(Ly)}{\sum L} = \frac{80.78}{22.65} = 3.57 \text{ in.}$$

Because the distance y_{cg} is less than the half-depth of 4.0 in., the neutral axis is closer to the compression flange, and therefore the maximum stress occurs in the tension flange. The maximum compressive stress can be computed as follows:

$$f = 50\left(\frac{3.57}{8 - 3.57}\right) = 40.29 \text{ ksi}$$

Since this stress is less than the assumed value, another trial is required.

Second approximation: Assuming that $f = 40.29$ ksi,

$$p_{max} = 40.29 \text{ ksi}$$

$$\sqrt{\frac{p_{max}}{p_{cr}}} = \sqrt{\frac{40.29}{9.02}} = 2.11 > 0.673$$

$$b = w\sqrt{\frac{p_{cr}}{p_{max}}}\left(1 - 0.22\sqrt{\frac{p_{cr}}{p_{max}}}\right) = 4.842 \text{ in.}$$

Element	Effective Length L (in.)	Distance from Top Fiber y (in.)	Ly (in.^{2})	Ly^2 (in.^{3})
1	1.92	7.95	15.23	121.08
2	0.75	7.86	5.9	46.37
3	14.83	4.0000	59.32	237.28
4	0.75	0.14	0.11	−0.02
5	4.84	0.05	0.24	−0.01
Total	23.09		80.80	404.76

$$y_{cg} = \frac{80.8}{23.09} = 3.5 \text{ in.}$$

Check the effectiveness of the web: Using Section B2.3 of the AISI Specification the effectiveness of the web element can be checked as follows. From Fig. 47.30(c),

$$f_1 = 50(3.2075/4.50) = 35.64 \text{ ksi (compression)}$$

$$f_2 = -50(4.2075/4.50) = -46.75 \text{ (tension)}$$

$$\psi = f_2/f_1 = -1.312$$

$$k = 4 + 2(1-\psi)^3 + 2(1-\psi)$$

$$= 33.341$$

$$\frac{h}{t} = \frac{7.415}{0.105} = 70.62 < 200 \qquad \text{OK}$$

$$\lambda = \frac{1.052}{\sqrt{33.341}}(70.62)\sqrt{\frac{35.64}{29500}}$$

$$= 0.447 < 0.673$$

$$b_e = h = 7.415 \text{ in.}$$

$$b_1 = b_e/(3-\psi) = 1.72 \text{ in.}$$

Since $\psi < -0.236$,

$$b_2 = b_e/2 = 3.7075 \text{ in.}$$

$$b_1 + b_2 = 5.4275 \text{ in.}$$

Because the computed value of $(b_1 + b_2)$ is greater than the compression portion of the web (3.2075 in.), the web element is fully effective.

Moment of inertia and section modulus: The moment of inertia based on line elements is

$$2I_3' = 2\left(\frac{1}{12}\right)(7.415)^3 = 67.95$$

$$\sum(Ly^2) = 404.76$$

$$I_z' = 67.95 + 404.76 = 472.71 \text{ in.}^3$$

$$-\left(\sum L\right)(y_{cg})^2 = 23.0942(3.5)^2 = 282.9 \text{ in.}^3$$

$$I_x' = 189.81 \text{ in.}^3$$

The actual moment of inertia is

$$I_x = I_x't = 189.81(0.105) = 19.93 \text{ in.}^4$$

The section modulus relative to the extreme tension fiber is

$$S_x = 19.93/4.50 = 4.43 \text{ in.}^3$$

Nominal and allowable moments: The nominal moment for section strength is

$$M_n = S_eF_y = S_xF_y = (4.43)(50) = 221.50 \text{ in.-kips}$$

The allowable moment is

$$M_a = M_n/\Omega_f = 221.50/1.67 = 132.63 \text{ in.-kips}$$

47.4 Members Subject to Axial Load

Axial loading is a very common and very important type of loading, and the requirements to deal with this type of loading in cold-formed steel members vary according to the type of loading—tension or compression—and the geometry and use of the member. Due to the thinness of the walls in cold-formed steel sections and the variety of different cross-sectional shapes that can be produced, types of behavior not commonly found in traditional hot-rolled members can occur, and these must be recognized and taken into account in design. Codes provide design methods to deal with the various phenomena associated with thin-walled sections in a fairly simple manner.

Short Struts

Local buckling must be taken into consideration in the analysis of members in compression. We have seen in previous chapters how individual elements are dealt with in this regard. In the case of complete sections subjected to compression, we must take into account the possibility of local buckling in all elements of a cross section. We initially consider a short length of member that is acted upon by compressive loads as shown in Fig. 47.31.

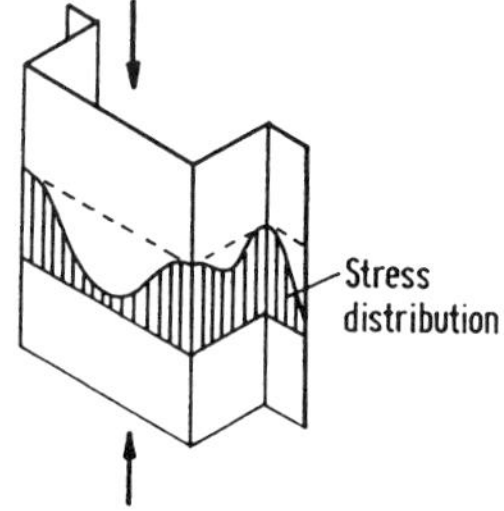

FIGURE 47.31 Short length of compressed member.

Due to the compressive loads, each element of the cross section can suffer local buckling. We therefore consider each flat element in turn; find the effective width, and hence effective area, of the element; and sum these, together with the areas of the corners, to obtain the total effective area A_{eff}. The ratio of the effective cross-sectional area to the full cross-sectional area A is denoted as Q:

$$Q = \frac{A_{\text{eff}}}{A} \tag{47.31}$$

The factor Q was adopted partly because it described the actual situation in a cross section (e.g., effective and ineffective portions) more realistically.

The load capacity of a short strut under uniform compression is given by the product of the effective area (A_{eff}) and the yield stress (Y_s):

$$P_{cs} = QAY_s \tag{47.32}$$

Flexural Buckling

Euler Buckling

We have seen how short, uniformly compressed members behave and how the effects of local buckling must be taken into account in design analysis. For long members under compression different modes of failure arise, due to overall buckling. We first of all consider buckling due to flexure, or *Euler buckling*. Euler buckling occurs when a long slender member (i.e., a column), is compressed. The elastic buckling load, or Euler load, for such a column under pin-end conditions is very well known as

$$P_E = \frac{\pi^2 EI}{\ell^2} \tag{47.33}$$

where I is the relevant second moment of area, E is the elasticity modulus, and ℓ is the column length.

By writing $I = Ar^2$, where r is the radius of gyration of the cross section corresponding to I, Eq. (47.33) can be put in terms of the critical, or Euler buckling, stress p_E as follows:

$$p_E = \frac{\pi^2 E}{(\ell/r)^2} \tag{47.34}$$

As the length of the column increases, the critical stress to cause Euler buckling decreases, so that for a very long column, Euler buckling occurs at extremely low stress levels. In the case of local buckling, we have seen that the local buckling stress is relatively unaffected by length. Thus, for long columns the effects of local buckling do not arise, and in determining the Euler load for such a column we do not need to take local buckling into account.

Effective Lengths

If the ends of a column are not pinned but subject to some other degree of fixity, Eq. (47.33) and (47.34) do not apply directly but must be modified to take the actual end conditions into account. In design, this is very often accomplished using the effective length concept, in which the actual column length L is replaced by an effective length L_E in these equations. The effective length of a column is normally taken as the distance between the points of contraflexure in a buckled column. Values for the effective length as a proportion of the actual length between supports are given in AISI Specification and BS 5950: Part 5 for a number of conditions of column support.

The ratio of effective length to the relevant radius of gyration of a column is termed the *slenderness ratio*. Maximum permitted values of the slenderness ratios of columns are given in Codes for different types of member. For members that normally act in tension but may be subject to load reversal due to the action of wind, high slenderness ratios are permitted. For members subjected to loads other than wind loads, the maximum slenderness ratio is given as 180 in BS 5950, and AISI stipulates a maximum value of 200.

In the design analysis of columns the complete range of slenderness ratios must be accounted for. We have seen that for short columns local buckling is important and Euler buckling is of little consequence, while for long columns Euler buckling assumes the highest significance and local buckling has little effect. For short members that are fully effective, failure occurs when the load reaches the *squash load* $Y_s \times A$. If local buckling is present, this load is modified, due to the local buckling effects, to $Y_s \times A_{\text{eff}}$, or QY_sA. If the slenderness ratio is greater than a fixed value, Euler buckling occurs, and the failure load decreases with increase in slenderness ratio.

Real columns are, of course, not perfect, and column imperfections cause some bending to occur even in very short members, thus hastening yield in these members and causing failure at loads less than the Euler load. It is imperative that the effects of imperfections are accounted for in the design analysis.

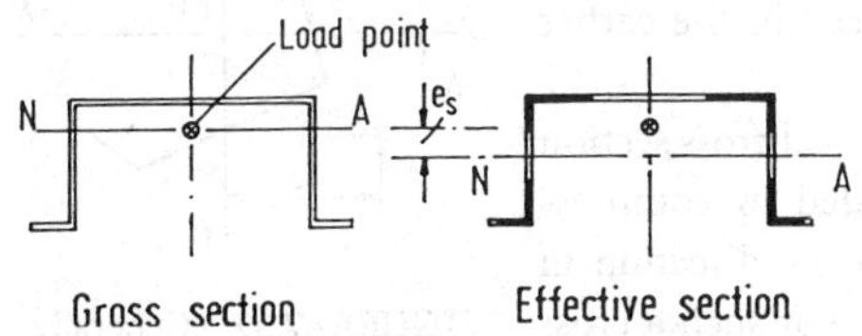

FIGURE 47.32 Neutral axis shift for locally buckled cross section.

Effects of Neutral Axis Shift

If we examine the gross cross section and the effective cross section together, as illustrated in Fig. 47.32, we can see that the effects of local buckling have been not only to alter the column's effective area but also to change the geometry, since some elements have become more ineffective than others. Because of this, the neutral axis of the effective cross section moves from its original position as local buckling progresses. If the loading is applied at the centroid of the full cross section, then it becomes eccentric to the centroid of the effective cross section, thus inducing bending in the member.

It is therefore evident that any section that is not doubly symmetric, and is subject to loads that induce local buckling effects, is likely to incur bending in addition to axial load if the loading is applied through its centroid. The degree of bending incurred depends on the distance by which the effective neutral axis is displaced from its initial position, and this in turn depends on the degree of local buckling undergone by the member. Since this bending has the effect of reducing the column load capacity, and since the magnitude of the neutral axis shift increases with load, it should make for conservative estimates of load capacity if the neutral axis shift is determined on the basis of the short strut load P_{cs}.

If the neutral axis of the effective section is displaced by an amount e_s from that of the gross cross section, the moment produced by a load applied through the original neutral axis is the product of load P and displacement e_s. To take the combination of axial load and moment into account a simple linear interaction formula is used of the form

$$\frac{M}{M_c} + \frac{P}{P_c} = 1 \tag{47.35}$$

where M_c is the moment capacity in the absence of axial load, determined as illustrated in the previous chapter, and P_c is the failure load of the column under uniform compression. At the ultimate load of the member P_c', the moment acting is $P_c' \times e_s$. Equation (47.34) becomes

$$\frac{P_c' e_s}{M_c} + \frac{P_c'}{P_c} = 1 \tag{47.36}$$

The full effects of neutral axis shift will not be incurred in practice for columns that are not, in fact, pin-ended. If the effective length of a column is less than the full length between supports, then any accurate assessment of the effects of neutral axis shift is complex, and there is as yet no satisfactory solution to this question. Experimental results suggest that for completely fixed ends the effects of neutral axis shift may be completely neglected in assessing the column capacity.

Torsional-Flexural Buckling

Theoretical Basis

Apart from local buckling, perhaps the major difference in behavior between hot-rolled steel and cold-formed steel structural members is to be found in the relative susceptibility of the latter to **torsional-flexural buckling**. Designers in hot-rolled steel do not come across this phenomenon to a great extent, partly because hot-rolled steel sections are generally thicker and more compact than cold-formed steel sections, but more generally because of the greater variety of sectional shapes that are designed in cold-formed steel. When dealing with members of arbitrary cross section, a more general theoretical approach must be adopted than that used in the earlier sections of this chapter.

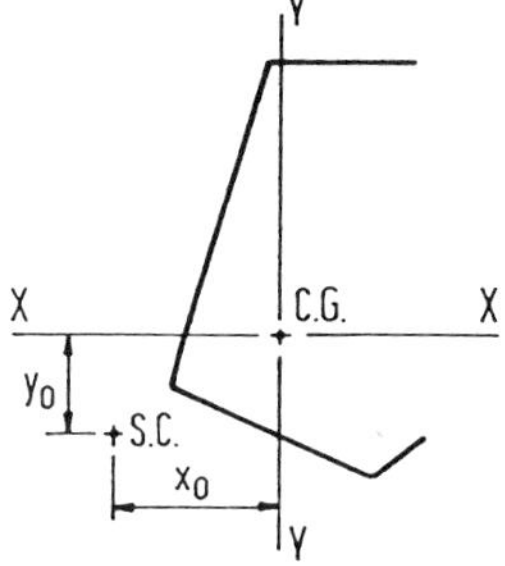

FIGURE 47.33 Generally unsymmetrical cross section.

Consider a member having a generally unsymmetrical cross section as depicted in Fig. 47.33. If this member is loaded in compression, it is not possible to determine by inspection the direction in which the cross section will move during buckling. For such a cross section, on the basis of classical theory, which is detailed in Murray [1984] and Allen and Bulson [1980], the deflections of the member will have components in the x and y directions, and twisting will also occur about the shear center, or center of twist. Indeed, if precise analysis of the situation were to be carried out, it would be found that distortion, or change in shape of the cross section, is also a distinct possibility in thin-walled sections, but this complicates the analysis considerably.

The application of classical theory to cold-formed steel sections has been researched extensively at Cornell University in the U.S. [Chajes and Winter, 1965 Peköz and Celebi, 1969], and Yu [1985] gives a thorough summary of the design approach based on this work. If we consider that the section of Fig. 47.33 is loaded through its centroid, and axes x–x and y–y are the principal axes, then the buckling load, which in the general case is due to a combination of bi-axial flexural load and twisting and is thus denoted P_{TF}, may be obtained from the following equation

$$\frac{I_c}{A}(P_{TF} - P_{EY})(P_{TF} - P_{EX})(P_{TF} - P_T) - P_{TF}^2\{y_o^2(P_{TF} - P_{EX}) + x_o^2(P_{TF} - P_{EY})\} = 0 \tag{47.37}$$

In this equation I_c is the polar second moment of area with respect to the shear center of the section, P_{EX} and P_{EY} are the critical loads for buckling about the x and y axes, respectively, and P_T is the torsional buckling load. The dimensions x_o and y_o are the distances between the centroid of the section and its shear center measured in the x and y directions, respectively. The smallest root of the equation gives the value of P_{TF} of interest, and this is always less than or equal to the smallest value of the individual critical loads.

If the member has simple support conditions, as normally defined, at its ends, then P_{EX} and P_{EY} are simply the Euler loads for buckling about the x–x and y–y axes, respectively. The torsional buckling load P_T, however, is not fully described by the commonly accepted "simple support" conditions, and closer examination of the support conditions must be carried out to define this load.

P_T is defined by the following equation:

$$P_T = \frac{1}{r_o^2}\left(GJ + k\pi^2 E\frac{C_w}{L_E^2}\right) \tag{47.38}$$

where G is the shear modulus for the material, J is the torsion constant for the section, C_w is the *warping constant* for the section, and r_o is the polar radius of gyration about the shear center, given by

$$r_o = (r_x^2 + r_y^2 + x_o^2 + y_o^2)^{1/2} \tag{47.39}$$

where r_x and r_y are radii of gyration corresponding to the x and y axes.

The constant k depends on the degree of warping restraint afforded by the end connections of the column. If warping is completely unrestrained, then $k = 1$. There is, as can be seen, an analogy with the Euler load for flexural buckling, in which a fully fixed column has a buckling load four times as great as the corresponding simply supported column.

Warping Restraint

We can see that a column that is nominally simply supported, or simply supported with regard to flexural buckling, may exhibit a wide range of variation in buckling load under the more general torsional-flexural buckling situation. If the walls of such a column are very thin, then for column lengths of commercial applicability, the torsion constant J, which is equal to $\Sigma bt^3/3$, can become very small in comparison with C_w/L_E^2, and the degree of warping restraint becomes very important.

The degree of warping restraint by different types of end connections is not easy to quantify, as this depends on a wide range of factors. However, warping of the ends is often prevented by the end plates, and the effects of torsional-flexural buckling are often minimized. This assumption of full warping restraint is therefore optimistic for the general case.

In the AISI Specification and in the various design codes that are based on it, the opposite view is taken. Here it is assumed that for safe design no account should be taken of the effects of connections on warping restraint, and so torsional-flexural buckling is based on $k = 1$. This gives much lower design loads for many cases and results in torsional-flexural buckling being the governing design criterion for a much wider range of section shapes than is otherwise the case.

Members under Combined Bending and Compression

Under practical conditions structural members are very often subjected to combinations of bending and axial loading. The interaction of the different effects must be taken into account when this occurs. When the axial loading is compressive, the different types of buckling that can occur for beams and columns may interact with each other, and this must be guarded against. Even in the case of members subjected to combinations of bending and tension, there are possibilities for buckling of some form to occur, and this must be taken into consideration. In this section, however, we shall consider combinations of compressive forces and bending.

Under hypothetical simplified conditions, in which buckling effects are absent and a member behaves perfectly elastically until failure occurs when the maximum stress reaches yield point, the effects of each different type of loading that acts on a member can simply be added together to produce an equation governing the capacity under simultaneous applications of moment M and axial load P of the form

$$\frac{P}{P_s} + \frac{M}{M_y} = 1 \tag{47.40}$$

where P_s is the load capacity in the absence of moments and M_y is the moment capacity in the absence of axial loads.

If buckling possibilities are negligible, and if the material has any postyield capacity, then this type of equation tends to give conservative results for the load-carrying capacity of the member under combined loadings. If, on the other hand, there is any tendency for the member to undergo buckling, this type of equation can underestimate the degree of interaction and give nonconservative estimates of carrying capacity.

The AISI recommends that the axial force and bending moments shall satisfy the following interaction equations:

$$\frac{P}{P_a} + \frac{C_{mx}M_x}{M_{ax}\alpha_x} + \frac{C_{my}M_y}{M_{ay}\alpha_y} \le 1.0 \tag{47.41}$$

$$\frac{P}{P_{a0}} + \frac{M_x}{M_{ax}} + \frac{M_y}{M_{ay}} \le 1.0 \tag{47.42}$$

When $P/P_a \le 0.15$, the following formula may be used in lieu of the two preceding formulas:

$$\frac{P}{P_a} + \frac{M_x}{M_{ax}} + \frac{M_y}{M_{ay}} \le 1.0 \tag{47.43}$$

where

P = applied axial load

M_x and M_y = applied moments with respect to the centroidal axes of the effective section determined for the axial load alone

P_a = allowable axial load determined in accordance with Sec. C4

P_{a0} = allowable axial load determined in accordance with Sec. C4, with $F_n = F_y$

M_{ax} and M_{ay} = allowable moments about the centroidal axes determined in accordance with Sec. C3

$1/\alpha_x, 1/\alpha_y$ = magnification factors

$= 1/[1 - \Omega_c P/P_{\text{cr}}]$

Ω_c = factor of safety used in determining P_a

$P_{\text{cr}} = \dfrac{\pi^2 E I_b}{(K_b L_b)^2}$

I_b = moment of inertia of the full, unreduced cross section about the axis of bending

L_b = actual unbraced length in the plane of bending

K_b = effective length factor in the plane of bending

C_{mx}, C_{my} = coefficients, whose value shall be taken from the AISI Specifications

BS 5950 considers two different possibilities that should be examined in dealing with the interaction of axial compression and bending.

1. The capacity of the member may be exceeded locally at discrete points at which the maximum loadings occur.
2. The overall capacity of the member to resist buckling due to the action of the combined loadings may be exceeded.

The first possibility is likely to occur in members subjected largely to bending with some additional axial loading. At points of maximum moment, as indicated in Fig. 47.34, the effects of the axial loading on the ultimate condition must be considered. This situation can be adequately covered by a linear interaction equation, and in BS 5950: Part 5 the following expression is used to check the local capacity of a member subjected to axial load F_c and bending moments M_x and M_y about the x and y axes respectively:

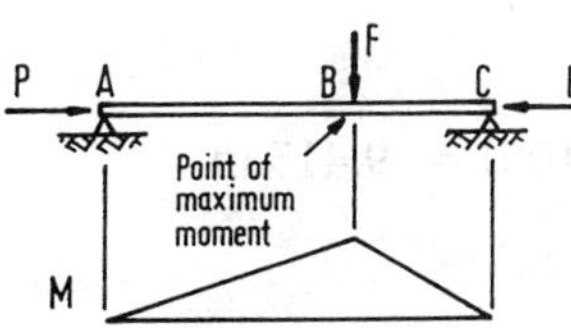

FIGURE 47.34 Moment variation along a member with axial and transverse loading.

$$\frac{F_c}{P_{cs}} + \frac{M_x}{M_{cx}} + \frac{M_y}{M_{cy}} \leq 1 \qquad (47.44)$$

where P_{cs} is the axial load capacity in the absence of moments, and M_{cx} and M_{cy} are the moment capacities if the member is subjected only to bending about the relevant axis in the absence of all other load actions.

This equation should be satisfied at discrete points on the member where the local load or moment magnitudes are at their peak. It is worthy of mention here that this equation takes account of local buckling effects, since the calculated load and moment capacities are determined on the basis of the methods already described in this book.

The second possibility that must be considered is that the overall buckling capacity of the member may be attained by a combination of loads. The possible types of buckling, in addition to local buckling, are flexural buckling or torsional-flexural buckling of members loaded largely in compression, and lateral-torsional buckling of members loaded largely in bending. A suitable overall buckling capacity check should consider all of these possibilities. BS 5950: Part 5 prescribes two interaction equations to deal with the overall buckling capacity check; the particular equation to be used in any given situation depends on whether or not lateral-torsional buckling is a possibility.

Members under Combined Bending and Tension

In the case of members subjected to combinations of bending and tension, there is not in general any need for an overall buckling capacity check, as the application of tensile loads has no tendency to increase the possibility of overall buckling. There are, of course, possibilities for the effects of local buckling to be present in the member, since not all elements of a cross section will be in tension under the combination of loads.

In BS 5950: Part 5, an interaction equation of the same form as Eq. (47.40) is used to check the local capacity at discrete points on a member. The relevant equation is

$$\frac{F_t}{P_t} + \frac{M_x}{M_{cx}} + \frac{M_y}{M_{cy}} \leq 1 \tag{47.45}$$

with

$$\frac{M_x}{M_{cx}} \leq 1 \qquad \frac{M_y}{M_{cy}} \leq 1$$

In this equation F_t and P_t are the applied tensile load and the tensile capacity of the member, respectively. M_{cx} and M_{cy} are the moment capacities computed in the absence of any other loading. Since M_{cx}, for example, is evaluated on the basis of an effective cross section, local buckling is automatically taken into account in the interaction equation.

Example 47.4. Compute the allowable axial load for the square tubular column shown in Fig. 47.35. Assume that the yield point of steel is 40 ksi and $K_x L_x = K_y L_y = 10$ ft.

Solution. The solution given below is based on the AISI Code.

Since the square tube is a doubly symmetric closed section, it will not be subject to torsional-flexural buckling.

Sectional properties of full section:

$$b = 10.00 - 2(R + t) = 10.00 - 2(0.1875 + 0.105) = 9.415 \text{ in.}$$

$$A = 4(9.415 \times 0.105 + 0.0396) = 4.113 \text{ in.}^2$$

$$I_x = I_y = 2(0.105)\left[\frac{1}{12}(9.415)^3 + 9.415\left(5 - \frac{0.105}{2}\right)^2\right]$$
$$+ 4(0.0396)(5.0 - 0.1373)^2$$
$$= 66.75 \text{ in.}^4$$

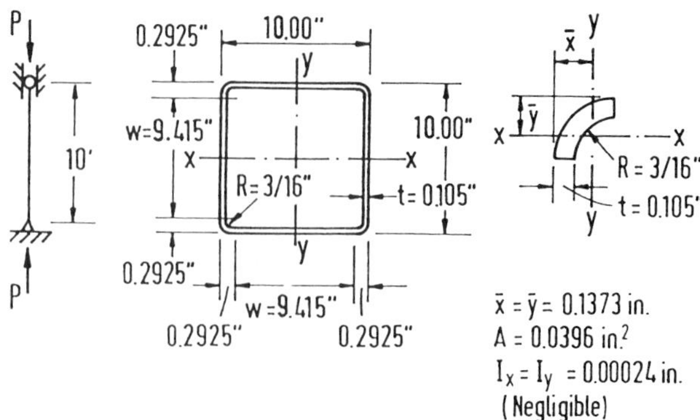

FIGURE 47.35 Column for Example 47.4.

$$r_x = r_y = \sqrt{\frac{I_x}{A}} = \sqrt{\frac{66.75}{4.113}} = 4.029 \text{ in.}$$

Nominal buckling stress, F_n: The elastic flexural buckling stress F_e is computed as follows:

$$\frac{KL}{r} = \frac{10 \times 12}{4.029} = 29.78 < 200 \qquad \text{OK}$$

$$F_e = \frac{\pi^2 E}{(KL/r)^2} = \frac{\pi^2(29500)}{(29.78)^2} = 328.3 \text{ ksi}$$

Since $F_e > F_y/2 = 20$ ksi,

$$\begin{aligned} F_n &= F_y(1 - F_y/4F_e) \\ &= 40[1 - 40/(4 \times 328.3)] \\ &= 38.78 \text{ ksi} \end{aligned}$$

Effective area, A_{eff}: Because the given square tube is composed of four stiffened elements, the effective width of stiffened elements subjected to uniform compression can be computed by using $k = 4.0$:

$$b/t = 9.415/0.105 = 89.67$$

$$p_{\text{cr}} = \frac{4\pi^2 E}{12(1-\nu^2)} \frac{1}{(b/t)^2} = 13.26 \text{ ksi}$$

$$\sqrt{\frac{f_{\max}}{p_{\text{cr}}}} = \sqrt{\frac{40}{13.26}} = 1.737$$

Since $\sqrt{f_{\max}/p_{\text{cr}}} > 0.637$,

$$\begin{aligned} b_{\text{eff}} &= b\sqrt{\frac{p_{\text{cr}}}{f_{\max}}}\left(1 - 0.22\sqrt{\frac{p_{\text{cr}}}{f_{\max}}}\right) \\ &= 4.73 \text{ in.} \end{aligned}$$

The effective area is $(4.73 \times 0.105 + 0.0396) = 2.145 \text{ in.}^2$

Nominal and allowable loads:

$$P_n = A_e F_n = (2.145)(38.78) = 83.18 \text{ kips}$$

$$P_a = P_n/\Omega_c = 83.18/1.92 = 43.32 \text{ kips}$$

$\Omega_c = 1.92$ is used because the section is not fully effective.

47.5 Connections for Cold-Formed Steelwork

All of the connection methods applicable to hot-rolled sections, such as bolting and welding, are also applicable to cold-formed steel sections at the thicker end of the range. In the case of thinner cold-formed steel sections, an extremely wide assortment of proprietary fasteners and fastening techniques exists. This wide range raises problems in setting realistic and reliable approaches to defining connection strength and the connection properties are generally evaluated on the basis of testing.

Types of Fasteners

Davies [Rhodes, 1991] listed fasteners in three main groupings, as shown in Table 47.2. The selection of the most suitable type of fastener for a given application is governed by several factors:

1. Load-bearing requirements: strength, stiffness, deformation capacity
2. Economic requirements: number of fasteners required, cost of labor and materials, skill required in fabrication, design life, maintenance, ability to be dismantled
3. Durability: resistance to aggressive environments
4. Watertightness
5. Appearance

Although structural engineers tend to be primarily concerned with the most economic way of meeting the load-bearing requirements, in many applications other factors may be equally important.

Bolts

The use of bolts to connect cold-formed steel components is similar to the practice in hot-rolled construction, because of the thinness of the material and the relatively small size of the components. However, the main design considerations tend to be end distance and bearing. With regard to bearing, it is worthy of note that hot-rolled steel design codes tend to have significantly different design treatments of this than cold-formed steel codes. This may be partially due to the different behavior of thin material, but it is mainly due to the adoption of different philosophies by the writers of cold-formed steel and hot-rolled steel codes. In British codes in particular, the cold-formed steel rules are based on strength design, while the hot-rolled steel rules are based on limiting slip. Table 47.3, from Davies [Rhodes, 1991] shows the comparison between permissible bearing stresses in bolted connections according to various codes.

Self-Tapping Screws

Self-tapping screws fall into two distinct types depending on whether or not they require a predrilled hole. Conventional self-tapping screws require a predrilled hole and fall into a number of subgroups depending on the type of thread, head, and washer. **Self-drilling,** self-tapping screws drill their own hole and form their own thread in a single operation. There are two basic types, depending on the thickness of the base material. Both types of screws are usually combined with washers, which serve to increase the load-bearing capacity, the sealing ability, or both.

Blind Rivets

Blind rivets are normally used for fastening two thin sheets of material together when access is available from one side only. These are installed in a single operation; for example, by pulling a mandrel, which forms a head on the blind side of the rivet and expands the rivet shank. Fastening is completed when the mandrel either pulls through or breaks off. This type of fastener generally requires a strength considerably in excess of the sheet material to minimize possibilities of brittle failure.

Table 47.2 Typical Applications of Different Types of Fasteners

Thin to Thin	Thin to Thick or Thin to Hot-Rolled	Thick to Thick or Thick to Hot-Rolled
Self-drilling, self-tapping screws	Self-drilling, self-tapping screws	Bolts
Blind rivets	Fired pins	Arc welds
Single flare Vee welds	Bolts	
Spot welds	Arc puddle welds	
Lock seaming		

Table 47.3 Permissible Bearing Stress in Bolted Connections

Code	Bearing Stress	Basis
Addendum No. 1 (April 1975) to BS449	$1.4U_s$	Permissible stress
BS 5950: Part 1 (hot-rolled)	$0.64(U_s + Y_s)$	Limit state
BS 5950: Part 5 (cold-formed):		Limit state
(a) $t \leq 1$ mm	$2.1Y_s$	
(b) 1 mm $\leq t \leq$ 3 mm	$(1.65 + 0.45t)Y_s$	
(c) 3 mm $\leq t \leq$ 8 mm	$3.0Y_s$	
Note: Values in BS 5950: Part 5 reduced by 25% unless two washers used		
AISI Specification (August 1986):		Permissible stress
(a) With washers	$1.35U_s$ to $1.50U_s$	
(b) Without washers	$1.00U_s$ to $1.35U_s$	
European Recommendations:		Limit state
(a) $t \leq 1$ mm	$2.1Y_s$	
(b) 3 mm $\leq t \leq$ 6 mm	$4.3Y_s$	

Where Y_s = yield stress of fastened sheet, U_s = ultimate stress of fastened sheet.

Fired Pins

These pins, as the name suggests, are fired through thin material into thicker base material to form a connection. Two different methods of firing the pins are commonly used: powder actuation and pneumatic actuation. In the former, the pins are fired from a tool that contains an explosive cartridge, and in the latter, compressed air is used as the firing agent. Fired pins generally provide a very tight grip and a very good sealing capability.

Spot Welds

Electrical resistance spot welds are a widely used method of connecting thin sheets. In the U.K. these are governed by BS 1140, *General Requirements for Spot Welding of Light Assemblies in Mild Steel,* and in BS 5950: Part 5 and the AISI Specification, recommendations regarding weld sizes and capacities are provided. Further information on loads in resistance spot welds has been given by Baehre and Berggren [1973].

Arc Welds

Conventional fillet and butt welds are applicable to cold-formed steel sections if the material is relatively thick, and in such cases design considerations are the same as for hot-rolled sections. In the case of thinner material a wide range of special weld types are used in the U.S., but not outside the U.S. Guidance on making these special welds has been given by O.W. Blodgett [1978]. Expressions for evaluation of the ultimate loads for various types of these connections are available from a large testing and analysis program carried out by Peköz and McGuire [1979].

Assemblies of Fasteners

The available information on the the behavior of assemblies of fasteners in light-gauge steelwork suggests that the performance of such assemblies is not in any way inferior to that of similar connection assemblies in hot-rolled members. Connections in cold-formed members, however, do tend to be more flexible and ductile than in hot-rolled members, and herein lies a bonus that is well appreciated in some areas of cold-formed steel usage. The ductility afforded by such connections can provide the "plastic plateau" behavior that often occurs in hot-rolled sections, thus allowing redistribution of moments in the postyield range. A prime example of this in the U.K. is the wide use of "sleeves" in purlin design. These sleeves are used to connect purlins in a "semirigid" fashion, which produces a moment distribution close to the ideal at the point of failure.

47.6 Sheeting and Decking

There has recently been a substantial increase in the usage of profiled sheeting and decking, accompanied by a corresponding development of the design principles applicable to this type of construction. While only a few years ago 3-inch corrugated profiles or simple trapezoidal profiles were the norm, today there is a multiplicity of different shapes available. Profiles can be obtained in a wide variety of colors, surface coatings, and other options, with a life to first maintenance of up to 25 years. A large number of structural improvements have been developed involving the use of thinner material with increased yield strength in highly stiffened profiles. It is most probable, however, that in this area improvements in aesthetics, utility, and durability outweigh the structural improvements.

Profiles for Roof Sheeting

Cold roof construction is the term applied to construction in which profiled steel sheeting is used as the outer waterproof skin of a roof, with insulation placed inside this skin. It is important that the skin provide optimum resistance to moisture penetration, and the side laps should occur near the crests of the corrugations. With modern fasteners (e.g., self-drilling, self-tapping screws with neoprene washers), it is quite permissible to fix the sheeting to the roof purlins through the troughs. These fasteners satisfactorily prevent moisture penetration while providing various structural benefits in the fixity provided to the purlins over through-crest fixings.

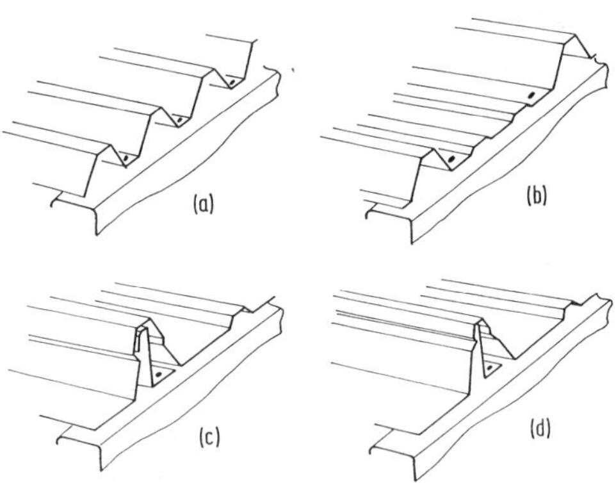

FIGURE 47.36 Typical roof sheeting profiles.

Modern profiles can be grouped into four main types, as illustrated in Fig. 47.36. For the trapezoidal profile shown in Fig. 47.36(a), it is found that to obtain the most economic use of material for a given span and loading using standard analysis procedures, the thickness of material and web slope angle that result are impractical, the thickness usually being very small and the web slope being very shallow. Fairly wide variations in the trapezoidal geometry result in relatively small variations in the economy, and the design of this type of sheeting is generally governed by practical considerations. In the profile of 47.36(b), again a fairly widely used shape of profile, the trapezoidal shape is altered to incorporate intermediate stiffeners in the lower flanges. In order to reduce or eliminate the requirement to penetrate the skin with fasteners, *concealed-fix* or *standing seam* sheeting has been developed over a number of years, Figure 47.36(c) and (d). The attachment in these systems is generally by means of clips, which connect mechanically to the sheeting and by screws to the purlin, with the mechanical connections to the sheeting often incorporated in the standing seam, eliminating penetration of the sheeting. In this type of system the purlin-sheeting connections are less useful in providing stability to the purlin than the more direct screw attachment and may necessitate close consideration of purlin design in such circumstances.

Profiles for Roof Decking

Profiled steel decking, supporting a built-up finish including insulation and waterproofing, is often termed "warm roof" construction. In this type of construction there is a tendency toward longer spans, more complex profiles, and the use of higher-strength steels. A typical profile resulting from these tendencies is shown in Fig. 47.37. The profile depth is of the order of 4 in. and the material

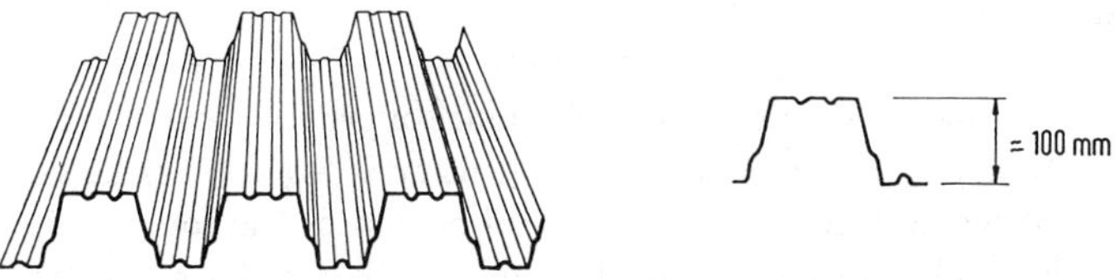

FIGURE 47.37 Typical long-span decking profiles.

yield strength of the order of 50 ksi. These profiles are suitable for spans of up to 30 ft and can thus eliminate the need for purlins. In general, stiffeners are rolled into both the flanges and the webs, as flanges and webs are slender and require stiffeners for efficiency.

There are now a large number of different design specifications dealing with roof decking. In Europe calculation procedures for complex shapes, as shown in Fig. 47.37, have been developed in Sweden for incorporation into the Swedish code of practice [Höglund, 1980], and these procedures form the basis of the European Recommendations [ECCS, 1983] as well as the national standards of some European countries, for example DIN 18807 [1987]. These procedures have been verified by a very substantial number of tests [Baehre and Fick, 1982].

Wall Coverings

Wall cladding carries significant wind loading and must satisfy structural considerations. However, since the appearance of a building is very substantially dependent on the sheeting, aesthetics are very often the primary consideration. Typical wall cladding profiles are shown in Fig. 47.38. The variety of profiles available, the wide range of coatings and colors, and the possibility of orienting the cladding in any desired direction provide substantial scope for imagination and artistry in the use of modern wall coverings in building design.

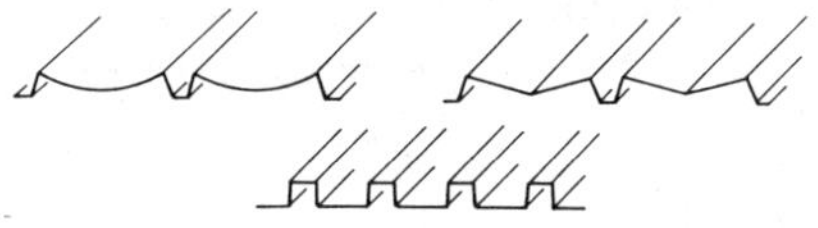

FIGURE 47.38 Typical wall cladding profiles.

Composite Panels

Composite panels, having outer faces of thin steel (or other material such as aluminum) and an internal core of foamed plastic, such as polyurethane or expanded polystyrene, are now widely used in construction (Fig. 47.39). The design criteria for composite panels are complex and are not discussed here. However, this type of construction must be mentioned because of its current growth and future potentialities. Foam-filled composite construction utilizes the favorable attributes of skin and core materials in combination. The core connects the metal skins, ensures that they are kept a distance apart to provide flexural capacity for the panel, and also provides a degree of resistance to local buckling of the skins. The skins protect the core from accidental damage and from the elements. Foam-filled panels combine high load-carrying capacity with low weight, attractive appearance, and other made-to-order attributes, such as extremely high thermal insulation capacity, to provide very good solutions to a variety of construction problems.

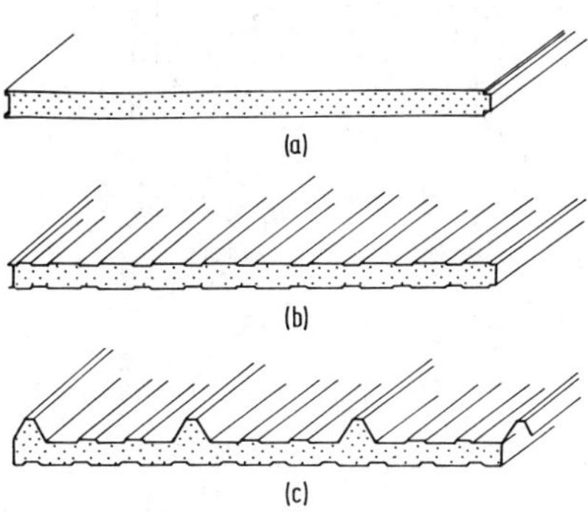

FIGURE 47.39 Typical sandwich panels. (a) Panel with flat faces. (b) Panel with lightly profiled faces. (c) Panel with profiled face.

47.7 Storage Racking

The term *storage racking* covers an extremely wide range of products that have as their purpose the storage of material in a secure and easily accessible manner. Storage racking systems range from small shelving systems to extremely large rack clad buildings. The components used in storage racking are largely of cold-formed steel, and the storage racking industry is one of the major users of cold-formed steel.

The forerunners of the modern storage structures were slotted angle products, first introduced in the 1930s. These consisted of cold-formed angle sections with perforations to provide a simple, flexible means whereby the designer could produce storage systems in a variety of shapes and configurations. By this means, simple shelving could be provided for warehouses of any size. From these beginnings natural evolution led to the present situation. It was found that for many systems bolted connections could be beneficially replaced by clips or other proprietory connections, and a wide variety of alternative connection methods was developed for different systems. While perforated angles had much in their favor for some purposes, particularly for smaller storage systems, the requirements for bracing in these torsionally weak members meant that for many purposes other cross sections were superior. Nowadays slotted angles are still widely used for shelving systems, but in larger storage racking systems other perforated members, generally monosymmetric, are used as columns, while normally unperforated members are used as beams.

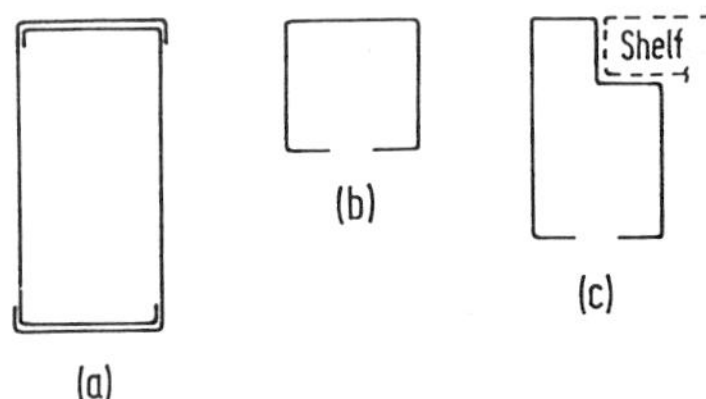

FIGURE 47.40 Typical beam profiles. (a) Box. (b) Open. (c) Stepped open.

Components

Beams

In pallet racking the most common types of beams used are boxed beams made from two interlocking lipped channels, as shown in Fig. 47.40(a), and open section beams, Fig. 47.40(b), which may be stepped to receive the edge of a shelf as shown in Fig. 47.40(c). The boxed beam is generally more structurally efficient than the open beam, but for lighter loads the economical aspects can favor the open section. Over the normal range of spans used in storage racking the limiting factor for the beams used is most often the bending capacity. The most common configuration of beam loading is that shown in Fig. 47.41 with two pallets in place.

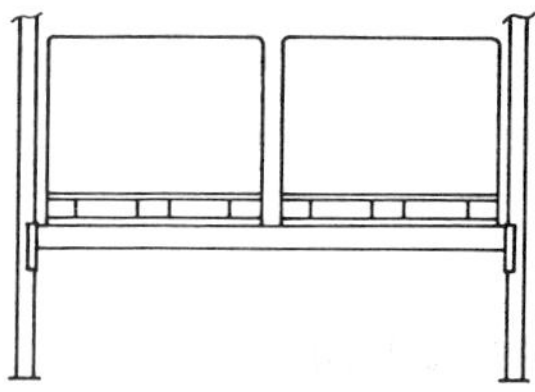

FIGURE 47.41 Two-pallet beam loading.

Uprights

Figure 47.42 shows typical perforated upright profiles used in storage racking at the present time. These are, in the main, developments from the simple lipped channel section. The developments generally involve the addition of more bends, which can be used to assist the attachment of bracing members as well as to enhance the structural capacity of the uprights. The bracing members, often of channel section, connect through the rear, circular, holes of the section while the beams are connected, via beam end connectors, through the perforations on the front face. The loads carried by the uprights are spread into the

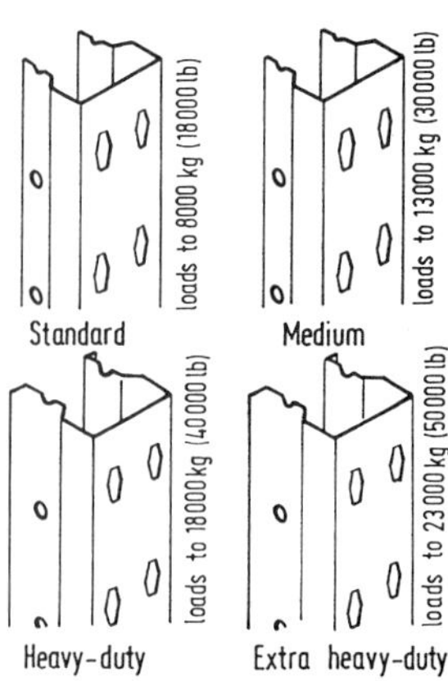

FIGURE 47.42 Typical upright profiles.

floor through base plates. These are normally thin plates, about 3 mm thick, which are bolted to the upright and to the floor.

Beam End Connectors

The beams used in storage racking are generally connected to the uprights via beam end connectors. These are generally made from hot-rolled material and are welded to the ends of the beam to provide a means of connecting the beam to the upright at any of a large number of positions through the perforations in the uprights. The connectors have hooks, which fit through the upright perforations to provide a semirigid connection between beam and upright. In evaluation of the behavior of beams and uprights, the connection behavior is very important. The moment-rotation behavior of beam end connectors is normally obtained on the basis of testing, and the strength and stiffness values so obtained are used in design analysis of the structure.

Design Codes of Practice

In Europe storage racking systems are generally considered to lie outside the scope of the normal codes of practice for structural design. This is largely because of the presence of perforations in uprights and other components. In the U.K. the Storage Equipment Manufacturers Association devised and published its own code of practice in 1980 [SEMA, 1980]. In Europe the Fédération Européenne de la Manutention (FEM) is working on the preparation of a code of practice that will most probably have currency throughout Europe, superseding the SEMA code in the U.K. and similar codes in other countries [FEM, 1986]. This code is being written for limit state design, and in its final form will be compatible with Eurocode 3.

Defining Terms

Beam: A straight or curved structural member, primarily supporting loads applied at right angles to the longitudinal axis.

Blind rivet: Mechanical fastener capable of joining workpieces together where access to the assembly is limited to one side only.

Buckle: To kink, wrinkle, bulge, or otherwise lose original shape as a result of elastic or inelastic strain.

Buckling load: The load at which a compressed element, member, or frame collapses in service or buckles in a loading test.

Cladding: Profiled sheet for walls.

Cold-formed steel structural members: Shapes that are manufactured by press-braking blanks sheared from sheets, cut length of coils, or plates or by roll-forming cold- or hot-rolled coils or sheets.

Composite flooring slab: Floor in which the structural bearing capacity is formed by the cooperation of concrete and floor decking (as reinforcement).

Effective length: The equivalent or effective length that, in the buckling formula for a hinged-end column, results in the same elastic critical load as for the framed member or other compression element under consideration at its theoretical critical load.

Effective width: Reduced width of a compression element in a flexural or compression members computed to account for local buckling when the **flat width–to–thickness ratio** exceeds a certain limit.

Flat width: Width of the straight portion of the element excluding the bent portion of the section.

Flat width–to–thickness ratio: Ratio of the **flat width** of an element to its thickness.

Flexural buckling: Buckling of a column due to flexure.

Limit state: Condition beyond which a structure would cease to be fit for its intended use.

Local buckling: Buckling of the walls of elements of a section characterized by a formation of waves or ripples along the member. It modifies the capacity of cross sections.

Multiply stiffened element: An element stiffened between webs, or between a web and a stiffened edge, by means of intermediate stiffeners parallel to the direction of stress.

Safety factor: A ratio of the stress (or strength) at incipient failure to the computed stress (or strength) at design load.

Self-drilling screw: Screw that drills its own hole and forms its mating thread in one operation.

Self-tapping screw: Screw that taps a counter thread in a prepared hole.

Sheeting: Profiled sheet for floors, roofs, or walls.

Stiffened or partially stiffened element: A flat compression element in which both edges parallel to the direction of stress are stiffened either by a web, flange, stiffening lip, or intermediate stiffener.

Tension-field action: Postbuckling behavior of a plate girder panel under shear force, during which diagonal compressive stresses cause the web to form diagonal waves.

Thickness: Thickness of base steel in cold-formed sections.

Torsional buckling: Buckling of a column by twisting.

Torsional-flexural buckling: A mode of buckling in which compression members can bend and twist simultaneously. This type of buckling mode is critical, in particular when the shear center of the section does not coincide with the centroid.

Unstiffened element: A flat compression element that is stiffened at one of the two edges parallel to the direction of stress.

Web crippling: Failure mode of the web of a beam caused by the combination of a bending moment and a concentrated load.

Yield point: The maximum stress recorded in a tensile or compressive test of steel specimen prior to entering the plastic range.

Yield strength: In a tension or compression test, the stress at which there is a specified amount of measured deviation from an extension of the initial linear stress-strain plot.

References

AISI (American Iron and Steel Institute). 1986. *Specification for the Design of Cold-Formed Steel Structural Members.* American Iron and Steel Institute.

AISI (American Iron and Steel Institute). 1991. *Load and Resistance Factor Design (LRFD) Specification for Cold-Formed Steel Structural Members.* American Iron and Steel Institute.

Allen, H. G. and Bulson, P. S., 1980. *Background to Buckling.* McGraw-Hill, New York.

Baehre, R. and Berggren, L. Joints in sheet metal panels. 1973. *National Swedish Building Res., Doc. D8.*

Baehre, R. and Fick, K. 1982. Berechnung und Bemessung von Trapezprofilen mit Erläuterungen zur DIN 18807. *Berichte der Versuchsanstalt für Stahl, Holz und Stein,* Universität Fridericiana Karlsruhe.

Blodgett, O. W. 1978. Report on proposed standards for sheet steel structural welding. In *Proc. 4th Int. Specialty Conf. Cold-Formed Steel Struct.* University of Missouri—Rolla, pp. 923–949.

British Standards Institution. 1975. *Specification for the Use of Cold Formed Steel in Building.* Addendum No. 1 to BS 449 (1969).

British Standards Institution. 1985. *Structural Use of Steelwork in Building.* BS 5950: Part 1.

British Standards Institution. 1987. *Structural Use of Steelwork in Building.* BS 5950: Part 5.

Canadian Standards Association. 1989. *Cold-Formed Steel Structural Members.* CAN3-S136-M89.

Chajes, A. and Winter, G. *J. Structural Div.* 1965. Torsional-flexural buckling of thin-walled members. *ASCE.* 91.

Desmond, T. P. Peköz, T., and Winter, G. 1981. Edge stiffeners for thin-walled members. *J. Structural Div., ASCE.* 107(ST2): 329–353.

DIN 18807. 1987. *Trapezprofile im Hochbau* (Trapezoidal Profiled Sheeting in Building). Deutsche Norm (German Standard).

Eurocode 3. 1989. *Design of Steel Structures, Draft Issue 2.*

ECCS (European Convention for Constructional Steelwork). 1983. *European Recommendations for the Design of Profiled Sheeting,* European Convention for Constructional Steelwork. ECCS-TC7-1983.

FEM (Fédération Européenne de la Manutention). 1986. Draft edition. *Recommendations for the Design of Steel Static Pallet Racking.*

Hancock, G. J. 1988. *Cold-Formed Steel Structures.* Australian Institute of Steel Construction, Sydney.

Hetrakul, N. and Yu, W. W. 1980. Cold-formed steel I-beams subjected to combined bending and web crippling. In *Thin-Walled Structures,* ed. J. Rhodes and A. C. Walker, Granada Publishing.

Höglund, T. Design of trapezoidal sheeting provided with stiffeners in the flanges and webs. *Swedish Counc. for Building Res., Doc. D28,* 1980.

Megson, T. H. G. 1975. *Analysis of Thin-Walled Members.* Intertext.

Metal Roof Deck Association. 1970. *Code of Design and Technical Requirements for Light Gauge Metal Roof Decks.* Metal Roof Deck Association, London.

Murray, N. W. 1984. *Introduction to the Theory of Thin Walled Structures.* Oxford Press.

Peköz, T. and Celebi, N. 1969. Torsional-flexural buckling of thin-walled sections under eccentric load. *Cornell Univ. Eng. Res. Bull.* 69-1.

Peköz, T. and McGuire, W. 1979. Welding of sheet steel. *Private Report to AISI,* Cornell University, Ithaca, NY.

Rhodes, J. 1983. Buckling and failure of edge stiffened plates. In *Collapse,* ed. J. M. T. Thompson and G. W. Hunt, Cambridge University Press, Cambridge, U.K.

Rhodes, J. 1971. *Design of Cold-Formed Steel Members.* Elsevier Applied Science.

Rhodes, J. 1987. Cold formed steel sections—state of the art in Great Britain. In *Steel Structures, Advances, Design and Construction,* ed. R. Narayanan, Elsevier Applied Science, London, pp. 406–418.

Rockey, K. C. 1967. Shear buckling of thin-walled sections. In *Thin-walled Structures,* ed. A.H. Chilver Chatto and Windus, London.

SEMA (Storage Equipment Manufacturers Association). 1980. *Interim Code of Practice for the Design of Static Racking.* Storage Equipment Manufacturers Association.

Swedish Code for Light-Gauge Metal Structures. 1982. National Swedish Committee on Regulations for Steel Structures. English translation.

Winter, G. 1947. Four papers on the performance of thin-walled steel structures. *Cornell Univ. Eng., Exp. Stn. Reprint No. 33.*

Winter, G. and Pian, R. H. J. 1946. Crushing strength of thin steel webs. *Cornell Bulletin,* 35.

Yu, Wei Wen. 1991. *Cold-Formed Steel Design.* John Wiley & Sons, Chichester.

Zetlin, L. 1955. Elastic instability of flat plates subjected to partial edge loads. *J. Structural Div., ASCE.* 81.

For Further Information

Guide to Stability Design Criteria for Metal Structures, by Theodore V. Galambos, is the most comprehensive treatment of stability design of metal structures. It covers a wide range of topics including postbuckling strength of unstiffened and stiffened plates, plate girders, and thin-walled metal girders.

Cold-Formed Steel in Tall Buildings, edited by W. W. Yu *et al.,* is a state-of-the-art design guide devoted to the most efficient and economical use of cold-formed steel in high-rise building.

Background to Buckling, by H. G. Allen and P. S. Bulson, provides a coherent account of the buckling problem, including an analytical treatment for studying the stability of plates and shells. Provides a valuable aid to forging the vital links between research, analysis, and design.

Plated Structures—Stability and Strength, by R. Narayanan, deals with various aspects related to stability problems of plated structures.

The Stressed Skin Design of Steel Buildings, by E. R. Bryan, and *Manual of Stressed Skin Diaphragm Design,* by E. R. Bryan and J. M. Davies, provide design methods and recommendations for roof decking and wall cladding profiles.

The *Proceedings of the International Conference on Cold-Formed Steel Structures,* edited by W. W. Yu *et al.* and available in a number of volumes, extensively covers all aspects of cold-formed steel structures.

Perhaps the best-known bridge in the world, the Golden Gate Bridge has adorned San Francisco Bay since its completion in 1937. Completion of the bridge required surmounting numerous obstacles of design and construction caused by the rugged weather and topography of the Golden Gate straits. It is a remarkable feat of engineering and art.

The finished product was built to endure these same harsh elements for ages to come. It features the largest bridge cables ever made, containing 80,000 miles of wire with a tensile strength of 200 million pounds; the bridge floor swings out as much as six feet under the brunt of 70-mile-per-hour winds; the roadway moves up and down with the cables' expansions and contractions during the Bay's rapid changes in temperature; and the entire structure is undergoing a seismic upgrade to withstand a maximum credible earthquake of 8.2 on the Richter scale.

During more than 50 years of operation, the bridge has been closed only three times due to weather conditions, and it suffered no damage during the devastating Loma Prieta quake of 1989. Carrying more than 40 million vehicles per year, the Golden Gate Bridge has made possible the growth of a major metropolis around the San Francisco Bay. (Photo courtesy of the American Society of Civil Engineers.)

48

Structural Concrete Design

Amy Grider
Purdue University

Julio A. Ramirez
Purdue University

Young Mook Yun
Purdue University

At this point in the history of development of reinforced and prestressed concrete it is necessary to reexamine the fundamental approaches to design of these composite materials. Structural engineering is a worldwide industry. Designers from one nation or a continent are faced with designing a project in another nation or continent. The decades of efforts dedicated to harmonizing concrete design approaches worldwide have resulted in some successes but in large part have

0-8493-8953-4/95/$0.00 + $.50

led to further differences and numerous different design procedures. It is this abundance of different design approaches, techniques, and code regulations that justifies and calls for the need for a unification of design approaches throughout the entire range of structural concrete, from plain to fully prestressed [Breen, 1991].

The effort must begin at all levels: university courses, textbooks, handbooks, and standards of practice. Students and practitioners must be encouraged to think of a single continuum of structural concrete. Based on this premise, this chapter on concrete design is organized to promote such unification. In addition, effort will be directed at dispelling the present unjustified preoccupation with complex analysis procedures and often highly empirical and incomplete sectional mechanics approaches that tend to both distract the designers from fundamental behavior and impart a false sense of accuracy to beginning designers. Instead, designers will be directed to give careful consideration to overall structure behavior, remarking the adequate flow of forces throughout the entire structure.

48.1 Properties of Concrete and Reinforcing Steel

The designer needs to be knowledgeable about the properties of concrete, reinforcing steel, and prestressing steel. This part of the chapter summarizes the material properties of particular importance to the designer.

Properties of Concrete

Workability is the ease with which the ingredients can be mixed and the resulting mix handled, transported, and placed with little loss in homogeneity. Unfortunately, workability cannot be measured directly. Engineers therefore try to measure the consistency of the concrete by performing a slump test.

The slump test is useful in detecting variations in the uniformity of a mix. In the slump test, a mold shaped as the frustum of a cone, 12 in. (305 mm) high with an 8 in. (203 mm) diameter base and 4 in. (102 mm) diameter top, is filled with concrete (ASTM Specification C143). Immediately after filling, the mold is removed and the change in height of the specimen is measured. The change in height of the specimen is taken as the slump when the test is done according to the ASTM Specification.

A well-proportioned workable mix settles slowly, retaining its original shape. A poor mix crumbles, segregates, and falls apart. The slump may be increased by adding water, increasing the percentage of fines (cement or aggregate), entraining air, or by using an admixture that reduces water requirements; however, these changes may adversely affect other properties of the concrete. In general, the slump specified should yield the desired consistency with the least amount of water and cement.

Concrete should withstand the weathering, chemical action, and wear to which it will be subjected in service over a period of years; thus, durability is an important property of concrete. Concrete resistance to freezing and thawing damage can be improved by increasing the watertightness, entraining 2 to 6% air, using an air-entraining agent, or applying a protective coating to the surface. Chemical agents damage or disintegrate concrete; therefore, concrete should be protected with a resistant coating. Resistance to wear can be obtained by use of a high-strength, dense concrete made with hard aggregates.

Excess water leaves voids and cavities after evaporation, and water can penetrate or pass through the concrete if the voids are interconnected. Watertightness can be improved by entraining air or reducing water in the mix, or it can be prolonged through curing.

Volume change of concrete should be considered, since expansion of the concrete may cause buckling and drying shrinkage may cause cracking. Expansion due to alkali-aggregate reaction can be avoided by using nonreactive aggregates. If reactive aggregates must be used, expansion may be reduced by adding pozzolanic material (e.g., fly ash) to the mix. Expansion caused by heat of hydration of the cement can be reduced by keeping cement content as low as possible; using Type IV cement; and chilling the aggregates, water, and concrete in the forms. Expansion from temperature increases can be reduced by using coarse aggregate with a lower coefficient of thermal expansion. Drying shrinkage can be reduced by using less water in the mix, using less cement, or allowing adequate moist curing. The addition of pozzolans, unless allowing a reduction in water, will increase drying shrinkage. Whether volume change causes damage usually depends on the restraint present; consideration should be given to eliminating restraints or resisting the stresses they may cause [MacGregor, 1992].

Strength of concrete is usually considered its most important property. The compressive strength at 28 days is often used as a measure of strength because the strength of concrete usually increases with time. The compressive strength of concrete is determined by testing specimens in the form of standard cylinders as specified in ASTM Specification C192 for research testing or C31 for field testing. The test procedure is given in ASTM C39. If drilled cores are used, ASTM C42 should be followed.

The suitability of a mix is often desired before the results of the 28-day test are available. A formula proposed by W. A. Slater estimates the 28-day compressive strength of concrete from its 7-day strength:

$$S_{28} = S_7 + 30\sqrt{S_7} \tag{48.1}$$

where

S_{28} = 28-day compressive strength, psi, and
S_7 = 7-day compressive strength, psi.

Strength can be increased by decreasing water-cement ratio, using higher strength aggregate, using a pozzolan such as fly ash, grading the aggregates to produce a smaller percentage of voids in the concrete, moist curing the concrete after it has set, and vibrating the concrete in the forms. The short-time strength can be increased by using Type III portland cement, accelerating admixtures, and by increasing the curing temperature.

The stress-strain curve for concrete is a curved line. Maximum stress is reached at a strain of 0.002 in./in., after which the curve descends.

The modulus of elasticity, E_c, as given in ACI 318-89 (Revised 92), *Building Code Requirements for Reinforced Concrete* [ACI Committee 318, 1992], is:

$$E_c = w_c^{1.5} 33\sqrt{f_c'} \qquad \text{lb/ft}^3 \text{and psi} \tag{48.2a}$$

$$E_c = w_c^{1.5} 0.043\sqrt{f_c'} \qquad \text{kg/m}^3 \text{and MPa} \tag{48.2b}$$

where

w_c = unit weight of concrete, and
f_c' = compressive strength at 28 days.

Tensile strength of concrete is much lower than the compressive strength—about $7\sqrt{f_c'}$ for the higher-strength concretes and $10\sqrt{f_c'}$ for the lower-strength concretes.

Creep is the increase in strain with time under a constant load. Creep increases with increasing water-cement ratio and decreases with an increase in relative humidity. Creep is accounted for in design by using a reduced modulus of elasticity of the concrete.

Lightweight Concrete

Structural lightweight concrete is usually made from aggregates conforming to ASTM C330 that are usually produced in a kiln, such as expanded clays and shales. Structural lightweight concrete has a density between 90 and 120 lb/ft^3 (1440–1920 kg/m^3).

Production of lightweight concrete is more difficult than normal-weight concrete because the aggregates vary in absorption of water, specific gravity, moisture content, and amount of grading of undersize. Slump and unit weight tests should be performed often to ensure uniformity of the mix. During placing and finishing of the concrete, the aggregates may float to the surface. Workability can be improved by increasing the percentage of fines or by using an air-entraining admixture to incorporate 4 to 6% air. Dry aggregate should not be put into the mix, because it will continue to absorb moisture and cause the concrete to harden before placement is completed. Continuous water curing is important with lightweight concrete.

No-fines concrete is obtained by using pea gravel as the coarse aggregate and 20 to 30% entrained air instead of sand. It is used for low dead weight and insulation when strength is not important. This concrete weighs from 105 to 118 lb/ft^3 (1680–1890 kg/m^3) and has a compressive strength from 200 to 1000 psi (1–7 MPa).

A porous concrete made by gap grading or single-size aggregate grading is used for low conductivity or where drainage is needed.

Lightweight concrete can also be made with gas-forming or foaming agents which are used as admixtures. Foam concretes range in weight from 20 to 110 lb/ft^3 (320–1760 kg/m^3). The modulus of elasticity of lightweight concrete can be computed using the same formula as normal concrete. The shrinkage of lightweight concrete is similar to or slightly greater than for normal concrete.

Heavyweight Concrete

Heavyweight concretes are used primarily for shielding purposes against gamma and x-radiation in nuclear reactors and other structures. Barite, limonite and magnetite, steel punchings, and steel shot are typically used as aggregates. Heavyweight concretes weigh from 200 to 350 lb/ft^3 (3200 to 5600 kg/m^3) with strengths from 3200 to 6000 psi (22–41 MPa). Gradings and mix proportions are similar to those for normal weight concrete. Heavyweight concretes usually do not have good resistance to weathering or abrasion.

High-Strength Concrete

Concretes with strengths in excess of 6000 psi (41 MPa) are referred to as high-strength concretes. Strengths up to 18,000 psi (124 MPa) have been used in buildings.

Admixtures such as superplasticizers, silica fume, and supplementary cementing materials such as fly ash improve the dispersion of cement in the mix and produce workable concretes with lower water-cement ratios, lower void ratios, and higher strength. Coarse aggregates should be strong, fine-grained gravel with rough surfaces.

For concrete strengths in excess of 6000 psi (41 MPa), the modulus of elasticity should be taken as

$$E_c = 40{,}000\sqrt{f_c'} + 1.0 \times 10^6 \tag{48.3}$$

t A:wq A noindent where f_c' = compressive strength at 28 days, psi [ACI Committee 36].

The shrinkage of high-strength concrete is about the same as that for normal concrete.

Reinforcing Steel

Concrete can be reinforced with welded wire fabric, deformed reinforcing bars, and prestressing tendons.

Welded wire fabric is used in thin slabs, thin shells, and other locations where space does not allow the placement of deformed bars. Welded wire fabric consists of cold drawn wire in orthogonal patterns—square or rectangular and resistance-welded at all intersections. The wires may be smooth (ASTM A185 and A82) or deformed (ASTM A497 and A496). The wire is specified

Table 48.1 Wire and Welded Wire Fabric Steels

AST Designation	Wire Size Designation	Minimum Yeild Stress,[a] f_y		Minimum Tensile Strength	
		ksi	MPa	ksi	MPa
A82-79 (cold-drawn wire) (properties apply when material is to be used for fabric)	W1.2 and larger[b]	65	450	75	520
	Smaller than W1.2	56	385	70	480
A185-79 (welded wire fabric)	Same as A82; this is A82 material fabricated into sheet (so-called "mesh") by the process of electric welding.				
A496-78 (deformed steel wire) (properties apply when material is to be used for fabric)	D1–D31[c]	70	480	80	550
A497-79	Same as A82 or A496; this specification applies for fabric made from A496, or from a combination of A496 and A82 wires.				

[a]The term "yield stress" refers to either *yield point,* the well-defined deviation from perfect elasticity, or *yield strength,* the value obtained by a specified offset strain for material having no well-defined yield point.

[b]The W number represents the nominal cross-sectional area in square inches multiplied by 100, for smooth wires.

[c]The D number represents the nominal cross-sectional area in square inches multiplied by 100, for deformed wires.

Source: Wang and Salmon, 1985.

by the symbol W for smooth wires or D for deformed wires followed by a number representing the cross-sectional area in hundredths of a square inch. On design drawings it is indicated by the symbol WWF followed by spacings of the wires in the two 90° directions. Properties for welded wire fabric are given in Table 48.1.

The deformations on a deformed reinforcing bar inhibit longitudinal movement of the bar relative to the concrete around it. Table 48.2 gives dimensions and weights of these bars. Reinforcing bar steel can be made of billet steel of grades 40 and 60 having minimum specific yield stresses of 40,000 and 60,000 psi, respectively (276 and 414 MPa) (ASTM A615) or low-alloy steel of grade

Table 48.2 Reinforcing Bar Dimensions and Weights

Bar Number	Nominal Dimensions: Diameter (in.)	Diameter (mm)	Area (in.^{2})	Area (cm^2)	Weight (lb/ft)	Weight (kg/m)
3	0.375	9.5	0.11	0.71	0.376	0.559
4	0.500	12.7	0.20	1.29	0.668	0.994
5	0.625	15.9	0.31	2.00	1.043	1.552
6	0.750	19.1	0.44	2.84	1.502	2.235
7	0.875	22.2	0.60	3.87	2.044	3.041
8	1.000	25.4	0.79	5.10	2.670	3.973
9	1.128	28.7	1.00	6.45	3.400	5.059
10	1.270	32.3	1.27	8.19	4.303	6.403
11	1.410	35.8	1.56	10.06	5.313	7.906
14	1.693	43.0	2.25	14.52	7.65	11.38
18	2.257	57.3	4.00	25.81	13.60	20.24

Table 48.3 Standard Prestressing Strands, Wires, and Bars

Tendon Type	Grade f_{pu} ksi	Nominal Dimension Diameter in.	Nominal Dimension Area in.^{2}	Weight plf
Seven-wire strand	250	1/4	0.036	0.12
	270	3/8	0.085	0.29
	250	3/8	0.080	0.27
	270	1/2	0.153	0.53
	250	1/2	0.144	0.49
	270	0.6	0.215	0.74
	250	0.6	0.216	0.74
Prestressing wire	250	0.196	0.0302	0.10
	240	0.250	0.0491	0.17
	235	0.276	0.0598	0.20
Deformed prestressing bars	157	5/8	0.28	0.98
	150	1	0.85	3.01
	150	1 1/4	1.25	4.39
	150	1 3/8	1.58	5.56

Source: Collins and Mitchell, 1991.

60, which is intended for applications where welding and/or bending is important (ASTM A706). Presently, grade 60 billet steel is the most predominately used for construction.

Prestressing tendons are commonly in the form of individual wires or groups of wires. Wires of different strengths and properties are available with the most prevalent being the 7-wire low-relaxation strand conforming to ASTM A416. ASTM A416 also covers a stress-relieved strand, which is seldom used in construction nowadays. Properties of standard prestressing strands are given in Table 48.3. Prestressing tendons could also be bars; however, this is not very common. Prestressing bars meeting ASTM A722 have been used in connections between members.

The modulus of elasticity for non-prestressed steel is 29,000,000 psi (200,000 MPa). For prestressing steel, it is lower and also variable, so it should be obtained from the manufacturer. For 7-wire strands conforming to ASTM A416, the modulus of elasticity is usually taken as 27,000,000 psi (186,000 MPa).

48.2 Proportioning and Mixing Concrete

Proportioning Concrete Mix

A concrete mix is specified by the weight of water, sand, coarse aggregate, and admixture to be used per 94-pound bag of cement. The type of cement (Table 48.4), modulus of the aggregates, and maximum size of the aggregates (Table 48.5) should also be given. A mix can be specified by the weight ratio of cement to sand to coarse aggregate with the minimum amount of cement per cubic yard of concrete.

Table 48.4 Types of Portland Cement*

Type	Usage
I	Ordinary construction where special properties are not required
II	Ordinary construction when moderate sulfate resistance or moderate heat of hydration is desired
III	When high early strength is desired
IV	When low heat of hydration is desired
V	When high sulfate resistance is desired

*According to ASTM C150.

In proportioning a concrete mix, it is advisable to make and test trial batches because of the many variables involved. Several trial batches should be made with a constant water-cement ratio but varying ratios of aggregates to obtain the desired workability with the least cement. To obtain results similar to those in the field, the trial batches should be mixed by machine.

Table 48.5 Recommended Maximum Sizes of Aggregate*

Minimum Dimension of Section, in.	Maximum Size, in., of Aggregate for Reinforced-Concrete Beams, Columns, Walls	Heavily Reinforced Slabs	Lightly Reinforced or Unreinforced Slabs
5 or less	. . .	¾–1 ½	¾–1 ½
6–11	¾–1 ½	1 ½	1 ½–3
12–29	1 ½–3	3	3–6
30 or more	1 ½–3	3	6

**Concrete Manual.* U.S. Bureau of Reclamation.

When time or other conditions do not allow proportioning by the trial batch method, Table 48.6 may be used. Start with mix B corresponding to the appropriate maximum size of aggregate. Add just enough water for the desired workability. If the mix is undersanded, change to mix A; if oversanded, change to mix C. Weights are given for dry sand. For damp sand, increase the weight of sand 10 lb, and for very wet sand, 20 lb, per bag of cement.

Admixtures

Admixtures may be used to modify the properties of concrete. Some types of admixtures are set accelerators, water reducers, air-entraining agents, and waterproofers. Admixtures are generally helpful in improving quality of the concrete. However, if admixtures are not properly used, they could have undesirable effects; it is therefore necessary to know the advantages and limitations of the proposed admixture. The ASTM Specifications cover many of the admixtures.

Set accelerators are used (1) when it takes too long for concrete to set naturally, such as in cold weather, or (2) to accelerate the rate of strength development. Calcium chloride is widely used as a set accelerator. If not used in the right quantities, it could have harmful effects on the concrete and reinforcement.

Table 48.6 Typical Concrete Mixes*

Maximum Size of Aggregate, in.	Mix Designation	Bags of Cement per yd^3 of Concrete	Aggregate, lb per Bag of Cement — Sand: Air-Entrained Concrete	Sand: Concrete Without Air	Gravel or Crushed Stone
½	A	7.0	235	245	170
	B	6.9	225	235	190
	C	6.8	225	235	205
¾	A	6.6	225	235	225
	B	6.4	225	235	245
	C	6.3	215	225	265
1	A	6.4	225	235	245
	B	6.2	215	225	275
	C	6.1	205	215	290
1 ½	A	6.0	225	235	290
	B	5.8	215	225	320
	C	5.7	205	215	345
2	A	5.7	225	235	330
	B	5.6	215	225	360
	C	5.4	205	215	380

**Concrete Manual.* U.S. Bureau of Reclamation.

Water reducers lubricate the mix and permit easier placement of the concrete. Since the workability of a mix can be improved by a chemical agent, less water is needed. With less water but the same cement content, the strength is increased. Since less water is needed, the cement content could also be decreased, which results in less shrinkage of the hardened concrete. Some water reducers also slow down the concrete set, which is useful in hot weather and in integrating consecutive pours of the concrete.

Air-entraining agents are probably the most widely used type of admixture. Minute bubbles of air are entrained in the concrete, which increases the resistance of the concrete to freeze-thaw cycles and the use of ice-removal salts.

Waterproofing chemicals are often applied as surface treatments, but they can be added to the concrete mix. If applied properly and uniformly, they can prevent water from penetrating the concrete surface. Epoxies can also be used for waterproofing. They are more durable than silicone coatings, but they may be more costly. Epoxies can also be used for protection of wearing surfaces, patching cavities and cracks, and glue for connecting pieces of hardened concrete.

Mixing

Materials used in making concrete are stored in batch plants that have weighing and control equipment and bins for storing the cement and aggregates. Proportions are controlled by automatic or manually operated scales. The water is measured out either from measuring tanks or by using water meters.

Machine mixing is used whenever possible to achieve uniform consistency. The revolving drum–type mixer and the countercurrent mixer, which has mixing blades rotating in the opposite direction of the drum, are commonly used.

Mixing time, which is measured from the time all ingredients are in the drum, "should be at least 1.5 minutes for a 1-yd^3 mixer, plus 0.5 min for each cubic yard of capacity over 1 yd^3" [ACI 304-73, 1973]. It also is recommended to set a maximum on mixing time since overmixing may remove entrained air and increase fines, thus requiring more water for workability; three times the minimum mixing time can be used as a guide.

Ready-mixed concrete is made in plants and delivered to job sites in mixers mounted on trucks. The concrete can be mixed en route or upon arrival at the site. Concrete can be kept plastic and workable for as long as 1.5 hours by slow revolving of the mixer. Mixing time can be better controlled if water is added and mixing started upon arrival at the job site, where the operation can be inspected.

48.3 Flexural Design of Beams and One-Way Slabs

Reinforced Concrete Strength Design

The basic assumptions made in flexural design are:

1. Sections perpendicular to the axis of bending that are plane before bending remain plane after bending.
2. A perfect bond exists between the reinforcement and the concrete such that the strain in the reinforcement is equal to the strain in the concrete at the same level.
3. The strains in both the concrete and the reinforcement are assumed to be directly proportional to the distance from the neutral axis (ACI 10.2.2) [ACI Committee 318, 1992].
4. Concrete is assumed to fail when the compressive strain reaches 0.003 (ACI 10.2.3).
5. The tensile strength of concrete is neglected (ACI 10.2.5).
6. The stresses in the concrete and reinforcement can be computed from the strains using stress-strain curves for concrete and steel, respectively.

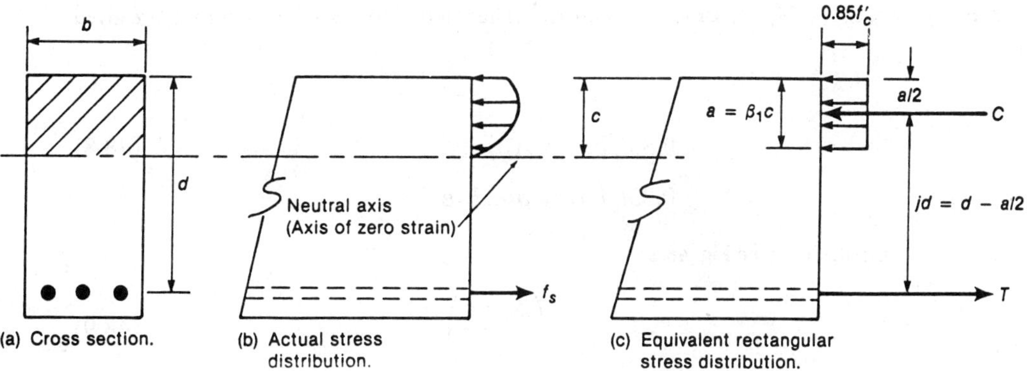

FIGURE 48.1 Stresses and forces in a rectangular beam. (*Source:* MacGregor, 1992.)

7. The compressive stress-strain relationship for concrete may be assumed to be rectangular, trapezoidal, parabolic, or any other shape that results in prediction of strength in substantial agreement with the results of comprehensive tests (ACI 10.2.6). ACI 10.2.7 outlines the use of a rectangular compressive stress distribution which is known as the Whitney rectangular stress block. For other stress distributions see *Reinforced Concrete Mechanics and Design* by James G. MacGregor [1992].

Analysis of Rectangular Beams with Tension Reinforcement Only

Equations for M_n and ϕM_n: Tension Steel Yielding. Consider the beam shown in Fig. 48.1. The compressive force, C, in the concrete is

$$C = (0.85f_c')ba \tag{48.3}$$

The tension force, T, in the steel is

$$T = A_s f_y \tag{48.4}$$

For equilibrium, $C = T$, so the depth of the equivalent rectangular stress block, a, is

$$a = \frac{A_s f_y}{0.85 f_c' b} \tag{48.5}$$

Noting that the internal forces C and T form an equivalent force-couple system, the internal moment is

$$M_n = T(d - a/2) \tag{48.6}$$

or

$$M_n = C(d - a/2)$$

ϕM_n is then

$$\phi M_n = \phi T(d - a/2) \tag{48.7}$$

or

$$\phi M_n = \phi C(d - a/2)$$

where $\phi = 0.90$.

Equation for M_n and ϕM_n: Tension Steel Elastic. The internal forces and equilibrium are given by:

$$\begin{aligned} C &= T \\ 0.85f_c'ba &= A_s f_s \\ 0.85f_c'ba &= \rho bdE_s\varepsilon_s \end{aligned} \tag{48.8}$$

From strain compatibility (see Fig. 48.1),

$$\varepsilon_s = \varepsilon_{cu}\left(\frac{d-c}{c}\right) \tag{48.9}$$

Substituting ε_s into the equilibrium equation, noting that $a = \beta_1 c$, and simplifying gives

$$\left(\frac{0.85f_c'}{\rho E_s\varepsilon_{cu}}\right)a^2 + (d)a - \beta_1 d^2 = 0 \tag{48.10}$$

which can be solved for a. Equations (48.6) and (48.7) can then be used to obtain M_n and ϕM_n.

Reinforcement Ratios. The reinforcement ratio, ρ, is used to represent the relative amount of tension reinforcement in a beam and is given by

$$\rho = \frac{A_s}{bd} \tag{48.11}$$

At the balanced strain condition the maximum strain, ε_{cu}, at the extreme concrete compression fiber reaches 0.003 just as the tension steel reaches the strain $\varepsilon_y = f_y/E_s$. The reinforcement ratio in the balanced strain condition, ρ_b , can be obtained by applying equilibrium and compatibility conditions. From the linear strain condition, Fig. 48.1,

$$\frac{c_b}{d} = \frac{\varepsilon_{cu}}{\varepsilon_{cu}+\varepsilon_y} = \frac{0.003}{0.003 + \dfrac{f_y}{29{,}000{,}000}} = \frac{87{,}000}{87{,}000 + f_y} \tag{48.12}$$

The compressive and tensile forces are:

$$\begin{aligned} C_b &= 0.85f_c'b\beta_1 c_b \\ T_b &= f_y A_{sb} = \rho_b bdf_y \end{aligned} \tag{48.13}$$

Equating C_b to T_b and solving for ρ_b gives

$$\rho_b = \frac{0.85f_c'\beta_1}{f_y}\left(\frac{c_b}{d}\right) \tag{48.14}$$

which on substitution of Eq. (48.12) gives

$$\rho_b = \frac{0.85f_c'\beta_1}{f_y}\left(\frac{87{,}000}{87{,}000 + f_y}\right) \tag{48.15}$$

ACI 10.3.3 limits the amount of reinforcement in order to prevent nonductile behavior:

$$\max\rho = 0.75\rho_b \tag{48.16}$$

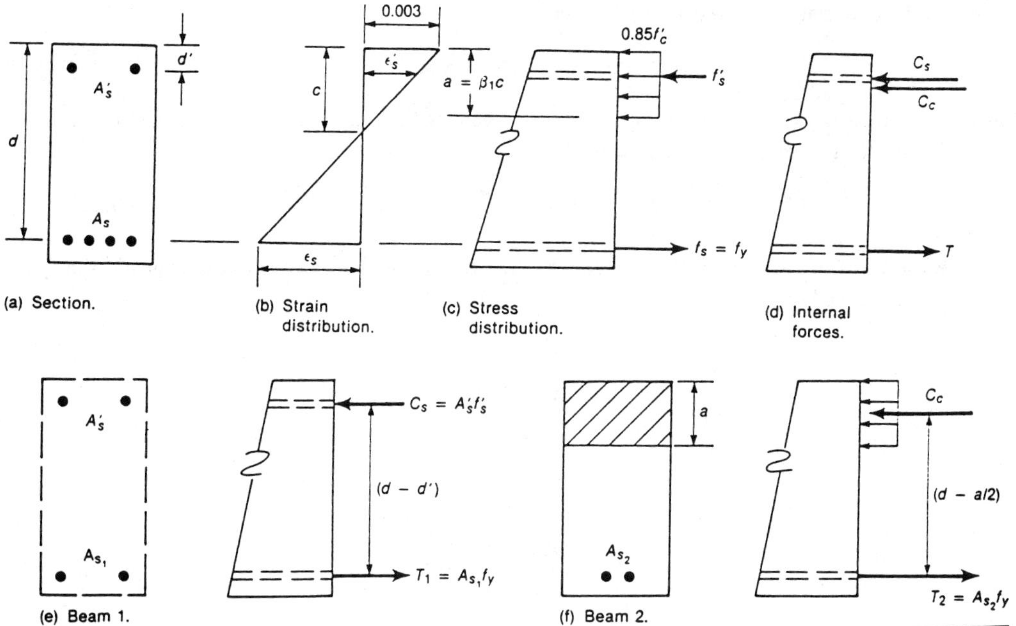

FIGURE 48.2 Strains, stresses, and forces in beam with compression reinforcement. (*Source:* MacGregor, 1992.)

ACI 10.5 requires a minimum amount of flexural reinforcement:

$$\rho_{\min} = \frac{200}{f_y} \tag{48.17}$$

Analysis of Beams with Tension and Compression Reinforcement

For the analysis of doubly reinforced beams, the cross section will be divided into two beams. Beam 1 consists of the compression reinforcement at the top and sufficient steel at the bottom so that $T_1 = C_s$; beam 2 consists of the concrete web and the remaining tensile reinforcement, as shown in Fig. 48.2.

Equation for M_n: Compression Steel Yields. The area of tension steel in beam 1 is obtained by setting $T_1 = C_s$, which gives $A_{s1} = A_s'$. The nominal moment capacity of beam 1 is then

$$M_{n1} = A_s' f_y (d - d') \tag{48.18}$$

Beam 2 consists of the concrete and the remaining steel, $A_{s2} = A_s - A_{s1} = A_s - A_s'$. The compression force in the concrete is

$$C = 0.85 f_c' ba \tag{48.19}$$

and the tension force in the steel for beam 2 is

$$T = (A_s - A_s') f_y \tag{48.20}$$

The depth of the compression stress block is then

$$a = \frac{(A_s - A_s') f_y}{0.85 f_c' b} \tag{48.21}$$

Therefore, the nominal moment capacity for beam 2 is

$$M_{n2} = (A_s - A_s')f_y(d - a/2) \tag{48.22}$$

The total moment capacity for a doubly reinforced beam with compression steel yielding is the summation of the moment capacity for beam 1 and beam 2; therefore,

$$M_n = A_s' f_y(d - d') + (A_s - A_s')f_y(d - a/2) \tag{48.23}$$

Equation for M_n: Compression Steel Does Not Yield. Assuming that the tension steel yields, the internal forces in the beam are

$$\begin{aligned} T &= A_s f_y \\ C_c &= 0.85 f_c' b a \\ C_s &= A_s'(E_s \varepsilon_s') \end{aligned} \tag{48.24}$$

where

$$\varepsilon_s' = \left(1 - \frac{\beta_1 d'}{a}\right)(0.003) \tag{48.25}$$

From equilibrium, $C_s + C_c = T$ or

$$0.85 f_c' b a + A_s' E_s \left(1 - \frac{\beta_1 d'}{a}\right)(0.003) = A_s f_y \tag{48.26}$$

This can be rewritten in quadratic form as

$$(0.85 f_c' b)a^2 + (0.003 A_s' E_s - A_s f_y)a - (0.003 A_s' E_s \beta_1 d') = 0 \tag{48.27}$$

where a can be calculated by means of the quadratic equation. Therefore, the nominal moment capacity in a doubly reinforced concrete beam where the compression steel does not yield is

$$M_n = C_c\left(d - \frac{a}{2}\right) + C_s(d - d') \tag{48.28}$$

Reinforcement Ratios. The reinforcement ratio at the balanced strain condition can be obtained in a similar manner as that for beams with tension steel only. For compression steel yielding, the balanced ratio is

$$(\rho - \rho')_b = \frac{0.85 f_c' \beta_1}{f_y}\left(\frac{87{,}000}{87{,}000 + f_y}\right) \tag{48.29}$$

For compression steel not yielding, the balanced ratio is

$$\left(\rho - \frac{\rho' f_s'}{f_y}\right)_b = \frac{0.85 f_c' \beta_1}{f_y}\left(\frac{87{,}000}{87{,}000 + f_y}\right) \tag{48.30}$$

The maximum and minimum reinforcement ratios as given in ACI 10.3.3 and 10.5 are

$$\rho_{\max} = 0.75 \rho_b \tag{48.31}$$

$$\rho_{\min} = \frac{200}{f_y}$$

Prestressed Concrete Strength Design

Elastic Flexural Analysis

In developing elastic equations for prestress, the effects of prestress force, dead load moment, and live load moment are calculated separately, and then the separate stresses are superimposed, giving

$$f = -\frac{F}{A} \pm \frac{Fey}{I} \pm \frac{My}{I} \tag{48.32}$$

where $(-)$ indicates compression and $(+)$ indicates tension. It is necessary to check that the stresses in the extreme fibers remain within the ACI-specified limits under any combination of loadings that may occur. The stress limits for the concrete and prestressing tendons are specified in ACI 18.4 and 18.5 [ACI Committee 318, 1992].

ACI 18.2.6 states that the loss of area due to open ducts shall be considered when computing section properties. It is noted in the commentary that section properties may be based on total area if the effect of the open duct area is considered negligible. In pretensioned members and in post-tensioned members after grouting, section properties can be based on gross sections, net sections, or effective sections using the transformed areas of bonded tendons and nonprestressed reinforcement.

Flexural Strength

The strength of a prestressed beam can be calculated using the methods developed for ordinary reinforced concrete beams, with modifications to account for the differing nature of the stress-strain relationship of prestressing steel compared with ordinary reinforcing steel.

A prestressed beam will fail when the steel reaches a stress f_{ps}, generally less than the tensile strength f_{pu} . For rectangular cross sections the nominal flexural strength is

$$M_n = A_{ps} f_{ps} d - \frac{a}{2} \tag{48.33}$$

where

$$a = \frac{A_{ps} f_{ps}}{0.85 f_c' b} \tag{48.34}$$

The steel stress f_{ps} can be found based on strain compatibility or by using approximate equations such as those given in ACI 18.7.2. The equations in ACI are applicable only if the effective prestress in the steel, f_{se}, which equals P_e/A_{ps}, is not less than $0.5\,f_{pu}$. The ACI equations are as follows.

(a) For members with bonded tendons:

$$f_{ps} = f_{pu}\left(1 - \frac{\gamma_p}{\beta_1}\left[\rho \frac{f_{pu}}{f_c'} + \frac{d}{d_p}(\omega - \omega')\right]\right) \tag{48.35}$$

If any compression reinforcement is taken into account when calculating f_{ps} with Eq. (48.35), the following applies:

$$\left[\rho_p \frac{f_{pu}}{f_c'} + \frac{d}{d_p}(\omega - \omega')\right] \geq 0.17 \tag{48.36}$$

and

$$d' \leq 0.15 d_p$$

(b) For members with unbonded tendons and with a span-to-depth ratio of 35 or less:

$$f_{ps} = f_{se} + 10{,}000 + \frac{f'_c}{100\rho_p} \leq \begin{Bmatrix} f_{py} \\ f_{se} + 60{,}000 \end{Bmatrix} \tag{48.37}$$

(c) For members with unbonded tendons and with a span-to-depth ratio greater than 35:

$$f_{ps} = f_{se} + 10{,}000 + \frac{f'_c}{300\rho_p} \leq \begin{Bmatrix} f_{py} \\ f_{se} + 30{,}000 \end{Bmatrix} \tag{48.38}$$

The flexural strength is then calculated from Eq. (48.33). The design strength is equal to ϕM_n, where $\phi = 0.90$ for flexure.

Reinforcement Ratios

ACI requires that the total amount of prestressed and nonprestressed reinforcement be adequate to develop a factored load at least 1.2 times the cracking load calculated on the basis of a modulus of rupture of $7.5\sqrt{f'_c}$.

To control cracking in members with unbonded tendons, some bonded reinforcement should be uniformly distributed over the tension zone near the extreme tension fiber. ACI specifies the minimum amount of bonded reinforcement as

$$A_s = 0.004A \tag{48.39}$$

where A is the area of the cross section between the flexural tension face and the center of gravity of the gross cross section. ACI 19.9.4 gives the minimum length of the bonded reinforcement.

To ensure adequate ductility, ACI 18.8.1 provides the following requirement:

$$\begin{Bmatrix} \omega_p \\ \omega_p + \left(\dfrac{d}{d_p}\right)(\omega - \omega') \\ \omega_{pw} + \left(\dfrac{d}{d_p}\right)(\omega_w - \omega'_w) \end{Bmatrix} \leq 0.36\beta_1 \tag{48.40}$$

ACI allows each of the terms on the left side to be set equal to 0.85 a/d_p in order to simplify the equation.

When a reinforcement ratio greater than $0.36\beta_1$ is used, ACI 18.8.2 states that the design moment strength shall not be greater than the moment strength based on the compression portion of the moment couple.

48.4 Columns under Bending and Axial Load

Short Columns under Minimum Eccentricity

When a symmetrical column is subjected to a concentric axial load, P, longitudinal strains develop uniformly across the section. Because the steel and concrete are bonded together, the strains in the concrete and steel are equal. For any given strain it is possible to compute the stresses in the concrete and steel using the stress-strain curves for the two materials. The forces in the concrete and steel are equal to the stresses multiplied by the corresponding areas. The total load on the column is the sum of the forces in the concrete and steel:

$$P_o = 0.85f'_c(A_g - A_{st}) + f_yA_{st} \tag{48.41}$$

To account for the effect of incidental moments, ACI 10.3.5 specifies that the maximum design axial load on a column be, for spiral columns,

$$\phi P_{n(\max)} = 0.85\phi[.85f_c'(A_g - A_{st}) + f_y A_{st}] \tag{48.42}$$

and for tied columns,

$$\phi P_{n(\max)} = 0.80\phi[.85f_c'(A_g - A_{st}) + f_y A_{st}] \tag{48.43}$$

For high values of axial load, ϕ values of 0.7 and 0.75 are specified for tied and spiral columns, respectively (ACI 9.3.2.2b) [ACI Committee 318, 1992].

Short columns are sufficiently stocky such that slenderness effects can be ignored.

Short Columns under Axial Load and Bending

Almost all compression members in concrete structures are subjected to moments in addition to axial loads. Although it is possible to derive equations to evaluate the strength of columns subjected to combined bending and axial loads, the equations are tedious to use. For this reason, interaction diagrams for columns are generally computed by assuming a series of strain distributions, each corresponding to a particular point on the interaction diagram, and computing the corresponding values of P and M. Once enough such points have been computed, the results are summarized in an interaction diagram. For examples on determining the interaction diagram, see *Reinforced Concrete Mechanics and Design* by James G. MacGregor [1992] or *Reinforced Concrete Design* by Chu-Kia Wang and Charles G. Salmon [1985].

Figure 48.3 illustrates a series of strain distributions and the resulting points on the interaction diagram. Point A represents pure axial compression. Point B corresponds to crushing at one face and zero tension at the other. If the tensile strength of concrete is ignored, this represents the onset of cracking on the bottom face of the section. All points lower than this in the interaction diagram represent cases in which the section is partially cracked. Point C, the farthest right point, corresponds to the balanced strain condition and represents the change from compression failures for higher loads and tension failures for lower loads. Point D represents a strain distribution where the reinforcement has been strained to several times the yield strain before the concrete reaches its crushing strain.

The horizontal axis of the interaction diagram corresponds to pure bending where $\phi = 0.9$. A transition is required from $\phi = 0.7$ or 0.75 for high axial loads to $\phi = 0.9$ for pure bending. The change in ϕ begins at a capacity ϕP_a, which equals the smaller of the balanced load, ϕP_b , or $0.1f_c'A_g$. Generally, ϕP_b exceeds $0.1f_c'A_g$ except for a few nonrectangular columns.

ACI Publication SP-17A(85), *A Design Handbook for Columns*, contains nondimensional interaction diagrams as well as other design aids for columns [ACI Committee 340, 1990].

Slenderness Effects

ACI 10.11 describes an approximate slenderness-effect design procedure based on the moment magnifier concept. The moments are computed by ordinary frame analysis and multiplied by a moment magnifier that is a function of the factored axial load and the critical buckling load of the column. The following gives a summary of the moment magnifier design procedure for slender columns in frames.

1. *Length of column.* The unsupported length, l_u, is defined in ACI 10.11.1 as the clear distance between floor slabs, beams, or other members capable of giving lateral support to the column.
2. *Effective length.* The effective length factors, k, used in calculating δ_b shall be between 0.5 and 1.0 (ACI 10.11.2.1). The effective length factors used to compute δ_s shall be greater than

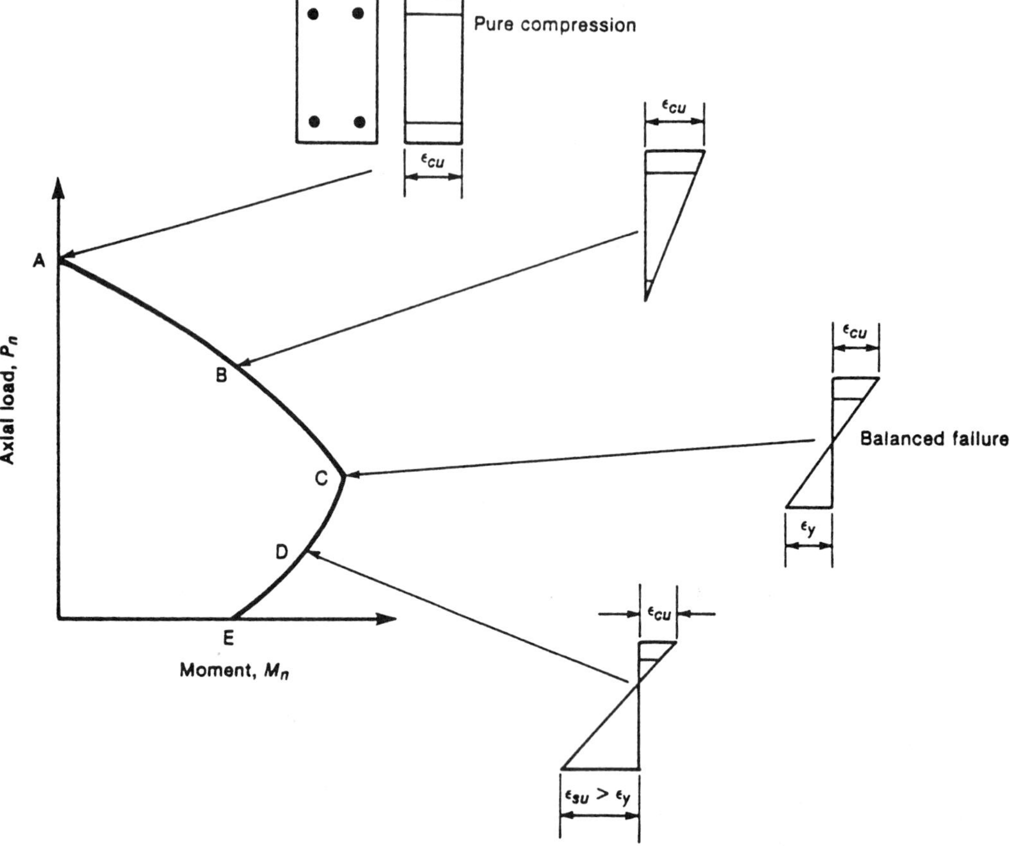

FIGURE 48.3 Strain distributions corresponding to points on interaction diagram.

1 (ACI 10.11.2.2). The effective length factors can be estimated using ACI Fig. R10.11.2 or using ACI Equations (A)–(E) given in ACI R10.11.2. These two procedures require that the ratio, ψ, of the columns and beams be known:

$$\psi = \frac{\sum(E_c I_c / l_c)}{\sum(E_b I_b / l_b)} \tag{48.44}$$

In computing ψ it is acceptable to take the EI of the column as the uncracked gross $E_c I_g$ of the columns and the EI of the beam as 0.5 $E_c I_g$.

3. *Definition of braced and unbraced frames.* The ACI Commentary suggests that a frame is braced if either of the following are satisfied:
   (a) If the stability index, Q, for a story is less than 0.04, where

$$Q = \frac{\sum P_u \Delta_u}{H_u h_s} \leq 0.04 \tag{48.45}$$

   (b) If the sum of the lateral stiffness of the bracing elements in a story exceeds six times the lateral stiffness of all the columns in the story.
4. *Radius of gyration.* For a rectangular cross section r equals 0.3 h, and for a circular cross section r equals 0.25 h. For other sections, r equals $\sqrt{I/A}$.

5. *Consideration of slenderness effects.* ACI 10.11.4.1 allows slenderness effects to be neglected for columns in braced frames when

$$\frac{kl_u}{r} < 34 - 12\frac{M_{1b}}{M_{2b}} \tag{48.46}$$

ACI 10.11.4.2 allows slenderness effects to be neglected for columns in unbraced frames when

$$\frac{kl_u}{r} < 22 \tag{48.47}$$

If kl_u/r exceeds 100, ACI 10.11.4.3 states that design shall be based on second-order analysis.
6. *Minimum moments.* For columns in a braced frame, M_{2b} shall be not less than the value given in ACI 10.11.5.4. In an unbraced frame ACI 10.11.5.5 applies for M_{2s}.
7. *Moment magnifier equation.* ACI 10.11.5.1 states that columns shall be designed for the factored axial load, P_u, and a magnified factored moment, M_c, defined by

$$M_c = \delta_b M_{2b} + \delta_s M_{2s} \tag{48.48}$$

where M_{2b} is the larger factored end moment acting on the column due to loads causing no appreciable sidesway (lateral deflections less than $l/1500$) and M_{2s} is the larger factored end moment due to loads that result in an appreciable sidesway. The moments are computed from a conventional first-order elastic frame analysis. For the above equation, the following apply:

$$\delta_b = \frac{C_m}{1 - P_u/\phi P_c} \geq 1.0 \tag{48.49}$$

$$\delta_s = \frac{1}{1 - \sum P_u/\phi \sum P_c} \geq 1.0$$

For members braced against sidesway, ACI 10.11.5.1 gives $\delta_s = 1.0$.

$$C_m = 0.6 + 0.4\frac{M_{1b}}{M_{2b}} \geq 0.4 \tag{48.50}$$

The ratio M_{1b}/M_{2b} is taken as positive if the member is bent in single curvature and negative if the member is bent in double curvature. Equation (48.50) applies only to columns in braced frames. In all other cases, ACI 10.11.5.3 states that $C_m = 1.0$.

$$P_c = \frac{\pi^2 EI}{(kl_u)^2} \tag{48.51}$$

where

$$EI = \frac{E_c I_g/5 + E_s I_{se}}{1 + \beta_d} \tag{48.52}$$

or, approximately

$$EI = \frac{E_c I_g/2.5}{1 + \beta_d} \tag{48.53}$$

When computing δ_b,

$$\beta_d = \frac{\text{Axial load due to factored dead load}}{\text{Total factored axial load}} \tag{48.54}$$

When computing δ_s,

$$\beta_d = \frac{\text{Factored sustained lateral shear in the story}}{\text{Total factored lateral shear in the story}} \tag{48.55}$$

If δ_b or δ_s is found to be negative, the column should be enlarged. If either δ_b or δ_s exceeds 2.0, consideration should be given to enlarging the column.

Columns under Axial Load and Biaxial Bending

The nominal ultimate strength of a section under biaxial bending and compression is a function of three variables, P_n, M_{nx}, and M_{ny}, which may also be expressed as P_n acting at eccentricities $e_y = M_{nx}/P_n$ and $e_x = M_{ny}/P_n$ with respect to the x and y axes. Three types of failure surfaces can be defined. In the first type, S_1, the three orthogonal axes are defined by P_n, e_x, and e_y; in the second type, S_2, the variables defining the axes are $1/P_n, e_x$, and e_y; and in the third type, S_3, the axes are P_n, M_{nx}, and M_{ny}. In the presentation that follows, the Bresler reciprocal load method makes use of the reciprocal failure surface S_2, and the Bresler load contour method and the PCA load contour method both use the failure surface S_3.

Bresler Reciprocal Load Method

Using a failure surface of type S_2, Bresler proposed the following equation as a means of approximating a point on the failure surface corresponding to prespecified eccentricities e_x and e_y:

$$\frac{1}{P_{ni}} = \frac{1}{P_{nx}} + \frac{1}{P_{ny}} - \frac{1}{P_0} \tag{48.56}$$

where

P_{ni} = nominal axial load strength at given eccentricity along both axes
P_{nx} = nominal axial load strength at given eccentricity along x axis
P_{ny} = nominal axial load strength at given eccentricity along y axis
P_0 = nominal axial load strength for pure compression (zero eccentricity)

Test results indicate that Eq. (48.56) may be inappropriate when small values of axial load are involved, such as when P_n/P_0 is in the range of 0.06 or less. For such cases the member should be designed for flexure only.

Bresler Load Contour Method

The failure surface S_3 can be thought of as a family of curves (load contours) each corresponding to a constant value of P_n. The general nondimensional equation for the load contour at constant P_n may be expressed in the following form:

$$\left(\frac{M_{nx}}{M_{ox}}\right)^{\alpha_1} + \left(\frac{M_{ny}}{M_{oy}}\right)^{\alpha_2} = 1.0 \tag{48.57}$$

where

$M_{nx} = P_n e_y; M_{ny} = P_n e_x$
M_{ox} = M_{nx} capacity at axial load P_n when M_{ny} (or e_x) is zero
M_{oy} = M_{ny} capacity at axial load P_n when M_{nx} (or e_y) is zero

The exponents α_1 and α_2 depend on the column dimensions, amount and arrangement of the reinforcement, and material strengths. Bresler suggests taking $\alpha_1 = \alpha_2 = \alpha$. Calculated values of α vary from 1.15 to 1.55. For practical purposes, α can be taken as 1.5 for rectangular sections and between 1.5 and 2.0 for square sections.

PCA (Parme–Gowens) Load Contour Method

This method has been developed as an extension of the Bresler load contour method in which the Bresler interaction equation (48.57) is taken as the basic strength criterion. In this approach, a point on the load contour is defined in such a way that the biaxial moment strengths M_{nx} and M_{ny} are in the same ratio as the uniaxial moment strengths M_{ox} and M_{oy},

$$\frac{M_{ny}}{M_{nx}} = \frac{M_{oy}}{M_{ox}} = \beta \tag{48.58}$$

The actual value of β depends on the ratio of P_n to P_0 as well as the material and cross-sectional properties, with the usual range of values between 0.55 and 0.70. Charts for determining β can be found in ACI Publication SP-17A(85), *A Design Handbook for Columns* [ACI Committee 340, 1990].

Substituting Eq. (48.58) into Eq. (48.57),

$$\begin{aligned}\left(\frac{\beta M_{ox}}{M_{ox}}\right)^{\alpha} + \left(\frac{\beta M_{oy}}{M_{oy}}\right)^{\alpha} &= 1\\ 2\beta^{\alpha} &= 1\\ \beta^{\alpha} &= 1/2\\ \alpha &= \frac{\log 0.5}{\log \beta}\end{aligned} \tag{48.59}$$

thus,

$$\left(\frac{M_{nx}}{M_{ox}}\right)^{\log 0.5/\log\beta} + \left(\frac{M_{ny}}{M_{oy}}\right)^{\log 0.5/\log\beta} = 1 \tag{48.60}$$

For more information on columns subjected to biaxial bending, see *Reinforced Concrete Design* by Chu-Kia Wang and Charles G. Salmon [1985].

48.5 Shear and Torsion

Reinforced Concrete Beams and One-Way Slabs Strength Design

The cracks that form in a reinforced concrete beam can be due to flexure or a combination of flexure and shear. Flexural cracks start at the bottom of the beam, where the flexural stresses are the largest. Inclined cracks, also called *shear cracks* or *diagonal tension cracks*, are due to a combination of flexure and shear. Inclined cracks must exist before a shear failure can occur.

Inclined cracks form in two different ways. In thin-walled I-beams in which the shear stresses in the web are high while the flexural stresses are low, a web-shear crack occurs. The inclined cracking shear can be calculated as the shear necessary to cause a principal tensile stress equal to the tensile strength of the concrete at the centroid of the beam.

In most reinforced concrete beams, however, flexural cracks occur first and extend vertically in the beam. These alter the state of stress in the beam and cause a stress concentration near the tip of the crack. In time, the flexural cracks extend to become flexure-shear cracks. Empirical equations have been developed to calculate the flexure-shear cracking load, since this cracking cannot be predicted by calculating the principal stresses.

In the ACI Code, the basic design equation for the shear capacity of concrete beams is as follows:

$$V_u \leq \phi V_n \tag{48.61}$$

where V_u is the shear force due to the factored loads, ϕ is the strength reduction factor equal to 0.85 for shear, and V_n is the nominal shear resistance, which is given by

$$V_n = V_c + V_s \tag{48.62}$$

where V_c is the shear carried by the concrete and V_s is the shear carried by the shear reinforcement.

The torsional capacity of a beam as given in ACI 11.6.5 is as follows:

$$T_u \leq \phi T_n \tag{48.63}$$

where T_u is the torsional moment due to factored loads, ϕ is the strength reduction factor equal to 0.85 for torsion, and T_n is the nominal torsional moment strength given by

$$T_n = T_c + T_c \tag{48.64}$$

where T_c is the torsional moment strength provided by the concrete and T_s is the torsional moment strength provided by the torsion reinforcement.

Design of Beams and One-Way Slabs Without Shear Reinforcement: for Shear

The critical section for shear in reinforced concrete beams is taken at a distance d from the face of the support. Sections located at a distance less than d from the support are designed for the shear computed at d.

Shear Strength Provided by Concrete. Beams without web reinforcement will fail when inclined cracking occurs or shortly afterwards. For this reason the shear capacity is taken equal to the inclined cracking shear. ACI gives the following equations for calculating the shear strength provided by the concrete for beams without web reinforcement subject to shear and flexure:

$$V_c = 2\sqrt{f_c'}b_w d \tag{48.65}$$

or, with a more detailed equation:

$$V_c = \left(1.9\sqrt{f_c'} + 2500\rho_w \frac{V_u d}{M_u}\right) b_w d \leq 3.5\sqrt{f_c'}b_w d \tag{48.66}$$

The quantity $V_u d/M_u$ is not to be taken greater than 1.0 in computing V_c where M_u is the factored moment occurring simultaneously with V_u at the section considered.

Combined Shear, Moment, and Axial Load. For members that are also subject to axial compression, ACI modifies Eq. (48.65) as follows (ACI 11.3.1.2):

$$V_c = 2\left(1 + \frac{N_u}{2000A_k}\right)\sqrt{f_c'}b_w d \tag{48.67}$$

where N_u is positive in compression. ACI 11.3.2.2 contains a more detailed calculation for the shear strength of members subject to axial compression.

For members subject to axial tension, ACI 11.3.1.3 states that shear reinforcement shall be designed to carry total shear. As an alternative, ACI 11.3.2.3 gives the following for the shear strength of members subject to axial tension:

$$V_c = 2\left(1 + \frac{N_u}{500A_g}\right)\sqrt{f_c'}b_w d \tag{48.68}$$

where N_u is negative in tension. In Eq. (48.67) and (48.68) the terms $\sqrt{f_c'}$, N_u/A_g, 2000, and 500 all have units of psi.

Combined Shear, Moment, and Torsion. For members subject to torsion, ACI 11.3.1.4 gives the equation for the shear strength of the concrete as the following:

$$V_c = \frac{2\sqrt{f_c'}b_w d}{\sqrt{1 + (2.5C_t T_u/V_u)^2}} \tag{48.69}$$

where

$$T_u \geq \phi(0.5\sqrt{f_c'}\sum x^2 y)$$

Design of Beams and One-Way Slabs Without Shear Reinforcements: for Torsion. ACI 11.6.1 requires that torsional moments be considered in design if

$$T_u \geq \phi(0.5\sqrt{f_c'}\sum x^2 y) \tag{48.70}$$

Otherwise, torsion effects may be neglected.

The critical section for torsion is taken at a distance d from the face of support, and sections located at a distance less than d are designed for the torsion at d. If a concentrated torque occurs within this distance, the critical section is taken at the face of the support.

Torsional Strength Provided by Concrete. Torsion seldom occurs by itself; bending moments and shearing forces are typically present also. In an uncracked member, shear forces as well as torques produce shear stresses. Flexural shear forces and torques interact in a way that reduces the strength of the member compared with what it would be if shear or torsion were acting alone. The interaction between shear and torsion is taken into account by the use of a circular interaction equation. For more information, refer to *Reinforced Concrete Mechanics and Design* by James G. MacGregor [1992].

The torsional moment strength provided by the concrete is given in ACI 11.6.6.1 as

$$T_c = \frac{0.8\sqrt{f_c'}x^2 y}{\sqrt{1 + (0.4V_u/C_t T_u)^2}} \tag{48.71}$$

Combined Torsion and Axial Load. For members subject to significant axial tension, ACI 11.6.6.2 states that the torsion reinforcement must be designed to carry the total torsional moment, or as an alternative modify T_c as follows:

$$T_c = \frac{0.8\sqrt{f_c'}x^2 y}{\sqrt{1 + (0.4V_u/C_t T_u)^2}}\left(1 + \frac{N_u}{500A_g}\right) \tag{48.72}$$

where N_u is negative for tension.

Design of Beams and One-Way Slabs without Shear Reinforcement:

Minimum Reinforcement. ACI 11.5.5.1 requires a minimum amount of web reinforcement to be provided for shear and torsion if the factored shear force V_u exceeds one half the shear strength provided by the concrete ($V_u \geq 0.5\phi V_c$) except in the following:

(a) Slabs and footings
(b) Concrete joist construction
(c) Beams with total depth not greater than 10 inches, 2 ½ times the thickness of the flange, or ½ the width of the web, whichever is greatest

The minimum area of shear reinforcement shall be at least

$$A_{v(\min)} = \frac{50 b_w s}{f_y} \quad \text{for} \quad T_u < \phi(0.5\sqrt{f_c'} \sum x^2 y) \tag{48.73}$$

When torsion is to be considered in design, the sum of the closed stirrups for shear and torsion must satisfy the following:

$$A_v + 2A_t \geq \frac{50 b_w s}{f_y} \tag{48.74}$$

where A_v is the area of two legs of a closed stirrup and A_t is the area of only one leg of a closed stirrup.

Design of Stirrup Reinforcement for Shear and Torsion

Shear Reinforcement. Shear reinforcement is to be provided when $V_u \geq \phi V_c$, such that

$$V_s \geq \frac{V_u}{\phi} - V_c \tag{48.75}$$

The design yield strength of the shear reinforcement is not to exceed 60,000 psi.

When the shear reinforcement is perpendicular to the axis of the member, the shear resisted by the stirrups is

$$V_s = \frac{A_v f_y d}{s} \tag{48.76}$$

If the shear reinforcement is inclined at an angle α, the shear resisted by the stirrups is

$$V_s = \frac{A_v f_y (\sin\alpha + \cos\alpha) d}{s} \tag{48.77}$$

The maximum shear strength of the shear reinforcement is not to exceed $8\sqrt{f_c'} b_w d$ as stated in ACI 11.5.6.8.

Spacing Limitations for Shear Reinforcement. ACI 11.5.4.1 sets the maximum spacing of vertical stirrups as the smaller of $d/2$ or 24 inches. The maximum spacing of inclined stirrups is such that a 45° line extending from midheight of the member to the tension reinforcement will intercept at least one stirrup.

If V_s exceeds $4\sqrt{f_c'} b_w d$, the maximum allowable spacings are reduced to one half of those just described.

Torsion Reinforcement. Torsion reinforcement is to be provided when $T_u \geq \phi T_c$, such that

$$T_s \geq \frac{T_u}{\phi} - T_c \tag{48.78}$$

The design yield strength of the torsional reinforcement is not to exceed 60,000 psi.

The torsional moment strength of the reinforcement is computed by

$$T_s = \frac{A_t \alpha_t x_1 y_1 f_y}{s} \tag{48.79}$$

where

$$\alpha_t = [0.66 + 0.33(y_t/x_t)] \geq 1.50 \tag{48.80}$$

where A_t is the area of one leg of a closed stirrup resisting torsion within a distance s. The torsional moment strength is not to exceed 4 T_c as given in ACI 11.6.9.4.

Longitudinal reinforcement is to be provided to resist axial tension that develops as a result of the torsional moment (ACI 11.6.9.3). The required area of longitudinal bars distributed around the perimeter of the closed stirrups that are provided as torsion reinforcement is to be

$$A_l \geq 2A_t \frac{(x_1 + y_1)}{s}$$

$$A_l \geq \left[\frac{400xs}{f_y} \left(\frac{T_u}{T_u + \dfrac{V_u}{3C_t}} \right) = 2A_t \right] \left(\frac{x_1 + y_1}{s} \right) \tag{48.81}$$

Spacing Limitations for Torsion Reinforcement. ACI 11.6.8.1 gives the maximum spacing of closed stirrups as the smaller of $(x_1 + y_1)/4$ or 12 inches.

The longitudinal bars are to be spaced around the circumference of the closed stirrups at not more than 12 inches apart. At least one longitudinal bar is to be placed in each corner of the closed stirrups (ACI 11.6.8.2).

Design of Deep Beams

ACI 11.8 covers the shear design of deep beams. This section applies to members with $l_n/d < 5$ that are loaded on one face and supported on the opposite face so that compression struts can develop between the loads and the supports. For more information on deep beams, see *Reinforced Concrete Mechanics and Design*, 2nd ed. by James G. MacGregor [1992].

The basic design equation for simple spans deep beams is

$$V_u \leq \phi(V_c + V_s) \tag{48.82}$$

where V_c is the shear carried by the concrete and V_s is the shear carried by the vertical and horizontal web reinforcement.

The shear strength of deep beams shall not be taken greater than

$$V_n = 8\sqrt{f_c'} b_w d \quad \text{for} \quad l_n/d < 2$$

$$V_n = \frac{2}{3}\left(10 + \frac{l_n}{d}\right)\sqrt{f_c'} b_w d \quad \text{for} \quad 2 \leq l_n/d \leq 5 \tag{48.83}$$

Design for shear is done at a critical section located at 0.15 l_n from the face of support in uniformly loaded beams, and at the middle of the shear span for beams with concentrated loads. For both cases, the critical section shall not be farther than d from the face of the support. The shear reinforcement required at this critical section is to be used throughout the span.

The shear carried by the concrete is given by

$$V_c = 2\sqrt{f_c'}b_w d \tag{48.84}$$

or, with a more detailed calculation,

$$V_c = \left(3.5 - 2.5\frac{M_u}{V_u d}\right)\left(1.9\sqrt{f_c'} + 2500\rho_w\frac{V_u d}{M_u}\right)b_w d \le 6\sqrt{f_c'}b_w d \tag{48.85}$$

where

$$\left(3.5 - 2.5\frac{M_u}{V_u d}\right) \le 2.5 \tag{48.86}$$

In Eqs. (48.85) and (48.86) M_u and V_u are the factored moment and shear at the critical section.

Shear reinforcement is to be provided when $V_u \ge \phi V_c$ such that

$$V_s = \frac{V_u}{\phi} - V_c \tag{48.87}$$

where

$$V_s = \left[\frac{A_v}{s}\left(\frac{1 + l_n/d}{12}\right) + \frac{A_{vh}}{s_2}\left(\frac{11 - l_n/d}{12}\right)\right]f_y d \tag{48.88}$$

where A_v and s are the area and spacing of the vertical shear reinforcement and A_{vh} and s_2 refer to the horizontal shear reinforcement.

ACI 11.8.9 and 11.8.10 require minimum reinforcement in both the vertical and horizontal sections as follows:

(a) vertical direction

$$A_v \ge 0.0015 b_w s \tag{48.89}$$

where

$$s \le \begin{Bmatrix} d/5 \\ 18 \text{ in.} \end{Bmatrix} \tag{48.90}$$

(b) horizontal direction

$$A_{vh} \ge 0.0025 b_w s_2 \tag{48.91}$$

where

$$s_2 \le \begin{Bmatrix} d/3 \\ 18 \text{ in.} \end{Bmatrix} \tag{48.92}$$

Prestressed Concrete Beams and One-Way Slabs Strength Design

At loads near failure, a prestressed beam is usually heavily cracked and behaves similarly to an ordinary reinforced concrete beam. Many of the equations developed previously for design of web reinforcement for nonprestressed beams can also be applied to prestressed beams.

Shear design is based on the same basic equation as before,

$$V_u \leq \phi(V_c + V_s)$$

where $\phi = 0.85$.

The critical section for shear is taken at a distance $h/2$ from the face of the support. Sections located at a distance less than $h/2$ are designed for the shear computed at $h/2$.

Shear Strength Provided by the Concrete

The shear force resisted by the concrete after cracking has occurred is taken as equal to the shear that caused the first diagonal crack. Two types of diagonal cracks have been observed in tests of prestressed concrete.

1. Flexure-shear cracks, occurring at nominal shear V_{ci}, start as nearly vertical flexural cracks at the tension face of the beam, then spread diagonally upward toward the compression face.
2. Web shear cracks, occurring at nominal shear V_{cw}, start in the web due to high diagonal tension, then spread diagonally both upward and downward.

The shear strength provided by the concrete for members with effective prestress force not less than 40% of the tensile strength of the flexural reinforcement is

$$V_c = \left(0.6\sqrt{f_c'} + 700\frac{V_u d}{M_u}\right)b_w d \leq 2\sqrt{f_c'}b_w d \tag{48.93}$$

V_c may also be computed as the lesser of V_{ci} and V_{cw}, where

$$V_{ci} = 0.6\sqrt{f_c'}b_w d + V_d + \frac{V_i M_{cr}}{M_{\max}} \geq 1.7\sqrt{f_c'}b_w d \tag{48.94}$$

$$M_{cr} = \left(\frac{I}{y_t}\right)\left(6\sqrt{f_c'} + f_{pc} - f_d\right) \tag{48.95}$$

$$V_{cw} = \left(3.5\sqrt{f_c'} + 0.3f_{pc}\right)b_w d + V_p \tag{48.96}$$

In Eqs. (48.94) and (48.96) d is the distance from the extreme compression fiber to the centroid of the prestressing steel or $0.8h$, whichever is greater.

Shear Strength Provided by the Shear Reinforcement

Shear reinforcement for prestressed concrete is designed in a similar manner as for reinforced concrete, with the following modifications for minimum amount and spacing.

Minimum Reinforcement. The minimum area of shear reinforcement shall be at least

$$A_{v(\min)} = \frac{50 b_w s}{f_y} \quad \text{for} \quad T_u < \phi\left(0.5\sqrt{f_c'}\Sigma x^2 y\right) \tag{48.97}$$

or

$$A_{v(\min)} = \frac{A_{ps} f_{pu} s}{80 f_y d} \sqrt{\frac{d}{b_w}} \tag{48.98}$$

Spacing Limitations for Shear Reinforcement. ACI 11.5.4.1 sets the maximum spacing of vertical stirrups as the smaller of (3/4)h or 24 in. The maximum spacing of inclined stirrups is such that a 45° line extending from midheight of the member to the tension reinforcement will intercept at least one stirrup.

If V_s exceeds $4\sqrt{f_c'} b_w d$, the maximum allowable spacings are reduced to one-half of those just described.

48.6 Development of Reinforcement

The development length, l_d, is the shortest length of bar in which the bar stress can increase from zero to the yield strength, f_y. If the distance from a point where the bar stress equals f_y to the end of the bar is less than the development length, the bar will pull out of the concrete. Development lengths are different for tension and compression.

Development of Bars in Tension

ACI Fig. R12.2 gives a flow chart for determining development length. The steps are outlined below.

The basic tension development lengths have been found to be (ACI 12.2.2). For no. 11 and smaller bars and deformed wire:

$$l_{db} = \frac{0.04 A_b f_y}{\sqrt{f_c'}} \tag{48.99}$$

For no. 14 bars:

$$l_{db} = \frac{0.085 f_y}{\sqrt{f_c'}} \tag{48.100}$$

For no. 18 bars:

$$l_{db} = \frac{0.125 f_y}{\sqrt{f_c'}} \tag{48.101}$$

where $\sqrt{f_c'}$ is not to be taken greater than 100 psi.

The development length, l_d, is computed as the product of the basic development length and modification factors given in ACI 12.2.3, 12.2.4, and 12.2.5. The development length obtained from ACI 12.2.2 and 12.2.3.1 through 12.2.3.5 shall not be less than

$$\frac{0.03 d_b f_y}{\sqrt{f_c'}} \tag{48.102}$$

as given in ACI 12.2.3.6.

The length computed from ACI 12.2.2 and 12.2.3 is then multiplied by factors given in ACI 12.2.4 and 12.2.5. The factors given in ACI 12.2.3.1 through 12.2.3.3 and 12.2.4 are required, but the factors in ACI 12.2.3.4, 12.2.3.5, and 12.2.5 are optional.

The development length is not to be less than 12 inches (ACI 12.2.1).

Development of Bars in Compression

The basic compression development length is (ACI 12.3.2)

$$l_{db} = \frac{0.02 d_b f_y}{\sqrt{f_c'}} \geq 0.003 d_b f_y \tag{48.103}$$

The development length, l_d, is found as the product of the basic development length and applicable modification factors given in ACI 12.3.3.

The development length is not to be less than 8 inches (ACI 12.3.1).

Development of Hooks in Tension

The basic development length for a hooked bar with $f_y = 60{,}000$ psi is (ACI 12.5.2)

$$l_{db} = \frac{1200 d_b}{\sqrt{f_c'}} \tag{48.104}$$

The development length, l_{dh}, is found as the product of the basic development length and applicable modification factors given in ACI 12.5.3.

The development length of the hook is not to be less than 8 bar diameters or 6 inches (ACI 12.5.1).

Hooks are not to be used to develop bars in compression.

Splices, Bundled Bars, and Web Reinforcement

Splices

Tension Lap Splices. ACI 12.15 distinguishes between two types of tension lap splices depending on the amount of reinforcement provided and the fraction of the bars spliced in a given length—see ACI Table R12.15.2. The splice lengths for each splice class are as follows:

Class A splice : $1.0 l_d$

Class B splice : $1.3 l_d$

where l_d is the tensile development length as computed in ACI 12.2 without the modification factor for excess reinforcement given in ACI 12.2.5. The minimum splice length is 12 inches.

Lap splices are not to be used for bars larger than no. 11 except at footing to column joints and for compression lap splices of no. 14 and no. 18 bars with smaller bars (ACI 12.14.2.1). The center-to-center distance between two bars in a lap splice cannot be greater than one-fifth the required lap splice length with a maximum of 6 inches (ACI 12.14.2.3). ACI 21.3.2.3 requires that tension lap splices of flexural reinforcement in beams resisting seismic loads be enclosed by hoops or spirals.

Compression Lap Splices. The splice length for a compression lap splice is given in ACI 12.16.1 as

$$l_s = 0.0005 f_y d_b \quad \text{for} \quad f_y \leq 60{,}000 \text{ psi} \tag{48.105}$$

$$l_s = (0.0009 f_y - 24) d_b \quad \text{for} \quad f_y > 60{,}000 \text{ psi} \tag{48.106}$$

but not less than 12 inches. For f_c' less than 3000 psi, the lap length must be increased by one-third.

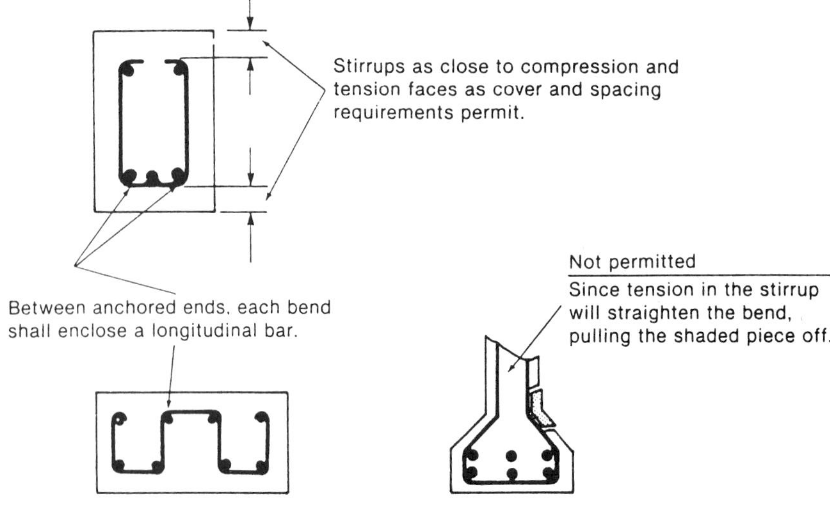
Stirrups as close to compression and tension faces as cover and spacing requirements permit.
Between anchored ends, each bend shall enclose a longitudinal bar.
Not permitted
Since tension in the stirrup will straighten the bend, pulling the shaded piece off.

(a) General requirements.

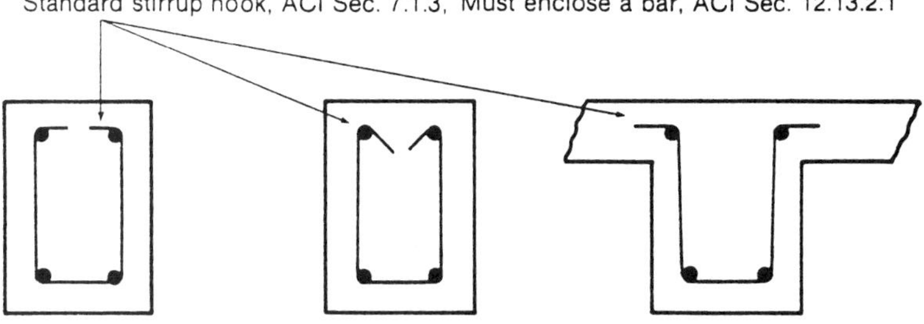
Standard stirrup hook, ACI Sec. 7.1.3, Must enclose a bar, ACI Sec. 12.13.2.1

(b) Stirrup anchorage requirements for No. 5 and smaller bars as per ACI Secs. 7.1.3 and 12.13.2.1.

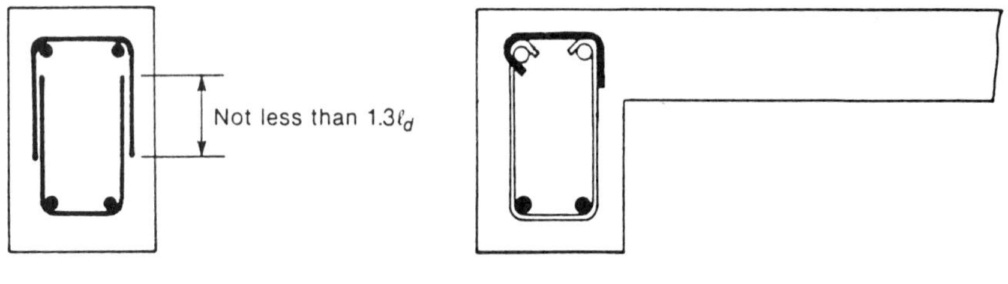
Not less than $1.3\ell_d$

(c) Stirrup anchorage as per ACI Sec. 12.13.5.

(d) Two piece closed stirrup —Beams with torsion or compression reinforcement. ACI Secs. 7.11 and 11.6.7.3.

FIGURE 48.4 Stirrup detailing requirements. (*Source:* Wang and Salmon, 1985.)

When different size bars are lap spliced in compression, the splice length is to be the larger of:

1. Compression splice length of the smaller bar, or
2. Compression development length of larger bar.

Compression lap splices are allowed for no. 14 and no. 18 bars to no. 11 or smaller bars (ACI 12.16.2).

End-Bearing Splices. End-bearing splices are allowed for compression only where the compressive stress is transmitted by bearing of square cut ends held in concentric contact by a suitable device. According to ACI 12.16.4.2 bar ends must terminate in flat surfaces within $1\frac{1}{2}^\circ$ of right angles to the axis of the bars and be fitted within 3° of full bearing after assembly. End-bearing splices are only allowed in members containing closed ties, closed stirrups, or spirals.

Welded Splices or Mechanical Connections. Bars stressed in tension or compression may be spliced by welding or by various mechanical connections. ACI 12.14.3, 12.15.3, 12.15.4, and 12.16.3 govern the use of such splices. For further information see *Reinforced Concrete Design*, by Chu-Kia Wang and Charles G. Salmon [1985].

Bundled Bars

The requirements of ACI 12.4.1 specify that the development length for bundled bars be based on that for the individual bar in the bundle, increased by 20% for a three-bar bundle and 33% for a four-bar bundle. ACI 12.4.2 states that "a unit of bundled bars shall be treated as a single bar of a diameter derived from the equivalent total area" when determining the appropriate modification factors in ACI 12.2.3 and 12.2.4.3.

Web Reinforcement

ACI 12.13.1 requires that the web reinforcement be as close to the compression and tension faces as cover and bar-spacing requirements permit. The ACI Code requirements for stirrup anchorage are illustrated in Fig. 48.4.

(a) ACI 12.13.3 requires that each bend away from the ends of a stirrup enclose a longitudinal bar, as seen in Fig. 48.4(a).

(b) For no. 5 or D31 wire stirrups and smaller with any yield strength and for no. 6, 7, and 8 bars with a yield strength of 40,000 psi or less, ACI 12.13.2.1 allows the use of a standard hook around longitudinal reinforcement, as shown in Fig. 48.4(b).

(c) For no. 6, 7, and 8 stirrups with f_y greater than 40,000 psi, ACI 12.13.2.2 requires a standard hook around a longitudinal bar plus an embedment between midheight of the member and the outside end of the hook of at least $0.014 d_b f_y / \sqrt{f_c'}$.

(d) Requirements for welded wire fabric forming U stirrups are given in ACI 12.13.2.3.

(e) Pairs of U stirrups that form a closed unit shall have a lap length of $1.3 l_d$ as shown in Fig. 48.4(c). This type of stirrup has proven unsuitable in seismic areas.

(f) Requirements for longitudinal bars bent to act as shear reinforcement are given in ACI 12.13.4.

48.7 Two-Way Systems

Definition

When the ratio of the longer to the shorter spans of a floor panel drops below 2, the contribution of the longer span in carrying the floor load becomes substantial. Since the floor transmits loads in two directions, it is defined as a *two-way system*, and flexural reinforcement is designed for both directions. Two-way systems include *flat plates, flat slabs, two-way slabs,* and *waffle slabs* (see Fig. 48.5). The choice between these different types of two-way systems is largely a matter of the architectural layout, magnitude of the design loads, and span lengths. A flat plate is simply a slab of uniform thickness supported directly on columns, generally suitable for relatively light loads. For larger loads and spans, a flat slab becomes more suitable with the column capitals and drop panels providing higher shear and flexural strength. A slab supported on beams on all sides of each floor panel is generally referred to as a two-way slab. A waffle slab is equivalent to a two-way joist system or may be visualized as a solid slab with recesses in order to decrease the weight of the slab.

Design Procedures

The ACI Code [ACI Committee 318, 1992] states that a two-way slab system "may be designed by any procedure satisfying conditions of equilibrium and geometric compatibility if shown that the design strength at every section is at least equal to the required strength.... and that all serviceability conditions, including specified limits on deflections, are met" (p. 204). There are a number of possible approaches to the analysis and design of two-way systems based on elastic theory, limit analysis, finite element analysis, or combination of elastic theory and limit analysis. The designer is permitted by the ACI Code to adopt any of these approaches provided that all safety and serviceability criteria are satisfied. In general, only for cases of a complex two-way system or unusual loading would a finite element analysis be chosen as the design approach. Otherwise, more practical design approaches are preferred. The ACI Code details two procedures—the *direct design method* and the *equivalent frame method*—for the design of floor systems with or without beams. These procedures were derived from analytical studies based on elastic theory in conjunction with aspects of limit analysis and results of experimental tests. The primary difference between the direct

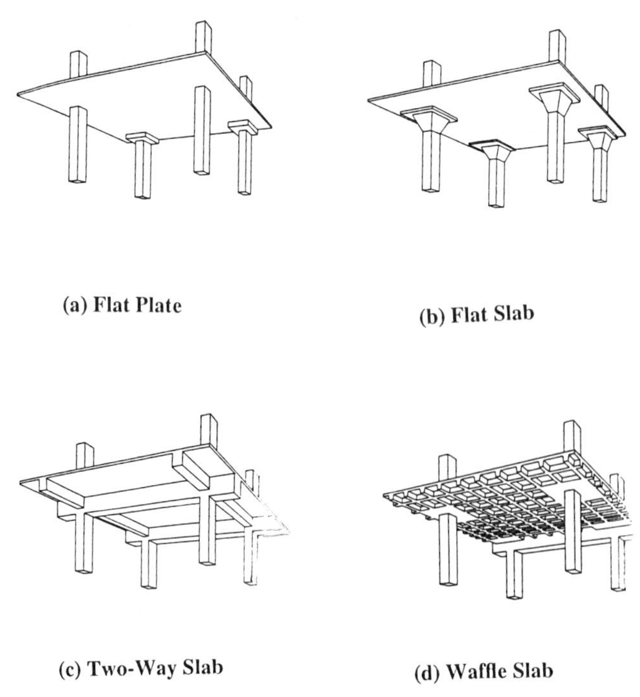

FIGURE 48.5 Two-way systems.

design method and equivalent frame method is in the way moments are computed for two-way systems.

The *yield-line theory* is a limit analysis method devised for slab design. Compared to elastic theory, the yield-line theory gives a more realistic representation of the behavior of slabs at the ultimate limit state, and its application is particularly advantageous for irregular column spacing. While the yield-line method is an upper-bound limit design procedure, *strip method* is considered to give a lower-bound design solution. The strip method offers a wide latitude of design choices and it is easy to use; these are often cited as the appealing features of the method.

Some of the earlier design methods based on moment coefficients from elastic analysis are still favored by many designers. These methods are easy to apply and give valuable insight into slab behavior; their use is especially justified for many irregular slab cases where the preconditions of the direct design method are not met or when column interaction is not significant. Table 48.7 lists the moment coefficients taken from method 2 of the 1963 ACI Code. As in the 1989 code, two-way slabs are divided into column strips and middle strips as indicated by Fig. 48.6, where l_1 and l_2 are the center-to-center span lengths of the floor panel. A column strip is a design strip with a width on each side of a column centerline equal to $0.25l_2$ or $0.25l_1$, whichever is less. A middle strip is a design strip bounded by two column strips. Taking the moment coefficients from Table 48.7, bending moments per unit width M for the middle strips are computed from the formula

$$M = (\text{Coef.})wl_s^2$$

Table 48.7 Elastic Moment Coefficients for Two-Way Slabs

	Short Span						Long Span, All Span Ratios
	Span Ratio, Short/Long						
Moments	1.0	0.9	0.8	0.7	0.6	0.5 and less	
Case 1—Interior panels							
Negative moment at:							
Continuous edge	0.033	0.040	0.048	0.055	0.063	0.083	0.033
Discontinuous edge	—	—	—	—	—	—	—
Positive moment at midspan	0.025	0.030	0.036	0.041	0.047	0.062	0.025
Case 2—One edge discontinuous							
Negative moment at:							
Continuous edge	0.041	0.048	0.055	0.062	0.069	0.085	0.041
Discontinuous edge	0.021	0.024	0.027	0.031	0.035	0.042	0.021
Positive moment at midspan	0.031	0.036	0.041	0.047	0.052	0.064	0.031
Case 3—Two edges discontinuous							
Negative moment at:							
Continuous edge	0.049	0.057	0.064	0.071	0.078	0.090	0.049
Discontinuous edge	0.025	0.028	0.032	0.036	0.039	0.045	0.025
Positive moment at midspan:	0.037	0.043	0.048	0.054	0.059	0.068	0.037
Case 4—Three edges discontinuous							
Negative moment at:							
Continuous edge	0.058	0.066	0.074	0.082	0.090	0.098	0.058
Discontinuous edge	0.029	0.033	0.037	0.041	0.045	0.049	0.029
Positive moment at midspan	0.044	0.050	0.056	0.062	0.068	0.074	0.044
Case 5—Four edges discontinuous							
Negative moment at:							
Continuous edge	—	—	—	—	—	—	—
Discontinuous edge	0.033	0.038	0.043	0.047	0.053	0.055	0.033
Positive moment at midspan	0.050	0.057	0.064	0.072	0.080	0.083	0.050

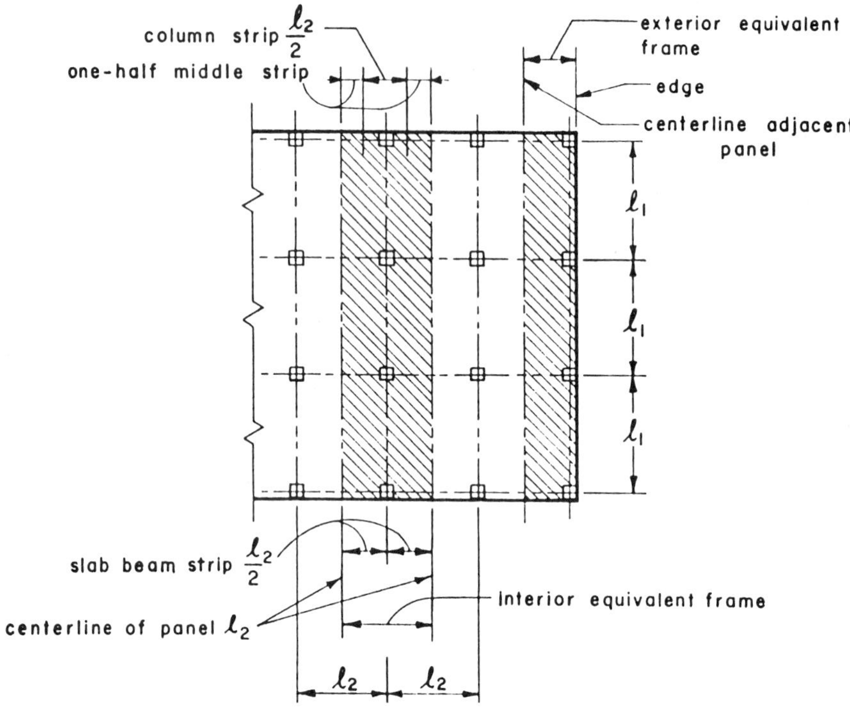

FIGURE 48.6 Definitions of equivalent frame, column strip, and middle strip. (*Source:* ACI Committee 318, 1992.)

where w is the total uniform load per unit area and l_s is the shorter span length of l_1 and l_2. The average moments per unit width in the column strip is taken as two-thirds of the corresponding moments in the middle strip.

Minimum Slab Thickness and Reinforcement

ACI Code Section 9.5.3 contains requirements to determine minimum slab thickness of a two-way system for deflection control. For slabs without beams, the thickness limits are summarized by Table 48.8, but thickness must not be less than 5 in. for slabs without drop panels or 4 in. for slabs with drop panels. In Table 48.8 l_n is the length of clear span in the long direction and α is the ratio of flexural stiffness of beam section to flexural stiffness of a width of slab bounded laterally by centerline of adjacent panel on each side of beam.

For slabs with beams, it is necessary to compute the minimum thickness h from

Table 48.8 Minimum Thickness of Two-Way Slabs without Beams

Yield Stress f_y, psi[1]	Exterior Panels		
	Without Edge Beams	With Edge Beams[2]	Interior Panels
	Without Drop Panels		
40,000	$l_n/33$	$l_n/36$	$l_n/36$
60,000	$l_n/30$	$l_n/33$	$l_n/33$
	With Drop Panels		
40,000	$l_n/36$	$l_n/40$	$l_n/40$
60,000	$l_n/33$	$l_n/36$	$l_n/36$

[1]For values of reinforcement yield stress between 40,000 and 60,000 psi minimum thickness shall be obtained by linear interpolation.

[2]Slabs with beams between columns along exterior edges. The value of α for the edge beam shall not be less than 0.8.

Source: ACI Committee 318, 1992.

$$h = \frac{l_n\left(0.8 + \dfrac{f_y}{200{,}000}\right)}{36 + 5\beta\left[\alpha_m - 0.12\left(1 + \dfrac{1}{\beta}\right)\right]} \tag{48.108}$$

but not less than

$$h = \frac{l_n\left(0.8 + \dfrac{f_y}{200{,}000}\right)}{36 + 9\beta} \tag{48.109}$$

and need not be more than

$$h = \frac{l_n\left(0.8 + \dfrac{f_y}{200{,}000}\right)}{36} \tag{48.110}$$

where β is the ratio of clear spans in long-to-short direction and α_m is the average value of α for all beams on edges of a panel. In no case should the slab thickness be less than 5 in. for $\alpha_m < 2.0$ or less than 3 ½ in. for $\alpha_m \geq 2.0$.

Minimum reinforcement in two-way slabs is governed by shrinkage and temperature controls to minimize cracking. The minimum reinforcement area stipulated by the ACI Code shall not be less than 0.0018 times the gross concrete area when grade 60 steel is used (0.0020 when grade 40 or grade 50 is used). The spacing of reinforcement in two-way slabs shall exceed neither two times the slab thickness nor 18 in.

Direct Design Method

The direct design method consists of a set of rules for the design of two-way slabs with or without beams. Since the method was developed assuming simple designs and construction, its application is restricted by the code to two-way systems with a minimum of three continuous spans, successive span lengths that do not differ by more than one-third, columns with offset not more than 10% of the span, and all loads are due to gravity only and uniformly distributed with live load not exceeding three times dead load. The direct design method involves three fundamental steps: (1) determine the total factored static moment; (2) distribute the static moment to negative and positive sections; and (3) distribute moments to column and middle strips and to beams, if any. The total factored static moment M_o for a span bounded laterally by the centerlines of adjacent panels (see Fig. 48.6) is given by

$$M_o = \frac{w_u l_2 l_n^2}{8} \tag{48.111}$$

In an interior span, $0.65M_o$ is assigned to each negative section and $0.35M_o$ is assigned to the positive section. In an end span, M_o is distributed according to Table 48.9. If the ratio of dead load to live load is less than 2, the effect of pattern loading is accounted for by increasing the positive moment following provisions in ACI Section 13.6.10. Negative and positive moments are then proportioned to the column strip following the percentages in Table 48.10, where β_t is the ratio of the torsional stiffness of edge beam section to flexural stiffness of a width of slab equal to span length of beam. The remaining moment not resisted by the column strip is proportionately assigned to the corresponding half middle strip. If beams are present, they are proportioned to

Table 48.9 Direct Design Method—Distribution of Moment in End Span

	(1)	(2)	(3)	(4)	(5)
		Slab with Beams	Slab without Beams between Interior Supports		
	Exterior Edge Unrestrained	between All Supports	Without Edge Beam	With Edge Beam	Exterior Edge Fully Restrained
Interior negative-factored moment	0.75	0.70	0.70	0.70	0.65
Positive-factored moment	0.63	0.57	0.52	0.50	0.35
Exterior negative-factored moment	0	0.16	0.26	0.30	0.65

Source: ACI Committee 318, 1992.

resist 85% of column strip moments. When $(\alpha l_2/l_1)$ is less than 1.0, the proportion of column strip moments resisted by beams is obtained by linear interpolation between 85% and zero. The shear in beams is determined from loads acting on tributary areas projected from the panel corners at 45 degrees.

Equivalent Frame Method

For two-way systems not meeting the geometric or loading preconditions of the direct design method, design moments may be computed by the equivalent frame method. This is a more general method and involves the representation of the three-dimensional slab system by dividing it into a series of two-dimensional "equivalent" frames (Fig. 48.6). The complete analysis of a two-way system consists of analyzing the series of equivalent interior and exterior frames that span longitudinally and transversely through the system. Each equivalent frame, which is centered on a column line and bounded by the center lines of the adjacent panels, comprises a horizontal slab-beam strip and equivalent columns extending above and below the slab beam (Fig. 48.7). This structure is analyzed as a frame for loads acting in the plane of the frame, and the moments obtained at critical sections across the slab-beam strip are distributed to the column strip, middle strip, and beam in the same manner as the direct design method (see Table 48.10). In its original development, the equivalent frame method assumed that analysis would be done by moment distribution. Presently, frame analysis is more easily accomplished in design practice with computers using general purpose programs based on the direct stiffness method. Consequently, the equivalent frame method is now often used as a method for modeling a two-way system for computer analysis.

Table 48.10 Proportion of Moment to Column Strip in Percent

Interior Negative-Factored Moment				
ℓ_2/ℓ_1		0.5	1.0	2.0
$(\alpha_1\ell_2/\ell_1) = 0$		75	75	75
$(\alpha_1\ell_2/\ell_1) \geq 1.0$		90	75	45
Positive-Factored Moment				
$(\alpha_1\ell_2/\ell_1) = 0$	$B_t = 0$	100	100	100
	$B_t \geq 2.5$	75	75	75
$(\alpha_1\ell_2/\ell_1) \geq 1.0$	$B_t = 0$	100	100	100
	$B_t \geq 2.5$	90	75	45
Exterior Negative-Factored Moment				
$(\alpha_1\ell_2/\ell_1) = 0$		60	60	60
$(\alpha_1\ell_2/\ell_1) \geq 1.0$		90	75	45

Source: ACI Committee 318, 1992.

For the different types of two-way systems, the moment of inertias for modeling the slab-beam element of the equivalent frame are indicated in Fig. 48.8. Moments of inertia of slab beams are based on the gross area of concrete; the variation in moment of inertia along the axis is taken into account, which in practice would mean that a node would be located on the computer model where

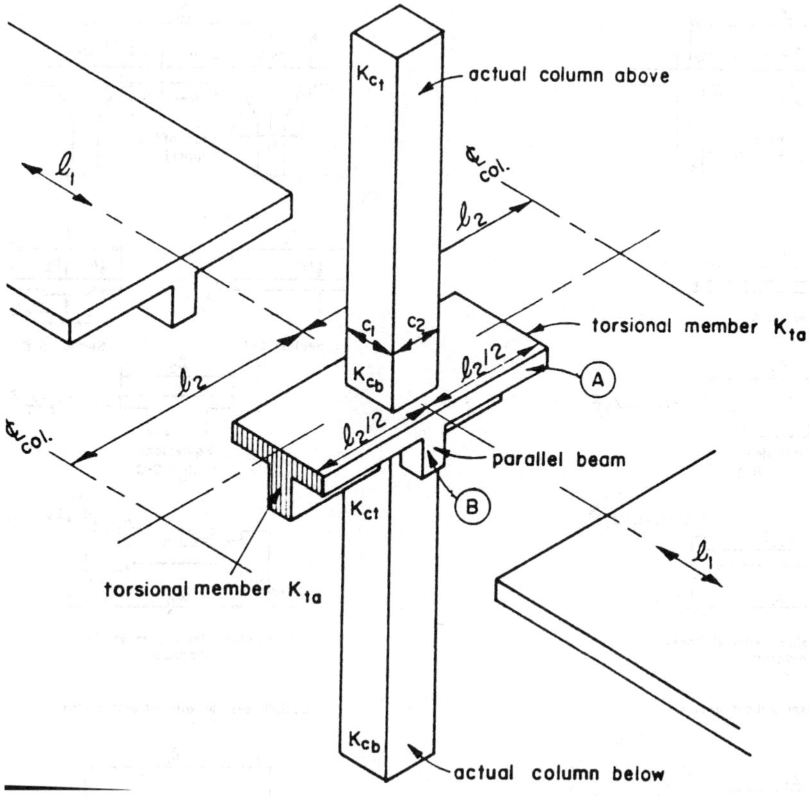

FIGURE 48.7 Equivalent column (columns plus torsional members).

a change of moment of inertia occurs. To account for the increased stiffness between the center of the column and the face of column, beam, or capital, the moment of inertia is divided by the quantity $(1 - c_2/l_2)^2$, where c_2 and l_2 are measured transverse to the direction of the span. For column modeling, the moment of inertia at any cross section outside of joints or column capitals may be based on the gross area of concrete, and the moment of inertia from the top to bottom of the slab-beam joint is assumed infinite.

Torsion members (Fig. 48.7) are elements in the equivalent frame that provide moment transfer between the horizontal slab beam and vertical columns. The cross section of torsional members are assumed to consist of the portion of slab and beam having a width according to the conditions depicted in Fig. 48.9. The stiffness K_t of the torsional member is calculated by the following expression:

$$K_t = \sum \frac{9E_{cs}C}{l_2\left(1 - \dfrac{c_2}{l_2}\right)^3} \tag{48.112}$$

where E_{cs} is the modulus of elasticity of the slab concrete and the torsional constant C may be evaluated by dividing the cross section into separate rectangular parts and carrying out the following summation:

$$C = \sum \left(1 - 0.63\frac{x}{y}\right)\frac{x^3 y}{3} \tag{48.113}$$

where x and y are the shorter and longer dimension, respectively, of each rectangular part. Where beams frame into columns in the direction of the span, the increased torsional stiffness K_{ta} is

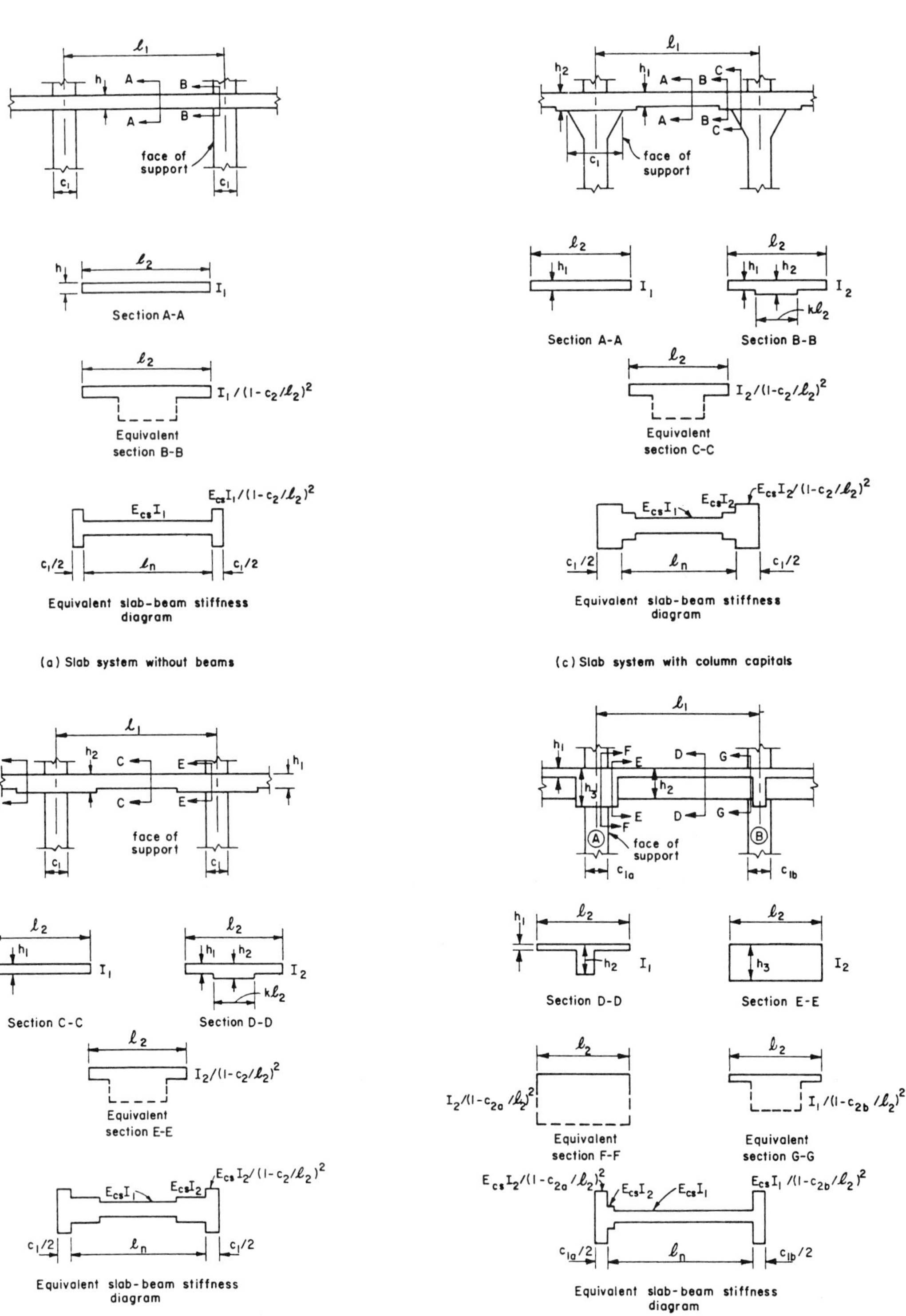

FIGURE 48.8 Slab-beam stiffness by equivalent frame method. (*Source:* ACI Committee 318, 1992.)

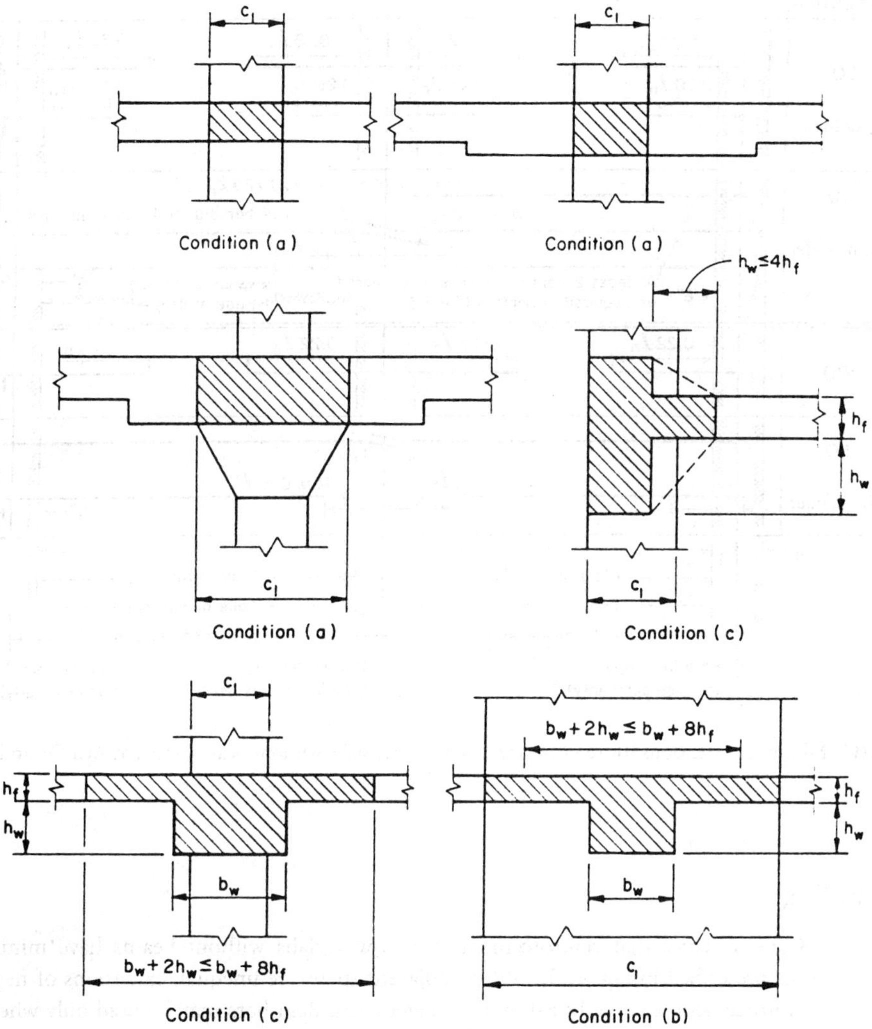

FIGURE 48.9 Torsional members. (*Source:* ACI Committee 318, 1992.)

obtained by multiplying the value K_t obtained from Eq. (48.112) by the ratio of (a) moment inertia of slab with such beam, to (b) moment of inertia of slab without such beam. Various ways have been suggested for incorporating torsional members into a computer model of an equivalent frame. The model implied by the ACI Code is one that has the slab beam connected to the torsional members, which are projected out of the plane of the columns. Others have suggested that the torsional members be replaced by rotational springs at column ends or, alternatively, at the slab-beam ends. Or, instead of rotational springs, columns may be modeled with an equivalent value of the moment of inertia modified by the equivalent column stiffness K_{ec} given in the commentary of the code. Using Fig. 48.7, K_{ec} is computed as

$$K_{ec} = \frac{K_{ct} + K_{cb}}{1 + \dfrac{K_{ct} + K_{cb}}{K_{ta} + K_{ta}}} \tag{48.114}$$

where K_{ct} and K_{cb} are the top and bottom flexural stiffnesses of the column.

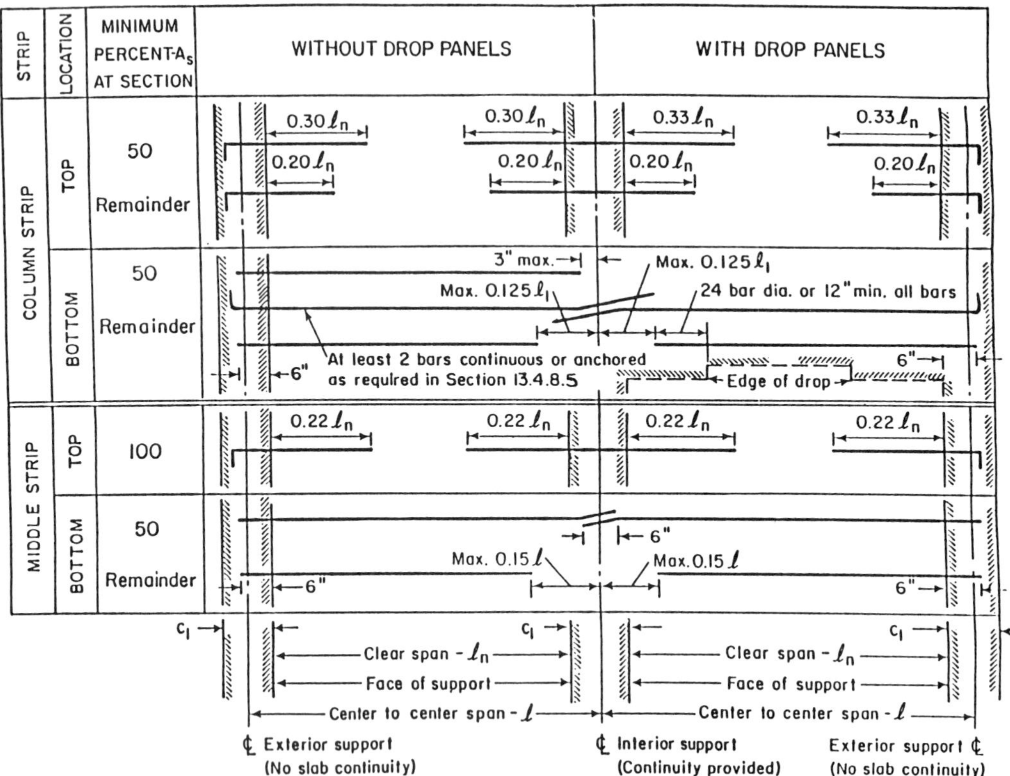

FIGURE 48.10 Minimum extensions for reinforcement in two-way slabs without beams. (*Source:* ACI Committee 318, 1992.)

Detailing

The ACI Code specifies that reinforcement in two-way slabs without beams have minimum extensions as prescribed in Fig. 48.10. Where adjacent spans are unequal, extensions of negative moment reinforcement shall be based on the longer span. Bent bars may be used only when the depth-span ratio permits use of bends 45 degrees or less. And at least two of the column strip bottom bars in each direction shall be continuous or spliced at the support with Class A splices or anchored within support. These bars must pass through the column and be placed within the column core. The purpose of this "integrity steel" is to give the slab some residual capacity following a single punching shear failure.

The ACI Code requires drop panels to extend in each direction from centerline of support a distance not less than one-sixth the span length, and the drop panel must project below the slab at least one-quarter of the slab thickness. The effective support area of a column capital is defined by the intersection of the bottom surface of the slab with the largest right circular cone whose surfaces are located within the column and capital and are oriented no greater than 45 degrees to the axis of the column.

48.8 Frames

A structural frame is a three-dimensional structural system consisting of straight members that are built monolithically and have rigid joints. The frame may be one bay long and one story high—such as portal frames and gable frames—or it may consist of multiple bays and stories. All members of frame are considered continuous in the three directions, and the columns participate with the beams in resisting external loads.

Consideration of the behavior of reinforced concrete frames at and near the ultimate load is necessary to determine the possible distributions of bending moment, shear force, and axial force that could be used in design. It is possible to use a distribution of moments and forces different from that given by linear elastic structural analysis if the critical sections have sufficient ductility to allow redistribution of actions to occur as the ultimate load is approached. Also, in countries that experience earthquakes, a further important design is the ductility of the structure when subjected to seismic-type loading, since present seismic design philosophy relies on energy dissipation by inelastic deformations in the event of major earthquakes.

Analysis of Frames

A number of methods have been developed over the years for the analysis of continuous beams and frames. The so-called classical methods—such as application of the theorem of three moments, the method of least work, and the general method of consistent deformation—have proved useful mainly in the analysis of continuous beams having few spans or of very simple frames. For the more complicated cases usually met in practice, such methods prove to be exceedingly tedious, and alternative approaches are preferred. For many years the closely related methods of slope deflection and moment distribution provided the basic analytical tools for the analysis of indeterminate concrete beams and frames. In offices with access to high-speed digital computers, these have been supplanted largely by matrix methods of analysis. Where computer facilities are not available, moment distribution is still the most common method. Approximate methods of analysis, based either on an assumed shape of the deformed structure or on moment coefficients, provide a means for rapid estimation of internal forces and moments. Such estimates are useful in preliminary design and in checking more exact solutions, and in structures of minor importance may serve as the basis for final design.

Slope Deflection

The method of slope deflection entails writing two equations for each member of a continuous frame, one at each end, expressing the end moment as the sum of four contributions: (1) the restraining moment associated with an assumed fixed-end condition for the loaded span, (2) the moment associated with rotation of the tangent to the elastic curve at the near end of the member, (3) the moment associated with rotation of the tangent at the far end of the member, and (4) the moment associated with translation of one end of the member with respect to the other. These equations are related through application of requirements of equilibrium and compatibility at the joints. A set of simultaneous, linear algebraic equations results for the entire structure, in which the structural displacements are unknowns. Solution for these displacements permits the calculation of all internal forces and moments.

This method is well suited to solving continuous beams, provided there are not very many spans. Its usefulness is extended through modifications that take advantage of symmetry and antisymmetry, and of hinge-end support conditions where they exist. However, for multistory and multibay frames in which there are a large number of members and joints, and which will, in general, involve translation as well as rotation of these joints, the effort required to solve the correspondingly large number of simultaneous equations is prohibitive. Other methods of analysis are more attractive.

Moment Distribution

The method of moment distribution was developed to solve problems in frame analysis that involve many unknown joint displacements. This method can be regarded as an iterative solution of the slope-deflection equations. Starting with fixed-end moments for each member, these are modified in a series of cycles, each converging on the precise final result, to account for rotation and translation of the joints. The resulting series can be terminated whenever one reaches the degree of accuracy required. After obtaining member-end moments, all member stress resultants can be obtained by use of the laws of statics.

Matrix Analysis

Use of matrix theory makes it possible to reduce the detailed numerical operations required in the analysis of an indeterminate structure to systematic processes of matrix manipulation, which can be performed automatically and rapidly by computer. Such methods permit the rapid solution of problems involving large numbers of unknowns. As a consequence, less reliance is placed on special techniques limited to certain types of problems; powerful methods of general applicability have emerged, such as the matrix displacement method. Account can be taken of such factors as rotational restraint provided by members perpendicular to the plane of a frame. A large number of alternative loadings may be considered. Provided that computer facilities are available, highly precise analyses are possible at lower cost than for approximate analyses previously employed.

Approximate Analysis

In spite of the development of refined methods for the analysis of beams and frames, increasing attention is being paid to various approximate methods of analysis. There are several reasons for this. Prior to performing a complete analysis of an indeterminate structure, it is necessary to estimate the proportions of its members in order to know their relative stiffness upon which the analysis depends. These dimensions can be obtained using approximate analysis. Also, even with the availability of computers, most engineers find it desirable to make a rough check of results—using approximate means—to detect gross errors. Further, for structures of minor importance, it is often satisfactory to design on the basis of results obtained by rough calculation.

Provided that points of inflection (locations in members at which the bending moment is zero and there is a reversal of curvature of the elastic curve) can be located accurately, the stress resultants for a framed structure can usually be found on the basis of static equilibrium alone. Each portion of the structure must be in equilibrium under the application of its external loads and the internal stress resultants. The use of approximate analysis in determining stress resultants in frames is illustrated using a simple rigid frame in Fig. 48.11.

ACI Moment Coefficients

The ACI Code [ACI Committee 318, 1992] includes moment and shear coefficients that can be used for the analysis of buildings of usual types of construction, span, and story heights. They are given in ACI Code Sec. 8.3.3. The ACI coefficients were derived with due consideration of several factors: a maximum allowable ratio of live to dead load (3:1); a maximum allowable span difference (the larger of two adjacent spans not exceed the shorter by more than 20%); the fact that reinforced concrete beams are never simply supported but either rest on supports of considerable width, such as walls, or are built monolithically like columns; and other factors. Since all these influences are considered, the ACI coefficients are necessarily quite conservative, so that actual moments in any particular design are likely to be considerably smaller than indicated. Consequently, in many reinforced concrete structures, significant economy can be effected by making a more precise analysis.

Limit Analysis

Limit analysis in reinforced concrete refers to the redistribution of moments that occurs throughout a structure as the steel reinforcement at a critical section reaches its yield strength. Under working loads, the distribution of moments in a statically indeterminate structure is based on elastic theory,

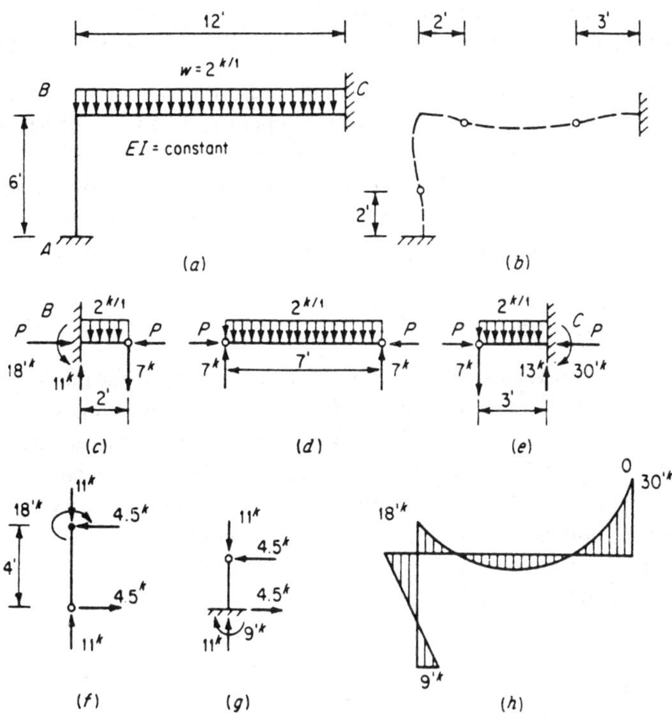

FIGURE 48.11 Approximate analysis of rigid frame. (*Source:* Nilson and Winter, 1992.)

and the whole structure remains in the elastic range. In limit design, where factored loads are used, the distribution of moments at failure when a mechanism is reached is different from that distribution based on elastic theory. The ultimate strength of the structure can be increased as more sections reach their ultimate capacity. Although the yielding of the reinforcement introduces large deflections, which should be avoided under service, a statically indeterminate structure does not collapse when the reinforcement of the first section yields. Furthermore, a large reserve of strength is present between the initial yielding and the collapse of the structure.

In steel design the term *plastic design* is used to indicate the change in the distribution of moments in the structure as the steel fibers, at a critical section, are stressed to their yield strength. Limit analysis of reinforced concrete developed as a result of earlier research on steel structures. Several studies had been performed on the principles of limit design and the rotation capacity of reinforced concrete plastic hinges.

Full utilization of the plastic capacity of reinforced concrete beams and frames requires an extensive analysis of all possible mechanisms and an investigation of rotation requirements and capacities at all proposed hinge locations. The increase of design time may not be justified by the limited gains obtained. On the other hand, a restricted amount of redistribution of elastic moments can safely be made without complete analysis and may be sufficient to obtain most of the advantages of limit analysis.

A limited amount of redistribution is permitted under the ACI Code, depending upon a rough measure of available ductility, without explicit calculation of rotation requirements and capacities. The ratio ρ/ρ_b—or in the case of doubly reinforced members, $(\rho - \rho')/\rho_b$—is used as an indicator of rotation capacity, where ρ_b is the balanced steel ratio. For singly reinforced members with $\rho = \rho_b$, experiments indicate almost no rotation capacity, since the concrete strain is nearly equal to ε_{cu} when steel yielding is initiated. Similarly, in a doubly reinforced member, when $\rho - \rho' = \rho_b$, very little rotation will occur after yielding before the concrete crushes. However, when ρ or $\rho - \rho'$ is low, extensive rotation is usually possible. Accordingly, ACI Code Sec. 8.3 provides as follows:

> Except where approximate values for moments are used, it is permitted to increase or decrease negative moments calculated by elastic theory at supports of continuous flexural members for any assumed loading arrangement by not more than $20[1 - (\rho - \rho')/\rho_b]$ percent. The modified negative moments shall be used for calculating moments at sections within the spans. Redistribution of negative moments shall be made only when the section at which moment is reduced is so designed that ρ or $\rho - \rho'$ is not greater than $0.5\rho_b$. [1992]

Design for Seismic Loading

The ACI Code contains provisions that are currently considered to be the minimum requirements for producing a monolithic concrete structure with adequate proportions and details to enable the structure to sustain a series of oscillations into the inelastic range of response without critical decay in strength. The provisions are intended to apply to reinforced concrete structures located in a seismic zone where major damage to construction has a high possibility of occurrence, and are designed with a substantial reduction in total lateral seismic forces due to the use of lateral load-resisting systems consisting of ductile moment-resisting frames. The provisions for frames are divided into sections on flexural members, columns, and joints of frames. Some of the important points stated are summarized below.

Flexural Members

Members having a factored axial force not exceeding $A_g f_c'/10$, where A_g is gross section of area (in.^{2}), are regarded as flexural members. An upper limit is placed on the flexural steel ratio ρ. The maximum value of ρ should not exceed 0.025. Provision is also made to ensure that a minimum quantity of top and bottom reinforcement is always present. Both the top and the bottom steel are to have a steel ratio of at least $200/f_y$, with the steel yield strength f_y in psi throughout the length of the member. Recommendations are also made to ensure that sufficient steel is present to allow for unforeseen shifts in the points of contraflexure. At column connections, the positive moment capacity should be at least 50% of the negative moment capacity, and the reinforcement should be continuous through columns where possible. At external columns, beam reinforcement should be terminated in the far face of the column using a hook plus any additional extension necessary for anchorage.

The design shear force V_e should be determined from consideration of the static forces on the portion of the member between faces of the joints. It should be assumed that moments of opposite sign corresponding to probable strength M_{pr} act at the joint faces and that the member is loaded with the factored tributary gravity load along its span. Figure 48.12 illustrates the calculation. Minimum web reinforcement is provided throughout the length of the member, and spacing should not exceed $d/4$ in plastic hinge zones and $d/2$ elsewhere, where d is effective depth of member. The stirrups should be closed around bars required to act as compression reinforcement and in plastic hinge regions, and the spacing should not exceed specified values.

Columns

Members having a factored axial force exceeding $A_g f_c'/10$ are regarded as columns of frames serving to resist earthquake forces. These members should satisfy the conditions that the shortest cross-sectional dimension—measured on a straight line passing through the geometric centroid—should not be less than 12 in. and that the ratio of the shortest cross-sectional dimension to the perpendicular dimension should not be less than 0.4. The flexural strengths of the columns should satisfy

$$\sum M_e \geq (6/5) \sum M_g \tag{48.115}$$

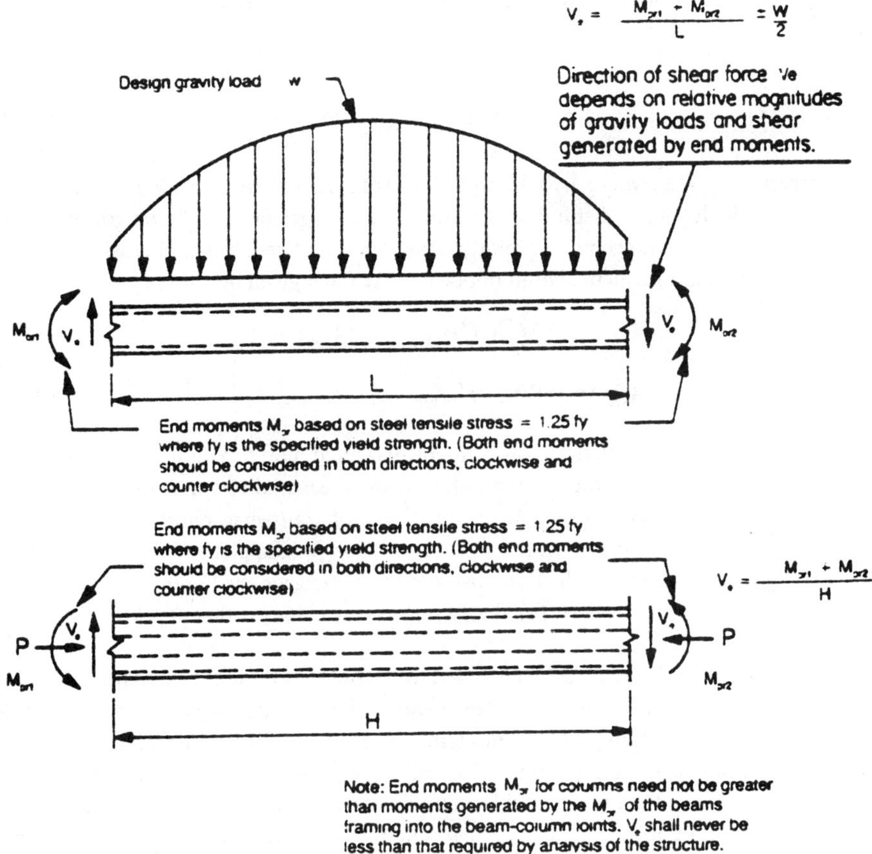

$V_e = \frac{M_{pr1} + M_{pr2}}{L} \pm \frac{w}{2}$
Design gravity load w
Direction of shear force Ve depends on relative magnitudes of gravity loads and shear generated by end moments.
M_{pr1}
V_e
M_{pr2}
L
End moments M_{pr} based on steel tensile stress = 1.25 fy where fy is the specified yield strength. (Both end moments should be considered in both directions, clockwise and counter clockwise)
End moments M_{pr} based on steel tensile stress = 1.25 fy where fy is the specified yield strength. (Both end moments should be considered in both directions, clockwise and counter clockwise)
$V_e = \frac{M_{pr1} + M_{pr2}}{H}$
P
H
Note: End moments M_{pr} for columns need not be greater than moments generated by the M_{pr} of the beams framing into the beam-column joints. V_e shall never be less than that required by analysis of the structure.

FIGURE 48.12 Design shears for girders and columns. (*Source:* ACI 318, 1992.)

where $\sum M_e$ is sum of moments, at the center of the joint, corresponding to the design flexural strength of the columns framing into that joint and where $\sum M_g$ is sum of moments, at the center of the joint, corresponding to the design flexural strengths of the girders framing into that joint. Flexural strengths should be summed such that the column moments oppose the beam moments. Eq. (48.115) should be satisfied for beam moments acting in both directions in the vertical plane of the frame considered. The requirement is intended to ensure that plastic hinges form in the girders rather than the columns.

The longitudinal reinforcement ratio is limited to the range of 0.01 to 0.06. The lower bound to the reinforcement ratio refers to the traditional concern for the effects of time-dependent deformations of the concrete and the desire to have a sizable difference between the cracking and yielding moments. The upper bound reflects concern for steel congestion, load transfer from floor elements to column in low-rise construction, and the development of large shear stresses. Lap splices are permitted only within the center half of the member length and should be proportioned as tension splices. Welded splices and mechanical connections are allowed for splicing the reinforcement at any section, provided not more than alternate longitudinal bars are spliced at a section and the distance between splices is 24 in. or more along the longitudinal axis of the reinforcement.

If Eq. (48.115) is not satisfied at a joint, columns supporting reactions from that joint should be provided with transverse reinforcement over their full height to confine the concrete and provide lateral support to the reinforcement. Where a spiral is used, the ratio of volume of spiral reinforcement to the core volume confined by the spiral reinforcement, ρ_s, should be at least that given by

$$\rho_s = 0.45 \frac{f_c'}{f_y}\left(\frac{A_g}{A_c} - 1\right) \tag{48.116}$$

but not less than $0.12 f_c'/f_{yh}$, where A_c is the area of core of spirally reinforced compression member measured to outside diameter of spiral in in.^{2} and f_{yh} is the specified yield strength of transverse reinforcement in psi. When rectangular reinforcement hoop is used, the total cross-sectional area of rectangular hoop reinforcement should not be less than that given by

$$A_{sh} = 0.3(sh_c f_c'/f_{yh})[(A_g/A_{ch}) - 1] \tag{48.117}$$

$$A_{sh} = 0.09 sh_c f_c'/f_{yh} \tag{48.118}$$

where s is the spacing of transverse reinforcement measured along the longitudinal axis of column, h_c is the cross-sectional dimension of column core measured center-to-center of confining reinforcement, and A_{sh} is the total cross-sectional area of transverse reinforcement (including crossties) within spacing s and perpendicular to dimension h_c. Supplementary crossties, if used, should be of the same diameter as the hoop bar and should engage the hoop with a hook. Special transverse confining steel is required for the full height of columns that support discontinuous shear walls.

The design shear force V_e should be determined from consideration of the maximum forces that can be generated at the faces of the joints at each end of the column. These joint forces should be determined using the maximum probable moment strength M_{pr} of the column associated with the range of factored axial loads on the column. The column shears need not exceed those determined from joint strengths based on the probable moment strength M_{pr}, of the transverse members framing into the joint. In no case should V_e be less than the factored shear determined by analysis of the structure (Fig. 48.12).

Joints of Frames

Development of inelastic rotations at the faces of joints of reinforced concrete frames is associated with strains in the flexural reinforcement well in excess of the yield strain. Consequently, joint shear force generated by the flexural reinforcement is calculated for a stress of $1.25 f_y$ in the reinforcement.

Within the depth of the shallowed framing member, transverse reinforcement equal to at least one-half the amount required for the column reinforcement should be provided where members frame into all four sides of the joint and where each member width is at least three-fourths the column width. Transverse reinforcement as required for the column reinforcement should be provided through the joint to provide confinement for longitudinal beam reinforcement outside the column core if such confinement is not provided by a beam framing into the joint.

The nominal shear strength of the joint should not be taken greater than the forces specified below for normal weight aggregate concrete:

$20\sqrt{f_c'}A_j$ for joints confined on all four faces

$15\sqrt{f_c'}A_j$ for joints confined on three faces or on two opposite faces

$12\sqrt{f_c'}A_j$ for others

where A_j is the effective cross-sectional area within a joint in a plane parallel to plane of reinforcement generating shear in the joint (see Fig. 48.13). A member that frames into a face is considered to provide confinement to the joint if at least three-quarters of the face of the joint is covered by the framing member. A joint is considered to be confined if such confining members

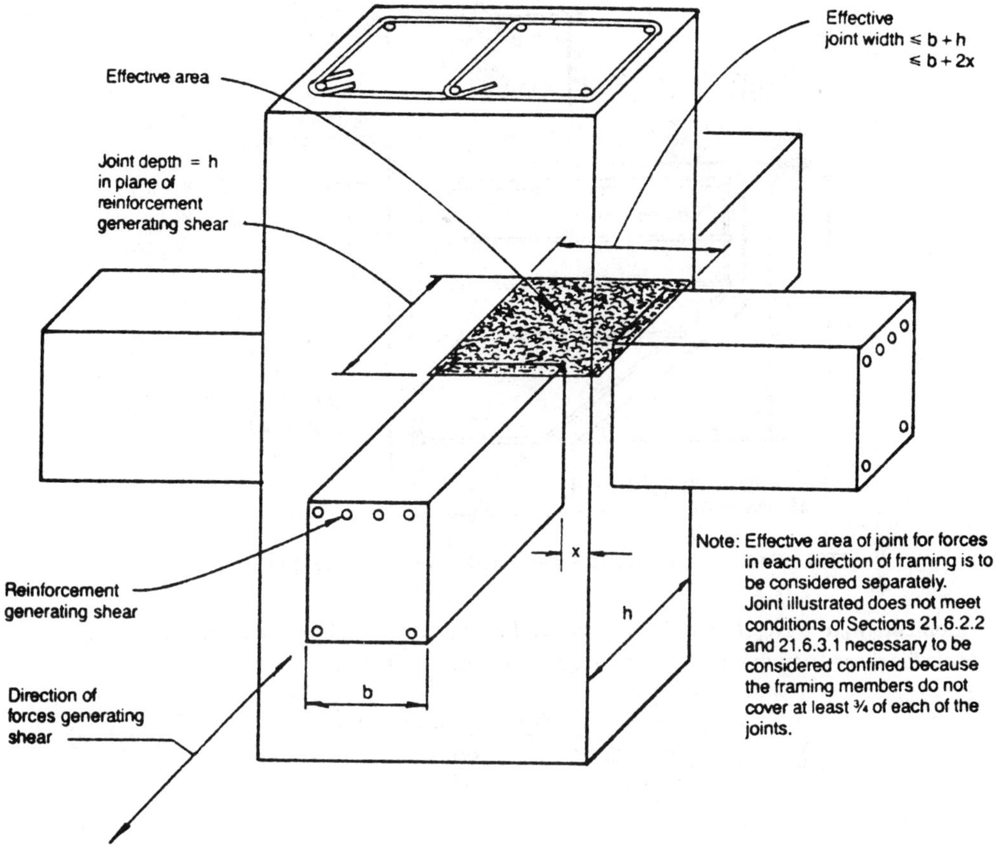

FIGURE 48.13 Effective area of joint. (*Source:* ACI Committee 318, 1992.)

frame into all faces of the joint. For lightweight-aggregate concrete, the nominal shear strength of the joint should not exceed three-quarters of the limits given above.

Details of minimum development length for deformed bars with standard hooks embedded in normal and lightweight concrete and for straight bars are contained in ACI Code Sec. 21.6.4.

48.9 Brackets and Corbels

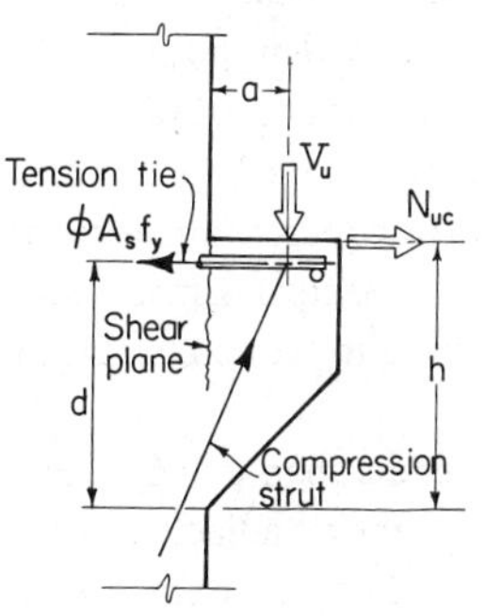

FIGURE 48.14 Structural action of a corbel. (*Source:* ACI Committee 318. 1992.)

Brackets and corbels are cantilevers having shear span to depth ratio, a/d, not greater than unity. The shear span a is the distance from the point of load to the face of support, and the distance d shall be measured at face of support (see Fig. 48.14).

The corbel shown in Fig. 48.14 may fail by shearing along the interface between the column and the corbel, by yielding of the tension tie, by crushing or splitting of the compression strut, or by localized bearing or shearing failure under the loading plate.

The depth of a bracket or corbel at its outer edge should be less than one-half of the required depth d at the support. Reinforcement should consist of main tension bars with area A_s and shear reinforcement with area A_h (see Fig. 48.15 for notation). The area of primary tension reinforcement A_s should be made equal to the greater of $(A_f + A_n)$ or $(2A_{vf}/3 + A_n)$, where A_f is the flexural reinforcement required to

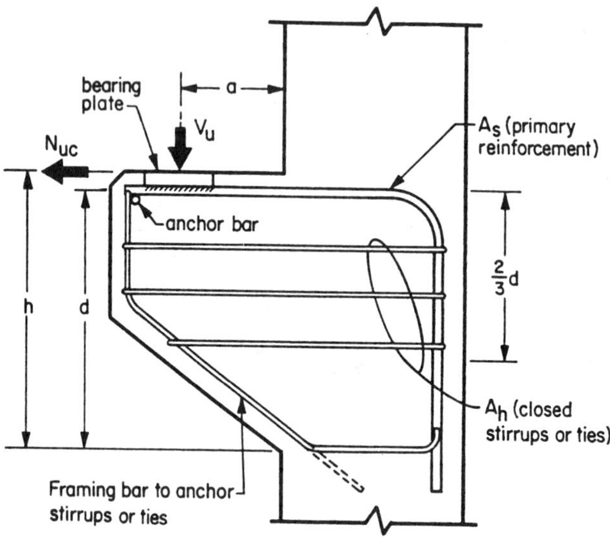

FIGURE 48.15 Notation used. (*Source:* ACI Committee 318, 1992.)

resist moment $[V_u a + N_{uc}(h - d)], A_n$ is the reinforcement required to resist tensile force N_{uc}, and A_{vf} is the shear-friction reinforcement required to resist shear V_u:

$$A_f = \frac{M_u}{\phi f_y j d} = \frac{V_u a + N_{uc}(h - d)}{\phi f_y j d} \tag{48.119}$$

$$A_n = \frac{N_{uc}}{\phi f_y} \tag{48.120}$$

$$A_{vf} = \frac{V_u}{\phi f_y \mu} \tag{48.121}$$

In the above equations, f_y is the reinforcement yield strength; ϕ is 0.9 for Eq. (48.119) and 0.85 for Eqs. (48.120) and (48.121). In Eq. (48.119), the lever arm jd can be approximated for all practical purposes in most cases as $0.85d$. Tensile force N_{uc} in Eq. (48.120) should not be taken less than $0.2V_u$ unless special provisions are made to avoid tensile forces. Tensile force N_{uc} should be regarded as a live load even when tension results from creep, shrinkage, or temperature change. In Eq. (48.121), $V_u/\phi(= V_n)$ should not be taken greater than $0.2f_c' b_w d$ nor $800 b_w d$ in pounds in normal-weight concrete. For "all-lightweight" or "sand-lightweight" concrete, shear strength V_n should not be taken greater than $(0.2 - 0.07a/d)f_c' b_w d$ nor $(800 - 280a/d)b_w d$ in pounds. The coefficient of friction μ in Eq. (48.121) should be 1.4λ for concrete placed monolithically, 1.0λ for concrete placed against hardened concrete with surface intentionally roughened, 0.6λ for concrete placed against hardened concrete not intentionally roughened, and 0.7λ for concrete anchored to as-rolled structural steel by headed studs or by reinforcing bars, where λ is 1.0 for normal weight concrete, 0.85 for "sand-lightweight" concrete, and 0.75 for "all-lightweight" concrete. Linear interpolation of λ is permitted when partial sand replacement is used.

The total area of closed stirrups or ties A_h parallel to A_s should not be less than $0.5(A_s - A_n)$ and should be uniformly distributed within two-thirds of the depth of the bracket adjacent to A_s.

At front face of bracket or corbel, primary tension reinforcement A_s should be anchored in one of the following ways: (a) by a structural weld to a transverse bar of at least equal size; weld to be designed to develop specified yield strength f_y of A_s bars; (b) by bending primary tension bars A_s back to form a horizontal loop, or (c) by some other means of positive anchorage. Also, to ensure development of the yield strength of the reinforcement A_s near the load, bearing area of load on bracket or corbel should not project beyond straight portion of primary tension bars A_s, nor project beyond interior face of transverse anchor bar (if one is provided). When corbels are designed to resist horizontal forces, the bearing plate should be welded to the tension reinforcement A_s.

48.10 Footings

Footings are structural members used to support columns and walls and to transmit and distribute their loads to the soil in such a way that (a) the load bearing capacity of the soil is not exceeded, (b) excessive settlement, differential settlement, and rotations are prevented, and (c) adequate safety against overturning or sliding is maintained. When a column load is transmitted to the soil by the footing, the soil becomes compressed. The amount of settlement depends on many factors, such as the type of soil, the load intensity, the depth below ground level, and the type of footing. If different footings of the same structure have different settlements, new stresses develop in the structure. Excessive differential settlement may lead to the damage of nonstructural members in the buildings, even failure of the affected parts.

Vertical loads are usually applied at the centroid of the footing. If the resultant of the applied loads does not coincide with the centroid of the bearing area, a bending moment develops. In this case, the pressure on one side of the footing will be greater than the pressure on the other side, causing higher settlement on one side and a possible rotation of the footing.

If the bearing soil capacity is different under different footings—for example, if the footings of a building are partly on soil and partly on rock—a differential settlement will occur. It is customary in such cases to provide a joint between the two parts to separate them, allowing for independent settlement.

Types of Footings

Different types of footings may be used to support building columns or walls. The most commonly used ones are illustrated in Fig. 48.16(a–g). A simple file footing is shown in Fig. 48.16(h).

For walls, a spread footing is a slab wider than the wall and extending the length of the wall [Fig. 48.16(a)]. Square or rectangular slabs are used under single columns [Fig. 48.16(b–d)]. When two columns are so close that their footings would merge or nearly touch, a combined footing [Fig. 48.16(e)] extending under the two should be constructed. When a column footing cannot project in one direction, perhaps because of the proximity of a property line, the footing may be helped out by an adjacent footing with more space; either a combined footing or a strap (cantilever) footing [Fig. 48.16(f)] may be used under the two.

For structures with heavy loads relative to soil capacity, a mat or raft foundation [Fig. 48.16(g)] may prove economical. A simple form is a thick, two-way-reinforced-concrete slab extending under the entire structure. In effect, it enables the structure to float on the soil, and because of its rigidity it permits negligible differential settlement. Even greater rigidity can be obtained by building the raft foundation as an inverted beam-and-girder floor, with the girders supporting the columns. Sometimes, also, inverted flat slabs are used as mat foundations.

Design Considerations

Footings must be designed to carry the column loads and transmit them to the soil safely while satisfying code limitations. The design procedure must take the following strength requirements into consideration:

- The area of the footing based on the allowable bearing soil capacity
- Two-way shear or punching shear
- One-way shear
- Bending moment and steel reinforcement required

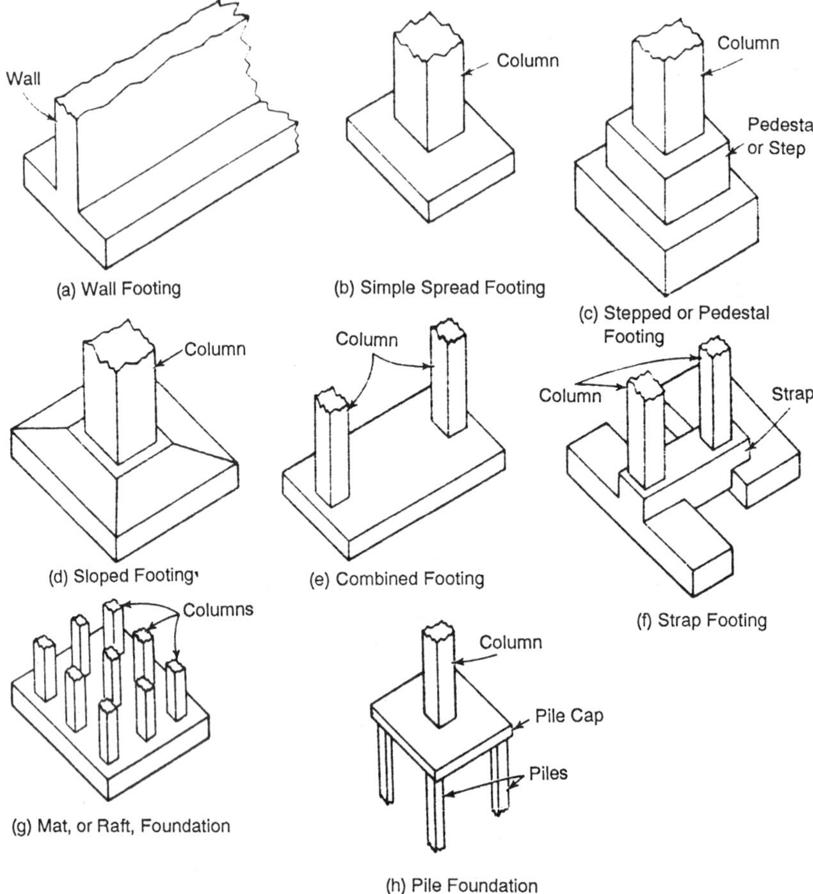

FIGURE 48.16 Common types of footings for walls and columns. (*Source:* ACI Committee 340, 1990.)

- Dowel requirements
- Development length of bars
- Differential settlement

These strength requirements will be explained in the following sections.

Size of Footings

The required area of concentrically loaded footings is determined from

$$A_{req} = \frac{D + L}{q_a} \tag{48.122}$$

where q_a is allowable bearing pressure and D and L are, respectively, unfactored dead and live loads. Allowable bearing pressures are established from principles of soil mechanics on the basis of load tests and other experimental determinations. Allowable bearing pressures q_a under service loads are usually based on a safety factor of 2.5 to 3.0 against exceeding the ultimate bearing capacity of the particular soil and to keep settlements within tolerable limits. The required area of footings under the effects of wind W or earthquake E is determined from the following:

$$A_{req} = \frac{D + L + W}{1.33q_a} \quad or \quad \frac{D + L + E}{1.33q_a} \tag{48.123}$$

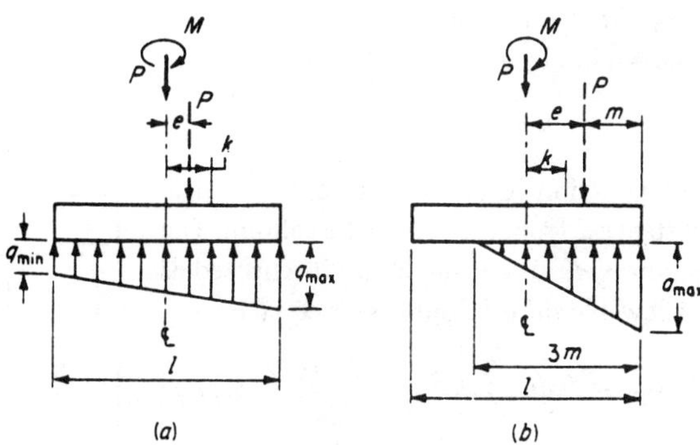

FIGURE 48.17 Assumed bearing pressures under eccentric footings. (*Source:* Wang and Salmon, 1985.)

It should be noted that footing sizes are determined for unfactored service loads and soil pressures, in contrast to the strength design of reinforced concrete members, which utilizes factored loads and factored nominal strengths.

A footing is eccentrically loaded if the supported column is not concentric with the footing area or if the column transmits—at its juncture with the footing—not only a vertical load but also a bending moment. In either case, the load effects at the footing base can be represented by the vertical load P and a bending moment M. The resulting bearing pressures are again assumed to be linearly distributed. As long as the resulting eccentricity $e = M/P$ does not exceed the kern distance k of the footing area, the usual flexure formula

$$q_{\text{max,min}} = \frac{P}{A} + \frac{Mc}{I} \tag{48.124}$$

permits the determination of the bearing pressures at the two extreme edges, as shown in Fig. 48.17(a). The footing area is found by trial and error from the condition $q_{\text{max}} \leq q_a$. If the eccentricity falls outside the kern, Eq. (48.124) gives a negative value for q along one edge of the footing. Because no tension can be transmitted at the contact area between soil and footing, Eq. (48.124) is no longer valid and bearing pressures are distributed as in Fig. 48.17(b).

Once the required footing area has been determined, the footing must then be designed to develop the necessary strength to resist all moments, shears, and other internal actions caused by the applied loads. For this purpose, the load factors of the ACI Code apply to footings as to all other structural components.

Depth of footing above bottom reinforcement should not be less than 6 in. for footings on soil, nor less that 12 in. from footings on piles.

Two-Way Shear (Punching Shear)

ACI Code Sec. 11.12.2 allows a shear strength V_c of footings without shear reinforcement for two-way shear action as follows:

$$V_c = \left(2 + \frac{4}{\beta_c}\right)\sqrt{f_c'}b_o d \leq 4\sqrt{f_c'}b_o d \tag{48.125}$$

where β_c is the ratio of long side to short side of rectangular area, b_o is the perimeter of the critical section taken at $d/2$ from the loaded area (column section), and d is the effective depth of footing. This shear is a measure of the diagonal tension caused by the effect of the column load on the footing. Inclined cracks may occur in the footing at a distance $d/2$ from the face of the column on all sides. The footing will fail as the column tries to punch out part of the footing, as shown in Fig. 48.18.

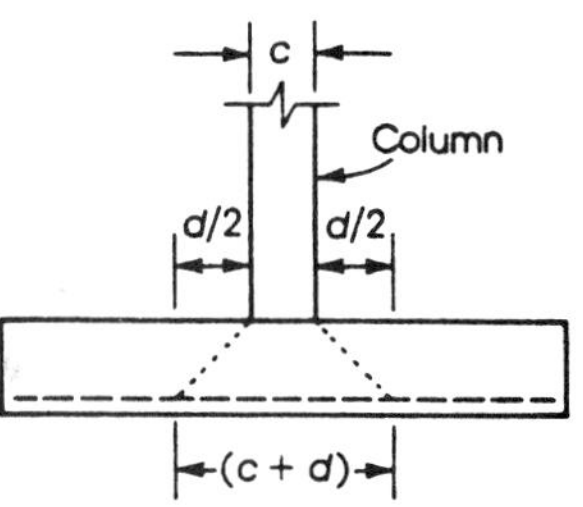

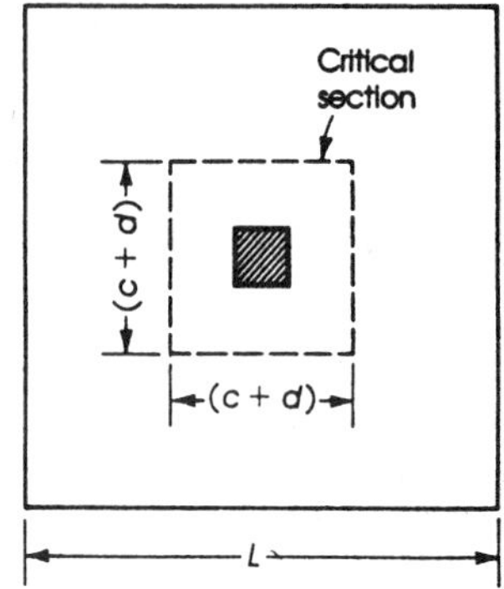

FIGURE 48.18 Punching shear (two-way). (*Source:* MacGregor, 1992.)

One-Way Shear

For footings with bending action in one direction, the critical section is located at a distance d from the face of the column. The diagonal tension at section m-m in Fig. 48.19 can be checked as is done in beams. The allowable shear in this case is equal to

$$\phi V_c = 2\phi \sqrt{f_c'} bd \qquad (48.126)$$

where b is the width of section m-m. The ultimate shearing force at section m-m can be calculated as follows:

$$V_u = q_u b \left(\frac{L}{2} - \frac{c}{2} - d \right) \qquad (48.127)$$

where b is the side of footing parallel to section m-m.

Flexural Reinforcement and Footing Reinforcement

The theoretical sections for moment occur at face of the column (section n-n, Fig. 48.20). The bending moment in each direction of the footing must be checked and the appropriate reinforcement must be provided. In square footings the bending moments in both directions are equal. To determine the reinforcement required, the depth of the footing in each direction may be used. As the bars in one direction rest on top of the bars in the other direction, the effective depth d varies with the diameter of the bars used. The value of d_{min} may be adopted.

The depth of footing is often controlled by the shear, which requires a depth greater than that required by the bending moment. The steel reinforcement in each direction can be calculated in the case of flexural members as follows:

$$A_s = \frac{M_u}{\phi f_y (d - a/2)} \qquad (48.128)$$

The minimum steel percentage requirement in flexural members is equal to $200/f_y$. However, ACI Code Sec. 10.5.3 indicates that for structural slabs of uniform thickness, the minimum area and maximum spacing of steel in the direction of bending should be as required for shrinkage and temperature reinforcement. This last minimum steel reinforcement is very small and a higher minimum reinforcement ratio is recommended, but not greater than $200/f_y$.

The reinforcement in one-way footings and two-way footings must be distributed across the entire width of the footing. In the case of two-way rectangular footings, ACI Code Sec. 15.4.4 specifies that in the long direction the total reinforcement must be placed uniformly within a band width equal to the length of the short side of the footing according to

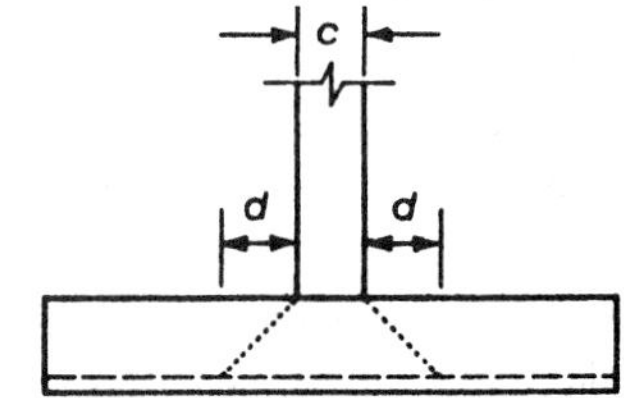

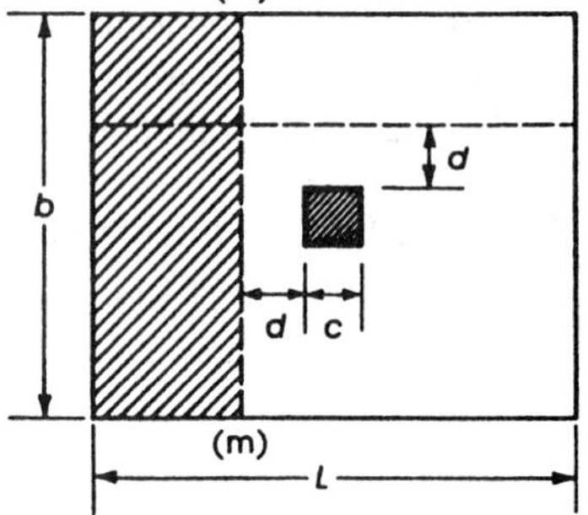

FIGURE 48.19 One-way shear. (*Source:* MacGregor, 1992.)

$$\frac{\text{Reinforcement band width}}{\text{Total reinforcement in short direction}} = \frac{2}{\beta + 1} \tag{48.129}$$

where β is the ratio of the long side to the short side of the footing. The band width must be centered on the centerline of the column (Fig. 48.21). The remaining reinforcement in the short direction must be uniformly distributed outside the band width. This remaining reinforcement percentage should not be less than required for shrinkage and temperature.

When structural steel columns or masonry walls are used, the critical sections for moments in footings are taken at halfway between the middle and the edge of masonry walls, and halfway between the face of the column and the edge of the steel base place (ACI Code Sec. 15.4.2).

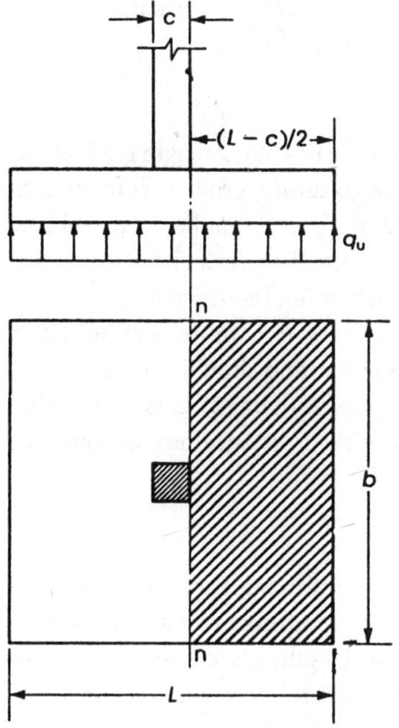

FIGURE 48.20 Critical section of bending moment. (*Source:* MacGregor, 1992.)

Bending Capacity of Column at Base

The loads from the column act on the footing at the base of the column, on an area equal to the area of the column cross section. Compressive forces are transferred to the footing directly by bearing on the concrete. Tensile forces must be resisted by reinforcement, neglecting any contribution by concrete.

Forces acting on the concrete at the base of the column must not exceed the bearing strength of concrete as specified by the ACI Code Sec. 10.15:

$$N = \phi(0.85f_c'A_1) \tag{48.130}$$

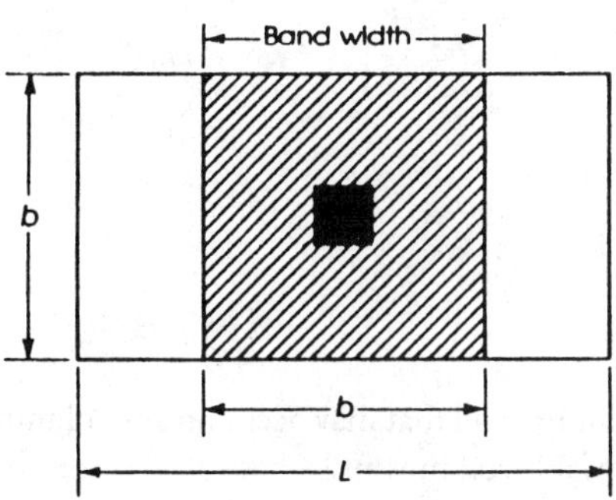

FIGURE 48.21 Band width for reinforcement distribution. (*Source:* MacGregor, 1992.)

where ϕ is 0.7 and A_1 is the bearing area of the column. The value of the bearing strength given in Eq. (48.130) may be multiplied by a factor $\sqrt{A_2/A_1} \leq 2.0$ for bearing on footings when the supporting surface is wider on all sides other than the loaded area. Here A_2 is the area of the part of the supporting footing that is geometrically similar to and concentric with the loaded area (Fig. 48.22). Since $A_2 > A_1$, the factor $\sqrt{A_2/A_1}$ is greater than unity, indicating that the allowable bearing strength is increased because of the lateral support from the footing area surrounding the column base. If the calculated bearing force is greater than N or the modified one with $r\sqrt{A_2/A_1}$, reinforcement must be provided to transfer the excess force. This is achieved by providing dowels or extending the column bars into the footing. If the calculated bearing force is less than either N or the modified one with $r\sqrt{A_2/A_1}$, then minimum reinforcement must be provided. ACI Code Sec. 15.8.2 indicates that the minimum area of the dowel reinforcement is at least $0.005A_g$ but not less than 4 bars, where A_g is the gross area of the column section of

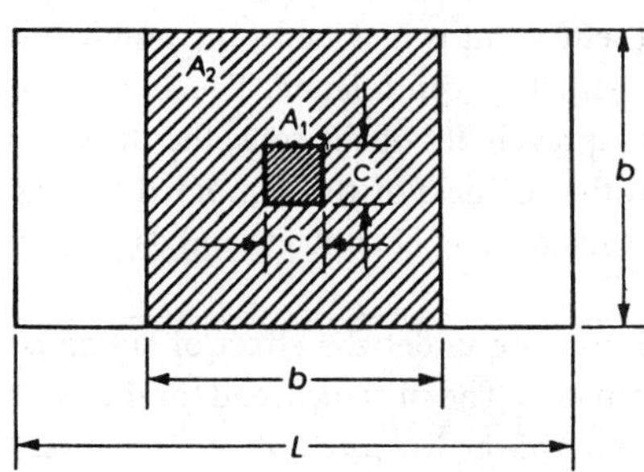

FIGURE 48.22 Bearing areas on footings. $A_1 = c^2, A_2 = b^2$. (*Source:* MacGregor, 1992.)

the supported member. The minimum reinforcement requirements apply to the case in which the calculated bearing forces are greater than N or the modified one with $r\ \sqrt{A_2/A_1}$.

Dowels on Footings

It was explained earlier that dowels are required in any case, even if the bearing strength is adequate. The ACI Code specifies a minimum steel ratio $\rho = 0.005$ of the column section as compared to $\rho = 0.01$ as minimum reinforcement for the column itself. The minimum number of dowel bars needed is four; these may be placed at the four corners of the column. The dowel bars are usually extended into the footing, bent at their ends, and tied to the main footing reinforcement.

ACI Code Sec. 15.8.2 indicates that #14 and #18 longitudinal bars, in compression only, may be lap-spliced with dowels. Dowels should not be larger than #11 bar and should extend (1) into supported member a distance not less than the development length of #14 or 18″ bars or the splice length of the dowels—whichever is greater, and (2) into the footing a distance not less than the development length of the dowels.

Development Length of the Reinforcing Bars

The critical sections for checking the development length of the reinforcing bars are the same as those for bending moments. Calculated tension or compression in reinforcement at each section should be developed on each side of that section by embedment length, hook (tension only) or mechanical device, or a combination thereof. The development length for compression bar is

$$l_d = 0.02 f_y d_b \sqrt{f_c'} \tag{48.131}$$

but not less than $0.0003 f_y d_b \geq 8$ in. For other values, refer to ACI Code, Chapter 12. Dowel bars must also be checked for proper development length.

Differential Settlement

Footings usually support the following loads:

- Dead loads from the substructure and superstructure
- Live loads resulting from materials or occupancy
- Weight of materials used in backfilling
- Wind loads

Each footing in a building is designed to support the maximum load that may occur on any column due to the critical combination of loadings, using the allowable soil pressure.

The dead load, and maybe a small portion of the live load, may act continuously on the structure. The rest of the live load may occur at intervals and on some parts of the structure only, causing different loadings on columns. Consequently, the pressure on the soil under different loadings will vary according to the loads on the different columns, and differential settlement will occur under the various footings of one structure. Since partial settlement is inevitable, the problem is defined by the amount of differential settlement that the structure can tolerate. The amount of differential settlement depends on the variation in the compressibility of the soils, the thickness of the compressible material below foundation level, and the stiffness of the combined footing and superstructure. Excessive differential settlement results in cracking of concrete and damage to claddings, partitions, ceilings, and finishes.

For practical purposes it can be assumed that the soil pressure under the effect of sustained loadings is the same for all footings, thus causing equal settlements. The sustained load (or the usual load) can be assumed equal to the dead load plus a percentage of the live load, which occurs very frequently on the structure. Footings then are proportioned for these sustained loads to produce the same soil pressure under all footings. In no case is the allowable soil bearing capacity to be exceeded under the dead load plus the maximum live load for each footing.

Wall Footings

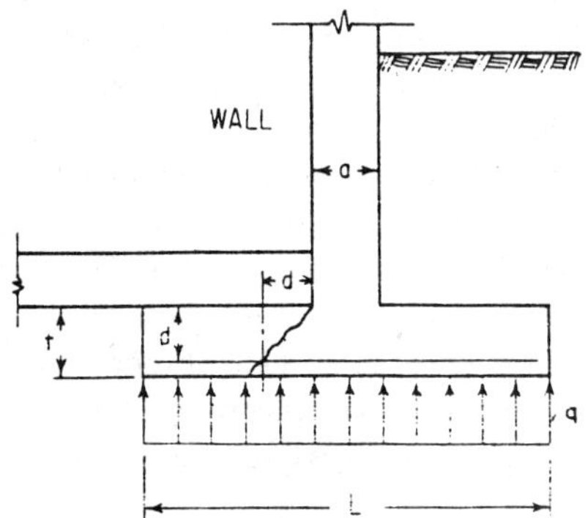

FIGURE 48.23 Reinforced-concrete wall footing. (*Source:* Wang and Salmon, 1985.)

The spread footing under a wall [Fig. 48.16(a)] distributes the wall load horizontally to preclude excessive settlement. The wall should be so located on the footings as to produce uniform bearing pressure on the soil (Fig. 48.23), ignoring the variation due to bending of the footing. The pressure is determined by dividing the load per foot by the footing width.

The footing acts as a cantilever on opposite sides of the wall under downward wall loads and upward soil pressure. For footings supporting concrete walls, the critical section for bending moment is at the face of the wall; for footings under masonry walls, halfway between the middle and edge of the wall. Hence, for a one-foot-long strip of symmetrical concrete-wall footing, symmetrically loaded, the maximum moment, ft-lb, is

$$M_u = \frac{1}{8} q_u (L - a)^2 \qquad (48.132)$$

where q_u is the uniform pressure on soil (lb/ft^2), L is the width of footing (ft), and a is wall thickness (ft).

For determining shear stresses, the vertical shear force is computed on the section located at a distance d from the face of the wall. Thus,

$$V_u = q_u \left(\frac{L - a}{2} - L \right) \qquad (48.133)$$

The calculation of development length is based on the section of maximum moment.

Single-Column Spread Footings

The spread footing under a column [Fig. 48.16(b–d)] distributes the column load horizontally to prevent excessive total and differential settlement. The column should be located on the footing so as to produce uniform bearing pressure on the soil, ignoring the variation due to bending of the footing. The pressure equals the load divided by the footing area.

In plan, single-column footings are usually square. Rectangular footings are used if space restrictions dictate this choice or if the supported columns are of strongly elongated rectangular cross section. In the simplest form, they consist of a single slab [Fig. 48.16(b)]. Another type is that of Fig. 48.16(c), where a pedestal or cap is interposed between the column and the footing slab; the pedestal provides for a more favorable transfer of load and in many cases is required in order to provide the necessary development length for dowels. This form is also known as a *stepped footing*. All parts of a stepped footing must be poured in a single pour in order to provide monolithic action. Sometimes sloped footings like those in Fig. 48.16(d) are used. They require less concrete than stepped footings, but the additional labor necessary to produce the sloping surfaces (formwork, etc.) usually makes stepped footings more economical. In general, single-slab footings [Fig. 48.16(b)] are most economical for thicknesses up to 3 ft.

The required bearing area is obtained by dividing the total load, including the weight of the footing, by the selected bearing pressure. Weights of footings, at this stage, must be estimated and usually amount to 4 to 8% of the column load, the former value applying to the stronger types of soils.

Once the required footing area has been established, the thickness h of the footing must be determined. In single footings the effective depth d is mostly governed by shear. Two different types of shear strength are distinguished in single footings: two-way (or punching) shear and one-way (or beam) shear. Based on the Eqs. (48.125) and (48.126) for punching and one-way shear strength, the required effective depth of footing d is calculated.

Single-column footings represent, as it were, cantilevers projecting out from the column in both directions and loaded upward by the soil pressure. Corresponding tension stresses are caused in both these directions at the bottom surface. Such footings are therefore reinforced by two layers of steel, perpendicular to each other and parallel to the edge. The steel reinforcement in each direction can be calculated using Eq. (48.128). The critical sections for development length of footing bars are the same as those for bending. Development length may also have to be checked at all vertical planes in which changes of section or of reinforcement occur, as at the edges of pedestals or where part of the reinforcement may be terminated.

When a column rests on a footing or pedestal, it transfers its load to only a part of the total area of the supporting member. The adjacent footing concrete provides lateral support to the directly loaded part of the concrete. This causes triaxial compression stresses that increase the strength of the concrete, which is loaded directly under the column. The design bearing strength of concrete must not exceed the one given in Eq. (48.130) for forces acting on the concrete at the base of column and the modified one with $r\sqrt{A_2/A_1}$ for supporting area wider than the loaded area. If the calculated bearing force is greater than the design bearing strength, reinforcement must be provided to transfer the excess force. This is done either by extending the column bars into the footing or by providing dowels, which are embedded in the footing and project above it.

Combined Footings

Spread footings that support more than one column or wall are known as *combined footings*. They can be divided into two categories: those that support two columns, and those that support more than two (generally large numbers of) columns.

In buildings where the allowable soil pressure is large enough for single footings to be adequate for most columns, two-column footings are seen to become necessary in two situations: (1) if columns are so close to the property line that single-column footings cannot be made without projecting beyond that line, and (2) if some adjacent columns are so close to each other that their footings would merge.

When the bearing capacity of the subsoil is low so that large bearing areas become necessary, individual footings are replaced by continuous strip footings, which support more than two columns and usually all columns in a row. Mostly, such strips are arranged in both directions, in which case a grid foundation is obtained, as shown in Fig. 48.24. Such a grid foundation can be done by single footings because the individual strips of the grid foundation represent continuous beams whose moments are much smaller than the cantilever moments in large single footings that project far out from the column in all four directions.

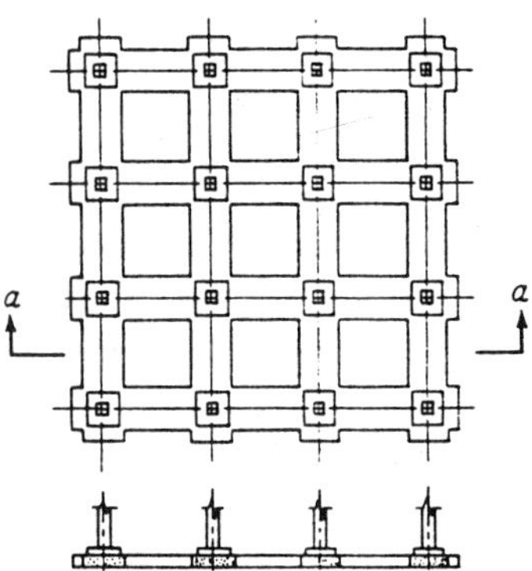

FIGURE 48.24 Grid foundation. (*Source:* Wang and Salmon, 1985.)

For still lower bearing capacities, the strips are made to merge, resulting in a mat foundation, as shown in Fig. 48.25. That is, the foundation consists of a solid reinforced concrete slab under the entire building. In structural action such a mat is very similar to a flat slab or a flat plate, upside down—that is, loaded upward by the bearing pressure and downward by the concentrated column reactions. The mat foundation evidently develops the maximum available bearing area under the building. If the soil's capacity is so low

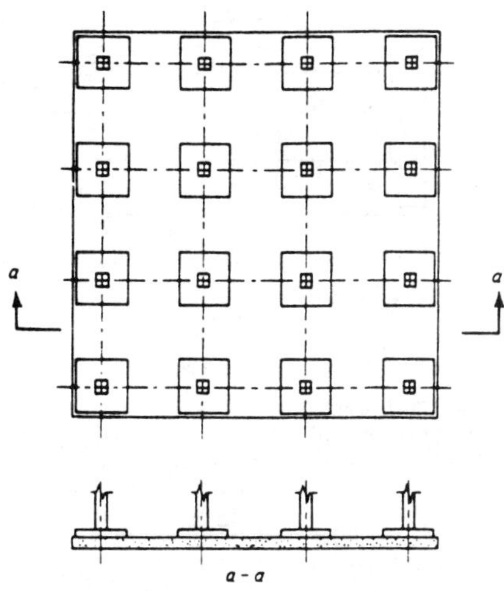

FIGURE 48.25 Mat foundation. (*Source:* Wang and Salmon, 1985.)

that even this large bearing area is insufficient, some form of deep foundation, such as piles or caissons, must be used.

Grid and mat foundations may be designed with the column pedestals—as shown in Figs. 48.24 and 48.25—or without them, depending on whether or not they are necessary for shear strength and the development length of dowels. Apart from developing large bearing areas, another advantage of grid and mat foundations is that their continuity and rigidity help in reducing differential settlements of individual columns relative to each other, which may otherwise be caused by local variations in the quality of subsoil, or other causes. For this purpose, continuous spread foundations are frequently used in situations where the superstructure or the type of occupancy provides unusual sensitivity to differential settlement.

Two-Column Footings

The ACI Code does not provide a detailed approach for the design of combined footings. The design, in general, is based on an empirical approach. It is desirable to design combined footings so that the centroid of the footing area coincides with the resultant of the two column loads. This produces uniform bearing pressure over the entire area and forestalls a tendency for the footings to tilt. In plan, such footings are rectangular, trapezoidal, or T shaped, the details of the shape being arranged to produce coincidence of centroid and resultant. The simple relationships of Fig. 48.26 facilitate the determination of the shapes of the bearing area [Fintel, 1985]. In general, the distances m and n are given, the former being the distance from the center of the exterior column to the property line and the latter the distance from that column to the resultant of both column loads.

Another expedient, which is used if a single footing cannot be centered under an exterior column, is to place the exterior column footing eccentrically and to connect it with the nearest interior column by a beam or strap. This strap, being counterweighted by the interior column load, resists the tilting tendency of the eccentric exterior footings and equalizes the pressure under it. Such foundations are known as *strap, cantilever*, or *connected footings*.

The strap may be designed as a rectangular beam spacing between the columns. The loads on it include its own weight (when it does not rest on the soil) and the upward pressure from the footings. Width of the strap usually is selected arbitrarily as equal to that of the largest column plus 4 to 8 inches so that column forms can be supported on top of the strap. Depth is determined by the maximum bending moment. The main reinforcing in the strap is placed near the top. Some of the steel can be cut off where not needed. For diagonal tension, stirrups normally will be needed near the columns (Fig. 48.27). In addition, longitudinal placement steel is set near the bottom of the strap, plus reinforcement to guard against settlement stresses.

The footing under the exterior column may be designed as a wall footing. The portions on opposite sides of the strap act as cantilevers under the constant upward pressure of the soil. The interior footing should be designed as a single-column footing. The critical section for punching shear, however, differs from that for a conventional footing. This shear should be computed on a section at a distance $d/2$ from the sides and extending around the column at a distance $d/2$ from its faces, where d is the effective depth of the footing.

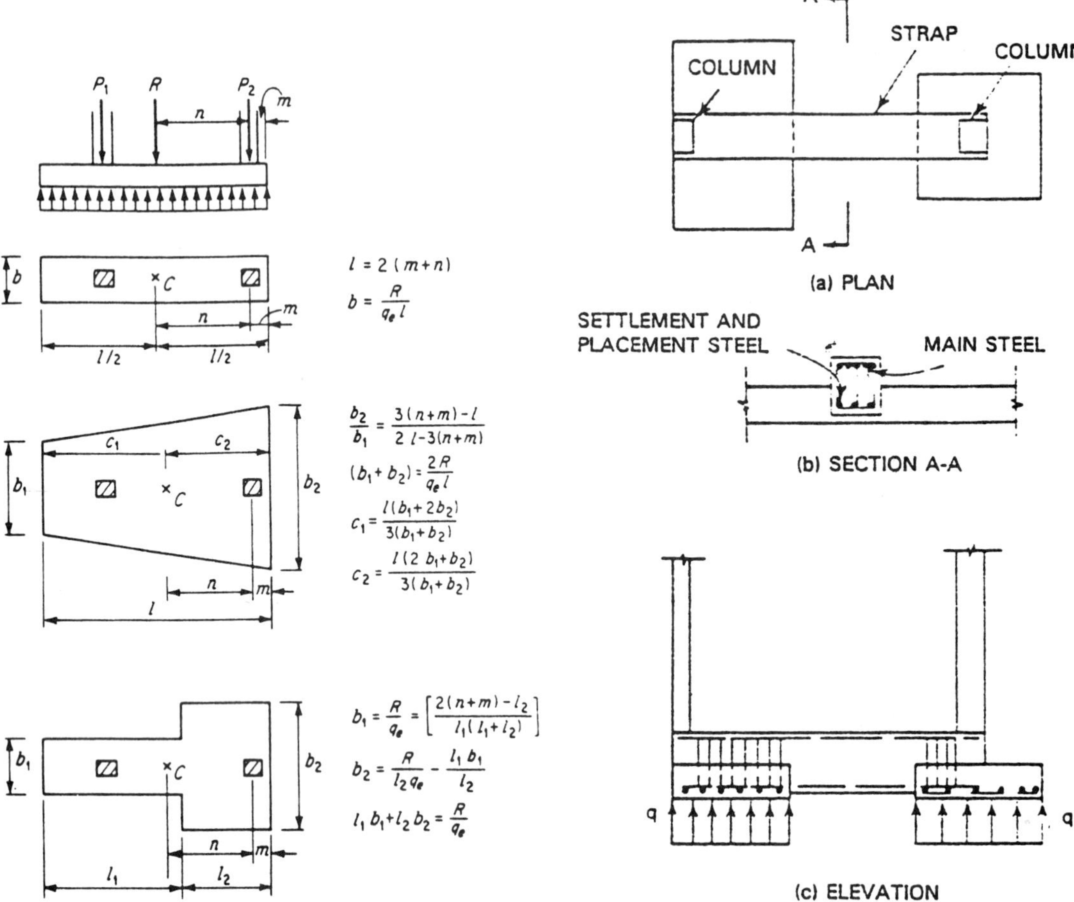

FIGURE 48.26 Two-column footings. (*Source:* Fintel, 1985.)

FIGURE 48.27 Strap (cantilever) footing. (*Source:* Fintel, 1985.)

Strip, Grid, and Mat Foundations

In the case of heavily loaded columns, particularly if they are to be supported on relatively weak or uneven soils, continuous foundations may be necessary. They may consist of a continuous strip footing supporting all columns in a given row or, more often, of two sets of such strip footings intersecting at right angles so that they form one continuous grid foundation (Fig. 48.24). For even larger loads or weaker soils the strips are made to merge, resulting in a mat foundation (Fig. 48.25).

For the design of such continuous foundations it is essential that reasonably realistic assumptions be made regarding the distribution of bearing pressures, which act as upward loads on the foundation. For compressible soils it can be assumed in first approximation that the deformation or settlement of the soil at a given location and the bearing pressure at that location are proportional to each other. If columns are spaced at moderate distances and if the strip, grid, or mat foundation is very rigid, the settlements in all portions of the foundation will be substantially the same. This means that the bearing pressure, also known as *subgrade reaction*, will be the same provided that the centroid of the foundation coincides with the resultant of the loads. If they do not coincide, then for such rigid foundations the subgrade reaction can be assumed as linear and determined from statics in the same manner as discussed for single footings. In this case, all loads—the downward column loads as well as the upward-bearing pressures—are known. Hence, moments and shear

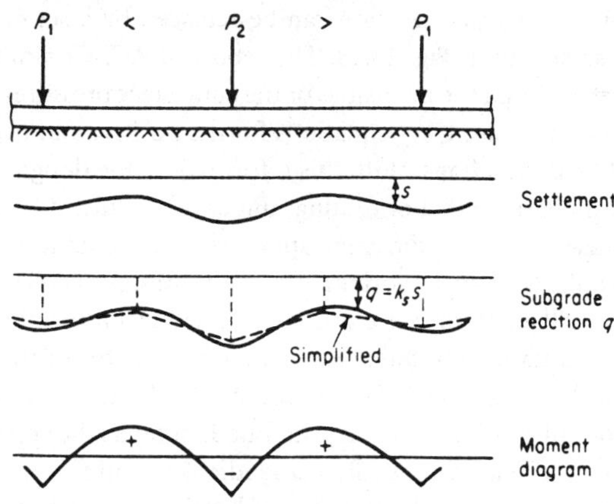

FIGURE 48.28 Strip footing. (*Source:* Fintel, 1985.)

forces in the foundation can be found by statics alone. Once these are determined, the design of strip and grid foundations is similar to that of inverted continuous beams, and design of mat foundations is similar to that of inverted flat slabs or plates.

On the other hand, if the foundation is relatively flexible and the column spacing large, settlements will no longer be uniform or linear. For one thing, the more heavily loaded columns will cause larger settlements, and thereby larger subgrade reactions, than the lighter ones. Also, since the continuous strip or slab midway between columns will deflect upward relative to the nearby columns, soil settlement—and thereby the subgrade reaction— will be smaller midway between columns than directly at the columns. This is shown schematically in Fig. 48.28. In this case the subgrade reaction can no longer be assumed as uniform. A reasonably accurate but fairly complex analysis can then be made using the theory of beams on elastic foundations.

A simplified approach has been developed that covers the most frequent situations of strip and grid foundations [ACI Committee 436, 1966]. The method first defines the conditions under which a foundation can be regarded as rigid so that uniform or overall linear distribution of subgrade reactions can be assumed. This is the case when the average of two adjacent span lengths in a continuous strip does not exceed 1.75/λ, provided also that the adjacent span and column loads do not differ by more than 20% of the larger value. Here,

$$\lambda = \sqrt[4]{\frac{k_s b}{3E_c I}} \tag{48.134}$$

where

$k_s = Sk_s'$

k_s' = coefficient of subgrade reaction as defined in soils mechanics, basically force per unit area required to produce unit settlement, kips/ft^3

b = width of footing, ft

E_c = modulus of elasticity of concrete, kips/ft^2

I = moment of inertia of footing, ft^4

S = shape factor, being $[(b + 1)/2b]^2$ for granular soils and $(n + 0.5)/1.5n$ for cohesive soils, where n is the ratio of longer to shorter side of strip

If the average of two adjacent spans exceeds 1.75/λ, the foundation is regarded as flexible. Provided that adjacent spans and column loads differ by no more than 20%, the complex curvilinear distribution of subgrade reaction can be replaced by a set of equivalent trapezoidal reactions, which are also shown in Fig. 48.28. The report of ACI Committee 436 contains fairly simple equations for determining the intensities of the equivalent pressures under the columns and at the middle of the spans and also gives equations for the positive and negative moments caused by these equivalent subgrade reactions. With this information, the design of continuous strip and grid footings proceeds similarly to that of footings under two columns.

Mat foundations likewise require different approaches, depending on whether they can be classified as rigid or flexible. As in strip footings, if the column spacing is less than $1/\lambda$, the structure may be regarded as rigid, the soil pressure can be assumed as uniformly or linearly distributed, and the design is based on statics. On the other hand, when the foundation is considered flexible as defined above, and if the variation of adjacent column loads and spans is not greater than 20%, the same simplified procedure as for strip and grid foundations can be applied to mat foundations. The mat is divided into two sets of mutually perpendicular strip footings of width equal to the distance between midspans, and the distribution of bearing pressures and bending moments is carried out for each strip. Once moments are determined, the mat is in essence treated the same as a flat slab or plate, with the reinforcement allocated between column and middle strips as in these slab structures.

This approach is feasible only when columns are located in a regular rectangular grid pattern. When a mat that can be regarded as rigid supports columns at random locations, the subgrade reactions can still be taken as uniform or as linearly distributed and the mat analyzed by statics. If it is a flexible mat that supports such randomly located columns, the design is based on the theory of plates on elastic foundation.

Footings on Piles

If the bearing capacity of the upper soil layers is insufficient for a spread foundation, but firmer strata are available at greater depth, piles are used to transfer the loads to these deeper strata. Piles are generally arranged in groups or clusters, one under each column. The group is capped by a spread footing or cap that distributes the column load to all piles in the group. Reactions on caps act as concentrated loads at the individual piles, rather than as distributed pressures. If the total of all pile reactions in a cluster is divided by area of the footing to obtain an equivalent uniform pressure, it is found that this equivalent pressure is considerably higher in pile caps than for spread footings. Thus, it is in any event advisable to provide ample rigidity—that is, depth for pile caps—in order to spread the load evenly to all piles.

As in single-column spread footings, the effective portion of allowable bearing capacities of piles, R_a, available to resist the unfactored column loads is the allowable pile reaction less the weight of footing, backfill, and surcharge per pile. That is,

$$R_e = R_a - W_f \tag{48.135}$$

where W_f is the total weight of footing, fill, and surcharge divided by the number of piles.

Once the available or effective pile reaction R_e is determined, the number of piles in a concentrically loaded cluster is the integer next larger than

$$n = \frac{D + L}{R_e} \tag{48.136}$$

The effects of wind and earthquake moments at the foot of the columns generally produce an eccentrically loaded pile cluster in which different piles carry different loads. The number and location of piles in such a cluster is determined by successive approximation from the condition that the load on the most heavily loaded pile should not exceed the allowable pile reaction R_a. Assuming a linear distribution of pile loads due to bending, the maximum pile reaction is

$$R_{\max} = \frac{P}{n} + \frac{M}{I_{pg}/c} \tag{48.137}$$

where P is the maximum load (including weight of cap, backfill, etc.), M is the moment to be resisted by the pile group, both referred to the bottom of the cap, I_{pg} is the moment of inertia of the entire pile group about the centroidal axis about which bending occurs, and c is the distance from that axis to the extreme pile.

Piles are generally arranged in tight patterns, which minimizes the cost of the caps, but they cannot be placed closer than conditions of deriving and of undisturbed carrying capacity will permit. AASHTO requires that piles be spaced at least 2 ft 6 in. center to center and that the distance from the side of a pile to the nearest edge of the footing be 9 in. or more.

The design of footings on piles is similar to that of single-column spread footings. One approach is to design the cap for the pile reactions calculated for the factored column loads. For a concentrically loaded cluster this would give $R_u = (1.4D + 1.7L)/n$. However, since the number of piles was taken as the next larger integer according to Eq. (48.137), determining R_u in this manner can lead to a design where the strength of the cap is less than the capacity of the pile group. It is therefore recommended that the pile reaction for strength design be taken as

$$R_u = R_e \times \text{Average load factor} \tag{48.138}$$

where the average load factor is $(1.4D + 1.7L)/(D + L)$. In this manner the cap is designed to be capable of developing the full allowable capacity of the pile group.

As in single-column spread footings, the depth of the pile cap is usually governed by shear. In this regard both punching and one-way shear need to be considered. The critical sections are the same as explained earlier under "Two-Way Shear (Punching Shear)" and "One-Way Shear." The difference is that shears on caps are caused by concentrated pile reactions rather than by distributed bearing pressures. This poses the question of how to calculate shear if the critical section intersects the circumference of one or more piles. For this case the ACI Code accounts for the fact that pile reaction is not really a point load, but is distributed over the pile-bearing area. Correspondingly, for piles with diameters d_p, it stipulates as follows:

> Computation of shear on any section through a footing on piles shall be in accordance with the following:
>
> (a) The entire reaction from any pile whose center is located $d_p/2$ or more outside this section shall be considered as producing shear on that section.
>
> (b) The reaction from any pile whose center is located $d_p/2$ or more inside the section shall be considered as producing no shear on that section.
>
> (c) For intermediate portions of the pile center, the portion of the pile reaction to be considered as producing shear on the section shall be based on straight-line interpolation between the full value at $d_p/2$ outside the section and zero at $d_p/2$ inside the section [1992].

In addition to checking punching and one-way shear, punching shear must be investigated for the individual pile. Particularly in caps on a small number of heavily loaded piles, it is this possibility of a pile punching upward through the cap which may govern the required depth. The critical perimeter for this action, again, is located at a distance $d/2$ outside the upper edge of the pile. However, for relatively deep caps and closely spaced piles, critical perimeters around adjacent piles may overlap. In this case, fracture, if any, would undoubtedly occur along an outward-slanting surface around both adjacent piles. For such situations the critical perimeter is so located that its length is a minimum, as shown for two adjacent piles in Fig. 48.29.

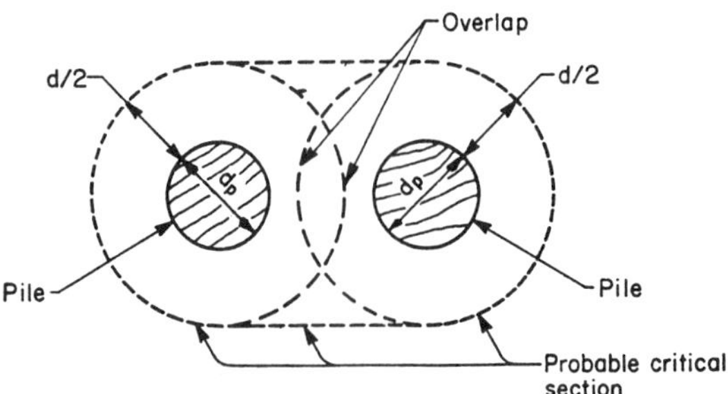

FIGURE 48.29 Modified critical section for shear with overlapping critical perimeters.

48.11 Walls

Panel, Curtain, and Bearing Walls

As a general rule, the exterior walls of a reinforced concrete building are supported at each floor by the skeleton framework, their only function being to enclose the building. Such walls are called *panel walls*. They may be made of concrete (often precast), cinder concrete block, brick, tile blocks, or insulated metal panels. The thickness of each of these types of panel walls will vary according to the material, type of construction, climatological conditions, and the building requirements governing the particular locality in which the construction takes place. The pressure of the wind is usually the only load that is considered in determining the structural thickness of a wall panel, although in some cases exterior walls are used as diaphragms to transmit forces caused by horizontal loads down to the building foundations.

Curtain walls are similar to panel walls except that they are not supported at each story by the frame of the building; rather, they are self supporting. However, they are often anchored to the building frame at each floor to provide lateral support.

A bearing wall may be defined as one that carries any vertical load in addition to its own weight. Such walls may be constructed of stone masonry, brick, concrete block, or reinforced concrete. Occasional projections or pilasters add to the strength of the wall and are often used at points of load concentration. Bearing walls may be of either single or double thickness, the advantage of the latter type being that the air space between the walls renders the interior of the building less liable to temperature variation and makes the wall itself more nearly moistureproof. On account of the greater gross thickness of the double wall, such construction reduces the available floor space.

According to ACI Code Sec. 14.5.2 the load capacity of a wall is given by

$$\phi P_{nw} = 0.55\phi f_c' A_g \left[1 - \left(\frac{kl_c}{32h}\right)^2\right] \tag{48.139}$$

where

ϕP_{nw} = design axial load strength
A_g = gross area of section, in.^{2}
l_c = vertical distance between supports, in.
h = thickness of wall, in.
ϕ = 0.7

and where the effective length factor k is taken as 0.8 for walls restrained against rotation at top or bottom or both, 1.0 for walls unrestrained against rotation at both ends, and 2.0 for walls not braced against lateral translation.

In the case of concentrated loads, the length of the wall to be considered as effective for each should not exceed the center-to-center distance between loads; nor should it exceed the width of the bearing plus 4 times the wall thickness. Reinforced concrete bearing walls should have a thickness of at least 1/25 times the unsupported height or width, whichever is shorter. Reinforced concrete bearing walls of buildings should be not less than 4 in. thick.

Minimum ratio of horizontal reinforcement area to gross concrete area should be 0.0020 for deformed bars not larger than #5—with specified yield strength not less than 60,000 psi or 0.0025 for other deformed bars—or 0.0025 for welded wire fabric not larger than W31 or D31. Minimum ratio of vertical reinforcement area to gross concrete area should be 0.0012 for deformed bars not larger than #5—with specified yield strength not less than 60,000 psi or 0.0015 for other deformed bars—or 0.0012 for welded wire fabric not larger than W31 or D31. In addition to the minimum reinforcement requirement, not less than two #5 bars shall be provided around all window and door openings. Such bars shall be extended to develop the bar beyond the corners of the openings but not less than 24 in.

Walls more than 10 in. thick should have reinforcement for each direction placed in two layers parallel with faces of wall. Vertical and horizontal reinforcement should not be spaced further apart than three times the wall thickness, or 18 in. Vertical reinforcement need not be enclosed by lateral ties if vertical reinforcement area is not greater than 0.01 times gross concrete area, or where vertical reinforcement is not required as compression reinforcement.

Quantity of reinforcement and limits of thickness mentioned above are waived where structural analysis shows adequate strength and stability. Walls should be anchored to intersecting elements such as floors, roofs, or to columns, pilasters, buttresses, and intersecting walls, and footings.

Basement Walls

In determining the thickness of basement walls, the lateral pressure of the earth, if any, must be considered in addition to other structural features. If it is part of a bearing wall, the lower portion may be designed either as a slab supported by the basement and floors or as a retaining wall, depending upon the type of construction. If columns and wall beams are available for support, each basement wall panel of reinforced concrete may be designed to resist the earth pressure as a simple slab reinforced in either one or two directions. A minimum thickness of 7.5 in. is specified for reinforced concrete basement walls. In wet ground a minimum thickness of 12 in. should be used. In any case, the thickness cannot be less than that of the wall above.

Care should be taken to brace a basement wall thoroughly from the inside (1) if the earth is backfilled before the wall has obtained sufficient strength to resist the lateral pressure without such assistance, or (2) if it is placed before the first-floor slab is in position.

Partition Walls

Interior walls used for the purpose of subdividing the floor area may be made of cinder block, brick, precast concrete, metal lath and plaster, clay tile, or metal. The type of wall selected will depend upon the fire resistance required; flexibility of rearrangement; ease with which electrical conduits, plumbing, etc., can be accommodated; and architectural requirements.

Shear Walls

Horizontal forces acting on buildings—for example, those due to wind of seismic action—can be resisted by a variety of means. Rigid-frame resistance of the structure, augmented by the

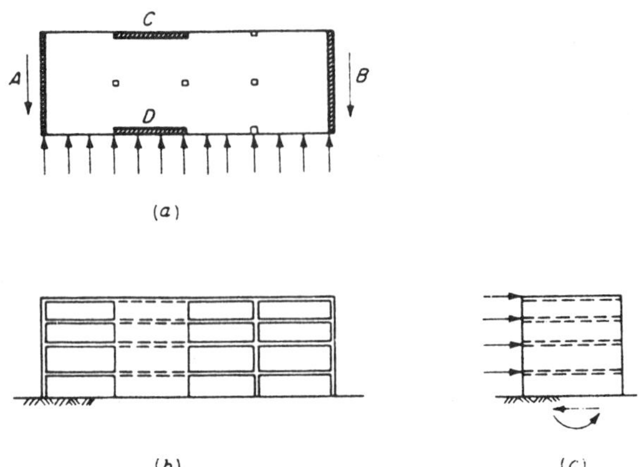

FIGURE 48.30 Building with shear walls subject to horizontal loads: (a) typical floor; (b) front elevation; (c) end elevation.

contribution of ordinary masonry walls and partitions, can provide for wind loads in many cases. However, when heavy horizontal loading is likely—such as would result from an earthquake—reinforced concrete shear walls are used. These may be added solely to resist horizontal forces; alternatively, concrete walls enclosing stairways or elevator shafts may also serve as shear walls.

Figure 48.30 shows a building with wind or seismic forces represented by arrows acting on the edge of each floor or roof. The horizontal surfaces act as deep beams to transmit loads to vertical resisting elements A and B. These shear walls, in turn, act as cantilever beams fixed at their base to carry loads down to the foundation. They are subjected to (1) a variable shear, which reaches maximum at the base, (2) a bending moment, which tends to cause vertical tension near the loaded edge and compression at the far edge, and (3) a vertical compression due to ordinary gravity loading from the structure. For the building shown, additional shear walls C and D are provided to resist loads acting in the log direction of the structure.

The design basis for shear walls, according to the ACI Code, is of the same general form as that used for ordinary beams:

$$V_u \leq \phi V_n \tag{48.140}$$

$$V_n = V_c + V_s \tag{48.141}$$

Shear strength V_n at any horizontal section for shear in plane of wall should not be taken greater than $10\sqrt{f_c'}h_d$. In this and all other equations pertaining to the design of shear walls, the distance of d may be taken equal to $0.8l_w$. A larger value of d, equal to the distance from the extreme compression face to the center of force of all reinforcement in tension, may be used when determined by a strain compatibility analysis.

The value of V_c, the nominal shear strength provided by the concrete, may be based on the usual equations for beams, according to ACI Code. For walls subjected to vertical compression,

$$V_c = 2\sqrt{f_c'}hd \tag{48.142}$$

and for walls subjected to vertical tension N_u,

$$V_c = 2\left(1 + \frac{N_u}{500A_g}\right)\sqrt{f_c'}hd \tag{48.143}$$

where N_u is the factored axial load in pounds, taken negative for tension, and A_g is the gross area of horizontal concrete section in square inches. Alternatively, the value of V_c may be based on a more detailed calculation, as the lesser of

$$V_c = 3.3\sqrt{f_c'}hd + \frac{N_u d}{4l_w} \tag{48.144}$$

or

$$V_c = \left[0.6\sqrt{f_c'} + \frac{l_w(1.25\sqrt{f_c'} + 0.2N_u/l_w h)}{M_u/V_u - l_w/2}\right]hd \tag{48.145}$$

Eq. (48.144) corresponds to the occurrence of a principal tensile stress of approximately $4\sqrt{f_c'}$ at the centroid of the shear-wall cross section. Eq. (48.145) corresponds approximately to the occurrence of a flexural tensile stress of $6\sqrt{f_c'}$ at a section $l_w/2$ above the section being investigated. Thus the two equations predict, respectively, web-shear cracking and flexure-shear cracking. When the quantity $M_u/V_u - l_w/2$ is negative, Eq. (48.145) is inapplicable. According to the ACI Code, horizontal sections located closer to the wall base than a distance $l_w/2$ or $h_w/2$, whichever less, may be designed for the same V_c as that computed at a distance $l_w/2$ or $h_w/2$.

When the factored shear force V_u does not exceed $\phi V_c/2$, a wall may be reinforced according to the minimum requirements given in Sec. 12.1. When V_u exceeds $\phi V_c/2$, reinforcement for shear is to be provided according to the following requirements.

The nominal shear strength V_s provided by the horizontal wall steel is determined on the same basis as for ordinary beams:

$$V_s = \frac{A_v f_y d}{s_2} \tag{48.146}$$

where A_v is the area of horizontal shear reinforcement within vertical distance s_2, (in.^{2}), s_2 is the vertical distance between horizontal reinforcement, (in.), and f_y is the yield strength of reinforcement, psi. Substituting Eq. (48.146) into Eq. (48.141), then combining with Eq. (48.140), one obtains the equation for the required area of horizontal shear reinforcement within a distance s_2:

$$A_v = \frac{(V_u - \phi V_c)s_2}{\phi f_y d} \tag{48.147}$$

The minimum permitted ratio of horizontal shear steel to gross concrete area of vertical section, ρ_n, is 0.0025 and the maximum spacing s_2 is not to exceed $l_w/5$, $3h$, or 18 in.

Test results indicate that for low shear walls, vertical distributed reinforcement is needed as well as horizontal reinforcement. Code provisions require vertical steel of area A_h within a spacing s_1, such that the ratio of vertical steel to gross concrete area of horizontal section will not be less than

$$\rho_n = 0.0025 + 0.5\left(2.5 - \frac{h_w}{l_w}\right)(\rho_h - 0.0025) \tag{48.148}$$

nor less than 0.0025. However, the vertical steel ratio need not be greater than the required horizontal steel ratio. The spacing of the vertical bars is not to exceed $l_w/3$, $3h$, or 18 in.

Walls may be subjected to flexural tension due to overturning moment, even when the vertical compression from gravity loads is superimposed. In many but not all cases, vertical steel is provided, concentrated near the wall edges, as in Fig. 48.31. The required steel area can be found by the usual methods for beams.

The ACI Code contains requirements for the dimensions and details of structural walls serving as part of the earthquake-force resisting systems. The reinforcement ratio, ρ_v ($= A_{sv}/A_{cv}$; where A_{cv} is the net area of concrete section bounded by web thickness and length of section in the direction of shear force considered, and A_{sv} is the projection on A_{cv} of area of distributed shear reinforcement crossing the plane of A_{cv}), for structural walls should not be less than 0.0025 along the longitudinal and transverse axes. Reinforcement provided for shear strength should be continuous and should be distributed across the shear plane. If the design shear force does not exceed $A_{cv}\sqrt{f_c'}$, the shear reinforcement may conform to the reinforcement ratio given in Sec. 12.1. At least two curtains of reinforcement should be used in a wall if the in-plane factored shear force assigned to the wall exceeds $2A_{cv}\sqrt{f_c'}$. All continuous reinforcement in structural walls should be anchored or spliced in accordance with the provisions for reinforcement in tension for seismic design.

Proportioning and details of structural walls that resist shear forces caused by earthquake motion is contained in the ACI Code Sec. 21.7.3.

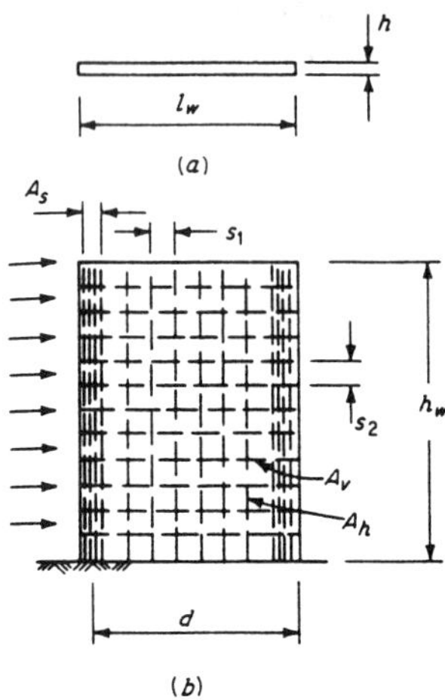

FIGURE 48.31 Geometry and reinforcement of typical shear wall: (a) cross section; (b) elevation.

References

ACI Committee 318. 1992. *Building Code Requirements for Reinforced Concrete and Commentary, ACI 318-89 (Revised 92) and ACI 318R-89 (Revised 92)* (347 pp.). Detroit, MI.

ACI Committee 340. 1990. *Design Handbook in Accordance with the Strength Design Method of ACI 318-89.* Volume 2, SP-17 (222 pp.).

ACI Committee 363. 1984. State-of-the-art report on high strength concrete. *ACI J. Proc.* 81(4):364–411.

ACI Committee 436. 1966. Suggested design procedures for combined footings and mats. *J. ACI.* 63:1041–1057.

Breen, J. E. 1991. Why structural concrete? *IASE Colloq. Struct. Concr.* Stuttgart, pp. 15–26.

Collins, M. P. and Mitchell, D. 1991. *Prestressed Concrete Structures,* 1st ed. Prentice Hall, Englewood Cliffs, N.J.

Fintel, M. 1985. *Handbook of Concrete Engineering.* 2nd ed. Van Nostrand Reinhold, New York.

MacGregor, J. G. 1992. *Reinforced Concrete Mechanics and Design,* 2nd ed. Prentice Hall, Englewood Cliffs, N.J.

Nilson, A. H. and Winter, G. 1992. *Design of Concrete Structures,* 11th ed. McGraw-Hill, New York.

Standard Handbook for Civil Engineers, 2nd ed. McGraw-Hill, New York.

Wang, C.-K., and Salmon, C. G. 1985. *Reinforced Concrete Design,* 4th ed. Harper Row, New York.

49

Structural Reliability

David V. Rosowsky
Clemson University

49.1 Introduction

Definition of Reliability

Reliability and **reliability-based design (RBD)** are terms which are increasingly being associated with the design of civil engineering structures. While the subject of reliability may not typically be treated explicitly in the undergraduate civil engineering curriculum, some basic knowledge of the concepts of structural reliability can be useful in understanding the development and bases for many modern design codes (including the AISC, ACI, AASHTO, and other codes).

Reliability simply refers to some probabilistic measure of satisfactory (or safe) performance, and as such, may be viewed as a complementary function of the probability of **failure**.

$$\text{Reliability} = fcn(1 - P_{\text{failure}}) \tag{49.1}$$

When we talk about the reliability of a structure (or member or system), we are referring to the *probability* of "safe performance" for a particular **limit state**. A limit state can refer to ultimate failure (such as collapse) or a condition of unserviceability (such as excessive vibration, deflection, or cracking). The treatment of structural loads and resistances using probability (or reliability) theory, and of course the theories of structural analysis and mechanics, has led to the development of the latest generation of probability-based, reliability-based, and **limit states design** codes.

0-8493-8953-4/95/$0.00 + $.50

If the subject of structural reliability is generally not treated in the undergraduate civil engineering curriculum, why include a basic (introductory) treatment in this handbook? Besides providing some insight into the bases for modern codes, it is likely that future generations of structural codes and specifications will rely more and more on probabilistic methods and reliability analyses. The treatment of structural analysis, structural design, and probability and statistics that is provided in most curricula permits this introduction to structural reliability without the need for more advanced study. This section by no means contains a complete treatment of the subject, nor does it contain a complete review of probability theory. At this point in time, structural reliability is usually treated at the graduate level. However, it is likely that as reliability-based design becomes more accepted and more prevalent, additional material will appear in the undergraduate curriculum.

Introduction to Reliability-Based Design Concepts

The concept of reliability-based design is most easily illustrated in Fig. 49.1. As shown in the figure, we consider the acting load and the structural resistance to be random variables. There is also the possibility of a resistance (or strength) that is inadequate for the acting load (or conversely, that the load exceeds the available strength). This possibility is indicated by the region of overlap on Fig. 49.1 in which realizations of the load and resistance variables lead to failure. The objective of RBD is to ensure that the probability of this condition is acceptably small. Of course, the "load" can refer to any appropriate structural, service, or environmental loading (actually, its *effect*), and the resistance can refer to any limit state "capacity" (flexural strength, bending stiffness, maximum tolerable deflection, etc.). If we formulate the simplest expression for the probability of failure (P_f) as

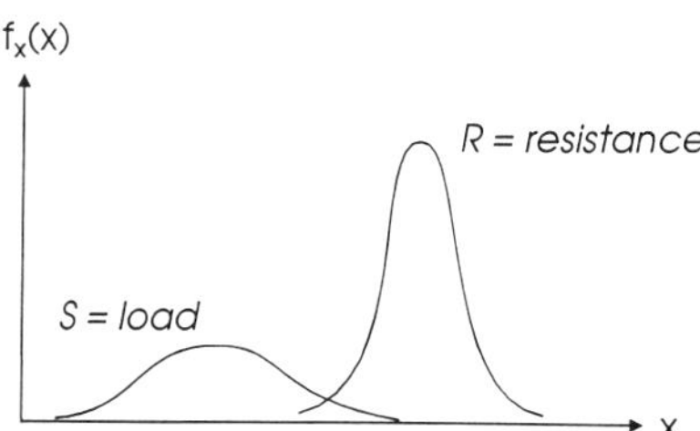

Figure 49.1 Basic concept of structural reliability.

$$P_f = P[(R - S) < 0] \tag{49.2}$$

we need only ensure that the units of the resistance (R) and the load (S) are consistent. We can then use probability theory to estimate these limit state probabilities.

Since RBD is intended to provide (or ensure) uniform and acceptably small failure probabilities for similar designs (limit states, materials, occupancy, etc.), these "acceptable" levels must be predetermined. This is the responsibility of code-development groups and is based largely on previous experience (i.e., calibration to previous design philosophies such as **allowable stress design** for steel) and engineering judgment. Finally, with information describing the statistical variability of the loads and resistances, and the target probability of failure (or target reliability) established, factors for codified design can be evaluated for the relevant load and resistance quantities (again, for the particular limit state being considered). This results, for instance, in the familiar form of design checking equations

$$\phi R_n \geq \sum_i \gamma_i Q_{n,i} \tag{49.3}$$

referred to as **load and resistance factor design (LRFD)** in the U.S. and in which R_n is the **nominal** (or design) **resistance** and Q_n is the **nominal load effect**. The factors γ_i and ϕ in Eq. (49.3) are the load and resistance factors, respectively. This will be described in more detail in later sections. Additional information on this subject may be found in a number of available texts [Melchers, 1987; Ang and Tang, 1975b].

49.2 Basic Probability Concepts

This section presents an introduction to basic probability concepts. Only a sufficient presentation of topics to permit the discussion of reliability theory and applications that follows is included herein. For additional information and a more detailed presentation, the reader is referred to a number of widely used textbooks [e.g., Benjamin and Cornell, 1970; Ang and Tang, 1975a].

Random Variables and Distributions

Random variables can be classified as being either *discrete* or *continuous*. Discrete random variables can assume only discrete values, whereas continuous random variables can assume any value within a range (which may or may not be bounded from above or below). In general, the random variables considered in structural reliability analyses are continuous, though some important cases exist where one or more variables are discrete (i.e., the number of earthquakes in a region). A brief discussion of both discrete and continuous random variables is presented here, but the reliability analysis (theory and applications) sections which follow will focus mainly on continuous random variables.

The *relative frequency* of a variable is described by its probability mass function (PMF), denoted $p_X(x)$ if it is discrete, or its probability density function (PDF), denoted $f_X(x)$ if it is continuous. (A histogram is an example of a PMF, whereas its continuous analog, a smooth function, would represent a PDF.) The *cumulative frequency* (for either a discrete or continuous random variable) is described by its cumulative distribution function (CDF), denoted $F_X(x)$. (See Fig. 49.2.)

There are three basic axioms of probability which serve to define valid probability assignments and provide the basis for probability theory.

1. The probability of an event is bounded by 0 and 1 (corresponding to the cases of zero probability and certainty, respectively).

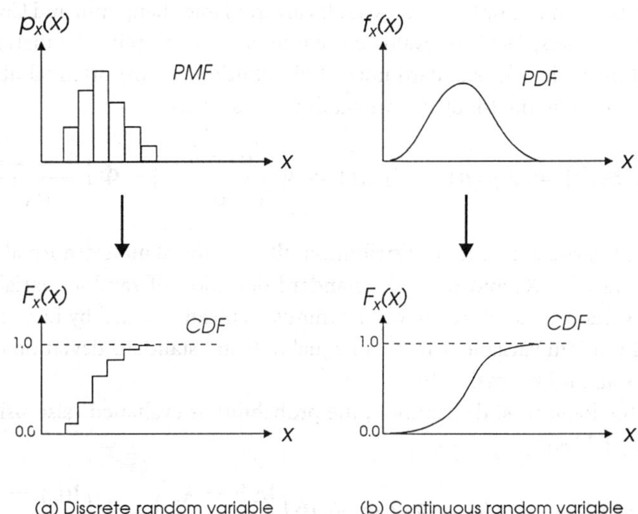

FIGURE 49.2 Sample probability functions.

2. The sum of all possible outcomes in a sample space must equal 1 (a statement of *collectively exhaustive* events).
3. The probability of the union of two *mutually exclusive* events is the sum of the two individual event probabilities, $P[A \cup B] = P[A] + P[B]$.

The PMF or PDF, describing the relative frequency of the random variable, can be used to evaluate the probability that a variable takes on a value within some range.

$$P[a < X_{\text{discr}} \leq b] = \sum_{a}^{b} p_X(x) \tag{49.4}$$

$$P[a < X_{\text{cts}} \leq b] = \int_{a}^{b} f_X(x)\, dx \tag{49.5}$$

The CDF is used to describe the probability that a random variable is less than or equal to some value. Thus, there exists a simple integral relationship between the PDF and the CDF. For example, for a continuous random variable,

$$F_X(a) = P[X \leq a] = \int_{-\infty}^{a} f_X(x)\, dx \tag{49.6}$$

There are a number of common distribution forms. The probability functions for these distribution forms are given in Table 49.1.

An important class of distributions for reliability analysis is based on the statistical theory of *extreme values*. Extreme value distributions are used to describe the distribution of the largest or smallest of a set of independent and identically distributed random variables. This has obvious implications for reliability problems in which we may be concerned with the largest of a set of 50 annual-extreme snow loads, or the smallest (lowest) concrete strength from a set of 100 cylinder tests, for example. There are three important extreme value distributions (referred to as type I, II, and III, respectively), which are also included in Table 49.1. Additional information on the derivation and application of extreme value distributions may be found in various texts [e.g., Ang and Tang, 1975b; Melchers, 1987].

In most cases, the solution to the integral of the probability function [see Eqs. (49.5) and (49.6)] is available in closed-form. The exceptions are two of the more common distributions, the normal and lognormal distributions. For these cases, tables are available [Benjamin and Cornell, 1970; Ang and Tang, 1975a; Melchers, 1987] to evaluate the integrals. To simplify the matter and eliminate the need for multiple tables, the standard normal distribution is most often tabulated. In the case of the normal distribution, the following probability is evaluated:

$$P[a < X \leq b] = F_X(b) - F_X(a) = \Phi\left(\frac{b - \mu_x}{\sigma_x}\right) - \Phi\left(\frac{a - \mu_x}{\sigma_x}\right) \tag{49.7}$$

where $F_X(\cdot)$ is the particular normal distribution, $\Phi(\cdot)$ is the standard normal CDF, μ_x is the mean of random variable X, and σ_x is the standard deviation of random variable X. Since the standard normal variate is therefore the variate minus its mean, divided by its standard deviation, it too is a normal random variable with mean equal to 0 and standard deviation equal to 1. Table 49.2 presents the standard normal CDF.

In the case of the lognormal distribution, the probability is evaluated (also using the standard normal probability tables):

$$P[a < Y \leq b] = F_Y(b) - F_Y(a) = \Phi\left(\frac{\ln b - \lambda_y}{\xi_y}\right) - \Phi\left(\frac{\ln a - \lambda_y}{\xi_y}\right) \tag{49.8}$$

Table 49.1 Common Distribution Forms and Their Parameters

Distribution	PMF or PDF	Parameters	Mean and Variance
Binomial	$p_X(x) = \binom{n}{x} p^x (1-p)^{n-x}$	p	$E[X] = np$
	$x = 0, 1, 2, \ldots, n$		$\text{Var}[X] = np(1-p)$
Geometric	$p_X(x) = p(1-p)^{x-1}$	p	$E[X] = 1/p$
	$x = 0, 1, 2, \ldots$		$\text{Var}[X] = (1-p)/p^2$
Poisson	$p_X(x) = \frac{(vt)^x}{x!} e^{-vt}$	v	$E[X] = vt$
	$x = 0, 1, 2, \ldots$		$\text{Var}[X] = vt$
Exponential	$f_X(x) = \lambda e^{-\lambda x}$	λ	$E[X] = 1/\lambda$
	$x \geq 0$		$\text{Var}[X] = 1/\lambda^2$
Gamma	$f_X(x) = \frac{v(vx)^{k-1} e^{-vx}}{\Gamma(k)}$	v, k	$E[X] = k/v$
	$x \geq 0$		$\text{Var}[X] = k/v^2$
Normal	$f_X(x) = \frac{1}{\sqrt{2\pi}\sigma} \exp\left[-\frac{1}{2}\left(\frac{x-\mu}{\sigma}\right)^2\right]$	μ, σ	$E[X] = \mu$
	$-\infty < x < \infty$		$\text{Var}[X] = \sigma^2$
Lognormal	$f_X(x) = \frac{1}{\sqrt{2\pi}\zeta x} \exp\left[-\frac{1}{2}\left(\frac{\ln x - \lambda}{\zeta}\right)^2\right]$	λ, ζ	$E[X] = \exp\left(\lambda + \frac{1}{2}\zeta^2\right)$
	$x \geq 0$		$\text{Var}[X] = E^2[X]\left(\exp\left(\zeta^2\right) - 1\right)$
Uniform	$f_X(x) = \frac{1}{b-a}$	a, b	$E[X] = \frac{(a+b)}{2}$
	$a < x < b$		$\text{Var}[X] = \frac{1}{12}(b-a)^2$
Extreme type-I	$f_X(x) = \alpha \exp\left[-\alpha(x-u) - e^{-\alpha(x-u)}\right]$	α, u	$E[X] = u + \frac{\gamma}{\alpha}$ $(\gamma \cong 0.5772)$
(largest)	$-\infty < x < \infty$		$\text{Var}[X] = \frac{\pi^2}{6\alpha^2}$
Extreme type-II	$f_X(x) = \frac{k}{x}\left(\frac{u}{x}\right)^k e^{-(u/x)^k}$	k, u	$E[X] = u\Gamma\left(1 - \frac{1}{k}\right)$ $(k > 1)$
(largest)	$x \geq 0$		$\text{Var}[X] = u^2\left[\Gamma\left(1 - \frac{2}{k}\right) - \Gamma^2\left(1 - \frac{1}{k}\right)\right]$ $(k > 2)$
Extreme type-III	$f_X(x) = \frac{k}{w-\varepsilon}\left(\frac{x-\varepsilon}{w-\varepsilon}\right)^{k-1} \exp\left[-\left(\frac{x-\varepsilon}{w-\varepsilon}\right)^k\right]$	k, w, ε	$E[X] = \varepsilon + (u-\varepsilon)\Gamma\left(1 + \frac{1}{k}\right)$
(smallest)	$x \geq \varepsilon$		$\text{Var}[X] = (u-\varepsilon)^2\left[\Gamma\left(1 + \frac{2}{k}\right) - \Gamma^2\left(1 + \frac{1}{k}\right)\right]$

where $F_Y(\cdot)$ is the particular lognormal distribution, $\Phi(\cdot)$ is the standard normal CDF, and λ_y and ξ_y are the lognormal distribution parameters related to μ_y (mean of random variable Y) and V_y (coefficient of variation of random variable Y) by the following:

$$\lambda_y = \ln \mu_y - \tfrac{1}{2}\xi_y^2 \tag{49.9}$$

$$\xi_y^2 = \ln(V_y^2 + 1) \tag{49.10}$$

Note that for relatively low coefficients of variation ($V_y \approx 0.3$ or less), Eq. (49.10) suggests the approximation $\xi \approx V_y$.

Table 49.2 Complementary Standard Normal Table, $\Phi(-\beta) = 1 - \Phi(\beta)$

β	$\Phi(-\beta)$	β	$\Phi(-\beta)$	β	$\Phi(-\beta)$	β	$\Phi(-\beta)$
.00	.50000+00	.60	.2743E+00	1.20	.1151E+00	1.80	.3593E−01
.01	.4960E+00	.61	.2709E+00	1.21	.1131E+00	1.81	.3515E−01
.02	.4920E+00	.62	.2676E+00	1.22	.1112E+00	1.82	.3438E−01
.03	.4880E+00	.63	.2643E+00	1.23	.1093E+00	1.83	.3363E−01
.04	.4840E+00	.64	.2611E+00	1.24	.1075E+00	1.84	.3288E−01
.05	.4801E+00	.65	.2578E+00	1.25	.1056E+00	1.85	.3216E−01
.06	.4761E+00	.66	.2546E+00	1.26	.1038E+00	1.86	.3144E−01
.07	.4721E+00	.67	.2514E+00	1.27	.1020E+00	1.87	.3074E−01
.08	.4681E+00	.68	.2483E+00	1.28	.1003E+00	1.88	.3005E−01
.09	.4641E+00	.69	.2451E+00	1.29	.9853E−01	1.89	.2938E−01
.10	.4602E+00	.70	.2420E+00	1.30	.9680E−01	1.90	.2872E−01
.11	.4562E+00	.71	.2389E+00	1.31	.9510E−01	1.91	.2807E−01
.12	.4522E+00	.72	.2358E+00	1.32	.9342E−01	1.92	.2743E−01
.13	.4483E+00	.73	.2327E+00	1.33	.9176E−01	1.93	.2680E−01
.14	.4443E+00	.74	.2297E+00	1.34	.9012E−01	1.94	.2619E−01
.15	.4404E+00	.75	.2266E+00	1.35	.8851E−01	1.95	.2559E−01
.16	.4364E+00	.76	.2236E+00	1.36	.8691E−01	1.96	.2500E−01
.17	.4325E+00	.77	.2207E+00	1.37	.8534E−01	1.97	.2442E−01
.18	.4286E+00	.78	.2177E+00	1.38	.8379E−01	1.98	.2385E−01
.19	.4247E+00	.79	.2148E+00	1.39	.8226E−01	1.99	.2330E−01
.20	.4207E+00	.80	.2119E+00	1.40	.8076E−01	2.00	.2275E−01
.21	.4168E+00	.81	.2090E+00	1.41	.7927E−01	2.01	.2222E−01
.22	.4129E+00	.82	.2061E+00	1.42	.7780E−01	2.02	.2169E−01
.23	.4090E+00	.83	.2033E+00	1.43	.7636E−01	2.03	.2118E−01
.24	.4052E+00	.84	.2005E+00	1.44	.7493E−01	2.04	.2068E−01
.25	.4013E+00	.85	.1977E+00	1.45	.7353E−01	2.05	.2018E−01
.26	.3974E+00	.86	.1949E+00	1.46	.7215E−01	2.06	.1970E−01
.27	.3936E+00	.87	.1922E+00	1.47	.7078E−01	2.07	.1923E−01
.28	.3897E+00	.88	.1894E+00	1.48	.6944E−01	2.08	.1876E−01
.29	.3859E+00	.89	.1867E+00	1.49	.6811E−01	2.09	.1831E−01
.30	.3821E+00	.90	.1841E+00	1.50	.6681E−01	2.10	.1786E−01
.31	.3783E+00	.91	.1814E+00	1.51	.6552E−01	2.11	.1743E−01
.32	.3745E+00	.92	.1788E+00	1.52	.6426E−01	2.12	.1700E−01
.33	.3707E+00	.93	.1762E+00	1.53	.6301E−01	2.13	.1659E−01
.34	.3669E+00	.94	.1736E+00	1.54	.6178E−01	2.14	.1618E−01
.35	.3632E+00	.95	.1711E+00	1.55	.6057E−01	2.15	.1578E−01
.36	.3594E+00	.96	.1685E+00	1.56	.5938E−01	2.16	.1539E−01
.37	.3557E+00	.97	.1660E+00	1.57	.5821E−01	2.17	.1500E−01
.38	.3520E+00	.98	.1635E+00	1.58	.5705E−01	2.18	.1463E−01
.39	.3483E+00	.99	.1611E+00	1.59	.5592E−01	2.19	.1426E−01
.40	.3446E+00	1.00	.1587E+00	1.60	.5480E−01	2.20	.1390E−01
.41	.3409E+00	1.01	.1562E+00	1.61	.5370E−01	2.21	.1355E−01
.42	.3372E+00	1.02	.1539E+00	1.62	.5262E−01	2.22	.1321E−01
.43	.3336E+00	1.03	.1515E+00	1.63	.5155E−01	2.23	.1287E−01
.44	.3300E+00	1.04	.1492E+00	1.64	.5050E−01	2.24	.1255E−01
.45	.3264E+00	1.05	.1469E+00	1.65	.4947E−01	2.25	.1222E−01
.46	.3228E+00	1.06	.1446E+00	1.66	.4846E−01	2.26	.1191E−01
.47	.3192E+00	1.07	.1423E+00	1.67	.4746E−01	2.27	.1160E−01
.48	.3156E+00	1.08	.1401E+00	1.68	.4648E−01	2.28	.1130E−01
.49	.3121E+00	1.09	.1379E+00	1.69	.4551E−01	2.29	.1101E−01
.50	.3085E+00	1.10	.1357E+00	1.70	.4457E−01	2.30	.1072E−01
.51	.3050E+00	1.11	.1335E+00	1.71	.4363E−01	2.31	.1044E−01
.52	.3015E+00	1.12	.1314E+00	1.72	.4272E−01	2.32	.1017E−01
.53	.2981E+00	1.13	.1292E+00	1.73	.4182E−01	2.33	.9903E−02
.54	.2946E+00	1.14	.1271E+00	1.74	.4093E−01	2.34	.9642E−02
.55	.2912E+00	1.15	.1251E+00	1.75	.4006E−01	2.35	.9387E−02
.56	.2877E+00	1.16	.1230E+00	1.76	.3920E−01	2.36	.9138E−02
.57	.2843E+00	1.17	.1210E+00	1.77	.3836E−01	2.37	.8894E−02
.58	.2810E+00	1.18	.1190E+00	1.78	.3754E−01	2.38	.8656E−02
.59	.2776E+00	1.19	.1170E+00	1.79	.3673E−01	2.39	.8424E−02

Table 49.2 (continued)

β	$\Phi(-\beta)$	β	$\Phi(-\beta)$	β	$\Phi(-\beta)$
2.40	.8198E−02	3.01	.1306E−02	3.62	.1473E−03
2.41	.7976E−02	3.02	.1264E−02	3.63	.1417E−03
2.42	.7760E−02	3.03	.1223E−02	3.64	.1363E−03
2.43	.7549E−02	3.04	.1183E−02	3.65	.1311E−03
2.44	.7344E−02	3.05	.1144E−02	3.66	.1261E−03
2.45	.7143E−02	3.06	.1107E−02	3.67	.1212E−03
2.46	.6947E−02	3.07	.1070E−02	3.68	.1166E−03
2.47	.6756E−02	3.08	.1035E−02	3.69	.1121E−03
2.48	.6569E−02	3.09	.1001E−02	3.70	.1077E−03
2.49	.6387E−02	3.10	.9676E−02	3.71	.1036E−03
2.50	.6210E−02	3.11	.9354E−02	3.72	.9956E−04
2.51	.6037E−02	3.12	.9042E−02	3.73	.9569E−04
2.52	.5868E−02	3.13	.8740E−02	3.74	.9196E−04
2.53	.5703E−02	3.14	.8447E−02	3.75	.8837E−04
2.54	.5543E−02	3.15	.8163E−02	3.76	.8491E−04
2.55	.5386E−02	3.16	.7888E−02	3.77	.8157E−04
2.56	.5234E−02	3.17	.7622E−02	3.78	.7836E−04
2.57	.5085E−02	3.18	.7363E−02	3.79	.7527E−04
2.58	.4940E−02	3.19	.7113E−02	3.80	.7230E−04
2.59	.4799E−02	3.20	.6871E−02	3.81	.6943E−04
2.60	.4661E−02	3.21	.6636E−02	3.82	.6667E−04
2.61	.4527E−02	3.22	.6409E−02	3.83	.6402E−04
2.62	.4396E−02	3.23	.6189E−02	3.84	.6147E−04
2.63	.4269E−02	3.24	.5976E−02	3.85	.5901E−04
2.64	.4145E−02	3.25	.5770E−02	3.86	.5664E−04
2.65	.4024E−02	3.26	.5570E−02	3.87	.5437E−04
2.66	.3907E−02	3.27	.5377E−02	3.88	.5218E−04
2.67	.3792E−02	3.28	.5190E−02	3.89	.5007E−04
2.68	.3681E−02	3.29	.5009E−03	3.90	.4804E−04
2.69	.3572E−02	3.30	.4834E−03	3.91	.4610E−04
2.70	.3467E−02	3.31	.4664E−03	3.92	.4422E−04
2.71	.3364E−02	3.32	.4500E−03	3.93	.4242E−04
2.72	.3264E−02	3.33	.4342E−03	3.94	.4069E−04
2.73	.3167E−02	3.34	.4189E−03	3.95	.3902E−04
2.74	.3072E−02	3.35	.4040E−03	3.96	.3742E−04
2.75	.2980E−02	3.36	.3897E−03	3.97	.3588E−04
2.76	.2890E−02	3.37	.3758E−03	3.98	.3441E−04
2.77	.2803E−02	3.38	.3624E−03	3.99	.3298E−04
2.78	.2718E−02	3.39	.3494E−03	4.00	.3162E−04
2.79	.2635E−02	3.40	.3369E−03	4.10	.2062E−04
2.80	.2555E−02	3.41	.3248E−03	4.20	.1332E−04
2.81	.2477E−02	3.42	.3131E−03	4.30	.8524E−05
2.82	.2401E−02	3.43	.3017E−03	4.40	.5402E−05
2.83	.2327E−02	3.44	.2908E−03	4.50	.3391E−05
2.84	.2256E−02	3.45	.2802E−03	4.60	.2108E−05
2.85	.2186E−02	3.46	.2700E−03	4.70	.1298E−05
2.86	.2118E−02	3.47	.2602E−03	4.80	.7914E−06
2.87	.2052E−02	3.48	.2507E−03	4.90	.4780E−06
2.88	.1988E−02	3.49	.2415E−03	5.00	.2859E−06
2.89	.1926E−02	3.50	.2326E−03	5.10	.1694E−06
2.90	.1866E−02	3.51	.2240E−03	5.20	.9935E−07
2.91	.1807E−02	3.52	.2157E−03	5.30	.5772E−07
2.92	.1750E−02	3.53	.2077E−03	5.40	.3321E−07
2.93	.1695E−02	3.54	.2000E−03	5.50	.1892E−07
2.94	.1641E−02	3.55	.1926E−03	6.00	.9716E−09
2.95	.1589E−02	3.56	.1854E−03	6.50	.3945E−10
2.96	.1538E−02	3.57	.1784E−03	7.00	.1254E−11
2.97	.1489E−02	3.58	.1717E−03	7.50	.3116E−13
2.98	.1441E−02	3.59	.1653E−03	8.00	.6056E−15
2.99	.1395E−02	3.60	.1591E−03	8.50	.9197E−17
3.00	.1350E−02	3.61	.1531E−03	9.00	.1091E−18

Moments

Random variables are characterized by their distribution form (i.e., probability functions) and their *moments.* These values may be thought of as shifts and scales for the distribution and serve to uniquely define the probability function. In the case of the familiar normal distribution, there are two moments: the mean and the standard deviation. The mean describes the central tendency of the distribution (the normal distribution is a symmetric distribution), while the standard deviation is a measure of the dispersion about the mean value. Given a set of n data points, the sample mean and the sample variance (the square of the sample standard deviation) are computed as

$$m_x = \frac{1}{n}\sum_i X_i \tag{49.11}$$

$$\hat{\sigma}_x^2 = \frac{1}{n-1}\sum_i (X_i - m_x)^2 \tag{49.12}$$

Many common distributions are two-parameter distributions and, while not necessarily symmetric, are completely characterized by their first two moments (see Table 49.1). The population mean, or first moment, of a continuous random variable is computed as

$$\mu_x = E[X] = \int_{-\infty}^{+\infty} x f_X(x)\,dx \tag{49.13}$$

where $E[X]$ is the expected value of X. The population variance (the square of the population standard deviation) of a continuous random variable is computed as

$$\sigma_x^2 = \mathrm{Var}[X] = E[(X - \mu_x)^2] = \int_{-\infty}^{+\infty} (x - \mu_x)^2 f_X(x)\,dx \tag{49.14}$$

The population variance can also be expressed in terms of expectations as

$$\sigma_x^2 = E[X^2] - E^2[X] = \int_{-\infty}^{+\infty} x^2 f_X(x)\,dx - \left(\int_{-\infty}^{+\infty} x f_X(x)\,dx\right)^2 \tag{49.15}$$

The coefficient of variation (COV) is defined as the ratio of the standard deviation to the mean, and therefore serves as a nondimensional measure of variability.

$$\mathrm{COV} = V_X = \frac{\sigma_x}{\mu_x} \tag{49.16}$$

In some cases, higher-order (> 2) moments exist, and these may be computed similarly as

$$\mu_x^{(n)} = E[(X - \mu_x)^n] = \int_{-\infty}^{\infty} (x - \mu_x)^n f_X(x)\,dx \tag{49.17}$$

where $\mu_x^{(n)}$ is the nth moment of random variable X. Often, it is more convenient to define the probability distribution in terms of its *parameters.* These parameters can be expressed as functions of the moments (see Table 49.1).

Concept of Independence

The concept of statistical independence is very important in structural reliability as it often permits great simplification of the problem. While not all random quantities in a reliability analysis may be assumed independent, it is certainly reasonable to assume (in most cases) that loads and resistances are statistically independent. Often, the assumption of independent loads (actions) can be made as well.

Two events A and B are statistically independent if the outcome of one in no way affects the outcome of the other. Therefore, two random variables X and Y are statistically independent if information on one variable's probability of taking on some value in no way affects the probability of the other random variable taking on some value. One of the most significant consequences of this statement of independence is that the joint probability of occurrence of two (or more) random variables can be written as the product of the individual marginal probabilities. Therefore, if we consider two events (A is probability that an earthquake occurs and B is probability that a hurricane occurs), and we assume these occurrences are statistically independent in a particular region, the probability of both an earthquake *and* a hurricane occurring is simply the product of the two probabilities:

$$P[A \text{ and } B] = P[A \cap B] = P[A]P[B] \tag{49.18}$$

Similarly, if we consider resistance (R) and load (S) to be continuous random variables, and assume independence, we can write that the probability of R being less than or equal to some value r *and* the probability that S exceeds some value s (i.e., failure) as

$$\begin{aligned} P[P \leq r \cap S > s] &= P[R \leq r]P[S > s] = P[R \leq r](1 - P[S \leq s]) \\ &= F_R(r)(1 - F_S(s)) \end{aligned} \tag{49.19}$$

Additional implications of statistical independence will be discussed in later sections. The treatments of *dependent* random variables, including issues of correlation, joint probability, and conditional probability, are beyond the scope of this introduction, but may be found in any elementary text [e.g., Benjamin and Cornell, 1970; Ang and Tang, 1975a].

Examples

Some relatively simple examples are presented here. These examples serve to illustrate some important elements of probability theory and introduce the reader to some basic reliability concepts in structural engineering and design.

Example 49.1. The Richter magnitude of an earthquake, given that it has occurred, is assumed to be exponentially distributed. For a particular region in southern California, the exponential distribution parameter (λ) has been estimated to be 2.23. What is the probability that a given earthquake will have a magnitude greater than 5.5?

$$\begin{aligned} P[M > 5.5] &= 1 - P[M \leq 5.5] = 1 - F_X(5.5) \\ &= 1 - [1 - e^{-5.5\lambda}] \\ &= e^{-2.23 \times 5.5} = e^{-12.265} \\ &\approx 4.71 \times 10^{-6} \end{aligned}$$

Given that two earthquakes have occurred in this region, what is the probability that *both* of their magnitudes were greater than 5.5?

$$\begin{aligned} P[M_1 > 5.5 \cap M_2 > 5.5] &= P[M_1 > 5.5]P[M_2 > 5.5] \quad \text{(assumed independence)} \\ &= (P[M > 5.5])^2 \qquad \text{(identically distributed)} \\ &= (4.71 \times 10^{-6})^2 \\ &\approx 2.22 \times 10^{-11} \qquad \text{(very small!)} \end{aligned}$$

Example 49.2. Consider the cross section of a reinforced concrete column with 12 reinforcing bars. Assume the load-carrying capacity of each of the 12 reinforcing bars (R_i) is normally distributed with a mean of 100 kN and a standard deviation of 20 kN. Assume that the load-carrying capacity of the concrete itself is $r_c = 500$ kN (deterministic), and that the column is subjected to a known load of 1500 kN. What is the probability that this column will fail?

First, we can compute the mean and standard deviation of the column's total load-carrying capacity.

$$E[R] = m_R = r_c + \sum_{i=1}^{12} E[R_i] = 500 + 12(100) = 1700\ \text{kN}$$

$$\text{Var}[R] = \sigma_R^2 = \sum_{i=1}^{12} \sigma_{R_i}^2 = 12(20)^2 = 4800\ \text{kN}^2 \qquad \therefore \sigma_R = 69.28\ \text{kN}$$

Since the total capacity is the sum of a number of normal variables, it too is a normal variable (central limit theorem). Therefore, we can compute the probability of failure as the probability that the load-carrying capacity R is less than the load of 1500 kN.

$$P[R < 1500] = F_R(1500) = \Phi\left(\frac{1500 - 1700}{69.28}\right) = \Phi(-2.89) \approx 0.00193$$

Example 49.3. The moment capacity (M) of the simply supported beam ($l = 10'$) shown in Fig. 49.3 is assumed to be normally distributed with a mean of 25 ft-kips and a coefficient of variation of 0.20. Failure occurs if the maximum moment exceeds the moment capacity. If only the concentrated load (P) is applied at midspan, what is the failure probability?

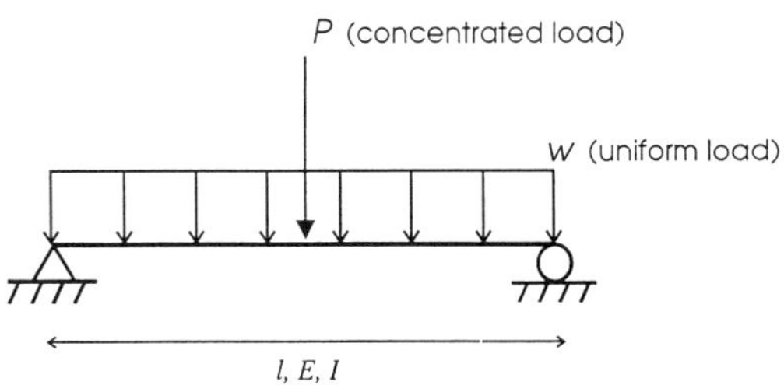

FIGURE 49.3 Simply supported beam.

$$M_{\max} = \frac{Pl}{4} = \frac{4(10')}{4} = 10\ \text{ft-kips}$$

$$P_f = P[M < M_{\max}] = F_M(10) = \Phi\left(\frac{10 - 25}{5}\right) = \Phi(-3.0) \approx 0.00135$$

If only the uniform load (w) is applied along the entire length of the beam, what is the failure probability?

$$M_{\max} = \frac{wl^2}{8} = \frac{1(10')^2}{8} = 12.5\ \text{ft-kips}$$

$$P_f = P[M < M_{\max}] = F_M(12.5) = \Phi\left(\frac{12.5 - 25}{5}\right) = \Phi(-2.5) \approx 0.00621$$

If the beam is subjected to both P and w simultaneously, what is the probability that the beam performs safely?

$$M_{\max} = \frac{Pl}{4} + \frac{wl^2}{8} = 10 + 12.5 = 22.5 \text{ ft-kips}$$

$$P_f = P[M < M_{\max}] = F_M(22.5) = \Phi\left(\frac{22.5 - 25}{5}\right) = \Phi(-0.5) \approx 0.3085$$

$$\therefore P(\text{safety}) = P_s = (1 - P_f) = 0.692$$

Note that this failure probability is *not* simply the sum of the two individual failure probabilities computed previously. Finally, for design purposes, suppose we want a probability of safe performance $P_s = 99.9\%$, for the case of the beam subjected to the uniform load (w) only. What value of $w_{\max}$ (i.e., maximum allowable uniform load for design) should we specify?

$$M_{\text{allow}} = \frac{w_{\max}(l^2)}{8} = w_{\max}\left(\frac{10^2}{8}\right) = 12.5(w_{\max})$$

$$\text{Goal: } P[M > 12.5w_{\max}] = 0.999$$

$$1 - F_M(12.5w_{\max}) = 0.999$$

$$1 - \Phi\left(\frac{12.5w_{\max} - 25}{5}\right) = 0.999$$

$$\therefore \Phi^{-1}(1.0 - 0.999) = \left(\frac{12.5w_{\max} - 25}{5}\right)$$

$$\therefore w_{\max} = \frac{(-3.09)(5) + 25}{12.5} \approx 0.76 \text{ kips/ft}$$

Example 49.4. The total annual snowfall for a particular location is modeled as a normal random variable with a mean of 60 inches and a standard deviation of 15 inches. What is the probability that in any given year the total snowfall in that location is between 45 and 65 inches?

$$P[45 < S \leq 65] = F_S(65) - F_S(45) = \Phi\left(\frac{65 - 60}{15}\right) - \Phi\left(\frac{45 - 60}{15}\right)$$

$$= \Phi(0.33) - \Phi(-1.00) = \Phi(0.33) - [1 - \Phi(1.00)]$$

$$= 0.629 - (1 - 0.841) \approx 0.47 \quad \text{(about 47\%)}$$

What is the probability the total annual snowfall is at least 30 inches in this location?

$$1 - F_S(30) = 1 - \Phi\left(\frac{30 - 60}{15}\right)$$

$$= 1 - \Phi(-2.0) = 1 - [1 - \Phi(2.0)]$$

$$= \Phi(2.0) \approx 0.977 \quad \text{(about 98\%)}$$

Suppose for design we want to specify the 95th-percentile snowfall value (i.e., a value which has a 5% *exceedance probability*). Estimate the value of $S_{.95}$.

$$P[S > S_{.95}] \equiv 0.05, \qquad P[S < S_{.95}] = .95, \qquad \Phi\left(\frac{S_{.95} - 60}{15}\right) = 0.95$$

$$\begin{aligned} \therefore S_{.95} &= [15 \times \Phi^{-1}(.95)] + 60 \\ &= (15)(1.64) + 60 = 84.6 \text{ inches} \quad \text{(so, specify 85 inches)} \end{aligned}$$

Now, assume the total annual snowfall is lognormally distributed (rather than normally) with the same mean and standard deviation as before. Recompute $P[45'' \leq S \leq 65'']$. First, we obtain the lognormal distribution parameters:

$$\xi^2 = \ln(V_S^2 + 1) = \ln\left(\left(\frac{15}{60}\right)^2 + 1\right) = 0.061$$

$$\xi = 0.246 \qquad (\approx 0.25 = V_S; \text{ OK for } V \approx 0.3 \text{ or less})$$

$$\lambda = \ln(m_S) - 0.5\xi^2 = \ln(60) - 0.5(.061) = 4.064$$

Now, using these parameters, recompute the probability:

$$\begin{aligned} P[45 < S_{LN} \leq 65] &= F_S(65) - F_S(45) = \Phi\left(\frac{\ln(65) - 4.06}{0.25}\right) - \Phi\left(\frac{\ln(45) - 4.06}{0.25}\right) \\ &= \Phi(0.46) - \Phi(-1.01) = \Phi(0.46) - [1 - \Phi(1.01)] \\ &= 0.677 - (1 - 0.844) \approx 0.52 \quad \text{(about 52\%)} \end{aligned}$$

Note that this is slightly higher than the value obtained assuming the snowfall was normally distributed (47%). Finally, again assuming the total annual snowfall to be lognormally distributed, recompute the 5% exceedance limit (i.e., the 95th-percentile value):

$$P[S < S_{.95}] = .95$$

$$\Phi\left(\frac{\ln(S_{.95}) - 4.06}{0.25}\right) = 0.95$$

$$\begin{aligned} \therefore \ln(S_{.95}) &= \left[.25 \times \Phi^{-1}(.95)\right] + 4.06 \\ &= (.25)(1.64) + 4.06 = 4.47 \end{aligned}$$

$$\therefore S_{.95} = \exp(4.47) \approx 87.4 \text{ inches} \quad \text{(specify 88 inches)}$$

Again, this value is slightly higher than the value obtained assuming the total snowfall was normally distributed (about 85 inches).

49.3 Basic Reliability Problem

A complete treatment of structural reliability theory is not included in this chapter. However, a number of texts are available (in varying degrees of difficulty) on this subject [Ang and Tang, 1975b; Melchers, 1987; Thoft-Christensen and Baker, 1982; Ditlevsen, 1981]. For the purpose of an introduction, an elementary treatment of the basic (two-variable) reliability problem is provided in the following sections.

Basic *RS* Problem

As described previously, the simplest formulation of the failure probability problem may be written as follows:

$$P_f = P[R < S] = P[R - S < 0] \tag{49.20}$$

where R is resistance and S is load. The simple function, $g(\mathbf{X}) = R - S$, where $\mathbf{X}$ is the vector of basic random variables, is termed the *limit state function.* It is customary to formulate this limit state function such that the condition $g(\mathbf{X}) < 0$ corresponds to failure, while $g(\mathbf{X}) > 0$ corresponds to a condition of safety. The *limit state surface* corresponds to points where $g(\mathbf{X}) = 0$ (the term *surface* implies that it is possible to have problems involving more than two random variables). For the simple two-variable case, if the assumption can be made that the load and resistance quantities are statistically independent, and that the population statistics can be estimated by the sample statistics, the failure probabilities for the cases of normal or lognormal variates (R, S) are given by

$$P_{f_{(N)}} = \Phi\left(\frac{0 - m_M}{\hat{\sigma}_M}\right) = \Phi\left(\frac{m_S - m_R}{\sqrt{\hat{\sigma}_S^2 + \hat{\sigma}_R^2}}\right) \tag{49.21}$$

$$P_{f_{(LN)}} = \Phi\left(\frac{0 - m_M}{\hat{\sigma}_M}\right) = \Phi\left(\frac{\lambda_S - \lambda_R}{\sqrt{\xi_S^2 + \xi_R^2}}\right) \tag{49.22}$$

where $M = R - S$ is the safety margin (or limit state function). The concept of a safety margin and the reliability index β is illustrated in Fig. 49.4. Here, it can be seen that the reliability index β corresponds to the distance (specifically, the number of standard deviations) the mean of the safety margin is away from the origin (recall that $M = 0$ corresponds to failure). The most common, generalized definition of reliability is the second-moment reliability index β, which derives from this simple two-dimensional case, and is related (approximately) to the failure probability by

$$\beta \approx \Phi^{-1}(1 - P_f) \tag{49.23}$$

where $\Phi^{-1}(\cdot)$ is the inverse standard normal cumulative distribution function. Table 49.2 can also be used to evaluate this function. (In the case of normal variates, Eq. (49.23) is "exact." Additional discussion of the reliability index β may be found in any of the texts cited previously.) To gain a "feel" for relative values of the reliability index β, the corresponding failure probabilities are shown in Table 49.3. Based on Eqs. (49.21) through (49.23), for the case of R and S both distributed normal or lognormal, expressions for the reliability index are given by

$$\beta_{(N)} = \frac{m_M}{\hat{\sigma}_M} = \frac{m_R - m_S}{\sqrt{\hat{\sigma}_R^2 + \hat{\sigma}_S^2}} \tag{49.24}$$

$$\beta_{(LN)} = \frac{m_M}{\hat{\sigma}_M} = \frac{\lambda_R - \lambda_S}{\sqrt{\xi_R^2 + \xi_S^2}} \tag{49.25}$$

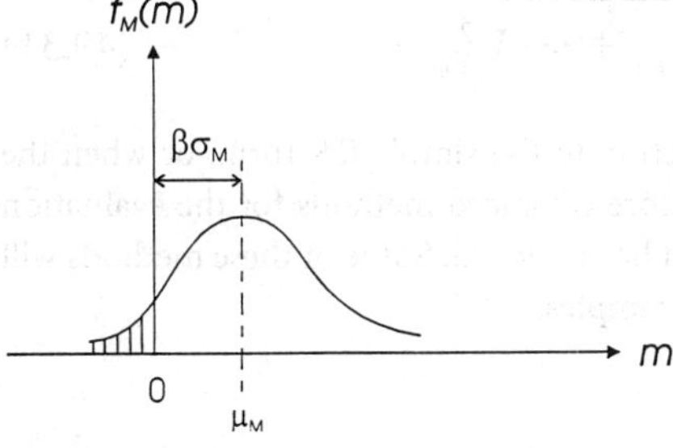

FIGURE 49.4 Safety margin concept, $M = R - S$.

Table 49.3 Failure Probabilities and Corresponding Reliability Values

Probability of Failure, P_f	Reliability Index, β
.5	0.00
.1	1.28
.01	2.32
.001	3.09
10^{-4}	3.71
10^{-5}	4.75
10^{-6}	5.60

For the less generalized case where R and S are not necessarily both distributed normal or lognormal (but are still independent), the failure probability may be evaluated by solving the convolution integral shown in Eq. (49.26a) or (49.26b) either numerically or by simulation:

$$P_f = P[R < S] = \int_{-\infty}^{+\infty} F_R(x) f_S(x)\, dx \tag{49.26a}$$

$$P_f = P[R < S] = \int_{-\infty}^{+\infty} [1 - F_S(x)] f_R(x)\, dx \tag{49.26b}$$

Again, the second-moment reliability is approximated as $\beta = \Phi^{-1}(1 - P_f)$. Additional methods for evaluating β (for the case of multiple random variables and more complicated limit state functions) are presented in subsequent sections.

More Complicated Limit State Functions Reducible to *RS* Form

It may be possible that what appears to be a more complicated limit state function (i.e., more than two random variables) can be reduced, or simplified, to the basic RS form. Three points may be useful in this regard:

1. If the coefficient of variation of one random variable is very small relative to the other random variables, it may be able to be treated as a deterministic quantity.
2. If multiple, statistically independent random variables (X_i) are taken in a summation function ($Z = aX_1 + bX_2 + \cdots$), and the random variables are assumed to be normal, the summation can be replaced with a single normal random variable (Z) with moments

$$E[Z] = aE[X_1] + bE[X_2] + \cdots \tag{49.27}$$

$$\mathrm{Var}[Z] = \sigma_z^2 = a^2\sigma_{x_1}^2 + b^2\sigma_{x_2}^2 + \cdots \tag{49.28}$$

3. If multiple, statistically independent random variables (Y_i) are taken in a product function ($Z' = cY_1Y_2\ldots$), and the random variables are assumed to be lognormal, the product can be replaced with a single lognormal random variable (Z') with moments (shown here for the case of the product of two variables):

$$E[Z'] = E[Y_1]E[Y_2] \tag{49.29}$$

$$\mathrm{Var}[Z'] = \mu_{Y_1}^2\sigma_{Y_2}^2 + \mu_{Y_2}^2\sigma_{Y_1}^2 + \sigma_{Y_1}^2\sigma_{Y_2}^2 \tag{49.30}$$

Note that the last term in Eq. (49.30) is very small if the coefficients of variation are small. In this case, and more generally, for the product of n random variables, the coefficient of variation of the product may be expressed as

$$V_Z \approx \sqrt{V_{Y_1}^2 + V_{Y_2}^2 + \cdots V_{Y_n}^2} \tag{49.31}$$

When it is *not* possible to reduce the limit state function to the simple RS form, or when the random variables are not both normal or lognormal, more advanced methods for the evaluation of the failure probability (and hence the reliability) must be employed. Some of these methods will be described in the next section after some illustrative examples.

Examples

The following examples all contain limit state functions which are in, or can be reduced to, the form of the basic RS problem. Note that in all cases, the random variables are all either normal or lognormal. Additional information suggesting when such distribution assumptions may be reasonable (or acceptable) is also provided in these examples.

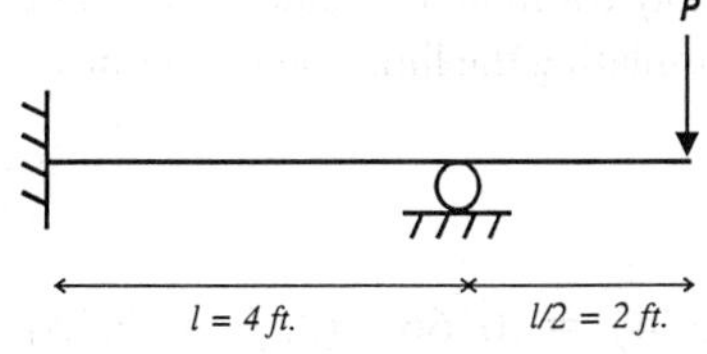

FIGURE 49.5 Cantilever beam subject to point load.

Example 49.5. Consider the statically indeterminate beam shown in Fig. 49.5, subjected to a concentrated load, P. The moment capacity M_{cap} is a random variable with a mean of 20 ft-kips and a standard deviation of 4 ft-kips. The load P is a random variable with a mean of 4 kips and a standard deviation of 1 kip. Compute the second moment reliability index assuming P and M_{cap} are normally distributed and statistically independent.

$$M_{\max} = \frac{Pl}{2}$$

$$P_f = P\left[M_{cap} < \frac{Pl}{2}\right] = P\left[M_{cap} - \frac{Pl}{2} < 0\right] = P[M - 2P < 0]$$

Here, the failure probability is expressed in terms of $R - S$ where $R = M$ and $S = 2P$. Now, we compute the moments of the safety margin given by $M = R - S$:

$$\begin{aligned} m_M &= E[M] = E[R - S] = E[R] - E[S] \\ &= m_{M_{cap}} - 2m_P = 20 - 2(4) = 12 \text{ ft-kips} \end{aligned}$$

$$\begin{aligned} \hat{\sigma}_M^2 &= \text{Var}[M] = \text{Var}[R] + \text{Var}[S] \\ &= \hat{\sigma}_{M_{cap}}^2 + (2)^2\hat{\sigma}_P^2 = (4)^2 + 4(1)^2 = 20(\text{ft-kips})^2 \end{aligned}$$

Finally, we can compute the second moment reliability index β as

$$\beta = \frac{m_M}{\hat{\sigma}_M} = \frac{m_R - m_S}{\sqrt{\hat{\sigma}_R^2 + \hat{\sigma}_S^2}} = \frac{12}{\sqrt{20}} \approx 2.68$$

(The corresponding failure probability is therefore $P_f \approx \Phi(-\beta) = \Phi(-2.68) \approx 0.00368$.)

Example 49.6. When designing a building, the total force acting on the columns must be considered. For a particular design situation, the total column force may consist of components of dead load (self-weight), live load (occupancy), and wind load, denoted D, L, and W, respectively. It is reasonable to assume that these variables are statistically independent, and here we will further assume them to be normally distributed with the following moments:

Variable	Mean (m)	Std. Dev. (σ)
D	4.0 kips	0.4 kips
L	8.0 kips	2.0 kips
W	3.4 kips	0.7 kips

If the column has a strength which is assumed to be deterministic, $R = 20$ kips, what is the probability of failure and the corresponding second moment reliability index β?

First, we compute the moments of the combined load, $S = D + L + W$:

$$m_S = m_D + m_L + m_W = 4.0 + 8.0 + 3.4 = 15.4 \text{ kips}$$

$$\hat{\sigma}_S = \sqrt{\hat{\sigma}_D^2 + \hat{\sigma}_L^2 + \hat{\sigma}_W^2} = \sqrt{(0.4)^2 + (2.0)^2 + (0.7)^2} = 2.16 \text{ kips}$$

Since S is the sum of a number of normal random variables, it is itself a normal variable. Now, since the resistance is assumed to be deterministic, we can simply compute the failure probability directly in terms of the standard normal CDF (rather than formulating the limit state function):

$$\begin{aligned} P_f &= P[S > R] = 1 - P[S < R] = 1 - F_S(20) \\ &= 1 - \Phi\left(\frac{20 - 15.4}{2.16}\right) = 1 - \Phi(2.13) \approx 1 - (.9834) = .0166 \quad (\therefore \beta = 2.13) \end{aligned}$$

If we were to formulate this in terms of a limit state function (of course, the same result would be obtained), we would have $g(\mathbf{X}) = R - S$, where the moments of S are given above and the moments of R are $m_R = 20$ kips and $\sigma_R = 0$. Now, if we assume the resistance R is a random variable (rather than being deterministic), with mean and standard deviation given by $m_R = 20$ kips and $\sigma_R = 2$ kips (i.e., COV $= 0.10$), how would this additional uncertainty affect the probability of failure (and the reliability)? To answer this, we analyze this as a basic RS problem, assuming normal variables, and making the reasonable assumption that the loads and resistance are independent quantities. Therefore, from Eq. (49.21):

$$\begin{aligned} P_f &= P[R - S < 0] = \Phi\left(\frac{m_S - m_R}{\sqrt{\hat{\sigma}_S^2 + \hat{\sigma}_R^2}}\right) = \Phi\left(\frac{15.4 - 20}{\sqrt{(2.16)^2 + (2)^2}}\right) \\ &= \Phi\left(\frac{-4.6}{\sqrt{8.67}}\right) = \Phi(-1.56) \approx 0.0594 \quad (\therefore \beta = 1.56) \end{aligned}$$

As expected, the uncertainty in the resistance serves to increase the failure probability (in this case, significantly), thereby decreasing the reliability.

Example 49.7. The fully plastic flexural capacity of a steel beam section is given by the product YZ, where Y is the steel yield strength and Z is the section modulus. Therefore, for an applied moment M, we can express the limit state function as $g(\mathbf{X}) = YZ - M$, where failure corresponds to the condition $g(\mathbf{X}) < 0$. Given the statistics shown below and assuming all random variables are lognormally distributed (this assures nonnegativity of the load and resistance variables), reduce this to the simple RS form and estimate the second moment reliability index.

Variable	Distribution	Mean	COV
Y	Lognormal	40 ksi	0.10
Z	Lognormal	50 in.^{3}	0.05
M	Lognormal	1000 in.-kips	0.20

First, we obtain the moments of R and S as follows. For $R = YZ$:

$$E[R] = m_R = m_Y m_Z = (40)(50) = 2000 \text{ in.-kips}$$

$$V_R = \text{COV} \approx \sqrt{V_Y^2 + V_Z^2} = 0.112 \quad \text{(since COVs are "small")}$$

For $S = M$:

$$E[S] = m_M = 1000 \text{ in.-kips}$$

$$V_S = \text{COV} = V_M = 0.20$$

Now, we can compute the lognormal parameters (λ and ξ) for R and S:

$$\xi_R \approx V_R = 0.112 \quad \text{(since small COV)}$$

$$\lambda_R = \ln m_R - \tfrac{1}{2}\xi_R^2 = \ln(2000) - \tfrac{1}{2}(.112)^2 = 7.595$$

$$\xi_S \approx V_S = 0.20 \quad \text{(since small COV)}$$

$$\lambda_S = \ln m_S - \tfrac{1}{2}\xi_S^2 = \ln(1000) - \tfrac{1}{2}(.2)^2 = 6.888$$

Finally, the second moment reliability index β is computed:

$$\beta_{\text{LN}} = \frac{\lambda_R - \lambda_S}{\sqrt{\xi_R^2 + \xi_S^2}} = \frac{7.595 - 6.888}{\sqrt{(.112)^2 + (.2)^2}} \approx 3.08$$

Since the variability in the section modulus Z is very small ($V_Z = 0.05$), we could choose to neglect it in the reliability analysis (i.e., assume Z deterministic). Still assuming variables Y and M to be lognormally distributed, and using Eq. (49.25) to evaluate the reliability index, we obtain $\beta = 3.17$. If we further assumed Y and M to be normal (instead of lognormal) random variables, the reliability index computed using Eq. (49.24) would be $\beta = 3.54$. This illustrates the relative error one might expect from (a) assuming certain variables with low COVs to be essentially deterministic (i.e., 3.17 vs. 3.08), and (b) assuming the incorrect distributions, or simply using the normal distribution when more statistical information is available suggesting another distribution form (i.e., 3.54 vs. 3.08).

Example 49.8. Consider the simply supported beam shown in Fig. 49.3, subjected to a uniform load w (only) along its entire length. Assume that, in addition to w being a random variable, the member properties E and I are also random variables. (The length, however, may be assumed to be deterministic.) Formulate the limit state function for excessive deflection (assume a maximum allowable deflection of $l/360$ where l = length of the beam) and then reduce it to the simple RS form. (Setup *only*.)

$$\delta_{\max} = \frac{5wl^4}{384EI}$$

$$P_f = P[\delta_{\text{allow}} - \delta_{\max} < 0]$$

The failure probability is in the RS form ($R = \delta_{\text{allow}}$ and $S = \delta_{\max}$), but we still have to express the limit state function in terms of the basic variables.

$$\begin{aligned} g(\mathbf{X}) &= \frac{l}{360} - \frac{5wl^4}{384EI} < 0 \quad \text{(for failure)} \\ &= \frac{EI}{360} - \frac{5wl^3}{384} < 0 \\ &= \frac{384}{360}(EI) - 5wl^3 < 0 \\ &= 1.067(EI) - 5l^3(w) < 0 \end{aligned}$$

Note that the limit state function is now expressed in the simple RS form, with $R = EI$ and $S = w$. If E and I are assumed to be lognormally distributed, their product EI is also a lognormal random variable, and the moments can be computed as in the previous example. Finally, if the uniform load w can be assumed lognormal as well, the second moment reliability index could be computed (also as done in the previous example).

49.4 Generalized Reliability Problem

Introduction

As discussed previously, the simple two-variable (RS) case in which R and S are assumed to be independent, identically distributed (normal or lognormal) random variables permits a closed-form solution to the failure probability. However, such a two-variable simplification of the limit state is often unavailable for structural reliability problems. Furthermore, the joint probability function for the random variables in the limit state equation are seldom known precisely, due to limited data. Even if the basic variables are mutually statistically independent and all marginal density functions are known, it is often impractical (or impossible) to perform the numerical integration of the multidimensional convolution integral over the failure domain. In this section, a number of widely used techniques for evaluating structural reliability under general conditions are presented.

FORM/SORM Techniques

First-order second-moment (**FOSM**) methods were the first techniques used to evaluate structural reliability. The name refers to the way in which the limit state is linearized (first-order) and the way in which the method characterizes the basic variables (second-moment). Later, more advanced methods were developed to include information about the complete distributions characterizing the random variables. These advanced FOSM techniques became known as first-order reliability methods (**FORM**). Finally, among the most recent developments has been the refined curve-fitting of the limit state surface in the analysis, giving rise to the so-called second-order reliability methods (**SORM**). Details of these reliability analysis techniques may be found in the literature [Melchers, 1987; Ang and Tang, 1975b; Thoft-Christensen and Baker, 1982].

When the simple limit state (safety margin) is defined by $M = R - S$, we have already seen that the reliability index β can be expressed (see Fig. 49.4):

$$\beta = \frac{\mu_M}{\sigma_M} = \frac{E[g(\mathbf{X})]}{SD[g(\mathbf{X})]} \tag{49.32}$$

where $E[g(\mathbf{X})]$ and $SD[g(\mathbf{X})]$ are the mean and standard deviation of the limit state function, respectively. Therefore, for the simple RS case, β is the distance from the mean of the safety margin ($\mu_M = \mu_R - \mu_S$) to the origin in units of standard deviations of M. This is illustrated in Fig. 49.1. In this simple second-moment formulation, no mention is made of the underlying probability distributions. The reliability index β depends only on measures of central tendency and dispersion of the margin of safety M for the limit state function.

For the more general case where the number of random variables may be greater than two, the limit state surface may be nonlinear, and the random variables may not be normal, a number of iterative solution techniques have been developed. These techniques are all very similar, differing mainly in the approach taken to a minimization problem. One general procedure is presented at the end of this section. Other approaches may be found in the literature [Ellingwood *et al.*, 1980; Ang and Tang, 1975b; Melchers, 1987; Thoft-Christensen and Baker, 1982]. What follows is a summary of the mathematics behind the formulation of FORM techniques. It is not necessary to fully understand the development of these methods, and those wishing to skip over this material can go directly to the algorithm provided later in this section.

To simplify the presentation herein, the basic variables X_i are assumed to be statistically independent and therefore uncorrelated. This assumption, as discussed earlier, is often reasonable for structural reliability problems. Further, it can be shown that weak correlation (i.e., $\rho < 0.2$, where ρ is the correlation coefficient) can generally be neglected and strong correlation (i.e., $\rho > 0.8$) can be considered to imply fully dependent variables. Additional discussion of correlated variables in FORM/SORM may be found in Melchers [1987] and Thoft-Christensen and Baker [1982]. The limit state function, expressed in terms of the basic variables X_i, is first transformed to reduced variables u_i having zero mean and unit standard deviation:

$$u_i = \frac{X_i - \mu_{X_i}}{\sigma_{X_i}} \tag{49.33}$$

A transformed limit state function can then be expressed in terms of the reduced variables:

$$g_1(u_1, \ldots, u_n) = 0 \tag{49.34}$$

with failure now being defined as $g_1(\mathbf{u}) < 0$. The space corresponding to the reduced variables can be shown to have rotational symmetry, as indicated by the concentric circles of equiprobability shown on Fig. 49.6. The reliability index β is now defined as the shortest distance between the limit state surface $g_1(\mathbf{u}) = 0$ and the origin in reduced variable space (see Fig. 49.6). The point $(u_1^*, \ldots, u_n^*)$ on the limit state surface which corresponds to this minimum distance is referred to as the *checking* (or *design*) point and can be determined by simultaneously solving the following set of equations:

$$\alpha_i = \frac{\partial g_1/\partial u_i}{\sqrt{\sum_i (\partial g_1/\partial u_i)^2}} \tag{49.35}$$

$$u_i^* = -\alpha_i \beta \tag{49.36}$$

$$g_1(u_1^*, \ldots, u_n^*) = 0 \tag{49.37}$$

searching for the direction cosines α_i which minimize β. The partial derivatives in Eq. (49.35) are evaluated at the reduced space design point $(u_1^*, \ldots, u_n^*)$. This procedure, and Eqs. (49.35)–(49.37),

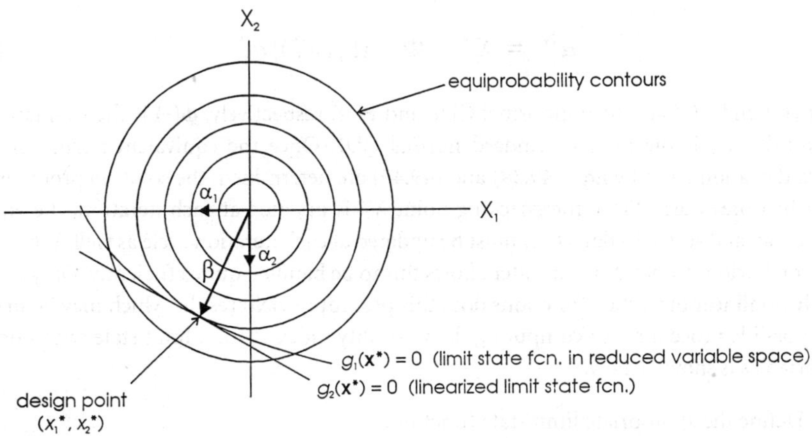

FIGURE 49.6 Formulation of reliability analysis in reduced variable space. (After Ellingwood, B., Galambos, T. V., MacGregor, J. C., and Cornell, C. A. 1980. *Development of a Probability Based Load Criterion for American National Standard A58.* NBS Special Publication SP577, National Bureau of Standards, Washington, D.C.

result from linearizing the limit state surface (in reduced space) and computing the reliability as the shortest distance from the origin in reduced space to the limit state hyperplane. It may be useful at this point to compare Figs. 49.4 and 49.6 to gain some additional insight into this technique.

Once the convergent solution is obtained, it can be shown that the checking point in the original random variable space corresponds to the points

$$X_i^* = \mu_{X_i}\left(1 - \alpha_i \beta V_{X_i}\right) \tag{49.38}$$

such that $g(X_1^*, \ldots, X_n^*) = 0$. These variables will correspond to values in the upper tails of the probability distributions for load variables and the lower tails for resistance (or geometric) variables.

The formulation described above provides an exact estimate of the reliability index β for cases in which the basic variables are normal and in which the limit state function is linear. In other cases, the results are only approximate. Because many structural load and resistance quantities are known to be nonnormal, it seems reasonable that information on distribution type be incorporated into the reliability analysis. This is especially true since the limit state probabilities can be affected significantly by different distributions' tail behaviors. Methods which include distribution information are known as full-distribution methods or advanced FOSM methods. One commonly used technique is described below.

Because of the ease of working with normal variables, the objective here is to transform the nonnormal random variables into equivalent normal variables and then perform the analysis for a solution of the reliability index, as described previously. This transformation is accomplished by approximating the true distribution by a normal distribution at the value corresponding to the design point on the failure surface. By fitting an equivalent normal distribution at this point, we are forcing the best approximation to be in the tail of interest of the particular random variable. The fitting is accomplished by determining the mean and standard deviation of the equivalent normal variable such that, at the value corresponding to the design point, the cumulative probability and the probability density of the actual (nonnormal) and the equivalent normal variable are equal. (This is the basis for the so-called Rackwitz–Fiessler algorithm.) These moments of the equivalent normal variable are given by

$$\sigma_i^N = \frac{\phi\left(\Phi^{-1}\left(F_i(X_i^*)\right)\right)}{f_i(X_i^*)} \tag{49.39}$$

$$\mu_i^N = X_i^* - \Phi^{-1}\left(F_i(X_i^*)\right)\sigma_i^N \tag{49.40}$$

where $F_i(\cdot)$ and $f_i(\cdot)$ are the nonnormal CDF and PDF, respectively, $\phi(\cdot)$ is the standard normal PDF, and $\Phi^{-1}(\cdot)$ is the inverse standard normal CDF. Once the equivalent normal mean and standard deviation given by Eqs. (49.39) and (49.40) are determined, the solution proceeds exactly as described previously. Since the checking point X_i^* is updated at each iteration, the equivalent normal mean and standard deviation must be updated at each iteration cycle as well. While this can be rather laborious by hand, a computer allows this to be handled quite efficiently. Only in the case of highly nonlinear limit state functions does this procedure yield results which may be in error.

One possible procedure for computing the reliability index β for a limit state with nonnormal basic variables is shown below:

1. Define the appropriate limit state function.
2. Make an initial guess at the reliability index β.
3. Set the initial checking point values $X_i^* = \mu_i$ for all i variables.
4. Compute the equivalent normal mean and standard deviation for nonnormal variables.
5. Compute the partial derivatives $(\partial_g / \partial X_i)$ evaluated at the design point X_i^*.

6. Compute the direction cosines α_i as

$$\alpha_i = \frac{(\partial g/\partial X_i)\sigma_i^N}{\sqrt{\sum_i \left((\partial g/\partial X_i)\sigma_i^N\right)^2}} \tag{49.41}$$

7. Compute the new values of design point X_i^* as

$$X_i^* = \mu_i^N - \alpha_i \beta \sigma_i^N \tag{49.42}$$

8. Repeat steps 4 through 7 until estimates of α_i stabilize (usually fast).
9. Compute the value of β such that $g(X_1^*, \ldots, X_n^*) = 0$.
10. Repeat steps 4 through 9 until the value for β converges. (This normally occurs within five cycles or less, depending on the nonlinearity of the limit state function.)

As with the previous procedure, this method is easily programmed on a computer. Many spreadsheet programs and other numerical analysis software packages have considerable statistical capabilities, and therefore can be used to perform these types of analyses. This procedure can also be modified to estimate a design parameter (i.e., a section modulus) such that a specific target reliability is achieved. Other procedures are presented elsewhere in the literature [Ang and Tang, 1975b; Melchers, 1987; Thoft-Christensen and Baker, 1982; Ellingwood *et al.*, 1980] including a somewhat different technique in which the equivalent normal mean and standard deviation are used directly in the reduction of the variables to standard normal form (i.e., u_i-space). Additional information on second-order (SORM) techniques may be found in the literature [Chen and Lind, 1983; Der Kiureghian and Liu, 1986].

Monte Carlo Simulation

An alternative to integration of the relevant joint probability equation over the domain of random variables corresponding to failure is to use Monte Carlo simulation (MCS). While FORM/SORM techniques are approximate in the case of nonlinear limit state functions or with nonnormal random variables (even when advanced FORM/SORM techniques are used), MCS offers the advantage of providing an exact solution to the failure probability. The potential disadvantage of MCS is the amount of computing time needed, especially when very small probabilities of failure are being estimated. Still, as computing power continues to increase and with the development and refinement of variance reduction techniques, MCS is becoming more accepted and more utilized, especially for the analysis of increasingly complicated structural systems. Variance reduction techniques (VRTs) such as importance sampling, stratified sampling, and Latin hypercube sampling can often be used to significantly reduce the number of simulations required to obtain reliable estimates of the failure probability.

A brief description of Monte Carlo simulation is presented here. Additional information may be found elsewhere [Rubinstein, 1981; Melchers, 1987]. The concept behind MCS is to generate sets of realizations of the random variables in the limit state function (with the assumed known probability distributions) and record the number of times the resulting limit state function is less than zero (i.e., failure). The estimate of the probability of failure (P_f) then is simply the number of failures divided by the total number of simulations (N). Clearly, the accuracy of this estimate increases as N increases, and a larger number of simulations are required to reliably estimate smaller failure probabilities. Table 49.4 presents the number of simulations required to obtain three different confidence intervals on the estimate of P_f for some typical values in structural reliability analyses.

The generation of random variates is a relatively simple task (provided the random variables may be assumed independent) and requires only (1) that the relevant CDF is invertible (or in

Table 49.4 Approximate Number of Simulations Required for Given Confidence Intervals ($\alpha \times 100\%$) on Reliability Index

$\beta \pm \varepsilon$	$\alpha = 0.90$ ($k = 1.64$)	$\alpha = 0.95$ ($k = 1.96$)	$\alpha = 0.99$ ($k = 2.58$)
1.5±.10	1,000	1,400	2,500
1.5±.05	4,000	5,700	9,800
1.5±.01	100,000	142,000	246,000
2.0±.10	2,000	3,000	5,100
2.0±.05	8,200	12,000	20,500
2.0±.01	240,000	342,000	592,000
3.0±.10	18,000	25,600	44,300
3.0±.05	75,000	107,000	186,000
3.0±.01	2,270,000	3,240,000	5,610,000

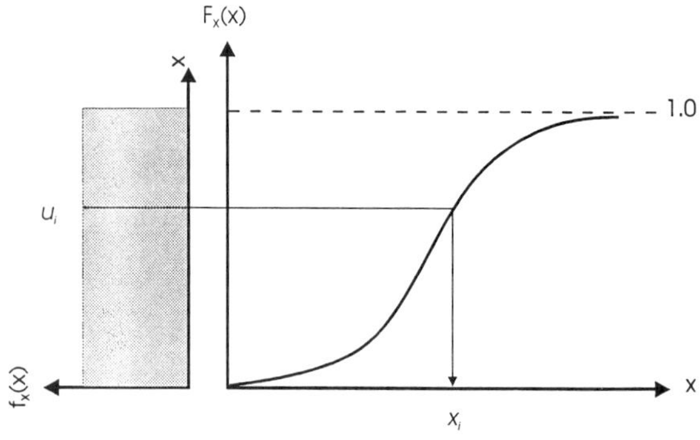

FIGURE 49.7 Random variable simulation.

the case of normal and lognormal variates, numerical approximations exist for the inverse CDF), and (2) that a uniform random number generator is available. (Random number generators for other distributions would further simplify the simulation analysis.) The generation of correlated variates is not described here, but information may be found in the literature [Melchers, 1987; Thoft-Christensen and Baker, 1982; Der Kiureghian and Liu, 1986]. As shown in Fig. 49.7, the value of the CDF for random variable X is (by definition) uniformly distributed on {0, 1}. Therefore, if we generate a uniform {0, 1} deviate and substitute this into the inverse of the CDF of interest (with the relevant parameters or moments), we obtain a realization of a variate with this CDF. For example, consider the generation of an exponential variate with parameter λ. The CDF is expressed as follows:

$$F_X(x) = 1 - \exp(-\lambda x) \tag{49.43}$$

If we substitute u_i (a uniform {0, 1} deviate) for $F_X(x)$ and invert the CDF to solve for x_i, we obtain

$$x_i = -\frac{1}{\lambda}\ln(1 - u_i) \tag{49.44}$$

Here, x_i is an exponential variate with parameter λ. As another example, consider the normal distribution, for which no closed-form expression exists for the CDF *or* its inverse. The generalized normal CDF can be written as a function of the standard normal CDF as

$$F_X(x) = \Phi\left(\frac{x - \mu_x}{\sigma_x}\right) \tag{49.45}$$

Therefore, an expression for a generalized normal variate would be

$$x_i = \mu_x + \sigma_x \Phi^{-1}(u_i) \tag{49.46}$$

where μ_x and σ_x are the mean and standard deviation, respectively, u_i is the uniform {0, 1} deviate, and $\Phi^{-1}(\cdot)$ is the inverse standard normal CDF. While not available in closed-form, numerical approximations for $\Phi^{-1}(\cdot)$ (i.e., in the form of algorithms or subroutines) are available [e.g., Ellingwood *et al.*, 1980].

MCS is a very powerful tool for the solution of a wide variety of problems. Improvements in efficiency over "crude" or "direct" MCS can be realized by improved algorithmic design (programming) and by the utilization of VRTs. Monte Carlo techniques can also be used for the simulation of discrete and continuous random processes.

49.5 System Reliability

Introduction

While most structural codes in the U.S. specify that structures be designed on a member-by-member basis, most elements within a structure are actually performing as part of an often complicated structural *system.* Interest in characterizing the performance and safety of structural systems has led to an increased interest in the area of system reliability. The classical theories of series and parallel system reliability are well developed and have been applied to the analysis of such complicated structural systems as nuclear power plants and offshore structures. In the following sections, a brief introduction to system reliability is presented along with some examples. This subject within the broad field of structural reliability is relatively new, and advances in the theory and application of system reliability concepts to civil engineering design can be expected in the coming years.

Basic Systems

The two types of systems in classical theory are the *series* (or *weakest link*) system and the *parallel* system. The literature is replete with formulations for the reliability of these systems, including the possibility of correlated element strengths (e.g., Thoft-Christensen and Baker, 1982; Thoft-Christensen and Morotsu, 1986). The relevant limit state is defined by the system type. For a series system, the system limit state is taken by definition to correspond to the first member failure, hence the name weakest link. In the case of the strictly parallel system, the system limit state is taken by definition to correspond to the failure of *all* members. Formulations for the system reliability of a parallel system in which the load-deformation behavior of the members is assumed to be ductile or brittle are well developed [Thoft-Christensen and Morotsu, 1986]. In all cases, the system reliabilities are expressed in terms of the component (or member) reliabilities.

Classical system reliability theory has been able to be extended somewhat to model more complicated systems using combinations of series and parallel systems. These formulations, however, are still subject to limitations with regard to possible load sharing (distribution of load among components of the system) and time-dependent effects, such as degrading member resistances.

Introduction to Classical System Reliability Theory

For a system limit state defined by $g(x_1, \ldots, x_m) = 0$, where x_i are the basic variables, the failure probability is computed as the integral over the failure domain ($g(\mathbf{X}) < 0$) of the joint probability density function of $\mathbf{X}$. In general, the failure of any system can be expressed as a union or intersection of events. For example, the failure of an ideal series (or weakest link) system may be expressed as

$$F_{\text{sys}} = F_1 \cup F_2 \cup \cdots \cup F_m \tag{49.47}$$

where $\cup$ denotes the Boolean "or" operator and F_i is the ith component (element) failure event. A statically determinate truss is modeled as a series system since the failure of the truss corresponds to the failure of any single member. Both first-order and second-order (which includes information on the joint probability behavior) bounds have been developed to express the system failure probability as a function of the individual element failure probabilities. These formulations are well developed and presented in the literature [Melchers, 1987; Ang and Tang, 1975b; Ditlevsen, 1981; Thoft-Christensen and Morotsu, 1986].

The failure of a strictly parallel system may be expressed as

$$F_{\text{sys}} = F_1 \cap F_2 \cap \cdots \cap F_m \tag{49.48}$$

where $\cap$ denotes the Boolean "and" operator. Such is the case for the classical Daniels system of parallel, ductile rods or cables subject to equal deformation. In this case, system failure corresponds to the failure of *all* members or elements. First- and second-order bounds are also available for this system idealization [e.g., Grigoriu, 1989]. Furthermore, bounds which account for possible dependence of failure modes (modal correlation) have been developed [Ang and Tang, 1975b]. If the parallel system is composed of brittle elements, the analysis may be further complicated by having to account for load redistribution following member failure. This total failure may therefore be the result of progressive element failures.

Returning again to the two fundamental system types, series and parallel, we can examine the probability distributions for the strength of these systems as functions of the distributions of the strengths of the individual members (elements). In the simple structural idealization of a series system of n elements (for which the characterization of the member failures as brittle or ductile is irrelevant since system failure corresponds to first-member failure), the distribution function for the system strength R_{sys} can be expressed as

$$F_{R_{sys}}(r) = 1 - \prod_{i=1}^{n} \left(1 - F_{R_i}(r)\right) \tag{49.49}$$

when the individual member strengths are assumed independent. In Eq. (49.49), $F_{R_i}(r)$ is the CDF for the individual member resistance. If the n individual member strengths are also *identically distributed* (i.e., have the same parent distribution, $F_R(r)$, with the same moments), Eq. (49.49) can be simplified to

$$F_{R_{sys}}(r) = 1 - (1 - F_R(r))^n \tag{49.50}$$

In the case of the idealized parallel system of n elements, the system failure depends on whether the member behavior is perfectly brittle or perfectly ductile. In the simple case of the parallel system with n perfectly ductile elements, the system strength is given by

$$R_{sys} = \sum_{i=1}^{n} R_i \tag{49.51}$$

where R_i is the strength of element i. The central limit theorem [see Benjamin and Cornell, 1970; Ang and Tang, 1975a] suggests that as the number of members in this system gets large, the system strength approaches a normal random variable, regardless of the distributions of the individual member strengths. When the member behavior is perfectly brittle, the system behavior depends on the degree of indeterminacy (redundancy) of the system and the ability of the system to redistribute loads to other members. For some applications, it may be reasonable to model structures idealized as parallel systems with brittle members as series systems, if the brittle failure of one member is likely to overload the remaining members. The issue of correlated member strengths (and correlated failure modes) is beyond the scope of this introduction, but information may be found in Thoft-Christensen and Baker [1982], Thoft-Christensen and Morotsu [1986], and Ang and Tang [1975b].

It is appropriate at this point to present the simple first-order bounds for the two fundamental systems. Additional information on the development and application of these as well as the second-order bounds may be found in the literature cited previously. The first-order bounds for a series system are given by

$$\max_{i=1}^{n}\{P_{f_i}\} \leq P_{f_{sys}} \leq 1 - \left(\prod_{i=1}^{n}(1 - P_{f_i})\right) \tag{49.52}$$

where P_{f_i} is the failure probability for member (element) i. The first-order bounds for a parallel system are given by

$$\prod_{i=1}^{n} P_{f_i} \leq P_{f_{\text{sys}}} \leq \min_{i=1}^{n}\{P_{f_i}\} \tag{49.53}$$

Improved (second-order) bounds (the first-order bounds are often too broad to be of practical use) which include information on the joint probability behavior (i.e., member or modal correlation) have been developed and are described in the literature.

Classical system reliability theory, as briefly introduced above, is limited in that it cannot account for more complicated load-deformation behavior and the time dependencies associated with load redistribution following (brittle) member failure. Generalized formulations for the reliability of systems which are neither strictly series nor strictly parallel are not available. Analyses of these systems are often based on combined series and parallel system models in which the complete system is modeled as some arrangement of these classical subsystems. These solutions tend to be problem-specific and still do not address any possible time-dependent or load-sharing issues.

Redundant Systems

A redundant (indeterminate) system may be defined as having some overload capacity following the failure of an element. The level of redundancy (or degree of indeterminacy) refers to the number of element failures which can be tolerated without the system failing. The reliability of such a structure depends on the nature (type) of redundancy. The level of redundancy dictates how many members can fail prior to collapse, and therefore answers the question, Would the failure of member j lead to impending collapse? Furthermore, load-deformation behavior of the individual members specifies whether or not the limit states are load-path dependent. For ductile element behavior (i.e., the Daniels system), the limit state is effectively load-path independent, implying that the order of member failures is not significant. For a system of brittle elements, however, the limit state may be load-path dependent. In this case, the performance of the system is related to the load redistribution behavior following member failure, and hence the order (or relative position) of member failures becomes important. The parallel-member system model with brittle elements (i.e., perfectly elastic load-deformation behavior) is appropriate for (and has been used to model) a wide range of redundant structural systems, including floors, roofs, and wall systems.

Examples

Three examples are described in this section. The first example considers a series system in which the elements are considered to represent different modes of failure. Modal failure analysis is often treated using the concepts of system reliability [e.g., Ang and Tang, 1975b]. Here, the structure being considered (actually, the simply supported beam element, Fig. 49.3) may fail in any one of three different modes: flexure, shear, or excessive deflection. (The last mode corresponds to a serviceability-type limit state rather than an ultimate strength type.) The failure of the structural element is assumed to occur when any one of these limit states is violated. For simplicity, the modal failure probabilities are assumed to be uncorrelated. [For information on handling correlated failure modes, see Ang and Tang, 1975b.] In other words, the element (system) fails when it fails in flexure or it fails in shear or it experiences excessive deflection:

$$F_{\text{sys}} = F_M \cup F_V \cup F_\delta \tag{49.54}$$

If, for example, the probabilities of moment, shear, and deflection failure, respectively, are given by $F_M = 0.0015$, $F_V = 0.002$, and $F_\delta = 0.005$, the first-order bounds shown in Eq. (49.52) result in the following:

$$0.005 \leq P_{f_{\text{sys}}} \leq 1 - (1 - 0.0015)(1 - 0.002)(1 - 0.005)$$

$$0.005 \leq P_{f_{\text{sys}}} \leq 0.0085 \tag{49.55}$$

This corresponds to a range for β of $2.39 \leq \beta_{\text{sys}} \leq 2.58$. The second example considers a strictly parallel system of five cables supporting a load (see Fig. 49.8). In this case, the system failure corresponds to the condition where the cable system can no longer carry *any* load. Therefore, *all* of the cables must have failed for the system to have failed. In this simple example, the issue of load-redistribution following the failure of one of the cables is not addressed; however, this problem has been studied extensively [e.g., Hohenbichler and Rackwitz, 1983]. Here, the five cable strengths are assumed to be statistically independent, and the system failure probability is the probability that P is large enough to fail all of the cables simultaneously:

$$F_{\text{sys}} = F_1 \cap F_2 \cap \cdots \cap F_5 \tag{49.56}$$

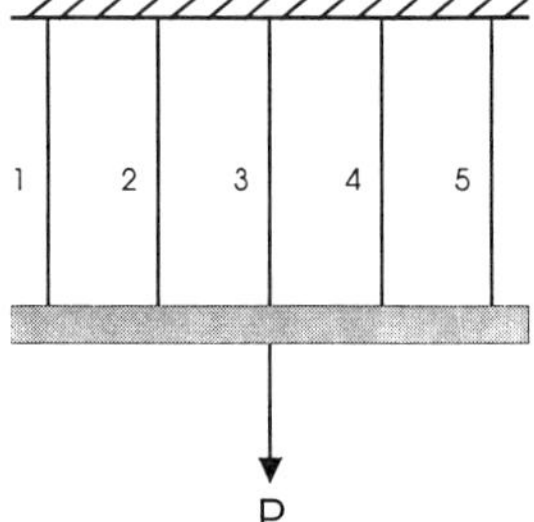

FIGURE 49.8 Five-element parallel system.

If, for example, the probability of failure of an individual cable is 0.001, and the cable strengths are assumed to be independent, identically distributed random variables, the first-order bounds on the system failure probability given by Eq. (49.53) become

$$(0.001)^5 \leq P_{f_{\text{sys}}} \leq 0.001 \tag{49.57}$$

Here, the lower bound corresponds to the case of perfectly uncorrelated member strengths (i.e., independent cable failures), while the upper bound corresponds to the case of perfect correlation. These first-order bounds, as indicated by Eq. (49.57), become very wide with increasing n. Here, information on correlation can be important in computing narrower and more useful bounds.

Finally, as a third example, a combined (series and parallel) system is considered. In this case, the event probabilities correspond to the failure probabilities of different components required for a "safe shutdown" of a nuclear power plant. While these events are assumed to be independent, their arrangement (see Fig. 49.9) forms a combined series-parallel system. In this case, the three subsystems are arranged in series: subsystem A is a series system, and subsystems B and C are parallel systems. In this case, the system failure probability is given by

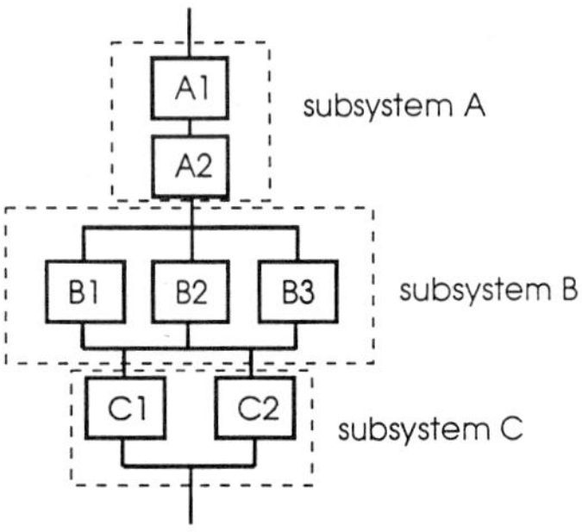

$$F_{\text{sys}} = F_A \cup F_B \cup F_C \tag{49.58}$$

FIGURE 49.9 Safe shutdown of a nuclear power plant.

or, expressed in terms of the individual component failure probabilities,

$$F_{\text{sys}} = [F_{A_1} \cup F_{A_2}] \cup [F_{B_1} \cap F_{B_2} \cap F_{B_3}] \cup [F_{C_1} \cap F_{C_2}] \tag{49.59}$$

49.6 Reliability-Based Design (Codes)

Introduction

This section will provide a brief introduction to reliability-based design concepts in civil engineering, with specific emphasis on structural engineering design. Since the 1970s, the theories of probability and statistics and reliability have provided the bases for modern structural design codes and specifications. Thus, probabilistic codes have been replacing previous deterministic-format codes in recent years. Reliability-based design procedures are intended to provide more predictable levels of safety and more "risk-consistent" (i.e., design-to-design) structures, while utilizing the most up-to-date statistical information on material strengths, as well as structural and environmental loads. An excellent discussion of RBD in the U.S. as well as other countries is presented in Melchers [1987]. Other references are also available which deal specifically with probabilistic code development in the United States [Ellingwood *et al.*, 1980; Ellingwood *et al.*, 1982; Galambos *et al.*, 1982]. The following sections provide some basic information on the application of reliability theory to aspects of reliability-based design.

Calibration and Selection of Target Reliabilities

Calibration refers to the linking of new design procedures to existing design philosophies. Much of the need for calibration arises from making any new code changes acceptable to the engineering and design communities. For purely practical reasons, it is undesirable to make drastic changes in the procedures for estimating design values (for example) or in the overall formats of design checking equations. If such changes are to be made, it is impractical and uneconomical to make them often. Hence, code development is an often slow process, involving many years and many revisions. The other justification for code calibration has been the notion that previous design philosophies (e.g., ASD) have resulted in safe designs (or designs with acceptable levels of performance), and that therefore these previous levels of safety should serve as benchmarks in the development of new specifications or procedures (e.g., LRFD).

The actual process of calibration is relatively simple. For a given design procedure (e.g., ASD for steel beams in flexure), estimate the reliability based on the available statistical information on the loads and resistances, and the governing checking equation. This becomes the target reliability and is used in order to develop the appropriate load and resistance factors for the new procedure. In the development of LRFD for both steel and wood, for example, the calibration process revealed an inconsistency in the reliability levels for different load combinations. As this was undesirable, a single target reliability was selected and the new LRFD procedures were able to correct this problem. For more information on code calibration, the reader is referred to the literature [Melchers, 1987; Ellingwood *et al.*, 1980; Galambos *et al.*, 1982].

Material Properties and Design Values

The basis for many design values encountered in structural engineering design is now probabilistic. Earlier design values were often based on mean values of member strength, for example, with the **factor of safety** intended to account for all forms of uncertainty, including material property variability. Later, as more statistical information became available, as people became more aware of the concept of relative uncertainty, and with the use of probabilistic methods in code development, characteristic values were selected for use in design. The characteristic values (referred to as nominal or design values in most specifications) are generally selected from the lower tail of the distribution describing the material property (see Fig. 49.10). Typically, the 5th-percentile value (that value below which 5% of the probability density lies) is selected as the nominal resistance (i.e., nominal strength), though in some cases, a different percentile value may be selected. While this value may serve as the starting point for establishing the design value, modifications are often needed to account for such things as size effects, system effects, or (in the case of wood) moisture content effects. The bases for the design resistance values for specifications in the U.S. are described in the literature [e.g., Galambos and Ravindra, 1978; Ellingwood *et al.,* 1980; MacGregor *et al.,* 1983]. An excellent review of resistance modeling and a summary of statistical properties for structural elements are presented in Melchers [1987]. Table 49.5 [Ellingwood *et al.,* 1980] presents some typical resistance statistics for concrete and steel members. Additional statistics are available, along with statistics for masonry, aluminum, and wood members in Ellingwood *et al.,* [1980] as well. The mean values are presented in ratio to their nominal (or design) values, m_R/R_n. In addition, the coefficient of variation, V_R, and the probability distribution are listed in Table 49.5.

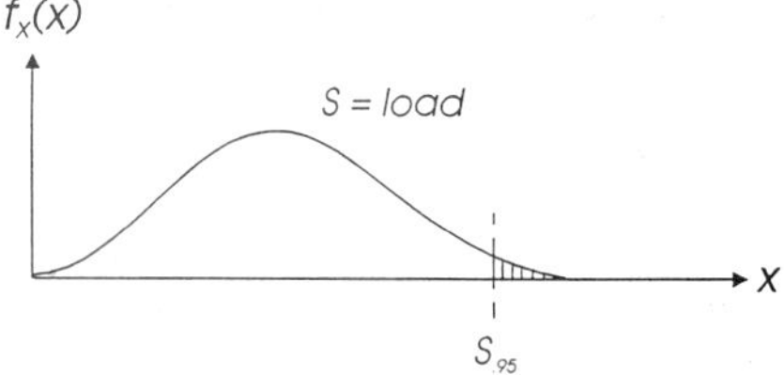

(a) design load (e.g., 5% exceedence probability)

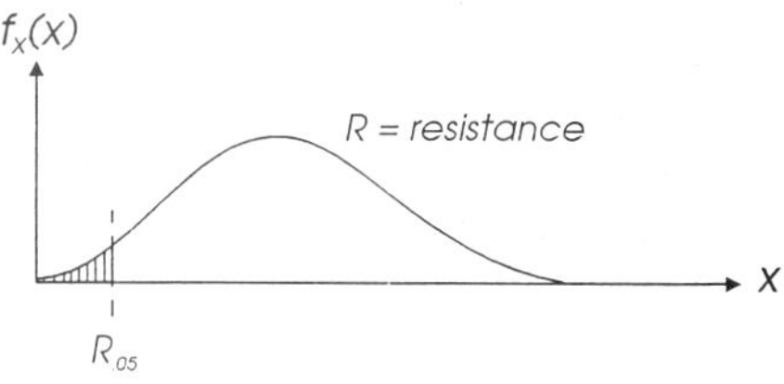

(b) nominal resistance (e.g., 5th-percentile)

FIGURE 49.10 Typical specification of design (nominal) load and resistance values.

Design Loads and Load Combinations

The selection of design load values, such as those found in the ASCE 7-88 Standard [1990] (formerly the ANSI A58.1 Standard), *Minimum Design Loads for Buildings and Other Structures,* is also largely probability-based. Though somewhat more complicated than the selection of design resistance values as described above, the concept is quite similar. Of course, greater complexity is introduced since we may be concerned with both spatial and temporal variations in the load effects. In addition, because of the difficulties in conducting load surveys and the large amount of variability associated with naturally occurring phenomena giving rise to many structural and environmental loadings, there is a high degree of uncertainty associated with these quantities. A number of load surveys have been conducted, and the valuable data collected have formed the basis for many of our design values [e.g., Chalk and Corotis, 1980; Ellingwood and Culver, 1977; Ellingwood and Redfield, 1982; Harris *et al.,* 1981]. When needed, such as in the case where data are not available or able to be collected with any reasonable amount of effort, this information is supplemented by engineering judgment and expert opinion. Therefore, design load values are based on (1) statistical information, such as load survey data; and (2) engineering judgment, including past experience, and scenario analysis. As shown in Fig. 49.10, the design load value can be visualized as some characteristic value in the upper tail of the distribution describing the load. For example, the 95th-percentile wind speed is that value of wind speed which has a 5% $(1 - .95)$ exceedance probability. Probabilistic load modeling represents an extensive area of research, and a significant amount of work is reported on in the literature [Melchers, 1987; Wen, 1990]. A summary of load statistics is presented in Table 49.6.

In most codes, a number of different load combinations are suggested for use in the appropriate checking equation format. For example, the ASCE 7-88 Standard (formerly ANSI A58.1) recommends the following load combinations (1990)[1]:

[1]Note that there have been some changes in these combinations in ASCE 7-93, released in 1994.

Table 49.5 Typical Resistance Statistics for Concrete and Steel Members

	Type of Member	m_R/R_n	V_R
Concrete Elements			
	Continuous one-way slabs	1.22	0.16
Flexure	Two-way slabs	1.12–1.16	0.15
Reinforced concrete	One-way pan joists	1.13	0.14
	Beams, grade 40, $f_c' = 5$ ksi	1.14–1.18	0.14
	Beams, grade 60, $f_c' = 5$ ksi	1.01–1.09	0.08–0.12
	Overall values	1.05	0.11
Flexure	Plant precast pretensioned	1.06	0.08
Prestressed concrete	Cast-in-place post-tensioned	1.04	0.10
Axial load and flexure	Short columns, compression	0.95–1.05	0.14–0.16
	Short columns, tension	1.05	0.12
	Slender columns, compression	1.10	0.17
	Slender columns, tension	0.95	0.12
Shear	Beams with $a/d < 2.5$, $\rho_w = 0.008$		
	No stirrups	0.93	0.21
	Minimum stirrups	1.00	0.19
	Moderate stirrups	1.09	0.17
Hot-Rolled Steel Elements			
	Tension member, yield	1.05	0.11
	Tension member, ultimate	1.10	0.11
	Compact beam, uniform moment	1.07	0.13
	Compact beam, continuous	1.11	0.13
	Elastic beam, LTB	1.03	0.12
	Inelastic beam, LTB	1.11	0.14
	Beam-columns	1.07	0.15
	Plate-girders, flexure	1.08	0.12
	Plate-girders, shear	1.14	0.16
	Compact composite beams	1.04	0.14
	Fillet welds	0.88	0.18
	ASS bolts in tension, A325	1.20	0.09
	ASS bolts in tension, A490	1.07	0.05
	HSS bolts in shear, A325	0.60	0.10
	HSS bolts in shear, A490	0.52	0.07

Source: Ellingwood, B., Galambos, T. V., MacGregor, J. C., and Cornell, C. A. 1980. *Development of a Probability Based Load Criterion for American National Standard A58.* NBS Special Publication SP577, National Bureau of Standards, Washington, D.C.

$$
\begin{aligned}
U &= 1.4D_n \\
U &= 1.2D_n + 1.6L_n \\
U &= 1.2D_n + 1.6S_n + (0.5L_n \text{ or } 0.8W_n) \\
U &= 1.2D_n + 1.3W_n + 0.5L_n \\
U &= 1.2D_n + 1.5E_n + (0.5L_n \text{ or } 0.2S_n) \\
U &= 0.9D_n - (1.3W_n \text{ or } 1.5E_n)
\end{aligned} \tag{49.60}
$$

where D_n, L_n, S_n, W_n, and E_n are the nominal (design) values for dead load, live load, snow load, wind load, and earthquake load, respectively. A similar set of load combinations may be found in both the ACI and AISC Specifications, though in the case of the ACI code the load factors (developed earlier) are slightly different. These load combinations were developed in order to ensure essentially equal exceedance probabilities for all combinations, U. A discussion of the bases for these load combinations may be found in Ellingwood *et al.* [1980]. A comparison of LRFD with other countries' codes may be found in Melchers [1987].

Table 49.6 Typical Load Statistics

Load Type	Mean-to-Nominal	COV	Distribution
Dead load	1.05	0.10	Normal
Live load			
Sustained component	0.30	0.60	Gamma
Extraordinary component	0.50	0.87	Gamma
Total (max., 50 years)	1.00	0.25	Type-I
Snow load (annual max.)			
General site (northeast U.S.)	0.20	0.87	Lognormal
Wind load			
50-year maximum	0.78	0.37	Type-I
Annual maximum	0.33	0.59	Type-I
Earthquake load	0.5–1.0	0.5–1.4	Type-II

One important tool used in the development of the load combinations is Turkstra's rule [Turkstra, 1972; Turkstra and Madsen, 1980], developed as an alternative to more complicated load combination analysis. This rule says that, in effect, the maximum of a combination of two or more load effects will occur when one of the loads is at its maximum value while the other loads take on their "instantaneous" or "arbitrary point-in-time" values. Therefore, if n time-varying loads are being considered, there are at least n corresponding load combinations which would need to be considered. This rule may be written generally as

$$\max\{Z\} = \max_i \left[\max_T X_i(t) + \sum_{\substack{j=1 \\ j\neq i}}^{n} X_j(t) \right] \tag{49.61}$$

where $\max\{Z\}$ is the maximum combined load, $X_i(t)$, $i = 1, \ldots, n$ are the time-varying loads being considered in combination, and t is time. In the equation above, the first term in the brackets represents the maximum in the lifetime (T) of load X_i, while the second term is the sum of all other loads at their point-in-time values. This approximation may be unconservative in some cases where the maximum load effect occurs as a result of the combination of multiple loads at near maximum values. However, in most cases, the probability of this occurring is small, and thus Turkstra's rule has been shown to be a good approximation for most structural load combinations [Wen, 1977].

Evaluation of Load and Resistance Factors

Recall that for the generalized case of nonnormal random variables, the following expression was developed [see Eq. (49.42)]:

$$X_i^* = \mu_i^N - \alpha_i \beta \sigma_i^N \tag{49.62}$$

If we further define the design point value X_i^* in terms of a nominal (design) value X_n:

$$X_i^* = \gamma_i X_n \tag{49.63}$$

where γ_i is the partial factor on load X_i (or the inverse of the resistance factor). Therefore, for the popular LRFD format in the U.S., in which the design equation has the form

$$\phi R_n \geq \sum_i \gamma_i X_{n,i} \tag{49.64}$$

the load factors may be computed as

$$\gamma_i = \frac{\mu_i^N - \alpha_i \hat{\beta} \sigma_i^N}{X_{n,i}} \tag{49.65}$$

and the resistance factor is given by

$$\phi = \frac{R_n}{\mu_i^N - \alpha_i \hat{\beta} \sigma_i^N} \tag{49.66}$$

In Eqs. (49.65) and (49.66), α_i is the direction cosine from the convergent iterative solution for random variable i, β is the convergent reliability index (i.e., the target reliability), and $X_{n,i}$ and R_n are the nominal load and resistance values, respectively.

Additional information on the evaluation of load and resistance factors based on FORM/SORM techniques, as well as comparisons between different code formats, may be found in the literature [Ang and Tang, 1975b; Ellingwood *et al.*, 1980; Melchers, 1987].

Acknowledgments

The author is grateful for the comments and suggestions provided by James T. P. Yao of Texas A&M University and Theodore V. Galambos of the University of Minnesota. In addition, discussions with Bruce Ellingwood at Johns Hopkins University were very helpful in preparing this chapter.

Defining Terms

Allowable stress design, working stress design: A method of proportioning structures such that the computed elastic stress does not exceed a specified limiting stress.

Calibration: A process of adjusting the parameters in a new standard to achieve approximately the same reliability as exists in a current standard or specification.

Factor of safety: A factor by which a designated limit state force or stress is divided to obtain a specified allowable value.

Failure: A condition where a limit state is reached.

FORM/SORM (FOSM): First/second-order reliability methods, first-order second-moment reliability methods. Methods which involve (1) a first (second)-order Taylor series expansion of the limit state surface, and (2) computing a notional reliability measure which is a function only of the means and variances (the first two moments) of the random variables. (Advanced FOSM methods include full distribution information as well as any possible correlations of random variables.)

Limit state: A criterion beyond which a structure or structural element is judged to be no longer useful for its intended function (serviceability limit state) or beyond which it is judged to be unsafe (ultimate limit state).

Limit states design: A design method which aims at providing safety against a structure or structural element being rendered unfit for use.

Load factor: A factor by which a nominal load effect is multiplied to account for the uncertainties inherent in the determination of the load effect.

LRFD: Load and resistance factor design, a design method which uses load factors and resistance factors in the design format.

Nominal load effect: Calculated using a nominal load. The nominal load frequently is determined with reference to a probability level; for example, 50-year mean recurrence interval wind speed used in calculating the wind load for design.

Nominal resistance: Calculated using nominal material and cross-sectional properties and a rationally developed formula based on an analytical and/or experimental model of limit state behavior.

Reliability: A measure of relative safety of a structure or structural element.

Reliability-based design (RBD): A design method which uses reliability (probability) theory in the safety checking process.

Resistance factor: A factor by which the nominal resistance is multiplied to account for the uncertainties inherent in its determination.

Source: Ellingwood, B., Galambos, T. V., MacGregor, J. C., and Cornell, C. A. 1980. *Development of a Probability Based Load Criterion for American National Standard A58.* NBS Special Publication SP577, National Bureau of Standards, Washington, D.C.

References

Ang, A. H.-S. and Tang, W. H. 1975a. *Probability Concepts in Engineering Planning and Design, Volume I: Basic Principles.* John Wiley & Sons, New York.

Ang, A. H.-S. and Tang, W. H. 1975b. *Probability Concepts in Engineering Planning and Design, Volume II: Reliability.* John Wiley & Sons, New York.

ASCE. 1990. *Minimum Design Loads for Buildings and Other Structures.* ASCE 7-88, American Society of Civil Engineers, New York.

Benjamin, J. R. and Cornell, C. A. 1970. *Probability, Statistics, and Decision for Civil Engineers.* McGraw-Hill, New York.

Chalk, P. and Corotis, R. B. 1980. A probability model for design live loads. *J. Struct. Div., ASCE.* 106(10):2017–2033.

Chen, X. and Lind, N. C. 1983. Fast probability integration by three-parameter normal tail approximation. *Struct. Saf.* 1(4):269–276.

Der Kiureghian, A. and Liu, P. L. 1986. Structural reliability under incomplete probability information. *J. Eng. Mech., ASCE.* 112(1):85–104.

Ditlevsen, O. 1981. *Uncertainty Modelling.* McGraw-Hill, New York.

Ellingwood, B. and Culver, C. G. 1977. Analysis of live loads in office buildings. *J. Struct. Div., ASCE.* 103(8):1551–1560.

Ellingwood, B., Galambos, T. V., MacGregor, J. C., and Cornell, C. A. 1980. *Development of a Probability Based Load Criterion for American National Standard A58.* NBS Special Publication SP577, National Bureau of Standards, Washington, D.C.

Ellingwood, B., MacGregor, J. G., Galambos, T. V., and Cornell, C. A. 1982. Probability based load criteria: Load factors and load combinations. *J. Struct. Div., ASCE.* 108(5):978–997.

Ellingwood, B. and Redfield, R. 1982. Ground snow loads for structural design. *J. Struct. Eng., ASCE.* 109(4):950–964.

Galambos, T. V., Ellingwood, B., MacGregor, J. G., and Cornell, C. A. 1982. Probability based load criteria: Assessment of current design practice. *J. Struct. Div., ASCE.* 108(5):959–977.

Galambos, T. V. and Ravindra, M. K. 1978. Properties of steel for use in LRFD. *J. Struct. Div., ASCE.* 104(9):1459–1468.

Grigoriu, M. 1989. Reliability of Daniels systems subject to Gaussian load processes. *Struct. Saf.,* 6(2–4):303–309.

Harris, M. E., Corotis, R. B., and Bova, C. J. 1981. Area-dependent processes for structural live loads. *J. Struct. Div., ASCE.* 107(5):857–872.

Hohenbichler, M. and Rackwitz, R. 1983. Reliability of parallel systems under imposed uniform strain. *J. Eng. Mech. Div., ASCE,* 109(3):896–907.

MacGregor, J. G., Mirza, S. A., and Ellingwood, B. 1983. Statistical analysis of resistance of reinforced and prestressed concrete members. *J. Am. Concr. Inst.* 80(3):167–176.

Melchers, R. E. 1987. *Structural Reliability: Analysis and Prediction.* Ellis Horwood, New York.

Rubinstein, R. Y. 1981. *Simulation and the Monte Carlo Method.* John Wiley & Sons, New York.

Thoft-Christensen, P. and Baker, M. J. 1982. *Structural Reliability Theory and Its Applications.* Springer-Verlag, Berlin.

Thoft-Christensen, P. and Morotsu, Y. 1986. *Application of Structural Systems Reliability Theory.* Springer-Verlag, Berlin.

Turkstra, C. J. 1972. *Theory of Structural Design Decisions, Solid Mechanics Study No. 2.* University of Waterloo, Ontario, Canada.

Turkstra, C. J. and Madsen, H. O. 1980. Load combinations in codified structural design. *J. Struct. Div., ASCE.* 106(12):2527–2543.

Wen, Y.-K. 1977. Statistical combinations of extreme loads. *J. Struct. Div., ASCE.* 103(6):1079–1095.

Wen, Y.-K. 1990. *Structural Load Modeling and Combination for Performance and Safety Evaluation.* Elsevier, Amsterdam.

For Further Information

Melchers [1987] provides one of the best overall presentations of structural reliability, both its theory and applications. Ang and Tang [1975b] also provides a good summary. For a more advanced treatment, refer to Ditlevsen [1981] or Thoft-Christensen and Baker [1982] or Thoft-Christensen and Morotsu [1986].

The International Conference on Structural Safety and Reliability (ICOSSAR) is held every four years, and the proceedings from these conferences include short papers on a wide variety of state-of-the-art topics in structural reliability. Many of these papers are authored by researchers as well as practicing engineers. The conference proceedings may be found in the engineering libraries at most universities. A number of other conferences, including periodic specialty conferences cosponsored by ASCE, also include sessions pertaining to reliability.

The Ev-K2 survey project on top of Mount Everest rates as the most precise and comprehensive ever to have been carried out on the roof of the world. The summit level of the world's highest peak was measured using Leica instruments in September 1992. On the 21st of April, 1993, after long computations which took gravimetric, meteorological, and astronomical factors into account, the new value for the summit level was communicated to the international press, and the world was taken by surprise. Mount Everest was only 8846.10 meters high, 2.03 meters less than indicated on existing maps. For the first time its height was determined from two sides of the mountain, in Nepal and China, respectively, and by two methods: GPS, and angle and distance measurement.

This "measurement of the century" has created new world records galore: the first satellite survey of Everest using a GPS system located on the summit; the first trigonometric survey of the mountain with the most precise electronic theodolites and laser distance-measuring equipment; the first survey signal to involve double-faced reflector prisms on the summit; the first simultaneous summit height determination from Nepal and China; the first time that snow samples for environmental analysis were collected from various heights; and the largest-ever international survey team to visit the Himalayas, with Italian, Chinese, Nepalese, and French scientists and mountaineers. (Photo courtesy of Leica Inc.)

VII

Surveying Engineering

Edward M. Mikhail
Purdue University

SURVEYING IS ONE OF THE OLDEST ACTIVITIES of the civil engineer, and remains a primary component of civil engineering. It is also one field that continues to undergo phenomenal changes due to technological developments in digital imaging and satellite positioning. These modern surveying tools are revolutionizing not only regular surveying engineering tasks but are also impacting a myriad of applications in a variety of fields where near-real-time positioning is of great value.

Surveying and engineering are closely related professional activities. The area of surveying and mapping is in many countries a discipline by itself, and taken in total, it is almost as broad in scope as civil engineering. In the U.S., surveying engineering has been historically allied to civil engineering,

> Engineering surveying is defined as those activities involved in the planning and execution of surveys for the location, design, construction, operation, and maintenance of civil and other engineered projects. Such activities include the preparation of survey and related mapping specifications; execution of photogrammetric and field surveys for the collection of required data, including topographic and hydrographic data; calculation, reduction, and plotting of survey data for use in engineering design; design and provision of horizontal and vertical control survey networks; provision of line and grade and other layout work for construction and mining activities; execution and certification of quality control measurements during construction; monitoring of ground and structural stability, including alignment observations, settlement levels, and related reports and certifications; measurement of material and other quantities for inventory, economic assessment, and cost accounting purposes; execution of as-built surveys and preparation of related maps and plans and profiles upon completion of construction; and analysis of errors and tolerances associated with the measurement, field layout, and mapping or other plots of survey measurements required in support of engineering projects. Engineering surveying may be regarded as a specialty within the broader professional practice of engineering and, with the exception of boundary, right-of-way, or other cadastral surveying, includes all surveying and mapping activities required to support the sound conception, planning, design, construction, maintenance, and operation of engineered projects. Engineering surveying does not include surveys for the retracement of existing land ownership boundaries or the creation of new boundaries.

Modern surveying engineering encompasses several specialty areas, each of which requires substantial knowledge and training in order to attain proper expertise. The most primary area perhaps is plane surveying because it is so widely applied in engineering and surveying practice. In plane surveying, we consider the fundamentals of measuring distance, angle, direction, and elevation. These measured quantities are then used to determine position, slope, area, and volume—the basic parameters of civil engineering design and construction. Plane surveying is applied in civil engineering projects of limited areal extent, where the effects of the earth's curvature are negligible relative to the positional accuracy required for the project.

Geodesy, or higher surveying, is an extensive discipline dealing with mathematical and physical aspects of modeling the size and shape of the earth, and its gravity field. Since the launch of earth-orbiting satellites, geodesy has become a truly three-dimensional science. Terrestrial and space geodetic measurement techniques, and particularly the relatively new technique of satellite surveying using the Global Positioning System (GPS), are applied in geodetic surveying. GPS surveying has not only revolutionized the art of navigation but has brought about an efficient positioning technique for a variety of users, prominent among them the engineering community. GPS has had a profound impact on the fundamental problems of determining relative and absolute positions on the earth, including improvements in speed, timeliness, and accuracy. It is safe to say that any geometry-based data collection scheme profits to some degree from the full constellation of 24 GPS satellites. In addition to the obvious applications in geodesy, surveying, and photogrammetry, the use of GPS is applied in civil engineering areas such as transportation (truck and emergency vehicle monitoring, intelligent vehicle and highway systems, etc.) and structures (monitoring of deformation of structures such as water dams). Even in other areas such as forestry and agriculture (crop yield management) GPS provides the geometric backbone of modern (geographic) information systems.

Photogrammetry and remote sensing encompass all activities involved in deriving qualitative and quantitative information about objects and environments from their images. Such imagery may be acquired at close range, from aircraft, or from satellites. In addition to large-, medium-, and small-scale mapping, many other applications such as resource management and environmental assessment and monitoring rely on imageries of various types. Close-range applications include such tasks as accident reconstruction, mapping of complex piping systems, and shape determina tion for parabolic antennas. Large-scale mapping (including the capture of data on infrastructure) remains the primary civil engineering application of photogrammetry. Recent evolution toward working with digital imagery has brought about the increasing acceptance of the digital orthophoto to augment or supplant the planimetric map. Digital image processing tools offer the probability of great increases in mapping productivity through automation. For small- and medium-scale mapping, the increasing availability of satellite image data offers an alternative to chemical photography. Commercially available satellite data with spatial resolutions of 1–3 meters, proposed for the near future, would have a profound impact on all mapping activities within civil engineering. Inclusion of GPS in photogrammetric and remote sensing acquisition platforms will lead to substantial improvements in accuracy, timeliness, and economy.

For centuries, maps have provided layered information in graphical form and have been used as legal documents and as tools to support decision making for applications such as urban planning. Recently, geographic information systems (GIS) have broadened the role played by all types of maps to encompass a total system of hardware, software, and procedures designed to capture, manage, manipulate, and produce information in a spatial context. GIS applications are broad indeed; they include land record management, base mapping, infrastructure maintenance, facilities management, and many others.

A driving force behind the move toward integrating mapping and other spatially oriented data has been the various utility industries and municipalities who need to plan and manage their infrastructure facilities and property assets. This automated mapping/facilities management, or AM/FM, concept is being successfully used today by many cities, counties, and utility industries, who may have embarked on the transition as much as 15 years ago. Successful practitioners of GIS can satisfy the needs of a broad spectrum of users with a single system, minimizing the duplication of resources required to support historically independent user groups.

All other components of Surveying Engineering contribute to the construction of a GIS. The range of survey methods, from classical to modern geodesic and space-based technologies, provide the required reference framework. Digital mapping provides an efficient technology to populate the GIS with spatial information. Remote sensing techniques applied to the earth and its environment provide the various thematic layers of information.

Scope of This Section of the Handbook

The scope of Section VII, Surveying Engineering, in this handbook is to present the reader with the basic information involved in the performance of different surveying engineering projects. As was mentioned earlier, this is a discipline of many areas, each of which will be covered in a separate chapter. The underlying mathematical concepts used by the different areas of surveying engineering are covered first in Chapter 50, followed by four chapters covering, in sequence, plane surveying, geodesy, photogrammetry and remote sensing, and geographic information systems. Of particular importance is the topic on measurements, their errors, and least squares adjustment of redundant data. Since surveying is fundamentally a measurement science, all phases are covered: preanalysis (design), data acquisition (observations), data preprocessing, data adjustment, and postadjustment analysis of the results (quality assessment). Each engineering surveying project must properly execute these phases.

50

General Mathematical and Physical Concepts

Edward M. Mikhail
Purdue University

0-8493-8953-4/95/$0.00 + $.50

50.1 Coordinate Systems

Two-Dimensional Systems

Figure 50.1 depicts two commonly used coordinate systems, one polar (r, θ) and the other *Cartesian* or *rectangular* (x_1, x_2). A point p can be located either by the angle θ, measured from the reference direction x_1, and range r from the reference point 0, or by its two distances from two perpendicular axes, x_1, x_2. The relationships between the two systems are given by

$$\begin{aligned} x_1 &= r\cos\theta \\ x_2 &= r\sin\theta \end{aligned} \tag{50.1}$$

$$\begin{aligned} r &= (x_1^2 + x_2^2)^{1/2} \\ \theta &= \tan^{-1}(x_2/x_1) \end{aligned} \tag{50.2}$$

Three-Dimensional Coordinate Systems

Figure 50.2 shows two systems of three dimensional coordinates: *spherical* (α, β, r) and Cartesian or rectangular (x_1, x_2, x_3). The Cartesian system depicted in Fig. 50.2 is *right-handed,* since a right-threaded screw rotated by an angle less than 90° from $+x_1$ to $+x_2$ would advance in the direction of $+x_3$. The relations between these two systems are as follows:

$$\begin{aligned} r' &= r\cos\beta \\ x_1 &= r'\cos\alpha \\ x_2 &= r'\sin\alpha \end{aligned}$$

$$\begin{aligned} x_1 &= r\cos\alpha\cos\beta \\ x_2 &= r\sin\alpha\cos\beta \\ x_3 &= r\sin\beta \end{aligned} \tag{50.3}$$

$$\begin{aligned} r &= (x_1^2 + x_2^2 + x_3^2)^{1/2} \\ \alpha &= \tan^{-1}(x_2/x_1) \\ \beta &= \sin^{-1}(x_3/r) \end{aligned} \tag{50.4}$$

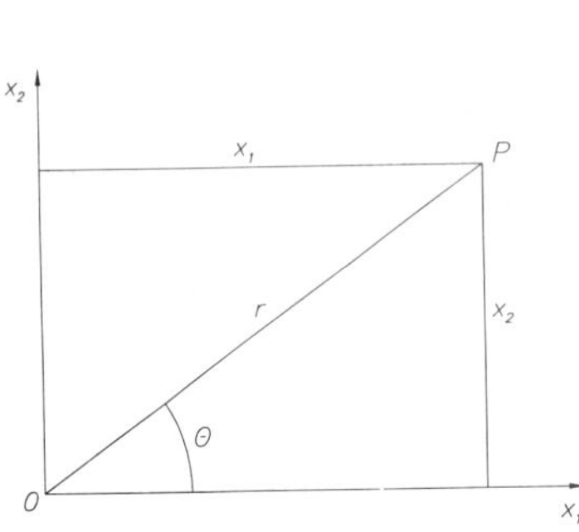

FIGURE 50.1 Two-dimensional coordinate systems.

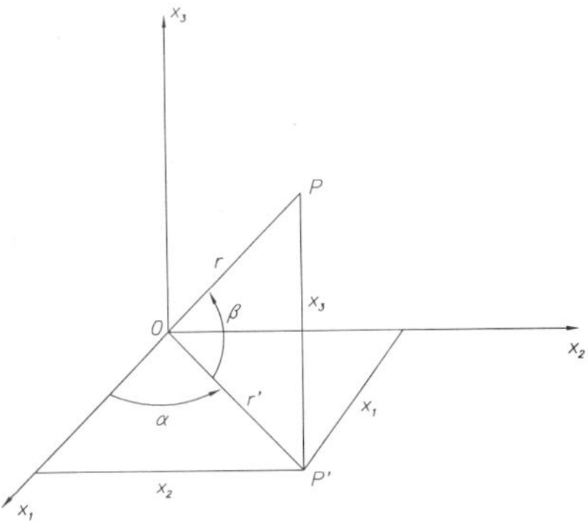

FIGURE 50.2 Three-dimensional right-handed Cartesian coordinate system.

One example of spherical coordinates consists of latitude ϕ, and longitude λ, and earth radius R, when the earth is considered as a sphere. A more accurate representation of the earth is an ellipsoid of revolution about its minor (shorter) axis. Coordinates referred to the earth ellipsoid, and other figures, are given in Chapter 52, "Geodesy."

Examples of rectangular systems include the geocentric and local space rectangular (LSR) coordinate systems. The geocentric system has its origin at the center of the earth, Z through the North Pole, X through the point of zero longitude, and the Y completing a right-handed system. The LSR varies with location, with its XY plane either tangent or secant to the ellipsoid and passing through a selected point λ_0, ϕ_0, and its Z along the local zenith.

50.2 Plane Geometry

Distance between Two Points

The distance between two points P_1 and P_2, with coordinates X_1, Y_1 and X_2, Y_2, is given by

$$L_{12} = \sqrt{(X_2 - X_1)^2 + (Y_2 - Y_1)^2} \tag{50.5}$$

Slope of a Line between Two Points

The slope of any line in a plane is the tangent of the angle it makes with the X axis. The slope angle is the counterclockwise angle from $+X$ axis to the line in a *specified direction.* If the slope angle for the line P_1P_2 is θ_{12}, which is $< 90°$, and the slope angle for the line P_2P_1 is θ_{21}, which is $> 180°$, then

$$\theta_{21} = \theta_{12} + 180° \tag{50.6}$$

The slope m_{12} of the line P_1P_2 is

$$m_{12} = \tan\theta_{12} = \frac{Y_2 - Y_1}{X_2 - X_1} \tag{50.7}$$

and that for the line P_2P_1 is

$$m_{21} = \tan\theta_{21} = \frac{Y_1 - Y_2}{X_1 - X_2} \tag{50.8}$$

Note that $m_{21} = \tan\theta_{21} = \tan(\theta_{12} + 180°) = \tan\theta_{12}$.

Azimuth

Azimuth is a clockwise angle with a magnitude between 0 and 360°. It is measured either from North (or the $+Y$ axis) or from South (or the $-Y$ axis). Thus, for any line, α_N and α_S designate azimuths from North and azimuth from South, respectively. One azimuth angle is obtained from the other by simply adding 180° and dropping 360° whenever the sum exceeds 360°. Thus

$$\begin{aligned} \alpha_S &= \alpha_N + 180° \\ \alpha_N &= \alpha_S + 180° \end{aligned} \tag{50.9}$$

Bearing

Bearing is another form of expressing the direction of a line in surveying. It is always an *acute* angle, with a magnitude between 0 and 90°, and is a positive quantity. The bearing is the angle the line

Table 50.1 Conversion of Bearing to Azimuth

Bearing	α_N, deg	α_s, deg
$N\beta°E$	β	$180+\beta$
$S\beta°E$	$180-\beta$	$360-\beta$
$S\beta°W$	$\beta+180$	β
$N\beta°W$	$360-\beta$	$180-\beta$

Table 50.2 Conversion of Azimuth to Bearing

α_N, deg	α_s, deg	β	Bearing
0–90	180–270	α_N or $\alpha_s - 180°$	$N\beta°E$
90–180	270–360	$180° - \alpha_N$ or $360° - \alpha_s$	$S\beta°E$
180–270	0–90	$\alpha_N - 180°$ or α_s	$S\beta°W$
270–360	90–180	$360° - \alpha_N$ or $180° - \alpha_s$	$N\beta°W$

Source: Tables 50.1 and 50.2 are from Anderson, J. M., and Mikhail, E. M. 1985. *Introduction to Surveying*, p. 665. McGrawHill, Inc., New York, NY. With permission.

makes with either N (for North) or S (for South). The quadrant is indicated by specifying whether the angle is on the east or west side of the meridian (Y axis). Thus, the bearing angle is preceded by either N or S and succeeded by either E or W. If β designates a bearing, Table 50.1 shows how to convert bearing to azimuth. Table 50.2 shows how to convert azimuth to bearing.

Relation between Azimuth and Slope Angle

Considering the commonly used azimuth from North, α_N, the slope angle θ is obtained from

$$\theta = 90° - \alpha_N \tag{50.10}$$

Internal and External Division of a Line Segment

Let points I and E divide a line segment in the proportion s_1/s_2, where I is an internal and E is an external point. If (x_1, y_1) and (x_2, y_2) are the coordinates of the ends of the line segment, the coordinates of I and E are given by

$$\begin{aligned} X_I &= \frac{s_1 X_2 + s_2 X_1}{s_1 + s_2} \\ Y_I &= \frac{s_1 Y_2 + s_2 Y_1}{s_1 + s_2} \end{aligned} \tag{50.11}$$

$$\begin{aligned} X_E &= \frac{s_1 X_2 - s_2 X_1}{s_1 - s_2} \\ y_E &= \frac{s_1 Y_2 - s_2 Y_1}{s_1 - s_2} \end{aligned} \tag{50.12}$$

Equation of a Line with Known *X* and *Y* Intercepts

$$\begin{aligned} \frac{X}{X_0} + \frac{Y}{Y_0} &= 1 \\ Y_0 X + X_0 Y &= X_0 Y_0 \end{aligned} \tag{50.13}$$

where X_0 and Y_0 are the X and Y intercepts, respectively.

General Equation of a Line

$$aX + bY + c = 0 \tag{50.14}$$

Since two points define a line, only two of the three coefficients a, b, c are independent. This can be shown by dividing by c, or

$$\frac{a}{c}X + \frac{b}{c}Y + 1 = 0 \quad \text{or} \quad a'X + b'Y + 1 = 0$$

The two intercepts are obtained from Eq. (50.14) as

$$\begin{aligned} X \text{ intercept} &= -\frac{c}{a} \\ Y \text{ intercept} &= -\frac{c}{b} \end{aligned} \tag{50.15}$$

and the slope of the line is given by

$$m = \tan\theta = -\frac{a}{b} \tag{50.16}$$

Lines Parallel to the Axes

The equation of a line parallel to the X axis is simply

$$Y = k_1 \tag{50.17}$$

where k_1 is the distance of the line from the X axis. Similarly, the equation of a line parallel to the Y axis is

$$X = k_2 \tag{50.18}$$

where k_2 is the distance between the Y axis and the line.

Equation of a Line with Given Slope and *Y* Intercept

$$Y = mX + k \tag{50.19}$$

where m is the slope and k is the Y intercept.

Equation of a Line with a Given Slope Passing through a Given Point

$$Y - Y_P = m(X - X_p) \tag{50.20}$$

where m is the slope and X_p, Y_p are the coordinates of the point. In terms of azimuth α, the equation becomes

$$Y - Y_p = (X - X_p)\cot\alpha \tag{50.21}$$

Equation of a Line Joining Two Points

The equation of a line passing through two points P_1 and P_2 with coordinates X_1, Y_1 and X_2, Y_2 is given by

$$\frac{Y - Y_1}{X - X_1} = \frac{Y_2 - Y_1}{X_2 - X_1} \tag{50.22}$$

Equation of a Line with Given Length and Slope of the Perpendicular from Origin

$$X \cos \theta + Y \sin \theta = p \tag{50.23}$$

where p is the length of the perpendicular from the origin to the line, and θ is the angle it makes with the X axis.

Perpendicular Distance from the Origin to a Line

From the general form of the equation of a line, Eq. (50.14), the length of the perpendicular from the origin to the line is given by

$$p = \left| \frac{c}{\sqrt{a^2 + b^2}} \right| \tag{50.24}$$

Perpendicular Distance from a Point to a Line

$$s = \left| \frac{aX_1 + bY_1 + c}{\sqrt{a^2 + b^2}} \right| \tag{50.25}$$

where X_1, Y_1 are the coordinates of the point, and the line is given by the general equation $aX + bY + c = 0$.

Equation of a Line through a Point and Parallel to Another Line

$$X \cos \theta + Y \sin \theta = p + s \tag{50.26}$$

in which $X \cos \theta + Y \sin \theta = p$ is the equation of the given line and s is the perpendicular distance between the two lines. The value of s cannot be taken as its absolute value; its proper sign must be determined. This is done by realizing that the value of the left-hand side of Eq. (50.14) will always be positive for all points on one side of the line and negative for all points on the other side. (The value is of course zero for points falling on the line.) Thus,

$$\frac{a}{\sqrt{a^2 + b^2}} X + \frac{b}{\sqrt{a^2 + b^2}} Y = -\frac{c}{\sqrt{a^2 + b^2}} + \frac{aX_1 + bY_1 + c}{\sqrt{a^2 + b^2}}$$

which when clearing fractions becomes

$$aX + bY - (aX_1 + bY_1) = 0 \tag{50.27}$$

This is the equation sought, in which a, b belong to the given line and X_1, Y_1 are the coordinates of the given point.

Equation of a Line through a Point and Perpendicular to Another Line

The given line has the general equation $aX + bY + c = 0$, and the given point P has coordinates X_p, Y_p. The slope of the given line is

$$m = -\frac{a}{b}$$

The slope of the line perpendicular to the given line is b/a. Thus,

$$Y - Y_p = \frac{b}{a}(X - X_p)$$

or

$$bX - aY - (bX_p - aY_p) = 0 \tag{50.28}$$

Angle between Two Lines

The angle γ, between two lines is given by

$$\gamma = \theta_2 - \theta_1 \tag{50.29}$$

where θ_1 and θ_2 are the slope angles of the two lines. Then

$$\tan\gamma = \tan(\theta_2 - \theta_1)$$

or

$$\tan\gamma = \frac{\tan\theta_2 - \tan\theta_1}{1 + \tan\theta_1 \tan\theta_2}$$

With the line slopes $m_1 = \tan\theta_1$ and $m_2 = \tan\theta_2$,

$$\tan\gamma = \frac{m_2 - m_1}{1 + m_1 m_2} \tag{50.30}$$

If $\tan\gamma = 0$ or $m_1 = m_2$, the two lines are parallel. On the other hand, if $\tan\gamma = \infty$ or $m_1 m_2 = -1$, the two lines are perpendicular to each other.

Point of Intersection of Two Lines

If the point of intersection of the two lines is q, its coordinates X_q, Y_q satisfy their equations. Then, to get X_q, Y_q we simultaneously solve the two equations of the lines:

$$X_q = \frac{c_1 b_2 - c_2 b_1}{a_1 b_2 - a_2 b_1}$$
$$Y_q = \frac{c_1 a_2 - c_2 a_1}{a_1 b_2 - a_2 b_1} \tag{50.31}$$

Equation of a Circle

Given a circle of radius r and center coordinates X_c, Y_c, its equation is

$$(X - X_c)^2 + (Y - Y_c)^2 = r^2 \tag{50.32}$$

If the circle's center is the origin of the coordinate system, $X_c = Y_c = 0$, its equation reduces to

$$X^2 + Y^2 = r^2 \tag{50.33}$$

Equation (50.32) may be expanded to the form

$$X^2 + Y^2 + 2dX + 2eY + f = 0 \tag{50.34}$$

which represents the general form of the equation of a circle. It contains three coefficients, d, e, f, which represent three geometric elements such as the radius and the two coordinates of its center.

Intersection of a Line and a Circle

In general, a straight line intersects a circle in two points. Given the equations of a circle and a line,

$$X^2 + Y^2 + 2dX + 2eY + f = 0$$

$$Y = mX + k$$

we can eliminate Y and get a general *quadratic* equation in X,

$$AX^2 + BX + C = 0$$

Its two roots are in general given by

$$X_1 = \frac{-B + \sqrt{B^2 - 4AC}}{2A} \quad \text{and} \quad X_2 = \frac{-B - \sqrt{B^2 - 4AC}}{2A}$$

for each of which a value for r is obtained from the equation of the line.

In addition to the case of two points of intersection, two other situations are possible, depending upon the quantity under the radical, $B^2 - 4AC$. If this quantity is zero, or $B^2 = 4AC$, then $X_1 = X_2$ and the line is tangent to the circle at one point. In the second situation, $B^2 < 4AC$, the quantity is negative, which means that one cannot take the square root. (This is usually referred to as the *imaginary* solution to the quadratic equation.) In this case, the line misses the circle and no intersection takes place.

Areas

Circle.

$$A = \pi r^2 = \frac{\pi d^2}{4} \tag{50.35}$$

where r and d are the radius and diameter of the circle, respectively.

Sector.

$$A = \frac{1}{2} ra = \frac{\pi r^2 \theta^\circ}{360^\circ} \tag{50.36}$$

where r is the radius, a is the arc length, and θ is the angle at the center in degrees.

Segment (Less than a Semicircle).

$$A = \frac{1}{2} ra - \frac{1}{2} r^2 \sin\theta \tag{50.37}$$

or

$$A = \frac{\pi r^2 \theta^\circ}{360^\circ} - \frac{1}{2} r^2 \sin\theta \tag{50.38}$$

where r, a, and θ are as defined above. If the chord length c is given, then

$$\sin\frac{\theta}{2} = \frac{c}{2r} \tag{50.39}$$

Triangle. For a *general* triangle, the area T is given by

$$T = \tfrac{1}{2}bh \qquad (h \text{ is perpendicular to } b) \tag{50.40}$$

or

$$T = \tfrac{1}{2}bc\sin A \tag{50.41}$$

or

$$T = \sqrt{s(s-a)(s-b)(s-c)} \tag{50.42}$$

with

$$s = \tfrac{1}{2}(a+b+c) \tag{50.43}$$

where a, b, c are the lengths of the sides, h is the height of the triangle (which is perpendicular to the base b), and A, B, C are the interior angles opposite to the side lengths a, b, c, respectively. Other useful relations for a plane triangle are

$$A + B + C = 180^\circ \tag{50.44a}$$

$$\frac{\sin A}{a} = \frac{\sin B}{b} = \frac{\sin C}{c} \tag{50.44b}$$

$$a^2 = b^2 + c^2 - 2bc\cos A \tag{50.44c}$$

$$b^2 = c^2 + a^2 - 2ac\cos B \tag{50.44d}$$

$$c^2 = a^2 + b^2 - 2ab\cos C \tag{50.44e}$$

$$\tan\frac{A}{2} = \frac{1}{s-a}\sqrt{\frac{(s-a)(s-b)(s-c)}{s}} \tag{50.44f}$$

$$\cos\frac{A}{2} = \sqrt{\frac{s(s-a)}{bc}} \tag{50.44g}$$

$$\cos\frac{B}{2} = \sqrt{\frac{s(s-b)}{ca}} \tag{50.44h}$$

$$\cos\frac{C}{2} = \sqrt{\frac{s(s-c)}{ab}} \tag{50.44i}$$

$$\frac{a-b}{a+b} = \frac{\tan(A-B)/2}{\tan(A+B)/2} \tag{50.44j}$$

$$h = \frac{2}{b}\sqrt{s(s-a)(s-b)(s-c)} \tag{50.44k}$$

For an *equilateral* triangle, where sides $a = b = c$ and angles $A = B = C = 60°$, then the area T becomes

$$T = \frac{a^2\sqrt{3}}{4} \qquad \left(\text{with } h = \frac{a\sqrt{3}}{2}\right) \tag{50.45}$$

Square.

$$A = a^2 \tag{50.46}$$

where a is the side length.

Rectangle.

$$A = ab \tag{50.47}$$

where a and b are its width and length.

Parallelogram. Let a and b be the sides, h the altitude upon side b, C the acute angle, and A the area. Then

$$A = bh = ab \sin C \tag{50.48}$$

Trapezoid. If b_1 and b_2 are the parallel sides, and h is the altitude between them, then the area A is

$$A = \tfrac{1}{2}h(b_1 + b_2) \tag{50.49}$$

For an *isosceles* trapezoid, if a is the length of one of the two nonparallel sides, and C is the acute angle between a and b_2, then

$$A = \tfrac{1}{2}a(b_1 + b_2)\sin C \tag{50.50}$$

50.3 Three-Dimensional Geometry

Distance between Two Points

The distance between two points (X_1, Y_1, Z_1) and (X_2, Y_2, Z_2) is

$$d_{12} = [(X_1 - X_2)^2 + (Y_1 - Y_2)^2 + (Z_1 - Z_2)^2]^{1/2} \tag{50.51}$$

Equation of a Plane

The general equation of a plane is given by

$$AX + BY + CZ + D = 0 \tag{50.52}$$

Only three of the four coefficients A, B, C, D are independent, since we can divide by D and get

$$\frac{A}{D}X + \frac{B}{D}Y + \frac{C}{D} + 1 = 0$$

or

$$EX + FY + GZ + 1 = 0 \tag{50.53}$$

Three noncollinear points determine a plane by writing three linear equations and solve them for E, F, G. When D in Eq. (50.52) is zero, or when no 1 is in Eq. (50.53), the plane passes through the origin. (For vector representation of a plane, see "Planes and Lines" in section 50.4.)

Equation of a Straight Line

Since a straight line is the intersection of two planes, and since a plane is expressed by one linear equation, a straight line in three-dimensional space is represented by two linear equations. The two equations of a line passing through two points (X_1, Y_1, Z_1) and (X_2, Y_2, Z_2) are given by

$$\frac{X - X_1}{X_2 - X_1} = \frac{Y - Y_1}{Y_2 - Y_1} = \frac{Z - Z_1}{Z_2 - Z_1} \tag{50.54}$$

For vector representation of a straight line, see "Planes and Lines" in the following section.

Equation of a Sphere

If (X_c, Y_c, Z_c) represents the center of the sphere and R is its radius, its equation is given by

$$(X - X_c)^2 + (Y - Y_c)^2 + (Z - Z_c)^2 = R^2 \tag{50.55}$$

50.4 Vector Algebra

Definitions

A *vector* is an entity which has a magnitude and direction. In two- and three-dimensional spaces, it is a directed line segment from one point to another. The projections of the vector on the x_1, x_2, and x_3 axes are a_1, a_2, and a_3 and are called the vector *components.* It is represented by a column:

$$\mathbf{a} = \begin{bmatrix} a_1 \\ a_2 \\ a_3 \end{bmatrix}$$

The *length* of the vector is designated by $|\mathbf{a}|$ and is given by

$$|\mathbf{a}| = (a_1^2 + a_2^2 + a_3^2)^{1/2} \tag{50.56}$$

A vector's *direction* is given either by the angles α, β, γ it makes With the axes or by their cosines. The latter are called *direction cosines* and are given by

$$\cos\alpha = \frac{a_1}{|\mathbf{a}|} \qquad \cos\beta = \frac{a_2}{|\mathbf{a}|} \qquad \cos\gamma = \frac{a_3}{|\mathbf{a}|} \tag{50.57}$$

It is evident that

$$\cos^2\alpha + \cos^2\beta + \cos^2\gamma = 1 \tag{50.58}$$

Generalizing a vector to n dimensions, we write

$$\mathbf{a} = \begin{bmatrix} a_1 \\ a_2 \\ \vdots \\ a_n \end{bmatrix}$$

Vector Operations

Equality.

$$\mathbf{a} = \mathbf{b} \qquad \text{when } a_1 = b_1, a_2 = b_2, \cdots, a_n = b_n$$

Addition/Subtraction.

$$\mathbf{c} = \mathbf{a} \pm \mathbf{b} \qquad \text{or} \qquad c_1 = a_1 \pm b_1, c_2 = a_2 \pm b_2, \cdots, c_n = a_n \pm b_n$$

$$\mathbf{a} + \mathbf{b} = \mathbf{b} + \mathbf{a}$$

$$(\mathbf{a} + \mathbf{b}) + \mathbf{c} = \mathbf{a} + (\mathbf{b} + \mathbf{c})$$

Multiplication by a Scalar. A *scalar* is a quantity which has magnitude but no direction, such as mass, temperature, time, etc., and will be designated by a lowercase Greek letter.

$$\lambda\mathbf{a} = \begin{bmatrix} \lambda a_1 \\ \lambda a_2 \\ \vdots \\ \lambda a_n \end{bmatrix}$$

$$\lambda(\mu\mathbf{a}) = (\lambda\mu)\mathbf{a} = \mu(\lambda\mathbf{a}) \tag{50.59}$$

$$(\lambda + \mu)\mathbf{a} = \lambda\mathbf{a} + \mu\mathbf{a}$$

$$\lambda(\mathbf{a} + \mathbf{b}) = \lambda\mathbf{a} + \lambda\mathbf{b}$$

$$|\lambda\mathbf{a}| = \lambda|\mathbf{a}|$$

Any vector $\mathbf{a}$ is reduced to a unit vector $\mathbf{a}^\circ$ when dividing its components by its length, which is a scalar, or $\mathbf{a}^\circ = \mathbf{a}/|\mathbf{a}|$. The components of $\mathbf{a}^\circ$ are the direction cosines of $\mathbf{a}$. Unit vectors along the coordinate axes are called *base* or *basis vectors* and are given by

$$\mathbf{i} = \begin{bmatrix} 1 \\ 0 \\ 0 \end{bmatrix} \qquad \mathbf{j} = \begin{bmatrix} 0 \\ 1 \\ 0 \end{bmatrix} \qquad \mathbf{k} = \begin{bmatrix} 0 \\ 0 \\ 1 \end{bmatrix} \tag{50.60}$$

Any vector in 3-space is uniquely expressed as

$$\mathbf{a} = a_1\mathbf{i} + a_2\mathbf{j} + a_3\mathbf{k} \tag{50.61}$$

The right-handed system introduced in section 50.1 can be generalized for three vectors **a, b, c**. If they are not coplanar, and they have the same initial point, then they are said to form a *right-handed* system if a right-threaded screw rotated through an angle *less than 180°* from **a** to **b** would advance in the direction **c**.

Vector Products

Dot (or Scalar) Product.

$$\mathbf{a} \cdot \mathbf{b} = \sum_{p=1}^{n} a_p b_p = a_1 b_1 + a_2 b_2 + \cdots + a_n b_n \tag{50.62}$$

This is also called the *inner product.* It is a scalar and has the following properties:

$$\mathbf{a}\cdot\mathbf{b} = \mathbf{b}\cdot\mathbf{a}$$
$$\mathbf{a}\cdot(\mathbf{b}+\mathbf{c}) = \mathbf{a}\cdot\mathbf{b}+\mathbf{a}\cdot\mathbf{c}$$
$$\lambda(\mathbf{a}\cdot\mathbf{b}) = (\lambda\mathbf{a})\cdot\mathbf{b} = \mathbf{a}\cdot(\lambda\mathbf{b}) = (\mathbf{a}\cdot\mathbf{b})\lambda \tag{50.63}$$
$$\mathbf{i}\cdot\mathbf{i} = \mathbf{j}\cdot\mathbf{j} = \mathbf{k}\cdot\mathbf{k} = 1$$
$$\mathbf{i}\cdot\mathbf{j} = \mathbf{j}\cdot\mathbf{k} = \mathbf{k}\cdot\mathbf{i} = 0$$

The dot product of a vector with itself is equal to the square of its length, or

$$\mathbf{a}\cdot\mathbf{a} = a_1^2 + a_1^2 + \cdots a_n^2 = |\mathbf{a}|^2 \tag{50.64}$$

If θ is the angle between two vectors **a** and **b** (in two- or three-dimensional space), it can be shown that

$$\mathbf{a}\cdot\mathbf{b} = |\mathbf{a}|\,|\mathbf{b}|\cos\theta \tag{50.65}$$

It follows that if **a** is perpendicular to **b**, then $\mathbf{a}\cdot\mathbf{b} = 0$.

Cross (or Vector) Product. $\mathbf{a}\times\mathbf{b}$ (read "**a** cross **b**") is another vector **c**, which is perpendicular to both **a** and **b** and in a direction such that **a**, **b**, **c** (in this order) form a right-handed system. The length of **c** is given by

$$|\mathbf{c}| = |\mathbf{a}\times\mathbf{b}| = |\mathbf{a}|\,|\mathbf{b}|\sin\theta \tag{50.66}$$

where θ is the angle between **a** and **b**. This quantity is the area of the parallelogram determined by **a** and **b**. If $\mathbf{a} = a_1\mathbf{i} + a_2\mathbf{j} + a_3\mathbf{k}$, and $\mathbf{b} + b_1\mathbf{i} + b_2\mathbf{j} + b_3\mathbf{k}$, then **c** is given by the determinant

$$\mathbf{c} = \mathbf{a}\times\mathbf{b} = \begin{vmatrix} \mathbf{i} & \mathbf{j} & \mathbf{k} \\ \mathbf{a}_1 & \mathbf{a}_2 & \mathbf{a}_3 \\ b_1 & b_2 & b_3 \end{vmatrix} \tag{50.67}$$

It has the following properties

$$\mathbf{a}\times\mathbf{b} = -(\mathbf{b}\times\mathbf{a})$$
$$\mathbf{a}\times(\mathbf{b}+\mathbf{c}) = \mathbf{a}\times\mathbf{b}+\mathbf{a}\times\mathbf{c} \quad \text{(observing the order)}$$
$$\mathbf{a}\cdot(\mathbf{a}\times\mathbf{c}) = 0$$
$$|\mathbf{a}\times\mathbf{b}|^2 = |\mathbf{a}|^2\,|\mathbf{b}|^2 - (\mathbf{a}\cdot\mathbf{b})^2 \tag{50.68}$$
$$\mathbf{i}\times\mathbf{i} = \mathbf{j}\times\mathbf{j} = \mathbf{k}\times\mathbf{k} = 0$$
$$\mathbf{i}\times\mathbf{j} = \mathbf{k}; \mathbf{j}\times\mathbf{k} = \mathbf{i}; \mathbf{k}\times\mathbf{i} = \mathbf{j}$$

For two nonzero vectors, if $\mathbf{a}\times\mathbf{b} = \mathbf{0}$, then **a** and **b** are parallel.

Scalar Triple Product.

$$\mathbf{a}\times\mathbf{b}\cdot\mathbf{c} = \begin{vmatrix} a_1 & a_2 & a_3 \\ b_1 & b_2 & b_3 \\ c_1 & c_2 & c_3 \end{vmatrix} \tag{50.69}$$

is a scalar which is equal to the volume of the parallelepiped determined by **a, b, c.** If it is zero, then the three vectors are coplanar. It has the following properties:

$$\mathbf{a} \times \mathbf{b} \cdot \mathbf{c} = \mathbf{b} \times \mathbf{c} \cdot \mathbf{a} = \mathbf{c} \times \mathbf{a} \cdot \mathbf{b}$$

$$\mathbf{a} \times \mathbf{b} \cdot \mathbf{c} = \mathbf{a} \cdot \mathbf{b} \times \mathbf{c}$$

Planes and Lines

If $\mathbf{p_0}$ is a given point in a plane, $\mathbf{n}$ is a nonzero vector normal to the plane, and $\mathbf{p}$ is any point in the plane, then the equation of the plane takes the form

$$(\mathbf{p} - \mathbf{p_0}) \cdot \mathbf{n} = 0 \qquad \text{or} \qquad \mathbf{p} \cdot \mathbf{n} - \mathbf{p_0} \cdot \mathbf{n} = 0 \tag{50.70}$$

Let $\mathbf{n} = A\mathbf{i} + B\mathbf{j} + C\mathbf{k}$, $\mathbf{p_0} = X_0\mathbf{i} + Y_0\mathbf{j} + Z_0\mathbf{k}$, and $\mathbf{p} = X\mathbf{i} + Y\mathbf{j} + Z\mathbf{k}$. Then Eq. (50.70) becomes

$$A(X - X_0) + B(Y - Y_0) + C(Z - Z_0) = 0 \tag{50.71}$$

or

$$AX + BY + CZ + D = 0$$

where $D = -(AX_0 + BY_0 + CZ_0)$. Two planes are parallel when they have a common normal vector $\mathbf{n}$, and are perpendicular when their normals are, or $\mathbf{n_1} \cdot \mathbf{n_2} = 0$.

If $\mathbf{p_0}$ represents a given point on a line, $\mathbf{p}$ any other point on the line, and $\mathbf{v}$ is a given nonzero vector parallel to the line, then

$$\mathbf{p} = \mathbf{p_0} + \lambda\mathbf{v} \tag{50.72}$$

is an equation of the line. In component form, it yields three scalar equations describing the parametric form (λ is the *running parameter*):

$$\begin{aligned} X &= X_0 + \lambda v_x \\ Y &= Y_0 + \lambda v_y \\ Z &= Z_0 + \lambda v_z \end{aligned} \tag{50.73}$$

If λ is eliminated, one gets the usual two-equation form of a straight line in space; see Eq. (50.54).

50.5 Matrix Algebra

Definitions

A *matrix* is a group of numbers or scalar functions collected in two-dimensional (rectangular) array. A matrix is designated by a boldface capital Roman letter. Thus, an $m \times n$ matrix can be symbolically written as

$$\underset{m,n}{\mathbf{A}} = \begin{bmatrix} a_{11} & a_{12} & \cdots & a_{1n} \\ a_{21} & a_{22} & & a_{2n} \\ \vdots & & \ddots & \vdots \\ a_{m1} & a_{m2} & \cdots & a_{mn} \end{bmatrix}$$

Types of Matrices

Square Matrix. This is a matrix in which the number of rows equals the number of columns. In this case, $\underset{m,m}{A}$ is a square matrix of *order m*. The *principal* (or *main*) diagonal of a square matrix is composed of all elements a_{ij} for which $i = j$.

Row Matrix.

$$\underset{1,n}{\mathbf{a}} = \begin{bmatrix} a_1 & a_2 & \cdots & a_n \end{bmatrix}$$

Column Matrix or Vector.

$$\underset{m,1}{\mathbf{b}} = \begin{bmatrix} b_1 \\ b_2 \\ \vdots \\ b_m \end{bmatrix}$$

Diagonal Matrix.

$$\mathbf{D} = \begin{bmatrix} d_{11} & 0 & \cdots & 0 \\ 0 & d_{22} & & 0 \\ \vdots & & \ddots & \vdots \\ 0 & 0 & \cdots & d_{mm} \end{bmatrix}$$

That is, $d_{ij} = 0$ for all $i \neq j$.

Scalar Matrix.

$$\mathbf{A} = \begin{bmatrix} a & 0 & \cdots & 0 \\ 0 & a & & 0 \\ \vdots & & \ddots & \vdots \\ 0 & 0 & \cdots & a \end{bmatrix}$$

$$a_{ij} = 0 \qquad \text{for all } i \neq j$$

$$a_{ij} = a \qquad \text{for all } i = j$$

Identity or Unit Matrix.

$$\mathbf{I} = \begin{bmatrix} 1 & 0 & \cdots & 0 \\ 0 & 1 & & 0 \\ \vdots & & \ddots & \vdots \\ 0 & 0 & \cdots & 1 \end{bmatrix}$$

$$a_{ij} = 0 \qquad \text{for all } i \neq j$$

$$a_{ij} = 1 \qquad \text{for all } i = j$$

Null Matrix. A *null* or *zero matrix* is a matrix whose elements are all zero. It is denoted by a boldface zero, **0.**

Upper Triangular Matrix.

$$\mathbf{U} = \begin{bmatrix} u_{11} & u_{12} & \cdots & u_{1m} \\ 0 & u_{22} & & u_{2m} \\ \vdots & & \ddots & \vdots \\ 0 & 0 & \cdots & u_{mm} \end{bmatrix}$$

with $u_{ij} = 0$ for $i > j$.

Lower Triangular Matrix.

$$\mathbf{L} = \begin{bmatrix} l_{11} & 0 & \cdots & 0 \\ l_{21} & l_{22} & & 0 \\ \vdots & & \ddots & \vdots \\ l_{m1} & l_{m2} & \cdots & l_{mm} \end{bmatrix}$$

where $l_{ij} = 0$ for $i > j$.

Basic Matrix Operations

Two matrices **A** and **B** are *equal* if they are of the *same dimensions* and each element $a_{ij} = b_{ij}$ for all *i* and *j*. The sum of two matrices **A** and **B** is possible only if they are of equal dimensions, and the elements of the resulting matrix **C** are $c_{ij} = a_{ij} + b_{ij}$ for all *i*, *j*. The following relations apply to addition (and subtraction) of matrices:

$$\begin{aligned} \mathbf{A} + \mathbf{B} &= \mathbf{B} + \mathbf{A} \\ \mathbf{A} + (\mathbf{B} + \mathbf{C}) &= (\mathbf{A} + \mathbf{B}) + \mathbf{C} = \mathbf{A} + \mathbf{B} + \mathbf{C} \\ \mathbf{A} + (-\mathbf{A}) &= \mathbf{0} \end{aligned} \tag{50.74}$$

with **0** being the zero or null matrix, and $-\mathbf{A}$ is the matrix composed of $-a_{ij}$ as elements.

Multiplication of a matrix by a scalar α results in another $\mathbf{B} = \alpha\mathbf{A}$ whose elements are $b_{ij} = \alpha a_{ij}$ for all i and j.

The following relations hold for scalar multiplication (λ, μ are scalars):

$$\begin{aligned} \lambda(\mathbf{A} + \mathbf{B}) &= \lambda\mathbf{A} + \lambda\mathbf{B} \\ (\lambda + \mu)\mathbf{A} &= \lambda\mathbf{A} + \mu\mathbf{A} \\ \lambda(\mathbf{AB}) + (\lambda\mathbf{A})\mathbf{B} &= \mathbf{A}(\lambda\mathbf{B}) \\ \lambda(\mu\mathbf{A}) &= (\lambda\mu)\mathbf{A} \end{aligned} \tag{50.75}$$

The product of two matrices is another matrix. The two matrices must be *conformable for multiplication,* i.e., the number of columns of the first matrix must equal the number of rows of the second matrix. Thus, if **A** is an $m \times q$ matrix and **B** is a $q \times n$ matrix, the product **AB,** *in that order,* is another matrix **C** with *m* rows (as in **A**) and *n* columns (as in **B**). Each element c_{ij} in **C** is obtained by multiplying each one of the *q* elements in the *i*th row in **A** by the corresponding element in the *j*th column in **B** and adding. Algebraically, this is written as

$$c_{ij} = a_{i1}b_{1j} + a_{i2}b_{2j} + \cdots + a_{iq}b_{qj} = \sum_{k=1}^{q} a_{ik}b_{kj} \tag{50.76}$$

To illustrate matrix multiplication:

$$\underset{2,1}{\mathbf{C}} = \underset{2,3}{\mathbf{A}}\,\underset{3,1}{\mathbf{B}} = \begin{bmatrix} 1 & 0 & 2 \\ 2 & 1 & 0 \end{bmatrix} \begin{bmatrix} 1 \\ 5 \\ 3 \end{bmatrix}$$
$$= \begin{bmatrix} (1 \times 1) + (0 \times 5) + (2 \times 3) \\ (2 \times 1) + (1 \times 5) + (0 \times 3) \end{bmatrix} = \begin{bmatrix} 7 \\ 7 \end{bmatrix}$$

Matrix multiplication is *not* commutative; that is, in general $\mathbf{FG} \neq \mathbf{GF}$, even if the dimensions of the matrices allow multiplication in both directions (e.g., $m \times n$ and $n \times m$, or square matrices). The following relationships regarding matrix multiplication hold:

$$\begin{aligned} \mathbf{AI} &= \mathbf{IA} = \mathbf{A} && \text{in which } \mathbf{I} \text{ is the unit or identity matrix} \\ \mathbf{AB} &\neq \mathbf{BA} \\ \mathbf{A}(\mathbf{BC}) &= (\mathbf{AB})\mathbf{C} = \mathbf{ABC} && \text{(associative law)} \\ \mathbf{A}(\mathbf{B}+\mathbf{C}) &= \mathbf{AB} + \mathbf{AC} && \text{(distributive laws)} \\ (\mathbf{A}+\mathbf{B})\mathbf{C} &= \mathbf{AC} + \mathbf{BC} && \text{(distributive laws)} \end{aligned} \tag{50.77}$$

$\mathbf{AB} = \mathbf{0}$ is possible without either $\mathbf{A}$ or $\mathbf{B}$ equaling $\mathbf{0}$. Also, $\mathbf{AB} = \mathbf{AC}$ does *not* imply $\mathbf{B} = \mathbf{C}$.

The *transpose* of the $m \times n$ matrix $\mathbf{A}$ is an $n \times m$ matrix formed from $\mathbf{A}$ by interchanging rows and columns such that the ith row of $\mathbf{A}$ becomes the ith column of the transposed matrix. We denote the transpose of $\mathbf{A}$ by $\mathbf{A}^{\mathrm{T}}$. If $\mathbf{B} = \mathbf{A}^{\mathrm{T}}$, it follows that $b_{ij} = a_{ji}$ for all i and j. The following relationships apply to the transpose of a matrix:

$$\begin{aligned} (\mathbf{A}+\mathbf{B})^{\mathrm{T}} &= \mathbf{A}^{\mathrm{T}} + \mathbf{B}^{\mathrm{T}} \\ (\mathbf{AB})^{\mathrm{T}} &= \mathbf{B}^{\mathrm{T}}\mathbf{A}^{\mathrm{T}} \quad \text{(note reverse order)} \\ (\alpha\mathbf{A})^{\mathrm{T}} &= \alpha\mathbf{A}^{\mathrm{T}} \\ (\mathbf{A}^{\mathrm{T}})^{\mathrm{T}} &= \mathbf{A} \end{aligned} \tag{50.78}$$

A square matrix $\mathbf{A}$ is *symmetric* if $\mathbf{A}^{\mathrm{T}} = \mathbf{A}$. Diagonal, scalar, and identity matrices are symmetric, since each is equal to its transpose. For any matrix $\mathbf{A}$ (not necessarily square), both $\mathbf{AA}^{\mathrm{T}}$ and $\mathbf{A}^{\mathrm{T}}\mathbf{A}$ are symmetric. If $\mathbf{B}$ is a symmetric matrix of suitable dimensions, then for any matrix $\mathbf{A}$, both $\mathbf{ABA}^{\mathrm{T}}$ and $\mathbf{A}^{\mathrm{T}}\mathbf{BA}$ are also symmetric.

If $\mathbf{a}$ is a column matrix (or vector), then $\mathbf{a}^{\mathrm{T}}\mathbf{a}$ is a positive scalar which is equal to the sum of the squares of its elements; for example, the square of the vector's length.

A square matrix $\mathbf{A}$ is *skew-symmetric* if $\mathbf{A}^{\mathrm{T}} = -\mathbf{A}$ and $a_{ij} = -a_{ji}$ for all i,j. For any square matrix $\mathbf{A}$, the matrix $(\mathbf{A} + \mathbf{A}^{\mathrm{T}})$ is symmetric and $(\mathbf{A} - \mathbf{A}^{\mathrm{T}})$ is skew-symmetric.

The *trace* of a *square* matrix is the scalar which is equal to the sum of its main diagonal elements. It is denoted by tr $(\mathbf{A})$; thus $\operatorname{tr}(\underset{n,n}{\mathbf{A}}) = a_{11} + a_{22} + \cdots a_{nn}$. The following are properties of the trace:

$$\begin{aligned} \operatorname{tr}(\mathbf{A}) &= \operatorname{tr}(\mathbf{A}^{\mathrm{T}}) \\ \operatorname{tr}(\lambda\mathbf{A}) &= \lambda \operatorname{tr}(\mathbf{A}^{\mathrm{T}}) \\ \operatorname{tr}(\mathbf{A}+\mathbf{B}) &= \operatorname{tr}(\mathbf{A}) + \operatorname{tr}(\mathbf{B}) \\ \operatorname{tr}(\mathbf{AB}) &= \operatorname{tr}(\mathbf{BA}) \\ \operatorname{tr}(\mathbf{FAF}^{-1}) &= \operatorname{tr}(\mathbf{A}) \quad (\mathbf{F} \text{ nonsingular matrix}) \end{aligned} \tag{50.79}$$

Matrix Inverse

Division of matrices is not defined. Instead, the *inverse* of a *square* matrix **A**, *if it exists,* is the unique matrix $\mathbf{A}^{-1}$ with the following property:

$$\mathbf{A}\mathbf{A}^{-1} = \mathbf{A}^{-1}\mathbf{A} = \mathbf{I} \tag{50.80}$$

where **I** is the identity matrix.

The properties of the inverse are

$$\begin{aligned} (\mathbf{AB})^{-1} &= \mathbf{B}^{-1}\mathbf{A}^{-1} \quad \text{(note reverse order)} \\ (\mathbf{A}^{-1})^{-1} &= \mathbf{A} \\ (\mathbf{A}^{\mathrm{T}})^{-1} &= (\mathbf{A}^{-1})^{\mathrm{T}} \\ (\lambda\mathbf{A})^{-1} &= \tfrac{1}{\lambda}\mathbf{A}^{-1} \end{aligned} \tag{50.81}$$

A square matrix which has an inverse is called *nonsingular,* whereas a matrix which does not have an inverse is called *singular.*

It was stated previously that **AB** can equal **0** without either $\mathbf{A} = \mathbf{0}$ or $\mathbf{B} = \mathbf{0}$. If, however, either **A** or **B** is nonsingular, then the other matrix must be a null matrix. Hence, the product of two nonsingular matrices cannot be a null or zero matrix.

Associated with each *square* matrix **A** is a unique scalar called the *determinant* of **A.** It is denoted either by det **A** or by $|\mathbf{A}|$. Thus, for

$$\mathbf{A} = \begin{bmatrix} 3 & 1 \\ 1 & 2 \end{bmatrix}$$

the determinant is expressed as

$$|\mathbf{A}| = \begin{vmatrix} 3 & 1 \\ 1 & 2 \end{vmatrix}$$

The determinant of order n (for an $n \times n$ square matrix) can be defined in terms of determinants of order $n - 1$ and less. The determinant of a 1×1 matrix is defined as the value of that one element, i.e., for $\mathbf{A} = [a_{11}]$, $|\mathbf{A}| = \det\mathbf{A} = a_{11}$.

If **A** is an $n \times n$ matrix, and one row and one column of **A** are deleted, the resulting matrix is an $(n - 1) \times (n - 1)$ *submatrix* of **A**. The determinant of such a submatrix is called a *minor* of **A**, and it is designated by m_{ij}, where i and j correspond to the deleted row and column, respectively. More specifically, m_{ij} is known as the *minor of the element* a_{ij} in **A**. Thus, each element of **A** has a minor.

The *cofactor* c_{ij} of an element a_{ij} is defined as

$$c_{ij} = (-1)^{i+j} m_{ij} \tag{50.82}$$

The determinant of an $n \times n$ matrix **A** can now be defined as

$$|\mathbf{A}| = a_{11}c_{11} + a_{12}c_{12} + \cdots + a_{1n}c_{1n} \tag{50.83}$$

which states that the determinant of **A** is the sum of the products of the elements of the first row of **A** and their corresponding cofactors. (It is equally possible to define $|\mathbf{A}|$ in terms of any other row or column, but for simplicity we used the first row.) On the basis of this definition, the 2×2 matrix

$$\mathbf{A} = \begin{bmatrix} a_{11} & a_{12} \\ a_{21} & a_{22} \end{bmatrix}$$

has cofactors $c_{11} = |a_{22}| = a_{22}$ and $c_{12} = -|a_{21}| = -a_{21}$, and the determinant of **A** is

$$|\mathbf{A}| = a_{11}c_{11} + a_{12}c_{12} = a_{11}a_{22} - a_{12}a_{21}$$

The *cofactor matrix* **C** of a matrix **A** is the square matrix of the same order as **A** in which each element a_{ij} is replaced by its cofactor c_{ij}.

The *adjoint matrix* of **A**, denoted by adj **A**, is the transpose of its cofactor matrix, i.e.,

$$\text{adj}\,\mathbf{A} = \mathbf{C}^{\mathrm{T}} \tag{50.84}$$

It can be shown that

$$\mathbf{A}(\text{adj}\,\mathbf{A}) = (\text{adj}\,\mathbf{A})\mathbf{A} = |\mathbf{A}|\,\mathbf{I} \tag{50.85}$$

Comparison of Eqs. (50.80) and (50.85) leads directly to a procedure for evaluating the inverse from the adjoint matrix, namely,

$$\mathbf{A}^{-1} = \frac{\text{adj}\,\mathbf{A}}{|\mathbf{A}|} \tag{50.86}$$

It is easy to show that for a 2 × 2 matrix, the adjoint matrix is simply

$$\begin{bmatrix} a_{22} & -a_{12} \\ -a_{21} & a_{11} \end{bmatrix}$$

A square matrix is called *orthogonal* if its inverse is equal to its transpose, or $\mathbf{A}^{-1} = \mathbf{A}^{\mathrm{T}}$. Thus a matrix **M** is orthogonal when

$$\mathbf{M}^{\mathrm{T}}\mathbf{M} = \mathbf{M}\mathbf{M}^{\mathrm{T}} = \mathbf{I} \tag{50.87}$$

The columns of an orthogonal matrix are mutually orthogonal vectors of unit length. Also,

$$|\mathbf{M}| = \pm 1 \tag{50.88}$$

When $|\mathbf{M}| = +1$, **M** is called "proper orthogonal"; otherwise it is termed "improper orthogonal." The product of two orthogonal matrices is also an orthogonal matrix.

Matrix Inverse by Partitioning

Let **A** be an $n \times n$ square nonsingular matrix whose inverse is to be evaluated. We *partition* **A** in the form

$$\mathbf{A} = \begin{matrix} & s & m & \\ & \left[\begin{matrix} \mathbf{A}_{11} \\ \mathbf{A}_{21} \end{matrix}\right. & \left.\begin{matrix} \mathbf{A}_{12} \\ \mathbf{A}_{22} \end{matrix}\right] & \begin{matrix} s \\ m \end{matrix} \end{matrix}$$

where $\mathbf{A}_{11}$ is $s \times s$, $\mathbf{A}_{12}$ is $s \times m$, $\mathbf{A}_{21}$ is $m \times s$, $\mathbf{A}_{22}$ is $m \times m$, and $m + s = n$. The inverse $\mathbf{A}^{-1}$ exists, and we shall denote it, in the correspondingly partitioned form, by

$$\mathbf{A}^{-1} = \mathbf{B} = \begin{bmatrix} \mathbf{B}_{11} & \mathbf{B}_{12} \\ \mathbf{B}_{21} & \mathbf{B}_{22} \end{bmatrix}$$

From the basic definition of an inverse we have $\mathbf{A}\mathbf{A}^{-1} = \mathbf{A}\mathbf{B} = \mathbf{I}$, or in the partitioned form,

$$\begin{bmatrix} \mathbf{A}_{11} & \mathbf{A}_{12} \\ \mathbf{A}_{21} & \mathbf{A}_{22} \end{bmatrix}\begin{bmatrix} \mathbf{B}_{11} & \mathbf{B}_{12} \\ \mathbf{B}_{21} & \mathbf{B}_{22} \end{bmatrix} = \begin{bmatrix} \mathbf{I}_s & \mathbf{0} \\ \mathbf{0} & \mathbf{I}_m \end{bmatrix}$$

which, when multiplied out, leads to four matrix equations in the four $\mathbf{B}_{ij}$ submatrices as unknowns, the solution of which, when $\mathbf{A}_{11}^{-1}$ exists, is given by

$$\begin{aligned} \mathbf{B}_{11} &= \mathbf{A}_{11}^{-1} - \mathbf{A}_{11}^{-1}\mathbf{A}_{12}\mathbf{B}_{21} \\ \mathbf{B}_{12} &= -\mathbf{A}_{11}^{-1}\mathbf{A}_{12}\mathbf{B}_{22} \\ \mathbf{B}_{21} &= -\mathbf{B}_{22}\mathbf{A}_{21}\mathbf{A}_{11}^{-1} \\ \mathbf{B}_{22} &= (\mathbf{A}_{22} - \mathbf{A}_{21}\mathbf{A}_{11}^{-1}\mathbf{A}_{12})^{-1} \end{aligned} \tag{50.89}$$

Alternatively, when $\mathbf{A}_{22}^{-1}$ exists, the solution is

$$\begin{aligned} \mathbf{B}_{11} &= [\mathbf{A}_{11} - \mathbf{A}_{12}\mathbf{A}_{22}^{-1}\mathbf{A}_{21}]^{-1} \\ \mathbf{B}_{12} &= -\mathbf{B}_{11}\mathbf{A}_{12}\mathbf{A}_{22}^{-1} \\ \mathbf{B}_{21} &= -\mathbf{A}_{22}^{-1}\mathbf{A}_{21}\mathbf{B}_{11} \\ \mathbf{B}_{22} &= \mathbf{A}_{22}^{-1} - \mathbf{A}_{22}^{-1}\mathbf{A}_{21}\mathbf{B}_{12} \end{aligned} \tag{50.90}$$

If $\mathbf{A}$ is originally a symmetric matrix, then $\mathbf{A}_{21} = \mathbf{A}_{12}^{\mathrm{T}}$ are correspondingly $\mathbf{B}_{21} = \mathbf{B}_{21}^{\mathrm{T}}$.

The *rank* of a matrix is the order of the largest nonzero determinant that can be formed from the elements of the matrix by appropriate deletion of rows or columns (or both). Thus, a matrix is said to be of *rank m* if and only if it has *at least one nonsingular submatrix of order m*, but has no nonsingular submatrix of order more than m. A nonsingular matrix of order n has a rank n. A matrix with zero rank has elements that must all be zero.

The inverse $\mathbf{A}^{-1}$ is defined only for square matrices, and it exists when the rank of $\mathbf{A}$ is equal to its order. A more general inverse may be defined for rectangular matrices with arbitrary rank. It is called the *generalized inverse*, denoted by $\mathbf{A}^{-}$, and satisfies the relation

$$\mathbf{A}\mathbf{A}^{-}\mathbf{A} = \mathbf{A} \tag{50.91}$$

This condition is not sufficient to define a unique $\mathbf{A}^{-}$. Additional conditions may be imposed on $\mathbf{A}^{-}$, such as

$$\begin{aligned} \mathbf{A}^{-}\mathbf{A}\mathbf{A}^{-} &= \mathbf{A}^{-} \\ (\mathbf{A}\mathbf{A}^{-})^{\mathrm{T}} &= \mathbf{A}\mathbf{A}^{-} \\ (\mathbf{A}^{-}\mathbf{A})^{\mathrm{T}} &= \mathbf{A}^{-}\mathbf{A} \end{aligned} \tag{50.92}$$

If we impose all four conditions in Eqs. (50.91) and (50.92), the inverse is called the *pseudo inverse* or the Moore-Penrose inverse, and is denoted by $\mathbf{A}^{+}$.

The Eigenvalue Problem

For a square matrix $\mathbf{A}$ of order n, we seek a nonzero vector $\mathbf{x}$ and a scalar λ such that

$$\mathbf{A}\mathbf{x} = \lambda\mathbf{x} \tag{50.93}$$

which is called the "eigenvalue problem." A solution λ_0 and $\mathbf{x}_0$ to this problem is called an *eigenvalue* (proper value, characteristic value) and the corresponding *eigenvector* (proper vector, characteristic vector) of the matrix $\mathbf{A}$. An eigenvector, if one exists, can be determined only to a scalar multiplication, for if λ_0, $\mathbf{x}_0$ satisfy Eq. (50.93), then λ_0, $\alpha\mathbf{x}_0$, where α is an arbitrary scalar, will also.

Equation (50.93) can be rewritten as

$$(\mathbf{A} - \lambda\mathbf{I})\mathbf{x} = \mathbf{0} \tag{50.94}$$

which represents a set of homogenous linear equations. For a nontrivial solution to this set the following condition must be satisfied:

$$|\mathbf{A} - \lambda\mathbf{I}| = 0 \tag{50.95}$$

Equation (50.95) represents a real polynomial equation of degree n:

$$b_n(-\lambda)^n + b_{n-1}(-\lambda)^{n-1} + \cdots + b_0 = 0 \tag{50.96}$$

where

$$\begin{aligned}
b_n &= 1 \\
b_{n-1} &= a_{11} + a_{22} + \cdots + a_{nn} = \sum_{i=1}^{n} a_{ii} = \mathrm{tr}(\mathbf{A}) = \text{trace of } \mathbf{A} \\
&\vdots \\
b_{n-r} &= \text{sum of all principal minors of order } r \text{ of } \mathbf{A} \\
&\vdots \\
b_0 &= |\mathbf{A}| = \text{determinant of } \mathbf{A}
\end{aligned} \tag{50.97}$$

Equation (50.96) is called the *characteristic equation* of **A**, or the *eigenvalue equation.* The matrix $(\mathbf{A} - \lambda\mathbf{I})$ is called the *characteristic matrix.* There are n roots for Eq. (50.96), counting multiplicity. These are the n eigenvalues of **A**, $\lambda_1, \lambda_2, \ldots, \lambda_n$. For an eigenvalue λ_i, we solve the set of (homogeneous) linear equations $(\mathbf{A} - \lambda_i\mathbf{I})\mathbf{x} = \mathbf{0}$ to determine the components of the corresponding eigenvector $\mathbf{x}_i$. In general, λ_i and $\mathbf{x}_i$ are either real or complex numbers and vectors, respectively.

If the matrix **A** is *symmetric,* then

1. The eigenvalues are real.
2. The eigenvectors are all mutually orthogonal; that is,

$$\mathbf{x}_i^{\mathrm{T}}\mathbf{x}_j = \mathbf{x}_j^{\mathrm{T}}\mathbf{x}_i = 0$$

Bilinear and Quadratic Forms

If **A** is a square matrix of order n and **x** and **y** are two arbitrary n-vectors, then the scalar

$$u = \mathbf{x}^{\mathrm{T}}\mathbf{A}\mathbf{y} \tag{50.98}$$

is called a *bilinear form.* If, however, the matrix **A** is also *symmetric,* then

$$v = \mathbf{x}^{\mathrm{T}}\mathbf{A}\mathbf{x} \tag{50.99}$$

is called a *quadratic form* with the kernel **A**.

The matrix **A** is called *positive definite* if $v > 0$ for all $\mathbf{x} \neq \mathbf{0}$, and we write $\mathbf{A} > \mathbf{0}$. If $v \geq 0$ for all **x** and there exists a nonzero vector **x** for which equality holds, we say **A** is *positive semidefinite* (or *nonnegative definite*) and write $\mathbf{A} \geq \mathbf{0}$. There are corresponding definitions for *negative definite* (or *nonpositive definite*). If there exist vectors $\mathbf{x}_1$ and $\mathbf{x}_2$ such that $\mathbf{x}_1^{\mathrm{T}}\mathbf{A}\mathbf{x}_1 > 0$ and $\mathbf{x}_2^{\mathrm{T}}\mathbf{A}\mathbf{x}_2 < 0$, we say **A** is *indefinite.*

For a positive definite matrix **A** it is necessary and sufficient that

$$a_{11} > 0, \quad \begin{vmatrix} a_{11} & a_{12} \\ a_{21} & a_{22} \end{vmatrix} > 0, \quad \cdots, \quad |\mathbf{A}| > 0$$

A quadratic form represents, in general, a conic section of some kind. Considering the two-dimensional case for simplicity, we write

$$\mathbf{x}^{\mathrm{T}}\mathbf{A}\mathbf{x} = b \quad \text{with } \mathbf{A} \text{ symmetric} \tag{50.100}$$

or

$$a_{11}x_1^2 + 2a_{12}x_1x_2 + a_{22}x_2^2 = b$$

which is the equation of an ellipse.

50.6 Coordinate Transformations

Linear Transformations

A general *linear transformation* of a vector **x** to another vector **y** takes the form

$$\mathbf{y} = \mathbf{M}\mathbf{x} + \mathbf{t} \tag{50.101}$$

Each element of the **y** vector is a linear combination of the elements of **x** plus a translation or shift represented by an element of the **t** vector. The matrix **M** is called the *transformation matrix*, which is in general rectangular, and **t** is called the translation vector. For our use we restrict **M** to being square nonsingular; thus, the inverse relation exists, or

$$\mathbf{x} = \mathbf{M}^{-1}(\mathbf{y} - \mathbf{t}) \tag{50.102}$$

in which case it is called *affine transformation.* Although both Eqs. (50.101) and (50.102) apply to higher-dimension vectors, we will limit our discussions, without loss of generality, to the more practical two- and three-dimensional spaces, where the elements of the transformations can be depicted geometrically.

Two-Dimensional Linear Transformations

There are six *elementary* transformations, each representing a single effect, which are geometrically represented in Fig. 50.3. Initially, four vectors (1,3) (1,5), (3,3) (3,5) representing the corners of a square (solid lines in Fig. 50.3) are referred to the x_1, x_2 coordinate system. Each of the six elementary transformations operates on the square, and the resulting y_1, y_2 coordinates are plotted to show the effect on the location, orientation, size, and shape of the square after the transformation (dashed lines in Fig. 50.3). In displaying the effects of the transformations, we either display the new figure (dashed lines) in the same coordinate system, or we change the coordinate system. It is easier for the student to visualize these transformations if the new figure is drawn without changing the coordinate system, which we did in Fig. 50.3. However, as we discuss each elementary transformation, we will comment on the second interpretation when appropriate.

1. **Translation**

$$\mathbf{y} = \mathbf{x} + \mathbf{t} \qquad \text{where } \mathbf{M} = \mathbf{I} \tag{50.103}$$

The square is shifted 3 units in x_1 direction and 1 unit in x_2 direction, as shown in Fig. 50.3(a). Alternatively, the solid square remains and the coordinate axes shifted (in the opposite direction and shown in dashed lines).

2. **Uniform Scale**

$$\mathbf{y} = \mathbf{M}\mathbf{x} \qquad \mathbf{M} = \mathbf{U} = \begin{bmatrix} u & 0 \\ 0 & u \end{bmatrix} = u\mathbf{I} \tag{50.104}$$

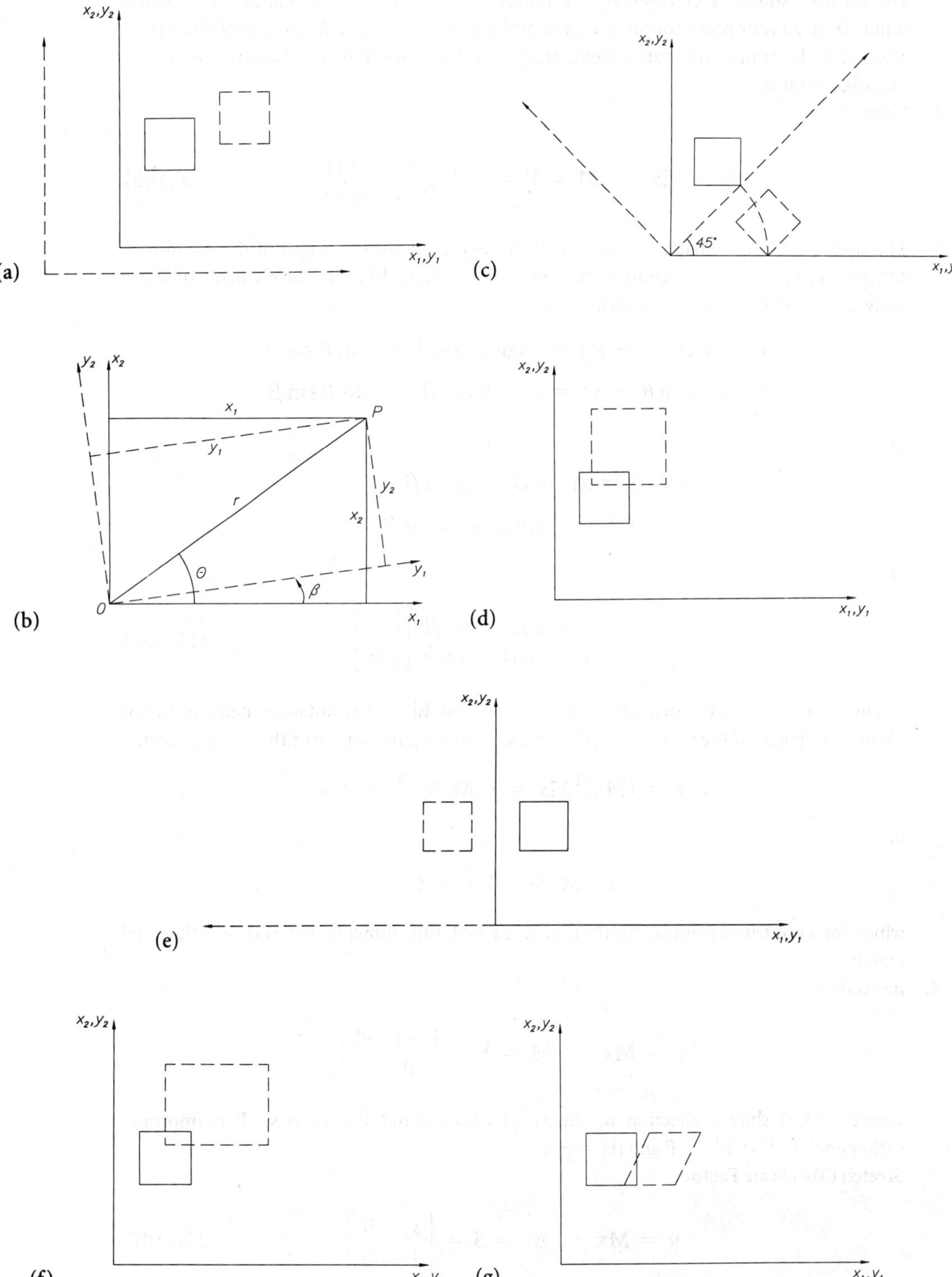

FIGURE 50.3 (a) Translation. (b) Uniform scale. (c) Rotation. (d) Rotation of a two-dimensional coordinate system. (e) Reflection. (f) Stretch (nonuniform scale). (g) Skew (nonperpendicularity of axes).

The (dotted) square is enlarged by the uniform scale u (= 1.5 in Fig. 50.3(b)), which results from all four point coordinate pairs multiplied by u. Alternatively, the solid square is referred to the same coordinate system, except that the units along the axes are now $1/u$ of the original units.

3. **Rotation**

$$\mathbf{y} = \mathbf{Mx} \qquad \mathbf{M} = \mathbf{R} = \begin{bmatrix} \cos\beta & \sin\beta \\ -\sin\beta & \cos\beta \end{bmatrix} \tag{50.105}$$

The square retains its shape, but is rotated through β about the origin of the coordinate system. In Fig. 50.3(c), the coordinate system is also rotated (45°). The elements of **R** are derived from Fig. 50.3(d) as follows:

$$\begin{aligned} y_1 &= r\cos(\theta - \beta) = r\cos\theta\cos\beta + r\sin\theta\sin\beta \\ y_2 &= r\sin(\theta - \beta) = r\sin\theta\cos\beta - r\cos\theta\sin\beta \end{aligned}$$

or

$$\begin{aligned} y_1 &= x_1\cos\beta + x_2\sin\beta \\ y_2 &= -x_1\sin\beta + x_2\cos\beta \end{aligned}$$

or

$$\begin{bmatrix} y_1 \\ y_2 \end{bmatrix} = \begin{bmatrix} \cos\beta & \sin\beta \\ -\sin\beta & \cos\beta \end{bmatrix} \begin{bmatrix} x_1 \\ x_2 \end{bmatrix} \tag{50.106}$$

The matrix **R** is proper orthogonal, $\mathbf{R}^{-1} = \mathbf{R}^{\mathrm{T}}$ and $|\mathbf{R}| = +1$. Rotation matrices do not change the length of the vector, i.e., $|\mathbf{x}| = |\mathbf{y}|$. Considering the square of the vector length,

$$\mathbf{y}^{\mathrm{T}}\mathbf{y} = (\mathbf{Mx})^{\mathrm{T}}\mathbf{Mx} = \mathbf{x}^{\mathrm{T}}\mathbf{M}^{\mathrm{T}}\mathbf{Mx}^{\mathrm{T}} = \mathbf{x}^{\mathrm{T}}\mathbf{x}$$

or

$$\mathbf{x}^{\mathrm{T}}(\mathbf{M}^{\mathrm{T}}\mathbf{M} - \mathbf{I})\mathbf{x} = \mathbf{0}$$

which for a nontrivial solution means that $\mathbf{M}^{\mathrm{T}}\mathbf{M} = \mathbf{I}$, thus showing that **M** is an orthogonal matrix.

4. **Reflection**

$$\mathbf{y} = \mathbf{Mx} \qquad \mathbf{M} = \mathbf{F} = \begin{bmatrix} -1 & 0 \\ 0 & 1 \end{bmatrix}$$

Figure 50.3(e) shows reflection of the x_1 axis (i.e., about the x_2 axis). **F** is improper orthogonal, $\mathbf{F}^{-1} = \mathbf{F}^{\mathrm{T}} = \mathbf{F}$ and $|\mathbf{F}| = -1$.

5. **Stretch (Two Scale Factors)**

$$\mathbf{y} = \mathbf{Mx} \qquad \mathbf{M} = \mathbf{S} = \begin{bmatrix} s_1 & 0 \\ 0 & s_2 \end{bmatrix} \tag{50.107}$$

The square is transformed into a rectangle as shown in Fig. 50.3(f), in which

$$\mathbf{S} = \begin{bmatrix} 2 & 0 \\ 0 & 1.5 \end{bmatrix}$$

6. **Skew (Shear)**

$$\mathbf{y} = \mathbf{Mx} \qquad \mathbf{M} = \mathbf{K} = \begin{bmatrix} 1 & k \\ 0 & 1 \end{bmatrix} \tag{50.108}$$

The square is transformed into a parallelogram as shown in Fig. 50.3(g), where

$$\mathbf{K} = \begin{bmatrix} 1 & 0.5 \\ 0 & 1 \end{bmatrix}$$

From these elementary transformations several affine transformations may be constructed using various sequences. The following are two of the commonly used transformations in photogrammetry.

Four-Parameter Transformation.

$$\begin{bmatrix} y_1 \\ y_2 \end{bmatrix} = \begin{bmatrix} u & 0 \\ 0 & u \end{bmatrix} \begin{bmatrix} \cos\beta & \sin\beta \\ -\sin\beta & \cos\beta \end{bmatrix} \begin{bmatrix} x_1 \\ x_2 \end{bmatrix} + \begin{bmatrix} t_1 \\ t_2 \end{bmatrix} \tag{50.109a}$$

or

$$\begin{aligned} y_1 &= ux_1\cos\beta + ux_2\sin\beta + t_1 \\ y_2 &= -ux_1\sin\beta + ux_2\cos\beta + t_2 \end{aligned} \tag{50.109b}$$

or

$$\begin{aligned} y_1 &= ax_1 + bx_2 + c \\ y_2 &= -bx_1 + ax_2 + d \end{aligned} \tag{50.109c}$$

or

$$\begin{bmatrix} y_1 \\ y_2 \end{bmatrix} = \begin{bmatrix} a & b \\ -b & a \end{bmatrix} \begin{bmatrix} x_1 \\ x_2 \end{bmatrix} + \begin{bmatrix} c \\ d \end{bmatrix} \tag{50.109d}$$

The inverse transformation is given by

$$\begin{bmatrix} x_1 \\ x_2 \end{bmatrix} = \frac{1}{u} \begin{bmatrix} \cos\beta & -\sin\beta \\ \sin\beta & \cos\beta \end{bmatrix} \begin{bmatrix} y_1 - c \\ y_2 - d \end{bmatrix} \tag{50.109e}$$

or

$$\begin{bmatrix} x_1 \\ x_2 \end{bmatrix} = \frac{1}{a^2 + b^2} \begin{bmatrix} a & -b \\ b & a \end{bmatrix} \begin{bmatrix} y_1 - c \\ y_2 - d \end{bmatrix} \tag{50.109f}$$

This transformation has four parameters: a uniform scale, a rotation, and two translations. It is a conformal transformation.

Six-Parameter Transformation.

$$\begin{bmatrix} y_1 \\ y_2 \end{bmatrix} = \begin{bmatrix} s_1 & 0 \\ 0 & s_2 \end{bmatrix} \begin{bmatrix} 1 & k \\ 0 & 1 \end{bmatrix} \begin{bmatrix} \cos\beta & \sin\beta \\ -\sin\beta & \cos\beta \end{bmatrix} \begin{bmatrix} x_1 \\ x_2 \end{bmatrix} + \begin{bmatrix} t_1 \\ t_2 \end{bmatrix} \tag{50.110a}$$

or

$$\begin{bmatrix} y_1 \\ y_2 \end{bmatrix} = \begin{bmatrix} a & b \\ d & e \end{bmatrix} \begin{bmatrix} x_1 \\ x_2 \end{bmatrix} + \begin{bmatrix} c \\ f \end{bmatrix} \tag{50.110b}$$

The six parameters of this transformation consist of two scales, one skew factor (lack of perpendicularity of the axes), one rotation, and two shifts. The inverse transformation is given by

$$\begin{bmatrix} x_1 \\ x_2 \end{bmatrix} = \frac{1}{ae - bd} \begin{bmatrix} e & -b \\ -d & a \end{bmatrix} \begin{bmatrix} y_1 - c \\ y_2 - f \end{bmatrix} \tag{50.110c}$$

Three-Dimensional Linear Transformations

As in the two-dimensional case, affine transformation in three dimensions can be factored out in several elementary transformations: translation, uniform scale, nonuniform scale, rotations, reflections, etc. Consideration, however, is limited to the seven-parameter transformation, which is composed of a uniform scale change, three translations, and three rotations.

We first consider rotations in three-dimensional space.

Rotations of a Three-Dimensional Coordinate System. There are three elementary rotations, one about each of the three axes. They are frequently performed in sequence one after the other. A set of three of these is as follows, where $\mathbf{x}$ is the original system, $\mathbf{x}'$ is once rotated, and $\mathbf{x}''$ is twice rotated:

1. β_1 about x_1 axis, positive rotation advances $+x_2$ to $+x_3$
2. β_2 about x_2' axis, positive rotation advances $+x_3'$ to $+x_1'$
3. β_3 about x_3'' axis, positive rotation advances $+x_1''$ to $+x_2''$

Each of the three elementary rotations is represented in matrix form by

$$\begin{bmatrix} x_1' \\ x_2' \\ x_3' \end{bmatrix} = \begin{bmatrix} 1 & 0 & 0 \\ 0 & \cos\beta_1 & \sin\beta_1 \\ 0 & -\sin\beta_1 & \cos\beta_1 \end{bmatrix} \begin{bmatrix} x_1 \\ x_2 \\ x_3 \end{bmatrix} = \mathbf{M}_{\beta_1} \begin{bmatrix} x_1 \\ x_2 \\ x_3 \end{bmatrix} \tag{50.111a}$$

where x_1, x_2, x_3 are the coordinates before rotation and x_1', x_2', x_3' are the coordinates after rotation. Similarly, a rotation of $+\beta_2$ about the x_2' axis and $+\beta_3$ about the x_3'' axis are given by

$$\begin{bmatrix} x_1'' \\ x_2'' \\ x_3'' \end{bmatrix} = \begin{bmatrix} \cos\beta_2 & 0 & -\sin\beta_2 \\ 0 & 1 & 0 \\ \sin\beta_2 & 0 & \cos\beta_1 \end{bmatrix} \begin{bmatrix} x_1' \\ x_2' \\ x_3' \end{bmatrix} = \mathbf{M}\beta_2 \begin{bmatrix} x_1 \\ x_2 \\ x_3 \end{bmatrix} \tag{50.111b}$$

$$\begin{bmatrix} y_1 \\ y_2 \\ y_3 \end{bmatrix} = \begin{bmatrix} x_1''' \\ x_2''' \\ x_3''' \end{bmatrix} = \begin{bmatrix} \cos\beta_3 & \sin\beta_3 & 0 \\ -\sin\beta_3 & \cos\beta_3 & 0 \\ 0 & 0 & 1 \end{bmatrix} \begin{bmatrix} x_1'' \\ x_2'' \\ x_3'' \end{bmatrix} = \mathbf{M}\beta_3 \begin{bmatrix} x_1'' \\ x_2'' \\ x_3'' \end{bmatrix} \tag{50.111c}$$

The three rotations in Eqs. (50.111) are often referred to as *elementary* rotations, since they may be used to construct any required set of sequential rotations. By successive substitution, the total rotation matrix is obtained:

$$\mathbf{y} = \mathbf{x}''' = \mathbf{M}_{\beta_3}\mathbf{M}_{\beta_2}\mathbf{M}_{\beta_1}\mathbf{x} = \mathbf{M}\mathbf{x} \tag{50.112}$$

in which $\mathbf{M}$ is now a function of the three rotation angles $\beta_1, \beta_2, \beta_3$. The most commonly used set of sequential rotations (in photogrammetry) is given the symbols ω, ϕ, κ where $\omega \equiv \beta_1$; $\phi \equiv \beta_2$; $\kappa \equiv \beta_3$. In this case, the matrix $\mathbf{M}$ which rotates the object coordinate system (X, Y, Z) parallel to the photo coordinate system (x, y, z) is given by

$$\mathbf{M} = \left[\begin{array}{c|c|c} \cos\phi\cos\kappa & \cos\omega\sin\kappa + \sin\omega\sin\phi\cos\kappa & \sin\omega\sin\kappa - \cos\omega\sin\phi\cos\kappa \\ -\cos\phi\sin\kappa & \cos\omega\cos\kappa - \sin\omega\sin\phi\sin\kappa & \sin\omega\cos\kappa + \cos\omega\sin\phi\sin\kappa \\ \sin\phi & -\sin\omega\cos\phi & \cos\omega\cos\phi \end{array}\right] \tag{50.113}$$

in which ω is about the X axis, ϕ is about the once-rotated Y axis, and κ is about the twice-rotated Z axis. The matrix $\mathbf{M}$ is orthogonal, since $\mathbf{M}_\omega$, $\mathbf{M}_\phi$, and $\mathbf{M}_\kappa$ are each orthogonal.

Seven-Parameter Transformation. This transformation contains seven parameters: a uniform scale change u, three rotations $\beta_1, \beta_2, \beta_3$, and three translations t_1, t_2, t_3. It takes the general form

$$\mathbf{y} = u\mathbf{M}\mathbf{x} + \mathbf{t} \tag{50.114}$$

The orthogonal matrix $\mathbf{M}$ is a function of only three independent parameters, in this case the angles $\beta_1, \beta_2, \beta_3$. This transformation is useful for different applications, such as absolute orientation, model connection, etc.

The orthogonal matrix $\mathbf{M}$ may be constructed by other methods besides sequential rotations. Two such methods follow.

Constructing* M *by One Rotation about a Line. This is also often referred to as the *solid body rotation*. Given a three-dimensional object in two different orientations, there exists a line in space about which the object may be rotated by a finite angle to change it from one orientation to the other. If the said line has λ, μ, ν as direction cosines and the angle of rotation is designated by α, the rotation matrix is given by

$$\mathbf{M} = \left[\begin{array}{c|c|c} \lambda^2(1-\cos\alpha)+\cos\alpha & \lambda\mu(1-\cos\alpha)-\nu\sin\alpha & \lambda\nu(1-\cos\alpha)+\mu\sin\alpha \\ \lambda\mu(1-\cos\alpha)+\nu\sin\alpha & \mu^2(1-\cos\alpha)+\cos\alpha & \mu\nu(1-\cos\alpha)-\lambda\sin\alpha \\ \lambda\nu(1-\cos\alpha)-\mu\sin\alpha & \mu\nu(1-\cos\alpha)+\lambda\sin\alpha & \nu^2(1-\cos\alpha)+\cos\alpha \end{array}\right] \tag{50.115}$$

A Purely Algebraic Derivation of M. The following skew-symmetric matrix contains only three parameters a, b, c:

$$\mathbf{S} = \begin{bmatrix} 0 & -c & b \\ c & 0 & -a \\ -b & a & 0 \end{bmatrix} \tag{50.116a}$$

An orthogonal matrix $\mathbf{M}$ can be obtained from $\mathbf{S}$ using

$$\mathbf{M} = (\mathbf{I}+\mathbf{S})(\mathbf{I}-\mathbf{S})^{-1} = (\mathbf{I}-\mathbf{S})^{-1}(\mathbf{I}+\mathbf{S}) \tag{50.116b}$$

in which $\mathbf{I}$ is the identity matrix. Then

$$\begin{aligned} \mathbf{M} &= (\mathbf{I}-\mathbf{S})^{-1}(\mathbf{I}+\mathbf{S}) \\ &= \frac{1}{1+a^2+b^2+c^2}\begin{bmatrix} 1+a^2-b^2-c^2 & 2ab-2c & 2ac+2b \\ 2ab+2c & 1-a^2+b^2-c^2 & 2bc-2a \\ 2ac-2b & 2bc+2a & 1-a^2-b^2+c^2 \end{bmatrix} \end{aligned} \tag{50.117}$$

Nonlinear Transformations

In addition to the linear transformations discussed so far, we use nonlinear transformations both in two and three dimensions. In two dimensions we have the following two transformations:

Eight-Parameter Transformation. The equations

$$\begin{aligned} y_1 &= \frac{a_1x_1 + b_1x_2 + c_1}{a_0x_1 + b_0x_2 + 1} \\ y_2 &= \frac{a_2x_1 + b_2x_2 + c_2}{a_0x_1 + b_0x_2 + 1} \end{aligned} \tag{50.118a}$$

represent the projective transformation from the $\mathbf{x}$ to the $\mathbf{y}$ coordinate systems, with the eight transformation parameters being $a_0, b_0, a_1, \ldots c_2$. Its inverse is given by

$$
\begin{aligned}
x_1 &= \frac{(c_1 - y_1)(b_0 y_2 - b_2) - (c_2 - y_2)(b_0 y_1 - b_1)}{(a_0 y_1 - a_1)(b_0 y_2 - b_2) - (a_2 y_2 - a_2)(b_0 y_1 - b_1)} \\
x_2 &= \frac{(a_0 y_1 - a_1)(c_2 - y_2) - (a_0 y_2 - a_2)(c_1 - y_1)}{(a_0 y_1 - a_1)(b_0 y_2 - b_2) - (a_0 y_2 - a_2)(b_0 y_1 - b_1)}
\end{aligned}
\qquad (50.118b)
$$

These equations describe the central projectivity between two planes.

Two-Dimensional General Polynomials.

$$
\begin{aligned}
y_1 &= a_0 + a_1 x_1 + a_2 x_2 + a_3 x_1 x_2 + a_4 x_1^2 + a_5 x_2^2 + \cdots \\
y_2 &= b_0 + b_1 x_1 + b_2 x_2 + b_3 x_1 x_2 + b_4 x_1^2 + b_5 x_2^2 + \cdots
\end{aligned}
\qquad (50.119a)
$$

These polynomials can obviously be extended to higher powers in x_1, x_2. A special case of these is the conformal form given in the following section.

Two-Dimensional Conformal Polynomials. The conformal property preserves the angles between intersecting lines after the transformation. If we impose the two conditions

$$
\frac{\partial y_1}{\partial x_1} = \frac{\partial y_2}{\partial x_2} \quad \text{and} \quad \frac{\partial y_1}{\partial x_2} = -\frac{\partial y_2}{\partial x_1} \qquad (50.119b)
$$

on the general polynomials in Eqs. (50.119a), we get

$$
\begin{aligned}
y_1 &= A_0 + A_1 x_1 + A_2 x_2 + A_3(x_1^2 - x_2^2) + A_4(2x_1 x_2) + \cdots \\
y_2 &= B_0 - A_2 x_1 + A_1 x_2 - A_4(x_1^2 - x_2^2) + A_3(2x_1 x_2) + \cdots
\end{aligned}
\qquad (50.119c)
$$

Note that the first three terms after the equal signs are the same as those in the four-parameter transformation given in Eq. (50.109c). Equation (50.119c) can also be derived using complex numbers by writing

$$
(y_1 + y_2 i) = (a_0 + b_0 i) + (a_1 + b_1 i)(x_1 + x_2 i) + (a_3 + b_3 i)(x_1 + x_2 i)^2 + \cdots
$$

in which $i = \sqrt{-1}$. Expanding and equating y_1 to the real part and y_2 to the imaginary part (multiplier of i) on the right-hand side leads to Eq. (50.119c).

Three-Dimensional General Polynomials.

$$
\begin{aligned}
y_1 &= a_0 + a_1 x_1 + a_2 x_2 + a_3 x_3 + a_4 x_1^2 + a_5 x_2^2 + a_6 x_1 x_2 + a_7 x_2 x_3 + a_8 x_1 x_3 + \cdots \\
y_2 &= b_0 + b_1 x_1 + b_2 x_2 + b_3 x_3 + b_4 x_1^2 + b_5 x_2^2 + b_6 x_1 x_2 + b_7 x_2 x_3 + b_8 x_1 x_3 + \cdots \\
y_3 &= c_0 + c_1 x_1 + c_2 x_2 + c_3 x_3 + c_4 x_1^2 + c_5 x_2^2 + c_6 x_1 x_2 + c_7 x_2 x_3 + c_8 x_1 x_3 + \cdots
\end{aligned}
\qquad (50.120a)
$$

We can extend these polynomials to higher order. Unlike the two-dimensional case, conformal transformation does *not* exist in three dimensions beyond the first-order (or linear) case given by the seven-parameter transformation, Eq. (50.114). A close approximation, which exists for only second-degree terms, is derived by imposing conditions similar to those in Eq. (50.119b) on every pair of coordinates in Eq. (50.120a). This makes the projections of the three-space onto each of the three planes conformal. Thus, imposing the following on the general polynomials in Eq. (50.120a)

$$\frac{\partial y_1}{\partial x_1} = \frac{\partial y_2}{\partial x_2} = \frac{\partial y_3}{\partial x_3}$$
$$\frac{\partial y_1}{\partial x_2} = -\frac{\partial y_2}{\partial x_1}; \quad \frac{\partial y_2}{\partial x_3} = -\frac{\partial y_3}{\partial x_2}; \quad \frac{\partial y_1}{\partial x_3} = -\frac{\partial y_3}{\partial x_1} \tag{50.120b}$$

leads to

$$\begin{aligned} y_1 &= A_0 + Ax_1 + Bx_2 - Cx_3 + E(x_1^2 - x_2^2 - x_3^2) + 0 + 2Gx_3x_1 + 2Fx_1x_2 + \cdots \\ y_2 &= B_0 - Bx_1 + Ax_2 + Dx_3 + F(-x_1^2 + x_2^2 - x_3^2) + 2Gx_2x_3 + 0 + 2Ex_1x_2 + \cdots \\ y_3 &= C_0 + Cx_1 - Dx_2 + Ax_3 + G(-x_1^2 - x_2^2 + x_3^2) + 2Fx_2x_3 + 2Ex_3x_1 + 0 + \cdots \end{aligned} \tag{50.120c}$$

50.7 Linearization of Nonlinear Functions

Frequently, the equations expressing the geometric and physical conditions of a problem are nonlinear, which makes their direct solution difficult and uneconomical. We linearize these equations using series expansion, usually Taylor's series, which in general is given by the following for $y = f(x)$:

$$y = f(x^0) + \left.\frac{df}{dx}\right|_{x^0} \Delta x + \frac{1}{2!}\left.\frac{d^2y}{dx^2}\right|_{x^0} (\Delta x)^2 + \cdots + \frac{1}{n!}\left.\frac{d^ny}{dx^n}\right|_{x^0} (\Delta x)^n + \cdots \tag{50.121}$$

This gives the value of y at $(x^0 + \Delta x)$, given the value of the function $f(x^0)$ at x^0. Equation (50.121) includes still higher-order terms, and therefore we usually drop the second- and higher-order terms and use the approximation

$$y \approx f(x^0) + \left.\frac{dy}{dx}\right|_{x^0} \Delta x \approx y^0 + j\Delta x \tag{50.122}$$

with obvious correspondence in terms.

The technique of linearization is demonstrated in Fig. 50.4. The curve represents the original nonlinear function $f(x)$, whereas the straight line represents the linearized form, Eq. (50.122).

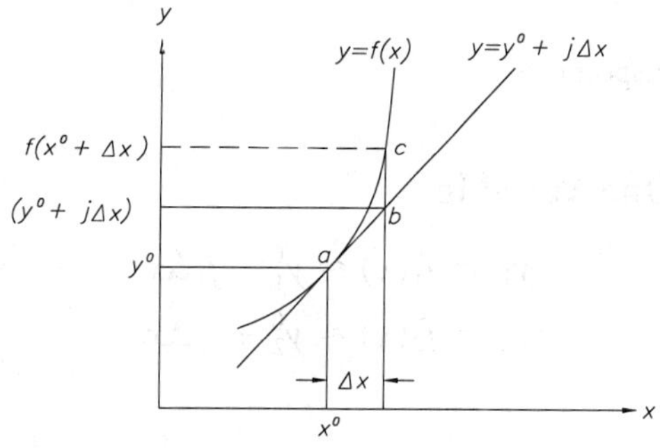

FIGURE 50.4 Linearization.

That line is tangent to the curve at the given point a, (x^0, y^0). When Δx is given (or evaluated), the value of the function would be approximated by point b, whose ordinate is $(y^0 + j\Delta x)$, and the exact value from the nonlinear function is point c, with ordinate $f(x^0 + \Delta x)$. The error arising from using the linear form is the line segment bc.

One Function of Two Variables

$$
\begin{aligned}
y &= f(x_1, x_2) \\
&= f(x_1^0, x_2^0) + \left.\frac{\partial y}{\partial x_1}\right|_{x_1^0,x_2^0} \Delta x_1 + \left.\frac{\partial y}{\partial x_2}\right|_{x_1^0,x_2^0} \Delta x_2 \\
&+ \frac{1}{2!}\left.\frac{\partial^2 y}{\partial x_1^2}\right|_{x_1^0,x_2^0} (\Delta x_1)^2 + \frac{1}{2!}\left.\frac{\partial^2 y}{\partial x_2^2}\right|_{x_1^0,x_2^0} (\Delta x_2)^2 \\
&+ \left.\frac{\partial y}{\partial x_1}\right|_{x_1^0,x_2^0} \left.\frac{\partial y}{\partial x_2}\right|_{x_1^0,x_2^0} (\Delta x_1)(\Delta x_2) + \cdots
\end{aligned}
\tag{50.123}
$$

For the linearized form, Eq. (50.123) is truncated to

$$y = y^0 + j_1 \Delta x_1 + j_2 \Delta x_2 \tag{50.124}$$

where

$$y^0 = f(x_1^0, x_2^0) \qquad j_1 = \left.\frac{\partial y}{\partial x_1}\right|_{x_1^0,x_2^0} \qquad j_2 = \left.\frac{\partial y}{\partial x_2}\right|_{x_1^0,x_2^0}$$

Equation (50.124) can be rewritten in matrix form as

$$y = y^0 + [j_1 \quad j_2]\begin{bmatrix} \Delta x_1 \\ \Delta x_2 \end{bmatrix}$$

or

$$y = y^0 + \mathbf{J}_{yx}\Delta\mathbf{x} \tag{50.125}$$

where

$$\mathbf{J}_{yx} = \frac{\partial y}{\partial \mathbf{x}} = \begin{bmatrix} \dfrac{\partial y}{\partial x_1} & \dfrac{\partial y}{\partial x_2} \end{bmatrix}$$

is the Jacobian of y with respect to $\mathbf{x}$.

Two Functions of One Variable

$$
\begin{aligned}
y_1 &= f_1(x) \approx y_1^0 + j_1 \Delta x \\
y_2 &= f_2(x) \approx y_2^0 + j_2 \Delta x
\end{aligned}
\tag{50.126}
$$

or

$$\mathbf{y} = \mathbf{y}^0 + \mathbf{J}_{yx}\Delta x$$

with

$$\begin{aligned} y_1^0 &= f_1(x^0) \\ y_2^0 &= f_2(x^0) \\ \mathbf{J}_{yx} &= [j_1 \quad j_2]^{\mathrm{T}} = \left[\left.\frac{dy_1}{dx}\right|_{x^0} \quad \left.\frac{dy_2}{dx}\right|_{x^0} \right]^{\mathrm{T}} \end{aligned}$$

Two Functions of Two Variables Each

$$\begin{aligned} y_1 &= f_1(x_1, x_2) \approx y_1^0 + j_{11}\Delta x_1 + j_{12}\Delta x_2 \\ y_2 &= f_2(x_1, x_2) \approx y_2^0 + j_{21}\Delta x_1 + j_{22}\Delta x_2 \end{aligned} \tag{50.127a}$$

or

$$\begin{bmatrix} y_1 \\ y_2 \end{bmatrix} \approx \begin{bmatrix} y_1^0 \\ y_2^0 \end{bmatrix} + \begin{bmatrix} j_{11} & j_{12} \\ j_{21} & j_{22} \end{bmatrix} \begin{bmatrix} \Delta x_1 \\ \Delta x_2 \end{bmatrix} \tag{50.127b}$$

or

$$\mathbf{y} = \mathbf{y}^0 + \mathbf{J}_{yx}\mathbf{\Delta x} \tag{50.127c}$$

where

$$\mathbf{y}^0 = \begin{bmatrix} y_1^0 \\ y_2^0 \end{bmatrix} = \begin{bmatrix} f_1(x_1^0, x_2^0) \\ f_2(x_1^0, x_2^0) \end{bmatrix}$$

and

$$\mathbf{J}_{xy} = \frac{\partial \mathbf{y}}{\partial \mathbf{x}} = \begin{bmatrix} \dfrac{\partial y_1}{\partial x_1} & \dfrac{\partial y_1}{\partial x_2} \\ \dfrac{\partial y_2}{\partial x_1} & \dfrac{\partial y_2}{\partial x_2} \end{bmatrix}$$

evaluated at x_1^0, x_2^0.

General Case of *m* Functions of *n* Variables

$$\begin{aligned} y_1 &= f_1(x_1, x_2, \ldots, x_n) \\ y_2 &= (x_1, x_2, \ldots, x_n) \\ &\vdots \qquad \vdots \\ y_m &= f_m(x_1, x_2, \ldots, x_n) \end{aligned} \tag{50.128a}$$

With the auxiliaries,

$$\mathbf{y}^0 = \begin{bmatrix} y_1^0 \\ y_2^2 \\ \vdots \\ y_m^0 \end{bmatrix} = \begin{bmatrix} f_1(x_1^0, x_2^0, \ldots, x_n^0) \\ f_2(x_1^0, x_2^0, \ldots, x_n^0) \\ \vdots \qquad \vdots \\ f_m(x_1^0, x_2^0, \ldots, x_n^0) \end{bmatrix}$$

$$\mathbf{J}_{yx} = \frac{\partial \mathbf{y}}{\partial \mathbf{x}} = \begin{bmatrix} \frac{\partial y_1}{\partial x_1} & \frac{\partial y_1}{\partial x_2} & \cdots & \frac{\partial y_1}{\partial x_n} \\ \vdots & & \ddots & \vdots \\ \frac{\partial y_m}{\partial x_1} & \frac{\partial y_m}{\partial x_2} & \cdots & \frac{\partial y_m}{\partial x_n} \end{bmatrix} \text{ evaluated at } \mathbf{x}^0$$

$$\mathbf{\Delta x} = \begin{bmatrix} \Delta x_1 \\ \Delta x_2 \\ \vdots \\ \Delta x_n \end{bmatrix}$$

the linearized form of Eq. (50.128a) becomes

$$\mathbf{y} \approx \mathbf{y}^0 + \mathbf{J}_{yx}\mathbf{\Delta x} \tag{50.128b}$$

which represents the general form, with $\mathbf{y}$, $\mathbf{y}^0$ being $m \times 1$ vectors, $\mathbf{J}$ an $m \times n$ Jacobian matrix, and $\mathbf{\Delta x}$ an $n \times 1$ vector. Equations (50.122), (50.125), and (50.127c) are special cases of Eq. (50.128b).

Differentiation of a Determinant

The partial derivative of a $p \times p$ determinant with respect to a scalar is composed of the sum of p determinants, each having the elements of only one row or one column replaced by their derivatives. Thus, given the determinant $d = |\mathbf{D}_1 \ \mathbf{D}_2 \cdots \mathbf{D}_p|$ in which $\mathbf{D}_i, i = 1, 2 \ldots p$ represents its p columns, then

$$\frac{\partial d}{\partial x} = \left|\frac{\partial \mathbf{D}_1}{\partial x}\mathbf{D}_2 \cdots \mathbf{D}_p\right| + \left|\mathbf{D}_1\frac{\partial \mathbf{D}_2}{\partial x} \cdots \mathbf{D}_\mathbf{p}\right| + \cdots + \left|\mathbf{D}_1\mathbf{D}_2 \cdots \frac{\partial \mathbf{D}_p}{\partial x}\right| \tag{50.129}$$

An expression similar to Eq. (50.129) can be written in which rows instead of the columns of d are partially differentiated.

Differentiation of a Quotient

The partial derivative of $g = U/W$ with respect to a variable x is given by

$$\frac{\partial g}{\partial x} = \frac{1}{W}\left[\frac{\partial U}{\partial x} - \frac{U}{W}\frac{\partial W}{\partial x}\right] \tag{50.130}$$

Both U and W can be general functions, including determinants, of several variables.

50.8 Map Projections

Map projection is concerned with the theory and techniques of proper representation of the curved earth surface on the plane of a map. When the map is of such large scale as to represent a very limited area, the earth curvature is insignificant, and field survey measurements can be directly represented on the map. On the other hand, as the surface area of the earth gets larger, this curvature becomes significant and must be dealt with. The earth is an ellipsoid and is also sometimes approximated by a sphere; neither of these surfaces can accurately be developed into a plane. Therefore, all map projection methods must by necessity contain some distortion. Various methods are selected to fit best the shape of the area to be mapped and to minimize the effects of particular distortions.

Locations on the earth are represented by meridians of longitude, λ, and parallels of latitude, ϕ. On the map these are represented by scaled linear distances X, Y, using the dimensions of the earth ellipsoid and selected criteria which the specific map projection must satisfy. These are obtained from transformation equations taking the general functional form of

$$\begin{aligned} X &= f_x(\lambda, \phi) \\ Y &= f_y(\lambda, \phi) \end{aligned} \tag{50.131}$$

Although all modern map projections are performed by computer programs, several are based on geometric projection of the earth onto one of three surfaces: a plane, a cylinder, or a cone. It is clear that the cylinder and cone are chosen because they can be developed into a plane—that of the map. When a plane is used, it is tangent to the earth's surface at a point and the projection center is either the center of the earth, as in the *gnomonic* projection shown in Fig. 50.5(a), or the point diametrically opposite to the tangent point, as in the *stereographic* projection shown in Fig. 50.5(b). If the projection lines are perpendicular to the plane, we have an *orthographic* projection, Fig. 50.5(c).

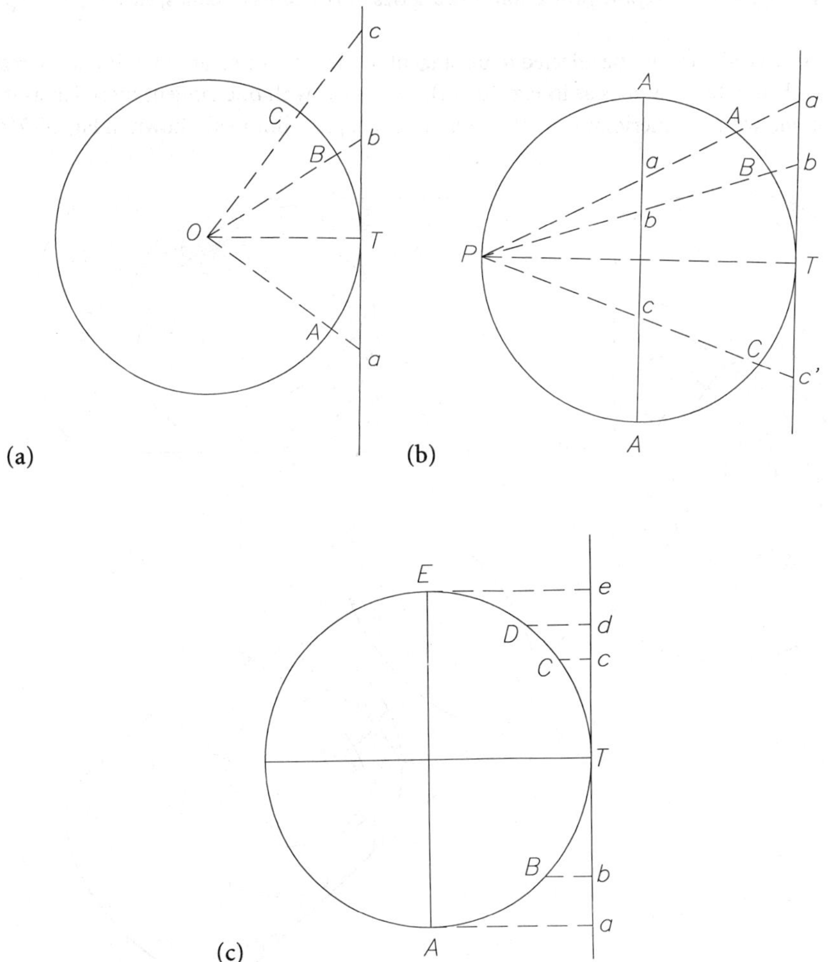

FIGURE 50.5 (a) Gnomonic projection. (b) Stereographic projection. (c) Orthographic projection.

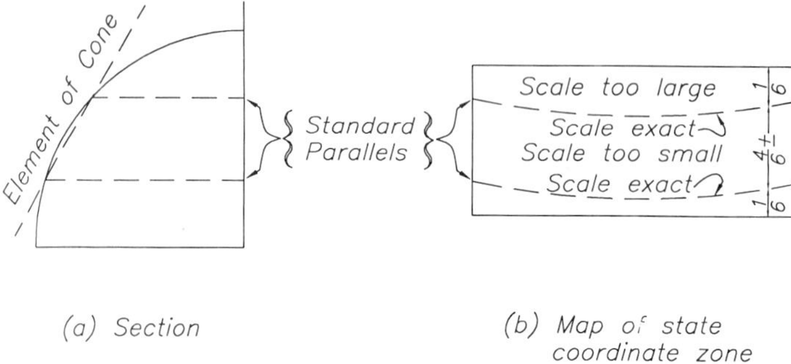

FIGURE 50.6 Lambert conformal conic projection. (*Source:* Davis, R. E., Foote, F. S., Anderson, J. M., and Mikhail, E. M. 1981. *Surveying: Theory and Practice,* 6th ed., p. 570. McGraw-Hill, New York, NY. With permission.)

A cone is usually selected with its axis coincident with the earth's polar axis. It may be tangent to the earth at one small circle, called *standard parallel,* or intersect it in two standard parallels. When developed, the scale will be true (i.e., without any distortion or error) at the standard parallels; see Fig. 50.6. Polyconic projections use a series of frustums of cones, each from a separate cone.

Like a cone, a cylinder may be selected to be tangent to the earth or secant to it. It may be *regular,* with its axis being the polar axis as in Fig. 50.7(a); *transverse,* with one tangent meridian as in Fig. 50.7(b); *secant,* with two meridians of intersection; or *oblique* cylindrical, shown in Fig. 50.7(c).

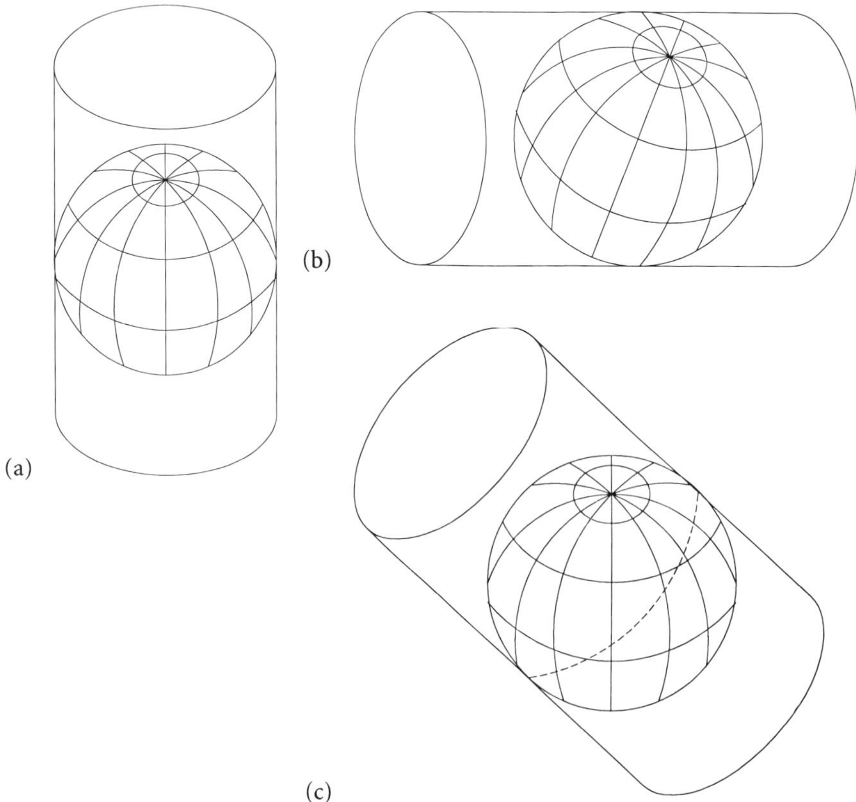

FIGURE 50.7 (a) Vertical cylinder. (b) Transverse horizontal cylinder. (c) Oblique cylinder.

Equation (50.131) will produce a *perfect* map—without any distortions—if it satisfies *all* of the following conditions:

1. All distances and areas have correct relative magnitudes.
2. All azimuths and angles are correctly represented on the map.
3. All great circles are shown on the map as straight lines.
4. Geodetic longitudes and latitudes are correctly shown on the map.

No one map projection can satisfy *all* these conditions. However, each class can satisfy some selected conditions. The following are four classes:

1. **Conformal or orthomorphic projection** results in a map showing the correct angle between any pair of short intersecting lines, thus making small areas appear in correct *shape*. As the scale varies from point to point, the shapes of larger areas are incorrect.
2. An **equal-area projection** results in a map showing all areas in proper relative *size*, although these areas may be much out of shape and the map may have other defects.
3. In an **equidistant projection** distances are correctly represented from one central point to other points on the map.
4. In an **azimuthal projection** the map shows the correct *direction* or azimuth of any point relative to one central point.

For conformal mapping, a new latitude, ψ, called the *isometric latitude,* is used in place of ϕ, where

$$\psi = \ln\left[\left(\frac{1 - e\sin\phi}{1 + e\sin\phi}\right)^{e/2} \tan\left(\frac{\pi}{4} + \frac{\phi}{2}\right)\right] \tag{50.132}$$

in which $e^2 = (a^2 - b^2)/a^2$, with a, b being the semimajor and semiminor axes of the earth ellipsoid, respectively. Then, Eq. (50.131) is replaced by

$$\begin{aligned} X &= f_1(\lambda, \psi) \\ Y &= f_2(\lambda, \psi) \end{aligned} \tag{50.133}$$

In order for the mapping in Eq. (50.133) to be conformal, the following Cauchy-Riemann conditions must be satisfied:

$$\frac{\partial X}{\partial \lambda} = \frac{\partial Y}{\partial \psi} \quad \text{and} \quad \frac{\partial X}{\partial \psi} = -\frac{\partial Y}{\partial \lambda} \tag{50.134}$$

Two commonly used conformal projections are the Lambert conformal conic projection and the transverse Mercator projection. A figure of the former is shown in Fig. 50.6, where the projection cone intersects the ellipsoid in two standard parallels. It is very widely used in the U.S., particularly as a State Plane Coordinate System for those states, or zones thereof, with greater east-west extent than north-south. The transverse Mercator projection is shown in Fig. 50.8, where the cylinder is either tangent or secant to the ellipsoid. When it is tangent, the scale at the central meridian is 1:1. But when it is not, the scale at the central meridian is less than 1:1, as shown in Fig. 50.8. The central meridian is the origin of the map X coordinate, while the origin of the map Y coordinate is the equator. This projection is used as a State Plane Coordinate System for states with greater north-south extent.

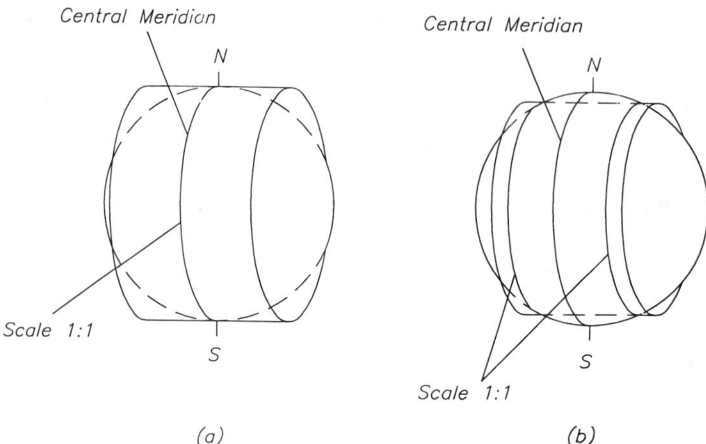

FIGURE 50.8 Transverse Mercator projection. (a) Cylinder with one standard line. (b) Cylinder with two standard lines.

An extensively used map projection system is the *Universal* Transverse Mercator, or UTM, schematically shown in Fig. 50.9. It is in 6° wide zones, with the scale at each central meridian of a zone being 0.9996. A false easting for each central meridian is 500,000 m. A transverse Mercator projection with 3° wide zones is possible, where the scale at the central meridian is improved to 0.9999.

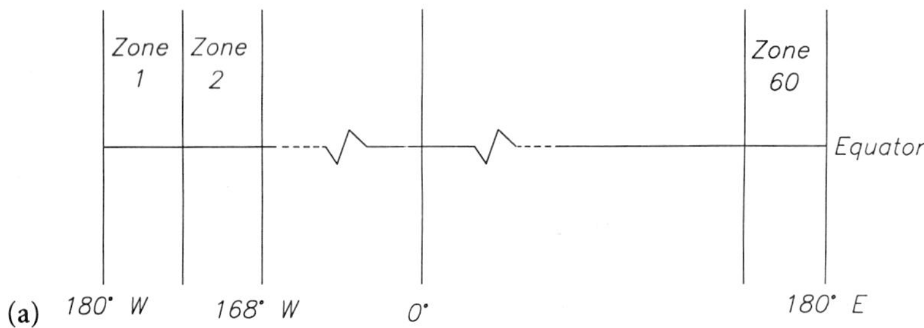

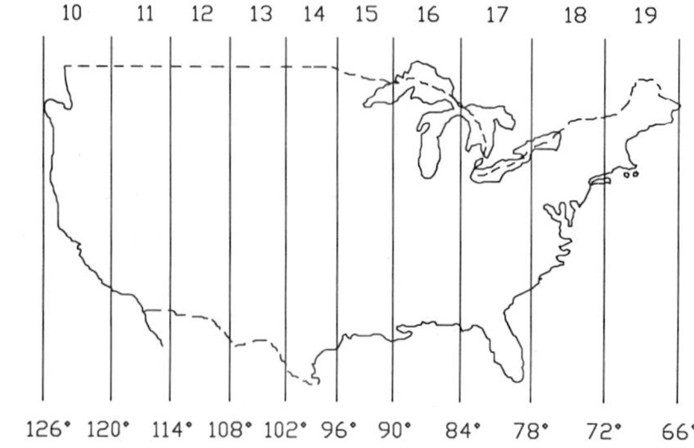

FIGURE 50.9 (a) Universal transverse Mercator zones. (b) UTM zones in the United States.

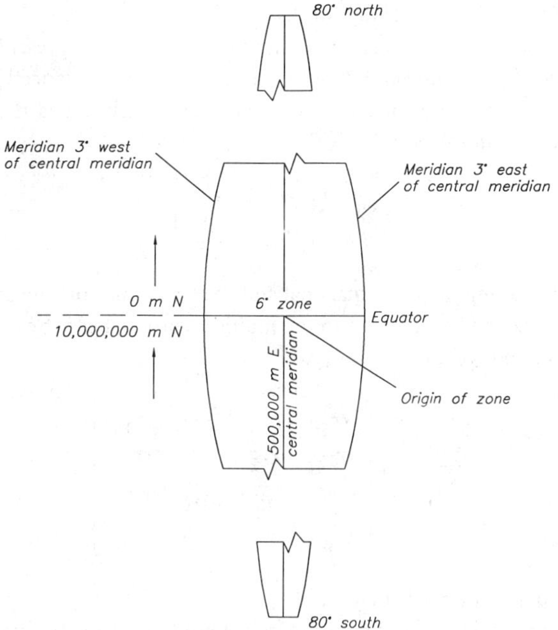

FIGURE 50.9 (continued) (c) X and Y coordinates of the origin of a UTM grid zone. (*Source:* U.S. Army Field Manual, *Map Reading,* FM 21-26.)

50.9 Observational Data Adjustment

Mathematical Model for Adjustment

In surveying engineering, measurements are rarely used directly as the required information. They are frequently used in subsequent operations to derive other quantities, often computationally, such as directions, lengths, relative positions, areas, shapes, and volumes. The relationships applied in the computational effort are the mathematical representations of the geometric and physical conditions of the problem, which, together with the quality of the measurements, are called the *mathematical model.*

This mathematical model is composed of two parts: a functional model and a stochastic model. The *functional model* is the part which describes the geometric or physical characteristics of the survey problem and the resulting mathematical relationships. The *stochastic model* is the part of the mathematical model that describes the statistical properties of all the elements involved in the functional model. It designates which parts are the observables, which are constants, and which are unknown parameters to be estimated in the **adjustment**. It also provides the information necessary to properly describe the quality of the observations to be used in the adjustment.

As a simple example, consider the size and shape of a plane triangle. While the shape depends only on angles, its size requires at least one side. Therefore, this model has three angular elements, the interior angles, and three linear (or distance) elements, the triangle sides. Two angles and one side will be the minimum number of measurements necessary to uniquely fix the triangle. If more measurements than these three are obtained, *redundancy* will exist, thus leading to inconsistency, which is resolved through an adjustment technique. Once the number of measurements is decided upon, the required set of independent condition equations can be written to express the functional model. The stochastic model will denote those elements (of the total of six) which are observed, and the quality of the observations.

The a priori quality of an observed angle or distance is usually expressed by a *standard deviation,* σ, or its square, the *variance,* σ^2. Correlation between observations is represented by the *covariance.* Thus, for two observables, or *random variables,* say $\tilde{x}$ and $\tilde{y}$, the variances, σ_x^2 and σ_y^2, and the covariance, σ_{xy}, are collected in a single *square symmetric* matrix called the *variance-covariance matrix,* or simply the *covariance matrix:*

$$\boldsymbol{\Sigma} = \begin{bmatrix} \sigma_x^2 & \sigma_{xy} \\ \sigma_{xy} & \sigma_y^2 \end{bmatrix} \tag{50.135}$$

where the variances are along the main diagonal and the covariance off the diagonal. The concept of the covariance matrix can be extended to the multidimensional case by considering n random variables $\tilde{x}_1, \tilde{x}_2, \ldots, \tilde{x}_n$ and writing

$$\boldsymbol{\Sigma}_{xx} = \begin{bmatrix} \sigma_1^2 & \sigma_{12} & \cdots & \sigma_{1n} \\ \sigma_{12} & \sigma_2^2 & & \sigma_{2n} \\ \vdots & & \ddots & \vdots \\ \sigma_{1n} & \sigma_{2n} & \cdots & \sigma_n^2 \end{bmatrix} \tag{50.136}$$

which is an $n \times n$ square symmetric matrix.

Often in practice, the variances and covariances are not known in absolute terms but only to a scale factor. The scale factor is given the symbol σ_0^2 and is termed the *reference variance,* although other names, such as "variance factor" and "variance associated with weight unity," have also been used. The square root σ_0, of σ_0^2 is called the "reference standard deviation" and was classically known as the "standard error of unit weight." The relative variances and covariances are called *cofactors* and are given by

$$\mathbf{q}_{ii} = \frac{\sigma_i^2}{\sigma_0^2} \quad \text{and} \quad \mathbf{q}_{ij} = \frac{\sigma_{ij}}{\sigma_0^2} \tag{50.137}$$

Collecting the cofactors in a square symmetric matrix produces the *cofactor matrix* **Q**, with the obvious relationship with the covariance matrix.

$$\mathbf{Q} = \frac{1}{\sigma_0^2}\boldsymbol{\Sigma} \tag{50.138}$$

When **Q** is nonsingular, its inverse is called the *weight matrix* and designated by **W**; thus,

$$\mathbf{W} = \mathbf{Q}^{-1} = \sigma_0^2\boldsymbol{\Sigma}^{-1} \tag{50.139}$$

If σ_0^2 is equal to 1, or, in other words, if the covariance matrix is known, the weight matrix becomes its inverse.

Design/Preanalysis

Engineering *design* is more frequently known as *preanalysis* in surveying engineering. It refers to the task of determining the observations to be made and their required accuracy so that the required accuracy of the final product is met. This is usually done by an iterative procedure, often using interactive graphics, and applying the established mathematical model of the problem. The physical limitations of the project, such as visibility and accessibility problems, are first imposed on the design. Next, what is considered to be an adequate set of measurements, with suitable accuracy estimates, is input in a design program to estimate the required unknown parameters and their expected accuracy. The overall accuracy of the estimated parameters is reflected by the posterior covariance matrix $\sum$. (This is why preanalysis is equivalent to *covariance analysis* used in the mathematical literature.) From $\sum$, individual confidence measures, such as **error ellipses** and **ellipsoids,** are computed and compared to the design requirements. If they are too large, additional observations or measurements of increased quality are attempted and the process repeated. On the other hand, if they are too small (i.e., too good), reduced observations or observations of decreased quality (using less expensive equipment/techniques) are attempted instead. The procedure is iterated until an optimum design results.

Data Acquisition

The results of preanalysis provide the information required to set up the specifications for data acquisition, particularly the quantities to be measured and their required accuracy. This leads to deciding on equipment to be used and observational techniques to be applied. It is important that the selected instruments be properly calibrated and in good working order, and that the procedures specified be rigorously followed. Good field practices must be followed to minimize blunders. In fact, all field activities are carefully planned and monitored so that the following operation, preprocessing of data, can be effectively carried out to yield the most suitable data entering the adjustment.

Data Preprocessing

Preprocessing of survey data involves the elimination of *blunders* and the correction for all known systematic errors. The resulting preprocessed measurements should have essentially nothing but random errors before they are used in the adjustment. Any uncorrected systematic errors known to still exist in the measurements must then be modeled mathematically and accounted for during the adjustment.

Because of the significant cost of the field operations, it is becoming more of a requirement to perform preprocessing in the field so as to minimize the cost of any remeasurement. Progressively more sophisticated preprocessing programs are available for use with portable computers in conjunction with electronic data collectors.

There are several techniques for checking the observations which apply to the different types of surveys, such as triangulation, trilateration, traverse, leveling, etc. These techniques depend on the shape of the network and various minimum combinations of observations.

Data Adjustment

When redundant measurements exist, an *adjustment* of the observed data becomes necessary in order to resolve the inconsistency between the observations and the model. As an illustration, consider the size and shape of a plane triangle in which the three interior angles A, B, C and two sides a, b (opposite to A and B, respectively) are measured. Suppose that we are interested in the length of the side c. We obviously have more information than needed, since any one side, a or b, and any two angles suffice to solve the triangle and hence determine c. However, due to random measurement errors, every combination selected would be expected to lead to a slightly different value for c. A choice from all possible combinations would be essentially arbitrary. More important is the fact that one should take advantage of using *all* the available information in one operation, in which each measurement contributes relative to its role in the mathematical model and commensurate with its quality (variance) relative to the quality of all other measurements. One of the most commonly used techniques of adjustment of redundant survey data is the method of **least squares**.

Least Squares Adjustment

Although least squares is an estimation procedure, it has been traditionally referred to in surveying as an *adjustment* technique. This stems from the fact that after the adjustment the original observations are replaced by a set of *adjusted* observations that *are* consistent with the model. Thus, after least squares adjustment the five observations in the triangle example are replaced by a consistent set, $\hat{A}$, $\hat{B}$, $\hat{C}$, $\hat{a}$, $\hat{b}$, in that *any* minimum combination of these would yield the *same* triangle solution. Each adjusted observation is the sum of the original measurement and a *residual*, ν, which is calculated in the adjustment.

The method of least squares is based on the observational residuals. If all the residuals in a given set of n observations, **l** are denoted by the vector $\boldsymbol{\nu}$, and the weight matrix associated with these observations is **W**, the least squares criterion is given by

$$\phi = \boldsymbol{\nu}^{\mathrm{T}}\mathbf{W}\boldsymbol{\nu} \rightarrow \text{minimum} \tag{50.140}$$

Note that ϕ is a scalar, for which a minimum is obtained by equating to zero its partial derivatives with respect to $\boldsymbol{\nu}$. In Eq. (50.140) the weight matrix of the observations, **W**, may be full, implying that the observations are correlated. If the observations are uncorrelated, **W** will be a diagonal matrix, and the criterion simplifies to

$$\phi = \sum_{i=1}^{n} w_i \nu_i^2 = w_1 \nu_1^2 + w_2 \nu_2^2 + \cdots w_n \nu_n^2 \rightarrow \text{minimum} \tag{50.141}$$

which says that the sum of the weighted squares of the residuals is a minimum. Another and simpler case involves observations which are uncorrelated and of equal weight (precision), for which $\mathbf{W} = \mathbf{I}$, and ϕ becomes

$$\phi = \sum_{i=1}^{n} \nu_1^2 = \nu_i^2 + \nu_2^2 + \cdots + \nu_n^2 \rightarrow \text{minimum} \tag{50.142}$$

The case covered by Eq. (50.142) is the oldest and may have accounted for the name "least squares," since it seeks the "least" sum of the squares of the residuals.

If we denote by n_0 the minimum number of independent variables needed to determine the selected model uniquely, then the *redundancy*, r, is given by

$$r = n - n_0 \tag{50.143}$$

As illustrations, consider the following examples.

1. The shape of a plane triangle is uniquely determined by a minimum of two interior angles, or $n_0 = 2$. If three interior angles are measured, then, with $n = 3$, redundancy is $r = 1$.
2. The size and shape of a plane triangle require a minimum of three observations, at least one of which is the length of one side, or $n_0 = 3$. If three interior angles and two side lengths are available, then with $n = 5$ the redundancy is $r = 2$.

After the redundancy r is determined, the adjustment proceeds by writing equations that relate the model variables in order to reflect the existing redundancy. Such equations will be referred to either as *condition equations* or simply as *conditions*. The number of independent conditions, c, will be equal to r if only observational variables and constants are involved. In many situations, however, additional unknown variables, called *parameters*, are carried in the adjustment. In such a case, if the number of unknown parameters is u, then a total of

$$c = r + u \tag{50.144}$$

independent condition equations in terms of both the n observations and u parameters must be written. In order for the parameters to be functionally independent, their number, u, should not exceed the minimum number of variables, n_0, necessary to specify the model. Hence, the following relation must be satisfied:

$$0 \leq u \leq n_0 \tag{50.145}$$

Similarly, for the formulated condition equations to be independent, their number, c, should not be larger than the total number of observations n. Hence,

$$r \leq c \leq n \tag{50.146}$$

Techniques of Least Squares

Although there is only one least squares criterion, there are several techniques by which least squares may be applied. Regardless of which technique is applied, the results of an adjustment of a given set of measurements associated with a specified model *must* be the same. The choice of a technique, therefore, is mostly a matter of convenience and computational economy. The first technique is called *adjustment of observations only*. The condition equations take the form

$$\underset{r,n}{\mathbf{A}}\ \underset{n,1}{\mathbf{v}} = \underset{r,1}{\mathbf{f}} \tag{50.147}$$

For linear adjustment problems the vector $\mathbf{f}$ is given by

$$\mathbf{f} = \mathbf{d} - \mathbf{Al} \tag{50.148}$$

in which $\mathbf{d}$ is a vector of numerical constants and $\mathbf{l}$ is the vector of given numerical values of the measurements.

Let the cofactor matrix of the observations be denoted by $\mathbf{Q}(= \mathbf{W}^{-1})$. Then

$$\mathbf{k} = (\mathbf{AQA}^{\mathrm{T}})^{-1}\mathbf{f} = \mathbf{Q}_e^{-1}\mathbf{f} = \mathbf{W}_e\mathbf{f} \tag{50.149}$$

$$\mathbf{v} = \mathbf{QA}^{\mathrm{T}}\mathbf{k} \tag{50.150}$$

$$\hat{\mathbf{l}} = \mathbf{l} + \mathbf{v} \tag{50.151}$$

Error propagation:

$$\mathbf{Q}_{vv} = \mathbf{QA}^{\mathrm{T}}\mathbf{W}_e\mathbf{AQ} \tag{50.152}$$

$$\mathbf{Q}_{\hat{l}\hat{l}} = \mathbf{Q} - \mathbf{Q}_{vv} \tag{50.153}$$

$$\hat{\sigma}_0^2 = \frac{\mathbf{v}^{\mathrm{T}}\mathbf{W}\mathbf{v}}{r} \tag{50.154}$$

The second technique is called *adjustment of indirect observations.* The condition equations are of the general (nonlinear) form:

$$\mathbf{l} + \mathbf{v} + \mathbf{F}(\mathbf{x}) = \mathbf{0} \tag{50.155}$$

When linearized at approximations x^0 for the $u = n_0$ parameters x, they become

$$\mathbf{v} + \mathbf{B\Delta} = \mathbf{f} \tag{50.156}$$

where $\mathbf{\Delta}$ is a vector of unknown parameter corrections, $\mathbf{B} = \frac{\partial \mathbf{F}}{\partial \mathbf{x}}\big|_{x^0}$ is an n by u coefficient matrix, and $\mathbf{f} = -\mathbf{F}(\mathbf{x^0}) - \mathbf{l}$. With $\mathbf{W} = \mathbf{Q}^{-1}(= \mathbf{\Sigma}^{-1}$ if $\sigma_0 = 1)$ as the weight matrix of the observations, then

$$\mathbf{N} = \mathbf{B}^{\mathrm{T}}\mathbf{WB} \tag{50.157}$$

$$\mathbf{t} = \mathbf{B}^{\mathrm{T}}\mathbf{W}\mathbf{f} \tag{50.158}$$

$$\mathbf{\Delta} = \mathbf{N}^{-1}\mathbf{t} \tag{50.159}$$

$$\hat{\mathbf{x}} = \mathbf{x}^0 + \Sigma\mathbf{\Delta} \qquad \text{(iterate)} \tag{50.160}$$

Error propagation:

$$\mathbf{Q}_{\hat{x}\hat{x}} = \mathbf{Q}_{\Delta\Delta} = \mathbf{N}^{-1} \tag{50.161}$$

$$\mathbf{Q}_{vv} = \mathbf{B}\mathbf{N}^{-1}\mathbf{B}^{\mathrm{T}} \tag{50.162}$$

$$\mathbf{Q}_{\hat{l}\hat{l}} = \mathbf{Q} - \mathbf{Q}_{vv} \tag{50.163}$$

$$\hat{\sigma}_0^2 = \frac{\mathbf{v}^{\mathrm{T}}\mathbf{W}\mathbf{v}}{r} \tag{50.164}$$

In the preceding technique, the largest number allowed for the parameters, $u = n_0$, must be carried in the adjustment so that each condition equation contains one and only one observation. In many applications, it is more economical to carry in the adjustment only the parameters of interest, which are fewer in number: $u < n_0$. This technique is called *combined adjustment of observations and parameters.* The general (nonlinear) condition equations are of the form

$$\mathbf{F}(\mathbf{l}, \mathbf{x}) = \mathbf{0} \tag{50.165}$$

which in linearized form becomes

$$\mathbf{A}\mathbf{v} + \mathbf{B}\mathbf{\Delta} = \mathbf{f} \tag{50.166}$$

where

$$\mathbf{A} = \left.\frac{\partial \mathbf{F}}{\partial \mathbf{l}}\right|_{l^0, x^0} \qquad \mathbf{B} = \left.\frac{\partial \mathbf{F}}{\partial \mathbf{x}}\right|_{l^0, x^0}$$

and

$$\mathbf{f} = -\left[\mathbf{F}(\mathbf{l}^0, \mathbf{x}^0) + \mathbf{A}(\mathbf{l} - \mathbf{l}^0)\right] \tag{50.167}$$

Then

$$\mathbf{Q}_e = \mathbf{A}\mathbf{Q}\mathbf{A}^{\mathrm{T}} \tag{50.168}$$

$$\mathbf{W}_e = \mathbf{Q}_e^{-1} = (\mathbf{A}\mathbf{Q}\mathbf{A}^{\mathrm{T}})^{-1} \tag{50.169}$$

$$\mathbf{N} = \mathbf{B}^{\mathrm{T}}\mathbf{W}_e\mathbf{B} \tag{50.170}$$

$$\mathbf{t} = \mathbf{B}^{\mathrm{T}}\mathbf{W}_e\mathbf{f} \tag{50.171}$$

$$\mathbf{\Delta} = \mathbf{N}^{-1}\mathbf{t} \tag{50.172}$$

$$\hat{\mathbf{x}} = \mathbf{x}^0 + \Sigma\mathbf{\Delta} \qquad \text{(iterate)} \tag{50.173}$$

$$\mathbf{v} = \mathbf{Q}\mathbf{A}^{\mathrm{T}}\mathbf{W}_e(\mathbf{f} - \mathbf{B}\mathbf{\Delta}) \tag{50.174}$$

$$\hat{\sigma}_0^2 = \frac{\mathbf{v}^{\mathrm{T}}\mathbf{W}\mathbf{v}}{r} \tag{50.175}$$

Error propagation:

$$\mathbf{Q}_{\hat{x}\hat{x}} = \mathbf{Q}_{\Delta\Delta} = \mathbf{N}^{-1} \tag{50.176}$$

$$\mathbf{Q}_{w} = \mathbf{Q}\mathbf{A}^{\mathrm{T}}\left(\mathbf{W}_e - \mathbf{W}_e\mathbf{B}\mathbf{Q}_{\Delta\Delta}\mathbf{B}^{\mathrm{T}}\mathbf{W}_e\right)\mathbf{A}\mathbf{Q} \tag{50.177}$$

$$\mathbf{Q}_{\hat{l}\hat{l}} = \mathbf{Q} - \mathbf{Q}_{vv} \tag{50.178}$$

The linear set $\mathbf{N}\mathbf{\Delta} = \mathbf{t}$ is called the *normal equations.*

In the three techniques above, the parameters **x** are functionally independent. In many surveying engineering applications, functions may exist between the parameters in the adjustment. Examples include points on geometric forms (lines, planes, surfaces), known distances, angles, etc., to be fixed in the adjustment. These functions are called *constraint equations* or simply *constraints,* to distinguish them from conditions. They are characterized by not containing any observations. The technique when constraints exist is called *adjustment with functional constraints;* of course, the condition equations to be combined with the constraints can take any of the three situations discussed. It is more general to consider the combined adjustment technique, thus:

$$\mathbf{F}(\mathbf{l}, \mathbf{x}) = \mathbf{0} \tag{50.179}$$

$$\mathbf{G}(\mathbf{x}) = \mathbf{0} \tag{50.180}$$

are the general (nonlinear) conditions and constraints. The linearized form is

$$\mathbf{A}\mathbf{v} + \mathbf{B}\boldsymbol{\Delta} = \mathbf{f} \tag{50.181}$$

$$\mathbf{C}\boldsymbol{\Delta} = \mathbf{g} \tag{50.182}$$

where $\mathbf{C} = \frac{\partial \mathbf{G}}{\partial \mathbf{x}}\big|_{\mathbf{x}^0}$ and $\mathbf{g} = -\mathbf{G}(\mathbf{x}^0)$

$$\mathbf{N} = \mathbf{B}^{\mathrm{T}}\mathbf{W}_e\mathbf{B} \qquad \mathbf{t} = \mathbf{B}^{\mathrm{T}}\mathbf{W}_e\mathbf{f} \tag{50.183}$$

$$\mathbf{M} = \mathbf{C}\mathbf{N}^{-1}\mathbf{C}^{\mathrm{T}} \tag{50.184}$$

$$\mathbf{k}_c = \mathbf{M}^{-1}(\mathbf{g} - \mathbf{C}\mathbf{N}^{-1}\mathbf{t}) \tag{50.185}$$

$$\boldsymbol{\Delta} = \mathbf{N}^{-1}(\mathbf{t} + \mathbf{C}^{\mathrm{T}}\mathbf{k}_c) \tag{50.186}$$

$$\mathbf{x} = \mathbf{x}^0 + \boldsymbol{\Sigma\Delta} \qquad \text{(iterate)} \tag{50.187}$$

$$\mathbf{Q}_{\Delta\Delta} = \mathbf{N}^{-1}(\mathbf{I} - \mathbf{C}^{\mathrm{T}}\mathbf{M}^{-1}\mathbf{C}\mathbf{N}^{-1}) \tag{50.188}$$

The quantities v, $\hat{\sigma}_0^2$, $\mathbf{Q}_{vv}$ and $\mathbf{Q}_{\hat{l}\hat{l}}$ are as given by Eqs. (50.174), (50.175), (50.177), and (50.178), respectively.

Another technique is to perform the adjustment sequentially. For a fixed number of parameters, both the inverse of the normal equations coefficient matrix **N** and the constant terms vector **t** are sequentially updated due to either addition or deletion of a set of condition equations. The relations for *sequential adjustment* are as follows:

$$\mathbf{A}_i\mathbf{v}_i + \mathbf{B}_i\boldsymbol{\Delta} = \mathbf{f}_i \qquad \text{with } \mathbf{Q}_i \tag{50.189}$$

are the conditions to be added or subtracted. Let $\mathbf{N}_{I-1}$, $\mathbf{t}_{I-1}$ and $\mathbf{N}_I$, $\mathbf{t}_I$ designate the matrices *before* and *after*, respectively, incorporating the effects of Eq. (50.189). Then

$$\mathbf{N}_I^{-1} = \mathbf{N}_{I-1}^{-1}\left[\mathbf{I} \mp \mathbf{B}_i^{\mathrm{T}}(\mathbf{Q}_{e_i} \pm \mathbf{B}_i\mathbf{N}_{I-1}^{-1}\mathbf{B}_i^{\mathrm{T}})^{-1}\mathbf{B}_i\mathbf{N}_{I-1}^{-1}\right] \tag{50.190}$$

$$\mathbf{t}_I = \mathbf{t}_{I-1} \pm \mathbf{B}_i^{\mathrm{T}}\mathbf{W}_{e_i}\mathbf{f}_i \tag{50.191}$$

in which $\mathbf{Q}_{e_i} = \mathbf{A}_i\mathbf{Q}_i\mathbf{A}_i^{\mathrm{T}}$ and $\mathbf{W}_{e_i} = \mathbf{Q}_{e_i}^{-1}$. In Eqs. (50.190) and (50.191) the *upper signs* refer to condition *addition* and the *lower signs* to condition *deletion.*

Finally, a very flexible technique is used in which *all the variables in the mathematical model are considered as observables.* This requires a priori estimates for all model variables, but more importantly, estimates of their *a priori weights.* The a priori weights provide the mechanism for effectively distinguishing between different groups of variables in the model. Very large weights leave a variable essentially as a constant in the adjustment, while very small or even zero weight leaves it as an unknown parameter that can freely adjust. This is referred to as the *unified least squares* technique and applies to *all* of the techniques presented above. As an example, we consider the combined adjustment of observations and parameters with the linear (or linearized) conditions: $\mathbf{Av} + \mathbf{B\Delta} = \mathbf{f}$. Let $\mathbf{x}$ be the prior estimates of the parameters and $\mathbf{W}_{xx}$ its corresponding prior weight matrix. The solution given by Eqs. (50.168) to (50.178) essentially apply, except that

$$\mathbf{\Delta} = (\mathbf{N} + \mathbf{W}_{xx})^{-1}(\mathbf{t} - \mathbf{W}_{xx}\mathbf{f}_x) \tag{50.192}$$

$$\mathbf{f}_x = \mathbf{x}^0 - \mathbf{x} \tag{50.193}$$

$$\mathbf{Q}_{\Delta\Delta} = (\mathbf{N} + \mathbf{W}_{xx})^{-1} \tag{50.194}$$

Assessment of Adjustment Results

After least squares adjustment it is quite important in surveying to analyze the results and provide a statement regarding the quality of the estimates. This operation is often referred to as *postadjustment analysis*, which applies various statistical techniques.

Test on Reference Variance

The first test is on the estimated reference variance, $\hat{\sigma}_0^2$. Let the a priori reference variance be σ_0^2; let r be the degrees of freedom (redundancy) in the adjustment, and assume that the residuals v_i are normally distributed. The statistic $r\hat{\sigma}_0^2/\sigma_0$ has a χ^2 distribution with r degrees of freedom. The two-tailed $100(1 - \alpha)$ confidence region for σ_0^2 is given by

$$(r\hat{\sigma}_0^2/\chi_{r,\alpha/2}^2) < \sigma_0^2 < (r\hat{\sigma}_0^2/\chi_{r,1-\alpha/2}^2) \tag{50.195}$$

If σ_0^2 is incorrect or the mathematical model used is improper or incomplete (does not adequately account for systematic errors), then σ_0^2 will fall outside this interval.

Test for Blunders or Outliers

If v_i is the ith residual and σ_{v_i} is its standard deviation, then

$$\bar{v}_i = v_i/\sigma_{v_i} \tag{50.196}$$

is called the *standardized residual*. Frequently, the effort involved in computing $\mathbf{\Sigma}_{vv}$ is quite extensive, and therefore an approximate estimate of σ_{v_i} may be obtained from

$$\hat{\sigma}_{v_i} = [(n - u)/n]^{1/2}\hat{\sigma}_0\sigma_{\ell_i}/\sigma_0 \tag{50.197}$$

in which n is the number of observations, u the number of parameters (thus $n - u = r$, the redundancy), and σ_{ℓ_i} is the a priori standard deviation of observation ℓ_i. When σ_0^2 is known, $\bar{v}_i$ has a probability density function, or pdf, $N(0, \sigma_{v_i^2})$ and

$$\bar{v}_i = \left|\frac{v_i}{\sigma_{v_i}}\right| < N_{1-\alpha/2} \tag{50.198}$$

If σ_0^2 is not known, then

$$\bar{v}_i = |v_i/\hat{\sigma}_{v_i}| < \tau_{r,1-\alpha/2} \tag{50.199}$$

in which $\hat{\sigma}_{v_i}$ is computed from Eq. (50.197), and τ_r has a Tau pdf with r degrees of freedom. If r is large, as in surveying, photogrammetric, or geodetic nets with extensive observations, τ_r may be replaced by Student t_r pdf or even normal pdf.

Confidence Region for Estimated Parameters

The covariance matrix for the parameters as evaluated from the least squares is given by (see, for example, Eq. (50.161))

$$\mathbf{\Sigma}_{\hat{x}\hat{x}} = \sigma_0^2 N^{-1} \tag{50.200}$$

It can be shown that a region of constant probability is bounded by a u-dimensional hyperellipsoid centered at $\hat{\mathbf{x}}$, if the parameters are assumed to have a multivariate normal pdf. The function

$$k^2 = (\mathbf{x} - \hat{\mathbf{x}})^T \Sigma_{\hat{x}\hat{x}}^{-1} (\mathbf{x} - \hat{\mathbf{x}}) \tag{50.201}$$

describes the hyperellipsoid. The quadratic k^2 has a χ_u^2 distribution, with the probability for a point estimate being

$$P(\chi_u^2 < k^2) = 1 - \alpha$$

For the two-dimensional case (error ellipses), typical values are

P	0.394	0.500	0.900	0.950	0.990
k	1.000	1.177	2.146	2.447	3.035

and for the three-dimensional case, they are

P	0.199	0.500	0.900	0.950	0.990
k	1.000	1.538	2.500	2.700	3.368

When $k = 1$, we usually call it the *standard* region, *standard error ellipse* (for 2-D), or *standard error ellipsoid* (for 3-D). The standard regions for several dimensions are

Dimension	1	2	3	4	5	6
P	0.683	0.394	0.199	0.090	0.037	0.014

Given Σ, for example for a point in a plane, the semimajor axis, a, and semiminor axis, b, of the standard error ellipse are computed from the eigenvalues and eigenvectors (see the discussion of "Basic Matrix Operations," in section 50.5). If one is interested in the 90% confidence region (i.e., significance level of $\alpha = 0.10$), the a, b are multiplied by 2.146. For the standard regions, there is a 0.683 probability that an adjusted point falls in a one-dimensional interval, a 0.394 probability that it falls inside the standard error ellipse, and only a 0.199 probability that it falls within the standard error ellipsoid.

Applications in Surveying Engineering

Level Net

Let l_{ij} represent the observed difference in elevation between two points whose (unknown) elevations are x_i and x_j. If l_{ij} is from point i to point j, then the condition equation is given by

$$x_i + l_{ij} + v_{ij} - x_j = 0$$

or

$$v_{ij} + x_i - x_j = -l_{ij} \tag{50.202}$$

This condition equation is in the form of adjustment of indirect observations. At least one benchmark (a point of known elevation) is needed for any given level net.

Traverse

There are two conditions, one for a measured angle α_i, and one for a measured distance d_{ij}. If α_i is at station i from the line $i - 1$ to i clockwise to the line from i to $i + 1$, then

$$\alpha_i = A_{i,i+1} - A_{i,i-1}$$

or

$$v_i + \tan^{-1}\left(\frac{x_{i-1} - x_i}{y_{i-1} - y_i}\right) - \tan^{-1}\left(\frac{x_{i+1} - x_i}{y_{i+1} - y_i}\right) = -\alpha_i \tag{50.203}$$

where **A** represents the azimuth and $(x, y)_{i-1,i,i+1}$ are the coordinates of the three points involved. The distance condition is given by

$$v_{ij} - [(x_i - x_j)^2 + (y_i - y_j)^2]^{1/2} = -d_{ij} \tag{50.204}$$

Again, this is in the form of adjustment of indirect observations. If any of the points involved is a fixed point (i.e., with known coordinates), its coordinates are not carried as unknown parameters in the adjustment.

Trilateration

This is the operation in which *only* distances are measured in the network. Therefore, the only condition used is that given by Eq. (50.204).

Triangulation

This is the operation in which chains of triangles are connected together. The fundamental measurement is the angle, and the adjustment unit is the triangle. The single condition for one triangle is

$$\sum l_i - \pi = 0 \tag{50.205}$$

where l_i represents all the measured angles inside a single triangle. For a quadrilateral, there is another condition called the *side condition*, the composition of which can be found in the literature.

Defining Terms

Adjustment of observations: The mathematical technique used to resolve the inconsistency between the measurements collected and the underlying mathematical model when *redundancy* exists, or when the measurements exceed the minimum necessary to uniquely define the model.

Azimuthal projection: A map projection that yields a map where correct *direction* or *azimuth* of any point relative to one central point is shown.

Conformal or orthomorphic projection: A map projection technique that preserves *angles* between short intersecting lines, thus making small areas appear in their correct *shape* on the map. Scale varies from point to point, and thus larger areas are incorrect.

Equal-area projection: A map projection technique that results in a map showing all areas in proper relative *size*; however, they may be distorted in *shape.*

Equidistant projection: A map projection technique in which distances are correctly represented from one central point to other points on the map.

Error ellipses and ellipsoids: Confidence regions about estimated survey points in two dimension (ellipses) or three dimensions (ellipsoids) for specified probabilities.

Error propagation: The general technique of determining the covariance matrix of a set of quantities, which are estimated from functions of another set of quantities of known values and covariance matrix.

Least squares adjustment: The most common adjustment technique used in surveying engineering; it is based on minimizing the sum of the weighted squares of the observational residuals.

Map projection: The theory and techniques of proper representation of the curved earth surface on the plane of a map.

References

Anderson, J. M. and Mikhail, E. M. 1985. *Introduction to Surveying.* McGraw-Hill, New York.

Davis, R. E., Foote, F. S., Anderson, J. M., and Mikhail, E.M. 1981. *Surveying: Theory and Practice,* 6th ed. McGraw-Hill, New York.

Krakiwsky, E. J., ed. 1983. *Papers for the CIS Adjustment and Analysis Seminars.* The Canadian Institute of Surveying, Ottawa.

Mikhail, E. M. 1976. *Observations and Least Squares.* University Press of America, Lanham, MD.

Mikhail, E. M. and Gracie, G. 1981. *Analysis and Adjustment of Survey Measurements.* Van Nostrand Reinhold, New York.

Vanicek, P. and Krakiwsky, E. J. 1982. *Geodesy: The Concepts.* North-Holland, Amsterdam.

Further Information

Articles on advances in the general field, and particularly in observational data adjustment, can be found in the following periodicals:

Bulletin Géodésique, published by Springer International
Manuscripta Geodetica, published by Springer International
Photogrammetric Engineering and Remote Sensing, published by the American Society for Photogrammetry and Remote Sensing, Bethesda, MD
Surveying and Land Information Systems, published by the American Congress on Surveying and Mapping, Bethesda, MD
The Photogrammetric Record, published by The Photogrammetric Society, London, England
Photogrammetria, Journal of the International Society for Photogrammetry and Remote Sensing, published by Elsevier, Amsterdam, The Netherlands
Geomatica, Journal of the Canadian Institute of Geomatics, Ottawa, Canada

51

Plane Surveying

Steven D. Johnson
Purdue University

Wesley G. Crawford
Purdue University

51.1 Introduction

The roots of surveying are contained in plane-surveying techniques that have developed since the very first line was measured. Though the fundamental processes haven't changed, the technology used to make the measurements has improved tremendously. The basic methods of distance measurement, angle measurement, determining elevation, etc., are becoming easier, faster, and more accurate.

Even though electronic measurement is becoming commonplace, distances are still being taped. While many transits and optical theodolites are still being used for layout purposes, electronic instruments are rapidly becoming the instruments of choice by the engineer for measurement of angles. Although establishing elevations on the building site is still being performed using levels, the laser is everywhere. It is the responsibility of the engineer in the field to choose the measuring method and the technology that most efficiently meet the accuracy needed.

The requirements of construction layout are more extensive than ever, with the complex designs of today's projects. The engineer must be exact in measurement procedures and must check and double-check to ensure that mistakes are eliminated and errors are reduced to acceptable limits to meet the tolerances required.

This chapter is directed to the engineer who will be working in the field on a construction project. Often that individual is called the *field engineer,* and that term will be used throughout this chapter.

0-8493-8953-4/95/$0.00 + $.50

The field engineer should be aware of fundamental measuring techniques as well as advances in the technology of measurement.

Although measurement concepts and calculation procedures are introduced and discussed in this chapter, only the basics are presented. It is the responsibility of the field engineer to review other sources for more detailed information to become more competent in these procedures. See the further information and references at the end of this chapter.

51.2 Distance Measurement

Distance may be measured by indirect and direct measurement procedures. Indirect methods include odometers, optical rangefinders, and tacheometry. Direct methods include pacing, taping, and electronic distance measurement.

When approximate distances are appropriate, pacing can be used over short distances and odometers or optical rangefinders can be used over longer or inaccessible distances. These methods can yield accuracies in the range of 1 part in 50 to 1 part in 100 over modest distances. The accuracy of pacing and optical distance measuring methods decreases rapidly as the distance increases. However, these approximate methods can be useful when checking for gross blunders, narrowing search areas in the field, and making preliminary estimates for quantities or future surveying work. Applications where these lower-order accuracy methods can be used to obtain satisfactory distance observations occur regularly in surveying, construction and engineering, forestry, agriculture, and geology.

Tacheometry

Tacheometry uses the relationship between the angle subtended by a short base distance perpendicular to the bisector of the line and the length of the bisecting line. Stadia and subtense bar are two tacheometric methods capable of accuracies in the range of 1 part in 500 to 1 part in 1000.

In the stadia method the angle is fixed by the spacing of the stadia cross hairs (i) on the telescope reticule and the focal length (f) of the telescope. Then the distance on the rod (d) is observed, and the distance is computed by

$$\frac{D}{d} = \frac{f}{i}$$

$$D = d\left(\frac{f}{i}\right)$$

When the telescope is horizontal, as in a level, a horizontal distance is obtained. When the telescope is inclined, as in a transit or theodolite, a slope distance is obtained. Stadia may be applied with level or transit, plane table and alidade, or self-reducing tacheometers.

The subtense bar is a tacheometric method in which the base distance, d, is fixed by the length of the bar and the angle, α, is measured precisely using a theodolite. The horizontal distance to the midpoint of the bar is computed by

$$\tan\frac{\alpha}{2} = \frac{d/2}{D}$$

$$D = \frac{d}{2\tan\left(\frac{\alpha}{2}\right)}$$

Since a horizontal angle is measured, this method yields a horizontal distance and no elevation information is obtained.

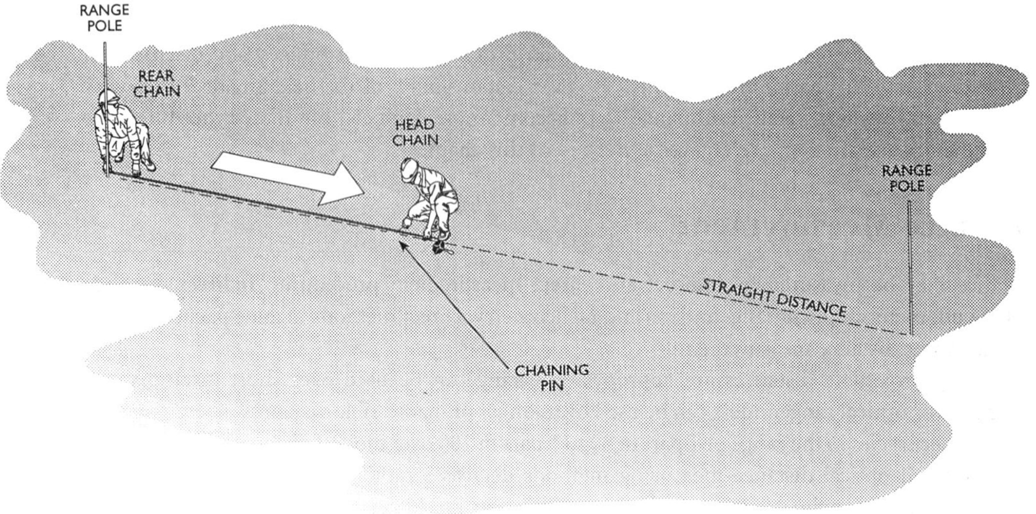

FIGURE 51.1 Taping on level terrain.

Taping

Taping is a direct method of measuring distance in which a tape of known length and graduated at intervals throughout its length is stretched along the line to be measured. Figure 51.1 illustrates taping on level terrain. Low-order accuracy measurements can be obtained using tapes made of cloth or fiberglass material. These tapes are available in a variety of lengths, and they may be graduated in feet, meters, or both. Accurate taped distances are obtained using calibrated metal ribbon tapes made of tempered steel. Tapes made of invar steel for temperature stability are also available. Table 51.1 summarizes the sources of errors and the procedures required to measure distances accurately.

Accuracy in taping requires a **calibrated** tape that has been compared to a length standard under known conditions. A tape calibration report will include the true length of the tape between marked intervals on the tape. The tape is calibrated when it is fully supported throughout its length at a specified pull or tension applied to the tape and at a specified temperature of the tape. The calibration report should also include the cross-sectional area of the tape, the weight of the tape per unit length, the modulus of elasticity for the tape material, and the coefficient of thermal expansion for the tape material. When a distance is observed in the field and the conditions of tape support, alignment, temperature, and tension are recorded, the **systematic errors** affecting the measurement can be corrected and a true distance obtained. The tape correction formulas are summarized in the following example.

Example 51.1. A typical 100-foot surveyor's steel tape is calibrated while supported throughout its length. The calibration report lists the following:

True length, 0-ft to 100-ft mark $L_0 = 99.98$ ft

Other length intervals may also be reported (and the tape characteristic constants):

Tension	$P_0 = 25$ lb
Temperature	$T_0 = 68°$F
Cross-sectional area	$A = 0.003$ in.^{2}
Weight	$w = 0.01$ lb/ft
Coefficient of thermal expansion	$\alpha = 0.00000645/°$F
Modulus of elasticity	$E = 29(10^6)$ lb/in.^{2}

Table 51.1 Taping Errors and Procedures

Error	Source	Type	Makes Tape	Importance	Procedure to Eliminate
Tape length	Instrumental	Systematic	Long or short	Direct impact to recorded measurement, always check.	Calibrate tape and apply adjustment.
Temperature	Natural	Systematic or random	Long or short	For a chain standardized at 68 degrees, 0.01 per 15 degrees per 100 foot chain.	Observe temperature of chain, calculate and apply adjustment.
Tension	Personal	Systematic or random	Long or short	Often not important.	Apply the proper tension. When in doubt, PULL HARD!
Tape not level	Personal or natural	Systematic	Short	Negligible on slopes less than 1%; must be calculated for greater than 1%.	Break chain by using a hand level and plumb bob to determine horizontal; or correct by formula for slope or elevation difference.
Alignment	Personal	Systematic	Short	Minor if less than 1 foot off of line in 100 foot; major if 2 or more feet off of line.	Stay on line; or determine amount off of line and calculate adjustment by formula.
Sag	Natural or personal	Systematic	Short	Large impact on recorded measurement.	Apply proper tension; or calculate adjustment by formula.
Plumbing	Personal	Random	Long or short	Direct impact to recorded measurement.	Avoid plumbing if possible; or plumb at one end of chain only.
Interpolation	Personal	Random	Long or short	Direct impact to recorded measurement.	Check and recheck any measurement that requires estimating a reading.
Improper marking	Personal	Random	Long or short	Direct impact to recorded measurement.	Mark all points so that they are distinct.

Table 51.2 Tape Corrections for Systematic Errors

Systematic Error	Correction Formula	0.01-ft Correction Caused by
Tape length	$C_l = \dfrac{L_{\text{tape}} - 100.00}{100.00}$	0.01 per 50-ft tape length
Tension	$C_p = \dfrac{(P - P_0)L}{AE}$	±9 lb
Temperature	$C_t = \alpha(T - T_0)L$	±15°F
Sag	$C_s = \dfrac{-w^2 L_s^3}{24P^2}$	−5 lb in 100-ft tape length
Alignment	$C_a = -\dfrac{h}{2L}$	±1.4 ft

Systematic errors in taping, the correction formula, and the change required to cause a 0.01-foot tape correction in a full 100-foot length of the example tape are summarized in Table 51.2. When a distance is measured directly by holding the tape horizontal, the effect of alignment error is present in both the alignment of the tape along the direction of the line to be measured and the vertical alignment necessary to keep the tape truly horizontal.

Normal taping procedures require that the tape be held horizontal for each interval measured. When taping on sloping terrain or raising the tape to clear obstacles, a plumb bob is required to transfer the tape mark to the ground. Steadying the hand-held tape and plumb bob will be the largest source of **random error** in normal taping. Accuracies of 1 part in 2000 to 1 part in 5000 can be expected.

Precise taping procedures eliminate the use of the plumb bob by measuring a slope distance between fixed marks that either are on the ground or are supported on tripods or taping stands. Taping the slope distance, S, will require that the horizontal distance, H, be computed by

$$H = S \cos \alpha = S \sin z$$

where α = **vertical** angle of the tape and z = **zenith** angle of the tape, or by

$$H = \sqrt{S^2 - h^2}$$

where h = difference in elevation between the ends of the tape. If all tape corrections are carefully applied, accuracies of 1 part in 10,000 to 1 part in 20,000 can be obtained.

Electronic Distance Measurement

Distance can be measured electronically if the velocity and travel time of electromagnetic energy propagated along a survey line are determined. Terrestrial electronic distance measurement instruments (EDMI) measure the travel time by comparing the phase of the outgoing measurement signal to the phase of the signal returning from the remote end of the line. The phase difference is thus a function of the double path travel of the measurement signal. The distance is given by

$$D = \frac{VT}{2}$$

where V = the velocity of electromagnetic energy in the atmosphere and T = the double path travel time determined using the phase difference.

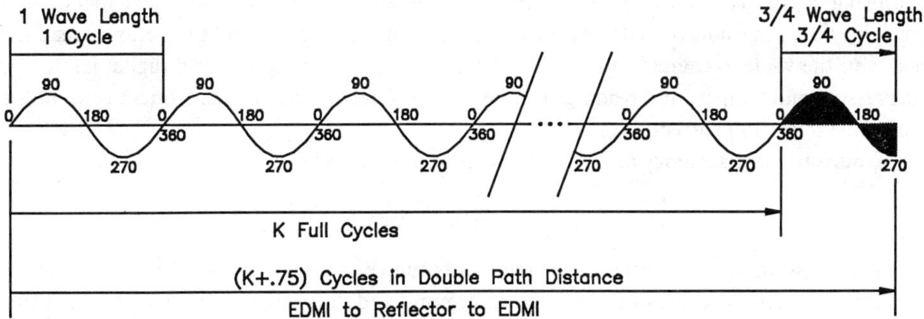

FIGURE 51.2 Phase shift principle of distance measurement.

If $\Delta\phi$ is the phase difference observed for the fine or shortest wavelength measurement signal, then the total travel time is found from the equation

$$T = k\left(\frac{1}{f}\right) + \frac{\Delta\phi}{2\pi}\left(\frac{1}{f}\right)$$

where f = the frequency of the measurement signal and k = the integer number of full cycles in the double path distance. The integer k is ambiguous for the fine measurement since the phase difference only determines the fractional part of the last cycle in the double path distance, as illustrated in Fig. 51.2. The value of k can be determined by measuring the phase difference of one or more coarse- or long-wavelength signals to resolve the distance to the nearest full cycle of the fine wavelength.

EDM instruments may be classified by type of energy used to carry the measurement signal or by the maximum measurement range of the system. Visible (white) light, infrared light, laser (red) light, and microwaves have been used as carrier energy.

The velocity of electromagnetic energy in the atmosphere is given by the expression

$$V = \frac{c}{n}$$

where c is the velocity in a vacuum, 299,792,158 m/s, and n is the atmospheric index of **refraction** for the conditions at the time of observation. The atmospheric index of refraction is a function of the wavelength of the electromagnetic energy propagated and the existing conditions of atmospheric temperature, pressure, and water vapor pressure. For EDMI using visible, laser, or infrared light carrier wavelengths, the effect of water vapor pressure is negligible, and it is often ignored. For EDMI using microwave carrier wavelengths, the effect of water vapor pressure is more significant.

Refraction causes a scale error in the observation. The correction is usually expressed in terms of parts per million, ppm, of the distance measured.

$$D_{\text{corrected}} = D + S\left(\frac{D}{10^6}\right)$$

The ppm correction for electro-optical instruments can be expressed in the form

$$S = A + \frac{Bp}{t + 273.2}$$

where p is the atmospheric pressure and t is the atmospheric temperature. The constants A and B are functions of the wavelength of the carrier and the precise frequency used for the highest-resolution measuring signal. The values of A and B can be obtained from the instrument manufacturer. Typically, the ppm value, S, is determined graphically using pressure and temperature read in the field, and the value is entered into the EDM instrument. On many modern digital instruments, the pressure and temperature readings can be entered directly, and the instrument computes and applies the refraction correction.

The measurement accuracy of an EDM instrument is expressed as

$$\sigma = \pm(c + s_{\text{ppm}})$$

In this expression, c is a constant that represents the contribution of uncertainty in the offset between the instruments measurement reference point and the geometric reference point centered over the survey station. The ppm term is a distance-dependent contribution representing the uncertainty caused by measurement frequency drift and atmospheric refraction modeling. An EDM instrument should be checked periodically to verify that it is operating within its specified error tolerance, σ. A quick check can be done by measuring a line of known length or at least a line that has been measured previously to see if the observed distance changes. When a more rigorous instrument calibration is warranted, use a calibrated base line and follow the procedures recommended in *NOAA Technical Memorandum NOS NGS-10* [Fronczek, 1980]. EDMI should be sent to the manufacturer for final calibration and adjustment.

51.3 Elevation Measurement

Elevation is measured with respect to a **datum** surface that is everywhere perpendicular to the direction of gravity. The datum surface most often chosen is called the *geoid.* The geoid is an equipotential surface that closely coincides with **mean sea level**. Elevations measured with respect to the geoid are called orthometric heights. The relationships between the mean sea level geoid, a level surface, and a horizontal line at a point are illustrated in Fig. 51.3.

Benchmark (BM)

A benchmark is best described as a permanent, solid point of known elevation. Benchmarks can be concrete **monuments** with a brass disk in the middle, iron stakes driven into the ground, or railroad spikes driven into a tree, etc.

Turning Point (TP)

A turning point is a point used in the differential leveling process to temporarily transfer the elevation from one setup to the next.

Types of Instruments

Elevation may be measured by several methods. Some of these methods measure elevation directly, whereas some measure the difference in elevation from a reference benchmark to the point to be

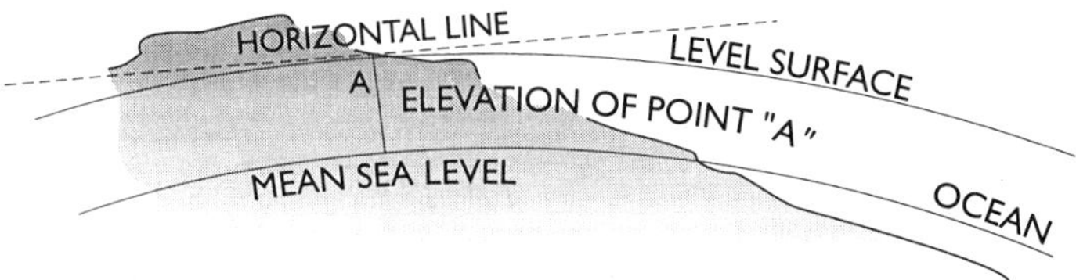

FIGURE 51.3 Elevation datum and level surface.

determined. Many types of instruments have been used for leveling purposes. The instruments have ranged from water troughs and hoses to instruments such as the barometer, dumpy level, automatic level, and laser. The field engineer should realize how each of these can be used to determine elevations for vertical control and construction.

Altimeters

Surveying altimeters are precise aneroid barometers that are graduated in feet or meters. As the altimeter is raised in elevation, the barometer senses the atmospheric pressure drop. The elevation is read directly on the face of the instrument. Although the surveying altimeter may be considered to measure elevation directly, best results are obtained if a difference in elevation is observed by subtracting readings between a base altimeter kept at a point of known elevation and a roving altimeter read at unknown points in the area to be surveyed. The difference in altimeter readings is a better estimate of the difference in elevation because local weather changes, temperature, and humidity that affect altimeter readings are canceled in the subtraction process. By limiting the distance between base and roving altimeters, accuracies of 3 to 5 feet are possible. Other survey configurations utilizing low and high base stations, and/or leap-frogging roving altimeters, can yield good results over large areas.

Level Bubble Instruments

Level bubble instruments include the builder's transit level, the transit, and the dumpy level. Each of these instruments contains a level vial with a bubble that must be centered to be used for leveling. Each instrument consists of three main components: a four-screw leveling head, a level vial attached to the telescope, and a telescope for magnification of the objective.

Instruments that use a level bubble to orient the axes to the direction of gravity depend on the bubble's sensitivity for accuracy. Level bubble sensitivity is defined as the central angle subtended by an arc of one division on the bubble tube. The smaller the angle subtended, the more sensitive the bubble is to dislevelment. A bubble division is typically 2 millimeters long, and bubble sensitivity typically ranges from 60 seconds to 1 second.

Builder's Level. The builder's level is one of the most inexpensive and versatile instruments that is used by field engineers. In addition to being able to perform leveling operations, it can be used to turn angles, and the scope can be tilted for sighting. Many residential builders use this instrument because it serves their purpose of laying out a building. See Fig. 51.4.

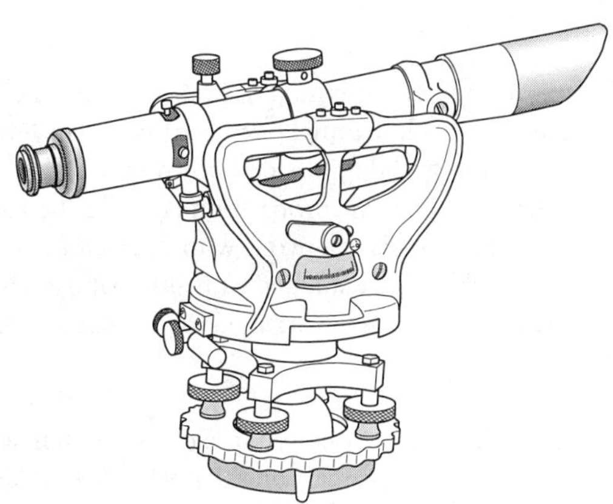

FIGURE 51.4 Builder level.

Transit. Although the primary functions of the transit are for angle measurement and layout, it can also be used for leveling because there is a bubble attached to the telescope. In fact, many construction companies who don't have both an angle measuring instrument and a dumpy or automatic level will do all of their leveling work with a transit. Some people prefer to use the transit for leveling because they are comfortable with its operation. However, the field engineer should be aware that the transit may not be as sensitive and stable as a quality level.

Dumpy Level. The engineer's dumpy level shown in Fig. 51.5 has been the workhorse of leveling instruments for more than 150 years. It has been used extensively for many of the great railroad, canal, bridge, tunnel, building, and harbor projects for the last century and a half. Even with advancements in other leveling instruments such as the automatic level and the laser, the dumpy is still the instrument of choice for a number of persons in construction because of its stability. On any type of project where there is going to be a great deal of vibration—such as pile driving, heavy-equipment usage, or high-rise construction—the dumpy may be the best choice for a leveling instrument.

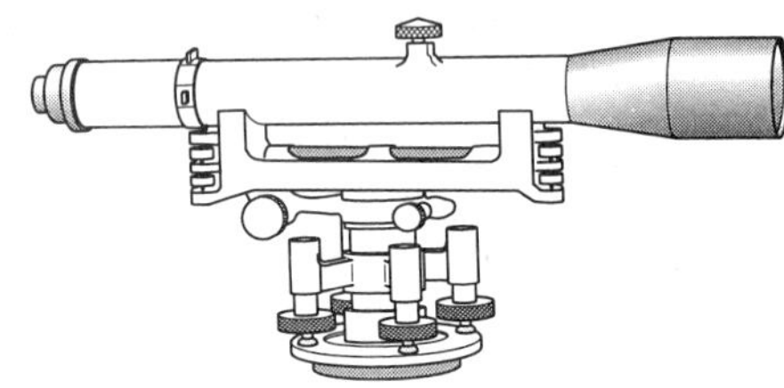

FIGURE 51.5 Engineer's dumpy level.

Automatic Compensator Instruments

Compensator instruments illustrated in Fig. 51.6 were developed about 50 years ago. Although each manufacturer may have developed a unique compensator, all compensators serve the same purpose—maintaining a fixed relationship between the line of sight and the direction of gravity. If the instrument is in adjustment, the line of sight will be maintained as a horizontal line. The operation of a compensator is illustrated in Fig. 51.7. Compensator instruments are extremely fast to set up and level. An experienced person can easily have an automatic level ready for a backsight in less than ten seconds, compared to a minute or more with a bubble-based leveling system. Compensators are available in several styles. Some are constructed by suspending a prism on wires. Some have a prism that is contained within a magnetic dampening system. Note that the instrument must be manually leveled to within the working range of the compensator by centering a bull's-eye bubble.

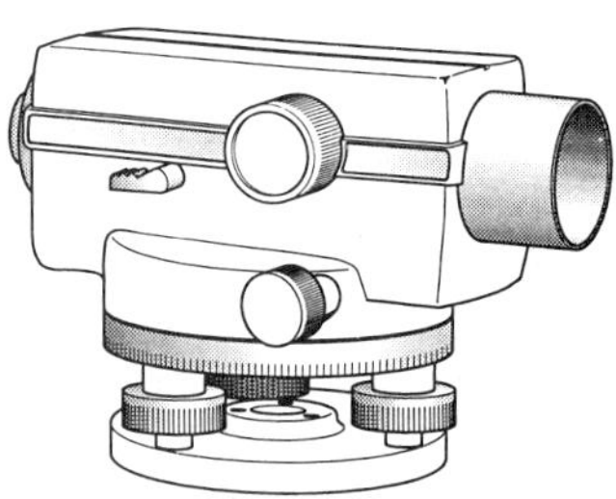

FIGURE 51.6 Automatic level.

Laser Levels

A laser level uses a laser beam directed at a spinning optical reflector. The reflector is oriented so that the rotating laser beam sweeps out a horizontal reference plane. The level rod is equipped with a sensor to detect the rotating beam. By sliding the detector on the rod, a vertical reading can be obtained. Laser levels are especially useful on construction sites. As the laser beam continuously sweeps out a constant elevation reference plane, anyone with a detector can get a rod reading to set or check grade elevations. Laser levels can be equipped with automatic leveling devices to maintain level orientation. The spinning optics can also be oriented to produce a vertical reference plane.

Digital Levels

Digital levels are electronic levels that can be used to more quickly obtain a rod reading and make the reading process more reliable. The length scale on the level rod is replaced by a bar code. The digital level senses the bar code pattern and compares it to a copy of the code held in its internal memory. By matching the bar code pattern, a rod-reading length can be obtained.

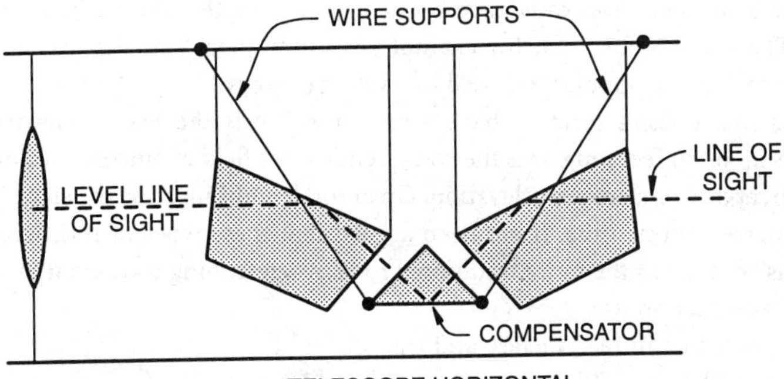

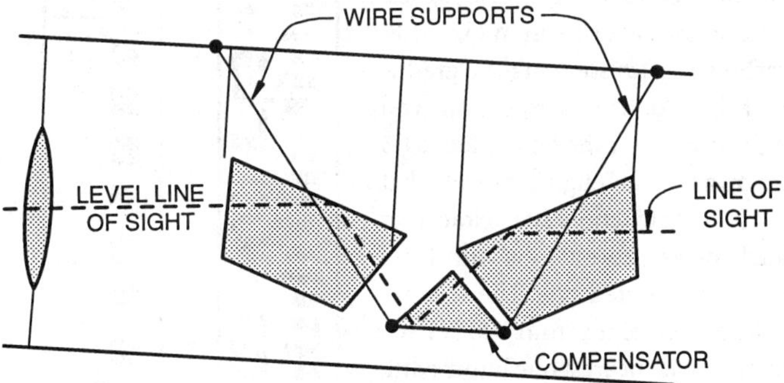

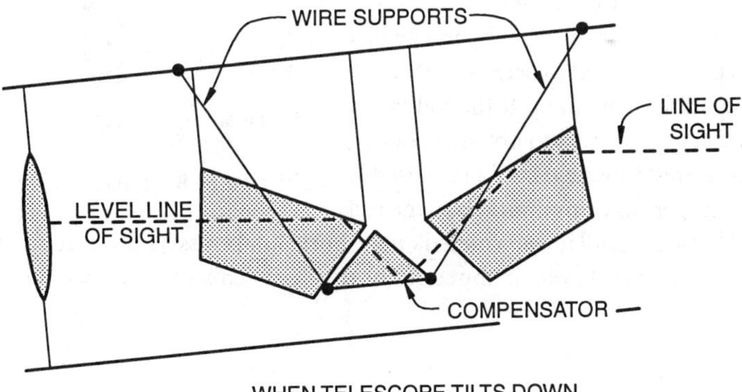

FIGURE 51.7 Leveling the line of sight by compensator.

Level Rods

In addition to the leveling instrument, a level rod is required to be able to transfer elevations from one point to another. The level rod is a graduated length scale affixed to a rod and held vertically on a turning point or benchmark. The scale is read by the person at the instrument. The reading taken is the vertical distance from the point to the line of sight.

Level rods are available in many sizes, shapes, and colors. They are made of wood, fiberglass, metal, or a combination of these materials. There are one-piece rods, two-piece rods, three-piece rods, six-piece rods, etc. Some have a square cross section and others are round or oval. Some are less than 10 feet long, while others are up to 30 feet long. Practically whatever type of rod a field engineer needs is available. Level rods have been named after the cities where they have been manufactured. The Philadelphia rod, for example, is a two-piece rod that can be extended to approximately 13 feet. It is a popular rod used in leveling surveys.

A popular rod that doesn't seem to have a proper name is the telescoping rod. Because a telescoping rod is 25 or 30 feet long, it is the rod of choice for field engineers working on projects where there is a great deal of change in elevation. Great rod lengths increase the elevation difference that can be transferred at one time. It has been argued that these types of rods wear rapidly and therefore aren't as accurate as the more traditional rods. If telescoping rods are well cared for, they are excellent for construction use.

Level rods are graduated in feet, inches, and fractions; feet, tenths, and hundredths; or meters and centimeters. The method of representing units of measurement onto the face of the rod also varies. Typical level rod graduations are shown in Fig. 51.8. The field engineer should, after studying the graduations on the face of the rod, be able to use any rod available. The rod illustrated on the left in Fig. 51.8 shows the markings on a typical "engineer's rod" that is widely used on the construction site. Note that the rod is graduated to the nearest foot with large numbers that are usually painted red. The feet are then graduated to the nearest tenth from 1 to 9. Each tenth is then graduated to the nearest one hundredth, which is the width of the smallest mark on the rod.

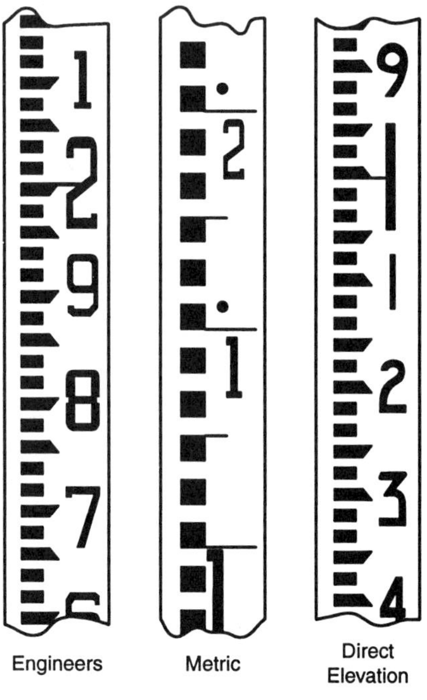

FIGURE 51.8 Typical level rod graduations.

Rod Targets

Rod targets are useful devices for several purposes. The rod target can be used as a target by the person looking through the instrument to help locate the rod when visibility conditions are poor. If the rod target has a vernier, it can also be used to obtain a reading on the rod to the nearest thousandth of a foot. In this case the instrument person communicates to the rod person to move the target until it has centered on the horizontal cross hair. The rod person can then read the rod and the vernier to obtain thousandths of a foot. This accuracy is sometimes required on very precise leveling work.

Rod Levels

Rod levels are used to keep the level rod plumb while the reading is being taken. Rod levels are simple devices made of metal or plastic and have a bull's-eye bubble attached. They are held along the edge of the level rod while the level rod is moved until the bubble is centered. If the rod level is in proper adjustment, the rod is plumb. When a rod level is not used, the rod person should slowly rock the rod through the vertical position. The instrument person watches the cross hair appear to move on the rod and records the lowest reading—the point when the rod was vertical.

Turning Point Pin

If a solid natural point is not available during the leveling process to be used as a turning point, then the field engineer will have to create one. This is accomplished by carrying something such as a railroad spike, piece of rebar, wooden stake, or plumb bob that can be inserted solidly into the ground. The rod is placed on top of the solid point while the foresight and backsight readings are taken. These solid points are removed after each backsight to be used the next time a solid turning point is needed.

Fundamental Relationships

All level instruments are designed around the same fundamental relationships and lines. The principal relationships among these lines are described as follows:

- The **axis of the level bubble** (or compensator) should be perpendicular to the vertical axis.
- The **line of sight** should be parallel to the axis of the level bubble.

When these instrument adjustment relationships are true and the instrument is properly set up, the line of sight will sweep out a horizontal plane that is perpendicular to gravity at the instrument location. However, several effects must be considered if the instrument is to be used for differential leveling.

Earth Curvature

The curved shape of the earth means that the level surface through the telescope will depart from the horizontal plane through the telescope as the line of sight proceeds to the horizon. This effect makes actual level rod readings too large by

$$C = 0.0239D^2$$

where D is the sight distance in thousands of feet.

Atmospheric Refraction

The atmosphere refracts the horizontal line of sight downward, making the level rod reading smaller. The typical effect of refraction is equal to about 14% of the effect of earth curvature. Thus, the combined effect of curvature and refraction is approximately

$$(C - r) = 0.0206D^2$$

Instrument Adjustment

If the geometric relationships defined above are not correct in the leveling instrument, the line of sight will slope upward or downward with respect to the horizontal plane through the telescope. The method of testing the line of sight of the level to ensure that it is horizontal is called the *two-peg test.* It requires setting up the level exactly between two points about 200 feet apart and subtracting readings taken on the points to determine the true difference in elevation between them. The instrument is then moved and placed adjacent to one of the points, and rod readings are again taken on the two points. If the line of sight of the instrument is truly horizontal, the difference in elevation obtained from the two setups will be the same. If the line of sight is inclined, the difference in elevation obtained from the two setups will not be equal. Either the instrument must be adjusted, or the slope of the line of sight must be calculated. The slope is expressed as a collimation factor, C, in terms of rod-reading correction per unit sight distance. It may be applied to each sight by

$$\text{Corrected rod reading} = \text{Rod reading} + \left(C_{\text{Factor}} \cdot D_{\text{Sight}}\right)$$

In ordinary differential leveling discussed next, these effects are canceled in the field procedure by always setting up so that the backsight distance and foresight distance are equal. The errors are canceled in the subtraction process. If long unequal sight distances are used, the rod readings should be corrected for curvature and refraction and for collimation error.

Ordinary Differential Leveling

Determining or establishing elevations is, at times, the most essential activity of the field engineer. Elevations are needed to set slope stakes, grade stakes, footings, anchor bolts, slabs, decks, sidewalks,

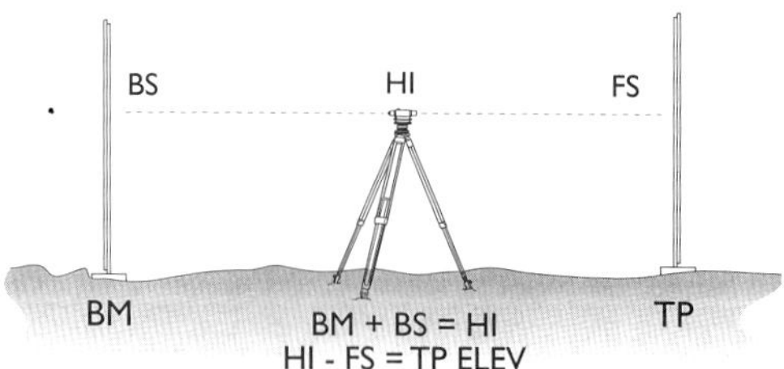

FIGURE 51.9 Differential leveling.

curbs, etc. Just about everything located on the project requires elevation. Differential leveling is the process used to determine or establish those elevations.

Differential leveling is a very simple process based on the measurement of vertical distances from a horizontal line. Elevations are transferred from one point to another through the process of using a leveling instrument to read a rod held vertically on, first, a point of known elevation and, then, on the point of unknown elevation. Simple addition and subtraction are used to calculate the unknown elevations.

A single-level setup is illustrated in Fig. 51.9. A backsight reading is taken on a rod held on a point of known elevation. That elevation is transferred vertically to the line of sight by reading the rod and then adding the known elevation and the backsight reading. The elevation of the line of sight is the height of instrument, HI. By definition, the line of sight is horizontal; therefore, the line of sight elevation can then be transferred down to the unknown elevation point by turning the telescope to the foresight and reading the rod. The elevation of the foresight station is found by subtracting the rod reading from the height of instrument. Note that the difference in elevation from the backsight station to the foresight station is determined by subtracting the foresight rod reading from the backsight rod reading.

A level route consists of several level setups, each one carrying the elevation forward to the next foresight using the differential-leveling method. Figure 51.10 shows a short level route and illustrates the typical format used in the field for differential level notes.

Leveling Closure

Level route **closure** is obtained by taking the last foresight on a benchmark of known elevation. If a second benchmark is not available near the end of the level route, the route should be looped back to the starting benchmark to obtain closure. At the closing benchmark,

$$\text{Closure} = \text{Computed elevation} - \text{Known elevation}$$

Since differential leveling is usually performed with approximately equal setup distances between turning points, the level route is adjusted by distributing the closure equally to each setup:

$$\text{Adjustment} = \left(-\frac{\text{Closure}}{n}\right)\text{per setup}$$

where n is the number of setups in the route. When a network of interconnected routes is surveyed, a **least-squares** adjustment is warranted.

Precise Leveling

Precise leveling methods are required for engineering work that requires extreme accuracy. The process requires that special equipment and methods be used. Instruments used in precise leveling are specifically designed to obtain a high degree of accuracy in leveling. Improved optics in the

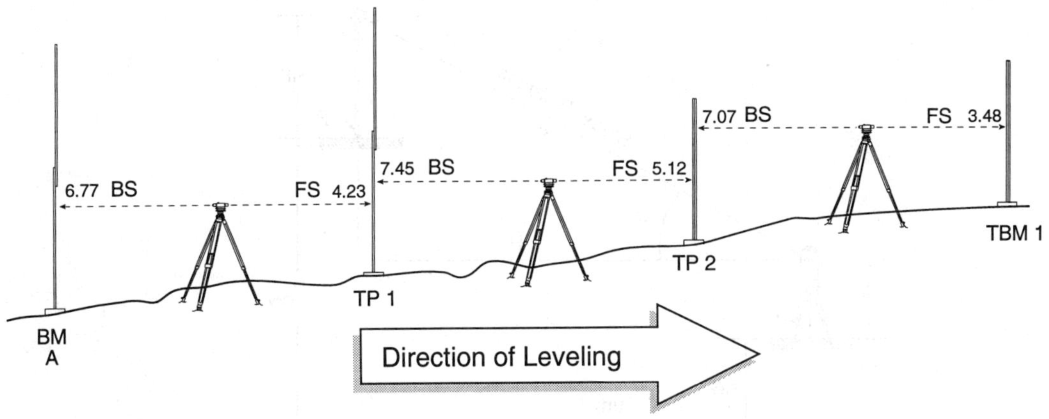

BM. A TO. TBM. 1

POINT	BS	HI	FS	SS	ELEVATION
BM. A					5280.00
INST. 1	6.77	5286.77			
TP. 1			4.23		5282.54
INST. 2	7.45	5289.99			
TP. 2			5.12		5284.87
INST. 3	7.07	5291.94			
TBM 1			3.48		5288.46

FIGURE 51.10 Level route and field note form.

telescope, improved level sensitivity, and carefully calibrated rod scales are all incorporated into the differential-leveling process. Methods have been developed to ensure that mistakes are eliminated and errors are minimized.

Typically when performing precise leveling, a method of leveling called *three-wire leveling* is used. This involves reading the center cross hair as well as the upper and lower "stadia" cross hairs. The basic process of leveling is the same, except that the three cross hair readings are averaged to improve the **precision** of each backsight and foresight value.

Another method that can be used to improve the precision of the level rod reading involves using an optical micrometer on the telescope. The optical micrometer is a rotating parallel-plate prism attached in front of the objective lens of the level. The prism enables the observer to displace the line of sight parallel with itself and set the horizontal cross hair exactly on the nearest rod

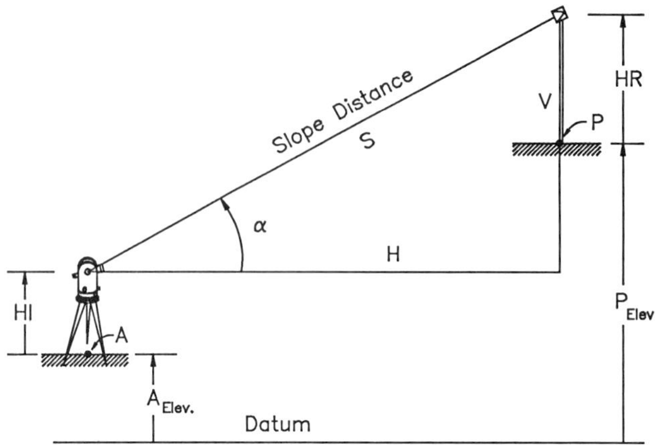

FIGURE 51.11 Trigonometric leveling in plane surveying.

graduation. The observer adds the middle cross hair rod reading and the displacement reading on the micrometer to obtain a rod reading precise to the nearest 0.1 millimeter.

When long precise level routes are surveyed, it is necessary to account for the fact that the level surfaces converge as the survey proceeds north. The correction to be applied for convergence of level surfaces at different elevations can be calculated by

$$\text{Correction} = -0.0053 \sin 2\phi H \Delta\phi_{\text{rad}}$$

where ϕ is the latitude at the beginning point, H is the elevation at the beginning point, and $\Delta\phi$ is the change in latitude from beginning point to end point in radians.

Trigonometric Leveling

Trigonometric leveling is a method usually applied when a total station is used to measure the slope distance and the vertical angle to a point. This method is illustrated in Fig. 51.11. Assuming the total station is set up on a station of known elevation, the elevation of the unknown station is

$$V = S \sin \alpha$$

$$P_{\text{Elev}} = A_{\text{Elev}} + HI + V - HR$$

The precision of trigonometric elevations is determined by the uncertainty in the vertical angle measurement and the uncertainty in the atmospheric refraction effects. For long lines the effects of earth curvature and atmospheric refraction must be included.

51.4 Angle Measurement

Horizontal and Vertical Angles

The angular orientation of a line is expressed in terms of horizontal angles and vertical angles. Survey angles illustrated in Fig. 51.12 are defined by specifying the plane that contains the angle, the reference line or plane where the angle starts, the direction of turning, and the terminal line or plane where the angle ends.

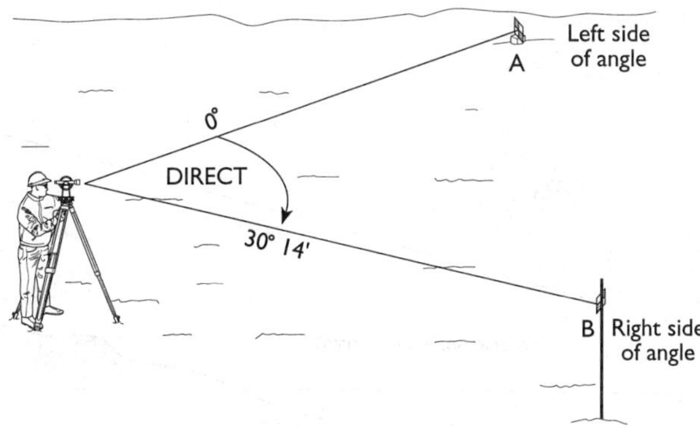

FIGURE 51.12 Horizontal survey angle.

Field angles that are measured in the horizontal plane include clockwise angles, counterclockwise angles, and deflection angles. The clockwise angle or "angle to the right," shown in Fig. 51.13, is measured from a backsight line, clockwise in the horizontal plane, to a foresight line. All transits or theodolites have horizontal circles marked to read clockwise angles. Transits used to stake out projects typically include counterclockwise markings to facilitate laying out angles to either side of a reference line. A deflection angle, also shown in Fig. 51.13, is measured from the prolongation of the backsight line either left or right to the foresight line. It is commonly used in route centerline surveys.

Vertical angles are measured in the vertical plane as shown in Fig. 51.14. An instrument with a vertical circle graduated to measure vertical angles reads 0° when the instrument is level and the telescope is in the horizontal plane. The vertical angle increases to +90° above the horizontal and to −90° below the horizontal plane. Many instruments today have vertical circles graduated to measure the zenith angle. Zero degrees is at the zenith point directly overhead on the vertical axis. The zenith angle will increase from 0° at the zenith to 90° at the horizon. The zenith angle is greater than 90° if the telescope is pointing below the horizon. The circle graduations continue to 180° at the **nadir,** to 270° at the horizon with the telescope reversed, and to 360° closing the circle at the zenith.

Direction Angles

Direction angles are horizontal angles from a north-south reference meridian to a survey line. Direction angles that are computed from field angles include bearings and azimuths. A bearing, as

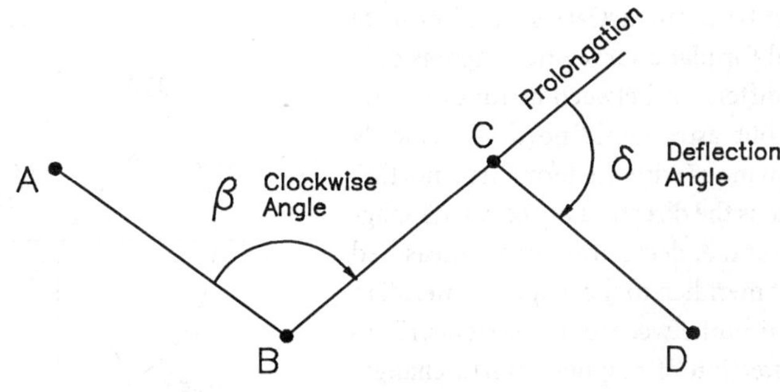

FIGURE 51.13 Horizontal field angles.

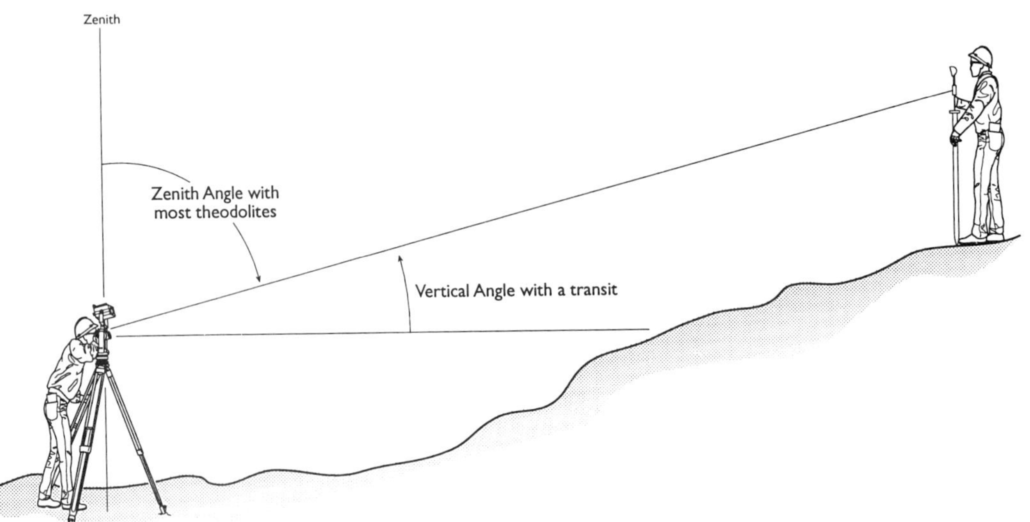

FIGURE 51.14 Vertical field angles.

shown in Fig. 51.15, is the acute angle from the reference meridian to the survey line. It always includes the quadrant designation. An azimuth, as shown in Fig. 51.16, is the angle from the north end of the reference meridian clockwise to the survey line. It can have a magnitude from 0° to 360°, and a quadrant designation is not necessary.

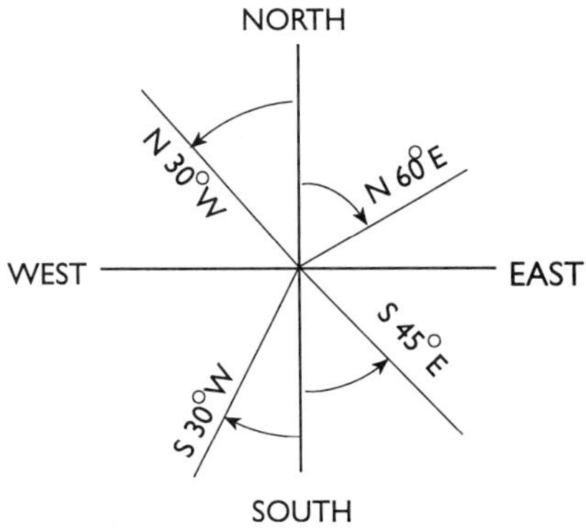

FIGURE 51.15 Bearing direction angles.

Several different meridians can be used as a reference for direction. The direction of north at a point can be defined as any of the following:

- Astronomic north
- Geodetic north
- Magnetic north
- Grid north
- Assumed north

The first three types of meridians converge to their respective north poles on the earth. Therefore, they do not form the basis for a Cartesian coordinate system to be used for plane surveying. There is only a small angular difference between astronomic and geodetic north, but astronomic north is typically taken to be synonymous with the term "true north."

Magnetic north is the direction of the earth's magnetic field at a point. A declination angle measured from the geodetic meridian to the magnetic meridian defines the relationship between the two systems. However, since the direction of magnetic north changes with time, proper use of magnetic north must account for variation in the magnetic declination angle.

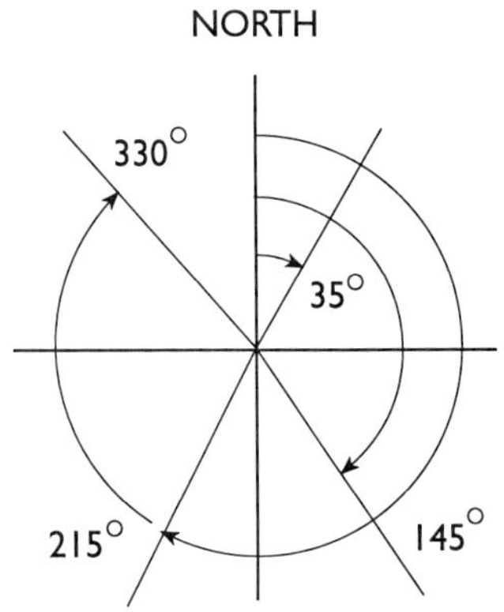

FIGURE 51.16 Azimuth direction angles.

Grid north is used in plane surveying. The grid should be defined on an appropriate map projection, such as the state plane coordinate systems available in each state. Then a defined relationship exists between the geodetic and grid meridians.

Assumed north may be used for preliminary surveys, but it is not recommended for permanent work. If the stations on the ground are lost or destroyed, the basis of direction is lost and the survey cannot be retraced.

Types of Instruments

There are many types of angle-measuring instruments used in surveying and construction. They include transits, theodolites, digital theodolites, and total stations. Even though the names are different, they look different, and the operation is slightly different, they all are used for the same purpose—angle measurement and layout.

Although there are a number of differences in the construction and features of transits, optical theodolites, and digital theodolites, the major difference is in the method of reading angles. Instruments are often classified according to the smallest interval, or so-called least count, that can be read directly in the instrument. The least count may range from 1 minute for a construction transit to 0.1 seconds for a first-order theodolite. Least counts from 10 seconds to 1 second are common in surveying and engineering instruments. The field engineer should become familiar with the instrument being used on the project.

Transit

The transit was developed to its present form during the 1800s. It has been used on construction projects from railroads across the wild west to the skyscrapers of the modern city. A good, solid, and reliable instrument, the transit is still used by many contractors today. However, optical theodolite technology, and now digital electronic technology, have passed it by. The major companies who manufactured transits have dropped them from their product line. They are slower to use than modern instruments, and they are not as easily adaptable to having an EDMI attached to them. Its major contribution to the field engineer today is that it is a good tool for learning the fundamentals of angle measurement. Its parts are exposed, making it easier to see what is going on in the manipulation of the clamps during the angle measurement process. By understanding the transit, one can easily move on to any other type of angle-measuring equipment.

Repeating Optical Theodolite

The repeating theodolite contains the same upper and lower clamp system as the transit; however, the reading of the angle is different because of the optical-reading capability. *Optical theodolite* is a term that was originally applied in Europe to instruments similar to the transit. However, as instrument technology progressed, *theodolite* became synonymous with a style of instrument that was enclosed, used a magnified optical system to read the angles, had a detachable tribrach with an optical plummet, used a three-screw leveling system, and was more precise than the transit. These features have made it much easier to use than the transit. The better optical theodolites have been "delicate" workhorses since they were introduced. That is, if they are properly cared for, they seem as though they will last forever because of their excellent construction and quality materials. However, they must be handled gently and carefully. A typical optical theodolite may have as many as 20 prisms or lenses as part of the optical angle–reading system. With a sharp bump, these can get out of alignment, which may render the instrument unusable. As with any surveying instrument, the theodolite cannot be exposed to inclement weather because of the optical system.

Scale-Reading Optical Theodolite. The typical scale-reading theodolite has a glass circle with a simple scale that is read directly. The scale is read where it is intersected by the degree readings from the circle. See Fig. 51.17 for an example scale reading. Simply read the degree that shows up

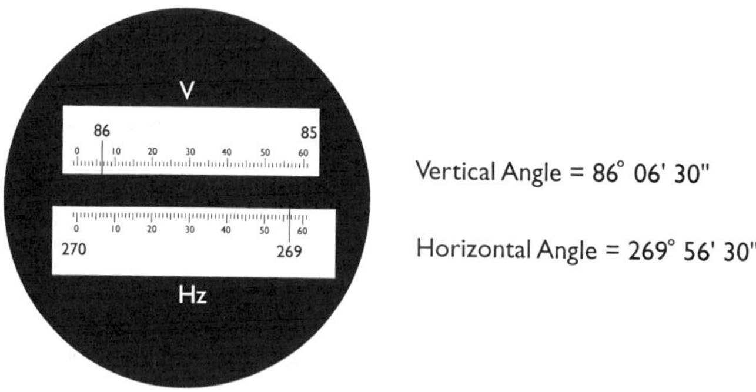

FIGURE 51.17 Scale angle reading display.

in the window and observe where the degree index mark intersects the scale. Both the horizontal circle and the vertical circle are generally observed at the same time.

Micrometer-Reading Optical Theodolite. The micrometer-reading instrument also has a glass circle, but it does not have a scale. An adjustable micrometer is used to precisely read the circle and subdivide the degree intervals into minutes and seconds. See Fig. 51.18 for an example micrometer reading. The operator points the instrument and then uses the micrometer to align the degree index marks. The readings from the degree window and the micrometer window are added together to obtain the angle.

Directional Optical Theodolite

The directional theodolite is different from other instruments because it does not have a lower motion clamp and tangent screw. Directions (circle readings) are observed and recorded, and then the directions are subtracted to obtain the angle between the backsight and foresight lines of sight. This type of system has generally been used only on the most precise instruments. Zero is usually not set on the instrument. A micrometer is used to optically read the circle. Optical theodolites are excellent instruments, but, just like the transit, they are being surpassed by the technology of electronics.

Digital Theodolite

Digital electronics has recently entered the area of surveying instruments. Angles are no longer read optically, they are displayed on a screen in degree, minute, and second format. The micrometer has been replaced with electronic sensors that determine the angle quickly and precisely. The digital theodolite has the appearance of an optical theodolite in size and overall shape. The telescope, the clamping system, the tribrach, and the optical plummet are the same. Only the angle-measuring

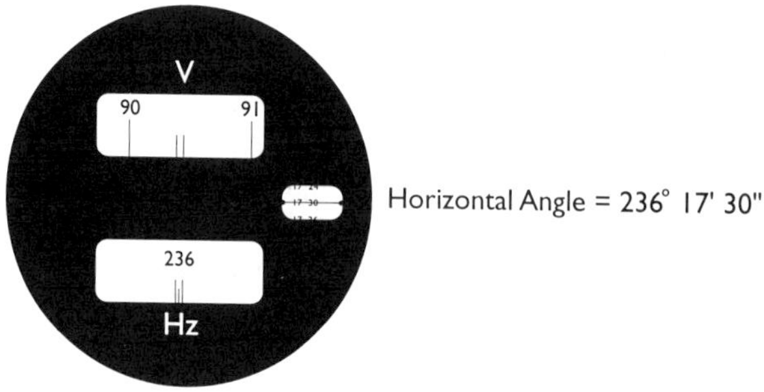

FIGURE 51.18 Micrometer angle reading display.

and -reading system is different. Digital theodolites are easy to read and fewer reading and recording blunders occur in the field. Because it is electronically based, the digital theodolite is like other electronic equipment—it either works or it doesn't work. If a circuit goes out or the battery is not charged, the instrument is unusable.

Most digital theodolites are designed to be interfaced with top-mounted EDMIs. This essentially has the impact of turning them into what are commonly called *semitotal stations*, which measure distances and angles and can be connected to a data collector for recording measurements. Sighting by both the instrument telescope and the EDMI scope are accomplished separately with this configuration.

Total Stations

The electronic total station is the ultimate in surveying measurement instruments. It is a combination digital theodolite and EDMI that allows the user to measure distances and angles electronically, calculate coordinates of points, and attach an electronic field book to collect and record the data. Since the total station is a combination digital theodolite and EDMI, its cost is quite high compared to a single instrument. However, its capabilities are simply phenomenal in comparison to the way surveyors and engineers had to measure just a few years ago. Total stations with data collectors are especially effective when a large number of points are to be located in the field, as in topographic mapping. The data collector can be used to transfer the points to a computer for final map preparation. Conversely, complex projects can be calculated on a computer in the office and the data uploaded to the data collector. The data collector is then taken to the field and connected to the total station, where hundreds of points can be rapidly established by radial layout methods.

Instrument Components

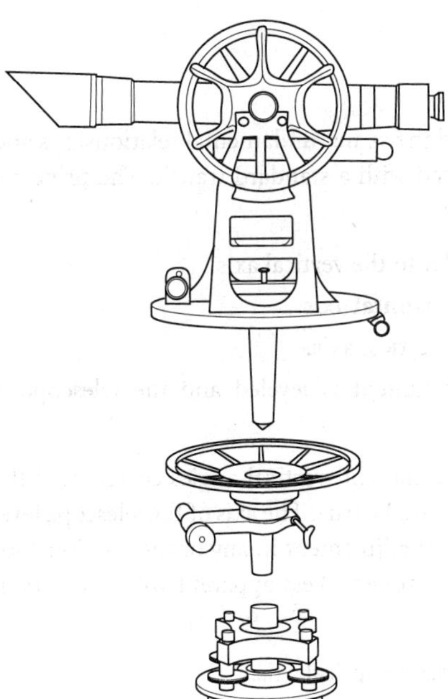

FIGURE 51.19 Transit or theodolite components.

Although there are differences between the transit, optical theodolite, and electronic theodolite, the construction and operation of all these instruments is basically the same. The three major components of any instrument are illustrated, using a transit, in Fig. 51.19. The upper plate assembly or the alidade, the lower plate assembly or the horizontal circle, and the leveling head are shown.

Alidade Assembly

The alidade assembly consists of the telescope, the vertical circle, the vertical clamp and vertical tangent, the standards or structure that holds everything together, the verniers, plate bubbles, telescope bubble, and the upper tangent screw. A spindle at the bottom of the assembly fits down into a hollow spindle on the horizontal circle assembly.

Horizontal Circle Assembly

The horizontal circle assembly comprises the horizontal circle, the upper clamp that clamps the alidade and horizontal circle together, and the hollow spindle that accepts the spindle from the alidade and fits into the leveling head.

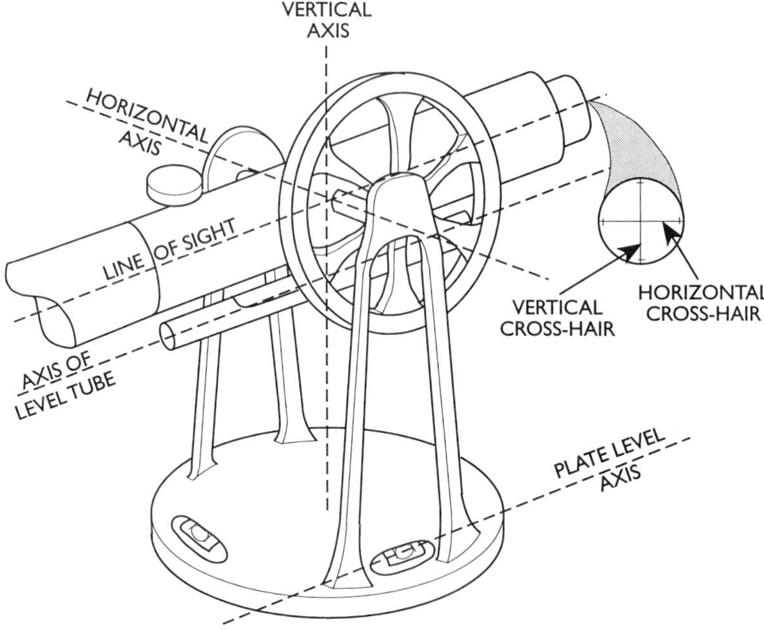

FIGURE 51.20 Principal adjustment relationships.

Leveling Head

The leveling head is the foundation that attaches the instrument assemblies to the tripod. It consists of leveling screws, the lower clamp that clamps the horizontal circle and the leveling head together, and a threaded bracket for attaching to the tripod.

Fundamental Relationships

All transit or theodolite instruments are designed around the same fundamental relationships and lines. These lines are shown in Fig. 51.20, again illustrated with a standard transit. The principal relationships between these lines are described as follows:

- The axis of the plate level(s) should be perpendicular to the vertical axis.
- The line of sight should be perpendicular to the **horizontal axis.**
- The horizontal axis should be perpendicular to the vertical axis.
- The vertical circle should read 0° when the instrument is leveled and the telescope is horizontal.

These four relationships are necessary to measure horizontal and vertical angles correctly. If the instrument will be used as a level, then the following must also be true: The axis of the telescope level should be parallel to the line of sight. Incorrect instrument adjustment in any of the relationships above will be readily apparent during field use of the instrument. Less apparent will be errors in the following relationships:

- The line of sight should be coincident with the telescope optical axis.
- The circles should be mounted concentrically on the axis of rotation.
- The circles should be accurately graduated throughout their circumference.
- The optical plummet should correctly position the instrument vertically over the field station.
- The vertical crosshair should lie in a vertical plane perpendicular to the horizontal axis.

Instrument Operation

The fundamental principle in using any transit or theodolite is the principle of reversion. That is, all operations should be performed in pairs, once with the telescope in the direct position and again with the telescope inverted on the horizontal axis or the reverse position. The correct value is the average of the two observations. Instrument operators should always use the double-centering ability of the instrument to eliminate the instrumental errors listed above. All of the principal instrument adjustment errors—except the vertical axis not being truly vertical—will be compensated for by averaging direct and reversed pairs of observations.

Prolonging a Straight Line

Prolonging a straight line from a backsight station through the instrument station to set a foresight station is a basic instrument operation that illustrates the principle of reversion. Often referred to as double centering, the steps to be performed are outlined as follows:

- Set up and level on the instrument station. With the telescope in the direct position, sight to the backsight station and clamp all horizontal motions.
- Rotate the telescope on the horizontal axis to the reverse position. Sight and set a point P_1 in the foresight direction as illustrated in Fig. 51.21.
- Revolve the instrument on the vertical axis and sight the backsight station again. The telescope will be in the reverse position.
- Rotate the telescope on the horizontal axis to the direct position. Sight and set a point P_2 in the foresight direction.
- Set the final point, P, on the true extension of the backsight line at the midpoint between P_1 and P_2.

If the stations are on nearly level terrain, the instrument error apparent in the distance between P_1 and P_2 is the line of sight not being perpendicular to the horizontal axis. When an instrument is severely out of adjustment, it is inconvenient to use it in the field. The instrument should be cleaned and adjusted periodically by a qualified instrument service technician.

Horizontal Angles by Repetition

The clamping system is probably the most important feature on an instrument because it determines the angle measurement procedures that can be used with the instrument. Transits and some optical theodolites have two horizontal motion clamps. They are commonly called the *upper motion* and *lower motion*. These types of instruments are called *repeating instruments* since the angle can be repeated and accumulated on the instrument circle. A procedure for measuring angles by repetition is outlined below.

1. Set the horizontal angle to read zero with the upper clamp and upper tangent screws. This is for convenience and any initial angle can be used.
2. Point on the backsight station with the telescope in the direct position using the lower motion. Check the circle reading and record the value in the notes.

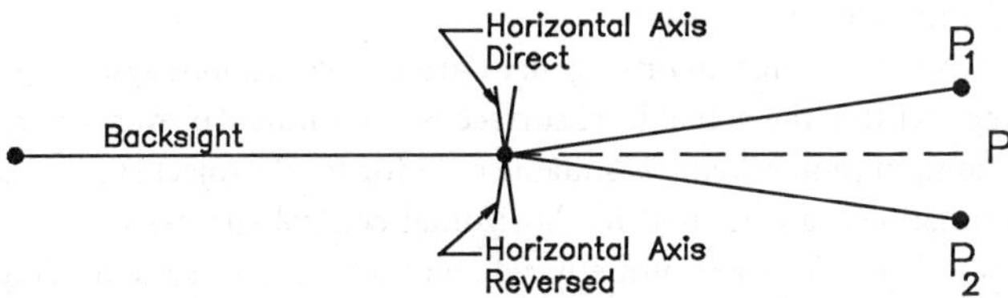

FIGURE 51.21 Double-centering to prolong a straight line.

3. Loosen the upper clamp (the lower clamp remains fixed), turn the instrument to the foresight station, and point using the upper motion clamp and upper tangent screw.
4. Note the circle reading. An approximate value of the angle can be obtained from this first turning.
5. Loosen the lower clamp, keeping the angle reading on the circle, and repeat steps 2 and 3. The number of repetitions must be an even number with half of the turnings with the telescope in the direct position and half of the turnings with the telescope in the reversed position.
6. With each turning, the circle reading will be incremented by the value of the angle. Thus, the average angle can be determined quickly by dividing the total angle read from the circle by the number of times the angle was measured.

Horizontal Angles by Direction

Many theodolites have only the upper horizontal motion clamp and the fine adjustment screw for pointing. A backsight cannot be made while holding an angle on the circle because there is no lower-motion fine adjustment for pointing. These types of instruments are called *direction instruments* since the circle reading is a clockwise direction angle from an arbitrary orientation of the 0° mark on the circle. A procedure for measuring angles by reading directions is outlined below.

1. Set the horizontal circle to read approximately 0° when pointed toward the backsight point with the telescope in the direct position.
2. Point on the backsight station with the telescope in the direct position using the upper-motion clamp and tangent screw. Read the horizontal circle and record the value in the notes.
3. Loosen the upper clamp and turn the instrument to point on the foresight station using the upper-motion clamp and tangent screw. Read the horizontal circle and record the value in the notes.
4. The clockwise angle is the difference between the foresight reading and the backsight reading.
5. Invert the telescope to the reversed position and repeat steps 2, 3, and 4. The circle readings should be exactly 180° from the telescope direct readings if there are no instrument adjustment, pointing, and reading errors. Of course the direct and reversed angles should agree closely and be averaged.
6. The direct and reversed pointings and the average angle constitute one position. Observe several positions and average the results to improve the precision of the final average angle. Advance the horizontal circle approximately between positions in order to distribute the readings around the circle.

51.5 Plane Survey Computations

Plane surveys use a three-dimensional Cartesian coordinate system as shown in Fig. 51.22. The horizontal components of distance, angle, and direction are assumed to be in the plane defined by the X and Y axes. The vertical components are along the Z axis or "up axis" in a vertical plane perpendicular to the horizontal XY plane.

Plane surveys referenced to such an orthogonal Cartesian coordinate system ignore the effect of earth curvature, the fact that the actual level surface is perpendicular everywhere to the direction of gravity. Such a computation scheme is suitable only for local project surveys of limited extent. Plane survey computations can be used for horizontal control surveys over a large areal extent if all observations and positions are properly referenced to a grid system using an appropriate map projection. State plane coordinate systems, defined for each state, are an example of this technique.

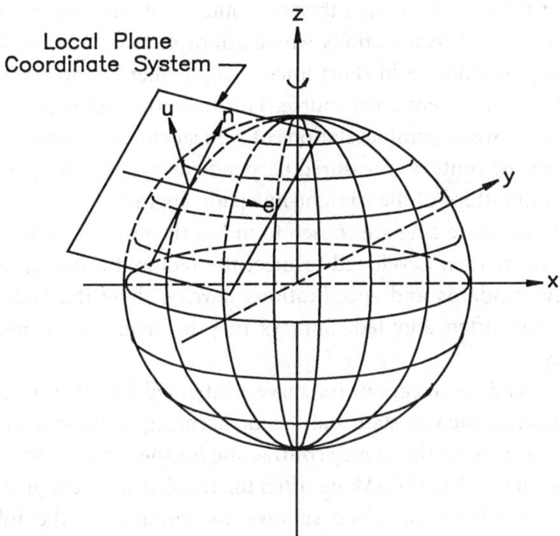

FIGURE 51.22 Local tangent plane coordinate system.

Traverse

A traverse is an efficient and flexible method of field surveying used to connect points of interest and establish horizontal and/or vertical coordinate reference values for project control. A traverse consists of interconnected straight lines along a traverse route. The straight lines meet at angle points that must be permanently marked by a traverse station monument. The length of each line is measured by field survey and then reduced to the horizontal reference plane. At each traverse station a horizontal angle is measured that will relate the directions of each line to one another in the horizontal plane. The elevation of each traverse station may also be determined if required for the purpose of the survey. Elevation can be determined by trigonometric leveling as part of the traverse measurements or by differential leveling as a separate survey operation.

Traverses are characterized as open or closed. An open traverse has no check on computed direction or position at the end of the route. It is used only for preliminary and uncontrolled work. Whenever traverse stations are to be used to control subsequent engineering or surveying work, the route should be designed to close on the beginning traverse station or to close on another known point of equal or higher-order accuracy. The closed linear traverse between known control stations is preferred since it can best detect length measurement errors that affect the scale of a survey.

Traverse design is influenced by the purpose of the survey and the terrain through which the survey must progress. The overriding factor in determining where to locate traverse stations is the purpose of the station. In general, a traverse station becomes a reference system control point that is used for one or more of the following purposes:

1. Provide control for mapping by field survey or photogrammetric methods.
2. Provide control for construction layout.
3. Locate property boundaries.
4. Connect lines within the traverse and secondary traverses that may be added to the survey network.

The traverse station must be located so that it is accessible for its intended purpose. Secondary to the purpose of the survey, the traverse station should be located in an area where the monument will be stable and undisturbed for the intended useful life of the station.

The traverse survey route is flexible and generally follows the path of least resistance so that clear lines of sight from traverse station to traverse station are obtained. However, several guidelines should be kept in mind when planning a traverse route. First, avoid lines of sight that pass close to the intervening terrain between stations where atmospheric refraction will seriously degrade measurement accuracy. Second, avoid short lines of sight where setup errors and pointing errors can be the dominant measurement error source. Third, the traverse route should proceed along a generally straight path between terminal stations of a linear closed traverse. A useful rule of thumb is that the deviation of the route, as measured perpendicular to the straight-line path, should not exceed approximately one-third of the straight-line path length.

Accuracy standards for traversing are chosen to match the purpose of the survey. Specifications for the survey process are then developed to meet the required accuracy standards. For control surveys the accuracy standards and specifications developed by the Federal Geodetic Control Committee (FGCC) are often adopted. Surveys may be required to meet **first-**, **second-**, or **third-order accuracy.**

Surveying standards and specifications for surveys intended for other purposes may be available from a variety of sources, such as state statutes or licensing board rules regulating the type of survey, professional societies, or the agency contracting for the survey. For example, standards and specifications developed by ALTA/ACSM are often the requirements adopted for property surveys.

Traverse computations for local plane surveys are outlined in the following conventional procedure:

1. Draw a complete sketch of the traverse.
2. Compute the angular closure. If angle closure is equal to or less than allowable limit, adjust the angles; if angle closure is not acceptable, remeasure angles. The allowable closure is typically specified in the following form:

$$\text{Allowable closure} = c = k\sqrt{n}$$

where

c = allowable error in a series of measurements
k = a value specified for the accuracy order of the survey
n = number of angles measured

3. Compute the direction (azimuths or bearings) of all lines.
4. Compute the latitude and departure (relative northing and relative easting components) for each course. Set up the computations in tabular form.

$$\text{Latitude} = L\cos\alpha$$
$$\text{Departure} = L\sin\alpha$$

5. Compute the traverse misclosure.

$$c = \sqrt{(\text{Latitude error})^2 + (\text{Departure error})^2}$$

6. Compute the traverse precision ratio.

$$\text{Precision} = \frac{\text{Closure error}}{\text{Perimeter}} = \frac{1}{?}$$

7. If the precision ratio is equal to or better than the allowable precision specified for the accuracy order of the survey, then distribute the error of closure throughout the traverse by an appropriate rule or by a least-squares adjustment.
8. Compute the coordinates of the traverse stations using the balanced latitudes and departures.
9. Determine the adjusted traverse course lengths and directions using an inverse computation from the adjusted coordinates or latitudes and departures.
10. Compute the area of a closed loop traverse.

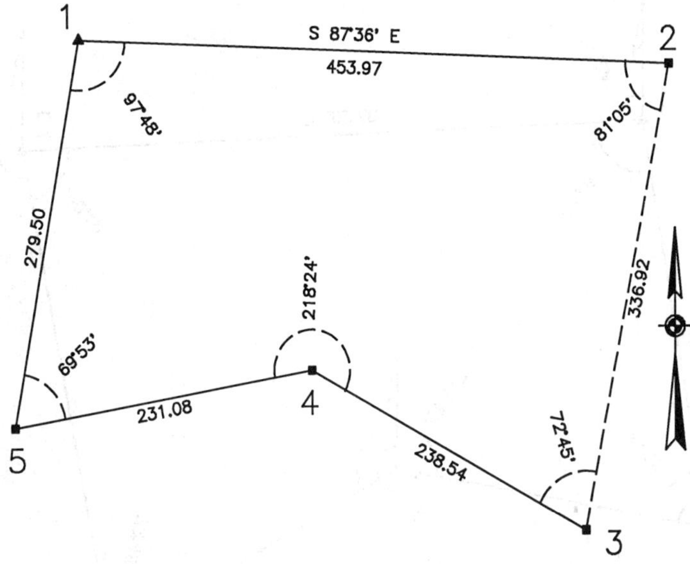

FIGURE 51.23 Sketch of sample traverse.

The traverse shown in Fig. 51.23 is used to illustrate conventional plane survey traverse computations. The measured horizontal distance of each line is given, and horizontal angles are measured as interior angles on the closed loop traverse. The orientation of the traverse is defined by the azimuth given for line 1-2. The coordinate system is defined by known coordinates at station of 1000.00N, 500.00E.

1. Draw a complete sketch.
2. Compute the angular closure.

$$\begin{aligned}\text{Field angle closure} &= (97°48' + 81°05' + 72°45' + 218°24' + 69°53') \\ &\quad - (5-2)180° \\ &= 539°55' - 540°00' \\ &= .05'\end{aligned}$$

Assuming the angular error of closure is less than the maximum allowable, apply an equal correction to each angle.

$$\text{Correction per angle} = \frac{05'}{5 \text{ angles}} = 0°01' \text{ to be added to each angle}$$

3. Compute bearings or azimuths. Use the adjusted angles. In this example the azimuth of each line is computed beginning from the given azimuth for line 1-2.

   Azimuth 1 to 2 = given = 92°24′
   Azimuth 2 to 3 = 92°24′ + 180° − 81°06′ = 191°18′
   Azimuth 3 to 4 = 191°18′ + 180° − 72°46′ = 298°32′
   Azimuth 4 to 5 = 298°32′ + 180° − 218°25′ = 260°07′
   Azimuth 5 to 1 = 260°07′ − 180° − 69°54′ = 10°13′
   Azimuth 1 to 2 = 10°13′ + 180° − 97°49′ = 92°24′

   Note that the azimuth of the first line is computed at the end to verify that the computations close on the given azimuth. The adjusted angles and azimuths are shown in Fig. 51.24.

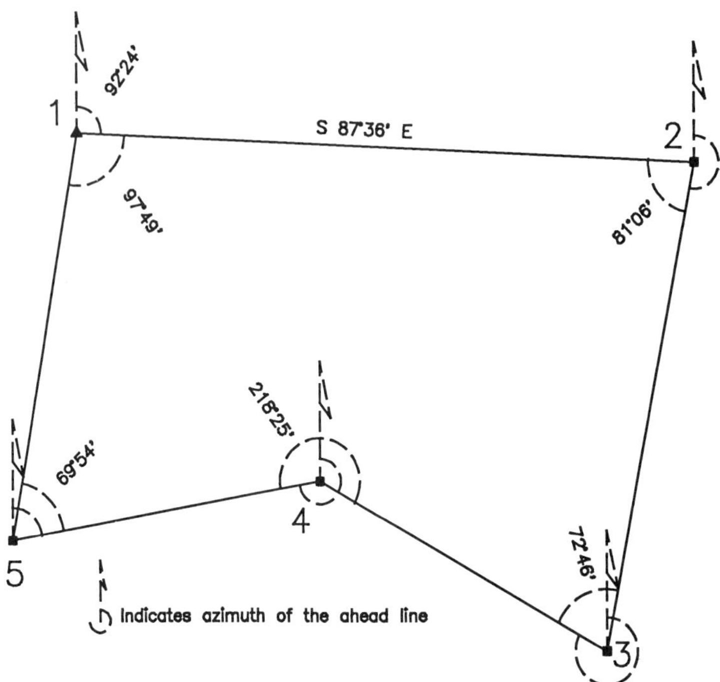

FIGURE 51.24 Adjusted angles and azimuths.

4. Compute the latitude and departure of each course, as shown in Table 51.3.
5. Compute the traverse misclosure.

$$c = \sqrt{(-0.051)^2 + (-0.089)^2} = 0.102 \text{ ft}$$

6. Compute the traverse precision.

$$\text{Precision} = \frac{0.102}{1540.01} = \frac{1}{15{,}100}$$

7. Distribute the error of closure. We will assume this precision is sufficient for the particular survey we are carrying out. Adjust the traverse by the compass rule:

$$C_{\text{lat}} = \frac{+0.051}{1540.01}(\text{Course length})$$

$$C_{\text{dep}} = \frac{+0.089}{1540.01}(\text{Course length})$$

Table 51.3 Computed Latitudes and Departures

Course	Length (L)	Azimuth α	Latitude $= L\cos\alpha$	Departure $= L\sin\alpha$
1-2	453.97	92°24′	−19.010	453.572
2-3	336.92	191°18′	−330.389	−66.018
3-4	238.54	298°32′	113.943	−209.567
4-5	231.08	260°07′	−39.663	−227.651
5-1	279.50	10°13′	275.068	49.575
Sum =	1540.01	Error =	−0.051	−0.089
		Correction	+0.051	+0.089

Table 51.4 Compass Rule Adjustment

	Applying Corrections		Balanced		Coordinates	
Station	Latitude	Departure	Latitude	Departure	North (Y)	East (X)
1					1000.00	500.00
	−19.010	453.572				
	+.015	+.026	−18.995	+453.598	−19.00	+453.60
2					981.00	953.60
	−330.389	−66.018				
	+.011	+.020	−330.378	−65.998	−330.38	−66.00
3					650.62	887.60
	113.943	−209.567				
	+.008	+.014	+113.951	−209.553	+113.95	−209.55
4					764.57	678.05
	−39.663	−227.651				
	+.008	+.013	−39.655	−227.638	−39.65	−227.64
5					724.92	450.41
	275.068	49.575				
	+.009	+.016	+275.077	+49.591	+275.08	+49.59
1					1000.00	500.00
		Total =	0.000	0.000	Check	Check

Table 51.5 Adjusted Lengths and Azimuths

	Coordinates		Adjusted	Adjusted
Station	North (Y)	East (X)	Length	Azimuth
1	1000.00	500.00		
			454.00	92°23′53″
2	981.00	953.60		
			336.90	191°17′50″
3	650.62	887.60		
			238.53	298°32′12″
4	764.57	678.05		
			231.07	260°07′05″
5	724.92	450.41		
			279.51	10°13′10″
1	1000.00	500.00		

8. Compute the balanced latitudes and departures by applying the corrections, and then calculate coordinates for all of the traverse stations, as shown in Table 51.4.
9. Adjusted traverse line lengths and directions are computed by an inverse computation using the adjusted latitudes and departures or adjusted coordinates, as shown in Table 51.5.

$$\text{Length} = \sqrt{Dep^2 + Lat^2}$$

$$= \sqrt{(X_j - X_i)^2 + (Y_j - Y_i)^2}$$

$$\text{Azimuth}_{ij} = \tan^{-1}\left[\frac{Dep_{ij}}{Lat_{ij}}\right] = \tan^{-1}\left[\frac{X_j - X_i}{Y_j - Y_i}\right]$$

10. Compute the area. Traverse areas are typically computed using the coordinate method, as shown in Table 51.6.

$$\text{Area} = \frac{\left|\sum_1^n X_i Y_{i-1} - \sum_1^n X_i Y_{i+1}\right|}{2}$$

The traverse area equals one-half of the difference between the totals of the double-area columns.

Table 51.6 Coordinate Method for Area

	Coordinates		Double Area	
Station	North (Y)	East (X)	↗	↘
1	1000.00	500.00	490,500	
2	981.00	953.60	620,431	953,600
3	650.62	887.60	678,632	870,736
4	764.57	678.05	491,532	441,153
5	724.92	450.41	450,410	344,370
1	1000.00	500.00		362,460
	Total =		2,731,505	2,972,319

$$\text{Traverse area} = \left| \frac{2{,}731{,}505 - 2{,}972{,}319}{2} \right| = 120{,}407\ \text{ft}^2 = 2.764\ \text{acres}$$

Partitioning Land

Partitioning land is a problem that can usually be classified according to one of two types of dividing line—a line of known direction or a line through a known point. A preliminary line is often required that satisfies the given condition. Then the line is translated parallel to itself in the first condition or pivoted about the known point in the second condition to obtain the required area.

As an illustration of these methods, partition the adjusted traverse of the previous section so that 65,000 square feet lie west of a true north line. A preliminary cutoff line bearing true north can be constructed through station 4, as shown in Fig. 51.25. The area west of the preliminary line is 50,620 square feet. The line is translated true east a distance X so that 14,380 square feet will be added to the west parcel. The parcel added is a trapezoid as shown in Fig. 51.25, and the area can be expressed by

$$14{,}380 = 227.97X - \tfrac{1}{2}X(X \tan \theta_1) + \tfrac{1}{2}X(X \tan \theta_2)$$

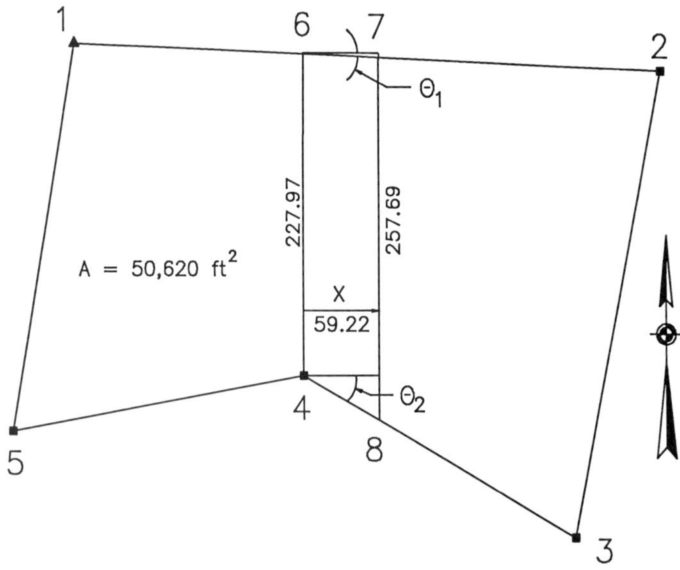

FIGURE 51.25 Partition by sliding a line.

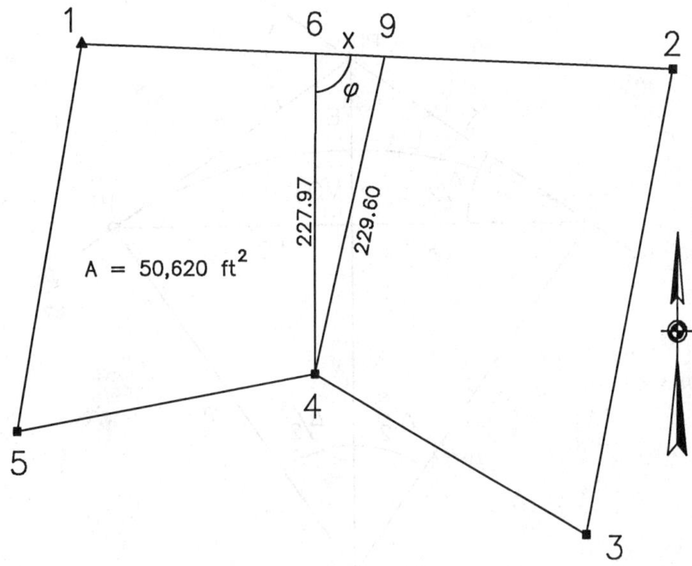

FIGURE 51.26 Partition by pivoting a line.

where $\theta_1 = 2°23'53''$ and $\theta_2 = 28°32'12''$. Rearranging this expression results in a quadratic equation, and the solution for X is found to be 59.22 feet. Then the following distances can be determined for the final cutoff line:

6-7 = 59.27 feet
4-8 = 67.41 feet
7-8 = 257.69 feet

Next, partition the adjusted traverse so that 55,000 square feet lie west of a line through station 4. The same preliminary cutoff line can be used. The line is pivoted easterly about station 4 so that 4380 square feet is added to the west parcel. The parcel added is a triangle as shown in Fig. 51.26, and the area can be expressed by

$$4380 = \tfrac{1}{2}227.97X \sin\phi$$

where $\phi = 87°36'07''$. The solution for X is found to be 38.46 feet, and the final cutoff line is

$$\text{4-9} = 229.60 \text{ feet}, 9°38'05'' \text{ azimuth}$$

51.6 Horizontal Curves

Horizontal curves are used in route projects to provide a smooth transition between straight-line tangent sections. These curves are simple circular curves. The components of a circular curve are illustrated in Fig. 51.27. The design of a circular curve requires that two curve elements be defined and then the remaining elements are determined by computation. In the typical case the intersection angle of the tangents is determined by field survey and the curve radius is chosen to meet design specifications such as vehicle speed and minimum sight lengths. The principal relationships between curve elements are

$$T = R\tan\frac{\Delta}{2}$$

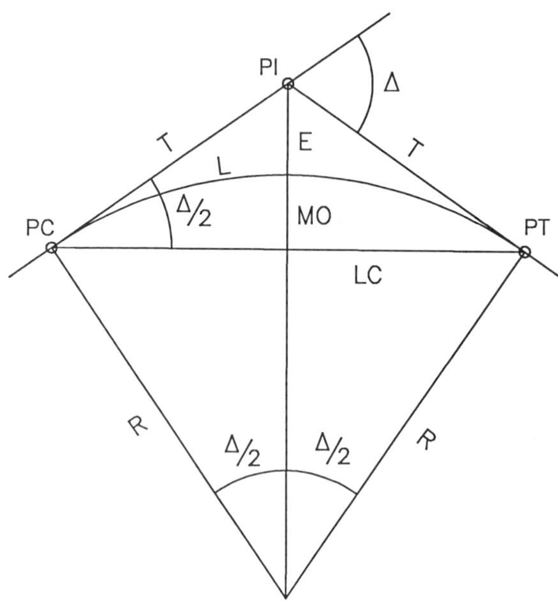

FIGURE 51.27 Circular horizontal curve components.

$$LC = \frac{1}{2} R \sin \frac{\Delta}{2}$$

$$E = R\left(\sec \frac{\Delta}{2} - 1\right)$$

$$MO = R\left(1 - \cos \frac{\Delta}{2}\right)$$

$$L = R \frac{\pi}{180^\circ} \Delta$$

Important angle relationships used when solving circular curve problems are illustrated in Fig. 51.28.

When a circular curve is staked out in the field, the most common method is to lay off deflection angles at the PC station. Field layout notes are prepared for a theodolite set up at the PC. A backsight is taken along the tangent line to the PI, and foresights are made to specific stations on the curve using the computed deflection angles. If a total station instrument is used the chord distance from the PC to the curve station can be used to set the station. If a tape is used the chord distance measured from the previous station on the curve is intersected with the line of sight to locate the curve station.

The following example illustrates both types of distances used to lay out a horizontal curve. Determine the field information necessary to stake out the horizontal curve shown in Fig. 51.29. First determine the stationing of the PC and PT. The tangent distance is

$$T = R \tan\left(\frac{\Delta}{2}\right) = 572.96 \ \tan 40^\circ 00' = 503.46 \text{ ft}$$

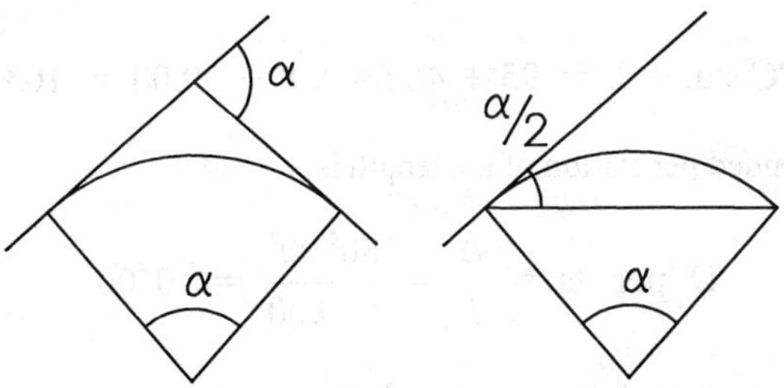

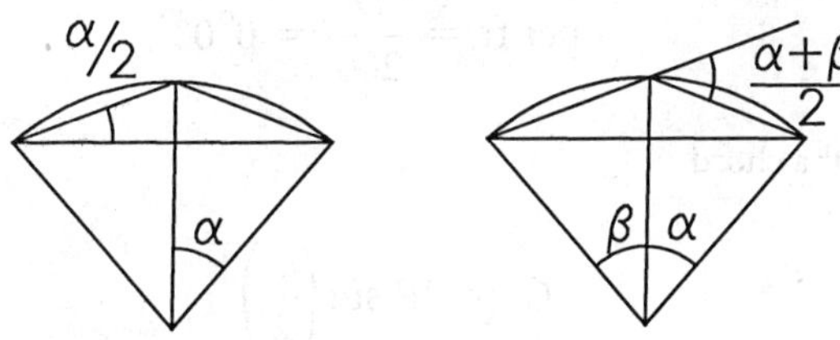

FIGURE 51.28 Angle relationships for circular curves.

so the PC station is

$$\text{PC sta.} = \text{PI sta.} - T = 100 + 50.00 - 5 + 03.46 = 95 + 46.54 \text{ ft}$$

The arc length of the curve is

$$L = R\left(\frac{\pi}{180^\circ}\right)\delta = 800.00 \text{ ft}$$

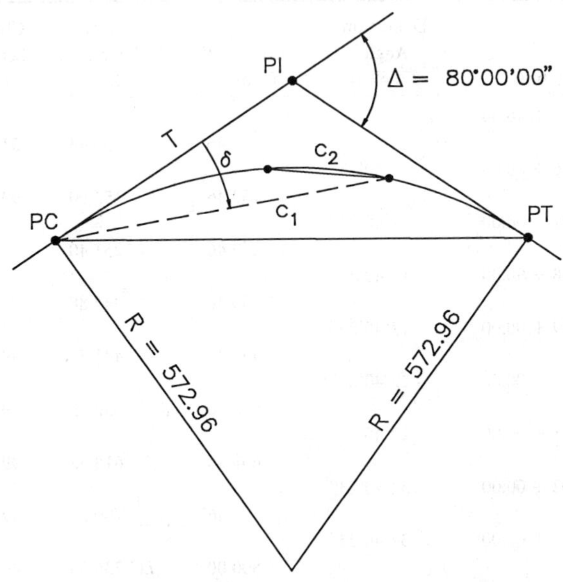

FIGURE 51.29 Horizontal curve layout.

so the PT station is

$$\text{PT sta.} = \text{PC sta.} + L = 95 + 46.54 + 8 + 00.00 = 103 + 46.54 \text{ ft}$$

The central angle subtended per station of arc length is

$$D^\circ \text{per sta} = \frac{\Delta}{L} = \frac{80^\circ 00'}{800.00} = 10^\circ 00'$$

The deflection angle is one-half of the central angle or

$$\delta \text{ per sta} = \frac{D}{2} = 5^\circ 00'$$

$$\delta \text{ per ft} = \frac{D}{200} = 0^\circ 03'$$

Recall the equation for a chord

$$C = 2R \sin\left(\frac{d}{2}\right)$$

where d is the central angle of the chord. The full stations falling on the curve are determined, and the values in Table 51.7 are computed for an instrument set up on the PC and oriented by a backsight on the PI. Note that C_1 denotes the chord measured from the PC and that C_2 denotes the chord measured from the previous station on the curve.

Alternative methods for staking out a curve include tangent offsets, chord offsets, middle ordinates, and radial staking out from the radius point of the curve or from the PI station.

Table 51.7 Deflection Angles and Chords for Layout of a Horizontal Curve

Station	Deflection Angle δ	Arc Length from PC	Chord Length C_1	Chord Length C_2
PC 95 + 46.54				
		53.46	53.44	53.44
96 + 00.00	2°40′23″			
		153.46	153.00	99.87
97 + 00.00	7°40′23″			
		253.46	251.40	99.87
98 + 00.00	12°40′23″			
		353.46	347.88	99.87
99 + 00.00	17°40′23″			
		453.46	441.72	99.87
100 + 00.00	22°40′23″			
		553.46	532.19	99.87
101 + 00.00	27°40′23″			
		653.46	618.62	99.87
102 + 00.00	32°40′23″			
		753.46	700.33	99.87
103 + 00.00	37°40′23″			
		L 800.00	*LC* 736.58	46.53
PT103 + 46.54	Δ/240°00′00″			

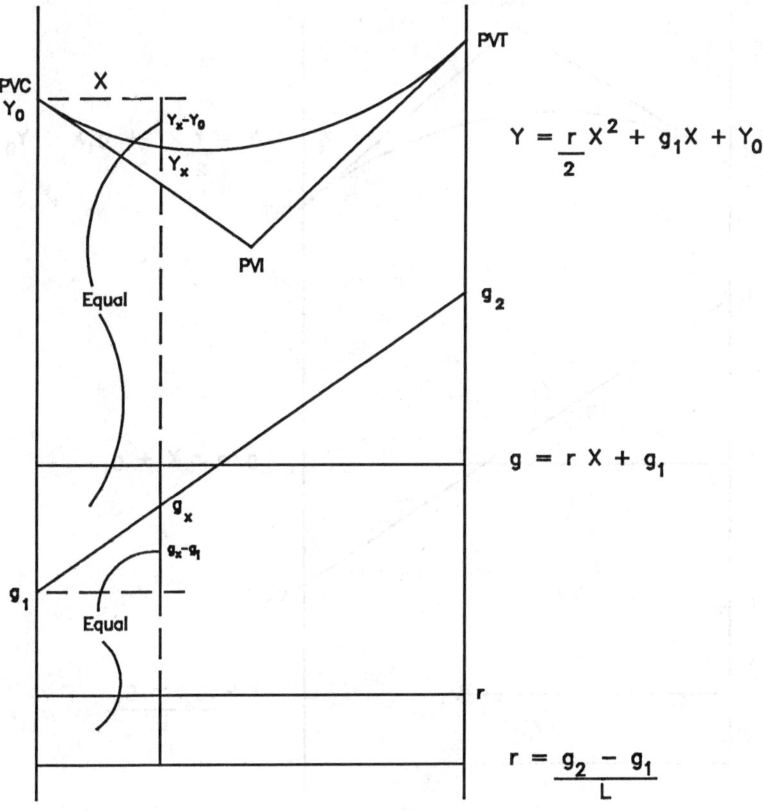

FIGURE 51.30 Sag vertical curve.

51.7 Vertical Curves

Vertical curves are used in vertical alignments of route projects to provide a smooth transition between grade lines. These curves are usually equal-tangent parabolic curves. The point of vertical intersection, PVI, of the entrance and exit grade lines always occurs at the midpoint of the length of curve. The length of curve and all station distances are measured in the horizontal plane. Figures 51.30 and 51.31 illustrate the geometry of a sag and crest vertical curve, respectively. Note the following relationships between the curves in Figs. 51.30 and 51.31. The top curve is the elevation curve. The middle curve is the grade curve; it is the derivative of the elevation curve. The bottom curve is the rate of change of grade curve; it is the derivative of the grade curve. The rate of change of grade is always a constant, r, for a parabolic vertical curve.

Since the curves are related by the derivative/integration operation, the change in an ordinate value on one curve is equal to the area under the next lower curve. Therefore, in Fig. 51.30, the change in elevation on the curve from the PVC to a point at a distance X on the curve is equal to the trapezoidal area under the grade curve

$$Y_x - Y_0 = X\left(\frac{g_x + g_1}{2}\right)$$

The change in grade over the same interval X is the rectangular area under the rate of change of grade curve.

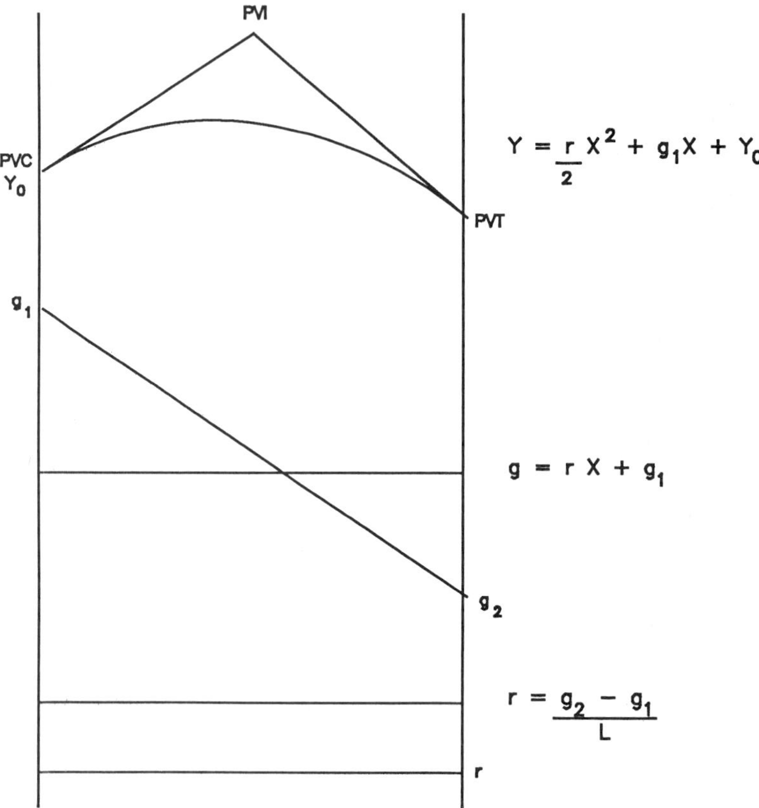

FIGURE 51.31 Crest vertical curve.

$$g_x - g_1 = Xr$$

These relationships can be used to calculate design values for vertical curves.

For example, suppose it is necessary to find the elevation and station of the high point for the crest curve in Fig. 51.31. Let X equal the distance to the high point that occurs at the point where the grade curve crosses the zero line. The change in grade must be equal to the area under the rate of change of grade curve.

$$g_{\text{high}} = 0$$

$$0 - g_1 = X(-r)$$

$$X = \frac{-g_1}{-r}$$

Note that r is negative for a crest curve. Then the station of the high point is

$$P_{\text{high}} = \text{PC} + X$$

The elevation of the high point can be found by evaluating the elevation equation at the known value of X or by calculating the area under the grade curve:

$$Y_x - Y_0 = X\left(\frac{g_1 + 0}{2}\right)$$

$$Y_x = Y_0 + X\left(\frac{g_1}{2}\right)$$

The design of a vertical curve involves choosing the length of the curve that will satisfy design speed considerations, earthwork considerations, and sometimes geometric constraints. As an example of a constrained design, suppose a PVC is located at station 150 + 40.00 and elevation 622.45 feet. The grade of the back tangent is −3.00% and the grade of the forward tangent is −7.00%. It is required that a vertical curve between these tangents must pass through station 152 + 10.00 at elevation 619.05 feet. Since this is a sag vertical curve, refer to Fig. 51.30. The value of r can be expressed as

$$r = \frac{(+7) - (-3)}{L} = +\frac{10}{L}$$

The distance X from the PVC to the known station on the curve is

$$X = (152 + 10.00) - (150 + 40.00) = 1.70000 \text{ stations}$$

The grade at the known station is then found from

$$g_x = (-3) + \left(\frac{10}{L}\right)1.7 = \frac{17}{L} - 3$$

The change in elevation from the PVC to the known station can be set equal to the trapezoidal area under the grade curve.

$$619.05 - 622.45 = 1.70000\left[\frac{\left(\frac{17}{L} - 3\right) + (-3)}{2}\right]$$

Solving this expression for L, we obtain

$$L = 8.5000 \text{ stations} = 850.00 \text{ ft}$$

The design elevation at each full station along this curve can be evaluated from the parabolic equation. First the station and elevation of the PVI is

$$\text{PVI sta} = \text{PVC sta} + \frac{L}{2} = 150 + 40.00 + 4 + 25.00 = 154 + 65.00 \text{ ft}$$

$$\text{PVI elev} = \text{PVC elev} + g_1\frac{L}{2} = 622.45 + (-3)(4.25) = 609.70 \text{ ft}$$

and the station and elevation of the PVT is

$$\text{PVT sta} = \text{PVI sta} + \frac{L}{2} = 154 + 65.00 + 4 + 25.00 = 158 + 90.00 \text{ ft}$$

$$\text{PVT elev} = \text{PVI elev} + g_1\frac{L}{2} = 609.70 + (+7)(4.25) = 639.45 \text{ ft}$$

Then the elevation of any point on the curve is found from

$$Y = \left(\frac{r}{2}\right)X^2 + g_1X + Y_0 = \left(\frac{10}{17}\right)X^2 - 3X + 622.45$$

A tabular solution for each full station along the curve is given in Table 51.8.

Table 51.8 Vertical Curve Elevations

Station	X sta	$(r/2)X^2$	g_1X	Elevation Y ft
PVC 150 + 40.00				Y_0 622.45
	0.6000	0.21	−1.80	
151 + 00.00				620.86
	1.6000	1.50	−4.80	
152 + 00.00				619.15
	2.6000	3.98	−7.80	
153 + 00.00				618.83
	3.6000	7.62	−10.80	
154 + 00.00				619.27
	4.6000	12.45	−13.80	
155 + 00.00				621.10
	5.6000	18.45	−16.80	
156 + 00.00				624.10
	6.6000	25.62	−19.80	
157 + 00.00				628.27
	7.6000	33.98	−22.80	
158 + 00.00				633.63
	8.5000	42.50	−25.50	
PVT 158 + 90.00				639.45

51.8 Volume

The determination of volume is necessary before a project begins, throughout the project, and at the end of the project. In the planning stages, volumes are used to estimate project costs. After the project is started, volumes are determined so the contractor can receive partial payment for work completed. At the end, volumes are calculated to determine final quantities that have been removed or put in place to make final payment. The field engineer is often the person who performs the field measurements and calculations to determine these volumes. Discussed here are the fundamental methods used by field engineers.

General

To compute volumes, field measurements must be made. This typically involves determining the elevations of points in the field by using a systematic approach to collect the needed data. If the project is a roadway, cross-sectioning is used to collect the data that are needed to calculate volume.

If the project is an excavation for a building, borrow-pit leveling will be used to determine elevations of grid points to calculate the volume. Whatever the type of project, the elevation and the location of points will need to be determined. It is the responsibility of the field engineer to determine the most efficient method of field measurement to collect the data.

However, it should be mentioned that sometimes volumes can be determined using no field measurements at all. In some situations the contractor may be paid for the number of truckloads removed. Keeping track of the number of trucks leaving the site is all that may be necessary. However, this isn't a particularly accurate method since the soil that is removed expands or swells and takes up a larger amount of space than the undisturbed soil. Depending on how the project is bid, it is sometimes accurate enough.

Field Measurements for Volume Computations

Measurements for volume are nothing more than applying basic distance and elevation measurements to determine the locations and elevations of points where the volume is to be determined.

It usually is not practical to take the time to collect data everywhere there is a slight change in elevation. Therefore, it must be understood that volume calculations do not give exact answers. Typically, approximations must be made and averages determined. The field engineer will analyze the data and make decisions that result in the best estimate of the volume.

Area

The key to volume calculation is the determination of area. Most volume calculation formulas contain within them the formula for an area, which is simply multiplied by the height to determine the volume. For instance, the area of a circle is pi times the radius squared. The volume of a cylinder is the area of the circle times the height of the cylinder. If an area can be determined, it is generally easy to determine the volume.

Counting Squares

Approximation is possible by plotting the figure to scale on cross-sectional paper and counting the squares. Each square represents *x* number of square feet. Incomplete squares along the edges of the cross section are visually combined and averaged.

Planimeter

The electromechanical digital planimeter is a quick method of determining the area of irregularly shaped figures. The irregular shape is drawn to scale and the planimeter is used to trace the outline of the shape. Inputting a scale factor into the planimeter results in a digital readout of the area.

Geometric Formula

Although a shape at first may seem irregular, it is often possible to break it into smaller regular shapes such as squares, rectangles, triangles, trapezoids, etc., that will allow the use of standard geometric formulas to determine the area. This method may be cumbersome because of all the shapes that may need to be calculated.

Cross Section Coordinates

If cross-sectional field data are available, use of this data is the recommended method of calculating volume. Once understood, this process is fast and the most accurate way of determining area. Cross section data collected on a project represent elevation and location information for points on the ground. These points can be used as coordinates to determine area.

Volume Computations—Road Construction

In road construction the shape of the ground must be changed to remove the ups and downs of the hills and valleys for the planned roadway. Often mountains of dirt must be moved to create a gentle grade for the roadway. Payment for the removal and placement of dirt is typically on a unit cost basis. That is, the contractor will be paid per cubic yard of soil and will receive a separate price per cubic yard of rock. It can be seen that accurate determination of the volume moved is critical to the owner and to the contractor. Each wants an accurate volume so payment for the work is correct.

For road projects, cross sections of the ground elevations are measured at the beginning of the project, during the project, and at the end of the project. Comparisons between final cross sections and original cross sections are used to determine the volume moved. Areas of the cross sections are most easily determined by using the elevations of the points and their locations from the centerline (coordinates).

The average end area method uses the end areas of adjacent stations along a route and averages them. Refer to Fig. 51.32. This average is then multiplied by the distance between the two end areas

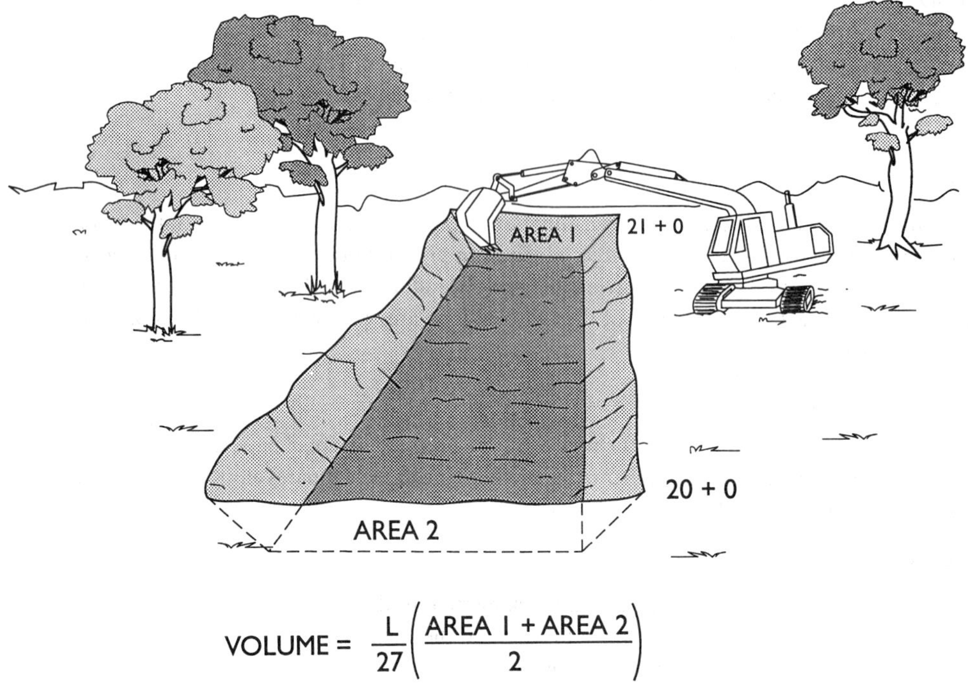

FIGURE 51.32 Average end area method.

to obtain the volume between them. In formula form the process is as follows:

$$\text{Volume} = \frac{L}{27}\left(\frac{\text{Area 1} + \text{Area 2}}{2}\right)$$

where L represents the distance between the cross-sectional end areas being used in the formula, and 27 represents the number of cubic feet in one cubic yard. Dividing cubic feet by 27 converts to cubic yards.

Volume Computations—Building Excavation

A method known as *borrow-pit leveling* can be used effectively to determine volume on building projects. A grid is established by the field engineer and elevations on the grid points are determined both before the excavation begins and when the work is complete.

The borrow-pit method uses a grid and the average depth of the excavation to determine the volume. Before the excavating begins the field engineer creates a grid over the entire area where the excavation is to occur. Elevation data are collected at each of the grid points and recorded for future reference. At any time during the excavating, the field engineer can reestablish the grid and determine new elevations for each of the field points. Using the average height formula shown below, the volume of soil removed from each grid area can be determined. Refer to Fig. 51.33. The smaller the grid interval, the more accurate the volume.

Summary

Only two general methods of calculating volumes have been presented here. There are many others that are very specific for the particular situation. For example, when determining volumes along a

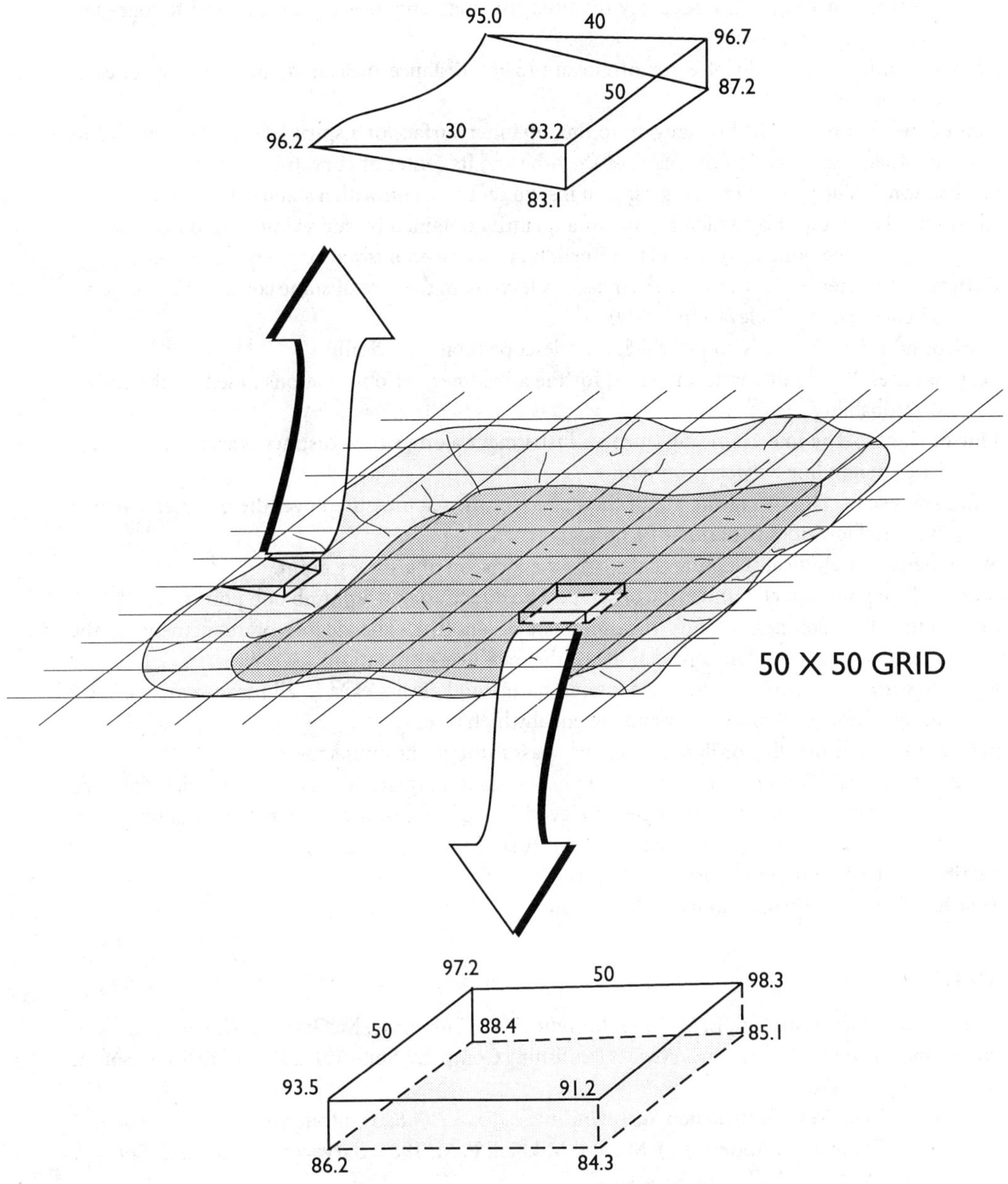

FIGURE 51.33 Borrow-pit method.

roadway, there is a constant transition from cut to fill to cut to fill, etc. To more accurately compute the volume, a prismoidal formula is used. The field engineer should check with textbooks that discuss in detail route surveying and earthwork for additional information.

Defining Terms

Accuracy: Refers to the degree of perfection obtained in measurements. It is a measure of the closeness to the true value.

Accuracy (first-order): The highest accuracy required for engineering projects such as dams, tunnels, and high-speed rail system.

Accuracy (second-order): The accuracy required for large engineering projects such as highways, interchanges, and short tunnels.

Accuracy (third-order): The accuracy required for small engineering projects and topographic mapping control.

Accuracy ratio: The ratio of error of closure to the distance measured for one or a series of measurements.

Axis of level bubble: The line tangent to the top inner surface of a spirit level at the center of its graduated scale, and in the plane of the tube and its center of curvature.

Calibration: The process of comparing an instrument or chain with a standard.

Closure: The amount by which a value of a quantity obtained by surveying fails to agree with a value (of the same quantity) determined. It is also called *misclosure* or *error of closure.*

Datum: A reference elevation such as mean sea level or, in the case of some construction projects, a benchmark with elevation 100.00.

Horizontal axis: The axis about which the telescope rotates vertically.

Least squares: A mathematical method for the adjustment of observations based on the theory of probability.

Line of sight: The line extending from an instrument along which distinct objects can be seen. The straight line between two points.

Mean sea level: The average height of the surface of the sea measured over the complete cycle of high and low tides (a period of 18.6 years).

Monument: A physical structure that marks the location of a survey point.

Nadir: The point directly under the observer. The direction that a plumb bob points.

Precision: The closeness of one measurement to another. The degree of refinement in the measuring process. The repeatability of the measuring operation.

Random errors: Errors that are accidental in nature and always exist in all measurements. They follow the laws of probability and are equally high or low.

Refraction: The bending of light rays as they pass through the atmosphere.

Systematic error: Those errors that occur in the same magnitude and the same sign for each measurement of a distance, angle, or elevation. Can be eliminated by mechanical operation of the instrument or by mathematical formula.

Vertical: The direction in which gravity acts.

Zenith: The point directly above a given point on earth.

References

Anderson, J. M. and Mikhail, E. M. 1985. *Introduction of Surveying.* McGraw-Hill, New York.

Professional Surveyor. American Surveyors Publishing Company, Suite 501, 2300 Ninth Street South, Arlington, VA.

Crawford, W. G. 1994. *Construction Surveying and Layout. P.O.B.*, Canton, MI.

Davis, R. E., Foote, F. S., Anderson, J. M., and Mikhail, E. M. 1981. *Surveying: Theory and Practice,* 6th ed. McGraw-Hill, New York.

Fronczek, C. J. 1980. *NOAA Technical Memorandum NOS NGS-10.* National Geodetic Information Branch, NOAA, Silver Springs, MD.

Federal Geodetic Control Committee. 1984. *Standards and Specifications for Geodetic Control Networks.* National Geodetic Information Branch, NOAA, Silver Springs, MD.

National Geodetic Survey. 1986. *Geodetic Glossary.* National Geodetic Information Branch, NOAA, Silver Springs, MD.

Point of Beginning. Business News Publishing Company, Troy, MI.

Wolf, P. R. and Brinker, R. C. 1994. *Elementary Surveying,* 9th ed. HarperCollins, New York.

For Further Information

The material here is intended only as an overview of plane surveying. There are many textbooks dedicated completely to the various aspects of surveying. The authors recommend the following books. For a more complete presentation of surveying theory, consult *Elementary Surveying*, 9th ed., by Wolf and Brinker, HarperCollins, 1992; or *Surveying: Theory and Practice*, 6th ed., by Davis, Foote, Anderson, and Mikhail, McGraw-Hill, 1981. For illustrated step-by-step descriptions of performing field work, consult: *Construction Surveying and Layout*, Wesley G. Crawford, P.O.B. Publishing, 1994.

To obtain detailed information on the capabilities of various instruments and software, P.O.B. Publishing prepares the trade magazine *P.O.B.*, and American Surveyors Publishing Company prepares the trade magazine *Professional Surveyor.* Each of these publications conducts annual reviews of theodolites, total stations, EDMIs, data collectors, GPSs, and software. These listings allow the reader to keep up to date and compare "apples to apples" when analyzing equipment.

Survey control information, software, and many useful technical publications are available from the National Geodetic Survey (NGS). The address is

National Geodetic Survey Division
National Geodetic Information Branch, N/CG17
1315 East-West Highway, Room 9218
Silver Spring, MD 20910-3282

52

Geodesy

B. H. W. van Gelder
Purdue University

52.1 Introduction

This chapter covers the basic mathematical and physical aspects of modeling the size and shape of the earth and its gravity field. Terrestrial and space geodetic measurement techniques are reviewed. Extra attention is paid to the relatively new technique of satellite surveying using the Global Positioning System (GPS). GPS surveying has not only revolutionized the art of navigation but also brought about an efficient positioning technique for a variety of users, engineers not the least. It is safe to say that any geometry-based data collecting scheme profits in some sense from the full constellation of 24 GPS satellites. Except for the obvious applications in geodesy, surveying, and photogrammetry, the use of GPS is applied in civil engineering areas such as transportation (truck

0-8493-8953-4/95/$0.00 + $.50

and emergency vehicle monitoring, intelligent vehicle and highway systems, etc.) and structures (monitoring of deformation of such structures as water dams). Even in areas as forestry and agriculture (crop yield management), GPS provides the geometric backbone to the (geographic) information systems.

Modern geodetic measurement techniques, using signals from satellites orbiting the earth, necessitate a new look at the science of geodesy. Classical measurement techniques divided the theoretical problem of mapping small or large parts of the earth into a horizontal issue and a vertical issue. Three-dimensional measurement techniques "solve" the geodetic problem at once. However, careful interpretation of these 3-D results is still warranted, probably even more than previously. This chapter will center around this three-dimensional approach. Less attention has been devoted to classical issues such as the computation of a geodesic on an ellipsoid of revolution. Although the latter issue still has some importance, the reader is referred to textbooks as listed at the end of this chapter.

More than in classical texts, 3-D polar (spherical) coordinate representations are used, because the fundamental issues pertaining to various geodetic models are easier to illustrate by spherical coordinates than by ellipsoidal coordinates. Moreover, the increased influence of the satellite techniques in everyday surveying revives the use of 3-D polar coordinate representations, because the 3-D location of a point is equally accurately represented by Cartesian, spherical, or ellipsoidal coordinates.

Throughout this chapter all coordinate frames are treated as right-handed orthogonal trihedrals. Because this also applies to curvilinear coordinates, the well-known geographic coordinates of latitude and longitude are presented in the following order:

1. The longitude (positive in East direction), λ
2. The latitude, ψ or ϕ
3. The height, h

In short, $\{\lambda, \psi, h\}$ or $\{\lambda, \phi, h\}$. Local Cartesian and curvilinear coordinates are treated in a similar fashion.

52.2 Coordinate Representations

For a detailed discussion on coordinate frames and transformations, the reader is referred to Chapter 50.

2-D

In surveying and mapping, 2-D frames are widely used. The different representations are all dependent, because only two numbers suffice to define the location of a point in 2-space. Cartesian frames consist of two often perpendicular reference axes, denoted as x and y, or e (easting) and n (northing). Points in 2-D are equally well represented by polar coordinates r (distance from an origin) and α (polar angle, counted positive counterclockwise from a reference axis).

We have

$$\begin{aligned} x &= r\cos\alpha \\ y &= r\sin\alpha \end{aligned} \tag{52.1}$$

The polar coordinates $\{r, \alpha\}$ are expressed in terms of the Cartesian counterparts by

$$\begin{aligned} r &= \sqrt{x^2 + y^2} \\ \alpha &= \arctan(y/x) \end{aligned} \tag{52.2}$$

3-D

3-D Cartesian Coordinates

There are various ways to represent points in a three-dimensional space. One of the well-known is the representation by the so-called Cartesian coordinates x, y, z: we represent the position of a point A through three distances x, y, z (coordinates) to three perpendicular planes, respectively the yz-, xz-, and xy-planes. The intersecting lines between the three planes are the perpendicular coordinate axes. The position of point A is thought to be represented by the vector **x** with elements $\{x, y, z\}$:

$$\mathbf{x} = \begin{pmatrix} x \\ y \\ z \end{pmatrix} \tag{52.3}$$

3-D Polar Coordinates: Spherical

We may want to represent the position of these points with respect to a sphere with radius R. We make use of so-called spherical coordinates. The earth's radius is about $R = 6371.0$ km.

The sphere is intersected by two perpendicular planes, both of which pass through the center O of the sphere: a reference *equatorial plane* (perpendicular to the rotation axis) and a reference *meridian plane* (through the rotation axis). The angle between the vector and the reference equatorial plane is called *latitude*, ψ. The angle between the reference meridian plane (through Greenwich) and the (local) meridian plane through A is called *longitude*, λ (positive east). The distance to the surface of the sphere we call *height*, h. Consequently, the position of a point A is represented by $\{\lambda, \psi, h\}$ or $\{\lambda, \psi, r\}$ or $\{\lambda, \psi, R + h\}$; see also Fig. 52.1.

3-D Polar Coordinates: Ellipsoidal

The earth is flattened at the poles, and the average ocean surface has about the shape of an ellipsoid. For this reason, ellipsoidal coordinates are more often used in geodesy than spherical coordinates. We express the coordinates with respect to an ellipsoid of revolution with an equatorial semimajor axis a and a polar semiminor axis b. The semimajor axis thus represents the equatorial radius, and b is the distance between the ellipsoidal origin and the poles. The equation of such an ellipsoid of revolution is

$$\frac{x^2}{a^2} + \frac{y^2}{a^2} + \frac{z^2}{b^2} = 1 \tag{52.4}$$

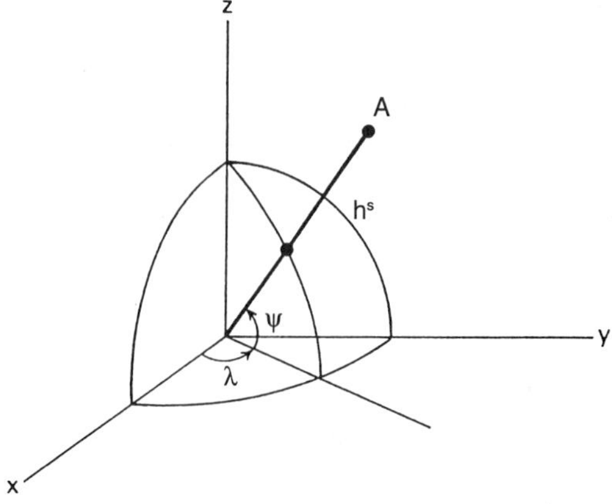

FIGURE 52.1 (Geographic) spherical coordinates: longitude λ, latitude ψ, and height h.

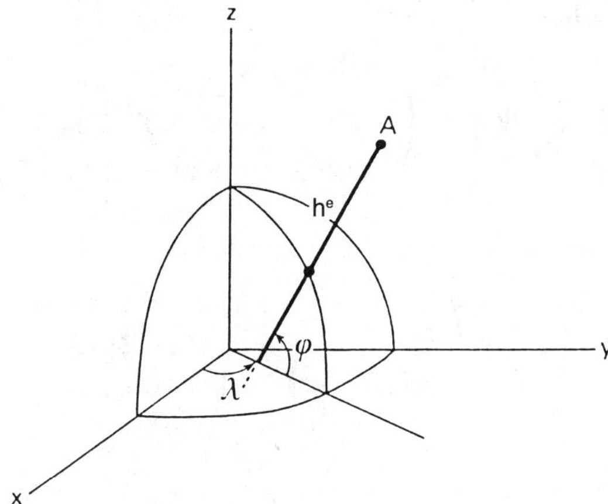

FIGURE 52.2 (Geodetic) ellipsoidal coordinates: longitude λ, latitude ϕ, and height h.

For the earth we have a semimajor axis $a = 6378.137$ km and a semiminor axis $b = 6356.752$ km. This means that the poles are about 21.4 km closer to the center of the earth than the equator.

The flattening of the earth is expressed by f and the (first) eccentricity by e:

$$f = \frac{a-b}{a} \qquad (\approx 1/298.257) \tag{52.5}$$

$$e^2 = \frac{a^2-b^2}{a^2} \qquad (\approx 0.00669438) \tag{52.6}$$

See also Fig. 52.2.

Coordinate Transformations

We have to distinguish between two classes of transformations:

- Transformations between dissimilar coordinate representations. An example would be the transformation between Cartesian coordinates and curvilinear coordinates such as the ellipsoidal (geodetic) coordinates.
- Transformations between similar coordinate frames. An example is the relation between geocentric Cartesian coordinates and topocentric Cartesian coordinates.

The latter group is to be discussed subsequently in this section after we consider transformations between dissimilar coordinate representations.

Transformations of Different Kind

If the xy-plane coincides with the equator plane and the xz plane with the reference meridian plane, then we have:

From Spherical to Cartesian.

$$\begin{pmatrix} x \\ y \\ z \end{pmatrix} = (R+h) \begin{pmatrix} \cos\psi\cos\lambda \\ \cos\psi\sin\lambda \\ \sin\psi \end{pmatrix} \tag{52.7}$$

From Cartesian to Spherical.

$$\begin{pmatrix} \lambda \\ \psi \\ h \end{pmatrix} \begin{matrix} = \\ = \\ = \end{matrix} \begin{pmatrix} \arctan(y/x) \\ \arctan(z/\sqrt{x^2+y^2}) \\ \sqrt{x^2+y^2+z^2} - R \end{pmatrix} \tag{52.8}$$

From Ellipsoidal to Cartesian.

$$\begin{pmatrix} x \\ y \\ z \end{pmatrix} = \begin{pmatrix} [N & +h] \cos\phi \cos\lambda \\ [N & +h] \cos\phi \sin\lambda \\ [N(1-e^2) & +h] \cos\phi \end{pmatrix} \tag{52.9}$$

with

$$N = \frac{a}{W} \tag{52.10}$$

and

$$W = \sqrt{1 - e^2 \sin^2\phi} \tag{52.11}$$

In these equations the variable N has a distinct geometric significance: it is the radius of curvature in the prime vertical plane. This plane goes through the local normal and is perpendicular to the meridian plane. In other words, N describes the curvature of the curve obtained through intersection of the prime vertical plane and the ellipsoid. The curve formed through intersection of the meridian plane and the ellipsoid is given by M, see Fig. 52.3.

The varying radius of curvature M of the elliptic meridian is given by

$$M = \frac{a(1-e^2)}{W^3} \tag{52.12}$$

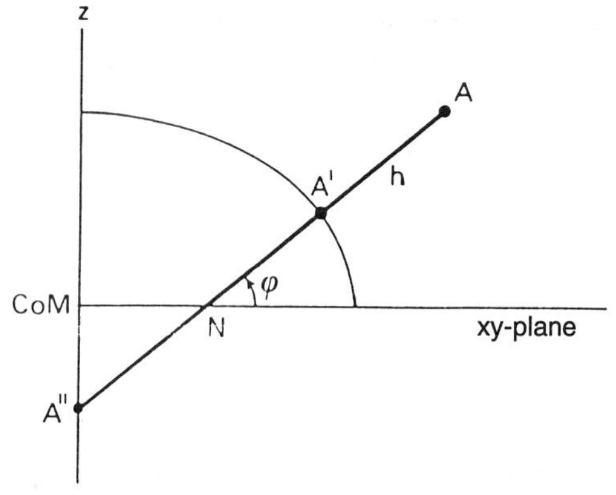

FIGURE 52.3 Meridian plane through point A.

From Cartesian to Ellipsoidal.

$$\lambda = \arctan(y/x) \tag{52.13}$$

The geodetic latitude ϕ can be obtained by the following iteration scheme, starting with an approximate value for the geodetic latitude ϕ_0:

$$\begin{aligned} \phi_0 &= \arctan\left[z/\sqrt{x^2+y^2}\right] \\ N_0 &= a/\sqrt{1-e^2\sin^2\phi_0} \\ \phi &= \arctan\left[(z + N_0 e^2 \sin\phi_0)/\sqrt{x^2+y^2}\right] \end{aligned} \tag{52.14}$$

If $|\phi - \phi_0| > \epsilon$ then set ϕ_0 equal to ϕ and go back to computation of N. After the iteration, h can be computed directly:

$$h = \sqrt{[x^2 + y^2 + (z + Ne^2 \sin\phi)^2]} - N \tag{52.15}$$

Through more cumbersome expressions an analytical solution for the geodetic latitude ϕ as function of the Cartesian coordinates $\{x, y, z\}$ is possible.

Transformations of Same Kind

Orthogonal Transformations: Translation and Rotation. When groups of points are known in their relative position with respect to each other, the use of Cartesian coordinates ($3n$ in total) becomes superfluous. As a matter of fact there are six degrees of freedom, since the position of the origin with respect to the group is arbitrary, as is the orientation of the frame axes. Two groups of identical points or, for that matter, one and the same group of points expressed in two arbitrary but different coordinate frames, may be represented by the following orthogonal transformation:

$$\mathbf{x}' = \mathbf{R}\mathbf{x} + \mathbf{t}' \tag{52.16}$$

or

$$\mathbf{x}' = \mathbf{R}(\mathbf{x} - \mathbf{t}) \tag{52.17}$$

The vector $\mathbf{t}$ represents a translation. In Eq. (52.16) $\mathbf{t}'$ represents the vector of the old origin in the new frame ($\mathbf{x}' - \mathbf{t}' = \mathbf{R}\mathbf{x}$); in Eq. (52 .17) $\mathbf{t}$ represents the coordinates of the origin of the new frame in the old coordinate frame. The relation between the two translation vectors is represented by

$$\mathbf{t}' = -\mathbf{R}\mathbf{t} \tag{52.18}$$

The rotation matrix describes the rotations around the frame axes. In Eq. (52.16) $\mathbf{R}$ describes a rotation around axes through the origin of the $\mathbf{x}$-frame; in Eq. (52.17) rotations around axes through the origin of the $\mathbf{x}'$-frame. We define the sense of rotations as follows: the argument angle of the rotation matrix is taken positive if one views the rotation as counterclockwise from the positive end of the rotation axis looking back to the origin. For an application relating coordinates in a local frame to coordinates in a global frame, see section 52.3.

In the above equations we assume three consecutive rotations, first around the z-axis with an argument angle γ, then around the y-axis around an argument angle β, and finally around the x-axis with the argument angle α. So we have

$$\mathbf{R} = \mathbf{R}_1(\alpha)\mathbf{R}_2(\beta)\mathbf{R}_3(\gamma) \tag{52.19}$$

$$\mathbf{R} = \begin{pmatrix} \cos\beta\cos\gamma & \cos\beta\sin\gamma & -\sin\beta \\ -\cos\alpha\sin\gamma + \sin\alpha\sin\beta\cos\gamma & \cos\alpha\cos\gamma + \sin\alpha\sin\beta\sin\gamma & \sin\alpha\cos\beta \\ \sin\alpha\sin\gamma + \cos\alpha\sin\beta\cos\gamma & -\sin\alpha\cos\gamma + \cos\alpha\sin\beta\sin\gamma & \cos\alpha\cos\beta \end{pmatrix} \tag{52.20}$$

One outcome of these orthogonal transformations is an inventory of variables that are invariant under these transformations. Without any proof, these include lengths, angles, sizes and shapes of figures, volumes, important quantities for the civil/survey engineer.

Similarity Transformations: Translation, Rotation, Scale. In the previous section we saw that the relative location of n points can be described by fewer than $3n$ coordinates: $3n - 6$ quantities (for instance, an appropriate choice of distances and angles) are necessary but also sufficient. Exceptions have to be made for so-called critical configurations such as four points in a plane. The 6 is nothing else than the six degrees of freedom supplied by the orthogonal transformation: three translations and three rotations.

A simple but different reasoning leads to the same result. Imagine a tetrahedron in 3-D. The four corner points are connected by six distances. These are exactly the six necessary but sufficient quantities to describe the form and shape of the tetrahedron. These six sides determine this figure completely in size and shape. A fifth point will be positioned by another three distances to any three of the four previously mentioned points. Consequently, a field of n points (in 3-D) will be necessarily but sufficiently described by $3n - 6$ quantities. We need these types of reasoning in 3-D geometric satellite geodesy.

If we just consider the *shape* of a figure spanned by n points (we are not concerned any more about the *size* of the figure) then we need even one quantity fewer than $3n - 6$ (i.e., $3n - 7$): we are ignoring, in addition to the position and orientation of the figure, now also the scale. This constitutes just the addition of a seventh parameter to the six-parameter orthogonal transformation: the scale parameter σ. So we have

$$\mathbf{x}' = \boldsymbol{\sigma}\mathbf{R}\mathbf{x} + \mathbf{t}' \tag{52.21}$$

or

$$\mathbf{x}' = \sigma\mathbf{R}(\mathbf{x} - \mathbf{t}) \tag{52.22}$$

Here also the vector $\mathbf{t}$ represents a translation. In Eq. (52.21) $\mathbf{t}'$ represents the vector of the old origin in the new scaled and rotated frame ($\mathbf{x}' - \mathbf{t}' = \sigma\mathbf{R}\mathbf{x}$); in Eq. (52.22) $\mathbf{t}$ represents the coordinates of the origin of the new frame in the old coordinate frame. The relation between the two translation vectors is represented by

$$\mathbf{t}' = -\sigma\mathbf{R}\mathbf{t} \tag{52.23}$$

One outcome of these similarity transformations is an inventory of invariant variables under these transformations. Without any proof, these include length ratios, angles, shapes of figures, and volume ratios, which are important quantities for the civil/survey engineer. The reader is referred to Leick and van Gelder [1975] for other important properties.

Curvilinear Coordinates and Transformations

One usually prefers to express coordinate differences in terms of the curvilinear coordinates on the sphere or ellipsoid or even locally rather than in terms of the Cartesian coordinates. This approach also facilitates the study of effects due to changes in the adopted values for the reference ellipsoid (so-called *datum transformations*).

Curvilinear Coordinate Changes in Terms of Cartesian Coordinate Changes

Differentiating the transformation formulas in which the Cartesian coordinates are expressed in terms of the ellipsoidal coordinates [see Eq. (52.9)], we obtain a differential formula relating the Cartesian total differentials $\{dx, dy, dz\}$ as a function of the ellipsoidal total differentials $\{d\lambda, d\phi, dh\}$.

$$\begin{pmatrix} dx \\ dy \\ dz \end{pmatrix} = \mathbf{J}\begin{pmatrix} d\lambda \\ d\phi \\ dh \end{pmatrix} \tag{52.24}$$

The projecting matrix $\mathbf{J}$ is nothing else than the Jacobian of partial derivatives:

$$\mathbf{J} = \frac{\partial(x, y, z)}{\partial(\lambda, \phi, h)} \tag{52.25}$$

Carrying out the differentiation, one finds

$$\mathbf{J} = \begin{bmatrix} -(N+h)\cos\phi\sin\lambda & -(M+h)\sin\phi\cos\lambda & \cos\phi\cos\lambda \\ (N+h)\cos\phi\cos\lambda & -(M+h)\sin\phi\sin\lambda & \cos\phi\sin\lambda \\ 0 & (M+h)\cos\phi & \sin\phi \end{bmatrix} \tag{52.26}$$

On inspection, this Jacobian $\mathbf{J}$ is simply a product of a rotation matrix $\mathbf{R}(\lambda, \phi)$ and a metric matrix $\mathbf{H}(\phi, h)$ [Soler, 1976]:

$$\mathbf{J} = \mathbf{RH} \tag{52.27}$$

or, in full,

$$\mathbf{R} = \begin{bmatrix} -\sin\lambda & -\sin\phi\cos\lambda & \cos\phi\cos\lambda \\ \cos\lambda & -\sin\phi\sin\lambda & \cos\phi\sin\lambda \\ 0 & \cos\phi & \sin\phi \end{bmatrix} \tag{52.28}$$

$$\mathbf{H} = \begin{bmatrix} (N+h)\cos\phi & 0 & 0 \\ 0 & (M+h) & 0 \\ 0 & 0 & 1 \end{bmatrix} \tag{52.29}$$

It turns out that the rotation matrix $\mathbf{R}(\lambda, \phi)$ relates the local $\{e, n, u\}$ frame to the geocentric $\{x, y, z\}$ frame; see further the discussion of "Earth-fixed" coordinates in section 52.3. The metric matrix $\mathbf{H}(\phi, h)$ relates the curvilinear coordinates longitude, latitude, and height in radians/meters to the curvilinear coordinates all expressed in meters.

The formulas just given are the simple expressions relating a small arc distance ds to the corresponding small angle $d\alpha$ through the radius of curvature. The radius of curvature for the longitude component is equal to the radius of the local parallel circle, which in turn

$$\begin{pmatrix} d\lambda_m \\ d\phi_m \\ dh_m \end{pmatrix} = \mathbf{H}(\phi, h) \begin{pmatrix} d\lambda_{\text{rad}} \\ d\phi_{\text{rad}} \\ dh_m \end{pmatrix} \tag{52.30}$$

equals the radius of curvature in the prime vertical plane times the cosine of the latitude.

The power of this evaluation is the more apparent if one realizes that the inverse Jacobian, expressing the ellipsoidal total differentials $\{d\lambda, d\phi, dh\}$ as a function of the Cartesian total differentials $\{dx, dy, dz\}$, is easily obtained, whereas an analytic solution expressing the geodetic ellipsoidal coordinates in terms of the Cartesian coordinates is extremely difficult to obtain. So, we have

$$\begin{pmatrix} d\lambda \\ d\phi \\ dh \end{pmatrix} = \mathbf{J}^{-1} \begin{pmatrix} dx \\ dy \\ dz \end{pmatrix} \tag{52.31}$$

With the relationship in Eq. (52.27) $\mathbf{J}^{-1}$ becomes simply

$$\mathbf{J}^{-1} = (\mathbf{RH})^{-1} = \mathbf{H}^{-1}\mathbf{R}^{\mathrm{T}} \tag{52.32}$$

or, in full,

$$\mathbf{J}^{-1} = \begin{bmatrix} \dfrac{-\sin\lambda}{(N+h)\cos\phi} & \dfrac{\cos\lambda}{(N+h)\cos\phi} & 0 \\ \dfrac{-\sin\phi\cos\lambda}{M+h} & \dfrac{-\sin\phi\sin\lambda}{M+h} & \dfrac{\cos\phi}{M+h} \\ \cos\phi\cos\lambda & \cos\phi\sin\lambda & \sin\phi \end{bmatrix} \tag{52.33}$$

This equation gives a simple analytic expression for the inverse Jacobian, whereas the analytic expression for the original function is virtually impossible.

Curvilinear Coordinate Changes Due to a Similarity Transformation

Differentiating Eq. (52.21) with respect to the similarity transformation parameters $\alpha, \beta, \gamma, t'_x, t'_y, t'_z, \sigma$, one obtains

$$\begin{pmatrix} dx \\ dy \\ dz \end{pmatrix}_7 = \mathbf{J}_7 \begin{pmatrix} d\alpha \\ d\beta \\ d\gamma \\ dt'_x \\ dt'_y \\ dt'_z \\ d\sigma \end{pmatrix} \tag{52.34}$$

with

$$\mathbf{J}_7 = \frac{\partial(x, y, z)}{\partial(\alpha, \beta, \gamma, t'_x, t'_y, t'_z, \sigma)} \tag{52.35}$$

The Jacobian $\mathbf{J}_7$ is a matrix that consists of seven column vectors

$$\mathbf{J}_7 = [\mathbf{j}_1 | \mathbf{j}_2 | \mathbf{j}_3 | \mathbf{j}_4 | \mathbf{j}_5 | \mathbf{j}_6 | \mathbf{j}_7] \tag{52.36}$$

with

$$\begin{aligned} \mathbf{j}_1 &= \sigma \mathbf{L}_1 \mathbf{R}_1(\alpha) \mathbf{R}_2(\beta) \mathbf{R}_3(\gamma) \mathbf{x} \\ &= \sigma \mathbf{L}_1 \mathbf{R} \mathbf{x} \\ &= \mathbf{L}_1 \mathbf{x}' \end{aligned} \tag{52.37}$$

$$\mathbf{j}_2 = \sigma \mathbf{R}_1(\alpha) \mathbf{L}_2 \mathbf{R}_2(\beta) \mathbf{R}_3(\gamma) \mathbf{x} \tag{52.38}$$

$$\begin{aligned} \mathbf{j}_3 &= \sigma \mathbf{R}_1(\alpha) \mathbf{R}_2(\beta) \mathbf{R}_3(\gamma) \mathbf{L}_3 \mathbf{x} \\ &= \sigma \mathbf{R} \mathbf{L}_3 \mathbf{x} \end{aligned} \tag{52.39}$$

$$[\mathbf{j}_4 | \mathbf{j}_5 | \mathbf{j}_6 |] = \mathbf{I} \qquad (3 \times 3) \text{ identity matrix} \tag{52.40}$$

$$\begin{aligned} \mathbf{j}_7 &= \mathbf{R}_1(\alpha) \mathbf{R}_2(\beta) \mathbf{R}_3(\gamma) \mathbf{x} \\ &= \mathbf{R} \mathbf{x} \\ &= (\mathbf{x}' - \mathbf{t}')/\sigma \end{aligned} \tag{52.41}$$

since

$$\partial \mathbf{R}_1 / \partial \alpha = \mathbf{L}_1 \mathbf{R}_1(\alpha) = \mathbf{R}_1(\alpha) \mathbf{L}_1 \tag{52.42}$$

$$\partial \mathbf{R}_2 / \partial \beta = \mathbf{L}_2 \mathbf{R}_2(\beta) = \mathbf{R}_2(\beta) \mathbf{L}_2 \tag{52.43}$$

$$\partial \mathbf{R}_3 / \partial \gamma = \mathbf{L}_3 \mathbf{R}_3(\gamma) = \mathbf{R}_3(\gamma) \mathbf{L}_3 \tag{52.44}$$

and

$$\mathbf{L}_1 = \begin{pmatrix} 0 & 0 & 0 \\ 0 & 0 & 1 \\ 0 & -1 & 0 \end{pmatrix} \tag{52.45}$$

$$\mathbf{L}_2 = \begin{pmatrix} 0 & 0 & -1 \\ 0 & 0 & 0 \\ 1 & 0 & 0 \end{pmatrix} \tag{52.46}$$

$$\mathbf{L}_3 = \begin{pmatrix} 0 & 1 & 0 \\ -1 & 0 & 0 \\ 0 & 0 & 0 \end{pmatrix} \tag{52.47}$$

The advantage of these **L** matrices is that in many instances the derivative matrix (product) can be written as the original matrix pre- or postmultiplied by the corresponding **L**-matrix [Lucas, 1963].

Curvilinear Coordinate Changes Due to a Datum Transformation

Differentiating Eq.(52.9) with respect to the semimajor axis a and the flattening f, one obtains

$$\begin{pmatrix} dx \\ dy \\ dz \end{pmatrix}_{a,f} = \mathbf{J}_{a,f} \begin{pmatrix} da \\ df \end{pmatrix} \tag{52.48}$$

with (see Soler and van Gelder [1987])

$$\mathbf{J}_{a,f} = \frac{\partial(x,y,z)}{\partial(a,f)} \tag{52.49}$$

and

$$\mathbf{J}_{a,f} = \begin{bmatrix} \cos\phi\cos\lambda/W & a(1-f)\sin^2\phi\cos\phi\cos\lambda/W^3 \\ \cos\phi\sin\lambda/W & a(1-f)\sin^2\phi\cos\phi\sin\lambda/W^3 \\ (1-e^2)\sin\phi/W & (M\sin^2\phi - 2N)(1-f)\sin\phi \end{bmatrix} \tag{52.50}$$

See also Soler and van Gelder [1987] for the second-order derivatives.

Curvilinear Coordinate Changes Due to a Similarity and a Datum Transformation

The curvilinear effects of a redefinition of the coordinate frame due to a similarity transformation and due to a datum transformation are computed by adding the Eqs. (52.34) and (52.48) and substituting them into

$$\begin{pmatrix} d\lambda \\ d\phi \\ dh \end{pmatrix} = \mathbf{J}^{-1} \begin{pmatrix} dx \\ dy \\ dz \end{pmatrix} \tag{52.51}$$

with

$$\begin{pmatrix} dx \\ dy \\ dz \end{pmatrix} = \begin{pmatrix} dx \\ dy \\ dz \end{pmatrix}_7 + \begin{pmatrix} dx \\ dy \\ dz \end{pmatrix}_{a,f} \tag{52.52}$$

52.3 Coordinate Frames Used in Geodesy and Some Additional Relationships

Earth-Fixed

Earth-Fixed Geocentric

From the moment satellites were used to study geodetic aspects of the earth, one had to deal with modeling the motion of the satellite (a point mass) around the earth's center of mass (CoM). The formulation of the equations of motion is the easiest when referred to the CoM. This point became almost naturally the origin of the coordinate frame in which the earth-bound observers were situated. For the orientation of the x- and z-axes, see the discussion of "3-D Polar Coordinates: Spherical" in section 52.2 and the discussion of "Polar Motion" in this section.

Earth-Fixed Topocentric Cartesian

An often used local frame is the earth-fixed topocentric coordinate frame. The origin resides at the position of the observer's instrument. Although in principle arbitrary, one often chooses the x axis pointing east, the y axis pointing north, and the z axis pointing up. This e, n, u frame is again a right-handed frame. With respect to the direction of the local z- or u-axis, various choices are possible: the u axis coincides with the negative direction of the local gravity vector (the first axis of a leveled theodolite) or along the normal perpendicular to the surface of the ellipsoid.

An Important Relationship Using an Orthogonal Transformation

The transformation formulas between a geocentric coordinate frame and a local coordinate frame are (see also Fig. 52.4):

$$\begin{pmatrix} x \\ y \\ z \end{pmatrix} = \mathbf{R}_3\left(-\lambda_a - \frac{\pi}{2}\right)\mathbf{R}_1\left(+\phi_a - \frac{\pi}{2}\right)\begin{pmatrix} e \\ n \\ u \end{pmatrix} + \begin{pmatrix} [N_a + h_a]\cos\phi_a\cos\lambda_a \\ [N_a + h_a]\cos\phi_a\sin\lambda_a \\ [N_a(1-e^2) + h_a]\sin\phi_a \end{pmatrix} \tag{52.53}$$

One should realize that this transformation formula is of the orthogonal type shown in Eq. (52.16):

$$\mathbf{x}' = \mathbf{R}\mathbf{x} + \mathbf{t}' \tag{52.54}$$

whereby

$\mathbf{x}'$ = the geocentric Cartesian vector

$\mathbf{R} = \mathbf{R}_3(-\lambda_a - \pi/2)\mathbf{R}_1(\phi_a - \pi/2)$

$\mathbf{t}'_a$ = the location of a in the (new) geocentric frame, and is equal to

$$\mathbf{t}'_a = \begin{pmatrix} [N_a + h_a]\cos\phi_a\cos\lambda_a \\ [N_a + h_a]\cos\phi_a\sin\lambda_a \\ [N_a(1-e^2) + h_a]\sin\phi_a \end{pmatrix} \tag{52.55}$$

The rotation matrix $\mathbf{R} = \mathbf{R}_3(-\lambda_a - \pi/2)\mathbf{R}_1(\phi_a - \pi/2)$ is given in section 52.2 as Eq. (52.28).

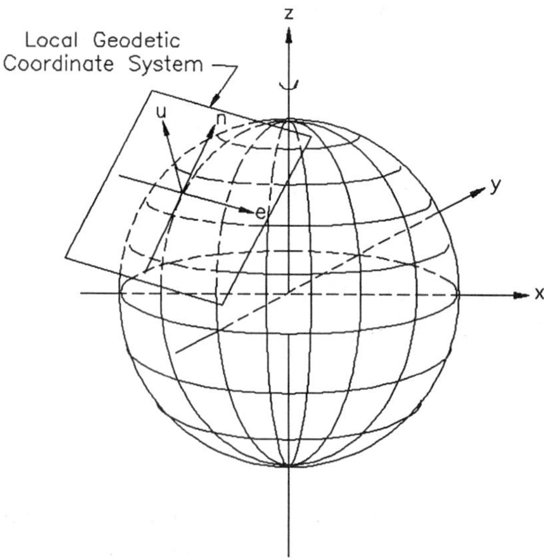

FIGURE 52.4 A geocentric and a local Cartesian coordinate frame.

Given the geocentric coordinates $\{x, y, z\}$ of an arbitrary point (e.g., a satellite) and one wants to compute the local coordinates $\{e, n, u\}$ of that point the local frame being centered at a, then one obtains for the inverse relationship

$$\begin{pmatrix} e \\ n \\ u \end{pmatrix} = \mathbf{R}_1\left(-\phi_a + \frac{\pi}{2}\right)\mathbf{R}_3\left(+\lambda_a + \frac{\pi}{2}\right)\left[\begin{pmatrix} x \\ y \\ z \end{pmatrix} - \begin{pmatrix} [N_a + h_a]\cos\phi_a \cos\lambda_a \\ [N_a + h_a]\cos\phi_a \sin\lambda_a \\ [N_a(1 - e^2) + h_a]\sin\phi_a \end{pmatrix}\right] \tag{52.56}$$

The rotation matrix $\mathbf{R}_1(-\phi_a + \pi/2)\mathbf{R}_3(+\lambda_a + \pi/2)$ is the transpose of the matrix given in section 52.2 as Eq. (52.28).

Earth-Fixed Topocentric Spherical

Satellites orbit the earth at finite distances. For such purposes as visibility calculations, one relates the local e, n, u coordinates to local spherical coordinates El (elevation or altitude angle), Az (azimuth: clockwise positive from the north), and Sr (slant range to the object):

$$\begin{pmatrix} e \\ n \\ u \end{pmatrix} = \mathrm{Sr}\begin{pmatrix} \cos \mathrm{El} \sin \mathrm{Az} \\ \cos \mathrm{El} \cos \mathrm{Az} \\ \sin \mathrm{El} \end{pmatrix} \tag{52.57}$$

The inverse relationships are

$$\begin{pmatrix} \mathrm{El} \\ \mathrm{Az} \\ \mathrm{Sr} \end{pmatrix} = \begin{pmatrix} \arctan[u/\sqrt{e^2 + n^2}] \\ \arctan(e/n) \\ \sqrt{e^2 + n^2 + u^2} \end{pmatrix} \tag{52.58}$$

Note again that the El, Az, Sr form themselves a right-handed (curvilinear) frame.

Some Important Relationships Using Similarity and Datum Transformations

Increasing measurement accuracies and improved insights in the physics of the earth often cause reference frames to be reviewed. For instance, if coordinates of a station are given in an old frame, then with current knowledge of similarity transformation parameters relating the old $\mathbf{x}$ frame to the new $\mathbf{x}'$ frame, the new coordinates can be computed according to

$$\mathbf{x}' = \sigma \mathbf{R}\mathbf{x} + \mathbf{t}' \tag{52.59}$$

In many instances the translation and rotation transformation parameters are small, and the scale parameter σ deviates little from 1, so we introduce the following new symbols:

$$\begin{aligned} \sigma &= 1 + \delta\sigma \\ \alpha &= \delta\epsilon \\ \beta &= \delta\psi \\ \gamma &= \delta\omega \\ t'_x &= \Delta x \\ t'_y &= \Delta y \\ t'_z &= \Delta z \end{aligned} \tag{52.60}$$

Neglecting second-order effects, the rotation matrix $\mathbf{R}$ can be written as the sum of an identity matrix $\mathbf{I}$ and a skew-symmetric matrix $\boldsymbol{\delta}\mathbf{R}$:

$$\mathbf{R} = \mathbf{R}_1(\delta\epsilon)\mathbf{R}_2(\delta\psi)\mathbf{R}_3(\delta\omega) \tag{52.61}$$

$$= \mathbf{I} + \boldsymbol{\delta}\mathbf{R} \tag{52.62}$$

with

$$\boldsymbol{\delta}\mathbf{R} = \begin{pmatrix} 0 & \delta\omega & -\delta\psi \\ -\delta\omega & 0 & \delta\epsilon \\ \delta\psi & -\delta\epsilon & 0 \end{pmatrix} \tag{52.63}$$

Equation (52.59) becomes

$$\mathbf{x}' = (1 + \delta\sigma)(\mathbf{I} + \boldsymbol{\delta}\mathbf{R})\mathbf{x} + \boldsymbol{\Delta}\mathbf{x} \tag{52.64}$$

or, neglecting second-order effects,

$$\mathbf{x}' = \mathbf{x} + \mathbf{dx} \tag{52.65}$$

with

$$\mathbf{dx} = \delta\sigma\mathbf{x} + \boldsymbol{\delta}\mathbf{Rx} + \boldsymbol{\Delta}\mathbf{x} \tag{52.66}$$

See section 52.8 for a variety of parameter sets relating the various reference frames and/or datum values.

Inertial and Quasi-Inertial

Inertial Geocentric Coordinate Frame

For the derivation of the equations of motion of point masses in space we need so-called inertial frames. These are frames where Newton's laws apply. These frames are nonrotating, where point masses either have uniform velocity or are at rest. Popularly speaking, in these frames the stars, or better extragalactic points/quasars, are "fixed" (i.e., not moving in a rotational sense). Since the stars are at such large distances from the earth, it is often sufficient in geodetic astronomy to consider the inertial directions. Instead of the inertial coordinates of the stars we consider the vector **d**, consisting of the three direction cosines. One has to realize that these direction cosines are dependent on only two angles. Consequently, only two direction cosines contain independent information, because the three direction cosines squared sum up to 1.

The two angles are (see Fig. 52.5)

α right ascension
δ declination

$$\begin{pmatrix} X \\ Y \\ Z \end{pmatrix} = \ell\mathbf{d} \tag{52.67}$$

with

$$\mathbf{d} = \begin{pmatrix} \cos\delta & \cos\alpha \\ \cos\delta & \sin\alpha \\ \sin\delta \end{pmatrix} \tag{52.68}$$

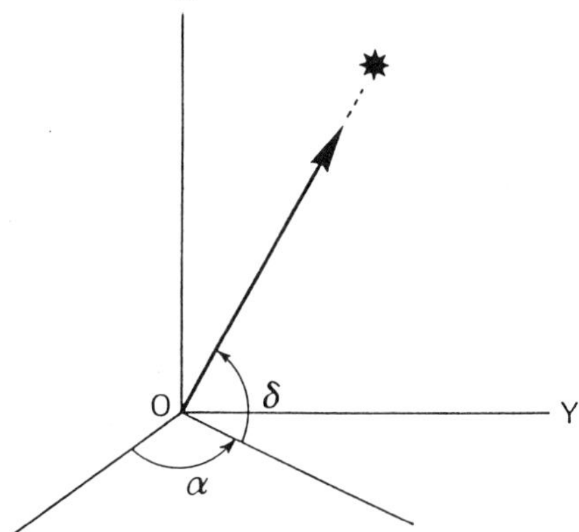

FIGURE 52.5 Direction to a satellite or star: right ascension, α, and declination, δ.

The right ascension is counted counterclockwise positive from the X-axis being defined as the intersection of the earth's equatorial plane and the plane of the earth's orbit around the Sun. One of the points of intersection is called the *vernal equinox:* it is that point in the sky among the stars where the sun appears as viewed from Earth at the beginning of spring in the northern hemisphere. The declination is counted from the equatorial plane in the same manner as the latitude; see Fig. 52.6.

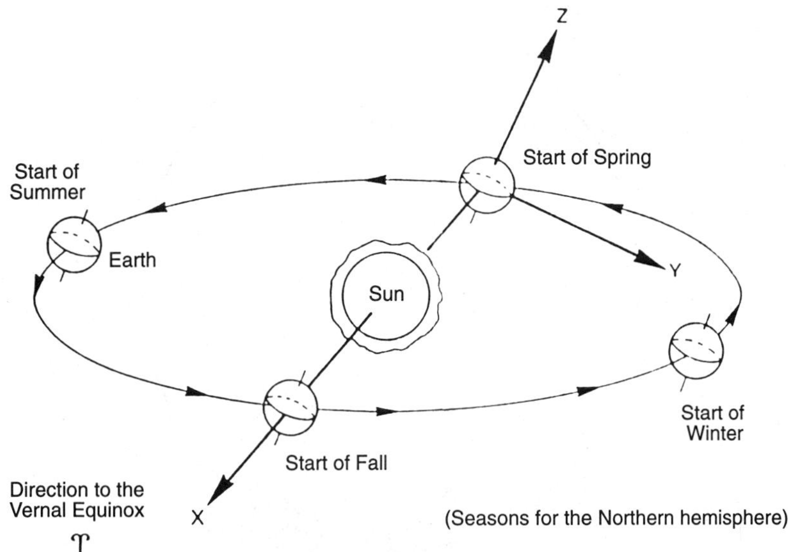

FIGURE 52.6 The (quasi-)inertial reference frame with respect to sun and earth.

Quasi-Inertial Coordinate Frame

In the previous section the position of the origin was not defined yet: the origin of the inertial frame is not to coincide with the center of mass of the earth; since the earth itself orbits around the sun, the center of mass of the earth is subject to accelerations. Similarly, the center of mass of the sun and all its planets is rotating around the center of our galaxy, and the galaxy experiences gravitational forces from other galaxies. A continuation of this reasoning will improve the quality of "inertiality" of the coordinate frame, but the practical application for the description of the motion of earth-orbiting satellites has been completely lost.

In section 52.5 a practical solution is presented: in orientation the frame is as inertial as possible, but the origin has been chosen to coincide with the earth's center of mass. Such frames are called *quasi-inertial* frames. The apparent forces caused by the (small) accelerations of the origin have to be accounted for later.

Relation between Earth-Fixed and Inertial

Satellite equations of motion are easily dealt with in an inertial frame, but we observers are likely to model our positions and relatively slow velocities in an earth-fixed frame. The relationship between these two frames has to be dealt with.

Time and Sidereal Time

The diurnal rotation of the earth is given by its average angular velocity (see also section 52.9):

$$\omega_e = 7.292115 \times 10^{-5} \text{ rad/s} \tag{52.69}$$

This inertial angular velocity results in an average day length of

$$T = \frac{2\pi}{\omega_e} = 86164.1 \text{ s} \tag{52.70}$$

This *sidereal day,* based on the earth's spin rotation with respect to the fixed stars, deviates from our 24 × 60 × 60 = 86,400-s day by 3 min, 55.9 seconds. That is why we see an arbitrary star

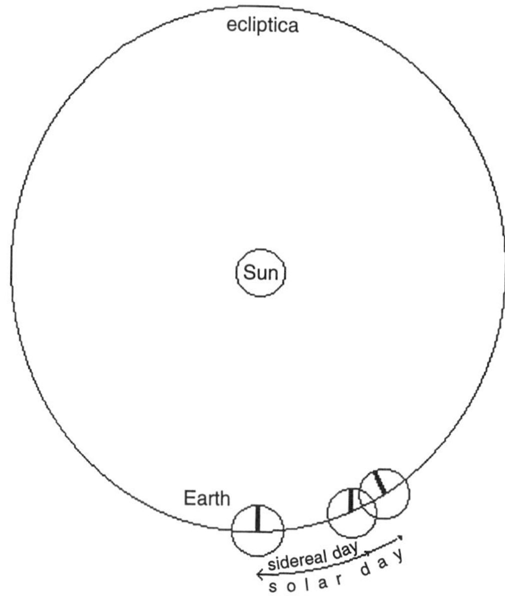

FIGURE 52.7 The earth's orbit around the sun: sidereal day and solar day.

constellation in the same position in the sky each day about four minutes earlier. Our daily lives are based on the earth's spin with respect to the sun, the *solar day.* Since the earth advances about 1 degree per day in its orbit around the sun, the earth has not completed a full spin with respect to the sun when it completes one full turn with respect to the stars; see Fig. 52.7.

For practical purposes the angular velocity must include the effect of precession; see the next section. We have

$$\omega_e^* = 7.2921158553 * 10^{-5} \text{ rad/s} \tag{52.71}$$

The angle between the vernal equinox and the Greenwich meridian as measured along the equator is *Greenwich apparent sidereal time* (GAST). This angle increases in time by ω_e^* per second. With the help of the formula of Newcomb we are able to compute GAST [IERS, 1992]:

$$\text{GAST}(t) = a + bT_u + cT_u^2 + dT_u^3 + \text{ee} \tag{52.72}$$

with

$$\begin{aligned} a &= 18 \text{ h } 41 \text{ min } 50.54841 \text{ s} \\ b &= 8{,}640{,}184.812866 \text{ s/century} \\ c &= 0.093104 \text{ s/century}^2 \\ d &= -0.0000062 \text{ s/century}^3 \\ \text{ee} &= \text{the equation of the equinoxes} \end{aligned} \tag{52.73}$$

T_u is measured in Julian centuries of 36,525 universal days since 1.5 January 2000 (JD_0 = 2,451,545.0). This means that T_u, until the year 2000, is negative. T_u can be computed from

$$T_u = \frac{(\text{JD} - 2{,}451{,}545.0)}{36{,}525} \tag{52.74}$$

when the Julian day number, JD, is given.

Polar Motion

Polar motion, or on a geological time scale "polar wandering," represents the motion of the earth's spin axis with respect to an earth-fixed frame. Polar motion changes our latitudes, since if the z-axis were chosen to coincide with the instantaneous position of the spin axis, our latitudes would change continuously.

Despite the earth's nonelastic characteristics, excitation forces keep polar motion alive. Polar motion is the motion of the instantaneous rotation axis, or celestial ephemeris pole (CEP), with respect to an adopted reference position, the Conventional Terrestrial Pole (CTP) of the Conventional Terrestrial Reference Frame (CTRF). The adopted reference position, or Conventional International Origin (CIO), was the main position of the CEP between 1900 and 1905. Since then the mean CEP has drifted about 10 meters away from the CIO in a direction of longitude 280°.

The transformation from the somewhat earth-fixed frame $\mathbf{x}_{\text{CEP}}$ to $\mathbf{x}_{\text{CTRF}}$ is

$$\mathbf{x}_{\text{CTRF}} = \mathbf{R}_2(-x_p)\mathbf{R}_1(-y_p)\mathbf{x}_{\text{CEP}} \tag{52.75}$$

The position of the pole is expressed in local curvilinear coordinates that have their origin in the CIO. The angle x_p (radians!) increases along the Greenwich meridian south, whereas the angle y_p increases along the meridian $\lambda = 270°$ south; see Fig. 52.8. As we saw in the previous subsection the earth rotates daily around its (moving) CEP axis. Expanding the transformation of Eq. (52.75) but now also including the sidereal rotation of the earth we obtain the following relationship between an inertial reference frame and the CTRF:

$$\mathbf{x}_{\text{CEP}} = \mathbf{R}_3(\text{GAST})\mathbf{x}_{\text{in}} \tag{52.76}$$

This relationship is shown in Fig. 52.9. Combining Eqs. (52.75) and (52.76), we have

$$\mathbf{x}_{\text{CTRF}} = \mathbf{R}_2(-x_p)\mathbf{R}_1(-y_p)\mathbf{R}_3(\text{GAST})\mathbf{x}_{\text{in}} \tag{52.77}$$

or, in short,

$$\mathbf{x}_{\text{CTRF}} = \mathbf{R}_S\mathbf{x}_{\text{in}} \tag{52.78}$$

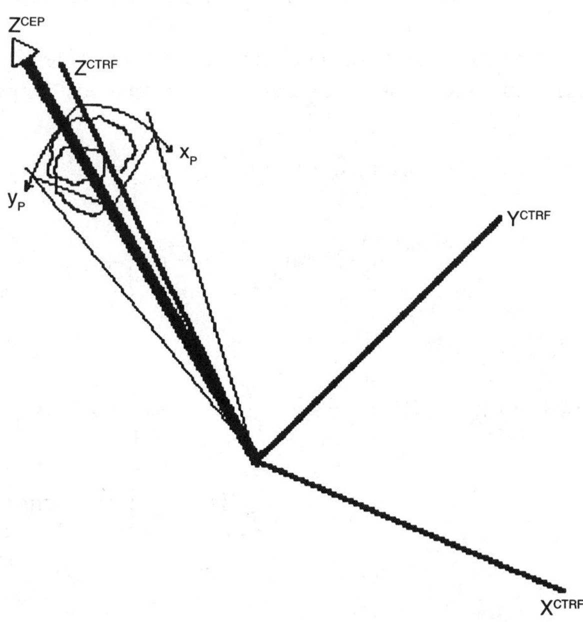

FIGURE 52.8 Local curvilinear pole coordinates x_p, y_p with respect to the Conventional Terrestrial Reference Frame.

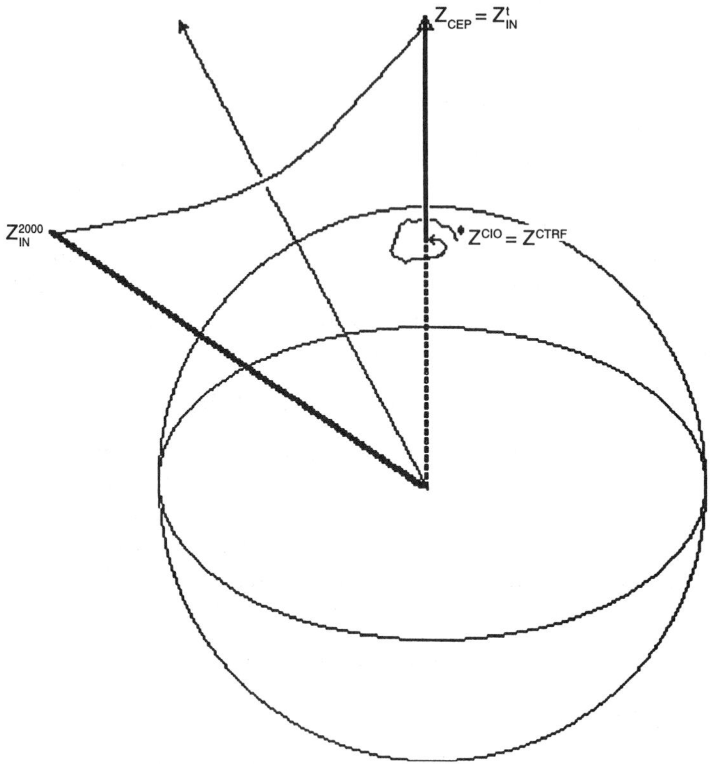

FIGURE 52.9 Relationship between the quasi-earth-fixed frame $\mathbf{x}_{CEP}$ and the inertial frame $\mathbf{x}_{in}$.

where $\mathbf{R}_S$ represents the combined earth rotation due to polar motion and diurnal rotation (length of day).

Two main frequencies make up the polar motion: the Chandler wobble of 435 days (14 months, more or less) and the annual wobble of 365.25 days. A prediction model for polar motion is

$$\begin{aligned} x_p(t) &= x_p(t_0) + \dot{x}_p(t - t_0) + A_s^x \sin\left[\frac{2\pi}{T_1}(t - t_0)\right] + A_c^x \cos\left[\frac{2\pi}{T_1}(t - t_0)\right] \\ &\quad + C_s^x \sin\left[\frac{2\pi}{T_2}(t - t_0)\right] + C_c^x \cos\left[\frac{2\pi}{T_2}(t - t_0)\right] \end{aligned} \tag{52.79}$$

$$\begin{aligned} y_p(t) &= y_p(t_0) + \dot{y}_p(t - t_0) + A_s^y \sin\left[\frac{2\pi}{T_1}(t - t_0)\right] + A_c^y \cos\left[\frac{2\pi}{T_1}(t - t_0)\right] \\ &\quad + C_s^y \sin\left[\frac{2\pi}{T_2}(t - t_0)\right] + C_c^y \cos\left[\frac{2\pi}{T_2}(t - t_0)\right] \end{aligned} \tag{52.80}$$

where

T_1 = 365.25 days
T_2 = 435 days

Similarly, the ever-increasing angle GAST has to be corrected for seasonal variations in ω. These variations are in the order of ± 1 ms. A prediction model for length-of-day variations is

$$\delta\text{UT1} = a \sin 2\pi t + b \cos 2\pi t + c \sin 4\pi t + d \cos 4\pi t \tag{52.81}$$

with

$$
\begin{aligned}
a &= +0.0220 \text{ s} \\
b &= -0.0120 \text{ s} \\
c &= -0.0060 \text{ s} \\
d &= +0.0070 \text{ s} \\
t &= 2000.000 + [(\text{MJD} - 51544.03)/365.2422]
\end{aligned}
$$

With Eq. (52.72), we have

$$\text{GAST} = \text{GAST}_{(t)} + \delta\text{UT1} \tag{52.82}$$

Due to the flattening, f, the $\mathbf{x}_{\text{in}}$ in Eq. (52.77) is not star-fixed but would be considered a proper (quasi-)inertial frame if we froze the frame at epoch t. So, Eq. (52.78) is more properly expressed as

$$\mathbf{x}_{\text{CTRF}} = \mathbf{R}_S \mathbf{x}_{\text{in}}^t \tag{52.83}$$

The motions of the CEP with respect to the stars are called *precession* and *nutation.*

Precession and Nutation

Precession and nutation represent the motion of the earth's spin axis with respect to an inertial frame.

To facilitate comparisons between observations of spatial objects (satellites, quasars, stars) that may be made at different epochs, a transformation is carried out between the inertial frame at epoch t (CEP at t) and the mean position of the inertial frame at an agreed-upon reference epoch t_0. The reference epoch is again 1.5 January 2000, for which $JD_0 = 2{,}451{,}545.0$; see the discussion of "Time and Sidereal Time" in this section. At this epoch we define the Conventional Inertial Reference Frame. So we have

$$\mathbf{x}_{\text{in}}^t = \mathbf{R}_{\text{NP}} \mathbf{x}_{\text{in}}^{2000} \tag{52.84}$$

All masses in the ecliptic plane (earth and sun) and close to it (planets) exert a torque on the tilted equatorial bulge of the earth. The result is that the CEP describes a cone with its half top angle equal to the obliquity ϵ: precession. The period is about 25,800 years. The individual orbits of the planets and moon cause deviations with respect to this cone: nutation. The largest effect is about 9 seconds of arc, with a period of 18.6 years, caused by the inclined orbit of the moon.

The transformation $\mathbf{R}_{\text{NP}}$ is carried out in two steps. The mean position of the CEP is first updated for precession to a mean position at epoch t (now):

$$\mathbf{x}_{\text{in}}^{t,\text{mean}} = \mathbf{R}_P \mathbf{x}_{\text{in}}^{2000,\text{mean}} \tag{52.85}$$

Subsequently, the mean position of the CEP at epoch t is transformed to the true position of the CEP at epoch t due to nutation:

$$\mathbf{x}_{\text{in}}^{t,\text{true}} = \mathbf{R}_N \mathbf{x}_{\text{in}}^{t,\text{mean}} \tag{52.86}$$

Combining all transformations, we have

$$\mathbf{x}_{\text{CTRF}} = \mathbf{R}_S \mathbf{R}_N \mathbf{R}_P \mathbf{x}_{\text{CIRF}} \tag{52.87}$$

$\mathbf{x}_{\text{CTRF}}$ is identical to $\mathbf{x}_{\text{earth-fixed}}^{\text{CIO}}$

$\mathbf{x}_{\text{CIRF}}$ is identical to $\mathbf{x}_{\text{star-fixed}}^{2000}$

The rotation matrices $\mathbf{R}_P$ and $\mathbf{R}_N$ depend on the obliquity ϵ, longitude of sun, moon, etc. We have

$$\mathbf{R}_P = \mathbf{R}_3(-z)\mathbf{R}_2(\theta)\mathbf{R}_3(-\zeta) \tag{52.88}$$

with (see, e.g., IERS [1992])

$$\zeta = 2\,306''.218\,1T_u + 0''.301\,88T_u^2 + 0''.017\,998T_u^3$$
$$\theta = 2\,004''.310\,9T_u - 0''.426\,65T_u^2 - 0''.041\,833T_u^3$$
$$z = 2\,306''.218\,1T_u + 1''.094\,68T_u^2 + 0''.018\,203T_u^3$$

and

$$\mathbf{R}_N = \mathbf{R}_1(-\epsilon - \Delta\epsilon)\mathbf{R}_3(-\Delta\psi)\mathbf{R}_1(\epsilon) \tag{52.89}$$

with

$$\epsilon = 84\,381''.448 - 46''.815\,0T_u - 0''.018\,203T_u^2 + 0''.001\,813T_u^3$$

For the nutation in longitude, $\Delta\psi$, and the nutation in obliquity, $\Delta\epsilon$, a trigonometric series expansion is available consisting of $106 \times 2 \times 2$ parameters and five arguments: mean anomaly of the moon, the mean anomaly of the sun, mean elongation of the moon from the sun, the mean longitude of the ascending node of the moon, and the difference between the mean longitude of the moon and the mean longitude of the ascending node of the moon. There are 2×2 constants in each term: a sine coefficient, a cosine coefficient, a time-invariant coefficient, and a time-variant coefficient.

The reader is referred to the earth orientation literature for more detail on these transformations, such as Mueller [1969], Moritz and Mueller [1988], and IERS [1992].

52.4 Mapping

Introduction

The art of mapping is referred to a technique that maps information from an n-dimensional space $\mathbb{R}_n$ to an m-dimensional space $\mathbb{R}_m$. Often information belonging to a high-dimensional space is mapped to a low-dimensional space. In other words, one has

$$n < m \tag{52.90}$$

In geodesy and surveying one may want to map a three-dimensional world onto a two-dimensional world. In photogrammetry, an aerial photograph can be viewed as a mapping procedure as well: a two-dimensional photo of the three-dimensional terrain. In least squares adjustment we map an n-dimensional observation space onto a u-dimensional parameter space. In this section we restrict our discussion to the mapping

$$\mathbb{R}_n \rightarrow \mathbb{R}_m \tag{52.91}$$

with

$$n = 3 \tag{52.92}$$

$$m = 2 \tag{52.93}$$

A three-dimensional earth, approximated by a sphere or, better, by an ellipsoid of revolution, cannot be mapped onto a two-dimensional surface, which is flat at the start (plane) or can be made to be flat (the surface of a cylinder or a cone), without distorting the original relative positions in $\mathbb{R}_3$. Any figure or, better, the relative positions between an arbitrary number of points on a sphere

Table 52.1 Errors in Length dS (km), Angle $d\alpha$ (arcseconds), and Height dh (km)[a]

Diameter, D	Length, S	dS		$d\alpha$		dh
km	km	m	ppm	arcseconds	ppm	m
0.100	0.050	−0.000	−0.000	0.000	0.000	0.000
0.200	0.100	−0.000	−0.000	0.000	0.000	0.001
0.500	0.250	−0.000	−0.000	0.000	0.000	0.005
2.000	1.000	−0.000	−0.003	0.001	0.005	0.078
5.000	2.500	−0.000	−0.019	0.007	0.032	0.491
20.000	10.000	−0.003	−0.308	0.110	0.509	7.848
50.000	25.000	−0.048	−1.925	0.688	3.184	49.050
200.000	100.000	−3.080	−30.797	11.002	50.937	784.790
500.000	250.000	−48.123	−192.494	68.768	318.372	4904.409
2000.000	1000.000	−3084.329	−3084.329	1101.340	5098.796	78,319.621

[a] Depending on the diameter D (km) of a project or distance $S = (D/2)$ from the center of the project for an equilateral triangle with sides of S km.

or ellipsoid, will also be distorted when mapped onto a plane, cylinder, or cone. The distortions of the figure (or part thereof) will increase with the area. Likewise, the mapping will introduce distortions that will become larger as the extent of the area to be mapped increases.

If one approximates (maps) a sphere of the size of the earth, the radius R being

$$R \approx 6371.000 \text{ km} \tag{52.94}$$

onto a plane tangent in the center of one's engineering project of diameter D km, one finds increased errors in lengths, angles, as well as heights the further one gets away from the center of the project. Table 52.1 lists these errors in distance dS, angle $d\alpha$, and height dh, if one assumes the following case: one measures in the center of the project one angle of 60° and two equal distances of S km. In the plane assumption we would find in the two terminal points of both lines: two equal angles of 60° and a distance between them of exactly S km. Basically we have an equilateral triangle. In reality, on the curved, spherical earth we would measure angles larger than 60° and a distance between them shorter than S km. The error dh shows how the Earth curves away from underneath the tangent plane in the center of the project.

The table shows that errors in length and angle of larger than 1 part per million start to occur for project diameters larger than 20 km. Height differences obtained through leveling would be accurate enough, but vertical angles would start deviating from 90° by 1 arcsecond per 30 m. Within an area of 20 km fancy mapping procedures would not be needed to avoid errors of 1 ppm. The trouble starts if one wants to map an area of the size of the state of Indiana. A uniform strict mapping procedure has to be adhered to if one wants to work in one consistent system of mapping coordinates. In practice, a state of the size of Indiana is actually divided into two regions to keep the distortions within bounds. One may easily reduce the errors by a factor of 2 by making the plane not tangent to the sphere but by lowering the plane from the center of the project by such an amount that the errors of dS in the center of the area are equal but of opposite sign to the errors dS at the border of the area (see Fig. 52.10). The U.S. State Plane Coordinate Systems are based on this practice.

As an alternative one may choose the mapping plane to be cylindrical or conical so that the mapping plane "follows" the earth's curvature at least in one direction (see Figs. 52.11 and 52.12). Also with these alternatives distortions are reduced even further by having the cylindrical or conical surface not tangent to the sphere or ellipsoid but intersecting the surface to be mapped just below the tangent point. After the mapping the cylinder or cone can be "cut" and be made into a 2-dimensional map. A cylinder and a cone are called *developable surfaces.*

The notion of distortion of a figure is applied to different elements of a figure. If a (spherical) triangle is mapped, one may investigate how the length of a side or the angle between two sides is distorted in the mapping plane.

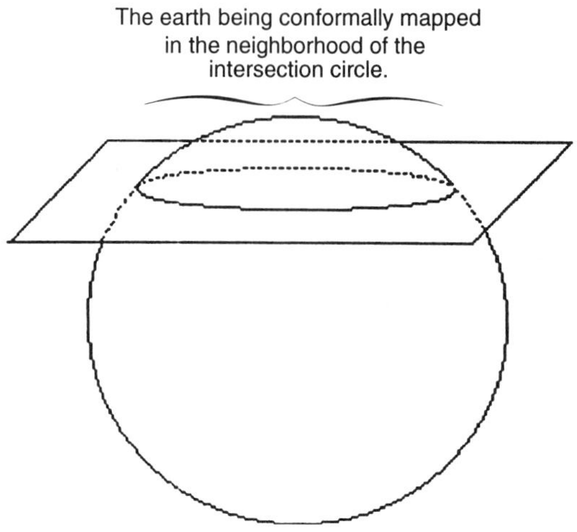

The earth being conformally mapped
in the neighborhood of the
intersection circle.

FIGURE 52.10 Mapping a sphere to a plane or lowered plane.

Two Worlds

We live in $\mathbb{R}_3$, which is to be mapped into $\mathbb{R}_2$. In $\mathbb{R}_3$ we need three quantities to position ourselves; $\{x, y, z\}$, $\{\lambda, \psi, h\}_R$, or $\{\lambda, \phi, h\}_{a,f}$, see section 52.2. In $\mathbb{R}_2$ we need only two quantities, such as two Cartesian mapping coordinates $\{X, Y\}$ or two polar mapping coordinates $\{r, \alpha\}$.

$$\text{Real world} \quad \longrightarrow \quad \text{Mapped world}$$

$$\left.\begin{matrix} \{x, y, z\} \\ \text{or} \\ \{\lambda, \psi, h\}_R \\ \text{or} \\ \{\lambda, \phi, h\}_{a,f} \end{matrix}\right\} \longrightarrow \left\{\begin{matrix} \{X, Y\} \\ \text{or} \\ \{r, \alpha\} \end{matrix}\right.$$

The mapping M may be written symbolically as

$$\{r, \alpha\} = M\{x, y, z\} \tag{52.95}$$

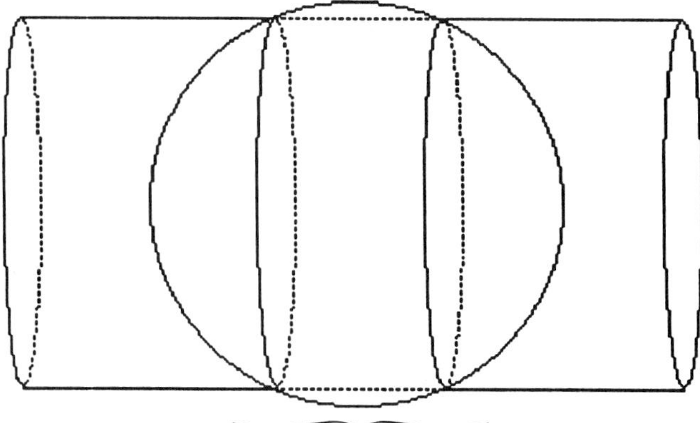

The earth being conformally mapped
in the neighborhood of two
intersection circles.

FIGURE 52.11 A cylinder trying to follow the earth's curvature.

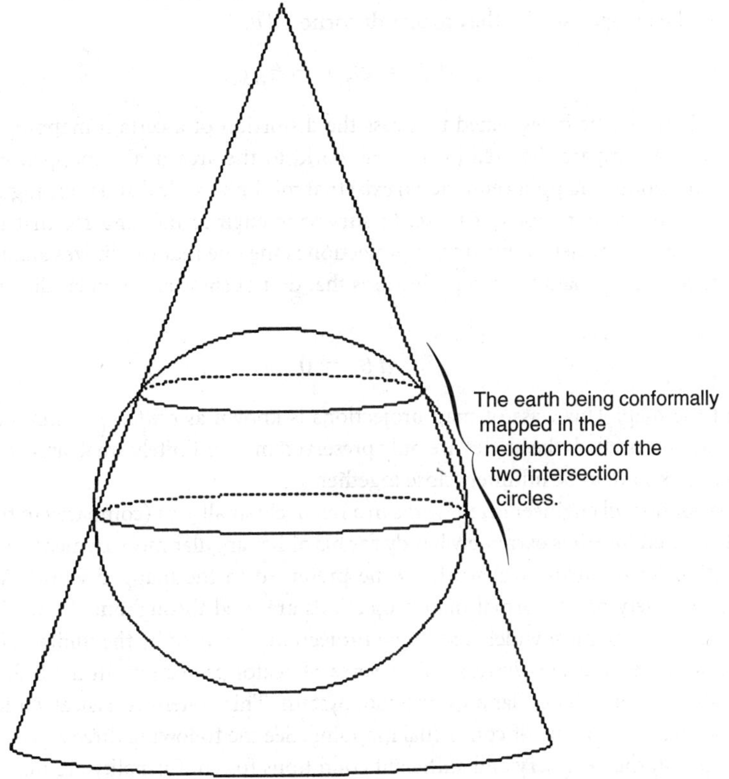
The earth being conformally
mapped in the
neighborhood of the
two intersection
circles.

FIGURE 52.12 A cone trying to follow the Earth's curvature.

or

$$\{X, Y\} = M'\{\lambda, \phi, h\}_{a,f} \tag{52.96}$$

M represents a mere mapping "prescription" how the $\mathbb{R}_3$ world is condensed into $\mathbb{R}_2$ information. Note that the computer era made it possible to "store" the $\mathbb{R}_3$ world also digitally in $\mathbb{R}_3$ (a file with three numbers per point). Only at the very end, if that information is to be presented on a map or computer screen, do we map to $\mathbb{R}_2$.

The analysis of distortion is, in this view, the mere comparison of corresponding geometrical elements in the real world and in the mapped world. A distance $s(x, y, z)$ between points i and j in the real world is compared to a distance $S = S(X, Y)$ in the mapped world. A scale distortion may be defined as the ratio

$$\sigma = \frac{S(X_i, Y_i, X_j, Y_j)}{s(x_i, y_i, z_i, x_j, y_j, z_j)} \tag{52.97}$$

or, for infinitely small distances,

$$\sigma = \frac{dS(dX_{ij}, dY_{ij})}{ds(dx_{ij}, dy_{ij}, dz_{ij})} \tag{52.98}$$

where

$$x_{ij} = x_j - x_i \quad \text{and so forth} \tag{52.99}$$

Similarly, an angle θ_{jik} in point i to points j and k in the real world is being compared to an angle Θ_{jik} in the mapped world, the angular distortion $d\,\theta$:

$$d\,\theta = \Theta_{jik} - \theta_{jik} \tag{52.100}$$

Other features may be investigated to assess the distortion of a certain mapping; for example, one may want to compare the area in the real world to the area in the mapped world. Various mapping prescriptions (mapping equations) exist that minimize scale distortion, angular distortion, area distortion, or combinations of these. In surveying engineering, and for that matter in civil engineering, the most widely applied map projection is the one that minimizes angular distortion. Moreover, there exists a class of map projections that do not show any angular distortion; in other words,

$$d\,\theta = 0 \tag{52.101}$$

throughout the map! This class of map projections is known as *conformal map projections.* One word of caution is needed: the angles are only preserved in an infinitely small area. In other words, the points $i,\ j,\ k$ have to be infinitely close together.

The surveyor or civil engineer often works in a relatively small area (compared to the dimensions of the earth). Therefore, it is extremely handy for his or her angular measurements from theodolite or total station, made in the real world, to be preserved in the mapped world. As a matter of fact, any of a variety of conformal map projections are used throughout the world by national mapping agencies. The most widely used map projection, also used by the military, is a conformal map projection, the so-called *Universal* Transverse Mercator projection. In the U.S. all states have adopted some sort of a State Plane Coordinate System. This system is basically a local reference frame based on a certain type of conformal mapping. See the following three subsections.

Without proof, the necessary and sufficient conditions for conformality are that the real world coordinates p and q and the mapping coordinates X and Y fulfill the Cauchy-Riemann equations or conditions:

$$\begin{aligned} \frac{\partial p}{\partial X} &= +\frac{\partial q}{\partial Y} \\ \frac{\partial p}{\partial Y} &= -\frac{\partial q}{\partial X} \end{aligned} \tag{52.102}$$

Purposely, the real world variables p and q have not been identified. First we want to enforce the natural restriction on $p,\ q,\ X,\ Y$: they have to be isometric coordinates. The mapping coordinates are often isometric by definition; however, the real world coordinates, as the longitude λ and the spherical latitude ψ (or geodetic latitude ϕ), are not isometric.

For instance, one arcsecond in longitude expressed in meters is very latitude-dependent, and moreover, is not equal to one arcsecond in latitude in the very same point; see Table 52.2.

Conformal Mapping Using Cartesian Differential Coordinates

In principle, we have four choices to map from $\mathbb{R}_3$ to $\mathbb{R}_2$:

A1: 3-D Cartesian → 2-D Cartesian
A2: 3-D Cartesian → 2-D Curvilinear

B1: 3-D Curvilinear → 2-D Cartesian
B2: 3-D Curvilinear → 2-D Curvilinear

Although all four modes have known applications, we treat an example in this section with the B1 mode of mapping, whereas the following subsection deals with an example from mode B2.

Table 52.2 Radius of Curvature in the Meridian M, Radius of Curvature in the Prime Vertical N, and Metric Equivalence of One Arcsecond in Ellipsoidal/Spherical Longitude λ (Meters) and in Ellipsoidal/Spherical Latitude ϕ/ψ (Meters) as a Function of Geodetic Latitude ϕ and Spherical Latitude ψ

			$1''$ in λ	$1''$ in ϕ	$1''$ in λ	$1''$ in ψ
ϕ/ψ	M	N	Ellipsoid		Sphere	
00.0	6,335,439	6,378,137	30.922	30.715	30.887	30.887
10.0	6,337,358	6,378,781	30.455	30.724	30.418	30.887
20.0	6,342,888	6,380,636	29.069	30.751	29.025	30.887
30.0	6,351,377	6,383,481	26.802	30.792	26.749	30.887
40.0	6,361,816	6,386,976	23.721	30.843	23.661	30.887
50.0	6,372,956	6,390,702	19.915	30.897	19.854	30.887
60.0	6,383,454	6,394,209	15.500	30.948	15.444	30.887
70.0	6,392,033	6,397,072	10.607	30.989	10.564	30.887
80.0	6,397,643	6,398,943	05.387	31.017	05.364	30.887
90.0	6,399,594	6,399,594	00.000	31.026	00.000	30.887

Ellipsoidal values for WGS84: $a = 6{,}378{,}137$ m, $1/f = 298.257\,223\,563$.
Spherical values: $R = 6{,}371{,}000$ m.

From Eqs. (52.29) and (52.30) we have

$$\begin{aligned} d\lambda_m &= (N+h)\cos\phi d\lambda_{\text{rad}} \\ d\phi_m &= (M+h)d\phi_{\text{rad}} \end{aligned} \tag{52.103}$$

A line element (small distance) in the real world (on the ellipsoid, $h = 0$) is

$$ds^2 = d\lambda_m^2 + d\phi_m^2 \tag{52.104}$$

A line element (small distance) in the mapped world (on paper) is

$$dS^2 = dX^2 + dY^2 \tag{52.105}$$

Equation (52.104) leads to

$$ds^2 = N^2\cos^2\phi d\lambda_{\text{rad}}^2 + M^2 d\phi_{\text{rad}}^2 \tag{52.106}$$

or

$$ds^2 = N^2\cos^2\phi\left(d\lambda_{\text{rad}}^2 + \frac{M^2}{N^2\cos^2\phi}d\phi_{\text{rad}}^2\right) \tag{52.107}$$

Since we want to work with isometric coordinates in the real world (note that the mapping coordinates $\{X, Y\}$ are already isometric), we introduce the new variable dq. In other words, Eq. (52.107) becomes

$$ds^2 = N^2\cos^2\phi(d\lambda^2 + dq^2) \tag{52.108}$$

So, we have

$$dq = \frac{M}{N\cos\phi}d\phi_{\text{rad}} \tag{52.109}$$

Upon integration of Eq. (52.109) we obtain the isometric latitude q:

$$q = \ln\left[\tan\left(\frac{\pi}{4} + \frac{\phi}{2}\right)\left(\frac{1 - e\sin\phi}{1 + e\sin\phi}\right)^{e/2}\right] \tag{52.110}$$

The isometric latitude for a sphere ($e = 0$) becomes simply

$$q = \ln \tan\left(\frac{\pi}{4} + \frac{\psi}{2}\right) \tag{52.111}$$

Equating the variables from Eqs. (52.105) and (52.108) we have the mapping M between Cartesian mapping coordinates $\{X, Y\}$ and curvilinear coordinates $\{\lambda, q\}$ for both the sphere and the ellipsoid. Note that $\{\lambda, q\}$ play the role of the isometric coordinates $\{p, q\}$ in the Cauchy-Riemann equations.

So the mapping equations are simply,

$$\begin{aligned} X &= \sigma\lambda \\ Y &= \sigma q \end{aligned} \tag{52.112}$$

or, for the sphere,

$$\begin{aligned} X &= \sigma\lambda \\ Y &= \sigma \ln \tan\left(\frac{\pi}{4} + \frac{\psi}{2}\right) \end{aligned} \tag{52.113}$$

Equations (52.113) represent the conformal mapping equations from the sphere to $\mathbb{R}_2$. These are the formulas of the well-known Mercator projection (cylindrical type). Inspection of the linear scale σ reveals that this factor depends on the term $M/(N \cos\phi)$ for the ellipsoid or $1/\cos\phi$ for the sphere. This means that the linear distortion is only latitude-dependent. In order to minimize this distortion, we simply apply this mapping to regions that are elongated in the longitudinal direction, where the linear distortion is constant. In case we want to map an arbitrarily oriented elongated region, we simply apply a coordinate transformation. In the subsection on "Coordinate Transformations and Conformal Mapping" we will perform such a coordinate transformation on these mapping equations.

So far, we have mapped a spherical quadrangle $d\lambda, d\psi$ or ellipsoidal "quadrangle" $d\lambda, d\phi$ to a planar quadrangle dX, dY; see Fig. 52.13.

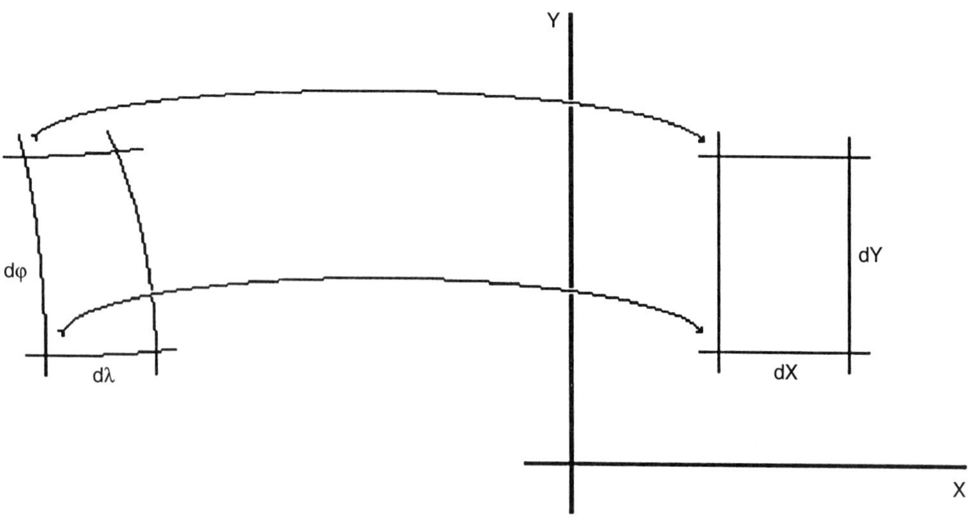

FIGURE 52.13 Spherical/ellipsoidal quadrangle mapped onto a planar Cartesian quadrangle.

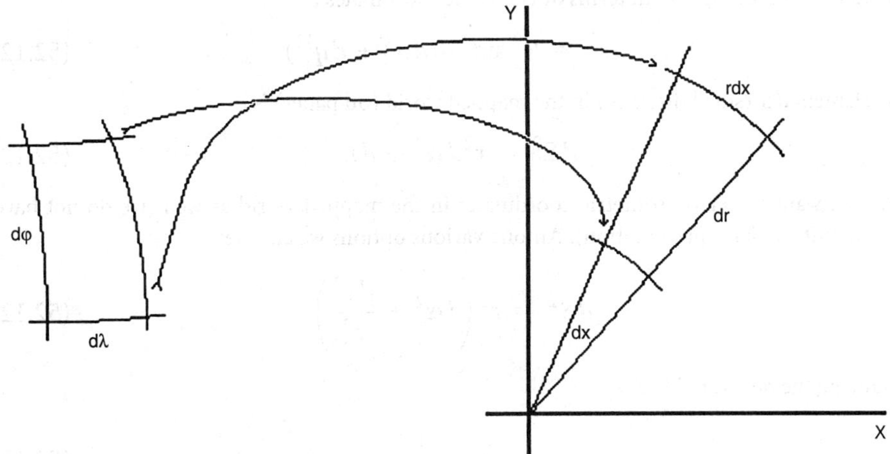

FIGURE 52.14 Spherical/ellipsoidal quadrangle mapped onto a planar polar quadrangle.

Conformal Mapping Using Polar Differential Coordinates

The alternative is to map an ellipsoidal/spherical quadrangle onto a polar quadrangle. Figure 52.14 shows that one option is to look for a mapping between the polar mapping coordinates $\{r, \alpha\}$ and the real world coordinates $\{\lambda, \phi\}$ or $\{\lambda, \psi\}$.

The similarity of roles played by the radius r and the latitude ϕ or ψ is more apparent if we view the *colatitude* θ, since this real world coordinate radiates from one point as the radius r does:

$$\theta = \frac{\pi}{2} - \psi \qquad (\text{or } \phi) \tag{52.114}$$

Considering now the line element ds on a sphere, we have

$$ds^2 = R^2 \sin^2 \theta d\lambda_{\text{rad}}^2 + R^2 d\theta_{\text{rad}}^2 \tag{52.115}$$

or

$$ds^2 = R^2 \sin^2 \theta \left(d\lambda_{\text{rad}}^2 + \frac{d\theta_{\text{rad}}^2}{\sin^2 \theta} \right) \tag{52.116}$$

Introducing the new variable dq', we have

$$dq' = \frac{d\theta_{\text{rad}}}{\sin \theta} \tag{52.117}$$

Integration of Eq. (52.117) gives the isometric colatitude q'

$$q' = \int dq = \int \frac{d\theta}{\sin \theta} \tag{52.118}$$

or

$$q' = -\ln\left(\cot \frac{\theta}{2}\right) \tag{52.119}$$

A line element on the sphere in terms of isometric coordinates is

$$ds^2 = R^2 \sin^2\theta(d\lambda^2 + dq'^2) \tag{52.120}$$

A line element dS (small distance) in the mapped world (on paper) is

$$dS^2 = r^2 d\alpha^2 + dr^2 \tag{52.121}$$

Now we want to derive isometric coordinates in the mapped world as well (we do not have a Cartesian but a polar representation). Among various options we choose

$$dS^2 = r^2\left(d\alpha^2 + \frac{dr^2}{r^2}\right) \tag{52.122}$$

Introducing the new variable $d\rho$,

$$d\rho = \frac{dr}{r} \tag{52.123}$$

Upon integration of Eq. (52.123), we obtain the isometric radius ρ:

$$\rho = \int d\rho = \int \frac{dr}{r} = \ln r + c_r \tag{52.124}$$

or

$$r = e^{\rho} \qquad r^2 = e^{2\rho} \tag{52.125}$$

The line element dS becomes with the new variable

$$dS^2 = e^{2\rho}(d\alpha^2 + d\rho^2) \tag{52.126}$$

The line element ds was

$$ds^2 = \mathbf{R}^2 \sin^2\theta(d\lambda^2 + dq'^2) \tag{52.127}$$

So the mapping equations are simply

$$\begin{aligned} \alpha &= \sigma\lambda \\ \rho &= \sigma q' \end{aligned} \tag{52.128}$$

or

$$\begin{aligned} \alpha &= \sigma\lambda \\ \ln r &= \sigma\left[-\ln\left(\cot\frac{\theta}{2}\right)\right] \end{aligned} \tag{52.129}$$

If one appropriately chooses the integration constant c_r, Eq. (52.126), to be

$$c_r = \ln 2 \tag{52.130}$$

the mapping Eqs. (52.129) become

$$\begin{aligned} \alpha &= \sigma\lambda \\ \ln r &= \sigma\left[\ln 2 - \ln\left(\cot\frac{\theta}{2}\right)\right] = \ln\left(2\tan\frac{\theta}{2}\right) \end{aligned} \tag{52.131}$$

or

$$\begin{aligned} \alpha &= \sigma\lambda \\ r &= 2\sigma \tan\frac{\theta}{2} \end{aligned} \tag{52.132}$$

Equations (52.132) represent also conformal mapping equations from the sphere to $\mathbb{R}_2$. They are the formulas of the well-known stereographic projection (planar type). Other choices of integration constants and integration interval would have led to the Lambert conformal projection (conical type).

The approaches laid out in this subsection and the preceding one are the theoretical basis of the U.S. State Plane Coordinate Systems. The reader is referred to Stem [1991] for the formulas of the ellipsoidal equivalents.

Coordinate Transformations and Conformal Mapping

The two examples treated in the preceding two subsections can be treated for any arbitrary curvilinear coordinates. The widely used Transverse Mercator projection for the sphere is easily derived using a simple coordinate transformation. Rather than having the origin of the (co)latitude variable at the pole, we define a similar pole at the equator, and the new equator will be perpendicular to the old equator.

Having mapping poles at the equator leads to transverse types of conformal mapping. If the pole is neither at the north pole nor at the equator, we obtain oblique variants of conformal mapping.

For the Transverse Mercator, the new equator may pass through a certain (old) longitude λ_0. For a UTM projection this λ_0 has specified values; for a State Plane Coordinate System the longitude λ may define the central meridian in a particular (part of the) state.

Two successive rotations will bring the old **x**-frame to the new **x**′-frame (see section 52.2):

$$\mathbf{x}' = \mathbf{R}_1(\pi/2)\mathbf{R}_3(\lambda_0)\mathbf{x} \tag{52.133}$$

The original **x**-frame expressed in curvilinear coordinates is

$$\mathbf{x} = R\begin{pmatrix} \cos\psi\cos\lambda \\ \cos\psi\sin\lambda \\ \sin\psi \end{pmatrix} \tag{52.134}$$

The new **x**’-frame expressed in curvilinear coordinates is

$$\mathbf{x}' = R\begin{pmatrix} \cos\psi'\cos\lambda' \\ \cos\psi'\sin\lambda' \\ \sin\psi' \end{pmatrix} \tag{52.135}$$

Multiplying out the rotations in Eqs. (52.135) we get

$$\mathbf{x}' = R\begin{pmatrix} \cos\psi'\cos\lambda' \\ \cos\psi'\sin\lambda' \\ \sin\psi' \end{pmatrix} = R\begin{pmatrix} x\cos\lambda_0 + y\sin\lambda_0 \\ z \\ x\sin\lambda_0 - y\cos\lambda_0 \end{pmatrix} \tag{52.136}$$

Substituting Eq. (52.134) into Eq. (52.136) and dividing by R, we obtain

$$\begin{pmatrix} \cos\psi'\cos\lambda' \\ \cos\psi'\sin\lambda' \\ \sin\psi' \end{pmatrix} = \begin{pmatrix} \cos\psi\cos\lambda\cos\lambda_0 + \cos\psi\sin\lambda\sin\lambda_0 \\ \sin\psi \\ \cos\psi\cos\lambda\sin\lambda_0 - \cos\psi\sin\lambda\cos\lambda_0 \end{pmatrix} \tag{52.137}$$

which directly leads to

$$
\begin{aligned}
\tan\lambda' &= \frac{\sin\psi}{\cos\psi\,\cos(\lambda-\lambda_0)} = \frac{\tan\psi}{\cos(\lambda-\lambda_0)} \\
\tan\psi' &= \frac{\cos\psi\sin(\lambda-\lambda_0)}{\sqrt{\cos^2\psi\cos^2(\lambda-\lambda_0)+\sin^2\psi}} = \frac{\sin(\lambda-\lambda_0)}{\sqrt{\cos^2(\lambda-\lambda_0)+\tan^2\psi}}
\end{aligned}
\tag{52.138}
$$

The new longitudes λ' and latitudes ψ' are subjected to a (Normal) Mercator projection according to

$$
\begin{aligned}
X' &= \sigma\lambda' \\
Y' &= \sigma q' = \sigma\ln\left[\tan\left(\frac{\pi}{4}+\frac{\psi'}{2}\right)\right]
\end{aligned}
\tag{52.139}
$$

The Transverse Mercator projection with respect to the central meridian λ_0 is obtained by a simple rotation about -90 degrees. The final mapping equations are

$$
\begin{pmatrix} X \\ Y \\ 0 \end{pmatrix} = \mathbf{R}_3\left(-\frac{\pi}{2}\right)\begin{pmatrix} X' \\ Y' \\ 0 \end{pmatrix} = \begin{pmatrix} -Y' \\ X' \\ 0 \end{pmatrix}
\tag{52.140}
$$

When we start with ellipsoidal curvilinear coordinates, we cannot apply this procedure directly. However, when we follow a two-step procedure—mapping the ellipsoid conformal to the sphere, and then using the "rotated conformal" mapping procedure as just described—the treatise in this section will have a more general validity.

For more details on conformal projections using ellipsoidal coordinates, the reader should consult Maling [1993], Stem [1991], and others.

52.5 Basic Concepts in Mechanics

Equations of Motion of a Point Mass in an Inertial Frame

To understand the motion of a satellite around the earth, we resort to two fundamental laws of physics: Isaac Newton's second law (the law of inertia) and Newton's law of gravitation.

Law of Inertia

The Second Law of Newton (the Law of Inertia) is as follows:

$$
F = ma
\tag{52.141}
$$

The mass of a point mass m is the constant ratio that experimentally exists between the force $\mathbf{F}$, acting on that point mass, and the acceleration that is the result of that force.

The acceleration $\mathbf{a}$, the velocity $\mathbf{v}$, and the distance $\mathbf{s}$ are related as follows:

$$
a = \frac{dv}{dt} = \frac{d^2s}{dt^2}
\tag{52.142}
$$

Equation (52.141) can be written in vector form (see Fig. 52.15):

$$
\mathbf{F} = m\mathbf{a} \quad \text{or} \quad \begin{pmatrix} F_x \\ F_y \\ F_z \end{pmatrix} = m\begin{pmatrix} \mathbf{a}_x \\ \mathbf{a}_y \\ \mathbf{a}_z \end{pmatrix} = m\begin{pmatrix} \ddot{X} \\ \ddot{Y} \\ \ddot{Z} \end{pmatrix} = m\ddot{\mathbf{X}}
\tag{52.143}
$$

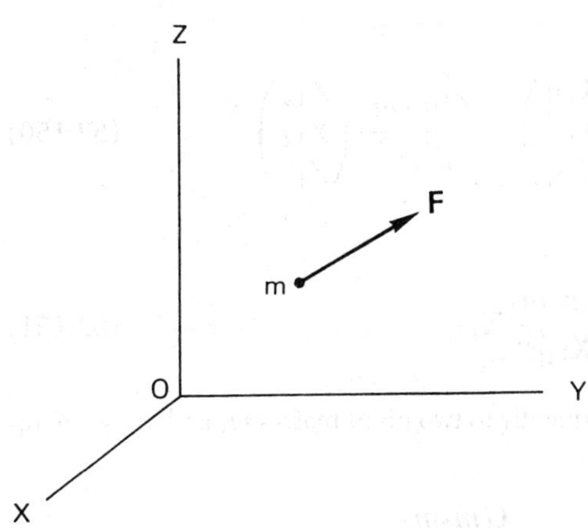

FIGURE 52.15 The acceleration of a point mass m.

with

$$\ddot{X} = \frac{d^2X}{dt^2} \quad \text{and similarly for } \ddot{Y} \text{ and } \ddot{Z}. \tag{52.144}$$

Law of Gravitation

Until now we did not mention the cause of the force **F** acting on point mass m. If this force is caused by the presence of a second point mass, then Newton's law of gravitation says

$$F = G\frac{m_1 m_2}{|\mathbf{X}_{12}|^2} \tag{52.145}$$

Two point masses m_1 and m_2 attract each other with a force that is proportional to the masses of each point mass and inversely proportional to the square of the distance between them, $|\mathbf{X}_{12}|$.

$$G = 6.67259 \pm 0.00085 \times 10^{-11}\text{m}^3\,\text{kg}^{-1}\,\text{s}^{-2} \tag{52.146}$$

is the gravitation constant (e.g., Cohen and Taylor [1988]), and

$$|\mathbf{X}_{12}|^2 = (X_2 - X_1)^2 + (Y_2 - Y_1)^2 + (Z_2 - Z_1)^2 \tag{52.147}$$

The force **F** will be written in vector form (see Fig. 52.16):

$$\mathbf{F}_{12} = \begin{pmatrix} F_{12X} \\ F_{12Y} \\ F_{12Z} \end{pmatrix} = |\mathbf{F}_{12}| \begin{pmatrix} \sin\alpha \\ \sin\beta \\ \sin\gamma \end{pmatrix} = |\mathbf{F}_{12}| \begin{pmatrix} X_{12}/|\mathbf{X}_{12}| \\ Y_{12}/|\mathbf{X}_{12}| \\ Z_{12}/|\mathbf{X}_{12}| \end{pmatrix} \tag{52.148}$$

with

$$X_{12} = X_2 - X_1 \qquad \text{and so forth.} \tag{52.149}$$

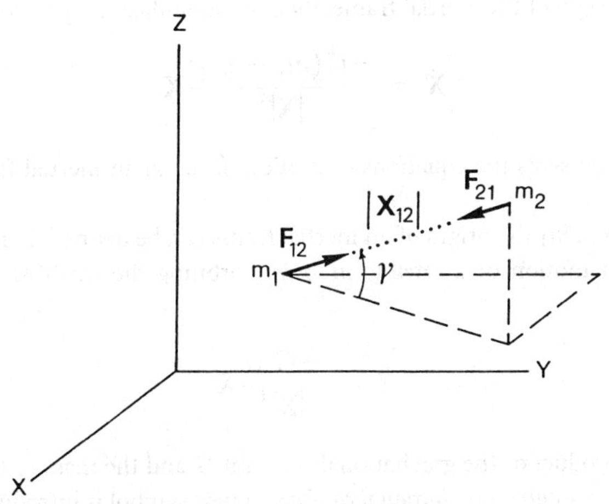

FIGURE 52.16 The attracting force $\mathbf{F}_{12}$ between two point masses m_1 and m_2.

Substituting Eq. (52.148) into Eq. (52.145),

$$\mathbf{F}_{12} = \frac{Gm_1m_2}{|\mathbf{X}_{12}|^2}\begin{pmatrix} X_{12}/|\mathbf{X}_{12}| \\ Y_{12}/|\mathbf{X}_{12}| \\ Z_{12}/|\mathbf{X}_{12}| \end{pmatrix} = \frac{Gm_1m_2}{|\mathbf{X}_{12}|^3}\begin{pmatrix} X_{12} \\ Y_{12} \\ Z_{12} \end{pmatrix} \tag{52.150}$$

or

$$\mathbf{F}_{12} = \frac{Gm_1m_2}{|\mathbf{X}_{12}|^3}\mathbf{X}_{12} \tag{52.151}$$

Equations (52.143) and (52.151) applied subsequently to two point masses m_1 and m_2, yield, for m_1:

$$\mathbf{F}_1 = m_1\ddot{\mathbf{X}}_1 = \mathbf{F}_{12} = \frac{Gm_1m_2}{|\mathbf{X}_{12}|^3}\mathbf{X}_{12} \tag{52.152}$$

or

$$\ddot{\mathbf{X}}_1 = \frac{Gm_2}{|\mathbf{X}_{12}|^3}\mathbf{X}_{12} \tag{52.153}$$

For m_2:

$$\mathbf{F}_2 = m_2\ddot{\mathbf{X}}_2 = \mathbf{F}_{21} = -\mathbf{F}_{12} = \frac{-Gm_1m_2}{|\mathbf{X}_{12}|^3}\mathbf{X}_{12} \tag{52.154}$$

or

$$\ddot{\mathbf{X}}_2 = \frac{-Gm_1}{|\mathbf{X}_{12}|^3}\mathbf{X}_{12} \tag{52.155}$$

By subtracting Eq. (52.153) from Eq. (52.155), we get

$$\ddot{\mathbf{X}}_{12} = \ddot{\mathbf{X}}_2 - \ddot{\mathbf{X}}_1 = \frac{-G(m_1 + m_2)}{|\mathbf{X}_{12}|^3}\mathbf{X}_{12} \tag{52.156}$$

If m_1 resides in the origin of the inertial frame, then the subindices may be omitted

$$\ddot{\mathbf{X}} = \frac{-G(m_1 + m_2)}{|\mathbf{X}|^3}\mathbf{X} \tag{52.157}$$

Equation (52.157) represents the equations of motion of m_2 in an inertial frame centered at m_1; see Fig. 52.17.

It will be clear that in m_1 the origin of an inertial frame can be defined if and only if $m_1 \gg m_2$. For the equations of motion of a satellite $m = m_2$ orbiting the earth $M = m_1$, Eq. (52.157) simplifies to

$$\ddot{\mathbf{X}} = \frac{-GM}{|\mathbf{X}|^3}\mathbf{X} \tag{52.158}$$

In Eq. (52.158) the product of the gravitational constant G and the mass of the earth M appears. For this product, the *geocentric gravitational constant,* a new symbol is introduced:

$$\mu = GM \tag{52.159}$$

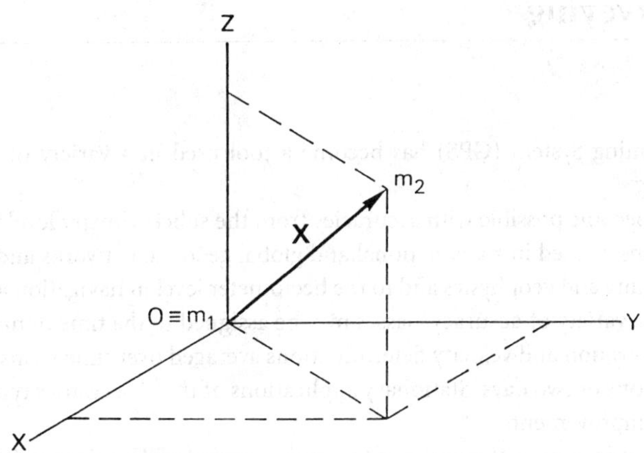

FIGURE 52.17 The inertial frame centered in m_1.

Potential

The equations of motion around a mass M,

$$\ddot{\mathbf{X}} = \frac{-\mu}{|\mathbf{X}|^3}\mathbf{X} \tag{52.160}$$

can be obtained through the definition of a scalar function V:

$$V = V(X, Y, Z) = \frac{\mu}{|\mathbf{X}|} = \frac{\mu}{(X^2 + Y^2 + Z^2)^{1/2}} \tag{52.161}$$

The partial derivatives of V with respect to $\mathbf{X}$ are

$$\frac{\partial V}{\partial X} = -\frac{\mu}{|\mathbf{X}|^2}\frac{\partial|\mathbf{X}|}{\partial X} = -\frac{\mu}{|\mathbf{X}|^3}X \tag{52.162}$$

$$\frac{\partial V}{\partial Y} = -\frac{\mu}{|\mathbf{X}|^2}\frac{\partial|\mathbf{X}|}{\partial Y} = -\frac{\mu}{|\mathbf{X}|^3}Y \tag{52.163}$$

$$\frac{\partial V}{\partial Z} = -\frac{\mu}{|\mathbf{X}|^2}\frac{\partial|\mathbf{X}|}{\partial Z} = -\frac{\mu}{|\mathbf{X}|^3}Z \tag{52.164}$$

Consequently, the equations of motion may be written as

$$\ddot{\mathbf{X}} = \begin{pmatrix} \dfrac{\partial V}{\partial X} \\ \dfrac{\partial V}{\partial Y} \\ \dfrac{\partial V}{\partial Z} \end{pmatrix} \equiv \text{grad} V \equiv \nabla V \tag{52.165}$$

V has physical significance: it is the potential of a point mass of negligible mass in a gravity field of a point mass with sizable mass M at a distance $|\mathbf{X}|$.

52.6 Satellite Surveying

Introduction

The Global Positioning System (GPS) has become a tool used in a variety of fields within and outside engineering.

Positioning has become possible with accuracies from the subcentimeter level for high-accuracy geodetic applications as used in state, national and global geodetic networks and for deformation analysis in engineering and geophysics and to the hectometer level in navigation applications. As in the space domain, a variety of accuracy classes may be assigned to the time domain: GPS provides a means to obtain position and velocity determinations averaged over time spans from subseconds (instantaneous) to one or two days. Stationary applications of the observatory type are used in GPS tracking for orbit improvement.

First of all, the physics and mathematics of the space segment will be given (without derivations).

Numerical Solution of Three Second-Order Differential Equations

Equation (52.158) in section 52.5 represented the equations of motion of a satellite expressed in vector form. Written in the three Cartesian components, we have

$$\begin{pmatrix} \dfrac{d^2X}{dt^2} \\ \dfrac{d^2Y}{dt^2} \\ \dfrac{d^2Z}{dt^2} \end{pmatrix} = \frac{-GM}{(X^2+Y^2+Z^2)^{3/2}} \begin{pmatrix} X \\ Y \\ Z \end{pmatrix} \tag{52.166}$$

or

$$\begin{aligned} \ddot{X} &= C[X/(X^2+Y^2+Z^2)^{3/2}] \\ \ddot{Y} &= C[Y/(X^2+Y^2+Z^2)^{3/2}] \\ \ddot{Z} &= C[Z/(X^2+Y^2+Z^2)^{3/2}] \end{aligned} \tag{52.167}$$

with $C = GM$. Rather than solving for three second-order differential equations (DEs), we make a transformation to six first-order DEs. Introduce three new variables U, V, W:

$$U = \dot{X} = \frac{dX}{dt}; \qquad V = \dot{Y} = \frac{dY}{dt}; \qquad W = \dot{Z} = \frac{dZ}{dt} \tag{52.168}$$

Equations (52.162) through (52.165) yield the following six DEs:

$$\begin{aligned} U &= \dot{X} \\ V &= \dot{Y} \\ W &= \dot{Z} \\ \dot{U} &= C[X/(X^2+Y^2+Z^2)^{3/2}] \\ \dot{V} &= C[Y/(X^2+Y^2+Z^2)^{3/2}] \\ \dot{W} &= C[Z/(X^2+Y^2+Z^2)^{3/2}] \end{aligned} \tag{52.169}$$

Integration of Eq. (52.169) results in six constants of integration. One is free to choose at an epoch t_0 six variables, $\{X_0, Y_0, Z_0, U_0, V_0, W_0\}$ or $\{X_0, Y_0, Z_0, \dot{X}_0, \dot{Y}_0, \dot{Z}_0\}$. These six starting values determine uniquely the orbit of m around M. In other words, if we know the position $\{X_0, Y_0, Z_0\}$ and its velocity $\{\dot{X}_0, \dot{Y}_0, \dot{Z}_0\}$ at an epoch t_0, then we are able to determine the position and velocity of m at any other epoch t by numerical integration of Eqs. (52.169).

Analytical Solution of Three Second-Order Differential Equations

The differential equations of an earth-orbiting satellite can also be solved analytically. Without derivation, the solution is presented in computational steps in terms of transformation formulas.

In history the solution to the motion of planets around the sun was found before its explanation. Through the analysis of his own observations and those made by Tycho Brahe, Johannes Kepler discovered certain regularities in the motions of planets around the sun and formulated the following three laws:

1. (formulated 1609) The orbit of each planet around the sun is an ellipse. The sun is in one of the two focal points.
2. (formulated 1609) The sun–planet line sweeps out equal areas in equal time periods.
3. (formulated 1611) The ratio between the square of a planet's orbital period and the third power of its average distance from the sun is constant.

Kepler's third law leads to the famous equation

$$n^2 a^3 = GM \tag{52.170}$$

in which n is the average angular rate and a the semimajor axis of the orbital ellipse.

In 1665–1666 Newton formulated his more fundamental laws of nature (which were only published after 1687) and showed that Kepler's laws follow from them.

Orientation of the Orbital Ellipse

In a (quasi-)inertial frame the ellipse of an earth-orbiting satellite has to be positioned: the focal point will coincide with the center of mass (CoM) of the earth. Instead of picturing the ellipse itself we project the ellipse on a celestial sphere centered at the CoM. On the celestial sphere we also project the earth's equator (see Fig. 52.18).

The orientation of the orbit ellipse requires three orientation angles with respect to the inertial frame XYZ: two for the orientation of the plane of the orbit (Ω and I); one for the orientation of the ellipse in the orbital plane in terms of the point of closest approach, the perigee (ω).

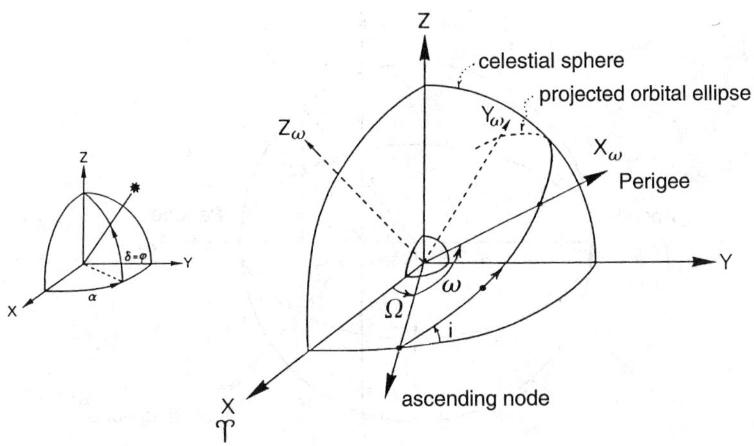

FIGURE 52.18 Celestial sphere with projected orbit ellipse and equator.

- Ω represents the right ascension (α) of the ascending node. The ascending node is the (projected) point where the satellite rises above the equator plane.
- I represents the inclination of the orbital plane with respect to the equator plane.
- ω represents the *argument of perigee:* the angle from the ascending node (in the plane of the orbit) to the perigee (for planets: the perihelion), which is that point where the satellite (planet) approaches the closest the earth (sun), or more precisely, the center of mass of the earth (sun).

We define now another reference frame $\mathbf{X}_\omega$ of which the $X_\omega Y_\omega$-plane coincides with the orbit plane. The X_ω-axis points to the perigee, and the origin coincides with the earth's CoM ($\equiv$ focal point ellipse $\equiv$ origin of $\mathbf{X}$-frame). The relationship between the inertial frames $\mathbf{X}_\mathbf{I}$ and $\mathbf{X}_\omega$ is

$$\mathbf{X}_I = \mathbf{R}_{I\omega}\mathbf{X}_\omega \tag{52.171}$$

in which

$$\mathbf{R}_{I\omega} = \mathbf{R}_3(-\Omega)\mathbf{R}_1(-I)\mathbf{R}_3(-\omega) \tag{52.172}$$

and

$$\mathbf{X}_\omega = \begin{pmatrix} X_\omega \\ Y_\omega \\ Z_\omega = 0 \end{pmatrix} \tag{52.173}$$

Reference Frame in the Plane of the Orbit

Now we know the orientation of the orbital ellipse, we have to define the size and shape of the ellipse and the position of the satellite along the ellipse at a certain epoch t_0.

Similarly to the earth's ellipsoid, discussed in section 52.2, we define the ellipse by a semimajor axis a and eccentricity e. In orbital mechanics it is unusual to describe the shape of the orbital ellipse by its flattening.

The position of the satellite in the orbital $X_\omega Y_\omega$-plane is depicted in Fig. 52 .19. In the figure the auxiliary circle enclosing the orbital ellipse reveals the following relationships:

$$X_\omega = r_\omega \cos\nu = a(\cos E - e) \tag{52.174}$$

$$Y_\omega = r_\omega \sin\nu = a\sqrt{1-e^2}\sin E \tag{52.175}$$

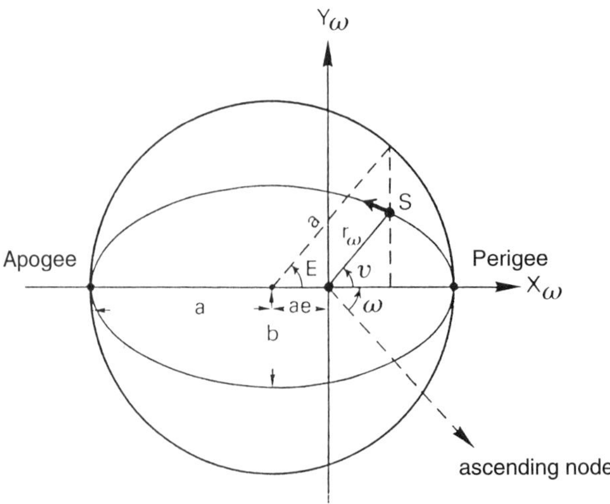

FIGURE 52.19 The position of the satellite S in the orbital plane.

in which

$$r_\omega = a(1 - e\cos E) \tag{52.176}$$

In Fig. 52.19 and the Eqs. (52.174) through (52.176),

a = the semimajor axis of the orbital ellipse
b = the semiminor axis of the orbital ellipse
e = the eccentricity of the orbital ellipse, with

$$e^2 = \frac{a^2 - b^2}{a^2} \tag{52.177}$$

ν = the true anomaly, sometimes denoted by an f
E = the eccentric anomaly

The relation between the true and the eccentric anomalies can be derived to be

$$\tan\left(\frac{E}{2}\right) = \sqrt{\frac{1-e}{1+e}}\tan\left(\frac{\nu}{2}\right) \tag{52.178}$$

Substitution of Eqs. (52.174) through (52.176) into (52.171) gives

$$\mathbf{X}_I = \begin{pmatrix} X \\ Y \\ Z \end{pmatrix} = \mathbf{R}_3(-\Omega)\mathbf{R}_1(-I)\mathbf{R}_3(-\omega)\begin{pmatrix} a(\cos E - e) \\ a\sqrt{1-e^2}\sin E \\ 0 \end{pmatrix} \tag{52.179}$$

In Eq. (52.179) the Cartesian coordinates are expressed in the six so-called Keplerian elements: a, e, I, Ω, ω, and E. Paraphrasing an earlier remark: If we know the position of the satellite at an epoch t_0 through $\{a, e, I, \Omega, \omega$, and $E_0\}$, we are capable of computing the position of the satellite at an arbitrary epoch t through the Eq. (52.179) if we know the relationship in time between E and E_0. In other words, how does the angle E increase with time?

We define an auxiliary variable (angle) M, which increases linearly in time with the mean motion $n(= (GM/a^3)^{1/2})$ according to Kepler's third law. The angle M, the *mean anomaly,* may be expressed as function of time by

$$M = M_0 + n(t - t_0) \tag{52.180}$$

Through *Kepler's equation*

$$M = E - e\sin E \tag{52.181}$$

the (time) relationship between M and E is given. Kepler's equation is the direct result of the enforcement of Kepler's second law ("equal area law").

Combining Eqs. (52.180) and (52.181) gives an equation that expresses the relationship between a given eccentric anomaly E_0 (or M_0, or ν_0) at an epoch t_0 and the eccentric anomaly E at an arbitrary epoch t:

$$E - E_0 = e(\sin E - \sin E_0) + n(t - t_0) \tag{52.182}$$

Transformation from Keplerian to Cartesian Orbital Elements

So far, the position vector $\{X, Y, Z\}$ of the satellite has been expressed in terms of the Keplerian elements. The transformation is complete when we express the velocity vector $\{\dot{X}, \dot{Y}, \dot{Z}\}$ in terms of those Keplerian elements. Differentiating Eq. (52.171) with respect to time we get

$$\dot{\mathbf{X}}_I = \mathbf{R}_{I\omega}\dot{\mathbf{X}}_\omega + \dot{\mathbf{R}}_{I\omega}\mathbf{X}_\omega \tag{52.183}$$

Since we consider the two-body problem with $m_1 = M \gg m = m_2$, the orientation of the orbital ellipse is time-independent in the inertial frame. This means that the orientation angles I, Ω, ω are time independent as well:

$$\dot{\mathbf{R}}_{I\omega} = [0] \tag{52.184}$$

Equation (52.183) simplifies to

$$\dot{\mathbf{X}}_I = \mathbf{R}_{I\omega}\dot{\mathbf{X}}_{\omega} \tag{52.185}$$

Differentiating Eqs. (52.174) through (52.176) with respect to time, we have

$$\dot{X}_{\omega} = -a\dot{E}\sin E \tag{52.186}$$

$$\dot{Y}_{\omega} = a\dot{E}\sqrt{1-e^2}\cos E \tag{52.187}$$

$$\dot{r}_{\omega} = a\dot{E}e\sin E \tag{52.188}$$

The remaining variable $\dot{E}$ is obtained through differentiation of Eq. (52.181):

$$\dot{E} = \frac{n}{1 - e\cos E} \tag{52.189}$$

Now all transformation formulas express the Cartesian orbital elements (state vector elements) in terms of the six Keplerian elements:

$$[\mathbf{X}_I \,\vdots\, \dot{\mathbf{X}}_I] = \mathbf{R}_{I\omega}[\mathbf{X}_{\omega} \,\vdots\, \dot{\mathbf{X}}_{\omega}] \tag{52.190}$$

or

$$[\mathbf{X}_I \,\vdots\, \dot{\mathbf{X}}_I] = \begin{bmatrix} X & \dot{X} \\ Y & \dot{Y} \\ Z & \dot{Z} \end{bmatrix} = \tag{52.191}$$

$$= \mathbf{R}_3(-\Omega)\mathbf{R}_1(-I)\mathbf{R}_3(-\omega)\begin{bmatrix} a(\cos E - e) & -a\dot{E}\sin E \\ a\sqrt{1-e^2}\sin E & a\dot{E}\sqrt{1-e^2}\cos E \\ 0 & 0 \end{bmatrix} \tag{52.192}$$

Transformation from Cartesian to Keplerian Orbital Elements

To compute the inertial position of a satellite in a central force field, it is simpler to perform a time update in the Keplerian elements than in the Cartesian elements. The time update takes place through Eqs. (52.180) through (52.182).

Schematically the following procedure is to be followed:

$t_0 : \{X, Y, Z, \dot{X}, \dot{Y}, \dot{Z}\}$

$\downarrow$ Converson to Keplerian elements, this subsection

$t_0 : \{a, e, I, \Omega, \omega, E_0\}$

$\downarrow$ Equation of Kepler, Eq. (52.182)

$t_1 : \{a, e, I, \Omega, \omega, E_1\}$

$\downarrow$ Conversion to Cartesian elements, previous subsection

$t_1 : \{X, Y, Z, \dot{X}, \dot{Y}, \dot{Z}\}$

The conversion from Keplerian elements to state vector elements has been treated in the previous subsection. In this section the somewhat more complicated conversion from position and velocity vector to Keplerian representation will be described. Basically we "invert" Eq. (52.192) by solving for the six elements $\{a, e, I, \Omega, \omega, E\}$ in terms of the six state vector elements.

First, we introduce another reference frame $\mathbf{X_u}$:

X_I, Y_I, Z_I:	inertial reference frame	(X_I-axis $\rightarrow$ vernal equinox)
$X_\omega, Y_\omega, Z_\omega$:	orbital reference frame	(X_ω-axis $\rightarrow$ perigee)
X_u, Y_u, Z_u:	orbital reference frame	(X_u-axis $\rightarrow$ satellite)

The $\mathbf{X_u}$-frame is defined similarly to the $\mathbf{X_\omega}$-frame, except that the $\mathbf{X_u}$-axis continuously points to the satellites (see Fig. 52.20). Thus

$$\mathbf{X}_u = \begin{pmatrix} X_u \\ Y_u \\ Z_u \end{pmatrix} = \begin{pmatrix} X_u \\ 0 \\ 0 \end{pmatrix} \tag{52.193}$$

The angle in the orbital plane enclosed by the X_u-axis and the direction to the ascending node is called u, the *argument of latitude,* with

$$u = \omega + \nu \tag{52.194}$$

As in the discussion of the orientation of the orbit ellipse, the following relationships hold:

$$\mathbf{X_\omega} = \mathbf{R}_{\omega I}\mathbf{X}_I \tag{52.195}$$

$$\mathbf{X}_u = \mathbf{R}_{uI}\mathbf{X}_I \tag{52.196}$$

$$\mathbf{X}_u = \mathbf{R}_{u\omega}\mathbf{X}_\omega = \mathbf{R}_{u\omega}\mathbf{R}_{\omega I}\mathbf{X}_I \tag{52.197}$$

Figure 52.20 reveals that

$$\mathbf{R}_{\omega I} = \mathbf{R}_3(\omega)\mathbf{R}_1(I)\mathbf{R}_3(\Omega) \tag{52.198}$$

$$\mathbf{R}_{uI} = \mathbf{R}_3(u)\mathbf{R}_1(I)\mathbf{R}_3(\Omega) \tag{52.199}$$

$$= \mathbf{R}_3(\nu + \omega)\mathbf{R}_1(I)\mathbf{R}_3(\Omega) \tag{52.200}$$

$$= \mathbf{R}_3(\nu)\mathbf{R}_3(\omega)\mathbf{R}_1(I)\mathbf{R}_3(\Omega) \tag{52.201}$$

$$= \mathbf{R}_3(\nu)\mathbf{R}_{\omega I} = \mathbf{R}_{u\omega}\mathbf{R}_{\omega I} \tag{52.202}$$

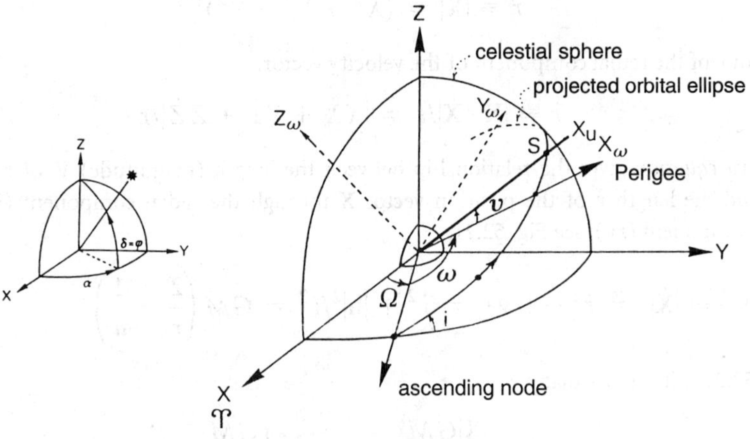

FIGURE 52.20 The orbital reference frame $\mathbf{X_u}$.

We define a vector **h** perpendicular to the orbital plane according to

$$\mathbf{h} \equiv \mathbf{X} \times \dot{\mathbf{X}} = |\mathbf{h}|\mathbf{w} = |\mathbf{h}| \begin{pmatrix} \sin\Omega \sin I \\ -\cos\Omega \sin I \\ \cos I \end{pmatrix} \tag{52.203}$$

in which **w** is the unit vector along **h**. Consequently,

$$\mathbf{h} = \begin{pmatrix} h_1 \\ h_2 \\ h_3 \end{pmatrix} = \begin{pmatrix} Y\dot{Z} - Z\dot{Y} \\ Z\dot{X} - X\dot{Z} \\ X\dot{Y} - Y\dot{X} \end{pmatrix} \tag{52.204}$$

h represents the angular momentum vector (vector product of position vector and velocity vector). The Keplerian elements Ω and I follow directly from Eqs. (52.203) and (52.204):

$$\tan\Omega = \frac{h_1}{-h_2} \tag{52.205}$$

$$\tan I = \frac{\sqrt{h_1^2 + h_2^2}}{h_3} \tag{52.206}$$

From

$$\mathbf{R}_3(-u)\mathbf{X}_u = \mathbf{R}_1(I)\mathbf{R}_3(\Omega)\mathbf{X}_1 \tag{52.207}$$

it follows that

$$X_u \cos u = X \cos\Omega + Y \sin\Omega \tag{52.208}$$

$$X_u \sin u = -X \cos I \sin\Omega + Y \cos I \cos\Omega + Z \sin I \tag{52.209}$$

and

$$\tan u = \frac{-X \cos I \sin\Omega + Y \cos I \cos\Omega + Z \sin I}{X \cos\Omega + Y \sin\Omega} \tag{52.210}$$

Before determining the third Keplerian element defining the orientation of the orbit (ω) from the argument of latitude (u), we define the following quantities:

- Length r of the radius vector **X**:

$$r \equiv |\mathbf{X}| = (X^2 + Y^2 + Z^2)^{1/2} \tag{52.211}$$

- Length $\dot{r}$ of the radial component of the velocity vector:

$$\dot{r} \equiv |\mathbf{X} \cdot \dot{\mathbf{X}}|/r = |X\dot{X} + Y\dot{Y} + Z\dot{Z}|/r \tag{52.212}$$

The *Vis-Viva equation* gives the relationship between the length (magnitude) V of the velocity vector $\dot{\mathbf{X}}$ and the length r of the position vector **X** through the radial component ($\dot{r}$) and the tangential component ($r\dot{v}$); see Fig. 52.21.

$$V^2 \equiv |\dot{\mathbf{X}}|^2 = \dot{r}^2 + (r\dot{v})^2 = \dot{r}^2 + |\mathbf{h}|^2/r^2 = GM\left(\frac{2}{r} - \frac{1}{a}\right) \tag{52.213}$$

From Eq.(52.213) it follows that

$$a = \frac{|\mathbf{X}|GM}{2GM - |\mathbf{X}|V^2} = \frac{rGM}{2GM - rV^2} \tag{52.214}$$

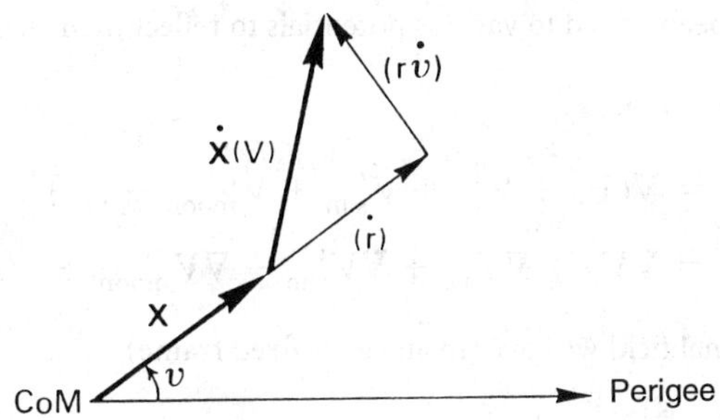

FIGURE 52.21 Illustration of the Vis-Viva equation.

In a similar manner one arrives at

$$1 - e^2 = \frac{|\mathbf{h}|^2}{aGM} \tag{52.215}$$

From Eqs. (52.188), (52.189), (52.176), and (52.170) the eccentric anomaly E may be computed:

$$\tan E = \frac{\sin E}{\cos E} = \left(\frac{r\dot{r}}{e\sqrt{GMa}}\right) \Big/ \left(\frac{a - r}{ae}\right) \tag{52.216}$$

The mean anomaly M is given by

$$M = E - e \sin E \tag{52.217}$$

The true anomaly ν follows from Eqs. (52.174) and (52.175):

$$\tan \nu = \frac{\sqrt{1 - e^2} \sin E}{\cos E - e} \tag{52.218}$$

after which finally the last Keplerian element, ω, is determined through Eq. (52.194):

$$\omega = u - \nu \tag{52.219}$$

Orbit of a Satellite in a Noncentral Force Field

The equations of motion for a real satellite are more difficult than reflected by Eq. (52.165). First of all, we do not deal with a central force field: the earth is not a sphere, nor does it have a radial symmetric density. Secondly, we deal with other forces, chiefly the gravity of the moon and the sun, atmospheric drag, and solar radiation pressure. Equation (52.165) gets a more general meaning if we suppose that the potential function is generated by the sum of the forces acting on the satellite:

$$V = V_c + V^t_{\mathrm{nc}} + V^t_{\mathrm{sun}} + V^t_{\mathrm{moon}} + \dots \tag{52.220}$$

with the central part of the earth's gravitational potential

$$V_c = \mu/|\mathbf{X}| \tag{52.221}$$

and the noncentral and time-dependent part of the earth's gravitational field

$$V^t_{\mathrm{nc}} = [\text{see Eq. (52.224)}] \tag{52.222}$$

and so forth.

The superscript t has been added to various potentials to reflect their time variance with respect to the inertial frame.

The equations of motion to be solved are

$$\begin{aligned}\ddot{\mathbf{X}} &= \nabla(V_c + V^t_{\text{nc}} + V^t_{\text{sun}} + V^t_{\text{moon}} + \ldots) \\ &= \nabla V_c + \nabla V^t_{\text{nc}} + \nabla V^t_{\text{sun}} + \nabla V^t_{\text{moon}} + \ldots\end{aligned} \tag{52.223}$$

For the earth's gravitational field we have (in an earth-fixed frame)

$$V_c + V_{\text{nc}} = \frac{\mu}{r}\left[1 + \sum_{l=1}^{\infty}\sum_{m=0}^{l}\left(\frac{a_e}{r}\right)^l (C_{lm}\cos m\lambda + S_{lm}\sin m\lambda)P_{lm}(\sin\phi)\right] \tag{52.224}$$

With Eq. (52.224) one is able to compute the potential at each point $\{\lambda, \phi, r\}$ necessary for the integration of the satellite's orbit. The coefficients C_{lm} and S_{lm} of the spherical harmonic expansion are in the order of 10^{-6} except for C_{20} ($l = 2, m = 0$), which is about 10^{-3}. This has to do with the fact that the earth's equipotential surface at mean sea level can be best approximated by an ellipsoid of revolution. One has to realize that the coefficients C_{lm}, S_{lm} describe the shape of the potential field and not the shape of the physical earth, despite a high correlation between the two. $P_{lm}(\sin\phi)$ are the associated Legendre functions of the first kind, of degree l and order m; a_e is some adopted value for the semimajor axis (equatorial radius) of the earth. See section 52.8 for values of a_e, $\mu(= GM)$, and $C_{20}(= -J_2)$.

The equatorial radius a_e, the geocentric gravitational constant GM, and the dynamic form factor J_2 characterize the Earth by an ellipsoid of revolution of which the surface is an equipotential surface.

Restricting ourselves to the central part ($\mu = GM$) and the dynamic flattening ($C_{20} = -J_2$), then Eq. (52.224) becomes

$$V_c + V_{\text{nc}} = \frac{\mu}{r}\left[1 + \frac{J_2 a_e^2}{2r^2}(1 - 3\sin^2\phi)\right] \tag{52.225}$$

with

$$\sin\phi = (\sin\delta =)\frac{z}{r} \tag{52.226}$$

in which ϕ is the latitude and δ the declination, see Fig. 52.20.

A solution of the DE [by substitution of Eq. (52.225) in Eq. (52.165)]

$$\ddot{\mathbf{X}} = \nabla(V_c + V_{20}) \tag{52.227}$$

in a closed analytical expression is not possible. The solution expressed in Keplerian elements shows periodic perturbations and some dominant secular effects. An approximate solution using only the latter effects is (position only):

$$\mathbf{X}_I = \mathbf{R}_3[-(\Omega_0 + \dot{\Omega}\Delta t)]\mathbf{R}_1(-I)\mathbf{R}_3[-(\omega_0 + \dot{\omega}\Delta t)]\mathbf{X}_\omega \tag{52.228}$$

with

$$\Delta t = t - t_0 \tag{52.229}$$

with

$$\dot{\Omega} = -\frac{3}{2}\frac{J_2 a_e^2}{a^2(1-e^2)^2}n\cos I \tag{52.230}$$

$$\dot{\omega} = \frac{3}{2} \frac{J_2 a_e^2}{a^2(1-e^2)^2} n \left(2 - \frac{5}{2}\sin^2 I\right) \tag{52.231}$$

$$n = n_0 \left[1 + \frac{3}{2} \frac{J_2 a_e^2 \sqrt{1-e^2}}{a^2(1-e^2)^2} \left(1 - \frac{3}{2}\sin^2 I\right)\right] \tag{52.232}$$

$$n_0 = \sqrt{\frac{GM}{a^3}} \tag{52.233}$$

Whenever	we have an
$I = 0°$	equatorial orbit
$0° < I < 90°$	direct orbit
$I = 90°$	polar orbit
$90° < I < 180°$	retrograde orbit
$I = 180°$	retrograde equatorial orbit

Equation (52.230) shows that the ascending node of a direct orbit slowly drifts to the west. For a satellite at a height of about 150 km above the earth's surface the right ascension of the ascending node decreases with about 9° per day. For satellites used in geodesy/geodynamics, such as STARLETTE (a = 7340 km, I = 50°) and LAGEOS-1 (a = 12,270 km, I = 110°), these values are −4° per day and $+\frac{1}{3}^{\circ}$ per day, respectively.

The satellites belonging to the Global Positioning System have an inclination of about 55°. Their nodal regression rate is about −0.04187 degrees per day.

The Global Positioning System (GPS)

Introduction

The Navstar GPS space segment consists of one Block I satellite, nine Block II, and fifteen Block IIA satellites as of October, 1994 [*GPS World,* 1994]. This means that the full Block II satellite constellation, in six orbital planes at a height of about 20,000 kilometers, is completed. With this number of satellites, 3-D positioning is possible every hour of the day. However, care must be exercised, since an optimum configuration for 3-D positioning is not available on a full day's basis.

In the meantime, GPS receivers, ranging in cost between $300 and $20,000, are readily available. Over 100 manufacturers are marketing receivers, and the prices are still dropping. Magazines such as *G.I.M., GPS World, P.O.B.,* and *Professional Surveyor* publish regularly on the latest models; see, for example, the recent GPS equipment surveys in Reilly [1994] and in Chan and Schorr [1994].

GPS consumer markets have been rapidly expanded. In the areas of land, marine, and aviation navigation, of precise surveying, of electronic charting, and of time transfer the deployment of GPS equipment seems to have become indispensable. This holds for military as well as civilian users.

Positioning

Several classes of positioning are recognized:

- SPS: Standard Positioning Service
- PPS: Precise Positioning Service

In terms of positional accuracies one has to distinguish between SPS, with and without selective availability (SA), on the one hand and PPS on the other. Selective availability deliberately introduces clock errors and ephemeris errors in the data being broadcast by the satellite. The current accuracy of the (civil) signal without SA is in the order of 20–40 meters. With SA implemented, SPS accuracy is degraded to 100 m. At the time of this writing (October 1994) the satellite constellation guarantees 100-m SPS accuracy to civilian and commercial users 95% of the time.

In several applications, GPS receivers are interfaced with other positioning systems such as inertial navigation systems (INS), hyperbolic systems, or even automatic braking systems (ABS) in cars.

GPS receivers in combination with various equipment are able to answer such general questions as [Wells and Kleusberg, 1990]

Absolute positioning	Where am I?
	Where are you?
Relative positioning	Where am I with respect to you?
	Where are you with respect to me?
Orientation	Which way am I heading?
Timing	What time is it?

All of these questions may refer to either an observer at rest (*Static* positioning) or one in motion (*Kinematic* positioning). The questions may be answered immediately (*Real-time* processing, often misnamed *DGPS*, for Differential GPS) or after the fact (*Batch* processing). The various options are summarized in Table 52.3 [Wells and Kleusberg, 1990]. Similarly, the accuracies for time dissemination are summarized in Table 52.4 [Wells and Kleusberg, 1990].

Table 52.3 Accuracy of Various GPS Positioning Modes

Absolute positioning:	
SPS with SA	100 m
SPS without SA	40 m
PPS	20 m
Relative/Differential positioning:	
Differential SPS	10 m
Carrier-smoothed code	2 m
Ambiguity-resolved carrier	10 cm
Surveying between fixed points	1 mm to 10 cm

Note: Accuracy of differential models depends on inter-receiver distance!

Table 52.4 Accuracy of Various GPS Time Dissemination Modes

Time and time interval:	
With SA	500 ns
Without SA, correct position	100 ns
Common mode, common view	<25 ns

Limiting Factors

Physics of the environment, instruments, broadcast ephemeris, and the relative geometry between orbits and network all form limiting factors on the final accuracy of the results. *Dilution of precision* (DOP) is used as a scaling factor between the observational accuracy and positioning accuracy.

The atmosphere of the earth changes the speed and the geometrical path of the electromagnetic signals broadcast by the GPS satellites. In the uppermost part of the atmosphere (the ionosphere) charged particles vary in number spatially as well as temporally. The so-called ionospheric refraction errors may amount to several tens of meters. Since this effect is frequency-dependent, the first-order effect can be largely eliminated by the use of dual-frequency receivers. The lower part of the atmosphere (the troposphere) causes refraction errors of several meters. Fortunately, the effect can be modeled rather well by measuring the atmospheric conditions at the measuring site.

GPS instruments are capable of measuring one or a combination of the following signals:

C/A code	with an accuracy of a few meters
P code	with an accuracy of a few decimeters
Carrier phase	with an accuracy of a few millimeters

In addition to this measurement noise, receiver clock errors are present and have to be modeled as to-be-solved-for parameters. It is this synchronization parameter between satellite time and receiver time that makes it necessary to have at least four satellites in view in order to get a 3-D fix.

Because of the high frequency of the GPS signals, multipath effects may hamper the final accuracy; the signal arriving at the receiver through a reflected path may be stronger than the direct signal. By careful antenna design and positioning of the antenna, multipath effects are reduced. The phase center of the antenna needs to be carefully calibrated with respect to a geometric reference point on the antenna assembly. However, because of the varying inclination angle of the incoming electromagnetic signals, effects of a moving phase center may be present at all times.

Information on the orbit of the satellite, as well as the orbital geometry relative to the network/receiver geometry, influences the overall positioning accuracy. The information the satellite broadcasts on its position and velocity is necessarily the result of a process of prediction. This causes the broadcast ephemeris to be contaminated with extrapolation errors. Typical values are

Radial error	about 5 m
Across-track error	about 10 m
Along-track errors	about 15 m

Also, the on-board satellite clock is not free of errors. Orbital and satellite clock errors can be largely taken care of by careful design of the functional model.

As mentioned before, deliberate contamination of the broadcast ephemeris and satellite time degrades the system to an accuracy of several tens of meters. PPS users are able to use the P code on two frequencies and have access to the SA code. Consequently, they are capable of eliminating the ionospheric effects and of removing deliberately introduced orbital and satellite clock errors. The resulting measurement error will be about 5 m. SPS users are able to use the C/A code (on one frequency only). They do not have access to the SA code. Consequently, the ionospheric error can only be roughly modeled, and these users are stuck with the deliberate errors. The resulting measurement error may be as large as 50 m.

Translocation techniques, such as having a stationary receiver continuously supporting the other (roving) receivers, will reduce the measurement error to well below the 10 meters for SPS users.

Differencing techniques applied to the carrier phase measurements are successfully used to eliminate a wide variety of errors, provided the receivers are not too far apart. In essence, two close-by receivers are influenced almost equally by (deliberate) orbital errors and by part of the atmosphere error. Differencing of the measurements of both receivers will cancel a large portion of the first-order effects of these errors.

Modeling and the GPS Observables

Developing well-chosen functional models **F**, relating the GPS measurements **L** to the modeled parameters **X**, enables users to fit GPS perfectly to their needs. A wide class of applications, from monitoring the subsidence of oil rigs in the open sea to real-time navigating vehicles collecting geoinformation, belong to the range of possibilities that have been opened up by the introduction of GPS.

In satellite geodesy, one traditionally modeled the state of the satellite: a vector combining the positional ($\mathbf{X}$) and the velocity ($\dot{\mathbf{X}}$) information. Although a known fact in the area of navigation, nowadays GPS provides geodesists, or geoscientists in general, with a tool by which the state of the observer, also in terms of position ($\mathbf{x}$) and velocity ($\dot{\mathbf{x}}$), can be determined with high accuracy and often in real time.

The GPS satellite geodetic model has evolved to

$$\mathbf{L} = \mathbf{F}(\mathbf{X}, \dot{\mathbf{X}}, \mathbf{x}, \dot{\mathbf{x}}, \mathbf{p}, t) \tag{52.234}$$

with

L = C/A code, P code, or carrier phase observations, $i = 1, \ldots, n$

X = 3-D position of the satellite at epoch t

$\dot{X}$ = 3-D velocity of the satellite at epoch t

x = 1-D, 2-D or 3-D position of the observer at epoch t

$\dot{x}$ = 1-D, 2-D or 3-D velocity of the observer at epoch t

p = vector of modeled (known or unknown) parameters, $j = 1, \ldots, u$

t = epoch of measurement taking

Various differencing operators $\mathbf{D}^{\mathbf{k}}$, up to order 3, are applied to the original observations in order to take full benefit of the GPS measurements. The difference operator $\mathbf{D}^{\mathbf{k}}$ may be applied in the observation space spanned by the vector **L**, Eq. (52.235), or the $\mathbf{D}^{\mathbf{k}}$ operator may be applied in the parameter space **x**, Eq. (52.236):

$$\mathbf{D}^{\mathbf{k}}[\mathbf{L}] = \mathbf{D}^{\mathbf{k}}[\mathbf{F}(\mathbf{X}, \dot{\mathbf{X}}, \mathbf{x}, \dot{\mathbf{x}}, \mathbf{p}, t)] \tag{52.235}$$

$$\mathbf{L} = \mathbf{F}[\mathbf{X}, \dot{\mathbf{X}}, \mathbf{D}^{\mathbf{1}}(\mathbf{x}, \dot{\mathbf{x}}), \mathbf{p}, t] \tag{52.236}$$

The latter method is sometimes referred to as *delta positioning.*

This is a difficult way of saying that one may either construct so-called derived observations from the original observations by differencing techniques or model the original observations, compute parameters (e.g., coordinates) in this way, and subsequently start a differencing technique on the results obtained from the roving receiver and the base receiver.

Pseudo Ranging. We restrict the discussion to the C/A-based *pseudo-range observables.* The ranges are called *pseudo* because this technique is basically a one-way ranging technique with two independent clocks: the offset δt_E^S between the satellite clock S and the receiver clock E yields one additional parameter to be solved for. Writing the observation equation in the earth-fixed reference frame, we have

$$\text{pr} = \sqrt{(x^S - x_E)^2 + (y^S - y_E)^2 + (z^S - z_E)^2} - c\,\delta t_E^S \tag{52.237}$$

Inspection of the partials

$$\frac{\partial \text{pr}}{\partial x^S} = \frac{x^S - x_E}{\text{pr}} = -\frac{\partial \text{pr}}{\partial x_E} \tag{52.238}$$

$$\frac{\partial \text{pr}}{\partial y^S} = \frac{y^S - y_E}{\text{pr}} = -\frac{\partial \text{pr}}{\partial y_E} \tag{52.239}$$

$$\frac{\partial \text{pr}}{\partial z^S} = \frac{z^S - z_E}{\text{pr}} = -\frac{\partial \text{pr}}{\partial z_E} \tag{52.240}$$

$$\frac{\partial \text{pr}}{\partial (\delta t_E^S)} = -c \tag{52.241}$$

reveals that

- The coordinates of the stations are primarily obtained in a frame determined by the satellites, or better, by their broadcast ephemeris.
- Partials evaluated for neighboring stations are practically identical, so the coordinates of one station need to be adopted.

Phase (Carrier Wave) Differencing. For precise engineering applications the phase of the carrier wave is measured. Two wavelengths are available in principle:

$$\begin{aligned} L_1: \quad \lambda_1 &= \frac{c}{f_1} \quad \text{with} \quad f_1 = 1.57542 \text{ GHz} \\ &= 19.0 \text{ cm} \end{aligned} \tag{52.242}$$

and

$$\begin{aligned} L_2: \quad \lambda_2 &= \frac{c}{f_2} \quad \text{with} \quad f_2 = 1.22760 \text{ GHz} \\ &= 24.4 \text{ cm} \end{aligned} \tag{52.243}$$

For phase measurements the following observation equation can be set up:

$$\text{Range} = \Phi + N\lambda_l; \quad l = 1, \ldots, 2 \tag{52.244}$$

or

$$\Phi_E^S = \sqrt{(x^S - x_E)^2 + (y^S - y_E)^2 + (z^S - z_E)^2} - N_E^S \lambda_l \tag{52.245}$$

where Φ_E^S is the phase observable in a particular S–E combination, and N_E^S is the integer multiple of wavelengths in the range: the *ambiguity*. Phase measurements can be done with probably 1% accuracy. This yields an observational accuracy—in case the ambiguity N can be properly determined—in the millimeter range!

Using the various differencing operators on the phase measurements:

$\mathbf{D}^{k=1}$ yields single differences:

- Between receiver differences, $\Delta\Phi$, eliminating or reducing satellite-related errors
- Between satellite differences, $\nabla\Phi$, eliminating or reducing receiver-related errors
- Between epoch differences, $\delta\Phi$, eliminating phase ambiguities per satellite/receiver combination

$\mathbf{D}^{k=2}$ yields double differences:

- Between receiver/satellite differences, $\Delta\nabla\Phi$, eliminating or reducing satellite- and receiver-related errors, and so forth

$\mathbf{D}^{k=3}$ yields triple differences:

- Between epoch/receiver/satellite differences, $\delta\nabla\Delta\Phi$, eliminating or reducing satellite/receiver-related errors and ambiguities

Receivers that use carrier wave observations have, in addition to the electronic components that do the phase measurements, a counter as well, which counts the complete cycles between selected epochs. GPS analysis software uses the triple differences to detect and possibly repair cycle slips occurring during loss of lock.

Design specifications and receiver selection depend on the specific project accuracy requirements. In the U.S. the Federal Geodetic Control Committee has adopted the specifications [FGCC, 1989] given in Tables 52.5 and 52.6.

GPS Receivers

A variety of receivers are on the market. Basically, they can be grouped in four classes, listed in Table 52.7. The types of observations of the first three types of receivers are subjected to models that can be characterized as *geometric* models. The position of the satellite is considered to be known, mostly the information taken from the broadcast ephemeris. The known positions are of course not errorless. First of all, the positions are predicted; thus, they contain errors because of an extrapolation process in time. Secondly, the positions being broadcast may be corrupted by intentional errors (due to SA, see previous paragraphs). Differencing techniques are capable of eliminating most of the error if the separation between base station and roving receiver is not too large.

Millimeter-accurate observations from geodesy-grade receivers are often subjected to analysis through models of the *dynamic* type. Software packages containing dynamic models are very elaborate and allow for some kind of orbit improvement estimation process.

Reilly [1994] and Chan and Schorr [1994] list overviews of recent GPS receivers and supporting software packages.

Table 52.5 Guidelines for GPS Field Survey Procedures

	AA	A	B	C
Geometric Relative Positioning Standards	AA 0.01	A 0.1	B 1.0	1, 2-I&II, 3 10, 20, 50, 100
Two frequency observations (1 and L2) required[a]*:* daylight observations[b]	Y	Y	Y	op
Recommended number of receivers observing simultaneously, not less than	5	5	4	3
Satellite observations: RDOP values during observing session (meters/cycle)[d]				
Period of observing session (observing span), not less than (min) (4 or more simultaneous satellite observations)[e]				
Triple difference processing[f]	na	na	240	60–120
Other processing techniques[g]				
General requirements[h]	240	240	120	30–60
Continuous and simultaneous between all receivers, period not less than[i]	180	120	60	20–30
Data sampling rate—maximum time interval between observations (sec)	15	30	30	15–30
Minimum number of quadrants from which satellite signals are observed	4	4	3	3 or 2[j]
Maximum angle above horizon for obstructions[u] (degrees)	10	15	20	20–40
Independent occupations per station[k]				
Three or more (percent of *all* stations, not less than)	80	40	20	10
Two or more (percent of stations, not less than)				
New stations	100	80	50	30
Vertical control stations	100	100	100	100
Horizontal control stations	100	75	50	25
Two or more for each station of "station-pairs"[l]	Y	Y	Y	Y

nr—not required; na—not applicable; op—optional.

[a] If two-frequency observations cannot be obtained, it is possible that an alternative method for estimating the ionospheric refraction correction would be acceptable, such as modeling the ionosphere using two-frequency data obtained from other sources. Or, if observations are during darkness, single-frequency observations may be acceptable depending on the expected magnitude of the ionospheric refraction error.

[b] When spacing between any two stations occupied during an observing session is more than 50 km, two-frequency observations may need to be considered for accuracy standards of order 2 or higher.

[c] Multiple baseline processing techniques.

[d] Studies are under way to investigate the relationship of geometric dilution of precision (GDOP) values to the accuracy of the baseline determinations. Initial results of these studies indicate a possible correlation. It appears the best results may be achieved when the GDOP values are changing during the observing session.

[e] The number of satellites that are observed simultaneously cannot be less than the number specified for more than 25% of the specified period for each observing session.

[f] Absolute minimum criterion is 100% of specified period.

[g] "Other" includes processing carrier phase data using single, double, nondifferencing, or other comparable precise relative positioning processing techniques.

[h] The times for the observing span are conservative estimates to ensure that the data quantity and quality will give results that will meet the desired accuracy standard.

[i] Absolute minimum criterion for the data collection observing span is that period specified for an observing session that includes continuous and simultaneous observations. Continuous observations are data collected that do not have any breaks involving *all* satellites; occasional breaks for individual satellites caused by obstructions are acceptable, but must be minimized. A set of observations for each measurement epoch is considered simultaneous when it includes data from at least 75% of the receivers participating in the observing session.

[j] Satellites should pass through quadrants diagonally opposite each other.

[k] Two or more independent occupations for the stations of a network are specified to help detect instrument and operator errors. Operator errors include those caused by antenna centering and height offset blunders. When a station is occupied during two or more sessions, back to back, the antenna/tripod will be reset and replumbed between sessions to meet the criteria for an independent occupation. To separate biases caused by receiver and/or antenna equipment problems from operator-induced blunders, a calibration test may need to be performed.

[l] Redundant occupations are required when pairs of intervisible stations are established to meet azimuth requirements, when the distance between the station pair is less than 2 km, and when the order is 2 or higher.

Table 52.5 (continued) Guidelines for GPS Field Survey Procedures

	AA	A	B	C
	AA	A	B	1, 2-I&II, 3
Geometric Relative Positioning Standards	0.01	0.1	1.0	10, 20, 50, 100
Master or fiducial stations[m]				
Required, yes or no[n]	Y	Y	Y	op
If yes, minimum number	4	3	2	—
Repeat baseline measurements, about equal number in N-S and E-W directions, minimum not less than [percent of total independently (nontrivial) determined base lines]	25	15	5	5
Loop closure, requirements when forming loops for postanalyses				
Baselines from independent observing sessions, not less than	3	3	2	2
Baselines in each loop, total not more than	6	8	10	10
Loop length, generally not more than (km)	2000	300	100	100
Baselines not meeting criteria for inclusion in any loop, not more than [percent of all independent nontrivial lines[o]]	0	5	20	30
Stations not meeting criteria for inclusion in any loop, not more than (percent of all stations)	0	5	10	15
Direct connections are required: Between any adjacent (NGRS and/or new GPS) stations (new or old, GPS or non-GPS) located near or within project area, when spacing is *less* than (km)	30	10	5	3
Antenna setup				
Number of antenna phase center height measurements per session, not less than	3[p]	3[p]	2	2
Independent plumb point check required[q]	Y	Y	Y	op
Photograph (closeup) and/or pencil rubbing required for each mark occupied	Y	Y	Y	Y
Meteorological observations				
Per observing session, not less than	3[r]	3[r]	2[s]	2[s] or op
Sampling rate (measurement interval), not more than (min)	30	30	60	60
Water vapor radiometer measurements required at selected stations?	op	op	N	N
Frequency standard warm-up time (hr)[t]				
Crystal	12	12	*u*	*u*
Atomic	—	1	*t*	*t*

[m] Master or fiducial stations are those that are continuously monitored during a sequence of sessions, perhaps for the complete project. These could be sites with permanently tracking equipment in operation where the data are available for use in processing with data collected with the mobile units.

[n] If simultaneous observations are to be processed in the session or network for baseline determinations while adjusting one or more components of the orbit, then two or more master stations shall be established.

[o] For each observing session there are $r - 1$ independent baselines, where r is the number of receivers collecting data simultaneously during a session; e.g., if there were 10 sessions and 4 receivers used in each session, 30 independent baselines would be observed.

[p] A measurement will be made both in meters and feet, at the beginning, midpoint, and end of each station occupation.

[q] To ensure that the antenna was centered accurately with the optical plummet over the reference point on the marker, when specified, a heavyweight plumb bob will be used to check that the plumb point is within specifications.

[r] Measurements of station pressure (in millibars), relative humidity, and air temperature (in °C) will be recorded at the beginning, midpoint, and end depending on the period of the observing session.

[s] Report only unusual weather conditions, such as major storm fronts passing over the sites during the data collection period. This report will include station pressure, relative humidity, and air temperature.

[t] The amount of warm-up time required is very instrument dependent. It is important to follow the manufacturer's specifications.

[u] An obstruction is any object that would effectively block the signal arriving from the satellite. These include buildings, trees, fences, humans, vehicles, etc.

Source: FGCC, 1989.

Table 52.6 Office Procedures for Classifying GPS Relative Positioning Networks Independent of Connections to Existing Control

Geometric Relative Positioning Standards	AA 0.01	A 0.1	B 1.0	1 10	2-I 20	2-II 50	3 100
Ephemerides							
Orbit accuracy, minimum (ppm)	0.008	0.05	0.5	5	10	25	50
Precise ephemerides required?	Y[a]	Y[a]	Y	op	op	N	N
Loop closure analyses[b]—when forming loops, the following are minimum criteria:							
Baselines in loop from independent observations not less than	4	3	2	2	2	2	2
Baselines in each loop, total not more than	6	8	10	10	10	15	15
Loop length, not more than (km)	2000	300	100	100	100	100	100
Baselines not meeting criteria for inclusion in any loop, not more than (percent of all independent lines)	0	0	5	20	30	30	30
In any component (X, Y, Z), "maximum" misclosure not to exceed (cm)	10	10	15	25	30	50	100
In any component (X, Y, Z), "maximum" misclosure, in terms of loop length, not to exceed (ppm)	0.2	0.2	1.25	12.5	25	60	125
In any component (X, Y, Z), "average" misclosure, in terms of loop length, not to exceed (ppm)	0.09	0.09	0.9	8	16	40	80
Repeat baseline differences							
Baseline length, not more than (km)	2000	2000	500	250	250	100	50
In any component (X, Y, Z), "maximum" not to exceed (ppm)	0.01	0.1	1.0	10	20	50	100

[a] The precise ephemerides is presently limited to an accuracy of about 1 ppm. The accuracy has been improved to about 0.1 ppm. It is unlikely orbital coordinate accuracies of 0.01 ppm will be achieved in the near future. Thus to achieve precisions approaching 0.01 ppm, it will be necessary to collect data simultaneously with continuous trackers or fiducial stations. Then all the data are processed in a session or network solution mode where the initial orbital coordinates are adjusted while solving for the baselines. In this method of processing the carrier phase data, the coordinates at the continuous trackers are held fixed.

[b] Between any combination of stations, it must be possible to form a loop through three or more stations which never passes through the same station more than once.

Source: FGCC, 1989.

GPS Base Station

GPS, like most other classical survey techniques, has to be applied in a differential mode if one wants to obtain reliable relative positional information. This implies that for most applications of GPS in geodesy, surveying and mapping, photogrammetry, GIS, and so forth, one has to have at least two GPS receivers at one's disposal. If one of the receivers occupies a known location during an acceptable minimum period, than one may obtain accurate coordinates for the second receiver *in the same frame.* In surveying/geodesy applications one should preferably include three stations with known horizontal coordinates and at least four with known vertical (orthometric) heights. In most GIS applications one receiver is left at one particular site. This station serves as a so-called *base station.*

GIS, Heights, and High-Accuracy Reference Networks

In order to reduce influences from satellite-related errors and atmospheric conditions in geodesy, surveying, and Geographic Information Systems (GIS) applications, GPS receivers are operated in a

Table 52.7 Accuracy Grades of (Civilian/Commercial) GPS Receivers

Navigation grade	40–100 m	C/A code, in stand-alone mode
Mapping (GIS) grade	2–5 m	C/A code, in differenced mode
Surveying grade	1–2 cm (within 10 km)	C/A code plus phase, differenced
Geodesy grade	5–15 mm over any distance	C/A plus P codes plus phase, differenced

differential mode. Whenever the roving receiver is not too far from the base station receiver, "errors at high altitudes (satellite and atmosphere)" are more or less canceled if the "fix from the field" is differenced with the "fix from the base."

Washington editor of *GPS World,* Hale Montgomery, writes, "As a peripheral industry, the reference station business has grown almost into an embarrassment of riches, with stations proliferating nationwide and sometimes duplicating services." William Strange, Chief Geodesist at the National Geodetic Survey (NGS), is quoted as saying, "Only about 25 full-service, fixed stations would be needed to cover the entire United States" [Montgomery, 1993]. A group of the interagency Federal Geodetic Control subcommittee has compiled a list of about 90 operating base stations on a more or less permanent basis. If all GIS/GPS base stations being planned or in operation are included, the feared proliferation will be even larger. From the point of view of the U.S. tax-paying citizen, "duplication of services" of work may be wasteful, although decentralization of services may often be more cost-effective than all-encompassing projects run by even more all-encompassing agencies.

From the geodetic point of view, the duplication of services (base station–generated fixes in the field) will proliferate the coordinate fields and the reference frames they supposedly are tied to. The loss of money and effort in the years to come while trying to make sense out of these most likely not-matching point fields may be far larger than the money lost in "service duplicating" base stations. If we are not careful, the "coordinate-duplicating" base stations will create a chaos among GIS-applying agencies.

Everyone is convinced about the necessity to collect GIS data in one *common frame.* Formerly, the GIS community was satisfied with positions of the 3- to 5-m accuracy. Manufacturers are aggressively marketing GPS/GIS equipment with 0.5-m accuracy (price tag $20,000 per receiver, and remember, you need at least two). The increased demand for accuracy requires that a reference frame be in place that lasts at least two decades. This calls for a consistent reference of one, probably two orders of magnitude more accurate than presently available.

The accuracy of the classical horizontal control was in the order of 1 part in 100,000 (1 cm over 1 km). GPS is a survey tool with an accuracy of 1 part per 1,000,000 (1 cm over 10 km). Many U.S. states put new High-Accuracy Reference Networks (HARN) in place to accommodate the accuracy of GPS surveys. Even for GIS applications where 0.5-m accuracies are claimed for the roving receivers, one may speak of 1 ppm surveys whenever those rovers operate at a distance of 500 km from their base station.

It should not be forgotten that GPS is a geometric survey tool yielding results in terms of earth-fixed coordinate differences x_{ij}. From these coordinate differences expressed in curvilinear coordinates we obtain, at best, somewhat reproducible ellipsoidal height differences. These height differences are *not* easily converted to orthometric height differences of equal accuracy. The latter height differences are of interest in engineering and GIS applications; see further section 52.7.

52.7 Gravity Field and Related Issues

One-Dimensional Positioning: Heights and Vertical Control

One of the most accurate measurements surveyors are able to make are the determinations of height differences by spirit leveling. Since a leveling instrument's line of sight is tangent to the potential surface, one may say that leveling actually determines the height differences with respect to equipotential surfaces. If one singles out one particular equipotential surface at mean sea level (the so-called *geoid*), then the heights a surveyor determines are actually orthometric heights; Fig. 52.22. Leveling in a closed loop is not a check on the actual height differences in a metrical sense but in a potential sense: the distance between equipotential surfaces varies due to local gravity variations.

FIGURE 52.22 Orthometric heights.

In spherical approximation the potential at a point A is

$$V = -\frac{GM}{r} = -\frac{GM}{R+h} \tag{52.246}$$

The gravity is locally dependent on the change in potential per height unit, or

$$\frac{dV}{dr} = g = \frac{GM}{r^2} \tag{52.247}$$

The potential dV difference between two equipotential surfaces is

$$dV = g\,dr \tag{52.248}$$

Consequently, if one levels in a loop, one has

$$\sum dV = \sum g\,dr = 0 \tag{52.249}$$

or

$$\oint dV = \oint g\,dr = 0 \tag{52.250}$$

This implies that for each metrically leveled height difference dr, one has to multiply this difference by the local gravity. Depending on the behavior of the potential surfaces in a certain area and the diameter of one's project, one has to "carry along a gravimeter" while leveling. The variations of local gravity vary depending on the geology of the area. Variations in the order $10^{-7}g$ may yield errors as large as 10 mm for height differences in the order of several hundred meters. For precise leveling surveys (≤ 0.1 mm/km) gravity observations must be made with an interval of:

2 to 3 km	in relatively "flat" areas
1 to 2 km	in hilly terrain
0.5 to 1.5 km	in mountainous regions

Table 52.8 FGCC Vertical Control Accuracy Standards (Differential Leveling)

Class	b^1
First Order	
I	< 0.5
II	0.7
Second Order	
I	1.0
II	1.3
Third Order	
	2.0

[1] $b = S/\sqrt{d}$ [mm km$^{-1/2}$]

S = standard deviation of elevation difference between control points (in mm)

d = approximate horizontal distance along leveled route (in km)

For more design criteria on leveling and gravity surveys, see FGCC [1989] and Table 52.8.

GPS surveys yield, at best, ellipsoidal height differences. These are rather meaningless from the engineering point of view. Therefore, extreme caution should be exercised when GPS height information, even after correction for geoidal undulations, is to be merged with height information from leveling. For two different points i and j,

$$h_i = H_i + N_i \tag{52.251}$$

$$h_j = H_j + N_j \tag{52.252}$$

Subtracting Eq. (52.251) from (52.252), we find the ellipsoidal height differences h_{ij} (from GPS) in terms of the orthometric height differences H_{ij} (from leveling) and the geoidal height differences N_{ij} (from gravity surveys):

$$h_{ij} = H_{ij} + N_{ij} \tag{52.253}$$

where

$$\begin{aligned} h_{ij} &= h_j - h_i \\ H_{ij} &= H_j - H_i \\ N_{ij} &= N_j - N_i \end{aligned} \tag{52.254}$$

For instance, with the National Geodetic Survey's software program GEOID93 package, geoidal height differences are as accurate as 10 cm over 100 km for the conterminous U.S. For GPS leveling this means that GPS may compete with third-order leveling as long as the stations are more than 5 km apart.

In principle any equipotential surface can act as a vertical datum such as the National Geodetic Vertical Datum of 1929 (NGVD29). Problems may arise merging GPS heights, gravity surveys, and orthometric heights referring to NGVD29. Heights referring to the NGVD88 datum will be more suitable for use with GPS surveys. In the U.S. about 600,000 vertical control stations are in existence.

Two-Dimensional Positioning: East/North and Horizontal Control

In classical geodesy the measurements in height had to be separated (leveling) from the horizontal measurements (directions, angles, azimuths, distances). To allow for the curvature of the earth and the varying gravity field, the horizontal observations were reduced first to the geoid, taking into account the orthometric heights. Subsequently, if it was desired to take advantage of geometrical properties between the once-reduced horizontal observations, the observations had to be reduced once more, but now from the geoid to the ellipsoid. An ellipsoid approximates the geoid up to 0.01%; the variations of the geoid are nowhere larger than 150 m. On the ellipsoid, which is a precise mathematical figure, one could check, for instance, whether the sum of the three angles equaled a prescribed value.

So far, geodesists relied on a biaxial ellipsoid of revolution. A semimajor axis a and a semiminor axis b define the dimensions of the ellipsoid. Rather than this semiminor axis, one specifies the *flattening* of the ellipsoid

$$f = \frac{a-b}{a} \approx \frac{1}{298.257} \tag{52.255}$$

For a semimajor axis of about 6378.137 km, it implies that the semiminor axis is 6378.137/298.257 $\approx$ 22 km shorter than a.

Distance measurements need to be reduced to the ellipsoid. Angular measurements made with theodolites, total stations, or other instruments need to be corrected for several effects:

- The direction of local gravity does not coincide with the normal to the ellipsoid. The direction of the first axis of the instrument coincides with the direction of the local gravity vector. Notwithstanding this effect, the earth's curvature causes nonparallelism of first axes of one arcsecond for each 30 m.
- The targets aimed at generally do not reside on the ellipsoid.

The non-coincidence of the gravity vector and the normal is called "deflection of the vertical." Proper knowledge of the behavior of the local geopotential surfaces is needed for proper distance and angle reductions. Consult Vaniček and Krakiwsky [1982], for example, for the mathematical background of these reductions. The FGCC adopted the following accuracy standards for horizontal control using classical geodetic measurement techniques, see Table 52.9. In the U.S. over 270,000 horizontal control stations exist.

Table 52.9 FGCC Horizontal Control Accuracy Standards (Classical Techniques)

First Order	
	1:100,000 (10 mm/km)
Second Order	
Class I	1:50,000 (20 mm/km)
Class II	1:20,000 (50 mm/km)
Third Order	
Class I	1:10,000 (100 mm/km)
Class II	1:5,000 (200 mm/km)

Three-Dimensional Positioning: Geocentric Positions and Full 3-D Control

Modern 3-D survey techniques, most noticeably GPS, allow for immediate 3-D relative positioning. 3-D coordinates are equally accurate expressed in ellipsoidal, spherical, or Cartesian coordinates. Care should be exercised in properly labeling curvilinear coordinates as spherical (*geographic*) or ellipsoidal (*geodetic*). Table 52.10 shows the large discrepancies between the two. At the mid-latitudes they may differ by more than 11′. This could result in a north-south error of 20 kilometers. While merging GIS data sets, one should be aware of the meaning LAT/LON in any instance.

Despite its 3-D characteristics, networks generated by GPS are the weakest in the vertical component, not only because of the lack of physical significance of GPS's determined heights, as described in the preceding subsection, but also because of the geometrical distribution of satellites with respect to the vertical: no satellite signals are received from "below the network." This lopsidedness makes the vertical the worst, determined component in 3-D.

Because of the inclination of the GPS satellites, there are places on earth, most notoriously the mid-latitudes, where there is not an even distribution of satellites in the azimuth sense. For instance, in the northern mid-latitudes we never have as many satellites to the north as we have to the south; see, for example, Santerre [1991]. This makes the latitude the second-best-determined curvilinear coordinate. For space techniques the FGCC has proposed the classification in Table 52.11.

Table 52.10 Geographic (Spherical) Latitude as a Function of Geodetic Latitude

Geodetic Latitude ° ′ ″	Geographic Latitude ° ′ ″	Geodetic Minus Geographic Latitude ° ′ ″
00 0 0.000	00 00 00.000	00 00 00.000
10 0 0.000	09 56 03.819	00 03 56.181
20 0 0.000	19 52 35.868	00 07 24.132
30 0 0.000	29 50 01.089	00 09 58.911
40 0 0.000	39 48 38.198	00 11 21.802
50 0 0.000	49 48 37.402	00 11 22.598
60 0 0.000	59 49 59.074	00 10 00.926
70 0 0.000	69 52 33.576	00 07 26.424
80 0 0.000	79 56 02.324	00 03 57.676
90 0 0.000	90 00 00.000	00 00 00.000

Table 52.11 FGCC 3-D Accuracy Standards (Space System Techniques)

AA Order (Global)	3 mm + 1:100,000,000 (1 mm/100 km)
A Order (Primary)	5 mm + 1:10,000,000 (1 mm/10 km)
B Order (Secondary)	8 mm + 1:1,000,000 (1 mm/km)
C Order (Dependent)	10 mm + 1:100,000 (10 mm/km)

52.8 Reference Systems and Datum Transformations

Geodetic Reference Frames

There has always been a necessity, in a variety of fields but especially in engineering, to rely on a set of adopted parameter values that describe various quantities related to the earth to the current state of the art of accuracy. Questions asked to geodesists are:

- What is the best estimate of the equatorial radius of the earth?
- What is the mass of the earth?
- How fast does the earth rotate?
- How flattened is the earth?

The values adopted at the General Assemblies of the International Association of Geodesy (IAG) of Madrid (1924) and Stockholm (1930) were valid for a long time:

$$a = 6{,}378.388 \text{ km} \quad \text{and} \quad f = 1/297.0$$

This was the so-called International Ellipsoid of the American geodesist J. F. Hayford (1909). The appropriate "International Gravity Formula" was

$$\gamma = 978.0490(1 + 0.005\,288\,4 \sin^2 \phi - 0.000\,005\,9 \sin 2\phi) \text{ cm s}^{-2}(\text{gal})$$

This formula for γ describes the gravity on the surface of the International Ellipsoid as function of the geodetic (ellipsoidal) latitude ϕ. After some years these values turned out to be not accurate enough, and the Soviet geodesist F. N. Krassovsky proposed a new ellipsoid (1943):

$$a = 6378.245 \text{ km} \quad \text{and} \quad f = 1/298.3$$

Geodetic Reference System 1967

In Hamburg (1964) the International Astronomical Union (IAU), after advice of the International Union of Geodesy and Geophysics, adopted values of three variables:

a	the equatorial radius of the earth
GM	the geocentric gravitational constant of the earth including the atmosphere
J_2	the dynamical form factor of the earth

The General Assembly of the IAG in Lucerne (1967) followed this proposal, resulting in the so-called

Geodetic Reference System 1967

$$\begin{aligned} a &= 6{,}378{,}160 \text{ m} \\ GM &= 398{,}603 \times 10^9 \text{ m}^3 \text{ s}^{-2} \\ J_2 &= 10{,}827 \times 10^{-7} \end{aligned}$$

It was agreed upon that the semiminor axis of the reference ellipsoid be parallel to the direction of the rotation axis as defined by the "Conventional International Origin." In addition, the reference meridian would be parallel to the zero meridian, as resulted from the adopted values for the longitudes of various observatories by the Bureau International de l'Heure (BIH) in Paris.

A fourth parameter completed the Geodetic Reference System (GRS) 1967, the (inertial or sidereal) angular velocity of the earth:

$$\omega = 0.000\,072\,921\,151\,467 \text{ rad/s}$$

The duration of one turn of the earth around its spin axis is then

$$\text{Length of day} = \frac{2\pi}{\omega} \approx 86{,}164.1 \text{ s} \tag{52.256}$$

This is not equal to $24 \times 60 \times 60$ s $= 86{,}400$ seconds, as explained in section 52.3.

Officially, the XVth General Assembly of the IAG in Moscow, 1971, approved and adopted these values [IAG, 1971].

Geodetic Reference System 1980

Accurate results of modern geodetic and astronomic measurement techniques necessitated the replacement of the GRS 1967 within thirteen years by the

Geodetic Reference System 1980

$$\begin{aligned} a &= 6378\,137 \text{ m} \\ GM &= 3986\,05 \times 10^{8} \text{ m}^3 \text{ s}^{-2} \\ J_2 &= 108\,263 \times 10^{-8} \\ \omega &= 7292\,115 \times 10^{-11} \text{ rad s}^{-1} \end{aligned}$$

The GRS 1980 was adopted at the XVIIth General Assembly of the IUGG in Canberra, December 1979 [IAG, 1980]. Agreements with respect to the orientation of the reference ellipsoid were not altered.

1983 Best Values

During the XVIIIth General Assembly of the IUGG in Hamburg, August 1983, the IAG adopted a resolution that stated that the GRS 1980 did not need to be replaced, but that the following values were (at the time) the most representative [IAG, 1984]:

1983 Best Values

Velocity of light in vacuum	$c =$	$(299{,}792{,}458 \pm 1.2)$ m s^{-1}
Gravitational constant (Newton)	$G =$	$(6{,}673 \pm 1) \times 10^{-14}$ m^3 s^{-2} kg^{-1}
Angular velocity of the earth	$\omega =$	$7{,}292{,}115 \times 10^{-11}$ rad s^{-1}
Geocentric gravitational constant of the earth including the atmosphere	$GM =$	$(39{,}860{,}044 \pm 1) \times 10^{7}$ m^3 s^{-2}
Geocentric gravitational constant of atmosphere	$GM_A =$	$(35 \pm 0.3) \times 10^{7}$ m^3 s^{-2}
Second-degree harmonic coefficient without permanent deformation due to tides	$J_2 =$	$(1{,}082{,}629 \pm 1) \times 10^{-9}$
Equatorial radius of the earth	$a =$	$(6{,}378{,}136 \pm 1)$ m
Equatorial gravity	$\gamma_e =$	$(978{,}032 \pm 1) \times 10^{-5}$ m s^{-2}
Flattening	$1/f =$	$(298{,}257 \pm 1) \times 10^{-3}$
Potential of the geoid	$W_0 =$	$(6{,}263{,}686 \pm 2) \times 10$ m^2 s^{-2}
Triaxial ellipsoid (rounded values):		
Equatorial flattening	$1/f_1 =$	$90{,}000$
Longitude semimajor axis	$\lambda_1 =$	$345°$ (east)

1987 Best Values and Secular Changes

In Hamburg a special study group (SSG 5.100), whose chairman was B. H. Chovitz, got the task to evaluate the status of the GRS 1980 and to prepare possible recommendations to the XIXth General Assembly of the IUGG in Vancouver, August, 1987. The result of the four-year study was that the GRS 1980 still did not need replacement [IAG, 1988a], but that a series of "1987 Best Values and Secular Changes" could be forwarded. The interesting conclusion was that, from this moment on, two values had to be recognized for many parameters: a time-invariant component and the first derivative with respect to time [IAG, 1988b].

In the meantime the velocity of light had been raised to a physical constant, adopting the value as listed in the previous section [Cohen and Taylor, 1988]:

$$c = 299{,}792{,}458 \text{ m s}^{-1} \qquad \text{exactly}$$

This means that the meter is now a derived parameter through the adopted length of a second.

1987 Best Values

Angular velocity of the earth (rounded)	$\omega = 7.292115 \times 10^{-5}$ rad s^{-1}
Geocentric gravitational constant of the earth including the atmosphere	$GM = (3{,}986{,}004.40 \pm .03) \times 10^{8}$ m^3 s^{-2}
Equatorial radius of the earth	$a = (6{,}378{,}136 \pm 1)$ m
Second-degree spherical harmonic coefficient without permanent effect due to tides	$J_2 = (1{,}082{,}626 \pm 2) \times 10^{-9}$

1987 Best Secular Changes

Decrease of angular velocity due to tides	$d\omega_{T}/dt = (-6.0 \pm 0.3) \times 10^{-22}$ rad/s^{-2}
Decrease of angular velocity due to other causes	$d\omega_{NT}/dt = (+1.4 \pm 0.3) \times 10^{-22}$ rad/s^{-2}
Total decrease in angular velocity	$d\omega/dt = (-4.6 \pm 0.4) \times 10^{-22}$ rad/s^{-2}
Relative change of the gravitational constant	$(dG/dt)/G = (0 \pm 1) \times 10^{-11}$ yr^{-1}
Change of the earth's mass	$dM/dt = 0$ kg yr^{-1}
Change of the earth's radius	$da/dt = 0$ mm yr^{-1}
Decrease in the second-degree harmonic coefficient	$dJ_2/dt = (-2.8 \pm 0.3) \times 10^{-11}$ yr^{-1}

It is interesting to note that we do not zero in on the earth's semimajor axis in an arbitrary manner. The earth seems to "shrink" under our increasingly accurate estimates of its equatorial radius, as shown in Table 52.12.

Table 52.12 Decrease of the Semimajor Axis in Recent Decades

International Ellipsoid 1924/Hayford 1909	6,378,388 m
Krassovsky 1943	6,378,245 m
Geodetic Reference System 1967	6,378,160 m
Geodetic Reference System 1980	6,378,137 m
Best Values 1983/1987	6,378,136 m

World Geodetic System 1984

The World Geodetic System 1984 (WGS) and its three predecessors (WGS 60, WGS 66, and WGS 72), developed by the American Department of Defense/Defense Mapping Agency (DoD/DMA), are also reference *systems* in the real sense of the word. The WGS 84 consists of a reference coordinate frame, an ellipsoidal gravity formula, an ellipsoid, an earth's gravity model, a geoid, and a series of transformation parameters and formulas between various geodetic reference frames and datum values [DMA, 1988].

WGS 84 Coordinate Frame

The WGS 84 coordinate frame is based on the frames that were originally developed for the U.S. Navy's Doppler/Transit System. One of its most recent reference frames, known as NSWC 9Z-2, had the following known problems:

- Its geocentricity or, better non-geocentricity
- The misorientation of the zero meridian as compared to the zero meridian adopted by the BIH
- A scale error of almost 1 ppm

These problems can be expressed in terms of specific values:

- The origin with respect to the center of mass of the earth, as accurately determined by satellite laser ranging (SLR), was 4.5 m too high.
- The x-axis of the NSWC frame was 0.814 arcseconds to the east of the BIH zero meridian—an error of about 24 m at the equator.
- The scale of NSWC 9Z-2 was 0.6×10^{-6} too large.

The WGS 84 corrected these problems. The transformation formulas (see the discussion of earth-fixed frames in section 52.3) from $\mathbf{X}_{\text{NSWC 9Z}-2}$ to $\mathbf{X}_{\text{WGS 84}}$ become

$$\mathbf{x}_{\text{WGS 84}} = \sigma \mathbf{R}_3(\gamma)\mathbf{x}_{\text{NSWC 9Z}-2} + \mathbf{t} \tag{52.257}$$

with

$$\sigma = 1. + \delta\sigma = 1. - 0.000\,000\,6 \tag{52.258}$$

$$\gamma = 0''.814 \tag{52.259}$$

$$\mathbf{t} = \begin{pmatrix} 0 \\ 0 \\ 4.5\text{ m} \end{pmatrix} \tag{52.260}$$

WGS 84 Ellipsoid

The origin of the WGS 84 coordinate frame is also the center of the WGS 84 ellipsoid, so that the symmetry axis of the ellipsoid of revolution coincides with the z-axis of the WGS 84 frame. The values for the standard WGS 84 earth are

WGS 84 Ellipsoid

Semimajor axis	$a =$	$(6{,}378{,}137 \pm 2)$ m
Geocentric gravitational constant of the earth including the atmosphere	$GM =$	$(3{,}986{,}005 \pm 0.6) \times 10^8\ \text{m}^3\ \text{s}^{-2}$
Geocentric gravitational constant of atmosphere	$GM_A =$	$(3.5 \pm 0.1) \times 10^8\ \text{m}^3\ \text{s}^{-2}$

Geocentric gravitational constant of the earth without atmosphere	$GM_{-A} = (3{,}986{,}001.5 \pm 0.6) \times 10^8\ \text{m}^3\ \text{s}^{-2}$
Normalized second-degree zonal gravitational coefficient without permanent deformation due to tides	$\bar{C}_{2,0}^{-T} = (-484.166\ 85 \pm .00130) \times 10^{-6}$
Normalized second-degree zonal gravitational coefficient with permanent deformation due to tides ($J_2 = -\sqrt{5}\bar{C}_{2,0}$)	$\bar{C}_{2,0}^{+T} = (-484.171\ 01 \pm .00130) \times 10^{-6}$
Angular velocity of the earth	$\omega = (7{,}292{,}115 \pm .1500) \times 10^{-11}\ \text{rad s}^{-1}$
Angular velocity of the earth in a precessing frame	$\omega* = (7{,}292{,}115.8553 \times 10^{-11} + 4.3 \times 10^{-15}\ T_u)\ \text{rad s}^{-1}$ with T_u in Julian centuries since January 1.5, 2000; see Eq. (52.74)

The velocity of light is the same as in GRS1980; see section 52.8.

Since the WGS 84 ellipsoid is a geocentric equipotential ellipsoid, the flattening f of this reference ellipsoid is a derived constant. It means that f can be computed from the listed "fundamental" parameters; see Heiskanen and Moritz [1968, sections 2-7 through 2-10] or IAG [1971].

The flattening f has the value:

$$\text{flattening (ellipticity)} \qquad f = 1/298.257\ 223\ 563$$

WGS 84 Ellipsoidal Gravity Formula

The normal gravity γ along the surface of the WGS 84 ellipsoid is given by

$$\gamma = 978{,}032.677\ 14 \times (1. + 0.001\ 931\ 851\ 386\ 39 \times \sin^2 \phi)/$$
$$(1. - 0.006\ 694\ 379\ 990\ 13 \times \sin^2 \phi)^{1/2}\ \text{milligal}$$

where 1 milligal $= 0.001\ \text{cm s}^{-2}$.

WGS 84 Earth Gravitational Model (EGM)

The WGS 84 Earth Gravitational Model is a gravity model with harmonic coefficients up to degree and order 180. For civilian users, the coefficients above degree and order 18 are classified.

WGS 84 Geoid

As with the EGM, for civilian users the geoid based on coefficients above degree and order 18 is classified.

In general, the geoidal undulations (geoidal heights) are accurate to about 2 to 6 m. For 55% of the earth's coverage the accuracy is 2 to 3 m, for 93% 2 to 4 m.

IERS Standards 1992

The International Earth Rotation Service adopted a standard reference system to be used in earth rotation analysis. The system includes not only numerical values for a set of variables, but also sets of adopted equations reflecting certain physical models. The reader is referred for more details to IERS [1992].

Table 52.13 Transformation Parameters among Several Global Reference Frames

Coordinate System (Datum) (1)	Δx (m) (2)	Δy (m) (3)	Δr (m) (4)	$\delta \varepsilon$ (arc sec) (5)	$\delta \psi$ (arc sec) (6)	$\delta \omega$ (arc sec) (7)	δs (ppm) (8)
NWL-9D→WGS-72	0	0	0	0	0	−0.26	−0.827
NWL-9D→WGS-84	0	0	+4.5	0	0	−0.814	−0.6
WGS-72→WGS-84	0	0	+4.5	0	0	−0.554	+0.227
BTS87→NWL-9D	+0.071	−0.509	−4.666	−0.0179	+0.005	+0.8073	+0.583
BTS87→WGS-84	+0.071	−0.509	−0.166	−0.0179	+0.005	−0.0067	−0.107
BTS87→VLBI (NGS)	−0.089	+0.143	−0.016	+0.0043	−0.0093	+0.0033	+0.009
BTS87→SLR (GSFC)	0.000[a]	0.000[a]	0.000[a]	+0.0018	−0.0062	+0.0075	0.000[a]
WGS-84→WGS-84 (GPS)	+0.026	−0.006	+0.093	+0.001	0.000	+0.002	−0.128

[a] These values were held fixed [i.e., the BTS87 frame origin and scale are assumed to be defined through satellite laser ranging SLR (GSFC)].

Source: Soler and Hothem, 1988.

Datum and Reference Frame Transformations

In section 52.3 the transformation formulas are described relating coordinates given in two different reference frames. Also the adoption of new ellipsoidal parameters (so-called datum transformations) causes the geodetic coordinates to change. Tables 52.13, 52.14, and 52.15, from such sources as Soler and Hothem [1988] and DMA [1988], list the values for various ellipsoids and reference frame transformations.

Table 52.14 Parameters of Some Adopted Reference Ellipsoids

Coordinate System (Datum) (1)	Reference Ellipsoid Used (2)	a (m) (3)	$1/f$ (4)
AGD	AN (or SA-69)	6,378,160	298.25
ED-79	International	6,378,388	297
GEM-8	GEM-8	6,378,145	298.255
GEM-9 (or GEM-10)	GEM-9 (or GEM-10)	6,378,140	298.255
GEM-10B	GEM-10B	6,378,138	298.257
GEM-T1	GEM-T1	6,378,137	298.257
NAD-27	Clarke 1866	6,378,206.4	294.9786982
NAD-83	GRS-80	6,378,137	298.257222101
NWL-9D = NSWC-9Z2	WGS-66	6,378,145	298.25
SA-69	SA-69 (or AN)	6,378,160	298.25
WGS-72	WGS-72	6,378,135	298.26
WGS-84	WGS-84	6,378,137	298.257223563

Note: AGD = Australian geodetic datum; AN = Australian national; ED = European datum; GEM = Goddard earth model; GRS = geodetic reference system; NAD = North American datum; NSWC = Naval surface warfare center; NWL = Naval Weapons Laboratory; SA = South American; WGS = world geodetic system.

Source: Soler and Hothem, 1988.

Table 52.15 Transformation Parameters, Local Geodetic Systems to WGS 84

	Reference Ellipsoids and Parameter Differences			Transformation Parameters		
Local Geodetic Systems	Name	Δa(m)	$\Delta f \times 10^4$	ΔX(m)	ΔY(m)	ΔZ(m)
Adindan	Clark 1880	−112.145	−0.54750714	−162	−12	206
Afgooye	Krassovsky	−108	0.00480795	−43	−163	45
Ain El ABD 1970	International	−251	−0.14192702	−150	−251	−2
Anna 1 Astro 1965	Australian National	−23	−0.00081204	−491	−22	435
ARC 1950	Clarke 1880	−112.145	−0.54750714	−143	−90	−294
ARC 1960	Clarke 1880	−112.145	−0.54750714	−160	−8	−300
Ascension Island 1958	International	−251	−0.14192702	−207	107	52
Astro Beacon "E"	International	−251	−0.14192702	145	75	−272

Table 52.15 (continued) Transformation Parameters, Local Geodetic Systems to WGS 84 *(continues)*

	Reference Ellipsoids and Parameter Differences			Transformation Parameters		
Local Geodetic Systems	Name	$\Delta a(m)$	$\Delta f \times 10^4$	$\Delta X(m)$	$\Delta Y(m)$	$\Delta Z(m)$
Astro B4 Sorol Atoll	International	−251	−0.14192702	114	−116	−333
Astro Dos 71/4	International	−251	−0.14192702	−320	550	−494
Astronomic Station 1952	International	−251	−0.14192702	124	−234	−25
Australian Geodetic 1966	Australian National	−23	−0.00081204	−133	−48	148
Australian Geodetic 1984	Australian National	−23	−0.00081204	−134	−48	149
Bellevue (IGN)	International	−251	−0.14192702	−127	−769	472
Bermuda 1957	Clarke 1866	−69.4	−0.37264639	−73	213	296
Bogota Observatory	International	−251	−0.14192702	307	304	−318
Campo Inchauspe	International	−251	−0.14192702	−148	136	90
Canton Astro 1966	International	−251	−0.14192702	298	−304	−375
Cape	Clarke 1880	−112.45	−0.54750714	−136	−108	−292
Cape Canaveral	Clarke 1866	−69.4	−0.37264639	−2	150	181
Carthage	Clarke 1880	−112.145	−0.54750714	−263	6	431
Chatham 1971	International	−251	−0.14192702	175	−38	113
Chua Astro	International	−251	−0.14192702	−134	229	−29
Corrego Alegre	International	−251	−0.14192702	−206	172	−6
Djakarta (Batavia)	Bessel 1841	739.845	0.10037483	−377	681	−50
DOS 1968	International	−251	−0.14192702	230	−199	−752
Easter Island 1967	International	−251	−0.14192702	211	147	111
European 1950	International	−251	−0.14192702	−87	−98	−121
European 1979	International	−251	−0.14192702	−86	−98	−119
Gandajika Base	International	−251	−0.14192702	−133	−321	50
Geodetic Datum 1949	International	−251	−0.14192702	84	−22	209
Guam 1963	Clarke 1866	−69.4	−0.37264639	−100	.248	259
GUX 1 Astro	International	−251	−0.14192702	252	−209	−751
Hjorsey 1955	International	−251	−0.14192702	−73	46	−86
Hong Kong 1963	International	−251	−0.14192702	−156	−271	−189
India	Everest	860.655	0.28361368	214	836	303
				289	734	257
Ireland 1965	Modified Airy	796.811	0.11960023	506	−122	611
ISTS 073 Astro 1969	International	−251	−0.14192702	208	−435	−229
Johnson Island 1961	International	−251	−0.14192702	191	−77	−204
Kandawala	Everest	860.655	0.28361368	−97	787	86
Kerguelen Island	International	−251	−0.14192702	145	−187	103
Kertau 1948	Modified Everest	832.937	0.28361368	−11	851	5
L.C. 5 Astro	Clarke 1866	−69.4	−0.37264639	42	124	147
Liberia 1964	Clarke 1880	−112.145	−0.54750714	−90	40	88
Luzon	Clarke 1866	−69.4	−0.37264639			
Philippines				−133	−77	−51
Mindanao Island				−133	−79	−72
Mahe 1971	Clarke 1880	−112.145	−0.54750714	41	−220	−134
Marco Astro	International	−251	−0.14192702	−289	−124	60
Massawa	Bessel 1841	739.845	0.10037483	639	405	60
Merchich	Clarke 1880	−112.145	−0.54750714	31	146	47
Midway Astro 1961	International	−251	−0.14192702	912	−58	1227
Minna	Clarke 1880	−112.145	−0.54750714	−92	−93	122
Nahrwan	Clarke 1880	−112.145	−0.54750714			
Masirah Island				−247	−148	369
United Arab Emirates				−249	−156	381
Saudi Arabia				−231	−196	482
Naparima, BWI	International	−251	−0.14192702	−2	374	172
North American 1927	Clarke 1866	−69.4	−0.37264639			
Mean Value (Conus)				−8	160	176
Alaska				−5	135	172
Bahamas				−4	154	178
San Salvador Island				1	140	165
Canada				−10	158	187

Table 52.15 (continued) Transformation Parameters, Local Geodetic Systems to WGS 84

	Reference Ellipsoids and Parameter Differences			Transformation Parameters		
Local Geodetic Systems	Name	$\Delta a(m)$	$\Delta f \times 10^4$	$\Delta X(m)$	$\Delta Y(m)$	$\Delta Z(m)$
Canal Zone				0	125	201
Caribbean				−7	152	178
Central America				0	125	194
Cuba				−9	152	178
Greenland				11	114	195
Mexico				−12	130	190
North American 1983	GRS	0	−0.00000016	0	0	0
Observatorio 1966	International	−251	−0.14192702	−425	−169	81
Old Egyptian	Helmert 1906	−63	0.00480795	−130	110	−13
Old Hawaiian	Clarke 1866	−69.4	−0.37264639	61	−285	−181
Oman	Clarke 1880	−112.145	−0.54750714	−346	−1	224
Ordnance Survey of Great Britain 1936	Airy	573.604	0.11960023	375	−111	431
Pico De Las Nieves	International	−251	−0.14192702	−307	−92	127
Pitcairn Astro 1967	International	−251	−0.14192702	185	165	42
Provisional South Chilean 1963	International	−251	−0.14192702	16	196	93
Provisional South American 1956	International	−251	−0.14192702	−288	175	−376
Puerto Rico	Clarke 1866	−69.4	−0.37264639	11	72	−101
Qatar National	International	−251	−0.14192702	−128	−283	22
Qornoq	International	−251	−0.14192702	164	138	−189
Reunion	International	−251	−0.14192702	94	−948	−1262
Rome 1940	International	−251	−0.14192702	−225	−65	9
Santo (DOS)	International	−251	−0.14192702	170	42	84
Sao Braz	International	−251	−0.14192702	−203	141	53
Sapper Hill 1943	International	−251	−0.14192702	−355	16	74
Schwarazeck	Bessel 1841	653.135	0.10037483	616	97	−251
South American 1969	South American 1969	−23	−0.00081204	−57	1	−41
South Asia	Modified Fischer 1960	−18	0.00480795	7	−10	−26
Southeast Base	International	−251	−0.14192702	−499	−249	314
Southwest Base	International	−251	−0.14192702	−104	167	−38
Timbalai 1948	Everest	860.655	0.28361368	−689	691	−46
Tokyo	Bessel 1841	739.845	0.10037483	−128	481	664
Tristan Astro 1968	International	−251	−0.14192702	−632	438	−609
Viti Levu 1916	Clarke 1880	−112.145	−0.54750714	51	391	−36
Wake-Eniwetok 1960	Hough	−133	−0.14192702	101	52	−39
Zanderij	International	−251	−0.14192702	−265	120	−358

Source: DMA, 1988.

References

Chan, L. and Schorr, J. 1994. GPS world receiver survey. *GPS World*. 5(1):38–56.

Cohen, E. R. and Taylor, B. N. 1988. The fundamental physical constants. *Physics Today*. 41(8):9–13.

DMA (Defense Mapping Agency). 1988. Department of Defense World Geodetic System. Its definition and relationships with local geodetic systems. *DMA Technical Report 8350.2* [revised: 1 March 1988].

FGCC (Federal Geodetic Control Committee). 1989. *Geometric Geodetic Accuracy Standards and Specifications for using GPS Relative Positioning Techniques,* Version 5.0. Federal Geodetic Control Committee, Rockville, MD.

GPS World, 1994. Satellite almanac overview. *GPS World*. 5(10):60.

Heiskanen, W. A. and Moritz, H. 1967. *Physical Geodesy*. W. H. Freeman and Company, San Francisco/London.

IAG (International Association of Geodesy). 1971. Geodetic Reference System 1967. Publication spéciale No. 3. Paris.

IAG. 1980. Geodetic Reference System 1980 (compiled by H. Moritz). In *The Geodesist's Handbook 1980. Bulletin Géodésique.* 54(3):395–405.

IAG. 1984. Geodetic Reference System 1980 (compiled by H. Moritz). In *The Geodesist's Handbook 1984. Bulletin Géodésique.* 58(3):388–398.

IAG. 1988a. Geodetic Reference System 1980 (compiled by H. Moritz). In *The Geodesist's Handbook 1988. Bulletin Géodésique.* 62(3):348–358.

IAG. 1988b. Parameters of common relevance of astronomy, geodesy, and geodynamics (compiled by B. H. Chovitz, former president of SSG 5.100). In *The Geodesist's Handbook 1988. Bulletin Géodésique.* 62(3):359–367.

IERS (International Earth Rotation Service). 1992. *IERS Standards* (1992). ed. D. D. McCarthy. IERS Technical Note 12. Central Bureau of the IERS. Observatoire de Paris.

Leick, A. and van Gelder, B. H. W. 1975. On similarity transformations and geodetic network distortions based on Doppler satellite coordinates. *Reports of the Department of Geodetic Science, No. 235.* The Ohio State University, Columbus, OH.

Lucas, J. 1963. Differentiation of the orientation matrix by matrix multipliers. *Photogrammetric Engineering,* 29(4):708–715.

Maling, D. H. 1993. *Coordinate Systems and Map Projections.* Pergamon Press, Oxford/New York/Seoul/Tokyo.

Montgomery, H. 1993. City streets, airports, and a station roundup. *GPS World,* 4(2):16–19.

Moritz, H. and Mueller, I. I. 1988. *Earth Rotation: Theory and Observation.* Frederick Ungar Publishing Co., New York.

Mueller, I. I. 1969. *Spherical and Practical Astronomy, As Applied to Geodesy.* Frederick Ungar Publishing Co., New York.

Reilly, J. P. 1994. *P.O.B.* 1994 GPS equipment survey. *P.O.B.* 19(5):75–86.

Santerre, R. 1991. Impact of GPS satellite sky distribution. *Manuscripta Geodaetica,* 61(1):28–53.

Soler, T. 1976. On differential transformations between Cartesian and curvilinear (geodetic) coordinates. *Reports of the Department of Geodetic Science, No. 236.* The Ohio State University, Columbus, OH.

Soler, T. and Hothem, L. D. 1988. Coordinate systems used in geodesy: basic definitions and concepts. *J. Surveying Eng.* 114(2):84–97.

Soler, T. and van Gelder, B. H. W. 1987. On differential scale changes and the satellite Doppler z-shift. *Geophys. J. Roy. Astron. Soc.* 91:639–656.

Stem, J. E. 1991. State Plane Coordinate System of 1983. *NOAA Manual NOS NGS 5.* Rockville, MD.

Vaniček, P. and E. J. Krakiwsky, 1982. *Geodesy: the Concepts.* North-Holland Publishing Company, Amsterdam/New York/Oxford.

Wells, D. (ed.) and Kleusberg, A. 1990. GPS: a multipurpose system. *GPS World.* 1(1):60–63.

For Further Information

Textbooks and Reference Books

For additional reading and more background, from the very basic to the advanced level, in geodesy, satellite geodesy, physical geodesy, mechanics, orbital mechanics, and relativity, the reader is referred to the following textbooks (in English):

Bomford, G. 1980. *Geodesy.* Clarendon Press, Oxford.

Escobal, P.R. 1976. *Methods of Orbit Determination.* John Wiley & Sons, Inc. New York/London/Sydney.

FGCC, 1984. *Standards and Specifications for Geodetic Control Networks.* Reprint version February 1991. Federal Geodetic Control Committee. Rockville, MD.

Goldstein, H. 1965. *Classical Mechanics.* Addison-Wesley Publishing Company, Inc., Reading, MA/Palo Alto/London/Dallas/Atlanta.

Hofmann-Wellenhof, B., Lichtenegger, H., and Collins, J. 1993. *GPS: Theory and Practice.* Springer-Verlag, Vienna/New York.

Jeffreys, Sir H. 1970. *The Earth: Its Origin, History and Physical Constitution.* Cambridge University Press, Cambridge/London/New York/ Melbourne.

Kaula, W. M. 1966. *Theory of Satellite Geodesy: Applications of Satellites to Geodesy.* Blaisdell Publishing Company, Waltham, MA/ Toronto/New York.

Lambeck, K. 1988. *Geophysical Geodesy: The Slow Deformations of the Earth.* Clarendon Press, Oxford.

Leick, A. 1990. *GPS: Satellite Surveying.* John Wiley & Sons, Inc., New York/Chichester/Brisbane/Toronto/Singapore.

Melchior, P. 1978. *The Tides of the Planet Earth.* Pergamon Press, Oxford/New York/Toronto/Sydney/Paris/Frankfurt.

Moritz, H. 1990. *The Figure of the Earth: Theoretical Geodesy and the Earth's Interior.* Wichmann, Karlsruhe.

Munk, W. H. and MacDonald, G. J. F. 1975. *The Rotation of the Earth: A Geophysical Discussion.* Cambridge University Press, Cambridge/London/New York/Melbourne.

NATO (North Atlantic Treaty Organization). 1988. Standardization agreement on NAVSTAR Global Positioning System (GPS), system characteristics—preliminary draft. *STANAG 4294* [revision: 15 April 1988].

Seeber, G. 1993. *Satellite Geodesy: Foundations, Methods, and Applications.* Walter de Gruyter, Berlin/New York.

Soffel, M. H. 1989. *Relativity in Astrometry, Celestial Mechanics and Geodesy.* Springer-Verlag, Berlin/Heidelberg/New York/London/Paris/Tokyo.

Torge, W. 1991. *Geodesy.* Walter de Gruyter, Berlin/New York.

Wells, D. (ed.) 1986. *Guide to GPS Positioning.* Canadian GPS Associates, Fredericton, New Brunswick, Canada.

Journals and Organizations

The latest results from research in geodesy are published in two international magazines under the auspices of the International Association of Geodesy, both published by Springer-Verlag, Berlin/Heidelberg/New York:

- *Bulletin Géodésique*
- *Manuscripta Geodaetica*

Geodesy- and geophysics-related articles can be found in

- American Geophysical Union: EOS and Journal of Geophysical Research, Washington, D.C.
- Royal Astronomical Society: Geophysical Journal International, London

Kinematic GPS-related articles can be found in

- Institute of Navigation: *Navigation*
- American Society of Photogrammetry and Remote Sensing: *Photogrammetric Engineering & Remote Sensing*

Many national mapping organizations publish journals in which recent results in geodesy/surveying/mapping are documented:

- American Congress of Surveying and Mapping: *Surveying and Land Information Systems; Cartography and Geographic Information Systems*
- American Society of Civil Engineers: *Journal of Surveying Engineering*
- Deutscher Verein für Vermessungswesen: *Zeitschrift für Vermessungswesen,* Konrad Wittwer Verlag, Stuttgart

- The Canadian Institute of Geomatics: *Geomatica*
- The Royal Society of Chartered Surveyors: *Survey Review*
- Institution of Surveyors of Australia: *Australian Surveyor*

Worth special mention are the following trade magazines:

- *GPS World,* published by Advanstar Communications, Eugene, OR
- *P.O.B.* (Point of Beginning), published by P.O.B. Publishing Company, Canton, MI
- *Professional Surveyor,* published by American Surveyors Publishing Company, Inc., Arlington, VA
- *G.I.M.* (*Geomatics Info Magazine*), published by Geodetical Information & Trading Centre bv., Lemmer, the Netherlands

National mapping organizations as the U.S. National Geodetic Survey (NGS) regularly make geodetic software available (free and at cost). Information can be obtained from

National Geodetic Survey
Geodetic Services Branch
National Ocean Service, NOAA
1315 East-West Highway, Station 8620
Silver Spring, MD 20910-3282

53

Photogrammetry and Remote Sensing

J. S. Bethel
Purdue University

53.1 Basic Concepts in Photogrammetry

The term *photogrammetry* refers to the measurement of photographs and images for the purpose of making inferences about the size, shape, and spatial attributes of the objects appearing in the images. The term *remote sensing* refers to the analysis of photographs and images for the purpose of extracting the best interpretation of the image content. Thus the two terms are by no means mutually exclusive and each one includes some aspects of the other. However, the usual connotations designate geometric inferences as photogrammetry and radiometric inferences as remote sensing. Classically, both photogrammetry and remote sensing relied on photographs, that is, silver halide emulsion products, as the imaging medium. In recent years digital images or computer resident images have taken on an increasingly important role in both photogrammetry and remote sensing. Thus many of the statements to be found herein will refer to the general term *images* rather than to the more restrictive term *photographs.* The characteristic of photogrammetry and remote sensing

0-8493-8953-4/95/$0.00 + $.50

that distinguishes them from casual photography is the insistence on a thorough understanding of the sensor geometry and its radiometric response.

The predominant type of imaging used for civil engineering applications is the traditional 23 centimeter format frame aerial photograph. Photogrammetric and remote sensing techniques can be equally applied to satellite images and to close-range images acquired from small format cameras. The main contributions of photogrammetry and remote sensing to civil engineering include topographic mapping, orthophoto production, planning, environmental monitoring, database development for geographic information systems (GIS), resource inventory and monitoring, and deformation analysis.

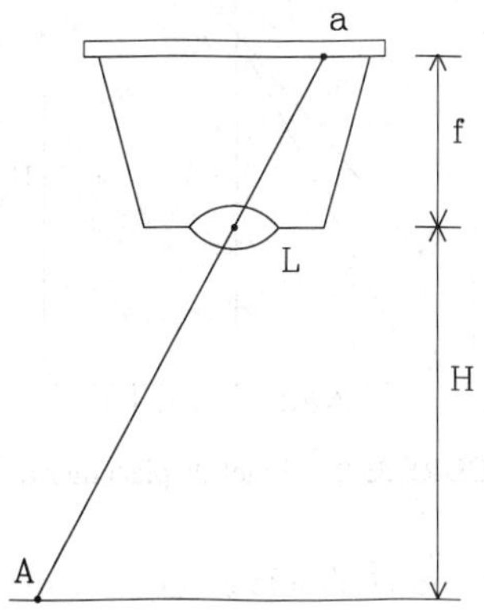

FIGURE 53.1 Typical aerial camera geometry.

Scale and Coverage

A typical aerial camera geometry is shown in Fig. 53.1. The perspective geometry of an ideal camera would dictate that each object point, A, would lie along a line containing the corresponding image point, a, and the perspective center, L. H represents the flying height above the terrain; f is the focal length or principal distance of the camera or sensor. In the ideal case that the camera axis is strictly vertical and the terrain is a horizontal plane surface, the *scale* of the image can be expressed as a fraction:

$$\text{Scale} = \frac{\text{Image distance}}{\text{Object distance}} \tag{53.1}$$

The units of measure should be the same for the numerator and the denominator so the fraction becomes a unitless ratio. Usually one forces the numerator to be 1, and this is often called the representative fraction or scale. In practice, units are sometimes introduced, but this should be discouraged. For instance, "one inch equals one hundred feet" should be converted as

$$\frac{1 \text{ inch}}{100 \text{ feet}} = \frac{1 \text{ inch}}{1200 \text{ inches}} = \frac{1}{1200} \tag{53.2}$$

This scale could also be represented in the form 1:1200. By the geometry shown in Fig. 53.1, the scale may also be expressed as a ratio of focal length and flying height above the terrain:

$$\text{Scale} = \frac{f}{H} \tag{53.3}$$

In practice, images never fulfill the ideal conditions stated above. In the general case of tilted photographs and terrain which is not horizontal, such a simple scale determination is only approximate. In fact the scale may be different at every point, and further the scale at each point may be a function of direction. Because of this one often speaks of a *nominal scale* which may apply to a single image or to a strip or block of images.

From Eq. 53.3 it is clear that high altitude imagery, having large H, yields a small scale ratio compared with lower altitude imagery, for which H is smaller. Thus one speaks of small scale imagery from high altitude and large scale imagery from low altitude. The area covered by a fixed size image will of course be inversely related to scale. Small scale images would cover a larger area than large scale images.

Relief and Tilt Displacement

If the ideal conditions of horizontal terrain and vertical imagery were fulfilled, a scale could be determined for the image and it could be used as a map. Terrain relief and image tilt are always present to some extent, however, and this prevents the use of raw images or photographs as maps. It is common practice to modify images to remove tilt displacement (this is referred to as *rectification*); such images can be further modified to remove relief displacement (this is referred to as *differential rectification*). Such a differentially rectified image is referred to as an *orthophoto.*

The concept of relief displacement is shown schematically in Fig. 53.2. If the flagpole were viewed from infinity its image would be a point and there would be no relief displacement. If it is viewed from a finite altitude its image will appear to "lay back" on the adjacent terrain. The displacement vector in the image is aa'. The magnitude of this displacement vector will depend on the height of the object, the flying height, and its location in the image. Such displacements in the image are always radial from the image of the *nadir point.* The nadir point is the point exactly beneath the perspective center. In a vertical photograph this would also coincide with the image *principal point,* or the foot of the perpendicular from the perspective center. Assuming vertical imagery, the height of an object can be determined by relief displacement to be

$$h = \frac{dH}{r} \tag{53.4}$$

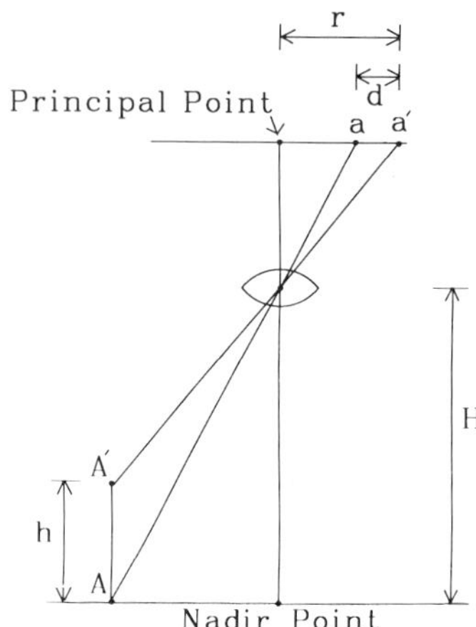

FIGURE 53.2 Relief displacement.

d and r should be measured in the same units in the image, and h and H should be in the same object space units. Equation (53.4) could be used to obtain approximate heights for discrete objects with respect to the surrounding terrain. The same displacement occurs with the terrain itself, if it is higher than the reference plane. However, because the terrain is a continuous surface this displacement is not obvious as in the case of the flagpole.

Image tilt also creates image point displacements that would not be present in a vertical image. Extreme image tilts are sometimes introduced on purpose for *oblique photography.* For nominally vertical images, tilts are usually kept smaller than three degrees. Figure 53.3 illustrates some of these concepts in a sequence of sketches. Figure 53.3(a) depicts a tilted image of a planimetrically orthogonal grid draped over a terrain surface, Fig. 53.3(b) shows the image with tilt effects removed, and Fig. 53.3(c) shows the image with relief displacement effects removed. Only after the steps to produce image (c) can one use the image as a map. Prior to this there are severe systematic image displacements.

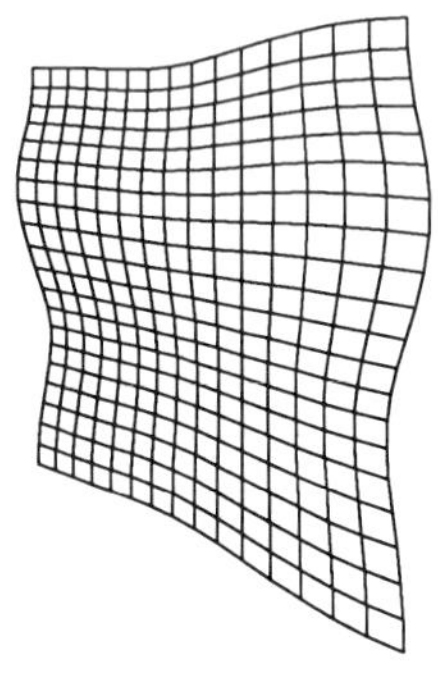

(a) Tilted Perspective Image

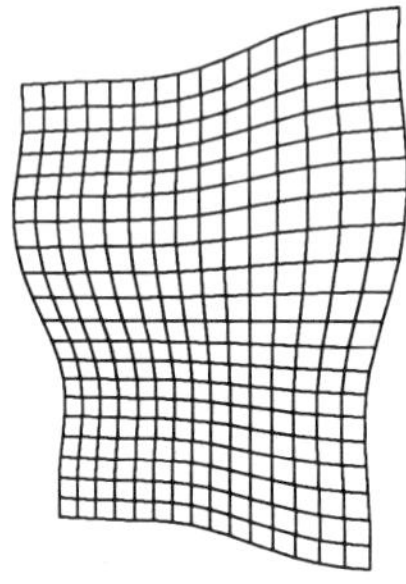

(b) Rectified Perspective Image (Equivalent Vertical Photograph)

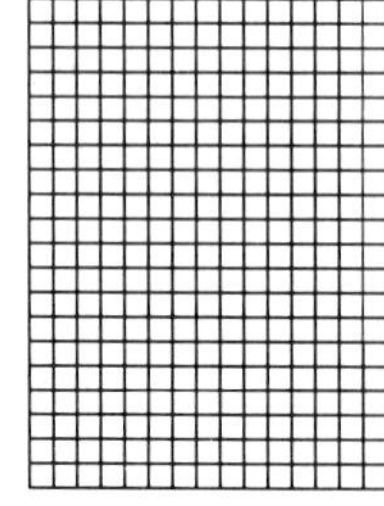

(c) Differentially Rectified Image (Orthophotograph)

FIGURE 53.3 Image rectification sequence.

Parallax and Stereo

Parallax is defined as the apparent shift in the position of an object, caused by a shift in the position of the viewer. Alternately closing one eye and then the other will demonstrate the concept of parallax, as near objects appear to shift, whereas far objects will appear stationary. This effect is the primary mechanism by which we achieve binocular depth perception, or stereo vision. This effect is exploited in photogrammetry when two overlapping photographs are taken. Within the overlap area, objects are imaged from two different exposure positions and parallax is evidenced in the resulting images. Parallax measurements can be used for approximate height computations in the following way using the geometry in Fig. 53.4. Two vertical overlapping photographs are taken at the same altitude, and an image coordinate system is set up within each image with the x axis parallel to the flight line, and with origin at the principal point in each image. For a given point, the parallax is defined as

$$p = x_{\text{left}} - x_{\text{right}} \tag{53.5}$$

The dimension H can be computed from

$$H = \frac{fB}{p} \tag{53.6}$$

B represents the base, or distance between the exposure stations. As is evident from Eq. (53.6) the distance H is inversely related to parallax, so that large parallax yields small H. Equation (53.6) is most often used with a pair of nearby points to determine a difference in elevation rather than the absolute elevation itself. This is given by the following equation, in which b is the image dimension of the base, and Δp is the difference in parallax.

$$\Delta H = \frac{H \Delta p}{b} \tag{53.7}$$

Figure 53.4 also illustrates the concept of B/H or *base-height ratio.* For a given focal length, format size, and overlap there is a corresponding B/H value for some average elevation in the area of interest. Large B/H yields strong geometry for elevation determination and small B/H yields weak geometry. For typical aerial photography for engineering mapping one would have $f = 152$ mm, $H = 732$ m, $B = 439$ m, overlap $= 60\%$, and $B/H = 0.6$.

The viewing geometry as opposed to the taking geometry for this imagery yields a $b/h = 0.3$. The difference between the taking geometry and the viewing geometry causes *vertical exaggeration.*

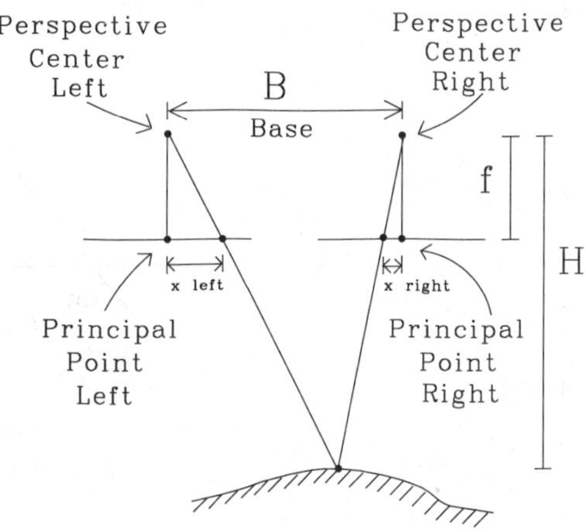

FIGURE 53.4 Parallax geometry.

That is features appear stretched in the z-dimension compared to the planimetric dimension. This does not affect measurements, but only the viewer's perception of depth.

53.2 Sensors and Platforms

Cameras

Aerial cameras for engineering mapping use standard aerial film with 23 cm width. The emulsion type is usually panchromatic (black and white) but color film is becoming more popular. For some interpretation tasks, particularly involving vegetation, color infrared film is also used. This should not be confused with thermal infrared imaging, which requires a special sensor and cannot be directly captured on film. Resolution of camera components can be, individually, in the range of 140 line pairs per millimeter. But taken as a system the resolution would be in the range of 30 line pairs per millimeter. To maintain high resolution at low altitude, modern cameras employ image motion compensation to translate the film in synchronism with the apparent ground motion while the shutter is open. The ratio of image velocity in the image plane, v, to camera velocity over the ground, V, for a nominally vertical photograph is given by

$$\frac{v}{V} = \text{Scale} = \frac{f}{H} \tag{53.8}$$

A vacuum system holds the film firmly in contact with the platen during exposure, releasing after exposure for the film advance. During a photo flight, the camera is typically rotated to be parallel with the flight direction, rather than the aircraft body, in case there is a cross wind. This prevents *crab* between the adjacent photographs.

Calibration of aerial cameras should be executed every two to three years. Such a calibration determines the resolution, shutter characteristics, fiducial mark coordinates, principal point location, and a combination of focal length and lens distortion parameters that permits the modeling of the system to have central perspective geometry. Specifications for engineering mapping should always include a requirement for up-to-date camera calibration. A typical radial lens distortion curve is given in Fig. 53.5. The horizontal axis is radial distance from the principal point in millimeters, and the vertical axis is distortion at the image plane in micrometers.

Terrestrial cameras for photogrammetry usually employ a smaller film format than the standard aerial case; 70 mm film is a popular size for terrestrial photogrammetry. Whereas aerial cameras

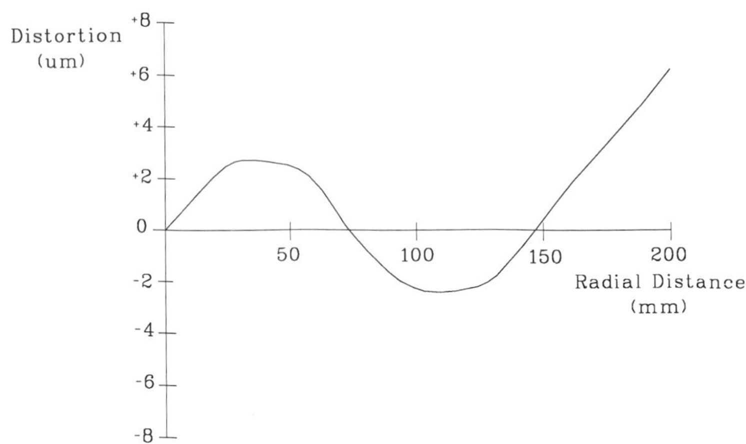

FIGURE 53.5 Lens distortion curve.

are at fixed, infinity focus, terrestrial cameras are typically used at a variety of focus positions. Thus calibration is more difficult since the lens components, and hence their image-forming attributes, are changing from one focus position to another. Often a camera manufacturer will provide principal distance and lens distortion data for each of several focus positions. Alternatively, the user may wish to model such behavior with additional parameters in a block adjustment with self-calibration.

Before leaving the subject of photogrammetric cameras, mention should be made of the possibility to use an electronic image sensor in place of hardcopy film. In the case of large scale, aerial, engineering mapping photography, sensors of a size and resolution to duplicate the functionality of film are not available, and will not be for many years into the future. In the case of small format cameras, it is a different matter. Area sensors with 2048 × 2048 picture elements are available, with an element size of 10 micrometers. The total area of such an array would be therefore about 20 mm by 20 mm. This begins to approach the format size and resolution that could be usable for terrestrial photogrammetry. The direct capture of a digital image has advantages when feature extraction or enhancement by image processing are to be employed.

Scanners

Satellite imagers necessarily employ electronic sensors rather than film-based image capture because of the difficulty of retrieving film packets versus the relative ease of transmitting digital data by telemetry. A variety of sensor technologies is employed in satellite imagers, and the first of these, mechanical scanners, is discussed here. The MSS, multispectral scanner, on Landsats 1 to 5 is shown schematically in Fig. 53.6. For this instrument the mirror scans across the track of the satellite. The focal plane elements consist of light-gathering fiber-optic bundles which carry the radiation to PMT, photomultiplier tubes, or photodiode detectors. The *IFOV, instantaneous field of view,* determined by detector size, telescope magnification, and altitude, is 83 m. The scan rate and the sampling rate determine the *pixel size*. For the MSS it is 83 m in the along-track dimension and 68 m in the across-track dimension. The MSS has detectors in 4 bands covering the visible and near infrared portions of the spectrum. The TM, thematic mapper, on Landsats 4 and 5 has a similar design but enhanced performance compared to the MSS. Data recording takes place on both swings of the mirror, and there is a wider spectral range among the seven bands. The pixel size is 30 m at ground scale.

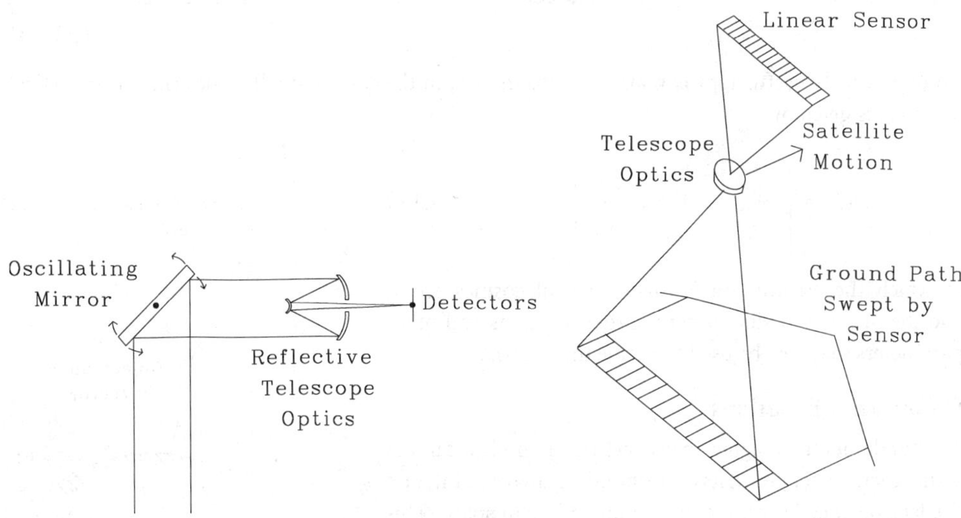

FIGURE 53.6 Schematic of mechanical scanner.

FIGURE 53.7 Pushbroom linear sensor.

Pushbroom Linear Sensors

Linear CCDs, charge-coupled devices, are gaining wide acceptance for satellite imagers. A typical linear array is 2000 to 8000 elements in length. Longer effective pixel widths may be obtained by combining multiple arrays end to end (or equivalently using optical beamsplitters). The individual elements in the sensor collect incident illumination during the integration period. The resulting charges are then transferred toward one end of the chip, and emerge as an analogue signal. This signal must be digitized for transmission or storage. Pixel sizes are usually in the 5 to 15 micrometer range. Since telescope optics have far superior resolution at the image plane, a purposeful defocusing of the optical system is performed to prevent aliasing. (See Fig. 53.7.)

53.3 Mathematics of Photogrammetry

Condition Equations

Using photogrammetry to solve spatial position problems inevitably leads to the formation of equations which link the observables to the quantities of interest. It is extremely rare that one would directly observe these quantities of interest. The form of the condition equations will reflect the nature of the observations, such as, 2-D image coordinates or 3-D model coordinates, and the particular problem that one wishes to solve, such as space resection or relative orientation.

Preliminaries

The 3×3 rotation that is often employed in developing photogrammetric condition equations is a function of three independent quantities. These quantities are usually the sequential rotations ω, ϕ, and κ, about the x, y, and z axes, respectively. The usual order of application is

$$M = M_\kappa M_\phi M_\omega \tag{53.9}$$

The elements are given by

$$M = \begin{bmatrix} \cos\phi\cos\kappa & \cos\omega\sin\kappa + \sin\omega\sin\phi\cos\kappa & \sin\omega\sin\kappa - \cos\omega\sin\phi\cos\kappa \\ -\cos\phi\sin\kappa & \cos\omega\cos\kappa - \sin\omega\sin\phi\sin\kappa & \sin\omega\cos\kappa + \cos\omega\sin\phi\sin\kappa \\ \sin\phi & -\sin\omega\cos\phi & \cos\omega\cos\phi \end{bmatrix} \tag{53.10}$$

An occasionally useful approximation to this matrix, in the case of small angles (i.e., near vertical imagery) is given by

$$M \approx \begin{bmatrix} 1 & \kappa & -\phi \\ -\kappa & 1 & \omega \\ \phi & -\omega & 1 \end{bmatrix} \tag{53.11}$$

in which the assumption is made that all cosines are 1, and the product of sines is zero. Other rotations and other parameters may also be used to define this matrix.

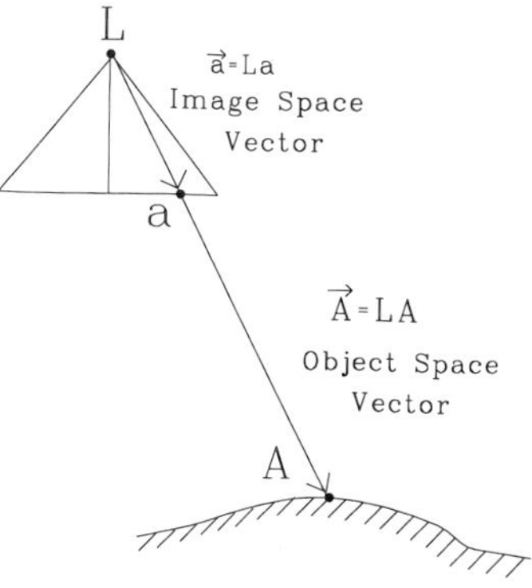

FIGURE 53.8 Collinearity geometry.

Collinearity Equations

The fundamental imaging characteristic of an ideal camera is that each object point, its corresponding image point, and the lens perspective center all lie along a line in space. This can be expressed in the following way, referring to Fig. 53.8,

$$\vec{a} = k \vec{A} \tag{53.12}$$

This equation is valid only if the two vectors are expressed in the same coordinate system. The image space vector is inevitably expressed in an image coordinate system,

$$\vec{a} = \begin{bmatrix} x - x_0 \\ y - y_0 \\ -f \end{bmatrix} \tag{53.13}$$

while the object space vector is expressed in an object space coordinate system (shifted to the camera perspective center):

$$\vec{A} = \begin{bmatrix} X - X_L \\ Y - Y_L \\ Z - Z_L \end{bmatrix} \tag{53.14}$$

One thus needs to scale, rotate, and translate one of these two coordinate systems until they are coincident. This transformation is usually applied to the object space vector and is expressed as follows:

$$\begin{bmatrix} x - x_0 \\ y - y_0 \\ -f \end{bmatrix} = kM \begin{bmatrix} X - X_L \\ Y - Y_L \\ Z - Z_L \end{bmatrix} \tag{53.15}$$

Eliminating the scale parameter k yields the classical form of the collinearity equations:

$$\begin{aligned} x - x_0 &= -f \left[\frac{m_{11}(X - X_L) + m_{12}(Y - Y_L) + m_{13}(Z - Z_L)}{m_{31}(X - X_L) + m_{32}(Y - Y_L) + m_{33}(Z - Z_L)} \right] \\ y - y_0 &= -f \left[\frac{m_{21}(X - X_L) + m_{22}(Y - Y_L) + m_{33}(Z - Z_L)}{m_{31}(X - X_L) + m_{32}(Y - Y_L) + m_{33}(Z - Z_L)} \right] \end{aligned} \tag{53.16}$$

Examples of particular problems for which the collinearity equations are useful include *space resection* (camera exterior orientation unknown, object points known, observed image coordinates given, usually implying a single image), *space intersection* (camera exterior orientations known, object point unknown, observed image coordinates given, usually implying a single object point), and *bundle block adjustment* (simultaneous resection and intersection, multiple images and multiple points). This equation is nonlinear and a linear approximation is usually made if we attempt to solve for any of the variables as unknowns. This dictates an iterative solution.

Coplanarity Equation

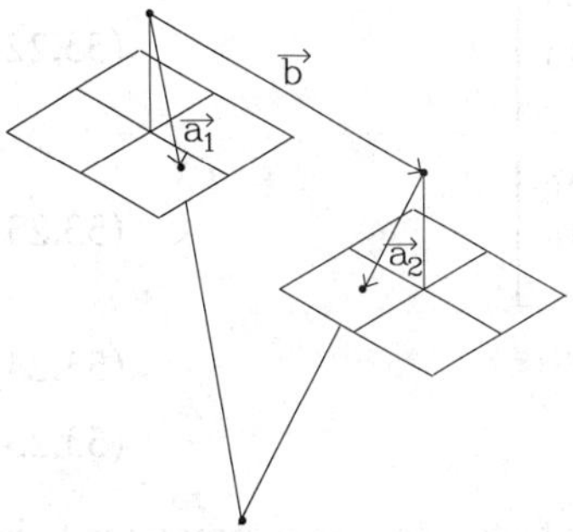

The phrase *conjugate image points* refers to multiple image instances of the same object point. If we consider a pair of properly oriented images, and the pair of rays defined by two conjugate image points, then this pair of rays together with the base vector between the perspective centers should define a plane in space. The *coplanarity* condition enforces this geometrical configuration. This is done by forcing these three vectors to be coplanar, which is in turn guaranteed by setting the triple scalar product to zero. An alternative explanation is that the parallelepiped defined by the three vectors as edges has zero volume. Figure 53.9 illustrates this geometry. The left vector is given by

FIGURE 53.9 Coplanarity geometry.

$$a_1 = \begin{bmatrix} u_1 \\ v_1 \\ w_1 \end{bmatrix} = M_1^t \begin{bmatrix} x - x_0 \\ y - y_0 \\ -f \end{bmatrix}_1 \quad (53.17)$$

The right vector is given by

$$a_2 = \begin{bmatrix} u_2 \\ v_2 \\ w_2 \end{bmatrix} = M_2^t \begin{bmatrix} x - x_0 \\ y - y_0 \\ -f \end{bmatrix}_2 \quad (53.18)$$

and the base vector is given by

$$b = \begin{bmatrix} b_x \\ b_y \\ b_z \end{bmatrix} = \begin{bmatrix} X_{L_2} - X_{L_1} \\ Y_{L_2} - Y_{L_1} \\ Z_{L_2} - Z_{L_1} \end{bmatrix} \quad (53.19)$$

The coplanarity condition equation is the above-stated triple scalar product

$$F = \begin{vmatrix} b_x & b_y & b_z \\ u_1 & v_1 & w_1 \\ u_2 & v_2 & w_2 \end{vmatrix} = 0 \quad (53.20)$$

The most prominent application for which the coplanarity equation is used is relative orientation. The equation is nonlinear and a linear approximation is usually made in order to solve for any of the variables as unknowns. This dictates an iterative solution.

Scale Restraint Equation

If photograph one is relatively oriented to photograph two, and photograph two is relatively oriented to photograph three, there is no guarantee that photograph one and photograph three are also relatively oriented. There are several methods to enforce this condition, and among them the most robust would be the *scale restraint* condition. This states that the intersection of conjugate rays from the three photographs should in fact occur at a single point. From Fig. 53.10 we see the three rays, $\mathbf{a}_i$, and two "mismatch" vectors, $\mathbf{d}_i$:

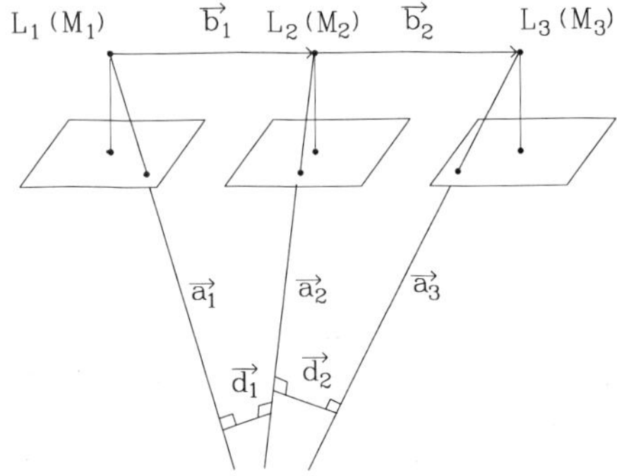

FIGURE 53.10 Scale restraint geometry.

$$\mathbf{a}_1 = \begin{bmatrix} a_{1_x} \\ a_{1_y} \\ a_{1_z} \end{bmatrix} = \mathbf{M}_1^t \begin{bmatrix} x_1 - x_0 \\ y_1 - y_0 \\ -f \end{bmatrix} \quad (53.21)$$

$$\mathbf{a}_2 = \begin{bmatrix} a_{2_x} \\ a_{2_y} \\ a_{2_z} \end{bmatrix} = \mathbf{M}_2^t \begin{bmatrix} x_2 - x_0 \\ y_2 - y_0 \\ -f \end{bmatrix} \quad (53.22)$$

$$\mathbf{a}_3 = \begin{bmatrix} a_{3_x} \\ a_{3_y} \\ a_{3_z} \end{bmatrix} = \mathbf{M}_3^t \begin{bmatrix} x_3 - x_0 \\ y_3 - y_0 \\ -f \end{bmatrix} \quad (53.23)$$

$$\mathbf{d}_1 = \mathbf{a}_1 \times \mathbf{a}_2 \quad (53.24)$$

$$\mathbf{d}_2 = \mathbf{a}_2 \times \mathbf{a}_3 \quad (53.25)$$

The scale restraint equation itself forces the independent scale factors for the common ray to be equal:

$$F = \frac{\begin{vmatrix} a_{1_x} & d_{1_x} & b_{1_x} \\ a_{1_y} & d_{1_y} & b_{1_y} \\ a_{1_z} & d_{1_z} & b_{1_z} \end{vmatrix}}{\begin{vmatrix} a_{1_x} & d_{1_x} & a_{2_x} \\ a_{1_y} & d_{1_y} & a_{2_y} \\ a_{1_z} & d_{1_z} & a_{2_z} \end{vmatrix}} + \frac{\begin{vmatrix} b_{2_x} & d_{2_x} & a_{3_x} \\ b_{2_y} & d_{2_y} & a_{3_y} \\ b_{2_z} & d_{2_z} & a_{3_z} \end{vmatrix}}{\begin{vmatrix} a_{2_x} & d_{2_x} & a_{3_x} \\ a_{2_y} & d_{2_y} & a_{3_y} \\ a_{2_z} & d_{2_z} & a_{3_z} \end{vmatrix}} = 0 \tag{53.26}$$

This equation is used primarily in the analytical formation of strips by successive relative orientation of image pairs in the strip. Common points in any photo triplet (that is, two adjacent models) would be subjected to the scale restraint condition. This equation is nonlinear and would require linear approximation for practical use in the given application.

Linear Feature Equations

It can happen, particularly in close-range photogrammetry, that the object space parameters of a straight line feature are to be determined. If at the same time stereo observation is either unavailable or difficult because of convergence or scale, then it becomes helpful if one can observe the feature monoscopically on each image without the need for conjugate image points. This can be elegantly accomplished with a condition equation which forces the ray associated with an observed image coordinate to pass through a straight line in object space. For each point in each photograph, a condition equation of the following kind may be written:

$$\begin{vmatrix} \rho_x & \rho_y & \rho_z \\ \beta_x & \beta_y & \beta_z \\ LC_x & LC_y & LC_z \end{vmatrix} = 0 \tag{53.27}$$

In this equation the vector ρ is the object space vector from the observed image point, the vector β is the vector along the straight line in object space, and LC is the vector from the perspective center to the point on the line closest to the origin. The six linear feature parameters must be augmented by two constraints which fix the magnitude of β to 1, and guarantee that β and C are orthogonal. A variation on this technique is the case where the object space feature is a circle in space. In this case each point on each image contributes an equation of the form

$$\left\| (L - C) - \frac{(L - C)\eta}{\rho\eta}\rho \right\| = r \tag{53.28}$$

where ρ and $\boldsymbol{L}$ have the same meaning as before. η represents the normal vector to the circle plane, r represents the circle radius, and C represents the circle center. In this case the normal vector must be constrained to unit magnitude. As in every case described here these equations are nonlinear in the variables of interest and when we solve for them, the equations must be approximated using the Taylor series.

Block Adjustment

The internal geometry of a block of overlapping photographs or images may be sufficient to determine relative point positions, but for topographic mapping and feature extraction, one needs to tie this block to a terrestrial coordinate system. It would be possible to provide field survey determined coordinates for every point, but this would be prohibitively expensive. Thus arises the need to simultaneously tie all the photographs to each other, as well as to a sparse network of terrestrial control points. This process is referred to as *block adjustment*. The minimum amount of control necessary would be seven coordinate components, i.e., two horizontal (X, Y) points and three vertical (Z) points, or two complete control points (X, Y, Z) and one point with only vertical (Z). In practice, of course, one usually provides control in excess of the minimum, the redundancy providing increased confidence in the results.

Block Adjustment by Bundles

Block adjustment by bundles is the most mathematically rigorous way to perform this task. Observations consist of 2-D photograph image coordinates, usually read from a comparator or analytical plotter, transformed to the principal point origin, and refined for all known systematic errors. These systematic errors, described below, consist of at least lens distortion and atmospheric refraction, although at low altitude refraction may be considered negligible. Each image point, i, on each image, j, contributes two collinearity condition equations of the form

$$\begin{aligned} x_i - x_0 &= -f\left[\frac{m_{11}(X_i - X_{L_j}) + m_{12}(Y_i - Y_{L_j}) + m_{13}(Z_i - Z_{L_j})}{m_{31}(X_i - X_{L_j}) + m_{32}(Y_i - Y_{L_j}) + m_{33}(Z_i - Z_{L_j})}\right] \\ y_i - y_0 &= -f\left[\frac{m_{21}(X_i - X_{L_j}) + m_{22}(Y_i - Y_{L_j}) + m_{33}(Z_i - Z_{L_j})}{m_{31}(X_i - X_{L_j}) + m_{32}(Y_i - Y_{L_j}) + m_{33}(Z_i - Z_{L_j})}\right] \end{aligned} \tag{53.29}$$

in which (x_i, y_i) are the observed image coordinates, transformed into a coordinate system defined by the camera fiducial coordinates. The variables (x_0, y_0) represent the position of the principal point and f represents the focal length or principal distance. The last three variables would often be considered as fixed constants from a camera calibration report, or they may be carried as unknowns in the adjustment. The variables (X_i, Y_i, Z_i) represent the object coordinates of the point i. They may be known or partially known if point i is a control point, or they may be unknown if point i is a pass point. The variables (X_{Lj}, Y_{Lj}, Z_{Lj}) represent the coordinates of the exposure station or perspective center of image j. Each image, j, also has an associated orientation matrix M_j whose elements are shown in the equation. If there are n points observed on m images, the total number of condition equations will be $2nm$. The total number of unknowns will be $3n + 6m$ − (Number of fixed coordinate components). With the advent of GPS in the photo aircraft, control may be introduced not only at the object points but also at the exposure stations. If the solution to the overdetermined problem is carried out by normal equations, the form of these equations is shown in Fig. 53.11.

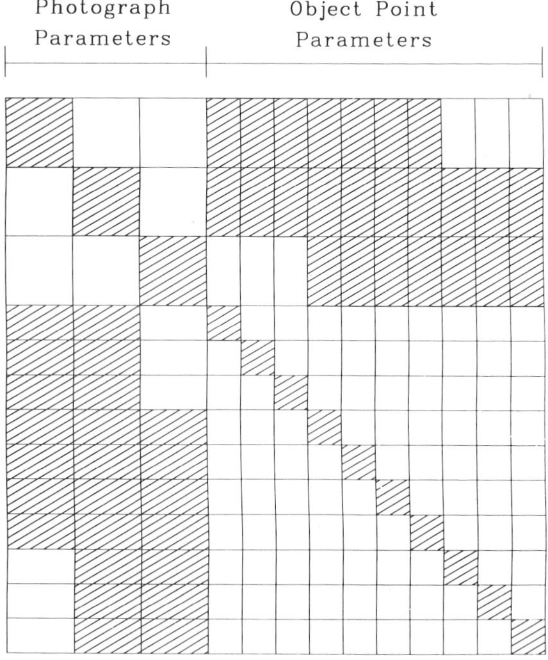

FIGURE 53.11 Organization of normal equations for bundle adjustment.

Block Adjustment by Models

Block adjustment by models is necessary if the original observations of the photographs are made in stereo with an observed (x, y, z) model coordinate for each control point and pass point. Together with the model coordinates of the perspective centers, all model points are simultaneously transformed to the object (or control) coordinate system. There is usually redundant information to specify this transformation, so a least squares estimation is necessary. Practitioners have used comparator-derived image coordinates to compute model coordinates, and then subjected these derived model coordinates to an independent model block adjustment. This practice should be discouraged in favor of using the image coordinates directly in a bundle adjustment as described above. For the independent model block adjustment, each point, i, in model (stereo pair), j, will be related to the object space coordinates by a seven-parameter transformation unique to each model. The seven parameters include scale, s, rotations, Ω, Φ, and K, and translations T_x, T_y, T_z. For each point, i, in each model, j, the following three equations can be written:

$$\begin{bmatrix} F_1 \\ F_2 \\ F_3 \end{bmatrix} = -\begin{bmatrix} X \\ Y \\ Z \end{bmatrix}_i + s_j M_j \begin{bmatrix} x \\ y \\ z \end{bmatrix}_i + \begin{bmatrix} T_x \\ T_y \\ T_z \end{bmatrix}_j = \begin{bmatrix} 0 \\ 0 \\ 0 \end{bmatrix} \tag{53.30}$$

The uppercase coordinate vector represents the coordinate system of the control points, the lowercase vector represents the model coordinates, and the matrix M contains the rotation parameters. Only for control points will the (X, Y, Z) values be known; for all other points these will be unknown parameters solved for in the block adjustment. For n points and m models the total number of condition equations would be $3nm$. The total number of unknown parameters would be $3n + 7m -$ (Number of fixed coordinate components).

Strip Formation and Block Adjustment by Polynomials

Strip formation and block adjustment by polynomials assumes that the input data are xyz model coordinates. These would usually come directly from a relatively oriented model in an analogue stereoplotter. They could also come from analytical computation from image coordinates. If this is the case it would be preferable to use the image coordinates directly in a bundle block adjustment. A similar comment was made with regard to the independent model block adjustment. The strip formation consists of linking successive models (including perspective center coordinates) by seven parameter transformations, and then transforming each new model into the strip system based on the first model. This process is illustrated in Fig. 53.12. If a single strip is sufficiently short, say five models or less, the strip can be fitted to the control points by a global seven parameter transformation. This is given in the following equation:

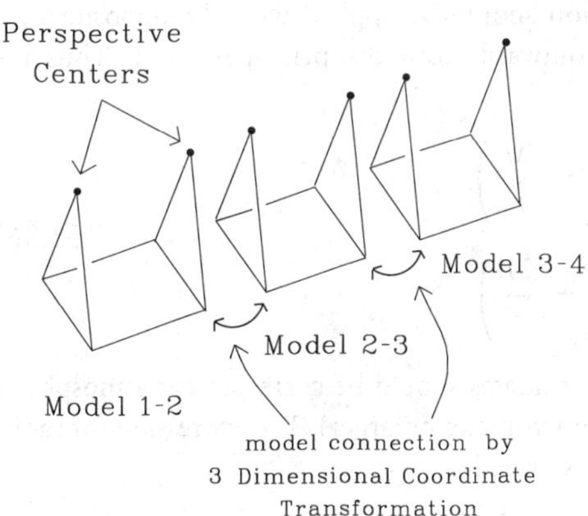

FIGURE 53.12 Model connection and strip formation.

$$\begin{bmatrix} E \\ N \\ h \end{bmatrix} = sM \begin{bmatrix} x_m \\ y_m \\ z_m \end{bmatrix} + \begin{bmatrix} t_x \\ t_y \\ t_z \end{bmatrix} \tag{53.31}$$

If the strip is longer than this, then because of adverse error propagation, artificial bends and bows will be present in the strip coordinates and polynomials present a way to model such effects. This technique is primarily of historical interest since its computational efficiency is no longer a compelling attribute. Both model coordinates and ground coordinates are transformed into an "axis of flight" coordinate system centered within the strip. Then conformal polynomials are used to transform the strip planimetric coordinates into the control system:

$$x' = x + a_1 + a_3x - a_4y + a_5(x^2 - y^2) - 2a_6xy + a_7(x^3 - 3xy^2) - a_8(3x^2y - y^2) + \cdots \tag{53.32}$$

$$y' = y + a_2 + a_4x + a_3y + a_6(x^2 - y^2) + 2a_5xy + a_7(3x^2y - y^3) + a_8(x^3 - 3x^2y) + \cdots \tag{53.33}$$

The vertical control points are used in the following polynomial:

$$\begin{aligned} z' &= z + b_0 - 2b_2x + 2b_1y + c_1x^2 + c_2x^3 + c_3x^4 \\ &+ d_1xy + d_2x^y + d_3x^3y + d_4x^4y + e_1y^2 + e_2xy^2 \end{aligned} \tag{53.34}$$

The number of terms is selected based on the quantity of control points and the length of the strip. Following the polynomial estimation, the points are transformed back into the original control coordinate system.

Image Coordinate Refinement

Raw stage coordinates from a comparator or analytical plotter must undergo a number of transformations and refinements before being used in further photogrammetric processing such as relative orientation or bundle block adjustment. Firstly the stage coordinates are transformed into the coordinate system defined by the camera *fiducial marks* or registration marks. This is usually done with a four or six parameter transformation. They are then shifted to the *principal point of autocollimation* by the principal point offsets (x_0, y_0). Following this they are corrected for radial lens distortion based on their position with respect to the *principal point of best symmetry*. The radial lens distortion is provided as part of the calibration of the camera either in the form of a table, a graph, or a polynomial function. A sample radial lens distortion graph is shown in Fig. 53.5. The usual form for polynomial lens distortion functions is given by

$$\Delta r = k_0r + k_1r^3 + k_2r^5 + k_3r^7 \tag{53.35}$$

in which the radial distance r is the distance from the symmetry point mentioned above. If a distortion table or function is given, the correction should be applied with the opposite sign. Conventionally, "+" indicates radial distortion outward from the principal point. Thus the correction equations would be

$$\begin{aligned} x_c &= x\left(1 - \frac{\Delta r}{r}\right) \\ y_c &= y\left(1 - \frac{\Delta r}{r}\right) \end{aligned} \tag{53.36}$$

Following lens distortion correction the image coordinates should be corrected for atmospheric refraction (if it is significant, i.e., on the order of a micrometer or larger). The expression for radial image displacement due to atmospheric refraction is given by

$$d_r = K\left(r + \frac{r^3}{f^2}\right) \tag{53.37}$$

where r is the radial distance from the principal point of autocollimation and f is the focal length. The value for K is a function of the camera altitude and terrain elevation, and is given according to the ARDC Model Atmosphere (Air Research and Development Command of the U.S. Air Force):

$$K = \left[\frac{2410H}{H^2 - 6H + 250} - \frac{2410h}{h^2 - 6h + 250}\left(\frac{h}{H}\right)\right] \times 10^{-6} \tag{53.38}$$

where H is the flying height in kilometers above sea level and h is the terrain height, also in kilometers above sea level. The displacement due to atmospheric refraction is always radially outward, therefore the correction is always radially inward. The same correction formulas may be used as in the case of lens distortion, replacing Δr by dr. Some practitioners have advocated handling earth curvature effects by modifying the image coordinates. This is to be discouraged. A better solution is to ensure that the object space coordinate system is truly cartesian, and then the "problem" disappears. See the following section for a discussion of this.

Object Space Coordinate Systems

Geodetic Coordinates ϕ, λ, h

Geodetic coordinates, that is, latitude, longitude, and height, are the most fundamental way to represent the position of a point in space with respect to a terrestrial ellipsoid. However, photogrammetric condition equations are usually expressed in terms of rectangular, cartesian coordinates. Thus, for the purpose of providing a reference frame for photogrammetric computations, one would usually transform ϕ, λ, and h into a rectangular system.

Space Rectangular Coordinates

Geocentric space rectangular coordinates may be derived from geodetic coordinates in such a way that the Z axis is parallel with the axis of rotation of the ellipsoid, and the X axis passes through the meridian of Greenwich in the plane of the equator. The Y axis is constructed so that the system is right-handed. In the following, h is assumed to be the ellipsoid height; if geoid height is given, it must be modified by the local geoid separation. The equations transforming geodetic coordinates into geocentric space rectangular coordinates are given by

$$\begin{aligned} X &= (N + h)\cos\phi\cos\lambda \\ Y &= (N + h)\cos\phi\sin\lambda \\ Z &= [N(1 - e^2) + h]\sin\phi \end{aligned} \tag{53.39}$$

where N, the radius of curvature in the prime vertical is given by

$$N = \frac{a}{\sqrt{1 - e^2\sin^2\phi}} \tag{53.40}$$

where a is the semimajor axis of the ellipsoid, b is the semiminor axis, and e is the eccentricity given by

$$e = \sqrt{\frac{(a^2 - b^2)}{a^2}} \tag{53.41}$$

The inverse transformation cannot be given in a closed form. One can solve for ϕ, λ, and h by choosing an initial approximation and proceeding iteratively by the conventional newton method,

$$X_{i+1} = X_i - J^{-1}F(X_i) \tag{53.42}$$

or

$$\begin{bmatrix} \phi_{i+1} \\ \lambda_{i+1} \\ h_{i+1} \end{bmatrix} = \begin{bmatrix} \phi_i \\ \lambda_i \\ h_i \end{bmatrix} - \begin{bmatrix} \dfrac{\partial F_1}{\partial \phi} & \dfrac{\partial F_1}{\partial \lambda} & \dfrac{\partial F_1}{\partial h} \\ \dfrac{\partial F_2}{\partial \phi} & \dfrac{\partial F_2}{\partial \lambda} & \dfrac{\partial F_2}{\partial h} \\ \dfrac{\partial F_3}{\partial \phi} & \dfrac{\partial F_3}{\partial \lambda} & \dfrac{\partial F_3}{\partial h} \end{bmatrix}^{-1} \begin{bmatrix} F_1(\phi_i, \lambda_i, h_i) \\ F_2(\phi_i, \lambda_i, h_i) \\ F_3(\phi_i, \lambda_i, h_i) \end{bmatrix} \tag{53.43}$$

and the three functions are the ones given in Eq. (53.39). One possible difficulty with the use of geocentric space rectangular coordinates is the large magnitude of the coordinate values. If one wished to maintain point precision to the nearest millimeter, 10 significant digits would have to be carried in the coordinates and single precision floating point computations would be insufficient. An alternative is the local space rectangular system, which is just the geocentric space rectangular coordinates, rotated so that Z' passes through a local point, Y' is in the meridian plane, and X' is constructed for a right-handed system. The LSR, or local space rectangular coordinates, are given by

$$\begin{bmatrix} X' \\ Y' \\ Z' \end{bmatrix}_{\text{LSR}} = M \left[\begin{bmatrix} X \\ Y \\ Z \end{bmatrix}_{\text{GSR}} - \begin{bmatrix} T_X \\ T_Y \\ T_Z \end{bmatrix} \right] \tag{53.44}$$

where GSR refers to the geocentric space rectangular coordinate vector, the T-vector is the translation to the local origin, and the rotation matrix M is given by

$$M = \begin{bmatrix} -\sin\lambda & \cos\lambda & 0 \\ -\sin\Phi\cos\lambda & -\sin\Phi\sin\lambda & \cos\Phi \\ \cos\Phi\cos\lambda & \cos\Phi\sin\lambda & \sin\Phi \end{bmatrix} \tag{53.45}$$

Map Projection Coordinates

The common map projections used to express terrestrial control points are the lambert conformal conic and the transverse mercator. In the U.S., each state has a state plane coordinate system utilizing possibly multiple zones of these projections. Globally, there is the UTM, or universal transverse mercator system, in which the globe is divided into 60 zones of width six degrees. Zone 1 is from 180 degrees west to 174 degrees west, and the zone numbering proceeds eastward until the globe is covered at zone 60. The zones are limited to ± 80 degrees latitude, and the scale factor at the central meridian is 0.9996. All of the above map projection coordinates have a common deficiency when used in photogrammetry. The XY coordinate is with respect to the developed projection surface, but the height coordinate is usually with respect to a sea level datum. Thus the system is not cartesian. Over a very small region one could neglect the curved Z-reference, but over any substantial project area the nonorthogonality of the coordinate system will present itself in photogrammetric computations as a so-called earth curvature effect. The best approach to handling this situation is to either transform all control points into an LSR system described above, or construct a local tangent plane system from the map projection coordinates, modifying the height component as follows:

$$h_{tp} = h_{sl} - \frac{D^2}{2}R \tag{53.46}$$

where the subscript tp refers to the tangent plane system, sl refers to the sea level system, D is the distance from a project centered tangent point, and R is the nominal earth radius. Following all photogrammetric computations, that is, block adjustment, the heights can be corrected back into the sea level system for use by compilers and engineers.

53.4 Instruments and Equipment

Stereoscopes

Stereo viewing is possible with the unaided eyes if conjugate imagery is placed at a spacing approximately equal to the eye base, at a comfortable distance in front of the eyes. Prolonged viewing at such a distance may produce eye fatigue and therein lies the value of a stereoscope. A simple *lens stereoscope* allows the eyes to focus comfortably at infinity, thus permitting longer working sessions. For frame photographs, the overlapping pair should be laid out with the flight lines coincident and the spacing adjusted for comfortable viewing. Only a small portion of a standard 23 cm photograph overlap area can be viewed in this way, and some bending of the paper prints may be necessary to access the full model area. A *mirror stereoscope,* being larger, permits viewing almost an entire overlap area, at a necessarily smaller scale. Approximate elevations can be read via a *parallax bar* and the associated 3-D measuring mark. Some modern softcopy stereo viewing systems employ nothing more than a simple mirror stereoscope to view conjugate imagery presented in split-screen mode on a video monitor.

Monocomparator, Stereocomparator, Point Marker

Both of the comparator instruments have been largely superseded by the analytical plotter, which is really nothing more than a computer-controlled stereocomparator. In any case, a monocomparator is a single two-axis stage with a measuring microscope and a coordinate readout, preferably with an accuracy of 1 or 2 micrometers. A stereocomparator is a pair of two-axis stages which permit stereo viewing by a pair of measuring microscopes, and simultaneous coordinate readout of two pairs of (XY) coordinates. Accuracy levels should be comparable to that mentioned for the monocomparator. Both of these comparator instruments are used chiefly for *aerial triangulation, bridging,* or *control extension.* In this process, all *control points* and *pass points* are read for all photographs in a *strip* or *block.* The photos are then linked by geometric condition equations and tied to the ground coordinate system, thus producing ground coordinates for all observed pass points. These pass points may then be used for individual model setups in a stereo restitution instrument. If pass points are desired in an area of the photograph without identifiable detail points, artificial emulsion marks or "pug points" are introduced by a *point marker* or "pug." These marks are typically 40 to 80 micrometer diameter drill holes in the photograph emulsion, sized to be compatible with the stereo measuring device.

Stereo Restitution: Analog, Analytical, Softcopy

Early instruments for map compilation consisted of optical projectors and a small viewing screen with a means to direct the image from one projector to the left eye and from the other projector to the right eye. This binocular separation was effected by *anaglyph* (red and blue filters), by mechanical shutter, and by polarization. Analogue instruments in use today employ exclusively mechanical projection in which a collection of gimbals, space rods, and cardan joints emulate the optical light paths. All analogue instruments must provide a way to re-create the inner camera geometry by positioning the principal point (via the fiducial marks) and setting the principal distance or focal length. These steps constitute the interior orientation. A procedure is also necessary to reestablish the *relative orientation* of the photographs at instant of exposure. This is accomplished by clearing *y-parallax,* or *y* displacement in model space between the projected images, in at least five points spaced throughout the model. For the point layout in Fig. 53.13 the sequence of steps for two-projector relative orientation is as follows:

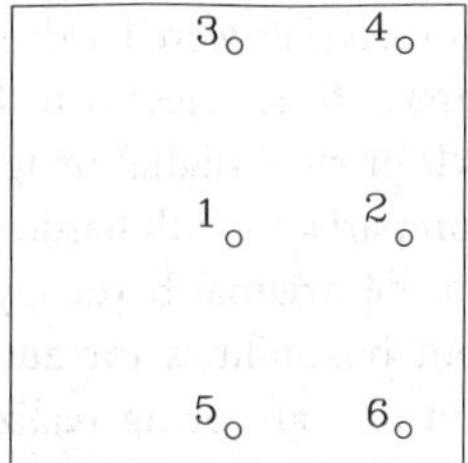

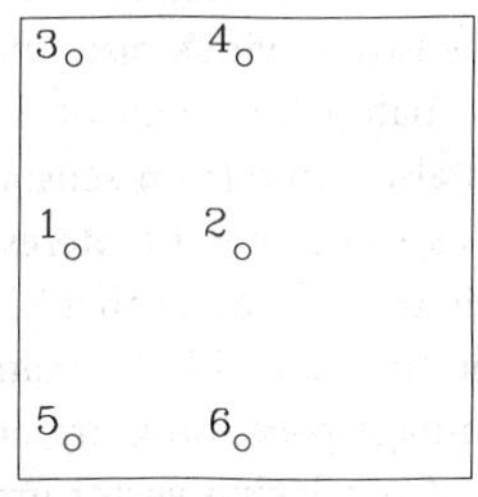

FIGURE 53.13 Point layout for relative orientation.

1. Clear at point 1 with kappa-right
2. Clear at point 2 with kappa-left
3. Clear at point 3 with phi-right
4. Clear at point 4 with phi-left
5. Clear at point 5 with omega-left or omega-right
6. Check for no parallax at point 6

If there is parallax at point 6, the procedure is repeated until no parallax is seen at point 6. Convergence can be speeded up by overcorrecting by about half at step 5. When this is complete, the entire model should be free of y-parallax. If there is visible parallax at other points, it could be due to uncompensated lens distortion, excessive film deformation, or other factors. Following relative orientation comes the *absolute orientation,* in which the relation is established between the model coordinates and the ground coordinates, defined by control points in the model. In the past this was done by physically orienting a map manuscript to a mechanical tracing device. This physical procedure would involve scaling, by adjusting the base components, and leveling, by adjusting either common rotation elements or combinations of projector rotations and corresponding base components. Now it is done analytically by computing the parameters of the three-dimensional similarity transformation between the model and ground coordinates. The computed rotations would then be introduced into the instrument as before. This computationally assisted absolute orientation requires that the instrument be fitted with position encoders for coordinate readout of xyz model coordinates. Accuracies on the order of 5 micrometers are typically seen for this task. Schematic depictions of an optical and a mechanical stereo restitution instrument are shown in Fig. 53.14(a) and (b). In addition to map compilation of planimetry and elevation data, an analog stereo instrument can also be used to collect model coordinates for independent model aerial triangulation. This requires an additional step of determining the model coordinates of the perspective center, which is necessary to link adjacent models in a strip.

All of the functions of an analogue instrument can be duplicated and usually exceeded in an *analytical plotter.* Such a device, shown schematically in Fig. 53.14(c), consists of two computer-controlled stages, a viewing stereomicroscope, operator controls for three-axis motion, and a suite of computer software to automate and assist in all of the desired operations. Interior orientation consists of measuring the fiducial marks and introducing the calibrated camera parameters. Relative orientation consists of measuring conjugate points and computing the five orientation parameters. Absolute orientation consists of measuring the control points and computing the seven parameter transformation as above. Of course, the two steps of relative and absolute orientation can be combined in a two-photo bundle solution using the collinearity equations as described previously. In addition to conventional map compilation, analytical plotters are well suited to aerial triangulation and block adjustment, digital elevation model collection, cross-section and profile collection, and terrestrial or close-range applications. Stage accuracies are typically 1 or 2 micrometers. Today one would always have a CAD system connected to the instrument for direct digitizing of features into the topographic or GIS database.

The most recent variant on the stereo viewer/plotter is the *softcopy* stereo system. Here the stereo images are presented to the operator on a computer video monitor. This is shown schematically in Fig. 53.14(d). In the two previous cases the input materials were hardcopy film transparencies. In this case the input material is a pair of digital image files. These usually come from digitized photographs, but can also come from sensors which provide digital image data directly, such as SPOT. Softcopy stereo systems present interesting comparisons with hardcopy–based instruments. Spatial resolution may be inferior to that visible in the original hardcopy image, depending on the resolution and performance of the scanner, but possibilities for automation and operator assistance by digital image processing are abundant and are being realized today. In addition, the complicated task of overlaying vector graphics onto the stereo images in a hardcopy instrument becomes a simple task in a softcopy environment. This can be enormously beneficial for

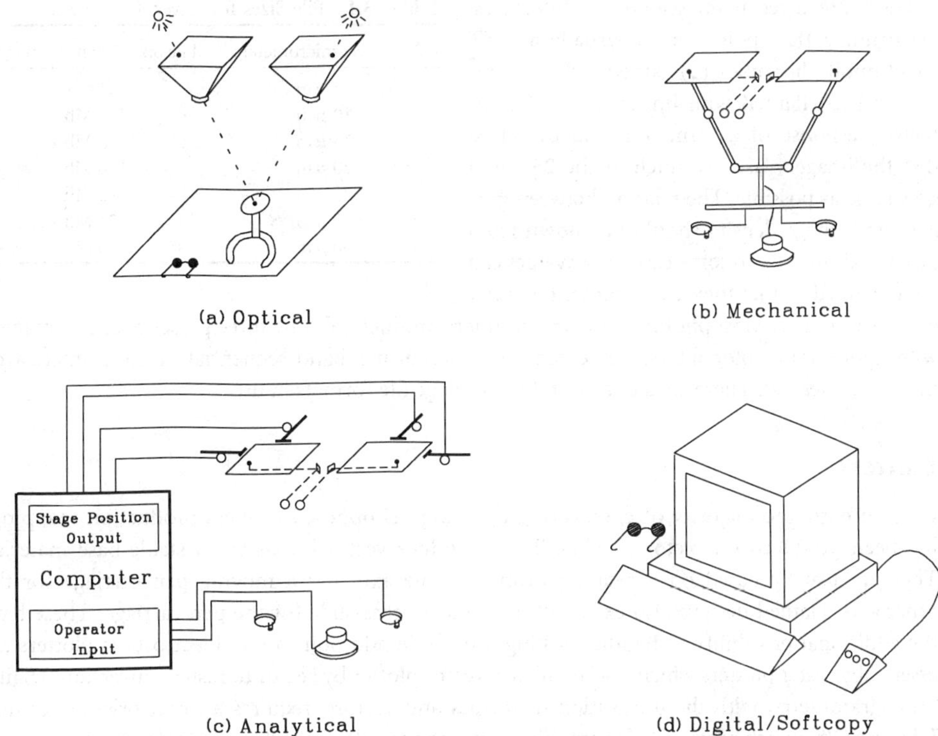

FIGURE 53.14 Schematic diagrams of stereo restitution instruments.

editing and completeness checking. The orientation aspect of a softcopy system is very similar to the analytical plotter in that all computations for orientation parameters are done via computer from image coordinate input. The dramatic impact of softcopy systems will not be apparent until specialized image processing tools for feature extraction and height determination are improved to the point that they can reliably replace manual operation for substantial portions of the map compilation task. A few definitions are now presented to encourage standardized terminology. *Digital mapping* refers to the collection of digital map data into a CAD or GIS system. This can be done from any type of stereo device: analog, analytical, or softcopy. *Digital photogrammetry* refers to any photogrammetric operations performed on digital images. *Softcopy photogrammetry* is really synonymous with digital photogrammetry, with the added connotation of softcopy stereo viewing.

Scanners

With the coming importance of digital photogrammetry, scanners will play an important role in the conversion of hard copy photograph transparencies into digital form. For aircraft platforms, and therefore for the majority of large scale mapping applications, film-based imaging is still preferred because of the high resolution and the straightforward geometry of non–time-dependent imagery. Thus arises the need for scanning equipment to make this hard copy imagery available to digital photogrammetric workstations. To really capture all of the information in high-performance film cameras, a pixel size of about 5 micrometers would be needed. Because of large resulting file sizes, many users are settling for pixel sizes of 12 to 30 micrometers. For digital orthophotos, sometimes an even larger size is used. Table 53.1 shows the relation between pixel size and file size for a 230 mm square image assuming no compression and assuming that each pixel is quantized to one of 256 gray levels (8 bits).

Table 53.1 File Sizes for Given Pixel Sizes

Pixel Size, micrometers	File Size, 230 mm Image
5 μm	2.1 Gb
10 μm	530 Mb
15 μm	235 Mb
20 μm	130 Mb
25 μm	85 Mb
50 μm	21 Mb
100 μm	5 Mb

There are three main scanner architectures: (1) drum with a point sensor, usually a PMT (photomultiplier tube); (2) flatbed with area sensor; and (3) flatbed with linear sensor. Radiometric response of a scanner is usually set so that the imagery uses as much of the 256 level gray scale as possible. The relation between gray level and image density should be known from system calibration. In some cases, gray values can be remapped so that they are linear with density or transmittance. Most photogrammetric scanners produce color by three passes over the imagery with appropriate color filters. These can be recorded in a band sequential or band interleaved manner as desired. There are a large number of image file formats in use.

Plotters

Until recently the majority of engineering mapping (as opposed to mass production mapping) has been produced on vector plotters. These produce vector linework on stable base material. They are usually based on a rotating drum for one axis and a moving pen carriage for the other axis. Flatbed designs also exist with a two-axis cross-slide for the pen carriage. These have the additional possibility to handle scribing directly in addition to ink. Electrostatic plotters are essentially raster plotters which may emulate a vector plotter by vector to raster conversion. Digital photogrammetry, with the integration of images and vectors, requires a raster-oriented device. Likewise GIS, which often calls for graphic presentations with area fills or raster layers, may require a raster device.

53.5 Photogrammetric Products

Topographic Maps

The classical product of photogrammetric compilation is a *topographic map*. A topographic map consists typically of planimetric features such as roads, buildings, and waterways, as well as terrain elevation information usually in the form of contours. In the past these were manually drafted in ink or scribed onto scribecoat material. Today they are recorded directly into a CAD or GIS system. U.S. National Map Accuracy Standards (NMAS) dictate an accuracy of planimetric features as well as contour lines, which is tied to hard copy scale. In the digital environment, such standards may need to be revised to reflect the increasingly prominent role of the (scaleless) digital map representation. The following is a summary of the Office of Management and Budget standards:

1. Horizontal accuracy. For maps with publication scale greater than 1:20,000, not more than 10% of the "well-defined" points tested shall be in error by more than 1/30th of an inch at publication scale. For maps with publication scale less than 1:20,000, the corresponding tolerance is 1/50th of an inch.
2. Vertical accuracy. Not more than 10% of the elevations tested shall be in error by more than one-half contour interval. Allowances are made for contour line position errors as above.
3. Any testing of map accuracy should be done by survey systems of a higher order of accuracy than that used for the map compilation.
4. Published maps meeting these accuracy requirements shall note this fact in their legends, as follows: "This map complies with national map accuracy standards."
5. Published maps whose errors exceed these limits shall omit from their legends any mention of compliance with accuracy standards.

6. When a published map is a considerable enlargement of a map designed for smaller-scale publication, this fact shall be stated in the legend. For example, "This map is an enlargement of a 1:20,000 scale map."

Other commonly accepted accuracy standards are as follows. Reference grid lines and control point positions should be within 1/100th of an inch of their true position. Ninety percent of spot elevations should be accurate to within one-fourth contour interval, and the remaining 10% shall not be in error by more than one-half contour interval.

Image Products

Image products from photogrammetry include uncontrolled mosaics, controlled mosaics, rectified enlargements, and orthophotos. Mosaics are collections of adjoining photograph enlargements mated in a way to render the join lines as invisible as possible. In the case of controlled mosaics, the photographs are enlarged in a rectifier, using control points to remove the effects of camera tilt, and to bring all enlargements for the mosaic to a common scale. Uncontrolled mosaics are similar except that no tilt removal is done, enlargement scales are less accurately produced, and continuity is attempted by the careful matching of image features. In the past all mosaicking has been done with paper prints, glue, and considerable manual dexterity. If the photographs are scanned, or if the imagery is originally digital, then the mosaicking process can be entirely digital. Digital techniques allow great flexibility for such tasks as tone matching, scaling, rectifying, and vector/annotation addition. Orthophotos are photographs which have been differentially rectified to remove both tilt displacements as well as relief displacements. A well-produced orthophoto can meet horizontal map accuracy standards, and can function as a planimetric map. Individual orthophotos can be further merged into orthophoto mosaics. Digital techniques for orthophoto production are also becoming very popular because of the flexibility mentioned above. People who are not mapping specialists seem to have a particularly easy time interpreting an orthophoto, compared to an abstract map with point, line, and area symbology that may be unfamiliar. With digital orthophoto generation, it is particularly effective to overlay contour lines on the imagery.

Digital Elevation Models

The concept of digitally recording discrete height points to characterize the topographic surface has been in practice for a number of years. Names and acronyms used to describe this concept include *DEM, digital terrain model; DEM, digital elevation model; DHM, digital height model;* and *DTED, digital terrain elevation data*. The philosophy behind this concept is that one obtains sufficient digital data to describe the terrain, and then generates graphic products such as contours or profiles only as a means for visualizing the terrain. This is in contrast to the conventional practice of recording a contour map and having the contours be the archival record which represents the landforms. The advantage of the DEM approach is that the height database can be used for several different applications such as contour generation, profile/cross-section generation, automated road design, and orthophoto interpolation control. Potential pitfalls in this approach mostly revolve around decisions to balance the conflicting demands of accurate terrain representation versus fast data collection and reasonable file sizes. There are basically two alternatives to consider when collecting such data: random data points selected to describe the terrain, or a regular grid of points, with interval selected to describe the terrain.

Random Data Points

Random data points in a DEM may be used directly in a *TIN, triangulated irregular network,* or they may be used to interpolate a regular grid via a variety of interpolation methods. The TIN may be created by a number of algorithms which are producing the equivalent *Dirichlet tessellation*, *Thiesson Polygons*, or the *Delauney triangulation*. One of the simpler methods is the basic Watson algorithm, described by the following steps:

1. Create three fictitious points such that the defined triangle includes all of the data points.
2. Pick a point.
3. Find all of the triangles whose "circumcircle" (the circle passing through triangle vertices) contains the point.
4. The union of all triangles in step 3 forms an "insertion polygon."
5. Destroy all internal edges in the insertion polygon, and connect the current point with all vertices of the polygon.
6. Go to step 2, until no more points are left.
7. When done, eliminate any triangle with a vertex consisting of one of the initial three fictitious points.

To enforce a breakline, one can overlay the breakline on the preliminary TIN, and introduce new triangles as required by the breakline. Interpolation within a TIN usually means locating the required triangular plane facet, and evaluating the plane for Z as a function of X and Y. Contour line generation is particularly easy within the triangular plane facets, with all lines being straight and parallel. Connecting contour lines between facets requires a searching and concatenation operation.

Interpolation in a random point DEM not organized as a TIN can be carried out by various *moving surface* methods, or by *linear prediction*. An example of a moving surface model would be a moving "tilted" plane:

$$z = a_0 + a_1 x + a_2 y \tag{53.47}$$

One equation is written for each point within a certain radius, possibly with weighting inversely related to distance from desired position, and the three parameters are solved for, thereby allowing an estimate of the interpolated height. Such moving surface models can be higher-order polynomials in two dimensions, as dictated by point density and terrain character. With linear prediction, an elevation is interpolated as follows:

$$z_0 = \sigma^t_{zz_0} \Sigma^{-1}_{zz} z \tag{53.48}$$

where z is an $n \times 1$ vector representing height points in the vicinity of the point to be interpolated, Σ_{zz} is an $n \times n$ matrix of covariances between the reference points, usually based on distance, and σ_{zz_0} is an $n \times 1$ vector representing the covariances between the point to be interpolated and the reference points, again based on distance. Breaklines can be enforced in these methods by not allowing reference points on one side of the break to influence interpolations on the other side of the break.

Gridded Data Points

Points can be easily collected directly in a regular grid by programming an analytical stereo instrument to move in this pattern. Grids may also be "created" by interpolating at grid "posts" using random data as outlined above. Interpolation within a regular grid could be done by the methods outlined for random points, but is more often done by bilinear interpolation. Bilinear interpolation can be done by making two linear interpolations along one axis, followed by a single linear interpolation along the other axis. Alternatively, one could solve uniquely for the following four parameters a, b, c, d at the grid cell corners, and evaluate the equation at the unknown point. x and y can be local grid cell coordinates.

$$z = a + bx + cy + dxy \tag{53.49}$$

Contouring is also relatively straightforward in a gridded DEM by linear interpolation along the edges of each grid cell. An ambiguity may arise when there is the same elevation on all four sides of the grid cell.

Geographic Information Systems and Photogrammetry

Historically, map data collected over the same region for different purposes was typically stored separately with no attempts at registration and integration. Likewise, textual attribute data describing the map features were likely kept in yet another storage location. Increasingly, the trend, particularly for municipalities, is to coregister all of this diverse map and attribute data within a *geographic information system*, or GIS. The photogrammetric process typically plays a very important role in creating the *land base*, or *land fabric*, based on well-defined geodetic control points, to which all other GIS data layers are registered. Photogrammetry also plays a role in making periodic updates to the GIS land base in order to keep it current with new land development, subdivision, or construction. The photogrammetric compiler may also be involved in collecting facilities features and tagging them with attributes or linking them to existing attribute records in a facilities database.

53.6 Digital Photogrammetry

The term *digital photogrammetry* refers to photogrammetric techniques applied to digital imagery. The current trend in favor of digital photogrammetry has been driven by the enormous increase in computer power and availability, and the advances in image capture and display techniques.

Data Sources

Digital image data for photogrammetry can come from a number of sources. Satellite imagers transmit data directly in digital form. Primary among the satellite sources would be SPOT and LANDSAT, each with a series of spacecraft and sensors to provide continuous availability of imagery. Digital cameras for airborne or terrestrial use have up to now been a minor contributor as source imagery for photogrammetry. This is principally due to the limited size and density of sensor element packing for area sensors, and to the adverse restitution capabilities with linear sensors using motion-induced scanning. The predominant source of digital imagery for engineering photogrammetry is the scan conversion of conventional film transparencies on high-accuracy scanning systems. Such image data is typically collected as 8-bit (256 level) gray values for monochrome imagery, or as 3×8-bit red, green, and blue components for color imagery. Monochrome digital images are stored as a *raster*, or rectangular grid format. For fast access they may also be *tiled*, or subdivided into subgrids. Color images may be stored in *band sequential* or *band interleaved* order.

Digital Image Processing Fundamentals

Sampling

Digital image data is generated by sampling from a continuous source of incident radiation. In the case of direct imaging, the source is truly continuous; in the case of sampling from a silver-halide emulsion image, the source is usually of sufficiently higher resolution as to be effectively continuous. The *sampling theorem* states that for a band-limited spatial signal with period T corresponding to the highest frequency present, we must sample it with a period no greater than $T/2$ in order to be able to reconstruct it without error. If there are frequencies present which are higher than can be reconstructed, *aliasing* will occur and may be visible in the reconstructed image as *moiré patterns* or other effects. If this is the case, one should purposely blur or defocus the image until the maximum frequencies present are consistent with the sampling interval. This creates an antialiasing filter. The sampled image will be referred to as $F(x, y)$. The gray values recorded in the digital image may represent estimates of well-defined photometric quantities such as irradiance, density, or transmittance. They may also be a rather arbitrary quantization of the range of irradiance levels reaching the sensor.

Histogram Analysis and Modification

The *histogram* of a digital image is a series of counts of frequency of occurrence of individual gray levels or ranges of gray levels. It corresponds to the probability function or probability density function of a random variable. Any operation on an image, A,

$$B(x, y) = f[A(x, y)] \tag{53.50}$$

in which the output gray level at a point is a function strictly of the input gray level is referred to as a *point operation* and will result in a modification of the histogram. If an image operation is a function of the position as well as the input gray level,

$$B(x, y) = f[A(x, y), x, y] \tag{53.51}$$

This may be referred to as a *spatially variant point operation.* It will result in a modification of the histogram that is position dependent. This would be used to correct for a sensor which has a position-dependent response. To illustrate a point operation, function $f(D)$ in Fig. 53.15 maps an input gray level D_A to an output gray level D_B. We assume for simplicity that f is monotone increasing or decreasing, and therefore has an inverse. The output histogram, $H_B(D)$, as a function of the input histogram, $H_A(D)$, and of the function, f, is

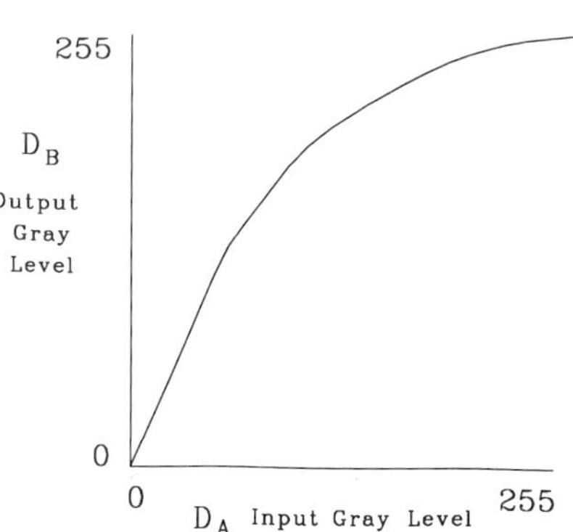

FIGURE 53.15 Gray level remapping.

$$H_B(D) = \frac{H_A[f^{-1}(D)]}{|df/dD|} \tag{53.52}$$

This is analogous to the transformation of distributions based on a functional relation between random variables. Other common applications of histogram modification include contrast stretching and brightness shifting, that is, gain and offset.

Resampling, Geometric Image Manipulation

Whenever the geometric form of an image requires modification, *resampling* is necessary. This is really just two-dimensional interpolation. Examples of such geometric modifications are rotation, rectification, and epipolar line extraction. If a simple change of scale is desired, magnification can be accomplished by *pixel replication,* and minification can be accomplished by *pixel aggregation,* or more quickly (with aliasing) by *subsampling*. For the resampling task, the fastest method is *nearest neighbor,* which assigns the gray level of the nearest pixel center to the point being interpolated. The most common method would be *bilinear* interpolation. This math model was described under digital elevation models. Less common would be interpolation by higher-order polynomials, such as cubic polynomials. As an example, the process to generate a digital orthophoto can be easily summarized in this context. A grid is defined in object space, which represents the locations of the "output" pixels. The elevation for these points is observed directly or interpolated from a digital terrain model. Each of these (XYZ) object points is passed through the collinearity equations using the known orientation of an image, and image coordinates are obtained. These image coordinates are transformed into the row/column coordinate space of the digital image. If the output spacing is on the order of the digital image pixel spacing, bilinear interpolation can be used to interpolate the orthophoto pixel gray level. If the output spacing is much greater than the image spacing, the output gray value should be obtained by averaging over all image pixels within the larger pixel coverage, thus avoiding aliasing problems.

Filtering

For continuous functions in one dimension, *convolution* is expressed as

$$y(t) = f * g = \int_{-\infty}^{\infty} f(\tau)g(t-\tau)\,d\tau \tag{53.53}$$

For continuous functions in two dimensions, convolution is given by

$$h(x,y) = f * g = \int_{-\infty}^{\infty}\int_{-\infty}^{\infty} f(u,v)g(x-u,y-v)\,du\,dv \tag{53.54}$$

For discrete two-dimensional functions, that is, images, convolution is given by

$$H(i,j) = F * G = \sum_{m}\sum_{n} F(m,n)G(i-m,j-n) \tag{53.55}$$

Image enhancements and feature exaggeration can be introduced by convolving the following 3×3 kernels with an image. For an edge parallel to azimuth 90 degrees,

$$G = \begin{bmatrix} 1 & 1 & 1 \\ 1 & -2 & 1 \\ -1 & -1 & -1 \end{bmatrix} \tag{53.56}$$

Edge parallel to azimuth 135 degrees,

$$G = \begin{bmatrix} 1 & 1 & 1 \\ -1 & -2 & 1 \\ -1 & -1 & 1 \end{bmatrix} \tag{53.57}$$

General edge sharpening via a Laplacian,

$$G = \begin{bmatrix} -1 & -1 & -1 \\ -1 & 8 & -1 \\ -1 & -1 & -1 \end{bmatrix} \tag{53.58}$$

Edge parallel to azimuth 0 degrees using a Sobel operator,

$$G = \begin{bmatrix} 1 & 0 & -1 \\ 2 & 0 & -2 \\ 1 & 0 & -1 \end{bmatrix} \tag{53.59}$$

Low pass filtering,

$$G = \frac{1}{9}\begin{bmatrix} 1 & 1 & 1 \\ 1 & 1 & 1 \\ 1 & 1 & 1 \end{bmatrix} \tag{53.60}$$

Matching Techniques

Digital image matching represents one of the means by which digital photogrammetry can potentially far exceed the productivity of conventional photogrammetry. Point mensuration and stereo compilation are tasks requiring skilled operators if done manually, and few means are available to speed up the manual process. Point mensuration and stereo "perception" by image matching will increase in speed with each advance in computer technology, including parallel processing. Point mensuration occurs at numerous stages in the photogrammetric restitution process, such as interior orientation (fiducial marks), relative orientation (parallax points), absolute orientation (signalized control points), and aerial triangulation (pass points and signalized control points). In the case of signalized points, a strategy would be as follows.

1. Derive a rotation-independent detection template for the target type. Convolve this template with the image(s) under study, and, by thresholding, obtain the approximate locations of the points.
2. With a fine-pointing template estimate the target position in the image, while simultaneously modeling rotation, scale, affinity, and radiometry.

The search criterion can be the maximization of a correlation coefficient, or the minimization of a sum of squares of residuals. The correlation coefficient is given by

$$C_{uv} = \frac{\Sigma(u_i - \overline{u})(v_i - \overline{v})}{\left[\Sigma(u_i - \overline{u})^2 \Sigma(v_i - \overline{v})^2\right]^{1/2}} \tag{53.61}$$

where u and v represent image and template, or vice versa, and

$$\overline{u} = \Sigma\frac{u_i}{N} \quad \text{and} \quad \overline{v} = \Sigma\frac{v_i}{N} \tag{53.62}$$

The other significant application area of digital image matching is digital elevation model extraction. Three techniques will be described for this task: (1) vertical line locus, (2) least squares matching, and (3) epipolar matching.

Vertical Line Locus

Vertical line locus (VLL) is used to estimate the elevation at a single point, appearing in two photographs. An initial estimate is required of the object space elevation of the point to be estimated. A "search range" is then established in the Z-dimension, extending above and below the initial estimate. This search range is then subdivided into "test levels." At each test level, a matrix of points is defined surrounding the point to be estimated, all in a horizontal plane. All points in this matrix are then projected back into each of the two photographs, and via interpolation in the digital image, a corresponding matrix of gray levels is determined for each photograph, at each level. When the test level most nearly coincides with the actual terrain elevation at the point, the match between the pair of gray level matrices should be the maximum. Some measure of this match is computed, often the correlation coefficient, and the elevation corresponding to the peak in the match function is the estimated elevation. Variants on this procedure involving an iterative strategy of progressively finer elevation intervals, variable sized matrix, and "matrix shaping" based on estimated terrain slope can all be implemented, yielding more accurate results at the expense of more computing effort.

Least Squares Matching

In the usual least squares matching (LSM) approach, one assumes that the two images are the same except for an affine geometric relationship and a radiometric relationship with a gain and offset. One takes the gray values from one image of the pair to be observations and then computes geometric and radiometric transformation parameters in order to minimize the sum of squares of the discrepancies (residuals) with the second image. In the following equations, g will represent the first image and h will represent the second image. The affine geometric relationship is defined by

$$\begin{aligned} x_h &= a_1 x_g + a_2 y_g + a_3 \\ y_h &= b_1 x_g + b_2 y_g + b_3 \end{aligned} \tag{53.63}$$

The linearized equation relating the gray levels is

$$g(x,y) = h(x,y) + \frac{\partial h}{\partial x}dx + \frac{\partial h}{\partial y}dy + h(x,y)\,dk_1 + dk_2 \tag{53.64}$$

Taking differentials of Eq. (53.63),

$$\begin{aligned} dx_h &= x_g\,da_1 + y_g\,da_2 + da_3 \\ dy_h &= x_g\,db_1 + y_g\,db_2 + db_3 \end{aligned} \tag{53.65}$$

making the following substitution for compact notation:

$$h_x = \frac{\partial h(x,y)}{\partial x} \quad \text{and} \quad h_y = \frac{\partial h(x,y)}{\partial y} \tag{53.66}$$

adding a residual to g, and substituting the differentials into Eq. (53.64), we obtain the condition equation to be used for the least squares estimation,

$$\begin{aligned} g + v_g - h - h_x x_g\,da_1 - h_x y_g\,da_2 - h_x\,da_3 - h_y x_g\,db_1 \\ - h_y y_g\,db_2 - h_y\,db_3 - h\,dk_1 - dk_2 = 0 \end{aligned} \tag{53.67}$$

In matrix form,

$$v + B\boldsymbol{\Delta} = f \tag{53.68}$$

the condition equation becomes

$$v_g + [-h_x x_g \quad -h_x y_g \quad -h_x \quad -h_y x_g \quad -h_y y_g \quad -h_y \quad -h \quad -1]\begin{bmatrix} da_1 \\ da_2 \\ da_3 \\ db_1 \\ db_2 \\ db_3 \\ dk_1 \\ dk_2 \end{bmatrix} = h - g \tag{53.69}$$

One such equation may be written for each pixel in the area surrounding the point to be matched. The usual solution of the least squares problem yields the parameter estimates, in particular the shift parameters to yield the "conjugate" image point. If the parameters are not small, it may be necessary to resample for a new "shaped" image, h, and solve repeated iterations until the parameter estimates are sufficiently small.

Epipolar Matching

Both of the previous techniques involved matching areas, albeit small ones. In epipolar matching, the two matched signals are one-dimensional. Also of interest is the degree to which these methods make use of a priori knowledge of the image orientation. In the VLL technique we make use of this information. In LSM we do not. In epipolar matching this information is used.

In theory, the use of this information should further restrict the solution space and thereby yield faster solutions. In the epipolar technique, one takes the two planes defined by the two images to be matched, and a third plane defined by the two perspective centers and intersecting the two photograph planes. This third plane is referred to as an epipolar plane (there is an infinite number of them). The intersection of this epipolar plane with each of the photograph planes defines two lines, one in each of the photographs. If the orientation is correct, it is guaranteed that each point on one of the lines has a conjugate point on the other line. Thus to search for this conjugate point one needs only to search in one dimension. The practical steps necessary to implement this technique would be as follows.

1. For each photograph, determine two points in object space which lie in the epipolar plane.
2. For each photograph, project the two corresponding points into the photograph plane and solve for the parameters defining the resulting line in image space.
3. Resample each digital image at an appropriate interval along these lines.
4. Determine a match point interval in image space. Select points in one image at this spacing along the epipolar line.
5. For each match point in one image, bracket it by some pixels on either side of it. Then search the epipolar line in the other image for the best match "element." The match criterion is usually the above-mentioned cross-correlation function, but it can be any objective function which measures the degree of "sameness" between the elements.
6. Create in this way an irregularly spaced line of points in object space.
7. Rotate the epipolar plane about the photograph baseline by a small amount, and repeat the process. This will create an irregular grid of XYZ points in object space.
8. Use these points directly to form a TIN, or interpolate a regular grid as described earlier.

Up to now, digital image matching has been most effective and reliable when used on small and medium scale imagery. Large scale images, showing the fine structure of the terrain along with individual trees, buildings, and other man-made objects, may contain steep slopes and vertical planes (i.e., not a smooth, continuous surface). This generally interferes with the simple matching algorithms described here. For large scale engineering mapping, more research needs to be done to develop more robust matching methods.

53.7 Photogrammetric Project Planning

Project planning usually starts with analysis of the specifications for the maps or image products to be produced. Requirements for these final products generally place fairly rigid constraints on the choices available to the planner. The constraints arise because the specified accuracies must be achieved and on the other hand the process must be economical.

Flight Planning

For conventional topographic mapping in the U.S., a compilation system (camera, field survey system, stereoplotter, operators, etc.) is often characterized by a *C-factor,* used as follows:

$$\text{Flying height} = C\text{-factor} \times \text{Contour interval} \tag{53.70}$$

This C-factor can be in the range of 150 to 250 depending on the quality of the equipment and the skill of the personnel doing the compilation. Thus if one knows from the specifications that a 1 meter contour interval is required and the C-factor is 2000, the maximum flying height (above the terrain) would be 2000 meters. One could always be conservative and fly at a lower height, resulting in more photographs at a larger scale. Table 53.2 gives a series of representative parameters for some of the traditional mapping scales used in the U.S. Conventional forward overlap between successive photographs is 60%. Conventional side overlap is 30%. In Table 53.2 the enlargement factor is the factor from negative scale to map scale. Figure 53.16 shows the geometry for determining the base, B, or forward gain, from the ground dimension of the photograph coverage, W, and the forward overlap as a fraction of W (60% is a common value). The following equation gives B as described ($OL = 0.6$, the forward overlap fraction):

$$B = (1 - OL) \times W \tag{53.71}$$

Table 53.2 Typical Flight Parameters (Length Units: Feet)

Map Scale	Ratio	Cont. Intvl.	Neg. Scale	Engl. Factor	H	C-Factor	W	Fwd. Gain	Flt. Line Spc.
1″ = 50′	600	1	4,200	7 ×	2,100	2,100	3,150	1,260	2,205
1″ = 100′	1,200	2	7,800	6.5×	3,900	1,950	5,850	2,340	4,095
1″ = 200′	2,400	5	12,000	5 ×	6,000	1,200	9,000	3,600	6,300
1″ = 400′	4,800	5	16,800	3.5×	8,400	1,680	12,600	5,040	8,820
1″ = 1000′	12,000	10	30,000	2.5×	15,000	1,500	22,500	9,000	15,750
1″ = 2000′	24,000	10	38,400	1.6×	19,200	1,920	28,800	11,520	20,160
1″ = 4000′	48,000	20	57,600	1.2×	28,800	1,440	43,200	17,280	30,240

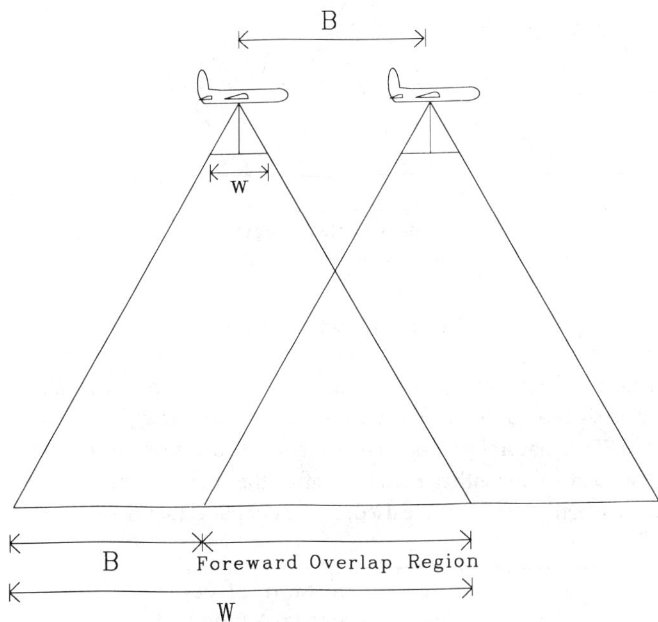

FIGURE 53.16 Forward overlap geometry.

The photograph coverage, W, is related to the actual photograph dimension, w, by the scale

$$W = \text{Scale} \times w \tag{53.72}$$

The photograph dimension, w, is approximately 23 cm for standard aerial film. An analogous situation is shown in Fig. 53.17 for determining S, the distance between flight lines, from W and from the side overlap as a fraction of W (30% is a common value). The following equation gives $S(SL = 0.3$, the side overlap fraction):

$$S = (1 - SL) \times W \tag{53.73}$$

Control Points

Horizontal control points (XY or NE) allow proper scaling of the model data collected from the stereo photographs. Vertical control points (Z or h) allow the proper terrain slopes to be preserved. Classical surveying techniques determined these two classes of control separately and the distinction is still made. Even with GPS-derived control points, which are inherently 3-D, the elevation or Z-dimension is often disregarded because we lack an adequate geoid undulation

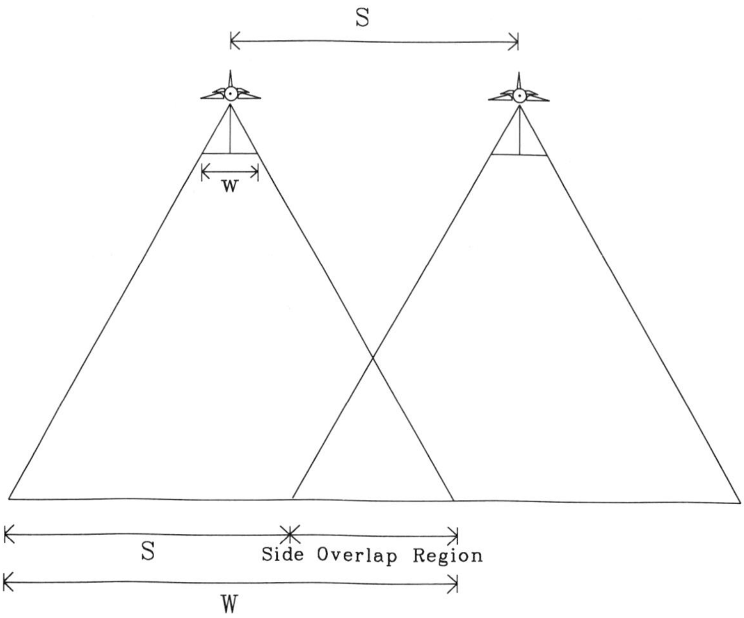

FIGURE 53.17 Side overlap geometry.

map to convert ellipsoid heights to orthometric (sea level) referenced heights. Control points may be targeted or signalized before the photo flight so that they are visible in the images, or one may use *photo-ID* or natural points such as manhole covers or sidewalk corners. Coordinates are determined by field survey either before or after the flight, though usually after in the case of photo-ID points. If artificial targets are used, the center panel (at image scale) should be modestly

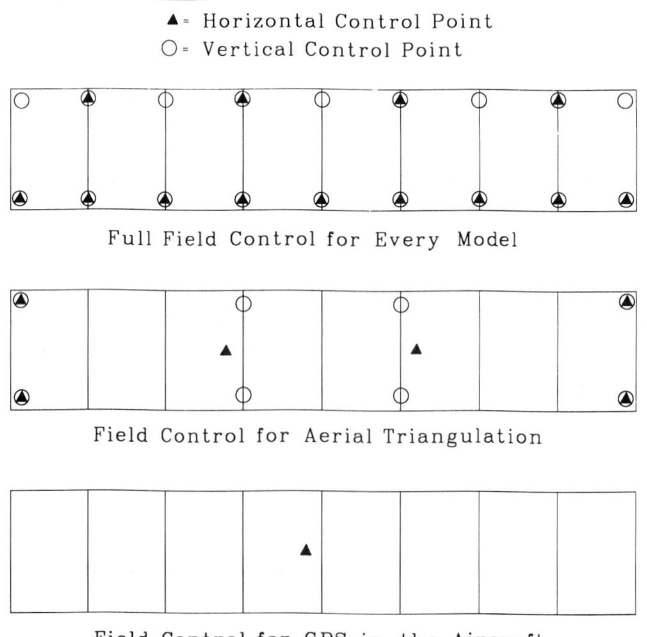

FIGURE 53.18 Control point distributions.

larger than the measuring mark in the stereocomparator or plotter. There should also be two or more prominent legs radiating from the target center panel to allow easy identification in the photographs. Each stereo model requires an absolute minimum of two horizontal control points and three vertical control points. These could all be established by field techniques but this would usually be too expensive. More commonly a sparser network of control points is established, and the needed control in between the field control is determined by bridging or aerial triangulation. When aerial triangulation is used, artificial marking or pugging of the film diapositives is necessary to create points for the operator to use for absolute orientation. When GPS is used in the aircraft to determine exposure station coordinates independently, then in theory, assuming one has photo strips crossing at near right angles, no ground control is needed. Without crossing strips, at least one control point is necessary. In practice the best number to use will probably fall between 1 and the larger number necessary for ground control based aerial triangulation. These control point requirements are shown in Fig. 53.18.

53.8 Close Range Metrology

Close range metrology using photogrammetry offers some unique advantages and presents some equally unique challenges. The advantages are (1) it is a noncontact measurement method, (2) photography can be acquired relatively quickly in dangerous or contaminated areas, (3) the photographs represent an archival record in case any dimensions are disputed, and (4) the camera can be brought to the object, rather than the other way around (as with a coordinate measuring machine). Accuracies obtainable using close range photogrammetry (as in aerial photogrammetry) would fall in the range of 1/5,000 to 1/100,000 of the object distance.

Equipment

Cameras are typically small format, usually with 70 mm film size. Stereometric cameras are mounted rigidly in pairs, so that fewer parameters are required in solving for the orientations. The cameras used would preferably be metric, although with very special handling, and much extra effort, nonmetric cameras can be used for some applications. There should be at least some sort of fiducial marks in the focal plane, preferably a reseau grid covering the image area. Film flatness can be a severe problem in nonmetric cameras. Ideally there should be a vacuum back permitting the film to be flattened during exposure. There should be detents in the focus settings to prevent accidental movement or slippage in the setting. Calibration will be carried out at each of these focus settings, resulting in a set of principal distances and lens distortion curves for each setting. These are often computed using added parameters in the bundle adjustment, performed by the user, rather than sending the camera to a testing laboratory. Lighting can be a problem in close range photogrammetry. For some objects and surfaces which have very little detail and texture, some sort of "structured lighting" is used to create a texture. To obtain strong geometry for the best coordinate accuracy, highly convergent photography is often used. This may prevent conventional stereo viewing, leading to all photo observations being made in monoscopic view. This introduces an additional requirement for good targeting, and can make advantageous the use of feature equations rather than strictly point equations.

Applications

Close range photogrammetry has been successfully used for tasks such as mapping of complex piping systems, shape determination for parabolic antennas, mating verification for ship hull sections, architectural/restoration work, accident reconstruction, and numerous medical/dental applications.

53.9 Remote Sensing

Remote sensing is considered here in its broad sense, including the photogrammetric aspects of using nonframe imagery from spaceborne or airborne platforms. A thorough treatment must include the metric aspects of the image geometry and the interpretive and statistical aspects of the data available from these sources.

Data Sources

Following is a partial listing of sensors, with associated platform and image descriptions. These sensors provide imagery which could be used to support projects in civil engineering.

1. MSS, multispectral scanner, Landsat 1-5, altitude 920 km, rotating mirror, telescope focal length 0.826 m, IFOV (instantaneous field of view) 83 m on the ground, gray levels 64, image width 2700 pixels, 4 spectral bands: 0.4–1.0 micrometers
2. TM, thematic mapper, Landsat 4-5, altitude 705 km, rotating mirror, telescope focal length 1.22 m, IFOV 30 m on the ground, gray levels 256, image width 6000 pixels, 6 spectral bands: 0.4–0.9, 1.5–1.7, 2.1–2.4, 10.4–12.5 micrometers
3. SPOT, SPOT 1-3, panchromatic (multispectral not described), altitude 822 km, pushbroom, telescope focal length 1.082, IFOV 10 m on the ground, gray levels 256, image width 6000 pixels, 1 panchromatic band 0.55–0.75 micrometers plus 3 multispectral bands, off-nadir viewing capability for stereo

Geometric Modeling

Coordinate Systems

The two primary coordinate systems needed for describing elliptical orbits as occupied by imaging sensor platforms are the XYZ system and the $q_1q_2q_3$ system shown in Fig. 53.19. The XYZ system has the XY plane defined by the earth equatorial plane, Z oriented with the spin axis of the earth, and with X through the vernal equinox or the first point of Aries. The second system has q_1q_2

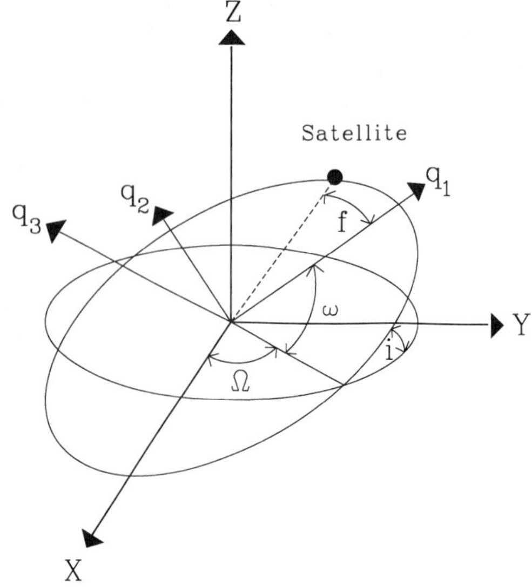

FIGURE 53.19 Orbital parameters.

in the orbit plane with origin at one focus of the ellipse, as shown in Fig. 53.19. q_1 is along the semimajor axis, $q_2 q_3$ are as described above to define a right-handed coordinate system.

Orbital Mechanics

The gravitational attraction between the earth and an orbiting satellite is

$$F = \frac{GMm}{r^2} \tag{53.74}$$

where m is the mass of the satellite, r is the geocentric distance to the satellite, and GM is the earth's gravitational constant:

$$GM = \mu = 3{,}986{,}005 \times 10^8 \mathrm{m}^3 \mathrm{s}^{-2}$$

with $\mathbf{r}$ the position vector of the satellite,

$$\bar{\mathbf{r}} = -\frac{G(M+m)\mathbf{r}}{r^3} \simeq -\frac{GM\mathbf{r}}{r^3} \tag{53.75}$$

Expressing the position vector in 3-D coordinates and the position in the orbital plane in polar coordinates, extracting an expression for the acceleration, and further manipulating the expressions yields a differential equation whose solution is a general conic section in polar coordinates:

$$r = \frac{p}{1 + e\cos(f)} \tag{53.76}$$

where f is the true anomaly as shown in Fig. 53.19, p is the semi–latus rectum, and e is the eccentricity. For $e > 1$ the orbit is a hyperbola, for $e = 1$ the orbit is a parabola, and for $e < 1$ the orbit is an ellipse. The use of the term *anomaly* as a synonym for angle is a vestige of Ptolemaic misconceptions, originally implying an angle which did not vary linearly with time. The true anomaly, f, the eccentric anomaly, E, and the mean anomaly, M, are related by Kepler's equation:

$$M = E - e\sin E \tag{53.77}$$

and

$$\tan\left(\frac{f}{2}\right) = \sqrt{\frac{1+e}{1-e}}\tan\left(\frac{E}{2}\right) \tag{53.78}$$

The point on an elliptical orbit furthest from the earth is the *apogee;* the point closest is the *perigee.* At perigee the true anomaly is zero; at apogee it is 180 degrees. An orbit is specified when the satellite position and velocity vectors, $\mathbf{r}$ and $\dot{\mathbf{r}}$, are given at a particular time. A more common set of six parameters to specify an orbit are the classical Keplerian elements $\{a, e,, T_0, \Omega, i, \text{ and } \omega\}$, where a is the semimajor axis, e is the eccentricity, T_0 is the time of perigee passage, Ω is the right ascension of the ascending node (shown in Fig. 53.19), i is the inclination, and ω is the argument of the perigee.

In order to obtain the Kepler elements from the position and velocity vectors at a given time, the following equations can be used. The angular momentum vector is

$$\mathbf{h} = \begin{bmatrix} h_x \\ h_y \\ h_z \end{bmatrix} = \mathbf{r} \times \dot{\mathbf{r}} \tag{53.79}$$

This yields

$$\Omega = \tan^{-1}\left(\frac{h_x}{-h_y}\right) \tag{53.80}$$

$$i = \tan^{-1}\left(\frac{\sqrt{h_x^2 + h_y^2}}{h_z}\right) \tag{53.81}$$

With v representing the velocity,

$$a = \frac{r}{2 - (rv^2/\mu)} \tag{53.82}$$

$$e = \sqrt{1 - \frac{h^2}{\mu a}} \tag{53.83}$$

Obtaining the eccentric anomaly from

$$\cos E = \frac{a - r}{ae} \tag{53.84}$$

the true anomaly is found from

$$f = \tan^{-1}\left(\frac{\sqrt{1 - e^2}\sin E}{\cos E - e}\right) \tag{53.85}$$

Defining an intermediate coordinate system in the orbit plane such that p_1 is along the nodal line,

$$p = \begin{bmatrix} p_1 \\ p_2 \\ p_3 \end{bmatrix} = R_1(i)R_3(\Omega)r \tag{53.86}$$

$$\omega + f = \tan^{-1}\left(\frac{p_2}{p_1}\right) \tag{53.87}$$

T_0 can be obtained from

$$M = \sqrt{\frac{\mu}{a^3}}(t - T_0) \tag{53.88}$$

To go in the other direction, from the Kepler elements to the position and velocity vectors, begin with Eq. (53.88) to obtain the mean anomaly, then solve Kepler's equation, Eq. (53.77), numerically for the eccentric anomaly, E. The true anomaly, f, is found by Eq. (53.78). The magnitude of the position vector in the orbit plane is

$$r = \frac{a(1 - e^2)}{1 + e\cos(f)} \tag{53.89}$$

the position vector is

$$r = \begin{pmatrix} q_1 \\ q_2 \\ q_3 \end{pmatrix} = \begin{pmatrix} r\cos(f) \\ r\sin(f) \\ 0 \end{pmatrix} \tag{53.90}$$

the velocity vector is

$$v = \sqrt{\frac{\mu}{p}}\begin{pmatrix} -\sin(f) \\ e + \cos(f) \\ 0 \end{pmatrix} \tag{53.91}$$

where p, the semi–latus rectum, is given by

$$p = a(1 - e^2) \tag{53.92}$$

The rotation matrix relating the orbit plane system and the vernal equinox, spin-axis system is given by

$$r_{XYZ} = Rr_{q_1 q_2 q_3} \tag{53.93}$$

where R is

$$R = R_3(-\Omega)R_1(-i)R_3(-\omega) \tag{53.94}$$

This R can be used to transform both the position and velocity vectors above into the XYZ "right ascension" coordinate system defined above.

Platform and Sensor Modeling

If we consider a linear sensor on an orbiting platform, with the sensor oriented such that the long dimension is perpendicular to the direction of motion (i.e., pushbroom), we can construct the imaging equations in the following way. Each line of imagery, corresponding to one integration period of the linear sensor, can be considered a separate perspective image, having only one dimension. Each line would have its own perspective center and orientation parameters. Thus, what may appear to be a static image frame is, in fact, a mosaic of many tiny "framelets." These time dependencies within the image "frame" are the result of normal platform translational velocity, angular velocity, and additional small, unpredictable velocity components. The instantaneous position and orientation of a platform coordinate system is shown in Fig. 53.20.

The equations which relate an object point, $(X_m, Y_m, Z_m)^t$, and the time-varying perspective center, $(X_s(t), Y_s(t), Z_s(t))^t$ are of the form

$$\begin{bmatrix} 0 \\ y \\ -f \end{bmatrix} = k \times R_{III} \times R_{II} \times R_I \times \begin{bmatrix} X_m - X_s(t) \\ Y_m - Y_s(t) \\ Z_m - Z_s(t) \end{bmatrix} \tag{53.95}$$

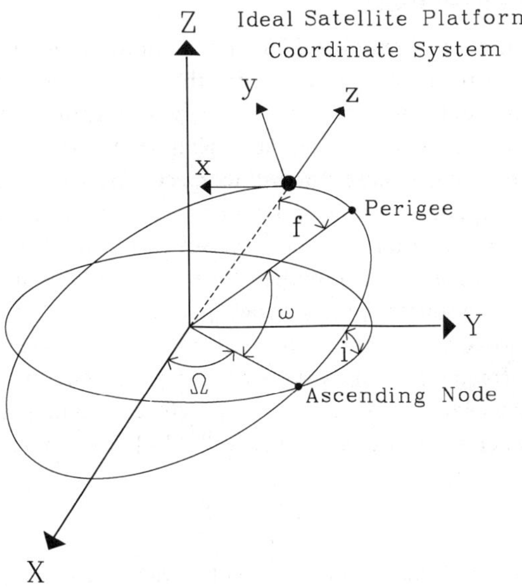

FIGURE 53.20 Imaging satellite orientation.

R_{III} is a platform to sensor rotation matrix which should be fixed during a normal frame. This could accommodate off-nadir view angle options, for instance. R_{II} transforms the ideal platform to the actual platform and the time dependency in its angles would probably take the form of low degree polynomial functions. R_I represents the time dependent rotation from a space-fixed geocentric coordinate system to the instantaneous platform coordinate system. In a manner similar to the previous section, this matrix would be the composition of the rotations,

$$R_I = R_z(\pi) \times R_y\left(\frac{\pi}{2} - \omega - f(t)\right) \times R_x\left(i - \frac{\pi}{2}\right) \times R_z(\Omega) \tag{53.96}$$

with the orbit elements as described in the previous section. The cartesian coordinates on the right side of Eq. (53.95) are with respect to a space-fixed coordinate system (i.e., not rotating with the earth). In order to transform a point in a terrestrial, earth-fixed coordinate system, a coordinate rotation, R_0, is necessary to account for the earth's rotation, polar motion, precession, and nutation. For $(X_e, Y_e, Z_e)^t$ in an earth-fixed system,

$$\begin{bmatrix} X_m \\ Y_m \\ Z_m \end{bmatrix} = R_0 \times \begin{bmatrix} X_e \\ Y_e \\ Z_e \end{bmatrix} \tag{53.97}$$

Analogous to the decomposition of conventional orientation in photogrammetry into exterior and interior components, we could construct here a decomposition of the imaging Eq. (53.95) into a platform component and a sensor component. Eliminating the nuisance scale parameter in Eq. (53.95) would yield two equations per object point, per image "frame." Unknown parameters would consist of a user-selected subset of the platform and sensor orientation parameters, some time varying, the orbit parameters, and the object point parameters. These equation systems could be formed for a single image (resection), a set of sequential images along the orbit path (strip), or an arbitrary collection of adjacent frames (block).

The imminent arrival of commercially available satellite data in the 1–3 meter pixel range means that this kind of image modeling will become even more important in the future as digital sensors continue to slowly supplant film-based photography.

Interpretive Remote Sensing

Remote sensing, as the term is usually employed, implies the study of images to identify the objects and features in them, rather than merely to determine their size or position. In this way, remote sensing is a natural successor to the activity described by the slightly outdated and more restrictive term *photo interpretation.* Remote sensing almost always implies digital data sources and processing techniques. It often implies multiple sensors in distinct spectral bands, all simultaneously imaging the same field of view. Along with the multispectral concept, there is no restriction that sensors respond only within the range of visible radiation. Indeed, many remote sensing tasks rely on the availability of a wide range of spectral data. Witness the recent move toward "hyperspectral" data, which may have hundreds of distinct, though narrow, spectral bands.

Systems have been deployed with sensitivities in the ultraviolet, visible, near infrared, thermal infrared, and microwave frequencies. Microwave systems are unique in that they can be *active* (providing source radiation as well as detection) whereas all of the others listed are strictly *passive* (detection only). Active microwave systems include both SLAR, side-looking airborne radar, and synthetic aperture radar.

Multispectral Analysis

Multispectral image data can be thought of as a series of coregistered image planes, each representing the scene reflectance in a discrete waveband. At each pixel location, the values from the spectral bands constitute an n-dimensional data vector for that location. Consider an example with two

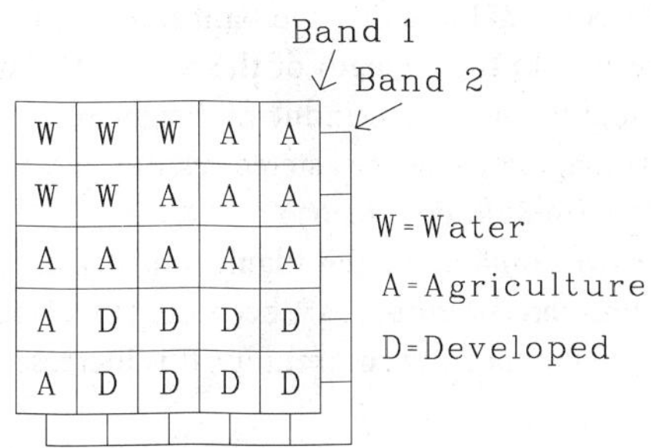

FIGURE 53.21 Training sample for supervised classification.

spectral bands in which the scene has been classified a priori into three categories: water (W), agricultural land (A), and developed land (D), as shown in Fig. 53.21. From this *training sample* one can (assuming normality) construct a mean vector and a variance/covariance matrix for each class. New observation vectors could then be assigned to one of the defined classes by selecting the minimum of the Mahalanobis function,

$$D^2 = (X - \mu_i)^t \Sigma_i^{-1} (X - \mu_i) + \ln |\Sigma_i| \tag{53.98}$$

where μ_i and Σ_i are evaluated for each category. Clustering of feature classes is shown in Fig. 53.22. This is an example of a maximum likelihood classifier, using *supervised* classification. Other techniques can be used such as discriminant functions and Bayesian classification. In addition to statistical and heuristic approaches to pattern recognition, one can also employ syntactic methods, wherein a feature is identified by its context or relationship to adjacent or surrounding features.

Change Detection

In order to detect changes on a portion of the earth's surface it is necessary to have imagery at different times. In addition, when possible, it is desirable to minimize apparent, but spurious, differences due to season, view attitude, and time of day. In order to make a proper pixel-by-pixel comparison, both images should be resampled and brought to the same map projection surface. This is best done by using identifiable ground control points (road intersections, etc.) and a modeling of the sensor and platform positions and orientations. Less effective registration is sometimes done using two-dimensional polynomial functions and a "rubber sheet" transformation to the control points. Once the scenes have been brought to a common coordinate base, a classification is made individually on each scene. A pixel-by-pixel comparison is made on the classification results and a "change" layer can be generated where differences are encountered.

Microwave Remote Sensing

Microwave radiation has the useful characteristic of penetrating clouds and other weather conditions which are opaque to visible wavelengths. Aircraft-based imaging radars are thus able to acquire imagery under less restrictive flight conditions than, for example, conventional photography. The geometry of the radar image is fundamentally different from the near perspective geometry of other sensor systems. In both real aperture side-looking airborne radar and synthetic aperture radar, imagery is presented as a succession of scan lines perpendicular to the flight direction. The

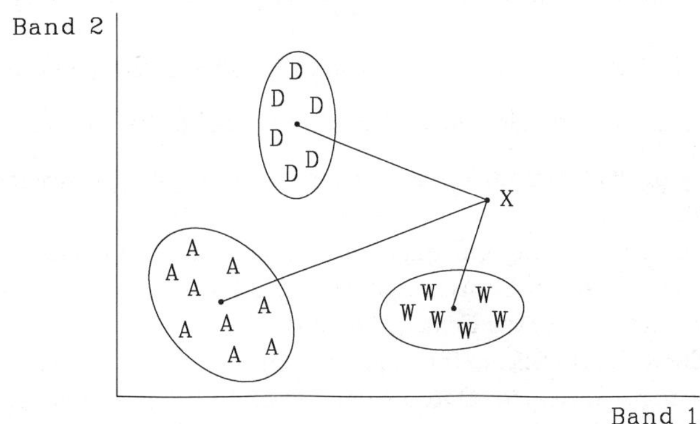

FIGURE 53.22 Clustering of two band data in scatter diagram.

features displayed along the scan lines have position proportional to either the slant range or the ground range, and the image density or gray level is related to the strength of the return. Radar imagery can be acquired of the same area from two flight trajectories, inducing height-related parallax into the resulting stereo pair. With the proper imaging equations, stereo restitution and feature compilation can be carried out, giving rise to the term *radargrammetry*.

Until recently, remote sensing activities were the exclusive domain of a few highly industrialized countries. Now a number of countries and even some commercial ventures are becoming involved in systems to provide remote sensing imagery. For the civil engineering community, this increased availability of data can be very beneficial.

References

Burnside, C. D. 1985. *Mapping from Aerial Photographs.* John Wiley & Sons, New York.

Colwell, R. N., ed. 1983. *Manual of Remote Sensing.* American Society of Photogrammetry and Remote Sensing, Bethesda, MD.

Ebner, H., Fritsch, D., and Heipke, C. 1991. *Digital Photogrammetric Systems.* Wichman, Karlsruhe.

Faugeras, O. 1993. *Three-Dimensional Computer Vision.* MIT Press, Cambridge, MA.

Karara, H. M., ed. 1989. *Non-Topographic Photogrammetry,* 2nd ed. American Society of Photogrammetry and Remote Sensing, Bethesda, MD.

Kraus, K. 1993. *Photogrammetry.* Dummler Verlag, Bonn.

Leick, A. 1990. *GPS Satellite Surveying.* John Wiley & Sons, New York.

Moffitt, F. H. and Mikhail, E. M. 1980. *Photogrammetry,* 3rd ed. Harper & Row, New York.

Pease, C. B. 1991. *Satellite Imaging Instruments.* Ellis Horwood, Chichester.

Richards, J. A. 1993. *Remote Sensing Digital Image Analysis: An Introduction,* 2nd ed. Springer-Verlag, New York.

Slama, C. C., ed. 1980. *Manual of Photogrammetry,* 4th ed. American Society of Photogrammetry and Remote Sensing, Bethesda, MD.

Wolf, P. R. 1983. *Elements of Photogrammetry.* McGraw-Hill, New York.

For Further Information

The following journals are useful sources of information:

Photogrammetric Engineering and Remote Sensing, American Society of Photogrammetry and Remote Sensing, Bethesda, MD.

The Photogrammetric Record, The Photogrammetric Society, London.

ISPRS Journal of Photogrammetry and Remote Sensing (formerly *Photogrammetria*), Elsevier Science Publishers B. V., Amsterdam, The Netherlands.

CISM Journal (formerly *Canadian Surveyor*), Canadian Institute of Surveying and Mapping, Ottawa, Canada.

Journal of Surveying Engineering, American Society of Civil Engineers, New York.

The following organizations provide valuable technical and reference data:

American Society of Photogrammetry and Remote Sensing, 5410 Grosvenor Lane, Suite 210, Bethesda, MD 20814

American Congress on Surveying and Mapping, 5410 Grosvenor Lane, Bethesda, MD 20814

AM/FM International, 14456 E. Evans Ave., Aurora, CO 80014

U.S. Geological Survey, EROS Data Center, Sioux Falls, SD 57198

U.S. Geological Survey, Earth Science Information Center, 53. National Center, Reston, VA 22092

SPIE, The International Society for Optical Engineering, P.O. Box 10, Bellingham, WA 98227

American Society of Civil Engineers, 345 East 47th Street, New York

54

Geographic Information Systems

Jolyon D. Thurgood
Leica, Inc.

J. S. Bethel
Purdue University

54.1 Introduction

Background

Information about our world has been depicted on maps of various forms for many centuries. During the golden age of exploration maps showed critical paths of navigation in the known world, as well as strategic political boundaries and information about settlements and natural resources. Over the past 300 years the art of cartography has been complemented by the development of scientific methods of surveying and related technologies, which have enabled increasingly accurate and complete representation of both physical and cultural features.

Most recently, computer-related advances have led to a revolution in the handling of geographic information.

First of all, raw spatial information can be gathered and processed much more efficiently and quickly, based on technologies such as analytical and digital photogrammetry, global positioning systems (GPS), and satellite remote sensing.

0-8493-8953-4/95/$0.00 + $.50

Second, it has become possible to automate drafting and map production techniques to replace manual drafting procedures.

Third, instead of providing simply a graphical representation through paper maps, it has become possible to model the real world in a much more structured fashion, and to use that spatial model as the basis for comprehensive and timely analyses. Based on this model, it is possible to geographically reference critical data generated by government agencies and private enterprises, and using information modeling and management techniques, to query large amounts of spatially integrated data, and to do so across multiple departments and users, thereby sharing common elements.

The result of this revolution is that there has been a very rapid growth in the production and manipulation of geographic information, to the extent that many organizations can fulfill their production and operational goals only through the use of a geographic information system (GIS).

A geographic information system may be defined as an integrated system designed to collect, manage, and manipulate information in a spatial context. The geographic component, the various technologies involved, and the approach to information modeling set a GIS apart from other types of information systems. A geographic information system provides an abstract model of the real world, stored and maintained in a computerized system of files and databases in such a way as to facilitate recording, management, analysis, and reporting of information. It can be more broadly stated that a geographic information system consists of a set of software, hardware, processes, and organization that integrates the value of spatial data. Various authors provide more detailed definitions of geographic information systems [Antenucci *et al.*, 1991; Dueker, 1987; Parker, 1988].

Early automated mapping systems used interactive computer graphics to generate, display, and edit cartographic elements using computer-aided drafting (CAD) techniques, more or less emulating the manual processes previously used. Over the past decade, more advanced techniques designed to more comprehensively integrate geometric (graphic) elements with associated nongraphic elements (attributes) and designed specifically for map-based and geographic data have resulted in more powerful and flexible implementation of GIS. Continuous mapping, in which a seamless geographic database system replaces map sheets or arbitrary facets earlier used, and the manipulation of geographic information as spatial objects or features are two aspects of most recent geographic information systems that provide users with more intuitive and realistic models of the real world. Also, the integration of vector-based graphics, imagery, and other cell-based information has provided increasingly powerful visualization and analysis capabilities.

Applications

At a broad range of scales, maps have become increasingly important as legal documents that convey land ownership and jurisdictional boundaries, as tools to support decision making (for example, in urban planning), and as a means of visualizing multiple levels of information on political, social, and ecological issues, for example, in thematic mapping of demographic data.

It is estimated that typically 70 to 80% of information maintained by government agencies may be geographically referenced. In addition to directly specifying spatial location on basemap information, such elements as taxpayer identifier, home-owner address, phone numbers, and parcel numbers may be used as the spatial key. Perhaps for this reason, GIS is often seen as the means to promote information sharing, more efficient information management and maintenance, and as a key to providing better and more timely services in a competitive environment. In addition, GIS applications are often both graphics- and database-intensive and provide strong visualization capabilities. The GIS offers the power to process large amounts of various types of information, but also to present results in a powerful graphical medium: The most common standard product of a GIS is for the time being still the printed map, but it is likely to be a cartographic product customized for a specific task or analysis, as opposed to a standard map series product.

In general, a GIS can provide the following information on geographic elements or features: location, characteristics, logical and geometric relationships with other features, and dependencies on other features. This information can generally be used as the basis for tabular reports, standard and custom map output plots, spatial decision support, trend analysis, as well as output to other potential users and analyses. A geographic information system may be accessed from a single PC, a local area network (LAN) of UNIX workstations, or through a virtual, wide area network (WAN) of distributed information.

Standard or common components of a GIS that enable full implementation of such tasks include drafting, data entry, polygon processing and network analysis, spatial querying, and application development tools (macro language, programming libraries).

From earliest times, maps have been used to establish land ownership. One of the first application areas for modern GIS has been in the area of property ownership and records. Within a municipality, the assessor's office or appraisal district is normally responsible for the identification, listing, and appraisal of parcels of real estate and personal property. A GIS provides real benefits to such an office by allowing accurate and complete appraisals, based on access not just to property attributes such as lot size and building square footage, but also to spatial information, such as the comparison of similar properties within a neighborhood. Once the complete map base has been established in digital form and linked to the nongraphic attribute database system, such tasks as property transactions and applications for building permits can be performed efficiently and without a lengthy manual, and often bureaucratic, delay.

Such a parcel-based land information system can provide the basis for a much more sophisticated GIS. For example, within an urban environment, various boundaries define school, library, fire department, sewer and water supply districts, special business zones, and other special tax assessment districts. The allocation of real estate taxes for a given property may be determined by overlaying all of these special districts with property boundaries. Done manually, it is a cumbersome process, and one that makes redistricting—that is, changing the boundaries of any of the constituent districts—a complex process. Polygon processing within a GIS provides the means to perform such an overlay and to determine very quickly how the various tax components apply to one or many properties. The same function can be used to provide answers to discussions regarding proposed changes to these districts, for example, to examine the impact on a city's tax base by annexing an adjacent unincorporated business region. The GIS therefore offers benefits in two areas—first in new capabilities, and second in its ability to produce results in a timely manner: two months of visual inspection and transcription can be replaced by one hour of computer time.

Another key application area is one based on linear networks, such as those defining transportation routes, or an electricity distribution network. In the area of transportation the GIS provides the ability to model individual road elements and intersections and to analyze routes between any two points within an urban street network. Such a network trace can be used in conjunction with emergency services planning to identify the shortest path to a hospital or to examine the average response time to a call to the fire department. By extending the GIS data structure to incorporate one-way streets, turn restrictions in a downtown area, and rush-hour speed statistics, a sophisticated, multipurpose model of the transportation network may be derived. This model can be designed and optimized exclusively for emergency response activities or for planning purposes only—for example, to examine commuting patterns and traffic congestion projections.

The geographic information system provides the ability to completely model utility networks, such as those supplying water, power, and telecommunications to large numbers of consumers. Such a system may operate at a variety of scales, modeling service connections to consumers, service districts, as well as detailed facilities inventories and layouts, such as transformers, valves, conduits, and schematic diagrams. The GIS then becomes a key element at many levels: in customer support (to respond to service failure), in maintenance and daily operations (to identify work requirements and assess inventories), and in planning (to respond to projected needs). It provides the link

GIS Overview

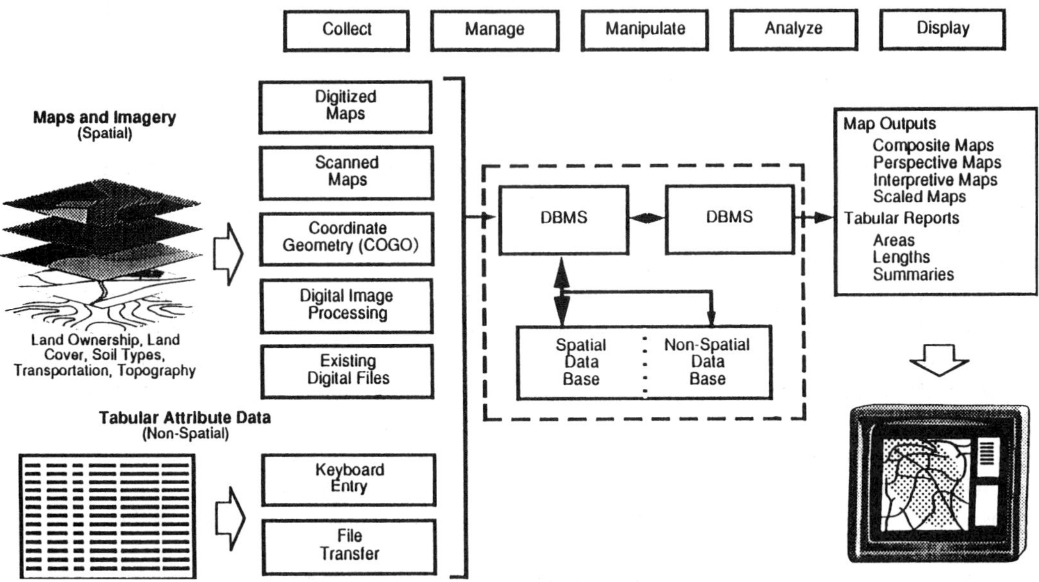

FIGURE 54.1 Overview of a geographic information system(GIS).

between many information systems, including engineering, planning, and customer billing, which can increase overall performance and operational efficiency.

These simple examples identify the key elements of a geographic information system: a base model that identifies spatial features and spatial relationships, a set of descriptors that can be used to discriminate and identify individual elements, and a set of functional processes and tools that operate against all information components. This structure is also shown in Fig. 54.1. Typical bases for application areas are shown in Fig. 54.2.

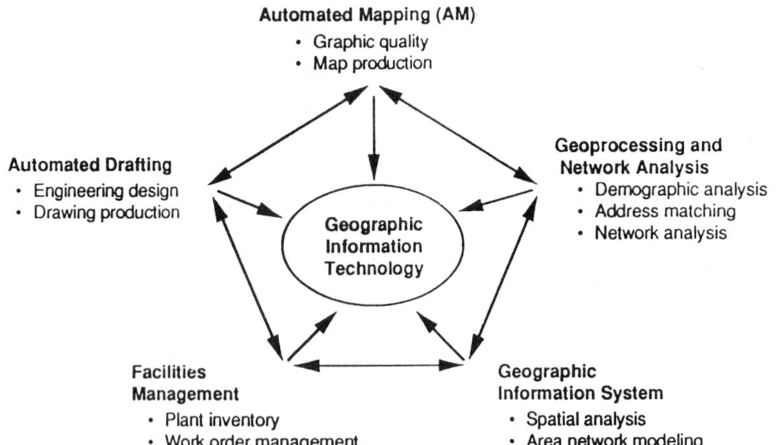

FIGURE 54.2 Application areas based on geographic information technology. (*Source:* Antenucci *et al.* 1991. *Geographic Information Systems: A Guide to the Technology.* Van Nostrand Reinhold, New York.)

54.2 Geographic Information Components

Geography provides a reference for a very large amount of information commonly stored and maintained in information systems. In many cases this reference or key is not necessarily absolute position or a set of geographic coordinates, but an indirect key to location such as street address, parcel or property identifier, phone number, or taxpayer number. One of the distinguishing properties of a geographic information system is the ability to tie such keys to a common geographic base, such as a basemap containing streets, property information, and so on. Some common keys for geographic information are identified in Fig. 54.3.

In fact, the geographic information system is often seen as a focal point for various types of graphical and nongraphical data collections.

Initially, we can look at these components separately.

Geometry (Graphics)

As previously discussed, GIS has generally evolved from computer graphics systems that allowed the graphical representation conventionally depicted on map products to be modeled as layers that can be displayed, edited, and otherwise manipulated by means of specialized software. Today's CAD systems still provide the same type of structure. In such a GIS, graphical elements forming a logical grouping or association are stored on distinct layers or even in separate files. A final graphical display or map output is formed by switching on or off the appropriate layers of information and assigning to each layer a predefined cartographic representation designed for the scale of map or specific application. The symbology or line style to be used is traditionally stored with the layer definition, although it is also normally possible to define special representations for specific graphical elements.

Spatial location is typically stored directly or indirectly within the graphical component of a geographic information system. In earlier systems a complete GIS project stretching across many map sheet boundaries would be stored still in the form of distinct tiles or facets, each representing a drawing with its original spatial extent. Absolute spatial location in a reference coordinate system, such as a state plane coordinate system, would be obtained by interpreting for each drawing part a local transformation (a two-dimensional conformal transformation typically) applied to drawing coordinates. A librarian system would allow such transformations to be applied transparently to the human operator or indeed to applications interested only in absolute spatial location.

More recent systems provide a more seamless continuous map base, where geometric components are stored in a single reference coordinate system that models the real-world representation of geographic features. Any trimming or splitting of the database to map sheet boundaries or other artificial tiling systems are more typically completely hidden from anyone but the project manager or system administrator, as depicted in Fig. 54.4.

Key	Type	Usage
absolute location	X,Y(,Z) coordinates	survey monuments
relative location	distance, bearing	property boundaries
parcel identifier	alphanumeric	land ownership transactions
street address	alphanumeric	ownership
street segment ID	alphanumeric	traffic engineering
manhole ID	alphanumeric	water/wastewater utilities
zip code	numeric	demographics
telephone number	numeric	pizza delivery

FIGURE 54.3 Keys to geographic information.

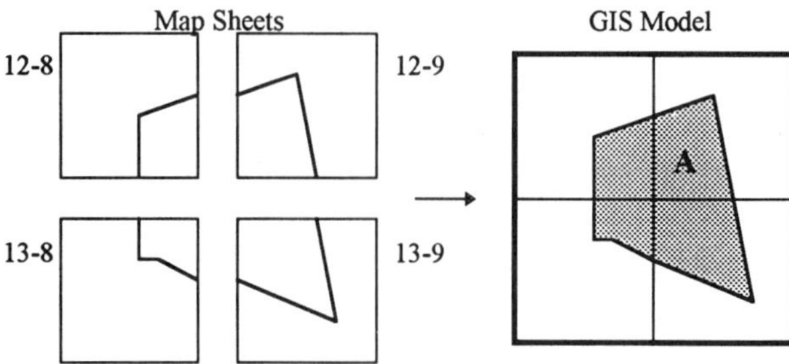

FIGURE 54.4 Map sheets in a GIS. A diagram showing how map sheets stored in separate data files are referenced within a geographic information system in order to model a continuous spatial extent. Polygon A can be accessed as a single geographic feature, regardless of where individual segments of boundary lines are stored.

Attributes (Nongraphics)

As already mentioned, the critical distinguishing property of a GIS lies in its ability to relate various types of information within a spatial context. Such a context is provided directly through absolute or relative location or through nongraphic characteristics, also referred to as attributes. Such attributes may act as direct or indirect keys that allow the analysis of otherwise unrelated sets of information.

We should also distinguish between internally defined attributes or keys that typically maintain the link between the geometry component and the nongraphic database system. This key is often the weakest link within a GIS, especially if geometry and attributes are stored and maintained in distinct file or database systems.

Attribute information is typically held in a relational database management system. GIS applications typically allow the retrieval of nongraphic and geographic information in a linked fashion. For example, it is possible to point to a specific location in a graphical map display, to select and identify an individual spatial feature, and to retrieve or update attribute information relating to this feature. Conversely, it is possible to select spatial elements on the basis of attribute matching and to display the selection in a graphical form such as a thematic map. A variety of attribute types are commonly used, as shown in Fig. 54.5.

Attribute Type	Typical Size / Storage Requirements	GIS Feature Example
short integer	2 bytes	land use class
long integer	4 bytes	land value
floating point real	4 bytes	area
double precision real	8 bytes	centroid coordinate
fixed length text	32 bytes	street name
variable length text	80 bytes per line	property description
bulk	30 kbytes for compressed 256x256	scanned image of building
date	6 bytes	land purchase date

FIGURE 54.5 Typical attribute types.

Vector Element	Parameters	Typical Size / Storage Requirements	GIS Feature Example
node / point	(x,y)	16 bytes (double precision floating point)	centroid of parcel
line	(x_1,y_1), (x_2,y_2)	32 bytes	service connection between house and utility main
line string / polyline	(x_1,y_1), ... , (x_n,y_n)	20 points require 320 bytes	stream
circular arc / circle	(x_0,y_0),radius,θ_{start},θ_{end}	40 bytes per arc	curved portions of roadway
spline	B_0,B_1,B_2,B_3 coefficients for each segment	20 segments require 640 bytes	contour line
complex / composite line	combinations of the above	combinations of the above	oddly-shaped parcel boundary
polygon / surface	(x_1,y_1), ..., (x_n,y_n)	4-sided polygon requires 64 bytes	boundary of municipal service district
Triangulated Irregular Network (TIN) element	(x_1,y_1,y_1),(x_2,y_2,z_2), (x_3,y_3,z_3)	500 triangles require 3600 bytes	digital terrain model (DTM)

FIGURE 54.6 Vector data components (2-D, 2.5-D applications).

Vector

The tremendous growth in comprehensive geographic information systems over the past decade reflects the implementation of vector-based GIS software systems that can handle a variety of point, line, and polygon geometries. Based on these structures, as in Fig. 54.6, all geometric information previously depicted on hard copy maps could be modeled, stored, and manipulated. Although in earlier systems it was common to hold a third dimension (the Z coordinate or elevation) only as a nongraphic attribute of the geometric entity, geometric structures in a modern GIS typically hold all three coordinates for each primitive component. The following geometric structures, also shown in Fig. 54.5, are normally available:

1. Point or node elements, each containing X, Y or X, Y, Z coordinates.
2. Line elements, containing or referencing beginning and end nodes, along with a set of intermediate points, sometimes referred to as shape points. In addition to straight-line connections between shape points, other primitive line types may be modeled, including circular arcs defined by three successive points or parametrically (including complete circles) or B-spline connections between all points in a single entity. Such flexibility in line primitives allows more efficient modeling of certain types of spatial features: for example, the outline of a building would be defined by digitizing corner points (vertices) only; a street centerline may be defined using the appropriate parametric geometry, whereas a digitized contour line may be held for cartographic purposes with a B-spline connection.
3. Polygons, surface or area primitives, are closed polyline elements used to represent enclosed areas of the earth's surface, such as property boundaries, lakes, and so on. A GIS typically allows for the formation of polygon boundaries based on one or more line primitives and also maintains special information about "islands" or "holes" within the enclosed area. Such structures allow the appropriate modeling of, for example, a pond on an island

on a large inland lake. As this example might indicate, more important than the geometric primitives themselves is the overall data model, including vector-based topology between elements.

Raster

Some of the earliest computer-based GISs were based on grid-cell information. Before the definition of more complex structures and the availability of computing power to support their processing, the simplest way to perform a spatial analysis between multiple layers of information was to subdivide each layer into a grid of small cells, with an associated numeric cell value which denoted a class or set value representing the characteristic of that location. By performing simple Boolean operations between the grids representing each characteristic, useful results could be obtained.

At the same time that raster-based GIS products have been incorporating more vector graphics and database structures, vector-based products are providing support for raster information. It is now possible to display scanned engineering drawing, maps, digital orthophoto products, and other raster-based imagery, coregistered with vector map information in a common geographic reference system. This can be seen as part of a broader trend to incorporate both vector- and raster-based structures in a single software system where spatial analyses may be performed using the most appropriate technique. For example, cartographic modeling based on raster data allows the simple analysis of such properties as adjacency and proximity, but processing requirements increase according to the spatial precision or resolution required. Future software is likely to allow rule-based conversion of information (from raster to vector, and vector to raster) prior to manipulation, dependent on the specific output requirements.

Topology

The power of a geographic information system lies not only in the data that are held within the system, but also in the data model that provides the fundamental structure or framework for the data. A basic component of a vector-based GIS is a set of topologic structures that allows the appropriate modeling of points, lines, and polygons.

Start- and end-node points of linear elements are often stored as distinct structures, thereby allowing the analysis of adjacent or connected entities through such logical connections, as opposed to simply a graphical operation. For example, two lines representing road segments that cross each other may intersect at an intersection point stored explicitly as an intersection, as opposed to an overpass.

Arc-node topology allows the formation of geometric structures that model elements such as a street centerline that contains straight-line segments, circular, and spiral curves. In addition, aspects such as tangency conditions between connected line elements are also considered a part of the topology. During geometric processing, such as interactive editing, all topology properties would normally be retained or at least restored after processing.

Polygonal geometry stores its own set of topology, including the element of closure between beginning and end points of an enclosing boundary. References to islands or holes within the boundary are also maintained as part of the topology for that feature.

Software tools to create and maintain polygon- and linear-based topology are common required elements of a GIS.

In a polygon coverage—such as a continuous coverage of soils, land use, or similar—it is often of value to store such area classifications with line-polygon topology that allows the analysis of shared boundaries. In such a structure, linear elements that form boundaries between adjacent polygons are referenced by both polygons. Such a structure also allows common geometry to be stored only once, allowing efficient storage and maintenance of related information. In a modern GIS it is possible for single geometric elements to reference not just two areas within a single

Topologic Type	Description	Properties / Operations	Graphical Representation
Node / point	discrete x,y(,z) identifying unique feature	position / proximity	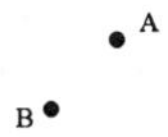 A B
Line	straight-line (two points), arcs / circles, splines	continuity / connectivity nodes at all intersections and at endpoints	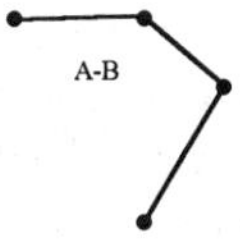 A-B
Polygon / surface	series of connected line segments	areal closure / adjacency	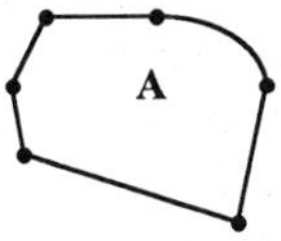 A
Spaghetti	sequence of connected points, with no topologic checking	length (line segments may cross/intersect each other)	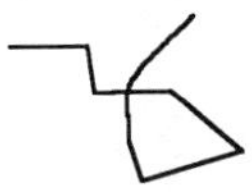

FIGURE 54.7 Node-line-polygon topology.

classification (for example, adjacent areas of soil), but any real-world elements that refer to the same geometry—for example, a property boundary that also forms part of the boundary of municipal jurisdictions (tax districts, school districts) as well as forming a right-of-way boundary.

Another important topologic structure is the composite or complex feature, which allows the grouping of logically related graphical and nongraphic elements. This extends the power of a GIS beyond simple relationships between a single spatial element and one set of attributes to a more sophisticated modeling of real-world elements. Through the use of composite features, a logical spatial feature may be defined—such as a school district composed of a school district boundary, a set of school facilities (classrooms, playing fields, administrative offices), school bus routes, and residential catchment areas—that permits more sophisticated modeling of spatial elements.

More sophisticated topologic structures involve those that allow the formation of a single seamless geographic base structure, where graphic or geometric components cross tile or district boundaries that are stored in distinct files or database tables, but referenced. In such systems users may access large or small spatial features.

Topologic structures such as those identified in Fig. 54.7 form one of the most critical aspects of a GIS since they determine how efficiently certain operations or analyses can be performed. (Network analysis depends on connectivity between line elements; polygon analysis requires handling of closed polylines, islands, and so on.)

54.3 Modeling Geographic Information

In this section various methods of modeling spatial information are described. The method chosen has broad implications on the scope and application of GIS.

Layer-Based Approaches

The traditional method of classifying information in a geographic information system derived from the ability to graphically distinguish various layers or levels of data, also affected by the practical limitations of available computing capacity (for example, limits of 256 layers or 32 colors in a palette). Compared to previous means of producing hard copy maps, though, these restrictions did not prevent reasonable modeling of geographic data. Rather, it allowed the storage, maintenance, and manipulation of many more levels of information than previously possible through manual drafting or overlay means.

In a layer-based approach, the ability to distinguish different types of information relies on a common definition and usage of layers, the schema or data dictionary. Such a schema, agreed upon by all potential users, defines that graphic primitives (lines, arcs, point symbols, text labels) common to a well-defined set of spatial features are stored in separate and distinct layers. The layer specifications determine how information once collected can be used, since in a graphics system the layer may be the only level at which logical data may be segregated. For example, text labels annotating parcel numbers may be stored in layer 25, and parcel boundaries themselves (the graphical primitives) in layer 24: By separating text and pure graphics, it is then possible to display either graphics, text, or both. In practice, separate layers of information may in fact reside in separate files on disk, physically as well as logically distinct. The layer-based model is closely analogous to logical map overlays or use of color separates as used in map production. Figure 54.8 shows typical graphical layers in a GIS.

Relational Approaches

As a first step in the evolution from a purely graphical system, it was recognized that links to additional nongraphic attributes in a GIS would radically increase the utility and power of such an information system. Although early GISs provided an ability to graphically model and depict cartographic information (much in the same way that paper maps had done previously), such systems became extremely cumbersome and limiting when trying to really apply the power of computers to selective retrieval and analysis of spatial information. This resulted in the development and introduction during the 1980s of a GIS with two distinct components: the graphical database or file system, and an associated nongraphic database system. Early systems used proprietary database structures to store and reference nongraphic information, but the more rapid growth and acceptance of GIS in the past few years has come with the use of standard commercial relational database management systems (RDBMS). Spatial information systems using such an approach are also referred to as *geo-relational* or *all-relational* GIS, the latter term denoting that at least some of the geographic or graphic components are also stored in a relational model.

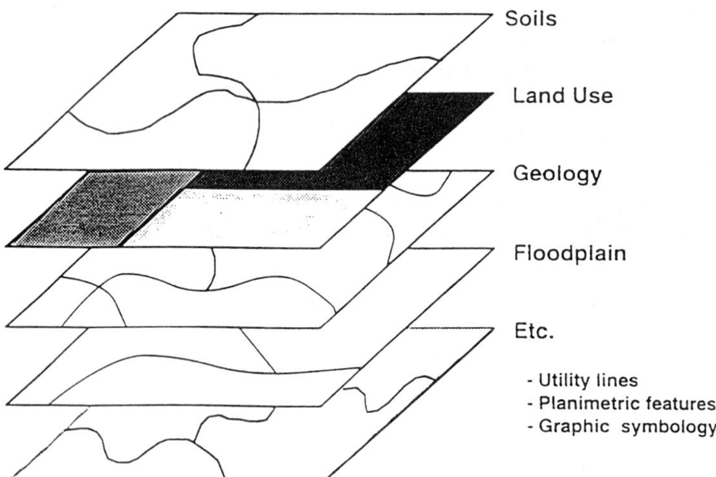

FIGURE 54.8 Layers in a geographic information system. (*Source:* ESRI. 1992. *ARC/INFO: GIS Today and Tomorrow.* ESRI, Redlands, CA.)

The simplest relational model used in GIS consists of a graphics-based component that carries with each geometric primitive (point, line, or polygon) a unique internal identifier or tag, which is the means to associate the geometry with additional nongraphic information defining the characteristics of the spatial feature. The geometry identifier is therefore the common key upon which the power of the relational GIS depends. Although the use of such identifiers is often hidden from the casual user, its role is critical: the means that a GIS software system uses to establish and maintain this vital link between graphics and nongraphics determines the practical potential of the system.

Once linked using the relational model, commonly used and widely understood techniques and tools may be applied in all areas of spatial data management, including data collection, update of nongraphic information, extraction, and reporting. Apart from the key link between geometry (graphics) and primary attribute (nongraphics) tables, it becomes a simple matter to extend the model through additional tables and keys. Common elements in such tables may be used as additional keys through table joins to incorporate the spatial context to a variety of nongraphic information. For example, property information including property addresses can be joined with a table containing address and telephone number in order to provide direct retrieval of phone numbers based on property and location. The relational model can also be used to associate sets of geographic features with common characteristics, such as those properties lying in a specific school district, thereby extending the spatial and logical models as appropriate.

In addition, certain relational terminology can be applied directly in the spatial domain: for example, a polygon overlay can be viewed simply as a "spatial join."

With the acceptance of standards within the RDBMS industry and with extensions of commercial products to provide transparent access and manipulation of information in a broadly distributed network of computing platforms, the geo-relational model has fit well for those GIS projects seeking to play a key role in a multidepartmental or enterprise-wide information technology environment. In such situations the GIS project manager is often happy to take advantage of these standard commercial products and in turn is able to concentrate on issues of data management specific to the geographic nature of the information.

As already mentioned, it is also possible to model geometry and topology using relational database technology. In such GIS, sometimes referred to as all-relational, such information as start and end nodes, pointers between line entities, and polygon elements may be stored in the form of relational tables. A purely relational approach sometimes adds tremendous computing overhead for simple operations in a GIS (for example, geometric editing of polygonal areas) and may require additional levels of information to be stored in a hybrid fashion to improve performance. Figure 54.9 shows the geo-relational model.

Object-Oriented Approaches

Object-oriented approaches in data modeling were applied early on in computer-aided design (CAD) products, in such areas as construction and manufacturing. Such a model allowed an architect to apply object-oriented rules to form walls with certain properties and optional or mandatory components (for example, windows and doors). Once a certain type of (standard) window had been modeled, it could be introduced and reused wherever appropriate.

Over the past few years, object-oriented techniques have been applied to GISs in at least two major areas: (1) spatial data modeling, and (2) spatial analysis or programming tools.

In any geographic information system the database and related software attempt to create an abstract model of the real world (as in Fig. 54.10) containing spatially dispersed information. The closer this abstract comes to modeling true characteristics of the physical and human-made world, the more powerful can be the application of computing power to interpret and provide decision support based on the geographic information. We discussed in a previous section the various components of spatial information: An object-oriented approach to spatial data modeling allows

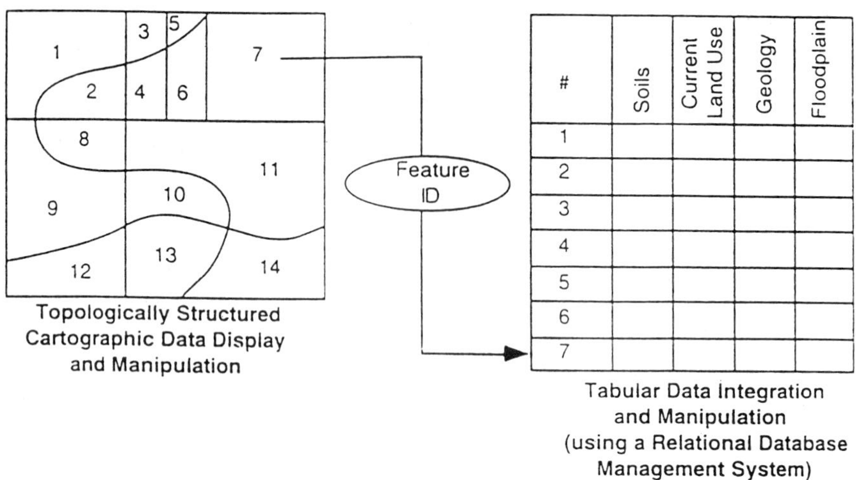

FIGURE 54.9 The geo-relational model. (*Source:* ESRI. 1992. *ARC/INFO: GIS Today and Tomorrow.* ESRI, Redlands, CA.)

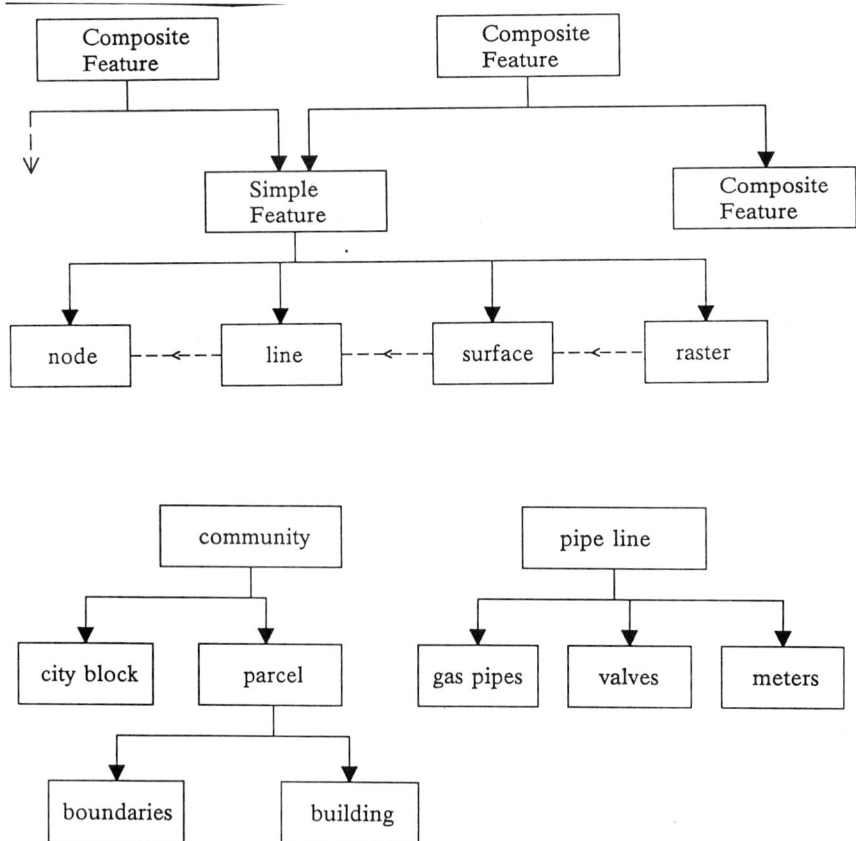

FIGURE 54.10 A GIS data model.

all related components to be encapsulated in a single object or feature definition, along with rules that govern how the object may be manipulated or how components may be related or dependent on components of other objects. The terms *feature oriented* or *feature based* are also used to denote software systems in which the modeled entity is intended to closely reflect the real-world element or object. In such an example a land-ownership record may be treated as a single spatial object or feature containing one or more boundary lines with surface (polygon) topology, additional nongraphic characteristics, and logical associations. Relationships based on common boundaries (logical adjacency) or common ownership may be accommodated. The object-oriented paradigm allows for the modeling of sophisticated groupings of geographic and logically related elements, such as a "census tract" composed of a collection of "parcel" objects, or a "gas distribution network" composed of various subelements such as "pipes" and "valves." Once such elements are defined, the data dictionary contains not just a list of element classes, but the rule base for their use and application. This permits, at least in theory, the population of highly structured geographic data sets in which many real-world relationships and constraints are retained.

Many GIS software products now claim to be object oriented or feature oriented. To a certain degree, such GISs have introduced levels of object-oriented user interfaces or modeling techniques, which may mask a layer-based or fully relational model that the system uses internally. For the time being at least, there are few GIS products that claim to be fully object oriented in both data modeling and data storage: This situation will probably change as object-oriented database management systems (OODBMS) become more widespread.

Extended Relational Model

Although the object-oriented paradigm carries many advantages—especially in areas of accurately modeling real-world phenomena or features and ability to develop or customize geo-related applications based on reusable blocks of code—the relational model carries with it many positive aspects when considering implementation of GIS projects. These relate to existing (or legacy) information systems that are to be somehow incorporated into or integrated with geographic information, the flexibility and increasing power of commercial RDBMS in handling large amounts of distributed data, and the use of industry-standard products and techniques in accessing and manipulating such information. The use of *structured query language* (SQL) and products such as Oracle and Informix has become widespread in large and small organizations, to the extent that large commercial companies may choose to standardize on the use of a specific RDBMS product. Commercial database products are constantly being extended to address such areas as multimedia (additional and custom data types), the object-component paradigm, as well as specific requirements for spatial data. In this way the growth of OODBMS is being countered by the major RDBMS vendors, many of whom are in the best position to package and offer the best of familiar (relational) technology with additional features addressing these more specialized requirements.

The geographic information system places a considerable burden on data management, from the aspects of both modeling real geographic features and the various types of spatially distributed data. A GIS is often seen as providing the common interface or natural reference system to which attribute, vector, raster, video, sound, and other multimedia data are to be attached. Such an information system was beyond the scope of purely relational database management technology a few years ago, but now several RDBMS provide the facilities to develop such support in an extended relational model.

Security and Information Sharing in a GIS

Early GIS projects were restricted to single users or single departments where data were gathered and processed by at most a small group of individuals. In such implementations little attention was paid to the role of making information secure or of handling multiple user transactions.

However, in many organizations today, the GIS project implementation carries a significant weight as a means of integrating or linking multiple agencies and departments based on (spatial) elements of common interest or value. The general problems of data duplication, inconsistencies, and inaccuracy associated with uncontrolled access to all information within the database system can no longer be ignored. In fact, the introduction of multiuser security to spatially related data elements allows the full benefit of a GIS.

First of all, the layers or classes of information stored in a GIS are typically of interest to more than one user or group of users. However, one user or department is responsible for the creation and maintenance of a single class of data. The goal must be to eliminate duplication both of spatial geometry and of attribute information.

As an example, in a large municipal environment, the private land ownership unit—the "parcel"—may have many fields associated with it. The same parcel is referenced by many departments, but each department views the fields differently. Most importantly, only one department—for example, the department of public works—is empowered with the creation of new property boundaries. This department must work closely with the appraisal district to allow the assignment of a unique property identifier and street address, to allow further information to be applied correctly. Figure 54.11 shows additional examples.

Data inconsistencies can be minimized by the use of controlled procedures and standards for all information handling. However, a database system must also provide controls by allowing tables and basic elements to be accessed in a read-only mode, for read-write, or to be completely restricted.

CLASS / LAYER	**Users**		
attributes	**Public Works / Engineering**	**Assessor**	**Planning**
PARCEL			
boundary	write	read	read
identifier	read	write	read
street address	write	read	read
land value	-	write	read
improvement value	-	write	read
total value	-	write	read
date of last sale	-	write	read
owner name	-	write	read
owner address	-	write	read
zoning	read	read	write
tax district	read	read	write
STREET			
centerline	write	read	read
right-of-way	write	read	read
name	write	read	read
TAX DISTRICT			
boundary	-	write	read
name	-	write	read
taxes	-	write	read

FIGURE 54.11 Security in a multiuser GIS. This diagram shows read/write access privileges to classes of information for three different groups of users (departments) in a multidepartmental GIS.

When dealing with geographic data, the example above shows that although one user may require write access to geometric components of a spatial object, other users require write access to specific attribute fields for which they are responsible. For example, the assessor is responsible for assigning taxable values to individual properties, whereas the planner may be responsible for the zoning for specific properties. A good GIS provides a unified database management system that permits such combinations for geometric and attribute components. In addition, many GISs that rely on the relational model support and make use of security characteristics of the underlying relational database management system. In this situation individual tables and table records may be locked for read-only access or restricted completely.

In many GISs access to spatial data is managed in conjunction with a checkout procedure, whereby a group of information corresponding to the area of interest (a map sheet for example) is made available to one user only for updating and restricted to read-only access for all others. A transaction-oriented information system often takes this to a more elemental level, in which any database transaction locks out a small set of data immediately prior to the transaction. An AM/FM system often aims for this level of security, in which changes to facilities status and geometric elements are critical. This issue of concurrency—where in theory no two users may access the same data elements for update at the same time—can be achieved only with a sophisticated management of graphic and attribute elements.

54.4 Building and Maintaining a GIS

This section discusses the primary issues in building a geographic information system.

Reference Coordinate Systems

A geographic information system requires the definition and application of a reference coordinate system. The choice of a reference system is typically based on accuracy requirements, geographic scope of the proposed GIS, and the predominantly used reference system for source data, as shown by examples in Fig. 54.12.

For example, engineering and property maps within a small municipality are typically referenced to a local grid based on state plane coordinates. An information system that is to be applied for work at national or international levels may require the use of a true geographic reference system (latitude, longitude) to provide true seamlessness and consistency across the entire area of interest. It should be noted that the reference coordinates for the GIS database storage often differ from a local working coordinate system. For example, data entry may derive from large-scale maps referenced to state plane coordinates. In this situation the digitizing and data validation operations take place in state plane, but, once complete, all information is transformed and stored in

Type	Extent	Source Data
Latitute/Longitude (geographic coordinates)	global	field geodetic / world maps
Universal Transverse Mercator (UTM)	worldwide zones, each 6 degrees of latitude wide	national / transcontinental
US State Plane Coordinate System	State / portion of State	state / county maps and documents
Project-specific	Projectwide (engineering / construction)	airport / building site

FIGURE 54.12 Reference systems for source data.

the project's primary coordinate system. Information may later be extracted and presented in a variety of local coordinate systems, without corrupting the primary data storage.

Consideration of Scale

In a traditional mapping system, consideration of precision of data is closely tied to final map scale. In theory, a GIS requires a much broader view of both data accuracy and precision, because it is open to data at a large range of scales and accuracies. Unfortunately, most GISs do not in practice allow the maintenance of data quality and accuracy information. Furthermore, many applications of GISs involve the merging and integration of spatial data of differing resolutions and accuracy. The results of such operations are often used in decision support without regard to their statistical reliability. For example, the results of overlaying parcel boundary information derived from 1″ to 50′ scale property maps and soil classification boundaries derived from 1″ to 1000′ orthophotos may be used to determine the taxable value of farm properties.

It is therefore critical that the overall design of a GIS takes into account the accuracy requirements as a function of the application intent.

Data Sources

A variety of data sources are available for input to a GIS, as depicted in Figs. 54.13 and 54.14. They can be classified most simply as follows:

- Local maps and related documents
- Existing local information systems
- Commercially available information
- Government sources

GIS implementation is often the driving force behind conversion of existing maps and other documents and records.

Source	Source Scale	Accuracy
Field Survey, total station		.02 m
Field Survey, GPS Carrier Phase Differential		.05 m
Photogrammetry, large scale	1:4,000	0.3 m
Digitized Engineering Map	1:600	0.5 m
Photogrammetry, medium scale	1:16,000	1.5 m
Photogrammetry, small-scale	1:40,000	3 m
Field Survey, GPS Pseudo-range Differential		5 m
Digitized USGS Quad sheet	1:24,000	15 m
Bureau of the Census TIGER files	1:24,000	20 m
Remote Sensing, SPOT panchromatic, rigorous model		20 m
Field Survey, GPS pseudo-range single receiver		50 m
Remote Sensing, Landsat TM panchromatic, rigorous model		60 m
USGS Digital Line Graph	1:2million	100 m

FIGURE 54.13 Typical data sources. This table indicates typical source materials and methods of spatial data collection, with resultant accuracies.

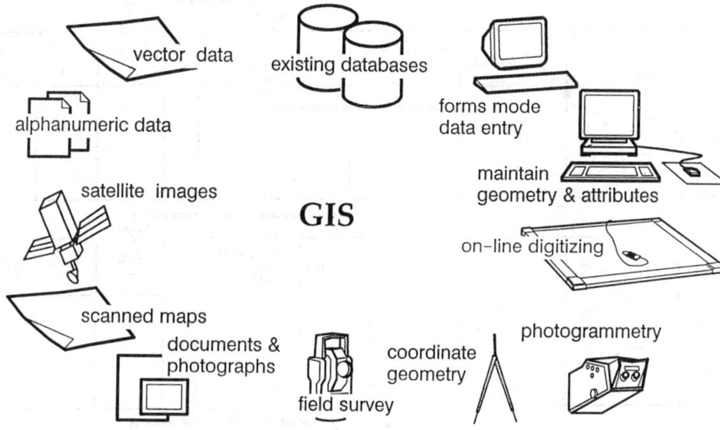

FIGURE 54.14 Data sources.

Data Entry and Processing

The data entry process may be reviewed in terms of:

- Interactive or semiautomated conversion of existing map documents
- Interactive update of attribute information through use of forms
- Batch loading of digital files

Interactive digitizing of existing maps has been superseded in part by a semiautomated process in which documents are scanned, displayed as a registered backdrop to vector data, and converted as necessary. Software that provides line-following and recognition capabilities, augmented by text recognition and a rules-based means to associate printed text labels as descriptors for adjacent graphic elements, may be applied successfully to conversion projects involving a large number of consistent map documents.

In addition, a key component to GIS database implementation is the linking of existing information systems. This is achieved as part of the overall database schema definition by identifying the key that provides the link between the core GIS and the existing data records. For example, the digitizing of parcel maps may include the interactive input of parcel identifier. Once a spatial feature containing the parcel number has been created, existing databases also containing the parcel number can be incorporated and accessed through the GIS. Alternatively, the parcel number may be used as a key to load additional attributes through a batch import process or as a guide to interactive form entry.

Structure/Topology

The data entry process normally includes facilities to check overall consistency and validity of data. The level to which these facilities operate is determined by the source data and the means of entry.

For digitizing operations and import of digital files representing vector geometry, it is critical that the input geometry defines a record that is consistent with both nongraphic descriptors and other graphic elements. For processing of vector line data, operations that determine intersections with adjacent geometry, extend lines to provide polygon closure, or eliminate overshoots are typically applied. Since these operations alter geometric components, they must be used with care and often use a correction tolerance determined by the ultimate accuracy requirements. This operation is often referred to as "building and cleaning," or "feature assembly." In addition, the process of "conflation" allows the merging of line data from disparate sources to minimize data storage and

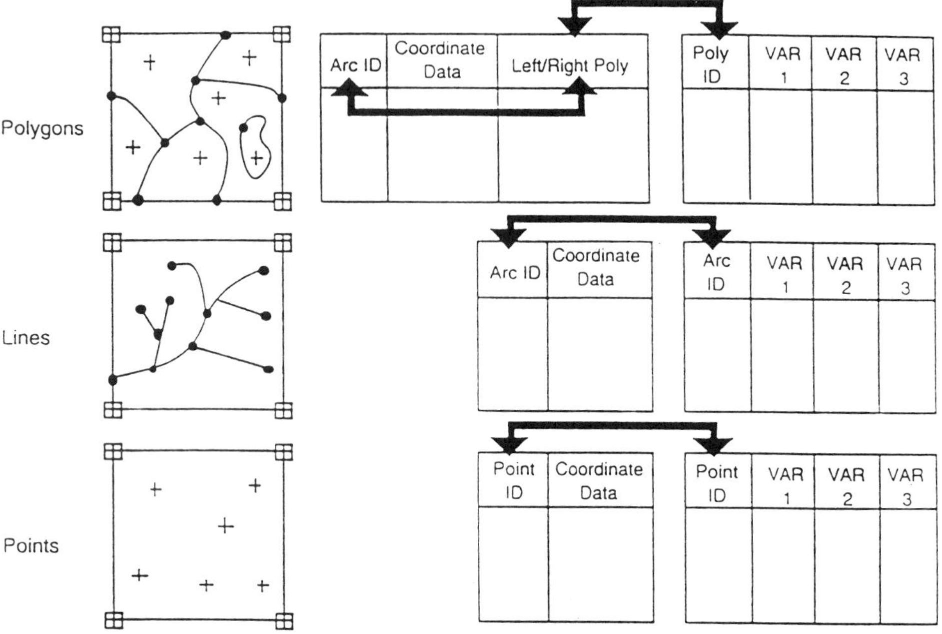

FIGURE 54.15 Database and geometry links. (*Source:* ESRI. 1992. *ARC/INFO: GIS Today and Tomorrow.* ESRI, Redlands, CA.)

permit more consistent assignment of descriptive attributes. For example, street centerlines derived from small-scale TIGER data may be conflated to highway alignment data derived directly from the local engineering department. This process actually "moves" inaccurate TIGER geometry to the more accurate engineering data, but then allows the use of the other TIGER data, such as street names and address ranges. Figure 54.15 identifies database and geometry linkages used to build a GIS.

Maintenance Operations

Once the primary spatial database has been created, it is important that maintenance operations may be applied without corrupting either geometric or attribute components. This is the most critical aspect to GIS editing as compared to a graphics-driven CAD system, in which the graphical elements and representation are most critical. For this reason, maintenance operations are most often performed on a primitive spatial object with both graphic and attribute components accessible. Nongraphic updates and edits may take place at a graphics workstation or at a simple alphanumeric screen, where operators update form entries.

54.5 Spatial Analysis

Beyond the implementation of GIS purely for mapping and map production, its key value lies in the ability to perform analysis based on spatial location and relationships between spatial elements. Such spatial analysis is augmented by the incorporation of external database records and application models.

Database Operations

We will now consider three groups of database operations associated with spatial information systems.

The first is simple querying and identification of individual geographic features and all related information. These may be nongraphic queries that return information based on nongraphic characteristics or graphical queries that incorporate a spatial extent to limit the extent of a querying process. Address matching, in which a nongraphic query provides street locations and/or property identification, is a fundamental operation in linking address as a reliable key between geographic and nongraphic information. Conversely, a graphical selection (for example, by pointing to a display) can be used to return a set of nongraphic information or a report. Querying and extraction operations are described later in this chapter.

The second group of operations are based on pure spatial analysis, based on geometric properties of individual elements. These apply to point, line, and polygon data as seen in Fig. 54.16. These include the generation of polygonal buffers around points, lines, or polygons and the overlay of multiple layers of polygonal information to provide a "spatial join" to be used in further processing and analysis. Properties such as proximity, distance, overlapping, and containment are included in this group.

The third group of operations rely on point, line, and polygon topology to provide the basis for analysis. Such properties as adjacency, connectivity, and composed of are included in this group. Linear network analysis provides the basis for shortest path calculations in transportation, distribution, and collection applications. Figure 54.16 identifies some commonly used spatial database operations.

Of course, the typical GIS application may use many database operations, spatial and nonspatial, to provide the desired answer. As an example, the simple question, "How many people live within 1/2 mile of the proposed light rail line?", is typically answered by proceeding through the following steps, each one a distinct database operation:

- Identify the light rail line (spatial).
- Create a 1/2-mile corridor/buffer based on the location of the rail line.
- Overlay the rail corridor against census block information.
- Transfer demographic information from census blocks to results of overlay.
- Produce summary statistics for population from the overlay.

Coupling to External Analyses/Applications

As previously discussed, the geographic information provides a powerful, common context for many different applications. In many cases the geographic key provides a much more significant means of interpretation or analysis than a simple numeric or textual key (for example, location versus street address). The structure of a GIS lends itself to linking with additional software and procedures that can be considered "external" applications. Examples are:

- Groundwater pollution models
- Statistical analysis
- Traffic engineering analyses

Data Interchange Standards and Formats

In practice, the exchange of data between geographic information systems has been inefficient and incomplete, except in situations where both systems share a common implementation. Two departments within a single organization, for example, may exchange information in a format recognized only by those departments. Such a practice results in the widespread exchange of information in commercially proprietary formats, such as those of Arc/Info coverage files or Intergraph design files. When geographic data is to be transferred to other agencies or to be imported from other sources, it is common to resort to one of several widely used standards.

Topologic Type	Node / point	Line	Polygon / surface
Node / point	buffer / proximity 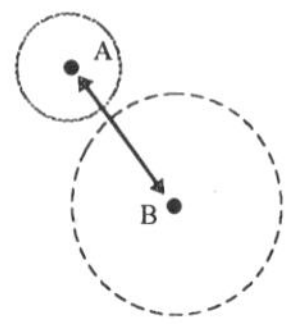	intersection  corridor / proximity 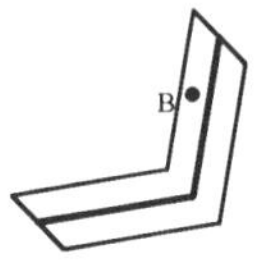	containment 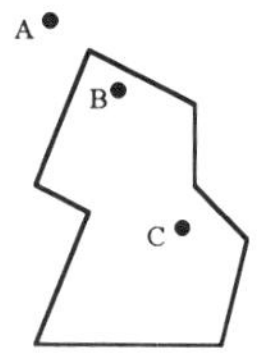
Line		intersection (create new nodes)  connectivity 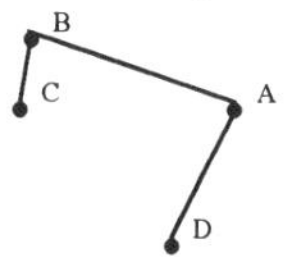	intersection 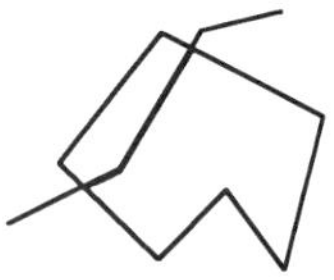
Polygon / surface			intersection 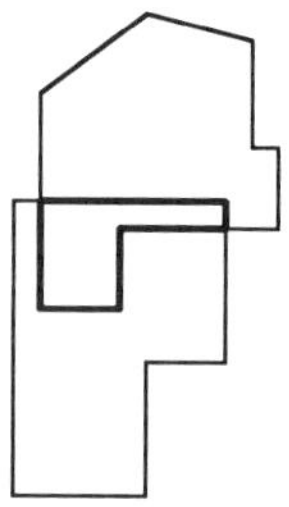

FIGURE 54.16 Spatial database operations. Typical operations involving two input geometries (row and column).

Digital Exchange Format (DXF), introduced by Autodesk in conjunction with its AutoCAD computer-aided drafting software product, is probably the most widely used means to exchange geographic data between CAD systems, desktop mapping systems, and even more sophisticated GISs. It is most effective as a means of encoding graphic elements but has severe limitations in the areas of handling topology, attributes, and more complex data structures. This means that the exchange of any DXF file must be accompanied by additional information describing the layering convention, symbology rules, and coordinate references. Such metadata is not normally incorporated in the DXF. Because spatial information must be resolved into simple graphic components and separate but related attribute information, both the source and target systems must provide additional utilities to ensure that all data is interpreted correctly.

Other formats, such as DLG (Digital Line Graph) and TIGER (Topologically Integrated Geographic Encoding and Referencing System), have been devised and introduced by federal government agencies as a means to deliver digital geographic data to the public, other agencies, and commercial users. They go beyond standards such as DXF in that they are designed to handle line and polygon topology, common data layer classifications, and real-world coordinate systems (as opposed to a graphics-based standard that often refers to "drawing units").

More recently, the Spatial Data Transfer Standard (SDTS) has been designed as a means to model, encode, and transfer geographic information. The SDTS goes far beyond previous standards, in that it provides a comprehensive means (1) to define spatial phenomena and spatial objects used to represent phenomena, and (2) to model spatial features in a hierarchical fashion. The exchanged data includes not only the raw geographic components, but also the definition of rules that associate the geometric, attribute and topologic elements, and any other metadata that further define the information. First practical implementations of SDTS, involving only a subset of the standard, have been developed for topologically structured vector data. This is referred to as the *vector profile.* Subsequent implementations or profiles will support raster- or grid-based data, such as imagery, digital orthophotos, and terrain models. The SDTS was approved as a Federal Information Processing Standard, Publication 173 (FIPS 173) in July 1992. Federal agencies such as the U.S. Geological Survey and the Bureau of the Census are now required to supply spatial information in an SDTS profile.

The use of structured query language (SQL) and other facilities associated with the relational database management system also provides the means to exchange spatially related data.

54.6 Information Extraction

The GIS has evolved over the past decade from a mainframe-based software system with distributed graphic terminals, through workstation clusters focusing high-performance graphics processors in a departmental fashion, to a more distributed organizational component. The corporate or enterprise-wide information system now typically includes a component dedicated to geographic data handling. Access to and extraction of information from a GIS takes place in various ways, depending on the nature and extent of the data required.

Displays and Reporting

As mentioned in the introduction to this chapter, the most standard product derived from a GIS remains a map. This map may be part of a standard map series, replacing the map product previously derived using manual drafting techniques, but increasingly common is a variety of custom maps and graphical displays that highlights the flexibility of a GIS in presentation of data.

For the typical GIS user the graphical display on a workstation or PC is the typical medium for interaction with the data. Layers or classes of information may be selected from a menu or list, for display at a variety of scales, along with attribute information as textual annotation or as a menu form. Specific layers and attribute information may be switched on or off based on scale of display or other criteria. Multiple windows may be used as a means to guide the user through large amounts of spatial information: For example, an overview window showing index maps or major features may be used as a constant reference to a larger-scale display in the current window.

In addition to information displays in standard cartographic forms (point symbols, line styles, and area fills and patterns), a thematic display of one class of information may be generated. Such displays rely on nongraphic characteristics such as "property value" or "soil type" or "road classification" to determine the cartographic representation. Thematic maps allow the efficient visualization of information. Also possible are more advanced representations, sometimes referred to as *cartograms,* such as the ability to size symbols based on attribute characteristics (for example, "tree height" or "drain size") or the ability to generate pie charts based on attributes (for example, "registered voters: Democrats" and "registered voters: Republicans").

Increasingly, the output from a GIS is not just a graphical product but a simple report or statistic. Spatial analysis can provide a set of information regarding a subset of spatial information. Thus, a letter to be sent to all property owners within a 1/2-mile buffer of a proposed industrial development may be the final product, based on spatial analysis and reporting.

Spatial Query Languages

In the geo-relational model a critical component in the GIS is the link between nongraphic and graphic elements. In such a model there are normally extensive sets of tools to allow access to, processing of, and retrieval of attribute information stored in tables. The *structured query language* (SQL) standard provides a common syntax for identifying and extracting specific table entries for further processing. Relational database management systems, such as Oracle and Ingres, are used broadly in conjunction with GIS components.

The critical nature of this link lies in the fact that once a database key is corrupted, the real value of the geographic information is lost and is difficult or impossible to recover. The second difficulty in the geo-relational model lies in the fact that the spatial component requires a distinct management system.

Commercial GIS products have attempted to introduce a geographic query language or spatial query language that incorporates many SQL components but is augmented by specific spatial operators. More recently, the ANSI SQL Committee has proposed spatial extensions to a query language. Commercial RDBMS vendors are also active in this area.

Figure 54.17 identifies some standard SQL operators, along with spatial operators as they might be used to form the basis for a spatial query language standard.

A where_clause may have Boolean expressions of the form:

```
EXPR [       is        ] CONDITION
             is not
null         EXPR
```

A CONDITION is:

```
[OPERATOR] [ EXPR]
in ( EXPR{, EXPR} )
```

Examples of OPERATOR are:

```
Comparision Operators:
             =, !=, > , >=, <, <=, range (between)
Pattern-matching Operators:
             like, match, not match
Spatial Operators:
             overlap, contain, near, left_of, adjacent to,
```

Examples of queries in a GIS:

```
select house where house.address = "5623 South Avalon" and house.zip_code = "43211"

select street where street.address_number range (5600,5700)

select house where house near (select river where river.name = "Colorado")

select house where house overlap (select soil_area where soil_area.type = "clay")
```

FIGURE 54.17 Query language. The where_clause forms part of a query language that is used to retrieve database records satisfying certain conditions specified. The where_clause consists of Boolean expressions that can be nested or used in combination, and that evaluate to TRUE or FALSE.

54.7 Applications

In this section we examine further some typical roles of geographic information systems.

Basemap and Infrastructure in Government

In an introductory example we discussed the use of GIS for handling land records and property ownership. In fact, this is but one role for geographic information management in local government. Typically, a set of basemap information—including property boundaries, highways and rights of way, public buildings, and other major transportation and drainage features—form the basis for a multidepartmental information system. During implementation and extension through multiple departments, additional attributes are added to these layers, and additional layers are created. In a sense the GIS allows the modeling in a rigorous, structured fashion of the annotated basemap that previously would have been found in individual agencies of municipal government.

In many situations a GIS is the sole means for providing required services, including meeting local, state, and federal legislative requirements. In Florida, for example, each community is required to submit a master plan that includes land-use zoning, transportation, and residential and commercial zoning plans. In addition, environmental impact regulations require the consideration of multiple levels of information (transportation, residences, zoning, water and wastewater, soils, vegetation, habitats) in producing reports and recommendations: A multilevel GIS is often the only means to meet planning requirements in a timely fashion.

Currently, federal legislation requires all communities to implement a management and control system for stormwater runoff. For large cities this requires major work not only in informing property owners and enforcing compliance with regulations, but also in compiling all of the background data required for analysis. This includes terrain-modeling data, as well as complete watershed and drainage network information. However, the timetable for compliance with these regulations requires a computer-based system of information for analysis and prediction.

In general, we can review the capabilities and advantages of a GIS as lying in the areas of:

- *Operation.* Being able to provide response and service requests effectively.
- *Planning.* Being able to anticipate and plan responses, for example, to emergencies.
- *Information-sharing.* Allowing multiple agencies to share cost of data collection and maintenance.
- *Decision support.* Providing qualified recommendations to management for review and action.

Facilities Management

For utility companies, such as those dealing with very extensive networks and facilities for the distribution of electricity and gas, a geographic information system is typically an integral part of a program aimed at improving operational efficiency, improvements in planning and engineering work, and higher service quality. These goals can be met by implementation of a GIS with powerful characteristics, which results in the ability to store and model spatial information at many levels: an upper management level that provides summaries updated on a regular basis; an intermediate level that provides all centralized management information to be used on a daily basis for business control and decision making; and the basic, "on-line" information level used for operational transactions.

A typical implementation is developed around a core facilities database that contains the following:

- Background cartographic, such as highway maps as political boundaries
- Operational facilities including all information on installations forming the distribution network and data on maintenance, breakdowns, and other incidents

- Network of facilities under construction, showing facilities not yet in service
- Planned network, containing planned facilities

The system provides the following capabilities based on this core database:

- Maintenance of all necessary network diagrams and work sheets
- Maintenance of all alphanumeric (descriptive) information regarding facilities and operations
- Alphanumeric queries on all networks and facilities
- Production of all necessary reports, drawing, plans, and maps

Development Tools (Means to Customize/Build New Applications)

There is always a compromise between implementing a GIS based on standard commercial products, while attempting to maintain the ability to customize a system to accommodate specific and changing requirements over time. There are several levels at which GIS products may offer such capabilities. We may consider these in terms of profiles of the various end-users:

- *The casual user,* with little programming experience
- *The GIS specialist,* with detailed knowledge of spatial data handling and operations
- *The GIS programmer,* with experience in spatial data handling and programming expertise

For the casual user of a GIS standard tools are typically available that allow repetitive or predictable operations to be stored as a macro or command file, which may then be invoked with a single command or menu operation. As commercial GIS products begin to reflect simplistic and intuitive graphical user interfaces—such as those consistent with Windows on a PC platform or X-Windows on a UNIX workstation—facilities such as the ability to "learn" a sequence of menu selections or command line arguments are those that most simply allow some limited customizing.

The GIS specialist who understands the full capabilities of a specific software product always uses this knowledge to extend its operational use through various combinations of basic functions and operations. For example, a well-known commercial GIS product now offers roughly 800 operations through command-line options and a much smaller number of graphical user interface components such as menus, submenus, and pull-down selection lists. The more advanced macro language permits the extension of simplistic sequences of operations to include flow control, user interaction, and menu components.

The GIS programmer uses knowledge of all programming components—including subroutine libraries supplied with the GIS, user interface libraries, and other programming tools—to develop a high-level application that can either be integrated with standard GIS operations or can completely replace the menus and options made available to the end-users of the information system. In this case the MIS department or an independent software developer may provide a completely customized interface for multiple departments within a large organization. For high levels of integration with other information systems, such programming is typically required. For example, GIS programming tools used in conjunction with high-level tools provided by Oracle relational database management system can provide the optimized functions and operational performance required.

54.8 Summary

This chapter has described the way in which map data is now often being incorporated directly into a computerized information system. A geographic information system typically attempts to link geometric, spatial information with related nongraphic or attribute data from one or many sources, thereby providing a direct spatial context for many databases. Since a large proportion of nongraphic data held by government agencies and certain types of private companies relate to geographic location and context, the introduction of GIS technology has been driven by such organizations in an attempt to make more efficient use of information and to allow more effective information management. GIS provides a more complete model of the real world than can simple (carto)graphic representations or tabular records or reports. In areas of local government and in private utility industries, the GIS approach is synonymous with information sharing, systems integration, and reengineering, in the name of providing more effective service in a competitive environment.

References

Antenucci, J., Brown, K., Croswell, P., Kevany, M., and Archer, H. 1991. *Geographic Information Systems: A Guide to the Technology.* Van Nostrand Reinhold, New York.

Aronoff, S. 1989. *Geographic Information Systems: A Management Perspective.* WDL Publications, Ottawa, ON.

Burrough, P. A. 1986. *Principles of Geographical Information Systems for Land Resources Assessment.* Oxford University Press, New York.

Calkins, H. W. and Tomlinson, R. F. 1984. *Basic Readings in Geographic Information Systems.* SPAD Systems, Ltd., Williamsville, NY.

Chance, A., Newell, R. G. and Theriault, D. G. 1990. An object oriented GIS: Issues and solutions. *Proceedings EGIS,* Vol.1. Amsterdam, Netherlands.

Cowen, D. J. 1988. GIS versus CAD versus DBMS: What are the differences? *Photogrammetric Engineering and Remote Sensing,* Vol. 54.

Date, G. J. 1987. *An Introduction to Database Systems.* Addison-Wesley, Reading, MA.

Digital Cartographic Data Standards Task Force. 1988. The proposed standard for digital cartographic data. *The American Cartographer,* Vol. 15.

Dueker, K. J. 1987. Geographic information systems and computer aided mapping. *Journal American Planning Association.* Summer 1987.

ESRI. 1990. *Understanding GIS: The ARC/Info Way.* ESRI, Redlands, CA.

ESRI. 1992. *ARC/INFO: GIS Today and Tomorrow.* ESRI, Redlands, CA.

Fletcher, D. 1987. Modeling GIS transportation networks. *Proceedings URISA 1988.* Los Angeles, CA.

GIS World. 1993. *GIS International Sourcebook, 1993.* GIS World, Fort Collins, CO.

Goodchild, M. F. 1988. A spatial analytical perspective on GIS. *International Journal of Geographical Information Systems,* Vol. 1.

Guptill, S. C. 1988. A process for evaluating GIS. *Proceedings GIS/LIS 1988.* San Antonio, TX.

Guttman, A. 1984. R-trees: A dynamic index structure for spatial searching. *ACM SIGMOD.*

Kilborn, K., Rifai, H. S., and Bedient, P. B. 1991. The integration of ground water models with geographic information systems. *Proceedings ACSM-ASPRS.*

Maguire, D., Goodchild, M. F., and Rhind, D. 1991. *Geographical Information Systems: Principles and Applications.* John Wiley & Sons, New York.

Montgomery, G. and Schuch, H. 1993. *Data Conversion in GIS.* GIS World, Fort Collins, CO.

Parker, H. D. 1988. The unique qualities of a geographic information system: A commentary. *Photogrammetric Engineering and Remote Sensing,* Vol. 54.

Tom, H. 1990. Geographic information systems standards: A federal perspective. *GIS World,* Vol. 3.

Tomlin, C. D. 1990. *Geographic Information Systems and Cartographic Modeling.* Prentice Hall, Englewood Cliffs, NJ.

The Dwight D. Eisenhower System of Interstate and Defense Highways is currently 99.8 percent finished at well over 42,000 miles. The project began in 1956 as a cooperative venture of the Federal Highway Administration and state highway agencies. The $130 billion interstate system has revolutionized surface transportation and has profoundly improved many aspects of American life, as well as helping to make the U.S. what President Clinton has called "the most mobile country in the world."

By standardizing highway design with features such as multiple travel lanes, full control of access, design for high speeds, and "forgiving roadsides," interstate planners and engineers made possible the fast, safe, and economical transport of people and goods to all parts of the nation. Today, the interstate system constitutes only 1 percent of U.S. total road mileage but carries fully 22 percent of all vehicular traffic. This is estimated to be some 510 billion vehicle miles per year. Above, Interstate 279 joins Interstate 579 in north Pittsburgh, Pennsylvania.

Often called the "greatest public works project in history," the interstate system has had far-reaching benefits over the past 40 years. The interstate system has boosted productivity and helped sustain a more-than-tenfold increase in the gross national product, stimulated the economy in many previously undeveloped areas, caused a steep decline in the highway fatality rate, enhanced access to jobs and housing throughout the country, linked a diverse nation both culturally and socially, spurred the development of auto tourism into a major industry and leisure activity along with the rise of businesses serving interstate travelers, and quickened national defense responsiveness. (Photo courtesy of the American Society of Civil Engineers.)

Transportation Engineering

Kumares C. Sinha
Purdue University

TRANSPORTATION ENGINEERING IS A MAJOR PART of the civil engineering profession. The importance of transportation engineering within the civil engineering discipline can be judged by the number of technical divisions in the American Society of Civil Engineers (ASCE) that are directly related to transportation. One-third of the 18 technical divisions within ASCE is devoted to transportation. In fact, a major component of ASCE's current thrust on research and technology application is highway-oriented.

Transportation plays a critical role in the U.S. economy; 20% of the U.S. gross national product is associated with transportation. About one of every seven U.S. workers is employed in some aspect of transportation. Most civil engineers engaged in the practice of transportation engineering are employed by state and local transportation agencies and associated contractors and consultants. A 1985 estimate by the Transportation Research Board indicated that there were 174,000 civil engineers in the U.S., and about 100,000 worked in transportation at least some of the time.

Transportation engineering, as practiced by civil engineers, primarily involves planning, design, construction, maintenance, and operation of transportation facilities. The facilities support air, highway, railroad, pipeline, water, and even space transportation. A specific indication of the subcomponents of the transportation engineering field with current importance to civil engineers can be obtained by examining the topics of the technical committees of the transportation-related divisions in ASCE. A review of descriptions of the scope of various committees indicates that while facility planning and design continue to be the core of the transportation engineering field, such areas as facility operations, management, and environmental considerations are of much current interest to civil engineers. In addition, the research and deployment of intelligent transportation systems, as well as the implementation of high-speed ground transportation systems, have gained wide attention in recent years.

In keeping with current needs and emerging interests, this section of the *Handbook* presents the basic principles and techniques for planning, design, operation, and management of highways, bridges, airports, and urban public transit systems. This section also presents materials on the rapidly developing areas of intelligent transportation systems and high-speed ground transportation systems.

Chapter 55 provides a detailed discussion on concepts and models used for both strategic (long-term) and tactical (short-term) planning processes. The primary thrust is to present a quantitative background on demand estimation for effective planning of surface transportation facilities.

The details of airport planning and design are given in Chapter 56. This chapter covers various aspects of airport planning, including air traffic control requirements, passenger terminal design, airport location, layout and design, and environmental considerations.

The chapter on high-speed ground transportation, Chapter 57, presents the planning requirements, design guidelines, and financing and policy issues. The lessons from Europe and Japan are also discussed. The details or urban transit systems are covered in Chapter 58, where procedures are discussed for operational planning, scheduling, and routing, patronage prediction and pricing, operations cost modeling, and system performance monitoring.

Aspects of structural design of pavements for highways and airports are dealt with in Chapter 59. The concept and methods of thickness design of both rigid and flexible pavements are presented. Highway geometric design fundamentals are given in Chapter 60, including design applications.

Principles of highway traffic operations are presented in Chapter 61, where the emphasis is on fundamental concepts and analytical techniques that can be applied to better understand traffic operating characteristics.

Potential applications of advanced technologies in the area of intelligent transportation systems (ITS) are examined in Chapter 62, where various components of ITS, along with the current status of operational tests and other field applications, are discussed. The concepts and principles of transportation infrastructure management are discussed in Chapter 63. Three specific systems are presented involving pavement, bridge, and highway maintenance management systems. Chapter 64 presents a discussion on environmental considerations in transportation planning and development. An overview to the environmental process is given, with emphasis on the physical impacts, particularly air quality and noise pollution.

The challenges and opportunities faced by the transportation engineering profession as it prepares to enter the 21st century are unique. These challenges cover a wide spectrum, including increasing traffic congestion at our highways and airports, continuing problems with transportation safety and environmental degradation of our communities, and ever more acute budget constraints. However, there are also opportunities offered by the timely application of technical innovations, such as intelligent transportation systems and high-speed ground transportation systems, to address these challenges. Major advances in these areas have the potential of opening new horizons in transportation engineering by developing new techniques and procedures while making substantial improvements in cost and productivity. This section of the *Handbook* provides a brief overview of the fundamentals of planning, design, operation, and management aspects of transportation engineering that will be useful not only to learn about the state of the art of transportation engineering in the U.S., but also to prepare for the future.

I gratefully acknowledge the cooperation of the authors, as well as the assistance of Dr. Ahmed E. Radwan of the University of Central Florida in getting the manuscripts reviewed. The help of the following reviewers is greatly appreciated: Haitham Al-Deek, University of Central Florida; Robert J. Behar, Carr Smith Associates; Jon D. Fricker, Purdue University; Shiou S. Kuo, University of Central Florida; Scot D. Leftwich, University of Central Florida; Joel P. Leisch, CH2M Hill-Leisch; Charles C. Liu, AEPCO, Inc.; Gary D. Logston, TAMS Consultants, Inc.; Samer Madanat, Purdue University; Patrick P. Martin, Indiana Dept. of Transportation; Sue McNeil, Carnegie Mellon University; Srinivas Peeta, Purdue University; Mitsuru Saito, City College of New York; Rocquin L. VanGuilder, TAMS Consultants, Inc.; and Robert Dorer, U.S. Department of Transportation.

55

Transportation Planning

David Bernstein
Massachusetts Institute of Technology

55.1 Introduction

Transportation plays an enormous role in our everyday lives. Each of us travels somewhere almost every day, whether it be to get to work or school, to go shopping, or for entertainment purposes. In addition, almost everything we consume or use has been transported at some point.

For a variety of reasons that are beyond the scope of the *Handbook*, many of the transportation services that affect our lives are provided by the public sector (rather than the private sector) and, hence, come under the aegis of civil engineering. This portion of *The Civil Engineering Handbook* deals with the role that *transportation planners* play in the provision of those services.

What Is Transportation Planning?

It is somewhat difficult to define **transportation planning** since the people who call themselves transportation planners are often involved in very different activities. For the purposes of the *Handbook* the easiest way to define transportation planning is by comparing it to other public sector activities related to the provision of transportation services. In general, these activities can be characterized as follows:

Management/Administration: Activities related to the transportation organization itself.
Operations/Control: Activities related to the provision of transportation services when the system is in a stable (or relatively stable) state.
Planning/Design: Activities related to changing the way transportation services are provided (i.e., state transitions).

Transportation planning activities are often characterized as being either strategic (i.e., with a fairly long time horizon) or tactical (i.e., with a fairly short time horizon).

Unfortunately, these definitions, in and of themselves, are not really enough to characterize transportation planning activities. To do so requires some concepts from systems theory.

0-8493-8953-4/95/$0.00 + $.50

A **system**, as defined by Hall and Fagen [1956], is a set of objects (the parameters of the system), their attributes, and the relationships between them. Any system can be described at varying levels of **resolution**. The resolution level of the system is, loosely speaking, defined by its elements and its environment. The environment is the set of all other systems, and the elements are treated as "black boxes"(i.e., the details of the elements are ignored; they are described in terms of their inputs and their outputs).

Thus, it is possible to talk about a variety of different transportation systems including (in increasing order of complexity):

1. Car
2. Driver + Car
3. Road + Driver + Car
4. Activities generating flows + Road + Driver + Car
5. Surveillance and control devices + Activities generating flows + Road + Driver + Car

Transportation planning is concerned with the fourth system listed above, treating the Road + Driver + Car subsystem as a black box. For example, transportation planners are interested in how activities generating flows interact with this black box to create congestion, and how congestion influences these activities. In contrast, automotive engineering is concerned with the vehicle as a system, human factors engineering is concerned with the Driver + Vehicle system, geometric design (Chapter 60) and infrastructure management (Chapter 63) are concerned with the Road or the Road + Car systems, and highway traffic operations (Chapter 61) and Intelligent Vehicle Highway Systems (Chapter 62) are concerned with the surveillance and control systems and how they interact with the Activities + Road + Driver + Car subsystem.

The Transportation Planning Process

The transportation planning process almost always involves the following six steps (in some form or another):

1. Identification of goals/objectives (anticipatory planning) or problems (reactive planning)
2. Generation of alternative methods of accomplishing these objectives or solving these problems
3. Determination of the impacts of the different alternatives
4. Evaluation of different alternatives
5. Selection of one alternative
6. Implementation

Some people have argued that this process is/should be completely "rational" or "scientific" and hence that the above steps are/should be performed in order (perhaps with a loop between evaluation of alternatives and generation of alternatives).[1] However, many others argue that the transportation planning process is not nearly this scientific. For example, Grigsby and Bernstein [1993] argue that there are a variety of factors that shape the transportation planning process:

Societal Setting: The laws, regulations, customs, and practices that distribute decision-making powers and that set limits on the process and on the range of alternatives.

Organizational Setting: The organization and administrative rules and practices that distribute decision-making powers and that set limits on the process and on the range of alternatives.

Planning Situation: The number of decision makers, the congruity and clarity of values, attitudes and preferences, the degree of trust among decision makers, the ability to forecast, time and

[1]This is sometimes called the *3C process:* continuing, comprehensive, and coordinated.

other resources available, quality of communications, size and distribution of rewards, and the permanency of relationships.

For these and other reasons, a variety of other "less-than-rational" descriptions of the planning processes have been presented. For example, Lindblom [1959] described what he called the "science of muddling through," in which planners build out from the current situation by small degrees rather than starting from the fundamentals each time. Etzioni [1967] described a mixed scanning approach which combines a detailed examination of some aspects of the "problem" with a truncated examination of others.

Fortunately, the exact process used has little impact on the day-to-day tasks that transportation planners are involved in. Transportation planners typically evaluate alternative proposals and sometimes generate alternative proposals. Hence, the transportation planner's job is primarily to determine the demand for the proposed alternatives (i.e., how the proposed alternatives affect the activities which generate flows).

Given that transportation planners are principally concerned with determining the demand changes that result from proposed projects, it would be natural to assume that they use the tools of the microeconomist (i.e., models of consumer and producer behavior). While this is true in some sense, the generic demand models used in microeconomics are usually not powerful enough to support the transportation planner. That is, for most applications it is not possible to reliably estimate the demand for a project/facility as a function of the attributes of that project/facility. This is because of the complex interactions that exist between different people and different facilities. Instead, transportation planners use a variety of different models depending on the specific decision they are trying to predict.

55.2 Transportation Planning Models

In order to determine the demand for a transportation project/facility the transportation planner must answer the following questions:

- Who travels?
- Why do they travel?
- Where do they travel?
- When do they travel?
- How do they travel?

The who and why questions are actually fairly easy to answer. In general, transportation planners need to distinguish between commuters, shoppers, holiday travelers, and business travelers. To answer the where, when, and how (and the aggregate question "how much") transportation planners develop theories and models of the decision-making processes that different travelers go through.

To do so, the transportation planner considers the following:

- The decision to travel
- The choice of a destination (and/or an origin)
- The choice of a mode
- The choice of a path (or route)
- The choice of a departure time

Models of the first four of these decisions are traditionally referred to as **trip generation, trip distribution, modal split**, and **traffic assignment** models. These types of models have been widely studied and applied. Departure-time choice has, for the most part, been ignored or handled in an ad hoc fashion.

It is important to observe that not all of these models need to be applied in all situations. In practice, the models used should depend on the time frame of the forecast being generated. For example, in the very short run, people are not likely to change where they live or where they work, but they may change their mode and/or path. Hence, when trying to predict the short-run reactions of commuters to a project, it does not make sense to run a trip generation or trip distribution model. However, it is important to run both the modal split and the traffic assignment models.

It is also important to note that it is often necessary to combine different models, and this can be done in one of two ways. Continuing the example above, if the choice of mode and path are tightly intertwined, then it may make sense to solve/run the two models simultaneously. If, on the other hand, people first choose a mode (based on some estimate of the costs on the two modes) and then choose the path on that mode, then it may make sense to solve/run the models sequentially. This will be discussed more fully below. For the time being, the models will be presented as if they are used sequentially.

The subsections that follow contain some of the more common models of each type. It is important to recognize at the outset that some of these models are very **disaggregate** while others are quite **aggregate** in nature. Disaggregate models consider the behavior of individuals (or sometimes households). They essentially consider the **choices** that individuals make among different **alternatives** in a given situation. Aggregate models, on the other hand, consider the decisions of a group in total. The groups themselves can be based either on geography (resulting in zonal models) or socioeconomic characteristics.[2]

Though each of the decisions that travelers make are modeled differently in the subsections that follow, it is important to realize that many of the techniques described in one subsection may be appropriate in others. In general, they are all models of how people make choices. Hence, they are applicable in a wide variety of different contexts (both inside and outside of transportation planning).

The Decision to Travel

In general, trip generation models relate the number of trips being taken to the characteristics of a "group" of travelers. The models themselves are usually statistical in nature. Zone-based models use aggregate data while household-based models use disaggregate data. These models typically fall into two groups: linear regression models and category analysis models. The output of a trip generation model is either **trip productions** (the number of trips originating from each location), **trip attractions** (the number of trips destined for each location), or both.

These models have, in general, received very little attention in recent years. That is, the techniques have not changed much in the past twenty years; only new parameters have been estimated. This is, in large part, because transportation planners have traditionally been concerned with congestion during the peak period, and it is relatively easy to model the decision to travel for work trips (i.e., everyone with a job takes a trip). However, this is beginning to change for several reasons:

- Congestion is increasingly occurring outside of the traditional morning and evening peaks. Hence, more attention needs to be given to nonwork trips.
- New technologies are changing the way in which people consider the decision to travel. The advent of **telecommuting** means that people may not commute to work every day. Similarly, **teleshopping** and **teleconferencing** can dramatically change the way people decide to take trips.

These trends have created a great deal of renewed interest in trip generation models.

[2]In some cases, disaggregate models are statistically estimated using aggregate data and knowledge of the distributional forms of the groups.

Linear Regression Models

In a linear regression model a statistical relationship is estimated between the number of trips and some characteristics of the zone or household. Typically, these models take the form

$$Y = \beta_0 + \beta_1 X_1 + \cdots + \beta_n X_n + \epsilon \tag{55.1}$$

where Y (called the dependent variable) is the number of trips, $X_1, X_2, \ldots, X_n$ (called the independent variables) are the n factors that are believed to affect the number of trips that are made, $\beta_1, \beta_2, \ldots, \beta_n$ are the coefficients to be estimated, and ϵ is an error term. Such models are often written in vector notation as $Y = \beta_0 + \beta X' + \epsilon$ where $\beta = (\beta_1, \ldots, \beta_n)$ and $X = (X_1, \ldots, X_n)$. Clearly, since $\beta_i = \partial Y / \partial X_i$, the coefficients represent the contribution of the independent variables to the magnitude of the dependent variable.

In a disaggregate model, Y is normally measured in trips (of different types) per household, whereas in an aggregate model it is measured in trips per zone. In general, the independent (or explanatory) variables should not be (linearly) related to each other, but should be highly correlated with the dependent variable. The selection of which dependent variables to include is part of the "art" of developing such models.

Such models are traditionally estimated using a technique known as *least squares estimation*. This technique determines the parameter estimates that minimize the sum of the squared differences between the observed and the expected values of the observations. It is described in almost every book on econometrics (see, for example, Theil [1971]).

In general, it is important to realize that linear regression models are much more versatile than one might immediately expect. In particular, observe that both the independent variables and the dependent variable can be transformed in nonlinear ways. For example, the model

$$\log(Y) = \beta_0 + \beta_1 \log(X_1) + \cdots + \beta_n \log(X_n) + \epsilon \tag{55.2}$$

can be estimated using ordinary least squares. In this case, the values of the coefficients can be interpreted as elasticities since

$$\frac{\partial \log(Y)}{\partial \log(X_i)} = \beta_i \Rightarrow \beta_i = \frac{\partial Y / Y}{\partial X_i / X_i} \tag{55.3}$$

Category Analysis Models

In category analysis, a mean trip rate is determined for different types (i.e., categories) of people and trips. The categories are typically based on social, economic, and demographic characteristics. The resulting models are nonparametric and have the following form (see, for example, Doubleday [1977]):

$$\Omega^p_{zc} = \frac{\sum_{r \in z} O^p_{rc}}{n_z} \tag{55.4}$$

where Ω^p_{zc} denotes the trip rate for people in category z for purpose p during time period c, O^p_{rc} denotes the number of trips by person r for purpose p during time period c, and n_z denotes the number of people in category z.

Models of this type are generally presented in tabular form as follows:

	Trip Type 1	Trip Type 2	...	Trip Type k
Category 1				
Category 2				
⋮				
Category m				

where the entries in the table would be the trip rates. These trip rates can then be used to predict future trip attractions and productions simply by predicting the number of people in each category and multiplying.

Origin and Destination Choice

Of course, each trip that a person takes must have an origin and a destination. For commuters, this origin/destination choice process is fairly long-term in nature. For morning trips to work, the origin is usually the person's place of residence and the destination is usually the place of work, and for evening trips from work it is exactly the opposite. Hence, for commuting trips the origin and destination choice processes are tantamount to the residential location and job choice processes. For shopping trips, the origin choice process is long-term in nature (i.e., the choice of a residence), but the destination choice process is very short-term in nature (i.e., where to go shopping for this particular trip). For holiday travel things are somewhat more confusing. However, in many cases we can treat holiday travel as if it involves short-term origin and short-term destination choices. For example, consider the holiday travel that occurs on Thanksgiving. You know that your family is going to get together, but where? Hence, the origin/destination choice process corresponds to determining where you will meet and, hence, who will be traveling from where and to where.

There are two widely used types of trip distribution models: gravity models and Fratar models. Gravity models are typically used to calculate a trip table from scratch, whereas Fratar models are used to adjust an existing trip table. Both types of models are aggregate in nature and use trip production and/or trip attractions to determine specific trip pairings (often called a **trip table**).

Gravity Models

The most popular models of origin/destination choice are collectively called *gravity models* (see, for example, Hua and Porell [1979], Erlander and Stewart [1989], and Sen and Smith [1994]). These models get their name because of their similarity to the Newtonian model of gravity. At the most basic level, these models assume that the movements of people tend to vary directly with the size of the attraction and inversely with the distance between the points of travel. So, for example, one could have a gravity model of the following kind:

$$T_{ij} = \alpha \frac{M_i M_j}{d_{ij}^2} \tag{55.5}$$

where T_{ij} denotes the number of trips between origin zone i and destination zone j, M_i denotes the population of zone i, M_j denotes the population of zone j, d_{ij} denotes the distance between i and j, and α is the so-called demographic gravitational constant.

Many models of this kind have been estimated and used over the years. However, they have also received a great deal of criticism. First, there is no particular reason to use d_{ij}^2 in the denominator; this seems to be carrying the Newtonian analogy farther than is justified. Second, there is no reason to use $M_i M_j$ in the numerator; it makes just as much sense to weight each of these terms (e.g., to use $w_i M_i^{\beta} u_j M_j^{\gamma}$). Finally, these models suffer from a small distance problem: as the distance between the origin and destination decreases, the number of trips increases without bound (i.e., as $d_{ij} \rightarrow 0$, $T_{ij} \rightarrow \infty$).

These criticisms led researchers to try many other forms of the gravity model. One of the more general specifications was given by Hua and Porell [1979]:

$$T_{ij} = A(i)B(j)F(d_{ij}) \tag{55.6}$$

where $A(i)$ and $B(j)$ are weighting functions and $F(d_{ij})$ is a distance deterrence function. Most of the variants of this model have differed in the form of the deterrence functions used. For example, the classical doubly constrained gravity model is given by

$$T_{ij} = A_i B_j O_i D_j f(c_{ij}) \tag{55.7}$$

where O_i is the number of trips originating at i, D_j is the number of trips destined for j, and A_i and B_j are defined as follows:

$$A_i = \left[\sum_j B_j D_j f(c_{ij})\right]^{-1} \tag{55.8}$$

$$B_j = \left[\sum_i A_i O_i f(c_{ij})\right]^{-1} \tag{55.9}$$

Though these variants have been motivated in a number of different ways, some formal and others more ad hoc (see, for example, Stouffer [1940], Niedercorn and Bechdolt [1969], and T. E. Smith [1975, 1976a, 1976b, 1988]), perhaps the most appealing to date are those based on the most probable state approach (see, for example, Wilson [1970] and Fisk [1985]).

In this approach, each individual is assumed to choose an origin and/or destination (the set of such choices are referred to as the **microstates** of the system). Any particular microstate will have associated with it a **macrostate,** which is simply the number of trips to and/or from each zone. A macrostate is feasible if it reproduces known properties referred to as **system states** (e.g., total cost of travel, total number of travelers). Letting $\mathcal{F}$ denote the set of feasible macrostates and $W(n)$ the number of microstates that are consistent with macrostate n, then the total number of possible microstates is given by

$$\Omega = \sum_{n \in \mathcal{F}} W(n) \tag{55.10}$$

Finally, if each microstate is equally likely to occur then the probability of a particular (feasible) macrostate is

$$P(n) = \frac{W(n)}{\Omega} \tag{55.11}$$

To develop specific gravity models using the most probable state approach one need simply derive an expression for $W(n)$ and then find the macrostate which maximizes (55.11). Fisk [1985] discusses several such models.

For shopping trips (from given origins), the following gravity model can be derived:

$$T_j = N \frac{D_j \exp(-\beta c_j)}{\sum_k D_k \exp(-\beta c_k)} \tag{55.12}$$

where T_j denotes the number of trips to destination j, N is the total number of travelers, D_j is the number of possible stores at destination j, and β is a parameter of the model. In general, β is expected to be negative.

For commuting trips, the following gravity model can be derived (assuming that the number of jobs is known and that one trip end is permitted per job):

$$T_j = \frac{D_j}{z^{-1} \exp(\beta c_j) + 1} \tag{55.13}$$

where D_j is the number of jobs at location j, and z^{-1} is found by substituting this expression into the equations defining the system states. For example, if the total number of travelers, N, and the total travel cost, C, are both known, z^{-1} would be obtained using

$$\sum_i n_i = N \tag{55.14}$$

$$\sum_i n_i c_i = C \tag{55.15}$$

These models are typically estimated using maximum likelihood techniques. These techniques attempt to find the value of the parameters that make the observed sample most likely. That is, a likelihood function is formed which represents the probability of the sample conditioned on the parameter estimates, and this likelihood function is then maximized using techniques from mathematical programming.

Fratar Models

A popular alternative to gravity models are Fratar models. While not as theoretically appealing, the Fratar model is sometimes used to adjust existing trip tables. The "symmetric" Fratar model, which is the only one presented here, requires that the number of trips from i to j equals the number of trips from j to i (i.e., $T_{ij} = T_{ji}$).

Letting T^0 denote the original trip table and O denote the future trip-end totals, this approach can be summarized as follows:

Step 0. Set the iteration counter to zero (i.e., $k = 0$).
Step 1. Calculate trip production totals. That is, set $P_i^k = \sum_j T_{ij}^k$.
Step 2. Set $k = k + 1$ and calculate the adjustment factors $f_i^k = O_i / P_i^{k-1}$ for all i. If $f_i^k \approx 1$ for all i then STOP.
Step 3. Set $N_{ij}^k = (T_{ij}^{k-1} f_j^k / \sum_n T_{in}^{k-1} f_n^k) O_i$.
Step 4. Set $T_{ij}^k = (T_{ij}^k + T_{ji}^k)/2$ and GOTO step 1.

Note that this algorithm does not always converge and that it cannot be used at all when the number of zones changes.

Mode Choice

Mode choice models are typically motivated in a disaggregate fashion. That is, the concern is with the choice process of individual travelers. As might be expected, there are many theories of individual choice that can be applied in this context.

One of the most successful theories of individual choice is the classical microeconomic theory of the consumer. This theory postulates that an individual chooses the consumption bundle that maximizes his or her utility given a particular budget. It assumes that the alternatives (i.e., the components of the consumption bundle) are continuously divisible. For example, it assumes that individuals can consume 0.317 units of good x, 5.961 units of good y, and 1.484 units of good z. As a result, it is not possible to directly apply this theory to the typical mode choice process in which travelers make discrete choices (e.g., whether to drive, take the bus, or walk).

Of course, one could modify the traditional theory of the consumer to incorporate discrete choices. In fact, such models have received a great deal of attention. The goal of these models is to impute the weights that an individual gives to different attributes of the alternatives based on the choices that are observed (again assuming that the individual chooses the alternative with the highest utility).

Unfortunately, however, these models do not always work well in practice. There are at least two reasons for this. First, individuals often select different alternatives when faced with (seemingly) identical choice situations. Second, individuals sometimes (seem to) make choices (or express preferences) that violate the **transitivity of preferences**. That is, they choose A over B, choose B over C, but choose C over A.

Two explanations have been given for these seeming inconsistencies. Some people, so-called random utility theorists, have argued that we (as observers) are unable to fully understand and measure all of the relevant factors that define the choice situation. Others, so-called constant utility theorists, have argued that decision makers actually behave based on choice probabilities. Both theories result in probabilistic models of choice rather than the deterministic models discussed thus far.

In the discussion that follows, a probabilistic model of choice will be motivated using random utility theory. However, it could just as easily have been motivated using constant utility theory. For the purposes of this *Handbook*, the end result would have been the same.

A General Probabilistic Model of Choice

Following the precepts of random utility theory, assume that individual n selects the mode with the highest utility but that these utilities cannot be observed with certainty. Then, from the analyst's perspective, the probability that individual n chooses mode i given choice set C_n is given by

$$P(i \mid C_n) = \text{Prob}\,[U_{in} \geq U_{jn}, \forall j \in C_n] \tag{55.16}$$

where U_{in} is the utility of mode i for individual n. In other words, the probability that n chooses mode i is simply the probability that i has the highest utility.

Now, since the analyst cannot observe the utilities with certainty they should be treated as random variables. In particular, assume that

$$U_{in} = V_{in} + \epsilon_{in} \tag{55.17}$$

where V_{in} is the systematic component of the utility and ϵ_{in} is the random component (i.e., the disturbance term). Combining (55.16) and (55.17) yields the following:

$$P(i \mid C_n) = \text{Prob}\,[V_{in} + \epsilon_{in} \geq V_{jn} + \epsilon_{jn}, \forall j \in C_n]. \tag{55.18}$$

Specific random utility models can now be derived by making assumptions about the joint probability distributions of the set of disturbances, $\{\epsilon_{jn}, j \in C_n\}$.

As with gravity models, these models are typically estimated using maximum likelihood techniques. In practice, it is generally assumed that the systematic utilities are linear functions of their parameters. That is,

$$V_{in} = \beta_1 x_{in1} + \beta_2 x_{in2} + \cdots + \beta_G x_{inG} \tag{55.19}$$

where x_{ing} is the gth attribute of alternative i for individual n, and β_{ing} is the "weight" of that attribute. However, as discussed above, this is not a very restrictive assumption.

Probit Models

Suppose that the disturbances are the sum of a large number of unobserved independent components. Then, by the central limit theorem, the disturbances would be normally distributed. The resulting model is called the *probit model.*

For the case of two alternatives, the (binary) probit model is given by

$$P(i \mid C_n) = \Phi\left(\frac{V_{in} - V_{jn}}{\sigma}\right) \tag{55.20}$$

where $\Phi(\cdot)$ denotes the cumulative normal distribution function and σ is the standard deviation of the difference in the error terms, $\epsilon_{jn} - \epsilon_{in}$. For more detail see Finney [1971] or Daganzo [1979].

Logit Models

Observe that the probit model above does not have a closed-form solution. That is, the probability is expressed in terms of an integral that must be evaluated numerically. This makes the probit model computationally burdensome. To get around this, a model has been developed which is probitlike but much more convenient. This model is called the *logit model.*

The logit model can be derived by assuming that the disturbances are independently and identically Type-I Extreme Value (i.e., Gumbel) distributed. That is,

$$F(\epsilon_{in}) = \exp[-\exp[-\mu(\epsilon_{in} - \eta]] \qquad \forall i, n \tag{55.21}$$

where $F(\epsilon_{in})$ denotes the cumulative distribution function of ϵ_{in}, μ is a positive scale parameter, and η is a location parameter.

With this assumption it is relatively easy to show that

$$P(i \mid C_n) = \frac{e^{\mu V_{in}}}{\sum_j e^{\mu V_{jn}}} \tag{55.22}$$

where j represents an arbitrary mode. For a more complete discussion see Domencich and McFadden [1975], McFadden [1976], Train [1984], and Ben-Akiva and Lerman [1985].

It is important to point out that, while widely used, the logit model has one serious limitation. To see this, consider the relative probabilities of two modes, i and k. It follows from (55.22) that

$$\frac{P(i \mid C_n)}{P(k \mid C_n)} = \frac{e^{\mu V_{in}} / \sum_j e^{\mu V_{jn}}}{e^{\mu V_{kn}} / \sum_j e^{\mu V_{jn}}} = \frac{e^{\mu V_{in}}}{e^{\mu V_{kn}}} = e^{\mu(V_{in} - V_{kn})} \tag{55.23}$$

Hence, the ratio of the choice probabilities for i and k is independent of all of the other modes. This property is known as **independence from irrelevant alternatives (IIA).**

Unfortunately, this property is problematic in some situations. Consider, for example, a situation in which there are two modes, automobile (A) and red bus (R). Assuming that that $V_{An} = V_{Rn}$ it follows from (55.22) that $P(A|C_n) = P(R|C_n) = 0.50$. Now, suppose a new mode is added, blue bus (B), that is identical to R except for the color of the vehicles. Then, one would still expect that $P(A \mid C_n) = P(\text{Bus} \mid C_n) = 0.50$ and hence that $P(R \mid C_n) = P(B \mid C_n) = 0.25$. However, in fact, it follows from (55.22) that $P(A \mid C_n) = P(R \mid C_n) = P(B \mid C_n) = 0.333$. Thus, the logit model would not properly predict the mode choice probabilities in this case. What is the reason? ϵ_{Rn} and ϵ_{Bn} are not independently distributed.

Nested Logit Models

In some situations, an individual's "choice" of mode is actually a series of choices. For example, when choosing between auto, bus, and train the person may also have to choose whether to walk or drive to the bus or train. This can be modeled in one of two ways. On the one hand, the choice set can be thought of as having five alternatives: auto, walk + bus, auto + bus, walk + train, auto + train. On the other hand, this can be viewed as a two-step process in which the person first chooses between auto, bus, and train, and then, if the person chooses bus or train, she must also choose between walk access and auto access.

The reason to use this second approach (i.e., multidimensional choice sets) is that some of the observed and some of the unobserved attributes of elements in the choice set may be equal across subsets of alternatives. Hence, the first approach may violate some of the assumptions of, say, the logit model. To correct for this it is common to use a nested logit model.

To understand the nested logit model, consider a mode and submode choice problem of the kind discussed above. Then, the utility of a particular choice of mode and submode (for a particular individual) is given by

$$U_{ms} = \tilde{V}_m + \tilde{V}_s + \tilde{V}_{ms} + \tilde{\epsilon}_m + \tilde{\epsilon}_s + \tilde{\epsilon}_{ms} \tag{55.24}$$

where $\tilde{V}_m$ is the systematic utility common to all elements of the choice set using mode m, $\tilde{V}_s$ is the systematic utility common to all elements of the choice set using submode s, $\tilde{V}_{ms}$ is the remaining systematic utility specific to the pair (m, s), $\tilde{\epsilon}_m$ is the unobserved utility common to all elements of the choice set using mode m, $\tilde{\epsilon}_s$ is the unobserved utility common to all elements of the choice set using submode s, and $\tilde{\epsilon}_{ms}$ is the remaining unobserved utility specific to the pair (m, s).

Now, assuming that $\tilde{\epsilon}_m$ has zero variance and $\tilde{\epsilon}_s$ and $\tilde{\epsilon}_{ms}$ are independent for all m and s, the terms $\tilde{\epsilon}_{ms}$ are independent and identically Gumbel distributed with scale parameter μ^m, and $\tilde{\epsilon}_s$ is distributed so that $\max_m U_{ms}$ is Gumbel distributed with scale parameter μ^s, then the choice probabilities can be represented as follows:

$$P(s) = \frac{e^{(\tilde{V}_s + V'_s)\mu^s}}{\sum_t e^{(\tilde{V}_t + V'_t)\mu^s}} \tag{55.25}$$

where the notation indicating the individual's choice set has been dropped for convenience, t denotes an arbitrary submode, and

$$V'_s = \frac{1}{\mu^m} \ln \sum_m e^{(\tilde{V}_m + \tilde{V}_{ms})\mu^s} \tag{55.26}$$

The conditional probability of choosing mode m given the choice of submode s is then given by

$$P(m \mid s) = \frac{e^{(\tilde{V}_{ms} + \tilde{V}_m)\mu^m}}{\sum_j e^{(\tilde{V}_{js} + \tilde{V}_j)\mu^m}} \tag{55.27}$$

where j is an arbitrary mode. That is, the conditional probabilities for this nested logit model are defined by a scaled logit model that omits the attributes that vary only across the submodes. Ben-Akiva [1973], Daly and Zachary [1979], Ben-Akiva and Lerman [1985], and Daganzo and Kusnic [1993] provide detailed discussions of these models.

Path Choice

While the shortest distance between any two points on a plane is described by a straight line, it is often impossible to actually travel that way. When using an automobile or bicycle you must, for the most part, use a path that travels along existing roads; when using a bus or train you must use a path that consists of different predefined route segments; even when flying you often must use a path that consists of different flight legs.

In some respects, it is pretty amazing that people are able to make path choices at all, given the enormous number of possible paths that can be used to travel from one point to another. Fortunately, people are able to make these choices and it is possible to model them.

The basic premise which underlies almost all path choice models is that people choose the "best" path available to them (where the "best" may be measured in terms of travel time, travel cost, comfort, etc.). Of course, in general, this assumption may fail to hold. For example, infrequent travelers may not have enough information to choose the best path and may, instead, choose the most obvious path. As another example, in some instances it may be too difficult to even calculate what the actual best path is, as is sometimes the case with complicated transit paths that involve numerous transfers or when a shopper needs to choose the best way to get from home to several destinations and back to home. Nonetheless, this relatively simplistic approach does seem to work fairly well in practice.[3]

Automobile Commuters

The most important thing to capture when modeling the path choices of automobile commuters is congestion. In other words, the path choice of one commuter affects the path choices of all other commuters. Hence, one can imagine that each day commuters choose a particular path, evaluate that path, and the next day choose a new path based on their past experiences. Given that the number of automobile commuters and the characteristics of the network are relatively constant from day to day, such an adjustment process might reasonably be expected to settle down at some point in time. Most models of automobile commuter path choice assume that this process does settle down and, in fact, only consider the final equilibrium point.

These models are typically set on a network comprised of a set of nodes N and a set of arcs (or links) A. Within this context, a path is just a sequence of links that a commuter can travel along from his/her origin to his/her destination. If arc a is a part of path k (connecting r and s) then $\delta^{rs}_{ak} = 1$; otherwise, $\delta^{rs}_{ak} = 0$. Most such models assume that the number of people traveling from each origin to each destination by automobile is known (i.e., the mode-specific trip table is known) and that each path uses a single link at most once.

The most popular behavioral theory of the path choices of automobile commuters was proposed by Wardrop [1952]. He postulated that, in practice, commuters will behave in such a way that "the journey times on all the routes actually used are equal, and less than those which would be experienced by a single vehicle on any unused route." When this situation prevails, Wardrop argued that "no driver can reduce his journey time by choosing a new route," and hence that this situation can be thought of as an equilibrium. Mathematically, this definition of a **Wardrop equilibrium** can be expressed as follows:

$$f^{rs}_k > 0 \Rightarrow c^{rs}_k = \min_{j \in \mathcal{K}_{rs}} c^{rs}_k \qquad \forall r \in \mathcal{R}, s \in \mathcal{S}, k \in \mathcal{K}_{rs} \tag{55.28}$$

where f^{rs}_k denotes the number of people traveling from origin r to destination s on path k, c^{rs}_k denotes the cost on path k (from r to s),[4] $\mathcal{K}_{rs}$ denotes the set of paths connecting r and s, $\mathcal{R}$ denotes the set of all origins, and $\mathcal{S}$ denotes the set of all destinations.

As it turns out, Wardrop was not completely correct in claiming that when (55.28) holds, no driver can reduce his or her travel cost by changing routes. This has led other researchers to define other notions of equilibrium that incorporate this latter idea explicitly. The first such definition was the **user equilibrium** concept proposed by Dafermos and Sparrow [1969] which requires that no portion of the flow on a path can reduce their costs by swapping to another path. A somewhat weaker definition of user equilibrium was proposed by Dafermos [1971] in which no small portion of the users on any path can reduce their travel costs by simultaneously switching to any other path connecting the same OD-pair. An even weaker definition was proposed by Bernstein and Smith [1994] which is closer in spirit to the notion of a Nash equilibrium in which there is no coordination. From a behavioral viewpoint, their definition makes no assertion about potential gains from simultaneous route shifts by any positive portion of the commuters. Rather, it simply asserts that no gains are possible for *arbitrarily* small shifts. A very different equilibrium concept

[3]It is important to note that many behavioral models consider idealized situations in which people make the best possible choice. For example, this is the basic assumption that underlies most of microeconomics. Though this assumption has received a great deal of criticism, as yet nobody has been able to propose as workable an alternative.

[4]Though Wardrop [1952] includes only travel time in his definition, it is clear that his ideas can easily be extended to include other costs as well.

was proposed by Heydecker [1986]. He says that **equilibrated path choices** exist when no portion of the flow on any path, p, can switch to any other path, r, connecting the same OD-pair without making the new cost on r at least as large as the new cost on p. We will ignore such differences here. In most cases of practical interest, the different definitions of user equilibrium and Wardrop equilibrium turn out to be identical.

To simplify the analysis, it is common to assume that commuters are infinitely divisible (i.e., that it makes sense to talk about fractions of commuters on a particular path). It is also common to assume that the cost on link a, which we denote by t_a, is a function only of the number of vehicles on arc a, which we denote by x_a. In this case, the cost functions are said to be separable, and the equilibrium can be found by solving the following nonlinear program:

$$\min \quad \sum_{a \in \mathcal{A}} \int_0^{x_a} t_a(\omega)\, d\omega \tag{55.29}$$

$$\text{s.t.} \quad \sum_{r \in \mathcal{R}} \sum_{s \in \mathcal{S}} \sum_{k \in \mathcal{K}_{rs}} f_k^{rs} \delta_{ak}^{rs} = x_a \qquad \forall a \in \mathcal{A} \tag{55.30}$$

$$\sum_{k \in \mathcal{K}_{rs}} f_k^{rs} = q_{rs} \qquad \forall r \in \mathcal{R}, s \in \mathcal{S} \tag{55.31}$$

$$f_k^{rs} \geq 0 \qquad \forall r \in \mathcal{R}, s \in \mathcal{S}, k \in \mathcal{K}_{rs} \tag{55.32}$$

where q_{rs} is the number of automobile commuters from r to s. The solution of this nonlinear program is an equilibrium because the Kuhn-Tucker conditions, which are both necessary and sufficient, are equivalent to the equilibrium conditions in (55.28). This result was first demonstrated by Beckmann *et al.* [1956]. This problem can be solved using a variety of different nonlinear programming algorithms (see, for example, LeBlanc, Morlok, and Pierskalla [1975], and Nguyen [1974, 1978]).

For cases where the arc cost functions are not separable we must instead solve a variational inequality problem in order to find the equilibrium.[5] In particular, letting H denote the set of all vectors $x = (x_a : a \in \mathcal{A})$ and $f = (f_k^{rs} : r \in \mathcal{R}, s \in \mathcal{S}, k \in \mathcal{K}_{rs})$ that satisfy

$$\sum_{r \in \mathcal{R}} \sum_{s \in \mathcal{S}} \sum_{k \in \mathcal{K}_{rs}} f_k^{rs} \delta_{ak}^{rs} = x_a \qquad \forall a \in \mathcal{A} \tag{55.33}$$

$$\sum_{k \in \mathcal{K}_{rs}} f_k^{rs} = q_{rs} \qquad \forall r \in \mathcal{R}, s \in \mathcal{S} \tag{55.34}$$

$$f_k^{rs} \geq 0 \qquad \forall r \in \mathcal{R}, s \in \mathcal{S}, k \in \mathcal{K}_{rs} \tag{55.35}$$

we must find vectors $(\bar{x}, \bar{f}) \in H$ that satisfy

$$\sum_{a \in \mathcal{A}} t_a(\bar{x})(x_a - \bar{x}_a) \geq 0 \tag{55.36}$$

for all $(x, f) \in H$. Fortunately, the solution to this variational inequality problem can be obtained in a variety of ways, one of which is to solve a sequence of nonlinear programs related to the one described above (see, for example, Dafermos and Sparrow [1969], Nagurney [1984, 1988], Harker and Pang [1990]).

[5]This is not, strictly speaking, true. When the cost functions are nonseparable but symmetric it is is still possible to develop a math programming formulation of the equilibrium problem. This is discussed more fully in Dafermos [1971], Abdulaal and LeBlanc [1979], and Smith and Bernstein [1993].

It is important to note that such equilibria are known to exist and be unique in most cases of practical interest (see Smith [1979] and Dafermos [1980]). It is also important to note that the assumption of perfect information can be relaxed and a stochastic version of the model developed (see, for example, Daganzo and Sheffi [1977], Sheffi and Powell [1982], and Smith [1988]). For a more complete discussion of these models see Friesz [1985], Sheffi [1985], Boyce *et al.* [1988], or Nagurney [1993].

Transit Travelers

The path choice problem faced by transit travelers is actually quite different from that faced by auto travelers. In particular, transit users must decide (based on a schedule, if one exists) how to best get from their origin to their destination using a group of vehicles traveling along predetermined routes. Of course, they make these choices knowing full-well that almost all aspects of transit service are stochastic (e.g., running times, vehicle arrival times, crowding, etc.).

Early models of transit path choice assumed that travelers essentially choose the path with the minimum expected cost. In the case of a tie (either on the entire path or a portion of the path), travelers are assumed to choose different routes in proportion to their frequency. Models of this kind are discussed by Dial [1967] and le Clercq [1972].

Recently, more attention has been given to how travelers might actually choose between multiple routes that service the same locations (whether they are intermediate points in the path or the actual origin and destination). These models assume that, because of the stochastic nature of vehicle departure and travel times, passengers will probably be willing to use several paths and will actually choose one based on the actual departure times of specific vehicles.

Chriqui and Robillard [1975] assume that travelers will first choose a set of routes they would be willing to use, and then actually choose the first vehicle that arrives which services one of the routes in that set. This model can be formalized as follows. Let n denote the number of routes providing service between two locations, let f_{W_i} denote the probability density function of the waiting times (for the next vehicle) on route i, let $\bar{F}_{W_i}$ denote the complement of the cumulative distribution function of the waiting times on route i, let $X = (x_1, \dots, x_n)$ denote the choice vector where $x_i = 1$ if route i is chosen and $x_i = 0$ otherwise, and let t_i denote the expected travel time after boarding a vehicle on route i. Then, following Hickman [1993], the problem of determining the optimal route set is given by

$$\min_{X} \quad \sum_{i=1}^{n} \int_{0}^{\infty} (z + t_i) \cdot x_i \cdot fw_i \prod_{j \neq i} \bar{F}w_j(z)^{x_j}\, dz \tag{55.37}$$

$$\text{s.t.} \quad \sum_{i=1}^{n} x_i \geq 1 \tag{55.38}$$

$$x_i \in \{0, 1\} \tag{55.39}$$

In this problem, the expression $x_i f_{W_i} \prod_{j \neq i} \bar{F}_{W_j}(z)^{x_j}$ denotes the probability that a vehicle on route i will arrive before any other vehicle in the choice set.

The solution technique proposed by Chriqui and Robillard [1975] is not guaranteed to find an optimal solution except when the waiting time distributions and in-vehicle travel times for all routes are identical and when the headways are exponentially distributed. Their heuristic proceeds as follows:

Step 0. Enumerate all of the possible routes. Set $k = 1$.
Step 1. Sort the routes by expected in-vehicle travel "cost" (e.g., time) letting route i denote the ith "cheapest" route.
Step 2. Let the initial guess at the choice set be given by $X^1 = (1, 0, \dots, 0)$ and let C^1 denote the expected travel cost associated with this choice set.

Step 3. Let the guess at iteration k be given by $X^k = (1_1, \ldots, 1_k, 0_{k+1}, \ldots, 0_n)$ where 1_i denotes a 1 in the ith position of the vector X and 0_i denotes a 0 in the ith position of the vector X. Calculate the expected cost of this choice set and denote it by C^k.

Step 4. If $C^k > C^{k-1}$ then STOP (the optimal choice set is given by X^{k-1}). Otherwise GOTO step 3.

This work is discussed and extended by Marguier [1981], Marguier and Ceder [1984], Janson and Ridderstolpe [1992], and Hickman [1993]. Other models of transit path choice are discussed in de Cea *et al.* [1988], Spiess and Florian [1989], and Nguyen and Pallottino [1988].

Departure-Time Choice

Traditionally, little attention has been given to the modeling of departure-time choice. Hence, this section will briefly discuss some of the approaches to modeling departure-time choice that have been proposed in the theoretical literature but, as yet, have not been widely implemented.

Automobile Commuters

In practice, the departure-time choices of automobile commuters are usually modeled very crudely. That is, the day is normally divided into several periods (e.g., morning peak, midday, evening peak, night) and a trip table is created for each period. Within-period departure-time choices are simply ignored.

The theoretical literature has considered two approaches for modeling within-period departure-time choice. The first approach makes use of the kinds of probabilistic choice models discussed above. These models, however, typically fail to consider congestion effects. The other approach accounts for congestion in a manner that is very similar to the path choice models described above. That is, this approach assumes that each person chooses the best departure-time given the behavior of all other commuters.

To understand this second approach, consider a simple example of the work-to-home commute in which each person chooses a departure time after 5:00 P.M. (denoted by $t = 0$) and before some time $\bar{t}$ in such a way that his or her cost is minimized given the behavior of all other commuters. Assuming that travel delays are modeled as a deterministic queuing process with service rate $1/\beta$ and the cost of departing at time t is given by

$$C(t) = \beta x(t) + \gamma t \tag{55.40}$$

where $x(t)$ is the size of the queue at time t and $\gamma < 1$ is a penalty for late departure, an equilibrium can be characterized as a departure pattern, h, that satisfies

$$C(t) = C(0) \qquad \forall t \in (0, \bar{t}] \tag{55.41}$$

$$x(0) + \int_0^{\bar{t}} h(w)\,dw = N \tag{55.42}$$

The first condition ensures that the costs are equal for all departure times, while the second ensures that everyone actually departs (where the total number of commuters is denoted by N).

In equilibrium, γN people will depart at exactly $t = 0$ (assuming that each individual member of this group will perceive the average cost for the entire group), and over the interval $(0, \beta N]$ the remaining commuters will depart at a rate of $(1 - \gamma)/\beta$.

To see that this is indeed an equilibrium, observe that as long as there are commuters in the queue throughout the period $[0, \bar{t}]$, the size of the queue at time t is given by

$$x(t) = x(0) + \int_0^{t} h(w)\,dw - 1/\beta t \tag{55.43}$$

Hence, the cost at time t is given by

$$C(t) = \beta[x(0) + \int_0^t h(w)\,dw - 1/\beta t] + \gamma t \quad (55.44)$$

Substituting for $x(0)$ and h yields $C(t) = \beta\gamma N + (1-\gamma)t - t + \gamma t = \beta\gamma N$ for $t \in (0, \bar{t}]$. And, since $x(0) = \gamma N$ it follows that $C(0) = \beta\gamma N$ and that the flow pattern is, in fact, an equilibrium.

These models are discussed in greater detail by Vickrey [1969], Hendrickson and Kocur [1981], Mahmassani and Herman [1984], Newell [1987] and Arnott *et al.* [1990a,b]. Stochastic versions are presented by Alfa and Minh [1979], de Palma *et al.* [1983], and Ben-Akiva *et al.* [1984].

Transit Travelers

Traditionally, transit models have assumed that (particularly when headways are relatively short) people depart from their homes (i.e., arrive at the transit stop) randomly. In other words, they assume that the interarrival times are exponentially distributed.

There has been some research, however, that has attempted to model departure time choices in more detail. This work is described by Jolliffe and Hutchinson [1975], Turnquist [1978], and Bowman and Turnquist [1981].

Combining the Models

The discussion above treated each of the different models in isolation. However, as mentioned at the outset, many of the decisions being modeled are actually interrelated. Hence, it is common practice to combine these models when they are actually applied.

The most obvious way to combine these models is to apply them sequentially. That is, obtain origin and/or destination totals from a trip generation model, use those totals as inputs to a trip distribution model and obtain a trip table, use the trips by origin-destination pair as inputs to a modal split model, and then assign the mode-specific trips to paths using an assignment model. Unfortunately, however, this process is not as "trouble free" as it might sound. For example, trip distribution models often have travel times as an input. What travel time should you use? Should you use a weighted average across different modes? Perhaps, but you have not yet modeled modal shares. In addition, since you have not yet modeled path choice you do not know what the travel times will be.

This has led many practitioners to apply the models sequentially but to do so iteratively, first guessing at appropriate inputs to the early models and then using the outputs from the later models as inputs in later iterations. Continuing the example above, you estimate travel times for the trip distribution model in the first iteration, then use the resulting trip table and an estimate of mode-specific travel times as an input to a modal split model. Next, you could use the output from the modal split model as an input to a traffic assignment model. Then, you could use the travel costs calculated by the traffic assignment model as inputs to the next iteration's trip distribution model, and so on.

Of course, one is naturally led to ask which approach is better. Unfortunately, there is no conclusive answer. Some people have argued that the simple sequential approach is an accurate predictor of observed behavior. In other words, they argue that the estimates of travel times that people use when choosing where to live and work often turn out to be inconsistent with the travel times that are actually realized. As a result, they are not troubled by the inconsistencies that arise using what is traditionally referred to as the "four-step process" (i.e., first trip generation, then trip distribution, then modal split, and finally traffic assignment).

Others have argued that the number of iterations should depend on the time frame of the analysis. That is, they believe that the iterative approach can be used to describe how these decisions are actually made over time. Hence, by iterating they believe that they can predict how the system will evolve over time.

Still others have argued that, while the iterative approach does not accurately describe how the system will evolve over time, it will eventually converge to the long-run equilibrium that is likely to be realized. That is, they believe that the trajectory of intermediate solutions is meaningless, but that the final solution (i.e., when the outputs across different iterations settle down) is a good predictor of the long-run equilibrium that will actually be realized.

Finally, others have argued that it makes sense to iterate until the outputs converge not because the final answer is likely to be a good predictor (since too many other things will change in the interim), but simply because it is internally consistent. They argue that it is impossible to compare the impacts of different projects otherwise.

Regardless of how you feel about the above debate, one thing is known for certain. There are more efficient ways of solving for the long-run equilibrium than iteratively solving each of the individual models until they converge. In particular, it is possible to solve most combinations of models simultaneously.

As an example, consider the problem of solving the combined mode and route choice problem, assuming that there are two modes (auto and train), that there is one train path for each OD-pair, that the two modes are independent (i.e., that neither mode congests the other), that the cost of the train is independent of the number of users of the train, and that the arc cost functions for auto are separable. Then, the combined model can be formulated as the following nonlinear program:

$$\min \quad \sum_{a \in \mathcal{A}} \int_0^{x_a} t_a(\omega)\, d\omega - \sum_{r \in \mathcal{R}, s \in \mathcal{S}} \int_0^{q_{rs}} \left[\frac{1}{\theta} \ln\left(\frac{\bar{q}_{rs}}{w} - 1 \right) + \hat{u}_{rs} \right] dw \tag{55.45}$$

$$\text{s.t.} \quad \sum_{r \in \mathcal{R}} \sum_{s \in \mathcal{S}} \sum_{k \in \mathcal{K}_{rs}} f_k^{rs} \delta_{ak}^{rs} = x_a \qquad \forall a \in \mathcal{A} \tag{55.46}$$

$$\sum_{k \in \mathcal{K}_{rs}} f_k^{rs} = q_{rs} \qquad \forall r \in \mathcal{R}, s \in \mathcal{S} \tag{55.47}$$

$$0 < q_{rs} < \bar{q}_{rs} \qquad \forall r \in \mathcal{R}, s \in \mathcal{S} \tag{55.48}$$

$$f_k^{rs} \geq 0 \qquad \forall r \in \mathcal{R}, s \in \mathcal{S}, k \in \mathcal{K}_{rs} \tag{55.49}$$

where $\bar{q}_{rs}$ denotes the total number of travelers on both modes and $\hat{u}_{rs}$ denotes the fixed transit travel cost.

Of course, there are far too many different combinations of the basic models to review them all here. Various different combinations of the traditional "four steps" are discussed by Tomlin [1971], Florian *et al.* [1975], Evans [1976], Florian [1977], Florian and Nguyen [1978], Sheffi [1985], and Safwat and Magnanti [1988]. There is also a considerable amount of activity currently being devoted to simultaneous models of route and departure-time choice. As these models are quite complicated, in general, they are beyond the scope of this *Handbook*. For auto commuters, see, for example, the deterministic models of Friesz *et al.* [1989], Smith and Ghali [1990], Bernstein *et al.* [1993], Friesz *et al.* [1993], Ran *et al.* [1993] and the stochastic models developed by Ben-Akiva *et al.* [1986] and Cascetta [1989]. For transit travel, see the models developed by Hendrickson and Plank [1984] and Sumi *et al.* [1990].

55.3 Applications and Example Calculations

In this section, several examples are presented and solved. Unfortunately, due to the complexity of some of the models and the ways in which they interact, these examples are not exhaustive.

The Decision to Travel

A number of different trip generation models have been developed over the years. This section contains examples of several.

The first example is a disaggregate regression model estimated by Douglas [1973]:

$$Y = -0.35^{*} + 0.63^{*}X_1 + 1.08^{*}X_2 + 1.88^{*}X_3 \tag{55.50}$$

where Y denotes the number of trips per household per day, X_1 denotes the number of people per household, X_2 denotes the number of employed people per household, and X_3 denotes the monthly income of the household (in thousands of U.K. pounds). The symbol * indicates that the estimate of the coefficient is significantly different from 0 at the 0.95 confidence level. As one example of how to use this model, observe that $\partial Y/\partial X_1 = 0.63$. Hence, this model says that, other things being equal, an additional unemployed member of the household would make (on average) 0.63 additional trips per day.

The second model is an example of a disaggregate category analysis model developed by Doubleday [1977]:

Type of Person		Total Trip Rate	Regular Trips	Nonregular Trips
Employed males	w/o a car	3.7	2.46	0.55
	with a car	5.7	2.80	1.38
Employed females	w/o a car	4.5	2.20	1.30
	with a car	6.0	2.39	2.13
Homemakers	w/o a car	4.1	—	3.25
	with a car	5.7	—	4.78
Retired persons	w/o a car	2.2	—	1.75
	with a car	4.1	—	3.16

where the numbers in the table are the number of trips per person per day. It should be relatively easy to see how such a model would be used in practice.

The third example is an aggregate regression model estimated by Keefer [1966] for the city of Pittsburgh:

$$Y = 3296.5 + 5.35X_1 + 291.9X_2 - 0.65X_3 - 22.31X_4 \tag{55.51}$$

where Y denotes the total number of automobile trips to shopping centers, X_1 denotes the number of work trips, X_2 denotes the distance of the shopping center from major competition (in tenths of miles), X_3 denotes the reported travel speed of trip makers (in miles per hour), and X_4 denotes the amount of floor space used for goods other than shopping and convenience goods (in thousands of square feet). The R^2 for this model is 0.920. What distinguishes this model from the disaggregate model above is that it does not focus on the individual household. Instead, it uses aggregate data and estimates the total number of automobile trips to shopping centers.

The final example is an aggregate category analysis model also developed by Keefer [1966]:

Land-Use Category	Square Feet (1000s)	Person Trips	Trips per 1000 sq. ft
Residential	2,744	6,574	2.4
Retail	6,732	54,833	8.1
Services	13,506	70,014	5.2
Wholesale	2,599	3,162	1.2
Manufacturing	1,392	1,335	1.0
Transport	1,394	5,630	4.0
Public buildings	31,344	153,294	4.9

where the numbers in the table are the total number of trips taken. Again, this is an aggregate model because, unlike the earlier category analysis model, it does not focus on the behavior of the individual. Instead, it is based on aggregate data about trip making.

Origin and Destination Choice

This section contains an example of an estimated gravity model and several iterations of an application of the Fratar model.

An Example of the Gravity Model

Putman [1983] presents an interesting example of a gravity model of commuter origin/destination choice. In this model

$$T_{ij} = \frac{L_i^{\delta} c_{ij}^{a} \exp(\beta c_{ij})}{\sum_k L_k^{\delta} c_{ik}^{a} \exp(\beta c_{ik})} \tag{55.52}$$

where T_{ij} denotes the number of commuting trips from i to j, L_i denotes the size of zone i, c_{ij} is the cost of traveling from i to j, and α, β, and δ are parameters.

He estimated this model for several cities (in slightly different years) and found the following:

City	α	β	δ
Gosford-Wyong, Australia	0.09	−0.03	−2.31
Melbourne, Australia	1.04	−0.06	0.13
Natal, Brazil	0.93	−0.01	0.48
Rio de Janeiro, Brazil	1.08	−0.01	0.60
Monclova-Frontera, Mexico	2.73	−0.14	0.24
Ankara, Turkey	0.64	−0.14	−0.31
Izmit, Turkey	0.90	−0.03	1.05

To see how this type of model would be applied, consider the following two-zone example in which $L_1 = 3000$, $L_2 = 1000$, $c_{11} = 2$, $c_{12} = 10$, $c_{21} = 7$, and $c_{22} = 3$, and suppose that these zones are in Melbourne, Australia. Then, for T_{11}, the numerator of (55.52) is given by $L_1^{\delta} c_{11}^{\alpha} \exp(\beta c_{11}) = 3000^{0.13} \cdot 2^{1.04} \cdot \exp(0.13 \cdot 2) = 2.83 \cdot 2.06 \cdot 0.89 = 5.16$. Continuing in this manner, the other numerators in (55.52) are given by 17.04 for $i = 1, j = 2$, 12.20 for $i = 2, j = 1$, and 6.43 for $i = 2, j = 2$. It then follows that

$$T_{11} = \frac{5.16}{5.16 + 17.04 + 12.20 + 6.43} = 698 \tag{55.53}$$

$$T_{12} = \frac{17.04}{5.16 + 17.04 + 12.20 + 6.43} = 2302 \tag{55.54}$$

$$T_{21} = \frac{12.20}{5.16 + 17.04 + 12.20 + 6.43} = 655 \tag{55.55}$$

$$T_{22} = \frac{6.43}{5.16 + 17.04 + 12.20 + 6.43} = 345 \tag{55.56}$$

Thus, the model predicts that there will be $698 + 655 = 1353$ trips to zone 1, and $2302 + 345 = 2647$ trips to zone 2.

In order to understand the sensitivity of this model, it is worth performing these same calculations using the parameters estimated for Ankara, Turkey. In this case, the resulting trip table is given by

$$T_{11} = 1567 \tag{55.57}$$

$$T_{12} = 1433 \tag{55.58}$$

$$T_{21} = 496 \tag{55.59}$$

$$T_{22} = 504 \tag{55.60}$$

An Example of the Fratar Model

Consider the following hypothetical example of the Fratar model in which there are four zones and the original trip table is given by

$$T^0 = \begin{bmatrix} 0.00 & 10.00 & 40.00 & 15.00 \\ 10.00 & 0.00 & 20.00 & 30.00 \\ 40.00 & 20.00 & 0.00 & 10.00 \\ 15.00 & 30.00 & 10.00 & 0.00 \end{bmatrix} \tag{55.61}$$

and the forecasted trip-end totals are given by

$$O = \begin{bmatrix} 130.00 \\ 140.00 \\ 225.00 \\ 90.00 \end{bmatrix} \tag{55.62}$$

In step 1, the production totals are calculated as

$$P^1 = \begin{bmatrix} 65.00 \\ 60.00 \\ 70.00 \\ 55.00 \end{bmatrix} \tag{55.63}$$

Then, in step 2, the factors are calculated as

$$f^1 = \begin{bmatrix} 2.00 \\ 2.33 \\ 3.21 \\ 1.64 \end{bmatrix} \tag{55.64}$$

Next, in step 3, the temporary (asymmetric) trip table is calculated as

$$N^1 = \begin{bmatrix} 0.00 & 17.19 & 94.73 & 18.08 \\ 20.99 & 0.00 & 67.48 & 51.53 \\ 125.85 & 73.41 & 0.00 & 25.74 \\ 20.43 & 47.68 & 21.89 & 0.00 \end{bmatrix} \tag{55.65}$$

Finally, the first iteration is concluded by calculating the symmetric trip table:

$$T^1 = \begin{bmatrix} 0.87 & 0.00 & 19.09 & 110.29 \\ 1.01 & 19.09 & 0.00 & 70.44 \\ 1.10 & 110.29 & 70.44 & 0.00 \\ 0.97 & 19.26 & 49.60 & 23.82 \end{bmatrix} \tag{55.66}$$

In step 1 of the second iteration, the trip-end totals are calculated as

$$P^2 = \begin{bmatrix} 148.64 \\ 139.13 \\ 204.55 \\ 92.68 \end{bmatrix} \quad (55.67)$$

Then in step 2 of iteration 2:

$$f^2 = \begin{bmatrix} 0.87 \\ 1.01 \\ 1.10 \\ 0.97 \end{bmatrix} \quad (55.68)$$

And in step 3 of iteration 2:

$$N^2 = \begin{bmatrix} 0.00 & 15.68 & 99.05 & 15.27 \\ 16.42 & 0.00 & 76.21 & 47.37 \\ 113.95 & 83.73 & 0.00 & 27.32 \\ 16.31 & 48.32 & 25.37 & 0.00 \end{bmatrix} \quad (55.69)$$

And, finally in step 4 of iteration 2:

$$T^2 = \begin{bmatrix} 0.00 & 16.05 & 106.50 & 15.79 \\ 16.05 & 0.00 & 79.97 & 47.85 \\ 106.50 & 79.97 & 0.00 & 26.35 \\ 15.79 & 47.85 & 26.35 & 0.00 \end{bmatrix} \quad (55.70)$$

In step 1 of iteration 3:

$$P^3 = \begin{bmatrix} 138.34 \\ 143.87 \\ 212.81 \\ 89.98 \end{bmatrix} \quad (55.71)$$

And, in step 2 of iteration 3:

$$f^3 = \begin{bmatrix} 0.94 \\ 0.97 \\ 1.06 \\ 1.00 \end{bmatrix} \quad (55.72)$$

Since all of these values are approximately equal to 1 the algorithm terminates at this point.

Mode Choice

Suppose the utility function for individual n is given by

$$V_{jn} = -t_j - \frac{5o_j}{Y_n} \quad (55.73)$$

where t_j is the travel time on mode j, o_j is the out-of-pocket cost on mode j, and Y_n is the income of individual n. Now, consider the following three modes:

Mode	t	o
Drive alone	0.50	2.00
Carpool	0.75	1.00
Bus	1.00	0.75

and consider this person's choices when her income was $15,000 and now that it is $30,000.

When her income was $15,000 the (systematic) utilities of the three modes were given by -1.17 for driving alone, -1.08 for carpooling, and -1.25 for taking the bus. Now that her income has increased to $30,000, the utilities have gone to -0.88 for driving alone, -0.92 for carpooling, and -1.13 for taking the bus. (Note: The utilities are negative because commuting itself decreases your overall utility.)

Using a deterministic choice model, one would conclude that this individual would choose the mode with the highest utility (i.e., the lowest disutility). In this case, when she earned $15,000 she would have carpooled, but now that she earns $30,000 she drives alone.

On the other hand, using a logit model with $\mu = 1$, the resulting probabilities are given by

$$P(i \mid C_n) = \frac{e^{V_{in}}}{\sum_j e^{V_{jn}}} \tag{55.74}$$

Hence, in this choice situation with $Y = 15$:

$$P(\text{Drive alone}) = \frac{0.31}{0.31 + 0.34 + 0.29} = \frac{0.31}{0.94} = 0.33 \tag{55.75}$$

Continuing in this way, one finds that

Mode	$P(i \mid C_n)$ for $Y = 15$	$P(i \mid C_n)$ for $Y = 30$
Drive alone	0.33	0.38
Carpool	0.36	0.34
Bus	0.31	0.28

Roughly speaking, this says that when her income was $15,000 she drove alone 33% of the time, carpooled 36% of the time, and took the bus 31% of the time. Now, however, she drives alone 38% of the time, carpools 34% of the time, and takes the bus 28% of the time.

Path Choice

This section contains an example of both highway path choice and transit path choice.

Highway Path Choice

A nice way to illustrate equilibrium path choice models is with a famous example called Braess's paradox. Consider the four-link network shown in Fig 55.1, where $t_1(x_1) = 50 + x_1$, $t_2(x_2) = 50 + x_2$, $t_3(x_3) = 10x_3$, and $t_4(x_4) = 10x_4$. Since these cost functions are separable, the following nonlinear program can be solved to obtain the equilibrium:

$$\min \quad \int_0^{x_1} (50 + \omega)\, d\omega + \int_0^{x_2} (50 + \omega)\, d\omega + \int_0^{x_3} (10\omega)\, d\omega + \int_0^{x_4} (10\omega)\, d\omega \tag{55.76}$$

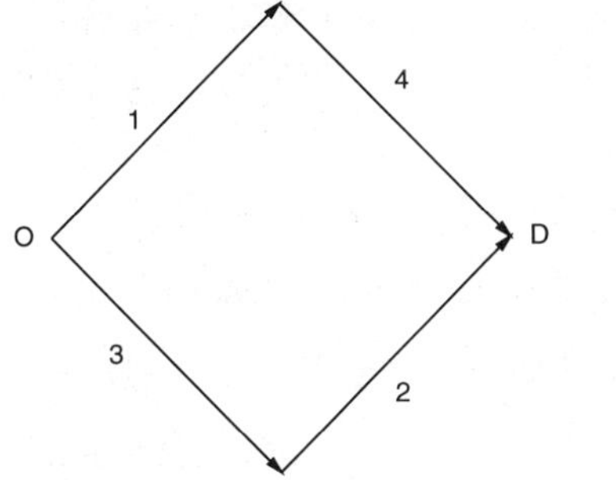

FIGURE 55.1 A four-link network.

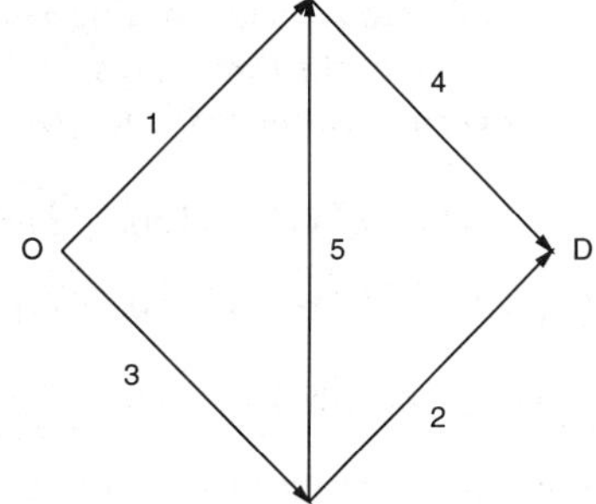

FIGURE 55.2 A five-link network.

$$\text{s.t.} \quad f_1 = x_1 \tag{55.77}$$
$$f_2 = x_2 \tag{55.78}$$
$$f_2 = x_3 \tag{55.79}$$
$$f_1 = x_4 \tag{55.80}$$
$$f_1 + f_2 = 6 \tag{55.81}$$
$$f_1 \geq 0 \tag{55.82}$$
$$f_2 \geq 0 \tag{55.83}$$

The solution to this problem is given by $x_1 = 3$, $x_2 = 3$, $x_3 = 3$, $x_4 = 3$. To verify that this is, indeed, an equilibrium, the costs on the two paths can be calculated as follows:

$$c_1^{OD} = t_1(x_1) + t_4(x_4) = (50 + 3) + (10 \cdot 3) = 83 \tag{55.84}$$
$$c_2^{OD} = t_3(x_3) + t_2(x_2) = (10 \cdot 3) + (50 + 3) = 83 \tag{55.85}$$

The total cost to all commuters is thus $(3 \cdot 83) + (3 \cdot 83) = 498$.

Now, suppose link 5 is added to the network as in Fig 55.2, where $t_5(x_5) = 10 + x_5$. Then, it follows that the following nonlinear program can be solved to obtain the new equilibrium:

$$\min \quad \int_0^{x_1} (50 + \omega)\, d\omega + \int_0^{x_2} (50 + \omega)\, d\omega + \int_0^{x_3} (10\omega)\, d\omega + \int_0^{x_4} (10\omega)\, d\omega + \int_0^{x_5} (10 + \omega)\, d\omega \tag{55.86}$$

$$\text{s.f.} \quad f_1 = x_1 \tag{55.87}$$
$$f_2 = x_2 \tag{55.88}$$
$$f_2 + f_3 = x_3 \tag{55.89}$$
$$f_1 + f_3 = x_4 \tag{55.90}$$
$$f_3 = x_5 \tag{55.91}$$
$$f_1 + f_2 + f_2 = 6 \tag{55.92}$$
$$f_1 \geq 0 \tag{55.93}$$
$$f_2 \geq 0 \tag{55.94}$$
$$f_3 \geq 0 \tag{55.95}$$

The solution to this problem is given by $x_1 = 2$, $x_2 = 2$, $x_3 = 4$, $x_4 = 4$, $x_5 = 2$ (with two commuters using each of the three paths). To verify that this is, indeed, an equilibrium, the costs on the three paths can be calculated as follows:

$$c_1^{OD} = t_1(x_1) + t_4(x_4) = (50 + 2) + (10 \cdot 4) = 92 \tag{55.96}$$

$$c_2^{OD} = t_3(x_3) + t_2(x_2) = (10 \cdot 4) + (50 + 2) = 92 \tag{55.97}$$

$$c_3^{OD} = t_3(x_3) + t_5(x_5) + t_4(x_4) = (10 \cdot 4) + (10 + 2) + (10 \cdot 4) = 92 \tag{55.98}$$

Now, however, the total cost to all commuters is $(2 \cdot 92) + (2 \cdot 92) + (2 \cdot 92) = 552$.

This example has received a great deal of attention because it illustrates that it is possible to increase total travel costs when you add a link to the network, and this seems counterintuitive. Of course, one is led to ask why people don't simply stop using path 3. The reason is that with 3 people on paths 1 and 2 (and hence with $x_1 = 3$, $x_2 = 3$, $x_3 = 3$, $x_4 = 3$) the cost on path 3 is given by

$$c_3^{OD} = t_3(x_3) + t_5(x_5) + t_4(x_4) = (10 \cdot 3) + (10 + 0) + (10 \cdot 3) = 70 \tag{55.99}$$

and hence people using paths 1 and 2 will want to switch to path 3. And, once they switch, even though their costs will go up they will not want to switch back. To see this, consider the equilibrium with the new link in place, and suppose someone on path 3 switches to path 1. Then, the resulting link volumes are $x_1 = 3$, $x_2 = 2$, $x_3 = 3$, $x_4 = 4$, and $x_5 = 1$. Hence

$$c_1^{OD} = t_1(x_1) + t_4(x_4) = (50 + 3) + (10 \cdot 4) = 93 \tag{55.100}$$

which is higher than the cost of 92 they would experience without switching.

Transit Path Choice

Consider an origin-destination pair that is serviced by four bus routes with the following characteristics:

Route	E [In-Vehicle Time]	E [Headway]	E [Travel Time]
A	20	5	22.5
B	10	30	25
C	30	30	45
D	35	25	47.5

The expected travel times in this table are calculated assuming that passengers arrive at the origin randomly and that the vehicle headways are randomly distributed. The expected waiting time for any particular route is half of the headway.

If one assumes that people simply choose the route with the lowest expected travel time, then it is clear that route A will be chosen. On the other hand, the Chriqui and Robillard [1975] model would predict that both routes A and B would be chosen. Their algorithm proceeds as follows.

In step 1 the routes are sorted based on their expected in-vehicle travel time. Hence, route B will be denoted by 1, route A will be denoted by 2, route C will be denoted by 3, and route D will be denoted by 4.

In step 2, the initial choice set is determined. In this case, $X^1 = (1, 0, 0, 0)$ and $C^1 = 25$.

In step 3, routes are iteratively added to this choice set until the expected travel time increases. So, in the first iteration the choice set is assumed to be $X^2 = (1, 1, 0, 0)$. To calculate the expected travel time for this choice set, observe that (given the above headways) 14 vehicles per hour from this choice set serve the OD-pair. Hence, the expected waiting time (for a randomly arriving passenger) is $4.29/2 = 2.14$ minutes. The expected travel time for this choice set is given by the probability-weighted travel times on the member routes. Hence, the expected travel time is

$(2/14) \cdot 10 + (12/14) \cdot 20 = 17.14 + 1.43 = 18.57$ minutes. Thus, the expected total travel time for this choice set, C^2, is 18.57 minutes. Since this is less than C^1, the algorithm continues.

In the second iteration the choice set is assumed to be $X^3 = (1, 1, 1, 0)$. Now, 16 vehicles per hour from this choice set serve the OD-pair. Hence, the expected waiting time is $3.75/2 = 1.875$ minutes. Further, the expected travel time is given by $[(2/16) \cdot 10] + [(12/16) \cdot 20] + [(2/16) \cdot 30] = 1.25 + 15 + 3.75 = 20$ minutes. Thus, the expected total travel time for this choice set, C^3, is 21.875 minutes. Since this is greater than C^2, the algorithm terminates.

Departure-Time Choice

As discussed above, the equilibrium departure pattern for a simple model of departure-time choice can be characterized as $x(0) = \gamma N, h(t) = (1 - \gamma)/\beta$ for $t \in (0, \beta N]$. Assuming $N = 10,000$, the service rate of the queue is 5000 vehicles/hr (i.e., $\beta = 1/5000$), and the late departure penalty is given by $\gamma = 0.1$, it follows that in equilibrium $x(0) = 1000$, the peak period ends at $\bar{t} = 2$ (i.e., lasts for 2 hours after 5:00 P.M.), and the departure rate during the peak period is 4500 vehicles/hr.

It also follows that the queue at time t is given by

$$x(t) = \gamma N - \frac{\gamma}{\beta} t, \qquad t \in [0, \beta N] \tag{55.101}$$

and hence that the queue is initially 1000 vehicles (at time $t = 0$) and decreases linearly at a rate of 500 vehicles/hr.

Combined Models

In this example, a hypothetical city is thinking about changing the fare on its transit line from \$1.50 to \$3.00 and would like to be able to predict how ridership and congestion levels will change. The network is shown in Fig 55.3. Node D is the single destination (the central business district) and node O is the single origin (the residential area). The solid line represents the highway link and the dotted line represents the transit link.

The Models

Highway travel times (in-vehicle) will be modeled using the following function recommended by the Bureau of Public Roads (BPR):

$$t_a = t_0 \left[1 + 0.15 \left(\frac{x_a}{k_a} \right)^4 \right] \tag{55.102}$$

where t_0 is the free-flow travel time (in minutes) and k_a is the practical capacity of link a.

Mode choice will be modeled using the following logit model:

$$P(T) = \frac{\exp(V_T)}{\exp(V_A) + \exp(V_T)} \tag{55.103}$$

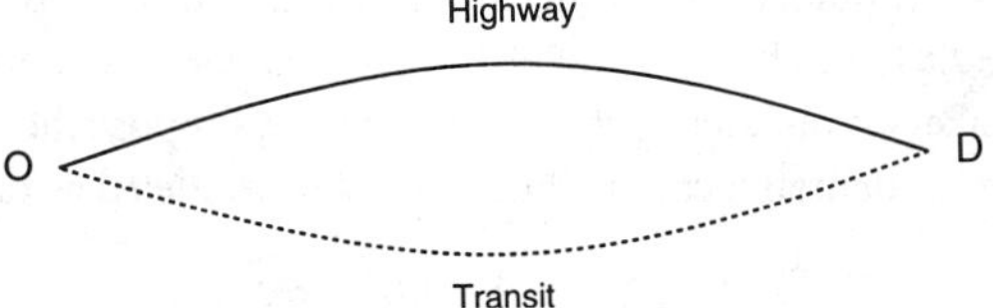

FIGURE 55.3 A multimodal network.

where $P(T)$ is the probability that a commuter chooses to go to work by transit (train), $P(A)$ is the probability that a commuter chooses to go to work by auto, V_T is the systematic component of the utility of transit, and V_A is the systematic component of the utility of auto.

Path choices obviously do not need to be modeled since there is only one path available to each mode. Hence, all of the commuters that choose a particular mode can simply be assigned to the single path for that mode.

The Data

There are 15,000 people, in total, commuting from node O to node D. Currently (i.e., with a transit fare of \$1.50) 12,560 people use transit and 2,440 people use the highway.

The systematic utilities for the logit model have been estimated as

$$V_A = 0.893 - 0.00897 \cdot i_A - 0.0308 \cdot o_A - 0.007 \cdot c_a \tag{55.104}$$

$$V_T = -0.00897 \cdot i_T - 0.0308 \cdot o_T - 0.007 \cdot c_t \tag{55.105}$$

where i is the in-vehicle travel time, o is the out-of-vehicle travel time, and c is the monetary cost per trip on that mode.

For transit, $i = 45$ (in minutes), $o = 10$, and $c = 150$ (cents). For the highway, $o = 5, c = 560$ (\$0.28 per mile times 20 miles), and, under current conditions, $i = 24.5$.

The practical capacity of the highway is 4,000, and the free-flow speed is 50 mph. Hence, since the highway is 20 miles long, the free-flow time, t_0, is 24 minutes.

Using the Models

It would seem like it should be relatively easy to use the logit model of mode choice to determine the impact of the fare increase. However, observe that this model requires the auto travel time as input and it is not clear what value should be used. Assuming that the highway will continue to operate at its current level of service, the travel time will be 24.5 minutes. Using this value, one finds that the systematic utilities are given by

$$V_A = 0.893 - 0.00897 \cdot 24.5 - 0.0308 \cdot 5 - 0.007 \cdot 560 = -3.4 \tag{55.106}$$

and

$$V_T = -0.00897 \cdot 45 - 0.0308 \cdot 10 - 0.007 \cdot 300 = -2.8 \tag{55.107}$$

Hence, the choice probabilities are given by $P_A = 0.3569$ and $P_T = 0.6431$. In other words, $0.3569 \cdot 15{,}000 = 5353$ people will use the highway and $0.6431 \cdot 15{,}000 = 9647$ people will use transit after the fare hike.

However, observe that these values are not consistent with the original assumption that the road would operate in near free-flow conditions. In particular, with 5353 highway users the travel time [calculated using (55.102)] will actually be 35.5 minutes, not 24 minutes. Hence, though people may make their initial choices based on free-flow speeds, they are likely to change their behavior in response to their incorrect estimate of the auto travel time.

If one believes that people will keep changing their paths until the travel time that is used as an input to the mode choice model is the same as the travel time that actually results, then it is necessary to solve the two models simultaneously. Doing so in this case, it turns out that $t_a = i_A = 33.38$, $V_A = -3.48$, $V_T = -2.81$, and hence that 5082 people will use auto and 9918 will use transit.

Of course, we could take this one step further. In particular, suppose there was another residential neighborhood, and that residential location choice could be modeled as follows:

$$n_j = N \cdot \frac{A_j \exp(-\theta c_j)}{\sum_i A_i \exp(-\theta c_i)} \tag{55.108}$$

where n_j is the number of commuters that choose to live in zone j, N is the total number of commuters, A_j is the "attractiveness" of the zone j, c_j is the commuting "cost" to the central business district, and θ is the cost sensitivity parameter. Then, it is easy to see that, since the cost of commuting has changed, the number of people living in each zone will also change (at least in the long run). Hence, one might want to simultaneously solve all three models.

Defining Terms

Aggregate models: Aggregate models consider the decisions of a group in total. The groups themselves can be based either on geography or socioeconomic characteristics.

Alternatives: The set of possible decisions that an individual can make.

Choice: The alternative that an individual selects in a given situation.

Disaggregate models: Disaggregate models consider the behavior of individuals (or sometimes households). They essentially consider the choices that individuals make among different alternatives in a given situation.

Equilibrated path choices: Path choices are equilibrated when no portion of the flow on any path, p, can switch to any other path, r, connecting the same OD-pair without making the new cost on r at least as large as the new cost on p.

Independence from irrelevant alternatives (IIA): The property that the ratio of the choice probabilities for i and k is independent of all of the other alternatives (within the context of probabilistic choice models).

Macrostate: The number of trips to and/or from each zone (within the context of a gravity model).

Management/administration: Activities related to the transportation organization itself.

Microstate: The set of such choices made by a group of individuals (within the context of a gravity model).

Modal split: This term is used to refer to both the process of modeling/predicting mode choices and the results of that process.

Operations/control: Activities related to the provision of transportation services when the system is in a stable (or relatively stable) state.

Organizational setting: The organization and administrative rules and practices that distribute decision-making powers and that set limits on the process and on the range of alternatives.

Planning/design: Activities related to changing the way transportation services are provided (i.e., state transitions).

Planning situation: The number of decision makers, the congruity and clarity of values, attitudes and preferences, the degree of trust among decision makers, the ability to forecast, time and other resources available, quality of communications, size and distribution of rewards, and the permanency of relationships.

Resolution: The resolution of a system is defined by how the system of interest is seen in relation to the environment (i.e., all other systems) and its elements which are treated as black boxes.

Societal setting: The laws, regulations, customs, and practices that distribute decision-making powers and that set limits on the process and on the range of alternatives.

System: A set of objects, their attributes, and the relationships between them.

System state: The known properties of the system (within the context of a gravity model).

Telecommuting: Using telecommunications technology (e.g., telephones, FAX machines, modems) to interact with coworkers in lieu of actually traveling to a central location.

Teleshopping: Using telecommunications technology (e.g., telephones, FAX machines, modem) to either acquire information about products or make purchases.

Traffic assignment: This term is used both to describe the process of modeling/predicting path choices and the results of that process.

Transitivity of preferences: Preferences are said to be transitive if whenever A is preferred to B and B is preferred to C it also follows that A is preferred to C.

Transportation planning: Activities related to changing the way transportation services are provided. Typical activities include the generation and evaluation of alternative proposals.

Trip attractions: The number of trips destined for a particular location.

Trip distribution: This term is used both to describe the process of modeling/predicting origin and destination choices and the result of that process.

Trip generation: Determining the number of trips that will originate from and terminate at each zone in the network. Trip generation models attempt to explain/predict the decision to travel.

Trip productions: The number of trips originating from a particular location.

Trip table: The number of trips traveling between each origin-destination pair.

User equilibrium: Several slightly different definitions of user equilibrium exist. The essence of these definitions is that no traveler can reduce his or her travel cost by unilaterally changing paths.

Wardrop equilibrium: A situation in which the cost on all of the paths between an origin and destination actually used are equal, and less than those which would be experienced by a single vehicle on any unused path.

References

Aashtiani, H. Z. 1979. *The Multi-Modal Traffic Assignment Problem.* Ph.D. Dissertation, Massachusetts Institute of Technology.

Aashtiani, H. Z. and Magnanti, T. L. 1981. Equilibria on a congested transportation network. *SIAM J. Algebraic Discrete Methods.* 2:213–226.

Abdulaal, M. and LeBlanc, L. J. 1979. Methods for combining modal split and equilibrium assignment models. *Transp. Sci.* 13:292–314.

Alfa, A. S. and Minh, D. L. 1979. A stochastic model for the temporal distribution of traffic demand—The peak hour problem. *Transp. Sci.* 13:315–324.

Arnott, R., de Palma, A., and Lindsey, R. 1990a. Economics of a bottleneck. *J. Urban Econ.* 27:111–130.

Arnott, R., de Palma, A., and Lindsey, R. 1990b. Departure time and route choice for the morning commute. *Transp. Res.* 24B:209–228.

Asmuth, R. L. 1978. *Traffic Network Equilibria.* Ph.D. Dissertation, Stanford University.

Beckmann, M., McGuire, C., and Winsten, C. 1956. *Studies in the Economics of Transportation.* Yale University Press, New Haven.

Ben-Akiva, M. 1973. *Structure of Passenger Travel Demand Models.* Ph.D. Dissertation, Massachusetts Institute of Technology.

Ben-Akiva, M., Cyna, M., and de Palma, A. 1984. Dynamic models of peak period congestion. *Transp. Res.* 18B:339–355.

Ben-Akiva, M. and Lerman, S. 1985. *Discrete Choice Analysis.* MIT Press, Cambridge.

Ben-Akiva, M., de Palma, A., and Kanaroglu, P. 1986. Dynamic models of peak period traffic congestion with elastic arrival rates. *Transp. Sci.* 20:164–181.

Bernstein, D., Friesz, T. L., Tobin, R. L, and Wie, B.-W. 1993. A Variational Control Formulation of the Simultaneous Route and Departure-Time Choice Equilibrium Problem. In *Proc. 12th Int. Symp. Theory Traffic Flow Transp.*

Bernstein, D. and Smith, T. E. 1994. Network equilibria with lower semicontinuous costs: With an application to congestion pricing. *Transp. Sci.* In press.

Bernstein, D. and Smith, T. E. 1993. Programmability of Discrete Network Equilibria. MIT Working Paper.

Boyce, D. E., LeBlanc, L. J., and Chon, K. S. 1988. Network equilibrium models of urban location and travel choices: A retrospective survey. *J. Reg. Sci.* 28:159–183.

Bowman, L. A. and Turnquist, M. A. 1981. Service frequency, schedule reliability, and passenger wait times at transit stops. *Transp. Res.* 15A:465–471.

Cascetta, E. 1989. A stochastic process approach to the analysis of temporal dynamics in transportation networks. *Transp. Res.* 23B:1–17.

Chriqui, C. and Robillard, P. 1975. Common bus lines. *Transp. Sci.* 9:115–121.

Dafermos, S. C. and Sparrow, F. T. 1969. The traffic assignment problem for a general network. *J. Res. Nat. Bur. Stand.* 73B:91–118.

Dafermos, S. C. 1971. An extended traffic assignment model with applications to two-way traffic. *Transp. Sci.* 5:366–389.

Dafermos, S. C. 1980. Traffic equilibrium and variational inequalities. *Transp. Sci.* 14:42–54.

Daganzo, C. F. 1979. *Multinomial Probit: The Theory and Its Application to Demand Forecasting.* Academic Press, New York.

Daganzo, C. F. and Sheffi, Y. 1977. On stochastic models of traffic assignment. *Transp. Sci.* 11:253–274.

Daganzo, C. F. and Kusnic, M. 1993. Two properties of the nested logit model. *Transp. Sci.* 27:395–400.

Daly, A. and Zachary, S. 1979. Improved multiple choice models. In *Determinants of Travel Choice,* eds. D. A. Hensher and M. Q. Dalvi. Prager, New York.

de Cea, J., Bunster, J. P., Zubieta, L., and Florian, M. 1988. Optimal strategies and optimal routes in public transit assignment models: An empirical comparison. *Traffic Eng. Control.* 29:520–526.

de Palma, A., Ben-Akiva, M., Lefevre, C., and Litinas, N. 1983. Stochastic equilibrium model of peak period traffic congestion. *Transp. Sci.* 17:430–453.

Devarajan, S. 1981. A note on network equilibrium and noncooperative games. *Transp. Res.* 15B:421–426.

Dial, R. B. 1967. Transit pathfinder algorithm. *Highway Res. Rec.* 205:67–85.

Domencich, T. and McFadden, D. 1975. *Urban Travel Demand—A Behavioral Analysis.* North-Holland, Amsterdam.

Doubleday, C. 1977. Some studies of the temporal stability of person trip generation models. *Transp. Res.* 11:255–263.

Douglas, A. A. 1973. Home-based trip end models—A comparison between category analysis and regression analysis procedures. *Transp.* 2:53–70.

Erlander, S. and Stewart, N. F. 1989. *The Gravity Model in Transportation.* VSP, Utrecht, Netherlands.

Etzioni, A. 1967. Mixed scanning: A "third" approach to decision-making. *Pub. Adm. Rev.* December.

Evans, S. 1976. Derivation and analysis of some models for combining trip distribution and assignment. *Transp. Res.* 10:37–57.

Finney, D. 1971. *Probit Analysis.* Cambridge University Press, Cambridge, England.

Fisk, C. 1985. Entropy and information theory: Are we missing something? *Environ. Plann.* 17A:679–687.

Florian, M., Nguyen, S., and Ferland, J. 1975. On the combined distribution-assignment of traffic. *Transp. Sci.* 9:43–53.

Florian, M. 1977. A traffic equilibrium model of travel by car and public transit modes. *Transp. Sci.* 11:166–179.

Florian, M. and Nguyen, S. 1978. A combined trip distribution, modal split and trip assignment model. *Transp. Res.* 4:241–246.

Friesz, T. L. 1985. Transportation network equilibrium, design and aggregation: Key developments and research opportunities. *Transp. Res.* 19A:413–427.

Friesz, T. L., Luque, F. J., Tobin, R. L., and Wie, B. W. 1989. Dynamic network traffic assignment considered as a continuous time optimal control problem. *Operations Res.* 37:893–901.

Friesz, T. L., Bernstein, D., Smith, T. E., Tobin, R. L., Wie, B. W. 1993. A variational inequality formulation of the dynamic network user equilibrium problem. *Operations Res.* 41:179–191.

Grigsby, W. and Bernstein, D. 1993. A new definition of planning. Fels Center Working Paper, University of Pennsylvania.

Hall, A. and Fagen, R. 1956. Definition of system. *Gen. Syst.* 1:18–28.

Harker, P. T. and Pang, J. S. 1990. Finite-dimensional variational inequality and complementarity problems. *Math. Programming.* 48:161–220.

Hendrickson, C. and Kocur, G. 1981. Schedule delay and departure time decisions in a deterministic model. *Transp. Sci.* 15:62–77.

Hendrickson, C. and Plank, E. 1984. The flexibility of departure times for work trips. *Transp. Res.* 18A:25–36.

Heydecker, B. G. 1986. On the definition of traffic equilibrium. *Transp. Res.* 20B:435–440.

Hickman, M. 1993. *Assessing the Impact of Real-Time Information on Transit Passenger Behavior.* Ph.D. Dissertation, Massachusetts Institute of Technology.

Hua, C.-I. and Porell, F. 1979. A critical review of the development of the gravity model. *Intl. Reg. Sci. Rev.* 4:97–126.

Jansson, K. and Ridderstolpe, B. 1992. A method for the route choice problem in public transport systems. *Transp. Sci.* 26:246–251.

Jolliffe, J. K. and Hutchinson, T. P. 1975. A behavioral explanation of the association between bus and passenger arrivals at a bus stop. *Transp. Sci.* 9:248–282.

Keefer, L. J. 1966. *Urban Travel Patterns for Airports, Shopping Centers and Industrial Plants.* National Cooperative Highway Research Project Report No. 24. Highway Research Board, Washington, D.C.

LeBlanc, L. J., Morlok, E. K., and Pierskalla, W. 1975. An efficient approach to solving the road network equilibrium traffic assignment problem. *Transp. Res.* 9:309–318.

le Clercq, F. 1972. A public transport assignment method. *Traffic Eng. Control.* 14:91–96.

Lindblom, C. 1959. The science of "muddling through." *Pub. Adm. Rev.* Spring.

Mahmassani, H. S. and Herman, R. 1984. Dynamic user equilibrium departure time and route choice on idealized traffic arterials. *Transp. Sci.* 18:362–384.

Marguier, P. H. J. 1981. *Optimal Strategies in Waiting for Common Bus Lines.* M.S. thesis, Massachusetts Institute of Technology.

Marguier, P. H. J. and Ceder, A. 1984. Passenger waiting strategies for overlapping bus routes. *Transp. Sci.* 18:207–230.

McFadden, D. 1976. The Mathematical Theory of Demand Models. In *Behavioral Travel Demand Models,* ed. P. Stopher and A. Meyburg. Lexington Books, Lexington, MA.

Nagurney, A. 1984. Comparative tests of multimodal traffic equilibrium methods. *Transp. Res.* 18B:469–485.

Nagurney, A. 1988. An equilibration scheme for the traffic assignment problem with elastic demands. *Transp. Res.* 22B:73–79.

Nagurney, A. 1993. *Network Economics.* Kluwer, Boston.

Newell, G. F. 1987. The morning commute for nonidentical travelers. *Transp. Sci.* 21:74–82.

Niedercorn, J. H. and Bechdolt, B. V. 1969. An economic derivation of the "gravity law" of spatial interaction. *J. Reg. Sci.* 9:273–281.

Nguyen, S. 1974. A unified approach to equilibrium methods for traffic assignment. In *Traffic Equilibrium Methods,* pp. 148–182. Lecture Notes in Economics and Mathematical Systems, Springer-Verlag, New York.

Nguyen, S. 1978. An algorithm for the traffic assignment problem. *Transp. Sci.* 8:203–216.

Nguyen, S. and Pallottino, S. 1988. Equilibrium traffic assignment for large scale transit networks. *Eur. J. Operational Res.* 37:176–186.

Putman, S. H. 1983. *Integrated Urban Models.* Pion, London.

Ran, B., Boyce, D. E., and LeBlanc, L. J. 1993. A new class of instantaneous dynamic user-optimal traffic assignment models. *Operations Res.* 41:192–202.

Rosenthal, R. W. 1973. The network problem in integers. *Networks.* 3:53–59.

Safwat, K. N. A. and Magnanti, T. L. 1988. A combined trip generation, trip distribution, modal split and trip assignment model. *Transp. Sci.* 18:14–30.

Sen, A. and Smith, T. E. 1994. *Gravity Models of Spatial Interaction Behavior.* Unpublished.

Sheffi, Y. 1985. *Urban Transportation Networks.* Prentice-Hall, Englewood Cliffs, NJ.

Sheffi, Y. and Powell, W. B. 1982. An algorithm for the equilibrium assignment problem with random link times. *Networks.* 12:191–207.

Smith, M. J. 1979. The existence, uniqueness, and stability of traffic equilibria. *Transp. Res.* 13B:295–304.

Smith, M. J. 1984. Two alternative definitions of traffic equilibrium. *Transp. Res.* 18B:63–65.

Smith, M. J. and Ghali, M. O. 1990. Dynamic traffic assignment and dynamic traffic control. In *Proc. 11th Int. Symp. Transp. Traffic Theory,* Elsevier, New York, pp. 273–290.

Smith, T. E. 1975. A choice theory of spatial interaction. *Reg. Sci. Urban Econ.* 5:137–176.

Smith, T. E. 1976a. Spatial discounting and the gravity hypothesis. *Reg. Sci. Urban Econ.* 6: 331–356.

Smith, T. E. 1976b. A spatial discounting theory of interaction preferences. *Environ. Plann.* 8A:879–915.

Smith, T. E. 1983. A cost-efficiency approach to the analysis of congested spatial-interaction behavior. *Environ. Plann.* 15A:435–464.

Smith, T. E. 1986. An axiomatic foundation for poisson frequency analyses of weakly interacting populations. *Reg. Sci. Urban Econ.* 16:269–307.

Smith, T. E. 1988. A cost-efficiency theory of dispersed network equilibria. *Environ. Plann.* 20A:231–266.

Smith, T. E. and Bernstein, D. 1993. Programmable Network Equilibria. In *Structure and Change in the Space Society,* ed. T. R. Lakshmanan and P. Nijkamp, pp. 91–130. Springer-Verlag, New York.

Spiess, H. and Florian, M. 1989. Optimal strategies: A new assignment model for transit networks. *Transp. Res.* 23B:83–102.

Stouffer, S. A. 1940. Intervening opportunities: A theory relating mobility and distance. *Am. Sociological Rev.* 5:845–867.

Sumi, T., Matsumoto, Y., and Miyaki, Y. 1990. Departure time and route choice of commuters on mass transit systems. *Transp. Res.* 24B:247–262.

Theil, H. 1971. *Principles of Econometrics.* John Wiley & Sons, New York.

Tomlin, J. A. 1971. A mathematical programming model for the combined distribution-assignment of traffic. *Transp. Sci.* 5:122–140.

Train, K. 1984. *Qualitative Choice Analysis: Theory, Economics, and an Application to Automobile Demand.* MIT Press, Cambridge.

Turnquist, M. A. 1978. A model for investigating the effects of service frequency and reliability on bus passenger waiting times. *Transp. Res. Rec.* 663:70–73.

Vickrey, W. 1969. Congestion theory and transport investment. *Am. Econ. Rev.* 56:251–260.

Wardrop, J. G. 1952. Some theoretical aspects of road traffic research. *Proc. Inst. Civ. Eng.* Part II. 1:325–378.

Wilson, A. G. 1970. *Entropy in Urban and Regional Planning.* Pion, London.

For Further Information

In addition to the references listed above, there are several introductory texts devoted to transportation planning, including *Fundamentals of Transportation Systems Analysis* by M. L. Manheim, *Introduction to Transportation Engineering and Planning* by E. K. Morlok, and *Fundamentals of Transportation Engineering* by C. S. Papacostas.

A variety of journals are also devoted (in whole or in part) to transportation planning, including the *Journal of Transport Economics and Policy, Transportation, Transportation Research,* the *Transportation Research Record,* and *Transportation Science.*

56

Airport Planning and Design

Robert K. Whitford
Purdue University

56.1 The Air Transportation System

Since World War II, air transportation has been one of the fastest-growing segments of the U.S. economy. In 1945 U.S. commercial airlines flew 5.3 billion revenue passenger miles (RPM), growing to 104.1 billion RPM in 1970, then to a phenomenal 480 billion RPM in 1992. U.S. air travel is expected to top 840 billion RPM by the year 2004 [FAA, 1993a]. Commercial and commuter air carriers increased their boardings from 312 million persons in 1982 to 471.6 million in 1992, an annual increase of passenger enplanements of about 4.2% per year. Aviation's present impact

0-8493-8953-4/95/$0.00 + $.50

on the economy has been no less spectacular, providing the basis for about 5.6% of the gross national product in 1989, with over 2 million persons directly employed in aviation [Wilbur Smith Associates, 1990]. The air transportation system, particularly its airports, is considered by many to be an engine for economic growth. Chicago's O'Hare airport alone accounted for adding an estimated $10.3 billion into the Chicago economy in 1985 [al Chalibi, 1993]. Aviation in the New York metro area alone was estimated to contribute about $30 billion to that economy in 1989 [Smith, 1990]. These significant contributions to the national and local economies are predicted to grow.

Civil Engineering and Airport Planning/Design

As the demand for air travel increases, so does the demand for airport capacity. In the last five to ten years, concern about capacity and the delay inherent in a system that operates close to saturation has caused the FAA to embark on a program to carefully examine the top 100 airports in the country and identify the needs for expanded capacity in the next 10 to 20 years [FAA, 1991]. Additional capacity is expected to be provided through a number of changes to the system. The primary focus at many airports is to provide more runways and/or high-speed exits. In addition, an increased number of reliever airports are planned, with improved instrument approach procedures, changes in limitations on runway spacing, provision for added on-site weather stations, and a more efficient air traffic control system.

Increased traffic and heavier aircraft place a demand on aprons. In addition, many airports face crowded conditions on the landside of their system, which will require terminal expansion or renovation, improved access by ground transportation, or increased parking.

Fundamentally, the airport is a point of connectivity in the transportation system. At the ends of a trip the airport provides for the change of mode from a ground to air mode or vice versa. As such, the airport is often analyzed using the schematic of Fig. 56.1, with the airport's *airside* consisting of approach airspace, landing aids, runway, taxiway, and apron all leading to the gate where the passenger (or cargo) passes through; and the airport's *landside*, consisting of the areas where the passenger (or cargo) is processed for further movement on land: the arrival and departure concourses, baggage handling, curbside, and access to parking lots, roads, and various forms of transit.

Most design aspects of the airport must reflect the composite understanding of several interrelated factors. Factors include aircraft performance and size, air traffic management, demand for safe and effective operation, the effects of noise on communities, and obstacles on the airways. All the disciplines of civil engineering are called into use in airport planning and design.

The Airport System

Figure 56.2 shows the top 100 airports in the U.S. in 1989 with a pattern that mirrors the population. As shown in Table 56.1, there are more than 16,000 airports in the U.S. Over 64% are privately owned and generally not lighted or paved. While there are many airports, only those that appear in a given state's aviation system plan are likely to involve the level of airport planning suggested here. These are usually the public airports, with some commercial operations such as air taxi or charter service that operate as a reliever airport, or where special conditions apply.

Focus on Planning

As part of an entire transportation system, airport planning must be broad, complete, and future oriented, because its design and operational features often exhibit strong interrelationships that reflect the long lead time of large investment decisions. The planning factors are as follows:

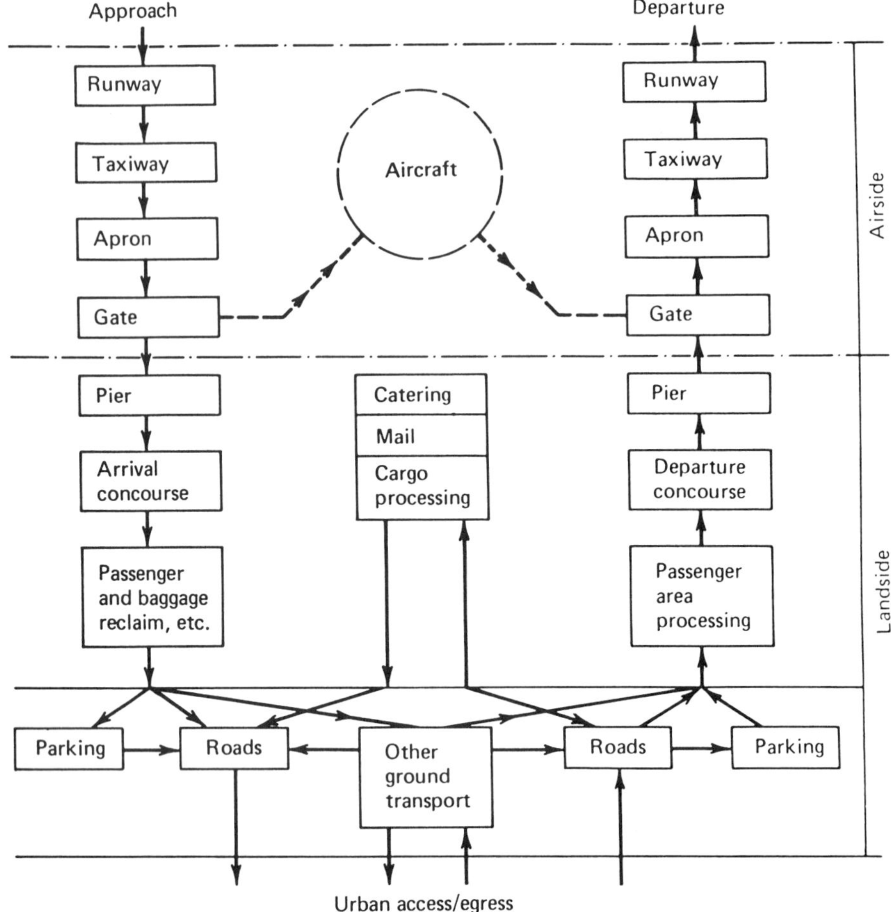

FIGURE 56.1 The airport system. (*Source:* Ashford, N., Stanton, H., and Moore, P. 1991. *Airport Operations.* Pitman, London.)

1. Demand for use of the airport by the community in both passengers and freight
2. Demand for airline use for hubbing
3. Operating characteristics, size, weight, and mix of potential aircraft using the airport
4. Meteorological and weather conditions at the airport
5. Volume, mix, and markets served by airlines and other aircraft operations
6. Constraints on navigation and navigable airspace
7. Environmental considerations associated with the community's land-use plan

Ownership and Management

Most public airports are owned by the municipal government(s) of the political jurisdiction(s) of the major markets which the airport serves. Where multiple jurisdictions are near airport boundaries or have significant use of the airport, an authority or board is set up with representatives from the involved jurisdictions, usually with some joint operating and funding arrangement. For example, the major airports around New York City—LaGuardia, John F. Kennedy International, and Newark International—are managed by the Port Authority of New York and New Jersey. The Port Authority also manages a general aviation airport at Teterboro and two heliports in the area encompassing

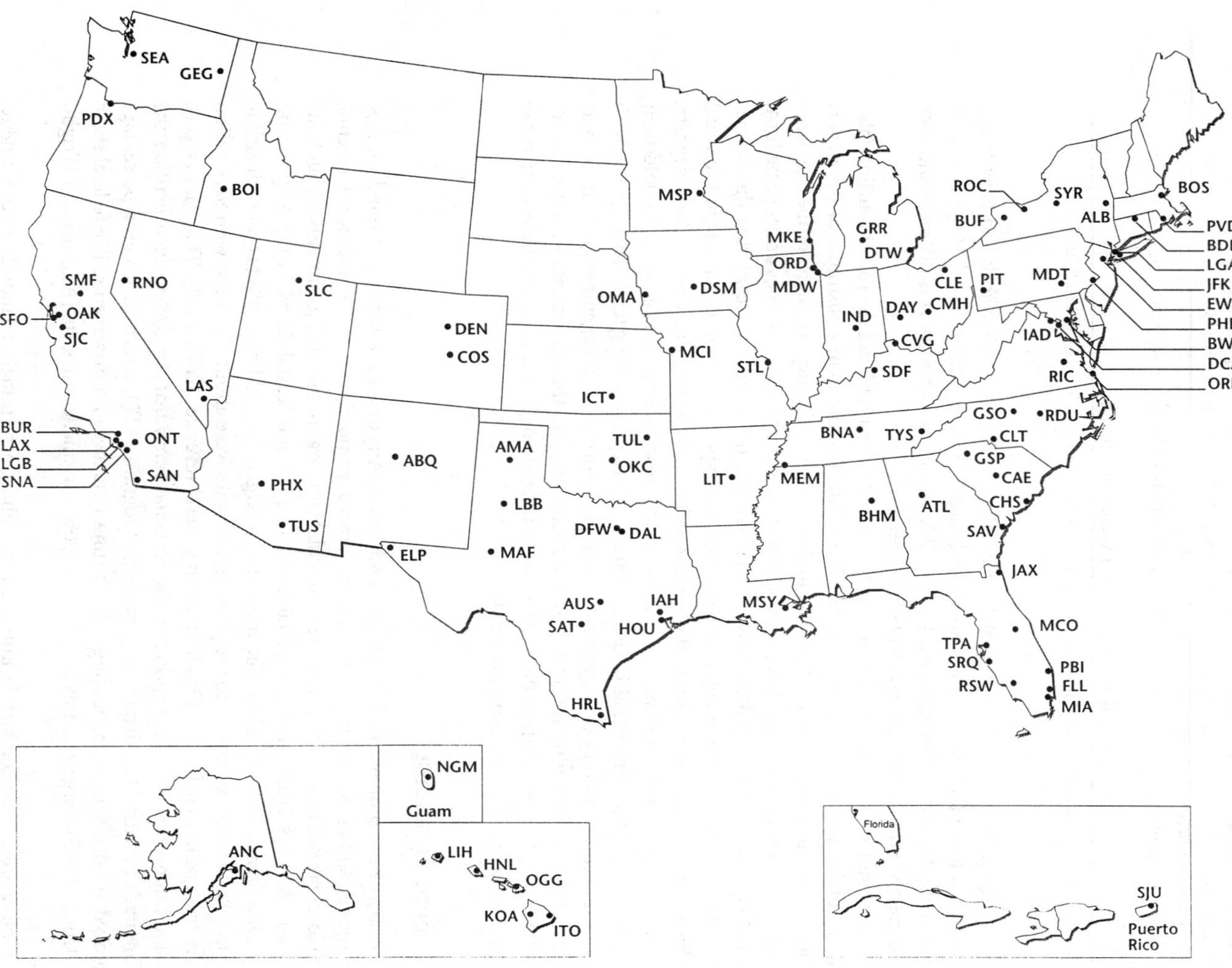

FIGURE 56.2 The top 100 airports in the U.S. (*Source:* FAA. 1991. *1991–1992 Aviation System Capacity Plan*, U.S. Department of Transportation Report DOT/FAA/ASC-91-1.)

Table 56.1 Airports in the U.S.

Airport Classification	Definition	Number (1990)
Large hubs	Enplanements > 1% of U.S. enplanements	30
Medium hubs	Enplanements between 1% and .25% of U.S. enplanements	41
Small hub/primary airports	Enplanements between .01% and 0.25% of U.S. enplanements	201
Other commercial airports	Enplanements between 2500 and 0.01% of U.S. enplanements	272
Reliever airports	Serves to relieve an airport with >250,000 enplanements	244
General aviation airports	Enplanements less than 2500	2,449
Other public-use airports	Not in the National Program Integrated System of Airports	2,588
Private airports		10,449
Heliports	Limited to vertical takeoff and landing aircraft	3600

Source: FAA. 1991b. *National Plan of Integrated Airport Systems.* Advisory Circular AC150/SC050-5.

about a 25 mile radius from the Statue of Liberty [Port Authority of New York and New Jersey, 1992]. On the other hand, the airports around Chicago (O'Hare, Midway, and Meigs Field) are managed by the Airport Authority of the City of Chicago. Thus each airport is different and each faces unique operational and management challenges.

It is important for the planner to know how the airport is financed and the role the airlines play in influencing the management of the airport. The airline is more than a customer of the airport since they often provide some of the financial underpinning of the airport. Many of the U.S. large and medium hubs have negotiated long-term agreements with the major airlines under some form of residual cost management [CBO, 1984]. (Residual cost management means that the airlines assume responsibility for paying any residual uncovered expenses the airport incurs in the year.) The airlines wield a considerable amount of power in the management decisions of these airports, because they are responsible for any cost excess and because they are always trying to hold their landing fees down. Other airports also have agreements that are not as long term. They operate with the more usual compensatory cost approach. (Compensatory cost management gives the airport management the responsibility for all airport cost accounts, and the agreements with the airlines are shorter term.) In this situation the local airport authority or board has more latitude to make plans more reflective of the community needs.

Investment Financing

Many airports raise money locally through bond issues. Airports have very good bond ratings. Where municipalities govern the airport, it is sometimes possible to raise additional revenue through local taxes. The federal government provides funding for airside investments through the Airport and Airways Trust Fund (first established in 1954 and reestablished in the Airport and Airways Development Act of 1970). The trust fund is largely funded by the 8% tax on each airline ticket. In 1988 the federal outlays for airports and airways were about $6 billion with $2.9 billion from the trust fund matched by $3 billion from general revenue [CBO, 1988]. That money goes for airside improvements. The airports in the *National Integrated Air System Plan* petition the FAA for funds through the Airport Improvement Program (AIP), which furnishes a percentage of approved airport navigation, landing aids, or runway/taxiway improvements. The federal share ranges from 75% for large and medium hubs to 90% for smaller airports [Ashford and Wright, 1992].

For their share in funding airside improvements and for terminal or landside improvements funding will usually come from the state and local governments through taxes and revenue bonds [CBO, 1984]. The Airport Safety and Capacity Expansion Act of 1990 allows airports to charge each passenger enplaning a passenger facility charge (PFC). The passenger facility charge provides the opportunity for airports to charge all users a fee not to exceed $3.00 for boarding at the airport. The DOT must approve applications for these funds, which are used for airside and terminal improvements but do not include improvements related to concessions or parking. The PFC was instituted to make it easier for airports to make improvements to airside or landside through direct user charge. Petitions to charge the PFC must be approved by the U.S. Department of Transportation.

56.2 The Airport Planning Process

There is a hierarchy of planning documents, beginning with the biannually published *National Plan of Integrated Airport Systems* (NPIAS) [FAA, 1991] that lists those public use airports where development is considered to be in the national interest and which are eligible for funding under the most recent congressional airport act. As suggested in Fig. 56.3, each state maintains a state system plan identifying their public use airports and indicating the needs for upgrading existing airports and development of new airports. The planning studies are partially funded by the FAA, usually with 90% from federal funds and 10% from state and local funds. The purpose of such planning is for the federal agencies in cooperation with regions and states to achieve an integrated plan facilitating further technical planning, refinements to transportation policy, integration of the various transportation modes, and multijurisdiction coordination.

The Master Plan

The individual airport master plan is the cornerstone of the continuing, comprehensive, and cooperative planning process [FAA, 1975]. It is a most exacting plan, generally prepared by the airport staff or consultants. It details long-range needs and implementation plans for the airport and is used by the airport's governing board or authority, the state, and the FAA in defining future funding requirements. The master plan reflects the complexity and size of the airport. A small, general aviation (GA) airport with 20,000 operations per year may require only a few pages and a short report indicating the airport's future needs. The state generally provides the forecast for such airports developed on a count of operations and on the number of aircraft based at the airport. (The number of operations at a small, nontowered GA airport is usually not well known. Some states use acoustical counters which are placed at the airport for a few weeks to monitor operations. Others make estimates based on surveys, fixed-base operator (FBO) counts, and other data.)

Large, sophisticated airports usually have ongoing studies involving several consultants and consisting of several volumes. For example, the master plan for Chicago's O'Hare Airport has some 19 volumes with over 6000 pages. Frequently, the master plan is aimed at solving a specific problem,

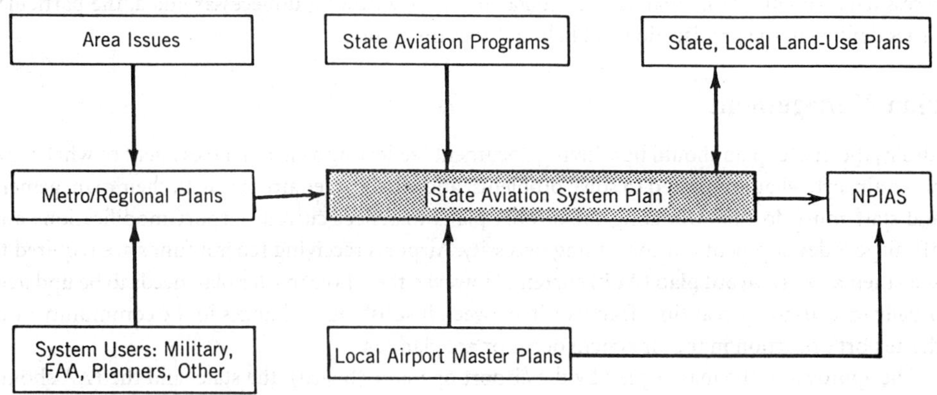

FIGURE 56.3 Planning relationships for a state aviation plan. (*Source:* FAA. 1975. *The Continuous Airport Planning Process.* Advisory Circular AC150/5050-5.)

Table 56.2 Steps in the Airport Master Planning Process

1.	**Decision** → A new master plan is needed (includes discussion of issues for airport)
	2. Developing the study grant application (includes scope)
	3. Consultant hired after agency coordination and approval
	4. Inventory of existing capability, capacity, and resources
	5. Forecast of demand
	6. Requirements analysis and concepts development
7.	**Decision** → New airport, or can we upgrade present airport?
	8. Site decision and planning
	9. Alternatives analysis
10.	**Decision** → Select approach desired from alternatives
	11. Detailed planning and preliminary engineering
	12. Financial plan (staged development)
	13. Implementation plan

Source: FAA. 1985. *Airport Master Plans.* Advisory Circular AC150/5070-6A.

such as repairing runways, evaluating obstructions, or improving the navigation or terminal landing aids. Physical improvements such as added or extended runways, taxiways, and apron expansion are also identified in the master plan.

The master planning process includes the steps indicated in Table 56.2. Each step involves some coordination with the FAA and the state. Public hearings may be a part of the process.

Airport Issues and Existing Conditions

Plans are not generated in a vacuum, nor are they generated if there are no issues. Almost every airport has some deficiency that the airport board or the community or some other airport stakeholder would like to see addressed. These issues can range from improving the capacity (and hence reducing the delay) to a desired improvement in the baggage-handling system. The study is undertaken by first identifying and gathering the issues obtained by examining prior studies and reports and by having in-depth discussions with the FAA region, the state aviation officials, the airport management, the air traffic controller, the airlines, the FBO, and others involved in the airport use.

Next, data are collected on the airport, the airspace infrastructure, and the nonaviation areas of airport land use. The data include an inventory of the existing physical plant, including an assessment of its condition and useful life, plus other relevant items such as land use surrounding the airport, financial data on the airport operation, community social and demographic data (to aid in forecasting), operational data on the airport, meteorological data, environmental data, ground access data, and air traffic management data. To avoid collecting unnecessary data, the particular issues defined in the preplanning will help to focus the efforts.

Plan Management

Ideally, the master plan should be a "living document" reflecting a current assessment of what exists at the airport, what is required to solve problems, and why. Larger airports with their management and staff must do this. Updating the airport plans to reflect current airport modifications and off-airport development is a continuing necessity. Airports receiving federal funds are required to keep their **airport layout plan (ALP)** current. However, the whole master plan needs to be updated, usually in a 10- to 20-year time frame or in between if substantive changes in the community or in the airport's function in the air system occur or are planned.

The approval of the master plan by the airport operator (board), the state, and the FAA should be done in a timely manner so that reimbursement for the consultant and FAA payments under federally assisted projects will be approved. The FAA approval of a given plan extends only to ensuring completion of the work elements specified in the grant agreement [FAA, 1985].

56.3 Forecasting Airport Traffic

Planning for an airport and building a credible airport investment program require that future traffic be forecast in a thorough, sensible manner. An overly optimistic forecast may cause premature investment costs and higher-than-needed operating costs; an overly conservative forecast will promote increased congestion with high levels of delay and potentially lost revenues.

In the exercise of its responsibility for investment planning, especially for the Airport and Airway Trust Fund and for future air traffic operations, the FAA has been forecasting overall traffic in the United States for a number of years [FAA, 1993]. The FAA also publishes forecasts of over 3600 airports in the U.S. that are eligible for AIP grants. The FAA forecasts are proven estimates weighing the inputs from many different sources [FAA, 1993]. Some important factors that need to be considered in the planning for a specific airport include the following:

- Unusual demographic factors existing in the community
- Geographic factors that will affect the amount of airplane use
- Changes in disposable income permitting some travelers to travel more
- Nearby airports whose operation may draw from the airport being planned
- Changes in how airlines use the airport (more hubbing, route changes, etc.)
- New local industry, meaning more jobs and more business travel
- New resort/convention industry(ies) or capacity that will bring vacation travelers

Forecasting traffic is generally handled differently for the large, medium, and small hubs than for small commercial, basic transport, general aviation airports. Figure 56.4 describes the flow of systems analysis on which much of the planning is based. Unless unusual conditions exist in an area, as is the case with very large urban areas like Chicago, the flow portrayed will determine the demand. The demand can be as simply stated as the percentage demand that "this Airport" enjoys when related to the national air system total demand. In more complex areas the demand forecast would be enriched by the addition of more detail about local economic conditions, about other transportation facilities, about the airline operations, and about aircraft to be used when demand changes.

Large, Medium, and Small Hubs

For the airports that have more than 0.05% of the national enplanements (255,000 in 1992) the forecasting is generally done by either comparing the airport in the context of the national airspace system using national statistics or using regression equations. Forecasting provides information about two important areas of design concern, namely, the prediction of passengers (enplanements) to aid in planning for terminal facilities and the anticipated number of operations (takeoffs and landings) needed for an appraisal of the adequacy of runways, taxiways, apron and air traffic control capability to handle the traffic without significant delay. The link between operations and enplanements is the capacity of the average aircraft (departing seats) coupled with the average passenger load factor as shown in Eq. (56.1).

$$\mathrm{DEP}_{\mathrm{A/C}} = \frac{\mathrm{OPS}_{\mathrm{A/C}}}{2} = \frac{\mathrm{ENP}}{\mathrm{SEATS}_{\mathrm{DEPART}} \times \mathrm{LF}} \tag{56.1}$$

where

$\mathrm{DEP}_{\mathrm{A/C}}$ = commercial aircraft departures
$\mathrm{OPS}_{\mathrm{A/C}}$ = commercial aircraft operations
$\mathrm{SEATS}_{\mathrm{DEPART}}$ = departing aircraft seats averaged over commercial aircraft departures
LF = average load factor or number of seats occupied
ENP = enplaning passengers

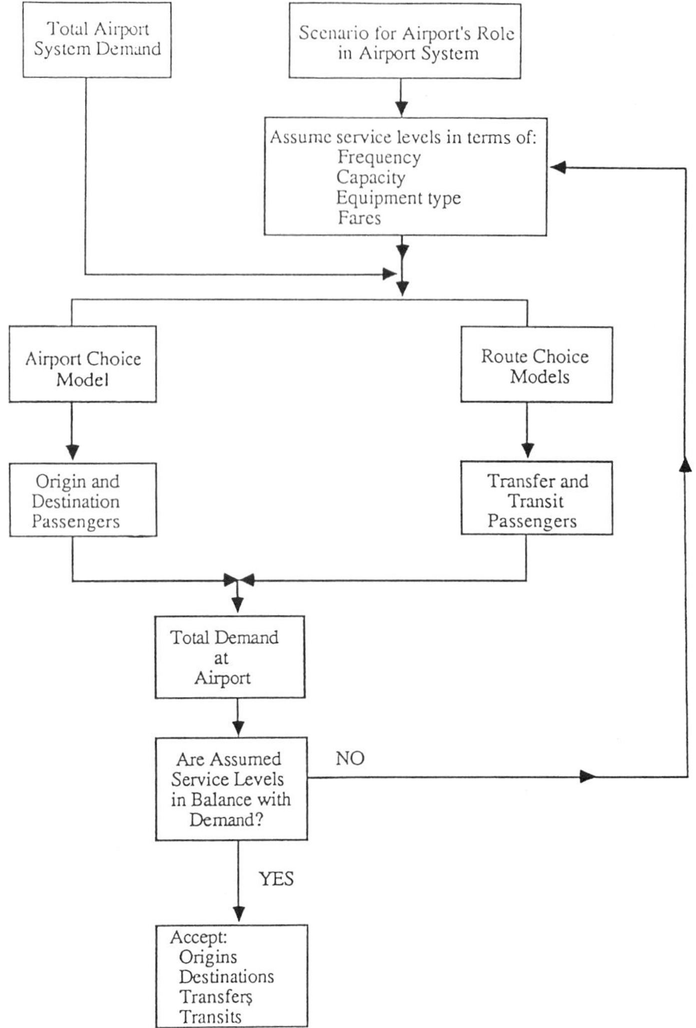

FIGURE 56.4 Flow chart of analysis for airport systems planning. (*Source:* FAA. 1975. *The Continuous Airport System Planning Process.* Advisory Circular AC150/5050-5.)

The analyst must carefully distinguish between passengers served (a number frequently used by many airport managers) and enplanements, a number of particular importance for the airlines and for terminal design. Equations (56.2) and (56.3) convert passenger data into enplanements. Origin-destination (O-D) passengers are those who either live in the local community or come into the local community for business or pleasure and are usually counted twice—each time they use the airport. Transferring passengers change from one airplane to another without leaving the terminal and are counted once.

$$\mathrm{ENP} = .5 \times \mathrm{PAX}_{\mathrm{O\text{-}D}} + \mathrm{PAX}_{\mathrm{TRANSFER}} \tag{56.2}$$

$$\mathrm{PAX}_{\mathrm{TOTAL}} = \mathrm{PAX}_{\mathrm{O\text{-}D}} + \mathrm{PAX}_{\mathrm{TRANSFER}} \tag{56.3}$$

where

$\mathrm{PAX}_{\mathrm{O\text{-}D}}$ = passengers passing through the airport who live/work/visit in the airport market area; a passenger is counted once when leaving and once when arriving

$PAX_{TRANSFER}$ = passengers who do not live/work in the airport market area and who are transferring from one aircraft to another

PAX_{TOTAL} = the total number of passengers served; often quoted by airports

One usually adequate method of forecasting is to decide that the airport use in the community (market area of the airport) will grow at the same rate as aviation across the U.S. and then use the present amount of the airport traffic as it reflects a percentage of that forecast for the U.S. provided by the FAA [FAA, 1993]. The FAA only guarantees their forecast for 11 years. National enplanement data beyond that date should use "unofficial" estimates available from the FAA Office of Policy and Plans.

Table 56.3 shows such a sample forecast for the hypothetical TBA airport (a medium hub). Note that the top part of the table indicates the historical data and their usual sources. If enplanements are not provided they can be calculated using Eqs. (56.2) and (56.3). In the TBA example when the enplanements are compared to national enplanements TBA has the history of being a 0.71% airport in the national airspace system. The planning wisdom is that the airport will stay at that level unless the community served by the airport is forecast to experience an unusual change in employment or economic capacity [FAA, 1985]. Growth or decline significantly different than the national statistics will be the cause for adjusting the simple forecast.

The past history and trends permit computation of several important planning factors such as percentage of the U.S. airport traffic, level of transfer passengers, departing seats, load factor, freight, and general aviation. These are shown in the heavier shaded portion of Table 56.3. It is assumed that departures will equal arrivals over a year and that the airlines will change the aircraft serving the airport to increase their capacity as the demand increases. The rest of the calculations, such as general aviation operations and freight operations, emanate from the planning factors.

When dramatic changes in employment in the community occur, historical data are used to determine the elasticity of a change in enplanements per change in jobs. From these data appropriate modifications to the spreadsheet of Table 56.3 are made to generate the forecast.

It may be necessary to review the variables used by the FAA for the development of their forecast and to alter the forecast if changes, for variables like disposable income, jobs, and population are vastly different from the national assumptions. (There are a number of references pertaining to the forecast methodology. However, the FAA includes in their forecast each year a list of the variables and their assumptions as to their growth.)

A number of consultants use regression equations in a manner similar to the FAA. However, the simple spreadsheet seems to offer as good a forecast. Since it is based on the FAA forecast, it should satisfy the FAA, which must approve the forecast as a part of its approval of the master plan.

Sometimes when a community projects a different economic pattern than is projected for the nation as a whole, the forecast must be developed using other variables. The FAA uses a regression equation with several variables, the most important being yield and disposable income. An example of the equation used in the planning of a small airport in Virginia [Ashford and Wright, 1992] is presented in Eq. (56.4) and an alternate one from the master plan update for Evansville Airport [HNTB, 1988] is shown in Eq. (56.5):

$$Ln\frac{E_i}{P_i} = 10.8 - 0.172F + 1.4Ln(Y_i) \tag{56.4}$$

where

E_i = predicted enplanements

P_i = population of market area of airport

F = average U.S. fare per mile or average yield per mile

Y_i = per capita income of the market area.

$$\text{ENP} = 2.2961 \times \text{EMP}^{1.126} \times \text{YIELD}^{-0.7306} \times \text{ACP}^{0.3317} \tag{56.5}$$

Table 56.3 Long-range Forecast for the TBA Airport

Usual Source of Data	Calc	Airlines	Airlines	FAA/Planning	Calc	Calc	FAA/ATC	Airlines	Airlines	FAA/ATC	Airport*	Calc	Carrier	Calc**	Calc	Calc
Year	O-D PAX	Interline Pax	Enplane-ments	Millions of Nat Enpla	Stature of Airport %	% Transfer/ Enplane	Commercial Departures	Departing Seats	Load Factor	Total Annual Ops.	Freight (tons)	Freight Ann. %	Capacity Freighter	Freight Ops.	GA Ops.	GA Ops/ Tot Ann Ops
1988	3,924,000	1,225,600	3,187,600	475.5	0.670%	38.4%	72,345	77.0	57.2%	161,709	3,000			226	16,793	10.38%
1989	3,904,900	1,356,000	3,308,450	480.4	0.689%	41.0%	75,678	79.0	55.3%	166,275	3,200	6.67%	26.50	242	14,677	8.83%
1990	4,123,600	1,356,000	3,417,800	497.9	0.686%	39.7%	76,980	83.0	53.5%	170,200	3,400	6.25%	26.50	257	15,983	9.39%
1991	4,137,000	1,343,000	3,411,500	487.0	0.701%	39.4%	79,300	84.0	51.2%	175,620	3,650	7.35%	26.50	275	16,745	9.53%
1992	4,368,000	1,382,270	3,566,270	503.6	0.708%	38.8%	81,456	85.0	51.5%	181,222	3,900	6.85%	26.50	294	18,016	9.94%
Projections					0.71 %	39 %			52 %			7.0 %				9.5 %
1993	4,491,247	1,435,727	3,681,350	518.5	0.71 %	39 %	81,844	86.5	52 %	179,577	4,173	7.0 %	27.0	309	15,580	9.5 %
1994	4,734,649	1,513,535	3,880,860	546.6	0.71 %	39 %	84,809	88	52 %	186,087	4,465	7.0 %	27.5	325	16,145	9.5 %
1995	4,971,988	1,589,406	4,075,400	574.0	0.71 %	39 %	87,081	90	52 %	191,082	4,778	7.0 %	28.0	341	16,578	9.5 %
2000	6,072,062	1,941,069	4,977,100	701.0	0.71 %	39 %	100,751	95	52 %	221,134	6,701	7.0 %	30.0	447	19,185	9.5 %
2005	7,224,108	2,309,346	5,921,400	834.0	0.71 %	39 %	111,640	102	52 %	245,135	9,398	7.0 %	32.0	587	21,267	9.5 %
2010	8,416,865	2,690,637	6,899,070	971.7	0.71 %	39 %	121,720	109	52 %	267,415	13,182	7.0 %	34.0	775	23,200	9.5 %
2015	9,294,326	2,971,137	7,618,300	1073.0	0.71 %	39 %	126,298	116	52 %	277,717	18,488	7.0 %	36.0	1027	24,094	9.5 %
2020	10,260,139	3,279,881	8,409,950	1184.5	0.71 %	39 %	130,427	124	52 %	287,130	25,930	7.0 %	38.0	1365	24,911	9.5 %

*This is the freight that is not carried in the belly of scheduled passenger aircraft.

**The average freight carried by freighters had been assumed at 26.5 tons (a small freighter can carry about 40 tons).

where

ENP = total passengers enplaned

EMP = regional employment

YIELD = air carrier yield

ACP = proportion of total possible passengers served by air carrier service (this factor depends on the number of passengers from the market area that use other airports)

The coefficients of the equations are determined from regression analysis and the average fare or yield is available from the FAA [FAA, 1993].

For the larger airports it is important to forecast the peak hour operations. Since there usually is little concern about capacity and delay until an airport with a single runway reaches approximately 35 operations an hour, the problem only surfaces in medium and large hubs. Figure 56.5, clearly labeled for planning purposes only, gives an indication of how peak hour operations are related to enplanements. For example, in the year 1993 the TBA airport might have as many as 45 operations in the peak hour; in 2020 that would be expected to grow to 55 to 60 in spite of a much larger growth in passengers. It is worth noting that the two variables, enplanements and operations, are linked by load factor and seats. So a doubling of enplanements may only result in a 30% to 40% increase in flight operations since airlines will tend to operate larger planes rather than fly more operations.

Small Commercial and General Aviation Airports

At the smaller airports, traffic is predominately general aviation (GA) traffic, which includes business flying. There may be a few commercial operations, air charter and air taxi operations, and

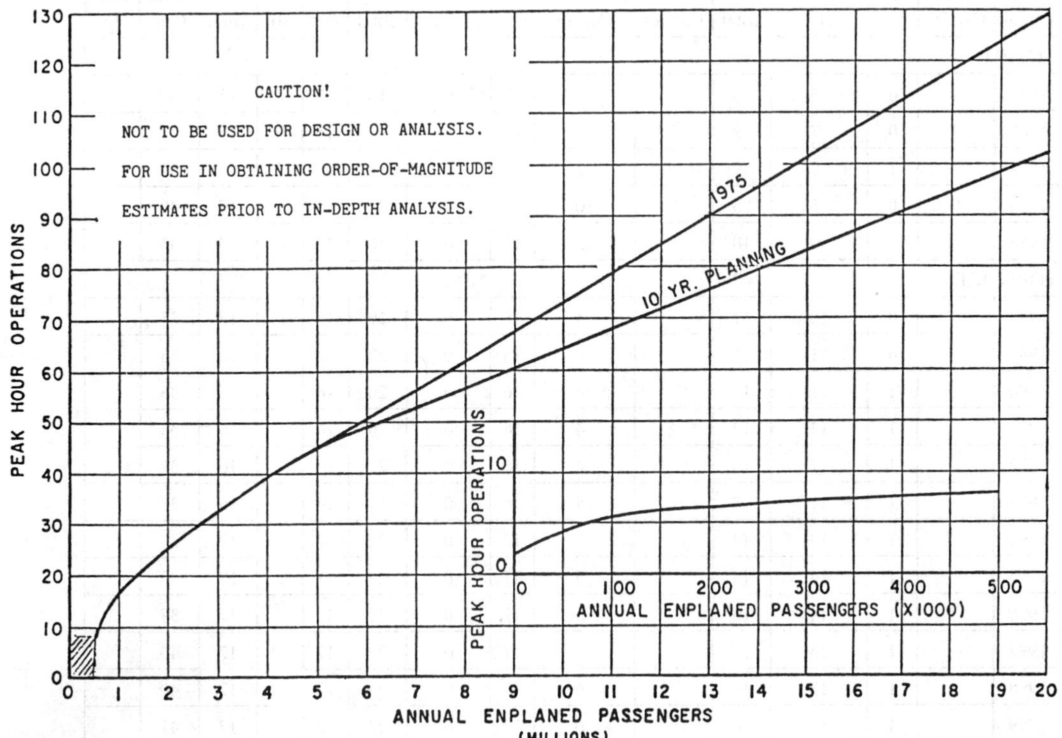

FIGURE 56.5 Estimated peak hour operations vs. annual enplaned passengers. (*Source:* FAA. 1988b. *Planning and Design Guidelines for Airport Terminal Facilities.* Advisory Circular AC 150/5360-13.)

even a few military. Each public-use airport in the state will have a history of the number of annual operations as a function of the based aircraft plus the number of annual operations expected from air carriers (usually commuter), air charter, air taxi, and military aviation.

Another important facet in the planning process for small airports is whether or not the airport is to be equipped for instrument approaches. Upgrades of navigation equipment or improved weather observation capability will generally increase the airport's percent time for landing, thus improving the airport's accessibility in inclement weather. Many commercial companies that use aircraft depend on airports that are not closed down every time there is low visibility or inclement weather.

To forecast future airport use it is essential that a history of operation be developed. For towered airports these data are available. However, for nontowered airports a count of traffic must be garnered by other means, such as acoustical counters on the runways. Critical aspects involved in forecasting for the smaller airports, listed in no particular order, include the following:

- The number of aircraft based at the airport, including mix or type
- The location of the airport and the weather data
- The instrument approach procedures and minimum altitudes
- Nearby airports and their relative appeal, including capability for landing
- Airport services and facilities, particularly the fixed base operator and T-hangars
- Touch-and-go operations, usually local operations

Table 56.4 Sample of Terminal Area Forecast Data Kept by FAA

REGION-STATE: ANN-WY　　LOCID: COD NONTOWERED　　BASED AIRCRAFT: 61
CITY: CODY　　AIRPORT: E. E. FAUST REGIONAL

					AIRCRAFT OPERATIONS									
	Enplanements (000)				Itinerant					Local				
Year Actual	Air Carrier	Air Taxi	Commercial	Total	Air Carrier	Air Taxi + Comm	GA	Military	Total	GA	Mil	Total	Total Ops	Inst. Ops
ACTUAL														
1982	1	0	3	5	2	1	15	0	18	15	0	15	33	
1983	2	0	7	9	1	1	15	0	17	15	0	15	32	
1984		1	8	9		2	15	0	17	15	0	15	32	
1985	0	0	5	5		10	15	0	25	15	0	15	40	
1986			9	10		3	15	0	18	15	0	15	33	
FORECAST								0						
1988		0	10	11		3	16	0	19	16		16	35	
1989		0	11	11		4	16	0	20	16		16	35	
1990		1	11	12		4	16	0	20	16		16	36	
1991		1	12	12		4	16	0	20	16		16	36	
1992		1	12	13		4	16	0	21	16		16	37	
1993		1	13	13		4	16	0	21	16		16	38	
1994		1	13	14		5	17	0	22	17		17	38	
1995		1	14	14		5	17	0	22	17		17	39	
1996		1	14	15		5	17	0	22	17		17	39	
1997		1	15	15		5	17	0	23	17		17	40	
1998		1	15	16		6	17	0	23	17		17	40	
1999		1	16	16		6	17	0	23	17		17	41	
2000		1	16	17		6	17		24	18		18	41	

Source: FAA. 1988. *Terminal Area Forecasts FY 1988–2000.* FAA-APO-88-3.

- Availability of mechanics and maintenance parts
- Level of air taxi, charter, and air carrier (usually commuter) operations
- Markets served and aircraft use

Once the existing GA operations have been determined, usually in the range of 300 to 500 annual operations per based aircraft, discussions with the airport manager, local businesspersons, and state aviation officials should be undertaken to provide added perspective on the rate of growth of based aircraft and of corporate/business operations. Local forecasts should be checked with the FAA forecast of air taxi and air charter growth. The FAA publishes forecasts for all airports in the *National Integrated Air System Plan*. A sample of the planning data presented on such airports is given in Table 56.4.

Sometimes it is necessary to develop a forecast that depends on the interaction of a number of GA airports. The Indianapolis metropolitan airport system is one such system. The area consists of eight counties involving 16 airports: one air carrier airport, seven relievers or potential relievers, five utility GA airports, and three basic GA airports. The objective of the plan is to look at the future of these airports, to decide the capabilities (e.g., runway length and width, instrumentation and lighting) each airport should have, and to decide if any new airports should be built in the area. A comprehensive inventory of each of the airports was made. A number of alternatives were addressed, including the possibility of developing new airports and closing down others. Here it is useful to indicate several questions that were analyzed:

- *How many based aircraft could be expected in the future?* Using a variety of national data and trends, projections of jobs and businesses in the Indianapolis area (including a new major airline maintenance facility at the airport), the number of single engine, multiengine, and turbine aircraft was projected. In spite of the general sluggishness of general aviation in the past ten years, there has been a higher growth than in the nation in this region of the country.
- *Where would the people who are pilots and/or own airplanes settle?* After examining a series of potential independent variables a regression equation involving population, households with

Table 56.5 Summary of 2012 Aviation Forecasts for the 1993 Indianapolis Airport System Plan

Airport	Based Aircraft Total	Based Aircraft Single	Based Aircraft Multi	Based Aircraft Turbine	Based Aircraft Rotor	Based Aircraft Other
1) Boone County	72	63	9	0	0	0
2) Eagle Creek Airpark	153	109	28	14	2	0
3) Franklin Flying Field	52	47	2	0	0	3
4) Greenwood Municipal	135	111	15	7	2	0
5) Hendricks County—new airport	93	67	15	9	2	0
6) Indianapolis International	88	0	10	71	7	0
7) Indianapolis Metropolitan	160	120	28	8	3	1
8) Indianapolis Mount Comfort	158	120	23	12	3	0
9) Indianapolis Speedway	0	0	0	0	0	0
10) Indianapolis Terry	82	57	4	6	0	15
11) McDaniel	15	12	0	0	0	3
12) Pope Field	26	20	1	0	0	5
13) Shelbyville Municipal	72	61	3	7	1	0
14) Sheridan	25	21	2	0	0	2
15) Westfield	34	33	0	0	0	1
16) Indianapolis Downtown Heliport	14	0	0	0	14	0
Sum of above airports	1179	841	140	134	34	30
All aircraft owned by residents (including businesses) of MSA	1294	946	142	134	42	30

Source: Indiana DOT. 1993. *Indianapolis Metropolitan Airport System Plan Update.* Prepared by TAMS and al Chalibi for Indiana Department of Transportation and the Indianapolis Airport Authority.

incomes in excess of $50,000, and number of airports within a 12-mile radius of the township (specific subarea in each county) was used to allocate the owners of aircraft to the region.

- *Where would these persons locate their aircraft?* The study examined the way present airports attract aircraft owners, considering several important attributes of a "good" airport (e.g., hangar capacity, fueling capability, mechanics, instrumented landing capability, cost of housing) and the convenience factor for the owner, primarily driving time from home to the airport. The transportation planning "intervening opportunity" model was used to allocate the aircraft as based aircraft to airports.
- *How many local operations, itinerant operations, and instrument approaches will there be?* The national average of local versus itinerant operation is expected to grow from 46 to 52% in the next 20 years. Instrument weather history and the landing capability of the airport were used to predict instrument landings.

Table 56.5 presents a summary of the findings, including a proposed airport in Hendricks County, Indiana.

56.4 Requirements Analysis—Capacity and Delay

Armed with the demand forecasts and having developed an inventory of the airport and reviewed its condition, the planning proceeds to determine the capability of the airport to accommodate the forecast demand. First is the determination of the capacity of the airport relative to the demand with special attention to the delay that will be incurred at peak times.

Capacity is used to denote the processing capability of a facility to serve its users over some period of time. For a facility to reach its maximum capacity there must be a continuous demand for service. At most facilities such a demand would result in large delays for the user and eventually become intolerable. To develop a facility where there was virtually no delay would require facilities that could not be economically justified. When a single runway serves arriving aircraft, the mean delay is given by Eq. (56.6) [Horonjeff and McKelvey, 1983].

$$W_a = \frac{\lambda_a \left[\sigma_a^2 + \left(1/\mu_a^2\right)\right]}{2\left[1 - (\lambda_a/\mu_a)\right]} \tag{56.6}$$

where

W_a = mean delay to arriving aircraft
λ_a = mean arrival rate, aircraft per unit time
μ_a = mean service rate, or reciprocal of the mean service time
σ_a = standard deviation of mean service time of arriving aircraft

For departing aircraft, Eq. (56.6) is used by replacing the subscript a with d. When aircraft share the same runway for landing and takeoff, arriving aircraft always have priority so the delay for arriving aircraft is the same as Eq. (56.6). The delay for departing aircraft is found by solving Eq. (56.7):

$$W_d = \frac{\lambda_d\left(\sigma_j^2 + j^2\right)}{2(1 - \lambda_d j)} + \frac{g\left(\sigma_f^2 + f^2\right)}{2(1 - \lambda_a f)} \tag{56.7}$$

where

W_d = mean delay to departing aircraft
λ_a = mean arrival rate, aircraft per unit time
λ_d = mean departure rate, aircraft per unit time

Table 56.6 Spacing Required for Safe Landing in IFR with Wake Vortex*

		Lead Aircraft				
		Heavy	Medium	Light	Type	Weight
Trailing	Heavy	4 n. mi.	3 n. mi.	3 n. mi.	D	≥300,000 lbs
Aircraft	Medium	5 n. mi.	4 n. mi.	3 n. mi.	C	>12,500 lbs <300,000 lbs
	Light	6 n. mi.	5 n. mi.	3 n. mi.	A&B	≤12,500 lbs

*To convert for VFR replace 3, 4, 5, and 6 n. mi. with 1.9, 2.7, 3.6, and 4.5 n. mi., respectively.
Source: FAA. 1978. *Parameters of Future ATC Systems Related to Airport Capacity and Delay.* Report FAA-EM-78-8A.

j = mean interval of time between two successive departures
σ_j = standard deviation of mean interval of time between two successive departures
g = mean rate at which gaps between successive aircraft occur
f = mean interval of time in which no departure can be released
σ_f = standard deviation of mean interval of time in which no departure can be released

During busy times the second term should approach zero if it is assumed that the aircraft are in a queue at the end of the runway [Horonjeff and McKelvey, 1983]. The following general rules for aircraft landing on a runway are important in the determination of capacity and delay.

- Two aircraft may not occupy the same runway at the same time.
- Arriving aircraft always have priority over departing aircraft.
- Departures may be released while the arriving aircraft is on approach providing it is two or more nautical miles from the threshold of the runway at the time of release.
- Spacing for successive landings incorporates wake vortex requirements for mixed aircraft landings as shown in Table 56.6.

In addition to separation on landing, the capacity is also a function of the configuration of runways, runway exit geometric design, landing speed, and braking ability. Air traffic control measures for noise abatement, heavy wind conditions, arriving and departing flight paths, and type of navigational aids also add complexity to the determination of capacity. Most significant is the safe spacing between successive aircraft.

To aid planning, capacity and delay may be estimated using the annual service volume (ASV) for the airport in combination with the annual demand. From the outset it is assumed that any airport configuration can be approximated by one of the eight depicted configurations of runways given in Fig. 56.6 with the note that crosswind runways do not significantly increase the ASV. The other assumptions for computing ASV are:

- Percent arrivals equal percent departures
- Full-length parallel taxiway with ample entrances and no taxiway crossing problems
- No airspace limitations that would adversely impact flight operations
- At least one runway equipped with an ILS and the ATC facilities to operate in a radar environment
- Operations occur within the ranges given in Table 56.7
- IFR weather conditions occur 10% of the time
- Roughly 80% of the time the runway configuration which produces the greatest hourly capacity is used

Example. Assume the TBA airport has two parallel runways separated by 1000 feet. The forecast from Table 56.3 indicates the requirement for 287,130 operations in the year 2020. The present demand is 183,000 operations. The aircraft mix during peak hours is derived from the anticipated peak hour aircraft traffic at TBA shown in Table 56.8.

No.	Runway-use Configuration	Mix Index %(C+3D)	Hourly Capacity Ops/Hr VFR	Hourly Capacity Ops/Hr IFR	Annual Service Volume Ops/Yr
1.		0 to 20	98	59	230,000
		21 to 50	74	57	195,000
		51 to 80	63	56	205,000
		81 to 120	55	53	210,000
		121 to 180	51	50	240,000
2.	 700' to 2499'*	0 to 20	197	59	355,000
		21 to 50	145	57	275,000
		51 to 80	121	56	260,000
		81 to 120	105	59	285,000
		121 to 180	94	60	340,000
3.	 2500'* to 4299'	0 to 20	197	62	355,000
		21 to 50	149	63	285,000
		51 to 80	126	65	275,000
		81 to 120	111	70	300,000
		121 to 180	103	75	365,000
4.	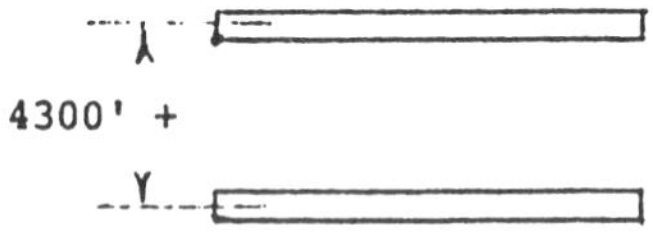 4300' +	0 to 20	197	119	370,000
		21 to 50	149	113	320,000
		51 to 80	126	111	305,000
		81 to 120	111	105	315,000
		121 to 180	103	99	370,000
5.	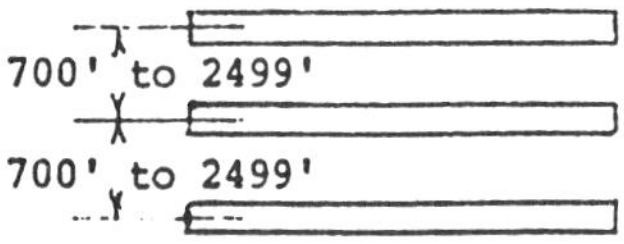 700' to 2499' 700' to 2499'	0 to 20	295	62	385,000
		21 to 50	213	63	305,000
		51 to 80	171	65	285,000
		81 to 120	149	70	310,000
		121 to 180	129	75	375,000
6.	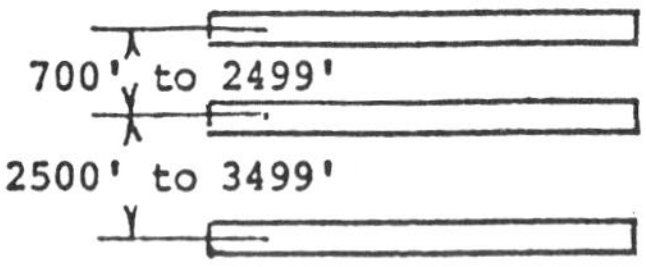 700' to 2499' 2500' to 3499'	0 to 20	295	62	385,000
		21 to 50	219	63	310,000
		51 to 80	184	65	290,000
		81 to 120	161	70	315,000
		121 to 180	146	75	385,000
7.	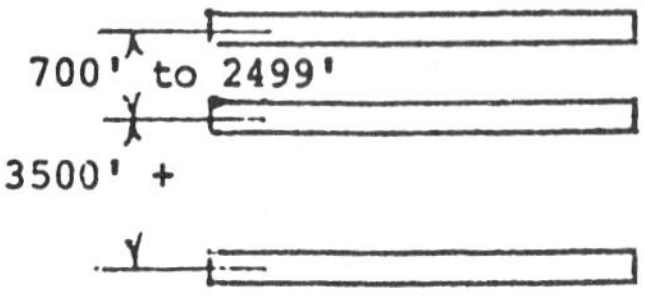 700' to 2499' 3500' +	0 to 20	295	119	625,000
		21 to 50	219	114	475,000
		51 to 80	184	111	455,000
		81 to 120	161	117	510,000
		121 to 180	146	120	645,000
8.	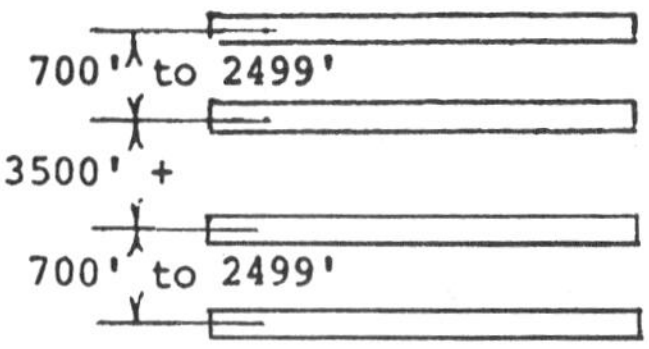 700' to 2499' 3500' + 700' to 2499'	0 to 20	394	119	715,000
		21 to 50	290	114	550,000
		51 to 80	242	111	515,000
		81 to 120	210	117	565,000
		121 to 180	189	120	675,000

FIGURE 56.6 Runway configurations capacity and ASV for long-range planning. (*Source:* FAA. 1983a. *Airport Capacity and Delay.* Advisory Circular AC150/5060-5. Incorporates change 1.)

Table 56.7 Chart for Calculating ASV for the Peak Month

Mix Index %(C + 3D)	Percent Arrivals	Percent Touch and Go	Demand Ratios: Annual Demand/ Average Daily Demand	Demand Ratios: Average Daily Demand/ Average Peak Hour Demand
0–20	50	0–50	290	9
21–50	50	0–40	300	10
51–80	50	0–20	310	11
81–120	50	0	320	12
121–180	50	0	350	14

Source: FAA. 1983a. *Airport Capacity and Delay.* Advisory Circular AC150/5060-5. Incorporates change 1.

Table 56.8 Aircraft Mix for Peak Hour Operation at TBA Airport

Aircraft Class	Typical Aircraft	Peak Hour Operations 1992		Peak Hour Operations 2020	
		VFR	IFR	VFR	IFR
A	Cessna, Piper, etc.	4 [11% T&G*] (10%)**	0	6 [15% T&G*] (9%)	0
B	Lear Jet, Shorts	8(22%)	4(14%)	18(27%)	10(18%)
C	DC-9, B-727, MD-80	22(61%)	22(79%)	35(51%)	35(65%)
D	DC-10, B-747, B-757	2(6%)	2(7%)	9(13%)	9(17%)

*T&G—touch and go is a training operation where aircraft come in for landing and once they touch down, increase power and take off again without stopping.

**Numbers in parentheses are the percent of the total operations.

$$MI = (\%C) + 3 \times (\%D) \tag{56.8}$$

The calculated mix index (MI) presently is $MI_{VFR} = 61 + 3(5) = 76$ and $MI_{IFR} = 79 + 2(7) = 93$, where VFR is *visual flight rules* and IFR is *instrument flight rules.* In 2020 it will grow to $MI_{VFR} = 90$ and $MI_{IFR} = 116$.

It is now appropriate to develop a delay specification. Let us assume that no more delay will be allowed in 2020 than the airport is now experiencing. The present mix index is 70 in VFR and is 94 in IFR. Using the VFR mix index, the ASV from Fig. 56.6 is 260,000 and for the number of annual operations from Table 56.3 of 181,222, the delay factor is indicated in Eq. (56.9).

$$DF = \frac{OPS}{ASV} = \frac{181,222}{260,000} = 0.7 \tag{56.9}$$

where

DF = delay factor

ASV = annual service volume read from Fig. 56.4

The average delay is read from Fig. 56.7 as between 0.6 and 1 minute. This is reasonable, but more than is good for the average delay over the day. Certainly during peak periods the actual delay may be five to ten times this average delay, and overall this airport is approaching a delay problem, although it is not yet major. Significant delay between 1.4 and 2.3 minutes average would accrue if one of the two runways had to be shut down for any period. An *average* delay of 2–4 minutes can become quite significant during the peak period.

To maintain an average delay in the range of 0.6 to 1 minute in the year 2020, an ASV of 287,130/0.7 = 410,185 is required. For the mix index of 90, runway configuration number 7 on Fig. 56.6 would satisfy the ASV. It would involve the addition of one parallel runway separated more than 3500 feet from the existing runways. The hourly capacity for VFR would be 161

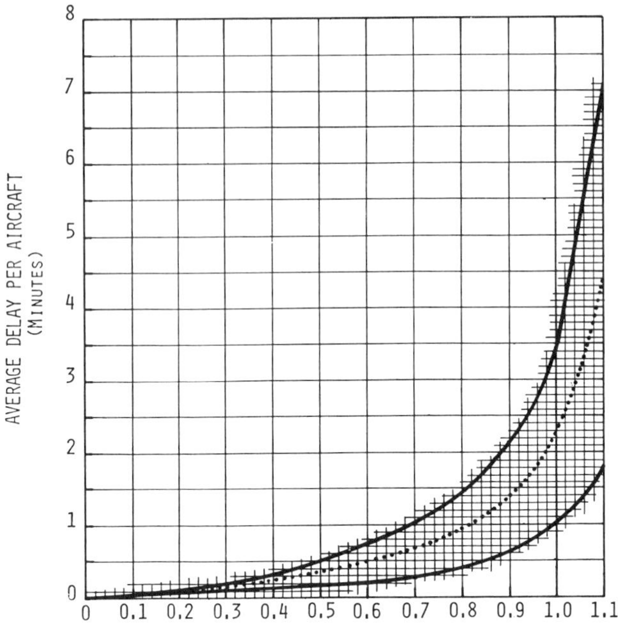

FIGURE 56.7 Average aircraft delay for long range planning. (*Source:* FAA. 1983a. *Airport Capacity and Delay.* Advisory Circular AC150/5060-5. Incorporates change 2.)

operations/hour and for IFR 117. This is well above the estimated operations of 68 and 54, respectively. The average delay in 2020 would be about 0.5 minutes.

Many simulation runs have been performed by the FAA to obtain better design data than the simplified annual service volume approach gives. Those simulation results are summarized in design curves like the one shown in Fig. 56.8, which is for TBA runway configuration needed in 2020. For the mix index of 90 the runway set will yield $C^* = 155$ with 50% arrivals, $T = 1$ with 15% touch and go and a mix index of 90, and $E = .89$ with one exit at 6000 feet from the threshold. This yields an hourly capacity of 138. While much below the 161 from the advanced planning charts, it is still far above the 68 aircraft expected during the peak hour in 2020. Similar curves are presented to calculate delay for arrivals and departures. For our example the delay during peak hour is 1.2 minutes for departing aircraft and 0.6 minutes for arriving aircraft. These are the peak delays so that when averaged over the day the delay will be less than 0.5 minutes.

This in-depth analysis approach of capacity and delay utilizes the charts as shown in Fig. 56.8 and can be found in FAA Advisory Circular AC 150/5060-5 entitled *Capacity and Delay*. In addition, National Technical Information Services (NTIS) has available capacity computer programs in FORTRAN. Personal computer programs are available to calculate capacity, delay, and ASV values which are more than adequate for planning [FAA, 1983].

56.5 Air Traffic Management

The second key aspect in the requirements analysis is to assess the capability of the airport to provide the traffic controls during poor weather flying conditions (IFR) as well as during good weather conditions (VFR). Except in airspace under positive control, VFR flying is based on a

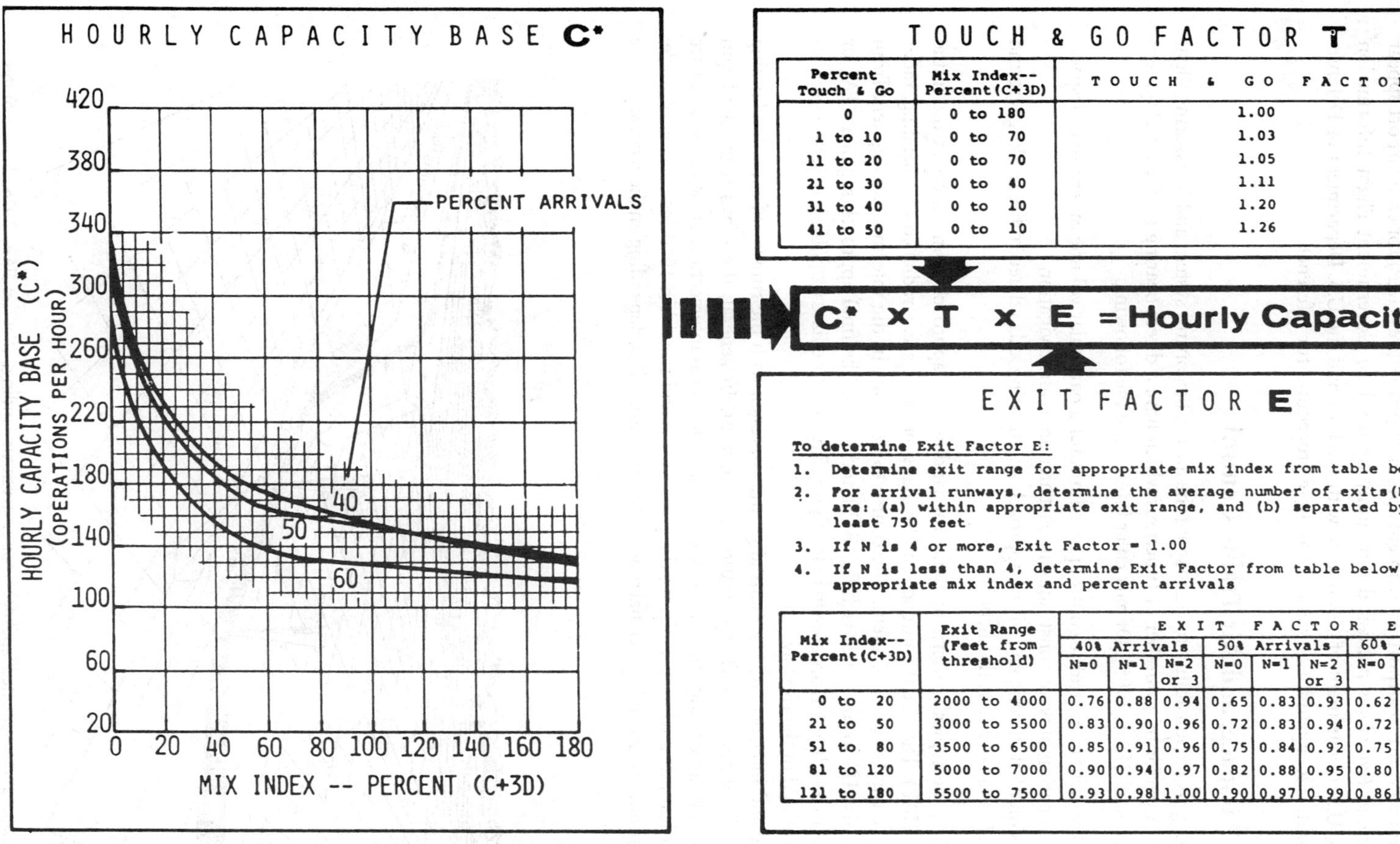

TOUCH & GO FACTOR T

Percent Touch & Go	Mix Index--Percent (C+3D)	TOUCH & GO FACTOR T
0	0 to 180	1.00
1 to 10	0 to 70	1.03
11 to 20	0 to 70	1.05
21 to 30	0 to 40	1.11
31 to 40	0 to 10	1.20
41 to 50	0 to 10	1.26

EXIT FACTOR E

To determine Exit Factor E:

1. Determine exit range for appropriate mix index from table below
2. For arrival runways, determine the average number of exits (N) which are: (a) within appropriate exit range, and (b) separated by at least 750 feet
3. If N is 4 or more, Exit Factor = 1.00
4. If N is less than 4, determine Exit Factor from table below for appropriate mix index and percent arrivals

Mix Index--Percent (C+3D)	Exit Range (Feet from threshold)	EXIT FACTOR E								
		40% Arrivals			50% Arrivals			60% Arrivals		
		N=0	N=1	N=2 or 3	N=0	N=1	N=2 or 3	N=0	N=1	N=2 or 3
0 to 20	2000 to 4000	0.76	0.88	0.94	0.65	0.83	0.93	0.62	0.79	0.90
21 to 50	3000 to 5500	0.83	0.90	0.96	0.72	0.83	0.94	0.72	0.83	0.93
51 to 80	3500 to 6500	0.85	0.91	0.96	0.75	0.84	0.92	0.75	0.84	0.92
81 to 120	5000 to 7000	0.90	0.94	0.97	0.82	0.88	0.95	0.80	0.87	0.94
121 to 180	5500 to 7500	0.93	0.98	1.00	0.90	0.97	0.99	0.86	0.95	0.98

FIGURE 56.8 Example of peak hour capacity determination charts for TBA airport in VFR. (*Source:* FAA. 1983a. *Airport Capacity and Delay.* Advisory Circular AC150/5060-5. Incorporates change 1.)

"pilot beware" or "see and be seen" approach to flying. General aviation pilots flying in VFR need only a functioning radio and altimeter. Commercial aircraft and many business aircraft are equipped with beacons, radar, and other equipment that permits them to fly in instrument weather and/or in controlled airspace. Capability for landing on a given runway and the use of navigation aids varies from airport to airport. Instrument approach procedures (IAP) for each airport appear in *U.S. Terminal Procedures* published bimonthly by the U.S. Government Flight Information Publications, U.S. Department of Commerce with the FAA and the U.S. Department of Defense. Every pilot with IFR capability carries a set of these procedures for reference.

Airways, Airspace, and Air Traffic Control

"In discharging its responsibility for managing the air traffic control system and in assuring flight safety, the FAA performs a number of functions which have a direct bearing on the development of the master plan" [FAA, 1985]. Of particular interest are the following:

1. Establishment of air traffic control procedures for a particular volume of terminal airspace
2. Determination of what constitutes an obstruction to air navigation
3. Provision of electronic and visual approach and landing aids related to the landing, ground control, and takeoff at the airport

In Fig. 56.9, the typical pattern of flight for landing, the approach commences at the initial approach fix (IAF). The initial approach can be made along an arc, a radial course, heading, radar vector, or by a combination of them. The course to be flown in the intermediate segment from intermediate fix (IF) to final approach fix (FAF) and during the final approach segment (FAF or outer marker to touchdown) are shown in the figure. The intermediate fix point is usually 5–9 miles from the threshold of the runway.

The initial and intermediate segments align the approach with the runway of intended landing and provide for initial aircraft stabilization and descent. In general, these two segments begin with signals from an en route navigation aid or the radio signal intersection of two aids. They are about 8 nautical miles wide, permitting the pilot to descend to within 1000 feet of any obstacle. The final approach segment is much narrower: 1 to 4 nautical miles, depending on the accuracy of the

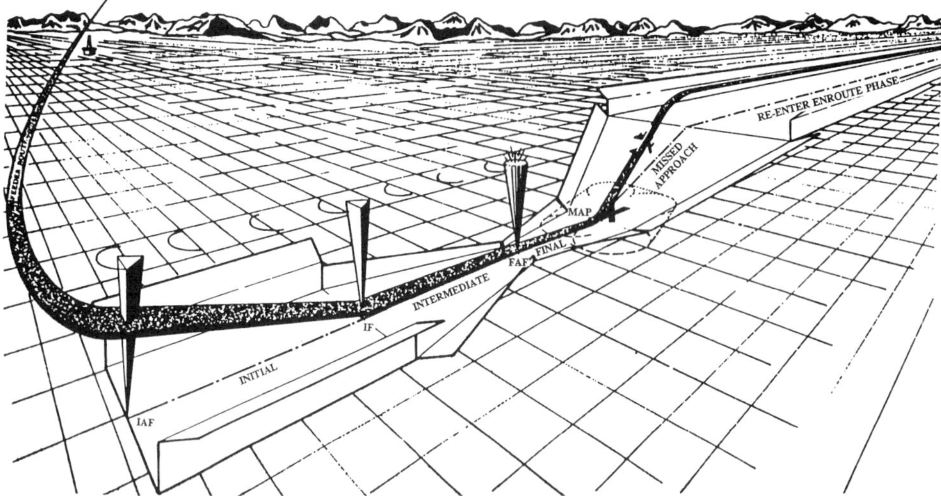

FIGURE 56.9 Markers and segmented approach for instrument landing. (*Source:* FAA. 1976. *United States Standard for Terminal Procedure (TERPS)*, 3d ed. FAA Handbook 8260.3B.)

Table 56.9 Aircraft Categories and Landing Speeds

Aircraft Category	1.3 Times the Stall Speed in Knots	Maximum Speed (Circling Approaches)	Typical Aircraft in This Category
A	<91 knots	90 knots	Small single engine
B	91–120 knots	120 knots	Small multiengine
C	121–140 knots	140 knots	Airline jet
D	141–165 knots	165 knots	Large jet/military jet
E*	>166 knots		Special military

*Category E is restricted to high-performance, special mission military aircraft and will not be addressed in this report.

Modified from FAA. 1976. *United States Standards for Terminal Instrument Procedures (TERPS)*, 3d ed. FAA Handbook, 8260.3B.

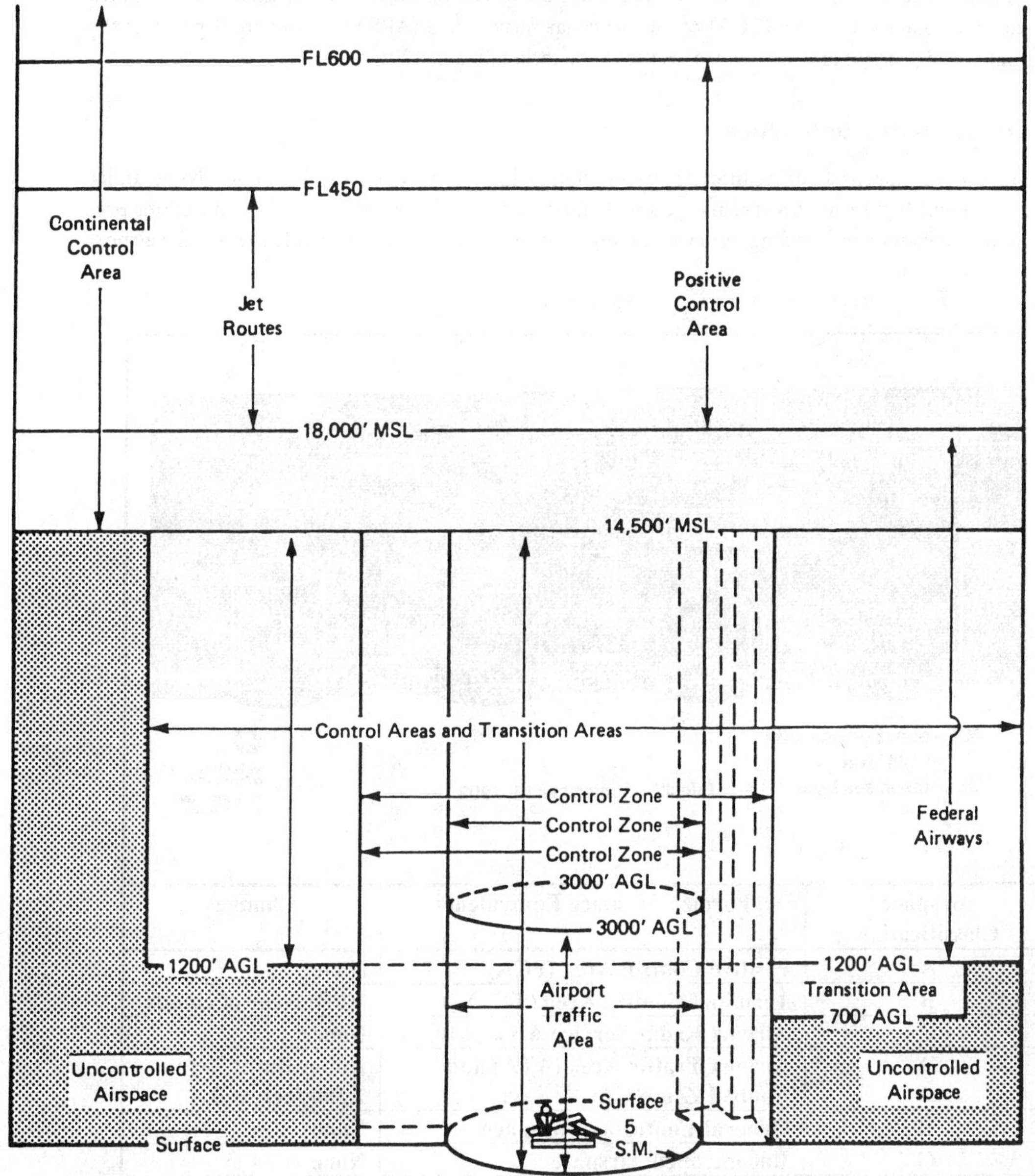

FIGURE 56.10 Controlled airspace. (*Source:* FAA. 1985. *Airport Master Plans.* Advisory Circular AC150/5070-6A.)

navigation aid being used. The missed approach segment transitions the pilot back to begin the approach again with 1000 feet of obstacle clearance.

The class of aircraft and amount of traffic play a significant role in determining the requirements for controlled airspace around the airport. All aircraft are categorized into one of five different approach speed categories (usually based on a landing speed of 1.3 times the aircraft's certificated stalling speed with the maximum certificated landing weight) called **aircraft approach categories.** These are listed in Table 56.9. The airspace with its safety or buffer zones is configured with these landing speeds in mind.

The volume of sky called "controlled airspace" shown in Fig. 56.10 gives the appearance of an upside-down wedding cake with the size dependent on the amount and nature of the traffic in the controlled zone. To maximize safety and efficiency, each aircraft within the terminal area controlled airspace volume will be under positive control by the air traffic controller when the weather is below minima. The aviation community has become used to calling these areas by their abbreviations, e.g., terminal control area (TCA) or airport radar service area (ARSA). Figure 56.11 presents the major categories of airspace control as they were reclassified in 1993.

Instrument Approaches

Instrument approach procedures (IAP) developed by the FAA for use by pilots flying under instrument flight rules provide navigational guidance to an airport when weather conditions preclude navigation and landing under visual flight conditions. If a pilot is unable to sight the airport

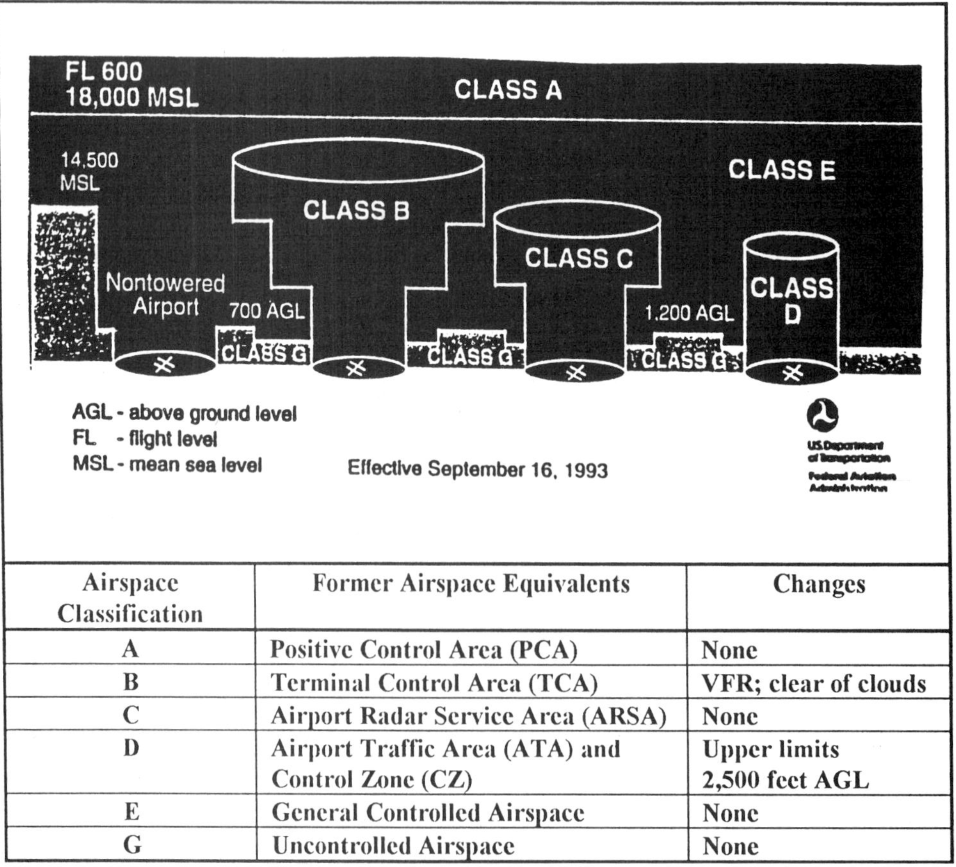

Airspace Classification	Former Airspace Equivalents	Changes
A	Positive Control Area (PCA)	None
B	Terminal Control Area (TCA)	VFR; clear of clouds
C	Airport Radar Service Area (ARSA)	None
D	Airport Traffic Area (ATA) and Control Zone (CZ)	Upper limits 2,500 feet AGL
E	General Controlled Airspace	None
G	Uncontrolled Airspace	None

FIGURE 56.11 Airspace reclassification. (*Source:* FAA. 1993d. *Classification of Airspace.* Brochure.)

visually while at the minimum en route altitude (MEA) permitted for VFR flight when traveling along an airway, the pilot must fly the instrument approach procedure developed specifically for the destination airport in order to land. Minimum en route altitude is defined as the lowest usable altitude on an airway with acceptable navigational signals and which meets obstacle clearance requirements. MEAs therefore vary for each airway at every airport, depending on navigation transmitter placement and local terrain elevation. Whenever the ceiling at an airport is below the MEA, pilots are required to conduct an instrument approach in order to complete their flight. There are three basic types of instrument approaches: circling, straight-in nonprecision, and straight-in precision, as briefly defined in Table 56.10.

A nonprecision approach typically uses an existing en route navigation aid (NAVAID), such as a VHF omnidirectional range (VOR) for guidance, and provides a path from that NAVAID to the airport. Precision approach procedures utilize the specially designed category I instrument landing system (ILS) or the newer microwave landing system (MLS). Both of these systems are specifically designed to provide highly accurate lateral and vertical guidance as shown in Fig. 56.12 to minima of 200 feet height above touchdown (HAT) and $\frac{1}{2}$-mile visibility. Two special ILS systems, category

Table 56.10 Definition of Instrument Approach Procedures

Circling approaches: If the navigational aid being utilized for the instrument approach does not line up within 30 degrees of any runway heading, the pilot must navigate to the general vicinity of the airport, and then circle to line up with the runway. This type of approach, known as a circling approach, is the least preferable of the three due to the fact that the pilot is not provided with any navigational assistance to line up with the runway of landing. A circling approach is basically akin to a nonprecision approach. Upon reaching the vicinity of the airport, the pilot must align the aircraft with the runway of intended landing. This approach may require extensive maneuvering just prior to landing, including an initial turn *away* from the airport. This approach is potentially more dangerous to execute and some aircraft operators either discourage or absolutely prohibit its use, particularly at night when ground references are less available. Instrument approaches may be specifically designed as circling approaches if navigation aids are unavailable for a straight-in approach. But most straight-in approaches can also be utilized as circling approaches to other runways located at the same airport. Circling approaches usually have higher minima than the straight-in approaches described below.

Straight-in nonprecision: Straight-in approaches are those that align the aircraft within 30 degrees of the landing runway. Nonprecision approaches provide lateral guidance only. During a nonprecision approach, the pilot navigates along a prescribed course until reaching a navigational fix known as the final approach fix (FAF). At this point the pilot initiates a descent to the lowest safe altitude, known as the minimum descent altitude (MDA). The pilot flies along this course until either reaching a predetermined point, or until a calculated period of time has elapsed. This point is known as the missed approach point (MAP). If the runway has not been sighted by the pilot before reaching the MAP, the pilot follows a procedure known as the missed approach that guides and climbs the aircraft back to a point where the approach can be initiated again, or the pilot can extend the flight to another airport where a landing is possible.

Vertical guidance is not provided to the pilot during a nonprecision approach; therefore, the lowest altitude to which a pilot may descend and the required in-flight visibility are fairly high. Usually, 300 to 900 feet is the lowest height above touchdown (HAT) to which the pilot may descend, and the required visibility is 1 to 2 miles. This type of approach is considered sufficiently accurate and safe for an airport that does not service a high level of commercial or essential air traffic.

Straight-in precision: A precision approach is similar to a nonprecision approach; the only difference is that the precision approach provides the pilot with electronic vertical guidance in addition to lateral guidance. A glidepath is transmitted from the ground and guides the aircraft on about a 3 degree descent path to the runway. The pilot simply follows the navigational directions during the descent. Since a glidepath is provided, the pilot need not level off at any minimum altitude. Precision approach procedures instead define a specific altitude at which the pilot must decide whether a landing can safely be conducted. This altitude is known as the decision height (DH). If the pilot has the runway, runway lights or approach lights in sight prior to, or upon reaching the decision height, the descent to the runway can be continued and the pilot may land the aircraft. If the runway or its associated lighting is not in sight, the pilot immediately begins to execute the missed approach instructions. Precision approaches utilize either instrument landing systems or microwave landing systems.

Modified from FAA. 1976. *United States Standard for Terminal Instrument Procedures (TERPS),* 3d ed. FAA Handbook 8260.3B.

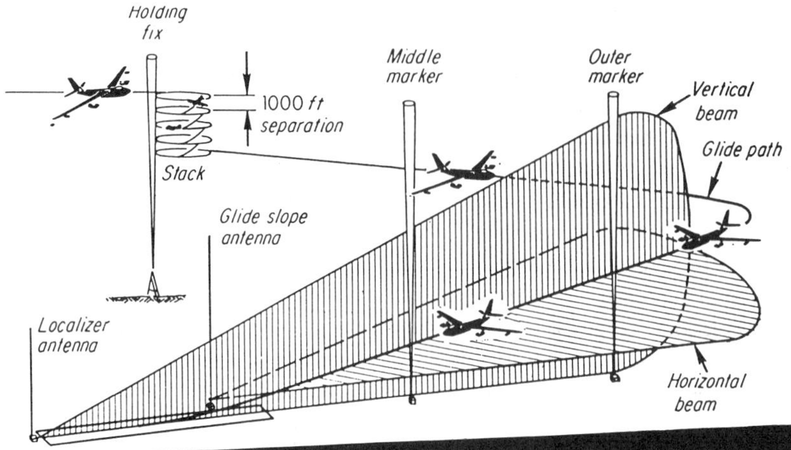

FIGURE 56.12 Instrument landing system. (*Source: Scientific American,* Dec. 1960.)

II and category III, will provide minima of 100-feet HAT with $\frac{1}{4}$-mile visibility and "all weather" landing minima, respectively. These two systems are very expensive and are usually installed only at the busiest commercial airports.

Circling approaches have been developed using both nonprecision and precision navigation aids, although nonprecision aids are most often used. Because a circling procedure does not align the aircraft with a specific runway but instead simply provides a path to the airport whereby the pilot decides which runway to land on and then circles to that runway, it is also sometimes considered as a visual approach.

A new navigation standard based on the use of the global positioning satellite (GPS) system, which became operational in 1992, should be available for aviation landing aid by 2000. With the proper receiver on board the aircraft, GPS signals will be able to be used to provide nonprecision approaches into any airport at little or no cost, other than the purchase of a low-cost satellite receiver on board the aircraft. Precision approaches have been demonstrated with GPS. The potential for GPS signal dropout during the critical landing phase is one of the limiting concerns.

Minimum Altitude Calculations

It is the final approach segment that is of most interest to the airport planner. In general terms, the lower a pilot is permitted to descend during the final approach, the greater the likelihood that a successful landing can be made. The more precise the navigation aid being used, the easier it is to "thread" a pilot around obstructions, and to authorize a lower final approach altitude.

The basic obstacle clearance distance and visibility requirements for a nonprecision straight-in instrument approach are 250 feet of obstacle clearance and one statute mile visibility. Therefore, if an airport located at 750 feet MSL has a 100-foot obstacle located along the final approach course, the instrument approach will mandate a minimum descent altitude (MDA) of 1100 feet (750-foot airport elevation + 100-foot actual obstacle height + 250-foot TERPS-mandated obstacle clearance height). The only methods that can be employed to reduce the 1100-foot MDA in this example are to utilize a more precise navigation aid (to navigate the pilot around the obstacle), develop an approach to a different runway with obstacles of lower height, or remove the obstruction (which may be impractical). The 250-foot basic obstacle clearance altitude may change depending on the type of navigation aid used at the airport, as indicated in Table 56.11.

Minimum Visibility

The visibility required during instrument approaches is a function of the aircraft's approach speed and the type of lighting associated with the landing runway. The standard visibility required for a nonprecision approach is one statute mile. The visibility value is designed so that when the pilot

Table 56.11 Obstacle Clearance Altitudes

Navigation Aid	Restriction	Obstacle Clearance Altitude
VOR	With a final approach fix	250 feet
VOR	Without a final approach fix	300 feet
Localizer (includes LDA and SDF)	Without glideslope	250 feet
Nondirectional beacon	With a final approach fix	300 feet
Nondirectional beacon	Without final approach fix	350 feet
Circling approach	None	300 feet
Circling approach	If an NDB approach without a final approach fix	350 feet

Modified from FAA. 1976. *United States Standard for Terminal Instrument Procedures (TERPS)*, 3d. ed. FAA Handbook 8260.3B.

sights the runway, a safe and controlled descent can be made to the runway. Higher minimum descent altitudes typically require higher visibility minima, since aircraft at those altitudes will need to sight the runway and begin a descent at a point more distant from the runway end. The basic 1-mile visibility for nonprecision approaches will be modified depending on the type of aircraft [FAA, 1976].

Required visibility can be reduced through the use of runway approach lights. In general, aircraft category A, B, and C visibility can be reduced to $\frac{3}{4}$ mile if fairly simple approach lights are installed. Visibility can be reduced even further to $\frac{1}{2}$ mile if higher quality approach landing systems with either sequenced flashers or runway alignment lights are installed (see section 56.7). Visibility minima for category D aircraft, with their higher landing speeds, can usually be reduced to an even 1 mile if any approach light systems are installed [FAA, 1976].

Precision Approach Minima

Precision approach minima are based on the type of approach, approach lighting, and runway lighting system. Because of the more accurate vertical and lateral navigation guidance the basic minima for a precision approach are a decision height of 200 feet HAT and a minimum visibility of $\frac{3}{4}$ mile for all categories of aircraft with the exceptions presented in Table 56.12.

Weather Effects

Since pilot altitude information during an instrument approach is derived from a barometric altimeter, it is crucial that when pilots are conducting instrument approaches with minimal obstacle clearance, the aircraft's altimeter be accurately set to the local barometric pressure. Inaccurate barometric pressure settings can result in inaccurate altitude measurement, which may reduce the aircraft's obstacle clearance during the approach. A certified and accurate barometric pressure measurement is available to the pilot at most airports. If such a measurement is not available at or within five miles of the airport, a barometric pressure reading from a nearby airport can be substituted, but the instrument approach descent altitude is adjusted upward to reflect the possibility that the pressure at the remote airport could be somewhat different from that at the airport of intended landing. The penalty for using barometric pressure from a remote site is an upward

Table 56.12 Visibility and Decision Height Exceptions

Runway and Approach Lighting	Decision Height in AGL	Minimum Visibility
None	200 feet	$\frac{3}{4}$ mile
Approach lights	200 feet	$\frac{1}{2}$ mile
Any of the above if no middle marker available	250 feet	$\frac{1}{2}$ mile $\frac{3}{4}$ mile without approach lights

Modified from FAA. 1976. *United States Standard for Terminal Instrument Procedures (TERPS)*, 3d. ed. FAA Handbook 8260.3B.

adjustment of 5 feet of altitude for every mile that the remote altimeter is distant from the main airport, after the first 5 miles.

Noncommercial operations do not have as many restrictions concerning the conduct of an instrument approach placed on them as do commercial operators. It is left up to the pilot to decide whether the minima exist when conducting the approach. If no weather reporting service is available at the airport, the pilot may very often conduct the instrument approach to "look and see" what the weather conditions are. If, in the pilot's preflight planning, the weather conditions at the destination airport appear to be unfavorable or are unknown, a pilot may not wish to risk the potential time lost to attempt an instrument approach at an airport without weather reporting. Thus the pilot may decide from the outset to fly to a more inconvenient airport with weather reporting, accepting the increased ground transportation time and cost, in order to eliminate the uncertainty and a possible unscheduled diversion to another airport.

Commercial operations require that weather observations be available at the airport during the times of arrival and departure. Previous to a recent technology change, weather reports were generated by human weather observers at the airport. Presently, automated weather observation and reporting stations, certified for airport use by the FAA, are available. They are the Automated Weather Observation System (AWOS III) and the Automated Surface Observation System (ASOS) developed by the National Weather Service. These systems permit the replacement of the human observer with the automated system, which is available 24 hours per day rather than being available only during certain hours of the day.

Navigational Aids

Aeronautical navigation aids (NAVAIDs) currently in use serve two purposes: as en route navigation aids or as instrument approach aids. In a few cases, they may do both. The master plan requires

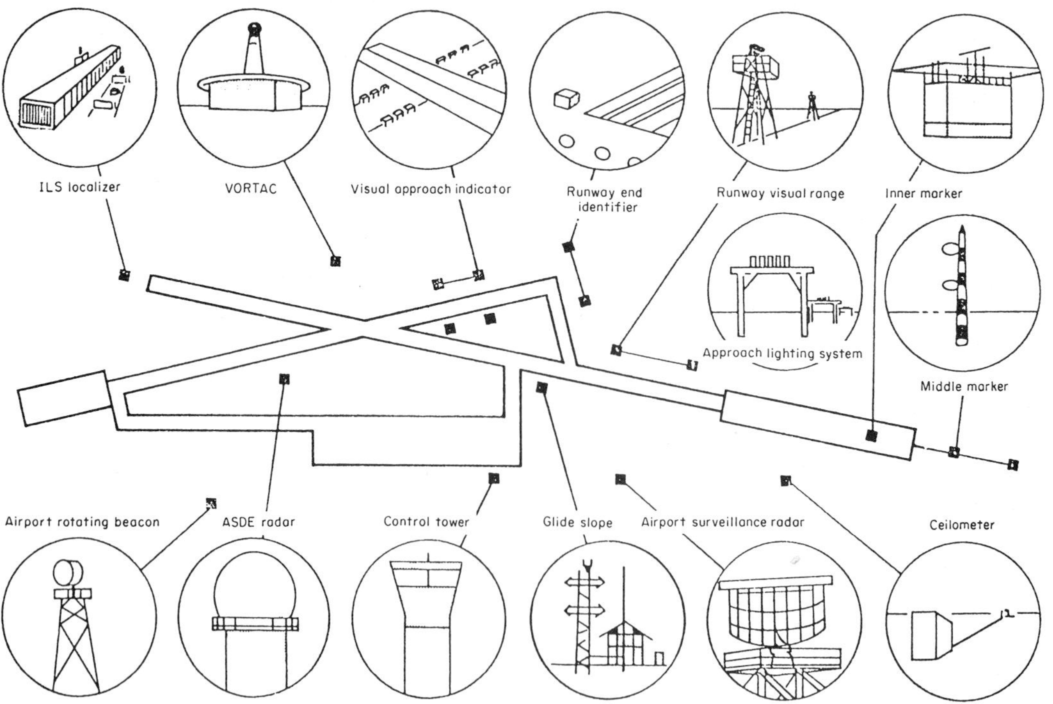

FIGURE 56.13 Relative location of terminal aids for approach to a runway system. (*Sources:* FAA. 1980. *Airport Design Standards—Site Requirements for Terminal Navigation Facilities.* Advisory Circular AC150/5300-2D; and FAA. 1989. *Airport Design.* Advisory Circular AC150/5300-13.)

Table 56.13 Example of Instrument Approach Capability for Airports

Instrument Approach	Airport Classification	Weather
Precision Ceiling 200 feet Visibility 1/2 mile	Primary Reliever Commercial GA Transport with > 24,000 annual OPS	AWOS III or ASOS
Nonprecision Ceiling 500 feet Visibility 1 mile	GA Transport with < 24,000 annual OPS GA Utility with > 12,000 annual OPS	All Part 135 operator airports AWOS III
Nonprecision Ceiling 1000 feet Visibility 1 mile	GA Utility with > 12,000 annual OPS GA Basic Utility	Part 135 Operations or Special Needs

Source: Purdue. 1993. *Instrument Approach and Weather Enhancement Plan.* Final Report on Contract 91-022-086 for Indiana Department of Transportation.

that an inventory of NAVAIDs be completed. TERPS [FAA 1976 with changes] clearly defines the capability of each of the many NAVAIDs. Figure 56.13 shows the general shape and relative location of navigational aids to the runway system.

Criteria for NAVAIDs and Weather Observation

Most public-use airports in any state plan should be considered for some instrument approach procedure. The master plan for a given airport must consider the procedures and weather observation capability that will be the best for the level and type of anticipated traffic. The criteria should be based on the number of annual instrument approach procedures that the airport could expect. For nontowered airports these data are not easily available. In lieu of complete AIA data, the number of annual operations provides an alternative measure, one which is usually forecast and is well understood.

For one state plan [Purdue University, 1993], it has been suggested that 24,000 annual operations provide the first level of separation between a precision approach and a nonprecision approach capability. The probability of various levels of IFR weather, when used with operations data, estimates that 24,000 annual operations (12,000 landings) would reflect conservatively about 650 annual instrument approaches [FAA, 1983a]. The probabilities of various weather conditions in the state suggest that converting from VFR (usually considered to be a 1000-foot ceiling and 3-mile visibility) to a precision instrument approach capability (MDA of 200 feet HAT and $\frac{1}{2}$-mile visibility) could increase possible airport use for an additional 30 to 40 days per year: a significant number of added operations (otherwise lost) for the airport.

Both the NAVAID and AWOS capabilities should be based on the number of annual itinerant operations (usually 40 to 60%) of the total general aviation operations, as shown in Table 56.13, where a precision and two levels of nonprecision IAPs are considered. Part 135 operations are commercial air taxi and air charter operations requiring weather observation.

Benefits to each airport community will accrue due to improved access to the airport from automated weather data: fewer abandoned flight plans due to questionable weather, fewer missed approaches, and increased airport utilization with its benefits to the economy of the community.

56.6 Passenger Terminal Requirements

For many airports the data reflecting present terminal size and capacity would be a part of the master plan inventory. Airports often need to plan for a new passenger terminal or for a major

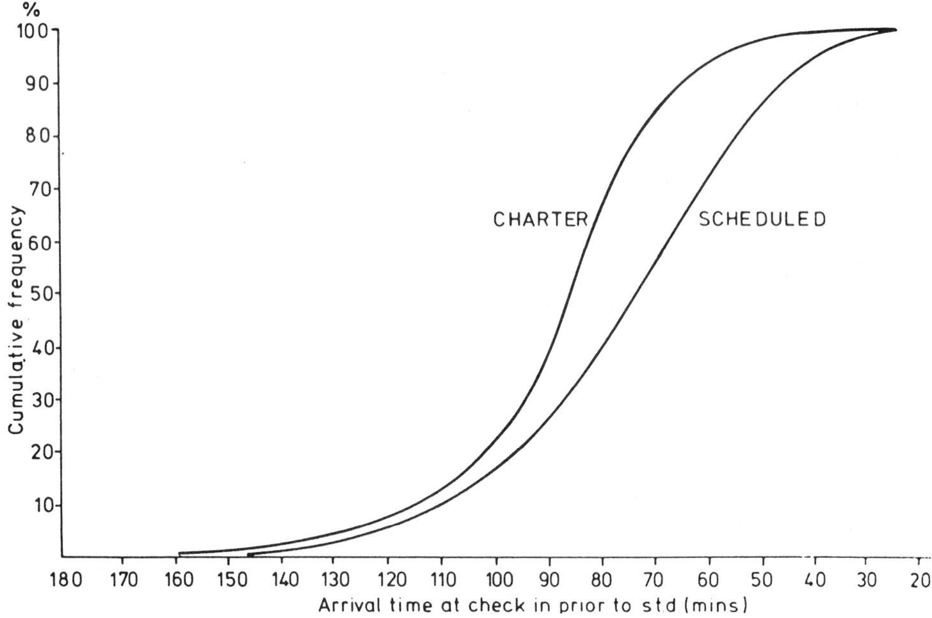

FIGURE 56.14 Typical arrival time for passengers. (*Source:* Ashford, N., Hawkins, N., O'Leary, M., Bennetts, D., and McGinity, P. 1976. Passenger behavior and design of airport terminals. *Transportation Research Record* #588.)

expansion of the existing one. Passenger terminal design should serve to accomplish the following functions:

- *Passenger processing* encompasses those activities associated with the air passenger's trip, such as baggage handling and transfer, ticket processing, and seating. Space is set aside for these activities.
- *Support facilities* for passengers, employees, airline crew and support staff, air traffic controllers, and airport management are provided in each airport. Airlines rent space for the crew to rest and prepare for their next flights.
- *Change mode of transportation* involves the local traveler who arrives by ground transport (car, subway, bus, etc.) and changes to the air mode. The origination-destination passengers require adequate access to the airport, parking, curbside for loading and unloading, and ticket/baggage handing.
- *Change of aircraft* usually occurs in the larger hubs as passengers change from one aircraft to another. While baggage and parking facilities are not needed for these persons, other amenities, such as lounges, good circulation between gates, and opportunities for purchasing food, are important.
- *Collection space* for passengers is necessary for effective air travel. The aircraft may hold from 15 to 400 passengers, each of whom arrives at the airport individually. Boarding passengers requires that the airport have holding or collecting areas adjacent to the airplane departure gate. Because different passengers will come at different times, as shown in Fig. 56.14, there should be amenities for the passenger, such as food, reading material, and seating lounges, as the group of passengers builds up to enplane. Likewise, the terminal provides the shift from group travel to individual travel and the handling of travelers' baggage when an aircraft arrives.

Passenger and Baggage Flow

Perhaps the greatest challenge for airport designers is the need for efficiency in the layout of the critical areas of flow and processing. The users of many airports experience sizable terminal delays

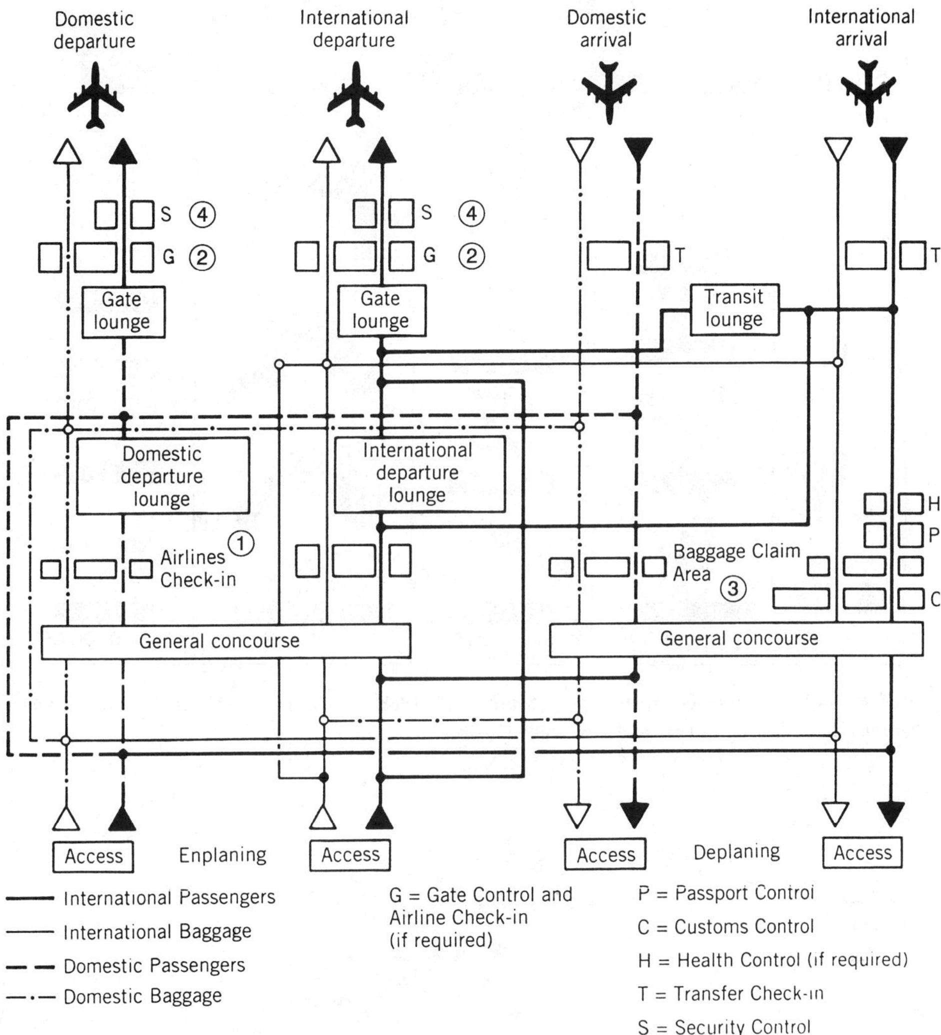

FIGURE 56.15 Passenger baggage flow system. (*Source:* Ashford, N. and Wright, P. 1992. *Airport Engineering*, p. 290. John Wiley & Sons, New York.)

because, under a heavy load, some areas of the terminal become saturated. Many airports designed some years ago were not prepared to handle the baggage from several heavy aircraft (e.g., DC-10, B-747, L1011, MD-81) all landing nearly simultaneously. Figure 56.15 shows the airport flow. The four potential terminal-related bottlenecks are noted in the figure.

1. Baggage/ticket check-in
2. Gate check-in and waiting area
3. Baggage retrieval area
4. Security checkpoints

Terminal Design Concepts

Several workable horizontal terminal configuration concepts are shown in Fig. 56.16. To accommodate growth, many airports have added space to the existing terminal. The new space may reflect a different design concept than the other parts of the terminal due in part to the airline's desires. The San Francisco airport layout shown in Fig. 56.17 provides an example of one terminal that grew and now employs several different gate configurations.

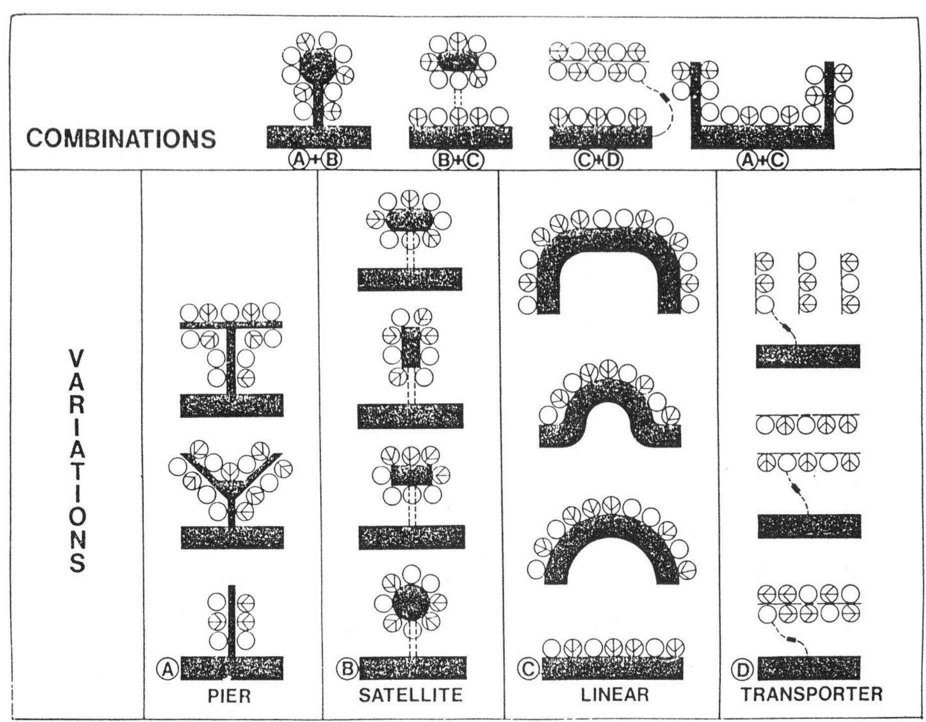

FIGURE 56.16 Terminal configurations. (*Source:* FAA. 1988b. *Planning and Design Guidelines for Airport Terminal Facilities.* Advisory Circular AC150/5360-13.)

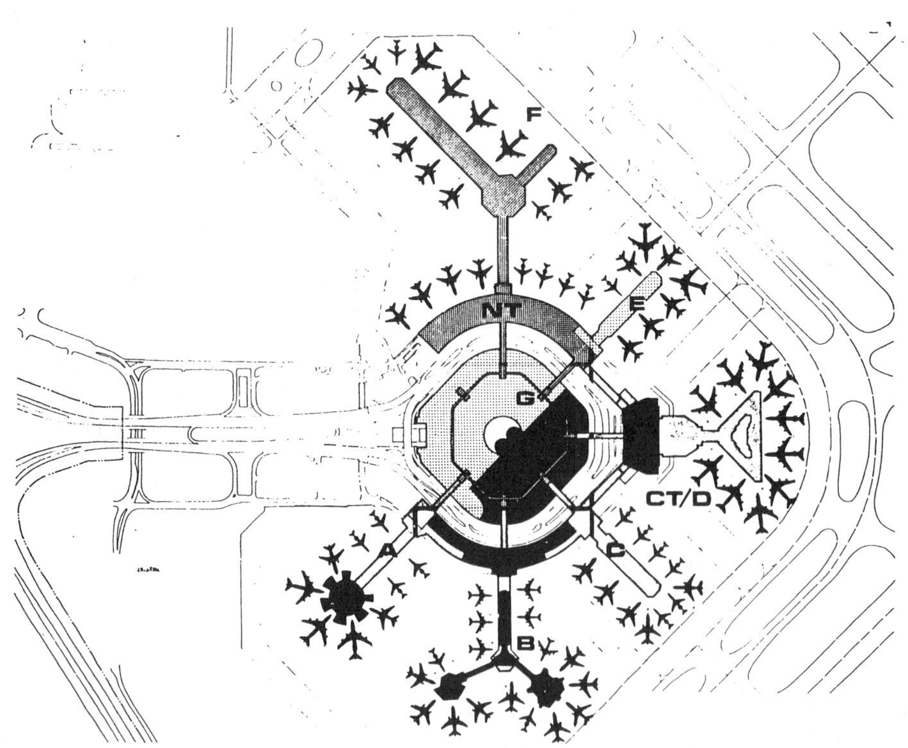

FIGURE 56.17 Layout of San Francisco Airport. (*Source:* San Francisco Airports Commission, circa 1981.)

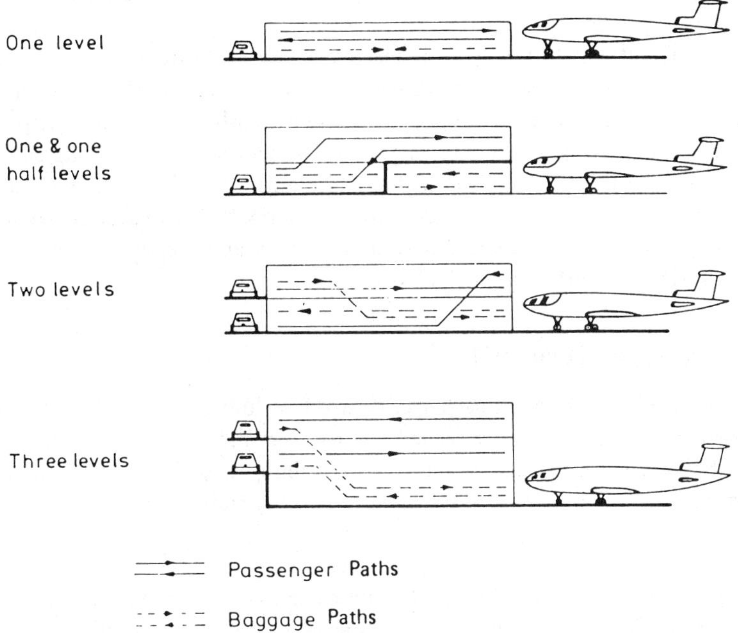

FIGURE 56.18 Vertical separation arrangements of passenger and baggage flows. (*Source:* FAA. 1988b. *Planning and Design Guidelines for Airport Terminal Facilities.* Advisory Circular AC150/5360-13.)

Airport Size by Annual Enplaned Passengers	CONCEPTS APPLICABLE	LINEAR	PIER	SATELLITE	TRANSPORTER	PHYSICAL ASPECTS OF CONCEPTS	SINGLE LEVEL CURB	MULTI LEVEL CURB	SINGLE LEVEL TERMINAL	MULTI LEVEL TERMINAL	SINGLE LEVEL CONNECTOR	MULTI LEVEL CONNECTOR	APRON LEVEL BOARDING	AIRCRAFT LEVEL BOARDING
FEEDER UNDER 25,000		X					X		X				X	
SECONDARY 25,000 TO 75,000		X					X		X				X	
75,000 TO 200,000		X					X		X		X		X	
200,000 TO 500,000		X	X				X		X		X		X	
PRIMARY OVER 75% PAX O/D 500,000 TO 1,000,000		X	X	X			X		X		X	X	X	X
OVER 25% PAX TRANSFER 500,000 TO 1,000,000		X	X	X			X		X		X	X	X	X
OVER 75% PAX O/D 1,000,000 TO 3,000,000			X	X	X		X	X		X	X	X	X	X
OVER 25% PAX TRANSFER 1,000,000 TO 3,000,000			X	X			X	X		X	X	X	X	X
OVER 75% PAX O/D OVER 3,000,000			X	X	X		X	X		X	X	X	X	X
OVER 25% PAX TRANSFER OVER 3,000,000			X	X			X	X		X	X	X		X

FIGURE 56.19 Matrix of concepts related to airport size. (*Source:* FAA. 1988b. *Planning and Design Guidelines for Airport Terminal Facilities.* Advisory Circular AC150/5360-13.)

There are also different vertical distribution concepts for passengers and aircraft. In many airports the passengers and baggage are handled on a single level. For others, the enplaning function is often separated from the deplaning function, especially where the curbside for departing passengers is on the upper level and the baggage claim and ground transportation for arriving passengers are reached on the lower level. Figure 56.18 shows four variations where the enplaning and deplaning passengers are separated as they enter the airport from the aircraft. The matrix shown in Fig. 56.19 indicates the type of terminal concept and separation that design experience has shown are most appropriate for various size airports.

Sizing the Passenger Terminal

The sizing of the terminal consists of passenger demand, including the anticipated requirements for transfer passengers; number of gates needed for boarding; and the anticipated aircraft size and mix. Three methodologies can assist the planner in determining the gross terminal size: the number of gates, the typical peak hour passenger, and the equivalent aircraft methods.

Size Estimate Using Gates

The number of gates can be crudely estimated by referring to the planning data given by the FAA in Fig. 56.20. The number of gates can be better estimated by noting the different types of aircraft that will be at the airport during the peak hour and including the dwell time for each at the gate. For planning purposes the large aircraft will be at the gate approximately 60 minutes. The medium jets like the DC-9s and B-727s will be at the gate for 35 to 50 minutes. However, for contingency planning, 50 minutes are usually allowed for noncommuter aircraft with less than 120 passengers

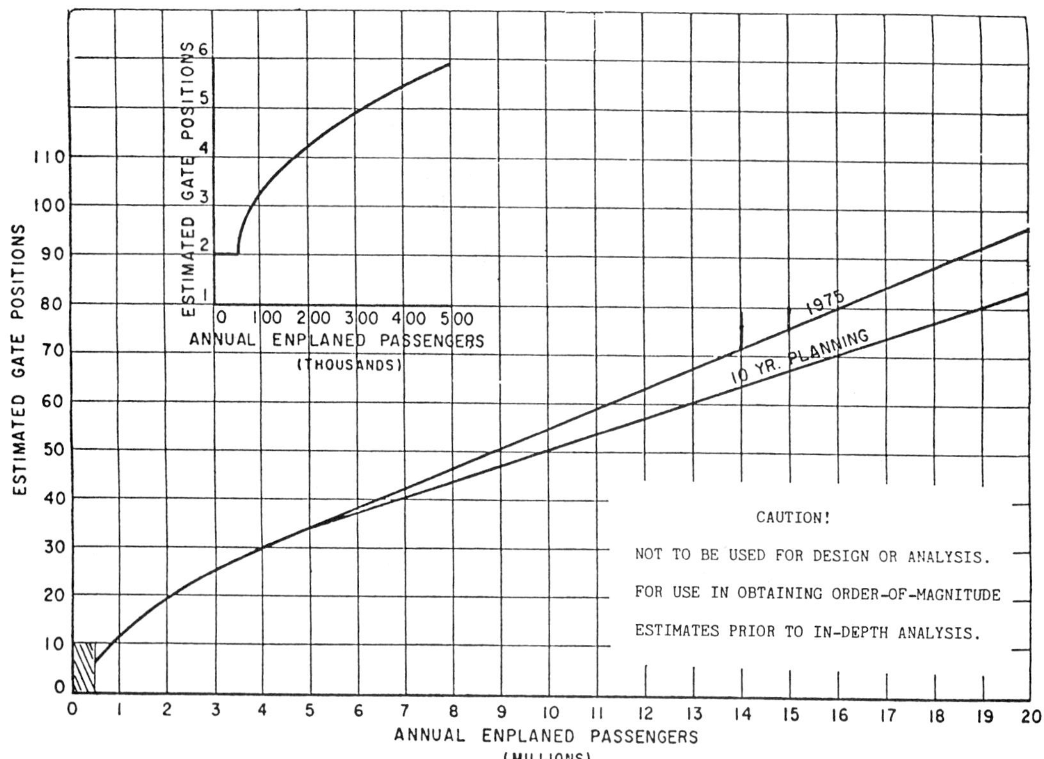

FIGURE 56.20 Planning curve to estimate the number of gates. (*Source:* FAA. 1988b. *Planning and Design Guidelines for Airport Terminal Facilities.* Advisory Circular AC150/5360-13.)

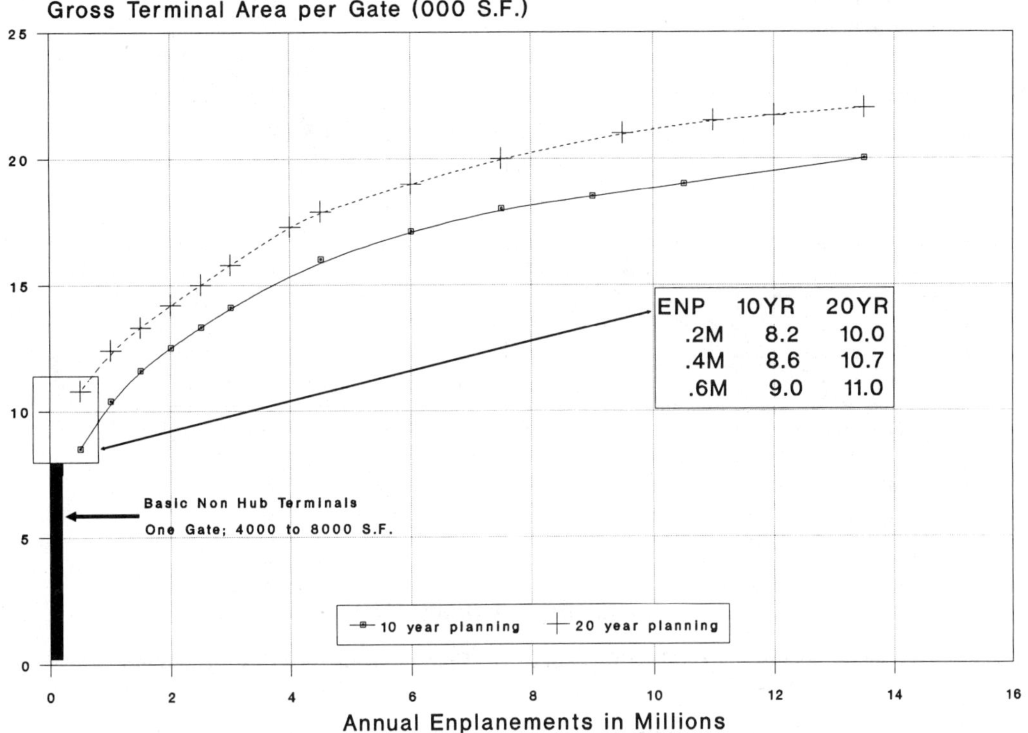

FIGURE 56.21 Using gates to estimate terminal space required. (*Source:* FAA. 1988b. *Planning and Design Guidelines for Airport Terminal Facilities.* Advisory Circular AC150/5360-13.)

Table 56.14 Calculation of Projected Overall Terminal Area for TBA Example Airport Using Number of Gates

Year	Annual Enplanements (Table 56.3)	# of Gates (Figure 56.20)	Area per Gate Square Feet (Figure 56.21)	Estimate of Size of Terminal (Square Feet)
1992 (present)	3,566,270	24	12,000 (Act)	360,000
2000	4,977,100	36	16,200	583,200
2020	8,409,950	45	20,500	922,500

and 1 hour is allowed for all other aircraft. This provides latitude for late (delayed) flights and/or the nonsharing of airline gates. The smaller commuter aircraft, usually with piston or turboprop engines, require about one gate for every three aircraft. The gross terminal area per gate is determined using the planning chart shown in Fig. 56.21. The results are indicated in Table 56.14.

Size Estimate Using Typical Peak Hour Passenger

Another method for sizing the terminal involves the use of the typical peak hour passenger (TPHP). The TPHP does not represent the maximum passenger demand of the airport. It is, however, well above the average demand and considers periods of high airport usage. The TPHP is computed using Eqs. (56.10a) for larger airports and (56.10b) for smaller airports (less than 500,000 annual enplanements). The curves in Fig. 56.22 show the small relative change in TPHP for airports that are entirely origin-destination (no hubbing) to airports where 50% of the enplanements transfer from

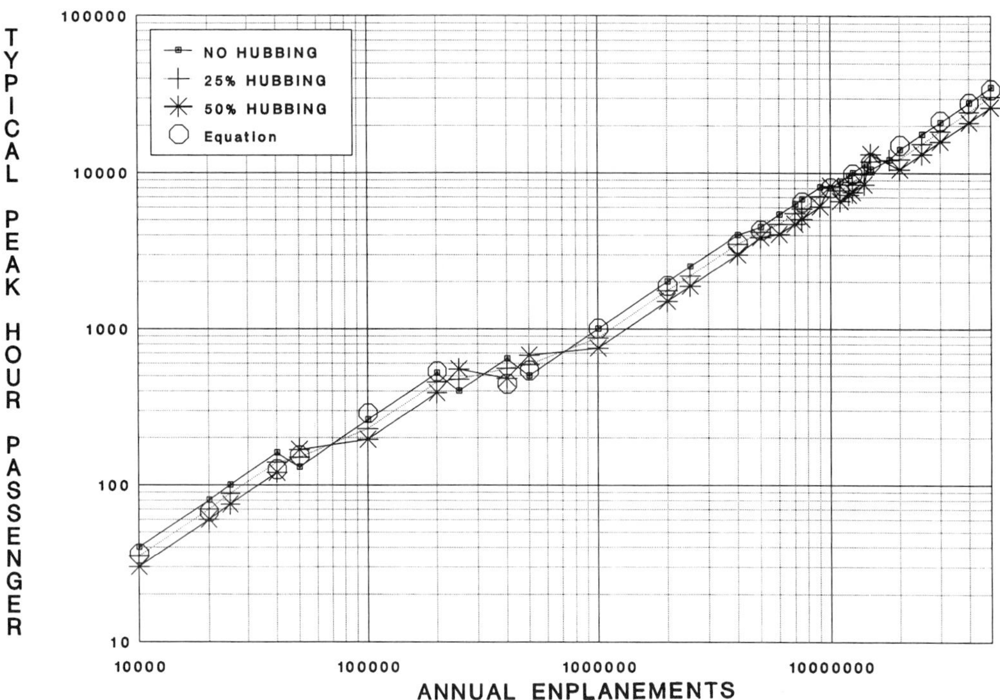

FIGURE 56.22 Typical peak hour passengers (TPHP) as a function of enplanements using FAA relationships.

one to another aircraft. The results are also plotted. For airports where annual enplanements exceed 500,000,

$$\text{TPHP} = .004\text{ENP}^{0.9} \tag{56.10a}$$

For airports where annual enplanements are less than 500,000,

$$\text{TPHP} = .009\text{ENP}^{0.9} \tag{56.10b}$$

where ENP equals annual enplanements.

One common measure used for long range planning is to estimate that 120 to 150 square feet will be required by each TPHP [Ashford and Wright, 1992]. (With an international component

Table 56.15 Calculating the Equivalent Aircraft Factor

							Forecast for 2020			
Type of Aircraft	No. of Aircraft Peak Hr.	Seat Range	Equiv. Aircraft Factor	Gates Req'd	EQA x Gates by Type	Departure Lounge (Sq. Feet)	No. of Aircraft Peak Hr.	Gates Req'd	EQA x Gates by Type	Departure Lounge (Sq. Feet)
B(a)	4	<80	0.6	2	1.2	2,400	8	3	1.8	2,100
C(b)	12	81–100	1.0	10	10.0	11,000	15	13	13.0	14,300
C(c)	10	111–160	1.4	10	14.0	15,000	20	20	28.0	30,000
D(d)	2	161–210	2.0	2	4.0	4,000	4	4	8.0	8,000
D(e)	0	211–280	2.5	—	—	—	3	3	7.5	8,100
D(f)	0	281–420	3.5	—	—	—	2	2	7.0	7,400
D(g)	0	421–500	4.6	—	—	—	—	—	—	—
	28		Total	24	34.1	31,400	52	45	65.3	69,900

Source: Computed from FAA. 1988b. *Planning and Design Guidelines for Airport Terminal Facilities.* Advisory Circular AC150/5360-13.

to the airport, this number increases to about 250 square feet per TPHP. The value of 150 square feet per TPHP is quoted by Ashford and Wright in *Airport Engineering*; its origin is not clear.) The current TPHP for TBA is about 3150, suggesting a terminal size of 473,200 square feet. In the year 2000 TPHP is estimated to be 4260, resulting in approximately 639,000 square feet. In 2020 a TPHP estimate of 6820 indicates a terminal size of 1,023,000 square feet.

Size Estimate Using the Equivalent Aircraft Factor

The FAA advisory circular presents a full range of design curves which are useful for preliminary layout and consideration of the adequacy of space by airport functional area, such as baggage claim. In using the FAA references there are two major areas of information about the airport needed: (1) the number of enplanements that are from the local community and (2) the number and types

Table 56.16 Example for Terminal Space Calculation

Function	How Determined	1992	2000	2020
Equivalent aircraft factor	Table 56.15	34.1	52.1	65.3
Gates	Table 56.15	24	36	45
TPHP	Eq. (56.10a)	3150	4260	6820
1. Departure lounge	Table 56.15	31,400 sq. ft.	51,200	69,900
2. Lobby and ticketing	FAA 1988b page 56	25,000 sq. ft.	40,000	45,000
3. Airline ticket operations	FAA 1988b page 65	7,200 sq. ft.	9,000	11,000
4. Airline space; crew, office, clubs	FAA 1988b page 69—5000 sq. ft. per peak hours aircraft departure	14,000 sq. ft.	21,000 sq. ft.	26,000 sq. ft.
5. Outbound baggage room	FAA 1988b page 67 (80% of the bag rooms)	17,000 sq. ft.	26,000	32,000
6. Baggage claim	60% arrivals with 50% in peak 20 minutes FAA 1988b page 86 for baggage claim frontage; FAA 1988b page 87 using T-shaped flat bed, dir. feed for area	360 feet of claim 11,000 sq. ft.	560 feet of claim 16,000 sq. ft.	750 feet of claim 21,000 sq. ft.
7. Lobby waiting area	FAA 1988b page 57 (seating for 20% TPHP)	12,000	16,000	24,000
8. Lobby for baggage claim	Two greeters plus one passenger, 20 minutes wait uses 21 square feet per person (see Table 56.18)	490 PAX 980 guest 30,800 sq. ft.	662 PAX 1314 guest 41,500 sq. ft.	1060 PAX 2120 guest 66,800 sq. ft.
9. Security	150 sq. ft. per station	1,200 sq. ft.	1500	1800
10. Food and beverage	FAA 1988b page 92 (assume 40–50% usage factor)	40,000 sq. ft.	44,000	52,000
11. Concessions	FAA 1988b page 93 (Upper value)	45,000 sq. ft.	60,000	80,000
12. Other circulation	Assume 80% of items 1 through 5	85,280 sq. ft.	130,500	163,900
13. HVAC, mechanical areas, structure	Use 25% of total	80,200 sq. ft.	114,200	148,400
	Total space required	401,100 sq. ft.	580,000	741,800
	Space per peak hour passenger	127.3 sq. ft. per TPHP	134.0	108.8

Source: Computed from FAA. 1988b. *Planning and Design Guidelines for Airport Terminal Facilities.* Advisory Circular AC150/5360-13, change 1.

Table 56.17 Comparison of Sizing Methods for the TBA Airport

	Method of Determination			
Year	Gates (sq. ft.)	TPHP (sq. ft.)	EQA (sq. ft.)	Recommended (sq. ft.)
1992	336,000	473,000	391,200	360,000(act)
2000	576,000	639,000	537,700	575,000
2020	945,000	1,023,000	710,800	900,000

Table 56.18 IATA Level of Service Space Standard for Airport Passenger Terminals

	Level of Service Standards in Square Feet per Occupant		
	Excellent	Good	Poor
Check-in queue area	19	15	11
Wait/circulate	29	20	11
Holding room	19	11	6.5
Bag claim area (no device)	21	17	13
Government inspection	15	11	6.5
Total	103	74	48

Source: International Air Transport Association. 1989. *Airport Terminals Reference Manual,* 7th ed. Montreal.

of aircraft that will use the airport in the peak hour, called the equivalent aircraft factor (EQA). The EQA for the TBA airport is shown in Table 56.15. It is based on the number of seats on arriving aircraft during the peak hour. Also shown is the departure lounge space, directly related to the EQA times the number of gates.

A terminal with a high level of hubbing results in a large number of passengers who will be changing aircraft rather than originating from the area. Thus, hubbing airports require reduced space for airline ticketing, baggage claim, curb access, and parking.

Table 56.16 gives a detailed breakdown of the area planning for a passenger terminal using the FAA design curves [FAA, 1988b]. The "how determined" column indicates how each number was computed. The estimate needed for baggage claim handling is percent arrivals (assumed during peak traffic to be 60%), the number of aircraft in the peak 20 minutes (assumed to be 50%), and the number of passengers and guests who will be getting baggage. It is assumed that 70% of arriving destination passengers will be getting baggage and each will have two guests. Use of the FAA Advisory Circular [FAA, 1988b] is indicated with a page number.

As shown in Table 56.17, the calculated space provides a range often useful in examining architect's renderings or developing preliminary cost estimates based on square foot cost standards. The International Air Transport Association (IATA) has established space requirements based on the level of service rated on a scale from excellent to poor for the major used portions of the airport. Given in Table 56.18, these data are useful in reviewing the terminal capabilities, capacities, and plans. The middle level is desirably the lowest level for peak operations. At the poor end, the system is at the point of breakdown.

Airport Airside Access

Parking of aircraft at the gate consists primarily of a "nose in" attitude requiring a pushback from the gate, or parking "parallel" to the terminal building. With the modern jetways, the parking space is usually governed by gate placement. The jetways themselves can be adjusted for aircraft door height from the ground and usually have sufficient extension capability to serve all the aircraft. Many airlines prefer boarding passengers on a Boeing 747 or other heavy aircraft through two doors. This requires two jetways for each gate destined to serve the heavy aircraft or for two gates. It also means that heavy aircraft will have special places to park at the gate. For the planning of

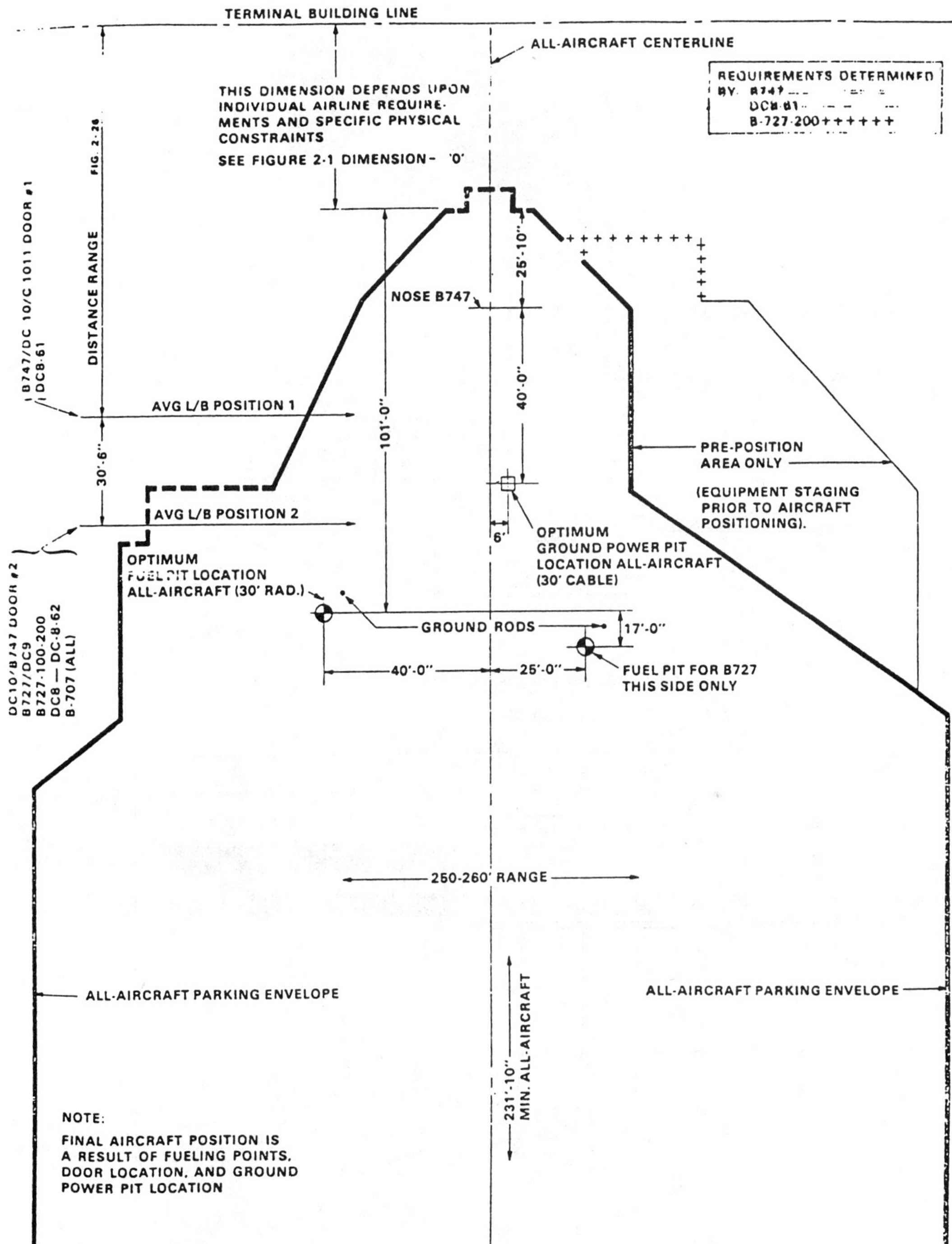

FIGURE 56.23 Aircraft and ground servicing parking envelope. (*Source:* Federal Aviation Administration. 1988b. *Planning and Design Guidelines for Airport Terminal Facilities.* Advisory Circular AC150/5360-13.)

the apron it is important to allow sufficient space to handle the expected aircraft according to the footprint shown in Fig. 56.23. Ease of aircraft movement to and from the taxiway dictates the space between aircraft parking areas.

The aircraft is unloaded, loaded, and serviced on the terminal apron. The spacing on the apron itself is determined by the physical dimensions of the aircraft and the parking configuration. Figure

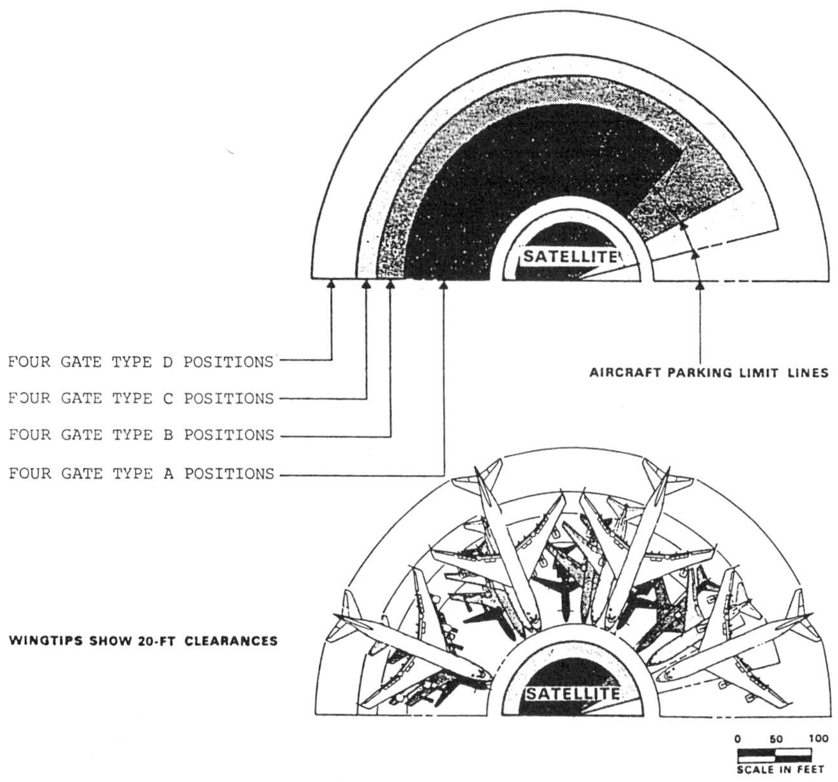

(a) Satellite push-out gate positioning

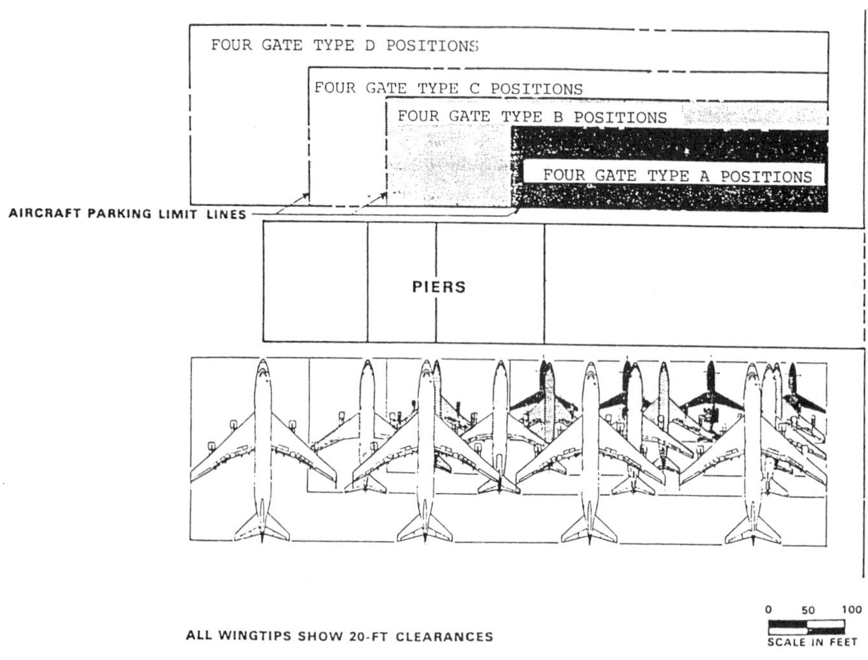

(b) Linear configuration push-out gate positioning

FIGURE 56.24 Configurations for parking at satellite or pier. (*Source:* (a) FAA. 1988b. *Planning and Design Guidelines for Airport Terminal Facilities.* Advisory Circular AC150/5360-13. (b) FAA. 1988b. *Planning and Design Guidelines for Airport Terminal Facilities.* Advisory Circular AC150/5360-13.)

Table 56.19 Apron Requirements for Parking and Aircraft Movement TBA Airport

Type of Aircraft	Space for Aircraft Movement and Parking (sq. ft.)	Aircraft in 2020	Apron Space Needed (sq. ft.)
Wide-bodied large-engine jet aircraft	160,000	9	1,440,000
Four-engine narrow-body jet aircraft	65,000	3	195,000
Three-engine narrow-body jet aircraft	43,000	17	731,000
Two-engine narrow-body jet aircraft	33,000	15	495,000
Two-engine turbojet aircraft	16,000	8	128,000
			2,989,000

56.24 shows the physical dimensions appropriate for pushout parking at either a satellite or a linear gate configuration. Apron dimensions are a function of the terminal concept chosen. However, for master planning where detailed geometry is not available, the total area is estimated by aircraft type. Table 56.19 presents the space numbers for aircraft movement and parking [Ashford and Wright, 1992] and extends them by the number of aircraft in the TBA example airport.

In the TBA airport example, for the 8.4 million annual enplanements in the year 2020, a total of just under three million square feet of apron area is required. This space allocation includes

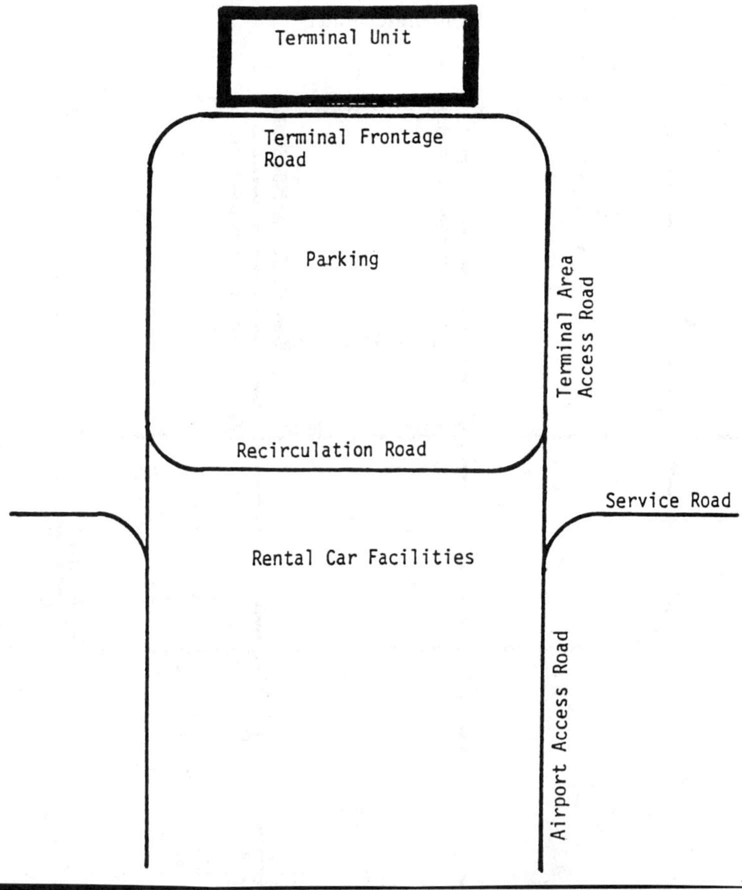

FIGURE 56.25 Four prominent ground access concepts. (*Source:* FAA. 1988b. *Planning and Design Guidelines for Airport Terminal Facilities.* Advisory Circular AC150/5360-13.) *(continues)*

adequate space for aircraft to move from the apron to the taxiway as well as space for aircraft to move freely when others are parked at the ramp.

Airport Landside Access

Access Planning

Planning for airport access, especially by highway, is best done in conjunction with the local or state highway departments, who will have the responsibility for maintaining efficient access and avoiding gridlock outside the airport. The access portion of the airport design and planning process would also take into account the potential for rail and special bus connections. The design of the roadways around the airport and for entering and leaving the airport will need to account for the heavy traffic flows that often occur near rush hour when local industry and airport traffic usually overlap. While these design aspects are covered in the highway design portion of the *Handbook*, Fig. 56.25 presents four of the more prominent layout options for airport access.

Terminal Curbside Dimensions

The curbside dimensions will depend on the anticipated mode of transportation that brings persons to the airport. For gross planning, 115 lineal feet per million originating passengers can be used [deNuefville, 1976]. For a more accurate estimate, the "dwell time" and length of each arriving vehicle at the curb must be determined. Since departing and arriving passengers exhibit different dwell times it is appropriate to consider them separately.

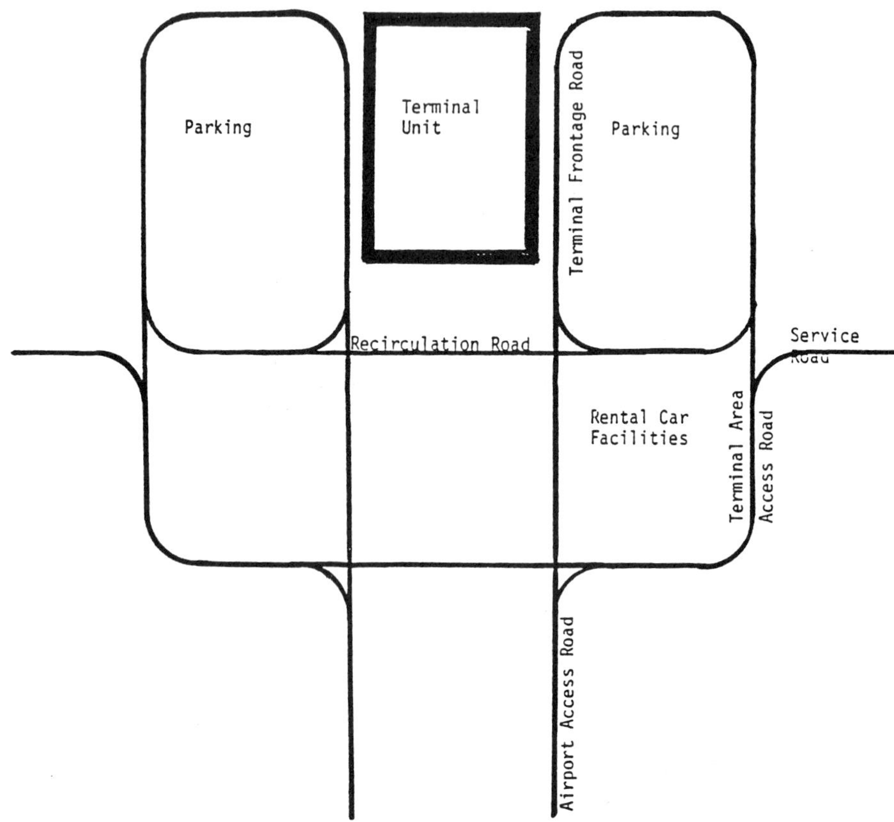

FIGURE 56.25 (continued) Four prominent ground access concepts.

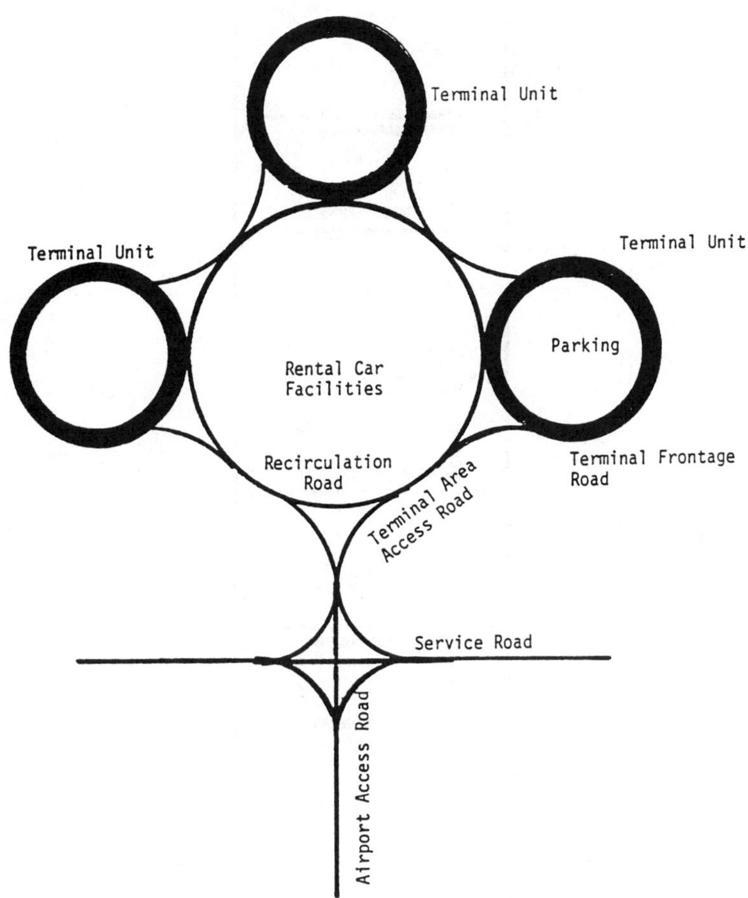

FIGURE 56.25 (continued) Four prominent ground access concepts. *(continues)*

Table 56.20 Curb Front Requirements from Fort Lauderdale–Hollywood Airport

		Average Dwell Time (s)		Curb front Required (feet-s)	
Mode	Length (ft)	Enplaning	Deplaning	Enplaning	Deplaning
Personal auto	26	130	170	3,380	4,420
Taxi	26	75	130	1,950	3,380
Limousine	36	180	400	6,480	14,400
Courtesy vehicle	46	80	180	3,680	8,280
Bus	46	270	400	12,420	18,400
Other	36	360	190	12,960	6,840

For example, Table 56.20 shows the average dwell times taken from data collected at the Fort Lauderdale–Hollywood airport. Table 56.21 then provides the curb length for the TBA airport in 2020 assuming that during the peak hour a TPHP of 1060 passengers arrives at the curb and 1060 depart (see Table 56.16). The mode split between and ridership in cars, taxis, buses, and courtesy cars would be as indicated.

Although theoretically one lineal foot of curb front can provide 3600 feet-seconds of curb front in one hour, it has been suggested that the practical capacity is about 70% of this number [Cherwony

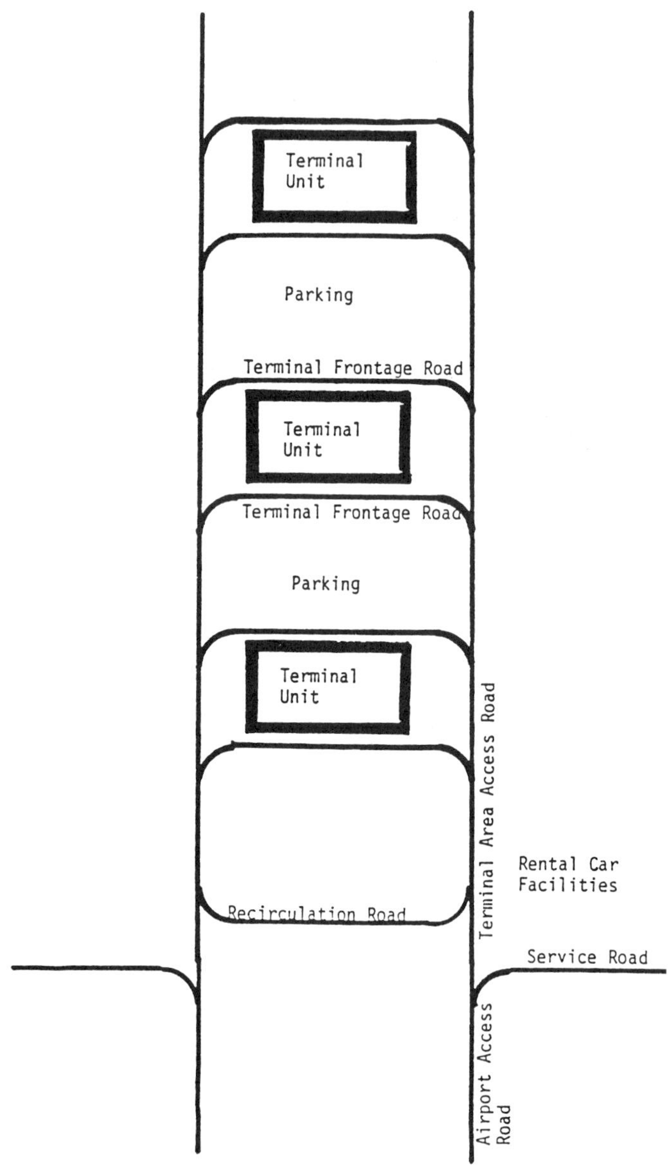

FIGURE 56.25 (continued) Four prominent ground access concepts.

Table 56.21 Example of Curb Front Design for TBA Airport

Mode	Enplaning			Deplaning		
	Passengers	Vehicles	Peak (ft-s)	Passengers	Vehicles	Peak (ft-s)
Personal auto	400	360	1,216,800	420	380	1,679,600
Taxi	100	100	195,000	100	100	338,000
Limousine	80	10	64,800	80	12	172,800
Courtesy vehicle	180	40	147,200	240	50	414,000
Bus	200	10	124,200	120	10	184,000
Other	100	10	129,600	100	12	82,000
			1,877,600			2,870,400

and Zabawski, 1983]. For the TBA airport in 2020, the curb necessary on the enplanement or departing level is 1,877,600/(0.7×3600) or about 745 feet, and for the deplaning curb front, 1139 feet.

Parking

Parking requirements for airports vary widely, depending on the nature of the airport and the manner in which people come to the airport. The long-term parking serves passengers who drive and park plus the employees on the site. The short-term lot accommodates well-wishers and greeters, visitors to the airport itself, and salespersons and is located next to the terminal. Separate lots for long-term and short-term parking should be provided when the total annual passenger volume exceeds 150,000 to 200,000 [FAA, 1988b].

High fees at the short-term lot relative to those for the long-term lot tend to discourage long-term parkers (more than three hours) from clogging short-term parking areas. The short-term lot can usually be sized on the basis of the originating peak hour passengers; one useful ratio is two short-term spaces for every seven originating peak hour passengers [Ashford and Wright, 1992]. Another rule of thumb is that the short-term parkers will require about 20% of the total parking space [FAA, 1988b].

The long-term lot requires a vastly different approach. The best way to develop the lot size is to obtain data from an airport similar to the one being designed, noting the time and day a car arrives, and its length of stay. From these data a simulation can be used to size the parking lot. The Institute of Air Transport surveyed 12 of the larger U.S. airports in 1979 and found that the parking ranged from 3.45 spaces per million annual originating enplanements for BWI to 0.86 at New York La Guardia. While this was a 1979 study and the parking at many airports has been upgraded it serve to indicate the disparity between airport parking facilities. Some cities have excellent transit connections to the airport which serve to relieve some of the pressure for long-term parking (at least for employees).

For preliminary planning, it would be safe to use 1.5 spaces for each TPHP originating passenger to size the total parking need. The land needed without a parking structure equates to 100 to 125 cars per acre. For TBA in 2020, the 6 million originating passengers would equate to 5040 TPHP resulting in an estimate of 7500 spaces or 60 to 75 acres of parking. The short-term lot would have about 1500 spaces with about 6000 allocated for long-term parking. Often the Achilles' heel of an airport, the parking lot is a good revenue producer and should be carefully managed. Shuttle buses provide courtesy transportation to the departure and arrival curbs for the convenience of the traveler.

56.7 Airport Site Determination and Considerations

It is often situations within 10 miles of the airport site that will have significant bearing on the success of an airport project. The airspace and associated ground tracks along the takeoff and landing corridors are critical not only to site location, but also for runway orientation, since they define:

- Where safe landing of aircraft for over 95% of the wind conditions must occur
- Where obstacles projecting into the flight path must be eliminated
- Where houses, buildings, and recreation sites could be subjected to unacceptable levels of aircraft noise

Siting of runways must seek to provide solutions to all three of these constraints. In addition, runways must avoid landing and takeoff paths that are over landfills and other areas that are prime bird habitats. In recognition of the severity of aircraft crashes when they occur in the vicinity of public assembly buildings, particularly schools, communities are encouraged to control the land

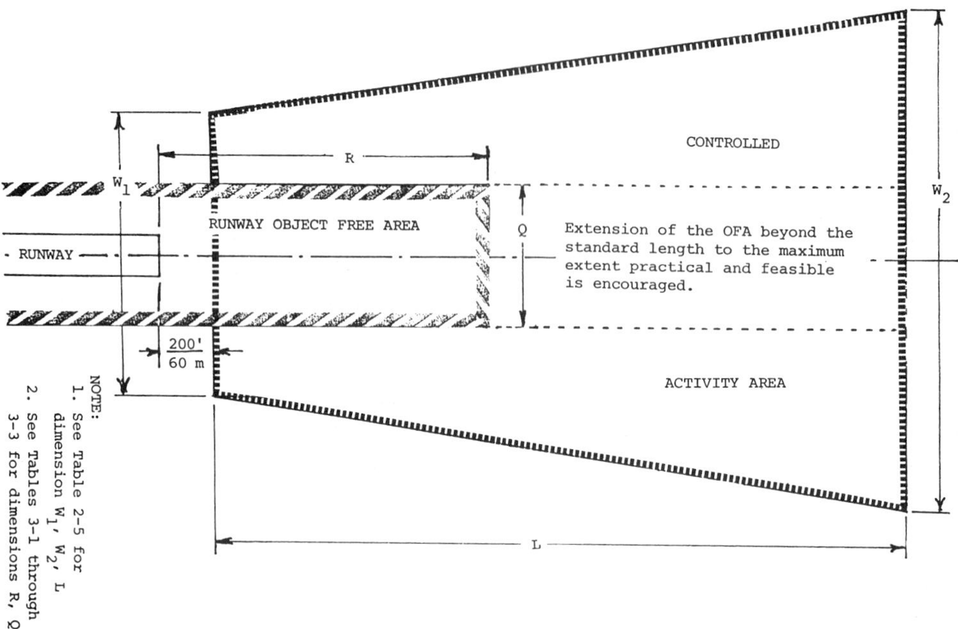

FIGURE 56.26 Runway protection zone. (*Source:* FAA. 1991c. *Airport Design.* Advisory Circular AC150/5300-13, change 1.)

use within three miles from the airport reference point (ARP) restricting the building of any such buildings [FAA, 1983a]. Other site considerations are the usual civil engineering concerns of soil condition, required grading and earthwork, wetlands, and suitable access connecting the airport with major business and industrial areas nearby.

Mandatory Control/Ownership

The land from the outer edge of the runway protection zone (RPZ) shown in Fig. 56.26 to the runway threshold is the minimum amount of land, beyond that associated with the runways themselves and the terminal, that should be in the possession (under direct control) of the airport management. If ownership of this area is not possible, then all activity in the trapezoidal area shown in Fig. 56.26 must be under total control of the airport. Sometimes special easements or other legal instruments are used to ensure positive control.

The more surrounding land the airport owns, especially land extending from the ends of major runways, the better the airport will be able to grow and expand to meet the ever growing demand for air travel while maintaining acceptable relationships with the community. As shown in Table 56.22, the RPZ for paved runways (formerly called the clearway) varies in size according to the approach capability (visual, nonprecision instrument, precision instrument) discussed in section 56.5.

For future expansion the best plan is to obtain land equivalent to the largest RPZ dimensions, which for precision approaches extend 2700 feet from the runway threshold. This could amount to a significant amount of land acquisition if the airport has not planned ahead. Even though the FAA will help fund purchase of land for safety improvements, obtaining the land around an existing airport is not always easy and can have as much neighborhood impact as the noise paths. While it is possible to fly special curved approaches during landing and takeoff to minimize noise, straight-in glide slopes are recommended as the safest.

Table 56.22 Runway Protection Zone Dimensions for Transport Airports (C&D Aircraft)

Runway Ends		Dimensions for Approach End			
Approach End	Opposite End	Length (ft) [m]	Inner Width (ft) [m]	Outer Width (ft) [m]	RPZ Area (Acres)
V	V, NP	1000 [300]	500 [150]	700 [210]	13.8
V	P	1000 [300]	1000 [300]	1100 [330]	24.1
NP	V, NP	1700 [510]	500 [150]	1010 [303]	29.5
NP	P	1700 [510]	1000 [300]	1425 [427.5]	47.3
P	V, NP, P	2500 [750]	1000 [300]	1750 [525]	78.9

V = visual approach; NP = nonprecision instrument approach (visibility > 3/4 statue mile); P = precision instrument approach

Source: FAA. 1991c. *Airport Design.* Advisory Council AC150/5300-13, change 1.

Obstacle Control

For the pilot on final approach the runway is an extension of the glide path. The length and slope of the glide path depend on the airport's traffic and the approach capability of the runway (visual, instrument nonprecision, or precision) landing system. The glide path for landing and taking off aircraft must be under the control of the airport to the extent that obstacles are avoided, navigation is facilitated, and landing is safe. Table 56.23 presents the dimensions of the approach surface for transport airports (C&D aircraft). The obstacles along the glide path pose a most severe situation. At a 50:1 slope, the distance from the end of the runway to clear a 200 foot (60 m) obstacle by 250 feet (75 m) is 22,500 feet (6850 m or 4.3 miles).

The airport is to be sited where it is free from obstructions that could be hazardous to aircraft taking off or landing at the airport. Imaginary surfaces are used to define the limits on potential obstacles on or near the glide slope. For takeoff these are also critical because it is required that a transport aircraft be able to take off successfully even if one engine is out. For aviation in the U.S. the imaginary surfaces are set forth in Part 77 of the Federal Aviation Regulations [FAA, 1975]. The imaginary surfaces are defined in the surface shown in Fig. 56.27. If the airport is ever to achieve precision instrument status, the precision instrument slope of 50:1 for 10,000 feet (3,000 m) followed by 40:1 for an additional 40,000 feet (12,000 m) should govern the land-use policies that restrict building and object heights. For nonprecision instrument landings and visual landings there is still need to control the obstacles out to at least 10,000 feet (3,000 m) at the landing slope of either 34:1 or 20:1, respectively.

Table 56.23 Approach Surface Dimensions for Transport Airport (C&D Aircraft)

Runway End		Approach Surface Dimensions			
Approach End	Opposite End	Length (ft) [m]	Inner Width (ft) [m]	Outer Width (ft) [m]	Slope Run:Rise
V	V, NP	5,000 [1500]	500 [150]	1,500 [450]	20:1
V	P	5,000 [1500]	1,000 [300]	1,500 [450]	20:1
NP	V, NP	10,000 [3000]	500 [150]	3,500 [1050]	34:1
NP	P	10,000 [3000]	500 [150]	3,500 [1050]	34:1
P	V, NP, P	10,000 [3000]	1,000 [300]	4,000 [1200]	50:1
		PLUS	+	+	
		40,000 [12,000]	4,000 [1200]	16,000 [4800]	40:1

V = visual approach; NP = nonprecision instrument approach (visibility > .75 statue mile); P = precision instrument approach

Source: FAA. 1991c. *Airport Design.* Advisory Circular AC150/5300-13, change 1.

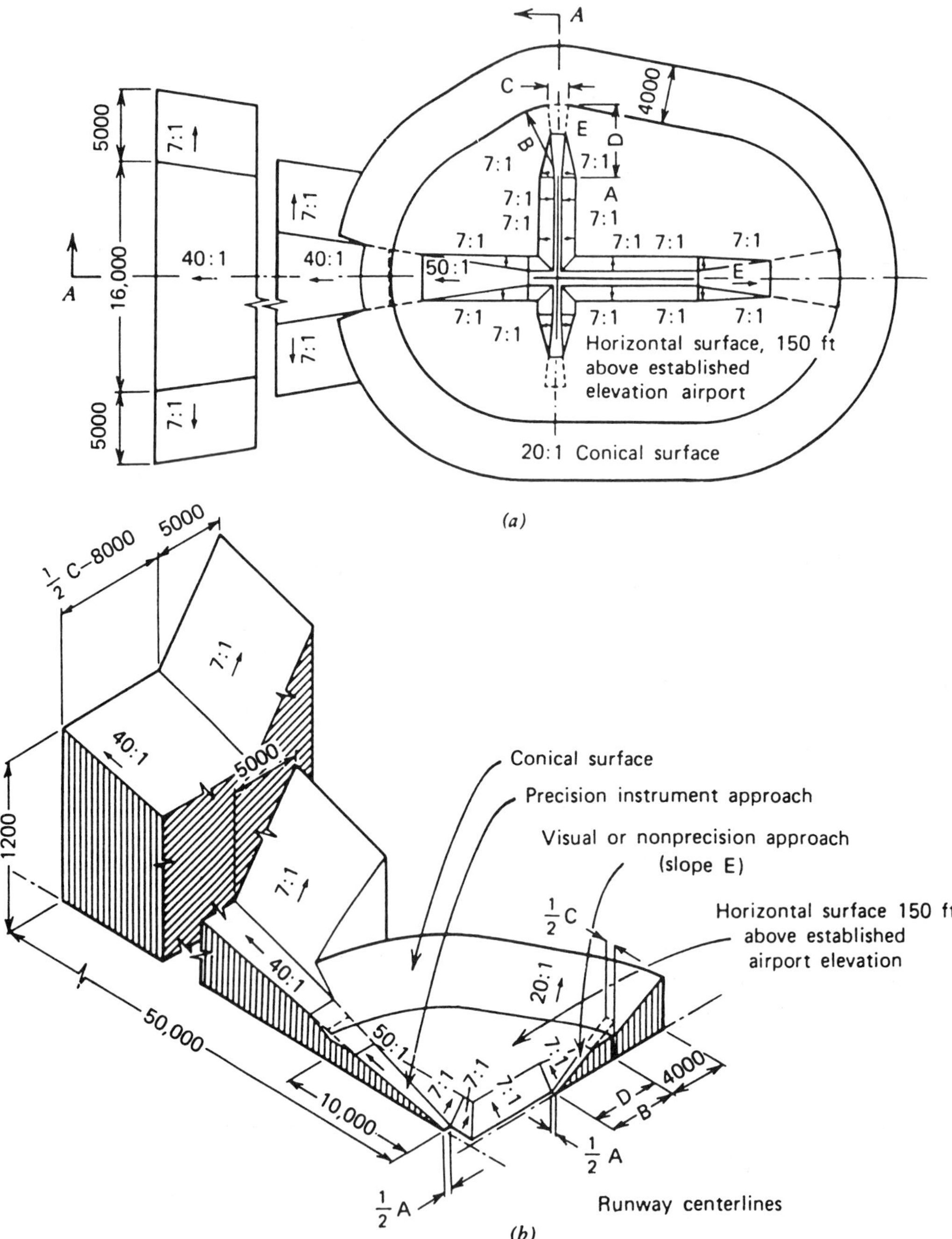

FIGURE 56.27 Imaginary surfaces used for obstacle control. All dimensions in feet. (*Source:* FAR 77.23. 1975.)

In terms of safety, the FAA has established object height requirements in the vicinity of the airport as follows:

> An object would be an obstruction to air navigation if of greater height than 200 feet (60 meters) above the ground at the site, or above the established airport elevation, which ever is higher (a) within 3 nautical miles (5.6 km) of the established reference point of an airport with its

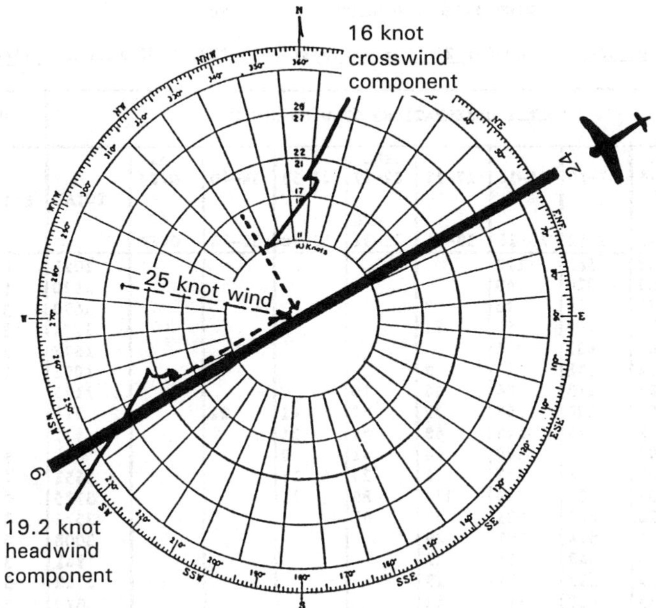

FIGURE 56.28 Headwind and crosswind components on a wind rose.

longest runway more than 3200 feet (975 m) in actual length and (b) that height increases in proportion of 100 feet (30 m) for each additional nautical mile from the airport reference point up to a maximum of 500 feet (150 m) [FAR Part 77.23(a)(2)].

Orientation for Winds

The orientation of the runway, in part, results from the physics of the aircraft. Airplanes operate best when they are flown heading into the wind, so the runway choice, if there is one, is always to land (or to take off) heading directly into the wind. Since the wind varies and the runway is fixed, this is usually not totally possible. Figure 56.28 shows an aircraft landing on runway 24 in a 25 knot wind blowing from 280 degrees azimuth.

Landing into the wind has also resulted in the convention for numbering runways where the runway number consists of the first two digits related to the azimuth of the runway rotated by 180 degrees to account for the direction of the wind. Thus the pilot landing on runway 24 will have the headwind component of 19.2 knots. The crosswind component of wind is 16 knots. The polar plot displaying these is called a wind rose.

The FAA standards, given in the U.S. code (CFR Title 14, Part 25), require that the airport must be able to accept landing (acceptable level of crosswind at 13 knots) along its runway(s) 95% of the time. When this cannot be accomplished with one runway, then the airport must add a "crosswind" runway. The two runways together then statistically eliminate unacceptable crosswinds to less than 5%. If possible, a ten-year sample of wind soundings taken hourly is used to establish a model of the wind velocity and direction. The wind data are then analyzed and placed in the appropriate cell, as shown in the wind rose of Fig. 56.29. Thus each cell shows the percentage of the time the wind has an amplitude and a direction indicated by the cell. The runways are then placed on the wind rose to analyze for minimum crosswinds in excess of the 13 knot criterion. For each orientation the cells outside the runway template are summed to determine if the 95% criterion has been met.

The rules relate the crosswind restriction only to the width of the runway, as indicated in Table 56.24. The crosswind restriction, for example, has been changed for basic transport aircraft to 20 knots [Ashford and Wright, 1992]. However, there is a trade-off between allowable crosswind and runway width for lighter planes which are difficult to control in heavy crosswinds. For example, a

WIND DIRECTION VERSUS WIND SPEED

STATION: Anywhere, USA HOURS: 24 Observations/Day PERIOD OF RECORD: 1964-1973

DIRECTION	HOURLY OBSERVATIONS OF WIND SPEED										AVERAGE SPEED	
	KNOTS 0-3	4-6	7-10	11-16	17-21	22-27	28-33	34-40	41 OVER	TOTAL	KNOTS	MPH
	MPH 0-3	4-7	8-12	13-18	19-24	25-31	32-38	39-46	47 OVER			
01	469	842	568	212						2091	6.2	7.1
02	568	1263	820	169						2820	6.0	6.9
03	294	775	519	73	9					1670	5.7	6.6
04	317	872	509	62	11					1771	5.7	6.6
05	260	861	437	106						1672	5.6	6.4
06	357	534	151	42	8					1092	4.9	5.6
07	369	403	273	84	36	10				1175	6.6	7.6
08	158	261	138	69	73	52	41	22		814	7.6	8.8
09	167	352	176	128	68	59	21			971	7.5	8.6
10	119	303	127	100	98	41	9			877	9.3	10.7
11	323	506	268	312	111	23	28			1651	7.9	9.1
12	610	1397	624	779	271	69	21			3779	8.3	9.6
13	472	1375	674	531	452	67				3571	8.4	9.7
14	647	1377	574	781	129					3008	6.2	7.1
15	338	1093	348	135	27					1941	5.6	6.4
16	560	1399	523	121	19					2622	5.5	6.3
17	587	883	469	128	12					2079	5.4	6.2
18	1046	1904	1068	297	83	10				4496	5.8	6.7
19	499	793	586	241	92					2211	6.2	7.1
20	371	946	615	243	64					2239	6.6	7.6
21	340	732	528	323	147	8				2078	7.6	8.8
22	479	768	603	231	115	38	19			2253	7,7	8.9
23	187	1008	915	413	192					2715	7.9	9.1
24	450	943	800	453	96	11	18			2779	7.2	8.2
25	351	899	752	297	102	21	9			2431	7.2	8.2
26	368	731	379	208	53					1739	6.3	7.2
27	411	748	469	232	118	19				1997	6.7	7.7
28	191	554	276	287	118					1426	7.3	8.4
29	271	642	548	479	143	17				2100	8.0	9.3
30	379	873	526	543	208	34				2563	8.0	9.3
31	299	643	597	618	222	19				2398	8.5	9.8
32	397	852	521	559	150	23				2510	7.9	9.1
33	236	721	324	238	48					1567	6.7	7.7
34	280	916	845	307	24					2372	6.9	7.9
35	252	931	918	487	23					2611	6.9	7.9
36	501	1568	1381	569	27					4046	7.0	8.0
00	7729									7720	0.0	0.0
TOTAL	21676	31828	19849	10437	3357	529	166	22		87864	6.9	7.9

FIGURE 56.29 Wind data and wind rose analysis. (*Source:* FAA. 1991c. *Airport Design.* Advisory Circular AC150/5300-13, change 1.)

200 foot wide runway gives the pilot of a light aircraft much more latitude for maintaining control in a heavy (20 knot) crosswind (provided the structural integrity of the aircraft is not exceeded) than for landing on a 75 foot wide runway. The acceptable practice for most airports has been to ensure that the runway configuration provides for a minimum of 95% against a 13 knot crosswind. Once the possible best directions of runways are established then other factors that impinge on direction obstacles and noise become critical.

Noise

Airport noise has restrained development, constrained operations, and restricted the expansion of many airports in the U.S. Its presence continues to plague airport managers and operators, who find it continually impinging on their desire to maintain good community relations. Aircraft primarily produce noise from their engines and from the flow of air over the aerodynamic surfaces. Jet-turbine driven aircraft produce considerably more noise than did their piston engine predecessors.

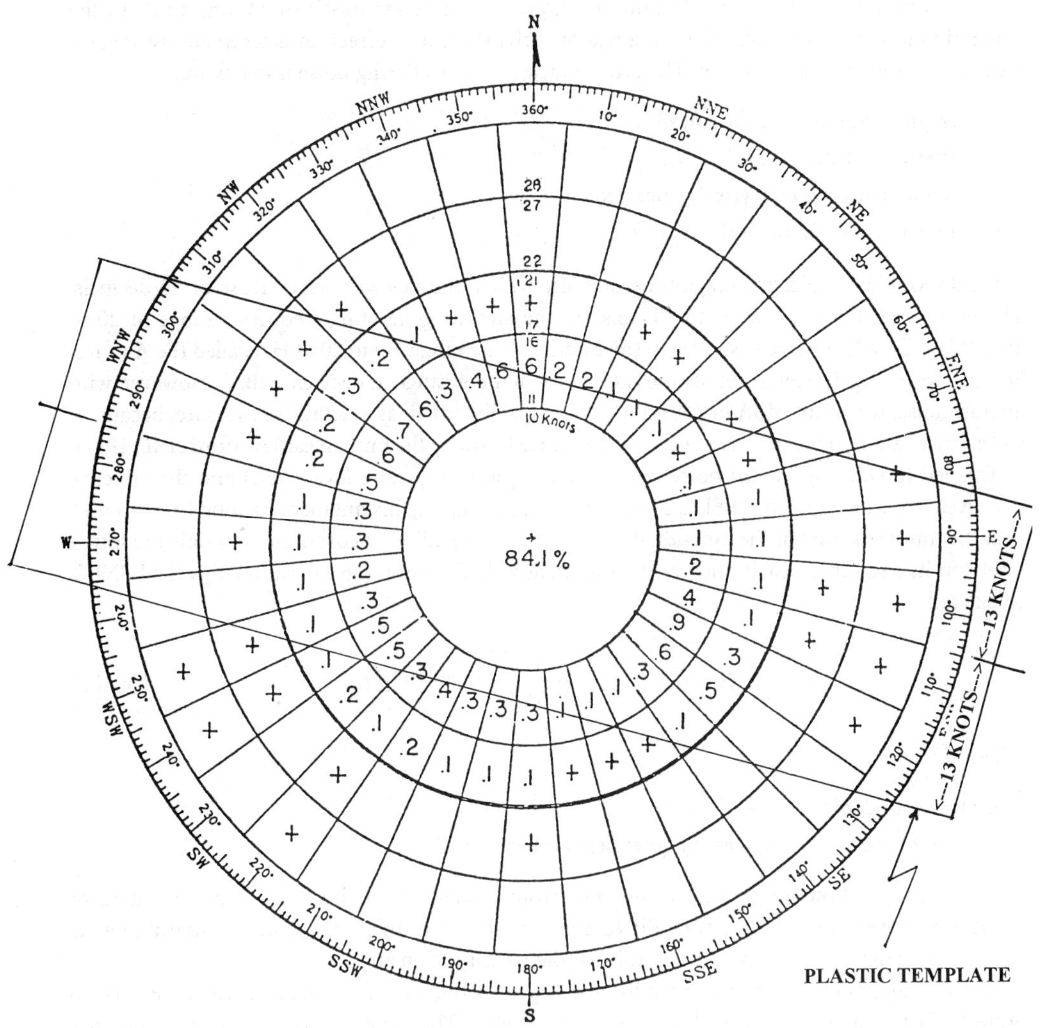

A runway at the airport represented by the wind data on the left that is oriented 105° - 285° (true) would have 2.72% of the winds exceeding the design crosswind/crosswind component of 13 knots.

FIGURE 56.29 (continued) Wind data and wind rose analysis.

Table 56.24 Runway Width and Allowable Crosswind for Landing

Runway Width W	Allowable Crosswind Component
$W < 75$ feet	10.5 knots
75 feet $\leq W < 100$ feet	13 knots
100 feet $\leq W < 150$ feet	16 knots
$W \geq 150$ feet	20 knots

Source: FAA. 1991c. *Airport Design.* Advisory Circular AC150/5300-13, change 1.

Noise from airports has evoked numerous lawsuits and excess media attention, much to the frustration of airport officials. Noise is a real disturbance and its effects and acceptability are best measured in the ears of the hearer. The critical factors in considering noise impacts are:

- Length or duration of the sound
- Repetition of the sound
- Predominant frequency(ies) generated
- Time of day when the noise occurs

Loudness is the subjective magnitude of noise which doubles with an increase of 10 decibels. The human ear is not sensitive to all noise in the aircraft generating frequency range of 20 to 20,000 Hz. Usually it perceives noise in the middle of the range, 50 to 2000 Hz, called the A range. Sound-measuring devices generally measure noise in the A range in decibels (dBA). However, with aircraft noise, the simple dBA or sound intensity was discarded, as a definitive measure, because it lacked correlation to the perceived noise disturbance heard by the human ear [Ashford *et al.*, 1991].

This led to two *single event* noise measures; the sound exposure level (SEL) and the effective perceived noise level (EPNL). SEL is computed by accumulating instantaneous sound levels in dBA over the time the sound of the individual event is detectable. EPNL incorporates not only the sound level, but its frequency distribution and duration as well. Equation (56.11) shows how the EPNL is calculated:

$$\text{EPNL} = 10\log\frac{1}{T}\int_0^T 10^{0.1L(t)}dt \tag{56.11}$$

where

$L(t)$ = sound level in dBA
T = 20 to 30 seconds to avoid quiet periods between aircraft

Since the irritation from noise comes not from a single event but from the integrated or cumulative measure of many events, EPNL and SEL, in and of themselves, are not useful metrics for modeling the impact from aircraft noise in the vicinity of an airport.

One of the models which has come to be accepted is the *noise exposure forecast* (NEF), which embeds EPNL in its definition [Ashford *et al.*, 1991]. The NEF has two different measures, depending on the time of day of the aircraft operation. Equation (56.12) indicates the NEF for day or night while Eq. (56.13) shows how the day and night measures are combined

$$\text{NEF} = \text{EPNL} + 10\log_{10}N - K \tag{56.12}$$

where

N = number of occurrences exceeding 80PNdB (peak level of noise from a Boeing 707 at full power at 12,000 ft altitude)
K = 88 for daytime operations (0700–2200)
K = 76 for nighttime operations (2200–0700)

$$\text{NEF}_{\text{day/night}} = 10\log_{10}\left(\text{antilog}\frac{\text{NEF}_{\text{day}}}{10} + \text{antilog}\frac{\text{NEF}_{\text{night}}}{10}\right) \tag{56.13}$$

More recently the FAA, airports, and community officials have adopted a cumulative noise measure based on SEL [FAA, 1983a]. Nighttime operations are weighted by a factor of 10 due to the additional disturbance from such operations. The measure is called the average day/night sound exposure or L_{DN}. Equation (56.14) indicates how L_{DN} is determined for each significant

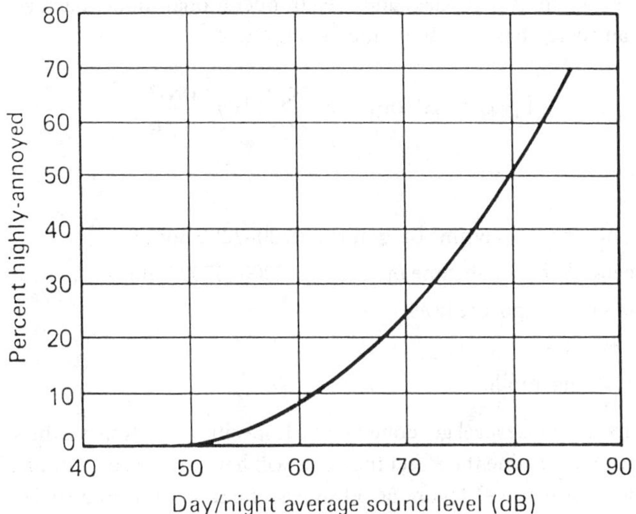

FIGURE 56.30 Degree of annoyance from noise observed in social surveys. (*Source:* Schultz, T. J., 1978. Synthesis of social surveys on noise annoyance. *Journal of Acoustical Society of America.* vol. 64.)

Table 56.25 Land Use for Various Levels of Airport Noise

Land-Use Zones	Noise Exposure Class	L_{DN}	NEF	HUD Noise Guidelines	Suggested Land-Use Controls	Recommended Land Use
A	Minimal	0 to 55	0 to 20	Acceptable	Normally requires no special consideration	Residential, cultural, public assembly, schools, resorts, mobile homes, parks, service
B	Low moderate	55 to 60	20 to 25	Normally acceptable	Some sound-reducing controls may be useful	Residential, hotels, apartments, business services, office complexes, light industry
BC	High moderate	60 to 65	25 to 30	Sometimes acceptable	Some sound-reducing controls may be useful	*See note*
C	Significant	65 to 75	30 to 40	Normally unacceptable	Noise easements required with strict land-use controls	Manufacturing, retail trade, construction services, refining, paper/pulp
D	Severe	>75	>40	Clearly unacceptable	Should be within the airport boundary; use of positive compatibility controls required	Highway right of way, motor vehicle transportation, rail transit, undeveloped area, heavy industry, farming

Note: Airport consultants pay special attention to areas impacted or potentially impacted by noise levels of NEF 25–30 and L_{DN} 60–65. Experience indicates that owners of property in these areas of noise transition from normally acceptable to normally unacceptable noise are frequently involved in noise litigation suits. The FAA has set L_{DN} limits at 65. Practitioners consider residential uses regardless of density as unacceptable land use below 60 dB. This is particularly true under the glide paths or tracks for landing or take-off. EPA in their levels document [EPA, 1974] suggests that the safe L_{DN} criterion for "health and welfare" is 55 dB. They set 60 dB with a 5 dB safety margin for outdoor noise in a residential neighborhood. (Adapted from FAA. 1983. *Airport Land Use Compatibility Planning.* Advisory Circular AC150/5050-6.)

noise intrusion for the ith aircraft class and the jth operational mode. Each single event (i, j) is then summed on an energy basis to obtain the total L_{DN}:

$$L_{DN} = 10\log_{10}\sum_i\sum_j(10)\frac{L_{DN}(i,j)}{10} \tag{56.14}$$

where

Ops_{day} are the number of daytime operations (0700–2200 hours)
Ops_{night} are the number of nighttime operations (2200–0700 hours)
SEL = average sound exposure level
i = ith aircraft class
j = jth operational mode

Having computed the noise level generated by each specific aircraft using the schedule of flights, it is then necessary to determine the effect the noise will have on the community. How much noise is too much? In what situations? Figure 56.30 shows one sample from a social survey indicating that below 50 dB on the day-night average sound level there is virtually no annoyance. Table 56.25 describes how communities and HUD have integrated the noise impacts into land-use planning recommendations (or regulations) in the community. While noise levels of L_{DN} below 65 dB are considered by some as acceptable, experience has indicated that airports would do well to plan their land acquisition program for L_{DN} below 60 dB or even 55 dB.

Integrated Noise Model

The computer software for determining the impact of noise around an airport is called the Integrated Noise Model (INM). Available for licensing from FAA Office of Environment and Energy, it can give the contours of equal noise exposure for any one of four different measures indicated in Table 56.26. The inputs are the airport elevation, the ambient temperature, runway geometry, the percentage use of each runway, number of operations during the day and at night, the expected aircraft in each time space, and the expected tracks of approach and takeoff in several altitude and distance segments.

Figure 56.31 shows a three-runway airport with the operational flight tracks that are to be used in computation of the noise. The noise along each track will differ depending on the number of aircraft in a day, the nighttime traffic, and the specific aircraft that are anticipated to fly each track. The model stores a database of existing aircraft by make, model number, the number of aircraft in a day, the nighttime traffic, and the specific aircraft that are anticipated to fly each track. The model stores a database of existing aircraft by make, model number, and engine type. Included are their

Table 56.26 Capabilities of Integrated Noise Model (INM)

Measure of Noise	Symbol	Description
Noise exposure forecast	NEF	Based on the effective perceived noise decibel (EPNL) as a unit of aircraft noise; nighttime operations are weighted by 16.7 per one operation
Equivalent sound level	SEL	Summation of aggregate noise environment A-weighted decibel units (dBA)
Day night average sound level	L_{DN}	Based on SEL with nighttime operations weighted by a 10 decibel penalty; see Eq. (56.13)
Time above threshold of A-weighted sound	TA	Time in minutes that a dBA level is exceeded in a 24-hour period

Source: FAA. 1992. *Integrated Noise Model.* Version 3, revision 1. DOT/FAA/EE-92/02.

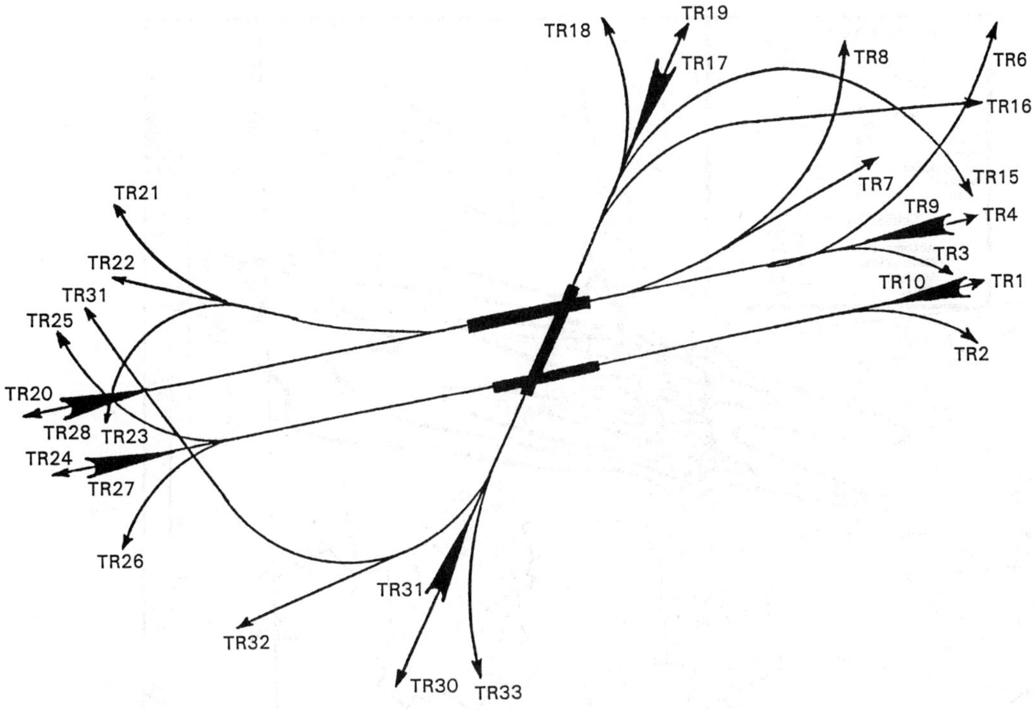

FIGURE 56.31 Input flight tracks to the integrated noise model.

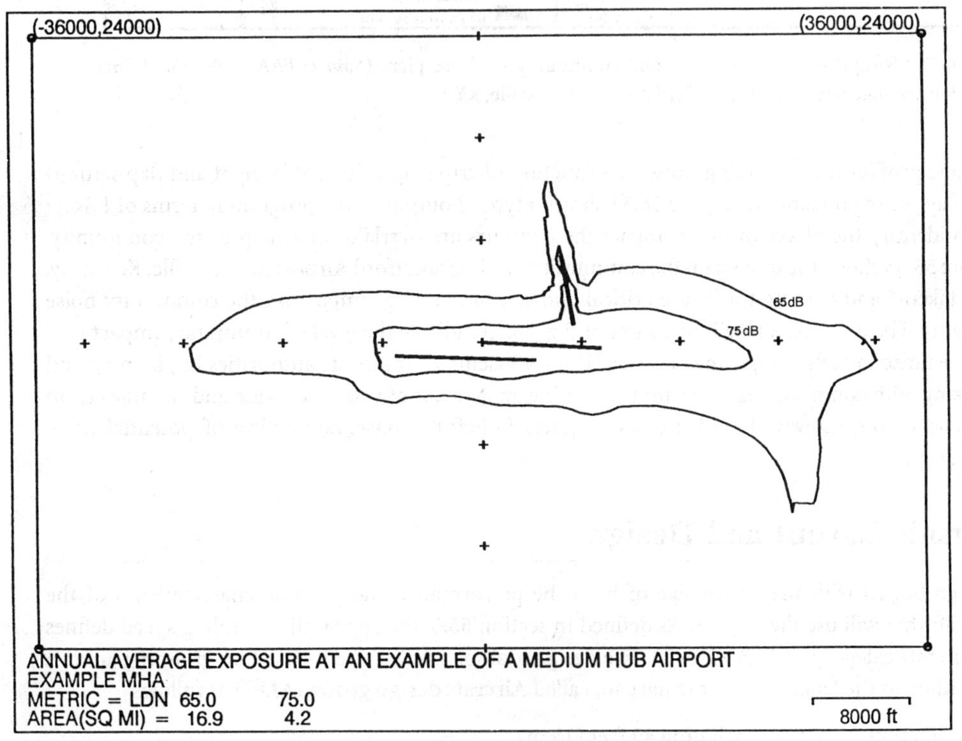

FIGURE 56.32 Noise contours for planning purposes.

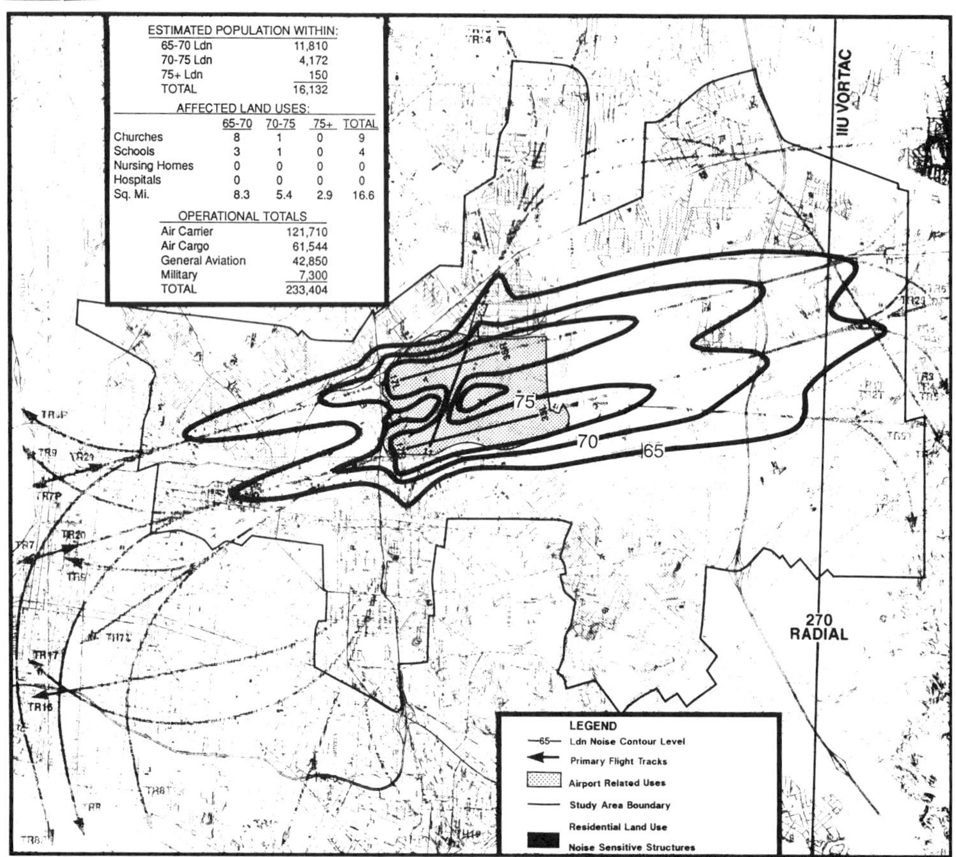

FIGURE 56.33 Sample of noise contours and community land-use plan. (*Source:* FAA. 1990. *Final Environmental Impact Statement,* Standiford Field Airport, Louisville, KY.)

altitude profiles for generating noise as a function of trip length (takeoff weight and flap setting) and flap setting on landing. Figure 56.32 shows a typical output of the program in terms of L_{DN}.

To identify the places of noise impact the contours are overlaid on a map of the community. Figure 56.33 shows the impact on the community for the Standiford Airport in Louisville, Kentucky. The takeoff and landing tracks are critical and can have a large impact on the community noise patterns. The data appearing in the inset of Fig. 56.33 indicate the level of community impact.

The noise models using either L_{DN} or NEF are essential for airport authorities in planning and working with communities. For simple planning an area about two miles wide and six miles from the end of the runway should provide a quick, hopefully conservative view of potential noise problems.

56.8 Airside Layout and Design

Design begins with the knowledge of both the performance and physical characteristics of the aircraft that will use the airport. As defined in section 56.5, the approach or landing speed defines an aircraft category as A, B, C, or D. The designation of aircraft size is based on grouping aircraft according to the length of their wingspan, called **Aircraft design group (ADG)** as follows:

Group I: up to but not including 49 feet (15 m)
Group II: 49 feet (15 m) up to but not including 79 feet (24 m)
Group III: 79 feet (24 m) up to but not including 118 feet (36 m)

Group IV: 118 feet (36 m) up to but not including 171 feet (52 m)
Group V: 171 feet (52 m) up to but not including 214 feet (65 m)
Group VI: 214 feet (65 m) up to but not including 262 feet (80 m)

The important physical characteristics of the aircraft affecting airport design are maximum takeoff weight (W), wingspan (A), length (B), tail height (C), wheel base (D), nose to centerline of main gear (E), undercarriage width (1.15 $\times$ main gear track, F); and line-of-sight/obstacle free zone at the nose of the aircraft. For reference these are presented for the Boeing 727 in Fig. 56.34 [FAA, 1991c].

Figure 56.35 displays a major problem faced by aircraft as they land and travel on the runway/taxiway/taxilane system. The pilot's view of the ground directly in front of the aircraft is obscured by the nose. This blind zone for the pilot is known as the object-free zone (OFZ) and is important for safe ground movement of aircraft. It affects the geometric design of the runway and taxiway. Table 56.27 shows the approach speed and physical characteristics for several specific aircraft.

MODEL	MAXIMUM TAKEOFF WEIGHT	MAXIMUM LANDING WEIGHT	A	B	C	D	E	F	G	J	K	M	N	P	TURN RADIUS
100	160,000 LB 72 575 KG	137,500 LB 62 369 KG	108'0" 32.92M	133'2" 40.59M	34'3" 10.44M	53'3" 16.23M	68'4" 20.83M	18'9" 5.72M	9'3" 2.82M	42'6" 12.95M	10'4" 3.15M	14'4" 4.37M	5'8" 1.72M	12'0" 3.66M	72'0" 21.95M
100-C	160,000 LB 72 575 KG	137,500 LB 62 369 KG	108'0" 32.92M	133'2" 40.59M	34'3" 10.44M	53'3" 16.23M	68'4" 20.83M	18'9" 5.72M	9'3" 2.82M	42'6" 12.95M	10'4" 3.15M	14'4" 4.37M	5'8" 1.72M	12'0" 3.66M	72'0" 21.95M
200	172,000 LB 78 018 KG	150,000 LB 68 039 KG	108'0" 32.92M	153'2" 46.68M	34'11" 10.65M	63'3" 19.28M	78'4" 23.88M	18'9" 5.72M	9'3" 2.82M	42'4" 12.90M	10'4" 3.15M	16'11" 5.16M	4'9" 1.44M	12'0" 3.66M	82'0" 24.99M

NOTE: OPTIONAL TAKEOFF AND LANDING WEIGHTS:

100	160,000 LB (72 575 KG) 142,500 LB (64 637 KG)	169,000 LB (76 657 KG) 142,500 LB (64 637 KG)			MAXIMUM TAKEOFF WEIGHT. MAXIMUM LANDING WEIGHT.
100C	160,000 LB (72 575 KG) 140,000 LB (63 503 KG)	169,000 LB (76 657 KG) 142,500 LB (64 637 KG)			MAXIMUM TAKEOFF WEIGHT. MAXIMUM LANDING WEIGHT.
200	184,800 LB (83 824 KG) 154,500 LB (70 080 KG)	190,500 LB (86 409 KG) 154,500 LB (70 080 KG)	197,000 LB (89 358 KG) 154,500 LB (70 080 KG)	209,500 LB (95 028 KG) 161,000 LB (73 028 KG)	MAXIMUM TAKEOFF WEIGHT. MAXIMUM LANDING WEIGHT.

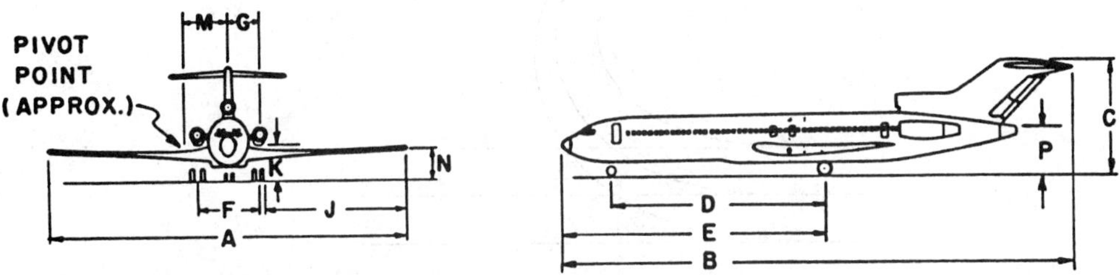

FIGURE 56.34 Sample aircraft dimensions (Boeing 727) for airport design. (*Source:* FAA. 1991c. *Airport Design.* Advisory Circular AC150/5300-13, change 1.)

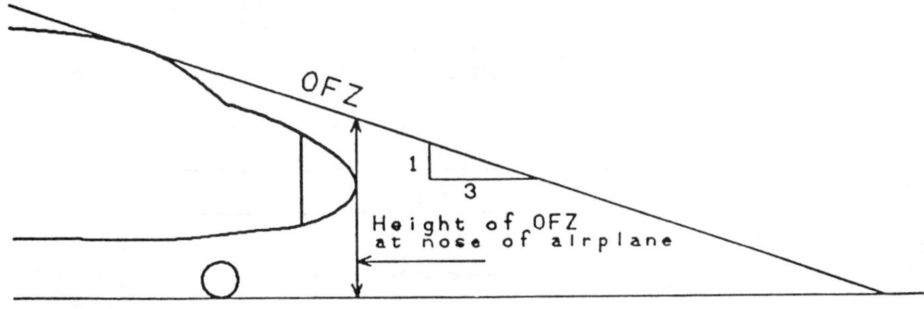

FIGURE 56.35 Object-free zone requirements as viewed from the cockpit. (*Source:* FAA. 1991c. *Airport Design.* Advisory Circular AC150/5300-13, change 1.)

Table 56.27 Aircraft Data Used by Design Program (Representative Sample)

Aircraft Make/Model	Airport Reference Code	Approx. Approach Speed (Knots)	Wingspan (Feet)	Length (Feet)	Tail Height (Feet)	Max. Takeoff Weight (Pounds)
Cessna-150	A-I	55	32.7	23.8	8.0	1,600
Beech King Air-B100	B-I	111	45.8	39.9	15.3	11,800
Gates Learjet 54-56	C-I	128	43.7	55.1	14.7	21,500
Dornier LTA	A-II	74	58.4	54.4	18.2	15,100
Shorts 360	B-II	104	74.8	70.8	23.7	26,453
Grumman Gulfstream III	C-II	136	77.8	83.1	24.4	68,700
DHC-8, Dash 8-300	A-III	90	90	84.3	24.6	41,100
Fairchild F-27	B-III	109	95.2	77.2	27.5	42,000
Boeing 727-200	C-III	138	108	153.2	34.9	209,500
Boeing 737-400	C-III	138	94.8	119.6	36.6	150,000
MDC-DC-9-50	C-III	132	93.3	133.6	28.8	121,000
Airbus 300-600	C-IV	135	147.1	177.5	54.7	363,763
Boeing 757	C-IV	135	124.8	155.3	45.1	255,000
Boeing 767-300	C-IV	130	156.1	180.3	52.6	350,000
MDC-DC-8-50	C-IV	133	142.4	150.8	43.3	325,000
MDC-DC-10-30	D-IV	151	165.3	181.6	58.6	590,000
MDC MD-11	D-IV	155	169.8	201.3	57.8	602,500
Boeing 747-200	D-V	152	195.7	231.8	64.7	833,000
Boeing 747-400	D-V	154	213.0	231.8	64.3	870,000
Lockheed C5A	C-VI	135	222.7	247.8	65.1	837,000

Source: FAA. 1991c. *Airport Design.* Advisory Circular AC150/5300-13, change 1.

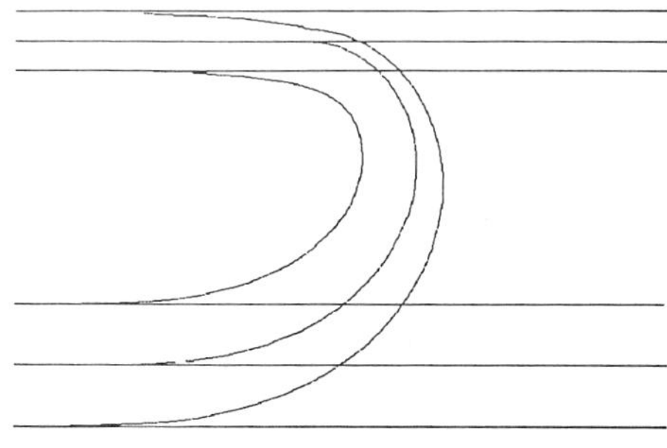

180-degree spiral double back exit taxiway

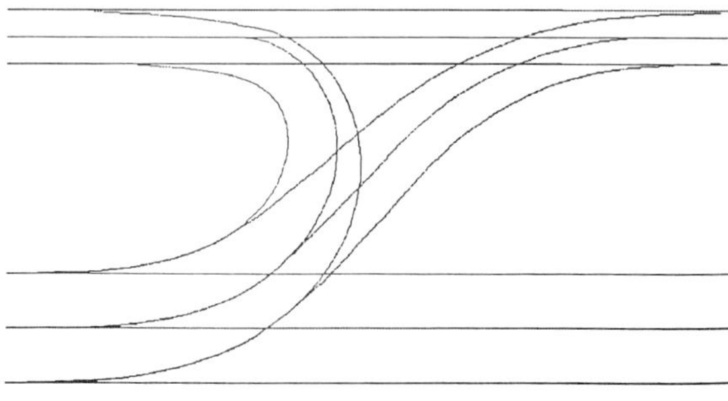

Acute-angled exit taxiway

FIGURE 56.36 Sample of steering and taxiway fillet design from airport design program.

Other input data to the computer program are the primary navigation capability, the altitude/elevation of the airport, and the mean temperature of the hottest month of the year. The program outputs include runway lengths, widths, and clearance standards. Outputs that develop taxiway design data, such as widths and clearance standards, steering angles on tangent section, circular curve layouts, spiral curve layouts, offset distances on taxiway intersections, offsets on exit taxiways, and the wing tip clearance on taxiways, are possible. The program has plotting capability for exit taxiways, taxiway intersections, or the curved track for wing tip clearance on taxiways as demonstrated in the spiral-double-back exit taxiway and acute-angled taxiway plotted in Fig. 56.36. The design program will also calculate the wind rose data. [See section 56.7.]

Runway Length

The length of the runway is determined by the aircraft, maximum takeoff weights, engine capabilities, landing and braking capabilities, flap settings, and required safety factors. For example,

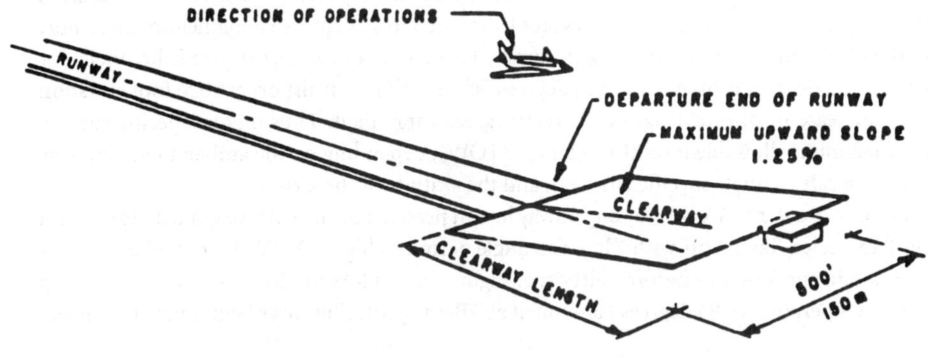

CLEARWAY

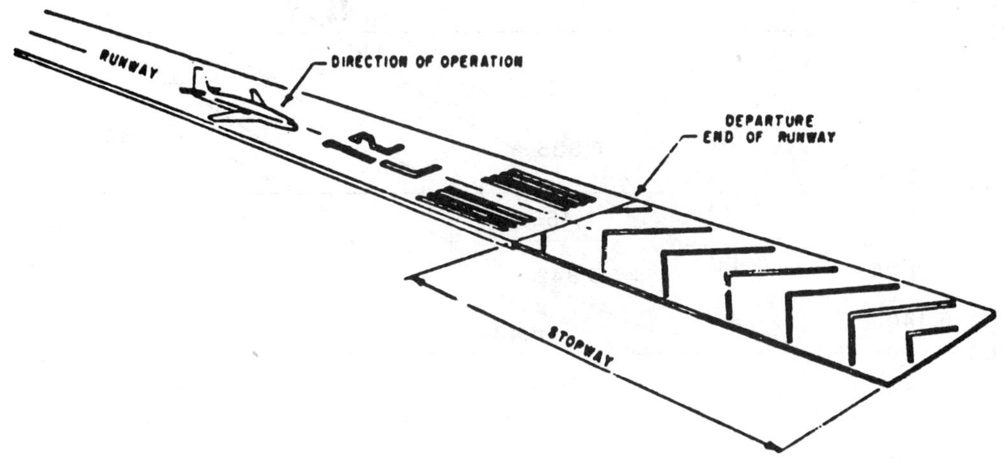

STOPWAY

FIGURE 56.37 View of the clearway and stopway. (*Source:* FAA. 1991c. *Airport Design.* Advisory Circular AC150/5300-13, change 1.)

the runway length for landing must be capable of permitting safe braking if touchdown occurs $\frac{1}{3}$ the length of the runway past the threshold. The runway must also be long enough to meet the obstacle-free capability to permit each aircraft to take off with one engine out. The stopping zone must include ample stopping distance in case the pilot chooses to abort takeoff just before rotating to become airborne (called stopway). As discussed, the runway safety areas are a must for airport control. Figure 56.37 shows the stopway, to prevent accidents at the end of the runway, and the clearway, also called the runway protection zone (RPZ).

The altitude of the airport and the temperature also have a significant impact on the airport runway length because lift capability is proportional to the air density, which diminishes as the altitude and temperature increase. Figure 56.38 illustrates how dramatic that change is for a Boeing 727-200 with JT8D-15 engine with a takeoff weight of 150,000 pounds and its wing flaps set at 20 degrees. The requirement for longer runways increases significantly as the altitude of the site above sea level increases. At an average temperature of 65 degrees Fahrenheit, the increase is from 4900 feet at sea level to 8660 feet at an altitude of 8000 feet or about 370 feet of added runway for each 1000 foot increase in altitude. The increase due to temperature, especially when the temperature is high, is equally dramatic. Going from 65 degrees Fahrenheit to 80 degrees Fahrenheit for an airport at 4000 foot elevation requires an increase in runway length of about 24 feet/degree Fahrenheit. For the shift from 95 degrees Fahrenheit to 110 degrees Fahrenheit for an airport at 4000 foot elevation the rate of increase in runway length is 58 feet/degrees Fahrenheit. Thus on any specific runway there is a maximum allowable takeoff weight (MATOW), depending on the ambient temperature, the specific aircraft (with its specific engines), and the altitude of the airport.

The advisory circular [FAA, 1990a] on runway length presents the takeoff weight data for several different flap angles. Taking off with a low flap angle permits a higher MATOW, but takes a longer runway to attain the speed to become airborne. Figure 56.39 plots the MATOW for various flap angles for a temperature of 90 degrees Fahrenheit at TBA airport. The curve beginning at the lower

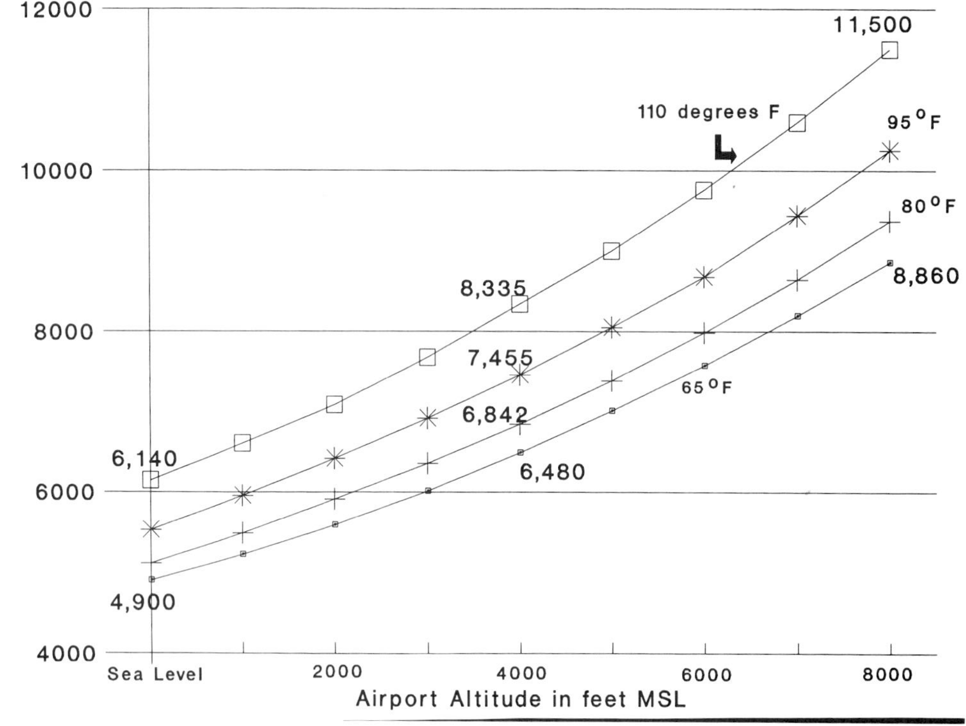

FIGURE 56.38 Change of required takeoff runway length due to temperature and altitude.

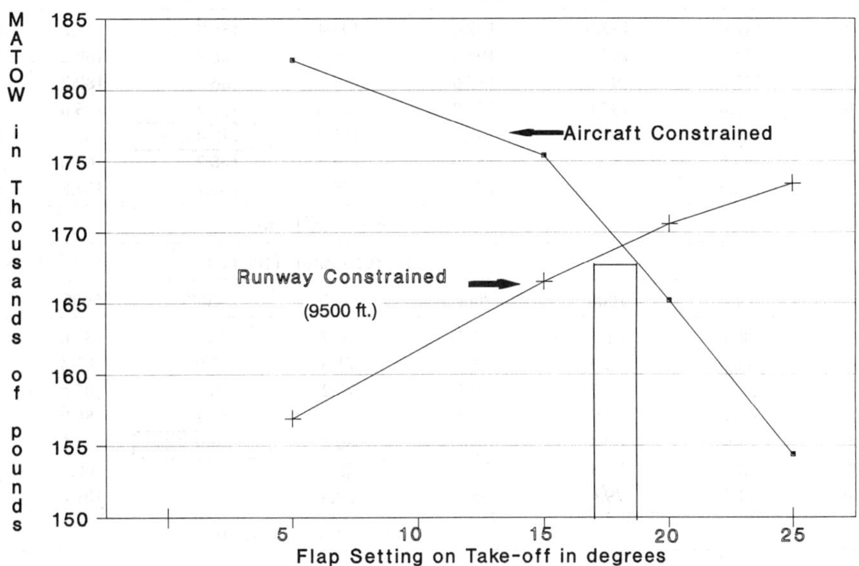

FIGURE 56.39 Takeoff weight as a function of flap angle setting.

left is constrained by the length of the 9500 foot (2900 m) runway while the curve beginning at the upper left is constrained only due to aircraft engine thrust capability of the JT8D-15 engines assuming sufficient runway length is available. A setting of flap angle at about 17 degrees will give the highest MATOW of 167,500 pounds (76,050 kg) for the day with a 90 degree Fahrenheit temperature.

The major operational constraint, when there is a weight limitation caused by a shorter than optimum runway, is the range that can be achieved. The 727-200 with JT8D-15 engines has an empty weight of 109,211 pounds and a structural payload weight of 40,339 pounds [FAA, 1990a]. Tables such as the example shown in Table 56.28 are available for most aircraft and for a range of flap angle settings for each aircraft. If the flaps are set at 15 degrees at the MRA airport on a 90 degree Fahrenheit day it can be seen that the MATOW should be 175,400 pounds (79,725 kg) as indicated by A on Table 56.28. By use of the reference factor of 86.9 (B on Table 56.28) and linear interpolation at the bottom portion (C), the runway would have to be 10,680 feet (3250 km). Since the runway is only 9500 feet (2900 m), interpolation would indicate a MATOW of 166,500 pounds (75,680 kg).

Using this value several different options of weight and range can be considered. These options are presented in Table 56.29 as the "Max Payload" case, the "1500 Mile Range" case, and the "50% Load Factor" case. The two critical numbers for all these cases are the fuel rate of 22 pounds/mile for this aircraft and the average weight of passengers with their luggage of 200 pounds per passenger. In the first case a full load would be determined by subtracting the structural payload weight of 40,339 pounds plus the operating empty weight of 109,211 pounds from the MATOW of 166,500 pounds. This leaves 16,950 pounds for fuel, which at 22 pounds per mile gives a range of 770 miles.

The next case is if the need is for a 1500 mile flight. After removing the operating empty weight and enough fuel for 1500 miles (33,000 pounds) the weight left for passengers and cargo is 24,289 pounds, which (if all of it is allotted to passengers) gives 141 passengers. The final case assumes a

Table 56.28 Aircraft Performance, Takeoff Boeing 727-200 JT8D-15 Engine, 15° Flaps

Maximum Allowable Takeoff Weight (1000 lb)								
Temp °F	Airport Elevation (1000 ft)							
	0	1000	2000	3000	4000	5000	6000	7000
50	197.0	197.0	197.0	193.4	186.9	180.2	173.5	166.9
60	197.0	197.0	197.0	193.4	186.9	180.2	173.5	166.9
70	197.0	197.0	197.0	193.4	186.9	180.2	173.5	166.9
80	197.0	197.0	196.2	189.0	182.2	175.8	169.5	163.4
90	197.0	195.8	188.7	181.9	**175.4 (A)**	169.0	162.9	156.8
100	194.5	187.7	181.0	174.5	168.2	162.0	156.0	150.2
110	186.5	179.7	173.1	166.8	160.7	154.8	149.0	143.4

Reference Factor								
Temp °F	Airport Elevation (1000 ft)							
	0	1000	2000	3000	4000	5000	6000	7000
50	58.6	61.6	65.7	70.7	76.5	83.0	90.2	97.9
60	58.4	62.0	66.3	71.3	77.0	83.4	90.5	98.4
70	59.4	63.5	68.1	73.2	78.9	85.4	92.6	100.6
80	61.4	65.9	70.8	76.2	82.2	88.9	96.4	104.9
90	64.5	69.4	74.7	80.5	**86.9 (B)**	94.0	102.0	111.0
100	68.7	73.9	79.6	85.9	92.9	100.7	109.4	119.0
110	74.0	79.4	85.5	92.5	100.3	109.0	118.5	129.0

Runway Length (1000 ft)									
Weight 1000 lb	Reference Factor								
	58	68	78	**86.9**	88	98	108	118	128
130	3.96	4.55	5.20		5.90	6.61	7.28	7.88	8.39
135	4.23	4.89	5.59		6.30	7.00	7.66	8.28	8.82
140	4.51	5.25	5.99		6.73	7.44	8.13	8.78	9.39
145	4.81	5.63	6.42		7.19	7.94	8.67	9.38	10.08
150	5.13	6.02	6.87		7.69	8.49	9.29	10.09	10.90
155	5.46	6.44	7.35		8.23	9.10	9.98	10.89	11.85
160	5.82	6.86	7.85		8.81	9.77	10.76	11.80	12.93
165	6.18	7.31	8.38	9.31	9.42	10.49	11.61	12.81	14.13
166.5				**9.50 (D)**					
170	6.57	7.77	8.92	9.95	10.08	11.26	12.53	13.92	
175	6.97	8.25	9.49	10.62	10.76	12.10	13.54		
175.4				**10.68 (C)**					
180	7.37	8.74	10.09	11.34	11.49	12.98			
185	7.83	9.25	10.71		12.25	13.92			
190	8.28	9.78	11.35		13.05				

Source: FAA. 1990a. *Runway Length Requirements Airport Design.* Advisory Circular AC150/5325-4A.

50% load factor of 81 passengers or 16,200 pounds, leaving 41,089 pounds for fuel. This amount of fuel would give a range of 1867 miles. The airlines will assign aircraft to meet the range/payload requirements of the markets they serve. It behooves the airport planner to make sure that the runway is long enough to serve the most distant markets that will attract airlines while also accounting for the hot summer weather.

The other runway length limitation is on landing, which usually requires less runway than for takeoff. Critical items are landing weight and flap settings. At TBA airport with a 90 degrees Fahrenheit temperature, the maximum allowable landing weight is 154,500 pounds (70,230 kg) with 30° flaps which would require 5720 feet (1750 m) of runway. Since the aircraft does not have the weight of fuel when landing, there is usually a good margin for landing.

Table 56.29 Range/Payload Calculation for 727-200 with JT8D-15 Engines and 15 Degree Flaps

Characteristic	Units/Notes	Max Payload Case	1500 Mile Range Case	50% Load Factor Case
Maximum allowable takeoff weight	Table 56.30 gives 175,400 lb as the maximum takeoff weight. However, that much weight requires a runway length of 10,680 feet. See Table 56.30 (C)			
Takeoff weight	Calculated using reference Factor (86.9) - Table 56.30 (D)	166,500 lb [76,050 kg]	166,500 lb [76,050 kg]	166,500 lb [76,050 kg]
Typical operating empty weight plus reserve	Given (1.25 hrs of fuel reserve required for domestic flight)	109,211 lb [49,650 kg]	109,211 lb [49,650 kg]	109,211 lb [49,650 kg]
Remaining for payload and fuel		57,289 lb [26,040 kg]	57,289 lb [26,040 kg]	57,289 lb [26,040 kg]
Passengers	Maximum = 162	162 (100% load factor)	141 (87% load factor)	81 (50% load factor)
Max. passengers 162	200 lbs [90 kg] per passenger	32,400 lb* [14,730 kg]	24,289 lb [11,040 kg]	16,200 lb [7365 kg]
Max belly air cargo	Fill to structural payload limit of 40,339 lb* [18,335 kg]	7,939 lb [3605 kg]	0 lb	0 lb
Amount of fuel		16,950 lb [7,700 kg]	33,000 lb [15,000 kg]	41,089 lb [18,675 kg]
Distance of market served range	Fuel rate given 22 lb/mile [6.2 kg/km]	770 miles [1240 km]	1500 miles [2415 km]	1867 miles [3005 km]

*Calculated from data in FAA. 1990a. *Runway Length Requirements for Airport Design.* Advisory Circular AC150/5325-4A.

"Declared distances" are distances the airport owner declares available and suitable for satisfying the airplane's takeoff distance, accelerate-stop distance, and landing distance requirements. The distances are:

- Takeoff run available (TORA)—the runway length declared available and suitable for the ground run of an airplane takeoff
- Takeoff distance available (TODA)—the TORA plus the length of any remaining runway and/or clearway (WY) beyond the far end of the TORA
- Accelerate-stop distance available (ASDA)—the runway plus stopway (SWY) length declared available and suitable for the acceleration and deceleration of an airplane aborting a takeoff
- Landing distance available (LDA)—the runway length declared available and suitable for a landing airplane

Runway and Taxiway Width and Clearance Design Standards

The FAA has developed a set of standard dimensions that determine runway width, separations between runways and taxiways, safety areas around runways and taxiways, shoulder width (possible areas of less than full-strength pavement), pads to deflect jet blast, object-free areas, and the like. These standards are a function of approach speed and aircraft size. Figure 56.40 presents the overall dimensions that are involved in parallel runways and taxiways while Table 56.30 shows the standards for airports that service aircraft in the approach speed category C&D. Figure 56.41 shows the plan view of major runway elements. The runway protection zone was shown in Fig. 56.26. There are similar data for airports serving approach categories A&B. These dimensions are all listed in the airport design computer program output [FAA, 1991c].

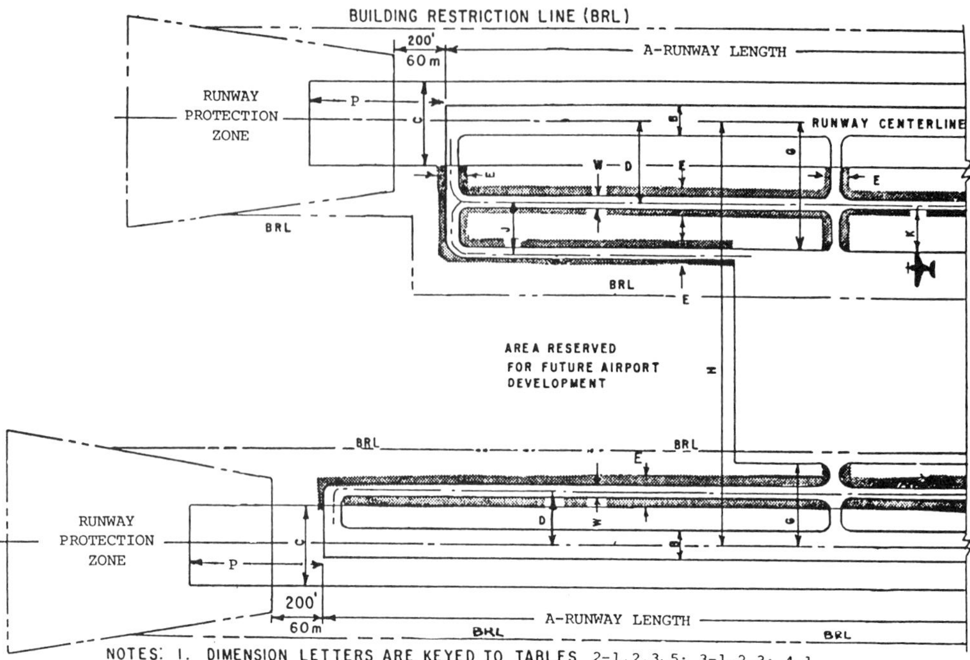

FIGURE 56.40 Runway and taxiway dimensions. (*Source:* FAA. 1991c. *Airport Design.* Advisory Circular AC150/5300-13, change 1.)

Runway Gradients

Longitudinal Gradient

The desire at any airport site is to have the runways and taxiways as level as possible, allowing for drainage with the design of the transverse grade. In many locations the grading for a perfectly level site would be too expensive when most aircraft can easily accept 1% grade. Where longitudinal grades are used, parabolic vertical curves are used for geometric design, as shown in Fig. 56.42. The penalty for gradients is to reduce the effective runway length by *10 feet per foot of difference between maximum and minimum* elevation of the runway [FAA, 1992]. Table 56.31 defines the gradients in terms of approach category.

For example, if the runway at TBA were 10,200 feet long but there was a differential between highest point and lowest point along the runway of 70 feet, the effective runway length for MATOW calculations would be 9500 (10,200 − 70 × 10) feet.

Line of Sight

The line of sight requirements also determine the acceptable profile of the runway. Any two points five feet above the runway centerline must be mutually visible for the entire runway or if on a parallel runway or taxiway for one-half of the runway. Likewise there needs to be a clear line of sight at the intersection of two runways, two taxiways, and taxiways that cross an active runway. Most line of sight requirements are within 800 to 1350 feet of the intersection, depending on the configuration.

Table 56.30 Separation Standards for Transport Airport Design (Approach Category C&D)

Design Item	Figs. 56.40 and 56.41 Dimen.	Airplane Design Group I	II	III	IV	V	VI
	For Runways						
Safety area width (ft)	C	500	500	500	500	500	500
Safety area length beyond runway end (ft)	P	1000	1000	1000	1000	1000	1000
Width (ft)	B	100	100	100	150	150	200
Shoulder width (ft)		10	10	20	25	35	40
Blast pad width (ft)		100	100	140	200	220	280
Object-free area width (ft)	Q	800	800	800	800	800	800
Object-free area length beyond RW end	R	1000	1000	1000	1000	1000	1000
Nonprecision instrument and visual runway centerline to:							
Parallel runway (simultaneous VFR ops**) (ft)	H	700	700	700	700	1200	1200
Taxiway/taxilane centerline (ft)	D	300	300	400	400	400 to 500*	600
Aircraft parking area (ft)	G	400	400	500	500	500	500
Precision instrument runway centerline to:							
Parallel runway (simultaneous IFR ops***)	H	4300	4300	4300	4300	4300	4300
Taxiway/taxiline centerline (ft)	D	400	400	400	400	400 to 500*	600
Aircraft parking area (ft)	G	500	500	500	500	500	500
	For Taxiways						
Safety area width (ft)	E	49	79	118	171	214	262
Width (ft)	W	25	35	50	75	75	100
Edge safety margin (ft)		5	7.5	10	15	15	20
Shoulder width		10	10	20	25	35	40
Object-free area width (ft)		89	131	186	259	320	386
Centerline to:							
Parallel taxiway/taxilane centerline (ft)	J	69	105	152	215	267	324
Fixed or movable object (ft)	K	44.5	65.5	93	129.5	160	193
	For Taxilanes						
Object-free area width (ft)		79	115	162	225	276	334
Centerline to:							
Parallel taxiway/taxiline centerline (ft)		64	97	140	198	245	298
Fixed or movable object (ft)		39.5	57.5	81	112.5	138	167

Note: *400 feet applies for airports from sea level to 1345 ft altitude; 450 feet from 1345 to 6560 feet; and 500 feet for altitudes greater than 6560 feet.

**Separations less than 2500 feet require wake turbulence procedures.

***Other separations are possible for simultaneous departures only; with radar 2500 feet; without radar 3500 feet.

Source: FAA. 1991c. *Airport Design.* Advisory Circular AC150/5300-13, change 1.

Transverse Gradients

The transverse gradients are important to ensure adequate drainage from the runways and the taxiways. The plan view shown in Fig. 56.41 indicates the typical gradients that are included in runways and taxiways. The chief concern is drainage and line of sight to adjacent runways/taxiways.

Drainage

Drainage on the airport surface is a prime requisite for operational safety and pavement durability. The drainage design is handled like most drainage for streets and highways. Avoidance of ponding and erosion of slopes that would weaken pavement foundations are critical for design. Because

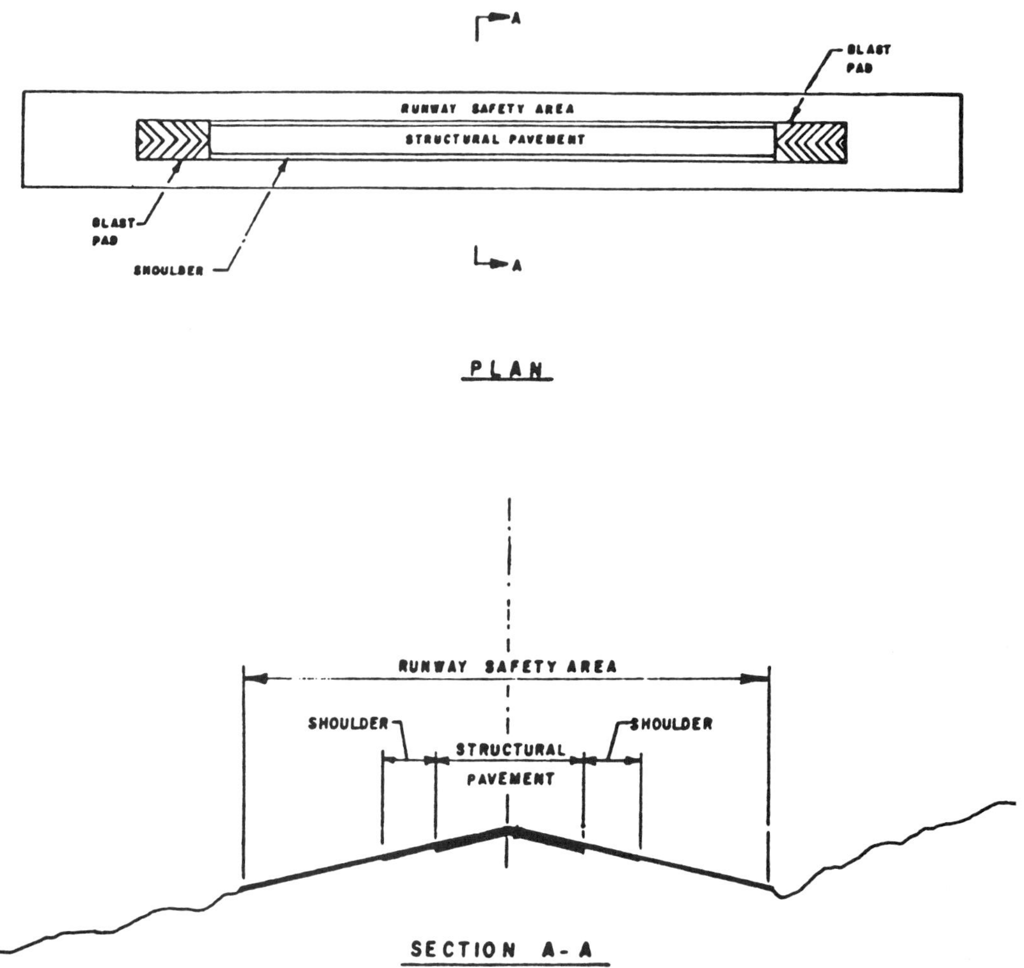

FIGURE 56.41 Plan and cross-section view of the runway elements. (*Source:* FAA. 1991c. *Airport Design.* Advisory Circular AC150/5300-13, change 1.)

of the need for quick and total water removal over the vast relatively flat airport surface an integrated drainage system is a must. Runoff is removed from the airport by means of surface gradients, ditches, inlets, an underground system of pipes, and retention ponds. Figure 56.43 shows one portion of an airport drainage system. Because of their large contiguous area, aprons are critical and must have an adequate sewer system. Runoff water treatment is required when there are fuel spills or during the winter when deicing chemical is used.

Lighting and Signing

Runway

Lighting and signing of the runway shown in Fig. 56.44 provide the pilot visual cues to ensure alignment with the runway, lateral displacement, and distance along the runway. Runway edge lights standing no more than 30 inches and no more than 10 feet from the runway edge are 200 feet or less apart and are white except for the last 2000 feet of runway, when they show yellow. Centerline lights are white and set two feet off the centerline of the runway, except for the last 3000 feet. In this area they are alternating red and white for 2000 feet and they are red 1000 feet from the

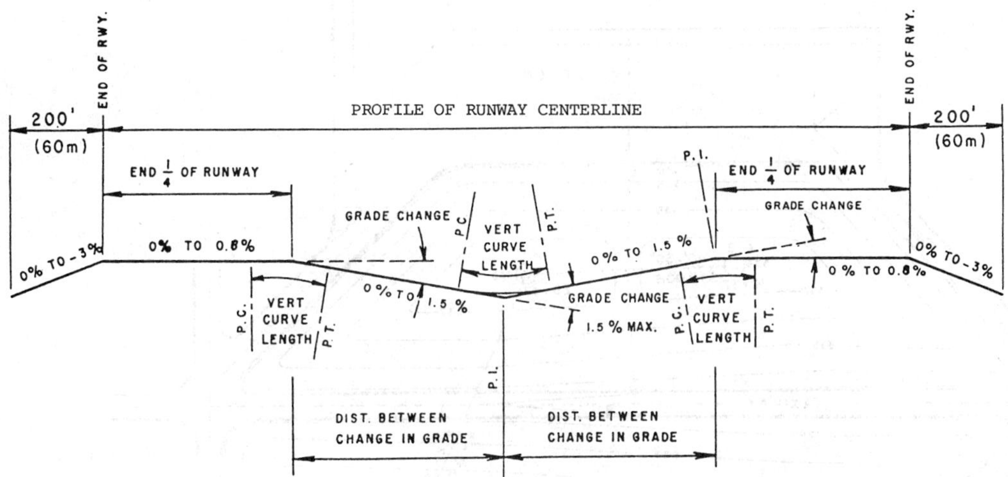

MINIMUM DISTANCE BETWEEN CHANGE IN GRADE = 1000' (300m) x SUM OF GRADE CHANGES (IN PERCENT).

MINIMUM LENGTH OF VERTICAL CURVES = 1000' (300m) x GRADE CHANGE (IN PERCENT)

FIGURE 56.42 Longitudinal grade criteria for airports (C&D approach criteria). (*Source:* FAA. 1991c. *Airport Design.* Advisory Circular AC150/5300-13, change 1.)

Table 56.31 Longitudinal Gradients for Runways and Taxiways

Design Item	Aircraft Served	Maximum Rate of Change	Vertical Curve	Remarks
Runway	A&B	±2%	300 feet/% change	Vertical curve not needed for changes less than 0.4%
Runway	C&D	±1.5%	1000 feet/% change	Grade on first and last $\frac{1}{4}$ of runway ±0.8%; vertical curve not needed for changes less than 0.4%
Runway	A&B	±2%	100 feet/% change	Elevation between taxiway and corresponding point on parallel runway
Taxiway	C&D	±1%	100 feet/% change	Taxiway or apron edge is 1.5% of shortest distance between the two
Apron	A&B	±2%		Consistent with drainage
Apron	C&D	±1%		Direct drainage away from building

Source: FAA. 1991c. *Airport Design.* Advisory Circular AC150/5300-13, change 1.

runway end. When aircraft are approaching the runway to land, the pilot determines the threshold because it is marked by a bar of green lights. However, those lights show red when aircraft approach the end of the runway from the other direction. As shown in Fig. 56.45, painted markings also indicate where the aircraft is relative to distance past the threshold. Exits, particularly high-speed exits, are clearly marked by signs placed at a distance of 1200 to 1500 feet before the exit.

Airfield

The airfield is marked with a variety of signs delineating the taxiways, stoplines, holding areas, and the like. Blue lights indicate taxiway edges. Stop bars before crossing or entering an active runway are yellow. There have been a number of accidents and near accidents on the ground, especially when the visibility is low. The FAA is experimenting with a new lighted stop bar. The controller

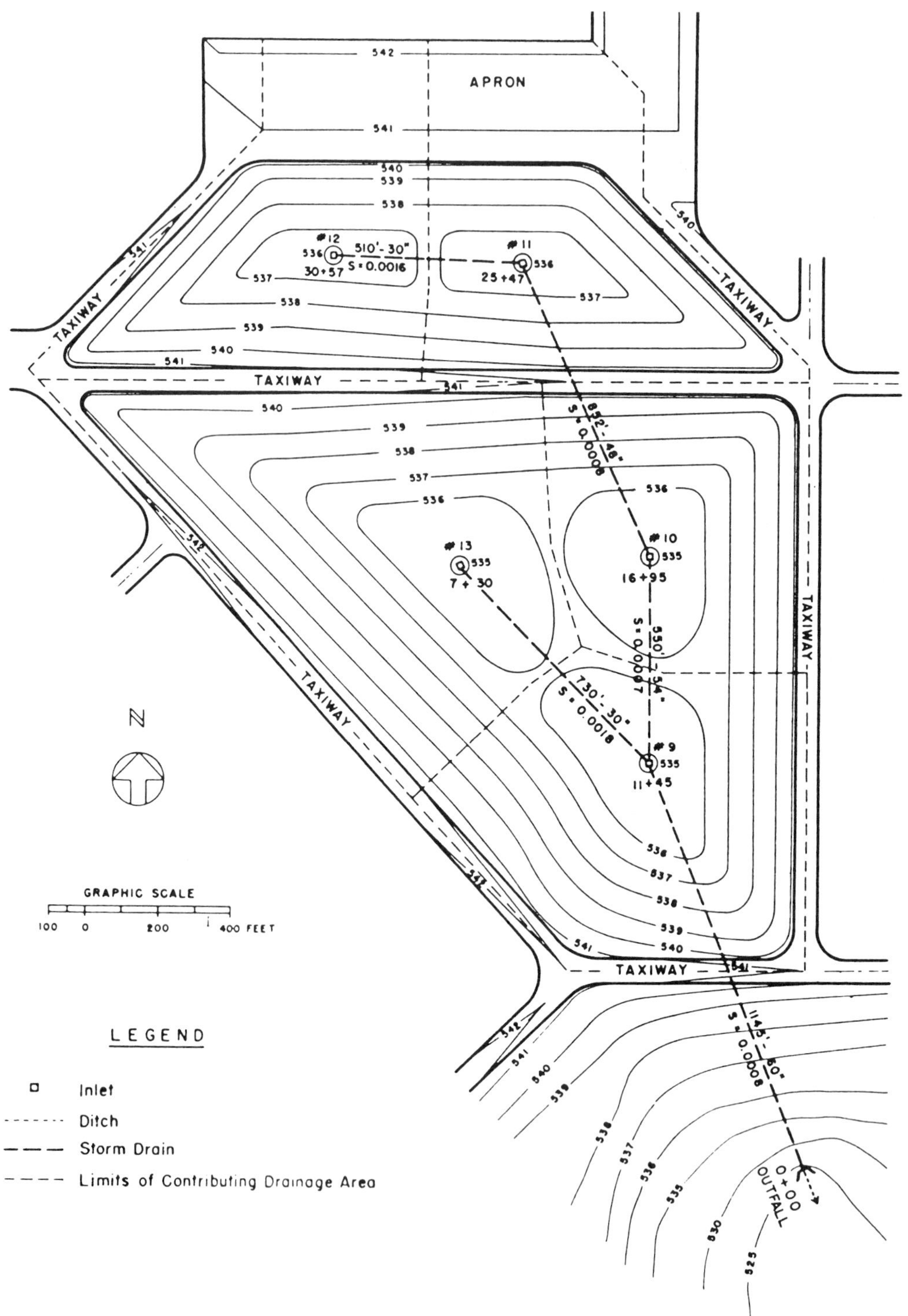

FIGURE 56.43 Portion of an airport showing drainage design. (*Source:* FAA. 1970. *Airport Drainage.* Advisory Circular AC150/5320-5B.)

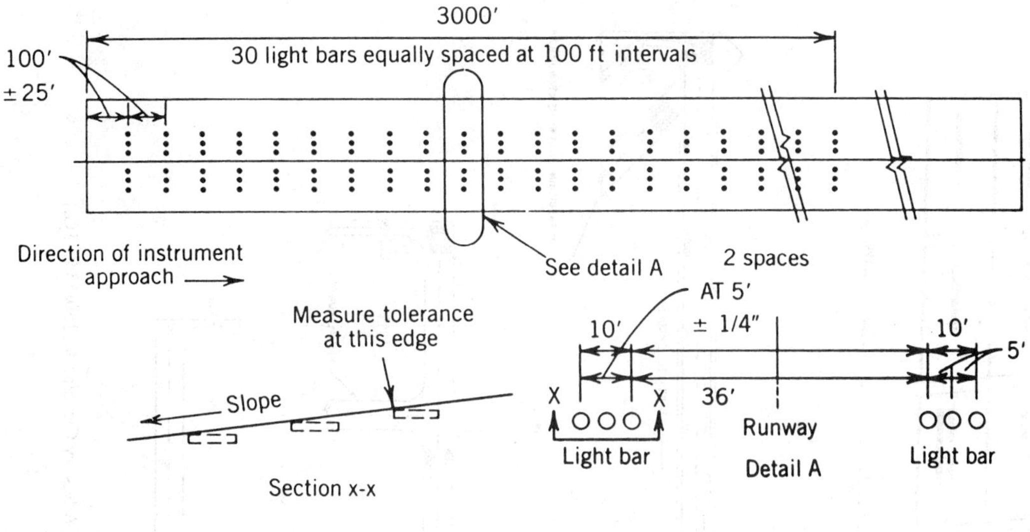

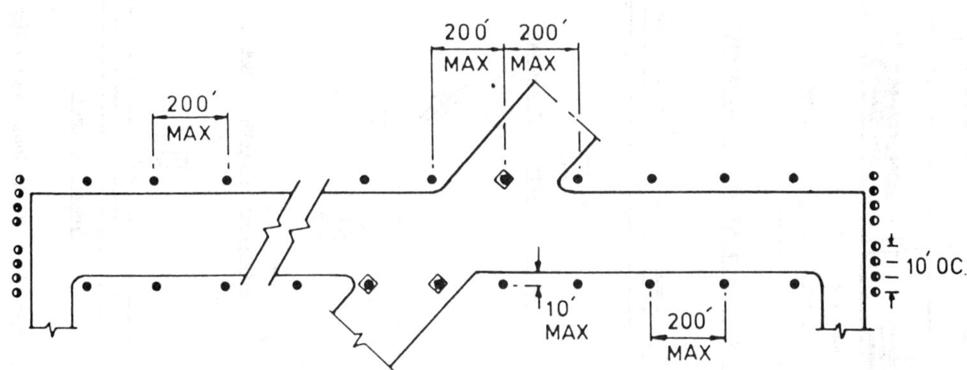

FIGURE 56.44 Runway lighting. (*Source:* FAA. 1993c. *Standards for Airport Marking.* Advisory Circular AC150/5340-1G.)

controls the lights. When the bar is lit there are now center lights ahead, creating a black hole effect. Once the aircraft is permitted on the runway the light bar is extinguished and the taxiway/runway lights are illuminated to guide the pilot onto the runway for takeoff [FAA, 1993b].

Typical airfield markings give the pilot directions to the ramp, parking areas, fuel, gates, areas for itinerant aircraft, ramps for military aircraft, cargo terminals, international terminals, and other airside functions. Visual cues also aid the pilot in docking the aircraft at the gate. Generally there is also an airline ground employee with handheld signal lights to direct the pilot as the aircraft pulls into the gate. Figure 56.46 shows the FAA's 1993 guide to airfield signs.

Approach to the Runway

The approach lighting system (ALS) dictates the navigation/approach capability. Light bars may extend as much as 3000 feet from the threshold along the aircraft's desired glide path. Lighting systems are available to provide runway glideslope cues indicating whether the pilot is above, below, right, or left of the hypothetical wire representing the proper descent trajectory. The visual approach slope indicator systems (VASIS) provide at the side of the runway red and white light bars.

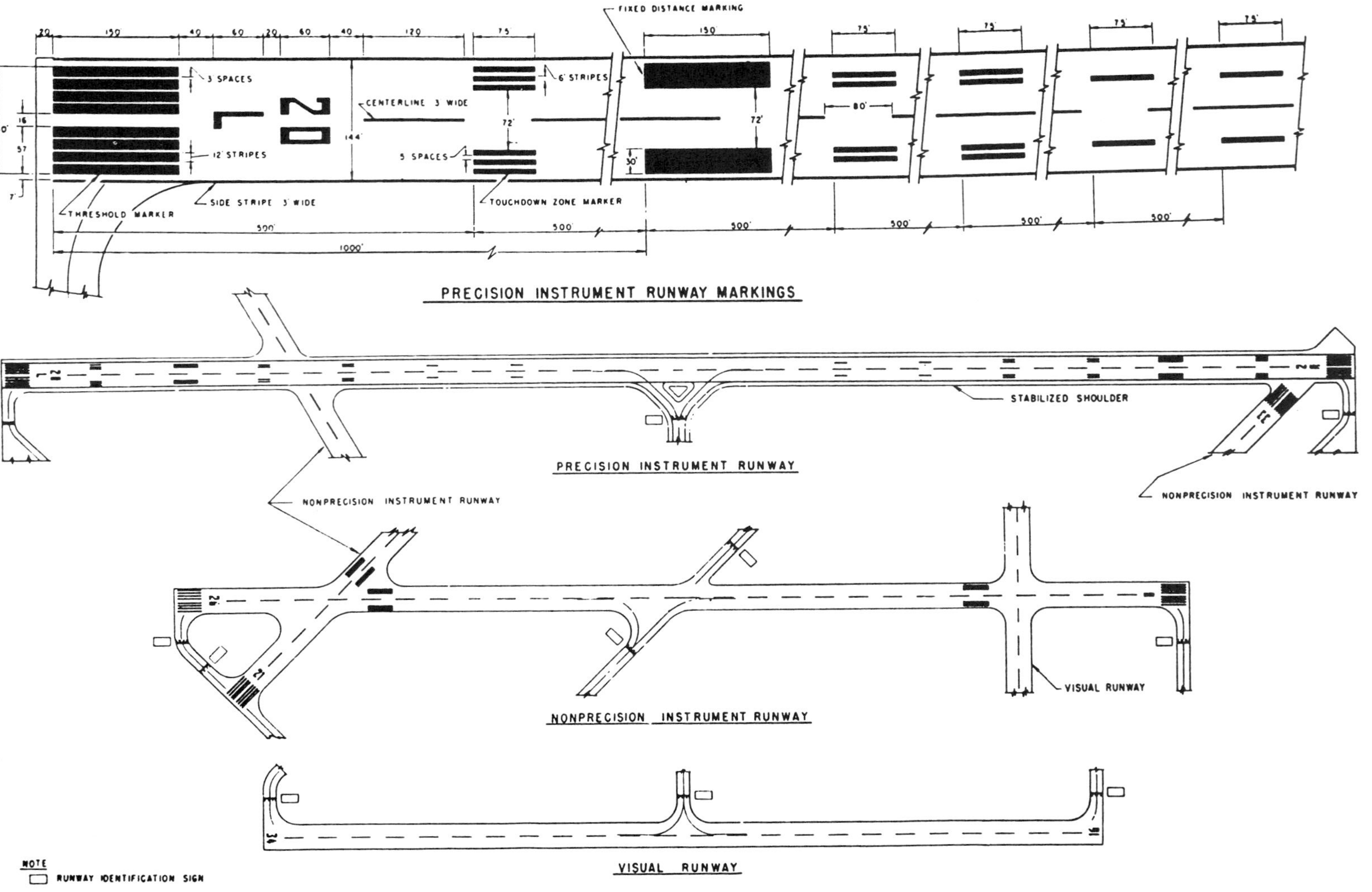

FIGURE 56.45 Marking along the runway. (*Source:* FAA. 1993c. *Standards for Airport Marking.* Advisory Circular AC150/5340-1G.)

GUIDE TO AIRFIELD SIGNS (U.S.)

SIGN and LOCATION	PILOT ACTION or SIGN PURPOSE
4-22 On Taxiways at Intersection with a Runway	Controlled Airport - **Hold** unless ATC clearance has been received. Uncontrolled Airport. - Proceed when **no** traffic conflict exists.
4-22 Runway/Runway Intersection	Taxiing - Same action as above. Taking Off or Landing - Disregard unless a "Land, Hold Short" clearance has been accepted. .
* 4-APCH Taxiway in Runway Approach or Departure Area	Controlled Airport - Hold when instructed by ATC. Uncontrolled Airport - Proceed when no traffic conflict exists.
* ILS ILS Critical Area	Hold when approaches are being made with visibility less than 2 miles or ceiling less than 800 feet.
Areas where Aircraft are Forbidden to Enter	Do not enter.
B Taxiway	Identifies taxiway on which aircraft is positioned.
22 Runway	Identifies runway on which aircraft is positioned.
* Edge of Protected Area for Runway	These signs are used on controlled airports to identify the boundary of the runway protected area. It is intended that pilots exiting this area would use this sign as a guide to judge when the aircraft is clear of the protected area.

Notes:
1. See the *Airman's Information Manual* for additional information on airfield signs.
2. The signs shown on this guide comply with FAA standards. In some cases ICAO's proposed sign standards differ with FAA's. The asterisk (*) in the left column denotes these cases so the pilot can be aware that some differences may be encountered outside the United Sates.

FIGURE 56.46 Guide to airfield signs. (*Source:* FAA. 1993c. *Standards for Airport Marking.* Advisory Circular AC150/5340-1G.) *(continues)*

The precision approach path indicator system (PAPI) provides upper and lower lights of red and white that in various combinations indicate whether the pilot is too low or too high. For example, an all-white bar indicates the aircraft is on a glide slope greater than 3.5 degrees, while an all-red bar is less than 2.5 degrees. Equal red and white indicates the aircraft is on the three degree glide slope.

Positioning along the glide path is accomplished by the use of light bars extending from the runway along the flight path. There are several different approach lighting systems, as suggested in Fig. 56.47. For precision approaches (Category I, II, or III) ILS, the high-intensity ALSF system is employed. The ALS consists of light bars 3000 feet from the threshold. From 3000 feet

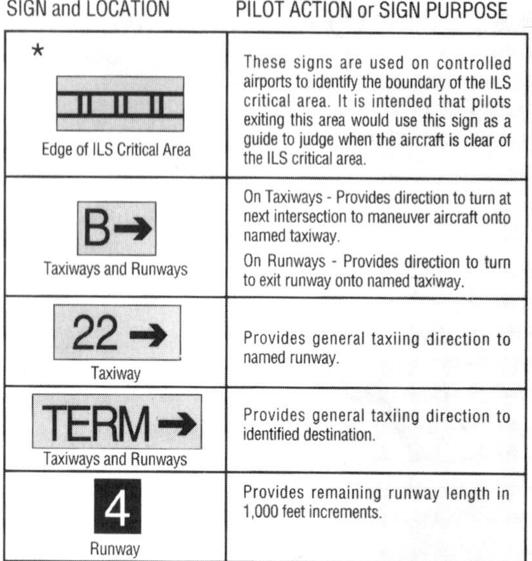

SIGN and LOCATION	PILOT ACTION or SIGN PURPOSE
* Edge of ILS Critical Area	These signs are used on controlled airports to identify the boundary of the ILS critical area. It is intended that pilots exiting this area would use this sign as a guide to judge when the aircraft is clear of the ILS critical area.
B→ Taxiways and Runways	On Taxiways - Provides direction to turn at next intersection to maneuver aircraft onto named taxiway. On Runways - Provides direction to turn to exit runway onto named taxiway.
22 → Taxiway	Provides general taxiing direction to named runway.
TERM → Taxiways and Runways	Provides general taxiing direction to identified destination.
4 Runway	Provides remaining runway length in 1,000 feet increments.

Arrangement of Signs at an Intersection

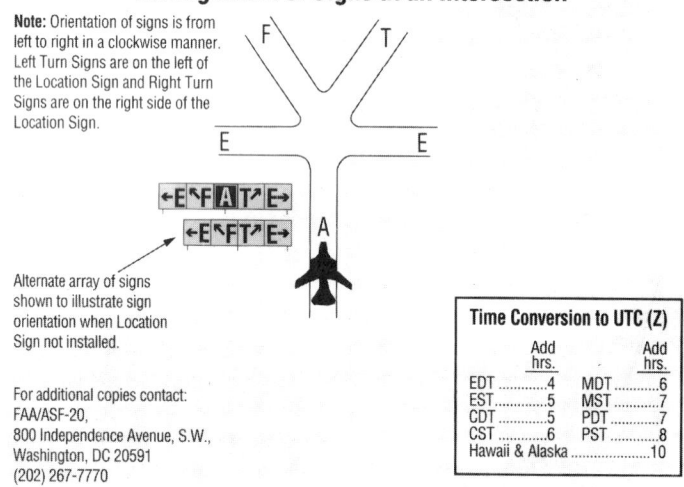

Note: Orientation of signs is from left to right in a clockwise manner. Left Turn Signs are on the left of the Location Sign and Right Turn Signs are on the right side of the Location Sign.

Alternate array of signs shown to illustrate sign orientation when Location Sign not installed.

For additional copies contact:
FAA/ASF-20,
800 Independence Avenue, S.W.,
Washington, DC 20591
(202) 267-7770

Time Conversion to UTC (Z)

	Add hrs.		Add hrs.
EDT	4	MDT	6
EST	5	MST	7
CDT	5	PDT	7
CST	6	PST	8
Hawaii & Alaska			10

FIGURE 56.46 (continued) Guide to airfield signs.

into 1000 feet the lights are a sequenced flasher that gives the appearance of a rolling ball leading to the runway centerline. From 1000 feet (inner marker) to the threshold there are white light bars in the center and bars of red lights on either side of the centerline spaced 100 feet apart. An extra light bar is placed at 500 feet to provide an added visual cue.

MALSR is a medium-intensity ALS with a runway alignment indicator light. It is the U.S. standard for ILS operations during Category I visibility minima. Five sequenced lights begin at 2400 feet from the threshold and extend to 1400 feet. Thereafter eight flashing light bars are installed along the extended runway centerline at 200 foot spacing extending to the threshold. Other medium-intensity approach lighting systems are for nonprecision approaches and consist of the white center marking bars sometimes augmented with the sequenced white flashers.

Runway Pavement Design

Pavement design methods are based on the gross weight of the aircraft. Since it is impracticable to develop design curves for each type of aircraft, composite aircraft are determined and loads

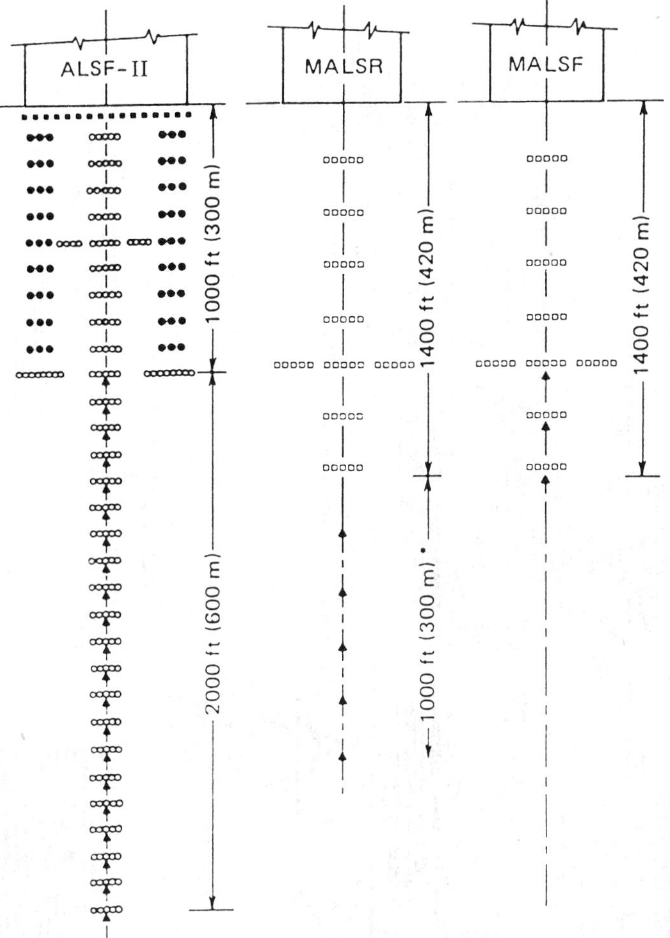

FIGURE 56.47 FAA approach light systems. (*Source:* FAA. 1993c. *Standards for Airport Marking.* Advisory Circular AC150/5340-1G.)

are converted from the actual aircraft to the design aircraft, the design aircraft being the one that requires the greatest thickness of pavement. The traffic forecast, which includes the mix of aircraft anticipated, is converted to a traffic forecast of equivalent annual departures.

The FAA advisory circular [FAA, 1978; AC 150/5320-6C CHG 2] presents a number of curves to be used to design the pavement thickness for both flexible and rigid pavements. The process is outlined in Chapter 59.

56.9 Airport Plans

Upon completion of the inventory, forecasting, requirements analysis, and site evaluation, the master planning proceeds to the synthesis of airside and landside concepts and plans. These include an airport layout plan and an approach and clear zone plan. Other plans could include the site plan, the access plan, and the environmental plan. The development of plans under the master planning process does not deal with detailed design plans.

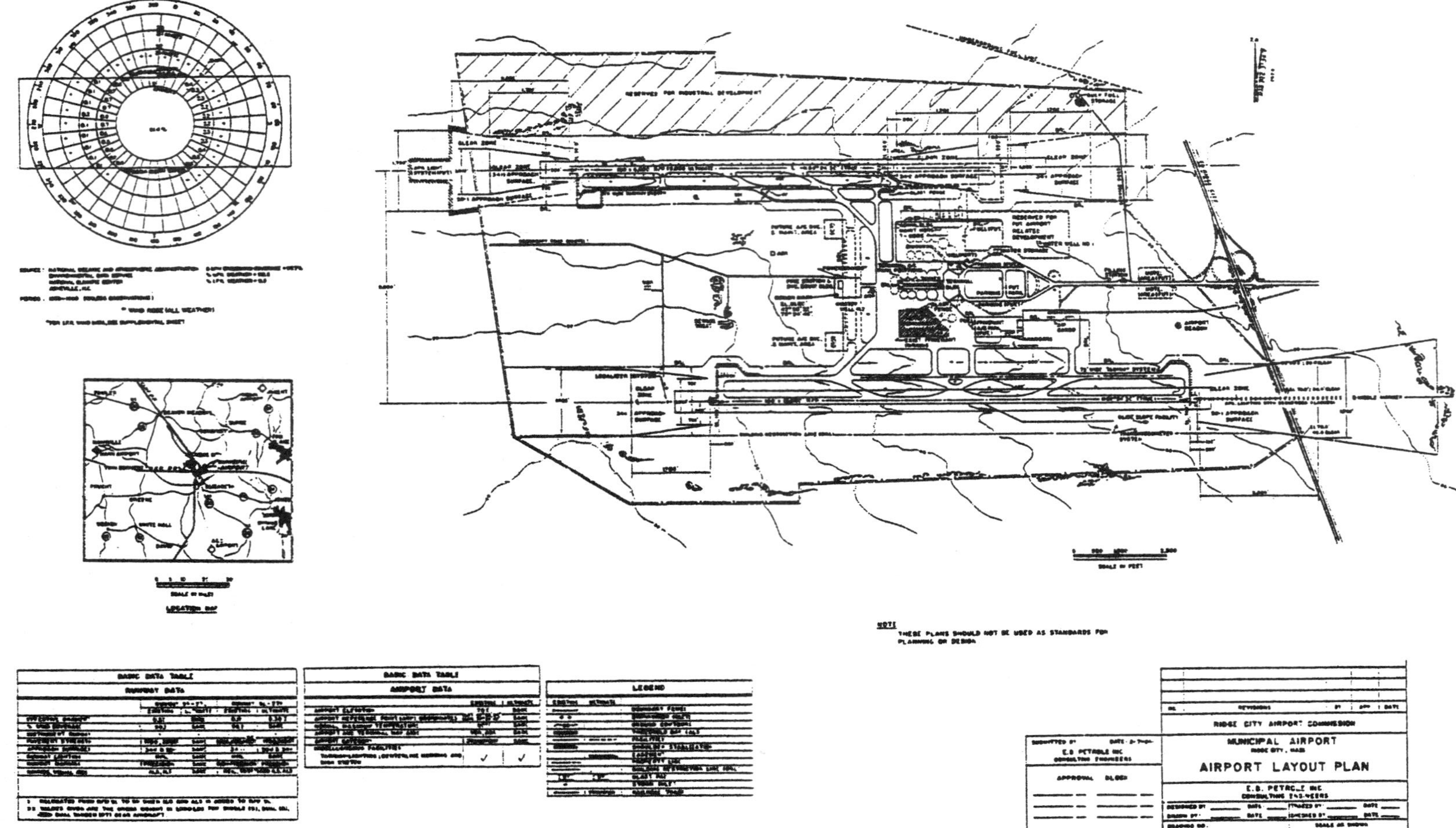

FIGURE 56.48 Sample airport layout plan. (*Source:* FAA. 1985. *Airport Master Plans.* Advisory Circular AC150/5070-6A.)

Airport Layout Plan

The airport layout plan (ALP) is a graphic representation to scale of existing and future airport facilities on the airport. An example is presented in Fig. 56.48. It will serve as the airport's public document, giving aeronautical requirements as well as pertinent clearance and dimensional data and relationships with the external area. The airfield configuration of runways, taxiways, aprons, and the terminal are shown schematically. The airport layout plan (usually $24'' \times 36''$ plate with minimum lettering of .120″) should include, as a minimum, the following:

- Airport layout details
    - Runways, taxiways, blast pads, stabilized shoulders, runway safety areas, buildings, NAVAIDs, parking areas, road lighting, runway marking, pipelines, fences, major drainage facilities, wind indicators, and beacon
    - Prominent features such as trees, streams, ponds, ditches, railroads, power lines, and towers
    - Revenue-producing nonaviation property
    - Areas reserved for future development, such as FBO facilities and fuel farms
    - Areas reserved for nonaviation development
    - Existing ground contours
    - Fueling facilities and tie-down areas
    - Airport boundaries
    - Clear zones and associated approach surfaces
    - Airport reference point
    - Latitude, longitude, and elevation of existing and ultimate runway ends and thresholds
    - True azimuth of the runways (measured from true north)
    - Pertinent dimensional data
- Location map depicting the airport with surrounding cities, railroads, major roads, and tall towers within 25 to 50 miles of the airport
- Vicinity map
- Basic data table on existing and future airport features, including elevation, reference point coordinates, magnetic variations, maximum daily temperature for the hottest month, airport and terminal navigational aids, runway identification, longitudinal gradients, percent wind coverage, instrument runways, pavement type, pavement strength in gross weight, type of main gear (single, dual, dual tandem), approach surfaces, runway lighting, runway marking, electronic and visual approach aids, and weather facilities
- Wind rose with runway orientation superimposed
- Designated instrumented runway [FAA, 1985]

Approach and Runway Clear Zone Plan

The approach and clear zone drawing permits the planner to determine how the airport will interface with the surrounding area in terms of safe flight. An example is presented in Fig. 56.49. It includes:

- Area under the imaginary surfaces defined in FAR Part 77 [FAR, 1975]
- Existing and ultimate approach slopes or slope protection established by local ordinance
- Runway clear zones and approach zones showing controlling objects in the airspace
- Obstructions which exceed the criteria
- Tall smokestacks, television towers, garbage dumps, landfills, or other bird habitats that could pose a hazard to flight

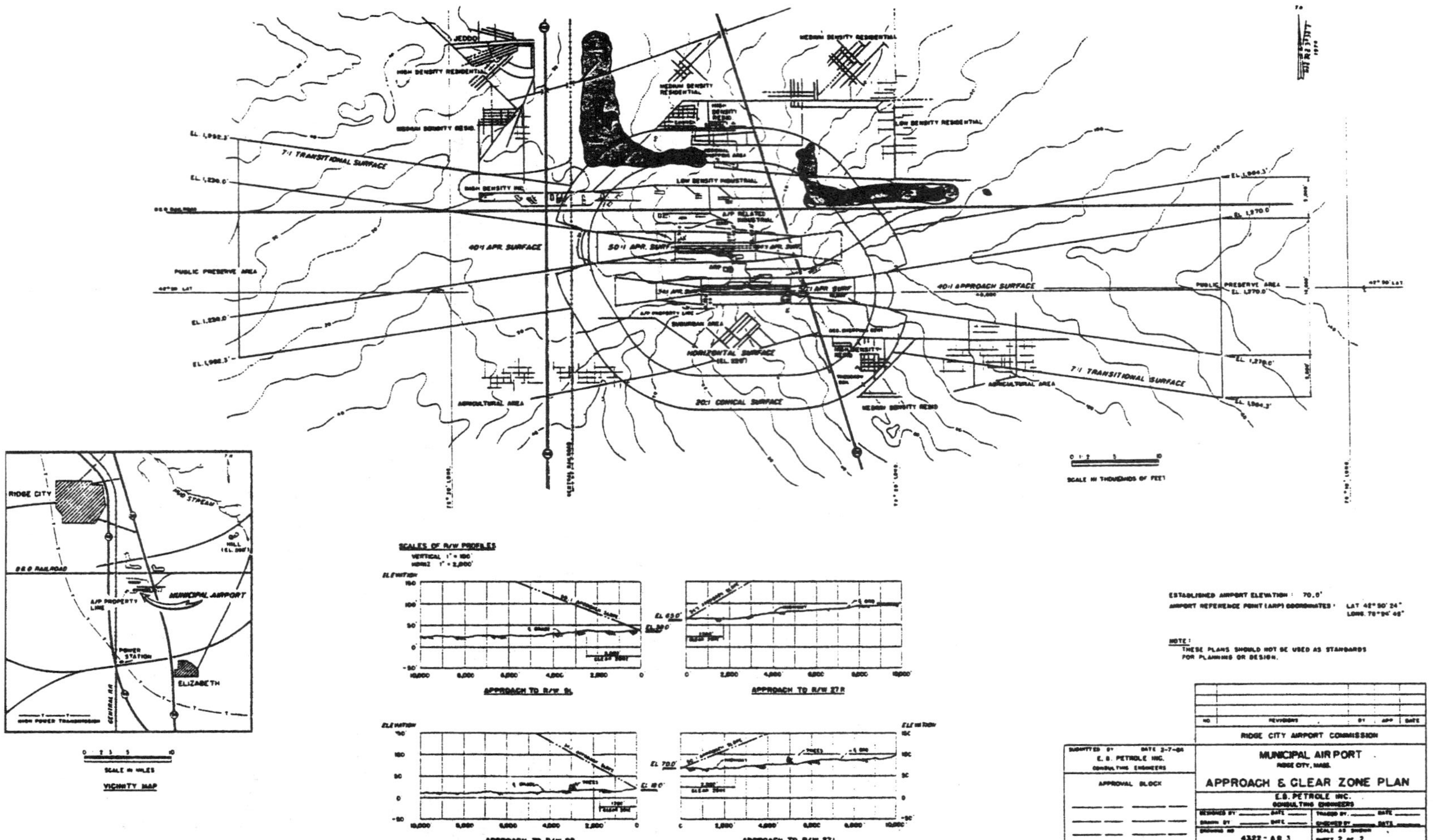

FIGURE 56.49 Sample runway and approach plan. (*Source:* FAA. 1985. *Airport Master Plans.* Advisory Circular AC150/5070-6A.)

Other Plans

Terminal Area Plan

The terminal area plan usually consists of a conceptual drawing showing the general plan for the terminal, including its possible expansion. Under some changes the terminal modification will have a major impact on the taxiway and apron and will be reflected in an altered ALP.

Noise Compatibility Plan

Using future airport traffic, noise contours should be generated to identify future impacts of noise in the community. It would include alternative takeoff tracks and operational constraints. The plan would identify buildings and other facilities that might potentially need to be moved or soundproofed.

56.10 Summary

The total airport system is the effective integration of both airside and landside systems to handle traveler requirements for airplane travel to and from distant points usually beyond the convenient range of automobile traffic or when time constraints require much higher-speed movement. The users of the airport include the traveler, the airlines (and their aircraft), flying enthusiasts, air freight forwarders, and air traffic controllers and other federal government representatives. The critical issues in airport design are:

- Complexities in design caused by the unique interaction of the aircraft performance and size with the engineering aspects of airport design
- Airport growth (terminal and runway) to account for the continued expansion of air travel demand, which is not expected to diminish in the next 50 years
- Integration of air traffic requirements into the design of the airport, particularly its operational capability in poor weather conditions
- Criteria for new and/or expanded sites for airports to increase capacity (minimize delay) while at the same time operating within the constraints imposed by noise and obstruction within the airways

The controlling document of any airport is the master plan, the outline of which was followed in this chapter.

There are other subjects that might have been treated here, such as:

- Design of an air cargo terminal
- Design of a heliport or vertiport
- Design of fuel farms and water supply
- Design of firefighting and rescue systems
- Design of snow and ice control

The FAA has provided definitive design guidelines for each of these items and many more. See FAA Advisory Circular AC 00-2.7 for a list of all the circulars that are available.

Defining Terms

Aircraft approach category: A grouping of aircraft based on 1.3 times their stall speed in their landing configuration at their maximum certificated landing weight. The categories are as follows:

A: Speed less than 91 knots
B: Speed 91 knots or more but less than 121 knots

C: Speed 121 knots or more but less than 141 knots
D: Speed 141 knots or more but less than 166
E: Speed 166 knots or more

Aircraft design group (ADG): A grouping of aircraft based on wingspan. The groups are as follows:

I: Up to but not including 49 feet (15 m)
II: 49 feet (15 m) up to but not including 79 feet (24 m)
III: 79 feet (24 m) up to but not including 118 feet (36 m)
IV: 118 feet (36 m) up to but not including 171 feet (52 m)
V: 171 feet (52 m) up to but not including 214 feet (65 m)
VI: 214 feet (65 m) up to but not including 262 feet (80 m)

Airport elevation: The highest point on an airport's usable runway expressed in feet above mean sea level.

Airport layout plan (ALP): The plan of an airport showing the layout of existing and proposed airport facilities.

Airport reference point (ARP): The latitude and longitude of the approximate center of the airport.

Blast fence: A barrier used to divert or dissipate jet blast or propeller wash.

Building restriction line (BRL): A line which identifies suitable building area locations on airports.

Clearway (WY): A defined rectangular area beyond the end of a runway cleared or suitable for use in lieu of runway to satisfy takeoff distance requirements.

Declared distances: The distances the airport owner declares available and suitable for satisfying the airplane's takeoff distance, accelerate-stop distance, and landing distance requirements.

Displaced threshold: The portion of pavement behind a displaced threshold may be available for takeoffs in either direction and landings from the opposite direction.

Hazard to air navigation: An object which, as a result of an aeronautical study, the FAA determines will have a substantial adverse effect upon the safe and efficient use of navigable airspace by aircraft, operation of air navigation facilities, or existing or potential airport capacity.

Inner-approach OFZ: The airspace above a surface centered on the extended runway centerline. It applies to runways with an approach lighting system.

Inner-transitional OFZ: The airspace above the surfaces located on the outer edges of the runway OFZ and the inner-approach OFZ. It applies to precision instrument runways.

Large airplane: An airplane of more than 12,500 pounds (5700 kg) maximum certificated takeoff weight.

Nonprecision instrument runway: A runway with an approved or planned straight-in instrument approach procedure which has no existing or planned precision instrument approach procedure.

Object: Includes, but is not limited to, aboveground structures, NAVAIDs, people, equipment, vehicles, natural growth, terrain, and parked aircraft.

Object free area (OFA): A two-dimensional ground area surrounding runways, taxiways, and taxilanes which is clear of objects except for objects whose location is fixed by function.

Obstacle-free zone (OFZ): The airspace defined by the runway OFZ and as appropriate, the inner-transitional OFZ, which is clear of object penetrations other than frangible NAVAIDs.

Obstruction to air navigation: An object of greater height than any of the heights or surfaces presented in Subpart C or FAR part 77. (Obstructions to air navigation are presumed to be hazards to air navigation until an FAA study has determined otherwise.)

Precision instrument runway: A runway with an existing or planned precision instrument approach procedure.

Relocated threshold: The portion of pavement behind a relocated threshold is not available for taking off or landing. It may be available for taxiing of aircraft.

Runway (RW): A defined rectangular surface on an airport prepared or suitable for the landing or takeoff of airplanes.

Runway blast pad: A surface adjacent to the ends of runways provided to reduce the erosive effect of jet blast and propeller wash.

Runway OFZ: The airspace above a surface centered on the runway centerline.

Runway protection zone (RPZ): An area off the runway end (formerly the clear zone) used to enhance the protection of people and property on the ground.

Runway safety area (RSA): A defined surface surrounding the runway prepared or suitable for reducing the risk of damage to airplanes in the event of an undershoot, overshoot, or excursion from the runway.

Runway type: A runway use classification related to its associated aircraft approach procedure.

Shoulder: An area adjacent to the edge of paved runways, taxiways, or aprons providing a transition between the pavement and the adjacent surface: support for aircraft running off the pavement; enhanced drainage; and blast protection.

Small airplane: An airplane of 12,500 pounds (5700 kg) or less maximum certificated takeoff weight.

Stopway (SWY): A defined rectangular surface beyond the end of a runway prepared or suitable for use in lieu of runway to support an airplane, without causing structural damage to the airplane, during an aborted takeoff.

Taxilane (IT): The portion of the aircraft parking area used for access between taxiways and aircraft parking positions.

Taxiway (TW): A defined path established for the taxiing of aircraft from one part of an airport to another.

Taxiway safety area (TSA): A defined surface alongside the taxiway prepared or suitable for reducing the risk of damage to an airplane unintentionally departing the taxiway.

Threshold (TH): The beginning of that portion of the runway available for landing. When the threshold is located at a point other than at the beginning of the pavement, it is referred to as either a displaced or a relocated threshold depending on how the pavement behind the threshold may be used.

Visual runway: A runway without an existing or planned straight-in instrument approach procedure.

Defining Acronyms

AC: Advisory Circular published by the FAA.

AGL: Above ground level.

AIP: Federal Aviation Administration Airport Improvement Program.

ALSF: Approach lighting system with sequenced flashing lights.

ALSF-1: Level 1 high-intensity approach lighting system.

ASOS: Automated surface observing system.

ASR: Airport surveillance radar.

ATCT: Air traffic control tower.

AWOS: Automated weather observing system.

CAT I ILS: Category I instrument landing system.

CAT II ILS: Category II instrument landing system.

CAT III ILS: Category III instrument landing system.

DH: Decision height.

DME: Distance measuring equipment.

FAA: Federal Aviation Administration.

FAF: Final approach fix.

FAR: Federal Aviation Regulations.
FBO: Fixed-base operator.
GPS: Global positioning system.
HAA: Height above airport elevation.
HAT: Height above touchdown.
HIRL: High-intensity runway lights.
IAP: Instrument approach procedure.
ICAO: International Civil Aviation Organization.
IFR: Instrument flight rules.
ILS: Instrument landing system.
LIRL: Low-intensity runway lights.
LOC: Localizer.
MALS: Medium-intensity approach lighting system.
MALSF: Medium-intensity approach lighting system with sequenced flashing lights.
MALSR: Medium-intensity approach lighting system with runway alignment indicator lights.
MAP: Missed approach point.
MDA: Minimum descent altitude.
MEA: Minimum en route altitude.
MIRL: Medium-intensity runway lights.
MLS: Microwave landing system.
MSL: Mean sea level.
NAVAID: Navigational aid.
NDB: Nondirectional radio beacon.
NDB-A: Circling approach utilizing NDB facility.
NOAA: National Oceanic and Atmospheric Administration.
NOTAM: Notice to airmen.
NPIAS: National Plan of Integrated Airport Systems.
NWS: National Weather Service.
ODALS: Omnidirectional approach lighting system.
PAPI: Precision approach path indicator.
PFC: Passenger facility charge.
RAP: Remote altimetry penalty.
RNAV: Area navigation.
RPM: Revenue passenger miles.
SASP: State aviation system plan.
SSAL: Simplified short approach lighting system.
SSALR: Simplified short approach lighting system with runway alignment indicator lights.
TERPS: Federal Aviation Administration's terminal instrument procedures.
TVOR: Terminal VOR.
VAPI: Visual approach path indicator.
VASI: Visual approach slope indicator.
VFR: Visual flight rules.
VHF: Very high frequency.
VOR: Very high-frequency omnidirectional range.
VOR-A: Circling approach utilizing VOR facility.
VORTAC: VOR and ultra-high-frequency tactical air navigation aid.

References

al Chalibi, M. 1993. *The Economic Impact of a Major Airport.* Urban Land Institute Research Paper # 622.

Ashford, N. and Wright, P. H. 1992. *Airport Engineering,* 3d ed. John Wiley & Sons, New York.

Ashford, N., Stanton, H., and Moore, P. 1991. *Airport Operations.* Pitman, London.

Cherwony, W. F. and Zabawski, F. 1983. Airport terminal curbfront planning. Presented at Transportation Research Board Annual Meeting, January 1983.

CBO (Congressional Budget Office). 1984. *Financing U.S. Airports in the 1980's.* U.S. Congress.

CBO. 1988. *The Status of the Airport and Airway Trust Fund.* U.S. Congress.

EPA. 1974. *Information on Levels of Environmental Noise Requisite to Protect Public Health and Welfare with an Adequate Margin of Safety.* EPA 550/9-74-004.

FAA (Federal Aviation Administration). 1993a. *Advisory Circular Checklist.* Advisory Circular AC00-2.7.

FAA. 1983a. *Airport Capacity and Delay.* Advisory Circular AC150/5060-5 (incorporates Change 1).

FAA. 1991c. *Airport Design.* Advisory Circular AC150/5300-13 Change 1.

FAA. 1970. *Airport Drainage.* Advisory Circular AC150/5320-5B.

FAA. 1985. *Airport Master Plans.* Advisory Circular AC150/5070-6A.

FAA. 1978. *Airport Pavement Design and Evaluation.* Advisory Circular AC150/5320-6C.

FAA. 1993b. *Aviation Forecasts; Fiscal Years 1993–2004.* FAA-APO-93-1.

FAA. 1991. *1991–92 Aviation System Capacity Plan.* U.S. Department of Transportation, DOT/FAA/ASC-91-1.

FAA. 1993d. *Classification of Airspace.* Brochure.

FAA. 1983c. *Establishment and Discontinuance Criteria for Automated Weather Observing Systems (AWOS).* Office of Aviation Policy and Plans, FAA-APO-83-6, May 1983.

FAA. 1990b. *Final Environmental Impact Statement.* Standiford Field Airport, Louisville, KY.

FAA. 1992. *Integrated Noise Model.* Version 3.10.

FAA. 1987. *Integrated Noise Model: User's Guide.* FAA-EE-81-17.

FAA. 1991b. *National Plan of Integrated Airport Systems.* AC150/5050-5.

FAA. 1983b. *Noise Control and Compatibility Planning for Airports.* Advisory Circular AC150/5020-1.

FAA. 1988b. *Planning and Design Guidelines for Airport Terminal Facilities.* Advisory Circular AC150/5360-13.

FAA. 1990a. *Runway Length Requirements for Airport Design.* Advisory Circular AC150/5325-4A.

FAA. 1993c. *Standards for Airport Markings.* Advisory Circular AC150/5340-1G.

FAA. 1988. *Terminal Area Forecasts FY 1988–2000.* FAA APO-88-3.

FAA. 1975. *The Continuous Airport System Planning Process.* Advisory Circular AC150/5050-5.

FAA. 1976. *United States Standard for Terminal Instrument Procedures (TERPS),* 3d ed. FAA Handbook 8260.3B.

Horonjeff, R. and McKelvey, F. 1994. *Planning and Design of Airports,* 4th ed. McGraw-Hill, New York.

HNTB. 1988. *Evansville Regional Airport; Master Plan Update.* Howard, Needles, Tammen & Bergendoff for Evansville Airport and State of Indiana DOT.

Indiana DOT. 1993. *Indianapolis Metropolitan Airport System Plan Update.* Prepared by TAMS and al Chalibi for Indiana Department of Transportation and the Indianapolis Airport Authority.

Port Authority of New York and New Jersey. 1992. *Airport Highlights.* Annual Report.

Purdue University. 1993. *Instrument Approach and Weather Facilities Enhancement Plan.* Final Report on Contract 91-022-086 for Indiana DOT (Aeronautics).

Transportation Research Board. 1989. *Aviation Forecasting Methodology; A Special Workshop.* TRB Circular 348.

United States Department of Commerce. 1993. *U.S. Terminal Procedures.* East Central Volume 2 of 3.

U.S. Code FAR. 1975. *Objects Affecting Navigable Airspace.* Federal Aviation Regulations Part 77, January 1975.

Wilbur Smith Associates and Partnership for Improved Air Travel. 1990. *The Economic Impact of Civil Aviation on the U.S. Economy.*

For Further Information

For further reading about planning, particularly the airport master plan, FAA Advisory Circular 150/5070-6A, *Airport Master Plans*, June 1985, is available from the Superintendent of Documents (SN 050-007-00703-5).

For further reference about capacity and delay, FAA Advisory Circular 150/5060-5, *Airport Capacity and Delay*, Sept. 1983, is available at no cost from U.S. Department of Transportation, Washington, D.C.

For further information about terminal design, FAA Advisory Circular 150/5360-13, *Planning and Design Guidelines for Airport Terminal Facilities*, April 1988, is available at no cost from the U.S. Department of Transportation, Washington, D.C.

For further information about site design, FAA Advisory Circular 150/5000, *Airport Design*, Sept. 1989, is available from the Superintendent of Documents (SN 050-007-853-8 plus Chg. #1 SN-050-007-929-1). Information on noise is found in FAA Advisory Circular 150/5020-1, *Noise Control and Compatibility Planning for Airports*, August 1983, available at no cost from the U.S. Department of Transportation, Washington, D.C.

Section 56.8 augments the airport design computer program available from the FAA discussed in Advisory Circular 150/5360-13, *Airport Design*. A diskette for PCs containing the airport design programs is available from your nearest FAA Airports Office.

For further design information about runway length for various aircraft consult FAA Advisory Circular 150/5325-4A, *Runway Length Requirements for Airport Design*, Jan. 1990, available at no cost from the U.S. Department of Transportation, Washington, D.C.

57

High-Speed Ground Transportation: Planning and Design Issues

Robert K. Whitford
Purdue University

Matthew G. Karlaftis
Purdue University

57.1 Introduction

Purpose

High-speed ground transportation (**HSGT**) usually refers to trains that achieve speeds higher than 200 kph (125 mph) in revenue service. There are two principal technologies for these and higher speeds:

1. Railroad trains that operate using steel-wheel-on-steel-rail technology and are powered by either electric locomotives or diesel-electric locomotives. Trains in this category are called high-speed rail (**HSR**).
2. Trains that are both suspended (levitated) and propelled by magnetic fields (MAGLEV).

The technologies that support the HSR trains operating at 300 to 400 kph (186 to 250 mph) are available now. These trains have been successful in revenue service in France, Germany, Spain, and Japan for several years. On the other hand, MAGLEV has not yet seen operational use and in many ways is still a promising but experimental technology. Most of this chapter will be devoted to HSR.

0-8493-8953-4/95/$0.00 + $.50

Further, since the U.S. is only beginning to plan for a major involvement of HSR in the late 1990s, most of this chapter will be devoted to the planning/systems issues associated with implementing this technology in the U.S. HSR is a complex system and has many technical facets beyond the scope of this chapter.

HSGT has many important civil engineering planning and design considerations. Somewhere between 65 and 80% of the investment cost is for civil-type facilities (track, right-of-way, stations, catenaries, bridges, etc.). System requirements that pertain to its introduction in the U.S. are under scrutiny as government and industry attempt to identify pertinent issues and problems. Of special concern are those areas of operation, such as safety and noise, that are considered in a different manner by non-U.S. manufacturers and operators whose history in passenger rail does not parallel that in the U.S.

Scope

A brief history sets the stage. Section 57.2 discusses pertinent systems issues such as market demand, corridor development, cost estimating, schedule, safety, noise, ride quality, and the like. Section 57.3 presents the specifications that pertain to a typical **TGV**-type trainset, while section 57.4 discusses the infrastructure. Section 57.5 identifies several track-train interactions. Section 57.6 discusses HSR examples from abroad, and section 57.7 discusses MAGLEV technology. A long list of references is provided to help the reader develop a better understanding of the work that must precede any implementation in the U.S. (The Volpe National Transportation Center in Cambridge, Mass., and the Federal Railroad Administration in Washington, D.C., are studying these and other issues.)

The HSR is often referred to as the TGV train. TGV (Train Grande Vitesse) is French, while Germany has a similar development operating in revenue service that is fully competitive with the TGV called the Intercity Express (**ICE**). Both systems have been submitted as competitors in response to a request by the Texas High-Speed Rail Authority for the Dallas–Fort Worth–Houston and Dallas–Fort Worth–San Antonio corridors.[1,2]

Brief History

High-speed trains began in the early 19th century when British engineers developed the steam railroad locomotives. These locomotives were the beginning of a long era in which varying technologies were pursued in order to achieve higher speeds coupled with operating and fuel efficiencies. Steam propulsion gave maximum speeds of about 160 kph (100 mph) in the early 20th century, and test speeds of 209 kph (130 mph) were attained by using diesel-electric and electric propulsion in the 1930s and 1940s. This change in technologies provided slightly improved performance and fuel efficiency, but it was not until the 1960s that breakthroughs in suspension, train-track dynamics, and other factors permitted an increase in train speeds by a factor of 2 or more.

On November 1, 1965, the Japanese introduced, as part of the regular train schedule, a standard-gauge railway service that reached a maximum speed of 209 kph (130 mph). The average speed of the train was 166 kph (103 mph). While the speed of the train was not a major breakthrough in terms of what had been previously achieved, the uniqueness was in the dedication of the right-of-way (ROW) to this system.

The ensuing decade brought much activity in the area of high-speed rail. The Pennsylvania railroad developed the "metro-liner" by placing traction motors on each axle. That train achieved a maximum speed of 251 kph (156 mph) in Princeton, New Jersey, in May 1967. "Metro-liner-like" coaches, pulled by AEM-7 locomotives operating at speeds up to 201 kph (125 mph) in revenue service, are the only relatively high-speed rail systems in the U.S. today.

At present, high-speed rail trains regularly operate in Japan, France, Germany, and Spain in revenue service at speeds between 240 and 300 kph (150 and 187 mph). The Japanese have been

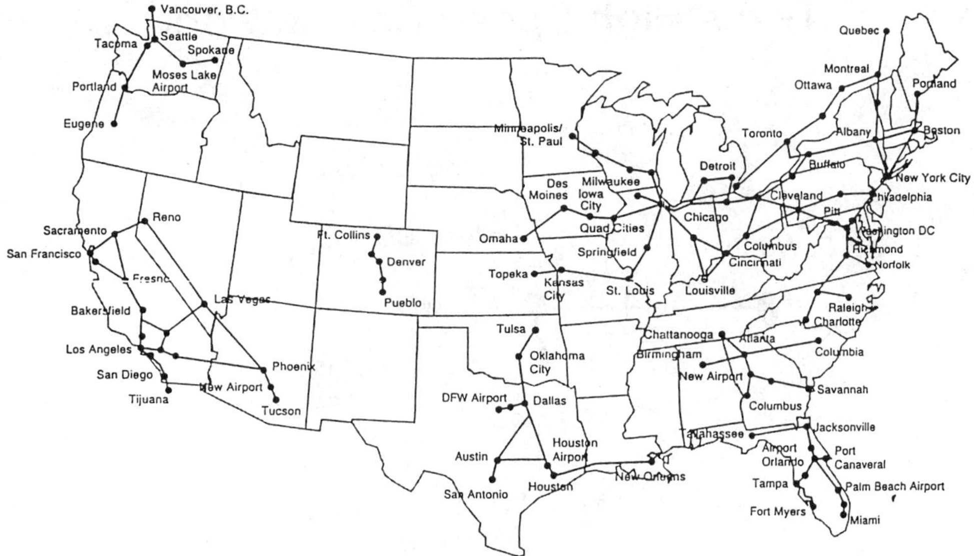

FIGURE 57.1 HSGT systems under study in the U.S. (*Source:* U.S. GAO, 1993.)

operating their Shinkansen ("Bullet train") between Osaka and Tokyo for over 30 years. The French initiated high-speed revenue service between Paris and Lyons with the TGV in 1981. The latest French TGV train operating in revenue service on the Atlantique High Speed Line serves the market between Paris and the Atlantic coast cities of LeMans and Tours. It achieves peak speeds of 320 kph (200 mph) in revenue service and has been tested at a sustained speed of over 530 kph (330 mph).

The U.S. has been slow to embrace the development of HSGT systems. This is probably due to a combination of several factors:

- The love of Americans for the automobile
- The advances of the airline industry
- The size of the U.S. land mass
- The skepticism of generating enough demand for a profitable venture in any but the highest population corridors
- The lack of a passenger railroad culture

Recent concern over gridlock and the failure of airports to expand to fully meet commuter demand, coupled with the successes in Europe, has brought about a resurgence of interest in high-speed passenger rail traffic. The HSR is most likely to succeed between cities with stage lengths in the range of 200 km (125 mi) to 400 km (250 mi). Figure 57.1 indicates the leading candidates for HSR or MAGLEV consideration in the U.S. HSGT systems could free capacity at some of the congested highways and airports, and moreover it is a mode of transportation that appears to be reasonably energy efficient when compared with other modes, and with electrification for its power source it is potentially less dependent on petroleum resources.[3,4]

57.2 Systems and Planning Issues

Because of its years of revenue service in Europe and Japan and because it has continued to achieve growth, HSR should be considered a mature technology that might be expected to find its way quickly and easily into operation in the U.S. That notion requires exploration in three major areas:

1. Where, how, and who will provide leadership to back high front-end costs in the face of widespread skepticism and competitive pressure? For example, the proposed map for the

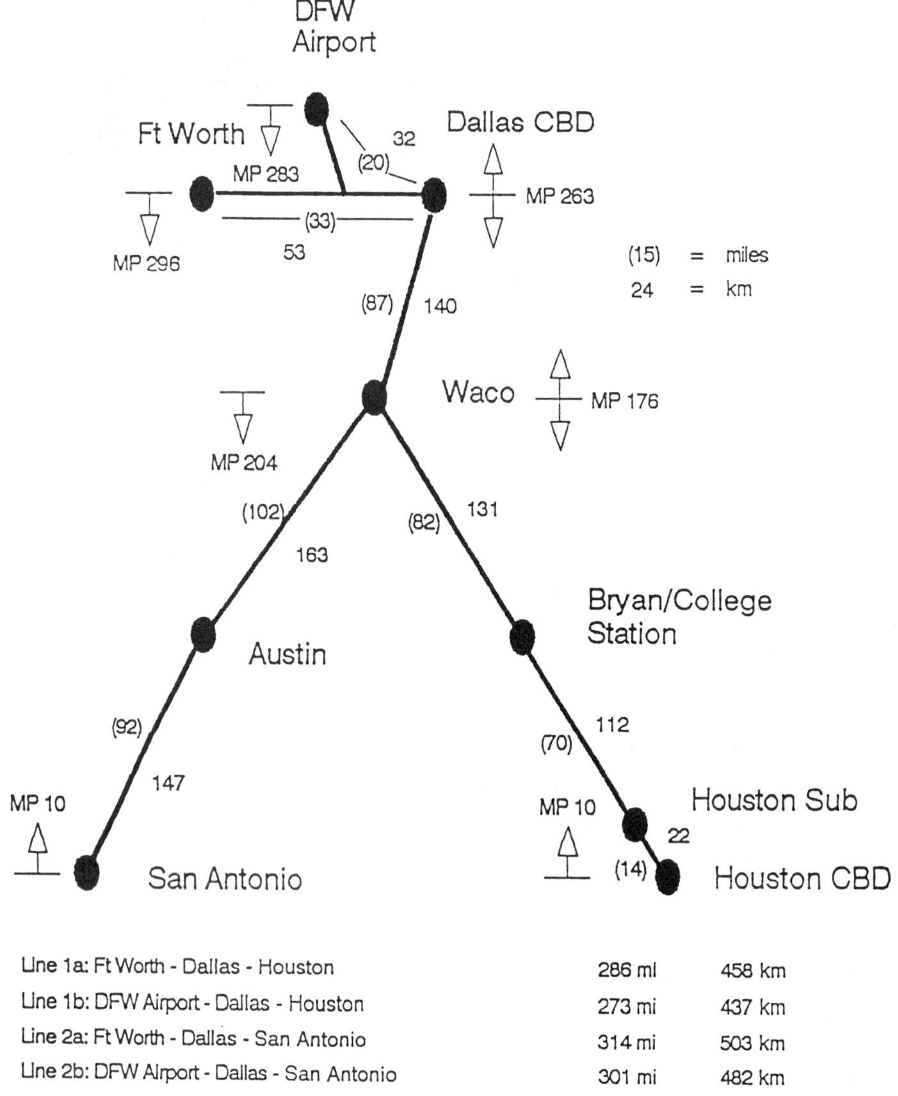

FIGURE 57.2 Proposed route for Texas GT HSR. (*Source:* Texas FasTrac, Inc. *Franchise Application.* Submitted to The Texas High-Speed Rail Authority, 1991.)

Texas HSR is shown in Fig. 57.2, with one 458-km (286-mi) leg connecting Dallas to Fort Worth to Houston and one 503-km (314-mi) leg connecting Fort Worth to Dallas to San Antonio. The Texas HSR train proposal in these corridors has received some very pointed opposition from a local airline that feels that this system will significantly cut into its business, which involves commuter lines between these same cities. The project is on hold because the necessary funds could not be raised for the next phase.

2. Can potential projects in the U.S. secure ROW with adequate space to meet the high-speed requirements of track geometry and provisions for electric power supply and catenary?

3. Will all of the safety concerns, many of which reflect the difference between American philosophy and that of Europe and Japan, be met without significantly reducing the speed of the trains?

In an HSGT system, average speed is usually considered to be the most important performance characteristic.[5] The requirements for high-speed vehicles in a specific corridor arise from the need to decrease travel time relative to other modes of transport or to increase corridor capacity. Although peak speed is considered the main travel-time-reducing parameter, average speed is critically important and is affected by the following:

- The HSR train's ability to negotiate curves
- The HSR train's ability to accelerate and decelerate quickly
- The number of station stops en route and dwell time at each station

Besides travel speed, there are a number of systems considerations and performance characteristics that should be used in evaluating HSGT systems. These include travel demand, corridor development, schedule performance, acceleration-deceleration, ride quality, noise, safety, energy conversion efficiency, and systemwide parameters.

Market Demand

The cost of the investment of most HSR systems is high and its introduction will require a thorough and positive market analysis. Since it impinges on the commuter air market and the automobile/bus vehicle market, its potential entry as a competitor may be difficult to judge. In addition, there will be a latent demand for TGV travel that will become part of the demand picture. Table 57.1 shows that the French have succeeded in shifting traffic from air and highway and in generating some new demand with the initiation of the Paris to Lyons train (with other nearby cities served by standard rail). The analysis of the growth in the Paris to Lyons market of 2.2 million passengers from 1980 (before TGV) to 1994 indicates that 40% of the passengers have been diverted from commuter air, 23% have been diverted from auto, and 37% are new or induced demand.[2]

How will the U.S. public respond to the placement of an HSR system in their area? Will they give up their autos for these trips? Is there adequate market in these days of electronic mail and teleconferencing? Will the public accept rail with trip times from origin to destination in the same

Table 57.1 Passenger and Traffic Demand for the French TGV Paris to Lyons Route

	Travel Demand,[a] Paris to				
	Dijon	Geneva	Lyons	Grenoble	Marseilles
1980 train	850	105	1470	435	720
1994 train (TGV)	1045	325	3670	710	1005
% increase	23%	209%	150%	38%	40%
1980 airplane		605	970	515	1480
1994 airplane (after TGV)		495	525	280	1195
% increase		−18%	−46%	−46%	−24%
Probable diversion from					
Plane		165	865	40	165
Road	50	20	500	65	50
Induced traffic	145	35	835	90	70

Source: Texas TGV. *Franchise Application.* (Submitted to The Texas High-Speed Rail Authority, 1991.)

[a] Thousands of passengers.

range for commuter air? These questions and more must be answered. Without adequate answers or without a strong infusion of capital from the federal government to help construct expensive facilities, HSR or MAGLEV will probably not reach any significant level of implementation in the U.S. For example, a study of the Chicago–Milwaukee–Twin Cities[6] corridor, in which fairly conservative demand models were developed and which studied three scenarios (125, 186, and 300 mph systems) concluded that the 125-mph HSR offered the best financial return for the least environmental cost. It also showed that the highest economic benefit per dollar invested was the most important feature in a public sector capital constrained investment program. The economic benefits achieved by the 125-mph HSR option are 80% of those achieved by the 185-mph HSR and 94% of those achieved by the 300-mph MAGLEV trains. The 185-mph TGV technology, while showing good financial return and the highest net economic benefit, had the highest environmental costs because of the potential woodland and farm severance problems associated with obtaining and using new right-of-way. The MAGLEV train showed good economic benefit, but its investment costs were so high that its overall economic performance was marginal. The key negative issue for MAGLEV is that its present early stage of development suggests further cost and performance risk.

Corridor Development

Any corridor to be used for HSR operations will require either an upgrade of existing track or new land acquisition coupled with the construction of straighter track, built with the stability in alignment required for these speeds. This may not be so easy, since the corridor will bisect many farms and create dead ends for many rural roads. The corridor would also require rerouting of utility service lines such as gas, water, sewer, telephone, and electricity.

At the speeds of the HSR, grade separation means that overpasses and underpasses will be required in many areas. For some farms it may be necessary to put an access tunnel under the roadbed in order for the farmer to get from the fields on one side of the track to those on the other. The 458 km between Fort Worth–Dallas and Houston is estimated to require 270 bridges over creeks, rivers, highways, and other railroads plus about 25,000 meters of elevated track, mostly in urban areas. There are also about one hundred forty-five 10 ft by 10 ft by 150 ft culverts for drainage and access along the same right-of-way.[1]

Small towns also pose a corridor problem. If the use of the present ROW is suggested, much of that ROW passes through small towns and other lightly populated areas. The choice of corridor, the placement of and approach to tunnels, and the design of other depressed or elevated ROW around these towns may significantly increase the cost. Most of these towns will not have a station, since the train cannot stop many times if it is to take advantage of the higher peak speed. Therefore, the economic development aspects that usually occur around stations will not take place and the town may see the HSR as a nuisance rather than a benefit.

Entering large urban areas may be easy or difficult depending on the existing roadbed. The TGV operates at reduced speed using existing track for entry to and exit from of Paris and Lyons.[7]

The corridor chosen may require some trade-off between the cost of wetlands remediation (if wetlands are in the path of the HSR corridor) and alternate routes that require less or no mitigation.

No doubt access of ROW for a corridor, whether through procurement, eminent domain, or condemnation, must occur early in the project, well before any construction is due to begin. With the corridor's disruption, or "severance," of the farms and landscape, the HSR will face the usual uphill battle from those who do not want the railroad. The NIMBY ("not in my backyard") syndrome will in all likelihood be prevalent.

Cost Estimating

The Transportation Research Board has developed a series of cost estimates based on several HSGT scenarios.[8] The scenarios compare several options against an "as is" railroad. The options for which data are presented in Tables 57.2 and 57.3 are as follows:

Table 57.2 Estimates for Six Alternatives of HSGT—300-mile Corridor

Alternative	Urban Track, 40 Miles, Cost/Mile	Suburban Track, 60 Miles, Cost/Mile	Rural Track, 200 Miles, Cost/Mile	Total System, 300 Miles, Total Cost
1	$915,000	$400,000	$235,000	$108,000,000
2	$1,660,000	$1,000,000	$740,000	$275,000,000
3	$7,645,000	$6,410,000	$5,260,000	$1,742,000,000
4	$7,680,000	$7,370,000	$8,340,000	$2,418,000,000
5	$8,110,000	$14,490,000	$10,495,000	$3,293,000,000
6	$31,115,000	$19,715,000	$14,055,000	$5,239,000,000

Source: High-Speed Surface Transportation Cost Estimate Report. Parsons Brinckerhoff for the Transportation Research Board, Washington, D.C.

1. *"As is" railroad* requiring typical class-3 track to be upgraded with the addition of block signaling and passing sidings to permit 125-kph (79-mph) passenger service.
2. *Low ROW/capital investment strategy* with top speeds of 175 kph (110 mph) and upgrades in track and cab signaling. ROW width sufficient for second track and major rehabilitation at stations.
3. *Intercity/shared ROW* with top speeds of 200 kph (125 mph) with electric propulsion, full double track, and concrete ties maintained to FRA class-6 standards. All high-speed crossings would be grade separated.
4. *Intercity/shared ROW/new bypass segment* with one to several bypass segments and both track geometry and signaling to permit top speed on the bypasses of 240 kph (150 mph).
5. *The TGV approach* with trains operating mostly on new ROW dedicated to the TGV with top speed operation at 290 to 320 kph (180 to 200 mph) range.
6. *New technology using MAGLEV* concepts with a top speed of 500 kph (320 mph).

The TRB report presents many assumptions on which the above numbers are based. However, these numbers should be sufficient for the purpose of planning. The costs include land, ROW preparation, utilities relocation, track construction, realignments, grade separations/enclosures, fencing, electrification, signaling, undergrade bridges, overhead bridges, tunnels, terminal, beltway stations, operations and maintenance (O&M), central control administration, and trainsets. The assumptions are given in sufficient detail so that a quick estimate can be made (Table 57.2). The costs for a 500-km (300-mile) system with 67 km (40 miles) of urban land, 100 km (60 miles) of suburban land, and 333 km (200 miles) of rural land are presented. The average cost per mile for the TGV (option 5) is about $6.84 million per kilometer ($11 million per mile).[8,9]

Schedule Performance

Providing that the ride quality is acceptable, the demand for the HSGT will depend on the perception that the consumers have as to its schedule performance and reliability. When comparing the HSGT with competing modes, the actual travel time for the user between origin and destination is critical. The line-haul speed, typical average speed, and time of trip are given in Table 57.3 for the six alternatives shown above.

Somewhat higher speeds than those quoted for alternatives 3 and 4 may be obtained if an active tilt train such as the X2000 or EGR 460 are used.[10–12] The speed of the train on curves will depend on the radius of the curve, the superelevation of the track (bank angle), and the allowable deficiency in the cant angle, the limits for which come from ride comfort (lateral forces on passengers). A train (such as alternatives 2, 3, and 4) that must regularly slow down for curves and then accelerate again will have a slower overall line-haul speed than one that can negotiate curves at or close to its regular speed. Improved acceleration and braking are aided by lower trainset weight achieved through the use of lightweight materials and appropriate construction techniques.

Table 57.3 Speeds and Trip Times for the Six Scenarios

Alternative	Urban Track, 40 Miles, Speed (kph)	Suburban Track, 60 Miles, Speed (kph)	Rural Track, 200 Miles, Speed (kph)	Total Trip	
				Total Time (Hours)	Avg. Speed (kph)
1	50	65	70	4.58	65
2	50	80	90	3.77	80
3	55	90	1090	3.39	88
4	55	90	130	2.93	102
5	60	100	150	2.60	115
6	100	170	230	1.62	185

Source: High-Speed Surface Transportation Cost Estimate Report. Parsons Brinckerhoff for the Transportation Research Board, Washington, D.C.

The number of stations and the dwell time in each station will also affect the overall time of travel between two points. Analysis of the TGV on one 400-km (250-mile) route with a maximum line-haul train speed of 300 kph (186 mph) and stops at four intermediate stations, each with a dwell time 3 minutes, took 118 minutes elapsed time from doors closing at the origin to opening at the final destination. The average speed was 204 kph (127 mph).

Safety

The potential severity of high-speed accidents can be offset by such factors as dedicated right-of-way, fully fenced corridors, automatic train control, and ROW maintenance all geared to reduce the overall risk. A large part of the research regarding HSGT systems has focused on safety issues.[13–16] Many of these issues deal with crash tests and with car designs that will minimize bodily injuries in case of a crash.[17–20] The existing HSGT systems in Europe are operating at a level of safety that is better than that of European conventional rail technology. Excellent system maintenance, highly automated train control, dedicated ROW, and fencing around the tracks have led to an amazingly safe accident record. For example, the Japanese Shinkasen line has had no fatal accidents in the past 18 years (through 1994), while having transported approximately 2 billion passengers.[4,21]

Higher-speed operations on existing lines pose an especially difficult problem. The track, ballast, and geometry are all designed to accommodate lower speeds. There will certainly be increased risk associated with the use of this track in high-speed operation. In fact, the use of an existing ROW such as in the northeast corridor is the most likely scenario to happen first in the U.S. With the large amounts of capital needed, it is clear that until the federal government decides to fund dedicated ROWs, the upgrading and increased use of existing ROWs will be the direction higher-speed rail will take in the U.S. With this technology comes the need to assess how ROW and equipment design, signal systems, onboard and wayside detectors of various sorts, grade crossings, ROW security, and so on contribute to enhancing the overall safety risk given the accident severity potential that inherently increases with speed. Such progress is achieved by using system elements other than speed to control risk at the same or lower levels than that currently accepted by the riding public.[22,23]

This leads to a careful design of an onboard monitoring system that will automatically slow a train that is starting to hunt. The elements with automation will include accelerometers to detect the onset of truck hunting, bearing temperature monitors, brake system sensors, and various wayside detectors.[7]

As speeds increase above 200 kph (125 mph) dynamic force control is the key factor in maintaining safety of operation. New inspection methodologies to move from the current accepted static geometry (even if loaded) measurements to more dynamic real-time monitoring of equipment forces (wheel) and the track response/interaction (rail) is needed for the HSR. The whole issue of maintaining track for ride comfort versus minimum track geometry standards must be addressed.[24,25]

Lighter weight but stronger materials will be required. One key to the success of TGV and ICE will be maintaining or reducing, not increasing, axle loads. Thus given the U.S.'s need to design for different collision scenarios (mixed freight/passenger operations) the challenge to be innovative will be even greater. (Even though the TGV and ICE in some cases do operate on existing/shared ROW as well as on dedicated lines, the type of freight equipment and thus accident scenarios are different than in the U.S.) Given some of the accidents France and Germany have had between "regular" passenger trains and freight trains and their resultant severity one could argue that such risks would not be acceptable to the U.S. riding public.

Studies have also been made of fire safety,[26] emergency preparedness,[27] control and communication,[28,29] and human factors and automation as applied to train control.[30] Finally, there is considerable concern over the effects on health of electromagnetic field (EMF) radiation. The **FRA,** through the Volpe National Transportation Systems Center (**VNTSC**) and the Environmental Protection Agency, has done considerable testing and analysis of this potential, both for the TGV-type train and the MAGLEV train. A series of 17 reports is available on the subject.[31]

Noise

Since HSGT is to be powered by electricity, air pollution is not a factor, leaving the major environmental considerations as severance of land, wetlands mitigation, and noise. The National Environmental Policy Act of 1969 requires that any project with federal government involvement be accompanied by an environmental impact statement. Certainly the choice of a corridor is critical as it may require wetlands mitigation, and the HSR noise can have an affect on the land use along the corridor. Although the HSR noise seems to be less than the noise associated with conventional rail operations,[32–34] there is sufficient noise, especially due to the high aerodynamic component, that either passive barriers or active noise-canceling systems may be necessary.[7]

The major noise sources in diesel operations are the engine and the interaction of the steel wheels and the steel rail. The noise levels from pre-1987 diesel locomotives vary from 67 dBA at idle to 89 dBA at full throttle when standing 100 feet from the locomotive.[32] The wheel/rail noise levels vary dramatically according to the type of wheel and track structure. Most irritating is the track squeal resulting from lateral sliding of the wheels. Wheel/rail noise is usually computed as a function of the speed of the train.

Likewise, the noise experienced by rapid rail transit is attributed to the electric engine, which is much quieter than its diesel-driven counterpart and the rail/wheel interaction. The noise level for the San Francisco Bay Area Rapid Transit System (**BART**) at 60 mph is approximately 83 dBA 50 feet from the train; the corresponding diesel noise is 97 dBA.[32]

The Japanese Shinkansen has been in operation since 1964 and has provided much noise data. The noise level measured at 15 feet from the train varies from 62 dBA at 118 mph to 76 dBA at 124 mph. The French TGV showed noise somewhat higher when operating on its Paris to Lyons route. However, 72 dBA was exceeded at only three homes along the route and the maximum noise measure 82 feet from the train was 97 dBA. The German ICE reported noise levels of 86 dBA at 11.5 feet from the train traveling 124 mph and 93 dBA at 186 mph.

Care in design will keep the noise at these relatively low levels. With noise barriers provided by trees and shrubs, or constructed walls, or depressed track, the noise of the HSGT should be well below any sound levels that would pose an annoyance to surrounding neighborhoods.

Ride Quality

In addition to the stress on performance the consumers will only ride the HSGT if as passengers they perceive it to be comfortable. Thus ride quality as experienced in the seat design and in the amenities is quite important. Although ride quality is subjective in nature, trainset appearance (both interior and exterior), lighting, sound levels, and air flow and temperature determine the appeal. Most of the European trains also provide places for small meetings, phone and fax service, and special workspaces including computer hookup and real-time trip-related status information.

Physical ride quality is determined by track design and alignment, car-body motion, and the design of the passenger seats. The track input comes from the track itself. Most of the high-speed lines maintain track to achieve lateral and vertical forces and acceleration levels low enough to ensure good ride quality. At present, **ISO** standards for ride quality do exist and are currently under review. From a vehicle point of view, dynamically balanced wheelsets, wheel profile, and low unsprung mass are considered the strongest influences on ride quality.[35] Low unsprung mass is essential for truck stability, while the suspension must be designed to minimize lateral and vertical movements of the car body. Wheel profile is important in maintaining safe levels of wheel–rail interaction forces to ensure a smooth and comfortable ride. In summary, ride comfort is very subjective and each railway authority develops its own criteria.

So important is ride quality that sensors and a computer are employed on many trains to give a real-time measure of ride quality and to make adjustments or signal the engineer as necessary. Such data may be used to determine portions of the track that need maintenance. Such a maintenance philosophy dramatically reduces the likelihood that unsafe levels will be encountered. The TGV uses truck-mounted accelerometers to detect truck hunting and requires immediate reduction in speed if such hunting is detected.

Energy Conversion

Energy use by the train is important because one of the goals of the HSGT systems is to conserve energy to minimize operating cost. Energy efficiency is a function of the propulsion system, gearing, and trainset design. The Federal Railroad Administration, through the Improved Passenger Equipment Evaluation Program (IPEEP), found that increased train weight leads to increased energy consumption in the range of 0.06 to 0.08 watt hours/seat-kilometer ton.[36]

The friction involved in steel-wheel-on-steel-rail technology is extremely low. The concern for HSGT is that at the very high speeds, significant additional energy losses will occur in bearing friction and from aerodynamics. Equation (57.1) is the generalized equation for the horsepower required to pull a train. Aerodynamic design is extremely important since the power to overcome aerodynamic effects increases by the cube of the increase in speed. Thus increasing speed from 60 mph to 180 mph requires 27 times the power and from 120 mph to 180 mph requires 3.4 times the power.[37] From the outset of the first TGV significant effort has gone into reducing aerodynamic drag.[7]

$$\mathrm{HP} = C_0 V(C_1 + C_2/W + C_3 V + C_4 V^2/W) \tag{57.1}$$

where

HP = horsepower
W = weight of the trainset
V = speed
C_0 = efficiency of the drive system
C_1, C_2 = friction between the rail and wheel
C_3 = rolling and bearing resistance
C_4 = aerodynamic coefficient

HSGT is not particularly energy efficient except when energy expended per passenger kilometer is compared with other modes. In addition, regenerative braking can be used if the power source is receptive to it. Energy input enables the train to accelerate in order to reach its line-haul speed. When it decelerates, some of the energy that would otherwise be dissipated as heat can be returned to the power source, thus reducing the energy needed to accelerate again.

Systemwide Parameters

HSGT vehicle performance can be seriously affected by certain other system parameters. For example, the location of the corridors is a critical factor in construction cost. Hilly or mountainous terrain, marshland, large numbers of crossings of rivers, creeks, drainage ditches, and roads all add to the alignment of the track system. The elimination of grade crossings is a must, and the communications and signaling systems must be designed to handle the high speed. Most HSGT lines are double track. The catenary will require more supports as track crosses mountains.

Automation is an important design requirement at these high speeds. The manner in which train operators (engineers, dispatchers, etc.) are trained, how the design of their workstation enhances their performance, and how emergencies are to be handled depend on the extent and nature of automation. In all likelihood it will be different from current intercity rail systems in the U.S. and abroad. Some lessons are available by examining BART, **WMATA, PATCO,** and other U.S. systems, and from the international front, where different uses of automation can be seen when comparing TGV with ICE operations. The similarity of these issues with those confronting the aviation industry increases as the use of automation increases.[30] Two obvious options for automation application are a highly automated system with a human-in-the-loop both heavily observed and managing a fully automated system; and human-out-of-the-loop, where the human operator is an observer of automatic systems with virtually no override capability, except to stop the train. Each system has its individual safety and design implications.[30]

The basic principles of operation of the signaling, communication, and control mechanisms, the extent and type of automatic train operation or control to be used, and the provisions for driver vigilance monitoring and override are beyond the scope of this chapter.

57.3 Trainset Specifications

There are several configurations that are often chosen for the trainset; however, as shown in Fig. 57.3, a trainset typically consists of the power car (engine) and 6 to 12 coaches and another power car. Table 57.4 gives the typical physical characteristics of the TGV and the ICE. Maximum speed in revenue service is between 290 and 340 kph (180 and 210 mph), but test runs on the TGV–Atlantique, which the French have built to more stringent specification, have posted test speeds greater than 510 kph (322 mph).[7,15]

57.4 Infrastructure Specifications and Design

The infrastructure that supports the HSR includes the track structure from the subbase, subballast, ballast section, ties, fasteners, rail, switches, turnouts and crossovers, rail anchors and tie pads, catenary and its supports, power substations, bridges, and tunnels. The specifications for the infrastructure TGV Southeastern and Atlantic routes are given in Table 57.5.[7,15]

The gauge of the track and the distance between the centers of the dual track are included in a specification. The amount of ballast determines the stiffness of the track/ballast/subgrade, taken as a combined subsystem under load. Ballast shoulder width is also important in maintaining adequate lateral track stability. Most roadbed has a minimum width of 14 m (46 ft).[1,7]

Geometric Design

Geometric design for the HSR is little different than good practice for geometric design of ROW was years ago, except that with the higher speed, more care is taken in design and the curves have

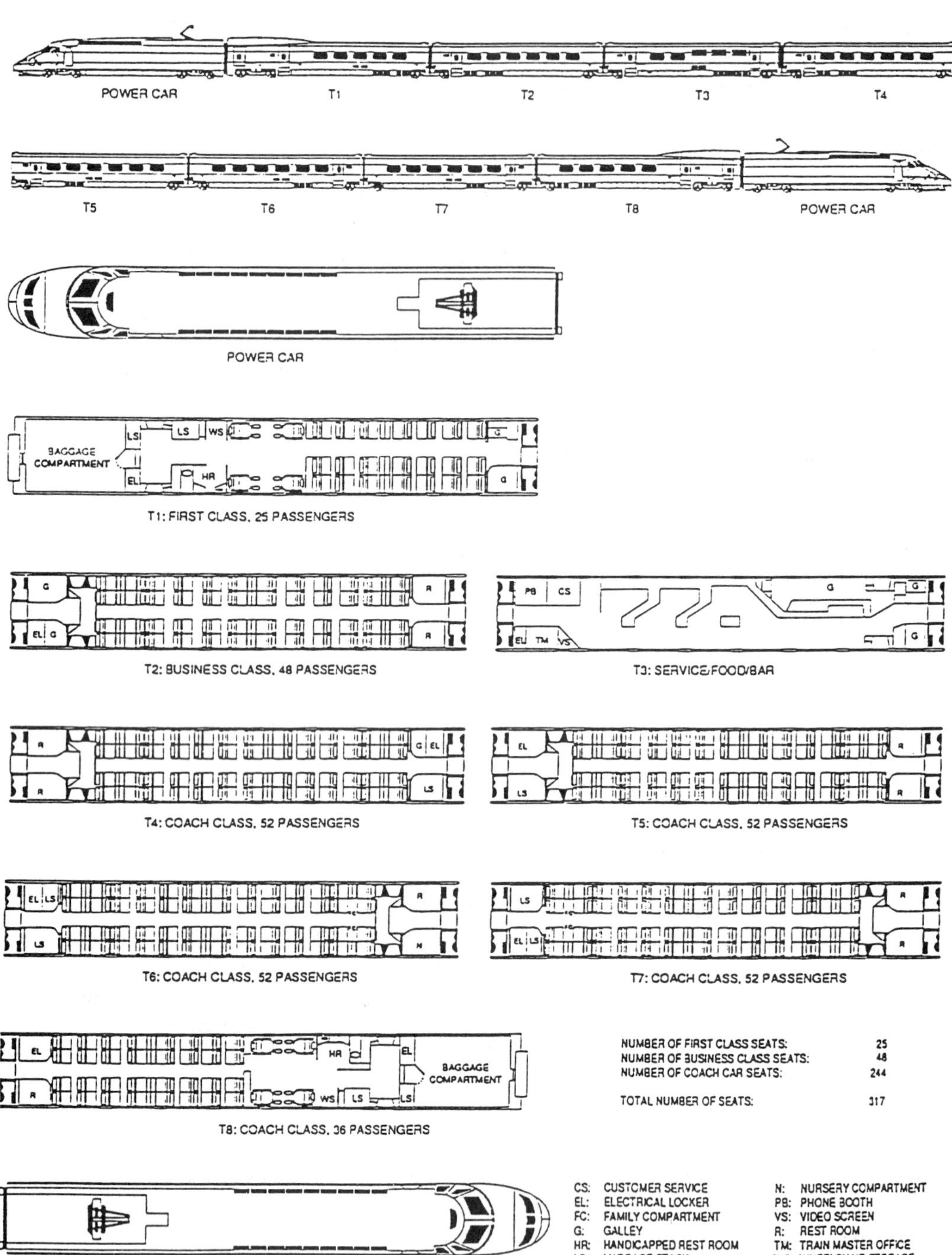

FIGURE 57.3 Typical TGV 1-8-1 trainset. (*Source:* Federal Railroad Administration.)

much larger radii. The critical elements in the design are the superelevation and the length of the transition spiral. As long as a safe speed is maintained, the performance on curves is dictated by ride comfort, which in turn is determined by the centrifugal force the passenger feels. Figure 57.4 shows how the centrifugal force acting on a passenger is developed. A curve that is banked properly (has the right superelevation) will have those forces canceling out.

Table 57.4 TGV Rolling Stock Characteristics for Southeastern and Atlantique Routes

	TGV-PSE	TGV-A
Consist configuration	1-8-1	1-10-1
Seating capacity	260 coach, 108 first	369 coach, 116 first
Fleet size	95 passenger, 2 mail	105
Consist dimensions:		
Length	200 m	237.6 m
Width	2.8 m	2.8 m
Height	4.1 m	4.1 m
Total weight	416 tonnes	490 tonnes
Operating speed	270 kph	300 kph
Total axles	26	30
Maximum axle load	16.25 tonnes	17.0 tonnes
Total powered axles	12	8
Unsprung mass	2.2 tonnes	2.2 tonnes
Transmission type	Sliding cardan shaft	Sliding cardan shaft
Traction		
Location	Body-mounted	Body-mounted
Type	dc THO 676	ac synch, 3-phase
Motor power	525 kW	1100 kW
Total power	6300 kW	8800 kW
Full-power speed range	109–175 mph	80–186 mph
Start-up tractive effort	212 kN	212 kN
Adhesion		
At start	0.12	0.16
At full speed	0.05	0.08
Brake system	Rheostatic; disc brakes on unpowered axles; tread brakes on all wheels	Rheostatic and tread brakes on powered axles; 4 discs on unpowered axles
Stop from max speed	3.6 km	3.6 km
Suspension		
Primary	Coil spring	Coil spring
Secondary	Coil spring	SR10 airbag
Articulation	Annular ring	Annular ring
Current collection	AMDE photograph	GPU pantograph
Truck	Y230 motorized; Y231 trailer	Y230A powered Y237 A&B trailer
Wheel size	36″ (920 mm)	36″ (920 mm)
Pressure-sealed	No	No

Source: Canadian Institute of Guided Ground Transport, GEC-Alsthom/SNCF TGV Baseline, Draft Report, Queen's University, 1992.

Table 57.5 TGV Infrastructure Characteristics for Southeastern and Atlantique Routes *(continues)*

	TGV-PSE	TGV-A
Line length	258 miles	193 miles
Line configuration	Full double track	Full double track
Design operating speed	168 mph	186 mph
	Track Geometry	
Horizontal curvature		
Minimum	10,660 ft	13,130 ft
Design	13,120 ft	20,000 ft
Vertical curvature		
Crest		
Minimum	39,370 ft	52,490 ft
Design	82,020 ft	82,020 ft
Trough		
Minimum	45,930 ft	45,930 ft
Design	82,020 ft	82,020 ft
Maximum gradient	3.5%	2.5%

Table 57.5 (continued) TGV Infrastructure Characteristics for Southeastern and Atlantique Routes

	TGV-PSE	TGV-A
	Parabolic Transitions	
Maximum superelevation	7.09°	7.09°
Unbalanced elevation		
Normal limit	4.53°	3.94°
Exceptional	n/a	5.125°
Exceptional (100 mph)	n/a	6.25°
Variation in superelevation on transition curves		
Normal	0.25°31′	0.22°31′
Exceptional	0.31°31′	0.27°31′
Exceptional at 100 mph	0.48°31′	0.48°31′
Rate of variation in unbalance elevation on transition curves		
Normal	1.19°246′	1.19°271′
Exceptional	1.97°246′	1.97°246′
Minimum spiral length	780′	987′
Minimum separation between transitions	500′	500′
Track gauge	4′8	4′8
Distance between track centers	4.2 m	4.2 m
	Track Structure	
Rail	UIC 60 (121 lb/yd) CWR	UIC 60 (121 lb/yd) CWR
Fasteners	Nabla double-curvature steel spring, 11 kN force with deflection of 0.32″; 0.36″ rubber pad with 1780 kN/in. stiffness	Nabla double-curvature steel spring, 11 kN force with deflection of 0.32″; 0.36″ rubber pad with 1780 kN/in. stiffness
Ties	U41 twin-block, 550 lb concrete, on 26″ centers	U41 twin-block, 550 lb concrete, on 26″ centers
Ballast	Minimum ballast depth 12″, clean crushed rock, with top 4″ of hard material	Minimum ballast depth 14″, clean crushed rock, with top layer of hard material
Crossovers	160 kph maximum at 25 km intervals	160 kph maximum at 25 km intervals
Turnouts	230 kph on deviated track, 270 kph on main line	230 kph on deviated track, 300 kph on main line
Signaling	Full CTC with in-cab signaling; current-coded track circuits; TVM-300 automatic train operation system with override train braking capacity	Full CTC with in-cab signaling; current-coded track circuits; TVM-300 automatic train operation system with override train braking capacity
Catenary	Power 2 × 25 kV/50 Hz; feeder/overhead system in phase opposition; OCS has 107 mm² reinforced contact wire at 16′9″ height	Power 2 × 25 kV/50 Hz; feeder/overhead system in phase opposition; OCS has 150 mm² reinforced contact wire at 16′9″ height
Substations	8 single-phase, 200 kV supply feed	5 single-phase, 220 kV supply feed
Bridges/flyovers	540 total; longest is 10 spans covering 419 m; all bridges carrying TGV line are ballast-deck	328 major bridges; ballast deck as for PSE; 488 culverts
Tunnels	None	Five bored tunnels; total length 6.3 miles; tunnel cross sections: double track: 125 mph–441 ft², 168 mph–764 ft²; single track: 168 mph–495 ft²

Source: Canadian Institute of Guided Ground Transport, GEC-Alsthom/SNCF TGV Baseline, Draft Report, Queen's University, 1992.

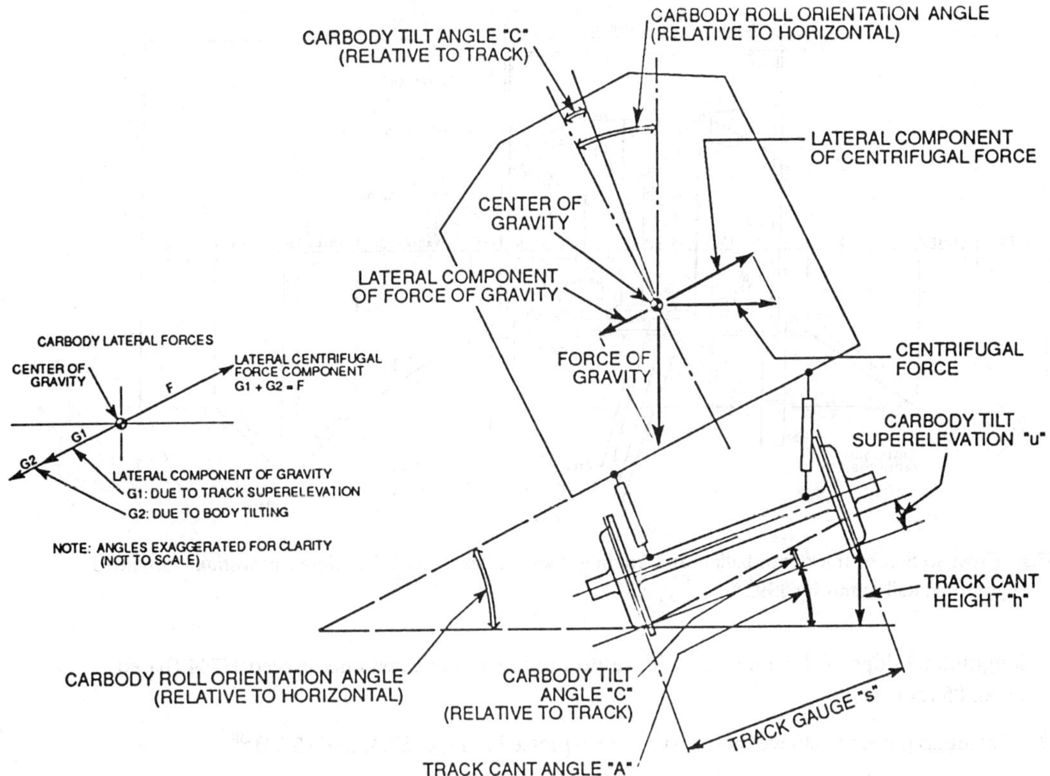

FIGURE 57.4 Effect of deliberate body tilting on forces acting on passengers. (*Source:* Federal Railroad Administration.)

Going from tangent track to curved track requires a spiral as the radius of the horizontal curve goes from infinity to a specific number. The spiral is not flat but must begin the run-in of the superelevation to meet that required for the curve. Likewise, as the track returns to tangent track there is a spiral and superelevation run-out as well.[37,38]

Equation (57.2) indicates how superelevation height difference, which is about 18 cm (7.1 in.), is derived:

$$E = 0.0007V^2D \tag{57.2}$$

where

E = superelevation distance in inches
V = design speed of vehicle along the curve in mph
D = degree of the curve

However, since the speed with which a train would actually traverse the curve is seldom the exact speed used in design, it is necessary to specify the deficiency in cant angle, which in turn indicates the degradation of ride quality.

In the U.S. the design of the spiral is dictated by the following quote from the American Railway Association Design Manual:[36]

> The desirable length of the spiral for tracks... should be such that when the passenger cars of average roll tendency are to be operated the rate of change of the unbalanced lateral acceleration acting on a passenger will not exceed 0.03 g's per second. Also the desirable length needed to limit possible racking and torsional forces produced should be such that the

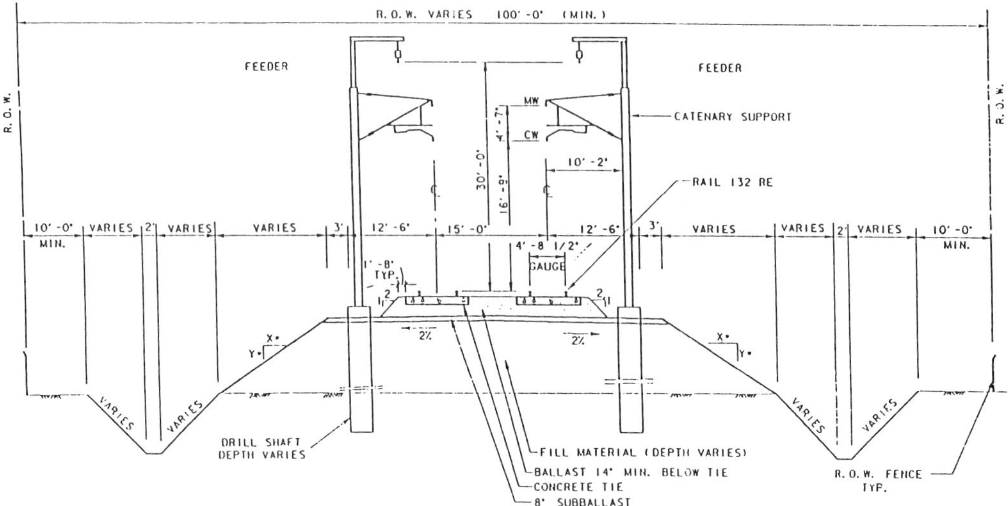

FIGURE 57.5 Cross section of track with ballast and grading. (*Source:* Texas TGV. *Franchise Application.* Submitted to The Texas High-Speed Rail Authority, 1992.)

longitudinal slope of the outer rail with respect to the inner rail will not exceed 1/744 (based on an 85 foot car).

The formulas given to achieve these results are expressed in Eqs. (57.3) and (57.4).[38]

$$L = 1.63E_u V \tag{57.3}$$

$$L = 62E_a \tag{57.4}$$

where

L = desirable minimum length of the spiral in feet
V = maximum train speed in mph
E_a = actual elevation in inches
E_u = unbalance elevation in inches

HSR track must provide accurate vehicle guidance at very high speeds and under various weather conditions, resist static forces, withstand extensive dynamic loading, and minimize the transmission of vibrations and noise.[39] A typical cross section of track used for the TGV train is shown in Fig. 57.5.[2]

Track and Ties

Rail stresses associated with energy absorption depend on the elastic properties of the track as a whole.[40] In the early stages of HSGT systems heavy rail was anticipated and a 142 pound per yard rail was developed. However, the development of this type of rail has proven to be unwarranted. Heavy rail may even cause problems to the elastic balance of the track due to its excessive stiffness.

In an effort to maintain the strict design requirements, the Japanese have utilized a "slab-track" design on their newer lines. This design incorporates the direct fastening, through elastometric rail fasteners, of the rail to a concrete slab. This approach is costly but provides some performance advantages. The French maintain excellent track alignment using dual-block concrete ties and spring-clip fasteners in their conventional tie-and-ballast approach.

Both the French and the Japanese use the lightweight rail tracks (121 pounds per yard) and ties spaced 23 to 26 inches apart. The French TGV track is maintained at a level of tolerance (±0.8 in.) that is four times more stringent than presently required in the U.S. under FRA Class 6 (110 mph) track standards.[21] The track geometry and lining are checked statically through laser positioning systems in both countries. It seems that the French use the onboard dynamic force measurements to align their tracks over the short term and the track geometry measurements over the long term.

High-speed rail also requires the use of improved fasteners that are able to both absorb lateral stress and account for the longitudinal continuity of the rail. Elastic fasteners, which allow for the rotation of the rail around the rail base edge and which are supported by the rigid shoulder of the base plate, reflect common practice.

The weight of the ties is needed to stabilize the track, and for this reason lightweight concrete, steel, or wood ties cannot be used in high-speed rail. In France, concrete two-block ties weighing 250 kg (550 pounds), which rely on more resultant lateral area and resisted tie sides (leading to lower weight), are preferred. Unprestressed concrete blocks can lead to very economical results.[21]

Ballast/Subgrade

An integral part of the system, the subgrade is a significant factor in overall track performance.[2]

- The subgrade should be constructed of materials that have an acceptable potential for shrink-swell.
- Active soils should be stabilized to reduce the potential for shrink-swell to an acceptable level.
- The shrink-swell potential should be controlled so that seasonal changes are minimized and long-term changes are adjusted using routine track releveling procedures.

One of the main advantages of supporting the rails on the ties and rock ballast/subballast system is that grade adjustments caused by active soils can be corrected by routine releveling techniques. This is, of course, not the case for slab track design, where periodic releveling would be extremely expensive. Further, the slab itself can increase and magnify the amount of moisture that will migrate with time beneath the slab. Where clay soils are prevalent, the subgrade will best be designed by removing the clay and replacing it with a compacted granular soil or by using in-place stabilization with materials such as liquid lime/flyash slurry.

> For the Atlantique the minimum ballast section is 35 cm of crushed rock laid in two stages beneath the ties. The ballast grading provides material 1# to 2# size. The bottom layer of normal stone is compacted, and the track placed on it. The top layer of harder material is then stabilized by vibration after the track has been lifted by tampers. A sub-ballast layer separates the ballast and the subgrade materials. Prior to the record setting run in May 1990, ballast cleaning was undertaken to ensure that there were no fine materials that could be blown up by the slipstream.[7]

Catenary

The catenary system is basically state of the art with special attention given to providing the following:[2]

- Ample tension on the contact wire to reduce uplift, thereby improving the collection capacity of the pantograph
- A more rigid suspension system to prevent swaying in lateral winds
- Use of flattened contact wire to reduce wear on both the contact wire and the current collector
- The spacing of support poles from 30 to 70 meters depending on the mechanical requirements of the system at any given location

57.5 Track-Train Interactions

Using Existing Right-of-Way

With the difficulty in procuring large portions of new right-of-way and with raising the new capital from private sources and with building new infrastructure, it becomes imperative that engineers find ways of better using the existing track. The principal problems are as follows:

- The tightness of curves built for slower trains will cause a loss of average speed and energy because of repeated acceleration and deceleration.
- The ride quality will suffer because the superelevation is designed for much slower trains.
- The track/ballast/subgrade may not have the width to provide the lateral stability needed.
- Grade separation does not exist and will require special provision for safety.
- The change to electrification of the line must be accommodated.

In any event, the roadbed will almost always require some rehabilitation and upgrading. Signaling will have to be upgraded to account for the higher speed. The potential of mixed freight/HSR operations may call for more siding and more frequent inspection and realignment. One solution is to depend more on the trainset. The result is trains with active tilting capability.

Tilt Trains[10–13]

The radii for curves that will accommodate the high speeds of a TGV or ICE are extremely high, so the use of existing track built with tighter curves and lower superelevation for slower trains would require excessive slowing and accelerating of a typical high-speed train. To maintain speed on tight curves, Asea Brown Boveri, Inc. (ABB), in cooperation with the Swedish State Railroads (SJ), developed a train using an active tilt mechanism. The train, known as X2000, is similar in its use of technology to FIATs, ETR 450, and ETR 460 trains.[11] They are both *active tilt* trains. The purpose of developing this technology was to significantly reduce trip times while using conventional track.

The X2000 features a self-steering radial truck which consists of a rigid frame in which two wheelsets are mounted in parallel. As the stiff truck travels through curves, the axles remain parallel to each other, exerting forces on the rails.[10] The higher the speed for a given curve radius, the greater the tendency of the wheels in the normal truck assembly to try to overturn the rail (rollover), or to climb over the rail (climbing). The solution developed by ABB was the self-steering radial truck. A soft chevron primary suspension system allows each truck to assume its natural radial position in each curve.[11] The result is a redistribution of forces exerted by the wheelsets. For example, the X2000 exerts no more force rounding a curve at 125 mph than does a regular train at 80 mph. The result of this capability is that it allows for significantly higher average speed.

The active carbody tilting system, shown in Fig. 57.6, was developed mainly for passenger comfort. Along with the increase of train speeds in the curves are associated lateral forces experienced by the passengers. By anticipating each curve and causing the carbodies to tilt inward at the appropriate angles, centrifugal forces are compensated for and passenger comfort is maintained. The X2000 is designed for speeds of 201 kph (125 mph) in Sweden and it has been tested at 250 kph (156 mph) in Germany.[10] The X2000 operated in revenue service in the Northeast corridor for several months. Figure 57.7 shows the train with the AMTRAK logo when it toured the U.S. in the summer of 1993. An active tilt train has high potential of becoming the first HSR system to operate in revenue service in the U.S.

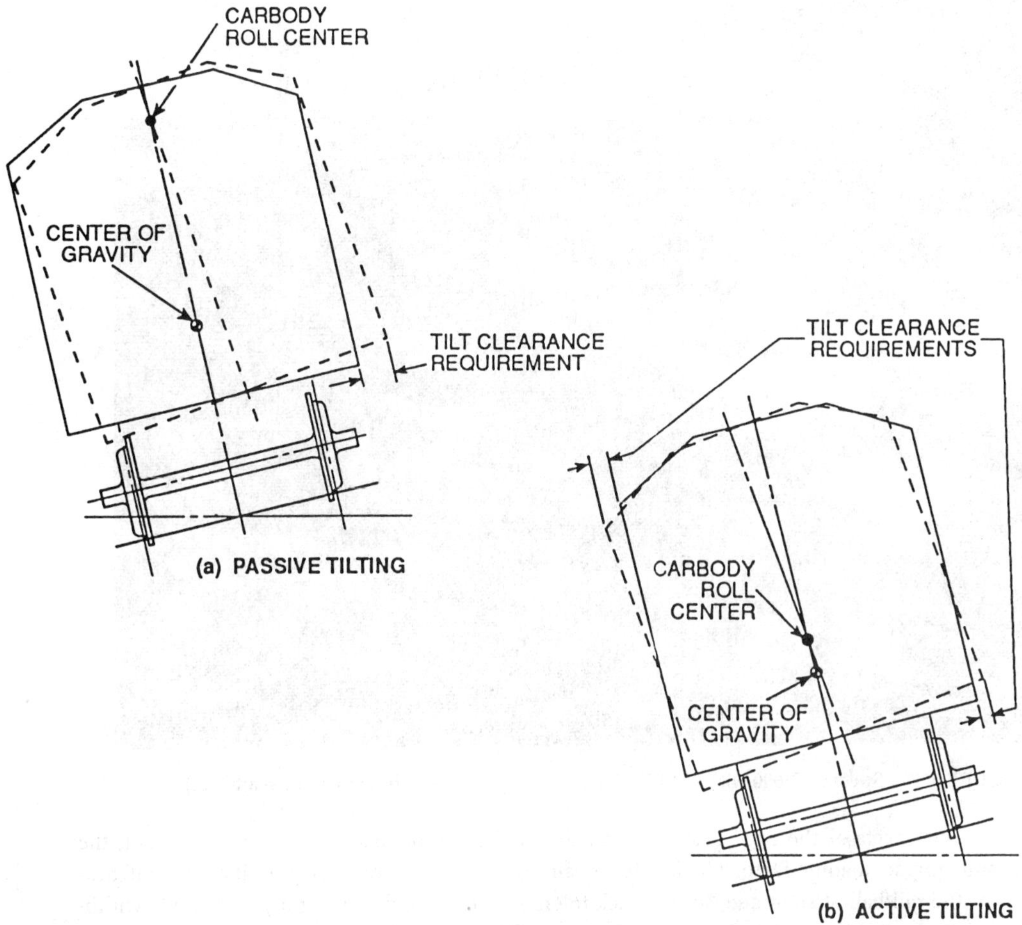

FIGURE 57.6 Location of carbody roll center and center of gravity for passive and active body tilting. (*Source:* Federal Railroad Administration.)

Train-Track Dynamics

The design of track components, special trackwork, ballast, subballast, subgrade, and acceptance of soils is controlled by the dynamic loading associated with track irregularities. Figure 57.8 shows the track–vehicle system for dynamic analysis of forces. The approach used to analyze the total vertical dynamic effects has been expressed as a function of the static loading. The total vertical dynamic impact is expressed by the coefficient I_t as indicated in Eq. (57.5). The value of the coefficient is given with respect to the impact of the vehicle design and the impact of the track design.[21]

$$I_t = f(I_v, I_s) \tag{57.5}$$

where

I_t = dynamic impact loading factor
I_s = factor for track stiffness
I_v = impact factor for the vehicle design

In the early days of railroads, the *vehicle component* dominated the combined dynamic impacts, forcing designers to focus more on improving vehicle design rather than track irregularities. The relatively one-sided improvement effort for vehicle design over track design through the years has

FIGURE 57.7 Swedish X2000 with AMTRAK logo touring the U.S. (Photo by R. K. Whitford.)

shifted the effects of the two coefficients on the total dynamic loading impact. At present, the dynamic impact loading factor (I_t) is affected almost entirely by track irregularities and stiffness (I_s) of the rail/ballast/subgrade, so the track tolerances are specified as tightly as possible within financially feasible limits.[24] Controlling the impact of dynamic loading is essential in HSR systems, and requires mainly uniformity of subgrade.

57.6 HSR Examples from Abroad

In the U.S., HSGT systems are still in the planning stages. However, in many other countries, HSGT has been in successful revenue service for several decades. The Japanese recognized their high-speed train's 30th anniversary in 1990. The train has captured around 80% of the travel market between Tokyo and Osaka (320 miles apart). The first line was built on a completely grade-separated and exclusive ROW. For the needs of the route, the Japanese built 66 tunnels and more than 3100 bridges. The original route was so successful that Japan now has almost 1300 miles of high-speed passenger lines. The Japanese system serves around 400,000 passengers a day and has an on-time arrival record of 99%.[4]

The TGV system has revolutionized passenger rail service in France and has spread throughout Europe with five more countries currently planning to use its technology. Besides France, Spain has been using the technology since 1992 under the name AVE (Alta Velosidad Español). Great Britain will use it as Eurostar, the train that will connect London with Paris and Brussels through the tunnel under the English Channel. Besides Belgium, Holland also plans to become part of the network on the PBKA (Paris–Brussels–Cologne–Amsterdam) route. This HSGT route will be fully operational by the year 2000. It should be noted that one of the main reasons that HSGT systems are so successful in Europe is that at speeds of 170 mph they are very competitive with air transport for the range of distances (usually less than 500 mi) common between European cities.

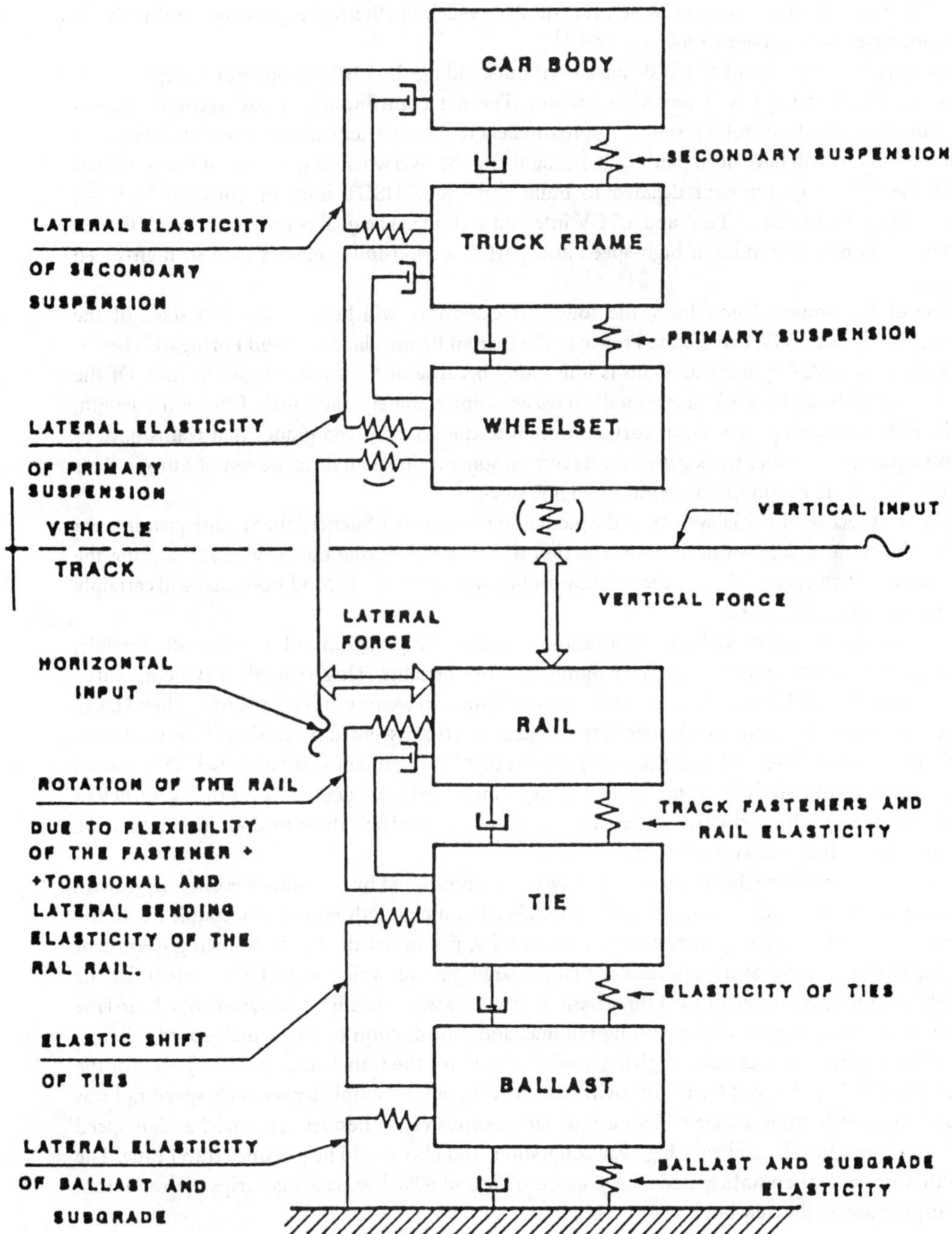

FIGURE 57.8 Track-train system for dynamic analysis of forces. (*Source:* U.S. Government Accounting Office.)

The catalyst for the development of HSGT systems in Europe was the resounding success this system had in France. In September 1981, the southeast TGV began operating. The length of the line was 259 mi, and has turned out to be both an operating and a financial success. From an operational standpoint, the TGV had the desired results. Trends in air service from Paris to southeast France declined, while there was no traffic growth on the Paris–Lyons expressway (traffic on the rest of the French expressways grew at a rate of 5% a year). Furthermore, passenger traffic in the TGV increased 90% between 1980 and 1990.[41]

Financially, the results of the TGV have been outstanding. In 1990 the operating surplus (revenue minus operating costs) was $503 million. The net contribution to cost recovery totaled $313 million, or 38% of total revenue. Approximately ten years after full operation of the line, its cost ($3 billion) will have been paid back. Recognizing the overwhelming success of the southeast TGV, the French government decided to build three new HSGT lines (a northern TGV, an extension of the southeast TGV, and a TGV interconnection). The goal is that by 1995 the French Railways will have 782 miles of high-speed line, part of a 3450-mile network used by high-speed trains.[41]

One of the proposed new lines, the southeast extension, will become the first stage of the "South-European" TGV, connecting France to the Iberian Peninsula (Spain and Portugal). The development of HSGT systems in Spain is interesting because of the mountainous terrain. Of the current Spanish rail network, horizontal curves are approximately one-third of the entire length, while 80% is made up by vertical curves. Moreover, due to the steep grades that trains have to climb, Spain uses a wider track gauge for its conventional trains than does the rest of Europe. This permits achieving maximum power of its engine stock.[42]

The great success of HSGT systems in the pioneering countries influenced the Spanish government in the direction of supporting the construction of a Madrid–Seville line as well as adopting the international track gauge, despite the severely problematic geographic conditions that will certainly increase construction costs.

VIA rail, Canada's national rail corporation, found that high-speed rail is technically feasible, financially attractive, and can result in significant user benefits. Three corridors (Quebec City–Windsor, Montreal–Ottawa–Toronto, and Toronto–Windsor) were examined regarding their HSGT system potential, but only the Quebec City–Windsor corridor seemed promising.[43] This corridor is about 700 miles long and contains a population of 15 million, approximately half of Canada's population. In this study VIA indicated that high-speed rail can succeed in capturing sufficient ridership over medium distances (250 to 350 miles) if it offers door-to-door times that are comparable with that of air transport.

There are cases where the need for HSGT systems is prompted by a country's economic growth. This is the case in South Korea, where the annual economic growth rate of 8%, coupled with the use of the existing highway infrastructure at capacity, prompted the South Korean government to implement a national transportation plan, a large part of which was HSGT system. In the feasibility study performed in 1984 three basic alternatives were examined: provision of a four-lane expressway, improvement of the existing rail line, and introduction of a new high-speed rail line. An efficiency index was calculated giving a value of 3.37 for the four-lane expressway, 4.87 for the high-speed rail, and only 1.0 for improving the existing rail. In simple terms, high-speed rail was found to be 45% more efficient than a four-lane expressway.[44] Besides efficiency, a high-speed rail system would help deal with highway congestion, and also would help reduce travel time. The South Korean planners anticipated 33,000 automobile and 8000 bus passenger trips to divert from the expressway to the new line.

57.7 Magnetic Levitation

Using magnetic levitation for suspension and propelling by means of electric fields is one technology that has been considered as a replacement for the conventional steel-wheel-on-steel-rail technology. The technology is referred to as MAGLEV.[45–49] Without friction, higher speeds are possible, but the system requires a specially designed guideway, often elevated. The "father of electromagnetic levitation," Herman Kemper, began his research on the subject in 1922, with a basic patent granted in 1934. The patent was proof of magnetic levitation and resulted in a model that could carry a load of 450 pounds. The research on magnetic levitation and its application to passenger transport has come a long way since, with the German government initiating an in-depth examination of the feasibility, safety, and planning issues of such a system in the 1970s.[45,48,49]

The German study pointed out that MAGLEV technology could be very successful, in terms of passenger traffic, for medium- and long-distance routes. Planners have calculated that the construction cost of a MAGLEV system would be about 30% higher than that of the steel-wheel-on-steel-rail system. Germany is taking a careful look at the prospect of initiating a MAGLEV line at the Hamburg–Berlin route, to accommodate the considerable increase in passenger traffic after the 1991 unification of Germany. It is worth noting that the interest from the industry in MAGLEV technology was such that private capital will be incorporated into the public infrastructure through the construction of the MAGLEV line.[46]

MAGLEV technology differs significantly from that of steel-wheel-on-steel-rail. The Transrapid (the name of the German MAGLEV system) vehicles are magnetically levitated and guided within a guideway (Fig. 57.9). They are propelled by synchronous linear motors along a guideway. Levitation forces are generated by magnets on the undercarriage, or levitation frame, below the guideway beam. Guidance magnets are also mounted on the undercarriage, but face the outer edges of the guideway, thus keeping the vehicle aligned with the guideway.[46] Each car has a series of levitation frames which align magnets and carry the levitation and guidance forces to the car body through pneumatic springs and links. When the levitation magnets are energized, the vehicle is lifted toward the guideway.[47] The magnets are energized with power provided by onboard batteries. The batteries are charged when the vehicle is moving at speeds greater than 75 mph.

Guideway structures are made up of steel or prestressed concrete beams. They can be elevated on piers up to 65 ft, or they can be elevated up to 130 ft with the use of special structures. Ground-level guideway is used in tunnels or in areas where an elevated guideway is undesirable. The switches, used to divert vehicles to different branch lines, are steel bending beams aligned elastically by a series of electromechanical or hydraulic actuators. This type of construction allows for a smooth ride during the switch, while switches with radii of 7500 ft allow for speeds of 125 mph in the branching position.

The accuracy of the guideway in the high speeds that the MAGLEV vehicles reach is extremely important. Accuracy and stability are achieved with the use of automated production techniques and construction of the piers and the foundations to appropriate specifications. The system is operated automatically, and its monitoring is achieved with the use of fiber-optic wave transmission between the vehicle and a central control center.

The MAGLEV system is not directly damaging to the environment, since it is electrically propelled and emits no pollutants. Any pollution caused by MAGLEV technology is the indirect effect of the power stations supplying the electric energy necessary for the system. The electromagnetic system necessary for propulsion, with magnets located beneath the guideway, results in minor magnetic fields in the vehicles or in the vicinity of the vehicles (10 to 30 mGauss at sea level).[46,47]

MAGLEV vehicles reach speeds of 250 to 300 mph, making safety a critical consideration.[49–52] The MAGLEV vehicles "wrap around" the guideway, essentially eliminating the danger of derailment. The automated operation and control minimize possible driver error. Furthermore, the fully separated guideway provides a natural barrier, preventing most types of conflict. Vehicle interior is designed to meet the fire protection standards set by the 1988 Air Transport Standards Act.[49] Because MAGLEV trains are elevated and there are no outside connections with the guideway, they are relatively quiet.[53,54]

Several sites in the U.S. have been examined for the possible implementation of a MAGLEV system, yet only the successful operation of a revenue system will prompt further development. On June 12, 1991, the state of Florida certified the Orlando-based Maglev Transit, Inc., to construct and operate the first commercial system based on MAGLEV technology in the U.S. The 14-mile-long system is to provide access from the airport to the Orlando tourist district. It is anticipated that the system will carry approximately 8 million passengers per year, at a top speed of 250 mph. It is also planned that arriving baggage will be checked through to the terminal at the tourist district.[45]

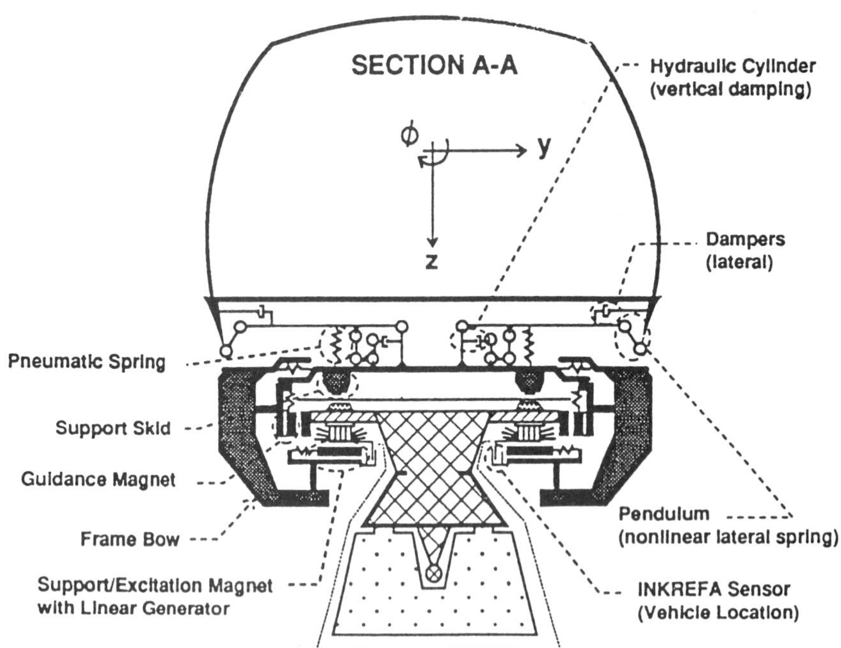

FIGURE 57.9 Transrapid TR-07 MAGLEV train. (*Source:* Galganski, R. A. *Safety of Highway Speed Guided Ground Transportation Systems—Collision Avoidance and Accident Survivability—Volume III: Accident Survivability*, DOT/FRA/ORD-93/02.III, March 1993.)

MAGLEV systems are convenient and attractive for development, in that they offer a comfortable, environmentally safe, and fast mode of transport. On the other hand, both the supporting structures and the equipment needed for electromagnetic levitation and propulsion are expensive, and along with the continuing concerns for safety make the use of such technology problematic.

Conclusion

Plans to introduce HSGT systems have been proposed for several corridors in the U.S. (Fig. 57.1). Two key issues that must be addressed before any project moves ahead are the adequacy of the cost-benefit analysis and identifying mechanisms to adequately fund the high-speed system.[55,56,57]

Questions often asked in terms of the adequacy of the cost-benefit analysis include the following:

- Is the HSGT system feasible from both the engineering and the socioeconomic perspective?
- Where will the market share for HSGT systems come from? Will travelers shift from existing modes of travel—automobile, mass transit, and aviation? How much? How often? What will be the impacts of this shift in mode of travel?
- Will the new system provide benefits that cannot be obtained from the existing infrastructure? Will the attracted ridership optimize revenue generation adequately so that operating expenses can be covered in the short run, and ultimately provide for system profitability?
- Can the system obtain the necessary approvals and permits from the regulatory and other agencies in a timely manner? Table 57.6 summarizes the major approval requirements before the HSR system can begin construction.[58] A portion of the process for new corridors will involve solutions to the severance problem.

The primary issue that must be resolved before HSGT systems can be developed in the U.S. is project funding.[9,58] The past policy has relied heavily on the private sector or user fees to develop many of the transportation systems in the U.S., yet little private investment is likely for HSGT without substantial federal government support. Private investors may fear that passenger demand will be overestimated and the fares collected will not cover costs. There is also the concern that other emerging technologies, such as tilt-rotor aircraft and video conferencing, may compete successfully with other forms of intercity travel, including HSR. The failure of the Texas High-Speed

Table 57.6 Major Approval Requirements for HSGT Systems in the U.S.

Level of Government Approvals	Conceptual Approvals	Detailed Approvals
Federal	Environmental assessment	Joint use of highways
	Section 4(f) recreation impact statement	Dredge and fill/(404) wetland permit(s)
	Environmental impact statement	Coast guard permit(s)
		Other federal permit(s)
State	Franchise certification	Dredge and fill
		Stormwater discharge
	Environmental certification	Water quality certification
		Historic and archaeological
	Financing package	Other federal permit(s)
		Financing package refinement
Local	Local government comprehensive plans	Noise ordinances
	Environmental or community impact statements	Zoning
		Other local approvals

Source: Mudd, C. B. Local, state and federal approvals in developing high-speed rail systems in the U.S. *Proceedings of the First International Conference on High-Speed Ground Transportation Systems I,* 1992, pp. 648–656.

Rail Authority to raise the necessary funds to continue the Dallas/Houston/San Antonio work is testimony to the critical nature of funding.

It is unlikely that the private sector will assume a major share of the risks associated with financing HSGT development. For HSGT systems to be developed in the U.S., the federal government will have to assume a substantial portion of the risk, making private investment through some sort of partnership arrangement possible.

With over 30 years of experience in France, Japan, and Germany, HSGT is a mature, useful technology that could meet a number of U.S. needs. There is substantial interest in HSGT systems because they provide a cost-effective means of intercity travel endorsed and widely accepted by passengers (as proven in the many countries where HSGT systems have been operating for many years), because they are an environmentally safe alternative, and because they can be a safe alternative to the capacity limitation and gridlock that surround the nation's busiest airports.

The vast area of the U.S. makes it unlikely that the HSGT systems will have as much success as the ones developed in Europe and Japan, unless a major and lasting oil crisis decreases automobile use to a significant extent. However, HSGT systems can efficiently and successfully serve corridors between markets that are located 200 to 500 km apart. These cities reflect long travel times for automobiles and are too close for anything more than commuter air travel.

HSGT will succeed only if the U.S. embraces technology developed and used by other countries and is willing to spend the money for the sizable investment required to create the infrastructure. It is one viable transportation mode that fits squarely in the unique market of intercity travel.

Defining Terms

Many abbreviations are in common use for railroad organizations and high-speed rail systems and their components. Note that some abbreviations, particularly those used for different train control systems (ATC, ATCS, ATP, etc.), may not have the same meaning for all users. Commonly accepted meanings are given.

AAR: Association of American Railroads

ASTREE: Automatization du Suivi en Temps (French onboard train control system).

ATC: Automatic train control. Systems that provide for automatic initiation of braking and/or other train control functions. ATP and ATO are subsystems of ATC.

ATO: Automatic train operation. A system of automatic control of train movements from start to stop. Customarily applied to rail rapid transit operations.

ATCS: Advanced train control systems. A specific project of the AAR to develop train control systems with enhanced capabilities.

ATP: Automatic train protection. Usually a comprehensive system of automatic supervision of train operator actions. Will initiate braking if speed limits or signal indications are not obeyed. All ATP systems are also ATC systems.

AWS: Automatic warning system. A simple cab signaling and ATC system used on British Rail.

BART: Bay Area Rapid Transit (San Francisco).

BN: Burlington Northern Railroad.

BR: British Rail.

CFR: Code of Federal Regulations.

CPU: Central processing unit (core unit of a microprocessor).

DB: Deutsche Bundesbahn—German Federal Railways.

DIN: Deutsche Institut for Normung—German National Standards Institute.

DLR: Docklands Light Railway, London, U.K.

EMI: Electromagnetic interference. Usually used in connection with the interference with signal control circuits caused by high-power electric traction systems.

FCC: Federal Communications Commission (U.S.).

FRA: Federal Railroad Administration of the U.S. Department of Transportation.

FTA: Federal Transit Administration.

HSGT: High-speed ground transportation.

HSR: High-speed rail.

HST: High-speed train—British Rail high-speed diesel-electric trainset.

ICE: Intercity Express. A high-speed trainset developed for German Federal Railways consisting of a locomotive at each end and approximately ten intermediate passenger cars.

ISO: International Standards Organization.

Intermittent: A term used in connection with ATC and ATO systems to describe a system that transmits instructions from track to train at discrete points rather than continuously.

JNR: Japanese National Railways. Organization formerly responsible for rail services in Japan, reorganized as the Japan Railways (JR) Group on April 1, 1987, comprising several regional railways, a freight business, and a Shinkansen holding company.

JR: Japan Railways—see JNR.

LCX: Leakage coaxial cables. LCX cables laid along a guideway can provide high-quality radio transmission between the vehicle and wayside. LCX is more reliable than airwave radio and can be used where airwaves cannot, for example, in tunnels.

LGV: Ligne a Grand Vitesse—newly built French high-speed lines. See also TGV.

LRC: Light, rapid, and comfortable. A high-speed tilt-body diesel-electric trainset developed in Canada.

LZB: Linienzugbeeinflussung. Comprehensive system of train control and automatic train protection developed by German Federal Railways.

MU: Multiple unit. A train on which all or most passenger cars are individually powered and no separate locomotive is used.

MARTA: Metropolitan Atlanta Rapid Transit Authority.

NBS: Neubaustrecken—newly built German Federal Railway high-speed lines.

NTSB: National Transportation Safety Board (U.S.).

PATCO: Port Authority Transit Corporation (Lindenwold line).

PSE: Paris Sud-Est. The high-speed line from Paris to Lyons on French National Railways.

RENFE: Rede Nacional de los Ferrocarriles Españoles—Spanish National Railways.

SBB: Schweizerische Bundesbahnen—Swiss Federal Railways.

SELTRAC: Moving-block signaling system developed by Alcatel, Canada.

SSI: Solid state interlocking.

SJ: Statens Jarnvagar—Swedish State Railways.

SNCF: Societe Nationale des Chemin de Fer Francais—French National Railways.

TGV: Train a Grand Vitesse—French High-Speed Train. Also used to refer to complete French high-speed train system.

UIC: Union Internationale de Chemins de Fer.

UMTA: Urban Mass Transportation Administration of the U.S. Department of Transportation. The name of this agency has now been changed to the Federal Transit Administration (FTA).

VNTSC: Volpe National Transportation Systems Center.

WMATA: Washington Metropolitan Area Transit Authority.

References

1. Texas TGV. *Franchise Application.* Chapter 5, "Description of Proposed Technology." Submitted to The Texas High-Speed Rail Authority, 1991.
2. Texas FasTrac, Inc. *Franchise Application.* Chapter 5, "Description of Proposed Technology." Submitted to The Texas High-Speed Rail Authority, 1991.
3. U.S. General Accounting Office, Report to the Chairman, Committee on Energy and Commerce. *High-Speed Ground Transportation—Issues Affecting Development in the United States.* House of Representatives, Nov. 1993.

4. Vranich, J. *Super-Trains: Solutions to America's Transportation Gridlock.* St. Martin's Press, New York, 1991.
5. Bachman, J. A. HSR vehicle performance characteristics. *Journal of Transportation Engineering,* 115:48–56, 1989.
6. TMS/Benesch. *Tri-State Study of High Speed Rail Service: Chicago–Milwaukee–Twin Cities Corridor.* For Illinois, Minnesota, and Wisconsin Departments of Transportation, 1991.
7. Canadian Institute of Guided Ground Transport, GEC-Alsthom/SNCF TGV Baseline, Draft Report, Queen's University, 1992.
8. Parsons Brinckerhoff Quade & Quade, Inc. *High-Speed Surface Transportation Cost Estimate Report.* TRB, April 1991.
9. Nassar, F. E. and Najafi, F. T. Development of simulation and cost models to compare HSR and MAGLEV systems. *Proceedings of the First International Conference on High-Speed Ground Transportation Systems I,* 1992, pp. 336–346, ASCE, New York.
10. Silien, J. S. The technology of X2000—ABB's high-speed tilting trains. *Proceedings of the First International Conference on High-Speed Ground Transportation Systems I,* 1992, pp. 735–743, ASCE, New York.
11. *Tilt Train Technology: A State of the Art Survey,* DOT/FRA-92/05, 1992.
12. Boon, C., *et al. High Speed Rail Tilt Train Technology—A State of the Art Survey,* DOT/FRA/ORD-92/02, May 1992.
13. Federal Railroad Administration. Safety relevant observations on the X2000 tilting train. *Moving America—New Directions, New Opportunities,* DOT/FRA/ORD-90-14.
14. Federal Railroad Administration. Safety relevant observations on the ICE high speed train. *Moving America—New Directions, New Opportunities,* DOT/FRA/ORD-90-04.
15. Federal Railroad Administration. Safety relevant observations on the TGV high speed train. *Moving America—New Directions, New Opportunities,* DOT/FRA/ORD-91-03, July 1991.
16. Argonne National Laboratory. *Safety of High Speed Guided Ground Transportation Systems, Estimate Report,* April 1991.
17. Bing, A. J. *Safety of Highway Speed Guided Ground Transportation Systems—Collision Avoidance and Accident Survivability—Volume I: Collision Threat,* DOT/FRA/ORD-93/02.I, March 1993.
18. Harrison, J., Bing, A. J., and Galganski, R. A. *Safety of Highway Speed Guided Ground Transportation Systems—Collision Avoidance and Accident Survivability—Volume II: Collision Avoidance,* DOT/FRA/ORD-93/02.II, March 1993.
19. Galganski, R. A. *Safety of Highway Speed Guided Ground Transportation Systems—Collision Avoidance and Accident Survivability—Volume III: Accident Survivability,* DOT/FRA/ORD-93/02.III, March 1993.
20. Bing, A. J. *Safety of Highway Speed Guided Ground Transportation Systems—Collision Avoidance and Accident Survivability—Volume IV: Proposed Specifications,* DOT/FRA/ORD-93/02.IV, March 1993.
21. Zicha, J. H. High-speed rail track design. *Journal of Transportation Engineering,* 115:68–73, 1989.
22. Little, A. D. and Brinckerhoff, P. *Safety of High Speed Rail Transportation Systems, Passenger Train and Freight Railroad Corridors.*
23. Hadden, J., Reich, S., and Bessoir, T. *Safety of High Speed Guided Ground Transportation Systems—Shared Right-of-Way Safety Issues,* DOT/FRA/ORD-92/13, September 1992.
24. Zicha, J. H. High-speed rail loading scenario and dynamic tuning of track. *Proceedings of the First International Conference on High-Speed Ground Transportation Systems I,* 1992, pp. 470–480, ACSE, New York.
25. Jenkins, H. Track maintenance for high-speed trains. *Bull. AREA,* latest publication.
26. Markos, S. H. *Safety of High Speed Guided Ground Transportation Systems, Emergency Preparedness Guidelines,* DOT/FRA/ORD-93/24, December 1993.

27. National Institute of Standards and Technology (NIST). *Fire Safety of Passenger Trains—A Review of U.S. and Foreign Approaches,* DOT/FRA/ORD-93/23, December 1993.
28. Reich, S. and Bessoir, T. *Safety of Vital Control and Communication Systems in Guided Ground Transportation—Analysis of Railroad Signaling System: Microprocessor Interlocking,* DOT/FRA-ORD-93/08, May 1993.
29. Leudeke, J. F. *Safety of High Speed Guided Ground Transportation Systems, Safety Verification/Validation Methodologies for Vital Computer Systems,* Battelle Institute, Columbus, OH.
30. MIT. *Safety of High Speed Guided Ground Transportation Systems, Human Factors and Automation,* Oct. 1994, DOT/FRA/ORD 94/24, Cambridge, MA.
31. FRA/Volpe Center. *High-speed Ground Transportation Bibliography.* Volpe National Transportation Systems Center, Cambridge, MA, August 1994.
32. Wayson, R. L. and Bowlby, W. Noise and air pollution of high-speed rail systems. *ASCE Journal of Transportation Engineering,* 115:20–36, 1989.
33. Taille, J. Y. The TGV network and the environment. *Proceedings of the First International Conference on High-Speed Ground Transportation Systems I,* 1992, pp. 136–145, ASCE, New York.
34. Hall, M. S. and Wayson, R. L. A combined model for HSGT traffic. *Proceedings of the First International Conference on High-Speed Ground Transportation Systems I,* 1992, pp. 176–188, ASCE, New York.
35. Isbell, T. S. Concurrent engineering planning in HSGT systems. *Proceedings of the First International Conference on High-Speed Ground Transportation Systems I,* 1992, pp. 457–467, ASCE, New York.
36. *AREA manual for railway engineering.* AREA, 1992, Washington, D.C.
37. Hay, W. W. *Railroad Engineering,* 2d ed. John Wiley & Sons, New York, 1982.
38. Wright, P. H. and Ashford, N. J. *Transportation Engineering Planning and Design,* 3d ed. John Wiley & Sons, New York, 1989.
39. Wakui, H. and Matsumoto, N. Dynamic study on new guideway structure for JR Maglev. *Proceedings of the First International Conference on High-Speed Ground Transportation Systems I,* 1992, pp. 487–496, ASCE, New York.
40. Hardgrove, M. S. and Mason, J. Estimating the ground transportation impacts of the MAGLEV system. *Proceedings of the First International Conference on High-Speed Ground Transportation Systems I,* 1992, pp. 199–208, ASCE, New York.
41. Mathieu, G. The French master plan for high-speed rail services. *Proceedings of the First International Conference on High-Speed Ground Transportation Systems I,* 1992, pp. 687–696, ASCE, New York.
42. Losada, M. The singularity of Spanish railway high speed. *Proceedings of the First International Conference on High-Speed Ground Transportation Systems I,* 1992, pp. 667–676, ASCE, New York.
43. Matyas, G. High-speed rail's prospects in Canada. *Proceedings of the First International Conference on High-Speed Ground Transportation Systems I,* 1992, pp. 677–686, ASCE, New York.
44. Cha, D. D. and Suh, S. Planning for the national high-speed rail network in Korea. *Proceedings of the First International Conference on High-Speed Ground Transportation Systems I,* 1992, pp. 697–706, ASCE, New York.
45. Witt, M. H. Application of the MAGLEV system in Germany. *Proceedings of the First International Conference on High-Speed Ground Transportation Systems I,* 1992, pp. 707–722, ASCE, New York.
46. Wackers, M. The Transrapid MAGLEV system. *Proceedings of the First International Conference on High-Speed Ground Transportation Systems I,* 1992, pp. 724–734, ASCE, New York.
47. Dickhart, W. W. and Pavlick, M. MAGLEV sky train. *Mechanical Engineering,* 106(1), 1984.
48. Sara, C. M. Plan for development and implementation of a domestic MAGLEV network. *Proceedings of the First International Conference on High-Speed Ground Transportation Systems I,* 1992, pp. 638–647, ASCE, New York.

49. Dorer, R. M. and Hathaway, W. T. *Safety of High Speed Magnetic Levitation Transportation Systems—Preliminary Safety Review of the Transrapid Maglev System,* DOT/FRA/ORD-90/09, May 1991.
50. Dorer, R. M. and Hathaway, W. T. *Safety of High Speed Magnetic Levitation Transportation Systems—German High-Speed MAGLEV Train Safety Requirements—Potential for Application in the United States,* Department of Transportation DOT/FRA/ORD-92/02, February 1992, Washington, D.C.
51. Arthur D. Little, Inc. *A Comparison of U.S. and Foreign Safety Regulations for Potential Application to MAGLEV Systems,* DOT/FRA/ORD-93/21, September 1993.
52. RW MSB Working Group. *Safety of High Speed Magnetic Levitation Transportation Systems—High-Speed MAGLEV Trains: German Safety Requirements RW-MSB,* DOT/FRA/ORD-92/01, January 1992.
53. ASCE Subcommittee on High-Speed Rail Systems. High-speed rail systems in the United States. *ASCE Journal of Transportation Engineering,* 111:79–94, 1985.
54. Hanson, C. E. Noise from high-speed MAGLEV transportation systems. *Proceedings of the First International Conference on High-Speed Ground Transportation Systems I,* 1992, pp. 146–155, ASCE, New York.
55. Brand, N. M. and Lucas, M. M. Operating and maintenance costs of the TGV high-speed rail system. *ASCE Journal of Transportation Engineering,* 115:37–47, 1989.
56. Pintag, G. Capital cost and operations of high-speed rail system in West Germany. *Journal of Transportation Engineering,* 115:57–67, 1989.
57. Harrison, J. A. High-speed surface transportation cost estimating. *Proceedings of the First International Conference on High-Speed Ground Transportation Systems I,* 1992, pp. 314–325, ASCE, New York.
58. Mudd, C. B. Local, state and federal approvals in developing high-speed rail systems in the U.S. *Proceedings of the First International Conference on High-Speed Ground Transportation Systems I,* 1992, pp. 648–656, ASCE, New York.

58

Urban Transit

Peter G. Furth
Northeastern University

58.1 Transit Modes

The principal transit modes are bus, light rail, and metro (heavy rail).

Bus

Bus is the most common transit mode, operating in every urban area in the U.S. Nearly all transit coaches in the U.S. are powered by diesel engines, although experimentation with alternative fuels has grown since the enactment of the Clean Air Act of 1990. The standard 40-ft coach can seat 40–55 passengers, depending on seating configuration. Smaller coaches are common in settings of lower passenger demand. Large, articulated coaches, seating 60–75, are common in some cities on high-volume routes.

0-8493-8953-4/95/$0.00 + $.50

Table 58.1 Priority Schemes for Bus

a. On freeways [often shared with other high occupancy vehicles (HOVs)]
   - Median HOV roadways
   - With-flow HOV lanes
   - Contraflow HOV lanes

b. On arterials and downtown streets
   - Bus lane (curb lane)
   - Express bus lane (inside lane—no stops)
   - Contraflow lane on one-way street
   - Bus-only street
   - Exemption from turning restrictions
   - Priority merge when departing from bus stop

c. At traffic signals, toll booths, ramp meters, and other bottlenecks
   - Timing signals to favor buses' progression
   - Signal preemption
   - Queue bypass lanes

The bus mode uses the existing road network. Therefore, its two greatest advantages are its low capital cost and its ability to access transit demand anywhere. Sharing the road with general traffic is also the bus's main weakness: Buses suffer traffic delays, and ride quality often suffers due to poor pavement quality. Priority schemes, such as those listed in Table 58.1, can be used to reduce traffic delays. The ultimate priority scheme is a *busway,* a bus-only roadway with grade separation that gives buses comparable level of service to that of rail lines [Bonsall, 1987]. Cities with busways include Pittsburgh, Ottawa, and Adelaide, Australia.

Light Rail

Streetcar was once the dominant transit mode, operating on tracks laid in city streets. Their replacement by buses, beginning around 1930 and largely completed by 1960, was due in part to the lower capital cost of bus systems and in part to the streetcar's inflexibility in mixed traffic. For example, to avoid being blocked by parked vehicles, tracks were usually laid in the inside lane of a multilane street, forcing passengers to board and alight in the middle of the street instead of at the curb. Few such streetcar operations remain in North America. Still surviving, and growing in number, are systems operating primarily on their own right-of-way, sometimes grade separated.

Light rail's main advantage over bus is its economy in carrying high passenger volumes, since rail cars are bigger and can be joined into trains. The economy of a single operator staffing a multicar train requires self-service fare collection, usually entailing ticket vending machines on platforms, validators (ticket-canceling machines) on platforms or on vehicles, and random fare inspection. Other advantages are that light rail produces no fumes, offers a higher-quality ride, and takes less space, which substantially lowers tunneling cost. The main disadvantage of light rail is the need for its own right-of-way, which can be compromised in small sections (e.g., a downtown transit mall) and at grade crossings, with a corresponding loss of speed. Given the right-of-way, another disadvantage of light rail versus bus is the large number of passengers who must transfer between rail and feeder bus, in contrast to a busway used as a trunk from which routes branch off covering a wide area.

Metro

Metro, or heavy-rail systems, are high-cost, high-capacity systems operating in an exclusive, grade-separated right-of-way, often in subway, but often elevated or at grade. Floors flush with the platforms and wide doors make for rapid boarding and alighting and easy accessibility for disabled persons. Modern systems feature automatic or nearly automatic control, allowing for small headways and more reliable operation.

The distinction between light rail and heavy rail is becoming blurred in intermediate systems such as those in Lille, France and Vancouver, Canada. They use small vehicles characteristic of light rail, but have high platform loading and automatic control characteristic of metro systems.

Other Modes

Trolleybus coaches resemble diesel coaches, except that they are powered by electric motors, drawing electric current from overhead wires. They are common in Europe and operate in several North American cities. Their virtually unlimited power enables them to accelerate more quickly and climb steep hills more easily than diesel buses, and because they need no transmission, their ride is smoother. Fumes are eliminated, improving the environment, and making it easier for them to operate in tunnels. The overhead wires, however, are sometimes seen as a detriment to the environment. Trolleybuses are not as flexible as diesel buses—they must be replaced by diesel buses when there is a detour, for example—but can maneuver around a blocked lane, making them better suited to mixed traffic than light rail. The cost of the power line network generally limits them to high-volume routes since the benefits of electrification are proportional to the number of passengers and vehicles on the route.

A more recent bus variation is the self-steering bus, guided by contact between a raised curb and small guide wheel that extends from the side of the bus. It allows a bus to operate in a smaller lane, lowering construction costs on elevated busways and in tunnels.

Commuter rail has long been used in older U.S. cities for long-distance commuting. Extensions are being built, and new systems have recently been opened in metropolitan Washington, Los Angeles, and in south Florida. They use standard locomotives (electric or diesel) pulling passenger cars on rail lines often shared with freight or long-distance passenger traffic. More than any other mode, it relies on auto access at the suburban end of the trip, and usually relies on metro for distribution at the downtown end. The design and operation of downtown terminals where many lines converge can be quite involved.

Numerous other transit modes operate in different cities. They include ferry boat, cable car, incline, and, more recently, downtown people mover.

Line Capacity

The line capacity of a transit line is the number of passengers that can be carried per hour in one direction past any point. Line capacity is often a major consideration of choice of mode for a new transit line. If there is no shortage of vehicles, the line capacity is constrained only by the headway (time interval) between vehicles:

$$\begin{aligned}\text{Line capacity}_{(\text{pass/h})} &= \text{Vehicle capacity}_{(\text{pass/veh})} \times \text{Train size}_{(\text{veh})} \\ &\quad \times 3600/\text{Minimum headway}_{(\text{s})}\end{aligned} \tag{58.1}$$

Vehicle capacity includes standees, according to the level of crowding deemed acceptable. Minimum headway is a function of safety and is governed primarily by vehicle interference at stations. For example, a metro with 6-car trains that can fit 240 people per car, if operating at a 4-min headway, can carry 21,600 pass/h in each direction; at a 2-min headway, line capacity is doubled. Typical line capacities for various modes are given in Table 58.2. By contrast, a freeway lane with headways of about 1.8 s and average occupancy of 1.2 persons carries only 2400 people per hour. If that lane were converted to a high-occupancy vehicle lane, with an average occupancy of 5 (95% carpools, 5% buses), then even with half the vehicular volume its passenger volume would more than double, and its line capacity would almost quadruple, since the minimum headway would only barely increase.

Table 58.2 Typical Line Capacities

Mode	Train Size	Minimum Headway	Occupancy	Line Capacity
Auto on freeway	1	1.8 s	1.2	2,400/h/lane
High-occupancy freeway lane (5% buses)	1	2 s	5	9,000/h
Bus-only freeway lane	1	4 s	40	36,000/h
Bus in arterial bus lane	1	30 s	65	7,800/h
Busway	1	20 s	60	10,800/h
Light-rail exclusive way	2	1 min	150	18,000/h
Metro	6	2 min	240	43,200/h
Commuter rail	10	10 min	275	13,750/h

Comparing Alternatives

Many studies have been done comparing transit alternatives. There is no clear consensus on the superiority or inferiority of any mode in a general sense. Any investment using federal funds requires an *alternatives analysis* that considers a no-build alternative (including low-cost transportation systems management improvements) and at least two different modes, with variations in alignment for each. Basic analysis is done using methods described in Chapter 55. Evaluation of alternatives is done following federal guidelines considering capital cost, operating cost, expected number of new passengers, benefits to existing passengers, and financing and political considerations [Zimmerman, 1989]. As a rule of thumb, construction cost for elevated exclusive guideway is 2 to 3 times greater than for at-grade, and subway is 4 to 10 times more costly than at-grade, creating a strong incentive to utilize existing rail rights-of-way and other alignments that avoid the need for tunneling or aerial construction.

58.2 The Transit Environment

Travel Patterns and Urban Form

Between about 1870 and 1940 streetcar (first horse drawn, later electric) was the primary mode of urban travel. Consequently, urban development during this period was oriented around streetcar use—dense development along radial streetcar lines, with a heavy concentration of commercial development in the central business district (CBD). Postwar development, in contrast, has been largely auto-oriented. At first homes, later stores, and finally employers became dispersed in large numbers in the suburbs. The travel patterns of streetcar-era development—many to one, or many to many along a linear corridor—lend themselves to the kind of demand concentration that transit can serve easily. The dispersed travel patterns of auto-oriented urban land use are far more difficult for transit to serve [Pushkarev and Zupan, 1977]. A transit system in an older city with a concentrated urban form faces far different problems than one serving a dispersed urban form. The former can face challenges such as how to carry the enormous demand; the main challenge in the latter is how to attract passengers.

Historically, as trip ends dispersed and income and auto ownership increased, transit ridership declined. "Captive" markets, primarily the carless poor and persons unable to drive, have shrunk. Within the "choice" market, discretionary trips (e.g., shopping trips that can be arranged to be done when an auto is available and where parking is free) were hardest hit. In cities with large downtown employment, parking is expensive, and the home-to-work commute market has held its own and in some cases grown. As a result, transit demand in large cities has become more and more peaked. The *peak-to-base ratio* (the number of buses in service during the A.M. peak period divided by the number in service during the base period) can be as great as 3:1. Passenger utilization is still more peaked because vehicles are more crowded during peak hours. This level of peaking hurts transit's economy because fixed facilities are underutilized and because of the costs inherent in starting and stopping service.

Transit Financing

Although transit first developed as a profitable private enterprise, inflation coupled with politically mandated caps on fares and competition from autos began to cripple the industry by the end of the First World War, leading to consolidation, disinvestment (e.g., abandoning streetcar lines, not replacing aging vehicles), public subsidy, and eventually public ownership. The chief reasons for transit's not being profitable are:

- *Low fare.* Political pressure has kept fares low for a variety of reasons, including accommodating poor riders and providing an alternative to autos which, transit proponents argue, are heavily subsidized.
- *Low demand and highly peaked demand.* These factors prevent transit from achieving economies of scale.
- *Social service.* It is politically mandated that service be offered on routes and at times of low demand.
- *High wages.* All large transit systems are unionized. When a large percentage of downtown workers use transit, the threat of a strike gives unions considerable bargaining power, which they have used effectively to negotiate high wages.
- *Restrictive work rules and various management problems.* These factors have also been blamed for transit's financial losses.

Capital costs are financed entirely by government subsidy. In the U.S. the Federal Transit Administration (FTA) usually covers 80%, with state and local government covering the remainder. For major construction projects, limited federal funds sometimes result in a smaller federal share. Public subsidies cover, as a national average, about 50% of operating costs in the U.S., although it varies a great deal from city to city. The federal contribution is small except in small cities; state and local government cover most of the operating deficit. Instability in the source of state and local funding is a cause of much uncertainty in management in many cities.

Transit Management

Ownership of public transit agencies is exercised through a board of directors whose members usually represent the local political constituencies (cities, counties, etc.) that subsidize it. Because the board members are political appointees, many have little knowledge about managing a transit system. And because they carry with them political views that sometimes conflict with one another, management can be politically charged, making the direction unstable. For example, urban board members may want to keep fares low, whereas suburban board members may be primarily concerned with reducing subsidies. Construction projects always entail significant political interest. Depending on how the balance of power changes, direction may change often.

The chief executive officer, usually called the *general manager*, is appointed by the board. The organization is usually divided functionally into departments such as transportation, maintenance, finance, administration, planning, real estate, and construction. Larger agencies may use a modal breakdown (bus, rail, etc.) as well, which may be under or above the functional division.

58.3 Fundamentals of Cyclical Operations

Most transit services are cyclical: Vehicles leave a depot, cycle over a given route, return, and then make another cycle. Routes that go back and forth between two terminals can still be considered cyclical, and either terminal can be considered the depot. This section looks at a route during a period of the day in which ridership and running times can be treated as constant.

Fundamental Operating Parameters and Relationships

Cycle time (c) is the time a vehicle uses to perform the cycle and wait for the next cycle. It consists of running time and layover (or recovery) time. On routes with two terminals, layover time is usually distributed between the two terminals. Its primary purpose is to serve as a buffer for run time delays, reducing the degree to which delays propagate from one cycle to the next. It also allows vehicle operators a rest. On short urban bus routes, layover is commonly 15 to 20% of running time; on longer routes or routes with less traffic congestion, 10% is typical. Layover may be further increased by schedule slack, as discussed later.

The service frequency (q) is the number of cycles per hour, the number of trips per hour passing a given point in a given direction. (*Trip* in transit terminology refers to vehicle trips, unless otherwise designated.) The reciprocal of frequency is headway (h), the time between successive trips.

If a route is operated in isolation, one can speak of the number of vehicles (n) operating on the route. In the rail context n is the number of trains. The fundamental relationship is

$$c = nh = n/q \tag{58.2a}$$

$$n = c/h = cq \tag{58.2b}$$

$$h = 1/q = c/n \tag{58.2c}$$

For example, to operate a route with a 40-min cycle at a 10-min headway (a frequency of 6/h) will require 4 vehicles. Or, given 6 vehicles and a 40-min cycle, the route can operate with a 6.67 min headway (a frequency of 9/h).

Fundamental Measures of Passenger Demand

An origin-destination (O-D) matrix—showing the number of passengers per hour traveling from one stop to another—is the fundamental descriptor of passenger demand on a route. (Further detail, such as fare category, usually does not matter for operations planning.) On routes operating between two terminals, it is best to divide the demand by direction, resulting in a triangular O-D matrix for each direction. The row and column totals represent the ons and offs (boardings and alightings) at each stop. The grand total is the boardings (b) on the route.

The volume profile is the passenger volume on each interstop segment on the route. On routes that empty out at a terminal, the volume profile is easily constructed stop by stop, beginning at the terminal, accumulating ons and deducting offs. On loop routes that do not empty out (e.g., a circumferential route), the volume on a segment is the sum of the demands in the O-D cells that involve travel over that segment. Once the volume on any one segment is calculated, the volume on the succeeding segments can be found by accumulating ons and deducting offs. The peak volume segment (also called *peak load point* or *peak point*) is the segment with the greatest volume; its volume is the peak volume (v^*). On bidirectional routes, the direction with the greater peak volume is the peak direction, and its peak volume is the route's peak volume. Another measure of demand is passenger-miles (or passenger-km), most easily calculated by multiplying the volume on each segment by the segment length and summing over all segments:

$$\text{pass-mi} = \sum(\text{Volume on segment } i)(\text{Length of segment } i) \tag{58.3}$$

The main quantifiable measures of service quality for transit passengers are travel time (in vehicle); waiting time, which is approximately half the headway when service is regular and headway is not too large; number of transfers; and level of crowding.

Basic Schedule Design

The simplest practical scheduling method involves three constraints in addition to the fundamental relationship [Eq. (58.2)]. First, the average vehicle load at the peak point, called *peak load* (l_p), must not exceed a design capacity (k), which depends on the size of the bus as well as standards of comfort and safety:

$$l_p = \frac{v^*}{q} \leq k \tag{58.4}$$

Second, the headway is usually restricted to a set of acceptable values, $\{h\}$. This set is based on four considerations: (1) whole minute headways are usually required because schedules are written in whole minutes (exception: some rail systems use the half minute or quarter minute as the basic unit); (2) multiples of 5 min are desired for long headways; (3) headways that repeat every hour (e.g., 12, 15, 20, 30 min) are desirable; and (4) there is a maximum headway, called a *policy headway*, that may not be exceeded, usually 60 min, sometimes 30 min or smaller in peak periods. For example, one set of acceptable headways might be

$$\{h_{\mathrm{I}}\} = \{1, 2, \ldots, 20, 25, 30, 35, 40, 45, 50, 55, 60\} \tag{58.5}$$

while a more restrictive set might be

$$\{h_{\mathrm{II}}\} = \{1, 2, \ldots, 10, 12, 15, 20, 30, 60\} \tag{58.6}$$

Third, the number of vehicles must be integer (unless the route is not operated in isolation, in which case scheduling must be done jointly for a number of routes, as discussed in section 58.5). The result of these last two constraints is to force additional slack into the schedule, both in the form of excess capacity and excess layover.

Schedule design usually begins with a given peak volume and a given minimum cycle time ($c_{\min}$) that accounts for running time and minimum necessary layover. The schedule design procedure that follows has as its primary objective minimizing fleet size (about the same as minimizing cost); its secondary objective, for a given fleet size, is to maximize service frequency (maximize service quality). In what follows, $[\ ]^+$ means round up and $[\ ]^-$ means round down. This procedure assumes that cycle times and headways are in minutes, while frequencies and passenger volumes are hourly.

Step 1. $h_{\max} = [k/(v^*/60)]^-$ (round down to next acceptable headway)
Step 2. $n = [c_{\min}/h_{\max}]^+$
Step 3. $h = [c_{\min}/n]^+$ (round up to next acceptable headway)
Step 4. Given n and h, determine the remaining parameters (c, q, l_p) using Eqs. (58.2) and (58.4). The difference between c and $c_{\min}$, called *schedule slack*, gets added to the layover.

The rounding involved in steps 1 and 2 can add substantially to operating cost. For example, consider a route for which $v^* = 260/h$, $c_{\min} = 51$ min, and $k = 50$. If one ignores rounding, the minimal service frequency is 260/50 = 5.2/h, the headway is 60/5.2 = 11.5 min, and the number of vehicles needed is 51/11.5 = 4.4. While this kind of analysis can be done in sketch planning, it does not produce a workable design. Following are two designs using the preceding procedure; their difference is that one uses set $\{h_{\mathrm{I}}\}$, which allows an 11-min headway, while the other uses $\{h_{\mathrm{II}}\}$, which does not.

Case	Set of Acceptable Headways	$h_{\max}$ Unrounded (min)	$h_{\max}$ (min)	n Unrounded	n	h (min)	c (min)	l_p
I	$\{h_{\mathrm{I}}\}$	11.5	11	4.64	5	11	55	47.7
II	$\{h_{\mathrm{II}}\}$	11.5	10	5.1	6	9	54	39.0

This example demonstrates the substantial effect of rounding. In case I rounding increased fleet requirements from the sketch planning value of 4.4 to 5. The extra resources consumed are manifest as slack in the cycle time (the final cycle time, 55 min, is 4 min greater than required) and in slack capacity (peak load, 47.7, is below the allowed capacity of 50). Case II, by not permitting an 11-min headway, requires more rounding, increasing the vehicle requirement to 6. However, the extra resources are not all wasted, but are partially converted into extra service as service frequency increases, reducing passenger waiting time and crowding. Case II also illustrates the role of step 3 in achieving the secondary objective of maximizing service level. Step 3 could have been omitted, leaving $h = h_{\max} = 10$ min, and the result would have been a viable design. However, the rounding involved in calculating n (step 2) made a better value of h (9 min vs. 10 min) possible without adding a vehicle.

Finally, a schedule with a large amount of slack time in the cycle begs for opportunities to adjust the minimum cycle time, either lowering it enough to save a vehicle (e.g., by eliminating a deviation or securing traffic improvements) or lengthening it by an amount less than or equal to the slack in an effort to attract new passengers (e.g., extending the route or adding a deviation) without increasing the fleet requirement.

Example 58.1 A downtown circulator on its own right-of-way with six stops is being planned. Stops, numbered clockwise, are 0.5 mi apart, and travel time (including dwell time at stops) is 1.5 min per segment, for a 9-min overall running time. There should be little or no layover because of the nature of the service; likewise, the only restriction on headway is that it be in quarter minutes. Vehicle design capacity is 40. Two alternative configurations are to be compared. Alternative A is service in the clockwise direction only. Alternative B is service in both directions; naturally, alternative B involves a greater construction cost.

Estimated P.M. peak demand is shown in the O-D matrix in Table 58.3(a). For alternative A the volume on segment 6-1 is the sum of the cells in the O-D matrix that involve travel over that segment; those cells are shaded in Table 58.3(a). Volume on the remaining segments is found by accumulating ons and subtracting offs, resulting in the volume profile shown in Table 58.3(b). Peak volume is seen to be 1333/h. Multiplying volume by segment length and by segment travel time results in passenger-miles and passenger-minute estimates.

Following the four steps for schedule design,

$$h_{\max} = \left[\frac{40}{(1333/60)}\right]^{-} = [1.80]^{-} = 1.75 \text{ min}$$

$$n = \left[\frac{9}{1.75}\right]^{+} = [5.14]^{+} = 6$$

$$h = \left[\frac{9}{6}\right]^{+} = [1.5]^{+} = 1.5 \text{ min}$$

$$c = (1.5)(6) = 9 \text{ min, and layover} = 0$$

$$q = 60/1.5 = 40/\text{h}$$

$$l_p = 1333/40 = 33.3$$

For alternative B demand must be split between the two directions. Assuming that passengers choose the shortest direction and that those traveling three stops split themselves evenly between the two directions, O-D matrices for the two routes are shown in Table 58.4(a). Volume profiles [Table 58.4(b)] are constructed as they were for alternative A, with the shaded cells in the corresponding O-D matrix constituting load on the first segment. The peak volumes are 517/h on the clockwise route, and 550/h on the counterclockwise route. Schedule design for the two routes is as follows:

Table 58.3 Demand Analysis, One-Directional Circumferential Route

a. Origin - Destination Matrix (passengers / hr)

FROM \ TO	1	2	3	4	5	6	TOTAL
1		150	150	300	150	150	900
2	100		50	100	50	50	350
3	100	33		100	50	50	333
4	200	67	67		67	67	467
5	100	33	33	33		50	250
6	100	33	33	33	33		233
TOTAL	600	317	333	567	350	367	2533

b. Volume Profile

SEGMENT AFTER STOP	6	1	2	3	4	5	6	TOTAL
OFF		600	317	333	567	350	367	2533
ON		900	350	333	467	250	233	2533
VOLUME	1000	1300	1333	1333	1233	1133	1000	
MILES		0.5	0.5	0.5	0.5	0.5	0.5	
PASS-MI		650	667	667	617	567	500	3667
MIN		1.5	1.5	1.5	1.5	1.5	1.5	
PASS-MIN		1950	2000	2000	1850	1700	1500	11000

Note: volume on first segment shown is sum of shaded cells in O-D matrix.

Clockwise route	Counterclockwise route
$h_{max} = \left[\frac{40}{517/60}\right]^- = [4.64]^- = 4.5$ min	$h_{max} = \left[\frac{40}{550/60}\right]^- = [4.36]^- = 4.25$ min
$n = \left[\frac{9}{4.5}\right]^+ = 2$	$n = \left[\frac{9}{4.25}\right]^+ = [2.12]^+ = 3$
$h = \left[\frac{9}{2}\right]^+ = 4.5$ min	$h = \left[\frac{9}{3}\right]^+ = 3$ min
$c = (2)(4.5) = 9$ min, layover $= 0$	$c = (3)(3) = 9$ min, layover $= 0$
$q = 60/4.5 = 13.33/\text{h}$	$q = 60/3 = 20/\text{h}$
$l_p = 517/13.33 = 38.8$	$l_p = 550/20 = 27.5$

A comparison between alternatives A and B is given in Table 58.5. Most of the table is self-explanatory. Averages in rows 3, 5, 8, and 10 are found by dividing the previous figure by total boardings (row 1). Average wait time in row 6 is taken to be half the headway; total wait (row 7) is the product of the average and the total boardings. In some alternatives analyses, wait time is weighted more heavily than travel time; this measure has not been taken in this example. In comparing the two examples one can see that alternative B requires one fewer vehicle and involves less passenger time overall. A decision between the two alternatives should consider other factors

Table 58.4 Demand Analysis, Two Circumferential Routes

CLOCKWISE ROUTE

a. Origin - Destination Matrix (passengers / hr)

FROM \ TO	1	2	3	4	5	6	TOTAL
1		150	150	150	0	0	450
2	0		50	100	25	0	175
3	0	0		100	50	25	175
4	100	0	0		67	67	233
5	100	17	0	0		50	167
6	100	33	17	0	0		150
TOTAL	300	200	217	350	142	142	1350

b. Volume Profile

SEGMENT AFTER STOP	6	1	2	3	4	5	6	TOTAL
OFF		300	200	217	350	142	142	1350
ON		450	175	175	233	167	150	1350
VOLUME	367	517	492	450	333	358	367	
MILES		0.5	0.5	0.5	0.5	0.5	0.5	
PASS-MI		258	246	225	167	179	183	1258
MIN		1.5	1.5	1.5	1.5	1.5	1.5	
PASS-MIN		775	737	675	500	538	550	3775

COUNTERCLOCKWISE ROUTE

a. Origin - Destination Matrix (passengers / hr)

FROM \ TO	1	2	3	4	5	6	TOTAL
1		0	0	150	150	150	450
2	100		0	0	25	50	175
3	100	33		0	0	25	158
4	100	67	67		0	0	233
5	0	17	33	33		0	83
6	0	0	17	33	33		83
TOTAL	300	117	117	217	208	225	1183

b. Volume Profile

SEGMENT AFTER STOP	1	6	5	4	3	2	1	TOTAL
OFF		225	208	217	117	117	300	1183
ON		83	83	233	158	175	450	1183
VOLUME	550	408	283	300	342	400	550	
MILES		0.5	0.5	0.5	0.5	0.5	0.5	
PASS-MI		204	142	150	171	200	275	1142
MIN		1.5	1.5	1.5	1.5	1.5	1.5	
PASS-MIN		612	425	450	512	600	825	3425

Note: volume on first segment is sum of shaded cells in O-D matrix.

Table 58.5 Comparison of Alternatives

	Alternative A	Alternative B Clockwise	Alternative B Counter-clockwise	Alternative B Total
1. Boardings (pass/h)	2,533	1350	1183	2,533
2. Passenger-miles (per h)	3,667	1258	1142	2,400
3. Average trip length (mi)	1.45			0.95
4. Travel time (pass-min/h)	11,000	3775	3425	7,200
5. Average ride time (min)	4.34			2.84
6. Average wait time (min)	0.75	2.25	1.5	
7. Total wait time (pass-min/h)	1,900	3038	1774	4,812
8. Average wait time (min)	0.75			1.90
9. Total pass-min per h	12,900			12,012
10. Average travel + ride time (min)	5.09			4.74
11. Vehicles needed	6	2	3	5

as well, including vehicle requirements in other periods, capital costs, operating statistics in other periods and in the future, and sensitivity to changes in the demand estimates.

58.4 Frequency Determination

Most bus, light rail, and metro routes operate at a constant headway over a time period. When demand is very low, routes follow a policy headway (H), the maximum headway allowed by system policy. When demand is very high, the headway is set so that average peak load equals or is just below the design capacity, as described in section 58.3.

In between very low and very high demand, there is no widely accepted method for setting frequencies. An optimization framework, based on a trade-off between operator cost and passenger waiting time, provides a rule that can be used to consistently set frequencies on routes. Let

$$(\text{OC}) = \text{operating cost per vehicle hour (\$/veh-h)}$$

$$(\text{VOT}) = \text{value of passenger wait time (\$/pass-h)}$$

Assuming that wait time is half the headway, the combined operator plus passenger cost per hour for route i is

$$(\text{OC})c_i q_i + 0.5(\text{VOT})b_i/q_i$$

Summing over all routes and then minimizing by setting to zero the derivative with respect to q_i yields the "square root rule":

$$q_i = \sqrt{\frac{0.5(\text{VOT})b_i}{(\text{OC})c_i}} \tag{58.7}$$

Incorporating policy headway and capacity constraints, the rule has two modifications:

- If the solution is below $60/H$, set $q_i = 60/H$ (use policy headway).
- If the solution is below v_i^*/k, set $q_i = v_i^*/k$ (make peak load equal capacity).

In addition, solutions must be rounded appropriately to satisfy integer constraints as described earlier.

The value of time can be explicitly specified as a matter of policy (one half the average wage is typical). Alternatively, it can be implicitly determined by a constraint on total operating cost per hour of system operation [$(\text{OC})\sum c_i q_i \leq$ Budget]; in which case the value of time will be that value for which the total operating cost when the q_i values are set by the constrained square root rule equals the budget [Furth and Wilson, 1981]. The framework can be further generalized by recognizing that demand is not fixed, but will respond to frequency changes, and so the generalized cost function should include the societal benefit of increased ridership and consumer surplus. Yet nearly the same result will be reached if demand is treated as constant because the goals of minimizing waiting time for existing passengers and trying to attract new passengers are so much in harmony. If the driving limitation is number of vehicles of varying types instead of an operating cost budget, the same framework applies, with OC = 1 and Budget equal to fleet size. A useful dynamic programming solution to the latter problem using a similar optimization framework is found in Hasselstrom [1981]; unlike the calculus-based models, it yields solutions that do not need to be rounded.

Figure 58.1 depicts graphically how the constrained square root rule applies to routes with varying levels of ridership in comparison with minimum cost scheduling, using typical values for OC ($60/veh-h), c_i (1h), H (60 min), k (60), VOT ($4/hour), and the ratio v_i^*/b_i (0.5). Part (a) shows how frequency varies with ridership; part (b) shows how peak load varies with frequency.

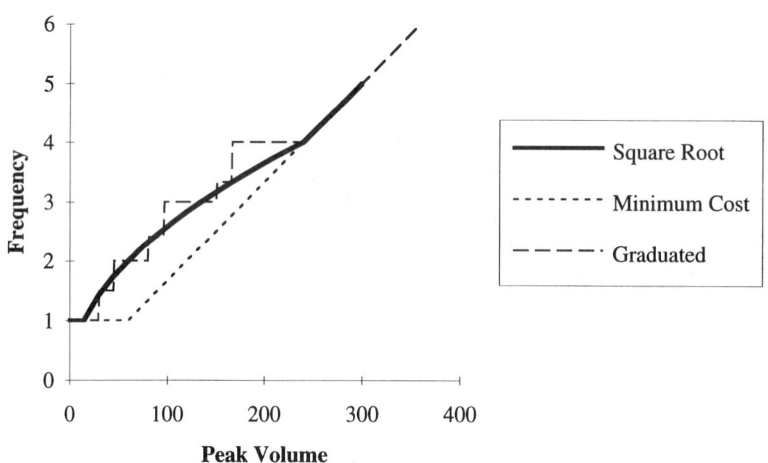

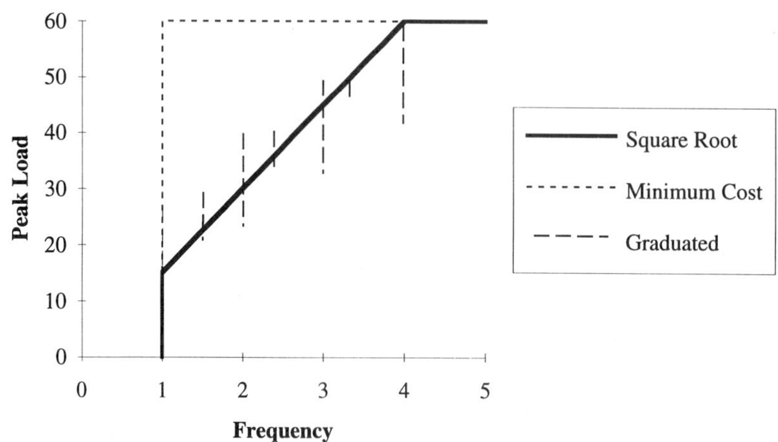

FIGURE 58.1 Frequency determination rules.

The curves have three ranges:

1. For low-volume routes, policy headway governs; frequency is independent of passenger volume; increases in ridership are simply absorbed by increasing load.
2. For intermediate-volume routes, the square root formula applies; frequency increases with passenger volume, though less than proportionally, because part of the passenger volume increase is absorbed by increasing the peak load. This intermediate portion is missing under minimum cost scheduling.
3. For high-volume routes, the load constraint governs; all routes are at capacity, and frequency is proportional to passenger volume.

Many transit systems claim to be minimizing operating cost subject only to policy headway and capacity constraints. If that were the case, the minimum cost curve would be followed. Yet in practice it is uncommon to find a route with a policy headway whose peak load is nearly equal to capacity. This is because the scheduling rules followed by most transit agencies—whether informal or formal—resemble the square root rule in that they include a transition between policy headway and capacity constrained. An example is a graduated peak load standard used by some agencies, also illustrated in Fig. 58.1. It states that maximum peak load (design capacity) decreases from a base value of 60 to 50, 40, and finally 30 on routes whose headway is above 15, 20, and 30 min, respectively. It is somewhat "saw-toothed" due to the requirement that, beyond 12 min, only certain headways are used. The fact that it closely parallels the constrained square root rule and at the same time is more readily understood by operations planners (for example, it avoids the troublesome value-of-time parameter) makes it a useful rule.

In rail systems, frequency determination has an additional dimension—train length. Whether train length should vary between peak and off-peak involves a trade-off in coupling costs as well as the usual operating costs and passenger convenience.

On many commuter rail routes and some express bus routes, cycle length is so long and demand so peaked that it makes little sense to speak of a constant service frequency within a time period. Scheduling in such cases is done at a more detailed level, tailoring departure time for each trip to passenger demand. An example is load-based scheduling for evening peak express service leaving a downtown terminal. One would construct a profile of cumulative passenger arrivals versus time, which may be quite irregular with numerous short peaks corresponding to common quitting times such as 4:00, 4:30, and 5:00. Departures are then scheduled whenever the cumulative arrivals since the last departure equals the desired vehicle load.

58.5 Scheduling and Routing

Desired headways and running times can vary throughout the day, usually with two peak periods when more vehicles and operators are needed than in the midday (base) period. In some cities the periods within which running time and/or headway change can be as small as 20 min. For this and other reasons, scheduling is far more complex than the fundamental case described in section 58.3.

Given the desired timetable and running times on a network of routes, scheduling is the task of creating vehicle and operator duties to perform the specified service. A basic overview of vehicle scheduling is given in the following discussions. Because scheduling both vehicles and operators is so complex, most transit agencies use automated scheduling. Nevertheless, it is still important for route and schedule designers to understand some basic scheduling paradigms. The final discussion of section 58.5 briefly describes route design.

Deficit Function Analysis

The number of vehicles needed to meet an arbitrary timetable (i.e., without any expectation of constant running times or headways) involving several terminals and routes can be found using *deficit functions* [Ceder and Stern, 1981]. The deficit at terminal i at time t, $d_i(t)$, is the cumulative

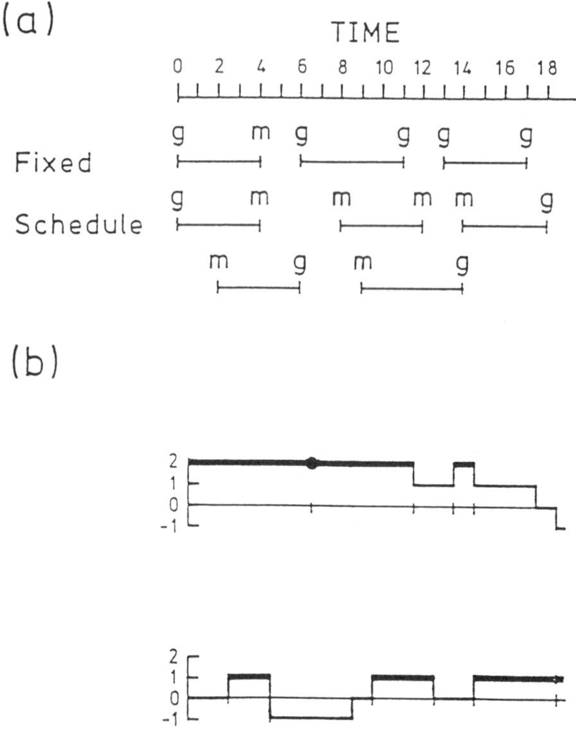

FIGURE 58.2 (a) A two-terminal fixed schedule between terminals g and m. (b) Deficit functions at g (above) and m.

number of departures from that terminal minus the cumulative number of arrivals at that terminal as of time t. Each departure from a terminal increases that terminal's deficit by 1; each arrival lowers it by 1. For a small network, deficit functions can be easily drawn, as in Fig. 58.2. Let $d_i^* =$ the greatest deficit occurring at terminal i; it represents that number of vehicles that must be on hand at that terminal at the start of the day to keep that terminal from running out of vehicles. The total vehicle requirement for the network is the sum of the d_i^* values. Once the vehicle requirement is known, vehicle duties can be easily constructed by simply chaining trips together, minimizing layover. For an isolated route with trips scheduled as round trips, another way of saying that the fleet requirement equals the peak deficit is to say that the fleet requirement equals the maximum number of departures occurring during any round-trip window.

Schedule adjustments to reduce the fleet requirement should naturally be aimed at reducing the peak deficit at the various terminals. Sometimes shifting a trip's time by a minute or two can reduce the peak deficit at one terminal without increasing it at another. Another such adjustment is to add to the schedule a deadhead (i.e., empty) trip that leaves one terminal after the time of its peak deficit and arrives at another terminal before the time of its peak deficit, reducing by one the peak deficit at the second terminal without increasing the peak deficit at the first. This kind of schedule analysis has proven particularly helpful in improving the efficiency of regional bus operations, where headways can be quite long, can vary greatly, and routes frequently do not operate as simple loops.

Network Analysis

A more comprehensive and flexible framework for analyzing schedules is network analysis. Each node in the network represents a terminal and a time (either a departure or arrival time). The network has five kinds of links:

1. *Service links*, representing trips in the timetable. These links have a minimum "flow" of one.
2. *Layover links*, going from one node to another node representing the same location at a later time. These links have no minimum flow.
3. *Deadhead links*, joining a node to another node representing a different location at a later time (the time must be enough later that the connection can be made). These links have no minimum flow.
4. *Source links*, leaving a central source node, representing vehicles entering service.
5. *Sink links*, going to a central sink node, representing vehicles leaving service.

An example network is shown in Fig. 58.3. The fleet requirement for the schedule is the minimum flow that must begin at the source node, filter through the network satisfying flow conservation at every node (inflow = outflow) and meeting the flow requirements of the service links, and return to the sink node. Mathematical programming solutions to this problem, including a linear programming approach, are well known. One intuitive way of looking for a solution is to examine cuts that divide the network. A valid cut (1) divides the network into two parts, an early part containing the source node and a late part containing the sink node; (2) intersects no link more than once; and (3) intersects links in such a way that the beginning of every intersected link lies in the early part of the network, and the end of every intersected link lies in the late part. The minimum number of vehicles needed is the greatest number of service links that can be intersected by a valid cut. In Fig. 58.3, two valid cuts are shown, one intersecting

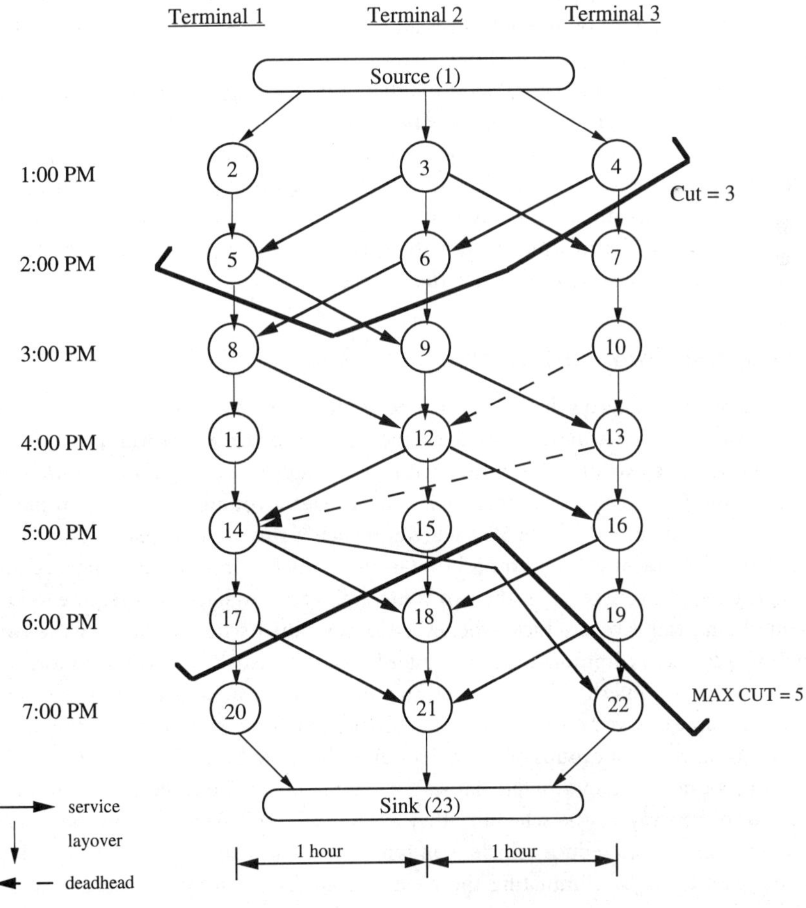

FIGURE 58.3 Network analysis.

3 service links, the other intersecting 5 service links. The reader can verify that the fleet requirement is indeed 5.

A refinement is to seek the minimum cost solution [Scott, 1984]. Costs are assessed to deadhead and layover links reflecting labor and vehicle costs. A flow of one is required on service arcs, so no cost need be applied. The cost of a source link should reflect the cost of deadheading from the garage, plus a substantial penalty for increasing the peak vehicle requirement. Sink links are assessed the cost of deadheading to the garage. The problem is now a "transshipment problem," which has several well-known solution algorithms including the *network simplex algorithm.* To apply transshipment model algorithms, it is necessary to consider the source and sink nodes as being connected to every other node.

Automated Scheduling

Vehicles can easily operate for 16 or 20 hours without a break, but operators' schedules are constrained by a variety of work rules, such as minimum number of paid hours (typically 8), maximum number of hours (typically 8.25 to 9), restrictions on the number of part-time operators, paid breaks after 5 hours of uninterrupted service, maximum spread (for a "split shift," i.e., a shift with an unpaid break in the middle, spread is the amount of time between beginning and end of the workday), pay premiums for time after 8 hours and for spread exceeding a certain amount, and the requirement that every shift end where it began. Most automated scheduling packages first do run cutting—using network optimization procedures described earlier to create efficient vehicle schedules—and then operator scheduling, using other optimization methods to split the vehicle schedules into pieces of about 4 hours and then match them into legal, efficient operator schedules.

There are several scheduling packages on the market that compete with one another and with manual scheduling. Although the cost of these packages can be high, benchmark tests have shown their ability to reduce operator labor costs by up to 3%, well justifying the investment [Blais *et al.*, 1990]. Scheduling software does a great deal of valuable bookkeeping and usually includes graphical interfaces that enable schedulers to easily make manual schedule adjustments. The software can also be integrated with other information systems, such as timetable publishing, payroll, work force management, and data collection, adding to their value.

Interlining and Through-Routing

Interlining means scheduling a vehicle to switch between routes. It can be done on an *ad hoc* basis, but it can also be done systematically in a scheme that can be called *cyclical interlining*. Cyclical interlining means that two or more routes with a common headway and a common terminal are scheduled jointly in such a way that each vehicle does a round trip on one route followed by a round trip on the other. Figure 58.4 illustrates such a situation. In part (a), the two routes are scheduled independently, and require a total of 5 vehicles. In part (b), they are interlined, with an aggregate cycle consisting of the two route cycles back to back, and require only 4 buses. Cyclical interlining can save a vehicle whenever the combined schedule slack of the two routes equals or exceeds their common headway. Interlining can also be done with more than two routes. If k routes are interlined, all having a common headway and common terminus, it is theoretically possible to save as many as $k - 1$ vehicles, if each route has a lot of schedule slack. In practice it is uncommon for groups of more than three routes to be cyclically interlined. Cyclically interlined routes can maintain separate names for the public, or the route combination can have a single name; either way, to the scheduler they are a single unit. While the primary motivation for cyclical interlining is to reduce vehicle requirements by eliminating schedule slack, interlining also benefits passengers by eliminating the need to transfer between the routes that are interlined.

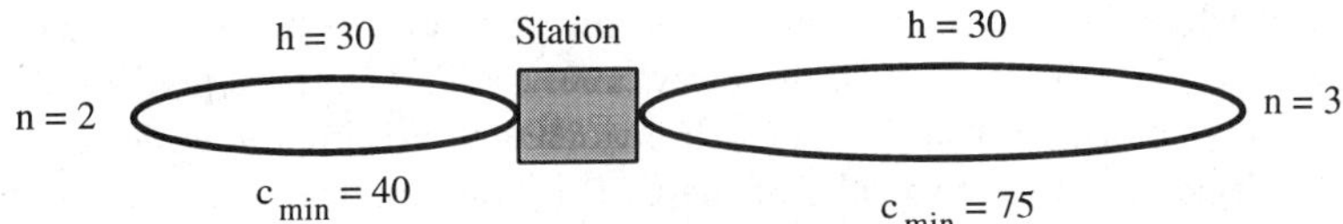

(a) scheduled independently, requiring 5 buses

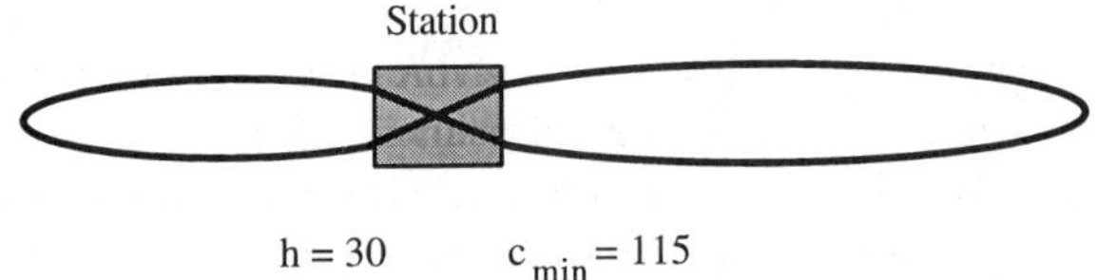

(b) interlined, requiring 4 buses

FIGURE 58.4 Cyclical interlining routing.

When two routes are interlined to save a vehicle, some freedom in choosing departure times is lost. For example, in Fig. 58.4(a), it is possible for the routes to have simultaneous departures. But in Fig. 58.4(b), once interlined and operated with 4 vehicles, they cannot have simultaneous departures. If there is a route 1 departure at 10:00, the same vehicle will depart on route 2 immediately after completing the route 1 cycle, that is, at 10:40, or up to 5 min later, since the interlined cycle has 5 min slack. Working backward with a 30-min headway, it is apparent that previous route 2 departures must lie in the time windows at 10:10–10:15, 9:40–9:45, etc. If the route 2 departures are scheduled outside this 5-min window, there will be no vehicle savings from interlining.

Through-routing is the practice of joining two routes that go to (roughly) opposite sides of the central business district (CBD) or a major terminal. In small cities, where it is common for all the radial routes to share a single, central CBD terminal, through-routing is the same as cyclical interlining. In cities with large CBDs, good CBD distribution can only be attained if routes extend more than halfway into the CBD, resulting in several CBD terminals and considerable cross-CBD traffic. In such a case joining two opposite radial routes is more than just interlining, since it also means sharing the cross-CBD link. The benefits of through-routing include the following: (1) Sharing the cross-CBD reduces the combined cycle time, potentially reducing vehicle requirements. Combining the cycles also tends to reduce the loss due to slack (as with cyclical interlining). (2) More extensive CBD distribution for passenger convenience, especially for reaching destinations such as universities and hospitals that are often located on the fringe of the CBD. (3) A smaller contribution to traffic congestion in the CBD. (4) Rail lines are nearly always through-routed to avoid the need for duplicate trackage, tunneling, etc. (5) Time-consuming (and, for rail systems, expensive) turnarounds in the CBD are eliminated. (6) Layovers in the CBD, where space is at a premium, are likewise eliminated.

Through-routing also has two disadvantages, however. First, eliminating CBD layovers means that departures from the CBD are less prone to be on time. This can cause severe crowding in the P.M. peak, when evenly spaced departures from the CBD are essential to keeping loads balanced. The longer routes are harder to control in other ways as well. For this reason some large cities avoid through-routing. Second, routes that are through-routed must have a common headway and common vehicle and train sizes at all times. They cannot easily be uncoupled when demand and, consequently, desired headways on the two sides of the CBD differ. If during a certain period of the day or at some time in the future the east side of a through-route requires an 8-min headway

(7.5 trips/hour) while the west side requires a 6-min headway (10 trips/hour), the combined route must be operated with the smaller headway, implying that 2.5 trips/hour on the east side will be made unnecessarily. When designing through-routed rail lines, one design objective is to balance the demands on the two sides of the CBD by varying line length and availability of parking.

Pulse (Timed Transfer) Systems

Perhaps the most disliked aspect of the passenger journey is the transfer from one route to another. Transfers cannot be eliminated; there is simply not enough demand to afford direct service for every desired trip. Pulse scheduling is a concept aimed at making transfers less onerous [Vuchic, 1981]. Routes are concentrated at a single terminal (in more complex versions of pulse scheduling, there are several major pulse points) and all have simultaneous departures at a common headway (usually every hour or half-hour). Vehicles from all the routes arrive shortly before a pulse, then depart together, leaving a small time window for passengers to transfer between routes, similar to operations at an airline hub. Route cycle times must be a multiple of the common headway, so routes should be laid out so that their running time is just under a multiple of the common headway to avoid excess schedule slack (which cannot be reduced by interlining due to the need for simultaneous departures). It is possible to compromise the ideal plan to accommodate routes of greater and less demand; for instance, high-demand routes could pulse every 15 min, while low-demand routes pulse every 30 min.

The main advantages of a pulse system are as follows: (1) The waiting time when transferring is small. (2) Compared to transferring on a street corner, the transfer environment is vastly improved. There is increased passenger security (due to not being alone), less anxiety about having missed a connection, and amenities can include concession stands and protected waiting areas. (3) The schedule is easy for passengers to remember. (4) The centralized terminal can increase transit's visibility and improve its image to its patrons and the community.

The chief disadvantages of a pulse system are as follows: (1) There is a cost for building and maintaining the transfer center. (2) There is increased need for space at the transfer center, which is usually located in the CBD or another major activity center where space is at a premium. In a pulse system, each route needs its own berth; if departures were staggered, routes could share berths. (3) The need for cycle time to be a multiple of the common headway can lead to considerable schedule slack, increasing vehicle and operator requirements. Schedule slack can be reduced by making routes more circuitous in an effort to reach more passengers, but the benefits are usually small. Interlining cannot save buses because departures must be simultaneous.

Pulse systems are most effective in small cities where most routes operate at long (30- or 60-min) headways; in larger-sized cities during evening hours; or in suburban areas, sometimes at rail stations, to facilitate bus-to-bus transfers. When demand is great enough that headways become smaller than 20 or 30 min, the transferring benefit of pulse scheduling becomes smaller and the operating cost impact too large to make pulse scheduling practical. Timed transfers can be also arranged for a pair or small subset of routes that share a common headway.

Operating Strategies for High-Demand Corridors

Corridors with high passenger demand create opportunities for differentiating service to better serve certain markets and to better match capacity to demand. Along low-demand routes there is little alternative to the standard local bus route that satisfies the fundamental needs of access and an acceptably small headway. As demand grows, one has the choice of simply increasing the frequency of the local route or employing other routing and scheduling strategies for bus service. The strategies described in this section apply equally to rail service, except to the extent that they rely on overtaking.

Express Service

A common strategy is to supplement local bus service with express service. Passengers obviously benefit from the reduced travel time, and operating cost can likewise be reduced due to higher speed if the load on the express route is sufficient. Express service in many cities attracts auto users via park-and-ride lots. The speed attainable depends, of course, on the available roadways. Priority treatments on highways, arterials, and in the downtown can make a great difference [Levinson *et al.*, 1975]. Express service in a high-volume corridor must offer a sufficient travel time savings over the local route and a small enough headway to "compete" successfully with the local route and thus capture its own market. Then the express and local services can be scheduled independently.

Zonal Express Service

When the demand for express service is great, rather than simply increase the frequency of the express route, it is often more cost effective to divide the corridor into zones and provide an express route for each zone [Turnquist, 1979]. Passengers in the outer zones will enjoy a further travel time savings (but a longer headway), and operating cost will be reduced as fewer vehicles have to cover the full length of the corridor. The same approach has been successfully applied to commuter rail [Salzborn, 1969].

Alternating Deadheading

During peak periods when there is a strong direction imbalance in travel demand, some routes, particularly express routes, deadhead (return empty) in the reverse direction. If offering reverse direction service is desirable but demand is still far below peak direction demand, and there exists a high-speed path for reverse direction deadheading, buses can sometimes be saved by having only a fraction of the trips return in service, with the remainder deadheading [Furth, 1985]. Because the deadheading trips must be coordinated with those returning in service (since they both will continue on the same peak direction route), a systematic coordination "mode" will be needed. The most effective mode is often 1:1, meaning that for every trip returning in service, one deadheads. As a rule of thumb, a bus can be saved if the time saved by deadheading equals two headways. The extent of alternating deadheading is limited by capacity and the policy headway constraints.

Restricted Zonal Service

Zonal design can be applied to radial local service as well as to express if the peak volume is large and the volume profile shows a steady increase from the outer end of the route and to just outside the downtown [Furth, 1986]. The corridor is divided into zones, each large enough to support a route with an acceptably small headway. Buses never leave the main route of the corridor (unless deadheading), but employ boarding and alighting restrictions between their zone and the downtown terminal. Inbound, buses let passengers board in their zone only, but let them alight anywhere. Outbound, alighting is restricted to the route's zone, while boarding is unrestricted. With this strategy, direct service is still offered between every pair of stops in the corridor, but there is only one zonal route that a passenger can take between any given origin and destination. Therefore the corridor O-D matrix can simply be split into the markets served by the different routes, and each route scheduled independently. In general, the advantage of restricted zonal service is that it allows the passenger-carrying capacity to increase along the route as the volume profile increases, reducing unused capacity in the outer zones and thereby saving vehicles. However, once an inbound bus enters the portion of its route in which boarding is restricted, alighting passengers cannot be replaced, resulting in some unused capacity in the inner zones. For this reason, restricted zonal service is effective only when the proportion of outer zone riders alighting before the downtown is small. Another disadvantage of the strategy is that with more routes, there is more unproductive slack due to rounding. Moreover, there can be problems in passenger understanding of and acceptance of the boarding and alighting restrictions, although the strategy has been used successfully in some cities for years.

Short-Turning

Short-turning, like restricted zonal service, means some buses traverse only the inner portion of the route, allowing provided capacity to more closely match demand [Furth, 1988]. Unlike restricted service, there are no boarding or alighting restrictions. In a two-route system (three-route systems are uncommon for practical reasons), passengers with either an origin or destination in the outer zone must use a bus serving the full route, while those whose trip lies entirely within the inner zone can use buses on either the full or short-turning route. Efficient operation demands that most of these "choice" passengers use the short-turning route. Unless a reduced fare can be offered on the short-turning route, the way to effect this choice is to coordinate the scheduling of the routes, having a short-turning bus lead a full route bus by a small time interval. For example, full route buses might pass the turnback point at 7:00, 7:10, 7:20, etc., while short-turning buses leave the turnback point at 6:58, 7:08, 7:18, etc. In this example the headway module is 10 min, divided into a "leader's headway" of 8 min and a "follower's headway" of 2 min. Each short-turning bus will therefore carry 8 minutes' worth of the choice market, while each full route bus carries only 2 minutes' worth of the choice market. This is an example of 1:1 schedule coordination. Other coordination modes are also possible. For example, if full route buses pass the turnback point at 7:00, 7:10, 7:20, etc., 1:2 coordination might have short-turning buses depart at 7:04, 7:08, 7:14, 7:18, 7:24, 7:28, etc. Then each full route bus will get two minutes' worth of the choice market, while each short-turning bus gets four minutes' worth. The challenge of design is to choose the turnback point(s) and headway module, and then split the headway module in such a way that the resulting split in the choice market gives each route a peak volume near, but not exceeding, design capacity.

Other strategies for high-volume corridors include *limited stop* and *skip-stop* service [Furth, 1985].

Route Design

For the most part, route design is done manually, using standard evaluation methods to choose between alternative routings. Standards have been developed in different agencies regarding route length, circuity, stop spacing, route spacing, and, of course, expected demand.

Efforts to automate route (and, for that matter, entire network) design have resulted in some useful models that are available as software packages [Hasselstrom, 1981; Babin *et al.*, 1982; Chapleau, 1986]. Some of these packages have the capacity to select route alignments, but their main contribution is in evaluating alternative networks. The investment in the software and in the network coding has limited their use to major investment analyses (e.g., design of a new rail line) and to systems with a large and dynamic ridership.

In newly developing areas, successful route layout can be very difficult if developers ignore transit and pedestrian access. Whenever possible, transit agencies and political authorities should try to influence developers to facilitate transit use by such measures as providing through streets that transit routes can use; providing walkways and pedestrian bridges for direct, easy pedestrian access to through streets; siting commercial buildings to allow for easy pedestrian access to through streets; and clustering high-density uses near potential transit routes.

58.6 Patronage Prediction and Pricing

Evaluation of proposed service or fare changes requires prediction of ridership and revenue impacts. The first requisite for such evaluation is methods for measuring current ridership, discussed in sections 58.8 and 58.9. Nearly all patronage-forecasting methods assume a knowledge of current ridership, and before/after studies, a valuable analysis technique, require little more than accurate measurement of actual ridership.

The amount of effort spent on predicting impacts of service changes should be proportionate to the cost of making the service change. For example, consider implementation of a new Saturday bus service requiring one bus for 8 hours a day. It is obviously not worth investing 100 hours in analysis to predict how successfully the service will be when operating it on a trial basis for 6 months would cost only 200 bus-hrs. For low-cost changes, a low-cost prediction method to screen out changes that are unlikely to succeed coupled with before/after evaluation is appropriate. On the other hand, for very large investments, full-scale modeling and demand-forecasting techniques, discussed in Chapter 55, are appropriate. For short-range transit planning, prediction methods can be divided into two groups: predicting ridership changes in response to service changes and predicting ridership on a new service.

Predicting Changes

The most commonly used method to predict ridership changes, particularly in response to fare changes, is elasticities. Fare elasticity is the relative change in demand divided by the relative change in fare, or

$$\varepsilon = \frac{\Delta\text{Demand}/\text{Demand}}{\Delta\text{Price}/\text{Price}} \tag{58.8}$$

Most fare elasticities measured from before/after studies lie between -0.10 and -0.70. The industry rule of thumb for many years was $\varepsilon = -0.30$; more recently, experience indicates that elasticity is closer to -0.2. Factors that lead to smaller elasticity (i.e., elasticity closer to 0) include high transit dependency, a predominance of work and school trips, and a low current fare. Opposite factors, such as a predominance of discretionary trips, lead to greater elasticity.

Predictions using fare elasticity can be made directly from Eq. (58.8), in which demand and price are entered at their base levels. For example, using the old rule-of-thumb elasticity, a fare increase from \$0.75 to \$1.00 will cause a relative ridership change of

$$\frac{\Delta\text{Demand}}{\text{Demand}} = -0.3\frac{\$0.25}{\$0.75} = -0.10$$

that is, a 10% drop. Of course, this simplistic example assumes that everybody pays full fare. Some market segments—those making transfers, those using passes, those getting discounts, for example—may experience different fare changes and may have different elasticities. It is therefore preferable to make predictions by market segment, or at least to base the prediction on average price rather than nominal fare.

Equation (58.8) represents one type of elasticity, the *shrinkage ratio*. There are some inherent inconsistencies in this form—for example, if fare in the previous example returns to \$0.75, predicted demand will not return to its original value. The *log-arc* elasticity is a theoretically consistent form. Log-arc elasticities are estimated from before/after data using the following equation:

$$\varepsilon = \frac{\ln(\text{New demand}/\text{Old demand})}{\ln(\text{New price}/\text{Old price})} \tag{58.9}$$

Use of the log-arc elasticity for making predictions is illustrated for the previous example:

$$\frac{\text{New demand}}{\text{Old demand}} = \left(\frac{\text{New price}}{\text{Old price}}\right)^{\varepsilon} = \left(\frac{\$1.00}{\$0.75}\right)^{-0.3} = 0.917 \tag{58.10}$$

implying an 8.3% drop in demand. The log-arc form gives different answers than the shrinkage ratio form, and the differences can be very large for large changes. There is a third type of elasticity as well—linear-arc elasticity—for which predictions are not materially different from predictions made using log-arc elasticity.

Although the log-arc elasticity is consistent with itself, it is not entirely consistent with reality—for example, it predicts infinite demand when price goes to zero. In reality, elasticity changes as price and demand change—demand becomes more elastic (elasticity increases in magnitude) as price becomes greater, and as the transit mode share becomes smaller. One way of facing this reality is to carefully select an elasticity from a catalog of elasticities [Mayworm *et al.*, 1980; Charles River Associates and Levinson, 1988], trying to best match it to current circumstances.

Another way to face the reality of varying elasticity is to use an incremental demand model that does not assume constant elasticity. The incremental logit method, an abbreviated form of the logit model described in Chapter 55 is such a method. It looks at transit demand as a share of the wider market that could use the transit service in question. The prediction formula is

$$\frac{\text{New demand}}{\text{Old demand}} = \frac{e^{(\text{coef})(\Delta\text{ price})}}{(\text{shr})e^{(\text{coef})(\Delta\text{ price})} + (1 - \text{shr})} \tag{58.11}$$

where shr = current transit share and coef = logit model coefficient for price. Logit model coefficients are not as widely catalogued as elasticities. A typical value is −0.6/$, so if transit's share of all trips that could use transit is 20%, the previous example leads to the prediction

$$\frac{\text{New demand}}{\text{Old demand}} = \frac{e^{-0.6(0.25)}}{(0.2)e^{-0.6(0.25)} + 0.8} = 0.89$$

implying a drop in demand of 11%.

The incremental logit method implies a point elasticity of

$$\varepsilon = (1 - \text{shr})(\text{coef})(\text{Current price}) \tag{58.12}$$

which for the example comes out to be $\varepsilon = (1 - 0.2)(-0.6)(0.75) = -0.36$. As Eq. (58.12) indicates, it is inherent in the incremental logit method for elasticity to change with both price and transit share. Equation (58.12) can also be used to determine a coefficient that is consistent with a given elasticity. The main drawbacks of the incremental logit model are in the need to specify a coefficient and to estimate the transit share.

Changes in service attributes such as headway can be evaluated in a similar manner to fare changes, using either elasticities or incremental logit. Elasticities with respect to attributes other than price are less well studied, and should be applied with care. The incremental logit model is suitable when the service change can be expressed as a change in a passenger's utility. For example, headway changes affect passenger waiting time, which is a part of most logit utility functions, with a typical coefficient of about −0.08/min. A change in headway from 18 to 12 min implies a drop in average waiting time of about 3 min, so the prediction will be

$$\frac{\text{New demand}}{\text{Old demand}} = \frac{e^{-0.08(-3)}}{(\text{shr})e^{-0.08(-3)} + (1 - \text{shr})}$$

which yields to an increase of 21% if the transit share is 20% (i.e., an increase from a share of 20% to a share of 24.2%).

Revenue Forecasting and Pricing

Revenue is simply ridership times average fare. Because different markets pay different fares and are affected differently by fare and service changes, revenue forecasting is best done by market segment. Market segmentation can be done along various lines, depending on the purpose. For example, the market can be segmented by type of fare paid (cash, pass, bulk purchase); by service (bus, metro, both bus and metro); by time of day (peak, off-peak, weekend); and by location (city, suburb, suburb to city). The matter is further complicated by the fact that some markets are fluid. For example, changes in pricing can make patrons switch from using a pass to paying cash, and so on.

Manipulating Eq. (58.9) leads to

$$\frac{\text{New revenue}}{\text{Old revenue}} = \frac{\text{New demand} \times \text{New price}}{\text{Old demand} \times \text{Old price}} = \left(\frac{\text{New price}}{\text{Old price}}\right)^{1+\varepsilon} \tag{58.13}$$

A desire is often expressed to increase revenue by *lowering* fares, the idea being that so many new riders will be attracted that revenues from them will more than make up the loss from current riders. As demonstrated by Eq. (58.13), this will not happen unless $|\varepsilon| > 1$, a situation almost never seen in transit. With typical low elasticities (around -0.2), fare increases will lead to revenue increases, although not as large (relatively) as the fare increase itself due to the loss in ridership.

Political and fiscal realities often lead planners to look for ways to increase revenue with the smallest possible attendant loss in ridership. The solution, in a theoretical sense, is to raise fares in the least elastic markets while holding steady or even lowering fares in the most elastic markets. This approach, called *price discrimination*, is widely practiced by the airlines. In transit, it has been the basis for peak/off-peak price differentials because peak period demand is less elastic, and for deep discounting for occasional riders, who are considered to be a relatively elastic market (i.e., willing to use transit more if offered a discount).

An example given in Table 58.6 illustrates this phenomenon. A uniform fare increase from \$1.00 to \$1.20 raises revenue by \$13.6 million, but at a ridership loss of 5.3%. A targeted fare increase, raising the fare to \$1.30 for the less elastic market (for argument's sake, peak period travelers) while lowering the fare to \$0.90 for the more elastic market, yields the same revenue increase with a ridership loss of only 1.5%.

Because transit services are subsidized, pricing determines in part how subsidies are distributed. There has been strong criticism that some pricing policies subsidize high-income users more than low-income users [Cervero, 1981]. Pricing efficiency has been hindered by practical difficulties with distance and time-based pricing. Some of these difficulties may be eliminated and new opportunities opened by advanced information technology that is beginning to appear in the industry.

Prediction for Service to New Markets

There is no well-defined method that is widely used for predicting patronage on a service to a market not currently served by transit. A common approach is to compare the new service to similar services in similar areas, with subjective adjustments made to account for the extent of dissimilarity. For example, a city might find that existing park and ride (P&R) express lots attract 12 riders per 1000 inhabitants living within a 3 mi × 3 mi square centered on the lot. This factor could be used for predicting ridership in a new market, with adjustments for population density, service frequency, income, distance from the city center, and so on.

Because predictions will necessarily be imprecise, it is important to monitor ridership and evaluate the new services at regular intervals, retaining services that meet established criteria for

Table 58.6 Varying Ridership Impacts of Fare Increases

		Current			Uniform Fare Increase			Targeted Fare Increase		
	Elasticity	Fare (\$)	Riders (Million)	Revenue (\$Million)	Fare (\$)	Riders (Million)	Revenue (\$Million)	Fare (\$)	Riders (Million)	Revenue (\$Million)
Market 1	−0.20	1.00	65.7	65.7	1.20	63.3	76.0	1.30	62.3	81.0
Market 2	−0.50	1.00	34.3	34.3	1.20	31.3	37.6	0.90	36.2	32.5
Total			100.0	100.0		94.7	113.6		98.5	113.6

new services. Such criteria might take a form such as "at least 10 passengers per service hour after 3 months," "at least 15 passengers per service hour after 6 months," and so on.

58.7 Operating Cost Models

Decisions regarding transit service often require estimates of the operating cost. The most simplistic way of estimating operating cost is based on a single factor, such as vehicle-miles. A transit agency can simply divide its annual operating costs by the annual vehicle-miles, resulting in a figure such as \$4.20/veh-mi, and multiply this by the change in vehicle-miles involved in a proposed change in service.

A more accurate model is based on several factors, such as vehicle-miles, vehicle-hours, peak vehicles (number of vehicles needed in the peak), trackage, and revenue. For bus service, a commonly used model has this form:

$$\Delta\text{Operating cost} = b_1(\Delta\text{vehicle-miles}) + b_2(\Delta\text{vehicle-hours}) + b_3(\Delta\text{peak vehicles})$$

where Δ means "change in" and b_1, b_2, and b_3 are coefficients estimated from accounting data. To estimate the coefficients, historical operating costs are allocated to one or more factors based on the most likely causal relationship. For example, costs for fuel, tires, and maintenance are allocated to factor 1, vehicle-miles. Labor costs for bus operators and supervisors are allocated to factor 2, vehicle-hours. Costs for insurance, space, and various overhead costs are usually allocated to factor 3, peak vehicles. Some costs may be allocated to more than one factor; for example, the cost of marketing may be allocated 50% to factor 1 and 50% to factor 3. The total cost allocated to each of various factors is then divided by the corresponding factor total (e.g., for factor 1, total vehicle-miles) to determine the coefficients b_1, b_2, and b_3 [Booz, Allen & Hamilton, 1981; Cherwony and Mundle, 1980].

It is generally conceded that peak period service is more expensive to operate than off-peak service. A schedule change that will require any additional vehicles operating during the peak can be substantially more costly than a schedule change (involving the same change in vehicle-miles and vehicle-hours) that would not require an extra vehicle during the peak, but would involve keeping a vehicle busy for longer during the off-peak. This phenomenon is partly reflected in cost models that include peak vehicles as a cost factor. Yet the typical models do not reflect the additional labor and maintenance costs associated with peaking (due to spread penalties, pull-outs, etc.). Efforts to better account for these factors have led to peak/off-peak cost models, which are theoretically appealing but not widely accepted because they lack an objective basis for allocating costs between peak and off-peak.

Most cost models are *fully allocated* models, in which the full annual operating cost is allocated to the various factors when estimating the coefficients. The result is estimates of *average* cost per factor unit instead of *marginal* cost, which is more appropriate for a model used to estimate the cost of service changes. Marginal cost coefficients can be estimated by eliminating from the accounting fixed costs and by using more complex models for various cost components. However, there is no widely accepted agreement on which costs are fixed since, in the long run, virtually all costs are variable, and so it is common to use simple average costs. The use of complex cost models for estimating labor costs of a schedule change has for the most part been made obsolete by the availability of scheduling software described in section 58.5, with which a detailed optimized schedule complete with actual operator costs and vehicle mileage can easily be obtained for any proposed service change.

58.8 Monitoring Operations, Ridership, and Service Quality

Effective transit management requires ongoing monitoring of what is actually happening: how close operations are to what was planned, how many riders there are, and the quality of the service delivered.

Operations Monitoring

Virtually all systems have supervisors or inspectors to oversee vehicle operators, see that service is operating as planned, authorize adjustments in response to disruptions, and maintain logs. In most transit systems, vehicles are equipped with two-way radios that can be used for obtaining information from operators as well as for sending messages. For many operations, these measures provide adequate monitoring.

In rail systems varying degrees of electronic monitoring are used. At the extreme end are automated systems that have constant communication between the vehicles and the central computer, and therefore constant monitoring of every vehicle's location, speed, and other attributes. Other systems maintain some degree of manual control, but still have constant communication with the central computer used to tell the operator desired speed or desired dwell time at a station. Either way, constant communication means that the location of every vehicle can be displayed on an electronic map, and various statistics such as actual headway at key points can be constantly monitored.

Systems lacking the facility for constant communication can use detectors located at key points along the track to monitor movements. If the detection system uses radio technology to identify a vehicle number (i.e., by sending a signal to a passing vehicle which is returned by an on-vehicle transponder), it is called an *automatic vehicle identification (AVI) system*. If it merely notes the presence of a passing vehicle, it is just a throughput detector. While either type of detector can monitor headway at detection points, AVI is vital for identifying service on various branches of a route, out-of-service vehicles, vehicles that have turned back, and individual vehicle running times.

Bus systems in mixed traffic do not afford the possibility of constant communication. Real-time monitoring can be done with a radio-based automatic vehicle location or automatic vehicle monitoring (AVL or AVM) system. A central radio tower polls every vehicle in turn once every polling cycle (typically one to three minutes) by sending out a signal to the effect of "bus number *xyz*, please respond." The bus radio then responds by sending back a stream of information that typically includes identifiers, location information, and alarm status (on or off) for various mechanical (e.g., oil pressure) and security alarms. In most existing systems location is ascertained using signposts located along routes that emit a weak radio signal that the bus radio receives as it passes the signpost. When polled, the bus sends back to the central tower the identification number of the signpost most recently passed, and the number of odometer "clicks" since passing the signpost (each click on a digital odometer is typically one axle revolution). Newer systems are being developed that rely on satellite-based systems rather than signposts for location information. AVL systems permit a central computer to display approximate location of vehicles and to calculate statistics such as schedule deviation. This kind of information can be used to radio instructions to an operator (e.g., slow down) or to suggest a service change, such as a turnback or placing a standby vehicle into service. It can be used to display real-time information to waiting passengers concerning vehicle arrival time. It is also valuable for locating vehicles when the operator activates a silent alarm, which can be of great importance in enhancing the security of operators who may have to drive near-empty buses at night in dangerous or isolated areas.

AVL systems are spreading slowly in the transit community because of their high costs and because of questions about how valuable location information is in practice. Besides cost, another limiting factor is the number of radio channels available. The following formula shows the relationship between the various parameters of an AVL system:

$$\text{Polling cycle} = \frac{\text{No. of vehicles}}{\text{No. of channels}} \times \text{Poll length} \tag{58.14}$$

For example, if there is only one channel and each poll takes 2 s, then monitoring 400 buses will require a polling cycle of 800 s (more than 13 min)—an unacceptably long time. In order to poll

400 vehicles every 120 s, 7 channels will be needed. In a large city, obtaining permission from the appropriate authorities to use 7 radio channels for an AVL system, in addition to the channels needed for voice communication, can be difficult. It is obviously desirable to reduce the poll length, which depends primarily on how many bits of information are sent during a poll.

Passenger Counting

Transit systems do not issue point-to-point tickets to all passengers as airlines do, and so they must rely on counts and samples. Common types of counts and samples are:

- *Farebox or driver counts.* In some cities, it is policy that every passenger is counted. The current generation of electronic fareboxes makes it possible to count passengers by numerous fare categories by vehicle, with a separate count for each trip.
- *Revenue.* Closely related to ridership, revenue is always counted systemwide, sometimes by farebox or turnstile, sometimes by time of day or route.
- *Ride checks.* An on-board checker records ons and offs by stop, as well as time at key points. Usually checkers are full-time employees of the transit system; sometimes temporary help is used. Hand-held electronic units with stored stop lists make collection and processing easier.
- *Point checks.* A wayside checker records the load of passing vehicles and the time. Accuracy can be questionable, and tinted windows and security considerations limit their use.
- *"No questions asked" surveys.* These involve distributing to each boarding passenger a card coded by origin stop (or segment) and collecting the cards as passengers alight, filing them by destination stop so that both boarding stop and alighting stop are known for each passenger [Stopher *et al.*, 1985]. Response rate is usually over 90%, and so the resulting O-D matrix is quite reliable provided the sample size is large enough.
- *Passenger surveys.* These surveys can request a variety of information used for planning and marketing, such as trip purpose, questions about travel habits (do you have a car? how frequently do you use transit?), trip origin and destination, transfers made, and customer satisfaction. Response rate can vary widely, sometimes as low as 20%. When the response rate is low, nonresponse bias becomes an issue that puts in question the validity of the expanded results. However, this is the only practical method to obtain much of this information and to learn about *linked* (or *revenue) trips*—passenger trips from their initial origin to final destination, including accessing the transit system and transfers between routes.

Other data sources include turnstile counts, ticket and pass sales, and transfer counts and surveys.

A few transit systems use automatic passenger counters, which detect and counts ons and offs by stop. Detection is based on either infrared beams across the doorway or instrumented treadle mats. Stop location has to be inferred from odometer and clock readings that are automatically recorded.

Most of the counts listed above are samples—they do not count every passenger, every day. Section 58.9 deals with making estimates from samples and determining their statistical validity and, conversely, determining the sample size needed to ensure a statistically valid estimate.

Service Standards

Effective management is assisted by adopting a set of measures of productivity, efficiency, and service quality that are regularly monitored. For most measures a minimally acceptable level is established based on management goals; this level is called a *performance standard* or a *service standard.* By comparing measures of performance with service standards, performance can be evaluated and, hopefully, improved [Wilson *et al.*, 1984].

Service standards at the system level are effective for monitoring the performance of an overall operating strategy. More helpful, however, are standards at the route level or route/direction/period

(R/D/P) level, which can be useful for monitoring individual services, evaluating service changes, and suggesting service improvements.

Some performance measures apply equally to transit and other service industries, such as absenteeism rate and measures of performance that relate to finance or data-processing functions. Likewise, service standards in the vehicle maintenance function are very helpful. This section concentrates on measures that are particular to the transportation function of transit.

One common group of service standards is productivity and economic standards. Productivity standards usually take the form of a ratio of a measure of utilization to a measure of input, such as passengers per vehicle-hour, passengers per vehicle-mile, passenger-miles per vehicle-mile (average payload). Routes that fail to meet standards may become candidates for elimination or remedial action with more careful monitoring. Comparing the performance of different routes and observing a route's performance over time can suggest improvements or be helpful in evaluating the effect of service or policy changes. Economic performance measures include revenue per vehicle-hour or per vehicle-mile, revenue per passenger (average fare), cost per vehicle-hour and per vehicle-mile, cost per passenger, and the ratio of revenue to operating cost, known as the *recovery ratio*. Some of these measures may be unavailable at the route level because of the difficulty in estimating route level revenue—for example, due to a high level of monthly pass usage—or from lack of a reliable route-costing model. Some systems have politically mandated recovery ratios; if they fail to meet the standard, they must either reduce operating costs or raise fares.

Efficiency standards are ratios of input to output, such as vehicle-hours per pay hour. Indicators of negative output likewise belong in this category, such as percent missed trips and accidents or breakdowns per vehicle-mile. Because efficiency measures are more under the control of the transit agency than are productivity measures, the agency can be held more accountable for them. In contrast, low productivity could be caused in part by bad management and in part by erosion in the demand for transit due to increased auto ownership or economic recession.

Service quality measures relate to the value or quality of the service as perceived by the passenger. They include relatively static measures that relate to the route network or the schedule. Examples are percentage of dwellings or jobs within 0.5 mi of a transit station; number of departures in the A.M. peak (e.g., on a commuter rail or ferry route); and vehicle-mi of service offered at night. Other measures require monitoring of actual service. The most direct measures of service quality are measures of crowding, on-time performance or waiting time, and travel time. Because the first two of these are strongly affected by randomness in operations, they will be elaborated upon in the following paragraphs. Measures of safety (injuries per 100,000 mi) and passenger satisfaction (complaints or survey results) can be important as well.

For services with a headway of 10 min or more, for which passengers tend to consult the timetable, on-time performance is measured against scheduled departure times. It is common to measure the percentage of trips that are early, on time, and late. "On time" is usually defined to be 0 to 5 min late. Early trips are especially cause for concern, since they are inexcusable and can leave a passenger stranded for a full headway.

For more frequent services, on-time performance is measured based on headway, since passengers are not usually aiming for a specific trip, but hope to enjoy a short wait. The average wait during a given headway, assuming that passengers arrive at random, is half the headway. But if headways are not regular—that is, if some headways are large while others are small—the average wait is not half the average headway; it is larger than that because, even if there are equal numbers of long headways and short headways, if passengers arrive at random, more will arrive during long headways than short, and thus more will experience a long wait. In fact, assuming that passengers arrive at random and can board the first vehicle that comes by, the average wait is given by the formula

$$\text{Average wait} = \frac{\text{Average headway}}{2}(1 + v_h^2) \tag{58.15}$$

where v_h = coefficient of variation of headway. A reduction in average wait can therefore be accomplished either by reducing the average headway or by reducing headway irregularity. The former remedy implies providing additional service, since average headway is the inverse of service frequency, while the latter requires only better control. For example, with perfectly regular headways, $v_h = 0$, so the average wait is half the average headway. But if vehicles come in bunches of two, with headways alternating between 0 and twice the average headway, $v_h = 1$ and the average wait equals the average headway. Likewise, if headways are so random that they follow the exponential probability distribution, $v_h = 1$ and the average wait equals the average headway. A good measure of service quality with respect to on-time performance, then, is average experienced wait, which can be divided into the scheduled wait time (half the scheduled headway) and the balance (increase due to missed trips and headway irregularities). Sometimes the balance is negative because extra service has been provided.

Another measure of on-time performance for frequent services is the percentage of passengers who wait one scheduled headway or less [Wilson *et al.*, 1992]. Assuming that passengers arrive at a steady rate during the period in question, if the scheduled headway is, say, 5 min, and all of the actual headways are 5 min or less, then 100% of the passengers wait less than a scheduled headway. Now suppose that, over the course of an hour, there is a 6-min headway, a 7-min headway, three 4-min headways, and seven 5-min headways. Then 3 minutes' worth of passengers (those arriving during the first minute of the 6-min headway and those arriving during the first two minutes of the 7-min headway) have to wait more than 5 min, so only 57/60 or 95% of the passengers wait less than a scheduled headway. A drawback of this measure is that it tends to be worse for lower headway services, when in fact having to wait more than one headway when the headway is very small (say, 3 min) is not as serious as extra waiting time when the headway is longer.

Crowding, like average waiting time, depends on both the frequency and the regularity of service, as well as on the passenger demand. However, because load variations are primarily due to headway variations, it is not usually the practice to measure variations in load (the airlines are an exception in this regard). Measuring average load will be sufficient to see that enough overall capacity is provided, and measuring headway variation will indicate whether there is sufficient control to keep headways regular and loads balanced.

However, to understand the passenger's experience, it is helpful to know that the average experienced load at a point is given by

$$\text{Average experienced load} = (\text{Average load}) \times (1 + v_{\text{load}}^2) \tag{58.16}$$

where v_{load} = coefficient of variation of load at that point. That average experienced load is greater than average load is clear if one considers two trips, one with a large load and the other with a small load. More passengers are on the first one, and so more than half the passengers experience the larger load. Therefore the average experienced load is greater than the average load of the two trips.

Common practice is to measure load at the peak point of the route, although that may not be the peak point for each individual trip. Moreover, load measured at a single point fails to distinguish between crowding that lasts for a long time and crowding that occurs on a brief segment only; on the other hand, load averaged over every route segment tends to hide crowding where it does occur. To further complicate things, the disutility to passengers due to crowding is highly nonlinear. As long as the load is less than the number of seats, there is very little disutility from increasing load. Once there are standees, the marginal effect of additional passengers becomes more and more severe as the crowding increases. No satisfactory measure of crowdedness has been accepted that reflects the passenger's viewpoint of all of these aspects of crowding.

Data Collection Program Design

Every transit agency has a data collection program, although with some it is more formalized than with others. This program consists of a set of data collection activities that are performed regularly;

a system of recording, processing, storing, and reporting the data; and a set of measures that are calculated and standards they are compared against [Furth *et al.*, 1985].

The design of the data collection program should first pay attention to data needs, which arise from (1) primary needs of various departments, such as scheduling, planning, budgeting, and marketing; (2) external reporting requirements; and (3) service standards. Second, methods of data collection should be determined. Where automated systems, such as electronic farebox systems, have been already installed, attention should be given to making full use of their data. Where automated systems have not been installed, a full range of methods, from manual to fully automated, can be considered. In many cases the most economical solution is technology-enhanced manual techniques, such as using hand-held devices. Third, for items requiring sampling, a sampling strategy and sample sizes must be determined to meet statistical accuracy requirements. Fourth, economies arising from overlapping needs should be identified. For example, a point checker at a single terminal can gather data to meet several needs, such as estimating average load and on-time performance on all the routes that pass that point. By stationing checkers simultaneously at the opposite ends of some of those routes, the same checker's data become useful for estimating running time. Finally, a schedule of data collection activities can be developed that meets the sample size requirements efficiently. It must be coupled with a system for obtaining counts that are not sampled (e.g., revenue counts or counts from turnstiles or the farebox system).

The plan for gathering the data must then be complemented with a plan for processing, reporting, and storing the data. Modern database and other data-processing software can greatly expedite this task. Efforts should be made to avoid "information overload" by reporting only what will actually be used. Regular communication between the users of the data and those responsible for gathering and processing the data is absolutely vital to keep the data collection program responsive, to correct errors, and to ensure the best use of the information. If data gatherers perceive that the data they are gathering is not being used, data quality will eventually deteriorate. On the other hand, if they are given rapid feedback concerning the value of the data, its quality will improve.

58.9 Ridership Estimation and Sampling

Ridership and Passenger-Miles Estimation

Knowing the current ridership is important for planning and scheduling service, estimating transit's benefits, monitoring service effectiveness, and meeting reporting requirements of funding agencies. All operators receiving federal operating assistance (which amounts to nearly all transit operators) must, at a minimum, meet the uniform reporting requirements of Section 15 of the Federal Transit Act (formerly called the Urban Mass Transportation Act of 1964), as amended. Although Section 15 deals mostly with accounting information, it also requires annual estimates of boardings and of passenger-miles for each mode with an accuracy of $\pm 10\%$ precision at the 95% confidence level.

In some transit systems all passengers are counted as a matter of course. However, in many systems—including most large bus systems, barrier-free rail systems and elsewhere—ridership must be estimated by sampling. In almost all systems, passenger-miles estimates must be made based on sampling. Estimates of other demand measures, including peak load and O-D matrices, also depend on sampling.

Direct Estimation with Simple Random Sampling

To estimate mean boardings per trip, mean load at a given point, or passenger-miles per trip, the item of interest can be measured for a sample of n trips and the mean calculated as

$$\bar{y} = \frac{1}{n} \sum y_i \tag{58.17}$$

where y_i = value for trip $i, \bar{y}$ is the sample mean, and n is the sample size. Simple random sampling is most easily accomplished by constructing a sampling frame (a list of all the trips) and using a random number sequence to select trips from that frame. "Trip" in this context can mean a one-way trip or a round trip; sampling by round trip is usually more efficient when checkers are involved since they almost always have to return to their starting point. If round trips are the basic sampling unit, the sampling frame is a list of round trips. One-way trips without a natural pair can either stand on their own or be linked to another round trip.

The sample variance is

$$s^2 = \frac{1}{n-1}\sum(y_i - \bar{y})^2 \tag{58.18}$$

The (absolute) tolerance and the (relative) precision of the estimate are

$$\text{Precision} = \frac{ts}{\bar{y}\sqrt{n}}, \qquad \text{Tolerance} = \frac{ts}{\sqrt{n}} \tag{58.19}$$

where t is the ordinate of the t-distribution for the desired confidence level with $n - 1$ degrees of freedom (t values are tabulated in most statistics texts). At the 95% confidence level, with a sample size of over 30, $t \approx 2.0$. For example, if $\bar{y} = 40, s = 24$, and $n = 64$, the tolerance at the 95% confidence level is 6 and the precision is 0.15, or $\pm 15\%$. This means that one can be 95% confident that the true mean lies within 15% of the estimated mean, that is, in the interval $[40 \pm 6]$. To expand the result to a system total such as total annual boardings, simply multiply the mean by the number of trips in the sampling frame. The tolerance expands likewise, whereas the precision remains unchanged.

To find the sample size necessary to achieve a given precision, these formulas can be reversed. Of course, an estimate of the coefficient of variation ($s/\bar{y}$) will be needed. It is best obtained from historical data. If historical data are unavailable, default coefficients of variation for various measures of interest are found in Furth *et al.* [1985].

For example, suppose we wish to estimate annual passenger-miles to a precision of $\pm 10\%$ at the 95% confidence level. A sample of round trips will be selected for which on/off counts at every stop will be done, and passenger-miles calculated using Eq. (58.3). How large a sample will be needed? The necessary sample size, reversing Eq. (58.17) and using $t = 2$, is

$$n = 2.0^2 \frac{(s/\bar{y})^2}{(0.10)^2}$$

Assuming that $(s/\bar{y})$ was estimated from previous data on round trips to be 1.05, $n = 440$ round trips will be needed. They should be spread throughout the entire year, doing 8 one week and 9 the next, with random sampling within the week. To randomly select 8 or 9 round trips within a week, determine N_{wk} = the number of round trips operated in that week (usually, it's 5 times the number of round trips on the weekday schedule plus the number of weekend round trips) and assign to each trip a sequence number from 1 to N_{wk}. Then select 8 or 9 random numbers between 1 and N_{wk}.

Quarterly estimates, based on only a quarter of the data, will have a precision twice as large as annual estimates, since precision is inversely proportional to the square root of sample size. To get an annual precision of $\pm 1\%$, 100 times more data would be needed than to achieve a precision of $\pm 10\%$.

Using Conversion Factors

When the variable to be estimated is closely related to another variable whose total is known, sampling can be done to estimate the ratio, or conversion factor, between the item of interest and the related or *auxiliary item* [Furth and McCollom, 1987]. For example, if total boardings is known from electronic farebox counts, passenger-miles can be estimated by first estimating average passenger-trip length (APL) and then expanding it by total boardings. APL is the ratio of passenger-miles (the item of interest) to boardings (the auxiliary item), which can be estimated from on/off counts made on a sample of trips. Or if total boardings is not known but total revenue is, a passenger-miles–to–revenue ratio can be estimated by measuring passenger-miles and cash revenue on a sample of trips. This technique can be far less costly than simple direct estimation.

Again, a simple random sample of n trips (either one-way or round trips can be the basic unit) is needed; for each sampled trip, both the item of interest (y) and the auxiliary item (x) are measured. The estimate of the ratio r is

$$r = \frac{\sum y_i}{\sum x_i} \tag{58.20}$$

and the estimated total is found by simple expansion:

$$Y_{\text{total}} = rX_{\text{total}} \tag{58.21}$$

The relative variance per sampled trip is

$$u_r^2 = \frac{1}{\bar{y}^2(n-1)} \sum (y_i - rx_i)^2 \tag{58.22}$$

and the precision of both the ratio and the expanded total is

$$\text{Precision} = \frac{tu_r}{\sqrt{n}} \tag{58.23}$$

This last formula can be inverted to determine necessary sample size, using historical data to determine u_r. For example, suppose electronic fareboxes count all boardings, and a sample of ride checks on round trips will be made to estimate the ratio of passenger-miles to boardings. From a historical sample of round trips, u_r is estimated [Eq. (58.22)] to be 0.6. The necessary sample size to achieve 10% precision at the 95% confidence level would be

$$n = \left(\frac{tu_r}{\text{Precision}}\right)^2 = \left(\frac{(2.0)(0.6)}{0.1}\right)^2 = 144 \tag{58.24}$$

The resulting sampling plan would probably call for 3 round trips per week.

Other Sampling Techniques

It is possible to meet Section 15 sampling requirements by following one of two specified sampling plans, using direct estimation (FTA Circular 2710.1) or ratio-to-revenue estimation (Circular 2710.4), avoiding the need to do statistical analysis. These plans require annually about 550 and 208 individually sampled one-way trips, respectively.

More advanced sampling techniques—including stratified sampling, cluster sampling, and multistage sampling—can be effective in improving precision or reducing necessary sampling size. For example, in estimating passenger-miles using a ratio to boardings, stratifying routes by length can be extremely effective. Cluster sampling can reduce costs by allowing for samples to be taken on efficient clusters of trips. On a system with only one or two lines (e.g., a light-rail system) and a high degree of precision specified, the sample size may be so great that every trip in the weekday schedule can be sampled at least once. Using two-stage sampling, the effect of variance between different scheduled trips can be eliminated due to the finite population correction, and only the variance between days for each scheduled trip will affect the precision. The same applies to a bus route that gets a so-called "100% ride check" in which every trip in the schedule is checked, usually on the same day. The result is still a sample, albeit one in which the trip-to-trip variation has been eliminated and only the day-to-day variation remains.

Estimating a Route-Level Origin-Destination Matrix

An O-D matrix for a route/direction/period (R/D/P) is a valuable input for designing service changes along a route such as short-turns, express service, or route restructuring. There are two general methods for estimating an R/D/P level O/D matrix: direct estimation and updating. Direct estimation is best done using the "no questions asked" survey on a sample of trips (see section 58.8) and simply expanding the results. This type of survey has a response rate near 100% and does not suffer from biases caused by low response rate in questionnaire-type surveys.

The simplest updating method relies on a sample of ride checks to obtain on and off totals at each stop for the R/D/P, which serve as row and column totals for the O/D matrix. Updating begins with a seed matrix, which can be an old O/D matrix, if available; a small-sample O/D matrix obtained through a questionnaire-type survey; or a matrix of propensities, as used in a gravity model [Ben-Akiva *et al.*, 1985]. Each row is balanced (factored proportionately to match its target row total); then each column is balanced likewise. The process is repeated iteratively until no more adjustments are needed. This method, known as either *biproportional method* or *iterative proportional fit*, is wholly equivalent to the doubly constrained gravity model.

References

Babin, A., Florian, M., James-Lefebvre, L., and Spiess, H. 1982. EMME/2: Interactive graphic method for road and transit planning. *Transportation Research Record.* 866:1–9.

Ben-Akiva, M., Macke, P., and Hsu, P. S. 1985. Alternative methods to estimate route level trip tables and expand on-board surveys. *Transportation Research Record.* 1037:1–11.

Blais, J. Y., Lamont, J., and Rousseau, J. M. 1990. The HASTUS vehicle and manpower scheduling system at the Societe de Transport de la Communaute Urbaine de Montreal. *Interfaces.* 20(1):26–42.

Bonsall, J. A. 1987. *Transitways—The Ottawa Experience.* OC Transpo, Ottawa.

Booz, Allen & Hamilton. 1981. *Bus Route Costing Procedures, Interim Report No. 2: Proposed Method.* Report no. UMTA-IT-09-9014-81-1. Urban Mass Transportation Administration.

Ceder, A. and Stern, H. 1981. Deficit function bus scheduling with deadheading trip insertions for fleet size reduction. *Transportation Science* 15:338–363.

Cervero, R. 1981. Efficiency and equity impacts of current transit fare policies. *Transportation Research Record.* 790:7–15.

Chapleau, R. 1986. *Transit Network Analysis and Evaluation With a Totally Different Approach Using MADITUC.* Ecole Polytechnique, Montreal.

Charles River Associates, Inc., and H. S. Levinson. *Characteristics of Urban Transportation Demand—An Update* (rev. ed.) 1988. UMTA Report no. DOT-T-88-18. U.S. Department of Transportation, Washington, D.C.

Cherwony, W. and Mundle, S. R. 1980. Transit cost allocation model development. *Transportation Engineering Journal of ASCE.* 106 (TE1):31–42.

Furth, P. G. Alternating deadheading in bus route operations. 1985. *Transportation Science.* 19:13–28.

Furth, P. G. Zonal route design for transit corridors. 1986. *Transportation Science.* 20:1–12.

Furth, P. G. Short-turning on transit routes. 1988. *Transportation Research Record.* 1108:42–52.

Furth, P. G., Attanucci, J. P., Burns, I., and Wilson, N. H. 1985. *Transit Data Collection Design Manual.* Report DOT-I-85-38. U.S. Department of Transportation, Washington, D.C.

Furth, P. G. and Day, F. B. Transit routing and scheduling strategies for heavy demand corridors. 1985. *Transportation Research Record.* 1011:23–26.

Furth, P. G., Killough, K. L., and Ruprecht, G. F. Cluster sampling techniques for estimating transit system patronage. 1988. *Transportation Research Record.* 1165:105–114.

Furth, P. G. and McCollom, B. Using conversion factors to lower transit data collection costs. 1987. *Transportation Research Record.* 1144:1–6.

Furth, P. G. and Wilson, N. H. 1981. Setting frequencies on bus routes: Theory and practice. *Transportation Research Record.* 818:1–7.

Gomez-Ibanez, J. A. and Meyer, J. R. 1993. *Going Private: The International Experience with Transit Privatization.* Brookings Institute, Washington, D.C.

Hasselstrom, D. 1981. *Public Transportation Planning: A Mathematical Programming Approach.* PhD Thesis. Department of Business Administration. University of Gothenburg, Sweden.

Lampkin, W. and Saalmans, P. D. 1967. The design of routes, service frequencies and schedules for a municipal bus undertaking: A case study. *Operations Research Quarterly.* 18(4):375–397.

Levinson, H. S., Adams, C. L., and Hoey, W. F. 1975. *Bus Use of Highways: Planning and Design Guidelines.* NCHRP Report 155. Transportation Research Board, Washington, D.C.

Mayworm, P., Lago, A. M., and McEnroe, J. M. 1980. *Patronage Impacts on Changes in Transit Fares and Services.* Report no. 1205-UT. U.S. Department of Transportation, Washington, D.C.

Metropolitan Transit Authority of Harris County (METRO). 1984. *Bus Service Evaluation Methods: A Review.* Report no. DOT-I-84-49. U.S. Department of Transportation, Washington, D.C.

Pickrell, D. H. 1983. *The Causes of Rising Transit Operating Deficits.* Report no. DOT-I-83-47. U.S. Department of Transportation, Washington, D.C.

Pickrell, D. H. 1989. *Urban Rail Transit Projects: Forecast Versus Actual Ridership and Costs.* Prepared by the Transportation Systems Center for UMTA, U.S. Government Printing Office, Washington, D.C.

Pushkarev, B. S. and Zupan, J. M. 1977. *Public Transportation and Land Use Policy.* Indiana University Press, Bloomington.

Salzborn, F. J. 1969. Timetables for a suburban rail transit system. *Transportation Science.* 3:297–316.

Scott, D. 1984. *A Method for Scheduling Urban Transit Vehicles Which Takes Account of Operation Labor Cost.* Publication #365. Centre de Recherche sur les Transports, Univ. de Montreal.

Stopher, P. R., Shillito, L., Grober, D. T., and Stopher, H. M. 1985. On-board bus surveys: No questions asked. *Transportation Research Record.* 1085:50–57.

TRB. 1980. *Bus Route and Schedule Planning Guidelines,* NCHRP Synthesis of Highway Practice 69. Transportation Research Board, Washington, D.C.

Turnquist, M. 1979. Zone scheduling of urban bus routes. *Transportation Engineering Journal* 105(1):1–12.

Vuchic, V. R. 1981. *Timed Transfer System Planning, Design, and Operation.* Prepared for UMTA University Research and Training Program. Report no. PA-11-0021. Department of Civil and Urban Engineering, University of Pennsylvania, Philadelphia.

Wilson, N. H., Bauer, A., Gonzalez S., and Shriver, J. 1984. Short range transit planning: Current practice and a proposed framework. Report DOT-I-84-44. U.S. Department of Transportation, Washington, D.C.

Wilson, N. H., Nelson, D., Palmere, A., Grayson, T., and Cederquist, C. 1992. Service quality monitoring for high-frequency transit lines. *Transportation Research Record.* 1349:3–11.

Zimmerman, S. L. 1989. UMTA and major investments: Evaluation process and results. *Transportation Research Record.* 1209:32–36.

For Further Information

Valuable comprehensive texts are

Canadian Transit Handbook (3rd edition), Canadian Urban Transit Association, 1993.

Gray, G. E. and L. A. Hoel (eds.), *Public Transportation* (2nd edition), Prentice Hall, 1992.

Vuchic, V. R., *Urban Public Transportation Systems and Technology,* Prentice Hall, 1981.

For a more thorough treatment of transit management, its political environment, and terminology, see

Altshuler, A., *The Urban Transportation System: Politics and Policy Innovation,* MIT Press, 1979.
Fielding, G. J., *Managing Public Transportation Strategically: A Comprehensive Approach to Strengthening Service and Monitoring Performance,* Jossey-Bass, 1987.
Smerk, G. M., *Mass Transit Management: A Handbook for Small Cities; Part 1: Goals, Support and Finance; Part 2: Management and Control; Part 3: Operations; Part 4: Marketing* (3rd ed., rev.), prepared by Indiana University Institute for Urban Transportation for UMTA, Report no. DOT-T-88-12, 1988.
White, P. R., *Public Transport, Its Planning, Management and Operation,* London: Hutchinson, 1986.
Gray, B. H. (ed.), *Urban Public Transportation Glossary,* Transportation Research Board, 1989.

More detail on advanced technology and software can be found in

Odoni, A. R., J. M. Rousseau, and N. H. Wilson, Models in urban and air transportation, in *Handbook on Operations Research and the Public Sector,* Elsevier, 1994.
Paixao, J. and J. R. Daduna (eds.), *Computer-Aided Transit Scheduling,* Springer-Verlag, 1994.
Rousseau, J. M. (ed.), *Computer Scheduling of Public Transport 2,* Elsevier Science Publishers B.V., North-Holland, 1985.
Wren, A. (ed.), *Computer Scheduling of Public Transport Urban Passenger Vehicle and Crew Scheduling,* Elsevier Science Publishers B.V., North-Holland, 1981.
U.S. Department of Transportation, *Advanced Public Transportation Systems: The State of the Art,* Report DOT-VNTSC-UMTA-91-2 and updated annually, U.S. Department of Transportation.

Many helpful conferences are held and reports published by the following organizations:

Federal Transit Administration, U.S. Department of Transportation
American Public Transportation Association
Canadian Urban Transit Association
Transportation Research Board

59

Highway and Airport Pavement Design

T. F. Fwa
National University of Singapore

59.1 Introduction

Pavements are designed and constructed to provide durable all-weather traveling surfaces for safe and speedy movement of people and goods with an acceptable level of comfort to users. These functional requirements of pavements are achieved through careful considerations in the following aspects during the design and construction phases: (a) selection of pavement type, (b) selection of materials to be used for various pavement layers and treatment of subgrade soils, (c) structural

0-8493-8953-4/95/$0.00 + $.50

thickness design for pavement layers, (d) subsurface drainage design for the pavement system, (e) surface drainage and geometric design, and (f) ridability of pavement surface.

The two major considerations in the structural design of highway and airport pavements are material design and thickness design. Material design deals with the selection of suitable materials for various pavement layers and mix design of bituminous materials (for flexible pavement) or portland cement concrete (for rigid and interlocking block pavements). These topics are discussed in other chapters of this handbook. This chapter presents the concepts and methods of pavement thickness design. As the name implies, *thickness design* refers to the procedure of determining the required thickness for each pavement layer to provide a structurally sound pavement structure with satisfactory performance for the design traffic over the selected design life. *Drainage design* examines the entire pavement structure with respect to its drainage requirements and incorporates facilities to satisfy those requirements.

59.2 Pavement Types and Materials

Flexible versus Rigid Pavement

Traditionally, pavements are classified into two categories, namely flexible and rigid pavements. The basis for classification is the way by which traffic loads are transmitted to the subgrade soil through the **pavement structure.** As shown in Fig. 59.1, a **flexible pavement** provides sufficient thickness for load distribution through a multilayer structure so that the stresses and strains in the subgrade soil layers are within the required limits. It is expected that the strength of subgrade soil would have a direct bearing on the total thickness of the flexible pavement. The layered pavement structure is designed to take advantage of the decreasing magnitude of stresses with depth.

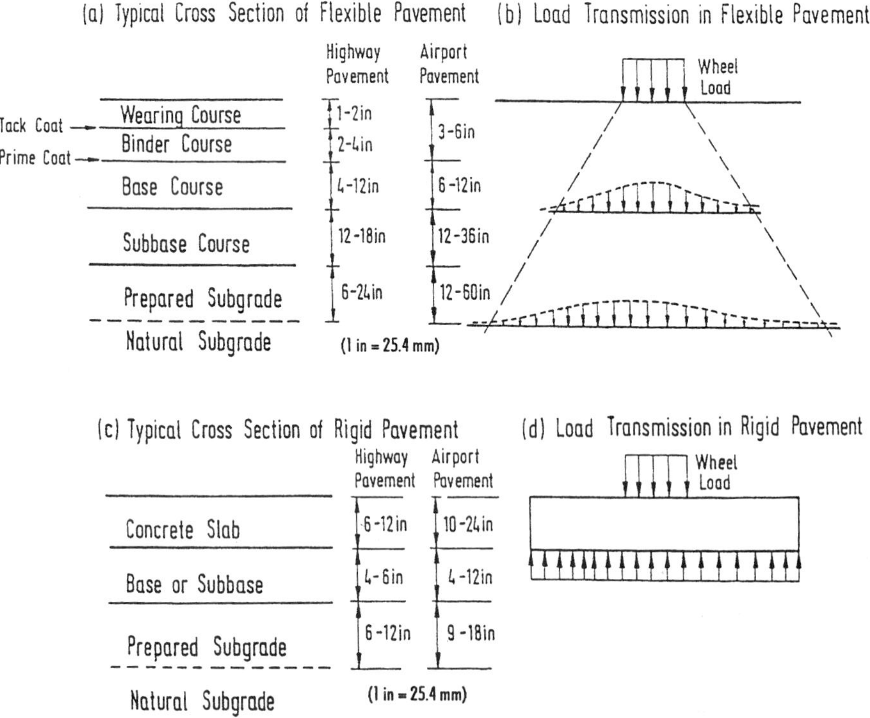

FIGURE 59.1 Flexible and rigid pavements.

A **rigid pavement**, by virtue of its rigidity, is able to effect a slab action to spread the wheel load over the entire slab area, as illustrated in Fig. 59.1. The structural capacity of the rigid pavement is largely provided by the slab itself. For the common range of subgrade soil strength, the required rigidity for a portland cement concrete slab (the most common form of rigid pavement construction) can be achieved without much variation in slab thickness. The effect of subgrade soil properties on the thickness of rigid pavement is therefore much less important than in the case of flexible pavement.

Layered Structure of Flexible Pavement

Surface Course

In a typical conventional flexible pavement, known as **asphalt pavement,** the surface course usually consists of two bituminous layers—a wearing course and a binder course. To provide a durable, watertight, smooth riding, and skid-resistant traveled surface, the wearing course is often constructed of dense-graded hot mix asphalt with polish-resistant aggregate. The binder course generally has larger aggregates and less asphalt. The composition of the bituminous mixtures and the nominal top size aggregates for the two courses is determined by the intended use, desired surface texture (for the case of wearing course), and layer thickness. A light application of tack coat of water-diluted asphalt emulsion may be used to enhance bonding between the two courses. Table 59.1 shows selected mix compositions listed in ASTM Standard Specification D3515 [1992]. Open-graded wearing courses, some with air void exceeding 20%, have also been used to improve skid resistance and reduce splash during heavy rainfall by acting as a surface drainage layer.

Base Course

Base and subbase layers of the flexible pavement make up a large proportion of the total pavement thickness needed to distribute the stresses imposed by traffic loading. Usually base course also serves as a drainage layer and provides protection against frost action. Crushed stone is the traditional material used for base construction to form what is commonly known as the *macadam base course*. In this construction, choking materials consisting of natural sand or the fine product resulting from crushing coarse aggregates are added to produce a denser structure with higher shearing

Table 59.1 Example Composition of Dense Bituminous Paving Mixtures

	Mix Designation and Nominal Maximum Size of Aggregate					
Sieve Size	2 in. (50 mm)	1 ½ in. (37.5 mm)	1 in. (25.0 mm)	3/4 in. (19.0 mm)	1/2 in. (12.5 mm)	3/8 in. (9.5 mm)
2 ½ in.	100	—	—	—	—	—
2 in.	90–100	90–100	100	—	—	—
1 ½ in.	—	90–100	100	—	—	—
1 in.	60–80	—	90–100	100	—	—
3/4 in.	—	56–80	—	90–100	100	—
1/2 in.	35–65	—	56–80	—	90–100	100
3/8 in.	—	—	—	56–80	—	90–100
No. 4	17–47	23–53	29–59	35–65	44–74	55–85
No. 8	10–36	15–41	19–45	23–49	28–58	32–67
No. 16	—	—	—	—	—	—
No. 30	—	—	—	—	—	—
No. 50	3–15	4–16	5–17	5–19	5–21	7–23
No. 100	—	—	—	—	—	—
No. 200	0–5	0–6	1–7	2–8	2–10	2–10

Note: Numbers in table refer to percent passing by weight.

Source: ASTM. 1992. ASTM Standard Specification D3515-84. *Annual Book of ASTM Standards.* Vol. 04.03—Road and Paving Materials; Travelled Surface Characteristics. With permission.

Table 59.2(a) Grading Requirements for Unbound Subbase and Base Materials—ASSHTO Designation M147-65 (1989)

	Grading: Percentage Passsing					
Sieve Size	A	B	C	D	E	F
50 mm	100	100				
25 mm	—	75–95	100	100	100	100
9.5 mm	30–60	40–75	50–85	60–100	—	—
4.75 mm	25–55	30–60	35–65	50–85	55–100	70–100
2 mm	15–40	20–45	25–50	40–70	40–100	55–100
425 μm	8–20	15–30	15–30	25–45	20–50	30–70
75 μm	2–8	5–20	5–15	5–20	6–20	8–25

Other requirements:

1. Coarse aggregate (> 2 mm) to have a percentage wear by Los Angeles test not more than 50.

2. Fraction passing 425 μm sieve to have a liquid limit not greater than 25% and a plasticity index not greater than 6%.

Source: AASHTO. 1989. AASHTO Designation M147–65. *AASHTO Standard Specifications for Transportation Materials and Methods of Sampling and Testing.* Copyright 1989 by the American Association of State Highway and Transportation Officials, Washington, D.C. Used by permission.

resistance. Such base courses are called by different names depending on the construction method adopted.

Dry-bound macadam is compacted by means of rolling and vibration that work the choking materials into the voids of larger stones. For water-bound macadam, after spreading of the choking materials, water is applied before the entire mass is rolled. Alternatively, a wet-mix macadam may be used by premixing crushed stone or slag with a controlled amount of water. The material is spread by a paving machine and compacted by vibrating roller. Table 59.2 shows specifications for unbound base and subbase materials specified by AASHTO and ASTM.

Granular base materials may be treated with either asphalt or cement to enhance load distribution capability. Bituminous binder can be introduced by spraying heated asphalt cement on consolidated and rolled crushed stone layer to form a penetration macadam road base. Alternatively, bituminous road bases can be designed and laid as in the case for bituminous surface courses. Cement-bound granular base material is plant mixed with an optimal moisture content for compaction. It is laid by paver and requires time for curing. Lean concrete base has also been used successfully under flexible pavements. Table 59.3 shows examples of grading requirements for these materials.

Table 59.2(b) ASTM Designation D2940–74 (Reapproved 1985)

	Grading: Percentage Passing	
Sieve Size	Bases	Subbases
50 mm	100	100
37.5 mm	95–100	90–100
19 mm	70–92	—
9.5 mm	50–70	—
4.75 mm	35–55	30–60
600 μm	12–25	—
75 μm	0–8	0–12

Other requirements:

1. Fraction passing the 75 μm sieve not to exceed 60% of the fraction passing the 600 μm sieve.

2. Fraction passing the 425 μm sieve shall have a liquid limit not greater than 25% and a plasticity index not greater than 4%.

Source: ASTM. 1992. ASTM Standard Specification D2940–74 (reapproved 1980), *Annual Book of ASTM Standards.* Vol. 04.03—Road and Paving Materials; Travelled Surface Characteristics. With permission.

Subbase Course

The subbase material is of lower quality than the base material in terms of strength, plasticity, and gradation, but is superior to the subgrade material in these properties. It may be compacted granular material or stabilized soil, thus allowing building up of sufficient thickness for the pavement structure at relatively low cost. On a weak subgrade it also serves as a useful working

Table 59.3 Requirements for Stabilized Base Courses

Specification	Cement Treated			Bituminous Treated		
	Class A	Class B	Class C	Class 1	Class 2	Lime Treated
(a) Stabilized Base Courses for Flexible Pavements						
Percent passing						
2 ½ in.	100	100	100			
3/4 in.	—	—	75–95			
No. 4	65–100	55–100	25–60			
No. 10	20–45	—	15–45			
No. 40	15–30	25–50	8–30			
No. 200	5–12	5–20	2–15			
7-day f_c (psi)	650–1000	300–650				
S (lb)	—	—	—	750 min	500 min	—
F (0.01 in.)	—	—	—	16 max	20 max	—
PI	12 max	—	—	6 max	6 max	6 max

Specification	Type A (Open Graded)	Type B (Dense Graded)	Type C (Cement Graded)	Type D (Lime Treated)	Type E (Bituminous Treated)	Type F (Granular)
(b) Base Materials for Concrete Pavement						
Percent passing						
1 ½ in.	100	100	100	—	—	100
3/4 in.	60–90	85–100	—	*	*	—
No. 4	35–60	50–80	65–100	—	—	65–100
No. 40	10–25	20–35	25–50	—	—	25–50
No. 200	0–7	5–12	5–20	—	—	0–15
(The minus No. 200 material should be held to a practical minimum)						
28-day f_c (psi)	—	—	400–750	100	—	—
S (lb)	—	—	—	—	500 min	—
F (0.01 in.)	—	—	—	—	20 max	—
Soil constants:						
LL	25 max	25 max	—	—	—	25 max
PI*	N.P.	6 max	10 max	—	6 max	6 max

Notes:
*To be determined by complete laboratory analysis, taking into consideration the ability of the stabilized mixture to resist under-slab erosion.

f_c = compressive strength as determined in unconfined compression tests on cylinders 4 inches in diameter and 4 inches high. Test specimens should contain the same percentage of portland cement and be compacted to the same density as achieved in construction.

S = Marshall stability.

F = Marshall flow.

PI = plasticity index performed on samples prepared in accordance with AASHTO Designation T-87 and applied to aggregate prior to mixing with the stablilizing admixture, except that, in the case of lime-treated base, the value is applied after mixing.

LL = liquid limit.

Source: AASHTO. 1972. *AASHTO Interim Guide for Design of Pavement Structures.* Copyright 1972 by the American Association of State Highway and Transportation Officials, Washington, D.C. Used by permission.

platform for constructing the base course. Examples of grading requirements for subbase materials are given in Table 59.2. The subbase course may be omitted if the subgrade soil satisfies the requirements specified for subbase material.

Prepared Subgrade

Most natural soils forming the roadbed for pavement construction require some form of preparation or treatment. The top layer of a specified depth is usually compacted to achieve a desired density.

The depth of compaction and the compacted density required depend on the type of soil, and magnitudes of wheel loads and tire pressures. For highway construction, compaction to 100% modified AASHTO density covering a thickness of 12 in. (300 mm) below the formation level is commonly done. Compaction depth of up to 24 in. (600 mm) may be required for heavily trafficked pavements. For example, in the case of cohesive subgrade, the Asphalt Institute [1981] requires a minimum of 95% of AASHTO T180 (Method D) density for the top 12 in. (300 mm), and a minimum of 90% for all fill areas below the top 12 in. (300 mm). For cohesionless subgrade the corresponding compaction requirements are 100% and 95% respectively.

Due to the higher wheel loads and tire pressures of aircraft, many stringent compaction requirements are found in airport pavement construction. Figure 59.2 shows an example of the compaction requirements recommended by the FAA [1978].

In some instances it may be economical to treat or stabilize poor subgrade materials and reduce the total required pavement thickness. Portland cement, lime, and bitumen have all been used successfully for this purpose. The choice of the method of stabilization depends on the soil properties, improvement expected, and the cost of construction.

Rigid Pavement

Rigid pavements constructed of portland cement concrete are mostly found in heavy-traffic highways and airport pavements. To allow for expansion, contraction, warping, or breaks in construction of the concrete slabs, joints are provided in **concrete pavements.** The joint spacing, which determines the length of individual slab panels, depends on the use of steel reinforcements in the slab. The *jointed plain concrete pavement* (JPCP), requiring no steel reinforcements and thus the least expensive to construct, is a popular form of construction. Depending on the thickness of the slab, typical joint spacings for plain concrete pavements are between 10 and 20 ft (3 and 6 m). For

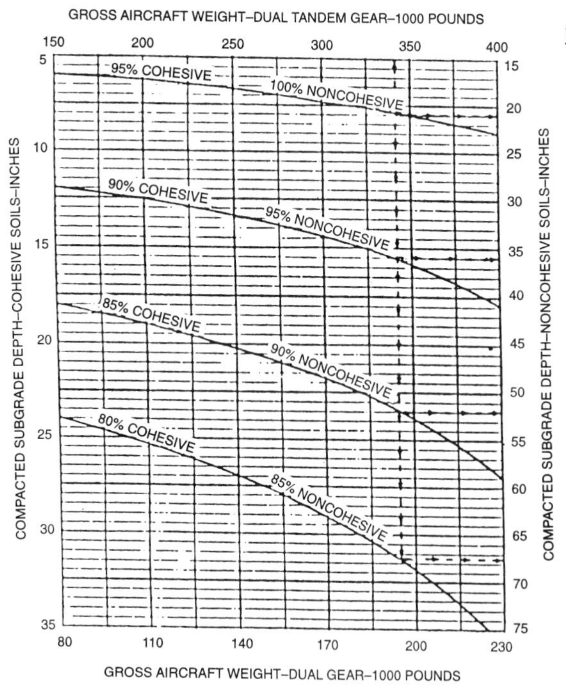

NOTES:

1. Curved lines denote depths below the finished subgrade above which densities should equal or exceed the indicated percentage of the maximum density at optimum moisture as determined by the FAA compaction control T-611.
2. For embankment areas the charted criteria should be met except that the minimum density of soils placed in fill should be 90% for cohesive and 95% for noncohesive, and for the top nine inches in fill should be not less than 95% for cohesive and 100% for noncohesive, of the T-611 density.
3. The subgrade in cut areas shall have natural densities shown or should (a) be compacted from the surface to achieve the required densities, (b) should be removed and replaced in which case the minimum densities for fills apply, or (c) when economics and grades permit, be covered with sufficient select or subbase material so that the uncompacted subgrade is at a depth where the in-place densities are satisfactory.
4. For swelling soils, reduced densities may be used.
5. 1 inch = 2.54 cm
   1 lb. = 0.454 kg

FIGURE 59.2 Subgrade compaction requirements for flexible airport pavements. (*Source:* Federal Aviation Administration. 1978. *Airport Pavement Design and Evaluation.* Advisory Circular AC No. 150/5320-6C, p. 41. With permission.)

slabs with joint spacing greater than 6 m, steel reinforcements have to be provided for crack control, giving rise to the use of *jointed reinforced concrete pavements* (JRCP) and *continuously reinforced concrete pavements* (CRCP). Continuously reinforced concrete pavements usually contain higher than 0.6% of steel reinforcement to eliminate the need to provide joints other than construction and expansion joints.

The base course for rigid pavement, sometimes called **subbase**, is often provided to prevent pumping (ejection of foundation material through cracks or joints resulting from vertical movement of slabs under traffic). The base course material must provide good drainage and be resistant to the erosive action of water. When dowel bars are not provided in short jointed pavements, it is common practice to construct cement-treated base to assist in load transfer across the joints.

Considerations for Highway and Airport Pavements

The two pavement types, flexible and rigid pavement, have been used for road and airport pavement construction. The choice of pavement type depends on the intended functional use of the pavement (such as operating speed and safety requirements), types of traffic loading, cost of construction, and maintenance consideration.

The main differences in design considerations for highway and airport pavements arise from the characteristics of traffic using them. Over the typical design life span of 10 to 20 years for flexible pavements, or 20 to 40 years for rigid pavements, a highway pavement will be receiving highly channelized wheel load applications in the millions. Consideration of the effects of load repetitions—such as cumulative permanent deformation, crack propagation, and fatigue failure—becomes important. The total number of load applications in the entire design life of a highway pavement must therefore be known for pavement structural design. In contrast, the frequency of aircraft loading on the airport pavement is much less. There are also the so-called wander effect of aircraft landing and taking off, and the large variation in the wheel assembly configurations and layout of different aircraft. These make wheel loading on airport pavements less channelized than on highway pavements. Identification of the most critical aircraft is therefore necessary for structural design of airport pavements.

Another important difference is in the magnitude of wheel loads. Airport pavements receive loads far exceeding those applied on the highway. An airport pavement may have to be designed to withstand equivalent single wheel loads of the order of 50 t (approximately 50 tons), whereas the maximum single wheel load allowed on the road pavement by most highway authorities is about 10 t (approximately 10 tons). Furthermore, the wheel tire pressure of an aircraft of about 1200 kPa (175 psi) is nearly twice the value of a normal truck tire. These differences greatly influence the material requirements for the pavements.

59.3 Traffic Loading Analysis for Highway Pavements

Although it is convenient to describe the design life of a pavement in years, it is the total traffic loading during service that determines the actual design life of the pavement. It is thus more appropriate to associate the design life of a pavement with the total design traffic loading. For example, a pavement designed for 20 years with an assumed traffic growth of 4% will reach the end of its design life sooner than 20 years if the actual traffic growth is higher than 4%.

The ultimate aim of traffic analysis for pavement design is to determine the magnitudes of wheel loads and the number of times each of these loads will be applied on the pavement during its design life. For highway pavements the computation of design traffic loading involves the following steps:

1. Estimation of expected initial year traffic volume
2. Estimation of expected annual traffic growth rate
3. Estimation of traffic stream composition
4. Computation of traffic loads

5. Estimation of directional split of design traffic loads
6. Estimation of design lane traffic loads

Information concerning the first two steps can be obtained from traffic surveys and forecasts based on historical trends or prediction using transportation models. The analyses required for the remaining steps are explained in the discussions that follow.

Traffic Stream Composition

The number of different types of vehicles—such as cars, buses, single-unit trucks, and multiple-unit trucks—expected to use the highways must be estimated. One may derive the vehicle type distribution from results of classification counts made on similar highway type within the same region or from general data compiled by highway authorities as illustrated in Table 59.4. However, as noted in the footnote of the table, individual situations may differ from the average values by 50% or more.

Traffic-Loading Computation

Two aspects of traffic loading are of concern in the structural design of highway pavements, namely, the number of applications and the magnitude of each load type. A traffic count survey that

Table 59.4 Asphalt Institute Data for Truck Loading Computation

	Average Trucks				
Truck Class	Interstate Rural	Other Rural	All Rural	All Urban	All System
(a) Average Distribution on Different Classes of Highways (U.S.)					
Single-unit trucks					
2 axle, 4 tire	39	58	47	61	49
2 axle, 6 tire	10	11	10	13	11
3 axle, or more	2	4	2	3	3
All single-unit	51	73	59	77	63
Multiple-unit trucks					
3 axle	1	1	1	1	1
4 axle	5	3	4	4	4
5 axle or more	43	23	36	18	32
All multiple-units	49	27	41	23	37
All trucks	100	100	100	100	100
(b) Average Truck Factors (TF) for Different Classes of Highways and Vehicles (U.S.)					
Single-unit trucks					
2 axle, 4 tire	0.02	0.02	0.03	0.03	0.02
2 axle, 6 tire	0.19	0.21	0.20	0.26	0.21
3 axle, or more	0.56	0.73	0.67	1.03	0.73
All single-unit	0.07	0.07	0.07	0.09	0.07
Multiple-unit trucks					
3 axle	0.51	0.47	0.48	0.47	0.48
4 axle	0.62	0.83	0.70	0.89	0.73
5 axle or more	0.94	0.98	0.95	1.02	0.95
All multiple-unit	0.93	0.97	0.94	1.00	0.95
All trucks	0.49	0.31	0.42	0.30	0.40

Note: Individual situations may differ from these average values by 50% or more.

Source: Asphalt Institute. 1983b. *Asphalt Technology and Construction Practices.* Educational Series ES-1, pp. J5–J7. With permission.

Table 59.5 Vehicle Classification by Axle Configuration

Vehicle Class	Axle Configuration	Total No. of Axles	Number of Single Axles	Number of Tandem Axles
1		2	2	
2		2	2	
3		2	2	
4		2	2	
5		3	3	
6		3	1	1
7		3	3	
8		4	4	
9		4	2	1
10		4	2	1
11		4	2	1
12		5	1	2
13		5	5	
14		6	4	1

Source: Fwa and Sinha. 1985. *A Routine Maintenance and Pavement Performance Relationship Model for Highways.* Reports JHRP-85-11, Purdue University, West Lafayette, IN. With permission.

classifies vehicles by axle configuration, as shown in Table 59.5, enables one to compute the number of repetitions by axle type (i.e., by single axle, tandem axle, and tridem axle). With this information, one must further subdivide each axle type by load magnitude to arrive at a traffic-loading table such as that illustrated in Table 59.6.

The combined loading effects of different axle types on pavements cannot be easily analyzed. In the late 1950s AASHO [Highway Research Board 1962] conducted the now well-known AASHO road test to provide, among other information, equivalency factors to convert one pass of any given single- or tandem-axle load to equivalent passes of an 18-kip (80 kN) single-axle load. The single-axle load of 18 kip (80 kN) was arbitrarily chosen in the AASHO road test as the standard axle with a damaging effect of unity. The equivalency factor, known as the equivalent single-axle load (ESAL) factor, was derived based on the relative damaging effects of various axle loads. Table 59.7 presents the ESAL factors of axle loads for different thicknesses of flexible pavements with a terminal serviceability index of 2.5. Table 59.8 presents the corresponding ESAL factors for rigid pavements.

Another approach to computing the combined effect of mixed traffic is to adopt the hypothesis of cumulative damage. For a given form of pavement damage the allowable number of repetitions by each vehicle type or load group is established separately. A damage ratio for vehicle type or load group i is defined as

$$D_i = (n_i / N_i) \tag{59.1}$$

Table 59.6 Examples of Axle-Load Data Presentation

Axle Load (kips)	No. Axles/Day	Axle Load (kips)	No. Axles/Day
Single Axle		Tandem Axle	
Less than 3	1438	9–11	2093
3–5	3391	11–13	1867
5–7	3432	13–15	1298
7–9	6649	15–17	1465
9–11	9821	17–19	1734
11–13	2083	19–21	1870
13–15	946	21–23	2674
15–17	886	23–25	2879
17–19	472	25–27	2359
19–21	299	27–29	2104
21–23	98	29–31	1994
		31–33	1779
		33–35	862
		35–37	659
		37–39	395
		39–41	46

Table 59.7 AASHTO Load Equivalency Factors for Flexible Pavements

Axle Load (kips)	Pavement Structural Number (*SN*) 1	2	3	4	5	6
	(a) Single Axles and p_t of 2.5					
2	.0004	.0004	.0003	.0002	.0002	.0002
4	.003	.004	.004	.003	.002	.002
6	.011	.017	.017	.013	.010	.009
8	.032	.047	.051	.041	.034	.031
10	.078	.102	.118	.102	.088	.080
12	.168	.198	.229	.213	.189	.176
14	.328	.358	.399	.388	.360	.342
16	.591	.613	.646	.645	.623	.606
18	1.00	1.00	1.00	1.00	1.00	1.00
20	1.61	1.57	1.49	1.47	1.51	1.55
22	2.48	2.38	2.17	2.09	2.18	2.30
24	3.69	3.49	3.09	2.89	3.03	3.27
26	5.33	4.99	4.31	3.91	4.09	4.48
28	7.49	6.98	5.90	5.21	5.39	5.98
30	10.3	9.5	7.9	6.8	7.0	7.8
32	13.9	12.8	10.5	8.8	8.9	10.0
34	18.4	16.9	13.7	11.3	11.2	12.5
36	24.0	22.0	17.7	14.4	13.9	15.5
38	30.9	28.3	22.6	18.1	17.2	19.0
40	39.3	35.9	28.5	22.5	21.1	23.0
42	49.3	45.0	35.6	27.8	25.6	27.7
44	61.3	55.9	44.0	34.0	31.0	33.1
46	75.5	68.8	54.0	41.4	37.2	39.3
48	92.2	83.9	65.7	50.1	44.5	46.5
50	112.	102.	79.	60.	53.	55.
	(b) Tandem Axles and p_t of 2.5					
2	.0001	.0001	.0001	.0000	.0000	.0000
4	.0005	.0005	.0004	.0003	.0003	.0002
6	.002	.002	.002	.001	.001	.001
8	.004	.006	.005	.004	.003	.003
10	.008	.013	.011	.009	.007	.006

Table 59.7 (continued) AASHTO Load Equivalency Factors for Flexible Pavements *(continues)*

Axle Load (kips)	Pavement Structural Number (SN)					
	1	2	3	4	5	6
	(b) Tandem Axles and p_t of 2.5					
12	.015	.024	.023	.018	.014	.013
14	.026	.041	.042	.033	.027	.024
16	.044	.065	.070	.057	.047	.043
18	.070	.097	.109	.092	.077	.070
20	.107	.141	.162	.141	.121	.110
22	.160	.198	.229	.207	.180	.166
24	.231	.273	.315	.292	.260	.242
26	.327	.370	.420	.401	.364	.342
28	.451	.493	.548	.534	.495	.470
30	.611	.648	.703	.695	.658	.633
32	.813	.843	.889	.887	.857	.834
34	1.06	1.08	1.11	1.11	1.09	1.08
36	1.38	1.38	1.38	1.38	1.38	1.38
38	1.75	1.73	1.69	1.68	1.70	1.73
40	2.21	2.16	2.06	2.03	2.08	2.14
42	2.76	2.67	2.49	2.43	2.51	2.61
44	3.41	3.27	2.99	2.88	3.00	3.16
46	4.18	3.98	3.58	3.40	3.55	3.79
48	5.08	4.80	4.25	3.98	4.17	4.49
50	6.12	5.76	5.03	4.64	4.86	5.28
52	7.33	6.87	5.93	5.38	5.63	6.17
54	8.72	8.14	6.95	6.22	6.47	7.15
56	10.3	9.6	8.1	7.2	7.4	8.2
58	12.1	11.3	9.4	8.2	8.4	9.4
60	14.2	13.1	10.9	9.4	9.6	10.7
62	16.5	15.3	12.6	10.7	10.8	12.1
64	19.1	17.6	14.5	12.2	12.2	13.7
66	22.1	20.3	16.6	13.8	13.7	15.4
68	25.3	23.3	18.9	15.6	15.4	17.2
70	29.0	26.6	21.5	17.6	17.2	19.2
72	33.0	30.3	24.4	19.8	19.2	21.3
74	37.5	34.4	27.6	22.2	21.3	23.6
76	42.5	38.9	31.1	24.8	23.7	26.1
78	48.0	43.9	35.0	27.8	26.2	28.8
80	54.0	49.4	39.2	30.9	29.0	31.7
82	60.6	55.4	43.9	34.4	32.0	34.8
84	67.8	61.9	49.0	38.2	35.3	38.1
86	75.7	69.1	54.5	42.3	38.8	41.7
88	84.3	76.9	60.6	46.8	42.6	45.6
90	93.7	85.4	67.1	51.7	46.8	49.7
	(c) Triple Axles and p_t of 2.5					
2	.0000	.0000	.0000	.0000	.0000	.0000
4	.0002	.0002	.0002	.0001	.0001	.0001
6	.0006	.0007	.0005	.0004	.0003	.0003
8	.001	.002	.001	.001	.001	.001
10	.003	.004	.003	.002	.002	.002
12	.005	.007	.006	.004	.003	.003
14	.008	.012	.010	.008	.006	.006
16	.012	.019	.018	.013	.011	.010
18	.018	.029	.028	.021	.017	.016
20	.027	.042	.042	.032	.027	.024
22	.038	.058	.060	.048	.040	.036
24	.053	.078	.084	.068	.057	.051

Table 59.7 (continued) AASHTO Load Equivalency Factors for Flexible Pavement

Axle Load (kips)	Pavement Structural Number (SN)					
	1	2	3	4	5	6
26	.072	.103	.114	.095	.080	.072
28	.098	.133	.151	.128	.109	.099
30	.129	.169	.195	.170	.145	.133
32	.169	.213	.247	.220	.191	.175
34	.219	.266	.308	.281	.246	.228
36	.279	.329	.379	.352	.313	.292
38	.352	.403	.461	.436	.393	.368
40	.439	.491	.554	.533	.487	.459
42	.543	.594	.661	.644	.597	.567
44	.666	.714	.781	.769	.723	.692
46	.811	.854	.918	.911	.868	.838
48	.979	1.015	1.072	1.069	1.033	1.005
50	1.17	1.20	1.24	1.25	1.22	1.20
52	1.40	1.41	1.44	1.44	1.43	1.41
54	1.66	1.66	1.66	1.66	1.66	1.66
56	1.95	1.93	1.90	1.90	1.91	1.93
58	2.29	2.25	2.17	2.16	2.20	2.24
60	2.67	2.60	2.48	2.44	2.51	2.58
62	3.09	3.00	2.82	2.76	2.85	2.95
64	3.57	3.44	3.19	3.10	3.22	3.36
66	4.11	3.94	3.61	3.47	3.62	3.81
68	4.71	4.49	4.06	3.88	4.05	4.30
70	5.38	5.11	4.57	4.32	4.52	4.84
72	6.12	5.79	5.13	4.80	5.03	5.41
74	6.93	6.54	5.74	5.32	5.57	6.04
76	7.84	7.37	6.41	5.88	6.15	6.71
78	8.83	8.28	7.14	6.49	6.78	7.43
80	9.92	9.28	7.95	7.15	7.45	8.21
82	11.1	10.4	8.8	7.9	8.2	9.0
84	12.4	11.6	9.8	8.6	8.9	9.9
86	13.8	12.9	10.8	9.5	9.8	10.9
88	15.4	14.3	11.9	10.4	10.6	11.9
90	17.1	15.8	13.2	11.3	11.6	12.9

Source: AASHTO. 1993. *AASHTO Guides for Design of Pavement Structures.*

Table 59.8 AASHTO Load Equivalency Factors for Rigid Pavements

Axle Load (kips)	Slab Thickness, D (inches)								
	6	7	8	9	10	11	12	13	14
	(a) Single Axles and p_t of 2.5								
2	.0002	.0002	.0002	.0002	.0002	.0002	.0002	.0002	.0002
4	.003	.002	.002	.002	.002	.002	.002	.002	.002
6	.012	.011	.010	.010	.010	.010	.010	.010	.010
8	.039	.035	.033	.032	.032	.032	.032	.032	.032
10	.097	.089	.084	.082	.081	.080	.080	.080	.080
12	.203	.189	.181	.176	.175	.174	.174	.173	.173
14	.376	.360	.347	.341	.338	.337	.336	.336	.336
16	.634	.623	.610	.604	.601	.599	.599	.599	.598
18	1.00	1.00	1.00	1.00	1.00	1.00	1.00	1.00	1.00
20	1.51	1.52	1.55	1.57	1.58	1.58	1.59	1.59	1.59

Table 59.8 (continued) AASHTO Load Equivalency Factors for Rigid Pavements *(continues)*

Axle Load (kips)	Slab Thickness, *D* (inches)								
	6	7	8	9	10	11	12	13	14
22	2.21	2.20	2.28	2.34	2.38	2.40	2.41	2.41	2.41
24	3.16	3.10	3.22	3.36	3.45	3.50	3.53	3.54	3.55
26	4.41	4.26	4.42	4.67	4.85	4.95	5.01	5.04	5.05
28	6.05	5.76	5.92	6.29	6.61	6.81	6.92	6.98	7.01
30	8.16	7.67	7.79	8.28	8.79	9.14	9.35	9.46	9.52
32	10.8	10.1	10.1	10.7	11.4	12.0	12.3	12.6	12.7
34	14.1	13.0	12.9	13.6	14.6	15.4	16.0	16.4	16.5
36	18.2	16.7	16.4	17.1	18.3	19.5	20.4	21.0	21.3
38	23.1	21.1	20.6	21.3	22.7	24.3	25.6	26.4	27.0
40	29.1	26.5	25.7	26.3	27.9	29.9	31.6	32.9	33.7
42	36.2	32.9	31.7	32.2	34.0	36.3	38.7	40.4	41.6
44	44.6	40.4	38.8	39.2	41.0	43.8	46.7	49.1	50.8
46	54.5	49.3	47.1	47.3	49.2	52.3	55.9	59.0	61.4
48	66.1	59.7	56.9	56.8	58.7	62.1	66.3	70.3	73.4
50	79.4	71.7	68.2	67.8	69.6	73.3	78.1	83.0	87.1
(b) Tandem Axles and p_t of 2.5									
2	.0001	.0001	.0001	.0001	.0001	.0001	.0001	.0001	.0001
4	.0006	.0006	.0005	.0005	.0005	.0005	.0005	.0005	.0005
6	.002	.002	.002	.002	.002	.002	.002	.002	.002
8	.007	.006	.006	.005	.005	.005	.005	.005	.005
10	.015	.014	.013	.013	.012	.012	.012	.012	.012
12	.031	.028	.026	.026	.025	.025	.025	.025	.025
14	.057	.052	.049	.048	.047	.047	.047	.047	.047
16	.097	.089	.084	.082	.081	.081	.080	.080	.080
18	.155	.143	.136	.133	.132	.131	.131	.131	.131
20	.234	.220	.211	.206	.204	.203	.203	.203	.203
22	.340	.325	.313	.308	.305	.304	.303	.303	.303
24	.475	.462	.450	.444	.441	.440	.439	.439	.439
26	.644	.637	.627	.622	.620	.619	.618	.618	.618
28	.855	.854	.852	.850	.850	.850	.849	.849	.849
30	1.11	1.12	1.13	1.14	1.14	1.14	1.14	1.14	1.14
32	1.43	1.44	1.47	1.49	1.50	1.51	1.51	1.51	1.51
34	1.82	1.82	1.87	1.92	1.95	1.96	1.97	1.97	1.97
36	2.29	2.27	2.35	2.43	2.48	2.51	2.52	2.52	2.53
38	2.85	2.80	2.91	3.03	3.12	3.16	3.18	3.20	3.20
40	3.52	3.42	3.55	3.74	3.87	3.94	3.98	4.00	4.01
42	4.32	4.16	4.30	4.55	4.74	4.86	4.91	4.95	4.96
44	5.26	5.01	5.16	5.48	5.75	5.92	6.01	6.06	6.09
46	6.36	6.01	6.14	6.53	6.90	7.14	7.28	7.36	7.40
48	7.64	7.16	7.27	7.73	8.21	8.55	8.75	8.86	8.92
50	9.11	8.50	8.55	9.07	9.68	10.14	10.42	10.58	10.66
52	10.8	10.0	10.0	10.6	11.3	11.9	12.3	12.5	12.7
54	12.8	11.8	11.7	12.3	13.2	13.9	14.5	14.8	14.9
56	15.0	13.8	13.6	14.2	15.2	16.2	16.8	17.3	17.5
58	17.5	16.0	15.7	16.3	17.5	18.6	19.5	20.1	20.4
60	20.3	18.5	18.1	18.7	20.0	21.4	22.5	23.2	23.6
62	23.5	21.4	20.8	21.4	22.8	24.4	25.7	26.7	27.3
64	27.0	24.6	23.8	24.4	25.8	27.7	29.3	30.5	31.3
66	31.0	28.1	27.1	27.6	29.2	31.3	33.2	34.7	35.7
68	35.4	32.1	30.9	31.3	32.9	35.2	37.5	39.3	40.5
70	40.3	36.5	35.0	35.3	37.0	39.5	42.1	44.3	45.9
72	45.7	41.4	39.6	39.8	41.5	44.2	47.2	49.8	51.7
74	51.7	46.7	44.6	44.7	46.4	49.3	52.7	55.7	58.0
76	58.3	52.6	50.2	50.1	51.8	54.9	58.6	62.1	64.8
78	65.5	59.1	56.3	56.1	57.7	60.9	65.0	69.0	72.3

Table 59.8 (continued) AASHTO Load Equivalency Factors for Rigid Pavements

Axle Load (kips)	Slab Thickness, D (inches)								
	6	7	8	9	10	11	12	13	14
80	73.4	66.2	62.9	62.5	64.2	67.5	71.9	76.4	80.2
82	82.0	73.9	70.2	69.6	71.2	74.7	79.4	84.4	88.8
84	91.4	82.4	78.1	77.3	78.9	82.4	87.4	93.0	98.1
86	102.	92.	87.	86.	87.	91.	96.	102.	108.
88	113.	102.	96.	95.	96.	100.	105.	112.	119.
90	125.	112.	106.	105.	106.	110.	115.	123.	130.
(c) Triple Axles and p_t of 2.5									
2	.0001	.0001	.0001	.0001	.0001	.0001	.0001	.0001	.0001
4	.0003	.0003	.0003	.0003	.0003	.0003	.0003	.0003	.0003
6	.001	.001	.001	.001	.001	.001	.001	.001	.001
8	.003	.002	.002	.002	.002	.002	.002	.002	.002
10	.006	.005	.005	.005	.005	.005	.005	.005	.005
12	.011	.010	.010	.009	.009	.009	.009	.009	.009
14	.020	.018	.017	.017	.016	.016	.016	.016	.016
16	.033	.030	.029	.028	.027	.027	.027	.027	.027
18	.053	.048	.045	.044	.044	.043	.043	.043	.043
20	.080	.073	.069	.067	.066	.066	.066	.066	.066
22	.116	.107	.101	.099	.098	.097	.097	.097	.097
24	.163	.151	.144	.141	.139	.139	.138	.138	.138
26	.222	.209	.200	.195	.194	.193	.192	.192	.192
28	.295	.281	.271	.265	.263	.262	.262	.262	.262
30	.384	.371	.359	.354	.351	.350	.349	.349	.349
32	.490	.480	.468	.463	.460	.459	.458	.458	.458
34	.616	.609	.601	.596	.594	.593	.592	.592	.592
36	.765	.762	.759	.757	.756	.755	.755	.755	.755
38	.939	.941	.946	.948	.950	.951	.951	.951	.951
40	1.14	1.15	1.16	1.17	1.18	1.18	1.18	1.18	1.18
42	1.38	1.38	1.41	1.44	1.45	1.46	1.46	1.46	1.46
44	1.65	1.65	1.70	1.74	1.77	1.78	1.78	1.78	1.79
46	1.97	1.96	2.03	2.09	2.13	2.15	2.16	2.16	2.16
48	2.34	2.31	2.40	2.49	2.55	2.58	2.59	2.60	2.60
50	2.76	2.71	2.81	2.94	3.02	3.07	3.09	3.10	3.11
52	3.24	3.15	3.27	3.44	3.56	3.62	3.66	3.68	6.68
54	3.79	3.66	3.79	4.00	4.16	4.26	4.30	4.33	4.34
56	4.41	4.23	4.37	4.63	4.84	4.97	5.03	5.07	5.09
58	5.12	4.87	5.00	5.32	5.59	5.76	5.85	5.90	5.93
60	5.91	5.59	5.71	6.08	6.42	6.64	6.77	6.84	6.87
62	6.80	6.39	6.50	6.91	7.33	7.62	7.79	7.88	7.93
64	7.79	7.29	7.37	7.82	8.33	8.70	8.92	9.04	9.11
66	8.90	8.28	8.33	8.83	9.42	9.88	10.17	10.33	10.42
68	10.1	9.4	9.4	9.9	10.6	11.2	11.5	11.7	11.9
70	11.5	10.6	10.6	11.1	11.9	12.6	13.0	13.3	13.5
72	13.0	12.0	11.8	12.4	13.3	14.1	14.7	15.0	15.2
74	14.6	13.5	13.2	13.8	14.8	15.8	16.5	16.9	17.1
76	16.5	15.1	14.8	15.4	16.5	17.6	18.4	18.9	19.2
78	18.5	16.9	16.5	17.1	18.2	19.5	20.5	21.1	21.5
80	20.6	18.8	18.3	18.9	20.2	21.6	22.7	23.5	24.0
82	23.0	21.0	20.3	20.9	22.2	23.8	25.2	26.1	26.7
84	25.6	23.3	22.5	23.1	24.5	26.2	27.8	28.9	29.6
86	28.4	25.8	24.9	25.4	26.9	28.8	30.5	31.9	32.8
88	31.5	28.6	27.5	27.9	29.4	31.5	33.5	35.1	36.1
90	34.8	31.5	30.3	30.7	32.5	34.4	36.7	38.5	39.8

Source: AASHTO, 1993. *AASHTO Guides for Design of Pavement Structures.* Copyright 1993 by the American Association of State Highway and Transportation Officials, Washington, D.C. Used by permission.

where n_i is the design repetitions and N_i the allowable repetitions. The total level of damage caused by the mixed traffic is computed as the sum of damage ratios of all vehicle types or load groups.

Example 59.1. This example involves computation of the ESAL contribution of a passenger car, a bus, and a combination truck. The axle loads of the three fully laden vehicles are given as follows:

Car. Front single axle = 2 kips; rear single axle = 2 kips.
Bus. Front single axle = 10 kips; rear single axle = 8 kips.
Truck. Front single axle = 12 kips; Middle single axle = 18 kips; rear tandem axle = 32 kips.

Assuming a terminal serviceability index of 2.5, the ESAL contributions of the three vehicles can be computed for a flexible pavement with structural number $SN = 5.0$ [see Eq. (59.17) for definition of SN], and a rigid pavement of slab thickness equal to 10 in.

For the ESAL on flexible pavement, Table 59.7 is used to obtain the ESAL factor for each axle. The ESAL contribution of the passenger car is $(0.0002 + 0.0002) = 0.0004$. The ESAL contribution of the bus is $(0.088 + 0.034) = 0.122$. The contribution of the truck is $(0.189 + 1.00 + 0.857) = 2.046$. Table 59.8 is used for the ESAL computation in the case of rigid pavement. The ESAL contributions are $(0.0002 + 0.0002) = 0.0004$ for the car, $(0.081 + 0.032) = 0.113$ for the bus, and $(0.175 + 1.00 + 1.50) = 2.675$ for the truck.

The ratios of ESAL contributions are (car):(bus):(truck) = 5012:305:1 for flexible pavement and 6688:283:1 for rigid pavement. It can be seen from this example that the damaging effects of a truck and a bus are, respectively, more than 5000 and 280 times that of a passenger car. This explains why passenger car volumes are often ignored in traffic loading computation for pavement design.

Example 59.2. This example involves ESAL computation based on axle load data. Calculate the total daily ESAL of the traffic-loading data of Table 59.6 for (a) a flexible pavement with structural number $SN = 5.0$ [see Eq. (59.17) for definition of SN], and (b) a rigid pavement with slab thickness of 10 in. The design terminal serviceability index for both pavements is 2.5.

The data in Table 59.6 are repeated in columns (1) and (2) of the following table. The ESAL factors in column (3) are obtained from Table 59.7 (second part) for $SN = 5.0$, and those in column (5) are obtained from Table 59.8 (second part) for slab thickness of 10 in. The ESAL contribution by each axle group is computed by multiplying its ESAL factor by the number of axles per day. The total ESAL of the traffic loading is 12,642 for the flexible pavement and 19,309 for the rigid pavement.

Axle Load (kips) (1)*	No. Axles per Day (2)	Flexible Pavement		Rigid Pavement	
		ESAL Factor (3)	ESAL (2) × (3)	ESAL Factor (5)	ESAL (2) × (5)
S2	1438	0.0002	0.2876	0.0002	0.2876
S4	3391	0.002	6.782	0.002	6.782
S6	3432	0.01	34.32	0.01	34.32
S8	6649	0.034	226.066	0.032	212.768
S10	9821	0.088	864.248	0.081	795.501
S12	2083	0.189	393.687	0.175	364.525
S14	946	0.36	340.56	0.338	319.748
S16	886	0.623	551.978	0.601	532.486
S18	472	1.00	472.00	1.00	472.00
S20	299	1.51	451.49	1.58	472.42
S22	98	2.18	213.64	2.38	233.24

*In column (1), the prefix S stands for single axle and T for tandem axle.

(continues)

Axle Load (kips) (1)*	No. Axles per Day (2)	Flexible Pavement ESAL Factor (3)	Flexible Pavement ESAL (2) × (3)	Rigid Pavement ESAL Factor (5)	Rigid Pavement ESAL (2) × (5)
T10	2093	0.007	14.651	0.012	25.116
T12	1867	0.014	26.138	0.025	46.675
T14	1298	0.027	35.046	0.047	61.006
T16	1465	0.047	68.855	0.081	118.665
T18	1734	0.077	133.518	0.132	228.888
T20	1870	0.121	226.27	0.204	381.48
T22	2674	0.18	481.32	0.305	815.57
T24	2879	0.26	748.54	0.441	1,269.639
T26	2359	0.364	858.676	0.62	1,462.58
T28	2104	0.495	1,041.48	0.85	1,788.4
T30	1994	0.658	1,312.052	1.14	2,273.16
T32	1779	0.857	1,524.603	1.5	2,668.5
T34	862	1.09	939.58	1.95	1,680.9
T36	659	1.38	909.42	2.48	1,634.32
T38	395	1.7	671.5	3.12	1,232.4
T40	46	2.08	95.68	3.87	178.02
			Total = 12,642.38		Total = 19,309.39

*In column (1), the prefix S stands for single axle and T for tandem axle.

Example 59.3. This example entails ESAL computation based on Asphalt Institute truck distribution and truck factors. Consider a two-lane rural highway with a design lane daily directional traffic of 2000 vehicles per day during the first year. The traffic growth factor is 4.5% and there are 16% trucks in the traffic. The total number of trucks in the 20-year design traffic is computed by the following geometric sum equation,

$$(2000 \times 365 \times 16\%) \cdot \frac{(1 + 0.01 \times 4.5)^{20} - 1}{0.01 \times 4.5} = 3{,}664{,}180$$

The following table summarizes the procedure for ESAL computation. The numbers of vehicles in column (2) are calculated based on the distribution of "other rural" in Table 59.4(a) and the truck factors in column (3) are for "other rural" given in Table 59.4(b).

Vehicle Type (1)	Number of Vehicles (2)	Truck Factor (3)	ESAL Contribution (4)
Single-unit trucks			
2-axle, 4-tire	2,125,220	0.02	45,200
2-axle, 6-tire	403.060	0.21	84,640
3-axle or more	146,570	0.73	107,000
All singles	2,674,850		236,840
Tractor semitrailers and combinations			
3-axle	36,640	0.48	17,590
4-axle	109,930	0.70	76,950
5-axle or more	842,760	0.95	800,620
All multiple units	989,330		895,160
			Total design ESAL = 1,132,000

Directional Split

It is common practice to report traffic volume of a highway to include flows for all lanes in both directions. To determine the design traffic loading on the design lane, one must split the traffic by direction and distribute the directional traffic by lanes. An even split assigning 50% of the traffic to each direction appears to be the norm. In circumstances where uneven split occurs, pavements are designed based on the heavier directional traffic loading.

Design Lane Traffic Loading

The design lane for pavement structural design is usually the slow lane (lane next to the shoulder in most cases) in which a large proportion of the directional heavy-vehicle traffic is expected to travel. Some highway agencies assign 100% of the estimated directional heavy-vehicle traffic to the design lane for the purpose of structural design. This leads to overestimation of traffic loading on roads with more than one lane in each direction. Studies have shown that—depending upon road geometry, traffic volume, and composition—as much as 50% of the directional heavy vehicles may not travel on the design lane [Fwa and Li 1994]. Figure 59.3 shows the lane-use distributions recommended by a number of organizations. It is noted that while most agencies apply lane-use factors to traffic volume, the factors in AASHTO recommendations are for lane distributions of ESAL. The latter tends to provide a better estimate of traffic loading in cases involving a higher concentration of heavily loaded vehicles in the slow lane.

Formula for Computing Total Design Loading

Depending on the information available, the computations of the design load for pavement structural design may differ slightly. Assuming that the initial-year total ESAL is known and a constant growth of ESAL at a rate of *r*% per annum is predicted, the design lane loading for an

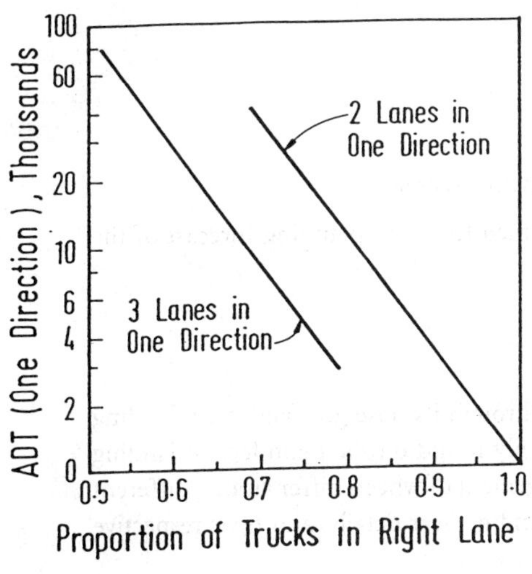

(b) Recommendation by Asphalt Institute [1991]

Number of Lanes per Direction	% Trucks in Design Lane
1	100%
2	90% (70–96%)
3 or more	80% (50–96%)

(c) Recommendation by AASHTO [1993]

Number of Lanes per Direction	% ESAL in Design Lane
1	100%
2	80–100%
3	60–80%
4	50–75%

Note: ESAL stands for equivalent 80 kN single axle load

FIGURE 59.3 Percentage of truck traffic in design lane.

analysis period of n years can be computed by the following equation:

$$(\text{ESAL})_T = (\text{ESAL})_0 \cdot \frac{(1 + 0.01r)^n - 1}{0.01r} \cdot f_D \cdot f_L \quad (59.2)$$

where

$(\text{ESAL})_T$ = total design lane ESAL for n years
$(\text{ESAL})_0$ = initial-year design lane ESAL
r = annual growth rate of ESAL in percent
f_D = directional split factor
f_L = lane-use distribution factor

For design methods that rely on damage ratio computation [see Eq. (59.1)], instead of computing cumulative ESAL, the total number of repetitions of each vehicle type or axle type is calculated.

Example 59.4. The daily ESAL computed in Example 59.2 is based on the initial traffic estimate for both directions of travel in an expressway with three lanes in each direction. The annual growth of traffic is estimated at 3%. The directional split is assumed to be 45 to 55%. The expressway is to be constructed of flexible pavement. Calculate the design lane ESAL for a service life of 15 years.

Adopting the AASHTO lane distribution factor shown in Fig. 59.3(b), we have $f_L = 0.7$. Given $f_D = 0.55$, $n = 15$, $i = 3$, and from Example 59.2 $(\text{ESAL})_0 = (12{,}642 \times 365)$, the total design lane ESAL is

$$\begin{aligned}(\text{ESAL})_T &= (12{,}642 \times 365) \cdot \frac{(1 + 0.01 \times 3)^{15} - 1}{0.01 \times 3} \cdot 0.7 \cdot 0.55 \\ &= 33.04 \times 10^6\end{aligned}$$

59.4 Traffic Loading Analysis for Airport Pavements

The procedure of traffic loading analysis for airport pavements differs slightly from that for highway pavements due to differences in traffic operations and functional uses of the pavements. The basic steps are:

1. Estimation of expected initial year traffic volume
2. Estimation of expected annual traffic growth rate
3. Estimation of traffic stream composition
4. Computation of traffic loading
5. Estimation of design traffic loading for different functional areas

Information concerning the first two steps is usually obtained from the planning forecast of the airport authority concerned.

Traffic Stream Composition

The weight of an aircraft is transmitted to the pavement through its nose gear and main landing gears. Figure 59.4 shows the wheel configurations commonly found on the main legs of landing gear of civil aircraft. Since the gross weight and exact arrangement of wheels differ among different aircraft, there is a need to identify the types of aircraft, landing gear details, and their respective frequencies of arrival for the purpose of pavement design.

Computation of Traffic Loading

For pavement design purposes the maximum takeoff weights of the aircraft are usually considered. It is also common to assume that 95% of the gross weight is carried by the main landing gears

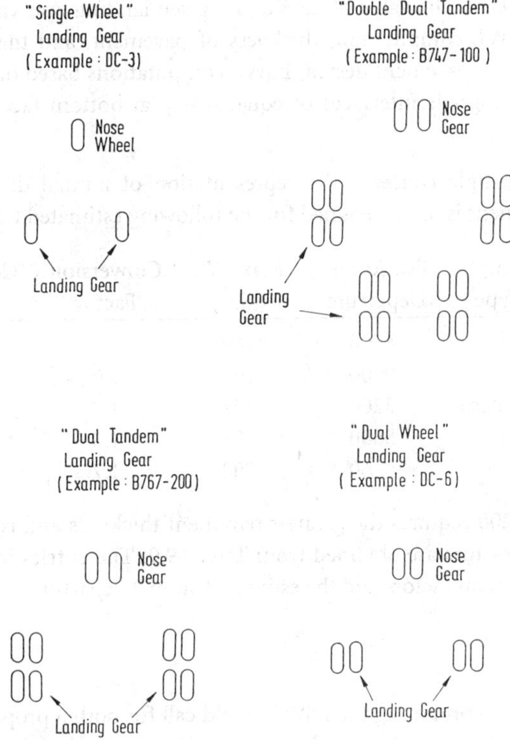

FIGURE 59.4 Typical wheel configurations of a main leg of aircraft landing gear.

and 5% by the nose gear. In the consideration of mixed traffic loading, both the equivalent load concept and Miner's hypothesis have been used. For example, the FAA method [Federal Aviation Administration 1978] converts the annual departure of all aircraft into the equivalent departures of a selected design aircraft using the factors in Table 59.9. In establishing the thickness design curves for flexible airport pavements, the concept of equivalent single-wheel load (ESWL) is adopted

Table 59.9 Conversion Factors for Computing Annual Departures

Aircraft Type	Design Aircraft	Conversion Factor F
Single wheel	Dual wheel	0.8
Single wheel	Dual tandem	0.5
Dual wheel	Dual tandem	0.6
Double dual tandem	Dual tandem	1.0
Dual tandem	Single wheel	2.0
Dual tandem	Dual wheel	1.7
Dual wheel	Single wheel	1.3
Double dual tandem	Dual wheel	1.7

Note: Multiply the annual departures of given aircraft type by the conversion factor to give annual departures in design aircraft landing gear.

Source: Federal Aviation Administration. 1978. *Airport Pavement Design and Evaluation.* Reprinted from FAA Advisory Circular. Report No. FAA/AC-150/5320-6C. 7 December 1978; NTIS Accession No. AD-A075 537/1.

by the FAA. The concept of ESWL is widely used in airport pavement design to assess the effect of multiple-wheel landing gears. The value of ESWL of a given landing gear varies with the control response selected for ESWL computation, thickness of pavement, and the relative stiffness of pavement layers. For airport pavement design, ESWL computations based on equal deflection (at surface or at pavement-subgrade interface) or equal stress (at bottom face of bound layer) are commonly used.

Example 59.5. This example concerns the representation of annual departures of designed aircraft. An airport pavement is to be designed for the following estimated traffic:

Aircraft	Landing Gear Type	Est. Annual Departures	Max. Wt. (kips)	Conversion Factor	Converted Annual Departures
727-100	Dual	4500	160	1.0	4500
727-200	Dual	9900	190.5	1.0	9900
707-320B	Dual tandem	3200	327	1.7	5440
DC-9-30	Dual	5500	108	1.0	5500
747-100	Dual DT	60	700	1.7	102

In this example the 727-200 requires the greatest pavement thickness and is therefore the design aircraft. The conversion factors are obtained from Table 59.9. The entries in the last column are the products of the conversion factors and the estimated annual departures.

Equal Stress ESWL

An elaborate procedure for computing the ESWL would call for both a proper analytical solution for the required stress produced by the wheel assembly of interest and a trial-and-error process to identify the magnitude of the single wheel that will produce identical stress. As this procedure is time consuming, simplified methods have been employed in practice.

Figure 59.5(a) presents a simplified procedure for estimating the equal subgrade stress ESWL of a set of dual wheels for flexible pavement design. With the assumed 45° spread of applied pressure,

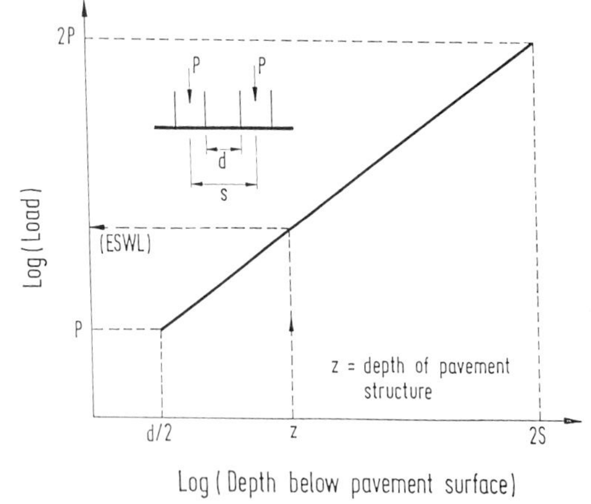

FIGURE 59.5 Computation of ESWL. (a) Equal subgrade stress ESWL of dual wheels for flexible pavement design. (*Source: Principles of Pavement Design* by E. J. Yoder and M. W. Witczak, 2nd edition, John Wiley & Sons, 1975, p. 138. With permission.)

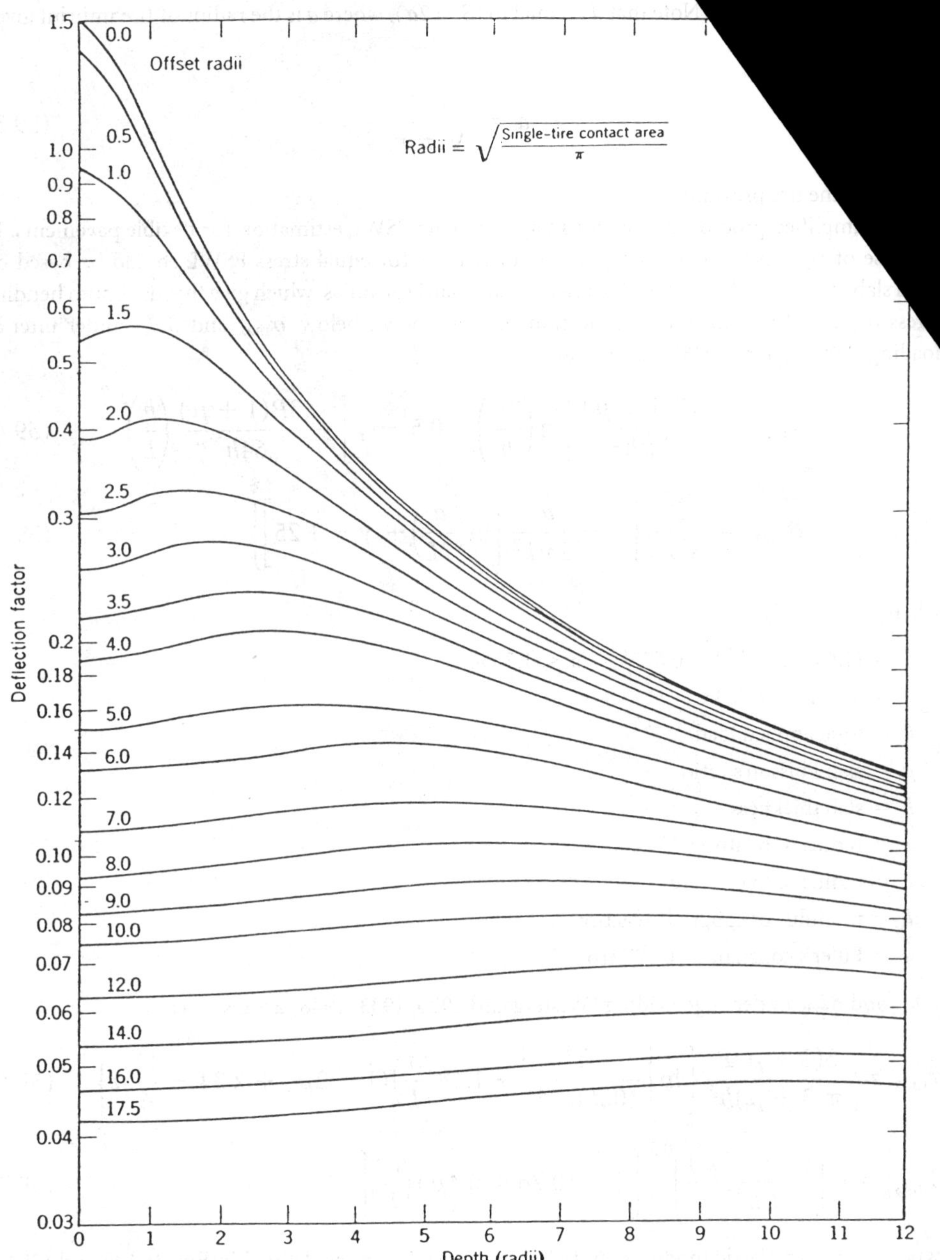

FIGURE 59.5 (continued) Computation of ESWL. (b) One-layer deflection factor for equal-deflection ESWL computation. (*Source: Principles of Pavement Design* by E. J. Yoder and M. W. Witczak, 2nd edition, John Wiley & Sons, 1975, p. 138. With permission.)

the ESWL is equal to one wheel load P if the pavement thickness is less than or equal to $d/2$, where d is the smallest edge-to-edge distance between the tire imprints of the dual wheels. The method further assumes that ESWL $= 2P$ for any pavement equal to or thicker than $2S$, where S is the center-to-center spacing of the dual wheels. For pavement thicknesses between $d/2$ and $2S$, ESWL is determined, as shown in Fig. 59.5(a), by assuming a linear log-log relationship between ESWL

$- 2a$), where a is the radius of tire imprint given

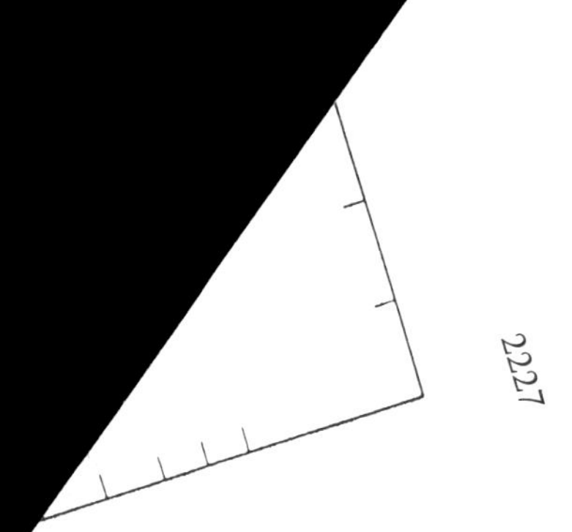

(59.3)

'L estimation for flexible pavements. In ual stress ESWL should be based on as, which give the maximum bending below. σ_{max} and δ_{max} under interior

$$\frac{\quad \mu)}{2\pi h^2}\left[\ln\left(\frac{2L}{b}\right) + 0.5 - \gamma\right] + \frac{3P(1+\mu)}{64h^2}\left(\frac{b}{L}\right)^2 \tag{59.4}$$

$$\delta_{max} = \frac{P}{8kL^2}\left\{1 + \frac{a^2}{2\pi L^2}\left[\ln\left(\frac{a}{2L}\right) + \gamma - 1.25\right]\right\} \tag{59.5}$$

where

$b = (1.6a^2 + h^2)^{0.5} - 0.675h, \quad a < 1.724h$
$= a, \quad a > 1.724h$
P = total applied load
μ = slab Poisson's ratio
h = slab thickness
L = radius of relative stiffness
a = radius of loaded area
k = modulus of subgrade reaction
γ = Euler's constant = 0.577216

σ_{max} and δ_{max} under edge loading [Westergaard 1926, 1933, 1948] are given as

$$\sigma_{max} = \frac{3(1+\mu)P}{\pi(3+\mu)h^2}\left\{\ln\left[\frac{Eh^3}{100k(a)^4}\right] + 1.18\left(\frac{a}{L}\right)(1+2\mu) + 2.34 - \frac{11}{6}\mu\right\} \tag{59.6}$$

$$\delta_{max} = P\left[\frac{(2+1.2\mu)}{Eh^3k}\right]^{0.5}\left\{1 - (0.76+0.4\mu)\left(\frac{a}{L}\right)\right\} \tag{59.7}$$

where E = slab elastic modulus, and all other variables are as defined in Eqs. (59.1) and (59.2). σ_{max} and δ_{max} under corner loading [Westergaard 1926] are given as

$$\sigma_{max} = \frac{3P}{h^2}\left(1 - \left(\frac{1.4142a}{L}\right)^{0.6}\right) \tag{59.8}$$

$$\delta_{max} = \frac{P}{kL^2}\left(1.1 - 0.88\left(\frac{1.4142a}{L}\right)\right) \tag{59.9}$$

Example 59.6. This example addresses equal subgrade-stress ESWL. The total load on a set of dual wheels is 45,000 lb. The tire pressure of the wheels is 185 psi. The center-to-center spacing of the wheels is 34 in. Calculate the equal-stress ESWL if the thickness of pavement structure is (a) $h = 30$ in., and (b) $h = 70$ in.

Load per wheel = 22,500 lb., radius of tire imprint $a = \sqrt{(22{,}500/185\pi)} = 6.22$ in., $S = 34$ in., $d = (S - 2a) = 21.56$. By means of a log-log plot as shown in Fig. 59.5(a), ESWL is determined to be 33,070 lb. for $h = 30$ in. For $h = 70$ in., since $h > 2S$, ESWL $= 2(22{,}500) = 45{,}000$ in.

Equal Deflection ESWL

Equal deflection ESWL can be derived by assuming either constant tire pressure or constant area of tire imprint. A simplified method for computing ESWL on flexible pavement, based on the Boussinesq one-layer theory [Boussinesq 1885], is presented in this section. It computes the ESWL of an assembly of n wheels by equating the surface deflection under the ESWL to the maximum surface deflection caused by the wheel assembly, that is,

$$\frac{(\mathrm{ESWL})^{0.5}}{\pi E}K = \frac{P^{0.5}}{\pi E}(K_1 + K_2 + \cdots + K_n)_{\max} \tag{59.10}$$

where

P = gross load on each tire of the wheel assembly

E = stiffness modulus of the soil

K, K_1, K_2, K_n = Boussinesq deflection factor given by Fig. 59.5(b)

With the assumption of constant tire pressure, Eq. (59.2) can be solved for ESWL by the following iterative procedure: (1) compute $(K_1 + K_2 + \cdots + K_n)_{\max}$ at the point of maximum surface deflection; (2) assume a, the radius of tire imprint for ESWL; (3) determine K from Fig. 59.5(b) with zero horizontal offset; (4) compute ESWL from Eq. (59.2); (5) calculate new $a = \sqrt{(P/\pi p)}$; and (6) if new a does not match the assumed a, return to step (3) and repeat the procedure with the new a until convergence.

Deflections of rigid pavements under loads are computed by means of Westergaard's theory [see Eqs. (59.4–9)] or more elaborate analysis using finite element method. Improved deflection computations using thick plate theory [Shi *et al.*, 1994; Fwa *et al.*, 1993] could also be used for the purpose of ESWL evaluation.

Example 59.7. Calculate the equal subgrade-deflection ESWL for the dual wheels in Example 59.6 for $h = 30$ in. Therefore, $(h/a) = 4.82$. For a point directly below one of the wheels, $(r/a)_1 = (34/6.22) = 5.47, K_1 = 0.15$ [from Fig. 59.5(b)]; and $(r/a)_2 = 0, K_2 = 0.31$. For the point on the vertical line midway between the two wheels, $(r/a)_1 = (r/a)_2 = 2.73$, and $K_1 = K_2 = 0.24$ [from Fig. 59.5(b)]. The critical $(K_1 + K_2) = 0.48$. The ESWL is obtained by trial and error as follows.

Trial a	(h/a)	K	ESWL by Eq. (59.10)	New a by Eq. (59.3)
6.5 in.	4.615	0.3177	51,361 lb.	9.40
8.0 in.	3.75	0.3865	34,704 lb.	7.73
7.85 in.	3.8217	0.3797	35,954 lb.	7.86

The ESWL is equal to 35,950 lb.

Critical Areas for Pavement Design

Runway ends, taxiways, aprons, and turnoff ramp areas receive a concentration of aircraft movements with maximum loads. They are designated as the critical areas for pavement design purposes. Reduced thickness may be used for other areas.

59.5 Thickness Design of Flexible Pavements

The thickness design of flexible pavement is a complex engineering problem involving a large number of variables. Most of the design methods in use today are largely empirical or semiempirical procedures derived from either full-scale pavement tests or performance monitoring of in-service pavements. This section presents the methods of the Asphalt Institute and AASHTO for flexible highway pavements and the FAA method for flexible airport pavements. A brief description of the development of the mechanistic approach to flexible pavement design is also presented.

AASHTO Design Procedure for Flexible Highway Pavements

The AASHTO design procedure [AASHTO 1993] was developed based on the findings of the AASHO road test [Highway Research Board 1962]. It defines pavement performance in terms of the *present serviceability index* (PSI), which varies from 0 to 5. The PSI of newly constructed flexible pavements and rigid pavements were found to be about 4.2 and 4.5, respectively. For pavements of major highways the end of service life is considered to be reached when PSI = 2.5. A terminal value of PSI = 2.0 may be used for secondary roads. Serviceability loss given by the difference of the initial and terminal serviceability is required as an input parameter. Pavement layer thicknesses are designed using the nomograph in Fig. 59.6. The design traffic loading in ESAL is computed by Eq. (59.2). Other input parameters are discussed in this section.

Reliability

The AASHTO guide incorporates in the design a reliability factor $R\%$ to account for uncertainties in traffic prediction and pavement performance. $R\%$ indicates the probability that the pavement designed will not reach the terminal serviceability level before the end of the design period. The AASHTO suggested ranges of $R\%$ are 85 to 99.9%, 80 to 99%, 80 to 95%, and 50 to 80% for urban interstates, principal arterials, collectors, and local roads, respectively. The corresponding ranges for rural roads are 80 to 99.9%, 75 to 95%, 75 to 95%, and 50 to 80%. The overall standard deviation, s_o, for flexible and rigid pavements developed at the AASHO road test are 0.45 and 0.35, respectively.

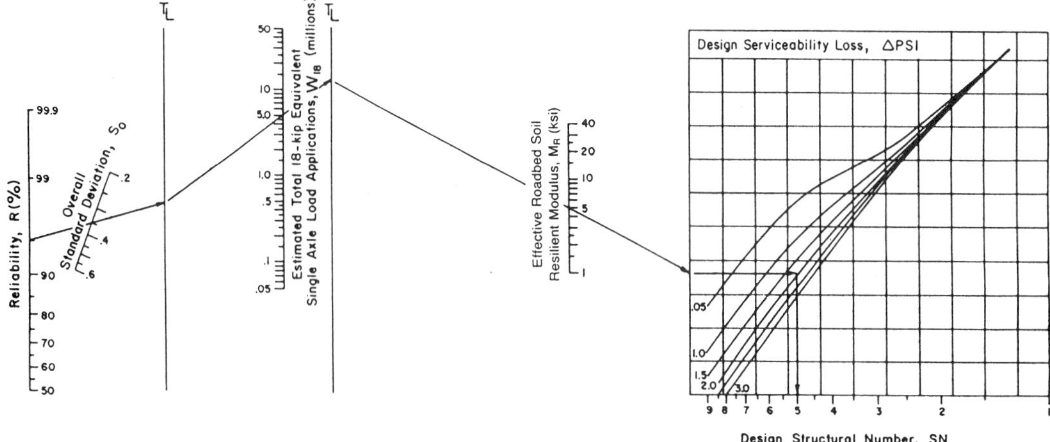

FIGURE 59.6 AASHTO design chart for flexible highway pavements. (*Source:* AASHTO. 1993. *AASHTO Guides for Design of Pavement Structures.* Copyright 1993 by the American Association of State Highway and Transportation Officials, Washington, D.C. Used by permission.)

Effective Roadbed Soil Resilient Modulus

Determination of Subgrade Resilient Modulus. The total pavement thickness requirement is a function of the resilient modulus, M_r, of subgrade soil. Methods for the determination of M_r for granular materials and fine-grained soils are described in AASHTO Test Method T274 [AASHTO 1989].

Since many laboratories are not equipped to perform the resilient modulus test for soils, it is common practice to estimate M_r through empirical correlation with other soil properties. Equation (59.11) is one such correlation suggested by AASHTO for fine-grained soils with soaked CBR of 10 or less.

$$M_r \text{ (psi)} = 1500 \times \text{CBR} \tag{59.11}$$

Other correlations are also found in the literature, such as in the work by Van Til *et al.* [1972].

Determination of Effective M_r. To account for seasonal variations of subgrade soil resilient modulus, AASHTO defines an effective roadbed soil M_r to represent the combined effect of all the seasonal modulus values. This effective M_r is a weighted value that would give the correct equivalent annual pavement damage for design purpose. The steps in computing the effective M_r are as follows:

1. Divide the year into equal-length time intervals each equal to the smallest season. AASHTO suggests that the smallest season should not be less than one-half month.
2. Estimate the relative damage u_f corresponding to each seasonal modulus by the following equation,

$$u_f = 1.18 \times 10^8 \times M_r^{-2.32} \tag{59.12}$$

   where M_r is expressed in 10^3 psi.
3. Sum up the u_f of all seasons and divide by the number of seasons to give the average seasonal damage.
4. Substitute the average seasonal damage into Eq. (59.12), and calculate M_r to arrive at the effective roadbed soil M_r.

Example 59.8. This example examines effective roadbed soil resilient modulus. The resilient moduli of a roadbed soil determined at 24 half-month intervals are 6000, 20,000, 20,000, 4000, 4500, 5000, 6000, 6000, 5000, 5000, 5000, 6000, 6000, 6500, 6500, 6500, 6500, 6500, 6000, 6000, 5500, 5500, 5500, and 6000. The total relative damage u computed by Eq. (59.12) is

$$\begin{aligned} u &= 0.2026 + 0.0124 + 0.0124 + 0.5189 + 0.3948 + 0.3092 + 0.2026 \\ &\quad + 0.2026 + 0.3092 + 0.3092 + 0.3092 + 0.2026 + 0.2026 + 0.1682 \\ &\quad + 0.1682 + 0.1682 + 0.1682 + 0.1682 + 0.2026 + 0.2026 + 0.2479 \\ &\quad + 0.2479 + 0.2479 + 0.2026 \\ &= 5.568 \end{aligned}$$

Mean $u = 0.232$. Applying Eq. (59.12) again, the effective M_r is 5655 psi.

Pavement Layer Modulus

Structural thicknesses required above other pavement layers are also determined based on their respective M_r values. For bituminous pavement layers M_r may be tested by the repeated load indirect tensile test described in ASTM Test D-4123 [ASTM, 1992]. Figure 59.7 shows a chart developed by Van Til *et al.* [1972] relating M_r of hot-mix asphalt mixtures to other properties.

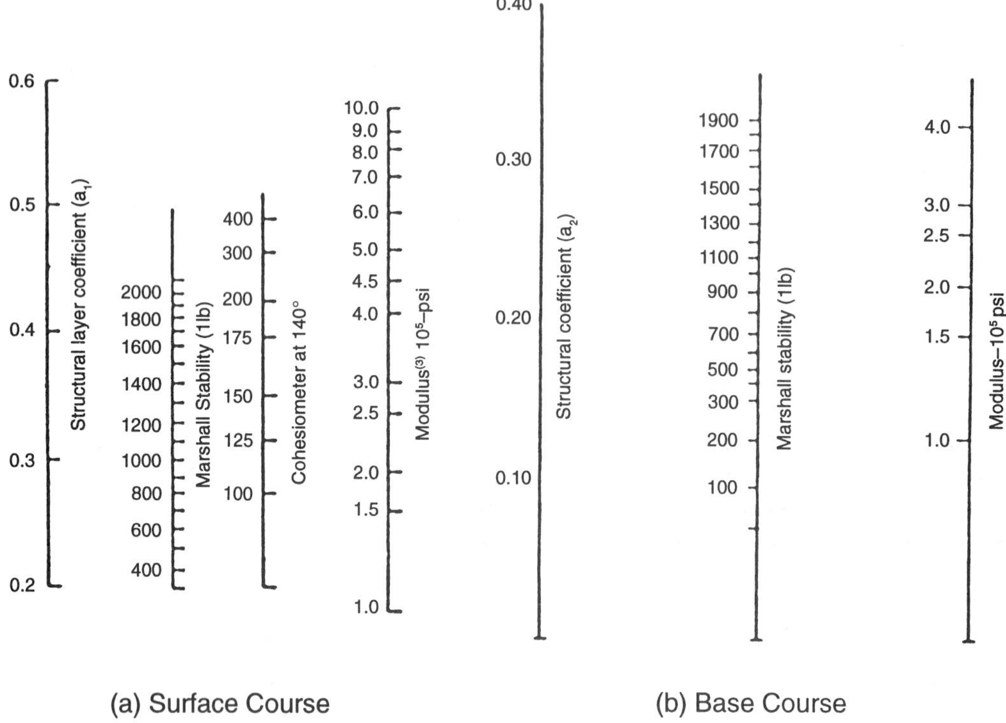

FIGURE 59.7 Correlation charts for estimating resilient modulus of asphalt concrete. (*Source:* Van Til *et al.* 1972. *Evaluation of AASHO Interim Guides for Design of Pavement Structures.* NCHRP Report 128. Highway Research Board. Washington, D.C.)

For unbound base and subbase materials M_r may be estimated from the following correlations:

$$M_r(\text{psi}) = 740 \times \text{CBR} \quad \text{for } \theta = 100\,\text{psi} \tag{59.13}$$

$$M_r(\text{psi}) = 440 \times \text{CBR} \quad \text{for } \theta = 30\,\text{psi} \tag{59.14}$$

$$M_r(\text{psi}) = 340 \times \text{CBR} \quad \text{for } \theta = 20\,\text{psi} \tag{59.15}$$

$$M_r(\text{psi}) = 250 \times \text{CBR} \quad \text{for } \theta = 10\,\text{psi} \tag{59.16}$$

where θ is the sum of principal stresses, $(\sigma_1 + \sigma_2 + \sigma_3)$.

Thickness Requirements

Using the input parameters described in the preceding sections, the total pavement thickness requirement is obtained from the nomograph in Fig. 59.6 in terms of structural number *SN*. *SN* is an index number equal to the weighted sum of pavement layer thicknesses, as follows:

$$SN = a_1 D_1 + a_2 D_2 m_2 + a_3 D_3 m_3 \tag{59.17}$$

where a_1, a_2, and a_3 are numbers known as layer coefficients; D_1, D_2, and D_3 are layer thicknesses; and m_2 and m_3 are layer drainage coefficients. *SN* can be considered a form of equivalent thickness, and layer coefficients and drainage coefficients are applied to actual pavement thicknesses to account for their structural and drainage properties, respectively.

Table 59.10 Base and Subbase Stress States

Asphalt Concrete Thickness (inches)	Roadbed Soil Resilient Modulus (psi)		
	3000	7500	15,000
(a) Stress State for Base Course			
Less than 2	201	25	30
2–4	10	15	20
4–6	5	10	15
Greater than 6	5	5	5

Asphalt Concrete Thickness (inches)	Stress State (psi)
(b) Stress State for Subbase (6–12 in.)	
Less than 2	10.0
2–4	7.5
Greater than 4	5.0

Source: AASHTO. 1993. *AASHTO Guide for Design of Pavement Structures.* Copyright 1993 by the American Association of State Highway and Transportation Officials, Washington, D.C. Used by permission.

Drainage coefficients are determined from Table 59.10. Coefficient a_1 can be estimated from Fig. 59.8. Coefficients a_2 and a_3 of granular base and subbase layers can be obtained from the following correlations:

$$a_2 = 0.249(\log_{10} M_r) - 0.977 \tag{59.18}$$

$$a_3 = 0.227(\log_{10} M_r) - 0.839 \tag{59.19}$$

where $M_r = k_1(\theta)^{k_2}$, θ is the stress state in psi, and k_1 and k_2 are regression constants. Recommended values are given in Table 59.11.

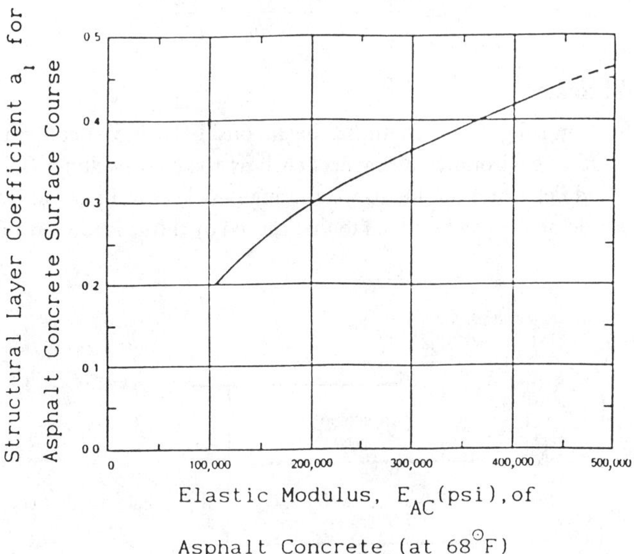

FIGURE 59.8 Chart for estimating structural layer coefficient of dense-graded asphalt concrete. (*Source:* AASHTO. 1993. *AASHTO Guides for Design of Pavement Structures.* Copyright 1993 by the American Association of State Highway and Transportation Officials, Washington, D.C. Used by permission.)

Table 59.11 Recommended m_i Value for Modifying Structural Layer Coefficient of Untreated Base and Subbase Materials in Flexible Pavements

	Percent of Time Pavement Structure is Exposed to Moisture Levels Approaching Saturation			
Quality of Drainage	Less than 1%	1–5%	5–25%	Greater than 25%
Excellent	1.40–1.35	1.35–1.30	1.30–1.20	1.20
Good	1.35–1.25	1.25–1.15	1.15–1.00	1.00
Fair	1.25–1.15	1.15–1.05	1.00–0.80	0.80
Poor	1.15–1.05	1.05–0.80	0.80–0.60	0.60
Very Poor	1.05–0.95	0.95–0.75	0.75–0.40	0.40

Source: AASHTO. 1993. *AASHTO Guides for Design of Pavement Structures.* Copyright 1993 by the American Association of State Highway and Transportation Officials, Washington, D.C. Used by permission.

The thicknesses of individual pavement layers are determined by means of the layer analysis concept depicted in Fig. 59.9. The total structural number required above the subgrade soil, denoted as SN_1, is determined from the nomograph in Fig. 59.6, with the effective roadbed soil M_r as input. SN_2 and SN_3 are determined likewise by replacing M_r with E_3 (stiffness modulus of subbase material) and E_2 (stiffness modulus of base course material), respectively. All pavement layer thicknesses are then derived by solving the following inequalities:

$$D_1 \geq \frac{SN_1}{a_1} \tag{59.20}$$

$$D_2 \geq \frac{SN_2 - a_1 D_1}{a_2 m_2} \tag{59.21}$$

$$D_3 \geq \frac{SN_3 - a_1 D_1 - a_2 D_2 m_2}{a_3 m_3} \tag{59.22}$$

Environmental Effects

The moisture effect on subgrade strength has been considered in the computation of effective roadbed soil M_r. Other environmental impacts such as roadbed swelling, frost heave, aging of asphalt mixtures, and deterioration due to weathering could result in considerable serviceability loss. This loss in serviceability can be added to that caused by traffic loading for design purposes.

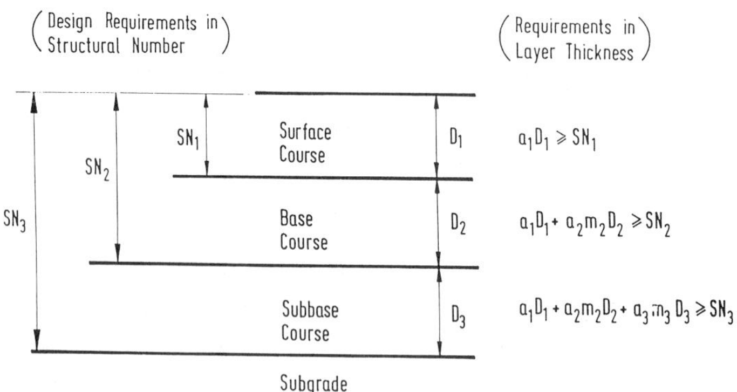

FIGURE 59.9 The concept of layer analysis.

Table 59.12(a) Minimum Thickness of Pavement Layers—AASHTO Thickness Requirements in Inches

Traffic, ESAL	Asphalt Concrete	Aggregate Base
Less than 50,000	1.0 (or surface treatment)	4
50,001–150,000	2.0	4
150,001–500,000	2.5	4
500,001–2,000,000	3.0	6
2,000,001–7,000,000	3.5	6
Greater than 7,000,000	4.0	6

Source: AASHTO. 1993. *AASHTO Guides for Design of Pavement Structures.* Copyright 1993 by the American Association of State Highway and Transportation Officials, Washington, D.C. Used by permission.

Table 59.12(b) Asphalt Institute Requirements

Traffic, ESAL	Asphalt Concrete Thickness
(a) Minimum Thickness of Asphalt Concrete on Aggregate Base	
Less than 10,000	1 in. (25 mm)
Less than 100,000	1.5 in. (40 mm)
Greater than 100,000	2 in. (50 mm)

Traffic (ESAL)	Asphalt Concrete Thickness: Type I Base	Asphalt Concrete Thickness: Type II and III Base
(b) Minimum Thickness of Asphalt Concrete over Emulsified Asphalt Bases		
$\leq 10^4$	1 in. (25mm)	2 in. (50 mm)
$\leq 10^5$	1.5 in. (40mm)	2 in. (50 mm)
$\leq 10^6$	2 in. (50mm)	3 in. (75 mm)
$\leq 10^7$	2 in. (50mm)	4 in. (100 mm)
$\geq 10^7$	2 in. (50mm)	5 in. (130 mm)

Source: Asphalt Institute. 1983b. *Asphalt Technology and Construction Practices.* Educational Series ES-1, p. J25. With permission.

Minimum Thickness Requirements

It is impractical to construct pavement layers less than a certain minimum thickness. AASHTO [1993] recommends minimum thicknesses for different layers as a function of design traffic, which are given in Table 59.12(a).

Example 59.9. On the subgrade examined in Example 59.8 is to be constructed a pavement to carry a design lane ESAL of 5×10^6. The elastic moduli of the surface, base, and subbase courses are respectively $E_1 = 360{,}000$ psi, $E_2 = 30{,}000$ psi, and $E_3 = 13{,}000$ psi. The drainage coefficients of the base and subbase courses are $m_2 = 1.20$ and $m_3 = 1.0$, respectively. The design reliability is 95% and the standard deviation s_o is 0.35. Provide a thickness design for the pavement if the initial serviceability level is 4.2 and the terminal serviceability level is 2.5.

For $R = 95\%$, $s_o = 0.35$, ESAL $= 5 \times 10^6$, $M_r = 5655$ psi, and $\Delta\text{PSI} = 4.2 - 2.5 = 1.7$, obtain $SN_3 = 5.0$ from Fig. 59.6. Repeat the procedure with $E_3 = 13{,}000$ to obtain $SN_2 = 3.8$, and $E_2 = 30{,}000$ to obtain $SN_1 = 2.7$. From Fig. 59.8, $a_1 = 0.40$. By Eq. (59.18), $a_2 = 0.249(\log 30{,}000) - 0.977 = 0.138$, and by Eq. (59.19), $a_3 = 0.227(\log 13{,}000) - 0.839 = 0.095$. The layer thicknesses are $D_1 = (2.7/0.4) = 6.75$ in.; $D_2 = \{3.8 - (0.4 \times 6.75)\}/(0.138 \times 1.20) = 6.64$ or 6.75 in.; and $D_3 = \{5.0 - (0.4 \times 6.75) - (0.138 \times 1.20 \times 6.75)\}/(0.095 \times 1.0) = 12.4$ or 12.5 in.

AI Design Procedure for Flexible Highway Pavements

The Asphalt Institute [1991] promotes the use of full-depth pavements in which asphalt mixtures are employed for all courses above the subgrade. Potential benefits of full-depth pavements derive from the higher load bearing and spreading capability and moisture resistance of asphalt mixtures as compared to unbound aggregates. Thickness design charts are provided for full-depth pavements, pavements with emulsified asphalt base, and untreated aggregate base. These charts are developed based on two design criteria: (1) maximum tensile strains induced at the underside of the lowest asphalt-bound layer, and (2) maximum vertical strains induced at the top of the subgrade layer. The design curves have incorporated the effects of seasonal variations of temperature and moisture on the subgrade and granular base materials.

Computation of Design ESAL

When detailed vehicle classification and weight data are available, the design lane ESAL is computed according to Eq. (59.2). The AASHTO ESAL factors for $SN = 5$ and terminal serviceability index $= 2.5$ are used. When such data are not available, estimates can be made based on the information in Table 59.4. The truck factor in the table refers to the total ESAL contributed by one pass of the truck in question.

Example 59.10. Calculate the design lane 20-year ESAL by the AI procedure for a 3-lane rural interstate with an initial directional AADT of 600,000. The predicted traffic growth is 3% per annum and the percent truck traffic is 16%.

From Fig. 59.3(b), the design lane is to be designed to carry 80% of the directional truck traffic. Total design lane truck volume $= 600{,}000 \times 16\% \times 80\% \times \{(1 + 0.03)^{20} - 1\}/0.03 = 2{,}063{,}645$. The total ESAL is computed as follows:

Truck Type	% Share from Table 59.4(a)	Truck Factor from Table 59.4(b)	ESAL Contribution (Col. 2 × Col. 3)
SU 2-axle, 4-tire	39%	0.02	16,096
SU 2-axle, 6-tire	10%	0.19	29,209
SU 3-axle or more	2%	0.56	23,113
MU 3-axle	1%	0.51	10,525
MU 4-axle	5%	0.62	63,973
MU 5-axle or more	43%	0.94	834,125
			Total ESAL = 987,041

Subgrade Resilient Modulus

The Asphalt Institute design charts require subgrade resilient modulus M_r as input. However, M_r can be estimated by performing the CBR test [ASTM Method D1883, 1992] or the R-value test [ASTM Method D2844, 1992] and applying the following relationships:

$$M_r\ (\text{MPa}) = 10.3\ \text{CBR} \quad \text{or} \quad M_r\ (\text{psi}) = 1500\ \text{CBR} \tag{59.23}$$

$$M_r\ (\text{MPa}) = 8.0 + 3.8\,R \quad \text{or} \quad M_r = 1155 + 555\,R \tag{59.24}$$

For each soil type, six to eight tests are recommended for the purpose of selecting the design subgrade resilient modulus by the following procedure: (1) arrange all M_r values in ascending order; (2) for each test value, compute y = percent of test values equal to or greater than it; (3) plot y against M_r; and (4) read from the plot the design subgrade strength at an appropriate percentile value. The design subgrade percentile value is selected according to the magnitude of design ESAL as follows: 60th percentile for ESAL $\leq 10^4$, 75th percentile if $10^4 <$ ESAL $< 10^6$, and 87.5th percentile if ESAL $\geq 10^6$.

Example 59.11. Eleven CBR tests on the subgrade for the pavement in Example 59.10 yield the following results: 7, 5, 7, 2, 8, 6, 5, 3, 4, 3, and 6. Determine the design subgrade resilient modulus by the AI method.

From Example 59.10, ESAL = 987,041, hence use 75th percentile according to Asphalt Institute recommendation. Next, arrange the test values in ascending order.

CBR	2	3	4	5	6	7	8
% ≥	100	91	73	64	45	27	9

The design CBR (75th percentile value) is 4%. By Eq. (59.23), the corresponding design $M_r = 1500 \times 4 = 6000$ psi.

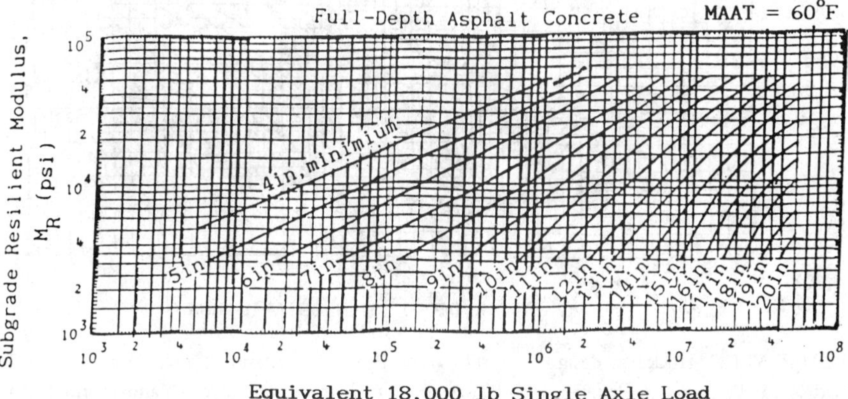

FIGURE 59.10 Thickness design curves for full-depth asphalt concrete. (*Source:* Asphalt Institute. 1991. *Thickness Design—Asphalt Pavements for Highways and Streets.* Manual Series MS-1. With permission.)

Pavement Thickness Requirements

Thickness requirement charts are developed for three different designs of pavement structure: (1) Fig. 59.10 for full-depth pavements, (2) Figs. 59.11–59.13 for pavements with emulsified asphalt base, and (3) Figs. 59.14 and 59.15 for pavements with untreated aggregate base. These charts are valid for mean annual air temperature (MAAT) of 60°F (15.5°C). Corresponding charts are also prepared by the Asphalt Institute for MAATs of 45°F (7°C) and 75°F (24°C). Type I, II, and III emulsified asphalt mixes differ in the aggregates used. Type I mixes are made with processed dense-graded aggregates; type II with semiprocessed, crusher-run, pit-run, or bank-run aggregates; and type III with sands or silty sands. The minimum thicknesses of full-depth pavements at different traffic levels are indicated in Fig. 59.10. The minimum thicknesses of asphalt concrete surface course for pavements with other base courses are given in Table 59.12(b).

Example 59.12. With the data in Examples 59.10 and 59.11, design the required thickness for (1) a full-depth pavement, (2) a pavement with type II emulsified base, and (3) a pavement with 12-in. aggregate base.

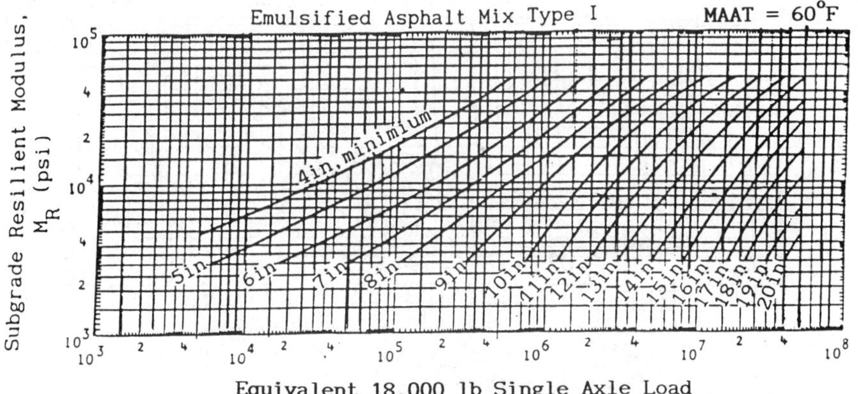

FIGURE 59.11 Thickness design curves for type I emulsified asphalt mix. (*Source:* Asphalt Institute. 1991. *Thickness Design—Asphalt Pavements for Highways and Streets.* Manual Series MS-1. With permission.)

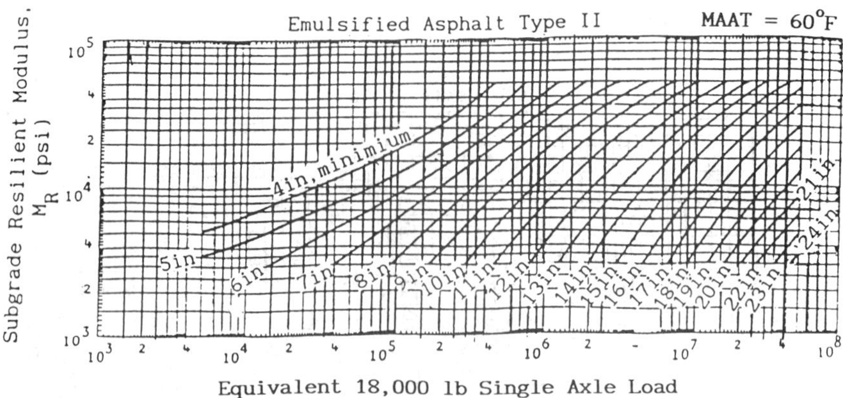

FIGURE 59.12 Thickness design curves for type II emulsified asphalt mix. (*Source:* Asphalt Institute. 1991. *Thickness Design—Asphalt Pavements for Highways and Streets.* Manual Series MS-1. With permission.)

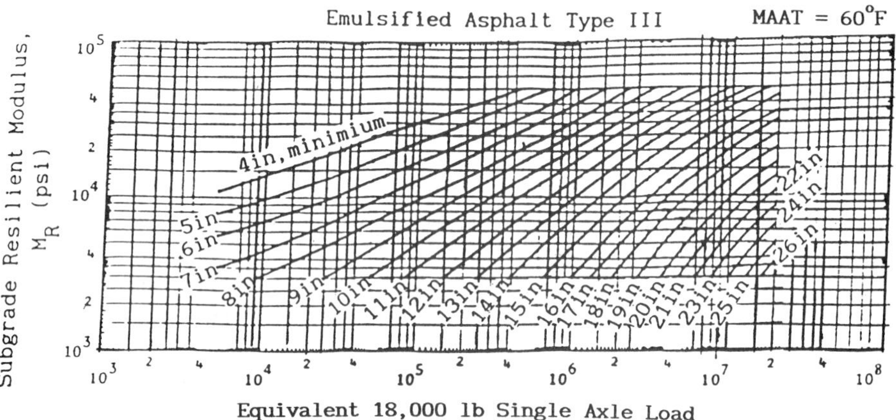

FIGURE 59.13 Thickness design curves for type III emulsified asphalt mix. (*Source:* Asphalt Institute. 1991. *Thickness Design—Asphalt Pavements for Highways and Streets.* Manual Series MS-1. With permission.)

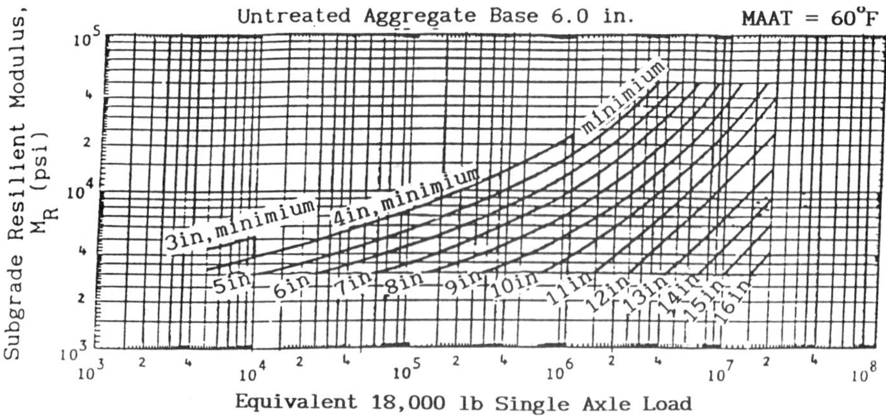

FIGURE 59.14 Thickness design curves for asphalt pavement with 6-in. untreated aggregate base. (*Source:* Asphalt Institute. 1991. *Thickness Design—Asphalt Pavements for Highways and Streets.* Manual Series MS-1. With permission.)

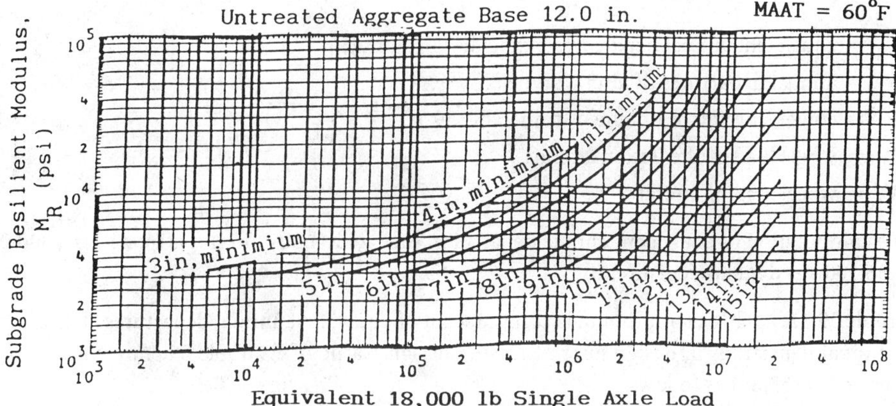

FIGURE 59.15 Thickness design curves for asphalt pavement with 12-in. untreated aggregate base. (*Source:* Asphalt Institute. 1991. *Thickness Design—Asphalt Pavements for Highways and Streets.* Manual Series MS-1. With permission.)

(1) With ESAL = 987,041 and M_r = 6000 psi, the required full-depth pavement thickness is 9.5 in. according to Fig. 59.10. (2) The total required thickness is 11.5 in., from Fig. 59.12. The minimum asphalt concrete surface course is 3 in. according to Table 59.12(b). Hence, the thickness of the emulsified base = 11.5 − 3 = 8.5 in. (3) The required thickness of asphalt concrete is 7.5 in. Use 2 in. of asphalt concrete surface according to Table 59.12(b) and (7.5 − 2) = 5.5 in. of asphalt base layer.

FAA Design Procedure for Flexible Airport Pavements

Based on the California bearing ratio (CBR) method of design, FAA [1978] developed—through test track studies and observations of in-service pavements—pavement thicknesses that are necessary to protect pavement layers with various CBR values from shear failure. In establishing the thickness requirements the equivalent single-wheel loads of wheel assemblies were computed based on deflection consideration. The design assumes that 95% of the gross aircraft weight is carried on the main landing gear assembly and 5% on the nose gear assembly. Generalized design curves are available for single, dual, and dual tandem main landing gear assemblies. Design curves for specific wide-body aircraft have also been developed.

Computation of Design Loading

The FAA design charts are based on the equivalent annual departures of a selected design aircraft. The annual departures are assumed to occur over a 20-year life. The following steps are involved in the selection of design aircraft and determination of equivalent annual departures.

1. Obtain forecasts of annual departures by aircraft type.
2. Determine for each aircraft type the required pavement thickness using the appropriate design curve with the forecast number of annual departures for that aircraft. The aircraft requiring the greatest pavement thickness is selected as the design aircraft.
3. Convert the annual departures of all aircraft to equivalent annual departures of the design aircraft by the following formula:

$$\log R_{\text{eq}} = \log(R_i \times F_i) \times \left\{ \frac{W_i}{W} \right\}^{0.5} \tag{59.25}$$

where

R_{eq} = equivalent annual departures by the design aircraft
R_i = annual departures of aircraft type i
F_i = conversion factor obtained from Table 59.9
W = wheel load of the design aircraft
W_i = wheel load of aircraft i

In the computation of equivalent annual departures, each wide-body aircraft is treated as a 300,000 lb (136,100 kg) dual tandem aircraft.

Example 59.13. This example entails computation of equivalent annual departures. The equivalent annual departures in design aircraft for the design traffic of Example 59.5 are computed by means of Eq. (59.25) as follows.

Aircraft	Single-Wheel Load W_i (lbs.)	$(R_1 \times F_i)$ from Example 59.5	R_{eq} by Eq. (59.25)
727-100	38,000	4500	2229
727-200	45,240	9900	9900
707-320B	38,830	5440	2890
DC-9-30	25,650	5500	655
747-100	35,625	102	61
	Total equivalent design annual departures = 15,735		

Pavement Thickness Requirements

Figures 59.16–59.22 are the FAA design charts for different aircraft types. The charts have incorporated the effects of load repetitions, landing gear assembly configuration, and the "wandering" (lateral distribution) effect of aircraft movements. With subgrade CBR, gross weight, and total equivalent annual departures of design aircraft as input, the total pavement thickness required can be read off from the appropriate chart. Each design chart also indicates the required thickness of bituminous surface course. The minimum base course thickness is obtained from Fig. 59.23.

FAA requires stabilized base and subbase courses to be used to accommodate jet aircraft weighing 100,000 lb or more. These stabilized courses may be substituted for granular courses using the equivalency factors in Table 59.13.

FAA [1978] suggests that the full design thickness T be used at critical areas where departing traffic will be using the pavement, $0.9T$ be used at areas receiving arriving traffic such as high-speed turnoffs, and $0.7T$ be used where traffic is unlikely. These reductions in thickness are applied to base and subbase courses. Figure 59.24 shows a typical cross section for runway pavements.

For pavements receiving high traffic volumes and exceeding 25,000 departures per annum, FAA requires that the bituminous surfacing be increased by 1 in. (3 cm) and the total pavement thickness be increased as follows: 104%, 108%, 110%, and 112% of design thickness (based on 25,000 annual departures) for annual departures of 50,000, 100,000, 150,000, and 200,000, respectively.

Example 59.14. For the design traffic in Example 59.13, determine the thickness requirements for a pavement with subgrade CBR = 5 and subbase CBR = 20.

The design aircraft has dual-wheel landing gear and a maximum weight of 190,500 lb. Figure 59.17 gives the total thickness requirement as 45 in. above subgrade and 18 in. above subbase. Minimum asphalt concrete surface for critical area is 4 in. Thickness of base = 18 − 4 = 14 in. Thickness of subbase = 45 − 4 − 14 = 27 in. Since the design aircraft weighs more than 100,000 lb, stabilized base and subbase are needed. Use bituminous base course with equivalency factor of 1.5 [see Table 59.13(a)] and cold-laid bituminous base course with equivalency factor

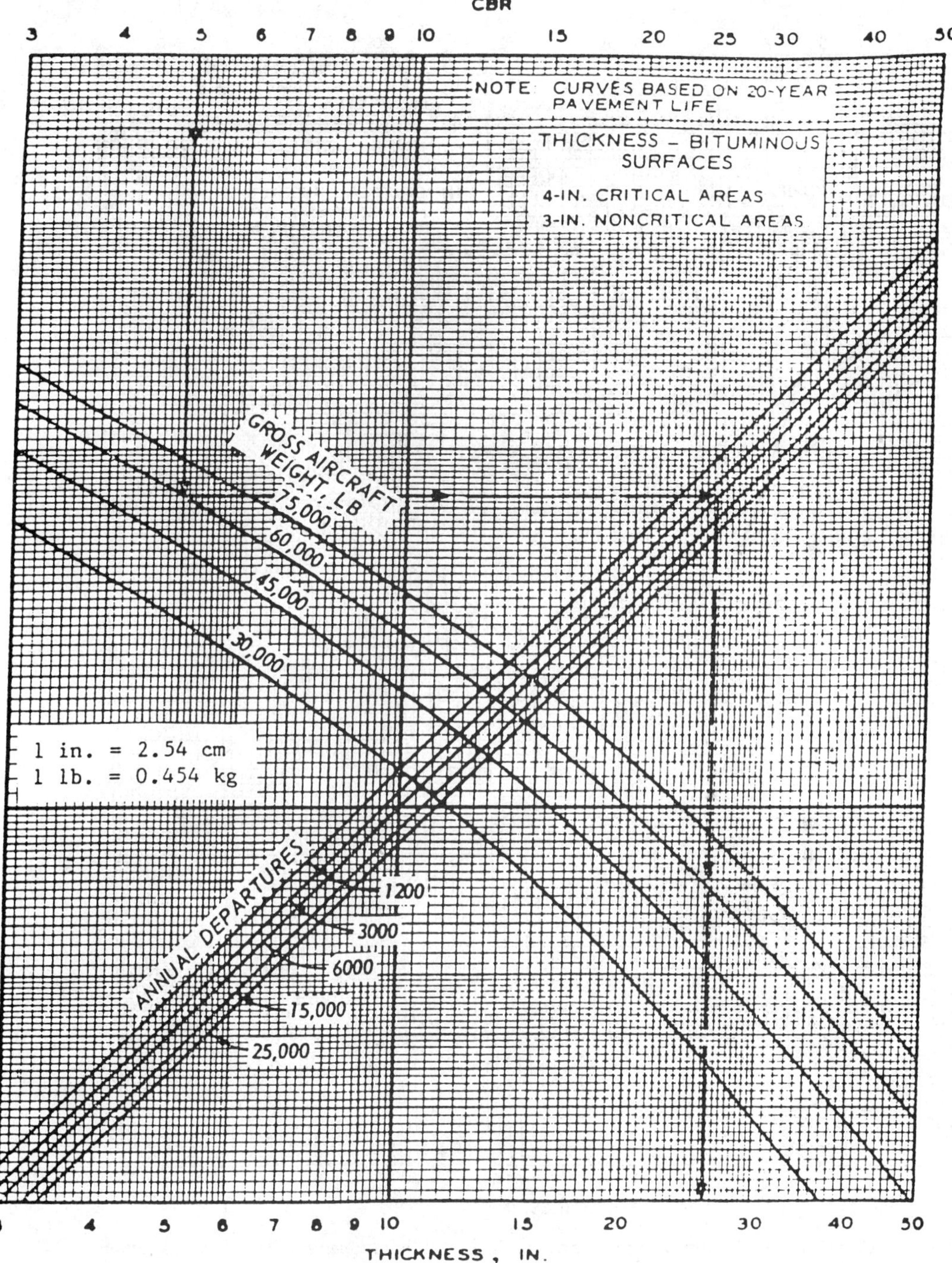

FIGURE 59.16 Critical area flexible pavement thickness for single-wheel gear. (*Source:* Federal Aviation Administration. 1978. *Airport Pavement Design and Evaluation.* Advisory Circular AC No. 150/5320-6C. With permission.)

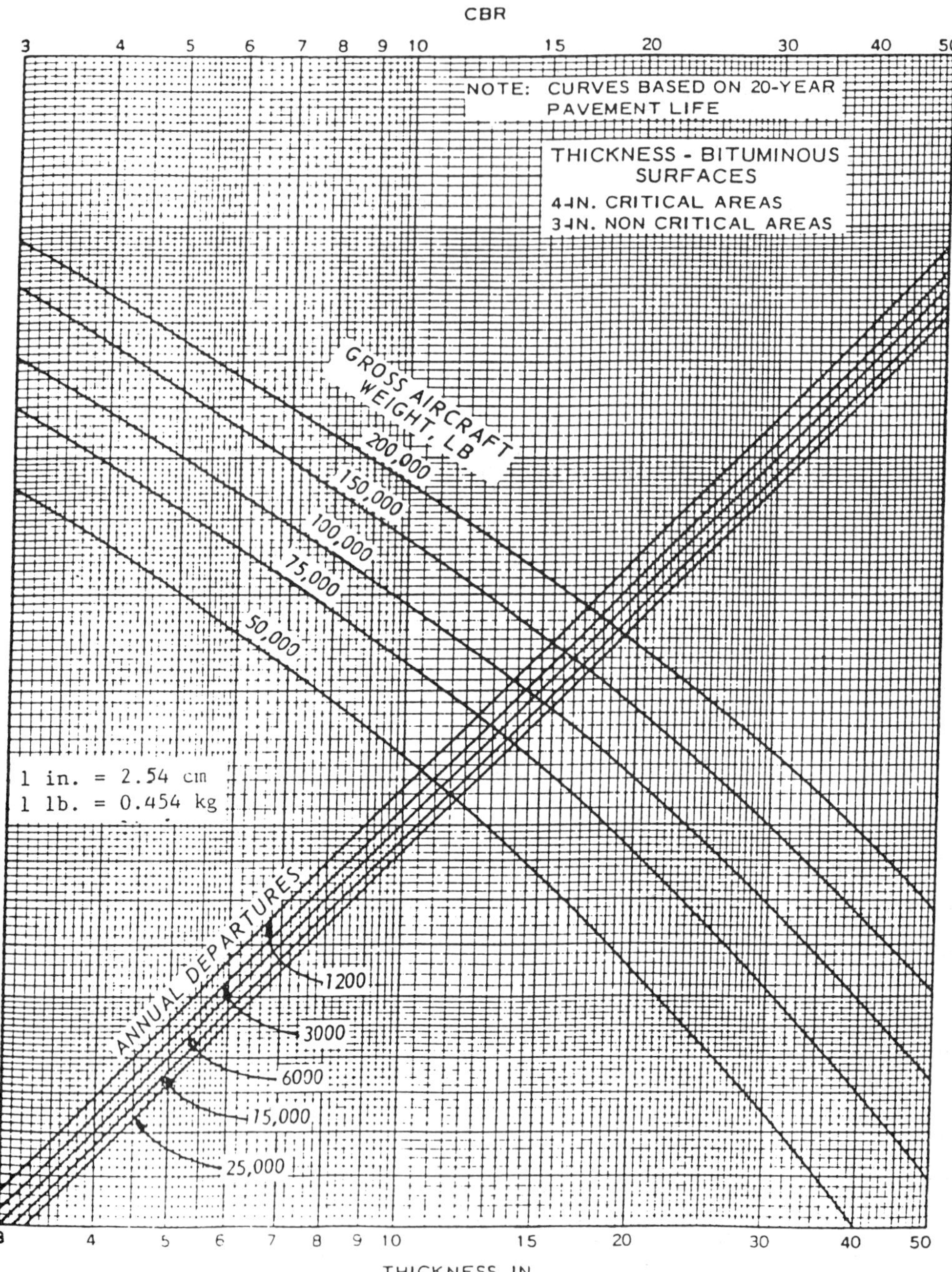

FIGURE 59.17 Critical area flexible pavement thickness for dual-wheel gear. (*Source:* Federal Aviation Administration. 1978. *Airport Pavement Design and Evaluation.* Advisory Circular AC No. 150/5320-6C. With permission.)

of 1.5 [see Table 59.13(b)]. The required stabilized base thickness = (14/1.5) = 9 in., and the required subbase thickness = (27/1.5) = 18 in., both for critical areas. In the case of noncritical areas, the asphalt concrete surface thickness is 3 in., and the corresponding base and subbase thicknesses are (9 × 0.9) = 8 in. and (18 × 0.9) = 16 in.

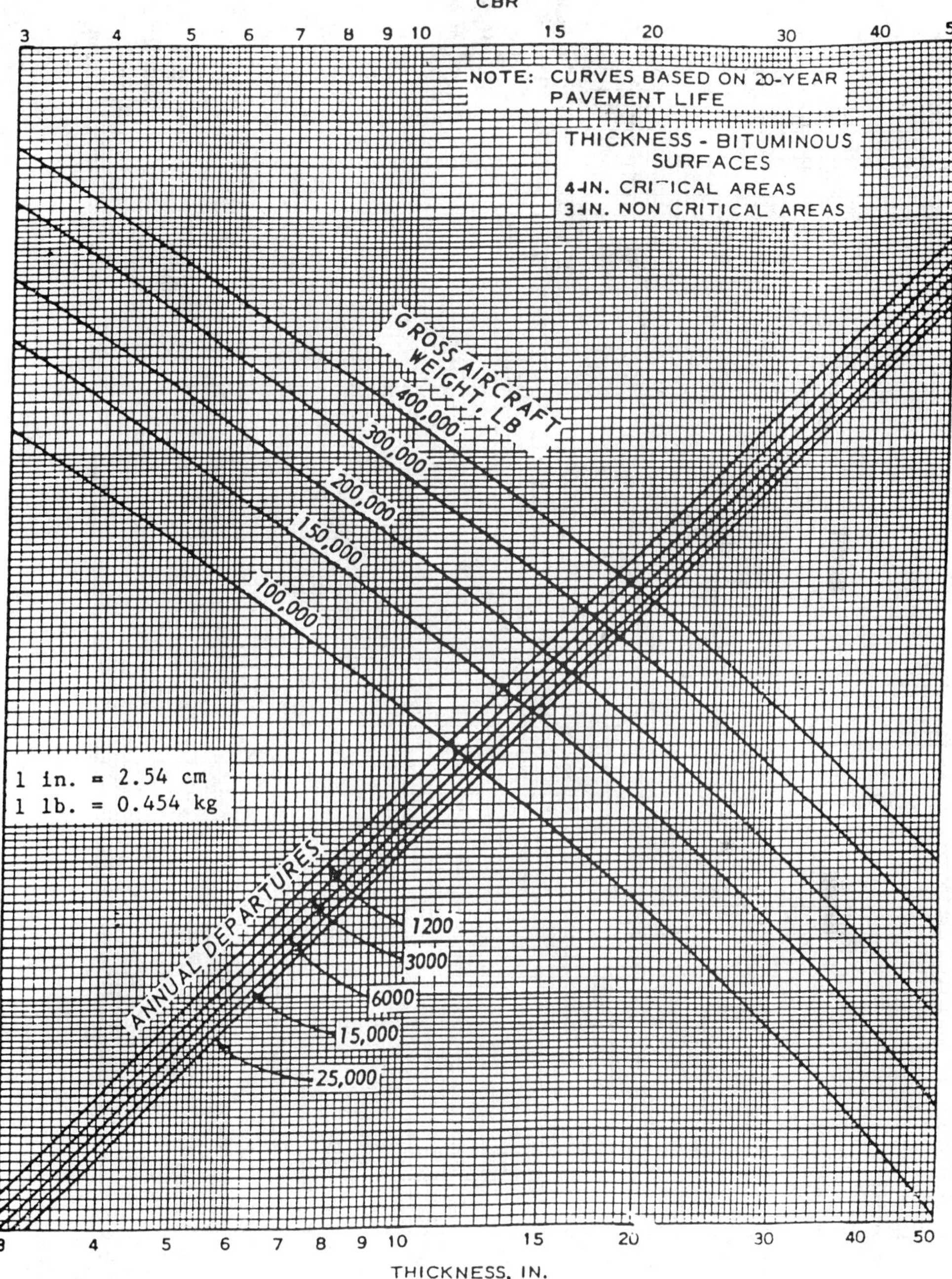

FIGURE 59.18 Critical area flexible pavement thickness for dual-tandem gear. (*Source:* Federal Aviation Administration. 1978. *Airport Pavement Design and Evaluation.* Advisory Circular AC No. 150/5320-6C. With permission.)

Mechanistic Approach for Flexible Pavement Design

The methods described in the preceding sections provide evidence of the continued effort and progress made by engineers towards adopting theoretically sound approaches with fundamental material properties in pavement design. For example, the Asphalt Institute method described is a complete revision that uses analyses based on elastic theory to generate pavement thickness

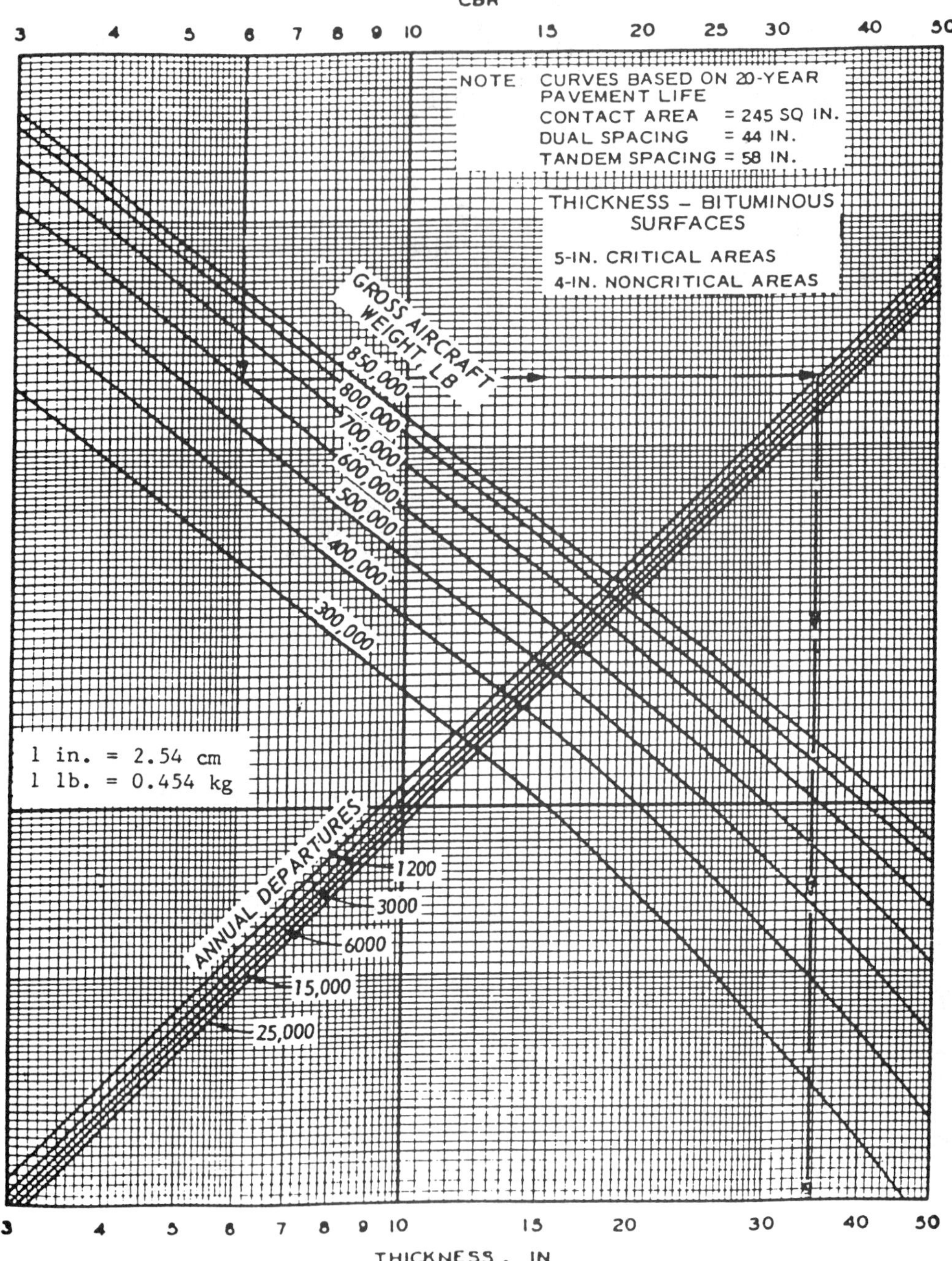

FIGURE 59.19 Critical area flexible pavement thickness for B-747-100, SR, 200B, 200C, and 200F. (*Source:* Federal Aviation Administration. 1978. *Airport Pavement Design and Evaluation.* Advisory Circular AC No. 150/5320-6C. With permission.)

requirements against two failure criteria: a fatigue cracking criterion for the asphalt layer and a rutting criterion for the subgrade.

More comprehensive mechanistic procedures, capable of handling the following aspects in pavement design, are available in the literature: (a) viscoelastic behavior of bituminous materials,

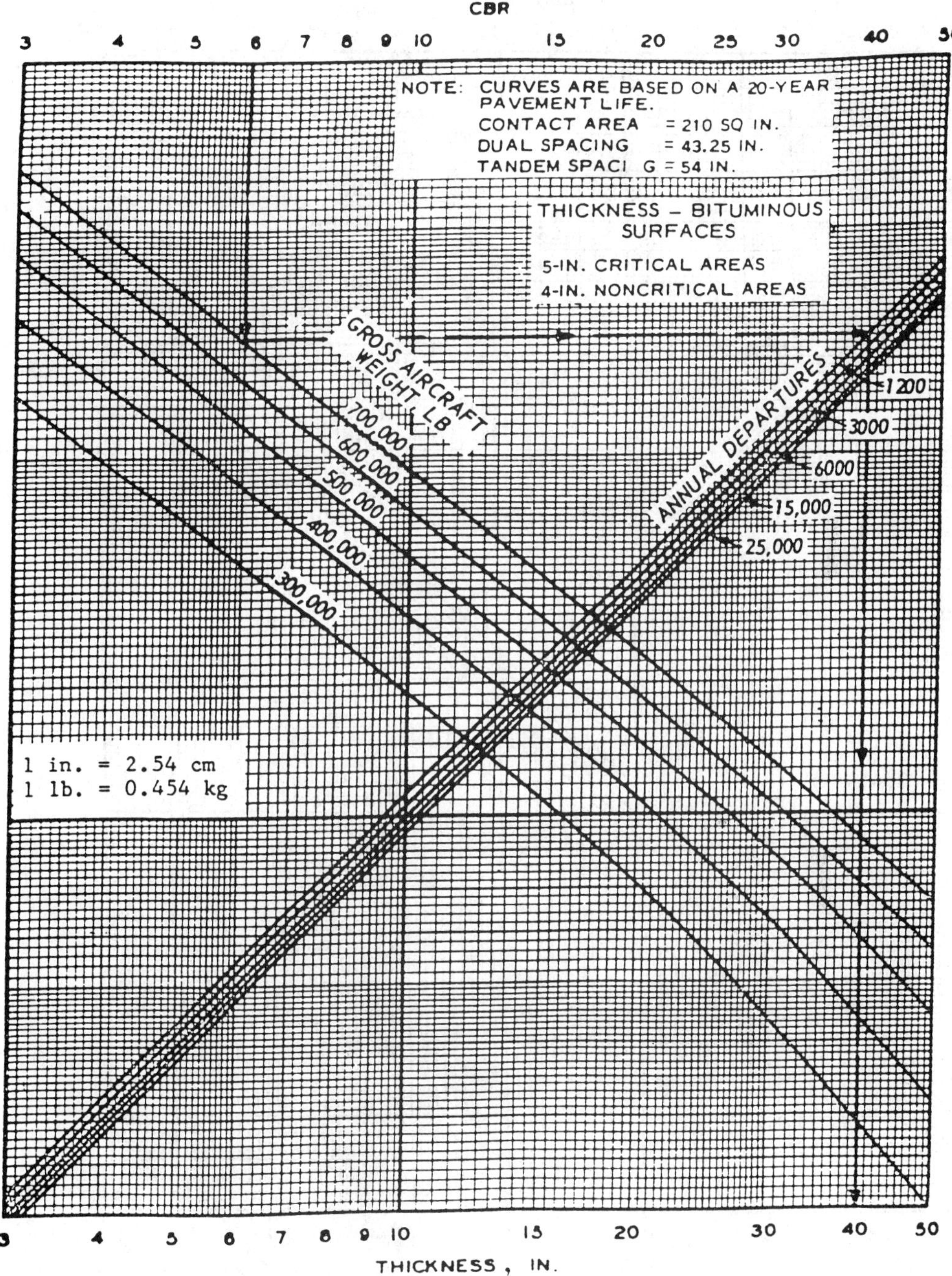

FIGURE 59.20 Critical area flexible pavement thickness for B-747-SP. (*Source:* Federal Aviation Administration. 1978. *Airport Pavement Design and Evaluation.* Advisory Circular AC No. 150/5320-6C. With permission.)

(b) nonlinear response of untreated granular and cohesive materials, (c) aging of bituminous materials, (d) material variabilities, (e) dynamic effect of traffic loading, (f) effect of mixed traffic loading, and (g) interdependency of the development of different distresses, including pavement roughness. Unfortunately, quantification of necessary material properties and analysis of pavement

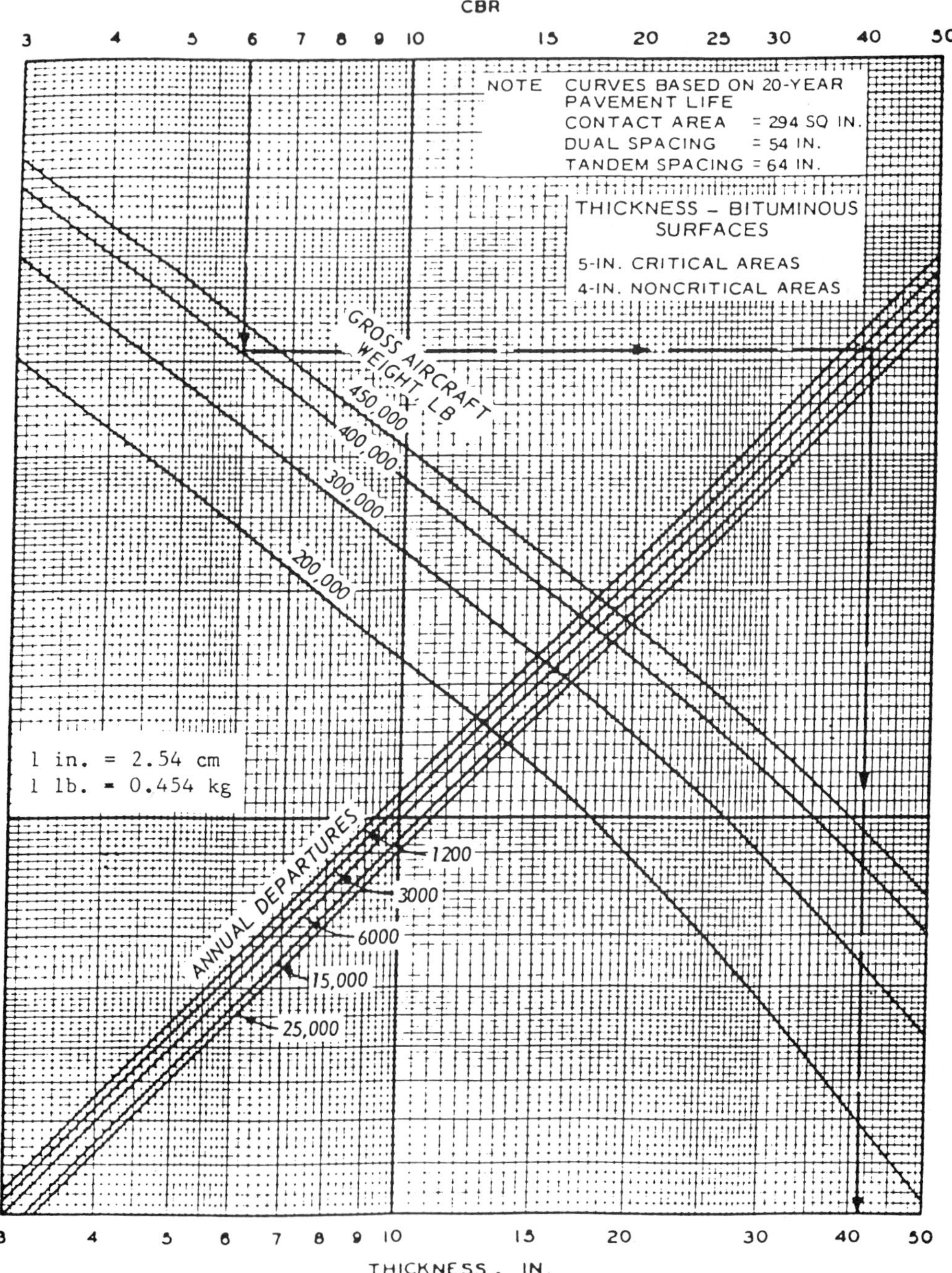

FIGURE 59.21 Critical area flexible pavement thickness for DC 10-10, 10CF. (*Source:* Federal Aviation Administration. 1978. *Airport Pavement Design and Evaluation.* Advisory Circular AC No. 150/5320-6C. With permission.)

response by these procedures are often complicated, time consuming, skill demanding, and costly. Although there are great potentials for these procedures when fully implemented, simplifications such as that adopted by the Asphalt Institute method will have to be applied for pavement design in practice.

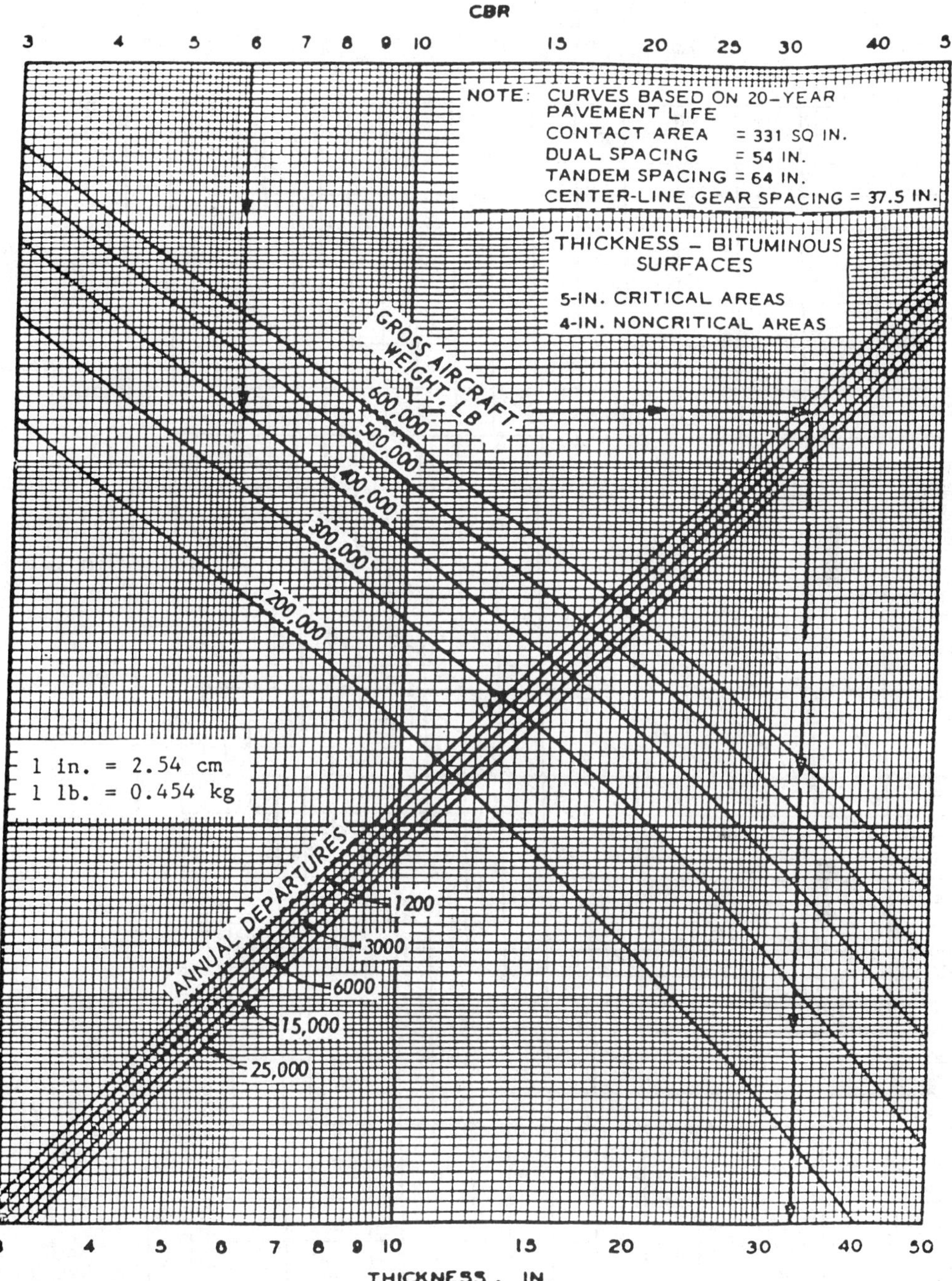

FIGURE 59.22 Critical area flexible pavement thickness for DC 10-30, 30CF, 40, and 40CF. (*Source:* Federal Aviation Administration. 1978. *Airport Pavement Design and Evaluation.* Advisory Circular AC No. 150/5320-6C. With permission.)

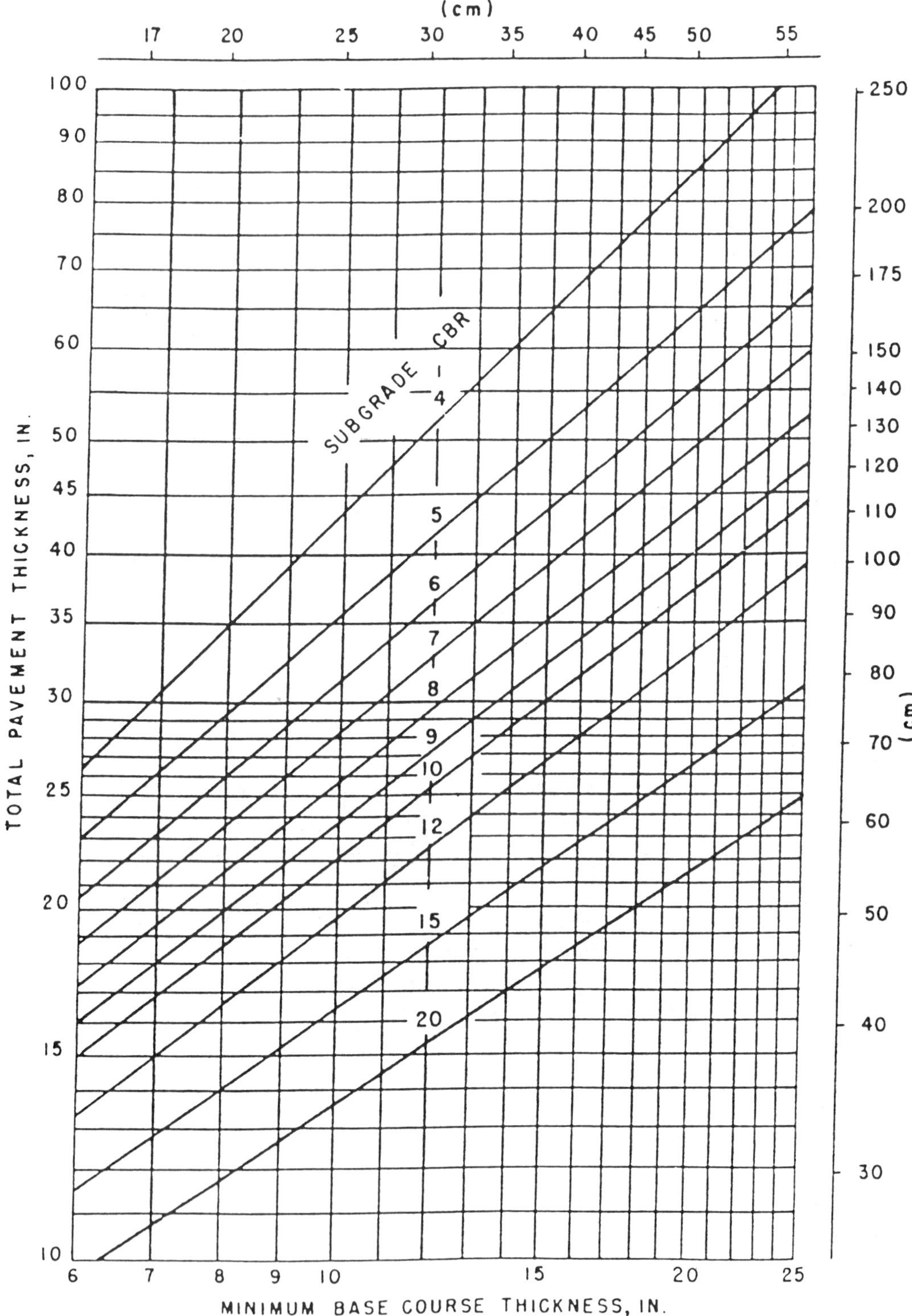

FIGURE 59.23 Minimum base course thickness requirements. (*Source:* Federal Aviation Administration. 1978. *Airport Pavement Design and Evaluation.* Advisory Circular AC No. 150/5320-6C, p. 53. With permission.)

Table 59.13 FAA Recommended Equivalency Factors for Stablilized Base and Subbase

Material	Equivalency Factor
(a) Equivalency Factors for Stabilized Base Course	
Bituminous surface course	1.2–1.6
Bituminous base course	1.2–1.6
Cold-laid bituminous base course	1.0–1.2
Mixed-in-place base course	1.0–1.2
Cement-treated base course	1.2–1.6
Soil cement base course	N/A
Crushed aggregate base course	1.0
Subbase course	N/A
(b) Equivalency Factors for Stabilized Subbase Course	
Bituminous surface course	1.7–2.3
Bituminous base course	1.7–2.3
Cold-laid bituminous base course	1.5–1.7
Mixed-in-place base course	1.5–1.7
Cement-treated base course	1.6–2.3
Soil cement base course	1.5–2.0
Crushed aggregate base course	1.4–2.0
Subbase course	1.0

Source: Federal Aviation Administration. 1978. *Airport Pavement Design and Evaluation.* Reprinted from FAA Advisory Circular. Report No. FAA/AC-150/5320-6C. 7 December 1978; NTIS Accession No. AD-A075 537/1.

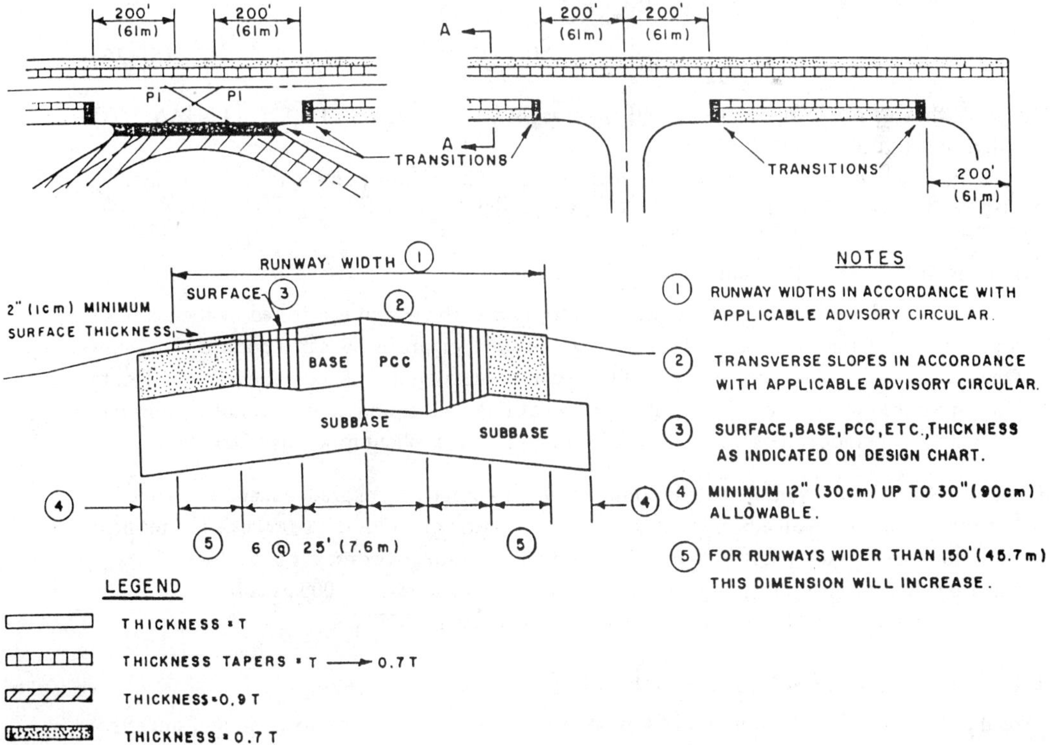

FIGURE 59.24 Typical plan and cross section for runway pavements. (*Source:* Federal Aviation Administration. 1978. *Airport Pavement Design and Evaluation.* Advisory Circular AC No. 150/5320-6C, p. 35. With permission.)

59.6 Structural Design of Rigid Pavements

Structural design of rigid pavements includes thickness and reinforcement designs. Two major forms of thickness design methods are being used today for concrete pavements. The first form is an approach that relies on empirical relationships derived from performance of full-scale test pavements and in-service pavements. The design procedure of AASHTO [1993] is an example. The second form develops relationships in terms of the properties of pavement materials, as well as load-induced and thermal stresses, and calibrates these relationships with pavement performance data. The PCA [1984] and the FAA [1978] methods of design adopt this approach. Thickness design procedures by AASHTO, PCA, and FAA are discussed in this section. Reinforcement designs by the AASHTO and FAA procedures will be presented.

AASHTO Thickness Design for Rigid Highway Pavements

The serviceability-based concept of the AASHTO design procedure for rigid pavements [AASHTO 1993] is similar to its design procedure for flexible pavements. Pavement thickness requirements are established from data of the AASHO road test [Highway Research Board 1962]. Input requirements such as reliability information and serviceability loss for design have been described in the section on AASHTO flexible pavement design. Details for other input requirements are described in this section.

Pavement Material Properties

The elastic modulus E_c and modulus of rupture S_c of concrete are required input parameters. E_c is determined by the procedure specified in ASTM C469. It could also be estimated using the following correlation recommended by ACI [1977]:

$$E_c \text{ (psi)} = 57{,}000(f_c)^{0.5} \tag{59.26}$$

where f_c is the concrete compressive strength in psi as determined by AASHTO T22, T140 [AASHTO 1989], or ASTM C39 [ASTM 1992].

S_c is the mean 28-day modulus of rupture determined using third-point loading as specified by AASHTO T97 [AASHTO 1989] or ASTM C39 [ASTM 1992].

Modulus of Subgrade Reaction

The value of modulus of subgrade reaction k to be used in the design is affected by the depth of bedrock and the characteristics of the subbase layer, if used. Figure 59.25 is first applied to account for the presence of subbase course and obtain the composite modulus of subgrade reaction. Figure 59.26 is next used to include adjustment for the depth of rigid foundation. It is noted from Fig. 59.25 that the subgrade soil property required for input is the resilient modulus M_r.

Example 59.15. This example entails computation of composite subgrade reaction. A concrete pavement is constructed on a 6-in.-thick subbase with elastic modulus of 20,000 psi. The resilient modulus of the subgrade soil is 7000 psi. The depth of subgrade to bedrock is 5 ft.

Entering Fig. 59.25, with D_{SB} = 6 in., E_{SB} = 20,000 psi, and M_r = 7000 psi, obtain k_∞ = 400 pci. With bedrock depth of 5 ft., composite k = 500 pci, from Fig. 59.26.

Effective Modulus of Subgrade Reaction

Like the effective roadbed soil resilient modulus M_r for flexible pavement design, an effective k is computed to represent the combined effect of seasonal variations of k. The procedure is identical to the computation of effective M_r, except that the relative damage u is now computed as

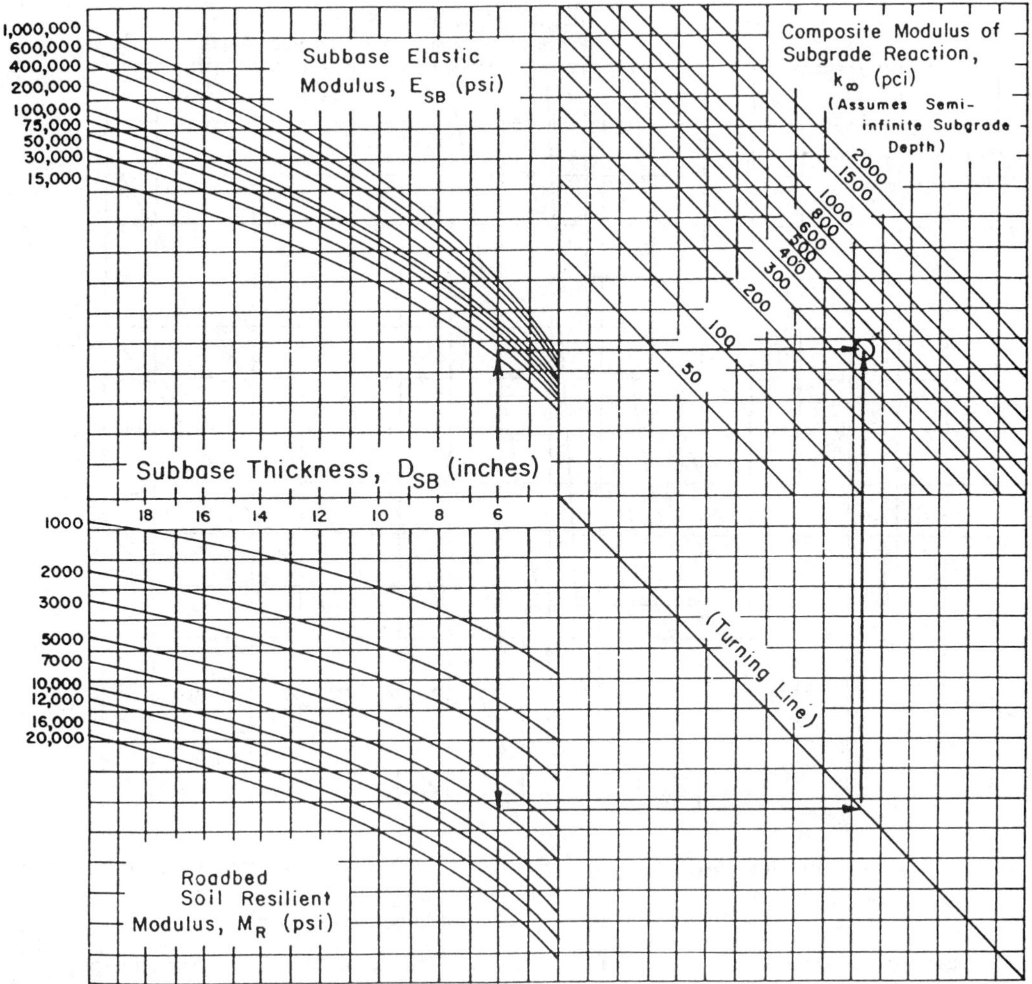

FIGURE 59.25 Chart for estimating composite k_∞. (*Source:* AASHTO. 1993. *AASHTO Guides for Design of Pavement Structures.* Copyright 1993 by the American Association of State Highway and Transportation Officials, Washington, D.C. Used by permission.)

$$u = (D^{0.75} - 0.39k^{0.25})^{3.42} \tag{59.27}$$

Instead of solving the above equation, u can be obtained from Fig. 59.27.

Depending on the type of subbase and subgrade materials, the effective k must be reduced according to Fig. 59.28 to account for likely loss of support by foundation erosion and/or differential soil movements. Suggested values of LS in Fig. 59.28 are given in Table 59.14.

Example 59.16. The value of composite k values determined at 1-month intervals are 400, 400, 450, 450, 500, 500, 450, 450, 450, 450, 450, and 450. Projected slab thickness is 10 in. and $LS = 1.0$. Determine effective k.

By means of Eq. (59.27) or Fig. 59.27, the relative damage for each k can be determined. Hence, total $u = 100 + 100 + 97 + 97 + 93 + 93 + 94 + 94 + 94 + 94 + 94 + 94 = 1144$. Average $u = 95.3$ and average $k = 470$ pci. Entering Fig. 59.28 with $LS = 1.0$ and $D = 10$ in., read effective $k = 150$ pci.

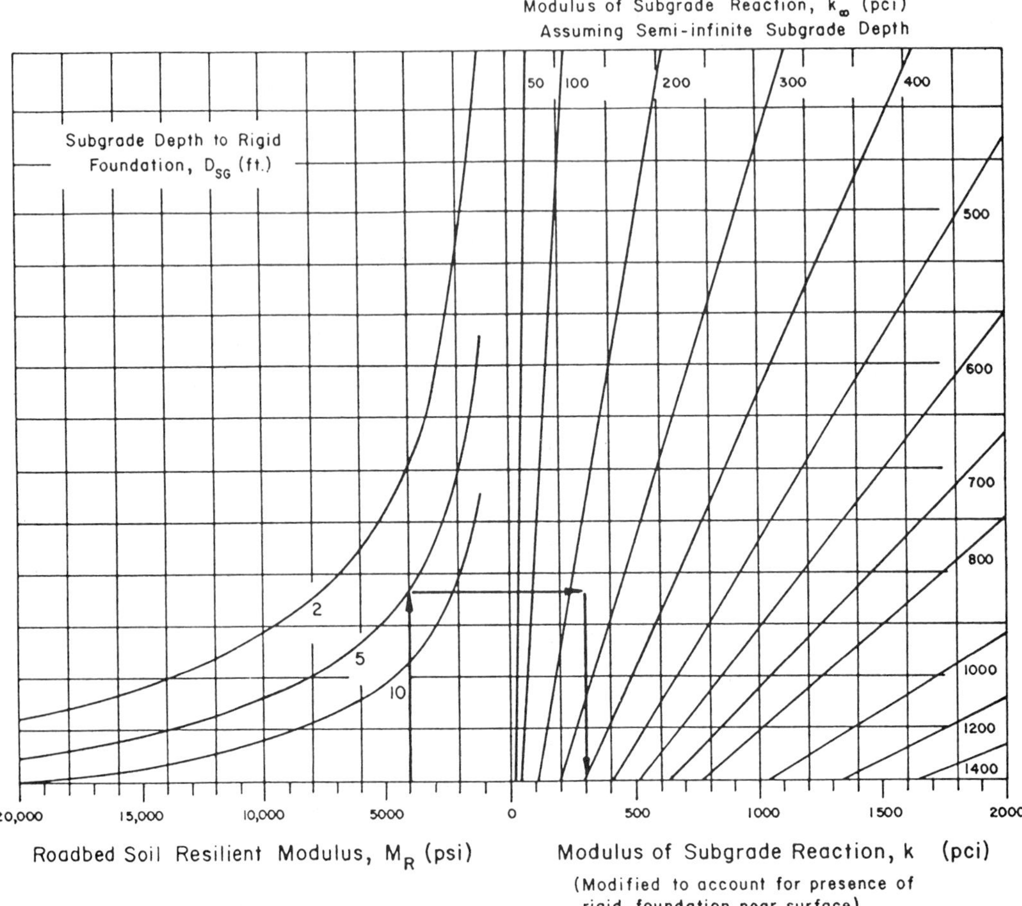

FIGURE 59.26 Chart for k as a function of bedrock depth. (*Source:* AASHTO. 1993. *AASHTO Guides for Design of Pavement Structures.* Copyright 1993 by the American Association of State Highway and Transportation Officials, Washington, D.C. Used by permission.)

Load Transfer Coefficient

Load transfer coefficient J is a numerical index developed from experience and stress analysis. Table 59.15 presents the J values for the AASHO road test conditions. Lower J values are associated with pavements with load transfer devices (such as dowel bars) and those with tied shoulders. For cases where a range of J values applies, higher values should be used with low k values, high thermal coefficients, and large variations of temperature. When dowel bars are used, the AASHTO guide recommends that the dowel diameter should be equal to the slab thickness multiplied by 1/8, with normal dowel spacing and length of 12 in. and 18 in., respectively.

Drainage Coefficient

To allow for changes in thickness requirement due to differences in drainage properties, pavement layers, and subgrade, a drainage coefficient C_d was included in the design. Setting $C_d = 1$ for conditions at the AASHO road test, Table 59.16 shows the C_d values for other conditions. The percentage of time during the year that the pavement structure would be exposed to moisture levels approaching saturation can be estimated from the annual rainfall and the prevailing drainage condition.

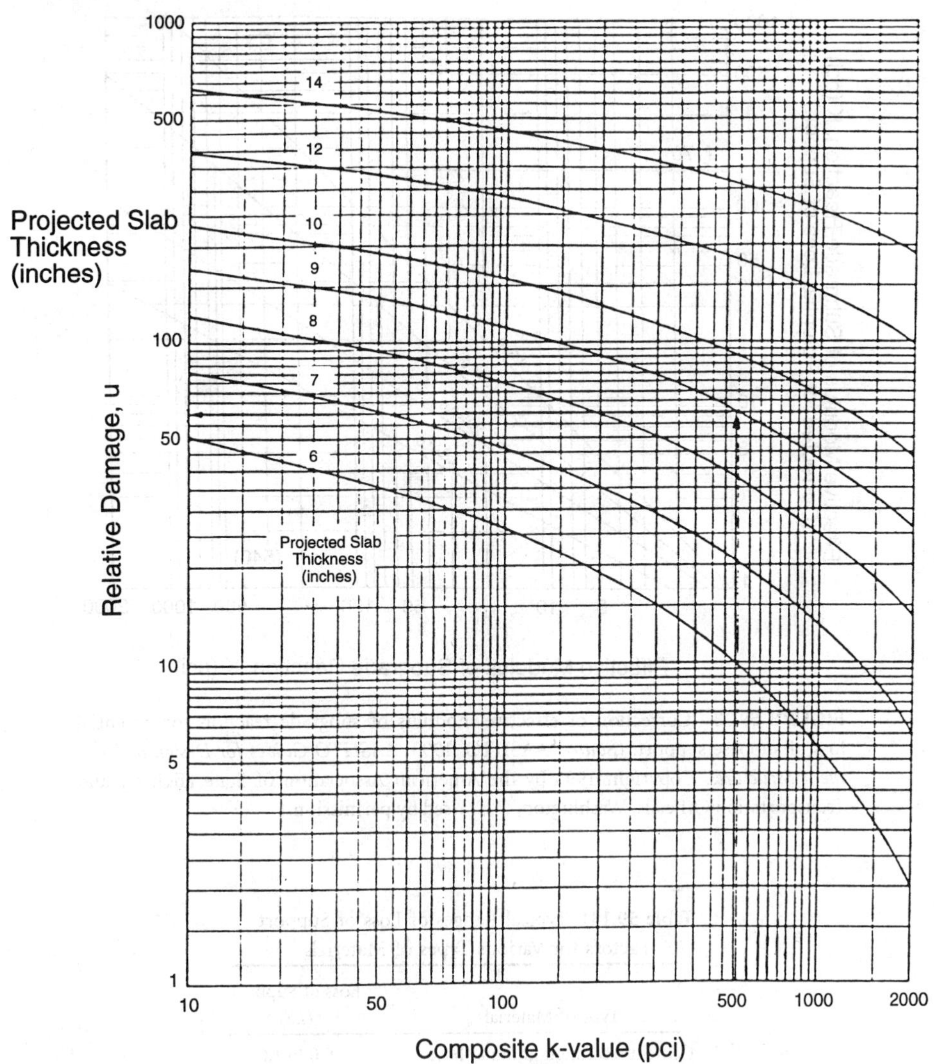

FIGURE 59.27 Chart for estimating relative damage to rigid pavements. (*Source:* AASHTO. 1993. *AASHTO Guides for Design of Pavement Structures.* Copyright 1993 by the American Association of State Highway and Transportation Officials, Washington, D.C. Used by permission.)

Thickness Requirement

The required slab thickness is obtained using the nomograph in Fig. 59.29. If the environmental effects of roadbed swelling and frost heave are important, they are considered in the same way as for flexible pavements.

Example 59.17. Apply AASHTO procedure to design a concrete pavement slab thickness for ESAL $= 11 \times 10^6$. The design reliability is 95%, with a standard deviation of 0.3. The initial and terminal serviceability levels are 4.5 and 2.5, respectively. Other design parameters are $E_c = 5 \times 10^6$, $S_c' = 650$ psi, $J = 3.2$, and $C_d = 1.0$.

Design PSI loss $= 4.5 - 2.5 = 2.0$. From Fig. 59.29, $D = 10$ in.

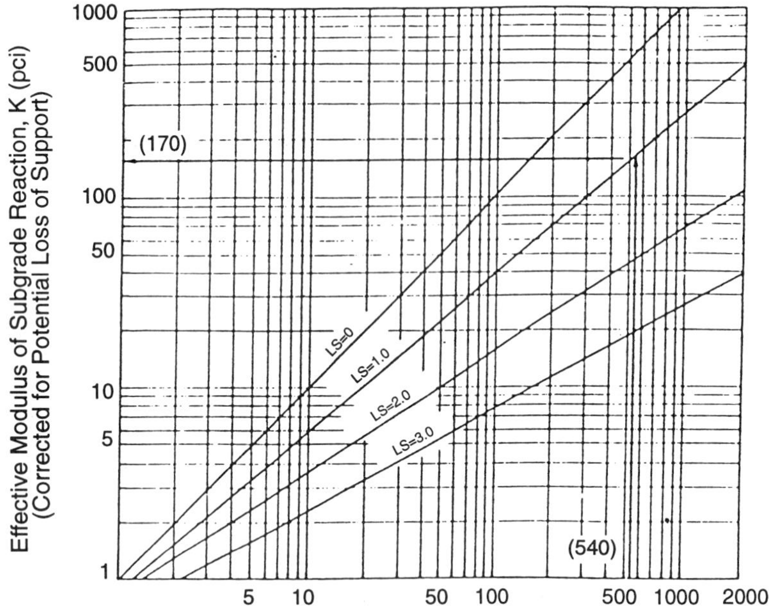

FIGURE 59.28 Correction of effective modulus of subgrade reaction for potential loss of subbase support. (*Source:* AASHTO. 1993. *AASHTO Guides for Design of Pavement Structures.* Copyright 1993 by the American Association of State Highway and Transportation Officials, Washington, D.C. Used by permission.)

Table 59.14 Typical Ranges of Loss of Support (*LS*) Factors for Various Types of Materials

Type of Material	Loss of Support (*LS*)
Cement treatment granular base (E = 1,000,000 to 2,000,000 psi)	0.0 to 1.0
Cement aggregate mixtures (E = 500,000 to 1,000,000 psi)	0.0 to 1.0
Asphalt treated base (E = 350,000 to 1,000,000 psi)	0.0 to 1.0
Bituminous stabilized mixtures (E = 40,000 to 300,000 psi)	0.0 to 1.0
Lime stabilized (E = 20,000 to 70,000 psi)	1.0 to 3.0
Unbound granular materials (E = 15,000 to 45,000 psi)	1.0 to 3.0
Fine-grained or natural subgrade materials (E = 3000 to 40,000 psi)	2.0 to 3.0

Source: AASHTO. 1993. *AASHTO Guides for Design of Pavement Structures.* Copyright 1993 by the American Association of State Highway and Transportation Officials, Washington, D.C. Used by permission.

Table 59.15 Recommended Load Transfer Coefficient for Various Pavement Types and Design Conditions

Shoulder	Asphalt		Tied P.C.C.	
Load Transfer Device	Yes	No	Yes	No
Pavement Type				
Plain Jointed and Jointed Reinforced	3.2	3.8–4.4	2.5–3.1	3.6–4.2
CRCP	2.9–3.2	N/A	2.3–2.9	N/A

Source: AASHTO. 1993. *AASHTO Guides for Design of Pavement Structures.* Copyright 1993 by the American Association of State Highway and Transportation Officials, Washington, D.C. Used by permission.

Table 59.16 Recommended Value of Drainage Coefficient, C_d, for Rigid Pavement Design

	Percent of Time Pavement Structure Is Exposed to Moisture Levels Approaching Saturation			
Quality of Drainage	Less than 1%	1–5%	5–25%	Greater than 25%
Excellent	1.25–1.20	1.20–1.15	1.15–1.10	1.10
Good	1.20–1.15	1.15–1.10	1.10–1.00	1.00
Fair	1.15–1.10	1.10–1.00	1.00–0.90	0.90
Poor	1.10–1.00	1.00–0.90	0.90–0.80	0.80
Very Poor	1.00–0.90	0.90–0.80	0.80–0.70	0.70

Source: AASHTO. 1993. *AASHTO Guides for Design of Pavement Structures.* Copyright 1993 by the American Association of State Highway and Transportation Officials, Washington, D.C. Used by permission.

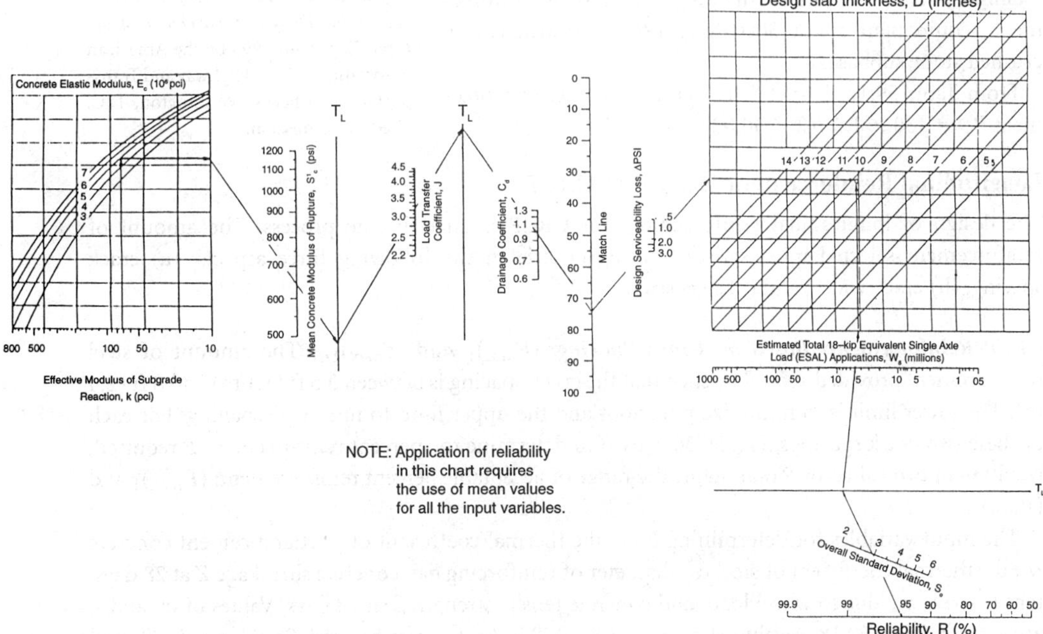

FIGURE 59.29 Rigid pavement thickness design chart. (*Source:* AASHTO. 1993. *AASHTO Guides for Design of Pavement Structures.* Copyright 1993 by the American Association of State Highway and Transportation Officials, Washington, D.C. Used by permission.)

AASHTO Reinforcement Design for Rigid Highway Pavements

Reinforcements are introduced in concrete pavements for the purpose of crack-width control. They are designed to hold cracks tightly closed so that the pavement remains as an integral structural unit. The amount of reinforcement required is a function of slab length (or joint spacing) and thermal properties of the pavement material.

Reinforcements are not required in jointed plain concrete pavements (JPCP) whose lengths are relatively short. As a rough guide proposed by AASHTO, the joint spacing (in feet) for JPCP should not greatly exceed twice the slab thickness (in inches), and the ratio of slab width to length should not exceed 1.25.

Reinforcement Design for JRCP

The percentage of steel reinforcement (either longitudinal or transverse reinforcement) required for jointed reinforced concrete pavement (JRCP) is given by

$$P_s = \frac{L \cdot F}{2f_s} \times 100\% \qquad (59.28)$$

where L = slab length in feet, F = friction factor between the bottom of slab and the top of underlying subbase or subgrade, and f_s = allowable working stress of steel reinforcement in psi. AASHTO's recommended values for F are given in Table 59.17. The allowable steel working stress is equal to 75% of the steel yield strength. For grade 40 and grade 60 steel, f_s is equal to 30,000 psi and 45,000 psi, respectively. For welded wire fabric, f_s is 48,750 psi. Equation (59.28) is also applicable to the design of transverse steel reinforcement for continuously reinforced concrete pavement (CRCP).

Table 59.17 Recommended Friction Factors

Type of Material beneath Slab	Friction Factor
Surface treatment	2.2
Lime stabilization	1.8
Asphalt stabilization	1.8
Cement stabilization	1.8
River gravel	1.5
Crushed stone	1.5
Sandstone	1.2
Natural subgrade	0.9

Source: AASHTO. 1993. *AASHTO Guides for Design of Pavement Structures.* Copyright 1993 by the American Association of State Highway and Transportation Officials, Washington, D.C. Used by permission.

Example 59.18. Determine the longitudinal steel reinforcement requirement for a 30-ft-long JRCP constructed on crushed stone subbase.

From Table 59.17, $F = 1.5$. Percentage of steel reinforcement $P_s = (30 \times 1.5)/(2 \times 30{,}000) = 0.075\%$.

Longitudinal Reinforcement Design for CRCP

The design of longitudinal reinforcement for CRCP is an elaborate process. The amount of reinforcement selected must satisfy limiting criteria in the following three aspects: (a) crack spacing, (b) crack width, and (c) steel stress.

CRCP Reinforcements Based on Crack Spacing: $(P_{\min})_1$ and $(P_{\max})_1$. The amount of steel reinforcement provided should be such that the crack spacing is between 3.5 ft (1.1 m) and 8 ft (2.4 m). The lower limit is to minimize punchout and the upper limit to minimize spalling. For each of these two crack spacings, Fig. 59.30 is used to determine the percent reinforcement P required, resulting in two values of P that define the range of acceptable percent reinforcement: $(P_{\max})_1$ and $(P_{\min})_1$.

The input variables for determining P are the thermal coefficient of portland cement concrete α_c, the thermal coefficient of steel α_s, diameter of reinforcing bar, concrete shrinkage Z at 28 days, tensile stress σ_w due to wheel load, and concrete tensile strength f_t at 28 days. Values of α_c and Z are given in Table 59.18. A value of $\alpha_s = 5.0 \times 10^{-6}$ in./in./°F may be used. Steel bars of 5/8- and 3/4-in. diameter are typically used, and the 3/4-in. bar is the largest practical size for crack-width control and bond requirements. The nominal diameter of a reinforcing bar, in inches, is simply the

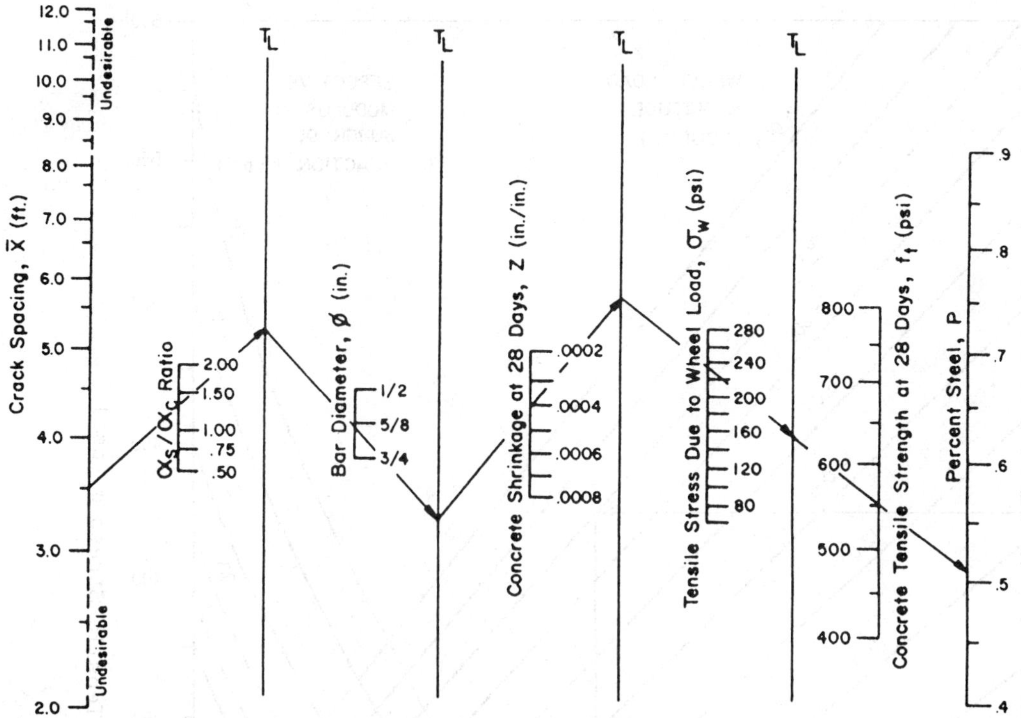

FIGURE 59.30 Minimum percent reinforcement to satisfy crack-spacing criteria. (*Source:* AASHTO. 1993. *AASHTO Guides for Design of Pavement Structures.* Copyright 1993 by the American Association of State Highway and Transportation Officials, Washington, D.C. Used by permission.)

Table 59.18 Shrinkage and Thermal Coefficient of Portland Cement Concrete

Indirect Tensile Strength (psi)	Shrinkage (in./in.)
(a) Approximate Relations between Shrinkage and Indirect Tensile Strength of Portland Cement Concrete	
300 (or less)	0.0008
400	0.0006
500	0.00045
600	0.0003
700 (or greater)	0.0002
Type of Coarse Aggregate	**Concrete Thermal Coefficient (10^{-6}/°F)**
(b) Recommended Value of the Thermal Coefficient of Concrete as a Function of Aggregate Types	
Quartz	6.6
Sandstone	6.5
Gravel	6.0
Granite	5.3
Basalt	4.8
Limestone	3.8

Source: AASHTO. 1993. *AASHTO Guides for Design of Pavement Structures.* Copyright 1993 by the American Association of State Highway and Transportation Officials, Washington, D.C. Used by permission.

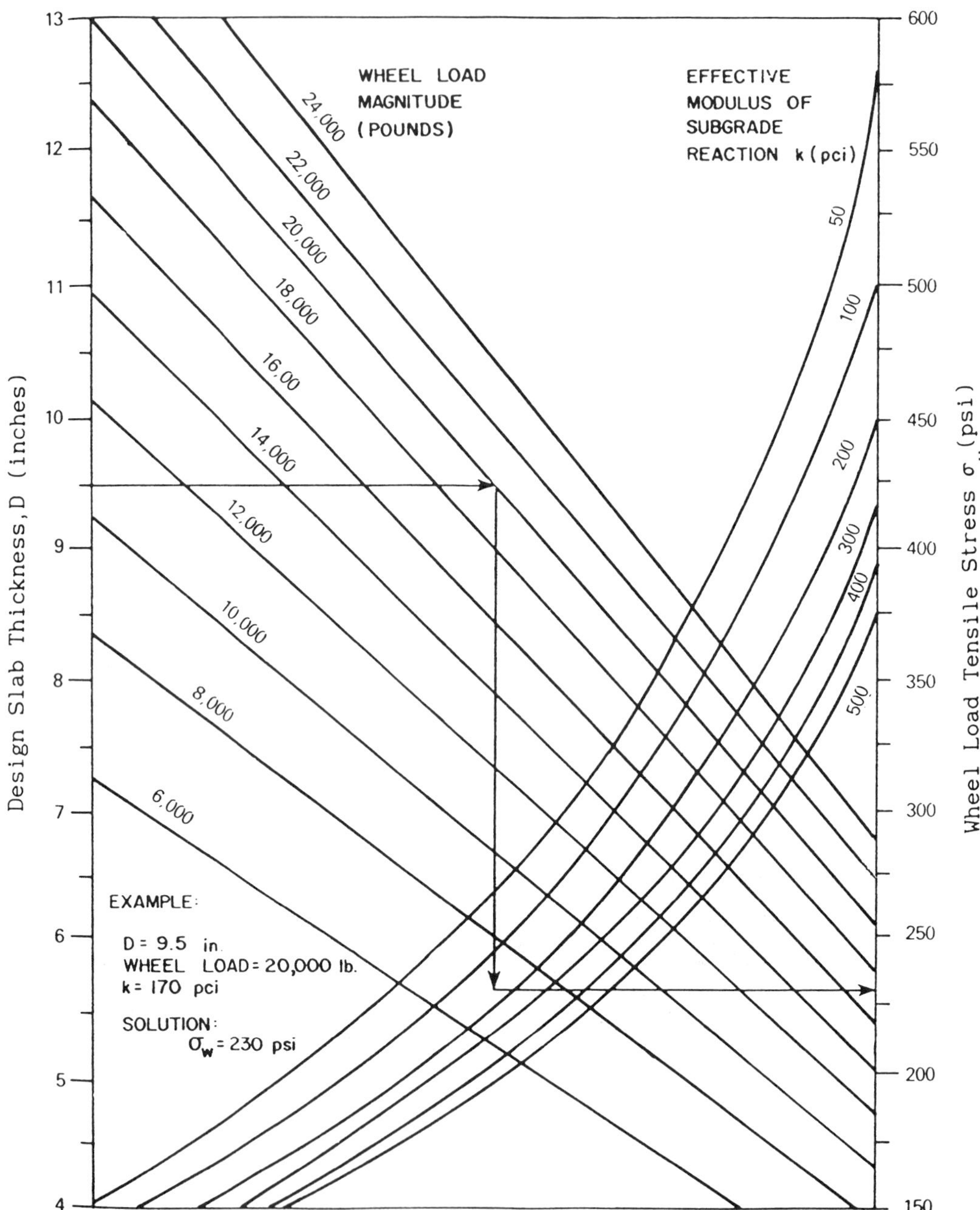

FIGURE 59.31 Chart for estimating wheel load tensile stress σ_w. (*Source:* AASHTO. 1993. *AASHTO Guides for Design of Pavement Structures.* Copyright 1993 by the American Association of State Highway and Transportation Officials, Washington, D.C. Used by permission.)

bar number divided by 8. Meanwhile, σ_w is the tensile stress developed during initial loading of the constructed pavement by either construction equipment or truck traffic. It is determined using Fig. 59.31 based on the design slab thickness, the magnitude of the wheel load, and the effective modulus of subgrade reaction. Likewise, f_t is the concrete indirect tensile strength determined by AASHTO T198 or ASTM C496. It can be assumed as 86% of the modulus of rupture S_c used for thickness design.

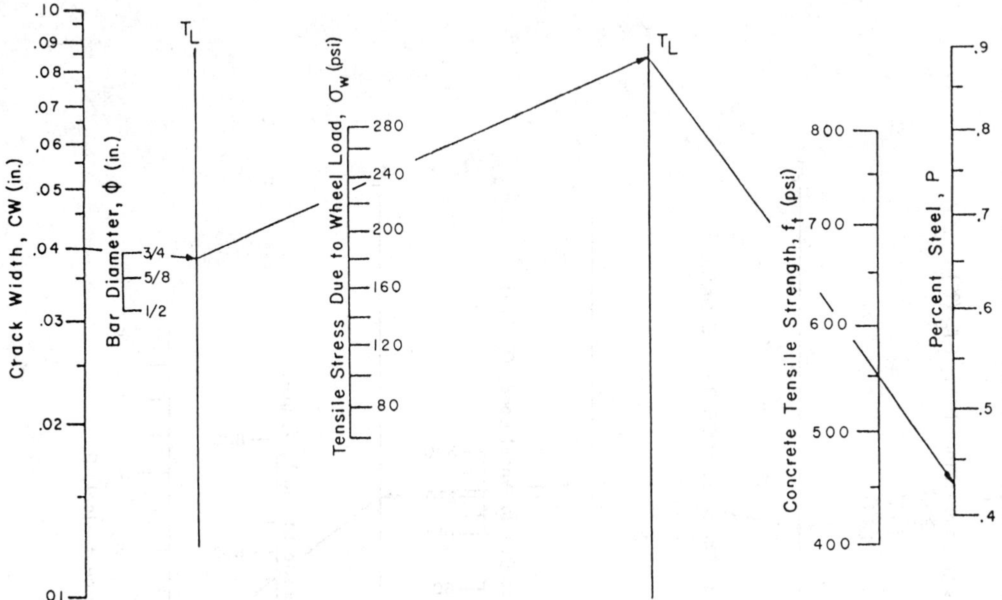

FIGURE 59.32 Minimum percent steel reinforcement to satisfy crack-width criteria. (*Source:* AASHTO. 1993. *AASHTO Guides for Design of Pavement Structures.* Copyright 1993 by the American Association of State Highway and Transportation Officials, Washington, D.C. Used by permission.)

***CRCP Reinforcements Based on Crack Width:* $(P_{\min})_2$.** Crack width in CRCP is controlled to within 0.04 in. (1.0 mm) to prevent spalling and water infiltration. The minimum percent steel $(P_{\min})_2$ that would produce crack widths of 0.04 in. can be determined from Fig. 59.32 with a selected bar size and input variables σ_w and f_t.

***CRCP Reinforcements Based on Steel Stress:* $(P_{\min})_3$.** To guard against steel fracture and excessive permanent deformation, a minimum amount of steel $(P_{\min})_3$ is determined according to Fig. 59.33. Input variables Z, σ_w, and f_t have been determined earlier. For the steel stress σ_s a limiting value equal to 75% of the ultimate tensile strength is recommended. Table 59.19 gives the allowable steel working stress for grade 60 steel meeting ASTM A615 specifications. The determination of $(P_{\min})_3$ also requires the computation of a design temperature drop given by

$$DT_D = T_H - T_L \tag{59.29}$$

where T_H is the average daily high temperature during the month the pavement is constructed and T_L is average daily low temperature during the coldest month of the year.

Reinforcement Design. Based on the three criteria discussed above, the design percent steel should fall within $P_{\max}$ and $P_{\min}$ given by

$$P_{\max} = (P_{\max})_1 \tag{59.30}$$

$$P_{\min} = \max\{(P_{\min})_1, (P_{\min})_2, (P_{\min})_3\} \tag{59.31}$$

If $P_{\max}$ is less than $P_{\min}$, a design revised by changing some of the input parameters is required. With $P_{\max}$ greater than $P_{\min}$, the number of reinforcing bars or wires required, N, is given by $N_{\min} \leq N \leq N_{\max}$ where $N_{\min}$ and $N_{\max}$ are computed by

$$N_{\min} = 0.01273 P_{\min} W_s D/\phi^2 \tag{59.32}$$

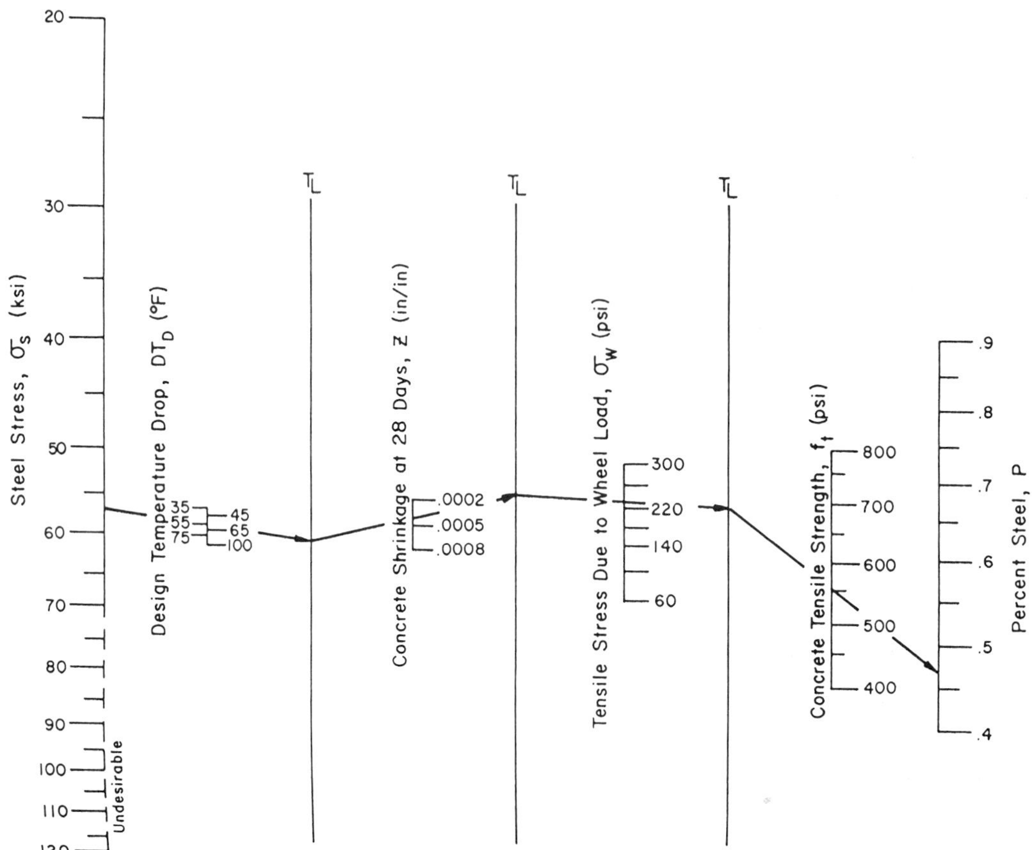

FIGURE 59.33 Minimum percent reinforcement to satisfy steel-stress criteria. (*Source:* AASHTO. 1993. *AASHTO Guides for Design of Pavement Structures.* Copyright 1993 by the American Association of State Highway and Transportation Officials, Washington, D.C. Used by permission.)

$$N_{\max} = 0.01273 P_{\max} W_s D/\phi^2 \tag{59.33}$$

in which W_s is the width of pavement in inches, D is the slab thickness in inches, and ϕ is the reinforcing bar or wire diameter in inches.

Example 59.19. This example concerns longitudinal steel reinforcement design for CRCP. Design data are $D = 9$ in., $\alpha_s = 5 \times 10^{-6}$ in./in./°F, $\alpha_c = 5.3 \times 10^{-6}$ for concrete with granite coarse aggregate (see Table 59.18), $f_t = 600$ psi at 28 days, $Z = 0.0003$ (see Table 59.18), maximum construction wheel load = 18,000 lb, effective $k = 100$ pci, and design temperature drop = 55°F.

Try steel bar diameter $\phi = 0.5$ in. (a) Crack spacing control. $\sigma_w = 240$ psi, from Fig. 59.31. With $(\alpha_s/\alpha_c) = 0.94$, obtain $(P_{\min})_1 < 0.4\%$ for $X = 8$ ft, and $(P_{\max})_2 = 0.49\%$ for $X = 3.5$ ft from Fig. 59.30. (b) Crack width control. Obtain $(P_{\min})_2 = 0.40\%$ from Fig. 59.32. (c) Steel stress control. $\sigma_s = 67$ psi, from Table 59.19. Obtain $(P_{\min})_3 = 0.45\%$. Hence,

Table 59.19 Allowable Steel Working Stress, ksi

Indirect Tensile Strength of Concrete at 28 days, psi	Reinforcing Bar Size		
	No. 4	No. 5	No. 6
300 (or less)	65	57	54
400	67	60	55
500	67	61	56
600	67	63	58
700	67	65	59
800 (or greater)	67	67	60

Source: AASHTO. 1993. *AASHTO Guides for Design of Pavement Structures.* Copyright 1993 by the American Association of State Highway and Transportation Officials, Washington, D.C. Used by permission.

overall (P_{min}) = 0.4% and (P_{max}) = 0.49%. For a pavement width of 12 ft, apply Eqs. (59.32) and (59.33), N_{min} = 29.7, N_{max} = 32.3. Use 30 numbers of 0.5-in. bars.

PCA Thickness Design Procedure for Rigid Highway Pavements

The thickness design procedure published by PCA [1984] was developed by relating theoretically computed values of stress, deflection, and pressure to pavement performance criteria derived from data of (1) major road test programs, (2) model and full-scale tests, and (3) performance of normally constructed pavements subject to normal mixed traffic.

Traffic-loading data in terms of axle-load distribution are obtained in the usual way as described earlier in this chapter. Each axle load is further multiplied by a load safety factor (LSF) according to the following recommendations: (1) LSF = 1.2 for interstate highways and other multilane projects with uninterrupted traffic flow and high volumes of truck traffic; (2) LSF = 1.1 for highways and arterial streets with moderate volumes of truck traffic; and (3) LSF = 1.0 for roads, residential streets, and other streets with small volumes of truck traffic. The flexural strength of concrete is determined by the 28-day modulus of rupture from third-point loading according to ASTM Test Method C78. Subgrade and subbase support is defined in terms of the modulus of subgrade reaction k.

The design procedure consists of a fatigue analysis and an erosion analysis, which are considered separately using different sets of tables and design charts. The final thickness selected must satisfy both analyses.

Fatigue Design

Fatigue design is performed with the aim to control fatigue cracking. The slab thickness based on fatigue design is the same for JRCP, for JPCP with doweled and undoweled joints, and for CRCP. This is because the most critical loading position is near mid-slab and the effect of joints is negligible. The presence of a tied concrete shoulder, however, must be considered since it significantly reduces the critical edge stress. The analysis is based on the concept of cumulative damage given by

$$D = \sum_{i=1}^{m} \frac{n_i}{N_i} \tag{59.34}$$

where m is the total number of axle load groups, n_i is the predicted number of repetitions for the ith load group, and N_i is the allowable number of repetitions for the ith load group.

The steps in the design procedure are:

1. Multiply the load of each design axle load group by the appropriate LSF.
2. Assume a trial slab thickness.
3. Obtain from Table 59.20(a) or (b) the equivalent stress for the input slab thickness and k, and calculate stress ratio factor as

$$\text{Stress ratio factor} = \frac{\text{Equivalent stress}}{\text{Concrete flexural strength}} \tag{59.35}$$

4. For each axle load i, obtain from Fig. 59.34 the allowable load repetitions N_i.
5. Compute D from Eq. (59.34). If D exceeds 1, select a greater trial thickness and repeat steps 3 through 5. The trial thickness is adequate if D is less than or equal to 1.

Erosion Design

PCA requires erosion analysis in pavement thickness design to control foundation and shoulder erosion, pumping, and faulting. Since the most critical deflection occurs at the corner, the presence

Table 59.20 Equivalent Stress for Fatigue Analysis

Slab Thickness	*k* of Subgrade-Subbase (pci) 50	100	150	200	300	500	700
(a) Equivalent Stress—No Concrete Shoulder (Single Axle/Tandem Axle)							
4 in.	825/679	726/585	671/542	634/516	584/486	523/457	484/443
4.5 in.	699/586	616/500	571/460	540/435	498/406	448/378	417/363
5 in.	602/516	531/436	493/399	467/374	432/349	390/321	363/307
5.5 in.	526/461	464/387	431/353	409/331	379/305	343/278	320/264
6 in.	465/416	411/348	382/316	362/296	336/271	304/246	285/232
6.5 in.	417/380	367/317	341/286	324/267	300/244	273/220	256/207
7 in.	375/349	331/290	307/262	292/244	271/222	246/199	231/186
7.5 in.	340/323	300/268	279/241	265/224	246/203	224/181	210/169
8 in.	311/300	274/249	255/223	242/208	225/188	205/167	192/155
8.5 in.	285/281	252/232	234/208	222/193	206/174	188/154	177/143
9 in.	264/264	232/218	216/195	205/181	190/163	174/144	163/133
9.5 in.	245/248	215/205	200/183	190/170	176/153	161/134	151/124
10 in.	228/235	200/193	186/173	177/160	164/144	150/126	141/117
10.5 in.	213/222	187/183	174/164	165/151	153/136	140/119	132/110
11 in.	200/211	175/174	163/155	154/143	144/129	131/113	123/104
11.5 in.	188/201	165/165	153/148	145/136	135/122	123/107	116/98
12 in.	177/192	155/158	144/141	137/130	127/116	116/102	109/93
12.5 in.	168/183	147/151	136/135	129/124	120/111	109/97	103/89
13 in.	159/176	139/144	129/129	122/119	113/106	103/93	97/85
13.5 in.	152/168	132/138	122/123	116/114	107/102	98/89	92/81
14 in.	144/162	125/133	116/118	110/109	102/98	93/85	88/78
(b) Equivalent Stress—Concrete Shoulder (Single Axle/Tandem Axle)							
4 in.	640/534	559/468	517/439	489/422	452/403	409/388	383/384
4.5 in.	547/461	479/400	444/372	421/356	390/338	355/322	333/316
5 in.	475/404	417/349	387/323	367/308	341/290	311/274	294/267
5.5 in.	418/360	368/309	342/285	324/271	302/254	276/238	261/231
6 in.	372/325	327/277	304/255	289/241	270/225	247/210	234/203
6.5 in.	334/295	294/251	274/230	260/218	243/203	223/188	212/180
7 in.	302/270	266/230	248/210	236/198	220/184	203/170	192/162
7.5 in.	275/250	243/211	226/193	215/182	201/168	185/155	176/148
8 in.	252/232	222/196	207/179	197/168	185/155	170/142	162/135
8.5 in.	232/216	205/182	191/166	182/156	170/144	157/131	150/125
9 in.	215/202	190/171	177/155	169/146	158/134	146/122	139/116
9.5 in.	200/190	176/160	164/146	157/137	147/126	136/114	129/108
10 in.	186/179	164/151	153/137	146/129	137/118	127/107	121/101
10.5 in.	174/170	154/143	144/130	137/121	128/111	119/101	113/95
11 in.	164/161	144/135	135/123	129/115	120/105	112/95	106/90
11.5 in.	154/153	136/128	127/117	121/109	113/100	105/90	100/85
12 in.	145/146	128/122	120/111	114/104	107/95	99/86	95/81
12.5 in.	137/139	121/117	113/106	108/99	101/91	94/82	90/77
13 in.	130/133	115/112	107/101	102/95	96/86	89/78	85/73
13.5 in.	124/127	109/107	102/97	97/91	91/83	85/74	81/70
14 in.	118/122	104/103	97/93	93/87	87/79	81/71	77/67

Source: Portland Cement Association. 1984. *Thickness Design for Concrete Highway and Street Pavements.* With permission.

of shoulder and the type of joint construction will both affect the analysis. The concept of cumulative damage as defined by Eq. (59.34) is again applied. The steps are:

1. Multiply the load of each design axle load group by LSF.
2. Assume a trial slab thickness.
3. Obtain from Table 59.21(a), (b), (c), or (d) the erosion factor for the input slab thickness and *k*.
4. For each axle load *i*, obtain from Fig. 59.35(a) or (b) the allowable load repetitions N_i.

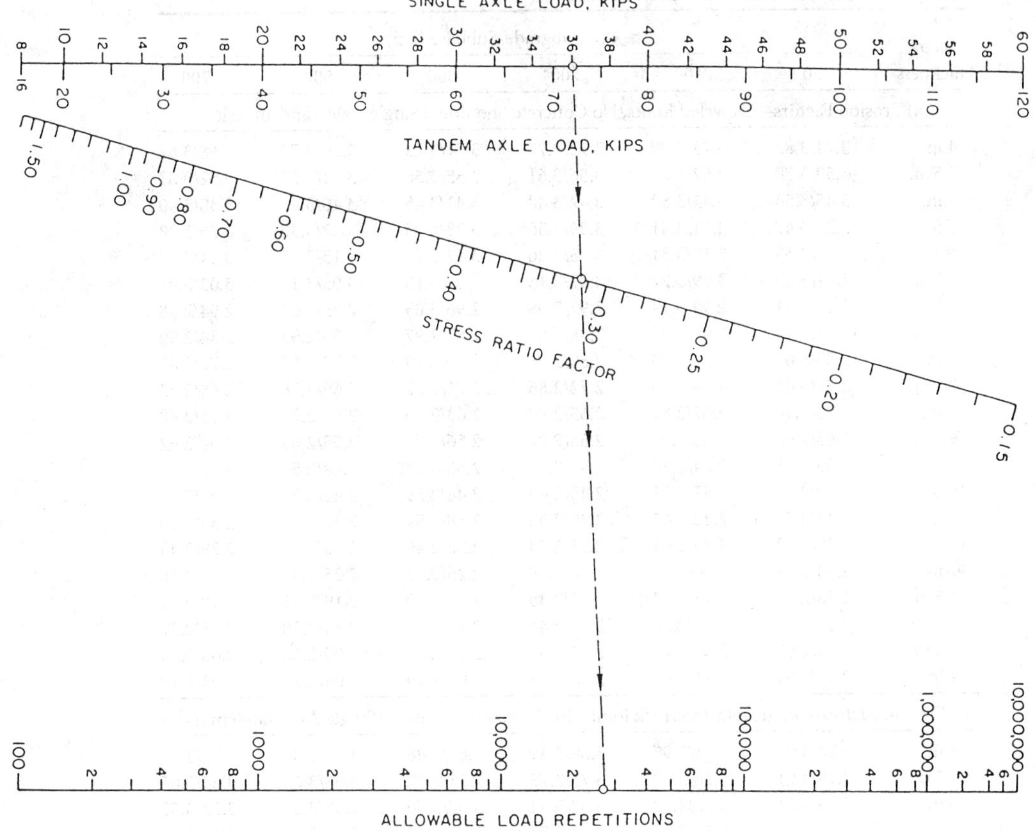

FIGURE 59.34 Allowable repetitions for fatigue analysis. (*Source:* Portland Cement Association. 1984. *Thickness Design for Concrete Highway and Street Pavements*, p. 15. With permission.)

5. Compute D from Eq. (59.34). If D exceeds 1, select a greater trial thickness and repeat steps 3 through 5. The trial thickness is adequate if D is less than or equal to 1.

Example 59.20. Determine the required slab thickness for an expressway with the design traffic shown in the table below. The pavement is to be constructed with doweled joint, but without concrete shoulder. Concrete modulus of rupture is 650 psi. The subgrade k is 130 pci.

Axle Load (kips)	Design Load (kips)	Design n	Fatigue N_1	Fatigue (n/N_1)	Erosion N_2	Erosion (n/N_2)
52T	62.4T	3,100	800,000	0.004	800,000	0.004
50T	60.0T	32,000	2,000,000	0.016	1,000,000	0.030
48T	57.6T	32,000	10,000,000	0.0032	1,200,000	0.027
46T	55.2T	48,000	unlimited	0	1,700,000	0.028
44T	52.8T	158,000	unlimited	0	2,000,000	0.079
42T	50.4T	172,000	unlimited	0	2,800,000	0.061
40T	48.0T	250,000	unlimited	0	3,500,000	0.071
30S	36.0S	3,100	25,000	0.124	1,700,000	0.002
28S	33.6S	3,100	70,000	0.044	2,200,000	0.001
26S	31.2S	9,300	200,000	0.045	3,000,000	0.002
24S	28.8S	545,000	800,000	0.682	5,000,000	0.033
22S	26.4S	640,000	10,000,000	0.064	9,000,000	0.071
				Total 0.982		0.41

Table 59.21 Erosion Factor for Erosion Analysis

Slab Thickness	k of Subgrade-Subbase (pci) 50	100	200	300	500	700
(a) Erosion Factors—Doweled Joints, No Concrete Shoulder (Single Axle/Tandem Axle)						
4 in.	3.74/3.83	3.73/3.79	3.72/3.75	3.71/3.73	3.70/3.70	3.68/3.67
4.5 in.	3.59/3.70	3.57/3.65	3.56/3.61	3.55/3.58	3.54/3.55	3.52/3.53
5 in.	3.45/3.58	3.43/3.52	3.42/3.48	3.41/3.45	3.40/3.42	3.38/3.40
5.5 in.	3.33/3.47	3.31/3.41	3.29/3.36	3.28/3.33	3.27/3.30	3.26/3.28
6 in.	3.22/3.38	3.19/3.31	3.18/3.26	3.17/3.23	3.15/3.20	3.14/3.17
6.5 in.	3.11/3.29	3.09/3.22	3.07/3.16	3.06/3.13	3.05/3.10	3.03/3.07
7 in.	3.02/3.21	2.99/3.14	2.97/3.08	2.96/3.05	2.95/3.01	2.94/2.98
7.5 in.	2.93/3.14	2.91/3.06	2.88/3.00	2.87/2.97	2.86/2.93	2.84/2.90
8 in.	2.85/3.07	2.82/2.99	2.80/2.93	2.79/2.89	2.77/2.85	2.76/2.82
8.5 in.	2.77/3.01	2.74/2.93	2.72/2.86	2.71/2.82	2.69/2.78	2.68/2.75
9 in.	2.70/2.96	2.67/2.87	2.65/2.80	2.63/2.76	2.62/2.71	2.61/2.68
9.5 in.	2.63/2.90	2.60/2.81	2.58/2.74	2.56/2.70	2.55/2.65	2.54/2.62
10 in.	2.56/2.85	2.54/2.76	2.51/2.68	2.50/2.64	2.48/2.59	2.47/2.56
10.5 in.	2.50/2.81	2.47/2.71	2.45/2.63	2.44/2.59	2.42/2.54	2.41/2.51
11 in.	2.44/2.76	2.42/2.67	2.39/2.58	2.38/2.54	2.36/2.49	2.35/2.45
11.5 in.	2.38/2.72	2.36/2.62	2.33/2.54	2.32/2.49	2.30/2.44	2.29/2.40
12 in.	2.33/2.68	2.30/2.58	2.28/2.49	2.26/2.44	2.25/2.39	2.23/2.36
12.5 in.	2.28/2.64	2.25/2.54	2.23/2.45	2.21/2.40	2.19/2.35	2.18/2.31
13 in.	2.23/2.61	2.20/2.50	2.18/2.41	2.16/2.36	2.14/2.30	2.13/2.27
13.5 in.	2.18/2.57	2.15/2.47	2.13/2.37	2.11/2.32	2.09/2.26	2.08/2.23
14 in.	2.13/2.54	2.11/2.43	2.08/2.34	2.07/2.29	2.05/2.23	2.03/2.19
(b) Erosion Factors—Aggregate-Interlock Joints, No Concrete Shoulder (Single Axle/Tandem Axle)						
4 in.	3.94/4.03	3.91/3.95	3.88/3.89	3.86/3.86	3.82/3.83	3.77/3.80
4.5 in.	3.79/3.91	3.76/3.82	3.73/3.75	3.71/3.72	3.68/3.68	3.64/3.65
5 in.	3.66/3.81	3.63/3.72	3.60/3.64	3.58/3.60	3.55/3.55	3.52/3.52
5.5 in.	3.54/3.72	3.51/3.62	3.48/3.53	3.46/3.49	3.43/3.44	3.41/3.40
6 in.	3.44/3.64	3.40/3.53	3.37/3.44	3.35/3.40	3.32/3.34	3.30/3.30
6.5 in.	3.34/3.56	3.30/3.46	3.26/3.36	3.25/3.31	3.22/3.25	3.20/3.21
7 in.	3.26/3.49	3.21/3.39	3.17/3.29	3.15/3.24	3.13/3.17	3.11/3.13
7.5 in.	3.18/3.43	3.13/3.32	3.09/3.22	3.07/3.17	3.04/3.10	3.02/3.06
8 in.	3.11/3.37	3.05/3.26	3.01/3.16	2.99/3.10	2.96/3.03	2.94/2.99
8.5 in.	3.04/3.32	2.98/3.21	2.93/3.10	2.91/3.04	2.88/2.97	2.87/2.93
9 in.	2.98/3.27	2.91/3.16	2.86/3.05	2.84/2.99	2.81/2.92	2.79/2.87
9.5 in.	2.92/3.22	2.85/3.11	2.80/3.00	2.77/2.94	2.75/2.86	2.73/2.81
10 in.	2.86/3.18	2.79/3.06	2.74/2.95	2.71/2.89	2.68/2.81	2.66/2.76
10.5 in.	2.81/3.14	2.74/3.02	2.68/2.91	2.65/2.84	2.62/2.76	2.60/2.72
11 in.	2.77/3.10	2.69/2.98	2.63/2.86	2.60/2.80	2.57/2.72	2.54/2.67
11.5 in.	2.72/3.06	2.64/2.94	2.58/2.82	2.55/2.76	2.51/2.68	2.49/2.63
12 in.	2.68/3.03	2.60/2.90	2.53/2.78	2.50/2.72	2.46/2.64	2.44/2.59
12.5 in.	2.64/2.99	2.55/2.87	2.48/2.75	2.45/2.68	2.41/2.60	2.39/2.55
13 in.	2.60/2.96	2.51/2.83	2.44/2.71	2.40/2.65	2.36/2.56	2.34/2.51
13.5 in.	2.56/2.93	2.47/2.80	2.40/2.68	2.36/2.61	2.32/2.53	2.30/2.48
14 in.	2.53/2.90	2.44/2.77	2.36/2.65	2.32/2.58	2.28/2.50	2.25/2.44
(c) Erosion Factors—Doweled Joints, Concrete Shoulder (Single Axle/Tandem Axle)						
4 in.	3.28/3.30	3.24/3.20	3.21/3.13	3.19/3.10	3.15/3.09	3.12/3.08
4.5 in.	3.13/3.19	3.09/3.08	3.06/3.00	3.04/2.96	3.01/2.93	2.98/2.91
5 in.	3.01/3.09	2.97/2.98	2.93/2.89	2.90/2.84	2.87/2.79	2.85/2.77
5.5 in.	2.90/3.01	2.85/2.89	2.81/2.79	2.79/2.74	2.76/2.68	2.73/2.65
6 in.	2.79/2.93	2.75/2.82	2.70/2.71	2.68/2.65	2.65/2.58	2.62/2.54
6.5 in.	2.70/2.86	2.65/2.75	2.61/2.63	2.58/2.57	2.55/2.50	2.52/2.45
7 in.	2.61/2.79	2.56/2.68	2.52/2.56	2.49/2.50	2.46/2.42	2.43/2.38
7.5 in.	2.53/2.73	2.48/2.62	2.44/2.50	2.41/2.44	2.38/2.36	2.35/2.31
8 in.	2.46/2.68	2.41/2.56	2.36/2.44	2.33/2.38	2.30/2.30	2.27/2.24
8.5 in.	2.39/2.62	2.34/2.51	2.29/2.39	2.26/2.32	2.22/2.24	2.20/2.18

Table 59.21 (continued) Erosion Factor for Erosion Analysis

Slab Thickness	k of Subgrade-Subbase (pci)					
	50	100	200	300	500	700
(c) Erosion Factors—Doweled Joints, Concrete Shoulder (Single Axle/Tandem Axle)						
9 in.	2.32/2.57	2.27/2.46	2.22/2.34	2.19/2.27	2.16/2.19	2.13/2.13
9.5 in.	2.26/2.52	2.21/2.41	2.16/2.29	2.13/2.22	2.09/2.14	2.07/2.08
10 in.	2.20/2.47	2.15/2.36	2.10/2.25	2.07/2.18	2.03/2.09	2.01/2.03
10.5 in.	2.15/2.43	2.09/2.32	2.04/2.20	2.01/2.14	1.97/2.05	1.95/1.99
11 in.	2.10/2.39	2.04/2.28	1.99/2.16	1.95/2.09	1.92/2.01	1.89/1.95
11.5 in.	2.05/2.35	1.99/2.24	1.93/2.12	1.90/2.05	1.87/1.97	1.84/1.91
12 in.	2.00/2.31	1.94/2.20	1.88/2.09	1.85/2.02	1.82/1.93	1.79/1.87
12.5 in.	1.95/2.27	1.89/2.16	1.84/2.05	1.81/1.98	1.77/1.89	1.74/1.84
13 in.	1.91/2.23	1.85/2.13	1.79/2.01	1.76/1.95	1.72/1.86	1.70/1.80
13.5 in.	1.86/2.20	1.81/2.09	1.75/1.98	1.72/1.91	1.68/1.83	1.65/1.77
14 in.	1.82/2.17	1.76/2.06	1.71/1.95	1.67/1.88	1.64/1.80	1.61/1.74
(d) Erosion Factors—Aggregate-Interlock Joints, Concrete Shoulder (Single Axle/Tandem Axle)						
4 in.	3.46/3.49	3.42/3.39	3.38/3.32	3.36/3.29	3.32/3.26	3.28/3.24
4.5 in.	3.32/3.39	3.28/3.28	3.24/3.19	3.22/3.16	3.19/3.12	3.15/3.09
5 in.	3.20/3.30	3.16/3.18	3.12/3.09	3.10/3.05	3.07/3.00	3.04/2.97
5.5 in.	3.10/3.22	3.05/3.10	3.01/3.00	2.99/2.95	2.96/2.90	2.93/2.86
6 in.	3.00/3.15	2.95/3.02	2.90/2.92	2.88/2.87	2.86/2.81	2.83/2.77
6.5 in.	2.91/3.08	2.86/2.96	2.81/2.85	2.79/2.79	2.76/2.73	2.74/2.68
7 in.	2.83/3.02	2.77/2.90	2.73/2.78	2.70/2.72	2.68/2.66	2.65/2.61
7.5 in.	2.76/2.97	2.70/2.84	2.65/2.72	2.62/2.66	2.60/2.59	2.57/2.54
8 in.	2.69/2.92	2.63/2.79	2.57/2.67	2.55/2.61	2.52/2.53	2.50/2.48
8.5 in.	2.63/2.88	2.56/2.74	2.51/2.62	2.48/2.55	2.45/2.48	2.43/2.43
9 in.	2.57/2.83	2.50/2.70	2.44/2.57	2.42/2.51	2.39/2.43	2.36/2.38
9.5 in.	2.51/2.79	2.44/2.65	2.38/2.53	2.36/2.46	2.33/2.38	2.30/2.33
10 in.	2.46/2.75	2.39/2.61	2.33/2.49	2.30/2.42	2.27/2.34	2.24/2.28
10.5 in.	2.41/2.72	2.33/2.58	2.27/2.45	2.24/2.38	2.21/2.30	2.19/2.24
11 in.	2.36/2.68	2.28/2.54	2.22/2.41	2.19/2.34	2.16/2.26	2.14/2.20
11.5 in.	2.32/2.65	2.24/2.51	2.17/2.38	2.14/2.31	2.11/2.22	2.09/2.16
12 in.	2.28/2.62	2.19/2.48	2.13/2.34	2.10/2.27	2.06/2.19	2.04/2.13
12.5 in.	2.24/2.59	2.15/2.45	2.09/2.31	2.05/2.24	2.02/2.15	1.99/2.10
13 in.	2.20/2.56	2.11/2.42	2.04/2.28	2.01/2.21	1.98/2.12	1.95/2.06
13.5 in.	2.16/2.53	2.08/2.39	2.00/2.25	1.97/2.18	1.93/2.09	1.91/2.03
14 in.	2.13/2.51	2.04/2.36	1.97/2.23	1.93/2.15	1.89/2.06	1.87/2.00

Source: Portland Cement Association. 1984. *Thickness Design for Concrete Highway and Street Pavements.* With permission.

Trial-and-error approach is needed by assuming slab thickness. Solution is shown only for slab thickness $h = 9.5$ in.

For an expressway LSF $= 1.2$. The design load is equal to ($1.2 \times$ axle load). From Table 59.20, equivalent stress for single axle is 206 and for tandem axle 192. The corresponding stress ratios are 0.317 and 0.295. N_1 for fatigue analysis is obtained from Fig. 59.34. From Table 59.21, erosion factor is 2.6 for single axle and 2.8 for tandem axle. N_2 for erosion analysis is obtained from Fig. 59.35(a). The results show that the design is satisfactory.

FAA Method for Rigid Airport Pavement Design

Both the thickness design and reinforcement design procedures by FAA [1978] are presented in this section.

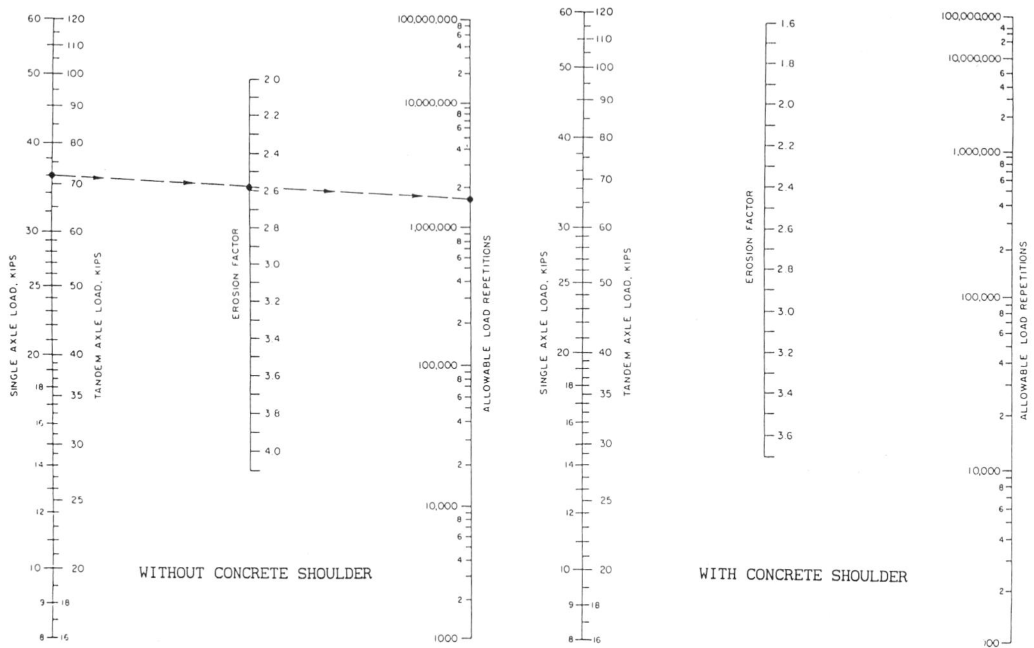

FIGURE 59.35 Allowable repetitions for erosion analysis. (*Source:* Portland Cement Association. 1984. *Thickness Design for Concrete Highway and Street Pavements.* With permission.)

FAA Thickness Design Procedure for Rigid Airport Pavements

The FAA thickness design method is based on the Westergaard analysis of an edge-loaded slab on a dense liquid foundation. The design curves in Figs. 59.36–59.42 have been developed with the assumption that the landing gear assembly is either tangent to a longitudinal joint or perpendicular to a transverse joint—whichever produces the largest stress. It is also assumed that 95% of the gross aircraft weight is carried on the main landing gear assembly. The design curves provide slab thickness T for the critical areas defined earlier in this chapter. The thickness of $0.9T$ for noncritical areas applies to the concrete slab thickness. As in the case for flexible pavement design, stabilized subbase is required to accommodate aircraft weighing 100,000 lb or more.

Design Loading. The same method of selecting a design aircraft and computing design annual departures is followed as for the FAA flexible airport pavement design.

Concrete Flexural Strength. The 28-day flexural strength of concrete is determined by ASTM Test Method C78. A 90-day flexural strength may be used. It can be taken to be 10% higher than the 28-day strength, except when high early strength cement or pozzolanic admixtures are used.

Foundation Modulus. The subgrade modulus k is determined by the test method specified in AASHTO T222. When a layer of subbase is used the design k is obtained from Fig. 59.43 for unstabilized subbase and Fig. 59.44 for stabilized subbase.

High Traffic Volumes. For airports with design traffic exceeding 25,000 annual departures, FAA suggests using thicker pavements as follows: 104%, 108%, 110%, and 112% of design thickness for 25,000 annual departures for annual departure levels of 50,000, 100,000, 150,000, and 200,000,

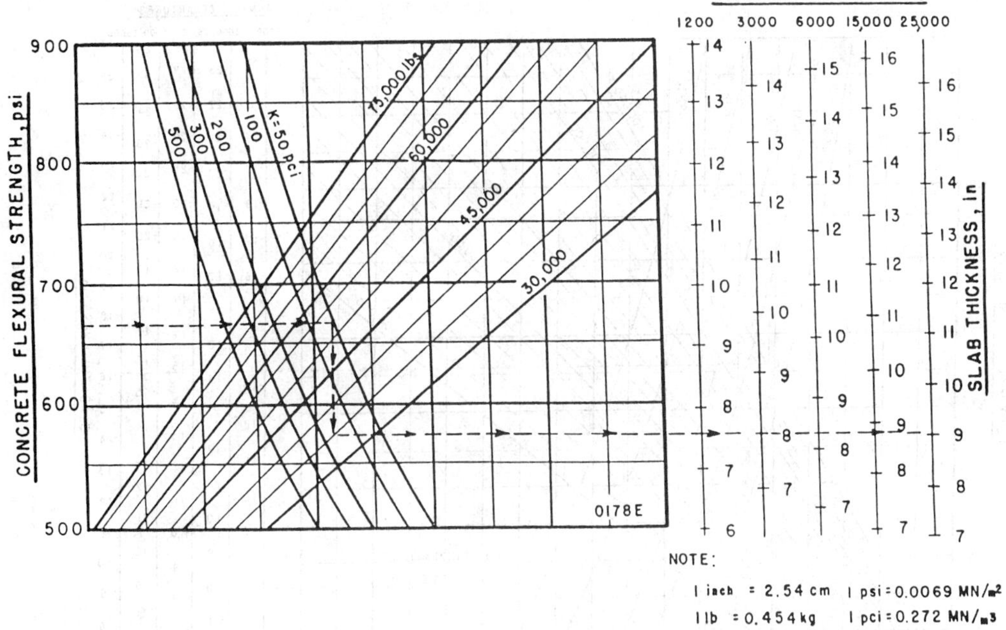

FIGURE 59.36 Rigid pavement thickness for single-wheel gear. (*Source:* Federal Aviation Administration. 1978. *Airport Pavement Design and Evaluation.* Advisory Circular AC No. 150/5320-6C. With permission.)

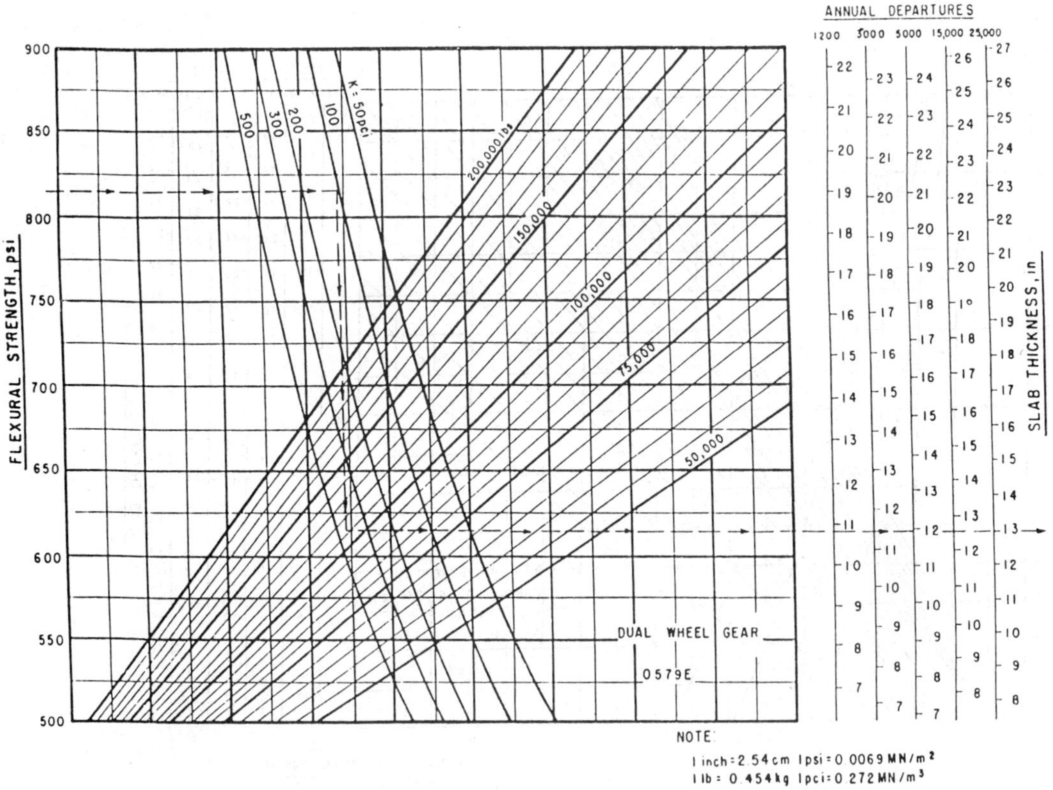

FIGURE 59.37 Rigid pavement thickness for dual-wheel gear. (*Source:* Federal Aviation Administration. 1978. *Airport Pavement Design and Evaluation.* Advisory Circular AC No. 150/5320-6C. With permission.)

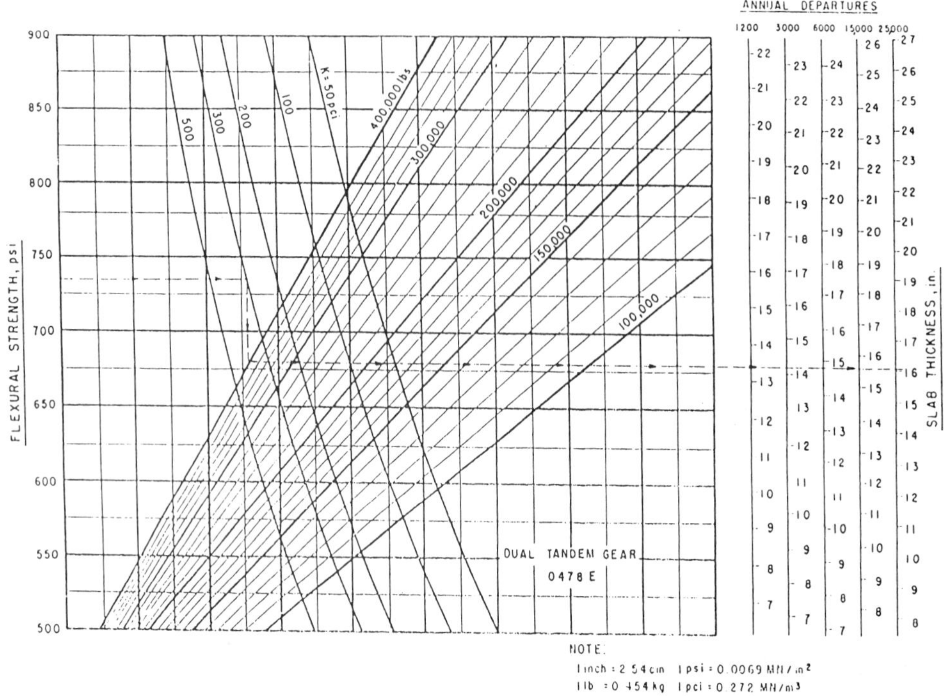

FIGURE 59.38 Rigid pavement thickness for dual-tandem gear. (*Source:* Federal Aviation Administration. 1978. *Airport Pavement Design and Evaluation.* Advisory Circular AC No. 150/5320-6C. With permission.)

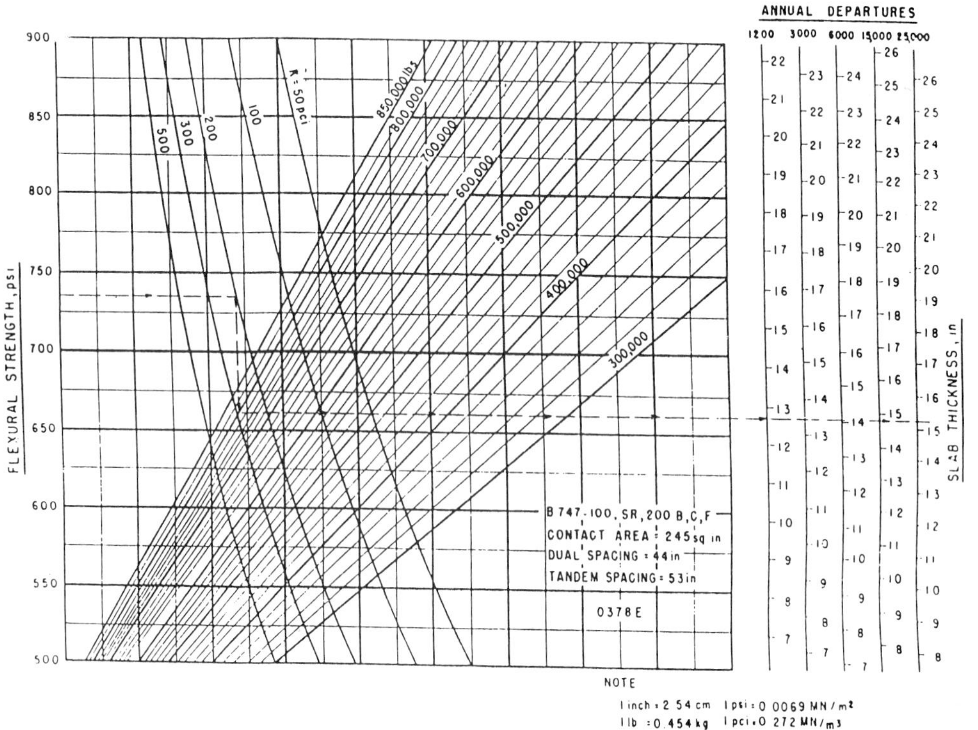

FIGURE 59.39 Rigid pavement thickness for B-747-100, SR, 200B, 200C, and 200F. (*Source:* Federal Aviation Administration. 1978. *Airport Pavement Design and Evaluation.* Advisory Circular AC No. 150/5320-6C. With permission.)

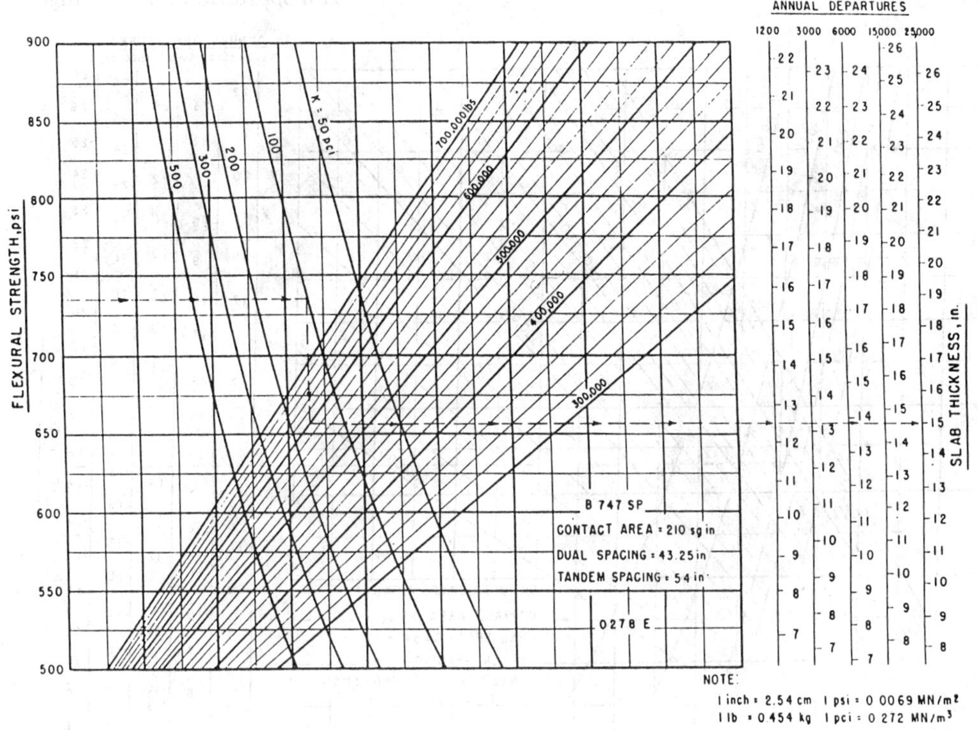

FIGURE 59.40 Rigid pavement thickness for B-747-SP. (*Source:* Federal Aviation Administration. 1978. *Airport Pavement Design and Evaluation.* Advisory Circular AC No. 150/5320-6C. With permission.)

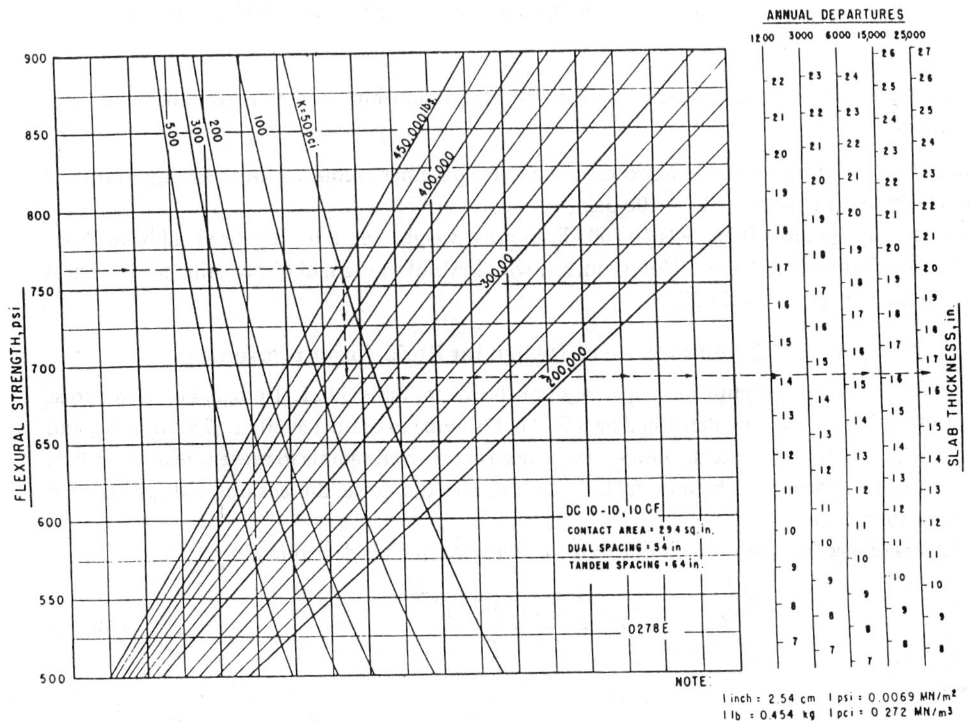

FIGURE 59.41 Rigid pavement thickness for DC 10-10, 10CF. (*Source:* Federal Aviation Administration. 1978. *Airport Pavement Design and Evaluation.* Advisory Circular AC No. 150/5320-6C. With permission.)

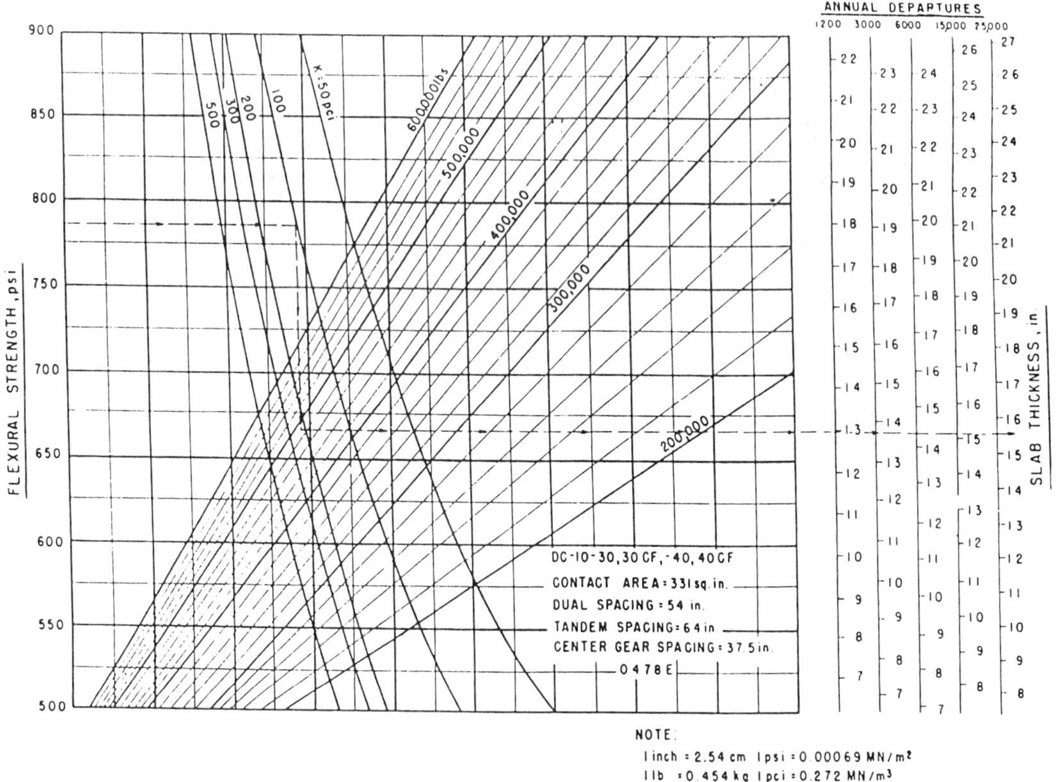

FIGURE 59.42 Rigid pavement thickness for DC 10-30, 30CF, 40, and 40CF. (*Source:* Federal Aviation Administration. 1978. *Airport Pavement Design and Evaluation.* Advisory Circular AC No. 150/5320-6C. With permission.)

respectively. This suggestion is based on a logarithmic relationship between percent thickness and departures.

Example 59.21. Determine the thickness of concrete pavement required for the design traffic of Example 59.13. The subgrade $k = 100$ pci.

Since the design aircraft exceeds 100,000 lb in gross weight, use 6 in. stabilized subbase. From Fig. 59.44, effective $k = 210$ pci. Using concrete with flexural strength of 650 psi, the slab thickness required is 18 in., from Fig. 59.37.

FAA Joint Spacing and Reinforcement Design for Rigid Airport Pavements

The recommended maximum joint spacings are shown in Table 59.22. Tie bars are used across longitudinal joints. They are deformed bars 5/8 in. (16 mm) in diameter, 30 in. (76 cm) long, and spaced 30 in. (76 cm) on centers. Dowel bars are used at transverse joints to prevent relative vertical displacement of adjacent slab ends. Table 59.23 indicates the dowel dimensions and spacings for various slab thicknesses.

The area of steel required for a reinforced concrete pavement is determined by

$$A_s = \frac{3.7(L)(Lt)^{0.5}}{f_s} \tag{59.36}$$

where

A_s = area of steel per foot width or length, square inches
L = length or width of slab, feet

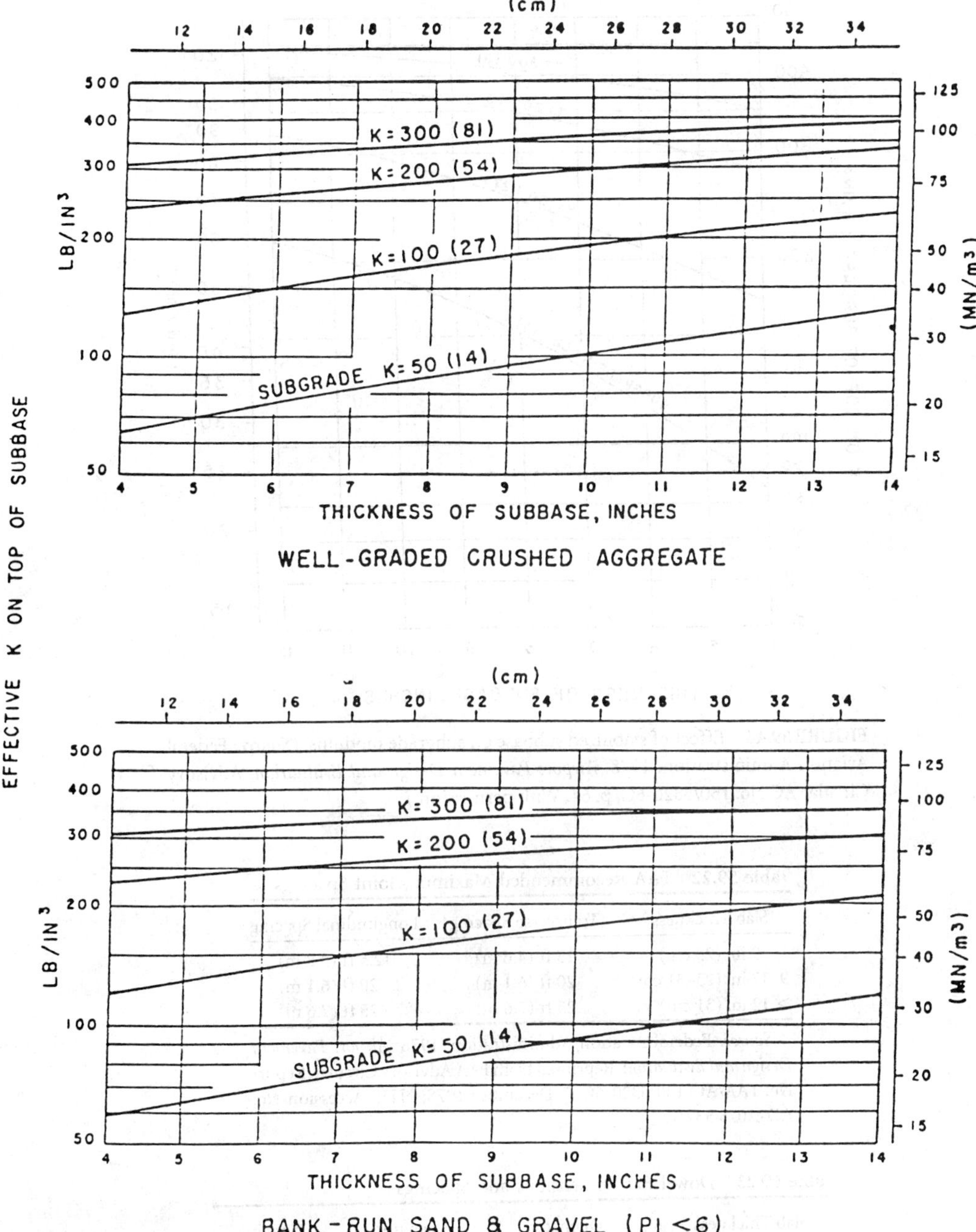

FIGURE 59.43 Effect of subbase on subgrade modulus. (*Source:* Federal Aviation Administration. 1978. *Airport Pavement Design and Evaluation.* Advisory Circular AC No. 150/5320-6C, p. 25. With permission.)

t = thickness of slab, inches

f_s = allowable tensile stress in steel, psi, taken as two-thirds of the yield strength of the steel

The minimum percentage of steel reinforcement is 0.05%. The maximum allowable slab length regardless of steel percentage is 75 ft (23 m).

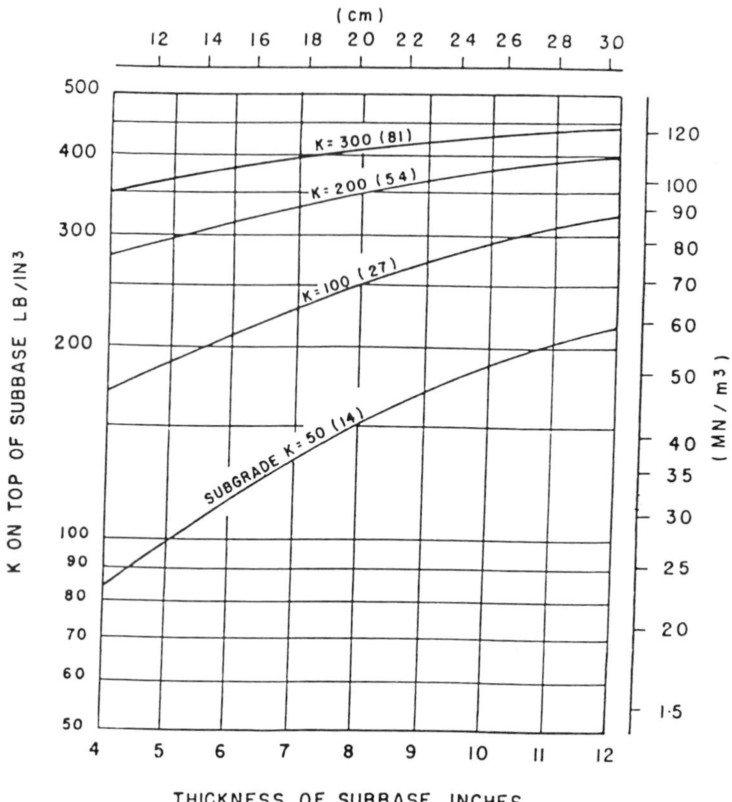

FIGURE 59.44 Effect of stabilized subbase on subgrade modulus. (*Source:* Federal Aviation Administration. 1978. *Airport Pavement Design and Evaluation.* Advisory Circular AC No. 150/5320-6C, p. 64. With permission.)

Table 59.22 FAA Recommended Maximum Joint Spacings

Slab Thickness	Transverse Spacing	Longitudinal Spacing
< 9 in. (23 cm)	15 ft (4.6 m)	12.5 ft (3.8 m)
9–12 in. (23–31 cm)	20 ft (6.1 m)	20 ft (6.1 m)
> 12 in. (31 cm)	25 ft (7.6 m)	25 ft (7.6 m)

Source: Federal Aviation Administration. 1978. *Airport Pavement Design and Evaluation.* Reprinted from FAA Advisory Circular. Report No. FAA/AC-150/5320-6C. 7 December 1978; NTIS Accession No. AD-A075 537/1.

Table 59.23 Dowel Bar Dimensions and Spacings

Slab Thickness	Diameter	Length	Spacing
6–7 in. (15–18 cm)	0.75 in. (20 mm)	18 in. (46 cm)	12 in. (31 cm)
8–12 in. (21–31 cm)	1 in. (25 mm)	19 in. (48 cm)	12 in. (31 cm)
13–16 in. (33–41 cm)	*1.25 in. (30 mm)	20 in. (51 cm)	15 in. (38 cm)
17–20 in. (43–51 cm)	*1.50 in. (40 mm)	20 in. (51 cm)	18 in. (46 cm)
21–24 in. (54–61 cm)	*2 in. (50 mm)	24 in. (61 cm)	18 in. (46 cm)

*Dowels may be a solid bar or high-strength pipe. High-strength pipe dowels must be plugged on each end with a tight-fitting plastic cap or with bituminous or mortar mix.

Source: Federal Aviation Administration. 1978. *Airport Pavement Design and Evaluation.* Reprinted from FAA Advisory Circular. Report No. FAA/AC-150/5320-6C. 7 December 1978; NTIS Accession No. AD-A075 537/1.

Example 59.22. An 18-in.-thick concrete airport pavement has a slab length of 50 ft. Determine the longitudinal steel requirement.

Using grade 60 steel, $f_s = \frac{2}{3}(60{,}000) = 40{,}000$ psi. By Eq. (59.36), $A_s = 3.7(50)\sqrt{(50)(18)/(40{,}000)} = 0.14$ in.^{2} per ft. This is equal to 0.016% steel, satisfying the minimum requirement of 0.05%.

59.7 Pavement Overlay Design

As a pavement reaches the end of its service life, a new span of service life can be provided by either a reconstruction or an application of **overlay** over the existing pavement. There are three common forms of overlay construction—namely, bituminous overlay on flexible pavement, bituminous overlay on concrete pavement, and concrete overlay on concrete pavement. The Asphalt Institute method of flexible overlay design for highway pavement, the Portland Cement Association method of concrete overlay design for highway pavement, and the Federal Aviation Administration method of overlay design for airport pavement are described in this section.

AI Design Procedure for Flexible Overlay on Flexible Highway Pavement

The Asphalt Institute [1983] presents two different approaches to flexible overlay design—one based on the concept of effective thickness and the other based on deflection analysis.

AI Effective Thickness Approach

This approach evaluates the so-called effective thickness T_e of the existing pavement and determines the required overlay thickness T_{OL} as

$$T_{OL} = T - T_e \tag{59.37}$$

where T is the required thickness of a new full-depth pavement if constructed on the existing subgrade, to be determined from Fig. 59.10.

The Asphalt Institute recommends two methods for evaluating effective pavement thickness. The first method involves the use of a conversion factor C based on the PSI (present serviceability index) of the existing pavement, plus the use of conversion factors E for converting various pavement layers into equivalent thickness of asphalt concrete. That is,

$$T_e = C \sum_{i=1}^{n} \{h_i E_i\} \tag{59.38}$$

where n is the total number of pavement layers. C is obtained from either line A or line B in Fig. 59.45. Line A assumes that the overlaid pavement would exhibit a reduced rate of change in PSI compared to before overlay. Line B represents a more conservative design assuming that the rate of change in PSI would remain unchanged after overlay. PSI is usually estimated from correlation with pavement roughness measurements. Equivalency factors E_i are obtained from Table 59.24.

Example 59.23. An old pavement has 3-in. asphalt surface course and 8.5-in. type II emulsified asphalt base (see Example 59.12). Its current PSI is 2.8. Provide an overlay to the pavement to carry the design traffic of Example 59.10.

With PSI = 2.8, $C = 0.75$ by line A of Fig. 59.45. Thickness of new full-depth asphalt pavement required is 9.5 in. (see Example 59.12). Equivalency factor of type II emulsified base is 0.83, from Table 59.24. Overlay thickness $T_e = 9.5 - 0.75\{(3 \times 1.0) + (8.5 \times 0.83)\} = 2$ in.

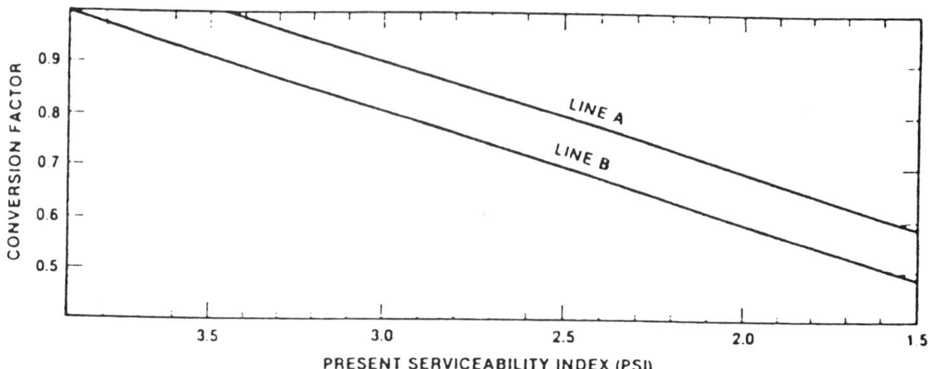

FIGURE 59.45 PSI-based conversion factors for determining effective thickness. (*Source:* Asphalt Institute. 1983a. *Asphalt Overlays for Highway and Street Rehabilitation.* Manual Series MS-17, p. 51. With permission.)

The second recommended method relies on component analysis that assigns conversion factors C_i from Table 59.25 to individual pavement layers based on their respective physical conditions. The effective thickness for the existing pavement structure is given by

$$T_e = \sum_{i=1}^{n} h_i C_i \tag{59.39}$$

where h_i is the layer thickness of layer i and n is the total number of layers in the existing pavement.

Table 59.24 Asphalt Institute Equivalency Factors for Converting Layers of Other Material Types to Equivalent Thickness of Asphalt Concrete

Material Type	Equivalency Factor E_i
Asphalt concrete	1.00
Type I emulsified asphalt base	0.95
Type II emulsified asphalt base	0.83
Type III emulsified asphalt base	0.57

Source: Asphalt Institute. 1983. *Asphalt Overlays for Highway and Street Rehabilitation.* MS-17, p. 52. With permission.

Example 59.24. For the old pavement in Example 59.23, it is observed that the asphalt concrete surface exhibits appreciable cracking, and the emulsified asphalt base has some fine cracking and slight deformation in the wheel paths. Design an overlay for the same traffic as in Example 59.23.

From Table 59.25, the conversion factors for the surface and base courses are both 0.6. $T_{OL} = 9.5 - \{(0.6 \times 3) + (0.6 \times 8.5)\} = 2.6$ in. Use 3 in.

AI Deflection-Based Approach

This approach is based on the correlation between wheel load, repetitions of wheel loads, and the magnitude of pavement rebound deflection. Rebound deflections are measured using the Benkelman beam on the outer wheel path at a minimum of 10 locations within the test section or a minimum of 20 measurements per mile (12 per km). The Benkelman beam is a 12-ft (3.66 m) beam pivoted at a point 8 ft (2.44 m) from the probe end. The probe is positioned at the test point between the dual tires of a rear wheel of a loaded truck that has an 18-kip (80 kN) load equally distributed on its two dual wheels of the rear axle. The amount of vertical rebound at the test point after the truck moves away is recorded as the rebound deflection.

The deflection measurements are used to determine a representative rebound deflection δ_r:

$$\delta_r = (\delta_m - 2s)Fc \tag{59.40}$$

where δ_m is the mean of rebound deflection measurements, s is the standard deviation, F is the temperature adjustment factor, and c is the critical period adjustment factor.

Table 59.25 Conversion Factors C for Determining Effective Thickness

Case	Description	Factor C
I	(a) Native subgrade in all cases (b) Improved subgrade predominantly granular material may contain some silt and clay but have P.I. of 10 or less. (c) Lime-modified subgrade constructed from high-plasticity soils; P.I. greater than 10.	0.0
II	Granular subbase or base, reasonably well-graded, hard aggregates with some plastic fines and CBR not less than 20. Use upper part of range if P.I. is 6 or less; lower part of range if P.I. is more than 6.	0.1–0.2
III	Cement or lime-fly ash stabilized subbases and bases constructed from low plasticity soils, P.I. of 10 or less.	0.2–0.3
IV	(a) Emulsified or cutback asphalt surfaces and bases that show extensive cracking, considerable raveling or aggregate degradation, appreciable deformation in the wheel paths, and lack of stablity. (b) Portland cement concrete pavements (including those under asphalt surfaces) that have been broken into small pieces 2 ft (0.6 m) or less in maximum dimension prior to overlay construction. Use upper part of range when subbase is present, lower part of range when slab is on subgrade. (c) Cement or lime-fly ash stablilzed bases that have developed pattern cracking, as shown by reflected surface cracks. Use upper part of range when cracks are narrow and tight, lower part of range with wide cracks, pumping, or evidence of instability.	0.3–0.5
V	(a) Asphalt concrete surface and base that exhibit appreciable cracking and crack patterns. (b) Emulsified or cutback asphalt surface and bases that exhibit some fine cracking, some raveling or aggregate degradation, and slight deformation in the wheel paths but remain stable. (c) Appreciably cracked and faulted portland cement concrete pavement (including such under asphalt surfaces) that cannot be efectively undersealed. Slab fragments, ranging in size from approximately 10 to 160 ft^2 (1 to 4 m^2), and have been well seated on the subgrade by heavy pneumatic-tired rolling.	0.5–0.7
VI	(a) Asphalt concrete surfaces and bases that exhibit some fine cracking, have small intermittent cracking patterns and slight deformation in the wheel paths but remain stable. (b) Emulsified or cutback asphalt surface and bases that are stable, generally uncracked, show no bleeding, and exhibit little deformation in the wheel paths. (c) Portland cement concrete pavements (including such under asphalt surfaces) that are stable and undersealed, have some cracking but contain no pieces smaller than about 10 ft^2(1 m^2).	0.7–0.9
VII	(a) Asphalt concrete, including asphalt concrete base, generally uncracked, and with little deformation in the wheel paths. (b) Portland cement concrete pavement that is stable, undersealed, and generally uncracked. (c) Portland cement concrete base, under asphalt surface, that is stable, nonpumping, and exhibits little reflected surface cracking	0.9–1.0

Source: Asphalt Institute. 1983a. *Asphalt Overlays for Highway and Street Rehabilitation.* Manual Series MS-17, pp. 54–55. With permission.

The factor c converts the measured deflection to the maximum deflection that would have occurred if the test were performed at the most critical time of the year. Numerically it is equal to the ratio of measured deflection to the corresponding deflection measurement if it were to be made during the critical period. It can be established from historical records or derived from engineering judgement when no record is available.

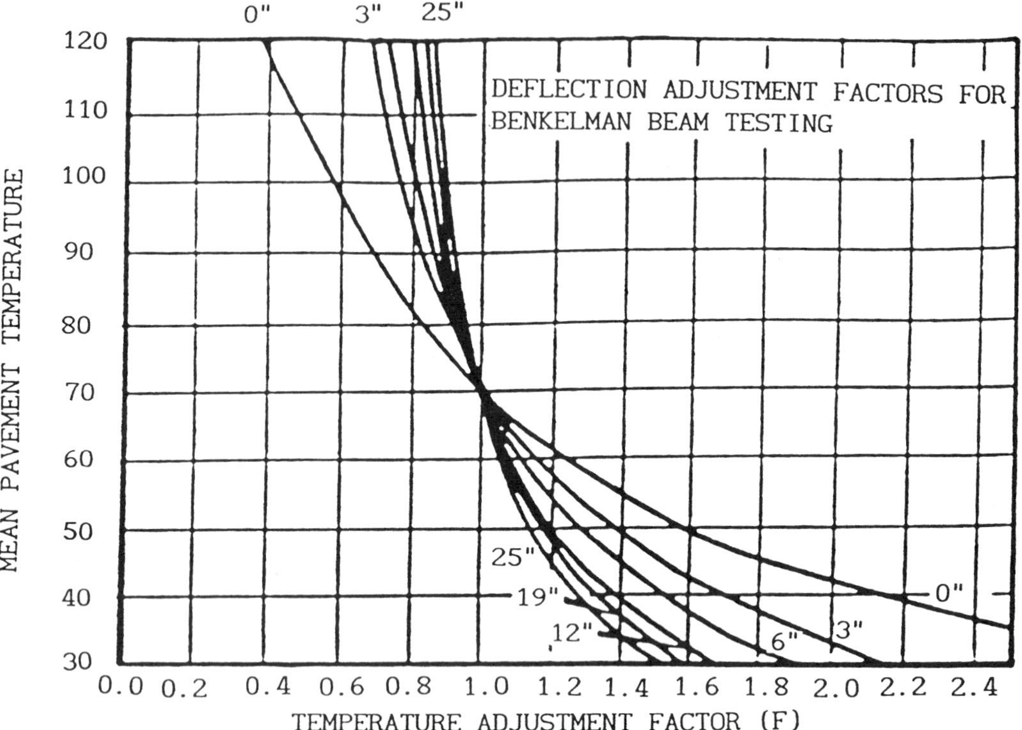

FIGURE 59.46 Chart for determining temperature correction factor *F*. (*Source:* Asphalt Institute. 1983a. *Asphalt Overlays for Highway and Street Rehabilitation.* Manual Series MS-17. With permission.)

F is determined from Fig. 59.46 with two inputs: thickness of untreated granular base and mean pavement temperature. The estimation of mean pavement temperature requires information of the pavement surface temperature at the time of test and the 5-day mean air temperature computed from the maximum and minimum air temperature for each of the 5 days prior to the date of deflection testing. Fig. 59.47 is used to obtain temperature at the middepth and bottom of the pavement. Next, the surface temperature, middepth temperature, and bottom temperature are averaged to provide the mean pavement temperature.

Having computed the representative rebound deflection, Fig. 59.48 is used to determine the required overlay thickness. The design ESAL is estimated by means of the procedure described under the heading of traffic-loading computation.

Example 59.25. Rebound deflection measurements made at 12 randomly selected locations on an old asphalt pavement using Benkelman beam produced the following net rebound deflections in in.: 0.038, 0.035, 0.039, 0.039, 0.039, 0.039, 0.044, 0.044, 0.037, and 0.036. The temperature of pavement surface was found to be 131°F. The extreme air temperatures in the previous 5 days are (88°F, 75°F), (86°F, 75°F), (90°F, 77°F), (88°F, 77°F), and (88°F, 75°F). The thickness of the asphalt layer is 6 in. The thickness of untreated granular base is 12 in. Determine the overlay thickness required to carry additional ESAL of 5×10^6.

Mean deflection $\delta_m = 0.0391$ in. and standard deviation $s = 0.0029$. Five-day mean air temperature = 81.9°F. From Fig. 59.47, pavement layer middepth temperature $T_1 = 105$°F, bottom temperature $T_2 = 100$°F. Mean pavement temperature = 112°F. From Fig. 59.46, $F = 0.82$.

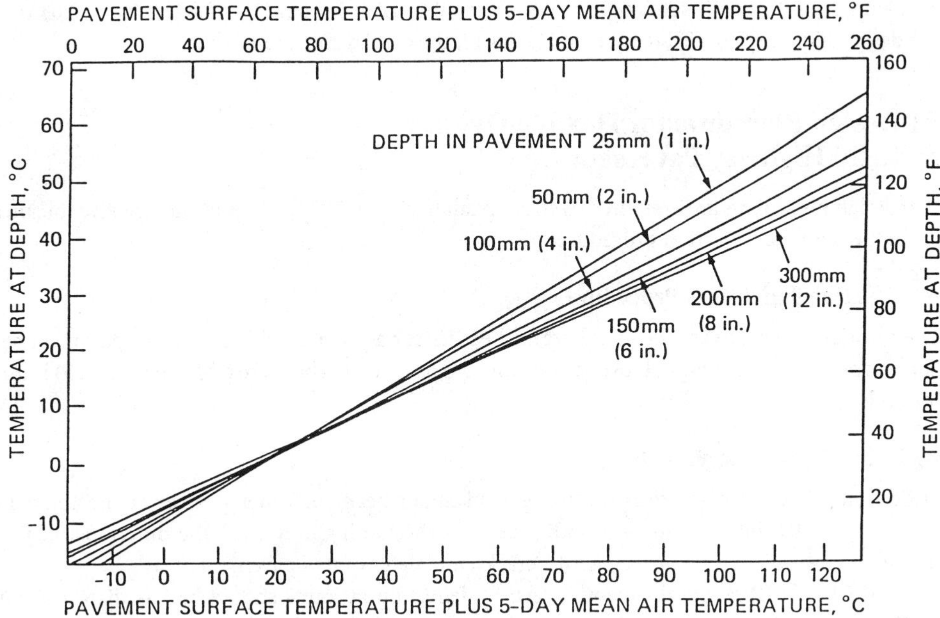

FIGURE 59.47 Estimation of pavement temperature. (*Source:* Asphalt Institute. 1983a. *Asphalt Overlays for Highway and Street Rehabilitation.* Manual Series MS-17. With permission.)

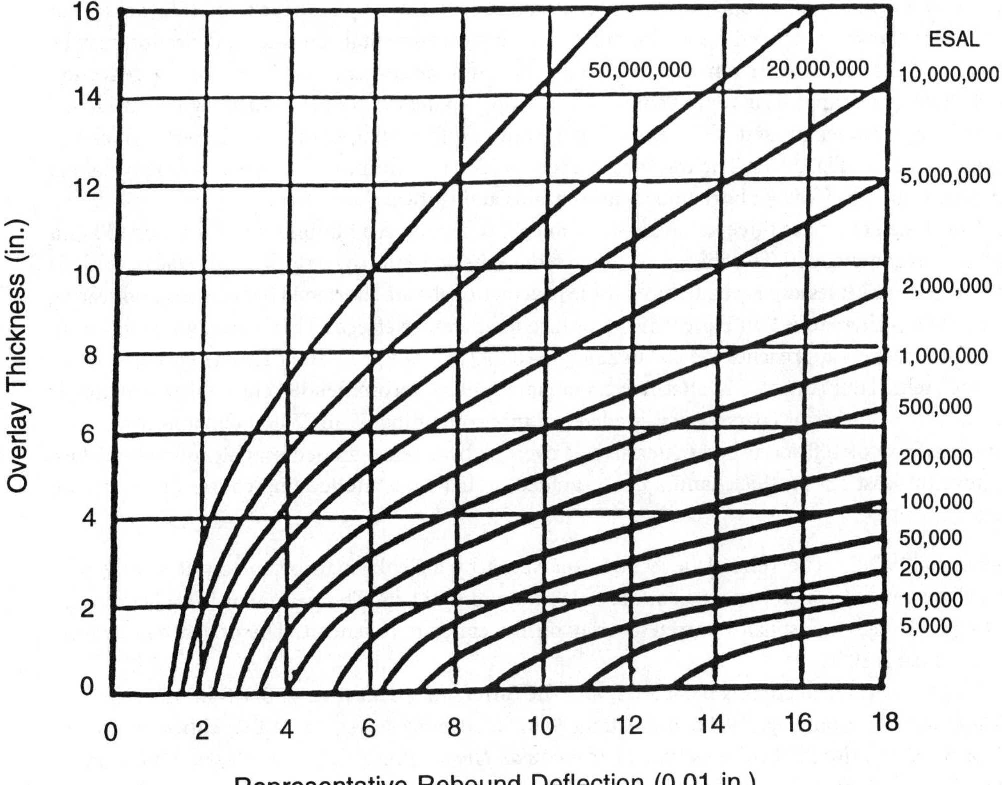

FIGURE 59.48 Design chart for overlay thickness. (*Source:* Asphalt Institute. 1983a. *Asphalt Overlays for Highway and Street Rehabilitation.* Manual Series MS-17. With permission.)

Assume a critical period factor of $c = 0.9$, $\delta_r = \{0.0391 + 2(0.0029)\} \cdot (0.82)(0.9) = 0.0331$ in. For design ESAL of 5×10^6, read from Fig. 59.48, the overlay thickness is 3 in.

AI Design Procedure for Flexible Overlay on Rigid Highway Pavement

Two design procedures are presented by the Asphalt Institute [1983], namely, the effective thickness procedure and the deflection procedure.

AI Effective Thickness Procedure

The component analysis procedure described earlier for asphalt overlay on flexible pavement also applies for the design of asphalt overlay on concrete pavement. The same table (Table 59.25) is used for both.

AI Deflection-Based Procedure

Deflection measurements are made using Benkelman beam or other devices at the following locations: (a) the outside edge on both sides of two-lane highways; (b) the outermost edge of divided highways; and (c) corners, joints, cracks, and deteriorated pavement areas.

For JPCP and JRCP the differential vertical deflection at joints should be less than 0.05 mm (0.002 in.), and the mean deflection should be less than 0.36 mm (0.014 in.). For CRCP, Dynaflect deflections of 15 to 23 μm (0.0006 to 0.0009 in.) or greater lead to excessive cracking and deterioration. Undersealing or stabilization is required when the deflection exceeds 15 μm (0.0006 in.).

Dense-graded asphalt concrete overlay can reduce deflections by 0.2% per mm (5% per in.) of thickness. However, depending on the mix type and environmental conditions, deflection may be as high as 0.4 to 0.5% per mm (10 to 12% per in.). If a reduction of 50% or more of deflection reduction is required, it is more economical to apply undersealing before overlay is considered. For a given slab length and mean annual temperature differential, the required overlay thickness is selected from Fig. 59.49. The thicknesses are provided to minimize reflective cracking by taking into account the effects of horizontal tensile strains and vertical shear stresses.

The design chart has three sections—A, B, and C. In section A a minimum thickness of 100 mm (4 in.) is recommended. This thickness should reduce the deflection by an estimated 20%. In sections B and C the thicknesses may be reduced if the pavement slabs are shortened by breaking and seating (denoted as alternative 2 in Fig. 59.49) to reduce temperature effects. This is recommended as an overlay thickness approaches the 200 to 225 mm (8 to 9 in.) range. Another alternative is the use of a crack relief layer (denoted as alternative 3 in Fig. 59.49). A recommended crack relief structure is a 3.5-in.-thick layer of coarse, open-graded hot mix containing 25 to 35% interconnecting voids and made up of 100% crushed material. It is overlain by a dense-graded asphalt concrete surface course (at least 1.5 in. thick) and a dense-graded asphalt concrete leveling course (at least 2 in. thick).

Example 59.26. The vertical deflections measured by a Benkelman beam test at a joint of a portland cement concrete pavement are 0.042 in. and 0.031 in. The pavement has a slab length of 40 ft. Design an asphalt concrete overlay on the concrete pavement. The design temperature differential is 80°F.

Mean vertical deflection is 0.0365 in., and the differential deflection is 0.009 in. *Alternative 1. Thick overlay*: From Fig. 59.49, more than 9 in. of overlay is required. Use either alternative 2 or 3. *Alternative 2. Break and seat to reduce slab length*: Break slab into 20-ft sections. From Fig. 59.49, 5.5 in. of overlay is required. For the overlaid pavement, mean vertical deflection $= 0.0365 - \{(5.5 \times 5\%) \times 0.0365\} = 0.0265 > 0.014$ in.; and vertical differential deflection $= 0.009 - \{(5.5 \times 5\%) \times 0.009\} = 0.0025 > 0.002$ in. Undersealing is needed. *Alternative 3. Crack*

TEMPERATURE DIFFERENTIAL* (°F)

Slab Length (Ft)	30	40	50	60	70	80	Slab Length (m)
10 or Less	100mm (4 in.)	100mm (4 in.)	100mm (4 in.)	100mm (4 in.)	100mm (4 in.)	100mm (4 in.)	3
15	100mm (4 in.)	100mm (4 in.)	100mm (4 in.)	100mm (4 in.)	100mm (4 in.)	100mm (4 in.)	4.5
20	100mm (4 in.)	100mm (4 in.)	100mm (4 in.)	100mm (4 in.)	125mm (5 in.)	140mm (5.5 in.)	6
25	100mm (4 in.)	100mm (4 in.)	100mm (4 in.)	125mm (5 in.)	150mm (6 in.)	175mm (7 in.)	7.5
30	100mm (4 in.)	100mm (4 in.)	125mm (5 in.)	150mm (6 in.)	175mm (7 in.)	200mm (8 in.)	9
35	100mm (4 in.)	115mm (4.5 in.)	150mm (6 in.)	175mm (7 in.)	215mm (8.5 in.)	Use Alternative 2 or 3	10.5
40	100mm (4 in.)	140mm (5.5 in.)	175mm (7 in.)	200mm (8 in.)	Use Alternative 2 or 3	Use Alternative 2 or 3	12
45	115mm (4.5 in.)	150mm (6 in.)	190mm (7.5 in.)	225mm (9 in.)	Use Alternative 2 or 3	Use Alternative 2 or 3	13.5
50	125mm (5 in.)	175mm (7 in.)	215mm (8.5 in.)	Use Alternative 2 or 3	Use Alternative 2 or 3	Use Alternative 2 or 3	15
60	150mm (6 in.)	200mm (8 in.)	Use Alternative 2 or 3	Use Alternative 2 or 3	Use Alternative 2 or 3	Use Alternative 2 or 3	18
	17	22	28	33	39	44	

TEMPERATURE DIFFERENTIAL* (°C)

FIGURE 59.49 Thickness of asphalt overlay on concrete pavement. (*Source:* Asphalt Institute. 1983a. *Asphalt Overlays for Highway and Street Rehabilitation.* Manual Series MS-17, p. 79. With permission.)

relief layer: Use 3.5-in. crack relief course with 1.5″ surface course and 2″ leveling course, giving a total of 7″ asphalt concrete courses. Similar procedure of deflection checks to those for alternative 2 indicate that undersealing is required.

PCA Design Procedure for Concrete Overlay on Concrete Highway Pavement

Depending on the bonding between the overlay and the existing pavement slab, concrete overlays can be classified into three types: bonded, unbonded, and partially bonded. *Bonded overlay* is achieved by applying a thin coating of cement grout before overlay placement. The construction of *unbonded overlay* involves the use of an unbonding medium at the surface of the existing pavement. Asphaltic concrete and sand asphalt are common unbonding media. *Partially bonded overlay* refers to a construction in which the overlay is placed directly on the existing pavement without the application of a bonding or unbonding medium.

Design of Unbonded Overlay

The procedure selects an overlay thickness that, under the action of an 18-kip (80-kN) single-axle load, would have an edge stress in the overlay equal to or less than the corresponding edge stress in an adequately designed new pavement under the same load. Design charts in Fig. 59.50 are provided for the following three cases:

Case 1. Existing pavement exhibiting a large amount of midslab and corner cracking; poor load transfer at cracks and joints.
Case 2. Existing pavement exhibiting a small amount of midslab and corner cracking; reasonably good load transfer across cracks and joints; localized repair performed to correct distressed slabs.
Case 3. Existing pavement exhibiting a small amount of midslab cracking; good load transfer across cracks and joints; loss of support corrected by undersealing.

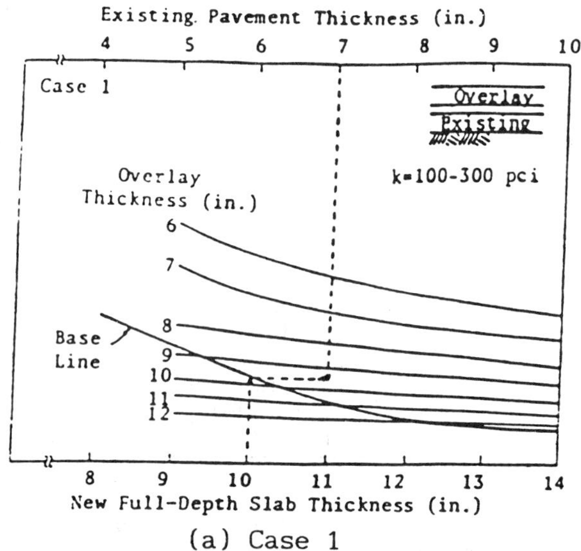

(a) Case 1

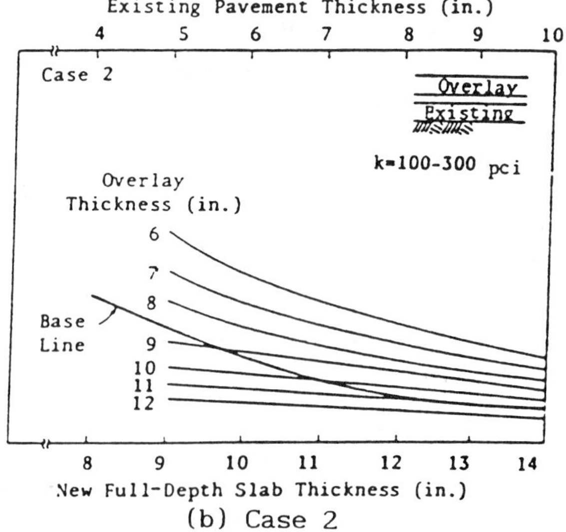

(b) Case 2

FIGURE 59.50 PCA design charts for unbonded overlays. (*Source:* Tayabji and Okamoto [1985], *Proceedings 3rd Int. Conf. on Concrete Pavement Design and Rehabilitation,* April 23–25, Purdue University, pp. 367–379. With permission.) *(continues)*

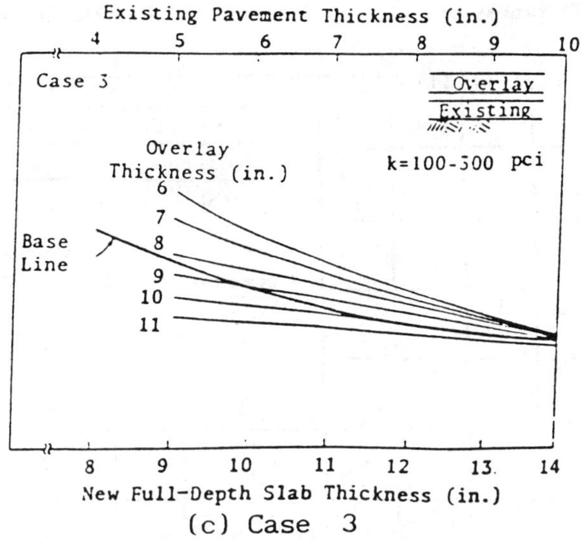

FIGURE 59.50 (continued) PCA design charts for unbonded overlays.

The design charts were obtained from computer analysis of pavements assuming modulus of elasticity of 5×10^6 psi (35 GPa) for overlays and 3×10^6 and 4×10^6 (21 and 28 GPa) for existing pavements. If a tied shoulder is provided the thickness of the overlay may be reduced by 1 in. (25 mm) subject to the minimum thickness requirement of 6 in. (150 mm).

Example 59.27. Design a concrete overlay for an existing 10-in.-thick concrete pavement if the required new single-slab thickness is 10 in.

For case 1 $T_{OL} = 9$ in., from Fig. 59.50(a). For case 2 $T_{OL} = 6$ in., from Fig. 59.50(b). For case 3, where $T_{OL} < 6$ in., from Fig. 59.50(c), use minimum 6 in.

Design of Bonded Overlay

The same structural equivalency concept as for unbonded overlay is adopted in the design for bonded overlay, except that the comparison is now made between the stress-to-strength ratios of the new and the overlaid pavements. The design chart (Fig. 59.51) has three curves for three different ranges of moduli of rupture S_c of the existing concrete. S_c may be estimated from the effective splitting tensile strength f_{te} as follows:

$$f_{te} = f_t - 1.65s \tag{59.41}$$

$$S_c = 0.9Af_{te} \tag{59.42}$$

where f_t is the average value of splitting tensile strength of cored specimens determined according to ASTM Test Method C496, s is the standard deviation of splitting tensile strength, and A is a regression constant ranging from 1.35 to 1.55. An A value of 1.45 is suggested in the absence of local information. One core should be taken every 300 to 500 ft (91 to 152 m) at midslab and about 2 ft (0.6 m) from the edge of outside lane. The 0.9 factor in Eq. (59.42) relates the strength of the concrete specimens to that near or at the edge.

Example 59.28. A 9-in.-thick concrete pavement is to be strengthened to match the capacity of a new 10-in. concrete pavement. What is the required thickness of bonded overlay if the existing concrete has a flexural strength of 450 psi?

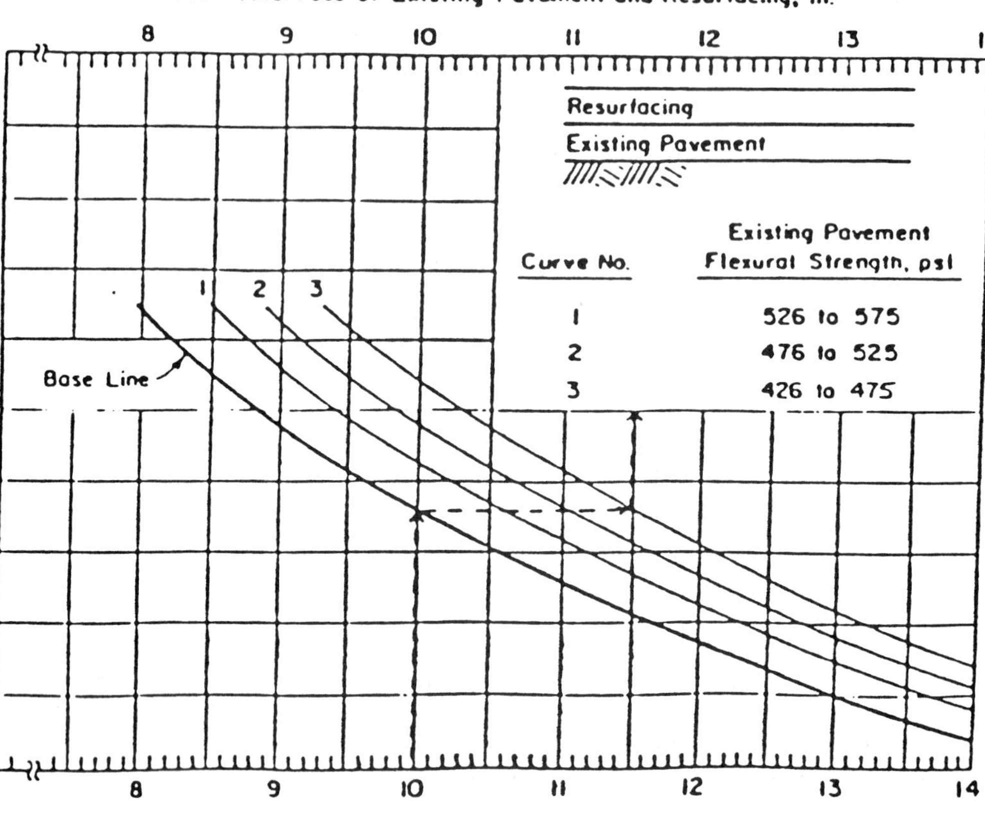

FIGURE 59.51 PCA design charts for bonded overlays. (*Source:* Tayabji and Okamoto [1985], *Proceedings 3rd Int. Conference on Concrete Pavement Design and Rehabilitation,* April 23–25, Purdue University, p. 379. With permission.)

Using curve 3 of Fig. 59.51, thickness of existing pavement plus overlay = 11.5 in. Overlay thickness = 11.5 − 9 = 2.5 in.

FAA Design Procedure for Flexible Overlay on Flexible Airport Pavement

The design method of FAA [1978] is similar in equivalent thickness concept to the PCA method that adopts the component analysis. The FAA equivalency factors are shown in Table 59.13. A high-quality material may be converted to a lower-quality material—for example, surfacing to base and base to subbase. A material may not be converted to a higher-quality material. The overlay thickness is equal to the difference between the total equivalent layer thicknesses of the existing pavement and the corresponding required layer thicknesses of a new pavement. The minimum overlay thickness allowed is 3 in. (75 mm).

Example 59.29. An existing asphalt concrete airport pavement has 4-in. bituminous surface course, 7-in. base course, and 14-in. subbase. The CBR of the subgrade is 8 and that of the subbase 12. Provide an overlay to strengthen the pavement for 6000 annual departures of a design aircraft (dual-wheel landing gear) with maximum weight of 100,000 lb.

From Fig. 59.17, a new pavement requires 30 in. total thickness based on subgrade CBR of 8 and a thickness of 17 in. above subbase based on subbase CBR of 12. Using 4 in. of asphalt concrete surface course, the base layer is (17 − 4) = 13 in. The deficiency of thickness of the existing pavement is all in the base layer. Assuming the existing asphalt concrete surface course can be converted to base at an equivalency ratio of 1.4 to 1 (see Table 59.13), the thickness of asphalt base required = (13 − 7)/1.4 = 4.3 in. An additional 0.3 in. of asphalt concrete base is needed. The total thickness of overlay = (4 in. of new surface course) + 0.3 = 4.3 in. Use 4.5 in. overlay.

FAA Design Procedure for Flexible Overlay on Concrete Airport Pavement

The equation for computing bituminous overlay thickness T is

$$T(\text{inches}) = 2.5(Fh - C_b h_e) \tag{59.43}$$

where

F = factor to be obtained from Fig. 59.52
h = single thickness of rigid pavement required for design condition, inches (use the exact value without rounding off)
C_b = condition factor for base pavement, $0.75 \leq C_b \leq 1.0$
h_e = thickness of existing rigid pavement, inches

The F factor is related to the degree of cracking that will occur in the base pavement. It has a value less than one, indicating that the entire single concrete slab thickness is not needed because a bituminous overlay pavement is allowed to crack and deflect more than a conventional rigid

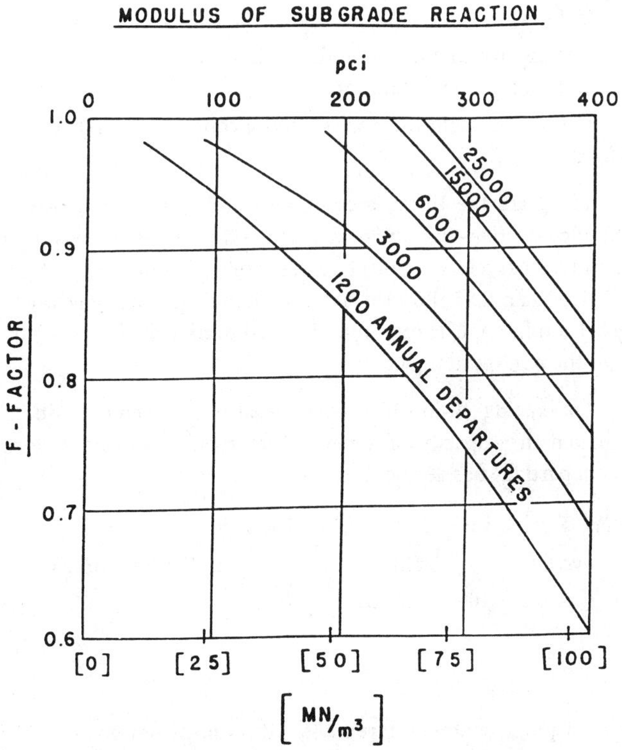

FIGURE 59.52 Graph for determination of F factor. (*Source:* Federal Aviation Administration. 1978. *Airport Pavement Design and Evaluation.* Advisory Circular AC No. 150/5320-6C, p. 105. With permission.)

pavement. C_b is an assessment of the structural integrity of the existing pavement. C_b is 1.0 when the existing slabs contain nominal initial cracking and 0.75 when the slabs contain multiple cracking.

Example 59.30. An existing 10-in. concrete pavement has a condition factor C_b of 0.8. The subgrade k is 200 pci. Provide a bituminous overlay to strengthen the pavement to be equivalent to a single rigid pavement thickness of 12 in. for a design traffic of 3000 annual departures.

From Fig. 59.52, $F = 0.92$. By Eq. (59.43), overlay thickness $t = 2.5\{(0.92 \times 12) - (0.8 \times 10)\} = 7.6$ in. Use 8-in.-thick overlay.

FAA Design Procedure for Concrete Overlay on Concrete Airport Pavement

The design of concrete overlay requires an assessment of the structural integrity of the existing pavement and the thickness of a new concrete pavement on the existing subgrade. The design equations are

$$\text{Unbonded overlay} \qquad T = (h^2 - C_r h_e^2)^{1/2} \tag{59.44}$$

$$\text{Partially bonded overlay} \qquad T = (h^{1.4} - C_r h_e^{1.4})^{1/1.4} \tag{59.45}$$

$$\text{Bonded overlay} \qquad T = h - h_e \tag{59.46}$$

where

C_r = 1.0 for existing pavement in good condition—some minor cracking evident but no structural defects

C_r = 0.75 for existing pavement containing initial corner cracks due to loading but no progressive cracking or joint faulting

C_r = 0.35 for existing pavement in poor structural condition—badly cracked or crushed and faulted joints.

The variables h and h_e are the thicknesses of new and existing pavements respectively. The use of partially bonded overlay—which is constructed directly on an existing pavement without debonding medium (such as a bituminous leveling course)—is not recommended for an existing pavement with C_r less than 0.75. Bonded overlays should be used only when the existing rigid pavement is in good condition. The minimum bonded overlay thickness is 3 in. For partially and unbonded overlays, the minimum thickness is 5 in.

Example 59.31. An existing 10-in.-thick concrete airport pavement with $C_r = 1.0$ is to be strengthened to match the capacity of a new 12-in. rigid pavement. Determine the required thickness of bonded, partially bonded, and unbonded overlays.

Unbonded overlay. $T = \sqrt{12^2 - 10^2} = 6.6$ in. Use 7 in.

Partially bonded overlay. $T = \sqrt[1.4]{12^{1.4} - 10^{1.4}} = 4.1$ in. Use 5 in. (min).

Bonded overlay. $T = 12 - 10 = 2$ in. Use 3 in. (min).

Defining Terms

Asphalt pavement (asphalt concrete pavement, bituminous pavement): The most common form of flexible pavement in which the surface course is constructed of asphaltic (or bituminous) mixtures.

Base course: The layer of selected material in a pavement structure placed between a subbase and a surface course.

Concrete pavement: The most common form of rigid pavement in which the top slab is constructed of portland cement concrete.

Flexible pavement: A pavement structure that does not distribute traffic load to the subgrade by means of slab action, but mainly through spreading of the load by providing sufficient thickness of the pavement structure.

Overlay: A new surface layer laid on an existing pavement to improve the latter's load-carrying capacity.

Pavement structure: A structure consisting of one or more layers of selected materials constructed on prepared subgrade to designed strength and thickness(es) for the purpose of supporting traffic.

Rigid pavement: A pavement structure that distributes traffic loads to the subgrade by means of slab action through its top layer of high-bending resistance.

Subbase: The layer of selected material in a pavement structure placed between the subgrade and the base or surface course.

Subgrade: The top surface of graded foundation soil, on which the pavement structure is constructed.

Surface course: The top layer of a pavement structure placed on the base course, the top surface of which is in direct contact with traffic loads.

References

AASHTO. 1972. *AASHTO Interim Guide for Design of Pavement Structures.* American Association of State Highway and Transportation Officials, Washington, D.C.

AASHTO. 1989. *Standard Specifications for Transportation Materials and Methods of Sampling and Testing.* Part I and II. American Association of State Highway and Transportation Officials, Washington, D.C.

AASHTO. 1993. *AASHTO Guides for Design of Pavement Structures.* American Association of State Highway and Transportation Officials, Washington D.C.

ACI. 1977. *Building Code Requirements for Reinforced Concrete.* American Concrete Institute, Detroit, MI.

Asphalt Institute. 1983a. *Asphalt Overlays for Highway and Street Rehabilitation.* Manual Series No. 17. Lexington, KY.

Asphalt Institute. 1983b. *Asphalt Technology and Construction Practices. Educational Series ES-1,* 2nd ed. Lexington, KY.

Asphalt Institute. 1991. *Thickness Design—Asphalt Pavements for Highways & Streets.* Manual Series No. 1. Lexington, KY.

ASTM. 1992. *Annual Books of ASTM Standards.* American Society for Testing and Materials, Philadelphia, PA.

Boussinesq, J. 1885. *Application des Potentiels a l'etude de l'equilibre et du Mouvement des Solids Elastiques.* Gauthier-Villars, Paris.

FAA. 1978. *Airport Pavement Design and Evaluation.* Advisory Circular AC No. 150/5320-6C. Federal Aviation Administration.

Fwa, T. F. and Li, S. 1994. Estimation of lane distribution of truck traffic for pavement design. Paper accepted for publication. *Journal of Transportation Engineering.*

Fwa, T. F., Shi, X. P., and Tan, S. A. 1993. *Load-Induced Stresses and Deflections in Concrete Pavement—Analysis by Rectangular Thick-Plate Model.* CTR Technical Report CTR-93-5. Centre for Transportation Research, Faculty of Engineering, National University of Singapore.

Fwa, T. F. and Sinha, K. C. 1985. *A Routine Maintenance and Pavement Performance Relationship Model for Highways.* Joint Highway Research Project Report JHRP-85-11. Purdue University, West Lafayette, IN.

Highway Research Board. 1962. *The AASHO Road Test, Report 5—Pavement Research.* HRB Special Report 61E. Washington, D.C.

PCA. 1984. *Thickness Design for Concrete Highway and Street Pavements.* Portland Cement Association, Skokie, IL.

Shi, S. P., Tan, S. A., and Fwa, T. F. 1994. Rectangular plate with free edges on a Pasternak foundation. *Journal of Engineering Mechanics.* 120(5):971–988.

Tayabji, S. D. and Okamoto, P. A. 1985. Thickness design of concrete resurfacing. *Proc. 3rd Int. Conf. on Concrete Pavement Design and Rehabilitation,* April 23–25, Purdue University, West Lafayette, IN, pp. 367–379.

Van Til, C. J., McCullough, B. F, Vallerga, B. A., and Hicks, R. G. 1972. *Evaluation of AASHO Interim Guides for Design of Pavement Structures.* NCHRP Report 128. Highway Research Board, Washington, D.C.

Westergaard, H. M. 1926. Stresses in concrete pavements computed by theoretical analysis. *Public Roads.* 7(2):25–35.

Westergaard, H. M. 1933. Analytical tools for judging results of structural tests of concrete pavements. *Public Roads.* 14(10).

Westergaard, H. M. 1948. New formulas for stresses in concrete pavements of airfield. *ASCE Transactions.* Vol. 113.

Yoder, E. J. and Witczak, M. W. 1975. *Principles of Pavement Design,* 2nd ed. John Wiley & Sons, New York.

For Further Information

A widely quoted reference to the basics of practical design of highway and airport pavements is *Principles of Pavement Design* by E. J. Yoder and M. W. Witczak. Although the described design methods by various agencies are outdated, the book is still a valuable reference on the requirements of pavement construction and design.

Detailed descriptions of pavement design methods, pavement material, and construction requirements by various organizations are available in their respective publications. The Asphalt Institute publishes a manual series addressing bituminous pavement-related topics—including thickness design, pavement rehabilitation and maintenance, pavement drainage, hot-mix design, and paving technology. Additional information concerning topics related to portland cement concrete pavement is found in publications by Portland Cement Association and American Concrete Institute.

The latest developments in various aspects of pavement design are reported in a number of technical journals in the field. The most important are the *Journal of Transportation Engineering,* published bimonthly by the American Society of Civil Engineers, and *Transportation Research Records* published by Transportation Research Board. There are about 40 issues of *Transportation Research Records* published each year recently, each collecting a group of technical papers addressing a specialized area of transportation engineering.

There are several major conferences that focus on highway and airport pavements. The International Conference on Structural Design of Asphalt Pavements has been held once every five years since 1962. The seventh conference, in 1992, was named *International Conference on Asphalt Pavements: Design, Construction and Performance* to reflect the added scope of the conference. The proceedings of the conferences document advances in areas of asphalt pavement technology. Another conference, the International Conference on Concrete Pavement Design and Rehabilitation, focuses on the development of concrete pavement technology. It has been organized once every four years by Purdue University since 1977. There is also the International Conference on the Bearing Capacity of Roads and Airfields, held at intervals of four years since 1982. Other related publications are the *Proceedings of the World Road Congress,* published by the Permanent International Association of Road Congress, and the *Proceedings of the Road Congress of the International Road Federation.*

60

Geometric Design

Said M. Easa
Lakehead University

60.1 Introduction

Geometric design of highways refers to the design of the visible dimensions of such features as curves, cross sections, sight distance, bicycle and pedestrian facilities, and intersections. The main objective of geometric design is to produce a highway with safe, efficient, and economic traffic operations, while maintaining aesthetic and environmental quality. Geometric design is influenced by vehicle, user, and traffic characteristics. The temporal changes of these characteristics make geometric design a dynamic field in which design standards are periodically updated to provide more satisfactory design.

Policies on highway geometric design in the U.S. are developed by the American Association of State Highway and Transportation Officials (AASHTO). These policies represent standards and guidelines agreed to by the state highway and transportation departments and the Federal Highway Administration (FHWA). Standards of highway geometric design are presented in *A Policy on Geometric Design of Highways and Streets* [AASHTO, 1990], which is based on many years of experience and research. Repeated citation of AASHTO throughout this chapter refers to this policy. The philosophy and approach to geometric design standards in a number of other countries are mostly similar to those of AASHTO.

This chapter discusses the fundamentals of highway geometric design and their applications, and is divided into three main sections: fundamentals of geometric design, basic design applications, and special design applications. It draws information mostly from the AASHTO policy and provides supplementary information on more recent developments. Since geometric design is a major component in both the preliminary location study and the final design of a proposed highway, it is helpful to describe first the highway design process.

0-8493-8953-4/95/$0.00 + $.50

Design Process

The design process of a proposed highway involves preliminary location study, environmental impact evaluation, and final design. This process normally relies on a team of professionals, including engineers, planners, economists, sociologists, ecologists, and lawyers. Such a team may have responsibility for addressing social, environmental, land use, and community issues associated with the highway development.

Preliminary Location Study

The preliminary location study involves collecting and analyzing data, locating feasible routes, determining preliminary horizontal and vertical alignments for each, and evaluating alternative routes to select the best route. The types of data required are related to the engineering, social and demographic, environmental, and economic characteristics of the area. Examples of such data are topography, land use pattern, wildlife types, and unit costs of construction. A preliminary-study report is prepared, and typically includes a general description of the proposed highway, description of alternative locations and designs, projected traffic volumes and estimated total costs, economic and environmental evaluation, and a recommended highway location. Before the project is approved, it is common to hold public hearings to discuss the preliminary study and environmental impacts.

Environmental Evaluation

Highway construction may impact the environment in a number of areas, including air quality, water quality, noise, wildlife, and socioeconomic. For example, highways may cause loss or degradation of a unique wildlife habitat and changes to migratory patterns. Socioeconomic impacts include displacement of people and businesses, removal of historically significant sites, and severance of the interpersonal ties of displaced residents to their former community. It is therefore essential that the environmental impacts of alternative highway locations be fully evaluated.

Provisions of the National Environment Policy Act of 1969 require that an environmental impact statement (EIS) be submitted for any project affecting the quality of the environment. The EIS must describe the environmental impacts of the proposed action, both positive and negative; probable unavoidable adverse environmental impacts; secondary environmental impacts, such as changes in the pattern of social and economic activities; analysis of short- and long-term impacts; irreversible and irretrievable commitments of resources; and public and minority involvement. Chapter 64 provides more details on the environmental process.

Final Design

The final design involves establishing the design details of the selected route, including final horizontal and vertical alignments, drainage facilities, and all items of construction. The design process has been revolutionized by advanced photogrammetric and computer techniques. For example, designers now can have a driver's-eye view of a proposed highway alignment displayed on a monitor and readily examine the effects of alignment refinements. Further details on the design process are found in Garber and Hoel [1988].

60.2 Fundamentals of Geometric Design

Geometric design involves a number of fundamentals and concepts that guide and control the manner in which a highway is designed. These include highway types, design controls, sight distance, and simple highway curves.

Highway Types

Classification of highways into functional types is necessary for communication among engineers, administrators, and the general public. The functional classification system facilitates grouping roads that require the same quality of design, maintenance, and operation. The system also facilitates the logical assignment of responsibility among different jurisdictions, and its structure of design standards is readily understood.

The basic highway functional types, adopted separately for urban and rural areas, are locals, collectors, arterials, and freeways. The two major considerations in the functional classification system are travel mobility and land access. Locals emphasize the land access function; collectors provide a balanced service for both functions; arterials emphasize the mobility function; and freeways provide optimum mobility. Design standards for local roads and streets, collector roads and streets, rural and urban arterials, and rural and urban freeways are addressed by AASHTO in Chapters V to VIII, respectively. Two additional references, published by the Institute of Transportation Engineers (ITE), discuss local street design [ITE, 1990] and urban arterial design [ITE, 1984].

Design Controls

The major controls that influence the geometric design of highways include design vehicle, driver characteristics, design volume, design speed, level of service, access control, and safety considerations. Other controls such as topography, aesthetics, environment, economics, and public concerns are important, but are reflected in either the preceding major controls or the preliminary location study.

Design Vehicle

A design vehicle is a selected motor vehicle whose weight, dimension, and operating characteristics are used to establish certain geometric design standards. The three general classes used in geometric design are passenger cars, trucks, and buses/recreational vehicles. The dimensions of 15 design vehicles within these general classes have been given by AASHTO. The design vehicle selected for geometric design is the largest vehicle likely to use the highway with considerable frequency. The design vehicle is directly used to determine the radii of traveled way at intersections and the width of **turning roadways**. The minimum turning path for a WB-40 design vehicle (intermediate tractor–semitrailer combination truck with an effective wheelbase of 40 ft) is shown in Fig. 60.1. The main dimensions affecting design are the minimum turning radius, the tread width, the wheel base, and the path of the inner rear tire. Other vehicle characteristics such as acceleration and braking capabilities, driver's-eye height, and vehicle headlight also affect many geometric design features.

Driver Characteristics

Driver characteristics affect highway geometric design in a number of ways. The design should focus a driver's attention on the safety-critical elements by providing clear sight lines and good visual quality. The design should take into account the longer reaction time required for complex decisions by providing adequate decision sight distance. On high-speed facilities, control and guidance activities should be simplified because speed reduces the visual field, restricts peripheral vision, and limits the time available to process information. Flat curving alignment that follows the natural contours of the terrain should be used along with well-spaced rest areas to avoid driver fatigue.

Design Volume

Design volume and composition determine the highway type, required roadway width, and other geometric features. The basic unit of measure is the **average daily traffic (ADT)**. This unit is used for selecting geometric design standards for local roads and streets. For other highways, the design-hour volume (DHV), a two-way volume, is used and is generally defined as the **30th highest-hour volume** of the year. The ratio of the DHV and the ADT, designated P, varies only slightly from year to

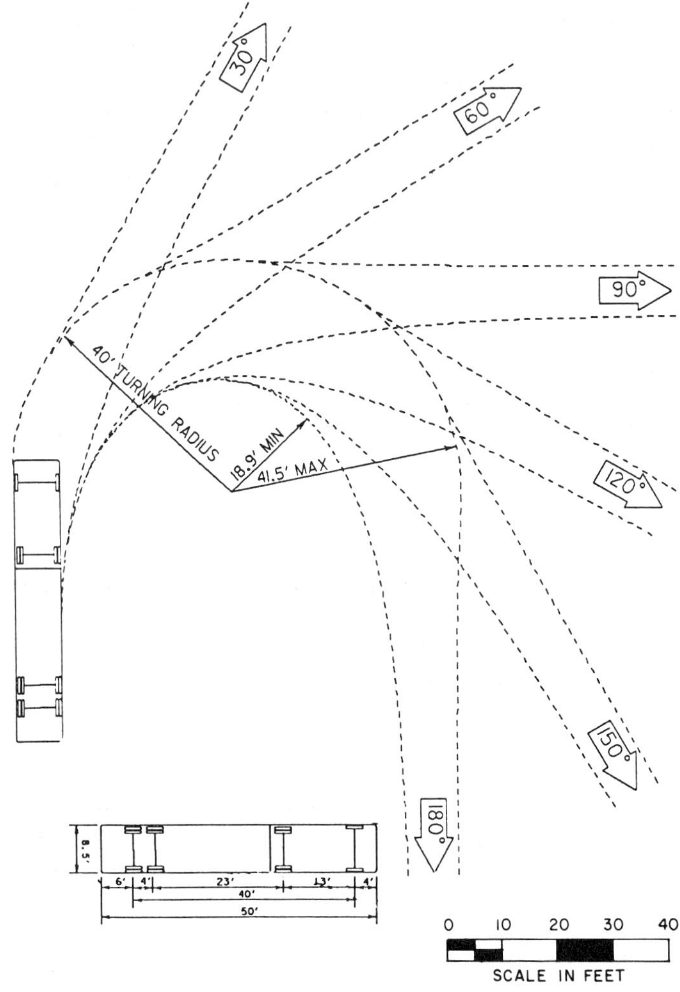
THIS TURNING TEMPLATE SHOWS THE TURNING PATHS OF THE AASHTO DESIGN VEHICLES. THE PATHS SHOWN ARE FOR THE LEFT FRONT OVERHANG AND THE OUTSIDE REAR WHEEL. THE LEFT FRONT WHEEL FOLLOWS THE CIRCULAR CURVE, HOWEVER, ITS PATH IS NOT SHOWN.
30°
60°
90°
120°
150°
180°
40' TURNING RADIUS
18.9' MIN
41.5' MAX
8.5'
6'
4'
23'
13'
4'
40'
50'
0 10 20 30 40
SCALE IN FEET

FIGURE 60.1 Minimum turning path for WB-40 design vehicle. (From *A Policy on Geometric Design of Highways and Streets.* Copyright 1990 by the American Association of State Highway and Transportation Officials, Washington, D.C. Used by permission.)

year. For design of a new highway, P can be determined using existing traffic volumes of similar highways.

To determine the DHV of a proposed highway, the ADT is forecast for the design year and multiplied by P. The typical range of P is 12 to 18% for rural highways and 8 to 12% for urban highways. For recreational routes, in practice the DHV is selected as 50% of the volume that occurs for only a few peak hours during the design year. Other volume characteristics required for the design year are the percentage of the two-way DHV occurring in the predominant direction (D) and the percentage of trucks in the design-hour volume (T).

Design Speed (*V*)

Design speed is a speed selected for design and correlation of the geometric elements of a highway. It is the maximum safe speed maintainable over a specified section of a highway when weather and traffic conditions are so favorable that the design features of the highway govern. Selection of design speed is influenced by type of highway, topography, density of land use, and driver expectations. Nearly all geometric design elements are directly or indirectly influenced by design speed.

Level of Service (LOS)

The required number of lanes of a highway depends on the DHV and the **level of service** intended for the design year. The *Highway Capacity Manual* (HCM) defines six levels of service ranging from A (free flow) to F (forced flow) with level of service E representing the capacity of the highway [TRB, 1985]. The HCM presents a thorough discussion on the level of service concept. Table 60.1 shows the AASHTO recommended design levels of service for different highway types. In heavily developed sections of metropolitan areas, conditions may necessitate the use of level of service D for freeways and arterials.

Access Control

Access control refers to the extent of roadside-interference regulations of public access to and from properties on the roadside. These regulations may be full control of access or partial control of access. Fully controlled access facilities (such as freeways) have no at-grade crossings and have access connections only with selected roads. With partial control of access, preference is given to through traffic to an extent, but there may be some at-grade crossings and driveway connections. Partial access control can be achieved by driveway permits, zoning restrictions, and frontage roads. The extent of access control is a significant factor in defining the functional type of a highway.

Safety Considerations

Safety is a major consideration in the design of nearly all elements of highway geometric design, including horizontal and vertical alignments, cross sections, roadsides, traffic control devices, and intersections. Safety must be reflected not only in new highway and major reconstruction projects, but also in the resurfacing, restoration, and rehabilitation projects. AASHTO stresses

Table 60.1 Guide for Design Levels of Service

	Type of Area and Appropriate Level of Service			
Highway Type	Rural Level	Rural Rolling	Rural Mountainous	Urban and Suburban
Freeway	B	B	C	C
Arterial	B	B	C	C
Collector	C	C	D	D
Local	D	D	D	D

Note: General operating conditions for levels of service:
A—free flow, with low volumes and high speeds
B—reasonably free flow, but speeds beginning to be restricted by traffic conditions
C—in stable flow zone, but most drivers restricted in freedom to select their own speed
D—approaching unstable flow, drivers have little freedom to maneuver
E—unstable flow, may be short stoppages

From *A Policy on Geometric Design of Highways and Streets.* Copyright 1990 by the American Association of State Highway and Transportation Officials, Washington, D.C. Used by permission.

the importance of establishing a safety evaluation program to identify safety hazards, evaluate the effectiveness of alternative improvements, and program available funds to the most effective uses.

Design Designation

Design designation, a summary of the major controls for which a highway is designed, is normally included in design contract drawings. The following is an example of a design designation for a new urban freeway:

Design year	=	2000
ADT (1990)	=	None
ADT (2000)	=	42,800
DHV	=	5140
D	=	65%
T	=	5%
V	=	60 mph
LOS	=	C
Control of access	=	Full

Sight Distance

Sight distance is the length of the roadway visible to the driver. It is a fundamental design element in the safe and efficient operation of a highway. Three basic types of sight distances must be considered in design: (1) stopping sight distance, applicable on all highways; (2) passing sight distance, applicable only on two-lane highways; and (3) decision sight distance, needed at complex locations.

Stopping Sight Distance (SSD)

Stopping sight distance is the distance that enables a driver to stop after seeing an object in the vehicle's path without hitting the object. The SSD in feet is computed by

$$\text{SSD} = 1.47tV + \frac{V^2}{30(f + G)} \tag{60.1}$$

where

V = initial speed prior to braking (mph)
t = perception-reaction time (2.5 s)
f = longitudinal friction factor between vehicle tires and pavement
G = percent of grade divided by 100 (positive for upgrade and negative for downgrade)

Design values of SSD for level grades ($G = 0$) are shown in Table 60.2, where f is based on wet pavements and relatively poor tires. The higher (desirable) values of the range correspond to an initial speed equal to the design speed. The lower (minimum) values assume that vehicles travel at the **average running speed** on wet pavements. The design values of Table 60.2 are based on tangent operations. On horizontal curves, SSD requirements will be larger because of reduced longitudinal friction. The *Manual of Geometric Design Standards for Canadian Roads* [RTAC, 1986] recommends increasing the tangent-based values by 5% on sharp horizontal curves.

Example 60.1. Compute the desirable SSD for a highway with a 60-mph design speed and $G = 0$. From Table 60.2, $f = 0.29$ and assumed initial speed $V = 60$ mph. From Eq. (60.1),

$$\text{SSD} = 1.47 \times 2.5 \times 60 + \frac{(60)^2}{30(0.29)} = 634.3 \text{ ft}$$

Table 60.2 Design Requirements for Stopping and Passing Sight Distances of AASHTO

	Stopping Sight Distance				Passing Sight Distance			
Design Speed (mph)	Assumed Initial Speed (mph)	Longitudinal Friction Factor f	Computed SSD[a] (ft)	Rounded for Design (ft)	Passed Vehicle Speed (mph)	Passing Vehicle Speed (mph)	Computed PSD[b] (ft)	Rounded for Design (ft)
20	20–20	0.40	106.8–106.8	125–125	20	30	790	800
30	28–30	0.35	177.6–196.0	200–200	26	36	1090	1100
40	36–40	0.32	267.3–313.7	275–325	34	44	1490	1500
50	44–50	0.30	376.8–461.5	400–475	41	51	1840	1800
60	52–60	0.29	501.9–634.3	525–650	47	57	2140	2100
70	58–70	0.28	613.6–840.6	625–850	54	64	2490	2500

[a] Values are computed using Eq. (60.1).
[b] Values are computed using Eq. (60.2).

which is the same as the computed value in Table 60.2. The minimum SSD is similarly computed using the average running speed (52 mph).

Passing Sight Distance (PSD)

Passing sight distance is the distance required for a vehicle to overtake a slower-moving vehicle safely on a two-lane highway. The AASHTO model is based on certain assumptions for traffic behavior, and considers PSD as the sum of four distances: (1) distance during perception, reaction, and acceleration of the passing vehicle to the encroachment point on the left lane; (2) distance while the passing vehicle occupies the left lane; (3) distance between the passing vehicle at the end of its maneuver and the opposing vehicle; and (4) distance traveled by an opposing vehicle for two-thirds of the time the passing vehicle occupies the left lane. Safe passing sight distances for various speed ranges based on field observations have been presented by AASHTO. The following best-fit relationship, based on AASHTO data, can be used for computing safe PSD:

$$\text{PSD} = 50v - 710 \tag{60.2}$$

where

PSD = safe passing sight distance (ft)
v = average speed of passing vehicle (mph), where v lies between 30 and 70 mph

Design values of PSD are shown in Table 60.2, where the computed values are based on Eq. (60.2). The speed of the passed vehicle is assumed to be the average running speed and the speed of the passing vehicle is 10 mph greater. The design values apply to a single passing only. The design requirements of SSD and PSD of Table 60.2 are for passenger cars.

Decision Sight Distance (DSD)

Decision sight distance is required at complex locations to enable drivers to maneuver their vehicles safely rather than to stop. It is the distance required for a driver to detect an unexpected hazard, recognize the hazard, decide on proper maneuvers, and execute the required action safely. Design values for DSD, developed from empirical data, are shown in Table 60.3. Since decision sight distance affords drivers sufficient length to maneuver their vehicles, its value is much greater than the stopping sight distance.

Examples of complex locations where provision of DSD is desirable include complex interchanges and intersections, toll plazas, lane drops, and areas where sources of information (such as signs, signals, and traffic control devices) compete. Where it is not feasible to provide DSD, designers

Table 60.3 Design Requirements for Decision Sight Distance

Design Speed (mph)	Decision Sight Distance for Avoidance Maneuver (ft)[a]				
	A	B	C	D	E
30	220	500	450	500	625
40	345	725	600	725	825
50	500	975	750	900	1025
60	680	1300	1000	1150	1275
70	900	1525	1100	1300	1450

[a] Avoidance Maneuver A: stop on rural road
Avoidance Maneuver B: stop on urban road
Avoidance Maneuver C: speed/path/direction change on rural road
Avoidance Maneuver D: speed/path/direction change on suburban road
Avoidance Maneuver E: speed/path/direction change on urban road

From *A Policy on Geometric Design of Highways and Streets.* Copyright 1990 by the American Association of State Highway and Transportation Officials, Washington, D.C. Used by permission.

should move the location or use suitable traffic control devices to provide advance warning of the conditions to be encountered. A recent study found that the application guidelines of DSD were too vague and recommended revised DSD values [McGee, 1989].

Design Heights for Sight Distances

The AASHTO design driver's-eye height and object height used for measuring various sight distances are shown in Table 60.4. Sight distances are measured from a 3.5-ft driver's-eye height to a 0.5-ft object height for SSD and DSD and a 4.25-ft object height for PSD. Recent research has suggested that SSD should be based on functional highway classifications, using different object heights, driver's reaction times, and pavement friction coefficients for different highway types [Neuman, 1989]. In Canadian standards, an appropriate object height for DSD depending on the prevailing conditions is used, with a zero object height needed in some circumstances [RTAC, 1986]. The driver's-eye height is 1.05 m and the object height is 0.38 m (taillight height) for SSD and 1.3 m for PSD.

Table 60.4 AASHTO Design Heights for Sight Distances

Sight Distance Type	Driver's-Eye Height, H_e (ft)	Object Height, H_o (ft)
SSD	3.5	0.5
PSD	3.5	4.25
DSD	3.5	0.5

Simple Highway Curves

Two basic curves are used for connecting straight (tangent) roadway sections in geometric design: a simple circular curve for horizontal alignment and a simple parabolic curve for vertical alignment. Other options include spirals, compound curves, and reverse circular curves for horizontal alignment, and unsymmetrical curves and reverse parabolic curves for vertical alignment. Details on the geometry of these curves can be found in Meyer and Gibson [1980].

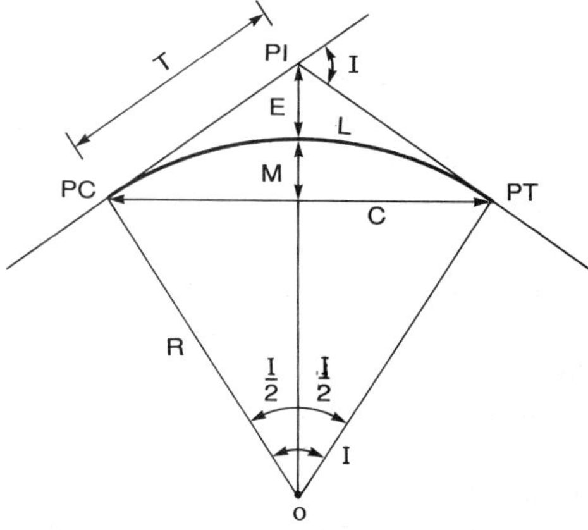

FIGURE 60.2 Geometry of simple horizontal curve.

Simple Horizontal Curves

A simple **horizontal curve** with radius R and deflection angle I is shown in Fig. 60.2. The basic elements required for laying out a horizontal curve are tangent distance T, external distance E, middle ordinate M, length of chord C, and curve length L. These elements can be easily computed in terms of R and I. The simple circular curve may be described by either the radius or the degree of curve. The degree of curve is defined as the central angle subtended by a 100-ft arc and is related to the curve radius by

$$D = \frac{5730}{R} \tag{60.3}$$

where

D = degree of curve (degrees)
R = radius of curve (ft)

Simple Vertical Curves

Vertical curves are normally parabolic. A simple vertical curve may be **crest vertical curve** or **sag vertical curve**, as illustrated in Fig. 60.3. In this figure, G_1 and G_2 are the grades of the first and second tangents (in percent), A is the absolute value of the algebraic difference in grades (in percent), L is the curve length measured in a horizontal plane, VPI is the vertical point of intersection, VPC is the vertical point of curvature, and VPT is the vertical point of tangent. For a simple vertical curve, VPI lies in the middle of the curve. The vertical curve parameter, K, is defined (with L in feet and A in percent) as

$$K = \frac{L}{A} \tag{60.4}$$

The elevations of the vertical curve, E_x, at horizontal distances, x, from VPC are required for laying out the vertical curve. If the elevation of VPC is E_{VPC}, then

$$E_x = E_{\text{VPC}} + \frac{G_1 x}{100} - \frac{x^2}{200K} \tag{60.5}$$

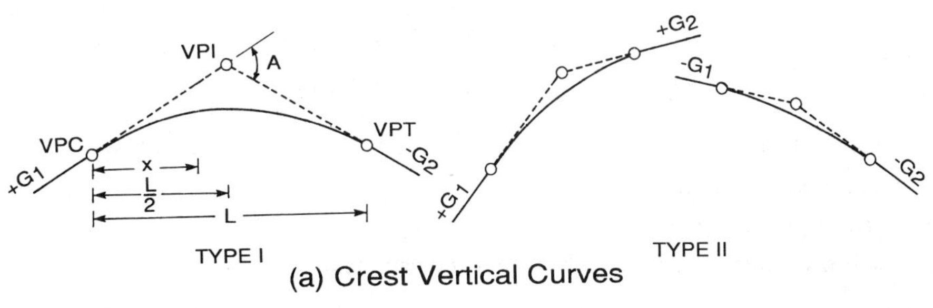

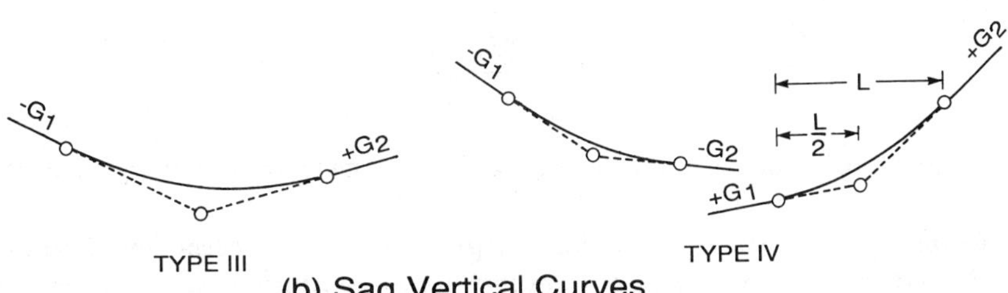

FIGURE 60.3 Types of vertical curves. (From *A Policy on Geometric Design of Highways and Streets.* Copyright 1990 by the American Association of State Highway and Transportation Officials, Washington, D.C. Used by permission.)

where E_x, x, and E_{VPC} are in feet. The location of the highest (lowest) point of the curve, x_{high}, is important for pavement drainage requirements. The lowest and highest points exist only for vertical curves of type I and type III, respectively (Fig. 60.3). Equating the first derivative of E_x with respect to x to zero gives

$$x_{high} = KG_1 \tag{60.6}$$

Chapter 51 presents more details on the layout of horizontal and vertical curves. The radius R of the horizontal curve and the parameter K of the vertical curve are determined based on highway design speed and other parameters, as discussed next.

60.3 Basic Design Applications

The basic elements of geometric design are horizontal alignment, vertical alignment, cross section, and intersection. The design of these elements mainly involves application of the fundamentals discussed in the previous section.

Horizontal Alignment

The horizontal alignment consists of straight roadway sections (tangents) connected by horizontal curves, which are normally circular curves with or without transition (spiral) curves. The basic design features of horizontal alignment include minimum radius, transition curves, **superelevation**, and sight distance. To understand how the minimum radius is determined, the radius-speed relationship is described first.

Radius-Speed Relationship

When a vehicle travels along a horizontal curve, it is forced radially outward by a centrifugal force. The centrifugal force is counterbalanced by the vehicle weight component related to the roadway superelevation and the friction force between the tire and pavement. From the law of mechanics,

$$R = \frac{V^2}{15(e + f)} \tag{60.7}$$

where

R = radius of curve (ft)
V = vehicle speed (mph)
e = rate of superelevation (ft/ft)
f = side friction factor between vehicle tires and pavement.

Substituting for R from Eq. (60.3) in Eq. (60.7) gives

$$D = \frac{85{,}950(e + f)}{V^2} \tag{60.8}$$

The minimum radius or maximum degree of curve is found based on limiting values of e and f.

Maximum Superelevation. The maximum superelevation, e_{max}, depends on climatic conditions, terrain, location (urban or rural), and frequency of slow-moving vehicles. For open highways, the maximum superelevation is 0.10 or 0.12 in areas without snow and ice; otherwise the maximum superelevation should be 0.08. A rate of 0.12 may also be used for low-volume gravel roads to facilitate cross drainage. A maximum rate of 0.04 or 0.06 is common in urban areas.

Table 60.5 AASHTO Design Side Friction Factors for Open Highways

Rural Highways and High-Speed Urban Streets		Low-Speed Urban Streets	
Design Speed (mph)	Side Friction Factor f	Design Speed (mph)	Side Friction Factor f
20	0.17	20	0.300
30	0.16	25	0.252
40	0.15	30	0.221
50	0.14	35	0.197
60	0.12	40	0.178
70	0.10	—	—

Maximum Side Friction. Maximum side friction factors for design are established by AASHTO based on field studies. Table 60.5 shows the recommended design values of f for open highways. The higher side friction factors used for low-speed urban streets reflect a tolerable degree of discomfort accepted by drivers and provide a margin of safety compared with actual conditions.

Minimum Radius. For open highways, the minimum radius, R_{min}, for a given design speed is calculated from Eq. (60.7) using the maximum superelevation and maximum side friction factor. When larger radii than R_{min} are used for a given design speed, the required superelevation is found based on a practical distribution of the superelevation rate over the range of curvature. For rural highways and high-speed urban streets, the recommended design superelevation rates for $e_{max} = 0.04$ are shown in Table 60.6. Similar tables for other e_{max} values are given by AASHTO. For low-speed urban streets, an accepted procedure is to compute the required superelevation rate with f equal to the maximum value. If the computed value of e is negative, superelevation will not be required (practically, superelevation is set equal to a minimum of 0.015).

For intersection curves and turning roadways, the minimum radius for various design speeds are shown in Table 60.7. The values are based on higher side friction factors and minimum rates of superelevation and are calculated from Eq. (60.7). If conditions allow more than this minimum superelevation, drivers will drive the curve more comfortably because of less friction or will travel at a higher speed. The following two examples illustrate the computations of minimum radius and superelevation rate for high-speed and low-speed urban streets, respectively.

Example 60.2. (a) Find the minimum radius and maximum degree of curve on a high-speed urban street with a 50-mph design speed and a 0.04 maximum superelevation. From Table 60.5, $f = 0.14$. From Eq. (60.7), the minimum radius is

$$R_{min} = \frac{(50)^2}{15(0.04 + 0.14)} = 926 \text{ ft}$$

From Eq. (60.8), the corresponding maximum degree of curve is

$$D_{max} = \frac{85{,}950(0.04 + 0.14)}{(50)^2} = 6.19^\circ$$

D_{max} can also be obtained from Table 60.6 as $6^\circ 00'$ (a rounded value), and the corresponding $R_{min} = 955$ ft.

(b) Find the required superelevation rate for a flatter curve on the above street with $D = 3^\circ$. From Table 60.6, for $V = 50$ mph and $D = 3^\circ$, the required superelevation rate is $e = 0.033$.

Table 60.6 Design Elements Related to Horizontal Curvature ($e_{max} = 0.04$)

		$V = 30$ mph			$V = 40$ mph			$V = 50$ mph			$V = 55$ mph			$V = 60$ mph		
			L (ft)			L (ft)			L (ft)			L (ft)			L (ft)	
D	R (ft)	e	Two Lanes	Four Lanes	e	Two Lanes	Four Lanes	e	Two Lanes	Four Lanes	e	Two Lanes	Four Lanes	e	Two Lanes	Four Lanes
0°15′	22,918	NC	0	0	NC	0	0	NC	0	0	NC	0	0	NC	0	0
0°30′	11,459	NC	0	0	NC	0	0	NC	0	0	NC	0	0	NC	0	0
0°45′	7,639	NC	0	0	NC	0	0	NC	0	0	RC	160	160	NC	175	265
1°00′	5,730	NC	0	0	NC	0	0	RC	150	225	0.021	160	160	0.023	175	265
1°30′	3,820	NC	0	0	RC	125	190	0.024	150	225	0.026	160	160	0.029	175	265
2°00′	2,865	RC	100	150	0.022	125	190	0.027	150	225	0.030	160	160	0.033	175	265
2°30′	2,292	RC	100	150	0.025	125	190	0.030	150	225	0.033	160	160	0.037	175	265
3°00′	1,910	0.020	100	150	0.027	125	190	0.033	150	225	0.036	160	160	0.039	175	265
3°30′	1,637	0.022	100	150	0.028	125	190	0.035	150	225	0.038	160	160	0.040	175	265
4°00′	1,432	0.024	100	150	0.030	125	190	0.037	150	225	0.039	160	160	$D_{max} = 3°45'$		
5°00′	1,146	0.026	100	150	0.033	125	190	0.039	150	225	$D_{max} = 4°45'$					
6°00′	955	0.028	100	150	0.035	125	190	0.040	150	225						
7°00′	819	0.030	100	150	0.037	125	190	$D_{max} = 6°00'$								
8°00′	716	0.031	100	150	0.039	125	190									
9°00′	637	0.033	100	150	0.040	125	190									
10°00′	573	0.034	100	150	0.040	125	190									
11°00′	521	0.035	100	150	$D_{max} = 10°00'$											
12°00′	477	0.036	100	150												
13°00′	441	0.037	100	150												
14°00′	409	0.038	100	150												
16°00′	358	0.039	100	150												
18°00′	318	0.040	100	150												
19°00′	302	0.040	100	150												
		$D_{max} = 19°00'$														

D = degree of curve
R = radius of curve
V = assumed design speed
e = rate of superelevation
L = minimum length of runoff (does not include tangent runout)

NC = normal crown section
RC = remove adverse crown, superelevate at normal crown slope
Note: Lengths rounded in multiples of 25 or 50 ft to permit simpler calculations.
In recognition of safety considerations, use of $e_{max} = 0.04$ should be limited to urban conditions.

From *A Policy on Geometric Design of Highways and Streets*. Copyright 1990 by the American Association of State Highway and Transportation Officials, Washington, D.C. Used by permission.

Table 60.7 Minimum Radii for Intersection Curves and Turning Roadways

Design Speed (mph)	Side Friction Factor f	Minimum Superelevation e	Minimum Radius (ft) Computed	Rounded for Design
10	0.38	0.00	18	25
15	0.32	0.00	47	50
20	0.27	0.02	92	90
25	0.23	0.04	154	150
30	0.20	0.06	231	230
35	0.18	0.08	314	310
40	0.16	0.09	426	430

Note: For design speeds of more than 40 mph, use values for open highway conditions.

From *A Policy on Geometric Design of Highways and Streets.* Copyright 1990 by the American Association of State Highway and Transportation Officials, Washington, D.C. Used by permission.

Example 60.3. (a) Find the minimum radius for a low-speed urban street with a 40-mph design speed and a 0.06 maximum superelevation rate. From Table 60.5, $f = 0.178$ and Eq. (60.7) gives

$$R_{\text{min}} = \frac{(40)^2}{15(0.06 + 0.178)} = 449 \text{ ft} \quad \text{(rounded to 450 ft)}$$

(b) Find the required superelevation rate for a flatter curve on the above street with $R = 530$ ft. With f equal to the maximum value, Eq. (60.7) becomes

$$530 = \frac{(40)^2}{15(e + 0.178)}$$

from which $e = 0.024$.

Transition (Spiral) Curves

A transition curve is a curve whose radius continuously changes. It provides a transition between a tangent and a circular curve (simple spiral) or between two circular curves with different radii (segmental spiral). For simple spirals, the degree of curve varies from zero at the tangent end to the degree of the circular curve at the curve end. For segmental spirals, the degree of curve varies from that of the first circular curve to that of the second circular curve. A transition curve has the following advantages:

1. It provides a natural, smooth path
2. It provides a length for attaining superelevation
3. It aids pavement widening on curves
4. It enhances the appearance of the highway

A practical method for determining the length of spiral is to use the length required for attaining superelevation.

Methods of Attaining Superelevation

The change in cross slope from a section with adverse crown removed to a fully superelevated section, or vice versa, is achieved over a highway length called *superelevation runoff.* The runoff depends on the design speed, superelevation rate, and pavement width. The minimum length of runoff for two-lane and four-lane highways is obtained from Table 60.6. The superelevation is attained by rotating a crowned pavement about the centerline, the inside edge, or the outside edge. Figure 60.4 shows the method of attaining superelevation for a curve to the right when the

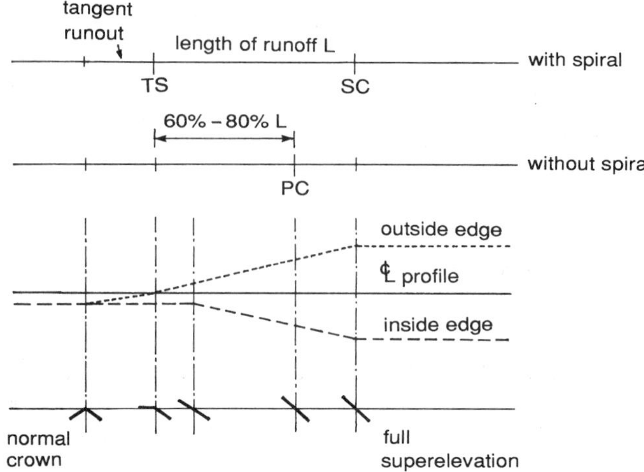

FIGURE 60.4 Attaining superelevation for a curve to the right by rotating pavement about its centerline.

pavement is rotated about its centerline. For spiraled circular curves, the length of spiral equals the superelevation runoff. The runoff starts at the tangent-spiral (TS) point and ends at the spiral-curve (SC) point. For unspiraled circular curves, the superelevation runoff is typically positioned such that 60 to 80% of the runoff is on the tangent and the remainder is on the curve. For safety and appearance, angular breaks should be rounded using vertical curves.

Sight Distance on Horizontal Curves

Sight obstacles such as walls, cut slopes, and buildings on the inside of horizontal curves may restrict the available sight distance. Figure 60.5 shows the geometry of lateral clearance and sight distance on a four-lane highway. The obstacle lies at the middle of the curve with a lateral clearance

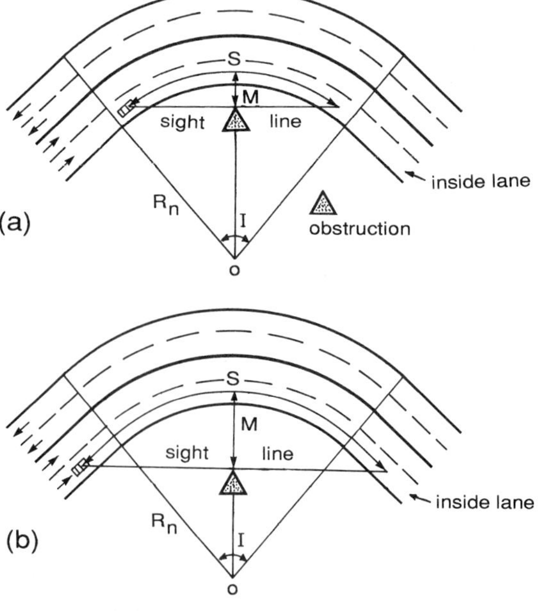

FIGURE 60.5 Lateral clearance on simple horizontal curve with middle obstacle: (a) $S \leq L$ and (b) $S > L$.

measured from the centerline of the inside lane. The required lateral clearance to satisfy a specified sight distance is computed [Olson *et al.*, 1984] by

$$M = R_n\left[1 - \cos\frac{28.65S}{R_n}\right], \quad S \leq L \tag{60.9}$$

$$M = \frac{L(2S - L)}{8R_n}, \quad S > L \tag{60.10}$$

where

M = lateral clearance (ft)
R_n = radius to the inside-lane centerline (ft)
S = sight distance, SSD or PSD (ft)
L = length of horizontal curve along the inside-lane centerline (ft).

When the obstacle lies near the ends of the curve, the lateral clearance needs will be less than the values computed by Eqs. (60.9) and (60.10). The lateral clearance needs for different obstacle locations have been established [Easa, 1991].

Example 60.4. A two-lane highway with 12-ft lanes has a horizontal curve designed for an 800-ft radius, a 50-mph design speed, and $I = 40°$. Find the required lateral clearance for a middle obstacle to satisfy SSD and PSD of AASHTO. Since the curve radius is typically given for the highway centerline, then $R_n = 800 - 6 = 794$ ft. The horizontal curve length is

$$L = \frac{R_n I \pi}{180} = \frac{794 \times 40 \times 3.14}{180} = 554 \text{ ft}$$

For SSD, from Table 60.2, SSD (desirable) = 475 ft. Since $S <$ L, from Eq. (60.9)

$$M = 794\left[1 - \cos\frac{28.65 \times 475}{794}\right] = 36 \text{ ft}$$

For PSD, from Table 60.2, PSD = 1800 ft. Since $S > L$, from Eq. (60.10)

$$M = \frac{554(2 \times 1800 - 554)}{8 \times 794} = 266 \text{ ft}$$

The required lateral clearance for PSD is clearly impractical because it would exceed the normal **right-of-way** line. Practical design for PSD occurs only for very flat curves.

Vertical Alignment

The vertical alignment consists of straight roadway sections (grades or tangents) connected by vertical curves. The grade line is laid out in the preliminary location study to reduce the amount of earthwork and to satisfy other constraints such as minimum and maximum grades. The basic design features of vertical alignment include grades, critical length of grade, climbing lanes, emergency escape ramps, and vertical curve length.

Grades

Maximum grades for different types of roads and design speeds have been established by AASHTO. Maximum grades of about 5% are considered appropriate for a 70-mph design speed. For a 30-mph design speed, maximum grades generally range from 7 to 12%, depending on topography. For intermediate design speeds, maximum grades lie between the above extremes. For low-volume rural highways, grades may be 2% steeper. Minimum grades are necessary to facilitate surface drainage. For uncurbed roads, the grade may be 0%, provided ditch grades are adequate. For curbed roads, the minimum grade is 0.3% but a 0.5% grade should be used if possible. For very flat terrains, a grade as low as 0.2% may be necessary.

Critical Length of Grade

The critical length of grade is the maximum length of a designated upgrade on which a loaded truck can operate without an unreasonable reduction in speed. A 10-mph reduction is used as a general guide, since accident rate increases significantly when the truck speed reduction is greater than this value. Figure 60.6 shows the critical length of grade for design for a typical design truck with a weight to horsepower ratio of 300 lb/hp and a 55-mph entering speed. For grades longer than the critical length, design adjustments to reduce grades or addition of a climbing lane should be considered.

Climbing Lanes

A climbing lane is an extra lane on the upgrade side of a two-lane highway for use by heavy vehicles whose speeds are significantly reduced on upgrades. Climbing lanes improve traffic operation and safety and are justified when the following three conditions are satisfied: (1) the grade length exceeds the critical length of grade or the upgrade level of service is E or F, (2) the upgrade traffic volume exceeds 200 vehicles per hour, and (3) the upgrade truck volume exceeds 20 vehicles per hour. A climbing lane normally begins where the speed of the design truck is reduced by 10 mph and ends when the design truck regains a speed equal to that at the start of the climbing lane. Details on the design of climbing lanes, including entrance and exit transition tapers, width, signing, and marking, are presented by AASHTO.

Emergency Escape Ramps

An emergency escape ramp is provided on a long, steep downgrade for use by heavy vehicles losing control because of brake failure (caused by heating or mechanical failure). The ramp allows these

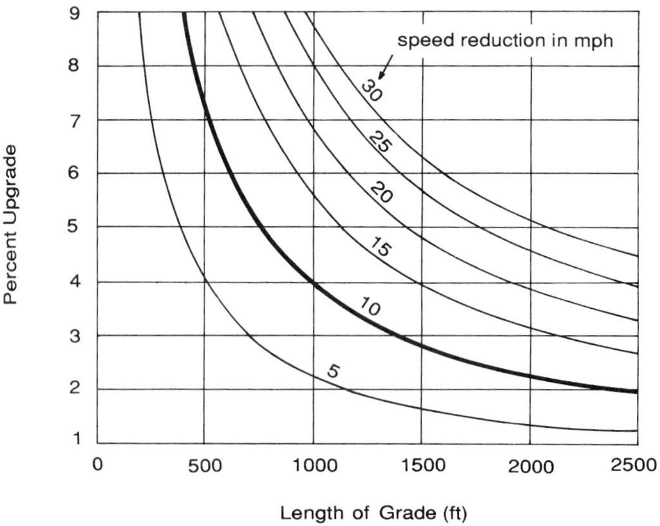

FIGURE 60.6 Critical lengths of grade for design, assumed typical heavy truck of 300 lb/hp, entering speed = 55 mph. (From *A Policy on Geometric Design of Highways and Streets.* Copyright 1990 by the American Association of State Highway and Transportation Officials, Washington, D.C. Used by permission.)

vehicles to decelerate and stop away from the main traffic stream. There are four basic types of emergency escape ramps: sandpile, descending grade, horizontal grade, and ascending grade. The rolling resistance on the ramps is supplied by the loose sand or an arresting bed of loose gravel. The ascending grade ramp provides a force of gravity opposite the vehicle movement and therefore its length can be shorter than the descending and horizontal grade ramps. Each ramp type is applicable to a particular topographic situation. The characteristics of emergency escape ramps used throughout the U.S. are described by Ballard [1983].

Vertical Curve Length

The length of a vertical parabolic curve, based on Eq. (60.4), is computed by

$$L = AK \tag{60.11}$$

where

L = length of vertical curve (ft)
A = algebraic difference in grades (%)
K = constant

For crest vertical curves, the constant K depends on the sight distance used for design, driver's-eye height H_e, and object height H_o. For sag vertical curves, the design is generally based on a headlight criterion, and the constant K depends on stopping sight distance, headlight height H (2 ft), and the upward divergence of the light beam from the longitudinal axis of the vehicle α (1°). The design K values for crest and sag vertical curves are shown in Table 60.8. These values are computed using the formulas shown in the table, where S equals the respective sight distance for crest curves and the SSD for sag curves. The heights H_e and H_o are given in Table 60.4. When the K value needed for design is greater than 167, pavement drainage near the highest (lowest) point, given by Eq. (60.6), must be more carefully designed. For small A, the length computed by Eq. (60.11) may be unrealistically small, and it is common practice to express the minimum curve length (in feet) as three times the design speed (in miles per hour).

Table 60.8 Design Rates of Vertical Curvature K for Crest and Sag Vertical Curves[a]

Design Speed (mph)	Crest Curves: Stopping[b] Sight Distance: Minimum	Crest Curves: Stopping[b] Sight Distance: Desirable	Crest Curves: Decision Sight Distance: A	B	C	D	E	Crest Curves: Passing Sight Distance[c]	Sag Curves: Stopping Sight Distance[b]: Minimum	Sag Curves: Stopping Sight Distance[b]: Desirable
20	10	10	—	—	—	—	—	210	20	20
30	30	30	40	190	160	190	300	400	40	40
40	60	80	90	400	270	400	520	730	60	70
50	110	160	190	720	430	610	790	1050	90	110
60	190	310	350	1270	760	1000	1230	1430	120	160
70	290	540	610	1750	910	1270	1580	2030	150	220

For crest curves: $K = \dfrac{S^2}{200\left(\sqrt{H_e} + \sqrt{H_o}\right)}$ For sag curves: $K = \dfrac{S^2}{200\,(H + S\tan\alpha)}$

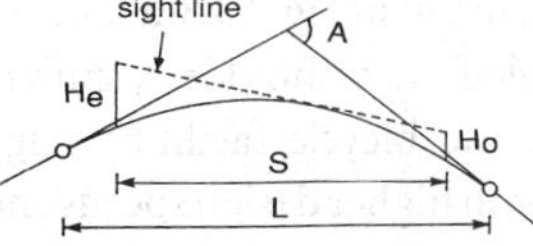

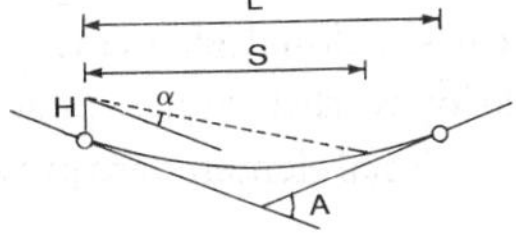

[a] K values for stopping and passing sight distances are adapted from AASHTO.
[b] Using computed values of SSD.
[c] Using rounded values of PSD.

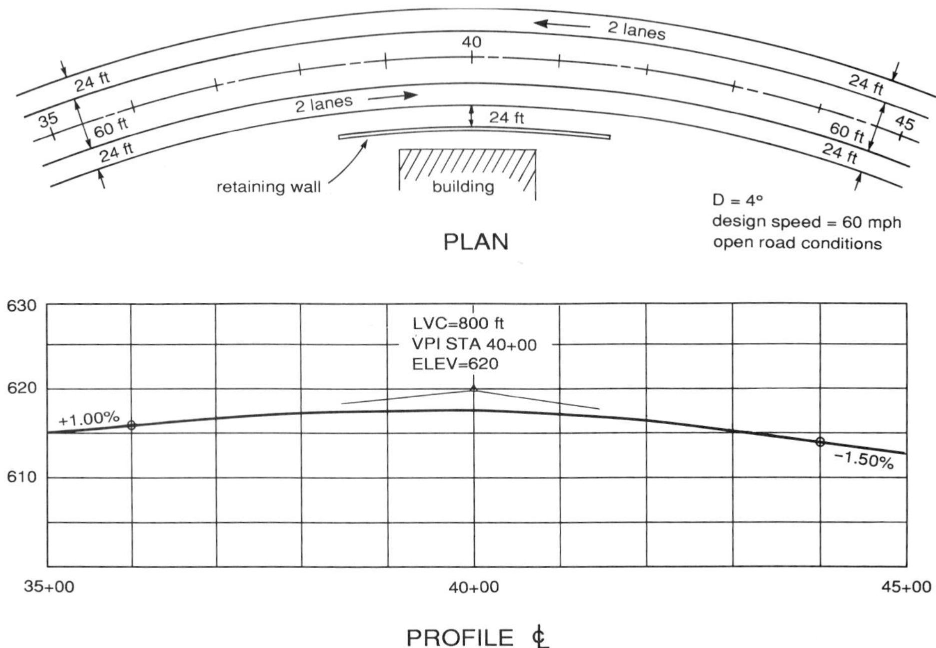

FIGURE 60.7 Example of sight distance analysis for combined horizontal and vertical curves. (*Source:* Jack E. Leisch & Associates. *Notes on Fundamentals of Highway Planning and Geometric Design.* Vol. 1, Evanston, IL.)

Example 60.5. A section of a four-lane highway with partial access control and a 60-mph design speed lies on combined horizontal curve ($D = 4°$) and crest vertical curve ($L = 800$ ft), as shown in Fig. 60.7. The length of the horizontal curve is greater than 800 ft. A retaining wall (5 ft high above the pavement) is required for a planned development near the highway. Determine the adequacy of the design for SSD. To check sight distance on the vertical curve, from Table 60.8, $K = 190$ (minimum) and 310 (desirable). For $A = 2.5\%$, the required lengths of the vertical curve, based on Eq. (60.11), are

$$L\,(\text{minimum}) = 2.5 \times 190 = 475\,\text{ft}$$
$$L\,(\text{desirable}) = 2.5 \times 310 = 775\,\text{ft}$$

Since the vertical curve is 800 ft long, sight distance on the vertical curve is adequate. To check sight distance on the horizontal curve, from Table 60.2, the required range of SSD is 525 to 650 ft. For $D = 4°, R = 1432.5$ ft from Eq. (60.3) and $R_n = 1432.5 - 30 - 18 = 1384.5$ ft. From Eq. (60.9) for $S \leq L$, the corresponding offsets are 25 ft and 38 ft, respectively. Since the distance from the retaining wall to the inside-lane centerline, given in the design, is $24 + 6 = 30$ ft, the minimum sight distance criterion is also met.

Cross Section Elements

Typical cross sections for rural highways and urban streets are shown in Fig. 60.8(a) and Fig. 60.8(b), respectively. The cross-sectional elements include the traveled way, shoulders, curbs, medians, sideslopes and backslopes, clear zones, pedestrian facilities, and bicycle facilities. Higher design standards for cross-sectional elements are provided for roads with higher design speeds and volumes.

Traveled Way

The main features of the traveled way are lane width and cross slope. Through-lane width ranges from 10 to 12 ft on most highways, with 12 ft being most common; auxiliary-lane width ranges

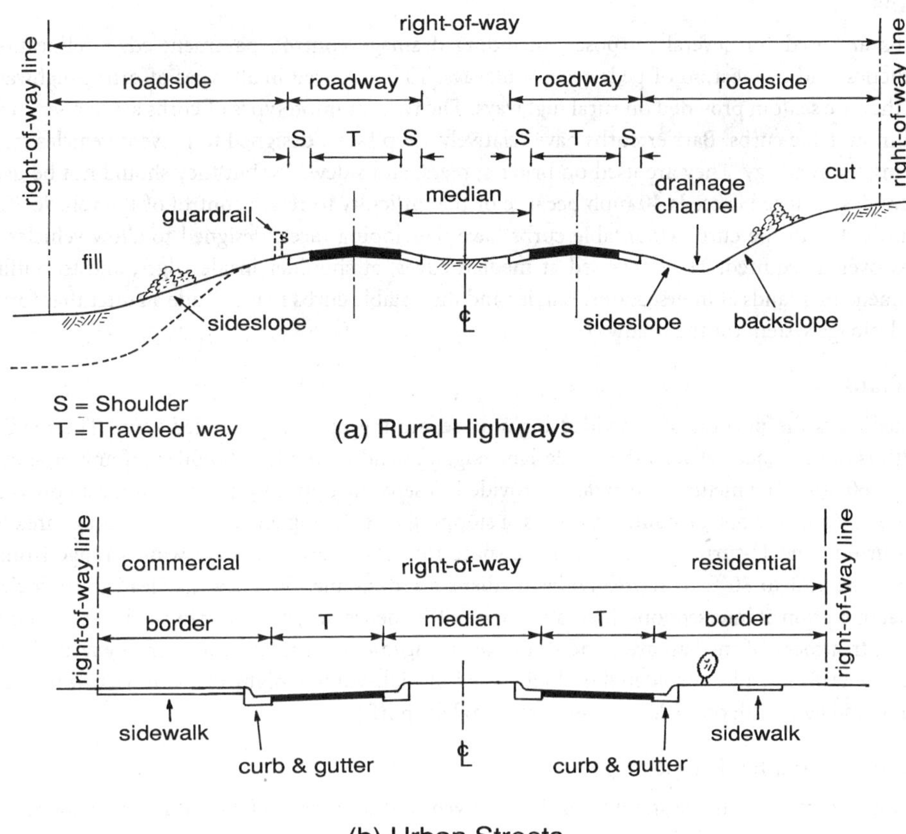

FIGURE 60.8 AASHTO typical cross sections for rural highways and urban streets.

from 8 to 12 ft. Auxiliary lanes are used for on-street parking in urban areas; for turning traffic at intersections; for passing slow vehicles on two-lane highways; and for entering, exiting, and weaving traffic on freeways. Safety increases with wider lanes up to a width of 12 ft but greater lane widths do not offer increased safety. Pavement widening on curves, to provide for vehicle off-tracking and additional lateral clearance, is an important design feature. To facilitate cross drainage, all highways are designed with a normal cross slope ranging from 1.5 to 2% for high-type pavement surface, 1.5 to 3% for intermediate-type surface, and 2 to 6% for low-type surface.

Shoulders

Shoulders vary in width from 2 ft to 12 ft and may be paved or unpaved. Some important advantages of shoulders are providing a space for emergency stopping, maintenance operations, and signs; improving sight distance and highway capacity; and providing a structural support for the pavement. Shoulder cross slopes depend on the type of shoulder construction. Bituminous shoulders should be sloped from 2 to 6%, gravel shoulders from 4 to 6%, and turf shoulders about 8%. The designer should pay attention to the difference between the pavement and shoulder cross slopes on horizontal curves. On the outside of a superelevated section, the difference should not exceed 8%. On the inside, the shoulder and pavement slopes are generally the same. Shoulder contrast is desirable and, when shoulders are used by bicycles, pavement edge lines should be designed as described by the *Manual on Uniform Traffic Control Devices* (MUTCD) [FHWA, 1988].

Curbs

Curbs are used for several purposes, including drainage control, pavement edge delineation, aesthetics, and delineation of pedestrian walkways. They are used in all types of urban highways, but they are seldom provided on rural highways. The two common types of curbs are barrier curbs and mountable curbs. Barrier curbs have relatively steep faces, designed to prevent vehicles from leaving the roadway. They are used on bridges, piers, and sidewalks, but they should not be used where design speed exceeds 40 mph because of the difficulty to retain control of the vehicle after an impact with the curb. Mountable curbs have flat sloping faces, designed to allow vehicles to cross over if required. They are used at median edges, at shoulder inside edges, and to outline channelizing islands at intersections. Barrier and mountable curbs may include a gutter that forms the drainage system for the roadway.

Medians

A median is the portion of a divided highway separating opposing traveled ways. The median width is the distance between the inside-lane edges, including the left shoulders, if any, as shown in Fig. 60.8(a). The median is primarily provided to separate opposing traffic, but it also provides a recovery area for out-of-control vehicles, a stopping area during emergencies, a storage area for left-turning and U-turning vehicles, and a space for future lanes. Median widths range from a minimum of 4 to 80 ft or more. Wider medians are desirable, but they may lead to inefficient signal operation at intersections. Medians can be either depressed, raised, or flushed, depending on width, treatment of median area, and drainage arrangements. A median barrier (a guardrail or a concrete wall) must be considered on high-speed or high-volume highways with narrow medians and on medians with obstacles or a sudden lateral dropoff.

Sideslopes and Backslopes

On fills, sideslopes provide stability for the roadway and serve as a safety feature by being part of a clear zone. In cuts, sideslopes and backslopes form the drainage channels. Sideslopes of 4:1 or flatter are desirable and can be used where heights of fill or cut are moderate. Sideslopes steeper than 3:1 on high fills generally require roadside barriers. Backslopes should be 3:1 or flatter to facilitate maintenance. Steeper backslopes should be evaluated for soil stability and traffic safety. In rock cuts, backslopes of 1:4 or vertical faces are commonly used. Rounding where slope planes intersect is an important element of safety and appearance.

Clear Zones

The clear zone is the unobstructed, relatively flat area outside the edge of the traveled way, including shoulder and sideslope, for the recovery of errant vehicles. The clear zone width depends on traffic volume, speed, and fill slope. Where a hazard potential exists on the roadside, such as high fill slopes, roadside barriers should be provided. The *Roadside Design Guide* [AASHTO, 1989] provides guidance for design of clear zones, roadside barriers, and sideslopes and backslopes.

Pedestrian Facilities

Pedestrian facilities include sidewalks, crosswalks, curb ramps for the handicapped, and grade separations. Sidewalks are usually provided in urban areas and in rural areas with high pedestrian concentrations such as schools, local businesses, and industrial plants. The width of sidewalks in residential areas varies from 4 to 8 ft. In commercial areas, the border (the area between the roadway and the right-of-way line) usually is devoted entirely to sidewalk, as shown in Fig. 60.8(b).

Pedestrian crosswalks, 6 to 8 ft wide, are provided at intersections and at midblocks. For guidance on pedestrian crosswalk marking refer to MUTCD. Curb ramps for the handicapped should be provided at all intersections that have curbs and sidewalks, and at midblock pedestrian crossings. Because these crossings are generally unexpected by drivers, warning signs and adequate visibility (by prohibiting parking) should be provided. Pedestrian grade separations are necessary when pedestrian and traffic volumes are high or where there is abnormal inconvenience to pedestrians, such as at freeways. The width of walkways for pedestrian overpasses is 8 ft minimum. Design issues on safe accommodation of pedestrians are addressed by Pietrucha and Opiela [1993].

Bicycle Facilities

Design of bicycle facilities is an important consideration in highway design. Design measures to enhance safety for bicycle traffic on existing highways include paved shoulders, wider outside traffic lanes (14 ft minimum), adjusting manhole covers to pavement surface, and providing a smooth riding surface. The highway system can also be supplemented by providing specifically designated bikeways. Important elements of bikeway design include design speed, bikeway width, superelevation, turning radii, grade, stopping sight distance, and vertical curves. The *Guide for Development of New Bicycle Facilities* [AASHTO, 1991] provides guidance for bikeway design.

Intersections

There are three general types of intersections: intersections at grade, grade separations without ramps, and interchanges. Selection of a specific intersection type depends on several factors, including highway classification, traffic volume, safety, topography, and highway user benefits. Selection guidelines based on highway classification are given in Table 60.9.

Intersections at Grade

The objective of intersection design is to reduce the severity of potential conflicts between vehicles, bicycles, and pedestrians. Intersection design is generally affected by traffic factors, physical

Table 60.9 Guide for Selection of Interchanges, Grade Separations, and Intersections Based on Classification

	Freeway	Arterial	Collector/Local
Rural			
Freeway	1	2	4
Arterial		6	7
Collector/local			8
Urban			
Freeway	1	3	5
Arterial		6	6 or 7
Collector/local			8

1. Interchange in all cases.
2. Normally interchange, but grade separation where traffic volume is light.
3. Normally interchange, but grade separation where interchange spacing is too close.
4. Normally grade separation or alternatively the collector/local may be closed.
5. Normally grade separation, but an interchange might be justified to
   - Relieve congestion
   - Serve high-density traffic generators
6. Normally intersection, but an interchange may be justified where
   - Capacity limitation causes serious delay
   - Injury and fatality rates are high
   - Cost is lower than an intersection
7. Normally intersection, or alternatively the collector/local may be closed.
8. Normally intersection, or alternatively one road may be closed.

factors, human factors, and economic factors. Examples of these factors are turning-movement design volumes, sight distance, perception-reaction time, and cost of improvements. The basic types of intersections and their variations are described by Neuman [1985]. The key features of intersection design include capacity analysis, alignment and profile, turning curve radius and width, channelization, median opening, traffic control devices, and sight distance. Details on these design features along with examples of good designs are given by AASHTO.

Intersection Sight Distance

Adequate sight distances must be provided at intersections at grade. The most common case is an intersection controlled by a stop sign on the minor roadway. The driver of a stopped vehicle on the minor roadway must have sufficient sight distance for safe departure (crossing, turning left, or turning right), even though an approaching vehicle on the major highway comes into view as the stopped vehicle begins to depart. The area of an obstructed sight distance is called the sight triangle. For the crossing maneuver shown in Fig. 60.9(a), the required sight distance along the

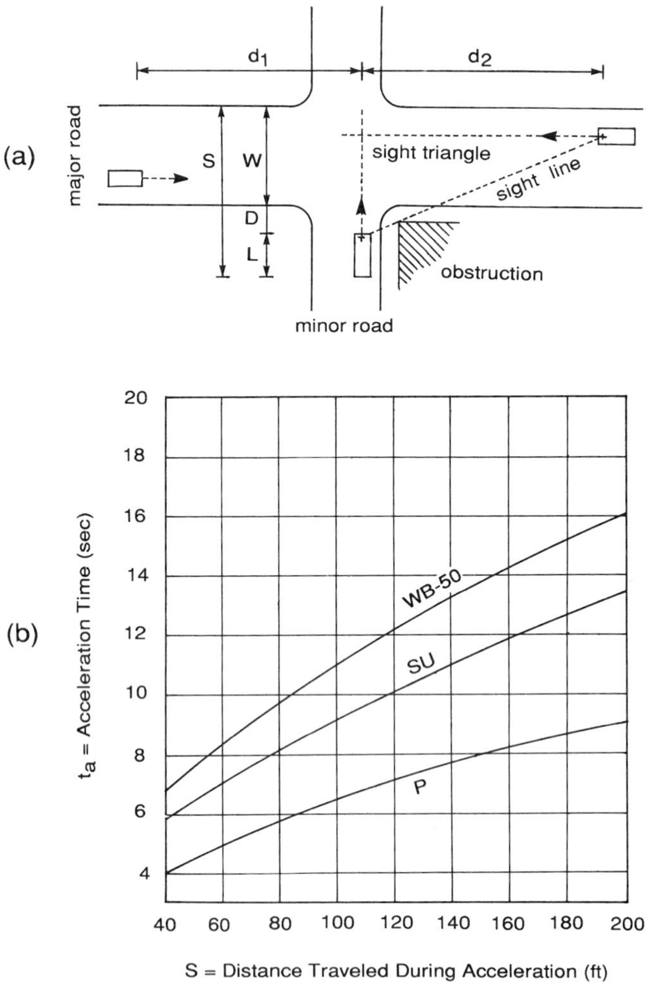

FIGURE 60.9 Crossing sight distance for intersection with stop control on the minor roadway: (a) sight triangle and (b) acceleration time t_a. (From *A Policy on Geometric Design of Highways and Streets.* Copyright 1990 by the American Association of State Highway and Transportation Officials, Washington, D.C. Used by permission.)

major highway is computed by

$$d = 1.47V(J + t_a) \tag{60.12}$$

where

d = d_1 or d_2, sight distance along the major highway (ft)
V = design speed on the major highway (mph)
J = sum of the perception time and the time required to actuate the clutch or an automatic shift (assumed 2 s)
t_a = time required to accelerate and traverse the distance S to clear the major highway pavement (s)

The distance S equals the sum of the distance from the near edge of pavement to the front of a stopped vehicle D (assumed 10 ft); the pavement width along the path of crossing vehicle W (ft); and the overall length of vehicle L (ft). The time t_a depends on the vehicle acceleration rate, and can be obtained from Fig. 60.9(b) for passenger cars (P), single unit trucks (SU), and large semitrailer combination trucks (WB-50) for level conditions.

The sight line of the sight triangle is measured from a 3.5-ft eye height to a 4.25-ft object height. The sight triangle is used to determine required building setbacks, to find whether an existing obstruction should be moved, and to determine appropriate traffic control measures if the obstruction cannot be moved. Sight distance needs for left and right turning maneuvers from a stopped position and for other types of intersection control are found in AASHTO. The AASHTO sight distance models for at-grade intersections and railroad crossings have been extended [Easa, 1993b; Mason *et al.*, 1989].

Example 60.6. A four-lane divided highway with a 24-ft median and a 50-mph design speed intersects with a minor road controlled by a stop sign. The sight distance along the major highway, restricted by an existing building that cannot be moved, is 780 ft. Check the adequacy of sight distance at this intersection for SU design vehicles ($L = 30$ ft). Assume 12-ft lanes and level conditions. First, compute $W = 4 \times 12 + 24 = 72$ ft and $S = 10 + 72 + 30 = 112$ ft. From Fig. 60.9(b), $t_a = 9.7$ s and Eq. (60.12) gives

$$d = 1.47 \times 50 \times (2 + 9.7) = 860 \text{ ft}$$

which is greater than 780 ft. Therefore, the available sight distance is not adequate. Since the obstruction cannot be moved, the speed on the major highway may be reduced. Substituting for $d = 780$ in Eq. (60.12) gives the required speed on the major highway as $V = 45.4$ mph, rounded to 45 mph.

Interchanges

Interchanges provide the greatest traffic safety and capacity. The basic types of interchanges are shown in Fig. 60.10. The trumpet pattern provides a loop ramp for accommodating the lesser left-turn volume. The three-leg directional pattern is justified when all turning movements are large. The one-quadrant interchange is provided because of topography, even though the volumes are low and do not justify the structure. Simple diamond interchanges are most common for major-minor highway intersections with limited right-of-way. A partial-cloverleaf interchange is normally dictated by site conditions and low turning volumes, while a full-cloverleaf interchange is adaptable to rural areas where the right-of-way is not prohibitive. An all-direction four-leg interchange is most common in urban areas where turning volumes are high.

In urban areas, where the interchanges are closely spaced, all interchanges should be integrated into a system design that includes the following aspects: (1) interchange spacing, (2) route continuity, (3) uniformity of exit and entrance patterns, (4) signing and marking, and (5) coordination of lane

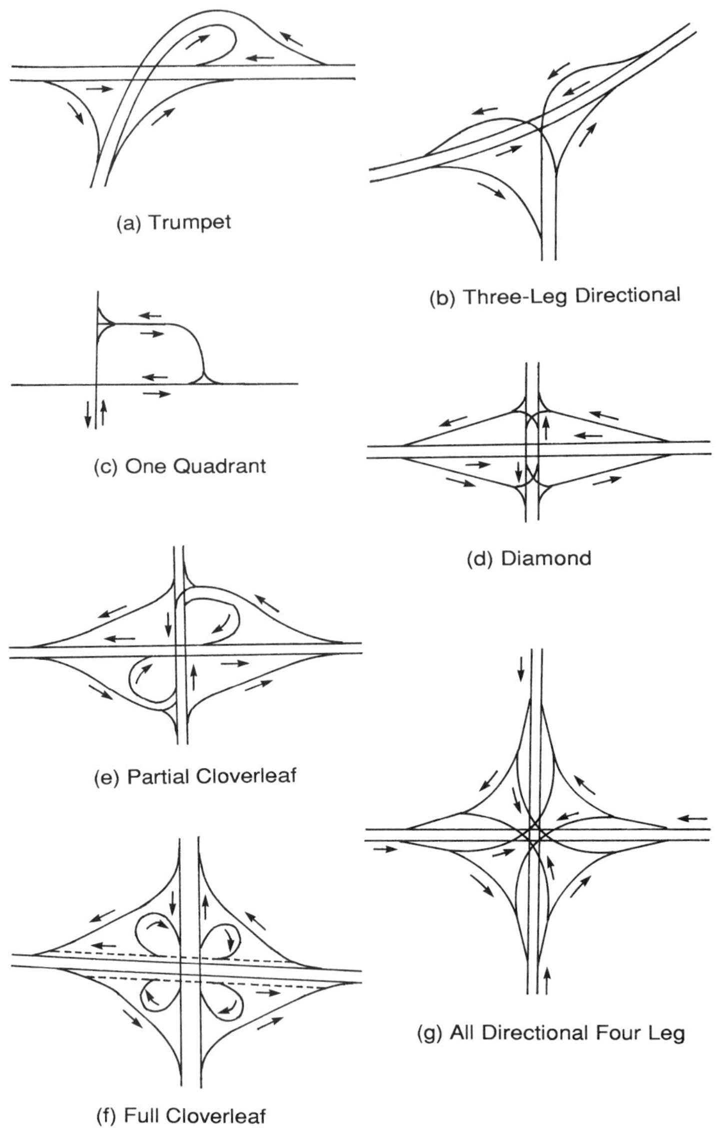

FIGURE 60.10 Basic types of interchanges. (From *A Policy on Geometric Design of Highways and Streets.* Copyright 1990 by the American Association of State Highway and Transportation Officials, Washington, D.C. Used by permission.)

balance and basic number of lanes. Details on these design aspects are presented by AASHTO and by Leisch [1993].

High-occupancy vehicle (HOV) roadways, designated for buses and car pools (typically with three or more persons per vehicle), have been incorporated into urban freeway corridors to improve traffic operations. There are three basic types of HOV lanes. The concurrent-flow type provides an HOV lane normally on the left lane of the roadway in both directions. The HOV lane is separated from the regular-use lanes with a buffer. The contra-flow type designates an HOV lane in the opposing direction of travel. The third type involves physically separating the HOV lane from the regular-use lanes. Details on the design of HOV facilities are given in the *Guide for the Design of High-Occupancy Vehicle and Public Transfer Facilities* [AASHTO, 1992a].

Aesthetic and General Considerations

Exact adherence to the preceding (specific) design standards does not guarantee obtaining a satisfactory and aesthetically pleasing design. A number of general guidelines should also be followed in practice for individual horizontal and vertical alignments, combined alignments, and cross sections and intersections.

Individual Alignments

For horizontal alignment, the alignment should be as directional as possible but should conform to the natural contours. The designer should avoid the use of minimum radius, small deflection angles, sharp curves at the end of long tangents, abrupt reversal in alignment, and **broken-back curves**. If small deflection angles cannot be avoided, curves should be sufficiently long to avoid the appearance of a kink. For compound curves, the ratio of the flatter radius to the shaper radius should not exceed 1.5. Transition curves should be used whenever possible.

For vertical alignment, a smooth-flowing profile is preferred to a profile with numerous breaks and short grades. The designer should avoid the hidden-dip type of profiles, broken-back curves, sag vertical curves in cuts, and crest vertical curves on fills. On long grades, the steepest grade should be placed at the bottom and the grades near the top of the ascent are lightened. Steep grades through important intersections should also be reduced to minimize potential hazards to turning vehicles.

Combined Alignments

Coordination of horizontal and vertical alignments can be achieved during both preliminary location study and final design. A change in horizontal alignment should be made at a sag vertical curve where the driver can readily see the change in direction. However, to avoid distorted appearance, sharp horizontal curves should not be introduced near the low point of the sag vertical curve. If combined horizontal and crest vertical curves cannot be avoided, the horizontal curve should lead the vertical curve. Providing adequate passing opportunities on two-lane highways should supersede other desirable combinations of horizontal and vertical alignments. Alignment coordination is aided by using computer perspectives. Figures 60.11 and 60.12 each show an example of poor and good practice.

Cross Sections and Intersections

Aesthetic features of cross section design include well-rounded sideslopes and backslopes, tapered piers and planting below elevated freeways, and spandrel arch bridges where excess vertical clearance is available. On depressed freeways, planting on median barriers, textured retaining walls with luminaire located on the top, and landscape planting on earth slopes aid roadway appearance. At interchanges, flat earth slopes and long transition grading between cut and fill slopes raise the aesthetic level of the area. Landscape development on roadsides and at interchanges is an important design feature for aesthetics and safety, and information on the subject is found in *Transportation Landscape and Environmental Design* [AASHTO, 1992b].

Design Aids

A number of design aids are used in geometric design, including physical aids and computer programs. These design aids provide greater flexibility and efficiency during various stages of highway design.

Physical Aids

Physical aids include templates and physical models. The following types of templates are used in geometric design: horizontal curve templates, vertical curve templates, turning vehicle templates, and sight distance templates. Horizontal curve templates may be circular curves, three-centered

(a)

(b)

FIGURE 60.11 Example of highways with (a) a long tangent-short curve design that slashes through the terrain and (b) the more desirable flowing alignment that conforms to the natural contours. (From *Manual of Geometric Design Standards for Canadian Roads.* Copyright 1986 by the Roads and Transportation Association of Canada, Ottawa, Ontario. Used by permission.)

(a)

(b)

FIGURE 60.12 Example of highway perspectives with (a) a short vertical curve imposed on a relatively long horizontal curve causing the appearance of a settlement of the roadway and (b) the more desirable alignment where the length of vertical curve is increased to nearly that of the horizontal curve. (From *Manual of Geometric Design Standards for Canadian Roads.* Copyright 1986 by the Roads and Transportation Association of Canada, Ottawa, Ontario. Used by permission.)

curves, or spiral curves, and vertical curve templates are parabolic. Both types of templates are used in preliminary location study and final design. Turning vehicle templates for design vehicles are used to establish the minimum vehicle path and minimum pavement width for different turning conditions [ITE, 1993].

Sight distance templates are used to measure the available sight distances graphically on highway plans and profiles, especially in preliminary location studies. Each template consists of four parallel horizontal lines. The top line represents the line of sight and the next three lines represent the profile elevations for object height, driver's-eye height, and opposing-vehicle height, respectively. Physical models are skeletonized, scaled structures easily adjusted to required changes in alignment (design models) or permanent structures used to illustrate the functional plan of a highway or interchange (presentation models). Examples of physical models are presented by AASHTO.

Computer Programs

Computer programs are available to carry out all (or specific) aspects of highway geometric design. Available microcomputer programs are regularly published by the Center for Microcomputers in Transportation [McTrans, 1993]. For example, Interactive Computer Assisted Highway Design (ICAHD) is a comprehensive rural highway design program for IBM microcomputers that is used for all geometric design aspects. A sister program, ICAHD Urban Design Model, allows for the design of urban streets and highways. AutoTURN is used for simulating vehicle turning path in a computer-aided design environment and SPUI is used for analyzing the geometric design of single-point urban interchanges.

Available mainframe computer programs are regularly published in the *Computer System Index* [AASHTO, 1987]. For example, Roadway Design System (RDS) is a popular program that is used in the preliminary location study and the final design. Besides alignment and cross-sectional designs, the program performs earthwork calculations, bridge design, and view perspectives. RDS is used by the Province of Manitoba and several states, including Idaho, Indiana, Iowa, and Michigan.

60.4 Special Design Applications

Special design applications include design of complex highway curves, sight distance needs for trucks, and design considerations for RRR projects. Design considerations for these special applications have been recently addressed in literature and, to some extent, they supplement the design aspects covered by AASHTO.

Design of Complex Highway Curves

Complex highway curves consist of two consecutive circular or parabolic curves. Sight distance characteristics of these curves are different from those of simple curves, and consequently, design requirements are different. Four types of complex curves are discussed: unsymmetrical vertical curves, **compound curves**, **reverse curves**, and highway sight-hidden dips.

Unsymmetrical Vertical Curves

Unsymmetrical vertical curves may be required on certain occasions because of vertical clearance or other controls. An unsymmetrical vertical curve consists of two consecutive (unequal) parabolic curves with a common tangent at VPI. The curve is described by the parameter Q, which is the ratio of the shorter curve to the total curve length. Design length requirements for unsymmetrical crest and sag vertical curves, based on AASHTO sight distance needs, have been developed. Table 60.10 shows the length requirements for crest and sag vertical curves for $Q = 0.4$. The length requirements for unsymmetrical vertical curves are much greater than those for simple vertical curves.

Table 60.10 Design Length Requirements for Unsymmetrical Crest and Sag Vertical Curves Based on SSD of AASHTO ($Q = 0.4$)

Algeb. Diff. in Grades	Crest Curves (ft)[a]						Sag Curves (ft)[b]					
	Design Speed (mph)						Design Speed (mph)					
(%)	20	30	40	50	60	70	20	30	40	50	60	70
2	60	90	120	160	460	720	60	90	120	150	180	210
4	60	90	260	680	1250	1770	60	160	340	560	820	1130
6	60	220	490	1090	1870	2650	140	300	600	970	1410	1920
8	100	340	690	1450	2490	3530	220	440	830	1310	1880	2580
10	150	450	860	1810	3120	4410	280	550	1010	1650	2380	3220

[a] Driver's-eye height = 3.5 ft, object height = 0.5 ft. (*Source:* Easa, S. M. 1991. Sight distance model for unsymmetrical crest curves, p. 46. *Transp. Res. Rec.* 1303:39–50.)

[b] Headlight height = 2 ft, $\alpha = 1°$. (*Source:* Easa, S. M. 1991. Sight distance models for unsymmetrical sag curves, p. 55. *Transp. Res. Rec.* 1303:51–62.)

Compound Curves

Compound horizontal curves are advantageous for turning roadways at intersections and interchanges and on open highways, and are more economical in mountainous terrains. A compound curve consists of two consecutive horizontal curves with different radii turning in the same direction. For obstacles within the sharper (flatter) arc, application of simple curve models, which ignore the flatter (sharper) arc, will underestimate (overestimate) lateral clearance needs. The exact lateral clearance needs for compound curves have been established [Easa, 1993a].

Reverse Curves

Reverse horizontal curves are useful in effecting a change of direction when conditions do not permit the use of simple curves. The alignment reversal reduces the needed lateral clearance on the first curve. Reverse vertical curves (a crest curve followed by a sag curve) are advantageous in hilly and mountainous terrains and their use is often necessary at interchange ramps. The alignment reversal of the sag curve improves the sight distance and consequently reduces the length requirements of the crest curve. Application of AASHTO standards to reverse curves will therefore be conservative.

Highway Sight-Hidden Dips

A sight-hidden dip (SHD) occurs when a sag curve follows a crest curve on a relatively straight horizontal alignment and contributes to passing maneuver accidents on two-lane highways. The start and end locations of the SHD on existing highways can be found [Easa, 1994]. These locations should be delineated on the highway using appropriate signing. In design, the SHD can be avoided by using flatter vertical curves or inserting a long tangent between the crest and sag curves. The AASHTO minimum curvatures for SSD of the crest and sag curves produce unsatisfactory SHD when the design speed and the algebraic difference in grades of the sag curve are large.

Sight Distance Needs for Trucks

Stopping and passing sight distance needs for trucks are shown in Table 60.11. For stopping sight distance, a truck with a conventional brake system and a worst-performing driver requires greater SSD and longer crest vertical curves than the AASHTO values [Harwood *et al.*, 1989]. In contrast, a truck with conventional brake system and a best-performing driver requires only slightly more SSD and shorter crest curve length. Trucks with antilock brake systems require less SSD and significantly shorter crest curve length. For passing sight distance, the needs for maneuvers involving trucks are greater than those involving only passenger cars [Harwood and Glennon, 1989]. The PSD criteria for a passenger car passing a passenger car, which are based on a different passing model, are much

Table 60.11 Design Requirements for Stopping and Passing Sight Distances for Trucks

	Stopping Sight Distance (ft)[a]			Passing Sight Distance (ft)[b]	
Design Speed (mph)	Worst-Performing Driver	Best-Performing Driver	Antilock Brake System	Passenger Car Passing Passenger Car	Passenger Car Passing Truck
20	150	125	125	325	350
30	300	250	200	525	575
40	500	375	325	700	800
50	725	525	475	875	1025
60	975	700	600	1025	1250
70	1275	900	775	1200	1450

[a] *Source:* Harwood, D. W., Glauz, W. D., and Mason, J. M., Jr. 1989. Stopping sight distance design for large trucks, p. 44. *Transp. Res. Rec.* 1208:36–46.

[b] *Source:* Harwood, D. W. and Glennon, J. C. 1989. Passing sight distance design for passenger cars and trucks, p. 64. *Transp. Res. Rec.* 1208:59–69.

shorter than the AASHTO design criteria. Important design issues on sight distance for trucks are found in Fitzpatrick *et al.* [1993].

Example 60.7. Check the SSD adequacy of the design in Example 60.5 for conventional trucks with best-performing drivers. Assume a 7.75-ft truck driver's-eye height and a 0.5-ft object height. From Table 60.11, the required stopping sight distance for $V = 60$ mph is 700 ft. Using the formula in Table 60.8 with $H_e = 7.75$ ft, $H_o = 0.5$ ft, and $S = 700$ ft, compute $K = 201$. From Eq. (60.11), the required length of the crest vertical curve is $L = 2.5 \times 201 = 503$ ft. Therefore, the vertical curve length is adequate.

For the horizontal curve, the required lateral clearance using Eq. (60.9) is 45 ft. This is greater than the available 30-ft lateral clearance. Therefore, the sight distance is restricted by the retaining wall and alternative improvements should be examined. This example illustrates an important fact: the greater truck-driver's-eye height offsets the larger SSD on crest vertical curves, but generally provides no comparable advantage for sight distance obstruction on horizontal curves.

Design Considerations for RRR Projects

Resurfacing, restoration, and rehabilitation (RRR) projects can improve highway safety by upgrading selected features of existing highways, such as minor widening of lanes, without the cost of full reconstruction. The AASHTO policy is intended for new highway and major reconstruction projects, but not for RRR projects. To address the need for design standards for these projects, a study by the Transportation Research Board examined the safety cost-effectiveness of geometric design standards and recommended several design practices, rather than minimum standards [TRB, 1987]. The recommendations fall into five categories: safety-conscious design process, design practices for key highway features, other design procedures and assumptions, planning and programming RRR projects, and safety research and training. Subsequent research on safety effectiveness of highway design features has been conducted and reported in six volumes; see, for example, Zegeer and Council [1992].

60.5 Economic Evaluation

The objective of economic evaluation is to rank alternative highway improvements so that a selection can be made. Economic evaluation of major improvements requires information about the highway costs and benefits. The costs consist of capital costs and maintenance and operating costs. Capital costs include engineering design, **right-of-way,** and construction. Maintenance and operating costs include roadside maintenance, snow removal, and lighting, and are incurred annually over the service life of the facility. The benefits are normally based on highway user benefits which are the savings in the costs of vehicle operation, travel time, and accidents. For local highway safety improvements, such as improving sight distance, the benefits normally are based on reduced accident costs.

To carry out an economic evaluation, the change in the value of money over the life of the facility must be considered. All costs and benefits of each alternative are combined into a single number (measure) using economic evaluation methods. These include net present value, benefit-cost ratio, equivalent uniform annual cost, and internal rate of return. It is stressed that the results of economic evaluation must be set alongside other strategic and nonmonetary considerations before the policymaker can reach a final decision. Further details on full project economic evaluation along with example calculations are presented in *A Manual on User Benefit Analysis of Highway and Bus Transit Improvements* [AASHTO, 1977].

A simplified method for estimating user cost savings for highway improvements is also available [TAC, 1993]. Simple look-up tables are presented for the following facility types: (1) two-lane highway, (2) two-lane highway with passing lane, (3) four-lane divided arterial highway, (4) signalized highway intersection, and (5) highway interchange. For each facility type, the tables give estimates of road user costs over a range of traffic levels. The tables are not intended to replace full project economic evaluation, but they provide a low-cost initial screening at the early stages of planning and designing highway improvements.

60.6 Summary: Key Considerations

The fundamentals of highway geometric design and their applications are presented in this chapter. The fundamentals include highway types, design controls, sight distance, and simple highway curves which influence the design of four basic highway elements: horizontal alignment, vertical alignment, cross section, and intersection. Recent information on the design of complex highway curves, sight distance needs for trucks, design considerations for RRR projects, economic evaluation, and design aids is presented.

Geometric design standards promote safety, efficiency, and comfort for the road users. However, strict application of these standards will not guarantee obtaining a good design. The following key ingredients are also required:

1. *Consistency.* Geometric design should avoid abrupt changes in standards. It should also provide positive guidance to the drivers to achieve safety and efficiency and should conform to driver expectations.
2. *Aesthetics.* Visual quality should be achieved in design by careful attention to coordinating horizontal and vertical alignments and to landscape developments. The process can be greatly aided by using computer perspectives and physical models.
3. *Engineering judgment.* Experience and skill of the designer are important in producing a good design. Considerable creativity is required in developing a design that addresses environmental and economic concerns.

Future developments in geometric design are likely to involve expert systems, intelligent vehicle-highway systems (IVHS), and reliability analysis. Expert systems utilize existing experience and knowledge in geometric design; vehicle and highway technologies of IVHS influence important design parameters; and reliability analysis quantifies the risk associated with the design. The dynamic nature of geometric design will aid these developments.

Defining Terms

Average daily traffic (ADT): The total traffic volume during a given time period (in whole days greater than one day and less than one year) divided by the number of days in that time period.

Average running speed: The distance divided by the average running time to traverse a segment of highway (time during which the vehicle is in motion).

Broken-back curve: An alignment in which a short tangent separates two horizontal or vertical curves turning in the same direction.

Crest vertical curve: A curve in the longitudinal profile of a road having a convex shape.

Compound curve: A curve composed of two consecutive horizontal curves turning in the same direction.

Horizontal curve: A curve in plan that provides a change of direction.

Level of service: A qualitative measure that describes operating conditions of a traffic stream and their perception by motorists and passengers.

Reverse curve: A curve composed of two consecutive horizontal or vertical curves turning in opposite directions.

Right-of-way: The land area (width) acquired for the provision of a highway.

Sag vertical curve: A curve in the longitudinal profile of a road having a concave shape.

Superelevation: The gradient across the roadway on a horizontal curve measured at right angles to the centerline from the inside to the outside edge.

30th highest-hour volume: The hourly volume that is exceeded by 29 hourly volumes during a designated year.

Turning roadway: A separate roadway to accommodate turning traffic at intersections or interchanges.

References

American Association of State Highway and Transportation Officials. 1992a. *Guide for the Design of High-Occupancy Vehicle and Public Transfer Facilities.* AASHTO, Washington, D.C.

American Association of State Highway and Transportation Officials. 1992b. *Transportation Landscape and Environmental Design.* AASHTO, Washington, D.C.

American Association of State Highway and Transportation Officials. 1991. *Guide for Development of New Bicycle Facilities.* AASHTO, Washington, D.C.

American Association of State Highway and Transportation Officials. 1990. *A Policy on Geometric Design of Highways and Streets.* AASHTO, Washington, D.C.

American Association of State Highway and Transportation Officials. 1989. *Roadside Design Guide.* AASHTO, Washington, D.C.

American Association of State Highway and Transportation Officials. 1987. *Computer System Index.* AASHTO, Washington, D.C.

American Association of State Highway and Transportation Officials. 1977. *A Manual on User Benefit Analysis of Highway and Bus Transit Improvements.* AASHTO, Washington, D.C.

Ballard, A. J. 1983. Current state of truck escape-ramp technology. *Transp. Res. Rec.* 923:35–42.

Easa, S. M. 1991. Lateral clearance to vision obstacles on horizontal curves. *Transp. Res. Rec.* 1303:22–32.

Easa, S. M. 1993a. Lateral clearance needs on compound horizontal curves. *J. Transp. Eng., ASCE.* 119(1):111–123.

Easa, S. M. 1993b. Should vehicle 15-percentile speed be used in railroad grade crossing design? *ITE J.* 63(8):37–46.

Easa, S. M. 1994. Design considerations for highway sight-hidden dips. *Transp. Res. A.* 28(1): 17–29.

Federal Highway Administration. 1988. *Manual on Uniform Traffic Control Devices.* U.S. Department of Transportation, Washington, D.C.

Fitzpatrick, K., Mason, J. M., Jr., and Harwood, D. W. 1993. Comparison of sight distance procedures for turning vehicles from a stop-controlled approach. *Transp. Res. Rec.* 1385:1–11.

Garber, N. J. and Hoel, L. A. 1988. *Traffic and Highway Engineering.* West Publishing, New York.

Harwood, D. W. and Glennon, J. C. 1989. Passing sight distance design for passenger cars and trucks. *Transp. Res. Rec.* 1208:59–69.

Harwood, D. W., Glauz, W. D., and Mason, J. M., Jr. 1989. Stopping sight distance design for large trucks.*Transp. Res. Rec.* 1208:36–46.

Institute of Transportation Engineers. 1984. *Guidelines for Urban Major Street Design.* ITE, Washington, D.C.

Institute of Transportation Engineers. 1990. *Guidelines for Residential Subdivision Street Design.* ITE, Washington, D.C.

Institute of Transportation Engineers. 1993. *Turning Vehicle Templates—Metric System.* Publication No. LP-022A, ITE, Washington, D.C.

Leisch, J. P. 1993. Operational considerations for systems of interchanges. *Transp. Res. Rec.* 1385:106–111.

Mason, J. M., Jr., Fitzpatrick, K., and Harwood, D. W. 1989. Intersection sight distance requirements for large trucks. *Transp. Res. Rec.* 1208:47–58.

McGee, H. W. 1989. Reevaluation of the usefulness and application of decision sight distance. *Transp. Res. Rec.* 1208:85–89.

McTrans—Center for Microcomputers in Transportation. 1993. *Software and Source Book.* Transportation Research Center, University of Florida, Gainesville, FL.

Meyer, C. F. and Gibson, D. W. 1980. *Route Surveying and Design.* International Textbook, Scranton, PA.

Neuman, T. R. 1985. *Intersection Channelization Design Guide.* NCHRP Report 279, National Research Council, Washington, D.C.

Neuman, T. R. 1989. New approach to design for stopping sight distance. *Transp. Res. Rec.* 1208:14–22.

Olson, P. L., Cleveland, D. E., Fancher, P. S., Kostyniuk, L. P., and Schneider, L. W. 1984. *Parameters Affecting Stopping Sight Distance.* NCHRP Report 270, National Research Council, Washington, D.C.

Pietrucha, M. T. and Opiela, K. S. 1993. Safe accommodation of pedestrians at intersections. *Transp. Res. Rec.* 1385:12–21.

Roads and Transportation Association of Canada (RTAC). 1986. *Manual of Geometric Design Standards for Canadian Roads.* RTAC, Ottawa, Ontario.

Transportation Association of Canada (TAC). 1993. *Highway User Cost Tables: A Simplified Method of Estimating User Cost Savings for Highway Improvements.* TAC, Ottawa, Ontario.

Transportation Research Board. 1985. *Highway Capacity Manual.* Special Report 209, National Research Council, Washington, D.C.

Transportation Research Board. 1987. *Designing Safer Roads—Practices for Resurfacing, Restoration, and Rehabilitation.* Special Report 214, National Research Council, Washington, D.C.

Zegeer, C. V. and Council, F. F. 1992. *Safety Effectiveness of Highway Design Features, Volume III: Cross Sections.* Federal Highway Administration, U.S. Department of Transportation, Washington, D.C.

For Further Information

American geometric design standards for urban and rural highways are found in *A Policy on Geometric Design of Highways and Streets* (1990) by AASHTO. Canadian geometric design standards are found in *Manual of Geometric Design Standards for Canadian Roads* (1986) by RTAC. These references contain design practices in universal use.

Geometric Design Projects for Highways (1993) by J. G. Schoon, published by the American Society of Civil Engineers, is particularly helpful for understanding the preliminary location study of a proposed highway. The book illustrates the design procedures with detailed case studies.

Available microcomputer programs for geometric design are described in *Software and Source Book*, published annually by the Center for Microcomputers in Transportation. For subscription information contact McTrans, Transportation Research Center, University of Florida, Gainesville, FL 32611-2083. Phone: (904) 392-0378; fax: (904) 392-3224.

A synthesis of past and ongoing research has been sponsored by the TRB committee on geometric design on the following topics: highway sight distance design issues, intersection and interchange design, and cross section and alignment design. Research on the first two topics has been published in *Transportation Research Records* 1208 and 1385, respectively. Research on the last topic will be published in a record in 1994.

Established in 1962 by the National Aeronautics and Space Administration (NASA), the John F. Kennedy Space Center is the nation's principal center for the testing and launching of space vehicles and payloads, including the space shuttle fleet. The center occupies 140,000 acres—one-fifth the size of Rhode Island—on Florida's central Atlantic coast. It consists of numerous uniquely engineered structures designed for processing space vehicles.

The centerpiece of the facility is Launch Complex 39, from which the powerful Saturn V Apollo rockets once blasted into space and which has been extensively modified for today's space shuttle operations. The complex features the 525-foot-tall Vehicle Assembly Building (VAB), the largest building in the world when completed in 1965. The structure is more than 200 feet taller than the Statue of Liberty. It covers eight acres and could easily fit the United Nations headquarters building through its gaping doors, which take 45 minutes to open. After the Shuttle undergoes maintenance at the Orbiter Processing Facility, it is towed to the VAB for attachment of its solid rocket boosters and external fuel tanks, and then transported by a mammoth crawler-transporter to Launch Pad 39A or 39B for liftoff.

The Kennedy Space Center—tourist mecca, employer of 18,000 workers, national wildlife refuge, and gateway to other worlds—is a permanent memorial to the crucial role of American civil engineers in the conquest of space. (Photo courtesy of the American Society of Civil Engineers.)

61

Highway Traffic Operations

Michael J. Cassidy
University of California, Berkeley

In pursuing a better understanding of vehicle movement on highway facilities, we must first concede that individual vehicle (i.e., motorist) behavior is variable and not always predictable. Fortunately, the goals and issues associated with the analysis of highway operation are generally directed at those aspects of traffic stream behavior that are *reproducible*. When assessing highway operation, the traffic engineer is typically concerned with designing and analyzing systems for *average* traffic stream conditions. As these *average* conditions are relatively insensitive to the detailed behavior of individual vehicles/motorists, one can evaluate highway traffic operation in an analytical manner.

61.1 Chapter Scope

Numerous published sources currently address a variety of important issues in traffic engineering. For example, there presently exist manuals for the capacity and level-of-service analysis of highway facilities [TRB, 1985]; manuals outlining the warrants and design standards for the implementation of traffic control devices [*MUTCD*, 1988]; references for designing and operating traffic signals [Kell and Fullerton, 1991]; handbooks addressing geometric standards for streets and highways [AASHTO, 1984]; and documents detailing the execution of traffic engineering studies [ITE, 1976].

Although not all the material presented in these existing documents is unanimously regarded as appropriate and proper, these published references serve to outline the current state of traffic engineering practice. The discussion concerning highway traffic operation presented in this chapter does not seek to replicate information already presented in these widely circulated documents. Rather, this chapter presents fundamental concepts and analytical techniques that can be applied toward better understanding of operating characteristics in a highway traffic stream. Such material can serve to aid and improve traffic engineering analysis and design.

0-8493-8953-4/95/$0.00 + $.50

We begin by defining and describing the three fundamental traffic stream parameters—flow, q; density, k; and speed, v—and their relationship on uninterrupted flow facilities. This initial discussion gives rise to descriptions of queueing and time-space diagrams—two important graphical techniques for assessing highway operating conditions. Discussion of traffic stream parameters is concluded with a description of time-mean and space-mean properties. The important distinctions between these two properties are illustrated.

The chapter next focuses on the presumed relationships between any two of the three aforementioned traffic stream parameters. Based upon such presumed relations, shock wave analysis, a technique for evaluating congested roadway operation, is presented.

Finally, the chapter highlights certain issues associated with signalized intersections, an interrupted highway facility. Discussion here is directed primarily at (1) understanding the fundamental dynamics of vehicle flow at traffic signals and (2) techniques for establishing signal timing plans to accommodate prevailing operating conditions.

The issues presented in this chapter reflect a summary of materials compiled from several sources. In particular, the author acknowledges G. F. Newell, University of California, for his work in enumerating many of the principles outlined in this chapter.

61.2 Traffic Flow: Observations at a Single Location

In the context of highway operations **flow**, q, is defined as the *rate* at which vehicles pass a fixed highway location. One can imagine an observer who records the arrival times of individual vehicles as they pass a specified point on the roadway. Graphically, this process can be represented by a cumulative vehicle arrival curve, illustrated in Fig. 61.1. This curve is a step function that increases by 1 each time a vehicle passes the fixed point of interest. If more than one vehicle simultaneously passes this point (i.e., the highway has multiple lanes), the step height would represent the appropriate positive integer.

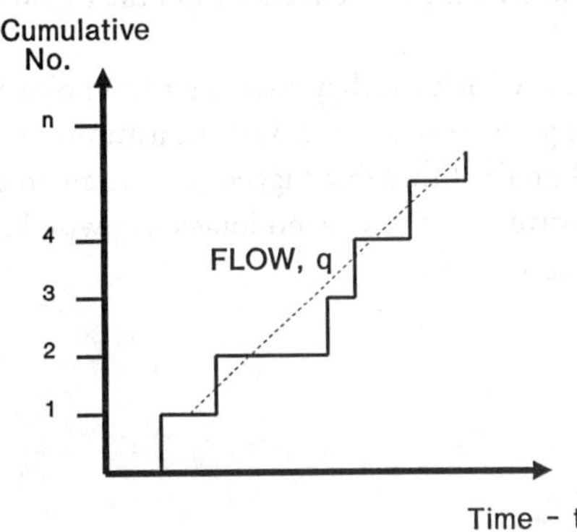

FIGURE 61.1 Vehicle counts past a fixed location over time.

Where the arrival curve in Fig. 61.1 exhibits fairly uniform step sizes, average flow can be represented as a single line drawn through the midpoints of the step function. The magnitude (i.e., value) of flow is reflected in the slope of this line in units of number of vehicles per time interval.

We note that average flow has the dimension of $(\text{time})^{-1}$. If the cumulative arrival curve in Fig. 61.1 depicts vehicles passing the fixed point one at a time, and if the specified time interval begins and ends with the passage of a (different) vehicle, then the time interval used for defining q becomes the sum of individual vehicle *interarrival* times.

We define the time duration between individual vehicle arrivals (i.e., step *widths* in Fig. 61.1) as **headways.** As q is now defined as the total number n of vehicles passing the fixed point, divided by the sum of the individual headways, $\sum_{j=1}^{n} h_j$,

$$q = \frac{n}{\sum_{j=1}^{n} h_j} = \frac{1}{\frac{1}{n}\sum_{j=1}^{n} h_j} \tag{61.1}$$

illustrating that flow is the inverse of average headway.

61.3 Cumulative Arrival Curves: Deterministic Queueing

The cumulative vehicle arrival curve depicted in Fig. 61.1 has value beyond merely illustrating the attributes of flow and headway. Suppose, for example, one were interested in evaluating operating conditions (e.g., travel times or delays) on some given section of highway extending from an upstream point x_1 to a downstream point x_2. One could, in theory, construct cumulative vehicle arrival curves at both locations x_1 and x_2, as presented in Fig. 61.2, in which the term $t_j(x_i)$ reflects the arrival time of vehicle j at location x_i, where $j = 1, 2, \ldots, n$ and $i = 1$ or 2.

The two curves in Fig. 61.2 have several meaningful interpretations:

- The vertical distance $Q(t)$ between the two curves at time t reflects the number of vehicles within the highway section (from x_1 to x_2).
- The horizontal distance W_j between the two curves at height j represents the travel time of the jth vehicle between points x_1 and x_2.
- Total travel time between points x_1 and x_2 is merely the sum of the individual W_j. Because each W_j can be viewed as an area of height 1 (or some appropriate positive integer) and width W_j, total travel time is the area between the cumulative arrival curves at x_1 and x_2. One can express this as the total travel time for a specified number of vehicles, n, by horizontally bounding the area between curves.
- Average travel time between points x_1 and x_2 is the total area divided by n, the total number of vehicles evaluated in the system.

The curves illustrated in Fig. 61.2 suggest that vehicles pass downstream point x_2 in the same sequence in which they pass upstream point x_1. Vehicle-overtaking maneuvers within the highway section x_1–x_2 would negate this ordered sequence.

If one could "tag" (e.g., sequentially number) individual vehicles as they pass upstream point x_1 and relay this information to the observer at downstream point x_2 (a not-too-realistic assumption), then arrival curves at x_2 could still be constructed based on the times that tagged (i.e., numbered) vehicles passed x_2. As illustrated in Fig. 61.3, the cumulative curve at x_2 is no longer stepwise in a consistently ascending manner where overtaking maneuvers occur.

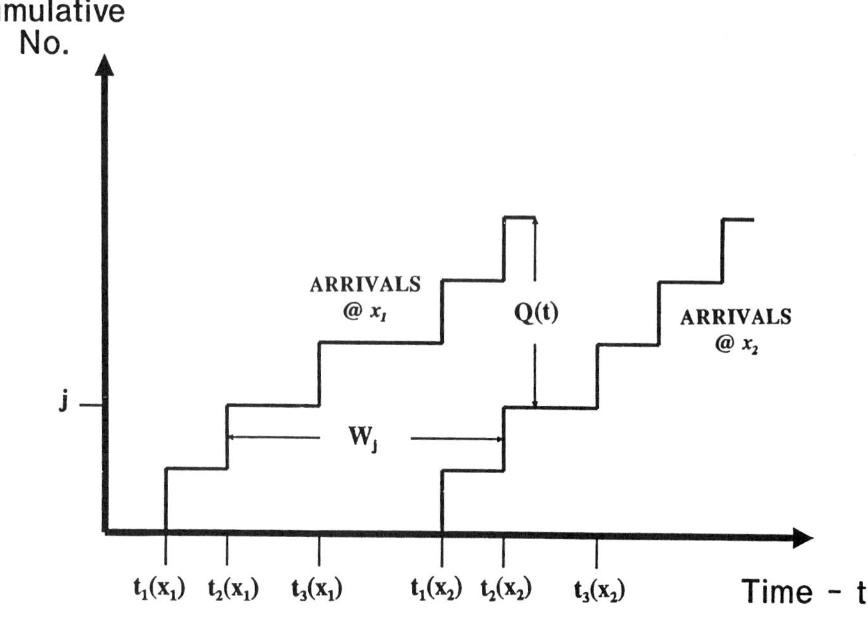

FIGURE 61.2 Upstream and downstream cumulative vehicle arrivals.

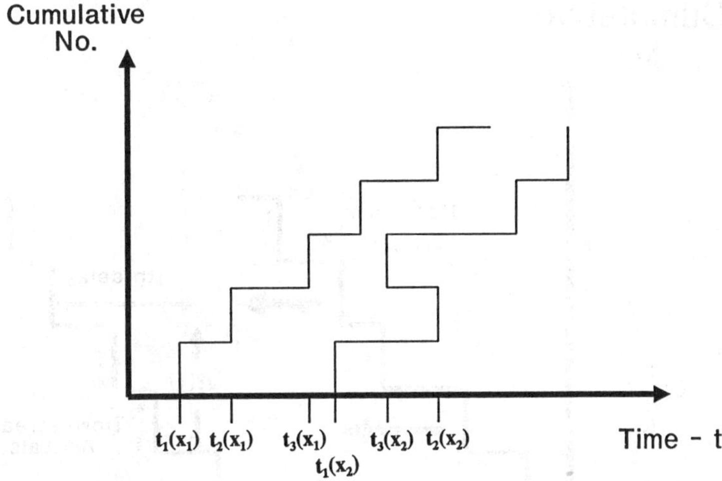

FIGURE 61.3 Upstream and downstream cumulative curves with vehicle overtaking.

More importantly, however, we recognize that we are typically more interested in total or average performance measures than in individual measures. These "global" or aggregate measures are independent of arrival sequence at points x_1 and x_2. For example, travel time is the difference between arrival times at downstream point x_2 and upstream point x_1. Total travel time is the sum of individual travel times past points x_1 and x_2 where, for now, we assume that vehicles maintain their identity (i.e., their sequential number assigned at x_1) as they pass downstream point x_2. Thus, total travel time for n vehicles can be expressed as

$$\sum_{j=1}^{n}[t_j(x_2) - t_j(x_1)] \tag{61.2}$$

This expression, however, is equivalent to

$$\sum_{j=1}^{n} t_j(x_2) - \sum_{j=1}^{n} t_j(x_1) \tag{61.3}$$

From Eq. (61.3) we note that total (and average) travel time is independent of arrival sequence past points x_1 and x_2. One can obtain total and average travel times by summing individual vehicle arrival times at upstream and downstream boundaries, regardless of overtaking activity. In the context of "viewing" the arrival curves in Fig. 61.2, one can interpret overtaking maneuvers within the highway section as merely an "exchange" of vehicle identities. What is essential is that both curves in Fig. 61.2 must represent the same collection of n vehicles.

One could horizontally translate the curve reflecting vehicle arrivals at upstream point x_1 to the right by a distance equal to the average free-flow (i.e., uncongested) travel time between points x_1 and x_2, as illustrated in Fig. 61.4. The remaining horizontal distance between the two curves at height j represents the *delay* imparted to the jth vehicle as a result of congestion (or interruption). The area between the curves in Fig. 61.4 reflects total *delay*.

Again referring to Fig. 61.4, the vertical distance between cumulative curves at time instant t is the number of (additional) vehicles between points x_1 and x_2 as a consequence of the delay. Specifically, the vertical distance in Fig. 61.4 represents the difference between the total number of vehicles that would have arrived at point x_2 in the absence of delay and the total number of vehicles that actually did arrive at x_2 by time t. This value is not equal to the number of vehicles actually

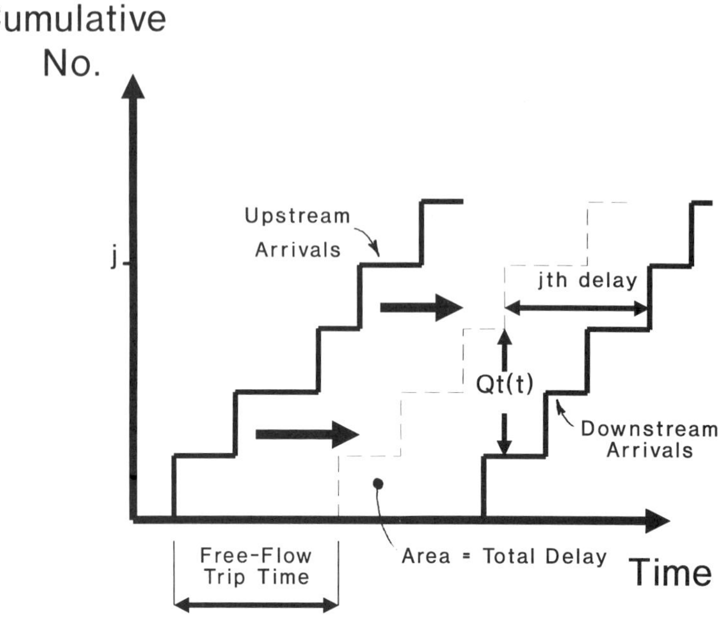

FIGURE 61.4 Translated arrival curve—queueing diagram.

"in queue" (i.e., in congestion) at t, which is a larger number, because at some instant t the actual queue will include vehicles that would not have yet been able to pass downstream point x_2, even in the absence of congestion.

One can readily derive the expression for the actual queue length $Q'(t)$ as

$$Q'(t) = Q_t(t)/\left[1 - \frac{Q(t) - Q_t(t)}{Q(t)}\right] \tag{61.4}$$

where $Q_t(t)$ represents the added number of vehicles, at time t, between x_1 and x_2 as a result of delay (i.e., $Q_t(t)$ represents the vertical distance between two cumulative arrival curves in Fig. 61.4), and $Q(t)$ represents the actual number of vehicles between x_1 and x_2 at time t (i.e., the vertical distance between arrival curves in Fig. 61.2).

Diagrams such as the one in Fig. 61.4 are referred to as deterministic **queueing diagrams**, and their applications in traffic engineering are numerous. Such diagrams are commonly used to evaluate delays and queueing activity at highway bottlenecks such as traffic signals, toll booths, and loading docks. If one does not know the precise arrival and departure times of individual vehicles within the queueing system, the step function curves depicted in Fig. 61.4 can be replaced with average flow rates. Thus, the traffic engineer is required to determine (or estimate) a priori only arrival and departure rates at the queueing system. Additional discussion of queueing diagrams is included later in this chapter.

61.4 Density: Observations at an Instant in Time

The "companion property" associated with individual vehicle arrival times past a fixed location, $t_j(x_i)$, is the count of vehicles between two points x_1 and x_2 at some instant in time t. As depicted in Fig. 61.5, one can construct a cumulative curve of the number of vehicles within roadway section x_2–x_1 (at time t) as a function of the physical distance between x_1 and x_2. Vehicle number is counted in the negative x direction, so the function exhibits a unit "step up" as x *decreases* past each vehicle location at instant t, $x_j(t)$.

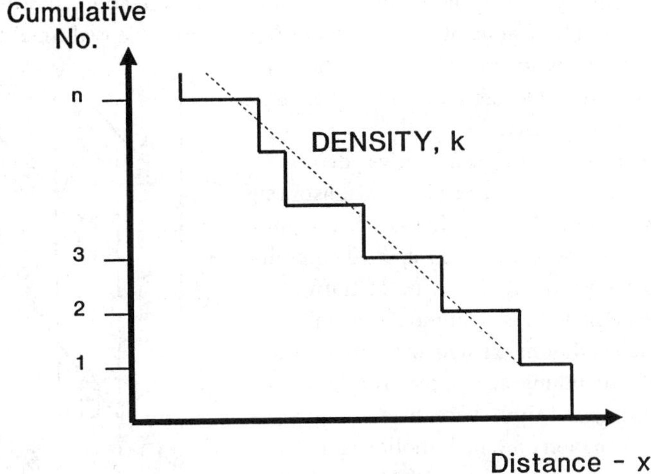

FIGURE 61.5 Vehicle counts over space at a fixed time.

The slope of the straight line through this function reflects **density**, k. The magnitude of density, which is measured over some specified spatial interval, is the absolute value of the (negative) slope as illustrated in Fig. 61.5. Density is measured in units of vehicles per distance.

If the specified spatial interval (over which density is measured) begins and ends with the presence of a vehicle, then the spatial interval used for defining density becomes the sum of the physical distances between individual vehicles. These individual distances, which are measured from identical points on consecutive vehicles (e.g., the front bumper), are termed **spacings** and are represented by the step *widths* in Fig. 61.5. As density, k, is defined as the total number of vehicles present in a specified highway section at instant t, n^*, divided by the sum of individual spacings, $\sum_{j=1}^{n^*} S_j^*$,

$$k = \frac{n^*}{\sum_{j=1}^{n^*} S_j^*} = \frac{1}{\frac{1}{n^*}\sum_{j=1}^{n^*} S_j^*} \tag{61.5}$$

illustrating that density is the inverse of average spacing.

61.5 Speed: Time-Space Trajectories

The traffic engineer is concerned with the *movement* of vehicles (and people) on a highway facility. The parameter **speed**, measured in units of distance per time, reflects the manner in which a vehicle changes location over time.

Because vehicles are typically constrained to move along a single channel (e.g., a road section), vehicle motion can typically be described on a two-dimensional plane depicting space and time. The manner in which these vehicle movements can be depicted is analogous to film (or video) images of a road section collected from an overhead satellite. Individual vehicles identified in a single film frame maintain their unique identities in subsequent frames in which they appear. Each vehicle exhibits a position or location on each film frame, as reflected by some identifiable reference point such as the front bumper. Unless a vehicle is stopped, its location will change in each subsequent film frame.

To interpret the motion of each vehicle, one can imagine that the film frames are projected onto a sheet of paper so that both space and time are represented, as shown in Fig. 61.6, where the channel (i.e., highway lane) projects forward in the vertical direction, and time is translated by placing consecutive film frames at a unit distance (representing elapsed time) to the left. Each vehicle is identified by a unique number. Smooth curves are drawn through identifiable reference points (e.g., front bumpers) on the "same" vehicle.

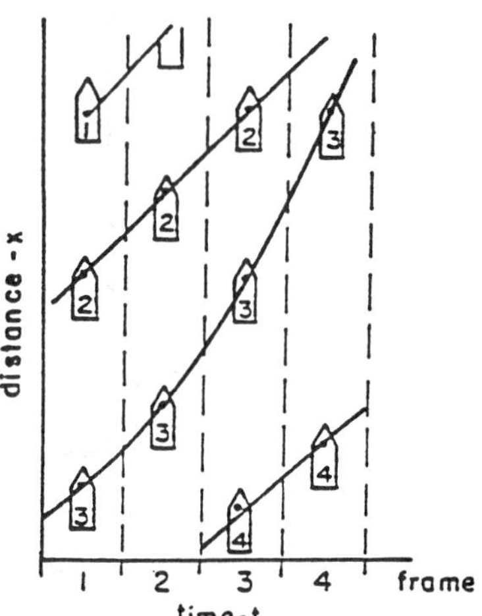

FIGURE 61.6 Construction of trajectories from film frames. *Source:* Daganzo, C. F. and Newell, G. F. Unpublished. *Notes on Introduction to Transportation Engineering.* University of California, Berkeley. With permission.

These smooth lines, termed **trajectories**, depict the motion of individual vehicles. The diagram illustrating these trajectories is referred to as a **time-space diagram.** Although time-space diagrams can be applied to evaluate vehicle motion on virtually any type of traffic flow facility, they are perhaps most commonly exploited to evaluate operating conditions on signalized arterials and to synchronize signal timing at adjacent traffic signals. The effective utilization of time-space diagrams does not require precise information on vehicle motion (e.g., "tracing" motion over consecutive aerial photographs). Rather, the traffic engineer is required only to "visualize" or estimate the manner in which vehicles travel over space and time.

Figure 61.7 depicts the one-directional movement of vehicles through two signalized intersections located at points y_1 and y_2. We note that the physical locations of the intersections do not change over time (the locations depicted by y_1 and y_2 might represent the stop bar or the middle of the intersections). The changing red and green times are represented by *solid walls* and *windows* defining when vehicles are, and are not, permitted to traverse the intersections.

In the time-space diagram, trajectory slope reflects vehicle speed. Thus, the downward bending, horizontal projection, and upward bending of individual trajectories in Fig. 61.7 represent vehicle deceleration, stop time, and acceleration, respectively.

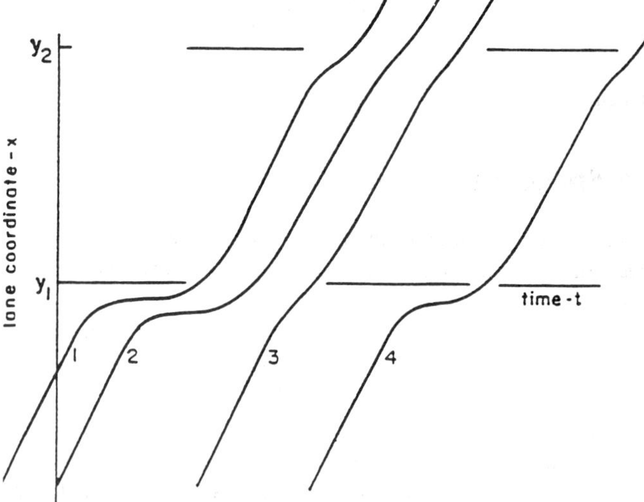

FIGURE 61.7 Vehicle trajectories through traffic signals. *Source:* Daganzo, C. F. and Newell, G. F. Unpublished. *Notes on Introduction to Transportation Engineering.* University of California, Berkeley. With permission.

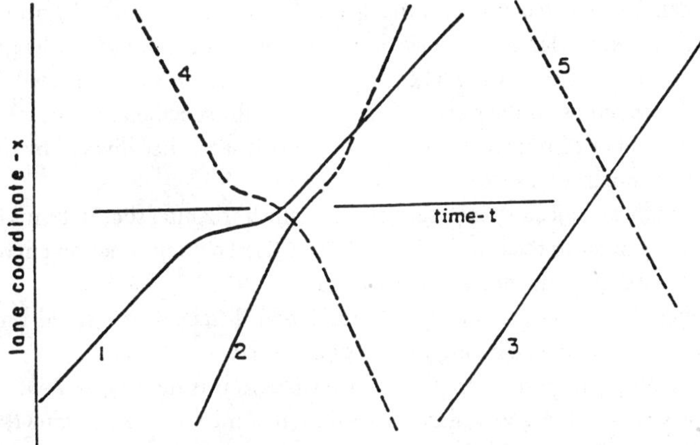

FIGURE 61.8 Trajectories for two-directional traffic. *Source:* Daganzo, C. F. and Newell, G. F. Unpublished. *Notes on Introduction to Transportation Engineering*. University of California, Berkeley. With permission.

Figure 61.8 illustrates two-directional traffic traveling through a single intersection; the dashed-line trajectories represent individual vehicles traveling in the opposing direction.

Where an individual vehicle exits a channel (e.g., changes lanes or turns onto a cross street), the behavior can be depicted by the sudden disappearance of the trajectory on the time-space plane and, if relevant, its reappearance on another time-space plane.

61.6 Time-Mean and Space-Mean Properties

The time-space diagram depicts a wealth of information for better understanding of traffic flow characteristics. Figure 61.9 illustrates simple vehicle trajectories moving in a single direction along some roadway. Recalling that flow is defined as the number of vehicles crossing some point

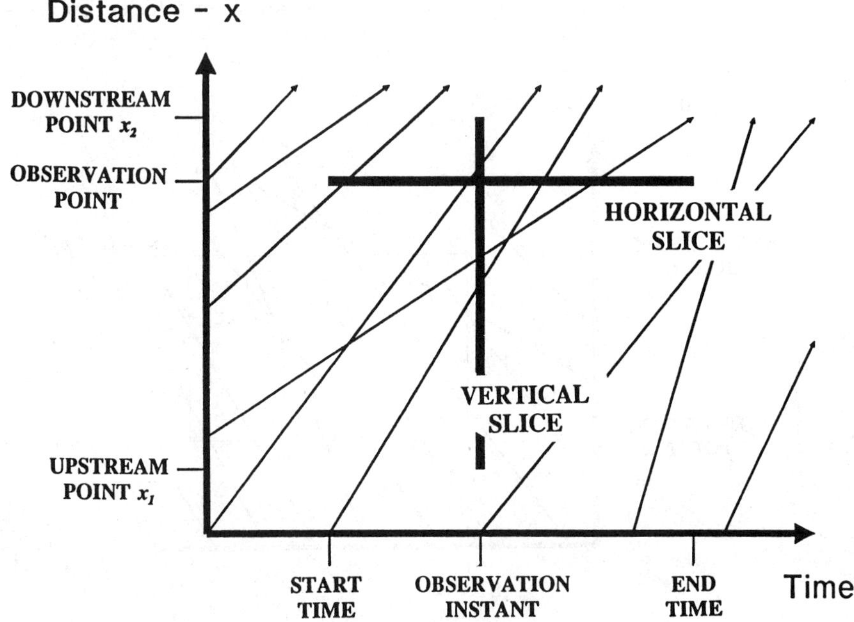

FIGURE 61.9 Interpreting operating characteristics using a time-space diagram.

over time, flow can be depicted on the time-space diagram by taking a horizontal slice in the time-space plane perpendicular to the space axis. The vertical location (i.e., height) of this slice represents the physical location on the highway where observations are collected. The horizontal boundaries of the slice represent the start and end times of the specified time interval. The value of flow is simply the number of trajectories crossing the horizontal slice divided by the time interval (i.e., the width of the horizontal slice).

One can visualize these trajectory intersections with the horizontal slice in terms of a cumulative vehicle arrival curve similar to the one depicted in Fig. 61.1. The step function curve increases by 1 each time a trajectory crosses the horizontal slice in Fig. 61.9.

Individual vehicle headways are merely the horizontal distances, measured on the time axis, between consecutive trajectories crossing the horizontal slice.

The companion property to flow on a time-space diagram is density, k. As density is defined as the number of vehicles in a given highway section at some specified instant t, density can be characterized as a vertical slice in the time-space plane perpendicular to the time axis. The location of the vertical slice (measured horizontally from the origin) reflects the instant in time when density is evaluated. The vertical boundaries of this slice represent the end points of the specified highway segment. The value of k is merely the number of trajectories crossing the vertical slice divided by the specified spatial interval (i.e., the height of the vertical slice measured on the space axis).

One can visualize these trajectory intersections with the vertical slice in terms of a cumulative curve of vehicle number at time t versus space similar to the one depicted in Fig. 61.5. The step function curve increases by 1 at each location where the trajectory crosses the vertical slice.

Individual vehicle spacings are merely the vertical distances, measured on the space axis, between consecutive trajectories crossing the vertical slice.

The properties depicted on the time-space diagram allude to issues in traffic engineering that are not always well understood. For example, when evaluating operating conditions on a given highway facility, the traffic engineer is actually interested in assessing characteristics over some specified physical length of highway section, during some time interval of interest. These dimensions of space and time can be represented by a rectangular area superimposed on a time-space diagram, as shown in Fig. 61.10, where we label the rectangular area of interest Area A. The actual spatial and temporal dimensions of Area A represent values judged to be appropriate for a given analysis. We recognize that the trajectories traveling through Area A represent the vehicles that actually travel

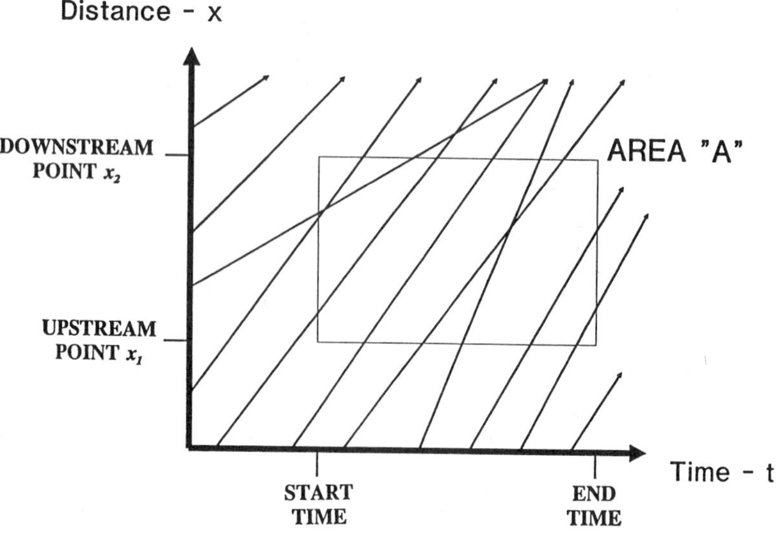

FIGURE 61.10 Evaluating traffic operation over time and space.

on the selected highway section during the time interval of interest. (The intersecting trajectories in Fig. 61.10 indicate overtaking maneuvers on the roadway section of interest.) Thus, the time and space boundaries of Area A must lie within a "homogeneous" traffic operating state. That is to say, the dimensions of Area A must be sufficiently small so that operating conditions (i.e., trajectories) within Area A do not reflect temporal or spatial changes as traffic demands change with time and flow patterns change with changing geometries over space.

The traffic engineer is interested in evaluating conditions prevailing within Area A (i.e., the selected highway segment and observation time period). The traffic engineer, for example, might wish to determine average speed prevailing in Area A or may be interested in identifying the percentage of a certain vehicle type (e.g., trucks) traveling in Area A.

The traffic engineer should recognize that such measures are generally not correctly obtained by computing algebraic averages of observations measured at a single point within the highway section (i.e., by taking a "horizontal slice" in Area A and algebraically averaging measured characteristics of interest as trajectories cross the horizontal slice). Because vehicles are traveling at different speeds, a disproportionally high percentage of faster-moving vehicles will be observed passing a single observation point on the roadway. The traffic stream measures of interest (e.g., speed, percent trucks, etc.) can only be correctly determined by averaging across all trajectories in Area A (or by observing a *random* sample thereof).

We recall that traffic engineering generally concerns itself with analyzing those collective (i.e., reproducible) behavior of vehicles within a traffic stream that are not sensitive to the behavior of individual motorists. In theory, then, the traffic engineer seeks to average flow over some spatial interval (i.e., across all points in Area A) and to average density over some temporal interval (i.e., across all time instants in Area A). By obtaining an average flow and an average density in Area A, $q(\mathrm{A})$ and $k(\mathrm{A})$, the average speed in Area A, $v(\mathrm{A})$, is computed from the fundamental equation of uninterrupted flow,

$$v(\mathrm{A}) = q(\mathrm{A})/k(\mathrm{A}) \tag{61.6}$$

This relationship holds for all vehicles exhibiting positive, uninterrupted motion.

To illustrate some of the common misconceptions concerning *time-mean* and *space-mean* properties, and to demonstrate the potential magnitudes of the resulting errors, we consider the following example.

Example. Suppose we are interested in evaluating certain operating conditions on a 1-mi long multilane highway section during a 1-h period.

For simplicity, we assume that the traffic stream consists of only three different vehicle types:

- "Fast" cars, which travel at constant speeds of 60 mph
- "Slow" cars, traveling at 50 mph
- Trucks, traveling at 45 mph

We also assume that the actual density of fast cars is 20 per mile; the density of slow cars is 10 per mile; and truck density is 10 per mile. (We would not actually obtain these values directly without studying a series of aerial photos.)

Finally, we assume that all vehicle types exhibit *stationary flow* conditions (i.e., when a certain vehicle type exits the downstream end of the highway section, it is immediately replaced by the same vehicle type entering the upstream end of the section).

If we were interested in measuring average speed within the 1-mi long highway section during the hour of interest, we might collect speed measurements at a specified point in the highway section, perhaps by using a radar gun.

As we would be collecting a measure (i.e., speed) past a fixed point over time, we would be obtaining observations based upon the *flow* of each vehicle type (recall that flow is defined as the number passing a point over time).

To determine what our measured average speed would be, we rely on the fundamental relationship expressed in Eq. (61.6), $v = q/k$ or $q = vk$. Therefore,

$$\begin{aligned} q(\text{fast cars}) &= 60 \text{ mph} \times 20 \text{ vpm} = 1{,}200 \text{ vph} \\ q(\text{slow cars}) &= 50 \text{ mph} \times 10 \text{ vpm} = 500 \text{ vph} \\ q(\text{trucks}) &= 45 \text{ mph} \times 10 \text{ vpm} = \underline{450 \text{ vph}} \\ q(\text{total}) &= 2{,}150 \text{ vph} \end{aligned}$$

Assume that we measure the speeds of all vehicles to pass the observation point during the hour and then make the mistake of computing the algebraic mean:

$$[1{,}200(60) + 500(50) + 450(45)]/2{,}150 = 54.5 \text{ mph}$$

This value is referred to as the *time-mean speed* (TMS), and it is *not* the true mean prevailing in the highway section during the observation period (i.e., it does not relate the value of average flow to average density).

We can obtain the *true* mean speed, known as the *space-mean speed* (SMS), by measuring the travel times of vehicles in our 1-mi long highway section, computing the average, and converting average travel time to average speed. For illustration purposes, again assume that we measure the travel times of all vehicles traversing the section in 1 h.

$$\begin{aligned} \text{Avg Travel Time} &= [1{,}200(1 \text{ mi}/60 \text{ mph}) + 500(1 \text{ mi}/50 \text{ mph}) \\ &\quad + 450(1 \text{ mi}/45 \text{ mph})]/2{,}150 = 0.019 \text{ hours} \\ \text{Avg Speed} &= 1 \text{ mi}/0.019 \text{ hours} = 53.8 \text{ mph} \end{aligned}$$

The difference between TMS and SMS increases with increasing speed variance in the traffic stream.

We concede that measuring speeds at a point (using a radar gun) is generally simpler and more practical than measuring travel times (or summed arrival times as previously discussed in this chapter). It is therefore worth noting that speed observations collected at a point can be converted to SMS by computing a *harmonic* mean rather than an algebraic mean. The formula for harmonic mean is simply

$$v(A) = \frac{1}{\frac{1}{n} \sum_{j=1}^{n} \frac{1}{v_j(x)}}$$

where $v_j(x)$ is the measured speed of vehicle j as it crosses the specified observation point x.

In similar fashion, if we were interested in obtaining the percentage of trucks in the traffic stream, collecting observations at a point over time would result in an errant measure.

$$\text{Percentage of trucks (at point)} = q(\text{trucks})/q(\text{total}) = 450/2150 = 21\%$$

The true percentage of trucks must be computed using average density, a spatial measure:

$$\begin{aligned} \text{Percentage of trucks} &= k(\text{trucks})/k(\text{total}) \\ &= [q(\text{trucks})/v(\text{trucks})]/[q(\text{total})/v(\text{average})] \\ &= (450/45)/(2150/53.8) = 25\% \end{aligned}$$

61.7 Relationships among q, v, k

Having discussed the relationship between the three traffic stream measures, flow, q; density, k; and speed, v; we briefly turn our attention to the relationships between any two of these three parameters. As early as 1934 [Greenshields, 1934], relationships between q and k, v and q, and v

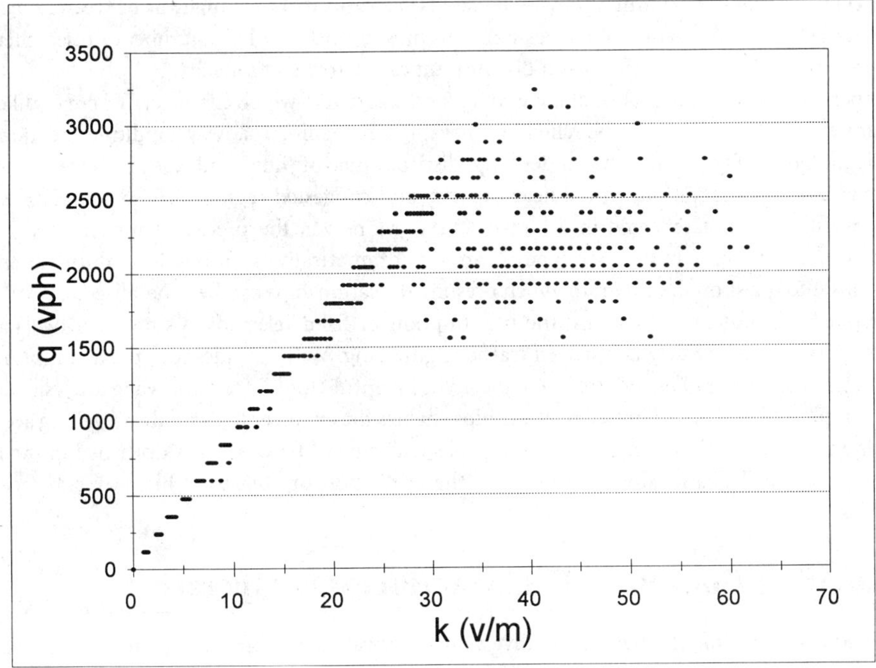

FIGURE 61.11 Scatterplot of flow versus density.

and k have been identified from empirical data. An example of the relationship between flow and density on an **uninterrupted-flow facility** (i.e., a freeway or highway segment) is illustrated in Fig. 61.11. The left-hand side of the scatterplot reflects the relationship between q and k under uncongested flow conditions (i.e., low densities). The right-hand side reflects operating states in congestion. The data points reflect observations collected over 30-second intervals.

Figure 61.11 clearly depicts a correlation between flow and density. These correlations have motivated the use of fundamental curves, such as the q-k curve illustrated in Fig. 61.12. Similar best-fit curves have been developed for flow versus speed and speed versus density, again based on relationships exhibited in scatterplots. Indeed, the existence of "some type" of relationship is perhaps indisputable. Less than certain, however, are the precise functional forms that best characterize these relationships.

For example, the extent to which increasing flow rates influence prevailing vehicle speeds remains a topic of considerable debate in the traffic engineering community. Moreover, there is

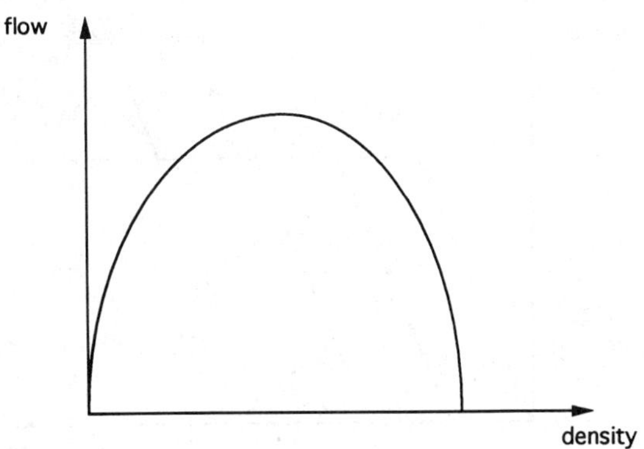

FIGURE 61.12 Flow-density characteristic curve.

some evidence that the maximum flow obtainable (i.e., **capacity**) on a highway or freeway section is greater before the formation of congestion. This presumption has led a number of researchers to construct functional relationships with discontinuities occurring at capacity.

Yet perhaps an even more relevant issue concerns the extent to which characteristic curves like the one presented in Fig. 61.12 can be relied on to evaluate prevailing roadway conditions adequately. As illustrated in Fig. 61.11, the underlying distributions of flow and speed create excessive "scatter" among observations in the near-capacity and congested regions. These *nonreproducible* characteristics are not reflected in a best-fit curve. Moreover, the precise characteristics of the relationships between any two parameters can vary dramatically from one location to the next. This calls into question the extent to which presumed relationships can be generalized.

Despite these potential concerns, the presumption of fixed relationships between two traffic stream parameters is widely adopted in traffic engineering. Without question, the assumption of such relationships provides the traffic engineer with helpful "tools" for addressing analysis needs. For example, reliance on presumed relationships enables the engineer to estimate prevailing average speeds and densities given only a known or projected value of flow. Also, as described in the next section, presumed relationships are vital for the evaluation of congested highway and freeway facilities.

61.8 Evaluating Congested Highway/Freeway Operation

As noted earlier in this chapter, cumulative vehicle arrival curves can be used for estimating the delay for a *specified* number of queued vehicles. As Fig. 61.4 shows, delay is graphically depicted by translating the upstream vehicle arrival curve to the right a distance equal to the undelayed trip time between the upstream and downstream observation points. Figure 61.13 depicts the implication of this horizontal translation on a time-space plane.

In effect, the translation assumes that vehicles travel at their free-flow speed until arriving at the roadway restriction, at which point the vehicles stop instantaneously and remain stopped until they exit the restriction. This depiction of trajectories is obviously not realistic. However, the delay estimate depends only on the "desired" and the actual exit times through the restriction. Thus, the delay estimate for a given vehicle, or group of vehicles, is independent of the trajectory characteristics.

The graphical construction of a queueing diagram as depicted in Fig. 61.4 results in a second "unrealistic" implication. Queued vehicles are assumed to occupy zero physical space on the roadway. This zero space is equivalent to assuming that vehicles wait in queue as if they were *vertically* stacked at the bottleneck entrance.

Once again, we emphasize that the presumed "vertical queueing" does not affect the delay estimate for a given vehicle or group of vehicles. What the queueing diagram does fail to represent,

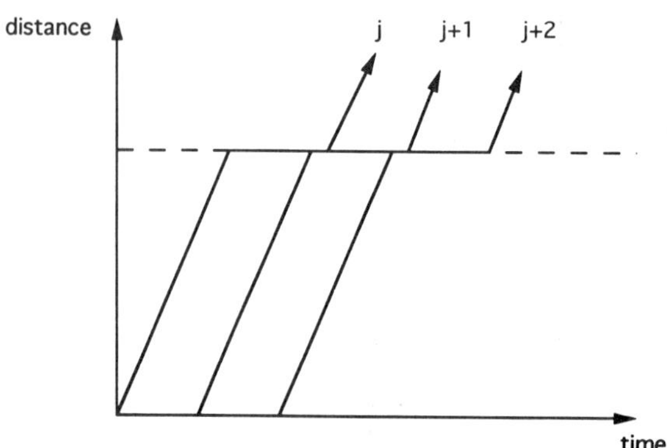

FIGURE 61.13 Delay depicted on time-space plane.

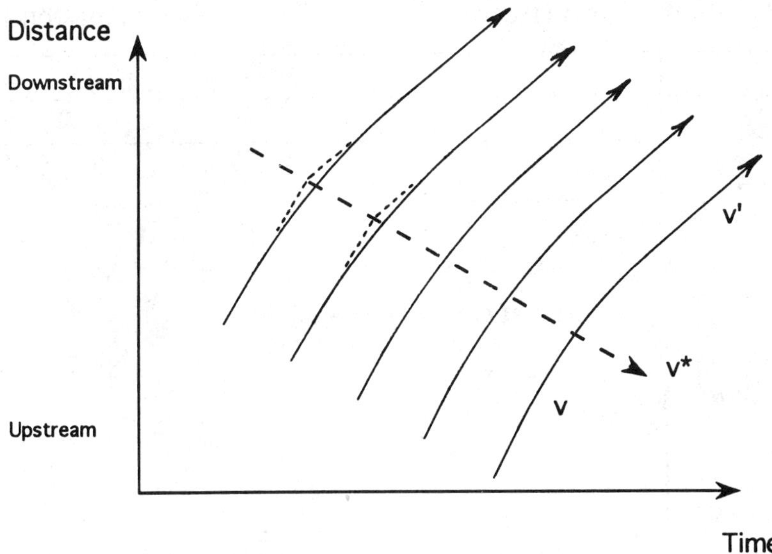

FIGURE 61.14 Shock wave propagation.

however, is the manner in which highway congestion actually propagates over time and space (because congested vehicles do indeed occupy physical space on the roadway).

A more illustrative example of congestion formation is presented in Fig. 61.14. The trajectories depict vehicles arriving from the uncongested region upstream and decelerating in response to downstream congestion. The dashed line separating the congested and uncongested region is termed a **shock wave.**

One can readily derive the velocity at which a shock wave travels through space as the difference in flows between the two regions divided by the difference in density states.

$$v^* = [q_A - q_B]/[k_A - k_B] \qquad (61.7)$$

where the subscripts A and B represent the differing traffic states. In Fig. 61.13, for example, A and B reflect the upstream uncongested region and the downstream congested region, respectively.

Shock wave analysis models the impacts of propagating discontinuities in traffic states (i.e., congestion) by assuming that the motion of highway traffic flow behaves like the kinematic wave motion of fluid flow [Lighthill and Whitham, 1955]. Shock wave analysis is suited to evaluating the impacts of propagating congestion, because the method recognizes that delayed vehicles do occupy physical space.

By extrapolating the asymptotes of the trajectories over the region of vehicle deceleration (shown as the dotted line in Fig. 61.14), one can graphically perform shock wave analysis by exploiting (1) a time-space diagram and (2) a presumed relationship between flow and density.

The application of shock wave analysis is next demonstrated by example.

Example: Application of Shock Wave Analysis. Consider the simple freeway system illustrated in Fig. 61.15. Vehicles arrive at the upstream source of this system, Point X_s, and travel downstream through the bottleneck caused by the lane reduction. We assume that the flow-density relationships for the upstream and downstream segments are represented by the parabolic functions illustrated in Fig. 61.16. Prior to some time t_0, arrival rate past X_s is q_1, where q_1 is less than the bottleneck capacity. At time t_0, however, demand at Point X_s rises to rate q_2, which is greater than the bottleneck capacity. Flow rate past X_s remains at q_2 until time t_1, after which flow past X_s returns to rate q_1.

Using the presumed q-k relationships and a time-space diagram, the propagation of congestion at the bottleneck entrance is examined.

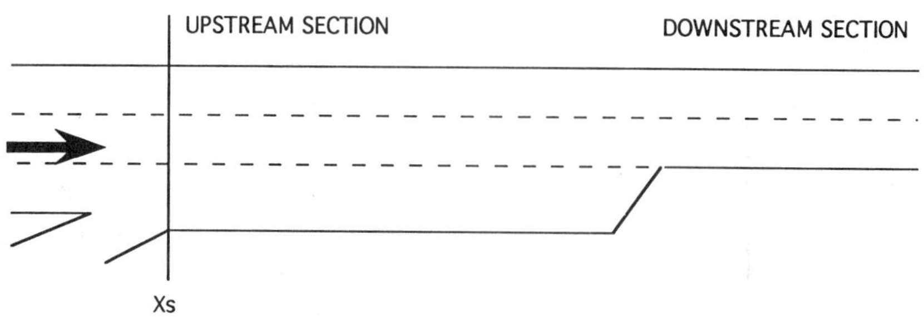

FIGURE 61.15 Simple freeway system.

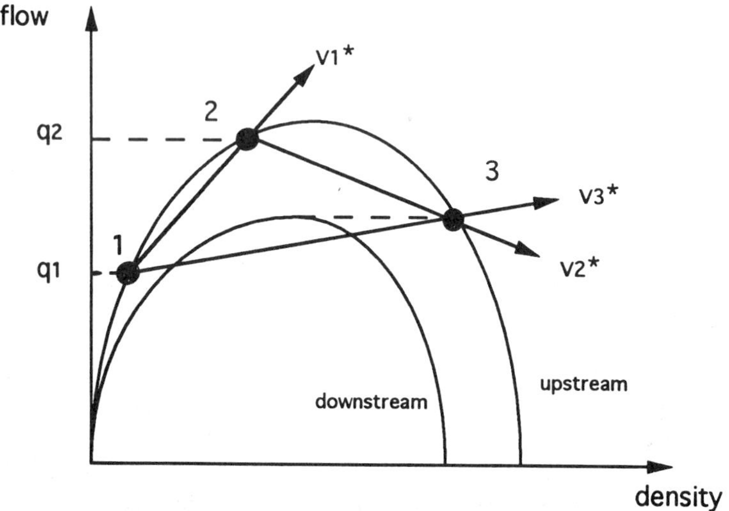

FIGURE 61.16 Shock wave propagation on q-k diagram.

- Prior to time t_0, prevailing flow and density conditions within the upstream freeway segment are represented by point 1 on Fig. 61.16.
- Following time t_0, flow rate past X_s rises to q_2, and the "new" traffic state is depicted by point 2 on the q-k diagram. The slope of the line connecting points 1 and 2 in Fig. 61.16 represents the speed of the "forward-moving" wave separating the two traffic states within the upstream freeway section. This wave velocity is also illustrated in Fig. 61.17.

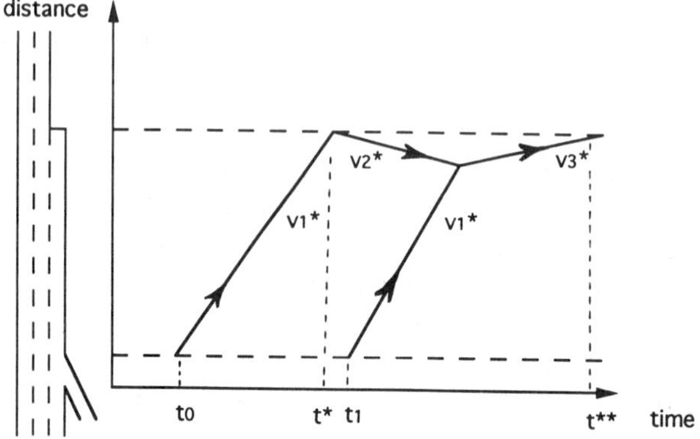

FIGURE 61.17 Shock wave propagation on time-space diagram.

- The discontinuity, moving forward at speed v_1^*, arrives at the bottleneck entrance (i.e., the lane drop), at time t^*, as illustrated in Fig. 61.17. At t^*, flow into the bottleneck enters at a rate equal to the bottleneck capacity. As the arrival rate q_2 is greater than this capacity, congestion forms at the bottleneck entrance. This congested state is represented by point 3 on Fig. 61.16. Note that flow within this congested region is equal to the bottleneck capacity. The speed at which congestion now propagates upstream of the bottleneck entrance is represented by the slope of line v_2^* in Figs. 61.16 and 61.17.
- Finally, when flow returns to rate q_1, the discontinuity between traffic state q_2 and q_1 again moves forward in the upstream section at speed v_1^*. When the reduced flow state q_1 arrives at the rear of the queue, congestion recovery begins. The discontinuity between congested and uncongested states moves forward at velocity v_3^*, as illustrated in Figs. 61.16 and 61.17. Congestion dissipates at time t^{**}, as shown in Fig. 61.17.

Once shock wave analysis has been used to identify congestion "evolution," the resulting delays can be readily estimated. The flow-density state within each region of the time-space diagram are defined by fixed points on the q-k fundamental curve (such as points 1, 2, and 3 in Fig. 61.16). Thus, resulting speeds (and delays) in all time-space regions are determined from the relationship $v = q/k$.

Our discussion on highway traffic operations has thus far been concerned with so-called uninterrupted facilities (i.e., roadways without control devices that interrupt the traffic stream). We now turn our attention to traffic signals, an **interrupted-flow facility.**

61.9 Traffic Signal Operation

To investigate the dynamics of vehicle flow at a signalized intersection, we again use a cumulative vehicle arrival curve. We consider a queue of through-moving vehicles in a single lane, which begins discharging after the initiation of the green phase. Figure 61.18 illustrates $n(t, x)$, the number of vehicles to pass point x to time t. Here, point x represents a location near the intersection (e.g., the stop bar), and time t is measured from the initiation of the green phase.

In Fig. 61.18, t_j represents the time the jth vehicle crosses point x (i.e., the time the vehicle enters the intersection). The time interval $t_1 - 0$ represents the elapsed time between the initiation of green and the first vehicle entry. This first headway is relatively long (perhaps 3 to 6 s at a typical intersection), because the first motorist must react to the green initiation and accelerate from a stopped position. The second headway, marked by the time interval $t_2 - t_1$, might typically

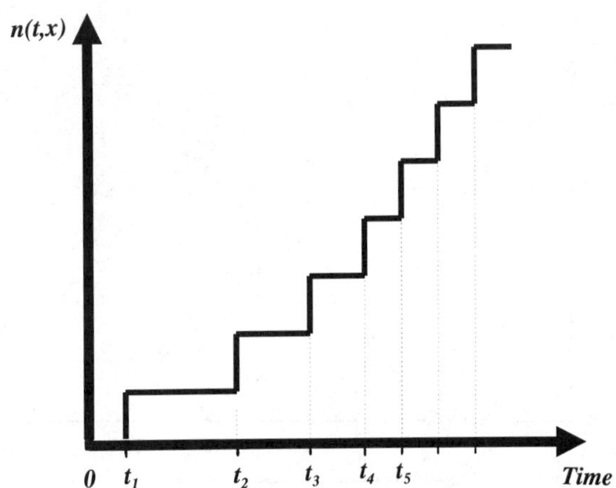

FIGURE 61.18 Queue discharge process.

be shorter than the first, as the reaction times for both motorists are shared. Likewise, the third headway is typically smaller than the second.

Typically, by the fourth, fifth, or sixth headway the vehicle interarrival times into the intersection stabilize and one headway becomes, on average, approximately equal to the next. This "fully accelerated" queue discharge headway is termed *saturation* headway, and its inverse is **saturation flow rate,** s. Saturation flow rate typically continues on the intersection approach until either

1. The vehicle queue has completely discharged into the intersection, at which point flow into the intersection becomes equal to arrival rate, q.
2. The green time is terminated, after which vehicle discharge ceases.

Suppose one were to collect repeated measurements over many cycles and compute an average of cumulative vehicle arrivals into the intersection:

$$\bar{n}(t, x) = \frac{1}{N} \sum_{k=1}^{N} n_k(t, x) \tag{61.8}$$

where

N = the number of observed cycles
$n_k(t, x)$ = the curve for the kth cycle

Then the average arrival curve $\bar{n}(t, x)$ would be a smooth, noninteger function, such as the one illustrated in Fig. 61.19.

In Fig. 61.19 the slope s represents saturation flow rate, while the subsequent slope q reflects the arrival rate following queue dissipation. The start-up lost time, ℓ_s, reflects the longer discharge headways exhibited by the first several vehicles in queue. The value of ℓ_s can be graphically represented as the interval between time 0 (the green initiation time) and the point where the extrapolation of slope s crosses the time axis. The intersection likewise exhibits an *ending* lost time, ℓ_e, caused by the cessation of flow into the intersection following the termination time of the green indication, G.

One can readily see from Fig. 61.19 that the duration of time, g, during which green is *effectively* available to a given movement of vehicles, can be expressed as

$$g = G + a - (\ell_s + \ell_e) \tag{61.9}$$

where a is the duration of the amber (i.e., yellow) **interval.**

As measuring ℓ_s and ℓ_e can be a tedious task, a commonly adopted assumption is that

$$a = \ell_s + \ell_e$$

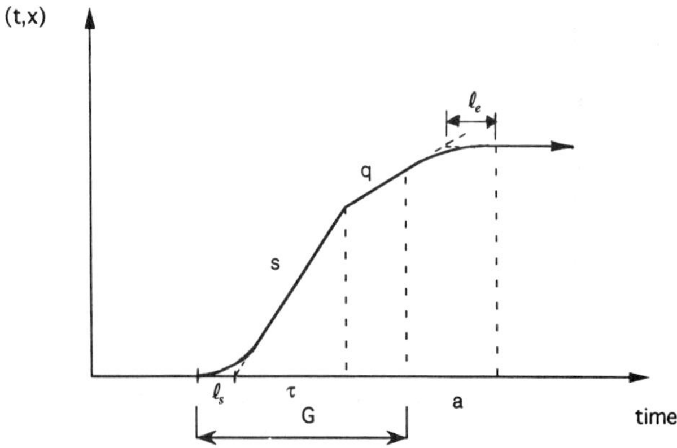

FIGURE 61.19 Discharge flow rate.

Thus, effective green time, g, is often approximated as being equal to the actual duration of the displayed green time, G.

The curve in Fig. 61.19 is valuable for more than just its depiction of traffic signal flow dynamics. The curve in Fig. 61.19 serves as the basis for developing equations for estimating average intersection delay.

If one imagines simultaneously collecting vehicle arrival times at a point x (e.g., the intersection stop bar) and some point u upstream of the queueing influence, the average cyclic arrival times can be depicted on the same graph, as illustrated in Fig. 61.20. Translating the arrival curve at upstream point u to the right by a distance equal to the average undelayed travel time, we obtain a queueing diagram for the movement.

From the queueing diagram depicted in Fig. 61.20, we introduce τ, the average time required to process the prevailing vehicle queue. We note that the average number of *queued* vehicles to enter the intersection per cycle is the product of the saturation flow rate and the discharge time, τ,

$$s\tau \tag{61.10}$$

Alternatively, we can express this number of queued vehicles as the number of arrivals since the onset of the previous red phase to time τ:

$$(C - g + \tau)q \tag{61.11}$$

where C is the *cycle length,* and red time is expressed as $C - g$. From the foregoing expressions, we determine

$$\tau = q(C - g)/(s - q) \tag{61.12}$$

We further note that the average fraction of vehicles delayed by signalization each cycle is the ratio

$$\tau s/qC \tag{61.13}$$

Having previously solved for τ, we compute the fraction of delayed vehicles as

$$\text{Fraction delayed} = (1 - g/C)/(1 - q/s) \tag{61.14}$$

Finally, we note that the average delay imparted to *all* arriving vehicles is merely the product of the fraction delayed and the average delay imparted to these queued vehicles. We recognize that the average delay in queue is one-half the red time.

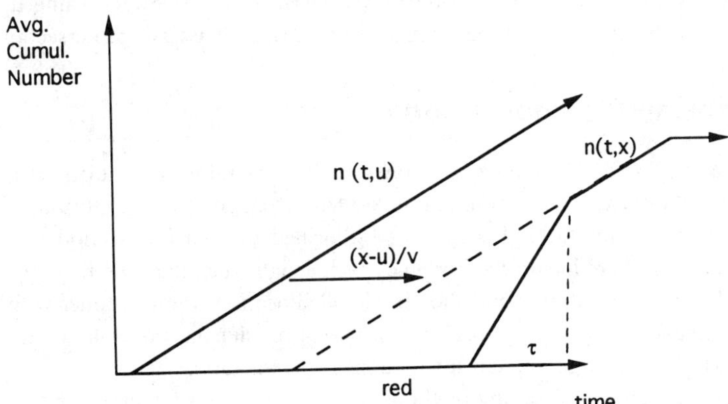

FIGURE 61.20 Average arrival and departure process.

Thus,

$$\bar{w} = \tfrac{1}{2}[1 - g/C]^2 C/(1 - q/s) \tag{61.15}$$

where $\bar{w}$ is the average delay imparted to intersection vehicles. Equation (61.15) signifies that the average intersection delay, assuming perfectly uniform vehicle arrivals, can be estimated knowing only the most basic details of signalization (g and C) and the average vehicle arrival and departure rates (q and s). The delay equation above is essentially identical to the delay equation in the *Highway Capacity Manual* (HCM) [TRB, 1985], with two exceptions:

1. The HCM employs an *adjustment factor* intended to account for the impacts created by arrivals that are not uniform (i.e., platooning effects caused by upstream signals).
2. The authors of the HCM have opted to use the average *stopped* delay, rather than average *approach* delay, for characterizing intersection operating quality. The former measure accounts only for the average amount of time vehicles spend completely stopped at the intersection, while the latter, estimated by Eq. (61.15), includes acceleration and deceleration delay on the approach. Thus, the HCM's delay equation incorporates an adjustment factor based on the rather crude assumption that 30% "extra" delay on an intersection approach results from acceleration and deceleration.

It is also important to note that Eq. (61.15) is sufficient for estimating average delays only under relatively low flow conditions (i.e., where the movement's volume-capacity ratio does not exceed 0.5). Under these low demand conditions, residual or "overflow" queues caused by random fluctuations in vehicle arrival rates are infrequent, and their deleterious impacts quickly disappear in subsequent cycles.

However, where flow rates increase, the frequency of residual queueing, and the persistence of the impacts, increase. Thus, a second, additive delay term is used for estimating average delay. The so-called *incremental* delay term included in the HCM represents only one of the overflow delay equations derived in an effort to model high-flow conditions [Hurdle, 1984].

What is perhaps most important to recognize about this second delay term is that the equation reflects a so-called *steady-state* assumption. This means that any delay prediction generated by the equation presumes that the intersection's operating conditions (e.g., arrival rates) have prevailed for a sufficiently long time that the intersection is no longer influenced by its previous operating state. Under higher-flow conditions this required *transition time* can be surprisingly large [Son *et al.*, 1994]. When an intersection has not completely transitioned from its previous state, actual prevailing delays can be substantially different from those predicted by equations. The steady-state assumption, in fact, is the reason why all incremental delay equations break down when the volume-capacity ratio approaches or exceeds 1.0.

It is therefore worth noting that field-measured delays, collected by recording upstream and downstream arrival times, represent the most accurate means for assessing intersection operation.

61.10 Signal Timing Considerations

As demonstrated in Eq. (61.15) and by intuition, signal timing influences the operating conditions (e.g., delay) prevailing at an intersection. In simplest terms, establishing signal timing plans involves determining the green durations that are to be allocated to each intersection movement. The techniques we present here identify a "suitable" cycle length (i.e., the time required to display a sequence of all signal indications) and the "efficient" allocation of green times within the cycle. The methods discussed herein are appropriate for fixed-time signalization, where the signal timing cannot be adjusted in response to cyclic fluctuations in arrival rates.

We also note that the signal timing methods we present are easy to apply and are commonly exploited. The methods are, for example, presented in the HCM.

The signal timing plan ultimately devised for a given intersection primarily depends on two considerations:

1. Prevailing traffic demands. As these average demands are time-variant, separate signal timing plans are often established for different times of day.
2. Signal phase plans. The manner in which right of way is allocated to conflicting movements can vary from simple two-phase plans to strategies that exclusively allocate green time to left-turn movements or extend right-of-way for given movements for two or more intervals. Many jurisdictions have established their own policies for developing phase plans. In the absence of local policies, the reader can refer to handbooks addressing the subject [Kell and Fullerton, 1991; TRB, 1985].

We now define several traffic stream measures relevant to signal timing.

Flow ratio, $(q/s)_j$, represents the ratio of vehicle arrival rate to queue discharge rate for movement j. These arrival and departure rates can be field-measured or assumed. The HCM, for example, devotes considerable discussion to the analytical estimation of s as a function of prevailing operating conditions.

Clearly, the amount of green time required for a jth movement will be proportional to the extent to which arrival rate q approaches discharge rate s.

Critical flow ratio, $(q/s)_{ci}$, represents the largest flow ratio occurring in the ith *phase* and is an indication of the required green time for phase interval i.

The variable Y is commonly used in the literature to represent the sum of the critical flow ratios prevailing in all intervals, $\sum_i (q/s)_{ci}$. The summation of individual flow ratios is demonstrated in an upcoming example.

Capacity, c_j, reflects the maximum rate of vehicles in movement j that can enter the intersection given the available green time and other relevant prevailing conditions. Intuitively, intersection capacity depends upon (1) the queue discharge rate and (2) the proportion of time that green is available to the movement. Thus,

$$c_j = s_j(g/C)_j \tag{61.16}$$

Using this expression for capacity, we define the volume-capacity ratio,

$$(q/c)_j = q_j/[s_j(g/C)_j] \tag{61.17}$$

Solving for g_j, the effective green time needed by movement j,

$$g_j = (q/s)_j(C/(q/c)_j) \tag{61.18}$$

Finally, we introduce X_c, the volume-capacity ratio of the overall intersection,

$$X_c = YC/(C - L) \tag{61.19}$$

where L is the total summed startup and ending lost time, ℓ_s and ℓ_e, occurring per cycle.

We note that the expression just obtained for X_c reflects the ratio of green time required per cycle (based on "critical" vehicle arrival and departure rates) to the total amount of green time actually available per cycle.

Having the expression for X_c, we can either estimate its value using a "desired" cycle length, C, or use the expression to solve for C given X_c. That is,

$$C = LX_c/[X_c - Y] \tag{61.20}$$

We now illustrate the application of these expressions, and further clarify certain issues, using an example.

Example. We assume that the flow ratios for each movement at the subject intersection have been measured or projected to be as illustrated in Fig. 61.21 and that the phase plan illustrated in Fig. 61.22 has been adopted. This phase plan exploits "overlaps." The westbound left turn movement in the first interval is continued in the second interval. Likewise, the westbound through movement is serviced in both interval 2 and interval 3. Thus, computing Y for this scenario is carried out as follows:

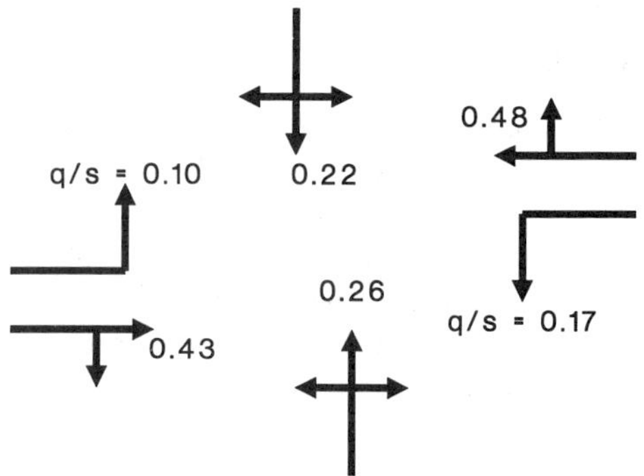

FIGURE 61.21 Example flow ratios.

Interval 1 will be extended for the duration required, on average, to serve the eastbound left turn. Thus, $(q/s)_{c1}$ is 0.10. The "remaining" flow ratio value for the westbound left turn movement is carried over to interval 2. When the westbound left turn demand is served, the third interval is initiated. Thus, $(q/s)_{c2}$ becomes 0.07. The remaining flow ratio is transferred to interval 3, and $(q/s)_{c3}$ is 0.43. Finally, $(q/s)_{c4}$ is 0.26.

The critical flow ratio for the intersection,

$$Y = 0.10 + 0.07 + 0.43 + 0.26 = 0.86$$

A value of Y less than 1.0 indicates that the intersection operates in an undersaturated manner, meaning, on average, that queues on all approaches can be served. Where a value of Y is greater than 1.0, oversaturated conditions (i.e., increased residual queueing) prevail. Such an operating state can be improved only by reducing arrival rate, by increasing discharge rate s (perhaps by adding approach lanes), or both.

To derive adequate signal timing parameters, we will rely upon a "desired" value for X_c. Convention dictates that, whenever possible, a cycle length should be sufficiently long to serve all vehicle queues (i.e., $X_c < 1.0$) without being excessively long, as Eq. (61.15) reveals that delay increases with cycle length. We therefore specify X_c to be 0.94.

We further assume that lost time per phase is equal to the amber time. We adopt an amber phase length of 4 seconds and note that the adopted phase plan interrupts all movements twice per cycle. Thus,

$$L = 4 \times 2 = 8 \text{ s/cycle}$$

Using Eq. (61.20), we determine cycle length:

$$C = LX_c/(X_c - Y) = 8 \times 0.94/(0.94 - 0.86) = 90 \text{ s}$$

To allocate green times to each interval, we use Eq. (61.18), and we adopt a conventional strategy that promotes a volume-capacity ratio for the *critical* movement in each interval to be equal to X_c.

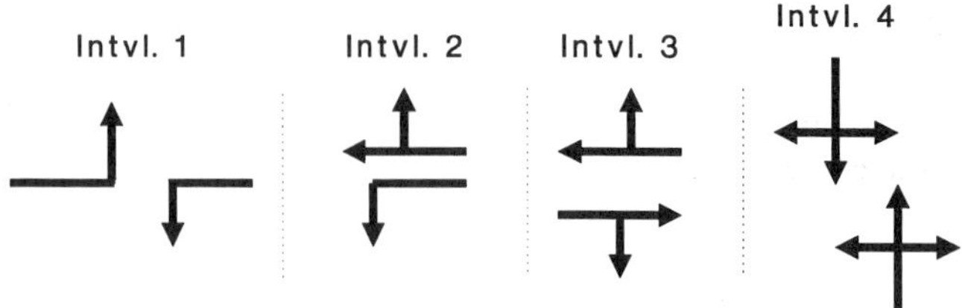

FIGURE 61.22 Example phase plan.

Thus,

$$
\begin{aligned}
g_1 = (q/s)_1(C/(q/c)_1) &= 0.10(90/0.94) = 9.5\,\text{s} \\
g_2 &= 0.07(90/0.94) = 6.5\,\text{s} \\
g_3 &= 0.43(90/0.94) = 41.0\,\text{s} \\
g_4 &= 0.26(90/0.94) = 25.0\,\text{s} \\
L &= \underline{8.0\,\text{s}} \\
C &= 90\ \ \text{s}
\end{aligned}
$$

Green time is thus established for each interval illustrated in Fig. 61.22 subject to the prevailing conditions depicted in Fig. 61.21.

Defining Terms

Capacity: The maximum rate at which vehicles can traverse a given facility.

Density: The number of vehicles occupying a specified length of highway section at an instant in time.

Flow: The rate at which vehicles arrive to a specified point on the highway.

Flow Ratio: The ratio of vehicle arrival rate to saturation flow rate at a signalized intersection.

Headway: The elapsed time between consecutive vehicle arrivals at some specified point on the highway.

Interrupted-Flow Facility: A facility deploying fixed causes of interruption (e.g., traffic signal or stop sign) to the traffic stream.

Interval: A signal display/indication.

Queueing Diagram: Graphical depiction of translated cumulative vehicle arrival curves.

Saturation Flow Rate: The maximum queue discharge rate at a signalized intersection.

Shock Wave: A propagating discontinuity between two traffic states.

Spacing: The physical distance between consecutive vehicles.

Speed: The rate of motion expressed as distance per time.

Time-Space Diagram: Graphical depiction of vehicle motion over time and space.

Trajectory: Vehicle motion over time and space.

Uninterrupted-Flow Facility: A facility without fixed causes of interruption to the traffic stream.

References

AASHTO (American Association of State Highway and Transportation Officials), *A Policy on Geometric Design of Highways and Streets,* Washington, D.C., 1984.

Daganzo, C. F. and Newell, G. F. Unpublished. *Notes on Introduction to Transportation Engineering.* University of California, Berkeley.

Greenshields, B. D. 1935. A study of traffic capacity. *Highway Res. Board Proc. 14.*

Hurdle, V.F. 1984. "Signalized intersection delay models—a primer for the uninitiated." *Transportation Research Record 971.* Transportation Research Board, National Research council, Washington, D.C.

ITE (Institute of Transportation Engineers). 1976. *Manual of Traffic Engineering Studies.* Arlington, VA.

Kell, J. H. and Fullerton, I. J. 1991. *Manual of Traffic Signal Design.* Institute of Transportation Engineers. Prentice Hall, Englewood Cliffs, NJ.

Lighthill, M. J. and Whitham, G. B. 1955. On kinematic waves. I: Flood movement in long rivers. II: A theory of traffic flow on long crowded roads. *Proc. Royal Society (London).*

MUTCD (Manual on Uniform Traffic Control Devices). 1988. U.S. DOT, FHWA, Washington, D.C.

Newell, G. F. Unpublished. "Notes on the Operation of Transportation Facilities," University of California, Berkeley.

Newell, G. F. 1988. *Theory of Highway Traffic Signals.* Institute of Transportation Studies, Working Paper UCB-ITS-WP-88-10, University of California, Berkeley. 1988.

Son, Y. T., Cassidy, M. J., and Madanat, S. M. 1994. Evaluating the appropriateness of the steady state assumption for a highway queueing system. In press. *ASCE J. of Transportation Eng.*

TRB (Transportation Research Board). 1985. *Highway Capacity Manual.* Special Report 209. Washington, DC.

For Further Information

The commonly adopted procedures for capacity and level of service analysis are detailed in the 1985 *Highway Capacity Manual,* published by the Transportation Research Board.

There also exist a number of references describing other aspects of traffic engineering, such as traffic control, traffic management, traffic studies, and design.

The *Manual on Uniform Traffic Control Devices,* published by the U.S. Department of Transportation, documents warrants and standards for the implementation of traffic control devices (e.g., signs, signals, pavement markers).

Suggested methods for designing and operating traffic signals are presented in the *Manual of Traffic Signal Design* by Kell and Fullerton.

The commonly adopted procedures for performing traffic engineering studies (e.g., speed studies, volume counts, origin-destination studies) are outlined in the *Manual of Traffic Engineering Studies* published by the Institute of Transportation Engineers.

Geometric standards for highway design are documented in *A Policy on Geometric Design of Highways and Streets* published by the American Association of State Highway and Transportation Officials.

62

Intelligent Transportation Systems*

Yorgos J. Stephanedes
University of Minnesota, Minneapolis

62.1 Introduction

The dilemma facing the American traveling public is that, while demands for mobility continue to increase, the available capacity of the roadway system is nearly exhausted. Except for fine tuning and relatively modest additions, the road system cannot be expanded in many areas. The only means left for increasing available travel capacity is to use the available capacity more effectively: Redirect traffic to avoid congestion, provide assistance to drivers and other travelers on planning and following optimal routes, increase the reliability of and access to public transportation, and refocus safety efforts on accident avoidance rather than merely minimizing the consequences of accidents [1].

Responding to this need, *the objective of intelligent transportation systems (ITS) is to use the available capacity more effectively, promote more efficient use of the existing highway and transportation network, make significant improvements in mobility, highway safety, and productivity by*

*This chapter contains substantial portions of text, largely unchanged, from references 1, 2, 4, 6, 7, 10, 19, and 20.

0-8493-8953-4/95/$0.00 + $.50

building transportation systems that draw upon advanced electronic technologies and control software, and decrease the environmental costs of travel. In particular, *ITS is the integrated application of advanced information processing and communications, sensing, display, and control technologies to surface transportation—both in the vehicle and on the highway [1,2].* It should be noted that the initial efforts in the use of advanced technologies in surface transportation were organized under the name of intelligent vehicle highway systems (IVHS). However, it was felt that the term IVHS was restrictive and in late 1994 it was changed to intelligent transportation systems (ITS). Some of the effects that ITS could have in transportation operations, safety, and productivity are described next.

62.2 Role of ITS in Tomorrow's Transportation Systems

Operations

The essence of ITS as it relates to transportation operations is the improved ability to manage transportation services as a result of the availability of accurate, real-time information and to greatly enhance control of traffic flow and individual vehicles. With ITS, decisions that individuals make as to time, mode, and route choices can be influenced by information that currently is not available when it is needed, or is incomplete, inconvenient, or inaccurate. For example, ITS technology would enable operators to detect incidents more quickly; to provide information immediately to the public on where the incident is located, its severity, its effect on traffic flow, and its expected duration; to change traffic controls to accommodate changes in flow brought about by the incident; and to provide suggestions on better routes and information on alternative means of transportation [2].

The availability of this information would also enable the development of new transportation control strategies. For example, to obtain recommended routing information, drivers will have to specify their origins and destinations. Knowledge of origin and destination information in real time will enable the development of traffic assignment models that will be able to anticipate when and where congestion will occur. Control strategies that integrate the operation of freeway ramp-metering systems, driver information systems, and arterial traffic signal control systems and that meter flow into bottleneck areas can be developed to improve traffic control. Eventually, perhaps toward the third decade of the 21st century, totally automated facilities may be built on which vehicles will be controlled by electronics in the highway [2].

Safety

Whereas many safety measures have been aimed at reducing the consequences of accidents (such as vehicle crashworthiness and forgiving roadside features), many ITS functions are directed toward the *prevention* of accidents. A premise of the European PROMETHEUS program, for example, is that 50% of all rear-end collisions and accidents at crossroads and 30% of head-on collisions could be prevented if the driver is given another half-second of advance warning and reacts correctly. Over 90% of these accidents could be avoided if drivers take the appropriate countermeasures one second earlier. ITS technologies that involve sensing and vehicle-to-vehicle communications will initially be designed to automatically warn the driver, providing enough lead time for evasive actions. The technologies may also assume some of the control functions that are now totally the responsibility of drivers, compensating for some of their limitations and enabling them to operate their vehicles closer together but safer. Potential safety dangers must, however, also be acknowledged. A key issue involves driver distraction and information overload from the various warning and display devices in the vehicle. Other issues include dangers resulting from system unreliability (for example, a warning or driver-aid system that fails to operate) and the incentive for risky driving that ITS technologies may provide [2].

Productivity

The availability of accurate, real-time information will be especially useful to operators of vehicle fleets, including transit/HOV, emergency, fire, and police services, as well as truck fleets. Operators will know where their vehicles are and how long a trip can be expected to take; thus, they will be able to advise on best routes to take and will be able to manage their fleets better. Further, there is great potential for productivity improvements in the area of regulation of commercial vehicles. Automating and coordinating regulatory requirements through application of ITS technologies can, for example, reduce delays currently incurred at truck weigh stations, reduce labor costs to the regulators, and minimize the frustration and costs of red tape to long-distance commercial vehicle operators. There is also potential for improved coordination among freight transportation modes; as an example, if the maritime and trucking industries were to use the same electronic container identifiers, freight-handling efficiencies would be greatly improved [2].

62.3 ITS Categories

By general agreement, ITS has been subdivided into six overlapping system areas, three focused on technology and three on applications [1]. Technology-oriented system areas are:

- Advanced traffic management systems (ATMS)
- Advanced traveler information systems (ATIS)
- Advanced vehicle control systems (AVCS)

Applications-oriented system areas are:

- Advanced public transportation systems (APTS)
- Commercial vehicle operations (CVO)
- Advanced rural transportation systems (ARTS)

Advanced Traffic Management Systems (ATMS)

ATMS address technologies to monitor, control, and manage traffic on streets and highways. ATMS technologies include [1]:

- Traffic management centers (TMCs) in major metropolitan areas to gather and report traffic information, and to control traffic movement to enhance mobility and reduce congestion through ramp, signal, and lane management, vehicle route diversion, etc.
- Sensing instrumentation along the highway system, which consists of several types of sensors, including magnetic loops and machine vision systems, that provide current information on traffic flow to the TMC
- Variable message signs that provide current information on traffic conditions to highway users and suggest alternate routes
- Priority control systems to provide safe travel for emergency vehicles when needed
- Programmable, directional traffic signal control systems
- Automated dispatch of tow, service, and emergency vehicles to accident sites

Advanced traffic management systems have six primary characteristics differentiating them from the typical traffic management system of today [2]. In particular, ATMS:

- Work in real time.
- Respond to changes in traffic flow. In fact, an ATMS will be one step ahead, predicting where congestion will occur based on collected origin-destination information.

- Include area-wide surveillance and detection systems.
- Integrate management of various functions, including transportation information, demand management, freeway ramp metering, and arterial signal control.
- Imply collaborative action on the part of transportation management agencies and jurisdictions involved.
- Include rapid response incident management strategies.

To implement ATMS, real-time traffic monitoring and data management capabilities are being developed, including advanced detection technology, such as image processing systems, automatic vehicle location and identification techniques, and the use of vehicles as probes. New traffic models are being created, including real-time dynamic traffic assignment models, real-time traffic simulation models, and corridor optimization techniques. The applicability of artificial intelligence and expert systems techniques is assessed, and applications such as rapid incident detection, congestion anticipation, and control strategy selection are being developed and tested [2].

Advanced Traveler Information Systems (ATIS)

ATIS address technologies to assist travelers with planning, perception, analysis, and decision making to improve the convenience and efficiency of travel. In the automobile ATIS technologies include [1]:

- On-board displays of maps and roadway signs (in-vehicle signing)
- On-board navigation and route guidance systems
- Systems to interpret digital traffic information broadcasts
- On-board traffic hazard warning systems (e.g., icy road warnings)

Outside the vehicle, ATIS technologies include [1]:

- Trip-planning services
- Public transit route and schedule information available on-line at home, office, kiosks, and transit stops

Advanced traveler information systems provide drivers with information on congestion and alternate routes, navigation and location, and roadway conditions through audio and visual means in the vehicle. This information can include incident location, location of fog or ice on the roadway, alternate routes, recommended speeds, and lane restrictions. ATIS provide information that assists in trip planning at home, at work, and by operators of vehicle fleets. ATIS also provide information on motorist services, such as restaurants, tourist attractions, and the nearest service stations and truck and rest stops (this has been called the "yellow pages function"). ATIS can include on-board displays that replicate warning or navigational roadside signs when they may be obscured during inclement weather or when the message should be changed, as when speed limits should be lowered on approaches to congested freeway segments or fog areas. An automatic mayday feature may also be incorporated, which would provide the capability to automatically summon emergency assistance and provide vehicle location [2].

A substantial effort is required to define the communications technology, architecture, and interface standards that will enable two-way, real-time communication between vehicles and a management center. Possibilities include radio data communications, cellular systems, roadside beacons used in conjunction with infrared or microwave transmissions or low-powered radio signals, and satellite communications. Software methods to fuse the information collected at the management center and format it for effective use by various parties must also be developed. These parties include commuters; other trip makers; commercial vehicle operators, both before they make a trip and en route; operators of transportation management systems; and police, fire, and emergency response services [2].

A number of critical human factors issues must also be investigated. These include identifying the critical pieces of information and the best way of conveying them to different individuals. The human factors issues also include a critical examination of in-vehicle display methods, among them: What are the roles of audio messages and in-vehicle displays? Where should visual displays be located and what technology should be used? What should be shown on visual displays? [2].

Advanced Vehicle Control Systems (AVCS)

AVCS address technologies to enhance the control of vehicles by facilitating and augmenting driver performance and, ultimately, relieving the driver of some tasks, through electronic, mechanical, and communications devices in the vehicle and in the roadway. AVCS technologies include [1]:

- Adaptive cruise control, which slows a cruise-controlled vehicle if it gets too close to a preceding vehicle
- Vision enhancement systems, which aid driver visibility in the dark or in adverse weather
- Lane departure warning systems, which help drivers avoid run-off-the-road crashes
- Automatic collision avoidance system, that is, automatic braking upon obstacle detection
- Automated highway systems (AHS), automatically controlling vehicles in special highway lanes to increase highway capacity and safety.

Whereas the other categories of ITS primarily serve to make traveling more efficient by providing more timely and accurate information about transportation, AVCS serve to greatly improve safety and potentially make dramatic improvements in highway capacity by providing information about changing conditions in the immediate environment of the vehicle, sounding warnings, and assuming partial or total control of the vehicle [2].

Early implementation of AVCS technologies may include a number of systems to aid with the driving task. These include hazard warning systems that sound an alarm or actuate a light when a vehicle moves dangerously close to an object, such as when backing up or when moving into the path of another vehicle when changing lanes. Infrared imaging systems may be implemented that enhance driver visibility at night. AVCS technologies also include adaptive cruise control and lane-keeping systems that automatically adjust vehicle speed and position within a lane through, for example, radar systems that detect the position and speed of a lead vehicle, or possibly through electronic transmitters in the pavement that detect the position of vehicles within the lane and send messages to a computer in the vehicle that has responsibility for partial control functions. As technology advances, lanes of traffic may be set aside exclusively for automated operation, known as platooning highway systems. Small groups of vehicles, perhaps up to 12 per platoon, will travel together at high speeds, maybe 105 kilometers per hour with very short headways, controlled through obstacle detection and automatic speed control and braking systems.

Within AVCS the ISTEA requires that the U.S. DOT develop an automated highway and vehicle system and establishes a goal of having a demonstration in place by 1997. Although it may be questionable if such a goal could be met, this is an exciting opportunity to dramatically advance the vehicle control aspects of ITS. The vision for the AHS program is to create a fully automated system that evolves from today's roads, beginning in selected corridors and routes; provides fully automated, "hands off" operation at better levels of performance than today in terms of safety, efficiency, and comfort; and allows equipped vehicles to operate in both urban and rural areas and on highways that are instrumented and not instrumented [7].

Although full deployment of an AHS is certainly a long-term goal, pursuit of this goal is extremely important. A new level of benefits could be realized with the complete automation of certain facilities. By eliminating human error, an automated highway could provide a nearly accident-free driving environment. In addition, the precise, automated control of vehicles on an automated vehicle-highway system could result in an increase of two to three times the capacity of present-day facilities, while encouraging use of more environmentally benign propulsion methods.

Initial AHS deployments might be on heavily traveled urban or interstate highway segments, and the automated lanes might be comparable to the high-occupancy vehicle lanes on today's highways. If successful, the AHS could evolve into a major advance of the nation's heavily traveled roadways or the interstate highway system [7].

However, much research and development work and testing are needed before such systems can be built and implemented. Perhaps the most important issues relate to the role of humans in the system—that is, public acceptability and how it is likely to affect system effectiveness. Other human factor issues include driver reaction to partial or full control—whether it will cause them to lose alertness or to drive more erratically. Another important area is AVCS reliability and the threat of liability [2].

Advanced Public Transportation Systems (APTS)

APTS address applications of ITS technologies to enhance the effectiveness, availability, attractiveness, and economics of public transportation. APTS strive to improve performance of the public transportation system at the unit level (vehicle and operator) and at the system level (overall coordination of facilities, provision of better information to users). APTS technologies include [1]:

- Fleet monitoring and dispatch management
- On-board displays for operators and passengers
- Real-time displays at bus stops
- Intelligent fare collection (e.g., using smart cards)
- Ride share and HOV information systems

Applications of ITS technologies could lead to substantial improvements in bus and paratransit operations in urban and rural areas. Dynamic routing and scheduling could be accomplished through on-board devices, communications with a fleet management center, and public access to a transportation information system containing information on routes, schedules, and fares. Automated fare collection systems could also be developed that would enable extremely flexible and dynamic fare structures and relieve drivers of fare collection duties [2].

Commercial Vehicle Operations (CVO)

CVO address applications of ITS technologies to commercial roadway vehicles (trucks, commercial fleets, intercity buses). Many CVO technologies—especially for interstate trucking—relate to the automated, no-stop-needed handling of the routine administrative tasks that have traditionally required stops and waiting in long lines: toll collection, road use calculation, permit acquisition, vehicle weighing, etc. Such automation can save time, reduce air pollution (most/worst emissions are produced during acceleration and deceleration), and increase the reliability of record keeping and fee collection. CVO technologies include [1]:

- Automatic vehicle identification (AVI)
- Weigh in motion (WIM)
- Automatic vehicle classification (AVC)
- Electronic placarding/bill of lading
- Automatic vehicle location (AVL)
- Two-way communications (TWC) between fleet operator and vehicles
- Automatic clearance sensing (ACS)

The application of ITS technologies holds great promise for improving the productivity, safety, and regulation of all commercial vehicle operations, including large trucks, local delivery vans, buses, taxis, and emergency vehicles. Faster dispatching, efficient routing, and more timely pick-ups and deliveries will be made possible, and this will have a direct effect on the quality and competitiveness of businesses and industries at both the national and the international level [2].

ITS technologies manifest themselves in numerous ways in commercial vehicle operations. For example, for long-distance freight operations, on-board computers will not only monitor the other systems of the vehicle, but could also function to analyze driver fatigue and provide communications between the vehicle and external sources and recipients of information. Applications include automatic processing of truck regulations (for example, commercial driver license information, safety inspection data, and fuel tax and registration data), thus avoiding the need to prepare redundant paperwork and leading to "transparent borders"; provision of real-time traffic information through advanced traveler information systems; proof of satisfaction of truck weight laws using weigh-in-motion scales, classification devices, and automated vehicle identification transponders; and two-way communication with fleet dispatchers using automatic vehicle location and tracking and in-vehicle text and map displays. Regulatory agencies would be able to take advantage of computerized record systems and target their weighing operations and safety inspections at those trucks that are most likely to be in violation [2].

Advanced Rural Transportation Systems (ARTS)

ARTS address applications of ITS technologies to rural needs, such as vehicle location, emergency signaling, and traveler information. The issues involved in implementing ITS in rural areas are significantly different from those in urban areas, even when services are similar. Rural conditions include low population density, fewer roads, low amount of congestion, sparse or unconventional street addresses, etc. Different technologies and/or communications techniques are needed in rural ITS to deal with those conditions. Safety is a major issue in ARTS; over half of all accidents occur on rural roads. ARTS technologies include [1]:

- Route guidance
- Two-way communications
- Automatic vehicle location (AVL)
- Automatic emergency signaling
- Incident detection
- Roadway edge detection

Application of ARTS technologies can address the needs of rural motorists who require assistance either because they are not familiar with the area in which they travel (tourists) or because they face extreme conditions such as weather, public works, and special events. Provision of emergency services is particularly important in rural areas. A state study [28] has determined that notification of spot hazardous conditions and collision avoidance at nonsignalized intersections are highly important issues that ARTS could address in the short term; long-term issues include construction zone assistance, transit applications, inclement weather trip avoidance and assistance, tourist en-route information and traffic control, and in-vehicle mayday devices.

62.4 Benefits of ITS

It is interesting to reflect on how technology has influenced transportation in the U.S. The steam engine improved travel by boat and railroad, resulting in coast-to-coast systems. The internal combustion engine freed the vehicle from a fixed guideway or waterway and encouraged the construction of farm-to-market roads as well as enabled travel by air. ITS will move transportation another step forward by providing traveler information and by operating traffic management and control systems. This quantum leap will have a major impact on today's lifestyle [6].

Attempting to quantify the benefits of widely deployed ITS technologies at this stage must be similar to what planners of the U.S. interstate highway system tried to do in the 1950s. It is impossible to anticipate all of the ways that applications of ITS technology may affect society, just as planners of the interstate highway system could not have anticipated all of its effects on American society. Recognizing the importance of the issue, however, Mobility 2000—an ad hoc coalition of industry; university; and federal, state, and local government participants whose work led to the establishment of IVHS America—addressed the potential benefits of applying ITS technology in the U.S. Numerous benefits were predicted for urban and rural areas and for targeted groups, such as elderly and disadvantaged travelers. Positive benefits were also found in regard to the environment. Some of the specific findings that were reported include the following [2]:

- Fully deployed ATMS/ATIS combinations can reduce congestion costs in urban areas from 20 to 40%. This was projected from initial simulation work and estimates for the Smart Corridor project in Los Angeles and has been confirmed from simulation results in the I-494 corridor in Minneapolis [29].
- It was estimated that the cost of delay in the U.S. in 1990 was approximately $100 billion. The value of time saved alone would therefore be at least $25 billion in 1990 and would grow substantially since total travel is expected to increase by about 50% by the year 2005.
- Unchecked traffic congestion is the single largest contributor to poor air quality and wasted fuel consumption. Reductions in traffic congestion will lead to improvements in these areas.
- By 2010 annual savings of approximately 11,500 lives and $22 billion in accident costs could be realized, based on an analysis that considered safety technology features, projected market penetrations, and estimated effectiveness in reducing various accident types.
- By 2020 a similar analysis estimated that annual savings of 33,500 lives and $65 billion in accident costs could be realized, as advanced vehicle control strategies achieve a large market penetration.
- Rural areas have the most to gain in relation to safety improvements, since 57% of fatal accidents occur in rural areas where collision speeds are likely to be higher.
- Older and disadvantaged drivers can benefit by having specific devices available to offset the slowing down of their capabilities; these devices could include infrared imaging, obstacle detection and warning systems, radar braking and steering override, and on-board replication of maps and signs.
- Motor carrier productivity can be significantly increased and fuel costs can be decreased through automated toll collection, the provision of real-time routing information and yellow pages services, automated processing of permits and licenses, and on-board computers that provide information on vehicle performance.

Assuming that the benefits and cost effectiveness are proven and the public exhibits a willingness to pay for these systems, it is estimated that the U.S. market alone for automotive electronics will amount to $28 billion annually by the year 2000, and that the U.S. highway infrastructure costs for these systems would total $30 billion through the year 2010. Thus, there will be a very substantial international market for ITS products and services that can be supplied by the private sector, contributing to economic growth [2].

Advanced technologies for managing traffic can provide significant benefits of reduced delay and improved safety. However, the acceptance and application of these technologies has proceeded more slowly than the incorporation of computer and communications technologies in "smart cars." The ATMS portion of ITS technologies is unique in that it is almost exclusively a public sector responsibility. The customer for these technologies is not the individual driver but the public agencies responsible for operating and maintaining the street systems. The streets and highways scheduled to be made "smart" with computer and communications technology are under the jurisdiction and control of state and local governments. In most areas, more than one jurisdiction of local government is involved. In addition, there might be more than one department of a state or local government involved in the ATMS program. In most cases the process of working out the institutional arrangements is a more daunting task than the installation of a technology [4].

Greater involvement by engineers who operate and maintain highway systems (city traffic engineers, in particular) can help to emphasize that the smart highway system includes urban arterials as well as freeways. These traffic engineers, most of whom had early training as civil engineers, might not have a complete knowledge of the computer and communications technologies available for intelligent highways. They do, however, have a unique understanding of traffic operations and human factors needed to ensure that new technologies address the highest-priority traffic problems [4].

Car owners will maintain the technology in the smart cars. The technology in the smart streets, however, will be maintained by public road agencies with jurisdictional responsibility. The benefits of advanced technology for traffic management systems cannot be realized over a long period of time unless adequate funds are available to these agencies for operations and maintenance [4].

62.5 U.S. Federal Actions to Deliver the ITS Program

The major efforts within the Federal Highway Administration (FHWA) and other U.S. DOT agencies have focused on implementing the provisions of the IVHS Act of 1991, which is an integral part of the Intermodal Surface Transportation Efficiency Act (ISTEA) signed into law on 18 December 1991. The major area of support from ISTEA is provision of a total of $659 million from fiscal years 1992 through 1997 in FHWA contract authority to carry out the goals of the ITS program. Much of this funding is to support the IVHS Corridors Program [7]. The Clinton administration has proposed increasing funding for ITS by $355 million [30].

Research and Development

Acceleration of the ITS research and development (R&D) program began in 1991, with funding of nearly $4 million of R&D activities. With passage of ISTEA the program was greatly expanded to about $24 million in 1992. The R&D program is designed to explore issues critical to ITS implementation and to provide needed insight into new technologies and applications that can be used as building blocks for emerging and future ITS applications. Several broad categories of research are pursued to advance the ITS program [7]:

Providing basic research tools and a knowledge base. These activities enhance overall research and analytical capabilities, rather than focus on any specific ITS application. Examples include data collection and analysis, computer modeling, creation of databases, and the development of research equipment and test facilities.

Creating a favorable environment for ITS applications. These activities focus on the broader, usually nontechnical, issues facing ITS evolution, such as the identification of economic/social/legal obstacles and ways to resolve them; cost-benefit studies/analyses of the impact of ITS; and the relationships among various participants, that is, federal government, state/local government, private industry, and travelers/users.

Defining potential opportunities for ITS applications. These projects identify specific highway and public transportation circumstances in which ITS could provide a solution to current problems and determine the feasibility of those concepts. Examples include issues such as fare collection, augmenting driver capabilities to avoid collisions, routing advice for travelers, and advanced traffic management.

Developing ITS applications. While this responsibility generally resides with the private sector, government also will play a role where appropriate. This includes the initial development of systems with high public benefit but low commercial potential that private industry might not pursue on its own, as well as the development of systems where government, rather than the private sector, is a substantial or primary user of the application.

Selection and initiation of research projects results form a formal program development process that takes into account input from the ITS community. U.S. DOT and other agencies are also turning to more innovative contracting mechanisms to help deliver a program of this size and complexity in a timely manner. For example, several agreements with the national laboratories have been initiated to pursue various research topics. These agreements take advantage of the wealth of applicable talent and sophisticated resources these organizations possess. Research consortia have been formed to develop ITS topics such as real-time traffic control strategies. These consortia offer the benefit of establishing partnerships with the private sector. The partnerships include significant cost sharing by the various partners, which leverages the federal funds available and accelerates the program [7].

Operational Tests

The U.S. DOT is supporting implementation of a comprehensive set of operational test projects designed to address key ITS technological and institutional issues. These projects are conducted as cooperative partnerships among U.S. DOT and a variety of public and private partners, including state and local governments, private companies, and universities. Funding, technical, and administrative responsibilities are shared among the partners. Operational tests are designed to evaluate applications of new technologies and system concepts, facilitating the transition from the realm of research and development into operational use. An operational test integrates existing technologies with new R&D products, as well as institutional and perhaps financial or regulatory arrangements, in order to evaluate one or usually more of these elements in an operational highway environment under "live" transportation conditions. Operational tests also provide more opportunities for R&D as the operational impact of proposed ITS concepts becomes known [7].

Evaluation of operational tests and the ITS program as a whole is a major federal responsibility. The ISTEA requires that operational tests have a written evaluation consistent with guidelines, criteria, and standard methodologies established by the U.S. DOT. In this manner it is ensured that a uniform basis is used in comparing the results of ITS operational tests [7].

Corridors Program

A large share of the funds provided in ISTEA for ITS is for the newly created ITS Corridors Program, with $501 million during the six-year period. The funds provided through the Corridors Program support advancement of ITS technologies and institutional issues, primarily through the implementation of operational tests of the various ITS services and functions. In addition, Corridors Program funds can support certain efforts in support of corridor plans, such as planning, feasibility, and conceptual studies. The corridor sites become national test beds for ITS and provide the public its first introduction to ITS. Over the long term, funds spent on operational tests at each corridor site, along with funds that the operational test money leverages, result in the establishment of an ITS infrastructure that supports continuing deployment of ITS technologies and services [7].

The ISTEA defines seven specific criteria to identify "priority corridors." Among the criteria are traffic density above the national average, severe or extreme ozone nonattainment, a variety of transportation facilities, and an inability to significantly expand capacity. Six sites meet these criteria: In the Northeast, along Interstate 95 from just north of Washington, D.C., to Connecticut; in the Midwest, Chicago and parts of Indiana and Wisconsin; the Houston, Texas, area; and, in California, Los Angeles, Orange, and San Diego Counties. An "other corridors" component of the program provides funding for ITS activities at sites that do not meet all of the ISTEA criteria, but possess the characteristics necessary to advance the national ITS program. Four such potential sites are southeast Florida, southeast Michigan, Minnesota, and Seattle, Washington. Over time additional corridors and areas will be considered, such as commercial vehicle-oriented and rural corridors. Further, additional operational test projects will be added at corridor sites, moving these test beds toward deployment of integrated ITS applications [7].

In addition to projects conducted by the Corridors Program participants, other interested parties are assured an opportunity to participate in the operational test program through presentation of project proposals for funding consideration. Each year in March U.S. DOT publishes an announcement in the *Federal Register* seeking offers from public and private entities to form partnerships to conduct ITS operational tests at sites other than those in the Corridors Program. The announcement identifies ITS technical or institutional areas in which tests are needed to supplement Corridors Program activities [7]. A complete listing of projects supported by the U.S. Department of Transportation is available from the FHWA. The location of a number of operational tests is marked on an accompanying map in Fig. 62.1 [5]. A small sampling of these is included in section 62.7.

ITS Issues

ITS System Architecture

A system architecture is a master plan—a structural outline—within whose dimensions a variety of real-world systems can be implemented. It describes how components interact to achieve total system goals. Ideally, a system architecture provides a modular design that allows the system to grow and change over time, to include new components, and to accommodate new technologies [1]. In the case of ITS an architecture would lay out the specific functions performed by in-vehicle equipment, by roadside devices, by traffic management facilities, or by information distribution centers [7].

Although all systems, regardless of function, have some sort of system architecture, there are compelling reasons for developing a thoughtful and forward-looking architecture before building the system rather than letting the architecture evolve alongside it. The still-emerging state of ITS affords a prime opportunity to develop a sound architecture that can be used to guide the implementation of an integrated ITS across the country. The benefits to ITS of careful architecture development, compatibility, and interoperability across areas are numerous [1,7]:

- Insurance of compatibility and interchangeability of components throughout the system, such that, for example, emergency-response and in-vehicle navigation systems can function consistently regardless of locale, anywhere in the country.

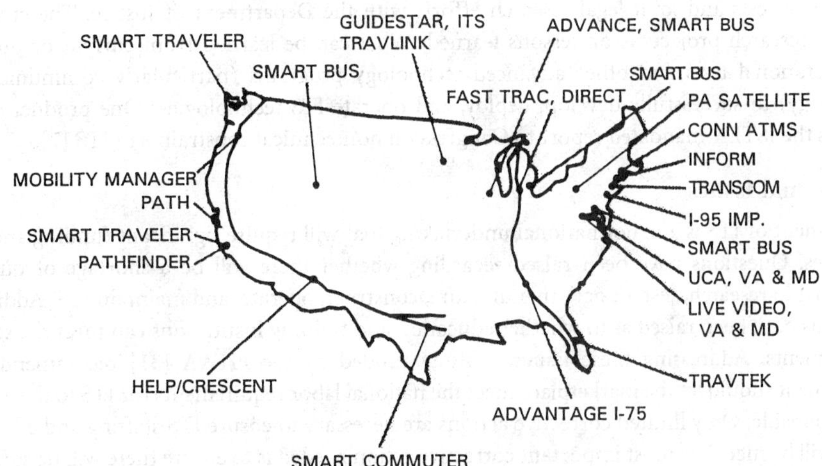

FIGURE 62.1 Some of the tests of ITS technologies taking place in the U.S. (*Source:* Costantino, James. *ITE.* Feb. 1993, p. 21. With permission.)

- Enforcement, permitting, and compliance systems can serve the widest possible community without requiring vehicles, particularly commercial vehicles, to carry multiple redundant systems.
- Avoidance of multiple standards proliferation, or the painful evolution of de facto standards, will reduce risk and cost for product manufacturers and consumers alike.
- Ability to "build in" capacity for growth and continual modernization of the system.
- Acceleration of actual installation of the system, since we will have a picture of the fully functional, completed product.

To this end the ITS community has begun a strong initiative to develop an embracing ITS system architecture. Such an architecture will identify the abstract components of the overall ITS system and, equally important, the critical interfaces among components. This initiative, which is led by the U.S. DOT, is expected to develop the system architecture for an integrated nationwide ITS by 1996. The program establishes a far-reaching, consensus-building effort to tie together the diverse interests of the many users of a nationwide ITS and allow their needs and concerns to be incorporated in the development process. Fifteen multidisciplinary teams initially participated in the proposal process to develop promising system concepts. Four of these teams were selected to work in parallel on competing systems architectures, one of which will eventually serve as a base for developing ITS services and products throughout the country [1,7,9].

Institutional Issues

The challenge of developing and implementing ITS is as much concerned with nontechnical issues as with technology. Successful system solutions must recognize the current transportation institutional environment, such as jurisdictional authorities and legal restrictions, so that deployment of ITS can naturally evolve from today's circumstances. Examples of these issues include [7]:

- Definition of true needs/expectations of all system users
- Social/societal issues such as privacy and equity
- Relationships between ITS and air quality requirements
- Legal considerations such as antitrust and liability
- Jurisdictional or implementation concerns such as public/private partnerships, multijurisdictional relationships, and privatization options

Institutional issues are studied using several research approaches including case studies of ITS operational tests and joint legal research efforts with the Department of Justice. The emphasis of most research projects is on lessons learned: What can be learned from current or previous ITS operational tests and other advanced technology programs (particularly communications technology) to successfully develop, deploy, and operate ITS technologies? One product of this effort is the ISTEA-mandated report to Congress on nontechnical constraints to ITS [7].

Educational Issues

Deployment of ITS is a major national undertaking that will require significant funding and staff resources. Questions have been raised regarding whether there will be a shortage of qualified personnel to research, test, design, manufacture, construct, operate, and maintain ITS. Additional questions have been raised as to whether education and training institutions can meet the staffing requirements. Addressing these issues, a study funded by the FHWA [31] recommends that government should let the marketplace meet the national labor requirements for ITS to the greatest extent possible. Only limited corrective actions are necessary to ensure ITS staffing and education needs will be met. The most important corrective action needed is to ensure there will be sufficient qualified staff employed at the regional and local level [31].

The study determined that there is a need for substantially broader, more integrated multidisciplinary college and university curricula to meet ITS needs. Civil engineering programs need to address the acquisition of broader technical and management knowledge including electrical and mechanical engineering, computer science, systems, and organizational skills. Similarly, non-civil engineers who wish to effectively address transportation problems in ITS need to acquire transportation knowledge, including basic civil and traffic engineering skills.

The study also noted that colleges and universities cannot meet all the education and training needs for ITS. Retraining programs, offered in-house by private firms and public agencies, will be essential to retooling people for jobs in ITS. The private sector has been successful in rapidly introducing training programs to respond to significant technological developments, whereas public vocational schools have generally been more concerned with providing basic education.

ITS User Services

In 1993 the ITS America National Program Plan introduced a set of ITS user services and subservices. This set presents ITS as 27 related services that are targeted toward the user. The ITS user services are summarized as follows:

Traveler services information
Pretrip travel information
Route guidance
En-route driver advisory
Emergency notification and personal security
Incident management
Travel demand management
Traffic control
Electronic payment services
Emergency vehicle management
Longitudinal collision avoidance
Lateral collision avoidance
Fully automated vehicle operation
Precrash restraint deployment
Impairment alert
Vision enhancement for crash avoidance
Intersection crash warning and control
Commercial vehicle preclearance
Automated roadside safety inspections
Commercial vehicle administrative processes
On-board safety monitoring
Commercial fleet management
Personalized public transit
Public transportation management
Public travel security
Ride matching and reservation
En-route transit advisory

62.6 An ITS State Team Effort

The establishment of a national ITS program has presented to state and local transportation engineers new opportunities for using advanced technologies to improve mobility during a time of continually increasing travel demands [25].

In Minnesota the establishment of the Institute for ITS Concepts has presented the opportunity for combining the capabilities of the ITS Laboratory of the new Institute with those of the Human Factors Laboratory and the I-394 test bed as well as other test beds in the Minneapolis/St. Paul metropolitan area. The Minnesota test beds are stretches of freeway, urban networks, or combinations of the two that have been instrumented to facilitate the implementation of ITS tools. This instrumentation has been installed to address a number of objectives, including:

Real-time detection of traffic flow
Real-time collection of ground truth information
Communications
Real-time traffic information
Real-time traffic management

In particular the test beds have been densely instrumented with magnetic loop and wide-area machine vision–based sensors that can collect real-time information on basic traffic flow variables such as volume, occupancy, and speed. Such information is communicated to a central computer over state-of-the-art communications networks. Traffic engineers use the central computer to automatically control the flow through metering or intersection timing and communicate appropriate messages to drivers by radio or VMS. The traffic flow data can also feed into incident management tools that can be alerted to respond to incidents. Ground truth information—for example, videotaping of queues and incidents—can also be collected to evaluate existing tools, such as ramp metering and incident detection algorithms, in the ITS Laboratory.

To take advantage of this unique opportunity, the University of Minnesota, the Minnesota DOT, and the City of Minneapolis have proposed a multilevel architecture for ATMS/ATIS that will be based on real-time information from the field, for example, instrumented intersections and freeway zones. An illustration of a prototype for this architecture, with the I-394 as an example test bed, is found in Fig. 62.2.

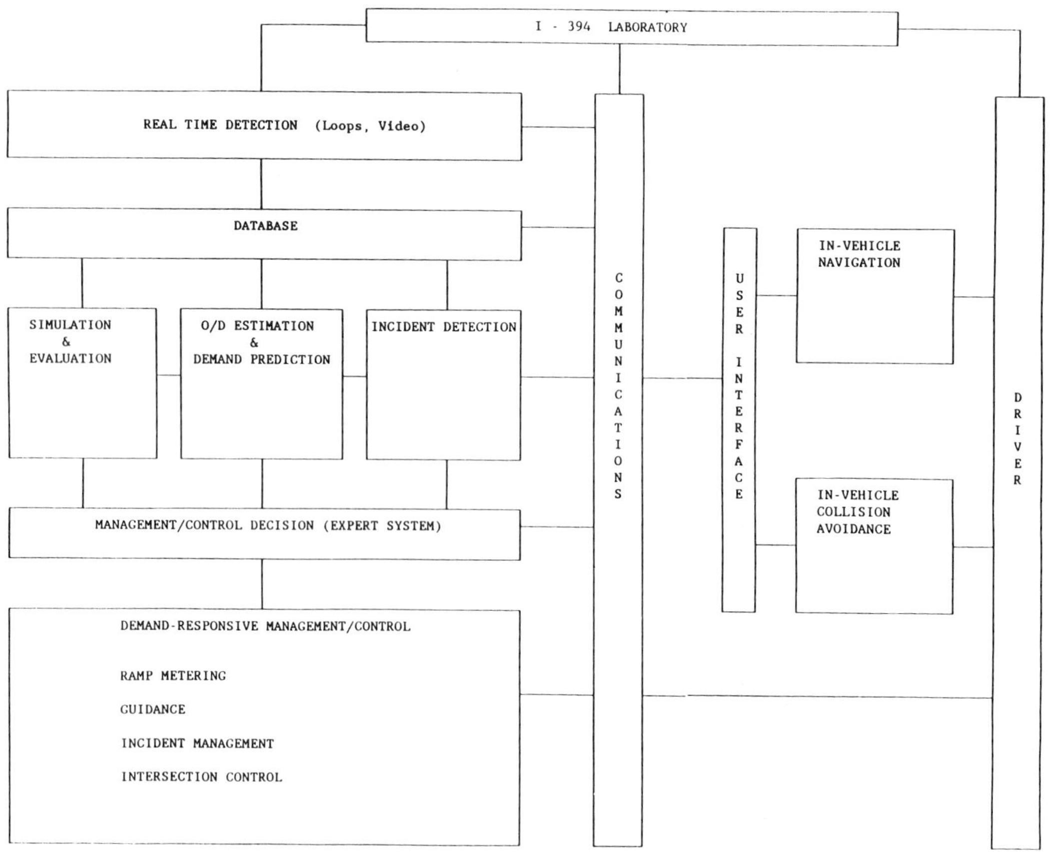

FIGURE 62.2 Advanced traffic management and driver information system.

As the figure indicates, the foundation of the ATMS/ATIS is the collection and use of real-time traffic data. This is being accomplished with a dense instrumentation of wide-area machine-vision stations (see the discussion of GuideStar in section 62.7) and classical loop detectors. The traffic data are stored in a dynamic database that can be interrogated both off-line and in real time by the ATMS. The I-394 information is also communicated to the Human Factors Laboratory in the ATIS for testing in-vehicle navigation and collision avoidance systems through appropriate user interfaces.

Off-line communication to the ATMS is, for instance, necessary for simulating and evaluating various management strategies prior to implementation. Real-time communication to the top layer of an integrated, adaptive/predictive control system [26] of the ATMS provides information necessary for dynamic estimation of origins/destinations, dynamic prediction of demand in the urban network and diversion at freeway ramp entrances [27], and real-time detection of incidents [19].

An expert system for traffic management assists the transfer of the information from the top layer of the adaptive/predictive control system to its lower layer. At the lower layer, real-time decisions are made for metering ramps, providing diversion guidance, managing incidents, and controlling intersections in freeway corridors and urban areas. Each of these real-time decisions represents a complete operation that can be analyzed further as a case study. For instance, the adaptive-predictive freeway zone control system, a prototype of which was tested on Interstate I-494 from December 1992 to February 1993 [29], is expected to help the Minnesota DOT operate and maintain approximately 500 ramp meters without the need to drastically increase staff. Automating the labor-intensive processes of developing new meter control strategies and updating these strategies in response to traffic requirements is the goal.

62.7 Case Study: Incident Management

Recurring Congestion

The most common cause of *recurring congestion* (congestion that routinely occurs at certain locations and during specific time periods) is excessive demand, the basic overloading of a facility that results in traffic stream turbulence. For instance, under ideal conditions, the capacity of a freeway is approximately 2000 to 2200 passenger cars per lane per hour. When the travel demand exceeds this number, an *operational bottleneck* will develop. An example is congestion associated with nonmetered freeway ramp access. If the combined volume of a freeway entrance ramp and the main freeway lanes creates a demand that exceeds the capacity of a section of freeway downstream from the ramp entrance, congestion will develop on the main lanes of the freeway, which will result in queuing upstream of the bottleneck. The time and location of this type of congestion can be predicted [18].

Another cause of recurring congestion is the reduced capacity created by a geometric deficiency such as a lane drop, difficult weaving section, or narrow cross section. The capacity of these isolated sections, called *geometric bottlenecks,* is lower than that of adjacent sections along the highway. When the demand upstream of the bottleneck exceeds the capacity of the bottleneck, congestion develops and queuing occurs on the upstream lanes. As above, the resulting congestion can also be predicted [18].

Nonrecurring Congestion: Incidents

Delay and hazards caused by random events constitute another serious highway congestion problem. Referred to as *temporary hazards* or *incidents,* they can vary substantially in character. Included in this category is any unusual event that causes congestion and delay [18]. According to FHWA estimates, incidents account for 60% of the vehicle-hours lost to congestion. Of the incidents that are recorded by police and highway departments, the vast majority, 80%, are vehicle disablements—cars and trucks that have run out of gas, had a flat tire, or been abandoned by their drivers. Of these, 80% wind up on the shoulder of the highway for an average of 15 to 30 minutes. During off-peak periods when traffic volumes are low, these disabled vehicles have little or no impact on traffic flow. But when traffic volumes are high, the presence of a stalled car or a driver changing a flat tire in the breakdown lane can slow traffic in the adjacent traffic lane, causing 100–200 vehicle-hours of delay to other motorists [10].

An incident that blocks one lane of three on a freeway reduces capacity in that sense of travel by 50%, and even has a substantial impact on the opposing sense of travel because of rubbernecking [12]. If traffic flow approaching the incident is high (near capacity), the resulting back-up can grow at a rate of about 8.5 miles per hour—that is, after one hour, the back-up will be 8.5 miles long [12,13]. Traffic also backs up on ramps and adjacent surface streets, affecting traffic that does not even intend to use the freeway. Observations in Los Angeles indicate that, in off-peak travel periods, each minute of incident duration results in 4–5 minutes of additional delay. In peak periods, the ratio is much greater [12,14]. Figure 62.3 shows a composite profile of recorded incidents, drawn from the limited research available on freeway incidents [10].

Accidents account for only 10% of reported incidents. Most are the result of minor collisions, such as side-swipes and slow-speed rear-end collisions [10]. 40% of accidents block one or two lanes of traffic. These often involve injuries or spills. Each such incident typically lasts 45–90 minutes, causing 1200–2500 vehicle-hours of delay [10,15]. It is estimated that major accidents make up 5 to 15% of all accidents and cause 2500–5000 vehicle-hours of delay per incident [10,16]. A very few of these major incidents, typically those involving hazardous materials, last 10–12 hours and cause 30,000–40,000 vehicle-hours of delay. These incidents are rare, but their impacts can be catastrophic and trigger gridlock [10]. To be sure, these statistics are location-specific and may differ across areas in the U.S.

Incident Management

Incident congestion can be minimized by detecting and clearing incidents as quickly as possible and diverting traffic before vehicles are caught up in the incident queue. About 30 cities have programs to manage freeway incidents and reduce incident congestion, but their scope and effectiveness vary widely [10,17]. Most major incidents are detected within 5–15 minutes; however, minor incidents may go unreported for 30 minutes or more [10]. Traffic information for incident detection is typically collected from loop detectors and includes occupancy and volume averaged at 20- to 60-second intervals usually across all lanes. Detector spacing along the freeway is half mile on the average. Certain systems in the U.S. and Canada (e.g., California I-880, Ontario's Queen Elizabeth Way) also use paired detectors to collect speed data. In Connecticut, in a demonstration project, overhead mounted radar detectors will return speed and volume data for incident detection. In Virginia a switch from loop to video detectors is underway [20].

Most systems use a California algorithm [32] for incident detection. With this algorithm incidents are detected by logically evaluating the variations in flow characteristics along the freeway and through time. In particular changes in the percentage of time that a vehicle is present between adjacent detectors (lane occupancy) are used to sense congestion and indicate that an incident has occurred. A computer calculates the difference in occupancy between adjacent detector stations, for example, spaced at half-mile intervals. At the end of each sampling period and when the relative percent change between the present occupancy and that of the preceding sample for the downstream detector exceeds a certain value, an alert is signaled automatically by the computer [18]. The original California algorithm is used in Minnesota, Ontario, and Virginia. The modified California algorithm #2, which additionally requires persistence of the incident alarm for two consecutive periods, is used in Los Angeles and Seattle. Algorithm #7 is used in tunnel locations in Seattle. Different algorithms are often employed depending on traffic conditions.

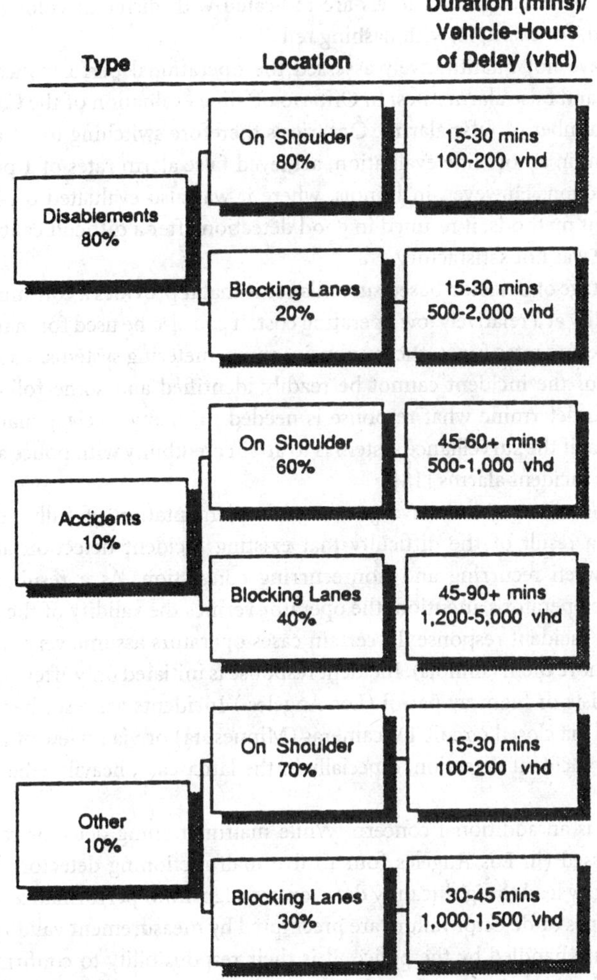

FIGURE 62.3 Composite profile of reported incidents by type. (*Source:* Cambridge Systematics Inc. Estimate based on case study interviews, incident management program records, and available studies. In *ITE*, March 1992, p. 17.)

In other cities locally developed algorithms have been implemented. In Connecticut a simple algorithm indicates an incident when speed drops below a threshold. In Illinois a Bayesian approach [33] employed the relative spatial occupancy difference as detection parameter. The approach focused on using the probability distributions of the detection parameter under incident and incident-free conditions for determining an optimal detection threshold. Because of excessive computer time requirements, the Bayesian algorithm was replaced by a simpler one that considers the occupancy difference between the upstream and downstream station; an incident is signaled if this difference continuously exceeds a threshold for five minutes. Although the 5-minute persistence test results in long response times, this is preferred to responding to frequent false alarms.

Thresholds for operational algorithms have been typically calibrated by trial and error (Los Angeles, Seattle), empirical experimentation on historical data (Illinois), and performance curves obtained from multiple runs of the respective algorithm on the data with incrementally changing thresholds (Ontario). In Los Angeles algorithms are frequently recalibrated, especially at locations that produce frequent incident alarms. Algorithm output consists of either textual description (Illinois, Virginia, Seattle) or color computer graphic maps (Los Angeles, Minnesota, Ontario).

In the latter case several congestion levels are indicated with different colors, and incidents are separately indicated, for example, with flashing red.

Most systems have not quantitatively assessed the operational performance of algorithms in terms of detection and false alarm rates. In Ontario off-line evaluation of the California algorithm produced a large number of false alarms. Ontario is therefore switching to a Canadian algorithm [34]; this algorithm, in an off-line evaluation, achieved false alarm rates of 1 per station every 64 hours at 75% detection. However, in Illinois, where it was also evaluated off-line as a potential substitute of current methods, it resulted in good detection after a difficult calibration period, but the false alarm rate was not satisfactory.

The main advantage of detector-based surveillance is that it provides a continuous network-wide monitoring capability at a relatively low operating cost. It can also be used for many other tasks such as establishing metering rates for traffic-responsive ramp-metering systems. Its main disadvantage is that the nature of the incident cannot be readily identified and some follow-up surveillance is often required to determine what response is needed. In many freeway management systems, television is essential if the surveillance system is to attain credibility with police and others charged with responding to incident alarms [18].

To date, high false alarm rates have prevented implementation of fully automated incident detection. This is a result of the difficulty that existing incident detection algorithms have in distinguishing between recurring and nonrecurring congestion. As a result, algorithm alarms typically trigger the operator's attention; the operator verifies the validity of the alarm and decides on the appropriate incident response. In certain cases operators assume very frequent alarms are false and tend to ignore them (Illinois). Incident response is initiated only after an incident has been reported by motorists or highway patrol (Los Angeles.) Incidents are elsewhere identified by the highway patrol and via closed circuit TV cameras (Minnesota) or via review of the raw data by the operator (Illinois). Incident detection, especially in the latter case, heavily relies on the operator's expertise.

Detector failure is an additional concern. While malfunctioning detector rates have not been systematically assessed (in Los Angeles four to five malfunctioning detectors are identified and repaired weekly), they lead to significantly deteriorated algorithm performance. In certain systems (Seattle) specific types of detector failure are preempted by measurement validity checks [20].

Most incidents are handled by the police. It is their responsibility to confirm the incident, assess what needs to be done, and summon help as needed. Communications about an incident are commonly handled directly by police dispatchers, but an increasing number of cities and states are building special-purpose traffic management centers to coordinate traffic and incident communications. Almost all urban areas have emergency response plans for catastrophic incidents, especially those involving hazardous materials; some have traffic diversion plans for major incidents, and a few have procedures for routine incidents. Traffic management teams representing state and local police and highway engineers are gaining in popularity as a way to develop response plans and traffic diversion routes [10].

Private tow-truck operators clear the great majority of freeway incidents. An effective clearance operation requires that the police officer on the scene diagnose the problem correctly, summon the right tow equipment, and make sure the operation is carried out with dispatch. This is seldom a problem in minor incidents, but is frequently a problem in major ones, especially those involving large trucks. Minor incidents are frequent, and police officers quickly develop the experience to deal with them. Large-truck incidents are much less frequent, and few police officers develop adequate experience unless they are assigned to special commercial-vehicle safety enforcement units or receive training in truck-clearance techniques. Most police departments do not provide specialized training in incident management; tow-truck rotation lists often turn out the wrong type of equipment; and most private tow operators lack the expertise to handle large truck accidents. These problems, often compounding, can double or triple the time required to clear an incident [10].

Incident recovery consists of three steps: (1) redirecting traffic flow at the site of the incident, (2) preventing more traffic from flowing into the area and getting trapped in the upstream queue,

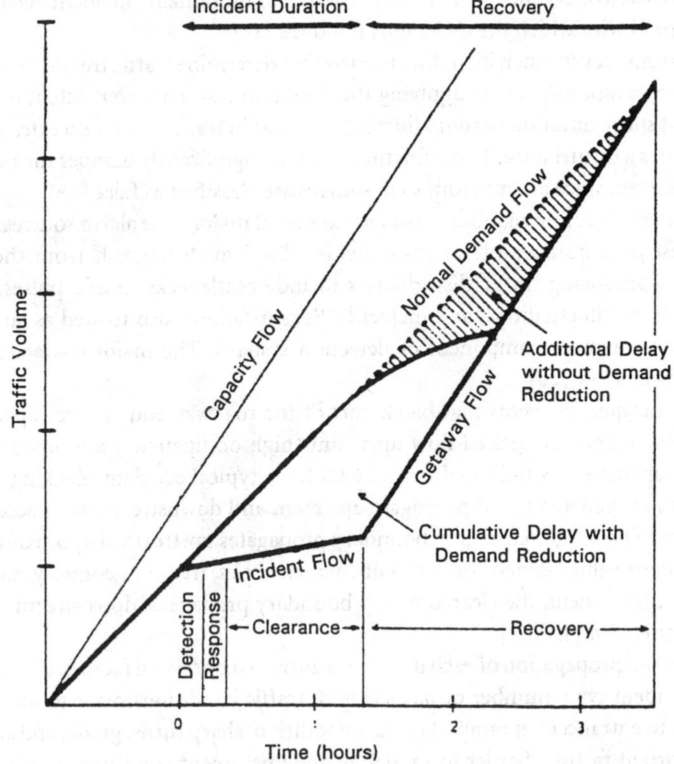

FIGURE 62.4 Schematic of traffic flow during an incident. (*Source:* Cambridge Systematics Inc. In *ITE*, March 1992, p. 17.)

and (3) preventing congestion from spilling across the metropolitan traffic network. Incident congestion can be minimized by these actions, but traffic management is the least developed element of most incident management programs. Police at the scene of an incident have little time to manage traffic flow around an incident; indeed, many police and fire departments encourage their officers to block traffic to prevent their being run down by drunken or unobservant drivers. In addition, radio traffic-reporting services often provide too little information, too late [10].

The dynamics of an incident are illustrated in Fig. 62.4. During an incident the queue continues to build until the incident is cleared and traffic flow is restored. The vehicle-hours of delay that accrue to motorists are represented in the exhibit by the area that lies between the normal flow rate and the lower incident flow rates. If the normal flow of traffic into the incident site is reduced by diverting traffic to alternate routes, the vehicle-hours of delay are minimized (shaded area.) If normal traffic flow is not diverted, additional vehicle-hours of delay (hatched area) are accrued [10].

Need for New Incident Detection Algorithm

The time saved by an incident management program depends on how well the four stages of an incident—detection, response, clearance, and recovery—are managed [10,14]. Fast and accurate detection of incidents can, therefore, substantially reduce the impact of incident congestion on freeway traffic. In particular, when an incident alarm is promptly signaled, traffic management plans can be adjusted in real time to produce the best control and guidance actions in freeway corridors. In addition, the incident management process is initiated as emergency vehicles can be promptly dispatched to clear the incident [19]. The key to minimizing incident-related congestion impacts is the speed with which the incident is removed [18].

Existing algorithms, even when including models that determine traffic trends, do not preprocess current traffic observations prior to applying the detection test. However, actual traffic measurements reveal that short-duration random fluctuations exist in traffic even if no external factor (e.g., incident) generates a disturbance. These fluctuations may significantly hamper the performance of a detection algorithm, since alarms from such sources are classified as false [19].

Effective incident detection requires consideration of all major false alarm sources. In particular, traffic flow presents a number of inhomogeneities, hard to distinguish from those driven by incidents. Events producing traffic disturbances include bottlenecks, traffic pulses, compression waves, random traffic fluctuations, and incidents. Sensor failure, also treated as an event, is only related to the measurement component of detection systems. The major characteristics of each event are described below [20].

Incidents are unexpected events that block part of the roadway and reduce capacity. Incidents create two traffic regimes, congested flow upstream (high occupancies) and uncongested downstream (low occupancies), as indicated in Fig. 62.5 for a typical accident blocking moving lanes. Two shock waves are generated and propagate upstream and downstream, each accompanying its respective regime. The congested-region boundary propagates upstream at approximately 16 km/h (10 mph), where the value depends on incident characteristics, freeway geometry, and traffic level. Downstream of the incident, the cleared region boundary propagates downstream at a speed that can reach 80 km/h (50 mph) [32].

The evolution and propagation of each incident is governed by several factors, the most important of which are incident type, number of lanes closed, traffic conditions prior to incident, incident location relative to entrance/exit ramps, lane drop/addition, sharp turns, grade, and sensor stations. Other less important factors, harder to model, include pavement condition, traffic composition, and driver characteristics.

Incident patterns vary depending on the nature of the incident and prevailing traffic conditions [32]. The most distinctive pattern occurs when the reduced capacity from incident blockage falls below oncoming traffic volume so that a queue develops upstream. This pattern, which is clearest when traffic is flowing freely prior to the incident, is typical when one or more moving lanes is

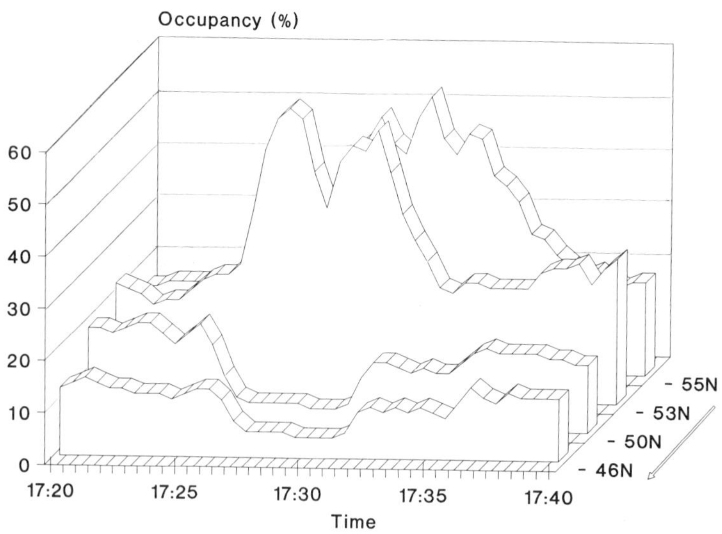

FIGURE 62.5 Incident pattern (I-35W North, 11/21/89).

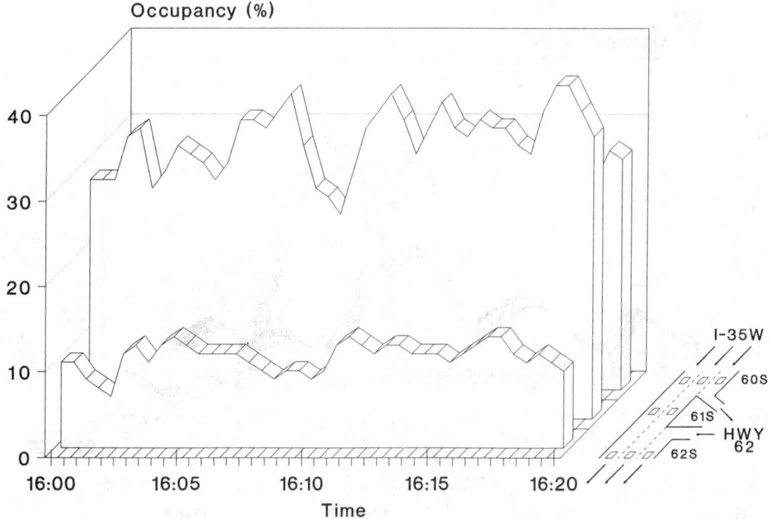

FIGURE 62.6 Occupancy measurements at bottleneck.

blocked following severe accidents (see Fig. 62.5 for an example of such an incident). The second pattern type occurs when the prevailing traffic condition is freely moving but the impact of the incident is not severe. This may result, for example, from lane blockage that still yields reduced capacity higher than the volume of incoming traffic. This situation may lead to missed detection, especially if the incident is not located near a detector. The third type characterizes incidents that do not create considerable flow discontinuity, as when a car stalls on the shoulder. These incidents usually do not create observable traffic shock waves and have limited or no noticeable impact on traffic operations. The fourth type of incident occurs in heavy traffic when a freeway segment is already congested. The incident generally leads to clearance downstream, but a distinguishable traffic pattern develops only after several minutes, except in a very severe blockage. This type of incident is often observed in secondary accidents at the congested region upstream of an incident in progress.

Bottlenecks are formed where the freeway cross section changes, for example, in lane drop or addition. While incidents have only temporary effect on occupancies, bottlenecks generally result in longer-lasting spatial density or occupancy discrepancies. A typical bottleneck situation is illustrated in Fig. 62.6. This figure presents occupancy measurements at three consecutive stations of a freeway segment involving lane drop between the first two and lane addition between the second and third stations. Under normal conditions the three stations operate at different average occupancy levels. This difference is more pronounced between stations 61S and 62S.

Traffic pulses are created by platoons of cars moving downstream. Such disturbances may be caused by a large entrance ramp volume as, for instance, a sporting event letting out. The observed pattern is an increase in occupancy in the upstream station followed by a similar increase in the downstream station (Fig. 62.7).

Compression waves occur in heavy, congested traffic, usually following a small disturbance and are associated with severe slow-down/speed-up vehicle speed cycles. Waves are typically manifested by a sudden, large increase in occupancy that propagates through the traffic stream in a direction counter to traffic flow (Fig. 62.8). Compression waves result in significantly high station occupancies of the same magnitude as in incident patterns.

Random fluctuations are often observed in the traffic stream as short-duration peaks of traffic occupancy. These fluctuations, although usually not high in magnitude, may form an incident pattern or obscure real incident patterns.

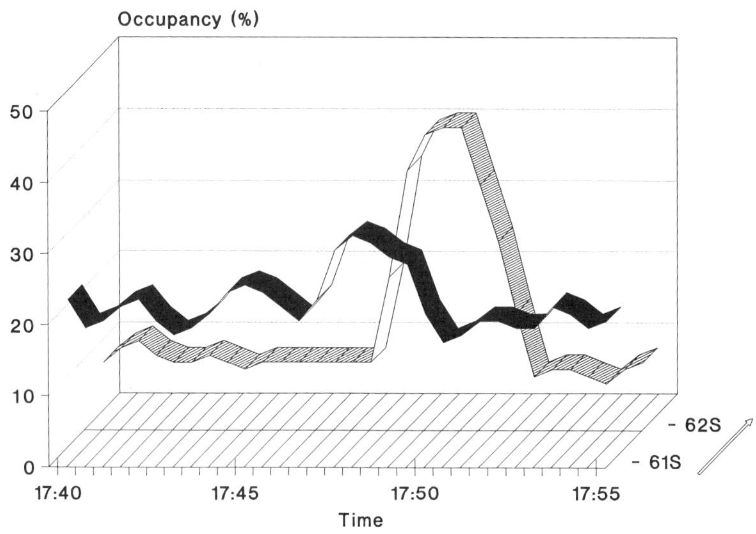

FIGURE 62.7 Traffic pulse (I-35W South, 11/13/89).

Detection system failures may be observed in several forms (Fig. 62.9). A specific pattern is observed with isolated high-magnitude impulses in 30-second volume/occupancy measurements, appearing simultaneously in several stations. These values are considered outliers or impulsive data noise [20].

To address the need for preprocessing traffic data and assess the improvement in detection performance that can result from such preprocessing, a new incident detection scheme has been developed at the University of Minnesota, with support by the Center for Transportation Studies and the National Science Foundation. The detection scheme is part of IDENTIFY (*I*ncident *De*tection *EN*hancements for *T*raffic *I*n *F*reewa*Y*s), a project evolving in two major directions, one employing filtering and a second focusing on neural network applications.

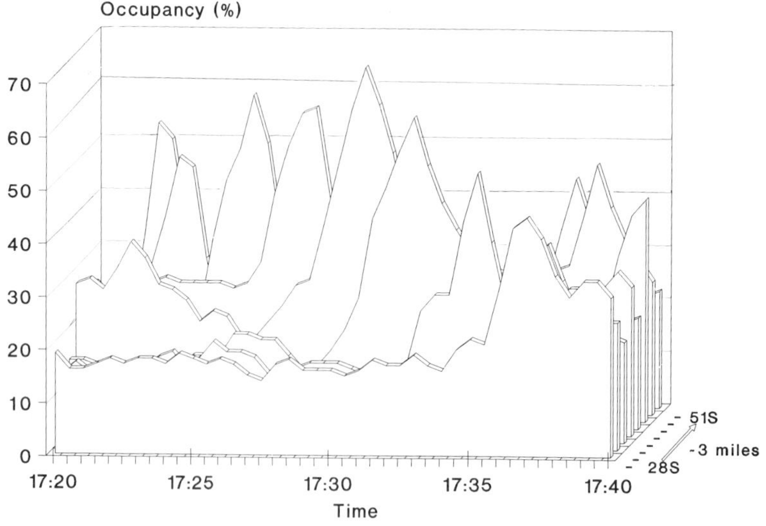

FIGURE 62.8 Compression wave (I-35W South, 11/16/89).

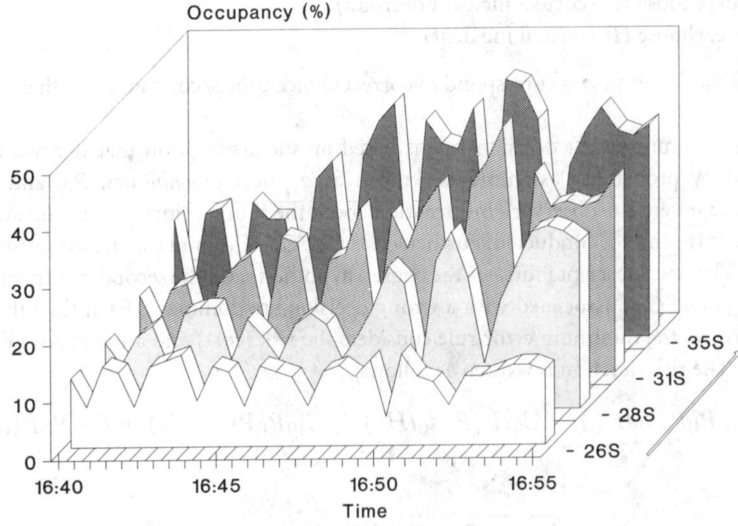

FIGURE 62.9 Erroneous measurements (I-35W South, 7/31/89).

Along the first direction DELOS (*DE*tection *LO*gic with *S*moothing) employs filtering of the traffic occupancy to determine whether a large deviation of current occupancy measurements from past values corresponds to a long-lasting incident or to a short-duration random fluctuation. A detection algorithm smoothes the data using, alternatively, the average, the statistical median, or exponential smoothing of measurements within moving windows. With random fluctuations filtered out, the algorithm examines the temporal change of the smoothed spatial occupancy difference to detect a major (in terms of severity and duration) traffic change that is most probably due to an incident. The algorithm and a set of algorithms from the literature have been tested in the Minneapolis–St. Paul freeway network and the Boulevard Peripherique in Paris, France. Test results indicate that the new algorithm yields performance superior to previous algorithms [19].

Formulation of Incident Detection Problem

Incident detection can be viewed as part of a statistical decision framework in which traffic observations are used to select the true hypothesis from a pair, that is, incident or no incident. Such a decision is associated with a level of risk and cost. The cost of a missed detection is expressed in terms of increased delays and the cost of a false alarm, in terms of incident management resources dispatched to the incident location [35]. The objective of incident detection is to minimize the overall cost.

To formulate the incident detection problem in a simple incident/no-incident environment, we observe the detector output that has a random character and seek to determine which of two possible causes, incident or normal traffic, produced it. The possible causes are assigned to a hypothesis—incident H_1 versus no-incident (normal traffic) H_0. Traffic information is collected in real time and processed through a detection test, in which a decision is made based on specific criteria. Traffic information, such as occupancy, represents the observation space. We can assume that the observation space corresponds to a set of N observations denoted by the observation vector $\mathbf{r}$. Following a suitable decision rule, the total observation space Z is divided into two subspaces, Z_1 and Z_0. If observation $\mathbf{r}$ falls within Z_1, the decision is d_1; otherwise, the decision is d_0.

To discuss suitable decision rules, we first observe that, each time the detection test is performed, four alternatives exist depending on the true hypothesis H_i and the actual decision d_i, where $i = 0, 1$:

1. H_0 true; choose H_0 (correct no-incident decision)
2. H_0 true; choose H_1 (false alarm)

3. H_1 true; choose H_1 (correct incident decision)
4. H_1 true; choose H_0 (missed incident)

The first and third alternatives correspond to correct choices; the second and fourth correspond to errors.

The Bayes minimum error decision rule is based on the assumption that the two hypotheses are governed by probability assignments, known as *a priori probabilities*, P_0, and P_1. These probabilities represent the observer's information about the sources (incident, no-incident) before the experiment (testing) is conducted. Further, costs C_{00}, C_{10}, C_{11}, and C_{01} are assigned to the four alternatives. The first subscript indicates the chosen hypothesis and the second, the true hypothesis. The costs, C_{10} and C_{01}, associated with a wrong decision are dominant. Each time the detection test is performed, the minimum error rule considers the risk (cost) and attempts to minimize the average risk. The risk function is written as follows:

$$R = C_{00}P_0P(d_0/H_0) + C_{01}P_1P(d_0/H_1) + C_{10}P_0P(d_1/H_0) + C_{11}P_1P(d_1/H_1)$$

or

$$R = \sum_i \sum_j C_{ij}P_jP(d_i/H_j), \qquad i, j = 0, 1 \tag{62.1}$$

where the conditional probabilities $P(d_i/H_j)$ result from integrating $p(\mathbf{r}/H_j)$, the conditional probability to observe the vector **r**, over Z_i, the observation subspace in which the decision is d_i. In particular, the probability of detection is

$$P(d_i/H_1) = P_D = \int_{Z1} p(\mathbf{r}/H_1)d\mathbf{r} \tag{62.2}$$

and the probability of false alarm is

$$P(d_1/H_0) = P_F = \int_{Z1} p(\mathbf{r}/H_0)d\mathbf{r} \tag{62.3}$$

Minimizing the average risk yields the *likelihood ratio* test:

$$\Lambda(\mathbf{r}) = \frac{p(\mathbf{r}/H_1)}{p(\mathbf{r}/H_0)} \underset{H_0}{\overset{H_1}{\gtrless}} \frac{(c_{10} - c_{00})P_0}{(c_{01} - c_{11})P_1} \tag{62.4}$$

where the second part in the inequality represents the test threshold, and the conditional and a priori probabilities can be estimated through time observations of incident and incident-free data. However, obtaining an optimal threshold requires realistic assignment of costs to each alternative. This is further impeded by the fact that incidents (or false alarms) are not alike in frequency, impact, and consequences. Therefore, an optimal threshold cannot practically be established. Previous attempts to use the Bayes decision rule [33,36] employed a simplified risk function to overcome the cost assignment issue and reduce the calibration effort. For instance, Levin and Krause [33] obtained a suboptimal threshold by maximizing the expression

$$R = P(d_0/H_0) + P(d_1/H_1)$$

using the relative spatial occupancy difference between adjacent stations as observation parameter.

An alternative procedure to the Bayes rule—applicable when assigning realistic costs or a priori probabilities is not feasible—is the Neyman–Pearson (NP) criterion. The NP criterion views the solution of the optimization of the risk function in Eq. (62.1) as a constrained maximization problem. This is necessitated by the fact that minimizing P_F and maximizing P_D are conflicting objectives. Therefore, one must be fixed while the other is optimized: Constrain $P_F \leq \alpha$ and design a test to maximize P_D under this constraint. Similarly to the minimum error criterion, the NP test results in a likelihood ratio test:

$$\Lambda(\mathbf{r}) \geq \lambda \tag{62.5}$$

where the threshold λ is a function of P_F only. Decreasing λ is equivalent to increasing Z_1, the region where the decision is d_1 (incident). Thus, both P_F and P_D increase as λ decreases. The Neyman–Pearson *lemma* [37] implies that the maximum P_D occurs at $P_F = \alpha$. The lemma holds since P_D is a nondecreasing function of P_F. In practical terms an NP procedure implies that, after an incident test has been designed, it is applied to a data set initially employing a high (restrictive) threshold, which results in low P_F. The threshold is incrementally reduced until P_F increases to the upper tolerable limit α. The corresponding P_D represents the detection success of the test at false alarm α. An NP procedure seems more applicable to incident detection than a minimum error procedure for two reasons. First, the only requirement is the constraint on P_F, which can easily be assessed by traffic engineers to a tolerable limit. Second, an NP procedure does not require separate threshold calibration since no optimal threshold, in the Bayesian sense, is sought. Instead, thresholds result from the desirable P_F.

The decision process is facilitated by the *likelihood ratio* $\Lambda(\mathbf{r})$. In signal detection practice $\Lambda(\mathbf{r})$ is replaced by a *sufficient statistic* $l(\mathbf{r})$ which is a simpler than $\Lambda(\mathbf{r})$ function of the data. The values of the *sufficient statistic* are then compared to appropriate thresholds to decide which hypothesis is true. In incident detection applications, however, the tests of an algorithm are designed empirically so that they only approximately can be considered *sufficient statistics.*

To illustrate the various terms in the sufficient statistic coordinate system, consider Fig. 62.10, where the typical conditional probability density functions $p(l/H_0)$ and $p(l/H_1)$ and cumulative conditional probabilities P_F and P_D are plotted. Clearly, both P_D and P_F are decreasing functions of the threshold. In the context of Fig. 62.10 an incident detection algorithm is expected to have good performance if it succeeds to minimize the overlapping of the density functions $p(l/H_0)$ and $p(l/H_1)$. The amount of overlapping depends mainly on the choice of the detection variable and the total number of false alarms that the algorithm produces. Different detection variables result in varying normalized distance of the two density functions. In Figs. 62.11(a) and (b) we illustrate the true incident and false alarm histograms across the values of the detection threshold resulting from the standard deviation and the median algorithms and data in the Twin Cities metropolitan area.

Existing Incident Detection Algorithms

Incident detection algorithms, depending on their logic, can be classified into three major categories [19]. In further discussing the relevant algorithms, define (with reference to Fig. 62.12) $o_i(t)$, $q_i(t)$, $v_i(t)$, and $x_i(t)$ as occupancy, flow, mean speed, and general variable, respectively, at detector station i at time interval t.

Comparative algorithms [38,39,34,40] establish rough incident patterns and attempt to recognize these patterns in traffic measurements by comparing detection variables to preselected thresholds. The algorithms consider single occupancy and volume measurements separated in time to detect incidents. The California algorithm [38] compares occupancies at neighboring detectors in the same time interval, and occupancies at different time intervals at the same detector. Occupancies are 1-minute values updated every 20 or 30 seconds. The decision rule detects an incident when the value at the left-hand side of each criterion below exceeds the corresponding threshold at time t:

$$o_i(t) - o_{i+1}(t) \geq D_1 \tag{62.6}$$

$$\frac{o_i(t) - o_{i+1}(t)}{o_i(t)} \geq D_2 \tag{62.7}$$

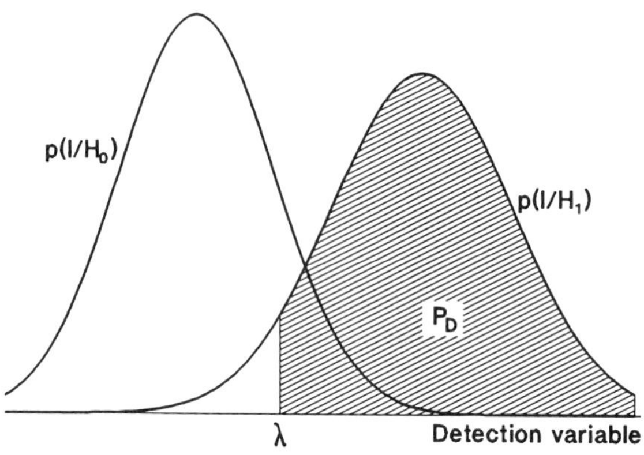

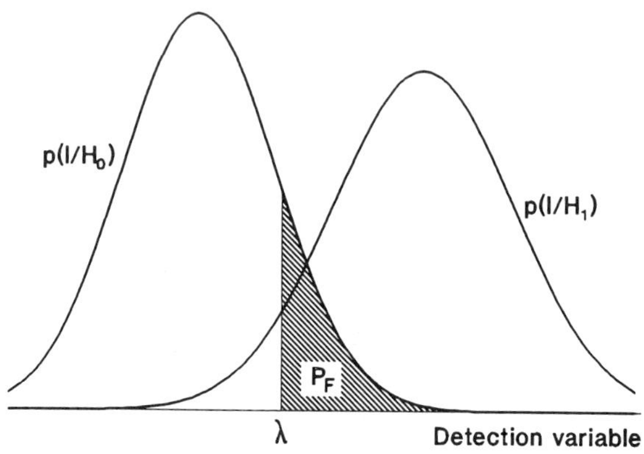

FIGURE 62.10 Conditional probability distributions in incident detection. (From *Transp. Res. J.*, Sept. 1993, p. 223.)

$$\frac{o_{i+1}(t-d) - o_{i+1}(t)}{o_{i+1}(t-d)} \geq D_3 \tag{62.8}$$

where D_1, D_2, D_3 are the test thresholds and $d = 2$ min is the time period for temporal tests. A modified version of the algorithm, which requires persistence of the incident alarm for two consecutive periods, was included in the evaluation tests below.

While most algorithms employ aggregate 30-second data from a pair of adjacent detectors, the HIOCC algorithm is based on 1-second occupancy data from a single detector [39]. The algorithm looks at consecutive occupancy outputs at the upstream detector and detects stationary or slow-moving cars when this occupancy remains high for a number of seconds. The algorithm assumes that such a pattern occurs after an incident. An algorithm developed in Canada [34] employs analysis of the traffic data in a two-dimensional space. In particular it separates the flow-occupancy diagram into three areas corresponding to different traffic states. The area boundaries are defined by the occupancy at maximum flow and a nonlinear equation of q_i as a function of o_i. Incidents are detected after observing changes of the traffic state in a short time period or a significant drop in speed.

In a different approach *time series algorithms* employ statistical or time series models to estimate the current traffic trend based on past traffic observations [41,42,43]. Incidents are detected after observing abrupt changes in the trend. Algorithms in this class have an advantage over comparative algorithms because they base their decision on more observations. On the other hand, while

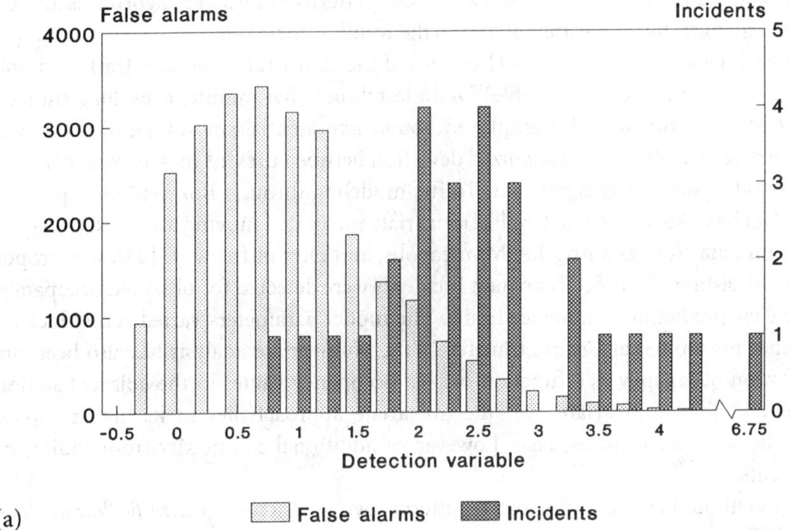

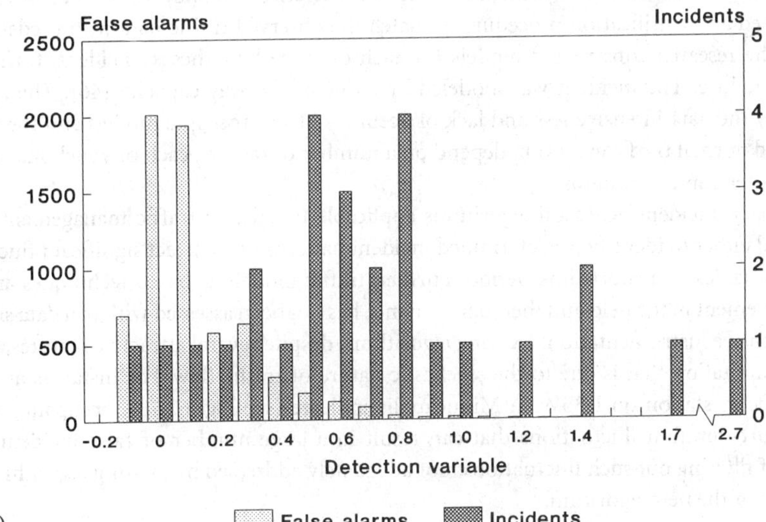

FIGURE 62.11 (a) Incident and false alarm histograms for the standard deviation algorithm. (b) Incident and false alarm histograms for the median algorithm. (From *Transp. Res. J.*, Sept. 1993, p. 231.)

Detector station i
Segment j
Segment j+1
Detector station i+1
k, q, v
o, q, v
Δx

FIGURE 62.12 Typical discretized freeway section. (From *Transp. Res. J.*, Sept. 1993, p. 223.)

comparative algorithms attempt to identify incident patterns, time series algorithms aim to detect abrupt traffic changes, but these are not always the result of incidents.

The standard deviation algorithm [41] employed the statistical mean of a traffic variable (e.g., volume, occupancy, or composite variable) for the last three to five minutes to evaluate the trend. The double exponential algorithm [42] employed double exponential smoothing of traffic variables. Incidents are detected when the normalized deviation between present measurement and expected value of the traffic variable is significant. Traffic modeling through Box–Jenkins type analysis of time series data has also been considered. A univariate integrated moving average (ARIMA) model, calibrated with data from Los Angeles, Minneapolis, and Detroit freeways [43], was proposed for short-term forecasting of traffic occupancy. Incidents were detected for observed occupancy values outside the 95% prediction confidence limits. The model parameters varied across locations and over time and this limited algorithm transferability. Multivariate analysis has also been proposed to model station occupancy as a function of past occupancy values in the adjacent stations. The improvement of the multivariate over the univariate approach lies in its higher capability to forecast traffic changes, at the expense, however, of additional complexity from adding variables and coefficients.

Detection methods based on anticipated patterns involve macroscopic *traffic flow modeling* with nonlinear differential equations to obtain traffic patterns for normal and incident conditions. Willsky *et al.* [44] considered Payne's model [45] in discretized freeway sections as in Fig. 62.12 and employed an identification procedure to match the observed traffic data with predetermined patterns. The research constructed models for each of three hypotheses: incident, traffic pulse, and sensor failure. The incident was modeled by reducing freeway capacity [46]. The approach is limited by the data intensiveness and lack of accuracy of macroscopic models and the fact that incident and normal traffic evolution depend on a number of factors, each of which may lead to a different pattern and hypothesis.

In summary, incident detection algorithms applicable to existing traffic management systems are designed either to identify predetermined incident patterns or to detect significant fluctuations of traffic variables in a short time period. Further, traffic modeling–based techniques are rather hard to implement in the field and their performance has not been assessed with real data since they require extensive instrumentation. Existing algorithms, despite several attractive features, share an important limitation that is due to the stochastic nature of traffic flow. For instance, as the data from a detector station on I-35W in Minneapolis indicate (see Fig. 62.13), occupancy typically exhibits short-duration fluctuations that can result in a large number of false incident alarms. The issue of filtering out such fluctuations, not adequately addressed by existing algorithms, is the major focus of the new algorithm.

New Algorithm

The new algorithm combines certain features of comparative and time series algorithms and seeks to increase the decision confidence that an alarm corresponds to an incident and not to unexpected traffic disturbances [47]. The algorithm aims to detect specific incident-related occupancy changes in time and space in a manner similar to the California algorithm. It also smoothes past detector output in a manner similar to the standard deviation algorithm. The major innovation of the algorithm is that, in addition to smoothing past values, it smoothes the current occupancy values. This feature results in the reduction of short-duration random traffic fluctuations that are frequently observed in congested flows.

To differentiate patterns between traffic states, the algorithm logic considers two time periods. The past period corresponds to the preincident period, and calculations within that period aim to assess the predominant traffic conditions before an incident. Such consideration is necessary since the impact of an incident depends on its severity and the traffic conditions prior to its occurrence. The upper part of Fig. 62.14 presents occupancy time series data at the upstream and downstream stations of an incident location (Minneapolis, I-35W, 11/16/89). The current period begins at the

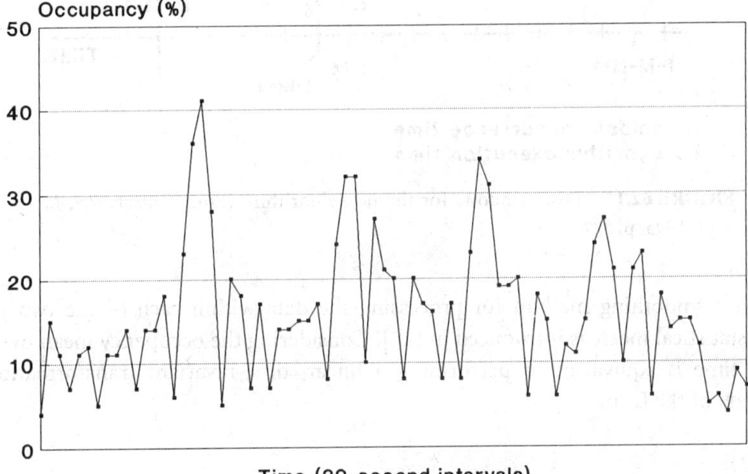

FIGURE 62.13 Occupancy measurements (I-35W, station 42S, 7/31/89, 16:00). (From *Transp. Res. J.*, Sept 1993, p. 225.)

end of the past period. Occupancy values in the current period assess current traffic conditions. If an incident is assumed to occur at time t_0, the past period is defined as the period from $t_0 - N$ to $t_0 - 1$—that is, $[t_0 - N, t_0 - 1]$—and the current as $[t_0, t_0 + M - 1]$ (Fig. 62.15), where M and N are determined below and M data values after the incident occurrence need to be known before making a decision.

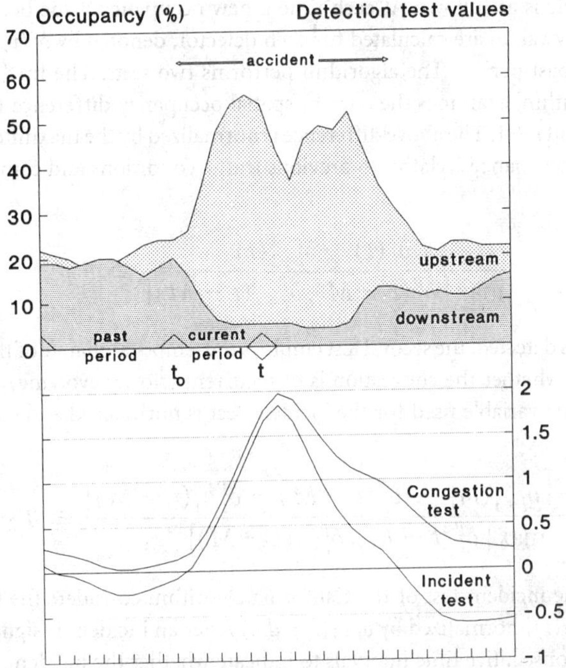

FIGURE 62.14 Occupancy and detection variable values during an accident (I-35W, stations 53N and 50N, 11/21/89, 17:20). (From *Transp. Res. J.*, Sept 1993, p. 226.)

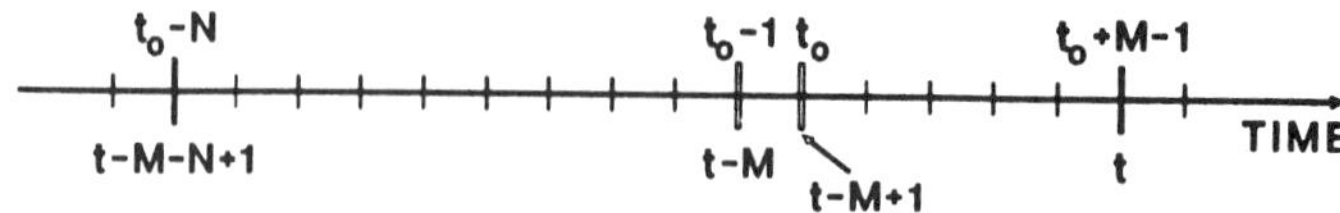

FIGURE 62.15 Data windows for the new algorithm. (From *Transp. Res. J.*, Sept. 1993, p. 227.)

The simplest smoothing method for processing the data within each of the two periods is through the statistical mean as introduced in [48]. Considering the occupancy mean over moving windows in time is equivalent to performing a linear, time-invariant transformation of the occupancy data of the form

$$\hat{o}_i(t) = \sum_{l=0}^{L-1} h(l) o_i(t-l) \tag{62.9}$$

where impulse response $h(l)$ is a rectangular pulse of width L and height $1/L$, and $L = M$ and N for the current and past period, respectively. The linear transformation, although effective in removing measurement fluctuations, distorts information-bearing edges (i.e., step-like changes that are due to incidents), possibly obscuring their information content. In such a case a nonlinear transformation can preserve sharp discontinuities in the data and can still filter out the undesirable fluctuations. Such a transformation employs the statistical median of the data window or exponential smoothing [48].

The detection logic is as follows. At each time a new occupancy value becomes available, two smoothed occupancy values are calculated for each detector, denoted by $\hat{o}_i^c(t)$ for the current and $\hat{o}_i^p(t-M)$ for the past period. The algorithm performs two tests. The first, in a way similar to the California algorithm, examines the current spatial occupancy difference to detect congestion between stations i and $i+1$. The above difference is normalized by the maximum of $\hat{o}_i^p(t-M)$ and $\hat{o}_{i+1}^p(t-M)$ to reflect changes relative to previous traffic conditions and compared to congestion test threshold T_1 :

$$\frac{\hat{o}_i^c(t) - \hat{o}_{i+1}^c(t)}{\max\left\{\hat{o}_i^p(t-M), \hat{o}_{i+1}^p(t-M)\right\}} \geq T_1 \tag{62.10}$$

After congestion is detected, the second test employs the temporal change of the spatial occupancy difference to decide whether the congestion is of recurrent (slowly evolving) type or has resulted from an incident. The variable used for the incident test is normalized and compared to incident test threshold T_2 :

$$\frac{\left[\hat{o}_i^c(t) - \hat{o}_{i+1}^c(t)\right] - \left[\hat{o}_i^p(t-M) - \hat{o}_{i+1}^p(t-M)\right]}{\max\left\{\hat{o}_i^p(t-M), \hat{o}_{i+1}^p(t-M)\right\}} \geq T_2 \tag{62.11}$$

(For comparison, the incident test of the California algorithm considers the temporal difference $[o_{i+1}(t-d) - o_{i+1}(t)]$ normalized by $o_{i+1}(t-d)$.) After an incident is signaled, the congestion test is employed at consecutive time intervals to indicate whether the incident is still in effect. The alarm is terminated at the first time interval that the congestion test fails.

The time series values of the congestion and incident tests for the incident in Fig. 62.14 are depicted in the lower part of the figure. The congestion test exhibits high values throughout the incident duration, whereas the incident test attains high positive values at the beginning of the incident and negative values at its end.

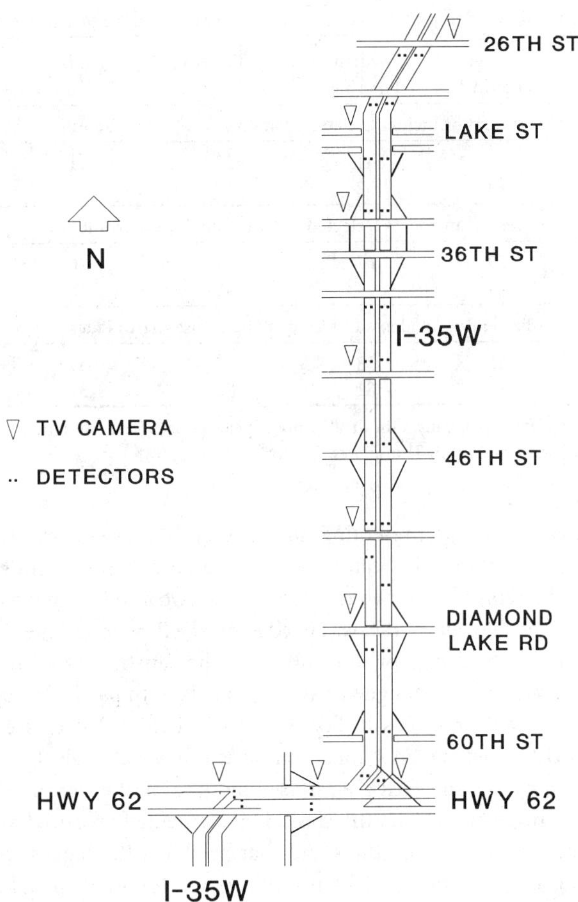

FIGURE 62.16 Study site in Minneapolis I-35W. (From *Transp. Res. J.*, Sept. 1993, p. 228.)

Test Results

A number of existing algorithms were tested with a set of actual field data. In particular, 140 hours of afternoon peak period traffic data from southbound I-35W in Minneapolis (Fig. 62.16) were collected through the MnDOT Traffic Management Center. The data cover a six-month period from June to December 1989. In the afternoon rush period the southbound direction of the freeway experiences recurrent congestion. The segment has three lanes along most of its length. It contains two major bottlenecks, one at a lane drop just before merging with highway 62, and another at an uphill section followed by shoulder elimination at the Minnehaha Creek bridge. The segment includes four entrance and five exit ramps. The data consist of 30-second volume and occupancy measurements from loop detectors forming 14 detector stations embedded along the road 0.3–0.7 miles apart. The 30-second data are averaged across lanes so that individual lane information is not available. During the period that traffic data were collected, 27 incidents were reported by the traffic engineer on duty. The set includes accidents and vehicle stalls in moving lanes and shoulder. Incidents cover a wide range of blockage severity, duration, and location relative to detectors. Detection of the incidents was performed manually by the traffic engineer, mostly through CCTV cameras installed along the freeway segment.

Detection performance evaluation was based on a Neyman–Pearson criterion. The procedure begins with the selection of α. For obtaining a comprehensive assessment of the algorithm detection

Table 62.1 Percent of Incidents Detected at Low, Moderate, and High False Alarm Rates

False Alarm Rate (number[1])	Average Filtering	Median Filtering	California Algorithm	Standard Deviation	Double Exponential
Percent of Incidents Detected at Low False Alarm Rates					
0.05% (1.5)	41	41	22	11	—
0.10% (3.0)	59	52	37	15	11
Percent of Incidents Detected at Moderate False Alarm Rates					
0.20% (5.9)	70	67	44	26	15
0.30% (8.9)	74	74	48	26	30
Percent of Incidents Detected at High False Alarm Rates					
0.40% (11.9)	81	78	67	33	37
0.50% (14.8)	85	81	70	37	48

[1]Number of false alarms per daily 2-hour afternoon peak period (test site length is 5.5 miles and includes 14 detector stations). From *Transp. Res. J.*, Sept. 1993, p. 229.)

performance, several values of α, representing different levels of false alarm rates (the change in notation from probabilities to rates indicates operations on a data sample rather than the whole incident population), were considered: two α values at low, two at moderate, and two at high false alarm rates. For each α, the false alarm rate was fixed at maxFAR $= \alpha$ and the maximum detection rate (DR) was determined for each algorithm subject to the constraint FAR $\leq \alpha$. To accomplish this, a number of experiments were designed based on the data. In particular, observing that both DR and FAR increase as the thresholds in Eqs. (62.10) and (62.11) decrease, the procedure to obtain the maximum DR under the FAR constraint is as follows. A high threshold is considered first and the corresponding DR and FAR are assessed by running the algorithm with the incident and incident-free data, respectively. The threshold is incrementally reduced and the experiment is repeated. The procedure ends at the threshold that produces the highest possible FAR $\leq \alpha$, and the corresponding DR is recorded. This procedure was followed for all algorithms in this evaluation.

Table 62.1 and Fig. 62.17 present the detection rates achieved by a number of existing and the new algorithm at three levels of false alarm rate (low, medium, high). The detection variable in the

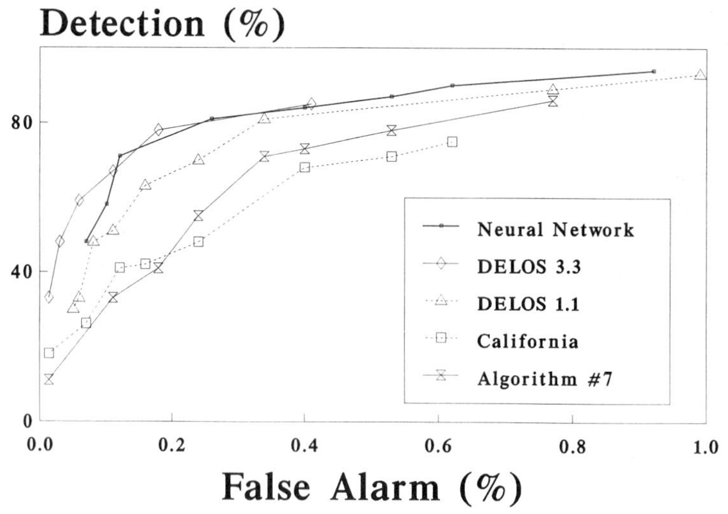

FIGURE 62.17 Algorithm performance comparison.

standard deviation and double exponential algorithms is the spatial occupancy difference between adjacent stations, a key variable in the California and the new algorithms. The actual number of alarms for this 5.5-mile freeway segment per daily two-hour peak period is also included in Table 62.1 to indicate the effectiveness of each algorithm in terms of operator load. The window sizes with the new algorithm were selected to best serve the trade-off between false alarm and mean algorithm response time as $M = 6$ and $N = 10$ for the average smoother and $M = 9$ and $N = 9$ for the median filter. Higher M values result in longer response times, and lower values may not provide adequate smoothing of traffic fluctuations.

Regarding existing algorithms, Table 62.2 indicates that algorithms developed to detect specific incident patterns yield higher performance over algorithms directed toward detection of abrupt traffic changes. The California algorithm detects roughly twice as many incidents as either the standard deviation or double exponential algorithm.

As the table also indicates, when tested with our incident data set the new algorithm—with either type of filtering (the average performs slightly better than the median)—accomplishes the highest detection rates among all examined algorithms at all six false alarm levels considered. More specifically, the new algorithm surpasses the detection rate of the California algorithm (the best performing examined algorithm) by 40 to 86%, 52 to 59%, and 16 to 22% at low, moderate, and high false alarm levels, respectively.

The improved performance of the new algorithm results from combining and enhancing positive features of previous algorithms, that is, California and standard deviation algorithms. In particular, the California algorithm seeks specific incident patterns by employing the downstream occupancy temporal reduction test. This test, however, is not always helpful since not all incidents result in observable occupancy reduction. The new algorithm replaces this test with the temporal change of spatial occupancy difference that is more easily observable. Further, unlike the California algorithm, which employs single occupancy observations in time, the new algorithm considers average occupancies over time windows, thus reducing the false alarm risk from random traffic fluctuations.

The algorithm is also similar to the standard deviation algorithm in that both use the same averaging technique to determine traffic trends. Further, filtering the current data increases the effectiveness of the new algorithm as the performance comparison in Table 62.1 indicates. The substantial difference in performance indicates that a large percentage of false incident alarms are produced by traffic fluctuations. The performance difference can be viewed in terms of the extent of overlapping between $p(l/H_0)$ and $p(l/H_1)$, discussed earlier. Considering, for instance, the actual histograms of incidents and false alarms across threshold values for two algorithms, the standard deviation and the new algorithm with median filtering [Figs. 62.11(a) and (b)], the higher performance of the new algorithm is explained by the smaller overlapping between false alarm and incident histograms. In particular, the new algorithm eliminates a large number of the alarms that the standard deviation algorithm produces and, in addition, better isolates the largest portion of incidents from the corresponding portion of false alarms.

Table 62.2 Algorithm Mean Response Time (minutes)

False Alarm Rate (%)	Average Filtering	Median Filtering	California Algorithm	Standard Deviation	Double Exponential
0.05	0.4	1.6	0.4	1.4	—
0.10	0.9	1.9	0.3	1.6	8.1
0.20	0.8	1.8	1.2	0.8	7.9
0.30	0.7	1.8	0.5	0.5	6.3
0.40	0.6	1.4	1.1	0.0	4.7
0.50	0.4	1.4	0.4	−0.6	4.4

From *Transp. Res. J.*, Sept. 1993, p. 232.

Table 62.3 Algorithm Performance Comparison

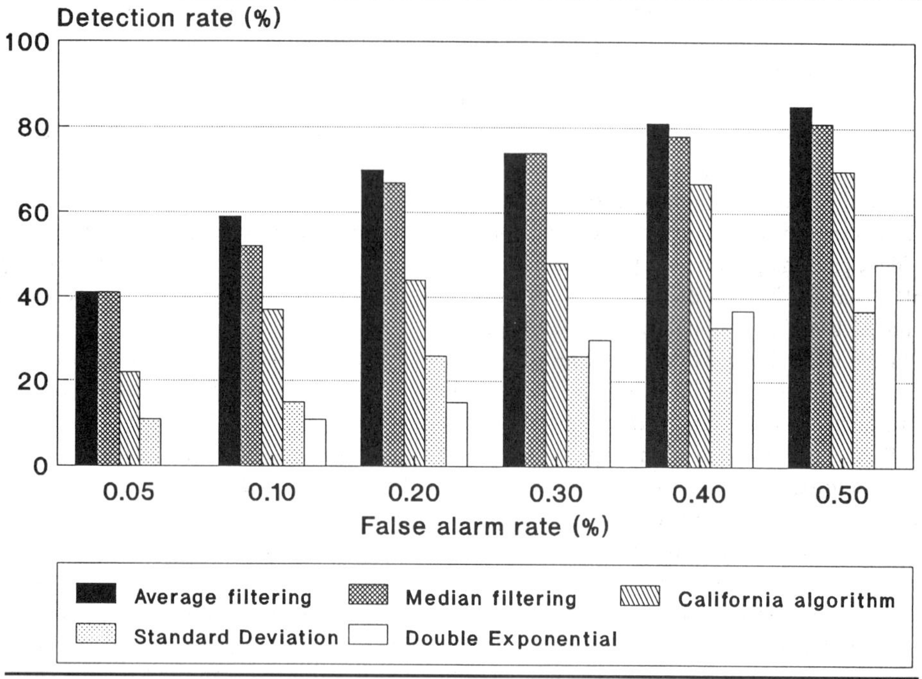

From *Transp. Res. J.*, Sept. 1993, p. 230.

Mean response time is an additional indicator of algorithm effectiveness. Table 62.2 presents the mean response time for each algorithm at the same FAR levels as in Tables 62.1 and 62.3. These values do not represent absolute response times but, rather, times relative to the engineer's detection, since the only information on incident occurrence comes from the engineer's log. A negative value implies that the algorithm detects the incident faster than the engineer. Response time ranges from −0.6 (standard deviation) to 8.1 (double exponential) minutes. The new algorithms exhibit response times from 0.4 to 0.9 minutes (average) and from 1.4 to 1.9 minutes (median)—well within the range established by previous algorithms. Because of the larger-size window for smoothing the current data, the median algorithm responds more slowly than the average.

62.8 Sampling of U.S. Operational Tests

ADVANCE

ADVANCE is a Chicago-area demonstration of ATIS and ATMS with focus on arterial traffic. The objective is to evaluate the performance of a large-scale dynamic route guidance system, the Mobile Navigation Assistant. The program seeks to relieve traffic congestion using:

- Alternative approaches for driver information systems
- Dynamic traffic information acquisition
- Incident detection, analysis, and forecasting

Several thousand private vehicles will be equipped with an in-vehicle route guidance system. Business listings, points of interest, and traffic information will also be available. The equipped vehicles will act as traffic probes, reporting congestion information on an exception basis. Loop detectors, video systems, and other traffic surveillance systems will also provide traffic information [1].

Advantage I-75

Advantage I-75 is a public-private partnership project along the Interstate 75 corridor, running from Florida to upstate Michigan, with an extension into Canada. Its objectives are to reduce congestion, increase efficiency, and enhance safety. Its initial project is an AVI application to facilitate motor carrier operations. Advantage I-75 features off-the-shelf technology and decentralized control, with each state retaining its constitutional and statutory authority relative to motor carriers and their operations [1].

Crescent Demonstration Project (HELP)

Crescent is a demonstration project of the Heavy Vehicle Electronic License Plate (HELP) program. The demonstration is being operated along Interstate 10 from the New Mexico/Texas state line, west through New Mexico, Arizona, and California, north from Los Angeles along Interstate 5 through California, Oregon, and Washington, and into British Columbia. Program participants also include the states of Alaska, Idaho, Iowa, Minnesota, Nevada, Pennsylvania, Virginia, the Port Authority of New York/New Jersey, and Transport Canada. Technologies being evaluated include:

- Weigh in motion (WIM)
- Automatic vehicle identification (AVI)
- Automatic vehicle classification (AVC)
- Automatic toll collection
- Speed and elapsed travel time monitoring
- Commercial license checking

Information will be processed at a central computer and used by both government and the trucking industry for regulatory, weight enforcement, and fleet management purposes. Crescent will include approximately 40 equipped sites and has about 2000 equipped vehicles in operation. The goal is to have a system in which a truck can drive through the entire network without having to stop at any weigh stations or ports of entry [1].

Fast-Trac

Fast-Trac (Faster and Safer Travel—Traffic Routing and Advanced Controls) is a five-year ATMS and ATIS demonstration project taking place in Oakland County, Michigan. The goal is to demonstrate the effectiveness of an integrated ATIS/ATMS system in improving mobility, reducing energy consumption, enhancing air quality, and reducing traffic accidents. The program will also evaluate and demonstrate requirements for the deployment of ITS. The integrated ITS to be demonstrated consists of an ATMS using a traffic management system developed in Australia and a wide-area video image–processing system developed in Minnesota. The system will also include an ATIS using technology for infrastructure-based real-time route guidance developed in Germany [1].

GuideStar

GuideStar is an ITS program developed jointly by the University of Minnesota Center for Transportation Studies and the Minnesota DOT to rapidly respond to growing congestion and safety problems on the highways. GuideStar ITS activities emphasize the gathering and distribution of traffic information for use by traffic managers and motorists. Activities also center around the automation of current traffic management and control practices. These activities are designed to improve driver decision making dramatically and provide a foundation for reaching GuideStar goals [3].

GuideStar is a cooperative effort that brings together a number of ongoing operational ATMS/ATIS efforts with other ITS projects and technologies throughout the state, such as the Minnesota Administrative Truck Center, a "one-stop" center that combines permit and other regulatory offices previously scattered over the Minneapolis–St. Paul area. GuideStar includes the Genesis and Travlink projects that provide traveler information, ACTS, and Roadstar I. The program includes [1]:

- Machine-vision research
- A regional traffic communications network
- Weigh In Motion research

GuideStar operates the Twin Cities Freeway Management System and has established a major laboratory on a freeway corridor along Interstate 394. The I-394 Laboratory is equipped with dense wide-area machine-vision instrumentation and cameras, and a densely located network of traditional traffic-management, driver-navigation, and data-collection equipment. Corridor activities include employment of machine vision to monitor traffic flow. Machine vision will also be used to identify congestion, incidents, and accidents by computer analysis of traffic data collected by television cameras (Fig. 62.18). Real-time information is disseminated to traffic managers and motorists and is used to test traffic control strategies and simulation models as well as to evaluate a full range of machine-vision capabilities [3].

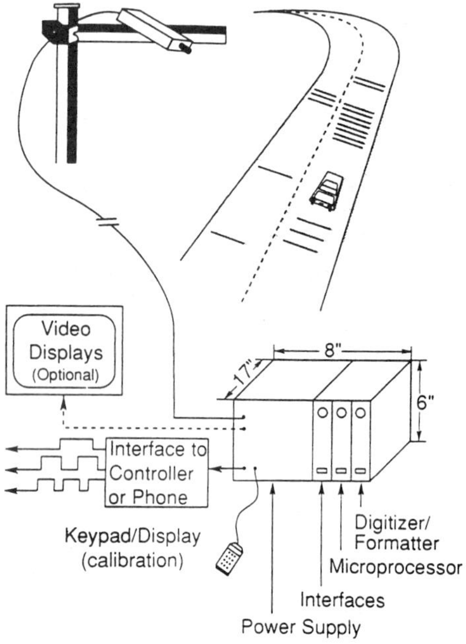

FIGURE 62.18 Video imaging system used in Minnesota's GuideStar project. (*Source: ITE*, March 1992, p. 44.)

The heart of the wide-area machine-vision instrumentation on I-394 is a patented device that was developed at the University of Minnesota with funding from FHWA, the Minnesota DOT, and the University of Minnesota. It is expected to become the centerpiece of a traffic communications network for monitoring 300 miles of freeways and major arterials in the Minneapolis/St. Paul area [24].

(Contact: R. Stehr, Minnesota DOT, St. Paul, Minnesota, 612/297-3532.)

ICTM

ICTM (Integrated Corridor Traffic Management) is a cooperative effort by the Minnesota DOT, Hennepin County, and the City of Bloomington. Objectives of the project are:

- Develop partnerships with potential participants.
- Develop strategies for traffic management from a corridor perspective.
- Install vehicle detection/surveillance on arterials parallel to and intersecting the freeway.
- Implement coordinated arterial traffic signal systems on intersecting and parallel arterials and integrate those systems with the freeway management system.

(Contact: L. Taylor, Minnesota DOT, St. Paul, Minnesota, 612/341-7291.)

INFORM

INFORM is a computerized traffic management and information system (ATMS/ATIS) operated by the New York State DOT. INFORM is designed to reduce travel time along the 40-mile central corridor of Long Island, in Queens, Nassau, and Suffolk counties, previously supported by project IMIS. The system gathers traffic information using citizens band radio monitors, closed-circuit television (CCTV), and roadway sensors in the Long Island Expressway, Northern State Parkway/Grand Central Parkway, Jericho Turnpike, and adjacent cross streets and parallel arterials. INFORM communicates traffic information to motorists through variable message signs (VMS), commercial radio broadcasts, and color-coded maps. INFORM also adjusts traffic signals and entrance ramp–metering signals in response to current traffic patterns [1].

Pathfinder

Pathfinder was a field test of in-vehicle urban freeway navigation and information systems sponsored by Caltrans and the private industry and aimed at improving traffic flow. The project, completed in June 1992, was divided into three stages. Stage one began in June 1990. Testing took place on the SMART Corridor, a 13-mile stretch of the Santa Monica Freeway between Los Angeles and Santa Monica. Twenty-five vehicles were equipped with a navigation system manufactured in Germany under license from the U.S. The cars received real-time information on accidents, congestion, highway construction, and alternate routes through a computerized map display or digital voice in the vehicle. A traffic management center (TMC) managed the communications, detecting traffic density and vehicle speeds (via detectors and using the Pathfinder vehicles as probes, see Fig. 62.19) and transmitting congestion information to equipped vehicles [1].

Stage two involved the use of hired drivers to travel between various origin and destination points to obtain data on differences in travel time with and without the equipment in operation. The drivers' choice of display mode, route selection, and diversion choices were also evaluated. In

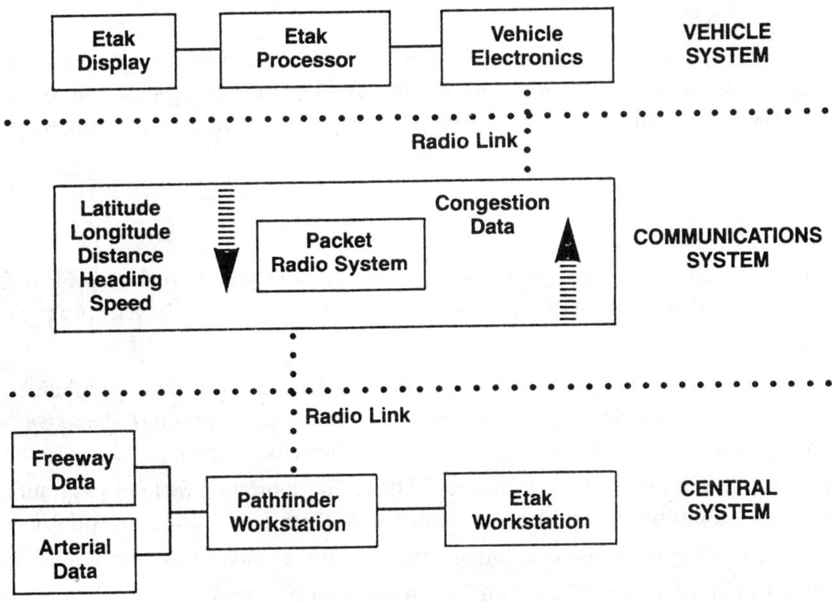

FIGURE 62.19 Components of Pathfinder system. (*Source: ITE*, Nov. 1990, p. 24.)

stage three, local commercial drivers used Pathfinder vehicles for regular business trips. The final stage was similar to stage one, in that destinations were chosen by the driver [1].

Smart Bus/Smart Vehicle

Smart Vehicle includes a series of operational tests that use different technologies to improve the efficiency and attractiveness of public transportation. Five Smart Vehicle programs have been sponsored by FTA [1]:

- *Ann Arbor.* Includes an on-board bus system capable of traffic signal preemption, a central control system, and a Smart Card fare collection/parking pass system.
- *Baltimore—MTA.* AVL technology provides bus location information to the public. Uses LORAN-C and radio and will be expanded to include GPS.
- *Chicago—CTA.* Preliminary studies have been done for an operational test of AVL, computer-assisted dispatch and control system, real-time passenger information signs, and traffic signal preemption for public buses.
- *Denver—RTD.* AVL technology provides bus location information to transit dispatchers. Uses GPS and will be expanded to include a transit information system with kiosks.
- *Portland.* Integrated fixed-route transit and contract taxi service information is provided to travelers and operators.

SMART Corridor

The SMART Corridor is a joint demonstration project sponsored by Caltrans, the Los Angeles Metropolitan Transportation Authority, the Los Angeles DOT, and the Los Angeles Police Department. Objectives include [1]:

- Congestion relief
- Accident reduction
- Reduced fuel consumption
- Improved air quality

Variable message signs, highway advisory radio (HAR), kiosks, videotex, and ramp metering are tested along 13 miles of the Santa Monica Freeway. Improved emergency response and coordinated interagency traffic management via voice communication and electronic data sharing will also be tested [1].

Smart Traveler

Smart Traveler includes a series of operational tests that examine ways to make reliable real-time information on ride sharing and public transit easily available. Four Smart Traveler programs have been sponsored by FTA [1]:

- *Bellevue, Washington.* Phase one advises private auto drivers of ride share possibilities using a prototype commuter information system and mobile communications. Phase two tests a prototype computer-based, interactive commuter information center.
- *California.* Test of the California Advanced Public Transportation Systems program in five sites. Audiotex and videotex are used for interactive ATIS services, including carpool matching.
- *Houston.* Development and evaluation of a real-time traffic and transit information system.
- *Minneapolis/St. Paul.* Test of "TravLink" automatic vehicle location and traveler information. Smart Card investigation.

SPIM

SPIM (St. Paul Incident Management) is a demonstration project in incident-based advanced traffic guidance and control. It seeks the application of CCTV and VMS and the integration of ramp meters and intersection signals. It features the application of off-the-shelf technology and the coordination of the activities of multiple jurisdictions in the St. Paul Metropolitan area.

The objective of the SPIM project is to provide traffic guidance and control during freeway incidents, to divert traffic off the problem scene and around the St. Paul CBD using alternate surface street routes, and to redirect the traffic back to the freeways. The project includes the installation of traffic surveillance and route guidance equipment, and the implementation of incident-based traffic control strategies. Included in the project are the following components:

- Communication between City of St. Paul and Traffic Management Center of the Minnesota DOT
- Traffic detection
- Traffic surveillance and guidance
- Incident-based signal-timing plans
- Integration of ramp meters and adjacent signals

(Contact: J. Wright, Minnesota DOT, St. Paul, Minnesota, 612/297-7166.)

TRANSCOM

TRANSCOM is a consortium of 14 transportation and public safety agencies in the New York and New Jersey area. It functions as an information clearinghouse for incident and construction information. Its three core projects are [1]:

- Construction coordination to avoid interagency scheduling conflicts
- Incident management planning to coordinate interagency response to incidents
- Operations information center to collect and disseminate real-time traffic information to member agencies, who then pass it on to travelers via HAR and variable message signs

TRANSCOM has initiated a cooperative effort to equip approximately 1000 commercial vehicles with transponders. Specific technologies include [1]:

- HAR
- Remote video surveillance
- Computer networking system
- AVI system for electronic toll collection
- Geographic information system (GIS) for data management

The test evaluates the use of transponder data to determine real-time traffic information such as speed, travel time, and the occurrence of incidents [1].

TRANSCOM is also involved in the EZ-Pass effort (an electronic toll collection system). TRANSCOM is funding a marketing study of the system, to explore its viability for advanced traffic management applications [1].

TravTek

TravTek was a three-year joint effort that included the participation of Florida DOT, the City of Orlando, and the American Automobile Association (AAA.) It conducted ATIS experiments with the objective of evaluating user acceptance and technical performance. TravTek employed ATIS technologies to maximize consumer use of traffic and service information. One hundred vehicles were equipped with a route guidance system and an in-vehicle TravTek device that provided real-time information on traffic congestion, as well as information on local events, hotels/motels, restaurants, attractions, and other points of interest [1].

The system included a traffic management center (TMC) that received information from a traffic information network and a TravTek information service center. AAA counselors provided assistance to TravTek drivers, answering questions, providing driving directions outside the automated coverage area, dispatching emergency road service, and other travel-related services. The TravTek information service center was also linked directly to the traffic information center to ascertain current vehicle location and provide daily special event and road restriction information to the traffic center for broadcast to the vehicles (see Fig. 62.20) [1].

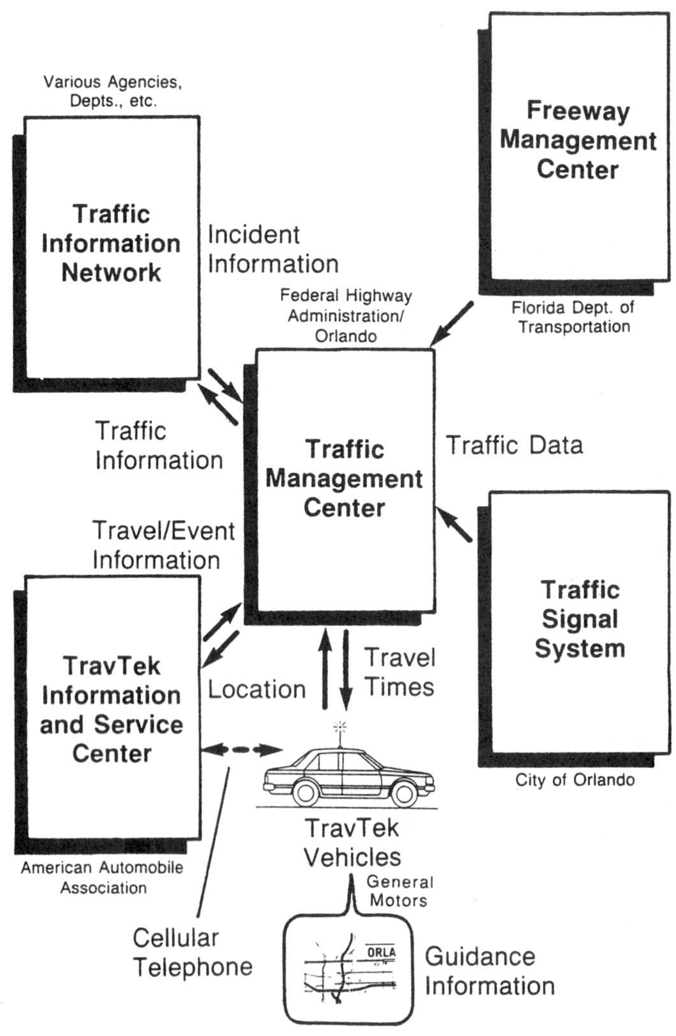

FIGURE 62.20 Components of TravTek system. (*Source: ITE,* Nov. 1990, p. 25.)

62.9 ITS Programs Outside the U.S.

Canada

Canadian ITS activities are carried out by both public and private organizations. The ITS Roundtable of the Transportation Association of Canada acts as a monitor for these activities. Canada is a participant in several North American operational field tests, including the Heavy Vehicle Electronic License Plate (HELP) Program, Advantage I-75, and the Wide Area Vehicle Monitoring (WAVM) Project. Canadian programs also include [1]:

AVION (vehicle monitoring automation)
COMPASS
MTIPS (Municipal Traffic Information Production System)
TravelGuide

Europe

In Europe government and industry work together to fund research and development of ITS services. ITS is called road transport informatics (RTI) or applications in transport telematics (ATT). European ITS standards activities are being addressed by Technical Committee 278 of CEN, the European Standards Organization. Current programs include [1]:

DRIVE
PROMETHEUS
EUREKA
ERTICO

Japan

The Japanese ITS program is extensive and involves close cooperation between industry and government. An ATMS/ATIS infrastructure has been installed in much of the country, and automakers offer navigation systems. Several government agencies have played a key role in the development and testing of ITS, including the Ministry of Construction, the Ministry of Posts and Telecommunications, the Ministry of International Trade and Industry, the Ministry of Transport, and the National Police Agency. Japanese ITS programs include [1]:

Japan Digital Road Map Association
Comprehensive Automobile Control System
Advanced Mobile Traffic Information/Communications Systems
Road-Automotive Communication System
Vehicle Information Communication System
Advanced Road Traffic System
Urban Traffic Control System
Super Smart Vehicle System
Personal Vehicle System
Advanced Traffic Information Service

References

1. TRB Committee on Intelligent Vehicle Highway Systems. 1993. Primer on Intelligent Vehicle Highway Systems. Transportation Research Circular 412, Transportation Research Board, National Research Council, Washington, D.C., ISSN 0097-8515, August.

2. Euler, G. W. 1990. Intelligent vehicle/highway systems: Definitions and applications. *ITE Journal.* 60(11):17–22.
3. Minnesota IVHS Planning Group. 1990. *GuideStar: Guiding the Future of Minnesota's Highways.* Minnesota Department of Transportation and University of Minnesota.
4. Beaubien, R. F. 1993. Deployment of intelligent vehicle-highway systems. *ITE Journal.* 63(2): 15–18.
5. Constantino, J. 1993. IVHS America two years later. *ITE Journal.* 63(2):19–22.
6. Kraft, W. H. 1993. IVHS and the transportation profession. *ITE Journal.* 63(2):23–25.
7. Carlson, E. D. 1993. Federal actions to deliver the IVHS program. *ITE Journal.* 63(2):26–30.
8. Davis, G. 1993. Private communication. University of Minnesota, Minneapolis, Minnesota, October.
9. Inside IVHS. 1993. FHWA awards contracts for IVHS architecture. *Intelligent Vehicle/Highway Systems Update.* 3(19):1–3. Waters Information Services, Inc., ISSN 1054-2647, September.
10. Grenzeback, L. R. and Woodle, C. E. 1992. The true costs of highway congestion. *ITE Journal.* 62(3):16–20.
11. Lindley, J. A. (1986). *Quantification of Urban Freeway Congestion and Analysis of Remedial Measures.* Federal Highway Administration Staff Rep. RD-871052, October. See also, Urban freeway congestion: Quantification of the problem and effectiveness of potential solutions. *ITE Journal.* 57:27–32.
12. Judycki, D. C. and Robinson, J. R. 1992. Managing traffic during nonrecurring congestion. *ITE Journal.* 62(3):21–26.
13. Morales, J. M. 1986. Analytical procedures for estimating freeway traffic congestion. *Public Roads.* 50(2):55–61.
14. Judycki, D. C. and Robinson, J. R. 1988. Freeway incident management. In *1988 Compendium of Technical Papers.* Institute of Transportation Engineers, Washington, D.C., pp. 161–165.
15. JHK & Associates. Estimate based on delay evaluation procedures in *Alternative Surveillance Concepts and Methods for Freeway Incident Management: Vol. 2—Planning and Trade-Off Analyses for Low-Cost Alternatives.* Washington, D.C.: Federal Highway Administration, March 1978; *A Freeway Management Handbook: Vol. 2—Planning and Design.* FHWA, May 1983; and modifications recommended by California Department of Transportation.
16. Recker, W. W. 1988. *An Analysis of the Characteristics and Congestion Impacts of Truck-Involved Freeway Accidents.* Institute of Transportation Studies, Irvine, CA.
17. Robinson, J. R. and McDade, J. D. 1990. *Incident Management Programs in the United States.* Office of Traffic Operations, FHWA, Washington, D.C.
18. Capelle, D. G. 1979. *Freeway Traffic Management.* Final Report, NCHRP Project 20-3D. Transportation Research Board, National Research Council, Washington, D.C.
19. Stephanedes, Y. J. and Chassiakos, A. P. 1993. Freeway incident detection through filtering. *Transportation Research, Part C: Emerging Technologies.* 1C(3):219–233. ISSN 0968-090X, September.
20. Chassiakos, A. P. and Stephanedes, Y. J. 1993. Smoothing algorithms for incident detection. *Transportation Research Record.* 1394:8–16. Transportation Research Board, National Research Council, Washington, D.C. ISBN 0-309-05463-X, August.
21. Ritchie, S. G. and Stack, R. 1993. A real-time expert system for freeway incident management in Orange County, California. *Proceedings, Fifth International Conference on Computing in Civil Engineering.* ASCE, New York.
22. Stephanedes, Y. J. and Liu, X. 1994. *Artificial Neural Networks for Freeway Incident Detection.* Presented at 73rd Annual Meeting of the Transportation Research Board, Washington, D.C., January.
23. Transportation Research Board. 1993. *IVHS-IDEA.* Program announcement, Washington, D.C., April.
24. French, R. L. 1990. Intelligent vehicle/highway systems in action. *ITE Journal.* 60(11):23–31.

25. Rowe, E. 1993. IVHS—Making it work, pulling it all together. *ITE Journal.* 63(2):45–48.
26. Stephanedes, Y. J. and Chang, K. 1993. Optimal control of freeway corridors. *Journal of Transportation Engineering.* 119(4):504–514. ASCE, ISSN 0733-947X, July/August.
27. Stephanedes, Y. J. and Kwon, E. 1993. Adaptive demand-diversion prediction for integrated control of freeway corridors. *Transportation Research, Part C: Emerging Technologies.* 1C(1). ISSN 0968-090X, March.
28. Castle Rock Consultants. 1994. *Rural IVHS Scoping Study.* Minnesota GuideStar, St. Paul, April.
29. Stephanedes, Y. J. 1994. Implementation of on-line zone control strategies for optimal ramp metering in the Minneapolis ring road. *Road Traffic Monitoring and Control,* no. 391. The Institution of Electrical Engineers, London, UK.
30. Executive Office of the President of the United States. 1993. *A Vision of Change for America.* U.S. Government Printing Office, Washington, D.C., 17 February.
31. Federal Highway Administration. 1993. *IVHS Staffing and Educational Needs.* Final Report. U.S. Department of Transportation, DTFH61-92-C-00145, Washington, D.C.
32. Payne, H. J. and Tignor, S. C. 1978. Freeway incident detection algorithms based on decision trees with states. *Transportation Research Record,* no. 682. TRB, Washington, D.C. pp. 30–37.
33. Levin, M. and Krause, G. M. 1978. Incident detection: A Bayesian approach. *Transportation Research Record.* (682):52–58.
34. Persaud, B. N., Hall, F. L., and Hall, L. M. 1990. Congestion identification aspects of the McMaster incident detection algorithm. *Transportation Research Record,* (1287):151–166.
35. Zografos, K. G. and Nathanail, T. 1991. An analytical framework for minimizing freeway incident response time. *Proc. 1991 ASCE Conf. on Applications of Advanced Technologies in Transportation Engineering.* Minneapolis, MN. pp. 157–161.
36. Tsai, J. and Case, E. R. 1979. Development of freeway incident detection algorithms by using pattern-recognition techniques. *Transportation Research Record.* (722):113–116.
37. Van Trees, H. 1968. *Detection, Estimation, and Modulation Theory Part I.* John Wiley & Sons, New York.
38. Payne, H. J., Helfenbein, E. D., and Knobel, H. C. 1976. *Development and Testing of Incident Detection Algorithms.* Final Report. FHWA, FH-11-8278, vol. 2, Washington, D.C.
39. Collins, J. F., Hopkins, C. M., and Martin, J. A. 1979. Automatic incident detection—TRRL algorithms HIOCC and PATREG. *TRRL Supplementary Report,* no. 526. Crowthorne, Berkshire, United Kingdom.
40. Masters, P. H., Lam, J. K., and Wong, K. 1991. Incident detection algorithms for COMPASS—An advanced traffic management system. *Proc. Vehicle Navigation & Information Systems Conference,* Part 1. Dearborn, Michigan, pp. 295–310.
41. Dudek, C. L. and Messer, C. J. 1974. Incident detection on urban freeways. *Transportation Research Record.* (495):12–24.
42. Cook, A. R. and Cleveland, D. E. 1974. Detection of Freeway Capacity-Reducing Incidents by Traffic-Stream Measurements. *Transportation Research Record.* (495):1–11.
43. Ahmed, S. A. and Cook, A. R. 1982. Discrete dynamic models for freeway incident detection systems. *Transportation Planning and Technology.* 7:231–242.
44. Willsky, A. S., Chow, E. Y., Gershwin, S. B., Greene, C. S., Houpt, P. K., and Kurkjan, A. L. 1980. Dynamic model-based techniques for the detection of incidents on freeways. *IEEE Transactions on Automatic Control.* 25(3):347–360.
45. Payne, H. J. 1971. Models of freeway traffic and control. *Math. Models Public Syst., Simulation Council Proc.* 28(1):51–61.
46. Cremer, M. 1981. Incident detection on freeways by filtering techniques. *Preprints 8th IFAC Congress.* Kyoto, vol. XVII, pp. 96–101.
47. Chassiakos, A. P. 1990. *An Improved Time-Space Incident Detection Algorithm.* M.Sc. Thesis. Dept. of Civil and Mineral Engineering, University of Minnesota, Minneapolis.

48. Stephanedes, Y. J. and Chassiakos, A. P. 1993. Application of filtering techniques for incident detection. *ASCE Journal of Transportation Engineering.* 119(1):13–26.

For Further Information

To gain additional knowledge in ITS, readers can contact ITS America, 400 Virginia Avenue, SW, Suite 800, Washington, D.C. 20024-2730, tel (202) 484-4847, fax (202) 484-3483. A biannual publication of the American Society of Civil Engineers (ASCE), *Applications of Advanced Technologies in Transportation Engineering,* which contains summaries from a respective biannual conference, provides insights into current ITS research by civil engineers. To obtain the most recent available copy, contact ASCE, United Engineering Center, 345 East 47th Street, New York, NY 10017-2398, tel. (212) 705-7496, fax (212) 980-4681.

For information on the DRIVE program of the European Community, contact Commission of the European Communities, Directorate General, Information Technologies and Industries, and Telecommunications, Rue de la Loi 200, B-1049 Brussels, Belgium, tel (32-2) 296-3591, fax (32-2) 296-2391. A good description of progress in European projects can be found in *Advanced Telematics In Road Transport,* Elsevier 1991, and *Advanced Transport Telematics,* DR400, Commission of the European Communities 1993, which contain summaries from project reviews.

63

Highway Infrastructure Management

Anton J. Kleywegt
Purdue University

Kumares C. Sinha
Purdue University

63.1 Introduction

There is increasing pressure on agencies around the world responsible for the provision and maintenance of transportation infrastructure to provide more and better facilities with less resources. To meet these demands efficiently, agencies can utilize tools such as advanced data collection equipment, databases, geographic information systems, modeling techniques, prioritization, and optimization. Increasing computer power makes it possible to routinely use techniques that were previously impractical. In many countries highway agencies are expected to use these techniques to improve budgeting and provision of service.

The U.S. Intermodal Surface Transportation Efficiency Act (ISTEA) of 1991 requires states to develop, establish, and implement systems to manage the following [U.S. Federal Register, 1993]:

1. Highway pavements of federal-aid highways (pavement management systems)
2. Bridges on and off federal-aid highways (bridge management systems)
3. Highway safety (safety management systems)
4. Traffic congestion (congestion management systems)

0-8493-8953-4/95/$0.00 + $.50

5. Public transportation facilities and equipment (public transportation management systems)
6. Intermodal transportation facilities and systems (intermodal management systems)

States must also implement traffic monitoring systems for highways and public transportation facilities and equipment.

This chapter discusses tools that can be used to develop these management systems, with the emphasis on pavement, bridge, and maintenance management systems. Although a maintenance management system is not required by the ISTEA of 1991, it is an important aid to improve the activities of a highway agency. Section 63.2 gives an overview of highway system management. Section 63.3 discusses data needs and data collection technologies. Section 63.4 gives a brief overview of some useful modeling techniques. In section 63.5 the analysis of highway life-cycle costs is discussed. Priority setting and optimization, which are useful techniques for strategy selection, are discussed in section 63.6. Section 63.7 is devoted to maintenance management systems, and section 63.8 concludes the chapter.

63.2 Highway System Management

Total Highway System Management—Facilities, Activities, and Objectives

In managing a highway system, the total system with its objectives, facilities and activities should be seen in perspective. A highway system consists of facilities, such as pavement, bridges, roadside elements, and traffic control devices. These facilities have different characteristics which influence their management, such as construction costs, deterioration rates, life spans, maintenance needs, and replacement costs. Different management issues are therefore important in managing different facilities. Overall, facilities all share the same funds and (to some extent) the same manpower, which makes their management interdependent.

To provide the facilities, a variety of activities have to be performed, such as planning, design, construction, operation, inspection, maintenance, improvement, and management. These activities influence each other, as well as other facilities (e.g., the geometric design influences the design of bridges and placing of traffic control devices). Activities are therefore highly interdependent.

Highway system management serves a number of objectives, such as provision of service, safety, condition preservation, cost reduction, environmental conservation, and other socioeconomic objectives. These objectives are to some extent complementary (e.g., improving level of service often also improves safety). Many times these objectives are in conflict (e.g., improving level of service increases agency costs). Ultimately the objective is to provide the best mix of highway facilities with the limited funds available, which again requires a global point of view in managing the highway system.

The ideal highway management system would therefore be a comprehensive, coordinated system, at the same time flexible and sensitive enough to adjust to changes in requirements. Due to a variety of reasons, such as the piecemeal development of different management systems, different characteristics and activities, organizational structures and budgets, and the need for less complexity, separate management systems, such as pavement, bridge, and maintenance management systems, were developed over the last two decades [Sinha and Fwa, 1989]. Where applicable, these three types will be discussed separately, always keeping in mind the need for a global point of view.

Components and Products of an Infrastructure Management System

An infrastructure management system should be able to provide the following products with the necessary accuracy and speed.

- A database with accurate, complete, and up-to-date data forms the core of any infrastructure management system. These data are used for all the other products. Types of data, data organization, and data collection and analysis are discussed in detail in sections 63.3 and 63.4 for each type of management system.

- An inventory of the infrastructure, with characteristics such as location, construction cost and date, maintenance history and cost, condition history, and service provided (e.g., traffic levels).
- Prediction of future condition of infrastructure elements, under different maintenance and rehabilitation (M&R) alternatives.
- M&R activities to be implemented—which activities, where, and when to be implemented.
- Impacts of different activities on objectives such as level of service, condition, safety, user cost, and the environment.
- Budget implications of strategies and activities—both immediate budget requirements and implications for future budgets.
- Project level analysis to trade off and choose the best of mutually exclusive alternatives.
- Prioritization of projects competing for the same limited funds.
- **Network level optimization** to optimally allocate scarce resources to maximize benefits for the network as a whole over multiple budget periods.
- **Programming** and scheduling of alternatives to be implemented.
- Generation of reports with different levels of detail for different facilities, geographic areas, and time periods, to facilitate better communication between maintenance personnel, engineers, managers, politicians, and outside organizations.

Flexibility and Applicability

Management system needs of agencies vary according to the type and extent of infrastructure the agency is responsible for, financial and manpower resources, political level of the agency (city, county, or state agency), organizational and budget structure, and experience with management systems and activities such as formal prioritization and optimization.

Management system needs also change with time as responsibilities shift, infrastructure elements change, organizations and budget compositions are restructured, and new technologies are developed. A management system should therefore be applicable to the type of agency it is intended to serve, and be flexible to change with changing requirements.

Database

Designing a good database forms the basis of a management system. A good database structure has the following properties:

- Minimum redundancy. A database is usually structured into different tables/files, each containing data on different characteristics, called fields, of different elements/transactions, called records. Redundancy is reduced by duplicating only the minimum amount of data in different files to enable the necessary cross-referencing. This is accomplished by normalization of the data structures and intelligent design of files and fields.
- Integrity. Due to the normalization of data structures, there will exist many cross-references between data entries in different files. Integrity means that consistency is enforced when cross-referenced data is added, changed, and deleted (e.g., a record in one file cannot be deleted if a record in another file is cross-referenced to it).
- Validity checks. This means that entered data are checked for reasonableness and accuracy, with corresponding warnings and opportunity to correct invalid data.
- Security. The rights of different users to modify the database structure, add, edit, delete, and read data have to be clearly spelled out and coded into the database schema.

- Uniform standards. The data dictionary has to provide uniform standards for allowable data types and the names and codes of items (e.g., pavement types, spelling and abbreviations, date formats).
- Networking. The database should be accessible to a network of users, who should be able to extract the data needed for different applications in whatever format required.

Many database management systems (DBMS) are available commercially. These have varying levels of sophistication and capabilities. The better DBMSs all have a programming language, with a compiler or at least interpreter, enabling the system developer to write additional application programs for the management system. A DBMS suited to the needs of the agency can therefore be selected, and used as a basis for implementing the infrastructure management system.

Choosing the correct hardware and software and designing a good database structure are such fundamental parts of implementing an infrastructure management system that the participation and careful thought of top management should be obtained. Even though a DBMS makes provision for changing the structure of the database after it has become operational, and therefore provides the necessary flexibility, these changes come at a cost. Especially if a large number of application programs have to be adjusted as well, these modifications can be expensive and time-consuming, and can cause many teething problems. It is important to do things right the first time, to build in the necessary flexibility, and to document the system well.

Location Referencing System

For spatially related data, such as highway condition data, a consistent **location referencing system** is essential. Many location referencing systems are in use, varying from milepost-based systems and systems based on distances from landmarks such as intersections, to more sophisticated systems such as latitude and longitude location and versions of the Gauss-Mercator projection system. Whichever system is chosen, the most important is consistent application, to make the location references of different facilities compatible.

There are a number of reasons for considering the more rigorous location referencing systems. Emerging new technologies such as global positioning systems (GPS) make it possible to determine the location of facilities and points of inspection to high accuracy in a fraction of a second using relatively unsophisticated equipment and a system of satellites. To use this capability, a good location referencing system must be in use, and/or a conversion routine must be developed between the referencing systems.

A large amount of geographically referenced data are available, such as the U.S. Census TIGER files [Federal Highway Administration, 1991]. These data can be incorporated into the agency's database if the referencing systems can be converted to each other. A good location referencing system is also essential to establish a **geographic information system (GIS)**.

Geographic Information Systems

A GIS is a database with additional capabilities to store, manipulate, display, and plot spatially referenced data, such as data on the condition of segments of a highway network, and to do spatial analyses of these data, such as computing and displaying indices for spatial entities having specified properties, for example, to compute remaining service lives for all highway links with a traffic load of more than 5000 vehicles per day, and to highlight those with less than 2 years remaining service life.

A large number of GIS packages are already commercially available. New technology is continuously being developed, and many new developments and improvements can be expected in the years to come. Important issues to be addressed when developing a GIS application are (1) a good location referencing system, (2) design of the database structure, (3) an efficient means of collecting and converting the huge amounts of data required by a GIS application, (4) choosing computer hardware that is powerful enough and has enough memory to store the data and do the large number of computations needed to process some types of requests in reasonable time, and (5) training of personnel.

63.3 Data Needs and Data Collection

Pavement Management Systems

The main purpose of a pavement management system (PMS) is to select and implement cost-effective pavement construction, rehabilitation, and maintenance strategies. At the network level, agencywide strategies are developed which will have the least total cost, or greatest benefit, over the selected analysis period. At the project level, detailed evaluation is made of costs and benefits associated with alternative activities for a particular pavement section or project over the analysis period [Haas *et al.*, 1994]. It should be noted that a PMS is applicable to both highways and airports.

Data Types

The data items to be collected and included in the database will depend on the management and analysis needs of the agency, which in turn will depend on the types of infrastructure, the available resources, and the organizational units that will use the data. Data needs of a typical pavement management system can be classified as follows [Sinha and Fwa, 1990].

Road Inventory Data. Road classification data include data items such as functional class of the road, identification codes, location, history of construction and rehabilitation, and geometrical characteristics such as divided/undivided roadway, number of lanes, lane widths, and shoulder widths.

Pavement data items include pavement type, layer thicknesses and materials, subgrade characteristics, overlays, and drainage.

Traffic Data. Traffic volume and composition data are obtained from traffic counts and surveys. Included are data on average daily traffic (ADT), traffic composition of different vehicle types, lane distribution, and directional distribution of different vehicle types.

Where available from weigh stations or surveys, data on the axle load distributions of the axle groups of different vehicle types should be included in the database. These data are useful for design procedures.

Accident Data. Traffic accident data can be useful to identify the causes of accidents (including dangerous features), to improve geometric designs, and to estimate accident cost impacts of different alternatives.

Pavement Condition Data. Pavement serviceability data are usually measured in terms of the roughness of the pavement. Rougher pavements have consequences for vehicle operating costs, such as tire costs and maintenance, as well as for passenger comfort. These effects also depend on the pavement type.

Many types of pavement surface **distress** data are collected. Data items depend on the pavement type. Table 63.1 shows surface distress types for pavements with asphalt concrete surfaces. Table 63.2 shows surface distress types for pavements with jointed portland cement concrete surfaces. Table 63.3 shows surface distress types for pavements with continuously reinforced concrete surfaces [Strategic Highway Research Program, 1993].

Skid resistance is an important pavement characteristic influencing safety. Skid resistance is measured by a coefficient of friction or a skid number, which depends on the amount of water on the pavement.

Table 63.1 Surface Distress Types for Pavements with Asphalt Concrete Surfaces

Distress Type	Unit of Measure
A. Cracking	
1. Fatigue cracking	Square meters
2. Block cracking	Square meters
3. Edge cracking	Meters
4a. Wheel-path longitudinal cracking	Meters
4b. Non–wheel-path longitudinal cracking	Meters
5. Reflection cracking at joints	
Transverse reflection cracking	Number, meters
Longitudinal reflection cracking	Meters
6. Transverse cracking	Number, meters
B. Patching and potholes	
7. Patch/patch deterioration	Number, square meters
8. Potholes	Number, square meters
C. Surface deformation	
9. Rutting	Millimeters
10. Shoving	Number, square meters
D. Surface defects	
11. Bleeding	Square meters
12. Polished aggregate	Square meters
13. Raveling	Square meters
E. Miscellaneous distress	
14. Lane-to-shoulder dropoff	Millimeters
15. Water bleeding and pumping	Number, meters

Source: Strategic Highway Research Program. 1993. *Distress Identification Manual for the Long-Term Pavement Performance Project.* National Research Council, Washington, D.C.

Table 63.2 Surface Distress Types for Pavements with Jointed Portland Cement Concrete Surfaces

Distress Type	Unit of Measure
A. Cracking	
1. Corner breaks	Number
2. Durability cracking ("D" cracking)	Number of slabs, square meters
3. Longitudinal cracking	Meters
4. Transverse cracking	Number, meters
B. Joint deficiencies	
5a. Transverse joint seal damage	Number
5b. Longitudinal joint seal damage	Number, meters
6. Spalling of longitudinal joints	Meters
7. Spalling of transverse joints	Number, meters
C. Surface defects	
8a. Map cracking	Number, square meters
8b. Scaling	Number, square meters
9. Polished aggregate	Square meters
10. Popouts	Number/square meter
D. Miscellaneous distress	
11. Blowups	Number
12. Faulting of transverse joints and cracks	Millimeters
13. Lane-to-shoulder dropoff	Millimeters
14. Lane-to-shoulder separation	Millimeters
15. Patch/patch deterioration	Number, square meters
16. Water bleeding and pumping	Number, meters

Source: Strategic Highway Research Program. 1993. *Distress Identification Manual for the Long-Term Pavement Performance Project.* National Research Council, Washington, D.C.

Table 63.3 Surface Distress Types for Pavements with Continuously Reinforced Concrete Surfaces

Distress Type	Unit of Measure
A. Cracking	
1. Durability cracking ("D" cracking)	Number, square meters
2. Longitudinal cracking	Meters
3. Transverse cracking	Number, meters
B. Surface defects	
4a. Map cracking	Number, square meters
4b. Scaling	Number, square meters
5. Polished aggregate	Square meters
6. Popouts	Number/square meter
C. Miscellaneous distresses	
7. Blowups	Number
8. Transverse construction joint deterioration	Number
9. Lane-to-shoulder dropoff	Millimeters
10. Lane-to-shoulder separation	Millimeters
11. Patch/patch deterioration	Number, square meters
12. Punchouts	Number
13. Spalling of longitudinal joints	Meters
14. Water bleeding and pumping	Number, meters
15. Longitudinal joint seal damage	Number, meters

Source: Strategic Highway Research Program. 1993. *Distress Identification Manual for the Long-Term Pavement Performance Project.* National Research Council, Washington, D.C.

Structural capacity, usually measured in terms of pavement structural characteristics, is an important indication of the ability of the pavement to carry different sizes and numbers of load applications, and therefore of the remaining life of the pavement.

Climatic and Environmental Data. Data on climatic conditions, such as rainfall, snowfall, and temperature variation, are useful to develop models to predict pavement deterioration. Pavement and subgrade characteristics such as drainage and thermal stresses should also be taken into account.

Productivity Data. Data on the resource requirements of standard activities are used to estimate the resource needs and costs of alternatives. Resource needs of activities include manpower, equipment, material, time, and money requirements.

Cost and Benefit Data. Alternatives are evaluated by comparing their costs and benefits. Cost data for maintenance, rehabilitation, and replacement activities are essential to estimate agency costs and budget requirements. For activities performed by the agency itself, productivity data can be combined with unit cost data to obtain estimates of costs of alternatives. For activities performed by contractors, previous contract prices can be adjusted to derive cost estimates.

Benefits include user and nonuser benefits. User benefits include reduction in vehicle operating costs, travel time, and accidents. Data regarding the different components of vehicle operating costs, as well as travel time and accident rates and severities, are used to estimate user benefits. These topics are covered in more detail in section 63.5.

Defining Pavement Segments

Data on pavements are usually collected and stored for discrete units called segments. For data to be compatible, it is necessary for these segments to be well defined. Two approaches for defining **pavement segments**, the equal length method and the uniform characteristics method, will be discussed.

Equal Length Pavement Segments. Equal length pavement segments are convenient, both for data collection purposes and for representation in the database. Pavement belonging to different links will usually be part of different segments, so that some segments might still have unequal lengths. For a **network level analysis**, the characteristics of equal length segments are uniform enough within each segment to obtain results of sufficient accuracy. For a **project level analysis**, more accuracy is required; therefore, shorter segment lengths or segments with uniform characteristics should be used.

Pavement Segments with Uniform Characteristics. There are a number of approaches to identify pavement segments with uniform characteristics. The pavement classification based approach, the pavement response based approach, and the cumulative difference approach will be discussed.

With the classification-based approach, pavement characteristics are chosen, and segments are identified so that the pavement characteristics are uniform within each segment. Usually these are characteristics that influence the deterioration of the segment and the type of rehabilitation action to be applied to the segment. Such pavement characteristics are pavement type, material types, layer thicknesses, subgrade type, highway classification, and traffic load. An advantage of this approach is that pavement segments delineated on this basis should retain their uniform characteristics over time, until major rehabilitation or reconstruction changes some properties. Figure 63.1 gives an example of pavement segment delineation with the classification-based approach.

With the response-based approach, a number of pavement response variables are chosen, and segments are identified so that the response variables remain fairly constant within each segment. Pavement response variables that can be chosen include roughness, rut depth, skid resistance, some surface distresses, and structural characteristics such as deflection. These variables are chosen according to their importance for predicting pavement deterioration and for determining the type

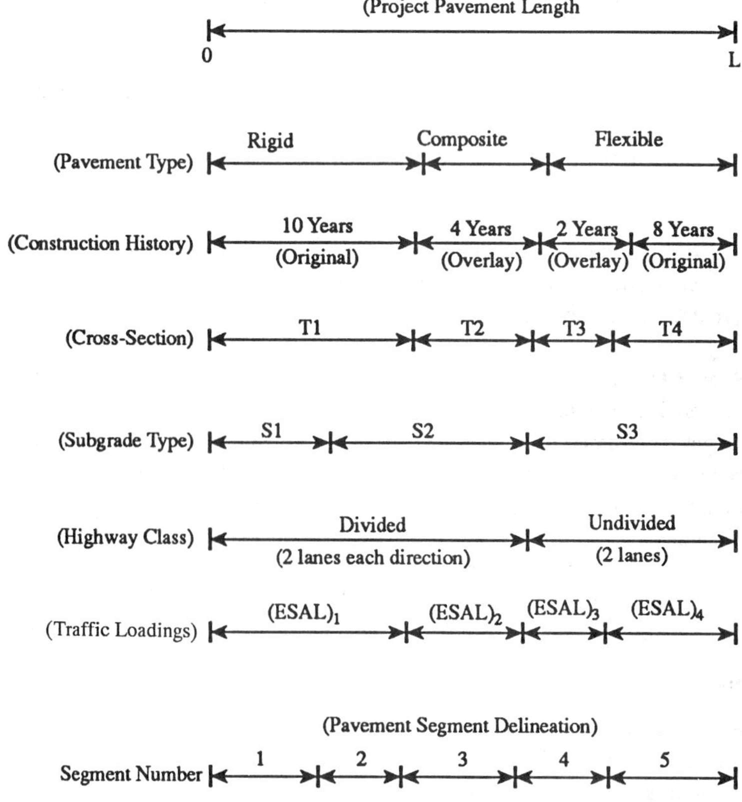

FIGURE 63.1 Pavement segment delineation with the classification-based approach.

of rehabilitation action to be implemented. Because these characteristics change over time, this approach has the disadvantage that segments might have to be redefined regularly.

Statistical tests can be used to test whether the characteristics of adjacent segments are different enough to classify them as separate segments. Assuming the characteristic is normally distributed in both segments with the same variance, the following t test can be used to test for statistically significant difference:

$$t = \frac{\overline{X}_1 - \overline{X}_2}{s_p\sqrt{\dfrac{1}{n_1} + \dfrac{1}{n_2}}}$$

$$s_p^2 = \frac{(n_1 - 1)s_1^2 + (n_2 - 1)s_2^2}{(n_1 - 1) + (n_2 - 1)}$$

where

$\overline{X}_1$ = average characteristic for segment 1
$\overline{X}_2$ = average characteristic for segment 2
s_p = pooled estimate of common standard deviation
n_1 = sample size of segment 1
n_2 = sample size of segment 2
s_1 = sample standard deviation of segment 1
s_2 = sample standard deviation of segment 2

The calculated t value can then be compared with the critical t value, $t_{k,\alpha/2}$, at the α level of significance, where $k = n_1 + n_2 - 2$ is the number of degrees of freedom. If $-t_{k,\alpha/2} < t < t_{k,\alpha/2}$, the characteristic is not significantly different between the two segments, and the segments can be combined. Figure 63.2 gives an example of pavement segment delineation with the response-based approach, the response used in this case being pavement deflection.

The cumulative difference approach identifies the boundary between adjacent segments as the point where the cumulative pavement characteristic versus distance function changes slope. It is recommended by AASHTO for pavement segment delineation. Figure 63.3(a) shows how the pavement characteristic changes with distance along the length of a pavement. The solid line in Fig. 63.3(b) shows the cumulative value of the characteristic along the length of the pavement. For a continuous function, the cumulative function is given by

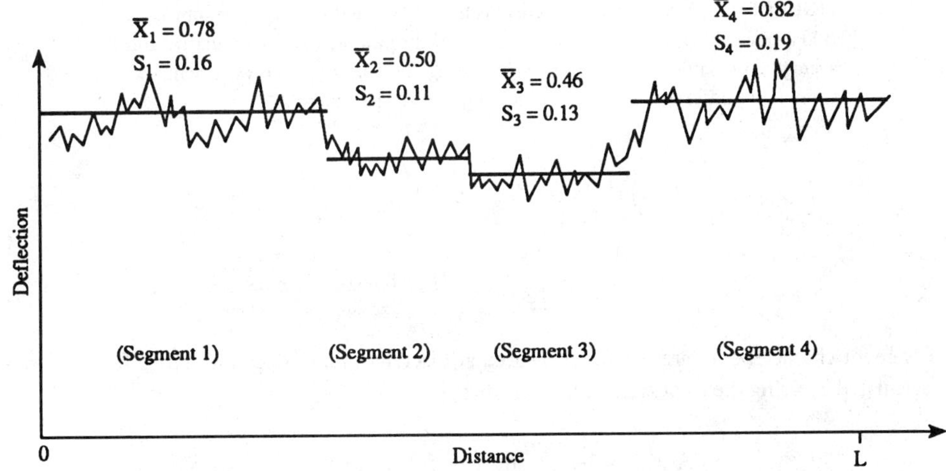

FIGURE 63.2 Pavement segment delineation with the response-based approach.

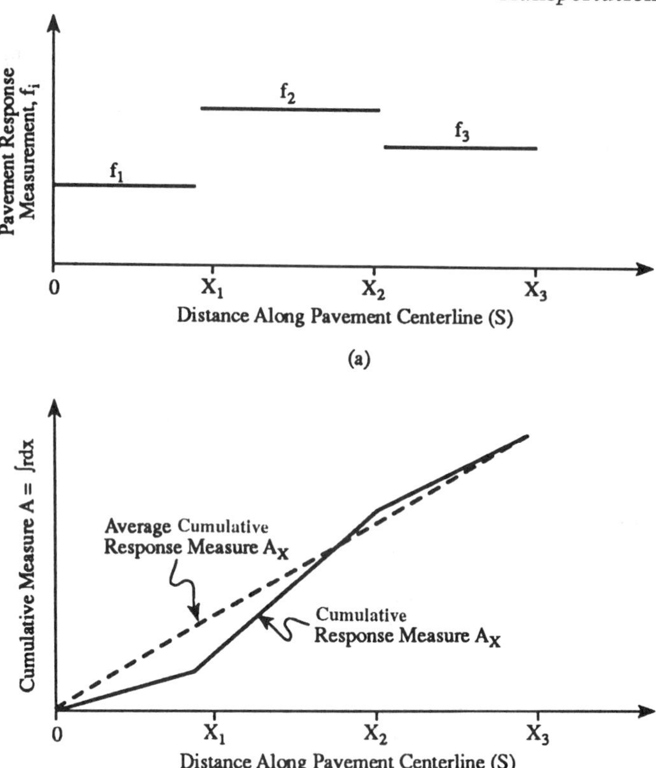

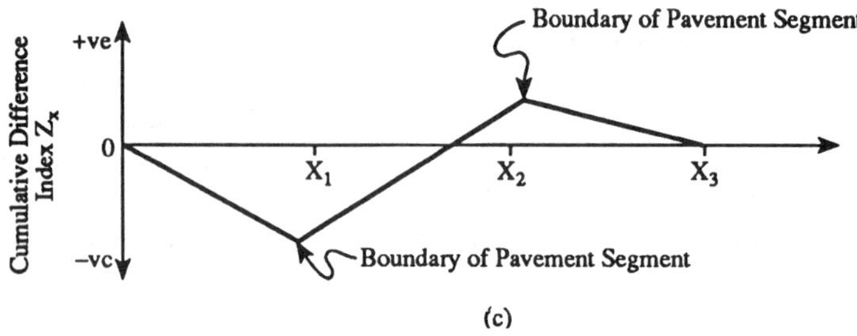

FIGURE 63.3 Pavement segment delineation by cumulative difference approach. (a) Pavement characteristic along the length of the pavement. (b) Cumulative and average cumulative characteristic value along the length of the pavement. (c) Cumulative difference index along the length of the pavement.

$$
\begin{aligned}
A(x) &= \int_0^x f_1(s)\,ds \quad \text{for } 0 \le x \le x_1 \\
&= \int_0^{x_1} f_1(s)\,ds + \int_{x_1}^{x} f_2(s)\,ds \quad \text{for } x_1 \le x \le x_2, \text{ etc.}
\end{aligned}
$$

Pavement characteristics are usually measured at discrete points. The cumulative function can be calculated by using the trapezium rule as follows:

$$
A(x) = \frac{1}{2}\sum_{i=1}^{n}(f_{i-1} + f_i)(x_i - x_{i-1})
$$

where

n = the total number of measurements
f_i = the characteristic value of the i th measurement
x_i = the distance along the pavement of the i th measurement

Figure 63.3(b) also shows a dashed line connecting the cumulative function values at the two endpoints. The slope of this line therefore represents the average characteristic value over the length of pavement. Let $\overline{A}(x)$ denote the cumulative value given by the dashed line. The cumulative difference index $Z(x)$ is then given by $Z(x) \equiv A(x) - \overline{A}(x)$.

Figure 63.3(c) shows a plot of $Z(x)$ versus distance. The boundaries of pavement segments are given by the points where the slope of the $Z(x)$ distance plot changes sign. The boundaries and their stability over time will depend on the characteristic chosen for segment delineation.

Data Collection Technologies

Specialized equipment are used to determine position, measure pavement friction, and determine ride quality or roughness, pavement distress, and structural properties.

Location Determination. Systems with mileposts or mile reference markers are common. The vehicle with which pavement condition data is collected has the ability to determine distance fairly accurately, and these relative distances, with the necessary corrections, are then used to determine the positions of points where condition data measurements are made. A newer development is global positioning systems, which can determine position automatically and accurately with radio signals and satellites. This technology holds even more promise when combined with geographic information systems.

Pavement Friction Measurement. The locked wheel friction tester is the most common device used to measure pavement friction. Disadvantages are low testing speed, high testing cost, and high wear and tear of equipment [Federal Highway Administration, 1991]. Alternative equipment in use or in development are the mu-meter, the locked wheel skid tester, the spin-up tester, and various devices for collecting and processing video and laser images of pavement texture, such as the Yandell Mee Texture Friction Device [Federal Highway Administration, 1989].

Pavement Roughness. Many different measures of roughness exist, and correspondingly many different types of equipment for measuring roughness. The most basic way of measuring pavement roughness is with a rod and level survey. Another manual method is the dipstick profiler, which makes recording of readings easier. Because use of manual devices is very time-consuming, they are mostly used to profile roughness calibration sections to calibrate other types of devices.

Profilographs are mostly used for construction quality control of portland cement concrete pavements. Response-type road roughness meters (RTRRM) measure the dynamic response of a mechanical system with specified characteristics traveling at a specified speed over the pavement, as a measure of pavement roughness. Disadvantages are that they do not measure pavement profile but dynamic response as a proxy, and that these devices have to be calibrated frequently to obtain consistent relationships between pavement characteristics and dynamic responses, which is an expensive process.

A number of profiling devices have been developed that more directly measure the profile of the pavement. Most of these devices consist of accelerometers, which are used to determine relative vertical displacement with time of the device, a distance measuring instrument to measure distance along the vehicle path, and an ultrasonic, optical, or laser system to measure the height of the device above the pavement surface. The **pavement profile** $p(x)$ at any point x along the vehicle path is then given by device vertical displacement $y(x)$ minus height of device above pavement surface $h(x)$, or $p(x) = y(x) - h(x)$. Raw collected data are noisy and must be smoothed or filtered first, and then processed further to convert the profile data to the required roughness indices. Software have been developed to do this [Federal Highway Administration, 1991].

Rut Depth. Rut depths are measured at highway speeds with basically the same technology used for profiling pavement surfaces. Most equipment types have a transverse rut bar with 3 or 5 ultrasonic or laser sensors. Laser sensors are more accurate and reliable than ultrasonic sensors, but are much more expensive. The heights between the sensors and the pavement surface are measured at regular intervals. The transverse profile and a measure of rutting can then be constructed using the height measurements for each cross section [Federal Highway Administration, 1991].

Various Surface Distresses. Most surface distress data are collected with visual inspections. These manual methods are time-consuming, expensive, and rely on subjective evaluations. Efficiency can be improved with clear and standardized manuals, training of inspectors, and inspection aids such as portable computers and specialized survey keyboards.

A number of technologies using laser, film, or video are being developed to improve inspections. Laser devices can detect some cracking, but are less reliable and repeatable, and do not produce visual records [Federal Highway Administration, 1991]. Film and video recordings can be used to detect distresses such as surface cracks, potholes, and rutting. Video equipment is easier to use, cheap, and the media can be reused.

Most of the effort in using these technologies is involved in processing the recorded images to extract distress data. Much research is being done to automate image processing. The technology and algorithms required for automatic image processing are complex, and many different approaches are currently being attempted [Federal Highway Administration, 1989].

Structural Capacity. Pavement structural carrying capacity determines the sizes of loads as well as the number of load applications that can be carried by a pavement, and is therefore an important factor influencing the remaining service life of a pavement. Deflection of the pavement under various static and dynamic loads is usually taken as a measure of pavement structural capacity. Pavement deflection is influenced by many factors, including size and duration of the load, pavement type, stiffness of the pavement, local defects such as joints and cracks, moisture, frost, temperature, and proximity of structures, which have to be taken into account when analyzing deflection data. Types of deflection equipment are static deflection equipment, steady state dynamic deflection equipment, and impulse deflection equipment.

Static deflection equipment measures pavement deflection under slowly applied loads. The best known device of this type is the Benkelman beam. It gives only a single deflection measurement, and therefore cannot be used to obtain the deflection profile of the pavement. Results are also not very repeatable. Other static deflection devices are the curvature meter and plate bearing test equipment. There are also automated beam deflection equipment, such as the La Croix Deflectograph and the Traveling Deflectometer.

Steady state dynamic deflection equipment apply a steady state sinusoidal force to the pavement after the application of a static preload. The change in deflection (vibration) is then measured and compared with the amplitude of the dynamic force. The Dynaflect and the more versatile Road Rater are well known devices of this type. Limitations of these equipment are the limited amplitudes of the dynamic loads compared with the static preloads.

Impulse deflection equipment applies an impulsive force to the pavement, usually with a falling weight. The size of the force can be varied by varying the drop height and the mass of the weight. The Dynatest Falling Weight Deflectometer is a widely used device of this type [Federal Highway Administration, 1989].

Bridge Management Systems

Like a PMS, a bridge management system (BMS) can also be developed for application at both network and project level. The primary data in a BMS include bridge inventory and condition ratings. Many of the approaches for data collection and analysis are the same for both PMS and BMS.

Data Types

Several basic data items required for a bridge management system database are given in the FHWA publication, *Recording and Coding Guide for the Structure Inventory and Appraisal of the Nation's Bridges* [Federal Highway Administration, 1988]. These are the basic data items state highway agencies have to collect to conform to the requirements of the National Bridge Inspection Standard (NBIS). These data items are reported to FHWA, and are used to make up the National Bridge Inventory (NBI) [Turner and Richardson, 1993].

It has been found that these data items are not adequate for bridge management systems. Data items to be collected for a BMS depend on the functions to be performed by the BMS. To provide a bridge inventory, data items such as location, number and length of spans, structural and material type of deck, superstructure and substructure, and age are needed. For condition prediction, data items such as element types, environmental conditions, traffic load, types and dates of maintenance and rehabilitation actions, and inspection results should be collected. To estimate agency costs, the above inventory and condition data, as well as data on the costs of M&R actions, are needed. To estimate user costs, data on the distribution of vehicle types, heights and weights on the different routes, vehicle operating costs, average lengths of detours, and bridge-related accident rates and costs are needed.

Data Collection

Some of these data items can be obtained from secondary data sources, such as traffic counts conducted for other purposes, a pavement management system's network data, or published information. The ideal is to directly link databases for different systems, such as for a PMS and a BMS, to have a single consistent network database. Other data items have to be collected only once, or at large intervals, such as most inventory data items. Data items related to implemented M&R actions are collected as the actions are performed. Condition data are collected with regular inspections.

Sampling Techniques

There are a number of basic statistical **sampling techniques** that can be used to design samples to obtain infrastructure condition data.

Simple Random Sampling

The population of infrastructure is divided into sampling units, such as pavement segments in pavement management systems, or bridges or bridge spans in bridge management systems. The units to be sampled are then selected in such a way that each unit has an equal probability of being chosen. This method is not preferred, because it is usually desirable to sample units with different characteristics in different proportions, and to cluster chosen units in such a way that they can be easily inspected together. These objectives give rise to other sampling techniques, such as stratified random sampling and cluster sampling.

Systematic Random Sampling

This technique entails selecting sampling units with a constant interval between them. For example, a highway may be divided into pavement segments, and it is required to sample every kth segment along the highway. The first segment to be sampled is drawn randomly from 1 to k, say the jth. The segments to be sampled are then segments number $j, j + k, j + 2k, j + 3k, j + 4k, \ldots$. This technique has the same disadvantages as simple random sampling.

Stratified Random Sampling

A number of characteristics are chosen according to which the sampling units are classified. These characteristics might represent the importance of different classes of units; for example, highway functional classification might be chosen, because it is desirable to sample proportionately more units from highways carrying more traffic than from less-trafficked highways. Sampling units are then selected randomly from each stratum. This approach provides more flexibility for sample design, and ensures that different classes of units will be included in the sample in the desired proportions. It can also be used to decrease the variance of condition estimators.

Single-Stage and Multistage Cluster Sampling

Where sampling units are distributed over a wide geographical area, as is typical for infrastructure, it is desirable to choose sampling units that are close to each other to decrease survey costs. This can be accomplished with cluster sampling. Units are grouped into clusters, usually on the basis of geographical proximity. The clusters to be included in the sample are then selected randomly. With single-stage cluster sampling, all the units in the chosen clusters are sampled. With multistage cluster sampling, subclusters, strata, or individual units within each chosen cluster are further randomly selected.

Combined Sampling Methods

The sampling techniques discussed above can be combined to produce desired sampling methods. Cluster sampling can be imbedded within stratified sampling, by selecting clusters to be sampled within each stratum. Stratified sampling can in turn be imbedded within cluster sampling, by grouping each cluster into a number of strata, and sampling from these strata in the desired proportions.

Sample Size Determination

Larger sample sizes are desirable for the sake of more complete information and more precise estimates of condition and other characteristics. Unfortunately, larger samples require more cost

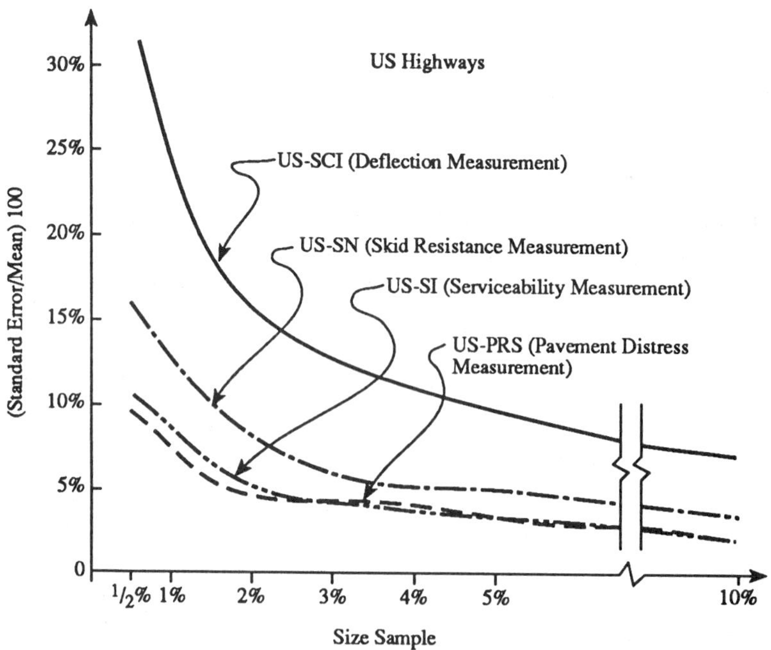

FIGURE 63.4 Sample coefficient of variation as a function of sample size for different condition measures. (*Source:* Mahoney, J. P. and Lytton, R. L. 1978. *Measurement of Pavement Performance Using Statistical Sampling Techniques.* Research report, 207-2, Texas Transportation Institute, Texas A&M University, College Station, TX.)

and time to obtain. With modern technology, it might even be possible to collect an excessive amount of data, where the storage and processing requirements become too large to justify the small increases in preciseness obtained with such large samples. The best sample size is therefore a trade-off between the preciseness required and the cost and time needed to collect and process the data. Factors influencing the best sample size are the amount of resources (funds, manpower, and equipment) available for surveys, available data collection and processing technology, the required precision, items on which data are collected, and the intended use of the data. Figure 63.4 shows the effect of sample size on the sample coefficient of variation for different data types, found by a study conducted in Texas [Mahoney and Lytton, 1978]. As can be seen, for small sample sizes, significant improvements are obtained when sample sizes are increased. When sample sizes are already large, improvements are negligible.

Sample sizes should be large enough to detect changes in characteristics at the required level of confidence, because the rate at which infrastructure condition changes is one of the most important factors in deciding when to schedule rehabilitation work. The following equations were suggested [Federal Highway Administration, 1987] to provide guidance for selecting sample sizes for detecting changes in condition:

$$m = \frac{Z_\alpha^2}{2(P_1 - P_2)^2}$$

$$n = \left(\frac{m}{m + N}\right)N$$

where

P_i = the proportion of infrastructure type that has condition worse than a specified level, at time period i

Z_α = value of standard normal random variable at the required level of confidence $(1 - \alpha)$

N = total number of infrastructure units (e.g., pavement segments)

n = number of units to be sampled to detect a change of $(P_1 - P_2)$ with the required level of confidence $(1 - \alpha)$.

A decision has to be made on the number of measurements to be taken for each sampling unit selected by the sampling method. This again involves a trade-off between precision and cost. If the characteristic to be measured is normally distributed within each sampling unit, the statistical relationship between precision and sample size is as follows:

$$(\overline{X} - \mu) = t_{k,\alpha/2}\left(\frac{s}{\sqrt{n}}\right)$$

where

$\overline{X}$ = sample mean of sampling unit

μ = true mean of sampling unit

$t_{k,\alpha/2}$ = value of t statistic with $k = n - 1$ degrees of freedom at the required level of confidence $(1 - \alpha)$

s = sample standard deviation of sampling unit

n = number of measurements taken on sampling unit

To obtain estimates of the average characteristic of the sampling unit which has error not exceeding $(\overline{X} - \mu)$ at least $(1 - \alpha)100\%$ of the time, n measurements per sampling unit are therefore required. Figure 63.5 shows a typical plot of error $(\overline{X} - \mu)$ versus number of measurements n. As can be seen, if n is small, precision improves dramatically with increasing n. If n is large, only small gains are obtained from increasing n.

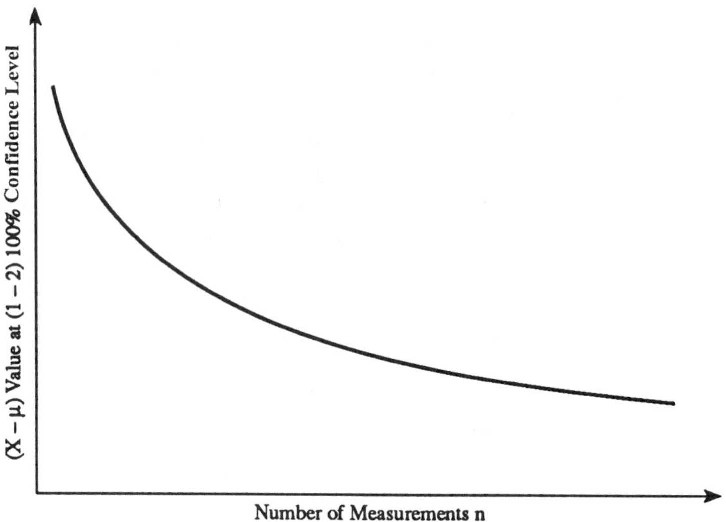

FIGURE 63.5 Typical precision characteristics of sample mean as a function of number of measurements.

Survey Frequency

Surveys should be conducted frequently enough to detect significant variation in infrastructure condition. Survey frequency is, similar to sample size, a trade-off between the cost of frequent surveys and the desirability of up-to-date and frequent data. Important factors determining survey frequency are the availability of resources, such as funds, manpower, and equipment, and the rate at which the condition of different components change.

The ideal is to adjust survey frequency to the changing rate of deterioration of infrastructure over its lifetime, with more frequent surveys when deterioration is rapid. This is illustrated in Fig. 63.6, with three profiles of condition versus time. Profile B indicates approximately linear deterioration with time, and in such a case a constant survey frequency would be good. In the case

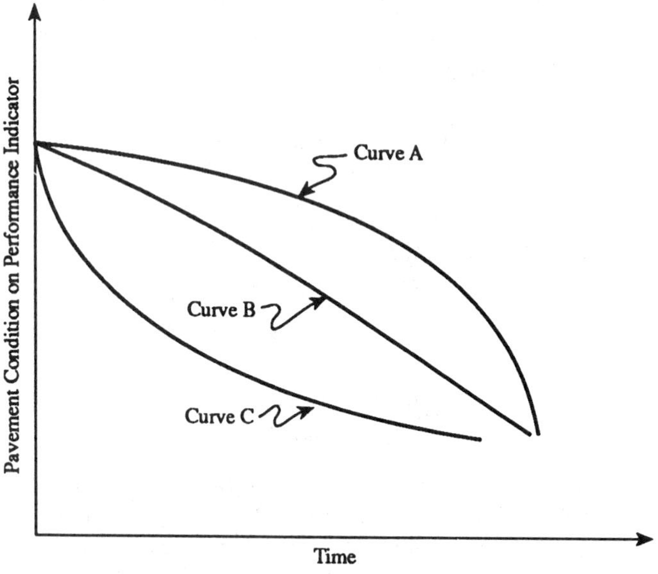

FIGURE 63.6 Possible pavement condition on performance indicator profiles.

Table 63.4 Examples of Survey Frequency and Sampling Rate for Pavement Data Collection

Data Type	Agency	Equipment/Method Used	Sampling Rate	Frequency
Roughness	Arizona	Mays meter	100%	Annually
	California	PCA meter	100%	Biennially
	Florida	Mays meter	100%	Annually
	Ontario	Subjective rating	100%	3 year or annually (25%)
Skid resistance	Arizona	Mu-meter	500 ft per mi	67% annually
	Florida	Locked-wheel trailer	35–40% interstate/primary	Annually
	New York	Skid trailer	Every 0.1 or 0.2 mi	Every 3 years
	Washington	Skid trailer	Every 1 mi section	Biennially
Structural capacity	Arizona	Dynaflect	3 tests per mi	40% annually
	Utah	Dynaflect	1 test per mi	50% annually
	Ontario	Dynaflect/Benkelman beam	Selective	20% annually
	Pennsylvania	Road rater	Pavement reaching terminal serviceability index	5% annually
Pavement distress	Arizona	Crack survey and visual comparison with photos	1000 ft^2 area every $\frac{1}{3}$ mi	40% annually
	California	Structural defects rated for extent and severity	300 ft per mi flexible pavement; every panel of rigid pavement with cracking	100% biennially
	Florida	Structural defects rated	100 ft section per mi	100% annually
	Utah	Detailed subjective evaluation	500 ft section per mi	50% annually
	Washington	Subjective evaluation	200 ft section per mi	100% biennially

Source: Transportation Research Board. 1981. *Collection and Use of Pavement Condition Data.* NCHRP Synthesis of Highway Practice 76.

of A, the deterioration rate increases with time, and with it survey frequency should also be increased. In the case of C, the deterioration rate is more rapid in the early stages, when survey frequency should be higher.

Most highway agencies establish a fixed frequency survey program. Table 63.4 gives examples of the sample sizes and survey frequencies of a number of North American highway agencies.

Data Quality Assurance

The quality of collected data can be improved through training of facility inspectors, carefully designed uniform procedures and measures, quality control and quality assurance programs, and better inspection manuals. Such quality assurance procedures were developed for the Pennsylvania DOT [Purris and Koretzky, 1988].

63.4 Data Analysis

Current Condition

Current condition data are obtained from regular condition surveys. For presentation and decision-making purposes condition data can be aggregated in a number of ways. One method is to derive condition indices, which take into account the extent as well as the severity of different distresses, by assigning appropriate weights. In this way indices can be constructed which aggregate data of the conditions of individual elements to obtain indices for larger elements, or for the facility as a whole, or even for a whole network of facilities. These indices are useful for presentation purposes, but not for detailed decision making.

Many condition evaluations are based on subjective ratings of inspectors. The theory of fuzzy sets utilizes less exact data items. Unlike classical set theory, where an element is either a member of a set or not, in fuzzy set theory degrees of membership are provided for. This seems to give a more realistic and flexible method to represent subjective condition evaluations.

Condition Prediction and Remaining Life

For decision making it is necessary to predict the conditions of the elements of facilities if different alternatives are implemented, to forecast the impacts of budgets and alternatives, and to select the best strategies. A number of techniques can be used for condition prediction, of which **regression analysis, Markov chains**, and **Bayesian estimation** will be discussed.

Regression Models

Regression analysis concerns the estimation of the parameters of equations with observed data. It has many applications, of which condition prediction is but one. Equations are estimated to predict the future conditions of facility elements as a function of the current condition, the age of the element, material types, maintenance practices, environmental conditions, traffic volume, and rehabilitation action taken. These predicted conditions are then used to estimate future agency and user costs, to evaluate different rehabilitation and replacement alternatives, to choose strategies under budget and other constraints, to predict the impacts of different budgets, and to plan work over the medium and longer term. To make forecasts, relevant data should be collected and be accurate and up-to-date.

A commonly used form of equation in regression analysis, due to the ease with which the parameters of such an equation can be estimated, is the linear regression equation. A linear regression equation can be stated as follows:

$$Y = \beta_0 + \beta_1 X_1 + \beta_2 X_2 + \cdots + \beta_k X_k + \epsilon$$

where

Y = dependent variable
X_j = explanatory or independent variable ($j = 1, 2, \ldots, k$)
β_j = unknown parameter to be estimated ($j = 1, 2, \ldots, k$)
ϵ = random error term

The dependent variable might, for example, be the future condition of a component, and explanatory variables might include the current condition, time since the previous major rehabilitation, the type of rehabilitation implemented, material type, and environmental conditions.

The random error term ϵ makes provision for the fact that the equation will never be a perfect representation of the underlying phenomenon. Certain statistical assumptions are made regarding these random errors. If these assumptions are violated, poor estimators might be obtained. Tests should be performed to test for violations of these assumptions, such as heteroscedasticity and autocorrelation. Figure 63.7 shows polynomial regression curves that were fitted to condition data of bridge decks, superstructures, and substructures in Indiana [Sinha *et al.*, 1991].

Different methods can be used to obtain parameter estimates that will make the equations fit the data as well as possible in some sense of goodness-of-fit. A simple and common method is ordinary least squares (OLS) or variations thereof, such as generalized least squares (GLS). Another more versatile method is maximum likelihood (ML).

The goodness-of-fit of the regression model can be evaluated in different ways. The most popular is the coefficient of determination, R^2, which measures the closeness of the equation to the data. As a single measure it has limited value.

Most real-world systems cannot usefully be modeled with a single equation alone. Realistic regression models are therefore often systems of simultaneous equations. Techniques for the estimation of simultaneous equation systems, such as two-stage least squares (2SLS), are discussed in the literature. Additional issues arise with the estimation of simultaneous equation systems, such as the identification problem, which has to be resolved before all parameters can be estimated.

The approach of latent variables considers the infrastructure "performance" or "condition" as a set of unobservable or latent variables, which depend on variables such as previous maintenance,

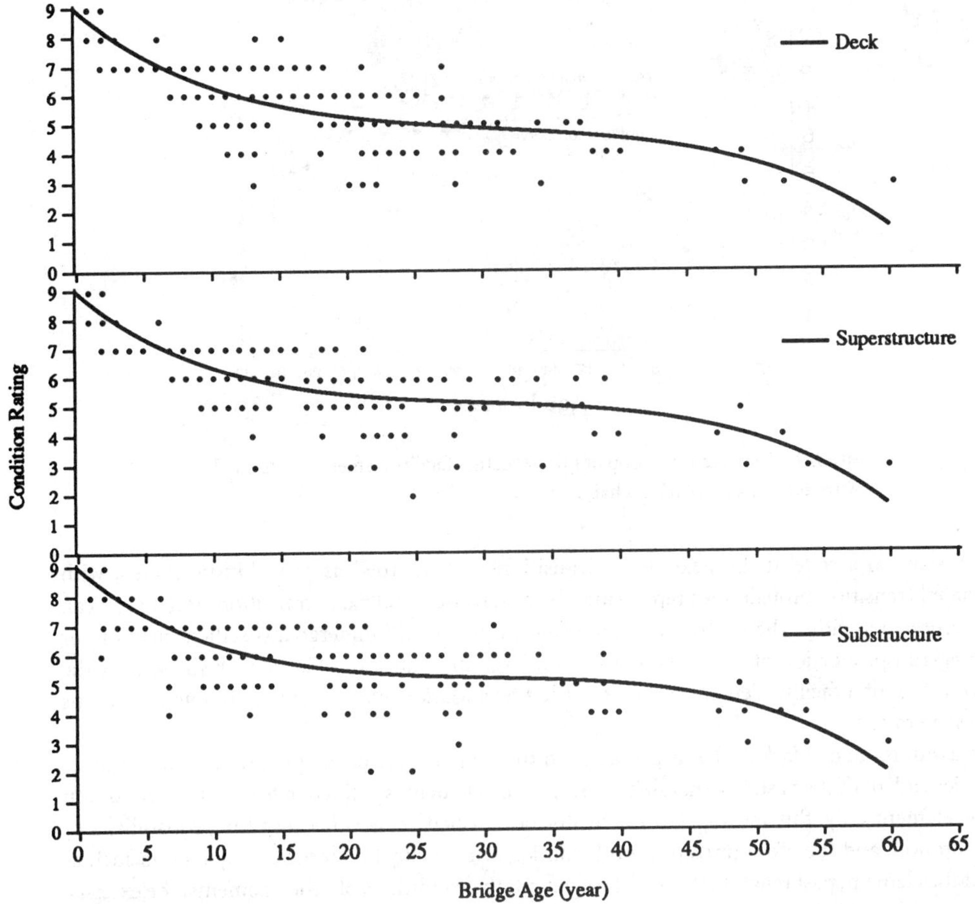

FIGURE 63.7 Bridge element deterioration curves estimated with regression analysis.

environmental conditions, and traffic load. The observed characteristics, such as the measured distresses, in turn simultaneously depend on the underlying latent variables. Because variables such as various distresses and structural capacity are measured with a large degree of error, the observed variables can be modeled as functions of the true values as well as stochastic measurement errors [Ramaswamy, 1989].

The model can also be enhanced by using lagged variables and by simultaneously modeling deterioration and maintenance. The last option is especially important, because deterioration tends to increase with decreasing maintenance, all other factors being constant, but maintenance tends to increase with increasing deterioration. If these two relationships are not modeled explicitly and simultaneously, the wrong model might be estimated. This wrong model might very well indicate that deterioration increases as maintenance increases, all other factors held constant, because the model that is estimated might be closer to maintenance as a function of deterioration than to deterioration as a function of maintenance.

Markov Chains

Markov chains can be used to model the deterioration process if the conditions of facility elements are classified into discrete states, for example, condition index represented by the numbers 1 to 9. The state of each element, or the proportion of elements that are in each state, can be measured during an inspection. A Markov chain describes a process that undergoes transitions from a state

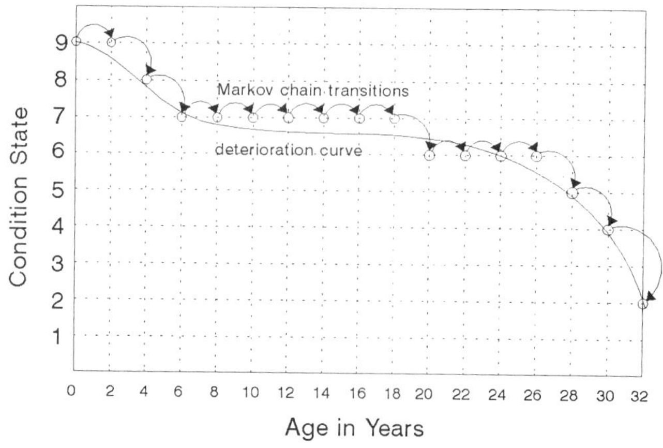

FIGURE 63.8 Deterioration of infrastructure facility element represented by transitions of a Markov chain.

at one stage to a state at the next stage. Transitions are regarded as probabilistic events, with associated transition probabilities represented by a transition matrix. A transition probability can be interpreted as either the probability that a single element will undergo a specific transition, or the long-run proportion of elements that will undergo the transition. Figure 63.8 shows how the deterioration of a facility element can be modeled by a Markov chain making discrete transitions from state to state.

An assumption of Markov chains is that given the present state of the process, the future states are independent of the past. In general, if the state of an element is defined in terms of the condition of the element only, this assumption might not be satisfied. This is because the probabilities of deterioration, and therefore transition probabilities, will usually be influenced by factors such as age of the element, past rehabilitation of the element, the conditions of other elements, the external forces such as traffic load applied, and others, and not only by the current condition of the element. To make better use of Markov chains for condition prediction, the states of an element therefore have to be defined in terms of the factors that have a significant influence on its deterioration, and not in terms of the element's current condition only.

The *latent Markov decision process* explicitly takes the uncertainty (e.g., due to measurement errors) associated with facility inspection into account, and incorporates this into a Markov decision process framework. It augments the definition of states to incorporate all information available up to each stage (all previous measured conditions and implemented actions). This causes the state space to grow very rapidly with the number of stages, which makes this method computationally very cumbersome. This approach is required to enable the recursive calculation of the conditional probabilities of the actual condition, given all information up to that stage.

With an appropriate cost function in terms of element condition and implemented action, the strategy selection problem can be formulated in terms of a dynamic program, to find the optimal strategy over a finite horizon with no budget constraints [Madanat, 1993].

Transition probabilities can be estimated in a variety of ways. The expected value method first partitions facilities into groups with homogeneous explanatory variables and into subgroups corresponding to age intervals. For each subgroup a deterioration model is estimated with regression analysis. A transition probability matrix is specified for each subgroup, and the values of the transition probabilities are found that minimize some measure of the differences between the expected conditions predicted by the regression model and those predicted by the transition matrix. Limitations are the assumption that condition ratings are continuously measurable on an interval scale, which is usually not the case; the fact that only expected values are used in obtaining the best fit and not probability distributions; and the maximum number of independent transition probabilities that can be estimated is equal to the length of the age interval. The frequency method uses the same partitioning as the expected value method, but estimates the transition probabilities by minimizing a measure of the differences between the observed frequencies and the frequencies predicted by the transition matrix. Limitations are the same as for the expected value method, except that more probabilities can be estimated. The ordered logit/probit method takes into account that condition ratings are discrete, ordinal variables. The true condition of the facility is regarded as an unobservable, continuous random variable, which depends on a number of explanatory variables. Associated with this variable are a number of thresholds, or boundaries, which determine how intervals of the continuous variable are mapped to discrete condition ratings. Procedures such as ordered logit or probit can be used to estimate the parameters of the deterioration model and the thresholds. These can in turn be used to estimate transition probabilities for different groups of facilities. This model gives more reasonable estimates of transition probabilities than the previous two methods [Madanat *et al.*, 1994]. In cases where sufficient data are not available, the subjective judgments of experts have been used to estimate transition probabilities. As data are collected through regular inspections, these initial estimates are updated and improved. One updating technique draws on the principles of Bayesian estimation.

Bayesian Estimation

Bayesian estimation can be used to update the estimated conditional probabilities of future condition states, given the current state, as additional data become available with inspections. Under suitable assumptions, the updated estimate (called posterior mean) can be shown to be a weighted average of the previous estimate (called prior mean) and the mean of the new data. The weights represent the value attached to the data from which the prior mean was estimated, relative to the new data. Usually, the relative numbers of observations are used as weights. If the prior mean was estimated from judgmental methods, it has to be valued as an equivalent number of observations, representing the amount of data on which the experts' judgments are based. With accurate data being collected, the effect of initial crude estimates fades away, and estimates are improved as new data become available.

63.5 Life-Cycle Cost Analysis

Estimation of Agency and User Impacts

The agency and user cost implications of alternatives are used in the comparison of alternatives for project-level decisions as well as in ranking and optimization routines for network-level decisions.

Agency costs, such as construction and maintenance costs, are more tangible and easier to estimate. Costs incurred by the public can be divided into user and nonuser costs. Usually only user costs are taken into account, because it is unclear to what extent the alternative actions taken by the agency can be regarded as the sole cause of nonuser costs, such as pollution, and because of the possibility of double counting of costs and benefits, as in the case of economic development effects.

Agency Costs

Agency costs include the resources, such as funds, manpower, equipment, and materials consumed in infrastructure-related activities, such as routine maintenance, rehabilitation, and replacement. To estimate agency costs, data should be recorded on actions performed on facility elements, the costs incurred, the conditions of the elements before and after the actions, and other relevant data items.

Routine Maintenance Costs

Routine maintenance costs can be estimated directly or indirectly. Directly, these costs can be estimated as a function of the material type, condition, environment, traffic load, highway classification, and other important factors for each element. Indirectly, these costs can be estimated by first estimating the quantity of different routine maintenance activities performed on a type of element per year, as a function of element condition, material type, traffic load, highway classification, environment, and other factors. The unit cost of each type of maintenance activity is also estimated as a function of material type, location, highway classification, and other factors. Together, the quantity of routine maintenance activities per year and their unit costs give an estimate of the routine maintenance costs. With the necessary data, regression analysis can be used to estimate both the quantity of work to be done and unit costs for each type of work.

Element Rehabilitation Costs

Element rehabilitation costs can be estimated for different types of elements and the different rehabilitation alternatives applicable for each element type. A good database or cost accounting system is essential to provide accurate and up-to-date cost estimates, broken down to individual element rehabilitation level. Data should also reflect the circumstances that might have a significant effect on rehabilitation costs, such as location, season, and the cost of handling traffic.

Element and Facility Replacement Costs

Replacement cost estimation also depends on accurate and up-to-date data, similar to that for rehabilitation costs. Replacement costs differ from rehabilitation costs because of different direct costs related to types and quantities of materials and work, as well as different indirect costs due to the contractor's preliminary costs, the cost of handling traffic, and different funding options.

User Costs

User costs include all additional costs incurred by road users over those costs that would have been incurred if the highway system had been in a specific predefined "ideal" state. User costs are therefore incurred if there is no facility in place and when a facility is in bad condition. User costs are also incurred during rehabilitation and replacement work.

User costs related to roads are different in nature from those related to bridges, due to the different implications of deficiencies.

User Costs Related to Roads

Road user costs make up by far the largest component of total transportation costs—approximately eight to ten times as much as highway agency costs [Chesher and Harrison, 1987]. Road users incur costs due to depreciation of vehicles, consumption of resources such as fuel, oil, and tires, vehicle maintenance, the value of the time spent by drivers, other crew members and passengers, and the inventory holding cost of goods in transit. Conventional practice is to combine estimates of amounts of different resources consumed by road transportation with prices to estimate total road user costs. The *highway design and maintenance (HDM) model* estimates amounts used of the following resources: fuel, lubricants, tires, maintenance parts, maintenance labor, crew time, depreciation, interest, overhead, passenger time, cargo holding, and miscellaneous costs.

The consumption of many of these resources, such as fuel, depreciation, crew time, passenger delays, and cargo holding, is influenced by operating speed. An important part of the HDMIII model is therefore the vehicle speed submodel. Explanatory variables taken into account include vehicle characteristics such as engine power, gross vehicle weight, aerodynamic drag coefficient, and braking capacity, as well as road characteristics, such as vertical gradient, rise and fall, horizontal curvature, superelevation, and roughness. The approach is to calculate limiting speeds based on engine power and vertical gradient, braking capacity, horizontal curvature, roughness, and a desired

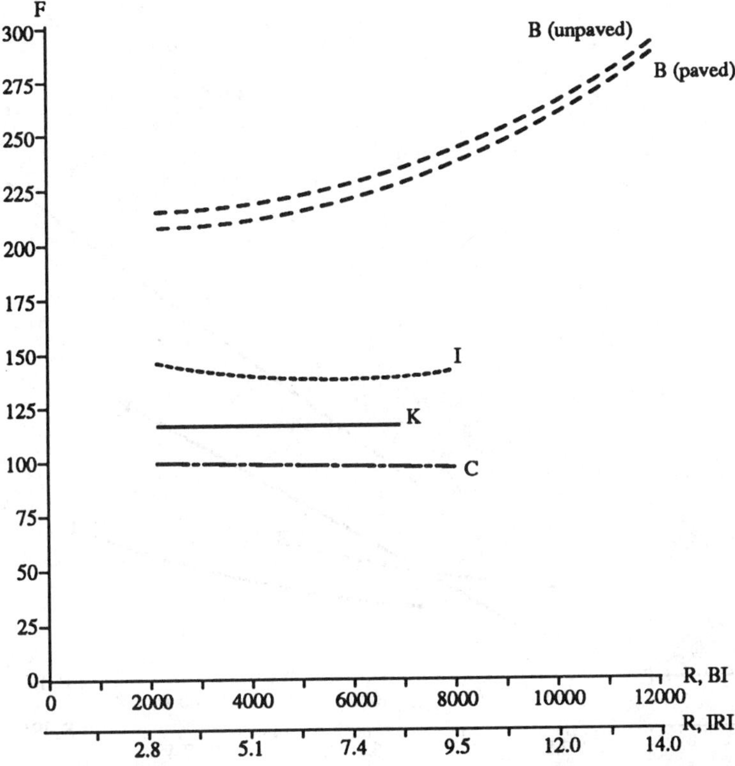

FIGURE 63.9 Passenger car fuel consumption versus road roughness. B: Brazil; C: Caribbean; I: India; K: Kenya; F = fuel consumption ($\delta/10^3$ km); R = roughness, BI = roughness measured by Bump Integrator (mm/km), IRI = International Roughness Index (m/km). (*Source:* Watanada, T., Harral, C. G., Paterson, W. D. O., Dhareshwar, A. M., Bhandari, A., and Tsunokawa, K. 1987. *The Highway Design and Maintenance Standards Model. Volume 1. Description of the HDM-III Model.* Transportation Department, World Bank, Washington, D.C.)

speed. The average speed for different vehicle types is then estimated as a probabilistic minimum of these speeds [Chesher and Harrison, 1987].

Fuel consumption can be estimated as a function of gross vehicle weight, engine power, speed, aerodynamic drag, vertical gradient, rise and fall, horizontal curvature, and roughness. Figure 63.9 shows passenger car fuel consumption as a function of road roughness. Explanatory variables used in the HDMIII model for estimating tire wear include roughness, gross vehicle weight, number of tires, and aerodynamic drag. Figure 63.10 shows passenger car tire consumption as a function of road roughness. Maintenance parts cost per vehicle per distance is estimated as a fraction of the cost of a new vehicle. Variables used include roughness, age in years, and distance traveled per year. Figure 63.11 shows passenger car maintenance parts consumption as a function of road roughness. Maintenance labor cost is then estimated as a function of maintenance parts cost and roughness. Lubricants consumption is influenced by roughness. Crew costs, passenger delays, and cargo holding costs are related to average speed. A variety of approaches can be used to incorporate depreciation and interest costs. If an interest accounting method is used to evaluate the cash flows of road users, including the replacement cost of new vehicles, depreciation and interest costs are taken into account by the method. If overhead cost is related to vehicle operating costs, it should also be taken into account, because then it can be influenced by the chosen M&R alternatives. The HDMIII model also identifies some miscellaneous costs [Chesher and Harrison, 1987].

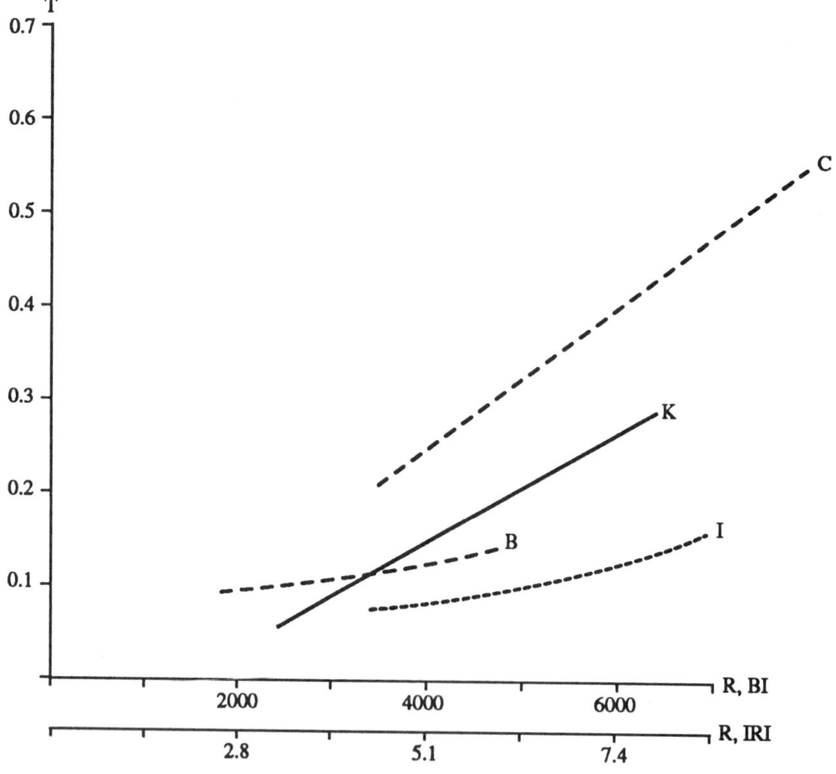

FIGURE 63.10 Passenger car tire consumption versus road roughness. (*Source:* Watanada, T., Harral, C. G., Paterson, W. D. O., Dhareshwar, A. M., Bhandari, A., and Tsunokawa, K. 1987. *The Highway Design and Maintenance Standards Model. Volume 1. Description of the HDM-III Model.* Transportation Department, World Bank, Washington, D.C.)

The above approach regards road user consequences of road condition from the supply side only. As road condition deteriorates, a number of simultaneous changes might take place. Road transportation costs increase, which might lead to a decrease in the quantity of road transportation demanded by shippers using the services of professional carriers, as well as agencies who provide their own transportation. Some substitution between different modes of transportation might take place. Carriers might modify equipment, buy different vehicles at replacement time, or change vehicle maintenance policies. These adjustments in transportation markets lead to consequences that are much wider than the changes in vehicle operating costs modeled above. A model of the cost-minimizing transportation firm that addresses some of these issues was suggested in the HDMIII model documentation [Chesher and Harrison, 1987].

User Costs Related to Bridges

Additional User Costs Due to Detours. User costs are incurred if vehicles have to take detours because of insufficient vertical clearance or load capacity. These costs will consist of additional vehicle operating costs as well as the value of the time lost. The additional congestion and pavement damage caused in the rest of the transportation system should also be taken into account, if this effect is likely to be significant.

The following approach can be used to estimate user costs for life-cycle cost analysis:

1. Estimate future traffic. With the necessary data, time series analysis of varying sophistication or regression analysis can be used. For simplified analysis the historic traffic growth rate along the same or a similar route can be extrapolated over the analysis period.

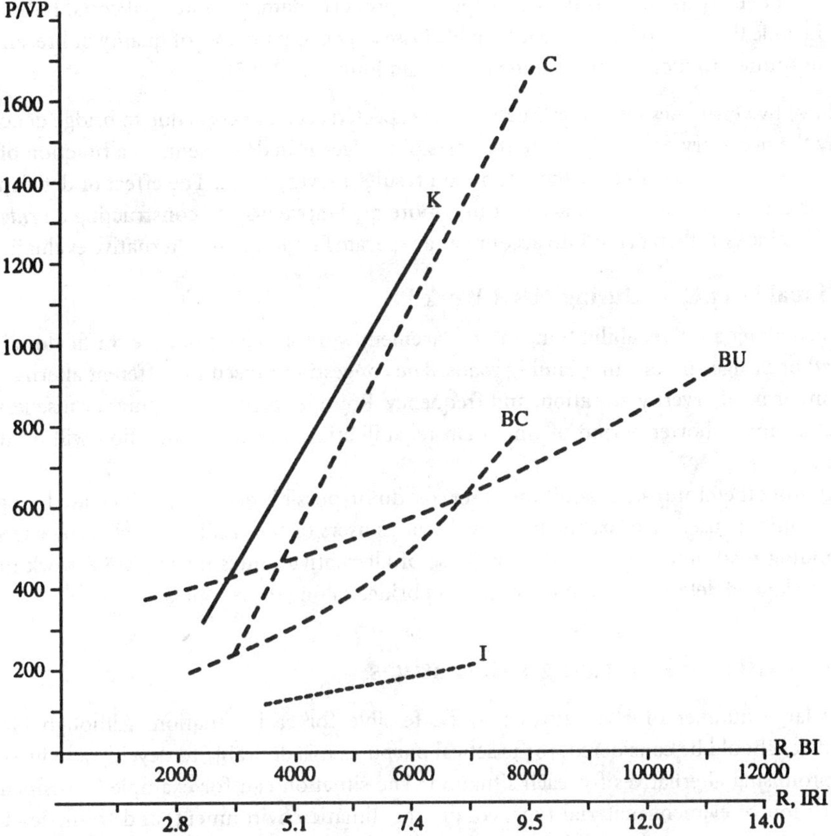

FIGURE 63.11 Passenger car maintenance parts consumption versus road roughness. (*Source:* Watanada, T., Harral, C. G., Paterson, W. D. O., Dhareshwar, A. M., Bhandari, A., and Tsunokawa, K. 1987. *The Highway Design and Maintenance Standards Model. Volume 1. Description of the HDM-III Model.* Transportation Department, World Bank, Washington, D.C.)

2. Predict load capacity. This can either be done directly using techniques such as regression analysis or Markov chains, or indirectly by using these techniques to predict the conditions of the applicable structural elements and then deriving the load capacity from these element conditions.
3. Estimate the distribution of different vehicle types on different routes, the distribution of vehicle weight and height for each vehicle type, the numbers of different types of vehicles detoured due to insufficient bridge load capacity or vertical clearance, the vehicle operating costs per distance for different types and weights of vehicles, and hence the additional vehicle operating costs and time costs due to bridge deficiencies.

Accident Costs. The following approach can be used to estimate the costs due to bridge-related accidents:

1. Estimate the expected rates of different types of accidents at each bridge as a function of its deficiencies. Accidents involving bridges are on average more serious than general vehicle accidents on the open highway [Chen and Johnston, 1987]. Highway accident statistics should therefore not be used to estimate rates of bridge-related accidents of different severity levels.
2. Estimate the costs of the different types of accidents related to bridges. This is usually done by separately considering the direct and indirect accident costs. Direct costs include the more

"tangible" costs, such as medical expenses, property damage, and legal costs. Indirect costs include the value of the more intangible losses, such as pain, loss of quality of life, and losses in future production and income [Chen and Johnston, 1987].

The above two estimates give an estimate of the expected accident costs due to bridge deficiencies. Usually the necessary data to estimate the costs of bridge-related accidents as a function of bridge deficiencies are not available, or the estimation results are very weak. The effect of deficiencies on accidents can also be taken into account in a more qualitative way by constructing a *bridge safety index*. This index is then taken into account as a separate factor during alternative evaluation.

Additional User Cost During M&R Work

Routine maintenance, rehabilitation, and replacement work usually influence traffic flow, both on the road or bridge and on surrounding roads. The congestion caused by different alternatives can differ in terms of severity, duration, and frequency. Routine maintenance might cause less severe congestion for a shorter period of time than rehabilitation, but this congestion will occur more frequently.

M&R work therefore causes additional user cost due to possible detours and increased congestion. These additional user costs are incurred by previous users of the facility as well as by users of the surrounding road network. The additional use of alternative routes during M&R work may also cause accelerated deterioration of the roads and bridges along these routes.

Identification of Promising Alternatives

A very large number of alternatives may be feasible for each situation. Although ideally all alternatives should be considered, for practical purposes it is desirable to develop a reduced list of more promising alternatives for each situation. The situation can for example be defined as the combination of element material type, condition, climatic environment, and traffic load. These alternatives are then further analyzed with their activity profiles and cash flows for project-level decisions, or with the promising alternatives of other bridge projects for network-level decisions.

The detail with which alternatives are formulated will depend on the level of detail at which the analyst wants to make distinctions between different alternatives, and the level of detail that the available data will permit. Results specifying specific actions for each facility can obviously not be expected if data of sufficient detail were not collected and the alternatives were not specified with the necessary level of detail.

Activity Profiles and Cash Flows

To evaluate the economic merit of each alternative, its activity profile and **cash flow** are constructed. For this the results of previous analyses are brought together. The current condition and ADT determine current agency and user costs, and which alternatives are currently feasible.

The condition prediction models are used to predict the condition for different alternatives. These are then used with the models for agency and user costs to estimate the costs associated with each activity profile, and thereby to derive each associated cash flow. The cash flow of each activity profile can then be analyzed with the techniques of interest accounting.

Criteria for selecting or ranking alternatives are net present value (NPV), equivalent uniform annual cost (EUAC), incremental benefit/cost ratio (B/C), and incremental internal rate of return (IRR). If the analysis is done correctly, all these criteria should lead to the same preferences of alternatives relative to each other. A large number of reference works describe the application of these techniques; see, for example, Grant *et al.* [1976].

A minimum acceptable rate of return has to be chosen to use these techniques. Theoretically, the chosen rate should be the rate of return that can be earned on projects or investments with a similar level of risk. The yield rate on an applicable type of long-term government bond, adjusted for inflation, can be used. The way in which inflation is to be taken into account is described in the literature.

If alternatives with different service lives have to be compared, suitable adjustments have to be made. One approach is to use the same cutoff date for all alternatives, after a fairly long analysis period. Differences between alternatives after this cutoff date are then represented by different residual values. Evaluation results should be insensitive to the residual values used. Another approach is to assume that each life cycle after replacement is repeated into perpetuity. If this approach is followed, growth in the ADT has to be addressed with care.

Direct and indirect costs can be taken into account in different ways. Agency as well as user costs, as influenced by the alternatives, should be included. Nonuser costs, such as those associated with air and noise pollution, aesthetics, and ecological disturbance, are difficult to relate to alternatives and to estimate. Most studies currently take these effects into account in a more qualitative way.

63.6 Priority Setting and Optimization

An important objective of infrastructure management is better decision making, leading to more effective utilization of scarce resources. To use the data and analysis results to make better decisions, a number of approaches and techniques have been developed. One approach is that of cash flow evaluation discussed above. This is useful for project-level decisions, where alternatives are mutually exclusive. For network-level decisions, where different projects are competing for limited resources, other approaches have to be used. The first approach discussed is that of priority setting, which is usually done by ranking projects according to some criteria to obtain a list of projects in priority order. Another approach is that of optimization, where the objective is maximized/minimized subject to constraints, by choosing the best values of the decision variables.

Priority Setting

Priority setting or ranking is usually used when there are multiple objectives, and objectives are too intangible to use the optimization methods. Many ranking methods have been developed to aid in priority setting. Most ranking methods develop a composite index or indices for each project. Projects are then ranked according to the values of these indices, which determines the priority of each project.

An example is the sufficiency rating calculated for bridges according to the FHWA's structure inventory and appraisal guide [Federal Highway Administration, 1988]. This makes provision for the calculation of indices reflecting the structural adequacy and safety, the serviceability and functional obsolescence, the essentiality for public use, and the overall sufficiency rating.

There are a number of methods that are based on pairwise comparisons. Concordance analysis has been used to select transit improvement alternatives and in bridge evaluation. Another method was developed using linear programming to estimate the weights of multiple attributes in constructing a composite criterion [Saito and Sinha, 1990]. The analytic hierarchy process is another pairwise comparison method and will be briefly discussed. A disadvantage of all these pairwise comparison methods is that the number of pairwise comparisons becomes very large as the number of alternatives increases.

Assignment of Relative Weights

The analytic hierarchy process (AHP) constructs a hierarchy and uses pairwise comparisons at each level of the hierarchy. System goals, objectives, criteria, and alternatives are related by the hierarchy. Relative weights are given to aspects on the same level in the hierarchy for measuring their contribution to an aspect on an adjacent higher level. For an infrastructure management system, the highest level might be the goal to maximize system effectiveness. The second level might consist of the objectives in terms of which achievement of the goal will be measured, such as facility condition, agency costs, user costs, safety, and external impacts. The third level might then consist of the criteria in terms of which each objective is measured. The criteria for user costs might consist of vehicle operating costs, value of travel time, and accident costs. The fourth level might then consist of individual alternative projects.

The above hierarchical structuring is very general, and similar structures are used in many ranking methods. What makes the AHP method different is the way in which the relative weights are derived. The aspects on each level are pairwise compared to produce relative weights, and these relative weights are arranged in a reciprocal matrix for each higher-level aspect. If the pairwise comparisons are fairly consistent, an eigenvector corresponding to the largest eigenvalue will give a set of relative weights for all the aspects. Alternatives can then be ranked according to these weights. More information on the AHP can be obtained in Saaty [1980].

Utility Functions

In infrastructure management, many alternatives might have to be compared. Making pairwise comparisons between all alternatives with respect to each criterion might therefore be an enormous task. Another approach is to develop utility functions for the characteristics, such as remaining service life, that will be impacted by alternatives. To compare alternative projects, the affected characteristics are thereby directly converted to utility points, without having to make pairwise comparisons between all alternatives [Saito and Sinha, 1990].

Optimization

A variety of optimization techniques can be used to determine an optimal set of projects to be implemented at different times on a network of facilities, subject to budget and other constraints.

Minimization of Life-Cycle Costs

An obvious, but somewhat brute force, approach is to do a life-cycle cost analysis for each facility or set of facilities in the system, for each promising alternative at each programming period. This reduces to continuing with routine maintenance until one of the rehabilitation or replacement alternatives is better than routine maintenance.

However, this approach does not find a "true global" optimum strategy, because of the following drawbacks:

1. It does not simultaneously take networkwide effects such as budget constraints into account.
2. At the point in time that an alternative is chosen, future choices are not yet determined. To choose the optimum alternative under these conditions, some simplifying assumptions about future alternatives usually have to be made.

Linear Programming

A widely used optimization technique is **linear programming** (LP). A linear program searches for the values of decision variables that maximize/minimize a linear objective function, subject to linear equality/inequality constraints, such as budget constraints. The decision variables should be continuous for practical purposes.

Maintenance optimization models have been formulated in such a way that they can be solved with linear programming. Decision variables might include expected discounted cost and the limiting probability that an element will be in a state and an action will be chosen [Golabi *et al.*, 1990, 1992].

Integer Linear Programming

In many management problems the decision variables are discrete, such as whether an alternative will be implemented ($x = 1$) or not ($x = 0$). Such problems can often be formulated as a linear program with integer variables, called an **integer linear program** (ILP). A simple example follows.

$$\min \sum_{t=1}^{T} \sum_{i=1}^{I} \sum_{a=1}^{A} c_{ait} x_{ait}$$

subject to

$$\sum_{i=1}^{I} \sum_{a=1}^{A} b_{ait} x_{ait} \leq B_t \qquad \forall t = 1, \ldots, T$$

$$\sum_{a=1}^{A} x_{ait} \leq 1 \qquad \forall i = 1, \ldots I, \; t = 1, \ldots, T$$

$$\sum_{u=t}^{t+\tau_a} x_{aiu} \leq 1 \qquad \forall a = 1, \ldots, A, \; i = 1, \ldots, I, \; t = 1, \ldots, T - \tau_a$$

$$x_{ait} = 0 \text{ or } 1 \qquad \forall a, i, t$$

Each decision variable x_{ait} indicates whether alternative a for each facility i for each programming period t is chosen ($x_{ait} = 1$) or not ($x_{ait} = 0$). The associated total agency and user costs are c_{ait}, and associated budget requirements are b_{ait}. The objective function minimizes the total costs over a finite time horizon T. The first constraint ensures that the budget limit B_t for each programming period t is not exceeded. Budget constraints can also be split between various sources and accounts. The second constraint ensures that at most one alternative is chosen for each facility in each programming period. The third constraint ensures that the same alternative a is not implemented more than once for each facility i during a time window with length τ_a. The time window depends on the alternative type (e.g., shorter time windows for routine maintenance than for replacement). Similar constraints can be formulated for mutually exclusive as well as for interdependent projects.

Linear and integer linear programming are very versatile. Formulations can be adjusted to changing needs and circumstances. The above formulation is just one simple example. Another advantage is that software for linear and integer linear programming is widely available and very powerful. Huge linear programs, with hundreds of thousands of decision variables, have been solved successfully.

However, the size of *integer* linear programs that are solvable in reasonable time is much more restricted than that for linear programs. Another restriction is that objective functions and constraints have to be linear functions of the decision variables. Nonlinear objective functions and constraints can also be handled with the techniques of nonlinear programming. This is computationally more demanding, and complicating issues might arise.

Dynamic Programming

The approach of **dynamic programming** is based on the *principle of optimality*. In the case of infrastructure management this means that optimal alternatives/policies over time consist of optimal subalternatives/subpolicies over shorter periods of time. This implies that optimal policies can be constructed by recursively finding optimal subpolicies for successive programming periods. This procedure can significantly reduce the computational effort to find optimal solutions.

Dynamic programming can be applied by doing the analysis over a finite, but fairly long, time horizon. At each stage a facility element can be in a number of different states. A terminal value/cost is assigned to each state at the end of the analysis period. A cost is also associated with being in each state and with the implementation of each alternative in each state.

The transition probabilities can be given by the transition matrix of a Markov chain, as long as the underlying assumptions of a Markov chain are satisfied, in which case it is called a *Markov decision process* (MDP). The optimal alternative can be calculated recursively for each state at each stage. However, constraints such as budget constraints cannot be incorporated into the optimization without large increases in computational complexity, which limits the usefulness of the procedure for practical scenarios. Markov decision processes have been suggested for a number of pavement and bridge management systems [Golabi *et al.*, 1990, 1992; Haas *et al.*, 1994].

If optimal alternatives are consistently implemented, the state of the system should move toward an optimal steady state. A useful analysis is to determine the optimal steady state and associated alternatives and costs. Because the system will currently not be in this optimal steady state, an associated problem is the optimal way of moving toward the optimal steady state. Both the optimal steady state problem and the optimal transition stage problem can be formulated as linear programs [Golabi *et al.*, 1990, 1992].

Network and Heuristic Methods

For large systems, such as infrastructure management systems, realistic optimization problems, without too many restrictive assumptions that reduce their usefulness, can become very large and computationally demanding to solve. **Heuristic optimization** procedures, which search for approximately optimal solutions, might be more practical.

Another popular approach to reduce computational effort is to formulate the problem as a network optimization model. Many efficient algorithms have been developed for network problems. Some types of problems lend themselves to network formulations, and with an appropriate formulation these efficient algorithms can be exploited to solve the problem [Garcia-Diaz and Liebman, 1980, 1983].

63.7 Maintenance Management Systems

Routine infrastructure maintenance consumes large amounts of resources. Many of the techniques and systems described above can be used to improve the effectiveness of routine maintenance. In the process the objectives are to better utilize the scarce resources available for routine maintenance and improve the coordination of maintenance and rehabilitation programs to account for the trade-off between (minor) routine maintenance activities and (major) rehabilitation.

Despite many similarities, a number of aspects make maintenance management different from other infrastructure management, such as pavement and bridge management. Maintenance management activities are performed on a more decentralized basis, such as at the district and subdistrict level of state highway authorities, whereas other infrastructure management tends to be more centralized. This leads to maintenance management having unique needs and addressing unique issues. An example is the additional need to simultaneously involve maintenance managers at the decentralized levels in the data collection as well as decision-making process, and to centrally coordinate all data collection, maintenance, and rehabilitation activities to improve the allocation of scarce resources.

A maintenance management system should make provision for routine maintenance needs assessment, workload and resource needs estimation, cost estimation, priority setting, optimization, and programming and scheduling of maintenance activities. To accomplish this, the infrastructure management database should incorporate the data necessary for maintenance management.

Database

For the sake of consistency and reduction of redundancy of data, the data needed for maintenance management should be incorporated into the overall infrastructure management database. In this way, highway network inventory data used for pavement management are available for maintenance management as well, and condition data collected as part of the routine maintenance process become available for pavement and bridge management as well. The data should be readily accessible to decentralized maintenance managers, for example through a computer network, or through the ability to transfer the necessary data files between decentralized offices and the central database.

Data needs unique to maintenance management include data regarding routine maintenance needs assessments; performance standards; unit costs of maintenance equipment, materials, and labor; and priority-rating weights.

Routine Maintenance Needs Assessment

The first step in maintenance needs assessment is an inspection or a condition survey of the infrastructure. For the sake of efficiency this should also be part of the agency's regular data collection program for the management of the specific infrastructure (e.g., for pavement management). All the data collection techniques discussed in section 63.3 are directly applicable.

The next step is to choose appropriate maintenance activities, given the collected condition and other data. The most appropriate treatment might depend on the condition, traffic load, climate, available resources, and competing maintenance requirements. One approach to choosing the best alternatives is to ignore some of these aspects, such as limited resources, for the time being, and to compile a short list of alternatives that have proved to be cost-effective for each given set of circumstances. Such a short list should preferably be standardized for all maintenance units of the highway agency. Scarce resources are taken into account during the priority setting and/or optimization phase of the management process.

Workload and Resource Needs Assessment

For the purpose of cost estimation, priority setting, optimization, programming and scheduling, and budgeting, it is necessary to estimate the resources needed to perform each of the identified maintenance alternatives. One approach is to first identify a measurement unit for each maintenance treatment. Next, estimates are obtained for the types and numbers of laborers needed, the number of labor-hours for each labor type or the number of crew days, the types and numbers of equipment needed, and the types and quantities of materials needed, per unit of the maintenance activity. These standard resource requirements should be incorporated into the database to enable automated resource need and cost estimation.

Cost Estimation

The quantity of the maintenance activity to be performed in its standard measurement units, the types and amounts of resources needed per unit of the maintenance activity, and the unit costs of each type of resource can be combined to obtain cost estimates for each type of maintenance activity. Unit costs should be incorporated in the database and kept up to date, to enable cost estimates to be automated and accurate.

Priority Setting and Optimization

Due to limited resources, usually it will not be possible to perform all the required maintenance activities. A scheme for priority setting and/or optimization is therefore needed, to effectively allocate scarce resources. The techniques discussed in section 63.6 can also be used for maintenance management. Typical constraints to be incorporated into mathematical programming models include minimum and maximum maintenance production requirements, manpower availability, equipment availability, materials availability, budget constraints, and constraints to coordinate rehabilitation and maintenance programs. One way of coordinating these programs is by suspending routine maintenance activities a certain time period before scheduled major rehabilitation work is to commence on the same facility.

Programming, Scheduling, and Budgeting

The results of the above phases of the maintenance management process provide the input for the programming and scheduling of maintenance activities at all levels, as well as for compiling budgets at the decentralized as well as centralized levels. All these results combined with the database provide for a powerful management system.

63.8 Conclusion

Highway management systems provide decision-making tools to improve transportation services with limited resources. The transportation needs and facilities of communities differ vastly, and so will the management systems of their highway agencies. Important factors influencing the type and scope of management systems are the nature of the facilities to be provided and maintained, as well as the available funds, equipment, and manpower. Flexible systems are therefore important.

Infrastructure management is not a single exercise, it is a way of life for the responsible agencies. Activities such as data collection, analysis, optimization, and strategy selection should be pursued continuously. Many of the technologies used in infrastructure management systems are regularly improved. As better techniques become available, systems should be updated. Better data collection equipment, computing capabilities, and modeling and optimization techniques are expected to be developed in the near future.

Defining Terms

Bayesian estimation: A statistical technique used to update estimates when new data become available.

Cash flow: The sequence of benefits and costs over time, associated with an alternative (not all of which need involve monetary transactions).

Distress: A deteriorated condition of an infrastructure facility. For definitions and descriptions of pavement distress types, see *Distress Identification Manual, Long-Term Pavement Performance Project, SHRP 1993*, or similar manuals.

Dynamic programming: The techniques used to formulate and solve optimization problems by recursively solving subproblems.

Geographic information system (GIS): A database with capabilities to manipulate, analyze, and display spatial data.

Heuristic optimization: Techniques used to find satisfactory, not necessarily optimal, solutions to optimization problems. Some heuristic techniques provide solutions with values that differ from the optimal solution value by not more than a provable bound.

Integer linear programming: The techniques used to formulate and solve optimization problems in which at least some decision variables are discrete, and the objective function as well as constraints are linear.

Linear programming: The techniques used to formulate and solve optimization problems in which the decision variables are continuous, and the objective function as well as constraints are linear.

Location referencing system: A system to uniquely describe the geographical location of a feature with respect to some projection or other position referencing system.

Markov chain: A stochastic process in which the present state of the system determines the future evolution of the process independently of the past.

Network level analysis: The evaluation of dependent or independent alternatives for a set of infrastructure facilities.

Network level optimization: The techniques used to formulate and solve optimization problems in such a way that the polynomial time algorithms developed for some network problems can be used.

Pavement profile: The vertical shape of the pavement surface.

Pavement segment: A length of pavement that forms the basic unit for which data is collected, stored, and analyzed, and for which alternatives are evaluated and decisions are taken.

Programming: The determination and scheduling of activities to be performed, and the resources needed to implement the program.

Project level analysis: The evaluation of mutually exclusive alternatives for a single infrastructure facility.

Regression analysis: The technique of mathematically estimating the parameters of functions describing a process, to fit the functions to data.

Sampling technique: The method used to choose units to include in the sample, with the objective of obtaining a statistically adequate, representative sample at reasonable cost.

References

Chen, C. and Johnston, D. W. 1987. *Bridge Management under a Level of Service Concept Providing Optimum Improvement Action, Time, and Budget Prediction.* Department of Civil Engineering, North Carolina State University, Raleigh, NC.

Chesher, A. and Harrison, R. 1987. *Vehicle Operating Costs. Evidence from Developing Countries* Transportation Department, World Bank, Washington, D.C.

Federal Highway Administration. 1987. *Highway Performance Monitoring System.* Washington, D.C.

Federal Highway Administration. 1988. *Recording and Coding Guide for the Structure Inventory and Appraisal of the Nation's Bridges.* U.S. Department of Transportation, Washington, D.C.

Federal Highway Administration. 1989. *Automated Pavement Condition Data Collection Equipment.* FHWA Pavement Division, Washington, D.C.

Federal Highway Administration. 1991. *An Advanced Course in Pavement Management Systems.* Washington, D.C.

Garcia-Diaz, A. and Liebman, J. S. 1980. An investment staging model for a bridge replacement problem. *Operations Res.* 28(3):736–753.

Garcia-Diaz, A. and Liebman, J. S. 1983. Optimal strategies for bridge replacement. *J. Transp. Eng.* 109(2):196–208.

Golabi, K., Thompson, P. D., and Jun, C. H. 1990. *Network Optimization System for Bridge Improvements and Maintenance.* Report to California Department of Transportation and FHWA, Cambridge Systematics/Optima.

Golabi, K., Thompson, P. D., and Hyman, W. A. 1992. *PONTIS Technical Manual, a Network Optimization System for Bridge Improvements and Maintenance.* Report to FHWA, Cambridge Systematics/Optima.

Grant, E. L., Ireson, W. G., and Leavenworth, R. S. 1976. *Principles of Engineering Economy.* John Wiley & Sons, New York.

Haas, R., Hudson, W. R., and Zaniewski, J. 1994. *Modern Pavement Management.* Krieger Publishing, Malabar, FL.

Jiang, Y. and Sinha, K. C. 1990. *The Development of Optimal Strategies for Maintenance, Rehabilitation and Replacement of Highway Bridges, Final Report Vol. 6: Performance Analysis and Optimization.* Joint Highway Research Project, Purdue University, West Lafayette, IN.

Madanat, S. 1993. Optimal infrastructure management decisions under uncertainty. *Transp. Res. C.* 1(1):77–88.

Madanat, S., Mishalani, R., and Wan Ibrahim, W. H. 1994. *Estimation of Infrastructure Transition Probabilities from Condition Rating Data.* School of Civil Engineering, Purdue University, West Lafayette, IN.

Mahoney, J. P. and Lytton, R. L. 1978. *Measurement of Pavement Performance Using Statistical Sampling Techniques.* Research report, 207-2, Texas Transportation Institute, Texas A&M University, College Station, TX.

Purvis, R. L. and Koretzky, H. P. 1988. *Bridge Safety Inspection Quality Assurance: Pennsylvania Department of Transportation. Transportation Research Record 1184.* Transportation Research Board, National Research Council, Washington, D.C., pp. 10–21.

Ramaswamy, R. 1989. *Estimation of Latent Pavement Performance from Damage Measurements.* Ph.D. Thesis, Department of Civil Engineering, MIT, Cambridge, MA.

Saaty, T. 1980. *The Analytic Hierarchy Process: Planning, Priority Setting, and Resource Allocation.* McGraw-Hill, New York.

Saito, M. and Sinha, K. C. 1990. *The Development of Optimal Strategies for Maintenance, Rehabilitation and Replacement of Highway Bridges, Final Report Vol. 5: Priority Ranking Method.* Joint Highway Research Project, Purdue University, West Lafayette, IN.

Sinha, K. C. and Fwa, T. F. 1989. *On the Concept of Total Highway Management.* Transportation Research Record 1229. Transportation Research Board, National Research Council, Washington, D.C., pp. 79–88.

Sinha, K. C. and Fwa, T. F. 1990. *Data Requirements for Pavement Management.* Unpublished report, School of Civil Engineering, Purdue University, West Lafayette, IN.

Sinha, K. C. and Fwa, T. F. 1993. A framework for systematic decision making in highway maintenance management. *Transportation Research Record 1409.* Transportation Research Board, National Research Council, Washington, D.C.

Sinha, K. C., Saito, M., Jiang, Y., Murthy, S., Tee, A. B., and Bowman, M. D. 1991. *The Development of Optimal Strategies for Maintenance, Rehabilitation and Replacement of Highway Bridges, Final Report Vol. 1: The Elements of the Indiana Bridge Management System.* Joint Highway Research Project, Purdue University, West Lafayette, IN.

Strategic Highway Research Program. 1993. *Distress Identification Manual for the Long-Term Pavement Performance Project.* National Research Council, Washington, D.C.

Transportation Research Board. 1981. *Collection and Use of Pavement Condition Data.* NCHRP Synthesis of Highway Practice 76.

Turner, D. S. and Richardson, J. A. 1993. Bridge management system data needs and data collection. *Transportation Research Circular 423.* Transportation Research Board, National Research Council, Washington, D.C., pp. 5–15.

U.S. Federal Register. 1993. FHWA, 23 CFR Parts 500 and 626; FTA, 49 CFR Part 614—Management and Monitoring Systems; Interim Final Rule. 58(229).

Watanada, T., Harral, C. G., Paterson, W. D. O., Dhareshwar, A. M., Bhandari, A., and Tsunokawa, K. 1987. *The Highway Design and Maintenance Standards Model. Volume 1. Description of the HDM-III Model.* Transportation Department, World Bank, Washington, D.C.

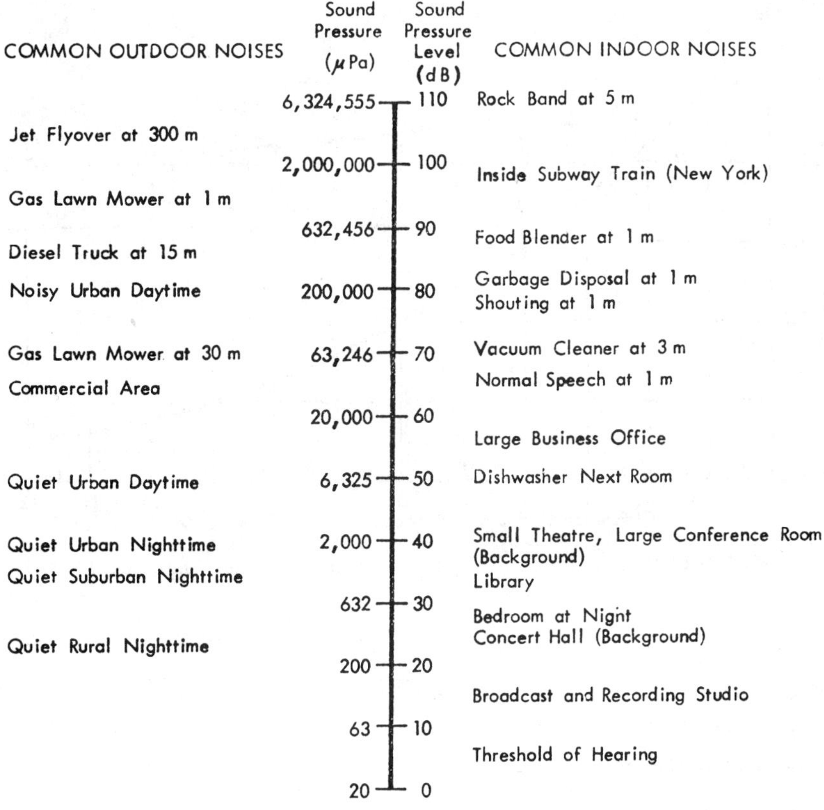

FIGURE 64.1 Typical noise levels. (*Source:* Ref. 2.)

The use of dB indicates the loudness is measured as a sound pressure level (SPL) and no longer just the sound pressure. Decibels do not add in a linear fashion, but logarithmically. This means that if the sound pressure is increased by a factor of two, an increase in the sound pressure *level* would only be 3 dB. Adding dB may be accomplished using a simple chart (Fig. 64.2)[2] or, to be more exact, the equation:

$$\mathrm{SPL_{total}} = 10\log_{10}\sum_{i=1}^{n} 10^{\mathrm{SPL}_{i/10}}$$

In outdoor situations, a change of greater than 3 dB is required to be noticeable. A change of 10 dB is generally perceived to be a doubling of the sound level. This means that a significant change in transportation patterns (vehicle volume, speed, mix, etc.) or alignment must occur for individuals to *objectively* determine a change in noise levels.

Frequency. The human ear can hear a large range of frequencies, or changes in the rate of pressure fluctuations in the air. The pressure changes per second, or oscillations per second, have the unit of hertz (Hz). The ear can detect a range of frequencies extending from about 20 Hz to 20,000 Hz. It is this difference in the rate of the pressure fluctuations that provides the tonal quality of the sound and permits identification of the source. A flute has a much higher frequency than a bass guitar and we are adept enough to easily tell the difference, just as we can discern aircraft sounds from the blowing wind.

Frequency, the wavelength of the sound wave, and the speed of sound are all related mathematically:

$$f = c/\lambda$$

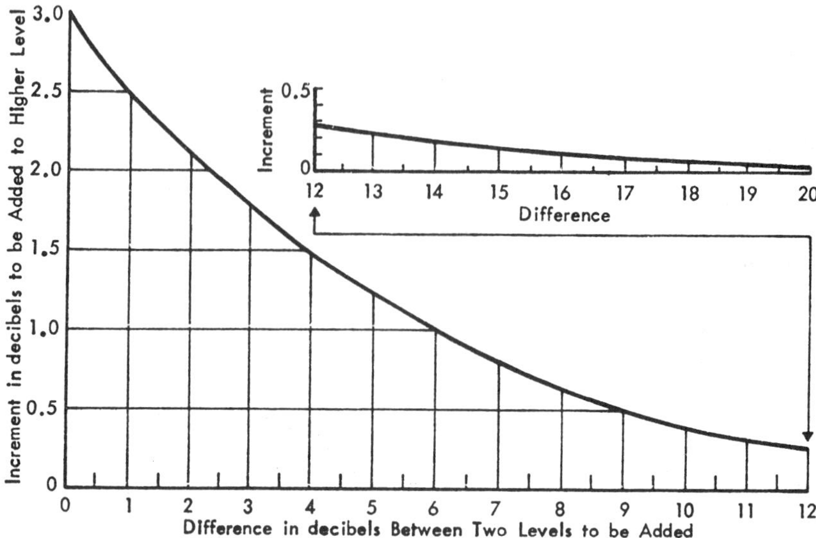

FIGURE 64.2 Chart for combining levels by decibel addition. (*Source:* Ref. 2.)

where

f = frequency (Hz)
c = speed of sound ($\approx$ 343 m/s)
λ = wavelength (distance)

The human ear does not detect all frequencies equally well. Low frequencies (less than 500 Hz) and higher frequencies (greater than 10,000 Hz) are not heard as well. This fact requires a sound to be described by more than just loudness—by including some description of the frequency spectra. The loudness of each frequency could be reported and evaluated, but this is not practical. Groups of frequencies, called *octave bands*, are used to describe sounds and provide a detailed description of the frequency components (see Table 64.1). However, in regard to transportation sounds, a broader approach is most often used. In this approach all frequency band contributions are first adjusted to approximate the way the ear hears each range, then the contributions are summed to a single number. Three common scales have been used. Figure 64.3 shows the A, B, and C weighting scales. The A scale is the way our ears respond to moderate sounds; the B scale is the response curve for more intense sound; and the C scale is the way our ears would respond to very loud sounds. The nonlinear response of the ear at low and high frequencies is quite apparent from these graphs. Most regulations and evaluations applicable to transportation analysis use the A scale.

Table 64.1 Octave Band and Center Frequencies

Octave Frequency Range (Hz)	Geometric Mean (or Center) Frequency of Band (Hz)
22–44	31.5
44–88	63
88–177	125
177–355	250
355–710	500
710–1,420	1,000
1,420–2,840	2,000
2,840–5,680	4,000
5,680–11,360	8,000
11,360–22,720	16,000

Source: Reference 2.

Duration. A firecracker may be loud, but it lasts only a fraction of a second. Traffic noise may not be as intense, but it is continual. Effective descriptors of how sound varies with time have been developed. Some of the more important descriptors in regards to transportation noise are maximum sound level, $L_{\max}(t)$; statistical sound levels, $L_{xx}(t)$; the equivalent sound level, $L_{\text{eq}}(t)$;

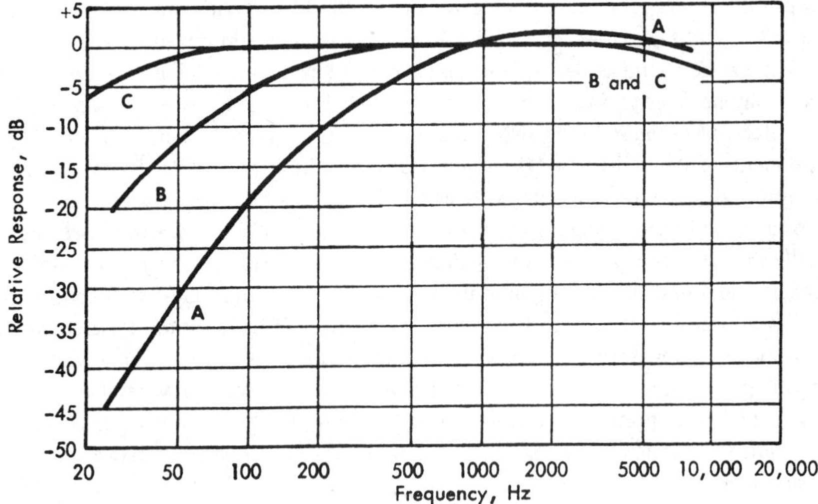

FIGURE 64.3 Frequency response curves. (*Source:* Ref. 2.)

and the day/night level, L_{dn}. In each of these descriptors L represents that each is a sound pressure level with units of dB. The (t) indicates each is given for a specific period of time. By definition, L_{dn} is a 24-hour metric.

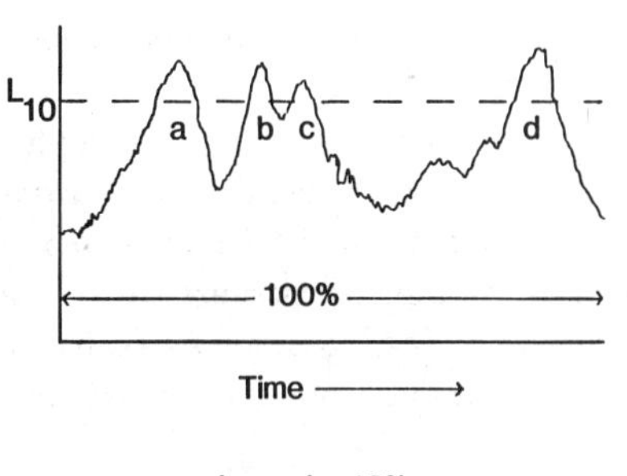

FIGURE 64.4 Graphical description of L_{10}.

L_{max} represents the maximum noise level that occurs during a defined time period. This allows for a more complete description of the noise when combined with loudness and frequency description. For example, 60 dB (A-weight) L_{max} (1h) defines the overall sound level, frequency weighting, and an indication of the noise changes during a defined time period.

More description is possible with statistical descriptors (L_{xx}). The subscript xx indicates the percentage of time that the listed level is exceeded. For instance, a reported sound level of 60 dB (A-weight) L_{10} (1h) would mean that a sound pressure level of 60 dB on an A-weighted scale was exceeded 10% of the time in a one-hour time period. The numeric value may be any fraction of the time, but L_{10}, L_{50}, and L_{90} are most commonly used (see Fig. 64.4). L_{90} is the sound pressure level exceeded 90% of the time and is commonly used as the background level.

L_{eq}, the equivalent sound pressure level, is a single number metric that represents the value of a nonvarying tone that contains the same acoustic energy as a varying tone over the same time period. One might think of L_{eq} as an average acoustic energy descriptor. It should be noted that the average energy is not an average of SPL over the time period because of the logarithmic nature of the dB. L_{eq} has the advantage of allowing different noises that occur during the same time period to be added. It has become the metric of choice in the U.S. for highway noise analysis.

The descriptor L_{dn}, the day/night level, is by definition a 24-hour metric. L_{dn} is sometimes shown as DNL. L_{dn} takes into account that duration is not the only important factor; the time of day the sound occurs is also important. L_{dn} consists of hourly L_{eq} (A-weighted) values, energy averaged over the entire 24-hour period, with a 10-dB (A-weight) penalty added to each hour during the time period from 10 P.M. until 7 A.M. The 10-dB penalty in effect requires the sound pressure to be 10 times lower in the nighttime hours.

Effects and Subjectivity. Until now we have discussed characteristics of sound that may be mathematically quantified. However, individuals have different responses to various sounds and, as such, whether the sound is desirable or considered noise is quite subjective. Rock music to one listener may be a refreshing sound, but to another listener only noise. Unwanted sound is commonly referred to as *noise*. Transportation noise is a common problem in urban areas. The first three components of sound discussed (loudness, frequency, duration) can be objectively described. Noise annoyance is subjective and criteria are usually based on attitudinal surveys.

It has been found that a single loud noise can result in an acute hearing loss, but this is not typically the case with transportation-related noise. Our mechanized societies tend to pose more of a chronic problem, resulting in reduced hearing ability after long-term exposure. In the short term, annoyance or irritation is of more importance. Transportation noise could lead to problems in our emotional well-being and cause increased tension by interfering with our sleep or causing disruption in our daily lives.[3] Studies have shown that noise prevents deep sleep cycles needed for complete refreshment, causes increased tension due to continual intrusion, inhibits communication ability, and reduces the learning abilities of students when excessive noise intrudes into the classroom.

Legislation and Regulations Affecting Transportation Noise

Federal legislation for noise pollution was passed in the 1960s and 1970s and is still in effect. The Housing and Urban Development Act of 1965, reinforced with the Noise Control Act of 1972, mandated the control of urban noise impact. The Control and Abatement of Aircraft Noise and Sonic Boom Act of 1968 led to noise standards being placed on aircraft. The Quiet Communities Act of 1978 better defined and added to the requirements of the Noise Control Act. This environmental legislation required noise pollution to be considered for all modes of transportation. Analysis methodologies and documentation requirements of noise impacts resulted.

To help ensure enforcement, EPA created the short-lived Office of Noise Abatement and Control that contributed significantly to determination of noise sources and determination of regulations. A desirable neighborhood goal of 55 dB (A-weight) L_{dn} was identified.[4]

Many discussions have surrounded the appropriate noise level and descriptor most applicable to various forms of transportation and land use. In the U.S. the Federal Highway Administration (FHWA) has defined procedures (23CFR772) that must be followed to predict the worst hour noise levels where human activity normally occurs. Included in these detailed procedures are the Noise Abatement Criteria for various land uses as shown in Table 64.2. The legislation states that when the Noise Abatement Criteria are approached or exceeded, noise mitigation must be considered.

Table 64.2 FHWA Noise Abatement Criteria

Activity Category	$L_{eq}(h)^1$	$L_{10}(h)^1$	Description of Activity Category
A	57 (exterior)	60 (exterior)	Lands on which serenity and quiet are of extraordinary significance and serve an important public need and where the preservation of those qualities is required if the area is to continue to serve its intended purpose
B	67 (exterior)	70 (exterior)	Picnic areas, recreation areas, playgrounds, active sports areas, parks, residences, motels, hotels, schools, churches, libraries, and hospitals
C	72 (exterior)	75 (exterior)	Developed lands, properties, or activities not included in categories A and B table
D	—	—	Undeveloped lands
E	52 (interior)	55 (interior)	Residences, hotels, motels, public meeting rooms, schools, churches, libraries, hospitals, and auditoriums

Source: 23 CFR Part 772.

If abatement is considered feasible (possible) and reasonable (cost effective), then abatement measures must be implemented. Abatement may not occur if it is infeasible or unreasonable even though the criteria are exceeded. This leads to the requirement that each project be documented and considered individually. In addition to the Noise Abatement Criteria, substantial increases also trigger abatement analysis for projects on new alignment or drastic changes to existing highways, even though the Noise Abatement Criteria are not exceeded.

Aircraft noise is also controlled by federal legislation. The Control and Abatement of Aircraft Noise and Sonic Boom Act of 1968 mandated noise emission limits on aircraft beginning in 1970. The standards for new aircraft created classifications of aircraft based on noise emissions called stage I, II, or III. The stage I (noisier) aircraft have all but been phased out in the U.S. New regulations, in the form of 14CFR91 (*Transition to An All Stage III Fleet Operating in the 48 Contiguous United States and District of Columbia*) and 14CFR161 (*Notice and Approval of Airport Noise and Access Restrictions*) call for the fast phase-in of the quieter stage III aircraft.

In 1979 the Aviation Safety and Noise Abatement Act placed more responsibility on local and regional airport authorities. The Airport Noise Control and Land Use Compatibility (ANCLUC) Planning process included in Part 150 of the Federal Aviation Regulations (FARs) allows federal funds to be allocated for noise abatement purposes. This process is often referred to as a "Part 150 study."

FAA has also implemented a program that requires computer modeling for environmental analysis and documentation. Impacts are defined to occur if the L_{dn} is predicted to be above 65 dB (A-weighted).

In response to a lawsuit by the Association of American Railroads, the Federal Railroad Administration (FRA) has released standards as 40CFR Part 201. Figure 64.5 presents these standards. The lawsuit was necessary to circumvent hindrances to interstate commerce caused by inconsistent local ordinances.

In addition to administration regulations of U.S. DOT, other criteria or regulations may be applicable such as the guidelines established by the Department of Housing and Urban Development (HUD) to protect housing areas. The HUD Site Acceptability Standards use L_{dn} (A-weighted) and are acceptable if less than 65 dB, normally unacceptable from 66 to 75 dB, and unacceptable if above 75 dB. In addition, state and/or local governments have also issued guidelines. The analyst should carefully review all applicable requirements before beginning any study.

Estimating Transportation Noise Impacts

At the heart of transportation noise prediction is the use of reference emission levels that are averages of noise levels and frequency spectra occurring from defined transportation sources for a specified distance and test condition. This level is then corrected for distance, environmental variables, transportation volumes, and other related parameters during the noise prediction process.

Most highway vehicle modeling in the U.S. is based upon a single pass-by of the defined vehicle type at a distance of 15 meters (50 feet) from the center of the vehicle track. In Europe 7.5 meters (25 feet) is more typical. Defined vehicle types are generally broken into automobiles and trucks with subcategories of each. The U.S. Federal Highway Administration uses the categories of cars, medium trucks, and heavy trucks. Frequency spectra considerations are usually taken into account by the use of A-weighting. The Federal Highway Administration national reference emission levels are shown in Fig. 64.6.[5] Note that as speed increases, so do the emission levels. These levels were based on in situ measurements.[6]

The reference levels must then be adjusted to the modeling conditions. Among these are geometric spreading (effects of distance), traffic volume adjustments, source characteristics, diffraction, and environmental adjustments.

Noise reduction occurs with increased distance from a source and is usually referred to as geometric spreading. The attenuation due to geometric spreading may be characterized by the geometry of the source. If noise is emitted from a single location, the source is referred to as a *point source*

SUMMARY OF RAILROAD NOISE STANDARDS
40 CFR PART 201

Noise Source	Operating Condition		Noise Metric	Meter Response	Meas't Location	Standard dB(A)
Railroad Cars	Speed ≤ 45 mph		L_{max}	Fast	100 Feet	88
	Speed > 45 mph		L_{max}	Fast	100 Feet	93
Active Retarders	Any		$L_{adj.ave. max.}$	Fast	Rec.Prop.	83
Car-Coupling	Any		$L_{adj.ave. max.}$	Fast	Rec.Prop.	92
Locomotive Load Cell Test Stands	Any		L_{90}*	Fast	Rec.Prop.	65
	or	(a) Primary Standard	L_{max}	Slow	100 Feet	78
		(b) If (a) Is Not Feasible	L_{90}*	Fast	Rec.Prop. >400 Feet	65

* L_{90} measurement must be validated by showing that L_{10}(Fast) – L_{99}(Fast) ≤ 4 dB(A).

SUMMARY OF RAILROAD NOISE STANDARDS
40 CFR PART 201
Locomotive Source Standards

Operating Condition	Noise Metric	Meter Response	Meas't Location	Locomotive Type: Non-Switchers Built On or Before 31 Dec 79	Locomotive Type: All Switchers;* Non-Switchers Built After 31 Dec 79
Stationary, Idle	L_{max}	Slow	100 Feet	73 dB(A)	70 dB(A)
Stationary, Non-Idle	L_{max}	Slow	100 Feet	93 dB(A)	87 dB(A)
Moving	L_{max}	Fast	100 Feet	96 dB(A)	90 dB(A)

* Switchers are in compliance if L_{90}(Fast) ≤ 65 dB(A) on receiving property. L_{90} measurement must be validated by showing that L_{10}(Fast) – L_{99}(Fast) ≤ 4 dB(A).

FIGURE 64.5 Railroad noise standards. (*Source:* 40 CFR Part 201.)

(see Fig. 64.7). A boat whistle, a locomotive at idle, or a single aircraft could be identified as a point source. If the point source is extruded in space, a line is formed and the source is referred to as a *line source* (see Fig. 64.7). Highway traffic may be modeled as either a moving point source or, for high-volume highways, a line source.

For a point source the sound energy spreads as the surface of a sphere ($4\pi r^2$). The intensity and the root-mean-square pressure decreases proportionally to the inverse of the square root of the distance from the source (inverse-square law). A definite relationship in dB can be derived, resulting in:

$$\Delta \text{SPL(dB)} = 10 \log_{10}(r_1/r_2)^2$$

where

ΔSPL(dB) = difference in SPL
r_1 = distance at point 1
r_2 = distance at point 2

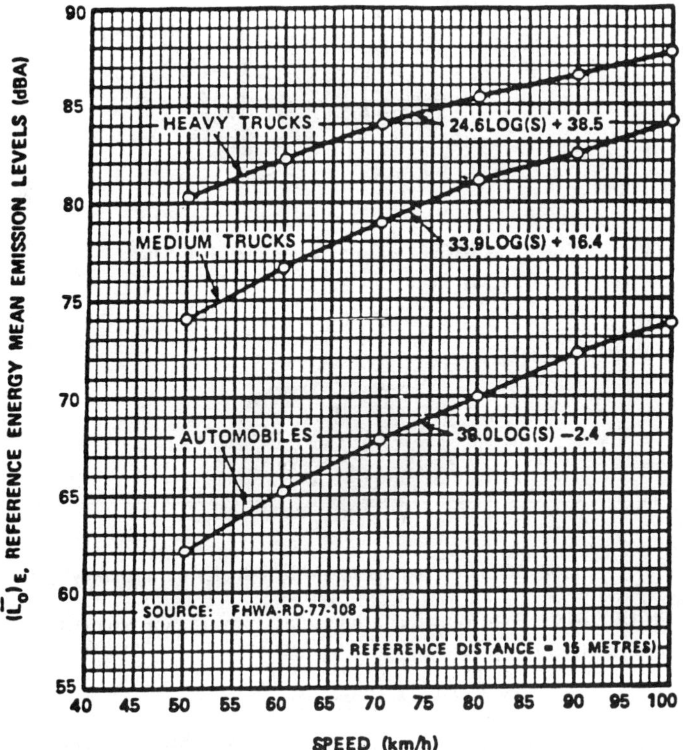

LEGEND:

1. **AUTOMOBILES: ALL VEHICLES WITH TWO AXLES AND FOUR WHEELS.**
2. **MEDIUM TRUCKS: ALL VEHICLES WITH TWO AXLES AND SIX WHEELS.**
3. **HEAVY TRUCKS: ALL VEHICLES WITH THREE OR MORE AXLES.**

FIGURE 64.6 Motor vehicle reference energy mean emission levels. (*Source:* 23 CFR Part 772.)

Consider when the distance—point source to receiver—is doubled:

$$\begin{aligned} \Delta\text{SPL(dB)} &= 10\log_{10}(1/2)^2 \\ &= 10\log_{10} 1/4 \\ &= -6\text{ dB} \end{aligned}$$

Then, for a point source, every time we double the distance we reduce the noise levels by 6 dB. This is the way we might expect noise to decrease from a stationary vehicle.

Geometric-spreading attenuation for a line source can be derived in a similar fashion to that for a point source. This time the energy is spread over the surface area of a cylinder. In addition, the line consists of an infinite number of closely spaced point sources, so only the spreading away from the source in a single plane must be considered. This means that the sound energy spreading is proportional to the circumference of a circle. The circumference of a circle is equal to $2\pi r$ and, using the same mathematical procedure as for a point source, a line source decreases as:

$$\Delta\text{SPL(dB)} = 10\log_{10}(r_1/r_2)$$

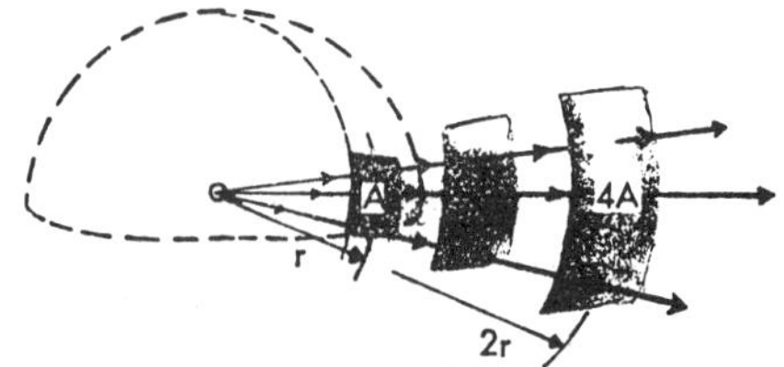

(A) POINT SOURCE: eg. single vehicle at distance

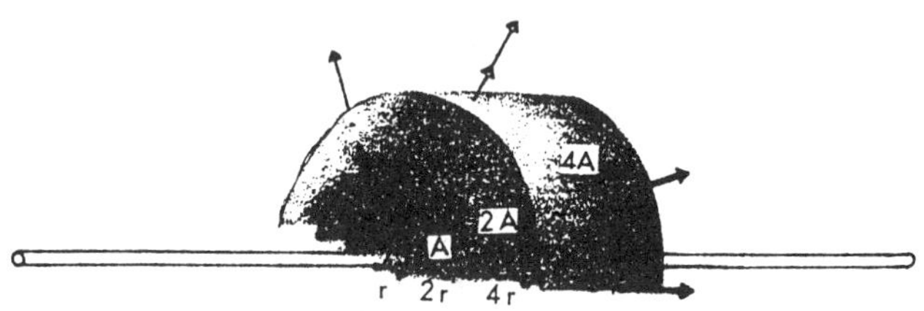

(B) LINE SOURCE: eg. many single vehicles on a roadway

FIGURE 64.7 Point source and line source propagation geometries. (*Source:* Ref. 2.)

For line sources the sound level decrease is proportional to the distance from the source, not the square of the distance. Solving as before for a doubling of distance, a decrease of 3 dB occurs with each doubling of distance. Accordingly, it could be expected that the sound level would decrease by 3 dB for each doubling of distance from a highway. However, the highway is not actually in free space but close to the earth's surface. As a result, the interaction of the sound wave with the surface causes excess attenuation above what would be expected from just geometric spreading. The excess attenuation effects are related to the type of soil, ground cover, and surface topography. Ground effects are difficult to predict. However, it has been determined that a value of 3 dB per doubling of distance is more typical of hard surfaces abutting the highway (e.g., water or pavement), and a value of approximately 4.5 dB for each doubling of distance has been found to be applicable for soft surfaces (e.g., vegetative coverings).

Since the reference levels are usually developed for a single-motor vehicle (e.g., locomotive, rail car, or aircraft), additional vehicles, the number of vehicles by type, the duration of each different vehicle event and the noise contribution of each vehicle allowing for vehicle path must be considered. Inclusion of these parameters permits a correct estimation of the overall noise for a defined transportation system.

The spatial relationship of the transportation source to the receiver not only determines the attenuation due to geometric spreading, but also determines characteristics of the noise path, such as obstructions to the sound path. Spatial relationships are usually accounted for by using an x, y, z Cartesian coordinate system. This permits distances—such as source to receiver, source to obstruction, obstruction to receiver—and other geometric relationships to be determined.

Obstructions in the noise path may cause diffraction or reflection of the sound (see Fig. 64.8). Diffraction, or the blocking of the sound, causes noise levels to be reduced. This area of decreased sound is called the *shadow zone*. Sound is attenuated the most immediately behind the object, and the attenuation decreases with distance behind the object as the wave reforms. Diffraction is the reason that properly designed highway noise barriers are effective. Obstructions may also reflect sound. This causes a redirection of the sound energy. The angle of incidence equals the angle of reflection.

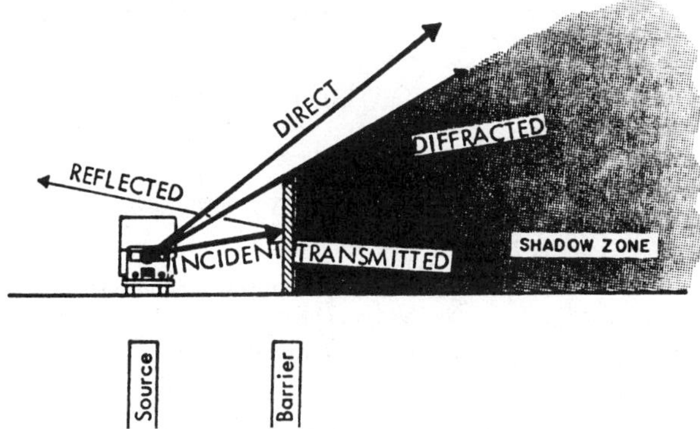

FIGURE 64.8 Effect of a barrier (shading indicates relative strength of sound energy). (*Source:* Ref. 2.)

Weather parameters may refract (bend) the sound waves causing reduced or increased noise levels according to the weather conditions. Figure 64.9 shows the effect of refraction that takes place when wind shear exists. Similar refraction is caused by temperature lapse rates (thermal vertical gradients). The upwind case occurs during normal lapse conditions (temperature decreases with height) and the downwind case would be expected when an inversion (temperature increase with height) occurs.

The refraction that occurs due to weather effects can be very significant but has been greatly ignored in many past models. It should be noted that, since atmospherical effects cause refraction of the sound wave, negative excess attenuation (amplification) may also occur (downwind and during inversions). New model development will most certainly contain adjustments for these effects.

These overall developed methodologies have led to regulatory models by various governmental entities for transportation noise specific to area of jurisdiction. For highway vehicles, the Federal Highway Administration (FHWA) has released such a model. Figure 64.10 shows the basic components of this model.[5] The FHWA model has been incorporated into a computer program called STAMINA. The latest release, STAMINA 2.0,[7] is in use by most state departments of transportation (DOT). The computer model includes many improvements over the basic model

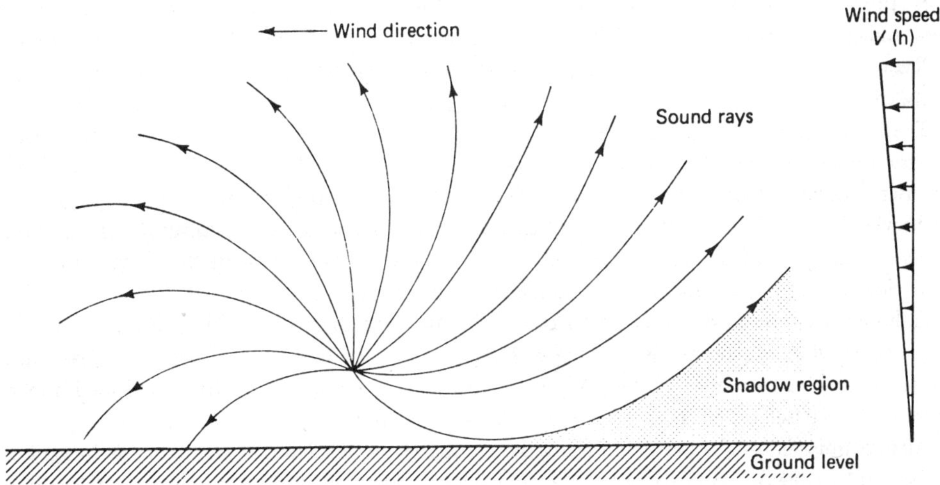

FIGURE 64.9 Refraction of sound waves due to wind speed changes with height.

$$
\begin{aligned}
L_{eq}(h)_i = (\overline{L_o})_{E_i} & \quad \text{reference energy mean emission level} \\
+10 \log \left(\frac{N_i \pi D_o}{S_i T}\right) & \quad \text{traffic flow adjustment} \\
+10 \log \left(\frac{D_o}{D}\right)^{1+\alpha} & \quad \text{distance adjustment} \\
+10 \log \left(\frac{\psi_\alpha(\phi_1, \phi_2)}{\pi}\right) & \quad \text{finite roadway adjustment} \\
+\Delta_s & \quad \text{shielding adjustment} \qquad (1)
\end{aligned}
$$

where

$L_{eq}(h)_i$ is the hourly equivalent sound level of the ith class of vehicles.

$(\overline{L_o})_{E_i}$ is the reference energy mean emission level of the ith class of vehicles.

N_i is the number of vehicles in the ith class passing a specified point during some specified time period (1 hour).

D is the perpendicular distance, in metres, from the centerline of the traffic lane to the observer.

D_o is the reference distance at which the emission levels are measured. In the FHWA model, D_o is 15 metres. D_o is a special case of D.

S_i is the average speed of the ith class of vehicles and is measured in kilometres per hour (km/h).

T is the time period over which the equivalent sound level is computed (1 hour).

α is a site parameter whose values depend upon site conditions.

ψ is a symbol representing a function used for segment adjustments, i.e., an adjustment for finite length roadways.

Δ_s is the attenuation, in dB, provided by some type of shielding such as barriers, rows of houses, densely wooded areas, etc.

FIGURE 64.10 FHWA highway traffic noise prediction model. (*Source:* Ref. 5.)

shown in Fig. 64.10, such as the ability to handle complex highway geometry, graphical presentations, and to a small degree atmospheric effects. The output of the model is L_{eq} and L_{10} (A-weighted) one-hour sound levels at user-defined locations. These values can then be compared to the Noise Abatement Criteria (previously discussed) as well as other state and/or local criteria.

STAMINA 2.0 is meant to be used for single diffraction, constant speed situations. To overcome these problem areas, methods such as that by Foss for multiple diffraction[8] or NCHRP 311 for stop-and-go conditions[9] are used. Other adjustments are also needed, which has led to work on a new computer model to replace STAMINA 2.0 that will be called the Traffic Noise Model and Software.

The FHWA computer model STAMINA 2.0 is meant to be run in conjunction with a sister program named OPTIMA.[7] This program reads an input file developed by STAMINA and allows the user to design effective noise barriers by changing barrier heights, eliminating barrier segments, and reviewing the effects on the overall noise levels at the specified receiver locations. In addition, a cost file is included to permit the user to determine the relative cost of the designed barrier scenarios, permitting consideration of both effectiveness and cost.

Although the noise analysis procedures for aircraft is similar to that for highways, it usually only occurs when the aircraft is near the ground. As with highway noise, expected increases and comparison to criteria are required. A level of 65 dB (A-weighted) L_{dn} is the established impact criteria and is based on yearly average operations at the airport.

Airport noise is usually predicted using the computer models NOISEMAP[10] or the Integrated Noise Model (INM).[11] Adjustments to reference levels are made as with highway models. While noise levels at user-defined locations can be predicted as with the FHWA model, the primary output

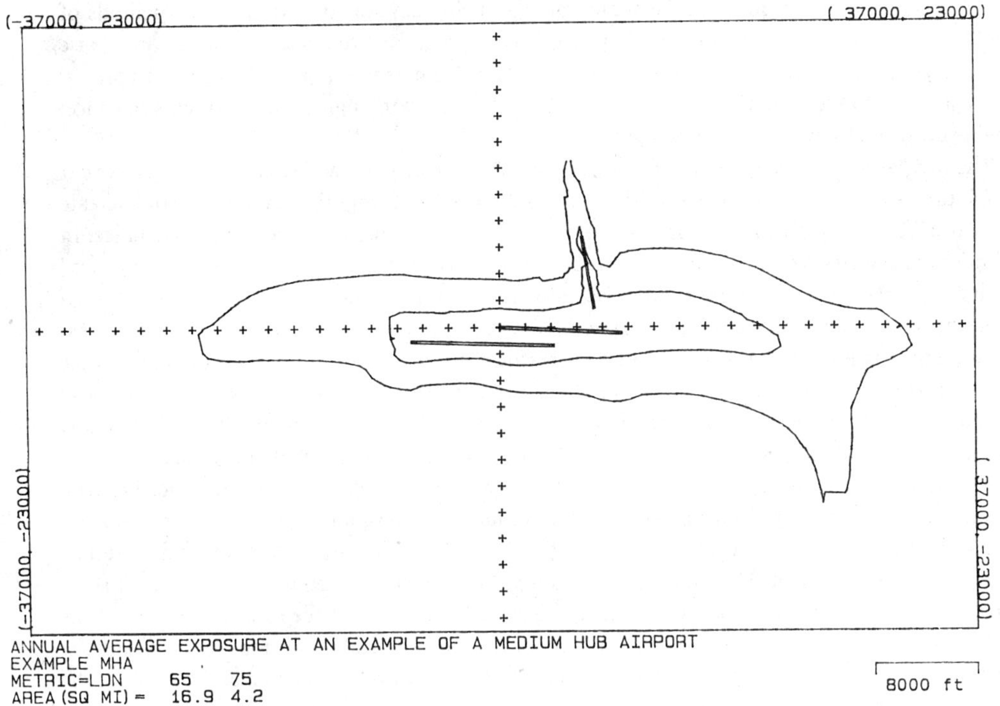

FIGURE 64.11 Typical noise contour plot from INM. (*Source:* Ref. 11.)

of these models are noise contours as shown in Fig. 64.11. Contours may be plotted for a specified level—typically the 65 dB L_{dn} FAA criteria—to determine areas defined to be impacted near the airport.

No models have been released by the Federal Railroad Administration, but it is generally agreed that rail noise can be predicted based upon a strong logarithmic relationship to speed:[12]

$$L_{max} = 30 \log(V) + C$$

where V is the speed of the train and C is a constant. In this situation, L_{max} is at a defined location for a single pass-by. Adjustments for site-specific conditions (e.g., distance) must be calculated as for highway vehicles. For the newer high-speed rail, a relationship of 40 log V appears to be appropriate.[13] Measurements at yards may be used for future predictions by "scaling" the levels according to expected future use. The Federal Transit Authority (FTA, previously UMTA) has released detailed prediction guidelines for urban rail noise.[14] An update of these guidelines is soon to be released by FTA and FRA.

Mitigation of Transportation Noise

Mitigation of transportation noise may occur at the source, in the propagation path, or at the receiver. Each has benefits and problems. Abatement of noise from highways has proven to be successful in all these locations.

Traffic noise is emitted from the tire/pavement interaction, the exhaust, and from the engine. At higher speeds the automobile noise is almost all from the tire/pavement interaction. For heavy trucks the noise from the exhaust and engine during acceleration are much more dominant than the tire noise. Accordingly, reduction of noise from various sources is needed to reduce noise impacts on the highway neighbor. Insulation in the engine compartment is used to reduce engine noise, mufflers are used for exhaust noise, and pavement type selection may reduce tire/pavement interaction and lessen the noise impact.

Noise abatement techniques at the vehicle are still being developed. Two of these methods of control are the use of contra-noise and open-graded asphalt for tire noise control. Contra-noise involves generating a wave form of equal amplitude but opposite phase to the cancel engine or exhaust noise. Implementation could consist of speakers used in engine compartments to cancel engine noise or electronic mufflers for exhaust noise.

The tire/pavement noise generation is a direct result of the friction and small impacts that occur as the tire rolls along the surface. Intuitively, it would seem that smoother pavement would create less noise. And while this is true, it leads to unsafe conditions due to reduced skid resistance. In addition, recent studies, particularly in Europe, have shown that open-graded asphalt ("popcorn" asphalt) can also result in a significant noise reduction. This is thought to occur because although more noise is generated, the rough surface diffuses the normally reflected wave and hence, less pressure fluctuations are transmitted. Highway traffic noise reduction may also occur at the source by using traffic management (e.g., reduce speeds or ban trucks). Near airports, noise may also be controlled at the source. FAR Part 36, along with the regulations discussed earlier, has greatly reduced the aircraft engine noise and is the most effective aircraft noise abatement measure.

The path may also be altered to reduce noise. As discussed earlier, increased distance between the source (motor vehicle) and the receiver (uses of land near or abutting the highway) results in reduced sound levels due to geometric spreading. It then follows that increased path distance results in traffic noise abatement. This abatement measure is possible if sufficient right-of-way widths are available. A greenbelt is established and the noise levels are reduced at the receiver locations. This option is very costly in large urban areas because of the many typical relocations required and the cost of land. As such, it is a costly option for highways or railways in urban areas, but may be the only feasible method for airports.

A more cost-effective abatement method for highways and railways may be achieved by requiring changes in the vertical or horizontal alignment. Another more cost-effective abatement measure that has been used extensively in North America, Europe, Japan, and Australia is the diffraction of the sound wave by noise barriers. A typical traffic noise barrier is shown by Fig. 64.12. The difference in noise levels with and without the wall is referred to as *insertion loss*. Insertion loss is typically calculated using Fresnel diffraction techniques that depend on the path length distance (difference between direct path and path over the obstacle). As the path length distance increases, so does the insertion loss. Highway neighbors often request vegetative plantings instead of noise barriers. Some work has been done to determine the insertion loss due to vegetation,[15] but this method is most often not very effective, except on a subjective basis.

FIGURE 64.12 Typical noise barrier for highway.

Sometimes it is not practical or feasible to mitigate traffic noise in the path, such as near airports. In these cases it may be possible to protect individuals by insulating buildings. A typical wood frame house, with the windows closed, attenuates the noise roughly 10 to 20 dB on an A-weighted scale. With increased insulation in the walls and roof, double-paned or thicker pane windows, acoustic vents, storm doors, and possible other changes, the house may provide a reduction of more than 40 dB (A-weighted). Many homes near airports are insulated in this manner using Part 150 funding. Public buildings, such as schools, have been insulated near highways using FHWA funding.

Other abatement measures have been defined by various agencies. The FAA has listed 37 abatement measures, shown in Table 64.3. In regard to rail noise abatement, methods of diffraction, reduction of rail squeal, dampening of track noise, and increased separation distance have all been applied. Noise barriers (diffraction) are quite efficient because of possible placement very near the guided path.

Table 64.3 FAA Noise Control Strategy Categories

Category Number	Description
1	State noise law
2	Local noise law or ordinance
3	Airport master plan
4	ANCLUC plan
5	Part 150 noise exposure map approved
6	Part 150 noise compatibility plan approved
7	Development of an EIS
8	Noise monitoring equipment: temporary or permanent
9	Restriction on ground runup
10	Limit on the number of operations by hour, day, month, year, or noise capacity
11	Preferential runway system
12	Runway restrictions imposed for specific aircraft type
13	Use restriction by aircraft type or class
14	Use restriction based on noise levels
15	Use restriction based on Part 36
16	Use restriction based on AC 36-3
17	Complete curfew
18	Arrivals and/or departures over a body of water
19	Displaced runway threshold
20	Rotational runway system
21	Maximum safe climb on takeoff
22	Takeoff thrust reduction
23	Reverse thrust limits
24	Flight training restriction
25	Weight or thrust limit
26	Informal flight operation restriction
27	Zoning
28	Purchase land for noise control
29	Use of capital improvements to direct development
30	Building codes and permits to control noise
31	Noise easements
32	Purchase assurance
33	Soundproofing programs
34	Noise use fees
35	Shift operations to a reliever airport
36	Local pattern restrictions
37	Navigational aid assisted departure

Source: FAA.

Cost may vary widely according to the mitigation method selected, materials available, location, and amount of abatement required. The Federal Highway Administration reports that the average noise barrier cost in the U.S., in 1986 dollars, is $12 per square foot.[16] Buying or insulation of homes near airports is also quite costly. Accordingly, abatement may be somewhat costly, but usually a small part of the overall project costs. It should be noted that a livable environment is the goal; not implementing mitigation measures may lead to drastic changes or cancellation of the project. As such, properly designed abatement would seem to be well worthwhile.

Transportation and Air Quality

Fundamental Concepts

General Considerations. Transportation sources are typically called *mobile sources.* Airports, parking lots, and other collectors of mobile sources are often referred to as *indirect sources.* From each source various amounts and types of air pollutants are emitted according to fuel type, vehicle mode, vehicle volumes, and various other vehicle specific factors. An air pollutant can be defined as a gas, liquid droplet, or solid particle dispersed in the air with sufficient concentration to cause an adverse impact on public health or welfare. Combustion exhaust gases from transportation sources include carbon monoxide (CO), nitrogen oxides (NO_x), volatile organic compounds (VOCs), and sulfur oxides (SO_x). The emitted VOCs are sometimes simply referred to as hydrocarbons (HC) and are emitted not only from the combustion process but from evaporated fuels as well.

Particulate matter (PM) is also created during combustion and includes small solids or liquid droplets (0.01–100 micrometers). Standards are now written in terms of particulate matter less than 10 micrometers in aerodynamic diameter, designated as PM_{10}.

Other pollutants not directly emitted from the source may form in the atmosphere using the directly emitted pollutants as precursors. These include nitrates, sulfates, and photochemical oxidants (ozone).

Common units of measurement for air pollutants include parts per million (ppm) and micrograms per cubic meter ($\mu g/m^3$). The unit ppm provides a measure of very small ratios of volume (or moles) of pollutant per volume (or moles) of air. One ppm represents one-millionth of the total volume. One $\mu g/m^3$ is a small amount of mass in a defined volume of air.

Although significant reductions of mobile source emissions have occurred, the EPA still reported that in 1991 transportation sources were the primary anthropogenic source of CO (>70%) and also contributed about 38%, 30%, and 21% of the NO_x, volatile organic carbons, and particulate matter, respectively.[17] Research is currently being conducted by FHWA to determine the overall contribution of diesel vehicles on urban PM_{10} concentrations. Lead emissions have been drastically reduced from mobile sources through the use of unleaded gasoline. Fuels in the U.S. are highly refined and sulfur is significantly removed. Accordingly, most SO_x emissions are attributed to nonmobile sources in the U.S.

Typically, efforts have been concentrated on a few regulated pollutants during transportation analysis. These include CO, HC, NO_x, and PM_{10}. However, regional analyses are now being done for ozone precursors and toxic compounds emitted by transportation. These types of analysis will become more important in the future.

The reader is encouraged to consult any general air pollution text for particular pollutant characteristics. Due to space limitation, a complete description cannot be included here.

Health and Public Welfare Effects. Air pollution is generally associated with respiratory damage (bronchitis, emphysema, pneumonia, and lung cancer) as well as irritation of the eyes, nose, and throat. Public welfare effects include damage to structures and materials, damage to crops and animal life, and atmospheric haze. Global effects from acid rain, global warming, and ozone depletion are also of concern. Standards are usually written as primary standards to protect human health and secondary standards to protect public welfare.

Air Quality Legislation and Regulations

Federal legislation relating to air quality and transportation began in the mid-1950s when the Air Pollution Control Act of 1955 was passed. This law gave authority to the Public Health Service and was amended in 1962 to include health effects of auto exhaust. In 1963 the original Clean Air Act (CAA) was passed and subsequently amended in 1965. The act authorized the Secretary of Health, Education and Welfare (HEW) to promulgate and enforce federal emission standards for new motor vehicles. The 1965 Motor Vehicle Control Act was passed in an effort to improve national vehicle emission standards to California standards by 1968. It is interesting that this legislation forced development of automotive emission controls and resulted in common use of such items as the positive crankcase ventilation (PCV) valve.

The Air Quality Control Act of 1967 charged the Secretary of HEW to publish air quality criteria. Unfortunately, criteria do not require compliance, as do standards. This made the federal legislation good in theory but unenforceable in practice. Major responsibilities were still at the state and local level, and evaluation methodologies varied widely.

In 1970 all federal air pollution control functions were transferred from HEW to the newly established Environmental Protection Agency (EPA). This gave direct authority for determining guidelines, methods, and standards to the EPA and forced state and local governments to look to EPA for guidance.

The Clean Air Act Amendments of 1970 (CAAA70) permitted federal intervention if state and local governments did not meet their responsibilities. This was the first air quality law that provided strong federal controls in the individual states. Section 110 led to the establishment of *state implementation plans* (SIPs) first released as 40CFR51 and 52. The SIP is a state-prepared document that completely outlines how the state will deal with air pollution problems.

Section 202 led to the promulgation of the National Ambient Air Quality Standards (NAAQSs). The dose an individual receives depends on the toxicity of a particular pollutant, the concentration of exposure, and the time of exposure. The NAAQSs were determined based on these three factors. Table 64.4 lists the NAAQSs promulgated by the U.S. EPA. Although revised, the NAAQSs are still used as the air quality evaluation criteria, including transportation projects.

Section 202 of the CAAA70 led to motor vehicle and aircraft emission standards, released as 40CFR85. A 90% reduction in CO and HC were originally required by 1975, with a similar reduction in nitrogen oxides (NO_x) required by 1976. This law directly led to the development of the catalytic oxidizer, which was included on cars manufactured after 1974. Unfortunately, events such as the Arab oil embargo of 1973 led to delays in implementation of emission standards. As a result the emission standards originally intended for the mid-1970s were not fully realized until the mid-to-late 1980s.

Another strong piece of environmental legislation passed in 1970—in regard to transportation—was the Federal Aid Highway Act of 1970 (FAHA70). This act required the secretary of the U.S. Department of Transportation—guided by discussions with the EPA administrator—to develop and issue guidelines governing the air quality impacts of highways. A strong planning process was implemented and transportation plans were required to be consistent with the SIP. To ensure consistency, *transportation control plans* (TCPs) and *transportation control measures* (TCMs) were required to improve air quality.

The CAA was amended again in 1977. In this law regions were classified as nonattainment areas (NAAs) if the NAAQSs were exceeded. With the NAA designations came the concept of sanctions (loss of federal funds if a good faith effort was not given to meet the NAAQSs),

Table 64.4 National Ambient Air Quality Standards

Pollutant	Averaging Time	Primary Standard	Secondary Standard
PM-10	Annual arithmetic mean	50 $\mu g/m^3$	same
	24-hour	150 $\mu g/m^3$	same
SO_2	Annual arithmetic mean	80 $\mu g/m^3$	
	24-hour	(0.14 ppm)	
		365 $\mu g/m^3$	
	3-hour		1300 $\mu g/m^3$
CO	8-hour	9 ppm	none
	1-hour	35 ppm	none
NO_2	Annual arithmetic mean	0.053 ppm	same
O_3	Maximum daily 1-hour average[a]	0.12 ppm	same
Pb	Maximum quarterly average	1.5 $\mu g/m^3$	same

Source: 40 CFR Part 50.

[a] The standard is attained when the expected number of days per calendar year with maximum hourly average concentrations above 0.12 ppm is equal to or less than 1, as determined according to Appendix H of the ozone NAAQS.

Note: The ambient standard for VOCs as a general category was rescinded by the EPA in 1983, but various source performance standards remain in effect.

and states hurried to ensure their SIPs were acceptable. The SIP had become a strong tool that allowed EPA to review and concur with state air quality policy. In June 1978, in direct response to the Section 108(e) of the CAAA77, EPA and the U.S. DOT issued joint Transportation–Air Quality Planning Guidelines.[18] These ambitious guidelines were issued with a national goal of attainment of the NAAQSs by 1982, with some extensions until 1987 (e.g., for O_3 and CO). SIP requirements, various agency coordination, and analysis/abatement methodologies were included in the guidelines. Unfortunately, the attainment dates were not met, which partially resulted in the CAAA90.

The CAAA90, signed by the president on 12 November 1990, contains strong transportation provisions. In fact, one entire title of the act that is related to mobile sources and other titles may have far-reaching effects on transportation issues.

Title I approaches nonattainment areas (NAAs) in a new way. Categories of nonattainment have been developed for O_3, CO, and PM_{10}. For example, ozone NAAs are rated from *transitional* to *extreme*. Each category allows different time schedules and levels of air pollution control measures to help ensure the NAAQSs will be attained. More heavily polluted areas have a greater amount of time to come into attainment, but must implement greater emission controls.

Title II (mobile sources) requires more strict emission controls, use of alternative fuels, off-road source (e.g., tractors and construction equipment) emission controls, and implementation of TCMs according to the severity of the air pollution problem in the area. Title II also required more strict tailpipe emission controls and includes: (1) phasing in tighter HC, CO, and NO_x tailpipe emission standards for cars and trucks, beginning with 1994 models (HC 0.25 g/mile, CO 3.4 g/mile, NO_x 0.4 g/mile); (2) requiring vehicle manufacturers to design for reducing evaporative HC emissions during refueling; (3) controlling fuel quality (e.g., volatility and sulfur content); (4) reformulated gasoline beginning in 1995 for most severely polluted ozone NAAs; (5) mandating oxygenated fuels during winter months for 41 moderate and serious carbon monoxide NAAs; and (6) establishing a clean fuel pilot program in Los Angeles, affecting 150,000 vehicles in 1996 and 300,000 in 1999, with stricter standards imposed in 2001.

Title III (toxics) and Title IV (acid rain) may also have long reaching effects on mobile sources as interpretation of the CAAA90 continues.

The CAAA90 directed the administrator of the EPA to "update the 1978 Transportation–Air Quality Planning Guidelines and publish guidance on the development and implementation of transportation and other measures necessary to demonstrate and maintain attainment of national ambient air quality standards (NAAQSs)." This resulted in the issuance of the updated Transportation and Air Quality Planning Guidelines,[19] which provide guidance to state and local government officials to assist them in planning for transportation-related emissions reductions that will contribute to the attainment and maintenance of NAAQS. Also in 1992, guidelines were released by EPA for evaluating "hot spot" intersections[20] for carbon monoxide concentrations.

The complex determination of when projects achieved SIP conformity led to a series of EPA and U.S. DOT conformity guidelines and memorandums.[21,22,23,24] The purpose of these documents was to provide guidance regarding the criteria and procedures to be followed by MPOs, other recipients of funds designated under Title 23 of the U.S. Code or the Federal Transit Act, and the U.S. DOT in making conformity determinations. The Final Conformity Rule was finally released in November 1993 by FHWA and EPA as 40CFR Parts 51 and 93. This 169-page document is still confusing to many air quality analysts and has resulted in controversy, including lawsuits. Conformity requires not exceeding established emission budgets based on established levels. Transportation projects generally are also required to show emission benefits.

The Intermodal Surface Transportation Efficiency Act (ISTEA) was signed by the president on 18 December 1991. The act included increased funding, flexibility for local projects, and additional metropolitan and statewide planning requirements. Emphasis is required on multimodal considerations, land use considerations, development decisions, and transportation-related air quality problem solving. As such, projects to increase single-occupancy-vehicle (SOV) capacity will not be easily approved in NAAs. In fact, for transportation management areas (TMAs) in NAAs, federal funds may not be programmed for any highway or transit project that will result in a significant increase in capacity for SOVs unless it is part of a congestion management system (CMS). This has led states to consider many nontraditional highway projects, such as high-occupancy-vehicle (HOV) lanes, across the country.

ISTEA also reinforces the CAAA90 requirements that transportation plans conform to the SIPs. MPOs must also coordinate the development of long-range TIPs to be in conformity with the SIP, and the TIP must have had an opportunity for public comment. Also, regardless of funding source, MPOs must now consider the emissions from *all* transportation projects. Consistency with the long-range transportation plan is required. This has provided MPOs with more flexibility, but at a cost of more planning requirements.

Modeling of Air Pollutants from Transportation

Dispersion of air pollutants is due to molecular diffusion, eddy diffusion, and random shifts in the instantaneous direction of the wind. Eddies are small, random swirls of air that transport parcels of air from one location to another, resulting in the dilution of pollutants much more rapidly than would occur from just diffusion. The end result is that, after emissions have been released, atmospheric dispersion determines the resulting concentrations. Because of the random nature of eddies and wind shifts, dispersion must be considered on a time-averaged rather than instantaneous basis. The wind also provides bulk transport of pollutants downwind. In general, the higher the wind speed is, the greater the mechanical mixing and the lower the concentrations as distance increases from the source.

In addition to the wind, solar insolation results in energy being absorbed by the ground, heating the surface. A temperature gradient (lapse rate) is formed and temperature decreases with height as the heat is transferred to the layer of air in contact with the ground. This causes the air to rise, and mixing occurs. This thermal mixing may be dominant during the daytime and results in an "unstable" atmosphere. Unstable air tends to disperse air pollution vertically—promoting mixing of pollutants—and results in lower concentrations. It has been found that unstable conditions generally exist when the temperature gradient, or lapse rate, is greater than 5.4°F for each 1000-foot change in elevation (0.98°C for every 100 meters of height change). Near this lapse rate, neutral conditions exist (mixing is neither hindered or helped), whereas atmospheric conditions with a lower lapse rate result in a stable atmosphere (mixing is hindered). When temperature increases with height, an inversion is formed and results in the greatest concentrations. These ideas are summarized in Fig. 64.13.

Based on this mechanical (due to the wind) and thermal mixing, atmospheric stability has been classified from A (very unstable) to F (very stable). The D stability class is considered neutral. This classification is called the Pasquill–Gifford atmospheric classification scheme and is shown in Fig. 64.14.

As the pollution is dispersed, the peak concentrations decline. However, the vertical and horizontal (perpendicular to the wind direction) distributions remain somewhat normally distributed or Gaussian.[25] The Gaussian model is one way to estimate dispersion for a nonreactive pollutant released steadily from a source at a defined downwind location. As such, modifications of this approach have been used extensively for modeling CO that has a half-life of over 40 days in the atmosphere. The general form of the equation for a point source is:

$$C = [Q/2\pi u \sigma_y \sigma_z][0.5(y/\sigma_y)^2][0.5\{(z+H)/\sigma_z\}^2 + 0.5\{(z-H)/\sigma_z\}^2]$$

where

C = concentration (mass/volume)

Q = emission rate (mass/time)

u = wind speed

σ_y, σ_z = standard deviation of dispersion in the y and z directions

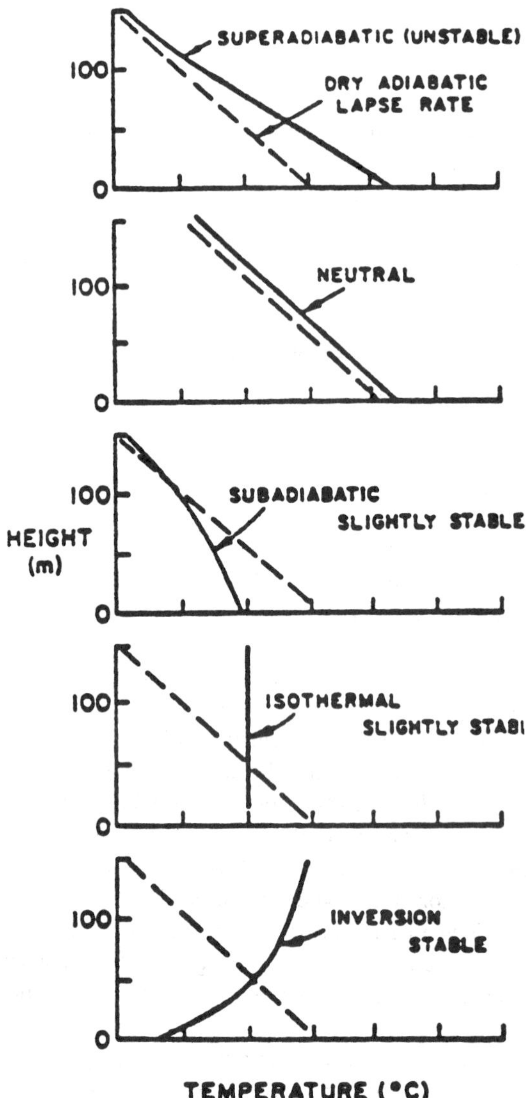

FIGURE 64.13 Graphical descriptions of lapse rates.

y = distance receiver is removed from the x axis
z = receptor height
H = source height

In the Gaussian equation y is parallel to the ground, perpendicular to the wind speed or x axis, and z is in the vertical, perpendicular to the x axis. This equation has been modified for approximating mobile sources as moving point sources, line sources, and area sources. These modifications result in various units for the emission rate.

Use of the Gaussian model requires knowledge of the atmospheric mixing, and the Pasquill–Gifford classes are commonly used. A good knowledge of the emission rate is also crucial to obtaining good performance of the dispersion model. With point sources, emission rates are relatively easy to obtain. With mobile sources, getting a good estimate of the average emission rate is much more difficult.

A - Extremely unstable conditions
B - Moderately unstable conditions
C - Slightly unstable conditions
D - Neutral conditions[a]
E - Slightly stable conditions
F - Moderately stable conditions

Surface Wind Speed (at 10m)	Daytime Conditions			Nighttime Conditions[b]	
	Incoming Solar Radiation[d]			Thin Overcast or $\geq$4/8 Cloudiness[c]	$\leq$3/8 Cloudiness
m/sec	Strong	Moderate	Slight		
<2	A	A-B	B	-	-
2-3	A-B	B	C	E	F
3-5	B	B-C	C	D	E
5-6	C	C-D	D	D	D
>6	C	D	D	D	D

[a]Applicable to heavy overcast, day or night.

[b]Nighttime conditions refers to the period one hour before sunset to one hour after sunrise

[c]The degree of cloudiness is defined as that fraction of the sky above the local apparent horizon which is covered by clouds.

[d]"Strong" incoming solar radiation corresponds to a solar altitude greater than 60° with clear skies; "moderate" insolation corresponds to a solar altitude of 35° to 60° with clear skies; and "slight" insolation corresponds to a solar altitude of 15° to 35° with clear skies.

FIGURE 64.14 Pasquill–Gifford stability classes. (*Source:* EPA, Pub. AP-26.)

Standardized emission databases for mobile sources are available. In the U.S. two major databases exist for highway vehicles. These are the EPA document, *Compilation of Air Pollutant Emission Factors, Volume 2, Mobile Sources,*[26] fourth edition, and the California specific emission factors. The national emission factors in the EPA document are commonly referred to as AP-42 from an abbreviation of the document number. Since these factors have numerous adjustments (mode, speed, type, etc.) a computer program has been promulgated by EPA to derive the proper emission factor. The computer program is called MOBILE and the latest version is called MOBILE5a[27]. The California factors also require adjustments and have been implemented in a model called EMFAC. The latest version is EMFAC7F.

Emissions factors are a measure of the source strength per unit of activity. As such, the units from the MOBILE model are mass emitted per distance in cruise conditions and mass emitted per time during idling. It should be noted that, in MOBILE, cruise is really a trip-generated model containing all four modes—cruise, idle, acceleration, and deceleration.

The FAA-approved aircraft emission factors can be obtained from AP42 or a computer database known as the FAA Emission Data (FAAED). Emission factors for water vessels and rail activities are included in AP-42 as well.

After emission data have been determined, two analyses are typically accomplished. The NEPA process usually requires the Gaussian dispersion modeling methodology to be used, so the pollutant concentrations may be predicted and compared to the NAAQSs, the existing concentrations, and various future alternative predictions (including the no-build alternative). Multiplication of emission factors times activity rates determines the total pollutant load (total mass emitted) and is compared to the emission budget, area-wide emissions, and the alternatives during the conformity determination process. Because of the complexity of the analysis procedure, many state agencies have written guidelines. Some examples have been included in the references.[28,29,30,31]

Because of the importance of weather on pollutant levels, historical meteorological data are used to evaluate dispersion. These data usually come from airports or other weather stations. Often, these stations are on the outskirts of the urban area and may not be suitable. For transportation impact analysis an alternative exists, especially for modeling the impacts of a future project. The approach is to use a set of "worst-case" meteorological conditions: lowest realistic wind speed, worst reasonable stability class, lowest (or highest) reasonable temperature, highest expected traffic and emissions, and closest reasonable receptor locations in the air quality models. This approach will result in the worst-case concentration that can reasonably be expected. If the predicted worst-case concentration does not violate the NAAQS, then it is safe to assume the project is not likely to exceed the standards under typical conditions.

Computer models that have been developed for highway projects (uninterrupted flow) include CALINE, HIWAY, and PAL. CALINE (California Line Source Model) has been updated several times. CALINE 4[32] is the latest version, but EPA maintains CALINE 3[33] as the approved model. HIWAY is another approved EPA line source Gaussian model; the latest version is HIWAY 2.[34] This model lacks some of the sophistication of CALINE 3, but is somewhat easier to use and can be manually applied if necessary. PAL is a model that includes *p*oint, *a*rea, and *l*ine source Gaussian dispersion algorithms. PAL is routinely applied in transportation projects such as parking lots or with the many considerations of different sources such as at airports. The current version is PAL 2.0.[35]

CALINE, HIWAY, and PAL are used in free-flow conditions. At intersections they lack the needed considerations for interrupted flows. At intersections other models are currently used. Recent EPA testing has identified three models that perform well: TEXIN 2,[36] CALINE 4, and CAL3QHC.[37] Each uses the Gaussian dispersion algorithms but also accounts for queuing, delays, excess emission due to modes, and cruise. CAL3QHC, based on the Highway Capacity Manual,[38] is the EPA-preferred model.

For times greater than one hour (e.g., eight hours) the error would be too great using dispersion models based on nonvarying volumes and weather conditions. In these cases persistence factors are used. A persistence factor is a number—less than one—that represents the ratio of the longer-term concentration to the short-term concentration:

$$\text{Persistence factor} = \frac{Ct_2}{Ct_1}$$

where Ct_2 = *long-term concentration* and Ct_1 = *short-term concentration*. Typical persistence factors for carbon monoxide near intersections—when predicting the eight-hour concentration based on the one-hour concentration—are in the range of 0.4 to 0.7.

Using these dispersion models, the first row of homes along a highway will be predicted to have greater concentrations than the second row of homes, and so on. During modeling, care must be taken to model those receptors close to the roadway, where normal human activity occurs and where the greatest concentrations of modeled pollutants (generally CO) will occur. If a violation of criteria or a standard occurs at these receptors, sites farther away must be modeled to determine the extent of the problem. Figure 64.15 shows a typical flow chart of events for this project level, or microscale modeling.

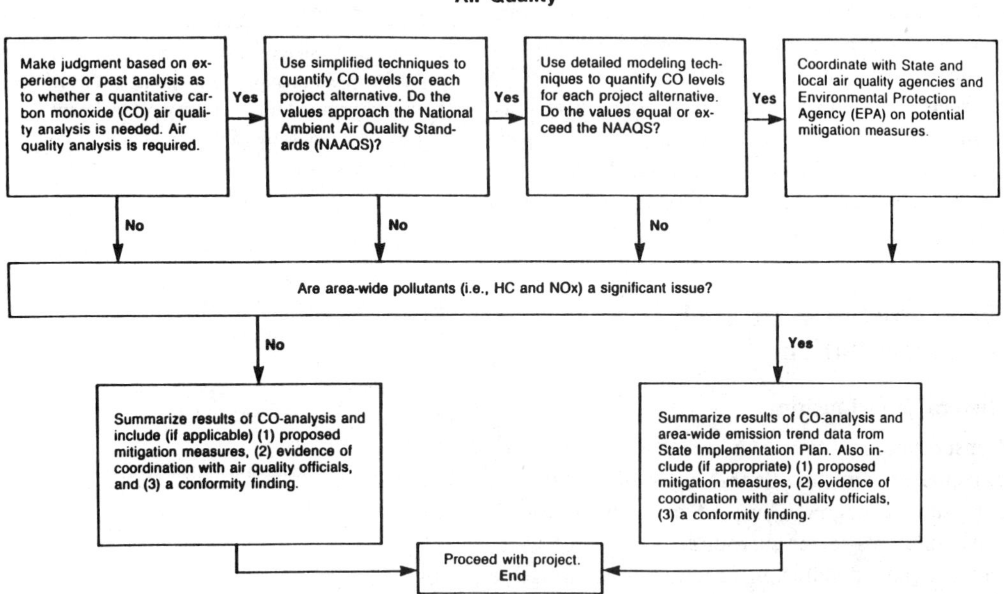

FIGURE 64.15 Generalized air quality evaluation flowchart. (*Source:* FHWA.)

Evaluation of air pollution from airports is similar to that of highways and is described in a series of documents.[39,40,41] As a first step, FAA Order 5050.4 uses an emission inventory to determine impacts.[42] If impacts are considered to occur, dispersion modeling is then required, again typically just for CO. Computer modeling is now quite common for airports. Originally, the airport vicinity air pollution model was used[43] on mainframe computers. This model has been replaced by the PC-based Emission and Dispersion Modeling System.[44]

No specific model has been issued for evaluating rail lines or yards. However, the Gaussian approach, coupled with AP-42 emission factors, would allow predictions to be made. Both rail systems and airports are compared with the NAAQSs during NEPA evaluations and are considered during conformity determinations.

Of course, secondary pollutants such as ozone form in the atmosphere. These pollutants reach maximum concentrations at long distances from the source, and Gaussian modeling is not applicable. For example, ozone forms from nitrogen dioxide in about two to three hours in an urban area. If the average wind speed is just five miles an hour, the peak ozone concentrations due to the highway would be ten to fifteen miles away. As such—and because of the numerous contributions and reactions from other emissions—large-scale regional models are used for these predictions and are not project-specific. A simple approach to regional modeling assumes that a defined volume (a "box") has complete mixing. Using this assumption, a simple mass balance can be used to predict concentrations in the box. The large regional model termed the urban airshed model (UAM) uses small, connected "boxes" to better define an entire area.

Abatement

At the Source. The most effective abatement for air pollution occurs at the source. In the U.S. emissions standards and test procedures have changed significantly since the first automobile emission standards were imposed in California in 1966. Standards for mobile sources were discussed earlier in this chapter. Control of fuel storage is also very significant. Stage I vapor recovery is accomplished if fueling vapors are collected during delivery of fuel by tanker truck, whereas stage II vapor recovery is defined to occur during the dispensing of fuel to the individual vehicles.

During Project Development. The CAAA90 required EPA to publish guidance on transportation control measures (TCMs). Table 64.5 shows a listing of the defined TCMs for highway projects. Although no such list exists for aircraft or rail operations, similar ideas can be applied. Problems at airports generally occur due to motor vehicle access. Problems for rail operations generally occur near the classification yards.

Table 64.5 Section 108(f) Transportation Control Measures

1. Trip-reduction ordinances
2. Vehicle use limitations/restrictions
3. Employer-based transportation management
4. Improved public transit
5. Parking management
6. Park and ride/fringe parking
7. Flexible work schedules
8. Traffic flow improvements
9. Areawide rideshare incentives
10. High-occupancy vehicle facilities
11. Major activity centers
12. Special events
13. Bicycling and pedestrian programs
14. Extended vehicle idling
15. Extreme cold starts
16. Voluntary removal of pre-1980 vehicles

Source: CAAA90.

Water Quality as Related to Transportation

General Information

Transportation systems may affect water quality or can interfere with the desirable use of a waterway. For example, highway, airport, or railroad runoff adds pollutants to the surrounding bodies of water and may cause flooding. Bridges may affect navigable waters. Construction may cause erosion.

Runoff refers to the volume and discharge rate of water occurring as overland flow from a highway, airport, or rail line immediately after a precipitation event. Hydrologic variables that affect runoff are precipitation amount, evaporation, transpiration, infiltration, and storage.

Approximately 7% of the land in the U.S. is classified as floodplain. *Floodplains* are low areas adjacent to streams, oceans, and lakes that are subject to flooding at least once in 100 years.[45] In the U.S. in the early 1980s about 90% of all losses from natural disasters were caused by floods. In addition to economic impacts, health and safety problems are evident. In the U.S. there are approximately 200 flood-related deaths per year. Transportation systems can exacerbate flooding conditions if the facility is not properly designed.

Construction, operation, and maintenance of transportation systems also contribute a variety of pollutants to runoff, such as solids, nutrients, heavy metals, oil and grease, pesticides, and bacteria. The extent to which the runoff affects the quality of the surrounding water requires adequate knowledge of quantity and quality of pollutants, their origin, and movement reactions within the system.

Water Quality Legislation and Regulations

Federal legislation for water-related activities in regard to transportation has been around since 1899 when the Rivers and Harbors Act was passed (Title 23 of the U.S. Code). This law, amended by the Department of Transportation Act of 1966, requires the U.S. Coast Guard to approve the plans for construction of any bridge over navigable waters. Accordingly, the required process (generally referred to as a Section 9 permit from the applicable portion of the act) protects navigation activities from being affected by other transportation modes.

In 1972 Section 404 was added to the Federal Water Pollution Control Act. This requires a permit (called a 404 permit) from the U.S. Army Corps of Engineers for any filling, dredging, or realignment of a waterway. For smaller projects that do not pass established threshold limits a general permit may be issued.

The Federal Water Pollution Act was changed in 1977 and issued as the Clean Water Act. This act reflected the desire to protect water quality and regulated the discharge of storm water from transportation facilities. Also included in this law was the option for the Corps of Engineers to transfer 404 permitting to the states.

The 404 permitting process also includes required assessments of potential wetland impacts. The amount of wetlands affected, the productivity (especially as related to endangered or protected species), overall relationship to regional ecosystems, and potential enhancements during the design of the project must all be considered. Executive Order 11990, "Protection of Wetlands," issued in 1977, required a public-oriented process to mitigate losses or damage to wetlands, as well as to preserve and enhance natural or beneficial values. This has led to a policy of wetlands being avoided and replacement required if destruction occurs. The Federal Highway Administration has released guidelines to help during this phase of the project.[46]

Modeling of Water Impacts

The rational formula has been used since 1851 to calculate the peak discharge flow rate and can be derived using a mass balance for precipitation rate and runoff rate. The rational equation is:

$$Q_P = CiA$$

where

Q_P = peak discharge
C = runoff coefficient
i = precipitation rate
A = watershed area

Typical units are used with conversion factors as required. Use of the equation must consider the following assumptions:

1. Rainfall intensity is constant over the time it takes to drain the watershed.
2. The runoff coefficient remains constant during the time of concentration.
3. The watershed area does not change.

These assumptions are reasonable for watersheds with short time of concentration (about 20 minutes) since the intensity is relatively constant for travel time below 20 minutes.

A detailed manual for flood analysis was developed by Davis[47] and is used by the U.S. Army Corps of Engineers. From basic hydrologic relationships of flow rates versus frequency and versus depth (stage), damages can be calculated as a function of flow rate.

Runoff water from transportation corridors, yards, parking lots, or airports has the potential to cause a pollution problem depending on the type and amount of pollutant present in runoff water and the ambient water quality characteristics of the receiving water it enters.[48] Suspended solids and associated pollutants in runoff discharges accumulate in localized areas close to the input sources, causing bioaccumulation of toxic materials in benthic organisms. Studies conducted in retention/detention ponds receiving highway runoff by Yousef *et al.*[49] indicate a decline in the number of benthic species present and the number of organisms in each specie. Also, species tolerant of pollutional loads dominated the bottom sediments in these ponds. Concentrations of dissolved pollutants may cause water quality criteria established for a particular receiving stream to be exceeded. The increased levels can produce visible impacts, fish kills, taste and odor problems, or alterations in the aquatic biological community.

Contaminants accumulate on roadway surfaces and medians from moving vehicles, highway construction and maintenance, natural contributions, and atmospheric fallout. The magnitude and pattern of accumulation varies with many factors including dry periods between rainfall events, sweeping practices, scour by wind and/or rainfall, type of pavement and grade, traffic volume, and adjacent land use. For example, Gupta *et al.*[50] identified the variables affecting the quality of highway runoff as follows:

a. Traffic (volume, speed, braking, type and age, etc.)
b. Climate (precipitation, wind, temperature, dust fall)

c. Maintenance (sweeping, mowing, repair, etc.)
d. Land use (residential, commercial, industrial, rural)
e. Percent impervious areas
f. Regulations (air emissions, littering laws, etc.)
g. Vegetation
h. Accidental spills

Particulates and other associated pollutants are generally attributed to atmospheric deposition, degradation, and traffic activities. Dust fall is a measure of the particulate matter in the 20–40 micrometer range that falls out of the atmosphere due to gravity. The quantity and quality of particulates vary greatly with land use and geographic location. Smith *et al.*[51] reported dust fall loads in the U.S. to approximately 0.23 g/m^2-d in the northern region, and 0.16–1.53 in the central region, 0.07–0.18 in the southern region, and 0.06–0.16 in the eastern region. He concluded that the dry areas of the central U.S. are dustier than the wet areas to the east. During the period from 1975 to 1981 emissions of particles from automobiles were estimated to have decreased about 20%.[52]

Predictive models for accumulation of particulate matter and associated pollutants have been developed with emphasis on that fraction of pollutant load that is available for wash-off. A simplified equation of a predictive model for highways by Gupta *et al.* can be expressed as follows:[50]

$$P = P_0 + K_1 H_L T$$

where

P = pollutant load after buildup
P_0 = initial surface pollutant load
K_1 = pollutant accumulation rate
H_L = highway length
T = time of accumulation

Wide variations were found to occur for K_1. Gupta reported K_1 could be estimated using

$$K_1 = 0.007(\text{ADT})^{0.89}$$

where ADT = average daily traffic.

Structured techniques have also been developed to evaluate wetlands. A methodology called WET[53] (Wetlands Evaluation Technique) has been promulgated by the FHWA and allows a comprehensive review of wetland impacts and mitigation.

Abatement of Water-Related Impacts

Storage of highway runoff water on highway right-of-way can provide both flood control and contaminant discharge abatement by permitting settling. To reduce flooding potential, the volume of storage is determined to be consistent with downstream flow rates for a specified return period and the available on-site land area for ponding of the runoff.

Erosion and sedimentation often occur during the construction phase of a transportation project and may also occur during operation. Silt fences, minimizing clearing, increased vegetative cover, embankments, rounding of slopes, water flow control, and on-site ponding have been used to mitigate erosion.

Stringent controls and regulations have been placed on point source discharges, such as sewage treatment plants and industrial wastewater outlets, in most of the industrialized countries of the world. However, a corresponding improvement in the receiving water quality of surface waters has not always been noticeable. During the last 20 years research into nonpoint sources has zeroed in on urban storm water runoff as a major pollution source and authors have reported that concentrations of certain constituents, such as heavy metals and nutrients, greatly exceed those found in secondary effluent discharges.[54,55,56] To mitigate these discharges, runoff from transportation systems can be transported in separate sewer systems, combined sewer systems, or held on site before allowing drainage directly into lakes, streams, rivers, and other surface water. Direct runoff discharge adversely impacts the water quality, and prior treatment is required by many regulatory agencies.[57] On-site treatment can add significant costs to a transportation project but may be extremely necessary in cases such as aircraft fuels spillage.

Energy Use

Fossil fuels are the primary source of energy for transportation. The automobile is the number one consumer. These fuels are becoming short in supply and are being used much faster than they are being formed.[58] The conservation of these fossil fuels is of extreme importance for many reasons (monetary, social, national defense, etc.). Unfortunately, transportation use of these fuels continues to increase, and it is expected that there will be more than one billion vehicles in use by the year 2030.[59] Many conservation techniques are possible, including vehicle technology improvements, ride sharing, traffic flow improvements, transportation systems management, and improved goods movement.[60] During project development, these techniques may be evaluated based on energy consumption. These estimates can be done directly by identifying vehicle movements and applying fuel consumption factors. However, many factors are difficult to determine, and in many cases surrogates must be used, such as "scaling" future values based on recorded fuel consumption or approximations using population or vehicle miles traveled (VMT). Each of these estimation procedures provides a method to determine which alternatives may result in a significant fuel savings.

It is also important to consider the fuel used during project construction. Fossil fuels are used in the materials (e.g., asphalt) for needed machinery (e.g., graders) and for transportation of supplies, equipment, and labor. The overall fuel savings of a project is related to the construction and operational fuel use.

During project development estimates may be required of the total fuel use. This will require fuel use rates as a minimum. Many such documents are available.[60]

Ecological Impacts

Transportation projects can have major impacts on ecological systems. Construction, physical removal of vegetation, compaction of soils, paving of surfaces, draining, and construction vehicle operation can all destroy needed habitats. Mowing, application of herbicides, accidental spills, vehicle operations, and human activity can interrupt the normal ecosystem cycle. Accordingly, these impacts must be considered during transportation planning.

The U.S. Fish and Wildlife Service has become quite involved during transportation planning since the Fish and Wildlife Coordination Act of 1958 [16USC 661-667(d)] required that the U.S. Fish and Wildlife Service be consulted when bodies of water are to be modified. In addition to this charge, the Fish and Wildlife Service is involved in operations affection coastal barrier resources (P.L. 97-348), endangered species (50CFR Part 402, Endangered Species Act), wetlands evaluations (discussed earlier), and wild or scenic rivers. The Endangered Species Act has caused many projects to be modified or stopped. Early coordination is required to avoid such problems.

In addition to these considerations, other ecological considerations are required during the transportation planning process. Coastal zone management must be considered if the project is located near the coastline (15CFR part 930) and requires involvement of the National Oceanic and Atmospheric Administration. The Farmland Protection Policy Act (7CFR Part 658) requires justification for taking such land and involvement by the Department of Agriculture. If the project is in a floodplain, the Federal Emergency Management Agency may become involved. If the project will affect hazardous waste areas, EPA must be brought in.

It becomes apparent that ecological impacts are very important considerations and requires involvement of many players during the transportation planning process. All of these impacts and the related analysis cannot be discussed here. The reader is encouraged to consult other references on these topics.

Sociological Concerns

NEPA requires conservation of national resources: national well-being and preservation of heritage certainly qualifies. During transportation planning, Section 106 of the Historic Preservation Act (36CFR Part 800) requires that historic properties be identified and protected. This requires coordination with the state historic preservation officer (SHPO), and sometimes with the Department of Interior and Advisory Council on Historic Preservation. Complete documentation is required by the SHPO to allow a determination of impact.

Socioeconomic impacts must also be considered. Generally these impacts include disruption of community cohesion, prevention of access to community facilities, general overall social or economic disruption, discrimination, and/or relocation. During planning, these items must be assessed and measures taken to mitigate any such impacts. Again the reader should consult other references as needed.

Aesthetics

The public often judges the quality of a transportation project by the visual impact. It is important to evaluate aesthetics from the point of view of both the traveler and the transportation neighbor. The FHWA has released a document to help during the planning process entitled "FHWA Visual Assessment Methodology." Listed are common mitigation measures, which include changes in horizontal and vertical alignment, landscaping, use of vegetation, litter pickup, and maintenance practices. To help with these considerations, another good guide has been issued by the American Association of State Highway Officials.[61] This document is also helpful in that it lists the many applicable laws related to project development.

64.3 Summation

This chapter was intended to present an overview to the environmental process. Information on basic principles, legislation/regulation, analysis, and abatement were presented for general impacts. A complete description would require much more discussion and, indeed, several volumes of text. Many laws and regulations not mentioned here come into play. It is hoped that the reader can use the basis provided here to build upon during environmental planning for transportation projects.

References

1. Martin, F. N. 1991. *Introduction to Audiology.* 4th ed. Prentice Hall, Englewood Cliffs, NJ.
2. Federal Highway Administration. 1980. *Fundamentals and Abatement of Highway Traffic Noise—Textbook and Training Course.* FHWA, Washington, D.C.
3. Bronzaft, A. L. 1989. *Public Health Effects of Noise.* Paper No. 89-101.3. Proceedings of the 82nd Annual Meeting of the Air & Waste Management Assn. Anaheim, CA.
4. U.S. Environmental Protection Agency. 1974. *Information on Levels of Environmental Noise Requisite to Protect Public Health and Welfare with an Adequate Margin of Safety.* EPA-550/9-74-004. U.S. Environmental Protection Agency, Washington, D.C.
5. Barry, T. M. and Reagan, J. A. 1978. *FHWA Highway Traffic Noise Prediction Model.* FHWA-RD-77-108. Federal Highway Administration, Washington, D.C.

6. Rickley, E. J., Ford, D. W., and Quinn, R. W. 1978. *Highway Measurements for Verification of Prediction Models.* Report No. DOT-TSC-OST-78-2. U.S. DOT Transportation System Center and Office of the Secretary, Cambridge, MA.
7. Bowlby, W., Higgins, J., and Reagan, J., eds. 1983. *Noise Barrier Cost Reduction Procedure, STAMINA 2.0/OPTIMA User's Manual.* FHWA-DP-58-1. Federal Highway Administration, Washington, D.C.
8. Foss, R. N. 1979. Double barrier noise attenuation and a predictive algorithm. *Noise Control Engineering.* 11(1):40–44.
9. Bowlby, W., Wayson, R. L., and Stammer, R. 1985. *Predicting Stop-and-Go Traffic Noise Levels.* National Cooperative Highway Research Program Report 311. Transportation Research Board, Washington, D.C.
10. Moulton, H. T. 1990. *Air Force Procedure for Predicting Aircraft Noise around Airbases: Noise Exposure Model (NOISEMAP) User's Manual.* AAMRL-TR-90-011. Armstrong Laboratory, Wright-Patterson AFB, OH.
11. Flythe, M. C. 1982. *INM, Integrated Noise Model, Version 3 User's Guide.* FAA-EE-81-17. Washington, D.C.
12. Hanson, C. E. 1976. *Environmental Noise Assessment of Railroad Electrification.* Inter-Noise '76 Proceedings. Washington, D.C.
13. Wayson, R. L. and Bowlby, W. 1989. Noise and air pollution of high-speed rail systems. *Journal of Transportation Engineering.* 115(1):20–36.
14. Saurenman, H., Nelson, J. T., and Wilson, G. P. 1982. *Handbook of Urban Rail Noise and Vibration Control.* UMTA-MA-06-099-82-1. U.S. Dept. of Transportation, Washington, D.C.
15. Harris, R. A. and Cohn, L. F. 1986. Use of vegetation for abatement of highway traffic noise. *Journal of Urban Planning and Development.* 111(1):34–38.
16. Federal Highway Administration. 1994. *Summary of Noise Barriers Constructed by December 31, 1992.* U.S. Dept. of Transportation, Washington, D.C.
17. U.S. Environmental Protection Agency. 1992. *National Air Quality and Emission Trends Report.* EPA Report No. 450/R-92-001. Office of Air Quality Planning and Standards, Research Triangle Park, NC.
18. *Code of Federal Regulations Part 770* (23CFR770), Chapter 23. Air quality guidelines. 1981, pp. 8429–8431).
19. U.S. Environmental Protection Agency. 1992. *The 1992 Transportation and Air Quality Planning Guidelines,* EPA 420/R-92-001. U.S. EPA, Research Triangle Park, NC.
20. U.S. Environmental Protection Agency. 1992. *Guidelines for Modeling Carbon Monoxide from Roadway Intersections.* EPA Report No. EPA-454/R-92-005. U.S. EPA Office of Air Quality Planning and Standards, Research Triangle Park, NC.
21. U.S. Environmental Protection Agency/U.S. Department of Transportation. 1991. *USEPA and USDOT Guidance for Determining Conformity of Transportation Plans, Programs, and Projects with Clean Air Act Implementation Plans During Phase 1 of the Interim Period.* Federal Highway Administration, Washington, D.C.
22. U.S. Department of Transportation. 1992. *Further Guidance on Conformity Determinations.* Memorandum released by Federal Highway Administration Office of Environment and Policy, July 27, 1992.
23. U.S. Department of Transportation. 1992. *Clarification of FHWA July 27, 1992, Memorandum to Regional Administrators.* Memorandum released by Federal Highway Administration Office of Environment and Policy, Oct. 9, 1992.
24. U.S. Environmental Protection Agency. 1993. Criteria and procedures for determining conformity to state or federal implementation plans, program, and projects funded or approved under Title 23 U.S.C. or the Federal Transit Act. *Federal Register* 58(6):3768–3798.
25. Pasquill, F. 1974. *Atmospheric Diffusion.* John Wiley & Sons, New York.

26. U.S. Environmental Protection Agency. 1985. *Compilation of Air Pollutant Emission Factors, Volume 2: Mobile Sources.* EPA Report No. AP-42-ED-4-VOL-2. Ann Arbor, MI.
27. U.S. Environmental Protection Agency. 1993. *User's Guide to MOBILE 5.0 (Mobile Source Emission Factor Model).* EPA-AA-TEB-93-01, Ann Arbor, MI.
28. California Department of Transportation. 1988. *Air Quality, Technical Analysis Notes.* California Department of Transportation, Office of Transportation Laboratory, Sacramento, CA.
29. Florida Department of Environmental Regulations. 1988. *Guidelines for Evaluating the Air Quality Impacts of Indirect Sources.* Final Draft. Florida Dept. of Environmental Regulation, Tallahassee, FL.
30. Florida Department of Transportation. 1991. Air quality analysis. *Project Development and Environmental Guidelines; Part 2: Analysis and Documentation.* Florida Dept. of Trans., Tallahassee, FL. Revised 1 Oct. 91.
31. New York State Department of Transportation. 1986. *New York State Department of Transportation's Project Environmental Guideline, PEG Transmittal #42.* Latest revision, #3, 1992 (Air Quality Analysis Procedures). New York State Department of Transportation, Environmental Analysis Bureau, Albany, NY.
32. Benson, P. 1984. *CALINE-4—A Dispersion Model for Predicting Air Pollutant Levels Near Roadways.* FHWA Report No. FHWA/CA/TL-84/15. Office of Transp. Lab., California Dept. of Trans., Sacramento, CA.
33. Benson, P. 1979. *CALINE-3—A Versatile Dispersion Model for Predicting Air Pollutant Levels Near Highways and Arterial Streets.* FHWA Report No. FHWA/CA/TL-79/23. Office of Transp. Lab., California Dept. of Trans., Sacramento, CA.
34. U.S. Environmental Protection Agency. 1980. *User's Guide to HIGHWAY-2: A Highway Air Pollution Model.* EPA 600/8-80-018. U.S. EPA, Research Triangle Park, NC.
35. Peterson, W. B. and Rumsey, E. D. 1987. *User's Guide for PAL 2.0—A Gaussian-Plume Algorithm for Point, Area, and Line Sources.* EPA 600/4-78-013, U.S. EPA, Research Triangle Park, NC.
36. Bullin, J. A., Korpics, J. J., and Hlavinka, M. W. 1986. *User's Guide to the TEXIN-2 Model: A Model for Predicting Carbon Monoxide Concentrations Near Intersections.* FHWA Report No. FHWA/TX-86/283-2. College Station, TX.
37. Schattanek, G. and Stranton, T. 1990. *CAL3QHC, A Modeling Methodology for Predicting Pollutant Concentrations Near Roadway Intersections.* User Guide. Parsons, Brinckerhoff, Quade & Douglas, New York.
38. Transportation Research Board. 1985. *The Highway Capacity Manual.* Special Report 209. Trans. Research Board, Washington, D.C.
39. Federal Aviation Administration. 1985. *Airport Environmental Handbook.* FAA Order 5050.4. Washington, D.C.
40. Federal Aviation Administration. 1982. *Air Quality Procedures for Civilian Airports and Air Force Bases.* FAA Report No. FAA-EE-82-21. Washington, D.C.
41. U.S. Environmental Protection Agency. 1973. *An Air Pollution Impact Methodology for Airports; Phase I.* EPA Report No. APTD-1470. NTIS, Springfield, VA.
42. Wayson, R. L. and Bowlby, W. 1988. Inventorying airport air pollutant emissions. *ASCE Journal of Transportation Engineering.* 114(1). American Society of Civil Engineers, New York.
43. Federal Aviation Administration. 1975. *Airport Vicinity Air Pollution Model User's Guide.* Report FAA-RD-75-230. Washington, D.C.
44. Segal, H. M. 1991. *EDMS—Microcomputer Pollution Model for Civilian Airports and Air Force Bases, User's Guide.* FAA Report No. FAA-EE-91-3. Air Force Report No. ESL-TR-91-31. Washington, D.C.
45. Water Resources Council, Hydrology Committee. 1981. Guidelines for Determining Flood Frequencies. *Bulletin 17B.* U.S. Water Resources Council, Washington, DC.
46. Federal Highway Administration. 1986. *Highways and Wetlands: Compensating Wetland Losses.* Report No. FHWA-IP-86-22. Washington, D.C.

47. Davis, S. 1988. *National Economic Development Procedures Manual: Urban Flood Damage.* U.S. Army Corps of Engineers. IWR Report 88-R-2. Fort Belvoir, VA.
48. Driscoll, E. D., Shelly, P. E., and Strecker, E. W. 1988. *Evaluation of Pollutant Impacts from Highway Stormwater Runoff: Analysis Procedure.* FHWA/RD-08/006. Federal Highway Administration, McLean, VA.
49. Yousef, Y.A., Lin, L., Sloat, J. V., and Kaye, K. 1991. *Maintenance Guidelines for Accumulated Sediments in Retention/Detention Ponds Receiving Highway Runoff.* FL-ER-47-91. Florida Dept. of Transportation, Tallahassee, FL.
50. Gupta, M. K., Agnew, R. W., Gruber, D., and Kreutzberger, W. A. 1981. *Constituents of Highway Runoff: Volume IV—Characteristics of Runoff from Operating Highways—Research Report.* FHWA/RD-81/045. Federal Highway Administration, Office of Research and Development, Washington, D.C.
51. Smith, R. M., Twiss, P. C., Krauss, R. K., and Brown, M. J. 1970. Dust deposition in relation to site, season and climatic variables. *Proceedings of Soil Sciences Soc. Am.* 34:112–117.
52. Cooper, C. D. and Alley, F. C. 1986. *Air Pollution Control: A Design Approach.* PWS, Boston, MA.
53. Adamus, P. R., Clairain, E. J., Smith, R. D., and Young, R. E. 1987. *Wetland Evaluation Techniques (WET).* Report No. FHWA-IP-88-029. Federal Highway Administration, McLean, VA.
54. Rimer, A. E., Nissen, J. A., and Reynolds, D.E. 1978. Characterization and impact of stormwater runoff from various land cover types. *Journal of the Water Pollution Control Federation.* 50:252.
55. Helsel, D. R., Kim, J., Grizzard, T. J., Randall, C., and Hoehn, R. C. 1979. Land use influences on metals storm drainage. *Journal of the Water Pollution Control Federation.* 51:709.
56. Sartor, J. D., Boyd, G. B., and Agardy, F. J. 1974. Water pollution aspects of street surface contaminants. *Journal of the Water Pollution Control Federation.* 46:458–167.
57. Wanielista, M. P., Yousef, Y. A., and Christopher, J. E. 1980. *Management of Runoff from Highway Bridges.* Florida Dept. of Transportation, Contract No. 99700-7198, Tallahassee, FL.
58. Davis, G. R. 1990. Energy for the planet earth. *Scientific American.* 263(3):55–62.
59. Bleviss, D. L. and Walzer, P. 1990. Energy for motor vehicles. *Scientific American.* 263(3):103–109.
60. Institute of Transportation Engineers. 1992. *Transportation Planning Handbook.* Prentice Hall, Englewood Cliffs, NJ.
61. American Association of State Highway Officials. 1991. *A Guide for Transportation Landscape and Environmental Design.* AASHO, Washington, D.C.

For Further Information

References have purposely been used quite heavily in this chapter, and the interested reader should obtain these references as needed. In addition, extensive literature in the form of preprints, proceedings, and journal articles is available through the U.S. DOT Administration, the U.S. EPA, the Air and Waste Management Association, the American Society of Civil Engineers, the Society of Automotive Engineers, and the Transportation Research Board. Discussion of most topics of interest can be obtained from these sources.

This 12.5-mile road project through the scenic Glenwood Canyon, located in the White River National Forest, is located 150 miles from Denver, Colorado. It forms the final link in Interstate 70 from Baltimore to Cove Fort, Utah. The $490 million project overcame a variety of design challenges and extreme physical constraints to create a four-lane highway. The project, which required a decade to plan and 12 years to construct, includes 39 bridges and viaducts; the first U.S. highway tunnels to use rock reinforcement to support performance; a retaining-wall system to fit four traffic lanes into a space barely fitting two previously; and numerous other technical innovations. (Photo courtesy of the Colorado Department of Transportation.)

Mathematics, Symbols, and Physical Constants

0-8493-8953-4/95/$0.00 + $.50

Mathematics, Symbols, and Physical Constants

Greek Alphabet

Greek letter	Greek name	English equivalent	Greek letter	Greek name	English equivalent
Α α	Alpha	a	Ν ν	Nu	n
Β β	Beta	b	Ξ ξ	Xi	x
Γ γ	Gamma	g	Ο ο	Omicron	ŏ
Δ δ	Delta	d	Π π	Pi	p
Ε ϵ	Epsilon	ĕ	Ρ ρ	Rho	r
Ζ ζ	Zeta	z	Σ σ ς	Sigma	s
Η η	Eta	ē	Τ τ	Tau	t
Θ θ ϑ	Theta	th	Υ υ	Upsilon	u
Ι ι	Iota	i	Φ ϕ φ	Phi	ph
Κ κ	Kappa	k	Χ χ	Chi	ch
Λ λ	Lambda	l	Ψ ψ	Psi	ps
Μ μ	Mu	m	Ω ω	Omega	ō

International System of Units (SI)

The International System of units (SI) was adopted by the 11th General Conference on Weights and Measures (CGPM) in 1960. It is a coherent system of units built from seven *SI base units,* one for each of the seven dimensionally independent base quantities: they are the meter, kilogram, second, ampere, kelvin, mole, and candela, for the dimensions length, mass, time, electric current, thermodynamic temperature, amount of substance, and luminous intensity, respectively. The definitions of the SI base units are given below. The *SI derived units* are expressed as products of powers of the base units, analogous to the corresponding relations between physical quantities but with numerical factors equal to unity.

In the International System there is only one SI unit for each physical quantity. This is either the appropriate SI base unit itself or the appropriate SI derived unit. However, any of the approved decimal prefixes, called *SI prefixes,* may be used to construct decimal multiples or submultiples of SI units.

It is recommended that only SI units be used in science and technology (with SI prefixes where appropriate). Where there are special reasons for making an exception to this rule, it is recommended always to define the units used in terms of SI units. This section is based on information supplied by IUPAC.

Definitions of SI Base Units

Meter—The meter is the length of path traveled by light in vacuum during a time interval of 1/299 792 458 of a second (17th CGPM, 1983).

Kilogram—The kilogram is the unit of mass; it is equal to the mass of the international prototype of the kilogram (3rd CGPM, 1901).

Second—The second is the duration of 9 192 631 770 periods of the radiation corresponding to the transition between the two hyperfine levels of the ground state of the cesium-133 atom (13th CGPM, 1967).

Ampere—The ampere is that constant current which, if maintained in two straight parallel conductors of infinite length, of negligible circular cross-section, and placed 1 meter apart in vacuum, would produce between these conductors a force equal to 2×10^{-7} newton per meter of length (9th CGPM, 1948).

Kelvin—The kelvin, unit of thermodynamic temperature, is the fraction 1/273.16 of the thermodynamic temperature of the triple point of water (13th CGPM, 1967).

Mole—The mole is the amount of substance of a system which contains as many elementary entities as there are atoms in 0.012 kilogram of carbon-12. When the mole is used, the elementary entities must be specified and may be atoms, molecules, ions, electrons, or other particles, or specified groups of such particles (14th CGPM, 1971).

Examples of the use of the mole:

1 mol of H_2 contains aboaut 6.022×10^{23} H_2 molecules, or 12.044×10^{23} H atoms
1 mol of HgCl has a mass of 236.04 g
1 mol of Hg_2Cl_2 has a mass of 472.08 g
1 mol of Hg_2^{2+} has a mass of 401.18 g and a charge of 192.97 kC
1 mol of $Fe_{0.91}S$ has a mass of 82.88 g
1 mol of e^- has a mass of 548.60 μg and a charge of –96.49 kC
1 mol of photons whose frequency is 10^{14} Hz has energy of about 39.90 kJ

Candela—The candela is the luminous intensity, in a given direction, of a source that emits monochromatic radiation of frequency 540×10^{12} hertz and that has a radiant intensity in that direction of (1/683) watt per steradian (16th CGPM, 1979).

Names and Symbols for the SI Base Units

Physical quantity	Name of SI unit	Symbol for SI unit
length	meter	m
mass	kilogram	kg
time	second	s
electric current	ampere	A
thermodynamic temperature	kelvin	K
amount of substance	mole	mol
luminous intensity	candela	cd

SI Derived Units with Special Names and Symbols

Physical quantity	Name of SI unit	Symbol for SI unit	Expression in terms of SI base units	
frequency[1]	hertz	Hz	s^{-1}	
force	newton	N	$m\ kg\ s^{-2}$	
pressure, stress	pascal	Pa	$N\ m^{-2}$	$= m^{-1}\ kg\ s^{-2}$
energy, work, heat	joule	J	N m	$= m^2\ kg\ s^{-2}$
power, radiant flux	watt	W	$J\ s^{-1}$	$= m^2\ kg\ s^{-3}$
electric charge	coulomb	C	A s	

Physical quantity	Name of SI unit	Symbol for SI unit	Expression in terms of SI base units	
electric potential, electromotive force	volt	V	$J\ C^{-1}$	$= m^2\ kg\ s^{-3}\ A^{-1}$
electric resistance	ohm	Ω	$V\ A^{-1}$	$= m^2\ kg\ s^{-3}\ A^{-2}$
electric conductance	siemens	S	Ω^{-1}	$= m^{-2}\ kg^{-1}\ s^3\ A^2$
electric capacitance	farad	F	$C\ V^{-1}$	$= m^{-2}\ kg^{-1}\ s^4\ A^2$
magnetic flux density	tesla	T	$V\ s\ m^{-2}$	$= kg\ s^{-2}\ A^{-1}$
magnetic flux	weber	Wb	V s	$= m^2\ kg\ s^{-2}\ A^{-1}$
inductance	henry	H	$V\ A^{-1}\ s$	$= m^2\ kg\ s^{-2}\ A^{-2}$
Celsius temperature[2]	degree Celsius	°C	K	
luminous flux	lumen	lm	cd sr	
illuminance	lux	lx	$cd\ sr\ m^{-2}$	
activity (radioactive)	becquerel	Bq	s^{-1}	
absorbed dose (of radiation)	gray	Gy	$J\ kg^{-1}$	$= m^2\ s^{-2}$
dose equivalent (dose equivalent index)	sievert	Sv	$J\ kg^{-1}$	$= m^2\ s^{-2}$
plane angle	radian	rad	1	$= m\ m^{-1}$
solid angle	steradian	sr	1	$= m^2\ m^{-2}$

1 For radial (circular) frequency and for angular velocity the unit rad s^{-1}, or simply s^{-1}, should be used, and this may not be simplified to Hz. The unit Hz should be used only for frequency in the sense of cycles per second.

2 The Celsius temperature θ is defined by the equation:

$$\theta/°C = T/K - 273.15$$

The SI unit of Celsius temperature interval is the degree Celsius, °C, which is equal to the kelvin, K. °C should be treated as a single symbol, with no space between the ° sign and the letter C. (The symbol °K, and the symbol °, should no longer be used.)

Units in Use Together with the SI

These units are not part of the SI, but it is recognized that they will continue to be used in appropriate contexts. SI prefixes may be attached to some of these units, such as milliliter, ml; millibar, mbar; megaelectronvolt, MeV; kilotonne, ktonne.

Physical quantity	Name of unit	Symbol for unit	Value in SI units
time	minute	min	60 s
time	hour	h	3600 s
time	day	d	86 400 s
plane angle	degree	°	(π/180) rad
plane angle	minute	′	(π/10 800) rad
plane angle	second	″	(π/648 000) rad
length	ångstrom[1]	Å	10^{-10} m
area	barn	b	$10^{-28}\ m^2$
volume	litre	l, L	$dm^3 = 10^{-3}\ m^3$
mass	tonne	t	$Mg = 10^3$ kg
pressure	bar[1]	bar	$10^5\ Pa = 10^5\ N\ m^{-2}$
energy	electronvolt[2]	eV (= $e \times$ V)	$\approx 1.60218 \times 10^{-19}$ J
mass	unified atomic mass unit[2,3]	u (= $m_a(^{12}C)/12$)	$\approx 1.66054 \times 10^{-27}$ kg

1 The ångstrom and the bar are approved by CIPM for "temporary use with SI units," until CIPM makes a further recommendation. However, they should not be introduced where they are not used at present.

2 The values of these units in terms of the corresponding SI units are not exact, since they depend on the values of the physical constants e (for the electronvolt) and N_A (for the unified atomic mass unit), which are determined by experiment.

3 The unified atomic mass unit is also sometimes called the dalton, with symbol Da, although the name and symbol have not been approved by CGPM.

Conversion Constants and Multipliers

Recommended Decimal Multiples and Submultiples

Multiples and submultiples	Prefixes	Symbols	Multiples and submultiples	Prefixes	Symbols
10^{18}	exa	E	10^{-1}	deci	d
10^{15}	peta	P	10^{-2}	centi	c
10^{12}	tera	T	10^{-3}	milli	m
10^{9}	giga	G	10^{-6}	micro	μ (Greek mu)
10^{6}	mega	M	10^{-9}	nano	n
10^{3}	kilo	k	10^{-12}	pico	p
10^{2}	hecto	h	10^{-15}	femto	f
10	deca	da	10^{-18}	atto	a

Conversion Factors—Metric to English

To obtain	Multiply	By
Inches	Centimeters	0.3937007874
Feet	Meters	3.280839895
Yards	Meters	1.093613298
Miles	Kilometers	0.6213711922
Ounces	Grams	$3.527396195 \times 10^{-2}$
Pounds	Kilograms	2.204622622
Gallons (U.S. Liquid)	Liters	0.2641720524
Fluid ounces	Milliliters (cc)	$3.381402270 \times 10^{-2}$
Square inches	Square centimeters	0.1550003100
Square feet	Square meters	10.76391042
Square yards	Square meters	1.195990046
Cubic inches	Milliliters (cc)	$6.102374409 \times 10^{-2}$
Cubic feet	Cubic meters	35.31466672
Cubic yards	Cubic meters	1.307950619

Conversion Factors—English to Metric*

To obtain	Multiply	By
Microns	Mils	**25.4**
Centimeters	Inches	**2.54**
Meters	Feet	**0.3048**
Meters	Yards	**0.9144**
Kilometers	Miles	**1.609344**
Grams	Ounces	28.34952313
Kilograms	Pounds	**0.45359237**
Liters	Gallons (U.S. Liquid)	**3.785411784**
Millimeters (cc)	Fluid ounces	29.57352956
Square centimeters	Square inches	**6.4516**
Square meters	Square feet	**0.09290304**
Square meters	Square yards	**0.83612736**
Milliliters (cc)	Cubic inches	**16.387064**
Cubic meters	Cubic feet	$2.831684659 \times 10^{-2}$
Cubic meters	Cubic yards	0.764554858

* Boldface numbers are exact; others are given to ten significant figures where so indicated by the multiplier factor.

Conversion Factors—General*

To obtain	Multiply	By
Atmospheres	Feet of water @ 4°C	2.950×10^{-2}
Atmospheres	Inches of mercury @ 0°C	3.342×10^{-2}
Atmospheres	Pounds per square inch	6.804×10^{-2}
BTU	Foot-pounds	1.285×10^{-3}
BTU	Joules	9.480×10^{-4}
Cubic feet	Cords	**128**
Degree (angle)	Radians	57.2958
Ergs	Foot-pounds	1.356×10^{7}
Feet	Miles	**5280**
Feet of water @ 4°C	Atmospheres	33.90
Foot-pounds	Horsepower-hours	1.98×10^{6}
Foot-pounds	Kilowatt-hours	2.655×10^{6}
Foot-pounds per min	Horsepower	3.3×10^{4}
Horsepower	Foot-pounds per sec	1.818×10^{-3}
Inches of mercury @ 0°C	Pounds per square inch	2.036
Joules	BTU	1054.8
Joules	Foot-pounds	1.35582
Kilowatts	BTU per min	1.758×10^{-2}
Kilowatts	Foot-pounds per min	2.26×10^{-5}
Kilowatts	Horsepower	0.745712
Knots	Miles per hour	0.86897624
Miles	Feet	1.894×10^{-4}
Nautical miles	Miles	0.86897624
Radians	Degrees	1.745×10^{-2}
Square feet	Acres	**43560**
Watts	BTU per min	17.5796

Temperature Factors

$$°F = 9/5\ (°C) + 32$$

$$\text{Fahrenheit temperature} = 1.8\ (\text{temperature in kelvins}) - 459.67$$

$$°C = 5/9\ [(°F) - 32)]$$

$$\text{Celsius temperature} = \text{temperature in kelvins} - 273.15$$

$$\text{Fahrenheit temperature} = 1.8\ (\text{Celsius temperature}) + 32$$

Conversion of Temperatures

From	To	
°Celsius	°Fahrenheit	$t_F = (t_C \times 1.8) + 32$
	Kelvin	$T_K = t_C + 273.15$
	°Rankine	$T_R = (t_C + 273.15) \times 18$
°Fahrenheit	°Celsius	$t_C = \dfrac{t_F - 32}{1.8}$
	Kelvin	$T_k = \dfrac{t_F - 32}{1.8} + 273.15$
	°Rankine	$T_R = t_F + 459.67$
Kelvin	°Celsius	$t_C = T_K - 273.15$
	°Rankine	$T_R = T_K \times 1.8$
°Rankine	°Fahrenheit	$t_F = T_R - 459.67$
	Kelvin	$T_K = \dfrac{T_R}{1.8}$

* Boldface numbers are exact; others are given to ten significant figures where so indicated by the multiplier factor.

Physical Constants

General

Equatorial radius of the earth = 6378.388 km = 3963.34 miles (statute).
Polar radius of the earth, 6356.912 km = 3949.99 miles (statute).
1 degree of latitude at 40° = 69 miles.
1 international nautical mile = 1.15078 miles (statute) = 1852 m = 6076.115 ft.
Mean density of the earth = 5.522 g/cm^3 = 344.7 lb/ft^3
Constant of gravitation $(6.673 \pm 0.003) \times 10^{-8}$ cm^3 gm^{-1} s^{-2}.
Acceleration due to gravity at sea level, latitude 45° = 980.6194 cm/s^2 = 32.1726 ft/s^2.
Length of seconds pendulum at sea level, latitude 45° = 99.3575 cm = 39.1171 in.
1 knot (international) = 101.269 ft/min = 1.6878 ft/s = 1.1508 miles (statute)/h.
1 micron = 10^{-4} cm.
1 ångstrom = 10^{-8} cm.
Mass of hydrogen atom = $(1.67339 \pm 0.0031) \times 10^{-24}$ g.
Density of mercury at 0°C = 13.5955 g/ml.
Density of water at 3.98°C = 1.000000 g/ml.
Density, maximum, of water, at 3.98°C = 0.999973 g/cm^3.
Density of dry air at 0°C, 760 mm = 1.2929 g/l.
Velocity of sound in dry air at 0°C = 331.36 m/s – 1087.1 ft/s.
Velocity of light in vacuum = $(2.997925 \pm 0.000002) \times 10^{10}$ cm/s.
Heat of fusion of water 0°C = 79.71 cal/g.
Heat of vaporization of water 100°C = 539.55 cal/g.
Electrochemical equivalent of silver 0.001118 g/s international amp.
Absolute wavelength of red cadmium light in air at 15°C, 760 mm pressure = 6438.4696 Å.
Wavelength of orange-red line of krypton 86 = 6057.802 Å.

π Constants

π = 3.14159 26535 89793 23846 26433 83279 50288 41971 69399 37511
$1/\pi$ = 0.31830 98861 83790 67153 77675 26745 02872 40689 19291 48091
π^2 = 9.8690 44010 89358 61883 44909 99876 15113 53136 99407 24079
$\log_e \pi$ = 1.14472 98858 49400 17414 34273 51353 05871 16472 94812 91531
$\log_{10} \pi$ = 0.49714 98726 94133 85435 12682 88290 89887 36516 78324 38044
$\log_{10} \sqrt{2\pi}$ = 0.39908 99341 79057 52478 25035 91507 69595 02099 34102 92128

Constants Involving *e*

e = 2.71828 18284 59045 23536 02874 71352 66249 77572 47093 69996
$1/e$ = 0.36787 94411 71442 32159 55237 70161 46086 74458 11131 03177
e^2 = 7.38905 60989 30650 22723 04274 60575 00781 31803 15570 55185
$M = \log_{10} e$ = 0.43429 44819 03251 82765 11289 18916 60508 22943 97005 80367
$1/M = \log_e 10$ = 2.30258 50929 94045 68401 79914 54684 36420 76011 01488 62877
$\log_{10} M$ = 9.63778 43113 00536 78912 29674 98645 –10

Numerical Constants

$\sqrt{2}$ = 1.41421 35623 73095 04880 16887 24209 69807 85696 71875 37695
$\sqrt[3]{2}$ = 1.25992 10498 94873 16476 72106 07278 22835 05702 51464 70151
$\log_e 2$ = 0.69314 71805 59945 30941 72321 21458 17656 80755 00134 36026
$\log_{10} 2$ = 0.30102 99956 63981 19521 37388 94724 49302 67881 89881 46211
$\sqrt{3}$ = 1.73205 08075 68877 29352 74463 41505 87236 69428 05253 81039
$\sqrt[3]{3}$ = 1.44224 95703 07408 38232 16383 10780 10958 83918 69253 49935
$\log_e 3$ = 1.09861 22886 68109 69139 52452 36922 52570 46474 90557 82275
$\log_{10} 3$ = 0.47712 12547 19662 43729 50279 03255 11530 92001 28864 19070

Symbols and Terminology for Physical and Chemical Quantities

Name	*Symbol*	*Definition*	*SI unit*
	Classical Mechanics		
mass	m		kg
reduced mass	μ	$\mu = m_1m_2/(m_1 + m_2)$	kg
density, mass density	ρ	$\rho = m/V$	kg m^{-3}
relative density	d	$d = \rho/\rho^{\theta}$	1
surface density	ρ_A, ρ_S	$\rho_A = m/A$	kg m^{-2}
specific volume	v	$v = V/m = 1/\rho$	m^3 kg^{-1}
momentum	$\boldsymbol{p}$	$\boldsymbol{p} = m\boldsymbol{v}$	kg m s^{-1}
angular momentum, action	$\boldsymbol{L}$	$\boldsymbol{L} = \boldsymbol{r} \times \boldsymbol{p}$	J s
moment of inertia	I, J	$I = \Sigma m_i r_i^2$	kg m^2
force	$\boldsymbol{F}$	$\boldsymbol{F} = \mathrm{d}\boldsymbol{p}/\mathrm{d}t = m\boldsymbol{a}$	N
torque, moment of a force	$\boldsymbol{T}$, $(\boldsymbol{M})$	$\boldsymbol{T} = \boldsymbol{r} \times \boldsymbol{F}$	N m
energy	E		J
potential energy	E_p, V, Φ	$E_\mathrm{p} = -\int \boldsymbol{F} \cdot \mathrm{d}\boldsymbol{s}$	J
kinetic energy	E_k, T, K	$E_\mathrm{k} = (1/2)mv^2$	J
work	W, w	$W = \int \boldsymbol{F} \cdot \mathrm{d}\boldsymbol{s}$	J
Hamilton function	H	$H(q, p) = T(q, p) + V(q)$	J
Lagrange function	L	$L(q, \dot{q}) = T(q, \dot{q}) - V(q)$	J
pressure	p, P	$p = F/A$	Pa, N m^{-2}
surface tension	γ, σ	$\gamma = \mathrm{d}W/\mathrm{d}A$	N m^{-1}, J m^{-2}
weight	G, (W, P)	$G = mg$	N
gravitational constant	G	$F = Gm_1m_2/r^2$	N m^2 kg^{-2}
normal stress	σ	$\sigma = F/A$	Pa
shear stress	τ	$\tau = F/A$	Pa
linear strain, relative elongation	ε, e	$\varepsilon = \Delta l/l$	1
modulus of elasticity, Young's modulus	E	$E = \sigma/\varepsilon$	Pa
shear strain	γ	$\gamma = \Delta x/d$	1
shear modulus	G	$\mathrm{G} = \tau/\gamma$	Pa
volume strain, bulk strain	θ	$\theta = \Delta V/V_0$	1
bulk modulus, compression modulus	K	$K = -V_0(\mathrm{d}p/\mathrm{d}V)$	Pa
viscosity, dynamic viscosity	η, μ	$\tau_{x,z} = \eta(\mathrm{d}v_x/\mathrm{d}z)$	Pa s
fluidity	ϕ	$\phi = 1/\eta$	m kg^{-1} s
kinematic viscosity	ν	$\nu = \eta/\rho$	m^2 s^{-1}
friction coefficient	μ, (f)	$F_\mathrm{frict} = \mu F_\mathrm{norm}$	1
power	P	$P = \mathrm{d}W/\mathrm{d}t$	W
sound energy flux	P, P_a	$P = \mathrm{d}E/\mathrm{d}t$	W
acoustic factors			
reflection factor	ρ	$\rho = P_\mathrm{r}/P_0$	1
acoustic absorption factor	α_a, (α)	$\alpha_\mathrm{a} = 1 - \rho$	1
transmission factor	τ	$\tau = P_\mathrm{tr}/P_0$	1
dissipation factor	δ	$\delta = \alpha_\mathrm{a} - \tau$	1

Elementary Algebra and Geometry

Fundamental Properties (Real Numbers)

$a+b=b+a$	Commutative Law for Addition
$(a+b)+c=a+(b+c)$	Associative Law for Addition
$a+0=0+a$	Identity Law for Addition
$a+(-a)=(-a)+a=0$	Inverse Law for Addition
$a(bc)=(ab)c$	Associative Law for Multiplication
$a\left(\frac{1}{a}\right)=\left(\frac{1}{a}\right)a=1,\ a\neq 0$	Inverse Law for Multiplication
$(a)(1)=(1)(a)=a$	Identity Law for Multiplication
$ab=ba$	Commutative Law for Multiplication
$a(b+c)=ab+ac$	Distributive Law

DIVISION BY ZERO IS NOT DEFINED

Exponents

For integers m and n

$$a^n a^m = a^{n+m}$$

$$a^n/a^m = a^{n-m}$$

$$(a^n)^m = a^{nm}$$

$$(ab)^m = a^m b^m$$

$$(a/b)^m = a^m/b^m$$

Fractional Exponents

$$a^{p/q} = (a^{1/q})^p$$

where $a^{1/q}$ is the positive qth root of a if $a>0$ and the negative qth root of a if a is negative and q is odd. Accordingly, the five rules of exponents given above (for integers) are also valid if m and n are fractions, provided a and b are positive.

Irrational Exponents

If an exponent is irrational, e.g., $\sqrt{2}$, the quantity, such as $a^{\sqrt{2}}$ is the limit of the sequence, $a^{1.4}, a^{1.41}, a^{1.414}, \ldots$.

Operations with Zero

$$0^m = 0;\ a^0 = 1$$

Logarithms

If x, y, and b are positive and $b \neq 1$

$$\log_b(xy) = \log_b x + \log_b y$$

$$\log_b(x/y) = \log_b x - \log_b y$$

$$\log_b x^p = p \log_b x$$

$$\log_b(1/x) = -\log_b x$$

$$\log_b b = 1$$

$$\log_b 1 = 0 \qquad \textit{Note}: b^{\log_b x} = x.$$

Change of Base ($a \neq 1$)

$$\log_b x = \log_a x \log_b a$$

Factorials

The factorial of a positive integer n is the product of all the positive integers less than or equal to the integer n and is denoted $n!$. Thus,

$$n! = 1 \cdot 2 \cdot 3 \cdot \ldots \cdot n.$$

Factorial 0 is defined: $0! = 1$.

Stirling's Approximation

$$\lim_{n \to \infty} (n/e)^n \sqrt{2\pi n} = n!$$

Binomial Theorem

For positive integer n

$$(x+y)^n = x^n + nx^{n-1}y + \frac{n(n-1)}{2!}x^{n-2}y^2 + \frac{n(n-1)(n-2)}{3!}x^{n-3}y^3 + \cdots + nxy^{n-1} + y^n.$$

Factors and Expansion

$$(a+b)^2 = a^2 + 2ab + b^2$$

$$(a-b)^2 = a^2 - 2ab + b^2$$

$$(a+b)^3 = a^3 + 3a^2b + 3ab^2 + b^3$$

$$(a-b)^3 = a^3 - 3a^2b + 3ab^2 - b^3$$

$$(a^2 - b^2) = (a-b)(a+b)$$

$$(a^3 - b^3) = (a-b)(a^2 + ab + b^2)$$

$$(a^3 + b^3) = (a+b)(a^2 - ab + b^2)$$

Progression

An *arithmetic progression* is a sequence in which the difference between any term and the preceding term is a constant (d):

$$a, a+d, a+2d, \ldots, a+(n-1)d.$$

If the last term is denoted $l\ [=a+(n-1)d]$, then the sum is

$$s = \frac{n}{2}(a+l).$$

A *geometric progression* is a sequence in which the ratio of any term to the preceding term is a constant r. Thus, for n terms

$$a, ar, ar^2, \ldots, ar^{n-1}$$

The sum is

$$S = \frac{a - ar^n}{1-r}$$

Complex Numbers

A complex number is an ordered pair of real numbers (a, b).

Equality: $(a,b) = (c,d)$ if and only if $a = c$ and $b = d$

Addition: $(a,b) + (c,d) = (a+c, b+d)$

Multiplication: $(a,b)(c,d) = (ac - bd, ad + bc)$

The first element (a, b) is called the *real* part; the second the *imaginary* part. An alternate notation for (a, b) is $a + bi$, where $i^2 = (-1, 0)$, and $i = (0, 1)$ or $0 + 1i$ is written for this complex number as a convenience. With this understanding, i behaves as a number, i.e., $(2-3i)(4+i) = 8 - 12i + 2i - 3i^2 = 11 - 10i$. The conjugate of $a + bi$ is $a - bi$ and the product of a complex number and its conjugate is $a^2 + b^2$. Thus, *quotients* are computed by multiplying numerator and denominator by the conjugate of the denominator, as illustrated below:

$$\frac{2+3i}{4+2i} = \frac{(4-2i)(2+3i)}{(4-2i)(4+2i)} = \frac{14+8i}{20} = \frac{7+4i}{10}$$

Polar Form

The complex number $x + iy$ may be represented by a plane vector with components x and y

$$x + iy = r(\cos\theta + i\sin\theta)$$

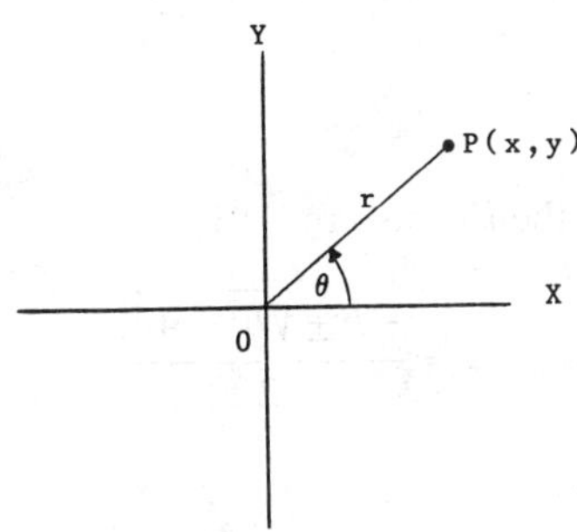

FIGURE 1 Polar form of complex number.

(see Figure 1). Then, given two complex numbers $z_1 = r_1(\cos\theta_1 + i\sin\theta_1)$ and $z_2 = r_2(\cos\theta_2 + i\sin\theta_2)$, the product and quotient are

product: $$z_1 z_2 = r_1 r_2[\cos(\theta_1 + \theta_2) + i\sin(\theta_1 + \theta_2)]$$

quotient: $$z_1/z_2 = (r_1/r_2)[\cos(\theta_1 - \theta_2) + i\sin(\theta_1 - \theta_2)]$$

powers: $$z^n = [r(\cos\theta + i\sin\theta)]^n = r^n[\cos n\theta + i\sin n\theta]$$

roots: $$z^{1/n} = [r(\cos\theta + i\sin\theta)]^{1/n}$$

$$= r^{1/n}\left[\cos\frac{\theta + k.360}{n} + i\sin\frac{\theta + k.360}{n}\right], \qquad k = 0, 1, 2, \ldots, n-1$$

Permutations

A permutation is an ordered arrangement (sequence) of all or part of a set of objects. The number of permutations of n objects taken r at a time is

$$p(n,r) = n(n-1)(n-2)\ldots(n-r+1)$$

$$= \frac{n!}{(n-r)!}$$

A permutation of positive integers is "even" or "odd" if the total number of inversions is an even integer or an odd integer, respectively. Inversions are counted relative to each integer j in the permutation by counting the number of integers that follow j and are less than j. These are summed to give the total number of inversions. For example, the permutation 4132 has four inversions: three relative to 4 and one relative to 3. This permutation is therefore even.

Combinations

A combination is a selection of one or more objects from among a set of objects regardless of order. The number of combinations of n different objects taken r at a time is

$$C(n,r) = \frac{P(n,r)}{r!} = \frac{n!}{r!(n-r)!}$$

Algebraic Equations

Quadratic

If $ax^2+bx+c=0$, and $a \neq 0$, then roots are

$$x=\frac{-b \pm \sqrt{b^2-4ac}}{2a}$$

Cubic

To solve $x^3+bx^2+cx+d=0$, let $x=y-b/3$. Then the *reduced cubic* is obtained:

$$y^3+py+q=0$$

where $p=c-(1/3)b^2$ and $q=d-(1/3)bc+(2/27)b^3$. Solutions of the original cubic are then in terms of the reduced cubic roots y_1, y_2, y_3:

$$x_1=y_1-(1/3)b \qquad x_2=y_2-(1/3)b \qquad x_3=y_3-(1/3)b$$

The three roots of the reduced cubic are

$$y_1=(A)^{1/3}+(B)^{1/3}$$

$$y_2=W(A)^{1/3}+W^2(B)^{1/3}$$

$$y_3=W^2(A)^{1/3}+W(B)^{1/3}$$

where

$$A=-\frac{1}{2}q+\sqrt{(1/27)p^3+\frac{1}{4}q^2},$$

$$B=-\frac{1}{2}q-\sqrt{(1/27)p^3+\frac{1}{4}q^2},$$

$$W=\frac{-1+i\sqrt{3}}{2}, \quad W^2=\frac{-1-i\sqrt{3}}{2}.$$

When $(1/27)p^3+(1/4)q^2$ is negative, A is complex; in this case A should be expressed in trigonometric form: $A=r(\cos\theta+i\sin\theta)$ where θ is a first or second quadrant angle, as q is negative or positive. The three roots of the reduced cubic are

$$y_1=2(r)^{1/3}\cos(\theta/3)$$

$$y_2=2(r)^{1/3}\cos\left(\frac{\theta}{3}+120°\right)$$

$$y_3=2(r)^{1/3}\cos\left(\frac{\theta}{3}+240°\right)$$

Geometry

Figures 2 to 12 are a collection of common geometric figures. Area (A), volume (V), and other measurable features are indicated.

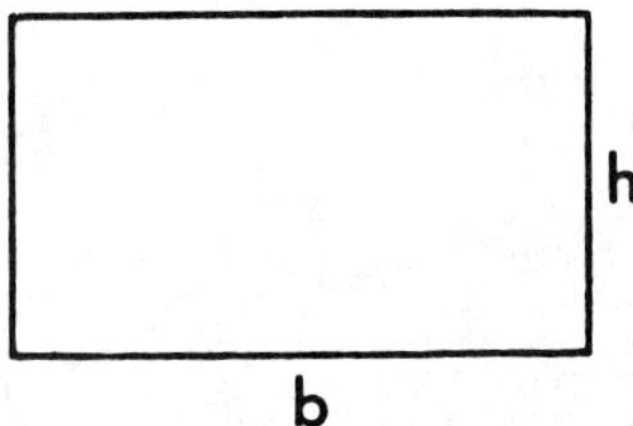

FIGURE 2 Rectangle. $A = bh$.

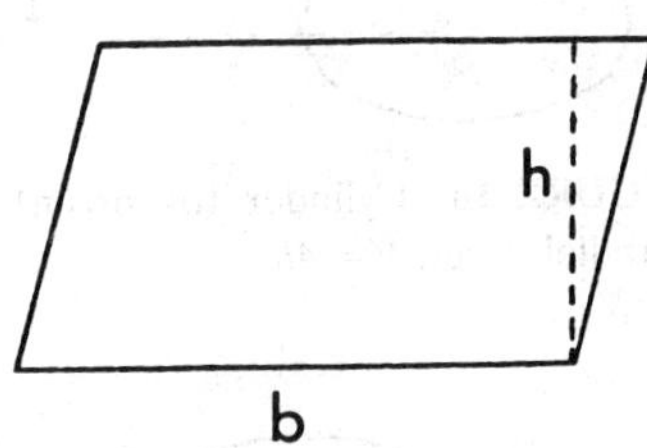

FIGURE 3 Parallelogram. $A = bh$.

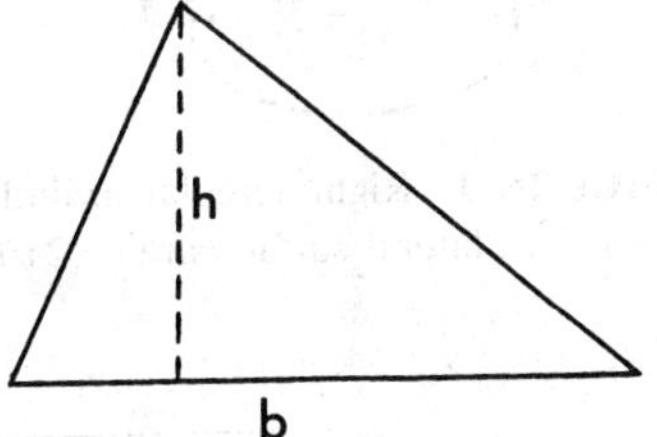

FIGURE 4 Triangle. $A = \frac{1}{2}bh$.

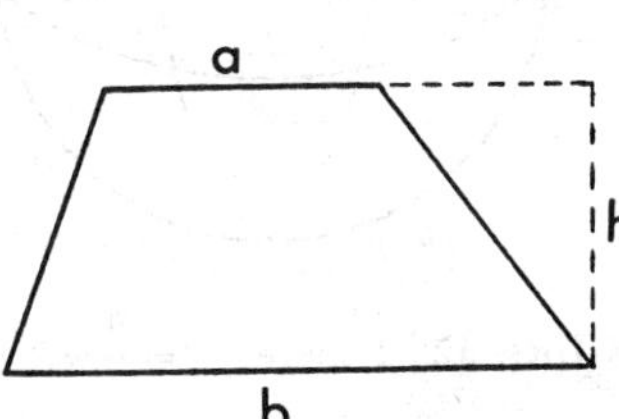

FIGURE 5 Trapezoid. $A = \frac{1}{2}(a + b)h$.

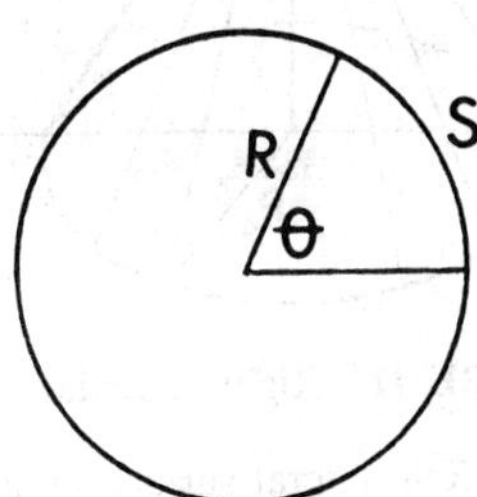

FIGURE 6 Circle. $A = \pi R^2$; circumference $= 2\pi R$; arc length $S = R\theta$ (θ in radians).

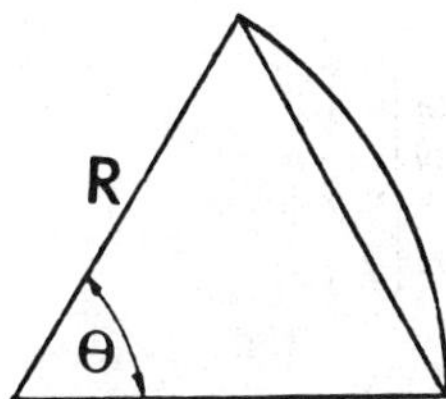

FIGURE 7 Sector of circle. $A_{\text{sector}} = \frac{1}{2}R^2\theta$; $A_{\text{segment}} = \frac{1}{2}R^2(\theta - \sin\theta)$.

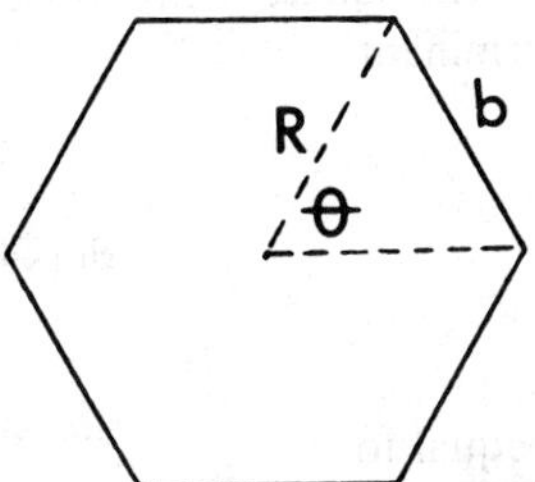

FIGURE 8 Regular polygon of n sides. $A = \frac{n}{4}b^2 \operatorname{ctn}\frac{\pi}{n}$; $R = \frac{b}{2}\csc\frac{\pi}{n}$.

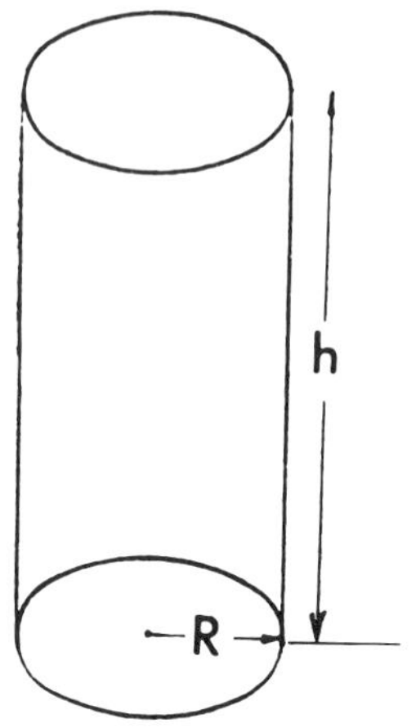

FIGURE 9 Right circular cylinder. $V = \pi R^2 h$; lateral surface area $= 2\pi Rh$.

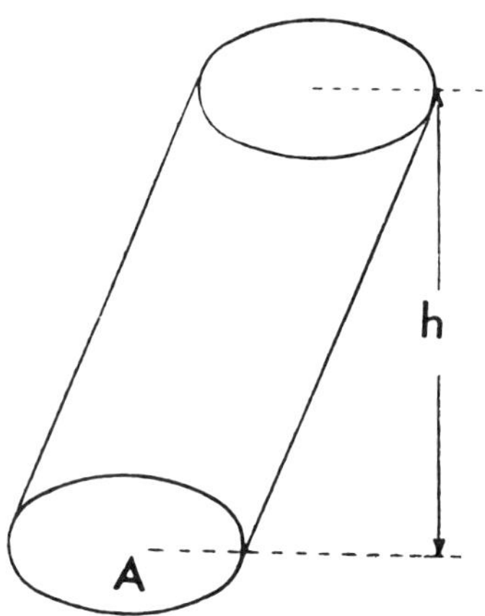

FIGURE 10 Cylinder (or prism) with parallel bases. $V = Ah$.

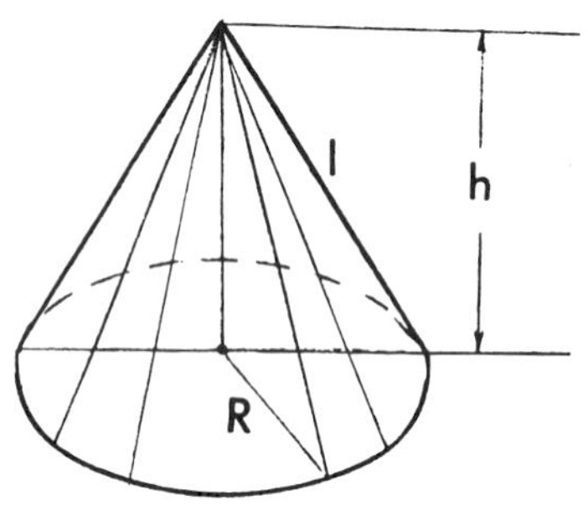

FIGURE 11 Right circular cone. $V = \frac{1}{3}\pi R^2 h$; lateral surface area $= \pi Rl = \pi R\sqrt{R^2 + h^2}$.

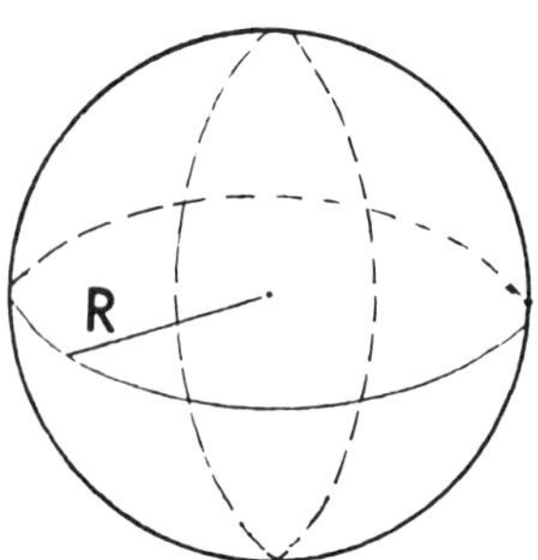

FIGURE 12 Sphere. $V = \frac{4}{3}\pi R^3$; surface area $= 4\pi R^2$.

Determinants, Matrices, and Linear Systems of Equations

Determinants

Definition. The square array (matrix) A, with n rows and n columns, has associated with it the determinant

$$\det A = \begin{vmatrix} a_{11} & a_{12} & \cdots & a_{1n} \\ a_{21} & a_{22} & \cdots & a_{2n} \\ \cdots & \cdots & \cdots & \cdots \\ a_{n1} & a_{n2} & \cdots & a_{nn} \end{vmatrix},$$

a number equal to

$$\sum (\pm) a_{1i} a_{2j} a_{3k} \ldots a_{nl}$$

where $i, j, k, \ldots, l$ is a permutation of the n integers $1, 2, 3, \ldots, n$ in some order. The sign is plus if the permutation is *even* and is minus if the permutation is *odd*. The 2×2 determinant

$$\begin{vmatrix} a_{11} & a_{12} \\ a_{21} & a_{22} \end{vmatrix}$$

has the value $a_{11}a_{22}-a_{12}a_{21}$ since the permutation (1,2) is even and (2,1) is odd. For 3×3 determinants, permutations are as follows:

1,	2,	3	even
1,	3,	2	odd
2,	1,	3	odd
2,	3,	1	even
3,	1,	2	even
3,	2,	1	odd

Thus,

$$\begin{vmatrix} a_{11} & a_{12} & a_{13} \\ a_{21} & a_{22} & a_{23} \\ a_{31} & a_{32} & a_{33} \end{vmatrix} = \begin{cases} +a_{11} \cdot a_{22} \cdot a_{33} \\ -a_{11} \cdot a_{23} \cdot a_{32} \\ -a_{12} \cdot a_{21} \cdot a_{33} \\ +a_{12} \cdot a_{23} \cdot a_{31} \\ +a_{13} \cdot a_{21} \cdot a_{32} \\ -a_{13} \cdot a_{22} \cdot a_{31} \end{cases}$$

A determinant of order n is seen to be the sum of $n!$ signed products.

Evaluation by Cofactors

Each element a_{ij} has a determinant of order $(n-1)$ called a *minor* (M_{ij}) obtained by suppressing all elements in row i and column j. For example, the minor of element a_{22} in the 3×3 determinant above is

$$\begin{vmatrix} a_{11} & a_{13} \\ a_{31} & a_{33} \end{vmatrix}$$

The cofactor of element a_{ij}, denoted A_{ij}, is defined as $\pm M_{ij}$, where the sign is determined from i and j:

$$A_{ij}=(-1)^{i+j}M_{ij}.$$

The value of the $n\times n$ determinant equals the sum of products of elements of any row (or column) and their respective cofactors. Thus, for the 3×3 determinant

$$\det A=a_{11}A_{11}+a_{12}A_{12}+a_{13}A_{13} \text{ (first row)}$$

or

$$= a_{11}A_{11} + a_{21}A_{21} + a_{31}A_{31} \text{ (first column)}$$

etc.

Properties of Determinants

a. If the corresponding columns and rows of A are interchanged, det A is unchanged.

b. If any two rows (or columns) are interchanged, the sign of det A changes.

c. If any two rows (or columns) are identical, det $A = 0$.

d. If A is triangular (all elements above the main diagonal equal to zero), $A = a_{11} \cdot a_{22} \cdot \ldots \cdot a_{nn}$:

$$\begin{vmatrix} a_{11} & 0 & 0 & \cdots & 0 \\ a_{21} & a_{22} & 0 & \cdots & 0 \\ \cdots & \cdots & \cdots & \cdots & \cdots \\ a_{n1} & a_{n2} & a_{n3} & \cdots & a_{nn} \end{vmatrix}$$

e. If to each element of a row or column there is added C times the corresponding element in another row (or column), the value of the determinant is unchanged.

Matrices

Definition. A matrix is a rectangular array of numbers and is represented by a symbol A or $[a_{ij}]$:

$$A = \begin{bmatrix} a_{11} & a_{12} & \cdots & a_{1n} \\ a_{21} & a_{22} & \cdots & a_{2n} \\ \cdots & \cdots & \cdots & \cdots \\ a_{m1} & a_{m2} & \cdots & a_{mn} \end{bmatrix} = [a_{ij}]$$

The numbers a_{ij} are termed *elements* of the matrix; subscripts i and j identify the element as the number in row i and column j. The order of the matrix is $m \times n$ ("m by n"). When $m = n$, the matrix is square and is said to be of order n. For a square matrix of order n the elements $a_{11}, a_{22}, \ldots, a_{nn}$ constitute the main diagonal.

Operations

Addition. Matrices A and B of the same order may be added by adding corresponding elements, i.e., $A + B = [(a_{ij} + b_{ij})]$.

Scalar multiplication. If $A = [a_{ij}]$ and c is a constant (scalar), then $cA = [ca_{ij}]$, that is, every element of A is multiplied by c. In particular, $(-1)A = -A = [-a_{ij}]$ and $A + (-A) = 0$, a matrix with all elements equal to zero.

Multiplication of matrices. Matrices A and B may be multiplied only when they are conformable, which means that the number of columns of A equals the number of rows of B. Thus, if A is $m \times k$ and B is $k \times n$, then the product $C = AB$ exists as an

$m \times n$ matrix with elements c_{ij} equal to the sum of products of elements in row i of A and corresponding elements of column j of B:

$$c_{ij} = \sum_{l=1}^{k} a_{il} b_{lj}$$

For example, if

$$\begin{bmatrix} a_{11} & a_{12} & \cdots & a_{1k} \\ a_{21} & a_{22} & \cdots & a_{2k} \\ \cdots & \cdots & \cdots & \cdots \\ a_{m1} & \cdots & \cdots & a_{mk} \end{bmatrix} \cdot \begin{bmatrix} b_{11} & b_{12} & \cdots & b_{1n} \\ b_{21} & b_{22} & \cdots & b_{2n} \\ \cdots & \cdots & \cdots & \cdots \\ b_{k1} & b_{k2} & \cdots & b_{kn} \end{bmatrix} = \begin{bmatrix} c_{11} & c_{12} & \cdots & c_{1n} \\ c_{21} & c_{22} & \cdots & c_{2n} \\ \cdots & \cdots & \cdots & \\ c_{m1} & c_{m2} & \cdots & c_{mn} \end{bmatrix}$$

then element c_{21} is the sum of products $a_{21}b_{11} + a_{22}b_{21} + \ldots + a_{2k}b_{k1}$.

Properties

$$A + B = B + A$$

$$A + (B + C) = (A + B) + C$$

$$(c_1 + c_2)A = c_1 A + c_2 A$$

$$c(A + B) = cA + cB$$

$$c_1(c_2 A) = (c_1 c_2)A$$

$$(AB)(C) = A(BC)$$

$$(A + B)(C) = AC + BC$$

$$AB \neq BA \text{ (in general)}$$

Transpose

If A is an $n \times m$ matrix, the matrix of order $m \times n$ obtained by interchanging the rows and columns of A is called the *transpose* and is denoted A^T. The following are properties of A, B, and their respective transposes:

$$(A^T)^T = A$$

$$(A + B)^T = A^T + B^T$$

$$(cA)^T = cA^T$$

$$(AB)^T = B^T A^T$$

A *symmetric* matrix is a square matrix A with the property $A = A^T$.

Identity Matrix

A square matrix in which each element of the main diagonal is the same constant a and all other elements zero is called a *scalar* matrix.

$$\begin{bmatrix} a & 0 & 0 & \cdots & 0 \\ 0 & a & 0 & \cdots & 0 \\ 0 & 0 & a & \cdots & 0 \\ \cdots & \cdots & \cdots & \cdots & \\ 0 & 0 & 0 & \cdots & a \end{bmatrix}$$

When a scalar matrix multiplies a conformable second matrix A, the product is aA; that is, the same as multiplying A by a scalar a. A scalar matrix with diagonal elements 1 is called the *identity*, or *unit* matrix and is denoted I. Thus, for any nth order matrix A, the identity matrix of order n has the property

$$AI = IA = A$$

Adjoint

If A is an n-order square matrix and A_{ij} the cofactor of element a_{ij}, the transpose of $[A_{ij}]$ is called the *adjoint* of A:

$$adj A = [A_{ij}]^T$$

Inverse Matrix

Given a square matrix A of order n, if there exists a matrix B such that $AB = BA = I$, then B is called the *inverse* of A. The inverse is denoted A^{-1}. A necessary and sufficient condition that the square matrix A have an inverse is $\det A \neq 0$. Such a matrix is called *nonsingular*; its inverse is unique and it is given by

$$A^{-1} = \frac{adj A}{\det A}$$

Thus, to form the inverse of the nonsingular matrix A, form the adjoint of A and divide each element of the adjoint by $\det A$. For example,

$$\begin{bmatrix} 1 & 0 & 2 \\ 3 & -1 & 1 \\ 4 & 5 & 6 \end{bmatrix} \text{ has matrix of cofactors } \begin{bmatrix} -11 & -14 & 19 \\ 10 & -2 & -5 \\ 2 & 5 & -1 \end{bmatrix},$$

$$\text{adjoint} = \begin{bmatrix} -11 & 10 & 2 \\ -14 & -2 & 5 \\ 19 & -5 & -1 \end{bmatrix} \text{ and determinant 27.}$$

Therefore,

$$A^{-1} = \begin{bmatrix} \frac{-11}{27} & \frac{10}{27} & \frac{2}{27} \\ \frac{-14}{27} & \frac{-2}{27} & \frac{5}{27} \\ \frac{19}{27} & \frac{-5}{27} & \frac{-1}{27} \end{bmatrix}.$$

Systems of Linear Equations

Given the system

$$\begin{array}{ccccccccc} a_{11}x_1 & + & a_{12}x_2 & +\cdots+ & a_{1n}x_n & = & b_1 \\ a_{21}x_1 & + & a_{22}x_2 & +\cdots+ & a_{2n}x_n & = & b_2 \\ \vdots & & \vdots & \vdots & \vdots & & \vdots \\ a_{n1}x_1 & + & a_{n2}x_2 & +\cdots+ & a_{nn}x_n & = & b_n \end{array}$$

a unique solution exists if det $A \neq 0$, where A is the $n \times n$ matrix of coefficients $[a_{ij}]$.

Solution by Determinants (Cramer's Rule)

$$x_1 = \begin{vmatrix} b_1 & a_{12} & \cdots & a_{1n} \\ b_2 & a_{22} & & \\ \vdots & \vdots & & \vdots \\ b_n & a_{n2} & & a_{nn} \end{vmatrix} \div \det A$$

$$x_2 = \begin{vmatrix} a_{11} & b_1 & a_{13} & \cdots & a_{1n} \\ a_{21} & b_2 & \cdots & & \cdots \\ \vdots & \vdots & & & \\ a_{n1} & b_n & a_{n3} & & a_{nn} \end{vmatrix} \div \det A$$

$$\vdots$$

$$x_k = \frac{\det A_k}{\det A},$$

where A_k is the matrix obtained from A by replacing the kth column of A by the column of b's.

Matrix Solution

The linear system may be written in matrix form $AX = B$ where A is the matrix of coefficients $[a_{ij}]$ and X and B are

$$X = \begin{bmatrix} x_1 \\ x_2 \\ \vdots \\ x_n \end{bmatrix} \qquad B = \begin{bmatrix} b_1 \\ b_2 \\ \vdots \\ b_n \end{bmatrix}$$

If a unique solution exists, det $A \neq 0$; hence A^{-1} exists and

$$X = A^{-1}B.$$

Trigonometry

Triangles

In any triangle (in a plane) with sides a, b, and c and corresponding opposite angles A, B, C,

$$\frac{a}{\sin A}=\frac{b}{\sin B}=\frac{c}{\sin C}. \qquad \text{(Law of Sines)}$$

$$a^2=b^2+c^2-2cb\cos A. \qquad \text{(Law of Cosines)}$$

$$\frac{a+b}{a-b}=\frac{\tan\frac{1}{2}(A+B)}{\tan\frac{1}{2}(A-B)}. \qquad \text{(Law of Tangents)}$$

$$\sin\frac{1}{2}A=\sqrt{\frac{(s-b)(s-c)}{bc}}, \qquad \text{where } s=\frac{1}{2}(a+b+c).$$

$$\cos\frac{1}{2}A=\sqrt{\frac{s(s-a)}{bc}}.$$

$$\tan\frac{1}{2}A=\sqrt{\frac{(s-b)(s-c)}{s(s-a)}}.$$

$$\text{Area}=\frac{1}{2}bc\sin A$$
$$=\sqrt{s(s-a)(s-b)(s-c)}.$$

If the vertices have coordinates $(x_1,y_1),(x_2,y_2),(x_3,y_3)$, the area is the *absolute value* of the expression

$$\frac{1}{2}\begin{vmatrix} x_1 & y_1 & 1 \\ x_2 & y_2 & 1 \\ x_3 & y_3 & 1 \end{vmatrix}$$

Trigonometric Functions of an Angle

With reference to Figure 13, $P(x,y)$ is a point in either one of the four quadrants and A is an angle whose initial side is coincident with the positive x-axis and whose terminal side contains the point $P(x,y)$. The distance from the origin $P(x,y)$ is denoted by r and is positive. The trigonometric functions of the angle A are defined as:

$$\begin{aligned}
\sin A &= \text{sine } A &&= y/r \\
\cos A &= \text{cosine } A &&= x/r \\
\tan A &= \text{tangent } A &&= y/x \\
\text{ctn } A &= \text{cotangent } A &&= x/y \\
\sec A &= \text{secant } A &&= r/x \\
\csc A &= \text{cosecant } A &&= r/y
\end{aligned}$$

z-Transform and the Laplace Transform

When $F(t)$, a continuous function of time, is sampled at regular intervals of period T the usual Laplace transform techniques are modified. The diagramatic form of a simple sampler together with its associated input-output waveforms is shown in Figure 35.

Defining the set of impulse functions $\delta_T(t)$ by

$$\delta_T(t) \equiv \sum_{n=0}^{\infty} \delta(t-nT)$$

the input-output relationship of the sampler becomes

$$F^*(t) = F(t)\cdot\delta_T(t)$$

$$= \sum_{n=0}^{\infty} F(nT)\cdot\delta(t-nT).$$

While for a given $F(t)$ and T the $F^*(t)$ is unique, the converse is not true.

For function $U(t)$ the output of the ideal sampler $U^*(t)$ is a set of values $U(kT)$, $k=0,1,2,\ldots$, that is,

$$U^*(t) = \sum_{k=0}^{\infty} U(t)\,\delta(t-kT)$$

The Laplace transform of the output is

$$\mathscr{L}\{U^*(t)\} = \int_0^{\infty} e^{-st}U^*(t)\,dt = \int_0^{\infty} e^{-st}\sum_{k=0}^{\infty} U(t)\delta(t-kT)\,dt$$

$$= \sum_{k=0}^{\infty} e^{-skT}\,U(kT)$$

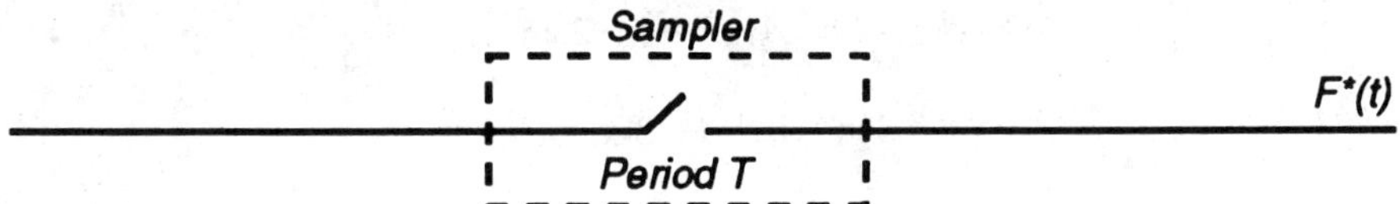

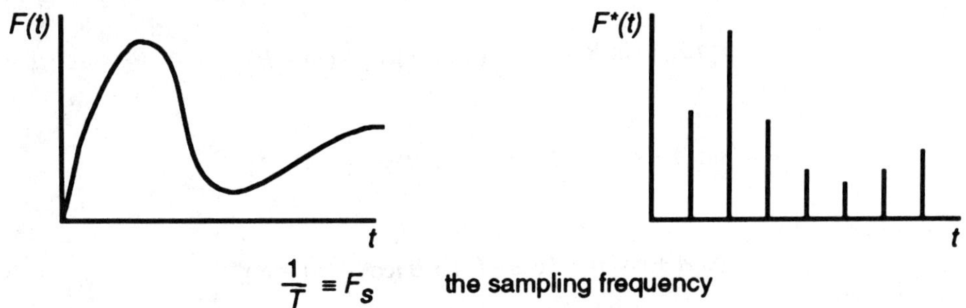

$\frac{1}{T} \equiv F_S$ the sampling frequency

FIGURE 35

$$\tan A = \frac{1}{\operatorname{ctn} A} = \frac{\sin A}{\cos A}$$

$$\csc A = \frac{1}{\sin A}$$

$$\sec A = \frac{1}{\cos A}$$

$$\operatorname{ctn} A = \frac{1}{\tan A} = \frac{\cos A}{\sin A}$$

$$\sin^2 A + \cos^2 A = 1$$

$$1 + \tan^2 A = \sec^2 A$$

$$1 + \operatorname{ctn}^2 A = \csc^2 A$$

$$\sin(A \pm B) = \sin A \cos B \pm \cos A \sin B$$

$$\cos(A \pm B) = \cos A \cos B \mp \sin A \sin B$$

$$\tan(A \pm B) = \frac{\tan A \pm \tan B}{1 \mp \tan A \tan B}$$

$$\sin 2A = 2 \sin A \cos A$$

$$\sin 3A = 3 \sin A - 4 \sin^3 A$$

$$\sin nA = 2 \sin(n-1)A \cos A - \sin(n-2)A$$

$$\cos 2A = 2\cos^2 A - 1 = 1 - 2\sin^2 A$$

$$\cos 3A = 4\cos^3 A - 3\cos A$$

$$\cos nA = 2\cos(n-1)A \cos A - \cos(n-2)A$$

$$\sin A + \sin B = 2 \sin\frac{1}{2}(A+B)\cos\frac{1}{2}(A-B)$$

$$\sin A - \sin B = 2 \cos\frac{1}{2}(A+B)\sin\frac{1}{2}(A-B)$$

$$\cos A + \cos B = 2 \cos\frac{1}{2}(A+B)\cos\frac{1}{2}(A-B)$$

$$\cos A - \cos B = -2 \sin\frac{1}{2}(A+B)\sin\frac{1}{2}(A-B)$$

$$\tan A \pm \tan B = \frac{\sin(A \pm B)}{\cos A \cos B}$$

$$\operatorname{ctn} A \pm \operatorname{ctn} B = \pm \frac{\sin(A \pm B)}{\sin A \sin B}$$

$$\sin A \sin B = \frac{1}{2}\cos(A - B) - \frac{1}{2}\cos(A + B)$$

$$\cos A \cos B = \frac{1}{2}\cos(A - B) + \frac{1}{2}\cos(A + B)$$

$$\sin A \cos B = \frac{1}{2}\sin(A + B) + \frac{1}{2}\sin(A - B)$$

$$\sin\frac{A}{2} = \pm\sqrt{\frac{1 - \cos A}{2}}$$

$$\cos\frac{A}{2} = \pm\sqrt{\frac{1 + \cos A}{2}}$$

$$\tan\frac{A}{2} = \frac{1 - \cos A}{\sin A} = \frac{\sin A}{1 + \cos A} = \pm\sqrt{\frac{1 - \cos A}{1 + \cos A}}$$

$$\sin^2 A = \frac{1}{2}(1 - \cos 2A)$$

$$\cos^2 A = \frac{1}{2}(1 + \cos 2A)$$

$$\sin^3 A = \frac{1}{4}(3\sin A - \sin 3A)$$

$$\cos^3 A = \frac{1}{4}(\cos 3A + 3\cos A)$$

$$\sin ix = \frac{1}{2}i(e^x - e^{-x}) = i \sinh x$$

$$\cos ix = \frac{1}{2}(e^x + e^{-x}) = \cosh x$$

$$\tan ix = \frac{i(e^x - e^{-x})}{e^x + e^{-x}} = i \tanh x$$

$$e^{x+iy} = e^x(\cos y + i \sin y)$$

$$(\cos x \pm i \sin x)^n = \cos nx \pm i \sin nx$$

Inverse Trigonometric Functions

The inverse trigonometric functions are multiple valued, and this should be taken into account in the use of the following formulas.

$$\sin^{-1} x = \cos^{-1}\sqrt{1-x^2}$$

$$= \tan^{-1}\frac{x}{\sqrt{1-x^2}} = \mathrm{ctn}^{-1}\frac{\sqrt{1-x^2}}{x}$$

$$= \sec^{-1}\frac{1}{\sqrt{1-x^2}} = \csc^{-1}\frac{1}{x}$$

$$= -\sin^{-1}(-x)$$

$$\cos^{-1} x = \sin^{-1}\sqrt{1-x^2}$$

$$= \tan^{-1}\frac{\sqrt{1-x^2}}{x} = \mathrm{ctn}^{-1}\frac{x}{\sqrt{1-x^2}}$$

$$= \sec^{-1}\frac{1}{x} = \csc^{-1}\frac{1}{\sqrt{1-x^2}}$$

$$= \pi - \cos^{-1}(-x)$$

$$\tan^{-1} x = \mathrm{ctn}^{-1}\frac{1}{x}$$

$$= \sin^{-1}\frac{x}{\sqrt{1+x^2}} = \cos^{-1}\frac{1}{\sqrt{1+x^2}}$$

$$= \sec^{-1}\sqrt{1+x^2} = \csc^{-1}\frac{\sqrt{1+x^2}}{x}$$

$$= -\tan^{-1}(-x)$$

Analytic Geometry

Rectangular Coordinates

The points in a plane may be placed in one-to-one correspondence with pairs of real numbers. A common method is to use perpendicular lines that are horizontal and vertical and intersect at a point called the *origin*. These two lines constitute the coordinate axes; the horizontal line is the x-axis and the vertical line is the y-axis. The positive direction of the x-axis is to the right whereas the positive direction of the y-axis is up. If P is a point in the plane one may draw lines through it that are perpendicular to the x- and y-axes (such as the broken lines of Figure 14). The lines intersect the x-axis at a point with coordinate x_1 and the y-axis at a point with coordinate y_1. We call x_1 the x-coordinate or *abscissa* and y_1 is termed the y-coordinate or *ordinate* of the point P. Thus, point P is associated with the pair of real numbers (x_1, y_1) and is denoted $P(x_1, y_1)$. The coordinate axes divide the plane into quadrants I, II, III, and IV.

Distance between Two Points; Slope

The distance d between the two points $P_1(x_1, y_1)$ and $P_2(x_2, y_2)$ is

$$d=\sqrt{(x_2-x_1)^2+(y_2-y_1)^2}$$

In the special case when P_1 and P_2 are both on one of the coordinate axes, for instance, the x-axis,

$$d=\sqrt{(x_2-x_1)^2}=|x_2-x_1|,$$

or on the y-axis,

$$d=\sqrt{(y_2-y_1)^2}=|y_2-y_1|.$$

The midpoint of the line segment P_1P_2 is

$$\left(\frac{x_1+x_2}{2}, \frac{y_1+y_2}{2}\right).$$

The slope of the line segment P_1P_2, provided it is not vertical, is denoted by m and is given by

$$m=\frac{y_2-y_1}{x_2-x_1}.$$

The slope is related to the angle of inclination α (Figure 15) by

$$m=\tan\alpha$$

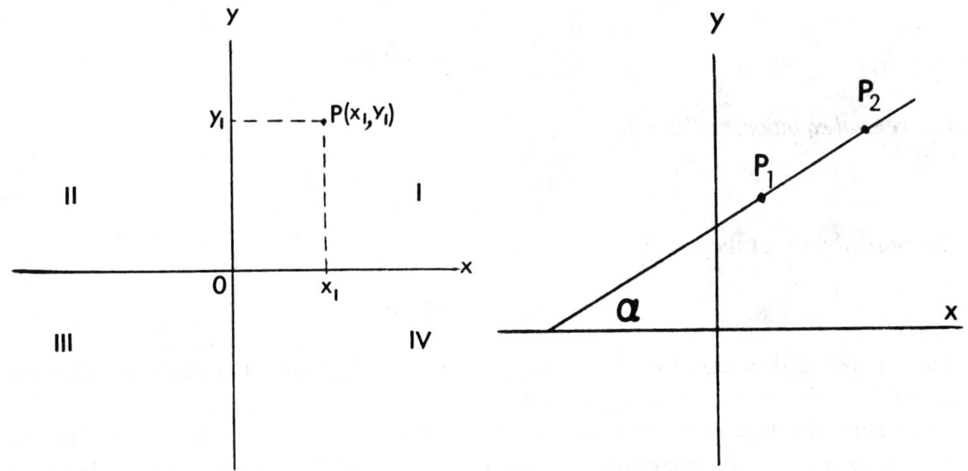

FIGURE 14 Rectangular coordinates.

FIGURE 15 The angle of inclination is the smallest angle measured counterclockwise from the positive x-axis to the line that contains P_1P_2.

Two lines (or line segments) with slopes m_1 and m_2 are perpendicular if

$$m_1 = -1/m_2$$

and are parallel if $m_1 = m_2$.

Equations of Straight Lines

A *vertical* line has an equation of the form

$$x = c$$

where $(c, 0)$ is its intersection with the x-axis. A line of slope m through point (x_1, y_1) is given by

$$y - y_1 = m(x - x_1)$$

Thus, a *horizontal line* (slope = 0) through point (x_1, y_1) is given by

$$y = y_1.$$

A nonvertical line through the two points $P_1(x_1, y_1)$ and $P_2(x_2, y_2)$ is given by either

$$y - y_1 = \left(\frac{y_2 - y_1}{x_2 - x_1}\right)(x - x_1)$$

or

$$y - y_2 = \left(\frac{y_2 - y_1}{x_2 - x_1}\right)(x - x_2).$$

A line with x-intercept a and y-intercept b is given by

$$\frac{x}{a} + \frac{y}{b} = 1 \qquad (a \neq 0, b \neq 0).$$

The *general equation* of a line is

$$Ax + By + C = 0$$

The *normal form* of the straight line equation is

$$x \cos\theta + y \sin\theta = p$$

where p is the distance along the normal from the origin and θ is the angle that the normal makes with the x-axis (Figure 16).

The general equation of the line $Ax + By + C = 0$ may be written in normal form by dividing by $\pm\sqrt{A^2 + B^2}$, where the plus sign is used when C is negative and the minus sign is used when C is positive:

$$\frac{Ax + By + C}{\pm\sqrt{A^2 + B^2}} = 0,$$

so that

$$\cos\theta = \frac{A}{\pm\sqrt{A^2+B^2}}, \qquad \sin\theta = \frac{B}{\pm\sqrt{A^2+B^2}}$$

and

$$p = \frac{|C|}{\sqrt{A^2+B^2}}.$$

Distance from a Point to a Line

The perpendicular distance from a point $P(x_1, y_1)$ to the line $Ax+By+C=0$ is given by d

$$d = \frac{Ax_1+By_1+C}{\pm\sqrt{A^2+B^2}}.$$

Circle

The general equation of a circle of radius r and center at $P(x_1, y_1)$ is

$$(x-x_1)^2+(y-y_1)^2=r^2.$$

Parabola

A parabola is the set of all points (x, y) in the plane that are equidistant from a given line called the *directrix* and a given point called the *focus*. The parabola is symmetric about a line that contains the focus and is perpendicular to the directrix. The line of symmetry intersects the parabola at its *vertex* (Figure 17). The eccentricity $e=1$.

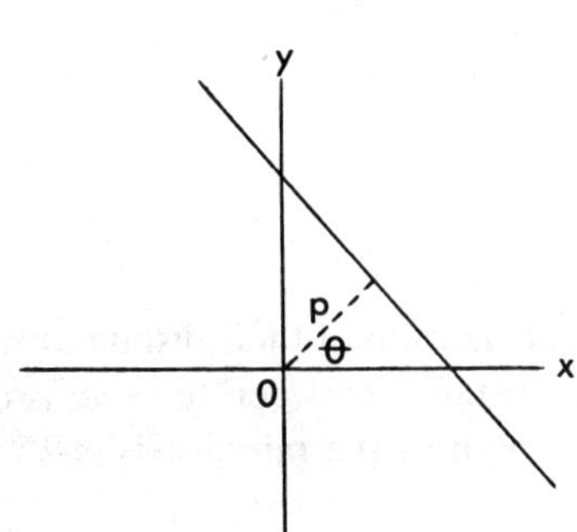

FIGURE 16 Construction for normal form of straight line equation.

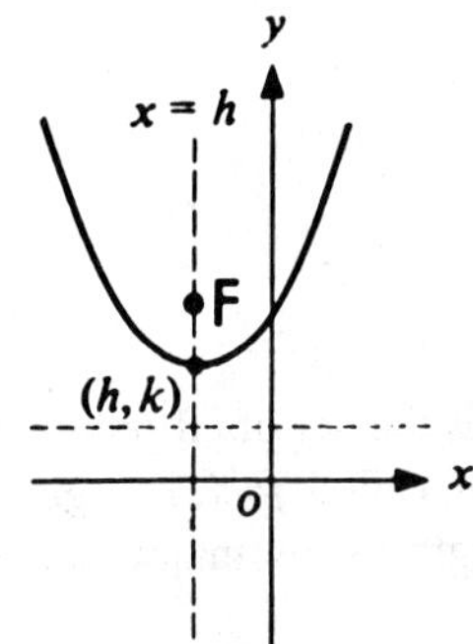

FIGURE 17 Parabola with vertex at (h,k). F identifies the focus.

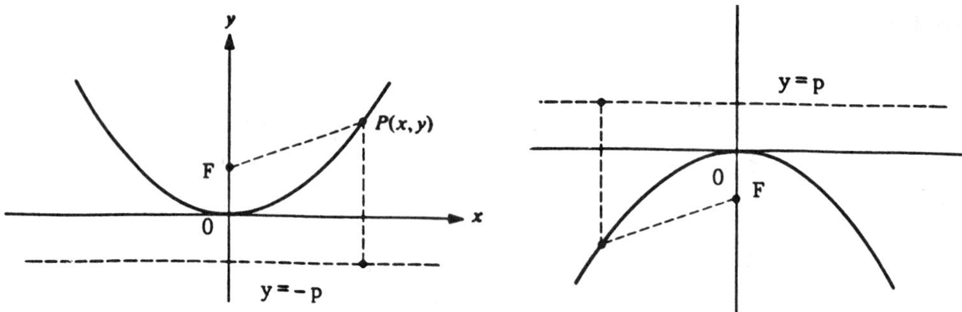

FIGURE 18 Parabolas with y-axis as the axis of symmetry and vertex at the origin. (Left) $y=\frac{x^2}{4p}$; (right) $y=-\frac{x^2}{4p}$.

The distance between the focus and the vertex, or vertex and directrix, is denoted by $p(>0)$ and leads to one of the following equations of a parabola with vertex at the origin (Figures 18 and 19):

$$y=\frac{x^2}{4p} \qquad \text{(opens upward)}$$

$$y=-\frac{x^2}{4p} \qquad \text{(opens downward)}$$

$$x=\frac{y^2}{4p} \qquad \text{(opens to right)}$$

$$x=-\frac{y^2}{4p} \qquad \text{(opens to left)}$$

For each of the four orientations shown in Figures 18 and 19, the coresponding parabola with vertex (h,k) is obtained by replacing x by $x-h$ and y by $y-k$. Thus, the parabola in Figure 20 has the equation

$$x-h=-\frac{(y-k)^2}{4p}.$$

Ellipse

An ellipse is the set of all points in the plane such that the sum of their distances from two fixed points, called *foci*, is a given constant $2a$. The distance between the foci is denoted $2c$; the length of the major axis is $2a$, whereas the length of the minor axis is $2b$ (Figure 21) and

$$a=\sqrt{b^2+c^2}.$$

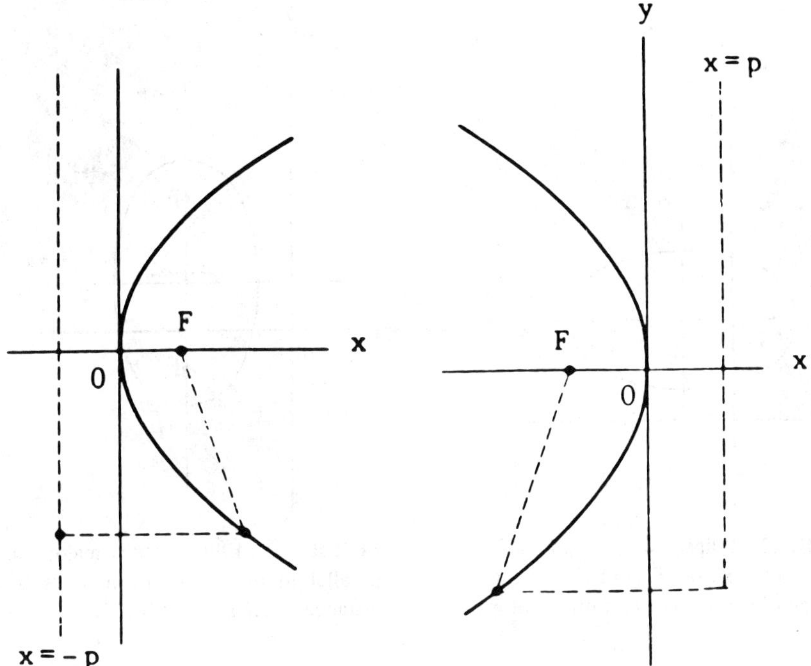

FIGURE 19 Parabolas with x-axis as the axis of symmetry and vertex at the origin. (Left) $x = \frac{y^2}{4p}$; (right) $x = -\frac{y^2}{4p}$.

The eccentricity of an ellipse, e, is <1. An ellipse with center at point (h, k) and major axis *parallel to the x-axis* (Figure 22) is given by the equation

$$\frac{(x-h)^2}{a^2} + \frac{(y-k)^2}{b^2} = 1.$$

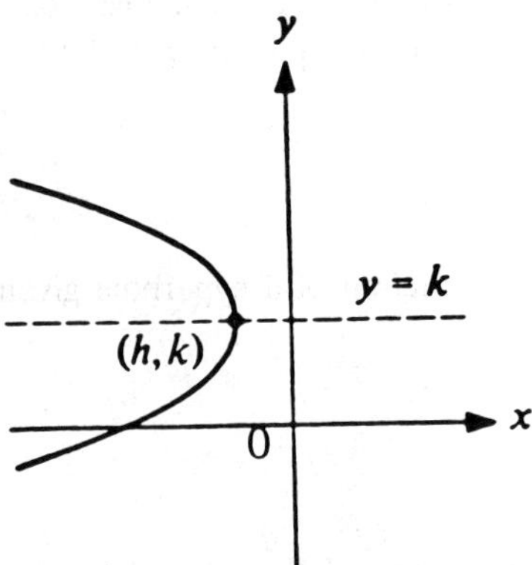

FIGURE 20 Parabola with vertex at (h,k) and axis parallel to the x-axis.

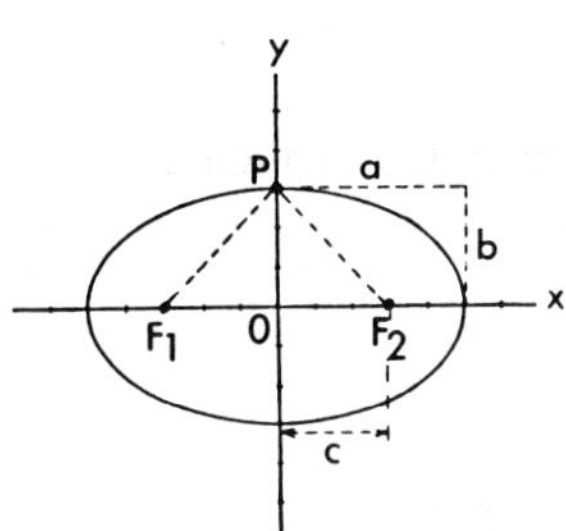

FIGURE 21 Ellipse; since point P is equidistant from foci F_1 and F_2 the segments F_1P and $F_2P = a$; hence $a = \sqrt{b^2 + c^2}$.

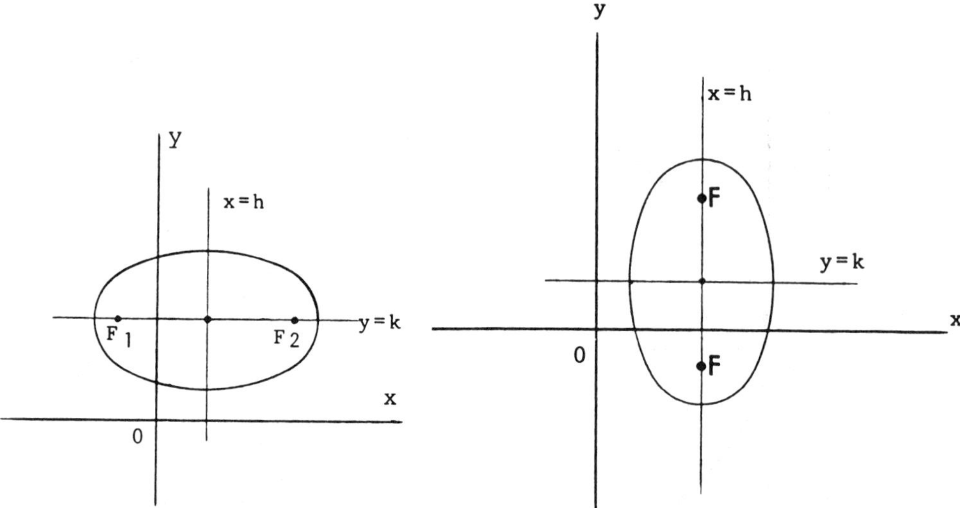

FIGURE 22 Ellipse with major axis parallel to the x-axis. F_1 and F_2 are the foci, each a distance c from center (h,k).

FIGURE 23 Ellipse with major axis parallel to the y-axis. Each focus is a distance c from center (h,k).

An ellipse with center at (h,k) and major axis parallel to the y-axis is given by the equation (Figure 23)

$$\frac{(y-k)^2}{a^2}+\frac{(x-h)^2}{b^2}=1.$$

Hyperbola $(e>1)$

A hyperbola is the set of all points in the plane such that the difference of its distances from two fixed points (foci) is a given positive constant denoted $2a$. The distance between the two foci is $2c$ and that between the two vertices is $2a$. The quantity b is defined by the equation

$$b=\sqrt{c^2-a^2}$$

and is illustrated in Figure 24, which shows the construction of a hyperbola given by the equation

$$\frac{x^2}{a^2}-\frac{y^2}{b^2}=1.$$

When the focal axis is parallel to the y-axis the equation of the hyperbola with center (h,k) (Figures 25 and 26) is

$$\frac{(y-k)^2}{a^2}-\frac{(x-h)^2}{b^2}=1.$$

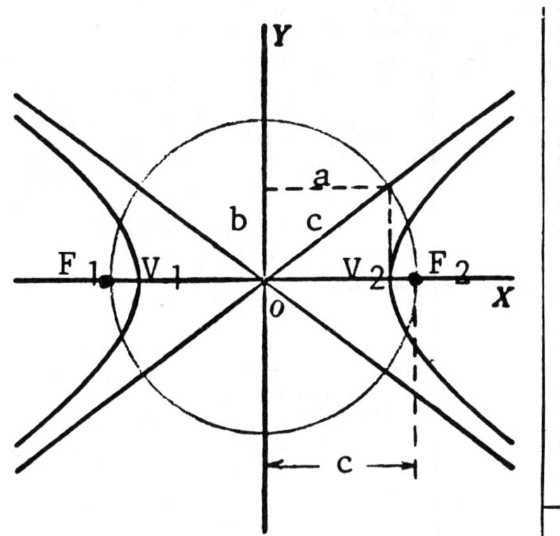

FIGURE 24 Hyperbola; V_1, V_2 = vertices; F_1, F_2 = foci. A circle at center 0 with radius c contains the vertices and illustrates the relation among a, b, and c. Asymptotes have slopes b/a and $-b/a$ for the orientation shown.

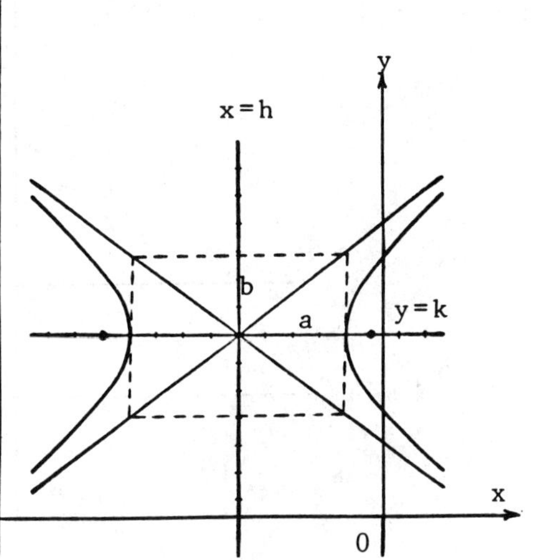

FIGURE 25 Hyperbola with center at (h,k): $\frac{(x-h)^2}{a^2} - \frac{(y-k)^2}{b^2} = 1$; slopes of asymptotes $\pm b/a$.

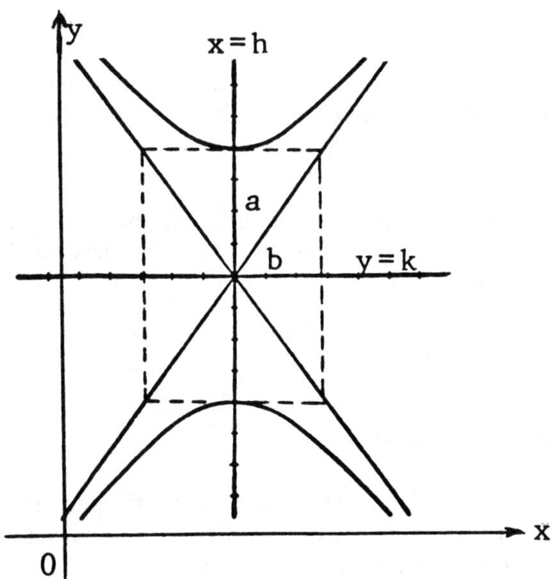

FIGURE 26 Hyperbola with center at (h,k): $\frac{(y-k)^2}{a^2} - \frac{(x-h)^2}{b^2} = 1$; slopes of asymptotes $\pm a/b$.

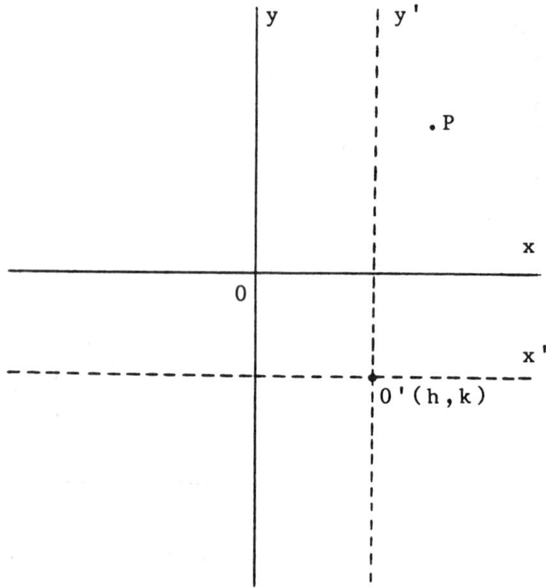

FIGURE 27 Translation of axes.

If the focal axis is parallel to the x-axis and center (h, k), then

$$\frac{(x-h)^2}{a^2} - \frac{(y-k)^2}{b^2} = 1$$

Change of Axes

A change in the position of the coordinate axes will generally change the coordinates of the points in the plane. The equation of a particular curve will also generally change.

Translation

When the new axes remain parallel to the original, the transformation is called a *translation* (Figure 27). The new axes, denoted x' and y', have origin $0'$ at (h, k) with reference to the x and y axes.

Series

Bernoulli and Euler Numbers

A set of numbers, $B_1, B_3, \ldots, B_{2n-1}$ (Bernoulli numbers) and $B_2, B_4, \ldots, B_{2n}$ (Euler numbers) appear in the series expansions of many functions. A partial listing follows; these are computed from the following equations:

$$B_{2n} - \frac{2n(2n-1)}{2!}B_{2n-2} + \frac{2n(2n-1)(2n-2)(2n-3)}{4!}B_{2n-4} - \cdots + (-1)^n = 0,$$

and

$$\frac{2^{2n}(2^{2n}-1)}{2n}B_{2n-1}=(2n-1)B_{2n-2}-\frac{(2n-1)(2n-2)(2n-3)}{3!}B_{2n-4}+\cdots+(-1)^{n-1}.$$

$$\begin{array}{ll} B_1 = 1/6 & B_2 = 1 \\ B_3 = 1/30 & B_4 = 5 \\ B_5 = 1/42 & B_6 = 61 \\ B_7 = 1/30 & B_8 = 1385 \\ B_9 = 5/66 & B_{10} = 50521 \\ & \\ B_{11} = 691/2730 & B_{12} = 2702765 \\ B_{13} = 7/6 & B_{14} = 199360981 \\ \vdots & \vdots \end{array}$$

Series of Functions

In the following, the interval of convergence is indicated, otherwise it is all x. Logarithms are to the base e. Bernoulli and Euler numbers (B_{2n-1} and B_{2n}) appear in certain expressions.

$$(a+x)^n=a^n+na^{n-1}x+\frac{n(n-1)}{2!}a^{n-2}x^2+\frac{n(n-1)(n-2)}{3!}a^{n-3}x^3+\cdots$$

$$+\frac{n!}{(n-j)!j!}a^{n-j}x^j+\cdots \qquad [x^2<a^2]$$

$$(a-bx)^{-1}=\frac{1}{a}\left[1+\frac{bx}{a}+\frac{b^2x^2}{a^2}+\frac{b^3x^3}{a^3}+\cdots\right] \qquad [b^2x^2<a^2]$$

$$(1\pm x)^n=1\pm nx+\frac{n(n-1)}{2!}x^2\pm\frac{n(n-1)(n-2)x^3}{3!}+\cdots \qquad [x^2<1]$$

$$(1\pm x)^{-n}=1\mp nx+\frac{n(n+1)}{2!}x^2\mp\frac{n(n+1)(n+2)}{3!}x^3+\cdots \qquad [x^2<1]$$

$$(1\pm x)^{\frac{1}{2}}=1\pm\frac{1}{2}x-\frac{1}{2\cdot4}x^2\pm\frac{1\cdot3}{2\cdot4\cdot6}x^3-\frac{1\cdot3\cdot5}{2\cdot4\cdot6\cdot8}x^4\pm\cdots \qquad [x^2<1]$$

$$(1\pm x)^{-\frac{1}{2}}=1\mp\frac{1}{2}x+\frac{1\cdot3}{2\cdot4}x^2\mp\frac{1\cdot3\cdot5}{2\cdot4\cdot6}x^3+\frac{1\cdot3\cdot5\cdot7}{2\cdot4\cdot6\cdot8}x^4\mp\cdots \qquad [x^2<1]$$

$$(1\pm x^2)^{\frac{1}{2}}=1\pm\frac{1}{2}x^2-\frac{x^4}{2\cdot4}\pm\frac{1\cdot3}{2\cdot4\cdot6}x^6-\frac{1\cdot3\cdot5}{2\cdot4\cdot6\cdot8}x^8\pm\cdots \qquad [x^2<1]$$

$$(1\pm x)^{-1}=1\mp x+x^2\mp x^3+x^4\mp x^5+\cdots \qquad [x^2<1]$$

$$(1\pm x)^{-2}=1\mp 2x+3x^2\mp 4x^3+5x^4\mp\cdots \qquad [x^2<1]$$

$$e^x=1+x+\frac{x^2}{2!}+\frac{x^3}{3!}+\frac{x^4}{4!}+\cdots$$

$$e^{-x^2} = 1 - x^2 + \frac{x^4}{2!} - \frac{x^6}{3!} + \frac{x^8}{4!} - \cdots$$

$$a^x = 1 + x \log a + \frac{(x \log a)^2}{2!} + \frac{(x \log a)^3}{3!} + \cdots$$

$$\log x = (x-1) - \frac{1}{2}(x-1)^2 + \frac{1}{3}(x-1)^3 - \cdots \qquad [0 < x < 2]$$

$$\log x = \frac{x-1}{x} + \frac{1}{2}\left(\frac{x-1}{x}\right)^2 + \frac{1}{3}\left(\frac{x-1}{x}\right)^3 + \cdots \qquad \left[x > \frac{1}{2}\right]$$

$$\log x = 2\left[\left(\frac{x-1}{x+1}\right) + \frac{1}{3}\left(\frac{x-1}{x+1}\right)^3 + \frac{1}{5}\left(\frac{x-1}{x+1}\right)^5 + \cdots\right] \qquad [x > 0]$$

$$\log(1+x) = x - \frac{1}{2}x^2 + \frac{1}{3}x^3 - \frac{1}{4}x^4 + \cdots \qquad [x^2 < 1]$$

$$\log\left(\frac{1+x}{1-x}\right) = 2\left[x + \frac{1}{3}x^3 + \frac{1}{5}x^5 + \frac{1}{7}x^7 + \cdots\right] \qquad [x^2 < 1]$$

$$\log\left(\frac{x+1}{x-1}\right) = 2\left[\frac{1}{x} + \frac{1}{3}\left(\frac{1}{x}\right)^3 + \frac{1}{5}\left(\frac{1}{x}\right)^5 + \cdots\right] \qquad [x^2 > 1]$$

$$\sin x = x - \frac{x^3}{3!} + \frac{x^5}{5!} - \frac{x^7}{7!} + \cdots$$

$$\cos x = 1 - \frac{x^2}{2!} + \frac{x^4}{4!} - \frac{x^6}{6!} + \cdots$$

$$\tan x = x + \frac{x^3}{3} + \frac{2x^5}{15} + \frac{17x^7}{315} + \cdots + \frac{2^{2n}(2^{2n}-1)B_{2n-1}x^{2n-1}}{(2n)!} \qquad \left[x^2 < \frac{\pi^2}{4}\right]$$

$$\operatorname{ctn} x = \frac{1}{x} - \frac{x}{3} - \frac{x^3}{45} - \frac{2x^5}{945} - \cdots - \frac{B_{2n-1}(2x)^{2n}}{(2n)!x} - \cdots \qquad [x^2 < \pi^2]$$

$$\sec x = 1 + \frac{x^2}{2!} + \frac{5x^4}{4!} + \frac{61x^6}{6!} + \cdots + \frac{B_{2n}x^{2n}}{(2n)!} + \cdots \qquad \left[x^2 < \frac{\pi^2}{4}\right]$$

$$\csc x = \frac{1}{x} + \frac{x}{3!} + \frac{7x^3}{3 \cdot 5!} + \frac{31x^5}{3 \cdot 7!} + \cdots + \frac{2(2^{2n+1}-1)}{(2n+2)!}B_{2n+1}x^{2n+1} + \cdots \qquad [x^2 < \pi^2]$$

$$\sin^{-1} x = x + \frac{x^3}{6} + \frac{(1 \cdot 3)x^5}{(2 \cdot 4)5} + \frac{(1 \cdot 3 \cdot 5)x^7}{(2 \cdot 4 \cdot 6)7} + \cdots \qquad [x^2 < 1]$$

$$\tan^{-1} x = x - \frac{1}{3}x^3 + \frac{1}{5}x^5 - \frac{1}{7}x^7 + \cdots \qquad [x^2 < 1]$$

$$\sec^{-1} x = \frac{\pi}{2} - \frac{1}{x} - \frac{1}{6x^3} - \frac{1 \cdot 3}{(2 \cdot 4)5x^5} - \frac{1 \cdot 3 \cdot 5}{(2 \cdot 4 \cdot 6)7x^7} - \cdots \qquad [x^2 > 1]$$

$$\sinh x = x + \frac{x^3}{3!} + \frac{x^5}{5!} + \frac{x^7}{7!} + \cdots$$

$$\cosh x = 1 + \frac{x^2}{2!} + \frac{x^4}{4!} + \frac{x^6}{6!} + \frac{x^8}{8!} + \cdots$$

$$\tanh x = (2^2-1)2^2 B_1 \frac{x}{2!} - (2^4-1)2^4 B_3 \frac{x^3}{4!} + (2^6-1)2^6 B_5 \frac{x^5}{6!} - \cdots \qquad \left[x^2 < \frac{\pi^2}{4}\right]$$

$$\operatorname{ctnh} x = \frac{1}{x}\left(1 + \frac{2^2 B_1 x^2}{2!} - \frac{2^4 B_3 x^4}{4!} + \frac{2^6 B_5 x^6}{6!} - \cdots\right) \qquad [x^2 < \pi^2]$$

$$\operatorname{sech} x = 1 - \frac{B_2 x^2}{2!} + \frac{B_4 x^4}{4!} - \frac{B_6 x^6}{6!} + \cdots \qquad \left[x^2 < \frac{\pi^2}{4}\right]$$

$$\operatorname{csch} x = \frac{1}{x} - (2-1)2B_1 \frac{x}{2!} + (2^3-1)2B_3 \frac{x^3}{4!} - \cdots \qquad [x^2 < \pi^2]$$

$$\sinh^{-1} x = x - \frac{1}{2}\frac{x^3}{3} + \frac{1\cdot 3}{2\cdot 4}\frac{x^5}{5} - \frac{1\cdot 3\cdot 5}{2\cdot 4\cdot 6}\frac{x^7}{7} + \cdots \qquad [x^2 < 1]$$

$$\tanh^{-1} x = x + \frac{x^3}{3} + \frac{x^5}{5} + \frac{x^7}{7} + \cdots \qquad [x^2 < 1]$$

$$\operatorname{ctnh}^{-1} x = \frac{1}{x} + \frac{1}{3x^3} + \frac{1}{5x^5} + \cdots \qquad [x^2 > 1]$$

$$\operatorname{csch}^{-1} x = \frac{1}{x} - \frac{1}{2\cdot 3x^3} + \frac{1\cdot 3}{2\cdot 4\cdot 5x^5} - \frac{1\cdot 3\cdot 5}{2\cdot 4\cdot 6\cdot 7x^7} + \cdots \qquad [x^2 > 1]$$

$$\int_0^x e^{-t^2}\,dt = x - \frac{1}{3}x^3 + \frac{x^5}{5\cdot 2!} - \frac{x^7}{7\cdot 3!} + \cdots$$

Error Function

The following function, known as the error function, erf x, arises frequently in applications:

$$\operatorname{erf} x = \frac{2}{\sqrt{\pi}} \int_0^x e^{-t^2}\,dt$$

The integral cannot be represented in terms of a finite number of elementary functions, therefore values of erf x have been compiled in tables. The following is the series for erf x:

$$\operatorname{erf} x = \frac{2}{\sqrt{\pi}}\left[x - \frac{x^3}{3} + \frac{x^5}{5\cdot 2!} - \frac{x^7}{7\cdot 3!} + \cdots\right]$$

There is a close relation between this function and the area under the standard normal curve (Table 1 in the Tables of Probability and Statistics). For evaluation it is convenient to use z instead of x; then erf z may be evaluated from the area $F(z)$ given in Table 1 by use of the relation

$$\operatorname{erf} z = 2F(\sqrt{2}\, z)$$

Example

$$\operatorname{erf}(0.5) = 2F[(1.414)(0.5)] = 2F(0.707)$$

By interpolation from Table 1, $F(0.707) = 0.260$; thus, $\operatorname{erf}(0.5) = 0.520$.

Series Expansion

The expression in parentheses following certain of the series indicates the region of convergence. If not otherwise indicated it is to be understood that the series converges for all finite values of x.

Binomial

$$(x+y)^n = x^n + nx^{n-1}y + \frac{n(n-1)}{2!}x^{n-2}y^2 + \frac{n(n-1)(n-2)}{3!}x^{n-3}y^3 + \cdots \quad (y^2 < x^2)$$

$$(1 \pm x)^n = 1 \pm nx + \frac{n(n-1)x^2}{2!} \pm \frac{n(n-1)(n-2)x^3}{3!} + \cdots \text{ etc.} \qquad (x^2 < 1)$$

$$(1 \pm x)^{-n} = 1 \mp nx + \frac{n(n+1)x^2}{2!} \mp \frac{n(n+1)(n+2)x^3}{3!} + \cdots \text{ etc.} \qquad (x^2 < 1)$$

$$(1 \pm x)^{-1} = 1 \mp x + x^2 \mp x^3 + x^4 \mp x^5 + \cdots \qquad (x^2 < 1)$$

$$(1 \pm x)^{-2} = 1 \mp 2x + 3x^2 \mp 4x^3 + 5x^4 \mp 6x^5 + \cdots \qquad (x^2 < 1)$$

Reversion of Series

Let a series be represented by

$$y = a_1x + a_2x^2 + a_3x^3 + a_4x^4 + a_5x^5 + a_6x^6 + \cdots \qquad (a_1 \neq 0)$$

to find the coefficients of the series

$$x = A_1y + A_2y^2 + A_3y^3 + A_4y^4 + \cdots$$

$$A_1 = \frac{1}{a_1} \qquad A_2 = -\frac{a_2}{a_1^3} \qquad A_3 = \frac{1}{a_1^5}(2a_2^2 - a_1a_3)$$

$$A_4 = \frac{1}{a_1^7}(5a_1a_2a_3 - a_1^2a_4 - 5a_2^3)$$

$$A_5 = \frac{1}{a_1^9}(6a_1^2a_2a_4 + 3a_1^2a_3^2 + 14a_2^4 - a_1^3a_5 - 21a_1a_2^2a_3)$$

$$A_6 = \frac{1}{a_1^{11}}(7a_1^3a_2a_5 + 7a_1^3a_3a_4 + 84a_1a_2^3a_3 - a_1^4a_6 - 28a_1^2a_2^2a_4 - 28a_1^2a_2a_3^2 - 42a_2^5)$$

$$A_7 = \frac{1}{a_1^{13}}(8a_1^4a_2a_6 + 8a_1^4a_3a_5 + 4a_1^4a_4^2 + 120a_1^2a_2^3a_4 + 180a_1^2a_2^2a_3^2 + 132a_2^6 - a_1^5a_7$$

$$-36a_1^3a_2^2a_5 - 72a_1^3a_2a_3a_4 - 12a_1^3a_3^3 - 330a_1a_2^4a_3)$$

Taylor

1. $$f(x) = f(a) + (x-a)f'(a) + \frac{(x-a)^2}{2!}f''(a) + \frac{(x-a)^3}{3!}f'''(a)$$

$$+ \cdots + \frac{(x-a)^n}{n!}f^{(n)}(a) + \cdots \text{ (Taylor's Series)}$$

(Increment form)

2. $$f(x+h) = f(x) + hf'(x) + \frac{h^2}{2!}f''(x) + \frac{h^3}{3!}f'''(x) + \cdots$$

$$= f(h) + xf'(h) + \frac{x^2}{2!}f''(h) + \frac{x^3}{3!}f'''(h) + \cdots$$

3. If $f(x)$ is a function possessing derivatives of all orders throughout the interval $a \leqq x \leqq b$, then there is a value X, with $a < X < b$, such that

$$f(b) = f(a) + (b-a)f'(a) + \frac{(b-a)^2}{2!}f''(a) + \cdots$$

$$+ \frac{(b-a)^{n-1}}{(n-1)!}f^{(n-1)}(a) + \frac{(b-a)^n}{n!}f^{(n)}(X)$$

$$f(a+h) = f(a) + hf'(a) + \frac{h^2}{2!}f''(a) + \cdots + \frac{h^{n-1}}{(n-1)!}f^{(n-1)}(a)$$

$$+ \frac{h^n}{n!}f^{(n)}(a+\theta h), \qquad b = a+h, 0 < \theta < 1.$$

or

$$f(x) = f(a) + (x-a)f'(a) + \frac{(x-a)^2}{2!}f''(a) + \cdots + (x-a)^{n-1}\frac{f^{(n-1)}(a)}{(n-1)!} + R_n,$$

where

$$R_n = \frac{f^{(n)}[a + \theta\cdot(x-a)]}{n!}(x-a)^n, \qquad 0 < \theta < 1.$$

The above forms are known as Taylor's series with the remainder term.

4. *Taylor's series for a function of two variables*

If $$\left(h\frac{\partial}{\partial x}+k\frac{\partial}{\partial y}\right)f(x,y)=h\frac{\partial f(x,y)}{\partial x}+k\frac{\partial f(x,y)}{\partial y};$$

$$\left(h\frac{\partial}{\partial x}+k\frac{\partial}{\partial y}\right)^2 f(x,y)=h^2\frac{\partial^2 f(x,y)}{\partial x^2}+2hk\frac{\partial^2 f(x,y)}{\partial x\partial y}+k^2\frac{\partial^2 f(x,y)}{\partial y^2}$$

etc., and if $\left(h\frac{\partial}{\partial x}+k\frac{\partial}{\partial y}\right)^n f(x,y)\Big|_{\substack{x=a\\y=b}}$ with the bar and subscripts means that after differentiation we are to replace x by a and y by b,

$$f(a+h,b+k)=f(a,b)+\left(h\frac{\partial}{\partial x}+k\frac{\partial}{\partial y}\right)f(x,y)\Big|_{\substack{x=a\\y=b}}+\cdots$$

$$+\frac{1}{n!}\left(h\frac{\partial}{\partial x}+k\frac{\partial}{\partial y}\right)^n f(x,y)\Big|_{\substack{x=a\\y=b}}+\cdots$$

MacLaurin

$$f(x)=f(0)+xf'(0)+\frac{x^2}{2!}f''(0)+\frac{x^3}{3!}f'''(0)+\cdots+x^{n-1}\frac{f^{(n-1)}(0)}{(n-1)!}+R_n,$$

where

$$R_n=\frac{x^n f^{(n)}(\theta x)}{n!},\qquad 0<\theta<1.$$

Exponential

$$e=1+\frac{1}{1!}+\frac{1}{2!}+\frac{1}{3!}+\frac{1}{4!}+\cdots$$

$$e^x=1+x+\frac{x^2}{2!}+\frac{x^3}{3!}+\frac{x^4}{4!}+\cdots \qquad \text{(all real values of } x\text{)}$$

$$a^x=1+x\log_e a+\frac{(x\log_e a)^2}{2!}+\frac{(x\log_e a)^3}{3!}+\cdots$$

$$e^x=e^a\left[1+(x-a)+\frac{(x-a)^2}{2!}+\frac{(x-a)^3}{3!}+\cdots\right]$$

Logarithmic

$$\log_e x=\frac{x-1}{x}+\frac{1}{2}\left(\frac{x-1}{x}\right)^2+\frac{1}{3}\left(\frac{x-1}{x}\right)^3+\cdots \qquad (x>\tfrac{1}{2})$$

$$\log_e x=(x-1)-\tfrac{1}{2}(x-1)^2+\tfrac{1}{3}(x-1)^3-\cdots \qquad (2\geq x>0)$$

$$\log_e x = 2\left[\frac{x-1}{x+1} + \frac{1}{3}\left(\frac{x-1}{x+1}\right)^3 \frac{1}{5}\left(\frac{x-1}{x+1}\right)^5 + \cdots\right] \qquad (x>0)$$

$$\log_e(1+x) = x - \tfrac{1}{2}x^2 + \tfrac{1}{3}x^3 - \tfrac{1}{4}x^4 + \cdots \qquad (-1 < x \leq 1)$$

$$\log_e(n+1) - \log_e(n-1) = 2\left[\frac{1}{n} + \frac{1}{3n^3} + \frac{1}{5n^5} + \cdots\right]$$

$$\log_e(a+x) = \log_e a + 2\left[\frac{x}{2a+x} + \frac{1}{3}\left(\frac{x}{2a+x}\right)^3 + \frac{1}{5}\left(\frac{x}{2a+x}\right)^5 + \cdots\right]$$

$$(a>0, -a<x<+\infty)$$

$$\log_e \frac{1+x}{1-x} = 2\left[x + \frac{x^3}{3} + \frac{x^5}{5} + \cdots + \frac{x^{2n-1}}{2n-1} + \cdots\right], \qquad -1<x<1$$

$$\log_e x = \log_e a + \frac{(x-a)}{a} - \frac{(x-a)^2}{2a^2} + \frac{(x-a)^3}{3a^3} - + \cdots, \qquad 0 < x \leqq 2a$$

Trigonometric

$$\sin x = x - \frac{x^3}{3!} + \frac{x^5}{5!} - \frac{x^7}{7!} + \cdots \qquad \text{(all real values of } x\text{)}$$

$$\cos x = 1 - \frac{x^2}{2!} + \frac{x^4}{4!} - \frac{x^6}{6!} + \cdots \qquad \text{(all real values of } x\text{)}$$

$$\tan x = x + \frac{x^3}{3} + \frac{2x^5}{15} + \frac{17x^7}{315} + \frac{62x^9}{2835} + \cdots$$

$$+ \frac{(-1)^{n-1}2^{2n}(2^{2n}-1)B_{2n}}{(2n)!}x^{2n-1} + \cdots, \qquad \left[x^2 < \frac{\pi^2}{4}, \text{ and } B_n \text{ represents the } n\text{th Bernoulli number.}\right]$$

$$\cot x = \frac{1}{x} - \frac{x}{3} - \frac{x^2}{45} - \frac{2x^5}{945} - \frac{x^7}{4725} - \cdots$$

$$- \frac{(-1)^{n+1}2^{2n}}{(2n)!}B_{2n}x^{2n-1} - \cdots, \qquad \left[x^2 < \pi^2, \text{ and } B_n \text{ represents the } n\text{th Bernoulli number.}\right]$$

Differential Calculus

Notation

For the following equations, the symbols $f(x)$, $g(x)$, etc., represent functions of x. The value of a function $f(x)$ at $x=a$ is denoted $f(a)$. For the function $y=f(x)$ the derivative of y with respect to x is denoted by one of the following:

$$\frac{dy}{dx}, \quad f'(x), \quad D_x y, \quad y'.$$

Higher derivatives are as follows:

$$\frac{d^2y}{dx^2} = \frac{d}{dx}\left(\frac{dy}{dx}\right) = \frac{d}{dx}f'(x) = f''(x)$$

$$\frac{d^3y}{dx^3} = \frac{d}{dx}\left(\frac{d^2y}{dx^2}\right) = \frac{d}{dx}f''(x) = f'''(x), \text{ etc.}$$

and values of these at $x = a$ are denoted $f''(a)$, $f'''(a)$, etc. (see Table of Derivatives).

Slope of a Curve

The tangent line at a point $P(x, y)$ of the curve $y = f(x)$ has a slope $f'(x)$ provided that $f'(x)$ exists at P. The slope at P is defined to be that of the tangent line at P. The tangent line at $P(x_1, y_1)$ is given by

$$y - y_1 = f'(x_1)(x - x_1).$$

The *normal line* to the curve at $P(x_1, y_1)$ has slope $-1/f'(x_1)$ and thus obeys the equation

$$y - y_1 = [-1/f'(x_1)](x - x_1)$$

(The slope of a vertical line is not defined.)

Angle of Intersection of Two Curves

Two curves, $y = f_1(x)$ and $y = f_2(x)$, that intersect at a point $P(X, Y)$ where derivatives $f_1'(X)$, $f_2'(X)$ exist, have an angle (α) of intersection given by

$$\tan \alpha = \frac{f_2'(X) - f_1'(X)}{1 + f_2'(X) \cdot f_1'(X)}.$$

If $\tan \alpha > 0$, then α is the acute angle; if $\tan \alpha < 0$, then α is the obtuse angle.

Radius of Curvature

The radius of curvature R of the curve $y = f(x)$ at point $P(x, y)$ is

$$R = \frac{\{1 + [f'(x)]^2\}^{3/2}}{f''(x)}$$

In polar coordinates (θ, r) the corresponding formula is

$$R = \frac{\left[r^2 + \left(\frac{dr}{d\theta}\right)^2\right]^{3/2}}{r^2 + 2\left(\frac{dr}{d\theta}\right)^2 - r\frac{d^2r}{d\theta^2}}$$

The *curvature* K is $1/R$.

Relative Maxima and Minima

The function f has a relative maximum at $x=a$ if $f(a) \geq f(a+c)$ for all values of c (positive or negative) that are sufficiently near zero. The function f has a relative minimum at $x=b$ if $f(b) \leq f(b+c)$ for all values of c that are sufficiently close to zero. If the function f is defined on the closed interval $x_1 \leq x \leq x_2$, and has a relative maximum or minimum at $x=a$, where $x_1 < a < x_2$, and if the derivative $f'(x)$ exists at $x=a$, then $f'(a)=0$. It is noteworthy that a relative maximum or minimum may occur at a point where the derivative does not exist. Further, the derivative may vanish at a point that is neither a maximum or a minimum for the function. Values of x for which $f'(x)=0$ are called "critical values." To determine whether a critical value of x, say x_c, is a relative maximum or minimum for the function at x_c, one may use the second derivative test

1. If $f''(x_c)$ is positive, $f(x_c)$ is a minimum
2. If $f''(x_c)$ is negative, $f(x_c)$ is a maximum
3. If $f''(x_c)$ is zero, no conclusion may be made

The sign of the derivative as x advances through x_c may also be used as a test. If $f'(x)$ changes from positive to zero to negative, then a maximum occurs at x_c, whereas a change in $f'(x)$ from negative to zero to positive indicates a minimum. If $f'(x)$ does not change sign as x advances through x_c, then the point is neither a maximum nor a minimum.

Points of Inflection of a Curve

The sign of the second derivative of f indicates whether the graph of $y=f(x)$ is concave upward or concave downward:

$$f''(x) > 0\text{: concave upward}$$

$$f''(x) < 0\text{: concave downward}$$

A point of the curve at which the direction of concavity changes is called a point of inflection (Figure 28). Such a point may occur where $f''(x)=0$ or where $f''(x)$ becomes infinite. More precisely, if the function $y=f(x)$ and its first derivative $y'=f'(x)$ are continuous in the interval $a \leq x \leq b$, and if $y''=f''(x)$ exists in $a < x < b$, then the graph of $y=f(x)$ for $a < x < b$ is concave upward if $f''(x)$ is positive and concave downward if $f''(x)$ is negative.

Taylor's Formula

If f is a function that is continuous on an interval that contains a and x, and if its first $(n+1)$ derivatives are continuous on this interval, then

$$f(x)=f(a)+f'(a)(x-a)+\frac{f''(a)}{2!}(x-a)^2+\frac{f'''(a)}{3!}(x-a)^3+\cdots+\frac{f^{(n)}(a)}{n!}(x-a)^n+R,$$

where R is called the *remainder*. There are various common forms of the remainder:

Lagrange's Form

$$R = f^{(n+1)}(\beta)\cdot\frac{(x-a)^{n+1}}{(n+1)!};\ \beta \text{ between } a \text{ and } x.$$

Cauchy's Form

$$R = f^{(n+1)}(\beta)\cdot\frac{(x-\beta)^{n}(x-a)}{n!};\ \beta \text{ between } a \text{ and } x.$$

Integral Form

$$R = \int_a^x \frac{(x-t)^n}{n!} f^{(n+1)}(t)\,dt.$$

Indeterminant Forms

If $f(x)$ and $g(x)$ are continuous in an interval that includes $x=a$ and if $f(a)=0$ and $g(a)=0$, the limit $\lim_{x\to a}(f(x)/g(x))$ takes the form "0/0", called an *indeterminant form*. *L'Hôpital's rule* is

$$\lim_{x\to a}\frac{f(x)}{g(x)} = \lim_{x\to a}\frac{f'(x)}{g'(x)}.$$

Similarly, it may be shown that if $f(x)\to\infty$ and $g(x)\to\infty$ as $x\to a$, then

$$\lim_{x\to a}\frac{f(x)}{g(x)} = \lim_{x\to a}\frac{f'(x)}{g'(x)}.$$

(The above holds for $x\to\infty$.)

Examples

$$\lim_{x\to 0}\frac{\sin x}{x} = \lim_{x\to 0}\frac{\cos x}{1} = 1$$

$$\lim_{x\to\infty}\frac{x^2}{e^x} = \lim_{x\to\infty}\frac{2x}{e^x} = \lim_{x\to\infty}\frac{2}{e^x} = 0$$

Numerical Methods

a. *Newton's method* for approximating roots of the equation $f(x)=0$: A first estimate x_1 of the root is made; then provided that $f'(x_1)\neq 0$, a better approximation is x_2

$$x_2 = x_1 - \frac{f(x_1)}{f'(x_1)}.$$

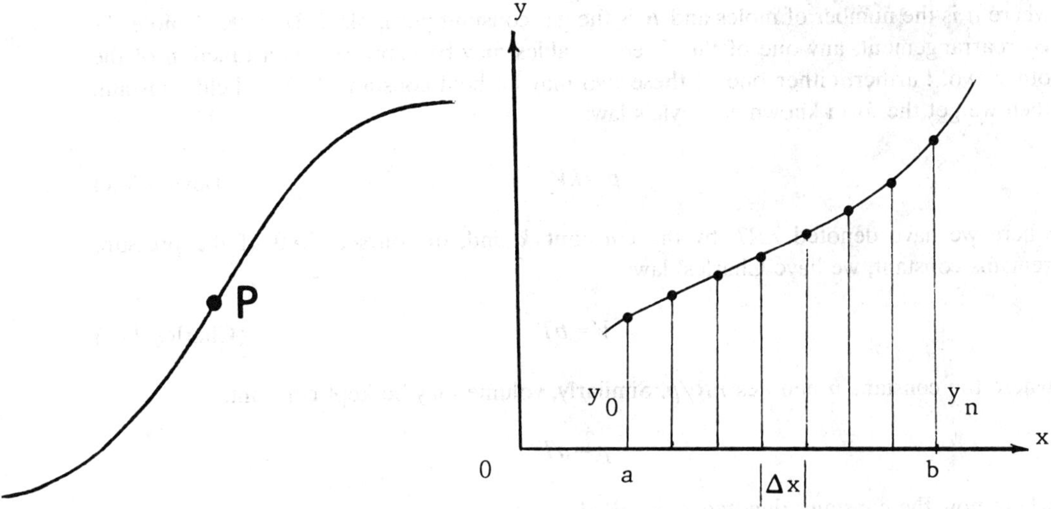

FIGURE 28 Point of inflection.

FIGURE 29 Trapezoidal rule for area.

The process may be repeated to yield a third approximation x_3 to the root:

$$x_3 = x_2 - \frac{f(x_2)}{f'(x_2)}.$$

provided $f'(x_2)$ exists. The process may be repeated. (In certain rare cases the process will not converge.)

b. *Trapezoidal rule for areas* (Figure 29): For the function $y = f(x)$ defined on the interval (a, b) and positive there, take n equal subintervals of width $\Delta x = (b-a)/n$. The area bounded by the curve between $x = a$ and $x = b$ (or definite integral of $f(x)$) is approximately the sum of trapezoidal areas, or

$$A \sim \left(\frac{1}{2}y_0 + y_1 + y_2 + \cdots + y_{n-1} + \frac{1}{2}y_n\right)(\Delta x)$$

Estimation of the error (E) is possible if the second derivative can be obtained:

$$E = \frac{b-a}{12} f''(c)(\Delta x)^2,$$

where c is some number between a and b.

Functions of Two Variables

For the function of two variables, denoted $z = f(x, y)$, if y is held constant, say at $y = y_1$, then the resulting function is a function of x only. Similarly, x may be held constant at x_1, to give the resulting function of y.

The Gas Laws

A familiar example is afforded by the ideal gas law that relates the pressure p, the volume V and the absolute temperature T of an ideal gas:

$$pV = nRT$$

where n is the number of moles and R is the gas constant per mole, 8.31 ($J \cdot {}^\circ K^{-1} \cdot mole^{-1}$). By rearrangement, any one of the three variables may be expressed as a function of the other two. Further, either one of these two may be held constant. If T is held constant, then we get the form known as Boyle's law:

$$p = kV^{-1} \qquad \text{(Boyle's law)}$$

where we have denoted nRT by the constant k and, of course, $V > 0$. If the pressure remains constant, we have Charles' law:

$$V = bT \qquad \text{(Charles' law)}$$

where the constant b denotes nR/p. Similarly, volume may be kept constant:

$$p = aT$$

where now the constant, denoted a, is nR/V.

Partial Derivatives

The physical example afforded by the ideal gas law permits clear interpretations of processes in which one of the variables is held constant. More generally, we may consider a function $z = f(x, y)$ defined over some region of the x-y-plane in which we hold one of the two coordinates, say y, constant. If the resulting function of x is differentiable at a point (x, y) we denote this derivative by one of the notations

$$f_x, \qquad \delta f/dx, \qquad \delta z/dx$$

called the *partial derivative with respect to x*. Similarly, if x is held constant and the resulting function of y is differentiable, we get the *partial derivative with respect to y*, denoted by one of the following:

$$f_y \qquad \delta f/dy \qquad \delta z/dy$$

Example

Given $z = x^4y^3 - y \sin x + 4y$, then

$$\delta z/dx = 4(xy)^3 - y \cos x;$$

$$\delta z/dy = 3x^4y^2 - \sin x + 4.$$

Integral Calculus

Indefinite Integral

If $F(x)$ is differentiable for all values of x in the interval (a, b) and satisfies the equation $dy/dx = f(x)$, then $F(x)$ is an integral of $f(x)$ with respect to x. The notation is $F(x) = \int f(x)\,dx$ or, in differential form, $dF(x) = f(x)\,dx$.

For any function $F(x)$ that is an integral of $f(x)$ it follows that $F(x)+C$ is also an integral. We thus write

$$\int f(x)\,dx = F(x) + C.$$

Definite Integral

Let $f(x)$ be defined on the interval $[a, b]$ which is partitioned by points $x_1, x_2, \ldots, x_j, \ldots, x_{n-1}$ between $a = x_0$ and $b = x_n$. The jth interval has length $\Delta x_j = x_j - x_{j-1}$, which may vary with j. The sum $\Sigma_{j=1}^{n} f(v_j)\Delta x_j$, where v_j is arbitrarily chosen in the jth subinterval, depends on the numbers $x_0, \ldots, x_n$ and the choice of the v as well as f; but if such sums approach a common value as all Δx approach zero, then this value is the definite integral of f over the interval (a, b) and is denoted $\int_a^b f(x)\,dx$. The *fundamental theorem of integral calculus* states that

$$\int_a^b f(x)\,dx = F(b) - F(a),$$

where F is any continuous indefinite integral of f in the interval (a, b).

Properties

$$\int_a^b [f_1(x) + f_2(x) + \cdots + f_j(x)]\,dx = \int_a^b f_1(x)\,dx + \int_a^b f_2(x)\,dx + \cdots + \int_a^b f_j(x)\,dx.$$

$$\int_a^b cf(x)\,dx = c\int_a^b f(x)\,dx, \text{ if } c \text{ is a constant.}$$

$$\int_a^b f(x)\,dx = -\int_b^a f(x)\,dx.$$

$$\int_a^b f(x)\,dx = \int_a^c f(x)\,dx + \int_c^b f(x)\,dx.$$

Common Applications of the Definite Integral

Area (Rectangular Coordinates)

Given the function $y = f(x)$ such that $y > 0$ for all x between a and b, the area bounded by the curve $y = f(x)$, the x-axis, and the vertical lines $x = a$ and $x = b$ is

$$A = \int_a^b f(x)\,dx.$$

Length of Arc (Rectangular Coordinates)

Given the smooth curve $f(x, y) = 0$ from point (x_1, y_1) to point (x_2, y_2), the length between these points is

$$L = \int_{x_1}^{x_2} \sqrt{1 + (dy/dx)^2}\,dx,$$

$$L = \int_{y_1}^{y_2} \sqrt{1 + (dx/dy)^2}\,dy.$$

Mean Value of a Function

The mean value of a function $f(x)$ continuous on $[a, b]$ is

$$\frac{1}{(b-a)}\int_a^b f(x)\,dx.$$

Area (Polar Coordinates)

Given the curve $r=f(\theta)$, continuous and non-negative for $\theta_1 \leq \theta \leq \theta_2$, the area enclosed by this curve and the radial lines $\theta=\theta_1$ and $\theta=\theta_2$ is given by

$$A=\int_{\theta_1}^{\theta_2}\frac{1}{2}[f(\theta)]^2\,d\theta.$$

Length of Arc (Polar Coordinates)

Given the curve $r=f(\theta)$ with continuous derivative $f'(\theta)$ on $\theta_1 \leq \theta \leq \theta_2$, the length of arc from $\theta=\theta_1$ to $\theta=\theta_2$ is

$$L=\int_{\theta_1}^{\theta_2}\sqrt{[f(\theta)]^2+[f'(\theta)]^2}\,d\theta.$$

Volume of Revolution

Given a function $y=f(x)$ continuous and non-negative on the interval (a,b), when the region bounded by $f(x)$ between a and b is revolved about the x-axis the volume of revolution is

$$V=\pi\int_a^b [f(x)]^2\,dx.$$

Surface Area of Revolution
(Revolution about the x-axis, between *a* and *b*)

If the portion of the curve $y=f(x)$ between $x=a$ and $x=b$ is revolved about the x-axis, the area A of the surface generated is given by the following:

$$A=\int_a^b 2\pi f(x)\{1+[f'(x)]^2\}^{1/2}\,dx$$

Work

If a variable force $f(x)$ is applied to an object in the direction of motion along the x-axis between $x=a$ and $x=b$, the work done is

$$W=\int_a^b f(x)\,dx.$$

Cylindrical and Spherical Coordinates

a. Cylindrical coordinates (Figure 30)

$$x=r\cos\theta$$

$$y=r\sin\theta$$

element of volume $dV=r\,dr\,d\theta\,dz$.

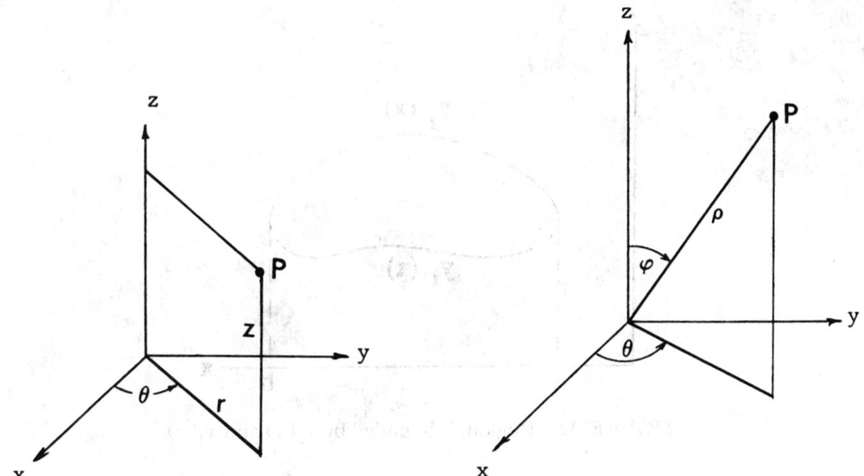

FIGURE 30 Cylindrical coordinates.

FIGURE 31 Spherical coordinates.

b. Spherical coordinates (Figure 31)

$$x = \rho \sin \phi \cos \theta$$

$$y = \rho \sin \phi \sin \theta$$

$$z = \rho \cos \phi$$

element of volume $dV = \rho^2 \sin \phi \, d\rho, d\phi \, d\theta$.

Double Integration

The evaluation of a double integral of $f(x, y)$ over a plane region R

$$\iint_R f(x, y) \, dA$$

is practically accomplished by iterated (repeated) integration. For example, suppose that a vertical straight line meets the boundary of R in at most two points so that there is an upper boundary, $y = y_2(x)$, and a lower boundary, $y = y_1(x)$. Also, it is assumed that these functions are continuous from a to b (see Figure 32). Then

$$\iint_R f(x, y) \, dA = \int_a^b \left(\int_{y_1(x)}^{y_2(x)} f(x, y) \, dy \right) dx$$

If R has left-hand boundary, $x = x_1(y)$, and a right-hand boundary, $x = x_2(y)$, which are continuous from c to d (the extreme values of y in R) then

$$\iint_R f(x, y) \, dA = \int_c^d \left(\int_{x_1(y)}^{x_2(y)} f(x, y) \, dx \right) dy$$

Such integrations are sometimes more convenient in polar coordinates, $x = r \cos \theta$, $y = r \sin \theta$; $dA = r dr d\theta$.

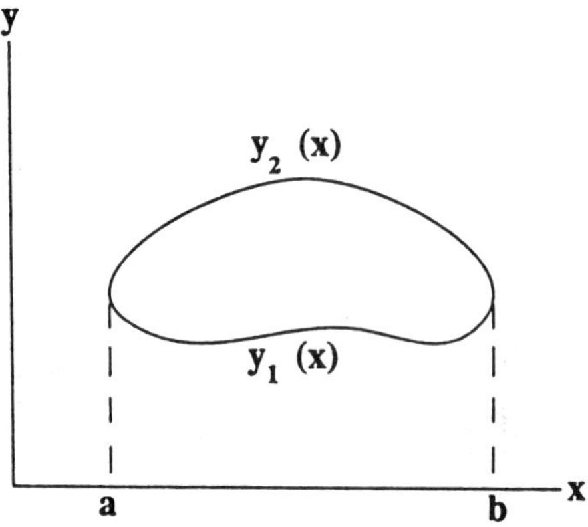

FIGURE 32 Region R bounded by $y_2(x)$ and $y_1(x)$.

Surface Area and Volume by Double Integration

For the surface given by $z=f(x,y)$, which projects onto the closed region R of the x–y-plane, one may calculate the volume V bounded above by the surface and below by R, and the surface area S by the following:

$$V=\iint_R z\,dA=\iint_R f(x,y)\,dx\,dy$$

$$S=\iint_R [1+(\delta z/\delta x)^2+(\delta z/\delta y)^2]^{1/2}\,dx\,dy$$

[In polar coordinates, (r,θ), we replace dA by $r\,dr\,d\theta$].

Centroid

The centroid of a region R of the x–y-plane is a point (x',y') where

$$x'=\frac{1}{A}\iint_R x\,dA; \qquad y'=\frac{1}{A}\iint_R y\,dA$$

and A is the area of the region.

Example. For the circular sector of angle 2α and radius R, the area A is αR^2; the integral needed for x', expressed in polar coordinates is

$$\iint x\,dA=\int_{-\alpha}^{\alpha}\int_0^R (r\cos\theta)r\,dr\,d\theta$$

$$=\left[\frac{R^3}{3}\sin\theta\right]_{-\alpha}^{+\alpha}=\frac{2}{3}R^3\sin\alpha$$

Centroids

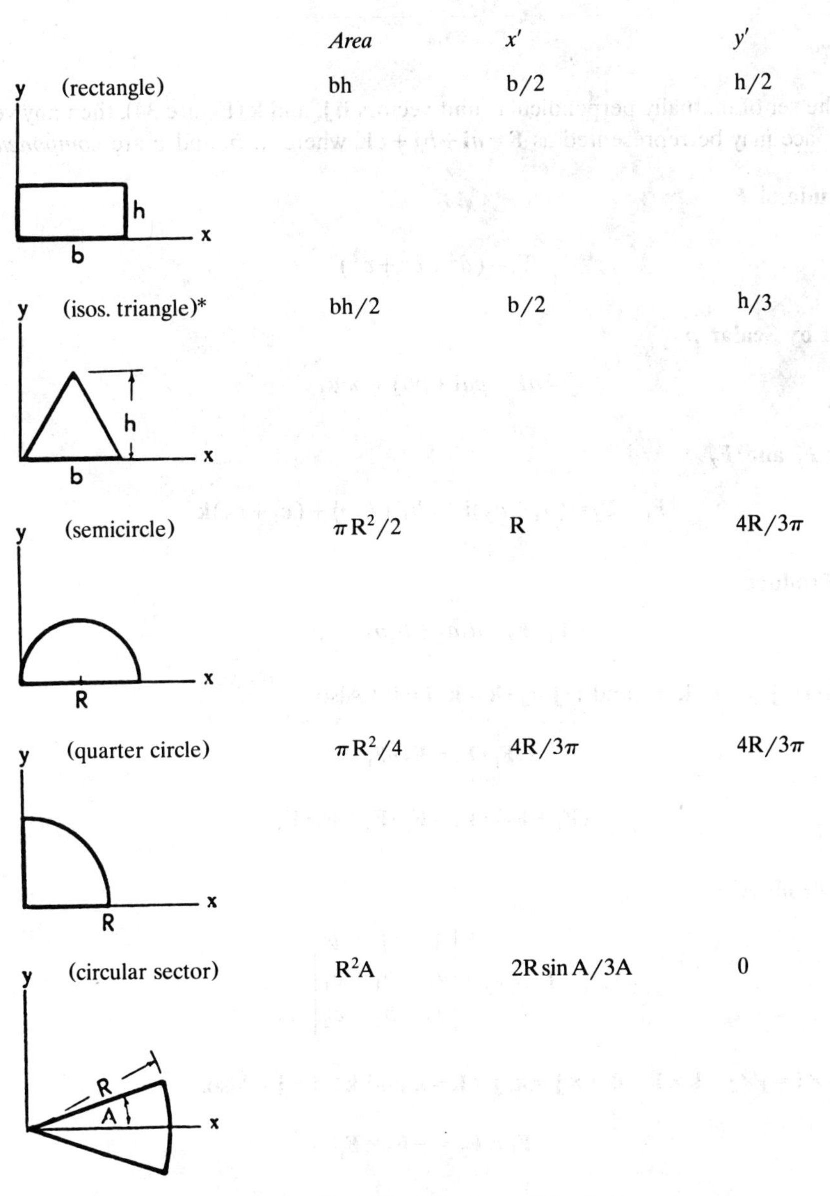

	Area	x'	y'
(rectangle)	bh	b/2	h/2
(isos. triangle)*	bh/2	b/2	h/3
(semicircle)	$\pi R^2/2$	R	$4R/3\pi$
(quarter circle)	$\pi R^2/4$	$4R/3\pi$	$4R/3\pi$
(circular sector)	R^2A	$2R\sin A/3A$	0

* $y' = h/3$ for any triangle of altitude h.

FIGURE 33

and thus,

$$x' = \frac{\frac{2}{3}R^3 \sin \alpha}{\alpha R^2} = \frac{2}{3}R\frac{\sin \alpha}{\alpha}.$$

Centroids of some common regions are shown in Figure 33.

Vector Analysis

Vectors

Given the set of mutually perpendicular unit vectors **i**, **j**, and **k** (Figure 34), then any vector in the space may be represented as $\mathbf{F} = a\mathbf{i} + b\mathbf{j} + c\mathbf{k}$, where a, b, and c are *components*.

Magnitude of *F*

$$|\mathbf{F}| = (a^2 + b^2 + c^2)^{\frac{1}{2}}$$

Product by Scalar *p*

$$p\mathbf{F} = pa\mathbf{i} + pb\mathbf{j} + pc\mathbf{k}.$$

Sum of F_1 and F_2

$$\mathbf{F}_1 + \mathbf{F}_2 = (a_1 + a_2)\mathbf{i} + (b_1 + b_2)\mathbf{j} + (c_1 + c_2)\mathbf{k}$$

Scalar Product

$$\mathbf{F}_1 \cdot \mathbf{F}_2 = a_1 a_2 + b_1 b_2 + c_1 c_2$$

(Thus, $\mathbf{i} \cdot \mathbf{i} = \mathbf{j} \cdot \mathbf{j} = \mathbf{k} \cdot \mathbf{k} = 1$ and $\mathbf{i} \cdot \mathbf{j} = \mathbf{j} \cdot \mathbf{k} = \mathbf{k} \cdot \mathbf{i} = 0$.) Also

$$\mathbf{F}_1 \cdot \mathbf{F}_2 = \mathbf{F}_2 \cdot \mathbf{F}_1$$

$$(\mathbf{F}_1 + \mathbf{F}_2) \cdot \mathbf{F}_3 = \mathbf{F}_1 \cdot \mathbf{F}_3 + \mathbf{F}_2 \cdot \mathbf{F}_3$$

Vector Product

$$\mathbf{F}_1 \times \mathbf{F}_2 = \begin{vmatrix} \mathbf{i} & \mathbf{j} & \mathbf{k} \\ a_1 & b_1 & c_1 \\ a_2 & b_2 & c_2 \end{vmatrix}$$

(Thus, $\mathbf{i} \times \mathbf{i} = \mathbf{j} \times \mathbf{j} = \mathbf{k} \times \mathbf{k} = \mathbf{0}$, $\mathbf{i} \times \mathbf{j} = \mathbf{k}$, $\mathbf{j} \times \mathbf{k} = \mathbf{i}$, and $\mathbf{k} \times \mathbf{i} = \mathbf{j}$.) Also,

$$\mathbf{F}_1 \times \mathbf{F}_2 = -\mathbf{F}_2 \times \mathbf{F}_1$$

$$(\mathbf{F}_1 + \mathbf{F}_2) \times \mathbf{F}_3 = \mathbf{F}_1 \times \mathbf{F}_3 + \mathbf{F}_2 \times \mathbf{F}_3$$

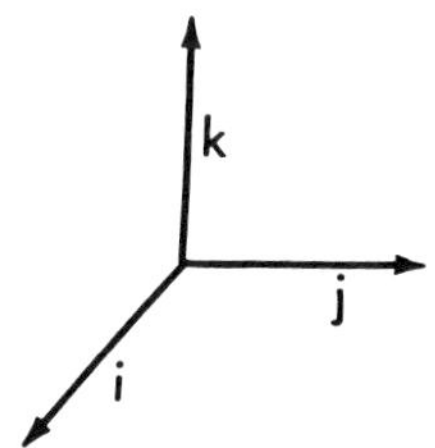

FIGURE 34 The unit vectors **i**, **j**, and **k**.

$$\mathbf{F}_1 \times (\mathbf{F}_2 + \mathbf{F}_3) = \mathbf{F}_1 \times \mathbf{F}_2 + \mathbf{F}_1 \times \mathbf{F}_3$$

$$\mathbf{F}_1 \times (\mathbf{F}_2 \times \mathbf{F}_3) = (\mathbf{F}_1 \cdot \mathbf{F}_3)\mathbf{F}_2 - (\mathbf{F}_1 \cdot \mathbf{F}_2)\mathbf{F}_3$$

$$\mathbf{F}_1 \cdot (\mathbf{F}_2 \times \mathbf{F}_3) = (\mathbf{F}_1 \times \mathbf{F}_2) \cdot \mathbf{F}_3$$

Vector Differentiation

If $\mathbf{V}$ is a vector function of a scalar variable t, then

$$\mathbf{V} = a(t)\mathbf{i} + b(t)\mathbf{j} + c(t)\mathbf{k}$$

and

$$\frac{d\mathbf{V}}{dt} = \frac{da}{dt}\mathbf{i} + \frac{db}{dt}\mathbf{j} + \frac{dc}{dt}\mathbf{k}.$$

For several vector functions $\mathbf{V}_1, \mathbf{V}_2, \ldots, \mathbf{V}_n$

$$\frac{d}{dt}(\mathbf{V}_1 + \mathbf{V}_2 + \cdots + \mathbf{V}_n) = \frac{d\mathbf{V}_1}{dt} + \frac{d\mathbf{V}_2}{dt} + \cdots + \frac{d\mathbf{V}_n}{dt},$$

$$\frac{d}{dt}(\mathbf{V}_1 \cdot \mathbf{V}_2) = \frac{d\mathbf{V}_1}{dt} \cdot \mathbf{V}_2 + \mathbf{V}_1 \cdot \frac{d\mathbf{V}_2}{dt},$$

$$\frac{d}{dt}(\mathbf{V}_1 \times \mathbf{V}_2) = \frac{d\mathbf{V}_1}{dt} \times \mathbf{V}_2 + \mathbf{V}_1 \times \frac{d\mathbf{V}_2}{dt}.$$

For a scalar valued function $g(x, y, z)$

(gradient) $$\text{grad}\, g = \nabla g = \frac{\delta g}{\delta x}\mathbf{i} + \frac{\delta g}{\delta y}\mathbf{j} + \frac{\delta g}{\delta z}\mathbf{k}.$$

For a vector valued function $\mathbf{V}(a, b, c)$, where a, b, c are each a function of x, y, and z,

(divergence) $$\text{div}\,\mathbf{V} = \nabla \cdot \mathbf{V} = \frac{\delta a}{\delta x} + \frac{\delta b}{\delta y} + \frac{\delta c}{\delta z}$$

(curl) $$\text{curl}\,\mathbf{V} = \nabla \times \mathbf{V} = \begin{vmatrix} \mathbf{i} & \mathbf{j} & \mathbf{k} \\ \frac{\delta}{\delta x} & \frac{\delta}{\delta y} & \frac{\delta}{\delta z} \\ a & b & c \end{vmatrix}$$

Also,

$$\operatorname{div}\operatorname{grad} g = \nabla^2 g = \frac{\delta^2 g}{\delta x^2} + \frac{\delta^2 g}{\delta y^2} + \frac{\delta^2 g}{\delta z^2}.$$

and

$$\operatorname{curl}\operatorname{grad} g = \mathbf{0}; \qquad \operatorname{div}\operatorname{curl}\mathbf{V} = 0;$$

$$\operatorname{curl}\operatorname{curl}\mathbf{V} = \operatorname{grad}\operatorname{div}\mathbf{V} - (\mathbf{i}\nabla^2 a + \mathbf{j}\nabla^2 b + \mathbf{k}\nabla^2 c).$$

Divergence Theorem (Gauss)

Given a vector function F with continuous partial derivatives in a region R bounded by a closed surface S, then

$$\iiint_R div\ \mathbf{F}\, dV = \iint_S \mathbf{n}\cdot\mathbf{F}\, dS,$$

where $\mathbf{n}$ is the (sectionally continuous) unit normal to S.

Stokes' Theorem

Given a vector function with continuous gradient over a surface S that consists of portions that are piecewise smooth and bounded by regular closed curves such as C, then

$$\iint_S \mathbf{n}\cdot\operatorname{curl}\mathbf{F}\, dS = \oint_C \mathbf{F}\cdot d\mathbf{r}$$

Planar Motion in Polar Coordinates

Motion in a plane may be expressed with regard to polar coordinates (r, θ). Denoting the position vector by $\mathbf{r}$ and its magnitude by r, we have $\mathbf{r} = r\mathbf{R}(\theta)$, where $\mathbf{R}$ is the unit vector. Also, $d\mathbf{R}/d\theta = \mathbf{P}$, a unit vector perpendicular to $\mathbf{R}$. The velocity and acceleration are then

$$\mathbf{v} = \frac{dr}{dt}\mathbf{R} + r\frac{d\theta}{dt}\mathbf{P};$$

$$\mathbf{a} = \left[\frac{d^2 r}{dt^2} - r\left(\frac{d\theta}{dt}\right)^2\right]\mathbf{R} + \left[r\frac{d^2\theta}{dt^2} + 2\frac{dr}{dt}\frac{d\theta}{dt}\right]\mathbf{P}.$$

Note that the component of acceleration in the $\mathbf{P}$ direction (transverse component) may also be written

$$\frac{1}{r}\frac{d}{dt}\left(r^2\frac{d\theta}{dt}\right)$$

so that in purely radial motion it is zero and

$$r^2 \frac{d\theta}{dt} = C \text{ (constant)}$$

which means that the position vector sweeps out area at a constant rate [see Area (Pola Coordinates) in the section entitled Integral Calculus].

Special Functions

Hyperbolic Functions

$$\sinh x = \frac{e^x - e^{-x}}{2} \qquad \operatorname{csch} x = \frac{1}{\sinh x}$$

$$\cosh x = \frac{e^x + e^{-x}}{2} \qquad \operatorname{sech} x = \frac{1}{\cosh x}$$

$$\tanh x = \frac{e^x - e^{-x}}{e^x + e^{-x}} \qquad \operatorname{ctnh} x = \frac{1}{\tanh x}$$

$$\sinh(-x) = -\sinh x \qquad \operatorname{ctnh}(-x) = -\operatorname{ctnh} x$$

$$\cosh(-x) = \cosh x \qquad \operatorname{sech}(-x) = \operatorname{sech} x$$

$$\tanh(-x) = -\tanh x \qquad \operatorname{csch}(-x) = -\operatorname{csch} x$$

$$\tanh x = \frac{\sinh x}{\cosh x} \qquad \operatorname{ctnh} x = \frac{\cosh x}{\sinh x}$$

$$\cosh^2 x - \sinh^2 x = 1 \qquad \cosh^2 x = \frac{1}{2}(\cosh 2x + 1)$$

$$\sinh^2 x = \frac{1}{2}(\cosh 2x - 1) \qquad \operatorname{ctnh}^2 x - \operatorname{csch}^2 x = 1$$

$$\operatorname{csch}^2 x - \operatorname{sech}^2 x = \operatorname{csch}^2 x \operatorname{sech}^2 x \qquad \tanh^2 x + \operatorname{sech}^2 x = 1$$

$$\sinh(x+y) = \sinh x \cosh y + \cosh x \sinh y$$

$$\cosh(x+y) = \cosh x \cosh y + \sinh x \sinh y$$

$$\sinh(x-y) = \sinh x \cosh y - \cosh x \sinh y$$

$$\cosh(x-y) = \cosh x \cosh y - \sinh x \sinh y$$

$$\tanh(x+y) = \frac{\tanh x + \tanh y}{1 + \tanh x \tanh y}$$

$$\tanh(x-y) = \frac{\tanh x - \tanh y}{1 - \tanh x \tanh y}$$

Laplace Transforms

The Laplace transform of the function $f(t)$, denoted by $F(s)$ or $L\{f(t)\}$, is defined

$$F(s)=\int_0^{\infty} f(t)e^{-st}\,dt$$

provided that the integration may be validly performed. A sufficient condition for the existence of $F(s)$ is that $f(t)$ be of exponential order as $t\rightarrow\infty$ and that it is sectionally continuous over every finite interval in the range $t\geq 0$. The Laplace transform of $g(t)$ is denoted by $L\{g(t)\}$ or $G(s)$.

Operations

$f(t)$	$F(s)=\int_0^{\infty} f(t)e^{-st}\,dt$
$af(t)+bg(t)$	$aF(s)+bG(s)$
$f'(t)$	$sF(s)-f(0)$
$f''(t)$	$s^2F(s)-sf(0)-f'(0)$
$f^{(n)}(t)$	$s^nF(s)-s^{n-1}f(0)-s^{n-2}f'(0)-\cdots-f^{(n-1)}(0)$
$tf(t)$	$-F'(s)$
$t^nf(t)$	$(-1)^nF^{(n)}(s)$
$e^{at}f(t)$	$F(s-a)$
$\int_0^t f(t-\beta)\cdot g(\beta)\,d\beta$	$F(s)\cdot G(s)$
$f(t-a)$	$e^{-as}F(s)$
$f\left(\frac{t}{a}\right)$	$aF(as)$
$\int_0^t g(\beta)\,d\beta$	$\frac{1}{s}G(s)$
$f(t-c)\delta(t-c)$	$e^{-cs}F(s), c>0$

where

$$\delta(t-c)=0 \text{ if } 0\leq t<c$$
$$=1 \text{ if } t\geq c$$

$f(t)=f(t+\omega)$ (periodic)	$\dfrac{\int_0^{\omega} e^{-s\tau}f(\tau)\,d\tau}{1-e^{-s\omega}}$

Table of Laplace Transforms

$f(t)$	$F(s)$		$f(t)$	$F(s)$	
1	$1/s$		$\sinh at$	$\frac{a}{s^2-a^2}$	
t	$1/s^2$		$\cosh at$	$\frac{s}{s^2-a^2}$	
$\frac{t^{n-1}}{(n-1)!}$	$1/s^n$	$(n=1,2,3,\ldots)$	$e^{at}-e^{bt}$	$\frac{a-b}{(s-a)(s-b)}$,	$(a \neq b)$
$\sqrt{t}$	$\frac{1}{2s}\sqrt{\frac{\pi}{s}}$		$ae^{at}-be^{bt}$	$\frac{s(a-b)}{(s-a)(s-b)}$,	$(a \neq b)$
$\frac{1}{\sqrt{t}}$	$\sqrt{\frac{\pi}{s}}$		$t \sin at$	$\frac{2as}{(s^2+a^2)^2}$	
e^{at}	$\frac{1}{s-a}$		$t \cos at$	$\frac{s^2-a^2}{(s^2+a^2)^2}$	
te^{at}	$\frac{1}{(s-a)^2}$		$e^{at} \sin bt$	$\frac{b}{(s-a)^2+b^2}$	
$\frac{t^{n-1}e^{at}}{(n-1)!}$	$\frac{1}{(s-a)^n}$	$(n=1,2,3,\ldots)$	$e^{at} \cos bt$	$\frac{s-a}{(s-a)^2+b^2}$	
$\frac{t^x}{\Gamma(x+1)}$	$\frac{1}{s^{x+1}}$,	$x>-1$	$\frac{\sin at}{t}$	$\mathit{Arc}\tan\frac{a}{s}$	
$\sin at$	$\frac{a}{s^2+a^2}$		$\frac{\sinh at}{t}$	$\frac{1}{2}\log_e\left(\frac{s+a}{s-a}\right)$	
$\cos at$	$\frac{s}{s^2+a^2}$				

z-Transform

For the real-valued sequence $\{f(k)\}$ and complex variable z, the z-transform, $F(z)=Z\{f(k)\}$ is defined by

$$Z\{f(k)\}=F(z)=\sum_{k=0}^{\infty} f(k)z^{-k}$$

For example, the sequence $f(k)=1$, $k=0,1,2,\ldots$, has the z-transform

$$F(z)=1+z^{-1}+z^{-2}+z^{-3}\cdots+z^{-k}+\cdots.$$

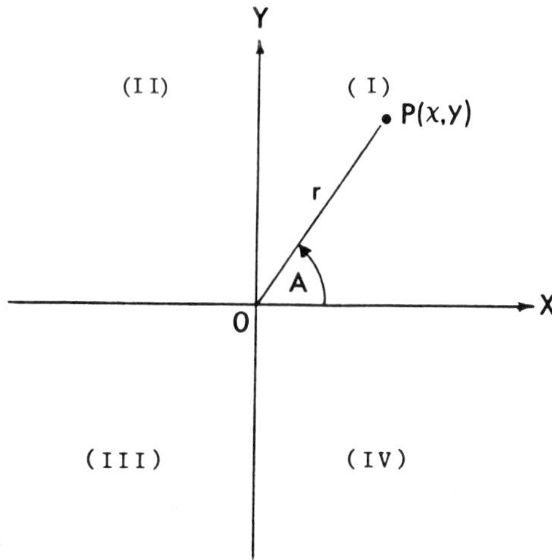

FIGURE 13 The trigonometric point. Angle A is taken to be positive when the rotation is counterclockwise and negative when the rotation is clockwise. The plane is divided into quadrants as shown.

Angles are measured in degrees or radians; $180° = \pi$ radians; 1 radian $= 180°/\pi$ degrees.

The trigonometric functions of 0°, 30°, 45°, and integer multiples of these are directly computed.

	0°	30°	45°	60°	90°	120°	135°	150°	180°
sin	0	$\frac{1}{2}$	$\frac{\sqrt{2}}{2}$	$\frac{\sqrt{3}}{2}$	1	$\frac{\sqrt{3}}{2}$	$\frac{\sqrt{2}}{2}$	$\frac{1}{2}$	0
cos	1	$\frac{\sqrt{3}}{2}$	$\frac{\sqrt{2}}{2}$	$\frac{1}{2}$	0	$-\frac{1}{2}$	$-\frac{\sqrt{2}}{2}$	$-\frac{\sqrt{3}}{2}$	-1
tan	0	$\frac{\sqrt{3}}{3}$	1	$\sqrt{3}$	∞	$-\sqrt{3}$	-1	$-\frac{\sqrt{3}}{3}$	0
ctn	∞	$\sqrt{3}$	1	$\frac{\sqrt{3}}{3}$	0	$-\frac{\sqrt{3}}{3}$	-1	$-\sqrt{3}$	∞
sec	1	$\frac{2\sqrt{3}}{3}$	$\sqrt{2}$	2	∞	-2	$-\sqrt{2}$	$-\frac{2\sqrt{3}}{3}$	-1
csc	∞	2	$\sqrt{2}$	$\frac{2\sqrt{3}}{3}$	1	$\frac{2\sqrt{3}}{3}$	$\sqrt{2}$	2	∞

Trigonometric Identities

$$\sin A = \frac{1}{\csc A}$$

$$\cos A = \frac{1}{\sec A}$$

Defining $z = e^{sT}$ gives

$$\mathscr{L}\{U^*(t)\} = \sum_{k=0}^{\infty} U(kT)z^{-k}$$

which is the z-transform of the sampled signal $U(kT)$.

Properties

Linearity: $Z\{af_1(k)+bf_2(k)\} = aZ\{f_1(k)\} + bZ\{f_2(k)\} = aF_1(z) + bF_2(z)$

Right-shifting property: $Z\{f(k-n)\} = z^{-n}F(z)$

Left-shifting property: $Z\{f(k+n)\} = z^{n}F(z) - \sum_{k=0}^{n-1} f(k)z^{n-k}$

Time scaling: $Z\{a^k f(k)\} = F(z/a)$

Multiplication by k: $Z\{kf(k)\} = -z\,dF(z)/dz$

Initial value: $f(0) = \lim_{z\to\infty}(1-z^{-1})F(z) = F(\infty)$

Final value: $\lim_{k\to\infty} f(k) = \lim_{z\to 1}(1-z^{-1})F(z)$

Convolution: $Z\{f_1(k)*f_2(k)\} = F_1(z)F_2(z)$

z-Transforms of Sampled Functions

$f(k)$	$Z\{f(kT)\} = F(z)$
1 at k; else 0	z^{-k}
1	$\dfrac{z}{z-1}$
kT	$\dfrac{Tz}{(z-1)^2}$
$(kT)^2$	$\dfrac{T^2 z(z+1)}{(z-1)^3}$
$\sin \omega kT$	$\dfrac{z \sin \omega T}{z^2 - 2z\cos \omega T + 1}$
$\cos \omega T$	$\dfrac{z(z-\cos \omega T)}{z^2 - 2z\cos \omega T + 1}$
e^{-akT}	$\dfrac{z}{z-e^{-aT}}$
kTe^{-akT}	$\dfrac{zTe^{-aT}}{(z-e^{-aT})^2}$
$(kT)^2 e^{-akT}$	$\dfrac{T^2 e^{-aT} z(z+e^{-aT})}{(z-e^{-aT})^3}$

$e^{-akT}\sin\omega kT$	$\dfrac{ze^{-aT}\sin\omega T}{z^2-2ze^{-aT}\cos\omega T+e^{-2aT}}$
$e^{-akT}\cos\omega kT$	$\dfrac{z(z-e^{-aT}\cos\omega T)}{z^2-2ze^{-aT}\cos\omega T+e^{-2aT}}$
$a^k\sin\omega kT$	$\dfrac{az\sin\omega T}{z^2-2az\cos\omega T+a^2}$
$a^k\cos\omega kT$	$\dfrac{z(z-a\cos\omega T)}{z^2-2az\cos\omega T+a^2}$

Fourier Series

The periodic function $f(t)$ with period 2π may be represented by the trigonometric series

$$a_0+\sum_1^\infty (a_n\cos nt+b_n\sin nt)$$

where the coefficients are determined from

$$a_0=\frac{1}{2\pi}\int_{-\pi}^{\pi} f(t)\,dt$$

$$a_n=\frac{1}{\pi}\int_{-\pi}^{\pi} f(t)\cos nt\,dt$$

$$b_n=\frac{1}{\pi}\int_{-\pi}^{\pi} f(t)\sin nt\,dt \qquad (n=1,2,3,\ldots)$$

Such a trigonometric series is called the Fourier series corresponding to $f(t)$ and the coefficients are termed Fourier coefficients of $f(t)$. If the function is piecewise continuous in the interval $-\pi\le t\le\pi$, and has left- and right-hand derivatives at each point in that interval, then the series is convergent with sum $f(t)$ except at points t_i at which $f(t)$ is discontinuous. At such points of discontinuity, the sum of the series is the arithmetic mean of the right- and left-hand limits of $f(t)$ at t_i. The integrals in the formulas for the Fourier coefficients can have limits of integration that span a length of 2π, for example, 0 to 2π (because of the periodicity of the integrands).

Functions with Period Other Than 2π

If $f(t)$ has period P the Fourier series is

$$f(t)\sim a_0+\sum_1^\infty\left(a_n\cos\frac{2\pi n}{P}t+b_n\sin\frac{2\pi n}{P}t\right),$$

where

$$a_0=\frac{1}{P}\int_{-P/2}^{P/2} f(t)\,dt$$

$$a_n=\frac{2}{P}\int_{-P/2}^{P/2} f(t)\cos\frac{2\pi n}{P}t\,dt$$

$$b_n=\frac{2}{P}\int_{-P/2}^{P/2} f(t)\sin\frac{2\pi n}{P}t\,dt.$$

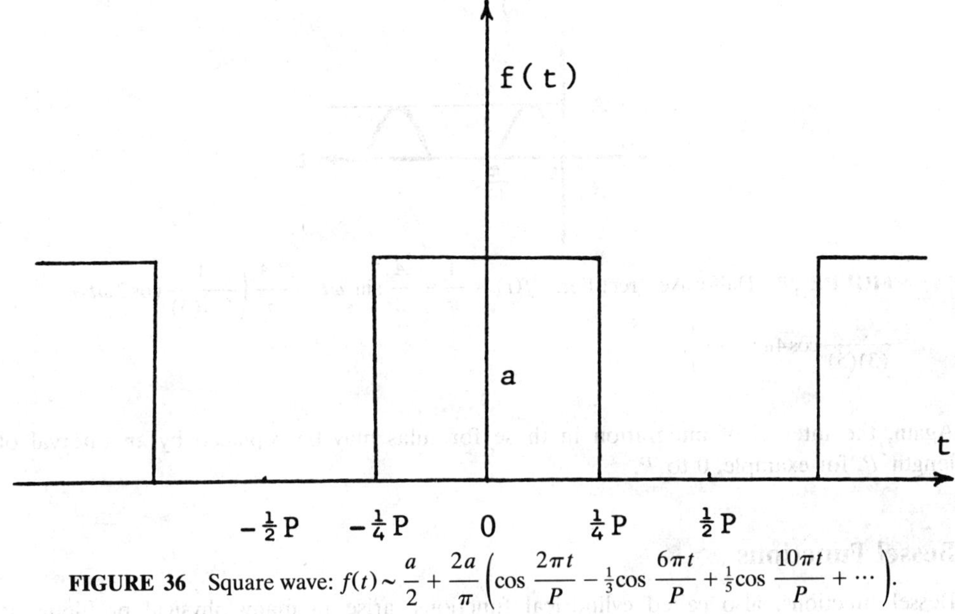

FIGURE 36 Square wave: $f(t) \sim \frac{a}{2} + \frac{2a}{\pi}\left(\cos\frac{2\pi t}{P} - \frac{1}{3}\cos\frac{6\pi t}{P} + \frac{1}{5}\cos\frac{10\pi t}{P} + \cdots\right)$.

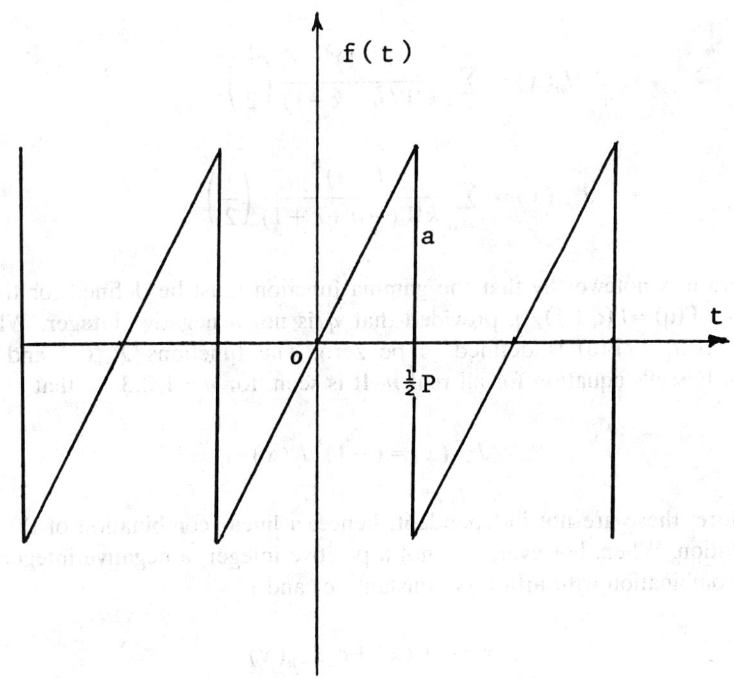

FIGURE 37 Sawtooth wave: $f(t) \sim \frac{2a}{\pi}\left(\sin\frac{2\pi t}{P} - \frac{1}{2}\sin\frac{4\pi t}{P} + \frac{1}{3}\sin\frac{6\pi t}{P} - \cdots\right)$.

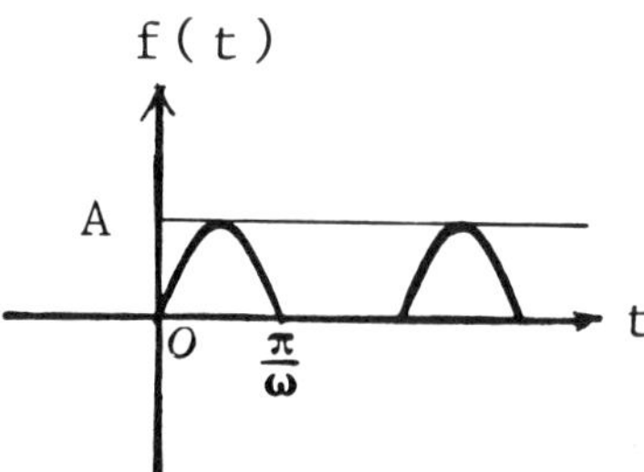

FIGURE 38 Half-wave rectifier: $f(t) \sim \frac{A}{\pi} + \frac{A}{2} \sin \omega t - \frac{2A}{\pi}\left(\frac{1}{(1)(3)} \cos 2\omega t + \frac{1}{(3)(5)} \cos 4\omega t + \cdots\right)$.

Again, the interval of integration in these formulas may be replaced by an interval of length P, for example, 0 to P.

Bessel Functions

Bessel functions, also called cylindrical functions, arise in many physical problems as solutions of the differential equation

$$x^2 y'' + xy' + (x^2 - n^2)y = 0$$

which is known as Bessel's equation. Certain solutions of the above, known as *Bessel functions of the first kind of order n*, are given by

$$J_n(x) = \sum_{k=0}^{\infty} \frac{(-1)^k}{k!\Gamma(n+k+1)} \left(\frac{x}{2}\right)^{n+2k}$$

$$J_{-n}(x) = \sum_{k=0}^{\infty} \frac{(-1)^k}{k!\Gamma(-n+k+1)} \left(\frac{x}{2}\right)^{-n+2k}$$

In the above it is noteworthy that the gamma function must be defined for the negative argument q: $\Gamma(q) = \Gamma(q+1)/q$, provided that q is not a negative integer. When q is a negative integer, $1/\Gamma(q)$ is defined to be zero. The functions $J_{-n}(x)$ and $J_n(x)$ are solutions of Bessel's equation for all real n. It is seen, for $n = 1,2,3,\ldots$ that

$$J_{-n}(x) = (-1)^n J_n(x)$$

and, therefore, these are not independent; hence, a linear combination of these is not a general solution. When, however, n is not a positive integer, a negative integer, nor zero, the linear combination with arbitrary constants c_1 and c_2

$$y = c_1 J_n(x) + c_2 J_{-n}(x)$$

is the general solution of the Bessel differential equation.

The zero order function is especially important as it arises in the solution of the heat equation (for a "long" cylinder):

$$J_0(x) = 1 - \frac{x^2}{2^2} + \frac{x^4}{2^2 4^2} - \frac{x^6}{2^2 4^2 6^2} + \cdots$$

while the following relations show a connection to the trigonometric functions:

$$J_{\frac{1}{2}}(x) = \left[\frac{2}{\pi x}\right]^{1/2} \sin x$$

$$J_{-\frac{1}{2}}(x) = \left[\frac{2}{\pi x}\right]^{1/2} \cos x$$

The following recursion formula gives $J_{n+1}(x)$ for any order in terms of lower order functions:

$$\frac{2n}{x} J_n(x) = J_{n-1}(x) + J_{n+1}(x)$$

Legendre Polynomials

If Laplace's equation, $\nabla^2 V = 0$, is expressed in spherical coordinates, it is

$$r^2 \sin\theta \frac{\delta^2 V}{\delta r^2} + 2r \sin\theta \frac{\delta V}{\delta r} + \sin\theta \frac{\delta^2 V}{\delta\theta^2} + \cos\theta \frac{\delta V}{\delta\theta} + \frac{1}{\sin\theta} \frac{\delta^2 V}{\delta\phi^2} = 0$$

and any of its solutions, $V(r, \theta, \phi)$, are known as *spherical harmonics*. The solution as a product

$$V(r, \theta, \phi) = R(r)\Theta(\theta)$$

which is independent of ϕ, leads to

$$\sin^2\theta\Theta'' + \sin\theta\cos\theta\Theta' + [n(n+1)\sin^2\theta]\Theta = 0$$

Rearrangement and substitution of $x = \cos\theta$ leads to

$$(1-x^2)\frac{d^2\Theta}{dx^2} - 2x\frac{d\Theta}{dx} + n(n+1)\Theta = 0$$

known as *Legendre's equation.* Important special cases are those in which n is zero or a positive integer, and, for such cases, Legendre's equation is satisfied by polynomials called Legendre polynomials, $P_n(x)$. A short list of Legendre polynomials, expressed in terms of x and $\cos\theta$, is given below. These are given by the following general formula:

$$P_n(x) = \sum_{j=0}^{L} \frac{(-1)^j (2n-2j)!}{2^n j!(n-j)!(n-2j)!} x^{n-2j}$$

where $L = n/2$ if n is even and $L = (n-1)/2$ if n is odd. Some are given below:

$P_0(x) = 1$

$P_1(x) = x$

$P_2(x) = \frac{1}{2}(3x^2 - 1)$

$P_3(x) = \frac{1}{2}(5x^3 - 3x)$

$P_4(x) = \frac{1}{8}(35x^4 - 30x^2 + 3)$

$P_5(x) = \frac{1}{8}(63x^5 - 70x^3 + 15x)$

$P_0(\cos\theta) = 1$

$P_1(\cos\theta) = \cos\theta$

$P_2(\cos\theta) = \frac{1}{4}(3\cos 2\theta + 1)$

$P_3(\cos\theta) = \frac{1}{8}(5\cos 3\theta + 3\cos\theta)$

$P_4(\cos\theta) = \frac{1}{64}(35\cos 4\theta + 20\cos 2\theta + 9)$

Additional Legendre polynomials may be determined from the *recursion formula*

$$(n+1)P_{n+1}(x) - (2n+1)xP_n(x) + nP_{n-1}(x) = 0 \qquad (n = 1, 2, \ldots)$$

or the *Rodrigues formula*

$$P_n(x) = \frac{1}{2^n n!}\frac{d^n}{dx^n}(x^2 - 1)^n$$

Laguerre Polynomials

Laguerre polynomials, denoted $L_n(x)$, are solutions of the differential equation

$$xy'' + (1 - x)y' + ny = 0$$

and are given by

$$L_n(x) = \sum_{j=0}^{n} \frac{(-1)^j}{j!} C_{(n,j)} x^j \qquad (n = 0, 1, 2, \ldots)$$

Thus,

$$L_0(x) = 1$$

$$L_1(x) = 1 - x$$

$$L_2(x) = 1 - 2x + \frac{1}{2}x^2$$

$$L_3(x) = 1 - 3x + \frac{3}{2}x^2 - \frac{1}{6}x^3$$

Additional Laguerre polynomials may be obtained from the recursion formula

$$(n+1)L_{n+1}(x) - (2n+1-x)L_n(x) + nL_{n-1}(x) = 0$$

Hermite Polynomials

The Hermite polynomials, denoted $H_n(x)$, are given by

$$H_0 = 1, \quad H_n(x) = (-1)^n e^{x^2} \frac{d^n e^{-x^2}}{dx^n}, \qquad (n = 1, 2, \ldots)$$

and are solutions of the differential equation

$$y'' - 2xy' + 2ny = 0 \qquad (n = 0, 1, 2, \ldots)$$

The first few Hermite polynomials are

$$H_0 = 1 \qquad H_1(x) = 2x$$

$$H_2(x) = 4x^2 - 2 \qquad H_3(x) = 8x^3 - 12x$$

$$H_4(x) = 16x^4 - 48x^2 + 12$$

Additional Hermite polynomials may be obtained from the relation

$$H_{n+1}(x) = 2xH_n(x) - H_n'(x),$$

where prime denotes differentiation with respect to x.

Orthogonality

A set of functions $\{f_n(x)\}$ $(n = 1, 2, \ldots)$ is orthogonal in an interval (a, b) with respect to a given weight function $w(x)$ if

$$\int_a^b w(x) f_m(x) f_n(x)\, dx = 0 \qquad \text{when } m \neq n$$

The following polynomials are orthogonal on the given interval for the given $w(x)$:

Legendre polynomials:	$P_n(x)$	$w(x) = 1$ $a = -1, b = 1$
Laguerre polynomials:	$L_n(x)$	$w(x) = \exp(-x)$ $a = 0, b = \infty$
Hermite polynomials:	$H_n(x)$	$w(x) = \exp(-x^2)$ $a = -\infty, b = \infty$

The Bessel functions *of order n*, $J_n(\lambda_1 x)$, $J_n(\lambda_2 x),\ldots,$ are orthogonal with respect to $w(x)=x$ over the interval $(0,c)$ provided that the λ_i are the positive roots of $J_n(\lambda c)=0$:

$$\int_0^c xJ_n(\lambda_j x)J_n(\lambda_k x)\,dx=0 \qquad (j\neq k)$$

where n is fixed and $n\geq 0$.

Statistics

Arithmetic Mean

$$\mu=\frac{\Sigma X_i}{N},$$

where X_i is a measurement in the population and N is the total number of X_i in the population. For a *sample* of size n the sample mean, denoted $\overline{X}$, is

$$\overline{X}=\frac{\Sigma X_i}{n}.$$

Median

The median is the middle measurement when an odd number (n) measurements is arranged in order; if n is even, it is the midpoint between the two middle measurements.

Mode

It is the most frequently occurring measurement in a set.

Geometric Mean

$$\text{geometric mean}=\sqrt[n]{X_1X_2\ldots X_n}$$

Harmonic Mean

The Harmonic mean H of n numbers $X_1, X_2,\ldots, X_n$, is

$$H=\frac{n}{\Sigma(1/Xi)}$$

Variance

The mean of the sum of squares of deviations from the mean (μ) is the population variance, denoted σ^2

$$\sigma^2=\Sigma(X_i-\mu)^2/N.$$

The sample variance, s^2, for sample size n is

$$s^2 = \Sigma(X_i - \bar{X})^2/(n-1).$$

A simpler computational form is

$$s^2 = \frac{\Sigma X_i^2 - \frac{(\Sigma X_i)^2}{n}}{n-1}$$

Standard Deviation

The positive square root of the population variance is the standard deviation. For a population

$$\sigma = \left[\frac{\Sigma X_i^2 - \frac{(\Sigma X_i)^2}{N}}{N}\right]^{1/2};$$

for a sample

$$s = \left[\frac{\Sigma X_i^2 - \frac{(\Sigma X_i)^2}{n}}{n-1}\right]^{1/2}.$$

Coefficient of Variation

$$V = s/\bar{X}.$$

Probability

For the sample space U, with subsets A of U (called "events"), we consider the probability measure of an event A to be a real-valued function p defined over all subsets of U such that:

$0 \leq p(A) \leq 1$
$p(U) = 1$ and $p(\Phi) = 0$
If A_1 and A_2 are subsets of U
$p(A_1 \cup A_2) = p(A_1) + p(A_2) - p(A_1 \cap A_2)$

Two events A_1 and A_2 are called mutually exclusive if and only if $A_1 \cap A_2 = \phi$ (null set). These events are said to be independent if and only if $p(A_1 \cap A_2) = p(A_1)p(A_2)$.

Conditional Probability and Bayes' Rule

The probability of an event A, given that an event B has occurred, is called the conditional probability and is denoted $p(A/B)$. Further

$$p(A/B) = \frac{p(A \cap B)}{p(B)}$$

Bayes' rule permits a calculation of *a posteriori* probability from given *a priori* probabilities and is stated below:

If $A_1, A_2, \ldots, A_n$ are n mutually exclusive events, and $p(A_1)+p(A_2)+\cdots+p(A_n)=1$, and B is any event such that $p(B)$ is not 0, then the conditional probability $p(A_i/B)$ for any one of the events A_i, *given that B has occurred* is

$$p(A_i/B)=\frac{p(A_i)p(B/A_i)}{p(A_1)p(B/A_1)+p(A_2)p(B/A_2)+\cdots+p(A_n)p(B/A_n)}$$

Example. Among 5 different laboratory tests for detecting a certain disease, one is effective with probability 0.75, whereas each of the others is effective with probability 0.40. A medical student, unfamiliar with the advantage of the best test, selects one of them and is successful in detecting the disease in a patient. What is the probability that the most effective test was used?

Let B denote (the event) of detecting the disease, A_1 the selection of the best test, and A_2 the selection of one of the other 4 tests; thus, $p(A_1)=1/5$, $p(A_2)=4/5$, $p(B/A_1)=0.75$ and $p(B/A_2)=0.40$. Therefore

$$p(A_1/B)=\frac{\frac{1}{5}(0.75)}{\frac{1}{5}(0.75)+\frac{4}{5}(0.40)}=0.319$$

Note, the *a priori* probability is 0.20; the outcome raises this probability to 0.319.

Binomial Distribution

In an experiment consisting of n independent trials in which an event has probability p in a single trial, the probability P_X of obtaining X successes is given by

$$P_X=C_{(n,X)}p^X q^{(n-X)}$$

where

$$q=(1-p) \text{ and } C_{(n,X)}=\frac{n!}{X!(n-X)!}.$$

The probability of between a and b successes (both a and b included) is $P_a+P_{a+1}+\cdots+P_b$, so if $a=0$ and $b=n$, this sum is

$$\sum_{X=0}^{n} C_{(n,X)}p^X q^{(n-X)}=q^n+C_{(n,1)}q^{n-1}p+C_{(n,2)}q^{n-2}p^2+\cdots+p^n=(q+p)^n=1.$$

Mean of Binomially Distributed Variable

The mean number of successes in n independent trials is $m=np$ with standard deviation $\sigma=\sqrt{npq}$.

Normal Distribution

In the binomial distribution, as n increases the histogram of heights is approximated by the bell-shaped curve (normal curve)

$$Y = \frac{1}{\sigma\sqrt{2\pi}} e^{-(x-m)^2/2\sigma^2}$$

where $m =$ the mean of the binomial distribution $= np$, and $\sigma = \sqrt{npq}$ is the standard deviation. For any normally distributed random variable X with mean m and standard deviation σ the probability function (density) is given by the above.

The *standard* normal probability curve is given by

$$y = \frac{1}{\sqrt{2\pi}} e^{-Z^2/2}$$

and has mean $= 0$ and standard deviation $= 1$. The total area under the standard normal curve is 1. Any normal variable X can be put into standard form by defining $Z = (X - m)/\sigma$; thus the probability of X between a given X_1 and X_2 is the area under the standard normal curve between the corresponding Z_1 and Z_2 (Table 1 in the Tables of Probability and Statistics). The standard normal curve is often used instead of the binomial distribution in experiments with discrete outcomes. For example, to determine the probability of obtaining 60 to 70 heads in a toss of 100 coins, we take $X = 59.5$ to $X = 70.5$ and compute corresponding values of Z from mean $np = 100\ \frac{1}{2} = 50$, and the standard deviation $\sigma = \sqrt{(100)(1/2)(1/2)} = 5$. Thus, $Z = (59.5 - 50)/5 = 1.9$ and $Z = (70.5 - 50)/5 = 4.1$. From Table 1, area between $Z = 0$ and $Z = 4.1$ is 0.5000 and between $Z = 0$ and $Z = 1.9$ is 0.4713; hence, the desired probability is 0.0287. The binomial distribution requires a more lengthy computation

$$C_{(100,60)}(1/2)^{60}(1/2)^{40} + C_{(100,61)}(1/2)^{61}(1/2)^{39} + \cdots + C_{(100,70)}(1/2)^{70}(1/2)^{30}.$$

Note that the normal curve is symmetric, whereas the histogram of the binomial distribution is symmetric only if $p = q = 1/2$. Accordingly, when p (hence q) differ appreciably from $1/2$, the difference between probabilities computed by each increases. It is usually recommended that the normal approximation not be used if p (or q) is so small that np (or nq) is less than 5.

Poisson Distribution

$$P = \frac{e^{-m} m^r}{r!}$$

is an approximation to the binomial probability for r successes in n trials when $m = np$ is small (< 5) and the normal curve is not recommended to approximate binomial probabilities (Table 2 in the Tables of Probability and Statistics). The variance σ^2 in the Poisson distribution is np, the same value as the mean. ***Example***: A school's expulsion rate is 5 students per 1000. If class size is 400, what is the probability that 3 or more will be expelled? Since $p = 0.005$ and $n = 400$, $m = np = 2$, and $r = 3$. From Table 2 we obtain for $m = 2$ and $r (= x) = 3$ the probability $p = 0.323$.

Tables of Probability and Statistics

TABLE 1: Areas Under the Standard Normal Curve

z	0.00	0.01	0.02	0.03	0.04	0.05	0.06	0.07	0.08	0.09
0.0	0.0000	0.0040	0.0080	0.0120	0.0160	0.0199	0.0239	0.0279	0.0319	0.0359
0.1	0.0398	0.0438	0.0478	0.0517	0.0557	0.0596	0.0636	0.0675	0.0714	0.0753
0.2	0.0793	0.0832	0.0871	0.0910	0.0948	0.0987	0.1026	0.1064	0.1103	0.1141
0.3	0.1179	0.1217	0.1255	0.1293	0.1331	0.1368	0.1406	0.1443	0.1480	0.1517
0.4	0.1554	0.1591	0.1628	0.1664	0.1700	0.1736	0.1772	0.1808	0.1844	0.1879
0.5	0.1915	0.1950	0.1985	0.2019	0.2054	0.2088	0.2123	0.2157	0.2190	0.2224
0.6	0.2257	0.2291	0.2324	0.2357	0.2389	0.2422	0.2454	0.2486	0.2517	0.2549
0.7	0.2580	0.2611	0.2642	0.2673	0.2704	0.2734	0.2764	0.2794	0.2823	0.2852
0.8	0.2881	0.2910	0.2939	0.2967	0.2995	0.3023	0.3051	0.3078	0.3106	0.3133
0.9	0.3159	0.3186	0.3212	0.3238	0.3264	0.3289	0.3315	0.3340	0.3365	0.3389
1.0	0.3413	0.3438	0.3461	0.3485	0.3508	0.3531	0.3554	0.3577	0.3599	0.3621
1.1	0.3643	0.3665	0.3686	0.3708	0.3729	0.3749	0.3770	0.3790	0.3810	0.3830
1.2	0.3849	0.3869	0.3888	0.3907	0.3925	0.3944	0.3962	0.3980	0.3997	0.4015
1.3	0.4032	0.4049	0.4066	0.4082	0.4099	0.4115	0.4131	0.4147	0.4162	0.4177
1.4	0.4192	0.4207	0.4222	0.4236	0.4251	0.4265	0.4279	0.4292	0.4306	0.4319
1.5	0.4332	0.4345	0.4357	0.4370	0.4382	0.4394	0.4406	0.4418	0.4429	0.4441
1.6	0.4452	0.4463	0.4474	0.4484	0.4495	0.4505	0.4515	0.4525	0.4535	0.4545
1.7	0.4554	0.4564	0.4573	0.4582	0.4591	0.4599	0.4608	0.4616	0.4625	0.4633
1.8	0.4641	0.4649	0.4656	0.4664	0.4671	0.4678	0.4686	0.4693	0.4699	0.4706
1.9	0.4713	0.4719	0.4726	0.4732	0.4738	0.4744	0.4750	0.4756	0.4761	0.4767
2.0	0.4772	0.4778	0.4783	0.4788	0.4793	0.4798	0.4803	0.4808	0.4812	0.4817
2.1	0.4821	0.4826	0.4830	0.4834	0.4838	0.4842	0.4846	0.4850	0.4854	0.4857
2.2	0.4861	0.4864	0.4868	0.4871	0.4875	0.4878	0.4881	0.4884	0.4887	0.4890
2.3	0.4893	0.4896	0.4898	0.4901	0.4904	0.4906	0.4909	0.4911	0.4913	0.4916
2.4	0.4918	0.4920	0.4922	0.4925	0.4927	0.4929	0.4931	0.4932	0.4934	0.4936
2.5	0.4938	0.4940	0.4941	0.4943	0.4945	0.4946	0.4948	0.4949	0.4951	0.4952
2.6	0.4953	0.4955	0.4956	0.4957	0.4959	0.4960	0.4961	0.4962	0.4963	0.4964
2.7	0.4965	0.4966	0.4967	0.4968	0.4969	0.4970	0.4971	0.4972	0.4973	0.4974
2.8	0.4974	0.4975	0.4976	0.4977	0.4977	0.4978	0.4979	0.4979	0.4980	0.4981
2.9	0.4981	0.4982	0.4982	0.4983	0.4984	0.4984	0.4985	0.4985	0.4986	0.4986
3.0	0.4987	0.4987	0.4987	0.4988	0.4988	0.4989	0.4989	0.4989	0.4990	0.4990

Source: R. J. Tallarida and R. B. Murray, *Manual of Pharmacologic Calculations with Computer Programs*, 2nd ed., New York: Springer-Verlag, 1987. With permission.

TABLE 2: Poisson Distribution

Each number in this table represents the probability of obtaining at least X successes, or the area under the histogram to the right of and including the rectangle whose center is at X.

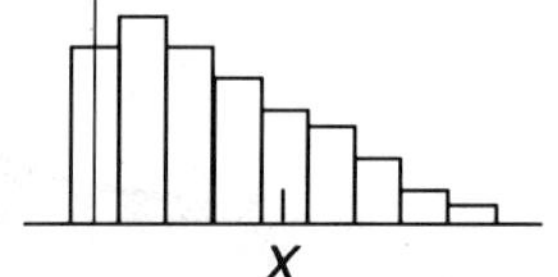

m	$X=0$	$X=1$	$X=2$	$X=3$	$X=4$	$X=5$	$X=6$	$X=7$	$X=8$	$X=9$	$X=10$	$X=11$	$X=12$	$X=13$	$X=14$
.10	1.000	.095	.005												
.20	1.000	.181	.018	.001											
.30	1.000	.259	.037	.004											
.40	1.000	.330	.062	.008	.001										
.50	1.000	.393	.090	.014	.002										
.60	1.000	.451	.122	.023	.003										
.70	1.000	.503	.156	.034	.006	.001									
.80	1.000	.551	.191	.047	.009	.001									
.90	1.000	.593	.228	.063	.013	.002									
1.00	1.000	.632	.264	.080	.019	.004	.001								
1.1	1.000	.667	.301	.100	.026	.005	.001								
1.2	1.000	.699	.337	.120	.034	.008	.002								
1.3	1.000	.727	.373	.143	.043	.011	.002								
1.4	1.000	.753	.408	.167	.054	.014	.003	.001							
1.5	1.000	.777	.442	.191	.066	.019	.004	.001							
1.6	1.000	.798	.475	.217	.079	.024	.006	.001							
1.7	1.000	.817	.507	.243	.093	.030	.008	.002							
1.8	1.000	.835	.537	.269	.109	.036	.010	.003	.001						
1.9	1.000	.850	.566	.296	.125	.044	.013	.003	.001						
2.0	1.000	.865	.594	.323	.143	.053	.017	.005	.001						
2.2	1.000	.889	.645	.377	.181	.072	.025	.007	.002						
2.4	1.000	.909	.692	.430	.221	.096	.036	.012	.003	.001					
2.6	1.000	.926	.733	.482	.264	.123	.049	.017	.005	.001					
2.8	1.000	.939	.769	.531	.308	.152	.065	.024	.008	.002	.001				
3.0	1.000	.950	.801	.577	.353	.185	.084	.034	.012	.004	.001				
3.2	1.000	.959	.829	.620	.397	.219	.105	.045	.017	.006	.002				
3.4	1.000	.967	.853	.660	.442	.256	.129	.058	.023	.008	.003	.001			
3.6	1.000	.973	.874	.697	.485	.294	.156	.073	.031	.012	.004	.001			
3.8	1.000	.978	.893	.731	.527	.332	.184	.091	.040	.016	.006	.002			
4.0	1.000	.982	.908	.762	.567	.371	.215	.111	.051	.021	.008	.003	.001		
4.2	1.000	.985	.922	.790	.605	.410	.247	.133	.064	.028	.011	.004	.001		
4.4	1.000	.988	.934	.815	.641	.449	.280	.156	.079	.036	.015	.006	.002	.001	
4.6	1.000	.990	.944	.837	.674	.487	.314	.182	.095	.045	.020	.008	.003	.001	
4.8	1.000	.992	.952	.857	.706	.524	.349	.209	.113	.056	.025	.010	.004	.001	
5.0	1.000	.993	.960	.875	.735	.560	.384	.238	.133	.068	.032	.014	.005	.002	.001

Source: H. L. Adler and E. B. Roessler, *Introduction to Probability and Statistics*, 6th ed., New York: W. H. Freeman, 1977. With permission.

TABLE 3: t-Distribution

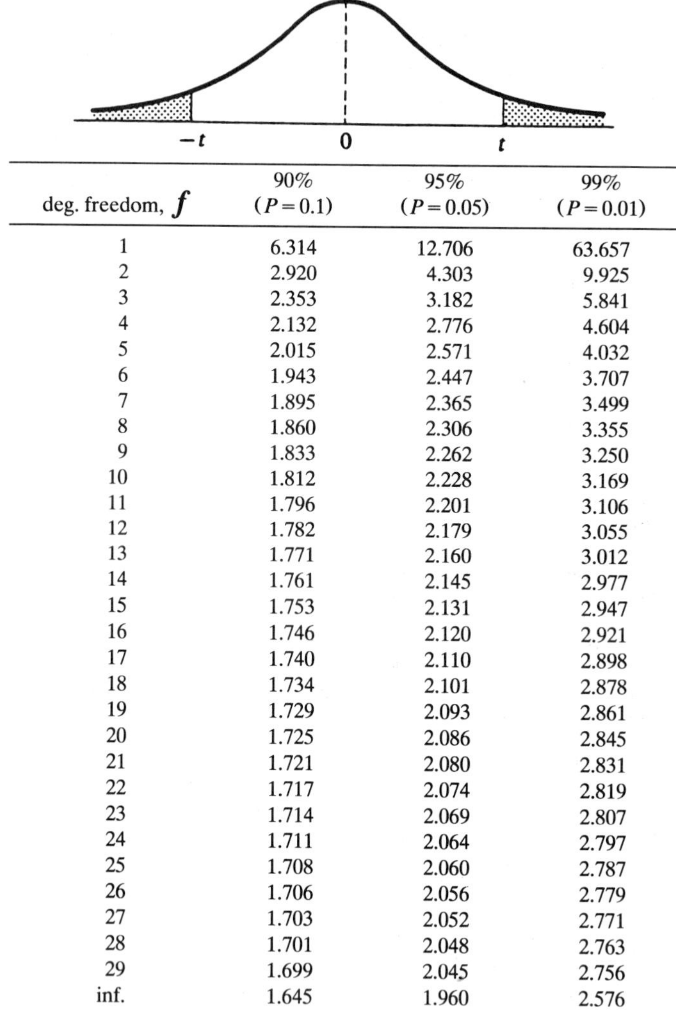

deg. freedom, f	90% ($P=0.1$)	95% ($P=0.05$)	99% ($P=0.01$)
1	6.314	12.706	63.657
2	2.920	4.303	9.925
3	2.353	3.182	5.841
4	2.132	2.776	4.604
5	2.015	2.571	4.032
6	1.943	2.447	3.707
7	1.895	2.365	3.499
8	1.860	2.306	3.355
9	1.833	2.262	3.250
10	1.812	2.228	3.169
11	1.796	2.201	3.106
12	1.782	2.179	3.055
13	1.771	2.160	3.012
14	1.761	2.145	2.977
15	1.753	2.131	2.947
16	1.746	2.120	2.921
17	1.740	2.110	2.898
18	1.734	2.101	2.878
19	1.729	2.093	2.861
20	1.725	2.086	2.845
21	1.721	2.080	2.831
22	1.717	2.074	2.819
23	1.714	2.069	2.807
24	1.711	2.064	2.797
25	1.708	2.060	2.787
26	1.706	2.056	2.779
27	1.703	2.052	2.771
28	1.701	2.048	2.763
29	1.699	2.045	2.756
inf.	1.645	1.960	2.576

Source: R. J. Tallarida and R. B. Murray, *Manual of Pharmacologic Calculations with Computer Programs*, 2nd ed., New York: Springer-Verlag, 1987. With permission.

TABLE 4: χ^2–Distribution

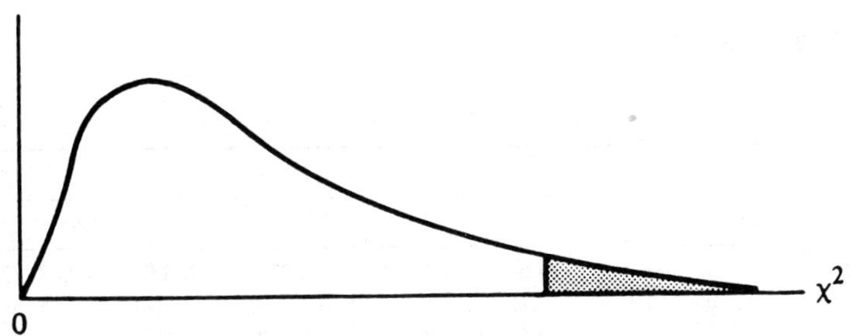

v	0.05	0.025	0.01	0.005
1	3.841	5.024	6.635	7.879
2	5.991	7.378	9.210	10.597
3	7.815	9.348	11.345	12.838
4	9.488	11.143	13.277	14.860
5	11.070	12.832	15.086	16.750
6	12.592	14.449	16.812	18.548
7	14.067	16.013	18.475	20.278
8	15.507	17.535	20.090	21.955
9	16.919	19.023	21.666	23.589
10	18.307	20.483	23.209	25.188
11	19.675	21.920	24.725	26.757
12	21.026	23.337	26.217	28.300
13	22.362	24.736	27.688	29.819
14	23.685	26.119	29.141	31.319
15	24.996	27.488	30.578	32.801
16	26.296	28.845	32.000	34.267
17	27.587	30.191	33.409	35.718
18	28.869	31.526	34.805	37.156
19	30.144	32.852	36.191	38.582
20	31.410	34.170	37.566	39.997
21	32.671	35.479	38.932	41.401
22	33.924	36.781	40.289	42.796
23	35.172	38.076	41.638	44.181
24	36.415	39.364	42.980	45.558
25	37.652	40.646	44.314	46.928
26	38.885	41.923	45.642	48.290
27	40.113	43.194	46.963	49.645
28	41.337	44.461	48.278	50.993
29	42.557	45.722	49.588	52.336
30	43.773	46.979	50.892	53.672

Source: J. E. Freund and F. J. Williams, *Elementary Business Statistics: The Modern Approach*, 2nd ed., Englewood Cliffs, N.J.: Prentice-Hall, 1972. With permission.

TABLE 5: Variance Ratio

	$F(95\%)$									
	n_1									
n_2	1	2	3	4	5	6	8	12	24	∞
1	161.4	199.5	215.7	224.6	230.2	234.0	238.9	243.9	249.0	254.3
2	18.51	19.00	19.16	19.25	19.30	19.33	19.37	19.41	19.45	19.50
3	10.13	9.55	9.28	9.12	9.01	8.94	8.84	8.74	8.64	8.53
4	7.71	6.94	6.59	6.39	6.26	6.16	6.04	5.91	5.77	5.63
5	6.61	5.79	5.41	5.19	5.05	4.95	4.82	4.68	4.53	4.36
6	5.99	5.14	4.76	4.53	4.39	4.28	4.15	4.00	3.84	3.67
7	5.59	4.74	4.35	4.12	3.97	3.87	3.73	3.57	3.41	3.23
8	5.32	4.46	4.07	3.84	3.69	3.58	3.44	3.28	3.12	2.93
9	5.12	4.26	3.86	3.63	3.48	3.37	3.23	3.07	2.90	2.71
10	4.96	4.10	3.71	3.48	3.33	3.22	3.07	2.91	2.74	2.54
11	4.84	3.98	3.59	3.36	3.20	3.09	2.95	2.79	2.61	2.40
12	4.75	3.88	3.49	3.26	3.11	3.00	2.85	2.69	2.50	2.30
13	4.67	3.80	3.41	3.18	3.02	2.92	2.77	2.60	2.42	2.21
14	4.60	3.74	3.34	3.11	2.96	2.85	2.70	2.53	2.35	2.13
15	4.54	3.68	3.29	3.06	2.90	2.79	2.64	2.48	2.29	2.07
16	4.49	3.63	3.24	3.01	2.85	2.74	2.59	2.42	2.24	2.01
17	4.45	3.59	3.20	2.96	2.81	2.70	2.55	2.38	2.19	1.96
18	4.41	3.55	3.16	2.93	2.77	2.66	2.51	2.34	2.15	1.92
19	4.38	3.52	3.13	2.90	2.74	2.63	2.48	2.31	2.11	1.88
20	4.35	3.49	3.10	2.87	2.71	2.60	2.45	2.28	2.08	1.84
21	4.32	3.47	3.07	2.84	2.68	2.57	2.42	2.25	2.05	1.81
22	4.30	3.44	3.05	2.82	2.66	2.55	2.40	2.23	2.03	1.78
23	4.28	3.42	3.03	2.80	2.64	2.53	2.38	2.20	2.00	1.76
24	4.26	3.40	3.01	2.78	2.62	2.51	2.36	2.18	1.98	1.73
25	4.24	3.38	2.99	2.76	2.60	2.49	2.34	2.16	1.96	1.71
26	4.22	3.37	2.98	2.74	2.59	2.47	2.32	2.15	1.95	1.69
27	4.21	3.35	2.96	2.73	2.57	2.46	2.30	2.13	1.93	1.67
28	4.20	3.34	2.95	2.71	2.56	2.44	2.29	2.12	1.91	1.65
29	4.18	3.33	2.93	2.70	2.54	2.43	2.28	2.10	1.90	1.64
30	4.17	3.32	2.92	2.69	2.53	2.42	2.27	2.09	1.89	1.62
40	4.08	3.23	2.84	2.61	2.45	2.34	2.18	2.00	1.79	1.51
60	4.00	3.15	2.76	2.52	2.37	2.25	2.10	1.92	1.70	1.39
120	3.92	3.07	2.68	2.45	2.29	2.17	2.02	1.83	1.61	1.25
∞	3.84	2.99	2.60	2.37	2.21	2.10	1.94	1.75	1.52	1.00

TABLE 5: Variance Ratio (continued)

F(99%)

	n_1									
n_2	1	2	3	4	5	6	8	12	24	∞
1	4,052	4,999	5,403	5,625	5,764	5,859	5,982	6,106	6,234	6,366
2	98.50	99.00	99.17	99.25	99.30	99.33	99.37	99.42	99.46	99.50
3	34.12	30.82	29.46	28.71	28.24	27.91	27.49	27.05	26.60	26.12
4	21.20	18.00	16.69	15.98	15.52	15.21	14.80	14.37	13.93	13.46
5	16.26	13.27	12.06	11.39	10.97	10.67	10.29	9.89	9.47	9.02
6	13.74	10.92	9.78	9.15	8.75	8.47	8.10	7.72	7.31	6.88
7	12.25	9.55	8.45	7.85	7.46	7.19	6.84	6.47	6.07	5.65
8	11.26	8.65	7.59	7.01	6.63	6.37	6.03	5.67	5.28	4.86
9	10.56	8.02	6.99	6.42	6.06	5.80	5.47	5.11	4.73	4.31
10	10.04	7.56	6.55	5.99	5.64	5.39	5.06	4.71	4.33	3.91
11	9.65	7.20	6.22	5.67	5.32	5.07	4.74	4.40	4.02	3.60
12	9.33	6.93	5.95	5.41	5.06	4.82	4.50	4.16	3.78	3.36
13	9.07	6.70	5.74	5.20	4.86	4.62	4.30	3.96	3.59	3.16
14	8.86	6.51	5.56	5.03	4.69	4.46	4.14	3.80	3.43	3.00
15	8.68	6.36	5.42	4.89	4.56	4.32	4.00	3.67	3.29	2.87
16	8.53	6.23	5.29	4.77	4.44	4.20	3.89	3.55	3.18	2.75
17	8.40	6.11	5.18	4.67	4.34	4.10	3.79	3.45	3.08	2.65
18	8.28	6.01	5.09	4.58	4.25	4.01	3.71	3.37	3.00	2.57
19	8.18	5.93	5.01	4.50	4.17	3.94	3.63	3.30	2.92	2.49
20	8.10	5.85	4.94	4.43	4.10	3.87	3.56	3.23	2.86	2.42
21	8.02	5.78	4.87	4.37	4.04	3.81	3.51	3.17	2.80	2.36
22	7.94	5.72	4.82	4.31	3.99	3.76	3.45	3.12	2.75	2.31
23	7.88	5.66	4.76	4.26	3.94	3.71	3.41	3.07	2.70	2.26
24	7.82	5.61	4.72	4.22	3.90	3.67	3.36	3.03	2.66	2.21
25	7.77	5.57	4.68	4.18	3.86	3.63	3.32	2.99	2.62	2.17
26	7.72	5.53	4.64	4.14	3.82	3.59	3.29	2.96	2.58	2.13
27	7.68	5.49	4.60	4.11	3.78	3.56	3.26	2.93	2.55	2.10
28	7.64	5.45	4.57	4.07	3.75	3.53	3.23	2.90	2.52	2.06
29	7.60	5.42	4.54	4.04	3.73	3.50	3.20	2.87	2.49	2.03
30	7.56	5.39	4.51	4.02	3.70	3.47	3.17	2.84	2.47	2.01
40	7.31	5.18	4.31	3.83	3.51	3.29	2.99	2.66	2.29	1.80
60	7.08	4.98	4.13	3.65	3.34	3.12	2.82	2.50	2.12	1.60
120	6.85	4.79	3.95	3.48	3.17	2.96	2.66	2.34	1.95	1.38
∞	6.64	4.60	3.78	3.32	3.02	2.80	2.51	2.18	1.79	1.00

Source: R. A. Fisher and F. Yates, *Statistical Tables for Biological, Agricultural and Medical Research*, London: The Lingman Group, Ltd. With permission.

Table of Derivatives

In the following table, a and n are constants, e is the base of the natural logarithms, and u and v denote functions of x.

1. $\frac{d}{dx}(a)=0$
2. $\frac{d}{dx}(x)=1$
3. $\frac{d}{dx}(au)=a\frac{du}{dx}$
4. $\frac{d}{dx}(u+v)=\frac{du}{dx}+\frac{dv}{dx}$
5. $\frac{d}{dx}(uv)=u\frac{dv}{dx}+v\frac{du}{dx}$
6. $\frac{d}{dx}(u/v)=\frac{v\frac{du}{dx}-u\frac{dv}{dx}}{v^2}$
7. $\frac{d}{dx}(u^n)=nu^{n-1}\frac{du}{dx}$
8. $\frac{d}{dx}e^u=e^u\frac{du}{dx}$
9. $\frac{d}{dx}a^u=(\log_e a)a^u\frac{du}{dx}$
10. $\frac{d}{dx}\log_e u=(1/u)\frac{du}{dx}$
11. $\frac{d}{dx}\log_a u=(\log_a e)(1/u)\frac{du}{dx}$
12. $\frac{d}{dx}u^v=vu^{v-1}\frac{du}{dx}+u^v(\log_e u)\frac{dv}{dx}$
13. $\frac{d}{dx}\sin u=\cos u\frac{du}{dx}$
14. $\frac{d}{dx}\cos u=-\sin u\frac{du}{dx}$
15. $\frac{d}{dx}\tan u=\sec^2 u\frac{du}{dx}$
16. $\frac{d}{dx}\text{ctn } u=-\csc^2 u\frac{du}{dx}$
17. $\frac{d}{dx}\sec u=\sec u\tan u\frac{du}{dx}$
18. $\frac{d}{dx}\csc u=-\text{csu ctn } u\frac{du}{dx}$
19. $\frac{d}{dx}\sin^{-1}u=\frac{1}{\sqrt{1-u^2}}\frac{du}{dx}, \quad (-\frac{1}{2}\pi\leq\sin^{-1}u\leq\frac{1}{2}\pi)$
20. $\frac{d}{dx}\cos^{-1}u=\frac{-1}{\sqrt{1-u^2}}\frac{du}{dx}, \quad (0\leq\cos^{-1}u\leq\pi)$
21. $\frac{d}{dx}\tan^{-1}u=\frac{1}{1+u^2}\frac{du}{dx}$
22. $\frac{d}{dx}\text{ctn}^{-1}u=\frac{-1}{1+u^2}\frac{du}{dx}$
23. $\frac{d}{dx}\sec^{-1}u=\frac{1}{u\sqrt{u^2-1}}\frac{du}{dx},$

    $(-\pi\leq\sec^{-1}u<-\frac{1}{2}\pi; 0\leq\sec^{-1}u<\frac{1}{2}\pi)$
24. $\frac{d}{dx}\csc^{-1}u=\frac{-1}{u\sqrt{u^2-1}}\frac{du}{dx},$

    $(-\pi<\csc^{-1}u\leq-\frac{1}{2}\pi; 0<\csc^{-1}u\leq\frac{1}{2}\pi)$
25. $\frac{d}{dx}\sinh u=\cosh u\frac{du}{dx}$
26. $\frac{d}{dx}\cosh u=\sinh u\frac{du}{dx}$
27. $\frac{d}{dx}\tanh u=\text{sech}^2 u\frac{du}{dx}$
28. $\frac{d}{dx}\text{ctnh } u=-\text{csch}^2 u\frac{du}{dx}$
29. $\frac{d}{dx}\text{sech } u=-\text{sech } u\tanh u\frac{du}{dx}$
30. $\frac{d}{dx}\text{csch } u=-\text{csch } u\text{ ctnh } u\frac{du}{dx}$
31. $\frac{d}{dx}\sinh^{-1}u=\frac{1}{\sqrt{u^2+1}}\frac{du}{dx}$
32. $\frac{d}{dx}\cosh^{-1}u=\frac{1}{\sqrt{u^2-1}}\frac{du}{dx}$
33. $\frac{d}{dx}\tanh^{-1}u=\frac{1}{1-u^2}\frac{du}{dx}$
34. $\frac{d}{dx}\text{ctnh}^{-1}u=\frac{-1}{u^2-1}\frac{du}{dx}$
35. $\frac{d}{dx}\text{sech}^{-1}u=\frac{-1}{u\sqrt{1-u^2}}\frac{du}{dx}$
36. $\frac{d}{dx}\text{csch}^{-1}u=\frac{-1}{u\sqrt{u^2+1}}\frac{du}{dx}$

Additional Relations with Derivatives

$\frac{d}{dt}\int_a^t f(x)\,dx = f(t) \quad \frac{d}{dt}\int_t^a f(x)\,dx = -f(t)$

If $x = f(y)$, then $\frac{dy}{dx} = \frac{1}{\frac{dx}{dy}}$

If $y = f(u)$ and $u = g(x)$, then $\frac{dy}{dx} = \frac{dy}{du} \cdot \frac{du}{dx}$ (chain rule)

If $x = f(t)$ and $y = g(t)$, then $\frac{dy}{dx} = \frac{g'(t)}{f'(t)}$, and $\frac{d^2y}{dx^2} = \frac{f'(t)g''(t) - g'(t)f''(t)}{[f'(t)]^3}$

(*Note*: exponent in denominator is 3.)

Integrals

Elementary Forms

1. $\int a\,dx = ax$

2. $\int a \cdot f(x)\,dx = a\int f(x)\,dx$

3. $\int \phi(y)\,dx = \int \frac{\phi(y)}{y'}\,dy$, where $y' = \frac{dy}{dx}$

4. $\int (u+v)\,dx = \int u\,dx + \int v\,dx$, where u and v are any functions of x

5. $\int u\,dv = u\int dv - \int v\,du = uv - \int v\,du$

6. $\int u\frac{dv}{dx}\,dx = uv - \int v\frac{du}{dx}\,dx$

7. $\int x^n\,dx = \frac{x^{n+1}}{n+1}$, except $n = -1$

8. $\int \frac{f'(x)\,dx}{f(x)} = \log f(x)$, $(df(x) = f'(x)\,dx)$

9. $\int \frac{dx}{x} = \log x$

10. $\int \frac{f'(x)\,dx}{2\sqrt{f(x)}} = \sqrt{f(x)}$, $(df(x) = f'(x)\,dx)$

11. $\int e^x\,dx = e^x$

12. $\int e^{ax}\,dx = e^{ax}/a$

13. $\int b^{ax}\,dx = \dfrac{b^{ax}}{a \log b}, \qquad (b > 0)$

14. $\int \log x\,dx = x \log x - x$

15. $\int a^{x} \log a\,dx = a^{x}, \qquad (a > 0)$

16. $\int \dfrac{dx}{a^2 + x^2} = \dfrac{1}{a} \tan^{-1} \dfrac{x}{a}$

17. $\int \dfrac{dx}{a^2 - x^2} = \begin{cases} \dfrac{1}{a} \tanh^{-1} \dfrac{x}{a} \\ \text{or} \\ \dfrac{1}{2a} \log \dfrac{a + x}{a - x}, \qquad (a^2 > x^2) \end{cases}$

18. $\int \dfrac{dx}{x^2 - a^2} = \begin{cases} -\dfrac{1}{a} \coth^{-1} \dfrac{x}{a} \\ \text{or} \\ \dfrac{1}{2a} \log \dfrac{x - a}{x + a}, \qquad (x^2 > a^2) \end{cases}$

19. $\int \dfrac{dx}{\sqrt{a^2 - x^2}} = \begin{cases} \sin^{-1} \dfrac{x}{|a|} \\ \text{or} \\ -\cos^{-1} \dfrac{x}{|a|}, \qquad (a^2 > x^2) \end{cases}$

20. $\int \dfrac{dx}{\sqrt{x^2 \pm a^2}} = \log(x + \sqrt{x^2 \pm a^2})$

21. $\int \dfrac{dx}{x\sqrt{x^2 - a^2}} = \dfrac{1}{|a|} \sec^{-1} \dfrac{x}{a}$

22. $\int \dfrac{dx}{x\sqrt{a^2 \pm x^2}} = -\dfrac{1}{a} \log\left(\dfrac{a + \sqrt{a^2 \pm x^2}}{x}\right)$

Forms Containing $(a + bx)$

For forms containing $a + bx$, but not listed in the table, the substitution $u = \dfrac{a + bx}{x}$ may prove helpful.

23. $\int (a + bx)^n\,dx = \dfrac{(a + bx)^{n+1}}{(n + 1)b}, \qquad (n \neq -1)$

24. $\int x(a + bx)^n\,dx = \dfrac{1}{b^2(n + 2)}(a + bx)^{n+2} - \dfrac{a}{b^2(n + 1)}(a + bx)^{n+1}, \qquad (n \neq -1, -2)$

25. $\int x^2(a+bx)^n\,dx = \frac{1}{b^3}\left[\frac{(a+bx)^{n+3}}{n+3} - 2a\frac{(a+bx)^{n+2}}{n+2} + a^2\frac{(a+bx)^{n+1}}{n+1}\right]$

26. $\int x^m(a+bx)^n\,dx = \begin{cases} \dfrac{x^{m+1}(a+bx)^n}{m+n+1} + \dfrac{an}{m+n+1}\int x^m(a+bx)^{n-1}\,dx \\ \text{or} \\ \dfrac{1}{a(n+1)}\Big[-x^{m+1}(a+bx)^{n+1} \\ \qquad + (m+n+2)\int x^m(a+bx)^{n+1}\,dx\Big] \\ \text{or} \\ \dfrac{1}{b(m+n+1)}\left[x^m(a+bx)^{n+1} - ma\int x^{m-1}(a+bx)^n\,dx\right] \end{cases}$

27. $\int \frac{dx}{a+bx} = \frac{1}{b}\log(a+bx)$

28. $\int \frac{dx}{(a+bx)^2} = -\frac{1}{b(a+bx)}$

29. $\int \frac{dx}{(a+bx)^3} = -\frac{1}{2b(a+bx)^2}$

30. $\int \frac{x\,dx}{a+bx} = \begin{cases} \dfrac{1}{b^2}[a+bx-a\log(a+bx)] \\ \text{or} \\ \dfrac{x}{b} - \dfrac{a}{b^2}\log(a+bx) \end{cases}$

31. $\int \frac{x\,dx}{(a+bx)^2} = \frac{1}{b^2}\left[\log(a+bx) + \frac{a}{a+bx}\right]$

32. $\int \frac{x\,dx}{(a+bx)^n} = \frac{1}{b^2}\left[\frac{-1}{(n-2)(a+bx)^{n-2}} + \frac{a}{(n-1)(a+bx)^{n-1}}\right], \quad n \neq 1,2$

33. $\int \frac{x^2\,dx}{a+bx} = \frac{1}{b^3}\left[\frac{1}{2}(a+bx)^2 - 2a(a+bx) + a^2\log(a+bx)\right]$

34. $\int \frac{x^2\,dx}{(a+bx)^2} = \frac{1}{b^3}\left[a+bx-2a\log(a+bx) - \frac{a^2}{a+bx}\right]$

35. $\int \frac{x^2\,dx}{(a+bx)^3} = \frac{1}{b^3}\left[\log(a+bx) + \frac{2a}{a+bx} - \frac{a^2}{2(a+bx)^2}\right]$

36. $\int \frac{x^2\,dx}{(a+bx)^n} = \frac{1}{b^3}\left[\frac{-1}{(n-3)(a+bx)^{n-3}} + \frac{2a}{(n-2)(a+bx)^{n-2}} - \frac{a^2}{(n-1)(a+bx)^{n-1}}\right], \quad n \neq 1,2,3$

37. $\int \frac{dx}{x(a+bx)} = -\frac{1}{a}\log\frac{a+bx}{x}$

38. $\int \frac{dx}{x(a+bx)^2} = \frac{1}{a(a+bx)} - \frac{1}{a^2}\log\frac{a+bx}{x}$

39. $\int \frac{dx}{x(a+bx)^3} = \frac{1}{a^3}\left[\frac{1}{2}\left(\frac{2a+bx}{a+bx}\right)^2 + \log\frac{x}{a+bx}\right]$

40. $\int \frac{dx}{x^2(a+bx)} = -\frac{1}{ax} + \frac{b}{a^2}\log\frac{a+bx}{x}$

41. $\int \frac{dx}{x^3(a+bx)} = \frac{2bx-a}{2a^2x^2} + \frac{b^2}{a^3}\log\frac{x}{a+bx}$

42. $\int \frac{dx}{x^2(a+bx)^2} = -\frac{a+2bx}{a^2x(a+bx)} + \frac{2b}{a^3}\log\frac{a+bx}{x}$

The Fourier Transforms

For a piecewise continuous function $F(x)$ over a finite interval $0 \leqq x \leqq \pi$, the *finite Fourier cosine transform* of $F(x)$ is

$$f_c(n) = \int_0^{\pi} F(x)\cos nx\,dx \qquad (n = 0, 1, 2, \ldots). \tag{1}$$

If x ranges over the interval $0 \leqq x \leqq L$, the substitution $x' = \pi x/L$ allows the use of this definition, also. The inverse transform is written

$$\bar{F}(x) = \frac{1}{\pi}f_c(0) + \frac{2}{\pi}\sum_{n=1}^{\infty} f_c(n)\cos nx \qquad (0 < x < \pi) \tag{2}$$

where $\bar{F}(x) = \dfrac{[F(x+0) + F(x-0)]}{2}$. We observe that $\bar{F}(x) = F(x)$ at points of continuity. The formula

$$f_c^{(2)}(n) = \int_0^{\pi} F''(x)\cos nx\,dx$$

$$= -n^2 f_c(n) - F'(0) + (-1)^n F'(\pi) \tag{3}$$

makes the finite Fourier cosine transform useful in certain boundary value problems.

Analogously, the *finite Fourier sine transform* of $F(x)$ is

$$f_s(n) = \int_0^{\pi} F(x)\sin nx\,dx \qquad (n = 1, 2, 3, \ldots) \tag{4}$$

and

$$\bar{F}(x) = \frac{2}{\pi} \sum_{n=1}^{\infty} f_s(n) \sin nx \qquad (0 < x < \pi) \tag{5}$$

Corresponding to (6) we have

$$f_s^{(2)}(n) = \int_0^{\pi} F''(x) \sin nx\, dx$$

$$= -n^2 f_s(n) - nF(0) - n(-1)^n F(\pi). \tag{6}$$

Fourier Transforms

If $F(x)$ is defined for $x \geqq 0$ and is piecewise continuous over any finite interval, and if

$$\int_0^{\infty} F(x)\, dx$$

is absolutely convergent, then

$$f_c(\alpha) = \sqrt{\frac{2}{\pi}} \int_0^{\infty} F(x) \cos(\alpha x)\, dx \tag{7}$$

is the *Fourier cosine transform* of $F(x)$. Furthermore,

$$\bar{F}(x) = \sqrt{\frac{2}{\pi}} \int_0^{\infty} f_c(\alpha) \cos(\alpha x)\, d\alpha. \tag{8}$$

If $\lim\limits_{x \to \infty} \dfrac{d^n F}{dx^n} = 0$, an important property of the Fourier cosine transform

$$f_c^{(2r)}(\alpha) = \sqrt{\frac{2}{\pi}} \int_0^{\infty} \left(\frac{d^{2r} F}{dx^{2r}} \right) \cos(\alpha x)\, dx$$

$$= -\sqrt{\frac{2}{\pi}} \sum_{n=0}^{r-1} (-1)^n a_{2r-2n-1} a^{2n} + (-1)^r \alpha^{2r} f_c(\alpha) \tag{9}$$

where $\lim\limits_{x \to 0} \dfrac{d^r F}{dx^r} = a_r$, makes it useful in the solution of many problems.

Under the same conditions,

$$f_s(\alpha) = \sqrt{\frac{2}{\pi}} \int_0^{\infty} F(x) \sin(\alpha x)\, dx \tag{10}$$

defines the *Fourier sine transform* of $F(x)$, and

$$\bar{F}(x)=\sqrt{\frac{2}{\pi}}\int_0^\infty f_s(\alpha)\sin(\alpha x)\,d\alpha. \tag{11}$$

Corresponding to (12) we have

$$f_s^{(2r)}(\alpha)=\sqrt{\frac{2}{\pi}}\int_0^\infty \frac{d^{2r}F}{dx^{2r}}\sin(ax)\,dx$$

$$=-\sqrt{\frac{2}{\pi}}\sum_{n=1}^{r}(-1)^n\alpha^{2n-1}a_{2r-2n}+(-1)^{r-1}\alpha^{2r}f_s(\alpha). \tag{12}$$

Similarly, if $F(x)$ is defined for $-\infty<x<\infty$, and if $\int_{-\infty}^{\infty}F(x)\,dx$ is absolutely convergent, then

$$f(\alpha)=\frac{1}{\sqrt{2\pi}}\int_{-\infty}^{\infty}F(x)e^{i\alpha x}\,dx \tag{13}$$

is the *Fourier transform* of $F(x)$, and

$$\bar{F}(x)=\frac{1}{\sqrt{2\pi}}\int_{-\infty}^{\infty}f(\alpha)e^{-i\alpha x}\,d\alpha. \tag{14}$$

Also, if

$$\lim_{|x|\to\infty}\left|\frac{d^nF}{dx^n}\right|=0 \qquad (n=1,2,\ldots,r-1),$$

then

$$f^{(r)}(\alpha)=\frac{1}{\sqrt{2\pi}}\int_{-\infty}^{\infty}F^{(r)}(x)e^{i\alpha x}\,dx=(-i\alpha)^r f(\alpha). \tag{15}$$

Finite Sine Transforms

$f_s(n)$	$F(x)$
1 $f_s(n)=\int_0^\pi F(x)\sin nx\,dx \quad (n=1,2,\ldots)$	$F(x)$
2 $(-1)^{n+1}f_s(n)$	$F(\pi-x)$
3 $\frac{1}{n}$	$\frac{\pi-x}{\pi}$

Finite Sine Transforms (continued)

	$f_s(n)$	$F(x)$
4	$\dfrac{(-1)^{n+1}}{n}$	$\dfrac{x}{\pi}$
5	$\dfrac{1-(-1)^n}{n}$	1
6	$\dfrac{2}{n^2}\sin\dfrac{n\pi}{2}$	$\begin{cases} x & \text{when } 0<x<\pi/2 \\ \pi-x & \text{when } \pi/2<x<\pi \end{cases}$
7	$\dfrac{(-1)^{n+1}}{n^3}$	$\dfrac{x(\pi^2-x^2)}{6\pi}$
8	$\dfrac{1-(-1)^n}{n^3}$	$\dfrac{x(\pi-x)}{2}$
9	$\dfrac{\pi^2(-1)^{n-1}}{n}-\dfrac{2[1-(-1)^n]}{n^3}$	x^2
10	$\pi(-1)^n\left(\dfrac{6}{n^3}-\dfrac{\pi^2}{n}\right)$	x^3
11	$\dfrac{n}{n^2+c^2}[1-(-1)^n e^{c\pi}]$	e^{cx}
12	$\dfrac{n}{n^2+c^2}$	$\dfrac{\sinh c(\pi-x)}{\sinh c\pi}$
13	$\dfrac{n}{n^2-k^2}(k\neq 0,1,2,\ldots)$	$\dfrac{\sinh k(\pi-x)}{\sin k\pi}$
14	$\begin{cases} \dfrac{\pi}{2} \text{ when } n=m \\ 0 \text{ when } n\neq m \end{cases} \quad (m=1,2,\ldots)$	$\sin mx$
15	$\dfrac{n}{n^2-k^2}[1-(-1)^n\cos k\pi] \quad (k\neq 1,2,\ldots)$	$\cos kx$
16	$\begin{cases} \dfrac{n}{n^2-m^2}[1-(-1)^{n+m}] & \text{when } n\neq m=1,2,\ldots \\ 0 & \text{when } n=m \end{cases}$	$\cos mx$
17	$\dfrac{n}{(n^2-k^2)^2}(k\neq 0,1,2,\ldots)$	$\dfrac{\pi\sin kx}{2k\sin^2 k\pi}-\dfrac{x\cos k(\pi-x)}{2k\sin k\pi}$
18	$\dfrac{b^n}{n}(\lvert b\rvert\leq 1)$	$\dfrac{2}{\pi}\arctan\dfrac{b\sin x}{1-b\cos x}$
19	$\dfrac{1-(-1)^n}{n}b^n \quad (\lvert b\rvert\leq 1)$	$\dfrac{2}{\pi}\arctan\dfrac{2b\sin x}{1-b^2}$

Finite Cosine Transforms

	$f_c(n)$	$F(x)$
1	$f_c(n) = \int_0^{\pi} F(x) \cos nx\, dx \quad (n = 0, 1, 2, \ldots)$	$F(x)$
2	$(-1)^n f_c(n)$	$F(\pi - x)$
3	0 when $n = 1, 2, \ldots$; $f_c(0) = \pi$	1
4	$\frac{2}{n} \sin \frac{n\pi}{2}$; $f_c(0) = 0$	$\begin{cases} 1 \text{ when } 0 < x < \pi/2 \\ -1 \text{ when } \pi/2 < x < \pi \end{cases}$
5	$-\frac{1 - (-1)^n}{n^2}$; $f_c(0) = \frac{\pi^2}{2}$	x
6	$\frac{(-1)^n}{n^2}$; $f_c(0) = \frac{\pi^2}{6}$	$\frac{x^2}{2\pi}$
7	$\frac{1}{n^2}$; $f_c(0) = 0$	$\frac{(\pi - x)^2}{2\pi} - \frac{\pi}{6}$
8	$3\pi^2 \frac{(-1)^n}{n^2} - 6 \frac{1 - (-1)^n}{n^4}$; $f_c(0) = \frac{\pi^4}{4}$	x^3
9	$\frac{(-1)^n e^c \pi - 1}{n^2 + c^2}$	$\frac{1}{c} e^{cx}$
10	$\frac{1}{n^2 + c^2}$	$\frac{\cosh c(\pi - x)}{c \sinh c\pi}$
11	$\frac{k}{n^2 - k^2}[(-1)^n \cos \pi k - 1] \quad (k \neq 0, 1, 2, \ldots)$	$\sin kx$
12	$\frac{(-1)^{n+m} - 1}{n^2 - m^2}$; $f_c(m) = 0 \quad (m = 1, 2, \ldots)$	$\frac{1}{m} \sin mx$
13	$\frac{1}{n^2 - k^2} \quad (k \neq 0, 1, 2, \ldots)$	$-\frac{\cos k(\pi - x)}{k \sin k\pi}$
14	0 when $n = 1, 2, \ldots$; $f_c(m) = \frac{\pi}{2} \quad (m = 1, 2, \ldots)$	$\cos mx$

Fourier Sine Transforms

	$F(x)$	$f_s(\alpha)$
1	$\begin{cases} 1 & (0<x<a) \\ 0 & (x>a) \end{cases}$	$\sqrt{\frac{2}{\pi}}\left[\frac{1-\cos\alpha}{\alpha}\right]$
2	$x^{p-1}(0<p<1)$	$\sqrt{\frac{2}{\pi}}\frac{\Gamma(p)}{\alpha^p}\sin\frac{p\pi}{2}$
3	$\begin{cases} \sin x & (0<x<a) \\ 0 & (x>a) \end{cases}$	$\frac{1}{\sqrt{2\pi}}\left[\frac{\sin[a(1-\alpha)]}{1-\alpha}-\frac{\sin[a(1+\alpha)]}{1+\alpha}\right]$
4	e^{-x}	$\sqrt{\frac{2}{\pi}}\left[\frac{\alpha}{1+\alpha^2}\right]$
5	$xe^{-x^2/2}$	$\alpha e^{-\alpha^2/2}$
6	$\cos\frac{x^2}{2}$	$\sqrt{2}\left[\sin\frac{\alpha^2}{2}C\left(\frac{\alpha^2}{2}\right)-\cos\frac{\alpha^2}{2}S\left(\frac{\alpha^2}{2}\right)\right]^*$
7	$\sin\frac{x^2}{2}$	$\sqrt{2}\left[\cos\frac{\alpha^2}{2}C\left(\frac{\alpha^2}{2}\right)+\sin\frac{\alpha^2}{2}S\left(\frac{\alpha^2}{2}\right)\right]^*$

*$C(y)$ and $S(y)$ are the Fresnel integrals

$$C(y)=\frac{1}{\sqrt{2\pi}}\int_0^y\frac{1}{\sqrt{t}}\cos t\,dt,$$

$$S(y)=\frac{1}{\sqrt{2\pi}}\int_0^y\frac{1}{\sqrt{t}}\sin t\,dt.$$

Fourier Cosine Transforms

	$F(x)$	$f_c(\alpha)$
1	$\begin{cases} 1 & (0<x<a) \\ 0 & (x>a) \end{cases}$	$\sqrt{\frac{2}{\pi}}\frac{\sin a\alpha}{\alpha}$
2	$x^{p-1}\ (0<p<1)$	$\sqrt{\frac{2}{\pi}}\frac{\Gamma(p)}{\alpha^p}\cos\frac{p\pi}{2}$
3	$\begin{cases} \cos x & (0<x<a) \\ 0 & (x>a) \end{cases}$	$\frac{1}{\sqrt{2\pi}}\left[\frac{\sin[a(1-\alpha)]}{1-\alpha}+\frac{\sin[a(1+\alpha)]}{1+\alpha}\right]$
4	e^{-x}	$\sqrt{\frac{2}{\pi}}\left(\frac{1}{1+\alpha^2}\right)$
5	$e^{-x^2/2}$	$e^{-\alpha^2/2}$
6	$\cos\frac{x^2}{2}$	$\cos\left(\frac{\alpha^2}{2}-\frac{\pi}{4}\right)$
7	$\sin\frac{x^2}{2}$	$\cos\left(\frac{\alpha^2}{2}+\frac{\pi}{4}\right)$

Fourier Transforms

	$F(x)$	$f(\alpha)$
1	$\dfrac{\sin ax}{x}$	$\begin{cases} \sqrt{\dfrac{\pi}{2}} & \lvert\alpha\rvert<a \\ 0 & \lvert\alpha\rvert>a \end{cases}$
2	$\begin{cases} e^{iwx} & (p<x<q) \\ 0 & (x<p, x>q) \end{cases}$	$\dfrac{i}{\sqrt{2\pi}}\dfrac{e^{ip(w+\alpha)}-e^{iq(w+\alpha)}}{(w+\alpha)}$
3	$\begin{cases} e^{-cx+iwx} & (x>0) \\ 0 & (x<0) \end{cases} \quad (c>0)$	$\dfrac{i}{\sqrt{2\pi}(w+\alpha+ic)}$
4	$e^{-px^2} \quad R(p)>0$	$\dfrac{1}{\sqrt{2p}}e^{-\alpha^2/4p}$
5	$\cos px^2$	$\dfrac{1}{\sqrt{2p}}\cos\left[\dfrac{\alpha^2}{4p}-\dfrac{\pi}{4}\right]$
6	$\sin px^2$	$\dfrac{1}{\sqrt{2p}}\cos\left[\dfrac{\alpha^2}{4p}+\dfrac{\pi}{4}\right]$
7	$\lvert x\rvert^{-p} \quad (0<p<1)$	$\sqrt{\dfrac{2}{\pi}}\dfrac{\Gamma(1-p)\sin\dfrac{p\pi}{2}}{\lvert\alpha\rvert^{(1-p)}}$
8	$\dfrac{e^{-a\lvert x\rvert}}{\sqrt{\lvert x\rvert}}$	$\dfrac{\sqrt{\sqrt{(a^2+\alpha^2)}+a}}{\sqrt{a^2+\alpha^2}}$
9	$\dfrac{\cosh ax}{\cosh \pi x} \quad (-\pi<a<\pi)$	$\sqrt{\dfrac{2}{\pi}}\dfrac{\cos\dfrac{a}{2}\cosh\dfrac{\alpha}{2}}{\cosh\alpha+\cos a}$
10	$\dfrac{\sinh ax}{\sinh \pi x} \quad (-\pi<a<\pi)$	$\dfrac{1}{\sqrt{2\pi}}\dfrac{\sin a}{\cosh\alpha+\cos a}$
11	$\begin{cases} \dfrac{1}{\sqrt{a^2-x^2}} & (\lvert x\rvert<a) \\ 0 & (\lvert x\rvert>a) \end{cases}$	$\sqrt{\dfrac{\pi}{2}}J_0(a\alpha)$
12	$\dfrac{\sin\left[b\sqrt{a^2+x^2}\right]}{\sqrt{a^2+x^2}}$	$\begin{cases} 0 & (\lvert\alpha\rvert>b) \\ \sqrt{\dfrac{\pi}{2}}J_0(a\sqrt{b^2-\alpha^2}) & (\lvert\alpha\rvert<b) \end{cases}$
13	$\begin{cases} P_n(x) & (\lvert x\rvert<1) \\ 0 & (\lvert x\rvert>1) \end{cases}$	$\dfrac{i^n}{\sqrt{\alpha}}J_{n+\frac{1}{2}}(\alpha)$

Fourier Transforms (continued)

	$F(x)$	$f(\alpha)$
14	$\begin{cases} \dfrac{\cos\left[b\sqrt{a^2-x^2}\right]}{\sqrt{a^2-x^2}} & (\lvert x\rvert<a) \\ 0 & (\lvert x\rvert>a) \end{cases}$	$\sqrt{\dfrac{\pi}{2}}\,J_0(a\sqrt{a^2+b^2})$
15	$\begin{cases} \dfrac{\cosh\left[b\sqrt{a^2-x^2}\right]}{\sqrt{a^2-x^2}} & (\lvert x\rvert<a) \\ 0 & (\lvert x\rvert>a) \end{cases}$	$\sqrt{\dfrac{\pi}{2}}\,J_0(a\sqrt{\alpha^2-b^2})$

The following functions appear among the entries of the tables on transforms.

Function	Definition	Name
$Ei(x)$	$\int_{-\infty}^{x} \frac{e^v}{v}\,dv$; or sometimes defined as $-Ei(-x)=\int_x^{\infty} \frac{e^{-v}}{v}\,dv$	
$Si(x)$	$\int_0^x \frac{\sin v}{v}\,dv$	
$Ci(x)$	$\int_{\infty}^{x} \frac{\cos v}{v}\,dv$; or sometimes defined as negative of this integral	
$erf(x)$	$\frac{2}{\sqrt{\pi}}\int_0^x e^{-v^2}\,dv$	Error function
$erfc(x)$	$1-erf(x)=\frac{2}{\sqrt{\pi}}\int_x^{\infty} e^{-v^2}\,dv$	Complementary function to error function
$L_n(x)$	$\frac{e^x}{n!}\frac{d^n}{dx^n}(x^n e^{-x}),\quad n=0,1,\ldots$	Laguerre polynomial of degree n

Numerical Methods

Solution of Equations by Iteration

Fixed-Point Iteration for Solving $f(x)=0$

Transform $f(x)=0$ into the form $x=g(x)$. Choose an x_0 and compute $x_1=g(x_0)$, $x_2=g(x_1)$, and in general

$$x_{n+1}=g(x_n), \qquad n=0,1,2,\ldots$$

Newton-Raphson Method for Solving $f(x) = 0$

f is assumed to have a continuous derivative f'. Use an approximate value x_0 obtained from the graph of f. Then compute

$$x_1 = x_0 - \frac{f(x_0)}{f'(x_0)}, \qquad x_2 = x_1 - \frac{f(x_1)}{f'(x_1)}$$

and in general

$$x_{n+1} = x_n - \frac{f(x_0)}{f'(x_n)}$$

Secant Method for Solving $f(x) = 0$

The secant method is obtained from Newton's method by replacing the derivative $f'(x)$ by the difference quotient

$$f'(x_n) = \frac{f(x_n) - f(x_{n-1})}{x_n - x_{n-1}}$$

Thus

$$x_{n+1} = x_n - f(x_n)\frac{x_n - x_{n-1}}{f(x_n) - f(x_{n-1})}$$

The secant method needs two starting values x_0 and x_1.

Method of Regula Falsi for Solving $f(x) = 0$

Select two starting values x_0 and x_1. Then compute

$$x_2 = \frac{x_0 f(x_1) - x_1 f(x_0)}{f(x_1) - f(x_0)}$$

If $f(x_0) \cdot f(x_2) < 0$, replace x_1 by x_2 in formula for x_2, leaving x_0 unchanged, and then compute the next approximation x_3; otherwise, replace x_0 by x_2, leaving x_1 unchanged, and compute the next approximation x_3. Continue in a similar manner.

Finite Differences

Uniform Interval *h*

If a function $f(x)$ is tabulated at a uniform interval h, that is, for arguments given by $x_n = x_0 + nh$, where n is an integer, then the function $f(x)$ may be denoted by f_n.

This can be generalized so that for all values of p, and in particular for $0 \leq p \leq 1$,

$$f(x_0 + ph) = f(x_p) = f_p,$$

where the argument designated x_0 can be chosen quite arbitrarily.

The following table lists and defines the standard operators used in numerical analysis.

Symbol	Function	Definition
E	Displacement	$Ef_p = f_{p+1}$
Δ	Forward difference	$\Delta f_p = f_{p+1} - f_p$
∇	Backward difference	$\nabla f_p = f_p - f_{p-1}$
⩚	Divided difference	
δ	Central difference	$of_p = f_{p+\frac{1}{2}} - f_{p-\frac{1}{2}}$
μ	Average	$uf_p = \frac{1}{2}(f_{p+\frac{1}{2}} + f_{p-\frac{1}{2}})$
Δ^{-1}	Backward sum	$\Delta^{-1}f_p = \Delta^{-1}f_{p-1} + f_{p-1}$
∇^{-1}	Forward sum	$\nabla^{-1}f_p = \nabla^{-1}f_{p-1} + f_p$
δ^{-1}	Central sum	$\delta^{-1}f_p = \delta^{-1}f_{p-1} + f_{p-\frac{1}{2}}$
D	Differentiation	$Df_p = \frac{d}{dx}f(x) = \frac{1}{h} \cdot \frac{d}{dp} f_p$
$I(=D^{-1})$	Integration	$If_p = \int^{x_p} f(x)\,dx = h\int^{p} f_p\,dp$
$J(=\Delta D^{-1})$	Definite integration	$Jf_p = h\int_p^{p+1} f_p\,dp$

I, Δ^{-1}, ∇^{-1} and δ^{-1} all imply the existence of an arbitrary constant which is determined by the initial conditions of the problem.

Where no confusion can arise the f can be omitted as, for example, in writing Δ_p for Δf_p.

Higher differences are formed by successive operations, e.g.,

$$\Delta^2 f_p = \Delta_p^2$$

$$= \Delta \cdot \Delta_p$$

$$= \Delta(f_{p+1} - f_p)$$

$$= \Delta_{p+1} - \Delta_p$$

$$= f_{p+2} - f_{p+1} - f_{p+1} + f_p$$

$$= f_{p+2} - 2f_{p+1} + f_p$$

Note that $f_p \equiv \Delta_p^0 \equiv \nabla_p^0 \equiv \delta_p^0$.

The disposition of the differences and sums relative to the function values is as shown (the arguments are omitted in these cases in the interest of clarity).

Calculus of Finite Differences

Forward difference scheme

Δ^{-2}_{-1}		f_{-2}		Δ^{2}_{-2}	
	Δ^{-1}_{-1}		Δ_{-2}		Δ^{3}_{-3}
Δ^{-2}_{0}		f_{-1}		Δ^{2}_{-2}	
	Δ^{-1}_{0}		Δ_{-1}		Δ^{3}_{-2}
Δ^{-2}_{1}		f_{0}		Δ^{2}_{-1}	
	Δ^{-1}_{1}		Δ_{0}		Δ^{3}_{-1}
Δ^{-2}_{2}		f_{1}		Δ^{2}_{0}	
	Δ^{-1}_{2}		Δ_{1}		Δ^{3}_{0}
Δ^{-2}_{3}		f_{2}		Δ^{2}_{1}	

Backward difference scheme

∇^{-2}_{-3}		f_{-2}		∇^{2}_{-1}	
	∇^{-1}_{-2}		∇_{-1}		∇^{2}_{0}
∇^{-2}_{-2}		f_{-1}		∇^{2}_{0}	
	∇^{-1}_{-1}		∇_{0}		∇^{3}_{1}
∇^{-2}_{-1}		f_{0}		∇^{2}_{1}	
	∇^{-1}_{0}		∇_{1}		∇^{3}_{2}
∇^{-2}_{0}		f_{1}		∇^{2}_{2}	
	∇^{-1}		∇_{2}		∇^{3}_{3}
∇^{-2}_{1}		f_{2}		∇^{2}_{3}	

Central difference scheme

δ^{-2}_{-2}		f_{-2}		δ^{2}_{-2}		δ^{4}_{-2}
	$\delta^{-1}_{-1\frac{1}{2}}$		$\delta_{-1\frac{1}{2}}$		$\delta^{3}_{-1\frac{1}{2}}$	
δ^{-2}_{-1}		f_{-1}		δ^{2}_{-1}		δ^{4}_{-1}
	$\delta^{-1}_{-\frac{1}{2}}$		$\delta_{-\frac{1}{2}}$		$\delta^{3}_{-\frac{1}{2}}$	
δ^{-2}_{0}		f_{0}		δ^{2}_{0}		δ^{4}_{0}
	$\delta^{-1}_{\frac{1}{2}}$		$\delta_{\frac{1}{2}}$		$\delta^{3}_{\frac{1}{2}}$	
δ^{-2}_{1}		f_{1}		δ^{2}_{1}		δ^{4}_{1}
	$\delta^{-1}_{1\frac{1}{2}}$		$\delta_{1\frac{1}{2}}$		$\delta^{3}_{1\frac{1}{2}}$	
δ^{-2}_{2}		f_{2}		δ^{2}_{2}		δ^{4}_{2}

In the forward difference scheme the subscripts are seen to move forward into the difference table and no fractional subscripts occur. In the backward difference scheme the subscripts lie on diagonals slanting backwards into the table while in the central difference scheme the subscripts maintain their position and the odd order subscripts are fractional.

All three however are merely alternative ways of labeling the same numerical quantities as any difference is the result of subtracting the number diagonally above it in the preceding column from that diagonally below it in the preceding column or, alternatively, it is the sum of the number diagonally above it in the subsequent column with that immediately above it in its own column.

In general $\Delta^{n}_{p-\frac{1}{2}n} \equiv \delta^{n}_{p} \equiv \nabla^{n}_{p+\frac{1}{2}n}$.

If a polynomial of degree r is tabulated exactly i.e., without any round-off errors, then the rth differences are constant.

The following table enables the simpler operators to be expressed in terms of the others:

	E	Δ	δ, μ	∇
E	—	$1+\Delta$	$1+\mu\delta+\frac{1}{2}\delta^2$	$(1-\nabla)^{-1}$
Δ	$E-1$	—	$\mu\delta+\frac{1}{2}\delta^2$	$\nabla(1-\nabla)^{-1}$
δ	$E^{\frac{1}{2}}-E^{-\frac{1}{2}}$	$\Delta(1+\Delta)^{-\frac{1}{2}}$	$2(\mu^2-1)^{\frac{1}{2}}$	$\nabla(1-\nabla)^{-\frac{1}{2}}$
∇	$-E^{-1}$	$\Delta(1+\Delta)^{-1}$	$\mu\delta-\frac{1}{2}\delta^2$	—
μ	$\frac{1}{2}(E^{\frac{1}{2}}+E^{-\frac{1}{2}})$	$\frac{1}{2}(2+\Delta)(1+\Delta)^{-\frac{1}{2}}$	$(1+\frac{1}{4}\delta^2)^{\frac{1}{2}}$	$\frac{1}{2}(2-\nabla)(1-\nabla)^{-\frac{1}{2}}$

In addition to the above there are other identities by means of which the above table can be extended, viz.,

$$E=e^{hD}=\Delta\nabla^{-1}$$

$$\mu=E^{-\frac{1}{2}}+\tfrac{1}{2}\delta=E^{\frac{1}{2}}-\tfrac{1}{2}\delta=\cosh(\tfrac{1}{2}hD)$$

$$\delta=E^{-\frac{1}{2}}\Delta=E^{\frac{1}{2}}\nabla=(\Delta\nabla)^{\frac{1}{2}}=2\sinh(\tfrac{1}{2}hD).$$

Note the emergence of Taylor's series from

$$\begin{aligned} f_p &= E^p f_0 \\ &= e^{phD} f_0 \\ &= f_0+phDf_0+\frac{1}{2!}p^2h^2D^2f_0+\cdots. \end{aligned}$$

Interpolation

Finite difference interpolation entails taking a given set of points and fitting a function to them. This function is usually a polynomial. If the graph of $f(x)$ is approximated over one tabular interval by a chord of the form $y=a+bx$ chosen to pass through the two points

$$(x_0, f(x_0)), \qquad (x_0+h, f(x_0+h))$$

the formula for the interpolated value is found to be

$$\begin{aligned} f(x_0+ph) &= f(x_0)+p[f(x_0+h)-f(x_0)] \\ &= f(x_0)+p\Delta f_0 \end{aligned}$$

If the graph of $f(x)$ is approximated over two successive tabular intervals by a parabola of the form $y=a+bx+cx^2$ chosen to pass through the three points

$$(x_0, f(x_0)), \qquad (x_0+h, f(x_0+h)), \qquad (x_0+2h, f(x_0+2h))$$

the formula for the interpolated value is found to be

$$f(x_0+ph)=f(x_0)+p[f(x_0+h)-f(x_0)]$$

$$+\frac{p(p-1)}{2!}[f(x_0+2h)-2f(x_0+h)+f(x_0)]$$

$$=f_0+p\Delta f_0+\frac{p(p-1)}{2!}\Delta^2 f_0$$

Using polynomial curves of higher order to approximate the graph of $f(x)$, a succession of interpolation formulas involving higher differences of the tabulated function can be derived. These formulas provide, in general, higher accuracy in the interpolated values.

Newton's Forward Formula

$$f_p=f_0+p\Delta_0+\frac{1}{2!}p(p-1)\Delta_0^2+\frac{1}{3!}p(p-1)(p-2)\Delta_0^3\cdots. \qquad 0\leqq p\leqq 1$$

Newton's Backward Formula

$$f_p=f_0+p\nabla_0+\frac{1}{2!}p(p+1)\nabla_0^2+\frac{1}{3!}p(p+1)(p+2)\nabla_0^3\cdots. \qquad 0\leqq p\leqq 1$$

Gauss' Forward Formula

$$f_p=f_0+p\delta_{\frac{1}{2}}+G_2\delta_0^2+G_3\delta_{\frac{1}{2}}^3+G_4\delta_0^4+G_5\delta_{\frac{1}{2}}^5\cdots. \qquad 0\leqq p\leqq 1$$

Gauss' Backward Formula

$$f_p=f_0+p\delta_{-\frac{1}{2}}+G_2^*\delta_0^2+G_3\delta_{-\frac{1}{2}}^3+G_4^*\delta_0^4+G_5\delta_{-\frac{1}{2}}^5\cdots \qquad 0\leqq p\leqq 1$$

In the above $G_{2n}=\binom{p+n-1}{2n}$

$$G_{2n}^*=\binom{p+n}{2n}$$

$$G_{2n+1}=\binom{p+n}{2n+1}$$

Stirling's Formula

$$f_p=f_0+\tfrac{1}{2}p(\delta_{\frac{1}{2}}+\delta_{-\frac{1}{2}})+\tfrac{1}{2}p^2\delta_0^2+S_3\left(\delta_{\frac{1}{2}}^3+\delta_{-\frac{1}{2}}^3\right)+S_4\delta_0^4+\cdots. \qquad -\tfrac{1}{2}\leqq p\leqq \tfrac{1}{2}$$

Steffenson's Formula

$$f_p = f_0 + \tfrac{1}{2}p(p+1)\delta_{\frac{1}{2}} - \tfrac{1}{2}(p-1)p\delta_{-\frac{1}{2}} + (S_3 + S_4)\delta^3_{\frac{1}{2}} + (S_3 - S_4)\delta^3_{-\frac{1}{2}} \cdots \quad -\tfrac{1}{2} \leqq p \leqq \tfrac{1}{2}.$$

In the above $S_{2n+1} = \dfrac{1}{2}\begin{pmatrix} p+n \\ 2n+1 \end{pmatrix}$

$$S_{2n+2} = \frac{p}{2n+2}\begin{pmatrix} p+n \\ 2n+1 \end{pmatrix}$$

$$S_{2n+1} + S_{2n+2} = \begin{pmatrix} p+n+1 \\ 2n+2 \end{pmatrix}$$

$$S_{2n+1} - S_{2n+2} = -\begin{pmatrix} p+n \\ 2n+2 \end{pmatrix}$$

Bessel's Formula

$$f_p = f_0 + p\delta_{\frac{1}{2}} + B_2\left(\delta_0^2 + \delta_1^2\right) + B_3\delta^3_{\frac{1}{2}} + B_4(\delta_0^4 + \delta_1^4) + B_5\delta^5_{\frac{1}{2}} + \cdots \qquad 0 \leqq p \leqq 1$$

Everett's Formula

$$f_p = (1-p)f_0 + pf_1 + E_2\delta_0^2 + F_2\delta_1^2 + E_4\delta_0^4 + F_4\delta_1^4 + E_6\delta_0^6 + F_6\delta_1^6 + \cdots \quad 0 \leqq p \leqq 1$$

The coefficients in the above two formulae are related to each other and to the coefficients in the Gaussian formulae by the identities

$$B_{2n} \equiv \tfrac{1}{2}G_{2n} \equiv \tfrac{1}{2}(E_{2n} + F_{2n})$$

$$B_{2n+1} \equiv G_{2n+1} - \tfrac{1}{2}G_{2n} \equiv \tfrac{1}{2}(F_{2n} - E_{2n})$$

$$E_{2n} \equiv G_{2n} - G_{2n+1} \equiv B_{2n} - B_{2n+1}$$

$$F_{2n} \equiv G_{2n+1} \equiv B_{2n} + B_{2n+1}$$

Also for $q \equiv 1-p$ the following symmetrical relationships hold:

$$B_{2n}(p) \equiv B_{2n}(q)$$

$$B_{2n+1}(p) \equiv -B_{2n+1}(q)$$

$$E_{2n}(p) \equiv F_{2n}(q)$$

$$F_{2n}(p) \equiv E_{2n}(q)$$

as can be seen from the tables of these coefficients.

Bessel's Formula (Unmodified)

$$f_p = f_0 + p\delta_{\frac{1}{2}} + B_2\left(\delta_0^2 + \delta_1^2\right) + B_3\delta^3_{\frac{1}{2}} + B_4(\delta_0^4 + \delta_1^4) + B_5\delta^5_{\frac{1}{2}} + B_6\left(\delta_0^6 + \delta_1^6\right) + B_7\delta^7_{\frac{1}{2}} + \cdots$$

Lagrange's Interpolation Formula

$$f(x)=\frac{(x-x_1)(x-x_2)\dots(x-x_n)}{(x_0-x_1)(x_0-x_2)\dots(x_0-x_n)}f(x_0)$$

$$+\frac{(x-x_0)(x-x_2)\dots(x-x_n)}{(x_1-x_0)(x_1-x_2)\dots(x_1-x_n)}f(x_1)$$

$$+\cdots+\frac{(x-x_0)(x-x_1)\dots(x-x_{n-1})}{(x_n-x_0)(x_n-x_1)\dots(x_n-x_{n-1})}f(x_n)$$

Newton's Divided Difference Formula

$$f(x)=f_0+(x+x_0)f[x_0,x_1]+(x-x_0)(x-x_1)f[x_0,x_1,x_2]$$

$$+\cdots+(x-x_0)(x-x_1)\dots(x-x_{n-1})f[x_0,x_1,\dots,x_n]$$

where

$$f[x_0,x_1]=\frac{f_1-f_0}{x_1-x_0},$$

$$f[x_0,x_1,x_2]=\frac{f[x_1,x_2]-f[x_0,x_1]}{x_2-x_0},$$

$$\dots$$

$$f[x_0,x_1,\dots,x_k]=\frac{f[x_1,x_2,\dots,x_k]-f[x_0,x_1,\dots,x_{k-1}]}{x_k-x_0}$$

The layout of a divided difference table is similar to that of an ordinary finite difference table.

$$\begin{array}{cccccc}
x_{-1} & f_{-1} & & \Delta^2_{-1} & & \Delta^4_{-1} \\
 & & \Delta_{-\frac{1}{2}} & & \Delta_{-\frac{1}{2}} & \\
x_0 & f_0 & & \Delta^2_0 & & \Delta^4_0 \\
 & & \Delta_{\frac{1}{2}} & & \Delta^3_{\frac{1}{2}} & \\
x_1 & f_1 & & \Delta^2_1 & & \Delta^4_1
\end{array}$$

where the Δ's are defined as follows:

$$\Delta^0_r\equiv f_r,\qquad \Delta_{r+\frac{1}{2}}\equiv(f_{r+1}-f_r)/(x_{r+1}-x_r),$$

and in general

$$\Delta^{2n}_r\equiv\left(\Delta^{2n-1}_{r+\frac{1}{2}}-\Delta^{2n-1}_{r-\frac{1}{2}}\right)/(x_{r+n}-x_{r-n})$$

and

$$\Delta^{2n+1}_{r+\frac{1}{2}}\equiv\left(\Delta^{2n}_{r+1}-\Delta^{2n}_r\right)/(x_{r+1+n}-x_{r-n}).$$

Iterative Linear Interpolation

Neville's modification of Aiken's method of iterative linear interpolation is one of the most powerful methods of interpolation when the arguments are unevenly spaced as no prior knowledge of the order of the approximating polynomial is necessary nor is a difference table required.

The values obtained are successive approximations to the required result and the process terminates when there is no appreciable change. These values are of course useless if a new interpolation is required when the procedure must be started afresh.

Defining
$$f_{r,s} \equiv \frac{(x_s-x)f_r-(x_r-x)f_s}{(x_s-x_r)}$$

$$f_{r,s,t} \equiv \frac{(x_t-x)f_{r,s}-(x_r-x)f_{s,t}}{(x_t-x_r)}$$

$$f_{r,s,t,u} \equiv \frac{(x_u-x)f_{r,s,t}-(x_r-x)f_{s,t,u}}{(x_u-x_r)},$$

the computation is laid out as follows:

$$\begin{array}{llllll}
x_{-1} & (x_{-1}-x) & f_{-1} & & & \\
 & & & f_{-1,0} & & \\
x_0 & (x_0-x) & f_0 & & f_{-1\,0,1} & \\
 & & & f_{0,1} & & f_{-1\,0\,1,2} \\
x_1 & (x_1-x) & f_1 & & f_{0,1,2} & \\
 & & & f_{1,2} & & \\
x_2 & (x_2-x) & f_2 & & &
\end{array}$$

As the iterates tend to their limit the common leading figures can be omitted.

Gauss' Trigonometric Interpolation Formula

This is of greatest value when the function is periodic, i.e., a Fourier series expansion is possible.

$$f(x) = \sum_{r=0}^{n} C_r f_r,$$

where $C_r = N_r(x)/N_r(x_r)$ and

$$N_r(x) = \left[\sin\frac{(x-x_0)}{2}\right]\left[\sin\frac{(x-x_1)}{2}\right]\cdots\left[\sin\frac{(x-x_{r-1})}{2}\right]\left[\sin\frac{(x-x_{r+1})}{2}\right]\cdots\left[\sin\frac{(x-x_n)}{2}\right].$$

This is similar to the Lagrangian formula.

Reciprocal Differences

These are used when the quotient of two polynomials will give a better representation of the interpolating function than a simple polynomial expression.

A convenient layout is as shown below:

$$
\begin{array}{cccccc}
x_{-1} & f_{-1} & & & & \\
 & & \rho_{-\frac{1}{2}} & & & \\
x_0 & f_0 & & \rho_0^2 & & \\
 & & \rho_{\frac{1}{2}} & & \rho_{\frac{1}{2}}^3 & \\
x_1 & f_1 & & \rho_1^2 & & \rho_1^4 \\
 & & \rho_{1\frac{1}{2}} & & \rho_{1\frac{1}{2}}^3 & \\
x_2 & f_2 & & \rho_2^2 & & \\
 & & \rho_{2\frac{1}{2}} & & & \\
x_3 & f_3 & & & &
\end{array}
$$

where

$$\rho_{r+\frac{1}{2}} \equiv \frac{x_{r+1}-x_r}{f_{r+1}-f_r}$$

and

$$\rho_r^2 \equiv \frac{x_{r+1}-x_{r-1}}{f_{r+\frac{1}{2}}-f_{r-\frac{1}{2}}}+f_r$$

In general

$$\rho_{r+\frac{1}{2}}^{2n+1} \equiv \frac{x_{r+n+1}-x_{r-n}}{\rho_{r+1}^{2n}-\rho_r^{2n}}+\rho_{r+\frac{1}{2}}^{2n-1}$$

$$\rho_r^{2n} \equiv \frac{x_{r+n}-x_{r-n}}{\rho_{r+\frac{1}{2}}^{2n-1}-\rho_{r-\frac{1}{2}}^{2n-1}}+\rho_r^{2n-2}.$$

The interpolation formula is expressed in the form of a continued fraction expansion.

The expansion corresponding to Newton's forward difference interpolation formula, in the sense of the differences involved, is

$$f(x)=f_0+\cfrac{(x-x_0)}{\rho_{\frac{1}{2}}+\cfrac{(x_2-x_1)}{\rho_1-f_0+\cfrac{(x-x_2)}{\rho_{1\frac{1}{2}}^3-\rho_{\frac{1}{2}}+\cfrac{(x_4-x_3)}{\rho_2^4-\rho_1^2+\cfrac{(x-x_4)}{\text{etc.}}}}}}$$

while that corresponding to Gauss' forward formula is

$$f(x)=f_0+\cfrac{(x-x_0)}{\rho_{\frac{1}{2}}+\cfrac{(x_2-x_1)}{\rho_0^2-f_0+\cfrac{(x_3-x_{-1})}{\rho_{\frac{1}{2}}^3-\rho_{\frac{1}{2}}+\cfrac{(x_4-x_2)}{\rho_0^4-\rho_0^2+\cfrac{(x-x_{-2})}{\text{etc.}}}}}}$$

Probability

Definitions

A sample space S associated with an experiment is a set S of elements such that any outcome of the experiment corresponds to one and only one element of the set. An event E is a subset of a sample space S. An element in a sample space is called a sample point or a simple event (unit subset of S).

Definition of Probability

If an experiment can occur in n mutually exclusive and equally likely ways, and if exactly m of these ways correspond to an event E, then the probability of E is given by

$$P(E) = \frac{m}{n}.$$

If E is a subset of S, and if to each unit subset of S a non-negative number, called its probability, is assigned, and if E is the union of two or more different simple events, then the probability of E, denoted by $P(E)$, is the sum of the probabilities of those simple events whose union is E.

Marginal and Conditional Probability

Suppose a sample space S is partitioned into rs disjoint subsets where the general subset is denoted by $E_i \cap F_j$. Then the marginal probability of E_i is defined as

$$P(E_i) = \sum_{j=1}^{s} P(E_i \cap F_j)$$

and the marginal probability of F_j is defined as

$$P(F_j) = \sum_{i=1}^{r} P(E_i \cap F_j)$$

The conditional probability of E_i, given that F_j has occurred, is defined as

$$P(E_i/F_j) = \frac{P(E_i \cap F_j)}{P(F_j)}, \qquad P(F_j) \neq 0$$

and that of F_j, given that E_i has occurred, is defined as

$$P(F_j/E_i) = \frac{P(E_i \cap F_j)}{P(E_i)}, \qquad P(E_i) \neq 0.$$

Probability Theorems

1. If ϕ is the null set, $P(\phi) = 0$.
2. If S is the sample space, $P(S) = 1$.

3. If E and F are two events

$$P(E \cup F) = P(E) + P(F) - P(E \cap F).$$

4. If E and F are mutually exclusive events,

$$P(E \cup F) = P(E) + P(F).$$

5. If E and E' are complementary events,

$$P(E) = 1 - P(E').$$

6. The conditional probability of an event E, given an event F, is denoted by $P(E/F)$ and is defined as

$$P(E/F) = \frac{P(E \cap F)}{P(F)},$$

where $P(F) \neq 0$.

7. Two events E and F are said to be independent if and only if

$$P(E \cap F) = P(E) \cdot P(F).$$

E is said to be statistically independent of F if $P(E/F) = P(E)$ and $P(F/E) = P(F)$.

8. The events $E_1, E_2, \ldots, E_n$ are called mutually independent for all combinations if and only if every combination of these events taken any number at a time is independent.

9. *Bayes Theorem.*
   If $E_1, E_2, \ldots, E_n$ are n mutually exclusive events whose union is the sample space S, and E is any arbitrary event of S such that $P(E) \neq 0$, then

$$P(E_k/E) = \frac{P(E_k) \cdot P(E/E_k)}{\sum_{j=1}^{n} \left[P(E_j) \cdot P(E/E_j)\right]}$$

Random Variable

A function whose domain is a sample space S and whose range is some set of real numbers is called a random variable, denoted by **X**. The function **X** transforms sample points of S into points on the x-axis. **X** will be called a discrete random variable if it is a random variable that assumes only a finite or denumerable number of values on the x-axis. **X** will be called a continuous random variable if it assumes a continuum of values on the x-axis.

Probability Function (Discrete Case)

The random variable **X** will be called a discrete random variable if there exists a function f such that $f(x_i) \geq 0$ and $\sum_i f(x_i) = 1$ for $i = 1, 2, 3, \ldots$ and such that for any event E,

$$P(E) = P[\mathbf{X} \text{ is in } E] = \sum_E f(x)$$

where $\sum_E$ means sum $f(x)$ over those values x_i that are in E and where $f(x) = P[\mathbf{X} = x]$. The probability that the value of $\mathbf{X}$ is some real number x, is given by $f(x) = P[\mathbf{X} = x]$, where f is called the probability function of the random variable $\mathbf{X}$.

Cumulative Distribution Function (Discrete Case)

The probability that the value of a random variable $\mathbf{X}$ is less than or equal to some real number x is defined as

$$\begin{aligned} F(x) &= P(\mathbf{X} \leq x) \\ &= \Sigma f(x_i), \qquad -\infty < x < \infty, \end{aligned}$$

where the summation extends over those values of i such that $x_i \leq x$.

Probability Density (Continuous Case)

The random variable $\mathbf{X}$ will be called a continuous random variable if there exists a function f such that $f(x) \geq 0$ and $\int_{-\infty}^{\infty} f(x)\,dx = 1$ for all x in interval $-\infty < x < \infty$ and such that for any event E

$$P(E) = P(\mathbf{X} \text{ is in } E) = \int_E f(x)\,dx.$$

$f(x)$ is called the probability density of the random variable $\mathbf{X}$. The probability that $\mathbf{X}$ assumes any given value of x is equal to zero and the probability that it assumes a value on the interval from a to b, including or excluding either end point, is equal to

$$\int_a^b f(x)\,dx.$$

Cumulative Distribution Function (Continuous Case)

The probability that the value of a random variable $\mathbf{X}$ is less than or equal to some real number x is defined as

$$\begin{aligned} F(x) &= P(\mathbf{X} \leq x), \qquad -\infty < x < \infty \\ &= \int_{-\infty}^{x} f(x)\,dx. \end{aligned}$$

From the cumulative distribution, the density, if it exists, can be found from

$$f(x) = \frac{dF(x)}{dx}.$$

From the cumulative distribution

$$\begin{aligned} P(a \leq \mathbf{X} \leq b) &= P(\mathbf{X} \leq b) - P(\mathbf{X} \leq a) \\ &= F(b) - F(a) \end{aligned}$$

Mathematical Expectation

Expected Value

Let **X** be a random variable with density $f(x)$. Then the expected value of **X**, $E(\mathbf{X})$, is defined to be

$$E(\mathbf{X}) = \sum_x xf(x)$$

if **X** is discrete and

$$E(\mathbf{X}) = \int_{-\infty}^{\infty} xf(x)\,dx$$

if **X** is continuous. The expected value of a function g of a random variable **X** is defined as

$$E[g(\mathbf{X})] = \sum_x g(x)\cdot f(x)$$

if **X** is discrete and

$$E[g(\mathbf{X})] = \int_{-\infty}^{\infty} g(x)\cdot f(x)\,dx$$

if **X** is continuous.

Positional Notation

In our ordinary system of writing numbers, the value of any digit depends on its position in the number. The value of a digit in any position is ten times the value of the same digit one position to the right, or one-tenth the value of the same digit one position to the left. Thus, for example,

$$173.246 = 1\times 10^2 + 7\times 10^1 + 3 + 2\times\frac{1}{10} + 4\times\frac{1}{10^2} + 6\times\frac{1}{10^3}.$$

There is no reason that a number other than 10 cannot be used as the *base*, or *radix*, of the number system. In fact, bases of 2, 8, and 16 are commonly used in working with digital computers. When the base used is not clear from the context, it is usually indicated as a parenthesized subscript or merely as a subscript. Thus

$$743_{(8)} = 7\times 8^2 + 4\times 8 + 3 = 7\times 64 + 4\times 8 + 3 = 448 + 32 + 3 = 483_{(10)}$$

$$1011.101_{(2)} = 1\times 2^3 + 0\times 2^2 + 1\times 2 + 1 + 1\times\tfrac{1}{2} + 0\times\tfrac{1}{4} + 1\times\tfrac{1}{8} = 11.625_{(10)}$$

Change of Base

In this section, it is assumed that all calculations will be performed in base 10, since this is the only base in which most people can easily compute. However, there is no logical reason that some other base could not be used for the computations.

To convert a number from another base into base 10:

Simply write down the digits of the number, with each one multiplied by its appropriate positional value. Then perform the indicated computations in base 10, and write down the answer.

For examples, see the two examples in the previous section.

To convert a number from base 10 into another base:

The part of the number to the left of the point and the part to the right must be operated on separately. For the integer part (the part to the left of the point):

a. Divide the number by the new base, getting an integer quotient and remainder.

b. Write down the remainder as the last digit of the number in the new base.

c. Using the quotient from the last division in place of the original number, repeat the above two steps until the quotient becomes zero.

For the fractional part (the part to the right of the point):

a. Multiply the number by the new base.

b. Write down the integral part of the product as the first digit of the fractional part in the new base.

c. Using the fractional part of the last product in place of the original number, repeat the above two steps until the product becomes an integer, or until the desired number of places have been computed.

Examples

These examples show a convenient method of arranging the computations.

1. Convert $103.118_{(10)}$ to base 8.

$$\begin{array}{lll} 8 & \underline{|103|} & 7 \\ 8 & \underline{|12|} & 4 \\ & 1 & \end{array} \qquad 147.074324\ldots \qquad \begin{array}{r} .118 \\ \underline{8} \\ .944 \\ \underline{8} \\ 7.552 \\ \underline{8} \\ 4.416 \\ \underline{8} \\ 3.328 \\ \underline{8} \\ 2.624 \\ \underline{8} \\ 4.992 \end{array}$$

The calculation of the fractional part could be carried out as far as desired. It is a non-terminating fraction which will eventually repeat itself.

$$103.118_{(10)} = 147.074324\ldots_{(8)}$$

The calculations may be further shortened by not writing down the multiplier and divisor at each step of the algorithm, as shown in the next example.

2. Convert $275.824_{(10)}$ to base 5.

$$\begin{array}{r|l|l} 5 & 275 & 0 \\ & 55 & 0 \\ & 11 & 1 \\ & 2 & \end{array} \qquad \begin{array}{r} .824 \\ 4.120 \\ 0.600 \\ 3.000 \end{array}$$

$$275.824_{(10)} = 2100.403_{(5)}$$

To convert from one base to another (neither of which is 10):

The easiest procedure is usually to convert first to base 10, and then to the desired base. However, there are two exceptions to this:

1. If computational facility is possessed in either of the bases, it may be used instead of base 10, and the appropriate one of the above methods applied.
2. If the two bases are different powers of the same number, the conversion may be done digit-by-digit to the base which is the common root of both bases, and then digit-by-digit back to the other base.

Example: Convert $127.653_{(8)}$ to base 16. (For base 16, the letters A–F are used for the digits $10_{(10)}$–$15_{(10)}$.)

The first step is to convert the number to base 2, simply by converting each digit to its binary equivalent:

$$127.653_{(8)} = 001 \quad 010 \quad 111 \quad \cdot \quad 110 \quad 101 \quad 011_{(2)}$$

Now by simply regrouping the binary number into groups of four binary digits, starting at the point, we convert to base 16:

$$127.653_{(8)} = 101 \quad 0111 \quad \cdot \quad 1101 \quad 0101 \quad 1_{(2)} = 57.\mathrm{D}58_{(16)}$$

Credits

Material in this section was reprinted from the following sources:

D. R. Lide, Ed., *CRC Handbook of Chemistry and Physics,* 73rd ed., Boca Raton, Fla.: CRC Press. 1992: International System of Units (SI), conversion constants and multipliers (conversion of temperatures), symbols and terminology for physical and chemical quantities, fundamental physical constants.

W. H. Beyer, Ed., *CRC Standard Mathematical Tables and Formulae,* 29th ed., Boca Raton, Fla.: CRC Press, 1991: Greek alphabet, conversion constants and multipliers (recommended decimal multiples and submultiples, metric to English, English to metric, general, temperature factors), physical constants, series expansion, integrals, the Fourier transforms, numerical methods, probability, positional notation.

R. J. Tallarida, *Pocket Book of Integrals and Mathematical Formulas,* 2nd ed., Boca Raton, Fla.: CRC Press, 1992: Elementary algebra and geometry; determinants, matrices, and linear systems of equations; trigonometry; analytic geometry; series; differential calculus; integral calculus; vector analysis; special functions; statistics; tables of probability and statistics; table of derivatives.

Associations and Societies

American Concrete Institute (ACI)

22400 West Seven Mile Road
P.O. Box 19159
Detroit, MI 48219

Tel. # (313) 532-2600 • Fax # (313) 533-4747

Founded in 1905, the American Concrete Institute (ACI) has grown into a chartered society with over 20,000 members worldwide. The ACI is a technical and educational nonprofit society dedicated to improving the design, construction, manufacture, and maintenance of concrete structures.

Among ACI's 20,000 members are structural designers, architects, civil engineers, educators, contractors, concrete craftsmen and technicians, representatives of materials suppliers, students, testing laboratories, and manufacturers from around the world. The 83 national and international chapters provide the membership with opportunities to network with their peers and keep in tune with the activities of ACI International.

Membership

Membership is open to individuals who work directly in, have an association with, or have an interest in concrete. All members are encouraged to participate in the activities of the ACI International, which include involvement on voluntary technical committees that develop ACI codes, standards, and reports. Various levels of membership exist to meet particular needs. Student memberships are available.

Publications

Concrete International. Published monthly. Covers Institute, chapter, and industry news. Several technical articles following a specific theme appear in each issue.

ACI Materials Journal. Published bimonthly. Describes research in materials and concrete, related ACI International standards, and committee reports.

ACI Structural Journal. Published bimonthly. Includes technical papers on structural design and analysis, state-of-the-art reviews on reinforced and structural elements, and the use and handling of concrete.

Other publications: ACI International makes available over 300 technical publications on concrete. Information is also available in computer software and compact disc formats. A free 72-page publications catalog describing what ACI International has to offer is available.

Other Activities

ACI International provides technical information in the form of high-quality conventions, seminars, and symposia.

American Iron and Steel Institute (AISI)

1101 17th Street, N.W.
Washington, DC 20036-4700

Tel. # (202) 452-7100

The American Iron and Steel Institute (AISI) was founded in 1908. The Institute is a nonprofit association of North American companies engaged in the iron and steel industry. AISI comprises 43 member companies that produce the full range of steel mill products. Also included are iron ore mining companies and member companies that produce raw steel, including integrated, electric furnace, and reconstituted mills. Member companies account for more than two-thirds of the raw steel produced in the U.S., most of the steel manufactured in Canada, and nearly two-thirds of the flat-rolled steel products manufactured in Mexico.

AISI has 230 associate members, including customers who distribute, fabricate, process, or consume steel. Also included are companies and representatives of organizations that supply the steel industry with materials, equipment, and services, as well as individuals associated with educational or research organizations.

American National Standards Institute (ANSI)

Corporate Headquarters
11 West 42nd Street
New York, NY 10036

Tel. # (212) 642-4900

Washington Office
655 15th Street, N.W.
Suite 300
Washington, D.C. 20005

Tel. # (202) 639-4090

Brussels Office
Avenue des Arts 50
1040 Brussels, Belgium

Tel. # 011-322-513-6892

Founded in 1918, the American National Standards Institute (ANSI) is a private, nonprofit membership organization that coordinates the U.S. voluntary consensus standards system and approves American National Standards. ANSI ensures that a single set of nonconflicting American National Standards are developed by ANSI-accredited standards developers and that all interests concerned have the opportunity to participate in the development process.

The Brussels office provides input on European standards activities, represents U.S. interests in international standardization, and disseminates information on draft standards.

The Washington office is a liaison with the administration, Congress, and relevant government authorities.

Membership

ANSI consists of approximately 1300 national and international companies, 30 government agencies, 20 institutional members, and 250 professional, technical, trade, labor, and consumer organizations. ANSI offers no individual membership. For more information on membership, write to the membership department at the headquarters office or call (212) 642-4948.

Publications

ANSI Reporter. Published monthly. Newsletter that updates members on major national and international standards activities. It also provides information on the activities of the European standards bodies, CEN and CENELEC.

Standards Action. Published biweekly. This newsletter outlines all national draft standards currently under consideration for approval as American National Standards and solicits comments from readers. Comments are also solicited on regional, international, and foreign standards. These comments are then reviewed as part of the development process.

Catalog of American National Standards. Published annually. Provides a complete listing of all ANSI-approved American National Standards. Supplements are also published.

American Railway Engineering Association

50 F Street, N.W.
Washington, DC 20001

Tel. # (202) 639-2190

During the latter half of the 19th century, railroads in North America underwent rapid growth and development. Engineering and maintenance-of-way department officers were faced with complex questions and needs for improved materials, designs, and procedures. On March 30, 1899, the American Railway Engineering Association (AREA) was formed. The purpose of the AREA was to study and report on problems in the maintenance-of-way and structures realm of railroading as practiced in North America.

Headquarters of the AREA was located in Chicago from its founding until 1979. The Association then moved its headquarters to Washington, D.C., to have a better liaison with the Association of American Railroads, the Federal Railroad Administration, and other related institutions. The need for closer contact with the U.S. federal government came with the advent of the track safety standards in 1971.

From its inception, AREA has dealt with technical challenges through committees. Currently there are 23 different committees. The result of the committee's work and study often becomes part of the *AREA Manual for Railway Engineering.* This manual is revised annually to make the latest in recommended practice information for railway engineering available to all interested parties.

The purpose of the AREA as it continues into the 21st century is "the advancement of knowledge pertaining to the scientific and economic location, construction, operation and maintenance of railways," as stated in its Constitution.

Membership

The basic qualifications for membership are five years of experience in the profession of maintaining, operating, constructing, or locating railways. Graduation from a recognized college or university with a degree in engineering is being taken as the equivalent to three years of experience. Today, the AREA has grown in membership to over 3800 members.

Publications

AREA Manual for Railway Engineering comprises the work of the association's committees. The *Manual* is revised annually to make the latest in recommended practice information for railway engineering available to all interested parties. The *Portfolio of Trackwork Plans* is also compiled and updated in some manner. The AREA publishes a bulletin five times a year and has a monthly section in *Railway Track & Structures Magazine.*

American Society of Civil Engineers (ASCE)

International Headquarters
345 East 47th Street
New York, NY 10017-2398

Tel. # (212) 705-7496 or (800) 548-ASCE

Washington Office
1015 15th Street, N.W., Suite 600
Washington, DC 20005-2605

Tel. # (202) 789-2200

Founded in 1852, the American Society of Civil Engineers (ASCE) is America's oldest national professional engineering society. The society has more than 115,000 individual members, including 6500 international members in 137 nations. Memberships consist of individual professional engineers rather than companies or organizations.

ASCE is organized geographically into 21 district councils, 83 sections, 143 branches, and 246 student chapters and clubs. The society is governed by a 28-member board and is headquartered in the United Engineering Center in New York City. A Washington, D.C., office is maintained for government relations.

ASCE maintains the Civil Engineering Research Foundation to focus national attention and resources on the research needs of the civil engineering profession. In addition, there are 25 technical divisions and councils which foster the development and advancement of the science and practice of engineering. ASCE has marked infrastructure renewal as a top national priority.

ASCE is the world's largest publisher of civil engineering information, publishing over 63,000 pages in 1994. Nearly 42% of the Society's yearly income is generated through publication sales.

Membership

Membership applicants must meet the requirements set in the Constitution of the ASCE. Various levels of membership exist to meet particular needs. Student memberships are available to students who meet the requirements of the Constitution. Various entrance fees and dues are required of the various levels of membership. Application materials may be requested from the ASCE Membership Services Department by mail or telephone 1-800-548-ASCE(2723).

Publications

Civil Engineering. Published monthly. This is the Society's official magazine and is mailed to all members of ASCE. The magazine contains articles of current interest in the various fields of civil engineering, news of a professional nature, and reports on the activities of ASCE and its members. Independently prepared papers may be sent directly to the editor of *Civil Engineering* at 345 East 47th Street, New York, NY 10017-2398.

ASCE News. Published monthly. Mailed to all members without charge. It concentrates on the activities of ASCE and its members, with the intent of promoting interest and participation in Society programs.

Worldwide Projects. Published quarterly. A copublication of ASCE and Intercontinental Media Inc., Westport, CT. Each issue provides engineers with articles giving insight into various topics related to international civil engineering projects and doing business outside the U.S.

Journals published: *Journal of Management in Engineering,* published bimonthly, and *Journal of Professional Issues in Engineering Education and Practice,* published quarterly, present professional and technical problems of broad interest and implications. They also publish significant reports of the Professional Activities Committee and its constituent committees.

Other publications: the Society also publishes transactions; standards; engineer, owner, and construction-related documents; the *Publications Information;* indexes; and newsletters. A civil engineering database is also available. For more information contact Carol Reese, manager, information products, ASCE, 345 East 47th Street, New York, NY 10017-2398 or telephone (212) 705-7520.

American Society for Testing and Materials (ASTM)

1916 Race Street
Philadelphia, PA 19103-1187

Tel. # (215) 299-5400 • Fax # (215) 997-9679

European Office:
27-29 Knowl Piece, Wilbury Way
Hitchin, Herts SG4 OSX, England

Tel. # 0462-437933 • Fax # 0462-433678

Founded in 1898, the American Society for Testing and Materials (ASTM) has grown into one of the largest voluntary standards development systems in the world. ASTM is a nonprofit organization that provides a forum for producers, users, ultimate consumers, and those having a general interest, such as representatives of government and academia, to meet on common ground and write standards for materials, products, systems, and services. From the work of 131 standards-writing committees, ASTM publishes standard test methods, specifications, practices, guides, classifications, and terminology. ASTM's standards development activities encompass metals, paints, plastics, textiles, petroleum, construction, energy, the environment, consumer products, medical services and devices, computerized systems, electronics, and many other areas. All technical research and testing is done voluntarily by over 35,000 technically qualified ASTM members located throughout the world.

Membership

ASTM members pay an annual administrative fee of $50.00 for individual membership and $350.00 for an organizational membership. The only other costs involved are the time and travel expenses of the committee members and the donated use of members' laboratory and research facilities.

Publications

Annual Book of ASTM Standards. A 70-volume set that includes standards and specs in the following subject areas:

- Iron and steel products
- Nonferrous metal products
- Metals test methods and analytical procedures

- Construction
- Petroleum products, lubricants, and fossil fuels
- Medical devices and services
- General methods and instrumentation
- Paints, related coatings, and aromatics
- Textiles
- Plastics
- Rubber
- Electrical insulation and electronics
- Water and environmental technology
- Nuclear, solar, and geothermal energy
- General products, chemical specialties and end-use products

Discounts applied when purchased as a complete set, or when purchased by complete sections. Volumes may also be purchased individually.

Standardization News. Published monthly.

Journals published: *Journal of Testing and Evaluation; Cement, Concrete, and Aggregates; Geotechnical Testing Journal; Journal of Composites Technology and Research;* and *Journal of Forensic Sciences.*

ASTM also publishes books containing reports on state-of-the-art testing techniques and their possible applications.

American Water Works Association

International Office
6666 West Quincy Avenue
Denver, CO 80235

Tel. # (303) 794-7711 • Fax # (303) 794-7310

Government Affairs Office
1401 New York Avenue, N.W.
Suite 640
Washington, DC 20005

Tel. # (202) 628-8303 • Fax # (202) 628-2846

The American Water Works Association (AWWA) was established in 1881 by 22 dedicated water supply professionals. Membership has grown to over 54,000 individuals and organizations. AWWA is an international, nonprofit, scientific, and educational association dedicated to improving drinking water for people everywhere. Today, AWWA has grown to be the largest organization of water supply professionals in the world, boasting members from virtually every country.

AWWA was formed to promote public health, safety, and welfare through the improvement of the quality and quantity of water delivered to the public, and through the development of public understanding. AWWA also takes an active role in shaping the water industry's direction through research, participation in legislative activities, development of products, procedural standards, and manuals of practice, and it educates the public on water issues to promote a spirit of cooperation between consumers and buyers.

Membership

Listed under individual memberships are active, affiliate, and student. Organization memberships include utility, municipal service subscriber, small water system, associate, consultant, contractor, technical service, and manufacturer's agent, distributor, or representative. The association is governed by a board of directors that establishes policy for the overall management and direction of association affairs.

Publications

AWWA is the world's major publisher of drinking water information. Its publications cover just about every area of interest in the water supply field. More than 500 titles are offered, covering all aspects of water resources, water quality, treatment and distribution, utility management, and employee training and safety.

Civil Engineering Research Foundation (CERF)

1015 15th Street N.W., Suite 600
Washington, DC 20005

Tel. # (202) 842-0555 • Fax # (202) 789-2943

The Civil Engineering Research Foundation (CERF) was created by the American Society of Civil Engineers and began operation in 1989 to advance the civil engineering profession through research. CERF is an industry-guided research organization which serves as a critical catalyst to help the design and construction industry and the civil engineer profession expedite the transfer of research results into practice through cooperative national programs. CERF integrates the efforts of industry, government, and academia in order to implement research that is beyond the capabilities of any single organization. CERF is an independent, nonprofit organization, but remains affiliated with ASCE.

Council on Tall Buildings and Urban Habitat

Lehigh University
13 East Packer Avenue
Bethlehem, PA 18015

Tel. # (215) 758-3515 • Fax # (215) 758-4522

The Council on Tall Buildings and Urban Habitat is an international organization sponsored by engineering, architectural, and planning professionals. The Council was founded in 1969 and was known as the "Joint Committee on Tall Buildings" until the name was changed in 1976 to its present form.

The Council was established to study and report on all aspects of the planning, design, construction, and operation of tall buildings. The Council is also concerned with the role of tall buildings in the urban environment and their impact thereon. However, the Council is not an advocate for tall buildings per se, but in those situations in which they are viable the Council seeks to encourage the use of the latest knowledge in their implementation.

Membership

Membership is available to associations, commercial organizations, individual members, and students. Membership is available to students at the rate of ten dollars per year. Membership fees vary for associations, commercial organizations, and individuals.

Publications

A major focus of the Council is the publication of a comprehensive monograph series for use by those responsible for tall building planning and design. The original five-volume *Monograph on the Planning and Design of Tall Buildings* was released between 1978 and 1981. This comprehensive source of tall building information is the only such reference tool now available to the high-rise specialist. The volumes are *Planning and Environmental Criteria for Tall Buildings, Tall Building Systems and Concepts, Tall Building Criteria and Loading, Structural Design of Tall Street Buildings,* and *Structural Design of Tall Concrete and Masonry Buildings.* These volumes are available as a set or sold separately. Updated monographs are continually added to the series in order to keep information current.

Structural Stability Research Council

Fritz Engineering Laboratory
13 East Parker Avenue
Lehigh University
Bethlehem, PA 18015-3191

Tel. # (215) 758-3522 • Fax # (215) 758-4522

The Structural Stability Research Council (formerly the Column Research Council) was founded in 1944 to review and resolve the conflicting opinions and practices that existed at the time with respect to solutions to stability problems and to facilitate and promote economical and safe design. Now, over 50 years later, the Council has broadened its scope within the field of structural stability, has become international in character, and continues to seek solutions to stability problems.

Membership

Various levels of membership exist for individuals. Organizations, companies, and firms concerned with investigation and design of metal and composite structures are invited by the Council to become sponsors, participating organizations, participating companies, or participating firms.

Publications

The Council maintains a library at its headquarters. Material from the library is available on request.

Transportation Research Board (TRB)

Cecil and Ida Green Building
2001 Wisconsin Avenue, N.W.
Washington, DC 20007

Mail Address
Transportation Research Board
2101 Constitution Avenue, N.W.
Washington, DC 20418

Tel. # (202) 334-2934 • Fax # (202) 334-2003

The Transportation Research Board (TRB) is a unit of the National Research Council, which serves the National Academy of Sciences and the National Academy of Engineering. The Board's purpose is to stimulate research concerning the nature and performance of transportation systems, to disseminate the information produced by the research, and to encourage the application of appropriate research findings. The board's program is carried out by more than 330 committees, task forces, and panels composed of more than 3900 administrators, engineers, social scientists, attorneys, educators, and other concerned with transportation; they serve without compensation.

The program is supported by state transportation and highway departments, modal administrations of the U.S. Department of Transportation, and others interested in the development of transportation.

In November 1920, after a series of preliminary meetings and conferences, the National Research Council created the Advisory Board on Highway Research. Four years later, the name was changed to the Highway Research Board. During the late 1960s the Highway Research Board expanded its scope to all modes of transportation. The name was again changed in 1974 to the Transportation Research Board to recognize its increased emphasis on a broadened approach to transportation problems and needs.

Today the Transportation Research Board devotes attention to all factors pertinent to the understanding, design, and function of systems for the safe and efficient movement of people and goods, including the following:

- Planning, design, construction, operation, safety, and maintenance of transportation facilities and their components
- Economics, financing, and administration of transportation facilities and services
- Interaction of transportation systems with one another and with the physical, economic, and social environment that they are designed to serve

Publications

One of the most important activities of the Transportation Research Board is the dissemination of current research results. The mainstay of the TRB publications program is the Transportation Research Record series. This series consists primarily of the papers delivered at the TRB annual meeting by authors from all over the world.

Ethics

The following code of ethics was adopted by the American Society of Civil Engineers on September 25, 1976. The code of ethics became effective on January 1, 1977. The ASCE has since amended this code on October 25, 1980, and April 17, 1993. The code of ethics shown below is in the most recent amended form.

The ASCE adopted the fundamental principles of the ABET Code of Ethics of Engineers as accepted by the Accreditation Board for Engineering and Technology, Inc. (ABET).

CODE OF ETHICS*

Fundamental Principles

Engineers uphold and advance the integrity, honor and dignity of the engineering profession by:

1. *using their knowledge and skill for the enhancement of human welfare;*
2. *being honest and impartial and serving with fidelity the public, their employers and clients;*
3. *striving to increase the competence and prestige of the engineering profession; and*
4. *supporting the professional and technical societies of their disciplines.*

Fundamental Canons

1. *Engineers shall hold paramount the safety, health and welfare of the public in the performance of their professional duties.*
2. *Engineers shall perform services only in areas of their competence.*
3. *Engineers shall issue public statements only in an objective and truthful manner.*
4. *Engineers shall act in professional matters for each employer or client as faithful agents or trustees, and shall avoid conflicts of interest.*
5. *Engineers shall build their professional reputation on the merit of their services and shall not compete unfairly with others.*
6. *Engineers shall act in such a manner as to uphold and enhance the honor, integrity, and dignity of the engineering profession.*
7. *Engineers shall continue their professional development throughout their careers, and shall provide opportunities for the professional development of those engineers under their supervision.*

*Published with permission of the American Society of Civil Engineers.

Guidelines to Practice Under the Fundamental Canons of Ethics

CANON 1. Engineers shall hold paramount the safety, health and welfare of the public in the performance of their professional duties.

a. Engineers shall recognize that the lives, safety, health and welfare of the general public are dependent upon engineering judgments, decisions and practices incorporated into structures, machines, products, processes and devices.
b. Engineers shall approve or seal only those design documents, reviewed or prepared by them, which are determined to be safe for public health and welfare in conformity with accepted engineering standards.
c. Engineers whose professional judgment is overruled under circumstances where the safety, health and welfare of the public are endangered, shall inform their clients or employers of the possible consequences.
d. Engineers who have knowledge or reason to believe that another person or firm may be in violation of any of the provisions of Canon 1 shall present such information to the proper authority in writing and shall cooperate with the proper authority in furnishing such further information or assistance as may be required.
e. Engineers should seek opportunities to be of constructive service in civic affairs and work for the advancement of the safety, health and well-being of their communities.
f. Engineers should be committed to improving the environment to enhance the quality of life.

CANON 2. Engineers shall perform services only in areas of their competence.

a. Engineers shall undertake to perform engineering assignments only when qualified by education or experience in the technical field of engineering involved.
b. Engineers may accept an assignment requiring education or experience outside of their own fields of competence, provided their services are restricted to those phases of the project in which they are qualified. All other phases of such project shall be performed by qualified associates, consultants, or employees.
c. Engineers shall not affix their signatures or seals to any engineering plan or document dealing with subject matter in which they lack competence by virtue of education or experience or to any such plan or document not reviewed or prepared under their supervisory control.

CANON 3. Engineers shall issue public statements only in an objective and truthful manner.

a. Engineers should endeavor to extend the public knowledge of engineering, and shall not participate in the dissemination of untrue, unfair or exaggerated statements regarding engineering.
b. Engineers shall be objective and truthful in professional reports, statements, or testimony. They shall include all relevant and pertinent information in such reports, statements, or testimony.
c. Engineers, when serving as expert witnesses, shall express an engineering opinion only when it is founded upon adequate knowledge of the facts, upon a background of technical competence, and upon honest conviction.

d. Engineers shall issue no statements, criticisms, or arguments on engineering matters which are inspired or paid for by interested parties, unless they indicate on whose behalf the statements are made.
e. Engineers shall be dignified and modest in explaining their work and merit, and will avoid any act tending to promote their own interests at the expense of the integrity, honor and dignity of thc profession.

CANON 4. Engineers shall act in professional matters for each employer or client as faithful agents or trustees, and shall avoid conflicts of interest.

a. Engineers shall avoid all known or potential conflicts of interest with their employers or clients and shall promptly inform their employers or clients of any business association, interests, or circumstances which could influence their judgment or the quality of their services.
b. Engineers shall not accept compensation from more than one party for services on the same project, or for services pertaining to the same project, unless the circumstances are fully disclosed to and agreed to, by all interested parties.
c. Engineers shall not solicit or accept gratuities, directly or indirectly, from contractors, their agents, or other parties dealing with their clients or employers in connection with work for which they are responsible.
d. Engineers in public service as members, advisors, or employees of a governmental body or department shall not participate in considerations or actions with respect to services solicited or provided by them or their organization in private or public engineering practice.
e. Engineers shall advise their employers or clients when, as a result of their studies, they believe a project will not be successful.
f. Engineers shall not use confidential information coming to them in the course of their assignments as a means of making personal profit if such action is adverse to the interests of their clients, employers or the public.
g. Engineers shall not accept professional employment outside of their regular work or interest without the knowledge of their employers.

CANON 5. Engineers shall build their professional reputation on the merit of their services and shall not compete unfairly with others.

a. Engineers shall not give, solicit or receive either directly or indirectly, any political contribution, gratuity, or unlawful consideration in order to secure work, exclusive of securing salaried positions through employment agencies.
b. Engineers should negotiate contracts for professional services fairly and on the basis of demonstrated competence and qualifications for the type of professional service required.
c. Engineers may request, propose or accept professional commissions on a contingent basis only under circumstances in which their professional judgments would not be compromised.
d. Engineers shall not falsify or permit misrepresentation of their academic or professional qualifications or experience.
e. Engineers shall give proper credit for engineering work to those to whom credit is due, and shall recognize the proprietary interests of others. Whenever possible, they shall name the person or persons who may be responsible for designs, inventions, writings or other accomplishments.
f. Engineers may advertise professional services in a way that does not contain misleading language or is in any other manner derogatory to the dignity of the profession. Examples of permissible advertising are as follows:

Professional cards in recognized, dignified publications, and listings in rosters or directories published by responsible organizations, provided that the cards or listings are consistent in size and content and are in a section of the publication regularly devoted to such professional cards.

Brochures which factually describe experience, facilities, personnel and capacity to render service, providing they are not misleading with respect to the engineer's participation in projects described.

Display advertising in recognized dignified business and professional publications, providing it is factual and is not misleading with respect to the engineer's extent of participation in projects described.

A statement of the engineers' names or the name of the firm and statement of the type of service posted on projects for which they render services.

Preparation or authorization of descriptive articles for the lay or technical press, which are factual and dignified. Such articles shall not imply anything more than direct participation in the project described.

Permission by engineers for their names to be used in commercial advertisements, such as may be published by contractors, material suppliers, etc., only by means of a modest, dignified notation acknowledging the engineers' participation in the project described. Such permission shall not include public endorsement of proprietary products.

g. Engineers shall not maliciously or falsely, directly or indirectly, injure the professional reputation, prospects, practice or employment of another engineer or indiscriminately criticize another's work.

h. Engineers shall not use equipment, supplies, laboratory or office facilities of their employers to carry on outside private practice without the consent of their employers.

CANON 6. Engineers shall act in such a manner as to uphold and enhance the honor, integrity, and dignity of the engineering profession.

a. Engineers shall not knowingly act in a manner which will be derogatory to the honor, integrity, or dignity of the engineering profession or knowingly engage in business or professional practices of a fraudulent, dishonest or unethical nature.

CANON 7. Engineers shall continue their professional development throughout their careers, and shall provide opportunities for the professional development of those engineers under their supervision.

a. Engineers should keep current in their specialty fields by engaging in professional practice, participating in continuing education courses, reading in the technical literature, and attending professional meetings and seminars.

b. Engineers should encourage their engineering employees to become registered at the earliest possible date.

c. Engineers should encourage engineering employees to attend and present papers at professional and technical society meetings.

d. Engineers shall uphold the principle of mutually satisfying relationships between employers and employees with respect to terms of employment including professional grade descriptions, salary ranges, and fringe benefits.

Index

Page on which term is defined is indicated in bold.

Page on which term is defined is indicated in bold.

B

Page on which term is defined is indicated in bold.

Page on which term is defined is indicated in bold.

C

Page on which term is defined is indicated in bold.

Page on which term is defined is indicated in bold.

Page on which term is defined is indicated in bold.

Page on which term is defined is indicated in bold.

D

Page on which term is defined is indicated in bold.

Page on which term is defined is indicated in bold.

Page on which term is defined is indicated in bold.

Page on which term is defined is indicated in bold.

F

Page on which term is defined is indicated in bold.

Page on which term is defined is indicated in bold.

G

Page on which term is defined is indicated in bold.

H

Page on which term is defined is indicated in bold.

Page on which term is defined is indicated in bold.

I

Page on which term is defined is indicated in bold.

J

Page on which term is defined is indicated in bold.

K

L

Page on which term is defined is indicated in bold.

M

Page on which term is defined is indicated in bold.

Page on which term is defined is indicated in bold.

Page on which term is defined is indicated in bold.

N

Page on which term is defined is indicated in bold.

O

P

Page on which term is defined is indicated in bold.

Page on which term is defined is indicated in bold.

Page on which term is defined is indicated in bold.

Q

Page on which term is defined is indicated in bold.

R

Page on which term is defined is indicated in bold.

S

Page on which term is defined is indicated in bold.

Page on which term is defined is indicated in bold.

Page on which term is defined is indicated in bold.

Page on which term is defined is indicated in bold.

Page on which term is defined is indicated in bold.

Page on which term is defined is indicated in bold.

T

Page on which term is defined is indicated in bold.

Page on which term is defined is indicated in bold.

U

V

Page on which term is defined is indicated in bold.

W

Page on which term is defined is indicated in bold.

Page on which term is defined is indicated in bold.

X

Y

Z